VRT A-Z (Part numbers from A to L) Volume 1a

Up – To – Date World's

Transistor

Thyristor

SMD Code

Diode

IC

Linear

Digital

Analog

Comparision Tables A...Z

Enlarged &
Updated Edition

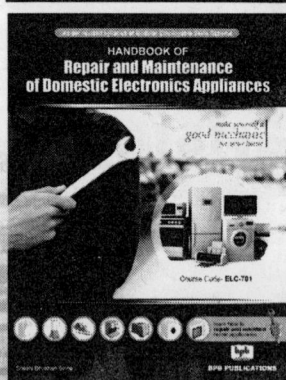

VRT A-Z (Part numbers from A to L) Volume 1a

Up – To – Date World's

Transistor

Thyristor

SMD Code

Diode

IC

Linear

Digital

Analog

Comparision Tables A...Z

Enlarged & Updated Edition

BPB PUBLICATIONS

B-14, CONNAUGHT PLACE, NEW DELHI-110001

ASIA-PACIFIC EDITION 2011
FIRST INDIAN EDITION 2011 REPRINT 2018
Copyright © BPB Publications, INDIA
ISBN : 978-81-8333-375-7

Original English Language Edition Published & Copyright © ECA –Electronic +GmbH, Germany.

Printed in India by arrangement with
ECA *Electronic + GmbH, USA / Tech Publications Singapore

Distributors:

BPB PUBLICATIONS
20,Ansari Road, Darya Ganj
New Delhi-110002
Ph: 23254990/23254991

DECCAN AGENCIES
4-3-329, Bank Street,
Hyderabad-500195
Ph: 24756967/24756400

BPB BOOK CENTRE
376 Old Lajpat Rai Market,
Delhi-110006 Ph: 23861747

MICRO MEDIA
Shop No. 5, Mahendra Chambers, 150
DN Rd. Next to Capital Cinema, V.T.
(C.S.T.) Station,
MUMBAI-400 001
Ph: 22078296/22078297

COMPUTER BOOK CENTRE
12, Shrungar Shopping Centre,
M.G.Road, BENGALURU–560001
Ph: 25587923/25584641

Published by Manish Jain for BPB Publications, 20, Ansari Road, Darya Ganj, New Delhi-110002 and Printed him at Vijeta Offset Printers, New Delhi.

TABLE OF CONTENTS

Important Notes on Use

In the following table section the most important semiconductor components – irrelevant as to whether it is dealing with transistors, diodes, thyristors integrated circuits etc. – are listed in strict alphabetical order. The individual types are provided with the most important relevant short data, and – as far as possible and as far as purposeful – suitable comparative types are named. This table is supplemented and extended with every new edition. Old types are **never** cancelled. Old tables should therefore always be removed (there is absolutely no risk of information loss) thus preventing valuable time being spent on searching through stacks of tables.

Column 1 ("Type")

The type designations correspond to those of the respective manufacturer documentations. They are sometimes stamped on the components themselves in a deviating or abbreviated form.

With several identical type designations of one kind or component (e. g. "BF 232") the abbreviation of the manufacturer is added in square brackets "[...]" to the type designation.

Selection identifications through added on letters or figures are only then considered if this is important in practice and with type comparison.

Column 2 ("Device")

Short definition of the semiconductor component.

Used abbreviations:

A/D-IC	Analog-to-digital converter
BiMOS-IC	Integrated circuit (Bipolar, + MOS technology)
CCD-IC	Charge-coupled device
CMOS-Logic	Digital-logic circuit (CMOS technology)
C-Di	Capacitance diode (Varactor, varicap)
Diac	3-layer-trigger diode, symmetrical
DIG-IC	Digital integrated circuit
DOC	Dual channel photocoupler
DPI	Dual channel photo interrupter
dRAM-IC	Read-write memory, dynamic
D/A-IC	Digital-to-analog converter
EAROM-IC	Memory, nonvolatile, alterable
ECL-Logic	Emitter-coupled logic
EEPROM-IC	Memory, nonvolatile, electrically erasable
EPROM-IC	Memory, nonvolatile, UV-erasable
ER	Emitter/detector
FIB IR LD	Infrared laser diode for fibre optics
FIB IR LD+MA	Infrared laser diode with monitor output for fibre optics
FIB IR LD+MD	Infrared laser diode with monitor diode for Fibre optics
FIB IR LD+MO	Infrared laser diode with modulator for fibre optics
FIB IR LD+NT	Infrared laser diode with NTC Resistor for fibre optics
FPLA	= PAL, field programmable
F-Thy	Fast thyristor
GaAs	Gallium arsenide
Ge-Di	Germanium diode
Ge-N	Germanium NPN transistor
Ge-P	Germanium PNP transistor
GTO-Thy	Gate turn off thyristor
Hybrid-IC	Integrated circuit (hybrid technology)
IC	Integrated circuit
IR LD	Infrared laser diode
IR LD+MA	Infrared laser diode with monitor output
IR LD+MD	Infrared laser diode with monitor diode
IRED	Infrared light emitting diode
IRED+MD	Light emitting diode
I/O-IC	Input/output IC for microcomputer
KOP-IC	Comparator (operational amplifier)
LIN-IC	Linear integrated circuit
LD	Laser diode
LDR	Photo Resistor
LED	Light emitting diode
LED/IRED	Light emitting diode and infrared light-emitting diode
MOS-...*	With integrated gate protection diode
MOS-FET-d	Metal oxide FET, depletion type
MOS-FET-e	Metal oxide FET, enhancement type
MOS-IC	Integrated circuit (MOS technology)
MOS-N/P-IGBT	Iso-gate bipolar transistor
MP	Matched emitter/detector
NMOS-IC	N-channel MOS-IC
N-FET	N-channel field-effect transistor
OC	Photocoupler
Opto	Optoelectronic component
OP-IC	Operational amplifier
PAL-IC	Programmable logic array
PC	Photo Element
PD	Photo Diode (for OPICs with Integrated Linear Amplifier)
PD ARRAY	Photo Diode Array
PI	Photo Interrupter
PIN-Di	PIN diode
PIN PD	PIN Photodiode (for OPICs with Integrated Linear Amplifier)
PMOS-IC	P-channel MOS-IC
PROM-IC	Electrically programmable ROM
PS	Photo reflective sensor
PT	Photo Transistor
PTR	Photo Triac
PUT	Programmable Unijunction transistor (UJT)
P-FET	P-channel field-effect transistor
PYRODET	Pyrodetector (Infrared Sensor)
QOC	Quad channel photocoupler
RCLED	Resonant cavity LED
RED LD	Red laser diode
RED LD+MD	Red laser diode with monitoring diode
Ref-Di	Reference diode (highly stable Z-diode)
ROM-IC	Read-only memory
SAS	Silicon asymmetrical switch
SBS	Silicon bilateral switch
Se-Di	Selenium diode
Si-Br	Silicon bridge rectifier
Si-Di	Silicon diode
Si-N	Silicon NPN transistor
Si-N-Darl	Silicon NPN Darlington transistor
Si-P	Silicon PNP transistor
Si-P-Darl	Silicon PNP Darlington transistor
Si-St	Silicon-stabi-diode (operation in forward direction)
sRAM-IC	Read-write memory, static
SUS	Silicon unilateral switch
Tetrode	P- + N-gate thyristor
Thermistor	Temperature depending resistor
Thy	Thyristor
Thy-Br	Thyristor bridge configuration
TOC	Triple channel photocoupler
Triac	Full-wave thyristor
Trigger-Di	4-layer trigger diode, asymmetrical
TTL-IC	Linear integrated circuit (TTL technology)
TTL-Logic	Digital logic circuit (transistor - transistor logic)
UJT	Unijunction transistor
Varistor	Voltage depending resistor (VDR)
Z-DI	Z-diode (operation in reverse direction)
Z-IC	Voltage regulator, voltage stabilizer
...+Di	With integrated damper diode
...+R	With integrated resistors
50Hz-Thy	Thyristor for mains operation
µC-IC	Single-chip microcomputer (MOS)
µP-IC	Microprocessor, CPU (MOS)

Column 3 ("Short description")

Short data or description of function of each type.

Used abbreviations:

A	Antenna and wideband amplifiers (CATV)
AFC	Automatic frequency control
AFT	Automatic fine tuning
AGC	Automatic gain control
ALC	Automatic level control
AM	RF application (AM range)
APC	Automatic phase control
ARI	Traffic information system (Germany)
Array	Arrangement of numerous elements in a single case
asym	Asymmetrical
AV	Audio/Video
A/W-Verst	Record/playback amplifier
B	DC forward current gain
Backward	Backward diode
Band-S	RF band switching
bidirektional	Bidirectional diode
Br	Bridge rectifier
Btx	Interactive video text
Camera	Video camera
CATV	Broad band cable amplifier
CB	CB-radio
CD	CD-player
Chopper	Chopper
contr. av.	Controlled avalanche
CPU	Central processing unit
CRT	Cathode ray tube
CTV	Color TV
Dem	Demodulator
Diskr	Discriminator
DMA	Direct memory access controller
Dual	Dual transistors for differential amplifiers or dual diode
E	Output stages
Equal	Equalizer
ESD	Electrostatic discharge
FB	Remote control
FED	Field effect diode
FIFO-IC	First-in First-out memory
FLT	Fluorescent tubes
FM	RF application (FM range)
FREDFET	V-MOS-FET with fast parasitic diode
F/V-Converter	Converter frequency to voltage
gep	Matched types
GI	Rectifier (general)
Gunn-Di	Gunn diode
HA	TV horizontal deflection stages
HF	RF application (general)
hi-beta	High current gain
hi-current	For high output current
hi-def	High definition
hi-power	High output power
hi-prec	High precision
hi-rel	High reliability
hi-res	High resolution
hi-speed	High speed
hi-volt	For high voltages
Horiz.	Horizontal
h-ohm	For high impedance demodulator circuits
Ib	Break-over current
Igt	Gate trigger current
Ih	Holding current
Impatt-Di	Impatt diode
Indic.	Indicator
IR	Infra-red
Ip	Peak point current
IPD	Intelligent power device
Is	Switching current
Iso	Insulated
Itsm	Surge current
Iv	Valley point current
kV-GI	High voltage rectifier
L	Power stages
LCD	Liquid crystal display
LED	Light-emitting diode
LIFO-IC	Last-in first-out memory
Limiter	Limiter
lo-drive	Low drive power required
lo-drop	Low voltage drop
LogL	Logic Level ($U_{th} \approx 0,8...2V$)
lo-power	Low power consumption
lo-sat	Low collector - emitter saturation voltage
lo-volt	For low voltages
M	Mixer stages
Min	Miniaturized
MMU	Memory management unit
Multipl	Frequency multiplier
NF	AF applications
Nix	Nixie driver (digital display tubes)
Noise suppr.	Noise suppression
n-ohm	For low impedance demodulator circuits
O	Oscillator stages
OFW/SAW-Filter	Surface acoustic wave filter
OP-Amp.	Operational amplifier
OSD	On-screen display
par	Parallel
PEP	Peak envelope power
PIP	Picture-in-picture
PLL	Phase-locked loop
PQ	RF-output power (transmitter transistor)
progr	Programmable
PS	Power supply
PWM	Pulse-width modulation
ra	Low noise
RadH	Designed for space aviation use (radiation hardened)
re	AGC stages
Recorder	Tape and cassette recorder
Reg	Regulator
S	Switching stages
Schottky	Schottky diode
ser	Serial
SHF	RF applications (>5 GHz)
SMD	Surface mounted device
SMPS, SN	Switch-mode power supplies
SS	Fast switching stages
SSB	Single sideband operation
stack	Rectifier stacks
sym	Symmetrical types
TAZ	Suppressor diode
tgq	Reset-time
Thy-Br	Thyristor bridge configuration
Thy-Modul	Arrangement of numerous thyristors in a single case
Ton	TV sound channel
Tr	Driver stages
Trigger-Di	4-layer trigger diode, asymmetrical
Tuning	RF tuning diode
Tunnel-Di	Tunnel diode
TV	Television applications
UART	Universal asynchronous receiver/transmitter
Ub	Break over voltage
Ucc, Us	Supply voltage
UHF	RF applications (>250 MHz)
Uni	General purpose types
US	Ultrasonics
USART	Universal synchronous/asynchronous receiver/transmitter
V	Pre/input stages
VA	TV vertical deflection types
VC	Video recorder
Vertik.	Vertical
VHF	RF applications (approx. 100...250MHz)
Vid	Video output stages

VIR	NTSC color correction
Vtx	Video text, teletext
V/F-Converter	Converter voltage to frequency
X-Ray-prot.	X-ray protection
ZF	IF stages
ZV	Integrated trigger amplification (Darlistor)
ß	Short-circuit current gain at 1 kHz
(eff)	r.m.s. value
(ss)	Peak value
(Ta = ...°)	Ambient temperature if not 25°C
(Tc =...°)	Case reference temperature if not 25°C (power types)
= [Typ]:	Identical with [type], however: [data, case, Pin-Code, etc.]
→	See under
-/...V	Only U_{CE0} known; otherwise always U_{CB0} or U_{CB0}/U_{CE0} (transistors)
...ns	Reverse recovery time (diodes)
.../-ns	Turn-on time (transistors)
.../...ns	Turn-on/turn-off time (transistors)
...µs	Circuit commutated turn-off time
	tq (thyristor)
... + Diac	With integrated diac
µComp	Microcomputer
≈ [Typ]	Similar to the named type (description see there)

Column 4 ("Fig.")

Related drawing number (figure) and pin assignment (appended letter) with discrete semiconductors. In the pin assignment table, the associated column ("Transistor", "FET", "Thyristor", "Diode", "Z-IC") must be observed. With ICs just a schematic drawing is provided without pin assignment. All drawings are situated at the end of the table. The degrees mark (°) means: case with additional heat sirik.

Column 5 ("Manufacturer")

The names of the manufacturers are abbreviated to save space. The complete name of each manufacturer is listed alphabetically on page XXII. We cannot vouch for completeness and availability.

If a number of manufacturers are named for a single type, data is given of only one, due to the data of a single type differing from one manufacturer to another slightly under unlike measurement conditions.

Column 6 ("Comparison types")

With the details on the alternate types, just a selection of the types "closely related" to the original type is given. They are preferably types which are not too difficult to obtain.

In practice there are usually many more alternate types applicable than there is room for in this table. The characterizing mark "+ +" after the last alternate type is to attract particular attention.

The **"parametric search"** in the online data base for semiconductor at **www.ecadata.de**, offering various abilities to find replacement types too. Furthermore you will be able to compare directly the parameters of the different types. Please read **page XXIV** for the **special offer for vrt book** users.

The "Selector tables" in the ECA data bases under **http://www.ecadata.de**, on DVD or as books conceal unsuspected furthering possibilities. These ECA data sources on DVD or at www.ecadata.de should by all means be consulted for an exact type comparison in critical cases.

If not otherwise mentioned in footnotes, the given alternate types have the same or better electrical data and can be applied without mechanical problems.

The same pin assignment has only been given attention to in the case of power types and ICs, as with small components the connections can be easily exchanged to fit. A check on any differing assignments is very much advised, and with this table is easy to carry out.

When a comparison type is followed by a maker's name (abbreviated) in square brackets, this means the data stems from this manufacturer and has not been verified by the author. In order not to exceed the capacity of this table, the detailed ECA data bases online or on DVD have partly had to be referred to. The incredible variety of the components and comparison possibilities cannot be fully shown in this scope with digital logic circuits, voltage stabilizers, Z-diodes, thyristors and operational amplifiers.

Footnote list on the comparison type
with comparison types in brackets ( ):

1	with integrated damper diode
2	damper diode between emitter and collector necessary externally
3	not insulated case
4	different case - possibly mechanical adaption necessary
5	the differing pin assignment possibly requires different assembly
6	alternate type is a little larger or with a different pin-spacing
7	alternate type with lower maximum ratings
8	lower transition frequency (f_T)
9	also with other gate trigger current
10	alternate IC is not pin-compatible or with a different pin-spacing
11	also with other gate trigger current and the alternate type is with different case
13	lower forward current gain (B, h_{FE}, ß, h_{fe})
15	No. 1 + 5
16	equivalent type with limited temperature range
17	higher saturation voltage (U_{CESat})
18	suitable resistor required externally

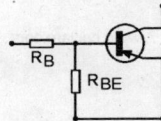

+ +   Many further comparison possibilities

Column 7 ("ECA Infos")

Further informations about this type like particularized data (data), original data sheets (pdf) or pinout (pinout) you will find in the ECA online data base for semiconductors at **http://www.ecadata.de**.

Please visit also our main page site **http://www.eca.de** for additional informations about our online data base.

In all critical application cases the relevant parameters of the comparison types are to be compared with those of the original in order to prevent any "surprises"!

Despite very careful research, errors excepted. There is no liability for failure in practice.

Abbreviations of manufacturers

++	More additional producer	
Acc	Accordion Electric Ltd. Co.	
Adv	Advanced Research Association	
Aeg	TEMIC TELEFUNKEN microelectronic	www.temic.de
Aei	AEI Semiconductors, Ltd.	
Akm	Asahi Kasei Microsystems	www.akm.com
Ald	Advanced Linear Devices Inc.	www.aldinc.com
Alp	Alpha Industries, Inc.	www.alphaind.com
Amc	Ampower Semiconductor Corp.	
Amd	Advanced Micro Devices GmbH	www.amd.com
Amd	Advanced Micro Devices Inc.	www.amd.com
Ame	Aksjeselskapet Mikro-Elektronik	
Amf	American Machine & Foundry	
Amp	Amperex Electronic Corp.	
Ams	American Microsystems, Inc.	www.amis.com
Anc	Ancom, Ishikawa Sangyo Ltd. Co.	
And	Analog Devices	www.analog.com
Ans	Analog Systems	
Apt	Advanced Power Technology	www.advancedpower.com
Apx	Apex Microtechnology Corp.	www.apexmicrotech.com
Atm	Atmel	www.atmel.com
Ava	Avantek, Inc.	
B&h	Bell & Howell Columbia Trading Ltd. Co.	
Bbc	BBC Brown Bovery, Ltd.	
Bec	Beckmann Instruments Inc.	
Bel	Bentron Elektronik GmbH	
Ben	Bendix Semiconductor Products	
Bog	Bogue Electric Manufacturing	
Bra	Bradley Semiconductor Corp.	
Bub	Burr-Brown Research Corp.	www.burr-brown.com
Cat	Catalyst Semiconductor	www.catsemi.com
Cbs	CBS electronics	
Cdi	Continental Device India, Ltd.	www.cdil.com
Cen	Central Semiconductor Corp. Devision	www.centralsemi.com
Chy	Cherry Semiconductor Corp.	www.cherry-semi.com
Cla	Clairex Technologies Inc.	www.clairex.com
Clv	Clevite Transistor	
Cml	Computer Labs	
Cod	CODI Corporation	
Com	Comlinear	
Cpc	C.P. Clare Transistor Corporation	www.cpclare.com
Cpd	Control Products Division Devar Inc.	
Crb	CR.BOX	
Cri	Crimson Semiconductor, Inc.	
Csr	CSR Industries, Inc.	
Ctr	Acrian Inc.	
Cys	Cypress Semiconductor Corp.	www.cypress.com
Dal	Dallas Semiconductor	
Dat	Datel Inc.	www.datel.com
Dat	Datel GmbH	www.datel.com
Dci	Dynamic Communications, Inc.	
Del	Delco Electronics	
Dic	Dickson Electronics Corp.	
Die	Diotec GmbH	www.diotec.com
Dio	Dionics, Inc.	
Dmc	Dynamic Measurement Corp.	
Dpm	Dense-Pac Microsystems, Inc.	www.dense-pac.com
Dsi	Discrete Semiconductors Industries	
Dtc	Diode Transistor Co., Inc.	
Ecg	ECG Philips	eu2.semiconductors.com
Edi	Electronic Designs, Inc. (EDI)	www.electronic-designs.com
Edl	EDAL Industries, Inc.	
Eiy	EI, Elektronskaja Industrija	
Ela	Élantec Semiconductor Inc.	www.elantec.com
Elc	Elcoma	
Ele	Electromation Co.	
Elx	Electroimpex	
Emi	Emihus Microcomponents Ltd.	
Ems	Enhanced Memory Systems Inc.	www.enhanced.com
Etc	Electronic Transistor Corporation	
EUR		
Exi	EXAR Corp.	www.exar.com
Exl	Exel Microelectronics, Inc. (Rohm Co. Ltd.)	www.rohm.com
Fag	Fagor Electrotecnica, S. Coop.	www.fagor.com
Fan	Fanon Transistor Corporation	
Fch	Fairchild Semiconductor GmbH	www.fairchildsemi.com
Fch	Fairchild Semiconductor	www.fairchildsemi.com
Fcs	Fairchild Semiconductor GmbH	www.fairchildsemi.com
Fcs	Fairchild Semiconductor	www.fairchildsemi.com
Fer	Ferranti GmbH	
Fer	Ferranti (Electronics) Semiconductors Ltd.	
Fjd	Fuji Denki Seizo Company	www.fujielectric.co.jp
Fjd	Fuji Electric GmbH	www.fujielectric.co.jp
Fjd	Fuji Semiconductor Inc.	www.fujisemiconductor.com
Fmi	FMI Inc.	
Fpm	Fine Products Microelectronics Corporation	
Fui	Fujitsu Mikroelektronik GmbH	www.fujitsu-ede.com
Fui	Fujitsu Ltd.	www.fujitsu.co.jp
Fuj	Fujitsu Ltd.	www.fujitsu.co.jp
Fuj	Fujitsu Mikroelektronik GmbH	www.fujitsu-ede.com
Gdc	General Diode Corporation	
Gen	Harris (General Electric) Semiconductor	www.semi.harris.com
Gie	General Instrument	www.gi.com
Gin	General Instrument, Optoelectronic Div.	www.gi.com
Gol	Goldstar (LG Electronics Deutschland)	www.goldstar.de
Gol	Goldstar (LG Electronics Inc.)	www.lg.co.kr
Gpd	Germanium Power Devices Corporation	
Gse	General Semiconductor	www.gensemi.com
Gsi	General Semiconductor	www.gensemi.com
GUS	UdSSR States located producer	
Haf	ASEA HAFO AB	
Hal	Halex Inc.	
Ham	HAMAMATSU PHOTONICS K.K.	www.hpk.co.jp
Ham	Hamamatsu Photonics Deutschland GmbH	www.hamamatsu.de
Har	Harris Semiconductor	www.harris.com
Har	Harris Semiconductor	www.harris.com
Hbc	Hybrid Systems Corp.	
Hei	EG&G Heimann Optoelectronics GmbH	www.heimann-opto.de
Hew	Hewlett Packard	www.hp.com/hp-comp
Hew	Hewlett-Packard GmbH	www.hp.com/hp-comp
Hfo	Halbleiterwerk Frankfurt/Oder	
Hit	Hitachi Europe GmbH	www.halsp.hitachi.com
Hit	Hitachi Europe Ltd.	www.halsp.hitachi.com
Hon	Honeywell AG Optoelectronics	www.honeywell.de
Hon	Honeywell International Inc.	www.honeywell.com
Hpa	Hewlett Packard	www.hp.com/hp-comp
Hsc	Helios Semiconductor	
Hsm	HI-SINCERITY MICROELECTRONICS	www.hsmc.com.tw
Hug	Hughes Aircraft Company	
Hun	Hyundai Electronics Industries Co., Ltd.	www.hei.co.kr
Hun	Hyundai Electronics Deutschland (HED)	www.hea.com
Idc	International Diode Corporation	
Idi	International Devices, Inc.	
Idr	I.P.R.S. Baneasa	
Idt	Integrated Device Technology (IDT)	www.idt.com
Idt	Integrated Device Technology GmbH	www.idt.com
Ilc	ILC Data Device Corp.	
Inf	Infineon Technologies AG (Siemens Semiconductor Group)	www.infineon.com
Inm	INMOS (siehe SGS-Thomson)	www.st.com
Inr	International Rectifier Corporation	www.irf.com
Int	Intel GmbH	www.intel.com
Int	Intel Corporation	www.intel.com
Isi	Harris Semiconductor	www.semi.harris.com
Iso	Isocom Inc. (Components, LTD.)	www.isocom.com
Iss	Integrated Silicon Solution, Inc.	www.issi.com
Ite	Intech/FMI Inc.	
Itl	Intersil Europe	www.intersil.com
Itl	Intersil Corporation Headquarters	www.intersil.com
Itr	Intronics Inc.	
Itt	ITT Intermetall (Micronas Semiconductors)	www.itt-sc.de
Ixy	IXYS Semiconductor GmbH	www.ixys.com
JAP	Unkown or several japanese Producter	
Jno	JENOPTIK Laserdiode GmbH	www.jold.de
Kec	KEC, Korea Electronics Co., Ltd.	www.kec.co.kr
Kem	Kemtron Electron Products	
Ker	Kertron, Inc.	
Khe	KH-Electronics	
Kmc	KMC-Semiconductor	
Kme	Kombinat Mikroelektronik Karl Marx Erfurt	
Ksw	KSW Electronics Corp.	
Lam	Lambda Semiconductors	
Lbd	Lambda Semiconductors	
Lic	Linear Technology Corporation	www.linear-tech.com
Lic	Linear Technology GmbH	www.linear-tech.com
Lif	Littelfuse	www.littelfuse.com
Lit	Taiwan Liton Electronic Co., Ltd.	www.liteon.com
Lrc	Leshan Radio Company Ltd.	www.lrc-china.com
Ltc	Linear Technology Corporation	www.linear-tech.com
Ltc	Linear Technology GmbH	www.linear-tech.com
Lte	Landsdale Transistor & Electronics, Inc.	
Ltt	Lignes Telegraphiques et Telephoniques (SGS-Thomson)	www.st.com
Luc	Lucas Electrical Co., Ltd.	
Mal	Mallory Distributor Products Co.	
Mat	Matsushita Electronics Corp.	www.mei.co.jp
Mat	Matsushita Electronics (Europe) GmbH	www.matsushita-europe.com
Max	Maxim Integrated Products Ltd.	www.maxim-ic.com
Mbl	M.B.L.E. (Philips Semiconductors)	eu2.semiconductors.com

Code	Manufacturer	Website
Mcc	Microchip Technology Inc.	www.microchip.com
Mcr	MICREL Semiconductor UK Lim.	www.micrel.com
Mdc	Microwave Diode Corp.	
Med	Meder Electronic AG	www.meder.com
Mhs	TemicTelefunken microelectronic GmbH	www.temic.de
Mic	Micro Electronics, Ltd.	
Min	Micron Technology, Inc.	
Mis	Mistral (SGS-Thomson)	www.st.com
Mit	Mitsubishi Electric Europe GmbH	www.mitsubishichips.com
Mit	Mitsubishi Electric Corporation	www.mitsubishichips.com
Miv	Microwave Associates	
Mmi	AMD (Advanced Micro Devices)	www.amd.com
Mod	Modular Devices Inc.	
Mos	Mostek (SGS-Thomson)	www.st.com
Mot	Motorola GmbH	www.mot-sps.com
Mot	Motorola Semiconductor Products Ltd.	www.mot-sps.com
Mpi	Micropac Industries Inc.	www.micropac.com
Mpi	Micropac Europe	www.micropac.com
Mps	Micro Power Systems	
Msc	Microsemi MicroPower Products	www.microsemi.com
Msp	MOSPEC SEMICONDUCTOR CORP.	www.mospec.com.tw
Msy	Microsystems	
Mts	Mitsumi Electric Co. Ltd.	www.mitsumi.com
Mul	Mullard Ltd. (Philips Semiconductors)	eu2.semiconductors.com
Mwa	Microwave Associates, Inc.	
Mws	Microwave Semiconductor Corporation	
Nac	National Aircraft Corporation	
Nae	NAE, Inc.	
Nas	North American Semiconductor Co., Inc.	
Nec	NEC Electronics (Germany) GmbH	www.nec.de
Nec	NEC Electronics Inc.,	www.ic.nec.co.jp
Neq	Nuclear Equipment Corp.	
New	Newmarket Transistors, Ltd.	
Nih	Nihon Inter Electronics Corporation	www.niec.co.jp
Njr	New Japan Radio Co., Ltd.	www.njr.com
Npp	National Power Products	
Nsc	National Semiconductor GmbH	www.national.com
Nsc	National Semiconductor	www.national.com
Nte	NTE Electronics Inc.	www.nteinc.com
Ntn	Hermann Koehler Elektrik GmbH & Co.	
Nuc	Nucleonic Products Co., Inc.	
Odc	Opto Diode Corporation	www.optodiode.com
Oiz	Oizumi Seisakusho	
Oki	OKI Electric Europe GmbH	www.oki.com
Oki	OKI Electric Industry Co. Ltd.	www.oki.com
old	Old unknown producer	
Omr	OMRON Europe B.V. (OMCE)	www.omron.de
Ons	ON Semiconductor Inc. www.onsemiconductor.com	
Opa	Opamp Labs Inc.	
Opt	Optical Electronics Inc.	
Org	Origin Electric Co., Ltd.	
Pai	Parametric Industries, Inc.	
Pan	Panasonic	www.panasonic.co.jp
Pan	Panasonic Deutschland GmbH	www.panasonic.de
Per	Pericom Semiconductor Corporation	www.pericom.com
Phb	Philco Radio Televisao, Ltda.	
Phc	Philco Corporation	
Phi	Philips Semiconductor	eu2.semiconductors.com
Pih	Piher Semiconductors	
Pls	Plessey Semiconductors, Ltd.	
Ply	Plessey Semiconductors, Ltd.	
Pmi	Precision Monolithics Inc.	
Poi	Power Innovations Ltd.	www.powinv.com
Ppc	PPC Products Corporation	
Ptc	Power Transistor Components/Allen-Bradley	
Pti	Power Tech, Inc.	www.power-tech.com
Pwi	Power Integrations, Inc.	www.powerint.com
Pwx	Powerex Inc.	www.pwrx.com
Qdc	Qualidyne Corporation	
Qse	Quality Semiconductor Inc.	www.qualitysemi.com
Qsi	Quality Semiconductor Inc.	www.qualitysemi.com
Qtc	Quality Technologies Corporation	
Qua	Quality Technologies Corporation	
Ray	Raytheon Halbleiter GmbH	www.raytheon.com
Ray	Raytheon Semiconductor Co.	www.raytheon.com
Rca	RCA Corporation (Harris Semiconductor)	www.semi.harris.com
Rec	Rectron Semiconductor	www.rectron.net
Ren	Renesas Technology Singapore Pte. Ltd.	www.renesas.com
Rfm	RF Micro Devices, Inc.	www.rfmd.com
Rhm	Rohm Co. Ltd.	www.rohm.co.jp
Rhm	Rohm Electronics GmbH	www.rohm.com
Riz	RIZ Radio Industrie Zagreb/Iskra Ljubljana	
Roc	Rochester Electronics Inc.	www.rocelec.com
Roe	Vishay Roederstein GmbH	www.vishay.com
Roh	Rohm Electronics GmbH	www.rohm.com
Roh	Rohm Co. Ltd.	www.rohm.co.jp
Roi	Roitner-Lasertechnik	
Rtc	R.T.C. La Radiotechnique-Compelec	
Sak	Sanken Electric Co., Ltd. (Allegro Micro Systems)	www.sanken-ele.co.jp
Sam	SAMSUNG Semiconductor Europe GmbH	www.samsungsemi.com
Say	Tokyo Sanyo Electric Co., Ltd.	www.semic.sanyo.co.jp
Say	SANYO Semiconductors (Europe) GmbH	www.sanyo.de
Sca	Semicoa	
Scn	Semicon Components Inc.	www.semicon.com
Seb	Semelab	
Sem	Semitronics Corporation	
Sen	Sensitron Semiconductor	
Seq	SEEQ Technology, Inc.	www.seeq.com
Ses	Sescosem (Thomson CSF)	
Sgs	SGS-THOMSON Microelectronics GmbH	www.st.com
Sha	Sharp Microelectronics	www.sharpmeg.com
Sha	Sharp Electronics (Europe) GmbH	www.sharp-eu.com
Shi	Shindengen Inc.	www.shindengen.com
Sie	Siemens AG	www.infineon.com
Sig	Signetics Corporation	www.signetics.com
Sii	Silonex Inc.	www.silonex.com
Sil	Silicon General Inc. (Linfinity Microelectronics)	www.linfinity.com
Sip	Sipex Corporation	www.sipex.com
Sip	SIPEX GmbH	www.sipex.com
Six	Siliconix GmbH (Temic Telefunken microelectronic GmbH)	www.siliconix.com
Skr	Semikron GmbH & Co. KG	www.semikron.de
Sld	Solid State Industries, Inc.	
Smi	Semitron, Ltd.	
Smo	Semi-Onics	
Smt	Semtech Corporation	www.semtech.com
Sol	Solitron Devices, Inc.	
Son	Sony Europa	www.sel.sony.com/semi/
Son	Sony Semiconductor Europe Limited	www.sel.sony.com/semi/
Sov	UdSSR	
Spc	Solid Power Corporation	
Spe	Space Power Electronic, Inc.	
Spr	Sprague Electric Co.	
Spt	Signal Processing Technologies	www.spt.com
Ssc	SSC, Silec - Semi - Conductors	
Ssi	Solid State Devices, Inc.	
Sss	Solid State Scientific, Inc.	
Sta	Standard Telephones & Cables	
Stc	Silicon Transistor Corp. (STC)	
Stl	Stanley Electric GmbH	www.stanleyelec.com
Stl	Stanley Electric Co.	www.stanleyelec.com
Stm	STMicroelectronics	www.st.com
Stw	Stow Laboratories, Inc.	
Stx	Supertex Inc.	www.supertex.com
Sty	Semiconductor Technology, Inc.	
Stz	ST Semicon, Inc.	
Syl	Syivania Semiconductors	
Sym	Symbol Semiconductor Inc.	
Tag	TAG Semiconductors	www.temic.de
Tcr	Teccor Electronics Inc.	www.teccor.com
Tcs	Telcom Semiconductor Inc. (Teledyne Components)	www.telcom-semi.com
Tdy	TELEDYNE Components	
Tel	Temic Telefunken Microelectronic GmbH	www.temic.de
Tem	Temic Telefunken Microelectronic GmbH	www.temic.de
Tes	Tesla Electronic Components (Ecimex)	
The	Theta-J Corporation	
Tho	SGS-THOMSON Mikroelektronik GmbH	www.st.com
Thr	Tree-Five System Inc.	
Tic	Transistor International Corporation	
Tix	Texas Instruments	www.ti.com
Tom	Thomson Bauelemente GmbH	www.tcs.thomson-csf.com
Tos	Toshiba Electronics Europe GmbH	www.toshiba.com
Tos	Toshiba Electronics U.K. Ltd.	www.toshiba.com
Toy	Toyo Denki Seizo Electronics Industry Corp.	
Tra	Transitron Electronic Corp.	
Trw	TRW Vertriebs GmbH	www.trw.com
Trw	TRW LSI Products Inc. (Semiconductors)	www.trw.com
Tsc	Teledyne Semiconductor	
Tsm	Tungsram GmbH	
Tun	Tung-Sol Electric	
Twa	Tokyo Wireless Apparatures	
Ucp	Unitra-Cemi (Polen)	
Unc	HI-Tech Co., Ltd (Unitra-Cemi)	www.hitech.com.pl
Ung	Ungarn	
Uni	Unitrode Corporation	www.unitrode.com
Uni	Unitrode Electronics GmbH	www.unitrode.com
Unz	Unizon	
Upi	UPI Semiconductor	
USA	USA located producer	
Usr	V/O Elektronzagranpostavka	

UTC	Unisonic Technologies Co., LTD.	www.unisonic.com.tw	Whs	Westinghouse Electric Corporation	
Vac	EG&G Vactec		Zel	Zeltex Inc.	
Val	Valvo (Philips Semiconductor)	eu2.semiconductors.com	Zex	Zetex Inc.	www.zetex.com
Vis	Vishay Intertechnology, Inc.	www.vishay.com	Zex	Zetex GmbH	www.zetex.com
Vtc	VTC Inc.	www.vtc.com	Zil	Zilog, Inc.	www.zilog.com
Wab	Walbern Devices, Inc.		Ztx	Zetex GmbH	www.zetex.com
Wea	West Ace				
Wes	Western Electric Co.				
Wfb	Werk für Fernsehelektronik Berlin				

Comparision Table
Reference table

Type	Device	Short Description	Fig.	Manu	Comparision Types	More at
A						
A...	...-P	→2SA.., z.B./e.g. "A748" = 2SA748→	Japantypen	JAP		
A...	...-P	→KSA.., z.B./e.g. "A 709"=KSA 709→	Samsung	Sam		
A...	...-P	→KTA.., z.B./e.g. "A 940"=KTA 940→	KEC	Kec		
A	Si-Di	=1SS400 (Typ-Code/Stempel/marking)	71(1,3mm)	Rhm	→1SS400	data
A	Si-P	=2SB1462 (Typ-Code/Stempel/marking)	35(1,6mm)	Mat	→2SB1462	data
A	Si-P	=2SB1462J (Typ-Code/Stempel/marking)	35(1,6mm)	Mat	→2SB1462	data
A	Si-P	=2SB766 (Typ-Code/Stempel/marking)	39	Mat	→2SB766	data
A	Si-N-Darl	=2SC2532 (Typ-Code/Stempel/marking)	35	Tos	→2SC2532	data
A	N-FET	=2SK1066 (Typ-Code/Stempel/marking)	35	Say	→2SK1066	data
A	GaAs-N-FET	=2SK1617 (Typ-Code/Stempel/marking)	52	Hit	→2SK1617	data
A	GaAs-FET	=2SK1619 (Typ-Code/Stempel/marking)	52	Tos	→2SK1619	data
A	MOS-N-FET-d	=3SK301 (Typ-Code/Stempel/marking)	44	Mat	→3SK301	data
A	MOS-N-FET-d	=3SK305 (Typ-Code/Stempel/marking)	44(2mm)	Mat	→3SK305	data
A	C-Di	=BA 892 (Typ-Code/Stempel/marking)	71(1,3mm)	Sie	→BA 892	data
A	C-Di	=HVS 303 (Typ-Code/Stempel/marking)	31a	Hit	→HVS 303	data
A	C-Di	=HVU 12 (Typ-Code/Stempel/marking)	71(1,7mm)	Hit	→HVU 12	data
A	Si-Di	=MA 2H735 (Typ-Code/Stempel/marking)	71(3,8mm)	Mat	→MA 2H735	data
A	Si-Di	=MA 2S111 (Typ-Code/Stempel/marking)	71(1,7mm)	Mat	→MA 2S111	data
A	Si-Di	=SB 01-05CP (Typ-Code/Stempel/marking)	35	Say	→SB 01-05CP	data
A	Si-Di	=SB 01-05Q (Typ-Code/Stempel/marking)	35(2mm)	Say	→SB 01-05Q	data
.A1	Si-Di	=1SS382 (Typ-Code/Stempel/marking)	44(2mm)	Tos	→1SS382	data
A 1	C-Di	=BA 591 (Typ-Code/Stempel/marking)	71(1,7mm)	Phi	→BA 591	data
A 1	Si-Di	=BAW 56 (Typ-Code/Stempel/marking)	35	Val,Mot,Sie	→BAW 56	data
A 1	Si-Di	=BAW 56L (Typ-Code/Stempel/marking)	35	Ons	→BAW 56L	
A 1	Si-Di	=BAW 56T (Typ-Code/Stempel/marking)	35(1,6mm)	Phi, Ons	→BAW 56T	data
A 1	Si-Di	=BAW 56W (Typ-Code/Stempel/marking)	35(2mm)	Phi, Ons	→BAW 56W	data
A 1	GaAs-N-FET-d	=CFY 10 (Typ-Code/Stempel/marking)	52	Sie	→CFY 10	data
A 1	GaAs-N-FET-d	=CFY 19-18 (Typ-Code/Stempel/marking)	51	Sie	→CFY 19	data
A 1	Si-N	=D70G.05T1 (Typ-Code/Stempel/marking)	39	Gen	→D70G.05T1	data
A 1	Si-Di	=DSE 015 (Typ-Code/Stempel/marking)	35(2mm)	Say	→DSE 015	data
A 1	Si-P+R	=FMA 1A (Typ-Code/Stempel/marking)	45	Rhm	→FMA 1A	data
A 1	Si-Di	=HN 2D01F (Typ-Code/Stempel/marking)	46	Tos	→HN 2D01F	data
A 1	Si-Di	=HSB 124SJ (Typ-Code/Stempel/marking)	35(2mm)	Hit	→HSB 124SJ	data
A 1	Si-Di	=HSM 124S (Typ-Code/Stempel/marking)	35	Hit	→HSM 124S	data
A 1	Z-Di	=MMSZ 10 (Typ-Code/marking)	71(2,7mm)	Ons	→MMSZ 2V4..56	
A 1	MOS-P-FET-e	=Si 2301DS (Typ-Code/Stempel/marking)	35	Six	→Si 2301DS	data
A 1	Si-P+R	=UMA 1N (Typ-Code/Stempel/marking)	45(2mm)	Rhm	→UMA 1N...11N	data
A 1	Si-Di	=1SS272 (Typ-Code/Stempel/marking)	44	Tos	→1SS272	data
A 1	Si-Di	=1SS308 (Typ-Code/Stempel/marking)	45	Tos	→1SS308	data
A 1	MOS-P-FET-e	=2SJ243 (Typ-Code/Stempel/marking)	35(1,6mm)	Nec	→2SJ243	data
A 1 p	Si-Di	=BAW 56 (Typ-Code/Stempel/marking)	35	Phi	→BAW 56	data
A 1 s	Si-Di	=BAW 56S (Typ-Code/Stempel/marking)	46(2mm)	Phi, Sie	→BAW 56S	data
A 1 s	Si-Di	=BAW 56 (Typ-Code/Stempel/marking)	35	Phi	→BAW 56	data
A 1 s	Si-Di	=BAW 56W (Typ-Code/Stempel/marking)	35(2mm)	Sie	→BAW 56W	data
A 1 t	Si-Di	=BAW 56 (Typ-Code/Stempel/marking)	35	Phi	→BAW 56	data
A 1 t	Si-Di	=BAW 56S (Typ-Code/Stempel/marking)	46(2mm)	Phi	→BAW 56S	data
A1X19X	LED	am ø5mm l:38lm Vf:2.5V If:350mA		Seo		data pdf pinout
A1X29X	LED	am ø5mm l:75lm Vf:2.5V If:700mA		Seo		data pdf pinout
A1X49X	LED	am ø5mm l:144lm Vf:2.5V If:1.4AmA		Seo		data pdf pinout
A 2	Si-Di	=BAT 18 (Typ-Code/Stempel/marking)	35	Phi	→BAT 18	data
A 2	GaAs-N-FET-d	=CFY 10-22 (Typ-Code/Stempel/marking)	51	Sie	→CFY 10	data
A 2	GaAs-FET	=CFY 30 (Typ-Code/Stempel/marking)	44	Sie	→CFY 30	data
A 2	Si-Di	=DSB 015 (Typ-Code/Stempel/marking)	35	Say	→DSB 015	data
A 2	Si-Di	=DSH 015 (Typ-Code/Stempel/marking)	35(2mm)	Say	→DSH 015	data
A 2	Z-Di	=EDZ 5.1B (Typ-Code/Stempel/marking)	71(1,3mm)	Rhm	→EDZ 5.1B	data
A 2	Si-P+R	=FMA 2A (Typ-Code/Stempel/marking)	45	Rhm	→FMA 2A	data
A 2	Si-Di	=HN 1D01F (Typ-Code/Stempel/marking)	46	Tos	→HN 1D01F	data
A 2	Si-Di	=HSM 221C (Typ-Code/Stempel/marking)	35	Hit	→HSM 221C	data
A 2	Si-Di	=IMBD 4148 (Typ-Code/Stempel/marking)	35	Gsi	→IMBD 4148	data
A 2	Si-Di	=IMBD 4448 (Typ-Code/Stempel/marking)	35	Gsi	→IMBD 4448	data
A 2	Si-P	=MBT 3906DW1 (Typ-Code/Stempel/marking	46(2mm)	Ons	→MBT 3906DW1	
A 2	Si-Di	=MMBD 2836 (Typ-Code/Stempel/marking)	35	Mot	→MMBD 2836	data
A 2	Z-Di	=MMSZ 11 (Typ-Code/marking)	71(2,7mm)	Ons	→MMSZ 2V4..56	
A 2	MOS-N-FET-e	=NTHD 5904 (Typ-Code/Stempel/marking)	≈8-MDIP	Ons	→NTHD 5904	
A 2	MOS-N-FET-e	=NTHS 5404 (Typ-Code/Stempel/marking)	≈8-MDIP	Ons	→NTHS 5404	
A 2	Si-Di	=PMBD 2836 (Typ-Code/Stempel/marking)	35	Phi	→PMBD 2836	
A 2	MOS-N-FET-e	=Si 2302DS (Typ-Code/Stempel/marking)	35	Six	→Si 2302DS	data
A 2	Z-Di	=UDZ 5.1B (Typ-Code/Stempel/marking)	71(1,7mm)	Rhm	→UDZ 2.0B...36B	data
A 2	Z-Di	=UDZS 5.1B (Typ-Code/Stempel/marking)	71(1,7mm)	Rhm	→UDZS 5.1B...10B	data
A 2	Si-P+R	=UMA 2N (Typ-Code/Stempel/marking)	45(2mm)	Rhm	→UMA 1N...11N	data
A 2	Si-Di	=ZC 833A (Typ-Code/Stempel/marking)	35	Fer	→ZC 833A	data
A 2	Si-Di	=1N4148W (Typ-Code/Stempel/marking)	71(2,7mm)	Gsi	→1N4148W	data
A 2	Si-Di	=1N4148WS (Typ-Code/Stempel/marking)	71(1,7mm)	Gsi	→1N4148WS	data
A 2	Si-Di	=1SS309 (Typ-Code/Stempel/marking)	45	Tos	→1SS309	data
A2V4 PH	Z-Di	=BZX 79-A2V4(Typ-Code/Stempel/marking)	31	Phi	→BZX 79-A...	data
A2V7 PH	Z-Di	=BZX 79-A2V7(Typ-Code/Stempel/marking)	31	Phi	→BZX 79-A...	data
A 2 X	Si-Di	=MMBD 2836 (Typ-Code/Stempel/marking)	35	Ons	→MMBD 2836	
A 3	PIN-Di	=BAP 64-03 (Typ-Code/Stempel/marking)	71(1,7mm)	Phi	→BAP 64-03	data
A 3	Si-Di	=BAS 16 (Typ-Code/Stempel/marking)	35	Fer	→BAS 16	data
A 3	Si-Di	=BAT 17 (Typ-Code/Stempel/marking)	35	Phi,Val	→BAT 17	data
A 3	Si-P+R	=FMA 3A (Typ-Code/Stempel/marking)	45	Rhm	→FMA 3A	data
A 3	Si-Di	=HN 1D02F (Typ-Code/Stempel/marking)	46	Tos	→HN 1D02F	data
A 3	Si-Di	=HN 1D02FU (Typ-Code/Stempel/marking)	46(2mm)	Tos	→HN 1D02FU	data
A 3	Si-Di	=KDS 120 (Typ-Code/Stempel/marking)	35(2mm)	Kec	→KDS 120	data

Type	Device	Short Description	Fig.	Manu	Comparision Types	More at
A 3	Si-Di	=KDS 120E (Typ-Code/Stempel/marking)	35(1,6mm)	Kec	→KDS 120E	data
A 3	Si-Di	=KDS 181 (Typ-Code/Stempel/marking)	35	Kec	→KDS 181	data
A 3	Si-Di	=MMBD 2835 (Typ-Code/Stempel/marking)	35	Mot	→MMBD 2835	data
A 3	Z-Di	=MMSZ 12 (Typ-Code/marking)	71(2,7mm)	Ons	→MMSZ 2V4...56	
A 3	MOS-P-FET-e	=NTHS 5441 (Typ-Code/Stempel/marking)	≈8-MDIP	Ons	→NTHS 5441	
A 3	Si-Di	=PMBD 2835 (Typ-Code/Stempel/marking)	35	Phi	→PMBD 2835	data
A 3	MOS-P-FET-e	=Si 2303DS (Typ-Code/Stempel/marking)	35	Six	→Si 2303DS	data
A 3	Si-P+R	UMA 3N (Typ-Code/Stempel/marking)	45(2mm)	Rhm	→UMA 1N...11N	data
A 3	Si-Di	=1PS300 (Typ-Code/Stempel/marking)	35(2mm)	Phi	→1PS300	data
A 3	Si-Di	=1S2835 (Typ-Code/Stempel/marking)	35	Nec	→1S2835	data
A 3	Si-Di	=1SS181 (Typ-Code/Stempel/marking)	35	Tos	→1SS181	data
A 3	Si-Di	=1SS300 (Typ-Code/Stempel/marking)	35(2mm)	Tos	→1SS300	data
A 3	Si-Di	=1SS306 (Typ-Code/Stempel/marking)	44	Tos	→1SS306	data
A 3	Si-Di	=1SS360 (Typ-Code/Stempel/marking)	35(1,6mm)	Tos	→1SS360	data
A 3	Si-Di	=1SS360F (Typ-Code/Stempel/marking)	35(1,6mm)	Tos	→1SS360F	data
A 3 p	Si-Di	=BAT 17 (Typ-Code/Stempel/marking)	35	Phi	→BAT 17	data
A 3 T	Si-Di	=1PS181 (Typ-Code/Stempel/marking)	35	Phi	→1PS181	data
A 3 t	Si-Di	=BAT 17 (Typ-Code/Stempel/marking)	35	Phi	→BAT 17	data
A3V0 PH	Z-Di	=BZX 79-A3V0(Typ-Code/Stempel/marking)	31	Phi	→BZX 79-A...	data
A3V3 PH	Z-Di	=BZX 79-A3V3(Typ-Code/Stempel/marking)	31	Phi	→BZX 79-A...	data
A3V6 PH	Z-Di	=BZX 79-A3V6(Typ-Code/Stempel/marking)	31	Phi	→BZX 79-A...	data
A3V9 PH	Z-Di	=BZX 79-A3V9(Typ-Code/Stempel/marking)	31	Phi	→BZX 79-A...	data
A 3 X	Si-Di	=MMBD 2835 (Typ-Code/Stempel/marking)	35	Ons	→MMBD 2835	
A3X18X	LED	am ⌀5mm I:38lm Vf:2.5V If:350mA		Seo		data pdf pinout
A3X28X	LED	am ⌀5mm I:68lm Vf:2.75V If:700mA		Seo		data pdf pinout
A4	Si-Di	=1SS383 (Typ-Code/Stempel/marking)	44(2mm)	Tos	→1SS383	data
A 4	Si-Di	=BAV 70 (Typ-Code/Stempel/marking)	35	Val,Mot,Sie	→BAV 70	data
A 4	Si-Di	=BAV 70L (Typ-Code/Stempel/marking)	35	Ons	→BAV 70L	
A 4	Si-Di	=BAV 70T (Typ-Code/Stempel/marking)	35(1,6mm)	Phi, Ons	→BAV 70T	data
A 4	Si-Di	=BAV 70W (Typ-Code/Stempel/marking)	35(2mm)	Phi, Ons	→BAV 70W	data
A 4	Si-Di	=DCF 015 (Typ-Code/Stempel/marking)	35(2mm)	Say	→DCF 015	data
A 4	Si-P+R	FMA 4A (Typ-Code/Stempel/marking)	45	Rhm	→FMA 4A	data
A 4	Si-Di	=HN 1D03F (Typ-Code/Stempel/marking)	46	Tos	→HN 1D03F	data
A 4	Si-Di	=HSB 2836 (Typ-Code/Stempel/marking)	35(2mm)	Hit	→HSB 2836	data
A 4	Si-Di	=HSM 2836C (Typ-Code/Stempel/marking)	35	Hit	→HSM 2836C	data
A 4	Si-Di	=MC 2836 (Typ-Code/Stempel/marking)	35	Mit	→MC 2836	data
A 4	Si-Di	=MC 2846 (Typ-Code/Stempel/marking)	35(2mm)	Mit	→MC 2846	data
A 4	Z-Di	=MMSZ 13 (Typ-Code/marking)	71(2,7mm)	Ons	→MMSZ 2V4...56	
A 4	MOS-P-FET-e	=NTHS 5443 (Typ-Code/Stempel/marking)	≈8-MDIP	Ons	→NTHS 5443	
A 4	MOS-N-FET-e	=Si 2304DS (Typ-Code/Stempel/marking)	35	Six	→Si 2304DS	data
A 4	Si-P+R	UMA 4N (Typ-Code/Stempel/marking)	45(2mm)	Rhm	→UMA 1N...11N	data
A 4	Si-Di	=1N4150W (Typ-Code/Stempel/marking)	71(2,7mm)	Gsi	→1N4150W	data
A 4	Si-Di	=1S2836 (Typ-Code/Stempel/marking)	35	Nec	→1S2836	data
A 4	Si-Di	=1SS319 (Typ-Code/Stempel/marking)	44	Tos	→1SS319	data
A 4 p	Si-Di	=BAV 70 (Typ-Code/Stempel/marking)	35	Phi	→BAV 70	data
A 4 s	Si-Di	=BAV 70S (Typ-Code/Stempel/marking)	46(2mm)	Sie	→BAV 70S	data
A 4 s	Si-Di	=BAV 70 (Typ-Code/Stempel/marking)	35	Sie	→BAV 70	data
A 4 s	Si-Di	=BAV 70W (Typ-Code/Stempel/marking)	35(2mm)	Sie	→BAV 70W	data
A 4 t	Si-Di	=BAV 70 (Typ-Code/Stempel/marking)	35	Phi	→BAV 70	data
A 4 t	Si-Di	=BAV 70S (Typ-Code/Stempel/marking)	46(2mm)	Phi	→BAV 70S	data
A4V3 PH	Z-Di	=BZX 79-A4V3(Typ-Code/Stempel/marking)	31	Phi	→BZX 79-A...	data
A4V7 PH	Z-Di	=BZX 79-A4V7(Typ-Code/Stempel/marking)	31	Phi	→BZX 79-A...	data
A 5	Si-Di	=1N4151W (Typ-Code/Stempel/marking)	71(2,7mm)	Gsi	→1N4151W	data
A 5	Si-Di	=1S2837 (Typ-Code/Stempel/marking)	35	Nec	→1S2837	data
A5	Si-Di	=1SS384 (Typ-Code/Stempel/marking)	44(2mm)	Tos	→1SS384	data
A 5	Si-Di	=1SS391 (Typ-Code/Stempel/marking)	44	Tos	→1SS391	data
A 5	PIN-Di	=BAP 51-03 (Typ-Code/Stempel/marking)	71(1,7mm)	Phi	→BAP 51-03	data
A 5	PUT	=BRY 61 (Typ-Code/Stempel/marking)	35	Val	→BRY 61	data
A 5	Si-P+R	FMA 5A (Typ-Code/Stempel/marking)	45	Rhm	→FMA 5A	data
A 5	Si-Di	=HN 2D02FU (Typ-Code/Stempel/marking)	46(2mm)	Tos	→HN 2D02FU	data
A 5	C-Di	=HVC 317B (Typ-Code/Stempel/marking)	71(1,3mm)	Hit	→HVC 317B	data
A 5	Si-Di	=MMBD2837 (Typ-Code/Stempel/marking)	35	Mot, Ons	→MMBD 2837	data
A 5	Z-Di	=MMSZ 15 (Typ-Code/marking)	71(2,7mm)	Ons	→MMSZ 2V4...56	
A 5	MOS-P-FET-e	=NTHS 5445 (Typ-Code/Stempel/marking)	≈8-MDIP	Ons	→NTHS 5445	
A 5	Si-Di	=PMBD2837 (Typ-Code/Stempel/marking)	35	Phi	→PMBD 2837	data
A 5	MOS-P-FET-e	=Si 2305DS (Typ-Code/Stempel/marking)	35	Six	→Si 2305DS	data
A05	Si-Di	=SMBYW02-50(Typ-Code/Stempel/marking)	71(6,4mm)	Tho	→SMBYW 02-...	
A 5	Z-Di	=UDZ 27B (Typ-Code/Stempel/marking)	71(1,7mm)	Rhm	→UDZ 2.0B...36B	data
A 5	Si-P+R	UMA 5N (Typ-Code/Stempel/marking)	45(2mm)	Rhm	→UMA 1N...11N	data
A 5 p	PUT	=BRY 61 (Typ-Code/Stempel/marking)	35	Val	→BRY 61	data
A 5 t	PUT	=BRY 61 (Typ-Code/Stempel/marking)	35	Val	→BRY 61	data
A5V1 PH	Z-Di	=BZX 79-A5V1(Typ-Code/Stempel/marking)	31	Phi	→BZX 79-A...	data
A5V6 PH	Z-Di	=BZX 79-A5V6(Typ-Code/Stempel/marking)	31	Phi	→BZX 79-A...	data
A 5 T6116...6118	PUT	=2N6116...6118	7a		→2N6116...6118	data
A 6	Si-Di	=1S2838 (Typ-Code/Stempel/marking)	35	Nec	→1S2838	data
A 6	Si-Di	=1SS399 (Typ-Code/Stempel/marking)	44	Tos	→1SS399	data
A 6	Si-Di	=BAS 16 (Typ-Code/Stempel/marking)	35	Mot,Sie,Phi	→BAS 16	data
A 6	Si-Di	=BAS 16D (Typ-Code/Stempel/marking)	71(2,7mm)	Gsi	→BAS 16D	data
A 6	Si-Di	=BAS 16H... (Typ-Code/Stempel/marking)	71(1,7mm)	Ons	→BAS 16H	data
A 6	Si-Di	=BAS 16T (Typ-Code/Stempel/marking)	35(1,6mm)	Phi	→BAS 16T	data
A 6	Si-Di	=BAS 16W (Typ-Code/Stempel/marking)	35(2mm)	Phi	→BAS 16W	data
A 6	Si-Di	=BAS 16WS (Typ-Code/Stempel/marking)	71(1,7mm)	Gsi	→BAS 16WS	data
A 6	Si-Di	=BAS 216 (Typ-Code/Stempel/marking)	71(2mm)	Phi	→BAS 216	data
A 6	Si-Di	=BAS 316 (Typ-Code/Stempel/marking)	71(1,7mm)	Phi	→BAS 316	data
A 6	Si-Di	=DCG 015 (Typ-Code/Stempel/marking)	35(2mm)	Say	→DCG 015	data

Type	Device	Short Description	Fig.	Manu	Comparision Types	More at
A 6	Si-P+R	=FMA 6A (Typ-Code/Stempel/marking)	45	Rhm	→FMA 6A	data
A 6	Si-Di	=HN 2S01F (Typ-Code/Stempel/marking)	46	Tos	→HN 2S01F	data
A 6	Si-Di	=HN 2S01FU (Typ-Code/Stempel/marking)	46(2mm)	Tos	→HN 2S01FU	data
A 6	Si-Di	=HSB 2838 (Typ-Code/Stempel/marking)	35(2mm)	Hit	→HSB 2838	data
A 6	Si-Di	=HSM 2838C (Typ-Code/Stempel/marking)	35	Hit	→HSM 2838C	data
A 6	Si-Di	=MC 2838 (Typ-Code/Stempel/marking)	35	Mit	→MC 2838	data
A 6	Si-Di	=MC 2848 (Typ-Code/Stempel/marking)	35(2mm)	Mit	→MC 2848	data
A 6	Si-Di	=MMBD 2838 (Typ-Code/Stempel/marking)	35	Mot	→MMBD 2838	data
A 6	MOS-N-FET-e	=NTHD 5902 (Typ-Code/Stempel/marking)	≈8-MDIP	Ons	→NTHD 5902	
A 6	MOS-N-FET-e	=NTHS 5402 (Typ-Code/Stempel/marking)	≈8-MDIP	Ons	→NTHS 5402	
A 6	Si-Di	=PMBD 2838 (Typ-Code/Stempel/marking)	35	Phi	→PMBD 2838	data
A 6	MOS-N-FET-e	=Si 2306DS (Typ-Code/Stempel/marking)	35	Six	→Si 2306DS	data
A 6	Si-P+R	=UMA 6N (Typ-Code/Stempel/marking)	45(2mm)	Rhm	→UMA 1N...11N	data
A 6 A	Si-P+R	=MMUN 2111 (Typ-Code/Stempel/marking)	35	Mot	→MMUN 2111	data
A 6 B	Si-P+R	=MMUN 2112 (Typ-Code/Stempel/marking)	35	Mot	→MMUN 2112	data
A 6 C	Si-P+R	=MMUN 2113 (Typ-Code/Stempel/marking)	35	Mot	→MMUN 2113	data
A 6 D	Si-P+R	=MMUN 2114 (Typ-Code/Stempel/marking)	35	Mot	→MMUN 2114	data
A 6 E	Si-P+R	=MMUN 2115 (Typ-Code/Stempel/marking)	35	Mot	→MMUN 2115	data
A 6 F	Si-P+R	=MMUN 2116 (Typ-Code/Stempel/marking)	35	Mot	→MMUN 2116	data
A 6 G	Si-P+R	=MMUN 2130 (Typ-Code/Stempel/marking)	35	Mot	→MMUN 2130	data
A 6 H	Si-P+R	=MMUN 2131 (Typ-Code/Stempel/marking)	35	Mot	→MMUN 2131	data
A 6 J	Si-P+R	=MMUN 2132 (Typ-Code/Stempel/marking)	35	Mot	→MMUN 2132	data
A 6 K	Si-P+R	=MMUN 2133 (Typ-Code/Stempel/marking)	35	Mot	→MMUN 2133	data
A 6 L	Si-P+R	=MMUN 2134 (Typ-Code/Stempel/marking)	35	Mot	→MMUN 2134	data
A 6 p	Si-Di	=PMBD 353 (Typ-Code/Stempel/marking)	35	Phi	→PMBD 353	data
A 6 p	Si-Di	=BAS 16 (Typ-Code/Stempel/marking)	35	Phi	→BAS 16	data
A 6 s	Si-Di	=BAS 16W (Typ-Code/Stempel/marking)	35(2mm)	Sie	→BAS 16W	data
A 6 s	Si-Di	=BAS 16 (Typ-Code/Stempel/marking)	35	Sie	→BAS 16	data
A 6 s	Si-Di	=BAS 16S (Typ-Code/Stempel/marking)	46(2mm)	Sie	→BAS 16S	data
A 6 t	Si-Di	=PMBD 353 (Typ-Code/Stempel/marking)	35	Phi	→PMBD 353	data
A6V2 PH	Z-Di	=BZX 79-A6V2(Typ-Code/Stempel/marking)	31	Phi	→BZX 79-A...	data
A6V8 PH	Z-Di	=BZX 79-A6V8(Typ-Code/Stempel/marking)	31	Phi	→BZX 79-A...	data
A 7	Si-Di	=BAS 321 (Typ-Code/Stempel/marking)	71(1,7mm)	Phi	→BAS 321	data
A 7	Si-Di	=BAV 99 (Typ-Code/Stempel/marking)	35	Mot,Val,Sie	→BAV 99	
A 7	Si-Di	=BAV 99L (Typ-Code/Stempel/marking)	35	Ons	→BAV 99L	
A 7	Si-Di	=BAV 99U (Typ-Code/Stempel/marking)	35(2mm)	Rhm	→BAV 99U	data
A 7	Si-Di	=BAV 99W (Typ-Code/Stempel/marking)	35(2mm)	Ons,Phi,Sie	→BAV 99W	data
A 7	Si-P+R	=FMA 7A (Typ-Code/Stempel/marking)	45	Rhm	→FMA 7A	data
A 7	Si-Di	=MC 2840 (Typ-Code/Stempel/marking)	35	Mit	→MC 2840	
A 7	Si-Di	=MC 2850 (Typ-Code/Stempel/marking)	35(2mm)	Mit	→MC 2850	data
A 7	MOS-P-FET-e	=NTHD 5903 (Typ-Code/Stempel/marking)	≈8-MDIP	Ons	→NTHD 5903	
A 7	MOS-P-FET-e	=Si 2307DS (Typ-Code/Stempel/marking)	35	Six	→Si 2307DS	data
A 7	Si-P+R	=UMA 7N (Typ-Code/Stempel/marking)	45(2mm)	Rhm	→UMA 1N...11N	data
A 7	Si-Di	=1SS 123 (Typ-Code/Stempel/marking)	35	Nec	→1SS 123	data
A 7	Si-Di	=1SS402 (Typ-Code/Stempel/marking)	44(2mm)	Tos	→1SS402	
A 7 p	Si-Di	=BAV 99 (Typ-Code/Stempel/marking)	35	Phi	→BAV 99	data
A 7 s	Si-Di	=BAV 99S (Typ-Code/Stempel/marking)	46(2mm)	Sie	→BAV 99S	data
A 7 s	Si-Di	=BAV 99 (Typ-Code/Stempel/marking)	35	Sie	→BAV 99	data
A 7 s	Si-Di	=BAV 99W (Typ-Code/Stempel/marking)	35(2mm)	Sie	→BAV 99W	data
A 7 t	Si-Di	=BAV 99 (Typ-Code/Stempel/marking)	35	Phi	→BAV 99	data
A 7 t	Si-Di	=BAV 756S (Typ-Code/Stempel/marking)	46(2mm)	Phi	→BAV 756S	data
A 7 T6027...6028	PUT	=2N6027...6028	7a		→2N6027...6028	data
A7V5 PH	Z-Di	=BZX 79-A7V5(Typ-Code/Stempel/marking)	31	Phi	→BZX 79-A...	data
A 8	PIN-Di	=BAP 50-03 (Typ-Code/Stempel/marking)	71(1,7mm)	Phi	→BAP 50-03	data
A 8	Si-Di	=BAS 19 (Typ-Code/Stempel/marking)	35	Val, Gsi	→BAS 19	data
A 8	Si-Di	=BAV 19W (Typ-Code/Stempel/marking)	71(2,7mm)	Gsi	→BAV 19W	data
A 8	Si-Di	=BAV 19WS (Typ-Code/Stempel/marking)	71(1,7mm)	Gsi	→BAV 19WS	data
A 8	Si-P+R	=FMA 8A (Typ-Code/Stempel/marking)	45	Rhm	→FMA 8A	data
A 8	Si-Di	=HSM 223C (Typ-Code/Stempel/marking)	35	Hit	→HSM 223C	data
A 8	MOS-N-FET-e	=Si 2308DS (Typ-Code/Stempel/marking)	35	Six	→Si 2308DS	data
A 8	Si-P+R	=UMA 8N (Typ-Code/Stempel/marking)	45(2mm)	Rhm	→UMA 1N...11N	data
A 8 A	Si-N+R	=MMUN 2211 (Typ-Code/Stempel/marking)	35	Mot	→MMUN 2211	data
A 8 B	Si-N+R	=MMUN 2212 (Typ-Code/Stempel/marking)	35	Mot	→MMUN 2212	data
A 8 C	Si-N+R	=MMUN 2213 (Typ-Code/Stempel/marking)	35	Mot	→MMUN 2213	data
A 8 D	Si-N+R	=MMUN 2214 (Typ-Code/Stempel/marking)	35	Mot	→MMUN 2214	data
A 8 E	Si-N+R	=MMUN 2215 (Typ-Code/Stempel/marking)	35	Mot	→MMUN 2215	data
A 8 F	Si-N+R	=MMUN 2216 (Typ-Code/Stempel/marking)	35	Mot	→MMUN 2216	data
A 8 G	Si-N+R	=MMUN 2230 (Typ-Code/Stempel/marking)	35	Mot	→MMUN 2230	data
A 8 H	Si-N+R	=MMUN 2231 (Typ-Code/Stempel/marking)	35	Mot	→MMUN 2231	data
A 8 J	Si-N+R	=MMUN 2232 (Typ-Code/Stempel/marking)	35	Mot	→MMUN 2232	data
A 8 K	Si-N+R	=MMUN 2233 (Typ-Code/Stempel/marking)	35	Mot	→MMUN 2233	data
A 8 L	Si-N+R	=MMUN 2234 (Typ-Code/Stempel/marking)	35	Mot	→MMUN 2234	data
A 8 M	Si-N+R	=MMUN 2235 (Typ-Code/Stempel/marking)	35	Ons	→MMUN 2235	data
A 8 R	Si-N+R	=MMUN 2238 (Typ-Code/Stempel/marking)	35	Ons	→MMUN 2238	data
A 8 U	Si-N+R	=MMUN 2241 (Typ-Code/Stempel/marking)	35	Ons	→MMUN 2241	data
A8V2 PH	Z-Di	=BZX 79-A8V2(Typ-Code/Stempel/marking)	31	Phi	→BZX 79-A...	data
A 9	Si-Di	=1SS294 (Typ-Code/Stempel/marking)	35	Tos	→1SS294	data
A 9	Si-Di	=1SS322 (Typ-Code/Stempel/marking)	35(2mm)	Tos	→1SS322	data
A 9	Ge-Di	=AAY 60 (Typ-Code/Stempel/marking)	35	Val	→AAY 60	data
A 9	Si-P+R	=FMA 9A (Typ-Code/Stempel/marking)	45	Rhm	→FMA 9A	data
A 9	Si-Di	=HSB 123 (Typ-Code/Stempel/marking)	35(2mm)	Hit	→HSB 123	data
A 9	Si-Di	=HSM 123 (Typ-Code/Stempel/marking)	35	Hit	→HSM 123	data
A 9	MOS-P-FET-e	=NTHD 5905 (Typ-Code/Stempel/marking)	≈8-MDIP	Ons	→NTHD 5905	
A 9	MOS-P-FET-e	=Si 2309DS (Typ-Code/Stempel/marking)	35	Six	→Si 2309DS	data
A 9	Si-P+R	=UMA 9N (Typ-Code/Stempel/marking)	45(2mm)	Rhm	→UMA 1N...11N	data

Type	Device	Short Description	Fig.	Manu	Comparision Types	More at
A 9 L	Si-N+R	=XN 1213 (Typ-Code/Stempel/marking)	45	Mat	→XN 1213	data
A9V1 PH	Z-Di	=BZX 79-A9V1(Typ-Code/Stempel/marking)	31	Phi	→BZX 79-A...	data
A 10	Si-P+R	=FMA 10A (Typ-Code/Stempel/marking)	45	Rhm	→FMA 10A	data
A10 PH	Z-Di	=BZX 79-A10 (Typ-Code/Stempel/marking)	31	Phi	→BZX 79-A...	data
A 10	Si-Di	=SMBYW02-100(Typ-Code/Stempel/marking)	71(6,4mm)	Tho	→SMBYW 02-...	data
A 10	Si-P+R	=UMA 10N (Typ-Code/Stempel/marking)	45(2mm)	Rhm	→UMA 1N...11N	data
A 11	Si-P+R	=FMA 11A (Typ-Code/Stempel/marking)	45	Rhm	→FMA 11A	data
A 11	Si-Di	=MMBD 1501A (Typ-Code/Stempel/marking)	35	Nsc	→MMBD 1501A	data
A11 PH	Z-Di	=BZX 79-A11 (Typ-Code/Stempel/marking)	31	Phi	→BZX 79-A...	data
A 11	Si-P+R	=UMA 11N (Typ-Code/Stempel/marking)	45(2mm)	Rhm	→UMA 1N...11N	data
A12 PH	Z-Di	=BZX 79-A12 (Typ-Code/Stempel/marking)	31	Phi	→BZX 79-A...	data
A 13	Si-Di	=1SS220 (Typ-Code/Stempel/marking)	35	Nec	→1SS220	data
A 13	Si-Di	=MMBD 1503A (Typ-Code/Stempel/marking)	35	Nsc	→MMBD 1503A	data
A13 PH	Z-Di	=BZX 79-A13 (Typ-Code/Stempel/marking)	31	Phi	→BZX 79-A...	data
A 14	Si-Di	=MMBD 1504A (Typ-Code/Stempel/marking)	35	Nsc	→MMBD 1504A	data
A 14	Si-Di	=1SS221 (Typ-Code/Stempel/marking)	35	Nec	→1SS221	data
A 14 A...U	Si-Di	GI, contr.av., 25...1000V, 1A, <6µs A=100, B=200, C=300, D=400, E=500, F=50, M=600, N=800, P=1000, U=25V	31a	Gen	BYW 52..56, 1N4245...4249, 1N5059...5062	data
A 15	Si-Di	=MMBD 1505A (Typ-Code/Stempel/marking)	35	Nsc	→MMBD 1505A	data
A 15	Si-Di	=SMBYW02-150(Typ-Code/Stempel/marking)	71(6,4mm)	Tho	→SMBYW 02-...	data
A 15	Si-Di	=1SS222 (Typ-Code/Stempel/marking)	35	Nec	→1SS222	data
A 15 A...U	Si-Di	GI, 25...800V, 3A, <3µs A=100, B=200, C=300, D=400, E=500V F=50, M=600, N=800, U=25V	31a	Gen	BY 251...255, BYW 17/.., 1N5624...5627	data
A15 PH	Z-Di	=BZX 79-A15 (Typ-Code/Stempel/marking)	31	Phi	→BZX 79-A...	data
A 16	C-Di	=ZC 934A (Typ-Code/Stempel/marking)	35	Ztx	→ZC 934A	data
A 16	Si-Di	=1SS223 (Typ-Code/Stempel/marking)	35	Nec	→1SS223	data
A16-H010...H012	Si-Di	→SSi A16-H010...H012	31a			data
A16 PH	Z-Di	=BZX 79-A16 (Typ-Code/Stempel/marking)	31	Phi	→BZX 79-A...	data
A 17	C-Di	=ZC 933A (Typ-Code/Stempel/marking)	35	Ztx	→ZC 933A	data
A 17	N-FET	=2SK436-A17 (Typ-Code/Stempel/marking)	35	Say	→2SK436	data
A 18	N-FET	=2SK436-A18 (Typ-Code/Stempel/marking)	35	Say	→2SK436	data
A18 PH	Z-Di	=BZX 79-A18 (Typ-Code/Stempel/marking)	31	Phi	→BZX 79-A...	data
A 19	N-FET	=2SK436-A19 (Typ-Code/Stempel/marking)	35	Say	→2SK436	data
A19-H025...H034	Si-Di	→SSi A19-H025...H034	31a			data
A 20	N-FET	=2SK1066-20 (Typ-Code/Stempel/marking)	35(2mm)	Say	→2SK1066	data
A 20	N-FET	=2SK436-A20 (Typ-Code/Stempel/marking)	35	Say	→2SK436	data
A20 PH	Z-Di	=BZX 79-A20 (Typ-Code/Stempel/marking)	31	Phi	→BZX 79-A...	data
A 20	Si-Di	=SMBYW02-200(Typ-Code/Stempel/marking)	71(6,4mm)	Tho	→SMBYW 02-...	data
A 21	N-FET	=2SK1066-21 (Typ-Code/Stempel/marking)	35(2mm)	Say	→2SK1066	data
A 21	N-FET	=2SK436-A21 (Typ-Code/Stempel/marking)	35	Say	→2SK436	data
A 22	N-FET	=2SK1066-22 (Typ-Code/Stempel/marking)	35(2mm)	Say	→2SK1066	data
A 22	N-FET	=2SK436-A22 (Typ-Code/Stempel/marking)	35	Say	→2SK436	data
A22 PH	Z-Di	=BZX 79-A22 (Typ-Code/Stempel/marking)	31	Phi	→BZX 79-A...	data
A24 PH	Z-Di	=BZX 79-A24 (Typ-Code/Stempel/marking)	31	Phi	→BZX 79-A...	data
A27 PH	Z-Di	=BZX 79-A27 (Typ-Code/Stempel/marking)	31	Phi	→BZX 79-A...	data
A 27	Si-N-Darl	=PXTA 27 (Typ-Code/Stempel/marking)	39	Phi	→PXTA 27	data
A30 PH	Z-Di	=BZX 79-A30 (Typ-Code/Stempel/marking)	31	Phi	→BZX 79-A...	data
A32 01	Si-Di	=ERA 32-01 (Typ-Code/Stempel/marking)	31a	Fjd	→ERA 32-...	data
A32 02	Si-Di	=ERA 32-02 (Typ-Code/Stempel/marking)	31a	Fjd	→ERA 32-...	data
A33 PH	Z-Di	=BZX 79-A33 (Typ-Code/Stempel/marking)	31	Phi	→BZX 79-A...	data
A36 PH	Z-Di	=BZX 79-A36 (Typ-Code/Stempel/marking)	31	Phi	→BZX 79-A...	data
A39 PH	Z-Di	=BZX 79-A39 (Typ-Code/Stempel/marking)	31	Phi	→BZX 79-A...	data
A43 PH	Z-Di	=BZX 79-A43 (Typ-Code/Stempel/marking)	31	Phi	→BZX 79-A...	data
A 44	Si-Di	=BAV 74 (Typ-Code/Stempel/marking)	35	Phi	→BAV 74	data
A 46	Si-Di	=BAR 46A (Typ-Code/Stempel/marking)	35	Tho	→BAR 46A	data
A47 PH	Z-Di	=BZX 79-A47 (Typ-Code/Stempel/marking)	31	Phi	→BZX 79-A...	data
A 51	Thy	=BRY 62 (Typ-Code/Stempel/marking)	44	Val	→BRY 62	data
A51 PH	Z-Di	=BZX 79-A51 (Typ-Code/Stempel/marking)	31	Phi	→BZX 79-A...	data
A56 PH	Z-Di	=BZX 79-A56 (Typ-Code/Stempel/marking)	31	Phi	→BZX 79-A...	data
A 61	Si-Di	=BAS 28 (Typ-Code/Stempel/marking)	44	Val	→BAS 28	data
A62 PH	Z-Di	=BZX 79-A62 (Typ-Code/Stempel/marking)	31	Phi	→BZX 79-A...	data
A 63	Si-P	=FZTA 63 (Typ-Code/Stempel/marking)	48	Ztx	→FZTA 63	data
A68 PH	Z-Di	=BZX 79-A68 (Typ-Code/Stempel/marking)	31	Phi	→BZX 79-A...	data
A75 PH	Z-Di	=BZX 79-A75 (Typ-Code/Stempel/marking)	31	Phi	→BZX 79-A...	data
A 80	Si-Di	=BAV 20W (Typ-Code/Stempel/marking)	71(2,7mm)	Gsi	→BAV 20W	data
A 81	Si-Di	=BAS 20 (Typ-Code/Stempel/marking)	35	Val, Gsi	→BAS 20	data
A 82	Si-Di	=BAS 21 (Typ-Code/Stempel/marking)	35	Val, Gsi	→BAS 21	data
A 82	Si-Di	=BAV 21W (Typ-Code/Stempel/marking)	71(2,7mm)	Gsi	→BAV 21W	data
A 91	Si-Di	=BAS 17 (Typ-Code/Stempel/marking)	35	Val	→BAS 17	data
A99	FET	Vs:±18V Vo:±10V f:4Mc f:150kc	7Y	Ite		data pinout
A100	HYBRID	Vs:±18V Vu:106dB Vo:±10V f:4Mc	7Y	Ite		data pinout
A 101	Si-P+R	=RA 101S (Typ-Code/Stempel/marking)	40	Say	→RA 101S	data
A101	HYBRID	Vs:±18V Vu:106dB Vo:±10V f:4Mc	7Y	Ite		data pdf pinout
A102	HYBRID	Vs:±18V Vu:106dB Vo:±10V f:4Mc	7Y	Ite		data pinout
A103	HYBRID	Vs:±18V Vu:106dB Vo:±10V f:4Mc	7Y	Ite		data pinout
A 104	Si-P+R	=RA 104S (Typ-Code/Stempel/marking)	40	Say	→RA 104S	data
A109	UNI	Vs:±18V Vu:90dB Vo:±13.2V Vi0:0.52mV	14D	Hfo		data pinout
A 109 C,D	OP-IC	Uni, Serie 109, ±18V, 0...+70°	DIP,DIC	Hfo	→Serie 109	data pinout
A110	COMP	Vs:-7/14V Vu:62dB Vo:TTL Vi0:0.92mV	14D	Hfo		data pinout

Type	Device	Short Description	Fig.	Manu	Comparision Types	More at
A 110 C,D	KOP-IC	Serie 110, 14/7V, 0...+70°	14-DIP,DIC	Hfo	→Serie 110	data pdf pinout
A122	HYBRID	Vs:±20V Vu:100dB Vo:±10V Vi0:1.5mV	9Y	Ite		data pinout
A123	HYBRID	Vs:±20V Vu:±10V Vi0:1.5mV	9Y	Ite		data pinout
A124	HYBRID	Vs:±20V Vu:100dB Vo:±10V Vi0:2mV	9Y	Ite		data pinout
A125	HYBRID	Vs:±20V Vu:100dB Vo:±10V Vi0:2mV	9Y	Ite		data pinout
A126	HYBRID	Vs:±18V Vu:100dB Vo:±10V Vi0:1mV	9Y	Ite		data pinout
A127	HYBRID	Vs:±18V Vu:100dB Vo:±10V Vi0:1mV	9Y	Ite		data pinout
A128	HYBRID	Vs:±18V Vu:108dB Vo:±10V Vi0:1mV	7Y	Ite		data pinout
A130	HYBRID	Vs:±18V Vu:100dB Vo:±10V f:20Mc	9Y	Ite		data pinout
A131	HYBRID	Vs:±18V Vu:100dB Vo:±10V f:20Mc	9Y	Ite		data pinout
A132	HYBRID	Vs:±18V Vu:100dB Vo:±10V Vi0:10mV	9Y	Ite		data pinout
A133	HYBRID	Vs:±18V Vu:100dB Vo:±10V Vi0:10mV	9Y	Ite		data pinout
A136	HYBRID	Vs:±20V Vu:100dB Vo:±10V Vi0:1mV	9Y	Ite		data pinout
A137	HYBRID	Vs:±20V Vu:100dB Vo:±10V Vi0:1mV	9Y	Ite		data pinout
A148	HYBRID	Vs:±18V Vu:104dB Vo:±10V f:15Mc	7Y	Ite		data pinout
A156	HYBRID	Vs:±18V Vu:104dB Vo:±10V f:6Mc	7Y	Ite		data pinout
A157	HYBRID	Vs:±18V Vu:104dB Vo:±10V f:15Mc	7Y	Ite		data pinout
A158	HYBRID	Vs:±18V Vu:104dB Vo:±10V f:15Mc	7Y	Ite		data pinout
A160	HYBRID	Vs:±60V Vu:106dB Vo:±50V f:6Mc	9Y	Ite		data pinout
A180	HYBRID	Vs:±18V Vu:110dB Vo:±10V Vi0:0.1mV	7Y	Ite		data pinout
A180A	HYBRID	Vs:±18V Vu:110dB Vo:±10V Vi0:1mV	7Y	Ite		data pinout
A183	HYBRID	Vs:±18V Vu:106dB Vo:±10V Vi0:3mV	9Y	Ite		data pinout
A185	HYBRID	Vs:±18V Vu:110dB Vo:±10V Vi0:0.25mV	7Y	Ite		data pinout
A186	HYBRID	Vs:±18V Vu:110dB Vo:±10V Vi0:0.1mV	7Y	Ite		data pinout
A188	HYBRID	Vs:±18V Vu:114dB Vi0:0.15mV f:0.4Mc	9Y	Ite		data pinout
A200	HYBRID	Vs:±18V Vu:60dB Vo:±10V f:0.05Mc	10Y	Ite		data pinout
A202	HYBRID	Vs:±18V Vu:60dB f:0.05Mc f:30kc	10Y	Ite		data pinout
A 202 D	LIN-IC	A/w-verst. + Alc f. Recorder	16-DIP	Hfo	TDA 1002A	pdf pinout
A203	HYBRID	Vs:±18V Vu:60dB f:0.05Mc f:30kc	10Y	Ite		data pinout
A 205 D	LIN-IC	NF-E, 20V, 2,2A, >4,5W(15V/4Ω)	12-DIP+a	Hfo	A 210E, TBA 810AS	
A 205 K	LIN-IC	=A 205D: +Kühlkoerper/heat sink	12-DIP+a°	Hfo	A 210K, (TBA 810AS)	
A 208 E,K	LIN-IC	NF-E, 3W		Hfo	(TBA 810)	
A 210 E	LIN-IC	NF-E, 20V, 2,5A, >5W(15V/4Ω)	12-DIP+a	Hfo	TBA 810AS	pdf pinout
A 210 K	LIN-IC	=A 210E: +Kühlkoerper/heat sink	12-DIP+a°	Hfo	(TBA 810AS)	
A 211 D	LIN-IC	NF-E, 15V, 1A, 1W(9V/8Ω)	14-DIP	Hfo	(TAA 611)[10]	pdf pinout
A212	HYBRID	Vs:±18V Vu:60dB Vo:±10V f:1.5Mc	10Y	Ite		data pinout
A213	HYBRID	Vs:±18V Vu:60dB Vo:±10V f:1.5Mc	10Y	Ite		data pinout
A214	HYBRID	Vs:±7.2V Vu:60dB f:15kkc f:20kc	10Y	Ite		data pinout
A 220 D	LIN-IC	Tv-ton-zf + NF-V	14-DIP	Hfo	SN 76620, TBA 120 S	pdf pinout
A221	HYBRID	Vu:160dB Vo:>±10V Vi0:±4mV	9Y	Ite		data pinout
A 223 D	LIN-IC	Tv-ton-zf + NF-V + VC-Signal	14-DIP	Hfo	SN 76622, TBA 120 U	pdf pinout
A 224 D	LIN-IC	→TBA 120T	16-DIP	Hfo	TBA 120T, SN 76623	pdf pinout
A225	HYBRID	Vs:±18V Vu:140dB Vo:±10V f:15Mc	7Y	Ite		data pinout
A 225 D	LIN-IC	FM-ZF, Dem., Afc	18-DIP	Hfo	TDA 1047	pdf pinout
A226	HYBRID	Vs:±18V Vu:140dB Vo:±10V f:15Mc	7Y	Ite		data pinout
A230	HYBRID	Vs:±18V Vu:140dB Vo:±10V f:1Mc	7Y	Ite		data pinout
A 230 D	LIN-IC	CTV-RGB-matrix	16-DIP	Hfo		
A 231 D	LIN-IC	Ctv-rgb-matrix	16-DIP	Hfo	(TBA530)	pdf pinout
A 232 D	LIN-IC	Ctv-rgb-matrix	16-DIP	Hfo	TDA 2532	pdf pinout
A233	HYBRID	Vs:±18V Vu:140dB Vo:±10V f:0.5Mc	7Y	Ite		data pinout
A240	HYBRID	Vs:±18V Vu:140dB Vo:±10V f:1Mc	7Y	Ite		data pinout
A 240 D	LIN-IC	TV-Video-ZF + Agc (PNP-tuner)	16-DIP	Hfo	TDA 440	pdf pinout
A241	HYBRID	Vs:±18V Vu:140dB Vo:±10V f:1Mc	7Y	Ite		data pinout
A 241 D	LIN-IC	CTV-video-ZF + Afc + Agc (pnp-tuner)	16-DIP	Hfo	TDA 2541	pdf pinout
A242	OP	Vs:±18V Vo:±10V Vi0:15µµV f:0.2Mc	7Y	Ite		data pinout
A 244 D	LIN-IC	AM-Empfänger/receiver	16-DIP	Hfo	TCA 440	pdf pinout
A245	HYBRID	Vs:±18V Vu:140dB Vo:±10V f:1Mc f:3kc	7Y	Ite		data pinout
A 250 D	LIN-IC	TV-HA-synchr. (f. Transistor-e)	14-DIP	Hfo	TBA 950	pdf pinout
A 252 D	LIN-IC	Tv-ha-synchr. (f. Thyristor-E)	14-DIP	Hfo	TBA 940(A)	
A 255 D	LIN-IC	TV-Horiz. + Vertik.-synchr. + Osc.	16-DIP	Hfo	TDA 2593	pdf pinout
A 270 D	LIN-IC	Ctv, Video-Signal	16-DIP	Hfo	TBA 970	pdf pinout
A 273 C,D	LIN-IC	2x NF-Stereo-Potentiometer (Vol.+Bal.)	16-DIP	Hfo	TCA 730	pdf pinout
A 274 D	LIN-IC	NF-Klangreg./AF DC tone-control	16-DIP	Hfo	TCA 740	pdf pinout
A 277 D	LIN-IC	LED-Display-Encoder, 12 LED	18-DIP	Hfo	(UAA 180, UL 1980)	pdf pinout
A 281 D	LIN-IC	AM-FM-ZF/IF-Empfängerschaltkreis/ Amplifier	14-DIP	Hfo	TAA 981	pdf pinout
A 283 D	LIN-IC	AM-V, AM/FM-ZF+NF, >0,3W(5,5V/8Ω)	16-DIP	Hfo	HA 12402, KA 22424, TA 7613AP, TDA 1083,	pdf pinout
A 290 D	LIN-IC	PLL-MPX-stereo-decoder	14-DIP	Hfo	CA 1310, LM 1310, MC 1310, SN 76115	pdf pinout
A 295 D	LIN-IC	CTV-SECAM-schalter/switch	16-DIP	Hfo	-	
A300	HYBRID	Vs:150V Vu:100dB Vo:115V Vi0:3mV	9Y	Ite		data pinout
A301	HYBRID	Vs:150V Vu:100dB Vo:115V Vi0:3mV	9Y	Ite		data pinout
A 301 D	LIN-IC	Schwellwertschalter/threshold switch	14-DIP	Hfo	-	pdf
A 301 V	LIN-IC	=A 301D: Fig.→	8-DIP	Hfo		pdf
A302	HYBRID	Vs:150V Vu:100dB Vo:115V Vi0:3mV	9Y	Ite		data pinout
A 302 D	LIN-IC	Schwellwertschalter/threshold switch	4-DIP	Hfo	TCA345	pdf pinout
A303	HYBRID	Vs:150V Vu:100dB Vo:115V Vi0:3mV	9Y	Ite		data pinout
A 306	Si-Di	=STTA 306B (Typ-Code/Stempel/marking)	30	Tho	→STTA 306B	data
A 321 G	LIN-IC	Camera-prozessor		Hfo	-	pdf
A440	HYBRID	Vs:±18V Vu:100dB Vo:±10V f:20Mc	9Y	Ite		data pinout
A 705 N	IC	High Power advanced Current regulator	43	Adm	-	pdf pinout
A 705 S	IC	High Power advanced Current regulator	43	Adm	-	pdf pinout
A 708 Y	IC	3 channels 20mA advanced Current regulator	43	Adm	-	pdf pinout
A 902 D	LIN-IC	Schwellwertschalter/threshold switch	4-DIP	Hfo	-	
A 910 D	LIN-IC	Transistorkombination f. Camera	14-DIP	Hfo	-	

Type	Device	Short Description	Fig.	Manu	Comparision Types	More at
A1005	HYBRID	Vs:±18V Vu:88dB Vo:±10 Vi0:6mV	7Y	Ite		data pinout
A1026	HYBRID	Vs:±18V Vu:108dB Vo:±10V Vi0:1mV	7Y	Ite		data pinout
A1027	HYBRID	Vs:±18V Vu:108dB Vo:±10V Vi0:1mV	7Y	Ite		data pinout
A1060	LDR	610nm Ro:33MOhm Vb:2V	TO18	Hei		data pinout
A 1524 D	LIN-IC	→TDA 1524A	18-DIP	Hfo	TDA 1524A	pdf pinout
A 1670 V,V1	LIN-IC	TV, VA-system, VA-E	15-SQL	Hfo	TDA 1670, TDA 1675	pdf pinout
A 1818 D	LIN-IC	Recorder-A/W-Verst. + Alc	20-DIP	Hfo	LM 1818	pdf pinout
A 2000 V	LIN-IC	2x AF/ NF- PAmp, 28V, 2.5A, 2x2, 8W(9V/ 2Ohm)	11-SQL	Hfo	A 2005V, TDA 2005	pdf
A 2005 V	LIN-IC	2x AF/NF-PAmp, 28V, 3.5A, 2x>9W(9V/2Ohm)	11-SQL	Hfo	TDA 2005	pdf
A 2014 DC	LIN-IC	Videosignalumschalter/video sign. sw.	8-DIP	Hfo	TEA 2014	pdf pinout
A 2030(H,V)	LIN-IC	NF-E, ±18V, 3,5A, >16W(±14V/4Ω)	17/5Pin	Hfo	TDA 2030(H,V)	pdf pinout
A 2525EL	IC	SMD, USB Power Control Switch, high-side, active high, −40... +85°C	MDIP	All		pdf pinout
A 2526EL	IC	SMD, dual USB Power Control Switch, high-side, active high, −40...+85°C	MDIP	All		pdf pinout
A 2535EL	IC	SMD, USB Power Control Switch, high-side, active low, −40... +85°C	MDIP	All		pdf pinout
A 2536EL	IC	SMD, dual USB Power Control Switch, high-side, active low, −40...+85°C	MDIP	All		pdf pinout
A 2918 SWH	IC	Dual full-bridge PWM motor driver	SIL	All	-	pdf pinout
A 2918 SWV	IC	Dual full-bridge PWM motor driver	SQL	All	-	pdf pinout
A 2927 SEB	IC	Dual full-bridge PWM motor driver	PLCC	All	-	pdf pinout
A3000	UNI	Vs:±18V Vu:100dB Vo:±10V Vi0:4mV	8A	Ite		data pinout
A3001	UNI	Vs:±18V Vu:100dB Vo:±10V Vi0:4mV	8A	Ite		data pinout
A3002	UNI	Vs:±18V Vu:100dB Vo:±10V Vi0:4mV	8A	Ite		data pinout
A3003	UNI	Vs:±18V Vu:100dB Vo:±10V Vi0:2mV	8A	Ite		data pinout
A 3048 DC	LIN-IC	IR-FB-empfänger/receiver	16-DIP	Hfo	TDA 3048	pdf pinout
A-3101 B	Hybrid-IC	NF-ra	≈7-SIP		-	
A-3103 C	Hybrid-IC	NF-ra	≈7-SIP		-	
A-3104 C	Hybrid-IC	NF-ra	≈7-SIP		-	
A-3133 B	Hybrid-IC	NF-ra	≈7-SIP		-	
A3240A	2xBIMOS	Vs:±18V Vi0:2mV f:4.5Mc	8D	Har	-	data pdf pinout
A 3340 ELHLT-T	Sensor	Chopper-Stabilized, Hall-effect Switch	QFP	All	-	pdf pinout
A 3340 EUA-T	Sensor	Chopper-Stabilized, Hall-effect Switch	QFP	All	-	pdf pinout
A 3340 LLHLT-T	Sensor	Chopper-Stabilized, Hall-effect Switch	QFP	All	-	pdf pinout
A 3340 LUA-T	Sensor	Chopper-Stabilized, Hall-effect Switch	QFP	All	-	pdf pinout
A 3423 EK-T	Sensor	Dual Channel Hall Effect Direction Detection Sensor	QFP	All	-	pdf pinout
A 3423 ELTR-T	Sensor	Dual Channel Hall Effect Direction Detection Sensor	QFP	All	-	pdf pinout
A 3423 LK-T	Sensor	Dual Channel Hall Effect Direction Detection Sensor	QFP	All	-	pdf pinout
A 3423 LLTR-T	Sensor	Dual Channel Hall Effect Direction Detection Sensor	QFP	All	-	pdf pinout
A 3501 D	LIN-IC	CTV-RGB + Video-E	28-DIP	Hfo	TDA 3501	pdf pinout
AAT 3510	µP-Periph	microprocessor reset circuit	FLP	Alt		pdf pinout
A 3510 D	LIN-IC	CTV-PAL-decoder	24-DIP	Hfo	TDA 3510	pdf
AAT 3511	µP-Periph	microprocessor reset circuit	FLP	Alt	-	pdf pinout
AAT 3512	µP-Periph	microprocessor reset circuit	FLP	Alt	-	pdf pinout
AAT 3513	µP-Periph	microprocessor reset circuit	FLP	Alt	-	pdf pinout
AAT 3514	µP-Periph	microprocessor reset circuit	FLP	Alt	-	pdf pinout
AAT 3515	µP-Periph	microprocessor reset circuit	FLP	Alt	-	pdf pinout
AAT 3516	µP-Periph	microprocessor reset circuit	FLP	Alt	-	pdf pinout
AAT 3517	µP-Periph	microprocessor reset circuit	FLP	Alt	-	pdf pinout
AAT 3518	µP-Periph	microprocessor reset circuit	FLP	Alt	-	pdf pinout
AAT 3519	µP-Periph	microprocessor reset circuit	FLP	Alt	-	pdf pinout
AAT 3520	Logic-IC	Microprocessor Reset Circuit	FLP	Alt		pdf pinout
A 3520 D	LIN-IC	CTV-SECAM-decoder	28-DIP	Hfo	TDA 3520	
AAT 3522	Logic-IC	Microprocessor Reset Circuit	FLP	Alt		pdf pinout
AAT 3524	Logic-IC	Microprocessor Reset Circuit	FLP	Alt		pdf pinout
AAT 3526	Logic-IC	Supervisory Circuit with Manual Reset	FLP	Alt		pdf pinout
AAT 3527	Logic-IC	Supervisory Circuit with Manual Reset	FLP	Alt		pdf pinout
AAT 3528	Logic-IC	Supervisory Circuit with Manual Reset	FLP	Alt		pdf pinout
AAT 3532	Logic-IC	Microprocessor Reset Circuit	FLP	Alt		pdf pinout
AAT 3560	Z-IC	Voltage Detector for portabel equipment	FLP	Alt		pinout
AAT 3562	Z-IC	Voltage Detector for portabel equipment	FLP	Alt		pinout
AAT 3564	Z-IC	Voltage Detector for portabel equipment	FLP	Alt		pinout
A 3904 ECGTR	IC	Low Voltage Voice Coil Motor Driver	QFP	All		pdf pinout
A 3904 ECW	IC	Low Voltage Voice Coil Motor Driver	QFP	All		pdf pinout
A 3904 EEWTR-T	IC	Low Voltage Voice Coil Motor Driver	QFP	All		pdf pinout
A 3951SB	LIN-IC	Full bridge PWM motor driver, Iout<2A	DIP	All	A3952SB	pdf pinout
A 3951SW	LIN-IC	Full bridge PWM motor driver, Iout<2A	SIL	All	A3952SW	pdf pinout
A 3953SB	IC	Full-bridge PWM motor drv., ±1.3A 50V	DIP	All	-	pdf pinout
A 3953SLB	IC	SMD, Full-bridge PWM motor drv., ±1.3A	MDIP	All	-	pdf pinout
A 3971 SLB	DMOS-IC	SMD, dual Dmos full-bridge driver for inductive loads, 2.5A 50V, -20...+85°C	24-MDIP	All	-	pdf pinout
A 3973SB	IC	Dual Dmos Full Bridge Microstepping PWM Motor Driver, ±1A 35V	DIP	All	-	pdf pinout
A 3973SLB	IC	SMD, Dual Dmos Full Bridge Microstepping PWM Motor Driver, ±1A 35V	MDIP	All		pdf pinout
A 3977 KED	IC	Microstepping motor drv., translator	PLCC	All	A3979	pdf pinout
A 3977 KLP	IC	SMD, Microstepp. motor drv, translator	28-SSMDIP	All	A3979	pdf pinout
A 3977 SED	IC	Microstepping motor drv., translator	PLCC	All	A3979	pdf pinout
A 3977 SLP	IC	SMD, Microstepp. motor drv., translator	28-SSMDIP	All	A3979	pdf pinout

Type	Device	Short Description	Fig.	Manu	Comparision Types	More at
A 3979 SLP	IC	SMD, Microstepp. motor drv, translator	28-SSMDIP	All		pdf pinout
A 4100 D	LIN-IC	AM-Radio, FM-ZF	22-DIP	Hfo	TDA 4100	pdf pinout
A 4401 KL-T	IC	Automotive Vacuum Fluorescent Display Power Supply	QFP	All	-	pdf pinout
A 4401 KLTR-T	IC	Automotive Vacuum Fluorescent Display Power Supply	QFP	All		pdf pinout
A 4510 D	LIN-IC	→TCA 4510	18-DIP	Hfo	TCA 4510	pdf pinout
A 4511 D	LIN-IC	→TCA 4511	18-DIP	Hfo	TCA 4511	pdf pinout
A 4555 DC	LIN-IC	→TDA 4555	28-DIP	Hfo	TDA 4555	pdf
A 4565 DC	LIN-IC	→TDA 4565	18-DIP	Hfo	TDA 4565	pdf pinout
A 4580 DC	LIN-IC	→TDA 4580	28-DIP	Hfo	TDA 4580	pdf pinout
A 4931 METTR-T	IC	3-Phase DC Motor Predriver	QFP	All	-	pdf pinout
A 5970 D	LIN/Z-IC	1.5A switch step down switching regulator	MDIP	Stm	-	pdf pinout
A 5972 D	LIN/Z-IC	2A switch step down switching regulator	MDIP	Stm	-	pdf pinout
A 5973 AD	LIN/Z-IC	2A switch step down switching regulator	MDIP	Stm	-	pdf pinout
A 5973 D	LIN/Z-IC	2.5A switch step down switching regulator	MDIP	Stm	-	pdf pinout
AU 6331 /LQFP	DIG-IC	Card Reader Controller	MP		-	pdf pinout
A 6810EA	DMOS-IC	DABiC-IV,10Bit serial input, latched source driver, -40°C to +85°C	DIP	All		pdf pinout
A 6810ELW	DMOS-IC	SMD, DABiC-IV,10Bit serial input, latched source driver, -40°C to +85°C	MDIP	All		
A 6810SA	DMOS-IC	DABiC-IV,10Bit serial input, latched source driver, -20°C to +85°C	DIP	All		pdf pinout
A 6810SLW	DMOS-IC	SMD, DABiC-IV,10Bit serial input, latched source driver, -20°C to +85°C	MDIP	All		pdf pinout
A 6902 D	LIN/Z-IC	Up to 1A step down switching regulator, adj. current limit	MDIP	Stm	-	pdf pinout
A7060	LDR	610nm Ro:33MOhm Vb:2V		Hei		data pinout
A 8285 SLB	IC	Lnb Supply & Control Voltage Reg.	MDIP	All	-	pdf pinout
A 8287 SLB	IC	Lnb Supply & Control Voltage Reg.	MDIP	All	-	pdf pinout
A 8481 EEJTR-T	IC	Dual Output Regulator, Single Inductor	QFP	All	-	pdf pinout
A 8504 EECTR-T	IC	Wled/rgb Backlight Driver, Medium Size LCDs	QFP	All	-	pdf pinout
A9060	LDR	610nm Ro:100MOhm Vb:2V		Hei		data pinout
A 9903	Diac	Ub=28...36V, Ib=0,4<1mA, Itsm=1A	31	Sie	D 32	data
A-DV 04	Hybrid-IC	Relaistreiber/relay driver		Fui		data
A-Q	Si-N	=2PD1820AQ (Typ-Code/Stempel/marking)	35(2mm)	Phi	→2PD1820A	data
A-R	Si-N	=2PD1820AR (Typ-Code/Stempel/marking)	35(2mm)	Phi	→2PD1820A	data
A-S	Si-N	=2PD1820AS (Typ-Code/Stempel/marking)	35(2mm)	Phi	→2PD1820A	data

AA

Type	Device	Short Description	Fig.	Manu	Comparision Types	More at
AA	Si-P	=2SA1415 (Typ-Code/Stempel/marking)	39	Say	→2SA1415	data
AA	Si-P	=2SA1864 (Typ-Code/Stempel/marking)	35(1,6mm)	Say	→2SA1864	data
AA	Si-N	=2SC4213A (Typ-Code/Stempel/marking)	35(2mm)	Tos	→2SC4213A	data
AA	Si-N	=2SD1366-AA (Typ-Code/Stempel/marking)	39	Hit	→2SD1366	data
AA	Si-P	=BCP 51 (Typ-Code/Stempel/marking)	48	Phi	→BCP 51	data
AA	Si-N	=BCW 60A (Typ-Code/Stempel/marking)	35	Mot,Phi,Sie Aeg,Fer,++	→BCW 60A	data
AA	Si-P	=BCX 51 (Typ-Code/Stempel/marking)	39	Mot,Sie,Val	→BCX 51	data
AA	Si-P	=CPH 3101 (Typ-Code/Stempel/marking)	35	Say	→CPH 3101	data
AA	Z-Di	=MA 5Z200-L (Typ-Code/Stempel/marking)	71(1,7mm)	Mat	→MA 5Z200-L	data
AA	MOS-N-FET-e	=MTSF 3N03HD(Typ-Code/Stempel/marking)	8-SSMDIP	Ons	→MTSF 3N03HD	
AA	Z-Di	=SMAJ 5.0CA (Typ-Code/Stempel/marking)	71(5x2,5)	Tho	→SMAJ ...	data
AA	Si-N+R	=XN 6214 (Typ-Code/Stempel/marking)	46	Mat	→XN 6214	data
AA	Si-N+R	=XP 6214 (Typ-Code/Stempel/marking)	46(2mm)	Mat	→XP 6214	data
AA	C-Di	=ZMV 829A (Typ-Code/Stempel/marking)	71(1,7mm)	Ztx	→ZMV 829	data
AA	Si-P	=µPA501T (Typ-Code/Stempel/marking)	45	Nec	→µPA501T	data
AA	Si-P	=µPA571T (Typ-Code/Stempel/marking)	45(2mm)	Nec	→µPA571T	data
AAp	Si-N	=BCW 60A (Typ-Code/Stempel/marking)	35	Phi	→BCW 60A	data
AAs	Si-N	=BCW 60A (Typ-Code/Stempel/marking)	35	Phi	→BCW 60A	data
AA 1	Si-N	=2SC4942-AA1(Typ-Code/Stempel/marking)	39	Nec	→2SC4942	data
AA 1 A3Q	Si-N+R	S, Rb=1k, Rbe=10kΩ, 60V, 0,1A, 0,25W	7c	Nec	DTC 113ZS, UN 4219	data
AA 1 A4M	Si-N+R	=AA 1A3Q: Rb=Rbe=10kΩ	7c	Nec	DTC 114ES, RN 1002, UN 4211, 2SC3402,++	data
AA 1 A4P	Si-N+R	=AA 1A3Q: Rb=10k, Rbe=47kΩ	7c	Nec	DTC 114YS, RN 1007, UN 4214, 2SC4048,++	data
AA 1 A4Z	Si-N+R	=AA 1A3Q: Rb=10kΩ, Rbe=-	7c	Nec	DTC 114TS, RN 1011, UN 4215, 2SC3860,++	data
AA 1 F4M	Si-N+R	=AA 1A3Q: Rb=Rbe=22kΩ	7c	Nec	DTC 124ES, RN 1003, UN 4212, 2SC3654,++	data
AA 1 F4N	Si-N+R	=AA 1A3Q: Rb=22k, Rbe=47kΩ	7c	Nec	DTC 124XS, KSR 1007, RN 1008	data
AA 1 F4Z	Si-N+R	=AA 1A3Q: Rb=22k, Rbe=-	7c	Nec	DTC 124TS, KSR 1011, UN 4217, 2SC4121	data
AA 1 L3M	Si-N+R	=AA 1A3Q: Rb=Rbe=4,7kΩ	7c	Nec	DTC 143ES, RN 1001, UN 421L, 2SC4363,++	data
AA 1 L3N	Si-N+R	=AA 1A3Q: Rb=4,7k, Rbe=10kΩ	7c	Nec	DTC 143XS, KSR 1005, UN 421F, 2SC4361	data
AA 1 L3Z	Si-N+R	=AA 1A3Q: Rb=4,7k, Rbe=-	7c	Nec	DTC 143TS, RN 1010, UN 4216, 2SC3901,++	data
AA 1 L4L	Si-N+R	=AA 1A3Q: Rb=47k, Rbe=22kΩ	7c	Nec	DTC 144WS, RN 1009, UN 421E, 2SC3401,++	data
AA 1 L4M	Si-N+R	=AA 1A3Q: Rb=Rbe=47kΩ	7c	Nec	DTC 144ES, RN 1004, UN 4213, 2SC3399,++	data
AA 1 L4Z	Si-N+R	=AA 1A3Q: Rb=47k, Rbe=-	7c	Nec	DTC 144TS, KSR 1012, UN 4210, 2SC3899,++	data
AA 2	Si-N	=2SC4942-AA2(Typ-Code/Stempel/marking)	39	Nec	→2SC4942	data
AA 3	Si-N	=2SC4942-AA3(Typ-Code/Stempel/marking)	39	Nec	→2SC4942	data
AA 103	Ge-Di	Uni, 70V, 10mA	31a	Eiy	AA 117, AA 118, AA 133, 1N54(A)	data
AA 110	Ge-Di	Dem, 22,5V, 15mA	31a	Eiy	AA 112, AA 116, AA 119, 1N60	data
AA 111	Ge-Di	Dem, h-ohm, 40V, 10mA	31a	Aeg	AA 113, AA 119, 1N60	data

Type	Device	Short Description	Fig.	Manu	Comparision Types	More at
AA 112	Ge-Di	Dem, n-ohm, 20V, 30mA	31a	EUR	AA 114, AA 116, 1N60	data
AA 113	Ge-Di	Dem, h-ohm, 65V, 25mA	31a	EUR	AA 113, 1N60	data
AA 114	Ge-Di	Dem, n-ohm, 30V, 40mA	31a	Tho	AA 116, 1N60	data
AA 115	Ge-Di	Dem, h-ohm, 45V, 35mA	31a	Sie	AA 113, AA 119, 1N60	data
AA 116 [EUR]	Ge-Di	Dem, n-ohm, 30V, 30mA	31a	EUR	AA 114, AA 138, 1N60	data
AA 116 [Eiy]	Ge-Di	Uni, 70V, 8mA	31a	Eiy	AA 117, AA 118, AA 133, 1N54(A)	data
AA 117	Ge-Di	Uni, 115V, 50mA	31a	EUR	AA 118, AA 133	data
AA 118	Ge-Di	Uni, 115V, 50mA	31a	EUR	AA 117, AA 133	data
AA 119	Ge-Di	Dem, h-ohm, 45V, 35mA	31a	EUR	AA 113, 1N60	data
AA 120 [Aei]	Ge-Di	Stabi, 0,26...0,32V(8,5mA)	2c	Aei	-	data
AA 120 [Eiy]	Ge-Di	Uni, 70V, 8mA	31a	Eiy	AA 117, AA 118, AA 133, 1N54(A)	data
AA 121 [Eiy]	Ge-Di	Uni, 70V, 10mA	31a	Eiy	AA 117, AA 118, AA 133, 1N54(A)	data
AA 121 [Tho]	Ge-Di	Dem, n-ohm, 25V, 30mA	31a	Tho	AA 114, AA 116, 1N60	data
AA 123	Ge-Di	Dem, n-ohm, 24V, 30mA	31a	Tho	AA 114, AA 116, 1N60	data
AA 127	Ge-Di	Uni, 95V, 15mA	31a	Eiy	AA 117, AA 118, AA 133	data
AA 129	Ge-Di	Ifm=20mA, Uf<0,23V(5mA)	1c	Phi	-	data
AA 130 [Tho]	Ge-Di	Dem, n-ohm, 15V, 20mA	31a	Tho	AA 112, AA 114, AA 116, AA 138, 1N60	data
AA 130 [Eiy]	Ge-Di	Uni, 115V, 15mA	31a	Eiy	AA 117, AA 118, AA 133	data
AA 131 [Eiy]	Ge-Di	Uni, 115V, 15mA	31a	Eiy	AA 117, AA 118, AA 133	data
AA 131 [Tho]	Ge-Di	Uni, n-ohm, 40V, 20mA	31a	Tho	AA 114, AA 116, AA 138, 1N60	data
AA 132	Ge-Di	Uni, 110V, 50mA	31a	Aeg,Phi,Tho	AA 133	data
AA 133	Ge-Di	=AA 132: 140V	31a	Aeg,Phi,Tho	1N39	data
AA 134	Ge-Di	=AA 132: 70V	31a	Aeg,Tho	AA 132, AA 133	data
AA 135	Ge-Di	Uni, n-ohm, 30V, 150mA	31a	Aeg, Tho	AA 136, AA 139	data
AA 136	Ge-Di	=AA 135: 60V	31a	Aeg	1N270	data
AA 137	Ge-Di	TV-AGC, n-ohm, 40V, 20mA	31a	Aeg, Tho	AA 114, AA 116, AA 138, 1N60	data
AA 138	Ge-Di	TV-Dem, 25V, 20mA	31a	Aeg, Tho	AA 112, AA 114, AA 116, 1N60	data
AA 139	Ge-Di	Uni, n-ohm, 20V, 200mA	31a	Aeg,Phi	1N270	data
AA 140 [Aeg]	Ge-Di	Dem, n-ohm, 32V, 20mA	9c	Aeg	AA 114, AA 116, AA 138, 1N60	data
AA 140 [Eiy]	Ge-Di	Uni, 150V, 10mA	31a	Eiy	1N39	data
AA 142	Ge-Di	=AA 140[Aeg]: Min	36d	Aeg	AA 114, AA 116, AA 138, 1N60	data
AA 143(S)	Ge-Di	Dem, n-ohm, 30V, 60mA	31a	Itt, Tho	AA 114, AA 116, AA 138, 1N60	data
AA 144	Ge-Di	Uni, 100V, 45mA	31a	Itt, Tho	AA 117, AA 118, AA 133	data
AA 8227	Dig. Audio	Dual Audio Power Amplifier	QIP_b	Bcd	-	pdf pinout
24 AA 1025 P	EEPROM-IC	1024K I2C Cmos Serial EEPROM	DIP	Mcp	-	pdf pinout
24 AA 1025 SM	EEPROM-IC	1024K I2C Cmos Serial EEPROM	DIP	Mcp	-	pdf pinout
AAA	N-FET	=MMBF 4856 (Typ-Code/Stempel/marking)	35	Mot	→MMBF 4856	data
AAQ	Si-N	=2SD1757K-Q (Typ-Code/Stempel/marking)	35	Rhm	→2SD1757K	data
AAR	Si-N	=2SD1757K-R (Typ-Code/Stempel/marking)	35	Rhm	→2SD1757K	data
AAR3220	R+	Io=>150mA P:500mW	3S	Alt		data pdf pinout
AAR3239	R+	Io=>500mA Vin:2.5V P:1.1W	8S	Alt		data pdf pinout
AAR3244	E2R	Io=>300mA V:0.6...3.6V Vin:3.6V	12S	Alt		data pdf pinout
AAS	Si-N	=2SD1757K-S (Typ-Code/Stempel/marking)	35	Rhm	→2SD1757K	data
AAs	Si-Di	=BAR 80 (Typ-Code/Stempel/marking)	44	Sie	→BAR 80	data
AAs	Si-P	=BCP 51M (Typ-Code/Stempel/marking)	≈45	Sie	→BCP 51M	data
AAT1106	S+	Io=600mA Vo:1.746...1.854V Vin:3.6V	5S	Alt		data pdf pinout
AAT1110	ES+	Io=800mA Vo:0.6...5.5V Vin:5.5V	8S	Alt		data pdf pinout
AAT1112	ES+	Io=1.5A Vo:0.6...5.5V P:2W	12SH	Alt		data pdf pinout
AAT1120	ES+	Io=500mA Vo:0.6...5.5V Vin:5.5V P:2W	8H	Alt		data pdf pinout
AAT1121	ES+	Io=250mA Vo:0.6...5.5V Vin:5.5V P:2W	8H	Alt		data pdf pinout
AAT1123	S+	Io=600mA Vin:3.6V P:625mW	8S	Alt		data pdf pinout
AAT1126	S+	Io=500mA Vin:3.6V P:625mW	5S,8S	Alt		data pdf pinout
AAT1140	ES+	Io=600mA Vo:0.6...5.5V Vin:5.5V	5S	Alt		data pdf pinout
AAT1141	S+	Io=600mA Vin:3.6V P:667mW	5S	Alt		data pdf pinout
AAT1142	S+	Io=800mA Vin:3.6V P:2W	12SH	Alt		data pdf pinout
AAT1143	S+	Io=400mA Vin:3.6V P:625mW	8S	Alt		data pdf pinout
AAT1145	S+	Io=1.5A Vin:3.6V P:2.2W	10H	Alt		data pdf pinout
AAT1146	S+	Io=400mA Vin:3.6V P:625mW	5S,8S	Alt		data pdf pinout
AAT1147	ES+	Io=400mA Vo:0.6...5.5V Vin:5.5V	8S	Alt		data pdf pinout
AAT1149	ES+	Io=400mA Vo:1.0...5.5V Vin:5.5V	6G,8S	Alt		data pdf pinout
AAT1150	S+	Io=1A P:667mW	8S	Alt		data pdf pinout
AAT1151	S+	Io=1A Vin:5V P:883mW	8S,16H	Alt		data pdf pinout
AAT1153	S+	Io=2A Vin:3.6V P:2.2W	10H	Alt		data pdf pinout
AAT1154	S+	Io=3A P:909mW	8S	Alt		data pdf pinout
AAT1155	S+	Io=2.5A P:833mW	8S	Alt		data pdf pinout
AAT1156	ES+	Io=700mA Vo:0.6...5.3V P:2W	16H	Alt		data pdf pinout
AAT1157	ES+	Io=1.2A Vo:0.6...5.3V P:2W	16H	Alt		data pdf pinout
AAT1160	ES+	Io=3A Vo:0.6...12.4V Vin:13.2V P:2.7W	16H	Alt		data pdf pinout
AAT1161	ES+	Io=3A Vo:0.6...12.4V Vin:13.2V P:2W	14H	Alt		data pdf pinout
AAT1162	ES+	Io=1.5A Vo:0.6...12.4V Vin:13.2V	16H	Alt		data pdf pinout
AAT1171	ES+	Io=600mA Vo:0.6...3.6V Vin:3.6V P:2W	12GH	Alt		data pdf pinout
AAT1210	ES+	Io=900mA Vo:3.3...18V Vin:3.6...18V	16H	Alt		data pdf pinout
AAT1217	S+	Vin:1.2...2.4V P:667mW	6S	Alt		data pdf pinout
AAT1230	ES+	Io=100mA Vo:...18V Vin:2.7...5.5V	16H,12S	Alt		data pdf pinout
AAT1230-1	ES+	Io=100mA Vo:...18V Vin:2.7...5.5V	12S	Alt		data pdf pinout

Type	Device	Short Description	Fig.	Manu	Comparision Types	More at
AAT1231	ES+	Io=50mA Vo:...24V Vin:2.7...5.5V	12S	Alt		data pdf pinout
AAT1231-1	ES+	Io=50mA Vo:...24V Vin:2.7...5.5V	12S	Alt		data pdf pinout
AAT1232	ES+	Io=100mA Vo:...24V Vin:2.7...5.5V	16H,12S	Alt		data pdf pinout
AAT1235	E5S	Io=100mA Vo:...24V Vin:2.7...5.5V P:2W	16H	Alt		data pdf pinout
AAT1236	E5S	Io=100mA Vo:...24V Vin:2.7...5.5V P:2W	16H	Alt		data pdf pinout
AAT1239-1	ES+	Io=30mA Vo:...40V Vin:2.7...5.5V	12S	Alt		data pdf pinout
AAT1265	S+	Io=400mA Vo:4.85..5.15V P:625mW	12S	Alt		data pdf pinout
AAT1270	E2S	Io=137mA Vo:...5.5V P:2W	14H	Alt		data pdf pinout
AAT1299	S+	Io=200mA Vo:3.1...3.4V	8S,10H	Alt		data pdf pinout
AAT2120	ES+	Io=500mA Vo:0.6...5.5V Vin:5.5V P:2W	8H	Alt		data pdf pinout
AAT2146	ES+	Io=600mA Vo:0.6...5.5V Vin:5.5V	8S	Alt		data pdf pinout
AAT2158	ES+	Io=1.5A Vo:0.6...5.5V Vin:5.5V P:2W	16H	Alt		data pdf pinout
AAT2500	MRS	Io=300mA Vin:3.9...5.5V P:2W	12H	Alt		data pdf pinout
AAT2500M	MRS	Io=600mA Vo:0.6...2.5V Vin:2.7...5.5V	12S	Alt		data pdf pinout
AAT2503	M2RS	Io=150mA Vo:0.9...5.5V Vin:5.5V P:2W	20H	Alt		data pdf pinout
AAT2504	M2RS	Io=300mA Vo:0.9...5.5V Vin:5.5V P:2W	20H	Alt		data pdf pinout
AAT2505	MRS	Io=400mA Vo:0.6...5.5V Vin:5.5V P:2W	12H	Alt		data pdf pinout
AAT2506	MRS	Io=400mA Vo:0.6...5.5V Vin:5.5V P:2W	12H	Alt		data pdf pinout
AAT2510	S++	Io=400mA Vin:2.7...5.5V P:2W	12H	Alt		data pdf pinout
AAT2511	E2S	Io=400mA Vo:0.6...2.5V Vin:2.7...5.5V	12H	Alt		data pdf pinout
AAT2512	S++	Io=400mA Vin:2.7...5.5V P:2W	12H	Alt		data pdf pinout
AAT2513	E2S	Io=600mA Vo:0.6...2.5V Vin:2.9...5.5V	16H	Alt		data pdf pinout
AAT2514	E2S	Io=600mA Vin:2.5...5.5V P:2.2W	10H	Alt		data pdf pinout
AAT2713	E2S	Io=600mA Vo:0.6...5.5V Vin:5.5V P:2W	16H	Alt		data pdf pinout
AAT2782	MR2S	Io=600mA Vo:0.6...5.5V Vin:5.5V P:2W	16H	Alt		data pdf pinout
AAT2783	MR2S	Io=400mA Vo:0.6...5.5V Vin:5.5V P:2W	16H	Alt		data pdf pinout
AAT2784	E3S	Io=1.5A Vo:0.6...5.5V Vin:2.7...5.5V	16H	Alt		data pdf pinout
AAT2785	MRS	Io=1.5A Vo:0.6...2.5V Vin:2.5...5.5V	16H	Alt		data pdf pinout
AAT2789	E2S	Io=800mA Vo:0.6...5.5V Vin:5.5V P:2W	16H	Alt		data pdf pinout
AAT3200	R+	Io=>250mA Vin:5.5V P:2W	3S	Alt		data pdf pinout
AAT3201	R+	Io=>150mA Vin:5.5V P:667mW	5S	Alt		data pdf pinout
AAT3215	R+	Io=>150mA Vin:5.5V P:526mW	5S,8S	Alt		data pdf pinout
AAT3216	R+	Io=>150mA Vin:5V P:526mW	5S,8S	Alt		data pdf pinout
AAT3218	R+	Io=>150mA Vin:4.5V P:526mW	5S,8S	Alt		data pdf pinout
AAT3221	R+	Io=>150mA Vin:4.5V P:667mW	5S,8S	Alt		data pdf pinout
AAT3222	R+	Io=>150mA Vin:4.5V P:667mW	5S	Alt		data pdf pinout
AAT3223	R+	Io=>250mA Vin:4.3V P:667mW	6S	Alt		data pdf pinout
AAT3236	R+	Io=>300mA Vin:4.6V P:526mW	6S,8S	Alt		data pdf pinout
AAT3237	R+	Io=>300mA Vin:Vout+1V P:667mW	6S,8S	Alt		data pdf pinout
AAT3238	R+	Io=>300mA Vin:Vout+1V P:667mW	6S,8S	Alt		data pdf pinout
AAT3258	R+	Io=>300mA Vin:3.4V P:833mW	6S	Alt		data pdf pinout
AAT4250	LOAD+	Vin:-0.3...6V Il:1.7A Ron:120mOhm	5S,8S	Alt		data pdf pinout
AAT4252A	2xLOAD+	Vin:-0.3...7V Il:1.8A Ron:87mOhm	12S	Alt		data pdf pinout
AAT4280	LOAD+	Vin:-0.3...6V Il:2.3A Ron:80mOhm	6S,8S	Alt		data pdf pinout
AAT4280A	LOAD+	Vin:-0.3...6V Il:2.3A Ron:80mOhm	6S	Alt		data pdf pinout
AAT4282A	2xLOAD+	Vin:-0.3...7V Il:3A Ron:56mOhm	8H	Alt		data pdf pinout
AAT4285	LOAD+	Vin:-0.3...14V Il:1.7A Ron:310mOhm	8S	Alt		data pdf pinout
AAT4290	5xLOAD+	Vin:-0.3...6V Il:250mA Ron:1.1Ohm	8S	Alt		data pdf pinout
AAT4291	3xLOAD+	Vin:-0.3...6V Il:250mA Ron:1.1Ohm	8S	Alt		data pdf pinout
AAT4295	3xLOAD+	Vin:6V Il:250mA Ron:1.9Ohm Ton:0.6µs	8S	Alt		data pdf pinout
AAT4296	5xLOAD+	Vin:-0.3...6V Il:250mA Ron:2.5Ohm	8S	Alt		data pdf pinout
AAT4297	6xLOAD+	Vin:6V Il:250mA Ron:1.9Ohm Ton:0.6µs	12S	Alt		data pdf pinout
AAT4298	6xLOAD+	Vin:-0.3...6V Il:250mA Ron:2.5Ohm	10S,12S	Alt		data pdf pinout
AAT4510A	LOAD+	Vin:-0.3...6V Ilim:1A Il:250mA	5S,8S	Alt		data pdf pinout
AAT4601	LOAD+	Vin:-0.3...6V Ilim:715mA Il:250mA	8S	Alt		data pdf pinout
AAT4601A	LOAD+	Vin:-0.3...6V Ilim:1A Il:250mA	8S	Alt		data pdf pinout
AAT4608	LOAD+	Vin:-0.3...6V Ilim:500mA Il:250mA	5S	Alt		data pdf pinout
AAT4610	LOAD+	Vin:-0.3...6V Ilim:1A Il:250mA	5S	Alt		data pdf pinout
AAT4611	LOAD+	Vin:-0.3...6V Ilim:1A Il:250mA	5S	Alt		data pdf pinout
AAT4612	LOAD+	Vin:-0.3...6V Ilim:1.5A Il:250mA	5S,8S	Alt		data pdf pinout
AAT4618	LOAD+	Vin:-0.3...6V Ilim:1.5A Il:250mA	5S,8S	Alt		data pdf pinout
AAT4625	LOAD+	Vin:-0.3...6V Ilim:1.5A Il:250mA	8S	Alt		data pdf pinout
AAT4650	2SEL+	Vin:-0.3...6V Il:250mA Ron:80mOhm	8S	Alt		data pdf pinout
AAT4651	2SEL+	Vin:-0.3...6V Il:250mA Ron:80mOhm	8S	Alt		data pdf pinout
AAT4670	2xLOAD+	Vin:-0.3...6V Ilim:1.25A Il:250mA	8S,12H	Alt		data pdf pinout
AAT4672	2SEL+	Vin:-0.3...6.5V Il:250mA Ron:120mOhm	12S	Alt		data pdf pinout
AAT4674	2SEL+	Vin:-0.3...6.5V Il:250mA Ron:120mOhm	12S	Alt		data pdf pinout
AAT4680	LOAD+	Vin:-0.3...6V Ilim:825mA Il:250mA	5S	Alt		data pdf pinout
AAT4682	2xLOAD+	Vin:-0.3...6V Ilim:825mA Il:250mA	6S	Alt		data pdf pinout
AAT4684	LOAD+	Vin:-0.3...30V Il:250mA Ron:100mOhm	12S	Alt		data pdf pinout
AAT 1270 IFO-T 1	IC	1A Step-up Current Regulator	QFN	Alt	-	pdf pinout
AAX	Si-N	=MMBTA42 (Typ-Code/Stempel/marking)	35	Kec	→MMBTA 42	data

AAY

Type	Device	Short Description	Fig.	Manu	Comparision Types	More at
AAY 10-120	Ge-Di	GI, 95V, 3,8A	32a	Phi	-	data
AAY 11	Ge-Di	S, 90V, 35mA	31a	Phi	AAY 49, AAZ 15, AAZ 17	data
AAY 12	Ge-Di	Uni, 100V, 115mA	31a	Phi	1N270	data
AAY 13	Ge-Di	S, 25V, 50mA, 500ns	31a	Itt	AAY 49, AAZ 15, AAZ 17	data
AAY 14	Ge-Di	Uni, 100V, 130mA	2c	Sie	1N270	data
AAY 15	Ge-Di	S, 30V, 190mA, 25ns	2c	Sie	AAZ 18	data
AAY 18	Ge-Di	4x Ge-Di, 55V, 30mA		Aeg	-	data

Type	Device	Short Description	Fig.	Manu	Comparision Types	More at
AAY 21	Ge-Di	S, 15V, 20mA, <12ns	31a	Mot, Phi	AAY 48	data
AAY 22	Ge-Di	Dem, n-ohm, 30V, 50mA	31a	Sie	AA 114, AA 116, 1N60	data
AAY 27	Ge-Di	Dem, S, 25V, 75mA, 15ns	31a	Sie	-	data
AAY 28	Ge-Di	Uni, S, 100V, 50mA, 100ns	31a	Sie	-	data
AAY 30	Ge-Di	S, 50V, 110mA, <150ns	31a	Phi	-	data
AAY 32	Ge-Di	S, 30V, 110mA, <50ns	31a	Phi	AAZ 18	data
AAY 33	Ge-Di	S, 15V, 100mA, <12ns	31a	Phi	-	data
AAY 34	Ge-Di	UHF, Q-Band-M, 26...40GHz	Koax	Mot, Phi	-	data
AAY 39(A)	Ge-Di	UHF, X-Band-M, 1...18GHz	Koax	Mot, Phi	-	data
AAY 40	Ge-Di	UHF, X-Band-M, 12GHz	Koax	Phi	-	data
AAY 41	Ge-Di	S, 30V, 300mA, 3500ns	31a	Aeg	(AA 139, 1N270)	data
AAY 42	Ge-Di	Uni, S, 70V, 75mA	31a	Phi	AA 136	data
AAY 43	Ge-Di	Ring-Dem (4xAAY 27)		Sie	-	data
AAY 46	Ge-Di	Dem, 2x2 Ge-Di, 70V, 50mA		Aeg	-	data
AAY 47	Ge-Di	S, 50V, 50mA, 200ns	31a	Tho	AAZ 15, AAZ 17, 1N276	data
AAY 48	Ge-Di	S, 12V, 50mA, <6ns	31a	Tho	-	data
AAY 49	Ge-Di	S, 40V, 150mA, 250ns	31a	Tho	AAZ 15, AAZ 17	data
AAY 50(R)	Ge-Di	UHF, X-Band-M, 12GHz	Koax	Phi	-	data
AAY 51(R)	Ge-Di	UHF, J-Band-M, 12...18GHz	Koax	Phi	-	data
AAY 52(R)	Ge-Di	UHF, J-Band-M, 12...18GHz	Koax	Phi	-	data
AAY 53	Ge-Di	UHF-Dem, 40V	Koax	Sie	-	data
AAY 54	Ge-Di	UHF-Dem, 40V	Koax	Sie	-	data
AAY 55	Ge-Di	UHF-Dem, 40V	Koax	Sie	-	data
AAY 56(R)	Ge-Di	UHF, S-Band-M, 4GHz	Koax	Phi	-	data
AAY 59(M)	Ge-Di	UHF, Q-Band-M, 26...40GHz	Koax	Phi	-	data
AAY 60	Ge-Di	SMD, Stabi, 0,15V(0,05mA)	35n	Phi	-	data

AAZ...AB

Type	Device	Short Description	Fig.	Manu	Comparision Types	More at
AAZ 10	Ge-Di	S, 30V, 30mA, 500ns	31a	Aeg	AAY 49, AAZ 15, AAZ 17	data
AAZ 12	Ge-Di	S, 30V, 220mA	1c	Phi	-	data
AAZ 13	Ge-Di	S, 8V, 30mA	31a	Mot, Phi	-	data
AAZ 14	Ge-Di	Ring-Dem, 4xGe-Di, 30V, 10mA		Aeg	-	data
AAZ 15	Ge-Di	S, 100V, 140mA, 350ns	31a	EUR	-	data
AAZ 17	Ge-Di	S, 75V, 140mA, <350ns	31a	Phi	AAZ 15	data
AAZ 18	Ge-Di	S, 30V, 180mA, <70ns	31a	Phi,Tho		data
AB	Si-P	=2SA1416 (Typ-Code/Stempel/marking)	39	Say	→2SA1416	data
AB	Si-N	=2SC4213B (Typ-Code/Stempel/marking)	35(2mm)	Tos	→2SC4213B	data
AB	Si-N	=2SD1101B (Typ-Code/Stempel/marking)	35	Hit	→2SD1101B	data
AB	Si-N	=2SD1366-AB (Typ-Code/Stempel/marking)	39	Hit	→2SD1366	data
AB	Si-N	=BCW 60B (Typ-Code/Stempel/marking)	35	Mot,Phi,Sie Aeg,Fer,++	→BCW 60B	data
AB	Si-P	=BCX 51-6 (Typ-Code/Stempel/marking)	39	Sie,Val,Mot	→BCX 51	data
AB	Z-Di	=MA 5Z200-H (Typ-Code/Stempel/marking)	71(1,7mm)	Mat	→MA 5Z200-H	data
AB	MOS-P-FET-e	=MTSF 1P02HD(Typ-Code/Stempel/marking)	8-SSMDIP	Ons	→MTSF 1P02HD	data
AB	C-Di	=ZMV 830A (Typ-Code/Stempel/marking)	71(1,7mm)	Ztx	→ZMV 830	data
AB 1	Si-N	=2SD2425-AB1(Typ-Code/Stempel/marking)	39	Nec	→2SD2425	data
AB 1 A3M	Si-N+R	S, Rb=Rbe=1kΩ, 30V, 0,7A, 0,75W	7c	Nec	-	data
AB 1 A4A	Si-N+R	=AB 1 A3M: Rb=-, Rbe=10kΩ	7c	Nec	-	data
AB 1 A4M	Si-N+R	=AB 1 A3M: Rb=10k, Rbe=10kΩ	7c	Nec	-	data
AB 1 F3P	Si-N+R	=AB 1 A3M: Rb=2,2k, Rbe=10kΩ	7c	Nec	-	data
AB 1 J3P	Si-N+R	=AB 1 A3M: Rb=3,3k, Rbe=10kΩ	7c	Nec	-	data
AB 1 L2Q	Si-N+R	=AB 1 A3M: Rb=0,47k, Rbe=4,7kΩ	7c	Nec	-	data
AB 1 L3N	Si-N+R	=AB 1 A3M: Rb=4,7k, Rbe=10kΩ	7c	Nec	-	data
AB 2	Si-N	=2SD2425-AB2(Typ-Code/Stempel/marking)	39	Nec	→2SD2425	data
AB 3	Si-N	=2SD2425-AB3(Typ-Code/Stempel/marking)	39	Nec	→2SD2425	data
ABG	Si-P	=2SA1312-GR (Typ-Code/Stempel/marking)	35	Tos	→2SA1312	data
ABL	Si-P	=2SA1312-BL (Typ-Code/Stempel/marking)	35	Tos	→2SA1312	data
ABp	Si-N	=BCW 60B (Typ-Code/Stempel/marking)	35	Phi	→BCW 60B	data
ABs	Si-N	=BCW 60B (Typ-Code/Stempel/marking)	35	Sie	→BCW 60B	data
ABt	Si-N	=BCW 60B (Typ-Code/Stempel/marking)	35	Phi	→BCW 60B	data
ABX	Si-N	=MMBTA43 (Typ-Code/Stempel/marking)	35	Kec	→MMBTA 43	data

AC

Type	Device	Short Description	Fig.	Manu	Comparision Types	More at
AC	Si-P	=2SA1417 (Typ-Code/Stempel/marking)	39	Say	→2SA1417	data
AC	Si-N	=2SD1101C (Typ-Code/Stempel/marking)	35	Hit	→2SD1101C	data
AC	Si-N	=2SD1366A-AC(Typ-Code/Stempel/marking)	39	Hit	→2SD1366A	data
AC	Si-P	=BCP 51-10 (Typ-Code/Stempel/marking)	48	Phi	→BCP 51	data
AC	Si-N	=BCW 60C (Typ-Code/Stempel/marking)	35	Mot,Phi,Sie Aeg,Fer,xx	→BCW 60C	data
AC	Si-P	=BCX 51-10 (Typ-Code/Stempel/marking)	39	Sie,Val,Mot	→BCX 51	data
AC	Z-Di	=MA 5Z220-L (Typ-Code/Stempel/marking)	71(1,7mm)	Mat	→MA 5Z220-L	data

Type	Device	Short Description	Fig.	Manu	Comparision Types	More at
AC	Si-N	=MT 3S06S (Typ-Code/Stempel/marking)	35(1,6mm)	Tos	→MT 3S06S	data
AC	Si-N	=MT 3S06T (Typ-Code/Stempel/marking)	35(1,6mm)	Tos	→MT 3S06T	data
AC	Si-N	=MT 3S06U (Typ-Code/Stempel/marking)	35(2mm)	Tos	→MT 3S06U	data
AC	Si-N	=MT 4S06 (Typ-Code/Stempel/marking)	44	Tos	→MT 4S06	data
AC	Si-N	=MT 4S06U (Typ-Code/Stempel/marking)	44(2mm)	Tos	→MT 4S06U	data
AC	MOS-N-FET-e	=MTSF 3N02HD(Typ-Code/Stempel/marking)	8-SSMDIP	Ons	→MTSF 3N02HD	
AC	Si-P	=RXT 2907A (Typ-Code/Stempel/marking)	39	Rhm	→RXT 2907A	data
AC	Z-Di	=SMAJ 10CA (Typ-Code/Stempel/marking)	71(5x2,5)	Tho	→SMAJ ...	data
AC	Si-N+R	=XN 1210 (Typ-Code/Stempel/marking)	45	Mat	→XN 1210	data
AC	Si-N+R	=XP 1210 (Typ-Code/Stempel/marking)	45(2mm)	Mat	→XP 1210	data
AC	C-Di	=ZMV 831A (Typ-Code/Stempel/marking)	71(1,7mm)	Ztx	→ZMV 831	data
AC 0V8 BGM	Triac	200V, 0,8A≈(Tc=60°C), Igt/Ih <10/<10mA	7a	Nec	MAC 94A-4, MAC 95A-4, MAC 96A-4	
AC 0V8 DGM	Triac	=AC 0V8BDGM: 400V	7a		MAC 94A-6, MAC 95A-6, MAC 96A-6	
AC 1 A3M	Si-N+R	S, Rb=Rbe=1kΩ, 20V, 2A, 0,75W	7c	Nec	-	data
AC 1 A4A	Si-N+R	=AC 1A3M: Rb=-, Rbe=10kΩ	7c	Nec	-	data
AC 01 DJM	Triac	400V, 1A≈(Tc=49°C), Igt/Ih <10/<10mA	39b	Nec	-	
AC 1 F2Q	Si-N+R	=AC 1A3M: Rb=0,22k, Rbe=2,2kΩ	7c	Nec	-	data
AC 1 F3M	Si-N+R	=AC 1A3M: Rb=Rbe=2,2kΩ	7c	Nec	-	data
AC 1 F3P	Si-N+R	=AC 1A3M: Rb=2,2k, Rbe=10kΩ	7c	Nec	-	data
AC 1 L2N	Si-N+R	=AC 1A3M: Rb=0,47k, Rbe=1kΩ	7c	Nec	-	data
AC 1 L2Q	Si-N+R	=AC 1A3M: Rb=0,47k, Rbe=4,7kΩ	7c	Nec	-	data
AC 2 A4A	Si-N+Di+R	S, Rb=10kΩ, 20V, ±3A, 0,75W, 140MHz	7c	Nec	-	data
AC 03 BGM...FGM L,R	Triac	200...600V, 3A≈(Tc=77°), Igt/Ih<15/=5mA L: >10V/μs, R: 4V/μs	13h§	Nec	TAG 137-...	data
AC 03 DJM(Z)	Triac	BGM=200V, DGM=400V, EGM=500V, FGM=600V 400V, 3A≈(Tc=110°C), Igt/Ih <12/=7mA	30h§	Nec	-	
AC 03 FJM(Z)	Triac	=AC 03DJM: 600V	30h§		-	
AC 04 BGM...FGM L,R	Triac	200...600V, 4A≈(Tc=80°), Igt/Ih<15/=5mA L: >10V/μs, R: 4V/μs BGM=200V, DGM=400V, EGM=500V, FGM=600V	13h§	Nec	TAG 137-...	
AC 4 DCM	Triac	=MAC 4 DCM (Typ-Code/Stempel/marking)	30	Ons	→MAC 4DCM	
AC 4 DCN	Triac	=MAC 4 DCN (Typ-Code/Stempel/marking)	30	Ons	→MAC 4DCN	
AC 4 DHM	Triac	=MAC 4 DHM (Typ-Code/Stempel/marking)	30	Ons	→MAC 4DHM	
AC 4 DLM	Triac	=MAC 4 DLM (Typ-Code/Stempel/marking)	30	Ons	→MAC 4DLM	
AC 05 DJM(Z)	Triac	400V, 5A≈(Tc=104°C), Igt/Ih <10/=10mA	30h§	Nec	-	
AC 05 FJM(Z)	Triac	=AC 05DJM: 600V	30h§		-	
AC 08 B	Triac	=MAC 08B (Typ-Code/Stempel/marking)	48	Ons	→MAC 08B	
AC 08 BGM...FGM L,R	Triac	200...600V, 8A≈(Tc=86°),Igt/Ih<75/=30mA L: >10V/μs, R: 4V/μs B..=200V, D..=400V, E..=500V, F...=600V	17h§	Nec	BT 158/..., MAC 222-.., TIC 225...,++	data
AC 08 BIM...FIM L,R	Triac	=AC 08BGM...FGM..: Iso	17h		-	data
AC 08 M	Triac	=MAC 08M (Typ-Code/Stempel/marking)	48	Ons	→MAC 08M	
AC 10 BGM...FGM L,R	Triac	200...600V,10A≈(Tc=86°),Igt/Ih<75/=30mA L: >10V/μs, R: 4V/μs B..=200V, D..=400V, E..=500V, F...=600V	17h§	Nec	BT 162/..., SC 146.., TIC 236...,++	data
AC 10 BIM...FIM L,R	Triac	=AC 10BGM...FGM..: Iso	17h		-	data
AC 12 BGM...FGM L,R	Triac	200...600V,12A≈(Tc=75°),Igt/Ih<50/=30mA L: >10V/μs, R: >1V/μs B..=200V, D..=400V, E..=500V, F...=600V	17h§	Nec	BT 162/..., SC 149.., TIC 236...,++	data
AC 12 BIM...FIM L,R	Triac	=AC 12BGM...FGM..: Iso	17h		-	data
AC 16 BGM...FGM L,R	Triac	200...600V,16A≈(Tc=80°),Igt/Ih<50/=40mA L: >10V/μs, R: >1V/μs B..=200V, D..=400V, E..=500V, F...=600V	17h§	Nec	MAC 15-..., SC 151.., TIC 246...,++	data
AC 16 BIF...FIF L,R	Triac	=AC 16BGM...FGM..: Iso	65h		-	data
AC 25 BIF...FIF L,R	Triac	200...600V,25A≈(Tc=75°),Igt/Ih<50/=40mA L: >10V/μs, R: >1V/μs BIF=200V, DIF=400V, EIF=500V, FIF=600V	65h§	Nec	MAC 515(A)-...	data
AC 105	Ge-P	NF-E, 40V, 1A, 0,4W, B=33	1a	Aeg	AC 128(K), AC 153(K), AC 188(K)	data
AC 106	Ge-P	=AC 105: B=50	1a	Aeg	AC 128(K), AC 153(K), AC 188(K)	data
AC 107	Ge-P	NF-V, ra, 15V, 10mA	1a	Phi	AC 125, AC 126, AC 151(r)	data
AC 108	Ge-P	NF-V/Tr, 20V, 50mA, ß=30...60	2a	Sie	AC 125, AC 126, AC 151	data
AC 109	Ge-P	=AC 108: ß=50...100	2a	Sie	AC 125, AC 126, AC 151	data
AC 110	Ge-P	=AC 108: ß=75...150	2a	Sie	AC 125, AC 126, AC 151	data
AC 113	Ge-P	NF-V/Tr, 26V, 50mA	2a	Aei	AC 125, AC 126, AC 151	data
AC 114	Ge-P	NF-V/Tr, 26V, 200mA	2a	Aei	AC 125, AC 126, AC 151	data
AC 115	Ge-P	NF-V/Tr, 26V, 50mA	2a	Aei	AC 125, AC 126, AC 151	data
AC 116	Ge-P	NF-Tr/E, 30V, 0,2A	3a	Aeg	AC 128K, AC 153K, (AC 188K)[7]	data
AC 117	Ge-P	NF-E, 32V, 1A, 1,1W	3a	Aeg	AC 128K, AC 153K, (AC 188K)[7]	data
AC 118	Ge-P	NF-E			AC 128, AC 153, AC 188	data
AC 119	Ge-P	NF-E	2a		AC 128, AC 153, AC 188	data
AC 120	Ge-P	NF-Tr/E, 20V, 0,3A, 0,6W	2a	Sie	AC 128, AC 153, AC 188	data
AC 121	Ge-P	NF-Tr/E, 20V, 0,3A, 0,9W	2a	Sie	AC 128, AC 153, AC 188	data
AC 122	Ge-P	NF-V, 30V, 0,2A	2a	Aeg,Tho	AC 125, AC 126, AC 151	data
AC 123	Ge-P	NF-Tr/E, 45V, 0,2A, 0,225W	3a	Aeg	AC 128K, AC 153K, ACY 24, ASY 48	data

Type	Device	Short Description	Fig.	Manu	Comparision Types	More at
AC 124	Ge-P	NF-E, 45V, 1A, 1,1W	3a	Aeg	(AC 128K, AC 153K, AC 188K)[7]	data
AC 125	Ge-P	NF-V/Tr, 32V, 0,2A, ß=80...170	2a	Mot,Phi,Tsm	AC 126, AC 151, ACY 24, ASY 48	data
AC 126	Ge-P	=AC 125: ß=130...300	2a	Mot,Phi,Tsm	AC 151, ACY 24, ASY 48	data
AC 127	Ge-N	NF-Tr/E, 32V, 0,5A, 0,34W	2a	Mot,Phi	AC 176, AC 187	data
AC 128	Ge-P	NF-E, 32V, 1A, 1W	2a	Mot,Phi,Tsm	AC 153, AC 188	data
AC 128 K	Ge-P	=AC 128:	3a		AC 153K, AC 188K	data
AC 129	Ge-P	Min, NF, 9V, 10mA	36a	Aeg	OC 57..60	data
AC 130	Ge-N	Tr, sym, 20V, 0,1A	2a	Phi	-	data
AC 131	Ge-P	NF-Tr/E, 30V, 1A, 0,75W	2a	Aeg	AC 128, AC 153, AC 188	data
AC 132	Ge-P	NF-Tr/E, 32V, 0,2A, 0,5W	2a	Mot,Phi	AC 128, AC 151, AC 153, AC 188	data
AC 134	Ge-P	NF-V/Tr, 32V, 0,2A, ß=45	2a	Sgs	AC 125, AC 126, AC 151	data
AC 135	Ge-P	NF-V/Tr, 32V, 0,2A, ß=110	2a	Sgs	AC 125, AC 126, AC 151	data
AC 136	Ge-P	NF-V/Tr, 40V, 0,2A, ß=110	2a	Sgs	AC 125, AC 126, AC 151	data
AC 137	Ge-P	NF-V/Tr, 32V, 0,05A, ß=170	2a	Sgs	AC 125, AC 126, AC 151	data
AC 138	Ge-P	NF-V/Tr, 25V, 1,2A, 1W	2a	Sgs	AC 128, AC 153, AC 188	data
AC 139	Ge-P	NF-Tr/E, 32V, 1A, 1W	2a	Sgs	AC 128, AC 153, AC 188	data
AC 139 K	Ge-P	=AC 139:)	3a		AC 128K, AC 153K, AC 188K	data
AC 141	Ge-N	NF-Tr/E, 32V, 1,2A, 1W	2a	Sgs	AC 176, AC 187	data
AC 141 K	Ge-N	=AC 141:	3a		AC 176K, AC 187K	data
AC 142	Ge-P	NF-Tr/E, 32V, 1,2A, 1W	2a	Sgs	AC 128, AC 153, AC 188	data
AC 142 K	Ge-P	=AC 142:	3a		AC 128K, AC 176K, AC 188K	data
AC 150	Ge-P	NF-V, ra, 30V, 50mA	2a	Aeg	AC 151r, ACY 32	data
AC 151(r)	Ge-P	NF-V/Tr, (ra), 32V, 0,2A	2a	Sie	AC 122, AC 125, AC 126, ACY 32	data
AC 152	Ge-P	NF-Tr/E, 32V, 0,5A, 0,9W	2a	Sie	AC 128, AC 153, AC 188	data
AC 153	Ge-P	NF-Tr/E, 32V, 2A, 1W	2a	Sie	AC 128, AC 188	data
AC 153 K	Ge-P	=AC 153:	3a		AC 128K, AC 188K	data
AC 154	Ge-P	NF-Tr/E, 26V, 0,5A, 0,2W	2a	Aei	AC 128, AC 153, AC 188	data
AC 155	Ge-P	NF-V/Tr, 26V, 50mA, B=20...68	2a	Aei	AC 125, AC 126, AC 151	data
AC 156	Ge-P	=AC 155: B=40...114	2a	Aei	AC 125, AC 126, AC 151	data
AC 157	Ge-N	NF-Tr/E, 26V, 0,5A, 0,2W	2a	Aei	AC 127, AC 176, AC 187	data
AC 160	Ge-P	NF-V, ra, 15V, 10mA	2a	Aeg	AC 151r, ACY 32	data
AC 161	Ge-P	NF-V, ra, 15V, 100mA	2a	Tho	AC 151r, ACY 32	data
AC 162	Ge-P	NF-V, 32V, 0,2A, ß=80...170	2a	Sie	AC 125, AC 126, AC 151	data
AC 163	Ge-P	=AC 162: ß=130...300	2a	Sie	AC 126, AC 151	data
AC 164	Ge-P	NF-V, 10V, 30mA	2a	Phi	AC 125, AC 126, AC 151	data
AC 165	Ge-P	NF-V/Tr, 32V, 50mA	2a	Aei	AC 125, AC 126, AC 151	data
AC 166	Ge-P	NF-Tr/E, 32V, 0,5A, 0,2W, B=52...315	2a	Aei	AC 128, AC 153, AC 188	data
AC 167	Ge-P	=AC 166: B=45..250	2a	Aei	AC 128, AC 153, AC 188	data
AC 168	Ge-N	NF-Tr/E, 32V, 0,5A, 0,2W	2a	Aei	AC 127, AC 176, AC 187	data
AC 169	Ge-P	Stabi, sym, 2V, 30mA	2a	Aei	-	data
AC 170	Ge-P	NF-V/Tr, 32V, 0,2A, ß=80...170	2a	Aeg	AC 125, AC 126, AC 151	data
AC 171	Ge-P	=AC 170: ß=130...300	2a	Aeg	AC 126, AC 151	data
AC 172	Ge-N	NF-V, ra, 32V, 10mA	2a	Phi,Sgs	(AC 127)	data
AC 173	Ge-P	NF-V/Tr, 32V, 0,3A, 0,2W	2a	Tho	AC 128, AC 151, AC 153, AC 188	data
AC 174	Ge-P	NF-E, 32V, 0,6A, 0,6W	2a		AC 128, AC 153, AC 188	data
AC 175	Ge-N	NF-Tr/E, 25V, 1A, 1,1W	3a	Aeg	AC 176K, AC 187K	data
AC 176	Ge-N	NF-Tr/E, 32V, 1A, 1W	2a	Phi,Sie,Tsm	AC 187	data
AC 176 K	Ge-N	=AC 176:	3a		AC 187K	data
AC 177	Ge-N	NF-Tr/E, 32V, 0,5A, 0,2W	2a	Aei	AC 128, AC 153, AC 188	data
AC 178	Ge-P	NF-Tr/E, 20V, 0,7A, 1,1W	3a	Aeg	AC 128K, AC 153K, AC 188K	data
AC 179	Ge-N	NF-Tr/E, 20V, 0,7A, 1,1W	3a	Aeg	AC 176K, AC 187K	data
AC 180	Ge-P	NF-Tr/E, 32V, 1,5A, 0,3W	2a	Tho	AC 128, AC 153, AC 188	data
AC 180 K,L	Ge-P	=AC 180: 0,44W	3a		AC 128K, AC 153K, AC 188K	data
AC 181	Ge-N	NF-Tr/E, 32V, 1,5A, 0,3W	2a	Tho	AC 176, AC 187	data
AC 181 K,L	Ge-N	=AC 181: 0,44W	3a		AC 176K, AC 187K	data
AC 182	Ge-P	NF-V/Tr, 32V, 0,15A	2a	Tho	AC 125, AC 126, AC 151	data
AC 183	Ge-N	NF-V/Tr, 32V, 0,15A	2a	Tho	AC 127, AC 176, AC 187	data
AC 184	Ge-P	NF-Tr/E, 32V, 0,5A, 0,16W	2a	Tho	AC 128, AC 153, AC 188	data
AC 185	Ge-N	NF-Tr/E, 32V, 0,5A, 0,16W	2a	Tho	AC 127, AC 176, AC 187	data
AC 186	Ge-N	NF-Tr/E, 30V, 0,7A, 0,75W	2a	Aeg	AC 127, AC 167, AC 187	data
AC 187	Ge-N	NF-Tr/E, 25V, 1A, 1W	2a	EUR	AC 176	data
AC 187 K	Ge-N	=AC 187:	3a		AC 176K	data
AC 188	Ge-P	NF-Tr/E, 25V, 1A, 1W	2a	EUR	AC 128, AC 153	data
AC 188 K	Ge-P	=AC 188:	3a		AC 128K, AC 153K	data
AC 190	Ge-N	Tr, sym		Sgs	-	data
AC 191	Ge-P	=AC 192: ra	2a	Sgs	AC 151r, ACY 32	data
AC 192	Ge-P	NF-V/Tr, 32V, 0,25A	2a	Sgs	AC 125, AC 126, AC 151	data
AC 193	Ge-P	NF-Tr/E, 32V, 1A, 1W	2a	Sgs	AC 128, AC 153, AC 188	data
AC 193 K	Ge-P	=AC 193:	3a		AC 128K, AC 153K, AC 188K	data
AC 194	Ge-N	NF-Tr/E, 32V, 1A, 1W	2a	Sgs	AC 176, AC 187	data
AC 194 K	Ge-N	=AC 194:	3a		AC 176K, AC 187K	data
AC 230	Ge-P	NF-V/Tr, 24V, 10mA	2a	Eiy	AC 125, AC 126, AC 151	data

Type	Device	Short Description	Fig.	Manu	Comparision Types	More at
AC 240	Ge-P	NF-V/Tr, 24V, 10mA, ß=30...50	2a	Eiy	AC 125, AC 126, AC 151	data
AC 241	Ge-P	=AC 240: ß=50...80	2a	Eiy	AC 125, AC 126, AC 151	data
AC 242	Ge-P	=AC 240: ß=80...150	2a	Eiy	AC 125, AC 126, AC 151	data
AC 250	Ge-P	NF-V/Tr, 32V, 50mA	2a	Eiy	AC 125, AC 126, AC 151	data
AC 251	Ge-P	NF-V/Tr, 32V, 125mA	2a	Eiy	AC 125, AC 126, AC 151	data
AC 330	Ge-N	NF-V/Tr, 24V, 10mA	2a	Eiy	AC 127	data
AC 340	Ge-N	NF-V/Tr, 24V, 10mA, ß=30...50	2a	Eiy	AC 127	data
AC 341	Ge-N	=AC 340: ß=50...80	2a	Eiy	AC 127	data
AC 342	Ge-N	=AC 340: ß=80...150	2a	Eiy	AC 127	data
AC 350	Ge-N	NF-V/Tr, 32V, 50mA	2a	Eiy	AC 127	data
AC 351	Ge-N	NF-V/Tr, 32V, 125mA	2a	Eiy	AC 127	data
AC 402	Ge-P	NF-E, 32V, 1,5A	2a	Elx	AC 128, AC 153, AC 188	data
AC 404	Ge-P	NF-Tr/E, 45V, 0,5A	2a	Elx	AC 128, AC 153, (AC 188)[7]	data
AC 502	Ge-P	NF, 16V, 0,1A, B=35...65	2a	Riz	AC 125, AC 126, AC 151	data
AC 503	Ge-P	NF, 16V, 0,1A, B=55...120	2a	Riz	AC 125, AC 126, AC 151	data
AC 504	Ge-P	NF, 16V, 0,1A, B=72..192	2a	Riz	AC 125, AC 126, AC 151	data
AC 508	Ge-P	NF, 16V, 0,1A, B=100...200	2a	Riz	AC 126, AC 151	data
AC 509	Ge-P	NF, 16V, 0,1A, B=100...200	2a	Riz	AC 126, AC 151	data
AC 515	Ge-P	NF, 16V, 0,1A, B=>60	2a	Riz	AC 126, AC 151	data
AC 516	Ge-P	NF, 16V, 0,1A, B=>95	2a	Riz	AC 125, AC 126, AC 151	data
AC 517	Ge-P	NF-Tr, 30V, 0,2A, B>45	2a	Riz	AC 125, AC 126, AC 151	data
AC 518	Ge-P	NF-Tr, 30V, 0,2A, B>85	2a	Riz	AC 125, AC 126, AC 151	data
AC 519	Ge-P	NF-Tr, 30V, 0,2A, B=25...45	2a	Riz	AC 125, AC 126, AC 151	data
AC 520	Ge-P	NF-Tr, 30V, 0,2A, B=35...65	2a	Riz	AC 125, AC 126, AC 151	data
AC 521	Ge-P	NF-Tr, 30V, 0,2A, B=55...120	2a	Riz	AC 125, AC 126, AC 151	data
AC 524	Ge-P	NF-Tr/E, 45V, 0,5A, B=20...42	2a	Riz	AC 128, AC 153, (AC 188)[7]	data
AC 525	Ge-P	NF-Tr/E, 45V, 0,5A, B=35...65	2a	Riz	AC 128, AC 153, (AC 188)[7]	data
AC 526	Ge-P	NF-Tr/E, 45V, 0,5A, B=55...90	2a	Riz	AC 128, AC 153, (AC 188)[7]	data
AC 527	Ge-P	NF-Tr/E, 45V, 0,5A, B=70...120	2a	Riz	AC 128, AC 153, (AC 188)[7]	data
AC 530	Ge-P	NF-V/Tr, 24V, 10mA, ß=20...40	2a	Eiy	AC 125, AC 126, AC 151	data
AC 540	Ge-P	NF-V/Tr, 24V, 10mA, ß=30...50	2a	Eiy	AC 125, AC 126, AC 151	data
AC 541	Ge-P	=AC 540: ß=50...80	2a	Eiy	AC 125, AC 126, AC 151	data
AC 542	Ge-P	=AC 540: ß=80...150	2a	Eiy	AC 125, AC 126, AC 151	data
AC 548	Ge-P	NF-Tr, 26V, 0,15A	2a	Eiy	AC 125, AC 126, AC 151	data
AC 549	Ge-P	NF-Tr, 26V, 0,3A	2a	Eiy	AC 125, AC 126, AC 151	data
AC 550	Ge-P	NF-V/Tr, 32V, 50mA	2a	Eiy	AC 125, AC 126, AC 151	data
AC 551(R)	Ge-P	NF-V/Tr, (ra), 32V, 125mA	2a	Eiy	AC 125, AC 126, AC 151	data
AC 552	Ge-P	NF-V/Tr, 60V, 125mA	2a	Eiy	ACY 24, ASY 48	data
AC 553	Ge-P	NF-Tr/E, 20V, 0,3A, 0,12W	2a	Eiy	AC 128, AC 153, AC 188	data
AC 554	Ge-P	NF-Tr/E, 26V, 0,3A, 0,12W	2a	Eiy	AC 128, AC 153, AC 188	data
AC 555	Ge-P	NF-Tr/E, 32V, 0,3A, 0,12W	2a	Eiy	AC 128, AC 153, (AC 188)[7]	data
AC 556	Ge-P	NF-Tr/E, 25V, 1A, 0,23W	2a	Eiy	AC 128, AC 153, AC 188	data
AC 556 K	Ge-P	=AC 556: 0,35W	3a		AC 128K, AC 153K, AC 188K	data
AC 558	Ge-N	NF-Tr/E, 25V, 1A, 0,23W	2a	Eiy	AC 176, AC 187	data
AC 558 K	Ge-N	=AC 558: 0,35W	3a		AC 176K, AC 187K	data
AC 570	Ge-P	NF-Tr/E, 70V, 0,5A, 0,15W, B=17...40	2a	Riz	ASY 48	data
AC 571	Ge-P	=AC 570: B=30...60	2a	Riz	ASY 48	data
AC 572	Ge-P	=AC 570: B=45...85	2a	Riz	ASY 48	data
AC 573	Ge-P	=AC 570: B=65...110	2a	Riz	ASY 48	data
AC 577	Ge-P	=AC 570: B=45...110	2a	Riz	ASY 48	data
AC 598	Ge-P	NF/S, 105V, 0,2A, 0,1W	2a	Riz	-	data
ACG	Si-P	=KTA1517-G (Typ-Code/Stempel/marking)	35	Kec	→KTA 1517	data
ACL	Si-N	=2SC3837K-L (Typ-Code/Stempel/marking)	35	Rhm	→2SC3837K	data
ACL	Si-N	=2SC4725-L (Typ-Code/Stempel/marking)	35(1,6mm)	Rhm	→2SC4725	data
ACL	Si-P	=KTA1517-L (Typ-Code/Stempel/marking)	35	Kec	→KTA 1517	data
ACM	Si-N	=2SC3837K-M (Typ-Code/Stempel/marking)	35	Rhm	→2SC3837K	data
ACM	Si-N	=2SC4725-M (Typ-Code/Stempel/marking)	35(1,6mm)	Rhm	→2SC4725	data
ACN	Si-N	=2SC3837K-N (Typ-Code/Stempel/marking)	35	Rhm	→2SC3837K	data
ACN	Si-N	=2SC4725-N (Typ-Code/Stempel/marking)	35(1,6mm)	Rhm	→2SC4725	data
ACO	Si-P	=2SA1313-O (Typ-Code/Stempel/marking)	35	Tos	→2SA1313	data
ACP	Si-N	=2SC3837K-P (Typ-Code/Stempel/marking)	35	Rhm	→2SC3837K	data
ACp	Si-N	=BCW 60C (Typ-Code/Stempel/marking)	35	Phi	→BCW 60C	data
ACQ	Si-N	=2SC3837K-Q (Typ-Code/Stempel/marking)	35	Rhm	→2SC3837K	data
ACs	Si-N	=BCW 60C (Typ-Code/Stempel/marking)	35	Sie	→BCW 60C	data
ACs	Si-N	=BF 620 (Typ-Code/Stempel/marking)	44(2mm)	Sie	→BF 620	data
ACS102-5T1	AC-Switch	AC On-/Off-Switch, 500V 0.2A, IGT<5mA	8-MDIP	Sgs	-	pdf pinout
ACS102-5TA	AC-Switch	AC On-/Off-Switch, 500V 0.2A, IGT<5mA	7	Sgs	-	pdf pinout
ACS108-5SA	AC-Switch	AC On-/Off-Switch, 500V 0.8A, IGT<10mA	7	Sgs	-	pdf pinout
ACS108-5SN	AC-Switch	SMD, AC On-/Off-Switch, 500V 0.8A	48	Sgs	-	pdf pinout
ACS110-7SB2	AC-Switch	AC On-/Off-Switch, 700V 1A, IGT<10mA	8-DIP	Sgs	-	pdf pinout
ACS110-7SN	AC-Switch	SMD, AC On-/Off-Switch, 700V 1A	48	Sgs	-	pdf pinout
ACS120-7SB	AC-Switch	SMD, AC On-/Off-Switch, 700V 2A	87	Sgs	-	pdf pinout
ACS120-7SFP	AC-Switch	AC On-/Off-Switch, 700V 2A, isolated	18	Sgs	-	pdf pinout
ACS120-7ST	AC-Switch	AC On-/Off-Switch, 700V 2A, IGT<10mA	18	Sgs	-	pdf pinout
ACS302-5T3	AC-Switch	SMD, 3x AC On-/Off-Switch, 500V 0.2A	20-MDIP	Sgs	-	pdf pinout
ACS402-5SB4	AC-Switch	SMD, 4x AC On-/Off-Switch, 500V 0.2A	20-DIP	Sgs	-	pdf pinout
ACST4-7(S,C)B	AC-Switch	SMD, AC Power Switch, 700V 4A	87	Sgs	-	pdf pinout
ACST4-7(S,C)FP	AC-Switch	AC Switch isolated, 700V 4A	18	Sgs	-	pdf pinout
ACST6-7S(R,T,FP)	AC-Switch	AC Switch, Ovp, 700V 6A, IGT<10mA	18	Sgs	-	pdf pinout
ACST6-7SG	AC-Switch	SMD, AC Switch, Ovp, 700V 6A IGT<10mA	18	Sgs	-	pdf pinout
ACST8-8C	AC-Switch	AC Switch isol., Ovp, 800V 8A IGT<30mA	18	Sgs	-	pdf pinout
ACt	Si-N	=BCW 60C (Typ-Code/Stempel/marking)	35	Phi	→BCW 60C	pdf pinout
ACVP 2205	IC	Ctv, Video-Prozessor		Itt	-	data

Type	Device	Short Description	Fig.	Manu	Comparision Types	More at
ACY...ACZ						
ACY	Si-P	=2SA1313-Y (Typ-Code/Stempel/marking)	35	Tos	→2SA1313	data
ACY 10	Ge-P	NF-V, 32V, 50mA, B>32	2a	Sie	AC 125, AC 126, AC 151	data
ACY 11	Ge-P	NF-V, 32V, 50mA, B>38	2a	Sie	AC 125, AC 126, AC 151	data
ACY 12	Ge-P	NF-V, ra, 32V, 50mA, B>38	2a	Sie	AC 151r, ACY 32	data
ACY 13	Ge-P	NF-V, 16V, 50mA, B>50	2a	Sie	AC 125, AC 126, AC 151	data
ACY 14	Ge-P	NF-V, 32V, 50mA, B>54	2a	Sie	AC 125, AC 126, AC 151	data
ACY 15	Ge-P	NF-V, ra, 32V, 50mA, B>54	2a	Sie	AC 151r, ACY 32	data
ACY 16	Ge-P	NF-Tr/E, 40V, 0,4A, 0,8W	3a	Aeg	AC 128K, AC 153K, ACY 33	data
ACY 17	Ge-P	NF/S, 70V, 0,5A, 0,26W, B=50...150	2a&	Phi	ASY 77	data
ACY 18	Ge-P	NF/S, 50V, 0,5A, 0,26W, B=40...120	2a&	Phi	ASY 77	data
ACY 19	Ge-P	NF/S, 50V, 0,5A, 0,26W, B=80...250	2a&	Phi	ASY 77	data
ACY 20	Ge-P	NF/S, 40V, 0,5A, 0,26W, B=80...145	2a&	Phi	AC 128, AC 153, ASY 77	data
ACY 21	Ge-P	NF/S, 40V, 0,5A, 0,26W, B=90...250	2a&	Phi	AC 128, AC 153, ASY 77	data
ACY 22	Ge-P	NF/S, 20V, 0,5A, 0,26W, B=30...300	2a&	Phi	AC 128, AC 153, AC 188, ASY 77	data
ACY 23	Ge-P	NF-V, 32V, 0,2A	2a	Sie	AC 125, AC 126, AC 151, ACY 32	data
ACY 24	Ge-P	NF-Tr/E, 70V, 0,3A, 0,53W	3a	Aeg	ASY 48	data
ACY 25	Ge-P	NF-V, 32V, 50mA	2a	Sie	AC 125, AC 126, AC 151	data
ACY 27	Ge-P	NF-V/Tr, 40V, 0,25A, 0,2W, ß=20...55	37a	Itt	AC 125, AC 126, AC 151, ACY 32	data
ACY 28	Ge-P	NF-V/Tr, 40V, 0,25A, 0,2W, ß=45...150	37a	Itt	AC 125, AC 126, AC 151, ACY 32	data
ACY 29	Ge-P	NF-V/Tr, ra, 40V, 0,25A, ß=45...145	37a	Itt	AC 151r, ACY 32	data
ACY 30	Ge-P	NF-V/Tr, 40V, 0,25A, 0,2W, ß=60...200	37a	Itt	AC 126, AC 151, ACY 32	data
ACY 31	Ge-P	NF-V/Tr, 50V, 0,25A, 0,2W, ß=>35	37/a	Itt	AC 125, AC 126, AC 151, ACY 32	data
ACY 32	Ge-P	=ACY 23: ra	2a	Sie	AC 151r	data
ACY 33	Ge-P	NF-Tr/E, 32V, 1A, 1,1W	2a	Sie	AC 128, AC 153, AC 188	data
ACY 34	Ge-P	NF, 30V, 50mA, ß=20...40	37a	Itt	AC 125, AC 126, AC 151	data
ACY 35	Ge-P	NF, 30V, 50mA, ß=30...75	37a	Itt	AC 125, AC 126, AC 151	data
ACY 36	Ge-P	NF, 30V, 250mA, B=30..90	37a	Itt	AC 125, AC 126, AC 151	data
ACY 38	Ge-P	NF-V, ra, 15V, 0,1A	2a	Tho	AC 151r, ACY 32	data
ACY 39	Ge-P	NF/S, 110V, 0,5A, 0,26W	2a&	Phi	2N2042, 2N2043	data
ACY 40	Ge-P	NF/S, 32V, 0,5A, 0,26W	2a&	Phi	AC 128, AC 153, ASY 77	data
ACY 41	Ge-P	NF/S, 21V, 0,5A, 0,26W	2a&	Phi	AC 128, AC 153, AC 188, ASY 77	data
ACY 44	Ge-P	NF/S, 50V, 0,5A, 0,26W	2a&	Phi	AC 128, AC 153, ASY 77	data
ACY 50	Ge-P	NF, 32V, 0,2A, 0,22W, B=30...150	2a	Eiy	AC 125, AC 126, AC 151, ACY 32	data
ACY 51(R)	Ge-P	NF, (ra), 32V, 0,2A, 0,22W, B=30...300	2a	Eiy	AC 125, AC 126, AC 151(r), ACY 32	data
ACY 52	Ge-P	NF, 60V, 0,2A, 0,22W, B=50...120	2a	Eiy	ASY 48	data
ACY 55	Ge-P	NF, 32V, 0,3A, 0,22W	2a	Eiy	AC 128, AC 153, AC 188	data
ACZ 10	Ge-P	NF-E, 70V, 0,3A, 0,4W	1a	Aeg	ASY 48	data
AD...ADX						
AD	Si-P	=2SA1418 (Typ-Code/Stempel/marking)	39	Say	→2SA1418	data
AD	Si-N	=2SD1366A-AD(Typ-Code/Stempel/marking)	39	Hit	→2SD1366A	data
AD	Si-P	=BCP 51-16 (Typ-Code/Stempel/marking)	48	Phi	→BCP 51	data
AD	Si-N	=BCW 60D (Typ-Code/Stempel/marking)	35	Mot,Phi,Sie Aeg,Fer,++	→BCW 60D	data
AD	Si-P	=BCX 51-16 (Typ-Code/Stempel/marking)	39	Sie,Val,Mot	→BCX 51	data
AD	Si-P	=CPH 3104 (Typ-Code/Stempel/marking)	35	Say	→CPH 3104	data
AD	Z-Di	=MA 5Z220-H (Typ-Code/Stempel/marking)	71(1,7mm)	Mat	→MA 5Z220-H	data
AD	Si-N	=MT 3S07S (Typ-Code/Stempel/marking)	35(1,6mm)	Tos	→MT 3S07S	data
AD	Si-N	=MT 3S07T (Typ-Code/Stempel/marking)	35(1,6mm)	Tos	→MT 3S07T	data
AD	Si-N	=MT 3S07U (Typ-Code/Stempel/marking)	35(2mm)	Tos	→MT 3S07U	data
AD	Si-N	=MT 4S07 (Typ-Code/Stempel/marking)	44	Tos	→MT 4S07	
AD	MOS-P-FET-e	=NTTS 2P02 (Typ-Code/Stempel/marking)	8-SSMDIP	Ons	→NTTS 2P02	
AD	Z-Di	=P4 SMA-6.8 (Typ-Code/Stempel/marking)	71(5x2,5)	Fag	→P4 SMA-6.8	data
AD	Si-P	=RXT 3906 (Typ-Code/Stempel/marking)	39	Rhm	→RXT 3906	data
AD	Si-P+R	=XN 1110 (Typ-Code/Stempel/marking)	45	Mat	→XN 1110	data
AD	Si-P+R	=XP 1110 (Typ-Code/Stempel/marking)	45(2mm)	Mat	→XP 1110	data
AD	C-Di	=ZMV 832A (Typ-Code/Stempel/marking)	71(1,7mm)	Ztx	→ZMV 832	data
AD	Z-Di	=SMBJ 5.0C (Typ-Code/Stempel/marking)	71(5x3,5)	Mop	→SMBJ ...	data
AD 1 A3M	Si-N+R	S, Rb=Rbe=1kΩ, 80V, 1A, 0,75W	7c	Nec	-	data
AD 1 A4A	Si-N+R	=AD 1A3M: Rb=0Ω, Rbe=10kΩ	7c	Nec	-	data
AD 1 A4M	Si-N+R	=AD 1A3M: Rb=Rbe=10kΩ	7c	Nec	-	data
AD 1 F2Q	Si-N+R	=AD 1A3M: Rb=0,22kΩ, Rbe=2,2kΩ	7c	Nec	-	data
AD 1 F3P	Si-N+R	=AD 1A3M: Rb=2,2kΩ, Rbe=10kΩ	7c	Nec	-	data
AD 1 L2Q	Si-N+R	=AD 1A3M: Rb=0,47kΩ, Rbe=4,7kΩ	7c	Nec	-	data
AD 1 L3N	Si-N+R	=AD 1A3M: Rb=4,7kΩ, Rbe=10kΩ	7c	Nec	-	data
AD 2 A3M...L3N	Si-N+R	=AD 1A3M...L3N: int. Z-Di(C→B), 60V	7c	Nec	-	data
AD004	FET	Vu:88dB Vo:±10V Vi0:20mV f:1Mc	8A	And		data pinout
AD101A	UNI	Vs:±22V Vu:104dB Vo:±14V Vi0:0.7mV	8A	And		data pinout
AD 103	Ge-P	NF-L, 50V, 15A, 22,5W		Sie	(AUY 29, 2N1549...1560)[4]	data
AD 104	Ge-P	NF-L, 65V, 10A, 22,5W		Sie	(AUY 21, 2N2526, 2N2289, 2N2292)[4]	data
AD 105	Ge-P	NF-L, 80V, 8A, 22,5W		Sie	(AUY 22, 2N2526, 2N2289, 2N2292)[4]	data
AD108	LO-POWER	Vs:±20V Vu:110dB Vo:±14V Vi0:0.7mV	8A	And		data pdf pinout
AD 130	Ge-P	NF-L, 32V, 3A, 30W	23a§	Gpd,Sie	AD 149, AL 102, 2N1539...1548	data
AD 131	Ge-P	=AD 130: 64V	23a§	Gpd,Sie	AL 102, 2N1540...1541, 2N1545...1546	data

15

Type	Device	Short Description	Fig.	Manu	Comparision Types	More at
AD 132	Ge-P	=AD 130: 80V	23a§	Gpd,Sie	AL 102, 2N1541...1542, 2N1546...1547	data
AD 133	Ge-P	NF/S-L, 50V, 15A, 36W	23a§	Gpd,Sie	AUY 29, 2N1549...1560	data
AD 134	Ge-P	NF/S-L, 65V, 10A, 22,5W	23a§	Sie	AUY 21, 2N2526, 2N2289, 2N2292	data
AD 135	Ge-P	NF/S-L, 80V, 8A, 22,5W	23a§	Sie	AUY 22, 2N2526, 2N2289, 2N2292	data
AD 136	Ge-P	S-L, 40V, 10A, 11W	2a§	Aeg,Gpd,Sie	AUY 18, AUY 35, AUY 36, 2SB627	data
AD 138(/50)	Ge-P	NF-L, 40(70)V, 8A, 30W	23a§	Aeg	AUY 21, AL 102, 2N2526, 2N2288...2293	data
AD 139	Ge-P	NF-L, 32V, 3,5A, 13W	22a§	Aeg,Phi	AD 162	data
AD 140	Ge-P	NF-L, 55V, 3A, 35W	23a§	Aeg,Gpd,Phi	AD 149, AL 102, AUY 19, 2N1540, 2N1545	data
AD 142	Ge-P	NF-L, 80V, 10A, 30W	23a§	Gpd,Sgs	AUY 22, AL 102, 2N2526, 2N2289, 2N2292	data
AD 143(R)	Ge-P	=AD 142: 40(32)V	23a§	Gpd,Sgs	AUY 21, AL 102, ASZ 16...17, 2N2288/91	data
AD 145	Ge-P	=AD 142: 20V	23a§	Sgs	AUY 21, AL 102, ASZ 16...17, 2N2288/91	data
AD 148	Ge-P	NF-L, 32V, 3,5A, 13,5W	22a§	Gpd,Sgs,Sie	AD 162	data
AD 149	Ge-P	NF-L, 50V, 3,5A, 27,5W	23a§	EUR	AL 102, AUY 19, ASZ 16...17, 2N1539...48	data
AD 150	Ge-P	=AD 149: 32V	23a§	Aeg,Gpd,Sie	AUY 21, AL 102, ASZ16...17, 2N1539..48	data
AD 152	Ge-P	NF-L, 45V, 1A, 6W	22a§	Aeg,Sgs	AD 162	data
AD 153	Ge-P	NF-L, 40V, 3A, 33W	23a§		AD 149, AL 102, AUY 19, 2N1539...1548	data
AD 155	Ge-P	NF-L, 25V, 1A, 6W	22a§	Aeg,Sgs	AD 162	data
AD 156	Ge-P	NF-L, 30V, 3A, 6W	22a§	Sie	AD 162	data
AD 157	Ge-N	NF-L, 30V, 3A, 6W	22a§	Sie	AD 161	data
AD 159	Ge-P	S-L, 40V, 8A, 9W	2a§	Aeg	AUY 18, AUY 35, AUY 36, 2SB627	data
AD 160	Ge-P	=AD 159: 30V	2a§	Aeg	AUY 18, AUY 35, AUY 36, 2SB627	data
AD 161	Ge-N	NF-L, 32V, 1A, 4W(Tc=70°)	22a§	EUR	-	data
AD 162	Ge-P	NF-L, 32V, 1A, 6W(Tc=63°)	22a§	EUR	-	data
AD 163	Ge-P	TV-VA, 100V, 3A, 30W	23a§	Gpd,Sie	AL 102, AUY 34, ASZ 15, ASZ 18	data
AD 164	Ge-P	NF-L, 25V, 1A, 6W	22a§	Aeg	AD 162	data
AD 165	Ge-N	NF-L, 25V, 1A, 5,3W	22a§	Aeg	AD 161	data
AD 166	Ge-P	NF-L, 60V, 5A, 27,5W	23a§	Sie	AL 102, AUY 21, 2N1540, 2N1545	data
AD 167	Ge-P	=AD 166: 75V	23a§	Sie	AL 102, AUY 22, 2N1541, 2N1546	data
AD 169	Ge-P	NF-L, 45V, 1A, 6W	22a§	Aeg	AD 162	data
AD201A	UNI	Vs:±22V Vu:94dB Vo:±10V Vio:2mV	8AD	And		data pinout
AD 202 J,K	LIN-IC	Trennverst./Isolation Amp, Ucc=±15V	38-DIP	And	-	data
AD 203 SN	LIN-IC	Trennverst./Isolation Amp, Ucc=±15V	38-DIP	And		pdf
AD 204 J,K	LIN-IC	Trennverst./Isolation Amp, Ucc=±15V	38-DIP	And	-	
AD208	LO-POWER	Vs:±20V Vu:94dB Vo:±13V Vio:2mV	8A	And		data pdf pinout
AD 208 AY,BY	LIN-IC	Trennverst./Isolation Amp, Ucc=±15V	38-SIP	And		
AD208A	LO-POWER	Vs:±20V Vu:98dB Vo:±13V Vio:0.5mV	8A	And		data pinout
AD 210 AN,BN	LIN-IC	Trennverst./Isolation Amp, Ucc=±15V	38-DIP	And	-	
AD 262	Ge-P	NF-L, 35V, 4A, 10W(Tc=60°)	22a§	Sgs	(AD 162)[7]	data
AD 263	Ge-P	=AD 262: 60V	22a§	Sgs	-	data
AD 301	Ge-P	NF-L, 30V, 3A, 30W	23a§	Elx	AD 149, AL 102, AUY 19, 2N1539...1548	data
AD301A	UNI	Vs:±18V Vu:104dB Vio:2mV	8AD	And		data pinout
AD 302	Ge-P	NF-L, 40V, 3A, 45W	23a§	Elx	AD 149, AL 102, AUY 19, 2N1539...1548	data
AD 303	Ge-P	NF-L, 60V, 3A, 45W	23a§	Elx	AL 102, AUY 19, 2N1540, 2N1545	data
AD 304	Ge-P	NF-L, 80V, 3A, 45W	23a§	Elx	AL 102, AUY 20, 2N1541, 2N1546	data
AD308	LO-POWER	Vs:±18V Vu:110dB Vo:±14V Vio:2mV	8A	And		data pinout
AD 312	Ge-P	NF-L, 40V, 6A, 45W	23a§	Elx	AL 102, AUY 28, 2N3611, 2N3613	data
AD 313	Ge-P	=AD 312: 60V	23a§	Elx	AL 102, AUY 28, 2N3612, 2N3614	data
AD 314	Ge-P	=AD 312: 80V	23a§	Elx	AL 102, AUY 28, 2N3615, 2N3617	data
AD 315	Ge-N	NF-L, 32V, 2,5A, 6W	22a§	Elx	AD 161	data
AD 325	Ge-P	NF-L, 100V, 10A, 45W	23a§	Elx	AL 102, AUY AUY 37, 2N2527, 2N2290/93	data
AD 362KD	A/D-IC	Precision S&H with 16 Channel MPX	DIP	And	-	pdf
AD 362SD	A/D-IC	Precision S&H with 16 Channel MPX	DIP	And	- -	pdf
AD 365	Ge-P	NF-L, 30V, 1,5A, 2W		Ucp	-	data
AD 366	Ge-P	=AD 365: 60V		Ucp	-	data
AD380	WIDEBAND	Vs:±15V Vo:±12V Vio:<2mV f:300Mc	12A	And		data pdf pinout
AD381	FET	Vs:±20V Vu:95.5dB Vo:±12V Vio:1mV	8A	And		data pinout
AD382	FET	Vs:±20V Vu:90dB Vo:±12V Vio:1mV	12A	And		data pinout
AD 412	Ge-P	NF-L, 24V, 1A	22a§	Eiy	AD 162	data
AD 415	Ge-P	NF-L, 32V, 2,5A, 6W	22a§	Eiy	AD 162	data
AD 420 AN	D/A-IC	16 Bit D/A-Conv., ser.Input, Vcc<32V Curr./Voltage Output, +5VRef., 3MBPS	24-DIP	And	-	pdf pinout
AD 420 AR	D/A-IC	SMD,16Bit D/A-Conv., ser.Inp., Vcc<32V Curr./Voltage Output, +5VRef., 3MBPS	24-MDIP	And		pdf pinout
AD 430	Ge-P	NF-L, 32V, 1,4A, 6W	23a§	Eiy	AD 149, AL 102, AUY 19, 2N2137, 2N2142	data
AD 431	Ge-P	NF-L, 32V, 2A,17W	23a§	Eiy	AD 149, AL 102, AUY 19, 2N2137, 2N2142	data
AD 432	Ge-P	NF-L, 32V, 1,5A, 20W	23a§	Eiy	AD 149, AL 102, AUY 19, 2N2137, 2N2142	data
AD 433	Ge-P	=AD 432: 45V	23a§	Eiy	AD 149, AL 102, AUY 19, 2N2138, 2N2143	data
AD 434	Ge-P	NF-L, 45V, 3A, 20W	23a§	Eiy	AD 149, AL 102, AUY 19, 2N2138, 2N2143	data
AD 436	Ge-P	NF-L, 40V, 3,5A, 30W	23a§	Eiy	AD 149, AL 102, AUY 19, 2N2138, 2N2143	data
AD 437	Ge-P	NF-L, 64V, 3A, 20W	23a§	Eiy	AL 102, AUY 19, 2N2139, 2N2144	data
AD 438	Ge-P	=AD 437: 60V	23a§	Eiy	AL 102, AUY 19, 2N2139, 2N2144	data
AD 439	Ge-P	=AD 437: 80V	23a§	Eiy	AL 102, AUY 20, 2N2141, 2N2146	data
AD 450	Ge-P	NF-L, 32V, 5A, 30W	23a§	Eiy	AL 102, AUY 28, 2N1539...1548	data

Type	Device	Short Description	Fig.	Manu	Comparision Types	More at
AD 451	Ge-P	NF-L, 32V, 5A, 30W	23a§	Eiy	AL 102, AUY 28, 2N1539...1548	data
AD 452	Ge-P	NF-L, 32V, 5A, 30W	23a§	Eiy	AL 102, AUY 28, 2N1539...1548	data
AD 453	Ge-P	NF-L, 35V, 5A, 30W	23a§	Eiy	AL 102, AUY 28, 2N1539...1548	data
AD 454	Ge-P	NF-L, 35V, 5A, 30W	23a§	Eiy	AL 102, AUY 28, 2N1539...1548	data
AD 455	Ge-P	NF-L, 35V, 5A, 30W	23a§	Eiy	AL 102, AUY 28, 2N1540...43, 2N1545...48	data
AD 456	Ge-P	NF-L, 50V, 5A, 30W	23a§	Eiy	AL 102, AUY 28, 2N1540...43, 2N1545...48	data
AD 457	Ge-P	NF-L, 50V, 5A, 30W	23a§	Eiy	AL 102, AUY 28, 2N1540...43, 2N1545...48	data
AD 458	Ge-P	NF-L, 80V, 5A, 30W	23a§	Eiy	AL 102, AUY 28, 2N1541...43, 2N1546...48	data
AD 459	Ge-P	NF-L, 80V, 5A, 30W	23a§	Eiy	AL 102, AUY 28, 2N1541...43, 2N1546...48	data
AD 460	Ge-P	NF-L, 32V, 6A, 30W	23a§	Eiy	AL 102, AUY 28, 2N1539...43, 2N1544...48	data
AD 461	Ge-P	NF-L, 32V, 6A, 30W	23a§	Eiy	AL 102, AUY 28, 2N1539...43, 2N1544...48	data
AD 462	Ge-P	NF-L, 32V, 6A, 30W	23a§	Eiy	AL 102, AUY 28, 2N1539...43, 2N1544...48	data
AD 463	Ge-P	NF-L, 35V, 6A, 30W	23a§	Eiy	AL 102, AUY 28, 2N1539...43, 2N1544...48	data
AD 464	Ge-P	NF-L, 35V, 6A, 30W	23a§	Eiy	AL 102, AUY 28, 2N1539...43, 2N1544...48	data
AD 465	Ge-P	NF-L, 35V, 6A, 30W	23a§	Eiy	AL 102, AUY 28, 2N1539...43, 2N1544...48	data
AD 466	Ge-P	NF-L, 50V, 6A, 30W	23a§	Eiy	AL 102, AUY 28, 2N1540...43, 2N1545...48	data
AD 467	Ge-P	NF-L, 50V, 6A, 30W	23a§	Eiy	AL 102, AUY 28, 2N1540...43, 2N1545...48	data
AD 468	Ge-P	NF-L, 70V, 6A, 30W	23a§	Eiy	AL 102, AUY 28, 2N1541...43, 2N1546...48	data
AD 469	Ge-P	NF-L, 70V, 6A, 30W	23a§	Eiy	AL 102, AUY 28, 2N1541...43, 2N1546...48	data
AD501	FET	Vu:100dB Vo:±12V Vi0:2mV f:4Mc	5Y	And		data pinout
AD502	UNI	Vs:±18V Vo:±14V Vi0:1mV	8A	And		data pinout
AD503	FET	Vs:±18V Vo:±10V f:1Mc f:100kc	8A	And		data pinout
AD504	HI-CMR	Vs:±18V Vu:108dB Vo:±13V Vi0:2.5mV	8A	And		data pinout
AD506	FET	Vo:±10V Vi0:3mV f:1Mc f:100kc	8A	And		data pinout
AD507	HI-SPEED	Vs:±20V Vu:98dB Vo:±12V Vi0:5mV	8A	And		data pinout
AD508	LO-POWER	Vs:±18V Vo:±13V Vi0:±2.5mV f:0.3Mc	8A	And		data pdf pinout
AD509	HI-SPEED	Vs:±15V Vu:78dB Vo:±10V Vi0:14mV	8A	And		data pinout
AD510	HI-PREC	Vs:±18V Vu:108dB Vo:±10V Vi0:0.1mV	8A	And		data pinout
AD511	FET	Vs:±18V Vo:±10V Vi0:1mV f:1Mc f:70kc	5Y	And		data pinout
AD511A	FET	Vs:±18V Vu:118dB Vo:±10V Vi0:2mV	5Y	And		data pinout
AD512	UNI	Vs:±18V Vu:94dB Vo:±10V Vi0:3mV	8A	And		data pinout
AD514	FET	Vs:±18V Vu:94dB Vo:±13V Vi0:20mV	8A	And		data pinout
AD515	HI-OHM	Vs:±18V Vu:86dB Vo:±12V Vi0:0.4mV	8A	And		data pdf pinout
AD515AJ	LO-POWER	Vs:±18V Vu:86dB Vo:±13V Vi0:<3mV	8A	And		data pdf pinout
AD515AK	LO-POWER	Vs:±18V Vu:86dB Vo:±13V Vi0:<1mV	8A	And		data pdf pinout
AD515AL	LO-POWER	Vs:±18V Vu:86dB Vo:±13V Vi0:<1mV	8A	And		data pdf pinout
AD516	UNI	V±:±15V Vu:66dB Vo:±12V Vi0:3mV	8A	And		data pinout
AD517	UNI	Vs:±18V Vu:120dB Vo:>±10V Vi0:0.15mV	8A	And		data pinout
AD518	HI-SPEED	Vs:±20V Vu:100dB Vo:±13V Vi0:4mV	8AD	And		data pinout
AD520	GAIN-SET	Vs:±18V Vu:0-60dB Vo:±10V f:0.125Mc	14D	And		data pinout
AD521	GAIN-SET	Vs:±18V Vu:0-60dB Vo:±10V Vi0:2mV	14D	And		data pdf pinout
AD 521 J,K,L,S	OP-IC	Gain-set(0,1...1000), ±18V, 0,04...>2MHz	14-DIP	And		data pinout
AD522	HYBRID	Vs:±18V Vu:0-60dB Vo:±10V Vi0:0.1mV	14D	And		data pdf pinout
AD522A	HYBRID	Vs:±18V Vu:0-60dB Vo:±10V Vi0:0.2mV	14D	And		data pinout
AD523	LO-BIAS	Vs:±18V Vu:88dB Vo:±10V Vi0:50mV	8A	And		data pinout
AD 524 (A,B,C,S)D	OP-IC	Gain-set(1...1000), ±18V, 0,025...1MHz	16-DIP	And	-	pdf
AD 524 (A,B,S)E	OP-IC	Gain-set(1...1000), ±18V, 0,025...1MHz	20-LCC	And	-	pdf
AD 524 AR-16	OP-IC	Gain-set(1...1000), ±18V, 0,025...1MHz	16-MDIP	And	-	pdf
AD526	OP	Vs:±16.5V Vo:±12V Vi0:0.25mV f:4Mc	16D	And	-	data pdf pinout
AD 526 (A,B,C,S)D	OP-IC	prog. gain- set(1... 16), ±16, 5V, 0, 35... 4MHz	16-DIC	And	-	pdf
AD526A	OP	Vs:±16.5V Vo:±12V Vi0:0.4mV f:4Mc	16D	And	-	data pdf pinout
AD 526 JN	OP-IC	Gain-set(1...16), ±16,5V, 0,35...4MHz	16-DIP	And	-	pdf
AD528	HI-SPEED	Vs:±20V Vu:100dB Vo:±10V Vi0:1mV	8A	And		data pinout
AD 532(J,K,S)D	LIN/OP-IC	OP-Amp, 2x Diff. Input, ±18V, S: ±22V	14-DIC			-
AD 532(J,K,S)H	LIN/OP-IC	OP-Amp, 2x Diff. Input, ±18V, S: ±22V	82			-
AD 533(J,K,L,S)D	LIN/OP-IC	Multiplier/Divider, ±18V, S: ±22V	14-DIC	And		-
AD 533(J,K,L,S)H	LIN/OP-IC	Multiplier/Divider, ±18V, S: ±22V	82			-
AD 534(J,K,L,S,T)D	LIN/OP-IC	Multiplier/Modulator, ±18V, 0...70°C S,T: ±22V -55...+125°C	14-DIC	And		pdf pinout
AD 534(S,T)E	LIN-IC	Multiplier/Modulat., ±22V, -55...+125°C	20-LCC	And		pdf pinout
AD 534(J,K,L,S,T)H	LIN/OP-IC	Multiplier/Modulator, ±18V, 0...70°C S,T: ±22V -55...+125°C	82	And		pdf pinout
AD 535(J,K)D	LIN/OP-IC	OP-Amp, Multiplier, ±18V	14-DIC	And		-
AD 535(J,K)H	LIN/OP-IC	OP-Amp, Multiplier, ±18V	82			-
AD 536A(...)N,D,Q	LIN-IC	RMS→DC Converter, ±3...18V	14-DIP,DIC	And, Max		pdf
AD 536A(...)H	LIN-IC	RMS→DC Converter, ±3...18V	82			pdf
AD 536A(...)WE	LIN-IC	SMD, RMS→DC Converter, ±3...18V	16-MDIP			pdf
AD 537(J,K,S)D	LIN-IC	Voltage/Frequ.-Converter, 5...36/±18V	14-DIC	And		pdf pinout
AD 537(J,K,S)H	LIN-IC	Voltage/Frequ.-Converter, 5...36/±18V	82	And		pdf pinout
AD 538(A,B,S)D	LIN-IC	Real Time Analogue Comput. Unit (acu) Multiplication,Division,Exponentation	18-DIC	And	-	pdf
AD 539(J,K,S)D	LIN-IC	2-Chan. Linear Multiplier/Divider, 2-Quadrant, 60MHz, ±4,5...16,5V	16-DIC	And		pdf pinout
AD 539 JN,KN	LIN-IC	2-Chan. Linear Multiplier/Divider, 2-Quadrant, 60MHz, ±4,5...16,5V	16-DIP	And		pdf pinout
AD 539 SE	LIN-IC	2-Chan. Linear Multiplier/Divider, 2-Quadrant, 60MHz, ±4,5...16,5V	20-LCC	And	-	pdf pinout
AD540	FET	Vs:±18V Vu:86dB Vo:±13V Vi0:50mV	8A	And		data pinout
AD 541	Ge-P	NF/S-L, 24V, 8A, 45W	38a§	Eiy	2N2078, 2N2082, 2N1980, 2N2491	data
AD 542	Ge-P	NF/S-L, 80V, 8A, 45W	38a§	Eiy	2N2075, 2N2079, 2N1982, 2N2492	data
AD542	FET	Vs:±18V Vo:±12V Vi0:<2mV f:1Mc	d,8A	And		data pdf pinout
AD544	FET	Vs:±18V Vo:±12V Vi0:<2mV f:2Mc	8A	And		data pdf pinout
AD 545	Ge-P	NF/S-L, 80V, 8A, 45W	38a§	Eiy	2N2076, 2N2080, 2N1982, 2N2491	data

Type	Device	Short Description	Fig.	Manu	Comparision Types	More at
AD545	FET	Vs:±18V Vo:±12V Vi0:<1mV f:700Mc	8A	And		data pinout
AD545A	FET	Vs:±18V Vo:±13V Vi0:<1mV f:1Mc	8A	And		data pinout
AD546	OP	Vs:±18V Vo:±12V Vi0:<2mV f:1Mc	8D	And		data pdf pinout
AD547	FET	Vs:±18V Vo:±12V Vi0:<1mV f:1Mc	8Ad	And		data pdf pinout
AD548	BIFET	Vs:±18V Vu:>100dB Vo:±13V Vi0:<2mV	d,8DSA	And		data pdf pinout
AD548A	BIFET	Vs:±18V Vu:>100dB Vo:±13V Vi0:<2mV	8ADS	And		data pdf pinout
AD548B	BIFET	Vs:±18V Vu:>100dB Vo:±13V Vi0:<0.5mV	8ADS	And		data pdf pinout
AD548C	BIFET	Vs:±18V Vu:>100dB Vo:±13V f:1Mc	8AD	And		data pinout
AD549	LO-BIAS	Vs:±18V Vo:±12V Vi0:0.5mV f:1Mc	d,8A	And		data pdf pinout
AD 558 JD	D/A-IC	8 Bit D/a-converter		And	(C 560D)	pdf
AD 562...	D/A-IC	12 Bit, hi-speed, multiplying	24-DIC	Mot		
AD 563 ...	D/A-IC	12 Bit, hi-speed	24-DIC	Mot		
AD 565 JN	D/A-IC	12 Bit D/a-converter	24-DIP	And	C 565D, µA 565...	
AD 565 ... N,D,Q	D/A-IC	12 Bit, hi-speed, int. 10V Reference	24-DIP,DIC	And, Max	C 565.., µA 565...	
AD 565 ... WG	D/A-IC	SMD, 12 Bit, hi-speed, int.10V Ref.	24-MDIP		-	
AD 566 ... N,D,Q	D/A-IC	12 Bit, hi-speed	24-DIP,DIC	And, Max	-	
AD 566 ... WG	D/A-IC	SMD, 12 Bit, hi-speed	24-MDIP		-	
AD 570 J,S	A/D-IC	8 Bit SAR A/D-Conv., Reference, Clock Tristate Out, +5V/-15V	18-DIC	And	-	pdf pinout
AD 571	A/D-IC	10-Bit A/D-Converter	18-DIP	And	C 571D, µA 571...	
AD 571 J,K,S	A/D-IC	10 Bit SAR A/D-Conv., +5...15V/-15V Clock, Reference, par.output, Tristate	18-DIC	And	C 571D, µA 571...	pdf pinout
AD 572 (A,B,S)D	A/D-IC	12 Bit SAR A/D-Conv., ser./par. Output Clock, +10VRef. temp.comp, +5/+15/-15V	32-DIC	And	-	pdf
AD 573 (J,K)D	A/D-IC	10 Bit SAR A/D-Converter, +5/-12...-15V	20-DIC	And	-	pdf pinout
AD 573 (J,K)N	A/D-IC	10 Bit SAR A/D-Converter, +5/-12...-15V	20-DIP	And	-	pdf pinout
AD 573 (J,K)P	A/D-IC	10 Bit SAR A/D-Converter, +5/-15V	20-PLCC	And	-	pdf pinout
AD 573 SD	A/D-IC	10 Bit SAR A/D-Conv.,-55...+125°C,±1LSB	20-DIC	And	-	pdf pinout
AD 574 A(J,K)E	A/D-IC	12 Bit SAR ADC, ±1/½LSB, +5V & ±12/15V CLK, Ref., 8/12Bit par.Out, 0...70°C	28-LCC	And	-	pdf pinout
AD 574 A(J,K,L)D	A/D-IC	12 Bit SAR ADC, ±1/½LSB, +5V & ±12/15V CLK, Ref., 8/12Bit par.Out, 0...70°C	28-DIC	And	-	pdf pinout
AD 574 A(J,K,L)N	A/D-IC	12 Bit SAR ADC, ±1/½LSB, +5V & ±12/15V CLK, Ref., 8/12Bit par.Out, 0...70°C	28-DIP	And	-	pdf pinout
AD 574 A(J,K,L)P	A/D-IC	12 Bit SAR ADC, ±1/½LSB, +5V & ±12/15V CLK, Ref., 8/12Bit par.Out, 0...70°C	28-PLCC	And	-	pdf pinout
AD 574 A(S,T,U)D	A/D-IC	12 Bit SAR ADC, ±1LSB, +5V & ±12/15V CLK, Ref., 8/12Bit par.Out, -55...125°C	28-DIC	And	-	pdf pinout
AD 578JN	A/D-IC	BiCMOS 12Bit, 6µs, Par/Ser Out, 0...+70°C	32-DIC	Max	-	pdf pinout
AD 578KN	A/D-IC	BiCMOS 12Bit, 4.5µs, Par/Ser Out, 0...+70°C	32-DIC	Max	-	pdf pinout
AD 578LN	A/D-IC	BiCMOS 12Bit, 3µs, Par/Ser Out, 0...+70°C	32-DIC	Max	-	pdf pinout
AD 578SD	A/D-IC	BiCMOS 12Bit, 6µs, Par/ Ser Out, 55... +125°C	32-DIC	Max	-	pdf pinout
AD 578TD	A/D-IC	BiCMOS 12Bit, 4.5µs, Par/Ser Out, -55...+125°C	32-DIC	Max	-	pdf pinout
AD 579JN	A/D-IC	BiCMOS 10Bit, 2.2µs, Par/Ser Out, 0...+70°C	32-DIC	Max	-	pdf pinout
AD 579KN	A/D-IC	BiCMOS 10Bit, 1.8µs, Par/Ser Out, 0...+70°C	32-DIC	Max	-	pdf pinout
AD 579TD	A/D-IC	BiCMOS 10Bit, 1.8µs, Par/Ser Out, -55...+125°C	32-DIC	Max	-	pdf pinout
AD580	R+	Io=10mA Vo:2.475..2.525V P:350mW	3A	And		data pdf pinout
AD580.H	R+	Io=10mA Vo:2.49..2.51V	3A	And		data pinout
AD 580 ... H	Ref-Z-IC	Prec. Voltage Ref., +2,5V ±10...75mV	2(+QM)	And, Max		data pdf
AD580J	R+	Io=10mA Vo:2.425..2.525V P:350mW	3A	And		data pdf pinout
AD580K	R+	Io=10mA Vo:2.45..2.55V P:350mW	3A	And		data pdf pinout
AD580KH	R+	Io=10mA Vo:2.425..2.525V	3A	And		data pinout
AD580L	R+	Io=10mA Vo:2.45..2.55V P:350mW	3A	And		data pdf pinout
AD580LH	R+	Io=10mA Vo:2.49..2.51V	3A	And		data pinout
AD 580 ... SA	Ref-Z-IC	=AD 580..H: SMD	8-MDIP			data pdf
AD580T	R+	Io=10mA Vo:2.45..2.55V P:350mW	3A	And		data pinout
AD581	R+	Io=10mA Vo:9.99...10.01V	3A	And		data pdf pinout
AD 581 ... H	Ref-Z-IC	Prec. Voltage Reference, +10V ±5..30mV	2(+QM)	And, Max	-	data pdf
AD 581 ... SA	Ref-Z-IC	=AD 581..H: SMD	8-MDIP		-	data pdf
AD583	SAMP&HOLD	Vs:±15V Vu:88dB Vo:±10V Vi0:8mV	14D	And		data pinout
AD 584 ... N	LIN-IC	Spg.-/Voltage-Ref., 10+7.5+5+2,5V	DIP	And	B 584X	data pdf
AD 584 ... H	LIN-IC	Prec. Voltage Ref., 10/7.5/5/2.5V	81	And	B 584X	data pdf
AD584J	R+	Io=10mA P:600mW	8AD	And		data pdf pinout
AD584K	R+	Io=10mA P:600mW	8AD	And		data pdf pinout
AD584L	R+	Io=10mA P:600mW	8A	And		data pdf pinout
AD 584 ... N,Q	Ref-Z-IC	=AD 584..H:	8-DIP,DIC		-	data
AD584S	R+	Io=10mA P:600mW	8A	And		data pdf pinout
AD 584 ... SA	Ref-Z-IC	=AD 584..H: SMD	8-MDIP		-	data
AD584T	R+	Io=10mA P:600mW	8A	And		data pdf pinout
AD586	R+	P:500mW	8DS	And		data pdf pinout
AD586J	R+	Vo:4.98..5.02V	d	And		data pdf pinout
AD 586 J(N,Q,R)	Z-Ref-IC	High Precision 5 V Reference	DIP	And		pdf pinout
AD587	R+	Vo:9.995..10.00V Vin:15V P:500mW	8DS	And		data pdf pinout
AD587J	R+	Vo:9.99...10.01V Vin:15V P:500mW	d,8DS	And		data pdf pinout
AD589	Z+	Io=5mA Vo:1.235...1.25V P:125mW	2A,8S	And		data pdf pinout
AD 589 JR	LIN-IC	Ref.-Spg.-Quelle/volt.refer., 1,235V	SDIP	And	B 589N	data pdf pinout
AD 589 N	LIN-IC	Ref.-Spg.-Quelle/volt.refer., 1,235V	7	And	B 589N	data pdf pinout
AD 592	Sensor	Precision IC Temperature Transducer		And		pdf pinout
AD 594	Sensor	Monolithic Thermocouple Amplifiers with	DIP	And	AD595	pdf pinout

Type	Device	Short Description	Fig.	Manu	Comparision Types	More at
AD 595	Sensor	Cold Junction Compensation. Monolithic Thermocouple Amplifiers with Cold Junction Compensation.	DIP	And	AD595	pdf pinout
AD611	JFET	Vs:±18V Vu:80V/mV Vi0:0.25mV f:2Mc	8A	And		data pinout
AD 620	LIN/OP-IC	Instrumentation Amplifier	DIP	And	-	pdf pinout
AD623B	RR-OP	Vs:±6V Vo:4.8/4.V Vi0:25µµV f:0.8Mc	8DS	And		data pdf pinout
AD 624 A,B,C,S	OP-IC	Gain-set(1...1000), ±18V, 0,025...1MHz	16-DIC	And	-	
AD 625 A,B,C,S	OP-IC	Gain-set(1...1000), ±18V, 25..650kHz	16-DIC	And	-	
AD 630(A,B,J,K,S)	LIN-IC	Hf Umschalter/Switch, ±5...16,5V	20-DIC,DIP	And	-	
AD 632(A,B,S,T)D	LIN/OP-IC	OP-Amp, Multiplier, ±18V, S,T: ±22V	14-DIC	And	-	
AD 632(A,B,S,T)H	LIN/OP-IC	OP-Amp, Multiplier, ±18V, S,T: ±22V	82	And	-	
AD 633 (...)N	LIN/OP-IC	4-Quadrant Analog Multiplier	DIP	And	-	pdf pinout
AD 633 (...)R	LIN/OP-IC	4-Quadrant Analog Multiplier	DIP	And	-	pdf pinout
AD 636(...)N,D,Q	LIN-IC	RMS→DC-Converter, +2/-2,5...±16,5V	14-DIP,DIC	And	-	pdf pinout
AD 636(...)H	LIN-IC	=AD 636...N,D,Q:	82	And		
AD 636(...)H	LIN-IC	RMS→DC-Converter, +2/-2,5...±16,5V	82	And	-	
AD 636(...)WE	LIN-IC	SMD, RMS→DC-Converter, ±2...±16,5V	16-MDIP	And	-	
AD 637(J,K,S)Q	LIN-IC	RMS/DC-Converter, ±3...18V	14-DIC	And	-	pdf pinout
AD 640(B,T)D	LIN-IC	Signal Amp, ±4,5...±7,5V	20-DIC	And	-	
AD 640(B,T)E	LIN-IC	Signal Amp, ±4,5...±7,5V	20-(P)LCC	And	-	
AD642	2xFET	Vs:±18V Vo:±12V Vi0:<2mV f:1Mc	d,8A	And		data pinout
AD644	2xBIFET	Vs:±18V Vi0:<2mV	8Ad	And		data pinout
AD645	FET	Vs:±18V Vu:130dB Vo:±11V Vi0:50µµV	8AD	And		data pdf pinout
AD647	2xBIFET	Vs:±18V Vi0:<1mV f:1Mc f:50kc	d,8A,20c	And		data pinout
AD648	2xBIFET	Vs:±18V Vu:>100dB Vo:±13V Vi0:<2mV	d,8DSA	And		data pinout
AD648A	2xBIFET	Vs:±18V Vu:>100dB Vo:±13V Vi0:<2mV	8AD	And		data pdf pinout
AD648B	2xBIFET	Vs:±18V Vu:>100dB Vo:±13V Vi0:<0.5mV	8AD	And		data pdf pinout
AD648C	2xBIFET	Vs:±18V Vu:>100dB Vo:±13V f:1Mc	8AD	And		data pinout
AD 650(A,B,J,K,S)	LIN-IC	V/F-Converter, ±9..±18V	14-DIC,DIP	And	-	
AD 651(A,B,S)Q	LIN-IC	V/F-Converter, ±6..±18V	16-DIC	And	-	
AD 652(J,K)P	LIN-IC	V/F-Converter, ±6...±18V	20-PLCC	And	-	
AD 652(A,B,S)Q	LIN-IC	V/F-Converter, ±6...±18V	16-DIC	And	-	
AD 654JR	LIN-IC	V/F-Converter, ±5...±18V	8-DIP	And	-	pdf pinout
AD 654(J)N	LIN-IC	V/F-Converter, ±5...±18V	8-DIP	And	-	pdf pinout
AD 674 B(A,B)R	A/D-IC	12 Bit SAR ADC, ±1/½LSB, +5V & ±12/15V CLK, Ref., 8/12Bit par.Out, tc=15µs	28-MDIP	And	-	pdf
AD 674 B(A,B,T)D	A/D-IC	12 Bit SAR ADC, ±1/½LSB, +5V & ±12/15V CLK, Ref., 8/12Bit par.Out, tc=15µs	28-DIC	And	-	pdf
AD 674 BJN,BKN	A/D-IC	12 Bit SAR ADC, ±1/½LSB, +5V & ±12/15V CLK, Ref., 8/12Bit par.Out, tc=15µs	28-DIP	And	-	pdf
AD 676 AD,BD	A/D-IC	16 Bit SAR A/D-Conv., S&H, +5V & ±12V par. Out, 100kSPS, --/1.5LSB, -40...85°	28-DIC	And	-	
AD 676 JD,KD	A/D-IC	16 Bit SAR A/D-Conv., S&H, +5V & ±12V par. Out, 100kSPS, --/1.5LSB, 0...70°	28-DIC	And	-	
AD 677 AD,BD	A/D-IC	16 Bit SAR A/D-Conv., S&H, +5V & ±12V ser. Out, 100kSPS, --/1.5LSB, -40...85°	16-DIC	And	-	
AD 677 JD,KD	A/D-IC	16 Bit SAR A/D-Conv., S&H, +5V & ±12V ser. Out, 100kSPS, --/1.5LSB, 0...70°	16-DIC	And	-	
AD 677 JN,KN	A/D-IC	16 Bit SAR A/D-Conv., S&H, +5V & ±12V ser. Out, 100kSPS, --/1.5LSB, 0...70°	16-DIP	And	-	
AD 677 JR,KR	A/D-IC	16 Bit SAR A/D-Conv., S&H, +5V & ±12V ser. Out, 100kSPS, --/1.5LSB, 0...70°	28-MDIP	And	-	
AD680	R+	Vo:2.495...2.505V Vin:5V P:500mW	8DS,3N	And		data pdf pinout
AD 684 (A,J,S)Q	LIN-IC	Quad, Sample&Hold Amplifier, ±12V	16-DIC	And	-	pdf pinout
AD688	R±	Vo:...±10V P:600mW	16D	And		data pdf pinout
AD 694	LIN-IC	4–20 mA Transmitter	DIP	And		pdf pinout
AD 701D...J	Ge-P	NF-L, 40...80V, 6A, 45W	23a§	Elx	AL 102, AUY 28, 2N1539...1548	data
AD 702A...Z	Ge-P	NF-L, 30...100V, 3...10A, 45W	23a§	Elx	AL 102, AUY 28, 2N1539...1548	data
AD704	4xBIFET	Vs:±18V Vu:2000V/mV Vo:±14V f:0.8Mc	14D,16S	And		data pdf pinout
AD704A	4xBIFET	Vs:±18V Vu:2000V/mV Vo:±14V f:0.8Mc	14D,16S	And		data pdf pinout
AD705	BIFET	Vs:±18V Vu:2000V/mV Vo:±14V f:0.8Mc	8DdS	And		data pdf pinout
AD705A	BIFET	Vs:±18V Vu:2000V/mV Vo:±14V f:0.8Mc	8D	And		data pdf pinout
AD706	2xBIFET	Vs:±18V Vu:2000V/mV Vo:±14V f:0.8Mc	8DdS	And		data pdf pinout
AD 706	LIN/OP-IC	Dual JFET OP, low power.	MDIP	And		pdf pinout
AD706A	2xBIFET	Vs:±18V Vu:2000V/mV Vo:±14V f:0.8Mc	8DS	And		data pdf pinout
AD707	LO-DRIFT	Vs:±22V Vo:±14V Vi0:30µµV f:0.9Mc	d,8DSA	And		data pinout
AD707A	LO-DRIFT	Vs:±22V Vo:±14V Vi0:30µµV f:0.9Mc	8ADS	And		data pinout
AD707B	LO-DRIFT	Vs:±22V Vo:±14V Vi0:10µµV f:0.9Mc	8AD	And		data pinout
AD707C	LO-DRIFT	Vs:±22V Vo:±14V Vi0:5µµV f:0.9Mc	8AD	And		data pinout
AD708	2xLO-OFFSET	Vs:±22V Vo:±14V Vi0:30µµV f:0.9Mc	d,8DA	And		data pinout
AD708A	2xLO-OFFSET	Vs:±22V Vo:±14V Vi0:30µµV f:0.9Mc	8AD	And		data pinout
AD708B	2xLO-OFFSET	Vs:±22V Vo:±14V Vi0:5µµV f:0.9Mc	8AD	And		data pinout
AD711	BIFET	Vs:±18V Vu:400V/mV Vo:±13V Vi0:0.3mV	8DSdA	And		data pdf pinout
AD711A	BIFET	Vs:±18V Vu:400V/mV Vo:±13V Vi0:0.3mV	8D	And		data pdf pinout
AD711B	BIFET	Vs:±18V Vu:400V/mV Vo:±13V Vi0:0.2mV	8AD	And		data pdf pinout
AD711C	BIFET	Vs:±18V Vu:400V/mV Vo:±13V Vi0:0.1mV	8AD	And		data pdf pinout
AD711SQ/883	BIFET	Vs:±18V Vu:400V/mV Vo:±13V Vi0:0.2mV	8D	And		data pdf pinout
AD711TQ/883	BIFET	Vs:±18V Vu:400V/mV Vo:±13V Vi0:0.2mV	8D	And		data pdf pinout
AD712	2xBIFET	Vs:±18V Vu:>100dB Vo:±13V Vi0:<3mV	8DSdA	And		data pdf pinout
AD712A	2xBIFET	Vs:±18V Vu:>100dB Vo:±13V Vi0:<3mV	d,8AD	And		data pdf pinout
AD712B	2xBIFET	Vs:±18V Vu:>100dB Vo:±13V Vi0:<1mV	8AD	And		data pdf pinout
AD712C	2xBIFET	Vs:±18V Vu:>100dB Vo:±13V Vi0:<0.5mV	8AD	And		data pdf pinout
AD713	4xBIFET	Vs:±18V Vu:400V/mV Vo:±13V Vi0:0.2mV	14Dd,16S	And		data pdf pinout
AD713A	4xBIFET	Vs:±18V Vu:400V/mV Vo:±13V Vi0:0.3mV	14D	And		data pdf pinout
AD717A	BIFET	Vs:±18V Vu:400V/mV Vo:±13V Vi0:0.3mV	8A	And		data pdf pinout

Type	Device	Short Description	Fig.	Manu	Comparision Types	More at
AD 736(A,B)N	LIN-IC	RMS/DC-Converter, +2,8/-3,2...±16,5V	8-DIP	And	-	
AD 736(A,B)Q	LIN-IC	RMS/DC-Converter, +2,8/-3,2...±16,5V	8-DIC	And	-	
AD 736(A,B)R	LIN-IC	SMD, RMS/DC-Conv., +2,8/-3,2...±16,5V	8-MDIP		-	
AD 737(A,B,J,K)N	LIN-IC	RMS/DC-Converter, +2,8/-3,2...±16,5V	8-DIP	And	-	
AD 737(A,B,J,K)Q	LIN-IC	RMS/DC-Converter, +2,8/-3,2...±16,5V	8-DIC	And	-	
AD 737(A,B,J,K)R	LIN-IC	SMD, RMS/DC-Conv., +2,8/-3,2...±16,5V	8-MDIP		-	
AD741	UNI	Vs:±15V Vu:93dB Vo:±10V Vi0:6mV	8A	And		data pdf pinout
AD 741(C)H	OP-IC	Uni, Serie 741	81	And	→Serie 741	data pinout
AD741C	UNI	Vs:±15V Vu:86dB Vo:±10V Vi0:7.5mV	8AD	And		data pdf pinout
AD741J	UNI	Vs:±15V Vu:94dB Vo:±10V Vi0:4mV	8AD	And		data pdf pinout
AD741K	UNI	Vs:±15V Vu:94dB Vo:±10V Vi0:3mV	8AD	And		data pdf pinout
AD741L	UNI	Vs:±15V Vu:94dB Vo:±10V Vi0:1mV	8AD	And		data pdf pinout
AD743	BIFET	Vs:±18V Vu:4000V/mV Vo:±13V f:4.5Mc	d,8D,16S	And		data pinout
AD743A	BIFET	Vs:±18V Vu:4000V/mV Vo:±13V f:4.5Mc	8D,16S	And		data pinout
AD743B	BIFET	Vs:±18V Vu:4000V/mV Vo:±13V f:4.5Mc	8D,16S	And		data pinout
AD744	BIFET	Vs:±18V Vu:400V/mV Vo:±13V Vi0:0.3mV	d,8DSA	And		data pinout
AD744A	BIFET	Vs:±18V Vu:400V/mV Vo:±13V Vi0:0.3mV	8AD	And		data pinout
AD744B	BIFET	Vs:±18V Vu:400V/mV Vo:±13V f:13Mc	8AD	And		data pinout
AD744C	BIFET	Vs:±18V Vu:400V/mV Vo:±13V f:13Mc	8AD	And		data pinout
AD745	BIFET	Vs:±18V Vu:4000V/mV Vo:±13V f:20Mc	d,8D,16S	And		data pinout
AD745A	BIFET	Vs:±18V Vu:4000V/mV Vo:±13V f:20Mc	8D,16S	And		data pinout
AD745B	BIFET	Vs:±18V Vu:4000V/mV Vo:±13V f:20Mc	8D	And		data pinout
AD746	2xBIFET	Vs:±18V Vu:300V/mV Vo:±13V Vi0:0.3mV	8DSd	And		data pdf pinout
AD746A	2xBIFET	Vs:±18V Vu:300V/mV Vo:±13V Vi0:0.3mV	8D	And		data pdf pinout
AD746B	2xBIFET	Vs:±18V Vu:300V/mV Vo:±13V f:13Mc	8D	And		data pdf pinout
AD 760 AQ	CMOS-D/A-IC	16/18 Bit, D/A-Converter, ±15V & +5V ser. & par. Inp., Self. Calibrating	28-DIC	And	-	data pdf pinout
AD 770 J,K	CMOS-A/D-IC	8 Bit, Video, >200 Msps, Ecl Level	40-DIP	And	-	
AD 774 B(A,B)R	A/D-IC	12 Bit SAR ADC, ±1/½LSB, +5V & ±12/15V CLK, Ref., 8/12Bit par.Out, tc=8µs	28-MDIP	And	-	pdf
AD 774 B(A,B,T)D	A/D-IC	12 Bit SAR ADC, ±1/½LSB, +5V & ±12/15V CLK, Ref., 8/12Bit par.Out, tc=8µs	28-DIC	And	-	pdf
AD 774 BJN,BKN	A/D-IC	12 Bit SAR ADC, ±1/½LSB, +5V & ±12/15V CLK, Ref., 8/12Bit par.Out, tc=8µs	28-DIP	And	-	pdf
AD779A	LO-NOISE	Vs:±18V Vu:20kV/mV Vo:±13V f:110Mc	8D	And		data pinout
AD780A	R+	Vo:2.995...3.005V Vin:5V P:500mW	8DS	And		data pdf pinout
AD780B	R+	Vo:2.999...3.001V Vin:5V P:500mW	8DS	And		data pdf pinout
AD780S	R+	Vo:2.995...3.005V Vin:5V P:500mW	8D	And		data pdf pinout
AD790	FAST-COMP	Vs:±18V Vi0:0.2mV	8DSd	And		data pdf pinout
AD790A	FAST-COMP	Vs:±18V Vi0:0.2mV	8D	And		data pdf pinout
AD790B	FAST-COMP	Vs:±18V Vi0:0.05mV	8D	And		data pdf pinout
AD795	LO-POWER	Vs:±18V Vu:120dB Vo:±11V Vi0:0.1mV	8DS	And		data pdf pinout
AD797A	LO-NOISE	Vs:±18V Vo:±13V Vi0:0.025mV f:110Mc	d,8DS	And		data pinout
AD797B	LO-NOISE	Vs:±18V Vo:±13V Vi0:0.01mV f:110Mc	8DS	And		data pinout
AD797S	LO-NOISE	Vs:±18V Vu:20kV/mV Vo:±13V f:110Mc	8DS	And		data pinout
AD801	UNI	Vs:±18V Vu:83.5dB Vo:±10V Vi0:5mV	8A	And		data pinout
AD810AN	LO-POWER	Vs:±18V Vu:100dB Vo:±12.9V Vi0:1.5mV	8D	And		data pinout
AD810AR	LO-POWER	Vs:±18V Vu:100dB Vo:±12.9V Vi0:1.5mV	8S	And		data pinout
AD811AC	HI-PERF	Vs:±18V Vo:±13V Vi0:0.5mV f:140Mc	d	And		data pdf pinout
AD811AE	HI-PERF	Vs:±18V Vo:±13V Vi0:0.5mV f:140Mc	20C	And		data pdf pinout
AD811AN	HI-PERF	Vs:±18V Vo:±13V Vi0:0.5mV f:140Mc	8D	And		data pdf pinout
AD811AQ	HI-PERF	Vs:±18V Vo:±13V Vi0:0.5mV f:140Mc	8D	And		data pdf pinout
AD811AR	HI-PERF	Vs:±18V Vo:±13V Vi0:0.5mV f:140Mc	16S,20S	And		data pdf pinout
AD811J	HI-PERF	Vs:±12V Vo:±12V Vi0:0.5mV f:140Mc	8S	And		data pdf pinout
AD811S	HI-PERF	Vs:±18V Vo:±12V Vi0:0.5mV f:140Mc	d,20c,8D	And		data pdf pinout
AD812AN	2xLO-POWER	Vs:±18V Vu:82dB Vo:±14V Vi0:2mV	8D	And		data pdf pinout
AD812AR	2xLO-POWER	Vs:±18V Vu:82dB Vo:±14V Vi0:2mV	8S	And		data pdf pinout
AD813AC	3xLO-POWER	Vs:±18V Vu:82dB Vo:±14V Vi0:2mV	d	And		data pdf pinout
AD813AN	3xLO-POWER	Vs:±18V Vu:82dB Vo:±14V Vi0:2mV	14D	And		data pdf pinout
AD813AR	3xLO-POWER	Vs:±18V Vu:82dB Vo:±14V Vi0:2mV	14S	And		data pinout
AD815AN	2xDIFF-DRV	Vs:±18V Vo:±11.7V Vi0:5mV f:40Mc	8S,24S	And		data pinout
AD815AV	2xDIFF-DRV	Vs:±18V Vo:±11.7V Vi0:5mV f:40Mc	15S	And		data pinout
AD815AY	2xDIFF-DRV	Vs:±18V Vo:±11.7V Vi0:5mV f:40Mc	15P	And		data pinout
AD816AV	2xDIFF-DRV	Vs:±18V Vi0:5mV f:10Mc	15S	And		data pinout
AD816AY	2xDIFF-DRV	Vs:±18V Vi0:5mV f:10Mc	15P	And		data pinout
AD817AN	HI-SPEED	Vs:±18V Vu:6V/mV Vo:±13.7V Vi0:0.5mV	8D	And		data pinout
AD817AR	HI-SPEED	Vs:±18V Vu:6V/mV Vo:±13.7V Vi0:0.5mV	8S	And		data pinout
AD818AN	LO-POWER	Vs:±18V Vu:9V/mV Vo:±13.7V Vi0:0.5mV	8D	And		data pinout
AD818AR	LO-POWER	Vs:±18V Vu:9V/mV Vo:±13.7V Vi0:0.5mV	8S	And		data pinout
AD820AN	LO-POWER	Vs:±18V Vi0:0.2mV f:1.5Mc f:240kc	8D	And		data pinout
AD820AQ	LO-POWER	Vs:±18V Vu:1000V/mV Vo:4.99/0V	8D	And		data pinout
AD820AR	LO-POWER	Vs:±18V Vi0:0.2mV f:1.5Mc f:240kc	8S	And		data pinout
AD820B	LO-POWER	Vs:±18V Vi0:0.3mV f:1.9Mc f:45kc	8DS	And		data pinout
AD820S	LO-POWER	Vs:±18V Vu:1000V/mV Vo:4.99/0V	8DS	And		data pinout
AD822AC	2xLO-POWER	Vs:±18V Vi0:0.4mV f:1.9Mc f:45kc	d	And		data pinout
AD822AN	2xLO-POWER	Vs:±18V Vi0:0.2mV f:1.5Mc f:240kc	8D	And		data pinout
AD822AQ	2xLO-POWER	Vs:±18V Vu:1000V/mV Vo:4.99/0V	9D	And		data pinout
AD822AR	2xLO-POWER	Vs:±18V Vi0:0.2mV f:1.5Mc f:240kc	8S	And		data pinout
AD822B	2xLO-POWER	Vs:±18V Vi0:0.3mV f:1.9Mc f:45kc	8DS	And		data pinout
AD822S	2xLO-POWER	Vs:±18V Vu:1000V/mV Vo:4.99/0V	8DS	And		data pinout
AD823AN	2xRR-OP	Vs:36V Vu:60V/mV Vo:±14.95V f:4Mkc	8D	And		data pdf pinout
AD823AR	2xRR-OP	Vs:36V Vu:60V/mV Vo:±14.95V f:4Mkc	8S	And		data pinout
AD824	4xRR-OP	Vs:±18V Vu:50V/mV Vi0:0.5mV f:2Mc	14SD	And		data pdf pinout
AD824AC	4xRR-OP	Vs:±18V Vu:50V/mV Vi0:0.5mV f:2Mc	d	And		data pdf pinout
AD824AN	4xRR-OP	Vs:±18V Vu:20V/mV Vi0:0.2mV f:2Mc	14D	And		data pinout

Type	Device	Short Description	Fig.	Manu	Comparision Types	More at
AD824AR	4xRR-OP	Vs:±18V Vu:20V/mV Vi0:0.2mV f:2Mc	14S,16S	And		data pinout
AD825AR	JFET	Vs:±18V Vu:76dB Vo:±13.3V Vi0:1mV	8S	And		data pinout
AD826AN	2xHI-SPEED	Vs:±18V Vu:6V/mV Vo:±13.7V Vi0:0.5mV	8D	And		data pdf pinout
AD826AR	2xHI-SPEED	Vs:±18V Vu:6V/mV Vo:±13.7V Vi0:0.5mV	8S	And		data pinout
AD827	2xHI-SPEED	Vs:±18V Vu:5.5V/mV Vo:±13.3V f:50Mc	8Dd,16S	And		data pdf pinout
AD828AN	2xLO-POWER	Vs:±18V Vu:6V/mV Vo:±13.7V Vi0:0.5mV	8D	And		data pdf pinout
AD828AR	2xLO-POWER	Vs:±18V Vu:6V/mV Vo:±13.7V Vi0:0.5mV	8S	And		data pinout
AD829AQ	HI-SPEED	Vs:±18V Vu:100V/mV Vo:±13.3V f:750Mc	8D	And		data pinout
AD829J	HI-SPEED	Vs:±18V Vu:100V/mV Vo:±13.3V f:750Mc	d,8DS	And		data pinout
AD829S	HI-SPEED	Vs:±18V Vu:100V/mV Vo:±13.3V f:750Mc	d,20c,8D	And		data pdf pinout
AD830	DIFF-OUT	Vs:±18V Vu:69dB Vo:±13.8V Vi0:±1.5mV	8DS	And		data pdf pinout
AD 834(J,S)Q	LIN/OP-IC	4 Quadr.-Multipl., 0...500MHz, ±4...9V	8-DIC	And		
AD 834(J,S)R	LIN-IC	SMD,4 Quadr.-Multipl., 500MHz, ±4...9V	8-MDIP	And	-	
AD840J	WIDEBAND	Vs:±18V Vu:130V/mV Vo:±10V Vi0:0.2mV	14D	And		data pinout
AD840K	WIDEBAND	Vs:±18V Vu:130V/mV Vo:±10V Vi0:0.1mV	14D	And		data pinout
AD840S	WIDEBAND	Vs:±18V Vu:130V/mV Vo:±10V Vi0:0.2mV	20C,14D	And		data pinout
AD841	WIDEBAND	Vs:±18V Vu:45V/mV Vo:>±10V Vi0:0.5mV	12A,14Dd	And		data pdf pinout
AD841J	WIDEBAND	Vs:±18V Vu:45V/mV Vo:>±10V Vi0:0.8mV	d,12A,14D	And		data pdf pinout
AD842A	FAST-SET	Vs:±18V Vu:25V/mV Vo:±11.5V Vi0:1mV	8D	And		data pinout
AD842J	WIDEBAND	Vs:±18V Vu:90V/mV Vo:>±10V Vi0:0.5mV	d,12A,14D	And		data pdf pinout
AD842K	WIDEBAND	Vs:±18V Vu:90V/mV Vo:>±10V Vi0:0.3mV	12A,14D	And		data pdf pinout
AD842S	WIDEBAND	Vs:±18V Vu:90V/mV Vo:>±10V Vi0:0.5mV	d,20c,12A	And		data pdf pinout
AD843B	FAST-SET	Vs:±18V Vu:30V/mV Vo:±11.5V f:34Mc	12A,8D	And		data pdf pinout
AD843J	FAST-SET	Vs:±18V Vu:25V/mV Vo:±11.5V Vi0:1mV	d,8D,16S	And		data pdf pinout
AD843K	FAST-SET	Vs:±18V Vu:30V/mV Vo:±11.5V f:34Mc	8D	And		data pdf pinout
AD843S	FAST-SET	Vs:±18V Vu:30V/mV Vo:±11.5V Vi0:1mV	d,20c,12A	And		data pdf pinout
AD844	HI-SLEW	Vs:±18V Vo:±11V Vi0:50µµV f:60Mc	d,8D,16S	And		data pdf pinout
AD 844(...)N,Q	OP-IC	hi-slew, ±4,5...18V, ±50mA, 2000/µs	8-DIP,DIC	And		data
AD845A	FET	Vs:±18V Vu:500V/mV Vo:>±12.5V f:16Mc	8D	And		data pdf pinout
AD845B	FET	Vs:±18V Vu:500V/mV Vo:>±12.5V f:16Mc	8D	And		data pdf pinout
AD845J	FET	Vs:±18V Vu:500V/mV Vo:>±12.5V f:16Mc	d,8D,16S	And		data pdf pinout
AD845K	FET	Vs:±18V Vu:500V/mV Vo:>±12.5V f:16Mc	8D	And		data pdf pinout
AD845S	FET	Vs:±18V Vu:500V/mV Vo:>±12.5V f:16Mc	d,8D	And		data pdf pinout
AD846	HI-PREC	Vs:±18V Vo:>±10V Vi0:25µµV f:80Mc	d,8D	And		data pinout
AD 846(...)N,Q	OP-IC	hi-slew, ±4,5...18V, 450/µs	8-DIP,DIC	And		data
AD847	HI-SPEED	Vs:±18V Vu:5.5V/mV Vo:>±10V f:50Mc	8DSd	And		data pdf pinout
AD848	HI-SPEED	Vs:±18V Vu:20V/mV Vo:>±10V Vi0:0.2mV	8DdS,20C	And		data pdf pinout
AD848S	HI-SPEED	Vs:±18V Vu:20V/mV Vo:>±10V Vi0:0.2mV	8D	And		data pinout
AD849A	HI-SPEED	Vs:±18V Vu:85V/mV Vo:>±10V Vi0:0.1mV	8D	And		data pdf pinout
AD849J	HI-SPEED	Vs:±18V Vu:85V/mV Vo:>±10V Vi0:0.3mV	d,8DS	And		data pinout
AD849S	HI-SPEED	Vs:±18V Vu:85V/mV Vo:>±10V Vi0:0.1mV	d,8D	And		data pinout
AD 974 AN,BN	A/D-IC	4 Chan./Kan. 16 Bit ADC, 200kSPS, +5V ser. Outp., Clock, INLmax=±3/±2 LSB	28-DIP	And	-	
AD 974 AR,BR	A/D-IC	4 Chan./Kan. 16 Bit ADC, 200kSPS, +5V ser. Outp., Clock, INLmax=±3/±2 LSB	28-MDIP	And	-	
AD 974 ARS,BRS	A/D-IC	4 Chan./Kan. 16 Bit ADC, 200kSPS, +5V ser. Outp., Clock, INLmax=±3/±2 LSB	28-SSMDIP	And	-	
AD 976 (A,B,C)N	A/D-IC	16 Bit SAR A/D-Conv., 100kSPS, +5V par. Outp., Clock, INLmax=±3/2/-- LSB	28-DIP	And	-	
AD 976 (A,B,C)R	A/D-IC	16 Bit SAR A/D-Conv., 100kSPS, +5V par. Outp., Clock, INLmax=±3/2/-- LSB	28-MDIP	And	-	
AD 976 (A,B,C)RS	A/D-IC	16 Bit SAR A/D-Conv., 100kSPS, +5V par. Outp., Clock, INLmax=±3/2/-- LSB	28-SSMDIP	And	-	
AD 976 A(A,B,C)N	A/D-IC	16 Bit SAR A/D-Conv., 200kSPS, +5V par. Outp., Clock, INLmax=±3/2/-- LSB	28-DIP	And	-	
AD 976 A(A,B,C)R	A/D-IC	16 Bit SAR A/D-Conv., 200kSPS, +5V par. Outp., Clock, INLmax=±3/2/-- LSB	28-MDIP	And	-	
AD 976 A(A,B,C)RS	A/D-IC	16 Bit SAR A/D-Conv., 200kSPS, +5V par. Outp., Clock, INLmax=±3/2/-- LSB	28-SSMDIP	And	-	
AD 977 (A,B,C)N	A/D-IC	16 Bit SAR A/D-Conv., 100kSPS, +5V ser. Outp., Clock, INLmax=±3/2/-- LSB	20-DIP	And	-	
AD 977 (A,B,C)R	A/D-IC	16 Bit SAR A/D-Conv., 100kSPS, +5V ser. Outp., Clock, INLmax=±3/2/-- LSB	20-MDIP	And	-	
AD 977 (A,B,C)RS	A/D-IC	16 Bit SAR A/D-Conv., 100kSPS, +5V ser. Outp., Clock, INLmax=±3/2/-- LSB	28-SSMDIP	And	-	
AD 977 A(A,B,C)N	A/D-IC	16 Bit SAR A/D-Conv., 200kSPS, +5V ser. Outp., Clock, INLmax=±3/2/-- LSB	20-DIP	And	-	
AD 977 A(A,B,C)R	A/D-IC	16 Bit SAR A/D-Conv., 200kSPS, +5V ser. Outp., Clock, INLmax=±3/2/-- LSB	20-MDIP	And	-	
AD 977 A(A,B,C)RS	A/D-IC	16 Bit SAR A/D-Conv., 200kSPS, +5V ser. Outp., Clock, INLmax=±3/2/-- LSB	28-SSMDIP	And	-	
AD 1202	Ge-P	NF-L, 45V, 1,5A, 13,5W	23a§	Tsm	AD 149, AL 102, AUY 19, 2N1539...1548	data
AD 1203	Ge-P	=AD 1202: 60V	23a§	Tsm	AL 102, AUY 19, 2N1540...34, 2N1545...48	data
AD1403	R+	Io=10mA	8D	And		data pinout
AD 1674 (A,B,J,K)R	A/D-IC	12 Bit SAR ADC, ±1/½LSB, +5V & ±12/15V CLK+Ref.+S&H, 8/12Bit par.Out, 100kSPS	28-MDIP	And	-	pdf
AD 1674 (A,B,T)D	A/D-IC	12 Bit SAR ADC, ±1/½LSB, +5V & ±12/15V CLK+Ref.+S&H, 8/12Bit par.Out, 100kSPS	28-DIC	And	-	pdf
AD 1674 JN,KN	A/D-IC	12 Bit SAR ADC, ±1/½LSB, +5V & ±12/15V CLK+Ref.+S&H, 8/12Bit par.Out, 100kSPS	28-DIP	And	-	pdf
AD 1990 ACPZ	Dig. Audio	Class-D Audio Power Amplifier	PLCC	And	-	pdf pinout
AD 1991 ASV	Dig. Audio	Class D/1-Bit Audio Power Output Stage	PLCC	And	-	pdf pinout
AD 1991 ASVRL	Dig. Audio	Class D/1-Bit Audio Power Output Stage	PLCC	And	-	pdf pinout
AD 1992 ACPZ	Dig. Audio	Class-D Audio Power Amplifier	PLCC	And	-	pdf pinout
AD 1994 ACPZ	Dig. Audio	Class-D Audio Power Amplifier	PLCC	And	-	pdf pinout

Type	Device	Short Description	Fig.	Manu	Comparision Types	More at
AD 1996 ACPZ	Dig. Audio	Class-D Audio Power Amplifier	PLCC	And	-	pdf pinout
AD 2020	A/D-IC	3 Digit A/D-Converter, Dual-Slope	16-DIP	And	C 520D	
ADNS-2030	CMOS-IC	Optical Mouse Sensor	DIP	Agt	-	pdf pinout
ADNS-2051	CMOS-IC	Optical Mouse Sensor	DIP	Agt	-	pdf pinout
AD2700	R+	Io=10mA Vin:15V P:400mW	14D	Max		data pinout
AD 2700 ... D	Ref-Z-IC	Prec. Voltage Ref., +10V, <10ppm/°C	14-DIC	And, Max	-	data
AD2701	R-	Io=10mA Vo:-9.995...-10.10V Vin:-15V	14D	Max		data pinout
AD 2701 ... D	Ref-Z-IC	Prec. Voltage Ref., -10V, <10ppm/°C	14-DIC	And, Max	-	data
AD2710	R+	Io=10mA Vo:9.999...V Vin:15V P:400mW	14D	Max		data pinout
AD 2710 ... D	Ref-Z-IC	Prec. Voltage Ref., +10V, <2ppm/°C	14-DIC	And, Max	-	data
AD3542	HI-OHM	Vs:±20V Vu:88dB Vo:±10V Vi0:20mV	8A	And		data pinout
AD3554	WIDEBAND	Vs:±20V Vu:106dB Vo:±11V Vi0:0.2mV	8O	Max		data pinout
AD3554A	WIDEBAND	Vs:±20V Vu:106dB Vo:±11V Vi0:0.5mV	8O	Max		data pinout
AD 3554(...)M	Hybrid-OP-IC	Wideband, JFET Inp, 40/±20V, >1000V/µs A,B: -25...+85°, M: -55...+125°	23/8Pin	Max	BB 3554...	data
AD 5241BR ...	Dig. Audio	SMD, 256 pos. dig. potentiometer, 10k/100k/1MOhm, I2C	MDIP	And	-	pdf pinout
AD 5241BRU ...	Dig. Audio	SMD, 256 pos. dig. potentiometer, 10k/100k/1MOhm, I2C	SSMDIP	And	-	pdf pinout
AD 5242BR ...	Dig. Audio	SMD, dual 256 pos. dig. potentiometer, 10k/100k/1MOhm, I2C	MDIP	And	-	pdf pinout
AD 5243 BRM10-RL7	IC	SMD, dual 256- position dig. potentiometer 10kOhm, I2C	VSMDIP	And		pdf pinout
AD 5243 BRM100-RL7	IC	SMD, dual 256- position dig. potentiometer 100kOhm, I2C	VSMDIP	And		pdf pinout
AD 5243 BRM2.5-RL7	IC	SMD, dual 256- position dig. potentiometer 2.5KOhm, I2C	VSMDIP	And		pdf pinout
AD 5243 BRM50-RL7	IC	SMD, 256- position dig. potentiometer 50kOhm, I2C	VSMDIP	And		pdf pinout
AD 5248 BRM10-RL7	IC	SMD, dual 256- position dig. potentiometer 10kOhm, I2C	VSMDIP	And		pdf pinout
AD 5248 BRM100-RL7	IC	SMD, dual 256- position dig. potentiometer 100kOhm, I2C	VSMDIP	And		pdf pinout
AD 5248 BRM2.5-RL7	IC	SMD, dual 256- position dig. potentiometer 2.5kOhm, I2C	VSMDIP	And		pdf pinout
AD 5248 BRM50-RL7	IC	SMD, dual 256- position dig. potentiometer 50kOhm, I2C	VSMDIP	And		pdf pinout
AD 5300 BRM	D/A-IC	SMD, 8 Bit D/A-Conv., ser. 3-wire Inp. Rail-to-Rail Outp., +2.7V...+5.5V	8-SSMDIP	And	-	data pdf pinout
AD 5300 BRT	D/A-IC	SMD, 8 Bit D/A-Conv., ser. 3-wire Inp. Rail-to-Rail Outp., +2.7V...+5.5V	6-SMDIP	And	-	data pdf pinout
AD 5301 BRM	D/A-IC	SMD, 8 Bit D/A-Conv., ser. 2-wire Inp. I²C, Rail-to-Rail Outp., +2.5V...+5.5V	8-SSMDIP	And	-	data pdf pinout
AD 5301 BRT	D/A-IC	SMD, 8 Bit D/A-Conv., ser. 2-wire Inp. I²C, Rail-to-Rail Outp., +2.5V...+5.5V	6-SMDIP	And	-	data pdf pinout
AD 5302 BRM	D/A-IC	SMD, Dual 8 Bit D/A-Conv.,+2.5V...+5.5V ser. 3-wire Input, Rail-to-Rail Out	10-VSMDIP	And	-	data pdf pinout
AD 5303 BRU	D/A-IC	SMD, Dual 8 Bit D/A-Conv.,+2.5V...+5.5V Rail-to-Rail Out, ser. Input/output	16-SSMDIP	And	-	data pdf pinout
AD 5304 BRM	D/A-IC	SMD, Quad 8 Bit D/A-Conv.,+2.5V...+5.5V Rail-to-Rail Out, ser. 3-wire Input	10-VSMDIP	And	-	data pdf pinout
AD 5305 BRM	D/A-IC	SMD, Quad 8 Bit D/A-Conv.,+2.5V...+5.5V Rail-to-Rail Out, ser. Input, I²C	10-VSMDIP	And	-	data pdf pinout
AD 5306 BRU	D/A-IC	SMD, Quad 8 Bit D/A-Conv.,+2.5V...+5.5V 2-wire Input, I²C, Rail-to-Rail Outp.	16-SSMDIP	And	-	data pdf pinout
AD 5307 BRU	D/A-IC	SMD, Quad 8 Bit D/A-Conv.,+2.5V...+5.5V ser. Input/output, Rail-to-Rail Output	16-SSMDIP	And	-	data pdf pinout
AD 5308 BRU	D/A-IC	SMD, Octal 8 Bit D/A-Conv.,+2.5...+5.5V ser. 3-wire Input, Rail-to-Rail Output	16-SSMDIP	And	-	data pdf pinout
AD 5310 BRM	D/A-IC	SMD, 10Bit D/A-Conv., ser. 3-wire Inp. Rail-to-Rail Outp., +2.7V...+5.5V	8-SSMDIP	And	-	data pdf pinout
AD 5310 BRT	D/A-IC	SMD, 10Bit D/A-Conv., ser. 3-wire Inp. Rail-to-Rail Outp., +2.7V...+5.5V	6-SMDIP	And	-	data pdf pinout
AD 5311 BRM	D/A-IC	SMD, 10Bit D/A-Conv., ser. 2-wire Inp. I²C, Rail-to-Rail Outp., +2.5V...+5.5V	8-SSMDIP	Agt	-	data pdf pinout
AD 5311 BRT	D/A-IC	SMD, 10Bit D/A-Conv., ser. 2-wire Inp. I²C, Rail-to-Rail Outp., +2.5V...+5.5V	6-SMDIP	And	-	data pdf pinout
AD 5312 BRM	D/A-IC	SMD, Dual 8 Bit D/A-Conv.,+2.5V...+5.5V ser. 3-wire Input, Rail-to-Rail Out	10-VSMDIP	And	-	data pdf pinout
AD 5313 BRU	D/A-IC	SMD, Dual 10Bit D/A-Conv.,+2.5V...+5.5V Rail-to-Rail Out, ser. Input/output	16-SSMDIP	And	-	data pdf pinout
AD 5314 BRM	D/A-IC	SMD, Quad 10Bit D/A-Conv.,+2.5V...+5.5V Rail-to-Rail Out, ser. 3-wire Input	10-VSMDIP	And	-	data pdf pinout
AD 5315 BRM	D/A-IC	SMD, Quad 10Bit D/A-Conv.,+2.5V...+5.5V Rail-to-Rail Out, ser. Input, I²C	10-VSMDIP	And	-	data pdf pinout
AD 5316 BRU	D/A-IC	SMD, Quad 10Bit D/A-Conv.,+2.5V...+5.5V 2-wire Input, I²C, Rail-to-Rail Outp.	16-SSMDIP	And	-	data pdf pinout
AD 5317 BRU	D/A-IC	SMD, Quad 10Bit D/A-Conv.,+2.5V...+5.5V ser. Input/output, Rail-to-Rail Output	16-SSMDIP	And	-	data pdf pinout
AD 5318 BRU	D/A-IC	SMD, Octal 10Bit D/A-Conv.,+2.5...+5.5V ser. 3-wire Input, Rail-to-Rail Output	16-SSMDIP	And	-	data pdf pinout
AD 5320 BRM	D/A-IC	SMD, 12Bit D/A-Conv., ser. 3-wire Inp. Rail-to-Rail Outp., +2.7V...+5.5V	8-SSMDIP	And	-	data pdf pinout
AD 5320 BRT	D/A-IC	SMD, 12Bit D/A-Conv., ser. 3-wire Inp. Rail-to-Rail Outp., +2.7V...+5.5V	6-SMDIP	And	-	data pdf pinout

Type	Device	Short Description	Fig.	Manu	Comparision Types	More at
AD 5321 BRM	D/A-IC	SMD, 12Bit D/A-Conv., ser. 2-wire Inp. I²C, Rail-to-Rail Outp., +2.5V...+5.5V	8-SSMDIP	And	-	data pdf pinout
AD 5321 BRT	D/A-IC	SMD, 12Bit D/A-Conv., ser. 2-wire Inp. I²C, Rail-to-Rail Outp., +2.5V...+5.5V	6-SMDIP	And	-	data pdf pinout
AD 5322 BRM	D/A-IC	SMD, Dual 8 Bit D/A-Conv.,+2.5V...+5.5V ser. 3-wire Input, Rail-to-Rail Out	10-VSMDIP	And	-	data pdf pinout
AD 5323 BRU	D/A-IC	SMD, Dual 12Bit D/A-Conv.,+2.5V...+5.5V Rail-to-Rail Out, ser. Input/output	16-SSMDIP	And	-	data pdf pinout
AD 5324 BRM	D/A-IC	SMD, Quad 12Bit D/A-Conv.,+2.5V...+5.5V Rail-to-Rail Out, ser. 3-wire Input	10-VSMDIP	And	-	data pdf pinout
AD 5325 BRM	D/A-IC	SMD, Quad 12Bit D/A-Conv.,+2.5V...+5.5V Rail-to-Rail Out, ser. Input, I²C	10-VSMDIP	And	-	data pdf pinout
AD 5326 BRU	D/A-IC	SMD, Quad 12Bit D/A-Conv.,+2.5V...+5.5V 2-wire Input, I²C, Rail-to-Rail Outp.	16-SSMDIP	And	-	data pdf pinout
AD 5327 BRU	D/A-IC	SMD, Quad 12Bit D/A-Conv.,+2.5V...+5.5V ser. Input/output, Rail-to-Rail Output	16-SSMDIP	And	-	data pdf pinout
AD 5328 BRU	D/A-IC	SMD, Octal 12Bit D/A-Conv.,+2.5...+5.5V ser. 3-wire Input, Rail-to-Rail Output	16-SSMDIP	And	-	data pdf pinout
AD 5330 BRU	D/A-IC	SMD, 8 Bit D/A-Conv., +2.5V...+5.5V par. Input, Rail-to-Rail Out	20-SSMDIP	And	-	data pdf pinout
AD 5331 BRU	D/A-IC	SMD, 10 Bit D/A-Conv., +2.5V...+5.5V par. Input, Rail-to-Rail Out	20-SSMDIP	And	-	data pdf pinout
AD 5332 BRU	D/A-IC	SMD, Dual 8 Bit D/A-Conv.,+2.5V...+5.5V par. Input, Rail-to-Rail Output	20-SSMDIP	And	-	data pdf pinout
AD 5333 BRU	D/A-IC	SMD, Dual 10Bit D/A-Conv.,+2.5V...+5.5V par. Input, Rail-to-Rail Output	24-SSMDIP	And	-	data pdf pinout
AD 5334 BRU	D/A-IC	SMD, Quad 8 Bit D/A-Conv.,+2.5V...+5.5V par. Input, Rail-to-Rail Output	24-SSMDIP	And	-	data pdf pinout
AD 5335 BRU	D/A-IC	SMD, Quad 10Bit D/A-Conv.,+2.5V...+5.5V par. Input, Rail-to-Rail Output	24-SSMDIP	And	-	data pdf pinout
AD 5336 BRU	D/A-IC	SMD, Quad 10Bit D/A-Conv.,+2.5V...+5.5V par. Input, Rail-to-Rail Output	28-SSMDIP	And	-	data pdf pinout
AD 5340 BRU	D/A-IC	SMD, 12 Bit D/A-Conv., +2.5V...+5.5V par. Input, Rail-to-Rail Out	24-SSMDIP	And	-	data pdf pinout
AD 5341 BRU	D/A-IC	SMD, 12 Bit D/A-Conv., +2.5V...+5.5V par. Input, Rail-to-Rail Out	20-SSMDIP	And	-	data pdf pinout
AD 5342 BRU	D/A-IC	SMD, Dual 12Bit D/A-Conv.,+2.5V...+5.5V par. Input, Rail-to-Rail Output	28-SSMDIP	And	-	data pdf pinout
AD 5343 BRU	D/A-IC	SMD, Dual 12Bit D/A-Conv.,+2.5V...+5.5V par. Input, Rail-to-Rail Output	20-SSMDIP	And	-	data pdf pinout
AD 5344 BRU	D/A-IC	SMD, Quad 12Bit D/A-Conv.,+2.5V...+5.5V par. Input, Rail-to-Rail Output	28-SSMDIP	And	-	data pdf pinout
AD 5410 BCPZ	D/A-IC	Single CH, 12-Bit, Current Source DAC	TSOP	And	-	pdf pinout
AD 5410 BREZ	D/A-IC	Single CH, 12-Bit, Current Source DAC	TSOP	And	-	pdf pinout
AD 5412 BCPZ	D/A-IC	Single CH, 12-Bit, Curr. Source & Volt. Out. DAC	TSOP	And	-	pdf pinout
AD 5412 BREZ	D/A-IC	Single CH, 12-Bit, Curr. Source & Volt. Out. DAC	TSOP	And	-	pdf pinout
AD 5420 BCPZ	D/A-IC	Single CH, 16-Bit, Ser.Inp., Curr. Source DAC	TSOP	And	-	pdf pinout
AD 5420 BREZ	D/A-IC	Single CH, 16-Bit, Ser.Inp., Curr. Source DAC	TSOP	And	-	pdf pinout
AD 5422 BCPZ	D/A-IC	Single CH, 16-Bit, Curr. Source+Volt. Output DAC	TSOP	And	-	pdf pinout
AD 5422 BREZ	D/A-IC	Single CH, 16-Bit, Curr. Source+Volt. Output DAC	TSOP	And	-	pdf pinout
AD 5516 ABC-(1...3)	D/A-IC	16 Chan./Kan., 12 Bit D/A-Conv., 20MHz ser. 3-wire Inp.,+2.7...5V/+5V/±5...±15V ..ABC-1/-2/-3: Vout=±2.5V/±5V/±10V	74-GRID	And	-	pdf
AD 5530, 5531 BRU	D/A-IC	SMD, 12/14 Bit D/a-converter ser. 3-wire Input, ±12/15V	16-SSMDIP	And	-	pdf
AD 5532 ABC-(1...5)	D/A-IC	32 Chan./Kan., 14 Bit D/A-Conv., 14MHz ser. Inp.,+2.7...5V/+5V/-5(+8)..±16,5V ..ABC-1/-2/-3/-5: Rout=0.5/0.5/500/1kΩ	74-GRID	And	-	pdf
AD 5532 BBC-1	D/A-IC	32 Chan./Kan., 14 Bit D/a-converter ser. 3-wire Inp., ser. Output, 14MHz Rout=0.5Ω, +2.7...5V/+5V/-5(+8)..±16,5V	74-GRID	And	-	data pdf
AD 5532HS ABC	D/A-IC	32 Chan./Kan., 14 Bit D/A-Conv., 30MHz ser. 3-wire Inp., Vout=±2,5V/0..+5V Rout=0.5Ω, +2.7...5V/+5V/±5..±12V	74-GRID	And	-	data pdf
AD5539	HI-FREQ	Vs:±10V Vu:52dB Vo:±3V Vi0:2mV	14D	And	-	data pdf pinout
AD 5543 BR,CR	D/A-IC	SMD, 16 Bit D/A-Converter, ±2/±1 LSB ser. 3-wire Input, Current Outp., +5V	8-MDIP	And	-	pdf
AD 5543 BRM	D/A-IC	SMD, 16 Bit D/A-Converter, ±2 LSB ser. 3-wire Input, Current Outp., +5V	8-SSMDIP	And	-	data pdf pinout
AD 5544 ARS	D/A-IC	SMD, Quad 16 Bit D/A-Conv., +5V ser. 3-wire Input, Curr.Out/Stromausg.	28-SSMDIP	And	-	data pdf pinout
AD 5545 BRU	D/A-IC	SMD, Dual 16 Bit D/A-Conv., +5V ser. 3-wire Input, Curr.Out/Stromausg.	16-SSMDIP	And	-	data pdf pinout
AD 5553 CRM	D/A-IC	SMD, 14 Bit D/A-Converter, ±1 LSB ser. 3-wire Input, Current Outp., +5V	8-SSMDIP	And	-	data pdf pinout
AD 5554 BRS	D/A-IC	SMD, Quad 14 Bit D/A-Conv., +5V ser. 3-wire Input, Curr.Out/Stromausg.	28-SSMDIP	And	-	data pdf pinout
AD 5555 CRU	D/A-IC	SMD, Dual 14 Bit D/A-Conv., +5V ser. 3-wire Input, Curr.Out/Stromausg.	16-SSMDIP	And	-	data pdf pinout

Type	Device	Short Description	Fig.	Manu	Comparision Types	More at
AD 5582, 5583 BRU	D/A-IC	SMD, Quad 12/10Bit D/A-Conv., par. Inp uni-/bipolar mode, +5V...12(15)V or ±5V	40-VSMDIP	And	-	pdf
AD 7147 ACPZ-REEL	DIG-IC	CapTouch Contr.1-Electrode Capacit. Sensors	TSOP	And		pdf pinout
AD 7224 ... N,R,Q	CMOS-D/A-IC	8 Bit, Double Buffered, Output Amp.	MDIP	And	-	pdf pinout
AD 7224 ... WN	CMOS-D/A-IC	SMD, 8 Bit, Double Buffered, Outp.Amp.	16-MDIP	And	-	pdf
AD 7225 ... E	CMOS-D/A-IC	8 Bit, Quad, Double Buff., Output Amp.	LCC	And	-	pdf pinout
AD 7225 ... N,R,Q	CMOS-D/A-IC	8 Bit, Quad, Double Buff., Output Amp.	24-DIP,DIC	And	-	pdf pinout
AD 7225 ... P	CMOS-D/A-IC	8 Bit, Quad, Double Buff., Output Amp.	PLCC	And	-	pdf pinout
AD 7225 ... WG	CMOS-D/A-IC	SMD, 8 Bit,Quad,Double Buff.,Outp.Amp.	24-MDIP	And	-	pdf
AD 7226 ... N,Q,R	CMOS-D/A-IC	8 Bit, Quad, Output Amplifier	20-DIP,DIC	And	-	pdf pinout
AD 7226 ... P	CMOS-D/A-IC	8 Bit, Quad, Output Amplifier	PLCC	And	-	pdf pinout
AD 7226 ... WP	CMOS-D/A-IC	SMD, 8 Bit, Quad, Output Amplifier	20-MDIP	And	-	pdf pinout
AD 7302 BN	D/A-IC	Dual, 8 Bit, D/A-Conv., +2.7V...+5.5V par. Input, Rail-to-Rail Out	20-DIP	And	-	data pdf pinout
AD 7302 BR	D/A-IC	SMD, Dual 8Bit D/A-Conv., +2.7V...+5.5V par. Input, Rail-to-Rail Out	20-MDIP	And	-	data pdf pinout
AD 7302 BRU	D/A-IC	SMD, Dual 8Bit D/A-Conv., +2.7V...+5.5V par. Input, Rail-to-Rail Out	20-SSMDIP	And	-	data pdf pinout
AD 7303 BN	D/A-IC	Dual 8 Bit D/A-Conv., +2.7V...+5.5V ser. 3-wire Input, Rail-to-Rail Out	8-DIP	And	-	data pdf pinout
AD 7303 BR	D/A-IC	SMD, Dual 8 Bit D/A-Conv.,+2.7V...+5.5V ser. 3-wire Input, Rail-to-Rail Out	8-MDIP	And	-	data pdf pinout
AD 7303 BRM	D/A-IC	SMD, Dual 8 Bit D/A-Conv.,+2.7V...+5.5V ser. 3-wire Input, Rail-to-Rail Out	8-SSMDIP	And	-	data pdf pinout
AD 7339 BS	A/D-D/A-IC	8 Bit, A/D-Conv. & Quad D/A-Conv., +5V T&H, 2 ser./ 2 par. Inp. DACs	52-QFP	And	-	data pdf pinout
AD 7357 BRUZ	A/D-IC	Diff. Input,4.25 MSPS, 14-Bit, SAR ADC	TSOP	And		pdf pinout
AD 7357 BRUZ-RL	A/D-IC	Diff. Input,4.25 MSPS, 14-Bit, SAR ADC	TSOP	And		pdf pinout
AD 7357 YRUZ	A/D-IC	Diff. Input,4.25 MSPS, 14-Bit, SAR ADC	TSOP	And		pdf pinout
AD 7357YRUZ-RL	A/D-IC	Diff. Input,4.25 MSPS, 14-Bit, SAR ADC	TSOP	And		pdf pinout
AD 7398,7399 BR	D/A-IC	SMD, Quad 12/10 Bit D/a-converter ser. 3-wire Inp., +3(+5,±5)V	16-MDIP	And		
AD 7398,7399 BRU	D/A-IC	SMD, Quad 12/10 Bit D/a-converter ser. 3-wire Inp., +3(+5,±5)V	16-SSMDIP	And		
AD 7400 YRWZ	IC	Isolated Sigma-Delta Modulator	TSOP	And		pdf pinout
AD 7400 YRWZ-REEL	IC	Isolated Sigma-Delta Modulator	TSOP	And		pdf pinout
AD 7440 BR(M,T)	A/D-IC	SMD, 10 Bit SAR A/D-Conv., 1MSPS diff. Input, ser. Interface, +3..5V	8-SSMDIP	And		pdf pinout
AD 7441(A,B) RM	A/D-IC	SMD, 10 Bit SAR A/D-Conv., 1MSPS pseudo diff. Inp., ser.Interf., +3..5V	8-SSMDIP	And		pdf pinout
AD 7441(A,B) RT	A/D-IC	SMD, 10 Bit SAR A/D-Conv., 1MSPS pseudo diff. Inp., ser.Interf., +3..5V	47	And		pdf pinout
AD 7450 AR,BR	A/D-IC	SMD, 12 Bit SAR A/D-Conv., 1MSPS diff. Input, ser. Interface, +3..5V	8-MDIP	And		pdf pinout
AD 7450 ARM,BRM	A/D-IC	SMD, 12 Bit SAR A/D-Conv., 1MSPS diff. Input, ser. Interface, +3..5V	8-SSMDIP	And		pdf pinout
AD 7450A BR(M,T)	A/D-IC	SMD, 12 Bit SAR A/D-Conv., 1MSPS diff. Input, ser. Interface, +3..5V	8-SSMDIP	And		pdf pinout
AD 7451(A,B) RM	A/D-IC	SMD, 12 Bit SAR A/D-Conv., 1MSPS pseudo diff. Inp., ser.Interf., +3..5V	8-SSMDIP	And		pdf pinout
AD 7451(A,B) RT	A/D-IC	SMD, 12 Bit SAR A/D-Conv., 1MSPS pseudo diff. Inp., ser.Interf., +3..5V	47	And		pdf pinout
AD 7452B RT	A/D-IC	SMD, 12 Bit SAR A/D-Conv., 555kSPS diff. Inp., ser.Interf., +3..5V	47	And		pdf pinout
AD 7453(A,B) RT	A/D-IC	SMD, 12 Bit SAR A/D-Conv., 555kSPS pseudo diff. Inp., ser.Interf., +3..5V	47	And		pdf pinout
AD 7457 BRT	A/D-IC	SMD, 12 Bit SAR A/D-Conv., 100kSPS pseudo diff. Inp., ser.Interf., +3..5V	47	And		pdf pinout
AD 7490 BCP	A/D-IC	SMD, 16-Chan. 12Bit SAR A/D-Conv., T&H MUX, ser.Interf., 1MSPS, +2.7...5.25V	32-LFCSP	And	-	pdf pinout
AD 7490 BRU	A/D-IC	SMD, 16-Chan. 12Bit SAR A/D-Conv., T&H MUX, ser.Interf., 1MSPS, +2.7...5.25V	28-SSMDIP	And	-	pdf pinout
AD 7510 DI E	CMOS-IC	4x Analog Switch, Udd=17V	20-(P)LCC	And		
AD 7510 DI N	CMOS-IC	4x Analog Switch, Udd=17V	16-DIP	And	SW-7510	
AD 7510 DI P	CMOS-IC	4x Analog Switch, Udd=17V	20-PLCC	And		
AD 7510 DI Q	CMOS-IC	4x Analog Switch, Udd=17V	...-DIC	And		
AD 7511 DI E	CMOS-IC	4x Analog Switch, Udd=17V	20-(P)LCC	And		
AD 7511 DI N	CMOS-IC	4x Analog Switch, Udd=17V	16-DIP	And	SW-7511	
AD 7511 DI P	CMOS-IC	4x Analog Switch, Udd=17V	20-PLCC	And		
AD 7511 DI Q	CMOS-IC	4x Analog Switch, Udd=17V	...-DIC	And		
AD 7510...7512 DI E	CMOS-IC	=AD 7510...7512DI N:	20-LCC			
AD 7512 DI E	CMOS-IC	2x Analog Switch, Udd=17V	20-(P)LCC	And		
AD 7512 DI N	CMOS-IC	2x Analog Switch, Udd=17V	14-DIP	And		
AD 7510...7512 DI P	CMOS-IC	=AD 7510...7512DI N:	20-PLCC			
AD 7512 DI P	CMOS-IC	2x Analog Switch, Udd=17V	20-PLCC	And		
AD 7510...7512 DI Q	CMOS-IC	=AD 7510...7512DI N:	...-DIC			
AD 7512 DI Q	CMOS-IC	2x Analog Switch, Udd=17V	...-DIC	And		
AD 7520 ...	CMOS-D/A-IC	10 Bit, Binary Multiplying, 600ns	16-DIC,DIP	Nsc		
AD 7520 ...N,D,Q	CMOS-D/A-IC	10 Bit, Io-power(20mW), Multiplying	16-DIP,DIC	And,Max,Nsc	-	
AD 7520 ... WE	CMOS-D/A-IC	SMD, 10 Bit, Io-power, Multiplying	16-MDIP			
AD 7521 ...	CMOS-D/A-IC	12 Bit, Binary Multiplying, 600ns	18-DIC,DIP	Nsc		
AD 7521 ... N,D,Q	CMOS-D/A-IC	12 Bit, Io-power(20mW), Multiplying	18-DIP,DIC	And,Max,Nsc	-	
AD 7521 ... WN	CMOS-D/A-IC	SMD, 12 Bit, Io-power, Multiplying	18-MDIP			
AD 7523 ... N	CMOS-D/A-IC	8 Bit, Multiplying	16-DIP	And, Max	-	

24

Type	Device	Short Description	Fig.	Manu	Comparision Types	More at
AD 7523 ... WE	CMOS-D/A-IC	SMD, 8 Bit, Multiplying	16-MDIP		-	
AD 7524 ... N,D,Q	CMOS-D/A-IC	8 Bit, Buffered, Multiplying	16-DIP,DIC	And, Max	-	
AD 7524 ... SE	CMOS-D/A-IC	SMD, 8 Bit, Buffered, Multiplying	16-MDIP		-	
AD 7528 ... N,D,Q	CMOS-D/A-IC	8 Bit, Dual, Buffered, Multiplying	20-DIP,DIC	And, Max	-	
AD 7528 ... WP	CMOS-D/A-IC	SMD ,8 Bit, Dual, Buffered, Multiplying	20-MDIP		-	
AD 7530 ...	CMOS-D/A-IC	10 Bit, Binary Multiplying, 600ns	16-DIC,DIP	Nsc	-	
AD 7530 ... N,D,Q	CMOS-D/A-IC	10 Bit, lo-power(20mW), Multiplying	16-DIP,DIC	And,Max,Nsc	-	
AD 7530 ... WE	CMOS-D/A-IC	SMD, 10 Bit, lo-power, Multiplying	16-MDIP		-	
AD 7531 ...	CMOS-D/A-IC	12 Bit, Binary Multiplying, 600ns	18-DIC,DIP	Nsc	-	
AD 7531 ... N,D,Q	CMOS-D/A-IC	12 Bit, lo-power(20mW), Multiplying	18-DIP,DIC	And,Max,Nsc	-	
AD 7531 ... WN	CMOS-D/A-IC	SMD, 12 Bit, lo-power, Multiplying	18-MDIP		-	
AD 7533 ... N,D,Q	CMOS-D/A-IC	10 Bit, lo-power(20mW), Multiplying	16-DIP,DIC	And, Max	-	
AD 7533 ... WE	CMOS-D/A-IC	SMD, 10 Bit, lo-power, Multiplying	16-MDIP		-	
AD 7541(A)..N,D,Q	CMOS-D/A-IC	12 Bit, lo-power, Multiplying	18-DIP,DIC	And, Max	-	
AD 7541 ... WN	CMOS-D/A-IC	SMD, 12 Bit, lo-power, Multiplying	18-MDIP		-	
AD 7542 ... N,D,Q	CMOS-D/A-IC	12 Bit, µP Compatible	16-DIP,DIC	And, Max	-	
AD 7542 ... WE	CMOS-D/A-IC	SMD, 12 Bit, µP Compatible	16-MDIP		-	
AD 7543 ... N,D,Q	CMOS-D/A-IC	12 Bit, Serial Input	16-DIP,DIC	And, Max	-	
AD 7543 ... WE	CMOS-D/A-IC	SMD, 12 Bit, Serial Input	16-MDIP		-	
AD 7545 ... N,D,Q	CMOS-D/A-IC	12 Bit, Buffered, Multiplying	20-DIP,DIC	And, Max	-	
AD 7545 ... WP	CMOS-D/A-IC	SMD, 12 Bit, Buffered, Multiplying	20-MDIP		-	
AD 7572(...)N,Q	CMOS-A/D-IC	12 Bit, hi-speed(5/12µs), µP Interface	24-DIP,DIC	Max	MAX 162... (3µs)	
AD 7572(...)WG	CMOS-A/D-IC	SMD, 12 Bit, hi-speed, µP Interface	24-MDIP		-	
AD 7574(...)N,Q	A/D-IC	8 Bit, 15µs, µP Interface	16-DIP	Max	MAX 150... (4µs)	
AD 7574(...)WN	A/D-IC	SMD, 8 Bit, 15µs, µP Interface	16-MDIP		-	
AD 7581(...)N,Q	CMOS-A/D-IC	8 Bit Data Acqu. System, 8-Ch. Input	28-DIP	Max	MAX 161...	
AD 7581(...)WI	CMOS-A/D-IC	SMD, 8 Bit Data Acqu.System,8-Ch. Inp.	28-MDIP		-	
AD 7621A CP	A/D-IC	16Bit diff. SAR A/ D- Conv. +2.5V par./ ser. Interf., 3MSPS	48-LFCSP	And		pdf
AD 7621A ST	A/D-IC	SMD, 16Bit diff. SAR A/D-Conv. +2.5V par./ser. Interf., int. Ref., 3MSPS	48-QFP	And		pdf
AD 7628 ... N,Q	CMOS-D/A-IC	8 Bit, Dual, Buffered, Multiplying	20-DIP,DIC	And, Max		
AD 7628 ... WP	CMOS-D/A-IC	SMD, 8 Bit, Dual, Buffered,Multiplying	20-MDIP			
AD 7650A CP	A/D-IC	16Bit SAR A/D-Converter, +5V par./ser. Interface, 570kSPS	48-LFCSP	And		pdf
AD 7650A ST	A/D 'C	SMD, 16Bit SAR A/D-Converter, +5V par./ser. Interface, 570kSPS	48-QFP	And		pdf
AD 7651 ACP,ACPRL	CMOS-A/D-IC	16 Bit SAR A/D-Conv., 100kSPS, +5V ser. 2-wire/par. Interface (3V/5V)	48-LFCSP	And		pdf
AD 7651 AST,ASTRL	CMOS-A/D-IC	16 Bit SAR A/D-Conv., 100kSPS, +5V ser. 2-wire/par. Interface (3V/5V)	48-QFP	And		pdf
AD 7652 ACP,ACPRL	CMOS-A/D-IC	16 Bit SAR A/D-Conv., 500kSPS, +5V ser. 2-wire/par. Interface (3V/5V)	48-LFCSP	And		pdf
AD 7652 AST,ASTRL	CMOS-A/D-IC	16 Bit SAR A/D-Conv., 500kSPS, +5V ser. 2-wire/par. Interface (3V/5V)	48-QFP	And		pdf
AD 7653 ACP,ACPRL	CMOS-A/D-IC	16 Bit SAR A/D-Conv., 1MSPS, +5V ser. 2-wire/par. Interface (3V/5V)	48-LFCSP	And		pdf
AD 7653 AST,ASTRL	CMOS-A/D-IC	16 Bit SAR A/D-Conv., 1MSPS, +5V ser. 2-wire/par. Interface (3V/5V)	48-QFP	And		pdf
AD 7654A CP	A/D-IC	Dual 2-Chan. 16Bit SAR A/D-Conv. MUX, par./ser. Interf., 500kSPS, +5V	48-LFCSP	And	-	pdf
AD 7654A ST	A/D-IC	SMD, Dual 2-Chan. 16Bit SAR A/D-Conv. MUX, par./ser. Interf., 500kSPS, +5V	48-QFP	And	-	pdf
AD 7655A CP	A/D-IC	4- Chan. 16Bit SAR A/ D- Conv. MUX, par./ ser. Interf., 1MSPS, +5V	48-LFCSP	And	-	pdf
AD 7655A ST	A/D-IC	SMD, 4-Chan. 16Bit SAR A/D-Conv. MUX, par./ser. Interf., 1MSPS, +5V	48-QFP	And	-	pdf
AD 7660 AST,ASTRL	CMOS-A/D-IC	16 Bit SAR A/D-Converter, 100kSPS, +5V ser. 2-wire / par. Output (3V/5V)	48-QFP	And	-	pdf
AD 7661A CP	A/D-IC	16Bit SAR A/D-Conv., 100kSPS unipolar Inp., par./ser. Interf., +5V	48-LFCSP	And	-	pdf
AD 7661A ST	A/D-IC	SMD, 16Bit SAR A/D-Conv., 100kSPS unipolar Inp., par./ser. Interf., +5V	48-QFP	And	-	pdf
AD 7663 ACP,ACPRL	CMOS-A/D-IC	16 Bit SAR A/D-Converter, 250kSPS, +5V ser. 2-wire / par. 8/16 Bit Out (3/5V)	48-CSP	And	-	pdf
AD 7663 AST,ASTRL	CMOS-A/D-IC	16 Bit SAR A/D-Converter, 250kSPS, +5V ser. 2-wire / par. 8/16 Bit Out (3/5V)	48-QFP	And	-	pdf
AD 7664 AST,ASTRL	CMOS-A/D-IC	16 Bit SAR A/D-Converter, 570kSPS, +5V ser. 2-wire / par. Output (3V/5V)	48-QFP	And	-	pdf
AD 7665 ACP,ACPRL	CMOS-A/D-IC	16 Bit SAR A/D-Converter, 570kSPS, +5V ser. 2-wire / par. 8/16 Bit Out (3/5V)	48-LFCSP	And	-	pdf
AD 7665 AST,ASTRL	CMOS-A/D-IC	16 Bit SAR A/D-Converter, 570kSPS, +5V ser. 2-wire / par. 8/16 Bit Out (3/5V)	48-QFP	And	-	pdf
AD 7666A CP	A/D-IC	16Bit SAR A/D-Conv., 500kSPS unipolar Inp., par./ser. Interf., +5V	48-LFCSP	And	-	pdf
AD 7666A ST	A/D-IC	SMD, 16Bit SAR A/D-Conv., 500kSPS unipolar Inp., par./ser. Interf., +5V	48-QFP	And	-	pdf
AD 7667A CP	A/D-IC	16Bit SAR A/D-Conv., 1MSPS unipolar Inp., par./ser. Interf., +5V	48-LFCSP	And	-	pdf
AD 7667A ST	A/D-IC	SMD, 16Bit SAR A/D-Conv., 1MSPS unipolar Inp., par./ser. Interf., +5V	48-QFP	And	-	pdf
AD 7671 ACP,ACPRL	CMOS-A/D-IC	16 Bit SAR A/D-Converter, 1MSPS, +5V ser. 2-wire / par. 8/16 Bit Out (3/5V)	48-LFCSP	And	-	pdf
AD 7671 AST,ASTRL	CMOS-A/D-IC	16 Bit SAR A/D-Converter, 1MSPS, +5V ser. 2-wire / par. 8/16 Bit Out (3/5V)	48-QFP	And	-	pdf

Type	Device	Short Description	Fig.	Manu	Comparision Types	More at
AD 7674A CP	A/D-IC	18Bit SAR A/D-Conv., 800kSPS diff. Inp., par./ser.Interf., +5V	48-LFCSP	And	-	pdf
AD 7674A ST	A/D-IC	SMD, 18Bit SAR A/D-Conv., 800kSPS diff. Inp., par./ser.Interf., +5V	48-QFP	And	-	pdf
AD 7675 ACP,ACPRL	CMOS-A/D-IC	16 Bit diff.SAR A/D-Conv., 100kSPS, 5V ser. 2-wire / par. 8/16 Bit Out (3/5V)	48-LFCSP	And	-	pdf
AD 7675 AST,ASTRL	CMOS-A/D-IC	16 Bit diff.SAR A/D-Conv., 100kSPS, 5V ser. 2-wire / par. 8/16 Bit Out (3/5V)	48-QFP	And	-	pdf
AD 7676 ACP,ACPRL	CMOS-A/D-IC	16 Bit diff.SAR A/D-Conv., 500kSPS, 5V ser. 2-wire / par. 8/16 Bit Out (3/5V)	48-LFCSP	And	-	pdf
AD 7676 AST,ASTRL	CMOS-A/D-IC	16 Bit diff.SAR A/D-Conv., 500kSPS, 5V ser. 2-wire / par. 8/16 Bit Out (3/5V)	48-QFP	And	-	pdf
AD 7677 ACP,ACPRL	A/D-IC	16 Bit diff.SAR A/D-Conv., 1MSPS, 5V ser. 2-wire / par. 8/16 Bit Out (3/5V)	48-LFCSP	And	-	pdf
AD 7677 AST,ASTRL	A/D-IC	16 Bit diff.SAR A/D-Conv., 1MSPS, 5V ser. 2-wire / par. 8/16 Bit Out (3/5V)	48-QFP	And	-	pdf
AD 7678A CP	A/D-IC	18Bit SAR A/D-Conv., 100kSPS diff. Inp., par./ser.Interf., +5V	48-LFCSP	And	-	pdf
AD 7678A ST	A/D-IC	SMD, 18Bit SAR A/D-Conv., 100kSPS diff. Inp., par./ser.Interf., +5V	48-QFP	And	-	pdf
AD 7679A CP	A/D-IC	18Bit SAR A/D-Conv., 570kSPS diff. Inp., par./ser.Interf., +5V	48-LFCSP	And	-	pdf
AD 7679A ST	A/D-IC	SMD, 18Bit SAR A/D-Conv., 570kSPS diff. Inp., par./ser.Interf., +5V	48-QFP	And	-	pdf
AD 7701 AN,BN	A/D-IC	16Bit Sigma- Delta A/ D- Conv., ±5V flexible ser.interf. Uart compatible	20-DIP	And	-	pdf pinout
AD 7701 AQ,BQ	A/D-IC	16Bit Sigma- Delta A/ D- Conv., ±5V flexible ser.interf. Uart compatible	20-DIC	And	-	pdf pinout
AD 7701 SQ,TQ	A/D-IC	16Bit Sigma-Delta A/D-Conv.,±5V, flex. ser.Interf. UART compat., -55...+125°C	20-DIC	And.	-	pdf pinout
AD 7701 AR,BR	A/D-IC	SMD, 16Bit Sigma-Delta A/D-Conv., ±5V flexible ser.interf. Uart compatible	20-MDIP	And	-	pdf pinout
AD 7701 ARS	A/D-IC	SMD, 16Bit Sigma-Delta A/D-Conv., ±5V flexible ser.interf. Uart compatible	28-SSMDIP	And	-	pdf pinout
AD 7705,7706 BN	CMOS-A/D-IC	2/3 Chan./Kan. 16 Bit A/D-Converter ser. 3-wire Outp., Sigma-Delta, +3/+5V	16-DIP	And	-	
AD 7705,7706 BR	CMOS-A/D-IC	SMD, 2/3 Chan./Kan. 16 Bit A/D-Conv. ser. 3-wire Outp., Sigma-Delta, +3/+5V	16-MDIP	And	-	
AD 7705,7706 BRU	CMOS-A/D-IC	SMD, 2/3 Chan./Kan. 16 Bit A/D-Conv. ser. 3-wire Outp., Sigma-Delta, +3/+5V	16-SSMDIP	And	-	
AD 7707 BR	CMOS-A/D-IC	SMD, 3 Chan./Kan. 16 Bit A/D-Conv. ser. 3-wire Outp., Sigma-Delta, +3(5)V	20-MDIP	And	-	pdf pinout
AD 7707 BRU	CMOS-A/D-IC	SMD, 3 Chan./Kan. 16 Bit A/D-Conv. ser. 3-wire Outp., Sigma-Delta, +3(5)V	20-SSMDIP	And	-	pdf pinout
AD 7708 BR	A/D-IC	SMD, 8/10 Chan./Kan. 16 Bit A/D-Conv. ser. 3-wire Outp., Sigma-Delta, +3(5)V	28-MDIP	And	-	pdf pinout
AD 7708 BRU	A/D-IC	SMD, 8/10 Chan./Kan. 16 Bit A/D-Conv. ser. 3-wire Outp., Sigma-Delta, +3(5)V	28-SSMDIP	And	-	pdf pinout
AD 7709 ARU,BRU	A/D-IC	SMD, 4 Chan./Kan. 16 Bit A/D-Converter ser. 3-wire Inp./Outp., Sigma-Delta switch. curr. sources, +3V(+5V)	24-SSMDIP	And	-	pdf
AD 7715 AN-(3,5)	CMOS-A/D-IC	16 Bit A/D-Converter, +3V/+5V ser-3-wire Out., ser.Inp, Sigma-Delta	16-DIP	And	-	
AD 7715 AR-(3,5)	CMOS-A/D-IC	SMD, 16 Bit A/D-Converter, +3V/+5V ser-3-wire Out., ser.Inp, Sigma-Delta	16-MDIP	And	-	
AD 7715 ARU-(3,5)	CMOS-A/D-IC	SMD, 16 Bit A/D-Converter, +3V/+5V ser-3-wire Out., ser.Inp, Sigma-Delta	16-SSMDIP	And	-	
AD 7718 BR	A/D-IC	SMD, 8/10 Chan./Kan. 24 Bit A/D-Conv. ser. 3-wire Outp., Sigma-Delta, +3(5)V	28-MDIP	And	-	pdf pinout
AD 7718 BRU	A/D-IC	SMD, 8/10 Chan./Kan. 24 Bit A/D-Conv. ser. 3-wire Outp., Sigma-Delta, +3(5)V	28-SSMDIP	And	-	pdf pinout
AD 7719 BR	A/D-IC	SMD, Dual 16/24 Bit A/D-Conv., +3V/+5V ser- 3-wire Out., ser.Inp, Sigma-Delta	28-MDIP	And	-	pdf pinout
AD 7719 BRU	A/D-IC	SMD, Dual 16/24 Bit A/D-Conv., +3V/+5V ser- 3-wire Out., ser.Inp, Sigma-Delta	28-SSMDIP	And	-	pdf pinout
AD 7731 BN	A/D-IC	3 diff. Chan./Kan. 24Bit A/D-Conv.,+5V ser- 3-wire Out., ser.Inp, Sigma-Delta	24-DIP	And	-	pdf pinout
AD 7731 BR	A/D-IC	SMD, 3 diff. Chan./Kan. 24Bit ADC, +5V ser- 3-wire Out., ser.Inp, Sigma-Delta	24-MDIP	And	-	pdf pinout
AD 7731 BRU	A/D-IC	SMD, 3 diff. Chan./Kan. 24Bit ADC, +5V ser- 3-wire Out., ser.Inp, Sigma-Delta	24-SSMDIP	And	-	pdf pinout
AD 7782 BRU	A/D-IC	SMD, 2 diff. Chan./Kan. 24 Bit ADC ser. 3-wire Outp., Sigma-Delta, +3V/5V	16-SSMDIP	And	-	pdf pinout
AD 7783 BRU	A/D-IC	SMD, 1 diff. Chan./Kan. 24 Bit ADC ser. 3-wire Out., Sigma-Delta 2 switch. current sources, +3V/+5V	16-SSMDIP	And	-	pdf pinout
AD 7801 BR	D/A-IC	SMD, 8 Bit D/ A- Conv., par. Input Rail- to- Rail Out, +2, 7V... +5, 5V	20-MDIP	And	-	data pdf pinout
AD 7801 BRU	D/A-IC	SMD, 8 Bit D/ A- Conv., par. Input Rail- to- Rail Out, +2, 7V... +5, 5V	20-SSMDIP	And	-	data pdf pinout
AD 7804 BN	D/A-IC	Quad, 10Bit D/A-Conv.,+3.0V...5.5V serial/serieller Input/eingang	16-DIP	And	-	data pdf pinout
AD 7804 BR	D/A-IC	SMD, Quad 10Bit D/A-Conv.,+3.0V...5.5V serial/serieller Input/eingang	16-MDIP	And	-	data pdf pinout
AD 7805 BN	D/A-IC	Quad, 10Bit D/A-Conv.,+3.0V...5.5V	28-DIP	And	-	data pdf pinout

Type	Device	Short Description	Fig.	Manu	Comparision Types	More at
		parallel/-er Input/eingang				
AD 7805 BR,CR	D/A-IC	SMD, Quad 10Bit D/A-Conv.,+3.0V...5.5V parallel/-er Input/eingang	28-MDIP	And	-	pdf
AD 7805 BRS	D/A-IC	SMD, Quad 10Bit D/A-Conv.,+3.0V...5.5V parallel/-er Input/eingang	28-SSMDIP	And	-	data pdf pinout
AD 7808 BN	D/A-IC	Octal, 10Bit D/A-Conv.,+3.0V...5.5V serial/serieller Input/eingang	24-DIP	And	-	data pdf pinout
AD 7808 BR	D/A-IC	SMD, Octal 10Bit D/A-Conv.,+3.0V...5.5V serial/serieller Input/eingang	24-MDIP	And	-	data pdf pinout
AD 7809 BST	D/A-IC	SMD, Octal 10Bit D/A-Conv.,+3.0V...5.5V parallel/-er Input/eingang	44-QFP	And	-	data pdf pinout
AD 7820(...)N,Q	CMOS-A/D-IC	8 Bit, hi-speed(1,34µs), Track/Hold	20-DIP,DIC	Max	MAX 150...	
AD 7820(...)WP	CMOS-A/D-IC	SMD, 8 Bit, hi-speed, Track/Hold	20-MDIP		-	
AD 7824(...)N,Q	CMOS-A/D-IC	8 Bit, 4-Ch. MPX, 2,5µs, Track/Hold	24-DIP,DIC	Max	MAX 154...	
AD 7824(...)WG	CMOS-A/D-IC	SMD, 8 Bit, 4-Ch. MPX, Track/Hold	24-MDIP		-	
AD 7828(...)N,Q	CMOS-A/D-IC	8 Bit, 8-Ch. MPX, 2,5µs, Track/Hold	28-DIP,DIC	Max	MAX 158...	
AD 7828(...)WI	CMOS-A/D-IC	SMD, 8 Bit, 8-Ch. MPX, Track/Hold	28-MDIP		-	
AD 7834 AN,BN	D/A-IC	Quad, 14 Bit D/A-Conv., +5V/±15V ser. Input, Lin. Error=±2/1LSB	28-DIP	And	-	pdf
AD 7834 AR,BR	D/A-IC	Quad, 14 Bit D/A-Conv., +5V/±15V ser. Input, Lin. Error=±2/1LSB	28-MDIP	And	-	pdf
AD 7834 SQ	D/A-IC	Quad, 14 Bit D/A-Conv., +5V/±15V ser. Inp., Lin.Err.=±2LSB, -55...+125°C	28-DIC	And	-	data pdf pinout
AD 7835 AP	D/A-IC	Quad, 14 Bit D/A-Conv., +5V/±15V par. Input, Lin. Error=±2LSB	44-PLCC	And	-	data pdf pinout
AD 7835 AS,BS	D/A-IC	Quad, 14 Bit D/A-Conv., +5V/±15V par. Input, Lin. Error=±2/1LSB	44-QFP	And	-	pdf
AD 7836 AS	D/A-IC	Quad, 14 Bit D/A-Conv., +5V/±15V par. Input, Lin. Error=±2LSB	44-QFP	And	-	data pdf pinout
AD 7839 AS	D/A-IC	Octal, 13 Bit D/A-Conv., +5V/±15V	44-QFP	And	-	data pdf pinout
AD 7841 AS,BS	D/A-IC	Octal, 14 Bit D/A-Conv., +5V/±15V par.Inp., DNL=-0.9/+2.0(AS) ±1LSB(BS)	44-QFP	And	-	pdf
AD 7849 AN,BN,CN	D/A-IC	14/16/16 Bit, D/A-Conv., ser. Input INL=±3/8/4 LSB, -40...85°C, ±15V & +5V	20-DIP	And	-	pdf
AD 7849 AR,BR,CR	D/A-IC	SMD, 14/16/16 Bit, D/A-Conv., ser. Inp INL=±3/8/4 LSB, -40...85°C, ±15V & +5V	20-MDIP	And	-	pdf
AD 7849 TQ	D/A-IC	16 Bit, D/A-Converter, ser. Input INL=±8 LSB, -55...125°C, ±15V & +5V	20-DIC	And	-	data pdf pinout
AD 7866 ARU,BRU	A/D-IC	SMD, Dual 2 Chan./Kan. 12 Bit SAR ADC ser. Out, 1MSPS, +2.7V/+5V	20-SSMDIP	And	-	
AD 7884 AN,BN	A/D-IC	16 Bit, 2-Pass-Flash A/D-Converter S&H, 16 Bit par. Outp., 166kSPS, ±5V	40-DIP	And		
AD 7884 AP,BP	A/D-IC	16 Bit, 2-Pass-Flash A/D-Converter S&H, 16 Bit par. Outp., 166kSPS, ±5V	44-PLCC	And		
AD 7885 AAP,ABP	A/D-IC	16 Bit, 2-Pass-Flash A/D-Converter S&H, 8 Bit par. Outp., 166kSPS, ±5V	44-PLCC	And		
AD 7885 AN,BN	A/D-IC	16 Bit, 2-Pass-Flash A/D-Converter S&H, 8 Bit par. Outp., 166kSPS, ±5V	28-DIP	And		
AD 7887 AR,BR	CMOS-A/D-IC	SMD, 2 Chan./Kan. 12 Bit SAR A/D-Conv. ser. 3-wire I/O, 125kSPS, +2.7V/+5V	8-MDIP	And		pdf
AD 7887 ARM	CMOS-A/D-IC	SMD, 2 Chan./Kan. 12 Bit SAR A/D-Conv. ser. 3-wire I/O, 125kSPS, +2.7V/+5V	8-SSMDIP	And		pdf pinout
AD 7888 AR,BR	CMOS-A/D-IC	SMD, 8 Chan./Kan. 12 Bit SAR A/D-Conv. ser. 3-wire I/O, 125kSPS, +2.7V/+5V	16-MDIP	And		pdf
AD 7888 ARU,BRU	CMOS-A/D-IC	SMD, 8 Chan./Kan. 12 Bit SAR A/D-Conv. ser. 3-wire I/O, 125kSPS, +2.7V/+5V	16-SSMDIP	And		pdf
AD 7904 BRU	A/D-IC	SMD, 4-Chan. 8Bit SAR A/D-Conv., MUX Sequencer, ser.Interf., +2.7...5.5V	16-SSMDIP	And	-	pdf pinout
AD 7908 BRU	A/D-IC	SMD, 8-Chan. 8Bit SAR A/D-Conv., MUX Sequencer, ser.Interf., +2.7...5.5V	20-SSMDIP	And	-	pdf pinout
AD 7910 ARM	A/D-IC	SMD, 10Bit SAR A/D-Conv., +2.35...5.25V Track&Hold, ser. Interface	8-SSMDIP	And	-	pdf pinout
AD 7911 A(RM,UJ)	A/D-IC	SMD, 2-Chan. 10Bit SAR A/D-Conv., T&H 250kSPS, ser. Interf., +2.35...5.25V	8-SSMDIP	And	-	pdf
AD 7912 A(RM,UJ)	A/D-IC	SMD, 2-Chan. 10Bit SAR A/D-Conv., T&H MUX, 1MSPS, ser. Interf., +2.35...5.25V	8-SSMDIP	And	-	pdf
AD 7914 BRU	A/D-IC	SMD, 4-Chan. 10Bit SAR A/D-Conv., MUX Sequencer, ser.Interf., +2.7...5.5V	16-SSMDIP	And	-	pdf pinout
AD 7918 BRU	A/D-IC	SMD, 8-Chan. 10Bit SAR A/D-Conv., MUX Sequencer, ser.Interf., +2.7...5.5V	20-SSMDIP	And	-	pdf pinout
AD 7920 (A,B)KS	A/D-IC	SMD, 12Bit SAR A/D-Conv., +2.35...5.25V Track&Hold, ser. Interface	46	And	-	pdf pinout
AD 7920 BRM	A/D-IC	SMD, 12Bit SAR A/D-Conv., +2.35...5.25V Track&Hold, ser. Interface	8-SSMDIP	And	-	pdf pinout
AD 7921 A(RM,UJ)	A/D-IC	SMD, 2-Chan. 12Bit SAR A/D-Conv., T&H 250kSPS, ser. Interf., +2.35...5.25V	8-SSMDIP	And	-	pdf
AD 7922 A(RM,UJ)	A/D-IC	SMD, 2-Chan. 12Bit SAR A/D-Conv., T&H MUX, 1MSPS, ser. Interf., +2.35...5.25V	8-SSMDIP	And	-	pdf
AD 7923 BRU	A/D-IC	SMD, 4-Chan. 12Bit SAR A/D-Conv., T&H MUX, ser.Interf., 200kSPS, +2.7...5.25V	16-SSMDIP	And	-	pdf pinout
AD 7924 BRU	A/D-IC	SMD, 4-Chan. 12Bit SAR A/D-Conv., MUX Sequencer, ser.Interf., +2.7...5.5V	16-SSMDIP	And	-	pdf pinout
AD 7927 BRU	A/D-IC	SMD, 8-Chan. 12Bit SAR A/D-Conv., T&H MUX, ser.Interf., 200kSPS, +2.7...5.25V	20-SSMDIP	And	-	pdf pinout
AD 7928 BRU	A/D-IC	SMD, 8-Chan. 12Bit SAR A/D-Conv., MUX	20-SSMDIP	And		pdf pinout

Type	Device	Short Description	Fig.	Manu	Comparision Types	More at
		Sequencer, ser.Interf., +2.7...5.5V				
AD 7933 BRU	A/D-IC	SMD, 4-Chan. 10Bit SAR A/D-Conv., T&H MUX, par.Interf., 1.5MSPS,+2.35...5.25V	28-SSMDIP	And	-	pdf pinout
AD 7934 BRU	A/D-IC	SMD, 4-Chan. 12Bit SAR A/D-Conv., T&H MUX, par.Interf., 1.5MSPS,+2.35...5.25V	28 SSMDIP	And	-	pdf pinout
AD 7934-6 BRU	A/D-IC	SMD, 4-Chan. 12Bit SAR A/D-Conv., T&H MUX, par.Interf., 625kSPS,+2.35...5.25V	28-SSMDIP	And	-	pdf pinout
AD 7935...36	A/D-IC	SMD, 8-Chan. 10/12Bit SAR A/D-Conv., MUX, par. Interf., 1.5MSPS,+2.7...5.25V	28-SSMDIP	And	-	pdf pinout
AD 7937 AR,BR	D/A-IC	SMD, Dual 12Bit D/A-Conv., +5...15V 8+4 Bit Loading, LC²MOS, -40...85°C	24-MDIP	And	-	pinout
AD 7938 BCP	A/D-IC	SMD, 8-Chan. 12Bit SAR A/D-Conv., T&H MUX, par.Interf., 1.5MSPS,+2.35...5.25V	32-LFCSP	And	-	pdf pinout
AD 7938 BSU	A/D-IC	SMD, 8-Chan. 12Bit SAR A/D-Conv., T&H MUX, par.Interf., 1.5MSPS,+2.35...5.25V	32-QFP	And	-	pdf pinout
AD 7938-6 BCP	A/D-IC	SMD, 8-Chan. 12Bit SAR A/D-Conv., T&H MUX, par.Interf., 625kSPS,+2.35...5.25V	32-LFCSP	And	-	pdf pinout
AD 7938-6 BSU	A/D-IC	SMD, 8-Chan. 12Bit SAR A/D-Conv., T&H MUX, par.Interf., 625kSPS,+2.35...5.25V	32-QFP	And	-	pdf pinout
AD 7939 BCP	A/D-IC	SMD, 8-Chan. 10Bit SAR A/D-Conv., T&H MUX, par.Interf., 1.5MSPS,+2.35...5.25V	32-LFCSP	And	-	pdf pinout
AD 7939 BSU	A/D-IC	SMD, 8-Chan. 10Bit SAR A/D-Conv., T&H MUX, par.Interf., 1.5MSPS,+2.35...5.25V	32-QFP	And	-	pdf pinout
AD 7940 BRJ	A/D-IC	SMD, 14Bit SAR A/D-Conv., +2.5...5.25V Track&Hold, ser. Interface, SOT-23	46	And	-	pdf pinout
AD 7940 BRM	A/D-IC	SMD, 14Bit SAR A/D-Conv., +2.5...5.25V Track&Hold, ser. Interface, SOT-23	8-SSMDIP	And	-	pdf pinout
AD 7942 BCP	A/D-IC	SMD, 14Bit, SAR A/D-Conv., +2...5.5V Track&Hold, ser. Interface, 250kSPS	10-QFN	And	-	pdf pinout
AD 7942 BRM	A/D-IC	SMD, 14Bit, SAR A/D-Conv., +2...5.5V Track&Hold. ser. Interface, 250kSPS	10-VSMDIP	And	-	pdf pinout
AD 7943 AN-B	D/A-IC	12Bit D/A-Conv.,ser.Interf., +3.3...5V	16-DIP	And	-	pdf pinout
AD 7943 ARS-B	D/A-IC	SMD,12Bit D/A-Conv., seriell, +3.3...5V	20-SSMDIP	And	-	pdf pinout
AD 7943 BN	D/A-IC	12Bit D/A-Conv.,ser.Interf., Iout, +5V	16-DIP	And	-	pdf pinout
AD 7943 BR	D/A-IC	SMD, 12Bit D/A-Conv.,ser.Interf., +5V	16-MDIP	And	-	pdf pinout
AD 7943 BRS	D/A-IC	SMD, 12Bit D/A-Conv.,ser.Interf., +5V	20-SSMDIP	And	-	pdf pinout
AD 7944 BCP	A/D-IC	SMD, 14Bit, diff. SAR A/D-Conv. T&H, ser. Interf., 250kSPS, +2...5.5V	10-QFN	And	-	pdf pinout
AD 7944 BRM	A/D-IC	SMD, 14Bit, diff. SAR A/D-Conv. T&H, ser. Interf., 250kSPS, +2...5.5V	10-VSMDIP	And	-	pdf pinout
AD 7945 AN-B	D/A-IC	12Bit D/A-Conv.,par.Interf., +3.3...5V	20-DIP	And	-	pdf pinout
AD 7945 ARS-B	D/A-IC	SMD,12Bit D/A-Conv.,parallel, +3.3...5V	20-SSMDIP	And	-	pdf pinout
AD 7945 BN	D/A-IC	12Bit D/A-Conv.,par.Interf., Iout, +5V	20-DIP	And	-	pdf pinout
AD 7945 BR	D/A-IC	SMD, 12Bit D/A-Conv.,par.Interf., +5V	20-MDIP	And	-	pdf pinout
AD 7945 BRS	D/A-IC	SMD, 12Bit D/A-Conv.,par.Interf., +5V	20-SSMDIP	And	-	pdf pinout
AD 7945 TQ	D/A-IC	12Bit D/A-Conv.,par.Interf., Iout, +5V	20-DIC	And	-	pdf pinout
AD 7946 BCP	A/D-IC	SMD, 14Bit, diff. SAR A/D-Conv. T&H, ser. Interf., 500kSPS, +2...5.5V	10-QFN	And	-	pdf pinout
AD 7946 BRM	A/D-IC	SMD, 14Bit, diff. SAR A/D-Conv. T&H, ser. Interf., 500kSPS, +2...5.5V	10-VSMDIP	And	-	pdf pinout
AD 7947 BCP	A/D-IC	SMD, 14Bit, diff. SAR A/D-Conv. T&H, ser. Interf., 500kSPS, +2...5.5V	QFN	And	-	pdf pinout
AD 7947 BRM	A/D-IC	SMD, 14Bit, diff. SAR A/D-Conv. T&H, ser. Interf., 500kSPS, +2...5.5V	10-VSMDIP	And	-	pdf pinout
AD 7948 AN-B	D/A-IC	12Bit D/A-Conv., par. 8Bit, +3.3...5V	20-DIP	And	-	pdf pinout
AD 7948 ARS-B	D/A-IC	SMD,12Bit D/A-Conv.,par.8Bit, +3.3...5V	20-SSMDIP	And	-	pdf pinout
AD 7948 BN	D/A-IC	12Bit D/A-Conv.,par.Interf. 8Bit, +5V	20-DIP	And	-	pdf pinout
AD 7948 BR	D/A-IC	SMD, 12Bit D/A-Conv., par. 8Bit, +5V	20-MDIP	And	-	pdf pinout
AD 7948 BRS	D/A-IC	SMD, 12Bit D/A-Conv., par. 8Bit, +5V	20-SSMDIP	And	-	pdf pinout
AD8001	LO-POWER	Vs:±6V Vo:±3.1V Vi0:2mV f:440Mc	d,8DS	And		data pdf pinout
AD8002A	2xLO-POWER	Vs:±6V Vo:±3.1V Vi0:2mV f:440Mc	8DS	And		data pdf pinout
AD8004A	4xLO-POWER	Vs:±6V Vo:±3.9V Vi0:5mV f:400Mc	14DS	And		data pdf pinout
AD8005A	FB-AMP	Vs:12.6V Vo:±3.9V Vi0:5mV f:30Mc	8DS,5S	And		data pdf pinout
AD8007C	HI-OHM	Vs:±18V Vu:86dB Vo:±10V Vi0:50mV	8A	And		data pinout
AD8009	FB-AMP	Vs:±6V Vo:±3.8V Vi0:2mV f:75Mc	8Sd	And		data pinout
AD8010A	HI-SPEED	Vs:±6V Vo:±3.0V Vi0:5mV f:60Mc	8D,16SS	And		data pdf pinout
AD8011	FB-AMP	Vs:12V Vo:±4.1V Vi0:±2mV f:25Mc	8DSd	And		data pdf pinout
AD8012A	2xLO-POWER	Vs:±6V Vo:±4V Vi0:±1.5mV f:40Mc	8S	And		data pdf pinout
AD8013A	3xVIDEO-AMP	Vs:13V Vi0:2mV f:50Mc	d,14DS	And		data pdf pinout
AD8014	LO-POWER	Vs:12V Vo:±4V Vi0:2mV f:20Mc	8S,5Sd	And		data pdf pinout
AD8015A	WIDEBAND	Vs:±5.5V f:240Mkc	d,8S	And		data pinout
AD8017A	2xHI-SPEED	Vs:±6V Vo:±5V Vi0:1.8mV f:70Mc	8S	And		data pdf pinout
AD8026A	4xHI-SPEED	Vs:14V Vo:0.2V Vi0:0.5mV f:12Mc	14S	And		data pinout
AD8031A	RR-OP	Vs:12V Vu:80dB Vo:±4.98V Vi0:±1mV	8DS,5S	And		data pdf pinout
AD8031B	RR-OP	Vs:12V Vu:80dB Vo:±4.98V Vi0:±0.5mV	8DS	And		data pdf pinout
AD8032A	2xRR-OP	Vs:12V Vu:80dB Vo:±4.98V Vi0:±1mV	8DS	And		data pdf pinout
AD8032B	2xRR-OP	Vs:12V Vu:80dB Vo:±4.98V Vi0:±0.5mV	8DS	And		data pdf pinout
AD8036	WIDEBAND	Vs:±6.5V Vu:56dB Vi0:±2mV f:350Mc	8DSd	And		data pdf pinout
AD8041	RR-OP	Vs:±6V Vu:99dB Vo:±4.95V Vi0:2mV	8DSd	And		data pinout
AD8042A	2xRR-OP	Vs:12V Vu:94dB Vo:±4.97V Vi0:3mV	8S	And		data pinout
AD8042AN(+3V)	2xRR-OP	Vs:12V Vu:100dB Vo:2.97/0V Vi0:3mV	8D	And		data pinout
AD8042AN(±5V)	2xRR-OP	Vs:12V Vu:94dB Vo:±4.97V Vi0:3mV	8D	And		data pinout
AD8042AR(+3V)	2xRR-OP	Vs:12V Vu:100dB Vo:2.97/0V Vi0:3mV	8S	And		data pinout
AD8044A	4xRR-OP	Vs:12V Vu:96dB Vo:±4.97V Vi0:1.4mV	14S	And		data pinout
AD8044AN(+3V)	4xRR-OP	Vs:12V Vu:92dB Vo:2.98/0V Vi0:1.5mV	14D	And		data pinout

Type	Device	Short Description	Fig.	Manu	Comparision Types	More at
AD8044AN(±5V)	4xRR-OP	Vs:12V Vu:96dB Vo:±4.97V Vi0:1.4mV	14D	And		data pinout
AD8044AR(+3V)	4xRR-OP	Vs:12V Vu:92dB Vo:2.98/0V Vi0:1.5mV	14S	And		data pinout
AD8047	FB-AMP	Vs:±6V Vu:62dB Vi0:1mV f:35Mc	8DSd	And		data pdf pinout
AD8048	FB-AMP	Vs:±6V Vu:68dB Vi0:1mV f:50Mc	8DSd	And		data pdf pinout
AD8051A	RR-OP	Vs:12V Vu:96dB Vo:±4.98V Vi0:1.8mV	8S,5S	And		data pdf pinout
AD8052A	2xRR-OP	Vs:12V Vu:96dB Vo:±4.98V Vi0:1.8mV	8S	And		data pinout
AD8054A	4xRR-OP	Vs:12V Vu:96dB Vo:±4.97V Vi0:1.8mV	14S	And		data pinout
AD8055A	FB-AMP	Vs:±6V Vo:±3.1V Vi0:3mV f:40Mc	8DS,5S	And		data pdf pinout
AD8056A	2xFB-AMP	Vs:±6V Vo:±3.1V Vi0:3mV f:40Mc	8DS	And		data pinout
AD8057A	HI-PERF	Vs:±6V Vo:±4V Vi0:1mV f:30Mc	8S,5S	And		data pdf pinout
AD8058A	2xHI-PERF	Vs:±6V Vo:±4V Vi0:1mV f:30Mc	8S	And		data pdf pinout
AD 8065 AR	LIN/OP-IC	145 MHz FastFET Op Amp	MDIP	And	-	pdf pinout
AD 8065 ART	LIN/OP-IC	145 MHz FastFET Op Amp	45	And	-	pdf pinout
AD 8066 AR	LIN/OP-IC	2x 145 MHz FastFET Op Amp	MDIP	And	-	pdf pinout
AD 8066 ARM	LIN/OP-IC	2x 145 MHz FastFET Op Amp	TSOP	And	-	pdf pinout
AD8072J	2xVIDEO-AMP	Vs:±6V Vo:3.7/0V Vi0:1.5mV f:10Mc	8DS	And		data pdf pinout
AD8073J	3xVIDEO-AMP	Vs:±6V Vo:3.7/0V Vi0:1.5mV f:10Mc	14DS	And		data pdf pinout
AD8079	2xBUFF	Vs:±6V Vo:±3.1V Vi0:10mV f:50Mc	8S	And		data pinout
AD 8221 AR	DIG-IC	Precision Instrumentation Amplifier	TSOP	And	-	pdf pinout
AD 8221 ARM	DIG-IC	Precision Instrumentation Amplifier	TSOP	And	-	pdf pinout
AD 8221 ARM-REEL	DIG-IC	Precision Instrumentation Amplifier	TSOP	And	-	pdf pinout
AD 8221 ARMZ	DIG-IC	Precision Instrumentation Amplifier	TSOP	And	-	pdf pinout
AD 8221 ARMZ-RL	DIG-IC	Precision Instrumentation Amplifier	TSOP	And	-	pdf pinout
AD 8221 AR-REEL	DIG-IC	Precision Instrumentation Amplifier	TSOP	And	-	pdf pinout
AD 8221 ARZ	DIG-IC	Precision Instrumentation Amplifier	TSOP	And	-	pdf pinout
AD 8221 ARZ-RL	DIG-IC	Precision Instrumentation Amplifier	TSOP	And	-	pdf pinout
AD 8221 BR	DIG-IC	Precision Instrumentation Amplifier	TSOP	And	-	pdf pinout
AD 8221 BR-REEL	DIG-IC	Precision Instrumentation Amplifier	TSOP	And	-	pdf pinout
AD 8231 ACPZ-RL	IC	Digit. Programma. Instrumentation Amp.	TSOP	And	-	pdf pinout
AD 8231 ACPZ-WP	IC	Digit. Programma. Instrumentation Amp.	TSOP	And	-	pdf pinout
AD 8270	IC	Precision, Dual-channel Difference Amp.	TSOP	And	-	pdf pinout
AD 8273	IC	Dual-channel Audio Difference Amp.	TSOP	And	-	pdf pinout
AD 8351A RM	HF-IC	SMD, diff. RF/IF(HF/ZF) Amp., +3...5.5V	10-VSMDIP	And	-	pdf pinout
AD 8352	IC	2 GHz Ultralow Distortion Differential RF/IF Amplifier	CSP	And	-	pdf pinout
AD8519A	RR-OP	Vs:±6V Vu:30V/mV Vo:±4.9V Vi0:0.6mV	8S,5S	And		data pdf pinout
AD8531A	SS-OP	Vs:7V Vu:80V/mV Vi0:<25mV f:3Mc	8S,5S	And		data pdf pinout
AD8532A	2xSS-OP	Vs:7V Vu:80V/mV Vi0:<25mV f:3Mc	8DS	And		data pdf pinout
AD8534A	4xSS-OP	Vs:7V Vu:80V/mV Vi0:<25mV f:3Mc	14DS	And		data pdf pinout
AD8541A	RR-OP	Vs:6V Vu:>300V/mV Vi0:1mV f:0.7Mc	8S,5S	And		data pdf pinout
AD8542A	2xRR-OP	Vs:6V Vu:>300V/mV Vi0:1mV f:0.7Mc	8DS	And		data pinout
AD8544A	4xRR-OP	Vs:6V Vu:>300V/mV Vi0:1mV f:0.7Mc	14DS	And		data pinout
AD8551A	RR-OP	Vs:6V Vu:120dB Vi0:1µµV f:1.5Mc	8DS	And		data pinout
AD8552A	2xRR-OP	Vs:6V Vu:120dB Vi0:1µµV f:1.5Mc	8DS	And		data pinout
AD8554A	4xRR-OP	Vs:6V Vu:120dB Vi0:1µµV f:1.5Mc	14DS	And		data pinout
AD 8555 ACP	LIN-IC	Sensor Signal Amp., +2.7...5.5V Zero-Drift, Single- Supply, Prog. Gain	16-LFCSP	And	-	pdf
AD 8555 AR	LIN-IC	SMD, Sensor Signal Amp., +2.7...5.5V Zero-drift, Single-Supply, Prog. Gain	8-MDIP	And	-	pdf pinout
AD8561	COMP	Vs:16V Vi0:1mV	8DS	And		data pdf pinout
AD8564	COMP	Vs:16V Vi0:1mV	16DS	And		data pdf pinout
AD 8591 ART	OP-IC	Cmos Single Supply Rail-to-Rail I/O Op Amp.	TSOP	And	-	pdf pinout
AD 8592 ARM	OP-IC	Cmos Single Supply Rail-to-Rail I/O Op Amp.	TSOP	And	-	pdf pinout
AD 8594 AR	OP-IC	Cmos Single Supply Rail-to-Rail I/O Op Amp.	TSOP	And	-	pdf pinout
AD 8594 ARU	OP-IC	Cmos Single Supply Rail-to-Rail I/O Op Amp.	TSOP	And	-	pdf pinout
AD 9002 AD,BD	CMOS-A/D-IC	8 Bit, Video, >125 Msps, Ecl Level	28-DIP	And	-	
AD 9006	CMOS-A/D-IC	6 Bit, Video, >470 Msps, Ecl Level	68-LCC	And	-	
AD 9012 AQ,BQ	CMOS-A/D-IC	8 Bit, Video, >75 Msps, TTL Level	28-DIP	And	-	
AD 9016 KZ	CMOS-A/D-IC	6 Bit, Video, >470 Msps, Ecl Level	68-LCC	And	-	
AD 9020 J,K	CMOS-A/D-IC	10 Bit, Video, >60 Msps, TTL Level	68-LCC	And	-	
AD 9028 J,K	CMOS-A/D-IC	8 Bit, Video, >300 Msps, Ecl Level	68-LCC	And	-	
AD 9060 J,K	CMOS-A/D-IC	10 Bit, Video, >75 Msps, Ecl Level	68-LCC	And	-	
AD 9280	CMOS-A/D-IC	8-Bit, 32 MSPS, 95 mW Cmos ADC	MDIP	And	AD876-8	pdf pinout
AD 9398	DIG-IC	Hdmi display interface	QFP	And	-	pdf pinout
AD 9521(J,K,S,T)E	LIN-IC	Log. Amp, 250MHz, 5,7...6,3V	20-(P)LCC	And	-	
AD 9521(J,K,S,T)H	LIN-IC	Log. Amp, 250MHz, 5,7...6,3V	81	And	-	
AD9610	WIDEBAND	Vs:±18V Vi0:±0.3mV f:>100Mc	12A	And		data pdf pinout
AD 9617	WIDEBAND	Vs:±7V Vi0:±3.8V Vi0:0.5mV f:190Mc	8DS	And		data pdf pinout
AD 9617 JN	OP-IC	wideband, ±5<7V, ±60mA, 1400/µs, 10ns	8-DIP	And		data pdf
AD 9617 JR	OP-IC	=AD 9617JN: SMD	8-MDIP	And		data pdf
AD 9618	WIDEBAND	Vs:±7V Vo:±3.7V Vi0:0.5mV f:160Mc	8DS	And		data pdf pinout
AD 9618 JN	OP-IC	wideband, ±5<7V, ±60mA, 1800/µs, 9ns	8-DIP	And		data pdf
AD 9618 JR	OP-IC	=AD 9618JN: SMD	8-MDIP	And		data pdf
AD9620	BUFF	Vs:±7V Vo:±2.4V Vi0:±2mV f:150Mkc	8Dd	And		data pdf pinout
AD9621	WIDEBAND	Vs:±6V Vu:56dB Vi0:±2mV f:350Mc	8DS	And		data pdf pinout
AD9622	WIDEBAND	Vs:±6V Vu:60dB Vi0:±2mV f:220Mc	8DS	And		data pdf pinout
AD9623	WIDEBAND	Vs:±6V Vu:69dB Vi0:±2mV f:270Mc	8DS	And		data pdf pinout
AD9624	WIDEBAND	Vs:±6V Vu:74dB Vi0:±2mV f:300Mc	8DS	And		data pdf pinout
AD9630	BUFF	Vs:±7V Vo:±3.6V Vi0:±3mV f:120Mkc	8DSd	And		data pdf pinout
AD9631	WIDEBAND	Vs:±6V Vu:52dB Vo:±3.9V Vi0:3mV	8DSd	And		data pdf pinout
AD9632	WIDEBAND	Vs:±6V Vu:52dB Vo:±3.9V Vi0:2mV	8DSd	And		data pdf pinout

Type	Device	Short Description	Fig.	Manu	Comparision Types	More at
AD9685B	COMP	Vs:±6V Vi0:±5mV	16D,10A	And	-	data pinout
AD 9685(B)D	KOP-IC	±6V, Iq=30mA, <3ns, -30...+85°	16-DIC	And	-	
AD 9685(B)H	KOP-IC	=AD 9685..D:	82		-	
AD9687B	2xCOMP	Vs:±6V Vi0:±5mV	16D	And	-	data pinout
AD 9687(B)D	KOP-IC	Dual, ±6V, Iq=30mA, <4ns, -30...+85°	16-DIC	And	-	
AD9696	FAST-COMP	Vs:±14V Vi0:1mV	10A,8DSF	And	-	data pinout
AD9698K	2xFAST-COMP	Vs:±14V Vi0:1mV	16DS	And	-	data pdf pinout
AD9698T	2xFAST-COMP	Vs:±14V Vi0:1mV	16DF	And	-	data pdf pinout
AD 9700 BD,BW	D/A-IC	8 Bit, Video, >100 Msps, Ecl Level	28-DIP	And	-	
AD 9701 BQ	D/A-IC	8 Bit, Video, >225 MHz, Ecl Level	22-DIP	And	-	
AD 9702 BD:ECL	D/A-IC	4 Bit, Video, >125 Msps, Ecl Level	24-DIP	And	-	pdf
AD 9702 BD:TTL	D/A-IC	4 Bit, Video, >75 Msps, TTL Level	24-DIP	And	-	
AD 9712 JP	D/A-IC	Video, 12Bit D/A-Conv. 100Msps, Ecl	28-PLCC	And	-	
AD 9712 JN	D/A-IC	Video, 12Bit D/A-Conv. 100Msps, Ecl	28-DIP	And	-	pdf
AD 9713 JP	D/A-IC	Video, 12Bit D/A-Conv. 80Msps, TTL	28-PLCC	And	-	pdf
AD 9713 JN	D/A-IC	Video, 12Bit D/A-Conv. 80Msps, TTL	28-DIP	And	-	pdf
AD 9780 BCPZ	D/A-IC	Dual 12-/14-/16-Bit, LVDS Interface 600 MSPS DACs	QFN	And	-	pdf pinout
AD 9781 BCPZ	D/A-IC	Dual 12-/14-/16-Bit, LVDS Interface 600 MSPS DACs	QFN	And	-	pdf pinout
AD 9783 BCPZ	D/A-IC	Dual 12-/14-/16-Bit, LVDS Interface 600 MSPS DACs	QFN	And	-	pdf pinout
AD 9880	IC	Analog/hdmi Dual Interface Display	QFP	And	-	pdf pinout
AD 9882A	IC	Dual Interface for Flat Panel Displays	QFP	And	-	pdf pinout
AD 9883A	IC	RGB Analog Interface, 110/140 MSPS für/for Flat Panel Displays	80-QFP	And	-	pdf pinout
AD 9884 A	IC	RGB analog flat panel interface	QFP	And	-	pdf pinout
AD 9885	IC	RGB analog flat panel interface	QFP	And	-	pdf pinout
AD 9885 A	IC	RGB analog flat panel interface	QFP	And	-	pdf pinout
AD 9887A	IC	Dual interface f. flat panel display	QFP	And	AD9887	pdf pinout
AD 9888	IC	RGB analog flat panel interface	QFP	And	-	pdf pinout
AD 9913	IC	Cmos Direct Digital Synthesizer	TSOP	And	-	pdf pinout
AD 9980	IC	8 Bit display interface	QFP	And	-	pdf pinout
AD 9981	IC	10 Bit display interface	QFP	And	-	pdf pinout
AD 7147 ACPZ-1REEL	DIG-IC	CapTouch Contr.1-Electrode Capacit. Sensors	TSOP	And	-	pdf pinout
AD 7400 YRWZ-REEL7	IC	Isolated Sigma-Delta Modulator	TSOP	And	-	pdf pinout
AD 8221 ARM-REEL7	DIG-IC	Precision Instrumentation Amplifier	TSOP	And	-	pdf pinout
AD 8221 ARMZ-R7	DIG-IC	Precision Instrumentation Amplifier	TSOP	And	-	pdf pinout
AD 8221 AR-REEL7	DIG-IC	Precision Instrumentation Amplifier	TSOP	And	-	pdf pinout
AD 8221 ARZ-R7	DIG-IC	Precision Instrumentation Amplifier	TSOP	And	-	pdf pinout
AD 8221 BR-REEL7	DIG-IC	Precision Instrumentation Amplifier	TSOP	And	-	pdf pinout
AD 8231 ACPZ-R 7	IC	Digit. Programma. Instrumentation Amp.	TSOP	And	-	pdf pinout
AD96685	FAST-COMP	Vs:±6.5V Vi0:1mV	10A,20C	And	-	data pdf pinout
AD 96685(B,T)Q	KOP-IC	±6,5V, Iq=30mA, <3ns, -25...+85°	16-DIC	And	-	data pinout
AD 96685(B)E	KOP-IC	=AD 96685...Q:	20-PLCC		-	data pinout
AD 96685(B)H	KOP-IC	=AD 96685..Q:	82		-	data pinout
AD96687	2xFAST-COMP	Vs:±6.5V Vi0:1mV	20C,16DS	And	-	data pdf pinout
AD 96687(B,T)Q	KOP-IC	Dual, ±6,5V, Iq=30mA, <3ns, -25...+85°	16-DIC	And	-	data pinout
AD 96687(B)E	KOP-IC	=AD 96687...Q:	20-PLCC		-	data pinout
AD 9780 BCPZRL 7	D/A-IC	Dual 12-/14-/16-Bit, LVDS Interface 600 MSPS DACs	QFN	And	-	pdf pinout
AD 9781 BCPZRL 7	D/A-IC	Dual 12-/14-/16-Bit, LVDS Interface 600 MSPS DACs	QFN	And	-	pdf pinout
AD 9783 BCPZRL 7	D/A-IC	Dual 12-/14-/16-Bit, LVDS Interface 600 MSPS DACs	QFN	And	-	pdf pinout
AD 73522-40	MOS-µC-IC	Dual Analog Front End, Flash based Dsp Microcomputer	QFN	And	-	pdf pinout
AD 73522-80	MOS-µC-IC	Dual Analog Front End, Flash based Dsp Microcomputer	QFN	And	-	pdf pinout
AD 7147ACPZ-500RL7	DIG-IC	CapTouch Contr.1-Electrode Capacit. Sensors	TSOP	And	-	pdf pinout
AD 7357BRUZ-500RL7	A/D-IC	Diff. Input,4.25 MSPS, 14-Bit, SAR ADC	TSOP	And	-	pdf pinout
AD 7357YRUZ-500RL7	A/D-IC	Diff. Input,4.25 MSPS, 14-Bit, SAR ADC	TSOP	And	-	pdf pinout
AD7147ACPZ-1500RL7	DIG-IC	CapTouch Contr.1-Electrode Capacit. Sensors	TSOP	And	-	pdf pinout
ADA 4937-2 YCPZ-RL	IC	Dual Distortion, Differential ADC Driver	QFN	And	-	pdf pinout
ADA 4938-2 ACPZ-RL	IC	Dual Distortion, Differential ADC Driver	QFN	And	-	pdf pinout
ADA 4937-2 YCPZ-R2	IC	Dual Distortion, Differential ADC Driver	QFN	And	-	pdf pinout
ADA 4937-2 YCPZ-R7	IC	Dual Distortion, Differential ADC Driver	QFN	And	-	pdf pinout
ADA 4938-2 ACPZ-R2	IC	Dual Distortion, Differential ADC Driver	QFN	And	-	pdf pinout
ADA 4938-2 ACPZ-R7	IC	Dual Distortion, Differential ADC Driver	QFN	And	-	pdf pinout
ADAU 1513 ACPZ	Dig. Audio	Class-D Audio Power Stage	TSOP	And	-	pdf pinout
ADAU 1513 ACPZ-RL	Dig. Audio	Class-D Audio Power Stage	TSOP	And	-	pdf pinout
ADAU 1513 ASVZ	Dig. Audio	Class-D Audio Power Stage	TSOP	And	-	pdf pinout
ADAU 1513 ASVZ-RL	Dig. Audio	Class-D Audio Power Stage	TSOP	And	-	pdf pinout
ADAU 1590 ACPZ	Dig. Audio	Class-D Audio Power Amplifier	TSOP	And	-	pdf pinout
ADAU 1590 ACPZ-RL	Dig. Audio	Class-D Audio Power Amplifier	TSOP	And	-	pdf pinout
ADAU 1590 ASVZ	Dig. Audio	Class-D Audio Power Amplifier	TSOP	And	-	pdf pinout
ADAU 1590 ASVZ-RL	Dig. Audio	Class-D Audio Power Amplifier	TSOP	And	-	pdf pinout
ADAU 1592 ACPZ	Dig. Audio	Class-D Audio Power Amplifier	TSOP	And	-	pdf pinout
ADAU 1592 ACPZ-RL	Dig. Audio	Class-D Audio Power Amplifier	TSOP	And	-	pdf pinout
ADAU 1592 ASVZ	Dig. Audio	Class-D Audio Power Amplifier	TSOP	And	-	pdf pinout
ADAU 1592 ASVZ-RL	Dig. Audio	Class-D Audio Power Amplifier	TSOP	And	-	pdf pinout
ADAU 1513 ACPZ-RL7	Dig. Audio	Class-D Audio Power Stage	TSOP	And	-	pdf pinout

Type	Device	Short Description	Fig.	Manu	Comparision Types	More at
ADAU 1513 ASVZ-RL7	Dig. Audio	Class-D Audio Power Stage	TSOP	And	-	pdf pinout
ADAU 1590 ACPZ-RL7	Dig. Audio	Class-D Audio Power Amplifier	TSOP	And	-	pdf pinout
ADAU 1590 ASVZ-RL7	Dig. Audio	Class-D Audio Power Amplifier	TSOP	And	-	pdf pinout
ADAU 1592 ACPZ-RL7	Dig. Audio	Class-D Audio Power Amplifier	TSOP	And	-	pdf pinout
ADAU 1592 ASVZ-RL7	Dig. Audio	Class-D Audio Power Amplifier	TSOP	And	-	pdf pinout
ADAV 804	Dig. Audio	Audio Codec For Recordable DVD	QFP	And	-	pdf pinout
ADB 1200 PCN	A/D-IC	12 Bit, Binary A/D Building Block	28-DIP	Nsc		
ADC 10D020 CIVS	CMOS-A/D-IC	Dual, 10 Bit, A/D-Converter, 20 MSPS Sample&Hold, +3V, 150mW	48-QFP	Nsc	-	data pdf pinout
ADC 12C065 CISQ	A/D-IC	SMD, 12-Bit, Cmos 65MSPS A/D Conv., S&H,LLP 1GHz Full Power Bw, single +3.3V		Nsc	-	pdf pinout
ADC 12C080 CISQ	A/D-IC	SMD, 12-Bit, Cmos 80MSPS A/D Conv., S&H,LLP 1GHz Full Power Bw, single +3.3V		Nsc	-	pdf pinout
ADC 12C095 CISQ	A/D-IC	SMD, 12-Bit, Cmos 95MSPS A/D Conv., S&H,LLP 1GHz Full Power Bw, single +3.3V		Nsc	-	pdf pinout
ADC 12C105 CISQ	A/D-IC	SMD, 12-Bit, Cmos 105MSPS A/D Conv., S&H, 1GHz Full Power Bw, single +3.3V	LLP	Nsc	-	pdf pinout
ADC 12DS065 CISQ	A/D-IC	SMD, 12-Bit, Cmos 65MSPS A/D Conv., S&H,LLP serial out, 1GHz, Bw, +3.3V		Nsc	-	pdf pinout
ADC 12DS080 CISQ	A/D-IC	SMD, 12-Bit, Cmos 80MSPS A/D Conv., S&H,LLP serial out, 1GHz, Bw, +3.3V		Nsc	-	pdf pinout
ADC 12DS095 CISQ	A/D-IC	SMD, 12-Bit, Cmos 95MSPS A/D Conv., S&H,LLP serial out, 1GHz, Bw, +3.3V		Nsc	-	pdf pinout
ADC 12DS105 CISQ	A/D-IC	SMD, 12-Bit, Cmos 105MSPS A/D Conv., S&H, serial out, 1GHz, Bw, +3.3V	LLP	Nsc	-	pdf pinout
ADC 14C065 CISQ	A/D-IC	SMD, 14-Bit, Cmos 65MSPS A/D Conv., S&H,LLP 1GHz Full Power Bw, single +3.3V		Nsc	-	pdf pinout
ADC 14C080 CISQ	A/D-IC	SMD, 14-Bit, Cmos 80MSPS A/D Conv., S&H,LLP 1GHz Full Power Bw, single +3.3V		Nsc	-	pdf pinout
ADC 14C095 CISQ	A/D-IC	SMD, 14-Bit, Cmos 95MSPS A/D Conv., S&H,LLP 1GHz Full Power Bw, single +3.3V		Nsc	-	pdf pinout
ADC 14C105 CISQ	A/D-IC	SMD, 14-Bit, Cmos 105MSPS A/D Conv., S&H, 1GHz Full Power Bw, single +3.3V	LLP	Nsc	-	pdf pinout
ADC 14DS065 CISQ	A/D-IC	SMD, 14-Bit, Cmos 65MSPS A/D Conv., S&H,LLP serial out, 1GHz, Bw, +3.3V		Nsc	-	pdf pinout
ADC 14DS080 CISQ	A/D-IC	SMD, 14-Bit, Cmos 80MSPS A/D Conv., S&H,LLP serial out, 1GHz, Bw, +3.3V		Nsc	-	pdf pinout
ADC 14DS095 CISQ	A/D-IC	SMD, 14-Bit, Cmos 95MSPS A/D Conv., S&H,LLP serial out, 1GHz, Bw, +3.3V		Nsc	-	pdf pinout
ADC 14DS105 CISQ	A/D-IC	SMD, 14-Bit, Cmos 105MSPS A/D Conv., S&H, serial out, 1GHz, Bw, +3.3V	LLP	Nsc	-	pdf pinout
ADC 207 LC,LM(-QL)	CMOS-A/D-IC	7 Bit, Flash A/D-Converter, S&H +5V, 20MHz, MC:0...70°C MM:-55...125°C	24-LCC	Dat	-	pdf
ADC 207 MC,MM(-QL)	CMOS-A/D-IC	7 Bit, Flash A/D-Converter, S&H +5V, 20MHz, MC:0...70°C MM:-55...125°C	18-DIC	Dat	-	pdf
ADC 208 MC	CMOS-A/D-IC	8 Bit, Flash A/D-Conv., S&H, 15 MHz	24-DIP	Dat		
ADC 228A MC,MM,883	A/D-IC	8 Bit, Flash A/D-Converter, ±15V +5V MC: 0...70°C MM,883: -55...125°C, 20MHz	24-DIC	Dat	-	pdf
ADC 228 MC,MM,883	A/D-IC .	8 Bit, Flash A/D-Converter, ±15V +5V MC: 0...70°C MM,883: -55...125°C, 20MHz	24-DIC	Dat	-	pdf
ADC 305-1	CMOS-A/D-IC	8 Bit, 2-Pass Flash A/D-Converter, S&H -40..85°C, 20MHz	24-DIP	Dat	-	data pdf pinout
ADC 305-3	CMOS-A/D-IC	SMD, 8 Bit, 2-Pass Flash A/D-Converter Sample&Hold, 20MHz, -40...85°C	24-MDIP	Dat	-	data pdf pinout
ADC 318(A)	A/D-IC	8 Bit, Flash A/D-Conv., S&H, -20...75°C ±5V or/oder +5V, 120MHz (A:140MHz)	48-QFP	Dat	-	pdf
ADC 321	CMOS-A/D-IC	8 Bit, Video A/D-Converter, S&H, 50MHz	32-QFP	Dat	-	data pdf pinout
ADC 0800 P(C)D	MOS-A/D-IC	8Bit, 3-State Out,-55...+125(C=0...+70°)	18-DIP	Nsc	-	pdf
ADC 0801...0805 LCN	CMOS-A/D-IC	8 Bit, SAR A/D-Converter, µP-komp. -40...85°C, ADC 0804 LCN: 0...70°C	20-DIP,DIC	Nsc	-	pdf
ADC 0801(-1)...	CMOS-A/D-IC	8 Bit, Access 135ns, 0,10% Accuracy	20-DIC,DIP	Nsc,Phi,Tix	-	
ADC 0802(-1)...	CMOS-A/D-IC	8 Bit, Access 135ns, 0,19% Accuracy	20-DIC,DIP	Nsc,Phi,Tix	-	
ADC 0802/0804 LCWM	CMOS-A/D-IC	SMD, 8 Bit, SAR A/D-Conv., µP-komp. 0...70°C	20-MDIP	Nsc	-	pdf
ADC 0803(-1)...	CMOS-A/D-IC	8 Bit, Access 135ns, 0,19% Accuracy	20-DIC,DIP	Nsc,Phi,Tix	-	pinout
ADC 0804(-1)...	CMOS-A/D-IC	8 Bit, Access 135ns, 0,39% Accuracy	20-DIC,DIP	Nsc,Phi,Tix	-	pinout
ADC 0804 LCJ	CMOS-A/D-IC	8 Bit, SAR A/D-Converter, µP-komp. -40...85°C, ±1Bit unadjusted	20-DIP,DIC	Nsc	-	data pdf pinout
ADC 0805(-1)...	CMOS-A/D-IC	8 Bit, Access 135ns, 0,39% Accuracy	20-DIC,DIP	Nsc,Phi,Tix	-	
ADC 0808 ...	CMOS-A/D-IC	8 Bit, 8-Channel Multiplexer	28-DIC,DIP	Nsc,Tix	TL 0808	
ADC 0808 (C)CJ	CMOS-A/D-IC	8 Bit, A/D-Converter, 8-Kan./Chan. MUX CJ: -55...125°C, CCJ:-40...85°C, ±½LSB	28-DIC	Nsc,Tix	TL 0808, MM 74C949	pdf
ADC 0808 CCN	CMOS-A/D-IC	8 Bit, A/D-Converter, 8-Kan./Chan. MUX -40...85°C, ±½LSB	28-DIP	Nsc,Tix	TL 0808, MM 74C949	data pdf pinout
ADC 0808 CCV	CMOS-A/D-IC	8 Bit, A/D-Converter, 8-Kan./Chan. MUX -40...85°C, ±½LSB	28-PLCC	Nsc,Tix	TL 0808, MM 74C949	data pdf pinout
ADC 0809 ...	CMOS-A/D-IC	8 Bit, 8-Channel Multiplexer	28-DIC,DIP	Nsc,Tix	TL 0809	
ADC 0809 CCN	CMOS-A/D-IC	8 Bit, A/D-Converter, 8-Kan./Chan. MUX -40...85°C, ±1LSB	28-DIP	Nsc,Tix	TL 0809, MM 74C949-1	data pdf pinout
ADC 0809 CCV	CMOS-A/D-IC	8 Bit, A/D-Converter, 8-Kan./Chan. MUX -40...85°C, ±1LSB	28-PLCC	Nsc,Tix	TL 0809, MM 74C949-1	data pdf pinout
ADC 0811	CMOS-A/D-IC	→TLC 541	20-DIP	Nsc	MC 145040, TLC 541	pdf pinout
ADC 0816 ...	CMOS-A/D-IC	8 Bit, 16-Channel Multiplexer	40-DIC,DIP	Nsc	-	
ADC 0816...0817 CCN	CMOS-A/D-IC	8 Bit, A/D-Converter, µP-komp., S&H 16 Kan./Chan. MUX	40-DIP	Nsc	-	pdf

Type	Device	Short Description	Fig.	Manu	Comparision Types	More at
ADC 0816 CCJ	CMOS-A/D-IC	8 Bit, A/D-Converter, µP-komp., ±½LSB 16 Kan./Chan. MUX, Sample&Hold	40-DIC	Nsc	MM 74C948	data pdf pinout
ADC 0817 ...	CMOS-A/D-IC	8 Bit, 16-Channel Multiplexer	40-DIC,DIP	Nsc	-	
ADC 0820 ...	CMOS-A/D-IC	8 Bit, <1,4µs, Track/Hold	20-DIC,DIP	Phi, Max	-	
ADC 0820 BCJ	CMOS-A/D-IC	8 Bit, A/D-Converter, T&H, ±½LSB, 25mW	20-DIC	Max	-	data pdf pinout
ADC 0820 BCM,CCM	CMOS-A/D-IC	SMD, 8 Bit, A/D-Conv., Track&Hold BCM:±½LSB, CCM:±1LSB, 25mW	20-MDIP	Max	-	pdf
ADC 0820 BCN,CCN	CMOS-A/D-IC	8 Bit, A/D-Converter, Track&Hold, 75mW BCN:±½LSB, CCN:±1LSB	20-DIP	Nsc, Max	-	pdf
ADC 0820 BCV	CMOS-A/D-IC	8 Bit, A/D-Converter, T&H, ±½LSB, 75mW	20-PLCC	Nsc, Max	-	data pdf pinout
ADC 0820 BCWM,CCWM	CMOS-A/D-IC	SMD, 8 Bit, A/D-Conv., Track&Hold BCWM:±½LSB, CCWM:±1LSB, 75mW	20-MDIP	Nsc, Max	-	pdf
ADC 0820 BJ,CJ	CMOS-A/D-IC	8 Bit, A/D-Converter, Track&Hold, 25mW BJ:±½LSB, CJ:±1LSB	20-DIC	Max	-	pdf
ADC 0820 CCJ	CMOS-A/D-IC	8 Bit, A/D-Converter, T&H, ±1LSB, 75mW	20-DIC	Nsc, Max	-	data pdf pinout
ADC 0820 CIWM	CMOS-A/D-IC	SMD, 8Bit, A/D-Conv., T&H, ±1LSB, 75mW	20-MDIP	Nsc, Max	-	data pdf pinout
ADC 0829	CMOS-A/D-IC	→TL 532	28-DIP	Nsc	MC 14442, TL 532	
ADC 0830	CMOS-A/D-IC	→TL 530	40-DIP	Nsc	MC 14444, TL 530	
ADC 0831..32 CCN	A/D-IC	8 Bit, A/D-Conv., ser.DataOut, ±1LSB ..31: 1 Kan., ..32: 2 Kan./Chan. MUX	8-DIP	Nsc	-	pdf
ADC 0831..32 CCWM	A/D-IC	SMD, 8 Bit, A/D-Conv., ser.Out, ±1LSB ..31: 1 Kan., ..32: 2 Kan./Chan. MUX	14-MDIP	Nsc	-	pdf
ADC 0831 A,B...	A/D-IC	8 Bit, 2-Channel Multiplexer	8-DIC,DIP	Nsc,Tix	-	
ADC 0832 A,B...	A/D-IC	8 Bit, 2-Channel Multiplexer	8-DIC,DIP	Nsc,Tix	-	
ADC 0832 CIWM	A/D-IC	SMD, 8 Bit, A/D-Converter, ser.DataOut 2 Kanal/Channel Multiplexer, ±1LSB	14-MDIP	Nsc	-	data pdf pinout
ADC 0833 ...	A/D-IC	8 Bit, 4-Channel Multiplexer	14-DIC,DIP	Nsc	-	
ADC 0834 A,B...	A/D-IC	8 Bit, 4-Channel Multiplexer	14-DIC,DIP	Nsc,Tix	-	
ADC 0834 BCN, CCN	A/D-IC	8 Bit, A/D-Converter, ser.DataOut 4 Kan./Chan. MUX, BCN:±½LSB CCN:±1LSB	14-DIP	Nsc	-	pdf
ADC 0834 CCWM	A/D-IC	SMD, 8 Bit, A/D-Converter, ser.DataOut 4 Kanal/Channel Multiplexer, ±1LSB	14-MDIP	Nsc	-	data pdf pinout
ADC 0838 A,B...	A/D-IC	8 Bit, 4-Channel Multiplexer	20-DIC,DIP	Nsc,Tix	-	
ADC 0838 BCV,CCV	A/D-IC	8 Bit, A/D-Converter, ser.DataOut 8 Kan./Chan. MUX, BCV:±½LSB CCV:±1LSB	20-PLCC	Nsc	-	pdf
ADC 0838 C(C,I)WM	A/D-IC	SMD, 8 Bit, A/D-Converter, ser.DataOut 8 Kanal/Channel Multiplexer, ±1LSB	20-MDIP	Nsc	-	pdf
ADC 0838 CCN	A/D-IC	8 Bit, A/D-Converter, ser.DataOut 8 Kanal/Channel Multiplexer, ±1LSB	20-DIP	Nsc	-	pdf
ADC 0844 BCJ,CCJ	CMOS-A/D-IC	8 Bit, A/D-Conv., 4 Kan./Chan. MUX BCJ: ±½ LSB, CCJ: ±1 LSB, -40...85°C	20-DIC	Nsc	-	pdf
ADC 0844 CCN	CMOS-A/D-IC	8 Bit, A/D-Conv., 4 Kan./Chan. MUX ±1 LSB, 0...70°C	20-DIP	Nsc	-	data pdf pinout
ADC 0848 BCN,CCN	CMOS-A/D-IC	8 Bit, A/D-Conv., 8 Kan./Chan. MUX BCN: ±½ LSB, CCN: ±1 LSB, 0...70°C	24-DIP	Nsc	-	pdf
ADC 0848 BCV,CCV	CMOS-A/D-IC	8 Bit, A/D-Conv., 8 Kan./Chan. MUX BCV: ±½ LSB, CCV: ±1 LSB, -40...85°C	28-PLCC	Nsc	-	pdf
ADC-908(...)P,X	CMOS-A/D-IC	8 Bit, 6µs, µcomp Interface	18-DIP,DIC	Pmi	AD 7574...	
ADC-908(...)S	CMOS-A/D-IC	SMD,8 Bit, 6µs, µcomp Interface	18-MDIP	Pmi	-	
ADC-910(...)T	A/D-IC	10 Bit, 6µs, µcomp Interface	28-DIC	Pmi	-	
ADC-912(...)P,W	CMOS-A/D-IC	12 Bit, 12µs, µcomp Interface	24-DIP,DIC	Pmi	-	
ADC-912(...)S	CMOS-A/D-IC	SMD,12 Bit, 12µs, µcomp Interface	24-MDIP	Pmi	-	
ADC-922(...)T	A/D-IC	12 Bit, <3µs, 10V Reference	28-DIC	Pmi	-	
ADC 1001 ...	A/D-IC	10 Bit, Access 170ns, 3-State Out.	20-DIC,DIP	Nsc	-	
ADC 1001 CCJ,CCJ-1	CMOS-A/D-IC	10 Bit, A/D-Conv., 2x 8Bit Data, ±1LSB CCJ: -40...85°C, CCJ-1: 0...70°C	20-DIC	Nsc	-	pdf
ADC 1005 BCJ,BCJ-1	CMOS-A/D-IC	10 Bit, A/D-Conv., 2x 8Bit Data, ±½LSB BCJ: -40...85°C, BCJ-1: 0...70°C	20-DIC	Nsc	-	pdf
ADC 1005 CCJ,CCJ-1	CMOS-A/D-IC	10 Bit, A/D-Conv., 2x 8Bit Data, ±1LSB CCJ: -40...85°C, CCJ-1: 0...70°C	20-DIC	Nsc	-	pdf
ADC 1021 ...	A/D-IC	10 Bit, Access 170ns, 3-State Out.	24-DIC,DIP	Nsc	-	
ADC 1061 CIN	A/D-IC	10 Bit, A/D-Converter, Track&Hold	20-DIP	Nsc	-	data pdf pinout
ADC 1061 CIWM	CMOS-A/D-IC	SMD, 10 Bit, A/D-Converter, Track&Hold	20-MDIP	Nsc	ADC 10061 CIWM	data pdf pinout
ADC 1080 ...	A/D-IC	12 Bit, Successive Approximation, ±18V	32-DIC,DIP	Nsc	-	
ADC 1173 CIJM	A/D-IC	SMD, 8 Bit, A/D-Converter, 15MSPS, S&H	24-MDIP	Nsc	-	data pdf pinout
ADC 1173 CIMTC	A/D-IC	SMD, 8 Bit, A/D-Converter, 15MSPS, S&H	24-SSMDIP	Nsc	-	data pdf pinout
ADC 1175-50 CIJM	A/D-IC	SMD, 8 Bit, A/D-Converter, 50MSPS, T&H	24-MDIP	Nsc	-	data pdf pinout
ADC 1175 CIJM	A/D-IC	SMD, 8 Bit, A/D-Converter, 20MHz, S&H	24-MDIP	Nsc	-	data pdf pinout
ADC 1175-50 CILQ	A/D-IC	SMD, 8 Bit, A/D-Converter, 50MSPS, T&H	24-LLP	Nsc	-	data pdf pinout
ADC 1175-50 CIMT	A/D-IC	SMD, 8 Bit, A/D-Converter, 50MSPS, T&H	24-SSMDIP	Nsc	-	data pdf pinout
ADC 1175 CIMTC	A/D-IC	SMD, 8 Bit, A/D-Converter, 20MHz, S&H	24-SSMDIP	Nsc	-	data pdf pinout
ADC 1210 ...	CMOS-A/D-IC	12 Bit, Successive Approximation, ±15V	24-DIC,DIP	Nsc	-	
ADC 1211 ...	CMOS-A/D-IC	12 Bit, Successive Approximation, ±15V	24-DIC,DIP	Nsc	-	
ADC 1241 BIJ,CIJ	CMOS-A/D-IC	12 Bit + Vorz./Sign, A/D-Conv., S&H par. Ausg./Out, self cal., -40...85°C	28-DIC	Nsc	ADC 12441 CIJ	pdf
ADC 1241 CMJ(/883)	CMOS-A/D-IC	12 Bit + Vorz./Sign, A/D-Conv., S&H par. Ausg./Out, self cal., MIL-STD-883	28-DIC	Nsc	ADC 12441 CMJ(/883)	pdf
ADC 1251 BIJ,CIJ	CMOS-A/D-IC	12 Bit + Vorz./Sign, A/D-Conv., S&H par. Ausg./Out, self cal., -40...85°C	24-DIC	Nsc		pdf
ADC 1251 CMJ(/883)	CMOS-A/D-IC	12 Bit + Vorz./Sign, A/D-Conv., S&H par. Ausg./Out, self cal., MIL-STD-883	24-DIC	Nsc		pdf
ADC 1280 ...	A/D-IC	12 Bit, Successive Approximation, ±18V	32-DIC,DIP	Nsc	-	
ADC 2300 E	A/D-IC	CTV-audio a/d-converter Europa	24-DIP	Itt	-	pdf
ADC 2300 J	A/D-IC	Ctv-audio-a/d-converter Japan	24-DIP	Itt	-	
ADC 2300 U	A/D-IC	Ctv-audio-a/d-converter USA	24-DIP	Itt	-	

Type	Device	Short Description	Fig.	Manu	Comparision Types	More at
ADC 2301 E	A/D-IC	Ctv-audio-a/d-converter Europa		Itt	-	
ADC 2310 E	A/D-IC	Ctv-audio-a/d-converter Europa	24-DIP	Itt	-	pdf
ADC 2311 E	A/D-IC	Audio	24-DIP	Itt	-	
ADC 2320 U	A/D-IC	Audio, f. Usa Btsc-standard	24-DIP	Itt	-	
ADC 08031 (B,C)IWM	A/D-IC	SMD, 8Bit, A/D-Conv., ser.DataOut, T&H	14-MDIP	Nsc	-	pdf
ADC 08031 BIN,CIN	A/D-IC	8 Bit, A/D-Converter, ser.DataOut, T&H	8-DIP	Nsc	-	pdf
ADC 08032 (B,C)IWM	A/D-IC	SMD, 8 Bit, 2-Kan./Chan. A/D-Converter ser.DataOut, Track&Hold	14-MDIP	Nsc	-	pdf
ADC 08034 (B,C)IWM	A/D-IC	SMD, 8 Bit, 4-Kan./Chan. A/D-Converter ser.DataOut, Track&Hold	14-MDIP	Nsc	-	pdf
ADC 08038 (B,C)IWM	A/D-IC	SMD, 8 Bit, 8-Kan./Chan. A/D-Converter ser.DataOut, Track&Hold	20-MDIP	Nsc	-	pdf
ADC 08038 BIN,CIN	A/D-IC	8 Bit, 8-Kan./Chan. A/D-Converter ser.DataOut, Track&Hold	20-DIP	Nsc	-	pdf
ADC 08060	A/D-IC	SMD, 8 Bit, 60MSPS, A/D-Converter, S&H	24-SSMDIP	Nsc	-	data pdf pinout
ADC 08061...62 BIN	CMOS-A/D-IC	A/D-Converter, S&H, 500ns, 100mW ...62: incl. 2-Kan./Chan. MUX	20-DIP	Nsc	-	pdf
ADC 08061...62 CIWM	CMOS-A/D-IC	SMD, A/D-Conv., S&H, 500ns, 100mW ...62: incl. 2-Kan./Chan. MUX	20-MDIP	Nsc	-	pdf
ADC 08100	A/D-IC	SMD, 8 Bit, 100MSPS, A/D-Converter,S&H	24-SSMDIP	Nsc	-	data pdf pinout
ADC 08131 CIWM	A/D-IC	SMD, 8 Bit, A/D-Converter, Track&Hold seriell DataOut, 1MHz, 8µs, 20mW	14-MDIP	Nsc	-	data pdf pinout
ADC 08134 CIWM	A/D-IC	SMD, 8Bit, 4-Kan./Chan. A/D-Conv., T&H seriell DataOut, 1MHz, 8µs, 20mW	14-MDIP	Nsc	-	data pdf pinout
ADC 08138 CIWM	A/D-IC	SMD, 8Bit, 8-Kan./Chan. A/D-Conv., T&H seriell DataOut, 1MHz, 8µs, 20mW	20-MDIP	Nsc	-	data pdf pinout
ADC 08161 CIWM	CMOS-A/D-IC	SMD, A/D-Conv., 500ns, Sample&Hold	20-MDIP	Nsc	-	data pdf pinout
ADC 08200	A/D-IC	SMD, 8 Bit, 200MSPS, A/D-Converter,S&H	24-SSMDIP	Nsc	-	data pdf pinout
ADC 08351 CILQ	CMOS-A/D-IC	SMD, 8 Bit A/D-Converter, 42MHz, S&H	24-LLP	Nsc	-	data pdf pinout
ADC 08351 CIMTC	CMOS-A/D-IC	SMD, 8 Bit A/D-Converter, 42MHz, S&H	20-SSMDIP	Nsc	-	data pdf pinout
ADC 08831...32 IM	CMOS-A/D-IC	SMD, 8 Bit, A/D-Conv., S&H, ser. Out 31: 1 Kan./Chan., 32: 2 Kan./Chan. MUX	8-MDIP	Nsc	-	pdf
ADC 08831...32 IMM	CMOS-A/D-IC	SMD, 8 Bit, A/D-Conv., S&H, ser. Out 31: 1 Kan./Chan., 32: 2 Kan./Chan. MUX	8-SSMDIP	Nsc	-	pdf
ADC 08831...32 IN	CMOS-A/D-IC	8 Bit, A/D-Conv., S&H, ser. Out/Ausg. 31: 1 Kan./Chan., 32: 2 Kan./Chan. MUX	8-DIP	Nsc	-	pdf
ADC 08831...32 IWM	CMOS-A/D-IC	SMD, 8 Bit, A/D-Conv., S&H, ser. Out 31: 1 Kan./Chan., 32: 2 Kan./Chan. MUX	14-MDIP	Nsc	-	pdf
ADC 10030 CIVT	CMOS-A/D-IC	10 Bit, 30 MSPS, A/D-Converter, S&H	32-QFP	Nsc	-	data pdf pinout
ADC 10061 CIWM	CMOS-A/D-IC	SMD, 10 Bit, A/D-Conv., Sample&Hold	20-MDIP	Nsc	ADC 10461 CIWM	data pdf pinout
ADC 10062 CIWM	CMOS-A/D-IC	SMD, 10 Bit, A/D-Conv., Sample&Hold 2 Kan./Chan. MUX	24-MDIP	Nsc	ADC 10462 CIWM	data pdf pinout
ADC 10064 CIWM	CMOS-A/D-IC	SMD, 10 Bit, A/D-Conv., Sample&Hold 4 Kan./Chan. MUX	28-MDIP	Nsc	ADC 10464 CIWM	data pdf pinout
ADC 10154 CIWM	CMOS-A/D-IC	SMD, 10 Bit + Vorz./Sign, A/D-Conv. 4 Kan./Chan. MUX, T&H, 2.5V Ref.	24-MDIP	Nsc	-	data pdf pinout
ADC 10158 CIN	CMOS-A/D-IC	SMD, 10 Bit + Vorz./Sign, A/D-Conv. 8 Kan./Chan. MUX, T&H, 2.5V Ref.	28-DIP	Nsc	-	data pdf pinout
ADC 10158 CIWM	CMOS-A/D-IC	SMD, 10 Bit + Vorz./Sign, A/D-Conv. 8 Kan./Chan. MUX, T&H, 2.5V Ref.	28-MDIP	Nsc	-	data pdf pinout
ADC 10221 CIVT	CMOS-A/D-IC	10 Bit, 15 MSPS, A/D-Converter, S&H	32-QFP	Nsc	-	data pdf pinout
ADC 10321 CIVT	CMOS-A/D-IC	10 Bit, 20 MSPS, A/D-Converter, S&H	32-QFP	Nsc	-	data pdf pinout
ADC 10461 CIWM	CMOS-A/D-IC	SMD, 10 Bit, A/D-Conv., Sample&Hold	20-MDIP	Nsc	ADC 10061 CIWM	data pdf pinout
ADC 10462 CIWM	CMOS-A/D-IC	SMD, 10 Bit, A/D-Conv., Sample&Hold 2 Kan./Chan. MUX	24-MDIP	Nsc	ADC 10062 CIWM	data pdf pinout
ADC 10464 CIWM	CMOS-A/D-IC	SMD, 10 Bit, A/D-Conv., Sample&Hold 4 Kan./Chan. MUX	28-MDIP	Nsc	ADC 10064 CIWM	data pdf pinout
ADC 10662 CIWM	CMOS-A/D-IC	SMD, 10 Bit, A/D-Conv., Sample&Hold 2 Kan./Chan. MUX	24-MDIP	Nsc	-	data pdf pinout
ADC 10664 CIWM	CMOS-A/D-IC	SMD, 10 Bit, A/D-Conv., Sample&Hold 4 Kan./Chan. MUX	28-MDIP	Nsc	-	data pdf pinout
ADC 10731 CIWM	CMOS-A/D-IC	SMD, 10 Bit + Vorz./Sign, A/D-Conv. S&H, ser. Out/Ausg., 2.5V Ref.	16-MDIP	Nsc	-	data pdf pinout
ADC 10732 CIWM	CMOS-A/D-IC	SMD, 10 Bit + Vorz./Sign, A/D-Conv. 2 Kan./Chan., S&H, ser. I/O, 2.5V Ref.	20-MDIP	Nsc	-	data pdf pinout
ADC 10734 CIMSA	CMOS-A/D-IC	SMD, 10 Bit + Vorz./Sign, A/D-Conv. 4 Kan./Chan., S&H, ser. I/O, 2.5V Ref.	20-SSMDIP	Nsc		data pdf pinout
ADC 10734 CIWM	CMOS-A/D-IC	SMD, 10 Bit + Vorz./Sign, A/D-Conv. 4 Kan./Chan., S&H, ser. I/O, 2.5V Ref.	20-MDIP	Nsc		data pdf pinout
ADC 10738 CIWM	CMOS-A/D-IC	SMD, 10 Bit + Vorz./Sign, A/D-Conv. 8 Kan./Chan., S&H, ser. I/O, 2.5V Ref.	24-MDIP	Nsc		data pdf pinout
ADC 12(HL)030 CIWM	A/D-IC	SMD, 8/12 Bit + Vorz./Sign, prog. A/D 2 Kan./Chan., S&H, ser. I/O, self cal. L: 3.3V Version, H: High Speed Version	16-MDIP	Nsc		pdf
ADC 12(HL)032 CIWM	A/D-IC	SMD, 8/12 Bit + Vorz./Sign, prog. A/D 2 Kan./Chan., S&H, ser. I/O, self cal. L: 3.3V Version, H: High Speed Version	20-MDIP	Nsc		pdf
ADC 12H034 CIMSA	A/D-IC	SMD, 8/12 Bit + Vorz./Sign, prog. A/D 4 Kan./Chan., S&H, ser. I/O, self cal.	24-SSMDIP	Nsc		data pdf pinout
ADC 12(H)034 CIN	A/D-IC	SMD, 8/12 Bit + Vorz./Sign, prog. A/D 4 Kan./Chan., S&H, ser. I/O, self cal.	24-DIP	Nsc		pdf
ADC 12(HL)034 CIWM	A/D-IC	SMD, 8/12 Bit + Vorz./Sign, prog. A/D 4 Kan./Chan., S&H, ser. I/O, self cal. L: 3.3V Version, H: High Speed Version	24-MDIP	Nsc		pdf
ADC 12(HL)038 CIWM	A/D-IC	SMD, 8/12 Bit + Vorz./Sign, prog. A/D 8	28-MDIP	Nsc		pdf

Type	Device	Short Description	Fig.	Manu	Comparision Types	More at
		Kan./Chan., S&H, ser. I/O, self cal. L: 3.3V Version, H: High Speed Version				
ADC 12040 CIVY	CMOS-A/D-IC	12 Bit, 40 MSPS, A/D-Converter, S&H	32-QFP	Nsc	-	data pdf pinout
ADC 12041 CIMSA	A/D-IC	SMD, 12 Bit + Vorz./Sign, A/D-Conv. Sample&Hold, pár. I/O, self cal.	28-SSMDIP	Nsc	-	data pdf pinout
ADC 12041 CIV	A/D-IC	12 Bit + Vorz./Sign, A/D-Converter Sample&Hold, par. I/O, self cal.	28-PLCC	Nsc		data pdf pinout
ADC 12048 CIV	A/D-IC	12 Bit + Vorz./Sign, A/D-Converter 8 Kan./Chan., S&H, par. I/O, self cal.	44-PLCC	Nsc		data pdf pinout
ADC 12048 CIVF	A/D-IC	12 Bit + Vorz./Sign, A/D-Converter 8 Kan./Chan., S&H, par. I/O, self cal.	44-QFP	Nsc		data pdf pinout
ADC 12062 (B,C)IVF	CMOS-A/D-IC	12 Bit, A/D-Converter, 1MHz, 75mW Sample&Hold, 2-Kanal Multiplexer	44-QFP	Nsc		pdf
ADC 12062 CIV	CMOS-A/D-IC	12 Bit, A/D-Converter, 1MHz, 75mW Sample&Hold, 2-Kanal Multiplexer	44-PLCC	Nsc		data pdf pinout
ADC 12L063 CIVY	CMOS-A/D-IC	12 Bit, A/D-Conv., S&H, 62MSPS, 3.3V	32-QFP	Nsc		data pdf pinout
ADC 12081 CIVT	CMOS-A/D-IC	12 Bit, 5 MHz, pipelined A/D-Converter self calibrating, Sample & Hold	32-QFP	Nsc		data pdf pinout
ADC 12130 CIN	A/D-IC	12 Bit + Vorz./Sign, prog. A/D-Conv. 2 Kan./Chan., S&H, ser. I/O, self cal.	16-DIP	Nsc		data pdf pinout
ADC 12130 CIWM	A/D-IC	SMD, 12 Bit + Vorz./Sign, prog. A/D 2 Kan./Chan., S&H, ser. I/O, self cal.	16-MDIP	Nsc		data pdf pinout
ADC 12132 CIMSA	A/D-IC	SMD, 12 Bit + Vorz./Sign, prog. A/D 2 Kan./Chan., S&H, ser. I/O, self cal.	20-SSMDIP	Nsc		
ADC 12138 CIMSA	A/D-IC	SMD, 12 Bit + Vorz./Sign, prog. A/D 8 Kan./Chan., S&H, ser. I/O, self cal.	28-SSMDIP	Nsc		data pdf pinout
ADC 12138 CIN	A/D-IC	12 Bit + Vorz./Sign, prog. A/D-Conv. 8 Kan./Chan., S&H, ser. I/O, self cal.	28-DIP	Nsc		data pdf pinout
ADC 12138 CIWM	A/D-IC	SMD, 12 Bit + Vorz./Sign, prog. A/D 8 Kan./Chan., S&H, ser. I/O, self cal.	28-MDIP	Nsc		data pdf pinout
ADC 12181 CIVT	CMOS-A/D-IC	12 Bit,10 MHz, pipelined A/D-Converter self calibrating, Sample & Hold	32-QFP	Nsc		data pdf pinout
ADC 12191 CIVT	CMOS-A/D-IC	12 Bit,10 MHz, pipelined A/D-Converter self calibrating, Sample & Hold	32-QFP	Nsc		data pdf pinout
ADC 12281 CIVT	CMOS-A/D-IC	12 Bit,20 MHz, pipelined A/D-Converter self calibrating, Sample & Hold	32-QFP	Nsc	-	data pdf pinout
ADC 12441 CIJ	CMOS-A/D-IC	12 Bit + Vorz./Sign, A/D-Conv., S&H par. Ausg./Out, self cal., -40...85°C	28-DIC	Nsc	ADC 1241 CIJ	data pdf pinout
ADC 12441 CMJ(883)	CMOS-A/D-IC	12 Bit + Vorz./Sign, A/D-Conv., S&H par. Ausg./Out, self cal., -55...125°C	28-DIC	Nsc	ADC 1241 CMJ(/883)	pdf
ADC 12451 CIJ	CMOS-A/D-IC	12 Bit + Vorz./Sign, A/D-Conv., S&H par. Ausg./Out, self cal., -40...85°C	24-DIC	Nsc		data pdf pinout
ADC 12451 CMJ	CMOS-A/D-IC	12 Bit + Vorz./Sign, A/D-Conv., S&H par. Ausg./Out, self cal., -55...125°C	24-DIC	Nsc		data pdf pinout
ADC 14061 CCVT	A/D-IC	14 Bit, 2,5 MSPS, A/D-Converter self calibrating, Sample & Hold	52-QFP	Nsc		data pdf pinout
ADC 14071 CIVBH	CMOS-A/D-IC	14 Bit, 7MSPS, A/D-Converter, S&H	48-QFP	Nsc		data pdf pinout
ADC 14161 CIVT	A/D-IC	14 Bit, 2,5 MSPS, A/D-Converter self calibrating, Sample & Hold	52-QFP	Nsc		data pdf pinout
ADC 16061 CCVT	A/D-IC	16 Bit, 2,5 MSPS, A/D-Converter self calibrating, Sample & Hold	52-QFP	Nsc		
ADC 30634	CMOS-A/D-IC	8 Bit, Flash A/D-Conv., S&H, 20MSPS	24-LCC	Dat	-	data pdf pinout
ADC 30720	CMOS-A/D-IC	8 Bit, Flash A/D-Conv., S&H, 20MSPS	24-DIP	Dat	-	data pdf pinout
ADCDS 1403 (EX)	A/D-IC	Imag. Sign. Proc., 3MPPS, ±5V +12V 14Bit, A/D-Conv., Cds, Sample&Hold 0...70°C (EX: -55...125°C)	40-DIP	Dat		pdf
ADCP 1100	CMOS-IC	Audio-a/d-converter, Filter, Interface	44-PLCC	Itt	-	
ADCV 0831 M6	A/D-IC	SMD, 8 Bit, A/D-Conv., ser.I/O, 3V	46	Nsc	-	data pdf pinout
ADD 8733 ACPZ-REEL	IC	Integrated TFT Panel Power Supply	TSOP	And	-	pdf pinout
ADE 3000	DIG/LIN-IC	LCD Display,DVI, ADC and Yuv Ports	QFP	Sgs	-	pdf pinout
ADE 3050	DIG/LIN-IC	LCD Display,Dvi, ADC and Yuv Ports	QFP	Sgs	-	pdf pinout
ADE 3100	DIG/LIN-IC	LCD Display, Dvi, ADC and Yuv Ports	QFP	Sgs	-	pdf pinout
ADE 3200	DIG/LIN-IC	LCD Display,Dvi, ADC and Yuv Ports	QFP	Sgs	-	pdf pinout
ADE 3250	DIG/LIN-IC	LCD Display,Dvi, ADC and Yuv Ports	QFP	Sgs	-	pdf pinout
ADE 3300	DIG/LIN-IC	LCD Display,Dvi, ADC and Yuv Ports	QFP	Sgs	-	pdf pinout
ADE 3800 SXL	IC	Analog LCD Display Engine for Xga, Sxga Resolutions	QFP	Stm	-	pdf pinout
ADE 3800 SXT	IC	Analog LCD Display Engine for Xga, Sxga Resolutions	QFP	Stm	-	pdf pinout
ADE 3800 XL	IC	Analog LCD Display Engine for XGA, SXGA Resolutions	QFP	Stm	-	pdf pinout
ADE 3800 XT	IC	Analog LCD Display Engine for Xga, Sxga Resolutions	QFP	Stm	-	pdf pinout
ADEL2020A	LO-POWER	Vs:±18V Vu:100dB Vo:±13V Vi0:1.5mV	20S	And		data pinout
ADF 7021 BCPZ	IC	High Perf. Narrow-band Transceiver IC	TSOP	And	-	pdf pinout
ADF 7021 BCPZ-RL	IC	High Perf. Narrow-band Transceiver IC	TSOP	And	-	pdf pinout
ADF 7021 BCPZ-RL7	IC	High Perf. Narrow-band Transceiver IC	TSOP	And	-	pdf pinout
ADG	Si-N	=KTC3911-G (Typ-Code/Stempel/marking)	35	Kec	→KTC 3911(S)	data
ADG 201 HS N	CMOS-IC	4x Analog Switch, Udd=44V	16-DIP	And	DG 201, LF 11201, SW-01	
ADG 201 HS E	CMOS-IC	4x Analog Switch, Udd=44V	20-(P)LCC	And	-	
ADG 201 HS P	CMOS-IC	4x Analog Switch, Udd=44V	20-PLCC	And	-	
ADG 201 HS Q	CMOS-IC	4x Analog Switch, Udd=44V	16-DIC	And		
ADG 772 BCPZ-REEL	CMOS-IC	Cmos Dual 2:1 Mux/Demux USB 2.0 USB 1.1	QFN	And	-	pdf pinout
ADG1204	MULT4	Vs:-16.5...16-5V Ron:Ohm at 15/-15V	12H,14S	And		data pdf pinout
ADG1408	MULT8	Vs:<13.2V Ron:Ohm Ton:245ns at 5/-5V	20C,16S	And		data pdf pinout

Type	Device	Short Description	Fig.	Manu	Comparision Types	More at
ADG1409	MULT4	Vs:<13.2V Ron:Ohm Ton:245ns at 5/-5V	20C,16S	And		data pdf pinout
ADG 2310	A/D-IC	Audio	24-DIP		-	
ADG 3308 BCPZ-REEL	DIG-IC	Bidirectional Logic Level Translators	QFN	And	-	pdf pinout
ADG 3308 BRUZ	DIG-IC	Bidirectional Logic Level Translators	QFN	And	-	pdf pinout
ADG 3308 BRUZ-REEL	DIG-IC	Bidirectional Logic Level Translators	QFN	And	-	pdf pinout
ADG 772 BCPZ-1REEL	CMOS-IC	Cmos Dual 2:1 Mux/Demux USB 2.0 USB 1.1	QFN	And	-	pdf pinout
ADG 772 BCPZ-REEL7	CMOS-IC	Cmos Dual 2:1 Mux/Demux USB 2.0 USB 1.1	QFN	And	-	pdf pinout
ADG3308BCBZ-1-REEL	DIG-IC	Bidirectional Logic Level Translators	QFN	And	-	pdf pinout
ADG3308BCBZ-REEL7	DIG-IC	Bidirectional Logic Level Translators	QFN	And	-	pdf pinout
ADG 3308BRUZ-REEL7	DIG-IC	Bidirectional Logic Level Translators	QFN	And	-	pdf pinout
ADG3308BCBZ-2-REEL	DIG-IC	Bidirectional Logic Level Translators	QFN	And	-	pdf pinout
ADG 3308BCBZ-1-RL7	DIG-IC	Bidirectional Logic Level Translators	QFN	And	-	pdf pinout
ADG 3308BCBZ-2-RL7	DIG-IC	Bidirectional Logic Level Translators	QFN	And	-	pdf pinout
ADL	Si-N	=2SC3838K-L (Typ-Code/Stempel/marking)	35	Rhm	→2SC3838K	data
ADL	Si-N	=2SC4726-L (Typ-Code/Stempel/marking)	35(1,6mm)	Rhm	→2SC4726	data
ADL	Si-N	=KTC3911-L (Typ-Code/Stempel/marking)	35	Kec	→KTC 3911(S)	data
ADL 5371 ACPZ-WP	IC	Rf quadrature modulator	QFN	And	-	pdf pinout
ADL 5371 EVALZ	IC	Rf quadrature modulator	QFN	And	-	pdf pinout
ADL 5373 ACPZ-WP	IC	Rf quadrature modulator	QFN	And	-	pdf pinout
ADL 5373 EVALZ	IC	Rf quadrature modulator	QFN	And	-	pdf pinout
ADL 5374 ACPZ-WP	IC	Rf quadrature modulator	QFN	And	-	pdf pinout
ADL 5374 EVALZ	IC	Rf quadrature modulator	QFN	And	-	pdf pinout
ADL 5531 ACPZ-WP	IC	20 MHz to 500 MHz IF Gain Blocks	TSOP	And	-	pdf pinout
ADL 5532 ACPZ-WP	IC	20 MHz to 500 MHz IF Gain Blocks	TSOP	And	-	pdf pinout
ADL 5534 ACPZ-WP	IC	20 MHz to 500 MHz IF Gain Blocks	TSOP	And	-	pdf pinout
ADL 5571	IC	2.5 GHz to 2.7 GHz WiMAX Power Amp.	TSOP	And	-	pdf pinout
ADL 5371 ACPZ-R 7	IC	Rf quadrature modulator	QFN	And	-	pdf pinout
ADL 5373 ACPZ-R 7	IC	Rf quadrature modulator	QFN	And	-	pdf pinout
ADL 5374 ACPZ-R 7	IC	Rf quadrature modulator	QFN	And	-	pdf pinout
ADL 5387 ACPZ-R7	IC	50 MHz to 2 GHz Quadrature Demodulator	TSOP	And	-	pdf pinout
ADL 5531 ACPZ-R 7	IC	20 MHz to 500 MHz IF Gain Blocks	TSOP	And	-	pdf pinout
ADL 5532 ACPZ-R7	IC	20 MHz to 500 MHz IF Gain Blocks	TSOP	And	-	pdf pinout
ADL 5534 ACPZ-R7	IC.	20 MHz to 500 MHz IF Gain Blocks	TSOP	And	-	pdf pinout
ADLH0032	FOLLOWER	Vs:±18V Vu:70dB Vo:±13V Vi0:5mV	12A	And		data pinout
ADLH0033	FOLLOWER	Vs:±20V Vo:±13V Vi0:10mV f:100Mc	12A	And		data pinout
ADM	Si-N	=2SC3838K-M (Typ-Code/Stempel/marking)	35	Rhm	→2SC3838K	data
ADM	Si-N	=2SC4726-M (Typ-Code/Stempel/marking)	35(1,6mm)	Rhm	→2SC4726	data
ADM501	HI-OHM	Vs:±15V Vu:87dB Vo:±10V Vi0:1mV		And		data pinout
ADM501A	HI-OHM	Vs:±15V Vu:87dB Vo:±10V Vi0:2mV		And		data pinout
ADM663AAN	R+	Io=100mA Vo:4.75...5.25V Vin:9V	8D	And		data pdf pinout
ADM663AAR	R+	Io=100mA Vo:4.75...5.25V Vin:9V	8S	And		data pdf pinout
ADM663AN	R+	Io=50mA Vo:4.75...5.25V Vin:9V	8D	And		data pinout
ADM663AR	R+	Io=50mA Vo:4.75...5.25V Vin:9V	8S	And		data pinout
ADM666AAN	R+	Io=100mA Vo:4.75...5.25V Vin:9V	8D	And		data pdf pinout
ADM666AAR	R+	Io=100mA Vo:4.75...5.25V Vin:9V	8S	And		data pdf pinout
ADM666AN	R+	Io=50mA Vo:4.75...5.25V Vin:9V	8D	And		data pinout
ADM666AR	R+	Io=50mA Vo:4.75...5.25V Vin:9V	8S	And		data pinout
ADM 3232 EARNZ	CMOS/TTL	Line Driver/Receiver	FLP	And	-	pdf pinout
ADM 3232 EARUZ	CMOS/TTL	Line Driver/Receiver	FLP	And	-	pdf pinout
ADM 3251 EARWZ	LIN-IC	Single-ch RS-232 Line Driver/Receiver	TSOP	And	-	pdf pinout
ADM3232EARNZ-REEL7	CMOS/TTL	Line Driver/Receiver	FLP	And	-	pdf pinout
ADM3232EARUZ-REEL7	CMOS/TTL	Line Driver/Receiver	FLP	And	-	pdf pinout
ADM 13305-4 ARZ	IC	Dual Processor Supervisors with Watchdog	TSOP	And	-	pdf pinout
ADM 13305-5 ARZ	IC	Dual Processor Supervisors with Watchdog	TSOP	And	-	pdf pinout
ADM 13307-4 ARZ	IC	Triple Processor Supervisors	TSOP	And	-	pdf pinout
ADM 13307-5 ARZ	IC	Triple Processor Supervisors	TSOP	And	-	pdf pinout
ADM 13305-18 ARZ	IC	Dual Processor Supervisors with Watchdog	TSOP	And	-	pdf pinout
ADM 13305-25 ARZ	IC	Dual Processor Supervisors with Watchdog	TSOP	And	-	pdf pinout
ADM 13305-33 ARZ	IC	Dual Processor Supervisors with Watchdog	TSOP	And	-	pdf pinout
ADM 13305-4ARZ-RL7	IC	Dual Processor Supervisors with Watchdog	TSOP	And	-	pdf pinout
ADM 13305-5ARZ-RL7	IC	Dual Processor Supervisors with Watchdog	TSOP	And	-	pdf pinout
ADM 13307-18 ARZ	IC	Triple Processor Supervisors	TSOP	And	-	pdf pinout
ADM 13307-25 ARZ	IC	Triple Processor Supervisors	TSOP	And	-	pdf pinout
ADM 13307-33 ARZ	IC	Triple Processor Supervisors	TSOP	And	-	pdf pinout
ADM 13307-4ARZ-RL7	IC	Triple Processor Supervisors	TSOP	And	-	pdf pinout
ADM 13307-5ARZ-RL7	IC	Triple Processor Supervisors	TSOP	And	-	pdf pinout
ADM13305-18ARZ-RL7	IC	Dual Processor Supervisors with Watchdog	TSOP	And	-	pdf pinout
ADM13305-25ARZ-RL7	IC	Dual Processor Supervisors with Watchdog	TSOP	And	-	pdf pinout
ADM13305-33ARZ-RL7	IC	Dual Processor Supervisors with Watchdog	TSOP	And	-	pdf pinout
ADM13307-18ARZ-RL7	IC	Triple Processor Supervisors	TSOP	And	-	pdf pinout
ADM13307-25ARZ-RL7	IC	Triple Processor Supervisors	TSOP	And	-	pdf pinout
ADM13307-33ARZ-RL7	IC	Triple Processor Supervisors	TSOP	And	-	pdf pinout
ADN	Si-N	=2SC3838K-N (Typ-Code/Stempel/marking)	35	Rhm	→2SC3838K	data
ADN	Si-N	=2SC4726-N (Typ-Code/Stempel/marking)	35(1,6mm)	Rhm	→2SC4726	data
ADO26	HI-OHM	Vs:±15V Vu:100dB Vo:±10V Vi0:70mV	7Y	Ite		data pinout
ADO27	HI-OHM	Vs:±15V Vu:100dB Vo:±10V Vi0:70mV	7Y	Ite		data pinout
ADO32	HI-OHM	Vs:±15V Vu:82dB Vo:±12V Vi0:1mV	7Y	Ite		data pinout
ADO39	HI-OHM	Vs:±15V Vu:97dB Vo:±11V Vi0:1mV	7Y	Ite		data pinout
ADO45	UNI	Vs:±15V Vu:98dB Vo:±10V f:1.5Mc	7Y	Ite		data pinout
ADO60	HI-SPEED	Vs:±15V Vu:120dB Vo:±10V Vi0:3mV	7Y	Ite		data pinout
ADO72	HYBRID	Vs:±15V Vu:113dB Vo:±12V Vi0:0.15mV	9Y	Ite		data pinout
ADOP07	HI-CMR	Vs:±22V Vo:±12.8V Vi0:60µµV f:600Mc	8DAS	And		data pinout
ADOP07A	HI-CMR	Vs:±22V Vo:±12.8V Vi0:30µµV f:600Mc	8DA	And		data pinout
ADOP27A	LO-NOISE	Vs:±18V Vu:1800V/mV Vo:±13.8V f:8Mc	8AD	And		data pinout
ADOP27B	LO-NOISE	Vs:±18V Vu:1800V/mV Vo:±13.8V f:8Mc	8AD	And		data pinout

Type	Device	Short Description	Fig.	Manu	Comparision Types	More at
ADOP27C	LO-NOISE	Vs:±18V Vu:1500V/mV Vo:±13.5V f:8Mc	8AD	And		data pinout
ADOP27E	LO-NOISE	Vs:±18V Vu:1800V/mV Vo:±13.8V f:8Mc	8AD	And		data pinout
ADOP27F	LO-NOISE	Vs:±18V Vu:1800V/mV Vo:±13.8V f:8Mc	8AD	And		data pinout
ADOP27G	LO-NOISE	Vs:±18V Vu:1500V/mV Vo:±13.5V f:8Mc	8AD	And		data pinout
ADOP37A	LO-NOISE	Vs:±18V Vu:125dB Vo:±13.5V Vi0:10μμV	8AD	And		data pinout
ADOP37B	LO-NOISE	Vs:±18V Vu:125dB Vo:±13.5V Vi0:20μμV	8AD	And		data pinout
ADOP37C	LO-NOISE	Vs:±18V Vu:123dB Vo:±13.5V Vi0:30μμV	8AD	And		data pinout
ADOP37E	LO-NOISE	Vs:±18V Vu:125dB Vo:±13.5V Vi0:10μμV	8AD	And		data pinout
ADOP37F	LO-NOISE	Vs:±18V Vu:125dB Vo:±13.5V Vi0:20μμV	8AD	And		data pinout
ADOP37G	LO-NOISE	Vs:±18V Vu:123dB Vo:±13.5V Vi0:30μμV	8AD	And		data pinout
ADP	Si-N	=2SC3838K-P (Typ-Code/Stempel/marking)	35	Rhm	→2SC3838K	data
ADp	Si-N	=BCW 60D (Typ-Code/Stempel/marking)	35	Phi	→BCW 60D	data
ADP501	HI-OHM	Vs:±15V Vu:87dB Vo:±10V Vi0:1mV		And		data pinout
ADP501A	HI-OHM	Vs:±15V Vu:87dB Vo:±10V Vi0:2mV		And		data pinout
ADP511	HI-OHM	Vs:±22V Vu:88dB Vo:±10V Vi0:2.5mV		And		data pinout
ADP511A	HI-OHM	Vs:±22V Vu:88dB Vo:±10V Vi0:6.5mV		And		data pinout
ADP 1828	Dig. Audio	Dual Audio Power Amplifier	QIP_b	Bcd	-	pdf pinout
ADP 3110 AKCPZ	CMOS-IC	2x Bootstrapped, 12V Mosfet Drv.	MLP	Ons	-	pdf pinout
ADP 3110 AKRZ	CMOS-IC	2x Bootstrapped, 12V Mosfet Drv.	MDIP	Ons	-	pdf pinout
ADP 3120 AJCPZ	CMOS-IC	2x Bootstrapped, 12V Mosfet Drv.	MLP	Ons	-	pdf pinout
ADP 3120 AJRZ	CMOS-IC	2x Bootstrapped, 12V Mosfet Drv.	MDIP	Ons	-	pdf pinout
ADP 3121 JCPZ	CMOS-IC	2x Bootstrapped, 12V Mosfet Drv.	MLP	Ons	-	pdf pinout
ADP 3121 JRZ	CMOS-IC	2x Bootstrapped, 12V Mosfet Drv.	MDIP	Ons	-	pdf pinout
ADP 3205	IC	Multiph. Imvp-iv Core Ctrl. f. CPU	LCC	And	-	pdf pinout
ADP 1828 YRQZ-R7	IC	Sync. Buck PWM, Step-Down, DC-to-DC Contr.	TSOP	And	-	pdf pinout
ADQ	Si-N	=2SC3838K-Q (Typ-Code/Stempel/marking)	35	Rhm	→2SC3838K	data
ADR 510 ART-R2	Z-Ref-IC	1.0 V Shunt Voltage Reference	TSOP	And	-	pdf pinout
ADR 510 ARTZ-R 2	Z-Ref-IC	1.0 V Shunt Voltage Reference	TSOP	And	-	pdf pinout
ADR 510 ART-REEL 7	Z-Ref-IC	1.0 V Shunt Voltage Reference	TSOP	And	-	pdf pinout
ADR 510 ARTZ-REEL7	Z-Ref-IC	1.0 V Shunt Voltage Reference	TSOP	And	-	pdf pinout
ADs	Si-N	=BCW 60D (Typ-Code/Stempel/marking)	35	Sie	→BCW 60D	data
ADS 112 MC, MM	A/D-IC	12 Bit, A/D-Converter, S&H, 1MHz MC: 0..70°C MM: -55...125°C	24-DIC	Dat	-	pdf
ADS 117 MC, MM	A/D-IC	12 Bit, A/D-Converter, S&H, 2MHz MC: 0..70°C MM: -55...125°C	24-DIC	Dat	-	pdf
ADS 118 AMC, AMM	A/D-IC	12 Bit, A/D-Converter, S&H, 5MHz AMC: 0..70°C AMM: -55...125°C	24-SMT	Dat	-	pdf
ADS 118 MC, MM	A/D-IC	12 Bit, A/D-Converter, S&H, 5MHz MC: 0..70°C MM: -55...125°C	24-DIC	Dat	-	pdf
ADS 119 GC, GM	A/D-IC	12 Bit, A/D-Converter, S&H, 10MHz GC: 0..70°C GM: -55...125°C	24-SMT	Dat	-	pdf
ADS 119 MC, MM	A/D-IC	12 Bit, A/D-Converter, S&H, 10MHz MC: 0..70°C MM: -55...125°C	24-DIC	Dat	-	pdf
ADS 230,231	CMOS-A/D-IC	12 Bit, 3-Step-Flash A/D-Conv., S&H ADS-230: 1MHz, ADS-231: 1.5MHz	44-PLCC	Dat	-	pdf
ADS 235..237 S	CMOS-A/D-IC	SMD, 12 Bit, 3-Step-Flash A/D-Conv. S&H, ..235S, ..236S: 5MHz ..237S: 9MHz	28-MDIP	Dat	-	pdf
ADS 238 Q	CMOS-A/D-IC	12 Bit, 3-Step-Flash A/D-Converter Sample&Hold, 20MHz	44-QFP	Dat	-	data pdf pinout
ADS 325 A	CMOS-A/D-IC	10 Bit, A/D-Converter, S&H, 20MHz	48-QFP	Dat	-	data pdf pinout
ADS 574 JE,KE	CMOS-A/D-IC	12 Bit, µP-komp. SAR A/D-Conv., S&H Lin.Error JE: ±1LSB KE: ±½LSB, +5V	28-DIP	Bub	-	pdf
ADS 574 JP,KP	CMOS-A/D-IC	12 Bit, µP-komp. SAR A/D-Conv., S&H Lin.Error JE: ±1LSB KE: ±½LSB, +5V	28-DIP	Bub	-	pdf
ADS 574 JU,KU	CMOS-A/D-IC	SMD,12Bit, µP-komp. SAR A/D-Conv., S&H Lin.Error JE: ±1LSB KE: ±½LSB, +5V	28-MDIP	Bub	-	pdf
ADS 774 JE,KE	CMOS-A/D-IC	12 Bit, µP-komp. SAR A/D-Conv., S&H Lin.Error JE: ±1LSB KE: ±½LSB, +5V	28-DIP	Bub	-	pdf
ADS 774 JP,KP	CMOS-A/D-IC	12 Bit, µP-komp. SAR A/D-Conv., S&H Lin.Error JE: ±1LSB KE: ±½LSB, +5V	28-DIP	Bub	-	pdf
ADS 774 JU,KU	CMOS-A/D-IC	SMD,12Bit, µP-komp. SAR A/D-Conv., S&H Lin.Error JE: ±1LSB KE: ±½LSB, +5V	28-MDIP	Bub	-	pdf
ADS 800...802 E	CMOS-A/D-IC	SMD, 12 Bit, pipelined A/D-Conv., +5V T&H, 800:40MHz 801:25MHz 802:10MHz	28-SSMDIP	Bub	-	pdf
ADS 800...802 U	CMOS-A/D-IC	SMD, 12 Bit, pipelined A/D-Conv., +5V T&H, 800:40MHz 801:25MHz 802:10MHz	28-MDIP	Bub	-	pdf
ADS 803...805 E	CMOS-A/D-IC	SMD, 12 Bit, pipelined A/D-Conv., T&H 803:5MHz 804:10MHz 805:20MHz,+5V sep. Drv. Supply:+3V/+5V	28-SSMDIP	Bub	-	pdf
ADS 803...805 U	CMOS-A/D-IC	SMD, 12 Bit, pipelined A/D-Conv., T&H 803:5MHz 804:10MHz 805:20MHz,+5V sep. Drv. Supply:+3V/+5V	28-MDIP	Bub	-	pdf
ADS 807 E	CMOS-A/D-IC	SMD, 12Bit, A/D-Conv., T&H, 53MHz, +5V	28-SSMDIP	Bub	-	data pdf pinout
ADS 808...809 Y	A/D-IC	12 Bit, pipelined A/D-Conv., T&H, +5V 70/80MHz, sep. Drv. Supply:+3V/+5V	48-QFP	Bub	-	pdf
ADS 820...821 E	CMOS-A/D-IC	SMD, 10Bit, A/D-Conv., Track&Hold, +5V 820: 20MHz, 821: 40MHz	28-SSMDIP	Bub	-	pdf
ADS 820...821 U	CMOS-A/D-IC	SMD, 10Bit, A/D-Conv., Track&Hold, +5V 820: 20MHz, 821: 40MHz	28-MDIP	Bub	-	pdf
ADS 822...823 E	CMOS-A/D-IC	SMD, 10 Bit, pipelined A/D-Conv., T&H 40/60MHz, +5V, sep. Drv. Supply:+3/+5V	28-SSMDIP	Bub	-	pdf
ADS 825...826 E	CMOS-A/D-IC	SMD, 10 Bit, pipelined A/D-Conv., T&H 40/60MHz, +5V, sep. Drv. Supply:+3/+5V	28-SSMDIP	Bub	-	pdf
ADS 828 E	CMOS-A/D-IC	SMD, 10 Bit, pipelined A/D-Conv., T&H	28-SSMDIP	Bub	-	data pdf pinout

Type	Device	Short Description	Fig.	Manu	Comparision Types	More at
ADS 830...831 E	CMOS-A/D-IC	75MHz, +5V. sep. Drv. Supply:-3V -5V SMD, 8 Bit. pipelined A D-Conv.. T&H 60/80MHz. +5V. sep. Drv. Supply:-3 -5V	20-SSMDIP	Bub	-	pdf
ADS 850 Y	A/D-IC	14 Bit, self cal. A D-Conv.. T&H. -5V 10MSPS. sep. Drv. Supply:-3V -5V	48-QFP	Bub	-	data pdf pinout
ADS 900...901 E	A/D-IC	10 Bit. pipelined A/D-Conv.. 20MHz T&H. +2.7...3.7V, ADS900E :integr. Ref.	28-SSMDIP	Bub	-	pdf
ADS 916 GC,GM	A/D-IC	14 Bit, Flash A/D-Conv.. S&H. 500kHz GC:0...70°C, GM:-55...125°C, ±12/15V	24-SMT	Dat	-	pdf
ADS 916 MC,MM	A/D-IC	14 Bit, Flash A/D-Conv.. S&H. 500kHz MC:0...70°C, MM:-55...125°C, ±12/15V	24-DIC	Dat	-	pdf
ADS 917 GC,GM	A/D-IC	14 Bit, Flash A/D-Conv., S&H. 1MHz GC:0...70°C, GM:-55...125°C, ±12/15V	24-SMT	Dat	-	pdf
ADS 917 MC,MM	A/D-IC	14 Bit, Flash A/D-Conv., S&H. 1MHz MC:0...70°C, MM:-55...125°C, ±12/15V	24-DIC	Dat	-	pdf
ADS 919 GC,GM	A/D-IC	14 Bit, Flash A/D-Conv., S&H, 2MHz GC:0...70°C, GM:-55...125°C, ±12/15V	24-SMT	Dat	-	pdf
ADS 919 MC,MM	A/D-IC	14 Bit, Flash A/D-Conv., S&H, 2MHz MC:0...70°C, MM:-55...125°C, ±12/15V	24-DIC	Dat	-	pdf
ADS 926/883	A/D-IC	14 Bit, Flash A/D-Conv., S&H, 500kHz bipolar input, MIL-STD-883, ±12/15V	24-DIC	Dat	-	data pdf pinout
ADS 926 GC,GM	A/D-IC	14 Bit, Flash A/D-Conv., S&H, 500kHz GC:0...70°C, GM:-55...125°C, ±12/15V	24-SMT	Dat	-	pdf
ADS 926 MC,MM	A/D-IC	14 Bit, Flash A/D-Conv., S&H, 500kHz MC:0...70°C, MM:-55...125°C, ±12/15V	24-DIC	Dat	-	pdf
ADS 927/883	A/D-IC	14 Bit, Flash A/D-Conv., S&H, 1MHz bipolar input, MIL-STD-883, ±12/15V	24-DIC	Dat	-	data pdf pinout
ADS 927 GC,GM	A/D-IC	14 Bit, Flash A/D-Conv., S&H, 1MHz GC:0...70°C, GM:-55...125°C, ±12/15V	24-SMT	Dat	-	pdf
ADS 927 MC,MM	A/D-IC	14 Bit, Flash A/D-Conv., S&H, 1MHz MC:0...70°C, MM:-55...125°C, ±12/15V	24-DIC	Dat	-	pdf
ADS 929/883	A/D-IC	14 Bit, Flash A/D-Conv., S&H, 2MHz bipolar input, MIL-STD-883, ±12/15V	24-DIC	Dat	-	data pdf pinout
ADS 929 GC,GM	A/D-IC	14 Bit, Flash A/D-Conv., S&H, 2MHz GC:0...70°C, GM:-55...125°C, ±12/15V	24-SMT	Dat	-	pdf
ADS 929 MC,MM	A/D-IC	14 Bit, Flash A/D-Conv., S&H, 2MHz MC:0...70°C, MM:-55...125°C, ±12/ 5V	24-DIC	Dat	-	pdf
ADS 930 MC,MM	A/D-IC	16 Bit, 3-Pass Flash A/D-Conv., S&H Inp. 0...-10V/-5...+5V, 500kHz MC: 0...70°C, MM: -55...125°C	40-DIC	Dat	-	pdf
ADS 931...933 MC,MM	A/D-IC	16 Bit, 2-Pass Flash A/D-Conv., S&H Inp. 0...-5.5V/±2.75V, ADS 93x: xMHz MC: 0...70°C, MM: -55...125°C	40-DIC	Dat	-	pdf
ADS 935 MC,MM	A/D-IC	16 Bit, 2-Pass Flash A/D-Conv., S&H Inp. 0...-5.5V/±2.75V, 5MHz MC: 0...70°C, MM: -55...125°C	40-DIC	Dat	-	pdf
ADS 937 MC,MM,883	A/D-IC	16 Bit, 2-Pass Flash A/D-Conv., S&H Inp. 0...-10V/-5...+5V, 1MHz MC: 0...70°C MM, 883: -55...125°C	32-DIC	Dat.	-	pdf
ADS 941...942 MC,ME	A/D-IC	14 Bit, 2-Pass Flash A/D-Conv., S&H Inp. 0...10V/-5...+5V, 941(942): 1(2)MHz MC: 0...70°C ME: -40...85°C	32-DIC	Dat	-	pdf
ADS 942A MC,ME	A/D-IC	14 Bit, 2-Pass Flash A/D-Conv., 2MHz Sample&Hold, Inp. 0...10V/-5V...+5V MC: 0...70°C ME: -40...85°C	32-DIC	Dat	-	pdf
ADS 943 GC,GM,G883	A/D-IC	14 Bit, 2-Pass Flash A/D-Conv., ±5V Inp. -2...+2V, Sample&Hold, 3MHz GC: 0...70°C GM, 883: -55...125°C	24-SMT	Dat	-	pdf
ADS 943 MC,MM,883	A/D-IC	14 Bit, 2-Pass Flash A/D-Converter Inp. -2...+2V, Sample&Hold, 3MHz MC: 0...70°C MM, 883: -55...125°C	24-DIC	Dat	-	pdf
ADS 944 MC,MM,883	A/D-IC	14 Bit, 2-Pass Flash A/D-Conv., 5MHz Sample&Hold, Inp. -1.25V...+1.25V MC: 0...70°C MM, 883: -55...125°C	32-DIC	Dat	-	pdf
ADS 945 (EX)	A/D-IC	14 Bit, 2-Pass Flash A/D-Conv., 10MHz S&H, Inp. ±1.25V, 0...70°C (-55...125°C)	76-DIC	Dat	-	pdf
ADS 946 GC,GM,G883	A/D-IC	14 Bit, 2-Pass Flash A/D-Conv., ±5V Inp. -2...+2V, Sample&Hold, 8MHz GC: 0...70°C GM, 883: -55...125°C	24-SMT	Dat	-	pdf
ADS 946 MC,MM,883	A/D-IC	14 Bit, 2-Pass Flash A/D-Converter Inp. -2...+2V, Sample&Hold, 8MHz MC: 0...70°C, MM, 883: -55...125°C	24-DIC	Dat	-	pdf
ADS 947 GC,GE	A/D-IC	14 Bit, 2-Pass Flash A/D-Conv., ±5V Inp. -2...+2V, Sample&Hold, 10MHz GC: 0...70°C GE: -40...100°C	24-SMT	Dat	-	pdf
ADS 947 MC,ME(-QL)	A/D-IC	14 Bit, 2-Pass Flash A/D-Conv., ±5V Inp. -2...+2V, Sample&Hold, 10MHz MC: 0...70°C ME(-QL): -40...100°C	24-DIC	Dat	-	pdf
ADS 950 MC,MM	A/D-IC	18 Bit, 2-Pass Flash A/D-Conv., S&H Inp. ±5V, MC: 0...70°C, MM: -55...125°C	32-DIC	Dat	-	pdf
ADS 951 MC,ME	A/D-IC	18 Bit, 2-Pass Flash A/D-Conv., S&H Inp. ±5V, MC: 0...70°C, ME: -40...110°C	32-DIC	Dat	-	pdf
ADS 1110	IC	16-Bit analog-to-digital Converter	TSOP	Bub	-	pdf pinout
ADS 1210 P	A/D-IC	24Bit, Delta-Sigma A/D-Conv., PGA ser. Ausg./Out, nur/single +5V	18-DIP	Bub	-	data pdf pinout

Type	Device	Short Description	Fig.	Manu	Comparision Types	More at
ADS 1210 U	A/D-IC	SMD, 24Bit, Delta-Sigma A/D-Conv., PGA ser. Ausg./Out, nur/single +5V	18-MDIP	Bub	-	data pdf pinout
ADS 1211 E	A/D-IC	SMD, 24Bit, Delta-Sigma A/D-Conv., PGA 4 Kan./Chan. MPX, ser. Ausg./Out, +5V	24-SSMDIP	Bub	-	data pdf pinout
ADS 1211 P	A/D-IC	24Bit, Delta-Sigma A/D-Conv., PGA 4 Kan./Chan. MPX, ser. Ausg./Out, +5V	24-DIP	Bub	-	data pdf pinout
ADS 1211 U	A/D-IC	SMD, 24Bit, Delta-Sigma A/D-Conv., PGA 4 Kan./Chan. MPX, ser. Ausg./Out, +5V	24-MDIP	Bub	-	data pdf pinout
ADS 1212 P	A/D-IC	24Bit, Delta-Sigma A/D-Conv., PGA ser. Ausg./Out, nur/single +5V	18-DIP	Bub	-	data pdf pinout
ADS 1212 U	A/D-IC	SMD, 24Bit, Delta-Sigma A/D-Conv., PGA ser. Ausg./Out, nur/single +5V	18-MDIP	Bub	-	data pdf pinout
ADS 1213 E	A/D-IC	SMD, 24Bit, Delta-Sigma A/D-Conv., PGA 4 Kan./Chan. MPX, ser. Ausg./Out, +5V	24-SSMDIP	Bub	-	data pdf pinout
ADS 1213 P	A/D-IC	24Bit, Delta-Sigma A/D-Conv., PGA 4 Kan./Chan. MPX, ser. Ausg./Out, +5V	24-DIP	Bub	-	data pdf pinout
ADS 1213 U	A/D-IC	SMD, 24Bit, Delta-Sigma A/D-Conv., PGA 4 Kan./Chan. MPX, scr. Ausg./Out, +5V	24-MDIP	Bub	-	data pdf pinout
ADS 1216 Y	A/D-IC	24Bit, Delta-Sigma A/D-Conv., PGA 8 Kan./Chan., ser. Ausg./Out, 2.7...5V	48-QFP	Bub	-	data pdf pinout
ADS 1218 Y	A/D-IC	24Bit, Delta-Sigma A/D-Conv., PGA 8 Kan./Chan., ser. Ausg./Out, Flashmem	48-QFP	Bub	-	data pdf pinout
ADS 1240 E	A/D-IC	SMD, 24 Bit, Delta-Sigma A/D-Conv., 4 Kan./Chan. MPX, ser. Ausg./Out	24-SSMDIP	Bub	-	data pdf pinout
ADS 1241 E	A/D-IC	SMD, 24 Bit, Delta-Sigma A/D-Conv., 8 Kan./Chan. MPX, ser. Ausg./Out	28-SSMDIP	Bub	-	data pdf pinout
ADS 1242	A/D-IC	SMD, 24 Bit, Delta-Sigma A/D-Conv., 4 Kan./Chan. MPX, ser. Ausg./Out	16-SSMDIP	Bub	-	data pdf pinout
ADS 1243	A/D-IC	SMD, 24 Bit, Delta-Sigma A/D-Conv., 8 Kan./Chan. MPX, ser. Ausg./Out	20-SSMDIP	Bub	-	data pdf pinout
ADS 1250 U	A/D-IC	SMD, 20Bit, Delta-Sigma A/D-Conv., PGA ser. Ausg./Out, nur/single +5V	16-MDIP	Bub	-	data pdf pinout
ADS 1251...52 U	A/D-IC	SMD, 24Bit, Delta-Sigma A/D-Conv., +5V ser. Ausg./Out, ...51:20kHz ...52:40kHz	8-MDIP	Bub	-	pdf
ADS 1253 E	A/D-IC	SMD, 24Bit, Delta-Sigma A/D-Conv., +5V 4 Kan./Chan.MPX, ser. Ausg./Out, 20kHz	16-SSMDIP	Bub	-	data pdf pinout
ADS 1254 E	A/D-IC	SMD, 24Bit, Delta-Sigma A/D-Conv., +5V 4 Kan./Chan. MPX, ser. Ausg./Out +1.8V	20-SSMDIP	Bub	-	data pdf pinout
ADS 1255 IDB	A/D-IC	SMD, 24Bit, Delta-Sigma A/D-Conv., +5V 2 Kan./Chan. MPX, ser. Ausg./Out +1.8V	20-SSMDIP	Bub	-	pdf pinout
ADS 1256 IDB	A/D-IC	SMD, 24Bit, Delta-Sigma A/D-Conv., +5V 8 Kan./Chan. MPX, ser. Ausg./Out +1.8V	28-SSMDIP	Bub	-	pdf pinout
ADS 1286 P(A,B,C)	A/D-IC	12 Bit, ser. A/D-Converter, S&H, 20kHz -40...85°C, Int. Lin. A:±2 B:±2 C:±1LSB	8-DIP	Bub	-	pdf
ADS 1286 P(K,L)	A/D-IC	12 Bit, ser. A/D-Converter, S&H, 20kHz 0...70°C, Int. Lin. P:±2 PK:±2 PL:±1LSB	8-DIP	Bub	-	pdf
ADS 1286 U(A,B,C)	A/D-IC	SMD, 12Bit, ser. A/D-Conv., S&H, 20kHz -40...85°C, Int. Lin. A:±2 B:±2 C:±1LSB	8-MDIP	Bub	-	pdf
ADS 1286 U(K,L)	A/D-IC	SMD, 12Bit, ser. A/D-Conv., S&H, 20kHz 0...70°C, Int. Lin. U:±2 UK:±2 UL:±1LSB	8-MDIP	Bub	-	pdf
ADS 1605 I	A/D-IC	SMD, 16Bit, Delta-Sigma A/D-Converter 5(10) MSPS, Vs: +5V, +3V, +2.7...5.25V	64-QFP	Bub	-	pdf
ADS 1606 I	A/D-IC	SMD, 16Bit, Delta-Sig. A/D-Conv., Fifo 5(10) MSPS, Vs: +5V, +3V, +2.7...5.25V	64-QFP	Bub	-	pdf
ADS 1625 I	A/D-IC	SMD, 18Bit, Delta-Sigma A/D-Converter 1.25 MSPS, Vs: +5V, +3V, +2.7...5.25V	64-QFP	Bub	-	pdf
ADS 1626 I	A/D-IC	SMD, 18Bit, Delta-Sig. A/D-Conv., Fifo 1.25 MSPS, Vs: +5V, +3V, +2.7...5.25V	64-QFP	Bub	-	pdf
ADS 2806...2807 Y	A/D-IC	Dual, 12 Bit, pipelined A/D-Conv., S&H ...06/07: 32/50MHz, Ausg./Out: +3V/+5V	64-QFP	Bub	-	pdf
ADS 5102 (C,I)PFB	CMOS-A/D-IC	10 Bit, A/D-Conv., S&H, +1.8V/+3.3V 65MSPS, ...C.: 0...70°C ...I.: -40...85°C	48-QFP	Tix	-	pdf
ADS 5103 (C,I)PFB	CMOS-A/D-IC	10 Bit, A/D-Conv., S&H, +1.8V/+3.3V 40MSPS, ...C.: 0...70°C ...I.: -40...85°C	48-QFP	Tix	-	pdf
ADS 5120	CMOS-A/D-IC	10 Bit, 8 Chan./Kan., A/D-Conv., S&H 40MSPS, +1.8V, I/O:+1.8V/+3.3V	257-GRID	Bub	-	data pdf
ADS 5237	IC	Dual, 10-Bit, 65MSPS, +3.3V A/D Conv.	QFP	Bub	-	pdf pinout
ADS 5270...72 IPFP	A/D-IC	SMD, 8-Chan. 12Bit A/D-Conv., +3.3V S&H, serial. LVDS Interf., 40-65 MSPS	80-QFP	And	-	pdf
ADS 5270 IPFP	A/D-IC	SMD, 8-Chan. 12Bit A/D-Conv., +3.3V S&H, serial. LVDS Interf., 40 MSPS	80-QFP	And	-	pdf
ADS 5273 IPFP	A/D-IC	SMD, 8-Chan. 12Bit A/D-Conv., +3.3V S&H, serial. LVDS Interf., 70 MSPS	80-QFP	And	-	pdf
ADS 5275...77 IPFP	A/D-IC	SMD, 8-Chan. 10Bit A/D-Conv., +3.3V S&H, serial. LVDS Interf., 40-65 MSPS	80-QFP	And	-	pdf
ADS 5421 Y	A/D-IC	14 Bit, pipelined A/D-Conv., T&H 40MHz, Ausg./Out: +5V/+3V	64-QFP	Bub	-	data pdf pinout
ADS 5422 Y	A/D-IC	14 Bit, pipelined A/D-Conv., T&H 62MSPS, Ausg./Out: +5V/+3V	64-QFP	Bub	-	data pdf pinout
ADS 5484	LIN-IC	16-Bit, 170/200-MSPS A/D-Converters	TSOP	Tix	-	pdf pinout
ADS 5485	LIN-IC	16-Bit, 170/200-MSPS A/D-Converters	TSOP	Tix	-	pdf pinout
ADS 5500I	A/D-IC	SMD, 14 Bit A/D-Conv., single +3,3V 125MSPS Sample&Hold	48-QFP	Bub	-	pdf
ADS 6142	IC	14-BITS, 125/105/80/65 MSPS ADC	QFP	Tix	-	pdf pinout
ADS 6143	IC	14-BITS, 125/105/80/65 MSPS ADC	QFP	Tix	-	pdf pinout

Type	Device	Short Description	Fig.	Manu	Comparision Types	More at
ADS 6144	IC	14-BITS, 125/105/80/65 MSPS ADC	QFP	Tix	-	pdf pinout
ADS 6145	IC	14-BITS, 125/105/80/65 MSPS ADC	QFP	Tix	-	pdf pinout
ADS 6222	IC	12-BIT, 125/105/80/65 MSPS ADC	QFP	Tix	-	pdf pinout
ADS 6223	IC	12-BIT, 125/105/80/65 MSPS ADC	QFP	Tix	-	pdf pinout
ADS 6224	IC	12-BIT, 125/105/80/65 MSPS ADC	QFP	Tix	-	pdf pinout
ADS 6225	IC	12-BIT, 125/105/80/65 MSPS ADC	QFP	Tix	-	pdf pinout
ADS 7800 AH,BH	CMOS-A/D-IC	12 Bit, SAR A/D-Conv., S&H, -40...85°C	24-DIC	Bub	-	pdf
ADS 7800 JP,KP	CMOS-A/D-IC	12 Bit, SAR A/D-Conv., S&H, 0...70°C	24-DIP	Bub	-	pdf
ADS 7800 JU,KU	CMOS-A/D-IC	SMD, 12Bit SAR A/D-Conv., S&H, 0...70°C	24-MDIP	Bub	-	pdf
ADS 7804 P(B)	CMOS-A/D-IC	12 Bit, SAR A/D-Conv., S&H, 10µs, +5V	28-DIP	Dat	-	pdf
ADS 7804 U(B)	CMOS-A/D-IC	SMD, 12 Bit, SAR A/D-Conv., S&H, +5V	28-MDIP	Dat	-	pdf
ADS 7805 P(B)	CMOS-A/D-IC	16 Bit, SAR A/D-Conv., S&H, 10µs, +5V	28-DIP	Dat	-	pdf
ADS 7805 U(B)	CMOS-A/D-IC	SMD, 16 Bit, SAR A/D-Conv., S&H, +5V	28-MDIP	Dat	-	pdf
ADS 7806 P(B)	CMOS-A/D-IC	12 Bit, SAR A/D-Conv., S&H, 35mW, +5V	28-DIP	Dat	-	pdf
ADS 7806 U(B)	CMOS-A/D-IC	SMD, 12 Bit, SAR A/D-Conv., S&H, +5V	28-MDIP	Dat	-	pdf
ADS 7807 P(B)	CMOS-A/D-IC	16 Bit, SAR A/D-Conv., S&H, 35mW, +5V	28-DIP	Dat	-	pdf
ADS 7807 U(B)	CMOS-A/D-IC	SMD, 16 Bit, SAR A/D-Conv., S&H, +5V	28-MDIP	Dat	-	pdf
ADS 7808 P(B)	CMOS-A/D-IC	12 Bit, SAR A/D-Conv., ser. Out, +5V	20-DIP	Bub	-	pdf
ADS 7808 U(B)	CMOS-A/D-IC	SMD,12Bit SAR A/D-Conv., ser. Out, +5V	20-MDIP	Bub	-	pdf
ADS 7809 P(B)	CMOS-A/D-IC	16 Bit, SAR A/D-Conv., ser. Out, +5V	20-DIP	Bub	-	pdf
ADS 7809 U(B)	CMOS-A/D-IC	SMD,16Bit SAR A/D-Conv., ser. Out, +5V	20-MDIP	Bub	-	pdf
ADS 7810 U(B)	CMOS-A/D-IC	SMD, 12 Bit, SAR A/D-Conv., S&H, ±5V Inp. ±10V, -40...85°C, ILE=±1(0.75)LSB	28-MDIP	Bub	-	pdf
ADS 7811 U	CMOS-A/D-IC	SMD, 16 Bit, SAR A/D-Conv., S&H, ±5V	28-MDIP	Bub	-	data pdf pinout
ADS 7812 P(B)	A/D-IC	12 Bit, SAR A/D-Conv., ser. Out, +5V 40kHz, mult.inp.ranges, ILE=±1(±½)LSB	16-DIP	Bub	-	pdf
ADS 7812 U(B)	A/D-IC	SMD, 12Bit, SAR A/D-Conv., ser.Out,+5V 40kHz, mult.inp.ranges, ILE=±1(±½)LSB	16-MDIP	Bub	-	pdf
ADS 7813 P(B)	A/D-IC	16 Bit, SAR A/D-Conv., ser. Out, +5V 40kHz, mult. inp. ranges, ILE=±3(2)LSB	16-DIP	Bub	-	pdf
ADS 7813 U(B)	A/D-IC	SMD, 16Bit, SAR A/D-Conv., ser.Out,+5V 40kHz, mult. inp. ranges, ILE=±3(2)LSB	16-MDIP	Bub	-	pdf
ADS 7815 U	CMOS-A/D-IC	SMD, 16 Bit, SAR A/D-Conv., S&H, ±5V	28-MDIP	Bub	-	data pdf pinout
ADS 7816 E(B,C)	A/D-IC	SMD,12Bit, SAR A/D-Conv., ser. Out,+5V 200kHz, unipolar, ILE= ±2 (±2,±1)LSB	8-SSMDIP	Bub	-	pdf
ADS 7816 P(B,C)	A/D-IC	12 Bit, SAR A/D-Conv., ser. Out, +5V 200kHz, unipolar, ILE= ±2 (±2,±1)LSB	8-DIP	Bub	-	pdf
ADS 7816 U(B,C)	A/D-IC	SMD,12Bit, SAR A/D-Conv., ser. Out,+5V 200kHz, unipolar, ILE= ±2 (±2,±1)LSB	8-MDIP	Bub	-	pdf
ADS 7817 E(B,C)	A/D-IC	SMD,12Bit, SAR A/D-Conv., ser. Out,+5V 200kHz, bipolar, ILE= ±2 (±2,±1)LSB	8-SSMDIP	Bub	-	pdf
ADS 7817 P(B,C)	A/D-IC	12 Bit, SAR A/D-Conv., ser. Out, +5V 200kHz, bipolar, ILE= ±2 (±2,±1)LSB	8-DIP	Bub	-	pdf
ADS 7817 U(B,C)	A/D-IC	SMD,12Bit, SAR A/D-Conv., ser. Out,+5V 200kHz, bipolar, ILE= ±2 (±2,±1)LSB	8-MDIP	Bub	-	pdf
ADS 7818 E(B)	A/D-IC	SMD,12Bit, SAR A/D-Conv., ser. Out,+5V 500kHz, unipolar, ILE= ±2(±1) LSB	8-SSMDIP	Bub	-	pdf
ADS 7818 P(B)	A/D-IC	12 Bit, SAR A/D-Conv., ser. Out, +5V 500kHz, unipolar, ILE= ±2(±1) LSB	8-DIP	Bub	-	pdf
ADS 7820 P(B)	CMOS-A/D-IC	12 Bit, SAR A/D-Conv., S&H, +5V Inp. 0..5V, -40...85°C, ILE=±1(±½)LSB	28-DIP	Bub	-	pdf
ADS 7820 U(B)	CMOS-A/D-IC	SMD, 12 Bit, SAR A/D-Conv., S&H, +5V Inp. 0..5V, -40...85°C, ILE=±1(±½)LSB	28-MDIP	Bub	-	pdf
ADS 7821 P(B)	CMOS-A/D-IC	16 Bit, SAR A/D-Conv., S&H, +5V Inp. 0..5V, -25..85°C, ILE=±4(3)LSB	28-DIP	Bub	-	pdf
ADS 7821 U(B)	CMOS-A/D-IC	SMD, 16 Bit, SAR A/D-Conv., S&H, +5V Inp. 0..5V, -25..85°C, ILE=±4(3)LSB	28-MDIP	Bub	-	pdf
ADS 7822 E(B,C)	A/D-IC	SMD, 12 Bit, A/D-Conv., ser.Out,+2.7V ILE=±2 (±1, ±0.75) LSB, -40...85°C	8-SSMDIP	Bub	-	pdf
ADS 7822 P(B,C)	A/D-IC	12 Bit, SAR A/D-Conv., ser.Out, +2.7V ILE=±2 (±1, ±0.75) LSB, -40...85°C	8-DIP	Bub	-	pdf
ADS 7822 U(B,C)	A/D-IC	SMD, 12 Bit, A/D-Conv., ser.Out, +2.7V ILE=±2 (±1, ±0.75) LSB, -40...85°C	8-MDIP	Bub	-	pdf
ADS 7823 E(B)	A/D-IC	SMD, 12Bit, A/D-Conv., S&H, I²C, 50kHz -40...85°C, ILE= ±2(1)LSB, +2.7V..5V	8-SSMDIP	Bub	-	pdf
ADS 7824 P(B)	CMOS-A/D-IC	4 Chan./Kan., 12Bit, SAR A/D-Conv.,+5V S&H, MUX, ser./par.Out, ILE=±1(±½)LSB	28-DIP	Bub	-	pdf
ADS 7824 U(B)	CMOS-A/D-IC	SMD, 4Chan./Kan., 12Bit, A/D-Conv.,+5V S&H, MUX, ser/par.Out, ILE= ±1(±½)LSB	28-MDIP	Bub	-	pdf
ADS 7825 P(B)	CMOS-A/D-IC	4 Chan./Kan., 16Bit, SAR A/D-Conv.,+5V S&H, MUX, ser./par.Out, ILE=±3(±2)LSB	28-DIP	Bub	-	pdf
ADS 7825 U(B)	CMOS-A/D-IC	SMD, 4Chan./Kan., 16Bit, A/D-Conv.,+5V S&H, MUX, ser/par.Out, ILE= ±3(±2)LSB	28-MDIP	Bub	-	pdf
ADS 7828 E(B)	A/D-IC	SMD, 8 Chan./Kan., 12 Bit, A/D-Conv. MUX, S&H, SAR, I²C, 50kHz, +2.7V...5V	16-SSMDIP	Bub	-	pdf
ADS 7830 I	A/D-IC	SMD, 8 Bit 8-Chan. SAR A/D-Conv., MUX 2.5V Ref., I²C, +2.7...5V	16-SSMDIP	And	-	pdf
ADS 7832 BN	CMOS-A/D-IC	4Chan./Kan., 12 Bit, A/D-Conv., 50kHz S&H, MUX, +3.3V or/oder +5V	28-PLCC	Bub	-	data pdf pinout
ADS 7832 BP	CMOS-A/D-IC	4Chan./Kan., 12 Bit, A/D-Conv., 50kHz S&H, MUX, +3.3V or/oder +5V	28-DIP	Bub	-	data pdf pinout
ADS 7834 E(B)	A/D-IC	SMD, 12Bit, A/D-Conv., 500kHz, +5V S&H, uni.Inp., ser. Out, ILE=±2(±1)LSB	8-SSMDIP	Bub	-	pdf
ADS 7834 P(B)	A/D-IC	12Bit, A/D-Converter, 500kHz, +5V S&H, uni.Inp., ser. Out, ILE=±2(±1)LSB	8-DIP	Bub	-	pdf

Type	Device	Short Description	Fig.	Manu	Comparision Types	More at
ADS 7835 E(B)	A/D-IC	SMD, 12Bit, A/D-Conv., 500kHz, +5V S&H, bi.Inp., ser. Out, ILE=±2(±1)LSB	8-SSMDIP	Bub	-	pdf
ADS 7835 P(B)	A/D-IC	12Bit, A/D-Converter, 500kHz, +5V S&H, bi.Inp., ser. Out, ILE=±2(±1)LSB	8-DIP	Bub	-	pdf
ADS 7841 E(B,S)	A/D-IC	SMD,12Bit, SAR A/D-Conv.,+2.7 or +5.5V 4Chan./4Kan., S&H, MUX, ser.Out E, ES: ±2LSB EB: ±1LSB	16-SSMDIP	Bub	-	pdf
ADS 7841 P(B)	A/D-IC	12Bit, SAR A/D-Conv.,+2.7 +5.5V 4Chan./4Kan., S&H, MUX, ser.Out	16-DIP	Bub	-	pdf
ADS 7842 E(B)	A/D-IC	SMD, 12Bit, SAR A/D-Conv., +2.7V...+5V 4Chan./Kan., MUX, S&H, par.Out	28-SSMDIP	Bub	-	pdf
ADS 7844 E(B),N(B)	A/D-IC	SMD, 12Bit, SAR A/D-Conv., +2.7V or 5V 8 Chan./Kan., MUX, S&H, ser. Out	20-SSMDIP	Bub	-	pdf
ADS 7852 Y(B)	A/D-IC	12 Bit, 8 Chan./Kan., SAR A/D-Conv. MUX, S&H, 500kHz, ±2(1)LSB, +5V	32-QFP	Bub	-	pdf
ADS 7861 E(B)	A/D-IC	SMD, 12 Bit, 2+2 Chan./Kan., A/D-Conv. simult. S&H, ser.Out, 500kHz, +5V	24-SSMDIP	Bub	-	pdf
ADS 7862 Y(B)	A/D-IC	SMD, 12 Bit, 2+2 Chan./Kan., A/D-Conv. simult. S&H, par.Out, 500kHz, +5V	32-QFP	Bub	-	pdf
ADS 7864 Y(B)	A/D-IC	SMD, 12 Bit, 3+3 Chan./Kan., A/D-Conv. simult. S&H, par.Out, 500kHz, +5V	48-QFP	Bub	-	pdf
ADS 7869 I	LIN-A/D-IC	SMD, Motor Control Front-End, diff.inp Sampling on 7 S/H Caps, ser. Interf. 3x 12Bit 12 Chan. 1MSPS A/D-Conv.	100-QFP	And	-	pdf
ADS 7890	A/D-IC	SMD, 14 Bit, SAR A/D-Conv., 1.25MSPS unipolar,pseudo diff.inp., ser.interf.	48-QFP	And	-	pdf pinout
ADS 8320 E(B)	A/D-IC	SMD, 16Bit SAR A/D-Conv., uni.Inp.,S&H ser.Out, ±0.018(0.012)%,100kHz, 2.7/5V	8-SSMDIP	Bub	-	pdf
ADS 8321 E(B)	A/D-IC	SMD, 16Bit, SAR A/D-Conv., bi.Inp.,S&H ser. Out, ±0.018(0.012)%, 100kHz, +5V	8-SSMDIP	Bub	-	pdf
ADS 8322 Y(B)	A/D-IC	16 Bit, SAR A/D-Conv., uni.Inp., S&H par. Out, ±8(6)LSB, 500kSPS, +5V	32-QFP	Bub	-	pdf
ADS 8323 Y(B)	A/D-IC	16 Bit, SAR A/D-Conv., bi.Inp., S&H par. Out, ±8(6)LSB, 500kSPS, +5V	32-QFP	Bub	-	pdf
ADS 8324 E(B)	A/D-IC	SMD, 14 Bit, SAR A/D-Conv., S&H Amp ser. Out, ±3(2)LSB, 50kHz, +1.8V	8-SSMDIP	Bub	-	pdf
ADS 8371 I(B)	A/D-IC	SMD, 16Bit SAR A/D-Conv., unipol. Inp. 750kSPS, ±2.5(1.5)% Lin., +3.3V & +5V	48-QFP	Tix	-	pdf
ADS 8381 I(B)	A/D-IC	SMD, 18Bit SAR A/D-Conv., unipol. Inp. 580kSPS, ±6(5)% Lin., +3.3V & +5V	48-QFP	Tix	-	pdf
ADS 8383 I(B)	A/D-IC	SMD, 18Bit SAR A/D-Conv., unipol. Inp. 500kSPS, ±10(7)% Lin., +3.3V & +5V	48-QFP	Tix	-	pdf
ADS 8422I	A/D-IC	SMD, 16Bit, 4MHz SAR ADC, int. 4.096V reference, S&H	QFP	Tix	-	pdf pinout
ADS 8422IB	A/D-IC	SMD, 16Bit, 4MHz SAR ADC, int. 4.096V reference, S&H	QFP	Tix	-	pdf pinout
ADS-CCD 1201 MC,MM	A/D-IC	12 Bit, A/D-Converter, S&H, 1.2MHz optimized for Ccd App's	24-DIC	Dat	-	pdf
ADS-CCD 1202 MC,MM	A/D-IC	12 Bit, A/ D- Converter, S&H, 2MHz optimized for Ccd App's	24-DIC	Dat	-	pdf
ADt	Si-N	=BCW 60D (Typ-Code/Stempel/marking)	35	Phi	→BCW 60D	data
ADT 7476 ARQZ	IC	Remote Thermal Contr., Volt.monitor	TSOP	And	-	pdf pinout
ADT 7476 ARQZ-REEL	IC	Remote Thermal Contr., Volt.monitor	TSOP	And	-	pdf pinout
ADT 6401 SRJZ-RL 7	LIN-IC	2.7 V to 5.5 V, Temperature Switches	TSOP	And	-	pdf pinout
ADT 6402 SRJZ-RL 7	LIN-IC	2.7 V to 5.5 V, Temperature Switches	TSOP	And	-	pdf pinout
ADT 7476ARQZ-REEL7	IC	Remote Thermal Contr., Volt.monitor	TSOP	And	-	pdf pinout
ADuC 841 BCP...	A/D-IC	8 Chan. 12 Bit A/D-Conv., T&H Dual D/A-Conv., 20MHz 8 Bit MCU (8051)	56-LFCSP	And	-	pdf
ADuC 841 BS...	A/D-IC	SMD, 8 Chan. 12 Bit A/D-Conv., T&H Dual D/A-Conv., 20MHz 8 Bit MCU (8051)	48-QFP	And	-	pdf
ADuC 842 BCP...	A/D-IC	8 Chan. 12 Bit A/D-Conv., T&H Dual D/A-Conv., 20MHz 8 Bit MCU (8051)	56-LFCSP	And	-	pdf
ADuC 842 BS...	A/D-IC	SMD, 8 Chan. 12 Bit A/D-Conv., T&H Dual D/A-Conv., 20MHz 8 Bit MCU (8051)	48-QFP	And	-	pdf
ADuC 843 BCP...	A/D-IC	8 Chan. 12 Bit A/D-Conv., T&H 20MHz 8 Bit MCU (8051)	56-LFCSP	And	-	pdf
ADuC 843 BS...	A/D-IC	SMD, 8 Chan. 12 Bit A/D-Conv., T&H 20MHz 8 Bit MCU (8051)	48-QFP	And	-	pdf
ADUC 7032 BSTZ-8L	A/D-IC	Integrated Precision Battery Sensor	TSOP	And	-	pdf pinout
ADUM 6132 ARWZ	LIN-IC	Isolated Half-Bridge Gate Driver	TSOP	And	-	pdf pinout
ADUM 6132 ARWZ-RL	LIN-IC	Isolated Half-Bridge Gate Driver	TSOP	And	-	pdf pinout
ADV 453 Kö40,Kö66	D/A-IC	8 Bit, 3Ch, Video, 40, 66Msps, TTL L.	40-DIP	And	-	
ADV 453 K..40, 66	D/A-IC	8 Bit, 3Ch, Video, 40, 66Msps, TTL L.	40-DIP	And	-	
ADV 476 KN50,KN66	D/A-IC	6 Bit, Video, 50, 66Msps, TTL Level	28-DIP	And	-	
ADV 7120 Kö50,Kö80	D/A-IC	8Bit, 3Ch, Video, 50, 80Msps, CMOS/TTL	40-DIP 44-PLCC	And	-	
ADV 7120 K..50, 80	D/A-IC	8Bit, 3Ch, Video, 50, 80Msps, CMOS/TTL	44-PLCC	And	-	
ADV 7120 K..50, 80	D/A-IC	8Bit, 3Ch, Video, 50, 80Msps, CMOS/TTL	40-DIP	And	-	
ADV 7121 K...,J...	D/A-IC	8 Bit, 3 Ch, Video, CMOS/TTL Level JN50,KN50: 50Msps, JN,KN80: 80Msps	40-DIP	And	-	
ADV 7122 K...,J...	D/A-IC	8 Bit, 3 Ch, Video, CMOS/TTL Level JN50,KN50: 50Msps, JN,KN80: 80Msps	44-PLCC	And	-	
ADX118	HI-SPEED	Vs:±15V Vu:88dB Vo:±12V Vi0:4mV	8A	And	-	data pinout
ADX218	HI-SPEED	Vs:±15V Vu:88dB Vo:±12V Vi0:4mV	8A	And	-	data pinout
ADX318	HI-SPEED	Vs:±15V Vu:88dB Vo:±12V Vi0:10mV	8A	And	-	data pinout

Type	Device	Short Description	Fig.	Manu	Comparision Types	More at
ADY...ADZ						
ADY 10	Ge-P	NF/S, 32V, 0,6A, 0,25W	2a	Gpd,Sie	2SB493	data
ADY 11	Ge-P	NF/S, 60V, 0,6A, 0,25W	2a	Gpd,Sie		data
ADY 12	Ge-P	NF/S, 32V, 0,6A, 0,25W	2a	Gpd,Sie	2SB493	data
ADY 13	Ge-P	NF/S, 60V, 0,6A, 0,25W	2a	Gpd,Sie	-	data
ADY 14	Ge-P	NF/S-L, 65V, 3A, 6W		Sie	(AL 102, AUY 19, 2N2140, 2N2145)[4]	data
ADY 15	Ge-P	NF/S-L, 80V, 3A, 6W		Sie	(AL 102, AUY 20, 2N2141, 2N2146)[4]	data
ADY 16	Ge-P	NF/S-L, 80V, 3A, 6W		Sie	(AL 102, AUY 20, 2N2141, 2N2146)[4]	data
ADY 18	Ge-P	NF/S-L, 60V, 15A, 45W	23a§	Aeg	2N1550...52, 2N1554...56, 2N1558...60	data
ADY 19	Ge-P	NF/S, 32V, 0,6A, 0,25W	2a	Sie	2SB493	data
ADY 20	Ge-P	NF/S, 60V, 0,6A, 0,25W	2a	Gpd,Sie	-	data
ADY 21	Ge-P	NF/S, 0,6A, 0,25W	2a	Gpd,Sie	-	data
ADY 22	Ge-P	NF-L, 30V, 10A, 40W	23a§	Itt	AUY 21, 2N2289, 2N2292, 2N2526	data
ADY 23	Ge-P	=ADY 22: 80V	23a§	Itt	AUY 37, 2N2289, 2N2292, 2N2526	data
ADY 24	Ge-P	=ADY 22: 80V	23a§	Itt	AUY 37, 2N2289, 2N2292, 2N2526	data
ADY 25	Ge-P	=ADY 22: 100V	23a§	Itt	AUY 37, 2N2290, 2N2293, 2N2527	data
ADY 26	Ge-P	NF/S-L, 80V, 25A, 100W	38a§	Gpd,Phi	2N1519, 2N1521, 2N1523	data
ADY 27	Ge-P	NF-L, 32V, 3,5A, 27,5W	23a§	Gpd,Sie	AD 149, AL 102, AUY 19, 2N2137, 2N2142	data
ADY 28	Ge-P	NF-L, 80V, 6A, 33W	23a§		AL 102, AUY 22, 2N3615, 2N3617	data
ADY 30	Ge-P	NF/S-L, 45V, 50A, 150W	38a§	Phi	2N4048...4053	data
ADY 31	Ge-P	NF/S-L, 60V, 160A, 85W	38a§	Phi	-	data
ADY 32	Ge-P	NF/S-L, 80V, 160A, 35W	38a§	Phi	-	data
ADZ 11	Ge-P	NF/S-L, 50V, 15A, 45W	38a§	Gpd,Phi	2N1980, 2N2077, 2N2081	data
ADZ 12	Ge-P	=ADZ 11: 80V	38a§	Gpd,Phi	2N1982, 2N2075, 2N2079, 2N2492	data
AE						
AE	Si-P	=BCP 52 (Typ-Code/Stempel/marking)	48	Phi	→BCP 52	data
AE	Si-N	=BCW 60E (Typ-Code/Stempel/marking)	35	Sie	→BCW 60E	data
AE	Si-P	=BCX 52 (Typ-Code/Stempel/marking)	39	Sie,Val,Mot	→BCX 52	data
AE	Si-P	=CPH 3105 (Typ-Code/Stempel/marking)	35	Say	→CPH 3105	data
AE	Z-Di	=MA 5Z240-L (Typ-Code/Stempel/marking)	71(1,7mm)	Mat	→MA 5Z240-L	data
AE	Si-N	=MT 3S04AS (Typ-Code/Stempel/marking)	35(1,6mm)	Tos	→MT 3S04AS	data
AE	Si-N	=MT 3S04AT (Typ-Code/Stempel/marking)	35(1,6mm)	Tos	→MT 3S04AT	data
AE	Si-N	=MT 3S04AU (Typ-Code/Stempel/marking)	35(2mm)	Tos	→MT 3S04AU	data
AE	Si-N	=MT 4S04A (Typ-Code/Stempel/marking)	44	Tos	→MT 4S04A	data
AE	MOS-P-FET-e	=NTTS 2P03 (Typ-Code/Stempel/marking)	8-SSMDIP	Ons	→NTTS 2P03	
AE	Z-Di	=P4 SMA-6.8A(Typ-Code/Stempel/marking)	71(5x2,5)	Fag	→P4 SMA-6.8A	data
AE	Z-Di	=SMAJ 5.0A (Typ-Code/Stempel/marking)	71(5x2,5)	Tho	→SMAJ ...	data
AE	Z-Di	=SMBJ 5.0CA (Typ-Code/Stempel/marking)	71(5x3,5)	Mop	→SMBJ ...	data
AE	C-Di	=ZMV 833A (Typ-Code/Stempel/marking)	71(1,7mm)	Ztx	→ZMV 833	data
AE	Si-P	=2SA1363-E (Typ-Code/Stempel/marking)	39	Mit	→2SA1363	data
AE	Si-P	=2SA1365-E (Typ-Code/Stempel/marking)	35	Mit	→2SA1365	data
AE	Si-P	=2SA1419 (Typ-Code/Stempel/marking)	39	Say	→2SA1419	data
AE	MOS-N-FET-d	=3SK219 (Typ-Code/Stempel/marking)	44	Mat	→3SK219	data
AE	MOS-N-FET-d	=3SK268 (Typ-Code/Stempel/marking)	44(2mm)	Mat	→3SK268	data
AEG	Si-P	=2SA1362-GR (Typ-Code/Stempel/marking)	35	Tos	→2SA1362	data
(AEG) 896	Se-Di	880/950V, 3,3mA	31a		E400C3,3	
(AEG) 972	Se-Di	1100/1200V, 3mA	31a		E500C3	
(AEG) 992	Se-Di	1300/1450V, 2,7mA	31a		E600C2,7	
(AEG) 1006	Se-Di	1600/1700V, 2,3mA	31a		E800C2,3	
AEL	Si-P	=2SA1362-BL (Typ-Code/Stempel/marking)	35	Tos	→2SA1362	data
AEL	Si-N	=2SC3839K-L (Typ-Code/Stempel/marking)	35	Rhm	→2SC3839K	data
AEM	Si-N	=2SC3839K-M (Typ-Code/Stempel/marking)	35	Rhm	→2SC3839K	data
AEN	Si-N	=2SC3839K-N (Typ-Code/Stempel/marking)	35	Rhm	→2SC3839K	data
AEO	Si-N	=KTC3883-O (Typ-Code/Stempel/marking)	35	Kec	→KTC 3883	data
AEP	Si-P	=2SB1424-P (Typ-Code/Stempel/marking)	39	Rhm	→2SB1424	data
AEP	Si-N	=2SC3839K-P (Typ-Code/Stempel/marking)	35	Rhm	→2SC3839K	data
AEQ	Si-P	=2SB1424-Q (Typ-Code/Stempel/marking)	39	Rhm	→2SB1424	data
AEQ	Si-N	=2SC3839K-Q (Typ-Code/Stempel/marking)	35	Rhm	→2SC3839K	data
AER	Si-P	=2SB1424-R (Typ-Code/Stempel/marking)	39	Rhm	→2SB1424	data
AER	Si-N	=KTC3883-R (Typ-Code/Stempel/marking)	35	Kec	→KTC 3883	data
AEs	Si-P	=BCP 52M (Typ-Code/Stempel/marking)	≈45	Sie	→BCP 52M	data
AEY	Si-P	=2SA1362-Y (Typ-Code/Stempel/marking)	35	Tos	→2SA1362	data
AF...AFX						
AF	Si-P	=2SA1363-F (Typ-Code/Stempel/marking)	39	Mit	→2SA1363	data
AF	Si-P	=2SA1365-F (Typ-Code/Stempel/marking)	35	Mit	→2SA1365	data
AF	Si-P	=2SA1575 (Typ-Code/Stempel/marking)	39	Say	→2SA1575	data
AF	MOS-N-FET-d	=3SK220 (Typ-Code/Stempel/marking)	44	Mat	→3SK220	data
AF	MOS-N-FET-d	=3SK270 (Typ-Code/Stempel/marking)	44(2mm)	Mat	→3SK270	data
AF	Si-N	=BCW 60FF (Typ-Code/Stempel/marking)	35	Sie,Val	→BCW 60FF	data
AF	Si-P	=BCX 52-6 (Typ-Code/Stempel/marking)	39	Sie,Val,Mot	→BCX 52	data
AF	Si-P	=CPH 3106 (Typ-Code/Stempel/marking)	35	Say	→CPH 3106	data
AF	Z-Di	=MA 5Z240-M (Typ-Code/Stempel/marking)	71(1,7mm)	Mat	→MA 5Z240-M	data
AF	Z-Di	=P4 SMA-7.5 (Typ-Code/Stempel/marking)	71(5x2,5)	Fag	→P4 SMA-7.5	data
AF	C-Di	=ZMV 834A (Typ-Code/Stempel/marking)	71(1,7mm)	Ztx	→ZMV 834	data

Type	Device	Short Description	Fig.	Manu	Comparision Types	More at
AFs	Si-N	=BCW 60FF (Typ-Code/Stempel/marking)	35	Sie,	→BCW 60FF	data
AF	Z-Di	=SMBJ 6.0C (Typ-Code/Stempel/marking)	71(5x3,5)	Mop	→SMBJ ...	data
AF 101	Ge-P	AM-V/M/O/ZF, 10MHz	1a	Aeg	AF 121, AF 127, AF 200	data
AF 102	Ge-P	VHF-V/M/O, 180MHz	1g	Phi	AF 106, AF 306	data
AF 105	Ge-P	AM/FM-ZF, 22MHz	1a	Aeg	AF 121, AF 126, AF 200	data
AF 106	Ge-P	VHF-V/M/O, 220MHz	5g	Aeg,Phi,Sie	AF 121, AF 306	data
AF 107	Ge-P	VHF-V/M/O, 330MHz	2a§	Sie	AF 109R	data
AF 108	Ge-P	VHF-V/M/O, 330MHz	2a§	Sie	AF 109R	data
AF 109(R)	Ge-P	VHF-V, re, 280MHz	5g	Aeg,Phi,Sie	AF 139, AF 239(S)	data
AF 110	Ge-P	ZF, re	5g	Sie,Phi	AF 121, AF 126, AF 200	data
AF 111	Ge-P	AM-V/M/ZF, 50MHz	2a	Itt	AF 127, AF 200	data
AF 112	Ge-P	AM/FM-M/O/ZF, 60MHz	2a	Itt	AF 126, AF 200	data
AF 113	Ge-P	FM-V/M/O/ZF, 80MHz	2a	Itt	AF 125, AF 200	data
AF 114	Ge-P	FM-V, 75MHz	1g	Phi,Sie	AF 124, AF 200	data
AF 115	Ge-P	FM-M, 75MHz	1g	Phi,Sie	AF 125, AF 200	data
AF 116	Ge-P	AM-V/M, FM-ZF, 75MHz	1g	Phi,Sie	AF 126, AF 200	data
AF 117	Ge-P	AM-V/M/ZF, 75MHz	1g	Phi,Sie	AF 127, AF 200	data
AF 118	Ge-P	HF/Vid, 70V, 175MHz	1g	Phi,Sie		data
AF 121	Ge-P	AM/FM-V/M/ZF, 270MHz	5k	Aeg,Phi	AF 200, AF 201	data
AF 121 S	Ge-P	TV-ZF, 32V, 270MHz	5k		AF 201, AF 202(S,L)	data
AF 122	Ge-P	VHF-V/M/O, 275MHz	2a	Aeg	AF 106, AF 109R, AF 306	data
AF 124	Ge-P	FM-V, 75MHz	5k	Phi,Sie	AF 200, AF 139, AF 239(S)	data
AF 125	Ge-P	FM-M, 75MHz	5k	Phi,Sie	AF 200, AF 139 ,AF 239(S)	data
AF 126	Ge-P	AM-V/M, FM-ZF, 75MHz	5k	Phi,Sie	AF 200, AF 139, AF 239(S)	data
AF 127	Ge-P	AM-V/M/ZF, 75MHz	5k	Phi,Sie	AF 200, AF 139, AF 239(S)	data
AF 128	Ge-P	Min, HF, 6MHz	36a	Aeg	(AF 127, AF 200)[6]	data
AF 129	Ge-P	FM-V, 150MHz	5k	Itt	AF 121, AF 124, AF 200	data
AF 130	Ge-P	FM-M, 150MHz	5k	Itt	AF 124, AF 200	data
AF 131	Ge-P	AM-V/M, 100MHz	5k	Itt	AF 125, AF 200	data
AF 132	Ge-P	AM/FM-ZF, 90MHz	5k	Itt	AF 126, AF 200	data
AF 133	Ge-P	AM-V/M/ZF, 100MHz	5k	Itt	AF 127, AF 200	data
AF 134	Ge-P	FM-V, 55MHz	5g	Aeg	AF 124, AF 200	data
AF 135	Ge-P	FM-M, 50MHz	5g	Aeg	AF 125, AF 200	data
AF 136	Ge-P	AM-V/M/O, 40MHz	5g	Aeg	AF 126, AF 200	data
AF 137	Ge-P	AM/FM-ZF, 35MHz	5g	Aeg	AF 126, AF 200	data
AF 138	Ge-P	AM/FM-ZF, re, 40MHz	5g	Aeg	AF 126, AF 200	data
AF 139	Ge-P	UHF-V/M/O, 550MHz	5g	EUR	AF 239(S)	data
AF 142	Ge-P	FM-V, 150MHz	1g	Sgs	AF 124, AF 200	data
AF 143	Ge-P	FM-M, 130MHz	1g	Sgs	AF 125, AF 200	data
AF 144	Ge-P	AM/FM-V/M/ZF, 130MHz	1g	Sgs	AF 126, AF 200	data
AF 146	Ge-P	AM/FM-V/M/ZF	1g	Sgs	AF 125, AF 200	data
AF 147	Ge-P	AM/FM-ZF	1g	Sgs	AF 126, AF 127, AF 200	data
AF 148	Ge-P	AM/FM-ZF	1g	Sgs	AF 126, AF 127, AF 200	data
AF 149	Ge-P	AM/FM-V/M/ZF	1g	Sgs	AF 126, AF 200	data
AF 150	Ge-P	AM-V/M/ZF	1g	Sgs	AF 126, AF 127, AF 200	data
AF 164	Ge-P	FM-V, 150MHz	4g	Sgs	AF 124, AF 200	data
AF 165	Ge-P	FM-M, 130MHz	4g	Sgs	AF 125, AF 200	data
AF 166	Ge-P	AM/FM-V/M/ZF, 130MHz	4g	Sgs	AF 126, AF 200	data
AF 168	Ge-P	FM-M/ZF	4g	Sgs	AF 125, AF 200	data
AF 169	Ge-P	AM-V/M/ZF	4g	Sgs	AF 126, AF 127, AF 200	data
AF 170	Ge-P	AM-V/M/O/ZF, 60MHz	4g	Sgs	AF 126, AF 127, AF 200	data
AF 171	Ge-P	AM-V/ZF	4g	Sgs	AF 126, AF 200	data
AF 172	Ge-P	AM-V/M/O/ZF, 60MHz	4g	Sgs	AF 127, AF 200	data
AF 178	Ge-P	VHF-V/M/O, 180MHz	5g	Aeg,Phi	AF 106, AF 306	data
AF 179	Ge-P	VHF-V/M/O, 270MHz	5g	Phi	AF 109R	data
AF 180	Ge-P	VHF-V, re, 250MHz	5g	Phi	AF 106, AF 109R, AF 306	data
AF 181	Ge-P	TV-ZF, re, 170MHz	5g	Aeg,Phi	AF 121, AF 200	data
AF 182	Ge-P	HF, Vid-Tr, >120MHz	4g	Tho	AF 121, AF 200, AF 201	data
AF 185	Ge-P	AM-V/M/O/ZF, 80MHz	5g	Phi	AF 106, AF 121, AF 200, AF 201, AF 306	data
AF 186	Ge-P	UHF-V/M/O	5g	Phi	AF 139, AF 239(S)	data
AF 187	Ge-P	NF/HF, 18V, 0,1A, 7MHz	2a	Tho	AC 125, AC 151, AF 127, AF 200	data
AF 188	Ge-P	NF/HF, 18V, 0,1A, 13MHz	2a	Tho	AC 126, AC 151, AF 127, AF 200	data
AF 189	Ge-P	NF/HF, 18V, 0,1A, 7MHz	2a	Tho	AC 125, AC 151, AF 127, AF 200	data
AF 190	Ge-P	NF/HF, 18V, 0,1A, 13MHz	2a	Tho	AC 126, AC 151, AF 127, AF 200	data
AF 192	Ge-N	NF/HF, sym, 10V, 0,1A	5g	Tho		data
AF 193	Ge-P	TV-ZF, 40MHz	2a	Tho	AF 126, AF 200	data
AF 194	Ge-P	FM-V, 110MHz	4g	Tho	AF 124, AF 200	data
AF 195	Ge-P	FM-M, 85MHz	4g	Tho	AF 125, AF 200	data
AF 196	Ge-P	AM/FM-V/M/ZF, 80MHz	4g	Tho	AF 126, AF 200	data
AF 197	Ge-P	AM-V/M/O/ZF, >60MHz	4g	Tho	AF 127, AF 200	data
AF 198	Ge-P	FM-ZF, re, >60MHz	4g	Tho	AF 127, AF 200	data

Type	Device	Short Description	Fig.	Manu	Comparision Types	More at
AF 200(U)	Ge-P	TV-ZF, re	5k	Sgs,Sie	AF 121	data
AF 201(U)	Ge-P	TV-ZF	5k	Sgs,Sie	AF 121	data
AF 202(L,S)	Ge-P	TV-ZF	5k	Sgs,Sie	AF 121(S)	data
AF 239(S)	Ge-P	UHF-V/M/O, 700MHz	5g	EUR	AF 279, AF 280, AF 379	data
AF 240(S)	Ge-P	UHF-M/O, 500MHz	5g	EUR	AF 239(S)	data
AF 250	Ge-P	UHF-V	5g	Phi	AF 239(S)	data
AF 251	Ge-P	UHF-V, re, 750MHz	9a	Aeg	AF 239(S)	data
AF 252	Ge-P	UHF-M, 650MHz	9a	Aeg	AF 239(S), AF 240	data
AF 253	Ge-P	VHF-V, re, 550MHz	9a	Aeg	AF 109R	data
AF 254	Ge-P	UHF	9a	Aeg	AF 239(S), AF 240	data
AF 256	Ge-P	VHF-V/M/O, >170MHz	9a	Aeg	AF 106, AF 306	data
AF 257	Ge-P	Min, VHF, >170MHz	36a	Aeg	(AF 109R)[6]	data
AF 260	Ge-P	AM/FM-ZF, 3MHz	2a	Eiy	AF 126, AF 200	data
AF 261	Ge-P	AM-V/M/O, 3MHz	2a	Eiy	AF 126, AF 127, AF 200	data
AF 263	Ge-P	VHF-V, re, 550MHz	9e	Aeg	AF 109R	data
AF 264	Ge-P	VHF-V/M/O, >170MHz	9e	Aeg	AF 106, AF 109R, AF 306	data
AF 265	Ge-P	S, 18V, 0,1A, 6MHz	2a	Eiy	AC 125, AC 151	data
AF 266	Ge-P	S, 18V, 0,1A, 8MHz	2a	Eiy	AC 126, AC 151	data
AF 267	Ge-P	UHF-V/M, 780MHz	24e	Phi	AF 279, AF 280, AF 379	data
AF 268	Ge-P	UHF	24e	Phi	AF 279, AF 280, AF 379	data
AF 269	Ge-P	UHF-M, 550MHz	24e	Phi	AF 279, AF 280, AF 379	data
AF 271	Ge-P	HF-V/M/O, 30MHz	2a	Eiy	AF 125, AF 126, AF 200, AF 201	data
AF 272	Ge-P	HF, 40MHz	2a	Eiy	AF 125, AF 126, AF 200, AF 201	data
AF 273	Ge-P	HF, 60MHz	2a	Eiy	AF 125, AF 126, AF 200, AF 201	data
AF 275	Ge-P	HF, 35MHz	2a	Eiy	AF 125, AF 126, AF 200, AF 201	data
AF 279(S)	Ge-P	UHF-V, 780MHz	24e	Aeg,Phi,++	AF 379	data
AF 280(S)	Ge-P	UHF-M/O, 550MHz	24e	Aeg,Phi,++	AF 279(S), AF 379	data
AF 280(III...VIII)	Ge-P	HF/ZF, 40MHz	2a	Eiy	AF 125, AF 126, AF 200, AF 201	data
AF 282(V...VIII)	Ge-P	HF, 80MHz	2a	Eiy	AF 125, AF 126, AF 200, AF 201	data
AF 284(IV...VIII)	Ge-P	HF, 80MHz	2d	Eiy	AF 125, AF 126, AF 200, AF 201	data
AF 289	Ge-P	UHF-V, re, 950MHz	24e	Sie	AF 279, AF 379	data
AF 290	Ge-P	UHF-M/O, 800MHz	24e	Sie	AF 279, AF 280, AF 379	data
AF 306	Ge-P	FM/VHF-V/M/O, 500MHz	7a	Phi,Sie	AF 106, AF 109R	data
AF 339	Ge-P	VHF-V, re, 750MHz	7a	Sie	AF 139, AF 239(S)	data
AF 367	Ge-P	UHF-V, 800MHz	24a	Phi	AF 279, AF 379	data
AF 369	Ge-P	UHF-M/O, 550MHz	24a	Phi	AF 279, AF 280, AF 379	data
AF 379	Ge-P	UHF-V, 1250MHz	24a	Phi,Sie	-	data
AF 387	Ge-P				-	
AF 439	Ge-P	VHF-V, 800MHz	24e	Sie	AF 279, AF 280, AF 379	data
AFL 2803R3 S	LIN/Z-IC	+3.3V, DC/DC Converter	FLP	Inr	-	pdf pinout
AFL 2805 D	LIN/Z-IC	±5V, DC/DC Converter	FLP	Inr	-	pdf pinout
AFL 2805 S	LIN/Z-IC	+5V, DC/DC Converter	FLP	Inr	-	pdf pinout
AFL 2807 S	LIN/Z-IC	+7V, DC/DC Converter	FLP	Inr	-	pdf pinout
AFL 2808 S	LIN/Z-IC	+8V, DC/DC Converter	FLP	Inr	-	pdf pinout
AFL 2809 S	LIN/Z-IC	+9V, DC/DC Converter	FLP	Inr	-	pdf pinout
AFL 2812 D	LIN/Z-IC	±12V, DC/DC Converter	FLP	Inr	-	pdf pinout
AFL 2812 S	LIN/Z-IC	+12V, DC/DC Converter	FLP	Inr	-	pdf pinout
AFL 2815 D	LIN/Z-IC	±12V, DC/DC Converter	FLP	Inr	-	pdf pinout
AFL 2815 S	LIN/Z-IC	+15V, DC/DC Converter	FLP	Inr	-	pdf pinout
AFL 2828 S	LIN/Z-IC	+28V, DC/DC Converter	FLP	Inr	-	pdf pinout
AFL 5003R3 S	LIN/Z-IC	3.3V, DC/DC Converter	FLP	Inr	-	pdf pinout
AFL 5005 D	LIN/Z-IC	±5V, DC/DC Converter	FLP	Inr	-	pdf pinout
AFL 5005 S	LIN/Z-IC	5V, DC/DC Converter	FLP	Inr	-	pdf pinout
AFL 5008 S	LIN/Z-IC	8V, DC/DC Converter	FLP	Inr	-	pdf pinout
AFL 5009 S	LIN/Z-IC	9V, DC/DC Converter	FLP	Inr	-	pdf pinout
AFL 5012 D	LIN/Z-IC	±12V, DC/DC Converter	FLP	Inr	-	pdf pinout
AFL 5012 S	LIN/Z-IC	12V, DC/DC Converter	FLP	Inr	-	pdf pinout
AFL 5015 D	LIN/Z-IC	±15V, DC/DC Converter	FLP	Inr	-	pdf pinout
AFL 5015 S	LIN/Z-IC	15V, DC/DC Converter	FLP	Inr	-	pdf pinout
AFL 5028 S	LIN/Z-IC	28V, DC/DC Converter	FLP	Inr	-	pdf pinout
AFL 12003R3 S	CMOS/LIN-IC	3.3V, Hybrid-High Reliability DC/DC Conv.	FLP	Inr	-	pdf pinout
AFL 12005 D	CMOS-Z-IC	±5V DC/DC Converter	FLP	Inr	-	pdf pinout
AFL 12005 S	CMOS/LIN-IC	5V, Hybrid-High Reliability DC/DC Conv.	FLP	Inr	-	pdf pinout
AFL 12007R5 S	CMOS/LIN-IC	7.5V, Hybrid-High Reliability DC/DC Conv.	FLP	Inr	-	pdf pinout
AFL 12008 S	CMOS/LIN-IC	8V, Hybrid-High Reliability DC/DC Conv.	FLP	Inr	-	pdf pinout
AFL 12009 S	CMOS/LIN-IC	9V, Hybrid-High Reliability DC/DC Conv.	FLP	Inr	-	pdf pinout
AFL 12012 D	CMOS-Z-IC	±12V DC/DC Converter	FLP	Inr	-	pdf pinout
AFL 12012 S	CMOS/LIN-IC	12V, Hybrid-High Reliability DC/DC Conv.	FLP	Inr	-	pdf pinout
AFL 12015 D	CMOS-Z-IC	±15V DC/DC Converter	FLP	Inr	-	pdf pinout
AFL 12015 S	CMOS/LIN-IC	15V, Hybrid-High Reliability DC/DC Conv.	FLP	Inr	-	pdf pinout
AFL 12028 S	CMOS/LIN-IC	28V, Hybrid-High Reliability DC/DC Conv.	FLP	Inr	-	pdf pinout
AFL 27003R3 S	LIN/Z-IC	+3.3V, DC/DC Converter	FLP	Inr	-	pdf pinout
AFL 27005 D	LIN/Z-IC	±5V, DC/DC Converter	FLP	Inr	-	pdf pinout
AFL 27005 S	LIN/Z-IC	+5V, DC/DC Converter	FLP	Inr	-	pdf pinout
AFL 27006 S	LIN/Z-IC	+6V, DC/DC Converter	FLP	Inr	-	pdf pinout
AFL 27009 S	LIN/Z-IC	+9V, DC/DC Converter	FLP	Inr	-	pdf pinout

Type	Device	Short Description	Fig.	Manu	Comparision Types	More at
AFL 27012 D	LIN/Z-IC	±12V, DC/DC Converter	FLP	Inr	-	pdf pinout
AFL 27012 S	LIN/Z-IC	+12V, DC/DC Converter	FLP	Inr	-	pdf pinout
AFL 27015 D	LIN/Z-IC	±15V, DC/DC Converter	FLP	Inr	-	pdf pinout
AFL 27015 S	LIN/Z-IC	+15V, DC/DC Converter	FLP	Inr	-	pdf pinout
AFL 27028 S	LIN/Z-IC	+28V, DC/DC Converter	FLP	Inr	-	pdf pinout
AFP	Si-N	=2SD1781K-P (Typ-Code Stempel marking)	35	Rhm	→2SD1781K	data
AFQ	Si-N	=2SD1781K-Q (Typ-Code Stempel marking)	35	Rhm	→2SD1781K	data
AFR	Si-N	=2SD1781K-R (Typ-Code Stempel marking)	35	Rhm	→2SD1781K	data
AFX	Si-P-Darl	=MMBTA 64 (Typ-Code Stempel marking)	35	Kec	→MMBTA 64	data

AFY...AK

Type	Device	Short Description	Fig.	Manu	Comparision Types	More at
AFY 10	Ge-P	VHF, 30V, 70mA, 350MHz	2a§	Sie	AFY 18, AFY 19	data
AFY 11	Ge-P	=AFY 10: 300MHz	2a§	Sie	AFY 18, AFY 19	data
AFY 12	Ge-P	VHF-V/M/O, 230MHz	5g	Aeg,Sgs,Sie	AF 106, AF 109R, AF 306	data
AFY 13	Ge-P	AM/FM-V/M/O, 50MHz	5g	Aeg	AF 125, AF 200, AF 201	data
AFY 14	Ge-P	HF-Tr/E, 60MHz	3a	Aeg	AFY 18, AFY 19	data
AFY 15	Ge-P	HF-Tr, 16MHz	2a	Aeg	AFY 18, AFY 19	data
AFY 16	Ge-P	UHF-V/M/O, 550MHz	5g	Aeg,Phi,Sie	AF 139, AF 239(S)	data
AFY 17	Ge-P	VHF, >250MHz	5g	Sie	AF 139, AF 239(S)	data
AFY 18	Ge-P	VHF-A, 30V, 100mA, 600MHz	2a§	Sie	-	data
AFY 19	Ge-P	FM/VHF-Tr/E, 32V, 250mA, 350MHz	2a§	Mot,Phi	-	data
AFY 20	Ge-P	HF, 20V, 0,1A, >2,5MHz	2a	Sie	AFY 18, AFY 19	data
AFY 21	Ge-P	=AFY 20: >5MHz	2a	Sie	AFY 18, AFY 19	data
AFY 22	Ge-P	=AFY 20: >5MHz	2a	Sie	AFY 18, AFY 19	data
AFY 23	Ge-P	=AFY 20: >10MHz	2a	Sie	AFY 18, AFY 19	data
AFY 24	Ge-P	=AFY 20: >10MHz	2a	Sie	AFY 18, AFY 19	data
AFY 25	Ge-P	UHF-V, 1800MHz	2d	Aeg	-	data
AFY 26	Ge-P	UHF-M/O, 1600MHz	2d	Aeg	-	data
AFY 29	Ge-P	AM/FM-ZF, 35MHz	5g	Aeg	AF 126, AF 200, AF 201	data
AFY 30	Ge-P	HF, 30V, 0,1A, >200MHz	2a§	Sie	AFY 18, AFY 19	data
AFY 31	Ge-P	HF, 30V, 0,1A, >200MHz	2a§	Sie	AFY 18, AFY 19	data
AFY 32	Ge-P	HF, 30V, 0,1A, >200MHz	2a§	Sie	AFY 18, AFY 19	data
AFY 33	Ge-P	HF/ZF, 20V, 70mA, >200MHz	2a§	Sie	AFY 18, AFY 19	data
AFY 34	Ge-P	UHF, 1500MHz	Koax	Sie	-	data
AFY 35	Ge-P	HF, 30V, 50mA, >300MHz	2a§	Sie	AFY 18, AFY 19	data
AFY 36	Ge-P	HF, 30V, 50mA, >300MHz	2a§	Sie	AFY 18, AFY 19	data
AFY 37	Ge-P	VHF/UHF-A, 32V, 20mA, 600MHz	5g	Sie	AF 239(S), AFY 40	data
AFY 38	Ge-P	VHF, ra, >190MHz	5g	Sie	AF 106, AF 109R, AF 306	data
AFY 39	Ge-P	VHF-A, 32V, 30mA, 500MHz	2k	Sie	AFY 18, AFY 37, AFY 40	data
AFY 40	Ge-P	VHF/VHF-A, 32V, 20mA, 700MHz	5g	Phi	AF 239(S), AFY 37	data
AFY 40 R	Ge-P	=AFY 40: 20V, 10mA, 600MHz	5g		AF 239(S), AFY 37	data
AFY 41	Ge-P	UHF-V/M/O, 650MHz	5g	Phi	AF 239(S), AFY 37, AFY 40	data
AFY 42(R)	Ge-P	UHF-V/M/O, 700MHz	5g	Sie	AF 239(S), AFY 37, AFY 40	data
AFY 60	Ge-P	HF, 4,5MHz	2a	Eiy	AF 126, AF 127, AF 200, AF 201	data
AFY 61	Ge-P	HF, 10MHz	2a	Eiy	AF 126, AF 127, AF 200, AF 201	data
AFY 66	Ge-P	HF, 18V, 0,1A, 8MHz	2a	Eiy	AFY 18, AFY 19	data
AFY 71	Ge-P	HF, 18V, 30MHz	2a	Eiy	AF 126, AF 127, AF 200, AF 201	data
AFY 75	Ge-P	HF, 18V, 35MHz	2a	Eiy	AF 126, AF 127, AF 200, AF 201	data
AFY 77	Ge-P	HF, 18V, 35MHz	2a	Eiy	AF 126, AF 127, AF 200, AF 201	data
AFZ 10	Ge-P	AM-Tr, 40V, 35MHz	1a	Aeg	AFY 18, AFY 19	data
AFZ 11	Ge-P	VHF, 140MHz	5k	Phi	AF 106, AF 109R, AF 306	data
AFZ 12	Ge-P	=AFZ 11: 180MHz	5k	Phi	AF 106, AF 109R, AF 306	data
AG	Si-P	=2SA1363-G (Typ-Code/Stempel/marking)	39	Mit	→2SA1363	data
AG	Si-P	=2SA1365-G (Typ-Code/Stempel/marking)	35	Mit	→2SA1365	data
AG	Si-P	=2SA1729 (Typ-Code/Stempel/marking)	39	Say	→2SA1729	data
AG	Si-P	=BCP 52-10 (Typ-Code/Stempel/marking)	48	Phi	→BCP 52	data
AG	Si-P	=BCX 52-10 (Typ-Code/Stempel/marking)	39	Sie,Val,Mot	→BCX 52	data
AG	Si-N	=BCX 70G (Typ-Code/Stempel/marking)	35	Mot,Phi,Sie Aeg,Fer,++	→BCX 70G	data
AG	Si-P	=CPH 3107 (Typ-Code/Stempel/marking)	35	Say	→CPH 3107	data
AG	Z-Di	=P4 SMA-7.5A(Typ-Code/Stempel/marking)	71(5x2,5)	Fag	→P4 SMA-7.5A	data
AG	C-Di	=ZMV 835A (Typ-Code/Stempel/marking)	71(1,7mm)	Ztx	→ZMV 835	data
AG	Z-Di	=SMBJ 6.0CA (Typ-Code/Stempel/marking)	71(5x3,5)	Mop	→SMBJ ...	data
AG 01	Si-Di	GI/S, 400V, 0,7A, <100ns	31a	Sak	BYV 26B...E, BYV 36B...D, BYX 92/400	data
AG 01 A	Si-Di	GI/S, 600V, 0,5A, <100ns	31a	Sak	BYV 26C...E, BYV 36C...D	data
AG 01 Y	Si-Di	GI/S, 70V, 1A, <100ns	31a	Sak	BYV 26B...D, BYV 36A...D, BYX 92/100,++	data
AG 01 Z	Si-Di	GI/S, 200V, 0,7A, <100ns	31a	Sak	BYV 26B...E, BYV 36A...D, BYX 92/200,++	data
AGP	Si-P	=2SA1797-P (Typ-Code/Stempel/marking)	39	Rhm	→2SA1797	data
AGp	Si-N	=BCX 70G (Typ-Code/Stempel/marking)	35	Phi	→BCX 70G	data
AGQ	Si-P	=2SA1797-Q (Typ-Code/Stempel/marking)	39	Rhm	→2SA1797	data
AGs	Si-N	=BCX 70G (Typ-Code/Stempel/marking)	35	Sie	→BCX 70G	data
AGt	Si-N	=BCX 70G (Typ-Code/Stempel/marking)	35	Phi	→BCX 70G	data

Type	Device	Short Description	Fig.	Manu	Comparision Types	More at
AGX	Si-P-Darl	=MMBTA 63 (Typ-Code/Stempel/marking)	35	Kec	→MMBTA 63	data
AH	Si-Di	=1SS345 (Typ-Code/Stempel/marking)	35	Say	→1SS345	data
AH	Si-P	=2SA1730 (Typ-Code/Stempel/marking)	39	Say	→2SA1730	data
AH	Si-P	=2SB1000-AH (Typ-Code/Stempel/marking)	39	Hit	→2SB1000	data
AH	Si-P	=BCP 53 (Typ-Code/Stempel/marking)	48	Phi, Ons	→BCP 53	data
AH	Si-P	=BCX 53 (Typ-Code/Stempel/marking)	39	Sie,Val,Mot	→BCX 53	data
AH	Si-N	=BCX 70H (Typ-Code/Stempel/marking)	35	Mot,Phi,Sie Fer	→BCX 70H	data
AH	Z-Di	=MA 5Z240-H (Typ-Code/Stempel/marking)	71(1,7mm)	Mat	→MA 5Z240-H	data
AH	Z-Di	=P4 SMA-8.2 (Typ-Code/Stempel/marking)	71(5x2,5)	Fag	→P4 SMA-8.2	data
AH	Z-Di	=SMBJ 6.5C (Typ-Code/Stempel/marking)	71(5x3,5)	Mop	→SMBJ ...	data
AH	Si-P+R	=XN 1101 (Typ-Code/Stempel/marking)	45	Mat	→XN 1101	data
AH	Si-P+R	=XP 1101 (Typ-Code/Stempel/marking)	45(2mm)	Mat	→XP 1101	data
AH	C-Di	=ZMV 930 (Typ-Code/Stempel/marking)	71(1,7mm)	Ztx	→ZMV 930	data
AH-10	Si-P	=BCP 53-10 (Typ-Code/Stempel/marking)	48	Ons	→BCP 53	data
AH-16	Si-P	=BCP 53-16 (Typ-Code/Stempel/marking)	48	Ons	→BCP 53	data
AH 173-P	IC	Pull-up Hall Effect Latch, High Temp.	TSOP	Dii	-	pdf pinout
AH 173-W	IC	Pull-up Hall Effect Latch, High Temp.	TSOP	Dii	-	pdf pinout
AH 175-P	IC	Hall Effect Latch For High Temperature	TSOP	Dii	-	pdf pinout
AH 175-W	IC	Hall Effect Latch For High Temperature	TSOP	Dii	-	pdf pinout
AH 180-FJ	IC	Micropower Omnipolar Hall-effect Sensor SWITCH	TSOP	Dii	-	pdf pinout
AH 180-P	IC	Micropower Omnipolar Hall-effect Sensor SWITCH	TSOP	Dii	-	pdf pinout
AH 180-SN	IC	Micropower Omnipolar Hall-effect Sensor SWITCH	TSOP	Dii	-	pdf pinout
AH 180-W	IC	Micropower Omnipolar Hall-effect Sensor SWITCH	TSOP	Dii	-	pdf pinout
AH 182-P	IC	low Power Hall Effect Switch	TSOP	Dii	-	pdf pinout
AH 182-W	IC	low Power Hall Effect Switch	TSOP	Dii	-	pdf pinout
AH 183-P	IC	low Power Hall Effect Switch	TSOP	Dii	-	pdf pinout
AH 183-W	IC	low Power Hall Effect Switch	TSOP	Dii	-	pdf pinout
AH 276 Q-P	IC	Complementary Output Hall Effect Latch	TSOP	Dii	-	pdf pinout
AH 277 A	IC	Complementary Output Hall Effect Latch	QFP	Bcd	-	pdf pinout
AH 284-P	IC	Hall-effect Smart Fan Motor Controller	TSOP	Dii	-	pdf pinout
AH 284-Y	IC	Hall-effect Smart Fan Motor Controller	TSOP	Dii	-	pdf pinout
AH 285-Y	IC	Hall-effect Smart Fan Motor Controller	TSOP	Dii	-	pdf pinout
AH 286-Y	IC	Hall-effect Smart Fan Motor Controller	TSOP	Dii	-	pdf pinout
AH 287-P	IC	High Voltage Hall-effect Smart Fan Motor CONTR.	TSOP	Dii	-	pdf pinout
AH 287-Y	IC	High Voltage Hall-effect Smart Fan Motor CONTR.	TSOP	Dii	-	pdf pinout
AH 288-Y	IC	High Voltage Hall-effect Smart Fan Motor Controller	TSOP	Dii	-	pdf pinout
AH 289-Y	IC	High Voltage Hall-effect Smart Fan Motor CONTR.	TSOP	Dii	-	pdf pinout
AH 291-P	IC	low Voltage Hall-effect Smart Fan Motor CONTR.	TSOP	Dii	-	pdf pinout
AH 291-W	IC	low Voltage Hall-effect Smart Fan Motor CONTR.	TSOP	Dii	-	pdf pinout
AH 291-Y	IC	low Voltage Hall-effect Smart Fan Motor CONTR.	TSOP	Dii	-	pdf pinout
AH 292-Y	IC	low Voltage Hall-effect Smart Fan Motor CONTR.	TSOP	Dii	-	pdf pinout
AH 293-W	IC	low Voltage Hall-effect Smart Fan Motor CONTR.	TSOP	Dii	-	pdf pinout
AH 293-Y	IC	low Voltage Hall-effect Smart Fan Motor CONTR.	TSOP	Dii	-	pdf pinout
AH 337-P	IC	Single Phase Hall Effect Switch	TSOP	Dii	-	pdf pinout
AH 337-W	IC	Single Phase Hall Effect Switch	TSOP	Dii	-	pdf pinout
AH 342	IC	Active High/low Complementary Output Hall-effect Latch	MSIP	Ana	-	pdf pinout
AH 373-P	IC	Internal Pull-up Hall Effect Latch	TSOP	Dii	-	pdf pinout
AH 373-W	IC	Internal Pull-up Hall Effect Latch	TSOP	Dii	-	pdf pinout
AH 375-P	IC	Single Phase Hall Effect Latch	TSOP	Dii	-	pdf pinout
AH 375-W	IC	Single Phase Hall Effect Latch	TSOP	Dii	-	pdf pinout
AH 1801-FJ	IC	Micropower, Ultra-sensitive Hall Effect SWITCH	TSOP	Dii	-	pdf pinout
AH 1801-SN	IC	Micropower, Ultra-sensitive Hall Effect SWITCH	TSOP	Dii	-	pdf pinout
AH 1801-W	IC	Micropower, Ultra-sensitive Hall Effect SWITCH	TSOP	Dii	-	pdf pinout
AH 1802-SN	IC	Ultra-sensitive Omnipolar Hall-effect Sensor Switch	TSOP	Dii	-	pdf pinout
AH 1802-W	IC	Ultra-sensitive Omnipolar Hall-effect Sensor Switch	TSOP	Dii	-	pdf pinout
AH 1883-Z	IC	Micropower, Hall Effect Switch	TSOP	Dii	-	pdf pinout
AH 1884-Z	IC	Micropower, Hall Effect Switch	TSOP	Dii	-	pdf pinout
AH 1885-Z	IC	Micropower, Hall Effect Switch	45	Dii	-	pdf pinout
AH 1886-Z	IC	Micropower, Hall Effect Switch	45	Dii	-	pdf pinout
AH 1887-Z	IC	Micropower, Hall Effect Switch	45	Dii	-	pdf pinout
AH 5009 CM	LIN-IC	SMD, 4x Analog Sw. 30V 30mA, 15V TTL	14-MDIP	Nsc	-	pdf pinout
AH 5009 CN	LIN-IC	4x Analog Sw. 30V 30mA, 150ns, 15V TTL	14-DIP	Nsc	-	pdf pinout
AH 5010 CM	LIN-IC	SMD, 4x Analog Sw., 30V 30mA, 5V TTL	14-MDIP	Nsc	-	pdf pinout
AH 5010 CN	LIN-IC	4x Analog Sw., 30V 30mA, 150ns, 5V TTL	14-DIP	Nsc	-	pdf pinout

Type	Device	Short Description	Fig.	Manu	Comparision Types	More at
AH 5011 CM	LIN-IC	SMD, 4x Analog Sw. 30V 30mA, 15V TTL	16-MDIP	Nsc	-	pdf pinout
AH 5011 CN	LIN-IC	4x Analog Sw. 30V 30mA, 150ns, 15V TTL	16-DIP	Nsc	-	pdf pinout
AH 5012 CM	LIN-IC	SMD, 4x Analog Sw. 30V 30mA, 5V TTL	16-MDIP	Nsc	-	pdf pinout
AH 5012 CN	LIN-IC	4x Analog Sw. 30V 30mA, 150ns, 5V TTL	16-DIP	Nsc	-	pdf pinout
AH 1803-SNG-7	IC	Micropower, Hall Effect Switch	TSOP	Dii	-	pdf pinout
AH 1803-WG-7	IC	Micropower, Hall Effect Switch	TSOP	Dii	-	pdf pinout
AHE 2805 S	LIN/Z-IC	+5V, DC/DC Converter	DIP	Inr	-	pdf pinout
AHE 2812 S	LIN/Z-IC	+12V, DC/DC Converter	DIP	Inr	-	pdf pinout
AHE 2815 S	LIN/Z-IC	+15V, DC/DC Converter	DIP	Inr	-	pdf pinout
AHN	Si-P	=2SA1759-N (Typ-Code/Stempel/marking)	39	Rhm	→2SA1759	data
AHP	Si-P	=2SA1759-P (Typ-Code/Stempel/marking)	39	Rhm	→2SA1759	data
AHP	Si-P	=2SB1197K-P (Typ-Code/Stempel/marking)	35	Rhm	→2SB1197K	data
AHp	Si-N	=BCX 70H (Typ-Code/Stempel/marking)	35	Phi	→BCX 70H	data
AHP 27005 D	LIN/Z-IC	±5V, DC/DC Converter	FLP	Inr	-	pdf pinout
AHP 27012 D	LIN/Z-IC	±12V, DC/DC Converter	FLP	Inr	-	pdf pinout
AHP 27015 D	LIN/Z-IC	±15V, DC/DC Converter	FLP	Inr	-	pdf pinout
AHQ	Si-P	=2SA1759-Q (Typ-Code/Stempel/marking)	39	Rhm	→2SA1759	data
AHQ	Si-P	=2SB1197K-Q (Typ-Code/Stempel/marking)	35	Rhm	→2SB1197K	data
AHR	Si-P	=2SB1197K-R (Typ-Code/Stempel/marking)	35	Rhm	→2SB1197K	data
AHs	Si-P	=BCP 53M (Typ-Code/Stempel/marking)	≈45	Sie	→BCP 53M	data
AHs	Si-N	=BCX 70H (Typ-Code/Stempel/marking)	35	Sie	→BCX 70H	data
AHt	Si-N	=BCX 70H (Typ-Code/Stempel/marking)	35	Phi	→BCX 70H	data
AHV 2805 D	LIN/Z-IC	±5V, DC/DC Converter	DIC	Inr	-	pdf pinout
AHV 2805 S	LIN/Z-IC	5V, DC/DC Converter	DIC	Inr	-	pdf pinout
AHV 2812 D	LIN/Z-IC	±12V, DC/DC Converter	DIC	Inr	-	pdf pinout
AHV 2812 S	LIN/Z-IC	12V, DC/DC Converter	DIC	Inr	-	pdf pinout
AHV 2812 T	LIN/Z-IC	+5V, ±12V, DC/DC Converter	DIC	Inr	-	pdf pinout
AHV 2815 D	LIN/Z-IC	±15V, DC/DC Converter	DIC	Inr	-	pdf pinout
AHV 2815 S	LIN/Z-IC	15V, DC/DC Converter	DIC	Inr	-	pdf pinout
AHV 2815 T	LIN/Z-IC	+5V, ±15V, DC/DC Converter	DIC	Inr	-	pdf pinout
AI	Si-P	=2SA1882 (Typ-Code/Stempel/marking)	39	Say	→2SA1882	data
AI	Z-Di	=MA 5Z270-L (Typ-Code/Stempel/marking)	71(1,7mm)	Mat	→MA 5Z270-L	data
AI	Si-N+R	=XN 1201 (Typ-Code/Stempel/marking)	45	Mat	→XN 1201	data
AI	Si-N+R	=XP 1201 (Typ-Code/Stempel/marking)	45(2mm)	Mat	→XP 1201	data
AI-	GaAs-N-FET-d	=3SK171 (Typ-Code/Stempel/marking)	44	Hit	→3SK171	data
AIC1520	LOAD+	Vin:7V II:250mA Ron:120mOhm	4S,3N	Aic		data pdf pinout
AIC1521	LOAD+	Vin:7V II:250mA Ron:120mOhm	8S	Aic		data pdf pinout
AIC1523	LOAD+	Vin:7V II:250mA Ron:120mOhm	8S	Aic		data pdf pinout
AIC1525	2xLOAD+	Vin:7V II:250mA Ron:70mOhm	8DS	Aic		data pdf pinout
AIC1526	2xLOAD+	Vin:7V II:250mA Ron:110mOhm	8DS	Aic		data pdf pinout
AIC1528	2xLOAD+	Vin:7V II:250mA Ron:110mOhm	8DS	Aic		data pdf pinout
AIC1529	4xLOAD+	Vin:7V II:250mA Ron:110mOhm	16DS	Aic		data pdf pinout
AIS	Si-P	=2SA1882-S (Typ-Code/Stempel/marking)	39	Say	→2SA1882	data
AIT	Si-P	=2SA1882-T (Typ-Code/Stempel/marking)	39	Say	→2SA1882	data
AIU	Si-P	=2SA1882-U (Typ-Code/Stempel/marking)	39	Say	→2SA1882	data
AIZ	Si-N	=KSC 3125 (Typ-Code/Stempel/marking)	35	Sam	→KSC 3125	data
AJ	Si-P	=2SA1724 (Typ-Code/Stempel/marking)	39	Say	→2SA1724	data
AJ	Si-P	=2SB1000-AJ (Typ-Code/Stempel/marking)	39	Hit	→2SB1000	data
AJ	Si-P	=BCX 53-6 (Typ-Code/Stempel/marking)	39	Sie,Val,Mot	→BCX 53	data
AJ	Si-N	=BCX 70J (Typ-Code/Stempel/marking)	35	Mot,Phi,Sie Fer	→BCX 70J	data
AJ	Si-P	=CPH 3109 (Typ-Code/Stempel/marking)	35	Say	→CPH 3109	data
AJ	Z-Di	=SMAJ 15CA (Typ-Code/Stempel/marking)	71(5x2,5)	Tho	→SMAJ ...	data
AJ	C-Di	=ZMV 931 (Typ-Code/Stempel/marking)	71(1,7mm)	Ztx	→ZMV 931	data
AJ 2	N-FET	=2SK67-AJ2 (Typ-Code/Stempel/marking)	≈35	Nec	→2SK67	data
AJ 3	N-FET	=2SK67-AJ3 (Typ-Code/Stempel/marking)	≈35	Nec	→2SK67	data
AJ 4	N-FET	=2SK67-AJ4 (Typ-Code/Stempel/marking)	≈35	Nec	→2SK67	data
AJ 5	N-FET	=2SK443-AJ5 (Typ-Code/Stempel/marking)	35	Say	→2SK443	data
AJ 5	N-FET	=2SK67-AJ5 (Typ-Code/Stempel/marking)	≈35	Nec	→2SK67	data
AJ 6	N-FET	=2SK443-AJ6 (Typ-Code/Stempel/marking)	35	Say	→2SK443	data
AJ 6	N-FET	=2SK67-AJ6 (Typ-Code/Stempel/marking)	≈35	Nec	→2SK67	data
AJ 7	N-FET	=2SK443-AJ7 (Typ-Code/Stempel/marking)	35	Say	→2SK443	data
AJ 7	N-FET	=2SK67-AJ7 (Typ-Code/Stempel/marking)	≈35	Nec	→2SK67	data
AJN	Si-P	=2SA1812-N (Typ-Code/Stempel/marking)	39	Rhm	→2SA1812	data
AJP	Si-P	=2SA1812-P (Typ-Code/Stempel/marking)	39	Rhm	→2SA1812	data
AJP	Si-N	=2SD1782K-P (Typ-Code/Stempel/marking)	35	Rhm	→2SD1782K	data
AJp	Si-N	=BCX 70J (Typ-Code/Stempel/marking)	35	Phi	→BCX 70J	data
AJQ	Si-P	=2SA1812-Q (Typ-Code/Stempel/marking)	39	Rhm	→2SA1812	data
AJQ	Si-N	=2SD1782K-Q (Typ-Code/Stempel/marking)	35	Rhm	→2SD1782K	data
AJR	Si-N	=2SD1782K-R (Typ-Code/Stempel/marking)	35	Rhm	→2SD1782K	data
AJs	Si-N	=BCX 70J (Typ-Code/Stempel/marking)	35	Sie	→BCX 70J	data
AJt	Si-N	=BCX 70J (Typ-Code/Stempel/marking)	35	Phi	→BCX 70J	data
AK	Si-P	=2SA1738 (Typ-Code/Stempel/marking)	35	Mat	→2SA1738	data
AK	Si-P	=2SA1740 (Typ-Code/Stempel/marking)	39	Say	→2SA1740	data
AK	Si-P	=2SA1806 (Typ-Code/Stempel/marking)	35(1,6mm)	Mat	→2SA1806	data
AK	Si-P	=2SB1000A-AK(Typ-Code/Stempel/marking)	39	Hit	→2SB1000A	data
AK	N-FET	=2SK2539 (Typ-Code/Stempel/marking)	35	Say	→2SK2539	data
AK	Si-P	=BCP 53-10 (Typ-Code/Stempel/marking)	48	Phi	→BCP 53	data
AK	Si-F	=BCX 53-10 (Typ-Code/Stempel/marking)	39	Sie,Val,Mot	→BCX 53	data
AK	Si-N	=BCX 70K (Typ-Code/Stempel/marking)	35	Sie,Phi,Fer	→BCX 70K	data
AK	Si-P	=CPH 3110 (Typ-Code/Stempel/marking)	35	Say	→CPH 3110	data
AK	Z-Di	=MA 5Z270-M (Typ-Code/Stempel/marking)	71(1,7mm)	Mat	→MA 5Z270-M	data
AK	Z-Di	=P4 SMA-8.2A (Typ-Code/Stempel/marking)	71(5x2,5)	Fag	→P4 SMA-8.2A	data
AK	Z-Di	=SMBJ 6.5CA (Typ-Code/Stempel/marking)	71(5x3,5)	Mop	→SMBJ ...	data
AK	C-Di	=ZMV 932 (Typ-Code/Stempel/marking)	71(1,7mm)	Ztx	→ZMV 932	data

Type	Device	Short Description	Fig.	Manu	Comparision Types	More at
AK 03	Si-Di	Schottky, S 20V, 1A, 100ns	31a	Sak	-	data
AK 04	Si-Di	Schottky, S, 40V, 1A, 100ns	31a	Sak	-	data
AK 06	Si-Di	Schottky, S, 60V, 0,7A, 100ns	31a	Sak	-	data
AK 6	N-FET	=2SK2539-6 (Typ-Code/Stempel/marking)	35	Say	→2SK2539	data
AK 7	N-FET	=2SK2539-7 (Typ-Code/Stempel/marking)	35	Say	→2SK2539	data
AK 8	N-FET	=2SK2539-8 (Typ-Code/Stempel/marking)	35	Say	→2SK2539	data
AK 09	Si-Di	Schottky, S, 90V, 0,7A, 100ns	31a	Sak	-	data
AK4117	IC	SMD, Low Power 192kHz Digital Audio Receiver, 24Bit, serial interface	SSMDIP	Akm	-	pdf pinout
AK 4324VF	A/D-D/A-IC	SMD, 96kHz Sampling 24Bit Delta-Sigma DAC	SSMDIP	Akm	-	pdf pinout
AK 4527BVQ	IC	SMD, high performance multi-channel audio codec, 3 wire serial and I2C bus	QFP	Akm	-	pdf pinout
AK 5365VQ	IC	SMD, 24Bit 96kHz ADC with 5 channel stereo input selector, PGA, ALC, I2C	QFP	Akm	-	pdf pinout
AK 8970N	IC	Electronic Compass, terrestrial magnetism detection, 2.5…3.6V	QFN	Akm	-	pdf pinout
AKBP(C) …	Si-Br	Brückengl./bridge rectifier, contr.av.		Gie	-	data
AKP	Si-P	=2SB1198K-P (Typ-Code/Stempel/marking)	35	Rhm	→2SB1198K	data
AKp	Si-N	=BCX 70K (Typ-Code/Stempel/marking)	35	Phi	→BCX 70K	data
AKQ	Si-P	=2SA1738-Q (Typ-Code/Stempel/marking)	35	Mat	→2SA1738	data
AKQ	Si-P	=2SA1806-Q (Typ-Code/Stempel/marking)	35(1,6mm)	Mat	→2SA1806	data
AKQ	Si-P	=2SB1198K-Q (Typ-Code/Stempel/marking)	35	Rhm	→2SB1198K	data
AKR	Si-P	=2SA1738-R (Typ-Code/Stempel/marking)	35	Mat	→2SA1738	data
AKR	Si-P	=2SA1806-R (Typ-Code/Stempel/marking)	35(1,6mm)	Mat	→2SA1806	data
AKR	Si-P	=2SB1198K-R (Typ-Code/Stempel/marking)	35	Rhm	→2SB1198K	data
AKs	Si-N	=BCX 70K (Typ-Code/Stempel/marking)	35	Sie	→BCX 70K	data
AKt	Si-N	=BCX 70K (Typ-Code/Stempel/marking)	35	Phi	→BCX 70K	data
AL						
AL	Si-P	=2SA1338 (Typ-Code/Stempel/marking)	35	Say	→2SA1338	data
AL	Si-P	=2SA1748 (Typ-Code/Stempel/marking)	35(2mm)	Mat	→2SA1748	data
AL	Si-P	=2SA1766 (Typ-Code/Stempel/marking)	39	Say	→2SA1766	data
AL	Si-P	=2SA1791 (Typ-Code/Stempel/marking)	35(1,6mm)	Mat	→2SA1791	data
AL	Si-P	=2SB1000A-AL(Typ-Code/Stempel/marking)	39	Hit	→2SB1000A	data
AL	Si-P	=BCP 53-16 (Typ-Code/Stempel/marking)	48	Phi	→BCP 53	data
AL	Si-P	=BCX 53-16 (Typ-Code/Stempel/marking)	39	Phi	→BCX 53	data
AL	Si-N	=BCX 70L (Typ-Code/Stempel/marking)	35	Sie	→BCX 70L	data
AL	Z-Di	=MA 5Z270-H (Typ-Code/Stempel/marking)	71(1,7mm)	Mat	→MA 5Z270-H	data
AL	Z-Di	=P4 SMA-9.1 (Typ-Code/Stempel/marking)	71(5x2,5)	Fag	→P4 SMA-9.1	data
AL	Z-Di	=SMBJ 7.0C (Typ-Code/Stempel/marking)	71(5x3,5)	Mop	→SMBJ …	data
AL	C-Di	=ZMV 933 (Typ-Code/Stempel/marking)	71(1,7mm)	Ztx	→ZMV 933	data
AL 01 Z	Si-Di	Gl/S, 200V, 1A, 50ns	31a	Sak	BYD 72D…G, EGP 10D, FE 1D	data
AL 100	Ge-P	NF-L, 130V, 10A, 30W	23a§	Sgs,Gpd	AUY 38, 2N2528	data
AL 101	Ge-P	=AL 100: 100V	23a§	Sgs	AUY 38, 2N2527, 2N2290, 2N2293	data
AL 102	Ge-P	NF-L, 130V, 6A, 30W	23a§	Sgs,Gpd	AL 100, AUY 38, 2N2528	data
AL102	LED	rd+gr ø4mm l:0.45mcd Vf:2.25V	TO18	Gus		data pinout
AL 103	Ge-P	=AL 102: 100V	23a§	Sgs,Gpd	AL 100…102, AUY 38, 2N2527, 2N2290/93	data
AL 112	Ge-P	NF-L, 130V, 6A, 10W	22a§	Sgs	-	data
AL 113	Ge-P	=AL 112: 100V	22a§	Sgs	-	data
AL336	LED	rd+yl+gr ø5mm l:10mcd Vf:2.8V	T1¾	Gus		data pinout
ALC 1001-131	CMOS-IC	SMD, TFT-LCD Prozessor, 5V, 30MHz	64-MP	Say	-	
ALC 1009-141	CMOS-IC	SMD, LCD Proz., Full Color, 5V, 30MHz	100-MP	Say	-	
ALC 1010-141	CMOS-IC	SMD, LCD Proz., Full Color, 3,3V, 33MH	100-MP	Say	-	
ALD1701	OP	Vs:±6V Vu:100V/mV Vo:±2.48V f:0.7Mc	8DSd	Ald		data pinout
ALD1701A	OP	Vs:±6V Vu:100V/mV Vo:±2.48V f:0.7Mc	8DS	Ald		data pinout
ALD1701B	OP	Vs:±6V Vu:100V/mV Vo:±2.48V Vi0:2mV	8D	Ald		data pinout
ALD1702	HI-PREC	Vs:±6V Vu:85V/mV Vo:±2.44V Vi0:4.5mV	8DS	Ald		data pinout
ALD1702A	HI-PREC	Vs:±6V Vu:85V/mV Vo:±2.44V Vi0:0.9mV	8DS	Ald		data pinout
ALD1702B	HI-PREC	Vs:±6V Vu:85V/mV Vo:±2.44V Vi0:2mV	8DS	Ald		data pinout
ALD1703	HI-PREC	Vs:±6V Vu:85V/mV Vo:±2.4V Vi0:10mV	8DSd	Ald		data pinout
ALD1704	FET	Vs:±6V Vu:150V/mV Vo:±4.95V f:2.1Mc	8DS	Ald		data pdf pinout
ALD1704A	FET	Vs:±6V Vu:150V/mV Vo:±4.95V f:2.1Mc	8DS	Ald		data pdf pinout
ALD1704B	FET	Vs:±6V Vu:150V/mV Vo:±4.95V Vi0:2mV	8DS	Ald		data pdf pinout
ALD1704Z	FET	Vs:±6V Vu:150V/mV Vo:±4.95V Vi0:10mV	d	Ald		data pinout
ALD1706	µ-POWER	Vs:±6V Vu:100V/mV Vo:±2.4V Vi0:4.5mV	8DSd	Ald		data pdf pinout
ALD1706A	µ-POWER	Vs:±6V Vu:100V/mV Vo:±2.4V Vi0:0.9mV	8DS	Ald		data pdf pinout
ALD1706B	µ-POWER	Vs:±6V Vu:100V/mV Vo:±2.4V Vi0:2mV	8D	Ald		data pdf pinout
ALD1712	HI-PREC	Vs:±6V Vu:85V/mV Vo:±2.44V Vi0:0.5mV	8DSd	Ald		data pdf pinout
ALD1712A	HI-PREC	Vs:±6V Vu:85V/mV Vo:±2.44V f:1.5Mc	8DS	Ald		data pdf pinout
ALD1712B	HI-PREC	Vs:±6V Vu:85V/mV Vo:±2.44V f:1.5Mc	8DS	Ald		data pdf pinout
ALD2301	2xCOMP	Vs:12V Vu:150V/mV Vi0:10mV	8DAS	Ald		data pdf pinout
ALD2301A	2xCOMP	Vs:12V Vu:150V/mV Vi0:2mV	8DA	Ald		data pdf pinout
ALD2301B	2xCOMP	Vs:12V Vu:150V/mV Vi0:5mV	8DAS	Ald		data pdf pinout
ALD2301C	2xCOMP	Vs:12V Vu:150V/mV Vi0:20mV	8DA	Ald		data pdf pinout
ALD2701	2xµ-POWER	Vs:±6V Vu:80V/mV Vo:±2.48V Vi0:10mV	8DS,14Sd	Ald		data pdf pinout
ALD2701A	2xµ-POWER	Vs:±6V Vu:100V/mV Vo:±2.48V Vi0:2mV	8DS	Ald		data pdf pinout
ALD2701B	2xµ-POWER	Vs:±6V Vu:100V/mV Vo:±2.48V Vi0:5mV	8D	Ald		data pdf pinout
ALD2702	2xHI-PREC	Vs:±6.6V Vu:50V/mV Vo:±2.44V Vi0:5mV	8DS	Ald		data pinout
ALD2702A	2xHI-PREC	Vs:±6.6V Vu:50V/mV Vo:±2.44V Vi0:1mV	8DS	Ald		data pinout

Type	Device	Short Description	Fig.	Manu	Comparision Types	More at
ALD2702B	2xHI-PREC	Vs:±6.6V Vu:50V/mV Vo:±2.44V Vi0:2mV	8DS	Ald		data pinout
ALD2704	2xFET	Vs:±6.6V Vu:50V/mV Vo:±4.95V Vi0:5mV	8DS	Ald		data pinout
ALD2704A	2xFET	Vs:±6.6V Vu:50V/mV Vo:±4.95V Vi0:1mV	8DS	Ald		data pinout
ALD2704B	2xFET	Vs:±6.6V Vu:50V/mV Vo:±4.95V Vi0:5mV	8DS	Ald		data pinout
ALD2706	2xµ-POWER	Vs:±6V Vu:80V/mV Vo:±2.4V Vi0:10mV	8DS,14Sd	Ald		data pinout
ALD2706A	2xµ-POWER	Vs:±6V Vu:100V/mV Vo:±2.4V Vi0:2mV	8DS	Ald		data pinout
ALD2706B	2xµ-POWER	Vs:±6V Vu:100V/mV Vo:±2.4V Vi0:5mV	8D	Ald		data pinout
ALD2711	2xHI-PREC	Vs:±6.3V Vu:100V/mV Vo:±2.48V	8DSd	Ald		data pdf pinout
ALD2711A	2xHI-PREC	Vs:±6.3V Vu:100V/mV Vo:±2.48V	8DS	Ald		data pdf pinout
ALD2711B	2xHI-PREC	Vs:±6.3V Vu:100V/mV Vo:±2.48V	8DS	Ald		data pdf pinout
ALD4201	SPST	Vs:3...12V Ron:Ohm Ton:150ns at 5/0V	16DS	Ald		data pdf pinout
ALD4202M	SPST	Vs:3...12V Ron:Ohm Ton:150ns at 5/0V	16DS	Ald		data pdf pinout
ALD4211	SPST	Vs:3...12V Ron:Ohm Ton:85ns at 5/0V	16DS	Ald		data pdf pinout
ALD4212	SPST	Vs:3...12V Ron:Ohm Ton:85ns at 5/0V	16DS	Ald		data pdf pinout
ALD4213	SPST	Vs:3...12V Ron:Ohm Ton:85ns at 5/0V	16DS	Ald		data pdf pinout
ALD4302	4xCOMP	Vs:±6V Vu:100V/mV Vi0:15mV	14DSd	Ald		data pdf pinout
ALD4302A	4xCOMP	Vs:±6V Vu:100V/mV Vi0:5mV	14DS	Ald		data pdf pinout
ALD4701	4xµ-POWER	Vs:±6V Vu:80V/mV Vo:±2.48V Vi0:10mV	14D,24S	Ald		data pdf pinout
ALD4701A	4xµ-POWER	Vs:±6V Vu:100V/mV Vo:±2.48V Vi0:2mV	14D,24S	Ald		data pdf pinout
ALD4701B	4xµ-POWER	Vs:±6V Vu:100V/mV Vo:±2.48V Vi0:5mV	14D,24S	Ald		data pdf pinout
ALD4706	4xµ-POWER	Vs:±6V Vu:50V/mV Vo:±2.40V Vi0:10mV	14D,24S	Ald		data pinout
ALD4706A	4xµ-POWER	Vs:±6V Vu:60V/mV Vo:±2.40V Vi0:2mV	14D,24S	Ald		data pinout
ALD4706B	4xµ-POWER	Vs:±6V Vu:60V/mV Vo:±2.40V Vi0:5mV	14D,24S	Ald		data pinout
ALG	Si-N	=KTC3875-G (Typ-Code/Stempel/marking)	35	Kec	→KTC 3875	data
ALL	Si-N	=KTC3875-L (Typ-Code/Stempel/marking)	35	Kec	→KTC 3875	data
ALM	Si-N	=2SC3802K-M (Typ-Code/Stempel/marking)	35	Rhm	→2SC3802K	data
ALN	Si-N	=2SC3802K-N (Typ-Code/Stempel/marking)	35	Rhm	→2SC3802K	data
ALO	Si-N	=KTC3875-O (Typ-Code/Stempel/marking)	35	Kec	→KTC 3875	data
ALP	Si-P	=2SA1900-P (Typ-Code/Stempel/marking)	39	Rhm	→2SA1900	data
ALP	Si-N	=2SC3802K-P (Typ-Code/Stempel/marking)	35	Rhm	→2SC3802K	data
ALQ	Si-P	=2SA1747-Q (Typ-Code/Stempel/marking)	35	Mat	→2SA1747	data
ALQ	Si-P	=2SA1748-Q (Typ-Code/Stempel/marking)	35(2mm)	Mat	→2SA1748	data
ALQ	Si-P	=2SA1791-Q (Typ-Code/Stempel/marking)	35(1,6mm)	Mat	→2SA1791	data
ALQ	Si-P	=2SA1900-Q (Typ-Code/Stempel/marking)	39	Rhm	→2SA1900	data
ALQ	Si-N	=2SC3802K-Q (Typ-Code/Stempel/marking)	35	Mat	→2SC3802K	data
ALR	Si-P	=2SA1747-R (Typ-Code/Stempel/marking)	35	Mat	→2SA1747	data
ALR	Si-P	=2SA1748-R (Typ-Code/Stempel/marking)	35(2mm)	Mat	→2SA1748	data
ALR	Si-P	=2SA1791-R (Typ-Code/Stempel/marking)	35(1,6mm)	Mat	→2SA1791	data
ALR	Si-P	=2SA1900-R (Typ-Code/Stempel/marking)	39	Rhm	→2SA1900	data
ALs	Si-N	=BFP 405 (Typ-Code/Stempel/marking)	44(2mm)	Sie	→BFP 405	data
ALY	Si-N	=KTC3875-Y (Typ-Code/Stempel/marking)	35	Kec	→KTC 3875	data
ALZ 10	Ge-P	NF/HF-E, 50V, 250mA, 40MHz		Aeg	(AC 128K, AC 153K)	data

AM

Type	Device	Short Description	Fig.	Manu	Comparision Types	More at
AM	Si-P	=2SA1896 (Typ-Code/Stempel/marking)	39	Say	→2SA1896	data
AM	Si-N	=2SC4562 (Typ-Code/Stempel/marking)	35(2mm)	Mat	→2SC4562	data
AM	Si-N	=2SC4656 (Typ-Code/Stempel/marking)	35(1,6mm)	Mat	→2SC4656	data
AM	Si-P	=BCP 52-16 (Typ-Code/Stempel/marking)	48	Phi	→BCP 52	data
AM	Si-P	=BCX 52-16 (Typ-Code/Stempel/marking)	39	Phi	→BCX 52	data
AM	Si-N	=BSS 64 (Typ-Code/Stempel/marking)	35	Mot, Sie	→BSS 64	data
AM	Si-P	=CPH 3112 (Typ-Code/Stempel/marking)	35	Say	→CPH 3112	data
AM	Z-Di	=MA 5Z300-L (Typ-Code/Stempel/marking)	71(1,7mm)	Mat	→MA 5Z300-L	data
AM	Si-N	=MMBT 3904T (Typ-Code/Stempel/marking)	35(1,6mm)	Ons	→MMBT 3904T	
AM	Si-N	=MMBT 3904W (Typ-Code/Stempel/marking)	35(2mm)	Ons	→MMBT 3904W	
AM	Z-Di	=P4 SMA-9.1A(Typ-Code/Stempel/marking)	71(5x2,5)	Fag	→P4 SMA-9.1A	
AM	Z-Di	=SMBJ 7.0CA (Typ-Code/Stempel/marking)	71(5x3,5)	Mop	→SMBJ ...	data
AM	C-Di	=ZMV 933A (Typ-Code/Stempel/marking)	71(1,7mm)	Ztx	→ZMV 933A	data
AMs	Si-N	=BSS 64 (Typ-Code/Stempel/marking)	35	Sie	→BSS 64	data
AM 01(A,B,Z)	Si-Di	Gl, Uni, 200...800V, 1A AM 01=400V, A=600V, B=800V, Z=200V	31a	Sak	BY 126...127, BY 133...134, 1N4003...07,++	data
AM 25 S10N	TTL-IC	Schottky-Interface, 4Bit Shifter	16-DIP	Amd	DS 2510DC	
AM 26 LS30CD	I/O-IC	=AM 26LS30CN: SMD	16-MDIP	Phi		
AM 26 LS30CN	I/O-IC	2x Line Driver, RS422, 0...+70°	16-DIP	Phi	-	
AM 26 LS30ID	I/O-IC	=AM 26LS30CN: SMD, -40...+85°	16-MDIP	Phi	-	
AM 26 LS30IN	I/O-IC	=AM 26LS30CN: -40...+85°	16-DIP	Phi	-	
AM 26 LS30MF,MN	I/O-IC	=AM 26LS30CN: -55...+125°	16-DIC,DIP	Phi	-	
AM 26 LS31CJ,DC	I/O-IC	4x Line Driver, RS422, 0...+70°	16-DIC	Mot,Phi,Sgs	SN 75172J, µA26LS31...	pinout
AM 26 LS31CN,PC	I/O-IC	=AM 26LS31CJ: Fig.→	16-DIP	Mot,Phi,Sgs	DL 2631D, SN 75172N, µA26LS31..., µA96172	pinout
AM 26 LS31D1,CD	I/O-IC	=AM 26LS31CJ: SMD	16-MDIP	Phi,Sgs		pinout
AM 26 LS31ID	I/O-IC	=AM 26LS31CJ: SMD, -40...+85°	16-MDIP	Phi	-	pinout
AM 26 LS31IN	I/O-IC	=AM 26LS31CJ: -40...+85°	16-DIP	Phi	-	pinout
AM 26 LS31MJ,MF,MN	I/O-IC	=AM 26LS31CJ: -55...+125°	16-DIC,DIP	Phi	-	pinout
AM 26 LS32CJ,D	I/O-IC	4x Line Receiver, RS422,423, 0...+70°	16-DIC	Mot,Sgs,Tos	SN 75173J, µA26LS32...	pinout
AM 26 LS32CN,P	I/O-IC	=AM 26LS32CJ: Fig.→	16-DIP	Mot,Phi,Sgs	DL 2632D, SN 75173N, µA26LS32...	pinout
AM 26 LS32D1,CD	I/O-IC	=AM 26LS32CJ: SMD	16-MDIP	Phi,Sgs		pinout
AM 26 LS32ID	I/O-IC	=AM 26LS32CJ: SMD, -40...+85°	16-MDIP	Phi	-	pinout
AM 26 LS32IN	I/O-IC	=AM 26LS32CJ: -40...+85°	16-DIP	Phi	-	pinout
AM 26 LS32MJ,MF,MN	I/O-IC	=AM 26LS32CJ: -55...+125°	16-DIC,DIP	Sgs		pinout
AM 26 LS33CJ	I/O-IC	4x Line Receiver, RS422,423, 0...+70°	16-DIC	Sgs		pinout
AM 26 LS33CN	I/O-IC	=AM 26LS33CJ: Fig.→	16-DIP	Phi,Sgs		pinout
AM 26 LS33D1,CD	I/O-IC	=AM 26LS33CJ: SMD	16-MDIP	Phi,Sgs		pinout
AM 26 LS33ID	I/O-IC	=AM 26LS33CJ: SMD, -40...+85°	16-MDIP	Phi	-	pinout
AM 26 LS33IN	I/O-IC	=AM 26LS33CJ: -40...+85°	16-DIP	Phi	-	pinout

Type	Device	Short Description	Fig.	Manu	Comparision Types	More at
AM 26 LS33MJ,MF,MN	I/O-IC	=AM 26LS33CJ: -55...+125°	16-DIC,DIP	Phi,Sgs	-	pinout
Am 26 S10N	TTL-IC	Schottky-Interface, 4x Bus-I/O	16-DIP	Amd	DS 2610DC, µA 9640	
AM 29 F400 AB...E	CMOS-sRAM-IC	4Mbit 5.0V, Sector Erase Flash Memory	TSOP	Max	-	pdf pinout
AM 29 F400 AB...F	CMOS-sRAM-IC	4Mbit 5.0V, Sector Erase Flash Memory	TSOP	Max	-	pdf pinout
AM 29 F400 AB...S	CMOS-sRAM-IC	4Mbit 5.0V, Sector Erase Flash Memory	MDIP	Max	-	pdf pinout
AM 29 F400 AT...E	CMOS-sRAM-IC	4x Mbit 5.0V, Sector Erase Flash Memory	TSOP	Max	-	pdf pinout
AM 29 F400 AT...F	CMOS-sRAM-IC	4Mbit 5.0V, Sector Erase Flash Memory	TSOP	Max	-	pdf pinout
AM 29 F400 AT...S	CMOS-sRAM-IC	4Mbit 5.0V, Sector Erase Flash Memory	MDIP	Max	-	pdf pinout
AM 97C09CN	LIN-IC	4x Analog Switch, Single Pole Single Throw switch, norm. open, Cmos version	14-DIP	Nsc	~AH5009CN,~AM9709CN	
AM 97C10CN	LIN-IC	4x Analog Switch, Single Pole Single Throw switch, norm. open, Cmos version	14-DIP	Nsc	~AH5010CN,~AM9710CN	
AM 97C11CN	LIN-IC	4x Analog Switch, Single Pole Single Throw switch, norm. open, Cmos version	16-DIP	Nsc	~AH5011CN,~AM9711CN	
AM 97C12CN	LIN-IC	4x Analog Switch, Single Pole Single Throw switch, norm. open, Cmos version	16-DIP	Nsc	~AH5012CN,~AM9712CN	
AM100	HYBRID	Vs:±15V Vu:114dB Vo:±10V f:18Mc	9Y	Dat		data pinout
Am101	UNI	Vs:±22V Vo:±14V Vi0:1mV	10F,8A	Amd		data pinout
AM101A	UNI	Vs:±22V Vu:104dB Vo:±14V Vi0:0.7mV	8A	Amd		data pinout
Am101A	UNI	Vs:±22V Vo:±14V Vi0:0.7mV	8A	Amd		data pinout
AM101B	UNI	Vs:±15V Vu:114dB Vo:±10V f:7Mc	9Y	Dat		data pinout
AM102	HI-SPEED	Vs:±15V Vu:112dB Vo:±10V f:45Mc	9Y	Dat		data pinout
AM103	HI-SPEED	Vs:±15V Vu:112dB Vo:±10V f:45Mc	9Y	Dat		data pinout
AM107	UNI	Vs:±22V Vo:±14V Vi0:0.7mV	10F,14D,8A	Amd		data pinout
Am108	LO-POWER	Vs:±20V Vi0:0.3mV	10F,14D,8A	Amd		data pinout
AM112	LO-POWER	Vs:±20V Vu:110dB Vo:±14V Vi0:0.7mV	14D,8A	Amd		data pinout
Am118	HI-SPEED	Vs:±13V Vi0:2mV f:15Mc	10F,14D,8A	Amd		data pinout
AM124	4xOP	Vs:±2.5V Vu:100dB Vo:0.75V Vi0:5mV	14DF	Amd		data pinout
AM201	UNI	Vs:±15V Vu:60dB Vo:±10V	10YF,8A	Dat		data pinout
AM201A	UNI	Vs:±22V Vu:104dB Vo:±14V Vi0:0.7mV	8A,10Y	Amd		data pinout
Am201A	UNI	Vs:±22V Vo:±14V Vi0:0.7mV	8A	Amd		data pinout
AM207	UNI	Vs:±22V Vu:104dB Vo:±14V Vi0:0.7mV	8A	Amd		data pinout
AM208	LO-POWER	Vs:±20V Vi0:0.3mV	10F,14D,8A	Amd		data pinout
AM212	LO-POWER	Vs:±20V Vu:110dB Vo:±14V Vi0:0.7mV	14D,8A	Amd		data pinout
AM216	LO-POWER	Vs:±20V Vu:86dB Vi0:10mV	10F,14D,8A	Amd		data pinout
AM216A	LO-POWER	Vs:±20V Vu:92dB Vi0:3mV	10F,14D,8A	Amd		data pinout
AM218	HI-SPEED	Vs:±20V Vo:±13V Vi0:2mV f:15Mc	10F,14D,8A	Amd		data pinout
AM224	4xOP	Vs:±2.5V Vu:100dB Vo:0.75V Vi0:5mV	14D	Amd		data pinout
AM301A	HI-VOLT	Vs:±120V Vu:120dB Vo:±110V Vi0:1mV	15Y,8A	Dat		data pinout
AM301B	HI-VOLT	Vs:±120V Vu:120dB Vo:±110V Vi0:1mV	15Y	Dat		data pinout
AM302	HI-VOLT	Vs:±150V Vu:120dB Vo:±120V Vi0:1mV	15Y	Dat		data pinout
AM303	FET	Vs:±150V Vo:±140V Vi0:±1mV f:5Mc	9Y	Dat		data pinout
AM307	UNI	Vs:±18V Vo:±14V Vi0:2mV	10F,14D,8A	Amd		data pinout
AM308	LO-POWER	Vs:±18V Vu:110dB Vo:±14V Vi0:2mV	14D,10F,8A	Amd		data pinout
Am308	LO-POWER	Vs:±15V Vo:±14V Vi0:0.3mV	10F,14D,8A	Amd		data pinout
Am312	LO-POWER	Vs:±18V Vo:±14V Vi0:2mV	10F,14D,8A	Amd		data pinout
AM316	LO-POWER	Vs:±20V Vu:86dB Vo:±13V Vi0:10mV	10F,14D,8A	Amd		data pinout
AM316A	LO-POWER	Vs:±20V Vi0:<3mV	10F,14D,8A	Amd		data pinout
Am318	HI-SPEED	Vs:±20V Vo:±13V Vi0:4mV f:15Mc	10F,14D,8A	Amd		data pinout
AM324	4xOP	Vs:±2.5V Vu:100dB Vo:0.75V Vi0:7mV	14D	Amd		data pinout
AM405-2	HI-SPEED	Vs:±15V Vu:78dB Vo:±10V f:20Mc	8A	Dat		data pinout
AM406-2	HI-SPEED	Vs:±15V Vu:98dB Vo:±10V f:100Mc	8A	Dat		data pinout
AM410-2	FET	Vs:±20V Vo:±11V Vi0:<1.5mV f:18Mc	8A	Dat		data pinout
AM411-2	FET	Vs:±20V Vo:±11V Vi0:<1.5mV f:50Mc	8A	Dat		data pinout
AM427-1	OP	Vs:±22V Vu:116dB Vo:>±11V f:>5Mc	8D	Dat		data pinout
AM427-2	OP	Vs:±22V Vu:116dB Vo:>±11V f:>5Mc	8A	Dat		data pinout
AM430	OP	Vs:±20V Vo:±10V Vi0:<75µµV f:2.5Mc	8A	Dat		data pinout
AM450-2	HI-SPEED	Vs:±20V Vo:±10V Vi0:<±4mV f:12Mc	8A	Dat		data pinout
AM452-2	HI-SPEED	Vs:±20V Vo:±10V Vi0:<±5mV f:20Mc	8A	Dat		data pinout
AM453-2	OP	Vs:±22V Vu:100V/mV Vo:±12V Vi0:0.5mV	8A	Dat		data pinout
AM460-2	HI-SPEED	Vs:±22.5V Vu:15V/mV Vo:±10V f:12Mc	8A	Dat		data pinout
AM462-1	HI-SPEED	Vs:±22.5V Vu:15V/mV Vo:±10V f:100Mc	14D	Dat		data pinout
AM462-2	HI-SPEED	Vs:±22.5V Vu:15V/mV Vo:±10V f:100Mc	8A	Amd		data pinout
AM464-2	HI-VOLT	Vs:±50V Vu:100V/mV Vo:±35V Vi0:<±6mV	8A	Dat		data pinout
AM470-2	OP	Vs:±22V Vu:300V/mV Vo:±12V Vi0:±1mV	8A	Dat		data pinout
AM490-2	COMP	Vs:±20V Vo:±10V Vi0:±20µµV f:3Mc	8A	Dat		data pinout
AM500	HI-SPEED	Vs:±18V Vo:>±10V Vi0:±0.5mV f:130Mc	14Y	Dat		data pinout
AM542	OP	Vs:±22V Vu:<1024dB Vo:>±10.5V	24Y	Dat		data pinout
AM543	OP	Vs:±16V Vu:<128dB Vo:>±11V f:7Mc	24Y	Dat		data pinout
AM551	3xPROG-GAIN	Vs:±18V Vo:>±11V Vi0:<±1mV f:0.6Mc	16D	Dat		data pinout
AM 685...	KOP-IC	→µA 685...		Amd	→µA 685...	
AM 687...	KOP-IC	→µA 687...		Amd	→µA 687...	
AM715	UNI	Vs:±15V Vu:78dB Vo:±10V Vi0:10mV	14D,10A	Amd		data pinout
AM723	ER+	Vo:2...37V P:850mW	14D,10A	Amd		data pinout
AM 723 HC,HM	Z-IC	+2...37V, 0,15A	82	Amd	...723...	data pinout
AM725	UNI	Vs:±15V Vu:88dB Vo:±12V Vi0:3.5mV	14D,8A	Amd		data pinout
Am725A	HI-GAIN	Vs:±22V Vo:>±12.5V Vi0:<0.5mV	8A	Amd		data pinout
AM725C	HI-GAIN	Vs:±22V Vo:±13.5V Vi0:0.5mV	8A	Amd		data pinout
AM741	UNI	Vs:±15V Vu:86dB Vo:±12V Vi0:7.5mV	14D,10F,8A	Amd		data pinout
AM 741 HC,HM	OP-IC	Uni, Serie 741	81	Amd	→Serie 741	data pinout
AM747	2xOP	Vs:±15V Vu:94dB Vo:±12V Vi0:7.5mV	14DF,10A	Amd		data pinout
AM 747 HC,HM	OP-IC	Dual, Serie 747	82	Amd	→Serie 747	data pinout
AM748	UNI	Vs:±15V Vu:94dB Vo:±12V Vi0:7.5mV	14D,10F,8A	Amd		data pinout
AM 748 HC	OP-IC	Uni, Serie 748	81	Amd	→Serie 748	data pinout
AM 1408...	D/A-IC	→DAC 1408...		Amd	→DAC 1408	

Type	Device	Short Description	Fig.	Manu	Comparision Types	More at
AM1435	WIDEBAND	Vs:±18V Vu:100dB Vo:±7V Vi0:±2mV	14Y	Dat		data pinout
AM 1458 H	OP-IC	Dual, Serie 158	81	Amd	→Serie 158	data
AM 1508...	D/A-IC	→DAC 1508...		Amd	→DAC 1508	
AM 1558 H	OP-IC	Dual, Serie 158	81	Amd	→Serie 158	data
AM1660	LO-POWER	Vs:±18V Vu:88dB Vo:±13V Vi0:2mV	14D,8A	Amd		data pinout
AM 2018	CMOS-IC	Progr. Gate Array		Amd	-	
AM 2064	CMOS-IC	Progr. Gate Array		Amd	-	
AM 2716B ... D ...	EPROM-IC	EPROM, 2048x8 Bit, TTL compatible In/out, +5V Vcc (±5% Tol.)	DIC	Amd	...2716...	pdf
AM 2716B ... L ...	EPROM-IC	EPROM, 2048x8 Bit, TTL compatible In/out, +5V Vcc (±5% Tol.)	LCC	Amd	...2716...	pdf
AM 2732B ... D ...	EPROM-IC	EPROM, 4096x8 Bit, TTL compatible In/out, +5V Vcc (±5% Tol.)	DIC	Amd	...2732...	pdf
AM 2732B ... L ...	EPROM-IC	EPROM, 4096x8 Bit, TTL compatible In/out, +5V Vcc (±5% Tol.)	LCC	Amd	...2732...	pdf
AM 2864 AE,BE-...	EEPROM-IC	8192 x 8 Bit, 5V, 200...350ns	28-DIP	Amd	...2864...	
AM 3020	CMOS-IC	Progr. Gate Array		Amd	-	
AM 3090	CMOS-IC	Progr. Gate Array		Amd	-	
AM 6012(A)D	D/A-IC	=AM 6012F: SMD	20-MDIP	Tho	-	pinout
AM 6012(A)PC	D/A-IC	=AM 6012F: Fig.→	20-DIP	Tho	-	pinout
AM 6012 F	D/A-IC	12 Bit, multiplying, 0...+70°	20-DIC	Phi	-	pdf pinout
AM 6651 B	IC	Motor Control Circuit	QFP	Bcd	-	pdf pinout
AM 6685...	KOP-IC	→μA 6685...		Amd	→μA 6685...	
AM 6687...	KOP-IC	→μA 6687...		Amd	→μA 6687...	
AM7650	HI-SPEED	Vs:±18V Vo:±4.85V Vi0:±0.7mV f:2Mc	14D,8AD	Dat		data pinout
AM 9709CN	LIN-IC	4x Analog Switch, Single Pole Single Throw switch, normally open	14-DIP	Nsc	AH5009CN	
AM 9710CN	LIN-IC	4x Analog Switch, Single Pole Single Throw switch, normally open	14-DIP	Nsc	AH5010CN	
AM 9711CN	LIN-IC	4x Analog Switch, Single Pole Single Throw switch, normally open	16-DIP	Nsc	AH5011CN	
AM 9712CN	LIN-IC	4x Analog Switch, Single Pole Single Throw switch, normally open	16-DIP	Nsc	AH5012CN	
AMA 2803R3 S	LIN/Z-IC	3.3V, Hybrid DC/DC Converter	FLP	Inr	-	pdf pinout
AMA 2805 D	LIN/Z-IC	±5V, Hybrid DC/DC Converter	FLP	Inr	-	pdf pinout
AMA 2805 S	LIN/Z-IC	5V, Hybrid DC/DC Converter	FLP	Inr	-	pdf pinout
AMA 2812 D	LIN/Z-IC	±12V, Hybrid DC/DC Converter	FLP	Inr	-	pdf pinout
AMA 2812 S	LIN/Z-IC	12V, Hybrid DC/DC Converter	FLP	Inr	-	pdf pinout
AMA 2815 D	LIN/Z-IC	±15V, Hybrid DC/DC Converter	FLP	Inr	-	pdf pinout
AMA 2815 S	LIN/Z-IC	15V, Hybrid DC/DC Converter	FLP	Inr	-	pdf pinout
AMC 317 SK/SJ	LIN-IC	1.2A / 3-terminal Adjustable regulator	48	Adm	-	pdf pinout
AMC 317 ST	LIN-IC	1.2A / 3-terminal Adjustable regulator	30	Adm	-	pdf pinout
AMC 317 T	LIN-IC	1.2A / 3-terminal Adjustable regulator	17	Adm	-	pdf pinout
AMC 358 DM	LIN-IC	Dual Operational Amplifiers	MDIP	Adm	LM358	pdf pinout
AMC 358 M	LIN-IC	Dual Operational Amplifiers	DIC	Adm	LM358	pdf pinout
AMC 393 DM	-IC	low Offset Voltage Dual Comparators	MDIP	Adm	LM393	pdf pinout
AMC 393 M	-IC	low Offset Voltage Dual Comparators	DIC	Adm	LM393	pdf pinout
AMC 431 DB	LIN-IC	Precision Adjustable Shunt Voltage Reference	35	Adm	TL431	pdf pinout
AMC 431 LP	LIN-IC	Precision Adjustable Shunt Voltage Reference	7	Adm	TL431	pdf pinout
AMC 431 PK	LIN-IC	Precision Adjustable Shunt Voltage Reference	48	Adm	TL431	pdf pinout
AMC 431 RDB	LIN-IC	Precision Adjustable Shunt Voltage Reference	35	Adm	TL431	pdf pinout
AMC 1112	LIN-IC	1A low dropout Positive Regulator, fixed 1.2V output	48	Adm	FAN1112	pdf pinout
AMC 1117	LIN-IC	1A low dropout Positive regulator	48	Adm	LT1117	pdf pinout
AMC 1117 B	LIN-IC	1A low dropout Positive regulator	48	Adm	LT1117	pdf pinout
AMC 2244 DM	LIN-IC	Dual Low-power 60MHz unity-gain stable Op Amplifiers	TSOP	Adm	-	pdf pinout
AMC 2244 N	LIN-IC	Dual Low-power 60MHz unity-gain stable Op Amplifiers	DIC	Adm	-	pdf pinout
AMC 2344 DM	LIN-IC	Triple Low-power 60MHz unity-gain stable Op Amplifiers	TSOP	Adm	-	pdf pinout
AMC 2344 N	LIN-IC	Triple Low-power 60MHz unity-gain stable Op Amplifiers	DIC	Adm	-	pdf pinout
AMC 2444 DM	LIN-IC	Quad Low-power 60MHz unity-gain stable Op Amplifiers	DIC	Adm	-	pdf pinout
AMC 2444 N	LIN-IC	Quad Low-power 60MHz unity-gain stable Op Amplifiers	DIC	Adm	-	pdf pinout
AMC 2576 DD	LIN-IC	3A step Down Voltage regulator	87	Adm	-	pdf pinout
AMC 2576 E/DM	LIN-IC	3A step Down Voltage regulator	TSOP	Adm	-	pdf pinout
AMC 2576 P(B)	LIN-IC	3A step Down Voltage regulator	86	Adm	-	pdf pinout
AMC 2596 DD	LIN-IC	150 KHz, 3A step Down Voltage regulator	87	Adm	-	pdf pinout
AMC 2596 P(B)	LIN-IC	150 KHz, 3A step Down Voltage regulator	86	Adm	-	pdf pinout
AMC 3100 -1.2	LIN-IC	1.5MHz, 600mA synchronous step-down Converter, 1.2V Fixed	45	Adm	-	pdf pinout
AMC 3100 -1.5	LIN-IC	1.5MHz, 600mA synchronous step-down Converter, 1.5V Fixed	45	Adm	-	pdf pinout
AMC 3100 -1.8	LIN-IC	1.5MHz, 600mA synchronous step-down Converter, 1.8V Fixed	45	Adm	-	pdf pinout
AMC 3100	LIN-IC	1.5MHz, 600mA synchronous step-down Converter, Adjustable	45	Adm	-	pdf pinout
AMC 3202 DM	IC	1.5A 280kHz Boost Regulators	43	Adm	-	pdf pinout

Type	Device	Short Description	Fig.	Manu	Comparision Types	More at
AMC 3842 BD	LIN-IC	Current Mode PWM Controller; Start-up = 16V, Hysteresis = 6V	MDIP	Adm	-	pdf pinout
AMC 3842 BDM	LIN-IC	Current Mode PWM Controller; Start-up = 16V, Hysteresis = 6V	MDIP	Adm	-	pdf pinout
AMC 3842 BM	LIN-IC	Current Mode PWM Controller; Start-up = 16V, Hysteresis = 6V	DIC	Adm	-	pdf pinout
AMC 3843 BD	LIN-IC	Current Mode PWM Controller; Start-up = 8.4V, Hysteresis = 0.8V	MDIP	Adm	-	pdf pinout
AMC 3843 BDM	LIN-IC	Current Mode PWM Controller; Start-up = 8.4V, Hysteresis = 0.8V	MDIP	Adm	-	pdf pinout
AMC 3843 BM	LIN-IC	Current Mode PWM Controller; Start-up = 8.4V, Hysteresis = 0.8V	DIC	Adm	-	pdf pinout
AMC 3844 BD	LIN-IC	Current Mode PWM Controller; Start-up = 16V, Hysteresis = 6V	MDIP	Adm	-	pdf pinout
AMC 3844 BDM	LIN-IC	Current Mode PWM Controller; Start-up = 16V, Hysteresis = 6V	MDIP	Adm	-	pdf pinout
AMC 3844 BM	LIN-IC	Current Mode PWM Controller; Start-up = 16V, Hysteresis = 6V	DIC	Adm	-	pdf pinout
AMC 3845 BD	LIN-IC	Current Mode PWM Controller; Start-up = 8.4V, Hysteresis = 0.8V	MDIP	Adm	-	pdf pinout
AMC 3845 BDM	LIN-IC	Current Mode PWM Controller; Start-up = 8.4V, Hysteresis = 0.8V	MDIP	Adm	-	pdf pinout
AMC 3845 BM	LIN-IC	Current Mode PWM Controller; Start-up = 8.4V, Hysteresis = 0.8V	DIC	Adm	-	pdf pinout
AMC 6821 SDBQ	IC	Intelligent Temp. Monitor, PWM Fan Contr.	TSOP	Bub	-	pdf pinout
AMC 7110 DB/DJ	BiMOS-IC	white/blue LED Driver For Li-ion Battery APPLICATION, 20mA, 3Ch	46	Adm	-	pdf pinout
AMC 7110 W	BiCMOS-IC	white/blue LED Driver For Li-ion Battery APPLICATION, 20mA, 3Ch.	MLP	Adm	-	pdf pinout
AMC 7111 DN	BiMOS-IC	white/blue LED Driver For Li-ion Battery APPLICATION, 20mA, 4Ch	TSOP	Adm	-	pdf pinout
AMC 7111 W	BiMOS-IC	white/blue LED Driver For Li-ion Battery APPLICATION, 20mA, 4Ch	MLP	Adm	-	pdf pinout
AMC 7113 DB	BiMOS-IC	white/blue LED Driver For Li-ion Battery APPLICATION, 15mA, 3Ch	46	Adm	-	pdf pinout
AMC 7114 DN	BiMOS-IC	white/blue LED Driver For Li-ion Battery APPLICATION, 15mA, 4Ch	TSOP	Adm	-	pdf pinout
AMC 7123 DN	BiMOS-IC	Flash light LED DRIVER; Max 120mA, 1Ch.	TSOP	Adm	-	pdf pinout
AMC 7124 DN	BiMOS-IC	Flash light LED DRIVER; Max 120mA, 3Ch.	TSOP	Adm	-	pdf pinout
AMC 7135	BiMOS-IC	350mA advanced Current regulator	48	Adm	-	pdf pinout
AMC 7140	LIN-IC	700mA High Voltage Adjustable Current regulator with ENABLE Control	87	Adm	-	pdf pinout
AMC 7150	LIN-IC	1.5A Power LED Driver	87	Adm	-	pdf pinout
AMC 7169 W	IC	500mA LED protector	43	Adm	-	pdf pinout
AMC 7585 SJ	LIN-IC	5A low dropout regulator	48	Adm	EZ1585B, LT1585A	pdf pinout
AMC 7585 ST(3)	LIN-IC	5A low dropout regulator	30	Adm	EZ1585B, LT1585A	pdf pinout
AMC 7585 T	LIN-IC	5A low dropout regulator	17	Adm	EZ1585B, LT1585A	pdf pinout
AMC 7635 DBFT	CMOS-IC	300mA Cmos low dropout regulator	45		-	pdf pinout
AMC 7637	LIN-IC	ultra low Iq 300mA low dropout regulator	35	Adm	-	pdf pinout
AMC 7660 DM	IC	DC - DC Converter Control Circuit	43	Adm	-	pdf pinout
AMC 7660 M	IC	DC - DC Converter Control Circuit	43	Adm	-	pdf pinout
AMC 78 L 05 DM	IC	100mA / 3-terminal 5V regulator	43	Adm	-	pdf pinout
AMC 78.L 05 LP	IC	100mA / 3-terminal 5V regulator	43	Adm	-	pdf pinout
AMC 78 L 05 PK	IC	100mA / 3-terminal 5V regulator	43	Adm	-	pdf pinout
AMC 79 L 05 DM	IC	100mA / 3-terminal negative Voltage regulator	43	Adm	-	pdf pinout
AMC 79 L 05 LP	IC	100mA / 3-terminal negative Voltage regulator	43	Adm	-	pdf pinout
AMC 8213 DW	IC	200MHz RGB Video Amp. System with OSD	43	Adm	-	pdf pinout
AMC 8213 N	IC	200MHz RGB Video Amp. System with OSD	43	Adm	-	pdf pinout
AMC 8878 DBT	IC	low NOISE 150mA low dropout regulator	43	Adm	-	pdf pinout
AMC 8879 DBT	IC	low NOISE 150mA low dropout regulator	43	Adm	-	pdf pinout
AMC 34063 ADM	LIN-IC	DC/DC Converter Control Circuits	MDIP	Adm	MC34063A	pdf pinout
AMC 34063 AM	LIN-IC	DC/DC Converter Control Circuits	DIC	Adm	MC34063A	pdf pinout
AMC 76381 PKF	CMOS-IC	450mA low dropout regulator	43	Adm	-	pdf pinout
AMC 76382 PKF	CMOS-IC	450mA low dropout regulator	43	Adm	-	pdf pinout
AMC 76383 SKF	CMOS-IC	450mA low dropout regulator	43	Adm	-	pdf pinout
AMC 76385 SKF	CMOS-IC	450mA low dropout regulator	43	Adm	-	pdf pinout
AMC 76386 DMF	CMOS-IC	Dual 450mA LDO regulator	43	Adm	-	pdf pinout
AMF 2803R3 S	LIN/Z-IC	+3.3V, Hybrid DC/DC Converter	FLP	Inr	-	pdf pinout
AMF 2805 D	LIN/Z-IC	±5V, Hybrid DC/DC Converter	FLP	Inr	-	pdf pinout
AMF 2805 S	LIN/Z-IC	+5V, Hybrid DC/DC Converter	FLP	Inr	-	pdf pinout
AMF 2807R5 S	LIN/Z-IC	+7.5V, Hybrid DC/DC Converter	FLP	Inr	-	pdf pinout
AMF 2812 D	LIN/Z-IC	±12V, Hybrid DC/DC Converter	FLP	Inr	-	pdf pinout
AMF 2812 S	LIN/Z-IC	+12V, Hybrid DC/DC Converter	FLP	Inr	-	pdf pinout
AMF 2815 D	LIN/Z-IC	±15V, Hybrid DC/DC Converter	FLP	Inr	-	pdf pinout
AMF 2815 S	LIN/Z-IC	+15V, Hybrid DC/DC Converter	FLP	Inr	-	pdf pinout
AMIS 30511	IC	2Ph., Micro-Stepping Motor Drv.	TSOP	Ons	-	pdf pinout
AMIS 30512	IC	2Ph., 0,8A, Micro-Stepping Motor Drv	TSOP	Ons	-	pdf pinout
AMIS 30521	IC	Micro-stepping Motor Drv.	QFN	Ons	-	pdf pinout
AMIS 30522	CMOS-IC	2Ph, Micro-Stepping Motor Drv.	QFN	Ons	-	pdf pinout
AMIS 30624 C6244	CMOS-IC	I2C Micro-stepping Motor Drv	MDIP	Ons	-	pdf pinout
AMIS 30624 C6245	CMOS-IC	I2C Micro-stepping Motor Drv	QFN	Ons	-	pdf pinout
AMIS 39100	CMOS-IC	8x High Side Drv., Protection	TSOP	Ons	-	pdf pinout

Type	Device	Short Description	Fig.	Manu	Comparision Types	More at
AMLM10	UNI	Vs:±20V Vu:94dB Vo:±12V Vi0:6mV	14D,10F,8A	Amd		data pinout
AMLM101	UNI	Vs:±20V Vu:94dB Vo:±12V Vi0:3mV	14D,10F,8A	Amd		data pinout
AMLM108	LO-POWER	Vs:±20V Vu:98dB Vo:±13V Vi0:1mV	14D,10F,8A	Amd		data pinout
AMLM112	LO-POWER	Vs:±20V Vu:94dB Vo:±13V Vi0:3mV	14D,10F,8A	Amd		data pinout
AMLM118	HI-SPEED	Vs:±20V Vu:92dB Vo:±12V Vi0:6mV	14D,10F,8A	Amd		data pinout
AMLM201	UNI	Vs:±20V Vu:94dB Vo:±12V Vi0:3mV	14D,10F,8A	Amd		data pinout
AMLM207	UNI	Vs:±20V Vu:94dB Vo:±12V Vi0:3mV	14D,10F,8A	Amd		data pinout
AMLM208	LO-POWER	Vs:±20V Vu:94dB Vo:±13V Vi0:1mV	14D,10F,8A	Amd		data pinout
AMLM212	LO-POWER	Vs:±20V Vu:94dB Vo:±13V Vi0:3mV	14D,8A	Amd		data pinout
AMLM216	LO-POWER	Vs:±20V Vu:86dB Vo:±13V Vi0:15mV	14D,8A	Amd		data pinout
AMLM216A	LO-POWER	Vs:±20V Vu:94dB Vo:±13V Vi0:6mV	14D,8A	Amd		data pinout
AMLM218	HI-SPEED	Vs:±20V Vu:92dB Vo:±12V Vi0:6mV	14D,8A	Amd		data pinout
AMLM301	UNI	Vs:±20V Vu:86dB Vo:±12V Vi0:10mV	8A	Amd		data pinout
AMLM301A	UNI	Vs:±15V Vu:88dB Vo:±12V Vi0:10mV	14D,8A	Amd		data pinout
AMLM307	UNI	Vs:±15V Vu:88dB Vo:±12V Vi0:10mV	14D,8A	Amd		data pinout
AMLM308	LO-POWER	Vs:±15V Vu:88dB Vo:±13V Vi0:10mV	14D,8A	Amd		data pinout
AMLM308A	LO-POWER	Vs:±15V Vu:88dB Vo:±13V Vi0:0.73mV	14D,8A	Amd		data pinout
AMLM311	COMP	Vs:±15V Vu:106dB Vi0:10mV	8A,14D	Amd		data pinout
AMLM312	LO-POWER	Vs:±15V Vu:88dB Vo:±13V Vi0:10mV	14D,8A	Amd		data pinout
AMLM316	LO-POWER	Vs:±20V Vu:88dB Vo:±13V Vi0:15mV	14D,8A	Amd		data pinout
AMLM316A	LO-POWER	Vs:±20V Vu:92dB Vo:±13V Vi0:6mV	14D,8A	Amd		data pinout
AMLM318	HI-SPEED	Vs:±20V Vu:86dB Vo:±12V Vi0:15mV	14D,8A	Amd		data pinout
AMLM319	COMP	Vs:±15V Vu:78dB Vi0:10mV		Amd		data pinout
AMLM339	COMP	Vs:±2.5V Vu:106dB Vi0:5mV		Amd		data pinout
AMN	Si-N	=2SC4018K-N (Typ-Code/Stempel/marking)	35	Rhm	→2SC4018K	data
AMp	Si-N	=BSS 64 (Typ-Code/Stempel/marking)	35	Phi	→BSS 64	data
AMP	Si-N	=2SC4018K-P (Typ-Code/Stempel/marking)	35	Rhm	→2SC4018K	data
AMP-01(...)X	OP-IC	Gain-set(0,1...10000), ±18V, 26...570kHz	18-DIC	Pmi	-	
AMP-01(...)S	OP-IC	=AMP-01...X: SMD	20-MDIP			
AMP-01(...)TC	OP-IC	=AMP-01...X:	28-LCC			
AMP-02(E,F)P	OP-IC	Gain-set(1...10000), ±18V, 200...1200kHz	8-DIP	Pmi	-	
AMP-02(E,F)S	OP-IC	=AMP-02...P: SMD	16-MDIP			
AMP-03(B,F,G)P	OP-IC	Unity-Gain, hi-prec, ±18V, 3MHz	8-DIP	Pmi	-	
AMP-03(B,F,G)J	OP-IC	=AMP-03..P:	81			
AMP-05(A,B,E,F)X	OP-IC	JFET, Gain-set(0,1..2000), ±18V	18-DIC	Pmi	-	
AMP310	OP	Vs:±18V Vo:>±10V	7Y	Ite		data pinout
AMQ	Si-N	=2SC4561-Q (Typ-Code/Stempel/marking)	35	Mat	→2SC4561	data
AMQ	Si-N	=2SC4562-Q (Typ-Code/Stempel/marking)	35(2mm)	Mat	→2SC4562	data
AMQ	Si-N	=2SC4656-Q (Typ-Code/Stempel/marking)	35(1,6mm)	Mat	→2SC4656	data
AMR	Si-N	=2SC4561-R (Typ-Code/Stempel/marking)	35	Mat	→2SC4561	data
AMR	Si-N	=2SC4562-R (Typ-Code/Stempel/marking)	35(2mm)	Mat	→2SC4562	data
AMR	Si-N	=2SC4656-R (Typ-Code/Stempel/marking)	35(1,6mm)	Mat	→2SC4656	data
AMR 2805 D	LIN-IC	±5V, Hybrid DC/DC Converter	FLP	Inr	-	pdf pinout
AMR 2805 S	LIN/Z-IC	5V, Radiation Tolerant DC/DC Converter	42	Inr	-	pdf pinout
AMR 2812 D	LIN-IC	±12V, Hybrid DC/DC Converter	FLP	Inr	-	pdf pinout
AMR 2812 S	LIN/Z-IC	12V Radiation Tolerant DC/DC Converter	42	Inr	-	pdf pinout
AMR 2815 D	LIN-IC	±15V, Hybrid DC/DC Converter	FLP	Inr	-	pdf pinout
AMR 2815 S	LIN/Z-IC	15V Radiation Tolerant DC/DC Converter	42	Inr	-	pdf pinout
AMR 2803R3 S	LIN/Z-IC	3.3V Radiation Tolerant DC/DC Converter	42	Inr	-	pdf pinout
AMs	Si-N	=BFP 420 (Typ-Code/Stempel/marking)	44(2mm)	Sie	→BFP 420	data
AMS	Si-P	=2SA1896-S (Typ-Code/Stempel/marking)	39	Say	→2SA1896	data
AMS 1086 CD-XX	LIN-IC	1.5A Fix Low Dropout Volt Reg.	30	Amo	-	pdf pinout
AMS 1086 CD	LIN-IC	1.5A Adj Low Dropout Volt Reg.	30	Amo	-	pdf pinout
AMS 1086 CM-XX	LIN-IC	1.5A Fix Low Dropout Volt Reg.	30	Amo	-	pdf pinout
AMS 1086 CM	LIN-IC	1.5A Adj Low Dropout Volt Reg.	30	Amo	-	pdf pinout
AMS 1086 CT-XX	LIN-IC	1.5A Fix Low Dropout Volt Reg.	17	Amo	-	pdf pinout
AMS 1086 CT	LIN-IC	1.5A Adj Low Dropout Volt Reg.	17	Amo	-	pdf pinout
AMIS 30623 C6239	CMOS-IC	Lin Micro-Stepping Motor Drv.	MDIP	Ons	-	pdf pinout
AMIS 30623 C623A	CMOS-IC	Lin Micro-Stepping Motor Drv.	QFN	Ons	-	pdf pinout
AMSSS725	UNI	Vs:±15V Vu:120dB Vo:±12.5V	8A	Amd		data pinout
AMSSS741	UNI	Vs:±15V Vu:94dB Vo:±12V Vi0:6mV	8A	Amd		data pinout
AMSSS747	2xOP	Vs:±15V Vu:94dB Vo:±12V Vi0:6mV	10A,14DF	Amd		data pinout
AMt	Si-N	=BSS 64 (Typ-Code/Stempel/marking)	35	Phi	→BSS 64	data
AMT	Si-P	=2SA1896-T (Typ-Code/Stempel/marking)	39	Say	→2SA1896	data
AMU 2480	NMOS-IC	CTV-Audio-Mischer/mixer f. D2-MAC	24-DIP	Itt	-	
AMU 2485	NMOS-IC	CTV-Audio-Mischer/mixer f. D2-MAC	24-DIP	Itt		pdf
AmZ 8121	LIN-IC	8 Bit Komparator	20-DIP		DL 8121D	
AmZ 8127	TTL-IC	System Clock f. 16 Bit µComp.	24-DIP		DL 8127D	

AN

Type	Device	Short Description	Fig.	Manu	Comparision Types	More at
AN	Si-P	=2SA1898 (Typ-Code/Stempel/marking)	39	Say	→2SA1898	data
AN	Si-N	=2SC2413-AN (Typ-Code/Stempel/marking)	≈35	Rhm	→2SC2413	data
AN	Si-N	=2SC2413K-N (Typ-Code/Stempel/marking)	35	Rhm	→2SC1413K	data
AN	Si-N	=2SC2532 (Typ-Code/Stempel/marking)	35	Tos	→2SC2532	data
AN	Si-N	=2SC4098-N (Typ-Code/Stempel/marking)	35(2mm)	Rhm	→2SC4098	data
AN	MOS-N-FET-e*	=2SK3287 (Typ-Code/Stempel/marking)	35	Hit	→2SK3287	data
AN	MOS-N-FET-e*	=2SK3289 (Typ-Code/Stempel/marking)	35(2mm)	Hit	→2SK3289	data
AN	Si-N	=BCW 60FN (Typ-Code/Stempel/marking)	35	Sie	→BCW 60FN	data
AN	Z-Di	=MA 5Z300-M (Typ-Code/Stempel/marking)	71(1,7mm)	Mat	→MA 5Z300-M	data
AN	Z-Di	=P4 SMA-10 (Typ-Code/Stempel/marking)	71(5x2,5)	Fag	→P4 SMA-10	data
AN	Z-Di	=SMBJ 7.5C (Typ-Code/Stempel/marking)	71(5x3,5)	Mop	→SMBJ ...	data
AN	Si-N	=XN 1509 (Typ-Code/Stempel/marking)	45	Mat	→XN 1509	data
AN	C-Di	=ZMV 934 (Typ-Code/Stempel/marking)	71(1,7mm)	Ztx	→ZMV 934	data

Type	Device	Short Description	Fig.	Manu	Comparision Types	More at
ANs	Si-N	=BCW 60FN (Typ-Code/Stempel/marking)	35	Sie	→BCW 60FN	data
AN 1 A3Q	Si-P+R	S, Rb=1k, Rbe=10kΩ, 60V, 0,1A, 0,25W	7c	Nec	DTA 113ZS, UN 4119	data
AN 1 A4M	Si-P+R	=AN 1A3Q: Rb=Rbe=10kΩ	7c	Nec	DTA 114ES, RN 2002, UN 4111, 2SA1348,++	data
AN 1 A4P	Si-P+R	=AN 1A3Q: Rb=10k, Rbe=47kΩ	7c	Nec	DTA 114YS, RN 2007, UN 4114, 2SA1564,++	data
AN 1 A4Z	Si-P+R	=AN 1A3Q: Rb=10k, Rbe=-	7c	Nec	DTA 114TS, RN 2011, UN 4115, 2SA1497,++	data
AN 1 F4M	Si-P+R	=AN 1A3Q: Rb=22k, Rbe=22kΩ	7c	Nec	DTA 124ES, RN 2003, UN 4112, 2SA1346,++	data
AN 1 F4N	Si-P+R	=AN 1A3Q: Rb=22k, Rbe=47kΩ	7c	Nec	DTA 124XS, KSR 2007, RN 2008	data
AN 1 F4Z	Si-P+R	=AN 1A3Q: Rb=22k, Rbe=-	7c	Nec	DTA 124TS, KSR 2011, UN 4117, 2SA1590	data
AN 1 L3M	Si-P+R	=AN 1A3Q: Rb=4,7k, Rbe=4,7kΩ	7c	Nec	DTA 143ES, RN 2001, UN 411L, 2SA1656,++	data
AN 1 L3N	Si-P+R	=AN 1A3Q: Rb=4,7k, Rbe=10kΩ	7c	Nec	DTA 143XS, KSR 2005, UN 411F, 2SA1654,++	data
AN 1 L3Z	Si-P+R	=AN 1A3Q: Rb=4,7k, Rbe=-	7c	Nec	DTA 143TS, RN 2010, UN 4116, 2SA1511,++	data
AN 1 L4L	Si-P+R	=AN 1A3Q: Rb=47k, Rbe=22kΩ	7c	Nec	DTA 144WS, RN 2009, UN 411E, 2SA1347,++	data
AN 1 L4M	Si-P+R	=AN 1A3Q: Rb=47k, Rbe=47kΩ	7c	Nec	DTA 144ES, RN 2004, UN 4113, 2SA1345,++	data
AN 1 L4Z	Si-P+R	=AN 1A3Q: Rb=47k, Rbe=-	7c	Nec	DTA 144TS, KSR 2012, UN 4110, 2SA1509	data
AN 77 L03	Z-IC	+3V, 0,1A	7b	Mat	...78L03...(TO-92)	pdf
AN 77 L04	Z-IC	+4V, 0,1A	7b	Mat	...78L04...(TO-92)	pdf
AN 77 L05	Z-IC	+5V, 0,1A	7b	Mat	...78L05...(TO-92)	pdf
AN 77 L06	Z-IC	+6V, 0,1A	7b	Mat	...78L06...(TO-92)	pdf
AN 77 L07	Z-IC	+7V, 0,1A	7b	Mat	...78L07...(TO-92)	pdf
AN 77 L08	Z-IC	+8V, 0,1A	7b	Mat	...78L08...(TO-92)	pdf
AN 77 L09	Z-IC	+9V, 0,1A	7b	Mat	...78L09...(TO-92)	pdf
AN 77 L10	Z-IC	+10V, 0,1A	7b	Mat	...78L10...(TO-92)	pdf
AN 77 L03M...L10M	Z-IC	=AN 77L03...L10: SMD	39b	Mat	...78Lxx...(SOT-89)	pdf
AN78L04	R+	Io=100mA Vo:3.84...4.16V Vin:9V	3N	Mat		data pdf pinout
AN 78 L04	Z-IC	+4V, 0,1A	7b	Mat	...78L04...(TO-92)	data pdf pinout
AN78L05	R+	Io=100mA Vo:4.8...5.2V Vin:10V	3N	Mat		data pdf pinout
AN 78 L05	Z-IC	+5V, 0,1A	7b	Mat	...78L05...(TO-92)	data pdf pinout
AN78L06	R+	Io=100mA Vo:5.76...6.24V Vin:11V	3N	Mat		data pdf pinout
AN 78 L06	Z-IC	+6V, 0,1A	7b	Mat	...78L06...(TO-92)	data pdf pinout
AN78L07	R+	Io=100mA Vo:6.72...7.28V Vin:12V	3N	Mat		data pdf pinout
AN 78 L07	Z-IC	+7V, 0,1A	7b	Mat	...78L07...(TO-92)	data pdf pinout
AN78L08	R+	Io=100mA Vo:7.7...8.3V Vin:14V	3N	Mat		data pdf pinout
AN 78 L08	Z-IC	+8V, 0,1A	7b	Mat	...78L08...(TO-92)	data pdf pinout
AN78L09	R+	Io=100mA Vo:8.64...9.35V Vin:15V	3N	Mat		data pdf pinout
AN 78 L09	Z-IC	+9V, 0,1A	7b	Mat	...78L09...(TO-92)	data pdf pinout
AN78L10	R+	Io=100mA Vo:9.6...10.4V Vin:16V	3N	Mat		data pdf pinout
AN 78 L10	Z-IC	+10V, 0,1A	7b	Mat	...78L10...(TO-92)	data pdf pinout
AN78L12	R+	Io=100mA Vo:11.5...12.5V Vin:19V	3N	Mat		data pdf pinout
AN 78 L12	Z-IC	+12V, 0,1A	7b	Mat	...78L12...(TO-92)	data pdf pinout
AN78L15	R+	Io=100mA Vo:14.4...15.6V Vin:23V	3N	Mat		data pdf pinout
AN 78 L15	Z-IC	+15V, 0,1A	7b	Mat	...78L15...(TO-92)	data pdf pinout
AN78L18	R+	Io=100mA Vo:17.3...18.7V Vin:27V	3N	Mat		data pdf pinout
AN 78 L18	Z-IC	+18V, 0,1A	7b	Mat	...78L18...(TO-92)	data pdf pinout
AN78L20	R+	Io=100mA Vo:19.2...20.8V Vin:29V	3N	Mat		data pdf pinout
AN 78 L20	Z-IC	+20V, 0,1A	7b	Mat	...78L20...(TO-92)	data pdf pinout
AN78L24	R+	Io=100mA Vo:23...25V Vin:33V P:630mW	3N	Mat		data pdf pinout
AN 78 L24	Z-IC	+24V, 0,1A	7b	Mat	...78L24...(TO-92)	data pdf pinout
AN 78 L03M...L24M	Z-IC	=AN 78L03...L24: SMD	39b	Mat	...78Lxx...(SOT-89)	pdf
AN78M05	R+	Io=500mA Vo:4.8...5.2V Vin:10V P:15W	3P	Mat		data pdf pinout
AN 78 M05 R	Z-IC	+5V, 0,5A	15/4Pin	Mat	-	
AN 78 M05	Z-IC	+5V, 0,5A	17b	Mat	...78M05...(TO-220)	data pinout
AN78M06	R+	Io=500mA Vo:5.75...6.25V Vin:11V P:15W	3P	Mat		data pdf pinout
AN 78 M06	Z-IC	+6V, 0,5A	17b	Mat	...78M06...(TO-220)	data pinout
AN78M07	R+	Io=500mA Vo:6.7...7.3V Vin:12V P:15W	3P	Mat		data pdf pinout
AN 78 M07	Z-IC	+7V, 0,5A	17b	Mat	...78M07...(TO-220)	data pinout
AN78M08	R+	Io=500mA Vo:7.7...8.3V Vin:14V P:15W	3P	Mat		data pdf pinout
AN 78 M08	Z-IC	+8V, 0,5A	17b	Mat	...78M08...(TO-220)	data pinout
AN 78 M08R	Z-IC	+8V, 0,5A	15/4Pin	Mat	-	
AN78M09	R+	Io=500mA Vo:8.65...9.35V Vin:15V P:15W	3P	Mat		data pdf pinout
AN 78 M09	Z-IC	+9V, 0,5A	17b	Mat	...78M09...(TO-220)	data pinout
AN 78 M09R	Z-IC	+9V, 0,5A	15/4Pin	Mat	-	
AN78M10	R+	Io=500mA Vo:9.6...10.4V Vin:16V P:15W	3P	Mat		data pdf pinout
AN 78 M10	Z-IC	+10V, 0,5A	17b	Mat	...78M10...(TO-220)	data pinout
AN78M12	R+	Io=500mA Vo:11.5...12.5V Vin:19V P:15W	3P	Mat		data pdf pinout
AN 78 M12 R	Z-IC	+12V, 0,5A	15/4Pin	Mat	-	
AN 78 M12	Z-IC	+12V, 0,5A	17b	Mat	...78M12...(TO-220)	data pinout
AN78M15	R+	Io=500mA Vo:14.4...15.6V Vin:23V P:15W	3P	Mat		data pdf pinout
AN 78 M15	Z-IC	+15V, 0,5A	17b	Mat	...78M15...(TO-220)	data pinout
AN78M18	R+	Io=500mA Vo:17.3...18.7V Vin:27V P:15W	3P	Mat		data pdf pinout
AN 78 M18	Z-IC	+18V, 0,5A	17b	Mat	...78M18...(TO-220)	data pinout
AN78M20	R+	Io=500mA Vo:19.2...20.8V Vin:29V P:15W	3P	Mat		data pdf pinout
AN 78 M20	Z-IC	+20V, 0,5A	17b	Mat	...78M20...(TO-220)	data pinout
AN78M24	R+	Io=500mA Vo:23...25V Vin:33V P:15W	3P	Mat		data pdf pinout
AN 78 M24	Z-IC	+24V, 0,5A	17b	Mat	...78M24...(TO-220)	data pinout
AN 78 M05F...M24F	Z-IC	=AN 78M05...M24: Iso	17b	Mat	...78Mxx...(TO-220Iso)	
AN78N04	R+	Io=300mA Vo:3.84...4.16V Vin:9V P:8W	3P	Mat		data pinout
AN 78 N04	Z-IC	+4V, 0,3A	14b	Mat	...78M04...(TO-126)	data pinout
AN78N05	R+	Io=300mA Vo:4.8...5.2V Vin:10V P:8W	3P	Mat		data pdf pinout
AN 78 N05	Z-IC	+5V, 0,3A	14b	Mat	...78M05...(TO-126)	data pinout
AN78N06	R+	Io=300mA Vo:5.75...6.25V Vin:11V P:8W	3P	Mat		data pdf pinout
AN 78 N06	Z-IC	+6V, 0,3A	14b	Mat	...78M06...(TO-126)	data pinout
AN78N07	R+	Io=300mA Vo:6.7...7.3V Vin:12V P:8W	3P	Mat		data pdf pinout
AN 78 N07	Z-IC	+7V, 0,3A	14b	Mat	...78M07...(TO-126)	data pinout
AN78N08	R+	Io=300mA Vo:7.7...8.3V Vin:14V P:8W	3P	Mat		data pdf pinout
AN 78 N08	Z-IC	+8V, 0,3A	14b	Mat	...78M08...(TO-126)	data pinout

Type	Device	Short Description	Fig.	Manu	Comparision Types	More at
AN78N09	R+	Io=300mA Vo:8.65...9.35V Vin:15V P:8W	3P	Mat		data pdf pinout
AN 78 N09	Z-IC	+9V, 0,3A	14b	Mat	... 78M09...(TO-126)	data pinout
AN78N10	R+	Io=300mA Vo:9.6...10.4V Vin:16V P:8W	3P	Mat		data pdf pinout
AN 78 N10	Z-IC	+10V, 0,3A	14b	Mat	... 78M10...(TO-126)	data pinout
AN78N12	R+	Io=300mA Vo:11.5...12.5V Vin:19V P:8W	3P	Mat		data pdf pinout
AN 78 N12	Z-IC	+12V, 0,3A	14b	Mat	... 78M12...(TO-126)	data pinout
AN78N15	R+	Io=300mA Vo:14.4...15.6V Vin:23V P:8W	3P	Mat		data pdf pinout
AN 78 N15	Z-IC	+15V, 0,3A	14b	Mat	... 78M15...(TO-126)	data pinout
AN78N18	R+	Io=300mA Vo:17.3...18.7V Vin:27V P:8W	3P	Mat		data pdf pinout
AN 78 N18	Z-IC	+18V, 0,3A	14b	Mat	... 78M18...(TO-126)	data pinout
AN78N20	R+	Io=300mA Vo:19.2...20.8V Vin:29V P:8W	3P	Mat		data pdf pinout
AN 78 N20	Z-IC	+20V, 0,3A	14b	Mat	... 78M20...(TO-126)	data pinout
AN78N24	R+	Io=300mA Vo:23...25V Vin:33V P:8W	3P	Mat		data pdf pinout
AN 78 N24	Z-IC	+24V, 0,3A	14b	Mat	... 78M24...(TO-126)	data pinout
AN79L04	R-	Io=100mA Vo:-3.84...-4.16V Vin:-9V	3N	Mat		data pdf pinout
AN 79 L04	Z-IC	-4V, 0,1A	7a	Mat	... 79L04...(TO-92)	data pdf pinout
AN79L05	R-	Io=100mA Vo:-4.8...-5.2V Vin:-10V	3N	Mat		data pdf pinout
AN 79 L05	Z-IC	-5V, 0,1A	7a	Mat	... 79L05...(TO-92)	data pdf pinout
AN79L06	R-	Io=100mA Vo:-5.76...-6.24V Vin:-11V	3N	Mat		data pdf pinout
AN 79 L06	Z-IC	-6V, 0,1A	7a	Mat	... 79L06...(TO-92)	data pdf pinout
AN79L07	R-	Io=100mA Vo:-6.72...-7.28V Vin:-12V	3N	Mat		data pdf pinout
AN 79 L07	Z-IC	-7V, 0,1A	7a	Mat	... 79L07...(TO-92)	data pdf pinout
AN79L08	R-	Io=100mA Vo:-7.7...-8.3V Vin:-14V	3N	Mat		data pdf pinout
AN 79 L08	Z-IC	-8V, 0,1A	7a	Mat	... 79L08...(TO-92)	data pdf pinout
AN79L09	R-	Io=100mA Vo:-8.65...-9.35V Vin:-15V	3N	Mat		data pdf pinout
AN 79 L09	Z-IC	-9V, 0,1A	7a	Mat	... 79L09...(TO-92)	data pdf pinout
AN79L10	R-	Io=100mA Vo:-9.6...-10.4V Vin:-16V	3N	Mat		data pdf pinout
AN 79 L10	Z-IC	-10V, 0,1A	7a	Mat	... 79L10...(TO-92)	data pdf pinout
AN 79 L12	Z-IC	-12V, 0,1A	7a	Mat	... 79L12...(TO-92)	data pdf pinout
AN79L15	R-	Io=100mA Vo:-14.4...-15.6V Vin:-23V	3N	Mat		data pdf pinout
AN 79 L15	Z-IC	-15V, 0,1A	7a	Mat	... 79L15...(TO-92)	data pdf pinout
AN79L18	R-	Io=100mA Vo:-17.3...-18.7V Vin:-27V	3N	Mat		data pdf pinout
AN 79 L18	Z-IC	-18V, 0,1A	7a	Mat	... 79L18...(TO-92)	data pdf pinout
AN79L20	R-	Io=100mA Vo:-19.2...-20.8V Vin:-29V	3N	Mat		data pdf pinout
AN 79 L20	Z-IC	-20V, 0,1A	7a	Mat	... 79L20...(TO-92)	data pdf pinout
AN79L24	R-	Io=100mA Vo:-23...-25V Vin:-33V	3N	Mat		data pdf pinout
AN 79 L24	Z-IC	-24V, 0,1A	7a	Mat	... 79L24...(TO-92)	data pdf pinout
AN 79 L04M...L24M	Z-IC	=AN 79L04...L24: SMD	39a	Mat	... 79Lxx...(SOT-89)	pdf
AN79M05	R-	Io=500mA Vo:-4.8...-5.2V Vin:-10V	3P	Mat		data pinout
AN 79 M05	Z-IC	-5V, 0,5A	17c	Mat	... 79M05...(TO-220)	data pinout
AN79M06	R-	Io=500mA Vo:-5.75...-6.25V Vin:-11V	3P	Mat		data pinout
AN 79 M06	Z-IC	-6V, 0,5A	17c	Mat	... 79M06...(TO-220)	data pinout
AN79M07	R-	Io=500mA Vo:-6.7...-7.3V Vin:-12V	3P	Mat		data pinout
AN 79 M07	Z-IC	-7V, 0,5A	17c	Mat	... 79M07...(TO-220)	data pinout
AN79M08	R-	Io=500mA Vo:-7.7...-8.3V Vin:-14V	3P	Mat		data pinout
AN 79 M08	Z-IC	-8V, 0,5A	17c	Mat	... 79M08...(TO-220)	data pinout
AN79M09	R-	Io=500mA Vo:-8.65...-9.35V Vin:-15V	3P	Mat		data pinout
AN 79 M09	Z-IC	-9V, 0,5A	17c	Mat	... 79M09...(TO-220)	data pinout
AN79M10	R-	Io=500mA Vo:-9.6...-10.4V Vin:-16V	3P	Mat		data pinout
AN 79 M10	Z-IC	-10V, 0,5A	17c	Mat	... 79M10...(TO-220)	data pinout
AN 79 M12	Z-IC	-12V, 0,5A	17c	Mat	... 79M12...(TO-220)	data pinout
AN79M15	R-	Io=500mA Vo:-14.4...-15.6V Vin:-23V	3P	Mat		data pinout
AN 79 M15	Z-IC	-15V, 0,5A	17c	Mat	... 79M15...(TO-220)	data pinout
AN79M18	R-	Io=500mA Vo:-17.3...-18.7V Vin:-27V	3P	Mat		data pinout
AN 79 M18	Z-IC	-18V, 0,5A	17c	Mat	... 79M18...(TO-220)	data pinout
AN79M20	R-	Io=500mA Vo:-19.2...-20.8V Vin:-29V	3P	Mat		data pinout
AN 79 M20	Z-IC	-20V, 0,5A	17c	Mat	... 79M20...(TO-220)	data pinout
AN79M24	R-	Io=500mA Vo:-23...-25V Vin:-33V P:15W	3P	Mat		data pinout
AN 79 M24	Z-IC	-24V, 0,5A	17c	Mat	... 79M24...(TO-220)	data pinout
AN 79 M05F...M24F	Z-IC	=AN 79M05...M24: Iso	17c	Mat	... 79Mxx...(TO-220Iso)	
AN79N04	R-	Io=300mA Vo:-3.84...-4.16V Vin:-9V	3P	Mat		data pinout
AN 79 N04	Z-IC	-4V, 0,3A	14c	Mat	... 79M04...(TO-126)	data pinout
AN79N05	R-	Io=300mA Vo:-4.8...-5.2V Vin:-10V P:8W	3P	Mat		data pinout
AN 79 N05	Z-IC	-5V, 0,3A	14c	Mat	... 79M05...(TO-126)	data pinout
AN79N06	R-	Io=300mA Vo:-5.75...-6.25V Vin:-11V	3P	Mat		data pinout
AN 79 N06	Z-IC	-6V, 0,3A	14c	Mat	... 79M06...(TO-126)	data pinout
AN79N07	R-	Io=300mA Vo:-6.7...-7.3V Vin:-12V P:8W	3P	Mat		data pinout
AN 79 N07	Z-IC	-7V, 0,3A	14c	Mat	... 79M07...(TO-126)	data pinout
AN79N08	R-	Io=300mA Vo:-7.7...-8.3V Vin:-14V P:8W	3P	Mat		data pinout
AN 79 N08	Z-IC	-8V, 0,3A	14c	Mat	... 79M08...(TO-126)	data pinout
AN79N09	R-	Io=300mA Vo:-8.65...-9.35V Vin:-15V	3P	Mat		data pinout
AN 79 N09	Z-IC	-9V, 0,3A	14c	Mat	... 79M09...(TO-126)	data pinout
AN79N10	R-	Io=300mA Vo:-9.6...-10.4V Vin:-16V	3P	Mat		data pinout
AN 79 N10	Z-IC	-10V, 0,3A	14c	Mat	... 79M10...(TO-126)	data pinout
AN 79 N12	Z-IC	-12V, 0,3A	14c	Mat	... 79M12...(TO-126)	data pinout
AN79N15	R-	Io=300mA Vo:-14.4...-15.6V Vin:-23V	3P	Mat		data pinout
AN 79 N15	Z-IC	-15V, 0,3A	14c	Mat	... 79M15...(TO-126)	data pinout
AN79N18	R-	Io=300mA Vo:-17.3...-18.7V Vin:-27V	3P	Mat		data pinout
AN 79 N18	Z-IC	-18V, 0,3A	14c	Mat	... 79N18...(TO-126)	data pinout
AN79N20	R-	Io=300mA Vo:-19.2...-20.8V Vin:-29V	3P	Mat		data pinout
AN 79 N20	Z-IC	-20V, 0,3A	14c	Mat	... 79M20...(TO-126)	data pinout
AN79N24	R-	Io=300mA Vo:-22.8...-25.2V Vin:-33V	3P	Mat		dcta pinout
AN 79 N24	Z-IC	-24V, 0,3A	14c	Mat	... 79M24...(TO-126)	data pinout
AN 90 B...(S)	LIN-IC	Transistor Arrays	...-(M)DIP	Mat	-	
AN 90 C...	LIN-IC	Transistor Arrays	-SIP	Mat	-	

Type	Device	Short Description	Fig.	Manu	Comparision Types	More at
AN 90 D21	LIN-IC	7x NPN-Darlington-Array	16-DIP	Mat	-	
AN 91 A10S	LIN-IC	Awb-system-interface	28-MDIP	Mat		
AN 91 A13S	LIN-IC	Awb-system-interface	28-MDIP	Mat		
AN 93 B06SCR	LIN-IC	SMD,Crt Monitor, RGB-Video-Verst./amp.	28-SMDIP	Mat		pdf
AN 96 A07K	LIN-IC	Crt Monitor, Dynamic Focus	28-SDIP	Mat		pdf
AN 101	LIN-IC	FM-noise suppr.	16-DIP	Mat	-	pdf
AN 103	LIN-IC	Cb-pll-hf	9-SIP	Mat		pdf
AN 115	LIN-IC	Stereo-decoder	14-DIP	Mat	BA 1310	pdf
AN 124	LIN-IC	NF-E, 4,4W(13V/4Ω)	9-SIP	Mat	AN 214(Q)	
AN 127	LIN-IC	NF-verst./AF amplifier		Mat	(OM 200)	
AN 179	LIN-IC	TV-Video	16-DIP	Mat	-	
AN 210	LIN-IC	AM/FM-ZF	14-DIP	Mat	-	
AN 211	LIN-IC	Stereo-decoder	14-DIP	Mat	(AN 362)	
AN 213	LIN-IC	NF-E, 40V, 15W(±16V/8Ω)	23/14-Pin	Mat	-	
AN 214(P,Q,R)	LIN-IC	NF-E, 18V, 1,2A, 4,4W(13V/4Ω)	9-SILP	Mat	-	
AN 215	LIN-IC	NF-V+E, 12V, 1A, >1W(6V/8Ω)	16-DILP	Mat	-	
AN 217(P)	LIN-IC	AM-V/M/O, AM/FM-ZF	16-DIP	Mat	-	
AN 219	LIN-IC	FM-Tuner	16-DIP	Mat	-	
AN 236	LIN-IC	CTV-hilfsträger/sub-carrier	16-DIP	Mat	-	
AN 239(Q)	LIN-IC	TV-Video + Ton-zf + Demod.	28-DIP	Mat	-	
AN 240(P,PD,PN)	LIN-IC	TV, Ton-ZF, Ucc=12V	14-DIP	Mat	→AN 241P(PD)	
AN 241(P,PD)	LIN-IC	TV-Ton-ZF, Ucc=24V	14-DIP	Mat	CA 3065, HA 1125, KA 2101, LA 1365, LM 3	
AN 245	LIN-IC	TV, Signal-ic		Mat	-	pinout
AN 246	LIN-IC	TV, Signal-ic	16-DIP	Mat	-	pdf
AN 247 P	LIN-IC	Tv-video-zf + Agc	16-DIP	Mat	-	
AN 252	LIN-IC	NF-E, 3W(13V/4Ω)	9-SILP	Mat	(AN 7140)[10]	
AN 253(P)	LIN-IC	AM/FM-ZF + NF-V	16-DIP	Mat	-	
AN 258	LIN-IC	FM-stereo-muting	14-DIP	Mat	-	
AN 259	LIN-IC	Am-tuner + Zf	14-DIP	Mat	-	
AN 260(P)	LIN-IC	AM-M/O, AM/FM-ZF	14-DIP	Mat	-	
AN 262(L)	LIN-IC	VC+recorder-A/W-Verst.	16-DIP	Mat	-	
AN 264	LIN-IC	2x NF-V-ra	14-DIP	Mat	(AN 7311)	
AN 270	LIN-IC	NF-V-ra	9-SIP	Mat	-	
AN 271	LIN-IC	Stereo-decoder	16-DIP	Mat	(AN 362)	
AN 272(U)	LIN-IC	NF-E, 34V, 2A, 5W(20V/8Ω)	10-DILP	Mat	-	
AN 274	LIN-IC	NF-E, 16V, 0,25A, 1,3W(10V/8Ω)	82	Mat	(AN 374P)	
AN 277	LIN-IC	AM-V/M/O, AM/FM-ZF	16-DIP	Mat	(AN 366P)	
AN 278	LIN-IC	FM-ZF	9-SIP	Mat	-	
AN 295	LIN-IC	TV-Horiz.+vertik.-synchr.+vertik.-e	24-DILP	Mat	-	
AN 301	LIN-IC	VC-Servo	16-DIP	Mat	-	
AN 302	LIN-IC	VC-Lumin.+ Video-AGC	16-DIP	Mat	-	
AN 303	LIN-IC	Vc-lumin.+ Noise Suppr.	18-DILP	Mat	-	
AN 304	LIN-IC	VC-FM-Limiter	14-DIP	Mat	-	
AN 305	LIN-IC	VC-Color-AGC	16-DIP	Mat	-	
AN 306	LIN-IC	VC-Color-APC	28-DIP	Mat	-	
AN 307	LIN-IC	VC-Color-AFC	28-DIP	Mat	-	
AN 313(U)	LIN-IC	2x NF-E, 2x3W(16V/8Ω)	16-DILP	Mat	-	
AN 315	LIN-IC	NF-E, 5,5W(13V/4Ω)	11-SILP	Mat	(AN 7154)[10]	
AN 316	LIN-IC	Vc, Drop-out-kompensation	16-DIP	Mat	-	
AN 318	LIN-IC	VC-Servo	28-DIP	Mat	-	
AN 320	LIN-IC	TV-AFT + Indicator	16-DIP	Mat	-	
AN 321	LIN-IC	TV-AFT	9-SIP	Mat	-	pdf
AN 325	LIN-IC	TV-AFT		Mat	-	
AN 331	LIN-IC	TV-signal-IC	16-DIP	Mat	-	
AN 0332 CG	MOS-IC	SMD, 8x N-Ch. Latchable Mosfet Array	16-MDIP	Stx	-	pdf pinout
AN 337	LIN-IC	VC-Chroma	28-DIP	Mat	-	
AN 340 P	LIN-IC	TV, Ton-ZF, Demod.	14-DIP	Mat	-	
AN 345(V)	LIN-IC	TV-Video	16-DIP	Mat	-	
AN 353	LIN-IC	AM/FM-ZF + Meter Driver	9-SIP	Mat	-	
AN 355	LIN-IC	TV-Ton-ZF+NF-E, 1,6W(16V/16Ω)	16-DIP+g	Phi	-	pdf pinout
AN360	OP	Vs:20V Vu:80dB	7l	Mat	-	data pdf pinout
AN 360	LIN-IC	NF-V-ra, Ucc=9V	7-SIP	Mat	AN 370	
AN 362	LIN-IC	Stereo-decoder(lärmpcienanz./lamp drv.)	16-DIP	Mat	-	
AN 362 L	LIN-IC	=AN 362: f.LED-Anz./LED driver	16-DIP	Mat	AN 362	
AN 363 N	LIN-IC	Stereo-decoder	16-DIP	Mat	BA 1320	
AN 366(P)	LIN-IC	AM-tuning, AM/FM-ZF	16-DIP	Mat	-	
AN 370	LIN-IC	NF-V-ra, (Ucc=35V)	7-SIP	Mat	-	
AN 374	LIN-IC	NF-E, 16V, 1A, 1W(9V/8Ω)	82	Mat	-	
AN 374 P	LIN-IC	=AN 374: Fig.→	7-SIP	Mat	-	
AN 377	LIN-IC	FM-ZF	16-DIP	Mat	-	
AN 380	LIN-IC	CTV-video + Chroma		Mat	(AN 5311)	
AN 603(N)	LIN-IC	Tachometer	14-DIP	Mat	-	
AN 605	LIN-IC	5x Schmitt-Trigger + 1x DC-Verst./amp.	14-DIC	Mat	-	
AN 606	LIN-IC	Video-Verstärker/amplifier	83	Mat	-	
AN 607	LIN-IC	=AN 607P: Fig.→	5	Mat	AN 608	
AN 607 P	LIN-IC	Wide bandwidth video amplifier IC	4-SIP	Pan	AN 608P	pdf pinout
AN 608	LIN-IC	=AN 608P: Fig.→	5	Mat	(AN 607)	
AN 608 P	LIN-IC	TV-Video, Wide Band Amp.	4-SIP	Mat	(AN 607P)	pdf
AN 610(P)	LIN-IC	Modulator	14-DIP	Mat	-	
AN 612	LIN-IC	Modulator	7-SIP	Mat	-	pdf
AN 614	LIN-IC	Camera, Video, Balance-modulator	7-SIP	Mat	-	pdf
AN 616	LIN-IC	Camera, Videosignal	14-DIP	Mat	-	
AN 829(P,Y)	LIN-IC	2x NF-Abschwächer/2x AF-Attenuator	14-DIP	Mat	-	
AN 829 S	LIN-IC	=AN 829: nur 1 Kanal/1 channel only	14-DIP	Mat	AN 829	
AN 915	LIN-IC	Transistor Array	14-DIP	Mat	-	

Type	Device	Short Description	Fig.	Manu	Comparision Types	More at
AN1081	JFET	Vs:±18V Vu:106dB Vo:±13.5V Vi0:2mV	8DS	Mat		data pinout
AN 1081	OP-IC	JFET, ±18V, -20...+75°	8-DIP	Mat	→Serie 080	data pdf pinout
AN 1081 S	OP-IC	=AN 1081: SMD	8-MDIP		-	data pdf pinout
AN1082	2xJFET	Vs:±18V Vu:106dB Vo:±13.5V Vi0:2mV	8DS	Mat		data pdf pinout
AN 1082	OP-IC	Dual, JFET, ±18V, -20...+75°	8-DIP	Mat	→Serie 080	data pinout
AN 1082 S	OP-IC	=AN 1082: SMD	8-MDIP		-	data pinout
AN1084	4xJFET	Vs:±18V Vu:106dB Vo:±13.5V Vi0:2mV	14D,18S	Mat		data pinout
AN 1084	OP-IC	Quad, JFET, ±18V, -20...+75°	14-DIP	Mat	→Serie 080	data pdf pinout
AN 1084 S	OP-IC	=AN 1084: SMD	18-MDIP		-	data pinout
AN1311	FAST-COMP	Vs:±18V Vu:106dB Vi0:2mV	8DS	Mat		data pinout
AN 1311	KOP-IC	hi-speed, ±18V, -20...+75°	8-DIP	Mat	→Serie 111	data pinout
AN 1311 S	OP-IC	=AN 1311: SMD	8-MDIP		-	data pinout
AN1319	2xFAST-COMP	Vs:36V Vu:40V/mV Vi0:2mV	14DS	Mat		data pinout
AN 1319	KOP-IC	Dual, hi-speed, ±18V, -20...+75°	14-DIP	Mat	→Serie 119	data pinout
AN 1319 S	OP-IC	=AN 1319: SMD	14-MDIP		-	data pinout
AN1324	4xOP	Vs:±16V Vu:100dB Vi0:2mV f:700Mc	14DS	Mat		data pdf pinout
AN 1324	OP-IC	Quad, Serie 124, ±15V, -20...+75°	14-DIP	Mat	→Serie 124	data pinout
AN 1324 NS	OP-IC	=AN 1324: SMD	14-MDIP		-	data pinout
AN1339	4xCOMP	Vs:36V Vu:200V/mV Vi0:2mV	14DS	Mat		data pdf pinout
AN 1339	KOP-IC	Quad, Serie 139, ±18V, -30...+85°	14-DIP	Mat	→Serie 139	data pinout
AN 1339 S	KOP-IC	=AN 1339: SMD	14-MDIP			data pinout
AN1358	2xOP	Vs:±16V Vu:100dB Vo:2mV f:700Mc	8DS	Mat		data pdf pinout
AN 1358	OP-IC	Dual, Serie 158, ±16V, -20...+75°	8-DIP	Mat	→Serie 158	data pinout
AN 1358 S	OP-IC	=AN 1358: SMD	8-MDIP		-	data pinout
AN1393	2xCOMP	Vs:36V Vu:200V/mV Vi0:1mV	8DS	Mat		data pdf pinout
AN 1393	KOP-IC	Dual, Serie 193, ±18V, -30...+85°	8-DIP	Mat	→Serie 193	data pinout
AN 1393 S	KOP-IC	=AN 1393: SMD	8-MDIP		-	data pinout
AN 1431	Z-IC	+2,5...36V, 0,1A	8-DIP	Mat	-	pdf
AN 1431 M	Z-IC	=AN 1431: SMD	39		-	pdf
AN 1431 T	Z-IC	=AN 1431: Fig.→	7		-	pdf
AN1458	2xOP	Vs:±18V Vu:106dB Vo:±14V Vi0:0.5mV	8DS	Mat		data pinout
AN 1458	OP-IC	Dual, Serie 158, ±18V, -20...+75°	8-DIP	Mat	→Serie 158	data pinout
AN 1458 S	OP-IC	=AN 1458: SMD	8-MDIP		-	data pinout
AN 1555 N	LIN-IC	Timer, -20...+75°	8-DIP	Mat	NE 555	pdf pinout
AN 1555 NS	LIN-IC	=AN 1555N: SMD	8-MDIP		NE 555D	pdf pinout
AN1741	UNI	Vs:±18V Vu:106dB Vo:±14V Vi0:0.5mV	8SD	Mat		data pinout
AN 1741	OP-IC	Uni, Serie 741, ±18V, -20...+75°	8-DIP	Mat	→Serie 741	data pdf pinout
AN1741S	UNI	Vs:±18V Vu:106dB Vo:±14V Vi0:0.5mV	8S	Pan		data pdf pinout
AN 1741 S	OP-IC	=AN 1741: SMD	8-MDIP		-	data pdf pinout
AN 1801	OP-IC		8-DIP	Mat	-	
AN 1801 S	OP-IC	=AN 1801: SMD	8-MDIP		-	
AN 1802	OP-IC	Dual	8-DIP	Mat	-	
AN 1802 S	OP-IC	=AN 1802: SMD	8-MDIP		-	
AN 1804	OP-IC	Quad	14-DIP	Mat	-	
AN 1804 S	OP-IC	=AN 1804: SMD	14-MDIP		-	
AN1833	2xOP	Vs:±18V Vo:±13.4V Vi0:0.3mV f:7Mc	8DS	Mat	-	data pinout
AN 1833	OP-IC	Dual, ±18V, -20...+75°, 6V/µs	8-DIP	Mat	-	data pinout
AN 1833 S	OP-IC	=AN 1833: SMD	8-MDIP		-	data pinout
AN 2010 S	LIN-IC	SMD, Ccd Correlated Double Sampling	16-MDIP	Mat	-	
AN 2011 S	LIN-IC	SMD, Ccd Correlated Double Sampling	16-MDIP	Mat	-	pdf
AN 2012 S	LIN-IC	SMD, Ccd Correlated Double Sampling	16-MDIP	Mat	-	
AN 2012 SB	LIN-IC	=AN 2012S:	16-SMDIP			
AN 2018 S	LIN-IC	SMD, Ccd Correlated Double Sampling	8-MDIP	Mat	-	pdf
AN 2020 S	LIN-IC	SMD, Camera, Dual Balanced Modulator	18-MDIP	Mat	-	pdf
AN 2034 FAP	LIN-IC	SMD, Ccd Camera, Signal-prozessor	32-MP	Mat	-	
AN 2035 FAQ	LIN-IC	SMD, Dig.CCD Camera, A/D Pre-process.	80-MP	Mat	-	
AN 2037 FAQ	LIN-IC	SMD, Dig.ccd Camera, A/D Pre-process.	80-MP	Mat	-	
AN 2038 FAQ	LIN-IC	SMD, Dig.ccd Camera, A/D Pre-process.	80-MP	Mat	-	
AN 2039 FAQ	LIN-IC	SMD, Dig.ccd Camera, A/D Pre-process.	80-MP	Mat	-	
AN 2042 SB	LIN-IC	SMD, Ccd Camera, Filter	16-SMDIP	Mat	-	
AN 2050 FB	LIN-IC	SMD, Ccd Camera, Signal-prozessor	44-MP	Mat	-	pdf
AN 2101 FH	LIN-IC	SMD, Camera, Signal-prozessor	80-MP	Mat	-	pdf
AN 2110 S	LIN-IC	SMD, Camera, Signal-prozessor	24-MDIP	Mat	-	pdf
AN 2125 FHS	LIN-IC	SMD, Camera, Signal-prozessor	128-MP	Mat	-	pdf
AN 2130	LIN-IC	Camera Agc	18-SIP	Mat	-	pdf pinout
AN 2131	LIN-IC	Camera Agc	18-SIP	Mat	-	pdf pinout
AN 2133	LIN-IC	Camera Agc	18-SIP	Mat	-	pdf pinout
AN 2140	LIN-IC	Camera Alc	18-SIP	Mat	-	
AN 2141	LIN-IC	Camera Alc	18-SIP	Mat	-	pdf pinout
AN 2145 NFHP	LIN-IC	SMD, Ccd Camera, Signal-prozessor	80-MP	Mat	-	pdf
AN 2146 FHP	LIN-IC	SMD, Ccd Camera, Signal-prozessor	80-MP	Mat	-	pdf
AN 2147 FHP	LIN-IC	SMD, Ccd Camera, Signal-prozessor	80-MP	Mat	-	pdf
AN 2150 S	LIN-IC	Ccd Camera, Signal-prozessor	28-MDIP	Mat	-	
AN 2153 S	LIN-IC	Camera, Signal-prozessor	42-MDIP	Mat	-	
AN 2154 FAP	LIN-IC	SMD, Camera, Signal-prozessor	48-MP	Mat	-	
AN 2163 FHP	LIN-IC	SMD, Camera, Signal-prozessor	80-MP	Mat	-	
AN 2210 S	LIN-IC	SMD, Camera, Video Output (ntsc)	24-MDIP	Mat	-	pdf
AN 2240	LIN-IC	Camera, NTSC-Signal	18-SIP	Mat	-	
AN 2241	LIN-IC	Camera, Ntsc-signal	18-SIP	Mat	-	
AN 2250 S	LIN-IC	SMD, Camera, Ccd Signal-prozessor	28-MDIP	Mat	-	
AN 2253 FAP	LIN-IC	SMD, Camera, Encoder	32-MP	Mat	-	
AN 2254 FAP	LIN-IC	SMD, Camera, Encoder (ntsc, PAL)	48-MP	Mat	-	
AN 2255 SB	LIN-IC	Camera, Fade	24-MDIP	Mat	-	
AN 2260 FAP	LIN-IC	SMD, Vc, Signal-prozessor (VHS, S-VHS)	48-MP	Mat	-	
AN 2276 S	LIN-IC	Ccd-camera, Signal-prozessor	32-MDIP	Mat	-	

Type	Device	Short Description	Fig.	Manu	Comparision Types	More at
AN 2310 S	LIN-IC	Camera, Color-signal	24-MDIP	Mat	-	
AN 2320 S	LIN-IC	Camera, Color-signal	24-MDIP	Mat	-	
AN 2330	LIN-IC	Camera, Color-signal	18-SIP	Mat	-	
AN 2331	LIN-IC	Camera, Color-signal	18-SIP	Mat	-	
AN 2340	LIN-IC	Camera, Color-signal	18-SIP	Mat	-	
AN 2341	LIN-IC	Camera, Color-signal	18-SIP	Mat	-	
AN 2350 S	LIN-IC	Ccd-camera, Signal-prozessor	28-MDIP	Mat	-	
AN 2354 S	LIN-IC	Camera, Color-signal-prozessor	24-MDIP	Mat	-	
AN 2355 FAP	LIN-IC	=AN 2355S:	32-MP		-	
AN 2355 S	LIN-IC	SMD, Camera, Color-signal-prozessor	24-MDIP	Mat	-	
AN 2365 S	LIN-IC	Camera, Balance	28-MDIP	Mat	-	
AN 2366 S	LIN-IC	Camera, Balance	28-MDIP	Mat	-	
AN 2373	LIN-IC	Differential Video-Verst./amplifier	14-DIP	Mat	-	
AN 2410 S	LIN-IC	Camera, Signal-prozessor	22-MDIP	Mat	-	
AN 2430	LIN-IC	Camera, Color-encoder	18-SIP	Mat	-	
AN 2431	LIN-IC	Camera, Color-encoder	18-SIP	Mat	-	
AN 2441 S	LIN-IC	SMD, Camera, Color-encoder (SECAM)	28-MDIP	Mat	-	
AN 2450 S	LIN-IC	Ccd-camera, Signal-prozessor	28-MDIP	Mat	-	
AN 2455 B	LIN-IC	Camera, Color-encoder (ntsc)	16-MDIP	Mat	-	
AN 2458 SH	LIN-IC	Ccd Camera, Color-encoder (ntsc/pal)	24-SSMDIP	Mat	-	pdf
AN 2460 FAP	LIN-IC	SMD, Vc, Signal-prozessor (vhs, S-vhs)	64-MP	Mat	-	
AN 2510 S	LIN-IC	SMD, Camera, Bildsucher/view Finder	24-MDIP	Mat	-	pdf pinout
AN 2512 S	LIN-IC	SMD, Camera, Bildsucher/view Finder	14-MDIP	Mat	-	pdf pinout
AN 2513 S	LIN-IC	SMD, Camera, Bildsucher/view Finder	16-MDIP	Mat	-	
AN 2514 S	LIN-IC	SMD, Camera, Bildsucher/view Finder	16-MDIP	Mat	-	pdf
AN 2515 S	LIN-IC	SMD, Camera, Bildsucher/view Finder	16-MDIP	Mat	-	pdf pinout
AN 2516 S	LIN-IC	SMD, Camera, Bildsucher/view Finder	18-MDIP	Mat	-	pdf
AN 2527 NFHP	LIN-IC	SMD, Camera, Color Signal Prozessor	48-MP	Mat	-	
AN 2560 S	LIN-IC	SMD, Camera, Batterie/battery	24-MDIP	Mat	-	
AN 2585 FAP	LIN-IC	SMD, Camera, Digital Auto-focus	32-MP	Mat	-	
AN 2661 NK	LIN-IC	Multi-laser-player, Signal-prozessor	30-SDIP	Mat	-	
AN 2662 K	LIN-IC	Multi-laser-player-prozessor	30-SDIP	Mat	-	
AN 2663 K	LIN-IC	Laser-disk, Signal-prozessor	24-SDIP	Mat	-	
AN 2663 S	LIN-IC	=AN 2662K: SMD	24-MDIP		-	
AN 2751 FAP	LIN-IC	SMD, Laser-disk, Signal-prozessor	48-MP	Mat	-	
AN 2870 FC	LIN-IC	SMD, Multi-laser-player-prozessor	48-MP	Mat	-	
AN 3122	LIN-IC	VHF-Converter	16-DIP	Mat	-	
AN 3126	LIN-IC	VHF-Modulator, Ucc=5V	14-DIP	Mat	-	
AN 3129 S	LIN-IC	SMD, VHF-Modulator, Ucc=5V	14-MDIP	Mat	-	
AN 3131	LIN-IC	HF-converter (secam)	14-QIP	Mat	-	
AN 3173 FB	LIN-IC	SMD, Vc, Digital-video-prozessor	48-MP	Mat	-	
AN 3183 FBP	LIN-IC	SMD, Vc, Digital-video-prozessor	64-MP	Mat	-	
AN 3220 K	LIN-IC	Vc, Aufn./Rec.-Vst./Amp.(4-Kopf/Head)	20-SDIP	Mat	-	
AN 3224 K	LIN-IC	Vc, Aufn./Rec.-Vst./Amp.(4-Kopf/Head)	20-SDIP	Mat	-	
AN 3231 K	LIN-IC	Vc, Signal-prozessor (VHS)	48-SDIP	Mat	-	
AN 3231(N)FA,FC	LIN-IC	=AN3231: SMD	48-MP		-	pdf pinout
AN 3232 FA,FB	LIN-IC	SMD, Vc, Signal-prozessor (S-VHS)	48-MP	Mat	-	
AN 3237 FA,FB	LIN-IC	SMD, Vc, Signal-prozessor (s-vhs)	48-MP	Mat	-	
AN 3247 NK	LIN-IC	Vc, Signal-prozessor	30-SDIP	Mat	-	
AN 3248 FAP	LIN-IC	=AN 3248NK: SMD	32-MP		-	
AN 3248 NK	LIN-IC	Vc, Luminanz Signal-prozessor	30-SDIP	Mat	-	pdf
AN 3268 NK	LIN-IC	Vc, Signal-prozessor	30-SDIP	Mat	-	
AN 3294 K	LIN-IC	Vc, I/O Interface (s-vhs)	30-SDIP	Mat	-	
AN 3296	LIN-IC	Vc, Synchr., Afc	16-DIP	Mat	-	pdf
AN 3296 S	LIN-IC	=AN 3296: SMD	16-MDIP		-	pdf
AN 3311 K	LIN-IC	Vc, Head Amp. (4-Kopf/Head), 5V	22-SDIP	Pan	-	pdf pinout
AN 3311 S	LIN-IC	SMD, Vc, Head Amp (4-Kopf/Head), 5V	22-MDIP	Pan	-	pdf pinout
AN 3316 K	LIN-IC	Vc, Hifi-audio-prozessor	22-SDIP	Pan	-	pdf pinout
AN 3327 K	LIN-IC	Vc, Hifi-audio-prozessor	22-SDIP	Mat	-	pdf pinout
AN 3328 S	LIN-IC	recording and playback amplifier IC	MDIP	Pan	-	pdf pinout
AN 3336 SB	LIN-IC	VCR Recording/playback Amplifier IC	36-SMDIP	Pan	-	pdf pinout
AN 3341 SC	LIN-IC	Vcr Recording/playback Amplifier IC	42-SMDIP	Pan	-	pdf pinout
AN 3346 FAS,FBP	LIN-IC	SMD, Vc, Cylinder Amp. (4-Kopf/Head)	48-MP	Mat	-	
AN 3347 FBP	LIN-IC	SMD, Vc, Cylinder Amp. (4-Kopf/Head)	48-MP	Mat	-	
AN 3348 FBP	LIN-IC	SMD,Vc,Camera, Verst./Amplifier(8Head)	64-MP	Mat	-	
AN 3358 SH	LIN-IC	recording and playback amplifier IC	42-SMDIP	Pan	-	pdf pinout
AN 3360 SB	LIN-IC	SMD, Vc, A/W-Verst./Amp. (4-Kopf/Head)	36-SMDIP	Mat	-	
AN 3361SB	LIN-IC	SMD, Vcr 4-head Rec/play amp, Vcc=12V	HSMDIP	Pan	-	pdf pinout
AN 3370 K	LIN-IC	Flying-erase IC	10-SIP	Pan	-	pdf pinout
AN 3371 SB	LIN-IC	2-head recording/playback amplifier IC	HSMDIP	Pan	-	pdf pinout
AN 3375 S	LIN-IC	recording/playback amplifier IC	MDIP	Pan	-	pdf pinout
AN 3380 NK	LIN-IC	Vc, FM-Audio, A/W-Verst./Amp. (2-Kopf)	22-SDIP	Mat	-	
AN 3383 K	LIN-IC	Vc, A/W-Verst./Amp. (4-Kopf/Head)	24-SDIP	Mat	-	
AN 3385 NK	LIN-IC	Vc, A/W-Verst./Amp. (4-Kopf/Head)	24-SDIP	Mat	-	
AN 3389 SB	LIN-IC	Recording/playback Amplifier IC	36-SMDIP	Pan	-	pdf pinout
AN 3398	LIN-IC	Vc, S-VHS	9-SIP	Mat	-	
AN 3398 S	LIN-IC	=AN 3398: SMD	8-MDIP		-	
AN 3450 FAP,FBP	LIN-IC	SMD, Vc, Signal-prozessor (vhs, HQ)	64-MP	Mat	-	
AN 3479 FBP	IC	Vcr Signal Processing IC	36-SMDIP	Pan	-	pdf pinout
AN 3493 K	LIN-IC	Vc, Lum./chroma Noise Red.(VHS, Hq)	30-SDIP	Mat	-	
AN 3495 K	LIN-IC	Vc, Lum./chroma Noise Red.(vhs, S-vhs)	30-SDIP	Mat	-	
AN 3495 S	LIN-IC	=AN 3495K: SMD	32-SMDIP		-	
AN 3497 SB	LIN-IC	SMD, Vc, Chroma Noise Reduction	16-SMDIP	Mat	-	pdf
AN 3501 NFBP	IC	processing IC for Vcr	36-SMDIP	Pan	-	pdf pinout
AN 3580 SB	LIN-IC	SMD, Vc, Video-Interface (vhs, S-vhs)	16-SMDIP	Mat	-	

Type	Device	Short Description	Fig.	Manu	Comparision Types	More at
AN 3592 K	LIN-IC	Vc, Jumping Correction (pal)	22-SDIP	Mat	-	pdf
AN 3592 S	LIN-IC	=AN 3592K: SMD	22-MDIP	Mat	-	pdf
AN 3594 K	LIN-IC	Vc, Color Signal Corr. (pal)	20-SDIP	Mat	-	
AN 3720 K,NK	LIN-IC	Vc, Slow/Still-Prozessor (2-Kopf/Head)	18-SDIP	Mat	-	
AN 3790 K	LIN-IC	Vc, Servo-interface	18-SDIP	Mat	-	
AN 3791 K	LIN-IC	Vc, X-value Shift	9-SIP	Mat	-	
AN 3792	LIN-IC	Vc, Cylinder Servo-interface	18-DIP	Mat	-	
AN 3793	LIN-IC	Vc, Cylinder Servo-interface	18-DIP	Mat	-	
AN 3794 N	LIN-IC	Vc, Capstan Servo-interface	18-DIP	Mat	-	
AN 3795 N	LIN-IC	Vc, Capstan Servo-interface (pal)	18-DIP	Mat	-	
AN 3811 NK	IC	Cylinder Hall motor driving IC for Vcr	36-SMDIP	Pan	-	pdf pinout
AN 3813 K	LIN-IC	Cylinder Motor Driver ICs vor Vcr	18-SDIP	Pan	-	pdf pinout
AN 3814 K	LIN-IC	Cylinder Motor Driver Ics vor Vcr	18-SDIP	Pan	-	pdf pinout
AN 3815 K	LIN-IC	Vc, Motor-Tr., 1,5A	18-SDIP	Mat	-	
AN 3816 SCR	LIN-IC	SMD, Vc, Motor-Tr., 1,5A	28-SMDIP	Mat	-	
AN 3821 K	LIN-IC	Vc, Capstan Motor Direct Drive, 1,5A	24-DILP	Mat	-	
AN 3826 NK	LIN-IC	Vc, Capstan Motor Drive, 1A	28-SDIP	Mat	-	
AN 3827 SB	LIN-IC	SMD, Vc, Capstan Motor Drive	36-MDIP	Mat	-	
AN 3830 K	LIN-IC	Vc, Reel Motor Direct Drive	24-DILP	Mat	-	
AN 3834 K	LIN-IC	Vc, Reel Motor Drive	24-DILP	Mat	-	
AN 3834 S	LIN-IC	SMD, Vc, Reel Motor Drive	24-MDIP	Mat	-	
AN 3840 SR	LIN-IC	Vc, Capstan Motor Drive	24-MDIP	Mat	-	
AN 3841 SR	LIN-IC	SMD, VC/DAT, Capstan Motor Drive, 1,5A	24-MDIP	Mat	-	pdf pinout
AN 3858SH	LIN-IC	SMD, 8mm Vcr 2-head Rec/play amp, Vcc=5V	VSMDIP	Pan	-	pdf pinout
AN 3860 A,SA	LIN-IC	SMD, Vc, Cylinder Motor Drive	32-SSMDIP	Mat	-	pdf pinout
AN 3890 FBS	LIN-IC	SMD, Vc,Camera, Capstan-motor Drive	36-MP	Mat	-	
AN 3891 FBP	LIN-IC	SMD, Multi Laser Player, Spindel-motor	42-MP	Mat	-	
AN 3893 NFHP	LIN-IC	SMD, Vc, Cylinder Motor Drive	48-MP	Mat	-	
AN 3900 NSC	LIN-IC	8mm-Camera,Stereo PLL, (ntsc, Pal, XT)	32-SMDIP	Mat	-	pdf pinout
AN 3915 S	LIN-IC	SMD, Vc,Video, Clock Regenrating, VCO	18-MDIP	Mat	-	
AN 3916	LIN-IC	Vc, Video Agc	14-SQP	Mat	-	pdf
AN 3917 S	LIN-IC	SMD, TV, VCO, Multiplier, Ucc=5V	16-MDIP	Mat	-	
AN 3920 K	LIN-IC	Vc, FM-audio	20-SDIP	Mat	-	
AN 3922 NK	LIN-IC	Vc, Fm-audio Signal-prozessor	20-SDIP	Mat	-	
AN 3922 NS	LIN-IC	=AN 3922NK: SMD	20-MDIP	Mat	-	
AN 3928 K	LIN-IC	Hifi-VC, Fm-audio Signal-prozessor	28-SDIP	Mat	-	
AN 3932 S	LIN-IC	Hifi-vc, FM Audio Signal-prozess.	32-SMDIP	Mat	-	
AN 3934 K	LIN-IC	Vc, Fm-audio Control Logic	24-SDIP	Mat	-	
AN 3935 NFHP	LIN-IC	SMD, Camera, FM Audio Signal Proz.	64-MP	Mat	-	
AN 3961 NFBPA	LIN-IC	SMD, Hifi-vc, Audio Signal-prozessor	64-MP	Mat	-	
AN 3963 NFBPA	LIN-IC	SMD, Hifi-vc, Audio Signal-prozessor	64-MP	Mat	-	
AN 3970 FBP	LIN-IC	SMD, Hifi-vc, Signal-prozessor (ntsc)	84-MP	Mat	-	
AN 3972 FB,FC	LIN-IC	SMD, Hifi-vc, FM Audio Noise Reduction	48-MP	Mat	-	
AN 3976 FBP	LIN-IC	SMD, Hifi-vc, Signal-prozessor (pal)	84-MP	Mat	-	
AN 3986 FBP	LIN-IC	SMD,8mm-Camera, Stereo Audio Prozessor	84-MP	Mat	-	pdf
AN 3986 FHP	LIN-IC	=AN 3986FBP:	80-MP	Mat	-	pdf
AN 3988 NFHP	LIN-IC	SMD,8mm-Camera, Stereo Audio Prozessor	64-MP	Mat	-	
AN 3990 K	LIN-IC	VC	18-SDIP	Mat	-	
AN 3991 K	LIN-IC	Vc, Signal-prozessor	20-SDIP	Mat	-	
AN 3991 NS	LIN-IC	=AN 3991K: SMD	20-MDIP	Mat	-	
AN 3993 K	LIN-IC	Vc, Signal-prozessor	22-SDIP	Mat	-	
AN 3994 NK	LIN-IC	Vc, Signal-prozessor	22-SDIP	Mat	-	
AN 3994 NS	LIN-IC	=AN 3994NK: SMD	22-MDIP	Mat	-	
AN 4136	OP-IC	Quad, Serie 124, ±20V, -20...+75°	14-DIP	Mat	→Serie 124	data
AN 4136 S	OP-IC	=AN 4136: SMD	14-MDIP	Mat	-	
AN4250	LO-POWER	Vs:±18V Vu:100dB Vo:±13V Vi0:0.5mV	8DS	Mat	-	data pdf pinout
AN 4250	OP-IC	lo-power, ±18V, -20...+75°	8-DIP	Mat	→Serie 250	data pinout
AN 4250 S	OP-IC	=AN 4250: SMD	8-MDIP	Mat	-	data pinout
AN4558	2xOP	Vs:±18V Vu:100dB Vo:±13V Vi0:0.5mV	8DS	Mat	-	data pdf pinout
AN 4558	OP-IC	Dual, Serie 158, ±18V, -20...+75°	8-DIP	Mat	→Serie 158	data pinout
AN 4558 S	OP-IC	=AN 4558: SMD	8-MDIP	Mat	-	data pinout
AN 5010	LIN-IC	TV-kanalwahl/channel selection	24-DIP	Mat	-	pdf
AN 5011	LIN-IC	Tv-kanalwahl/channel selection	18-DIP	Mat	-	
AN 5020	LIN-IC	FB-empf.-vorverst./preamplifier	9-SIP	Mat	-	
AN 5025 K	LIN-IC	FB-Empfänger/Receiver, Ucc=5V	10-SIP	Mat	-	
AN 5025 S	LIN-IC	=AN 5025K: SMD	14-MDIP	Mat	-	
AN 5026 K	LIN-IC	IR-FB-Empfänger/Receiver, Ucc=5V	10-SIP	Mat	-	pdf pinout
AN 5030	LIN-IC	TV, Tuner-steuerung/control	20-DIP	Mat	-	
AN 5031	LIN-IC	TV, Tuner-steuerung/control	20-DIP	Mat	-	
AN 5033	LIN-IC	TV, Tuner-steuerung/control	20-DIP	Mat	-	pdf pinout
AN 5035	LIN-IC	TV, Suchlauf/auto Search	9-SIP	Mat	-	
AN 5036	LIN-IC	TV, Tuner-steuerung/control	22-DIP	Mat	-	
AN 5043 SC	LIN-IC	SMD, TV, Tuner Band-umschaltung/switch	24-SMDIP	Mat	-	
AN 5070	LIN-IC	TV, Tuner-bandumschalt./band Switch	9-SIP	Mat	-	
AN 5071	LIN-IC	TV, Tuner-bandumschalt./band Switch	9-SIP	Mat	-	pdf
AN 5095 K	IC	Single chip IC with I2C bus interface	MDIP	Pan	-	pdf pinout
AN 5101 SC	LIN-IC	SMD, Tv/vc, Video+ton-zf, Agc, Afc	32-SMDIP	Mat	-	
AN 5111	LIN-IC	TV, Video-ZF, Afc, Agc	28-DIP	Mat	-	
AN 5112	LIN-IC	TV, Video-ZF, Agc, Afc	22-DIP	Mat	-	
AN 5120 N	LIN-IC	TV, Video-ZF, Agc, Demod.	16-DIP+g	Mat	-	
AN 5122	LIN-IC	TV, Video-ZF, Agc, Afc	22-DIP	Mat	-	
AN 5125	LIN-IC	Ctv, Video-ZF, Agc, Afc	22-DIP	Mat	-	pdf pinout
AN 5130	LIN-IC	TV, Video-ZF, Agc, Dem, neg.video-out	16-DIP+g	Mat	-	pdf pinout
AN 5132	LIN-IC	=AN 5130: Video-out=pos.	16-DIP+g	Mat	-	pdf pinout
AN 5135 K,NK	LIN-IC	Ctv, Video+ton-zf, Agc, Afc	28-SDIP	Mat	-	

Type	Device	Short Description	Fig.	Manu	Comparision Types	More at
AN 5136 K	LIN-IC	Ctv, Video+ton-zf, Agc, Afc	28-SDIP	Mat	-	
AN 5137 K	LIN-IC	Ctv, Video+ton-zf, Agc, Afc	28-SDIP	Mat	-	
AN 5138 NK	LIN-IC	Ctv/vc, Video+ton-zf, Agc, Afc	28-SDIP	Mat	-	pdf
AN 5150(N)	LIN-IC	TV, Video+ton-zf, AGC(neg.),HA/VA-O/Tr	28-DIP	Mat	-	pdf
AN 5151(N)	LIN-IC	TV, Video+ton-zf, Agc(pos.),Ha/va-o/tr	28-DIP	Mat	KA 2915	pdf pinout
AN 5156 K	LIN-IC	Ctv, Video+ton-zf, Sign.-proz.(ntsc)	42-SDIP	Mat	-	pdf pinout
AN 5160 NK	LIN-IC	Ctv, Video+ton-zf, Sign.-proz.(ntsc)	52-SDIP	Mat	-	pdf
AN 5163 K	LIN-IC	Ctv, Signal-prozessor (ntsc), I²C	52-SDIP	Mat	-	pdf
AN 5165 K	IC	A Single Chip IC for NTSC Color-TV	MDIP	Pan	-	pdf pinout
AN 5170 K	LIN-IC	VC/TV, Video+ton-zf, Agc, AFC(NTSC)	24-SDIP	Mat	-	pdf
AN 5177 NK	LIN-IC	Vc/tv, Video+ton-zf, Agc, Afc	30-SDIP	Mat	-	pdf pinout
AN 5179 K	LIN-IC	Vc/tv, Video+ton-zf, Agc, Afc	30-SDIP	Mat	-	
AN 5179 K	LIN-IC	Vc/tv, Video+ton-zf, Agc, Afc	30-SDIP	Mat	-	
AN 5182 K	LIN-IC	Vc/tv, Video+ton-zf, Agc, Afc	24-SDIP	Mat	-	
AN 5186 FB	IC	Vif/sif IC for car-tv	MDIP	Pan	-	pdf pinout
AN 5192 K	LIN-IC	Ctv, PAL/NTSC Signal-prozessor, I²C	64-SDIP	Mat	-	pdf
AN 5195 K	LIN-IC	Ctv, PAL/NTSC Signal-prozessor, I²C	64-SDIP	Mat	-	pdf
AN 5210	LIN-IC	TV, Ton-ZF, NF-E, 3,1W(24V/16Ω)	24-DILP	Mat	-	
AN 5215	LIN-IC	TV, Ton-ZF, Demod., Ucc=12V	7-SIP	Mat	-	pdf pinout
AN 5216	LIN-IC	TV, Ton-ZF, Demod.	12-SIP	Mat	-	pdf pinout
AN 5217	LIN-IC	TV, Ton-ZF, Demod.	12-SIP	Mat	-	pdf pinout
AN 5220	LIN-IC	TV, Ton-zf	14-DIP	Mat	AN 5221	pdf
AN 5221	LIN-IC	TV, Ton-zf	14-DIP	Mat	AN 5220	pdf
AN 5222	LIN-IC	TV, Ton-ZF, Demod., NF-V	14-DIP	Mat	-	
AN 5250	LIN-IC	TV, Ton-ZF, NF-E, 2W(17V/16Ω)	16-DIP+g	Mat	-	
AN 5255	LIN-IC	TV, Ton-ZF, NF-E	16-DIP+g	Mat	-	
AN 5256	LIN-IC	TV, Ton-ZF, NF-E, 2W(17V/16Ω)	16-DIP+g	Mat	-	
AN 5260	LIN-IC	TV, NF-V+E, 6,6W(24V/8Ω)	11-SILP	Mat	-	
AN 5262(N)	LIN-IC	TV, NF-V+Lautst.-Reg./DC Volume Contr.	7-SIP	Mat	-	pdf
AN 5265	LIN-IC	TV, NF-Endst./LF-PowAmp, 2.3W(18V/16Ω) Volume DC controlled	9-SIL	Pan	-	data pdf pinout
AN 5267	LIN-IC	TV, NF-V+E		Mat	-	
AN 5270	LIN-IC	NF-Endst./LF-PowAmp, 4.3W(18V/8Ω) Volume & Tone DC controlled	9-SIL	Pan	-	data pdf pinout
AN 5272	LIN-IC	TV, 2x NF- Endst./ LF- PowAmp 2x 4.0W(18V/ 8Ω), Volume DC controlled	12-SIL	Pan	-	data pdf pinout
AN 5273	LIN-IC	TV, 2x NF-Endst./LF-PowAmp, Mute 2x 4.0W(18V/8Ω), Volume DC contr.	12-SIL	Pan	-	data pdf pinout
AN 5274	LIN-IC	TV, 2x NF-Endst./LF-PowAmp, Mute 2x 4.0W(18V/8Ω), Volume DC contr.	12-SIL	Pan	-	data pdf pinout
AN 5275	LIN-IC	TV, 2x NF- Endst./ LF- PowAmp 2x 15W(32V/ 8Ω)	12-SIL	Pan	-	data pdf pinout
AN 5276	LIN-IC	TV, 2x NF- Endst./ LF- PowAmp 2x 5.0W(19V/ 8Ω), Volume DC controlled	12-SIL	Pan	-	data pdf pinout
AN 5277	LIN-IC	TV, 2x NF-Endst./LF-PowAmp, Mute 2x 10W(26V/8Ω)	12-SIL	Pan	-	data pdf pinout
AN 5278	LIN-IC	TV, NF-Endst./LF-PowAmp, 4.8W(18V/8Ω) Volume & Tone DC controlled	9-SIL	Pan	-	data pdf pinout
AN 5285 K	LIN-IC	TV,Vc, Sound Level Agc, Ucc=8,5...12,5V	10-SIP	Mat	-	pdf
AN 5290 S	IC	Antenna diversity IC for on-vehicle TV	MDIP	Pan	-	pdf pinout
AN 5295 NK	IC	3-ch. sound signal processing single chip IC	MDIP	Pan	-	pdf pinout
AN 5302 K	LIN-IC	Ctv, Signal-prozessor (ntsc)	52-SDIP	Mat	-	pdf
AN 5303 K	LIN-IC	Ctv, Signal-prozessor (ntsc)	52-SDIP	Mat	-	pdf
AN 5304 NK	LIN-IC	Ctv, Video/chroma (ntsc)	52-SDIP	Mat	-	pdf
AN 5306 NFBS	LIN-IC	SMD, Ctv, Video/Chroma/RGB (ntsc)	80-MP	Pan	-	pdf
AN 5308 NK	LIN-IC	Ctv, Signal-prozessor (ntsc), I²c-bus	64-SDIP	Mat	-	pdf
AN 5310	LIN-IC	Ctv, Video/chroma (ntsc)	28-DIP	Mat	-	pdf pinout
AN 5311	LIN-IC	Ctv, Video/Chroma-prozessor (ntsc)	28-DIP	Mat	-	pdf pinout
AN 5312	LIN-IC	Ctv, Video/chroma-prozessor (ntsc)	22-DIP	Mat	-	pdf pinout
AN 5313 NK	LIN-IC	Ctv, Video/chroma-prozessor (ntsc)	24-SDIP	Mat	-	
AN 5313 NS	LIN-IC	=AN 5313NK: SMD	24-MDIP			
AN 5314 K	LIN-IC	Ctv, Video/chroma-prozessor (ntsc)	24-SDIP	Mat	-	
AN 5314 S	LIN-IC	=AN 5314K: SMD	24-MDIP			
AN 5315	LIN-IC	Ctv, Video/chroma-prozessor (ntsc)	24-DIP	Mat	-	
AN 5316(N)	LIN-IC	Ctv, Video/chroma-prozessor (ntsc)	24-DIP	Mat	-	
AN 5318(A,N)	LIN-IC	Ctv, Video/chroma-prozessor (ntsc)	28-DIP	Mat	-	pdf pinout
AN 5320	LIN-IC	Ctv, NTSC-color-kompensation	14-DIP	Mat	-	pdf pinout
AN 5330	LIN-IC	Ctv, NTSC-VIR-Signal	24-DIP	Mat	-	
AN 5332 N	LIN-IC	Ctv, Video/chroma-prozessor (ntsc)	22-DIP	Mat	-	pdf
AN 5334 K	LIN-IC	Ctv, Video/synchr.-prozessor (ntsc)	52-SDIP	Mat	-	
AN 5337 K	LIN-IC	Ctv, Video/chroma/rgb (ntsc)	52-SDIP	Mat	-	pdf
AN 5340	LIN-IC	Ctv, Video, Synchr.	16-DIP	Mat	-	
AN 5342 FBP	LIN-IC	=AN 5342K: SMD	44-MP	Pan	-	pdf pinout
AN 5342 K	LIN-IC	Ctv, Horizontal Aperture Corr.	30-SDIP	Pan	-	pdf pinout
AN 5344 FBP	LIN-IC	SMD, Ctv, Signal-prozessor (ntsc)	34-MP	Pan	-	pdf pinout
AN 5352(N)	LIN-IC	Ctv, Video-interface f. Zeichen/char.	22-DIP	Mat	-	pdf
AN 5355	LIN-IC	Ctv, Rgb-interface f. Teletext	18-DIP	Mat	KA 6101	pdf pinout
AN 5356	LIN-IC	Ctv, Rgb-interface f. Teletext	18-DIP	Mat	KA 6102	pdf pinout
AN 5365 FBP	LIN-IC	SMD, Ctv/vc, Signal-prozessor (ntsc)	64-MP	Mat	-	
AN 5367 FB	IC	NTSC video, chroma signal processing circuit	MDIP	Pan	-	pdf pinout
AN 5370 S	LIN-IC	Ctv, Chroma-prozessor	22-MDIP	Mat	-	
AN 5371 S,NS	LIN-IC	SMD, Ctv, Chroma-prozessor f. LCD TV	22-MDIP	Mat	-	pdf pinout
AN 5372 S	LIN-IC	SMD,Ctv, Signal-p.(ntsc/pal) f. LCD TV	28-MDIP	Mat	-	pdf pinout

Type	Device	Short Description	Fig.	Manu	Comparision Types	More at
AN 5374 S	LIN-IC	SMD, Ctv, SECAM-PAL Transcoder	28-MDIP	Mat	-	
AN 5379 NS	LIN-IC	SMD, Ctv, Chroma-prozessor f. LCD TV	22-MDIP	Mat	-	pdf pinout
AN 5380 NK	LIN-IC	Ctv, RGB-Prozessor, I²c-bus Contr.	28-SDIP	Mat	-	
AN 5380 NS	LIN-IC	=AN 5380NK: SMD	28-MDIP		-	
AN 5385 K	IC	Contour correction IC	MDIP	Pan	-	pdf pinout
AN 5392 FBQ	IC	RGB processor IC	MDIP	Pan	-	pdf pinout
AN 5394 FB	IC	RGB processor IC, HDTV(J),wide-screen TV	MDIP	Pan	-	pdf pinout
AN 5395 FBP	IC	Contour correction IC for Hdtv	MDIP	Pan	-	pdf pinout
AN 5410	LIN-IC	Ctv, HA/VA-signal-prozesser	24-DIP	Mat	AN 5411	
AN 5411	LIN-IC	=AN 5410: X-Ray-Protect (Pin 9)	24-DIP	Mat	-	
AN 5415	LIN-IC	Ctv, HA/VA Signal-prozessor	18-DIP	Mat	-	pdf pinout
AN 5416	LIN-IC	Ctv, Ha/va Signal-prozesscr	18-DIP	Mat	-	pdf pinout
AN 5421(N)	LIN-IC	TV, Synchr.-signal, HA-O	9-SIP	Mat	-	
AN 5422 K	LIN-IC	TV/Monitor, Ha/va Signal-prozessor	22-SDIP	Mat	-	pdf
AN 5429	LIN-IC	TV, HA/VA-Synchr., 50Hz	18-DIP	Mat	-	
AN 5430	LIN-IC	=AN 5429: 60Hz	18-DIP	Mat	-	
AN 5431 N	LIN-IC	TV, HA/VA-synchr.	16-DIP	Mat	-	pdf pinout
AN 5435	LIN-IC	Ctv, Ha/va Signal-prozessor	18-DIP	Mat	-	
AN 5436(N)	LIN-IC	Ctv, Ha/va Signal-prozessor	18-DIP	Mat	KA 2134	
AN 5437 K	LIN-IC	Ctv, Ha/va Signal-prozessor	24-SDIP	Mat	-	
AN 5440	LIN-IC	TV, HA/VA-synchr./O	16-DIP	Mat	-	
AN 5441 S	IC	Deflection distortion correction IC	MDIP	Pan	-	pdf pinout
AN 5448	IC	Tilt correction IC	MDIP	Pan	-	pdf pinout
AN 5452	IC	IC for landing correction	MDIP	Pan	-	pdf pinout
AN 5491 K	IC	deflection processor IC	MDIP	Pan	-	
AN 5510	LIN-IC	TV, VA-E	11-SILP	Mat	-	
AN 5512	LIN-IC	TV, Vert. Deflection Output, Vcc=27.6V	9-SIL	Mat	KA 2131 [Pin10: n.c.]	pdf pinout
AN 5515	LIN-IC	TV, Vert. Deflection Output, Vcc=30V	7-SIL	Mat	-	pdf pinout
AN 5520	LIN-IC	TV, VA-E	11-SILP	Mat	-	pdf
AN 5521	LIN-IC	TV, VA-E	7-SIL	Mat	-	pdf pinout
AN 5530 K	LIN-IC	TV, HA-E	9-SIL	Mat	-	
AN 5531	LIN-IC	TV, HA-E	9-SIL	Mat	-	
AN 5532	LIN-IC	TV, HA-E, Ucc=12..24V	9-SIL	Mat	-	pdf pinout
AN 5534	LIN-IC	TV,Monitor, HA-E	12-SIL	Mat	-	
AN 5535	LIN-IC	TV, HA-E	12-SIL	Mat	-	
AN 5551	LIN-IC	TV,Monitor, Kissenentzerrung/pin cush.	9-SIP	Mat	-	pdf
AN 5560	LIN-IC	TV, 50- & 60Hz-Identifikation	7-SIP	Mat	-	pdf pinout
AN 5601 K	LIN-IC	Ctv, Chroma/RGB/synchr(PAL/NTSC)	42-SDIP	Mat	-	pdf pinout
AN 5607 NK	LIN-IC	Ctv, Chroma/RGB/synchr(PAL/NTSC), I²C	52-SDIP	Mat	-	pdf pinout
AN 5610 N	LIN-IC	Ctv, PAL/SECAM-Video/Chroma	16-DIP+g	Mat	-	
AN 5612	LIN-IC	Ctv, Pal/secam Chroma-/video-signal	18-DIP	Mat	-	pdf pinout
AN 5613	LIN-IC	Ctv, Pal/secam Chroma-/video Signal	18-DIP	Mat	-	pdf pinout
AN 5615	LIN-IC	TV, Video Signal-prozessor	12-SIP	Mat	-	
AN 5620 X	LIN-IC	Ctv, Chroma-signal-prozessor (pal)	16-DIP+g	Mat	TDA 5620	
AN 5622	LIN-IC	Ctv, Chroma-signal-prozessor (pal)	16-DIP+g	Mat	-	pdf pinout
AN 5625 N	LIN-IC	Ctv, Chroma-signal-proz. (pal/ntsc)	22-DIP	Mat	-	
AN 5630 N	LIN-IC	Ctv, Chroma-signal-prozessor (secam)	24-DIP	Mat	-	
AN 5633 K	LIN-IC	Ctv, Chroma-signal-proz. (SECAM/PAL)	28-SDIP	Mat	-	pdf
AN 5635 N	LIN-IC	Ctv, Chroma-signal-prozessor (secam)	24-DIP	Mat	-	pdf
AN 5635 NS	LIN-IC	=AN 5635N: SMD	28-MDIP		-	pdf
AN 5636 K	IC	Secam/pal signal conversion IC	MDIP	Pan	-	pdf pinout
AN 5637	LIN-IC	Ctv, Chroma-signal-prozessor (secam)	18-DIP	Mat	-	pdf
AN 5640	LIN-IC	Ctv, PAL/SECAM/NTSC-Identifikation	18-DIP	Mat	-	
AN 5641	LIN-IC	Ctv, Pal/secam/ntsc-identifikation	18-DIP	Mat	-	
AN 5650	LIN-IC	Ctv, Synchr.-signal-prozessor	16-DIP	Mat	-	pdf pinout
AN 5693 K	IC	Lumi.,Chroma and Sync. Sig. Processing IC	MDIP	Pan	-	pdf pinout
AN 5700	LIN-IC	TV, Tuner-bandumschalt./band Switch	9-SIP	Mat	-	
AN 5701(N)	LIN-IC	TV, Tuner-bandumschalt./band Switch	9-SIP	Mat	-	
AN 5702	LIN-IC	TV, Tuner-bandumschalt./band Switch	9-SIP	Mat	-	
AN 5703	LIN-IC	TV, Tuner-bandumschalt./band Switch	9-SIP	Mat	-	
AN 5707 NS	LIN-IC	SMD, Tuner-Steuerung/Control, Ucc=5V	28-MDIP	Mat	-	pdf
AN 5710	LIN-IC	TV, Video-ZF, Agc(pos.), Ucc=5,5V	9-SIP	Mat	AN 5712	pdf pinout
AN 5712	LIN-IC	TV, Video-ZF, Agc(pos.), Ucc=12V	9-SIP	Mat	-	
AN 5715 K	LIN-IC	TV, Video+ton-zf, Ucc=3,7...6V	24-DIP	Mat	-	pdf pinout
AN 5715 S	LIN-IC	=AN 5715K: SMD	24-MDIP		-	pdf pinout
AN 5720	LIN-IC	TV, Video-ZF, Demod., Ucc=5,5V	9-SIP	Mat	AN 5722	pdf pinout
AN 5722	LIN-IC	TV, Video-ZF, Demod., Ucc=12V	9-SIP	Mat	-	
AN 5730	LIN-IC	TV, Ton-ZF, Ucc=5,5V	7-SIP	Mat	AN 5732	
AN 5732	LIN-IC	TV, Ton-ZF, Ucc=12V	7-SIP	Mat	-	
AN 5733	LIN-IC	2x NF-Abschwächer/AF Attenuator	9-SIP	Mat	-	pdf
AN 5742	LIN-IC	TV, NF-E, 12V, 0,34W	9-SIP	Mat	-	
AN 5743	LIN-IC	TV, NF-E, 12V, 1,3W	9-SIL	Mat	-	
AN 5750	LIN-IC	TV, HA-Synchr.-Prozessor, Ucc=6V	9-SIP	Mat	-	pdf pinout
AN 5753	LIN-IC	TV, HA-Synchr., HA-Tr., Ucc=12V	9-SIP	Mat	-	pdf pinout
AN 5760	LIN-IC	TV, VA-Synchr., VA-E	9-SIP	Mat	-	pdf pinout
AN 5762	LIN-IC	TV, VA-E, 7"CRT, Ucc=12V	12-SIP	Mat	-	
AN 5763	LIN-IC	TV, VA-E, 12"CRT, Ucc=12V	12-SIL	Mat	-	pdf
AN 5764 N	IC	Horizontal picture position control IC	MDIP	Pan	-	pdf pinout
AN 5765	IC	Crt heater voltage control IC	MDIP	Pan	-	pdf pinout
AN 5766 K	LIN-IC	Crt, Kissenentzerrung/pin Cushion	22-SDIP	Mat	-	
AN 5767 K	IC	Synchronizing signal processing IC	SIP	Pan	-	pdf pinout
AN 5768	IC	Tilt correction IC	SILP	Pan	-	pdf pinout
AN 5769	IC	H/v convergence correctioi IC	SILP	Pan	-	pdf pinout
AN 5790(N)	LIN-IC	Crt Display-Tr., X-ray-prot.	12-SIP	Mat	KA 2135	pdf pinout

Type	Device	Short Description	Fig.	Manu	Comparision Types	More at
AN 5791	LIN-IC	Crt Display-Tr., Phase Shift	9-SIP	Mat	-	pdf
AN 5792	LIN-IC	Crt Display-Tr., X-ray-prot.	12-SIL	Mat	-	pdf
AN 5795 NK	LIN-IC	Crt,Monitor, HA/VA-Prozessor, ...130kHz	22-SDIP	Mat	-	pdf
AN 5817 NK	LIN-IC	Ctv, Stereo-Decoder, dbx System(USA)	42-SDIP	Mat	-	pdf pinout
AN 5819 K	IC	Sound multiplex demodulator IC, TV	SILP	Pan	-	pdf pinout
AN 5820	LIN-IC	Ctv, Stereo-zf/sub-channel Demod.	14-DIP	Mat	-	
AN 5821	LIN-IC	Ctv, Stereo-control-signal	14-DIP	Mat	-	
AN 5822	LIN-IC	Ctv, Stereo-matrix	14-DIP	Mat	-	
AN 5825	LIN-IC	TV, MPX-detector, Stereo-decoder	20-DIP	Mat	-	pdf
AN 5826 NK	LIN-IC	TV, Stereo-ton-prozessor	28-SDIP	Mat	-	
AN 5829 S	IC	Sound multiplex decoder IC, U.s. TV	SILP	Pan	-	pdf pinout
AN 5832 SA	IC	Silicon Monolithic Bipolar IC	SILP	Pan	-	pdf pinout
AN 5833 SA	IC	Silicon Monolithic Bipolar IC	SILP	Pan	-	pdf pinout
AN 5835	LIN-IC	Dual, Volume+tone-control	12-SIP	Mat	(AN 5836)	pdf
AN 5836	LIN-IC	=AN 5835: physiolog. Volume-Control	12-SIP	Mat	KA 2107	pdf pinout
AN 5837	LIN-IC	TV, FB-control-interface	9-SIP	Mat	-	
AN 5838	LIN-IC	TV, Fb-control-interface (sound MPX)	9-SIP	Mat	-	
AN 5850	LIN-IC	TV	16-DIP	Mat	-	pdf pinout
AN 5855 K	LIN-IC	TV, Audio/video-umschalter/switch	28-SDIP	Mat	-	
AN 5856 K	LIN-IC	Ctv, RGB-Umschalter/switch	28-SDIP	Mat	-	
AN 5858 K	LIN-IC	Ctv, Audio/video-umschalter/switch	42-SDIP	Mat	-	pdf
AN 5860	LIN-IC	Ctv, Analog-switch f. Rgb-interface	14-DIP	Mat	-	pdf pinout
AN 5860 S	LIN-IC	=AN 5860: SMD	24-MDIP		-	pdf pinout
AN 5862 K	LIN-IC	Ctv, Analog-switch f. Rgb-interface	13-SIP	Mat	-	pdf pinout
AN 5862 S	LIN-IC	=AN 5862K: SMD	18-MDIP	Mat	-	pdf pinout
AN 5867 K	LIN-IC	CTV/Monitor, Rgb-interface	28-SDIP	Mat	-	pdf pinout
AN 5868 NK	LIN-IC	CTV/Monitor, Crt-interface	28-SDIP	Mat	-	
AN 5870 K	IC	Wide bandwidth analog switch IC	SILP	Pan	-	pdf pinout
AN 5900	SMPS-IC	SMPS-Controller	9-SIP	Mat	-	pdf pinout
AN 5902 S	SMPS-IC	SMPS-Controller, Ucc=3,5...14V	16-MDIP	Mat	-	
AN 5905	SMPS-IC	Smps-controller	18-DIP	Mat	-	pdf pinout
AN 5905 S	SMPS-IC	=AN 5905: SMD	18-MDIP	Mat	-	pdf pinout
AN 6011	LIN-IC	Camera, Y-signal + Alc	18-DIP	Mat	-	
AN 6012	LIN-IC	Camera, Y-signal + Alc	18-DIP	Mat	-	
AN 6014	LIN-IC	Camera, Tracking + Agc	18-DIP	Mat	-	
AN 6015	LIN-IC	Camera, Tracking + Agc	18-DIP	Mat	-	
AN 6020	LIN-IC	Camera, Weißpegel/white level control	16-DIP	Mat	-	
AN 6022	LIN-IC	Camera, Konvergenz	18-DIP	Mat	-	
AN 6031	LIN-IC	Camera, Color-signal	18-DIP	Mat	-	
AN 6040	LIN-IC	Camera, Color-encoder	9-SIP	Mat	-	
AN 6041	LIN-IC	Camera, Dual Balanced Modulator	9-SIP	Mat	-	
AN 6045	LIN-IC	Camera, Color-encoder	22-DIP	Mat	-	
AN 6050	LIN-IC	Camera, Vidicon-signal	16-DIP	Mat	-	
AN 6055	LIN-IC	Camera, Fade-control, View-finder	18-DIP	Mat	-	
AN 6080 FHN	LIN-IC	Modulator IC, Cdma system cellular phone	4-SIP	Pan	AN 608P	pdf pinout
AN 6095 SH	LIN-IC	Reception IF + transm. quad. modul. IC	4-SIP	Pan	-	pdf pinout
AN 6096 FHN	LIN-IC	Transmission and reception IC	4-SIP	Pan	-	pdf pinout
AN 6105 FHN	LIN-IC	Quadrature demodulation IC	4-SIP	Pan	-	pdf pinout
AN 6107 SA	IC	Digital Communication IF Amplifier IC	78	Pan	-	pdf pinout
AN 6108 SA	IC	Digital Communication IF Amplifier IC	78	Pan	-	pdf pinout
AN 6130(N)	LIN-IC	Fm-noise suppr.	18-DIP	Mat	-	pdf pinout
AN 6132	LIN-IC	Fm-noise suppr.	18-DIP	Mat	-	
AN 6132 S	LIN-IC	=AN 6132: SMD	18-MDIP	Mat	-	
AN 6135	LIN-IC	Hifi-pop-noise-suppr.	9-SIP	Mat	AN 6136	pdf pinout
AN 6136	LIN-IC	Hifi-pop-noise-suppr.	9-SIP	Mat	-	pdf pinout
AN 6140	LIN-IC	Cb, NF-signal	16-DIP	Mat	-	
AN 6150	LIN-IC	Telefon, Sprechkreis/speech Network	16-DIP	Mat	-	pdf
AN 6151 K	LIN-IC	Telefon, Sprechkreis/speech Network	22-DIP	Mat	-	
AN 6152	LIN-IC	Telefon, Sprechkreis/speech Network	16-DIP	Mat	-	pdf
AN 6153 N	LIN-IC	Telefon, Sprechkreis/speech Network	16-DIP	Mat	-	
AN 6153 NS	LIN-IC	=AN 6153N: SMD	16-MDIP	Mat	-	
AN 6154 K	LIN-IC	Telefon, Sprechkreis/speech Network	24-SDIP	Mat	-	
AN 6154 S	LIN-IC	=AN 6154K: SMD	24-MDIP	Mat	-	
AN 6157 NK	LIN-IC	Telefon, Sprechkreis/speech Network	22-SDIP	Mat	-	
AN 6162 SC	LIN-IC	Telefon, FM-prozessor f. Cordless Tel.	32-MDIP	Mat	-	pdf pinout
AN 6164 K	LIN-IC	Telefon, Prozessor	28-SDIP	Mat	-	pdf pinout
AN 6164 S	LIN-IC	=AN 6164K: SMD	28-MDIP	Mat	-	pdf pinout
AN 6166 NK	IC	Cordless Telephone Speech Network IC	78	Pan	-	pdf pinout
AN 6167 S	IC	Cordless Telephone Speech Network IC	78	Pan	-	pdf pinout
AN 6167 SB	IC	Cordless Telephone Speech Network IC	78	Pan	-	pdf pinout
AN 6170	LIN-IC	Telefon, Glocke/tone Ringer	8-DIP	Mat	-	
AN 6170 S	LIN-IC	=AN 6170: SMD	8-MDIP	Mat	-	
AN 6171	LIN-IC	Telefon, Glocke/tone Ringer	14-DIP	Mat	-	
AN 6172	LIN-IC	Telefon, Glocke/tone Ringer	8-DIP	Mat	-	
AN 6175 K	IC	Hands-free Speech Network IC	78	Pan	-	pdf pinout
AN 6182 K	IC	Recording and playing amplifier IC	78	Pan	-	pdf pinout
AN 6182 S	IC	Recording and playing amplifier IC	78	Pan	-	pdf pinout
AN 6208	LIN-IC	Stereo-recorder	16-DIP	Mat	-	
AN 6209	LIN-IC	VC/Recorder, A/w-verst.	22-DIP	Mat	-	pdf pinout
AN 6209 S	LIN-IC	=AN 6209: SMD	22-MDIP	Mat	-	pdf pinout
AN 6210	LIN-IC	2x Mikrof.- + Line-Verst./Amp., Agc	28-DIP	Mat	-	pdf
AN 6212	LIN-IC	Recorder	16-DIP	Mat	-	
AN 6213	LIN-IC	Recorder	16-DIP	Mat	-	
AN 6214	LIN-IC	Recorder, Steuerung/control	16-DIP	Mat	-	
AN 6221 S	LIN-IC	Recorder, NF-V, Agc	20-MDIP	Mat	-	

Type	Device	Short Description	Fig.	Manu	Comparision Types	More at
AN 6230 S	LIN-IC	Recorder, NF-E(Kopfh./Earphone),Ucc=3V	18-MDIP	Mat	-	
AN 6246	LIN-IC	Recorder, Autoreverse-control	9-SIP	Mat	-	
AN 6247	LIN-IC	Recorder, Autoreverse-Control	7-SIP	Mat	-	
AN 6247	LIN-IC	Recorder, Autoreverse-control	7-SIP	Mat	-	
AN 6248	LIN-IC	Recorder, Autoreverse-Control	7-SIP	Mat	-	
AN 6248	LIN-IC	Recorder, Autoreverse-control	7-SIP	Mat	-	
AN 6249	LIN-IC	Recorder, Autoreverse-control	7-SIP	Mat	AN 6250	
AN 6250	LIN-IC	Recorder, Autoreverse-control	7-SIP	Mat	(AN 6249)	
AN 6251	LIN-IC	Recorder, Steuerung/control	24-DIP	Mat	-	
AN 6252	LIN-IC	Recorder, Steuerung/control	22-DIP	Mat	-	
AN 6256	LIN-IC	Recorder, Steuerung/control	16-DIP	Mat	-	
AN 6257	LIN-IC	Recorder, Steuerung/control	8-MDIP	Mat	-	
AN 6260	LIN-IC	Recorder, Suchlauf/program Selection	16-DIP+g	Mat	-	
AN 6262(N)	LIN-IC	Recorder, Pause Detect.	9-SIP	Mat	AN 6263	pdf
AN 6263(N)	LIN-IC	Recorder, Pause Detect.	9-SIP	Mat	AN 6262	pdf
AN 6270	LIN-IC	Recorder, Motor-steuerung/control	16-DIP+g	Mat	-	pdf pinout
AN 6280	LIN-IC	Zähler/counter, Treiber/driver	16-DIP	Mat	-	
AN 6290 S	LIN-IC	Recorder, dbx-kompander	20-MDIP	Mat	-	
AN 6291	LIN-IC	Recorder, Dual, dbx II, Noise Reduct.	22-DIP	Mat	-	pdf
AN 6291 S	LIN-IC	=AN 6291: SMD	22-MDIP	Mat	-	pdf
AN 6296	LIN-IC	Hifi-vc, Noise Reduction (vhs, 8mm)	22-DIP	Mat	-	
AN 6296 S	LIN-IC	=AN 6296: SMD	22-MDIP	Mat	-	
AN 6297 S	LIN-IC	Vc, Audio Noise Reduction (8mm)	20-MDIP	Mat	-	
AN 6298 NK	LIN-IC	Hifi-vc, Audio Noise Reduction (vhs)	28-SDIP	Mat	-	
AN 6298 NS	LIN-IC	=AN6298NK: SMD	28-MDIP	Mat	-	
AN 6300	LIN-IC	Vc, Video-signal	24-DIP	Mat	-	pdf pinout
AN 6306	LIN-IC	Vc, Video-signal	22-DIP	Mat	-	pdf pinout
AN 6306 S	LIN-IC	=AN 6306: SMD	22-MDIP	Mat	-	pdf pinout
AN 6307	LIN-IC	Vc, Video-kopfverst./head Amplifier	9-SIP	Mat	-	pdf pinout
AN 6308	LIN-IC	Vc, Analogschalter/analog Switch	8-DIP	Mat	-	pdf
AN 6308 S	LIN-IC	=AN 6308: SMD	8-MDIP	Mat	-	pdf
AN 6310	LIN-IC	VC	24-DIP	Mat	-	pdf pinout
AN 6320 N	LIN-IC	Vc, Kopf-verst./head Amplifier	14-DIP	Mat	-	pdf pinout
AN 6321	LIN-IC	Vc, Video-Playback	28-DIP+g	Mat	-	
AN 6326(N)	LIN-IC	Vc, Video-kopfverst./head Amplifier	18-DIP	Mat	-	
AN 6327	LIN-IC	Vc, Video-signal	22-DIP	Mat	-	
AN 6327 S	LIN-IC	=AN 6327: SMD	22-MDIP	Mat	-	
AN 6328	LIN-IC	Vc, Stoerunterdrückung/noise Cancellor	22-DIP	Mat	-	
AN 6328 S	LIN-IC	=AN 6328: SMD	22-MDIP	Mat	-	
AN 6330	LIN-IC	Vc, Kopf-verst./head Amplifier	14-DIP	Mat	-	
AN 6331	LIN-IC	Vc, Video-playback	28-DIP+g	Mat	AN 6332	
AN 6332	LIN-IC	Vc, Video-playback	28-DIP+g	Mat	-	pdf pinout
AN 6337	LIN-IC	VC	22-DIP	Mat	-	
AN 6337 S	LIN-IC	=AN 6337: SMD	22-MDIP	Mat	-	
AN 6340	LIN-IC	Vc, Servo	28-DIP	Mat	AN 6344	
AN 6341(N)	LIN-IC	Vc, Capstan-servo	16-DIP	Mat	-	pdf
AN 6342(N)	LIN-IC	Vc, Frequ.-teiler/divider	7-SIP	Mat	-	pdf
AN 6343	LIN-IC	Vc, Servo	18-DIP	Mat	-	
AN 6344	LIN-IC	Vc, Servo	28-DIP	Mat	-	pdf pinout
AN 6345	LIN-IC	Vc, Servo	16-DIP	Mat	-	
AN 6346(N)	LIN-IC	Vc, Servo (cylinder Interface)	18-DIP	Mat	-	
AN 6347	LIN-IC	Vc, Capstan-servo	20-DIP	Mat	-	
AN 6350	LIN-IC	Vc, Cylinder-servo	28-DIP	Mat	-	
AN 6352	LIN-IC	Vc, Pitch-control	12-SIP	Mat	-	
AN 6353	LIN-IC	Vc, Pitch-control	12-SIP	Mat	-	
AN 6354	LIN-IC	Vc, Servo	9-SIP	Mat	-	
AN 6356(N)	LIN-IC	Vc, Servo (cylinder Interface)	18-DIP	Mat	-	
AN 6357(N)	LIN-IC	Vc, Servo (capstan Interface)	20-DIP	Mat	-	
AN 6359(N)	LIN-IC	Vc, Servo (capstan Interface)	20-DIP	Mat	-	
AN 6360	LIN-IC	Vc, Color-signal	18-DIP	Mat	-	pdf pinout
AN 6360 S	LIN-IC	=AN 6360: SMD	22-MDIP	Mat	-	pdf pinout
AN 6361 N	LIN-IC	Vc, Color-apc	16-DIP	Mat	-	
AN 6361 NS	LIN-IC	=AN 6361: SMD	22-MDIP	Mat	-	
AN 6362	LIN-IC	Vc, Color-afc (ntsc,Pal)	18-DIP	Mat	-	pdf pinout
AN 6362 S	LIN-IC	=AN 6362: SMD	22-MDIP	Mat	-	pdf pinout
AN 6363	LIN-IC	Vc, Color-afc (pal)	20-DIP	Mat	-	pdf
AN 6363 S	LIN-IC	=AN 6363: SMD	22-MDIP	Mat	-	pdf
AN 6364 S	LIN-IC	Vc, Signal-discrim. (pal,Secam)	14-MDIP	Mat	-	
AN 6366 NK	LIN-IC	Vc, Color-signal-prozessor (ntsc)	22-SDIP	Mat	-	
AN 6366 NS	LIN-IC	=AN 6366NK: SMD	22-MDIP	Mat	-	
AN 6367 K,NK	LIN-IC	Vc, Color-signal-proz.(pal,SECAM,Ntsc)	22-SDIP	Mat	-	
AN 6367 S,NS	LIN-IC	=AN 6367K: SMD	22-MDIP	Mat	-	
AN 6368	LIN-IC	Vc, Color-signal-detector (pal,Secam)	14-DIP	Mat	-	
AN 6368 S	LIN-IC	=AN 6368: SMD	14-MDIP	Mat	-	
AN 6371	LIN-IC	Vc, Color-apc (pal)	16-DIP	Mat	-	pdf pinout
AN 6371 S	LIN-IC	=AN 6371: SMD	22-MDIP	Mat	-	pdf pinout
AN 6381 S	LIN-IC	Vc, Motor-interface (capstan)	14-MDIP	Mat	-	
AN 6386 K	LIN-IC	Vc, Cylinder-motor-servo	24-SDIP	Mat	-	
AN 6387	LIN-IC	Vc, Cylinder-motor-servo	24-SDIP	Mat	-	pdf
AN 6391 NK	LIN-IC	Vc, FM Audio-signal-prozessor	28-SDIP	Mat	-	
AN 6391 NS	LIN-IC	=AN6391NK: SMD	28-MDIP	Mat	-	
AN 6395	LIN-IC	Vc, FM-signal	9-SIP	Mat	-	
AN 6396	LIN-IC	Vc, A/W-verst./rec.p.b. Amplifier	22-DIP	Mat	-	
AN 6396 S	LIN-IC	=AN 6396: SMD	22-MDIP	Mat	-	
AN 6397	LIN-IC	Vc, Color-signal (secam)	24-DIP	Mat	-	

Type	Device	Short Description	Fig.	Manu	Comparision Types	More at
AN 6397 S	LIN-IC	=AN 6397: SMD	24-MDIP	Mat	-	
AN 6398	LIN-IC	Vc, Color-killer (secam)	20-DIP	Mat	-	
AN 6398 S	LIN-IC	=AN 6398: SMD	20-MDIP	Mat	-	
AN 6410	LIN-IC	NF-verst./amplifier f. Modulator	9-SIP	Mat	-	
AN 6425 K	LIN-IC	Telefon, Sprechkreis/speech Network	28-SDIP	Mat		pdf
AN 6426 K,NK	LIN-IC	Telefon, Hand-free Speech Network	42-SDIP	Mat		pdf pinout
AN 6426 K	LIN-IC	Telefon, Prozessor	42-SDIP	Mat		
AN 6448 NFBP	IC	Speech Network IC, Cross-point Switch	78	Pan	-	pdf pinout
AN 6454 SH	IC	Pager Direct Conv. Fsk Demod. Mixer IC	78	Pan	-	pdf pinout
AN 6480	LIN-IC	Auto-/car-telefon, Zf	18-DIP	Mat		pdf
AN6500	OP+V-REF	Vs:24V Vo:>3.5/0V Vi0:2mV	8DS	Mat		data pinout
AN 6500	OP-IC	int. Ref.-Volt., 24V, 160mA, -20…+75°	8-DIP	Mat		data pdf pinout
AN 6500 S	OP-IC	=AN 6500: SMD	8-MDIP	Mat		data pdf pinout
AN6501	OP+V-REF	Vs:24V Vo:>3.5/0V Vi0:2mV	7I	Mat	.	data pinout
AN 6501	OP-IC	=AN 6500: Fig.→	7-SIP	Mat		data pdf pinout
AN 6530	Z-IC	+5..30V, 0,5A	4-DIP+b	Mat		
AN 6531	Z-IC	=AN 6530: Fig.→	15/4Pin	Mat		pdf pinout
AN 6535	Z-IC	-5…-30V, 0,5A	15/4Pin	Mat		pdf
AN 6536	Z-IC	=AN 6535: Fig.→	6-DIP+g	Mat		pdf pinout
AN 6540	Z-IC	lo-drop, +8,5V, 0,24A, Adj. Rise Time	15/4Pin	Mat		pdf
AN 6541	Z-IC	lo-drop, +9V, 0,3…0,6A	17b	Mat	-	pdf
AN 6545	Z-IC	+5V, 0,15A	85/4Pin	Mat	-	pdf
AN 6545 SP	Z-IC	=AN 6545: Fig.→	≈5-SIL	Mat	-	pdf
AN 6546 SP	Z-IC	+5V	≈5-SIL	Mat		
AN 6548 S	Z-IC	+3,2V, Stand-by	8-MDIP	Mat		
AN6550	2xOP	Vs:±18V Vu:100dB Vo:±13V Vi0:0.5mV	9I	Mat		data pdf pinout
AN 6550	OP-IC	Dual, NF-Equal., ±12V, -20…+75°	9-SIP	Mat		data pinout
AN6551	2xOP	Vs:±18V Vu:100dB Vo:±13V Vi0:0.5mV	9I	Mat		data pdf pinout
AN 6551	OP-IC	Dual, NF-Equal., ±18V, -20…+75°	9-SIP	Mat	TA 75558S	data pinout
AN6552	2xOP	Vs:±18V Vu:100dB Vo:±13V Vi0:0.5mV	8DS	Mat		data pdf pinout
AN 6552	OP-IC	=AN 4558…	8-DIP	Mat	→AN 4558	data pinout
AN6553	2xOP	Vs:±18V Vu:100dB Vo:±13V Vi0:0.5mV	8DS	Mat		data pdf pinout
AN 6553	OP-IC	=AN 6552: verbessert/improved	8-DIP	Mat	NJM 4559…, TA 75559…, … µPC 4559…	data pinout
AN 6553 S	OP-IC	=AN 6553: SMD	8-MDIP	Mat	-	data pinout
AN6554	4xOP	Vs:±18V Vu:100dB Vo:±13V Vi0:0.5mV	14DS	Mat		data pdf pinout
AN 6554	OP-IC	Quad, NF, ±18V, -20…+75°	14-DIP	Mat	→Serie 124	data pinout
AN 6554 NS	OP-IC	=AN 6554: SMD	14-MDIP	Mat	-	data pinout
AN6555	2xOP	Vs:±18V Vu:100dB Vo:±13V Vi0:0.5mV	9I,8D	Mat		data pdf pinout
AN 6555	OP-IC	=AN 6556: Fig.→	9-SIP	Mat	TA 75559S	data pinout
AN6556	2xOP	Vs:±18V Vu:100dB Vo:±13V Vi0:0.5mV	8S	Mat		data pdf pinout
AN 6556	OP-IC	Dual, lo-noise, ±18V, -20…+75°	8-DIP	Mat		data pinout
AN 6556 S	OP-IC	=AN 6556: SMD	8-MDIP	Mat		data pinout
AN6557	2xOP	Vs:±18V Vu:100dB Vo:±14V Vi0:0.5mV	9I	Mat		data pdf pinout
AN 6557	OP-IC	=AN 6558: Fig.→	9-SIP	Mat		data pinout
AN6558	2xOP	Vs:±18V Vu:100dB Vo:±14V Vi0:0.5mV	8DS	Mat		data pdf pinout
AN 6558	OP-IC	Dual, lo-noise, ±18V, -20…+75°, 5V/µs	8-DIP	Mat		data pinout
AN 6558 S	OP-IC	=AN 6558: SMD	8-MDIP	Mat		data pinout
AN6561	2xOP	Vs:±16V Vu:100dB Vi0:2mV f:700Mc	9I	Mat		data pdf pinout
AN 6561(L)	OP-IC	Dual, ±15V, -20…+75°	9-SIP	Mat		data pinout
AN6562	2xOP	Vs:±16V Vu:100dB Vi0:2mV f:700Mc	8DS	Mat		data pdf pinout
AN 6562 …	OP-IC	=AN 1358…	8-DIP	Mat	→AN 1358	data pinout
AN6564	4xOP	Vs:±16V Vu:100dB Vi0:2mV f:700Mc	14DS	Mat		data pdf pinout
AN 6564 …	OP-IC	=AN 1324…	14-DIP	Mat	→AN 1324	data pinout
AN6567	2xOP	Vs:±9V Vu:100dB Vo:>3.3/0V Vi0:2mV	9I	Mat		data pdf pinout
AN 6567	OP-IC	=AN 6568: Fig.→	9-SIP	Mat		data pinout
AN6568	2xOP	Vs:±9V Vu:100dB Vo:>3.3/0V Vi0:2mV	8DS	Mat		data pdf pinout
AN 6568	OP-IC	Dual, hi-current, ±9V, -20…+75°	8-DIP	Mat		data pinout
AN 6568 S	OP-IC	=AN 6568: SMD	8-MDIP	Mat	-	data pinout
AN6570	UNI	Vs:±18V Vu:106dB Vo:±14V Vi0:0.5mV	8SD	Mat		data pinout
AN 6570 …	OP-IC	=AN 1741…	8-DIP	Mat	→AN 1741	data pdf pinout
AN6571	2xOP	Vs:±18V Vu:106dB Vo:±14V Vi0:0.5mV	9I	Mat		data pinout
AN 6571	OP-IC	=AN 1458: Fig.→	9-SIP	Mat	TA 75458S	data pinout
AN6572	2xOP	Vs:±18V Vu:106dB Vo:±14V Vi0:0.5mV	8D	Mat		data pinout
AN 6572	OP-IC	=AN 1458	8-DIP	Mat	→AN 1458	data pinout
AN6573	UNI	Vs:±18V Vu:106dB Vo:±14V Vi0:0.5mV	7I	Mat,Pan		data pdf pinout
AN 6573	OP-IC	=AN 1741: Fig.→	7-SIP	Mat	TA 7504S	data pdf pinout
AN6574	4xLO-NOISE	Vs:±18V Vu:110dB Vo:±13.4V Vi0:0.3mV	14DS	Mat		data pinout
AN 6574	OP-IC	Quad, lo-noise, ±18V, -20…+75°, 6V/µs	14-DIP	Mat		data pinout
AN 6574 S	OP-IC	=AN 6574: SMD	14-MDIP	Mat		data pinout
AN6581	2xFET	Vs:±18V Vu:106dB Vo:±14V Vi0:0.5mV	9I	Mat		data pinout
AN 6581	OP-IC	=AN 1082: Fig.→	9-SIP	Mat		data pinout
AN6583	FET	Vs:±18V Vu:106dB Vo:±13.5V Vi0:2mV	7I	Mat		data pinout
AN 6583	OP-IC	=AN 1081: Fig.→	7-SIP	Mat		data pinout
AN6592	2xLO-POWER	Vs:±18V Vu:100dB Vo:±13V Vi0:1mV	8DS	Mat		data pinout
AN 6592	OP-IC	Dual, lo-power, ±18V, -20…+75°	8-DIP	Mat		data pinout
AN 6592 S	OP-IC	=AN 6592: SMD	8-MDIP	Mat		data pinout
AN6593	UNI	Vs:±18V Vu:>96dB Vo:>±12V Vi0:6mV	9I	Mat		data pdf pinout
AN 6593	OP-IC	=AN 4250: Fig.→	9-SIP	Mat		data pinout
AN 6607 NS	LIN-IC	DAT-Rec., DC-Motor-Tr., 2Speed	16-MDIP	Mat		pdf pinout
AN 6608	LIN-IC	DC-Motor-Tr., 2 Speed	16-DIP+g	Mat		pdf
AN 6609 N	LIN-IC	Vc, DC-Motor-Tr., 2 Speed	16-DIP+g	Mat		pdf
AN 6610	LIN-IC	Motorregler/Speed Control, Uref=1,2V	14	Mat	TDA 1151	
AN 6611 S	LIN-IC	Recorder-steuerung/control	28-MDIP	Mat	-	
AN 6612	LIN-IC	Motorregler/Speed Control, Uref=1,32V	8-DIP	Mat	-	pdf
AN 6612 S	LIN-IC	=AN 6612: SMD	8-MDIP	Mat	-	pdf

Type	Device	Short Description	Fig.	Manu	Comparision Types	More at
AN 6631 S	LIN-IC	Direct-drive Motor Control	24-MDIP	Mat		
AN 6650	LIN-IC	Motorregler/Speed Control	8-DIP	Mat	UTC6650	pdf
AN 6650 S	LIN-IC	=AN 6650: SMD	8-MDIP	Mat		pdf pinout
AN 6651	LIN-IC	Motorregler/Speed Control, Uref=1,0V	85/4Pin	Mat	(KA 2407)	pdf
AN 6652	LIN-IC	Motorregler/Speed Control, Uref=1,25V	85/4Pin	Mat	-	pdf
AN 6654 S	LIN-IC	3V Motorregler/Speed Control	8-MDIP	Mat		
AN 6655 S	LIN-IC	Motorregler/Speed Control	16-MDIP	Mat		
AN 6656	LIN-IC	Motorregler/Speed Control	16-DIP	Mat		
AN 6656 S	LIN-IC	=AN 6656: SMD	16-MDIP	Mat		
AN 6657	LIN-IC	Motorregler/Speed Control, 4,5...14V	16-DIP	Mat		pdf
AN 6657 S	LIN-IC	=AN 6657: SMD	16-MDIP	Mat		pdf
AN 6659 S	LIN-IC	1,5V Motorregler/Speed Control	10-MDIP	Mat		
AN 6660	LIN-IC	Vc, DC-Motor-Tr., Ucc=4..20V	9-SIL(20mm	Mat		
AN 6660 K	LIN-IC	=AN 6660:	9-SIL(17mm	Mat		
AN 6662	LIN-IC	Vc, Tape-/cassette-loading Motor-tr.	10-SIP	Mat		
AN 6663 S	LIN-IC	SMD, Camera, Autofocus-motor-control	8-MDIP	Mat	-	pdf
AN 6663 SP	LIN-IC	=AN6663S: Mini-SIL	≤5-SIL	Mat	-	pdf
AN 6664 S	LIN-IC	SMD, Camera, Autofocus-motor-control	16-MDIP	Mat	-	
AN 6665 S	LIN-IC	Camera, DC-Motor-Tr.	20-MDIP	Mat	-	
AN 6666 S	LIN-IC	Camera, Dc-motor-tr.	28-MDIP	Mat	-	
AN 6667 S	LIN-IC	Camera, Dc-motor-tr.	18-MDIP	Mat	-	
AN 6676	LIN-IC	Vc, Servo (capstan)	24-DIP	Mat	-	
AN 6701	LIN-IC	Temp.-sensor	4-SIP	Mat		pdf
AN 6701 S	LIN-IC	=AN 6701: SMD	8-MDIP	Mat	-	pdf
AN 6721	IC	IGBT Drive IC	78	Pan	-	pdf pinout
AN 6751	LIN-IC	Stromüberwachung/Current Ctrl., Ucc=3V	8-DIP	Mat	-	
AN 6780	LIN-IC	Timer, -20...+75°	7-SIP	Mat	-	pdf
AN 6780 S	LIN-IC	=AN 6780: SMD	14-MDIP	Mat	-	pdf
AN 6781	LIN-IC	Timer, LED-Treiber/driver, 6 LED	16-DIP	Mat		
AN 6811	LIN-IC	Frequ.-Teiler/divider, :3/4/8/12/16	9-SIP	Mat	-	pdf pinout
AN 6817	LIN-IC	Frequ.-Teiler/divider, 1:252/1:256	8-DIP	Mat	-	
AN 6820	MOS-IC	=AN 6821: Fig.→	8-SIP	Mat	-	
AN 6821	MOS-IC	Frequ.-Teiler/divider 1:20	9-SIP	Mat	-	
AN 6823	LIN-IC	Stereo Prescaler	9-SIP	Mat	-	
AN 6855(T)	A/D-IC	4 Bit, hi-speed	16-DIP	Mat	-	
AN 6856	A/D-IC	6 Bit, hi-speed	24-DIP	Mat	-	
AN 6857(N)	A/D-IC	8 Bit, hi-speed	40-DIC	Mat	-	
AN 6870 N	LIN-IC	Dual FLT-Tr., 18 Dot Peak Hold	28-DIP	Mat	-	
AN 6873(N)	LIN-IC	FLT-Display, 8-Segment-Tr.	18-DIP	Mat	-	pdf
AN 6873 NS	LIN-IC	=AN 6873(N): SMD	18-MDIP	Mat	-	
AN 6875	LIN-IC	LED-Treiber/Driver, 5 LED, log.	9-SIP	Mat	-	
AN 6876	LIN-IC	=AN 6875: lin.	9-SIP	Mat	-	pdf pinout
AN 6877	LIN-IC	LED-Treiber/Driver, 7 LED, lin. Gnd connected to Heatsink/kühlfahne	16-DIP+g	Mat	-	pdf pinout
AN 6878	LIN-IC	LED-Treiber/Driver, 7 LED, log. Gnd connected to Heatsink/kühlfahne	16-DIP+g	Mat	-	pdf pinout
AN 6879	LIN-IC	LED-Decoder, 7 LED	16-DIP	Mat		
AN 6880	LIN-IC	Servo-motorregler/motor Control	7-SIP	Mat	-	
AN 6881	LIN-IC		9-SIP	Mat	-	
AN 6882	LIN-IC	LED-Treiber/Level Meter, 7 LED	16-DIP	Mat	-	
AN 6884	LIN-IC	LED-Treiber/Level Meter, 5 LED	9-SIP	Mat	BA 6124, KA 2284, LB 1403, KIA 6966S	pdf pinout
AN 6886	LIN-IC	Dual-Input LED-Decoder, 5 LED	14-DIP	Mat	(KIA 6976P)	
AN 6887	LIN-IC	Dual-Input LED-Decoder, 7 LED	16-DIP	Mat	-	
AN 6888	LIN-IC	Dual LED-Decoder, 2x5 LED	18-DIP	Mat	-	
AN 6889	LIN-IC	Dual LED-Decoder, 2x5 LED	18-DIP	Mat	-	
AN 6891	LIN-IC	LED-Decoder, 12 LED	18-DIP	Mat	-	
AN6912	4xCOMP	Vs:±18V Vu:200V/mV Vi0:2mV	14DS	Mat		data pinout
AN 6912(N)	KOP-IC	=AN 1339, =LM 2901	14-DIP	Mat	→AN 1339,→LM 2901	
AN 6912 S	KOP-IC	=AN 6912: SMD	14-MDIP	Mat		
AN6913	2xCOMP	Vs:36V Vu:200V/mV Vi0:1mV	9I	Mat		data pdf pinout
AN 6913(L)	KOP-IC	Dual, ±18V, -30...+85°	9-SIP	Mat		data pinout
AN6914	2xCOMP	Vs:36V Vu:200V/mV Vi0:1mV	8DS	Mat		data pdf pinout
AN 6914 ...	KOP-IC	=AN 1393, =LM 2903	8-DIP	Mat	→AN 1393,→LM 2903	data pinout
AN6915	2xCOMP	Vs:36V Vu:200V/mV Vi0:1mV	9I	Mat		data pdf pinout
AN 6915	KOP-IC	=AN 6916: Fig.→	9-SIP	Mat		data pinout
AN6916	2xCOMP	Vs:36V Vu:200V/mV Vi0:1mV	8DS	Mat		data pdf pinout
AN 6916	KOP-IC	Dual, 36V, -30...+85°	8-DIP	Mat		data pinout
AN 6916 S	KOP-IC	=AN 6916: SMD	8-MDIP	Mat		data pinout
AN6918	4xCOMP	Vs:36V Vu:200V/mV Vi0:1mV	14D	Mat		data pinout
AN 6918	KOP-IC	Quad, 36V, -30...+85°	14-DIP	Mat		data pinout
AN 6995	LIN-IC	Servo-control	22-DIP	Mat		
AN 7000	LIN-IC	AM/FM-Tuner, ZF, Stereo-decoder	28-DIP+g	Mat	-	
AN 7001	LIN-IC	AM/FM-Tuner, ZF, Stereo-decoder	28-DIP+g	Mat	-	pdf
AN 7002 K	LIN-IC	AM-Radio, 1Chip	22-DIP	Mat	-	pdf
AN 7002 S	LIN-IC	=AN 7002K: SMD	24-MDIP	Mat	-	
AN 7006 NS	LIN-IC	AM/FM-Radio, Tuner	28-MDIP	Mat	-	
AN 7007 S(U)	LIN-IC	AM/FM-Tuner, 1Chip	28-MDIP	Mat	-	
AN 7008 K	LIN-IC	AM-Radio, 1Chip, Ucc=1,5V	22-DIP	Mat	-	
AN 7009 S	LIN-IC	AM-Radio, 1Chip, Ucc=3V	24-MDIP	Mat	-	
AN 7014 K	LIN-IC	Recorder, 2x A/W-Verst., Mute	30-SDIP	Mat	-	
AN 7015 S	LIN-IC	Recorder, 2x A/W-Verst., Ucc=3V	22-MDIP	Mat	-	
AN 7016 NK	LIN-IC	Recorder, 2x A/W-Verst., Mute	30-SDIP	Mat	-	
AN 7017 S,SB	LIN-IC	FM/TV-V, Front-End, Ucc=1,5V	16-MDIP	Mat	-	pdf pinout
AN 7017 S	LIN-IC	SMD, FM/TV-V, Front-End, Ucc=1..2V	16-MDIP	Mat	-	pdf
AN 7024	LIN-IC	AM-Tuner, FM-ZF, MPX-stereo-decoder	18-SIP	Mat	-	

Type	Device	Short Description	Fig.	Manu	Comparision Types	More at
AN 7025 K	LIN-IC	AM-Tuner, FM-ZF, PLL-MPX-decoder	22-DIP	Mat	-	
AN 7025 S	LIN-IC	=AN 7025K: SMD	24-MDIP	Mat		
AN 7030 S	LIN-IC	R-dat, Kopf-verst./head-amp.	42-MDIP	Mat		
AN 7035 SC	LIN-IC	R-dat, Clock	32-MDIP	Mat		
AN 7060	LIN-IC	HiFi, NF-Treiber/Driver, Ucc=80V	9-SIP	Mat		pdf
AN 7062(N)	LIN-IC	2x HiFi-NF-V/Tr., Ucc=+74/-16V	18-DIP	Mat		pdf pinout
AN 7070	LIN-IC	NF-Treiber/Driver, ±40V, 1A	18-DILP	Mat		
AN 7071	LIN-IC	Vst.-schutzschaltg./amplif. Protect.	14-DIP	Mat		
AN 7072(N)	LIN-IC	NF-V, Mute	7-SIP	Mat		
AN 7074 K	LIN-IC	Hifi-verst./amplifier, Mute	13-SIP	Mat		
AN 7082 K	LIN-IC	Recorder, NF-V+Kopfh./Headph.-Amp., 3V	22-SDIP	Mat		
AN 7085 NS	LIN-IC	Recorder, NF-V+Kopfh./Headph.-Amp., 3V	20-MDIP	Mat		
AN 7086 S	LIN-IC	Recorder, NF-V+Kopfh./Headph.-Amp., 3V	24-MDIP	Mat		
AN 7090 FHQ	IC	Peripheral analog IC, audio	78	Pan	-	pdf pinout
AN 7100 S	LIN-IC	2x NF-E, 2x2mW(1,5V/150Ω), Ucc=1...3V	18-MDIP	Mat		
AN 7101 S	LIN-IC	2x NF-E, 2x12mW(3V/100Ω),Ucc=1,8..4,5V	14-MDIP	Mat		
AN 7103	LIN-IC	NF-E		Mat		
AN 7104	LIN-IC	NF-E		Mat		
AN 7105	LIN-IC	2x NF-V/E, 4,2...9V, 2x0,38W(6V/8Ω)	18-DIP	Mat		pdf pinout
AN 7106 K	LIN-IC	2x NF-V/E, 1,8...4,5V, 2x0,14W(3V/4Ω)	24-SDIP	Mat		pdf pinout
AN 7108	LIN-IC	Recorder, 2x NF-V/E(Kopfh./Headph.),3V	16-DIP	Mat	CXA 1034P, KA 22132	
AN 7109 S	LIN-IC	Recorder, 2x NF-V/E(Kopfh./Headph.),3V	28-MDIP	Mat		
AN 7110	LIN-IC	NF-E, 18V, 2A, 1,2W(9V/8Ω)	9-SIP	Mat	AN 7130, KIA 6278S	
AN 7111	LIN-IC	NF-E, 18V, 2A, 1,2W(9V/8Ω)	9-SIP	Mat	AN 7131, AN 7140	
AN 7112	LIN-IC	NF-E, 14V, 0,5A, 0,7W(9V/16Ω)	9-SIP	Mat	KA 2212, LA 4140, TA 7313(AP)	
AN 7114	LIN-IC	NF-E, 11V, 1,5A, 1W(6V/4Ω)	14-DIP+g	Mat	AN 7115	
AN 7115	LIN-IC	NF-E, 13V, 1,5A, 2,1W(9V/4Ω)	14-DIP+g	Mat		pdf pinout
AN 7116	LIN-IC	NF-E, 9V, 2A, 0,77W(6V/4Ω)	9-SIP	Mat	-	pdf pinout
AN 7117	LIN-IC	NF-E, 9V, 2A, 0,65W(6V/4Ω)	9-SIP	Mat		
AN 7118	LIN-IC	2x NF-E, 4,5V, 1A, 2x0,13W(3V/4Ω)	16-DIP	Mat		
AN 7118 S	LIN-IC	=AN 7118: SMD	18-MDIP	Mat		
AN 7120	LIN-IC	NF-E, 18V, 2A, 2,1W(9V/4Ω)	14-DIP+g	Mat		pdf pinout
AN 7124	LIN-IC	2x Aud. PAmp 3.1W, Standby, Muting	SIL	Pan	-	pdf pinout
AN 7125	LIN-IC	2x BTL Power Amp, 13.5W (12V,4Ohm)	SIL	Pan	-	
AN 7130	LIN-IC	NF-E, 18V, 3A, 4,2W(13V/4Ω)	9-SIL	Mat		
AN 7131	LIN-IC	NF-E, 24V, 4A, 5W(13V/4Ω)	9-SIL	Mat	AN 7140	
AN 7133 N	LIN-IC	2x NF-E, 24V, 6A, 2x5,8W(12V/3Ω)	23-SQL	Mat	-	
AN 7134 NR	LIN-IC	2x NF-E, 24V, 6A, 2x7,5W(15V/3Ω)	23-SQL	Mat	-	
AN 7139	LIN-IC	2x NF-E, 24V, 3,5A, 2x2,1W(9V/4Ω)	12-SIL	Mat	AN 7148	
AN 7140	LIN-IC	NF-E, 24V, 4A, 5W(13V/4Ω)	9-SIL	Mat	-	
AN 7141(N)	LIN-IC	NF-E, 3,8...18V, 2A, 1W(6V/4Ω)	9-SIP	Mat	-	
AN 7142	LIN-IC	2x NF-E, 3,8...18V, 4A, 2x1W(6V/4Ω)	16-DIP+g	Mat	-	
AN 7143	LIN-IC	2x NF-E, 4,8...24V, 2x>2W(9V/4Ω)	12-SIL	Mat	-	
AN 7145 H	LIN-IC	2x NF-E, 24V, 4A, 2x4,5W(16V/8Ω)	18-DILP	Mat	-	
AN 7145 L	LIN-IC	=AN7145 H: 20V, 4A, 2x1W(6V/4Ω)	18-DILP	Mat	-	
AN 7145 M	LIN-IC	=AN 7145 H: 20V, 4A, 2x2,4W(9V/4Ω)	18-DILP	Mat	-	
AN 7146 H	LIN-IC	2x NF-E, 24V, 4A, 2x4,5W(16V/8Ω)	18-DILP	Mat	AN 7145H	
AN 7146 M	LIN-IC	=AN 7146 H: 20V, 4A, 2x2,3W(9V/8Ω)	18-DILP	Mat	AN 7145M,H	
AN 7147(N)	LIN-IC	2x NF-E, 24V, 4A, 2x5,3W(12V/3Ω)	12-SIL	Mat	AN 7149(N)	pdf
AN 7148	LIN-IC	2x NF-E, 24V, 3,5A, 2x2,1W(9V/4Ω)	12-SIL	Mat	AN 7139	
AN 7149(N)	LIN-IC	Dual 5.3W Audio Power Amplifier	12-SIL	Pan	AN 7147(N)	pinout
AN 7150	LIN-IC	NF-E, 18V, 2,2A, 5,7W(13V/4Ω)	11-SIP	Mat	AN 7154	
AN 7151	LIN-IC	=AN 7150:spiegelb.Pinbel./rev.pinning	11-SIP	Mat	AN 7155	
AN 7154	LIN-IC	NF-E, 24V, 4A, 5,5W(13V/4Ω)	11-SIP	Mat	-	
AN 7155	LIN-IC	=AN 7154:spiegelb.Pinbel./rev.pinning	11-SILP	Mat	-	
AN 7156(N)	LIN-IC	2x NF-E, 24V, 4A, 2x5,5W(13V/4Ω)	12-SILP	Mat	-	
AN 7158(N)	LIN-IC	2x NF-E, 24V, 4A, 2x7,5W(16V/4Ω)	12-SILP	Mat	-	pdf
AN 7160	LIN-IC	NF-E, 24V, 4A, 17W(13V/4Ω)	12-SILP	Mat	-	pinout
AN 7161 N	LIN-IC	2x NF-E, 26V, 4A, 23W(15V/4Ω), BTL	12-SIL	Mat	-	
AN 7162 K	LIN-IC	2x NF-E, 20V, 4A, 14W(13V/4Ω), BTL	9-SIL	Mat	-	pdf
AN 7163	LIN-IC	2x NF-E, 24V, 4A, 17W(13V/4Ω), BTL	12-SIL	Mat	-	
AN 7164	LIN-IC	2x NF-E, 30V, 5A, 30W(21V/8Ω)	12-SILP	Mat	-	
AN 7166	LIN-IC	2x NF-E, 24V, 4A, 2x5,5W(13V/4Ω)	12-SILP	Mat	-	
AN 7168	LIN-IC	2x NF-E, 24V, 4A, 2x5,7W(13V/4Ω)	12-SILP	Mat	AN 7169	
AN 7169	LIN-IC	=AN 7168: rauscharm/low noise	12-SILP	Mat	-	
AN 7170	LIN-IC	NF-E, 35V, 4A, 18W(26V/4Ω)	11-SILP	Mat	-	
AN 7171 (N)K	LIN-IC	2x NF-E, 24V, 6A, 2x>9W(13V/4Ω),on=+5V	16-SQL	Mat	(AN 7173K)	
AN 7172 (N)K	LIN-IC	NF-E, 24V, 4A, 14W(13V/4Ω), BTL	9-SIP	Mat	-	
AN 7173 NK	LIN-IC	=AN 7171NK: Standby=0V, on=+5V	16-SQL	Mat	(AN 7171NK)	pdf pinout
AN 7174 K	LIN-IC	2x NF-E, 24V, 6A, 12,5W(13V/4Ω), BTL	16-SQL	Mat	-	pdf
AN 7177	LIN-IC	2x NF-E, 24V, 6A, 2x18W(13V/4Ω), BTL	23-SQL	Mat	-	
AN 7178	LIN-IC	2x NF-E, 18V, 4A, 2x5,7W(12V/3Ω)	12-SIP	Mat	AN 7168, AN 7169	
AN 7188 K	LIN-IC	2x NF-E, 24V, 9A, 2x18W(13V/4Ω), 2xBTL	16-SQL	Mat	-	
AN 7202 S	LIN-IC	FM-V, Front-End, Ucc=1,5V	10-MDIP	Mat	-	
AN 7203	LIN-IC	FM-V, Front-End, Ucc=3...5V	9-SIP	Mat	-	pdf pinout
AN 7204	LIN-IC	FM-V, Front-End, Ucc=5V	9-SIP	Mat	-	
AN 7205	LIN-IC	FM-V, Front-End, Ucc=3V	9-SIP	Mat	KA 22495, LA 1185, TA 7358AP, KIA 6058S	pdf pinout
AN 7208 SA	Dig. Audio	TV/FM front-end IC, 1.5V headphone stereo	78	Pan	-	pdf pinout
AN 7213	LIN-IC	FM-V, Front end	7-SIP	Mat	KA 2249	pdf
AN 7213 S	LIN-IC	=AN 7213: SMD	8-MDIP	Mat	-	
AN 7215	LIN-IC	FM-Tuner, Ucc=1,7...7V	7-SIP	Mat	-	
AN 7216	LIN-IC	=AN 7215: SMD	8-MDIP	Mat	-	
AN 7216 S	LIN-IC	=AN 7216: SMD	8-MSIP	Mat	-	
AN 7218	LIN-IC	AM-Tuner, AM/FM-ZF	16-DIP	Mat	-	

Type	Device	Short Description	Fig.	Manu	Comparision Types	More at
AN 7220	LIN-IC	AM-Tuner, AM/FM-ZF, Demod.	18-DIP	Mat	-	pdf
AN 7221	LIN-IC	=AN 7220: SMD	18-MDIP	Mat	-	
AN 7222(N)	LIN-IC	AM/FM-ZF, Afc, Agc, Demod.	18-DIP	Mat	-	pdf pinout
AN 7223	LIN-IC	AM-Tuner, AM/FM-ZF, Demod.	18-DIP	Mat	-	pdf
AN 7224	LIN-IC	AM-Tuner, AM/FM-ZF, Demod.	18-DIP	Mat	-	pdf
AN 7225	LIN-IC	AM/FM-ZF	8-MDIP	Mat	-	
AN 7230 S	LIN-IC	AM/FM-ZF, Demod., Ucc=1,5V	18-MDIP	Mat	-	
AN 7236	IC	FM-AM IF Amplifier Circuit	78	Pan	-	pdf pinout
AN 7238 S	LIN-IC	AM-Tuner, FM-ZF, Stereo-decoder	24-MDIP	Mat	-	
AN 7243 S	LIN-IC	Auto-/car-radio, FM-V, Front-End	14-MDIP	Mat	-	pdf
AN 7244	LIN-IC	FM-Mix, Osc, ZF, Agc	18-DIP	Mat	-	
AN 7244 S	LIN-IC	=AN 7244: SMD	18-MDIP	Mat	-	
AN 7246	LIN-IC	FM-ZF, Afc, Demod.	18-DIP	Mat	-	
AN 7246 S	LIN-IC	=AN 7246: SMD	18-MDIP	Mat	-	
AN 7248	LIN-IC	Auto-/car-radio, FM-ZF, Afc	18-DIP	Mat	-	
AN 7248 S	LIN-IC	=AN 7248: SMD	18-MDIP	Mat	-	
AN 7250 S	LIN-IC	AM-Tuner, FM, Agc, Demod.	18-MDIP	Mat	-	
AN 7254	LIN-IC	Auto-/car-radio, FM-V, FM Front-end	9-SIP	Mat	-	pdf
AN 7256	LIN-IC	Auto-/car-radio, FM-ZF, Demod.	18-SIP	Mat	-	pdf pinout
AN 7258	LIN-IC	Auto-/car-radio, FM-ZF, Demod.	18-DIP	Mat	-	
AN 7259 S	LIN-IC	Auto-/car-radio, FM-ZF, Demod., Mute	20-MDIP	Mat	-	pdf
AN 7260	LIN-IC	Auto-/car-radio, AM-Tuner, ZF, Demod.	18-DIP	Mat	-	
AN 7261 FBQ	IC	FM multiplex reception IC	78	Pan	-	pdf pinout
AN 7266	LIN-IC	AM-Tuner, AM/FM-ZF	18-DIP	Mat	-	
AN 7270	LIN-IC	FM-ZF, Afc, Demod.	16-DIP	Mat	-	
AN 7273	LIN-IC	AM-Tuner, AM/FM-ZF, Demod.	18-DIP	Mat	-	pdf
AN 7277	LIN-IC	FM-ZF, Afc, Demod.	18-DIP	Mat	-	
AN 7280 S	LIN-IC	Auto-/car-radio, FM-V, Front-End, Zf	20-MDIP	Mat	-	pdf
AN 7282 K	LIN-IC	Auto-/car-radio, AM-Tuner, Zf	22-DIP	Mat	-	
AN 7289 NFBQ	IC	FM-FE+AM IC for car radio	78	Pan	-	pdf pinout
AN 7291 SC	LIN-IC	FM-ZF, Noise-suppr., MPX-stereo-decod.	42-MDIP	Mat	-	pdf
AN 7293 NFBQ	Dig. Audio	Fm-fe+am IC for car radio	78	Pan	-	pdf pinout
AN 7298 FBP	Dig. Audio	Fm-fe+am IC for car radio	78	Pan	-	pdf pinout
AN 7299 S	IC	Antenna diversity IC for car TV	78	Pan	-	pdf pinout
AN 7310(N)	LIN-IC	=AN 7311: 16V	9-SIP	Mat	AN 7311, KIA 6225A	pdf pinout
AN7311	OP	Vs:18V Vu:90dB	9I	Mat	-	data pinout
AN 7311	LIN-IC	Auto-/car-radio, 2x NF-V, 18V	9-SIP	Mat	-	pdf
AN 7312	LIN-IC	Recorder, 2x AW-Verst., Alc	14-DIP	Mat	-	pdf
AN 7315	LIN-IC	2x NF-V, Ucc=1,6...4,5V	9-SIP	Mat	-	pdf pinout
AN 7315 S	LIN-IC	=AN 7315: SMD	14-MDIP	Mat	-	pdf pinout
AN 7316	LIN-IC	Recorder, 2x NF-V, Alc	16-DIP	Mat	-	pdf
AN 7320	LIN-IC	Recorder, Alc	7-SIP	Mat	-	pdf pinout
AN 7332 S	LIN-IC	Dual 4-Band Graphic Equalizer	24-MDIP	Mat	-	
AN 7333 K	LIN-IC	Dual 4-Band Graphic Equalizer	24-DIP	Mat	-	
AN 7333 S	LIN-IC	=AN 7333K: SMD	24-MDIP	Mat	-	pdf
AN 7337 N	LIN-IC	HiFi 7-Band Graphic Equalizer	20-DIP	Mat	-	pdf
AN 7345 K	LIN-IC	double cassette recorder, 2x rec./play. preamp	24-SDIP	Mat	-	pdf pinout
AN 7348 K	LIN-IC	double cassette recorder, 2x rec./play. preamp	24-SDIP	Mat	-	pdf pinout
AN 7350	LIN-IC	Differential-Verst./Ampl., Ucc=±35V	9-SIP	Mat	-	
AN 7351 K	LIN-IC	HiFi-Recorder, 2x A/W-Verst., Mute	42-SDIP	Mat	-	
AN 7351 SC	LIN-IC	=AN 7351K: SMD	42-MDIP	Mat	-	
AN 7352 S	IC	Playback Pre-amp, Stereo Cassette Deck	78	Pan	-	pdf pinout
AN 7353 S	IC	Record Equalizer Amp., Stereo Cas. Deck	78	Pan	-	pdf pinout
AN 7354 SC	IC	Dolby* B/c-type Noise Reduction	TSOP	Pan	-	pdf pinout
AN 7356 NSC	IC	Rec. and playback equalizer amp. IC	TSOP	Pan	-	pdf pinout
AN 7367 K	LIN-IC	Recorder, 2x dbx Ii, Noise Reduction	28-SDIP	Mat	-	pdf
AN 7368 K	LIN-IC	Recorder, 2x dbx Ii, Noise Reduction	28-SDIP	Mat	-	
AN 7370	LIN-IC	Recorder, Dolby C	28-DIP	Mat	-	
AN 7370 K	LIN-IC	=AN 7370: Fig.→	28-SDIP	Mat	-	
AN 7370 S	LIN-IC	=AN 7370: SMD	28-MDIP	Mat	-	
AN 7374 K	LIN-IC	Recorder, 2x Dolby B/C	28-SDIP	Mat	-	
AN 7375(N)	LIN-IC	Recorder, 2x Dolby B	18-DIP	Mat	-	pdf
AN 7375(N)S	LIN-IC	=AN 7375: SMD	18-MDIP	Mat	-	
AN 7381	LIN-IC	Dual Tone Control	9-SIP	Mat	-	pdf pinout
AN 7382	LIN-IC	Stereo-lautst.-reg./dc Volume Control	18-SIP	Mat	-	
AN 7384 N	LIN-IC	Stereo-lautst.-reg./dc Volume Control	16-DIP	Mat	-	pdf
AN 7395 K	Dig. Audio	Spatializer IC	TSOP	Pan	-	pdf pinout
AN 7395 S	Dig. Audio	Spatializer IC	TSOP	Pan	-	pdf pinout
AN 7396 K	Dig. Audio	Sound signal processing, Spatializer IC	TSOP	Pan	-	pdf pinout
AN 7397 K	Dig. Audio	Spatializer IC for I2C bus	TSOP	Pan	-	pdf pinout
AN 7397 S	Dig. Audio	Spatializer IC for I2C bus	TSOP	Pan	-	pdf pinout
AN 7399 S	Dig. Audio	Spatializer sound processor IC	TSOP	Pan	-	pdf pinout
AN 7410(N)	LIN-IC	FM-stereo, Mpx-decoder	16-DIP	Mat	BA1330, HA11227, KA2261, LA3361, TA7604	pdf
AN 7411	LIN-IC	Fm-stereo, Mpx-decoder	16-DIP	Mat	-	pdf
AN 7411 S	LIN-IC	=AN 7411: SMD	16-MDIP	Mat	-	
AN 7414	LIN-IC	Fm-stereo, Mpx-decoder	18-DIP	Mat	-	
AN 7415	LIN-IC	Fm-stereo, Mpx-decoder	16-DIP	Mat	-	pdf pinout
AN 7415 S	LIN-IC	=AN 7415: SMD	16-MDIP	Mat	-	pdf pinout
AN 7417	LIN-IC	Fm-stereo, Mpx-decoder	16-DIP	Mat	-	
AN 7418	LIN-IC	Fm-stereo, Mpx-decoder	18-DIP	Mat	-	
AN 7418 S	LIN-IC	=AN 7418: SMD	18-MDIP	Mat	-	
AN 7419	LIN-IC	Fm-stereo, Mpx-decoder	18-DIP	Mat	-	
AN 7420(N)	LIN-IC	Fm-stereo, Mpx-decoder	9-SIP	Mat	KA 2263(/N), TA 7343, KIA 6043S	pdf pinout

Type	Device	Short Description	Fig.	Manu	Comparision Types	More at
AN 7421	LIN-IC	Fm-stereo, Mpx-decoder	9-SIP	Mat	KA 2264, TA 7342	
AN 7463 S	LIN-IC	Auto-/car-radio, Fm-stereo, MPX-decod.	28-MDIP	Mat	-	
AN 7464 S	LIN-IC	Auto-/car-radio, Fm-stereo, Mpx-decod.	32-MDIP	Mat	-	
AN 7465 K	LIN-IC	Auto-/car-radio, Fm-stereo, Mpx-decod.	28-DIP	Mat	-	pdf
AN 7465 S	LIN-IC	=AN 7465K: SMD	28-MDIP	Mat	-	pdf
AN 7470	LIN-IC	Fm-stereo, Mpx-decoder	16-DIP	Mat	-	pdf pinout
AN 7472 S	LIN-IC	Fm-stereo, Mpx-decoder	28-MDIP	Mat	-	
AN 7510	Dig. Audio	1Wx2Ch., BTL Audio Power IC	DIP	Pan	-	pdf pinout
AN 7510 S	Dig. Audio	0.5W x2, BTL Audio Power IC	TSOP	Pan	-	pdf pinout
AN 7511	Dig. Audio	1.0W x2, BTL Audio Power IC	DIP	Pan	-	pdf pinout
AN 7511 S	Dig. Audio	0.5W x2, BTL Audio Power IC	TSOP	Pan	-	pdf pinout
AN 7512	Dig. Audio	1.0W x2, BTL Audio Power IC	DIP	Pan	-	pdf pinout
AN 7512 S	Dig. Audio	0.5W x2, BTL Audio Power IC	TSOP	Pan	-	pdf pinout
AN 7513	Dig. Audio	1.0W x2, BTL Audio Power IC	DIP	Pan	-	pdf pinout
AN 7513 S	Dig. Audio	0.5W x2, BTL Audio Power IC	TSOP	Pan	-	pdf pinout
AN 7522	Dig. Audio	3W BTL Audio Power IC	SIP	Pan	-	pdf pinout
AN 7523	Dig. Audio	3W BTL Audio Power Amp IC	SIP	Pan	-	pdf pinout
AN 77 L 00	IC	Low Power Loss Voltage Reg.(100mA Typ)	78	Pan	-	pdf pinout
AN 77 L 00 M	IC	Low Power Loss Voltage Reg.(100mA Typ)	78	Pan	-	pdf pinout
AN 7703	Z-IC	+3V, 1A	17b	Mat	... 7803... (TO-220)	pdf
AN 7704	Z-IC	+4V, 1A	17b	Mat	... 7804... (TO-220)	pdf
AN 7705	Z-IC	+5V, 1A	17b	Mat	... 7805... (TO-220)	pdf
AN 7706	Z-IC	+6V, 1A	17b	Mat	... 7806... (TO-220)	pdf
AN 7707	Z-IC	+7V, 1A	17b	Mat	... 7807... (TO-220)	pdf
AN 7708	Z-IC	+8V, 1A	17b	Mat	... 7808... (TO-220)	pdf
AN 7709	Z-IC	+9V, 1A	17b	Mat	... 7809... (TO-220)	pdf
AN 7710	Z-IC	+10V, 1A	17b	Mat	... 7810... (TO-220)	pdf
AN 7712	Z-IC	+12V, 1A	17b	Mat	... 7812... (TO-220)	pdf
AN 7715	Z-IC	+15V, 1A	17b	Mat	... 7815... (TO-220)	pdf
AN 7718	Z-IC	+18V, 1A	17b	Mat	... 7818... (TO-220)	pdf
AN 7720	Z-IC	+20V, 1A	17b	Mat	... 7820... (TO-220)	pdf
AN 7724	Z-IC	+24V, 1A	17b	Mat	... 7824... (TO-220)	pdf
AN 7703 F...7724 F	Z-IC	=AN 7703...7724: Iso	17b	Mat	... 78xx... (TO-220 Iso)	pdf
AN 7800	IC	3-pin Positive Output Volt. Reg. (1A Type)	TSOP	Pan	-	pdf pinout
AN 7800 F	IC	3-pin Positive Output Volt. Reg. (1A Type)	TSOP	Pan	-	pdf pinout
AN 78 L 00	IC	3-pin Positive Output Volt. Reg. (1A Type)	TSOP	Pan	-	pdf pinout
AN 78 L 00 M	IC	3-pin Positive Output Volt. Reg. (1A Type)	TSOP	Pan	-	pdf pinout
AN 78 M 00 R	IC	Positive Output Volt. Regul. Reset pin	TSOP	Pan	-	pdf pinout
AN 7800 R	IC	Positive Output Volt. Regul. Reset pin	TSOP	Pan	-	pdf pinout
AN7805	R+	Io=1A Vo:4.8...5.2V Vin:10V P:15W	3P	Mat		data pinout
AN 7805	Z-IC	+5V, 1A	17b	Mat	... 7805... (TO-220)	data pinout
AN 7805 R	Z-IC	+5V, 1A	15/4Pin	Mat		
AN7806	R+	Io=1A Vo:5.75...6.25V Vin:11V P:15W	3P	Mat		data pinout
AN 7806	Z-IC	+6V, 1A	17b	Mat	... 7806... (TO-220)	data pinout
AN7807	R+	Io=1A Vo:6.7...7.3V Vin:12V P:15W	3P	Mat		data pinout
AN 7807	Z-IC	+7V, 1A	17b	Mat	... 7807... (TO-220)	data pinout
AN7808	R+	Io=1A Vo:7.7...8.3V Vin:14V P:15W	3P	Mat		data pinout
AN 7808	Z-IC	+8V, 1A	17b	Mat	... 7808... (TO-220)	data pinout
AN7809	R+	Io=1A Vo:8.65...9.35V Vin:15V P:15W	3P	Mat		data pinout
AN 7809	Z-IC	+9V, 1A	17b	Mat	... 7809... (TO-220)	data pinout
AN 7809 R	Z-IC	+9V, 1A	15/4Pin	Mat	-	
AN7810	R+	Io=1A Vo:9.6...10.4V Vin:16V P:15W	3P	Mat		data pinout
AN 7810	Z-IC	+10V, 1A	17b	Mat	... 7810... (TO-220)	data pinout
AN7812	R+	Io=1A Vo:11.5...12.5V Vin:19V P:15W	3P	Mat		data pinout
AN 7812	Z-IC	+12V, 1A	17b	Mat	... 7812... (TO-220)	data pinout
AN 7812 R	Z-IC	+12V, 1A	15/4Pin	Mat	-	
AN7815	R+	Io=1A Vo:14.4...15.6V Vin:23V P:15W	3P	Mat		data pinout
AN 7815	Z-IC	+15V, 1A	17b	Mat	... 7815... (TO-220)	data pinout
AN7818	R+	Io=1A Vo:17.3...18.7V Vin:27V P:15W	3P	Mat		data pinout
AN 7818	Z-IC	+18V, 1A	17b	Mat	... 7818... (TO-220)	data pinout
AN7820	R+	Io=1A Vo:19.2...20.8V Vin:29V P:15W	3P	Mat		data pinout
AN 7820	Z-IC	+20V, 1A	17b	Mat	... 7820... (TO-220)	data pinout
AN7824	R+	Io=1A Vo:23...25V Vin:33V P:15W	3P	Mat		data pdf pinout
AN 7824	Z-IC	+24V, 1A	17b	Mat	... 7824... (TO-220)	data pinout
AN 7805 F...7824 F	Z-IC	=AN 7805...7824: Iso	17b	Mat	... 78xx... (TO-220 Iso)	pdf pinout
AN7905	R-	Io=1A Vo:-4.8...-5.2V Vin:-10V P:15W	3P	Mat		data pinout
AN 7905(T)	Z-IC	-5V, 1A	17c	Mat	... 7905... (TO-220)	data pdf pinout
AN7906	R-	Io=1A Vo:-5.75...-6.25V Vin:-11V P:15W	3P	Mat		data pinout
AN 7906(T)	Z-IC	-6V, 1A	17c	Mat	... 7906... (TO-220)	data pdf pinout
AN7907	R-	Io=1A Vo:-6.7...-7.3V Vin:-12V P:15W	3P	Mat		data pinout
AN 7907(T)	Z-IC	-7V, 1A	17c	Mat	... 7907... (TO-220)	data pdf pinout
AN7908	R-	Io=1A Vo:-7.7...-8.3V Vin:-14V P:15W	3P	Mat		data pinout
AN 7908(T)	Z-IC	-8V, 1A	17c	Mat	... 7908... (TO-220)	data pdf pinout
AN7909	R-	Io=1A Vo:-8.65...-9.35V Vin:-15V P:15W	3P	Mat		data pinout
AN 7909(T)	Z-IC	-9V, 1A	17c	Mat	... 7909... (TO-220)	data pdf pinout
AN7910	R-	Io=1A Vo:-9.6...-10.4V Vin:-16V P:15W	3P	Mat		data pinout
AN 7910(T)	Z-IC	-10V, 1A	17c	Mat	... 7910... (TO-220)	data pdf pinout
AN 7912(T)	Z-IC	-12V, 1A	17c	Mat	... 7912... (TO-220)	data pdf pinout
AN7915	R-	Io=1A Vo:-14.4...-15.6V Vin:-23V P:15W	3P	Mat		data pinout
AN 7915(T)	Z-IC	-15V, 1A	17c	Mat	... 7915... (TO-220)	data pdf pinout
AN7918	R-	Io=1A Vo:-17.3...-18.7V Vin:-27V P:15W	3P	Mat		data pinout

Type	Device	Short Description	Fig.	Manu	Comparision Types	More at
AN 7918(T)	Z-IC	-18V, 1A	17c	Mat	... 7918... (TO-220)	data pdf pinout
AN 7920	R-	Io=1A Vo:-19.2...-20.8V Vin:-29V P:15W	3P	Mat		data pinout
AN 7920(T)	Z-IC	-20V, 1A	17c	Mat	... 7920... (TO-220)	data pdf pinout
AN7924	R-	Io=1A Vo:-23...-25V Vin:-33V P:15W	3P	Mat		data pinout
AN 7924(T)	Z-IC	-24V, 1A	17c	Mat	... 7924... (TO-220)	data pdf pinout
AN 7905 F...7924 F	Z-IC	=AN 7905...7924(T): Iso	17c	Mat	... 79xx... (TO-220 Iso)	
AN 8002	Z-IC	lo-drop, +2V, 0,05A	7b	Mat	-	pdf
AN 8003	Z-IC	lo-drop, +3V, 0,05A	7b	Mat	-	pdf
AN 8004	Z-IC	lo-drop, +4V, 0,05A	7b	Mat	-	pdf
AN 8005	Z-IC	lo-drop, +5V, 0,05A	7b	Mat	-	pdf
AN 8006	Z-IC	lo-drop, +6V, 0,05A	7b	Mat	-	pdf
AN 8007	Z-IC	lo-drop, +7V, 0,05A	7b	Mat	-	pdf
AN 8008	Z-IC	lo-drop, +8V, 0,05A	7b	Mat	-	pdf
AN 8009	Z-IC	lo-drop, +9V, 0,05A	7b	Mat	-	pdf
AN 8010	Z-IC	lo-drop, +10V, 0,05A	7b	Mat	-	pdf
AN 8025	Z-IC	lo-drop, +2,5V, 0,05A	7b	Mat	-	pdf
AN 8028	IC	Ac/dc switching power supply	SIP	Pan	-	pdf pinout
AN 8035	Z-IC	lo-drop, +3,5V, 0,05A	7b	Mat	-	pdf
AN 8045	Z-IC	lo-drop, +4,5V, 0,05A	7b	Mat	-	pdf
AN 8050 S	Z-IC	Reset, +5V, 80mA, -5V, 80mA	18-	Mat	-	pdf
AN 8060	Z-IC	-4V, lo-drop, Reset	8-DIP	Mat	-	pdf
AN 8062	Z-IC	4V, lo-drop	8-DIP	Mat	-	
AN 8064 SP	Z-IC	4V, 0,15A, lo-drop	≈5-SIL	Mat	-	
AN 8066 SP	Z-IC	4V, 0,15A, lo-drop	≈5-SIL	Mat	-	
AN 8072 N	LIN-IC	5x Spannungsregler/Voltage Regulator	12-SIL	Mat	-	
AN 8079	Z-IC	5V, 0,1A, lo-drop, Reset	9-SIL	Mat	-	pdf
AN 8080 K	Z-IC	5V, 0,1A	20-SDIP	Mat	-	
AN 8083	SMPS-IC	DC-DC Converter	16-MDIP	Mat	-	
AN 8084 S	SMPS-IC	SMD, DC-DC Converter, 3,6...12V	10-MDIP	Mat	-	pdf
AN 8085	Z-IC	+8,5V, 0,05A, lo-drop	7b	Mat	-	pdf
AN 8002 M...8085 M	Z-IC	=AN 8002...8085: SMD	39b	Mat	-	pdf
AN 8090	SMPS-IC	Switching Regulator, 500kHz, 35V M51978, MB3769, HA16107, HA16108	16-DIP	Mat	(µPC1094, µPC1099, M51996, M51997)	pdf
AN 8090 S	SMPS-IC	=AN 8090: SMD	20-MDIP	Mat	-	pdf
AN 8120 K	A/D-IC	8 Bit, hi-speed	28-SDIP	Mat	-	
AN 8124 K	A/D-D/A-IC	BiCMOS, TV, 8 Bit, hi-speed	30-SDIP	Mat	-	
AN 8130 K	A/D-IC	BiCMOS, TV, 10 Bit, hi-speed	42-SDIP	Mat	-	
AN 8140 K	D/A-IC	BiCMOS, TV, 10 Bit, hi-speed	24-SDIP	Mat	-	pdf
AN 8140 S	D/A-IC	=AN 8140K: SMD	24-MDIP	Mat		pdf
AN 8146 FBP, FBQ	D/A-IC	SMD,BiCMOS, TV, 10Bit, 3-Ch., hi-speed	64-MP	Mat		pdf pinout
AN 8146 FBP	D/A-IC	BiCMOS, TV, 10Bit, 3Kan./Ch., hi-speed	64-MP	Mat		
AN 8201 S	LIN-IC	Floppy-disk, Stepping Motor Control	18-MDIP	Mat		
AN 8202 S	LIN-IC	Floppy-disk, Stepping Motor Control	18-MDIP	Mat		
AN 8210 NK	LIN-IC	Floppy-disk, Spindle Motor Control	24-DILP	Mat		
AN 8212 NK	LIN-IC	Floppy-disk, Spindle Motor Control	24-DILP	Mat		
AN 8230 K	LIN-IC	Floppy-disk, Motor Drive Control	28-SDIP	Mat		
AN 8231 K	LIN-IC	Floppy-disk, Motor Drv. Ctrl., lo-volt	28-SDIP	Mat		
AN 8231 S	LIN-IC	=AN 8230K: SMD	28-MDIP	Mat		
AN 8235 S	LIN-IC	3,5" Floppy-disk, Spindle Motor Ctrl.	16-MDIP	Mat		
AN 8236 S	LIN-IC	3,5" Floppy-disk, Spindle Motor Ctrl.	18-MDIP	Mat		
AN 8245 CR	LIN-IC	=AN 8245K: SMD	28-MDIC	Mat		
AN 8245 K	LIN-IC	Hard-disk, Spindle Motor Control	24-DILP	Mat		pdf pinout
AN 8250 N	LIN-IC	Schrittmotor-tr./stepping Motor Drive	16-DIP	Mat		
AN 8253 NS	LIN-IC	Floppy-disk, Stepping Motor Drive	18-MDIP	Mat		
AN 8254 S	LIN-IC	Floppy-disk, Stepping Motor Drive	16-MDIP	Mat		
AN 8261	LIN-IC	3-Ph. AC Motor Drive	18-DIP	Mat		pdf
AN 8267 S	LIN-IC	Lüfter/fan-motor Control	16-MDIP	Mat		pdf
AN 8270 K	LIN-IC	Video-disk, Motor-controller	24-DILP	Mat		
AN 8281 S	LIN-IC	Direct-drive Spindle Motor Control	24-MDIP	Mat		
AN 8290 NS	LIN-IC	Cd. Spindle Motor, PWM, Ucc=4,5...20V	24-MDIP	Mat		
AN 8315	LIN-IC	Hard-disk Interface	22-DIP	Mat		
AN 8320 NFA	LIN-IC	SMD, Dat-rec, Servo, Interface	48-MP	Mat		
AN 8340 UAS	LIN-IC	Hall-Verstärker, Ucc=4,5...10V	8-MDIP	Mat		
AN 8353 UB	LIN-IC		9-SIP	Mat		
AN 8356 S	LIN-IC	SMD, Vc, Barcode-scanner	16-MDIP	Mat		pdf
AN 8360 NK	LIN-IC	Bleibatterie/lead Battery Control	24-DIP	Mat	KA 7560	
AN 8374 S	LIN-IC	Cd, Servo-controller	42-MDIP	Mat		
AN 8375 S	LIN-IC	Cd, 3Kanal-/Channel PWM-Tr.	42-MDIP	Mat		
AN 8377 N	LIN-IC	Cd, 3Kanal-/Channel Linear-Tr.	16-DIP+g	Mat		pdf
AN 8387 S	LIN-IC	Cd, 2Kanal-/Channel Linear-Tr.	20-MDIP	Mat		pdf
AN 8806 SB	IC	Three-beam Method Head Amplifier IC	TSOP	Pan	-	pdf pinout
AN 8808 SB	IC	Three-beam Method Head Amplifier IC,CD	46	Pan	-	pdf pinout
AN 8812 SC	IC	4 Ch Linear Driver IC for CD/CD-ROM	46	Pan	-	pdf pinout
AN 8812 SCR	IC	4 Ch Linear Driver IC for CD/CD-ROM	46	Pan	-	pdf pinout
AN 8813 NSB	IC	4-channel driver IC,optical disk drive	46	Pan	-	pdf pinout
AN 8814 SB	IC	4-CH driver IC, optical disk drive	46	Pan	-	pdf pinout
AN 8816 SB	IC	4ch. Linear Driver IC for CD/CD-ROM	46	Pan	-	pdf pinout
AN 8819 NFB	IC	4Ch. Lin. Driver IC,DC-DC Conv. Control	46	Pan	-	pdf pinout
AN 8837 SB	IC	consumption CD-DA head amp. IC,3-beam system optical pick-up	46	Pan	-	pdf pinout
AN 8839 NSB	IC	CD-DA head amp.IC,3-beam system optical pick-up	46	Pan	-	pdf pinout
AN 8847 SB	IC	Head amp.IC for CD-ROM drive(<=32X speed)	46	Pan	-	pdf pinout
AN 8849 SB	IC	Head amp. IC for CD-ROM drive (<=24X)	46	Pan	-	pdf pinout

Type	Device	Short Description	Fig.	Manu	Comparision Types	More at
AN 8882 SB	IC	Head amp. IC for CD-ROM drive (<=32X)	46	Pan	-	pdf pinout
AN 8910 K	LIN-IC	DBS-system (noise Reduction)	18-DIP	Mat	-	
AN 8910 S	LIN-IC	=AN 8910K: SMD	24-MDIP	Mat	-	
AN 8914 K	LIN-IC	QPSK (quad Phase Shift Keying)	18-DIP	Mat	-	
AN 8914 S	LIN-IC	=AN 8914K: SMD	18-MDIP	Mat	-	
AN 8916 FBP	LIN-IC	SMD, TV, Bs/cs Tuner, QPSK	48-MP	Mat	-	
AN 8917 NFBP	LIN-IC	SMD, TV, Bs/cs Tuner, QPSK	48-MP	Mat	-	
AN 8920 K	LIN-IC	TV, Qpsk, QPR Demodulator	42-SDIP	Mat	-	
AN 8940 SB	LIN-IC	Dbs-system (noise Reduction)	36-MDIP	Mat	-	
AN 8946 SB	IC	FM demodula. IC for Bs/cs broadcasting	TSOP	Pan	-	pdf pinout
AN 8953 NFA	CMOS-IC	Silicon Monolithic Bi-cmos IC	TSOP	Pan	-	pdf pinout
AN 12941 A	Dig. Audio	Audio power IC for notebook PC	SIL	Pan	-	pdf pinout
AN 12943 A	Dig. Audio	Audio signal processing IC ,notebook	TSOP	Pan	-	pdf pinout
AN 12972 A	LIN-IC	Audio Power Amplifier, Agc	QFN	Pan	-	pdf
AN 12974 A	Dig. Audio	Stereo power amp, built- in Agc circuitry, Spatializer	SIL	Pan	-	pdf
AN 15524 A	IC	IC for Crt vertical deflection output	SIP	Pan	-	pdf pinout
AN 15525 A	IC	IC for Crt vertical deflection output	SIP	Pan	-	pdf pinout
AN 15526 A	IC	IC for Crt vertical deflection output	SIP	Pan	-	pdf pinout
AN 17000 A	Dig. Audio	Audio signal processing IC, notebook	TSOP	Pan	-	pdf pinout
AN 17020 A	Dig. Audio	Silicon Monolithic Bipolar IC	SIL	Pan	-	pdf pinout
AN 17020 B	Dig. Audio	Headphone amplifier IC	SIL	Pan	-	pdf pinout
AN 17813 A	Dig. Audio	Audio power amplifier IC	SIL	Pan	-	pdf pinout
AN 17821 A	LIN-IC	BTL 5.0W x2Ch. power amp.	SIL	Pan	-	pdf pinout
AN 17830A	LIN-IC	Audio PAmp, Mute, Standby, Vcc=17V	SIL	Mat	AN17831	pinout
AN 17831A	LIN-IC	Audio PAmp, Mute, Standby, Vcc=17V	SIL	Mat	-	pdf pinout
AN 17850 A	LIN-IC	70W 1xch Audio-amp, 10-33V	SIL	Mat	-	pdf pinout
ANi	Ge-P	Stabi	2a		(→AC 151)	
ANk	Ge-P	Stabi	2a		(→AC 151)	
ANm	Ge-N	Stabi	2a		(→AC 127)	
ANM	Si-N	=2SC4061K-M (Typ-Code/Stempel/marking)	35	Rhm	→2SC4061K	data
ANN	Si-N	=2SC4061K-N (Typ-Code/Stempel/marking)	35	Rhm	→2SC4061K	data
ANP	Si-N	=2SC4061K-P (Typ-Code/Stempel/marking)	35	Rhm	→2SC4061K	data
ANR	Si-P	=2SA1898-R (Typ-Code/Stempel/marking)	39	Say	→2SA1898	data
ANs	Si-N	=BFP 450 (Typ-Code/Stempel/marking)	44(2mm)	Sie	→BFP 450	data
ANS	Si-P	=2SA1898-S (Typ-Code/Stempel/marking)	39	Say	→2SA1898	data

AO

Type	Device	Short Description	Fig.	Manu	Comparision Types	More at
AO	Si-N	=2SC2880-O (Typ-Code/Stempel/marking)	39	Tos	→2SC2880	data
AO	Si-N	=2SC4210-O (Typ-Code/Stempel/marking)	35	Tos	→2SC4210	data
AO	Si-N	=BCW 60RA (Typ-Code/Stempel/marking)	35	Sie,Val	→BCW 60RA	data
AO	Si-N	=KTC4372-O (Typ-Code/Stempel/marking)	39	Kec	→KTC 4372	data
AO	Z-Di	=MA 5Z300-H (Typ-Code/Stempel/marking)	71(1,7mm)	Mat	→MA 5Z300-H	data
AO	Si-N	=XN 4509 (Typ-Code/Stempel/marking)	46	Mat	→XN 4509	data
AO	C-Di	=ZMV 934A (Typ-Code/Stempel/marking)	71(1,7mm)	Ztx	→ZMV 934A	data
AOs	Si-N	=BFP 490 (Typ-Code/Stempel/marking)	≈45	Sie	→BFP 490	data
A-OS 01	Hybrid-IC			Fui	-	
AOZ 1010 AI	IC	2A Simple Regulator	TSOP	Aos	-	pdf pinout
AOZ 1012 DI	IC	3A Simple Regulator	TSOP	Aos	-	pdf pinout
AOZ 1013 AI	IC	3A Simple Regulator	TSOP	Aos	-	pdf pinout
AOZ 1014 AI	IC	5A Simple Buck Regulator	TSOP	Aos	-	pdf pinout
AOZ 1014 DI	IC	5A Simple Buck Regulator	TSOP	Aos	-	pdf pinout
AOZ 1016 AI	IC	2A Simple Buck Regulator	TSOP	Aos	-	pdf pinout
AOZ 1017 AI	IC	3A Simple Regulator	TSOP	Aos	-	pdf pinout
AOZ 1018 AI	IC	2A Simple Regulator	TSOP	Aos	-	pdf pinout
AOZ 1300 AI	IC	Programm. 4A Power Distribution Switch	TSOP	Aos	-	pdf pinout
AOZ 8001 JI	IC	Ultra-low Capacitance TVS Diode Array	TSOP	Aos	-	pdf pinout
AOZ 8001 KI	IC	Ultra-low Capacitance TVS Diode Array	TSOP	Aos	-	pdf pinout
AOZ 8006 FI	IC	Ultra-low Capacitance TVS Diode Array	TSOP	Aos	-	pdf pinout
AOZ 8007 CI	IC *	Ultra-low Capacitance TVS Diode Array	TSOP	Aos	-	pdf pinout
AOZ 8007 FI	IC	Ultra-low Capacitance TVS Diode Array	TSOP	Aos	-	pdf pinout
AOZ 9004 BI	MOS-IC	1-Cell Battery Protection IC, Mosfet	TSOP	Aos	-	pdf pinout
AOZ 1320 DI-01	MOS-IC	Load Switch with Controlled Slew Rate	TSOP	Aos	-	pdf pinout
AOZ 1320 CI-02	MOS-IC	Load Switch with Controlled Slew Rate	TSOP	Aos	-	pdf pinout
AOZ 1320 DI-02	MOS-IC	Load Switch with Controlled Slew Rate	TSOP	Aos	-	pdf pinout
AOZ 1320 CI-03	MOS-IC	Load Switch with Controlled Slew Rate	TSOP	Aos	-	pdf pinout
AOZ 1320 DI-03	MOS-IC	Load Switch with Controlled Slew Rate	TSOP	Aos	-	pdf pinout
AOZ 1320 CI-04	MOS-IC	Load Switch with Controlled Slew Rate	TSOP	Aos	-	pdf pinout
AOZ 1320 DI-04	MOS-IC	Load Switch with Controlled Slew Rate	TSOP	Aos	-	pdf pinout
AOZ 1320 CI-05	MOS-IC	Load Switch with Controlled Slew Rate	TSOP	Aos	-	pdf pinout
AOZ 1320 DI-05	MOS-IC	Load Switch with Controlled Slew Rate	TSOP	Aos	-	pdf pinout
AOZ 1320 CI-06	MOS-IC	Load Switch with Controlled Slew Rate	TSOP	Aos	-	pdf pinout
AOZ 1320 DI-06	MOS-IC	Load Switch with Controlled Slew Rate	TSOP	Aos	-	pdf pinout
AOZ 9004 BI-01	MOS-IC	1-Cell Battery Protection IC, Mosfet	TSOP	Aos	-	pdf pinout
AOZ 9004 BI-02	MOS-IC	1-Cell Battery Protection IC, Mosfet	TSOP	Aos	-	pdf pinout
AOZ 9004 BI-03	MOS-IC	1-Cell Battery Protection IC, Mosfet	TSOP	Aos	-	pdf pinout
AOZ 9004 BI-04	MOS-IC	1-Cell Battery Protection IC, Mosfet	TSOP	Aos	-	pdf pinout
AOZ 9005 DI-00	MOS-IC	Single-cell Battery Protection IC	TSOP	Aos	-	pdf pinout
AOZ 9005 DI-01	MOS-IC	Single-cell Battery Protection IC	TSOP	Aos	-	pdf pinout
AOZ 1320 CI-01	MOS-IC	Load Switch with Controlled Slew Rate	TSOP	Aos	-	pdf pinout

Type	Device	Short Description	Fig.	Manu	Comparision Types	More at
AP..AS						
AP	Si-N	=2SC2413-AP (Typ-Code/Stempel/marking)	≈35	Rhm	→2SC2413	data
AP	Si-N	=2SC2413K-P (Typ-Code/Stempel/marking)	35	Rhm	→2SC2413K	data
AP	Si-N	=2SC4098-P (Typ-Code/Stempel/marking)	35(2mm)	Rhm	→2SC4098	data
AP	MOS-P-FET-e*	=2SJ575 (Typ-Code/Stempel/marking)	35	Hit	→2SJ575	data
AP	MOS-P-FET-e*	=2SJ576 (Typ-Code/Stempel/marking)	35(2mm)	Hit	→2SJ576	data
AP	Si-N	=BCW 60RB (Typ-Code/Stempel/marking)	35	Sie,Val	→BCW 60RB	data
AP	Z-Di	=MA 5Z330-L (Typ-Code/Stempel/marking)	71(1,7mm)	Mat	→MA 5Z330-L	data
AP	Z-Di	=P4 SMA-10A (Typ-Code/Stempel/marking)	71(5x2,5)	Fag	→P4 SMA-10A	data
AP	Z-Di	=SMBJ 7.5CA (Typ-Code/Stempel/marking)	71(5x3,5)	Mop	→SMBJ ...	data
AP 01 C	Si-Di	Gl, S, 1000V, 0,2A, 200ns	31a	Sak	BA 159, BYT 11/1000, BYT 52M, BYV 26E	data
AP 1 A3M	Si-P+R	S, Rb=Rbe=1kΩ, 25V, 0,7A, 0,75W	7c	Nec	-	data
AP 1 A4A	Si-P+R	=AP 1A3M: Rb=-, Rbe=10kΩ	7c	Nec	-	data
AP 1 A4M	Si-P+R	=AP 1A3M: Rb=10k, Rbe=10kΩ	7c	Nec	-	data
AP 1 F3P	Si-P+R	=AP 1A3M: Rb=2,2k, Rbe=10kΩ	7c	Nec	-	data
AP 1 J3P	Si-P+R	=AP 1A3M: Rb=3,3k, Rbe=10kΩ	7c	Nec	-	data
AP 1 L2Q	Si-P+R	=AP 1A3M: Rb=0,47k, Rbe=4,7kΩ	7c	Nec	-	data
AP 1 L3N	Si-P+R	=AP 1A3M: Rb=4,7k, Rbe=10kΩ	7c	Nec	-	data
AP 130-XXR	IC	300mA low dropout (LDO) linear Reg.	48	Dii	-	pdf pinout
AP 130-XXSA	IC	300mA low dropout (LDO) linear Reg.	48	Dii	-	pdf pinout
AP 130-XXW	IC	300mA low dropout (LDO) linear Reg.	48	Dii	-	pdf pinout
AP 130-XXY	IC	300mA low dropout (LDO) linear Reg.	48	Dii	-	pdf pinout
AP 130-XXYR	IC	300mA low dropout (LDO) linear Reg.	48	Dii	-	pdf pinout
AP 131-XXW	LIN-IC	300mA low dropout linear Regulator, SHUTDOWN	45	Dii	-	pdf pinout
AP 133-SN	LIN-IC	300mA, low dropout linear regulator	45	Dii	-	pdf pinout
AP 133-W	LIN-IC	300mA, low dropout linear regulator	45	Dii	-	pdf pinout
AP 139-XXW	CMOS/LIN-IC	300mA low-noice Cmos LDO	45	Dii	-	pdf pinout
AP 0332 CG	MOS-IC	SMD, 8x P-Ch. Latchable Mosfet Array	16-MDIP	Stx	-	pdf
AP 358 N	LIN-IC	low Power Dual Operational Amplifiers	TSOP	Dii	-	pdf pinout
AP 358 S	LIN-IC	low Power Dual Operational Amplifiers	TSOP	Dii	-	pdf pinout
AP 393 N	LIN-IC	lcw Offset Voltage Dual Comparators	TSOP	Dii	-	pdf pinout
AP 393 S	LIN-IC	low Offset Voltage Dual Comparators	TSOP	Dii	-	pdf pinout
AP 647	Si-Di	≈MR 31	31a		→MR 31	pdf pinout
AP 1084 D	IC	Positive ADJ. or Fixed-mode regulator	48	Dii	-	pdf pinout
AP 1084 K	IC	Positive ADJ. or Fixed-mode regulator	48	Dii	-	pdf pinout
AP 1084 T	IC	Positive ADJ. or Fixed-mode regulator	48	Dii	-	pdf pinout
AP 1086 D	IC	Positive Adjustable,Fixed-mode regulator	48	Dii	-	pdf pinout
AP 1086 K	IC	Positive Adjustable,Fixed-mode regulator	48	Dii	-	pdf pinout
AP 1086 T	IC	Positive Adjustable,Fixed-mode regulator	48	Dii	-	pdf pinout
AP 1115 XY	IC	Positive Adjustable,Fixed-mode regulator	48	Dii	-	pdf pinout
AP 1117 D	IC	1A low dropout Positive ADJ.,Fixed-mode regulator	48	Dii	-	pdf pinout
AP 1117 E	IC	1A low dropout Positive ADJ.,Fixed-mode regulator	48	Dii	-	pdf pinout
AP 1117 K	IC	1A low dropout Positive ADJ.,Fixed-mode regulator	48	Dii	-	pdf pinout
AP 1117 T	IC	1A low dropout Positive ADJ.,Fixed-mode regulator	48	Dii	-	pdf pinout
AP 1117 Y	IC	1A low dropout Positive ADJ.,Fixed-mode regulator	48	Dii	-	pdf pinout
AP 1120 S	IC	Dual 1A low dropout Positive regulator	TSOP	Dii	-	pdf pinout
AP 1121 AS	IC	Dual 1A low dropout Positive regulator	TSOP	Dii	-	pdf pinout
AP 1121 BS	IC	Dual 1A low dropout Positive regulator	TSOP	Dii	-	pdf pinout
AP 1122 D	IC	1A low dropout Positive regulator	48	Dii	-	pdf pinout
AP 1122 E	IC	1A low dropout Positive regulator	48	Dii	-	pdf pinout
AP 1122 Y	IC	1A low dropout Positive regulator	48	Dii	-	pdf pinout
AP 1184 S	IC	Positive Adjustable,Fixed-mode regulator	TSOP	Dii	-	pdf pinout
AP 1212 XS	IC	Dual USB High-side Power Switch	TSOP	Dii	-	pdf pinout
AP 1301	IC	Single Coil Fan Motor Full Wave Driver	TSOP	Ana	-	pdf pinout
AP 1302	IC	Single Coil Fan Motor Full Wave Driver	TSOP	Ana	-	pdf pinout
AP 1303	IC	Single Coil Fan Motor Full Wave Driver	TSOP	Ana	-	pdf pinout
AP 1304	IC	Single Coil Fan Motor Full Wave Driver	TSOP	Ana	-	pdf pinout
AP 1305	IC	Single Coil Fan Motor Full Wave Driver	TSOP	Ana	-	pdf pinout
AP 1306	LIN-IC	Two Coil Fan Motor Predriver	MDIP	Ana	-	pdf pinout
AP 1307	LIN-IC	Two Coil Fan Motor Predriver	MDIP	Ana	-	pdf pinout
AP 1308 SS	LIN-IC	Single Coil Fan Motor Full Wave Driver	MDIP	Ana	-	pdf pinout
AP 1308 TS	LIN-IC	Single Coil Fan Motor Full Wave Driver	TSOP	Ana	-	pdf pinout
AP 1501 A K 5	LIN-IC	150KHz, 5A PWM buck DC/DC Converter	30/5Pin	Dii	-	pdf pinout
AP 1501 A T 5	LIN-IC	150KHz, 5A PWM buck DC/DC Converter	17/5Pin	Dii	-	pdf pinout
AP 1501 A T 5 R	LIN-IC	150KHz, 5A PWM buck DC/DC Converter	86/5Pin	Dii	-	pdf pinout
AP 1501 A 33 K 5	LIN-IC	150KHz, 5A PWM buck DC/DC Converter	30/5Pin	Dii	-	pdf pinout
AP 1501 A 33 T 5	LIN-IC	150KHz, 5A PWM buck DC/DC Converter	17/5Pin	Ana	-	pdf pinout
AP 1501 A 33 T 5 R	LIN-IC	150KHz, 5A PWM buck DC/DC Converter	86/5Pin	Dii	-	pdf pinout
AP 1501 A 50 K 5	LIN-IC	150KHz, 5A PWM buck DC/DC Converter	30/5Pin	Dii	-	pdf pinout
AP 1501 A 50 T 5	LIN-IC	150KHz, 5A PWM buck DC/DC Converter	17/5Pin	Ana	-	pdf pinout
AP 1501 A 50 T 5 R	LIN-IC	150KHz, 5A PWM buck DC/DC Converter	86/5Pin	Dii	-	pdf pinout
AP 1501 A 12 K 5	LIN-IC	150KHz, 5A PWM buck DC/DC Converter	30/5Pin	Dii	-	pdf pinout
AP 1501 A 12 T 5	LIN-IC	150KHz, 5A PWM buck DC/DC Converter	17/5Pin	Dii	-	pdf pinout
AP 1501 A 12 T 5 R	LIN-IC	150KHz, 5A PWM buck DC/DC Converter	86/5Pin	Dii	-	pdf pinout
AP 1509-S	IC	150KHz, 2A PWM buck DC/DC Converter	TSOP	Dii	-	pdf pinout
AP 1510 S	LIN-IC	PWM Control 3A step-down Converter	TSOP	Dii	-	pdf pinout
AP 1511	MOS-IC	PWM Control 5A Step-down Converter	MDIP	Ana	-	pdf pinout

Type	Device	Short Description	Fig.	Manu	Comparision Types	More at
AP 1512 AK5	IC	50KHz, 3A PWM Buck DC/DC Converter	87	Ana	-	pdf pinout
AP 1512 AT5	IC	50KHz, 3A PWM Buck DC/DC Converter	86	Ana	-	pdf pinout
AP 1512 K5	IC	50KHz, 2A PWM Buck DC/DC Converter	87	Ana	-	pdf pinout
AP 1512 T5	IC	50KHz, 2A PWM Buck DC/DC Converter	86	Ana	-	pdf pinout
AP 1513 S	SMPS-IC	PWM Control 2A step-down Converter	TSOP	Dii	-	pdf pinout
AP 1520 S	LIN-IC	PWM Control 2A step-down Converter	TSOP	Dii	-	pdf pinout
AP 1530	LIN-IC	PWM Control 3A step-down Converter	MDIP	Dii	-	pdf pinout
AP 1533	LIN-IC	PWM Control 1.5A step-down Converter	MDIP	Ana	-	pdf pinout
AP 1534 S	LIN-IC	PWM Control 2A step-down Converter	TSOP	Dii	-	pdf pinout
AP 1580	LIN-IC	PWM Control 3A step-down Converter	MDIP	Ana	-	pdf pinout
AP 1581	LIN-IC	PWM Control 2.5A step-down Converter	MDIP	Ana	-	pdf pinout
AP 1603 W	IC	Step-up DC/DC Converter	TSOP	Dii	-	pdf pinout
AP 1604 SN	IC	Pwm/pfm Dual Mode step-down DC/DC Conv.	TSOP	Dii	-	pdf pinout
AP 1604 W	IC	Pwm/pfm Dual Mode step-down DC/DC Conv.	TSOP	Dii	-	pdf pinout
AP 1605 S	IC	Pwm/pfm Dual-mode step-down Switch. Reg.	TSOP	Dii	-	pdf pinout
AP 1609 S	IC	Pwm/pfm Dual Mode Step-up DC/DC Convert.	TSOP	Dii	-	pdf pinout
AP 1624	IC	PWM/ Pfm Dual Mode Step- up DC/ DC Controller	45	Ana	-	pdf pinout
AP 1635 S	IC	Pwm/pfm Dual Mode step-down DC/DC Conve	TSOP	Dii	-	pdf pinout
AP 1701 W	LIN-IC	3-PIN Microprocessor reset Circuits	43	Dii	-	pdf pinout
AP 1702 W	LIN-IC	3-PIN Microprocessor reset Circuits	43	Dii	-	pdf pinout
AP 1703 W	LIN-IC	3-PIN Microprocessor reset Circuits	43	Dii	-	pdf pinout
AP 1704 W	LIN-IC	3-PIN Microprocessor reset Circuits	43	Dii	-	pdf pinout
AP 2001 S	LIN-IC	Monolithic Dual Channel PWM Controller	TSOP	Dii	-	pdf pinout
AP 2004 S	LIN-IC	PWM buck Controller	TSOP	Dii	-	pdf pinout
AP 2008 S	LIN-IC	PWM buck Controller	TSOP	Dii	-	pdf pinout
AP 2010 S/TS	LIN-IC	Sine-wave Generator for Ccfl Supply	TSOP	Ana	-	pdf pinout
AP 2014 /AS	LIN-IC	synchronous PWM Controller	TSOP	Dii	-	pdf pinout
AP 3152 F	LIN-IC	120mA High efficiency White LED Driver	TSOP	Dii	-	pdf pinout
AP 393 AM 8	LIN-IC	low Offset Voltage Dual Comparators	TSOP	Dii	-	pdf pinout
AP 7115-W	LIN-IC	150mA low dropout linear Reg.,Shutdown	TSOP	Dii	-	pdf pinout
AP 7165-FN	LIN-IC	600mA low dropout regulator with POK	TSOP	Dii	-	pdf pinout
AP 7167-FN	LIN-IC	1.2A low dropout regulator with POK	TSOP	Dii	-	pdf pinout
AP 7173-FN	LIN-IC	1.5A dropout linear Reg. prog. SOFT-START	TSOP	Dii	-	pdf pinout
AP 7173-SP	LIN-IC	1.5A dropout linear Reg. prog. SOFT-START	TSOP	Dii	-	pdf pinout
AP 7201 FM	LIN-IC	Dual 150mA low dropout linear regulator	TSOP	Dii	-	pdf pinout
AP 78 L 05 S	LIN-IC	5V Output 3-terminal Positive regulator	30	Dii	-	pdf pinout
AP 78 L 05 V	LIN-IC	5V Output 3-terminal Positive regulator	30	Dii	-	pdf pinout
AP 78 L 05 Y	LIN-IC	5V Output 3-terminal Positive regulator	30	Dii	-	pdf pinout
AP 78 L 08 S	LIN-IC	8V Output 3-terminal Positive Regulators	30	Dii	-	pdf pinout
AP 78 L 08 V	LIN-IC	8V Output 3-terminal Positive Regulators	30	Dii	-	pdf pinout
AP 78 L 12 S	LIN-IC	12V Output 3-terminal Positive regulat.	30	Dii	-	pdf pinout
AP 78 L 12 V	LIN-IC	12V Output 3-terminal Positive regulat.	30	Dii	-	pdf pinout
AP 8910 A	IC	Voice Otp IC	DIP	Apl	-	pdf pinout
AP 8921 A	IC	Voice Otp IC	DIP	Apl	-	pdf pinout
AP 8942 A	CMOS-IC	Voice Otp IC	DIP	Apl	-	pdf pinout
AP 1184 K 5	IC	Positive Adjustable,Fixed-mode regulator	TSOP	Dii	-	pdf pinout
AP 1184 T 5	IC	Positive Adjustable,Fixed-mode regulator	TSOP	Dii	-	pdf pinout
AP 1186 K 5	IC	1.5A Positive ADJ. Fixed-mode regulator	TSOP	Dii	-	pdf pinout
AP 1186 T 5	IC	1.5A Positive ADJ. Fixed-mode regulator	TSOP	Dii	-	pdf pinout
AP 1501-K 5	IC	150KHz, 3A PWM buck DC/DC Converter	87	Dii	-	pdf pinout
AP 1501-T 5	IC	150KHz, 3A PWM buck DC/DC Converter	87	Dii	-	pdf pinout
AP 1501-T 5 R	IC	150KHz, 3A PWM buck DC/DC Converter	87	Dii	-	pdf pinout
AP 1506-K 5	LIN-IC	150KHz, 3A PWM buck DC/DC Converter	30	Dii	-	pdf pinout
AP 1506-T 5	LIN-IC	150KHz, 3A PWM buck DC/DC Converter	30	Dii	-	pdf pinout
AP 1506-T 5 R	LIN-IC	150KHz, 3A PWM buck DC/DC Converter	30	Dii	-	pdf pinout
AP 1507-D 5	LIN-IC	150KHz, 3A PWM buck DC/DC Converter	30	Dii	-	pdf pinout
AP 1512-K 5	LIN-IC	50KHz, 2A/3A PWM buck DC/DC Converter	TSOP	Dii	-	pdf pinout
AP 1601 M 8	LIN-IC	Step-up DC/DC Converter	TSOP	Dii	-	pdf pinout
AP 2280-1 FM	LIN-IC	1-CH slew rate Controlled load Switch	TSOP	Dii	-	pdf pinout
AP 2280-1 W	LIN-IC	1-CH slew rate Controlled load Switch	TSOP	Dii	-	pdf pinout
AP 2280-2 FM	LIN-IC	1-CH slew rate Controlled load Switch	TSOP	Dii	-	pdf pinout
AP 2280-2 W	LIN-IC	1-CH slew rate Controlled load Switch	TSOP	Dii	-	pdf pinout
AP 2281-1 W	LIN-IC	Single slew rate Contr. load Switch	TSOP	Dii	-	pdf pinout
AP 2281-3 W	LIN-IC	Single slew rate Contr. load Switch	TSOP	Dii	-	pdf pinout
AP 89085	IC	Voice Otp IC	DIP	Apl	-	pdf pinout
AP 89170	IC	Voice Otp IC	DIP	Apl	-	pdf pinout
AP 89341	IC	Voice Otp IC	DIP	Apl	-	pdf pinout
AP 1601 M 10	LIN-IC	Step-up DC/DC Converter	TSOP	Dii	-	pdf pinout
AP 34063 /N8	IC	Universal DC/DC Converter	TSOP	Ana	-	pdf pinout
AP 34063 /S8	IC	Universal DC/DC Converter	TSOP	Ana	-	pdf pinout
APA 0710 K/XA	IC	1.1W Mono Low-Volt. Audio Power Amp.	MDIP	Anp	-	pdf pinout
APA 0711 K/XA	IC	1.1W Mono Low-Volt. Audio Power Amp.	MDIP	Anp	-	pdf pinout
APA 0712 HA	IC	1.4W Mono Audio Low-Volt. Power Amp.	Anp	-	pdf pinout	
APA 0715 QB	IC	3W Mono Full Different. Audio PowerAmp	MLP	Anp	-	pdf pinout
APA 0715 X(A)	IC	3W Mono Full Different. Audio PowerAmp	MDIP	Anp	-	pdf pinout
APA 2010 (A)HA	IC	3W Mono Class-D Audio Power Amp.	GRID	Anp	-	pdf pinout
APA 2010 (A)QB	IC	3W Mono Class-D Audio Power Amp.	MLP	Anp	-	pdf pinout
APA 2020 K	IC	Stereo 2W Audio Power Amp.	MDIP	Anp	-	pdf pinout
APA 2020 R	IC	Stereo 2W Audio Power Amp.	TSOP	Anp	-	pdf pinout
APA 2030 R	IC	Stereo 2.6W Audio Amp., Gain Control	TSOP	Anp	-	pdf pinout
APA 2031 R	IC	Stereo 2.6W Audio Amp., Gain Control	TSOP	Anp	-	pdf pinout
APA 2035 R	IC	Stereo 2.6W Audio Amp., Gain Control	TSOP	Anp	-	pdf pinout

Type	Device	Short Description	Fig.	Manu	Comparision Types	More at
APA 2036	IC	Stereo 2.6W Audio Power Amp.	QFN	Anp	-	pdf pinout
APA 2037 QB	IC	3W 2Ch. Fully Differential Audio Power Amp.	QFN	Anp	-	pdf pinout
APA 2051 QB	IC	2.4W Stereo Audio Power Amp.	QFN	Anp	-	pdf pinout
APA 2057 AQB	IC	2.4W Stereo Audio Power Amp.	QFN	Anp	-	pdf pinout
APA 2057 AR	IC	2.4W Stereo Audio Power Amp.	TSOP	Anp	-	pdf pinout
APA 2058 QB	IC	2.4W 2Ch. Fully Differential Audio Power Amp.	QFN	Anp	-	pdf pinout
APA 2059 QB	IC	2.4W Stereo Fully Differential Audio Power Amp.	QFN	Anp	-	pdf pinout
APA 2065 J	IC	Stereo 2.7W Audio Power Amp.	DIP	Anp	-	pdf pinout
APA 2066 KA	IC	Stereo 1.8W Audio Power Amp.	MDIP	Anp	-	pdf pinout
APA 2068 KA	IC	Stereo 2.6W Audio Power Amp.	MDIP	Anp	-	pdf pinout
APA 2069 J	IC	Stereo 2.6W Audio Power Amp.	DIP	Anp	-	pdf pinout
APA 2070 J	IC	Stereo 2.6W Audio Power Amp.	DIC	Anp	-	pdf pinout
APA 2071 J	IC	Stereo 3.1W Non-inv. Audio Power Amp.	DIP	Anp	-	pdf pinout
APA 2120 R	IC	Stereo 2W Audio Power Amp.	TSOP	Anp	-	pdf pinout
APA 2121 R	IC	Stereo 2W Audio Power Amp.	TSOP	Anp	-	pdf pinout
APA 2176 AQB	IC	270mW Stereo Cap-Free Headphone Drv.	QFN	Anp	-	pdf pinout
APA 2176 O	IC	270mW Stereo Cap-Free Headphone Drv.	TSOP	Anp	-	pdf pinout
APA 2176 QB	IC	270mW Stereo Cap-Free Headphone Drv.	QFN	Anp	-	pdf pinout
APA 2178 HA	Dig. Audio	100mW Stereo Cap-Free Headphone Driver	GRID	Anp	-	pdf pinout
APA 2308 J	Dig. Audio	Class AB Stereo Headphone Driver	DIP	Anp	-	pdf pinout
APA 2308 K	Dig. Audio	Class AB Stereo Headphone Driver	DIP	Anp	-	pdf pinout
APA 2600 QB	Dig. Audio	2.8W Stereo Class-D Audio Power Amp.	QFN	Anp	-	pdf pinout
APA 2822 J	Dig. Audio	Dual Low-Voltage Power Amplifier	DIP	Anp	-	pdf pinout
APA 2822 K	Dig. Audio	Dual Low-Voltage Power Amplifier	DIP	Anp	-	pdf pinout
APA 3002 QB	Dig. Audio	12W Stereo Class-D Audio Power Amp.	QFN	Anp	-	pdf pinout
APA 3002 QCA	Dig. Audio	12W Stereo Class-D Audio Power Amp.	QFN	Anp	-	pdf pinout
APA 3010 KA	Dig. Audio	3W Mono Low-Voltage Audio Power Amp.	TSOP	Anp	-	pdf pinout
APA 3010 XA	Dig. Audio	3W Mono Low-Voltage Audio Power Amp.	TSOP	Anp	-	pdf pinout
APA 3011 KA	Dig. Audio	3W Mono Low-Voltage Audio Power Amp.	TSOP	Anp	-	pdf pinout
APA 3011 XA	Dig. Audio	3W Mono Low-Voltage Audio Power Amp.	TSOP	Anp	-	pdf pinout
APA 3012 KA	Dig. Audio	3W Mono Low-Voltage Audio Power Amp.	TSOP	Anp	-	pdf pinout
APA 3012 XA	Dig. Audio	3W Mono Low-Voltage Audio Power Amp.	TSOP	Anp	-	pdf pinout
APA 3541 J	Dig. Audio	Class AB Stereo Headphone Driver, Mute	DIP	Anp	-	pdf pinout
APA 3541 K	Dig. Audio	Class AB Stereo Headphone Driver, Mute	DIP	Anp	-	pdf pinout
APA 3544 J	Dig. Audio	Class AB Stereo Headphone Driver, Mute	DIP	Anp	-	pdf pinout
APA 3544 K	Dig. Audio	Class AB Stereo Headphone Driver, Mute	DIP	Anp	-	pdf pinout
APA 4558 K	LIN-IC	Dual Operational Amplifier	DIP	Anp	-	pdf pinout
APA 4800 J	Dig. Audio	Stereo 290mW 8W Speaker Driver	DIP	Anp	-	pdf pinout
APA 4800 K	Dig. Audio	Stereo 290mW 8W Speaker Driver	DIP	Anp	-	pdf pinout
APA 4801 J	Dig. Audio	Stereo 280mW 8 Ohm Speaker Driver, Mute	DIP	Anp	-	pdf pinout
APA 4801 K	Dig. Audio	Stereo 280mW 8 Ohm Speaker Driver, Mute	DIP	Anp	-	pdf pinout
APA 4838 R	Dig. Audio	Stereo 2.8W Audio Power Amplifier	TSOP	Anp	-	pdf pinout
APA 4863 KA	Dig. Audio	Stereo 2.2W Audio Power Amplifier	TSOP	Anp	-	pdf pinout
APA 4863 O	Dig. Audio	Stereo 2.2W Audio Power Amplifier	TSOP	Anp	-	pdf pinout
APA 4863 R	Dig. Audio	Stereo 2.2W Audio Power Amplifier	TSOP	Anp	-	pdf pinout
APA 4880 K	Dig. Audio	Stereo 330mW 8W Speak Driver With Mute	TSOP	Anp	-	pdf pinout
APA 4880 O	Dig. Audio	Stereo 330mW 8W Speak Driver With Mute	TSOP	Anp	-	pdf pinout
APA 4880 X	Dig. Audio	Stereo 330mW 8W Speak Driver With Mute	TSOP	Anp	-	pdf pinout
APC 2230	NMOS-IC	Ctv, NTSC-Signal-prozessor	40-DIP	Itt	-	
APD 203	Si-Di	≈BA 100	31a		→BA 100	
APL 1084-33	LIN-IC	low drop +3.3V/ 5A reg., pins Adj/ Out/ In/ Out	17		-	pdf
APL 3200 QA	LIN-IC	Dual-input, Usb/ac Adapter, 1-Cell Li+ Charger	QFN	Anp	-	pdf pinout
APL 3201 QA	LIN-IC	Li+ Battery Charger, Thermal Regulation	MLP	Anp	-	pdf pinout
APL 3202 (B)	LIN-IC	Li+ Battery Charger,Thermal Regulation	45	Anp	-	pdf pinout
APL 3203 AQB	LIN-IC	Li+ Charger Protection IC	MLP	Anp	-	pdf pinout
APL 3203 BQB	LIN-IC	Li+ Charger Protection IC	MLP	Anp	-	pdf pinout
APL 3204 AQB	LIN-IC	Li+ Charger Protection IC	MLP	Anp	-	pdf pinout
APL 3204 BQB	LIN-IC	Li+ Charger Protection IC	MLP	Anp	-	pdf pinout
APL 3205 AQB	LIN-IC	Li+ Charger Protection IC	MLP	Anp	-	pdf pinout
APL 3205 BQB	LIN-IC	Li+ Charger Protection IC	MLP	Anp	-	pdf pinout
APL 3206(A) QB	LIN-IC	Li+ Charger Protection IC, P-mosfet	MLP	Anp	-	pdf pinout
APL 3207 QA	LIN-IC	Li+ Battery Charger, Linear Regulator	MP	Anp	-	pdf pinout
APL 5508 -...A	LIN/Z-IC	1.5-4.5V Low IQ, 560mA Fix. Volt Reg.	35	Anp	-	pdf pinout
APL 5508 -...D	LIN/Z-IC	1.5-4.5V Low IQ, 560mA Fix. Volt Reg.	39	Anp	-	pdf pinout
APL 5508 R-...V	LIN/Z-IC	1.5-4.5V Low IQ, 560mA Fix. Volt Reg.	48	Anp	-	pdf pinout
APL 5508 -...V	LIN/Z-IC	1.5-4.5V Low IQ, 560mA Fix. Volt Reg.	48	Anp	-	pdf pinout
APL 5509 -...A	LIN/Z-IC	1.5-4.5V Low IQ, 560mA Fix. Volt Reg.	35	Anp	-	pdf pinout
APL 5509 -...D	LIN/Z-IC	1.5-4.5V Low IQ, 560mA Fix. Volt Reg.	39	Anp	-	pdf pinout
APL 5509 R-...V	LIN/Z-IC	1.5-4.5V Low IQ, 560mA Fix. Volt Reg.	48	Anp	-	pdf pinout
APL 5509 -...V	LIN/Z-IC	1.5-4.5V Low IQ, 560mA Fix. Volt Reg.	48	Anp	-	pdf pinout
APs	Si-N	=BFP 520 (Typ-Code/Stempel/marking)	44(2mm)	Sie	→BFP 520	data
APU 2400 E	NMOS-IC	Ctv, Audio-prozessor Europa	24-DIP	Itt	-	pinout
APU 2400 J	NMOS-IC	Ctv, Audio-prozessor Japan	24-DIP	Itt	-	pinout
APU 2400 K	NMOS-IC	Ctv, Audio-prozessor Korea	24-DIP	Itt	-	pinout
APU 2400 T	NMOS-IC	Ctv, Audio-prozessor Europa	24-DIP	Itt	-	pinout
APU 2400 U	NMOS-IC	Ctv, Audio-prozessor Usa	24-DIP	Itt	-	pinout
APU 2470	NMOS-IC	Ctv, Audio-prozessor D2-Mac	24-DIP	Itt	APU 2471	pdf
APU 2471	NMOS-IC	Ctv, Audio-prozessor	24-DIP	Itt	APU 2470	pinout
APX 321 SE	LIN-IC	-to-rail,Input/output,Single/dual/quad OP. AMP	45	Dii	-	pdf pinout

Type	Device	Short Description	Fig.	Manu	Comparision Types	More at
APX 321 W	LIN-IC	-to-rail,Input/output,Single/dual/quad OP. AMP	45	Dii	-	pdf pinout
APX 324 TS	LIN-IC	-to-rail,Input/output,Single/dual/quad OP. AMP	45	Dii	-	pdf pinout
APX 339 TS	LIN-IC	Rail-to-rail Input Dual/quad Comparators	45	Dii	-	pdf pinout
APX 358 S	LIN-IC	-to-rail,Input/output,Single/dual/quad OP. AMP	45	Dii	-	pdf pinout
APX 393 S	LIN-IC	Rail-to-rail Input Dual/quad Comparators	45	Dii	-	pdf pinout
APX 809 SA	LIN-IC	3-PIN Microprocessor reset Circuits	45	Dii	-	pdf pinout
APX 809 SR	LIN-IC	3-PIN Microprocessor reset Circuits	45	Dii	-	pdf pinout
APX 810 SA	LIN-IC	3-PIN Microprocessor reset Circuits	45	Dii	-	pdf pinout
APX 810 SR	LIN-IC	3-PIN Microprocessor reset Circuits	45	Dii	-	pdf pinout
APX 1117 D	LIN-IC	Positive Adj./fixed-mode regulator	45	Dii	-	pdf pinout
APX 1117 E	LIN-IC	Positive Adj./fixed-mode regulator	45	Dii	-	pdf pinout
APX 358 M 8	LIN-IC	-to-rail,Input/output,Single/dual/quad OP. AMP	45	Dii	-	pdf pinout
APX 393 M 8	LIN-IC	Rail-to-rail Input Dual/quad Comparators	45	Dii	-	pdf pinout
APX 7343 X	LIN-IC	Single-phase Full-wave Motor Driver	TSOP	Anp	-	pdf pinout
APX 823 W 5	LIN-IC	Processor Supervisory Circuits	45	Dii	-	pdf pinout
APX 824 W 5	LIN-IC	Processor Supervisory Circuits	45	Dii	-	pdf pinout
APX 825 A-W 6	LIN-IC	Processor Supervisory Circuits	45	Dii	-	pdf pinout
APX 9131 A/AT	LIN-IC	Hall Effect Micro Switch IC	43	Anp	-	pdf pinout
APX 9131 E	LIN-IC	Hall Effect Micro Switch IC	43	Anp	-	pdf pinout
APX 9132 A/AT	LIN-IC	Hall Effect Micro Switch IC	33	Anp	-	pdf pinout
APX 9142 E	LIN-IC	Hall Effect Sensor IC,Reverse Volt. Protec.	11	Anp	-	pdf pinout
APX 9262 N	LIN-IC	PWM Variable Speed Fan Motor Driver	TSOP	Anp	-	pdf pinout
APX 9263 N	LIN-IC	PWM Variable Speed Fan Motor Driver	TSOP	Anp	-	pdf pinout
APX 9266 X	LIN-IC	Single-phase Full-wave Motor Driver	TSOP	Anp	-	pdf pinout
APX 9267 X	LIN-IC	Single-phase Full-wave Motor Driver	TSOP	Anp	-	pdf pinout
APX 9268 X	LIN-IC	Single-phase Full-wave Motor Driver	TSOP	Anp	-	pdf pinout
APX 9270 N/R	LIN-IC	PWM Variable Speed Fan Motor Driver	TSOP	Anp	-	pdf pinout
APX 9280 O	LIN-IC	1-Phase Full-wave Motor Pre-driver	TSOP	Anp	-	pdf pinout
APX 9281 N	LIN-IC	Single-phase Full-wave Motor Pre-driver	TSOP	Anp	-	pdf pinout
APY 10(I,II)	Ge-Opto-Di	Foto-Di, 50V, Idark<5µA I=60, II=100nA/Lux	(13x2mmø)	Sie	-	
APY 11(I,II)	Ge-Opto-Di	Foto-Di, 25V, Idark<8µA I=60, II=100nA/Lux	(13x2mmø)	Sie	-	
APY 12(I,II,III)	Ge-Opto-Di	Foto-Di, 100V, 10mA, Idark<8µA I=100, II=180, III=220nA/Lux	2c	Sie	-	
APY 13(I,II,III)	Ge-Opto-Di	Foto-Di, 30V, 10mA, Idark<8µA I=100, II=180, III=220nA/Lux	2c	Sie	-	
AQ	Si-P	=2PB709Q (Typ-Code/Stempel/marking)	35	Phi	→2PB709	data
AQ	Si-P	=2SA1969 (Typ-Code/Stempel/marking)	39	Say	→2SA1969	data
AQ	Si-P	=2SB1218-Q (Typ-Code/Stempel/marking)	35(2mm)	Mat	→2SB1218	data
AQ	Si-P	=2SB1462-Q (Typ-Code/Stempel/marking)	35(1,6mm)	Mat	→2SB1462	data
AQ	Si-P	=2SB709-Q (Typ-Code/Stempel/marking)	35	Mat	→2SB709	data
AQ	Si-P	=2SB766-Q (Typ-Code/Stempel/marking)	39	Mat	→2SB766	data
AQ	Si-N	=2SC2413-AQ (Typ-Code/Stempel/marking)	≈35	Rhm	→2SC2413	data
AQ	Si-N	=2SC2413K-Q (Typ-Code/Stempel/marking)	35	Rhm	→2SC2413K	data
AQ	Si-N	=2SC4098-Q (Typ-Code/Stempel/marking)	35(2mm)	Rhm	→2SC4098	data
AQ	Z-Di	=P4 SMA-11 (Typ-Code/Stempel/marking)	71(5x2,5)	Fag	→P4 SMA-11	data
AQ	Z-Di	=SMBJ 8.0C (Typ-Code/Stempel/marking)	71(5x3,5)	Mop	→SMBJ ...	data
AQ	Si-N+R	=XN 121E (Typ-Code/Stempel/marking)	45	Mat	→XN 121E	data
AQ	Si-N+R	=XP 0121E (Typ-Code/Stempel/marking)	45(2mm)	Mat	→XP 0121E	data
AQ 1 A3M	Si-P+R	S, Rb=Rbe=1kΩ, 20V, 2A, 0,75W	7c	Nec	-	data
AQ 1 A4A	Si-P+R	=AQ 1A3M: Rb=-, Rbe=10kΩ	7c	Nec	-	data
AQ 1 F2Q	Si-P+R	=AQ 1A3M: Rb=0,22k, Rbe=2,2kΩ	7c	Nec	-	data
AQ 1 F3M	Si-P+R	=AQ 1A3M: Rb=2,2k, Rbe=2,2kΩ	7c	Nec	-	data
AQ 1 F3P	Si-P+R	=AQ 1A3M: Rb=2,2k, Rbe=10kΩ	7c	Nec	-	data
AQ 1 L2N	Si-P+R	=AQ 1A3M: Rb=0,47k, Rbe=1kΩ	7c	Nec	-	data
AQ 1 L2Q	Si-P+R	=AQ 1A3M: Rb=0,47k, Rbe=4,7kΩ	7c	Nec	-	data
AQ 2 A4A	Si-P+Di+R	S, Rb=10kΩ, 20V, ±3A, 0,75W, 140MHz	7c	Nec	-	data
AQ 25 Y	Si-N	≈BC 338			→BC 338	
AQO	Si-N	=KTC3880-O (Typ-Code/Stempel/marking)	35	Kec	→KTC 3880(S)	data
AQP	Si-P	=2SB1051K-P (Typ-Code/Stempel/marking)	35	Rhm	→2SB1051K	data
AQQ	Si-P	=2SB1051K-Q (Typ-Code/Stempel/marking)	35	Rhm	→2SB1051K	data
AQR	Si-P	=2SB1051K-R (Typ-Code/Stempel/marking)	35	Rhm	→2SB1051K	data
AQR	Si-N	=KTC3880-R (Typ-Code/Stempel/marking)	35	Kec	→KTC 3880(S)	data
AQY	Si-N	=KTC3880-Y (Typ-Code/Stempel/marking)	35	Kec	→KTC 3880(S)	data
AR	Si-P	=2PB709R (Typ-Code/Stempel/marking)	35	Phi	→2PB709	data
AR	Si-P	=2SA2009 (Typ-Code/Stempel/marking)	35(2mm)	Mat	→2SA2009	data
AR	Si-P	=2SA2011 (Typ-Code/Stempel/marking)	39	Say	→2SA2011	data
AR	Si-P	=2SB1218-R (Typ-Code/Stempel/marking)	35(2mm)	Mat	→2SB1218	data
AR	Si-P	=2SB1462-R (Typ-Code/Stempel/marking)	35(1,6mm)	Mat	→2SB1462	data
AR	Si-P	=2SB709-R (Typ-Code/Stempel/marking)	35	Mat	→2SB709	data
AR	Si-P	=2SB766-R (Typ-Code/Stempel/marking)	39	Mat	→2SB766	data
AR	Si-N	=2SC3338 (Typ-Code/Stempel/marking)	39	Hit	→2SC3338	data
AR	Si-N	=BCW 60RC (Typ-Code/Stempel/marking)	35	Sie,Val,Fer	→BCW 60RC	data
AR	Z-Di	=MA 5Z330-M (Typ-Code/Stempel/marking)	71(1,7mm)	Mat	→MA 5Z330-M	data
AR	Z-Di	=P4 SMA-11A (Typ-Code/Stempel/marking)	71(5x2,5)	Fag	→P4 SMA-11A	data
AR	Z-Di	=SMBJ 8.0CA (Typ-Code/Stempel/marking)	71(5x3,5)	Mop	→SMBJ ...	data
AR	Si-N+R	=XN 121F (Typ-Code/Stempel/marking)	45	Mat	→XN 121F	data

Type	Device	Short Description	Fig.	Manu	Comparision Types	More at
AR 1	Si-N	=BSP 40 (Typ-Code/Stempel/marking)	48	Phi	→BSP 40	data
AR 1	Si-N	=BSR 40 (Typ-Code/Stempel/marking)	39	Val	→BSR 40	data
AR 1 A3M	Si-P+R	S, Rb=Rbe=1kΩ, 60V, 1A, 0,75W	7c	Nec	-	data
AR 1 A4A	Si-P+R	=AR 1A3M: Rb=-, Rbe=10kΩ	7c	Nec	-	data
AR 1 A4M	Si-P+R	=AR 1A3M: Rb=10kΩ, Rbe=10kΩ	7c	Nec	-	data
AR 1 F2Q	Si-P+R	=AR 1A3M: Rb=0,22k, Rbe=2,2kΩ	7c	Nec	-	data
AR 1 F3P	Si-P+R	=AR 1A3M: Rb=2,2k, Rbe=10kΩ	7c	Nec	-	data
AR 1 L2Q	Si-P+R	=AR 1A3M: Rb=0,47k, Rbe=4,7kΩ	7c	Nec	-	data
AR 1 L3N	Si-P+R	=AR 1A3M: Rb=4,7k, Rbe=10kΩ	7c	Nec	-	data
AR 2	Si-N	=BSP 41 (Typ-Code/Stempel/marking)	48	Phi	→BSP 41	data
AR 2	Si-N	=BSR 41 (Typ-Code/Stempel/marking)	39	Val	→BSR 41	data
AR 3	Si-N	=BSP 42 (Typ-Code/Stempel/marking)	48	Phi	→BSP 42	data
AR 3	Si-N	=BSR 42 (Typ-Code/Stempel/marking)	39	Val	→BSR 42	data
AR 4	Si-N	=BSP 43 (Typ-Code/Stempel/marking)	48	Phi	→BSP 43	data
AR 4	Si-N	=BSR 43 (Typ-Code/Stempel/marking)	39	Val	→BSR 43	data
AR57G62B66BC-C	LED	rgb ø5mm I:4.5lm Vf:2V If:150mA		Roi		data pdf pinout
AR58G05B24BC-C	LED	rgb ø5mm I:6lm Vf:3.4V If:350mA		Roi		data pdf pinout
ARH 2805 S	LIN-IC	Radiation Hardened DC/DC Converters	42	Inr	-	pdf pinout
ARH 2812 S	LIN-IC	Radiation Hardened DC/DC Converters	42	Inr	-	pdf pinout
ARH 2815 S	LIN-IC	Radiation Hardened DC/DC Converters	42	Inr	-	pdf pinout
ARH 2802 R 5 S	LIN-IC	Radiation Hardened DC/DC Converters	42	Inr	-	pdf pinout
ARH 2803 R 3 S	LIN-IC	Radiation Hardened DC/DC Converters	42	Inr	-	pdf pinout
ARH 2805 R 2 S	LIN-IC	Radiation Hardened DC/DC Converters	42	Inr	-	pdf pinout
ARM 2812 T	IC	1 MEGA-RAD Hardened DC/DC Converter	42	Inr	-	pdf pinout
ARM 2815 T	IC	1 MEGA-RAD Hardened DC/DC Converter	42	Inr	-	pdf pinout
ART 2812 T	IC	Radiation Hardened DC/DC Converter	42	Inr	-	pdf pinout
ART 2815 T	IC	Radiation Hardened DC/DC Converter	42	Inr		pdf pinout
AS	Si-P	=2PB709S (Typ-Code/Stempel/marking)	35	Phi	→2PB709	data
AS	Si-P	=2SA1655 (Typ-Code/Stempel/marking)	35	Say	→2SA1665	data
AS	Si-P	=2SA2010 (Typ-Code/Stempel/marking)	35	Mat	→2SA2010	
AS	Si-P	=2SA2012 (Typ-Code/Stempel/marking)	39	Say	→2SA2012	data
AS	Si-P	=2SB1218-S (Typ-Code/Stempel/marking)	35(2mm)	Mat	→2SB1218	data
AS	Si-P	=2SB1462-S (Typ-Code/Stempel/marking)	35(1,6mm)	Mat	→2SB1462	data
AS	Si-P	=2SB709-S (Typ-Code/Stempel/marking)	35	Mat	→2SB709	data
AS	Si-P	=2SB766-S (Typ-Code/Stempel/marking)	39	Mat	→2SB766	data
AS	Si-N	=2SC3380 (Typ-Code/Stempel/marking)	39	Hit	→2SC3380	data
AS	Si-N	=BCW 60RD (Typ-Code/Stempel/marking)	35	Sie,Val	→BCW 60RD	data
AS	Z-Di	=MA 5Z330-H (Typ-Code/Stempel/marking)	71(1,7mm)	Mat	→MA 5Z330-H	data
AS	Z-Di	=P4 SMA-12 (Typ-Code/Stempel/marking)	71(5x2,5)	Fag	→P4 SMA-12	data
AS	Z-Di	=SMBJ 8.5C (Typ-Code/Stempel/marking)	71(5x3,5)	Mop	→SMBJ ...	data
AS	Z-IC	=TA 76431F (Typ-Code/Stempel/marking)	39	Tos	→TA 76431	
AS	Z-IC	=TA 76431F (Typ-Code/Stempel/marking)	39	Tos	→TA 76431	pdf
AS 01	Hybrid-IC	Vorverstärker/pre-amplifier		Fui		
AS 1	Si-N-Darl+Di	=BSP 50 (Typ-Code/Stempel/marking)	48	Phi, Sie	→BSP 50	data
AS 1	Si-N-Darl+Di	=BST 50 (Typ-Code/Stempel/marking)	39	Phi, Val	→BST 50	data
AS 01(A,Z)	Si-Di	GI, S, 200...600V, 0,6A, 1,5µs AS01=400V, A=600V, Z=200V	31a	Sak	BY 126...127, BY 133...134, 1N4003...07,++	data
AS 02	Hybrid-IC	Vorverstärker/pre-amplifier		Fui		
AS 2	Si-N-Darl+Di	=BSP 51 (Typ-Code/Stempel/marking)	48	Phi, Sie	→BSP 51	data
AS 2	Si-N-Darl+Di	=BST 51 (Typ-Code/Stempel/marking)	39	Phi, Val	→BST 51	data
AS 03	Hybrid-IC	Verstärker/amplifier		Fui		
AS 3	Si-N-Darl+Di	=BSP 52 (Typ-Code/Stempel/marking)	48	Phi, Sie	→BSP 52	data
AS 3	Si-N-Darl+Di	=BST 52 (Typ-Code/Stempel/marking)	39	Phi, Val	→BST 52	data
AS 04	Hybrid-IC	Verstärker/amplifier		Fui	-	
AS 05	Hybrid-IC	Photo-verstärker/amplifier		Fui	-	
AS 07	Hybrid-IC	Differential-verstärker/amplifier		Fui	-	
AS1117	Z-IC			Sip		pdf
AS 1117 R	Z-IC	800mA Low Dropout Regulator	17	Sip	-	pdf pinout
AS 1117 R-X	Z-IC	800mA Low Dropout Regulator	17	Sip	-	pdf pinout
AS 1117 S	Z-IC	800mA Low Dropout Regulator	17	Sip	-	pdf pinout
AS 1117 S-X	Z-IC	800mA Low Dropout Regulator	17	Sip	-	pdf pinout
AS 1117 T	Z-IC	800mA Low Dropout Regulator	17	Sip	-	pdf pinout
AS 1117 T-X	Z-IC	800mA Low Dropout Regulator	17	Sip	-	pdf pinout
AS 1117 U	Z-IC	800mA Low Dropout Regulator	17	Sip	-	pdf pinout
AS 1117 U-X	Z-IC	800mA Low Dropout Regulator	17	Sip	-	pdf pinout
AS 1117 M 1	Z-IC	800mA Low Dropout Regulator	17	Sip	-	pdf pinout
AS 1117 M 1-X	Z-IC	800mA Low Dropout Regulator	17	Sip	-	pdf pinout
AS 1117 M 3	Z-IC	800mA Low Dropout Regulator	17	Sip	-	pdf pinout
AS 1117 M 3-X	Z-IC	800mA Low Dropout Regulator	17	Sip	-	pdf pinout
ASG	Si-P	=KTA1504-G (Typ-Code/Stempel/marking)	35	Kec	→KTA 1504	data
ASO	Si-P	=KTA1504-O (Typ-Code/Stempel/marking)	35	Kec	→KTA 1504	data

ASY

Type	Device	Short Description	Fig.	Manu	Comparision Types	More at
ASY	Si-P	=KTA1504-Y (Typ-Code/Stempel/marking)	35	Kec	→KTA 1504	data
ASY 10	Ge-P	NF/S, 32V, 0,3A, B=30...50	2a	Sie	AC 125, AC 126, AC 151	data
ASY 11	Ge-P	=ASY 10: B=40...70	2a	Sie	AC 125, AC 126, AC 151	data
ASY 12	Ge-P	NF/S, 32V, 0,6A, 0,135W	2a	Itt	AC 128, AC 153, ASY 48, ASY 76	data
ASY 13	Ge-P	NF/S, 60V, 0,6A, 0,135W	2a	Itt	ASY 48, ASY 76	data
ASY 14	Ge-P	NF/S, 80V, 0,25A, 0,13W	2a	Itt		data

Type	Device	Short Description	Fig.	Manu	Comparision Types	More at
ASY 23	Ge-P	NF/S, 80V, 0,3A, 0,85W	2a§	Phi	-	data
ASY 24	Ge-P	NF/S, 50V, 0,25A, 0,1W	2a	Aeg	ASY 48, ASY 77	data
ASY 24 B	Ge-P	=ASY 24: 35V	2a		AC 125, AC 126, AC 151, ASY 48, ASY 76	data
ASY 25	Ge-P	NF/S, 32V, 0,3A, 0,15W	2a	Sie	AC 128, AC 153	data
ASY 26	Ge-P	NF/S, 30V, 0,2A, 0,15W	2a&	Aeg,Phi,++	ASY 48, ASY 76	data
ASY 27	Ge-P	NF/S, 25V, 0,2A, 0,15W	2a&	Aeg,Phi,++	ASY 48, ASY 76	data
ASY 28	Ge-N	NF/S, 30V, 0,2A, 0,15W	2a&	Aeg,Phi,++	ASY 73...75	data
ASY 29	Ge-N	NF/S, 25V, 0,2A, 0,15W	2a&	Aeg,Phi,++	ASY 73...75	data
ASY 30	Ge-P	NF/S, 50V, 0,25A, 0,2W	3a	Aeg	ASY 48, ASY 77	data
ASY 31	Ge-P	NF/S, 25V, 0,2A, 0,125W, B=30...80	1a	Phi	ASY 26, ASY 27, ASY 48, ASY 76	data
ASY 32	Ge-P	=ASY 31: B=50...150	1a	Phi	ASY 26, ASY 27, ASY 48, ASY 76	data
ASY 33	Ge-P	NF/S, 32V, 0,3A, 0,15W	2a	Sie	AC 125, AC 126, AC 151, ASY 48, ASY 76	data
ASY 37	Ge-P	NF/S, 64V, 0,3A, 0,15W	2a	Sie	ASY 48, ASY 77	data
ASY 48	Ge-P	NF/S, 64V, 0,3A, 0,22W	2a	Sie	ASY 77	data
ASY 49	Ge-P	NF/S, 100V, 0,25A, 0,09W	37a	Itt	-	data
ASY 50	Ge-P	NF/S, 20V, 0,5A, 0,06W	37a	Itt	AC 128, AC 153, AC 188	data
ASY 51	Ge-P	NF/S, 60V, 0,25A, 0,09W	37a	Itt	ASY 48, ASY 77	data
ASY 52	Ge-P	NF/S, 60V, 0,25A, 0,09W	37a	Itt	ASY 48, ASY 77	data
ASY 53	Ge-N	NF/S, 20V, 0,25A, 0,06W	37a	Itt	AC 127, AC 176, AC 187, ASY 29, ASY 73	data
ASY 54	Ge-P	NF/S, 30V, 0,5A, 0,2W	37a	Itt	AC 128, AC 153	data
ASY 54 N	Ge-P	=ASY 54: 0,15W	2a		AC 128, AC 153	data
ASY 55	Ge-P	NF/S, 20V, 0,5A, 0,2W	37a	Itt	AC 128, AC 153, AC 188	data
ASY 55 N	Ge-P	=ASY 55: 0,15W	2a		AC 128, AC 153, AC 188	data
ASY 56	Ge-P	NF/S, 20V, 0,2A, 0,2W, B=26...60	37a	Itt	AC 125, AC 126, AC 151	data
ASY 56 N	Ge-P	=ASY 56: 0,15W	2a		AC 125, AC 126, AC 151	data
ASY 57	Ge-P	=ASY 56: B=30...80	37a	Itt	AC 125, AC 126, AC 151	data
ASY 57 N	Ge-P	=ASY 57: 0,15W	2a		AC 125, AC 126, AC 151	data
ASY 58	Ge-P	=ASY 56: B=40...100	37a	Itt	AC 125, AC 126, AC 151	data
ASY 58 N	Ge-P	=ASY 58: 0,15W	2a		AC 125, AC 126, AC 151	data
ASY 59	Ge-P	=ASY 56: B=60...150	37a	Itt	AC 125, AC 126, AC 151	data
ASY 59 N	Ge-P	=ASY 59: 0,15W	2a		AC 125, AC 126, AC 151	data
ASY 60	Ge-P	NF/HF/S, sym, 20V, 0,25A, 0,2W	37a	Itt	-	data
ASY 61	Ge-N	NF/S, 30V, 0,25A, 0,1W	37a	Itt	AC 127, AC 176, ASY 28, ASY 73...75	data
ASY 62	Ge-N	NF/S, 20V, 0,25A, 0,1W	37a	Itt	AC 127, AC 176, AC 187, ASY 29, ASY 73	data
ASY 63	Ge-P	NF/S, 26V, 0,2W	37a	Itt	AC 125, AC 126, AC 151	data
ASY 63 N	Ge-P	=ASY 63: 0,075W	2a		AC 125, AC 126, AC 151	data
ASY 64	Ge-P	NF/S, 30V, 0,2W	37a	Itt	AC 125, AC 126, AC 151	data
ASY 66	Ge-P	NF/S, 30V, 0,2W	37a	Itt	AC 125, AC 126, AC 151	data
ASY 67	Ge-P	HF/S, 50V, 0,05A, 150MHz	5g	Phi	AFY 18, AFY 19	data
ASY 68	Ge-P	NF/S, 12V, 0,1A, 0,075W	2a	Sie	ASY 26...27	data
ASY 69	Ge-P	NF/S, 20V, 0,35A, 0,075W	2a	Sie	ASY 26...27	data
ASY 70	Ge-P	NF/S, 20V, 0,35A, 0,075W	2a	Sie	AC 128, AC 153, ASY 26, ASY 48, ASY 76	data
ASY 71	Ge-P	NF/S, 100V, 0,1A, 0,15W	37a	Phi	-	data
ASY 72	Ge-N	NF/S, 20V, 0,25A, 0,1W	37a	Itt	AC 127, AC 176, AC 187, ASY 29, ASY 75	data
ASY 73	Ge-N	NF/S, sym, 30V, 0,4A, 0,085W, >4MHz	2a&	Phi	-	data
ASY 74	Ge-N	=ASY 73: >6MHz	2a&	Phi	-	data
ASY 75	Ge-N	=ASY 73: >10MHz	2a&	Phi	-	data
ASY 76	Ge-P	NF/S, 40V, 0,5A, 0,16W	2a&	Phi	AC 128, AC 153, ASY 48	data
ASY 77	Ge-P	=ASY 76: 60V	2a&	Phi	ASY 48	data
ASY 78(T)	Ge-P	NF/S, 40V, 0,4A, 0,125W, 40MHz	2a	Tsm	-	data
ASY 80	Ge-P	NF/S, 40V, 0,5A, 0,16W	2a&	Phi	AC 128, AC 153, ASY 48, ASY 76	data
ASY 81	Ge-P	NF/S, 60V, 0,5A, 0,15W	2a&	Tho	ASY 48, ASY 77	data
ASY 82	Ge-P	NF/S, 26V, 0,5A, 0,2W, B=30...130	2a	Aei	AC 128, AC 153, AC 188	data
ASY 83	Ge-P	=ASY 82: B=70...320	2a	Aei	AC 128, AC 153, AC 188	data
ASY 84	Ge-P	NF/S, 40V, 0,5A, 0,2W, B=30...130	2a	Aei	AC 128, AC 153	data
ASY 85	Ge-P	=ASY 84: B=70...320	2a	Aei	AC 128, AC 153	data
ASY 86	Ge-N	NF/S, 16V, 0,5A, 0,2W, B=25...120	2a	Aei	AC 127, AC 176, AC 187	data
ASY 87	Ge-N	=ASY 86: B=60...295	2a	Aei	AC 127, AC 176, AC 187	data
ASY 88	Ge-N	NF/S, 26V, 0,5A, 0,2W, B=25...120	2a	Aei	AC 127, AC 176, AC 187	data
ASY 89	Ge-N	=ASY 88: B=60...295	2a	Aei	AC 127, AC 176, AC 187	data
ASY 90	Ge-P	NF/S, 40V, 0,25A, 0,185W	2a	Sgs	AC 125, AC 126, AC 151, ASY 48, ASY 76	data
ASY 91	Ge-P	=ASY 90: 25V	2a	Sgs	AC 125, AC 126, AC 151, ASY 48, ASY 76	data

ASZ

Type	Device	Short Description	Fig.	Manu	Comparision Types	More at
ASZ 10	Ge-P	NF/S, 50V, 0,25A, 0,15W	1a	Aeg	ASY 48, ASY 77	data
ASZ 11	Ge-P	NF/S, 20V, 0,2A, 0,125W, B>40	1a	Phi	AC 125, AC 126, AC 151, ASY 48, ASY 76	data
ASZ 12	Ge-P	=ASZ 11: B>60	1a	Phi	AC 125, AC 126, AC 151, ASY 48, ASY 76	data
ASZ 15	Ge-P	S-L, 100V, 8A, 30W	23a§	Gpd,Phi,Tsm	AUY 37, 2N2527	data
ASZ 16	Ge-P	S-L, 60V, 8A, 30W	23a§	Gpd,Phi,Tsm	AUY 21, 2N2526	data
ASZ 17	Ge-P	S-L, 60V, 8A, 30W	23a§	Gpd,Phi,Tsm	AUY 21, 2N2526	data
ASZ 18	Ge-P	S-L, 100V, 8A, 30W	23a§	Gpd,Phi,Tsm	AUY 37, 2N2527	data
ASZ 20	Ge-P	NF/S, 40V, 25mA, 100MHz	1g	Phi	(AF 202S/L)	data
ASZ 21	Ge-P	SS, 20V, 30mA, >300MHz, 50/80ns	2a	Phi	2N2635, 2N2955...2957	data
ASZ 23	Ge-P	SS, 30V, 0,1A, 0,05W	1g	Phi		data

Type	Device	Short Description	Fig.	Manu	Comparision Types	More at
ASZ 30	Ge-P	NF/S, 50V, 0,25A, 0,03W, 20MHz	1a	Aeg	ASY 48, ASY 77	data
ASZ 1015	Ge-P	S-L, 80V, 6A, 22,5W	23a§	Tsm	AUY 37, 2N2527	data
ASZ 1016	Ge-P	S-L, 60V, 6A, 22,5W	23a§	Tsm	AUY 21, 2N2526	data
ASZ 1017	Ge-P	S-L, 60V, 6A, 22,5W	23a§	Tsm	AUY 21, 2N2526	data
ASZ 1018	Ge-P	S-L, 80V, 6A, 22,5W	23a§	Tsm	AUY 37, 2N2527	data
AT						
AT	Si-P	=2SA2028 (Typ-Code/Stempel/marking)	35(2mm)	Mat	→2SA2028	
AT	Si-N	=2SC4066 (Typ-Code/Stempel/marking)	35	Say	→2SC4066	
AT	Si-N-Darl	=2SD1470 (Typ-Code/Stempel/marking)	39	Hit	→2SD1470	data
AT	Si-Di	=BAT 18-06 (Typ-Code/Stempel/marking)	35	Sie	→BAT 18-06	data
AT	Si-N	=BCW 60RE (Typ-Code/Stempel/marking)	35	Sie	→BCW 60RE	
AT	Z-Di	=P4 SMA-12A (Typ-Code/Stempel/marking)	71(5x2,5)	Fag	→P4 SMA-12A	data
AT	Z-Di	=SMBJ 8.5CA (Typ-Code/Stempel/marking)	71(5x3,5)	Mop	→SMBJ ...	data
AT 1	Si-N	=BSP 19 (Typ-Code/Stempel/marking)	48	Phi	→BSP 19	data
AT 1	Si-N	=BST 39 (Typ-Code/Stempel/marking)	39	Phi, Val	→BST 39	data
AT 2	Si-N	=BSP 20 (Typ-Code/Stempel/marking)	48	Phi	→BSP 20	data
AT 2	Si-N	=BST 40 (Typ-Code/Stempel/marking)	39	Phi, Val	→BST 40	data
AT 24C32	EEPROM-IC	(SMD), 32kBit(4kx8) Eeprom, I2C, 5V	DIP	Atm	...24C32...	pdf pinout
AT 24C64	EEPROM-IC	(SMD), 64kBit(8kx8) Eeprom, I2C, 5V	DIP	Atm	...24C64...	pdf pinout
AT 29C512	EEPROM-IC	512kB(64kx8) 5V Flash Memory	DIP	Atm	W29C512A, SST29EE512,	pdf pinout
AT 29 C 512 -70 JC	EEPROM-IC	512K 5-volt Only Flash Memory	PLCC	Atm	W29C512A, SST29EE512,	pdf pinout
AT 29 C 512-70 TC	EEPROM-IC	512K 5-volt Only Flash Memory	TSOP	Atm	W29C512A, SST29EE512,	pdf pinout
AT 120 A	Si-Di	=BAT 120A (Typ-Code/Stempel/marking)	48	Phi	→BAT 120A	data
AT 120 C	Si-Di	=BAT 120C (Typ-Code/Stempel/marking)	48	Phi	→BAT 120C	data
AT 120 S	Si-Di	=BAT 120S (Typ-Code/Stempel/marking)	48	Phi	→BAT 120S	data
AT 140 A	Si-Di	=BAT 140A (Typ-Code/Stempel/marking)	48	Phi	→BAT 140A	data
AT 140 C	Si-Di	=BAT 140C (Typ-Code/Stempel/marking)	48	Phi	→BAT 140C	data
AT 140 S	Si-Di	=BAT 140S (Typ-Code/Stempel/marking)	48	Phi	→BAT 140S	data
AT 160 A	Si-Di	=BAT 160A (Typ-Code/Stempel/marking)	48	Phi	→BAT 160A	data
AT 160 C	Si-Di	=BAT 160C (Typ-Code/Stempel/marking)	48	Phi	→BAT 160C	data
AT 160 S	Si-Di	=BAT 160S (Typ-Code/Stempel/marking)	48	Phi	→BAT 160S	data
AT 200	Ge-P	=AU 106	23a§	Sgs	→AU 106	data
AT 201	Ge-P	=AU 107	23a§	Sgs	→AU 107	data
AT 202	Ge-P	S-L, 100V, 3A, 3W	23a§	Sgs	AUY 34	data
AT 207	Ge-P	S-L, 60V, 10A, 9W	38a§	Sgs	2N2490..2492, 2N2079..2080, 2N1981	data
AT 208	Ge-P	S-L, 100V, 10A, 30W	23a§	Sgs	AUY 37, 2N2527	data
AT 209	Ge-P	NF/S, 40V, 0,25A, 0,18W	2a	Sgs	AC 125, AC 126, AC 151	data
AT 210	Ge-P	NF/S, 30V, 0,25A, 0,18W	2a	Sgs	AC 125, AC 126, AC 151	data
AT 216	Ge-P	S-L, 320V, 10A, 5W	23a§	Sgs	AU 106, AU 109, AU 111, AU 112, 2N5325	data
AT 270	Ge-P	=ASY 90	2a	Sgs	→ASY 90	data
AT 275	Ge-P	=ASY 91	2a	Sgs	→ASY 91	data
AT 431 UN	IC	1.24V low-Voltage Adj. Regulator	47	Aim	-	pdf pinout
AT 431 UN_GRE	IC	1.24V low-Voltage Adj. Regulator	47	Aim	-	pdf pinout
AT 431 Z	IC	1.24V low-Voltage Adj. Regulator	47	Aim	-	pdf pinout
AT 450	Ge-P	TV-HA, 420V, 10A, 5W	23a§	Sgs	-	data
AT 529	Si-N	NF/S-L, 140V, 8A	22a§	Sgs	BUX 59..60, BUW 64A..C, 2SC2334	data
AT 605	Si-N	=BU 126	23a§	Sgs	→BU 126	data
AT 1175 S	CMOS-IC	Video A/D Converter	TSOP	Aim	-	pdf pinout
AT 1175 S_GRE	CMOS-IC	Video A/D Converter	TSOP	Aim	-	pdf pinout
AT 1202 X	IC	High Voltage Linear Regulator	45	Aim	-	pdf pinout
AT 1202 X_GRE	IC	High Voltage Linear Regulator	45	Aim	-	pdf pinout
AT 1205 -BX	CMOS-IC	300mA Cmos LDO Regulator	46	Aim	-	pdf pinout
AT 1303 M_GRE	IC	Step-up DC/DC Converter	TSOP	Aim	-	pdf pinout
AT 1303 P_GRE	IC	Step-up DC/DC Converter	TSOP	Aim	-	pdf pinout
AT 1305 AP	IC	Pfm Step-up DC/DC Converter with LDO	45	Aim	-	pdf pinout
AT 1305 AP_GRE	IC	Pfm Step-up DC/DC Converter with LDO	45	Aim	-	pdf pinout
AT 1305 BP	IC	Pfm Step-up DC/DC Converter with LDO	45	Aim	-	pdf pinout
AT 1305 BP_GRE	IC	Pfm Step-up DC/DC Converter with LDO	45	Aim	-	pdf pinout
AT 1305 P	IC	PFM Step-up DC/DC Converter with LDO	45	Aim	-	pdf pinout
AT 1305 P_GRE	IC	Pfm Step-up DC/DC Converter with LDO	45	Aim	-	pdf pinout
AT 1306_GRE	IC	Step-up DC/DC Converter	TSOP	Aim	-	pdf pinout
AT 1308 X	IC	Step-up DC/DC Converter	TSOP	Aim	-	pdf pinout
AT 1308 X_GRE	IC	Step-up DC/DC Converter	TSOP	Aim	-	pdf pinout
AT 1309 X	IC	Micro-power Step up DC-DC Converter	45	Aim	-	pdf pinout
AT 1309 X_GRE	IC	Micro-power Step up DC-DC Converter	45	Aim	-	pdf pinout
AT 1310 X	IC	Micropower Charge Pump Converter	45	Aim	-	pdf pinout
AT 1310 X_PBF	IC	Micropower Charge Pump Converter	45	Aim	-	pdf pinout
AT 1312 AX	IC	Constant Current White-LEDs Driver	46	Aim	-	pdf pinout
AT 1312 AX_GRE	IC	Constant Current White-leds Driver	46	Aim	-	pdf pinout
AT 1312 BX	IC	Constant Current White-leds Driver	46	Aim	-	pdf pinout
AT 1312 BX_GRE	IC	Constant Current White-leds Driver	46	Aim	-	pdf pinout
AT 1312 CX	IC	Constant Current White-leds Driver	46	Aim	-	pdf pinout
AT 1312 CX_GRE	IC	Constant Current White-leds Driver	46	Aim	-	pdf pinout
AT 1313 X_GRE	IC	White-LED Driver	46	Aim	-	pdf pinout
AT 1314 S_GRE	IC	Step-down Current Regulator For LEDs	47	Aim	-	pdf pinout
AT 1325 R_GRE	IC	8-bit LED Sink Driver	46	Aim	-	pdf pinout
AT 1326 R_GRE	IC	16-bit LED Sink Driver	46	Aim	-	pdf pinout
AT 1362 AN	IC	Synchronous Buck Converter	47	Aim	-	pdf pinout

Type	Device	Short Description	Fig.	Manu	Comparision Types	More at
AT 1362 BN	IC	Synchronous Buck Converter	47	Aim	-	pdf pinout
AT 1366 BX	IC	Synchronous Buck Converter	47	Aim	-	pdf pinout
AT 1366 BX_GRE	IC	Synchronous Buck Converter	47	Aim	-	pdf pinout
AT 1366 X	IC	Synchronous Buck Converter	47	Aim	-	pdf pinout
AT 1366 X_GRE	IC	Synchronous Buck Converter	47	Aim	-	pdf pinout
AT 1367 AN	IC	Synchronous Buck Converter	47	Aim	-	pdf pinout
AT 1367 BN	IC	Synchronous Buck Converter	47	Aim	-	pdf pinout
AT 1367 CN	IC	Synchronous Buck Converter	47	Aim	-	pdf pinout
AT 1367 DN	IC	Synchronous Buck Converter	47	Aim	-	pdf pinout
AT 1368 AP	IC	1-CH Synchronous Buck PWM Controller	47	Aim	-	pdf pinout
AT 1368 AR	IC	1-CH Synchronous Buck PWM Controller	47	Aim	-	pdf pinout
AT 1368 BP	IC	1-CH Synchronous Buck PWM Controller	47	Aim	-	pdf pinout
AT 1368 BP_GRE	IC	1-CH Synchronous Buck PWM Controller	47	Aim	-	pdf pinout
AT 1368 BR	IC	1-CH Synchronous Buck PWM Controller	47	Aim	-	pdf pinout
AT 1369 AR	IC	2-CH Synchronous Buck PWM Controller	47	Aim	-	pdf pinout
AT 1369 AR_GRE	IC	2-CH Synchronous Buck PWM Controller	47	Aim	-	pdf pinout
AT 1369 BR	IC	2-CH Synchronous Buck PWM Controller	47	Aim	-	pdf pinout
AT 1369 BR_GRE	IC	2-CH Synchronous Buck PWM Controller	47	Aim	-	pdf pinout
AT 1380 AP	DIG-IC	Switching Regulator Controller	TSOP	Aim	-	pdf pinout
AT 1380 AP_GRE	DIG-IC	Switching Regulator Controller	TSOP	Aim	-	pdf pinout
AT 1380 AS	DIG-IC	Switching Regulator Controller	TSOP	Aim	-	pdf pinout
AT 1380 AS_GRE	DIG-IC	Switching Regulator Controller	TSOP	Aim	-	pdf pinout
AT 1380 P	IC	Switching Regulator Power Controller	47	Aim	-	pdf pinout
AT 1380 P_GRE	IC	Switching Regulator Power Controller	47	Aim	-	pdf pinout
AT 1380 S	IC	Switching Regulator Power Controller	47	Aim	-	pdf pinout
AT 1380 S_GRE	IC	Switching Regulator Power Controller	47	Aim	-	pdf pinout
AT 1382 AF	IC	4-Channel DC-DC Converter for DSC	47	Aim	-	pdf pinout
AT 1382 AF_GRE	IC	4-Channel DC-DC Converter for DSC	47	Aim	-	pdf pinout
AT 1382 AF_PBF	IC	4-Channel DC-DC Converter for DSC	47	Aim	-	pdf pinout
AT 1382 BF	IC	4-Channel DC-DC Converter for DSC	MP	Aim	-	pdf pinout
AT 1386 F	IC	6-Channel DC-DC Converter for DSC	MP	Aim	-	pdf pinout
AT 1386 F_GRE	IC	6-Channel DC-DC Converter for DSC	MP	Aim	-	pdf pinout
AT 1386 F_PBF	IC	6-Channel DC-DC Converter for DSC	MP	Aim	-	pdf pinout
AT 1387 AF	IC	6-Channel DC-DC Converter for DSC	MP	Aim	-	pdf pinout
AT 1387 AF_GRE	IC	6-Channel DC-DC Converter for DSC	MP	Aim	-	pdf pinout
AT 1387 AF_PBF	IC	6-Channel DC-DC Converter for DSC	MP	Aim	-	pdf pinout
AT 1388 AF	IC	6-Channel DC-DC Controller for DSC	MP	Aim	-	pdf pinout
AT 1388 AF_GRE	IC	6-Channel DC-DC Controller for DSC	MP	Aim	-	pdf pinout
AT 1388 AF_PBF	IC	6-Channel DC-DC Controller for DSC	MP	Aim	-	pdf pinout
AT 1388 F	IC	6-Channel DC-DC Controller for DSC	MP	Aim	-	pdf pinout
AT 1388 F_GRE	IC	6-Channel DC-DC Controller for DSC	MP	Aim	-	pdf pinout
AT 1388 F_PBF	IC	6-Channel DC-DC Controller for DSC	MP	Aim	-	pdf pinout
AT 1391 P	IC	1- Channel Step-down PWM Controller	MP	Aim	-	pdf pinout
AT 1391 P_GRE	IC	1- Channel Step-down PWM Controller	MP	Aim	-	pdf pinout
AT 1392 P	IC	1- Channel Synchronous Buck PWM Controller	MP	Aim	-	pdf pinout
AT 1392 P_GRE	IC	1- Channel Synchronous Buck PWM Controller	MP	Aim	-	pdf pinout
AT 1393 AF	IC	6-Channel DC-DC Converter IC for DSC	MP	Aim	-	pdf pinout
AT 1393 AF_GRE	IC	6-Channel DC-DC Converter IC for DSC	MP	Aim	-	pdf pinout
AT 1393 AF_PBF	IC	6-Channel DC-DC Converter IC for DSC	MP	Aim	-	pdf pinout
AT 1393 F	IC	6-Channel DC-DC Converter IC for DSC	MP	Aim	-	pdf pinout
AT 1393 F_GRE	IC	6-Channel DC-DC Converter IC for DSC	MP	Aim	-	pdf pinout
AT 1393 F_PBF	IC	6-Channel DC-DC Converter IC for DSC	MP	Aim	-	pdf pinout
AT 1395 F	IC	5-Channel DC-DC Converter for DSC	MP	Aim	-	pdf pinout
AT 1396 AN	IC	6-Channel DC-DC Converter for DSC	MP	Aim	-	pdf pinout
AT 1396 BN	IC	6-Channel DC-DC Converter for DSC	MP	Aim	-	pdf pinout
AT 1396 CN	IC	6-Channel DC-DC Converter for DSC	MP	Aim	-	pdf pinout
AT 1396 DN	IC	6-Channel DC-DC Converter for DSC	MP	Aim	-	pdf pinout
AT 1398 AN_GRE	MOS-IC	7-Channel DC-DC Converter	QFN	Aim	-	pdf pinout
AT 1398 BN_GRE	MOS-IC	7-Channel DC-DC Converter	QFN	Aim	-	pdf pinout
AT 1399 AN_GRE	MOS-IC	7-Channel DC-DC Converter	QFN	Aim	-	pdf pinout
AT 1399 CN_GRE	MOS-IC	7-Channel DC-DC Converter	QFN	Aim	-	pdf pinout
AT 1450 S	IC	Photoflash Capacitor Charger for DSC	TSOP	Aim	-	pdf pinout
AT 1450 S_GREE	IC	Photoflash Capacitor Charger for DSC	TSOP	Aim	-	pdf pinout
AT 1452 AP	IC	Photoflash Capacitor Charger for DSC	TSOP	Aim	-	pdf pinout
AT 1452 AP_GRE	IC	Photoflash Capacitor Charger for DSC	TSOP	Aim	-	pdf pinout
AT 1452 BP	IC	Photoflash Capacitor Charger for DSC	TSOP	Aim	-	pdf pinout
AT 1452 BP_GRE	IC	Photoflash Capacitor Charger for DSC	TSOP	Aim	-	pdf pinout
AT 1453 AP	IC	Photoflash Capacitor Charger for DSC	TSOP	Aim	-	pdf pinout
AT 1453 AP_GRE	IC	Photoflash Capacitor Charger for DSC	TSOP	Aim	-	pdf pinout
AT 1453 BP	IC	Photoflash Capacitor Charger for DSC	TSOP	Aim	-	pdf pinout
AT 1453 BP_GRE	IC	Photoflash Capacitor Charger for DSC	TSOP	Aim	-	pdf pinout
AT 1454 AM_GRE	DIG-IC	Photoflash Capacitor Charger	TSOP	Aim	-	pdf pinout
AT 1454 BM_GRE	DIG-IC	Photoflash Capacitor Charger	TSOP	Aim	-	pdf pinout
AT 1455 M	IC	Photoflash Capacitor Charger for DSC	TSOP	Aim	-	pdf pinout
AT 1455 M_GRE	IC	Photoflash Capacitor Charger for DSC	TSOP	Aim	-	pdf pinout
AT 1456 N_GRE	IC	Photoflash Capacitor Charger	TSOP	Aim	-	pdf pinout
AT 1457 M	IC	1-Cell Li-ion, Li-Pol Charge IC	TSOP	Aim	-	pdf pinout
AT 1458 N_GRE	DIG-IC	Battery Charger Management IC	QFN	Aim	-	pdf pinout
AT 1461 X_GRE	DIG-IC	Charge Management IC	QFN	Aim	-	pdf pinout
AT 1504 R	IC	Fet Bias Controller	TSOP	Aim	-	pdf pinout
AT 1504 R_GRE	IC	Fet Bias Controller	TSOP	Aim	-	pdf pinout
AT 1506 R	IC	Fet Bias Controller	TSOP	Aim	-	pdf pinout
AT 1506 R_GRE	IC	Fet Bias Controller	TSOP	Aim	-	pdf pinout

Type	Device	Short Description	Fig.	Manu	Comparision Types	More at
AT 1511 R	IC	Fet Bias Controller, Tone Detection	TSOP	Aim	-	pdf pinout
AT 1511 R_GRE	IC	Fet Bias Controller, Tone Detection	TSOP	Aim	-	pdf pinout
AT 1512 R	IC	Fet Bias Controller, Tone Detection	TSOP	Aim	-	pdf pinout
AT 1512 R_GRE	IC	Fet Bias Controller, Tone Detection	TSOP	Aim	-	pdf pinout
AT 1729 P	IC	3-Channel PWM Controller for LCD Bias	TSOP	Aim	-	pdf pinout
AT 1729 P_GRE	IC	3-Channel PWM Controller for LCD Bias	TSOP	Aim	-	pdf pinout
AT 1730 P	IC	2-Channel PWM Controller for LCD Bias	TSOP	Aim	-	pdf pinout
AT 1730 P_GRE	IC	2-Channel PWM Controller for LCD Bias	TSOP	Aim	-	pdf pinout
AT 1731 AP	IC	DC-DC Power IC for TFT Panel	TSOP	Aim	-	pdf pinout
AT 1731 AP_GRE	IC	DC-DC Power IC for TFT Panel	TSOP	Aim	-	pdf pinout
AT 1731 P	IC	DC-DC Power IC for TFT Panel	TSOP	Aim	-	pdf pinout
AT 1731 P_GRE	IC	DC-DC Power IC for TFT Panel	TSOP	Aim	-	pdf pinout
AT 1733 AN_GRE	IC	LCD Panel Power Supplies	TSOP	Aim	-	pdf pinout
AT 1733 N_GRE	IC	LCD Panel Power Supplies	TSOP	Aim	-	pdf pinout
AT 1735 N_GRE	IC	LCD Panel Power Supplies	TSOP	Aim	-	pdf pinout
AT 1741 D	IC	2-Channel PWM Controller	TSOP	Aim	-	pdf pinout
AT 1741 S	IC	2-Channel PWM Controller	TSOP	Aim	-	pdf pinout
AT 1741 S_CRE	IC	2-Channel PWM Controller	TSOP	Aim	-	pdf pinout
AT 1743 P	IC	2-Channel PWM Controller	TSOP	Aim	-	pdf pinout
AT 1743 P_GRE	IC	2-Channel PWM Controller	TSOP	Aim	-	pdf pinout
AT 1780 M_GRE	IC	Step-up Current-Mode PWM Controller	TSOP	Aim	-	pdf pinout
AT 1796-ADJ	IC	Step-down Switching Regulator	TSOP	Aim	-	pdf pinout
AT 1796-ADJ_PBF	IC	Step-down Switching Regulator	TSOP	Aim	-	pdf pinout
AT 1799 S	DIG-IC	Voltage Regulator	QFN	Aim	-	pdf pinout
AT 1799 S_GRE	DIG-IC	Voltage Regulator	QFN	Aim	-	pdf pinout
AT 3210	DIG-IC	16 GrayScales 160X160 STN LCD Controller	TSOP	Aim	-	pdf pinout
AT 5550M_GRE	DIG-IC	2-Channel Motor Driver	QFN	Aim	-	pdf pinout
AT 5550 N	DIG-IC	2-Channel Motor Driver	QFN	Aim	-	pdf pinout
AT 5556 F	DIG-IC	6-Channel DSC Motor Driver	QFN	Aim	-	pdf pinout
AT 5556 F_GRE	DIG-IC	6-Channel DSC Motor Driver	QFN	Aim	-	pdf pinout
AT 5556 F_PBF	DIG-IC	6-Channel DSC Motor Driver	QFN	Aim	-	pdf pinout
AT 5556 N_GRE	DIG-IC	6-Channel DSC Motor Driver	QFN	Aim	-	pdf pinout
AT 5557 N_GRE	IC	Serial Input 6CH Motor Driver for DSC	TSOP	Aim	-	pdf pinout
AT 5558 N_GRE	IC	5-Channel DSC Motor Driver	QFN	Aim	-	pdf pinout
AT 5561 D_PBF	IC	1-channel DC Motor Driver For Toy	QFN	Aim	-	pdf pinout
AT 5561 S_GRE	IC	1-channel DC Motor Driver For Toy	QFN	Aim	-	pdf pinout
AT 5562 D_PBF	IC	2-channel DC Motor Driver For Toy	QFN	Aim	-	pdf pinout
AT 5562 S_GRE	IC	2-channel DC Motor Driver For Toy	QFN	Aim	-	pdf pinout
AT 5566AN_GRE	DIG-IC	6CH motor driver IC	QFN	Aim	-	pdf pinout
AT 5566 N_GRE	DIG-IC	6CH motor driver IC	QFN	Aim	-	pdf pinout
AT 5608 S	IC	Bi-directional Motor Driver	QFN	Aim	-	pdf pinout
AT 5608 S_GRE	IC	Bi-directional Motor Driver	QFN	Aim	-	pdf pinout
AT 5609 S	IC	Bi-directional Motor Driver	QFN	Aim	-	pdf pinout
AT 5609 S_GRE	IC	Bi-directional Motor Driver	QFN	Aim	-	pdf pinout
AT 5654 H	IC	4-Channel BTL Motor Driver for CD-ROM	TSOP	Aim	-	pdf pinout
AT 5654 H_PBF	IC	4-Channel BTL Motor Driver for CD-ROM	TSOP	Aim	-	pdf pinout
AT 5658 H	IC	4-ch Motor Driver for CD/VCD Player	TSOP	Aim	-	pdf pinout
AT 5659 H	IC	5-ch Motor Driver for CD/VCD Player	TSOP	Aim	-	pdf pinout
AT 5660 H	IC	5-Channel Motor Driver for DVD/VCD	TSOP	Aim	-	pdf pinout
AT 5665 H	IC	5-Channel Motor Driver for DVD/VCD	TSOP	Aim	-	pdf pinout
AT 5669H	DIG-IC	5- Channel Motor Driver, Linear Regulator, 1 OPAMP	QFN	Aim	-	pdf pinout
AT5669H_PBF	DIG-IC	5- Channel Motor Driver, Linear Regulator, 1 OPAMP	QFN	Aim	-	pdf pinout
AT 5683 H	IC	4-channel BTL Motor Driver	TSOP	Aim	-	pdf pinout
AT 5801 F	IC	4-Channel Motor driver	TSOP	Aim	-	pdf pinout
AT 5801 F_GRE	IC	4-Channel Motor driver	TSOP	Aim	-	pdf pinout
AT 5802	IC	4-Channel driver + Power Controller	TSOP	Aim	-	pdf pinout
AT 5560 /DFN-6	IC	Vcm Driver for Mobile Phone	QFN	Aim	-	pdf pinout
AT 1117 -18V	IC	Positive Adjustable Regulator	48	Aim	-	pdf pinout
AT 1117 -18Y	IC	Positive Adjustable Regulator	48	Aim	-	pdf pinout
AT 1117 -25V	IC	Positive Adjustable Regulator	48	Aim	-	pdf pinout
AT 1117 -25Y	IC	Positive Adjustable Regulator	48	Aim	-	pdf pinout
AT 1117 -28V	IC	Positive Adjustable Regulator	48	Aim	-	pdf pinout
AT 1117 -28Y	IC	Positive Adjustable Regulator	48	Aim	-	pdf pinout
AT 1117 -33V	IC	Positive Adjustable Regulator	48	Aim	-	pdf pinout
AT 1117 -33Y	IC	Positive Adjustable Regulator	48	Aim	-	pdf pinout
AT 1117 -50V	IC	Positive Adjustable Regulator	48	Aim	-	pdf pinout
AT 1117 -50Y	IC	Positive Adjustable Regulator	48	Aim	-	pdf pinout
AT 1201-25X	IC	linear regulator	45	Aim	-	pdf pinout
AT 1201-25X_GREEN	IC	linear regulator	45	Aim	-	pdf pinout
AT 1201-25XR	IC	linear regulator	45	Aim	-	pdf pinout
AT 1201-25XR_GREEN	IC	linear regulator	45	Aim	-	pdf pinout
AT 1201-27X	IC	linear regulator	45	Aim	-	pdf pinout
AT 1201-27X_GREEN	IC	linear regulator	45	Aim	-	pdf pinout
AT 1201-27XR	IC	linear regulator	45	Aim	-	pdf pinout
AT 1201-27XR_GREEN	IC	linear regulator	45	Aim	-	pdf pinout
AT 1201-28X	IC	linear regulator	45	Aim	-	pdf pinout
AT 1201-28X_GREEN	IC	linear regulator	45	Aim	-	pdf pinout
AT 1201-28XR	IC	linear regulator	45	Aim	-	pdf pinout
AT 1201-28XR_GREEN	IC	linear regulator	45	Aim	-	pdf pinout
AT 1201-30X	IC	linear regulator	45	Aim	-	pdf pinout
AT 1201-30X_GREEN	IC	linear regulator	45	Aim	-	pdf pinout
AT 1201-30XR	IC	linear regulator	45	Aim	-	pdf pinout
AT 1201-30XR_GREEN	IC	linear regulator	45	Aim	-	pdf pinout

Type	Device	Short Description	Fig.	Manu	Comparision Types	More at
AT 1201-33X	IC	linear regulator	45	Aim	-	pdf pinout
AT 1201-33X_GREEN	IC	linear regulator	45	Aim	-	pdf pinout
AT 1201-33XR	IC	linear regulator	45	Aim	-	pdf pinout
AT 1201-33XR_GREEN	IC	linear regulator	45	Aim	-	pdf pinout
AT 1201-45X	IC	linear regulator	45	Aim	-	pdf pinout
AT 1201-45X_GREEN	IC	linear regulator	45	Aim	-	pdf pinout
AT 1201-45XR	IC	linear regulator	45	Aim	-	pdf pinout
AT 1201-45XR_GREEN	IC	linear regulator	45	Aim	-	pdf pinout
AT 1201-50X	IC	linear regulator	45	Aim	-	pdf pinout
AT 1201-50X_GREEN	IC	linear regulator	45	Aim	-	pdf pinout
AT 1201-50XR	IC	linear regulator	45	Aim	-	pdf pinout
AT 1201-50XR_GREEN	IC	linear regulator	45	Aim	-	pdf pinout
AT 1205 -15S	CMOS-IC	300mA Cmos LDO Regulator	46	Aim		pdf pinout
AT 1205 -15X	CMOS-IC	300mA Cmos LDO Regulator	46	Aim		pdf pinout
AT 1205 -18S	CMOS-IC	300mA Cmos LDO Regulator	46	Aim		pdf pinout
AT 1205 -18X	CMOS-IC	300mA Cmos LDO Regulator	46	Aim		pdf pinout
AT 1205 -25S	CMOS-IC	300mA Cmos LDO Regulator	46	Aim		pdf pinout
AT 1205 -25X	CMOS-IC	300mA Cmos LDO Regulator	46	Aim		pdf pinout
AT 1205 -30S	CMOS-IC	300mA Cmos LDO Regulator	46	Aim		pdf pinout
AT 1205 -30X	CMOS-IC	300mA Cmos LDO Regulator	46	Aim		pdf pinout
AT 1205 -33S	CMOS-IC	300mA Cmos LDO Regulator	46	Aim		pdf pinout
AT 1205 -33X	CMOS-IC	300mA Cmos LDO Regulator	46	Aim		pdf pinout
AT 1205 -50S	CMOS-IC	300mA Cmos LDO Regulator	46	Aim		pdf pinout
AT 1205 -50X	CMOS-IC	300mA Cmos LDO Regulator	46	Aim		pdf pinout
AT 1261-23 CL_GRE	IC	Current/voltage detectors	35	Aim		pdf pinout
AT 1261-23 CN_GRE	IC	Current/voltage detectors	35	Aim		pdf pinout
AT 1261-23 UL_GRE	IC	Current/voltage detectors	35	Aim		pdf pinout
AT 1261-23 UN_GRE	IC	Current/voltage detectors	35	Aim		pdf pinout
AT 1796-33	IC	Step-down Switching Regulator	TSOP	Aim		pdf pinout
AT 1796-33_PBF	IC	Step-down Switching Regulator	TSOP	Aim		pdf pinout
AT 1796-50	IC	Step-down Switching Regulator	TSOP	Aim		pdf pinout
AT 1796-50_PBF	IC	Step-down Switching Regulator	TSOP	Aim		pdf pinout
AT 5560 /SC-70	IC	Vcm Driver for Mobile Phone	QFN	Aim		pdf pinout
AT 29 C 020 32 J	EPROM-IC	2-megabit (256K x 8) 5-volt Only Flash Memory	PLCC	Atm	-	pdf pinout
AT 29 C 020 32 T	EPROM-IC	2-megabit (256K x 8) 5-volt Only Flash Memory	PLCC	Atm		pdf pinout
AT 29 C 256 28 T	EPROM-IC	256K (32K x 8) 5-volt Only Flash Memory	PLCC	Atm	-	pdf pinout
AT 29 C 256 32 J	EPROM-IC	256K (32K x 8) 5-volt Only Flash Memory	PLCC	Atm		pdf pinout
AT 29 C 512 -12 JC	EEPROM-IC	512K 5-volt Only Flash Memory	PLCC	Atm	W29C512A, SST29EE512,	pdf pinout
AT 29 C 512 -12 JI	EEPROM-IC	512K 5-volt Only Flash Memory	PLCC	Atm	W29C512A, SST29EE512,	pdf pinout
AT 29 C 512-12 TC	EEPROM-IC	512K 5-volt Only Flash Memory	TSOP	Atm	W29C512A, SST29EE512,	pdf pinout
AT 29 C 512-12 TI	EEPROM-IC	512K 5-volt Only Flash Memory	TSOP	Atm	W29C512A, SST29EE512,	pdf pinout
AT 29 C 512 -15 JC	EEPROM-IC	512K 5-volt Only Flash Memory	PLCC	Atm	W29C512A, SST29EE512,	pdf pinout
AT 29 C 512 -15 JI	EEPROM-IC	512K 5-volt Only Flash Memory	PLCC	Atm	W29C512A, SST29EE512,	pdf pinout
AT 29 C 512-15 TC	EEPROM-IC	512K 5-volt Only Flash Memory	TSOP	Atm	W29C512A, SST29EE512,	pdf pinout
AT 29 C 512-15 TI	EEPROM-IC	512K 5-volt Only Flash Memory	TSOP	Atm	W29C512A, SST29EE512,	pdf pinout
AT 29 C 512 -70 PC	EEPROM-IC	512K 5-volt Only Flash Memory	PLCC	Atm	W29C512A, SST29EE512,	pdf pinout
AT 29 C 512 -90 JC	EEPROM-IC	512K 5-volt Only Flash Memory	PLCC	Atm	W29C512A, SST29EE512,	pdf pinout
AT 29 C 512 -90 JI	EEPROM-IC	512K 5-volt Only Flash Memory	PLCC	Atm	W29C512A, SST29EE512,	pdf pinout
AT 29 C 512-90 TC	EEPROM-IC	512K 5-volt Only Flash Memory	TSOP	Atm	W29C512A, SST29EE512,	pdf pinout
AT 29 C 512-90 TI	EEPROM-IC	512K 5-volt Only Flash Memory	TSOP	Atm	W29C512A, SST29EE512,	pdf pinout
AT 29 C 020 32 P 6	EPROM-IC	2-megabit (256K x 8) 5-volt Only Flash Memory	PLCC	Atm		pdf pinout
AT 29 C 256 28 P 6	EPROM-IC	256K (32K x 8) 5-volt Only Flash Memory	PLCC	Atm	-	pdf pinout
AT 89 C 4051-12 PC	CMOS-ROM-IC	8-Bit Microcontroller with 4K Bytes Flash	QIP	Atm	-	pdf pinout
AT 89 C 4051-12 PA	CMOS-ROM-IC	8-Bit Microcontroller with 4K Bytes Flash	QIP	Atm		pdf pinout
AT 89 C 4051-12 PI	CMOS-ROM-IC	8-Bit Microcontroller with 4K Bytes Flash	QIP	Atm		pdf pinout
AT 89 C 4051-12 SA	CMOS-ROM-IC	8-Bit Microcontroller with 4K Bytes Flash	QIP	Atm	-	pdf pinout
AT 89 C 4051-12 SC	CMOS-ROM-IC	8-Bit Microcontroller with 4K Bytes Flash	QIP	Atm		pdf pinout
AT 89 C 4051-12 SI	CMOS-ROM-IC	8-Bit Microcontroller with 4K Bytes Flash	QIP	Atm	-	pdf pinout
AT 89 C 4051-24 PC	CMOS-ROM-IC	8-Bit Microcontroller with 4K Bytes Flash	QIP	Atm	-	pdf pinout
AT 89 C 4051-24 PI	CMOS-ROM-IC	8-Bit Microcontroller with 4K Bytes Flash	QIP	Atm	-	pdf pinout
AT 89 C 4051-24 SC	CMOS-ROM-IC	8-Bit Microcontroller with 4K Bytes Flash	QIP	Atm	-	pdf pinout
AT 89 C 4051-24 SI	CMOS-ROM-IC	8-Bit Microcontroller with 4K Bytes Flash	QIP	Atm	-	pdf pinout
AT 89 S 8252-24 AC	CMOS-IC	8-bit Microcontroller with 8K Bytes Flash	PLCC	Atm	-	pdf pinout
AT 89 S 8252-24 AI	CMOS-IC	8-bit Microcontroller with 8K Bytes Flash	PLCC	Atm	-	pdf pinout
AT 89 S 8252-24 JC	CMOS-IC	8-bit Microcontroller with 8K Bytes Flash	PLCC	Atm	-	pdf pinout
AT 89 S 8252-24 JI	CMOS-IC	8-bit Microcontroller with 8K Bytes Flash	PLCC	Atm	-	pdf pinout
AT 89 S 8252-24 PC	CMOS-IC	8-bit Microcontroller with 8K Bytes Flash	PLCC	Atm	-	pdf pinout

Type	Device	Short Description	Fig.	Manu	Comparision Types	More at
AT 89 S 8252-24 PI	CMOS-IC	8-bit Microcontroller with 8K Bytes Flash	PLCC	Atm	-	pdf pinout
AT 89 S 8252-24 QC	CMOS-IC	8-bit Microcontroller with 8K Bytes Flash	PLCC	Atm	-	pdf pinout
AT 89 S 8252-24 QI	CMOS-IC	8-bit Microcontroller with 8K Bytes Flash	PLCC	Atm	-	pdf pinout
ATA 6264-AL	IC	SMD, airbag power supply, diagnosis ...	QFP	Atm	-	pdf pinout
ATA 6824-PHQW	IC	SMD, high temp. H-bridge motor driver	QFN	Atm	-	pdf pinout
ATA 6827-PIQW	IC	SMD, 3x high temperature half-bridge driver, serial interface, PWM	QFN	Atm	-	pdf pinout
ATA 6832-PFQW	IC	SMD, 3x high temperature half-bridge driver, serial interface	QFN	Atm	-	pdf pinout
AT 89 C 52-12 AA	CMOS-PROM-IC	8-bit Microcontroller with 8K Bytes Flash	FLP	Atm	-	pdf pinout
AT 89 C 52-12 AC	CMOS-PROM-IC	8-bit Microcontroller with 8K Bytes Flash	FLP	Atm	-	pdf pinout
AT 89 C 52-12 AI	CMOS-PROM-IC	8-bit Microcontroller with 8K Bytes Flash	FLP	Atm	-	pdf pinout
AT 89 C 52-12 JA	CMOS-PROM-IC	8-bit Microcontroller with 8K Bytes Flash	FLP	Atm	-	pdf pinout
AT 89 C 52-12 JC	CMOS-PROM-IC	8-bit Microcontroller with 8K Bytes Flash	FLP	Atm	-	pdf pinout
AT 89 C 52-12 JI	CMOS-PROM-IC	8-bit Microcontroller with 8K Bytes Flash	FLP	Atm	-	pdf pinout
AT 89 C 52-12 PA	CMOS-PROM-IC	8-bit Microcontroller with 8K Bytes Flash	FLP	Atm	-	pdf pinout
AT 89 C 52-12 PC	CMOS-PROM-IC	8-bit Microcontroller with 8K Bytes Flash	FLP	Atm	-	pdf pinout
AT 89 C 52-12 PI	CMOS-PROM-IC	8-bit Microcontroller with 8K Bytes Flash	FLP	Atm	-	pdf pinout
AT 89 C 52-12 QA	CMOS-PROM-IC	8-bit Microcontroller with 8K Bytes Flash	FLP	Atm	-	pdf pinout
AT 89 C 52-12 QC	CMOS-PROM-IC	8-bit Microcontroller with 8K Bytes Flash	FLP	Atm	-	pdf pinout
AT 89 C 52-12 QI	CMOS-PROM-IC	8-bit Microcontroller with 8K Bytes Flash	FLP	Atm	-	pdf pinout
AT 89 C 52-16 AA	CMOS-PROM-IC	8-bit Microcontroller with 8K Bytes Flash	FLP	Atm	-	pdf pinout
AT 89 C 52-16 AC	CMOS-PROM-IC	8-bit Microcontroller with 8K Bytes Flash	FLP	Atm	-	pdf pinout
AT 89 C 52-16 AI	CMOS-PROM-IC	8-bit Microcontroller with 8K Bytes Flash	FLP	Atm	-	pdf pinout
AT 89 C 52-16 JA	CMOS-PROM-IC	8-bit Microcontroller with 8K Bytes Flash	FLP	Atm	-	pdf pinout
AT 89 C 52-16 JC	CMOS-PROM-IC	8-bit Microcontroller with 8K Bytes Flash	FLP	Atm	-	pdf pinout
AT 89 C 52-16 JI	CMOS-PROM-IC	8-bit Microcontroller with 8K Bytes Flash	FLP	Atm	-	pdf pinout
AT 89 C 52-16 PA	CMOS-PROM-IC	8-bit Microcontroller with 8K Bytes Flash	FLP	Atm	-	pdf pinout
AT 89 C 52-16 PC	CMOS-PROM-IC	8-bit Microcontroller with 8K Bytes Flash	FLP	Atm	-	pdf pinout
AT 89 C 52-16 PI	CMOS-PROM-IC	8-bit Microcontroller with 8K Bytes Flash	FLP	Atm	-	pdf pinout
AT 89 C 52-16 QA	CMOS-PROM-IC	8-bit Microcontroller with 8K Bytes Flash	FLP	Atm	-	pdf pinout
AT 89 C 52-16 QC	CMOS-PROM-IC	8-bit Microcontroller with 8K Bytes Flash	FLP	Atm	-	pdf pinout
AT 89 C 52-16 QI	CMOS-PROM-IC	8-bit Microcontroller with 8K Bytes Flash	FLP	Atm	-	pdf pinout
AT 89 C 52-20 AC	CMOS-PROM-IC	8-bit Microcontroller with 8K Bytes Flash	FLP	Atm	-	pdf pinout
AT 89 C 52-20 AI	CMOS-PROM-IC	8-bit Microcontroller with 8K Bytes Flash	FLP	Atm	-	pdf pinout
AT 89 C 52-20 JC	CMOS-PROM-IC	8-bit Microcontroller with 8K Bytes Flash	FLP	Atm	-	pdf pinout
AT 89 C 52-20 JI	CMOS-PROM-IC	8-bit Microcontroller with 8K Bytes Flash	FLP	Atm	-	pdf pinout
AT 89 C 52-20 PC	CMOS-PROM-IC	8-bit Microcontroller with 8K Bytes Flash	FLP	Atm	-	pdf pinout
AT 89 C 52-20 PI	CMOS-PROM-IC	8-bit Microcontroller with 8K Bytes Flash	FLP	Atm	-	pdf pinout
AT 89 C 52-20 QC	CMOS-PROM-IC	8-bit Microcontroller with 8K Bytes Flash	FLP	Atm	-	pdf pinout
AT 89 C 52-20 QI	CMOS-PROM-IC	8-bit Microcontroller with 8K Bytes Flash	FLP	Atm	-	pdf pinout
AT 89 C 52-24 AC	CMOS-PROM-IC	8-bit Microcontroller with 8K Bytes Flash	FLP	Atm	-	pdf pinout
AT 89 C 52-24 AI	CMOS-PROM-IC	8-bit Microcontroller with 8K Bytes Flash	FLP	Atm	-	pdf pinout
AT 89 C 52-24 JC	CMOS-PROM-IC	8-bit Microcontroller with 8K Bytes Flash	FLP	Atm	-	pdf pinout
AT 89 C 52-24 JI	CMOS-PROM-IC	8-bit Microcontroller with 8K Bytes Flash	FLP	Atm	-	pdf pinout
AT 89 C 52-24 PC	CMOS-PROM-IC	8-bit Microcontroller with 8K Bytes Flash	FLP	Atm	-	pdf pinout

Type	Device	Short Description	Fig.	Manu	Comparision Types	More at
AT 89 C 52-24 PI	CMOS-PROM-IC	8-bit Microcontroller with 8K Bytes Flash	FLP	Atm	-	pdf pinout
AT 89 C 52-24 QC	CMOS-PROM-IC	8-bit Microcontroller with 8K Bytes Flash	FLP	Atm	-	pdf pinout
AT 89 C 52-24 QI	CMOS-PROM-IC	8-bit Microcontroller with 8K Bytes Flash	FLP	Atm	-	pdf pinout
AT 89 C 55-16 AA	CMOS-IC	8-Bit Microcontroller with 20K Bytes Flash	FLP	Atm	-	pdf pinout
AT 89 C 55-16 JA	CMOS-IC	8-Bit Microcontroller with 20K Bytes Flash	FLP	Atm	-	pdf pinout
AT 89 C 55-16 PA	CMOS-IC	8-Bit Microcontroller with 20K Bytes Flash	FLP	Atm	-	pdf pinout
AT 89 C 55-16 QA	CMOS-IC	8-Bit Microcontroller with 20K Bytes Flash	FLP	Atm	-	pdf pinout
AT 89 C 55-24 AC	CMOS-IC	8-Bit Microcontroller with 20K Bytes Flash	FLP	Atm	-	pdf pinout
AT 89 C 55-24 AI	CMOS-IC	8-Bit Microcontroller with 20K Bytes Flash	FLP	Atm	-	pdf pinout
AT 89 C 55-24 JC	CMOS-IC	8-Bit Microcontroller with 20K Bytes Flash	FLP	Atm	-	pdf pinout
AT 89 C 55-24 JI	CMOS-IC	8-Bit Microcontroller with 20K Bytes Flash	FLP	Atm	-	pdf pinout
AT 89 C 55-24 PC	CMOS-IC	8-Bit Microcontroller with 20K Bytes Flash	FLP	Atm	-	pdf pinout
AT 89 C 55-24 PI	CMOS-IC	8-Bit Microcontroller with 20K Bytes Flash	FLP	Atm	-	pdf pinout
AT 89 C 55-24 QC	CMOS-IC	8-Bit Microcontroller with 20K Bytes Flash	FLP	Atm	-	pdf pinout
AT 89 C 55-24 QI	CMOS-IC	8-Bit Microcontroller with 20K Bytes Flash	FLP	Atm	-	pdf pinout
AT 89 C 55 WD-24AC	CMOS-IC	8-bit Microcontroller with 20K Bytes Flash	FLP	Atm	-	pdf pinout
AT 89 C 55 WD-24AI	CMOS-IC	8-bit Microcontroller with 20K Bytes Flash	FLP	Atm	-	pdf pinout
AT 89 C 55 WD-24JC	CMOS-IC	8-bit Microcontroller with 20K Bytes Flash	FLP	Atm	-	pdf pinout
AT 89 C 55 WD-24JI	CMOS-IC	8-bit Microcontroller with 20K Bytes Flash	FLP	Atm	-	pdf pinout
AT 89 C 55 WD-24PC	CMOS-IC	8-bit Microcontroller with 20K Bytes Flash	FLP	Atm	-	pdf pinout
AT 89 C 55 WD-24PI	CMOS-IC	8-bit Microcontroller with 20K Bytes Flash	FLP	Atm	-	pdf pinout
AT 89 C 55-33 AC	CMOS-IC	8-Bit Microcontroller with 20K Bytes Flash	FLP	Atm	-	pdf pinout
AT 89 C 55-33 JC	CMOS-IC	8-Bit Microcontroller with 20K Bytes Flash	FLP	Atm	-	pdf pinout
AT 89 C 55-33 PC	CMOS-IC	8-Bit Microcontroller with 20K Bytes Flash	FLP	Atm	-	pdf pinout
AT 89 C 55-33 QC	CMOS-IC	8-Bit Microcontroller with 20K Bytes Flash	FLP	Atm	-	pdf pinout
AT 89 C 55 WD-33AC	CMOS-IC	8-bit Microcontroller with 20K Bytes Flash	FLP	Atm	-	pdf pinout
AT 89 C 55 WD-33JC	CMOS-IC	8-bit Microcontroller with 20K Bytes Flash	FLP	Atm	-	pdf pinout
AT 89 C 55 WD-33PC	CMOS-IC	8-bit Microcontroller with 20K Bytes Flash	FLP	Atm	-	pdf pinout
AT 29 C010 A 32 J	EPROM-IC	1-Megabit (128K x 8) 5-volt Only Flash Memory	PLCC	Atm	-	pdf pinout
AT 29 C010 A 32 T	EPROM-IC	1-Megabit (128K x 8) 5-volt Only Flash Memory	PLCC	Atm	-	pdf pinout
AT 29 C 040 A 32 T	EPROM-IC	4- Megabit (512K x 8) 5- volt Only 256-Byte Sector Flash Memory	SDIP	Atm	-	pdf pinout
AT 29 C 512 -12 PC	EEPROM-IC	512K 5-volt Only Flash Memory	PLCC	Atm	W29C512A, SST29EE512,	pdf pinout
AT 29 C 512 -12 PI	EEPROM-IC	512K 5-volt Only Flash Memory	PLCC	Atm	W29C512A, SST29EE512,	pdf pinout
AT 29 C 512 -15 PC	EEPROM-IC	512K 5-volt Only Flash Memory	PLCC	Atm	W29C512A, SST29EE512,	pdf pinout
AT 29 C 512 -15 PI	EEPROM-IC	512K 5-volt Only Flash Memory	PLCC	Atm	W29C512A, SST29EE512,	pdf pinout
AT 29 C 512 -90 PC	EEPROM-IC	512K 5-volt Only Flash Memory	PLCC	Atm	W29C512A, SST29EE512,	pdf pinout
AT 29 C 512 -90 PI	EEPROM-IC	512K 5-volt Only Flash Memory	PLCC	Atm	W29C512A, SST29EE512,	pdf pinout
AT 29 C010 A 32 P6	EPROM-IC	1-Megabit (128K x 8) 5-volt Only Flash Memory	PLCC	Atm	-	pdf pinout
AT 29 C 040 A 32P6	EPROM-IC	4- Megabit (512K x 8) 5- volt Only 256-Byte Sector Flash Memory	SDIP	Atm	-	pdf pinout
AT 89 C 1051-12 PA	CMOS-ROM-IC	8-Bit Microcontroller with 1K Byte Flash	DIP	Atm	-	pdf pinout
AT 89 C 1051-12 PC	CMOS-ROM-IC	8-Bit Microcontroller with 1K Byte Flash	DIP	Atm	-	pdf pinout
AT 89 C 1051-12 PI	CMOS-ROM-IC	8-Bit Microcontroller with 1K Byte Flash	DIP	Atm	-	pdf pinout
AT 89 C 1051-12 SA	CMOS-ROM-IC	8-Bit Microcontroller with 1K Byte Flash	DIP	Atm	-	pdf pinout
AT 89 C 1051-12 SC	CMOS-ROM-IC	8-Bit Microcontroller with 1K Byte Flash	DIP	Atm	-	pdf pinout
AT 89 C 1051-12 SI	CMOS-ROM-IC	8-Bit Microcontroller with 1K Byte Flash	DIP	Atm	-	pdf pinout
AT 89 C 2051-12 PA	CMOS-ROM-IC	8-BitMicrocontroller with 2K Bytes Flash	DIP	Atm	-	pdf pinout
AT 89 C 2051-12 PC	CMOS-ROM-IC	8-BitMicrocontroller with 2K Bytes Flash	DIP	Atm	-	pdf pinout
AT 89 C 2051-12 PI	CMOS-ROM-IC	8-BitMicrocontroller with 2K Bytes Flash	DIP	Atm	-	pdf pinout
AT 89 C 2051-12 SA	CMOS-ROM-IC	8-BitMicrocontroller with 2K Bytes Flash	DIP	Atm	-	pdf pinout
AT 89 C 2051-12 SC	CMOS-ROM-IC	8-BitMicrocontroller with 2K Bytes Flash	DIP	Atm	-	pdf pinout
AT 89 C 2051-12 SI	CMOS-ROM-IC	8-BitMicrocontroller with 2K Bytes Flash	DIP	Atm	-	pdf pinout

Type	Device	Short Description	Fig.	Manu	Comparision Types	More at
AT 89 C 2051-24 PC	CMOS-ROM-IC	8-BitMicrocontroller with 2K Bytes Flash	DiP	Atm	-	pdf pinout
AT 89 C 2051-24 PI	CMOS-ROM-IC	8-BitMicrocontroller with 2K Bytes Flash	DIP	Atm	-	pdf pinout
AT 89 C 2051-24 SC	CMOS-ROM-IC	8-BitMicrocontroller with 2K Bytes Flash	DIF	Atm	-	pdf pinout
AT 89 C 2051-24 SI	CMOS-ROM-IC	8-BitMicrocontroller with 2K Bytes Flash	DIP	Atm	-	pdf pinout
ATN	Si-P	=2SA1812-N (Typ-Code/Stempel/marking)	39	Rhm	→2SA1812	data
ATP	Si-P	=2SA1812-P (Typ-Code/Stempel/marking)	39	Rhm	→2SA1812	data
AtQ	Si-N	=2PD1820AQ (Typ-Code/Stempel/marking)	35(2mm)	Phi	→2PD1820A	data
ATQ	Si-P	=2SA1812-Q (Typ-Code/Stempel/marking)	39	Rhm	→2SA1812	data
ATQ	Si-N	=2SC4326LK-Q(Typ-Code/Stempel/marking)	35	Rhm	→2SC4326LK	data
AtR	Si-N	=2PD1820AR (Typ-Code/Stempel/marking)	35(2mm)	Phi	→2PD1820A	data
ATR	Si-N	=2SC4326LK-R(Typ-Code/Stempel/marking)	35	Rhm	→2SC4326LK	data
ATR 2805 S	IC	Hybrid-high Reliability DC/DC Converter	42	Inr	-	pdf pinout
ATR 2812 D	IC	Hybrid-high Reliability DC/DC Converter	42	Inr	-	pdf pinout
ATR 2812 S	IC	Hybrid-high Reliability DC/DC Converter	42	Inr	-	pdf pinout
ATR 2815 D	IC	Hybrid-high Reliability DC/DC Converter	42	Inr	-	pdf pinout
ATR 2815 S	IC	Hybrid-high Reliability DC/DC Converter	42	Inr	-	pdf pinout
ATR 2803 R 3 S	IC	Hybrid-high Reliability DC/DC Converter	42	Inr	-	pdf pinout
AtS	Si-N	=2PD1820AS(Typ-Code/Stempel/marking)	35(2mm)	Phi	→2PD1820A	data
ATS	Si-N	=2SC4326LK-S(Typ-Code/Stempel/marking)	35	Rhm	→2SC4326LK	data
ATs	Si-N	=BF 540 (Typ-Code/Stempel/marking)	44(2mm)	Sie	→BF 540	data
ATS 137-P	LIN-IC	Single Hall Effect Switch	45	Dii	-	pdf pinout
ATS 137-W	LIN-IC	Single Hall Effect Switch	45	Dii	-	pdf pinout
ATS 177-P	LIN-IC	Single Output Hall Effect Latch	45	Dii	-	pdf pinout
ATS 610 LSC	Sensor	Dynamic, peak- detecting, diff. Hall-effect gear- tooth sensor		All	-	pdf pinout
ATS 612JSB	Sensor	Dynamic, self-cal., peak-detecting, diff. Hall-effect gear-tooth sensor		All		pdf pinout
ATS 632LSA	Sensor	Zero-speed, self-cal., non-oriented, Hall-effect gear-tooth sensor		All		pdf pinout
ATS 632LSC	Sensor	Zero-speed, self-cal., non-oriented, Hall-effect gear-tooth sensor		All		pdf pinout
ATS 640 JSB	Sensor	True zero speed, self. cal., diff. Hall-effect gear-tooth sensor, 2-wire		All		pdf pinout
ATS 2805 D	IC	High Reliability DC/DC Converter	42	Inr	-	pdf pinout
ATS 2805 S	IC	High Reliability DC/DC Converter	42	Inr	-	pdf pinout
ATS 2812 D	IC	High Reliability DC/DC Converter	42	Inr	-	pdf pinout
ATS 2812 S	IC	High Reliability DC/DC Converter	42	Inr	-	pdf pinout
ATS 2815 D	IC	High Reliability DC/DC Converter	42	Inr	-	pdf pinout
ATS 2815 S	IC	High Reliability DC/DC Converter	42	Inr	-	pdf pinout
ATS 2803 R 3 S	IC	High Reliability DC/DC Converter	42	Inr	-	pdf pinout
AT 89 S 53-24 AC	CMOS-IC	8-bit Microcontroller with 12K Bytes Flash	PLCC	Atm	-	pdf pinout
AT 89 S 53-24 AI	CMOS-IC	8-bit Microcontroller with 12K Bytes Flash	PLCC	Atm	-	pdf pinout
AT 89 S 53-24 JC	CMOS-IC	8-bit Microcontroller with 12K Bytes Flash	PLCC	Atm	-	pdf pinout
AT 89 S 53-24 JI	CMOS-IC	8-bit Microcontroller with 12K Bytes Flash	PLCC	Atm	-	pdf pinout
AT 89 S 53-24 PC	CMOS-IC	8-bit Microcontroller with 12K Bytes Flash	PLCC	Atm	-	pdf pinout
AT 89 S 53-24 PI	CMOS-IC	8-bit Microcontroller with 12K Bytes Flash	PLCC	Atm	-	pdf pinout
AT 89 S 53-33 AC	CMOS-IC	8-bit Microcontroller with 12K Bytes Flash	PLCC	Atm	-	pdf pinout
AT 89 S 53-33 JC	CMOS-IC	8-bit Microcontroller with 12K Bytes Flash	PLCC	Atm	-	pdf pinout
AT 89 S 53-33 PC	CMOS-IC	8-bit Microcontroller with 12K Bytes Flash	PLCC	Atm	-	pdf pinout
AT 89 S 8252-33 AC	CMOS-IC	8-bit Microcontroller with 8K Bytes Flash	PLCC	Atm	-	pdf pinout
AT 89 S 8252-33 JC	CMOS-IC	8-bit Microcontroller with 8K Bytes Flash	PLCC	Atm	-	pdf pinout
AT 89 S 8252-33 PC	CMOS-IC	8-bit Microcontroller with 8K Bytes Flash	PLCC	Atm	-	pdf pinout
AT 89 S 8252-33 QC	CMOS-IC	8-bit Microcontroller with 8K Bytes Flash	PLCC	Atm	-	pdf pinout

AU

Type	Device	Short Description	Fig.	Manu	Comparision Types	More at
AU	Si-P	=2SA2014 (Typ-Code/Stempel/marking)	39	Say	→2SA2014	data
AU	Si-P	=2SB804-AU (Typ-Code/Stempel/marking)	39	Nec	→2SB804	data
AU	Si-Di	=BAT 18-04 (Typ-Code/Stempel/marking)	35	Sie	→BAT 18-04	data
AU	Si-N	=BCX 70RG (Typ-Code/Stempel/marking)	35	Sie, Val	→BCX 70RG	data
AU	Z-Di	=P4 SMA-13 (Typ-Code/Stempel/marking)	71(5x2,5)	Fag	→P4 SMA-13	data
AU	Z-Di	=SMBJ 9.0C (Typ-Code/Stempel/marking)	71(5x3,5)	Mop	→SMBJ ...	data
AU 01(A,Z)	Si-Di	GI, S, 200...600V, 0,5A, 400ns AU01=400V, A=600V, Z=200V	31a	Sak	BA 157...159, BY 208/..., BY 407,++	data
AU 02(A,Z)	Si-Di	GI, S, 200...600V, 0,8A, 400ns AU02=400V, A=600V, Z=200V	31a	Sak	BYD 33J...M, BYT 11/600, BYT 52J...M,++	data
AU 101	Ge-P	TV-HA, 120V, 10A, 10W(Tc=75°)	23a§	Phi	AU 210	data
AU 102	Ge-P	TV-HA-Tr, 40V, 10A, 10W(Tc=75°)	23a§	Phi	AU 108	data
AU 103	Ge-P	TV-HA, 155V, 10A, 10W(Tc=60°)	23a§	Phi	AU 107, AU 110	data
AU 104	Ge-P	TV-HA, 185V, 12A, 15W(Tc=75°)	23a§	Phi	AU 107	data

Type	Device	Short Description	Fig.	Manu	Comparision Types	More at
AU 105	Ge-P	TV-HA, 130V, 10A, 27,5W	23a§	Sie	AU 110, AU 210	data
AU 106	Ge-P	TV-HA, 320V, 10A, 5W(Tc=55°)	23a§	Sgs, Gpd	AU 109, AU 111, AU 112, 2N5325	data
AU 107	Ge-P	TV-VA/HA, 200V, 10A, 30W	23a§	Sgs, Gpd	AU 113, AU 213, 2N5324	data
AU 108(F)	Ge-P	TV-HA-Tr, 100V, 10A, 30W	23a§	Sgs, Gpd	AU 210	data
AU 109	Ge-P	TV-HA, 320V, 10A, 15W	23a§	Sie	AU 106, AU 111, AU 112, 2N5325	data
AU 110	Ge-P	TV-HA, 160V, 10A, 30W	23a§	Sgs, Gpd	AU 107	data
AU 111	Ge-P	TV-HA, 320V, 10A, 5W(Tc=55°)	23a§	Sgs, Gpd	AU 106, AU 109, AU 112, 2N5325	data
AU 112	Ge-P	TV-HA, 320V, 10A, 5W(Tc=55°)	23a§	Sgs, Gpd	AU 106, AU 109, AU 111, 2N5325	data
AU 113	Ge-P	TV-HA, 250V, 10A, 5W(Tc=55°)	23a§	Sgs, Gpd	AU 213, 2N5324	data
AU 206	Ge-P	TV-HA, 320V, 10A, 5W(Tc=55°)	23a§	Sgs	AU 106, AU 109, AU 111, AU 112, 2N5325	data
AU 210	Ge-P	TV-HA, 140V, 10A, 5W(Tc=55°)	23a§	Sgs	AU 107, AU 110	data
AU 213	Ge-P	TV-HA, 250V, 10A, 5W(Tc=55°)	23a§	Sgs	AU 113, 2N5324	data
AU2901	4xCOMP	Vs:±18V Vu:100V/mV Vo:TTL Vi0:±2mV	14SD	Phi		data pdf pinout
AU 2901 D	KOP-IC	=LM 2901D: -40..+125°	14-MDIP	Phi	-	pinout
AU 2901 N	KOP-IC	=LM 2901N: -40..+125°	14-DIP	Phi		pinout
AU2902	4xLO-POWER	Vs:±16V Vu:100dB Vo:26/0V Vi0:±2mV	14SD	Phi		data pdf pinout
AU 2902 D	OP-IC	=LM 2902D: -40..+125°	14-MDIP	Phi	→LM 2902, LM 124	pinout
AU 2902 N	OP-IC	=LM 2902N: -40..+125°	14-DIP	Phi	→LM 2902, LM 124	pinout
AU2903	2xCOMP	Vs:±18V Vu:100V/mV Vo:TTL Vi0:±2mV	8SD	Phi		data pdf pinout
AU 2903 D	KOP-IC	=LM 2903D: -40..+125°	8-MDIP	Phi		pinout
AU 2903 N	KOP-IC	=LM 2903N: -40..+125°	8-DIP	Phi		pinout
AU2904	2xLO-POWER	Vs:±16V Vu:100V/mV Vo:>26/0V f:1Mc	8SD	Phi		data pdf pinout
AU 2904 D	OP-IC	=LM 2904D: -40..+125°	8-MDIP	Phi	→LM 2904, LM 158	pinout
AU 2904 N	OP-IC	=LM 2904N: -40..+125°	8-DIP	Phi	→LM 2904, LM 158	pinout
AU 6331 /LQFP	DIG-IC	Card Reader Controller	MP		-	pdf pinout

AUY...AX

Type	Device	Short Description	Fig.	Manu	Comparision Types	More at
AUY 10	Ge-P	NF/S-L, 70V, 0,7A, 6W(Tc=50°)	23a§	Phi	AUY 19, AUY 20	data
AUY 11	Ge-P	NF/S-L, 65V, 10A, 22,5W		Sie	(AUY 21, 2N2526)[4]	data
AUY 12	Ge-P	NF/S-L, 80V, 8A, 22,5W		Sie	(AUY 22, 2N2526)[4]	data
AUY 14	Ge-P	NF-L, 65V, 10A, 36,5W	23a§	Sie	AUY 21, 2N2289, 2N2292	data
AUY 15	Ge-P	NF-L, 65V, 10A, 36,5W	23a§	Sie	AUY 21, 2N2289, 2N2292	data
AUY 16	Ge-P	NF-L, 80V, 8A, 36,5W	23a§	Sie	AUY 22, 2N2289, 2N2292	data
AUY 17	Ge-P	NF-L, 80V, 8A, 36,5W	23a§	Sie	AUY 22, 2N2289, 2N2292	data
AUY 18	Ge-P	NF/S-L, 64V, 8A, 11W	2a§	Gpd,Sgs,Sie	AUY 35, AUY 36	data
AUY 19	Ge-P	NF/S-L, 64V, 3A, 30W	23a§	Gpd,Sgs,Sie	AL 102, AUY 20, 2N1531/36, 2N1541/46	data
AUY 20	Ge-P	=AUY 19: 80V	23a§	Gpd,Sgs,Sie	AL 102, AUY 28 2N1531/36, 2N1541/46	data
AUY 21	Ge-P	NF/S-L, 65V, 10A, 36W	23a§	Gpd,Sgs,Sie	ASZ 16, ASZ 17, 2N2526, 2N2289, 2N2292	data
AUY 22	Ge-P	NF/S-L, 80V, 8A, 36W	23a§	Gpd,Sgs,Sie	ASZ 15, ASZ 18, 2N2526, 2N2289, 2N2292	data
AUY 24	Ge-P	NF-L, 65V, 3A, 30W	23a§	Gpd, Sie	AL 102, AUY 19, 2N1530/35, 2N1540/45	data
AUY 26	Ge-P	NF-L, 80V, 3A, 30W	23a§	Gpd, Sie	AL 102, AUY 20, 2N1531/36, 2N1541/46	data
AUY 27	Ge-P	NF-L, 80V, 3A, 30W	23a§	Gpd, Sie	AL 102, AUY 20, 2N1531/36, 2N1541/46	data
AUY 28	Ge-P	NF/S-L, 90V, 6A, 30W	23a§	Aeg, Gpd	AL 102, ASZ 15, ASZ 18, 2N3616/18	data
AUY 29	Ge-P	NF/S-L, 50V, 15A, 36W	23a§	Gpd, Sie	2N1549...1560	data
AUY 30	Ge-P	NF/S-L, 100V, 10A, 33W	23a§	Tho	AUY 37, 2N2527, 2N2290, 2N2293	data
AUY 31	Ge-P	NF/S-L, 60V, 6A, 33W	23a§	Tho	AL 102, ASZ 16...17, AUY 21, 2N3612/14	data
AUY 32	Ge-P	NF/S-L, 80V, 3A, 33W	23a§	old	AL 102, AUY 20, 2N2141/46	data
AUY 33	Ge-P	=AUY 33: 60V	23a§	old	AL 102, AUY 19, 2N2139/44	data
AUY 34	Ge-P	NF/S-L, 100V, 3A, 30W	23a§	Gpd,Sie	AL 102, AUY 28, 2N1532/37, 2N1542/47	data
AUY 35	Ge-P	NF/S-L, 70V, 10A, 15W	2a§	Sgs	AUY 36	data
AUY 36	Ge-P	NF/S-L, 70V, 10A, 15W	2a§	Sgs	AUY 35	data
AUY 37	Ge-P	NF/S-L, 100V, 10A, 30W	23a§	Sgs	2N2527, 2N2290, 2N2293	data
AUY 38	Ge-P	S-L, 130V, 10A, 30W	23a§	Sgs	AL 100, 2N2528	data
AUZ 11(D)	Ge-P	NF/S-L, 50V, 1A, 6W		Aeg	(AD 162, AUY 18)[4]	data
AV	Si-P	=2SB804-AV (Typ-Code/Stempel/marking)	39	Nec	→2SB804	data
AV	Si-P	=2SA2015 (Typ-Code/Stempel/marking)	39	Say	→2SA2015	data
AV	Z-Di	=P4 SMA-13A (Typ-Code/Stempel/marking)	71(5x2,5)	Fag	→P4 SMA-13A	data
AV	Z-Di	=SMBJ 9.0CA (Typ-Code/Stempel/marking)	71(5x3,5)	Mop	→SMBJ ...	data
AV	C-Di	=SVC 208 (Typ-Code/Stempel/marking)	35	Say	→SVC 208	data
AV 03-03...-30	Z-Di	3...30V, ±5%, 10W	32a	Hit	BZX 98/.., BZY 93/.., 1N2970..89	data
AV 9107 C-03	LIN-IC	CPU Frequency Generator	DIC	Ics	-	pdf pinout
AV 9107 C-05	LIN-IC	CPU Frequency Generator	DIC	Ics		pdf pinout
AV 9107 C-10	LIN-IC	CPU Frequency Generator	DIC	Ics		pdf pinout
AV 9107 C-11	CMOS-IC	Green PC CPU Frequency Generator, 8 MHz	DIC	Ics		pdf pinout
AV 9107 C-13	LIN-IC	CPU Frequency Generator, 20/40 MHz	DIC	Ics		pdf pinout
AV 9107 C-17	LIN-IC	CPU Frequency Generator, 25/33 MHz	DIC	Ics		pdf pinout
AV 9107 C-19	LIN-IC	Frequency Generator for Fibre Channel Systems, 106.25 MHz clock	DIC	Ics		pdf pinout
AV 9107 C-20	LIN-IC	Frequency Generator for Fibre Channel Systems, 106.25 MHz clock	DIC	Ics		pdf pinout
AV 9107 C-21	LIN-IC	Frequency Generator for Fibre Channel Systems, 106.25 MHz clock	DIC	Ics	ICS9107C-21	pdf pinout
AV 9107 C-22	LIN-IC	Frequency Generator for Fibre Channel Systems, 106.25 MHz clock	DIC	Ics	ICS9107C-22	pdf pinout

Type	Device	Short Description	Fig.	Manu	Comparision Types	More at
AV 9110 -01	LIN-IC	Serially Programmable Frequency Generator, 5- 32 MHz	DIC	Ics	-	pdf pinout
AV 9110 -02	LIN-IC	Serially Programmable Frequency Generator, 5- 32 MHz	DIC	Ics	-	pdf pinout
AV 9154 A-04	LIN-IC	Low-Cost 16-Pin Frequency Generator	DIC	Ics		pdf pinout
AV 9154 A-10	LIN-IC	Low-Cost 16-Pin Frequency Generator	DIC	Ics		pdf pinout
AV 9154 A-26	LIN-IC	Low-Cost 16-Pin Frequency Generator	DIC	Ics		pdf pinout
AV 9154 A-27	LIN-IC	Low-Cost 16-Pin Frequency Generator	DIC	Ics		pdf pinout
AV 9154 A-39	LIN-IC	Low-Cost 16-Pin Frequency Generator	DIC	Ics	ICS9154A-39	pdf pinout
AV 9154 A-42	LIN-IC	Low-Cost 16-Pin Frequency Generator	DIC	Ics		pdf pinout
AV 9154 A-43	LIN-IC	Low-Cost 16-Pin Frequency Generator	DIC	Ics		pdf pinout
AV 9154 A-60	LIN-IC	Opti Notebook Frequency Generator	DIC	Ics		pdf pinout
AV 9155 C-01	LIN-IC	Low-Cost 20-Pin Frequency Generator, 16 MHz Bus Clk	DIC	Ics		pdf pinout
AV 9155 C-02	LIN-IC	Low-Cost 20-Pin Frequency Generator, 32 MHz Bus Clk	DIC	Ics		pdf pinout
AV 9155 C-23	LIN-IC	Low-Cost 20-Pin Frequency Generator, Includes PentiumÒ frequencies	DIC	Ics		pdf pinout
AV 9155 C-36	LIN-IC	Low-Cost 20-Pin Frequency Generator, Features a special 40 MHz Scsi clock	DIC	Ics		pdf pinout
AV 9155 C-44	LIN-IC	Low-Cost 20-Pin Frequency Generator	DIC	Ics		pdf pinout
AV 9170 -01	LIN-IC	Clock Synchronizer and Multiplier, 20-107Mhz at 5.0V, 20-66,7Mhz at 3,3V	DIC	Ics		pdf pinout
AV 9170 -02	LIN-IC	Clock Synchronizer and Multiplier, 5-26,75Mhz at 5.0V, 5-16,7Mhz at 3,3V	DIC	Ics		pdf pinout
AV 9170 -04	LIN-IC	Clock Synchronizer and Multiplier, 20-107Mhz at 5.0V, 20-66,7Mhz at 3,3V	DIC	Ics		pdf pinout
AV 9170 -05	LIN-IC	Clock Synchronizer and Multiplier, 5-26,75Mhz at 5.0V, 5-16,7Mhz at 3,3V	DIC	Ics		pdf pinout
AV 9170 -07	LIN-IC	Clock Multiplier	DIC	Ics		pdf pinout
AV 9172 -01	LIN-IC	Low Skew Output Buffer, Second source of GA1210E	DIC	Ics		pdf pinout
AV 9172 -03	LIN-IC	Low Skew Output Buffer, Clock doubler and buffer	DIC	Ics		pdf pinout
AV 9172 -07	LIN-IC	Low Skew Output Buffer, Clock buffer for 66 MHz input	DIC	Ics		pdf pinout
AV 9173 -01	LIN-IC	Video Genlock PLL, 1.25-75MHz	DIC	Ics		pdf pinout
AV 9173 -15	LIN-IC	Video Genlock PLL, 0.625-37.5 MHz	DIC	Ics		pdf pinout
AVS 08 ...	SMPS-IC	Control-IC+Triac(±500V, 5A~) f. SMPS	8-DIP+17I	Tho		
AVS 10 /P	SMPS-IC	Control-IC+Triac(±600V, 8A~) f. SMPS	8-DIP+17I	Tho	-	pdf pinout
AVS 10 /B	SMPS-IC	Control-IC+Triac(±600V, 8A~) f. SMPS	8-DIP+17I	Tho	-	pdf pinout
AVS 12 /P	SMPS-IC	Control-IC+Triac(±600V, 12A~) f. SMPS	8-DIP+17I	Tho	-	pdf pinout
AVS 12 /B	SMPS-IC	Control-IC+Triac(±600V, 12A~) f. SMPS	8-DIP+17I	Tho	-	pdf pinout
AVS 20 ...	SMPS-IC	Control-IC+Triac(±600V, 8A~) f. SMPS	8-DIP+17I	Tho	-	
AVS 200 ...	SMPS-IC	Control-IC+Triac(±800V, 8A~) f. SMPS	8-DIP+17I	Tho	-	
AVS 08..200 ...I	LIN-IC+Triac		17I		-	
AW	Si-P	=2SB804-AW (Typ-Code/Stempel/marking)	39	Nec	→2SB804	data
AW	Si-N	=BCX 70RH (Typ-Code/Stempel/marking)	35	Sie,Val	→BCX 70RH	data
AW	Z-Di	=P4 SMA-15 (Typ-Code/Stempel/marking)	71(5x2,5)	Fag	→P4 SMA-15	data
AW	Z-Di	=SMBJ 10C (Typ-Code/Stempel/marking)	71(5x3,5)	Mop	→SMBJ ...	data
AW 01-06...-33	Z-Di	6...33V, 5%, 1W	31a	Hit	BZV 85/..., BZW 22/..., BZX 61/..., ZPY...,+	data
AW 03-02...-05	Si-St	1,4...5,4V, 1W	31a	Hit		data
AX	Si-P	=2SA1739 (Typ-Code/Stempel/marking)	35(2mm)	Mat	→2SA1739	data
AX	Si-N	=BCX 70RJ (Typ-Code/Stempel/marking)	35	Sie,Val,Fer	→BCX 70RJ	data
AX	Z-Di	=P4 SMA-15A (Typ-Code/Stempel/marking)	71(5x2,5)	Fag	→P4 SMA-15A	data
AX	Z-Di	=SMAJ 10A (Typ-Code/Stempel/marking)	71(5x2,5)	Tho	→SMAJ ...	data
AX	Z-Di	=SMBJ 10CA (Typ-Code/Stempel/marking)	71(5x3,5)	Mop	→SMBJ ...	data
AX	Si-N+R	=XN 2212 (Typ-Code/Stempel/marking)	45	Mat	→XN 2212	data
AXQ	Si-P	=2SA1739-Q (Typ-Code/Stempel/marking)	35	Mat	→2SA1739	data
AXR	Si-P	=2SA1739-R (Typ-Code/Stempel/marking)	35	Mat	→2SA1739	data

AY...AZ

Type	Device	Short Description	Fig.	Manu	Comparision Types	More at
AY	Si-N	=2SC2880-Y (Typ-Code/Stempel/marking)	39	Tos	→2SC2880	data
AY	Si-N	=2SC3392 (Typ-Code/Stempel/marking)	35	Say	→2SC3392	data
AY	Si-N	=2SC4210-Y (Typ-Code/Stempel/marking)	35	Tos	→2SC4210	data
AY	Si-N	=BCX 70RK (Typ-Code/Stempel/marking)	35	Sie,Val	→BCX 70RK	data
AY	Si-N	=KTC4372-Y (Typ-Code/Stempel/marking)	39	Kec	→KTC 4372	data
AY	Z-Di	=P4 SMA-16 (Typ-Code/Stempel/marking)	71(5x2,5)	Fag	→P4 SMA-16	data
AY	Z-Di	=SMBJ 11C (Typ-Code/Stempel/marking)	71(5x3,5)	Mop	→SMBJ ...	data
AY-1-0212	IC	12 Note Top Octave Synthesizer	DIP		MO86B,M50242,S50242,50242,ECG-2043	pinout
AY-1-1007	IC	7 stage Divider	DIP		C10A29,C10130,PL4C07C	pinout
AY-1-5050	IC	7 stage Divider	DIP		SAJ-180,SA1005,TMS3612	pinout
AY-3-0214	IC	12 Note Top Octave Synthesizer	DIP		147196	pinout
AY-3-1232	LIN-IC	Digital-uhr/digital clock		Gie	-	
AY-3-8203	LIN-IC	US-FB-Decoder	40-DIP	Gie	-	
AY-3-8210	LIN-IC			Gie	-	
AY-3-8500	LIN-IC	TV-spielemodul/TV-game		Gie	-	
AY-3-8610	LIN-IC			Gie	-	

Type	Device	Short Description	Fig.	Manu	Comparision Types	More at
AY-3-8765	LIN-IC			Gie	-	
AY-3-9900	LIN-IC	Telefon, Codec, TTL-kompat.	24-DIP	Fer	ZNPCM 1	
AY-5-1203A	LIN-IC	TV-kanaleinbl./channel-# fade-in	28-DIP	Gie	-	
AY-5-8320	LIN-IC	Tv-kanaleinbl./channel-# fade-in	24-DIP	Gie	-	
AY-5-8322	LIN-IC	Digital-uhr/digital clock		Gie	-	
AY08A3-08...14	Triac	Triple, Triac 400-700V 0.8A -08:400V, -12:600V, -14:700V	16-DIP	Pwx	-	pdf
AY08A4-08...14	Triac	Quad, Triac 400-700V 0.8A -08:400V, -12:600V, -14:700V	16-DIP	Pwx	-	pdf
AY 101	Ge-Di	TV-Booster-Di, 150V, 15A(ss)	(32a)	Tho	-	data
AY 102	Ge-Di	TV-Booster-Di, 60/320V, 7A	23a§	Sgs	-	data
AY 103 K	Ge-Di	TV-Booster-Di, 60/320V, 7A	3(--AK)	Sgs	-	data
AY 104	Ge-Di	S, 90V, 5A(ss)	4(--AK)	Sgs	-	data
AY 105 K	Ge-Di	TV-Booster-Di, 80/250V, 5A(ss)	3(--AK)	Sgs	-	data
AY 106	Ge-Di	TV-Booster-Di, 60/200V, 7A	23a§	Sgs	-	data
AY 0438 /L	CMOS-IC	32-Segment Cmos LCD Driver	DIP	Mcp	-	pdf pinout
AY 0438 /P	CMOS-IC	32-Segment Cmos LCD Driver	DIP	Mcp	-	pdf pinout
AY-3-1014A	LIN-IC	Univ.asynchronous Receiver/transmitter	28-DIP	Gie	-	pdf pinout
AY-3-1015	LIN-IC	Univ.asynchronous Receiver/transmitter	28-DIP	Gie	-	pdf pinout
AY-3-8910	LIN-IC	Programmable Sound Generator	DIP	Gie	-	pdf pinout
AY-3-8912	LIN-IC	Programmable Sound Generator	DIP	Gie	-	pdf pinout
AY-3-8913	LIN-IC	Programmable Sound Generator	DIP	Gie	-	pdf pinout
AY-5-1013A	LIN-IC	Univ.asynchronous Receiver/transmitter	28-DIP	Gie	-	pdf pinout
AY-5-1224A	LIN-IC	4 Digit Clock Circuit	28-DIP	Gie	-	pdf pinout
AY-6-1013	LIN-IC	Univ.asynchronous Receiver/transmitter	28-DIP	Gie	-	pdf pinout
AZ	MOS-P-FET-e	=2SJ518 (Typ-Code/Stempel/marking)	39	Hit	→2SJ518	data
AZ	Si-N	=BCX 70RL (Typ-Code/Stempel/marking)	35	Sie	→BCX 70RL	data
AZ	Z-Di	=P4 SMA-16A (Typ-Code/Stempel/marking)	71(5x2,5)	Fag	→P4 SMA-16A	data
AZ	Z-Di	=SMBJ 11CA (Typ-Code/Stempel/marking)	71(5x3,5)	Mop	→SMBJ ...	data
AZ 23 C2V7...C51	Z-Di	SMD, Dual, 2,7...51V, 5%	35m	Aeg	-	data
AZ324	4xLO-POWER	Vs:20V Vu:100dB Vo:12V Vi0:2mV	14SD	Aac		data pdf pinout
AZ 324M	LIN-IC	SMD, 4 high gain and intern. freq. comp. OPamps, compatible with ...324...	MDIP	Bcd	...324	pdf pinout
AZ 324P	LIN-IC	4 high gain and intern. freq. comp. OPamps, compatible with ...324...	DIP	Bcd	...324	pdf pinout
AZ 384 XGM	IC	Green Mode PWM Controller	QFP	Bcd	-	pdf pinout
AZ 384 XGP	IC	Green Mode PWM Controller	QFP	Bcd	-	pdf pinout
AZ 386 M	IC	low Voltage Audio Power Amplifier	QFP	Bcd	-	pdf pinout
AZ 386 P	IC	low Voltage Audio Power Amplifier	QFP	Bcd	-	pdf pinout
AZ 431 K	LIN-IC	Adj. (2.5V-36V) Precision Shunt Reg.	45	Bcd		pdf pinout
AZ 431 M	LIN-IC	Adj. (2.5V-36V) Precision Shunt Reg.	MDIP	Bcd		pdf pinout
AZ 431 N	LIN-IC	Adj. (2.5V-36V) Precision Shunt Reg.	35	Bcd		pdf pinout
AZ 431 R	LIN-IC	Adj. (2.5V-36V) Precision Shunt Reg.	48	Bcd		pdf pinout
AZ 431 Z	LIN-IC	Adj. (2.5V-36V) Precision Shunt Reg.	7	Bcd		pdf pinout
AZ 2822 P	IC	Dual low Voltage Power Amplifier	QFP	Bcd		pdf pinout
AZ 2842 M	IC	Current Mode PWM Controller	QFP	Bcd		pdf pinout
AZ 2842 P	IC	Current Mode PWM Controller	QFP	Bcd	-	pdf pinout
AZ 2843 M	IC	Current Mode PWM Controller	QFP	Bcd		pdf pinout
AZ 2843 P	IC	Current Mode PWM Controller	QFP	Bcd		pdf pinout
AZ 2844 M	IC	Current Mode PWM Controller	QFP	Bcd	-	pdf pinout
AZ 2844 P	IC	Current Mode PWM Controller	QFP	Bcd		pdf pinout
AZ 2845 M	IC	Current Mode PWM Controller	QFP	Bcd		pdf pinout
AZ 2845 P	IC	Current Mode PWM Controller	QFP	Bcd	-	pdf pinout
AZ 3842 AM	IC	Current Mode PWM Controller	QFP	Bcd		pdf pinout
AZ 3842 AP	IC	Current Mode PWM Controller	QFP	Bcd		pdf pinout
AZ 3843 AM	IC	Current Mode PWM Controller	QFP	Bcd	-	pdf pinout
AZ 3843 AP	IC	Current Mode PWM Controller	QFP	Bcd		pdf pinout
AZ 3844 AM	IC	Current Mode PWM Controller	QFP	Bcd	-	pdf pinout
AZ 3844 AP	IC	Current Mode PWM Controller	QFP	Bcd	-	pdf pinout
AZ 3845 AM	IC	Current Mode PWM Controller	QFP	Bcd		pdf pinout
AZ 3845 AP	IC	Current Mode PWM Controller	QFP	Bcd		pdf pinout
AZ 6208 BM	IC	Reversible Motor Driver	TSOP	Bcd		pdf pinout
AZ 6208 BQ	IC	Reversible Motor Driver	TSOP	Bcd		pdf pinout
AZ 7500BM	IC	Pulse-width Modulation control circuit	SSMDIP	Bcd		pdf pinout
AZ 7500BP	IC	Pulse-width Modulation control circuit	DIC	Bcd	-	pdf pinout
AZ 7500CM	IC	Pulse-width Modulation control circuit	DIC	Bcd		pdf pinout
AZ 7500CP	IC	Pulse-width Modulation control circuit	DIC	Bcd		pdf pinout
AZ 34063 AM	IC	Step-down/step-up/inverting DC-DC Conv.	TSOP	Bcd		pdf pinout
AZ 34063 AP	IC	Step-down/step-up/inverting DC-DC Conv.	TSOP	Bcd		pdf pinout
AZ 34063 CM	IC	Step-down/step-up/inverting DC-DC Conv.	TSOP	Bcd		pdf pinout
AZ 34063 CP	IC	Step-down/step-up/inverting DC-DC Conv.	TSOP	Bcd		pdf pinout
AZ 34063 M	IC	Step-down/step-up/inverting DC-DC Conv.	TSOP	Bcd		pdf pinout
AZ 34063 P	IC	Step-down/step-up/inverting DC-DC Conv.	TSOP	Bcd		pdf pinout
AZ 2842 M 14	IC	Current Mode PWM Controller	QFP	Bcd		pdf pinout
AZ 2843 M 14	IC	Current Mode PWM Controller	QFP	Bcd		pdf pinout
AZ 2844 M 14	IC	Current Mode PWM Controller	QFP	Bcd		pdf pinout
AZ 2845 M 14	IC	Current Mode PWM Controller	QFP	Bcd		pdf pinout
AZ 3842 AM 14	IC	Current Mode PWM Controller	QFP	Bcd		pdf pinout
AZ 3843 AM 14	IC	Current Mode PWM Controller	QFP	Bcd		pdf pinout
AZ 3844 AM 14	IC	Current Mode PWM Controller	QFP	Bcd		pdf pinout
AZ 3845 AM 14	IC	Current Mode PWM Controller	QFP	Bcd		pdf pinout
AZ 5901 F 44	IC	4-channel driver and power controller	QFP	Bcd		pdf pinout
AZ 5954 M 28	IC	4-channel BTL driver IC, CDROM/DVD	QFP	Bcd		pdf pinout

Type	Device	Short Description	Fig.	Manu	Comparision Types	More at
AZ 6392 M 28	IC	4 Channel BTL Driver For VCD players	TSOP	Bcd	-	pdf pinout
AZ 9258 M 28	IC	4-Channel Motor Driver IC	TSOP	Bcd	-	pdf pinout
AZO	Si-P	=KTA1505-O (Typ-Code/Stempel/marking)	35	Kec	→KTA 1505	data
AZY	Si-P	=KTA1505-Y (Typ-Code/Stempel/marking)	35	Kec	→KTA 1505	data
АИ 101 А,Б,В,Г, Д,Е,Ж,И	GaAs-Di	Tunnel-Di, Ip=1...5mA, Up=0,16...0,18V 4...13pF, Rd=7...80Ω, Ip/Iv=5...6		Gus	-	
АИ 201 А,Б,В,Г, Д,Е,Ж,И,К,Л	GaAs-Di	Tunnel-Di, Ip=10...100mA,Up=0,18...0,33V 8...50pF, Rd=2,2...220Ω, Ip/Iv=10		Gus	-	
АИ 301 А,Б,В,Г	GaAs-Di	Tunnel-Di, Ip=2...10mA, Up=0,18V 12..50pF, Ip/Iv=8		Gus	-	
АИ 402 Б,Г,Е,И	GaAs-Di	Tunnel-Di, Up=0,6V, Irev=2...8mA		Gus	-	

B

Type	Device	Short Description	Fig.	Manu	Comparision Types	More at
B...	...-P	→2SB..., z.B./e.g. "B861" = 2SB861 →	Japantypen	JAP		
B...	...-P	→KSB..., z.B./e.g. "B1116"=KSB1116→	Samsung	Sam		
B...	...-P	→KTB..., z.B./e.g. "B 778"=KTB 778→	KEC	Kec		
B	Si-P	=2SB1218A (Typ-Code/Stempel/marking)	35(2mm)	Mat	→2SB1218A	data
B	Si-P	=2SB766A (Typ-Code/Stempel/marking)	39	Mat	→2SB766A	data
B	N-FET	=2SK1068 (Typ-Code/Stempel/marking)	35(2mm)	Say	→2SK1068	data
B	GaAs-FET	=2SK1325 (Typ-Code/Stempel/marking)	52	Tos	→2SK1325	data
B	GaAs-N-FET	=2SK1616 (Typ-Code/Stempel/marking)	52	Hit	→2SK1616	data
B	MOS-N-FET-d	=3SK302 (Typ-Code/Stempel/marking)	44	Mat	→3SK302	data
B	MOS-N-FET-d	=3SK306 (Typ-Code/Stempel/marking)	44(2mm)	Mat	→3SK306	data
B	Si-P	=BAS 16-03W (Typ-Code/Stempel/marking)	71(1,7mm)	Sie	→BAS 16-03W	data
B 0	Z-Di	=BZX 399-C4V3(Typ-Code/Stempel/marking)	71(1,7mm)	Phi	→BZX 399-...	data
B 0	C-Di	=HVC 350B (Typ-Code/Stempel/marking)	71(1,3mm)	Hit	→HVC 350B	data
B	Si-Di	=MA 2H736 (Typ-Code/Stempel/marking)	71(3,8mm)	Mat	→MA 2H736	data
B	Si-Di	=MA 2S728 (Typ-Code/Stempel/marking)	71(1,7mm)	Mat	→MA 728	data
B	Diode	=RB 520S-30 (Typ-Code/Stempel/marking)	71(1,3mm)	Rhm	→RB 520S-30	data
B	Si-Di	=S 5566B (Typ-Code/Stempel/marking)	31	Tos	→S 5566B	data
B	Si-Di	=SB 05-05CP (Typ-Code/Stempel/marking)	35	Say	→SB 05-05CP	data
B 1	MOS-N-FET-e	=2SK1824 (Typ-Code/Stempel/marking)	35(1,6mm)	Nec	→2SK1824	data
B 1	Si-Di	=BAS 40L... (Typ-Code/Stempel/marking)	35	Ons	→BAS 40L	data
B 1	Z-Di	=BZX 399-C1V8(Typ-Code/Stempel/marking)	71(1,7mm)	Phi	→BZX 399-...	data
B 1	Si-P	=D71G.05T1 (Typ-Code/Stempel/marking)	39	Gen	→D71G.05T1	data
B 1	Si-Di	=HSM 2692 (Typ-Code/Stempel/marking)	35	Hit	→HSM 2692	data
B 1	C-Di	=HVC 355B (Typ-Code/Stempel/marking)	71(1,3mm)	Hit	→HVC 355B	data
B 1	C-Di	=HVU 355B (Typ-Code/Stempel/marking)	71(1,7mm)	Hit	→HVU 355B	data
B 1	Si-P+R	=IMB 1A (Typ-Code/Stempel/marking)	46	Rhm	→IMB 1A	data
B 1	Z-Di	=MAZF 033 (Typ-Code/Stempel/marking)	71(1,7mm)	Mat	→MAZF 033	data
B 1	Z-Di	=MAZN 033 (Typ-Code/Stempel/marking)	71(1,3mm)	Mat	→MAZN 033	data
B 1	Si-P+R	=UMB 1N (Typ-Code/Stempel/marking)	46(2mm)	Rhm	→UMB 1N...11N	data
B 1 C	Si-Di	=MBRS 1100 (Typ-Code/Stempel/marking)	71(5x3,5)	Ons	→MBRS 1100	
B 1 L 3	Si-Di	=MBRA 130 (Typ-Code/Stempel/marking)	71(5x2,5)	Ons	→MBRA 130	
B 1 O	Si-N	=KSC 2715-O (Typ-Code/Stempel/marking)	35	Sam	→KSC 2715	data
B 1 R	Si-N	=KSC 2715-R (Typ-Code/Stempel/marking)	35	Sam	→KSC 2715	
B1X19X	LED	bl ø5mm I:11lm Vf:3.5V If:350mA		Seo		data pdf pinout
B1X29X	LED	bl ø5mm I:18lm Vf:3.5V If:700mA		Seo		data pdf pinout
B1X49X	LED	bl ø5mm I:30lm Vf:3.5V If:1.4AmA		Seo		data pdf pinout
B 1 Y	Si-N	=KSC 2715-Y (Typ-Code/Stempel/marking)	35	Sam	→KSC 2715	data
B 2	Si-N	=2SC1621-B2 (Typ-Code/Stempel/marking)	35	Nec	→2SC1621	data
B 2	Si-N	=2SC4175-B2 (Typ-Code/Stempel/marking)	35(2mm)	Nec	→2SC4175	data
B 2	Si-N	=BSV 52 (Typ-Code/Stempel/marking)	35	Mot,Phi,Tho Fer	→BSV 52	data
B 2	Z-Di	=BZX 399-C2V0(Typ-Code/Stempel/marking)	71(1,7mm)	Phi	→BZX 399-...	data
B 2	C-Di	=HVC 358B (Typ-Code/Stempel/marking)	71(1,3mm)	Hit	→HVC 358B	data
B 2	Si-P+R	=IMB 2A (Typ-Code/Stempel/marking)	46	Rhm	→IMB 2A	data
B 2	Z-Di	=MAZF 036 (Typ-Code/Stempel/marking)	71(1,7mm)	Mat	→MAZF 036	data
B 2	Z-Di	=MAZN 036 (Typ-Code/Stempel/marking)	71(1,3mm)	Mat	→MAZN 036	data
B 2	Si-Di	=MBR 0520 (Typ-Code/Stempel/marking)	71(2,7mm)	Mot	→MBR 0520	data
B 2	Si-Di	=SMBYT01-200(Typ-Code/Stempel/marking)	71(6,4mm)	Tho	→SMBYT 01-...	data
B 2	Si-P+R	=UMB 2N (Typ-Code/Stempel/marking)	46(2mm)	Rhm	→UMB 1N...11N	data
B 2 p	Si-N	=BSV 52 (Typ-Code/Stempel/marking)	35	Mot,Phi,Tho	→BSV 52	data
B 2 t	Si-N	=BSV 52 (Typ-Code/Stempel/marking)	35	Mot,Phi,Tho	→BSV 52	data
B2V4 PH	Z-Di	=BZX 79-B2V4(Typ-Code/Stempel/marking)	31	Phi	→BZX 79-B...	data
B2V7 PH	Z-Di	=BZX 79-B2V7(Typ-Code/Stempel/marking)	31	Phi	→BZX 79-B...	data
B 3	Si-Di	=1PS301 (Typ-Code/Stempel/marking)	35(2mm)	Phi	→1PS301	data
B 3	Si-Di	=1SS184 (Typ-Code/Stempel/marking)	35	Tos	→1SS184	data
B 3	Si-Di	=1SS301 (Typ-Code/Stempel/marking)	35(2mm)	Tos	→1SS301	data
B 3	Si-Di	=1SS361 (Typ-Code/Stempel/marking)	35(1,6mm)	Tos	→1SS361	data
B 3	Si-Di	=1SS361F (Typ-Code/Stempel/marking)	35(1,6mm)	Tos	→1SS361F	data
B 3	Si-N	=2SC1621-B3 (Typ-Code/Stempel/marking)	35	Nec	→2SC1621	data
B 3	Si-N	=2SC4175-B3 (Typ-Code/Stempel/marking)	35(2mm)	Nec	→2SC4175	data
B 3	Z-Di	=BZX 399-C2V2(Typ-Code/Stempel/marking)	71(1,7mm)	Phi	→BZX 399-...	data
B 3	Si-Di	=HSM 2694 (Typ-Code/Stempel/marking)	35	Hit	→HSM 2694	data
B 3	C-Di	=HVC 369B (Typ-Code/Stempel/marking)	71(1,3mm)	Hit	→HVC 369B	data
B 3	Si-P+R	=IMB 3A (Typ-Code/Stempel/marking)	46	Rhm	→IMB 3A	data
B 3	Si-Di	=KDS 121 (Typ-Code/Stempel/marking)	35(2mm)	Kec	→KDS 121	data
B 3	Si-Di	=KDS 121E (Typ-Code/Stempel/marking)	35(1,6mm)	Kec	→KDS 121E	data
B 3	Si-Di	=KDS 184 (Typ-Code/Stempel/marking)	35	Kec	→KDS 184	data
B 3	Z-Di	=MAZF 039 (Typ-Code/Stempel/marking)	71(1,7mm)	Mat	→MAZF 039	data
B 3	Z-Di	=MAZN 039 (Typ-Code/Stempel/marking)	71(1,3mm)	Mat	→MAZN 039	data

Type	Device	Short Description	Fig.	Manu	Comparision Types	More at
B 3	Si-Di	=MBR 0530 (Typ-Code/Stempel/marking)	71(2,7mm)	Mot	→MBR 0530	data
B 3	Si-Di	=MMBD 717 (Typ-Code/Stempel/marking)	35(2mm)	Ons	→MMBD 717	
B 3	Si-Di	=SMBYT01-300(Typ-Code/Stempel/marking)	71(6,4mm)	Tho	→SMBYT 01-...	data
B 3	Si-P+R	=UMB 3N (Typ-Code/Stempel/marking)	46(2mm)	Rhm	→UMB 1N...11N	data
B 03A-1B...	Si-Br				B...C300	
B 3 T	Si-Di	=1PS184 (Typ-Code/Stempel/marking)	35	Phi	→1PS184	data
B3V0 PH	Z-Di	=BZX 79-B3V0(Typ-Code/Stempel/marking)	31	Phi	→BZX 79-B...	data
B3V3 PH	Z-Di	=BZX 79-B3V3(Typ-Code/Stempel/marking)	31	Phi	→BZX 79-B...	data
B3V6 PH	Z-Di	=BZX 79-B3V6(Typ-Code/Stempel/marking)	31	Phi	→BZX 79-B...	data
B3V9 PH	Z-Di	=BZX 79-B3V9(Typ-Code/Stempel/marking)	31	Phi	→BZX 79-B...	data
B3X18X	LED	bl ø5mm l:11lm Vf:3.5V If:350mA		Seo		data pdf pinout
B3X28X	LED	bl ø5mm l:16lm Vf:4V If:700mA		Seo		data pdf pinout
B 4	Si-N	=2SC1621-B4 (Typ-Code/Stempel/marking)	35	Nec	→2SC1621	data
B 4	Si-N	=2SC4175-B4 (Typ-Code/Stempel/marking)	35(2mm)	Nec	→2SC4175	data
B 4	Si-N	=2SC4987-B4 (Typ-Code/Stempel/marking)	35(1,6mm)	Say	→2SC4987	data
B 4	Si-Di	=BAT 54W... (Typ-Code/Stempel/marking)	35(2mm)	Ons	→BAT 54W	
B 4	Si-N	=BSV 52R (Typ-Code/Stempel/marking)	35	Val,Tho,Fer	→BSV 52R	data
B 4	Z-Di	=BZX 399-C2V4(Typ-Code/Stempel/marking)	71(1,7mm)	Phi	→BZX 399-...	data
B 4	Si-Di	=HSM 2693A (Typ-Code/Stempel/marking)	35	Hit	→HSM 2693A	data
B 4	Si-P+R	=IMB 4A (Typ-Code/Stempel/marking)	46	Rhm	→IMB 4A	data
B 4	Si-N+P	=KTX 101U-Y (Typ-Code/Stempel/marking)	46(2mm)	Kec	→KTX 101U	data
B 4	Si-Di	=MBR 0540 (Typ-Code/Stempel/marking)	71(2,7mm)	Mot, Gsi	→MBR 0540	data
B 4	Si-Di	=SMBYT01-400(Typ-Code/Stempel/marking)	71(6,4mm)	Tho	→SMBYT 01-...	data
B 4	Si-P+R	=UMB 4N (Typ-Code/Stempel/marking)	46(2mm)	Rhm	→UMB 1N...11N	data
B4V3 PH	Z-Di	=BZX 79-B4V3(Typ-Code/Stempel/marking)	31	Phi	→BZX 79-B...	data
B4V7 PH	Z-Di	=BZX 79-B4V7(Typ-Code/Stempel/marking)	31	Phi	→BZX 79-B...	data
B 5	Si-N	=2SC4987-B5 (Typ-Code/Stempel/marking)	35(1,6mm)	Say	→2SC4987	data
B 5	Si-P	=BSR 12 (Typ-Code/Stempel/marking)	35	Phi	→BSR 12	data
B 5	Z-Di	=BZX 399-C2V7(Typ-Code/Stempel/marking)	71(1,7mm)	Phi	→BZX 399-...	data
B 5	C-Di	=HVC 372B (Typ-Code/Stempel/marking)	71(1,3mm)	Hit	→HVC 372B	data
B 5	Si-P+R	=IMB 5A (Typ-Code/Stempel/marking)	46	Rhm	→IMB 5A	data
B 5	Si-P+R	=UMB 5N (Typ-Code/Stempel/marking)	46(2mm)	Rhm	→UMB 1N...11N	data
B 5 A45 VI	Si-Di	Schottky-GI, 45V, 5A(Tc=115°)	17d	Kec	-	data
B 5 A45 VIC	Si-Di	=B 5A45VI: Dual, 5A(Tc=117°)	17p			data
B 5 A60 VI	Si-Di	=B 5A45VI: 60V, 5A(Tc=115°)	17p			data
B 5 A60 VIC	Si-Di	=B 5A45VI: Dual, 60V, 5A(Tc=117°)	17p			data
B 5 A90 VI	Si-Di	=B 5A45VI: 90V, 5A(Tc=115°)	17p			data
B 5 p	Si-P	=BSR 12 (Typ-Code/Stempel/marking)	35	Phi	→BSR 12	data
B5V1 PH	Z-Di	=BZX 79-B5V1(Typ-Code/Stempel/marking)	31	Phi	→BZX 79-B...	data
B5V6 PH	Z-Di	=BZX 79-B5V6(Typ-Code/Stempel/marking)	31	Phi	→BZX 79-B...	data
B 6	Si-P	=2SB815-6 (Typ-Code/Stempel/marking)	35	Say	→2SB815	data
B 6	Si-N	=2SC4987-B6 (Typ-Code/Stempel/marking)	35(1,6mm)	Say	→2SC4987	data
B 6	Si-Di	=BAT 54AL... (Typ-Code/Stempel/marking)	35	Ons	→BAT 54AL	
B 6	Z-Di	=BZX 399-C3V0(Typ-Code/Stempel/marking)	71(1,7mm)	Phi	→BZX 399-...	data
B 6	Si-P+R	=IMB 6A (Typ-Code/Stempel/marking)	46	Rhm	→IMB 6A	data
B 6	Si-N+P	=KTX 101U-GR(Typ-Code/Stempel/marking)	46(2mm)	Kec	→KTX 101U	data
B 6	Si-P+R	=UMB 6N (Typ-Code/Stempel/marking)	46(2mm)	Rhm	→UMB 1N...11N	data
B06 13	Si-Di	=ERB 06-13 (Typ-Code/Stempel/marking)	31a	Fjd	→ERB 06-...	data
B06 15	Si-Di	=ERB 06-15 (Typ-Code/Stempel/marking)	31a	Fjd	→ERB 06-...	data
B6V2 PH	Z-Di	=BZX 79-B6V2(Typ-Code/Stempel/marking)	31	Phi	→BZX 79-B...	data
B6V8 PH	Z-Di	=BZX 79-B6V8(Typ-Code/Stempel/marking)	31	Phi	→BZX 79-B...	data
B 7	Si-P	=2SB815-7 (Typ-Code/Stempel/marking)	35	Say	→2SB815	data
B 7	Z-Di	=BZX 399-C3V3(Typ-Code/Stempel/marking)	71(1,7mm)	Phi	→BZX 399-...	data
B 7	Si-P+R	=IMB 7A (Typ-Code/Stempel/marking)	46	Rhm	→IMB 7A	data
B 7	Si-P+R	=UMB 7 (Typ-Code/Stempel/marking)	46(2mm)	Rhm	→UMB 1N...11N	data
B7V5 PH	Z-Di	=BZX 79-B7V5(Typ-Code/Stempel/marking)	31	Phi	→BZX 79-B...	data
B 8	Si-Di	=BAT 54SW... (Typ-Code/Stempel/marking)	35(2mm)	Ons	→BAT 54SW	
B 8	Si-Di	=BAV 20WS (Typ-Code/Stempel/marking)	71(1,7mm)	Gsi	→BAV 20WS	data
B 8	Si-P	=BSR 12R (Typ-Code/Stempel/marking)	35	Phi	→BSR 12R	data
B 8	Z-Di	=BZX 399-C3V6(Typ-Code/Stempel/marking)	71(1,7mm)	Phi	→BZX 399-...	data
B 8	Si-P+R	=IMB 8A (Typ-Code/Stempel/marking)	46	Rhm	→IMB 8A	data
B 8	Si-P+R	=UMB 8N (Typ-Code/Stempel/marking)	46(2mm)	Rhm	→UMB 1N...11N	data
B8V2 PH	Z-Di	=BZX 79-B8V2(Typ-Code/Stempel/marking)	31	Phi	→BZX 79-B...	data
B 9	Si-Di	=1SS311 (Typ-Code/Stempel/marking)	35	Tos	→1SS311	data
B 9	Si-Di	=1SS397 (Typ-Code/Stempel/marking)	35(2mm)	Tos	→1SS397	data
B 9	Z-Di	=BZX 399-C3V9(Typ-Code/Stempel/marking)	71(1,7mm)	Phi	→BZX 399-...	data
B 9	Si-P+R	=IMB 9A (Typ-Code/Stempel/marking)	46	Rhm	→IMB 9A	data
B 9	Si-P+R	=UMB 9N (Typ-Code/Stempel/marking)	46(2mm)	Rhm	→UMB 1N...11N	data
B9V1 PH	Z-Di	=BZX 79-B9V1(Typ-Code/Stempel/marking)	31	Phi	→BZX 79-B...	data
B 10	N-FET	=2SK545-B10 (Typ-Code/Stempel/marking)	35	Say	→2SK545	data
B 10	Si-P+R	=IMB 10A (Typ-Code/Stempel/marking)	46	Rhm	→IMB 10A	data
B 10	Si-P+R	=UMB 10N (Typ-Code/Stempel/marking)	46(2mm)	Rhm	→UMB 1N...11N	data
B 10 A45V	Si-Di	Schottky-GI, 45V, 10A(Tc=135°), <35ns	17t§	Kec	-	data
B 10 A45VI	Si-Di	=B 10A45V: Iso, 10A(Tc=130°)	17d			data
B 10 A45 VIC	Si-Di	=B 10A45V: Iso, Dual, 10A(Tc=114°)	17p		-	data
B 10 A60 VIC	Si-Di	=B 10A45VI: Dual, 60V, 10A(Tc=114°)	17p		-	data
B 10 A90 VIC	Si-Di	=B 10A45VI: Dual, 90V, 10A(Tc=110°)	17p		-	data
B10 PH	Z-Di	=BZX 79-B10 (Typ-Code/Stempel/marking)	31	Phi	→BZX 79-B...	data
B 11	N-FET	=2SK545-B11 (Typ-Code/Stempel/marking)	35	Say	→2SK545	data
B 11	Si-P+R	=IMB 11A (Typ-Code/Stempel/marking)	46	Rhm	→IMB 11A	data
B11 PH	Z-Di	=BZX 79-B11 (Typ-Code/Stempel/marking)	31	Phi	→BZX 79-B...	data
B 11	Si-P+R	=UMB 11N (Typ-Code/Stempel/marking)	46(2mm)	Rhm	→UMB 1N...11N	data
B 12	N-FET	=2SK545-B12 (Typ-Code/Stempel/marking)	35	Say	→2SK545	data

87

Type	Device	Short Description	Fig.	Manu	Comparision Types	More at
B 12	Si-Di	=ERB 12-... (Typ-Code/Stempel/marking)	31a	Fjd	→ERB 12-...	data
B 12	Si-Di	=MBRS 120 (Typ-Code/Stempel/marking)	71(5x3,5)	Ons	→MBRS 120	
B 12	Si-N	=2SC3739-B12(Typ-Code/Stempel/marking)	35	Nec	→2SC3739	data
B 12	Si-N	=2SC4173-B12(Typ-Code/Stempel/marking)	35(2mm)	Nec	→2SC4173	data
B12 PH	Z-Di	=BZX 79-B12 (Typ-Code/Stempel/marking)	31	Phi	→BZX 79-B...	data
B 13	Si-Di	=MBRS 130 (Typ-Code/Stempel/marking)	71(5x3,5)	Ons	→MBRS 130	
B 13	Si-N	=2SC3739-B13(Typ-Code/Stempel/marking)	35	Nec	→2SC3739	data
B 13	Si-N	=2SC4173-B13(Typ-Code/Stempel/marking)	35(2mm)	Nec	→2SC4173	data
B13 PH	Z-Di	=BZX 79-B13 (Typ-Code/Stempel/marking)	31	Phi	→BZX 79-B...	data
B 14	Si-P+R	=IMB 14A (Typ-Code/Stempel/marking)	46	Rhm	→IMB 14A	data
B 14	Si-Di	=MBRA 140 (Typ-Code/Stempel/marking)	71(5x2,5)	Ons	→MBRA 140	
B 14	Si-Di	=MBRS 140 (Typ-Code/Stempel/marking)	71(5x3,5)	Ons	→MBRS 140	
B 14	Si-N	=2SC3739-B14(Typ-Code/Stempel/marking)	35	Nec	→2SC3739	data
B 14	Si-N	=2SC4173-B14(Typ-Code/Stempel/marking)	35(2mm)	Nec	→2SC4173	data
B 14 L	Si-Di	=MBRS 140L (Typ-Code/Stempel/marking)	71(5x3,5)	Ons	→MBRS 140L	
B 15	Si-N	=NTM 2222A (Typ-Code/Stempel/marking)	35	Nec	→NTM 2222A	data
B 15 A45V	Si-Di	Dual,Schottky, 45V, 15A(Tc=117°),<35ns	17p§	Kec	-	data
B 15 A45 VIC	Si-Di	=B 15A45V: Iso, 15A(Tc=105°)	17p		-	data
B 15 A60 VIC	Si-Di	=B 15A45V: Iso, 60V, 15A(Tc=114°)	17p		-	data
B15 PH	Z-Di	=BZX 79-B15 (Typ-Code/Stempel/marking)	31	Phi	→BZX 79-B...	data
B 16	Si-P+R	=IMB 16 (Typ-Code/Stempel/marking)	46	Rhm	→IMB 16	data
B16 PH	Z-Di	=BZX 79-B16 (Typ-Code/Stempel/marking)	31	Phi	→BZX 79-B...	data
B 17	Si-P+R	=IMB 17A (Typ-Code/Stempel/marking)	46	Rhm	→IMB 17A	data
B18 PH	Z-Di	=BZX 79-B18 (Typ-Code/Stempel/marking)	31	Phi	→BZX 79-B...	data
B 19	Si-Di	=MBRS 190 (Typ-Code/Stempel/marking)	71(5x3,5)	Ons	→MBRS 190	
B 20	MOS-N-FET-e	=BSN 20 (Typ-Code/Stempel/marking)	35	Gsi	→BSN 20	data
B 20 A45 VIC	Si-Di	Dual, Schottky-Gl, 45V, 20A(Tc=108°) <35ns	17p	Kec		data
B 20 A60 VIC	Si-Di	=B 20A45VIC: 60V	17p		-	data
B20 PH	Z-Di	=BZX 79-B20 (Typ-Code/Stempel/marking)	31	Phi	→BZX 79-B...	data
B 20	Si-Di	=SMBYW01-200(Typ-Code/Stempel/marking)	71(5x3,5)	Tho	→SMBYW 01-200	data
B 22	Si-N	=2SC3734-B22(Typ-Code/Stempel/marking)	35	Nec	→2SC3734	data
B22 PH	Z-Di	=BZX 79-B22 (Typ-Code/Stempel/marking)	31	Phi	→BZX 79-B...	data
B 23	Si-N	=2SC3734-B23(Typ-Code/Stempel/marking)	35	Nec	→2SC3734	data
B 24	Si-N	=2SC3734-B24(Typ-Code/Stempel/marking)	35	Nec	→2SC3734	data
B24 PH	Z-Di	=BZX 79-B24 (Typ-Code/Stempel/marking)	31	Phi	→BZX 79-B...	data
B 25	Si-N	=NTM 3904 (Typ-Code/Stempel/marking)	35	Nec	→NTM 3904	data
B 26	Si-N	=BF 570 (Typ-Code/Stempel/marking)	35	Phi	→BF 570	data
B27 PH	Z-Di	=BZX 79-B27 (Typ-Code/Stempel/marking)	31	Phi	→BZX 79-B...	data
B 30 A45V	Si-Di	Dual, Schottky-Gl, 45V, 30A(Tc=103°) <35ns	17p§	Kec	-	data
B 30 A45VIC	Si-Di	=B 30A45V: Iso, 30A(Tc=84°C)	17p			data
B 30 A45VN	Si-Di	=B 30A45V:	18p§			data
B30 PH	Z-Di	=BZX 79-B30 (Typ-Code/Stempel/marking)	31	Phi	→BZX 79-B...	data
B 32	Si-Di	=MBRS 320 (Typ-Code/Stempel/marking)	71(7x6mm)	Ons	→MBRS 320	
B32 01	Si-Di	=ERB 32-01 (Typ-Code/Stempel/marking)	31a	Fjd	→ERB 32-...	data
B32 02	Si-Di	=ERB 32-02 (Typ-Code/Stempel/marking)	31a	Fjd	→ERB 32-...	data
B 33	Si-Di	=MBRS 330 (Typ-Code/Stempel/marking)	71(7x6mm)	Ons	→MBRS 330	
B 33	Si-N	=2SC3735-B33(Typ-Code/Stempel/marking)	35	Nec	→2SC3735	data
B 33	Si-N	=2SC4176-B33(Typ-Code/Stempel/marking)	35(2mm)	Nec	→2SC4176	data
B33 PH	Z-Di	=BZX 79-B33 (Typ-Code/Stempel/marking)	31	Phi	→BZX 79-B...	data
B 34	Si-Di	=MBRS 340 (Typ-Code/Stempel/marking)	71(7x6mm)	Mot, Ons	→MBRS 340	data
B 34	Si-N	=2SC3735-B34(Typ-Code/Stempel/marking)	35	Nec	→2SC3735	data
B 34	Si-N	=2SC4176-B34(Typ-Code/Stempel/marking)	35(2mm)	Nec	→2SC4176	data
B 35	Si-N	=2SC3735-B35(Typ-Code/Stempel/marking)	35	Nec	→2SC3735	data
B 35	Si-N	=2SC4176-B35(Typ-Code/Stempel/marking)	35(2mm)	Nec	→2SC4176	data
B 36	Si-Di	=MBRS 360 (Typ-Code/Stempel/marking)	71(7x6mm)	Mot, Ons	→MBRS 360	data
B36 PH	Z-Di	=BZX 79-B36 (Typ-Code/Stempel/marking)	31	Phi	→BZX 79-B...	data
B37 08	Si-Di	=ERB 37-08 (Typ-Code/Stempel/marking)	31a	Fjd	→ERB 37-...	data
B37 10	Si-Di	=ERB 37-10 (Typ-Code/Stempel/marking)	31a	Fjd	→ERB 37-...	data
B38 04	Si-Di	=ERB 38-04 (Typ-Code/Stempel/marking)	31a	Fjd	→ERB 38-...	data
B38 05	Si-Di	=ERB 38-05 (Typ-Code/Stempel/marking)	31a	Fjd	→ERB 38-...	data
B38 06	Si-Di	=ERB 38-06 (Typ-Code/Stempel/marking)	31a	Fjd	→ERB 38-...	data
B39 PH	Z-Di	=BZX 79-B39 (Typ-Code/Stempel/marking)	31	Phi	→BZX 79-B...	data
B 42	Si-P	=2SB1475-B42(Typ-Code/Stempel/marking)	35(2mm)	Nec	→2SB1475	data
B 43	Si-P	=2SB1475-B43(Typ-Code/Stempel/marking)	35(2mm)	Nec	→2SB1475	data
B43 02	Si-Di	=ERB 43-02 (Typ-Code/Stempel/marking)	31a	Fjd	→ERB 43-...	data
B43 04	Si-Di	=ERB 43-04 (Typ-Code/Stempel/marking)	31a	Fjd	→ERB 43-...	data
B43 06	Si-Di	=ERB 43-06 (Typ-Code/Stempel/marking)	31a	Fjd	→ERB 43-...	data
B43 08	Si-Di	=ERB 43-08 (Typ-Code/Stempel/marking)	31a	Fjd	→ERB 43-...	data
B43 PH	Z-Di	=BZX 79-B43 (Typ-Code/Stempel/marking)	31	Phi	→BZX 79-B...	data
B 44	Si-P	=2SB1475-B44(Typ-Code/Stempel/marking)	35(2mm)	Nec	→2SB1475	data
B44 02	Si-Di	=ERB 44-02 (Typ-Code/Stempel/marking)	31a	Fjd	→ERB 44-...	data
B44 04	Si-Di	=ERB 44-04 (Typ-Code/Stempel/marking)	31a	Fjd	→ERB 44-...	data
B44 06	Si-Di	=ERB 44-06 (Typ-Code/Stempel/marking)	31a	Fjd	→ERB 44-...	data
B44 08	Si-Di	=ERB 44-08 (Typ-Code/Stempel/marking)	31a	Fjd	→ERB 44-...	data
B44 10	Si-Di	=ERB 44-10 (Typ-Code/Stempel/marking)	31a	Fjd	→ERB 44-...	data
B47 PH	Z-Di	=BZX 79-B47 (Typ-Code/Stempel/marking)	31	Phi	→BZX 79-B...	data
B 51	Si-N	=2SB736A-B51(Typ-Code/Stempel/marking)	35	Nec	→2SB736A	data
B51 PH	Z-Di	=BZX 79-B51 (Typ-Code/Stempel/marking)	31	Phi	→BZX 79-B...	data
B 52	Si-N	=2SB736A-B52(Typ-Code/Stempel/marking)	35	Nec	→2SB736A	data
B 53	Si-N	=2SB736A-B53(Typ-Code/Stempel/marking)	35	Nec	→2SB736A	data
B 54	Si-N	=2SB736A-B54(Typ-Code/Stempel/marking)	35	Nec	→2SB736A	data

Type	Device	Short Description	Fig.	Manu	Comparision Types	More at
B 55	Si-N	=2SB736A-B55(Typ-Code/Stempel/marking)	35	Nec	→2SB736A	data
B56 PH	Z-Di	=BZX 79-B56 (Typ-Code/Stempel/marking)	31	Phi	→BZX 79-B...	data
B060	BIFET	Vs:±18V Vi0:3-15mV	8D	Hfo		data pinout
B 060 D	OP-IC	→TL 060	8-DIP	Hfo	→TL 060	data pinout
B 060 SD,SG	OP-IC	=B 060D: SMD	8-MDIP	Hfo	→TL 060	pinout
B061	BIFET	Vs:±18V Vi0:3-15mV	8D	Hfo		data pinout
B 061 D	OP-IC	→TL 061	8-DIP	Hfo	→TL 061	data pinout
B 061 SD,SG	OP-IC	=B 061D: SMD	8-MDIP	Hfo	→TL 061	pinout
B062	2xBIFET	Vs:±18V Vi0:3-15mV	8D	Hfo		data pinout
B 062 D	OP-IC	→TL 062	8-DIP	Hfo	→TL 062	data pinout
B62 PH	Z-Di	=BZX 79-B62 (Typ-Code/Stempel/marking)	31	Phi	→BZX 79-B...	data
B 062 SD,SG	OP-IC	=B 062D: SMD	8-MDIP	Hfo	→TL 062	pinout
B064	4xBIFET	Vs:±18V Vi0:3-15mV	14D	Hfo		data pinout
B 064 D	OP-IC	→TL 064	14-DIP	Hfo	→TL 064	data pinout
B 064 SD,SG	OP-IC	=B 064D: SMD	14-MDIP	Hfo	→TL 064	pinout
B066	BIFET	Vs:±18V Vi0:3-15mV	8D	Hfo		data pinout
B 066 D	OP-IC	→TL 066	8-DIP	Hfo	→TL 066	data pinout
B 066 SD,SG	OP-IC	=B 066D: SMD	8-MDIP	Hfo	→TL 066	pinout
B68 PH	Z-Di	=BZX 79-B68 (Typ-Code/Stempel/marking)	31	Phi	→BZX 79-B...	data
B75 PH	Z-Di	=BZX 79-B75 (Typ-Code/Stempel/marking)	31	Phi	→BZX 79-B...	data
B080	FET	Vs:±18V Vu:68dB Vo:±10V Vi0:16μμV	8D	Hfo		data pinout
B 080 D	OP-IC	→TL 080	8-DIP	Hfo	→TL 080	data pinout
B081	FET	Vs:±18V Vu:68dB Vo:±10V Vi0:16μμV	8D	Hfo		data pinout
B 81	Si-P	=BSR 12R (Typ-Code/Stempel/marking)	35	Val	→BSR 12R	data
B81-004	Si-Di	=ERB 81-004 (Typ-Code/Stempel/marking)	31a	Fjd	→ERB 81-...	data
B 081 D	OP-IC	→TL 081	8-DIP	Hfo	→TL 081	data pinout
B082	2xFET	Vs:±18V Vu:68dB Vo:±10V Vi0:16μμV	8D	Hfo		data pinout
B 082 D	OP-IC	→TL 082	8-DIP	Hfo	→TL 082	data pinout
B083	2xFET	Vs:±18V Vu:68dB Vo:±10V Vi0:16μμV	14D	Hfo		data pinout
B83 004	Si-Di	=ERB 83-004 (Typ-Code/Stempel/marking)	31a	Fjd	→ERB 83-...	data
B83 006	Si-Di	=ERB 83-006 (Typ-Code/Stempel/marking)	31a	Fjd	→ERB 83-...	data
B 083 D	OP-IC	→TL 083	14-DIP	Hfo	→TL 083	data pinout
B084	4xFET	Vs:±18V Vu:68dB Vo:±10V Vi0:16μμV	14D	Hfo		data pinout
B84 009	Si-Di	=ERB 84-009 (Typ-Code/Stempel/marking)	31a	Fjd	→ERB 84-...	data
B 084 D	OP-IC	→TL 084	14-DIP	Hfo	→TL 084	data pinout
B91 02	Si-Di	=ERB 91-02 (Typ-Code/Stempel/marking)	31a	Fjd	→ERB 91-...	data
B93 02	Si-Di	=ERB 93-02 (Typ-Code/Stempel/marking)	31a	Fjd	→ERB 93-...	data
B109	UNI	Vs:±18V Vu:90.7dB Vo:±13.2V	14D	Hfo		data pinout
B 109 C,D	OP-IC	=A 109D: verbessert/improved,-25...+85°	14-DIC,DIP	Hfo	→Serie 109	data pinout
B 110	Si-Di	=MBRS 1100 (Typ-Code/Stempel/marking)	71(5x3,5)	Mot	→MBRS 1100	data
B110	COMP	Vs:-7/14V Vu:60dB Vo:TTL Vi0:0.76mV	14D	Hfo		data pinout
B 110 C,D	KOP-IC	=A 110D: verbessert/improved,-25...+85°	14-DIC,DIP	Hfo	→Serie 110	data pdf pinout
B 130	Si-Di	=MBRS 130 (Typ-Code/Stempel/marking)	71(5x3,5)	Mot	→MBRS 130	data
B 140	Si-Di	=MBRS 140 (Typ-Code/Stempel/marking)	71(5x3,5)	Mot	→MBRS 140	data
B 150	Si-Di	=MBR 150 (Typ-Code/Stempel/marking)	31	Mot	→MBR 150	data
B 160	Si-Di	=MBR 160 (Typ-Code/Stempel/marking)	31	Mot	→MBR 160	data
B165	BIFET	Vs:36V Vi0:50mV		Hfo		data pinout
B 165 H,V	OP-IC	→L 165	86/5Pin	Hfo	→L 165	data pdf
B165H1	BIFET	Vs:28V Vi0:50mV		Hfo		data pinout
B165V1	BIFET	Vs:28V Vi0:50mV		Hfo		data pinout
B 170	Si-Di	=MBR 170 (Typ-Code/Stempel/marking)	31	Mot	→MBR 170	data
B176	PROG	Vs:±18V Vu:25kdB Vi0:6mV	14D	Hfo		data pinout
B 176 D	OP-IC	→μA 776	8-DIP	Hfo	→μA 776	data pinout
B177	PROG	Vs:±18V Vu:25kdB Vi0:6mV	14D	Hfo		data pinout
B 177 D	OP-IC	~B 176D	14-DIP	Hfo	→B 176	data pinout
B 180	Si-Di	=MBR 180 (Typ-Code/Stempel/marking)	31	Mot	→MBR 180	data
B 190	Si-Di	=MBR 190 (Typ-Code/Stempel/marking)	31	Mot	→MBR 190	data
B 211	Ge-Di	≈2x AA 119			→AA 119	
B 222 D	LIN-IC	Doppelgegentaktmischer/dual-pp-mixer	14-DIP	Hfo		
B 260 D	SMPS-IC	TV, SN-controller	16-DIP	Hfo	NE 5560, SE 5560, TDA 1060	
B 290 SD	LIN-IC	=A 290D: SMD	14-MDIP	Hfo		
B 303 D	LIN-IC	Näherungsschalter/proximity detector	14-DIP	Hfo	-	pdf
B 303 SF	LIN-IC	=A 303D: SMD	14-MDIP	Hfo	-	pdf
B 304 D	LIN-IC	Näherungsschalter/proximity detector	14-DIP	Hfo	-	pdf
B 304 SF	LIN-IC	=A 304D: SMD	14-MDIP	Hfo	-	pdf
B 305 D	LIN-IC	Näherungsschalter/proximity detector	14-DIP	Hfo	-	pdf
B 305 SF	LIN-IC	=A 305D: SMD	14-MDIP	Hfo	-	pdf
B 306 D	LIN-IC	Näherungsschalter/proximitiy detector	8-DIP	Hfo	-	pdf
B 306 SF	LIN-IC	=A 306D: SMD	8-MDIP	Hfo	-	pdf
B 308 D	LIN-IC	Telefonverstärker/telephone ampl.	14-DIP	Hfo	(TBA 830)	
B 315 D	LIN-IC	4x NPN-Trans.-Array, 20V, 0,5A	14-DIP	Hfo	(Q 2 T2222)	
B 315 E	LIN-IC	=B 315D: Fig.→	14-DIP+a	Hfo	-	
B 315 K	LIN-IC	=B 315D: Fig.→	14-DIP+a°	Hfo	-	
B 318 D	LIN-IC	Telefonverstärker/telephone ampl.	14-DIP	Hfo		
B 320	Si-Di	=MBRD 320 (Typ-Code/Stempel/marking)	30	Mot	→MBRD 320	data
B 325 D	LIN-IC	=B 315D: 30V	14-DIP	Hfo	(Q 2 T2222)	
B 325 E	LIN-IC	=B 315D: 30V	14-DIP+a	Hfo		
B 325 K	LIN-IC	=B 315D: 30V	14-DIP+a°	Hfo		
B 330	Si-Di	=MBRD 330 (Typ-Code/Stempel/marking)	30	Mot	→MBRD 330	data

Type	Device	Short Description	Fig.	Manu	Comparision Types	More at
B 331 G	LIN-IC	Hoergeräteverst./hearing aid ampl.	14-FLP	Hfo	-	
B 340	Si-Di	=MBR 340 (Typ-Code/Stempel/marking)	31	Ons	→MBR 340	
B 340	Si-Di	=MBRD 340 (Typ-Code/Stempel/marking)	30	Mot	→MBRD 340	data
B 340 D	LIN-IC	4x NPN-Trans.-Array, 20V, 10mA	14-DIP	Hfo	-	
B 341 D	LIN-IC	4x NPN-Trans.-Array, 20V, 10mA	14-DIP	Hfo	-	
B 342 D	LIN-IC	4x NPN-Trans.-Array, 20V, 10mA	14-DIP	Hfo	-	
B 350	Si-Di	=MBR 350 (Typ-Code/Stempel/marking)	31	Ons	→MBR 350	
B 350	Si-Di	=MBRD 350 (Typ-Code/Stempel/marking)	30	Mot	→MBRD 350	data
B 360	Si-Di	=MBR 360 (Typ-Code/Stempel/marking)	31	Ons	→MBR 360	
B 360	Si-Di	=MBRD 360 (Typ-Code/Stempel/marking)	30	Mot	→MBRD 360	data
B 360 D	LIN-IC	=B 315D: 90V	14-DIP	Hfo	(TPQ 2222, TPQ 3724)	
B 360 E	LIN-IC	=B 315D: 90V	14-DIP+a	Hfo		
B 360 K	LIN-IC	=B 315D: 90V	14-DIP+a°	Hfo		
B 380 D	LIN-IC	=B 315D: 100V	14-DIP	Hfo	(TPQ 2222, TPQ 3725)	pdf
B 380 E	LIN-IC	=B 315D: 100V	14-DIP+a	Hfo	-	
B 380 K	LIN-IC	=B 315D: 100V	14-DIP+a°	Hfo	-	
B 384 D	LIN-IC	Telecom, Telefonspg.-versorg./ps	20-DIP	Hfo	-	
B 385 D	LIN-IC	Telecom, Telefontestkreis/test circuit	16-DIP	Hfo	-	
B 386 D	LIN-IC	Telecom, Telefonspeisekreis/supply c.	20-DIP	Hfo	-	
B 390 D	LIN-IC	Motorregler/motor control	18-DIP	Hfo	-	
B 391 D	LIN-IC	Recorder, Motorprozessor	18-DIP	Hfo	-	pdf
B411	PREC-BIFET	Vs:±18V Vi0:160μμV f:2.5Mc	8D	Hfo		data pinout
B 411 DD	OP-IC	FET-V, ±18V, -10...+70°	8-DIP	Hfo	LF 411 ACN	data pinout
B 451 G	LIN-IC	Hall-IC, Ucc=4,75...27V	8-SIP	Hfo	SAS 251	
B 452 G	LIN-IC	Hall-IC, Ucc=4,75...18V	8-SIP	Hfo	S 251S5	
B 453 G	LIN-IC	Hall-IC, Ucc=4,75...5,25V	8-SIP	Hfo	SAS 251S4	
B 460 G	LIN-IC	Hall-IC, Ucc=4,75...18V	8-SIP	Hfo	-	
B 461 G	LIN-IC	Hall-IC, Sensor, Ucc=4,75...5,25V	4-SIP	Hfo	SAS 261S4	pdf
B 462 G	LIN-IC	=B 461: Ucc= 4,75...18V	4-SIP	Hfo	SAS 261	pdf
B 466 GA	LIN-IC	Hall-IC, Ucc=0,5...14,5V	8-SIP	Hfo	-	
B 467 GE	LIN-IC	Hall-IC, Ucc=4,5...28	4-SIP	Hfo	-	
(SSi)B 0510...0580	Si-Di	→SSi B 0510...0580	27c	Sie		data
B 511 N	LIN-IC	Temperatur-sensor	7	Hfo		pdf
B 555 D	LIN-IC	→NE 555N	8-DIP	Hfo	→NE 555N	
B 556 D	LIN-IC	→NE 556N	14-DIP	Hfo	→NE 556N	
B 584 X	LIN-IC	Spg.-/voltage-Ref., 10+7,5+5+2,5V		Hfo	AD 584	
B 589 N	LIN-IC	Ref.-Spg.-Quelle/volt.refer., 1,235V	7	Hfo	AD 589	
B611	2xBIFET	Vs:±15V Vu:75dB Vo:TTL Vi0:<15mV	8D	Hfo		data pinout
B 611 D	OP-IC	→TCA 311A	6-DIP	Hfo	→TCA 311A	data pinout
B615	2xBIFET	Vs:±15V Vu:75dB Vo:TTL Vi0:<15mV	8D	Hfo		data pinout
B 615 D	OP-IC	→TCA 315A	6-DIP	Hfo	→TCA 315A	data pinout
B 620 T	Si-Di	=MBRD 620CT (Typ-Code/Stempel/marking)	30	Mot	→MBRD 620CT	data
B621	2xBIFET	Vs:±15V Vu:75dB Vo:TTL Vi0:<7.5mV	8DS	Hfo		data pinout
B 621 D	OP-IC	→TCA 321A	6-DIP	Hfo	→TCA 321A	data pinout
B 621 SC	OP-IC	=B 621D: SMD	8-MDIP	Hfo		data pinout
B625	2xBIFET	Vs:±15V Vu:75dB Vo:TTL Vi0:<7.5mV	8DS	Hfo		data pinout
B 625 D	OP-IC	→TCA 325A	6-DIP	Hfo	→TCA 325A	data pinout
B 625 SG	OP-IC	=B 625D: SMD	8-MDIP	Hfo		data pinout
B 630 T	Si-Di	=MBRD 630CT (Typ-Code/Stempel/marking)	30	Mot	→MBRD 630CT	data
B631	BIFET	Vs:±15V Vu:75dB Vo:TTL Vi0:<15mV	6D	Hfo		data pinout
B 631 D	OP-IC	→TCA 331A	6-DIP	Hfo	→TCA 331A	data pinout
B635	BIFET	Vs:±15V Vu:75dB Vo:TTL Vi0:<15mV	6D	Hfo		data pinout
B 635 D	OP-IC	→TCA 335A	6-DIP	Hfo	→TCA 335A	data pinout
B 640 T	Si-Di	=MBRD 640CT (Typ-Code/Stempel/marking)	30	Mot	→MBRD 640CT	data
B 650 T	Si-Di	=MBRD 650CT (Typ-Code/Stempel/marking)	30	Mot	→MBRD 650CT	data
B 654 D	LIN-IC	Servomotor Controller	14-DIP	Hfo	SN 28654	
B 660 T	Si-Di	=MBRD 660CT (Typ-Code/Stempel/marking)	30	Mot	→MBRD 660CT	data
B 735	Si-Di	=MBR 735 (Typ-Code/Stempel/marking)	17	Mot, Gie	→MBR 735	
B 745	Si-Di	=MBR 745 (Typ-Code/Stempel/marking)	17	Mot, Gie	→MBR 745	
B 745	Si-Di	=MBRF 745 (Typ-Code/Stempel/marking)	17	Mot	→MBRF 745	
B761	BIFET	Vs:±18V Vu:81.5dB Vi0:<6mV	6DS	Hfo		data pinout
B 761 D	OP-IC	→TAA 761A	6-DIP	Hfo	→TAA 761A	data pinout
B 761 SC	OP-IC	=B 761D: SMD	8-MDIP	Hfo		data pinout
B765	BIFET	Vs:±18V Vu:81.5dB Vi0:<6mV	6DS	Hfo		data pinout
B 765 D	OP-IC	→TAA 765A	6-DIP	Hfo	→TAA 765A	data pinout
B 765 SG	OP-IC	=B 765D: SMD	8-MDIP	Hfo	-	data pinout
(SSi)B 0810...0880	Si-Di	→SSi B 0810...0880	31a	Sie		data
B 835 L	Si-Di	=MBRD 835L (Typ-Code/Stempel/marking)	30	Mot	→MBRD 835L	data
B861	BIFET	Vs:±10V Vu:75dB Vi0:<10mV	6DS	Hfo		data pinout
B 861 D	OP-IC	→TAA 861A	6-DIP	Hfo	→TAA 861A	data pinout
B 861 SC	OP-IC	=B 861D: SMD	8-MDIP	Hfo		data pinout
B865	BIFET	Vs:±10V Vu:75dB Vi0:<10mV	6DS	Hfo		data pinout
B 865 D	OP-IC	→TAA 865A	6-DIP	Hfo	→TAA 865	data pinout
B 865 SG	OP-IC	=B 865D: SMD	8-MDIP	Hfo	-	data pinout
B 1035	Si-Di	=MBR 1035 (Typ-Code/Stempel/marking)	17	Mot, Gie	→MBR 1035	data
B 1035 CL	Si-Di	=MBRD 1035CTL (Typ-Code/Stempel/markin	30	Ons	→MBRD 1035CTL	
B 1045	Si-Di	=MBR 1045 (Typ-Code/Stempel/marking)	17	Mot, Gie	→MBR 1045	
B 1045	Si-Di	=MBRF 1045 (Typ-Code/Stempel/marking)	17	Mot	→MBRF 1045	data
B 1060	Si-Di	=MBR 1060 (Typ-Code/Stempel/marking)	17	Mot, Gie	→MBR 1060	data
B1060	LDR	610nm Ro:0.3MOhm Vb:2V	TO8	Hei		data pinout
B 1070	Si-Di	=MBR 1070 (Typ-Code/Stempel/marking)	17	Mot	→MBR 1070	data
B 1080	Si-Di	=MBR 1080 (Typ-Code/Stempel/marking)	17	Mot	→MBR 1080	data

Type	Device	Short Description	Fig.	Manu	Comparision Types	More at
B 1085	Ge-P	TV-HA	23a§		AU 106, AU 109, AU 111, AU 112, 2N5325	
B 1090	Si-Di	=MBR 1090 (Typ-Code/Stempel/marking)	17	Mot	→MBR 1090	data
B 1100	Si-Di	=MBR 1100 (Typ-Code/Stempel/marking)	31	Mot	→MBR 1100	data
B 1535	Si-Di	=MBR 1535CT (Typ-Code/Stempel/marking)	17	Mot, Gie	→MBR 1535CT	data
B 1545	Si-Di	=MBR 1545CT (Typ-Code/Stempel/marking)	17	Mot, Gie	→MBR 1545CT	data
B 1545	Si-Di	=MBRF 1545CT(Typ-Code/Stempel/marking)	17	Mot	→MBRF 1545CT	data
B 1545 T	Si-Di	=MBRB 1545CT(Typ-Code/Stempel/marking)	30	Mot	→MBRB 1545CT	data
B 1635	Si-Di	=MBR 1635 (Typ-Code/Stempel/marking)	17	Mot, Gie	→MBR 1635	data
B 1645	Si-Di	=MBR 1645 (Typ-Code/Stempel/marking)	17	Mot, Gie	→MBR 1645	data
B 2015	Si-Di	=MBR 2015CTL(Typ-Code/Stempel/marking)	17	Mot	→MBR 2015CTL	data
B 2030	Si-Di	=MBR 2030CTL(Typ-Code/Stempel/marking)	17	Mot	→MBR 2030CTL	data
B 2035	Si-Di	=MBR 2035CT (Typ-Code/Stempel/marking)	17	Mot, Gie	→MBR 2035CT	data
B 2045	Si-Di	=MBR 2045CT (Typ-Code/Stempel/marking)	17	Mot, Gie	→MBR 2045CT	data
B 2045	Si-Di	=MBRF 2045CT(Typ-Code/Stempel/marking)	17	Mot	→MBRF 2045CT	data
B 2060	Si-Di	=MBR 2060CT (Typ-Code/Stempel/marking)	17	Mot	→MBR 2060CT	data
B 2060	Si-Di	=MBRF 2060CT(Typ-Code/Stempel/marking)	17	Mot	→MBRF 2060CT	data
B 2060 T	Si-Di	=MBRB 2060CT(Typ-Code/Stempel/marking)	30	Mot	→MBRB 2060CT	data
B 2070	Si-Di	=MBR 2070CT (Typ-Code/Stempel/marking)	17	Mot	→MBR 2070CT	data
B 2080	Si-Di	=MBR 2080CT (Typ-Code/Stempel/marking)	17	Mot	→MBR 2080CT	data
B 2090	Si-Di	=MBR 2090CT (Typ-Code/Stempel/marking)	17	Mot	→MBR 2090CT	data
B 2510 B	Si-Di	→SSi B 2510...2580		Sie		
B 2515 L	Si-Di	=MBR 2515L (Typ-Code/Stempel/marking)	17	Mot	→MBR 2515L	data
B 2515 L	Si-Di	=MBRB 2515L (Typ-Code/Stempel/marking)	30	Mot	→MBRB 2515L	data
B 2535	Si-Di	=MBR 2535CT (Typ-Code/Stempel/marking)	17	Mot, Gie	→MBR 2535CT	data
B 2535 L	Si-Di	=MBR 2535CTL(Typ-Code/Stempel/marking)	17	Mot	→MBR 2535CTL	data
B 2535 L	Si-Di	=MBRB2535CTL(Typ-Code/Stempel/marking)	30	Mot	→MBRB 2535CTL	data
B 2545	Si-Di	=MBR 2545CT (Typ-Code/Stempel/marking)	17	Mot	→MBR 2545CT	data
B 2545	Si-Di	=MBRF 2545CT(Typ-Code/Stempel/marking)	17	Mot	→MBRF 2545CT	data
B 2545 T	Si-Di	=MBRB 2545CT(Typ-Code/Stempel/marking)	30	Mot	→MBRB 2545CT	data
B 2600 DG	SMPS-IC	SMPS Controller	18-DIP	Hfo	-	
B2761	2xBIFET	Vs:±15V Vu:80dB Vi0:<6mV	8DS	Hfo		data pinout
B 2761 D	OP-IC	→TAA 2761A	8-DIP	Hfo	→TAA 2761A	data pinout
B 2761 SC	OP-IC	=B 2761D: SMD	8-MDIP	Hfo	-	data pinout
B2765	2xBIFET	Vs:±15V Vu:80dB Vi0:<6mV	8DS	Hfo		data pinout
B 2765 D	OP-IC	→TAA 2765A	8-DIP	Hfo	→TAA 2765A	data pinout
B 2765 S	OP-IC	=B 2765D: SMD		Hfo	-	data pinout
B 2765 SG	OP-IC	=B 2765D: SMD	8-MDIP	Hfo	-	data pinout
B 2960 VG	SMPS-IC	DC-DC Schaltregler/switching regulator	15-SQL	Hfo	L 296 ·	
B 3030	Si-Di	=MBRB 3030CT(Typ-Code/Stempel/marking)	30	Mot	→MBRB 3030CT	data
B 3030 CTL	Si-Di	=MBRB3030CTL(Typ-Code/Stempel/marking)	30	Ons	→MBRB 3030CTL	data
B 3035	Si-Di	=MBR 3035PT (Typ-Code/Stempel/marking)	18	Mot	→MBR 3035PT	data
B 3035	Si-Di	=MBR 3035WT (Typ-Code/Stempel/marking)	16	Mot	→MBR 3035WT	data
B 3040 DA	LIN-IC	Treiber-sensor-kombi/driver sensor IC	28-DIP	Hfo	-	
B 3045	Si-Di	=MBR 3045PT (Typ-Code/Stempel/marking)	18	Mot	→MBR 3045PT	data
B 3045	Si-Di	=MBR 3045ST (Typ-Code/Stempel/marking)	17	Mot	→MBR 3045ST	data
B 3045	Si-Di	=MBR 3045WT (Typ-Code/Stempel/marking)	16	Mot	→MBR 3045WT	data
B 3100	Si-Di	=MBR 3100 (Typ-Code/Stempel/marking)	31	Ons	→MBR 3100	
B 3170 V	Z-IC	→LM 317	17I	Hfo	LM 317T	pdf
B 3171 V	Z-IC	→LM 317: +1,2...+57V	17I	Hfo	-	pdf
B 3370 V	Z-IC	→LM 337	17n	Hfo	LM 337T	pdf
B 3371 V	Z-IC	→LM 337: -1,2...-47V	17n	Hfo	-	pdf
B 3718 VC	LIN-IC	Stepper-Motor-Strg./ctrl., 45V, ±1,5A	15-SQL	Hfo	TEA 3718SP	
B 3870 D	LIN-IC	Telecom, NF-schaltung/AF circuit	28-DIP	Hfo	-	
B 3925 DD	LIN-IC	Motor-prozessor f. Floppy-disk (fdd)	20-DIP	Hfo	-	
B 4002 D	LIN-IC	Endstufentreiber/driver f. power outp.	16-DIP	Hfo	UAA 4002DP	
B 4015 L	Si-Di	=MBR 4015LWT(Typ-Code/Stempel/marking)	16	Mot	→MBR 4015LWT	data
B 4030	Si-Di	=MBRB 4030 (Typ-Code/Stempel/marking)	30	Mot	→MBRB 4030	data
B 4045	Si-Di	=MBR 4045PT (Typ-Code/Stempel/marking)	18	Mot	→MBR 4045PT	data
B 4045	Si-Di	=MBR 4045WT (Typ-Code/Stempel/marking)	16	Mot	→MBR 4045WT	data
B 4211 D	LIN-IC	Motorregler/motor speed control		Hfo	U 211B	
B4761	4xBIFET	Vs:±15V Vu:80dB Vi0:<6mV	14D	Hfo		data pinout
B 4761 D	OP-IC	→TAA 4761A	14-DIP	Hfo	→TAA 4761A	data pinout
B4765	4xBIFET	Vs:±15V Vu:80dB Vi0:<6mV	14D	Hfo		data pinout
B 4765 D	OP-IC	→TAA 4765A	14-DIP	Hfo	→TAA 4765A	data pinout
B 5000	Si-N	NF/HF-Tr/E, 35V, 3A, 25W(Tc=100°)			-	data
B 5025 L	Si-Di	=MBR 5025L (Typ-Code/Stempel/marking)	18	Mot	→MBR 5025L	data
B 5973 D	LIN-IC	2A step down switching regulator	MDIP	Stm	-	pdf pinout
B 6045	Si-Di	=MBR 6045PT (Typ-Code/Stempel/marking)	18	Mot	→MBR 6045PT	data
B 6045	Si-Di	=MBR 6045WT (Typ-Code/Stempel/marking)	16	Mot	→MBR 6045WT	data
B 7240 X	LIN-IC	12 Bit-Stromquelle/current source		Hfo	-	
B9060	LDR	610nm Ro:0.6MOhm Vb:2V		Hei		data pinout
B 10100	Si-Di	=MBR 10100 (Typ-Code/Stempel/marking)	17	Mot	→MBR 10100	data
B 16100	Si-Di	=MBR 16100CT (Typ-Code/Stempel/marking)	17	Ons	→MBR 16100CT	data
B 20100	Si-Di	=MBR 20100CT (Typ-Code/Stempel/marking)	17	Mot	→MBR 20100CT	data
B 20100	Si-Di	=MBRF20100CT(Typ-Code/Stempel/marking)	17	Mot	→MBRF 20100CT	data
B 20100 T	Si-Di	=MBRB20100CT(Typ-Code/Stempel/marking)	30	Mot	→MBRB 20100CT	data
B 20200	Si-Di	=MBR 20200CT (Typ-Code/Stempel/marking)	17	Mot	→MBR 20200CT	data
B 20200	Si-Di	=MBRF20200CT(Typ-Code/Stempel/marking)	17	Mot	→MBRF 20200CT	data
B 20200 T	Si-Di	=MBRB20200CT(Typ-Code/Stempel/marking)	30	Mot	→MBRB 20200CT	data

Type	Device	Short Description	Fig.	Manu	Comparision Types	More at
BA						
BA	Si-Di	=1SS154 (Typ-Code/Stempel/marking)	35	Tos	→1SS154	data
BA	Si-P	=2SA1865 (Typ-Code/Stempel/marking)	35(1,6mm)	Say	→2SA1865	data
BA	Si-P	=2SB1118 (Typ-Code/Stempel/marking)	39	Say	→2SB1118	data
BA	Si-P	=2SB1132 (Typ-Code/Stempel/marking)	39	Rhm	→2SB1132	data
BA	Si-N	=2SD1367-BA (Typ-Code/Stempel/marking)	39	Hit	→2SD1367	data
BA	Si-N	=BCP 54 (Typ-Code/Stempel/marking)	48	Phi	→BCP 54	data
BA	Si-P	=BCW 61A (Typ-Code/Stempel/marking)	35	Aeg,Phi,Sie Fer	→BCW 61A	data
BA	Si-N	=BCX 54 (Typ-Code/Stempel/marking)	39	Sie,Val,Mot	→BCX 54	data
BA	Z-Di	BZX 399-C4V7(Typ-Code/Stempel/marking)	71(1,7mm)	Phi	→BZX 399-...	data
BA	Si-N/P+R	=KRX 101U (Typ-Code/Stempel/marking)	45(2mm)	Kec	→KRX 101U	data
BA	Si-N/P+R	=KRX 201U (Typ-Code/Stempel/marking)	46(2mm)	Kec	→KRX 201U	data
BA	MOS-N-FET-e	=MTDF 1N02HD(Typ-Code/Stempel/marking)	8-SSMDIP	Ons	→MTDF 1N02HD	
BA	MOS-N-FET-e	=MTDF 2N06HD(Typ-Code/Stempel/marking)	8-SSMDIP	Ons	→MTDF 2N06HD	
BA	Si-N+R	=XN 2216 (Typ-Code/Stempel/marking)	45	Mat	→XN 2216	data
BA	Si-Di	=ZMS 2800 (Typ-Code/Stempel/marking)	71(1,7mm)	Ztx	→ZMS 2800	data
BA	Si-N	=µPA500T (Typ-Code/Stempel/marking)	45	Nec	→µPA500T	data
BA	Si-N	=µPA570T (Typ-Code/Stempel/marking)	45(2mm)	Nec	→µPA570T	data
BA	Si-N/P	=µPA674T (Typ-Code/Stempel/marking)	46(2mm)	Nec	→µPA674T	data
BAp	Si-P	=BCW 61A (Typ-Code/Stempel/marking)	35	Phi	→BCW 61A	data
BAs	Si-P	=BCW 61A (Typ-Code/Stempel/marking)	35	Sie	→BCW 61A	data
BA-S200	LIN-IC	Universal Building Block	24-DIC	Rhm	-	
BA 1 A3Q..L4Z	Si-N+R	=AA 1A3Q..L4Z: Fig.→	40c	Nec	-	data
BA 2 A3Q	Si-N+R	hi-beta, Rb=1k, Rbe=10kΩ, 60V, 0,1A	40c	Nec	-	data
BA 2 A4M	Si-N+R	=BA 2A3Q: Rb=10k, Rbe=10kΩ	40c	Nec	-	data
BA 2 A4P	Si-N+R	=BA 2A3Q: Rb=10k, Rbe=47kΩ	40c	Nec	-	data
BA 2 A4Z	Si-N+R	=BA 2A3Q: Rb=10k, Rbe=-	40c	Nec	-	data
BA 2 F4M	Si-N+R	=BA 2A3Q: Rb=22k, Rbe=22kΩ	40c	Nec	-	data
BA 2 F4N	Si-N+R	=BA 2A3Q: Rb=22k, Rbe=47kΩ	40c	Nec	-	data
BA 2 F4Z	Si-N+R	=BA 2A3Q: Rb=22k, Rbe=-	40c	Nec	-	data
BA 2 L3M	Si-N+R	=BA 2A3Q: Rb=Rbe=4,7kΩ	40c	Nec	-	data
BA 2 L3N	Si-N+R	=BA 2A3Q: Rb=4,7k, Rbe=10kΩ	40c	Nec	-	data
BA 2 L3Z	Si-N+R	=BA 2A3Q: Rb=4,7k, Rbe=-	40c	Nec	-	data
BA 2 L4L	Si-N+R	=BA 2A3Q: Rb=47k, Rbe=22kΩ	40c	Nec	-	data
BA 2 L4M	Si-N+R	=BA 2A3Q: Rb=47k, Rbe=47kΩ	40c	Nec	-	data
BA 2 L4Z	Si-N+R	=BA 2A3Q: Rb=47k, Rbe=-	40c	Nec	-	data
BA 03 FP	Z-IC	Iso, lo-drop, +3V, 1A	30b	Rhm	-	pdf
BA 3 L4Z	Si-N+R	Rb=47kΩ, int. Emitter-Di, 30V, 20mA	40c	Nec	-	data
BA 03 ST	Z-IC	Iso, lo-drop, on/off +3V, 1A	86/5Pin	Rhm	-	pdf
BA 03 T	Z-IC	Iso, lo-drop, +3V, 1A	17b	Rhm	-	pdf
BA 05 FP	Z-IC	Iso, lo-drop, +5V, 1A	30b	Rhm	-	pdf
BA 05 ST	Z-IC	Iso, lo-drop, on/off +5V, 1A	86/5Pin	Rhm	-	pdf
BA 05 T	Z-IC	Iso, lo-drop, +5V, 1A	17b	Rhm	-	pdf
BA 06 FP	Z-IC	Iso, lo-drop, +6V, 1A	30b	Rhm	-	pdf
BA 06 ST	Z-IC	Iso, lo-drop, on/off +6V, 1A	86/5Pin	Rhm	-	pdf
BA 06 T	Z-IC	Iso, lo-drop, +6V, 1A	17b	Rhm	-	pdf
BA 07 FP	Z-IC	Iso, lo-drop, +7V, 1A	30b	Rhm	-	pdf
BA 07 ST	Z-IC	Iso, lo-drop, on/off +7V, 1A	86/5Pin	Rhm	-	pdf
BA 07 T	Z-IC	Iso, lo-drop, +7V, 1A	17b	Rhm	-	pdf
BA 08 FP	Z-IC	Iso, lo-drop, +8V, 1A	30b	Rhm	-	pdf
BA 08 ST	Z-IC	Iso, lo-drop, on/off +8V, 1A	86/5Pin	Rhm	-	pdf
BA 08 T	Z-IC	Iso, lo-drop, +8V, 1A	17b	Rhm	-	pdf
BA 09 FP	Z-IC	Iso, lo-drop, +9V, 1A	30b	Rhm	-	pdf
BA 09 ST	Z-IC	Iso, lo-drop, on/off +9V, 1A	86/5Pin	Rhm	-	pdf
BA 09 T	Z-IC	Iso, lo-drop, +9V, 1A	17b	Rhm	-	pdf
BA 10 FP	Z-IC	Iso, lo-drop, +10V, 1A	30b	Rhm	-	pdf
BA 10 ST	Z-IC	Iso, lo-drop, on/off +10V, 1A	86/5Pin	Rhm	-	pdf
BA 10 T	Z-IC	Iso, lo-drop, +10V, 1A	17b	Rhm	-	pdf
BA 12 FP	Z-IC	Iso, lo-drop, +12V, 1A	30b	Rhm	-	pdf
BA 12 ST	Z-IC	Iso, lo-drop, on/off +12V, 1A	86/5Pin	Rhm	-	pdf
BA 12 T	Z-IC	Iso, lo-drop, +12V, 1A	17b	Rhm	-	pdf
BA 15 FP	Z-IC	Iso, lo-drop, +15V, 1A	30b	Rhm	-	pdf
BA 15 ST	Z-IC	Iso, lo-drop, on/off +15V, 1A	86/5Pin	Rhm	-	pdf
BA 15 T	Z-IC	Iso, lo-drop, +15V, 1A	17b	Rhm	-	pdf
BA 31 W12ST	Z-IC	Iso, Dual, lo-drop, 7V/1A, 5V/0,5A	86/5Pin	Rhm	-	
BA 33 FP	Z-IC	Iso, lo-drop, +3,3V, 1A	30b	Rhm	-	
BA 33 ST	Z-IC	Iso, lo-drop, on/off +3,3V, 1A	86/5Pin	Rhm	-	
BA 33 T	Z-IC	Iso, lo-drop, +3,3V, 1A	17b	Rhm	-	
BA 41 W12ST	Z-IC	Iso, Dual, lo-drop, 8V/1A, 5V/0,5A	86/5Pin	Rhm	-	pdf
BA 51 W12ST	Z-IC	Iso, Dual, lo-drop, 9V/1A, 5V/0,5A	86/5Pin	Rhm	-	pdf
BA 61 W12ST	Z-IC	Iso, Dual, lo-drop, 10V/1A, 5V/0,5A	86/5Pin	Rhm	-	pdf
BA 100	Si-Di	Uni, 60V, 90mA	31a	Phi,Tix	BA 128, 1N4148, 1N5194...5196, ++	data
BA 101(A,B,C)	C-Di	VHF-AFC	31a	Aeg	BA 125, BB 119, 1SV114, 1SV125	data
BA 102/A..D	C-Di	VHF-AFC	31a	Phi,Tho	BA 125, BB 119, 1SV114, 1SV125	data
BA 103	Si-Di	Uni, 6V, 0,2A	2c	Sie	BA 127, BA 187...188, BA 215, 1N4148, ++	data
BA 104	Si-Di	Uni, 100V, 0,19A	2c	Sie	BA 188...190, BAY 73, BAY 19...21, ++	data
BA 105	Si-Di	Uni, 300V, 0,15A	2c	Sie	BA 147/300, BAY 21, BAY 46, BAY 88, ++	data
BA 108	Si-Di	Uni, 50V, 0,19A	2c	Sie	BA 127, BA 188...190, BA 215, 1N4148, ++	data

Type	Device	Short Description	Fig.	Manu	Comparision Types	More at
BA 109	C-Di	VHF-AFC	31a	Phi	BA 125, BB 119, 1SV114, 1SV125	data
BA 110	C-Di	FM/VHF-AFC	31a	Itt	BA 111, BA 124, 1S2790, 1SV50	data
BA 110 G	C-Di	FM/VHF-Tuning	31a	Itt	BB 109, BB 143, MV 109, 1SV50	data
BA 111	C-Di	VHF-AFC	31a	Aeg,Itt,Tho	BA 125, BB 119, 1SV114, 1SV125	data
BA 112	C-Di	AFC	31a	Itt	(BA 111, BA 124)	data
BA 113	C-Di	AFC, 1400pF/2V		Itt	-	data
BA 114	Si-St	Stabi, 0,58...0,8V/3mA	31a	Phi	BA 216, BA 314, BA 315, 1N4148, ++	data
BA 115	Si-Di	S, 150V, 2mA	31a	Phi	BA 195, BAY 80, BAV 20, 1N3070, ++	data
BA 116	Si-Di	Dual, TV-AGC, 20/50V, 10mA	2q	Aei	-	data
BA 117	Si-St	Stabi, 0,2A, 0,72V/1mA	31a	Sie	BA 220, BZV49/C0V8, 1N4148, ++	data
BA 119	C-Di	FM-AFC	31a	Sie	BA 111, BA 124, 1S2790, 1SV50	data
BA 120	C-Di	VHF-AFC	31a	Sie	BA 125, BB 119, 1SV114, 1SV 125	data
BA 121	C-Di	VHF/UHF-AFC	31a	Aeg	BB 117, BB 417, 1SV89	data
BA 122	Si-Di	Uni, 100V, 90mA	31a	Phi	BA 147/100, 1N5195, 1N5606, ++	data
BA 123	C-Di	TV-HA-AFC, 1600...2400pF/2V	34a§	Itt	-	data
BA 124/50...65	C-Di	FM/VHF-AFC	31a	Aeg	BA 111, 1S2790, 1SV50	data
BA 125/35...50	C-Di	VHF-AFC	31a	Aeg,Tho	BB 119, 1SV114, 1SV125, 1SV 145...146	data
BA 126	Si-Di	Dual, TV-HA-Diskr.	79z	Aei	-	data
BA 127(D)	Si-Di	Uni, 60V, 0,2A	31a	Sie,Tix	BA 188...190, BA 215, BAY 45, 1N4148, ++	data
BA 128	Si-Di	Uni, 75V, 0,11A	31a	Fch,Sgs,Tix	BA 127, BA 147/50, BA 187...190, 1N4148++	data
BA 129	Si-Di	Uni, 200V, 0,225A	31a	Fch,Sgs,Tix	BAY 20...21, BAY 46, BAY 88, BAW 52, ++	data
BA 130	Si-Di	FM/Vid-Dem, 30V, 75mA	31a	Fch,Sgs,Tix	(BA 128, BA 222, 1N4148, 1N5194, ++)	data
BA 131	Si-Di	S, 600V, 0,3A	31a	Sie	BA 199/550, BA 158...159, BY 204/8...10	data
BA 132	Si-Di	=BA 131: 800V	31a	Sie	BA 159, BY 204/8...10	data
BA 133(F)	Si-Di	=BA 131: 1000V	31a	Sie	BA 159, BY 204/10	data
BA 136(A)	C-Di	VHF-Band-S	31a	Phi,Sie,Tho	BA 243, BA 283, BA 483...484, ++	data
BA 137	Si-Di	S, 150V, 0,1A, <50ns	31a	Sie	BA 196...198, BAV 20...21, BAY 80, 1N3070+	data
BA 138	C-Di	FM/VHF-Tuning	31a	Phi,Sie,Tho	BB 109, BB 143, MV 109, 1SV50, ++	data
BA 139	C-Di	UHF-Tuning	31a	Phi,Sie	BB 105A/B, BB 205A/B, BB 405A/B, ++	data
BA 140	C-Di	FM/VHF-Tuning	31a	Sie	BB 109, BB 143, MV 109, 1SV50, ++	data
BA 141	C-Di	VHF/UHF-Tuning	31a	Itt	BB 105A/B, BB 205A/B, BB 405A/B, ++	data
BA 142	C-Di	FM/VHF-Tuning	31a	Itt	BB 109, BB 143, MV 109, 1SV50, ++	data
BA 143 U	C-Di	UHF-Band-S	31a	Itt	BA 244, BA 282, BA 284, BA 482	data
BA 143 V	C-Di	VHF-Band-S	31a	Itt	BA 243, BA 283, BA 483...484, ++	data
BA 144	Si-Di	50V, 50mA	31a	Phi	-	data
BA 145	Si-Di	S, TV-Clamping, 350V, 0,3A	31a	Phi,Mot	BA 157...159, BA 245, BA 248, BY 206...207	data
BA 147/...	Si-Di	Uni, 25...200V, 0,15A	31a	Aeg,Tix	BAY 17...21, BAY 86...88, BAY 44...46, ++	data
BA 148	Si-Di	S, TV-Clamping, 350V, 0,3A	31a	Mot,Phi,Tho	BA 157...159, BA 245, BA 248, BY 206...207	data
BA 149/...	C-Di	VHF/UHF-Tuning	31a	Aeg	BB 105A/B, BB 205A/B, BB 405A/B, ++	data
BA 150/...	C-Di	FM-Tuning	31a	Aeg,Phi,Tho	BB 110, BB 203, 1SV68, 1SV84	data
BA 151	Si-Di	Stabi, 0,6V/0,05mA	31a	Aei,Tix	BA 216, BA 314, BA 315, 1N4148, ++	data
BA 152(A,P,PR)	C-Di	VHF-Band-S	31a	Tho	BA 243, BA 283, BA 483...484; ++	data
BA 153	Si-Di	Dual, 40V, 25mA	7e	Tix	-	data
BA 154	Si-Di	Uni, 50V, 30mA	31a	Phi,Tix	BA 128, BA 222, BA 147/50, 1N4148, ++	data
BA 155	Si-Di	Uni, 150V, 0,1A	31a	Fch,Phi,Tix	BA 147/150, BA 189...190, 1N5606...07, ++	data
BA 156	Si-St	Stabi, 30mA, <0,7V/3mA	31a	Phi,Tho	BA 216, BA 314, BA 315, 1N4148, ++	data
BA 157(GP)	Si-Di	S, TV-Clamping, 400V, 0,4A, <300ns	31a	Aeg,Itt,Mot Phi,Tho,++	BY 204/4, BY 206, BY 406, BY 208/600, ++	data
BA 158(GP)	Si-Di	=BA 157: 600V	31a	=BA 157	BY 204/8, BY 207, BY 407, BY 208/600, ++	data
BA 159(GP)	Si-Di	=BA 157: 1000V	31a	=BA 157	BY 204/10, BY 208/1000, BY 268, ++	data
BA 159 SGP	Si-Di	=BA 157: 1300V	31a	Gie	BY 231/1400, BY 268...269	data
BA 159 XGP	Si-Di	=BA 157: 1500V	31a	Gie	BY 231/1500, BY 269	data
BA 160	Si-Di	=BA 157: 1600V	31a	Itt	BY 269	data
BA 161	C-Di	VHF/UHF-Tuning	31a	Itt	BB 105A/B, BB 205A/B, BB 405A/B, ++	data
BA 162	C-Di	FM/VHF-Tuning	31a	Itt	BB 109, BB 143, MV 109, 1SV50	data
BA 163	C-Di	AM-Tuning	31a	Itt	BB 112, BB 130, BB 509, ++	data
BA 164	Si-St	Stabi, 75mA, 0,85V/10mA	31a	Sgs	BA 216, BA 314, BA 315, 1N4148, ++	data
BA 165(A)	C-Di	VHF-Band-S	31a	Tho	BA 243, BA 283, BA 483, BA 484, ++	data
BA 166	Si-Di	Uni, 20V, 50mA	31a	Gen,Tix	BA 128, BA 222, BA 147/50, 1N4148, ++	data
BA 167	Si-Di	S, Uni, 25V, 50mA, <200ns	31a	Gen,Tix	BAX 88, BAX 94, BAY 93, 1N4148, ++	data
BA 168	Si-Di	S, 15V, 50mA, <4ns	31a	Aei,Tix	BA 217, BA 317, BAY 71, 1N4148, ++	data
BA 169	Si-Di	20V, 75mA	31a	Tho	-	data
BA 170	Si-Di	S, Uni, 20V, 0,15A, 250ns	31a	Itt,Gen,Tix	BAW 21, BAX 15...17, 1N4148, ++	data
BA 171	Si-Di	=BA 170: 30V, 100ns	31a	Itt	BAW 21, BAX 15...17, 1N4148, ++	data
BA 172	Si-Di	=BA 170: 50V, 100ns	31a	Itt	BAW 21, BAX 15...17, 1N4148, ++	data
BA 173	Si-Di	S, TV-Clamping, 350V, 0,3A, <500ns	31a	Aeg	BA 157...159, BY 204/4, BY 206, BY 406,++	data
BA 174	Si-Di	Min, S, 35V, 0,115A, <35ns	36c	Aeg	BAS 20, BAX 90, BAX 94, 1N4148, ++	data

Type	Device	Short Description	Fig.	Manu	Comparision Types	More at
BA 175	Si-Di	Min, S, Uni, 75V, 0,25A, <300ns	36c	Aeg	BAS 20, BAX 90, BAX 94, 1N4148, ++	data
BA 176	Si-Di	Antennenschutzdiode/aerial protection	31a	Aeg	(BA 157...159, BA 199/..., BY 208/..., ++)	data
BA 177	C-Di	VHF-Band-S	31a	Aeg,Tho	BA 243, BA 283, BA 483...BA 484, ++	data
BA 178	C-Di	VHF-Band-S	71a(4mm)	Aeg	BA 243, BA 283, BA 483, BA 484, ++	data
BA 178 M...FP	Z-IC	=BA 178M...T:	30b	Rhm	... 78Mxx... (TO-251/252)	
BA 178 M05T	Z-IC	Iso, +5V, 0,5A	17b	Rhm	... 78M05.. (TO-220 Iso)	pdf
BA 178 M06T	Z-IC	Iso, +6V, 0,5A	17b	Rhm	... 78M06...(TO-220 Iso)	pdf
BA 178 M07T	Z-IC	Iso, +7V, 0,5A	17b	Rhm	... 78M07...(TO-220 Iso)	pdf
BA 178 M08T	Z-IC	Iso, +8V, 0,5A	17b	Rhm	... 78M08...(TO-220 Iso)	pdf
BA*178 M09T	Z-IC	Iso, +9V, 0,5A	17b	Rhm	... 78M09...(TO-220 Iso)	pdf
BA 178 M10T	Z-IC	Iso, +10V, 0,5A	17b	Rhm	... 78M10...(TO-220 Iso)	pdf
BA 178 M12T	Z-IC	Iso, +12V, 0,5A	17b	Rhm	... 78M12...(TO-220 Iso)	pdf
BA 178 M15T	Z-IC	Iso, +15V, 0,5A	17b	Rhm	... 78M15...(TO-220 Iso)	pdf
BA 178 M18T	Z-IC	Iso, +18V, 0,5A	17b	Rhm	... 78M18...(TO-220 Iso)	pdf
BA 178 M20T	Z-IC	Iso, +20V, 0,5A	17b	Rhm	... 78M20...(TO-220 Iso)	pdf
BA 178 M24T	Z-IC	Iso, +24V, 0,5A	17b	Rhm	... 78M24...(TO-220 Iso)	pdf
BA 179	Si-Di	Dual, 50V, 25mA	7e	Tix		data
BA 180(a,b,c)	Si-Di	Uni, 10V, 50mA	31a	Fch,Gen,Tix	BA 128, BA 222, BA 147/25, 1N4148, ++	data
BA 181(A,B,C)	Si-Di	=BA 180: 20V	31a	Fch,Gen,Tix	BA 128, BA 222, BA 147/25, 1N4148	data
BA 182	C-Di	VHF-Band-S	71a(4mm)	Aeg,Phi,Tho	BA 243, BA 283, BA 483, BA 484, ++	data
BA 184	Si-Di	S, Uni, 350V, 0,2A, 700ns	31a	Tix	BA 157...159, BY 204/4, BY 206, BY 406,++	data
BA 185	Si-Di	=BA 184: 450V	31a	Tix	BA 157...159, BY 204/4, BY 207, BY 407,++	data
BA 186	Si-Di	=BA 184: 500V	31a	Tix	BA 159...159, BY 204/8, BY 207, BY 407,++	data
BA 187	Si-Di	Uni, 50V, 0,2A	31a	Tix	BAY 18, BAY 44, BAY 86, BA 157...159, ++	data
BA 188	Si-Di	=BA 187: 100V	31a	Tix	BAY 19, BAY 45, BAY 87, BA 157...159, ++	data
BA 189	Si-Di	=BA 187: 150V	31a	Tix	BAY 20, BAY 45, BAY 88, BA 157...159, ++	data
BA 190	Si-Di	=BA 187: 200V	31a	Tix	BAY 20, BAY 46, BAY 88, BA 157...159	data
BA 191	Si-Di	Schottky-Di, S, UHF-M, 25V, 0,1A	71a(4mm)	Aeg	BAT 42...43, BAT 47, BAT 85, 1SS293	data
BA 192	Si-Di	Uni, 50V, 0,4A	31a	Tix	BA 157...159, BA 199/250, BY 208/600, ++	data
BA 193	Si-Di	=BA 192: 100V	31a	Tix	BA 157...159, BA 199/250, BY 208/600, ++	data
BA 194	Si-Di	=BA 192: 150V	31a	Tix	BA 157...159, BA 199/250, BY 208/600, ++	data
BA 195	Si-Di	S, Uni, 200V, 0,15A, <50ns	31a	Tix	BA 197, BAV 20, BAW 50, 1N3070, 1SS83	data
BA 196	Si-Di	S, Uni, 150V, 0,25A, <50ns	31a	Tix	BAV 20...21, BAW 50, 1SS83	data
BA 197	Si-Di	=BA 196: 200V	31a	Tix	BAV 20...21, BAW 50, 1SS83	data
BA 198	Si-Di	=BA 196: 250V	31a	Tix	BAV 21, 1SS83	data
BA 199/...	Si-Di	S, Uni, 250...550V, 0,4A, <1µs	31a	Tho,Tix	BA 157...159, BY 204/..., BY 208/..., ++	data
BA 200	Si-Di	S, 45V, 0,15A, 4ns	31a	Itt	BA 204, BAT 13, BAY 95, 1N3600, 1N4148	data
BA 201	Si-Di	=BA 200: 70V	31a	Itt	BAW 46, BAW 62, BAY 61, BAX 95, 1N4148	data
BA 202	Si-Di	=BA 200: 100V	31a	Itt	BAW 46, BAW 62, BAY 61, BAX 95, 1N4148	data
BA 203	Si-Di	=BA 200: 150V	31a	Itt	BAV 14	data
BA 204	Si-Di	S, 60V, 0,2A, <10ns	31a	Aeg	BAT 13, BAW 62, BAY 95, 1N3600, 1N4148++	data
BA 205	LIN-IC	Verstärker, Puffer/amplifier, buffer	16-DIP	Rhm	-	
BA 206	C-Di	AFC	31a	Tix	BA 121, BB 117, BB 417	data
BA 207	C-Di	AFC	31a	Tix	BA 121, BB 117, BB 417	data
BA 208	C-Di	AFC	31a	Tix	BA 121, BB 117, BB 417	data
BA 209	Si-Di	=1N4148	31a	Tix	→1N4148	data
BA 210	Si-Di	=1N4149	31a	Tix	→1N4149	data
BA 210	LIN-IC	Modulator	16-DIP	Rhm	-	
BA 211	Si-Di	=1N4446	31a	Tix	→1N4446	data
BA 212	Si-Di	=1N4447	31a	Tix	→1N4447	data
BA 213	Si-Di	=1N4448	31a	Tix	→1N4448	data
BA 214	Si-Di	=1N4449	31a	Tix	→1N4449	data
BA 215	Si-Di	Uni, 60V, 0,2A	31c	Tix	BA 127, BA 188, BAY 73, 1N4148, ++	data
BA 215	LIN-IC	FM-ZF	16-DIP	Rhm	-	
BA 216	Si-St	S, Stabi, 75mA, 0;58...0,8V/3mA, <4ns	31a	Fch,Phi	BA 217...219, BA 315...317, 1N4148, ++	data
BA 217	Si-Di	S, 30V, 75mA, <4ns	31a	Fch,Phi,Tix	BA 317...318, BAX 13, BAX 91, 1N4148, ++	data
BA 218	Si-Di	S, 50V, 75mA, <4ns	31a	Fch,Phi,Tix	BA 318, BAX 13, BAX 91, BAY 38, 1N4148	data
BA 219	Si-Di	S, 100V, 0,1A, <4ns	31a	Fch,Phi,Tix	BAW 47, BAX 96, 1N4148, 1SS115	data
BA 220	Si-St	S, Stabi, 0,2A, 0,64...0,7V/5mA, <4ns	31a	Phi,Tix	BA 221, BAY 74, 1N4154, 1N4148, ++	data
BA 221	Si-Di	S, Uni, 30V, 0,2A, <4ns	31a	Phi,Tix	BAW 62, BAX 95, BAY 74, 1N4154, 1N4148	data
BA 222	Si-Di	Uni, 50V, 75mA	31a	Phi	BA 128, BA 147/50, 1N4148, 1N5194, ++	data
BA 222	LIN-IC	CR Timer, Ucc=4,5...16V	7-SIP	Rhm	-	
BA 223-10...-70	Si-Di	GI, 1000...7000V, 1A...0,15A	31a	Edl		data
BA 223	LIN-IC	CR Timer, Ucc=4,5...16V	8-SIP	Rhm	-	
BA 223	LIN-IC	CR Timer, Ucc=4,5...16V	8-SIP	Rhm	-	
BA 224/...	Si-Di	S, Uni, 150...300V, 0,15A, <40ns	31a	Tho	BA 196...198, BAV 20..21	data
BA 224	LIN-IC	Timer	8-SIP	Rhm	-	
BA 224 F	LIN-IC	=BA 224: SMD	8-MDIP			
BA 225	Si-Di	SMD, Dual, 50V, 50mA, 6ns	35m(2mm)	Fer		data
BA 225	LIN-IC	Vc, CR Timer, 2x Multivibrator	8-DIP	Rhm	-	
BA 225 F	LIN-IC	=BA 225: SMD	8-MDIP	Rhm	-	
BA 226	Si-Di	=BA 225[Fer]:	35l(2mm)	Fer	-	data

Type	Device	Short Description	Fig.	Manu	Comparision Types	More at
BA 226	LIN-IC	Vc, CR Timer, 2x Multivibrator	8-DIP	Rhm	-	
BA 226 F	LIN-IC	=BA 226: SMD	8-MDIP	Rhm	-	
BA 227	Si-Di	Dual, 40V, 75mA	2	Tix	-	data
BA 228	Si-Di	Dual, 40V, 75mA	2	Tix	-	data
BA 235	LIN-IC	=BA 225: Fig.→	9-SIP	Rhm	-	
BA 236	LIN-IC	=BA 226: Fig.→	9-SIP	Rhm	-	
BA 243(A,S)	C-Di	VHF-Band-S	31a	Itt,Phi,++	BA 182, BA 283, BA 483, BA 484, ++	data
BA 244(A,S)	C-Di	VHF/UHF-Band-S	31a	Itt,Phi,++	BA 282, BA 284, BA 482	data
BA 245	Si-Di	S, TV, 350V, 0,4A, <300ns	31a	Tix	BA 157...159, BY 204/4, BY 206, BY 406	data
BA 248	Si-Di	S, TV, 350V, 0,4A, <500ns	31a	Tix	BA 157...159, BY 204/4, BY 206, BY 406	data
BA 277	C-Di	SMD, VHF-Band-S, 35V, <1,2pF, 0,7Ω	71a(1,3mm)	Phi		data
BA 280	Si-Di	Schottky-Di, UHF-M, 4V, 30mA	71a(4mm)	Phi	BA 480...481, BAR 19, BAT 29, 1SS88,++	data
BA 281	Si-Di	Dem, 50V, 0,2A	31a	Phi	(BA 127, BA 187, BA 215, 1N4148, ++)	data
BA 282	C-Di	VHF/UHF-Band-S	31a	Itt,Sie,++	BA 244, BA 284, BA 482	data
BA 283	C-Di	VHF-Band-S	31a	Itt,Sie,++	BA 243, BA 483, BA 484	data
BA 284	C-Di	VHF/UHF-Band-S	31a	Sie	BA 244, BA 282, BA 482	data
BA 301	LIN-IC	Uni, NF-V, Ucc=12V	7-SIP	Rhm	-	
BA 302	LIN-IC	NF-E, 1,8W(6V/8kΩ)	7-SIP	Rhm	-	
BA 306	LIN-IC	NF-Treiber/driver	7-SIP	Rhm		
BA 307	LIN-IC	NF-Verstärker/amplifier	7-SIP	Rhm	-	
BA 308	LIN-IC	NF-V-ra	7-SIP	Rhm		pdf
BA 311	LIN-IC	NF-V	7-SIP	Rhm	(BA 312)	
BA 312	LIN-IC	NF-V	7-SIP	Rhm	(BA 311)	
BA 313	LIN-IC	Recorder, A/W-Verst., ALC, Ucc=3...12V	9-SIP	Rhm	(BA 314)	
BA 314(A)	Si-St	Stabi, 0,68...0,76V/1mA	31a	Phi,Tix	BZX 62, BZX 84/C0V8, BZX 97/0V8	data
BA 314	LIN-IC	~BA 313	9-SIP	Rhm	(BA 313)	pdf
BA 315	Si-St	Stabi, S, 0,1A, 0,71..0,79V/10mA, <4ns	31a	Phi	BA 316, BA 220, BAY 94, 1N4148, ++	data
BA 316	Si-Di	S, Uni, 10V, 0,1A, <4ns	31a	Phi	BA 220, BA 221, BAY 94, 1N4148, ++	data
BA 317	Si-Di	=BA 316: 30V	31a	Phi	BAY 71, BAY 74, 1N4154, 1N4148, ++	data
BA 318	Si-Di	=BA 316: 50V	31a	Phi	BAX 80, BAY 38, BAY 95, 1N4148, ++	data
BA 318	LIN-IC	NF-V, + VU-meter	7-SIP	Rhm		
BA 319	Si-Di	S, Uni, 75V, 0,25A, <50ns	31a	Phi	BA 196, BAV 19, BAY 43, 1N4148, ++	data
BA 320	Si-Di	=BA 319: 100V	31a	Phi	BA 196, BAV 19, BAW 50	data
BA 328	LIN-IC	2x NF-V-ra, Ucc=8(6...16)V	8-SIP	Rhm	KA 2221, LA 3161, M5152L, TA 7375P	pdf pinout
BA 328 F	LIN-IC	=BA 328: SMD	8-MDIP		-	pdf pinout
BA 329	LIN-IC	2x NF-V-ra	9-SIP	Rhm	-	
BA 333	LIN-IC	Recorder, A/W-Verst, ALC, Ucc=2,5...16V	9-SIP	Rhm	KA 2220, LA 3210, TA 7137	pinout
BA 335	LIN-IC	Recorder, Band-suchlauf/music finder	9-SIP	Rhm		pdf
BA 336	LIN-IC	Recorder, Band-suchlauf/music finder	9-SIP	Rhm	BA 338	pdf
BA 337	LIN-IC	Recorder, Auto-reverse Controller	9-SIP	Rhm		
BA 338	LIN-IC	=BA 336: verbessert/improved	9-SIP	Rhm		pdf
BA 338 L	LIN-IC	=BA 336: verbessert/improved	9-SQP	Rhm		pdf
BA 340	LIN-IC	NF-V	8-SIP	Rhm		
BA 343	LIN-IC	2x Recorder-V, ALC, Ucc=8V	16-DIP	Rhm		
BA 379	PIN-Di	VHF/UHF, 30V, 20mA	71a(3mm)	Phi,Sie,Tho	BA 382, MPN 3401, MPN 3402	data
BA 382	PIN-Di	VHF/UHF, VHF-Band-S, 40V	71a(4mm)	Mot	MPN 3401, MPN 3402	data
BA 389	PIN-Di	1MHz...1GHz, 30V, 50mA	31a	Sie	BA 479	data
BA 401	LIN-IC	TV-Ton-ZF, FM-ZF	5-SIP	Rhm		
BA 401 F	LIN-IC	=BA 401: SMD	8-MDIP	Rhm	-	
BA 402	LIN-IC	TV-Ton-ZF, FM-ZF	7-SIP	Rhm		
BA 403	LIN-IC	FM-ZF + Demod.	7-SIP	Rhm	KA 2245, LA 1150, TA 7130P, μPC 1028H	pinout
BA 404	LIN-IC	FM-ZF	9-SIP	Rhm	KA 2244, TA 7303	
BA 423(A)	C-Di	AM-Band-S, 20V, 50mA	31a	Phi	BA 223	data
BA 423 (A)L	C-Di	=BA 423	72a(3,4mm)	Phi	-	data
BA 479(A,G,S)	PIN-Di	10...1000MHz, 30V, 50mA	71a(4mm)	Aeg	BA 389	data
BA 480	Si-Di	Schottky-Di, UHF-M, 4V, 30mA	31a	Phi	BAR 19, BAT 29, 1S1925	data
BA 481	Si-Di	Schottky-Di, UHF-M, 4V, 30mA	31a	Phi	BAR 19, BAT 29, 1S1925	data
BA 482	C-Di	VHF-Band-S, 35V, 0,1A, <1,2pF, <0,7Ω	31a	Phi,Tho	BA 243, BA 283, BA 483...484,++	data
BA 483	C-Di	VHF-Band-S, 35V, 0,1A, <1pF, <1,2Ω	31a	Phi,Tho	BA 243, BA 283, BA 483...484,++	data
BA 484	C-Di	VHF-Band-S, 35V, 0,1A, <1,6pF, <1,2Ω	31a	Phi,Tho	BA 243, BA 283, BA 483...484,++	data
BA 501	LIN-IC	NF-E, 18V, 1,5A, 4W(13V/4Ω)	8-SIL	Rhm	-	
BA 505	LIN-IC	Nf-treiber/driver	14-DIP+g	Rhm		
BA 505 G	Si-Di	VHF/CATV-Tuning		Sie		data
BA 511(A)	LIN-IC	NF-E, 4,5W(13V/4Ω)	10-SIL	Rhm	BA 532	
BA 514	LIN-IC	NF-V+E, 2W(9V/4Ω)	8-SIL	Rhm		
BA 515	LIN-IC	NF-E, 15V, 0,23W(3V/4Ω)	12-SQP	Rhm		
BA 516	LIN-IC	NF-E, 0,35W(9V/8Ω)	9-SIP	Rhm	BA 526, BA 527	
BA 518	LIN-IC	NF-E, 1,5W(12V/8Ω)	8-SIL	Rhm	BA 547	pdf
BA 521(S)	LIN-IC	NF-E, 5,8W(13V/4Ω)	10-SIL	Rhm	BA 532	
BA 524	LIN-IC	NF-E, 3,8W(12V/4Ω)	10-SIL	Rhm	-	
BA 526	LIN-IC	NF-E, 0,43W(6V/8Ω)	9-SIP	Rhm	BA 527	pdf
BA 527	LIN-IC	NF-E, 0,8W(6V/4Ω)	9-SIP	Rhm	KIA 6278S	pdf
BA 532	LIN-IC	NF-E, 5,8W(13V/4Ω)	10-SIL	Rhm		
BA 534	LIN-IC	NF-E, 2,3W(9V/4Ω)	10-SIL	Rhm	BA 524	pdf
BA 535	LIN-IC	2x NF-E, 20V, 2x4,8W(12V/4Ω)	12-SILP	Rhm	(BA 536)	pdf
BA 536	LIN-IC	2x NF-E, 2x4,5W(12V/4Ω)	12-SILP	Rhm	BA 5402(A)	
BA 546	LIN-IC	NF-E, 12V, 0,33W(6V/8Ω)	9-SIP	Rhm	BA 527	pdf

Type	Device	Short Description	Fig.	Manu	Comparision Types	More at
BA 547	LIN-IC	NF-E, 1,5W(12V/8Ω)	8-SIL	Rhm	BA 518	pdf
BA 555	LIN-IC	=NE 555	8-DIP	Rhm	→NE 555	
BA 567	LIN-IC	=LM 567CN	8-DIP	Rhm	→LM 567CN	pinout
BA 579 A	PIN-Di	SMD, Dual, VHF/UHF, 30V, 20mA	35m	Tho	-	data
BA 579 C(K)	PIN-Di	=BA 579A:	35l		-	data
BA 579 S	PIN-Di	=BA 579A:	35o		-	data
BA 582	C-Di	SMD, TV, VC, VHF-Band-S	71a(2,7mm)	Phi,Sie,Mot	1SS241	data
BA 585	PIN-Di	SMD, Abschw./attenu., 50V, 1..2000MHz SATV Ant. Switch	71a(2,7mm)	Sie	-	data
BA 586	PIN-Di	SMD, Abschw./attenuator, 50V, >1MHz	71a(2,7mm)	Sie	-	data
BA 591	C-Di	SMD, VHF-Band-S, 35V, <0,9pF, <0,7Ω	71a(1,7mm)	Phi	BA 592, MA 77, 1SS314, 1SS356	data
BA 592	C-Di	=BA 582:	71a(1,7mm)	Sie	BA 591, MA 77, 1SS314, 1SS356	data
BA 595	PIN-Di	=BA 585:	71a(1,7mm)	Sie	RN 731V	data
BA 596	PIN-Di	=BA 586:	71a(1,7mm)	Sie	-	data
BA 597	PIN-Di	SMD, Abschw./attenu., 50V, >10MHz	71a(1,7mm)	Sie	-	data
BA 604	Si-Di	SMD, Uni, S, 50V, 0,2A, <20ns	72a(3,4mm)	Aeg	-	data
BA 612	LIN-IC	5x PNP Darl.-Trans., 26V, 0,4A	14-DIP	Rhm	-	pdf
BA 614(A)	LIN-IC	6x PNP Darl.-Trans., 20V, 0,1A	14-DIP	Rhm	-	
BA 618	LIN-IC	7x LED+LCD-Tr, Ucc=10V, 0,1A	16-DIP	Rhm	-	pdf
BA 621	LIN-IC	Analog-schalter/switch	5-SIP	Rhm	-	
BA 631	LIN-IC	FM-intercom	14-DIP	Rhm	-	
BA 634	LIN-IC	T-flipflop + Reset	5-SIP	Rhm	-	
BA 634 F	LIN-IC	=BA 634: SMD	8-MDIP		-	
BA 653	LIN-IC	Berührungsschalter/touch sensor	16-DIP	Rhm	-	
BA 656	LIN-IC	LED-Linear-Meter, 5 LED	9-SIP	Rhm	(KA 2286, LB 1413)	
BA 658	LIN-IC	12P.-VU-Meter-Tr.(FLT display)	16-DIP	Rhm	-	pdf pinout
BA 664	LIN-IC	6x Uni-Tr, Ucc=20V, 0,1A	14-DIP	Rhm	-	
BA 668(A)	LIN-IC	12P.-VU-Meter-Tr.(FLT display)	18-DIP	Rhm	-	
BA 679(S)	PIN-Di	=BA 479(A,G,S): SMD	72a(3,4mm)	Aeg	BA 979	data
BA 681 A	LIN-IC	LED Linear Meter, 12 LED, Ucc=12V	18-DIP	Rhm	-	
BA 682	C-Di	=BA 282: SMD	72a(3,4mm)	Itt,Sie,++	BA 982	data
BA 682 A	LIN-IC	LED VU-Meter, 12 LED, Ucc=12V	18-DIP	Rhm	-	pdf
BA 682 AF	LIN-IC	=BA 682A: SMD	18-MDIP	Rhm	-	pdf
BA 683	C-Di	=BA 283: SMD	72a(3,4mm)	Itt,Sie,++	BA 983	data
BA 683 A	LIN-IC	LED Meter Driver, 12 LED, Ucc=12V	18-DIP	Rhm	-	pdf pinout
BA 684 A	LIN-IC	LED Linear-Meter, 8 LED	16-DIP	Rhm	-	
BA 685	LIN-IC	LED Linear-Meter, 5 LED	16-DIP	Rhm	-	
BA 689	LIN-IC	LED Linear Meter, 12 LED, Ucc=12V	18-DIP	Rhm	-	
BA 695	LIN-IC	AM/FM LED/LCD-Meter-Tr, 3 LED	9-SIP	Rhm	-	
BA704	R+	Io=10mA Vo:2.4..2.9V Vin:5.5V	3N	Roh	-	data pinout
BA 704	Z-IC	2,65V, 1,5mA	7b	Rhm	-	data pinout
BA707	R+	Io=10mA Vo:3...3.6V Vin:5.5V P:250mW	3N	Roh	-	data pinout
BA 707	Z-IC	3,3V, 1,8mA	7b	Rhm	-	data pinout
BA714	R+	Io=300µ Vo:3.05...3.55V Vin:4...7V	3l	Roh	-	data pinout
BA 714	Z-IC	3,3V, 50...250µA	9d	Rhm	-	data piriout
BA 715	OP-IC	Dual, Serie 158	9-SIP	Rhm	→Serie 158	data
BA718	2xOP	Vs:18V Vu:100dB Vo:±4.5V Vi0:2mV	9l	Roh	-	data pinout
BA 718	OP-IC	Dual	9-SIP	Rhm	-	data pinout
BA728	2xOP	Vs:18V Vu:100dB Vo:±4.5V Vi0:2mV	8DS	Roh	-	data pdf pinout
BA 728	OP-IC	Dual, lo-power, ±9V, -20..+75°	8-DIP	Rhm	-	data pinout
BA 728 F	OP-IC	=BA 728: SMD	8-MDIP		-	data pinout
BA 728 N	OP-IC	=BA 728: Fig.→	8-SIP		-	pdf
BA 735 F	OP-IC	Dual	8-MDIP	Rhm	-	
BA 779(S)	PIN-Di	=BA 479G(S): SMD	35e	Aeg	-	data
BA 779-2	PIN-Di	=BA 479(S): SMD, Dual	35o		-	data
BA 782	C-Di	SMD, VHF/UHF-Band-S, 35V, <1,25pF(3V) <0,7Ω(3mA, 50...1000MHz)	71a(2,7mm)	Gsi	-	data
BA 782 S	C-Di	=BA 782:	71a(1,7mm)		-	data
BA 783	C-Di	SMD, VHF/UHF-Band-S, 35V, <1,2pF(3V) <1,2Ω(3mA, 50...1000MHz)	71a(2,7mm)	Gsi	-	data
BA 783 S	C-Di	=BA 783:	71a(1,7mm)		-	data
BA 792	C-Di	SMD, VHF-Band-S, <1,1pF(3V), <0,7Ω	71a(2mm)	Phi	BA 592, HSU 277, MA 77, 1SS314, 1SS356	data
BA 802	LIN-IC	PLL Motor Controller	5-SIP	Rhm	-	
BA 806	LIN-IC	PLL Motor Controller	5-SIP	Rhm	-	
BA 820	LIN-IC	8 Bit Serial In Parallel Out Driver	16-DIP	Rhm	-	
BA 823	LIN-IC	8 Bit Serial In Parallel Out Driver	16-DIP	Rhm	-	pdf
BA 823 F	LIN-IC	=BA 823: SMD	16-MDIP		-	pdf
BA 829	LIN-IC	8 Bit Serial In Parallel Out Driver	18-DIP	Rhm	-	pdf
BA 841	LIN-IC	VC-Servo		Rhm	-	
BA 842	LIN-IC	Recordersteuerung/tape deck controller	40-DIP	Rhm	-	
BA 843	LIN-IC	Recorder, Tastenstrg./key Controller	16-DIP	Rhm	-	
BA 843 F	LIN-IC	=BA 843: SMD	16-MDIP		-	
BA 845	LIN-IC	Recorderspeicher/memory	24-DIP	Rhm	-	
BA 847	LIN-IC	VC-Servo, Bildschnitt/sp. playback	22-DIP	Rhm	-	
BA 852	LIN-IC	Vc, Servo Ctrl., FG System	16-DIP	Rhm	-	
BA 855(A)	LIN-IC	Vc, Servo, Special Playback(fine Slow)	20-DIP	Rhm	-	
BA 855 F,AF	LIN-IC	=BA 855: SMD	20-MDIP	Rhm	-	
BA 856	LIN-IC	Vc,Servo, Special Playback(fine Still)	16-DIP	Rhm	-	
BA 857	LIN-IC	Vc, Servo, Special Playback(fine Slow)	18-DIP	Rhm	-	

Type	Device	Short Description	Fig.	Manu	Comparision Types	More at
BA 857 F	LIN-IC	=BA 857: SMD	18-MDIP		-	
BA 860	LIN-IC	Vc, Servo, Motorregler/Speed Ctrl.	16-DIP	Rhm	-	pdf
BA 862	LIN-IC	Vc, Servo, Special Playback(fine Slow)	18-DIP	Rhm	-	
BA 866 F	LIN-IC	Vc, Servo, Special Playback(fine Slow)	18-MDIP	Rhm	-	
BA 867	LIN-IC	Vc, Servo, Special Playback(fine Slow)	18-DIP	Rhm	-	
BA 873	LIN-IC	Vc, Reel Sensor, Ucc=4,2...12V	16-DIP	Rhm	-	
BA 875	LIN-IC	Vc, Servo, Special Playback(fine Slow)	28-DIP	Rhm	-	
BA 875 F	LIN-IC	=BA 875: SMD	28-MDIP	Rhm	-	
BA 877 LS	LIN-IC	Vc, Servo, Special Playback(fine Slow)	24-SQP	Rhm	-	
BA 885	PIN-Di	=BA 585:	35e	Sie	-	data
BA 886	PIN-Di	=BA 586:	35e	Sie	-	data
BA 891	C-Di	SMD, VHF-Band-S, 35V, <0,9pF, 0,7Ω	71a(1,3mm)	Phi	-	data
BA 892	C-Di	SMD, VHF-Band-S, 35V, 0,6...1,1pF(3V)	71a(1,3mm)	Sie	-	data
BA 895	PIN-Di	=BA 585:	71a(1,6mm)	Sie	-	
BA 979(S)	PIN-Di	=BA 479G(S): SMD	72a(3,5mm)	Aeg	-	data
BA 980	PIN-Di	SMD,HF-Abschw./Attenuator, 10...1000MHz	72a(3,5mm)	Aeg	-	
BA 982	C-Di	=BA 282: SMD	72a(3,5mm)	Aeg	BA 682	data
BA 983	C-Di	=BA 283: SMD	72a(3,5mm)	Aeg	BA 683	data
BA 1282	C-Di	=BA 282: SMD	72a(1,9mm)	Aeg	-	data
BA 1283	C-Di	=BA 283: SMD	72a(1,9mm)	Aeg	-	data
BA 1310	LIN-IC	Stereo-decoder	14-DIP	Rhm	AN 115	pdf
BA 1310 F	LIN-IC	=BA 1310: SMD	14-MDIP		-	pdf
BA 1320	LIN-IC	FM-Multiplex Stereo-Decoder, Ucc=12V	16-DIP	Rhm	AN 363, µPC 1320	pdf pinout
BA 1320 F	LIN-IC	=BA 1320: SMD	16-MDIP		-	pdf
BA 1330	LIN-IC	=BA 1320: Ucc=3,6...6V	16-DIP	Rhm	AN7410, HA11227, KA2261, LA3361, TA7604	pdf
BA 1330 F	LIN-IC	=BA 1330: SMD	16-MDIP		-	pdf
BA 1332	LIN-IC	FM-Multiplex Stereo-Decoder,Ucc=3...14V	16-DIP	Rhm	BA 1320, BA 1330	pdf
BA 1332 F	LIN-IC	=BA 1332: SMD	16-MDIP		-	pdf
BA 1332 L	LIN-IC	=BA 1332: Fig.→	16-SQP		-	pdf
BA 1335	LIN-IC	FM-MPX Stereo-Decoder, Ucc=3,3...9V	16-DIP	Rhm	-	
BA 1335 F	LIN-IC	=BA 1335: SMD	16-MDIP		-	
BA 1350	LIN-IC	FM-MPX Stereo-Decoder, Ucc=6...12V	16-SQP	Rhm	-	
BA 1350 F	LIN-IC	=BA 1350: SMD	16-MDIP		-	
BA 1351	LIN-IC	=BA 1350: Fig.→	16-DIP	Rhm	-	
BA 1355	LIN-IC	FM-MPX Stereo-Decoder, Ucc=5...12V	16-DIP	Rhm	-	
BA 1355 F	LIN-IC	=BA 1355: SMD	16-MDIP		-	
BA 1356	LIN-IC	FM-MPX Stereo-Decoder, Ucc=9V	16-SQP	Rhm	-	
BA 1356 F	LIN-IC	=BA 1356: SMD	16-MDIP		-	
BA 1360	LIN-IC	FM-MPX Stereo-Decoder, Ucc=1,8...3V	16-SQP	Rhm	-	
BA 1360 F	LIN-IC	=BA 1360: SMD	16-MDIP		-	
BA 1362 F	LIN-IC	SMD, FM-MPX Stereo-Decoder, Ucc=1,5V	16-MDIP	Rhm	-	
BA 1362 FS	LIN-IC	=BA 1362F:	16-SMDIP	Rhm	-	
BA 1402	LIN-IC	FM Stereo-Radio, Ucc=1,5V	22-DIP	Rhm	-	
BA 1402 F	LIN-IC	=BA 1402: SMD	22-MDIP		-	
BA 1404	LIN-IC	FM Stereo-Sender/transmitter,Ucc=1,25V	18-DIP	Rhm	-	pdf
BA 1404 F	LIN-IC	=BA 1404: SMD	18-MDIP		-	pdf
BA 1405 F	LIN-IC	SMD, FM Stereo-Modulator, Ucc=4...6V	18-MDIP	Rhm	-	
BA 1407 AF	LIN-IC	=BA 1407AL: SMD	20-MDIP	Rhm	-	
BA 1407 AL	LIN-IC	TV, Multiplex-dem., Japan-standard	18-SQP	Rhm	-	
BA 1440	LIN-IC	AM/FM-ZF, FM Stereo-Dec., Ucc=3,5...7V	18-DIP	Rhm	-	
BA 1441	LIN-IC	AM/FM-ZF, FM-Stereo-Dec., Ucc=3,5...7V	18-DIP	Rhm	-	
BA 1442 A	LIN-IC	AM-Radio, FM-ZF-stereo-system	20-DIP	Rhm	-	pdf
BA 1443 A	LIN-IC	AM-Radio, Fm-zf-stereo-system	20-DIP	Rhm	-	
BA 1448 S	LIN-IC	AM-Radio, Fm-zf-stereo-system	24-SDIP	Rhm	-	pdf
BA 1449 F	LIN-IC	SMD, AM-Radio, Fm-zf-stereo-system	24-MDIP	Rhm	-	pdf
BA 1450 S	LIN-IC	SMD, AM-Radio, Fm-zf-stereo-system	24-SDIP	Rhm	-	pdf
BA 1602 L	LIN-IC	FM Intercom	18-SQP	Rhm	-	pdf
BA 1604	LIN-IC	Telecom, PLL-Ton-/tone-decoder	8-DIP	Rhm	→LM 567CN	pdf
BA 1604 F	LIN-IC	=BA 1604: SMD	8-MDIP		-	pdf
BA 1610	LIN-IC	Telecom, FSK Linear Modem	20-DIP	Rhm	-	
BA 1701	LIN-IC	VSC(Voice Compression System)prozessor	16-DIP	Rhm	-	pdf
BA 2266	LIN-IC	Radio-servo	18-DIP	Rhm	-	
BA 3112	LIN-IC	2x NF-V, Uni, Ucc=6...16V, Gv=10dB	8-SIP	Rhm	BA 3113, BA 3114, BA 3116	
BA 3113	LIN-IC	=BA 3112: Gv=15dB	8-SIP	Rhm	BA 3114, BA 3116	
BA 3114	LIN-IC	=BA 3112: Gp=20dB	8-SIP	Rhm	BA 3116	
BA 3116	LIN-IC	=BA 3112: Gv=30dB	8-SIP	Rhm	-	
BA 3118 L	LIN-IC	2x NF-Vst./amp., Ucc=4...16V, 6...20dB	18-SQP	Rhm	-	pdf
BA 3120	LIN-IC	Massefrei Verst./ground insul. Amp.	8-DIP	Rhm	-	
BA 3120 F	LIN-IC	=BA 3120: SMD	8-MDIP		-	
BA 3120 N	LIN-IC	=BA 3120: Fig.→	8-SIP		-	
BA 3121	LIN-IC	Massefrei Verst./ground insul. Amp.	8-DIP	Rhm	-	pdf
BA 3121 F	LIN-IC	=BA 3121: SMD	8-MDIP		-	pdf
BA 3121 N	LIN-IC	=BA 3121: Fig.→	8-SIP		-	pdf
BA 3124 F	LIN-IC	SMD, Cd, Line Output Muting, 4 Chan.	8-MDIP	Rhm	-	
BA 3126 F	LIN-IC	=BA 3126: SMD	14-MDIP		-	pdf
BA 3126 N	LIN-IC	2x Recorder Kopfumsch./head switch	9-SIP	Rhm	-	pdf
BA 3128 F	OP-IC	=BA 3128N: SMD	8-MDIP		-	
BA 3128 N	OP-IC	Audio Switched,±18V, 6,5MHz, -20...+75°	8-SIP	Rhm	-	
BA 3129	OP-IC	Dual, Audio Switched,±18V, -20...+75°	14-DIP	Rhm	-	
BA 3129 F	OP-IC	=BA 3129: SMD	14-MDIP		-	
BA 3131 FS	OP-IC	SMD, Triple, Audio Switched, ±18V	20-SMDIP	Rhm	-	pdf
BA3302	OP	Vs:14V Vu:90dB	8I	Roh	-	data pinout

Type	Device	Short Description	Fig.	Manu	Comparision Types	More at
BA 3302	LIN-IC	2x NF-V	8-SIP	Rhm	-	
BA 3304	LIN-IC	2x NF-V, Ucc=3V	16-SQP	Rhm	-	
BA 3304 F	LIN-IC	=BA 3304: SMD	16-MDIP		-	
BA 3306	LIN-IC	2x NF-V, ALC, Ucc=4,5...14V	9-SIP	Rhm	-	pdf
BA 3308	LIN-IC	2x NF-V, ALC, Ucc=4,5...14V	9-SIP	Rhm	KA 22241	pdf pinout
BA 3308 F	LIN-IC	=BA 3308: SMD	14-MDIP		-	pdf
BA 3310 N	LIN-IC	2x NF-V, ALC, Ucc=4...12V	10-SIP	Rhm	KA 22241	
BA 3311 L	LIN-IC	2x NF-V, ALC, Ucc=5...12V	12-SQP	Rhm	-	pdf
BA 3312 N	LIN-IC	2x NF-V, ALC, Ucc=4...12V	10-SIP	Rhm	KA 22242	
BA 3313 L	LIN-IC	2x NF-V, ALC-Detector, Ucc=5...12V	12-SQP	Rhm	-	pdf
BA 3314 F	LIN-IC	SMD, 2x NF-V, ALC-Detector, Ucc=5...12V	14-MDIP	Rhm	-	pdf
BA 3402	LIN-IC	2x Recorder-V f. Autoreverse, Ucc=8V	16-SQP	Rhm	-	
BA 3402 F	LIN-IC	=BA 3402: SMD	16-MDIP		-	
BA 3404 F	LIN-IC	2x Recorder-V f. Autoreverse, Ucc=3V	16-MDIP	Rhm	-	
BA 3404 FS	LIN-IC	=BA 3404F: 4,4 x 6,6mm	16-SMDIP		-	
BA 3404 L	LIN-IC	=BA 3404F: Fig.→	16-SQP		-	
BA 3406 AF	LIN-IC	=BA 3406AL: SMD	16-MDIP		-	
BA 3406 AL	LIN-IC	2x Recorder-V, Muting, Ucc=6...14V	16-SQP	Rhm	-	
BA 3408	LIN-IC	2x Recorder-V f. Autoreverse, Ucc=8V	16-DIP	Rhm	-	
BA 3408 F	LIN-IC	=BA 3408: SMD	16-MDIP		-	
BA 3410 AF	LIN-IC	SMD, Recorder, A/W-Verst., Ucc=3V	16-MDIP	Rhm	-	
BA 3412 K	LIN-IC	SMD, Recorder, 2x A/W-Verst., Ucc=3V	32-MP	Rhm	-	
BA 3413 F	LIN-IC	2x Recorder-V f. Autoreverse, Ucc=1,5V	16-MDIP	Rhm	-	
BA 3413 FS	LIN-IC	=BA 3413: 4,4 x 6,6mm	16-SMDIP			
BA 3414 L	LIN-IC	Transceiver Mikrofon-verst./amp.	16-SQP	Rhm	-	
BA 3416 BL	LIN-IC	2x Recorder-V, Ucc=3,5...12V	18-SQP	Rhm	-	pdf
BA 3420 AL	LIN-IC	Recorder, 2x A/W-Verst., Ucc=5...16V	18-SQP	Rhm	-	pdf
BA 3422 S	LIN-IC	Dual Recorder, A/W-Verst., Ucc=5...9V	32-SDIP	Rhm	-	
BA 3423 S	LIN-IC	Recorder, 2x A/W-Verst., Ucc=4,5...7V	32-SDIP	Rhm	-	pdf
BA 3424 F	LIN-IC	=BA 3424S: SMD	20-MDIP	Rhm	-	
BA 3424 FS	LIN-IC	=BA 3424S: SMD, 8,7 x 5,4mm	20-SMDIP	Rhm	-	
BA 3424 S	LIN-IC	2x Recorder-V f. Autoreverse, Ucc=9V	22-SDIP	Rhm	-	
BA 3426 S	LIN-IC	Double Recorder, System Preamp.,Ucc=9V	32-SDIP	Rhm	-	pdf
BA 3430 F	LIN-IC	=BA 3430S: SMD	24-MDIP		-	pdf
BA 3430 FS	LIN-IC	=BA 3430S: SMD	24-SMDIP		-	pdf
BA 3430 S	LIN-IC	2x Recorder-V, AMS, Ucc=7...18V	24-SDIP	Rhm	-	pdf
BA 3502 F	LIN-IC	SMD, 2x Recorder-V+E, Ucc=3V	24-MDIP	Rhm	KA 22131	
BA 3503 F	LIN-IC	SMD, 2x Recorder-V+E, Ucc=3V	24-MDIP	Rhm	-	
BA 3504 F	LIN-IC	SMD, 2x Recorder-V+E, Ucc=3V	24-MDIP	Rhm	-	
BA 3505 F	LIN-IC	SMD, 2x Recorder-V+E, Ucc=3V	24-MDIP	Rhm	-	
BA 3506 A	LIN-IC	2x Recorder-V+E, Ucc=3V	16-DIP	Rhm	-	
BA 3506 AF	LIN-IC	=BA 3506A: SMD	18-MDIP		-	
BA 3513 AF	LIN-IC	SMD, 2x Recorder-V+E, Ucc=3V	24-MDIP	Rhm	-	
BA 3513 AFS	LIN-IC	=BA 3513AF: 10 x 5,4mm	24-SMDIP		-	pdf
BA 3514 AF	LIN-IC	SMD, 2x Recorder-V+E, Ucc=3V	24-MDIP	Rhm	-	
BA 3516	LIN-IC	2x Recorder-V+E, Ucc=3V	16-DIP	Rhm	-	
BA 3516 F	LIN-IC	=BA 3516: SMD	18-MDIP		-	
BA 3518	LIN-IC	2x Recorder-V+E, Ucc=3V	16-DIP	Rhm	-	
BA 3518 F	LIN-IC	=BA 3518: SMD	16-MDIP		-	
BA 3519 F	LIN-IC	SMD, 2x Recorder-V+E, Ucc=3V	22-MDIP	Rhm	-	
BA 3519 FS	LIN-IC	=BA 3519: SMD, 10 x 5,4mm	24-SMDIP		-	
BA 3520	LIN-IC	2x Recorder-V+E, Ucc=3V	18-DIP	Rhm	-	
BA 3520 F	LIN-IC	=BA 3520: SMD	18-MDIP		-	
BA 3521	LIN-IC	2x Recorder-V+E, Ucc=3V	18-DIP	Rhm	-	
BA 3528 FP,AFP	LIN-IC	2x Recorder-V+E, Motor Ctrl., Ucc=3V	28-MDIP+b	Rhm	-	pdf
BA 3529 FP,AFP	LIN-IC	2x Recorder-V+E, Motor Ctrl., Ucc=3V	28-SMDIP+b	Rhm	-	pdf
BA 3558 K	LIN-IC	SMD, Cd, Endverst./post Amp., Filter	32-MP	Rhm	-	
BA 3570 F	LIN-IC	SMD, Kopfh.-verst./headphone Amp.	22-MDIP	Rhm	-	pdf
BA 3570 FS	LIN-IC	=BA 3570F: SMD	24-SMDIP		-	pdf
BA 3571 F	LIN-IC	SMD, Cd, Kopfh.-verst./headphone Amp.	20-MDIP	Rhm	-	pdf
BA 3571 FS	LIN-IC	=BA 3571F:	20-SMDIP		-	pdf
BA 3572 FS	LIN-IC	SMD, Kopfh.-verst./headphone Amp.	24-SMDIP	Rhm	-	
BA 3574 BFS	LIN-IC	SMD, Cd, Kopfh.-verst./headphone Amp.	20-SMDIP	Rhm	-	pdf
BA 3641 FV	LIN-IC	Recorder, Rec. Amp, ALC, Ucc=1,7...3,6V	20-SSMDIP	Rhm	-	pdf
BA 3702	LIN-IC	Recorder, Suchlaufspeicher/memory	16-DIP	Rhm	-	
BA 3703 F	LIN-IC	SMD, Recorder, Mute Detector, Ucc=9V	8-MDIP	Rhm	-	pdf
BA 3704	LIN-IC	Recorder, Suchlaufspeicher/memory	16-DIP	Rhm	-	
BA 3704 F	LIN-IC	=BA 3704: SMD	16-MDIP		-	
BA 3706	LIN-IC	Recorder, Suchlaufspeicher/memory	16-DIP	Rhm	-	
BA 3707	LIN-IC	Recorder, Band-suchlauf/music finder	9-SIP	Rhm	-	
BA 3708 F	LIN-IC	SMD, ~BA 3707, Ucc=3V	8-MDIP	Rhm	-	pdf
BA 3710	LIN-IC	Recorder, Band-suchlauf/music finder	16-DIP	Rhm	-	
BA 3711	LIN-IC	Recorder, Suchlaufspeicher/memory	16-DIP	Rhm	-	
BA 3712	LIN-IC	Recorder, Auto-reverse Controller	9-SIP	Rhm	-	
BA 3714 F	LIN-IC	SMD, Rec., Signal-Sensor, Ucc=1,5...3V	8-MDIP	Rhm	-	pdf
BA 3810 F	LIN-IC	SMD, Recorder-Controller, Ucc=3V	22-MDIP	Rhm	-	
BA 3812 F	LIN-IC	=BA 3812L: SMD	20-MDIP		-	
BA 3812 L	LIN-IC	Graphic Equalizer, 5 Point, Ucc=8V	18-SQP	Rhm	KA 22235	pdf
BA 3814 L	LIN-IC	Power on/off Controller	16-SQP	Rhm	-	
BA 3818 F	LIN-IC	SMD, Recorder, Comparator Array	8-MDIP	Rhm	-	
BA 3819 F	LIN-IC	SMD, Recorder, Comparator Array	8-MDIP	Rhm	-	
BA 3822 FS	LIN-IC	=BA 3822LS: SMD	24-SMDIP		-	pdf pinout
BA 3822 LS	LIN-IC	Graphic Equalizer, Stereo, 5 Points	24-SQP	Rhm	KA 22234	pdf pinout
BA 3823 LS	LIN-IC	Graphic Equalizer, Stereo, 5 Points	24-SQP	Rhm	-	pdf
BA 3824 LS	LIN-IC	Graphic Equalizer, Stereo, 5 Points	24-SQP	Rhm	-	pdf

Type	Device	Short Description	Fig.	Manu	Comparision Types	More at
BA 3826 S	LIN-IC	Band-pass Filter f. Spectrum Analyzer	18-SDIP	Rhm	-	pdf
BA 3828 F	LIN-IC	SMD, Recorder, Preset+comparator Array	22-MDIP	Rhm	-	
BA 3830 F	LIN-IC	=BA 3830S: SMD	18-MDIP		-	pdf
BA 3830 S	LIN-IC	Bandpass-filter f. Spectrum Analyzer	18-SDIP	Rhm	-	pdf
BA 3832 F	LIN-IC	SMD,Bandp.-filter f. Spectrum Analyzer	18-MDIP	Rhm	-	
BA 3834 F	LIN-IC	=BA 3834S: SMD	18-MDIP		-	pdf
BA 3834 S	LIN-IC	Bandpass-filter f. Spectrum Analyzer	18-SDIP	Rhm	-	pdf
BA 3835 F	LIN-IC	=BA 3835S: SMD	18-MDIP	.	-	pdf
BA 3835 S	LIN-IC	Bandpass-filter f. Spectrum Analyzer	18-SDIP	Rhm	-	pdf
BA 3837	LIN-IC	Vocal Fader f. Karaoke Systems	16-DIP	Rhm	-	pdf
BA 3837 F	LIN-IC	=BA 3837: SMD	16-MDIP		-	pdf
BA 3838 F	LIN-IC	SMD, Vocal Fader f. Karaoke Systems	16-MDIP	Rhm	-	pdf
BA 3840 KV	LIN-IC	SMD, Preset Graphic Equalizer	48-MP	Rhm	-	pdf
BA 3842 F	LIN-IC	SMD, Preset Equalizer Sound Ctrl.	24-MDIP	Rhm	-	pdf
BA 3853 AFS	LIN-IC	SMD, Dual Potentiometer (sound Ctrl.)	24-SMDIP	Rhm	-	pdf
BA 3870	LIN-IC	Recorder, Bass Boost System	20-DIP	Rhm	-	pdf
BA 3880 S	LIN-IC	Radio/TV++, Sound Prozessor, hi-def	22-SDIP	Rhm	-	pdf
BA 3884 F	LIN-IC	=BA 3884S: SMD	24-MDIP		-	pdf
BA 3884 S	LIN-IC	Radio/tv++, Sound Prozessor, hi-def	24-SDIP	Rhm	-	pdf
BA 3888 S	LIN-IC	Rec./vc, Sound Prozessor, hi-def	22-SDIP	Rhm	-	pdf
BA 3890 F	LIN-IC	SMD, Cd, Pulse Width Control	8-MDIP	Rhm	-	pdf
BA 3900	LIN-IC	Referenz-Spg.-/Ref.vltg. Generator	12-SILP	Rhm	-	
BA 3902	LIN-IC	Referenz-spg.-/ref.vltg. Generator	12-SILP	Rhm	-	pdf
BA 3904 A	LIN-IC	Referenz-spg.-/ref.vltg. Generator	12-SILP	Rhm	-	pdf
BA 3906	LIN-IC	Referenz-spg.-/ref.vltg. Generator	12-SILP	Rhm	-	pdf
BA 3908 B	LIN-IC	Radio, Stromversorgung/power Supply	12-SILP	Rhm	-	
BA 3910 B	LIN-IC	Referenz-Spg.-/Ref. Vltg. Generator	12-SILP	Rhm	-	pdf pinout
BA 3911	LIN-IC	Referenz-spg.-/ref. Vltg. Generator	12-SILP	Rhm	-	pdf
BA 3912	LIN-IC	Radio, Stromversorgung/power Supply	12-SILP	Rhm	-	
BA 3913	LIN-IC	Radio, Stromversorgung/power Supply	12-SILP	Rhm	-	
BA 3914 A	LIN-IC	Radio, Stromversorgung/power Supply	12-SILP	Rhm	-	
BA 3915 B	LIN-IC	Radio, Stromversorgung/power Supply	16-SILP	Rhm	-	pdf
BA 3916	LIN-IC	Radio, Stromversorgung/power Supply	12-SILP	Rhm	-	
BA 3918	LIN-IC	Radio, Stromversorgung/power Supply	12-SILP	Rhm	-	pdf
BA 3920	LIN-IC	Referenz-spg.-/ref.vltg. Generator	12-SILP	Rhm	-	
BA 3922	LIN-IC	Referenz-spg.-/ref.vltg. Generator	12-SILP	Rhm	-	
BA 3924	LIN-IC	Referenz-spg.-/ref.vltg. Generator	12-SILP	Rhm	-	pdf
BA 3926	LIN-IC	Referenz-spg.-/ref.vltg. Generator	12-SILP	Rhm	-	
BA 3928	LIN-IC	Referenz-spg.-/ref.vltg. Generator	12-SILP	Rhm	-	
BA 3930	LIN-IC	Rec./cd, Stromversorgung/power Supply	12-SILP	Rhm	-	
BA 3932	LIN-IC	Rec./cd, Stromversorgung/power Supply	12-SILP	Rhm	-	pdf
BA 3933	LIN-IC	Rec./cd, Stromversorgung/power Supply	12-SILP	Rhm	-	pdf
BA 3935	LIN-IC	Rec./cd, Stromversorgung/power Supply	12-SILP	Rhm	-	pdf
BA 3936	LIN-IC	Rec./cd, Stromversorgung/power Supply	12-SILP	Rhm	-	pdf
BA 3938	LIN-IC	Rec./cd, Stromversorgung/power Supply	12-SIL	Rhm	-	pdf
BA 3940 A	LIN-IC	Rec./cd, Stromversorgung/power Supply	12-SILP	Rhm	-	pdf
BA 3950 A	LIN-IC	Cd, Stromversorgung/power Supply	12-SILP	Rhm	-	pdf
BA 3960	Z-IC	1,5..21V/1,2A, 1,5..21V/0,9A	12-SIL	Rhm	-	pdf
BA 3962 A	LIN-IC	Rec./cd, Stromversorgung/power Supply	10-SIL	Rhm	-	
BA 3963	LIN-IC	Cd, Stromversorgung/power Supply	12-SILP	Rhm	-	pdf
BA 3990 F	LIN-IC	SMD, Cd, Ripple Filter, 38dB	8-MDIP	Rhm	-	pdf
BA 4110	LIN-IC	FM-ZF, Afc, Agc, Ucc=6...12V	16-SQP	Rhm	KA 22441, LA 1140	
BA 4111	LIN-IC	FM-ZF, Afc, Agc	16-DIP	Rhm		
BA 4111 F	LIN-IC	=BA 4111: SMD	16-MDIP			
BA 4112	LIN-IC	Telecom, FM-ZF, Schmalband/narrow band	16-DIP	Rhm		
BA 4113	LIN-IC	Telecom, FM-ZF, Schmalband/narrow band	18-DIP	Rhm		
BA 4114	LIN-IC	Telecom, FM-ZF, Schmalband/narrow band	16-DIP	Rhm		
BA 4116 FV	LIN-IC	SMD, Cordless Telephone, FM-ZF-Dem.	16-SSMDIP	Rhm	-	pdf
BA 4118 FV	LIN-IC	SMD, Pager, ZF-Dem., 455kHz	16-SSMDIP	Rhm	-	
BA 4210	LIN-IC	AM/FM-ZF, Ucc=2,5...6V	16-DIP	Rhm	-	
BA 4210 F	LIN-IC	=BA 4210: SMD	16-MDIP		-	
BA 4220	LIN-IC	AM/FM-ZF + Demod., Ucc=3...14V	16-DIP	Rhm	HA 12413, KA 2243	pdf pinout
BA 4220 F	LIN-IC	=BA 4220: SMD	16-MDIP			
BA 4222	LIN-IC	AM/FM-ZF	16-DIP	Rhm		
BA 4224	LIN-IC	AM-HF, FM-ZF	16-DIP	Rhm	-	
BA 4224 F	LIN-IC	=BA 4224: SMD	16-MDIP		-	
BA 4226	LIN-IC	AM/FM-ZF	16-DIP	Rhm	-	
BA 4228 F	LIN-IC	=BA 4228L: SMD	16-MDIP		-	
BA 4228 L	LIN-IC	AM-HF, AM/FM-ZF	16-SQP	Rhm	-	
BA 4230 AF	LIN-IC	SMD, AM/FM-ZF, Ucc=1,5V	18-MDIP	Rhm	-	
BA 4230 AFS	LIN-IC	=BA 4230AF:	20-SMDIP			
BA 4236 L	LIN-IC	AM/FM-ZF, Ucc=2,7...12V	18-SQP	Rhm	-	pdf
BA 4237 L	LIN-IC	AM/FM-ZF, Ucc=2,7...12V	18-SQP	Rhm	-	pdf
BA 4240 F	LIN-IC	=BA 4240L: SMD	18-MDIP			
BA 4240 L	LIN-IC	AM/FM-ZF, Ucc=3V	18-SQP	Rhm	-	
BA 4260	LIN-IC	AM-Tuner, FM-ZF	16-DIP	Rhm	KA 2247, LA 1260	
BA 4402(L)	LIN-IC	FM-V/Front End, Ucc=1,8...9V	9-SIP	Rhm	-	
BA 4403	LIN-IC	FM-V/Front End, Ucc=1,8...9V	7-SIP	Rhm	-	
BA 4404	LIN-IC	FM-V/Front End, Ucc=1,8...9V	9-SIP	Rhm	-	
BA 4405	LIN-IC	FM-V/Front End, Ucc=1,8...9V	7-SIP	Rhm	-	
BA 4408 F	LIN-IC	SMD, FM-V/Front End, Ucc=1,5V	14-MDIP	Rhm	-	
BA 4411	LIN-IC	FM-V/Front End, Ucc=2...8V	9-SIP	Rhm	-	
BA 4412	LIN-IC	FM-V/Front End, Ucc=2...8V	9-SIP	Rhm	-	
BA 4413	LIN-IC	FM-V/Front End, Ucc=2...8V	9-SIP	Rhm	-	
BA 4422 AN	LIN-IC	FM/TV-V/Front End, Ucc=2,4...5,5V	9-SIP	Rhm	KIA 6058S	

Type	Device	Short Description	Fig.	Manu	Comparision Types	More at
BA 4424 N	LIN-IC	FM/TV-V/Front End, Ucc=1,6...6V	9-SIP	Rhm	-	pdf
BA 4425 F	LIN-IC	SMD, FM/TV-V/Front End, Ucc=1,6...6V	8-MDIP	Rhm	-	pdf
BA 4510 F	OP-IC	SMD, Dual, ±5V; 9MHz, 5V/μs, -20...+75°	8-MDIP	Rhm	-	pdf
BA 4558	OP-IC	Dual, Serie 158, ±18V, -20...+75°	8-DIP	Rhm	→Serie 158	data pdf
BA 4558 F	OP-IC	=BA 4558: SMD	8-MDIP	Rhm	... 158..., ... 258..., ... 1458..., ... 4558...	pdf
BA 4558 N	OP-IC	=BA 4558: Fig.→	8-SIP	Rhm	... 158..., ... 258..., ..., 1458..., ...4558...	pdf
BA4560	2xOP	Vs:±18V Vu:100V/mV Vo:±14V Vi0:0.5mV	8D,9IA	Roh		data pinout
BA 4560	OP-IC	Dual, hi-current, ±18V, -20...+75°	8-DIP	Rhm	-	pdf
BA 4560 F	OP-IC	=BA 4560: SMD	8-MDIP	'	-	pdf
BA 4560 N	OP-IC	=BA 4560: Fig.→	8-SIP		-	pdf
BA 4900	LIN-IC	Radio, Stromversorgung/power Supply	12-SILP	Rhm	-	pdf
BA 4911	LIN-IC	car stereo, system reg. IC 5.0V/ 2x8.12V/ 7.9V/ 10.3V	SILP	Rhm	-	pdf pinout
BA 5101	LIN-IC	Recorder, A/W-Verst., ALC, Ucc=6...9V	16-SQP	Rhm	-	
BA 5102(A)	LIN-IC	Vc, Audio-a/w-verst., Logic	18-DIP	Rhm	KA 2983	pdf
BA 5102(A)F	LIN-IC	=BA 5102(A): SMD	18-MDIP			
BA 5102(A)L	LIN-IC	=BA 5102(A): Fig.→	18-SQP			
BA 5104 A	LIN-IC	Recorder, A/W-Verst., Ucc=6...9V	16-SQP	Rhm	(BA 5101)	
BA 5112 LS	LIN-IC	Vc, Audio-a/w-verst.	24-SQP	Rhm		
BA 5114 LS	LIN-IC	Vc, Audio-a/w-verst.	24-SQP	Rhm		
BA 5115	LIN-IC	Vc, Audio-a/w-verst.	18-DIP	Rhm		
BA 5115 L	LIN-IC	=BA 5115: Fig.→	18-SQP			
BA 5116	LIN-IC	Vc, Audio-a/w-verst.	20-DIP	Rhm		
BA 5117 L	LIN-IC	Vc, Audio-a/w-verst.	18-SQP	Rhm		
BA 5152 F	LIN-IC	SMD,2x Kopfh./Headphone Amp, Ucc=1,5V	16-MDIP	Rhm	-	pdf
BA 5204	LIN-IC	2x Kopfh.-Verst/Headphone Amp, 3V/32Ω	16-SQP	Rhm	-	
BA 5204 F	LIN-IC	=BA 5204: SMD	16-MDIP		-	pdf
BA 5206 F,BF	LIN-IC	SMD, 2x Kopfh./Headphone Amp, 3V/16Ω	16-MDIP	Rhm		
BA 5208 AF	IC	BTL Power Amplifier, Mute, 3V 400mW	16-SMDIP	Rhm	-	pdf pinout
BA 5210 FS	LIN-IC	SMD, Speaker/Headph. Amp, Ucc=2,5...6V	16-SMDIP'	Rhm	-	pdf
BA 5214	LIN-IC	2x Kopfh.-Verst/Headphone Amp, 3V/32Ω	16-DIP	Rhm	-	
BA 5302 A	LIN-IC	2x NF-E, 13V, 2x2,4W(9V/4Ω)	12-DIP+b	Rhm	BA 5304	
BA 5304	LIN-IC	2x NF-E, 2x3W(9V/3Ω)	12-DIP+b	Rhm	-	
BA 5402	LIN-IC	2x NF-E, 18V, 2x4,2W(12V/4Ω)	12-SILP	Rhm	BA 536, BA 5406	
BA 5402 A	LIN-IC	=BA 5402: rauscharm/low noise	12-SILP	Rhm	BA 536, BA 5406	
BA 5404	LIN-IC	NF-E, 20V, 0,36W(12V/32Ω)	9-SIP	Rhm	-	
BA 5406	LIN-IC	2x NF-E, 18V, 2x5W(12V/3Ω), Ucc=5...15V	12-SILP	Rhm	BA 536, BA 5402(A), KIA 6283K	pdf pinout
BA 5410	LIN-IC	2x NF-E, 20V, 2x5,2W(12V/3Ω)	10-SIL	Rhm	-	pdf
BA 5412	LIN-IC	NF-E, 24V, 2x5,4W(12V/3Ω)	12-SILP	Rhm	-	pdf pinout
BA 5413	LIN-IC	=BA 5412: Fig.→	12-SIL		-	pinout
BA 5415 A	LIN-IC	2x NF-E, 24V, 2x5,4W(12V/3Ω)	12-SILP	Rhm	-	pdf pinout
BA 5416	LIN-IC	2x NF-E, 24V, 2x5,4W(12V/3Ω)	12-SIL	Rhm	-	pdf pinout
BA 5939S	LIN-IC	MD/CD, 6 channel BTL motor driver	MDIP	Rhm	-	pdf pinout
BA 5983FM	LIN-IC	BTL-driver for Cd-rom, 4-ch.	TSOP	Rhm	-	pdf pinout
BA 5983FP	LIN-IC	BTL-driver for Cd-rom, 4-ch.	TSOP	Rhm	-	pdf pinout
BA 5995 FM	IC	4-channel BTL driver for CD/CD-ROM	DIP_K	Rhm	-	pdf pinout
BA 6104	LIN-IC	LED Linear-Meter, 5 LED, Ucc=12V	9-SIP	Rhm	-	
BA 6109	LIN-IC	Vc, Motor-Tr(reversible), 18V, 0,8A	10-SIL	Rhm	BA 6209	
BA 6110	OP-IC	Voltg. Controlled; 34V, -20...+70°	9-SIP	Rhm	-	pdf
BA 6110 FS	OP-IC	=BA 6110: SMD	16-SMDIP		-	pdf
BA 6118	LIN-IC	Motorsteuerung/motor control, 4,8V	14-DIP	Rhm	-	pdf
BA 6121	LIN-IC	VC-Logik (4 Sign.)	16-DIP	Rhm	-	
BA 6121 F	LIN-IC	=BA 6121: SMD	16-MDIP		-	
BA 6122 A	SMPS-IC	Vc PS, S-Reg, 5V, Stop-Func.	16-SQP	Rhm	-	pdf
BA 6122 AF	SMPS-IC	=BA 6122(A): SMD	16-MDIP		-	pdf
BA 6124	LIN-IC	LED VU-Meter, 5 LED, Ucc=6V	9-SIP	Rhm	AN 6884, KA 2284, LB 1403, KIA 6966S	pdf pinout
BA 6124 F	LIN-IC	=BA 6124: SMD	14-MDIP		-	pdf
BA 6125	LIN-IC	LED Linear-Meter, 5 LED, Ucc=6V	9-SIP	Rhm	KA 2287, LB 1433, (BA 6124)	pinout
BA 6129 F	LIN-IC	SMD, Reset IC + Battery Backup Func.	8-MDIP	Rhm	-	
BA 6135	LIN-IC	Vc, Sensor	14-DIP	Rhm	-	
BA 6137	LIN-IC	LED VU-Meter, 5 LED, Ucc=6V	9-SIP	Rhm	KA 2285, LB 1423, KIA 6966S, (BA 6144)	pdf pinout
BA 6138	LIN-IC	2x Kompress.-Vst./amp. f. Level-Meter	9-SIP	Rhm	-	pdf
BA 6138 F	LIN-IC	=BA 6138: SMD	14-MDIP		-	pdf
BA 6139(L)	LIN-IC	7 Bit FLT-Display-Tr, Ucc=24V	16-SQP	Rhm	-	
BA 6144	LIN-IC	LED VU-Meter, 5 LED, Ucc=12V	9-SIP	Rhm	(BA 6137)	pdf
BA 6146	LIN-IC	12P.-VU-Meter-Tr.(FLT display)	16-DIP	Rhm	-	pdf
BA 6148	LIN-IC	12P.-Level-Meter-Tr.(FLT Display)	16-DIP	Rhm	-	
BA 6149 LS	SMPS-IC	S-Reg, 6 Ausg.-Spg./output voltages	24-SQP	Rhm	-	pdf
BA 6154	LIN-IC	LED VU-Meter, 5 LED, Ucc=6V	9-SIP	Rhm	(BA 6137, BA 6144)	pdf
BA 6155 CH-2W	OPTO-IC	Photo-sensor, Schmitt-Trigger	Chip	Rhm	-	
BA 6161 F	SMPS-IC	=BA 6161N: SMD	8-MDIP		-	pdf
BA 6161 N	SMPS-IC	S-Reg f. Electr. Tuner, 30...45V	5-SIP	Rhm	-	pdf
BA 6162	LIN-IC	Reset IC + Battery Backup Function	8-MDIP	Rhm	-	pdf
BA 6162 F	LIN-IC	=BA 6162: SMD	8-DIP		-	pdf
BA 6191	LIN-IC	Cd, 2x Motor-Tr(reversible), 18V, 0,7A	12-SIL	Rhm	-	pdf
BA 6196 FP	LIN-IC	SMD, Cd, 4x Leistungs-Tr/power driver	28-SMDIP+b	Rhm	-	
BA 6199 FP	LIN-IC	SMD, Cd, 3x Leistungs-Tr/power driver	28-SMDIP+b	Rhm	-	
BA 6208	LIN-IC	Motor-Tr.(reversible), 18V, 0,5A	9-SIP	Rhm	-	data pdf pinout
BA 6208 F	LIN-IC	=BA 6208: SMD	8-MDIP	Rhm	-	data pdf pinout
BA 6209	LIN-IC	Vc, Motor-Tr.(reversible), 18V, 1,6A	10-SIL	Rhm	KA 8301	data pdf pinout
BA 6209 N	LIN-IC	=BA 6209: Fig.→	10-SIP	Rhm	-	data pdf pinout
BA 6212	LIN-IC	8x Uni-Tr, hi-current, 7V, 0,4A	20-DIP	Rhm	-	pdf
BA 6218	LIN-IC	Motor-Tr.(reversible), 18V, 0,7A	9-SIP	Rhm	-	data pdf pinout
BA 6219 B	LIN-IC	Vc, Motor-Tr.(reversible), 18V, 2,2A	10-SIL	Rhm	-	data pdf pinout
BA 6219 BFP-Y	LIN-IC	=BA 6219 B: SMD	25-SMDIP+k	Rhm	-	data pdf pinout

Type	Device	Short Description	Fig.	Manu	Comparision Types	More at
BA 6220	LIN-IC	Motorreg./el. Governor, Ucc=3,6...16V	8-DIP	Rhm	-	pdf
BA 6222	LIN-IC	Vc, Motor-Tr.(reversible), 18V, 2,2A	10-SIL	Rhm	-	data pdf pinout
BA 6227	LIN-IC	Motorregler/electr. Governor, Ucc=3V	8-DIP	Rhm		data pdf pinout
BA 6229	LIN-IC	Motor-Tr.(reversible), 23V, 1,2A	10-SIL	Rhm		data pdf pinout
BA 6235	LIN-IC	Motorregler/electr. Governor, Ucc=3V	8-DIP	Rhm	-	pdf
BA 6235 F	LIN-IC	SMD,Motorregler/electr.Governor,Ucc=3V	8-MDIP	Rhm	-	pdf
BA 6238 A	LIN-IC	Vc, Motor-Tr.(reversible), <20V, 1,6A	10-SIL	Rhm	KA 8305	data pdf pinout
BA 6238 AN	LIN-IC	=BA 6238A: Fig.→	10-SIP	Rhm	-	data pdf pinout
BA 6239 A	LIN-IC	Vc, Motor-Tr.(reversible), <20V, 1,2A	10-SIL	Rhm	-	data pdf pinout
BA 6239 AN	LIN-IC	=BA 6239A: Fig.→	10-SIP	Rhm	-	data pdf pinout
BA 6240	LIN-IC	Motorreg./el. Governor, Ucc=3,5...14V	8-DIP	Rhm	-	data pdf pinout
BA 6246	LIN-IC	2x Motor-Tr(reversible), <20V, 1A	10-SIL	Rhm		pdf pinout
BA 6246 N	LIN-IC	=BA 6246: Fig.→	10-SIP	Rhm		data pdf pinout
BA 6247	LIN-IC	2x Motor-Tr(reversible), <20V, 1A	10-SIL	Rhm		data pdf pinout
BA 6247 FP-Y	LIN-IC	=BA 6247: SMD	25-SMDIP+k	Rhm		data pdf pinout
BA 6247 N	LIN-IC	=BA 6247: Fig.→	10-SIP	Rhm		data pdf pinout
BA 6248	LIN-IC	2x Motor-Tr(reversible), 1,6A	10-SIL	Rhm		
BA 6249	LIN-IC	2x Motor-Tr(reversible), 18V, 1A	10-SIL	Rhm		
BA 6249 N	LIN-IC	=BA 6249: Fig.→	10-SIP	Rhm		
BA 6250	LIN-IC	7x PNP Darl.-Trans., 30V, 20mA	16-DIP	Rhm		pdf
BA 6250 F	LIN-IC	=BA 6250: SMD	16-MDIP			pdf
BA 6251	LIN-IC	7x PNP Darl.-Trans., 30V, 20mA	16-DIP	Rhm		pdf
BA 6251 F	LIN-IC	=BA 6251: SMD	16-MDIP			pdf
BA 6256	LIN-IC	6x Uni-Tr, Ucc=3...10V, 0,4A	16-DIP			
BA 6257	LIN-IC	7x PNP Darl.-Tr, Ucc=20V, 0,1A	16-DIP			
BA 6259 N	LIN-IC	2x Motor-Tr.(reversible), <20V, 1A	10-SIP	Rhm		data pdf pinout
BA 6266	LIN-IC	Hex Inverter, Ucc=5V	14-DIP	Rhm		pdf
BA 6266 F	LIN-IC	=BA 6266: SMD	14-MDIP		-	pdf
BA 6267	LIN-IC	Hex Buffer, Driver, Ucc=5V	14-DIP	Rhm	SN 7417	pdf
BA 6267 F	LIN-IC	=BA 6267: SMD	14-MDIP		-	pdf
BA 6268	LIN-IC	4x 2-Input NAND Buffer, Ucc=5V	14-DIP	Rhm	-	
BA 6280 AF	LIN-IC	SMD, Cd, PWM Motor Ctrl., Iq=0,5A	22-MDIP	Rhm	KA 9255D	
BA 6281 F	LIN-IC	SMD, Cd, PWM Motor Ctrl.	24-MDIP	Rhm	-	
BA 6283 N	LIN-IC	VC++, Motor-Tr.(reversible), <18V, 1A	9-SIP	Rhm		data pdf pinout
BA 6284 N	LIN-IC	VC++, Motor-Tr.(reversible), <18V, 1A	10-SIP	Rhm		data pdf pinout
BA 6285 FP	LIN-IC	SMD, Motor-Tr.(reversible), <18V, 1A	24-SMDIP+k	Rhm		data pdf pinout
BA 6285 FS	LIN-IC	=BA 6285FP: Fig.→	16-SMDIP	Rhm		data pdf pinout
BA 6286	LIN-IC	VC++, Motor-Tr.(reversible), <18V, 1A	10-SIL	Rhm		data pdf pinout
BA 6286 N	LIN-IC	=BA 6286: Fig.→	10-SIP	Rhm		data pdf pinout
BA 6287 F	LIN-IC	SMD, Motor-Tr.(reversible), <18V, 1A	8-MDIP	Rhm	-	data pdf pinout
BA 6288 FS	LIN-IC	SMD, Motor-Tr.(reversible), <18V, 1A	16-SMDIP	Rhm	-	data pdf pinout
BA 6289 F	LIN-IC	SMD, Motor-Tr.(reversible), <18V, 0,6A	8-MDIP	Rhm	BA 6417F	data pdf pinout
BA 6290 A	LIN-IC	Cd, 2x Leistungs-Tr/power driver	12-SILP	Rhm	KA 9257	pinout
BA 6292	LIN-IC	Cd, 2x Leistungs-Tr/power driver	12-SILP	Rhm	-	pdf
BA 6294	LIN-IC	Cd, 2x Leistungs-Tr/power driver	10-SIL	Rhm	-	
BA 6295 FP,AFP	LIN-IC	SMD, Cd, 2x Leistungs-Tr/power driver	28-SMDIP+b	Rhm	-	
BA 6296 FP	LIN-IC	SMD, Cd, 4x Leistungs-Tr/power driver	28-SMDIP+b	Rhm	-	
BA 6297 AFP	LIN-IC	SMD, Cd, 4x Leistungs-Tr/power driver	28-SMDIP+b	Rhm	-	pdf
BA 6298 FP	LIN-IC	SMD, Cd, 4x Leistungs-Tr/power driver	28-SMDIP+b	Rhm	-	
BA 6299 FP	LIN-IC	SMD, Cd, 4x Leistungs-Tr/power driver	28-SMDIP+b	Rhm	-	pdf
BA 6301	LIN-IC	Vc, Speed Servo Ctrl., Fg System	16-DIP	Rhm		pdf
BA 6301 F	LIN-IC	=BA 6301: SMD	16-MDIP			pdf
BA 6302(A)	LIN-IC	Vc, Speed Servo Ctrl., Fg System	16-DIP	Rhm		pdf
BA 6302(A)F	LIN-IC	=BA 6302: SMD	16-MDIP			pdf
BA 6303	LIN-IC	Vc, Speed Servo Ctrl., Fg System	16-DIP	Rhm		pdf
BA 6303 F	LIN-IC	=BA 6303: SMD	16-MDIP			pdf
BA 6304	LIN-IC	VC-drahtfernbed./wire remote ctrl.	14-DIP	Rhm		
BA 6305	LIN-IC	Vc, Servo, FG/CTL/DTP-verst./amp.	8-SIP	Rhm		pdf
BA 6305 F	LIN-IC	=BA 6305: SMD	8-MDIP			pdf
BA 6320	LIN-IC	Vc, Power Hold	16-SQP	Rhm		
BA 6321	LIN-IC	Vc, Speed Servo Ctrl., Fg System	16-DIP	Rhm		pdf
BA 6325 F	LIN-IC	VC(8mm), Sensor	24-MDIP	Rhm		
BA 6334	LIN-IC	VC/TV++, 16-Operation Mode Detector	9-SIP	Rhm		
BA 6340	LIN-IC	IR-FB, Encoder, Ucc=4,75...12,5V	8-SIP	Rhm		pdf
BA 6351 S	LIN-IC	Cd, Analog Servo	42-SDIP	Rhm		
BA 6353 S	LIN-IC	Cd, Analog Servo	42-SDIP	Rhm		
BA 6360	LIN-IC	Vc, Bandende/tape End Sensor	14-DIP	Rhm		
BA 6381 FS	LIN-IC	SMD, Cd, Vorverst./preamp.	16-MDIP	Rhm		
BA 6382 FS	LIN-IC	SMD, Cd, Vorverst./preamp.	-MDIP	Rhm		
BA 6392 FP	LIN-IC	SMD, Cd, 4x Leistungs-Tr/power driver	28-SMDIP+b	Rhm		pdf
BA 6393 FP	LIN-IC	SMD, Cd, 4x Leistungs-Tr/power driver	28-SMDIP+b	Rhm		
BA 6394 FP	LIN-IC	SMD, Cd, 4x Leistungs-Tr/power driver	28-SMDIP+b	Rhm		
BA 6395 AFP	LIN-IC	SMD, Cd, 5x Leistungs-Tr/power driver	28-SMDIP+b	Rhm		pdf
BA 6396 FP	LIN-IC	SMD, Cd, 4x Leistungs-Tr/power driver	28-SMDIP+b	Rhm		
BA 6397 FP	LIN-IC	SMD, Cd, 4x Leistungs-Tr/power driver	28-SMDIP+b	Rhm		pdf
BA 6398 FP	LIN-IC	SMD, Cd, 4x Leistungs-Tr/power driver	28-SMDIP+b	Rhm		pdf
BA 6399 FP	LIN-IC	SMD, Cd, 4x Leistungs-Tr/power driver	28-SMDIP+b	Rhm		
BA 6402 F	LIN-IC	=BA 6412: SMD	8-MDIP			pdf
BA 6404	LIN-IC	Fan Motor-Pre-Tr, 2-phase half-wave	8-DIP	Rhm		
BA 6404 F	LIN-IC	=BA 6404: SMD	8-MDIP			pdf
BA 6405	LIN-IC	Vc, Servo, FG/CTL-verst./amp.	14-DIP	Rhm		
BA 6405 F	LIN-IC	=BA 6405: SMD	14-MDIP			
BA 6406	LIN-IC	Fan Motor-Pre-Tr, 2-phase half-wave	8-DIP	Rhm		pdf
BA 6406 F	LIN-IC	=BA 6406: SMD	8-MDIP			pdf
BA 6407	LIN-IC	Fan Motor-Tr, 2-phase half-wave	8-DIP	Rhm		pdf

Type	Device	Short Description	Fig.	Manu	Comparision Types	More at
BA 6407 F,AF	LIN-IC	=BA 6407: SMD	8-MDIP		-	
BA 6408 FS	LIN-IC	Fan Motor-Tr, 2-phase half-wave	16-SMDIP		-	
BA 6411	LIN-IC	VC++, 2-Phase DD Motor-Tr, Ucc=9...18V	12-SIL	Rhm	KA 8304	pinout
BA 6411 FP	LIN-IC	=BA 6411: SMD	28-SMDIP+b		-	
BA 6412	LIN-IC	Fan Motor-Pre-Tr, 2-phase half-wave	8-DIP	Rhm	-	
BA 6413	LIN-IC	CD/VC++, 2-Phasen DD Motor-Tr	12-SIL	Rhm	-	
BA 6414 FP-Y	LIN-IC	Vc, 2-Phasen DD Motor-Tr, Ucc=8..20V	25-SMDIP+b	Rhm	-	
BA 6414 FS	LIN-IC	=BA 6414FP-Y:	24-SMDIP	Rhm	-	
BA 6417 F	LIN-IC	=BA 6289F: 1A	8-MDIP	Rhm	-	data pdf pinout
BA 6418 N	LIN-IC	Motor-Tr.(reversible), <18V, 0,7A	9-SIP	Rhm	-	data pdf pinout
BA 6419 F	LIN-IC	SMD, Motor-Tr(reversible), 15V, 0,7A	8-MDIP	Rhm	-	
BA 6424 FS	LIN-IC	SMD, Motor-Tr, single-phase full wave	16-SMDIP	Rhm	-	
BA 6431 F	LIN-IC	=BA 6431S: SMD	28-MDIP		-	
BA 6431 S	LIN-IC	Vc, 3-Phasen DD Motor-Tr (capstan)	24-DILP	Rhm	-	pdf
BA 6432 S	LIN-IC	Vc, 3-Phasen DD Motor-Tr (capstan)	24-DILP	Rhm	-	pdf
BA 6435 S	LIN-IC	Vc, 3-Phasen Motor-Tr (capstan)	24-DILP	Rhm	-	
BA 6436 P	LIN-IC	Vc, 3-Phasen Motor-Tr (capstan)	24-DILP	Rhm	-	
BA 6437 S	LIN-IC	Vc, 3-Phasen Motor-Tr (capstan)	24-DILP	Rhm	-	
BA 6438 S	LIN-IC	Vc, 3-Phasen Motor-Tr.(Capstan)	24-DILP	Rhm	-	data pdf pinout
BA 6439 P,S	LIN-IC	Vc, 3-Phasen Motor-Tr (capstan)	24-DILP	Rhm	-	
BA 6440 FP	LIN-IC	SMD, Vc, 3-Phasen Motor-Tr (capstan)	28-SMDIP+b	Rhm	-	
BA 6441 FP	LIN-IC	SMD, Vc, 3-Phasen Motor-Tr (capstan)	28-SMDIP+b	Rhm	-	
BA 6444 FP	LIN-IC	SMD, Vc, 3-Phasen Motor-Tr.(Capstan)	28-SMDIP+k	Rhm	-	data pdf pinout
BA 6446 FM	LIN-IC	=BA 6446FP	28-SMDIP+b		-	
BA 6446 FP	LIN-IC	SMD, Vc, 3-Phasen Motor-Tr (capstan)	28-SMDIP+b	Rhm	-	
BA 6450 F	LIN-IC	SMD, Vc, 3-Phasen DD Motor-Tr(Cylind.)	24-MDIP	Rhm	-	
BA 6453 FP	LIN-IC	SMD, Vc, 3-Phasen DD Motor-Tr	24-SMDIP	Rhm	-	
BA 6455 FS	LIN-IC	SMD, Vc, 3-Phasen Full Wave Motor-Tr	24-MDIP	Rhm	-	
BA 6456 FS	LIN-IC	SMD, Vc, 3-Phasen Full Wave Motor-Tr	24-MDIP	Rhm	-	
BA 6457 P	LIN-IC	SMD, Vc, 3-Phasen DD Mctor-Tr	24-SMDIP	Rhm	-	
BA 6458 FP-Y	LIN-IC	=BA 6458FP:	25-SMDIP+k	Rhm	-	data pdf pinout
BA 6458 FP	LIN-IC	SMD, Vc, 3-Phasen Motor-Tr, Ucc=5V	24-SMDIP	Rhm	-	
BA 6459 FS	LIN-IC	=BA 6459P,S: SMD	24-SMDIP		-	
BA 6459 P,S	LIN-IC	Vc, 3-Phasen DD Motor-Tr, Ucc=5V	24-DILP		-	
BA 6462 FP	LIN-IC	SMD, Vc, 3-Phasen Motor-Tr, Ucc=5V	24-SMDIP+b	Rhm	-	
BA 6463 FP-Y	LIN-IC	SMD, Vc, 3-Phasen Motor-Tr., 8..20V	25-SMDIP+k	Rhm	-	data pdf pinout
BA 6464 FP-Y	LIN-IC	SMD, Vc, 3-Phasen Motor-Tr, 7,5..20V	25-SMDIP+b	Rhm	-	
BA 6465 FP-Y	LIN-IC	SMD, Fdd, Spindle Motor Driver, Ucc=5V	25-SMDIP+b	Rhm	-	
BA 6470 FP	LIN-IC	SMD, Fdd, Spindle Motor Driver, Ucc=5V	24-SMDIP+b	Rhm	-	
BA 6471 FP	LIN-IC	SMD, Fdd, Spindle Motor Driver, Ucc=5V	24-SMDIP+b	Rhm	-	
BA 6472 FP	LIN-IC	SMD, Fdd, Spindle Motor Driver, Ucc=5V	24-SMDIP+b	Rhm	-	
BA 6473 FP-Y	LIN-IC	=BA 6473FP: Fig.→	25-SMDIP+b		-	
BA 6473 FP	LIN-IC	SMD, Fdd, Spindle Motor Driver, Ucc=5V	24-SMDIP+b	Rhm	-	
BA 6474 FP-Y	LIN-IC	=BA 6474FP: Fig.→	25-SMDIP+b		-	
BA 6474 FP	LIN-IC	SMD, Fdd, Spindle Motor Driver	24-SMDIP+b	Rhm	-	
BA 6475 FP	LIN-IC	SMD, Fdd, Spindle Motor Driver	24-SMDIP+b	Rhm	-	
BA 6476(A)FP-Y	LIN-IC	=BA 6476FP: Fig.→	25-SMDIP+b		-	
BA 6476 FP	LIN-IC	SMD, Fdd, Spindle Motor Drv, 4,2..6,5V	24-SMDIP+b	Rhm	-	
BA 6477 FS	LIN-IC	SMD, Fdd, Spindle Motor Driver, 5V	32-SMDIP	Rhm	-	data pdf pinout
BA 6479 AFP-Y	LIN-IC	Fdd, Spindle-Motor-Tr, 4,2..6,5V	25-SMDIP+b	Rhm	-	
BA 6480 K	LIN-IC	SMD,Fdd, Spindle-/Stepper-Motor-Tr, 5V	44-MP	Rhm	-	
BA 6482 AK	LIN-IC	SMD, Fdd, Spindle Motor Driver, Ucc=5V	32-MP	Rhm	-	
BA 6485 FP-Y	LIN-IC	SMD, Fdd, Spindle Motor Driver, Ucc=5V	25-SMDIP+b	Rhm	-	
BA 6486 FS	LIN-IC	SMD, Fdd, Spindle Motor Driver, 12V	32-SMDIP	Rhm	-	data pdf pinout
BA 6487 FP-Y	LIN-IC	SMD, Fdd, Spindle Motor Drv, Ucc=12V	25-SMDIP+b	Rhm	-	
BA 6488 FP-Y	LIN-IC	SMD, Fdd, Spindle Motor Drv, Ucc=12V	25-SMDIP+b	Rhm	-	
BA 6490 FS	LIN-IC	SMD, Fdd, Spindle Motor Driver, Ucc=5V	32-SMDIP	Rhm	-	
BA 6491 FS	LIN-IC	SMD, Fdd, Spindle Motor Driver, Ucc=5V	32-SMDIP	Rhm	-	
BA 6492 BFS	LIN-IC	SMD, Fdd, Spindle Motor Driver, Ucc=5V	32-SMDIP	Rhm	-	data pdf pinout
BA 6493 K	LIN-IC	SMD, Fdd, Spindle Motor Drv, 2,6..6,5V	32-MP	Rhm	-	
BA 6494 K	LIN-IC	SMD, Fdd, Spindle Motor Drv, 2,6..6,5V	32-MP	Rhm	-	
BA 6495 K	LIN-IC	SMD, Fdd, Spindle Motor Drv, 2,6..6,5V	32-MP	Rhm	-	
BA 6496 K	LIN-IC	SMD, Fdd, Spindle Motor Drv, 2,6..6,5V	32-MP	Rhm	-	
BA 6497 K	LIN-IC	SMD, Fdd, Spindle Motor Drv, 2,6..6,5V	32-MP	Rhm	-	
BA 6498 FS	LIN-IC	SMD, Fdd, Spindle Motor Drv, 2,6..6,5V	32-SMDIP	Rhm	-	
BA 6499 K	LIN-IC	SMD, Fdd, Spindle Motor Drv, 2,6..6,5V	32-MP	Rhm	-	
BA 6522	OP-IC	Dual, hi-speed	8-DIP	Rhm	-	
BA 6522 F	OP-IC	=BA 6522: SMD	8-MDIP		-	
BA 6527 A	LIN-IC	Telecom, Leak Breaker, Minute Sig.det.	8-SIP	Rhm	-	
BA 6562	LIN-IC	Telecom, Sprechkreis/speech network	16-DIP	Rhm	-	
BA 6562 F	LIN-IC	=BA 6562: SMD	18-MDIP		-	
BA 6564 A	LIN-IC	Telecom, Wecker/tone Ringer	8-DIP	Rhm	-	
BA 6564 AF	LIN-IC	=BA 6564A: SMD	8-MDIP		-	
BA 6565 A	LIN-IC	Telecom, Wecker/tone Ringer	8-DIP	Rhm	-	
BA 6565 AF	LIN-IC	=BA 6565A: SMD	8-MDIP		-	
BA 6566	LIN-IC	Telecom, Sprechkreis/speech network	18-DIP	Rhm	-	pdf
BA 6566 F	LIN-IC	=BA 6566: SMD	18-MDIP		-	pdf
BA 6566 FP	LIN-IC	=BA 6566: SMD	24-SMDIP+b		-	pdf
BA 6567 K	LIN-IC	SMD, Telecom, Sprechkreis/speech netw.	44-MP	Rhm	-	
BA 6569 AFP	LIN-IC	=BA 6569AS: SMD	24-SMDIP+b		-	pdf
BA 6569 FP	LIN-IC	=BA 6569S: SMD	24-SMDIP+b		-	pdf
BA 6569 S,AS	LIN-IC	Telecom, Sprechkreis/speech network	22-SDIP	Rhm	-	pdf
BA 6571 A	LIN-IC	Telecom, Hoehrerverst./speaker phone	28-DIP	Rhm	-	
BA 6571 AF	LIN-IC	=BA 6571A: SMD	28-MDIP		-	
BA 6580 DK	LIN-IC	SMD, Fdd, Read/write Amplifier	44-MP	Rhm	-	

Type	Device	Short Description	Fig.	Manu	Comparision Types	More at
BA 6587 K	LIN-IC	SMD, Fdd, Read/write Amplifier	44-MP	Rhm	-	
BA 6588 K	LIN-IC	SMD, Fdd, Read/write Amplifier	44-MP	Rhm	-	
BA 6589 K	LIN-IC	SMD, Fdd, Read/write Amp., Ucc=5+12V	44-MP	Rhm	-	
BA 6590 S	LIN-IC	Centronics Interface	42-SDIP	Rhm	-	
BA 6591 AF	LIN-IC	SMD,TV/camera, RGB-encoder f. NTSC/PAL	24-MDIP	Rhm	-	
BA 6600 K	LIN-IC	SMD, Fdd, Read/write Amplifier, Ucc=5V	44-MP	Rhm	-	
BA 6607 K	LIN-IC	SMD, Fdd, Read/write Amplifier, Ucc=5V	32-MP	Rhm	-	
BA 6608 K	LIN-IC	SMD, Fdd, Read/write Amplifier, Ucc=5V	32-MP	Rhm	-	
BA 6610 AK	LIN-IC	SMD, Fdd, Read/write Amplifier, Ucc=5V	32-MP	Rhm	-	
BA 6612 K	LIN-IC	SMD, Fdd, Read/write Amplifier, Ucc=5V	32-MP	Rhm	-	
BA 6722	SMPS-IC	2x S-Reg, 5,2V/80mA, 5V/80mA	10-SIL	Rhm	-	
BA 6780	DIG/LIN-IC	2 channel driver for CD changers	DIP	Rhm	-	pdf pinout
BA 6780 FP-Y	DIG/LIN-IC	2 channel driver for CD changers	DIP	Rhm	-	pdf pinout
BA 6790 FP	LIN-IC	SMD, Cd, 4x Leistungs-Tr/power driver	28-SMDIP+b	Rhm	-	pdf
BA 6795 FP	LIN-IC	SMD, Cd, 5x Leistungs-Tr/power driver	28-SMDIP+b	Rhm	-	pdf
BA 6796 FP	LIN-IC	SMD, Cd, 5x Leistungs-Tr/power driver	28-SMDIP+b	Rhm	-	pdf
BA 6798 S	LIN-IC	SMD, Cd, 4x Leistungs-Tr/power driver	24-DILP	Rhm	-	pdf
BA 6800 A	LIN-IC	16-P.-FLT VU-Meter-Tr, Peak Hold	28-DIP	Rhm	-	
BA 6800 AF	LIN-IC	=BA 6800A: SMD	28-MDIP		-	pdf
BA 6800 AS	LIN-IC	16-P.-FLT VU-Meter-Tr, Peak Hold	30-SDIP	Rhm	-	pdf
BA 6803 S	LIN-IC	16-P.-FLT VU-Meter-Tr, Peak Hold	30-SDIP	Rhm	-	pdf
BA 6805 A	LIN-IC	16-P.-FLT VU-Meter-Tr, Peak Hold	28-DIP	Rhm	-	
BA 6806 S	LIN-IC	16-P.-FLT VU-Meter-Tr, Peak Hold	30-SDIP	Rhm	-	
BA 6807	LIN-IC	Fan Motor-Tr, 2-phase half-wave	8-DIP	Rhm	-	
BA 6807 F	LIN-IC	=BA 6807: SMD	8-MDIP		-	
BA 6808 FS	LIN-IC	SMD, Fan Motor-Tr, 2-phase half-wave	16-SMDIP	Rhm	-	
BA 6809 F	LIN-IC	SMD, Fan Motor-Tr, 2-phase half-wave	8-MDIP	Rhm	-	
BA 6809 AF	LIN-IC	SMD, Fan Motor-Tr., 2-phase half-wave	8-MDIP	Rhm	-	data pdf pinout
BA 6810 F	LIN-IC	=BA 6810S: SMD	28-MDIP		-	pdf
BA 6810 S	LIN-IC	12P.-FLT VU-Meter-Tr, Peak Hold	30-SDIP	Rhm	-	pdf
BA 6811 F	LIN-IC	SMD, Fan Motor-Tr., 2-phase half-wave	8-MDIP	Rhm	-	data pdf pinout
BA 6812 FS	LIN-IC	SMD, Fan Motor-Tr, 2-phase half-wave	16-SMDIP	Rhm	-	
BA 6813 F	LIN-IC	SMD, Fan Motor-Tr., 2-phase half-wave	8-MDIP	Rhm	-	data pdf pinout
BA 6816 FS	LIN-IC	SMD, Fan Motor-Tr, 2-phase half-wave	16-SMDIP	Rhm	-	
BA 6817 F	LIN-IC	SMD, Fan Motor-Tr, 2-phase half-wave	8-MDIP	Rhm	-	data pdf pinout
BA 6818 FS	LIN-IC	SMD, Fan Motor-Tr., 2-phase half-wave	16-SMDIP	Rhm	-	data pdf pinout
BA 6819 AF	LIN-IC	SMD, Fan Motor-Tr., 2-phase half-wave	8-MDIP	Rhm	-	data pdf pinout
BA 6820 F	LIN-IC	SMD, 2x 12P.-LED VU-Meter, Ucc=5V	22-MDIP	Rhm	-	pdf
BA 6822 F	LIN-IC	=BA 6822S: SMD	22-MDIP		-	pdf
BA 6822 S	LIN-IC	2x 12P.-LED VU-Meter, Ucc=5V	22-SDIP	Rhm	-	pdf
BA 6825 FS	LIN-IC	Vc, 2-Phasen DD Motor-Tr., Ucc=8..20V	24-SMDIP	Rhm	-	data pdf pinout
BA 6826 FS	LIN-IC	Vc, 2-Phasen DD Motor-Tr., Ucc=8..20V	20-SMDIP	Rhm	-	data pdf pinout
BA 6827 FS	LIN-IC	Vc, 2-Phasen DD Motor-Tr, Ucc=8..20V	24-SMDIP	Rhm	-	
BA 6832 FS	LIN-IC	SMD, Fdd,Hdd, Voice Coil Motor Driver	24-SMDIP	Rhm	-	data pdf pinout
BA 6840 AFS	LIN-IC	SMD, Cd-rom, 3-Phasen Motor-Tr, 5V	20-SMDIP	Rhm	-	
BA 6840 BFP	LIN-IC	=BA 6840BFS: 1.7W	28-SMDIP+k	Rhm	-	data pdf pinout
BA 6840 BFP-Y	LIN-IC	=BA 6840BFS: 1.45W	25-SMDIP+k	Rhm	-	data pdf pinout
BA 6840 BFS	LIN-IC	=BA 6840AFS: verbessert/improved 0.93W	20-SMDIP	Rhm	-	data pdf pinout
BA 6842 BFS	LIN-IC	SMD, Cd-rom, 3-Phasen Motor-Tr., 5V 1W	32-SMDIP	Rhm	-	data pdf pinout
BA 6845 FS	LIN-IC	SMD, Fdd, Stepper Motor-Tr, 2,7..9V	16-SMDIP	Rhm	-	pdf
BA 6846 FS	LIN-IC	SMD, Fdd, Stepper Motor-Tr, 2,7..9V	16-SMDIP	Rhm	-	pdf
BA 6846 FV	LIN-IC	=BA 6846FS:	14-SSMDIP	Rhm	-	pdf
BA 6851 AFP-Y	LIN-IC	Optical Disk, Spindle-Motor-Tr, 5V	25-SMDIP+b	Rhm	-	
BA 6860 FS	LIN-IC	SMD, Camera, 3-Phasen Motor-Tr, 3...6V	32-SMDIP	Rhm	-	pdf
BA 6862 FS	LIN-IC	SMD, Camera, 3-Phasen Motor-Tr, 3...6V	32-SMDIP	Rhm	-	pdf
BA 6870 S	LIN-IC	Vc, 3-Phasen Motor-Tr.(Capstan)	24-DILP	Rhm	-	data pdf pinout
BA 6871 S,BS	LIN-IC	Vc, 3-Phasen Motor-Tr.(Capstan)	32-SDIP	Rhm	-	data pdf pinout
BA 6885 FP	LIN-IC	DC Motor-Tr, Ucc=6,5...28V, 1A	24-SMDIP+b	Rhm	-	
BA 6885 FS	LIN-IC	=BA 6885FP: Fig.→	16-SMDIP		-	
BA 6886	LIN-IC	DC Motor-Tr, Ucc=6,5...28V, 1A	10-SIP	Rhm	-	
BA 6886 N	LIN-IC	=BA 6886: Fig.→	10-SIL		-	
BA 6890 FP	LIN-IC	SMD, Cd, 4x Leistungs-Tr/power driver	28-SMDIP+b	Rhm	-	pdf
BA 6899 S	LIN-IC	SMD, Cd, 4x Leistungs-Tr/power driver	24-DILP	Rhm	-	
BA 6918	LIN-IC	Motor-Tr(reversible), <36V, 1A	10-SIL	Rhm	-	
BA 6918 N	LIN-IC	=BA 6918:	10-SIP		-	
BA 6919 FP-Y	LIN-IC	SMD, Motor-Tr(reversible), <36V, 1A	25-SMDIP+b	Rhm	-	
BA 6919 FS	LIN-IC	=BA 6919FP-Y:	16-SMDIP		-	
BA 6950 FS	LIN-IC	SMD, Motor-Tr.(reversible), <8V, 0,4A	16-SMDIP	Rhm	-	data pdf pinout
BA 6955 N	LIN-IC	VC++, Motor-Tr.(reversible), <18V, 1A	8-SIP	Rhm	-	data pdf pinout
BA 6970 FS	LIN-IC	Vc, 2-Phasen DD Motor-Tr., Ucc=8...16V	24-SMDIP	Rhm	-	data pdf pinout
BA 7001	LIN-IC	Audio,Vc, AV Switch, 2 Inputs	8-SIP	Rhm	-	
BA 7004	LIN-IC	Vc, Test-signal Generator	5-SIP	Rhm	-	pdf
BA 7004 F	LIN-IC	=BA 7004: SMD	8-MDIP		-	
BA 7005	LIN-IC	Vc, Modulator	16-SQP	Rhm	-	
BA 7007(L)	LIN-IC	Vc, SECAM-Diskriminator	16-SQP	Rhm	-	pdf pinout
BA 7007 F	LIN-IC	=BA 7007: SMD	16-MDIP		-	pdf pinout
BA 7021	LIN-IC	TV,Vc, Video Switch, 3 Inputs	9-SIP	Rhm	-	pdf
BA 7022 A	LIN-IC	Vc, 0,5H Skew Correction	22-DIP	Rhm	-	
BA 7023 L	LIN-IC	Vc, 0,5H Skew Detection	16-SQP	Rhm	-	
BA 7024	LIN-IC	Vc, Tsg Generator, Video Switch	7-SIP	Rhm	-	
BA 7025 L	LIN-IC	Vc/tv, Secam-diskriminator	18-SQP	Rhm	-	pdf
BA 7026 L	LIN-IC	VC++, Audio/video-signal Switch, mono	16-SQP	Rhm	-	
BA 7028 A	LIN-IC	Vc, Audio/video-signal Switch, stereo	22-DIP	Rhm	-	
BA 7032 L	LIN-IC	Vc, Chroma Skew Correction	16-SQP	Rhm	-	
BA 7036 LS	LIN-IC	Vc,Special Playback (noiseless Search)	24-SQP	Rhm	-	

Type	Device	Short Description	Fig.	Manu	Comparision Types	More at
BA 7039	LIN-IC	Vc(vhs), Auto Tracking Interface	16-DIP	Rhm	-	
BA 7041	LIN-IC	VCO, 18MHz, Ucc=4,5...5,5V	8-DIP	Rhm	-	
BA 7042	LIN-IC	Vco f. FM Intercom, Fsk Modem	8-DIP	Rhm	-	pdf
BA 7043 FS	LIN-IC	SMD, Vc(vhs), Auto Tracking Interface	20-SMDIP	Rhm	-	
BA 7044 N	LIN-IC	Vc, Testbild-/test pattern Generator	5-SIP	Rhm	-	
BA 7045	LIN-IC	Vc, NTSC→Pal Convertion	16-DIP	Rhm	-	pdf
BA 7045 FS	LIN-IC	=BA 7045: SMD	16-SMDIP		-	pdf
BA 7046	LIN-IC	Vc/tv, Synchr.-signal, Afc	8-DIP	Rhm	-	pdf
BA 7046 F	LIN-IC	=BA 7046: SMD	8-MDIP		-	pdf
BA 7047 S	LIN-IC	Vc(vhs), Auto Tracking Interface	22-SDIP	Rhm	-	
BA 7048 N	LIN-IC	Vc(vhs), Auto Tracking Interface	10-SIP	Rhm	-	
BA 7049 FS	LIN-IC	=BA 7049S: SMD	24-SMDIP		-	
BA 7049 S	LIN-IC	Vc, Chroma Signal Frequency Converter	24-SDIP	Rhm	-	
BA 7056 LS	LIN-IC	VC++, HiFi Audio Output Switch	24-SQP	Rhm	-	
BA 7057 S	LIN-IC	VC++, HiFi Audio Output Switch	22-SDIP	Rhm	-	
BA 7058 LS	LIN-IC	VC++, HiFi Audio Output Switch	24-SQP	Rhm	-	
BA 7062 F	LIN-IC	SMD, Tv/vc, Synchr.-signal, Afc	8-MDIP	Rhm	-	pdf
BA 7071 F	LIN-IC	SMD, Tv/vc, Synchr.-signal	8-MDIP	Rhm	-	pdf
BA 7072 N	LIN-IC	SMD, Tv/vc, Synchr.-signal	10-SIP	Rhm	-	
BA 7078 S	LIN-IC	Crt Display, hi-def, Synchr. Proz.	18-SDIP	Rhm	-	
BA 7082 F	LIN-IC	SMD, Sensivity Adj. Amp., Ucc=5V	16-MDIP	Rhm	-	
BA 7100	LIN-IC	Vc, Secam-diskriminator	16-DIP	Rhm	-	
BA 7101	LIN-IC	Vc, Lumin.-modulator	20-DIP	Rhm	-	
BA 7101 F	LIN-IC	=BA 7101: SMD	20-MDIP		-	
BA 7103	LIN-IC	Vc, Lumin.-modulator	20-DIP	Rhm	-	
BA 7103 F	LIN-IC	=BA 7103: SMD	20-MDIP		-	
BA 7106 LS	LIN-IC	Vc, Pal/secam Detect., Delay Line	24-SQP	Rhm	-	
BA 7107	LIN-IC	Vc(vhs), SECAM-Chroma Signal-proz.	28-DIP	Rhm	-	pdf
BA 7107 F	LIN-IC	=BA 7107: SMD	28-MDIP		-	pdf
BA 7107 S	LIN-IC	=BA 7107: Fig.→	30-SDIP		-	pdf
BA 7115 L	LIN-IC	Camera, Sucher-a/evf Deflection	18-SQP	Rhm	-	
BA 7116	LIN-IC	Vc, Lumin.-playback	20-DIP	Rhm	-	
BA 7116 F	LIN-IC	=BA 7116: SMD	20-MDIP		-	
BA 7122 F	LIN-IC	=BA 7122L: SMD	16-MDIP		-	
BA 7122 L	LIN-IC	Camera, FPN Elimination f. MOS Camera	16-SQP	Rhm	-	
BA 7125 L	LIN-IC	Camera, Sucher-a/evf Deflection	18-SQP	Rhm	-	
BA 7131 F	LIN-IC	SMD, TV,Vc, Video Switch, 2 Inputs	8-MDIP	Rhm	-	
BA 7135 L	LIN-IC	SMD, Camera, Sucher-a/evf Deflection	18-MDIP	Rhm	-	
BA 7145 F	LIN-IC	SMD, Camera, Sucher-a/evf Deflection	16-MDIP	Rhm	-	
BA 7148 F	LIN-IC	SMD,Camcorder, Sucher-a/evf Deflection	16-MDIP	Rhm	-	
BA 7149 F	LIN-IC	SMD,Camcorder, Sucher-a/evf Deflection	16-MDIP	Rhm	-	pdf
BA 7172 FS	LIN-IC	=BA 7172S: SMD	24-SMDIP		-	
BA 7172 S	LIN-IC	Vc, 2x Video-Signal-Verst./Amp.	22-SDIP	Rhm	-	
BA 7180 AS	LIN-IC	Vc, 2x Video-Signal-Verst./Amp.	22-SDIP	Rhm	-	pdf
BA 7180 FS,AFS	LIN-IC	=BA 7180AS: SMD	20-SMDIP	Rhm	-	
BA 7181 FS	LIN-IC	SMD, Vc, 2x Video-Signal-Verst./Amp.	20-SMDIP	Rhm	-	pdf
BA 7182 AS	LIN-IC	Vc, 2x Video-Signal-Verst./Amp.	22-SDIP	Rhm	-	pdf
BA 7184 S	LIN-IC	SMD, Vc, 4x Video-Signal-Verst./Amp.	24-SDIP	Rhm	-	pdf
BA 7207 AK	LIN-IC	=BA 7207AS: SMD	44-MP		-	pdf
BA 7207 S,AS	LIN-IC	Vc(vhs), Secam-chroma Signal-proz.	32-SDIP	Rhm	-	
BA 7212 S	LIN-IC	Vc, Video-signal A/w-verst.	22-SDIP	Rhm	-	
BA 7230 LS	LIN-IC	TV/camera/PC, Rgb-encoder f. NTSC	24-SQP	Rhm	-	pdf
BA 7232 FS	LIN-IC	SMD, TV, NTSC/PAL Color Encoder	24-SMDIP	Rhm	-	pdf
BA 7244 BS	LIN-IC	Vc, Video-signal A/w-verst.	32-SDIP	Rhm	-	
BA 7252 S	LIN-IC	Vc, Video-signal A/w-verst.	22-SDIP	Rhm	-	
BA 7253 S	LIN-IC	Vc, Video-signal A/w-verst.	22-SDIP	Rhm	-	
BA 7254 S	LIN-IC	Vc, Video-signal A/v-verst.	32-SDIP	Rhm	-	
BA 7258 BS,CS	LIN-IC	Vc, Luminanz Signal-prozessor	32-SDIP	Rhm	-	
BA 7258 AK	LIN-IC	=BA 7258AS: SMD	44-MP		-	
BA 7258 AS	LIN-IC	Vc, Luminanz Signal-prozessor	32-SDIP	Rhm	-	
BA 7266 F	LIN-IC	=BA 7266S: SMD	22-MDIP		-	
BA 7266 S	LIN-IC	Vc, Color Signal-prozessor	22-SDIP	Rhm	-	
BA 7267 F	LIN-IC	=BA 7267S: SMD	22-MDIP		-	
BA 7267 S	LIN-IC	Vc, Color Signal-prozessor	22-SDIP	Rhm	-	
BA 7274 S	LIN-IC	Vc, Video-signal A/w-verst.	32-SDIP	Rhm	-	pdf
BA 7277 S	LIN-IC	Vc, Video-signal A/w-verst.	32-SDIP	Rhm	-	
BA 7279 S	LIN-IC	Vc, Video-signal A/w-verst.	32-SDIP	Rhm	-	
BA 7280 AS	LIN-IC	Vc, Luminanz Signal-prozessor	32-SDIP	Rhm	-	
BA 7281 BS	LIN-IC	Vc, Luminanz Signal-prozessor	32-SDIP	Rhm	-	
BA 7288 K	LIN-IC	SMD, Vc, Luminanz Signal-prozessor	44-MP	Rhm	-	
BA 7307 S	LIN-IC	TV, NTSC Signal-prozessor, Filter	42-SDIP	Rhm	-	
BA 7356 S	LIN-IC	Tv/vc, Zf Signal-proz., Multistandard	22-SDIP	Rhm	-	pdf
BA 7357 S	LIN-IC	Tv/vc, Zf Signal-proz., Multistandard	22-SDIP	Rhm	-	pdf
BA 7358 S	LIN-IC	Tv/vc, Zf Signal-proz., Multistandard	22-SDIP	Rhm	-	
BA 7602	LIN-IC	TV,Vc, 3x AV Signal Switch, 2 Inputs	16-DIP	Rhm	-	pdf
BA 7602 F	LIN-IC	=BA 7602: SMD	16-MDIP		-	pdf
BA 7603	LIN-IC	TV,Vc, 3x Video Signal Switch, 2 Input	16-DIP	Rhm	-	pdf
BA 7603 F	LIN-IC	=BA 7603: SMD	16-MDIP		-	pdf
BA 7604 N	LIN-IC	TV,Vc, 2x AV Signal Switch	10-SIP	Rhm	-	pdf
BA 7605 N	LIN-IC	Vc,TV, 2x Video Switch, 2 Inputs	10-SIP	Rhm	-	pdf
BA 7606	LIN-IC	TV,Vc, 3x Video Signal Switch, 2 Input	16-DIP	Rhm	-	pdf
BA 7606 F	LIN-IC	=BA 7606: SMD	16-MDIP		-	pdf
BA 7607	LIN-IC	TV,Vc, 3x AV Signal Switch	16-DIP	Rhm	-	pdf
BA 7607 F	LIN-IC	=BA 7606: SMD	16-MDIP		-	pdf
BA 7608 N	LIN-IC	TV,Vc, 2x AV Signal Switch	10-SIP	Rhm	-	

Type	Device	Short Description	Fig.	Manu	Comparision Types	More at
BA 7609	LIN-IC	TV,Vc, 3x AV Signal Switch	16-DIP	Rhm	-	pdf
BA 7609 F	LIN-IC	=BA 7609: SMD	16-MDIP		-	pdf
BA 7611 Ñ,AN	LIN-IC	TV,Vc, Video Signal Switch, 3 Inputs	8-SIP	Rhm	-	
BA 7612 N	LIN-IC	TV,Vc, Video Signal Switch, 3 Inputs	8-SIP	Rhm	-	pdf
BA 7613 N	LIN-IC	TV,Vc, Video Signal Switch, 3 Inputs	8-SIP	Rhm	-	pdf
BA 7615 N	Hybrid-IC	Video signal switcher for VTR, TV	SIP	Roh	-	pdf pinout
BA 7622	LIN-IC	Tv/vc, 75Ω-Treiber/Driver, 3 Channel	8-DIP	Rhm	-	
BA 7622 F	LIN-IC	=BA 7622: SMD	8-MDIP		-	
BA 7623	LIN-IC	Tv/vc, 75Ω-Treiber/Driver, 3 Channel	8-DIP	Rhm	-	
BA 7623 F	LIN-IC	=BA 7623: SMD	8-MDIP		-	
BA 7625	LIN-IC	AV, 2x Video Signal Switch, 5 Inputs	16-DIP	Rhm	-	pdf
BA 7626	LIN-IC	AV, 2x Video Signal Switch, 5 Inputs	16-DIP	Rhm	-	pdf
BA 7626 F	LIN-IC	=BA 7626: SMD	16-MDIP		-	pdf
BA 7627 FV	LIN-IC	SMD, Vc, 3x AV Signal Switch, 2 Inputs	16-SSMDIP	Rhm	-	pdf
BA 7630 F	LIN-IC	=BA 7630S: SMD	28-MDIP		-	
BA 7630 S	LIN-IC	Vc, SECAM Switch f. Scramble TV(Video)	22-SDIP	Rhm	-	
BA 7631	LIN-IC	Vc, Switch f. Scramble TV (audio)	16-DIP	Rhm	-	pdf
BA 7631 F	LIN-IC	=BA 7631: SMD	16-MDIP		-	pdf
BA 7632 F,AF	LIN-IC	SMD, Vc, Switch f. Scramble TV (audio)	16-MDIP		-	
BA 7644 AN	LIN-IC	TV,Vc, AV Signal Switch, 4 Inputs	10-SIP	Rhm	-	
BA 7645 N	LIN-IC	TV,Vc, Video Signal Switch, 4 Inputs	10-SIP	Rhm	-	pdf
BA 7649 A	LIN-IC	ZV,Vc, Video Signal Switch, 5 Inputs	14-DIP	Rhm	-	pdf
BA 7649 AF	LIN-IC	=BA 4679A: SMD	14-MDIP		-	pdf
BA 7652 F,AF	LIN-IC	SMD, Vc, Video Switch, 3 Inputs	8-MDIP	Rhm	-	
BA 7653 F,AF	LIN-IC	SMD, Vc, Video Switch, 3 Inputs	8-MDIP	Rhm	-	
BA 7655 A	LIN-IC	Camcorder,Vc, Volt. Controlled Amp.	8-DIP	Rhm	-	pdf
BA 7655 AF	LIN-IC	=BA 7655: SMD	8-MDIP		-	pdf
BA 7657 F	LIN-IC	=BA 7657S: SMD	24-MDIP		-	pdf
BA 7657 S	LIN-IC	Crt/hdtv, Input Selector Switch	24-SDIP	Rhm	-	pdf
BA 7658 AFS	LIN-IC	SMD, Crt/hdtv, Input Selector Switch	32-SMDIP	Rhm	-	
BA 7700 K1	LIN-IC	SMD, Vc(vhs), HiFi Audio Signal-proz.	80-MP	Rhm	-	
BA 7703 K1	LIN-IC	SMD, Vc(vhs), HiFi Audio Signal-proz.	80-MP	Rhm	-	
BA 7705 K1	LIN-IC	SMD, Vc(vhs), HiFi Audio Signal-proz.	80-MP	Rhm	-	
BA 7706 KS	LIN-IC	SMD, Vc(vhs), HiFi Audio Signal-proz.	80-MP	Rhm	-	
BA 7710 S	LIN-IC	Vc, HiFi Audio FM Modem	30-SDIP	Rhm	-	
BA 7711 S	LIN-IC	Vc, HiFi Audio FM Modem	30-SDIP	Rhm	-	
BA 7720 S	LIN-IC	VC++, HiFi Audio Peak Noise Reduction	42-SDIP	Rhm	-	
BA 7721 S	LIN-IC	VC++, HiFi Audio Peak Noise Reduction	42-SDIP	Rhm	-	
BA 7725 FS	LIN-IC	=BA 7725S: SMD	20-SMDIP		-	pdf
BA 7725 S	LIN-IC	Audio Compression/expanding f. Karaoke	22-SDIP	Rhm	-	pdf
BA 7726 AFS	LIN-IC	=BA 7726AS: SMD	32-SMDIP		-	pdf
BA 7726 AS	LIN-IC	Audio Compression/expanding f. Karaoke	32-SDIP	Rhm	-	pdf
BA 7730 S	LIN-IC	VC++, HiFi Audio, Input/output Switch	32-SDIP	Rhm	-	
BA 7731 S	LIN-IC	VC++, HiFi Audio, Input/output Switch	32-SDIP	Rhm	-	
BA 7740 FS	LIN-IC	=BA 7740S: SMD	24-SMDIP		-	
BA 7740 S	LIN-IC	Vc, HiFi Audio A/W-Verst./Amp.	22-SDIP	Rhm	-	
BA 7743 FS	LIN-IC	Vc, HiFi Audio A/W-Verst./Amp.	24-SMDIP	Rhm	-	
BA 7745 FS	LIN-IC	SMD, Vc, HiFi Audio A/W-Verst./Amp.	32-SMDIP	Rhm	-	pdf
BA 7746 S	LIN-IC	Vc, HiFi Audio A/W-Verst./Amp.	24-SDIP	Rhm	-	pdf
BA 7750 AL	LIN-IC	Vc, Cue Detection	18-SQP	Rhm	-	
BA 7751 ALS	LIN-IC	Vc, Audio Signal-prozessor	24-SQP	Rhm	KA 8401	pdf
BA 7752 LS	LIN-IC	Vc, Audio Signal-prozessor	24-SQP	Rhm	-	pdf
BA 7755 A	LIN-IC	VC++, Audio Kopfumsch./head switch	5-SIP	Rhm	-	pdf
BA 7755 AF	LIN-IC	=BA 7755A: SMD	8-MDIP		-	pdf
BA 7757 BK	LIN-IC	SMD, Vc, Audio A/w-verst.	32-MP	Rhm	-	pdf
BA 7760	LIN-IC	Mikrofonverst./mike Amp. f. Karaoke	14-DIP	Rhm	-	pdf
BA 7760 F	LIN-IC	=BA 7760: SMD	14-MDIP		-	
BA 7765 AS	LIN-IC	VC++, Audio Signal-prozessor	32-SDIP	Rhm	-	
BA 7766 AS	LIN-IC	VC++, Audio Signal-prozessor	32-SDIP	Rhm	-	pdf
BA 7767 AS	LIN-IC	VC++, Audio Signal-prozessor	32-SDIP	Rhm	-	
BA 7780 KV	LIN-IC	SMD, Camera, Mikrofonverst./mike Amp.	64-MP	Rhm	-	pdf
BA 7781 K	LIN-IC	SMD, Camera, Mikrofonverst./mike Amp.	32-MP	Rhm	-	pdf
BA 7785 FS	LIN-IC	SMD, Speaker/Headph. Amp, Ucc=3..6V	20-SMDIP	Rhm	-	
BA 7786 FP-Y	LIN-IC	SMD, Speaker/Headph. Amp, Ucc=5V	25-SMDIP	Rhm	-	pdf
BA 7787 FS	LIN-IC	SMD, Speaker/Headph. Amp, Ucc=5V	20-SMDIP	Rhm	-	pdf
BA 7792 LS	LIN-IC	VC++, Audio Signal-prozessor	24-SPQ	Rhm	-	
BA 7795 FS	LIN-IC	=BA 7795LS: SMD	24-SMDIP		-	pdf
BA 7795 LS	LIN-IC	VC++, Audio Signal-prozessor	24-SQP	Rhm	-	pdf
BA 7796 FS	LIN-IC	SMD, VC++, Audio Signal-prozessor	24-SMDIP	Rhm	-	pdf
BA 7797 F	LIN-IC	SMD, VC++, Audio Signal-prozessor	18-MDIP	Rhm	-	pdf
BA 7798 FS	LIN-IC	SMD, VC++, Audio Signal-prozessor	18-SMDIP	Rhm	-	
BA 8201	LIN-IC	Telecom, Pseudo Inductance, Ucc=2...18V	8-DIP	Rhm	-	pdf
BA 8201 F	LIN-IC	=BA 8201: SMD	8-MDIP	Rhm	-	pdf
BA 8204	LIN-IC	Telecom, Klingelton/tone ringer	8-DIP	Rhm	-	pdf
BA 8204 F	LIN-IC	=BA 8204: SMD	8-MDIP		-	pdf
BA 8205	LIN-IC	Telecom, Klingelton/tone ringer	8-DIP	Rhm	-	
BA 8205 F	LIN-IC	=BA 8205: SMD	8-MDIP		-	
BA 8206	LIN-IC	Telecom, Klingelton/tone ringer	8-DIP	Rhm	-	pdf
BA 8206 F	LIN-IC	=BA 8206: SMD	8-MDIP		-	pdf
BA 8210 N	LIN-IC	Telecom, Sprechkreis/speech network	10-SIP	Rhm	-	
BA 8211 N	LIN-IC	Telecom, Sprechkreis/speech network	10-SIP	Rhm	-	
BA 8215	LIN-IC	Telecom, Sprechkreis/speech network	14-DIP	Rhm	-	
BA 8215 L	LIN-IC	=BA 8215: Fig.→	16-SQP		-	
BA 8216	LIN-IC	Telecom, Sprechkreis/speech network	14-DIP	Rhm	-	
BA 8221 AN	LIN-IC	Recorder, 2Ch. ALC, Ucc=12V	8-DIP	Rhm	-	

Type	Device	Short Description	Fig.	Manu	Comparision Types	More at
BA 8420	LIN-IC	Vc, Servo, Special Playback(fine Slow)	22-DIP	Rhm		
BA 8500	LIN-IC	Vc, Capstan-servo	28-DIP	Rhm		
BA 9101(B,S)	A/D-IC	8 Bit A/D-Converter	22-DIP	Rhm		
BA 9101(B,S)F	A/D-IC	=BA 9101(B,S): SMD	22-MDIP	Rhm		
BA 9201	D/A-IC	8 Bit, Input Data Latch, Ucc=5V	18-DIP	Rhm		
BA 9201 F	D/A-IC	=BA 9201: SMD	28-MDIP	Rhm		
BA 9211	D/A-IC	10 Bit, Reference Voltg. Supply	22-DIP	Rhm		
BA 9211 F	D/A-IC	=BA 9211: SMD	22-MDIP	Rhm		
BA 9221	D/A-IC	12 Bit, Digital Audio, Servo, ++	20-DIP	Rhm		
BA 9221 F	D/A-IC	=BA 9221: SMD	22-MDIP	Rhm		
BA 9700 A	SMPS-IC	S-Reg, DC-DC Converter, Ucc=3,55...24V	14-DIP	Rhm	-	pdf
BA 9700 AF	LIN-IC	=BA 9700A: SMD	14-MDIP			pdf
BA 9700 AFV	LIN-IC	=BA 9700A: SMD	14-SSMDIP			pdf
BA 9701	SMPS-IC	S-Reg, DC-DC Converter, Ucc=2,5V	8-DIP	Rhm		
BA 9701 F	LIN-IC	=BA 9701: SMD	8-MDIP			
BA 9702 FS	SMPS-IC	SMD, 3x S-Reg Controller, Ucc=3,6...23V	24-SMDIP	Rhm		
BA 9703 K	SMPS-IC	SMD, 3x S-Reg Controller, Ucc=3,6...18V	32-MP	Rhm		
BA 9704 N	SMPS-IC	DC-DC Converter f. Electr. Tuning	10-SIP	Rhm		
BA 9705 AK	SMPS-IC	SMD, 3x S-Reg Controller, Ucc=3,6...18V	32-MP	Rhm		
BA 9706 K	SMPS-IC	SMD, 3x S-Reg Controller, Ucc=3,6...18V	32-MP	Rhm		
BA 9707 KV	SMPS-IC	SMD, 4x S-Reg Controller, Ucc=3,5...12V	48-MP	Rhm		pdf
BA 9708 K	SMPS-IC	SMD, 3x S-Reg Controller, Ucc=3,5...12V	32-MP	Rhm		pdf
BA 9709 F	LIN-IC	SMD, Cd, Power Switch	8-MDIP	Rhm		
BA 9710 KV	SMPS-IC	SMD, 4x S-Reg Controller, Ucc=3,5...12V	48-MP	Rhm		pdf
BA 9741	LIN-IC	Two- Channel Switching regulator Contorller	MDIP	Utc		pdf pinout
BA 9743 AFV	SMPS-IC	SMD, 2x S-Reg Controller, Ucc=3,6...35V	16-SSMDIP	Rhm	-	pdf
BA 9755 S	LIN-IC	Multiscan Crt/hdtv, High Volt. Ctrl.	18-SDIP	Rhm		pdf
BA 9756 FS	LIN-IC	High-voltage control for Crt displays	HSMDIP	Rhm		pdf pinout
BA10324	4xOP	Vs:±16V Vu:>15V/mV Vo:±14V Vi0:<±9mV	14D	Roh		data pdf pinout
BA 10324(A)	OP-IC	Quad, Serie 124, ±16V, -40...+85°	14-DIP	Rhm	→Serie 124	data
BA 10324(A)F	OP-IC	=BA 10324(A): SMD	14-MDIP		... 124..., ... 224...	pdf
BA 10324 AFV	OP-IC	=BA 10324A: SMD	14-SMDIP		... 124..., ... 224...	pdf
BA 10339	KOP-IC	Quad, Serie 139, ±18V, -40...+85°	14-DIP	Rhm	→Serie 139	data pdf
BA 10339 F	KOP-IC	=BA 10339: SMD	14-MDIP		... 139..., ... 239...	pdf
BA 10339 FV	KOP-IC	=BA 10339: SMD	14-SSMDIP		... 139..., ... 239...	pdf
BA10358	2xOP	Vs:±16V Vu:<15V/mV Vo:28/0V Vi0:<5mV	8AD,9I	Roh		data pinout
BA 10358	OP-IC	Dual, Serie 158, ±16V, -40...+85°	8-DIP	Rhm	→Serie 158	data pdf
BA 10358 F	OP-IC	=BA 10358: SMD	8-MDIP		... 158..., ... 258..., ... 1458..., ... 1558...	pdf
BA 10358 N	OP-IC	=BA 10358: Fig.→	8-SIP		... 158..., ... 258..., ... 1458..., ... 1558...	pdf
BA 10393	KOP-IC	Dual, Serie 193, ±18V, -40...+85°	8-DIP	Rhm	→Serie 193	data pdf
BA 10393 F	KOP-IC	=BA 10393: SMD	8-MDIP		... 193..., ... 293...	pdf
BA 10393 N	KOP-IC	=BA 10393: Fig.→	8-SIP		... 193..., ... 293...	pdf
BA 12001	LIN-IC	7x PNP Darl.-Trans., 50V, 0,5A, B>1000	16-DIP	Rhm	-	pdf
BA 12002	LIN-IC	7x PNP Darl.-Trans., 50V, 0,5A, B>1000	16-DIP	Rhm	-	
BA 12003	LIN-IC	7x PNP Darl.-Trans., 50V, 0,5A, B>1000	16-DIP	Rhm	-	
BA 12004	LIN-IC	7x PNP Darl.-Trans., 50V, 0,5A, B>1000	16-DIP	Rhm	-	pdf
BA 13001	LIN-IC	6x PNP Darl, Ucc=3..8V, 0,5A, B>1000	16-DIP	Rhm	-	pdf
BA 13001 F	LIN-IC	=BA 13001: SMD	16-MDIP			
BA 13002	LIN-IC	6x Darl, Ucc=3..8V, 0,3A, B>1000	16-DIP	Rhm		
BA 13002 F	LIN-IC	=BA 13002: SMD	16-MDIP			pdf
BA 14741	OP-IC	Quad, NF, VCO, ±18V, -40...+85°	14-DIP	Rhm	-	pdf
BA 14741 F	OP-IC	=BA 14741: SMD	14-MDIP			
BA 15218	OP-IC	Dual,lo-noise,Serie 158,±18V, -20...75°	8-DIP	Rhm	→Serie 158	data pdf
BA 15218 F	OP-IC	=BA 15218: SMD	8-MDIP		... 158..., ... 258..., ... 1458..., ... 1558...	pdf
BA 15218 N	OP-IC	=BA 15218: Fig.→	8-SIP		... 158..., ... 258..., ... 1458..., ... 1558...	pdf
BA15532	2xLO-NOISE	Vs:±22V Vu:>10V/mV Vo:±13V Vi0:<5mV	8D	Roh		data pinout
BA 15532	OP-IC	Dual, ±21V, 20MHz, 8V/µs, -20...+75°	8-DIP	Rhm	NE 5532	pdf
BA 15532 F	OP-IC	=BA 15532: SMD	8-MDIP			pdf
BA 15532 N	OP-IC	=BA 15532: Fig.→	8-SIP		-	pdf
BA17805FP...17824FP	Z-IC	=BA 17805T...17824T:	30b	Rhm	... 7805...7824 (TO-251/252)	
BA 17805 T	Z-IC	Iso, +5V, 1A	17b	Rhm	... 7805... (TO-220 Iso)	pdf
BA 17806 T	Z-IC	Iso, +6V, 1A	17b	Rhm	... 7806... (TO-220 Iso)	pdf
BA 17807 T	Z-IC	Iso, +7V, 1A	17b	Rhm	... 7807... (TO-220 Iso)	pdf
BA 17808 T	Z-IC	Iso, +8V, 1A	17b	Rhm	... 7808... (TO-220 Iso)	pdf
BA 17809 T	Z-IC	Iso, +9V, 1A	17b	Rhm	... 7809... (TO-220 Iso)	pdf
BA 17810 T	Z-IC	Iso, +10V, 1A	17b	Rhm	... 7810... (TO-220 Iso)	pdf
BA 17812 T	Z-IC	Iso, +12V, 1A	17b	Rhm	... 7812... (TO-220 Iso)	pdf
BA 17815 T	Z-IC	Iso, +15V, 1A	17b	Rhm	... 7815... (TO-220 Iso)	pdf
BA 17818 T	Z-IC	Iso, +18V, 1A	17b	Rhm	... 7818... (TO-220 Iso)	pdf
BA 17820 T	Z-IC	Iso, +20V, 1A	17b	Rhm	... 7820... (TO-220 Iso)	pdf
BA 17824 T	Z-IC	Iso, +24V, 1A	17b	Rhm	... 7824... (TO-220 Iso)	pdf

BAL..BAQ

Type	Device	Short Description	Fig.	Manu	Comparision Types	More at
BAL 74	Si-Di	SMD, S, 50V, 0,25A, <4ns	35c	Phi,Sie,Tho	(BAR 74)[5]	data
BAL 74 W	Si-Di	=BAL 74: 85V, 0,175A	35c(2mm)			data
BAL 99	Si-Di	SMD, S, 70V, 0,25A, <6ns	35d	Phi,Sie,++	(BAR 99)[5]	data
BAL 99 W	Si-Di	=BAL 99: 85V, 0,15A	35d(2mm)		-	data
BAL 872	LIN-IC	Vc, Reel Motor Control	16-SQP	Rhm		
BAL 6309	LIN-IC	Vc, V-pulse Generator	16-SQP	Rhm		pdf
BAP	Si-P	=2SB1132-P (Typ-Code/Stempel/marking)	39	Rhm	→2SB1132	data

Type	Device	Short Description	Fig.	Manu	Comparision Types	More at
BAP 50-03	PIN-Di	SMD, Uni, 50V, 50mA, <40Ω(0,5mA) <5Ω(10mA/100MHz)	71a(1,7mm)	Phi	BA 595	data
BAP 50-04	PIN-Di	=BAP 50-03: Dual	35o		-	data
BAP 50-05	PIN-Di	=BAP 50-03: Dual	35p		-	data
BAP 51-03	PIN-Di	SMD, Uni, 60V, 60mA, <9Ω(0,5mA) <2,5Ω(10mA/100MHz)	71a(1,7mm)	Phi	BA 595	data
BAP 64-02	PIN-Di	SMD, 200V, 0,1A, 0,5pF, <40Ω(0,5mA) <3,8Ω(10mA/100MHz)	71a(1,3mm)	Phi	-	data
BAP 64-03	PIN-Di	=BAP 64-02:	71a(1,7mm)	Phi		data
BAP 64-04	PIN-Di	=BAP 64-02: Dual	35o	Phi		data
BAP 64-05	PIN-Di	=BAP 64-02: Dual	35p	Phi		data
BAQ	Si-P	=2SB1132-Q (Typ-Code/Stempel/marking)	39	Rhm	→2SB1132	data
BAQ 33	Si-Di	SMD, Pico-Ampere(Ir<3nA), 30V, 0,2A	72a(3,4mm)	Aeg	BAQ 133...135	data
BAQ 34	Si-Di	=BAQ 33: 60V	72a(3,4mm)	Aeg	BAQ 134...135	data
BAQ 35	Si-Di	=BAQ 33: 125V	72a(3,4mm)	Aeg	BAQ 135	data
BAQ 133	Si-Di	=BAQ 33:	72a(3,5mm)	Aeg	BAQ 33...35	data
BAQ 134	Si-Di	=BAQ 34:	72a(3,5mm)	Aeg	BAQ 34...35	data
BAQ 135	Si-Di	=BAQ 35:	72a(3,5mm)	Aeg	BAQ 35	data
BAQ 333	Si-Di	=BAQ 33:	72a(1,9mm)	Aeg	-	data
BAQ 334	Si-Di	=BAQ 34:	72a(1,9mm)	Aeg	-	data
BAQ 335	Si-Di	=BAQ 35:	72a(1,9mm)	Aeg	-	data
BAQ 800	PIN-Di	AM, HF-Abschwächer/Attenuator, >100kHz	31a	Phi	-	data
BAQ 806	PIN-Di	=BAQ 800: SMD	71a(5x2,5)		-	data

BAR

Type	Device	Short Description	Fig.	Manu	Comparision Types	More at
BAR	Si-P	=2SB1132-R (Typ-Code/Stempel/marking)	39	Rhm	→2SB1132	data
BAR 10	Si-Di	Schottky-Di, VHF/UHF-Dem, 20V, 35mA	31a	Tho	HSS 101, 1N5712, 1SS108	data
BAR 11	Si-Di	Schottky-Di, VHF/UHF-Dem, 15V, 20mA	31a	Tho	HSS 101, 1N5712, 1SS108	data
BAR 12-...	PIN-Di	1MHz...3GHz, 100V	31a	Sie	-	data
BAR 13-...	PIN-Di	=BAR 12:	71a(4mm)	Sie	-	data
BAR 14-...	PIN-Di	=BAR 12: SMD, Dual	35o	Sie	-	data
BAR 15-...	PIN-Di	=BAR 12: SMD, Dual	35l	Sie	-	data
BAR 16-...	PIN-Di	=BAR 12: SMD, Dual	35m	Sie	-	data
BAR 17-...	PIN-Di	=BAR 12: SMD	35e	Sie	-	data
BAR 18	Si-Di	SMD, Schottky-Di, 70V, 30mA	35e	Tho	1SS348	data
BAR 19	Si-Di	Schottky-Di, UHF-M, 4V, 30mA	31a	Tho	BA 480...481, BAT 19, BAT 29, 1S1925, ++	data
BAR 28	Si-Di	Schottky-Di, UHF-M, 70V, 15mA	31a	Tho	HSS 102, 1N5711	data
BAR 35	Si-Di	Schottky-Di, 5V	31a	Tho	-	data
BAR 42	Si-Di	SMD, Schottky-Di, 30V, 100mA	35e	Tho	BAT 54, BAT 64	data
BAR 43	Si-Di	SMD, Schottky-Di, 30V, 100mA	35e	Tho	BAT 54, BAT 64	data
BAR 43 A	Si-Di	=BAR 43: Dual	35m		BAT 54A, BAT 64-06	data
BAR 43 C	Si-Di	=BAR 43: Dual	35l		BAT 54C, BAT 64-05	data
BAR 43 S	Si-Di	=BAR 43: Dual	35o		BAT 54S, BAT 64-04	data
BAR 46	Si-Di	SMD, Schottky-Di, Uni, 100V	35e	Tho	-	data
BAR 46 A	Si-Di	=BAR 46: Dual	35m		-	data
BAR 60	PIN-Di	SMD, 3x PIN-Di, HF-Abschw., 100V, 0,1A	44	Sie	-	data
BAR 61	PIN-Di	SMD, 3x PIN-Di, HF-Abschw., 100V, 0,1A	44	Sie	-	data
BAR 63	PIN-Di	SMD, Uni, S, ...3GHz, 35V, 0,1A	35e	Sie	-	data
BAR 63-02W	PIN-Di	=BAR 63:	71a(1,3mm)		-	data
BAR 63-03W	PIN-Di	=BAR 63:	71a(1,7mm)		-	data
BAR 63-04	PIN-Di	=BAR 63: Dual	35o		-	data
BAR 63-04W	PIN-Di	=BAR 63: Dual	35o(2mm)		-	data
BAR 63-05	PIN-Di	=BAR 63: Dual	35l		-	data
BAR 63-05W	PIN-Di	=BAR 63: Dual	35l(2mm)		-	data
BAR 63-06	PIN-Di	=BAR 63: Dual	35m		-	data
BAR 63-06W	PIN-Di	=BAR 63: Dual	35m(2mm)		-	data
BAR 63-07	PIN-Di	=BAR 63: Dual	44s2		-	data
BAR 64	PIN-Di	SMD, Uni, S, ...3GHz, 200V, 0,1A	35e	Sie	-	data
BAR 64-02W	PIN-Di	=BAR 64:	71a(1,3mm)		-	data
BAR 64-03W	PIN-Di	=BAR 64:	71a(1,7mm)		-	data
BAR 64-04	PIN-Di	=BAR 64: Dual	35o		-	data
BAR 64-04W	PIN-Di	=BAR 64: Dual	35o(2mm)		-	data
BAR 64-05	PIN-Di	=BAR 64: Dual	35l		-	data
BAR 64-05W	PIN-Di	=BAR 64: Dual	35l(2mm)		-	data
BAR 64-06	PIN-Di	=BAR 64: Dual	35m		-	data
BAR 64-06W	PIN-Di	=BAR 64: Dual	35m(2mm)		-	data
BAR 64-07	PIN-Di	=BAR 64: Dual	44s2		-	data
BAR 65-02W	PIN-Di	=BAR 65:	71a(1,3mm)		-	data
BAR 65-03W	PIN-Di	SMD, Uni, S, 35V, 0,1A	71a(1,7mm)	Sie	-	data
BAR 65-07	PIN-Di	=BAR 65: Dual	44s2		-	data
BAR 66	PIN-Di	SMD, Dual, Surge Prot., 150V, 0,2A	35o	Sie	-	data
BAR 74	Si-Di	=BAL 74:	35e	Sie,Tho	(BAL 74)[5]	data
BAR 79	PIN-Di	60V, 5Ω(10mA), 1100Ω(10µA)	31a	Sie	-	data
BAR 80	Si-Di	SMD, HF-S, Shunt, 35V, 0,1A	44(AKAK)	Sie	-	data

Type	Device	Short Description	Fig.	Manu	Comparision Types	More at
BAR 81	Si-Di	SMD, HF-S, Shunt, 30V, 0,1A	44(AKAK)	Sie	-	data
BAR 81W	Si-Di	=BAR 81:	44(2mm)	Sie	-	data
BAR 99	Si-Di	=BAL 99:	35f	Sie,Tho	(BAL 99)[5]	data
BAR 223-10...-70	Si-Di	Gl, 1000...7000V, 0,5A..0,075A	31a	Edl		data

BAS

Type	Device	Short Description	Fig.	Manu	Comparision Types	More at
BAs	Si-N	=BCP 54M (Typ-Code/Stempel/marking)	≈45	Sie	→BCP 54M	data
BAS 11	Si-Di	Uni, contr.av., 300V, 0,35A, <1µs	31a	Phi	BYD 11G...M, BYX 57/...	data
BAS 12	Si-Di	=BAS 11: 400V	31a	Phi	BYD 11G...M, BYX 57/...	data
BAS 15	Si-Di	S, Uni, 50V, 75mA, <4ns	31a	Phi	BA 218, BAX 13, BAX 91, 1N4148, ++	data
BAS 16	Si-Di	SMD, SS, 85V, 0,25A, <6ns	35e	Phi,Sie,++	BAS 678, (BAL 99, BAR 99)[5]	data
BAS 16-02W	Si-Di	=BAS 16:	71a(1,3mm)		-	data
BAS 16-03W	Si-Di	=BAS 16:	71a(1,7mm)		MA 113	data
BAS 16 D	Si-Di	=BAS 16:	71a(2,7mm)	Gsi		data
BAS 16 H(T1)	Si-Di	=BAS 16:	71a(1,7mm)	Ons	MA 113	data
BAS 16 S	Si-Di	=BAS 16: Triple	46bg(2mm)		-	data
BAS 16 T	Si-Di	=BAS 16:	35e(1,6mm)		-	data
BAS 16 W	Si-Di	=BAS 16:	35e(2mm)		-	data
BAS 16 WS	Si-Di	=BAS 16:	71a(1,7mm)	Gsi	-	data
BAS 17	Si-St	SMD, Stabi, 0,75...0,83V/10mA	35e	Phi	BZX 84/C0V8	data
BAS 19	Si-Di	SMD, S, Uni, 120V, 0,2A, <50ns	35e	Phi,Sie,++	BAS 20, BAS 21	data
BAS 20	Si-Di	=BAS 19: 200V	35e	Phi,Sie,++	BAS 21	data
BAS 20 H(T1)	Si-Di	=BAS 19: 200V	71a(1,7mm)	Ons	BAV 20WS..21WS	
BAS 21	Si-Di	=BAS 19: 250V	35e	Phi,Sie,++	1SS250	data
BAS 21 H(T1)	Si-Di	=BAS 19: 250V	71a(1,7mm)	Ons	BAV 21WS	
BAS 21 L(T1)	Si-Di	=BAS 19: 250V	35e	Ons	1SS250	
BAS 21 SL(T1)	Si-Di	=BAS 19: Dual, 250V, 0,225mA	35o	Ons	-	
BAS 22	Si-Di	Schottky-Di, 4V, ...18GHz	Chip	Phi		data
BAS 23	Si-Di	Schottky-Di, 4V, ...18GHz	Chip	Phi		data
BAS 24	Si-Di	Schottky-Di, 4V, ...18GHz	Chip	Phi		data
BAS 25	Si-Di	Schottky-Di, 4V, ...18GHz	Chip	Phi		data
BAS 26-20	Si-Di	20V,	31a	Sie		data
BAS 27	Si-Di	SMD, S, 300V, 0,25A, <1µs	35e	Phi	-	data
BAS 28	Si-Di	=BAS 16: Dual	44s2	Phi, Sie	-	data
BAS 28W	Si-Di	=BAS 16: Dual	44s2(2mm)	Sie	-	data
BAS 29	Si-Di	SMD, S, 90V, 0,25A, <50ns	35e	Phi,Tho	BAS 19	data
BAS 31	Si-Di	=BAS 29: Dual	35o	Phi,Nsc,Tho	-	data
BAS 32(L)	Si-Di	SMD, SS, 75V, 0,2A, <4ns	72a(3,4mm)	Phi	LL 4148	data
BAS 33	Si-Di	Picoampere(Ir<3nA), 30V, 0,2A	31a	Aeg	BAS 45, BAY 135	data
BAS 34	Si-Di	=BAS 33: 60V	31a	Aeg	BAS 45, BAY 135	data
BAS 35	Si-Di	=BAS 29: Dual	35m	Phi, Nsc	-	data
BAS 40	Si-Di	SMD, Schottky, 40V, 0,12A, Uf<1V(40mA)	35e	Sie, Phi	1SS294, 1SS392	data
BAS 40 W	Si-Di	=BAS 40:	35e(2mm)		-	data
BAS 40-01	Si-Di	=BAS 40:	Chip		-	data
BAS 40-02	Si-Di	=BAS 40:	31a		-	data
BAS 40-03	Si-Di	=BAS 40:	71a		-	data
BAS 40-04	Si-Di	=BAS 40: Dual	35o		-	data
BAS 40-04L(T1)	Si-Di	=BAS 40L(T1): Dual	35o	Ons	-	
BAS 40-04W	Si-Di	=BAS 40: Dual	35o(2mm)		-	data
BAS 40-05	Si-Di	=BAS 40: Dual	35l		1SS392	data
BAS 40-05W	Si-Di	=BAS 40: Dual	35l(2mm)		1SS393	data
BAS 40-06	Si-Di	=BAS 40: Dual	35m		-	data
BAS 40-06L(T1)	Si-Di	=BAS 40L(T1): Dual	35m	Ons	-	
BAS 40-06W	Si-Di	=BAS 40: Dual	35m(2mm)		-	data
BAS 40-07	Si-Di	=BAS 40: Dual	44s2		-	data
BAS 40-07W	Si-Di	=BAS 40: Dual	44s2(2mm)		-	data
BAS 40 L(T1)	Si-Di	SMD, Schottky, 40V, Uf<1V(100mA)	35e	Ons	1SS294, 1SS392	data
BAS 45(A)	Si-Di	Picoampere(Ir<1nA), 125V, 0,225A A: 250mA	31a	Phi	BAY 135	data
BAS 45 (A)L	Si-Di	=BAS 45: SMD	72a(3,4mm)		-	data
BAS 46	Si-Di	Schottky-Di, X-Band-Dem, 2V		Phi	-	data
BAS 55	Si-Di	SMD, SS, 60V, 0,25A, <6ns	35e	Phi	BAS 16, BAS 678, (BAL 99, BAR 99)[5]	data
BAS 56	Si-Di	SMD, Dual, SS, 60V, 0,2A, <6ns	44s2	Phi	-	data
BAS 70	Si-Di	SMD, Schottky-Di, 70V, 0,07A	35e	Phi,Sie,Tho Ztx	-	data
BAS 70-01	Si-Di	=BAS 70:	Chip		-	data
BAS 70-02	Si-Di	=BAS 70:	31a		-	data
BAS 70-03	Si-Di	=BAS 70:	71a		-	data
BAS 70-04	Si-Di	=BAS 70: Dual	35o		-	data
BAS 70-04L(T1)	Si-Di	=BAS 70L: Dual	35o	Ons	-	
BAS 70-04S	Si-Di	=BAS 70: 2x Dual	46(2mm)		-	data
BAS 70-04W	Si-Di	=BAS 70: Dual	35o(2mm)		-	data
BAS 70-05	Si-Di	=BAS 70: Dual	35l		-	data
BAS 70-05W	Si-Di	=BAS 70: Dual	35l(2mm)		-	data
BAS 70-06	Si-Di	=BAS 70: Dual	35m		-	data
BAS 70-06S	Si-Di	=BAS 70: 2x Dual	46(2mm)		-	data

Type	Device	Short Description	Fig.	Manu	Comparision Types	More at
BAS 70-06W	Si-Di	=BAS 70: Dual	35m(2mm)		-	data
BAS 70-07	Si-Di	=BAS 70: Dual	44s2		-	data
BAS 70-07W	Si-Di	=BAS 70: Dual	44s2(2mm)		-	data
BAS 70 J	Si-Di	=BAS 70:	71a(1,7mm)		BAS 170W	data
BAS 70 L(T1)	Si-Di	SMD, Schottky, 70V, Uf<1V(15mA)	35e	Ons	1SS348	
BAS 70 W	Si-Di	=BAS 70:	35e(2mm)			data
BAS 78 A	Si-Di	S, Uni, 50V, 1A, 1µs	48a	Sie	-	data
BAS 78 B	Si-Di	=BAS 78A: 100V	48a	Sie	-	data
BAS 78 C	Si-Di	=BAS 78A: 200V	48a	Sie	-	data
BAS 78 D	Si-Di	=BAS 78D: 400V	48a	Sie	-	data
BAS 79 A...D	Si-Di	=BAS 78A...D: Dual	48p	Sie	-	data
BAS 81	Si-Di	=BAT 81: SMD	72a(3,4mm)	Aeg, Phi	-	data
BAS 82	Si-Di	=BAT 82: SMD	72a(3,4mm)	Aeg, Phi	-	data
BAS 83	Si-Di	=BAT 83: SMD	72a(3,4mm)	Aeg, Phi	-	data
BAS 85	Si-Di	SMD, Schottky-Di, 30V, 0,2A, <5ns	72a(3,4mm)	Aeg, Phi	TMMBAT 42...43, TMMBAT 48	data
BAS 86	Si-Di	SMD, Schottky-Di, 50V, 0,2A, <4ns	72a(3,4mm)	Phi	TMMBAT 49	data
BAS 116	Si-Di	SMD, Picoampere, 85V, 0,25A, Ir<5nA	35e	Sie, Mot	-	data
BAS 125	Si-Di	SMD, Schottky-Di, Clamping, 25V, 0,1A	35e	Sie	BAR 42...43	data
BAS 125 W	Si-Di	=BAS 125:	35e(2mm)		-	data
BAS 125-04	Si-Di	=BAS 125: Dual	35o		-	data
BAS 125-04W	Si-Di	=BAS 125: Dual	35o(2mm)		-	data
BAS 125-05	Si-Di	=BAS 125: Dual	35l		-	data
BAS 125-05W	Si-Di	=BAS 125: Dual	35l(2mm)		-	data
BAS 125-06	Si-Di	=BAS 125: Dual	35m		-	data
BAS 125-06W	Si-Di	=BAS 125: Dual	35m(2mm)		-	data
BAS 125-07	Si-Di	=BAS 125: Dual	44s2		-	data
BAS 125-07W	Si-Di	=BAS 125: Dual	44s2(2mm)		-	data
BAS 140W	Si-Di	SMD, Schottky-Di, Clamping, 40V, 0,12A	71a(1,7mm)	Sie	1SS357	data
BAS 170W	Si-Di	SMD, Schottky-Di, Clamping, 70V, 0,07A	71a(1,7mm)	Sie	BAS 70J	data
BAS 216	Si-Di	=BAS 16:	71a(2mm)	Phi	-	data
BAS 221	Si-Di	SMD, S, Uni, 250V, 0,3A, <50ns	71a(2mm)	Phi	-	data
BAS 281	Si-Di	=BAT 81: SMD	72a(3,5mm)	Aeg	-	data
BAS 282	Si-Di	=BAT 82: SMD	72a(3,5mm)	Aeg	-	data
BAS 283	Si-Di	=BAT 83: SMD	72a(3,5mm)	Aeg	-	data
BAS 285	Si-Di	=BAT 85: SMD	72a(3,5mm)	Aeg	-	data
BAS 286	Si-Di	=BAT 86: SMD	72a(3,5mm)	Aeg	-	data
BAS 316	Si-Di	SMD, SS, 85V, 0,25A, <4ns	71a(1,7mm)	Phi	BAS 16-03W, MA 113	data
BAS 321	Si-Di	SMD, S, Uni, 250V, 0,25A, <50ns	71a(1,7mm)	Phi	-	data
BAS 381	Si-Di	=BAT 81: SMD	72a(1,9mm)	Aeg	-	data
BAS 382	Si-Di	=BAT 82: SMD	72a(1,9mm)	Aeg	-	data
BAS 383	Si-Di	=BAT 83: SMD	72a(1,9mm)	Aeg	-	data
BAS 385	Si-Di	=BAT 85: SMD	72a(1,9mm)	Aeg	-	data
BAS 386	Si-Di	=BAT 86: SMD	72a(1,9mm)	Aeg	-	data
BAS 516	Si-Di	SMD, SS, 85V, 0,25A, <4ns	71a(1,3mm)	Phi	-	data
BAS 678	Si-Di	SMD, SS, 100V, 0,25A, <6ns	35e	Phi	-	data

BAT

Type	Device	Short Description	Fig.	Manu	Comparision Types	More at
BAT 10	Si-Di	Schottky-Di, X-Band-M	31a	Phi	-	data
BAT 11	Si-Di	Schottky-Di, X-Band-M	Chip	Phi	-	data
BAT 12	Si-Di	Schottky-Di, X-Band-Monitor		Phi	-	data
BAT 13	Si-Di	S, 60V, 0,2A, <10ns	9a	Aeg	BAW 62, BAX 95, BAY 61, 1N4148, 1N4151,+	data
BAT 14-...	Si-Di	Schottky-Di, S-Band-M	div.	Sie	-	data
BAT 15-...	Si-Di	Schottky-Di, S-Band-M	div.	Sie	-	data
BAT 16-...	Si-Di	Schottky-Di, 40V, 30mA	31a	Sie	-	data
BAT 17	Si-Di	SMD, Schottky-Di, S, UHF-M, 4V, 30mA	35e	Phi,Sie,Tho	1SS350	data
BAT 17 W	Si-Di	=BAT 17:	35e(2mm)			data
BAT 17 WS	Si-Di	=BAT 17:	71(1,7mm)	Gsi	HSR 276	data
BAT 17 W	Si-Di	=BAT 17:	71(2,7mm)	Gsi	HSU 276, HSU 88	data
BAT 17-04, DS	Si-Di	=BAT 17: Dual	35o			data
BAT 17-04W	Si-Di	=BAT 17: Dual	35o(2mm)			data
BAT 17-05	Si-Di	=BAT 17: Dual	35l			data
BAT 17-05W	Si-Di	=BAT 17: Dual	35l(2mm)			data
BAT 17-06	Si-Di	=BAT 17: Dual	35m			data
BAT 17-06W	Si-Di	=BAT 17: Dual	35m(2mm)			data
BAT 17-07	Si-Di	=BAT 17: Dual	44s2			data
BAT 18(DK)	Si-Di	SMD, VHF/UHF-S, 35V, 0,1A	35e	Phi,Sie,Tho	MA 57, 1SS153	data
BAT 18-04...06	Si-Di	Dual	35...		-	data
BAT 19	Si-Di	Schottky-Di, UHF-M, 10V, 30mA	31a	Tho	BAR 10...11, 1N5712, 1SS88, 1SS106	data
BAT 21	Si-Di	16-Di-Array, 60V, 0,4A, <20ns	14-DIP	Fer	-	data
BAT 22	Si-Di	16-Di-Array, 40V, 0,4A, <20ns	14-DIP	Fer	-	data
BAT 23	Si-Di	16-Di-Array, 60V, 0,3A, <20ns	14-DIP	Fer	-	data
BAT 24	Si-Di	16-Di-Array, 40V, 0,3A, <20ns	14-DIP	Fer	-	data
BAT 25	Si-Di	16-Di-Array, 60V, 0,4A, <20ns	14-DIP	Fer	-	data
BAT 26	Si-Di	16-Di-Array, 40V, 0,4A, <20ns	14-DIP	Fer	-	data

Type	Device	Short Description	Fig.	Manu	Comparision Types	More at
BAT 27	Si-Di	16-Di-Array, 60V, 0,3A, <20ns	14-DIP	Fer	-	data
BAT 28	Si-Di	16-Di-Array, 40V, 0,3A, <20ns	14-DIP	Fer	-	data
BAT 29	Si-Di	Schottky-Di, SS, UHF-M, 5V, 30mA	31a	Tho	BA 480..481, BAR 19, 1SS88, 1SS106	data
BAT 30	Si-Di	Schottky-Di, zero bias, ..25GHz	Chip	Sie	-	data
BAT 31	Si-Di	µWellen-Rausch-Di/µwave noise diode	Koax	Phi	-	data
BAT 32	Si-Di	Schottky-Di, ...18GHz, 6,5V, 50mA	51a/2Pin	Sie	-	data
BAT 34	Si-Di	5V, 0,05A	31a	Tho	-	data
BAT 35	Si-Di	=4x BAT 34	31a	Tho	-	data
BAT 36	Si-Di	5V, 0,1A	2c	Tho	-	data
BAT 37	Si-Di	=2x BAT 36	2c	Tho	-	data
BAT 38	Si-Di	Schottky-Di, ...40GHz	Koax	Phi	-	data
BAT 39(A)	Si-Di	Schottky-Di, 1...18GHz	Koax	Phi	-	data
BAT 40	Si-Di			Phi	-	data
BAT 41	Si-Di	Schottky-Di, 100V, 0,1A, <5ns	31a	Tho, Gsi	BAT 46	data
BAT 42	Si-Di	Schottky-Di, 30V, 0,2A, <5ns	31a	Itt,Tho,Gsi	BAT 48, BAT 85	data
BAT 43	Si-Di	Schottky-Di, 30V, 0,2A, <5ns	31a	Itt,Tho,Gsi	BAT 48, BAT 85	data
BAT 42W...43W	Si-Di	=BAT 42..43: SMD	71a(2,7mm)		-	data
BAT 45	Si-Di	Schottky-Di, SS, UHF-M, 15V, 30mA	31a	Tho	BAR 10...11, 1N5712	data
BAT 46	Si-Di	Schottky-Di, Uni, 100V, 0,15A	31a	Itt,Tho,Gsi	BAT 41	data
BAT 46 AW	Si-Di	=BAT 46: SMD, Dual	35m(2mm)	Tho	-	data
BAT 46 CW	Si-Di	=BAT 46: SMD, Dual	35l(2mm)	Tho	-	data
BAT 46 J	Si-Di	=BAT 46: SMD	71a(1,7mm)	Tho	-	data
BAT 46 SW	Si-Di	=BAT 46: SMD	35o(2mm)	Tho	-	data
BAT 46 W	Si-Di	=BAT 46: SMD	71a(2,7mm)	Itt, Gsi	-	data
BAT 46 W	Si-Di	=BAT 46: SMD	35e(2mm)	Tho	-	data
BAT 47	Si-Di	Schottky-Di, Uni, 20V, 0,35A	31a	Tho	BAT 42..43, BAT 85, MBR 030	data
BAT 48	Si-Di	Schottky-Di, Uni, 40V, 0,35A	31a	Tho, Gsi	BAT 86, MBR 040	data
BAT 49	Si-Di	Schottky-Di, 80V, 0,5A	31a	Tho	BYS 21-90, HRP 32, SB 180	data
BAT 50(R)	Si-Di	Schottky-Di, 8...12GHz	Koax	Phi	-	data
BAT 51(R)	Si-Di	Schottky-Di, 12...18GHz	Koax	Phi	-	data
BAT 52(R)	Si-Di	Schottky-Di, 12...18GHz	Koax	Phi	-	data
BAT 53	Si-Di	SMD, Schottky-Di, 10V, 10mA	35e	Tho		data
BAT 54	Si-Di	SMD, Schottky, SS, 30V, 0,2A, <5ns	35e	Aeg,Phi, ++	BAT 64	data
BAT 54 A	Si-Di	=BAT 54: Dual	35m		BAT 64-06	data
BAT 54 AL(T1)	Si-Di	=BAT 54L: Dual	35m	Ons	BAT 64-06	
BAT 54 AW	Si-Di	=BAT 54: Dual	35m(2mm)		BAT 64-06W	data
BAT 54 C	Si-Di	=BAT 54: Dual	35l		BAT 64-05	data
BAT 54 CW	Si-Di	=BAT 54: Dual	35l(2mm)		BAT 64-05W	data
BAT 54 H(T1)	Si-Di	=BAT 54L:	71a(1,7mm)	Ons	1PS 76SB10	
BAT 54 J	Si-Di	=BAT 54:	71a(1,7mm)		1PS 76SB10	
BAT 54 L(T1)	Si-Di	SMD, Schottky, SS, 30V, Uf<0,8V, <5ns	35e	Ons	BAT 64	
BAT 54 S	Si-Di	=BAT 54: Dual	35o		BAT 64-04	data
BAT 54 SL(T1)	Si-Di	=BAT 54L(T1): Dual	35o	Ons	BAT 64-04	
BAT 54 SW	Si-Di	=BAT 54: Dual	35o(2mm)		BAT 64-04W	data
BAT 54 SW(T1)	Si-Di	=BAT 54L(T1): Dual	35o(2mm)	Ons	BAT 64-04W	
BAT 54 T1	Si-Di	=BAT 54L:	71a(2,7mm)	Ons	BAT 43W	
BAT 54 W	Si-Di	=BAT 54:	35e(2mm)		-	data
BAT 54 W(T1)	Si-Di	=BAT 54L(T1):	35e(2mm)	Ons	-	
BAT 54 W	Si-Di	=BAT 54:	71(2,7mm)	Gsi	BAT 43W	data
BAT 59	Si-Di	Schottky-Di, 26..40GHz	Koax	Phi	-	data
BAT 60A,B	Si-Di	SMD, Schottky-Di, Uni, 10V, 3A	71a(1,7mm)	Sie	-	data
BAT 62	Si-Di	SMD, Dual, Schottky-Di, UHF, 40V, 20mA	44	Sie	-	data
BAT 62-02W	Si-Di	=BAT 62: 40mA	71a(1,3mm)		-	data
BAT 62-03W	Si-Di	=BAT 62: 40mA	71a(1,7mm)		-	data
BAT 62-07W	Si-Di	=BAT 62:	44s2(2mm)		-	data
BAT 63	Si-Di	SMD, Dual, Schottky-Di, 3V, 0,1A	44	Sie	-	data
BAT 63-03W	Si-Di	=BAT 63: SMD	71a(1,7mm)		-	data
BAT 64	Si-Di	SMD, Schottky-Di, SS, 30V, 0,2A, <5ns	35e	Sie	BAT 54	data
BAT 64 W	Si-Di	=BAT 64:	35e(2mm)		BAT 54W	data
BAT 64-04	Si-Di	=BAT 64: Dual	35o		BAT 54S	data
BAT 64-04W	Si-Di	=BAT 64: Dual	35o(2mm)		BAT 54SW	data
BAT 64-05	Si-Di	=BAT 64: Dual	35l		BAT 54C	data
BAT 64-05W	Si-Di	=BAT 64: Dual	35l(2mm)		BAT 54CW	data
BAT 64-06	Si-Di	=BAT 64: Dual	35m		BAT 54A	data
BAT 64-06W	Si-Di	=BAT 64: Dual	35m(2mm)		BAT 54AW	data
BAT 64-07	Si-Di	=BAT 64: Dual	44s2		BAT 74	data
BAT 64-07W	Si-Di	=BAT 64: Dual	44s2(2mm)		-	data
BAT 65	Si-Di	SMD, Schottky-Di, Clamping, 30V, 0,5A	71a(2,7mm)	Sie	MBR 0530	data
BAT 66	Si-Di	Schottky-Di, Clamping, 30V, 2A	48a	Sie	-	data
BAT 66-04	Si-Di	=BAT 66: Dual	48o		-	data
BAT 66-05	Si-Di	=BAT 66: Dual, 40V	48p		-	data
BAT 66-06	Si-Di	=BAT 66: Dual	48m		-	data
BAT 68	Si-Di	SMD,Schottky-Di,VHF/UHF-M,S, 8V, 0,13A	35e	Sie	-	data
BAT 68 W	Si-Di	=BAT 68:	35e(2mm)		-	data
BAT 68-03W	Si-Di	=BAT 68:	71a(1,7mm)		-	data

Type	Device	Short Description	Fig.	Manu	Comparision Types	More at
BAT 68-04	Si-Di	=BAT 68: Dual	35o		-	data
BAT 68-04W	Si-Di	=BAT 68: Dual	35o(2mm)		-	data
BAT 68-05	Si-Di	=BAT 68: Dual	35l		-	data
BAT 68-05W	Si-Di	=BAT 68: Dual	35l(2mm)		-	data
BAT 68-06	Si-Di	=BAT 68: Dual	35m		-	data
BAT 68-06W	Si-Di	=BAT 68: Dual	35m(2mm)		-	data
BAT 68-07	Si-Di	=BAT 68: Dual	44s2		-	data
BAT 68-07W	Si-Di	=BAT 68: Dual	44s2(2mm)		-	data
BAT 69	Si-Di	=BAT 66: SMD	71a(2,7mm)	Sie	-	data
BAT 70-05	Si-Di	SMD, Dual, Schottky, Clamp., 70V, 1,5A	48(A-AK)	Sie	-	data
BAT 74	Si-Di	SMD, Dual, Schottky-Di, 30V, 0,2A,<5ns	44s2	Phi	BAT 65-07	data
BAT 74 S	Si-Di	SMD, Dual, Schottky-Di, 30V, 0,2A,<5ns	46(2mm)	Phi	-	data
BAT 81(S)	Si-Di	Schottky-Di, SS, 40V, 30mA, <1ns	31a	Aeg, Phi	HSS 100, 1N6263	data
BAT 82(S)	Si-Di	=BAT 81: 50V	31a	Aeg, Phi	HSS 100, 1N6263	data
BAT 83(S)	Si-Di	=BAT 81: 60V	31a	Aeg, Phi	HSS 100, 1N6263	data
BAT 85(S)	Si-Di	Schottky-Di, SS, 30V, 0,2A, <5ns	31a	Aeg,Gsi,Phi	BAT 42...43	data
BAT 86(S)	Si-Di	Schottky-Di, SS, 50V, 0,2A, <4ns	31a	Phi, Gsi	BAT 49	data
BAT 114-099	Si-Di	SMD, Dual, Schottky-Di, 4V, 90mA	44	Sie	Pin-Code: A1K2A2K1	data
BAT 114-099R	Si-Di	Quad	44		-	data
BAT 120 A	Si-Di	SMD, Dual, Schottky-Di, 25V, 1A	48aa	Phi	-	data
BAT 120 C	Si-Di	=BAT 120A:	48ab		-	data
BAT 120 S	Si-Di	=BAT 120A:	48ac		-	data
BAT 140 A	Si-Di	SMD, Dual, Schottky-Di, 40V, 1A	48aa	Phi	-	data
BAT 140 C	Si-Di	=BAT 140A:	48ab		-	data
BAT 140 S	Si-Di	=BAT 140A:	48ac		-	data
BAT 160 A	Si-Di	SMD, Dual, Schottky-Di, 60V, 1A	48aa	Phi	-	data
BAT 160 C	Si-Di	=BAT 160A:	48ab		-	data
BAT 160 S	Si-Di	=BAT 160A:	48ac		-	data
BAT 165	Si-Di	SMD, Schottky-Di, Uni, 40V, 0,75A	71a(1,7mm)	Sie	-	data
BAT 240 A	Si-Di	SMD, Dual, Schottkyi, Uni, 250V, 0,4A	35o1	Sie	-	data
BAT 254	Si-Di	SMD, Schottky-Di, 30V, 0,2A, <5ns	71a(2mm)	Phi	-	data
BAT 720	Si-Di	SMD, Schottky-Di, 40V, 0,5A	35e	Phi	-	data
BAT 721	Si-Di	SMD, Schottky-Di, 40V, 0,2A	35e	Phi	-	data
BAT 721 A	Si-Di	=BAT 721: Dual	35m		-	data
BAT 721 C	Si-Di	=BAT 721: Dual	35l		-	data
BAT 721 S	Si-Di	=BAT 721: Dual	35o		-	data

BAV

Type	Device	Short Description	Fig.	Manu	Comparision Types	More at
BAV 10	Si-Di	SS, 75V, 0,3A, <6ns	31a	Phi,Tho,Tix	BAV 12, BAW 55, BAW 76, BAX 81	data
BAV 11	Si-Di	Schottky-Di, L-Band-M, 4V, 10mA	Koax	Tho	-	data
BAV 12	Si-Di	S, 90V, 0,35A, <10ns	31a	Tix	BAV 14, BAW 25, BAW 26, BAX 81	data
BAV 13	Si-Di	S, 50V, 0,4A, <10ns	31a	Tix	BAV 14, BAW 24, BAW 25, BAW 26	data
BAV 14	Si-Di	S, Uni, 120V, 0,5A, <10ns	31a	Tix	BAV 15, BAV 16	data
BAV 15	Si-Di	S, Uni, 140V, 0,5A, <20ns	31a	Tix	BAV 16	data
BAV 16	Si-Di	S, Uni, 150V, 0,5A, <15ns	31a	Tix	-	data
BAV 17	Si-Di	S, Uni, 25V, 0,25A, <50ns	31a	Itt,Phi,++	BA 196...198, BAW 48..50	data
BAV 18	Si-Di	=BAV 17: 60V	31a	Itt,Phi,++	BA 196...198, BAW 49..50	data
BAV 19	Si-Di	=BAV 17: 120V	31a	Itt,Phi,++	BA 196...198, BAW 50	data
BAV 20	Si-Di	=BAV 17: 200V	31a	Itt,Phi,++	BA 197...198, BAW 50	data
BAV 21	Si-Di	=BAV 17: 250V	31a	Itt,Phi,++	BA 198	data
BAV 19W..21W	Si-Di	=BAV 19..21: SMD	71a(2,7mm)	Gsi	-	data
BAV 19WS..21WS	Si-Di	=BAV 19..21: SMD	71a(1,7mm)	Gsi	BAS 20H..21H	data
BAV 22(R)	Si-Di	Schottky-Di, ...12GHz	Koax	Phi	-	data
BAV 23	Si-Di	SMD, Dual, S, 250V, 0,2A, <50ns	44s2	Phi	-	data
BAV 23 S	Si-Di	=BAV 23:	35o		-	data
BAV 24	Si-Di	S, 50V, 0,3A, <8ns	31a	Tix	BAV 10, BAW 54, BAW 55, BAW 76, 1N4150	data
BAV 25	Si-Di	Schottky-Di, S-Band-Dem, 4V, 10mA	Koax	Tho	-	data
BAV 26	Si-Di	Schottky-Di, S-Band-Dem, 4V, 10mA	Koax	Tho	-	data
BAV 27	Si-Di	Schottky-Di, C-Band-Dem, 4V, 10mA	Koax	Tho	-	data
BAV 28	Si-Di	Schottky-Di, C-Band-Dem, 4V, 10mA	Koax	Tho	-	data
BAV 29	Si-Di	Schottky-Di, X-Band-Dem, 4V, 10mA	Koax	Tho	-	data
BAV 30	Si-Di	Schottky-Di, X-Band-Dem, 4V, 10mA	Koax	Tho	-	data
BAV 31	Si-Di	Schottky-Di, L-Band-Dem, 4V, 10mA	Koax	Tho	-	data
BAV 32	Si-Di	Schottky-Di, L-Band-Dem, 4V, 10mA	Koax	Tho	-	data
BAV 33	Si-Di	Schottky-Di, S-Band-M, 4V, 10mA	Koax	Tho	-	data
BAV 34	Si-Di	Schottky-Di, S-Band-M, 4V, 10mA	Koax	Tho	-	data
BAV 35	Si-Di	Schottky-Di, C-Band-M, 4V, 10mA	Koax	Tho	-	data
BAV 36	Si-Di	Schottky-Di, C-Band-M, 4V, 10mA	Koax	Tho	-	data
BAV 37	Si-Di	Schottky-Di, X-Band-M, 4V, 10mA	Koax	Tho	-	data
BAV 38	Si-Di	Schottky-Di, X-Band-M, 4V, 10mA	Koax	Tho	-	data
BAV 39	Si-Di	Dual, 40V, 0,1A	2l	Tho	-	data
BAV 40	Si-Di	8-Di-Array, 60V, 0,3A, <6ns	14-DIP	Phi	-	data
BAV 41	Si-Di	8-Di-Array, 60V, 0,3A, <6ns	14-DIP	Phi	-	data
BAV 42	Si-Di	8-Di-Array, 60V, 0,3A, <6ns	14-DIP	Phi	-	data

Type	Device	Short Description	Fig.	Manu	Comparision Types	More at
BAV 44	Si-Di	S, Uni, 65V, 0,75A, <20ns	31a	Phi,Tho	-	data
BAV 45	Si-Di	Picoampere, 20V, 50mA, Ir<5pA(5V)	2c	Phi, Tho,++	BAS 33...34	data
BAV 45 A	Si-Di	=2x BAV 45	5u			data
BAV 46(D...F)	Si-Di	Schottky-Di, 1...18GHz	Koax	Phi	-	data
BAV 47	Si-Di	Pico-Ampere-Di, 45V, 50mA, Ir<5pA/20V	5t	Tix	BAS 33...34	data
BAV 48	Si-Di	=BAV 47: Ir<20pA/20V	5t	Tix	BAS 33...34	data
BAV 49	Si-Di	=BAV 47: 35V, Ir<100pA/20V	5t	Tix	BAS 33...34	data
BAV 50	Si-Di	16-Di-Array, 50V, 0,17A, <4ns	14-DIP	Sgs	-	data
BAV 50 D	Si-Di	=BAV 50:	16-DIP		-	data
BAV 51	Si-Di	Schottky-Di	Koax	Tho	-	data
BAV 53	Si-Di	Schottky-Di, 25V, 50mA	2c	Tho	BAT 42...43, BAT 85	data
BAV 53 A	Si-Di	=BAT 53:	31a		BAT 42...43, BAT 85	data
BAV 53 B	Si-Di	=BAT 53:	2c		BAT 42...43, BAT 85	data
BAV 54/...	Si-Di	S, 30...100V, 0,2A, <4ns	31a	Tho	BAW 62, BAW 76, BAX 95, 1N4148, ++	data
BAV 55	Si-Di	S, Uni, 150V, 0,22A, 50ns	31a	Sgs	BA 196...198, BAV 20...21, BAY 80, ++	data
BAV 56	Si-Di	S, 70V, <6ns		Tho	BAV 10, BAW 55, BAW 62, BAX 95, 1N4148++	data
BAV 65(A)	Si-Di	Schottky-Di, 25V, 0,1A	31a	Tho	BAT 42...43, BAT 85	data
BAV 66	Si-Di	Schottky-Di		Tho	-	data
BAV 67	Si-Di	S, 80V, 90mA, <6ns	31a	Sie	BA 219, BAW 46...47, 1N4148, ++	data
BAV 68	Si-Di	S, Uni, 180V, 0,25A, <50ns	31a	Phi	BA 197...198, BAV 20...21, BAW 50	data
BAV 69	Si-Di	=BAV 68: 240V	31a	Phi	BA 198, BAV 21	data
BAV 70	Si-Di	SMD, Dual, 70V, 0,25A, <6ns	35l	Phi,Sie,++	-	data
BAV 70 L(T1)	Si-Di	SMD, Dual, 70V, 0,2A, <6ns	35l	Ons	-	
BAV 70 S	Si-Di	=BAV 70: Quad	46bk(2mm)		-	data
BAV 70 T(T1)	Si-Di	=BAV 70:	35l(1,6mm)		-	data
BAV 70 W(T1)	Si-Di	=BAV 70:	35l(2mm)		-	data
BAV 71	Si-Di	Schottky-Di, 26...40GHz	Koax	Phi	-	data
BAV 72	Si-Di	Schottky-Di, 26...40GHz	Koax	Phi	-	data
BAV 74	Si-Di	SMD, Dual, 50V, 0,25A, <4ns	35l	Phi,Sie,++	BAV 70	data
BAV 74 L(T1)	Si-Di	SMD, Dual, 50V, 0,2A, <4ns	35l	Ons	BAV 70	
BAV 75	Si-Di	Schottky-Di, 8...12GHz	Koax	Phi	-	data
BAV 76	Si-Di	Schottky-Di, X-Band, 3V, 50mA	Koax	Tho	-	data
BAV 77	Si-Di	Schottky-Di, X-Band, 3V, 50mA	Koax	Tho	-	data
BAV 79	Si-Di	Schottky-Di, S-Band, 3V, 50mA	Koax	Tho	-	data
BAV 80	Si-Di	Schottky-Di, X-Band, 3V, 50mA	Koax	Tho	-	data
BAV 81	Si-Di	Schottky-Di, X-Band, 3V, 50mA	Koax	Tho	-	data
BAV 82	Si-Di	Schottky-Di, X-Band, 3V, 50mA	Koax	Tho	-	data
BAV 83	Si-Di	=2x BAV 82 gep.	Koax	Tho	-	data
BAV 84	Si-Di	Schottky-Di, L-Band, 25V, 10mA	31a	Tho	-	data
BAV 84 A	Si-Di	=BAV 84:	Koax		-	data
BAV 85	Si-Di	=BAV 84: 4 Dioden gep.	31a	Tho	-	data
BAV 85 A	Si-Di	=BAV 85:	Koax		-	data
BAV 86	Si-Di	Schottky-Di, X-Band, 3V, 50mA	Koax	Tho	-	data
BAV 87	Si-Di	=2x BAV 86 gep.	Koax	Tho	-	data
BAV 88	Si-Di	Schottky-Di, X-Band, 3V, 50mA	Koax	Tho	-	data
BAV 89	Si-Di	=2x BAV 88 gep.	Koax	Tho	-	data
BAV 90	Si-Di	Schottky-Di	31a	Tho	-	data
BAV 91	Si-Di	=2x BAV 90 gep.	31a	Tho	-	data
BAV 92	Si-Di	Schottky-Di, S-Band, 3V, 50mA	Koax	Tho	-	data
BAV 93	Si-Di	Schottky-Di, X-Band, 3V, 50mA	Koax	Tho	-	data
BAV 94	Si-Di	=2x BAV 93 gep.	Koax	Tho	-	data
BAV 95 A...D	Si-Di	Schottky-Di, X-Band-M	Koax	Phi	-	data
BAV 97	Si-Di	Schottky-Di, 1...18GHz	Koax	Phi	-	data
BAV 98	Si-Di	UHF-M, 18V	Chip	Aeg	-	data
BAV 99	Si-Di	SMD, Dual, 70V, 0,25A, <6ns	35o	Phi,Sie,++	-	data
BAV 99 L(T1)	Si-Di	SMD, Dual, 70V, 0,215A, <6ns	35o	Ons	-	
BAV 99 RW(T1)	Si-Di	=BAV 99:	35n(2mm)	Ons	-	
BAV 99 S	Si-Di	=BAV 99: Quad	46bk(2mm)		-	data
BAV 99 V	Si-Di	=BAV 99:	35o(2mm)		-	data
BAV 99 W(T1)	Si-Di	=BAV 99:	35o(2mm)		-	data
BAV 100	Si-Di	SMD, Uni, 60V, 0,25A, <50ns	72a(3,4mm)	Aeg,Gsi,Phi	BAV 200...203	data
BAV 101	Si-Di	=BAV 100: 120V	72a(3,4mm)	Aeg, Phi	BAV 201...203	data
BAV 102	Si-Di	=BAV 100: 200V	72a(3,4mm)	Aeg,Nsc,Phi	BAV 202...203	data
BAV 103	Si-Di	=BAV 100: 250V	72a(3,4mm)	Aeg,Nsc,Phi	BAV 203	data
BAV 105	Si-Di	SMD, SS, 60V, 0,3A, <6ns	72a(3,4mm)	Phi	-	data
BAV 170	Si-Di	=BAS 116: Dual	35l	Sie,Mot,Phi	-	data
BAV 199	Si-Di	=BAS 116: Dual	35o	Sie,Mot,Phi	-	data
BAV 199 L(T1,T3)	Si-Di	=BAS 116: Dual	35o	Ons	-	
BAV 199 W	Si-Di	=BAS 116: Dual	35o(2mm)	Phi	-	data
BAV 200	Si-Di	=BAV 100:	72a(3,5mm)	Aeg	BAV 100...103	data

Type	Device	Short Description	Fig.	Manu	Comparision Types	More at
BAV 201	Si-Di	=BAV 101:	72a(3,5mm)	Aeg	BAV 101...103	data
BAV 202	Si-Di	=BAV 102:	72a(3,5mm)	Aeg	BAV 102...103	data
BAV 203	Si-Di	=BAV 103:	72a(3,5mm)	Aeg	BAV 103	data
BAV 300	Si-Di	=BAV 100:	72a(1,9mm)	Aeg	-	data
BAV 301	Si-Di	=BAV 101:	72a(1,9mm)	Aeg	-	data
BAV 302	Si-Di	=BAV 102:	72a(1,9mm)	Aeg	-	data
BAV 303	Si-Di	=BAV 103:	72a(1,9mm)	Aeg	-	data
BAV 756 S	Si-Di	SMD, Quad, SS, 85V, 0,25A, <4ns	46bw(2mm)	Phi	-	data

BAW

Type	Device	Short Description	Fig.	Manu	Comparision Types	More at
BAW 10	Si-Di	Uni, 50V, 0,2A	31a	Tix	BA 157...159, BA 187...190, BAY 18..21, ++	data
BAW 11	Si-Di	=BAW 10: 100V	31a	Tix	BA 157...159, BA 188...190, BAY 20...21, ++	data
BAW 12	Si-Di	=BAW 10: 150V	31a	Tix	BA 157...159, BA 189...190, BAY 20..21, ++	data
BAW 13	Si-Di	=BAW 10: 200V	31a	Tix	BA 157...159, BA 190, BAY 20..21, ++	data
BAW 14	Si-Di	=BAW 10: 300V	31a	Tix	BA 157...159, BAY 21, BAY 46, ++	data
BAW 16	Si-Di	Uni, 150V, 0,2A	31a	Tix	BA 157...159, BA 189...190, BAY 20..21, ++	data
BAW 17	Si-Di	Uni, 200V, 0,2A	31a	Tix	BA 157...159, BA 190, BAY 21, ++	data
BAW 18	Si-Di	Uni, 150V, 0,2A	31a	Tix	BA 157...159, BA 189...190, BAY 20..21, ++	data
BAW 19	Varistor	VDR/Varistor, 75mA		Sgs	-	data
BAW 21(A,B)	Si-Di	S, contr.av., 70...90V, 0,4A, <300ns A=70V, B=90V	31a	Itt,Phi,Tho	BYX 57/..., (BA 173, BA 157...159, ++)	data
BAW 22	Si-Di	Ring-Dem., 50V, 0,1A	5y	Tho	-	data
BAW 23	Si-Di	=BAW 22: 30V	5y	Tho	-	data
BAW 24	Si-Di	SS, 50V, 0,6A, <6ns	31a	Aeg,Tix	BAW 25, BAW 26, BAW 27	data
BAW 25	Si-Di	SS, 50V, 0,6A, <6ns	31a	Aeg,Tix	BAW 26, BAW 27	data
BAW 26	Si-Di	SS, 75V, 0,6A, <6ns	31a	Aeg,Tix	BAW 27	data
BAW 27	Si-Di	SS, 75V, 0,6A, <6ns	31a	Aeg,Tix	BAW 26	data
BAW 28	Si-Di	SS, 30V, 0,4A, <6ns	31a	Sgs	BAV 13, BAV 14	data
BAW 29	Si-Di	Schottky-Di	31a	Sgs	-	data
BAW 31	Si-Di	Dual, 50V, 0,1A	5	Tho	-	data
BAW 32 A...E	Si-Di	Uni, 10...200V, 60mA A= 200V, B=150V, C=100V, D=50V, E=10V	31a	Tho	BA 147/..., BA 187...190, BAY 17..21, ++	data
BAW 33	Si-Di	S, Uni, 70V, 0,225A, <25ns	31a	Tix	BAV 19..21, BAY 43, BAW 50, 1N4148, ++	data
BAW 36	Si-Di	SS, Uni, 60V, 0,4A, <6ns	31a	Sgs	BAV 14, BAW 24..27	data
BAW 43	Si-Di	Uni, 125V, 0,3A	31a	Tix	BA 157...159, BA 194, BA 199/..., ++	data
BAW 45	Si-Di	SS, Uni, 20V, 0,1A, <6ns	31a	Sgs,Tix	BA 317, BAY 71, BAY 94, 1N4148, ++	data
BAW 46	Si-Di	SS, Uni, 75V, 0,12A, <6ns	31a	Sgs,Tix	BAW 62, BAW 76, BAX 95, 1N4148, ++	data
BAW 47	Si-Di	SS, Uni, 100V, 0,115A, <6ns	31a	Sgs,Tix	BA 202, BA 203, BA 219,	data
BAW 48	Si-Di	Dem, S, Uni, 50V, 0,3A, 35ns	31a	Sgs,Tix	BAX 78, BAW 49, BAW 50, 1N4148, ++	data
BAW 49	Si-Di	S, Uni, 100V, 0,35A, <60ns	31a	Sgs,Tix	BAW 50, BAX 12, BAY 72	data
BAW 50	Si-Di	S, Uni, 200V, 0,35A, <60ns	31a	Sgs,Tix	BAW 50	data
BAW 51	Si-Di	Uni, 80V, 0,3A	31a	Sgs,Tix	BA 157...159, BA 193...194, BAW 52, ++	data
BAW 52	Si-Di	Uni, 200V, 0,3A	31a	Sgs,Tix	BA 157...159, BA 199/..., BAS 11, ++	data
BAW 53	Si-Di	SS, Uni, 30V, 0,225A, <8ns	31a	Sgs,Tix	BA 204, BA 221, BAY 74, 1N4148, ++	data
BAW 54	Si-Di	SS, Uni, 50V, 0,3A, <6ns	31a	Sgs,Tix	BAV 10, BAV 24, BAW 55, BAW 76, 1N4150	data
BAW 55	Si-Di	SS, Uni, 75V, 0,3A, <6ns	31a	Sgs,Tix	BAV 10, BAV 12, BAW 76, BAX 81	data
BAW 56	Si-Di	SMD, Dual, SS, 85V, 0,25A, <4ns	35m	Phi,Sie,++	BAW 66, BAW 68	data
BAW 56 G	Si-Di	=BAW 56:	44s4	Sie	-	data
BAW 56 GT	Si-Di	=BAW 56:	44s2	Sie	-	data
BAW 56 L(T1)	Si-Di	SMD, Dual, SS, 70V, 0,2A, <6ns	35m	Ons	BAW 66, BAW 68	data
BAW 56 S	Si-Di	=BAW 56: Quad	46bj(2mm)		-	data
BAW 56 T(T1)	Si-Di	=BAW 56:	35m(1,6mm)	Phi, Ons	-	data
BAW 56 W(T1)	Si-Di	=BAW 56:	35m(2mm)	Phi, Ons	-	data
BAW 57	Si-Di	SMD, Array, 60V, 0,3A	14-MDIP	Tix	-	data
BAW 57 N	Si-Di	=BAW 57:	14-DIP		-	data
BAW 58	Si-Di	Min, S, Uni, 100V, 80mA, <50ns	36b	Sgs	BAV 19..21, BAY 43, BAY 80, 1N4148, ++	data
BAW 59	Si-Di	Min, SS, Uni, 40V, 60mA, <6ns	36b	Sgs	BA 217...218, BA 317, BAX 87, 1N4148, ++	data
BAW 60	Si-Di	SS, 15V, 50mA, <0,85ns	31a	Sgs	-	data
BAW 62	Si-Di	S, 75V, 0,2A, <4ns	31a	Gen,Phi,Tix	BAW 76, BAX 95, 1N4148..4149, ++	data
BAW 63	Si-Di	SMD, S, 60V, 0,2A, <4ns	35e(2mm)	Fer,Tix	(BAS 16)	data
BAW 63 A	Si-Di	=BAS 63: 30V	35e(2mm)		(BAS 16)	data
BAW 63 B	Si-Di	=BAS 63: 15V	35e(2mm)		(BAS 16)	data
BAW 64	Si-Di	=BAS 63: Dual	35l(2mm)	Fer,Tix	(BAV 70)	data
BAW 65	Si-Di	=BAS 63A: Dual	35l(2mm)	Fer,Tix	(BAV 70)	data
BAW 66	Si-Di	=BAS 63A: Dual	35m(2mm)	Fer,Tix	(BAW 56)	data
BAW 67	Si-Di	=BAS 63B: Dual	35l(2mm)	Fer,Tix	(BAV 70)	data
BAW 68	Si-Di	=BAS 63B: Dual	35m(2mm)	Fer,Tix	(BAW 56)	data
BAW 69	Si-Di	Schottky-Di, X-Band, 6V	Chip	Aeg	-	data
BAW 70	Si-Di	Schottky-Di, X-Band, 6V	Koax	Aeg	-	data
BAW 74	Si-Di	SMD, Dual, 50V	35m	Ztx	-	data
BAW 75	Si-Di	SS, 35V, 0,3A, <4ns	31a	Aeg,Sie,++	BAW 54...55, BAW 76, 1N4148, 1N4150	data
BAW 76	Si-Di	SS, 75V, 0,3A, <4ns	31a	Aeg,Sie,++	BAV 10, BAW 55, BAX 81, 1N4148	data
BAW 77	Si-Di	Uni, 120V, 0,1A	31a	Tix	BA 147/150, BA 157...159, BA 189...190, ++	data

Type	Device	Short Description	Fig.	Manu	Comparision Types	More at
BAW 78 A	Si-Di	SMD, Uni, 50V, 1A	39f	Sie	-	data
BAW 78 B	Si-Di	=BAW 78A: 100V	39f	Sie	-	data
BAW 78 C	Si-Di	=BAW 78A: 200V	39f	Sie	-	data
BAW 78 D	Si-Di	=BAW 78A: 400V	39f	Sie	-	data
BAW 78 M	Si-Di	=BAW 78A: 400V	≈45(AK--K)	Sie	-	data
BAW 79 A	Si-Di	=BAW 78A: Dual	39p	Sie	-	data
BAW 79 B	Si-Di	=BAW 78B: Dual	39p	Sie	-	data
BAW 79 C	Si-Di	=BAW 78C: Dual	39p	Sie	-	data
BAW 79 D	Si-Di	=BAW 78D: Dual	39p	Sie	-	data
BAW 84	Si-Di	S, Uni, 350V, 0,25A, 700ns	31a	Tix	BA 157...159, BY 204/..., BY 208/..., ++	data
BAW 85	Si-Di	=BAW 84: 450V	31a	Tix	BA 157...159, BY 204/..., BY 208/..., ++	data
BAW 86	Si-Di	=BAW 84: 500V	31a	Tix	BA 158...159, BY 204/..., BY 208/..., ++	data
BAW 90	Si-Di	Uni, 75V, 0,05A	2c	Tho	-	data
BAW 91	Si-Di	=BAW 90: Dual	2	Tho	-	data
BAW 92	Si-Di	=BAW 90: Dual	5	Tho	-	data
BAW 93	Si-Br	=BAW 90: Br	5	Tho	-	data
BAW 94	Si-Di	Dual, 30V, <2ns		Sgs	-	data
BAW 95 A...D	Si-Di	Schottky-Di, X-Band-M	Koax	Phi	-	data
BAW 96	Si-Di	Schottky-Di, 4V, 10mA	Koax	Tho	-	data
BAW 99	Si-Di	SMD, Dual, 25V, 0,05A, <4ns	Chip	Phi,Tho	-	data
BAW 100	Si-Di	=BAW 56:	44s4	Sie	MA 157A	data
BAW 101	Si-Di	SMD, Dual, 300V, 0,2A	44s2	Sie	-	data
BAW 156	Si-Di	=BAS 116: Dual	35m	Sie,Mot,Phi	-	data

BAX

Type	Device	Short Description	Fig.	Manu	Comparision Types	More at
BAX 11	Si-Di	UHF-Multiplier	Koax	Aeg	-	data
BAX 12(A)	Si-Di	S, Uni, contr.av., 90V, 0,4A, <60ns	31a	Phi,Tho,++	(BA 157...159, BAV 15...16)	data
BAX 13(A)	Si-Di	S, Uni, 50V, 75mA, <6ns	31a	Phi,Tho,++	BA 218, BA 318, BAX 91, BAY 38, 1N4148++	data
BAX 14	Si-Di	S, Uni, 40V, 0,5A, <30ns	31a	Phi,Tho	BAV 15, BAV 16, BAV 44	data
BAX 14 A	Si-Di	=BAX 14: <300ns	31a		BA 157...159, BY 204/..., BY 208/..., ++	data
BAX 15	Si-Di	S, Uni, 180V, 0,25A, <300ns	31a	Phi,Tix	BA 157...159, BA 173, BY 204/..., ++	data
BAX 16(A)	Si-Di	S, Uni, 150V, 0,2A, <120ns	31a	Phi,Tho,++	BA 157...159, BA 196...198, BAX 17, BAY 80	data
BAX 17(A)	Si-Di	S, Uni, 200V, 0,2A, <120ns	31a	Phi,Tho,Tix	BA 157...159, BA 197...198, BAV 20...21	data
BAX 18(A)	Si-Di	Uni, 75V, 0,5A	31a	Phi,Tix	BA 157...159, BY 402, BY 295/..., ++	data
BAX 20	Si-Di	S, Uni, 35V, 0,115A, <250ns	31a	Aeg	BAX 15...17, BAX 88, 1N4148, ++	data
BAX 21	Si-Di	=BAX 20: 75V	31a	Aeg	BA 157...159, BAX 15...17, 1N4148, ++	data
BAX 22	Si-Di	=BAX 20: 125V	31a	Aeg	BA 157...159, BA 173, BAX 15...17, ++	data
BAX 25	Si-Di	Schottky-Di, SS, 30V, 0,05A(ss),<0,5ns	31a	Aeg	-	data
BAX 26	Si-Di	Schottky-Di, SS, 30V, 0,1A(ss),<0,5ns	31a	Aeg	-	data
BAX 27	Si-Di	Schottky-Di, SS, 30V, 0,5A(ss),<0,5ns	31a	Aeg	-	data
BAX 28	Si-Di	3-Di-Array, 25V, 0,115A, <4ns	5r§	Sie	-	data
BAX 30	Si-Di	3-Di-Array, 25V, 0,115A, <4ns	5s&	Sie	-	data
BAX 32	Si-Di	RadH, 250V	31a	Sgs	-	data
BAX 33	Si-Di	Dual, SS, 30V, 0,115A, <2ns	42v	Sgs	-	data
BAX 34	Si-Di	Dual, SS, 30V, 0,115A, <2ns	42v	Sgs	-	data
BAX 35	Si-Di	Dual, SS, 30V, 0,115A, <2ns	42v	Sgs	-	data
BAX 36	Si-Di	Dual, S, 50V, 0,15A, <50ns	42v	Sgs	-	data
BAX 37	Si-Di	Dual, S, 50V, 0,15A, <50ns	42v	Sgs	-	data
BAX 38	Si-Di	Dual, S, 50V, 0,15A, <50ns	42v	Sgs	-	data
BAX 39	Si-Di	=BAX 33: 4-Di		Sgs		data
BAX 40	Si-Di	=BAX 34: 4-Di		Sgs		data
BAX 41	Si-Di	=BAX 35: 4-Di		Sgs		data
BAX 42	Si-Di	=BAX 36: 4-Di		Sgs		data
BAX 43	Si-Di	=BAX 37: 4-Di		Sgs		data
BAX 44	Si-Di	=BAX 38: 4-Di		Sgs		data
BAX 45	Si-Di	Dual, Logik-Gatter, 80V, 0,3A, <25ns	2n1	Sgs	-	data
BAX 46	Si-Di	=BAX 45	2n1	Sgs	-	data
BAX 47	Si-Br	=BAX 45: Br	5x	Sgs	-	data
BAX 48	Si-Br	=BAX 45: Br	5x	Sgs	-	data
BAX 49	Si-Di	=BAX 45: 6 Di	82	Sgs	-	data
BAX 50	Si-Di	=BAX 45: 8 Di	82	Sgs	-	data
BAX 51	Si-Di	=BAX 45: 16 Di	82	Sgs	-	data
BAX 52	Si-Br	=BAX 45: Br	5x	Sgs	-	data
BAX 53	Si-Br	=BAX 45: Br	5x	Sgs	-	data
BAX 54	Si-Di	=BAX 45: Ring-Dem	5y	Sgs	-	data
BAX 55	Si-Di	=BAX 45: Ring-Dem	5y	Sgs	-	data
BAX 56	Si-Di	Dual, Logik-Gatter, 60V, 0,3A, <4ns	2l	Sgs	-	data
BAX 57	Si-Di	=BAX 56:	2m	Sgs	-	data
BAX 58	Si-Di	=BAX 56:	2l	Sgs	-	data
BAX 59	Si-Di	=BAX 56:	2m	Sgs	-	data
BAX 60	Si-Di	=BAX 56: 3 Di	5r1	Sgs	-	data
BAX 61	Si-Di	=BAX 56: 3 Di	5s1	Sgs	-	data
BAX 62	Si-Di	=BAX 56: 3 Di	5r1	Sgs	-	data
BAX 63	Si-Di	=BAX 56: 3 Di	5s1	Sgs	-	data
BAX 64	Si-Di	=BAX 56: 4 Di	81cd	Sgs	-	data
BAX 65	Si-Di	=BAX 56: 4 Di	81cc	Sgs	-	data

Type	Device	Short Description	Fig.	Manu	Comparision Types	More at
BAX 66	Si-Di	=BAX 56: 5 Di	81cf	Sgs	-	data
BAX 67	Si-Di	=BAX 56: 5 Di	81ce	Sgs	-	data
BAX 68	Si-Di	=BAX 56: 6 Di	81ch	Sgs	-	data
BAX 69	Si-Di	=BAX 56: 6 Di	81cg	Sgs	-	data
BAX 70	Si-Di	=BAX 56: 7 Di	81ck	Sgs	-	data
BAX 71	Si-Di	=BAX 56: 7 Di	81cj	Sgs	-	data
BAX 72	Si-Di	=BAX 56: 8 Di	82db	Sgs	-	data
BAX 73	Si-Di	=BAX 56: 8 Di	82da	Sgs	-	data
BAX 74	Si-Di	S, 30V, <10ns	10	Tix	BA 317...318, BAY 38, BAY 71, 1N4148, ++	data
BAX 78	Si-Di	S, 55V, 0,3A, <20ns	31a	Phi,Tix	BA 318, BAY 38, BAY 69, 1N4148, ++	data
BAX 79	Si-Di	S, 50V, 0,4A, <4ns	31a	Sgs	BAV 13, BAV 14, BAW 24..27	data
BAX 80	Si-Di	S, 50V, 0,15A, <4ns	31a	Tix	BAX 80, BAW 62, BAY 95, 1N4148, ++	data
BAX 81	Si-Di	S, 90V, 0,35A, <6ns	31a	Tix	BAV 14	data
BAX 82	Si-Di	S, 50V, 0,25A, <6ns	31a	Tix	BAV 10, BAW 54...55, BAW 76, 1N4148, ++	data
BAX 83	Si-Di	S, 100V, 75mA, <10ns	31a	Tix	BA 202...203, BA 219, BAX 96, 1N4148, ++	data
BAX 84	Si-Di	S, 50V, 75mA, <6ns	31a	Tho,Tix	BA 218, BA 318, BAX 13, BAX 91, 1N4148++	data
BAX 85	Si-Di	=BAX 84: <15ns	31a	Tho,Tix	BA 218, BA 318, BAX 13, BAX 91, 1N4148++	data
BAX 86 A	Si-Di	S, 50V, 75mA, <8ns	31a	Tix	BA 218, BA 318, BAX 13, BAX 91, 1N4148++	data
BAX 86 B	Si-Di	=BAX 86A: <10ns	31a		BA 218, BA 318, BAX 13, BAX 91, 1N4148++	data
BAX 87	Si-Di	S, 40V, 75mA, 6ns	31a	Tix	BA 218, BA 318, BAX 13, BAX 91, 1N4148++	data
BAX 88	Si-Di	S, 20V, 75mA, <150ns	31a	Tix	BAX 15...17, BAY 92, BAW 21, 1N4148, ++	data
BAX 89 A	Si-Di	S, 45V, 75mA, <10ns	31a	Tix	BA 218, BA 318, BAX 13, BAY 38, 1N4148++	data
BAX 89 B	Si-Di	=BAX 89A: <20ns	31a		BA 218, BA 318, BAX 13, BAY 38, 1N4148++	data
BAX 89 C	Si-Di	=BAX 89A: 6ns	31a		BA 218, BA 318, BAX 13, BAY 38, 1N4148++	data
BAX 90 A	Si-Di	S, 45V, 75mA, <50ns	31a	Tix	BAX 80, BAX 85, BAX 94, 1N4148, ++	data
BAX 90 B	Si-Di	=BAX 90A: 50V	31a		BAX 80, BAX 85, BAX 94, 1N4148, ++	data
BAX 90 C	Si-Di	=BAX 90A: <4ns	31a		BA 218, BA 318, BAX 13, BAX 91, 1N4148++	data
BAX 91(A...C)	Si-Di	S, 50V, 75mA, <4ns	31a	Tix	BA 218, BA 318, BAX 13, BAX 91, 1N4148++	data
BAX 92	Si-Di	S, 50V, 75mA, <10ns	31a	Tix	BA 218, BA 318, BAX 13, BAX 91, 1N4148++	data
BAX 93	Si-Di	=BAX 92: <6ns	31a	Tix	BA 218, BA 318, BAX 13, BAX 91, 1N4148++	data
BAX 94	Si-Di	S, 50V, 75mA, <50ns	31a	Tix	BAX 85, BAX 89, BAX 90, 1N4148, ++	data
BAX 95	Si-Di	S, 75V, 0,2A, <4ns	31a	Tix	BAW 62, BAW 76, 1N4148...4149, ++	data
BAX 96(A...C)	Si-Di	S, 100V, 75mA, <4ns	31a	Tix	BA 219, 1N4148...4149, 1N4151	data
BAX 157	Si-Di	Uni, 400V, 0,5A, 400ns	31a	Idr	BY 204/4, BY 206, BY 406, BY 208/600, ++	data
BAX 158	Si-Di	=BAX 157: 600V	31a	Idr	BY 204/8, BY 207, BY 407, BY 208/600, ++	data
BAX 159	Si-Di	=BAX 157: 1000V	31a	Idr	BY 204/10, BY 208/1000, BY 268, ++	data
BAX 280	Si-Di	S, 1000V, 3,5A(Tc=80°), 55ns	48(-AKA)	Sie		data

BAY

Type	Device	Short Description	Fig.	Manu	Comparision Types	More at
BAY 14	Si-Di	Uni, 500V, 0,2A	34a	Aeg	BA 158...159, BAY 89, 1N4005...4007, ++	data
BAY 15	Si-Di	=BAY 14: 650V	34a	Aeg	BA 159, BAY 90, 1N4006...4007, ++	data
BAY 16	Si-Di	=BAY 14: 800V	34a	Aeg	BA 159, BAY 90, 1N4006...4007	data
BAY 17	Si-Di	Uni, 15V, 0,25A, 1µs	31a	Itt,Tho,Tix	BA 157...159, BAY 44...46, BAY 86...89, ++	data
BAY 18	Si-Di	=BAY 17: 60V	31a	Itt,Tho,Tix	BA 157...159, BAY 44...46, BAY 86...89, ++	data
BAY 19	Si-Di	=BAY 17: 120V	31a	Itt,Tho,Tix	BA 157...159, BAY 45...46, BAY 87...89, ++	data
BAY 20	Si-Di	=BAY 17: 180V	31a	Itt,Tho,Tix	BA 157...159, BAY 56, BAY 88...89, ++	data
BAY 21	Si-Di	=BAY 17: 350V	31a	Itt,Tho,Tix	BA 157...159, BAY 46, BAY 88...89, ++	data
BAY 23	Si-Di	Uni, 1000V, 80mA	31a	Itt,Tho	BAY 90...91, BY 203/12, 1N5181...5184	data
BAY 24	Si-Di	=BAY 23: 1500V	31a	Itt,Tho	BAY 91, BY 203/16, 1N5181...5184	data
BAY 25	Si-Di	=BAY 23: 2000V	31a	Itt,Tho	BAY 91, BY 203/20, 1N5181...5184	data
BAY 26	Si-Di	=BAY 23: 3000V	31a	Itt,Tho	HVG 3, 1N1733, 1N5181...5184	data
BAY 31	Si-Di	SS, 15V, 0,1A, 2,3ns	31a	Itt,Tix	BA 317...318, BAY 38, BAY 71, 1N4148, ++	data
BAY 32	Si-Di	Uni, 150V, 0,17A, <5µs	31a	Phi,Tix	BA 157...159, BA 147/150, BA 189...190, ++	data
BAY 33	Si-Di	Uni, S, 150V, 0,13A, <500ns	31a	Phi,Tix	BA 157...159, BAX 15...17, BA 173, ++	data
BAY 35	Si-Di	h-ohm, logarithm. Kennlinie	31a	Itt	-	data
BAY 36	Si-Di	SS, 30V, 0,1A, <10ns	31a	Tix	BA 317...318, BAY 69, BAY 71, 1N4148, ++	data
BAY 38	Si-Di	SS, 50V, 0,115A, <4ns	31a	Phi,Tho,Tix	BA 318, BAX 80, BAY 95, 1N4148, ++	data
BAY 39	Si-Di	S, 75V, 0,45A, <160ns	31a	Phi,Tix	BAV 15, BAV 16, BYX 57/...	data
BAY 40	Si-Di			Itt(SEL)	-	data
BAY 41	Si-Di	Uni, S, 40V, 0,225A, <15ns	31a	Sie,Tho,Tix	BA 204, BAX 82, BAX 95, BAY 61, 1N4148++	data
BAY 42	Si-Di	=BAY 41: 60V	31a	Sie,Tho,Tix	BAW 62, BAX 95, BAY 61, 1N4148, ++	data
BAY 43	Si-Di	=BAY 41: 80V	31a	Sie,Tho,Tix	BAV 14, 1N4148, 1N4149, 1N4151, ++	data
BAY 44	Si-Di	Uni, 50V, 0,25A	31a	Sie,Tho,Tix	BA 157...159, BAY 18...21, BAY 86, ++	data
BAY 45(DHD)	Si-Di	=BAY 44: 150V	31a	Sie,Tho,Tix	BA 157...159, BAY 20...21, BAY 88, ++	data
BAY 46	Si-Di	=BAY 44: 300V	31a	Sie,Tho,Tix	BA 157...159, BAY 21, BAY 88, ++	data
BAY 52	Si-Di	S, 15V, 0,1A, <15ns	31a	Itt(SEL)	BA 317...318, BAY 38, BAY 71, 1N4148, ++	data
BAY 60	Si-Di	SS, 25V, 0,115A, 4ns	31a	Phi,Sie,Tix	BA 317...318, BAY 38, BAY 71, 1N4148, ++	data
BAY 61	Si-Di	SS, 75V, 0,2A, <4ns	31a	Sie,Tix	BAW 62, BAW 76, BAX 95, 1N4148, ++	data
BAY 63	Si-Di	SS, 50V, 0,2A, 4ns	31a	Sie,Tho,Tix	BAW 62, BAW 76, BAX 95, 1N4148, ++	data
BAY 64	Si-Di	3-Di-Array, 20V, 0,1A	5	Phi,Tix	-	data
BAY 66	Si-Di	UHF-L, 100V, 0,4A(ss), 25GHz	Koax	Phi	-	data
BAY 67	C-Di	HF-Band-S	31a	Aeg,Tho,Tix	BA 243...244, BA 282...284, BA 482...484	data

Type	Device	Short Description	Fig.	Manu	Comparision Types	More at
BAY 68	Si-Di	SS, 35V, 0,115A, <10ns	31a	Aeg,Tho,Tix	BA 318, BAY 38, BAY 71, 1N4148, ++	data
BAY 69	Si-Di	=BAY 68: 60V	31a	Aeg,Tho,Tix	BA 318, BAX 80, BAY 38, 1N4148, ++	data
BAY 70	Si-Di	VHF-AFC, VHF-Tuning	31a	Aeg	BA 182, BA 243, BA 283, BA 483...484	data
BAY 71(DHD)	Si-Di	SS, 70V, 0,115A, <4ns	31a	Itt,Tho,++	BAW 62, BAW 76, BAX 95, 1N4148, ++	data
BAY 72	Si-Di	Uni, S, 125V, 0,375A, <400ns	31a	Fch,Tho,Tix	BA 157...159, BAV 15...16, BAW 50, ++	data
BAY 73	Si-Di	Uni, S, 125V, 0,225A, <3µs	31a	Fch,Tho,Tix	BA 157...159, BAY 20...21, BAY 45...46, ++	data
BAY 74	Si-Di	SS, 50V, 0,2A, <4ns	31a	Fch,Tho,Tix	BAW 62, BAX 95, BAY 95, 1N4148, ++	data
BAY 77	Si-Di	SS, 30V, 0,2A, 2,3ns	31a	Aeg	BAW 62, BAX 95, BAY 95, 1N4148, ++	data
BAY 78(M)	Si-Di	2x2 Si-Di, Dem, 75V, 0,1A		Aeg,Sgs	-	data
BAY 79	Si-Di	UHF-Multiplier-Di, 68...150GHz	Koax	Aeg	-	data
BAY 80	Si-Di	Uni, 150V, 0,25A, <50ns	31a	Aeg,Phi,Tho	BA 157...159, BA 196...198, BAV 20...21, ++	data
BAY 82	Si-Di	SS, 15V, 50mA, <0,75ns	31a	Fch,Sgs	FD 700, 1N4244, 1N4376	data
BAY 83	Si-Di			Sgs	-	data
BAY 84	Si-Di	SMD, Dual, 90V, 0,15A, <50ns	35o	Tho	BAS 31	data
BAY 85(S)	Si-Di	SMD, 300V, 0,15A, <40ns	35e	Tho	-	data
BAY 86	Si-Di	Uni, 600V, 0,25A, <3µs	31a	Aeg,Tix	BA 157...159, BAY 20...21, BAY 45, ++	data
BAY 87	Si-Di	=BAY 86: 120V	31a	Aeg,Tix	BA 157...159, BAY 20...21, BAY 45, ++	data
BAY 88	Si-Di	=BAY 86: 350V	31a	Aeg,Tix	BA 157...159, BAY 21, BAY 46, ++	data
BAY 89	Si-Di	Uni, 600V, 0,25A	31a	Aeg,Tix	BA 158...159, BY 133, BY 204/6, ++	data
BAY 90	Si-Di	=BAY 89: 1000V	31a	Aeg	BA 159, BY 133, BY 204/10, BYX 94...95,++	data
BAY 91	Si-Di	=BAY 89: 2000V	31a	Aeg	BY 203/20, BY 268, DM 516	data
BAY 91 A	Si-Di	=BAY 89: 1750V	31a		BY 203/20, BY 268, DM 516	data
BAY 92	Si-Di	Uni, S, 650V, 0,1A, <500ns	31a	Aeg,Tix	BA 158...159, BY 204/6.../10, BY 405, ++	data
BAY 93	Si-Di	S, 35V, 0,115A, <15ns	31a	Aeg,Tix,++	BA 317...318, BAY 38, BAY 69, 1N4148, ++	data
BAY 94	Si-Di	SS, 35V, 0,115A, <4ns	31a	Aeg,Mot,Tix	BA 317...318, BAY 38, BAY 69, 1N4148, ++	data
BAY 95	Si-Di	SS, 75V, 0,2A, <4ns	31a	Aeg,Mot,Tix	BAW 62, BAX 95, BAY 61, 1N4148, ++	data
BAY 96	Si-Di	UHF-L, 120V, 25GHz	32a§	Phi	-	data
BAY 97	Si-Di	SS, 25V, 0,1A, <4ns	31a	Sie,Tix	BA 317...318, BAY 38, BAY 69, 1N4148, ++	data
BAY 98	Si-Di	Uni, S, 150V, 0,2A, 50ns	31a	Sie,Tix	BA 157...159, BAX 16...17, BAY 80	data
BAY 99	Si-Di	Uni, S, 40V, 0,1A, <50ns	31a	Sie,Tix	BAX 80, BAX 94, BAY 41...43, 1N4148, ++	data
BAY 135	Si-Di	Picoampere, 140V, 0,2A, Ir<3nA(140V)	31a	Aeg	BAS 45	data

BB..BBZ

Type	Device	Short Description	Fig.	Manu	Comparision Types	More at
BB	Si-Di	=1SV128 (Typ-Code/Stempel/marking)	35	Tos	→1SV128	data
BB	Si-Di	=1SV237 (Typ-Code/Stempel/marking)	44	Tos	→1SV237	data
BB	Si-P	=2SB1119 (Typ-Code/Stempel/marking)	39	Say	→2SB1119	data
BB	Si-P	=2SB831-B (Typ-Code/Stempel/marking)	35	Hit	→2SB831	data
BB	Si-N	=2SD1367-BB (Typ-Code/Stempel/marking)	39	Hit	→2SD1367	data
BB	Si-N/P	=BC 846BPDW1(Typ-Code/Stempel/marking)	46(2mm)	Ons	→BC 846...PDW1	
BB	Si-P	=BCW 61B (Typ-Code/Stempel/marking)	35	Mot,Phi,Sie Aeg,Fer,++	→BCW 61B	data
BB	Si-N	=BCX 54-6 (Typ-Code/Stempel/marking)	39	Sie,Val,Mot	→BCX 54-6	data
BB	Z-Di	=BZX 399-C5V1(Typ-Code/Stempel/marking)	71(1,7mm)	Phi	→BZX 399-...	data
BB	Si-N/P+R	=KRX 103U (Typ-Code/Stempel/marking)	45(2mm)	Kec	→KRX 103U	data
BB	Si-N/P+R	=KRX 202U (Typ-Code/Stempel/marking)	46(2mm)	Kec	→KRX 202U	data
BB	MOS-N-FET-e	=MTDF 1N03HD(Typ-Code/Stempel/marking)	8-SSMDIP	Ons	→MTDF 1N03HD	data
BB	Si-Di	=SC 016-2 (Typ-Code/Stempel/marking)	71(5x2,5)	Fjd	→SC 016-...	data
BB	Si-Di	=ZMS 5800 (Typ-Code/Stempel/marking)	71(1,7mm)	Ztx	→ZMS 5800	data
BB 1 A3M...L3N	Si-N+R	=AB 1A3M...L3N: Fig.→	40c	Nec		data
BB 1...10	Si-Di	Uni, S, 250...1200V, 0,3A, 800ns BB1=250, BB2=350, BB4=600, BB6=800V BB8=1000, BB10=1200V	31a	Inr	BAY 89..91, BY 203/..., BY 208/..., ++	data
BB 3	UJT-P	Iv=1mA	5m§	Tho	-	data
BB 4 A	UJT-P	Iv=1mA	5m§	Tho	-	data
BB 4 B	UJT-P	Iv=8mA	5m§		-	data
BB 5 (A,B,C)	UJT-P	Iv=8mA	5m§	Tho	-	data
BB 11 A,B	UJT-P	Iv=2mA	5m§	Tho	-	data
BB 12	UJT-P	Iv=1mA	5m§	Tho	-	data
BB 14	UJT-P	Iv=25mA	5m§	Tho	-	data
BB 18	UJT-P	Iv=2mA	5m§	Tho	-	data
BB 100	C-Di	VHF-Tuning	31a	Tho	BB 105G, BB 106, BB 205G, BB 405G, ++	data
BB 100 G	C-Di	VHF-Tuning	31a		BB 105G, BB 106, BB 205G, BB 405G, ++	data
BB 101	C-Di	FM-Tuning	31a	Itt	BB 110, BB 203, MV 310, 1SV68, 1SV84	data
BB 102	C-Di	VHF-Tuning	31a	Aeg	BB 105G, BB 106, BB 205G, BB 405G, ++	data
BB 103	C-Di	FM-Tuning	31a	Sie,Tho	BB 110, BB 203, MV 310, 1SV68, 1SV84	data
BB 104	C-Di	Dual, FM-Tuning	12e	Phi,Sie,++	BB 204, BB 304, MV 104, 1SV55, 1SV109	data
BB 105 A,B	C-Di	UHF-Tuning	71a(4mm)	Phi,Sie,++	BB 121...122, BB 205A,B, BB 405A,B, ++	data
BB 105 G	C-Di	VHF-Tuning	71a(4mm)	Phi,Sie,++	BB 106, BB 205G, BB 405G, BB 505G, ++	data
BB 106	C-Di	VHF-Tuning (Euro-Band 1)	71a(4mm)	Phi,Tho	BB 209, BB 229, BB 309, BB 409	data
BB 107	C-Di	Dual, AM-Tuning	12p	Sie	BB 212, BB 312	data
BB 109	C-Di	FM/VHF-Tuning	31a	Phi,Sie,++	BB 143, MV 109, MV 2101..2115, 1SV50	data

Type	Device	Short Description	Fig.	Manu	Comparision Types	More at
BB 110(B,G)	C-Di	FM-Tuning	71a(4mm)	Aeg,Phi	BB 203, MV 310, 1SV68, 1SV84	data
BB 112	C-Di	AM-Tuning	7d	Phi,Sie	BB 130, BB 509, MV 1401, MV 1403...1405++	data
BB 113	C-Di	3-Di-AM-Tuning	5-SIP	Phi,Sie,++	BB 313, BB 413, MVAM-1	data
BB 117	C-Di	AFC	71a(4mm)	Phi	BA 121, BB 417, 1SV89	data
BB 119	C-Di	AFC, 15V, 20...25pF(4V), <1,5Ω	31a	Phi	BA 125, 1SV114, 1SV125	data
BB 121 A,B	C-Di	VHF/UHF-Tuning	31a	Itt,Tho	BB 105A,B, BB 205A,B, BB 405A,B, ++	data
BB 122	C-Di	VHF/UHF-Tuning	31a	Itt,Tho	BB 105A,B, BB 205A,B, BB 405A,B, ++	data
BB 125	C-Di	VHF/UHF-Tuning	31a	Idr	BB 105A,B, BB 205A,B, BB 405A,B, ++	data
BB 126	C-Di	VHF/UHF-Tuning	31a	Idr	BB 105A,B, BB 205A,B, BB 405A,B, ++	data
BB 130	C-Di	AM-Tuning, 450...550/12...21pF(1/28V)	7d	Phi	BB 112, BB 509, MV 1401, MV 1403...1405++	data
BB 131	C-Di	SMD,Sat-TV-Tuning,10,5/0,85pF(0,5/28V)	71a(1,7mm)	Phi	-	data
BB 132	C-Di	SMD, VHF-Tuning Band A(<160MHz) 60...75/2,3...2,75pF(0,5/28V)	71a(1,7mm)	Phi	-	data
BB 133	C-Di	SMD, VHF-Tuning Band B(<460MHz) 38...46/2,2...2,6pF(0,5/28V)	71a(1,7mm)	Phi	BB 639...640, MA 365...366, MA 371	data
BB 134	C-Di	SMD, TV-UHF-Tuning, 19/1,9pF(0,5/28V)	71a(1,7mm)	Phi	BB 535, MA 360, MA 363, MA 372	data
BB 135	C-Di	SMD, TV-UHF-Tuning, 19/1,9pF(0,5/28V)	71a(1,7mm)	Phi	BB 535, MA 360, MA 363, MA 372	data
BB 139	C-Di	VHF-Tuning, FCCOIRT, 5,2/29pF(25/3V)	31a	Fch,Itt,Tho	BB 106, BB 209, BB 229, BB 309, BB 409	data
BB 141 A,B	C-Di	VHF/UHF-Tuning, 30V, 0,6Ω(470MHz) A=2,18/11pF(25/3V), B=2,45/13pF(25/3V)	31a	Itt,Tho	BB 105A,B, BB 205A,B, BB 405A,B, ++	data
BB 141 [Philips]	C-Di	SMD, VCO, 6V, 4,2/2,38pF(1/4V)	71a(1,3mm)	Phi	-	data
BB 142	C-Di	VHF/UHF-Tuning, 30V, 2,5/12pF(25/3V)	31a	Itt,Tho	BB 105A,B, BB 205A,B, BB 405A,B, ++	data
BB 142 [Philips]	C-Di	SMD, VCO, 6V, 4,6/2,05pF(1/4V)	71a(1,3mm)	Phi	-	data
BB 143 A,B	C-Di	FM/VHF-Tuning, 30V, <0,7Ω(100MHz) A=5,6/9pF(25/10V), B=6,15/9pF(25/10V)	31a	Itt	BB 109, MV 109, MV 2101..2115, 1SV50	data
BB 143 [Philips]	C-Di	SMD, VCO, 6V, 5,3/2,25pF(1/4V)	71a(1,3mm)	Phi	-	data
BB 146	C-Di	SMD, TV-UHF-Tuning, extenden Band 35...43/1,7...2,1pF(0,5/28V)	71a(1,7mm)	Phi	-	data
BB 147	C-Di	SMD, TV-VHF-Tuning, extenden Band 92...112/2,4...2,8pF(0,5/28V)	71a(1,7mm)	Phi	-	data
BB 148	C-Di	SMD, TV-VHF/CATV-Tuning, ...460MHz 36,8...41,8/2,4...2,75pF(1/28V)	71a(1,7mm)	Phi	MA 355, MA 357, MA 365...366, MA 371	data
BB 149(A)	C-Di	SMD, TV-UHF-Tuning, 30V 18...19,5/1,9...2,25pF(1/28V) A: 18,22...21,26/1,951...2,225pF(1/28V)	71a(1,7mm)	Phi	BB 535, MA 360, MA 363, MA 372	data
BB 150	C-Di	SMD, TV-VHF/CATV-Tuning, ...460MHz 38...46/2,2...2,6pF(0,5/28V)	71a(1,7mm)	Phi	MA 355, MA 357, MA 365...366, MA 371	data
BB 151	C-Di	SMD, VCO, 10V, 19,1/9pF(0,5/4V)	71a(1,7mm)	Phi	-	data
BB 152	C-Di	SMD, TV-VHF-Tuning Band A(<160MHz) 52...62/2,48...2,89pF(1/28V)	71a(1,7mm)	Phi	-	data
BB 153	C-Di	SMD, TV-VHF-Tuning Band B(<460MHz) 34,65...42,35/2,361...2,754pF(1/28V)	71a(1,7mm)	Phi	MA 355, MA 357, MA 365...366, MA 371	data
BB 154	C-Di	SMD, TV-UHF-Tuning, 32V, <0,75Ω 18,5...21,25/1,9...2,2pF(1/28V)	71a(1,7mm)	Phi	BB 535, MA 360, MA 363, MA 372	data
BB 155	C-Di	SMD, Mobile Communication VCO, 30V 45,2...49,8/24,5...26,7pF(0,34/2,82V)	71a(1,7mm)	Phi	-	data
BB 156	C-Di	SMD, VCO, 10V, 16/4,8pF(1/7,5V)	71a(1,7mm)	Phi	-	data
BB 158	C-Di	SMD, TV-VHF-Tuning Band B(<460MHz) 36,8...41,8/2,4...2,75pF(1/28V)	71a(1,7mm)	Phi	BB 639...640, MA 365...366, MA 371	data
BB 159	C-Di	SMD, TV-UHF-Tuning 18...19,5/1,9...2,25pF(1/28V)	71a(1,7mm)	Phi	BB 535, MA 360, MA 363, MA 372	data
BB 164	C-Di	SMD, TV-VHF-Tuning Band A(<160MHz) 62...76/2,9...3,4pF(1/28V)	71a(1,7mm)	Phi	-	data
BB 178	C-Di	SMD, TV-VHF-Tuning Band B(<460MHz) 34,65...42,35/2,361...2,754pF(1/28V)	71a(1,3mm)	Phi	-	data
BB 179(B)	C-Di	SMD, TV-UHF-Tuning, 30V 18,22...21,26/1,951...2,225pF(1/28V) B: 18,22...20/1,9...2,25pF(1/28V)	71a(1,3mm)	Phi	-	data
BB 181	C-Di	SMD,Sat-TV-Tuning,10,5/0,85pF(0,5/28V)	71a(1,3mm)	Phi	-	data
BB 182	C-Di	SMD, TV-VHF-Tuning Band A(<160MHz) 52...62/2,48...2,89pF(1/28V)	71a(1,3mm)	Phi	-	data
BB 203	C-Di	=BB 103	31a	Sie	→BB 103	data
BB 204	C-Di	=BB 104:	7p	Phi,Sie,++	→BB 104	data
BB 205 A,B,G	C-Di	=BB 105 A,B,G	71a(4mm)	Phi,Sie,++	→BB 105A,B,G	data
BB 209(G)	C-Di	VHF-Tuning	71a(4mm)	Aeg,Phi,Sie	BB 106, BB 229, BB 309, BB 409, ++	data
BB 212(B)	C-Di	Dual, AM-Tuning	7p	Phi	BB 107, BB 312	data
BB 215	C-Di	SMD, UHF-Tuning	72a(3,4mm)	Phi	BB 621...623	data
BB 219	C-Di	SMD, VHF-Tuning	72a(3,4mm)	Phi	BB 629, BB 631	data
BB 221	C-Di	=BB 121 A,B	31a	Itt,Tho	→BB 121A,B	data
BB 222	C-Di	=BB 122	31a	Itt,Tho	→BB 122	data
BB 229	C-Di	=BB 209:	31a	Tho	→BB 209	data
BB 240	C-Di	SMD, VHF-Tuning, Band B, ...400MHz	72a(3,4mm)	Phi	-	data
BB 241	C-Di	SMD, VHF-Tuning, Band A, ...160MHz	72a(3,4mm)	Phi	-	data

117

Type	Device	Short Description	Fig.	Manu	Comparision Types	More at
BB 249	C-Di	SMD, VHF-Tuning, FCC, OIRT	72a(3,4mm)	Phi	-	data
BB 304(A)	C-Di	Dual, AM-Tuning, 32V, 42...47,5pF(2V) 20V	7p	Aeg,Phi,Sie	BB 104, BB 204, MV 104, 1SV55, 1SV109	data
BB 305 B	C-Di	UHF-Tuning	71a(4mm)	Mot	BB 105A,B, BB 205A,B, BB 405A, ++	data
BB 305 G	C-Di	VHF-Tuning	71a(4mm)	Mot	BB 105G, BB 205G, BB 405G, BB 505G, ++	data
BB 309	C-Di	VHF-Tuning	71a(4mm)	Sie	BB 106, BB 209, BB 229, BB 409, ++	data
BB 312	C-Di	Dual, AM-Tuning	7p	Sie	BB 107, BB 212	data
BB 313	C-Di	3-Di-AM-Tuning	5-SIP	Sie	BB 113, BB 413, MVAM-1	data
BB 314	C-Di	Dual, FM-Tuning	7p	Sie	BB 104, BB 204, BB 304, MV 104, 1SV 109	data
BB 319	C-Di	VHF-Tuning	31a	Tho	BB 106, BB 209, BB 229, BB 409, ++	data
BB 329 A,B	C-Di	VHF-Tuning	31a	Itt	BB 106, BB 209, BB 229, BB 409, ++	data
BB 404(A...E)	C-Di	SMD, Dual, FM-Tuning	35l	Aeg,Itt	BB 804, BB 814	data
BB 405 A,B	C-Di	UHF-Tuning	31a	Phi,Sie,Tho	BB 105A,B, BB 205A,B, BB 505B, ++	data
BB 405 G	C-Di	VHF-Tuning	31a	Phi,Sie,Tho	BB 105G, BB 205G, BB 505G, BB 609, ++	data
BB 406	C-Di	VHF-Tuning	31a	Tho	BB 106, BB 209, BB 229, BB 609, ++	data
BB 409	C-Di	VHF-Tuning, FCC, OIRT	31a	Sie	BB 106, BB 209, BB 229, BB 609, ++	data
BB 413	C-Di	3-Di-AM-Tuning	5-SIP	Sie	BB 113, BB 313, MVAM-1	data
BB 417	C-Di	VHF/UHF-AFC	31a	Phi	BA 121, BB 117, 1SV89	data
BB 419	C-Di	SMD, VHF-Tuning	71a(2,7mm)	Sie	BB 619...620, MA 329, MA335	data
BB 421	C-Di	UHF-Tuning	31a	Tho	BB 105A,B, BB 205A,B, BB 405A,B, ++	data
BB 422	C-Di	VHF-Tuning	31a	Tho	BB 105G, BB 205G, BB 405G, BB 505G, ++	data
BB 439	C-Di	=BB 419:	71a(1,7mm)	Sie	MA 361, MA 364	data
BB 501	C-Di	VHF/UHF-AFC	7c	Sie	BA 121, BB 117, BB 417, 1SV89	data
BB 502	C-Di	VHF-AFC	7c	Sie	BA 125, BB 119, 1SV114, 1SV125	data
BB 503	C-Di	=BB 501: SMD	35e	Sie,Tho	-	data
BB 503 DK	C-Di	=BB 501: SMD, Dual	35l	Tho	-	data
BB 504	C-Di	=BB 502: SMD	35e	Sie,Tho	-	data
BB 505 B	C-Di	UHF-Tuning	31a	Aeg,Sie	BB 105A,B, BB 205A,B, BB 405A,B, ++	data
BB 505 G	C-Di	VHF-Tuning	31a	Aeg,Sie	BB 105G, BB 205G, BB 405G, ++	data
BB 509	C-Di	AM-Tuning, 4,43MHz-AFC	7c	Itt	-	data
BB 510	C-Di	SMD, Dual, AM-Tuning	35m1	Itt	-	data
BB 512	C-Di	SMD, AM-Tuning	71a(2,7mm)	Sie	-	data
BB 515 B,G	C-Di	SMD, VHF/UHF-Tuning	71a(2,7mm)	Phi,Sie	BB 721	data
BB 521	C-Di	=BB 221: verbess./improved Linear.	31a	Itt	→BB 221	data
BB 523	C-Di	VHF/UHF-Tuning	31a	Itt	BB 105A,B, BB 205A,B, BB 405A,B, BB 505B	data
BB 525	C-Di	SMD, Tuning	71a(2,7mm)	Sie	-	data
BB 529	C-Di	=BB 329: verbess./improved Linear.	31a	Itt	→BB 329	data
BB 531	C-Di	VHF-Tuning	31a	Itt	BB 105G, BB 205G, BB 405G, BB 505G, ++	data
BB 535	C-Di	=BB 515:	71a(1,7mm)	Sie	MA 360, MA 372, 1SV214, 1SV223	data
BB 545	C-Di	=BB 525:	71a(1,7mm)	Sie	-	data
BB 555	C-Di	=BB 535:	71a(1,3mm)	Sie	-	data
BB 565	C-Di	=BB 545:	71a(1,3mm)	Sie	-	data
BB 601	C-Di	SMD, Satelliten-TV-Tuning	71a(2,7mm)	Itt	-	data
BB 609 A,B	C-Di	VHF-Tuning, OIRT, CATV	31a	Aeg,Sie	BB 106, BB 209, BB 229, BB 409, ++	data
BB 610	C-Di	VHF-Tuning	31a	Sie	BB 105G, BB 205G, BB 405G, BB 505G, ++	data
BB 619(A,B)	C-Di	SMD, VHF-TV/VC-Tuning, Extend. Band	71a(2,7mm)	Phi,Sie	BB 515B,G, BB 723, BB 729...731	data
BB 620	C-Di	SMD, VHF-TV/VC-Tuning, Hyperband	71a(2,7mm)	Phi,Sie	-	data
BB 621	C-Di	=BB 221: SMD	72a(3,4mm)	Aeg,Itt	BB 215, BB 623	data
BB 622	C-Di	=BB 222: SMD	72a(3,4mm)	Aeg,Itt	BB 215, BB 623	data
BB 623	C-Di	SMD, VHF/UHF-Tuning	72a(3,4mm)	Itt	BB 215, BB 621...622	data
BB 629	C-Di	=BB 329: SMD	72a(3,4mm)	Aeg,Itt	BB 219, BB 631	data
BB 631	C-Di	SMD, VHF-Tuning	72a(3,4mm)	Itt	BB 219, BB 629	data
BB 639(C)	C-Di	=BB 619:	71a(1,7mm)	Sie	MA 365, MA 374, 1SV231	data
BB 640	C-Di	=BB 620:	71a(1,7mm)	Sie	-	data
BB 644	C-Di	SMD, VHF-Tuning, 35V, 2,6/42pF(28/1V)	71a(1,7mm)	Sie	BB 439, MA 361, MA 364	data
BB 659(C)	C-Di	=BB 619:	71a(1,3mm)	Sie	-	data
BB 664	C-Di	=BB 644:	71a(1,3mm)	Sie	-	data
BB 669	C-Di	SMD, VHF-Tuning, Hyperband, 35V 2,7/56pF(28/1V)	71a(1,7mm)	Sie	-	data
BB 689	C-Di	=BB 669:	71a(1,3mm)	Sie	-	data
BB 709 A,B	C-Di	VHF-Tuning	71a(4mm)	Sie,Tho	BB 106, BB 209, BB 229, BB 409, ++	data
BB 721	C-Di	SMD, UHF-Tuning, 35V, 2,1...2,39(25V) 14...16,3pF(2V), <0,8Ω(470MHz)	71a(2,7mm)	Itt, Gsi	BB 723	data
BB 721 S	C-Di	=BB 721:	71a(1,7mm)	Gsi	-	data
BB 723	C-Di	SMD, VHF/UHF-Tuning	71a(2,7mm)	Itt	BB 721	data
BB 729	C-Di	SMD,VHF-Tuning, 32V, 2,38...2,93pF(28V) 30...33pF(28V), <0,8Ω(470MHz)	71a(2,7mm)	Itt, Gsi	BB 723, BB 730...731	data
BB 729 S	C-Di	=BB 729:	71a(1,7mm)	Gsi	-	data

Type	Device	Short Description	Fig.	Manu	Comparision Types	More at
BB 730	C-Di	SMD, VHF Tuning, 28V, 3,15...3,55(28V)	71a(2,7mm)	Itt	BB 723, BB 729, BB 731	data
BB 731	C-Di	SMD, VHF-Tuning(Hyperband), 32V 3,15..3,55/50pF(28/1V), <1Ω(300MHz)	71a(2,7mm)	Itt, Gsi	BB 723, BB 729...730	data
BB 731 S	C-Di	=BB 731:	71a(1,7mm)	Gsi	-	data
BB 801	C-Di	SMD, Satelliten-TV-Tuning	35a	Sie	-	data
BB 804(-0...-4)	C-Di	SMD, Dual, FM-Tuning	35l	Aeg,Phi,Sie	BB 404, BB 814	data
BB 809	C-Di	VHF-Tuning, OIRT, CATV	31a	Phi,Tho	BB 106, BB 209, BB 229, BB 409, ++	data
BB 811	C-Di	SMD, Satelliten-TV-Tuning, ..2GHz	71a(2,7mm)	Phi,Sie	-	data
BB 813	C-Di	SMD, Satelliten-TV-Tuning, ..2,5GHz	71a(2,7mm)	Sie	-	data
BB 814(-1,-2)	C-Di	SMD, Dual, FM-Tuning	35l	Aeg,Sie	BB 404, BB 804	data
BB 824	C-Di	SMD, Dual, Tuning	35l	Aeg		data
BB 831	C-Di	=BB 811:	71a(1,7mm)	Sie	1SV245	data
BB 833	C-Di	=BB 813:	71a(1,7mm)	Sie	1SV245	data
BB 835	C-Di	SMD, Satelliten-TV-Tuning, ..2,8GHz	71a(1,7mm)	Sie	1SV245	data
BB 857	C-Di	SMD, Satelliten-TV-Tuning, 35V	71a(1,3mm)	Sie	-	data
BB 901	C-Di	SMD, Sat-TV-Tuning, <1,055pF(28V)	35e	Phi	-	data
BB 909 A,B	C-Di	VHF-Tuning, CATV	31a	Phi,Tho	BB 106, BB 209, BB 229, BB 409, ++	data
BB 910	C-Di	VHF-Tuning, Band B, ..460MHz 2,4..2,7/>38pF(28/0,5V)	31a	Phi	-	data
BB 911(A)	C-Di	SMD, VHF-Tuning, Band A, ...160MHz	31a	Phi	-	data
BB 914	C-Di	SMD, Dual, FM-Tuning(Car), 20V 17,6...19,8/42,5...45pF(8/2V)	35l	Sie	BB 404, BB 804, BB 814	data
BB3553	FAST-BUFF	Vs:±20V Vo:±13V Vi0:<50mV f:300Mc	8O	Max		data pdf pinout
BB 3553 AM	Hybrid-OP-IC	±20V, 200MHz, 6000V/μs, -25..+85	23/8Pin	Max		data
BB3554	WIDEBAND	Vs:±20V Vu:106dB Vo:±11V Vi0:0.5mV	8O	Max		data pdf pinout
BB 3554(...)M	Hybrid-OP-IC	Wideband, JFET Inp, 40/±20V, >1000V/μs A,B: -25...+85°, M: -55...+125°	23/8Pin	Max	AD 3554...	data
BBA	Z-Di	=SMBJ 6.0CA (Typ-Code/Stempel/marking)	71(5x3,5)	Tho	→SMBJ ...	data
BBB	Z-Di	=SMBJ 6.5CA (Typ-Code/Stempel/marking)	71(5x3,5)	Tho	→SMBJ ...	data
BBC	Z-Di	=SMBJ 8.5CA (Typ-Code/Stempel/marking)	71(5x3,5)	Tho	→SMBJ ...	data
BBC C 106 B	50Hz-Thy	200V, 3,2A, Igt/lh<0,2/<5mA	17h§	Bbc	TIC 106, TAG 623, TAG 628, S 4060, MCR72	data
BBC C 106 B-1	50Hz-Thy	200V, 2A, Igt/lh<0,25/<2mA	13h§	Bbc	C 106, TAG 106, TAG 108, C 108, XO 403	data
BBC C 106 D	50Hz-Thy	400V, 3,2A, Igt/lh<0,2/<5mA	17h§	Bbc	TIC 106, TAG 623, TAG 628, S 4060, MCR72	data
BBC C 106 D-1	50Hz-Thy	400V, 2A, Igt/lh<0,25/<2mA	13h§	Bbc	C 106, TAG 106, TAG 108, C 108, XO 403	data
BBC C 106 E	50Hz-Thy	500V, 3,2A, Igt/lh<0,2/<5mA	17h§	Bbc	TIC 106, TAG 623, TAG 628, S 4060, MCR72	data
BBC C 106 E-1	50Hz-Thy	500V, 2A, Igt/lh<0,25/<2mA	13h§	Bbc	C 106, TAG 106, TAG 108, C 108, XO 403	data
BBC C 106 F	50Hz-Thy	600V, 3,2A, Igt/lh<0,2/<5mA	17h§	Bbc	TIC 106, TAG 623, TAG 628, S 4060, MCR72	data
BBC C 106 F-1	50Hz-Thy	600V, 2A, Igt/lh<0,25/<2mA	13h§	Bbc	C 106, TAG 106, TAG 108, C 108, XO 403	data
BBD	Z-Di	=SMBJ 10CA (Typ-Code/Stempel/marking)	71(5x3,5)	Tho	→SMBJ ...	data
BBE	Z-Di	=SMBJ 12CA (Typ-Code/Stempel/marking)	71(5x3,5)	Tho	→SMBJ ...	data
BBF	Z-Di	=SMBJ 13CA (Typ-Code/Stempel/marking)	71(5x3,5)	Tho	→SMBJ ...	data
BBG	Z-Di	=SMBJ 15CA (Typ-Code/Stempel/marking)	71(5x3,5)	Tho	→SMBJ ...	data
BBH	Z-Di	=SMBJ 18CA (Typ-Code/Stempel/marking)	71(5x3,5)	Tho	→SMBJ ...	data
BBI	Z-Di	=SMBJ 20CA (Typ-Code/Stempel/marking)	71(5x3,5)	Tho	→SMBJ ...	data
BBJ	Z-Di	=SMBJ 24CA (Typ-Code/Stempel/marking)	71(5x3,5)	Tho	→SMBJ ...	data
BBK	Z-Di	=SMBJ 26CA (Typ-Code/Stempel/marking)	71(5x3,5)	Tho	→SMBJ ...	data
BBL	Z-Di	=SMBJ 28CA (Typ-Code/Stempel/marking)	71(5x3,5)	Tho	→SMBJ ...	data
BBM	Z-Di	=SMBJ 30CA (Typ-Code/Stempel/marking)	71(5x3,5)	Tho	→SMBJ ...	data
BBN	Z-Di	=SMBJ 33CA (Typ-Code/Stempel/marking)	71(5x3,5)	Tho	→SMBJ ...	data
BBO	Z-Di	=SMBJ 58CA (Typ-Code/Stempel/marking)	71(5x3,5)	Tho	→SMBJ ...	data
BBp	Si-P	=BCW 61B (Typ-Code/Stempel/marking)	35	Phi	→BCW 61B	data
BBQ	Z-Di	=SMBJ 85CA (Typ-Code/Stempel/marking)	71(5x3,5)	Tho	→SMBJ ...	data
BBs	Si-Di	=BAR 81 (Typ-Code/Stempel/marking)	44	Sie	→BAR 81	data
BBs	Si-Di	=BAR 81W (Typ-Code/Stempel/marking)	44(2mm)	Sie	→BAR 81W	data
BBs	Si-P	=BCW 61B (Typ-Code/Stempel/marking)	35	Sie	→BCW 61B	data
BBS	Z-Di	=SMBJ 130CA (Typ-Code/Stempel/marking)	71(5x3,5)	Tho	→SMBJ ...	data
BBt	Si-P	=BCW 61B (Typ-Code/Stempel/marking)	35	Phi	→BCW 61B	data
BBT	Z-Di	=SMBJ 154CA (Typ-Code/Stempel/marking)	71(5x3,5)	Tho	→SMBJ ...	data
BBU	Si-N	=2SD2114K-U (Typ-Code/Stempel/marking)	35	Rhm	→2SD2114K	data
BBU	Z-Di	=SMBJ 170CA (Typ-Code/Stempel/marking)	71(5x3,5)	Tho	→SMBJ ...	data
BBV	Si-N	=2SD2114K-V (Typ-Code/Stempel/marking)	35	Rhm	→2SD2114K	data
BBV	Z-Di	=SMBJ 118CA (Typ-Code/Stempel/marking)	71(5x3,5)	Tho	→SMBJ ...	data
BBW	Si-N	=2SD2114K-W (Typ-Code/Stempel/marking)	35	Rhm	→2SD2114K	data
BBW	Z-Di	=SMBJ 48CA (Typ-Code/Stempel/marking)	71(5x3,5)	Tho	→SMBJ ...	data
BBY 30	C-Di	FM/VHF-Tuning	31a	Sie		data
BBY 31	C-Di	SMD, VHF/UHF-Tuning, 30V, 2,3/17,5pF (25/1V)	35e	Phi,Sie,Tho Ztx	-	data
BBY 39	C-Di	SMD, Dual, UHF/Sat-TV-Tuning 1,6...2/16,5pF(28/1V)	35l	Phi		data

Type	Device	Short Description	Fig.	Manu	Comparision Types	More at
BBY 40	C-Di	SMD, VHF-Tuning Band A(<160MHz) 4,3...6/26...32pF(25/3V)	35e	Phi, Ztx	-	data
BBY 42	C-Di	SMD, VHF-Tuning Band B(<460MHz) 2,4...3/>31pF(28/1V)	35c1,e	Phi	-	data
BBY 51	C-Di	SMD, Dual, Tuning, VCO, 7V 2,5...3,7/4,5...6,1pF(4/1V)	35l	Sie	-	data
BBY 51-02W	C-Di	=BBY 51:	71a(1,3mm)		-	data
BBY 51-03W	C-Di	=BBY 51:	71a(1,7mm)		-	data
BBY 51-07	C-Di	=BBY 51: 2,6...3,5/4,8...6pF(4/1V)	44s2		-	data
BBY 52	C-Di	SMD, Dual, Tuning, VCO, 7V 0,85...1,45/1,4...2,2pF(4/1V)	35l	Sie	-	data
BBY 52-02W	C-Di	=BBY 52:	71a(1,3mm)		-	data
BBY 52-03W	C-Di	=BBY 52:	71a(1,7mm)		-	data
BBY 53	C-Di	SMD, Dual, Tuning, VCO, 6V 1,85...3,1/4,8...5,8pF(6/1V)	35l	Sie	-	data
BBY 53-02W	C-Di	=BBY 53:	71a(1,3mm)		-	data
BBY 53-03W	C-Di	=BBY 53:	71a(1,7mm)		-	data
BBY 55-02W	C-Di	SMD, Dual, Tuning, VCO, 16V 5,5...6,5/14...16pF(10/2V)	71a(1,3mm)	Sie	-	data
BBY 55-03W	C-Di	=BBY 55:	71a(1,7mm)		-	data
BBY 56-02W	C-Di	SMD, Dual, Tuning, VCO, 10V 15,9...19/59...67pF(3/1V)	71a(1,3mm)	Sie	-	data
BBY 56-03W	C-Di	=BBY 56:	71a(1,7mm)		-	data
BBY 57-02W	C-Di	SMD, Dual, Tuning, VCO, 10V 4...5,5/16,5...18,6pF(4/1V)	71a(1,3mm)	Sie	-	data
BBY 57-03W	C-Di	=BBY 57:	71a(1,7mm)		-	data
BBY 58-02W	C-Di	SMD, Dual, Tuning, VCO, 10V 5,5...6,6/17,5...19,3pF(4/1V)	71a(1,3mm)	Sie	-	data
BBY 58-03W	C-Di	=BBY 58:	71a(1,7mm)		-	data
BBY 62	C-Di	SMD, Dual, UHF-Tuning, 30V 1,6...2/16,5pF(28/1V)	44s2	Phi	-	data
BBZ	Z-Di	=SMBJ 5.0CA (Typ-Code/Stempel/marking)	71(5x3,5)	Tho	→SMBJ ...	data

BC

Type	Device	Short Description	Fig.	Manu	Comparision Types	More at
BC	Si-P	=2SB1120 (Typ-Code/Stempel/marking)	39	Say	→2SB1120	data
BC	Si-P	=2SB1188 (Typ-Code/Stempel/marking)	39	Rhm	→2SB1188	data
BC	Si-P	=2SB831-C (Typ-Code/Stempel/marking)	35	Hit	→2SB831	data
BC	Si-N	=2SD1367-BC (Typ-Code/Stempel/marking)	39	Hit	→2SD1367	data
BC	Si-N	=BCP 54-10 (Typ-Code/Stempel/marking)	48	Phi	→BCP 54-10	data
BC	Si-P	=BCW 61C (Typ-Code/Stempel/marking)	35	Mot,Phi,Sie Aeg,Fer,++	→BCW 61C	data
BC	Si-N	=BCX 54-10 (Typ-Code/Stempel/marking)	39	Sie,Val,Mot	→BCX 54-10	data
BC	Z-Di	=BZX 399-C5V6(Typ-Code/Stempel/marking)	71(1,7mm)	Phi	→BZX 399-...	data
BC	Si-N/P+R	=KRX 203U (Typ-Code/Stempel/marking)	46(2mm)	Kec	→KRX 203U	data
BC	MOS-P-FET-e	=NTTD 1P02 (Typ-Code/Stempel/marking)	8-SSMDIP	Ons	→NTTD 1P02	
BC	Si-Di	=SC 016-4 (Typ-Code/Stempel/marking)	71(5x2,5)	Fjd	→SC 016-...	data
BC	Si-Di	=U 1BC44 (Typ-Code/Stempel/marking)	71(5x2,5)	Tos	→U 1BC44	data
BC	Si-Di	=ZMS 2811 (Typ-Code/Stempel/marking)	71(1,7mm)	Ztx	→ZMS 2811	data
BC 100	Si-N	Vid-E, 350V, 0,15A, 0,6W	2a§	Aeg	BF 259, BF 659, BFR 59, BFS 89, 2N5058	data
BC 101	Si-N	40V, 40mA, 0,3W, 30MHz	2a	Idr	BC 167, BC 182, BC 237, BC 547, ++	data
BC 107	Si-N	Uni, 50V, 0,1A, 0,3W, 300MHz	2a§	EUR	BC 167, BC 182, BC 237, BC 547, ++	data
BC 107 P	Si-N	=BC 107:	7a, 40e	Tix,Fer	→BC 107	data
BC 108	Si-N	=BC 107: 30V	2a§	EUR	BC 168, BC 183, BC 238, BC 548, ++	data
BC 108 P	Si-N	=BC 108:	7a, 40e	Tix,Fer	→BC 108	data
BC 109	Si-N	=BC 107: 30V, ra	2a§	EUR	BC 169, BC 184, BC 239, BC 549, ++	data
BC 109 P	Si-N	=BC 109:	7a, 40e	Tix,Fer	→BC 109	data
BC 110	Si-N	Uni, 80V, 50mA, 0,3W, 100MHz	2a§	Aeg,Sie	BC 546, 2SC1890	data
BC 111	Si-N	Min, NF, 20V, 50mA, 0,03W, >50MHz	2a	Phi	BC 122...123, BC 146	data
BC 112	Si-N	Min, NF, 20V, 50mA, 0,05W, 150MHz	36b	Phi	BC 122...123, BC 146	data
BC 113	Si-N	Uni, 30V, 50mA, 0,2W, 100MHz	8a	Nsc,Sgs	BC 168, BC 183, BC 238, BC 548, ++	data
BC 113 A	Si-N	=BC 113: 40V	8a		BC 167, BC 183, BC 237, BC 547, ++	data
BC 114	Si-N	=BC 133: ra	8a	Nsc,Sgs	BC 169, BC 184, BC 239, BC 549, ++	data
BC 114 A	Si-N	=BC 113: 40V, ra	8a		BC 184, BC 550	data
BC 115	Si-N	NF-Tr, 40V, 0,2A, 0,3W, 80MHz	8a	Nsc,Sgs	BC 167, BC 183, BC 237, BC 547, ++	data
BC 116(A)	Si-P	Uni, 45V, 0,5A, 0,3W, 200MHz	8a	Nsc,Sgs	BC 327, BC 636, BC 160...161, ++	data
BC 117	Si-N	NF, 120V, 50mA, 0,3W, >60MHz	8a	Sgs	BF 257, BF 297, 2SC1890A, 2SC2363, ++	data
BC 118	Si-N	Uni, 45V, 0,2W, >200MHz	8a	Nsc,Sgs	BC 167, BC 183, BC 237, BC 547, ++	data
BC 119	Si-N	NF-E, 60V, 1A, 0,8W, >40MHz	2a§	Sgs	BC 140...141, 2N1990, 2N2102, 2N2405, ++	data
BC 120	Si-N	NF-E, 60V, 1A, 0,8W, >40MHz	2a§	Sgs	BC 140...141, 2N1990, 2N2102, 2N2405, ++	data
BC 121	Si-N	Min, NF, ra, 5V, 75mA, 0,25W, 250MHz	(41c)	Sie	BC 122...123, BC 146	data
BC 122	Si-N	=BC 121: 30V	(41c)	Sie	BC 123, BC 146	data
BC 123	Si-N	=BC 121: 45V	(41c)	Sie		data
BC 125(A,B)	Si-N	NF-Tr, 50...60V, 0,5A, 0,3W, 350MHz	8a	Nsc,Sgs	BC 337, BC 637, BC 140...141, ++	data
BC 126(A)	Si-P	NF-Tr, 35...40V, 0,6A, 0,3W, 200MHz	8a	Nsc,Sgs	BC 327...328, BC 638, BC 160...161, ++	data

Type	Device	Short Description	Fig.	Manu	Comparision Types	More at
BC 127	Si-N	Min, NF, ra, 25V, 0,075W, 30MHz	36f	Sgs	BC 122...123	data
BC 128	Si-N	Min, NF, 25V, 0,1W, 30MHz	36f	Sgs	BC 122...123	data
BC 129	Si-N	Uni, 50V, 0,1A, 0,135W, 300MHz	2a	Aeg	BC 167, BC 182, BC 237, BC 547, ++	data
BC 130	Si-N	=BC 129: 30V	2a	Aeg	BC 168, BC 183, BC 238, BC 548, ++	data
BC 131	Si-N	=BC 129: 30V, ra	2a	Aeg	BC 169, BC 184, BC 239, BC 549, ++	data
BC 132(A)	Si-N	NF-V/Tr, 30...40V, 50mA, 0,4W, >40MHz	8a	Nsc,Sgs	BC 167, BC 183, BC 237, BC 547, ++	data
BC 134	Si-N	Uni, 45V, 0,2W, >400MHz	8a	Sgs	BC 167, BC 183, BC 237, BC 547, ++	data
BC 135(A)	Si-N	Uni, 45V, 0,2W, >200MHz	8a	Sgs	BC 167, BC 183, BC 237, BC 547, ++	data
BC 136	Si-N	NF-Tr, 60V, 0,5A, 0,3W, >60MHz	8a	Nsc,Sgs	BC 337, BC 637, BC 140...141, ++	data
BC 137	Si-P	NF-Tr, 60V, 0,5A, 0,3W, >60MHz	8a	Nsc,Sgs	BC 327, BC 638, BC 161, ++	data
BC 138	Si-N	NF-Tr/E, 60V, 1A, 0,8W, >40MHz	2a§	Sgs	BC 140...141, 2N3019...3020, ++	data
BC 139(A)	Si-P	NF-Tr/E, 40V, 0,5A, 0,7W	2a§	Sgs	BC 160...161, BC 303...304, 2N2303, ++	data
BC 140	Si-N	NF-Tr/E, 80V, 1A, 0,75W, >50MHz	2a§	EUR	BCX 40, 2N3019...3020, 2N4238...4239, ++	data
BC 141	Si-N	=BC 140: 100V	2a§	EUR	BCX 40, 2N3019...3020, 2N4239, ++	data
BC 142	Si-N	NF-Tr/E, 80V, 1A, 0,8W, 90MHz	2a§	Sgs,Tix	BC 140...141, BCX 40, 2N3019...3020, ++	data
BC 143	Si-P	NF-Tr/E, 60V, 1A, 0,7W, 170MHz	2a§	Sgs,Tix	BC 161, BC 461, BCX 60, 2N4236, ++	data
BC 144	Si-N	NF-Tr/E, 70V, 1A, 0,7W	2a§	Sgs	BC 140...141, BCX 40, 2N3019...3020, ++	data
BC 145	Si-N	NF-Tr, 120V, 0,1A, 0,3W, 80MHz	8a	Sgs	BF 257...259, BF 297...299, 2SC1890A, ++	data
BC 146	Si-N	Min, NF, ra, 20V, 50mA, 0,05W, 150MHz	7c	Aeg,Phi	BC 122...123	data
BC 147	Si-N	=BC 237:	11a	Aeg,Phi,Sie	→BC 237	data
BC 148	Si-N	=BC 238:	11a	Aeg,Phi,Sie	→BC 238	data
BC 149	Si-N	=BC 239:	11a	Aeg,Phi,Sie	→BC 239	data
BC 150	Si-N	Uni, 18V, 0,1A, 0,2W, 160MHz	7c	Aei	BC 168, BC 183, BC 238, BC 548, ++	data
BC 151	Si-N	Uni, 25V, 0,1A, 0,2W, 160MHz	7c	Aei	BC 168, BC 183, BC 238, BC 548, ++	data
BC 152	Si-N	Uni, 35V, 0,5A, 0,36W, 180MHz	7c	Aei	BC 337, BC 635, BC 637, BC 639, ++	data
BC 153	Si-P	NF-V, ra, 40V, 0,1A, 0,2W, 70MHz	8a	Nsc,Sgs	BC 214, BC 415...416, BC 560, ++	data
BC 154	Si-P	NF-V, ra, 40V, 0,1A, 0,2W, 70MHz	8a	Nsc,Sgs	BC 214, BC 415...416, BC 560, ++	data
BC 155	Si-N	Min, NF, 5V, 50mA, 0,105W, >50MHz	36c	Aeg	BC 121...123, BC 146	data
BC 156	Si-N	=BC 155: 0,05W	36c	Aeg	BC 121...123, BC 146	data
BC 157	Si-P	=BC 307:	11a	Aeg,Phi,Sie	→BC 307	data
BC 158	Si-P	=BC 308:	11a	Aeg,Phi,Sie	→BC 308	data
BC 159	Si-P	=BC 309:	11a	Aeg,Phi,Sie	→BC 309	data
BC 160	Si-P	NF-Tr/E, 40V, 1A, 0,75W, >50MHz	2a§	EUR	BC 460...461, BCX 60, 2N4234...4236, ++	data
BC 161	Si-P	=BC 160: 60V	2a§	EUR	BC 461, BCX 60, 2N4235...4236, ++ *	data
BC 167	Si-N	=BC 237:	7c	Aeg,Sie	→BC 237	data
BC 168	Si-N	=BC 238:	7c	Aeg,Sie	→BC 238	data
BC 169	Si-N	=BC 239:	7c	Aeg,Sie	→BC 239	data
BC 170	Si-N	Uni, 20V, 0,1A, 0,3W, 100MHz	7a	Itt,Nsc	BC 168, BC 183, BC 238, BC 548, ++	data
BC 171	Si-N	Uni, 50V, 0,1A, 0,3W, 250MHz	7a	Itt,Nsc	BC 167, BC 182, BC 237, BC 547, ++	data
BC 172	Si-N	=BC 171: 30V	7a	Itt,Nsc	BC 168, BC 183, BC 238, BC 548, ++	data
BC 173	Si-N	=BC 171: 30V, ra	7a	Itt,Nsc	BC 169, BC 184, BC 239, BC 549, ++	data
BC 174	Si-N	=BC 171: 64V	7a	Itt,Phi	BC 182, BC 190, BC 546	data
BC 174[Mot]	Si-N	=BC 174: Fig.→	7e	Mot		data
BC 175	Si-N	NF-Tr, 35V, 0,5A, 0,56W, 180MHz	7c	Aei	BC 337...338, BC 635, BC 637, BC 639, ++	data
BC 177	Si-P	Uni, 50V, 0,1A, 0,3W, 130MHz	2a§	EUR	BC 212, BC 257, BC 307, BC 557, ++	data
BC 177 P	Si-P	=BC 177:	7a, 40e	Tix,Fer	→BC 177	data
BC 178	Si-P	=BC 177: 30V	2a§	EUR	BC 213, BC 258, BC 308, BC 558, ++	data
BC 178 P	Si-P	=BC 178:	7a, 40e	Tix,Fer	→BC 178	data
BC 179	Si-P	=BC 177: 25V, ra	2a§	EUR	BC 214, BC 259, BC 309, BC 559, ++	data
BC 179 P	Si-P	=BC 179:	7a, 40e	Tix,Fer	→BC 179	data
BC 180	Si-N	NF-Tr, 45V, 0,5A, 0,36W, 180MHz	7c	Aei	BC 337, BC 635, BC 637, BC 639, ++	data
BC 181(A)	Si-P	Uni, 40V, 0,2A, 0,3W	7a	Tix	BC 213, BC 257, BC 307, BC 557, ++	data
BC 182	Si-N	=BC 183: 60V	7a	EUR	BC 174, BC 190, BC 546	data
BC 182 K	Si-N	=BC 182:	8a	Nsc	→BC 182	data
BC 182 L	Si-N	=BC 182:	7c	Nsc,Tix	→BC 182	data
BC 182 P	Si-N	=BC 182:	40e	Fer	→BC 182	data
BC 183	Si-N	Uni, 45V, 0,2A, 0,3W, 280MHz	7a	EUR	BC 167, BC 237, BC 547, ++	data
BC 183 K	Si-N	=BC 183:	8a	Nsc	→BC 183	data
BC 183 L	Si-N	=BC 183:	7c	Nsc,Tix	→BC 183	data
BC 183 P	Si-N	=BC 183:	40e	Fer	→BC 183	data
BC 184	Si-N	=BC 183: ra	7a	EUR	BC 413...414, BC 550	data
BC 184 K	Si-N	=BC 184:	8a	Nsc	→BC 184	data
BC 184 L	Si-N	=BC 184:	7c	Nsc,Tix	→BC 184	data
BC 184 P	Si-N	=BC 184:	40e	Fer	→BC 184	data
BC 185	Si-N	NF-Tr, 60V, 1A, 0,8W, 300MHz	2a§	Sgs	BC 140...141, BCX 40, 2N3019...3020, ++	data
BC 186	Si-P	Uni, 40V, 0,1A, 0,3W, >50MHz	2a§	Phi	BC 213, BC 257, BC 307, BC 557, ++	data
BC 187	Si-P	=BC 186: 30V	2a§	Phi	BC 213, BC 258, BC 308, BC 558, ++	data
BC 190	Si-N	Uni, 70V, 0,1A, 0,3W, 250MHz	2a§	Itt	BC 174, BC 546, 2SC1890	data
BC 192	Si-P	NF/S, 25V, 0,5A, 0,4W, >100MHz	2a§	Itt	BC 327...328, BC 636, BC 638, BC 640, ++	data
BC 194	Si-N	Min, NF, 40V, 0,8A, >250MHz	36c	Aeg	(BC 337, BC 635, BC 637, BC 639,++)[6]	data
BC 196	Si-P	Min, NF, 30V, 0,1A, 150MHz	36c	Aeg	BC 200, BC 202...203	data
BC 197	Si-N	=BC 198: 50V	36c	Aeg	BC 123	data
BC 198	Si-N	Min, NF, 30V, 0,1A, 0,05W, 300MHz	36c	Aeg	BC 122...123, BC 146	data
BC 199	Si-N	=BC 198: ra	36c	Aeg	BC 122...123, BC 146	data

Type	Device	Short Description	Fig.	Manu	Comparision Types	More at
BC 200	Si-P	Min, NF, ra, 20V, 50mA, 0,05W, 90MHz	7c	Aeg,Phi	BC 202...203	data
BC 201	Si-P	Min, NF, ra, 5V, 75mA, 0,25W, 80MHz	(41c)	Sie	BC 200	data
BC 202	Si-P	=BC 201: 30V	(41c)	Sie	-	data
BC 203	Si-P	=BC 201: 45V	(41c)	Sie	-	data
BC 204	Si-P	=BC 205: 50V	8a	Fch,Tho	BC 212, BC 257, BC 307, BC 557, ++	data
BC 205	Si-P	Uni, 25V, 0,1A, 0,2W, 200MHz	8a	Fch,Tho	BC 213, BC 258, BC 308, BC 558, ++	data
BC 206	Si-P	=BC 205: ra	8a	Fch,Tho	BC 214, BC 259, BC 309, BC 559, ++	data
BC 207	Si-N	=BC 208: 50V	8a	Fch,Tho	BC 167, BC 182, BC 237, BC 547, ++	data
BC 208	Si-N	Uni, 25V, 0,1A, 0,2W, 300MHz	8a	Fch,Tho	BC 168, BC 183, BC 238, BC 548, ++	data
BC 209	Si-N	=BC 208: ra	8a	Fch,Tho	BC 169, BC 184, BC 239, BC 549, ++	data
BC 210	Si-N	NF-Tr, 50V, 0,7A, 0,45W, >100MHz	2a§	Tho	BC 337, BC 637, BC 639	data
BC 211	Si-N	NF-Tr/E, 80V, 1A, 0,8W, 300MHz	2a§	Tho	BC 140...141, BCX 40, 2N3019...3020, ++	data
BC 211 A	Si-N	=BC 211: 100V	2a§		BC 141, BCX 40, 2N3019...3020, ++	data
BC 212	Si-P	=BC 213: 60V	7a	EUR	BC 256, BC 266, BC 556	data
BC 212 K	Si-P	=BC 212:	8a	Nsc	→BC 212	data
BC 212 L	Si-P	=BC 212:	7c	Nsc,Tix	→BC 212	data
BC 212 P	Si-P	=BC 212:	40e	Fer	→BC 212	data
BC 213	Si-P	Uni, 45V, 0,2A, 0,3W, 350MHz	7a	EUR	BC 257, BC 307, BC 557, ++	data
BC 213 K	Si-P	=BC 213:	8a	Nsc	→BC 213	data
BC 213 L	Si-P	=BC 213:	7c	Nsc,Tix	→BC 213	data
BC 213 P	Si-P	=BC 213:	40e	Fer	→BC 213	data
BC 214	Si-P	=BC 213: ra	7a	EUR	BC 415...416, BC 560	data
BC 214 K	Si-P	=BC 214:	8a	Nsc	→BC 214	data
BC 214 L	Si-P	=BC 214:	7c	Nsc,Tix	→BC 214	data
BC 214 P	Si-P	=BC 214:	40e	Fer	→BC 214	data
BC 215	Si-P	NF-Tr, 50V, 0,6A, 0,4W, 200MHz	2a§	Tho	BC 327, BC 638, BC 640	data
BC 216(A) [Riz]	Si-P	NF-Tr, 30...40V, 0,6A, 0,3W, 200MHz	2a§	Riz	BC 327, BC 636, BC 638, BC 640	data
BC 216(A,B) [SGS]	Si-N	NF, 45V, 20mA, 0,85W, 70MHz	2a§	Sgs	BC 167, BC 183, BC 237, BC 547, ++	data
BC 218(A)	Si-N	Uni, 30...45V, 0,1A, 0,3W, 350MHz	2a§	Riz	BC 167, BC 183, BC 237, BC 547, ++	data
BC 219	Si-N	NF-Tr, 60V, 0,5A, 0,8W, 40MHz	2a§	Riz	BC 140...141, BCX 40, 2N3019...3020, ++	data
BC 220	Si-N	Uni, 30V, 50mA, 0,2W, 80MHz	8a	Sgs	BC 168, BC 183, BC 238, BC 548, ++	data
BC 221	Si-P	NF-Tr, 30V, 0,5A, 0,3W, 150MHz	8a	Sgs	BC 327...328, BC 636, BC 638, BC 640, ++	data
BC 222	Si-N	NF-Tr, 30V, 0,5A, 0,3W, 250MHz	8a	Sgs	BC 337...338, BC 635, BC 637, BC 639, ++	date
BC 223	Si-N	NF-Tr, 50V, 0,4A, 0,36W	7a	Tix	BC 337, BC 637, BC 639	data
BC 224	Si-P	Uni, 30V, 30mA, 0,25W	7c	Tix	BC 213, BC 258, BC 308, BC 558, ++	data
BC 225	Si-P	NF, ra, 40V, 0,1A, 0,2W, 70MHz	8a	Sgs	BC 415...416, BC 560	data
BC 226 [SGS]	Si-N	NF-Tr, 40V, 0,1A, 0,8W, 40MHz	2a§	Sgs	BC 140...141, BC 337, BC 635, BC 637, ++	data
BC 226(A) [Riz]	Si-P	NF-Tr, 30...40V, 0,6A, 0,3W, 200MHz	2a§	Riz	BC 327...328, BC 636, BC 638, BC 640, ++	data
BC 231	Si-P	NF-Tr, 40V, 0,4A, 0,625W, 250MHz	7c	Tix	BC 327, BC 636, BC 638, BC 640, ++	data
BC 231 M	Si-P	=BC 231: 0,8W	2a§		BC 160...161, BCX 60, 2N2303, 2SA606,++	data
BC 232	Si-N	NF-Tr, 40V, 0,4A, 0,625W, 300MHz	7c	Tix	BC 337, BC 635, BC 637, BC 639, ++	data
BC 232 M	Si-N	=BC 232: 0,8W	2a§		BC 140...141, BCX 40, 2N3053, 2SC2108,++	data
BC 234(A)	Si-N	Uni, 30...45V, 0,1A, 0,3W, 350MHz	2a§	Riz	BC 167, BC 183, BC 237, BC 547, ++	data
BC 235(A)	Si-N	Uni, 30...45V, 0,1A, 0,3W, 350MHz	2a§	Riz	BC 167, BC 183, BC 237, BC 547, ++	data
BC 236	Si-N	Nix, 120V, 50mA, 0,3W	8a	Tho	BF 297..299, BF 422, BSS 38, BSX 21, ++	data
BC 237	Si-N	=BC 238: 50V	7a	EUR	BC 167, BC 182, BC 547, ++	data
BC 237 P	Si-N	=BC 237:	40e	Fer	→BC 237	data
BC 238	Si-N	Uni, 30V, 0,1A, 0,3W, 250MHz	7a	EUR	BC 168, BC 183, BC 548, ++	data
BC 238 P	Si-N	=BC 238:	40e	Fer	→BC 238	data
BC 239	Si-N	=BC 238: ra	7a	EUR	BC 169, BC 184, BC 549, ++	data
BC 239 P	Si-N	=BC 239:	40e	Fer	→BC 239	data
BC 238 [Riz]	Si-P	NF, -/28V, 0,5A, 0,7W, 200MHz	2a§	Riz	BC 160...161, BCX 60, 2N2303, 2SA606,++	data
BC 239 [Riz]	Si-P	NF, -/40V, 0,5A, 0,7W, 200MHz	2a§	Riz	BC 160...161, BCX 60, 2N2303, 2SA606,++	data
BC 250	Si-P	Uni, 20V, 0,1A, 0,3W, 180MHz	7a	Itt,Nsc	BC 213, BC 258, BC 308, BC 558, ++	data
BC 251	Si-P	=BC 252: 50V	7a	Itt,Nsc	BC 212, BC 257, BC 307, BC 557, ++	data
BC 252	Si-P	Uni, 30V, 0,1A, 0,3W, 130MHz	7a	Itt,Nsc	BC 213, BC 258, BC 308, BC 558, ++	data
BC 253	Si-P	=BC 252: ra	7a	Itt,Nsc	BC 214, BC 259, BC 309, BC 559, ++	data
BC 254 [Texas]	Si-N	Uni, 100V, 30mA, 0,25W	7c	Tix	BC 546, BF 297..299, BF 422, 2SC1890A,++	data
BC 254(A) [Riz]	Si-N	Uni, 18...25V, 0,1A, 0,3W, 120MHz	2a§	Riz	BC 168, BC 183, BC 238, BC 548, ++	data
BC 255 [Texas]	Si-N	=BC 254[Texas]: 0,625W	7c	Tix	→BC 254	data
BC 255(A) [Riz]	Si-N	Uni, 18...25V, 0,1A, 0,3W, 120MHz	2a§	Riz	BC 168, BC 183, BC 238, BC 548, ++	data
BC 256	Si-P	Uni, 64V, 0,1A, 0,3W, 130MHz	7a	Itt,Mot	BC 212, BC 266, BC 556	data
BC 257	Si-P	=BC 307:	7c	Sie	→BC 307	data
BC 258	Si-P	=BC 308:	7c	Sie	→BC 308	data
BC 259	Si-P	=BC 309:	7c	Sie	→BC 309	data
BC 260	Si-P	Uni, 20V, 0,1A, 0,3W, 180MHz	2a§	Itt	BC 213, BC 258, BC 308, BC 558, ++	data
BC 261	Si-P	=BC 262: 50V	2a§	Itt	BC 212, BC 257, BC 307, BC 557, ++	data
BC 262	Si-P	Uni, 30V, 0,1A, 0,3W, 130MHz	2a§	Itt	BC 213, BC 258, BC 308, BC 558, ++	data
BC 263	Si-P	=BC 262: ra	2a§	Itt	BC 241, BC 259, BC 309, BC 559, ++	data
BC 264	N-FET	NF-V, ra, 30V, Idss>2mA, Up<1,6V	7f	Phi,Mot,++	BF 244, BF 245, BFS 71, 2N3822, ++	data
BC 264 L	N-FET	=BC 264:	7e	Tix	→BC 264	data

Type	Device	Short Description	Fig.	Manu	Comparision Types	More at
BC 266	Si-P	=BC 261: 64V	2a§	Itt	BC 212, BC 256, BC 556	data
BC 267	Si-N	=BC 268: 50V	2a§	Sgs	BC 337, BC 637, BC 639, ++	data
BC 268	Si-N	Uni, 30V, 0,5A, 0,375W, 200MHz	2a§	Sgs	BC 337...338, BC 635, BC 637, BC 639, ++	data
BC 269	Si-N	=BC 268: ra	2a§	Sgs	BC 337...338, BC 635, BC 637, BC 639, ++	data
BC 270	Si-N	=BC 268: 20V	2a§	Sgs	BC 337...338, BC 635, BC 637, BC 639, ++	data
BC 271	Si-N	Uni, 25V, 1A, 0,3W, 225MHz	2a§	Sgs	BC 337...338, BC 635, BC 637, BC 639, ++	data
BC 272	Si-N	Uni, 45V, 1A, 0,3W, 150MHz	2a§	Sgs	BC 337, BC 635, BC 637, BC 639, ++	data
BC 274	Si-P	=BC 275: 50V	8a	Tho	BC 212, BC 257, BC 307, BC 557, ++	data
BC 275	Si-P	Uni, 30V, 0,1A, 0,33W, 130MHz	8a	Tho	BC 213, BC 258, BC 308, BC 558, ++	data
BC 276	Si-P	=BC 275: ra	8a	Tho	BC 214, BC 259, BC 309, BC 559, ++	data
BC 277	Si-N	=BC 278: 45V	8a	Tho	BC 167, BC 183, BC 237, BC 547, ++	data
BC 278	Si-N	Uni, 20V, 0,1A, 0,3W, 150MHz	8a	Tho	BC 168, BC 183, BC 238, BC 548, ++	data
BC 279	Si-N	=BC 278: ra	8a	Tho	BC 169, BC 184, BC 239, BC 549, ++	data
BC 280	Si-N	NF, ra, 45V, 0,1A, 0,36W, 130...250MHz	2a§	Sgs	BC 184, BC 413...414, BC 550	data
BC 281	Si-P	NF, ra, 45V, 0,2A, 0,36W, 130...250MHz	2a§	Sgs	BC 214, BC 415...416, BC 560	data
BC 282	Si-N	NF-Tr, 60V, 0,6A, 0,4W, 170MHz	2a§	Sgs	BC 337, BC 637, BC 639, ++	data
BC 283	Si-P	NF-Tr, 30V, 0,6A, 0,4W, 110MHz	2a§	Sgs	BC 327, BC 636, BC 638, BC 640, ++	data
BC 284	Si-N	Uni, 40V, 0,2A, 0,5W, 60MHz	2a§	Sgs	BC 167, BC 183, BC 237, BC 547, ++	data
BC 285	Si-N	Nix, Uni, 120V, 0,1A, 0,36W, 80MHz	2a§	Sgs	BF 297...299, BF 422, 2SC1890A, ++	data
BC 286	Si-N	NF-Tr/E, 70V, 1A, 0,8W, 100MHz	2a§	Sgs	BC 140...141, BCX 40, 2N3019...3020, ++	data
BC 287	Si-P	NF-Tr/E, 60V, 1A, 0,8W, 200MHz	2a§	Sgs	BC 161, BC 461, BCX 60, 2N4235...4236, ++	data
BC 288	Si-N	NF-E, 80V, 5A, 0,8W, 80MHz	2a§	Sgs	BFT 32...34, BDX 35...37, 2N5336...5339	data
BC 289	Si-N	Uni, 45V, 0,1A, 0,36W	2a§	Sgs	BC 167, BC 183, BC 237, BC 547, ++	data
BC 290	Si-N	=BC 289: ra	2a§	Sgs	BC 184, BC 413...414, BC 550, ++	data
BC 291	Si-P	Uni, 45V, 0,2A, 0,36W	2a§	Sgs	BC 213, BC 257, BC 307, BC 557, ++	data
BC 292	Si-P	=BC 291: ra	2a§	Sgs	BC 214, BC 415...416, BC 560, ++	data
BC 293	Si-N	TV-HA, 80V, 5A, 0,8W, 80MHz	2a§	Sgs	BFT 32...34, BDX 35...37, 2N5336...5339	data
BC 294	Si-P	NF-Tr, 60V, 0,5A, 0,6W	2a§	Sgs,Tix	BC 161, BC 303...304, BC 461, BCX 60	data
BC 295	Si-N	NF-V, 30V, 0,05A, 0,2W, 90MHz	8a	Sgs	BC 168, BC 183, BC 238, BC 548, ++	data
BC 297	Si-P	NF-Tr, 50V, 1A, 0,375W, 250MHz	2a§	Sgs	BC 327, BC 636, BC 638, BC 640, ++	data
BC 298	Si-P	=BC 297: 30V	2a§	Sgs	BC 327...328, BC 636, BC 638, BC 640, ++	data
BC 300	Si-N	NF-Tr/E, 120V, 0,5A, 0,85W, 120MHz	2a§	Sgs	2N1893(A), 2N2102, 2N3019...3020, ++	data
BC 301	Si-N	=BC 300: 90V	2a§	Sgs	BC 141, BCX 40, 2N1990, 2N2102, ++	data
BC 302	Si-N	=BC 300: 60V	2a§	Sgs	BC 140...141, BCX 40, 2N1990, 2N2102, ++	data
BC 303	Si-P	NF-Tr/E, 85V, 0,5A, 0,85W, 75MHz	2a§	Sgs	BC 461, BCX 60, 2N4236	data
BC 304	Si-P	=BC 303: 60V	2a§	Sgs	BC 161, BCX 60, BC 461, 2N4235...4236	data
BC 307	Si-P	=BC 308: 50V	7a	EUR	BC 212, BC 257, BC 557, ++	data
BC 307 P	Si-P	=BC 307:	40e	Fer	→BC 307	data
BC 308	Si-P	Uni, 30V, 0,1A, 0,3W, 130MHz	7a	EUR	BC 213, BC 258, BC 558, ++	data
BC 308 P	Si-P	=BC 308:	40e	Fer	→BC 308	data
BC 309	Si-P	=BC 308: 25V, ra	7a	EUR	BC 213, BC 259, BC 559, ++	data
BC 309 P	Si-P	=BC 309:	40e	Fer	→BC 309	data
BC 310	Si-N	NF-Tr/E, 70V, 1A, 0,8W, 90MHz	2a§	Sgs	BC 140...141, BCX 40, 2N3019...3020, ++	data
BC 311	Si-P	NF-Tr/E, 70V, 1A, 0,8W, 200MHz	2a§	Sgs	BC 161, BC 461, BCX 60, 2N4235...4236	data
BC 312	Si-N	NF, 100V, 0,15A, 0,8W	2a§	Sgs	BC 141, BC 300, BF 257...259, ++	data
BC 313	Si-P	NF-Tr/E, 60V, 1A, 0,8W, 200MHz	2a§	Tho	BC 161, BC 461, BCX 60, 2N4235...4236	data
BC 313 A	Si-P	=BC 313: 80V	2a§		BC 461, BCX 60, 2N4236	data
BC 314	Si-N	Uni, 120V, 0,05A, 0,18W, >50MHz	8a	Sgs	BF 297...299, BF 422, 2SC1890A, ++	data
BC 315	Si-P	NF-V, ra, 45V, 0,1A, 0,3W, >200MHz	7a	Tix	BC 212, BC 415...416, BC 560, ++	data
BC 317	Si-N	=BC 318: 50V	7e	Fch,Mot,++	BC 167, BC 182, BC 237, BC 547, ++	data
BC 318	Si-N	Uni, 40V, 0,15A, 0,35W, 280MHz	7e	Fch,Mot,++	BC 167, BC 183, BC 237, BC 547, ++	data
BC 319	Si-N	=BC 318: 30V, ra	7e	Fch,Mot,++	BC 169, BC 184, BC 239, BC 549, ++	data
BC 320	Si-P	=BC 321: 50V	7e	Fch,Mot,++	BC 212, BC 257, BC 307, BC 557, ++	data
BC 321	Si-P	Uni, 45V, 0,15A, 0,31W, 250MHz	7e	Fch,Mot,++	BC 213, BC 257, BC 307, BC 557, ++	data
BC 322	Si-P	=BC 321: 30V, ra	7e	Fch,Mot,	BC 214, BC 259, BC 309, BC 559, ++	data
BC 323	Si-N	TV-VA, 100V, 5A, 0,8W, 100MHz	2a§	Sgs	BFT 33...34, BDX 35...37, 2N5339...5339	data
BC 324	Si-N	TV-VA, 85V, 1A, 0,8W, 100MHz	2a§	Sgs	BC 140...141, BCX 40, 2N3019...3020, ++	data
BC 325	Si-P	NF, ra 60V, 50mA, 0,36W. B=40...120	2a§	Tix	BC 214, BC 415...416, BC 560, ++	data
BC 326	Si-P	=BC 325: B=100...500	2a§	Tix	BC 214, BC 415...416, BC 560, ++	data
BC 327	Si-P	NF-Tr, 50V, 0,8A, 0,625W, 100MHz	7a	EUR	BC 638, BC 640, 2SB647, ++	data
BC 327 A	Si-P	=BC 327: 60V	7a	Phi	BC 638, BC 640, 2SB647, ++	data
BC 327 L	Si-P	=BC 327:	7c	Tix	→BC 327	data
BC 327 P	Si-P	=BC 327:	40e	Fer	→BC 327	data
BC 328	Si-P	=BC 327: 30V	7a	EUR	BC 636, BC 638, BC 640, 2SB647, ++	data
BC 328 L	Si-P	=BC 328:	7c	Tix	→BC 328	data
BC 328 P	Si-P	=BC 328:	40e	Fer	→BC 328	data
BC 329	Si-N	NF-V, ra, 60V, 30mA, 0,25W	7a	Tix	BC 184, BC 413...414, BC 550, ++	data
BC 330	Si-N	=BC 329: 45V	7a	Tix	BC 184, BC 413...414, BC 550, ++	data
BC 331	Si-N	NF-V, ra, 60V, 30mA, 0,25W	7a	Tix	BC 184, BC 413...414, BC 550, ++	data
BC 332	Si-N	=BC 321: 45V	7a	Tix	BC 184, BC 413...414, BC 550, ++	data
BC 333	Si-N	Uni, 25V, 50mA, 0,31W, 50MHz	7e	Mot	BC 168, BC 183, BC 238, BC 548, ++	data
BC 334	Si-P	Uni, 25V, 50mA, 0,31W, 50MHz	7e	Mot	BC 213, BC 258, BC 308, BC 558, ++	data

Type	Device	Short Description	Fig.	Manu	Comparision Types	More at
BC 335	Si-N	=BC 333: ra	7e	Mot	BC 169, BC 184, BC 239, BC 549, ++	data
BC 336	Si-P	=BC 334: ra	7e	Mot	BC 214, BC 259, BC 309, BC 559, ++	data
BC 337	Si-N	NF-Tr, 50V, 0,8A, 0,625W, 100MHz	7a	EUR	BC 637, BC 639, 2SD667, ++	data
BC 337 A	Si-N	=BC 337: 60V	7a	Phi	BC 637, BC 639, 2SD667, ++	data
BC 337 L	Si-N	=BC 337:	7c	Tix	→BC 337	data
BC 337 P	Si-N	=BC 337:	40e	Fer	→BC 337	data
BC 338	Si-N	=BC 337: 30V	7a	EUR	BC 635, BC 637, BC 639, 2SD667, ++	data
BC 338 L	Si-N	=BC 338:	7c	Tix	→BC 338	data
BC 338 P	Si-N	=BC 338:	40e	Fer	→BC 338	data
BC 340	Si-N	NF-Tr/E, 40V, 0,5A, 0,8W, 100MHz	2a§	Itt	BC 140...141, BC 300...302, 2N1990, ++	data
BC 341	Si-N	=BC 340: 60V	2a§	Itt	BC 140...141, BC 300...302, 2N1990, ++	data
BC 342	Si-N	NF-Tr/E, 70V, 1A, 0,8W, 100MHz	2a§	Mot	BC 140...141, BCX 40, 2N3019...3020, ++	data
BC 343	Si-P	NF-Tr/E, 70V, 1A, 0,8W, 100MHz	2a§	Mot	BC 161, BCX 60, 2N4236, ++	data
BC 344	Si-N	=BC 342: 90V	2a§	Mot	BC 141, BCX 40, 2N3019...3020, ++	data
BC 345	Si-P	=BC 343: 90V	2a§	Mot	BC 461, BCX 60	data
BC 347	Si-N	Uni, 50V, 0,1A, 0,3W, >125MHz	7e	Mot	BC 167, BC 182, BC 237, BC 547, ++	data
BC 348	Si-N	=BC 347: 40V	7e	Mot	BC 167, BC 183, BC 237, BC 547, ++	data
BC 349	Si-N	=BC 347: 30V	7e	Mot	BC 168, BC 183, BC 238, BC 548, ++	data
BC 350	Si-P	Uni, 50V 0,1A, 0,3W, >125MHz	7e	Mot	BC 212, BC 257, BC 307, BC 557, ++	data
BC 351	Si-P	=BC 350: 40V	7e	Mot	BC 213, BC 257, BC 307, BC 557, ++	data
BC 352	Si-P	=BC 350: 30V	7e	Mot	BC 213, BC 258, BC 308, BC 558, ++	data
BC 354	Si-P	Uni, 30V, 0,2A, 0,31W, >200MHz	7e	Mot	BC 213, BC 258, BC 308, BC 558, ++	data
BC 355	Si-P	Uni, 30V, 0,2A, 0,31W, >200MHz	7e	Mot	BC 213, BC 258, BC 308, BC 558, ++	data
BC 357	Si-P	Uni, 25V, 0,1A, 0,31W, >125MHz	7e	Mot	BC 213, BC 258, BC 308, BC 558, ++	data
BC 358	Si-N	Uni, 25V, 0,1A, 0,31W, >125MHz	7e	Mot	BC 168, BC 183, BC 238, BC 548, ++	data
BC 360	Si-P	NF-Tr/E, 40V, 0,5A, 0,8W, 250MHz	2a§	Itt	BC 160...161, BC 303...304, BCX 60, ++	data
BC 361	Si-P	=BC 360: 60V	2a§	Itt	BC 161, BC 303...304, BCX 60, ++	data
BC 362	Si-P	NF-L, 50V, 2A, 8W, 100MHz	13g§	Mot	BD 518, BD 520, BD 526, BD 528, BD 530	data
BC 363	Si-P	=BC 362: 60V	13g§	Mot	BD 518, BD 520, BD 526, BD 528, BD 530	data
BC 364	Si-P	=BC 362: 80V	13g§	Mot	BD 520, BD 528, BD 530	data
BC 365	Si-N	NF-L, 50V, 2A, 8W, 150MHz	13g§	Mot	BD 517, BD 519, BD 525, BD 527, BD 529	data
BC 366	Si-N	=BC 365: 60V	13g§	Mot	BD 517, BD 519, BD 525, BD 527, BD 529	data
BC 367	Si-N	=BC 365: 80V	13g§	Mot	BD 519, BD 527, BD 529	data
BC 368	Si-N	NF-Tr, 25V, 1A, 0,8W, 65MHz	7c	Phi,Sie,++	BC 337...338, BC 635, BC 637, BC 639, ++	data
BC 369	Si-P	NF-Tr, 25V, 1A, 0,8W, 65MHz	7c	Phi,Sie,++	BC 327...328, BC 636, BC 638, BC 640, ++	data
BC 370	Si-P	NF-Tr, 20V, 0,5A, 0,375W, 150MHz	2a§	Sgs	BC 327...328, BC 636, BC 638, BC 640, ++	data
BC 371	Si-N	NF-Tr/E, 60V, 1A, 0,85W	2a§	Sgs	BC 140...141, BCX 40, 2N3019...3020, ++	data
BC 372	Si-N-Darl	100V 1A, 0,625W, >100MHz, B>10000	7e,7a	Mot,Ons,Ztx	(BC 879, BSR 52, 2SD1660, 2SD2067,++)[13]	data
BC 372 P	Si-N-Darl	=BC 372:	40e	Ztx	→BC 372	data
BC 373	Si-N-Darl	=BC 372: 80V	7e,7a	Mot, Ons	(BC 877, BC 879, BSR 51...52, 2SD1853,++)[13]	data
BC 375	Si-N	NF-Tr/E, 30V, 1A, 0,625W, 150MHz	7a	Phi	BC 337...338, BC 635, BC 637, BC 639, ++	data
BC 376	Si-P	NF-Tr/E, 30V, 1A, 0,625W, 100MHz	7a	Phi	BC 327...328, BC 636, BC 638, BC 640, ++	data
BC 377	Si-N	NF-Tr, 50V, 1A, 0,375W, 200MHz	2a§	Sgs	BC 337, BC 637, BC 639, ++	data
BC 378	Si-N	=BC 377: 30V	2a§	Sgs	BC 337...338, BC 635, BC 637, BC 639, ++	data
BC 381	Si-P	Uni, 40V, 0,2A, 0,625W	7a	Tix	BC 327, BC 636, BC 638, BC 640, ++	data
BC 382	Si-N	NF, ra, 50V, 0,1A, 0,3W, >150MHz	7a	Tix	BC 184, BC 413...414, BC 550, ++	data
BC 383	Si-N	=BC 382: 45V	7a	Tix	BC 184, BC 413...414, BC 550, ++	data
BC 384	Si-N	=BC 382: 45V, Ur<0,135µV	7a	Tix	BC 184, BC 413...414, BC 550, ++	data
BC 385	Si-N	Uni, 45V, 0,1A, 0,3W, >150MHz	7a	Tix	BC 167, BC 183, BC 237, BC 547, ++	data
BC 386	Si-N	Uni, 45V, 0,1A, 0,3W, >150MHz	7a	Tix	BC 167, BC 183, BC 237, BC 547, ++	data
BC 387	Si-N	NF-Tr, 35V, 0,6A, 0,31W	7e	Mot	BC 337, BC 635, BC 637, BC 639, :+	data
BC 388	Si-P	NF-Tr, 35V, 0,6A, 0,31W	7e	Mot	BC 327, BC 536, BC 638, BC 640, ++	data
BC 389	Si-N	Uni, -/45V, 0,2A, 0,3W, >150MHz, B>40	2a§	Sgs	BC 167, BC 183, BC 237, BC 547, ++	data
BC 390	Si-N	=BC 389: -/20V	2a§	Sgs	BC 168, BC 183, BC 238, BC 548, ++	data
BC 391	Si-N	=BC 389: -/20V, B>100	2a§	Sgs	BC 168, BC 183, BC 238, BC 548, ++	data
BC 393	Si-P	NF/Vid, 180V, 0,1A, 0,4W, 120MHz	2a§	Mot,Sgs	BF 423, BF 436...437, BF 491...493, ++	data
BC 394	Si-N	NF/Vid, 180V, 0,1A, 0,4W, 90MHz	2a§	Mot,Sgs	BF 298...299, BF 391...393, BF 423, ++	data
BC 395	Si-N	TV-VA-Tr, 80V, 0,5A, 0,3W, >60MHz	8a	Sgs	BC 140...141, BC 300...301, BC 639, ++	data
BC 396	Si-P	TV-VA-Tr, 80V, 0,5A, 0,3W	8a	Sgs	BC 303, BC 640	data
BC 397	Si-P	NF-Tr/E, 50V, 1A, 0,8W	2a§	Sgs	BC 161, BC 460...461, BCX 60, ++	data
BC 398	Si-N	NF-Tr/E, 60V, 1A, 0,8W	2a§	Sgs	BC 140...141, BC 440...441, BCX 40, ++	data
BC 399	Si-N	Min, NF, ra, 30V, 75mA, 75mW	36a	Sgs	BC 122...123	data
BC 400	Si-P	Uni, 90V, 50mA, 0,2W, 150MHz	8a	Sgs	BC 477, 2SA893(A), 2SA1017, SB715, ++	data
BC 404	Si-P	Uni, 80V, 0,15A, 0,36W, 150MHz	7a	Sgs	BC 477, BC 556, 2SA1136...37, 2SA1335,++	data
BC 405	Si-P	=BC 404: 60V, ra	7a	Sgs	BC 416, BC 560, 2SA1136...1137	data
BC 406	Si-P	=BC 404: 40V, ra	7a	Sgs	BC 214, BC 415...416, BC 560, ++	data
BC 407	Si-N	=BC 237:	8a	Phi	→BC 237	data
BC 408	Si-N	=BC 238:	8a	Phi	→BC 238	data
BC 409	Si-N	=BC 239:	8a	Phi	→BC 239	data
BC 410	Si-N	Uni, 30V, 50mA, 0,25W	8a	Sgs	BC 168, BC 183, BC 238, BC 548, ++	data
BC 411	Si-N	NF-Tr/E, 85V, 1A, 0,3W, 80MHz	8a	Sgs	BC 140...141, BCX 40, BC 639, ++	data
BC 412	Si-N	NF-Tr, 75V, 1A, 0,5W, 90MHz	2a§	Sgs	BC 639, 2SD667(A), 2SD774, 2SD1616A,++	data

124

Type	Device	Short Description	Fig.	Manu	Comparision Types	More at
BC 413	Si-N	Uni, ra, 45V, 0,1A, 0,3W, 250MHz	7a	EUR	BC 184, BC 550	data
BC 414	Si-N	=BC 413: 50V	7a	EUR	BC 184, BC 550	data
BC 415	Si-P	Uni, ra, 45V, 0,1A, 0,3W, 200MHz	7a	EUR	BC 214, BC 560	data
BC 416	Si-P	=BC 415: 50V	7a	EUR	BC 214, BC 560	data
BC 413P...416P	Si-N/P	=BC 413...416:	40e	Fer	→BC 413...416	data
BC 417	Si-P	=BC 307:	8a	Phi	→BC 307	data
BC 418	Si-P	=BC 308:	8a	Phi	→BC 308	data
BC 419	Si-P	=BC 309:	8a	Phi	→BC 309	data
BC 420	Si-P	Vid, 180V, 0,1A, 0,4W, 150MHz	2a§	Sgs	BF 423, BF 435...437, BF 491...493, ++	data
BC 424	Si-N	NF-Tr, 80V, 0,5A, 0,5W, >50MHz	7e	Mot	BC 639, 2N3700...3701, 2SD667	data
BC 425	Si-N	=BC 424: 60V	7e	Mot	BC 637, BC 639, 2N3700...3701, 2SD667	data
BC 426	Si-P	NF-Tr, 80V, 0,5A, 0,5W, >50MHz	7e	Mot	BC 640, 2SB647	data
BC 427	Si-P	=BC 426: 60V	7e	Mot	BC 638, BC 640, 2SB647	data
BC 429	Si-N	NF-Tr/E, 45V, 1A, 6W, >100MHz	14h§	Tix	BD 135, BD 137, BD 139, BD 226, ++	data
BC 429 A	Si-N	=BC 429: 60V	14h§		BD 137, BD 139, BD 228, BD 230, ++	data
BC 430	Si-P	NF-Tr/E, 45V, 1A, 6W, >100MHz	14h§	Tix	BD 136, BD 138, BD 140, BD 227, ++	data
BC 430 A	Si-P	=BC 430: 60V	14h§		BD 138, BD 140, BD 229, BD 231, ++	data
BC 431	Si-N	NF-Tr, 70V, 0,8A, 0,5W, 100MHz	7a	Aeg	BC 639, 2SD667, 2N3700...3701	data
BC 432	Si-P	NF-Tr, 70V, 0,8A, 0,5W, 100MHz	7a	Aeg	BC 640, 2SB647	data
BC 437	Si-N	=BC 438: 50V	9a	Hit	BC 167, BC 182, BC 237, BC 547, ++	data
BC 438	Si-N	Uni, 30V, 0,1A, 0,22W, 300MHz	9a	Hit	BC 168, BC 183, BC 238, BC 548, ++	data
BC 439	Si-N	=BC 438: ra	9a	Hit	BC 169, BC 184, BC 239, BC 549, ++	data
BC 440	Si-N	NF-Tr/E, 50V, 2A(ss), 1W, >50MHz	2a§	Sgs	BCX 40, 2N4237...4239, 2SC2214	data
BC 441	Si-N	=BC 440: 75V	2a§	Sgs	BCX 40, 2N4238...4239, 2SC2214	data
BC 445	Si-N	Uni, 60V, 0,3A, 0,625W, >100MHz	7a	Itt, Mot	BC 182, BC 546, BC 637, BC 639, ++	data
BC 446	Si-P	Uni, 60V, 0,3A, 0,625W, >100MHz	7a	Itt, Mot	BC 212, BC 556, BC 638, BC 640, ++	data
BC 447	Si-N	=BC 445: 80V	7a	Itt, Mot	BC 546, BC 639, 2SD667	data
BC 448	Si-P	=BC 446: 80V	7a	Itt, Mot	BC 556, BC 640, 2SB647	data
BC 449	Si-N	=BC 445: 100V	7a	Itt, Mot	BC 639, 2SD667, 2SC1670	data
BC 450	Si-P	=BC 446: 100V	7a	Itt, Mot	BC 640, 2SB647, 2SA840	data
BC 451	Si-N	Uni, 50V, 0,1A, 0,3W, >150MHz	7a	Tos	BC 167, BC 182, BC 237, BC 547, ++	data
BC 452	Si-N	=BC 451: 30V	7a	Tos	BC 168, BC 183, BC 238, BC 548, ++	data
BC 453	Si-N	=BC 451: 30V, ra	7a	Tos	BC 169, BC 184, BC 239, BC 549, ++	data
BC 454	Si-P	Uni, 50V, 0,1A, 0,3W, >150MHz	7a	Tos	BC 212, BC 257, BC 307, BC 557, ++	data
BC 455	Si-P	=BC 454: 30V	7a	Tos	BC 213, BC 258, BC 308, BC 558, ++	data
BC 456	Si-P	=BC 454: 30V, ra	7a	Tos	BC 214, BC 259, BC 309, BC 559, ++	data
BC 460	Si-P	NF-Tr/E, 50V, 2A(ss), 1W, >50MHz	2a§	Sgs	BCX 60, 2N4235...4236	data
BC 461	Si-P	=BC 460: 75V	2a§	Sgs	BCX 60, 2N4236	data
BC 462	Si-P	NF-Tr/E, 35V, 1,5A, 0,88W, 200MHz	11a	Phi	BC 636, BC 638, BC 640, 2SA966, ++	data
BC 463	Si-N	NF-Tr/E, 35V, 1,5A, 0,88W, 200MHz	11a	Phi	BC 635, BC 637, BC 639, 2SC2236, ++	data
BC 464	Si-P	=BC 462: 25V	11a	Phi	BC 636, BC 638, BC 640, 2SA966, ++	data
BC 465	Si-N	=BC 463: 25V	11a	Phi	BC 635, BC 637, BC 639, 2SC2236, ++	data
BC 467	Si-N	=BC 468: 50V	9b	Hit	BC 167, BC 182, BC 237, BC 547, ++	data
BC 468	Si-N	Uni, 30V, 0,1A, 0,22W, 300MHz	9b	Hit	BC 168, BC 183, BC 238, BC 548, ++	data
BC 469	Si-N	=BC 468: ra	9b	Hit	BC 169, BC 184, BC 239, BC 549, ++	data
BC 477	Si-P	Uni, 90V, 0,15A, 0,36W, 150MHz	2a§	Sgs	BC 556	data
BC 478	Si-P	=BC 477: 40V	2a§	Sgs	BC 213, BC 257, BC 307, BC 557, ++	data
BC 479	Si-P	=BC 477: 40V, ra	2a§	Sgs	BC 214, BC 415...416, BC 560	data
BC 485	Si-N	NF-Tr/E, 45V, 1A, 0,625W, 200MHz	7a	Mot, Nsc	BC 337, BC 635, BC 637, BC 639, ++	data
BC 486	Si-P	NF-Tr/E, 45V, 1A, 0,625W, 200MHz	7a	Mot, Nsc	BC 327, BC 636, BC 638, BC 640, ++	data
BC 487	Si-N	=BC 485: 60V	7a	Mot, Nsc	BC 637, BC 639, 2SD667, 2N3700...3701	data
BC 488	Si-P	=BC 486: 60V	7a	Mot, Nsc	BC 638, BC 640, 2SB647	data
BC 489	Si-N	=BC 485: 80V	7a	Mot, Nsc	BC 639, 2SD667, 2N3700...3701	data
BC 490	Si-P	=BC 486: 80V	7a	Mot, Nsc	BC 640, 2SB647	data
BC 507	Si-N	=BC 508: 70V	7a	Sgs	BC 174, BC 190, BC 546	data
BC 508	Si-N	Uni, 60V, 0,2A, 0,36W, 200MHz	7a	Sgs	BC 174, BC 182, BC 190, BC 546, ++	data
BC 509	Si-N	=BC 508: ra	7a	Sgs	BC 414, BC 550, 2SC2240, 2SC2459	data
BC 510	Si-N	=BC 508: 40V, ra	7a	Sgs	BC 184, BC 413...414, BC 550, ++	data
BC 512	Si-P	=BC 513: 50V	7a	Tix	BC 212, BC 257, BC 307, BC 557, ++	data
BC 513	Si-P	Uni, 30V, 0,2A, 0,3W, >200MHz	7a	Tix	BC 213, BC 258, BC 308, BC 558, ++	data
BC 514	Si-P	=BC 513: ra	7a	Tix	BC 214, BC 259, BC 309, BC 559, ++	data
BC 516(P)	Si-P-Darl	Uni, 40V, 0,4A, 0,625W, 250MHz, B>30k	7a	Phi,Tix,++	MPSA 75, (BC 876, BC 878, 2SA1555,++)[13]	data
BC 517(P)	Si-N-Darl	Uni, 40V, 0,4A, 0,625W, 220MHz, B>30k	7a	Phi,Tix,++	MPSA 25, (BC 875, BC 877, 2SC4017,++)[13]	data
BC 520	Si-N	Uni, ra, 60V, 0,05A, 0,625W, >100MHz	7e	Fch	BC 414, BC 550, 2SC2240, 2SC2459	data
BC 521	Si-N	=BC 520: 45V	7e	Fch	BC 184, BC 413...414, BC 550	data
BC 522	Si-N	=BC 520: 20V	7e	Fch	(BC 169, BC 184, BC 239, BC 549,++)[13]	data
BC 523	Si-N	=BC 520: 45V	7e	Fch	BC 184, BC 413...414, BC 550	data
BC 524	Si-N	Uni, ra, 45V, 0,1A, 0,625W	7e	Fch,Tix	BC 184, BC 413...414, BC 550	data
BC 525	Si-P	Uni, ra, 35V, 0,1A, 0,625W	7e	Fch	BC 214, BC 415...416, BC 560	data
BC 526	Si-P	Uni, 60V, 0,2A, 0,625W, >100MHz	7e	Fch	BC 212, BC 256, BC 266, BC 556	data
BC 527[Ucp]	Si-N	Uni, 45V, 0,05A, 0,3W, 150MHz	2a§	Ucp	BC 167, BC 182, BC 237, BC 547, ++	data
BC 527	Si-P	NF-Tr/E, 60V, 1A, 0,625W, >100MHz	7e	Fch,Mic	BC 638, BC 640, 2SB647	data
BC 528	Si-P	=BC 527: 80V	7e	Fch,Mic	BC 640, 2SB647	data
BC 528[Ucp]	Si-N	Uni, 45V, 0,05A, 0,3W, 150MHz	2a§	Ucp	BC 167, BC 182, BC 237, BC 547, ++	data

Type	Device	Short Description	Fig.	Manu	Comparision Types	More at
BC 529	Si-P	Uni, 50V, 0,2A, 0,625W, >100MHz	7e	Fch	BC 212, BC 556...557, BC 327, ++	data
BC 530	Si-P	Vid, 130V, 0,1A, 0,625W, >50MHz	7e	Fch	BF 398, BF 423, BF 435...437, 2SA1370, ++	data
BC 531	Si-P	=BC 530: 160V	7e	Fch	BF 423, BF 435...437, 2SA1370, ++	data
BC 532	Si-N	Vid, 160V, 0,1A, 0,625W, >50MHz	7e	Fch	BF 297...299, BF 422, 2SC3467, ++	data
BC 533	Si-N	=BC 532: 180V	7e	Fch	BF 298...299, BF 422, 2SC3467, ++	data
BC 534	Si-P	NF-Tr/E, 80V, 0,5A, 0,625W, >50MHz	7e	Fch	BC 640, 2SB647	data
BC 535	Si-N	NF-Tr/E, 80V, 0,5A, 0,625W, >50MHz	7e	Fch	BC 639, 2SD667	data
BC 537	Si-P	NF-Tr/E, 60V, 1A, 0,625W, >100MHz	7e	Fch	BC 637, BC 639, 2SD667	data
BC 538	Si-N	=BC 537: 80V	7e	Fch,Mic	BC 639, 2SD667	data
BC 546	Si-N	=BC 547: 80V	7a	EUR	2SC2240, 2SC2459, 2SC2674...75, 2SC3378	data
BC 547	Si-N	Uni, 50V, 0,1A, 0,5W, 300MHz	7a	EUR	BC 167, BC 182, BC 237, ++	data
BC 548	Si-N	=BC 547: 30V	7a	EUR	BC 168, BC 183, BC 238, ++	data
BC 549	Si-N	=BC 547: 30V, ra	7a	EUR	BC 169, BC 184, BC 239, ++	data
BC 550	Si-N	=BC 547: ra	7a	EUR	BC 184, BC 414, 2SC2240, 2SC2459	data
BC 551	Si-P	Uni, 50V, 0,1A, 0,5W	7a	Phi	BC 212, BC 257, BC 307, BC 557, ++	data
BC 556	Si-P	=BC 557: 80V	7a	EUR	2SA970, 2SA1049, 2SA1136	data
BC 557	Si-P	Uni, 50V, 0,1A, 0,5W, 150MHz	7a	EUR	BC 212, BC 257, BC 307, 2SB725, 2SA1137+	data
BC 558	Si-P	=BC 557: 30V	7a	EUR	BC 213, BC 258, BC 308, 2SB725, 2SA1137+	data
BC 559	Si-P	=BC 557: 30V, ra	7a	EUR	BC 214, BC 259, BC 309, ++	data
BC 560	Si-P	=BC 557: ra	7a	EUR	BC 214, BC 416, 2SA1136...1137	data
BC 582	Si-N	=BC 583: 50V	7a	Tix	BC 167, BC 182, BC 237, BC 547, ++	data
BC 583	Si-N	Uni, 30V, 0,2A, 0,3W, >150MHz	7a	Tix	BC 168, BC 183, BC 238, BC 548, ++	data
BC 584	Si-N	=BC 583: ra	7a	Tix	BC 169, BC 184, BC 239, BC 549, ++	data
BC 585	Si-N	Temp.-Sensor, 25V, 0,1A(ss), 0,35W	7a	Mot	(BC 168, BC 183, BC 238, BC 548)	data
BC 586	Si-P	Temp.-Sensor, 25V, 0,1A(ss), 0,35W	7a	Mot	(BC 213, BC 258, BC 308, BC 558)	data
BC 587	Si-P	60V, 1A, 4W(Tc=25°)	2a	Eiy	BC 161, BC 461, BCX 60, 2N4235...4236	data
BC 612	Si-P	Uni, 75V, 0,2A, 0,3W, >200MHz	7a	Tix	BC 556, 2SA1136...1137	data
BC 612 L	Si-P	=BC 612:	7c		→BC 612	data
BC 617	Si-N-Darl	NF-Tr/E, 50V, 1A, 0,625W,>150MHz,B>10k	7a	Phi,Tix,++	(BC 875, BC 877, BC 879, BSR 50...52,++)[13]	data
BC 618	Si-N-Darl	=BC 617: 80V	7a	Phi,Tix,++	(BC 877, BC 879, BSR 51...52,++)[13]	data
BC 635	Si-N	NF-Tr/E, 45V, 1A, 0,8W, 130MHz	7c	EUR	BC 537...538, 2N3700...3701, 2SD667, ++	data
BC 636	Si-P	NF-Tr/E, 45V, 1A, 0,8W, 50MHz	7c	EUR	BC 527...528, 2SB647, 2SA1013, ++	data
BC 637	Si-N	=BC 635: 60V	7c	EUR	BC 537...538, 2SD667, 2N3700...3701, ++	data
BC 638	Si-P	=BC 636: 60V	7c	EUR	BC 527...528, 2SB 647, 2SA1013, ++	data
BC 639	Si-N	=BC 635: 100V	7c	EUR	2SD667, 2N3700...3701, 2SC2383	data
BC 640	Si-P	=BC 636: 100V	7c	EUR	2SB647, 2SA1013	data
BC 650(S)	Si-N	Uni, ra, 30V, 0,2A, 0,625W, 300MHz	7e	Mot	BC 169, BC 184, BC 239, BC 549, ++	data
BC 651(S)	Si-N	=BC 650: 45V	7e	Mot	BC 184, BC 413...414, BC 550	data
BC 682	Si-N	Uni, 75V, 0,2A, 0,3W, >150MHz	7a	Tix	BC 174, BC 190, BC 546	data
BC 682 L	Si-N	=BC 682:	7c		→BC 682	data
BC 714	Si-P	Uni, ra, 45V, 0,2A, 0,3W, >200MHz	7a	Tix	BC 214, BC 415...416, BC 560, ++	data
BC 727	Si-P	NF-Tr/E, 50V, 1A, 0,625W, >100MHz	7e	Fch,Mic	BC 638, BC 640, 2SB647, ++	data
BC 728	Si-P	=BC 727: 30V	7e	Fch,Mic	BC 636, BC 638, BC 640, 2SB647, ++	data
BC 737	Si-N	NF-Tr/E, 50V, 1A, 0,625W, >100MHz	7e	Fch,Mic	BC 637, BC 639, 2SD667, 2N3700...3701, ++	data
BC 738	Si-N	=BC 737: 30V	7e	Fch,Mic	BC 635, BC 637, BC 639, 2SD667, ++	data
BC 807	Si-P	SMD, NF-Tr, 50V, 0,5A, 100MHz	35a	Phi,Sie,++	BCX 17, BCW 68	data
BC 807 R	Si-P	=BC 807:	35d		BCX 17R, BCW 68R	data
BC 807 U	Si-P	=2x BC 807:	46bh1	Sie	-	
BC 807-W	Si-P	=BC 807:	35a(2mm)	Phi	2SB1219A	
BC 808	Si-P	=BC 807: 30V	35a	Phi,Sie,++	BCX 17...18, BCW 67..68	data
BC 808 R	Si-P	=BC 808:	35d		BCX 17R...18R, BCW 67R...68R	data
BC 808-W	Si-P	=BC 808:	35a(2mm)	Phi	2SA1588, 2SB1219(A)	
BC 817	Si-N	SMD, NF-Tr, 50V, 0,5A, 200MHz	35a	Phi,Sie,++	BCX 19, BCW 65...66	data
BC 817 R	Si-N	=BC 817:	35d		BCX 19R, BCW 65R...66R	data
BC 817 U	Si-N	=2x BC 817:	46bh1	Sie	-	
BC 817 UPN	Si-N/P	=BC 817+BC 807:	46bh1	Sie	-	
BC 817-W	Si-N	=BC 817:	35a(2mm)	Phi	2SD1820A, 2SD1949	
BC 818	Si-N	=BC 817: 30V	35a	Phi,Sie,++	BCX 19...20, BCW 65...66	data
BC 818 R	Si-N	=BC 818:	35d		BCX 19R...20R, BCW 65R...66R	data
BC 818-W	Si-N	=BC 818:	35a(2mm)	Phi	2SC4097, 2SC4118, 2SD1820(A), 2SD1949	data
BC 827	Si-P	NF-Tr/E, 30V, 0,8A, 0,8W, 100MHz	7a	Sie	BC 327...328, BC 636, BC 638, BC 640, ++	data
BC 828	Si-P	=BC 827: 50V	7a	Sie	BC 327, BC 638, BC 640, 2SB647, ++	data
BC 837	Si-N	NF-Tr/E, 30V, 0,8A, 0,8W, 100MHz	7a	Sie	BC 337...337, BC 635, BC 637, BC 639, ++	data
BC 838	Si-N	=BC 837: 50V	7a	Sie	BC 337, BC 637, BC 639, 2SD667, ++	data
BC 846	Si-N	=BC 847: 80V	35a	Phi,Sie,++	BCV 71...72	data
BC 846 (B)DW1(T1)	Si-N	=2x BC 846(B):	46bh1(2mm)	Ons	BC 846(B)S	
BC 846(B)PDW1(T1)	Si-N/P	=BC 846(B) + BC 856(B):	46bh1(2mm)	Ons	BC 846(B)PN	
BC 846 (B)PN	Si-N/P	=BC 846(B) + BC 856(B):	46bh1(2mm)	Phi, Sie	BC 846(B)PDW1	
BC 846 R	Si-N	=BC 846:	35d		BCV 71R...72R	data
BC 846 (B)S	Si-N	=2x BC 846(B):	46bh1(2mm)	Phi, Sie	BC 846(B)DW1	data
BC 846 U	Si-N	=2x BC 846:	46bh1	Sie	-	
BC 846 UPN	Si-N/P	=BC 846 + BC 856:	46bh1	Phi, Sie	-	
BC 847	Si-N	SMD, Uni, 50V, 0,1A, 300MHz	35a	Phi,Sie,++	BCW 71...72, BCW 81	data
BC 847(B,C)DW1(T1)	Si-N	=2x BC 847(B,C):	46bh1(2mm)	Ons	BC 847(B,C)S	
BC847(B,C)PDW1(T1)	Si-N/P	=BC 847(B,C) + BC 857(B,C):	46bh1(2mm)	Ons	BC 847(B,C)PN	
BC 847 (B)PN	Si-N/P	=BC 847(B) + BC 857(B)	46bh1(2mm)	Phi, Sie	-	data
BC 847 R	Si-N	=BC 847:	35d		BCW 71R...72R, BCW 81R	data

Type	Device	Short Description	Fig.	Manu	Comparision Types	More at
BC 847 (B)S	Si-N	=2x BC 847(B)	46bh1(2mm)	Phi, Sie	BC 847(B,C)DW1	data
BC 848	Si-N	=BC 847: 30V	35a	Phi,Sie,++	BCW 31...33, BCW 71...72, BCW 81	data
BC 848(B,C)DW1(T1)	Si-N	=2x BC 848(B,C):	46bh1(2mm)	Ons	BC 847(B,C)S	
BC848(B,C)PDW1(T1)	Si-N/P	=BC 848(B,C) + BC 858(B,C):	46bh1(2mm)	Ons	BC 847(B,C)PN	
BC 848 R	Si-N	=BC 848:	35d		BCW 31R...33R, BCW 71R...72R, BCW 81R	data
BC 849	Si-N	=BC 847: 30V, ra	35a	Phi,Sie,++	BCF 32...33, BCF 81	data
BC 849 R	Si-N	=BC 849:	35d		BCF 32R...33R, BCF 81R	data
BC 850	Si-N	=BC 847: ra	35a	Phi,Sie,++	BCF 81	data
BC 846F...848F	Si-N	=BC 846...848:	35a(1,6mm)	Phi	2SC4617	data
BC 850 R	Si-N	=BC 850:	35d		BCF 81R	data
BC 846T...850T	Si-N	=BC 846...850:	35a(1,6mm)	Phi, Sie	2SC4617	data
BC 846W...850W	Si-N	=BC 846...850:	35a(2mm)	Phi, Sie	2SC4101...02, 2SC4117	data
BC 856	Si-P	=BC 857: 80V	35a	Phi,Sie,++	BCW 89	data
BC 856 (B)DW1(T1)	Si-P	=2x BC 856(B):	46bh1(2mm)	Ons	BC 856(B)S	
BC 856 R	Si-P	=BC 856:	35d		BCW 89R	data
BC 856 S	Si-P	=2x BC 856	46bh1(2mm)	Sie	BC 856(B)DW1	data
BC 856 U	Si-P	=2x BC 856:	46bh1	Sie	-	
BC 857	Si-P	SMD, Uni, 50V, 0,1A, 150MHz	35a	Phi,Sie,++	BCW 69...70, BCW 89	data
BC 857(B,C)DW1(T1)	Si-P	=2x BC 857(B,C):	46bh1(2mm)	Ons	BC 857(B)S	
BC 857 R	Si-P	=BC 857:	35d		BCW 69R...70R, BCW 89R	data
BC 857 (B)S	Si-P	=2x BC 857(B)	46bh1(2mm)	Phi, Sie	BC 857(B)DW1	data
BC 858	Si-P	=BC 857: 30V	35a	Phi,Sie,++	BCW 29...30, BCW 69...70, BCW 89,	data
BC 858(B,C)DW1(T1)	Si-P	=2x BC 858(B,C):	46bh1(2mm)	Ons	BC 857(B)S	
BC 858 R	Si-P	=BC 858:	35d		BCW 29R...30R, BCW 69R...70R, BCW 89R	data
BC 859	Si-P	=BC 857: 30V, ra	35a	Phi,Sie,++	BCF 29...30, BCF 70	data
BC 859 R	Si-P	=BC 859:	35d		BCF 29R...30R, BCF 70R	data
BC 860	Si-P	=BC 857: ra	35a	Phi,Sie,++	BCF 70	data
BC 856F...848F	Si-P	=BC 856...858:	35a(1,6mm)	Phi	2SB1462, 2SB1836	data
BC 860 R	Si-P	=BC 860:	35d		BCF 70R	data
BC 856T...860T	Si-P	=BC 856...860:	35a(1,6mm)	Phi, Sie	2SB1462, 2SB1836	data
BC 856W...860W	Si-P	=BC 856...860:	35a(2mm)	Phi, Sie	2SA1587, 2SA1587	data
BC 868	Si-N	SMD, NF-Tr/E, 25V, 1A, 60MHz	39b	Phi, Ztx	BCX 54...55, BCX 68	data
BC 869	Si-P	SMD, NF-Tr/E, 25V, 1A, 60MHz	39b	Phi, Ztx	BCX 51...52, BCX 69	data
BC 875	Si-N-Darl+Di	NF/S, 60V, 1A, 0,8W, 200MHz, B>2000	7c	Phi,Sie	BC 618, BSR 50...52	data
BC 876	Si-P-Darl+Di	NF/S, 60V, 1A, 0,8W, 200MHz, B>2000	7c	Phi,Sie	BSR 60...62	data
BC 877	Si-N-Darl+Di	=BC 875: 80V	7c	Phi,Sie	BC 618, BSR 51...52	data
BC 878	Si-P-Darl+Di	=BC 876: 80V	7c	Phi,Sie	BSR 61...62	data
BC 879	Si-N-Darl+Di	=BC 875: 100V	7c	Phi,Sie	BSR 52	data
BC 880	Si-P-Darl+Di	=BC 876: 100V	7c	Phi,Sie	BSR 62	data

BCF

Type	Device	Short Description	Fig.	Manu	Comparision Types	More at
BCF	Si-Di	=MBRM 120L (Typ-Code/Stempel/marking)	71(2mm)	Ons	→MBRM 120L	
BCF 29	Si-P	SMD, NF-V, ra, 32V, 0,1A, 150MHz	35a	Phi	BC 859...860, BCF 70	data
BCF 29 R	Si-P	=BCF 29:	35d		BC 859R...860R, BCF 70R	data
BCF 30	Si-P	SMD, NF-V, ra, 32V, 0,1A, 150MHz	35a	Phi	BC 859...860, BCF 70	data
BCF 30 R	Si-P	=BCF 30:	35d		BC 859R...860R, BCF 70R	data
BCF 32	Si-N	SMD, NF-V, ra, 32V, 0,1A, 300MHz	35a	Phi	BC 849...850, BCF 81	data
BCF 32 R	Si-N	=BCF 32:	35d		BC 849R...850R, BCF 81R	data
BCF 33	Si-N	SMD, NF-V, ra, 32V, 0,1A, 300MHz	35a	Phi	BC 849...850, BCF 81	data
BCF 33 R	Si-N	=BCF 33:	35d		BC 849R...850R, BCF 81R	data
BCF 70	Si-P	=BCF 30: 50V	35a	Phi	BC 860	data
BCF 70 R	Si-P	=BCF 70:	35d		BC 860R	data
BCF 81	Si-N	=BCF 33: 50V	35a	Phi	BC 850	data
BCF 81 R	Si-N	=BCF 81:	35d		BC 850R	data
BCG	Si-Di	=MBRM 130L (Typ-Code/Stempel/marking)	71(2mm)	Ons	→MBRM 130L	
BCJ	Si-Di	=MBRM 140 (Typ-Code/Stempel/marking)	71(2mm)	Ons	→MBRM 140	

BCP

Type	Device	Short Description	Fig.	Manu	Comparision Types	More at
BCP	Si-P	=2SB1188-P (Typ-Code/Stempel/marking)	39	Rhm	→2SB1188	data
BCp	Si-P	=BCW 61C (Typ-Code/Stempel/marking)	35	Phi	→BCW 61C	data
BCP 28	Si-P-Darl	=BCV 26: 1,5W	48j§	Sie	-	data
BCP 29	Si-N-Darl	=BCV 27: 1,5W	48j§	Sie	-	data
BCP 48	Si-P-Darl	=BCV 46: 1,5W	48j§	Sie	-	data
BCP 49	Si-N-Darl	=BCV 47: 1,5W	48j§	Sie	-	data
BCP 51	Si-P	SMD, NF-Tr, 45V, 1A, 1,5W, 50MHz	48j§	Phi,Sie,++	-	data
BCP 52	Si-P	=BCP 51: 60/60V	48j§	Phi,Sie,++	-	data
BCP 53	Si-P	=BCP 51: 100V	48j§	Phi,Sie,++	-	data
BCP 51M...53M	Si-P	=BCP 51...53: 1,7W(Tc=77°)	≈45(BCE-C)	Sie	-	data
BCP 54	Si-N	SMD, NF-Tr, 45V, 1A, 1,5W, 130MHz	48j§	Phi,Sie,++	-	data
BCP 55	Si-N	=BCP 54: 60V	48j§	Phi,Sie,++	-	data
BCP 56	Si-N	=BCP 54: 100V	48j§	Phi,Sie,++	-	data
BCP 54M...56M	Si-N	=BCP 54...56: 1,7W(Tc=77°)	≈45(BCE-C)	Sie	-	data
BCP 68	Si-N	=BCX 68: 1,5W	48j§	Phi, Sie,++	-	data
BCP 69	Si-P	=BCX 69: 1,5W	48j§	Phi, Sie,++	-	data
BCP 70 M	Si-P	SMD, NF-Tr/E, 32V, 3A, 100MHz	≈45(ECEBC)	Sie	-	data
BCP 71 M	Si-N	SMD, NF-Tr/E, 32V, 3A, 100MHz	≈45(ECEBC)	Sie	-	data
BCP 72 M	Si-P	SMD, NF-Tr/E, 10V, 3A, 100MHz	≈45(ECEBC)	Sie	-	data

Type	Device	Short Description	Fig.	Manu	Comparision Types	More at
BCQ	Si-P	=2SB1188-Q (Typ-Code/Stempel/marking)	39	Rhm	→2SB1188	data

BCR

Type	Device	Short Description	Fig.	Manu	Comparision Types	More at
BCR	Si-P	=2SB1188-R (Typ-Code/Stempel/marking)	39	Rhm	→2SB1188	data
BCR 1 AM-10	Triac	500V, 1A(Tc=56°), Igt<10mA	7e	Mit	ZO 104, ZO 101, ZO 106, TAG 204, TAG 205	data
BCR 1 AM-12	Triac	600V, 1A(Tc=56°), Igt<10mA	7e	Mit	ZO 104, ZO 101, ZO 106, TAG 204, TAG 205	data
BCR 1 AM-4	Triac	200V, 1A(Tc=56°), Igt<10mA	7e	Mit	ZO 104, ZO 101, ZO 106, TAG 204, TAG 205	data
BCR 1 AM-6	Triac	300V, 1A(Tc=56°), Igt<10mA	7e	Mit	ZO 104, ZO 101, ZO 106, TAG 204, TAG 205	data
BCR 1 AM-8	Triac	400V, 1A(Tc=56°), Igt<10mA	7e	Mit	ZO 104, ZO 101, ZO 106, TAG 204, TAG 205	data
BCR 3 AM-10	Triac	500V, 3A(Tc=71°), Igt<30mA	13h§	Mit	T 2323, ZO 410, T 2322, ZO 409	data
BCR 3 AM-12	Triac	600V, 3A(Tc=71°), Igt<30mA	13h§	Mit	T 2323, ZO 410, T 2322, ZO 409	data
BCR 3 AM-4	Triac	200V, 3A(Tc=71°), Igt<30mA	13h§	Mit	T 2323, ZO 410, T 2322, ZO 409	data
BCR 3 AM-8	Triac	400V, 3A(Tc=71°), Igt<30mA	13h§	Mit	T 2323, ZO 410, T 2322, ZO 409	data
BCR 6 AM-10	Triac	500V, 6A(Tc=103°), Igt<30mA	17h§	Mit	TAG 220, TAG 225, TAG 252, TAG 250	data
BCR 6 AM-12	Triac	600V, 6A(Tc=103°), Igt<30mA	17h§	Mit	TAG 220, TAG 225, TAG 252, TAG 250	data
BCR 6 AM-4	Triac	200V, 6A(Tc=103°), Igt<30mA	17h§	Mit	TAG 220, TAG 225, TAG 252, TAG 250	data
BCR 6 AM-6	Triac	300V, 6A(Tc=103°), Igt<30mA	17h§	Mit	TAG 220, TAG 225, TAG 252, TAG 250	data
BCR 6 AM-8	Triac	400V, 6A(Tc=103°), Igt<30mA	17h§	Mit	TAG 220, TAG 225, TAG 252, TAG 250	data
BCR 8 A-10	Triac	500V, 8A(Tc=92°), Igt<50mA	22a§	Mit	MAC 222, TIC 226, T 2802, BS 9-A	data
BCR 8 A-4	Triac	200V, 8A(Tc=92°), Igt<50mA	22a§	Mit	MAC 222, TIC 226, T 2802, BS 9-A	data
BCR 8 A-6	Triac	300V, 8A(Tc=92°), Igt<50mA	22a§	Mit	MAC 222, TIC 226, T 2802, BS 9-A	data
BCR 8 A-8	Triac	400V, 8A(Tc=92°), Igt<50mA	22a§	Mit	MAC 222, TIC 226, T 2802, BS 9-A	data
BCR 8CM-10	Triac	500V, 8A(Tc=105°), Igt<30mA	17h§	Mit	BTA 21, BT 158, T 2806, BT 162	data
BCR 8CM-12	Triac	600V, 8A(Tc=105°), Igt<30mA	17h§	Mit	BTA 21, BT 158, T 2806, BT 162	data
BCR 8CM-12	Triac	600V, 8A(Tc=105°), Igt<30mA	17h§	Mit	BTA 21, BT 158, T 2806, BT 162	data
BCR 8CM-4	Triac	200V, 8A(Tc=105°), Igt<30mA	17h§	Mit	BTA 21, BT 158, T 2806, BT 162	data
BCR 8CM-6	Triac	300V, 8A(Tc=105°), Igt<30mA	17h§	Mit	BTA 21, BT 158, T 2806, BT 162	data
BCR 8CM-8	Triac	400V, 8A(Tc=105°), Igt<30mA	17h§	Mit	BTA 21, BT 158, T 2806, BT 162	data
BCR 8DM-...	Triac	=BCR 8CM-...: 8A(Tc=94°)	17h§	Mit	BTA 21, BT 158, T 2806, BT 162	data
BCR 08 PN	Si-N/P+R	SMD, =BCR 108 + BCR 158	46bh1(2mm)	Sie	-	data
BCR 10A-2	Triac	100V, 10A(Tc=75°), Igt<50mA	76b§	Mit	-	data
BCR 10A-4	Triac	200V, 10A(Tc=75°), Igt<50mA	76b§	Mit	-	data
BCR 10A-6	Triac	300V, 10A(Tc=75°), Igt<50mA	76b§	Mit	-	data
BCR 10A-8	Triac	400V, 10A(Tc=75°), Igt<50mA	76b§	Mit	-	data
BCR 10AM-10	Triac	500V, 10A(Tc=100°), Igt<30mA	17h§	Mit	TAG 252, TAG 257	data
BCR 10AM-12	Triac	600V, 10A(Tc=100°), Igt<30mA	17h§	Mit	TAG 252, TAG 257	data
BCR 10AM-4	Triac	200V, 10A(Tc=100°), Igt<30mA	17h§	Mit	TAG 252, TAG 257	data
BCR 10AM-6	Triac	300V, 10A(Tc=100°), Igt<30mA	17h§	Mit	TAG 252, TAG 257	data
BCR 10AM-8	Triac	400V, 10A(Tc=100°), Igt<30mA	17h§	Mit	TAG 252, TAG 257	data
BCR 10B-2...-8	Triac	=BCR 10A-2...-8:	≈22	Mit		data
BCR 10C-2...-8	Triac	=BCR 10A-2...-8:	21b§	Mit	SC 245, BS 10-...A, TXD 99, SC 250	data
BCR 10CM-4...-12	Triac	=BCR 10AM-4...-12: 10A(Tc=103°)	17h§	Mit	TAG 252, TAG 257	data
BCR 10DM-4...-12	Triac	=BCR 10AM-4...-12: 10A(Tc=93°)	17h§	Mit	TAG 252, TAG 257	data
BCR 10EM-4...-12	Triac	=BCR 10AM-4...-12: 10A(Tc=80°)	17h§	Mit	TAG 252, TAG 257	data
BCR 10 PN	Si-N/P+R	SMD, =BCR 133 + BCR 183	46bh1(2mm)	Sie	-	data
BCR 12AM-10	Triac	500V, 12A(Tc=101°), Igt<30mA	17h§	Mit	TAG 257	data
BCR 12AM-12	Triac	600V, 12A(Tc=101°), Igt<30mA	17h§	Mit	TAG 257	data
BCR 12AM-4	Triac	200V, 12A(Tc=101°), Igt<30mA	17h§	Mit	TAG 257	data
BCR 12AM-6	Triac	300V, 12A(Tc=101°), Igt<30mA	17h§	Mit	TAG 257	data
BCR 12AM-8	Triac	400V, 12A(Tc=101°), Igt<30mA	17h§	Mit	TAG 257	data
BCR 12CM-4...-12	Triac	=BCR 12AM-4...-12: 12A(Tc=98°)	17h§	Mit	TAG 257	data
BCR 12DM-4...-12	Triac	=BCR 12AM-4...-12: 12A(Tc=84°)	17h§	Mit	TAG 257	data
BCR 12EM-4...-12	Triac	=BCR 12AM-4...-12: 12A(Tc=77°)	17h§	Mit	TAG 257	data
BCR 16A-10	Triac	500V, 16A(Tc=99°), Igt<30mA	76b§	Mit	-	data
BCR 16A-4	Triac	200V, 16A(Tc=99°), Igt<30mA	76b§	Mit	-	data
BCR 16A-6	Triac	300V, 16A(Tc=99°), Igt<30mA	76b§	Mit	-	data
BCR 16A-8	Triac	400V, 16A(Tc=99°), Igt<30mA	76b§	Mit	-	data
BCR 16AM-4...-10	Triac	=BCR 16A-4...-10:		Mit	-	data
BCR 16B-4...-10	Triac	=BCR 16A-4...-10:	≈22	Mit	-	data
BCR 16BM-4...-10	Triac	=BCR 16A-4...-10:		Mit	-	data
BCR 16C-4...-10	Triac	=BCR 16A-4...-10:	21b§	Mit	-	data
BCR 16CM-12	Triac	=BCR 16A-4...-10: 600V	17h§	Mit	-	data
BCR 16CM-4...-10	Triac	=BCR 16A-4...-10:	17h§	Mit	-	data
BCR 16DM-12	Triac	=BCR 16A-4...-10: 600V, 16A(Tc=79°)	17h§	Mit	-	data
BCR 16DM-4...-10	Triac	=BCR 16A-4...-10: 16A(Tc=79°)	17h§	Mit	-	data
BCR 16E-4...-10	Triac	=BCR 16A-4...-10:	≈22	Mit	-	data
BCR 16EM-4...-10	Triac	=BCR 16A-4...-10: 16A(Tc=73°)		Mit	-	data
BCR 16FM-4...-10	Triac	=BCR 16A-4...-10: 16A(Tc=73°)		Mit	-	data
BCR 16GM-4...-10	Triac	=BCR 16A-4...-10: 16A(Tc=73°)		Mit	-	data
BCR 16HM-4...-10	Triac	=BCR 16A-4...-10: 16A(Tc=82°)	65h	Mit	-	data
BCR 19 PN	Si-N/P+R	SMD, =BCR 119 + BCR 169	46bh1(2mm)	Sie	-	
BCR 20A-10	Triac	500V, 20A(Tc=98°), Igt<30mA	76b§	Mit	-	data
BCR 20A-4	Triac	200V, 20A(Tc=98°), Igt<30mA	76b§	Mit	-	data
BCR 20A-6	Triac	300V, 20A(Tc=98°), Igt<30mA	76b§	Mit	-	data
BCR 20A-8	Triac	400V, 20A(Tc=98°), Igt<30mA	76b§	Mit	-	data
BCR 20B-4...-10	Triac	=BCR 20A-4...-10:	≈22	Mit	-	data
BCR 20C-4...-10	Triac	=BCR 20A-4...-10:	21b§	Mit	-	data
BCR 20E-4...-10	Triac	=BCR 20A-4...-10:	≈22	Mit	-	data

Type	Device	Short Description	Fig.	Manu	Comparision Types	More at
BCR 22 PN	Si-N/P+R	SMD, =BCR 141 + BCR 191	46bh1(2mm)	Sie	-	data
BCR 25A-10	Triac	500V, 25A(Tc=92°), Igt<75mA	21b§	Mit	-	data
BCR 25A-4	Triac	200V, 25A(Tc=92°), Igt<75mA	21b§	Mit	-	data
BCR 25A-6	Triac	300V, 25A(Tc=92°), Igt<75mA	21b§	Mit	-	data
BCR 25A-8	Triac	400V, 25A(Tc=92°), Igt<75mA	21b§	Mit	-	data
BCR 25B-4...-10	Triac	=BCR 25A-4...-10:	≈22	Mit	-	data
BCR 30GM-10	Triac	500V, 30A(Tc=60°), Igt<50mA	65h	Mit	-	data
BCR 30GM-4	Triac	200V, 30A(Tc=60°), Igt<50mA	65h	Mit	-	data
BCR 30GM-6	Triac	300V, 30A(Tc=60°), Igt<50mA	65h	Mit	-	data
BCR 30GM-8	Triac	400V, 30A(Tc=60°), Igt<50mA	65h	Mit	-	data
BCR 35 PN	Si-N/P+R	SMD, =BCR 135 + BCR 185	46bh1(2mm)	Sie		data
BCR 39 PN	Si-N/P+R	SMD, =BCR 139 + BCR 189	46bh1(2mm)	Sie		
BCR 48 PN	Si-N/P+R	SMD, =BCR 148 + BCR 158	46bh1(2mm)	Sie	-	data
BCR 108	Si-N+R	SMD, Rb=2,2kΩ, Rbe=47kΩ, 50V, 0,1A	35a	Sie	DTC 123JK, KSR 1113, RN 1405, UN 221M	data
BCR 108 S	Si-N+R	=BCR 108: Dual	46bh1(2mm)		-	data
BCR 108 T	Si-N+R	=BCR 108:	35a(1,6mm)		DTC 123JE, RN 1105	
BCR 108 W	Si-N+R	=BCR 108:	35a(2mm)		DTC 123JU, RN 1305	data
BCR 112	Si-N+R	SMD, Rb=4,7kΩ, Rbe=4,7kΩ, 50V, 0,1A	35a	Sie	DTC 143EK, KSR 1101, RN 1401, 2SC4362	data
BCR 112 W	Si-N+R	=BCR 112:	35a(2mm)		DTC 143EU, RN 1301, UN 521L	
BCR 116	Si-N+R	SMD, Rb=4,7kΩ, Rbe=47kΩ, 50V, 0,1A	35a	Sie	DTC 143ZK, KSR 1114, RN 1406, 2SC4146	data
BCR 116 W	Si-N+R	=BCR 116:	35a(2mm)		DTC 143ZU, RN 1306	data
BCR 119	Si-N+R	SMD, Rb=4,7kΩ, Rbe=-, 50V, 0,1A	35a	Sie	DTC 143TK, KSR 1109, RN 1410, UN2216, ++	data
BCR 119 S	Si-N+R	=BCR 119: Dual	46bh1(2mm)		-	data
BCR 119 W	Si-N+R	=BCR 119:	35a(2mm)		DTC 143TU, RN 1310, UN 5216	
BCR 129	Si-N+R	SMD, Rb=10kΩ, Rbe=-, 50V, 0,1A	35a	Sie	DTC 114TK, KSR 1110, RN 1411, UN2215, ++	
BCR 129 S	Si-N+R	=BCR 129: Dual	46bh1(2mm)		-	
BCR 129 T	Si-N+R	=BCR 129T:	35a(1,6mm)		DTC 114TE, RN 1111, UN 9215	
BCR 133	Si-N+R	SMD, Rb=10kΩ, Rbe=10kΩ, 50V, 0,1A	35a	Sie	DTC 114EK, KSR 1102, RN 1402, UN 2211,++	data
BCR 133 S	Si-N+R	=BCR 133: Dual	46bh1(2mm)		-	data
BCR 133 T	Si-N+R	=BCR 133T:	35a(1,6mm)		DTC 114EE, RN 1102, UN 9211	
BCR 133 U	Si-N+R	=BCR 133: Dual	46bh1			
BCR 133 W	Si-N+R	=BCR 133:	35a(2mm)		DTC 114EU, UN 5211, 2SC4398	data
BCR 135	Si-N+R	SMD, Rb=10kΩ, Rbe=47kΩ, 50V, 0,1A	35a	Sie	DTC 114YK, KSR 1106, RN 1407, 2SC4047	data
BCR 135 S	Si-N+R	=BCR 135: Dual	46bh1(2mm)		-	data
BCR 135 T	Si-N+R	=BCR 135:	35a(1,6mm)		DTC 114YE, RN 1107, UN 9214	
BCR 135 W	Si-N+R	=BCR 135:	35a(2mm)		DTC 114YU, RN 1307, UN 5214	data
BCR 139	Si-N+R	SMD, Rb=22kΩ, Rbe=-, 50V, 0,1A	35a	Sie	DTC 124TK, KSR 1112, RN 1412, UN2217, ++	
BCR 141	Si-N+R	SMD, Rb=22kΩ, Rbe=22kΩ, 50V, 0,1A	35a	Sie	DTC 124EK, KSR 1103, RN 1403, UN 2212,++	data
BCR 141 S	Si-N+R	=BCR 141: Dual	46bh1(2mm)		-	data
BCR 141 T	Si-N+R	=BCR 141:	35a(1,6mm)		DTC 124EE, RN 1103, UN 9212	
BCR 141 W	Si-N+R	=BCR 141:	35a(2mm)		DTC 124EU, RN 1303, UN 5212, 2SC4397	data
BCR 142	Si-N+R	SMD, Rb=22kΩ, Rbe=47kΩ, 50V, 0,1A	35a	Sie	DTC 124XK, FA 1F4N, KSR 1107, RN 1408	data
BCR 142 T	Si-N+R	=BCR 142:	35a(1,6mm)		DTC 124XE, RN 1108	
BCR 142 W	Si-N+R	=BCR 142:	35a(2mm)		DTC 124XU, RN 1308	data
BCR 146	Si-N+R	SMD, Rb=47kΩ, Rbe=22kΩ, 50V, 0,1A	35a	Sie	DTC 144WK, KSR 1108, RN 1409, UN 221E,++	data
BCR 146 W	Si-N+R	=BCR 146:	35a(2mm)		DTC 144WU, RN 1309, UN 521E	data
BCR 148	Si-N+R	SMD, Rb=47kΩ, Rbe=47kΩ, 50V, 0,07A	35a	Sie	DTC 144EK, KSR 1104, RN 1404, UN 2213,++	data
BCR 148 S	Si-N+R	=BCR 148: Dual	46bh1(2mm)		-	data
BCR 148 T	Si-N+R	=BCR 148:	35a(1,6mm)		DTC 144EE, RN 1104, UN 9213	
BCR 148 U	Si-N+R	=BCR 148: Dual	46bh1			
BCR 148 W	Si-N+R	=BCR 148:	35a(2mm)		DTC 144EU, RN 1304, UN 5213	data
BCR 153	Si-N+R	SMD, Rb=2,2kΩ, Rbe=2,2kΩ, 50V, 0,1A	35a	Sie	DTC 123EK	
BCR 158	Si-P+R	SMD, Rb=2,2kΩ, Rbe=47kΩ, 50V, 0,1A	35a	Sie	DTA 123JK, KSR 2113, RN 2405, UN 211M	
BCR 158 T	Si-P+R	=BCR 158:	35a(1,6mm)		DTA 123JE, RN 2105	
BCR 158 W	Si-P+R	=BCR 158:	35a(2mm)		DTA 123JK, RN 2305	data
BCR 162	Si-P+R	SMD, Rb=4,7kΩ, Rbe=4,7kΩ, 50V, 0,1A	35a	Sie	DTA 143EK, KSR 2101, RN 2401, UN 211L,++	data
BCR 162 T	Si-P+R	=BCR 162:	35a(1,6mm)		DTA 143EE, RN 2101, UN 911L	
BCR 166	Si-P+R	SMD, Rb=4,7kΩ, Rbe=47kΩ, 50V, 0,1A	35a	Sie	DTA 143ZK, KSR 2114, RN 2406, 2SA1597	data
BCR 166 W	Si-P+R	=BCR 166:	35a(2mm)		DTA 143ZU, RN 2306	data
BCR 169	Si-P+R	SMD, Rb=4,7kΩ, Rbe=-, 50V, 0,1A	35a	Sie	DTA 143TK, KSR 2109, RN 2410, UN 2116,++	data
BCR 169 S	Si-P+R	=BCR 169: Dual	46bh1(2mm)		-	data
BCR 169 W	Si-P+R	=BCR 169:	35a(2mm)		DTA 143TU, RN 2310, UN 5116	
BCR 183	Si-P+R	SMD, Rb=10kΩ, Rbe=10kΩ, 50V, 0,1A	35a	Sie	DTA 114EK, KSR 2102, RN 2402, UN 2111,++	data
BCR 183 S	Si-P+R	=BCR 183: Dual	46bh1(2mm)		-	
BCR 183 T	Si-P+R	=BCR 183:	35a(1,6mm)		DTA 114EE, RN 2102, UN 9111	
BCR 183 U	Si-P+R	=BCR 183: Dual	46bh1			
BCR 183 W	Si-P+R	=BCR 183:	35a(2mm)		DTA 114EU, UN 5111, 2SA1678	
BCR 185	Si-P+R	SMD, Rb=10kΩ, Rbe=47kΩ, 50V, 0,1A	35a	Sie	DTA 114YK, KSR 2106, RN 2407, UN 2114,++	data
BCR 185 S	Si-P+R	=BCR 185: Dual	46bh1(2mm)		-	data
BCR 185 U	Si-P+R	=BCR 185: Dual	46bh1			
BCR 185 W	Si-P+R	=BCR 185:	35a(2mm)		DTA 114YU, RN 2307, UN 5114	data
BCR 189	Si-P+R	SMD, Rb=22kΩ, Rbe=-, 50V, 0,1A	35a	Sie	DTA 124TK, KSR 2111, RN2412, UN 2117, ++	
BCR 191	Si-P+R	SMD, Rb=22kΩ, Rbe=22kΩ, 50V, 0,1A	35a	Sie	DTA 124EK, KSR 2103, RN 2403, UN 2112,++	data
BCR 191 S	Si-P+R	=BCR 191: Dual	46bh1(2mm)		-	data

Type	Device	Short Description	Fig.	Manu	Comparision Types	More at
BCR 191 W	Si-P+R	=BCR 191:	35a(2mm)		DTA 124EU, RN 2303, UN 5112	
BCR 192	Si-P+R	SMD, Rb=22kΩ, Rbe=47kΩ, 50V, 0,1A	35a	Sie	DTA 124XK, FN 1F4N, KSR 2107, RN 2408	data
BCR 192 T	Si-P+R	=BCR 192:	35a(1,6mm)		DTA 124XE, RN 2108	
BCR 192 W	Si-P+R	=BCR 192:	35a(2mm)		DTA 124XU, RN 2308	
BCR 196	Si-P+R	SMD, Rb=47kΩ, Rbe=22kΩ, 50V, 0,1A	35a	Sie	DTA 144WK, KSR 2108, RN 2409, UN211E	data
BCR 196 W	Si-P+R	=BCR 196:	35a(2mm)		DTA 144WU, RN 2309, UN511E	data
BCR 198	Si-P+R	SMD, Rb=47kΩ, Rbe=47kΩ, 50V, 0,1A	35a	Sie	DTA 144EK, KSR 2104, RN 2404, UN 2113,++	data
BCR 198 S	Si-P+R	=BCR 198: Dual	46bh1(2mm)		-	data
BCR 198 W	Si-P+R	=BCR 198:	35a(2mm)		DTA 144EU, RN 2304, UN 5113, 2SA1676	data
BCR 400 R	Si-P/IC	SMD, Active Bias Controller	44	Sie	-	data
BCR 400 W	Si-P/IC	=BCR 400R:	44(2mm)		-	data
BCR 503	Si-N+R	SMD, Rb=2,2kΩ, Rbe=2,2kΩ, 50V, 0,5A	35a	Sie	DTD 123EK, RN 1222, UN 4221, 2SC3923, ++	data
BCR 505	Si-N+R	SMD, Rb=2,2kΩ, Rbe=10kΩ, 50V, 0,5A	35a	Sie	DTD 123YS, RN 1227, UN 4224, 2SC3922, ++	data
BCR 512	Si-N+R	SMD, Rb=4,7kΩ, Rbe=4,7kΩ, 50V, 0,5A	35a	Sie	DTD 143ES, RN 1223, UN 4222, 2SC3921, ++	data
BCR 519	Si-N+R	SMD, Rb=4,7kΩ, Rbe=0, 50V, 0,5A	35a	Sie	DTC 343TS, DTD 143TS	data
BCR 521	Si-N+R	SMD, Rb=1kΩ, Rbe=1kΩ, 50V, 0,5A	35a	Sie	DTD 113ES, RN 1221	data
BCR 523	Si-N+R	SMD, Rb=1kΩ, Rbe=10kΩ, 50V, 0,5A	35a	Sie	DTD 113ZS, RN 1018, RN 1226	data
BCR 533	Si-N+R	SMD, Rb=10kΩ, Rbe=10kΩ, 50V, 0,5A	35a	Sie	DTD 114ES, RN 1224, UN 4223, 2SC3920, ++	data
BCR 553	Si-P+R	SMD, Rb=2,2kΩ, Rbe=2,2kΩ, 50V, 0,5A	35a	Sie	DTB 123ES, RN 2222, UN 4121, 2SA1529, ++	data
BCR 555	Si-P+R	SMD, Rb=2,2kΩ, Rbe=10kΩ, 50V, 0,5A	35a	Sie	DTB 123YS, RN 2227, UN 4124, 2SA1528, ++	data
BCR 562	Si-P+R	SMD, Rb=4,7kΩ, Rbe=4,7kΩ, 50V, 0,5A	35a	Sie	DTB 143ES, RN 2223, UN 4122, 2SA1527, ++	data
BCR 569	Si-P+R	SMD, Rb=4,7kΩ, Rbe=0, 50V, 0,5A	35a	Sie	DTB 143TS	data
BCR 571	Si-P+R	SMD, Rb=1kΩ, Rbe=1kΩ, 50V, 0,5A	35a	Sie	DTB 113ES, RN 2221	data
BCR 573	Si-P+R	SMD, Rb=1kΩ, Rbe=10kΩ, 50V, 0,5A	35a	Sie	DTB 113ZS, RN 2226	data
BCR 583	Si-P+R	SMD, Rb=10kΩ, Rbe=10kΩ, 50V, 0,5A	35a	Sie	DTB 114ES, RN 2224, UN 4123, 2SA1526, ++	data
BCs	Si-P	=BCW 61C (Typ-Code/Stempel/marking)	35	Sie	→BCW 61C	data
BCt	Si-P	=BCW 61C (Typ-Code/Stempel/marking)	35	Phi	→BCW 61C	data

BCV

Type	Device	Short Description	Fig.	Manu	Comparision Types	More at
BCV	Si-Di	=MBRM 120E (Typ-Code/Stempel/marking)	71(2mm)	Ons	→MBRM 120E	
BCV 26	Si-P-Darl	SMD, 40V, 0,5A, 200MHz, B>20000	35a	Phi,Sie,Sgs	BCV 46	data
BCV 27	Si-N-Darl	SMD, 40V, 0,5A, 200MHz, B>20000	35a	Phi,Sie,Sgs	BCV 47	data
BCV 28	Si-P-Darl	=BCV 26:	39b	Phi, Sie,++	BST 60..62, 2SB1048	data
BCV 29	Si-N-Darl	=BCV 27:	39b	Phi, Sie,++	BST 50..52, 2SD1470, 2SD1511	data
BCV 46	Si-P-Darl	=BCV 26: 80V	35a	Phi,Sie,Sgs	-	data
BCV 47	Si-N-Darl	=BCV 27: 80V	35a	Phi,Sie,Sgs	-	data
BCV 48	Si-P-Darl	=BCV 46:	39b	Phi, Sie,++	BST 61..62	data
BCV 49	Si-N-Darl	=BCV 47:	39b	Phi, Sie,++	BST 51..52	data
BCV 61	Si-N	SMD, temp.-komp., 50V, 0,1A, 300MHz	44zc	Phi,Sie	-	data
BCV 62	Si-P	SMD, temp.-komp., 50V, 0,1A, 150MHz	44zc	Phi,Sie	-	data
BCV 63	Si-N	SMD, Schmitt-Trigger, ≈Darl	44w	Phi, Nsc	-	data
BCV 64	Si-P	SMD, Schmitt-Trigger, ≈Darl	44w	Phi, Nsc	-	data
BCV 65	Si-P+N	SMD, = BC 557+BC 547	44zj	Phi	-	data
BCV 71	Si-N	NF/S, 80V, 0,1A, 300MHz, B=110...220	35a	Fer,Phi,Sgs	BC 846, 2SC3324	data
BCV 71 R	Si-N	=BCV 71:	35d		BC 846R	data
BCV 72	Si-N	=BCV 71: B=200...450	35a	Fer,Phi,Sgs	BC 846, 2SC3324	data
BCV 72 R	Si-N	=BCV 72:	35d		BC 846R	data

BCW

Type	Device	Short Description	Fig.	Manu	Comparision Types	More at
BCW 10	Si-N	NF, 25V, 0,5A(ss), 0,3W, >150MHz	40e	Fer	BC 337...338, BC 635, BC 637, BC 639, ++	data
BCW 11	Si-P	NF, 25V, 0,5A(ss), 0,3W, >150MHz	40e	Fer	BC 327...328, BC 636, BC 638, BC 640, ++	data
BCW 12	Si-N	=BCW 10: 35V	40e	Fer	BC 337, BC 635, BC 637, BC 639, ++	data
BCW 13	Si-P	=BCW 11: 35V	40e	Fer	BC 327, BC 636, BC 638, BC 640, ++	data
BCW 14	Si-N	=BCW 10: 35V	40e	Fer	BC 337, BC 635, BC 637, BC 639, ++	data
BCW 15	Si-P	=BCW 11: 35V	40e	Fer	BC 327, BC 636, BC 638, BC 640, ++	data
BCW 16	Si-N	=BCW 10: 45V	40e	Fer	BC 337, BC 635, BC 637, BC 639, ++	data
BCW 17	Si-P	=BCW 11: 45V	40e	Fer	BC 327, BC 636, BC 638, BC 640, ++	data
BCW 18	Si-N	=BCW 10: 70V	40e	Fer	BC 639, 2SD667	data
BCW 19	Si-P	=BCW 11: 70V	40e	Fer	BC 640, 2SB647	data
BCW 20	Si-N	NF, ra, 30V, 0,3W, >30MHz	40e	Fer	BC 109, BC 184, BC 239, BC 549, ++	data
BCW 21	Si-P	NF, ra, 30V, 0,3W, >30MHz	40e	Fer	BC 214, BC 259, BC 309, BC 559, ++	data
BCW 22	Si-N	=BCW 20: 45V	40e	Fer	BC 184, BC 413...414, BC 550	data
BCW 23	Si-P	=BCW 20: 45V	40e	Fer	BC 214, BC 415...416, BC 560	data
BCW 24	Si-N	NF, ra, 45V, 30mA, 0,3W, >30MHz	2a§	Itt	BC 184, BC 413...414, BC 550	data
BCW 25	Si-N	Dual, 60V, 0,5A, 50MHz	81bs	Tix	2N2060(A), 2N2223(A)	data
BCW 26	Si-N	=2x BCW 25 gep.	81bs	Tix	2N2060(A), 2N2223(A)	data
BCW 27	Si-P	Uni, 150V, 0,1A, 0,625W	7d	Tix	BF 422, BF 398, BF 435...437, 2SA1019, ++	data
BCW 28	Si-P	Uni, 100V, 0,1A, 0,625W	7d	Tix	BF 422, BF 398, BF 435...437, 2SA1019, ++	data
BCW 29	Si-P	SMD, Uni, 30V, 0,1A, 150MHz, B=>120	35a	Phi,Sgs,++	BC 856...858, BCW 69...70, BCW 89, ++	data
BCW 29 R	Si-P	=BCW 29:	35d		BC 856R...858R, BCW 69R...70R, BCW 89R, ++	data
BCW 30	Si-P	=BCW 29: B>215	35a	Phi,Sgs,++	BC 856...858, BCW 69...70, BCW 89, ++	data
BCW 30 R	Si-P	=BCW 30:	35d		BC 856R...858R, BCW 69R...70R, BCW 89R, ++	data

Type	Device	Short Description	Fig.	Manu	Comparision Types	More at
BCW 31	Si-N	SMD, Uni, 30V, 0,1A, 300MHz, B>110	35a	Phi,Sgs,++	BC 846...848, BCW 71...72, BCW 81, ++	data
BCW 31 R	Si-N	=BCW 31:	35d		BC 846R...848R, BCW 71R...72R, BCW 81R, ++	data
BCW 32	Si-N	=BCW 31: B>200	35a	Phi,Sgs,++	BC 846...848, BCW 71...72, BCW 81, ++	data
BCW 32 R	Si-N	=BCW 32:	35d		BC 846R...848R, BCW 71R...72R, BCW 81R, ++	data
BCW 33	Si-N	=BCW 31: B>420	35a	Phi,Sgs,++	BC 846...848, BCW 71...72, BCW 81, ++	data
BCW 33 R	Si-N	=BCW 33:	35d		BC 846R...848R, BCW 71R...72R, BCW 81R, ++	data
BCW 34	Si-N	NF-Tr, 60V, 0,6A, 0,36W, >150MHz	2a§	Tix	BC 637, BC 639, BCX 22, BCX 24, ++	data
BCW 35	Si-P	NF-Tr, 60V, 0,6A, 0,36W, >150MHz	2a§	Tix	BC 638, BC 640, BCX 23, BCX 39, ++	data
BCW 36	Si-N	=BCW 34: 0,3W	7a	Tix	→BCW 34	data
BCW 37	Si-P	=BCW 35: 0,3W	7a	Tix	→BCW 35	data
BCW 38	Si-N	NF-Tr, 60V, 0,6A, 0,625W, >200MHz	7a	Tix	BC 637, BC 639, 2N3700...3701, 2SD667, ++	data
BCW 39	Si-N	=BCW 38: >250MHz	7a	Tix	BC 637, BC 639, 2N3700...3701, 2SD667, ++	data
BCW 44	Si-N	NF-Tr, 70V, 1A, 0,8W, 80MHz	2a§	Sgs	BC 140...141, BCX 40, 2N1990, 2N2102, ++	data
BCW 45	Si-P	NF-Tr, 70V, 1A, 0,8W, 200MHz	2a§	Sgs	BC 161, BCX 60, 2N4235...4236, ++	data
BCW 46	Si-N	=BCW 48: 80V	12a	Phi	BC 174, BC 190, BC 546, 2SC1890, ++	data
BCW 47	Si-N	=BCW 48: 50V	12a	Phi	BC 167, BC 182, BC 237, BC 547, ++	data
BCW 48	Si-N	Uni, 30V, 0,1A, 0,2W, 300MHz	12a	Phi	BC 168, BC 183, BC 238, BC 548, ++	data
BCW 49	Si-N	=BCW 48: ra	12a	Phi	BC 169, BC 184, BC 239, BC 549, ++	data
BCW 50	Si-N	NF, 120V, 0,5W, >50MHz	2a§	Sgs	BF 297...299, 2SC1890A, 2SC2240, ++	data
BCW 51	Si-N	NF/S, 50V, 0,2A, 0,3W, >200MHz	7a	Tix	BC 167, BC 182, BC 237, BC 547, ++	data
BCW 52	Si-P	NF/S, 50V, 0,2A, 0,3W, >200MHz	7a	Tix	BC 212, BC 257, BC 307, BC 557, ++	data
BCW 54	Si-N	NF, 64V, 0,1A, 0,3W, 300MHz, ß>125	2a§	Itt	BC 174, BC 190, BC 546, 2SC1890, ++	data
BCW 55	Si-N	=BCW 54: ß>240	2a§	Itt	BC 174, BC 190, BC 546, 2SC1890, ++	data
BCW 56	Si-P	=BCW 58: 80V	12a	Phi	BC 556, BC 477, 2SA893, 2SB715, ++	data
BCW 57	Si-P	=BCW 58: 50V	12a	Phi	BC 212, BC 257, BC 307, BC 557, ++	data
BCW 58	Si-P	Uni, 30V, 0,1A, 0,2W, 150MHz	12a	Phi	BC 213, BC 258, BC 308, BC 558, ++	data
BCW 59	Si-P	=BCW 58: ra	12a	Phi	BC 214, BC 259, BC 309, BC 559, ++	data
BCW 60	Si-N	SMD, Uni, 32V, 0,2A, 250MHz	35a	EUR	BC 846...848, BCW 31...33, BCX 70, ++	data
BCW 60 R	Si-N	=BCW 60:	35d		BC 846R...848R, BCW 31R...33R, BCX 70R, ++	data
BCW 61	Si-P	SMD, Uni, 32V, 0,2A, 180MHz	35a	EUR	BC 856...858, BCW 29...30, BCX 71, ++	data
BCW 61 R	Si-P	=BCW 61:	35d		BC 856R...858R, BCW 29R...30R, BCX 71R, ++	data
BCW 62	Si-P	=BCW 63: 60V	9c	Tix	BC 212, BC 256, BC 266, BC 556, ++	data
BCW 63	Si-P	Uni, 45V, 0,2A, 0,225W, >200MHz	9c	Tix	BC 213, BC 257, BC 307, BC 557, ++	data
BCW 64	Si-P	=BCW 63: ra	9c	Tix	BC 214, BC 415...416, BC 560	data
BCW 65	Si-N	SMD, NF/S, 60V, 0,8A, >100MHz	35a	Sie,Sgs,++	BCX 41	data
BCW 65 R	Si-N	=BCW 65:	35d		BCX 41R	data
BCW 66	Si-N	=BCW 65: 75V	35a	Sie,Sgs,++	BCX 41	data
BCW 66 R	Si-N	=BCW 66:	35d		BCX 41R	data
BCW 67	Si-P	SMD, NF/S, 45V, 0,8A, >100MHz	35a	Sie,Sgs,++	BCX 42	data
BCW 67 R	Si-P	=BCW 67:	35d		BCX 42R	data
BCW 68	Si-P	=BCW 67: 60V	35a	Sie,Sgs,++	BCX 42	data
BCW 68 R	Si-P	=BCW 68:	35d		BCX 42R	data
BCW 69	Si-P	SMD, Uni, 50V, 0,1A, 150MHz, B>120	35a	Phi,Sgs,++	BC 856...857, BCW 89	data
BCW 69 R	Si-P	=BCW 69:	35d		BC 856R...857R, BCW 89R	data
BCW 70	Si-P	=BCW 69: B>215	35a	Phi,Sgs,++	BC 856...857, BCW 89	data
BCW 70 R	Si-P	=BCW 70:	35d		BC 856R...857R, BCW 89R	data
BCW 71	Si-N	SMD, NF, 50V, 0,1A, 300MHz, B>110	35a	Phi,Sgs,++	BC 846...847, BCV 71...72	data
BCW 71 R	Si-N	=BCW 71:	35d		BC 846R...847R, BCV 71R...72R	data
BCW 72	Si-N	=BCW 71: B>200	35a	Phi,Sgs,++	BC 846...847, BCV 71...72	data
BCW 72 R	Si-N	=BCW 72:	35d		BC 846R...847R, BCV 71R...72R	data
BCW 73	Si-N	NF/S, 60V, 0,8A, 0,45W, >100MHz	2a§	Sie	BC 637, BC 639, BCX 73, 2N2221...2222, ++	data
BCW 74	Si-N	=BCW 73: 75V	2a§	Sie	BC 639, BCX 74, 2N2221A...2222A, ++	data
BCW 75	Si-P	NF/S, 45V, 0,8A, 0,45W, >100MHz	2a§	Sie	BC 638, BC 640, BCX 75, 2N2906...2907, ++	data
BCW 76	Si-P	=BCW 75: 60V	2a§	Sie	BC 640, BCX 76, 2N2906A...2907A, ++	data
BCW 77	Si-N	=BCW 73: 0,87W	2a§	Sie	BC 140...141, 2N2218...2219, ++	data
BCW 78	Si-N	=BCW 74: 0,87W	2a§	Sie	BC 140...141, 2N2218A...2219A, ++	data
BCW 79	Si-P	=BCW 75: 0,87W	2a§	Sie	BC 161, 2N2904...2905, ++	data
BCW 80	Si-P	=BCW 76: 0,87W	2a§	Sie	BC 161, 2N2904A...2905A, ++	data
BCW 81	Si-N	=BCW 71: B>420	35a	Phi,Sgs	BC 846...847, BCV 71...72	data
BCW 81 R	Si-N	=BCW 81:	35d		BC 846R...847R, BCV 71R...72R	data
BCW 82	Si-N	=BCW 83: 60V	9c	Tix	BC 182, BC 174, BC 190, BC 546, ++	data
BCW 83	Si-N	Uni, 45V, 0,2A, 0,225W, >150MHz	9c	Tix	BC 167, BC 183, BC 237, BC 547, ++	data
BCW 84	Si-N	=BCW 83: ra	9c	Tix	BC 184, BC 413...414, BC 550, ++	data
BCW 85	Si-P	NF/S, 90V, 0,2A, 0,3W, >200MHz	7a	Tix	BC 477, BC 556	data
BCW 86	Si-P	=BCW 85: 70V	7a	Tix	BC 477, BC 556, 2N2221A...2222A, ++	data
BCW 87	Si-N	NF, 45V, 0,1A, 0,225W, >125MHz	36e	Sie	BC 167, BC 183, BC 237, BC 547, ++	data
BCW 88	Si-P	NF, 45V, 0,1A, 0,225W, >180MHz	36e	Sie	BC 213, BC 257, BC 307, BC 557, ++	data
BCW 89	Si-P	=BCW 69: 80V	35a	Phi,Sgs,++	BC 856, 2SA1312, 2SA1325	data
BCW 89 R	Si-P	=BCW 89:	35d		BC 856R	data
BCW 90	Si-N	NF-Tr, 50V, 0,8A, 0,61W, >135MHz	7a	Tho	BC 337, BC 637, BC 639, 2N2221...2222, ++	data
BCW 90 K	Si-N	=BCW 90: 0,8W	7a°		→BCW 90	data
BCW 91	Si-N	=BCW 90: 70V	7a	Tho	BC 639, 2N2221A...2222A, ++	data
BCW 91 K	Si-N	=BCW 91: 0,8W	7a°		→BCW 91	data
BCW 92	Si-P	NF-Tr, 50V, 0,8A, 0,61W, >135MHz	7a	Tho	BC 327, BC 638, BC 640, 2N2906...2907, ++	data
BCW 92 K	Si-P	=BCW 92: 0,8W	7a°		→BCW 92	data
BCW 93	Si-P	=BCW 92: 70V	7a	Tho	BC 640, 2N2906A...2907A, ++	data
BCW 93 K	Si-P	=BCW 93: 0,8W	7a°		→BCW 93	data

Type	Device	Short Description	Fig.	Manu	Comparision Types	More at
BCW 94	Si-N	NF-Tr, 50V, 0,4A, 0,54W, 70MHz	7a	Tho	BC 337, BC 637, BC 639, 2N2221...2222, ++	data
BCW 94 K	Si-N	=BCW 94: 0,7W	7a°		→BCW 94	data
BCW 95	Si-N	=BCW 94: 70V	7a	Tho	BC 639, 2N2221A...2222A, ++	data
BCW 95 K	Si-N	=BCW 95: 0,7W	7a°		→BCW 95	data
BCW 96	Si-P	NF-Tr, 50V, 0,4A, 0,54W, 70MHz	7a	Tho	BC 327, BC 638, BC 640, 2N2906...2907, ++	data
BCW 96 K	Si-P	=BCW 96: 0,7W	7a°		→BCW 96	data
BCW 97	Si-P	=BCW 96: 70V	7a	Tho	BC 640, 2N2906A...2907A, ++	data
BCW 97 K	Si-P	=BCW 97: 0,7W	7a°		→BCW 97	data
BCW 98	Si-N	Min, 45V, 0,1A, 0,05W, >125MHz	9c	Aeg	BC 123	data
BCW 99	Si-P	Min, 45V, 0,1A, 0,05W, 200MHz	9c	Aeg	BC 203	data

BCX

Type	Device	Short Description	Fig.	Manu	Comparision Types	More at
BCX 10	Si-P	NF/S, 50V, 0,6A, 0,6W, 90MHz	2a§	Sgs	BC 161, BC 303...304, BCX 60, BSV 16...17	data
BCX 12	Si-N	NF/S, 125V, 0,8A, 0,625W, 100MHz	7a	Sie	2SC4488, 2SD1312, 2SD1616A, 2SD1857, ++	data
BCX 13	Si-P	NF/S, 125V, 0,8A, 0,625W, 120MHz	7a	Sie	2SA1708, 2SB984, 2SB1236, 2SB1456, ++	data
BCX 17	Si-P	SMD, NF-Tr, 50V, 0,5A, 100MHz	35a	Phi,Sgs,++	BC 807, BCW 68, BCX 42, 2SA1366	data
BCX 17 R	Si-P	=BCX 17:	35d		BC 807R, BCW 68R, BCX 42R	data
BCX 18	Si-P	=BCX 17: 30V	35a	Phi,Sgs,++	BC 807...808, BCW 67...68, 2SA1366	data
BCX 18 R	Si-P	=BCX 18:	35d		BC 807R...808R, BCW 67R...68R	data
BCX 19	Si-N	SMD, NF-Tr, 50V, 0,5A, 200MHz	35a	Phi,Sgs,++	BC 817, BCW 65...66, BCX 41, 2SC3441	data
BCX 19 R	Si-N	=BCX 19:	35d		BC 817R, BCW 65R...66R, BCX 41R	data
BCX 20	Si-N	=BCX 19: 30V	35a	Phi,Sgs,++	BC 817...818, BCW 65...66, 2SC3441	data
BCX 20 R	Si-N	=BCX 20:	35d		BC 817R...818R, BCW 65R...66R	data
BCX 21	Si-N-Darl	NF/S, 60V, 1A, 0,65W, 350MHz, B>2000	2a§	Phi,Fer	BSS 50...52, 2SD614...615, 2SD688	data
BCX 22	Si-N	NF-V/Tr, 125V, 0,8A, 0,45W, 100MHz	2a§	Phi,Sie	2N3700...3701, 2SC2235, 2SD667, ++	data
BCX 23	Si-P	NF-V/Tr, 125V, 0,8A, 0,45W, 100MHz	2a§	Phi,Sie	2N5400...5401, 2SA965, 2SB647, ++	data
BCX 24	Si-N	NF-V/Tr, 100V, 0,8A, 0,45W, 100MHz	2a§	Sie	BC 639, 2N3700...3701, 2SC2235, 2SD667,++	data
BCX 25	Si-N	NF/S, 60V, 0,2A, 0,35W, >100MHz	7a	Mot	BC 174, BC 182, BC 190, BC 546, ++	data
BCX 26	Si-P	NF/S, 60V, 0,2A, 0,35W, >100MHz	7a	Mot	BC 212, BC 256, BC 266, BC 556, ++	data
BCX 27	Si-N	=BCX 25: 80V	7a	Mot	BC 546, 2SC2240, 2SC2459, 2SC3245	data
BCX 28	Si-P	=BCX 26: 80V	7a	Mot	BC 556, 2SA970, 2SA1049, 2SA1285	data
BCX 29	Si-N	=BCX 25: 100V	7a	Mot	2SC2240, 2SC2459, 2SC3245	data
BCX 30	Si-P	=BCX 26: 100V	7a	Mot	2SA970, 2SA1049, 2SA1285	data
BCX 31	Si-N	NF, 100V, 0,5A, 0,83W, >80MHz	11a	Phi	BC 639, 2N3700...3701, 2SD667, ++	data
BCX 32	Si-N	NF, 80V, 1A, 0,83W, >80MHz	11a	Phi	BC 639, 2N3700...3701, 2SD667, ++	data
BCX 33	Si-N	=BCX 32: 60V	11a	Phi	BC 637, BC 639, 2N3700...3701, 2SD667, ++	data
BCX 34	Si-N	=BCX 32: 40V	11a	Phi	BC 635, BC 637, BC 639, 2SD667, ++	data
BCX 35	Si-P	NF, 80V, 0,6A, 0,83W, >80MHz	11a	Phi	BC 640, 2SA1013, 2SA1315, 2SB647, ++	data
BCX 36	Si-P	=BCX 35: 60V	11a	Phi	BC 638, BC 640, 2SA1315, 2SB647, ++	data
BCX 37	Si-P	=BCX 35: 40V	11a	Phi	BC 636, BC 638, BC 640, 2SB647, ++	data
BCX 38	Si-N-Darl	NF-Tr, 80V, 0,8A, 1W, B>500	40e	Ztx	BC 877, BC 879, BSR 51...52	data
BCX 39	Si-P	NF-V/Tr, 100V, 0,8A, 0,45W, 100MHz	2a§	Sie	BC 640, 2SB647, 2SA1013	data
BCX 40	Si-N	NF-Tr/E, 100V, 2A(ss), 1W, >50MHz	2a§	Sgs	BSS 15, 2N4239, 2N5320, 2SC2214	data
BCX 41	Si-N	SMD, NF, 125V, 0,8A, 100MHz	35a	Sie, Tho,++	-	data
BCX 41 R	Si-N	=BCX 41:	35d			data
BCX 42	Si-P	SMD, NF, 125V, 0,8A, 100MHz	35a	Sie,Tho	-	data
BCX 42 R	Si-P	=BCX 42:	35d			data
BCX 43	MOS-N-FET-e	150V, 1A	7	Tho	-	data
BCX 44	MOS-N-FET-e	90V, 0,5A	7	Tho	-	data
BCX 45	Si-N	NF/S, 45V, 1A, 0,625W, >100MHz	7a	Mot	BC 635, BC 637, BC 639	data
BCX 46	Si-P	NF/S, 45V, 1A, 0,625W, >60MHz	7a	Mot	BC 636, BC 638, BC 640	data
BCX 47	Si-N	=BCX 45: 60V	7a	Mot	BC 637, BC 639, 2SD667	data
BCX 48	Si-P	=BCX 46: 60V	7a	Mot	BC 638, BC 640, 2SB647	data
BCX 49	Si-N	=BCX 45: 80V	7a	Mot	BC 639, 2SD667, 2N3700...3701	data
BCX 50	Si-P	=BCX 46: 80V	7a	Mot	BC 640, 2SB647, 2SA1013	data
BCX 51	Si-P	SMD, NF-Tr, 45V, 1A, 50MHz	39b	Mot,Phi,Sie	2SA1364, 2SB804	data
BCX 52	Si-P	=BCX 51: 60V	39b	Mot,Phi,Sie	2SA1364, 2SB804	data
BCX 53	Si-P	=BCX 51: 100V	39b	Mot,Phi,Sie	2SB804, 2SB1025	data
BCX 54	Si-N	SMD, NF-Tr, 45V, 1A, 130MHz	39b	Mot,Phi,Sie	2SC3444, 2SD1005	data
BCX 55	Si-N	=BCX 54: 60V	39b	Mot,Phi,Sie	2SC3444, 2SD1005	data
BCX 56	Si-N	=BCX 54: 100V	39b	Mot,Phi,Sie	2SD1005, 2SD1418	data
BCX 58	Si-N	NF/S, 32V, 0,1A, 0,45W, 250MHz	7a	Phi,Sie,++	BC 183, BC 237, BC 547, 2N2221...2222, ++	data
BCX 59	Si-N	=BCX 58: 45V	7a	Phi,Sie,++	BC 182, BC 237, BC 547, 2N2221...2222, ++	data
BCX 60	Si-P	NF-Tr/E, 100V, 2A(ss), 1W, >50MHz	2a§	Sgs	BSS 17, 2N5322	data
BCX 68	Si-N	SMD, Uni, 25V, 1A, 65MHz	39b	Mot,Phi,Sie	BC 868, BCX 54...56, 2SC3439, 2SC3444	data
BCX 69	Si-P	SMD, Uni, 25V, 1A, 65MHz	39b	Mot,Phi,Sie	BC 869, BCX 51...53, 2SA1364, 2SA1369	data
BCX 70	Si-N	=BCW 60: 45V	35a	EUR	BC 846...847, BCW 71...72, BCW 81	data
BCX 70 R	Si-N	=BCX 70:	35d		BC 846R...847R, BCW 71R...72R, BCW 81R	data
BCX 71	Si-P	=BCW 61: 45V	35a	EUR	BC 856...857, BCW 69...70, BCW 89	data
BCX 71 R	Si-P	=BCX 71:	35d		BC 856R...857R, BCW 69R...70R, BCW 89R	data
BCX 73	Si-N	NF/S, 60V, 0,8A, 0,625W, >100MHz	7a	Mot,Sie	BC 637, BC 639, 2N2221...2222, ++	data
BCX 74	Si-N	=BCX 73: 75V	7a	Mot,Sie	BC 639, 2N2221A...2222A, ++	data
BCX 75	Si-P	NF/S, 60V, 0,8A, 0,625W, > 100MHz	7a	Mot,Sie	BC 638, BC 640, 2N2906...2907, ++	data
BCX 76	Si-P	=BCX 75: 75V	7a	Mot,Sie	BC 640, 2N2906A...2907A, ++	data
BCX 78	Si-P	NF/S, 32V, 0,1A, 0,45W, 200MHz	7a	Phi,Sie,++	BC 213, BC 307, BC 557, 2N2906...2907, ++	data

Type	Device	Short Description	Fig.	Manu	Comparision Types	More at
BCX 79	Si-P	=BCX 78: 45V	7a	Phi,Sie,++	BC 213, BC 307, BC 557, 2N2906...2907, ++	data
BCX 80	Si-N	NF-Tr, 50V, 0,75A, 0,5W, >120MHz	7a	Gen	BC 337, BC 637, BC 639, ++	data
BCX 81	Si-P	NF-Tr, 50V, 0,75A, 0,5W, >120MHz	7a	Gen	BC 327, BC 638, BC 640, ++	data
BCX 82	Si-N	Uni, ra, 100V, 0,1A, >75MHz, B>250	7a	Gen	2SC2240, 2SC2459	data
BCX 83	Si-N	=BCX 82: B>600	7a	Gen	2SC2240, 2SC2459	data
BCX 84	Si-P	NF-Tr, 60V, 0,5A, 0,5W, >60MHz	7a	Gen	BC 638, BC 640, 2SB647	data
BCX 85	Si-N	NF-Tr, 60V, 0,5A, 0,5W, >80MHz	7a	Gen	BC 637, BC 639, 2SD667	data
BCX 86	Si-N-Darl	NF, 25V, 0,5A, 0,5W, >80MHz, B>2000	7a	Gen	BC 517, BC 875, BC 877, BC 879	data
BCX 87	Si-P-Darl	NF, 25V, 0,5A, 0,5W, >80MHz, B>2000	7a	Gen	BC 516, BC 876, BC 878, BC 880	data
BCX 88	Si-N-Darl	=BCX 86: B>87000	7a	Gen	BC 517, (BC 875, BC 877, BC 879)	data
BCX 89	Si-P-Darl	=BCX 87: B>40000	7a	Gen	BC 516, (BC 876, BC 878, BC 880)	data
BCX 94	Si-N	NF-V/Tr, 100V, 0,8A, 0,45W, 100MHz	2a§	Sie	BC 639, 2N3700...3701, 2SD667	data
BCY						
BCY 10	Si-P	NF, 32V, 0,25A, 0,31W, 1,5MHz, B=15	1a	Phi	BC 327, BC 636, BC 638, BC 640, ++	data
BCY 11	Si-P	=BCY 10: 60V	1a	Phi	BC 638, BC 640, 2N2906A...2907A, ++	data
BCY 12	Si-P	=BCY 10: 60V, B=25	1a	Phi	BC 638, BC 640, 2N2906A...2907A, ++	data
BCY 13	Si-N	NF, 60V, 0,2A, 0,45W, 0,4MHz	2a§	Sie	BC 140...141, BC 637, BC 639, ++	data
BCY 14	Si-N	=BCY 14: 100V	2a§	Sie	BC 141, BC 639, 2SD667, ++	data
BCY 15	Si-N	=BCY 13: 0,3A	2a§	Sie	BC 140...141, BC 637, BC 639, ++	data
BCY 16	Si-N	=BCY 13: 100V, 0,3A	2a§	Sie	BC 141, BC 639, 2SD667, ++	data
BCY 17	Si-P	NF/S, 30V, 0,05A, 0,35W, 1,2MHz	2a	Sie,Tag	BC 213, BC 257, BC 307, BC 557, ++	data
BCY 18	Si-P	NF/S, 30V, 0,05A, 0,35W, 2MHz	2a	Sie,Tag	BC 213, BC 257, BC 307, BC 557, ++	data
BCY 19	Si-P	NF/S, 50V, 0,05A, 0,35W, 0,8MHz	2a	Sie,Tag	BC 212, BC 257, BC 307, BC 557, ++	data
BCY 20	Si-P	NF/S, 100V, 0,05A, 0,35W, 0,5MHz	2a	Sie,Tag	2N5400...5401, 2SB715...716, ++	data
BCY 21	Si-P	NF/S, 50V, 0,05A, 0,35W, 0,5MHz	2a	Tag	BC 212, BC 257, BC 307, BC 557, ++	data
BCY 22	Si-P	NF/S, 75V, 0,05A, 0,35W, 0,5MHz	2a	Tag	BC 256, BC 266, BC 556, ++	data
BCY 23	Si-P	NF/S, 30V, 0,05A, 0,35W, 1,5MHz	2a	Tag	BC 213, BC 258, BC 308, BC 558, ++	data
BCY 24	Si-P	NF/S, 30V, 0,05A, 0,35W, 1MHz	2a	Tag	BC 213, BC 258, BC 308, BC 558, ++	data
BCY 25	Si-P	NF/S, 30V, 0,05A, 0,35W, 2,5MHz	2a	Tag	BC 213, BC 258, BC 308, BC 558, ++	data
BCY 26	Si-P	NF/S, 30V, 0,05A, 0,35W, 0,6MHz	2a	Tag	BC 213, BC 258, BC 308, BC 558, ++	data
BCY 27	Si-P	NF/S, 30V, 0,05A, 0,275W, 1MHz	2a	Sie,Tag	BC 213, BC 258, BC 308, BC 558, ++	data
BCY 28	Si-P	NF/S, 30V, 0,05A, 0,275W, 1,5MHz	2a	Sie,Tag	BC 213, BC 258, BC 308, BC 558, ++	data
BCY 29	Si-P	NF/S, 60V, 0,05A, 0,275W, 0,5MHz	2a	Sie,Tag	BC 212, BC 256, BC 266, BC 556, ++	data
BCY 30	Si-P	NF/S, 64V, 0,05A, 0,25W, 1,2MHz	2a	Phi,Tag	BC 212, BC 256, BC 266, BC 556, ++	data
BCY 30 A	Si-P	=BCY 30: 0,1A, 0,6W, 7MHz	2a		BC 212, BC 256, BC 266, BC 556, ++	data
BCY 31	Si-P	NF/S, 64V, 0,05A, 0,25W, 1,7MHz	2a	Phi,Tag	BC 212, BC 256, BC 266, BC 556, ++	data
BCY 31 A	Si-P	=BCY 31: 0,1A, 0,6W, 7MHz	2a		BC 212, BC 256, BC 266, BC 556, ++	data
BCY 32	Si-P	NF/S, 64V, 0,05A, 0,25W, 2,5MHz	2a	Phi,Tag	BC 212, BC 256, BC 266, BC 556, ++	data
BCY 32 A	Si-P	=BCY 32: 0,1A, 0,6W, 7MHz	2a		BC 212, BC 256, BC 266, BC 556, ++	data
BCY 33	Si-P	NF/S, 32V, 0,05A, 0,25W, 1,5MHz	2a	Phi,Tag	BC 213, BC 258, BC 308, BC 558, ++	data
BCY 33 A	Si-P	=BCY 33: 0,1A, 0,6W, 7MHz	2a		BC 213, BC 258, BC 308, BC 558, ++	data
BCY 34	Si-P	NF/S, 32V, 0,05A, 0,25W, 2,4MHz	2a	Phi,Tag	BC 213, BC 258, BC 308, BC 558, ++	data
BCY 34 A	Si-P	=BCY 34: 0,1A, 0,6W, 7MHz	2a		BC 213, BC 258, BC 308, BC 558, ++	data
BCY 38	Si-P	NF, 32V, 0,25A, 0,35W, 1,5MHz	2a&	Phi,Tix	BC 328, BC 636, BC 638, BC 640, ++	data
BCY 39	Si-P	NF, 64V, 0,25A, 0,35W, 1,5MHz	2a&	Phi,Tix	BC 638, BC 640, 2N2904...2905, ++	data
BCY 40	Si-P	NF, 64V, 0,25A, 0,35W, 2,5MHz	2a&	Phi,Tix	BC 638, BC 640, 2N2904...2905, ++	data
BCY 42(P)	Si-N	Uni, 40V, 0,2A, 0,3W, >100MHz, B>45	2a§	Fer,Itt,Sgs	BC 167, BC 183, BC 237, BC 547, ++	data
BCY 43(P)	Si-N	=BCY 42: B>75	2a§	Fer,Itt,Sgs	BC 167, BC 183, BC 237, BC 547, ++	data
BCY 49	Si-P	NF, sym, 15V, 20mA, 0,25W	2a	Phi	-	data
BCY 50(r,i)	Si-N	NF-V, 10V, 0,1A, 0,3W, >60MHz i: Iso, r: ra	2a§ 2a	Itt	BC 169, BC 184, BC 239, BC 549, ++	data
BCY 51(r,i)	Si-N	=BCY 50: 30V, >50MHz i: Iso, r: ra	2a§ 2a	itt	BC 169, BC 184, BC 239, BC 549, ++	data
BCY 54	Si-P	NF, 50V, 0,25A, 0,35W, 2MHz	2a&	Phi	BC 327, BC 638, BC 640, 2N2904...2905, ++	data
BCY 55	Si-N	Dual, ra, 45V, 30mA, 0,3W, 80MHz	2x 2a°	Phi	-	data
BCY 56	Si-N	NF, ra, 45V, 0,1A, 0,3W, 250MHz	2a§	Phi,Sgs,Tix	BC 184, BC 413...414, BC 550, ++	data
BCY 57	Si-N	=BCY 56: 25V, 350MHz	2a§	Phi,Sgs,Tix	BC 169, BC 184, BC 239, BC 549, ++	data
BCY 58	Si-N	NF/S, 32V, 0,2A, 0,39W, 250MHz	2a§	EUR	BC 183, BC 548, BCX 58, 2N2221...2222, ++	data
BCY 58 P	Si-N	=BCY 58:	40e	Fer	→BCY 58	data
BCY 59	Si-N	=BCY 58: 45V	2a§	EUR	BC 182, BC 547, BCX 59, 2N2221...2222, ++	data
BCY 59 P	Si-N	=BCY 59:	40e	Fer	→BCY 59	data
BCY 65	Si-N	NF/S, 60V, 0,1A, 250MHz	2a§	EUR	BC 182, BC 546, 2N2221...2222, ++	data
BCY 65 P	Si-N	=BCY 65:	40e	Fer	→BCY 65	data
BCY 66	Si-N	NF, ra, 45V, 0,05A, 250MHz	2a§	Sie	BC 184, BC 413...414, BC 550, ++	data
BCY 67	Si-P	NF, ra, 45V, 0,05A, 180MHz	2a§	Sie	BC 214, BC 415...416, BC 560, ++	data
BCY 69	Si-N	NF, 20V, 0,1A, 0,3W, >150MHz	2a§	Tix,Tho	BC 168, BC 183, BC 238, BC 548, ++	data
BCY 70(P)	Si-P	NF/S, 50V, 0,2A, 0,35W, 450MHz	2a§	EUR	BCX 79, BCY 79, 2N2906...2907, ++	data
BCY 71(A,AP)	Si-P	=BCY 70, ra, 45V	2a§	EUR	BC 214, BC 415...416, BC 560, ++	data
BCY 72(P)	Si-P	=BCY 70: 30V	2a§	EUR	BCX 78, BCY 78, 2N2906...2907, ++	data
BCY 76	Si-N	Uni, 45V, 0,1A, 0,3W, 40MHz	2a§	Mic	BC 167, BC 182, BC 237, BC 547, ++	data
BCY 77	Si-P	NF/S, 60V, 0,1A, 0,35W, 180MHz	2a§	EUR	BC 212, BC 556, 2N2906A...2907A, ++	data
BCY 77 P	Si-P	=BCY 77:	40e	Fer	→BCY 77	data
BCY 78	Si-P	=BCY 77: 32V	2a§	EUR	BC 213, BC 558, BCX 78, 2N2906...2907, ++	data
BCY 78 P	Si-P	=BCY 78:	40e	Fer	→BCY 78	data

Type	Device	Short Description	Fig.	Manu	Comparision Types	More at
BCY 79	Si-P	=BCY 77: 45V	2a§	EUR	BC 213, BC 557, BCX 79, 2N2906..2907, ++	data
BCY 79 P	Si-P	=BCY 79:	40e	Fer	→BCY 79	data
BCY 85	Si-N	NF/S, 100V, 0,2A, 0,3W, >200MHz	7a	Tix	2SC2240, 2SC2459, 2SC3245	data
BCY 86	Si-N	=BCY 85: 80V	7a	Tix	BC 546, 2SC2240, 2SC2459, 2SC3245	data
BCY 87	Si-N	Dual, ra, 45V, 30mA, 0,15W, >50MHz	81br	Nsc,Phi,Tix	-	data
BCY 88	Si-N	Dual, ra, 45V, 30mA, 0,15W, >50MHz	81br	Nsc,Phi,Tix	-	data
BCY 89	Si-N	Dual, ra, 45V, 30mA, 0,15W, >50MHz	81br	Nsc,Phi,Tix	-	data
BCY 90(B)	Si-P	NF/S, 40V, 0,05A, 0,35W, 15MHz B: Iso (Fig. 2a)	2a§	Tag	BC 212, BC 257, BC 307, BC 557, ++	data
BCY 91(B)	Si-P	NF/S, 40V, 0,05A, 0,35W, 15MHz	2a	Tag	BC 212, BC 257, BC 307, BC 557, ++	data
BCY 92(B)	Si-P	NF/S, 40V, 0,05A, 0,35W, 15MHz	2a	Tag	BC 212, BC 257, BC 307, BC 557, ++	data
BCY 93(B)	Si-P	NF/S, 70V, 0,05A, 0,35W, 15MHz	2a	Tag	BC 212, BC 256, BC 266, BC 556, ++	data
BCY 94(B)	Si-P	NF/S, 70V, 0,05A, 0,35W, 15MHz	2a	Tag	BC 212, BC 256, BC 266, BC 556, ++	data
BCY 95(B)	Si-P	NF/S, 70V, 0,05A, 0,35W, 15MHz	2a	Tag	BC 212, BC 256, BC 266, BC 556, ++	data
BCY 96(B)	Si-P	NF/S, 90V, 0,05A, 0,35W, 15MHz	2a	Tag	BC 556, 2SA893, 2SA1016, 2SA1038, ++	data
BCY 97(B)	Si-P	NF/S, 90V, 0,05A, 0,35W, 15MHz	2a	Tag	BC 556, 2SA893, 2SA1016, 2SA1038, ++	data
BCY 98(B)	Si-P	NF/S, 40V, 0,05A, 0,35W, 15MHz	2a	Tag	BC 213, BC 257, BC 307, BC 557, ++	data
BCY 99(B)	Si-P	NF/S, 70V, 0,05A, 0,35W, 15MHz	2a	Tag	BC 212, BC 256, BC 266, BC 556, ++	data
BCZ						
BCZ 10	Si-P	NF, 25V, 0,05A, 0,21W, ß>15	1a	Phi,Mot	BC 213, BC 258, BC 308, BC 558, ++	data
BCZ 11	Si-P	=BCZ 10: ß>25	1a	Phi,Mot	BC 213, BC 258, BC 308, BC 558, ++	data
BCZ 12	Si-P	=BCZ 10: 60V, ß=15	1a	Phi,Mot	BC 212, BC 256, BC 266, BC 556, ++	data
BCZ 13	Si-P	Min, NF, 20V, 10mA, 85mW, ß>15	37a	Phi	(BC 213, BC 258, BC 308, BC 558,++)[6]	data
BCZ 14	Si-P	=BCZ 13: ß>30	37a	Phi	(BC 213, BC 258, BC 308, BC 558,++)[6]	data
BD						
BD	Si-Di	=1SS271 (Typ-Code/Stempel/marking)	35	Tos	→1SS271	data
BD	Si-P	=2SB1121 (Typ-Code/Stempel/marking)	39	Say	→2SB1121	data
BD	Si-P	=2SB1189 (Typ-Code/Stempel/marking)	39	Rhm	→2SB1189	data
BD	Si-N	=BCP 54-16 (Tyn-Code/Stempel/marking)	48	Phi	→BCP 54-16	data
BD	Si-P	=BCW 61D (Typ-Code/Stempel/marking)	35	Mot,Phi,Sie Aeg,Fer	→BCW 61D	
BD	Si-N	=BCX 54-16 (Typ-Code/Stempel/marking)	39	Sie,Val,Mot	→BCX 54-16	data
BD	Z-Di	=BZX 399-C6V2(Typ-Code/Stempel/marking)	71(1,7mm)	Phi	→BZX 399-...	data
BD	Si-N/P+R	=KRX 204U (Typ-Code/Stempel/marking)	46(2mm)	Kec	→KRX 204U	data
BD	Z-Di	=P4 SMA-18 (Typ-Code/Stempel/marking)	71(5x2,5)	Fag	→P4 SMA-18	data
BD	Si-Di	=SC 016-6 (Typ-Code/Stempel/marking)	71(5x2,5)	Fjd	→SC 016-...	data
BD	Si-Di	=ZHCS 400 (Typ-Code/Stempel/marking)	71(1,7mm)	Ztx	→ZHCS 400	data
BD	Z-Di	=SMBJ 12C (Typ-Code/Stempel/marking)	71(5x3,5)	Mop	→SMBJ ...	data
BD/1	Se-Di	Blitzschutz/Lightning Prot., 75V	12	Hfo	-	data
BD 04	Hybrid-IC	Plasma-anz.-treiber/plasma display dr.		Fui	-	
BD 106	Si-N	NF-L, 36V, 2,5A, 11,5W, 100MHz	22a§	Itt	BDW 25, BDX 25, 2SC1398, 2SC3252, ++	data
BD 107	Si-N	=BD 106: 64V	22a§	Itt	BDW 25, BDX 25, 2SC1398, 2SC3252, ++	data
BD 109	Si-N	NF/S-L, 60V, 3A, 18,5W, >30MHz	22a§	Sie	BDW 25, BDX 25, 2SC3252...3253, ++	data
BD 111	Si-N	S-L, TV-VA, 60V, 10A, 15W(Tc=75°)	23a§	Sgs	BD 245, BDY 90..92, 2N3055, 2SC3256, ++	data
BD 111 A	Si-N	=BD 111: 62,5W	23a	Sgs	BD 245, BDY 90..92, 2N3055, 2SC3256, ++	data
BD 112	Si-N	NF-L, 80V, 2A, 15W(Tc=75°), >30MHz	23a§	Sgs	BD 245B, BDY 90..92, 2N3055, 2SC3256, ++	data
BD 113	Si-N	NF-L, 60V, 10A, 15W(Tc=75°), >60MHz	23a§	Sgs	BD 245, BDY 90..92, 2N3055, 2SC3256, ++	data
BD 115	Si-N	NF, Vid, 245V, 0,15A, 0,8W, 145MHz	2a§	Phi,Tho	BF 381, BF 615, BFR 58..59, MPSU 10, ++	data
BD 116	Si-N	NF-L, 80V, 3A, 15W(Tc=75°), >30MHz	23a§	Sgs	BD 245B, BDY 90..92, 2N3055, 2SC3256, ++	data
BD 117	Si-N	NF-L, 100V, 5A, 30W(Tc=50°), 50MHz	23a§	Sgs	BD 245C, BDY 90..91, 2N3055, 2SC2681, ++	data
BD 118	Si-N	NF-L, 80V, 15W(Tc=75°), >30MHz	23a§	Sgs	BD 245B, BDY 90..92, 2N3055, 2SC3256, ++	data
BD 119	Si-N	NF/Vid-L, 300V, 0,4A, 6W(Tc=75°)	22a§	Sgs	2SC782, 2SC1929, 2SC2022, 2SC2354, ++	data
BD 120	Si-N	NF/Vid-L, 150V, 0,15A, 7,5W(Tc=75°)	22a§	Sgs	2SC1505...1507, 2SC1755...1757, 2SC1905,++	data
BD 121	Si-N	NF-L, 60V, 5A, 45W, 85MHz	23a§	Phi	BD 245A, BDY 90..92, 2N3055, 2SC3256, ++	data
BD 123	Si-N	=BD 121: 90V	23a§	Phi	BD 245C, BDY 90..91, 2N3055, 2SC2681, ++	data
BD 124(A)	Si-N	NF-L, 70V, 2A, 10W(Tc=100°), 120MHz	22a§	Phi	BDW 25, BDX 25, 2SC1398, 2SC3252, ++	data
BD 127	Si-N	NF/Vid-L, 300V, 0,5A, 17,5W(Tc=45°)	14h§	Aeg	BF 758, MJE 340, 2SC2899, 2SC3051, ++	data
BD 128	Si-N	=BD 127: 350V	14h§	Aeg	BF 759, MJE 340, 2SC2899, 2SC3051, ++	data
BD 129	Si-N	=BD 127: 400V	14h§	Aeg	2SC2899, 2SC3051, 2SC3425	data
BD 127...129	Si-N	=: 0,25A(ss), 16,5W	22a§	Aeg		
BD 130	Si-N	=2N3055	23a§	Rca,Sgs,Sie	→2N3055	data
BD 130 Y	Si-N	=BD 130: 60V	23a§		→2N3055	data
BD 131(A)	Si-N	NF-L, 70V, 3A, 15W(Tc=60°), >60MHz	14h§	Phi	BD 237, BD 441, BD 787, BD 789, ++	data
BD 132(A)	Si-P	NF-L, 45V, 3A, 15W(Tc=60°), >60MHz	14h§	Phi	BD 238, BD 440, BD 786, BD 788, ++	data
BD 133	Si-N	=BD 131: 90V	14h§	Phi	BD 237, BD 443, BD 791, BDX 35..37, ++	data
BD 134	Si-P	≈BD 238	14h§		→BD 238	
BD 135(H)	Si-N	NF-L, 45V, 1,5A, 12,5W(Tc=45°), >50MHz	14h§	EUR	BD 226, BD 375, BD 785	data
BD 135G	Si-N	=BD 135:	13j§	Gen	2SC1848, 2SC2483	data
BD 136(H)	Si-P	NF-L, 45V, 1,5A, 12,5W(Tc=45°), >50MHz	14h§	EUR	BD 227, BD 376, BD 786	data
BD 136 G	Si-P	=BD 136:	13j§	Gen	2SA887, 2SA1195	data
BD 137(H)	Si-N	=BD 135: 60V	14h§	EUR	BD 228, BD 377, BD 785	data
BD 137 G	Si-N	=BD 137:	13j§	Gen	2SC1848, 2SC2483	data
BD 138(H)	Si-P	=BD 136: 60V	14h§	EUR	BD 229, BD 378, BD 786	data

134

Type	Device	Short Description	Fig.	Manu	Comparision Types	More at
BD 138 G	Si-P	=BD 138:	13j§	Gen	2SA887, 2SA1195	data
BD 139(H)	Si-N	=BD 135: 100V	14h§	EUR	BD 230, BD 379, BD 791	data
BD 139 G	Si-N	=BD 139:	13j§	Gen	2SC2483	data
BD 140(H)	Si-P	=BD 136: 100V	14h§	EUR	BD 231, BD 380, BD 792	data
BD 140 G	Si-P	=BD 140:	13j§	Gen	2SA1195	data
BD 141	Si-N	NF-L, 140V, 8A, 117W, >0,8MHz	23a§	Sgs	BDX 11...12, BDX 51, 2N5634, 2SD551, ++	data
BD 142	Si-N	NF/S-L, 50V, 15A, 117W, 1,3MHz	23a§	Mot,Rca,Tho	BDW 51, BDX 13, BDX 60...61, 2N3055, ++	data
BD 144	Si-N	TV-VA, 800V, 0,25A, 8W(Tc=95°), 12MHz	23a§	Phi	2SC1101, 2SC3151, 2SC3533	data
BD 145	Si-N	TV-HA, 150V, 5A, 15W(Tc=100°), 100MHz	23a§	Phi	BU 104, BU 606...608	data
BD 148	Si-N	NF/S-L, 60V, 4A, 31W(Tc=45°), 1MHz	22a§	Sie	BD 243A...C, BD 543A...D, 2N3054, ++	data
BD 149	Si-N	=BD 148: 80V	22a§	Sie	BD 243B...C, BD 543B...D, 2N3054, ++	data
BD 150(A...C)	Si-N	Vid, 200...300V, 0,5A, 1W, 160MHz	2a§	Sgs	BF 382, BF 758...759, MPSU 10, ++	data
BD 151	Si-P	NF-L, 35V, 1A, 20W	14h§	Mot	BD 136, BD 227, BD 376, BD 786, ++	data
BD 152	Si-P	=BD 151: 50V	14h§	Mot	BD 138, BD 229, BD 378, BD 788, ++	data
BD 153	Si-N	NF-L, 35V, 1A, 20W	14h§	Mot	BD 135, BD 226, BD 375, BD 785, ++	data
BD 154	Si-N	=BD 153: 50V	14h§	Mot	BD 137, BD 228, BD 375, BD 785, ++	data
BD 155	Si-N	=BD 153: 70V	14h§	Mot	BD 139, BD 230, BD 377, BD 787, ++	data
BD 156	Si-P	=BD 151: 70V	14h§	Mot	BD 140, BD 231, BD 378, BD 790, ++	data
BD 157	Si-N	NF/Vid-L, 275V, 0,5A, 20W	14h§	Mot,Nsc,Tho	BF 758...759, MJE 340, 2SC2899, 2SC3051	data
BD 158	Si-N	=BD 157: 325V	14h§	Mot,Nsc,Tho	BF 759, MJE 340, 2SC2899, 2SC3051	data
BD 159	Si-N	=BD 157: 375V	14h§	Mot,Nsc,Tho	BF 759, MJE 340, 2SC2899, 2SC3051	data
BD 160	Si-N	TV-HA, 250V, 5A, 25W	23a§	Phi	BU 104, BU 606...608	data
BD 161	Si-N	NF/S-L, 90V, 4A, 19W(Tc=85°), 1,75MHz	22a§	Sgs	BD 243C, BD 543C...D, BD 953, 2N3054, ++	data
BD 162	Si-N	=BD 161: 40V	22a§	Sgs	BD 243, BD 543, BD 947, 2N3054, ++	data
BD 163	Si-N	=BD 161: 60V	22a§	Sgs	BD 243A...C, BD 543A...D, BD 949, 2N3054	data
BD 165	Si-N	NF-L, 45V, 1,5A, 20W, >3MHz	14h§	EUR	BD 175, BD 226, BD 233, BD 437, ++	data
BD 166	Si-P	NF-L, 45V, 1,5A, 20W, >3MHz	14h§	EUR	BD 176, BD 227, BD 234, BD 438, ++	data
BD 167	Si-N	=BD 165: 60V	14h§	EUR	BD 177, BD 228, BD 235, BD 439, ++	data
BD 168	Si-P	=BD 166: 60V	14h§	EUR	BD 178, BD 229, BD 236, BD 440, ++	data
BD 169	Si-N	=BD 165: 80V	14h§	EUR	BD 179, BD 230, BD 237, BD 441, ++	data
BD 170	Si-P	=BD 166: 80V	14h§	EUR	BD 180, BD 231, BD 238, BD 442, ++	data
BD 171	Si-N	NF-L, 100V, 0,5A, 20W, 6MHz	14h§	Aeg,Mot	BD 139, BD 237, MJE 340, 2SD1382, ++	data
BD 172	Si-N	=BD 171: 130V	14h§	Aeg,Mot	BD 524, MJE 340, 2SC2481, 2SC3117, ++	data
BD 173	Si-N	=BD 171: 170V	14h§	Aeg,Mot	BD 524, MJE 340, 2SC3117, 2SD669, ++	data
BD 175	Si-N	NF-L, 45V, 3A, 30W, >3MHz	14h§	Aeg,Mot,++	BD 185, BD 437, BD 785, 2N5190...5192, ++	data
BD 176	Si-P	NF-L, 45V, 3A, 30W, >3MHz	14h§	Aeg,Mot,++	BD 186, BD 438, BD 786, 2N5193...5195, ++	data
BD 177	Si-N	=BD 175: 60V	14h§	Aeg,Mot,++	BD 187, BD 439, BD 787, 2N5131...5192, ++	data
BD 178	Si-P	=BD 176: 60V	14h§	Aeg,Mot,++	BD 188, BD 440, BD 788, 2N5194...5195, ++	data
BD 179	Si-N	=BD 175: 80V	14h§	Aeg,Mot,++	BD 189, BD 441, BD 791, 2N5192, ++	data
BD 180	Si-P	=BD 176: 80V	14h§	Aeg,Mot,++	BD 190, BD 442, BD 792, 2N5195, ++	data
BD 181	Si-N	NF-L, 55V, 10A, 117W	23a§	Phi,Rca,Tho	BD 245, BD 311, BDY 39, 2N3055, ++	data
BD 182	Si-N	=BD 181: 70V, 15A	23a§	Phi,Rca,Tho	BD 745A, BD 315, BDY 39, 2N3055, ++	data
BD 183	Si-N	=BD 181: 85V, 15A	23a§	Phi,Rca,Tho	BD 745B, BD 317, BDY 39, 2N3055, ++	data
BD 184	Si-N	=BD 181: 95V, 15A	23a§	Phi,Rca,Tho	BD 745C, BD 317, BDY 39, 2N3055, ++	data
BD 185	Si-N	NF-L, 40V, 4A, 40W, >2MHz	14h§	Aeg,Mot	BD 437, BD 785, 2N5190	data
BD 186	Si-P	NF-L, 40V, 4A, 40W, >2MHz	14h§	Aeg,Mot	BD 438, BD 786, 2N5193	data
BD 187	Si-N	=BD 185: 55V	14h§	Aeg,Mot	BD 439, BD 785, 2N5191	data
BD 188	Si-P	=BD 186: 55V	14h§	Aeg,Mot	BD 440, BD 786, 2N5194	data
BD 189	Si-N	=BD 185: 70V	14h§	Aeg,Mot	BD 441, BD 787, 2N5192	data
BD 190	Si-P	=BD 186: 70V	14h§	Aeg,Mot	BD 442, BD 788, 2N5195	data
BD 191	Si-N	NF/S-L, 100V, 15A, 37,5W, >0,8MHz	22a§	Sgs	BD 743C, BD 911	data
BD 192	Si-N	=BD 191: 50V	22a§	Sgs	BD 743A, BD 907	data
BD 193	Si-N	=BD 191: 140V, 8A	22a§	Sgs	MJE 15030, 2SC4330, 2SD866	data
BD 195	Si-N	NF-L, 40V, 6A, 65W, >2MHz	16h§	Mot	BD 205, MJE 3055, 2N5983, 2SD499	data
BD 196	Si-P	NF-L, 40V, 6A, 65W, >2MHz	16h§	Mot	BD 206, MJE 2955, 2N5980, 2SB578	data
BD 197	Si-N	=BD 195: 55V	16h§	Mot	BD 205, MJE 3055, 2N5983, 2SD499	data
BD 198	Si-P	=BD 196: 55V	16h§	Mot	BD 206, MJE 2955, 2N5980, 2SB578	data
BD 199	Si-N	=BD 195: 70V	16h§	Mot	BD 207, MJE 3055, 2N5984, 2SD500	data
BD 200	Si-P	=BD 196: 70V	16h§	Mot	BD 208, MJE 2955, 2N5981, 2SB578	data
BD 201	Si-N	NF-L, 60/45V, 8A, 60W, >7MHz	17j§	Aeg,Phi,Rca	BD 243A, BD 543A, BD 707, BD 797, ++	data
BD 201 F	Si-N	=BD 201: Iso, >20W	17c	Phi	BDT 91F, BDX 77F, 2SD1411...12, 2SD1668++	data
BD 202	Si-P	NF-L, 60/45V, 8A, 60W, >7MHz	17j§	Aeg,Phi,Rca	BD 244A, BD 544A, BD 708, BD 798, ++	data
BD 202 F	Si-P	=BD 202: Iso, >20W	17c	Phi	BDT 92F, BDX 78F, 2SB1018...19, 2SB1135++	data
BD 203	Si-N	=BD 201: 60/60V	17j§	Aeg,Phi,Rca	BD 243A, BD 543A, BD 707, BD 797, ++	data
BD 203 F	Si-N	=BD 203: Iso, >20W	17c	Phi	BDT 91F, BDX 77F, 2SD1411...12, 2SD1668++	data
BD 204	Si-P	=BD 202: 60/60V	17j§	Aeg,Phi,Rca	BD 244A, BD 544A, BD 708, BD 798, ++	data
BD 204 F	Si-P	=BD 202: Iso, >20W	17c	Phi	BDT 92F, BDX 78F, 2SB1018...19, 2SB1135++	data
BD 205	Si-N	NF-L, 55V, 10A, 90W, 4MHz	16h§	Mot	MJE 3055, 2N5989, 2SD491	data
BD 206	Si-P	NF-L, 55V, 10A, 90W, 4MHz	16h§	Mot	MJE 2955, 2N5986, 2SB578	data
BD 207	Si-N	=BD 205: 70V	16h§	Mot	MJE 3055, 2N5990, 2SD491	data
BD 208	Si-P	=BD 206: 70V	16h§	Mot	MJE 2955, 2N5987, 2SB578	data
BD 213/45...80	Si-N	NF-L, 45...80V, 15A, 90W, >3MHz	18j§	Aeg	BD 245(A...C), BD 249(A...C), BD 745(A...C)	data
BD 214/45...80	Si-P	NF-L, 45...80V, 15A, 90W, >3MHz	18j§	Aeg	BD 246(A...C), BD 250(A...C), BD 746(A...C)	data
BD 215	Si-N	NF-L, 500/300V, 0,5A, 21,5W, 10MHz	22a§	Sgs	MJ4360...61, 2N4298...99, 2SC1078, 2SC1810	data
BD 216	Si-N	NF-L, TV-VA, 300V, 1A, 21,5W, 10MHz	22a§	Sgs	TIP47...48, 2N4296...97, 2SC867, 2SC2022...23	data
BD 220	Si-N	NF-L, 80/70V, 4A, 36W, >0,8MHz	17j§	Fch,Mic	BD 243B, BD 537, BD 539B, BD 951, ++	data

Type	Device	Short Description	Fig.	Manu	Comparision Types	More at ·
BD 221	Si-N	=BD 220: 60/40V	17j§	Fch,Mic	BD 243A, BD 535, BD 539A, BD 949, ++	data
BD 222	Si-N	=BD 220: 80/60V	17j§	Fch,Mic	BD 243B, BD 537, BD 539B, BD 951, ++	data
BD 223	Si-P	NF-L, 80/70V, 4A, 36W, >0,8MHz	17j§	Fch,Mic	BD 244B, BD 538, BD 540B, BD 952, ++	data
BD 224	Si-P	=BD 223: 60/40V	17j§	Fch,Mic	BD 244A, BD 536, BD 540A, BD 950, ++	data
BD 225	Si-P	=BD 223: 80/60V	17j§	Fch,Mic	BD 244B, BD 538, BD 540B, BD 952, ++	data
BD 226	Si-N	NF-L, 45V, 1,5A, 12,5W(Tc=62°), 125MHz	14h§	Phi	BD 135, BD 375, BD 785	data
BD 227	Si-P	NF-L, 45V, 1,5A, 12,5W(Tc=62°), 50MHz	14h§	Phi	BD 136, BD 376, BD 786	data
BD 228	Si-N	=BD 226: 60V	14h§	Phi	BD 137, BD 377, BD 785	data
BD 229	Si-P	=BD 227: 60V	14h§	Phi	BD 138, BD 378, BD 786	data
BD 230	Si-N	=BD 226: 100V	14h§	Phi	BD 139, BD 379, BD 791	data
BD 231	Si-P	=BD 227: 100V	14h§	Phi	BD 140, BD 380, BD 792	data
BD 232	Si-N	TV-HA-Tr, 500/300V, 0,25A, 15W, 20MHz	14h§	Phi,Mot	MJE 3439, 2SC2899, 2SC3051, 2SC3425	data
BD 232 G	Si-N	=BD 232: 15W(Tc=25°)	13j§	Gen	(MJE 3439, 2SC2899, 2SC3051, 2SC3425)[5]	data
BD 233	Si-N	NF-L, 45V, 2A, 25W, >3MHz	14h§	EUR	BD 175, BD 375, BD 437	data
BD 233 G	Si-N	=BD 233: 4A, 30W	17j§	Gen	BD 243, BD 533, BD 539, BD 947, ++	data
BD 234	Si-P	NF-L, 45V, 2A, 25W, >3MHz	14h§	EUR	BD 176, BD 376, BD 438	data
BD 234 G	Si-P	=BD 234: 4A, 30W	17j§	Gen	BD 244, BD 534, BD 540, BD 948, ++	data
BD 235	Si-N	=BD 233: 60V	14h§	EUR	BD 177, BD 377, BD 439	data
BD 235 G	Si-N	=BD 235: 4A, 30W	17j§	Gen	BD 243A, BD 535, BD 539A, BD 949, ++	data
BD 236	Si-P	=BD 234: 60V	14h§	EUR	BD 178, BD 378, BD 440	data
BD 236 G	Si-P	=BD 236: 4A, 30W	17j§	Gen	BD 244A, BD 536, BD 540A, BD 950, ++	data
BD 237	Si-N	=BD 233: 100V	14h§	EUR	BD 379, BD 443, 2SD1177...1178	data
BD 237 G	Si-N	=BD 237: 4A, 30W	17j§	Gen	BD 243C, BD 539C, BD 953, 2SD712, ++	data
BD 238	Si-P	=BD 234: 100V	14h§	EUR	ÞD 380, 2SB874...875	data
BD 238 G	Si-P	=BD 238: 4A, 30W	17j§	Gen	BD 244C, BD 540C, BD 954, 2SB682, ++	data
BD 239	Si-N	NF-L, 55V, 2A, 30W, >3MHz	17j§	EUR	BD 241, BD 243, BD 539A, BD 935, ++	data
BD 239 A	Si-N	=BD 239: 70V	17j§		BD 241A, BD 243A, BD 539B, BD 937, ++	data
BD 239 B	Si-N	=BD 239: 90V	17j§		BD 241B, BD 243B, BD 539C, BD 937, ++	data
BD 239 C	Si-N	=BD 239: 115V	17j§		BD 241C, BD 243C, BD 539D, BD 939, ++	data
BD 239 D	Si-N	=BD 239: 160V	17j§		BD 241D, BD 243D, 2SD759...760, 2SC2529	data
BD 239 E	Si-N	=BD 239: 180V	17j§		BD 241E, BD 243E, 2SD760, 2SC2660	data
BD 239 F	Si-N	=BD 239: 200V	17j§		BD 241F, BD 243F, 2SD760, 2SC2660	data
BD 240	Si-P	NF-L, 55V, 2A, 30W, >3MHz	17j§	EUR	BD 242, BD 244, BD 540A, BD 936, ++	data
BD 240 A	Si-P	=BD 240: 70V	17j§		BD 242A, BD 244A, BD 540B, BD 938, ++	data
BD 240 B	Si-P	=BD 240: 90V	17j§		BD 242B, BD 244B, BD 540C, BD 938, ++	data
BD 240 C	Si-P	=BD 240: 115V	17j§		BD 242C, BD 244C, BD 540D, BD 940, ++	data
BD 240 D	Si-P	=BD 240: 160V	17j§		BD 242D, BD 244D, 2SB719...720, 2SA1079	data
BD 240 E	Si-P	=BD 240: 180V	17j§		BD 242E, BD 244E, 2SB720, 2SA1133	data
BD 240 F	Si-P	=BD 240: 200V	17j§		BD 242F, BD 244F, 2SB720, 2SA1133	data
BD 241	Si-N	NF-L, 55V, 3A, 40W, >3MHz	17j§	EUR	BD 243, BD 539A, BD 543A, BD 935, ++	data
BD 241 A	Si-N	=BD 241: 70V	17j§		BD 243A, BD 539B, BD 543B, BD 937, ++	data
BD 241 B	Si-N	=BD 241: 90V	17j§		BD 243B, BD 539C, BD 543C, BD 937, ++	data
BD 241 BFI	Si-N	=BD 241B: Iso, 18W	17c	Tho	BD 937F, BDT 31AF, 2SC3851A, 2SD1586, ++	data
BD 241 BFP	Si-N	=BD 241B: Iso, 24W	17c	Tho	BD 937F, BDT 31AF, 2SC3851A, 2SD1586, ++	data
BD 241 C	Si-N	=BD 241: 115V	17j§		BD 243C, BD 539D, BD 543D, BD 939, ++	data
BD 241 D	Si-N	=BD 241: 160V	17j§		BD 243D, 2SD772A,B	data
BD 241 E	Si-N	=BD 241: 180V	17j§		BD 243E, 2SD772A,B	data
BD 241 F	Si-N	=BD 241: 200V	17j§		BD 243F, 2SD772A,B	data
BD 242	Si-P	NF-L, 55V, 3A, 40W, >3MHz	17j§	EUR	BD 244, BD 540A, BD 544A, BD 936, ++	data
BD 242 A	Si-P	=BD 242: 70V	17j§		BD 244A, BD 540B, BD 544B, BD 938, ++	data
BD 242 B	Si-P	=BD 242: 90V	17j§		BD 244B, BD 540C, BD 544C, BD 938, ++	data
BD 242 BFI	Si-P	=BD 242B: Iso, 18W	17c	Tho	BD 938F, BDT 32AF, 2SA1488A, 2SB1095, ++	data
BD 242 BFP	Si-P	=BD 242B: Iso, 24W	17c	Tho	BD 938F, BDT 32AF, 2SA1488A, 2SB1095, ++	data
BD 242 C	Si-P	=BD 242: 115V	17j§		BD 244C, BD 540D, BD 544D, BD 940, ++	data
BD 242 D	Si-P	=BD 242: 160V	17j§		BD 244D	data
BD 242 E	Si-P	=BD 242: 180V	17j§		BD 244E	data
BD 242 F	Si-P	=BD 242: 200V	17j§		BD 244F	data
BD 243	Si-N	NF-L, 55V, 6A, 65W, >3MHz	17j§	EUR	BD 543A, BD 797, BD 805	data
BD 243 A	Si-N	=BD 243: 70V	17j§		BD 543B, BD 799, BD 807	data
BD 243 B	Si-N	=BD 243: 90V	17j§		BD 543C, BD 801, BD 809	data
BD 243 C	Si-N	=BD 243: 115V	17j§		BD 543D, BD 801, 2SD866	data
BD 243 D	Si-N	=BD 243: 160V	17j§		BD 743F	data
BD 243 E	Si-N	=BD 243: 180V	17j§		-	data
BD 243 F	Si-N	=BD 243: 200V	17j§			data
BD 244	Si-P	NF-L, 55V, 6A, 65W, >3MHz	17j§	EUR	BD 544A, BD 798, BD 808	data
BD 244 A	Si-P	=BD 244: 70V	17j§		BD 544B, BD 800, BD 810	data
BD 244 B	Si-P	=BD 244: 90V	17j§		BD 544C, BD 802, BD 810	data
BD 244 C	Si-P	=BD 244: 115V	17j§		BD 544D, BD 802, 2SB870	data
BD 244 D	Si-P	=BD 244: 160V	17j§		BD 744F	data
BD 244 E	Si-P	=BD 244: 180V	17j§		-	data
BD 244 F	Si-P	=BD 244: 200V	17j§		-	data
BD 245	Si-N	NF-L, 55V, 10A, 80W, >3MHz	18j§	Tix	BD 249, BD 745A, BDV 91	data
BD 245 A	Si-N	=BD 245: 70V	18j§		BD 249A, BD 745B, BDV 93	data
BD 245 B	Si-N	=BD 245: 90V	18j§		BD 249B, BD 745C, BDV 95	data
BD 245 C	Si-N	=BD 245: 115V	18j§		BD 249C, BD 745D, 2SD1047	data
BD 245 D	Si-N	=BD 245: 160V	18j§		BD 249D, BD 745F, 2SC3263, 2SD1047	data
BD 245 E	Si-N	=BD 245: 180V	18j§		BD 249E, 2SC3263	data
BD 245 F	Si-N	=BD 245: 200V	18j§		BD 249F, 2SC3263	data
BD 246	Si-P	NF-L, 55V, 10A, 80W, >3MHz	18j§	Tix	BD 250, BD 746A, BDV 92	data
BD 246 A	Si-P	=BD 246: 70V	18j§		BD 250A, BD 746B, BDV 94	data
BD 246 B	Si-P	=BD 246: 90V	18j§		BD 250B, BD 746C, BDV 96	data
BD 246 C	Si-P	=BD 246: 115V	18j§		BD 250C, BD 746D, 2SB817	data

Type	Device	Short Description ·	Fig.	Manu	Comparision Types	More at
BD 246 D	Si-P	=BD 246: 160V	18j§		BD 250D, BD 746F, 2SA1294, 2SB817	data
BD 246 E	Si-P	=BD 246: 180V	18j§		BD 250E, 2SA1294	data
BD 246 F	Si-P	=BD 246: 200V	18j§		BD 250F, 2SA1294	data
BD 249	Si-N	NF-L, 55V, 25A, 125W, >3MHz	18j§	Mot,Tho,Tix	2SD1049	data
BD 249 A	Si-N	=BD 249: 70V	18j§		2SD1049	data
BD 249 B	Si-N	=BD 249: 90V	18j§		2SD1049	data
BD 249 C	Si-N	=BD 249: 115V	18j§		2SD1049	data
BD 249 D	Si-N	=BD 249: 160V	18j§		-	data
BD 249 E	Si-N	=BD 249: 180V	18j§		-	data
BD 249 F	Si-N	=BD 249: 200V	18j§		-	data
BD 250	Si-P	NF-L, 55V, 25A, 125W, >3MHz	18j§	Mot,Tho,Tix	-	data
BD 250 A	Si-P	=BD 250: 70V	18j§		-	data
BD 250 B	Si-P	=BD 250: 90V	18j§		-	data
BD 250 C	Si-P	=BD 250: 115V	18j§		-	data
BD 250 D	Si-P	=BD 250: 160V	18j§		-	data
BD 250 E	Si-P	=BD 250: 180V	18j§		-	data
BD 250 F	Si-P	=BD 250: 200V	18j§		-	data
BD 251	Si-N	NF-L, 40V, 3A, 20W(Tc=50°), 46MHz	23a§	Sgs	BD 245, BDY90...92, 2N3055, 2SC3256, ++	data
BD 253	Si-N	S-L, 350V, 4A, 100W, 25MHz	23a§	Tix	BU 526, BUW 71, BUX 16C, BUY 67, ++	data
BD 253 A	Si-N	=BD 253: 500V	23a§		BU 526, BUS 11, BUX 45, BUX 82, ++	data
BD 253 B	Si-N	=BD 253: 700V	23a§		BU 526, BUS 11, BUX 82, BUX 97, ++	data
BD 253 C	Si-N	=BD 253: 900V	23a§		BU 526, BUS 11A, BUX 83, 2SD802, ++	data
BD 254	Si-N	NF-L, 60V, 3A, 18,5W, 30MHz	22a§	Tho	BD 241A, BD 243A, BD 535, BD 935, ++	data
BD 255	Si-P	NF-L, 60V, 3A, 18,5W, 30MHz	22a§	Tho	BD 242A, BD 244A, BD 536, BD 936, ++	data
BD 257/45...100	Si-N	NF-L, 45...100V, 25A(ss), 125W, >3MHz	18j§	Aeg	BD 249(A...C)	data
BD 258/45...100	Si-P	NF-L, 45...100V, 25A(ss), 125W, >3MHz	18j§	Aeg	BD 250(A...C)	data
BD 260	Si-N	NF/S-L, 200V, 2A, 30W, >10MHz	22a§	Sgs	BUX 67, 2N3583, 2SC2023, 2SD610, ++	data
BD 261	Si-N	NF/S-L, 300V, 5A, 30W, >10MHz	22a§	Sgs	BUX 63, MJE 51T, 2N6234...35, 2SC2907, ++	data
BD 262	Si-P-Darl	NF-L, 60V, 4A, 36W, 7MHz, B>750	14h§	Phi	BD 678, BD 778, 2N6035	data
BD 262 A	Si-P-Darl	=BD 262: 80V	14h§		BD 680, BD 780, 2N6036	data
BD 262 B	Si-P-Darl	=BD 262: 100V	14h§		BD 682	data
BD 262 C	Si-P-Darl	=BD 262: 120V	14h§		BD 684	data
BD 262 L	Si-P-Darl	=BD 262: 45V	14h§		BD 678, BD 778, 2N6035	data
BD 263	Si-N-Darl	NF-L, 80V, 4A, 36W, 7MHz, B>750	14h§	Phi	BD 679, BD 779, 2N6039	data
BD 263 A	Si-N-Darl	=BD 263: 100V	14h§		BD 681	data
BD 263 B	Si-N-Darl	=BD 263: 120V	14h§		BD 683	data
BD 263 C	Si-N-Darl	=BD 263: 140V	14h§		-	data
BD 263 L	Si-N-Darl	=BD 263: 60V	14h§		BD 677, BD 777, 2N6038	data
BD 264	Si-P-Darl	NF-L, 60V, 4A, 40W, 7MHz, B>1000	17j§	Phi	BD 646, BD 898, BDW 24A, BDW 54A, ++	data
BD 264 A	Si-P-Darl	=BD 264: 80V	17j§		BD 648, BD 900, BDW 24B, BDW 54B, ++	data
BD 264 B	Si-P-Darl	=BD 264: 100V	17j§		BD 650, BD 902, BDW 24C, BDW 54C, ++	data
BD 264 L	Si-P-Darl	=BD 264: 45V	17j§		BD 644, BD 896, BDW 24, BDW 54, ++	data
BD 265	Si-N-Darl	NF-L, 80V, 4A, 40W, 7MHz, B>1000	17j§	Phi	BD 647, BD 899, BDW 23B, BDW 53B, ++	data
BD 265 A	Si-N-Darl	=BD 265: 100V	17j§		BD 649, BD 901, BDW 23C, BDW 53C, ++	data
BD 265 B	Si-N-Darl	=BD 265: 120V	17j§		BD 651, BDT 21, BDW 63D, 2SD1147	data
BD 265 L	Si-N-Darl	=BD 265: 60V	17j§		BD 645, BD 897, BDW 23A, BDW 53A, ++	data
BD 266	Si-P-Darl	NF-L, 60V, 8A, 60W, 7MHz, B>750	17j§	Phi	BD 646, BD 898, BDW 24A, BDW 64A, ++	data
BD 266 A	Si-P-Darl	=BD 266: 80V	17j§		BD 648, BD 900, BDW 24B, BDW 64B, ++	data
BD 266 B	Si-P-Darl	=BD 266: 100V	17j§		BD 650, BD 902, BDW 24C, BDW 64C, ++	data
BD 266 L	Si-P-Darl	=BD 266: 45V	17j§		BD 644, BD 896, BDW 24, BDW 64, ++	data
BD 267	Si-N-Darl	NF-L, 80V, 8A, 60W, 7MHz, B>750	17j§	Phi	BD 647, BD 899, BDW 23B, BDW 63B, ++	data
BD 267 A	Si-N-Darl	=BD 267: 100V	17j§		BD 649, BD 901, BDW 23C, BDW 63C, ++	data
BD 267 B	Si-N-Darl	=BD 267: 120V	17j§		BD 651, BDT 21, BDW 63D, 2SD1386	data
BD 267 L	Si-N-Darl	=BD 267: 60V	17j§		BD 645, BD 897, BDW 23A, BDW 63A, ++	data
BD 268	Si-P-Darl	NF-L, 80V, 10A, 75W	17j§	Phi	BDT 62A, BDT 64A, BDW 94B, BDX 34B	data
BD 268 A	Si-P-Darl	=BD 268: 100V	17j§		BDT 62B, BDT 64B, BDW 94C, BDX 34C	data
BD 269	Si-N-Darl	NF-L, 80V, 10A, 75W	17j§	Phi	BDT 63A, BDT 65A, BDW 93B, BDX 33B	data
BD 269 A	Si-N-Darl	=BD 269: 100V	17j§		BDT 63B, BDT 65B, BDW 93C, BDX 33C	data
BD 271	Si-N	NF-L, 55V, 4A, 36W, >2MHz	17j§	Phi	BD 243, BD 535, BD 539A, BD 949, ++	data
BD 272	Si-P	NF-L, 55V, 4A, 36W, >2MHz	17j§	Phi	BD 244, BD 536, BD 540A, BD 950, ++	data
BD 273	Si-N	=BD 271: 80V	17j§	Phi	BD 243B, BD 537, BD 539B, BD 951, ++	data
BD 274	Si-P	=BD 272: 80V	17j§	Phi	BD 244B, BD 538, BD 540B, BD 951, ++	data
BD 275	Si-N	=BD 271: 100V	17j§	Phi	BD 243C, BD 539C, BD 953, 2SD712, ++	data
BD 276	Si-P	=BD 272: 100V	17j§	Phi	BD 244C, BD 540C, BD 954, 2SB682, ++	data
BD 277	Si-P	NF/S-L, 45V, 7A, 70W, >10MHz	17j§	Phi,Rca	BD 244, BD 544A, BD 796, BD 806, ++	data
BD 278(A,AE)	Si-N	NF/S-L, 55V, 10A, 75W, >0,8MHz	17j§	Phi	BD 707, BD 805, BD 907	data
BD 279	Si-N-Darl	NF-L, -/40V, 2A, 10W, 75MHz, B>10000	13j§	Gen	-	data
BD 280	Si-P-Darl	NF-L, -/40V, 2A, 10W, 100MHz, B>10000	13j§	Gen	-	data
BD 281	Si-N	NF/S-L, 22V, 4A, 36W, >3MHz	14h§	Sgs	BD 185, BD 433, 2N5190	data
BD 282	Si-P	NF/S-L, 22V, 4A, 36W, >3MHz	14h§	Sgs	BD 186, BD 434, 2N5193	data
BD 283	Si-N	=BD 281: 32V	14h§	Sgs	BD 185, BD 435, 2N5190	data
BD 284	Si-P	=BD 282: 32V	14h§	Sgs	BD 186, BD 436, 2N5193	data
BD 285	Si-N	=BD 281: 45V	14h§	Sgs	BD 187, BD 437, 2N5191	data
BD 286	Si-P	=BD 282: 45V	14h§	Sgs	BD 188, BD 440, 2N5194	data
BD 287	Si-P	S-L, 30V, 12A, 36W, >50MHz	14h§	Sie	-	data
BD 288	Si-P	=BD 287: 45V	14h§	Sie	-	data
BD 291	Si-N	NF-L, 45V, 6A, 60W, >3MHz	≈14j§	Phi	(BD 243, BD 543A, BD 795, BD 805)[4]	data
BD 292	Si-P	NF-L, 45V, 6A, 60W, >3MHz	≈14j§	Phi	(BD 244, BD 544A, BD 798, BD 806)[4]	data

Type	Device	Short Description	Fig.	Manu	Comparision Types	More at
BD 293	Si-N	=BD 291: 60V	≈14j§	Phi	(BD 243A, BD 543A, BD 797, BD 807)[4]	data
BD 294	Si-P	=BD 292: 60V	≈14j§	Phi	(BD 244A, BD 544A, BD 800, BD 808)[4]	data
BD 295	Si-N	=BD 291: 80V	≈14j§	Phi	(BD 243B, BD 543B, BD 799, BD 809)[4]	data
BD 296	Si-P	=BD 292: 80V	≈14j§	Phi	(BD 244B, BD 544B, BD 800, BD 810)[4]	data
BD 301	Si-N	NF-L, TV-VA, 60V, 8A, 55W, >3MHz	17j§	Tho	BD 543A, BD 709, BD 797, BD 807	data
BD.302	Si-P	NF-L, TV-VA, 60V, 8A, 55W, >3MHz	17j§	Tho	BD 544A, BD 710, BD 798, BD 808	data
BD 303	Si-N	NF-L, TV-VA, 60V, 8A, 55W, >3MHz	17j§	Tho	BD 543A, BD 707, BD 797, BD 807	data
BD 303 A	Si-N	=BD 303: 80V	17j§		BD 543B, BD 709, BD 799, BD 809	data
BD 303 B	Si-N	=BD 303: 100V	17j§		BD 543C, BD 711, BD 801	data
BD 304	Si-P	NF-L, TV-VA, 60V, 8A, 55W, >3MHz	17j§	Tho	BD 544A, BD 708, BD 798, BD 806	data
BD 304 A	Si-P	=BD 304: 80V	17j§		BD 544B, BD 710, BD 800, BD 810	data
BD 304 B	Si-P	=BD 304: 100V	17j§		BD 544C, BD 712, BD 802	data
BD 306(A,B)	Si-N	NF-L, 36V, 2,5A, 10W, 100MHz	14h§	Itt	BD 233, BD 785, 2SC2877	data
BD 307(A,B)	Si-N	=BD 306: 64V	14h§	Itt	BD 235, BD 785, 2SD794(A)	data
BD 311	Si-N	NF-L, 60V, 10A, 150W, >4MHz	23a§	Mot,Phi	BD 315, BDW 21A, BDW 51A, 2N5877, ++	data
BD 312	Si-P	NF-L, 60V, 10A, 150W, >3MHz	23a§	Mot,Phi	BD 316, BDW 22A, BDW 52A, 2N5875, ++	data
BD 313	Si-N	=BD 311: 80V	23a§	Mot,Phi	BD 315, BDW 21B, BDW 51B, 2N5878, ++	data
BD 314	Si-P	=BD 312: 80V	23a§	Mot,Phi	BD 316, BDW 22B, BDW 52B, 2N5880, ++	data
BD 315	Si-N	NF-L, 80V, 16A, 200W, >1MHz	23a§	Mot	2N5629..5631	data
BD 316	Si-P	NF-L, 80V, 16A, 200W, >1MHz	23a§	Mot	2N6029..6031	data
BD 317	Si-N	=BD 315: 100V	23a§	Mot	2N5629..5631	data
BD 318	Si-P	=BD 316: 100V	23a§	Mot	2N6029..6031	data
BD 320(A,B,C)	Si-N-Darl	80V, 1A, 5W(Tc=25°), 80MHz, B>500	2a§	Fer	BSS 51..52, (BD 877, BDX 43)[4]	data
BD 321(A,B,C)	Si-N-Darl	=BD 320: 2A	2a§	Fer	2SC1888, 2SD406, 2SD614..615, 2SD688	data
BD 322(A,B,C)	Si-N-Darl	80V, 1A, 7,5W(Tc=25°), 80MHz, B>500	2a§	Fer	BSS 51..52, (BD 877, BDX 43)[4]	data
BD 323(A,B,C)	Si-N-Darl	=BD 322: 2A	2a§	Fer	2SC1888, 2SD406, 2SD614..615, 2SD688	data
BD 328	Si-N	5x Darl, 80V, 2A, 0,6W, 80MHz, B>1000	14-DIP	Fer		data
BD 329	Si-N	NF-L, 32V, 3A, 15W(Tc=45°), 130MHz	14h§	Phi,Sie	2SD882, 2SD1348, (BD 785, 2SD794(A))[17]	data
BD 330	Si-P	NF-L, 32V, 3A, 15W(Tc=45°), 100MHz	14h§	Phi,Sie	2SB772, 2SB986, (BD 766, 2SB744(A))[17]	data
BD 331	Si-N-Darl+Di	NF-L, 60V, 6A, 60W, >10MHz, B>750	≈14j§	Phi,Sgs	(BD 645, BD 897, BDW 23A, BDW 63A, ++)[4]	data
BD 332	Si-P-Darl+Di	NF-L, 60V, 6A, 60W, >10MHz, B>750	≈14j§	Phi,Sgs	(BD 646, BD 898, BDW 24A, BDW 64A, ++)[4]	data
BD 333	Si-N-Darl+Di	=BD 331: 80V	≈14j§	Phi,Sgs	(BD 647, BD 899, BDW 23B, BDW 63B, ++)[4]	data
BD 334	Si-P-Darl+Di	=BD 332: 80V	≈14j§	Phi,Sgs	(BD 648, BD 900, BDW 24B, BDW 64B, ++)[4]	data
BD 335	Si-N-Darl+Di	=BD 331: 100V	≈14j§	Phi,Sgs	(BD 649, BD 901, BDW 23C, BDW 63C, ++)[4]	data
BD 336	Si-P-Darl+Di	=BD 332: 100V	≈14j§	Phi,Sgs	(BD 650, BD 902, BDW 24C, BDW 64C, ++)[4]	data
BD 337	Si-N-Darl+Di	=BD 331: 120V	≈14j§	Phi,Sgs	(BD 651, BDT 21, BDW 63D, BDW 74D, ++)[4]	data
BD 338	Si-P-Darl+Di	=BD 332: 120V	≈14j§	Phi,Sgs	(BD 652, BDT 20, BDW 64D, BDW 74D, ++)[4]	data
BD 342	Si-N	NF-L, -/40V, 12A, 100W, 1,5MHz	23a§	Mot	BD 249, BD 315, BDW 51A, 2N5881, ++	data
BD 343	Si-P	NF-L, -/40V, 12A, 100W, 1,5MHz	23a§	Mot	BD 250, BD 316, BDW 52A, 2N5879, ++	data
BD 344	Si-P	NF-L, 60V, 1A, 20W, >50MHz	14h§	Nsc	BD 138, BD 229, BD 378, BD 786	data
BD 345	Si-N	NF-L, 60V, 1A, 20W, >50MHz	14h§	Nsc	BD 137, BD 228, BD 377, BD 785	data
BD 346	Si-P	NF-L, 60V, 8A, 60W, >4MHz	17j§	Nsc	BD 544A, BD 708, BD 798, BD 808, ++	data
BD 347	Si-N	NF-L, 60V, 8A, 60W, >4MHz	17j§	Nsc	BD 543A, BD 707, BD 797, BD 807, ++	data
BD 348	Si-P	NF-L, 80V, 1A, 20W, >50MHz	14h§	Nsc	BD 140, BD 231, BD 380, BD 792	data
BD 349	Si-N	NF-L, 80V, 1A, 20W, >50MHz	14h§	Nsc	BD 139, BD 230, BD 379, BD 791	data
BD 350	Si-P	NF-L, -/80V, 15A, 160W, >4MHz	23a§	Nsc	BD 316, BD 318, 2N6029..6031	data
BD 350 A	Si-P	=BD 350: -/70V	23a§		BD 316, BD 318, 2N6029..6031	data
BD 350 B	Si-P	=BD 350: -/60V	23a§		BD 316, BD 318, 2N6029..6031	data
BD 351	Si-N	NF-L, -/80V, 15A, 160W, >4MHz	23a§	Nsc	BD 315, BD 317, 2N5629..5631	data
BD 351 A	Si-N	=BD 351: -/70V	23a§		BD 315, BD 317, 2N5629..5631	data
BD 351 B	Si-N	=BD 351: -/60V	23a§		BD 315, BD 317, 2N5629..5631	data
BD 354(A,B,C)	Si-N	NF-L, -/40V, 3A, 12,5W(Tc=45°), >30MHz	22a§		MJE 15028, 2SC3252...3253	data
BD 355(A,B,C)	Si-P	NF-L, -/40V, 3A, 12,5W(Tc=45°), >30MHz	22a§		MJE 15029, 2SA1288...1289	data
BD 356	Si-P	NF-L, 20V, 5A, 20W, >50MHz	14h§	Sie	MJE 210	data
BD 357	Si-N	NF-L, 50V, 5A, 20W, >50MHz	14h§	Sie	MJE 200, 2SC2270, 2SD826	data
BD 358	Si-P	=BD 356: 8,3W	13h§	Sie	(MJE 210)	data
BD 359	Si-N	=BD 357: 8,3W	13h§	Sie	(MJE 200, 2SC2270, 2SD826)[4]	data
BD 361(A)	Si-N	NF-L, 32V, 3A, 15W	14h§	Mot	BD 175, BD 185, BD 435, 2N5190	data
BD 362(A)	Si-P	NF-L, 32V, 3A, 15W	14h§	Mot	BD 176, BD 186, BD 436, 2N5193	data
BD 363(A,B)	Si-N	NF-L, 60V, 6A, 75W	17j§	Sgs	BD 243A, BD 543A, BD 797, BD 807, ++	data
BD 364	Si-N	NF-L, 50V, 25A, 200W, >4MHz	23a§	Mot	MJ 802, 2N5885	data
BD 365	Si-P	NF-L, 50V, 25A, 200W, >4MHz	23a§	Mot	MJ 4502, 2N5883	data
BD 366	Si-N	=BD 364: 60V	23a§	Mot	MJ 802, 2N5885	data
BD 367	Si-P	=BD 365: 60V	23a§	Mot	MJ 4502, 2N5883	data
BD 368	Si-N	=BD 364: 80V	23a§	Mot	MJ 802, 2N5886	data
BD 369	Si-P	=BD 365: 80V	23a§	Mot	MJ 4502, 2N5884	data
BD 370 A	Si-P	NF-Tr/E, 45V, 1,5A, 2,5W, >50MHz	30e§	Nsc	(BC 362, BD 518, BD 526, BD 840)[4]	data
BD 370 B	Si-P	=BD 370A: 60V	30e§		(BC 363, BD 518, BD 526, BD 842)[4]	data
BD 370 C	Si-P	=BD 370A: 80V	30e§		(BC 364, BD 520, BD 528, BD 844)[4]	data
BD 370 D	Si-P	=BD 370A: 100V	30e§		(BD 530, BD 844)[4]	data
BD 371 A	Si-N	NF-Tr/E, 45V, 1,5A, 2,5W, >50MHz	30e§	Nsc	(BC 365, BD 517, BD 525, BD 839)[4]	data
BD 371 B	Si-N	=BD 371A: 60V	30e§		(BC 366, BD 517, BD 525, BD 841)[4]	data
BD 371 C	Si-N	=BD 371A: 80V	30e§		(BC 367, BD 519, BD 527, BD 843)[4]	data
BD 371 D	Si-N	=BD 371A: 100V	30e§		(BD 529, BD 843)[4]	data
BD 372(A..D)	Si-P	=BD 370A..D:	30c§	Nsc	→BD 370A..D	data
BD 373(A..D)	Si-N	=BD 371A..D:	30c§	Nsc	→BD 371A..D	data

Type	Device	Short Description	Fig.	Manu	Comparision Types	More at
BD 375	Si-N	NF/S-L, 50V, 2A, 25W, >50MHz	14h§	Nsc,Sgs	BD 177, BD 235, BD 785, 2SD1177...1178	data
BD 376	Si-P	NF/S-L, 50V, 2A, 25W, >50MHz	14h§	Nsc,Sgs	BD 178, BD 236, BD 786, 2SB874...875	data
BD 377	Si-N	=BD 375: 75V	14h§	Nsc,Sgs	BD 179, BD 237, BD 787, 2SD1177...1178	data
BD 378	Si-P	=BD 376: 75V	14h§	Nsc,Sgs	BD 180, BD 238, BD 788, 2SB874...875	data
BD 379	Si-N	=BD 375: 100V	14h§	Nsc,Sgs	BD 237, BD 791, 2SD1177...1178	data
BD 380	Si-P	=BD 376: 100V	14h§	Nsc,Sgs	BD 238, BD 792, 2SB874...875	data
BD 385	Si-N	NF-L, 60V, 1A, 10W, >350MHz	13h§	Mot	BD 827, BD 841	data
BD 386	Si-P	NF-L, 60V, 1A, 10W, >350MHz	13h§	Mot	BD 828, BD 842	data
BD 387	Si-N	=BD 385: 80V	13h§	Mot	BD 829, BD 843	data
BD 388	Si-P	=BD 386: 80V	13h§	Mot	BD 830, BD 844	data
BD 389	Si-N	=BD 385: 100V	13h§	Mot	BD 829, BD 843	data
BD 390	Si-P	=BD 386: 100V	13h§	Mot	BD 830, BD 844	data
BD 400	Si-N	TV-VA, 170V, 1A, 20W, 65MHz	14h§	Tix	2SC3117, 2SD669, (MJE 340)[7]	data
BD 401	Si-N	NF-L, 60/45V, 10A, 50W, 50MHz	17j§	Gen	BD 707, BD 743A, BD 807, BD 907	data
BD 402	Si-P	NF-L, 60/45V, 10A, 50W, 40MHz	17j§	Gen	BD 708, BD 744A, BD 808, BD 908	data
BD 403	Si-N	=BD 401: 60/60V	17j§	Gen	BD 707, BD 743A, BD 807, BD 907	data
BD 404	Si-P	=BD 402: 60/60V	17j§	Gen	BD 708, BD 744A, BD 808, BD 908	data
BD 410	Si-N	NF-L, Vid-L, 500V, 1A, 20W	14h§	Tix	BUV 93, 2SC3051	data
BD 411	Si-N-Darl	NF-L, 50V, 2A, 10W, B>25000	13g§	Mot	MPSU 45	data
BD 412	Si-N-Darl	=BD 411: B>15000	13g§	Mot	MPSU 45	data
BD 413	Si-P-Darl	NF-L, 50V, 2A, 10W, B>25000	13g§	Mot	MPSU 95	data
BD 414	Si-P-Darl	=BD 413: B>15000	13g§	Mot	MPSU 95	data
BD 415	Si-N	NF-L, 60V, 1A, 10W, >75MHz	13g§	Mot	BD 517, BD 525, (BD 385, BD 827)[5]	data
BD 416	Si-P	NF-L, 60V, 1A, 10W, >75MHz	13g§	Mot	BD 518, BD 526, (BD 386, BD 828)[5]	data
BD 417	Si-N	=BD 315: 80V	13g§	Mot	BD 519, BD 527, (BD 387, BD 829)[5]	data
BD 418	Si-P	=BD 416: 80V	13g§	Mot	BD 520, BD 528, (BD 388, BD 830)[5]	data
BD 419	Si-N	=BD 415: 100V	13g§	Mot	BD 529, (BD 389, BD 829)[5]	data
BD 420	Si-P	=BD 416: 100V	13g§	Mot	BD 530, (BD 390, BD 830)[5]	data
BD 421	Si-N-Darl	NF-L, -/100V, 2A(ss), 10W, >100MHz	13g§	Mot	(BD 879, BDX 44)[5]	data
BD 422	Si-N-Darl	=BD 421: -/80V	13g§	Mot	(BD 779, BD 877, BDX 43)[5]	data
BD 424	Si-N	TV-HA-Tr, 160V, 0,8A, 2,5W(Tc=100°)	13h§	Sie	BF 666...668, MJE 340, 2SC3117, (2SC2483)[5]	data
BD 429	Si-N	NF-L, 32V, 3A, 10W, 130MHz	13h§	Sie	BD 329, BD 785, 2SD1348	data
BD 430	Si-P	NF-L, 32V, 3A, 10W, 100MHz	13h§	Sie	BD 330, BD 786, 2SB986	data
BD 433	Si-N	NF-L, 22V, 4A, 36W, >3MHz	14h§	EUR	BD 185, 2N5190	data
BD 434	Si-P	NF-L, 22V, 4A, 36W, >3MHz	14h§	EUR	BD 186, 2N5193	data
BD 435	Si-N	=BD 433: 32V	14h§	EUR	BD 185, 2N5190	data
BD 436	Si-P	=BD 434: 32V	14h§	EUR	BD 186, 2N5193	data
BD 437	Si-N	=BD 433: 45V	14h§	EUR	BD 187, 2N5191	data
BD 438	Si-P	=BD 434: 45V	14h§	EUR	BD 188, 2N5194	data
BD 439	Si-N	=BD 433: 60V	14h§	EUR	BD 189, 2N5191	data
BD 440	Si-P	=BD 434: 60V	14h§	EUR	BD 190, 2N5194	data
BD 441	Si-N	=BD 433: 80V	14h§	EUR	BD 189, 2N5192	data
BD 442	Si-P	=BD 434: 80V	14h§	EUR	BD 190, 2N5195	data
BD 443	Si-N	NF/S-L, 120V, 3A, 30W, >0,8MHz	14h§	Mot	-	data
BD 443 A	Si-N	=BD 443: 170V	14h§		-	data
BD 450	Si-N	NF-L, 80V, 15A, 115W, >0,8MHz	23a§	Rca	BD 315, 2N5629...5631	data
BD 451	Si-N	=BD 450: 95V	23a§	Rca	BD 317, 2N5629...5631	data
BD 461	Si-N	NF-L, 35V, 4A, 30W	14h§	Tix	BD 185, BD 437, BD 785, 2N5190	data
BD 462	Si-P	NF-L, 35V, 4A, 30W	14h§	Tix	BD 186, BD 438, BD 786, 2N5193	data
BD 463	Si-N	NF-L, 35V, 4A, 30W	14h§	Tix	BD 185, BD 437, BD 785, 2B5190	data
BD 464	Si-P	NF-L, 35V, 4A, 30W	14h§	Tix	BD 186, BD 438, BD 786, 2N5193	data
BD 466 A	Si-P-Darl	NF-L, 30V, 1A, 8,5W, 170MHz, B>8000	14h§	Tix	BD 876, BDX 45...47	data
BD 466 B	Si-P-Darl	=BD 466A: 45V	14h§		BD 876, BDX 45...47	data
BD 477 A	Si-N-Darl	NF-L, 30V, 1A, 8,5W, 170MHz, B>8000	14h§	Tix	BD 875, BDX 42...44	data
BD 477 B	Si-N-Darl	=BD 477A: 45V	14h§		BD 875, BDX 42...44	data
BD 487	Si-P	S-L, 30V, 12A, 12,5W, >50MHz	13h§	Tix	-	data
BD 488	Si-P	=BD 487: 45V	13h§	Sie	-	data
BD 500	Si-P	NF-L, 60V, 10A, 75W, >5MHz	17j§	Rca	BD 708, BD 744A, BD 808, BD 908	data
BD 500 A	Si-P	=BD 500: 70V	17j§		BD 710, BD 744A, BD 810, BD 910	data
BD 500 B	Si-P	=BD 500: 90V	17j§		BD 712, BD 744B, BD 912	data
BD 501	Si-N	NF-L, 60V, 10A, 75W, >5MHz	17j§	Rca	BD 707, BD 743A, BD 807, BD 907	data
BD 501 A	Si-N	=BD 501: 70V	17j§		BD 709, BD 743A, BD 809, BD 909	data
BD 501 B	Si-N	=BD 501: 90V	17j§		BD 711, BD 743B, BD 911	data
BD 505	Si-N	NF-Tr/E, 30V, 2A, 10W, 250MHz	13g§	Mot	BD 515, BD 525, (BD 839)[5]	data
BD 506	Si-P	NF-Tr/E, 30V, 2A, 10W, 250MHz	13g§	Mot	BD 516, BD 526, (BD 840)[5]	data
BD 507	Si-N	=BD 505: 40V	13g§	Mot	BD 515, BD 525, (BD 839)[5]	data
BD 508	Si-P	=BD 506: 40V	13g§	Mot	BD 516, BD 526, (BD 840)[5]	data
BD 509	Si-N	=BD 505: 50V	13g§	Mot	BD 517, BD 525, (BD 841)[5]	data
BD 510	Si-P	=BD 506: 50V	13g§	Mot	BD 518, BD 526, (BD 842)[5]	data
BD 512	MOS-P-FET-e	V-MOS, NF-L, 60V, 1,5A, 10W	13b§	Itt	-	data
BD 515	Si-N	NF-Tr/E, 45V, 2A, 10W, 160MHz	13g§	Mot	BD 525, (BD 839)[5]	data
BD 516	Si-P	NF-Tr/E, 45V, 2A, 10W, 125MHz	13g§	Mot	BD 526, (BD 840)[5]	data
BD 517	Si-N	=BD 515: 60V	13g§	Mot	BD 525, (BD 841)[5]	data
BD 518	Si-P	=BD 516: 60V	13g§	Mot	BD 526, (BD 842)[5]	data
BD 519	Si-N	=BD 515: 80V	13g§	Mot	BD 527, (BD 843)[5]	data
BD 520	Si-P	=BD 516: 80V	13g§	Mot	BD 528, (BD 844)[5]	data

Type	Device	Short Description	Fig.	Manu	Comparision Types	More at
BD 522	MOS-N-FET-e	V-MOS, NF-L, 60V, 1,5A, 10W	13b§	Itt	-	data
BD 524	Si-N	TV-HA-Tr, 160V, 0,8A, 5W, 100MHz	14h§	Sie	2SC3051, 2SC3117, 2SC3425, 2SD669	data
BD 525	Si-N	NF-Tr/E, 60V, 2A, 10W, 150MHz	13g§	Mot	BD 517, (BD 841)[5]	data
BD 526	Si-P	NF-Tr/E, 60V, 2A, 10W, 100MHz	13g§	Mot	BD 518, (BD 842)[5]	data
BD 527	Si-N	=BD 525: 80V	13g§	Mot	BD 519, (BD 843)[5]	data
BD 528	Si-P	=BD 526: 80V	13g§	Mot	BD 518, (BD 844)[5]	data
BD 529	Si-N	=BD 525: 100V	13g§	Mot	(BD 843)[5]	data
BD 530	Si-P	=BD 526: 100V	13g§	Mot	(BD 844)[5]	data
BD 533(A)	Si-N	NF-L, 45V, 4A, 50W, >3MHz	17j§	Sie,Sgs,	BD 243, BD 539A, BD 543A, BD 947, ++	data
BD 533 FI	Si-N	=BD 533: Iso, 22W	17c	Sgs	BD 947F, 2SD1408, 2SC1667, 2SD2012,++	data
BD 533 FP	Si-N	=BD 533: Iso, 25W	17c	Sgs	BD 947F, 2SD1408, 2SC1667, 2SD2012,++	data
BD 534(A)	Si-P	NF-L, 45V, 4A, 50W, >3MHz	17j§	Sie,Sgs,++	BD 244, BD 540A, BD 544A, BD 950, ++	data
BD 534 FI	Si-P	=BD 534: Iso, 22W	17c	Sgs	BD 948F, 2SB1017, 2SA1488, 2SB1375,++	data
BD 534 FP	Si-P	=BD 534: Iso, 25W	17c	Sgs	BD 948F, 2SB1017, 2SA1488, 2SB1375,++	data
BD 535(A)	Si-N	=BD 533: 60V	17j§	Sie,Sgs,++	BD 243A, BD 539A, BD 543A, BD 949, ++	data
BD 536(A)	Si-P	=BD 534: 60V	17j§	Sie,Sgs,++	BD 244A, BD 540A, BD 544A, BD 950, ++	data
BD 537(A)	Si-N	=BD 533: 80V	17j§	Sie,Sgs,++	BD 243B, BD 539B, BD 543B, BD 951, ++	data
BD 538(A)	Si-P	=BD 534: 80V	17j§	Sie,Sgs,++	BD 244B, BD 540B, BD 544B, BD 952	data
BD 539	Si-N	NF-L, 40V, 5A, 45W, >3MHz	17j§	Tix	BD 243, BD 543, BD 795, BD 805	data
BD 539 A	Si-N	=BD 539: 60V	17j§		BD 243A, BD 543A, BD 797, BD 807	data
BD 539 B	Si-N	=BD 539: 80V	17j§		BD 243B, BD 543B, BD 799, BD 809	data
BD 539 C	Si-N	=BD 539: 100V	17j§		BD 243C, BD 543C, BD 801, 2SD866	data
BD 539 D	Si-N	=BD 539: 120V	17j§		BD 543D, 2SD866	data
BD 540	Si-P	NF-L, 40V, 5A, 45W, >3MHz	17j§	Tix	BD 244, BD 544, BD 796, BD 806	data
BD 540 A	Si-P	=BD 540: 60V	17j§		BD 244A, BD 544A, BD 798, BD 808	data
BD 540 B	Si-P	=BD 540: 80V	17j§		BD 244B, BD 544B, BD 800, BD 810	data
BD 540 C	Si-P	=BD 540: 100V	17j§		BD 244C, BD 544C, BD 802, 2SB870	data
BD 540 D	Si-P	=BD 540: 120V	17j§		BD 544D, 2SB870	data
BD 543	Si-N	NF-L, 40V, 8A, 70W, >3MHz	17j§	Tix	BD 705, BD 795, BD 805	data
BD 543 A	Si-N	=BD 543: 60V	17j§		BD 707, BD 797, BD 807	data
BD 543 B	Si-N	=BD 543: 80V	17j§		BD 709, BD 799, BD 809	data
BD 543 C	Si-N	=BD 543: 100V	17j§		BD 711, BD 801, 2SD866	data
BD 543 D	Si-N	=BD 543: 120V	17j§		2SD866	data
BD 544	Si-P	NF-L, 40V, 8A, 70W, >3MHz	17j§	Tix	BD 706, BD 796, BD 806	data
BD 544 A	Si-P	=BD 544: 60V	17j§		BD 708, BD 798, BD 808	data
BD 544 B	Si-P	=BD 544: 80V	17j§		BD 710, BD 800, BD 810	data
BD 544 C	Si-P	=BD 544: 100V	17j§		BD 712, BD 802, 2SB870	data
BD 544 D	Si-P	=BD 544: 120V	17j§		2SB870	data
BD 545	Si-N	NF-L, 40V, 15A, 85W, >3MHz	18j§	Tix	BD 249, BD 745	data
BD 545 A	Si-N	=BD 545: 60V	18j§		BD 249A, BD 745A	data
BD 545 B	Si-N	=BD 545: 80V	18j§		BD 249B, BD 745B	data
BD 545 C	Si-N	=BD 545: 100V	18j§		BD 249C, BD 745C	data
BD 545 D	Si-N	=BD 545: 120V	18j§		BD 745D	data
BD 546	Si-P	NF-L, 40V, 15A, 85W, >3MHz	18j§	Tix	BD 250, BD 746	data
BD 546 A	Si-P	=BD 546: 60V	18j§		BD 250A, BD 746A	data
BD 546 B	Si-P	=BD 546: 80V	18j§		BD 250B, BD 746B	data
BD 546 C	Si-P	=BD 546: 100V	18j§		BD 250C, BD 746C	data
BD 546 D	Si-P	=BD 546: 120V	18j§		BD 746D	data
BD 550	Si-N	NF-L, 130V, 7A, 150W, 5MHz	23a§	Rca	BDW 10...16, BUX 17, MJ 15015, 2SD665, ++	data
BD 550 A	Si-N	=BD 550: 200V	23a§		BDW 10, BUX 17, MJ 15015, 2SD665, ++	data
BD 550 B	Si-N	=BD 550: 275V	23a§		BUX 17A, 2SD555, 2SD583, 2SC1586	data
BD 561	Si-N	NF-L, 45V, 4A, 40W, >3MHz	14h§	Mot	BD 187, BD 437, 2B5191	data
BD 562	Si-P	NF-L, 45V, 4A, 40W, >3MHz	14h§	Mot	BD 188, BD 438, 2N5194	data
BD 566	Si-P-Darl	NF-L, -/60V, 10A, 50W, B>1000	17j§	Gen	BDT 62A, BDT 64A, BDW 94B, BDX 34B	data
BD 566 A	Si-P-Darl	=BD 566: -/80V	17j§		BDT 62B, BDT 64B, BDW 94C, BDX 34C	data
BD 567	Si-N-Darl	NF-L, -/60V, 10A, 50W, B>1000	17j§	Gen	BDT 63A, BDT 65A, BDW 93B, BDX 33B	data
BD 567 A	Si-N-Darl	=BD 567: -/80V	17j§		BDT 63B, BDT 65B, BDW 93C, BDX 33C	data
BD 575	Si-N	NF-L, 45V, 3A, 40W, >3MHz	15j§	Mot	BD 585, BD 241, BD 533, BD 933, ++	data
BD 576	Si-P	NF-L, 45V, 3A, 40W, >3MHz	15j§	Mot	BD 586, BD 242, BD 534, BD 934, ++	data
BD 577	Si-N	=BD 575: 60V	15j§	Mot	BD 587, BD 241A, BD 535, BD 935, ++	data
BD 578	Si-P	=BD 576: 60V	15j§	Mot	BD 588, BD 242A, BD 536, BD 936, ++	data
BD 579	Si-N	=BD 575: 80V	15j§	Mot	BD 589, BD 241B, BD 537, BD 937, ++	data
BD 580	Si-P	=BD 576: 80V	15j§	Mot	BD 590, BD 242B, BD 538, BD 938, ++	data
BD 581	Si-N	=BD 575: 100V	15j§	Mot	BD 591, BD 241C, BD 937, 2SD712, ++	data
BD 582	Si-P	=BD 576: 100V	15j§	Mot	BD 592, BD 242C, BD 938, 2SB682, ++	data
BD 585	Si-N	NF-L, 45V, 4A, 40W, >3MHz	15j§	Aeg,Mot	BD 595, BD 533, BD 539, BD 947, ++	data
BD 586	Si-P	NF-L, 45V, 4A, 40W, >3MHz	15j§	Aeg,Mot	BD 596, BD 534, BD 540, BD 948, ++	data
BD 587	Si-N	=BD 585: 60V	15j§	Aeg,Mot	BD 597, BD 535, BD 539A, BD 949, ++	data
BD 588	Si-P	=BD 586: 60V	15j§	Aeg,Mot	BD 598, BD 536, BD 540A, BD 950, ++	data
BD 589	Si-N	=BD 585: 80V	15j§	Aeg,Mot	BD 599, BD 537, BD 539B, BD 951, ++	data
BD 590	Si-P	=BD 586: 80V	15j§	Aeg,Mot	BD 600, BD 538, BD 540B, BD 952, ++	data
BD 591	Si-N	=BD 585: 100V	15j§	Aeg,Mot	BD 601, BD 539C, BD 953, 2SD712, ++	data
BD 592	Si-P	=BD 586: 100V	15j§	Aeg,Mot	BD 602, BD 540C, BD 954, 2SB682, ++	data
BD 595	Si-N	NF-L, 45V, 8A, 65W, >3MHz	15j§	Aeg,Mot	BD 605, BD 543, BD 795, BD 805, ++	data
BD 596	Si-P	NF-L, 45V, 8A, 65W, >3MHz	15j§	Aeg,Mot	BD 606, BD 544, BD 796, BD 806, ++	data
BD 597	Si-N	=BD 595: 60V	15j§	Aeg,Mot	BD 607, BD 543A, BD 797, BD 807, ++	data
BD 598	Si-P	=BD 596: 60V	15j§	Aeg,Mot	BD 608, BD 544A, BD 798, BD 808, ++	data
BD 599	Si-N	=BD 595: 80V	15j§	Aeg,Mot	BD 609, BD 543B, BD 799, BD 809, ++	data
BD 600	Si-P	=BD 596: 80V	15j§	Aeg,Mot	BD 610, BD 544B, BD 800, BD 810, ++	data
BD 601	Si-N	=BD 595: 100V	15j§	Aeg,Mot	BD 543C, BD 801, 2SD866	data

Type	Device	Short Description	Fig.	Manu	Comparision Types	More at
602	Si-P	=BD 596: 100V	15j§	Aeg,Mot	BD 544C, BD 802, 2SB869	data
605	Si-N	NF-L, 55V, 10A, 90W, >1,5MHz	15j§	Aeg,Mot	BD 743A, BD 805, BD 907	data
606	Si-P	NF-L, 55V, 10A, 90W, >1,5MHz	15j§	Aeg,Mot	BD 744A, BD 806, BD 908	data
607	Si-N	=BD 605: 70V	15j§	Aeg,Mot	BD 743A, BD 807, BD 909	data
608	Si-P	=BD 606: 70V	15j§	Aeg,Mot	BD 744A, BD 808, BD 910	data
609	Si-N	=BD 605: 80V	15j§	Aeg,Mot	BD 743B, BD 809, BD 909	data
610	Si-P	=BD 606: 80V	15j§	Aeg,Mot	BD 744B, BD 810, BD 910	data
611	Si-N	NF-L, 22V, 4A, 15W, >3MHz	13h§	Sie	BD 185, BD 433, 2N5190	data
612	Si-P	NF-L, 22V, 4A, 15W, >3MHz	13h§	Sie	BD 186, BD 434, 2N5193	data
613	Si-N	=BD 611: 32V	13h§	Sie	BD 185, BD 435, 2N5190	data
614	Si-P	=BD 612: 32V	13h§	Sie	BD 186, BD 436, 2N5193	data
615	Si-N	=BD 611: 45V	13h§	Sie	BD 187, BD 437, 2N5191	data
616	Si-P	=BD 612: 45V	13h§	Sie	BD 188, BD 438, 2N5194	data
617	Si-N	=BD 611: 60V	13h§	Sie	BD 189, BD 439, 2N5191	data
618	Si-P	=BD 612: 60V	13h§	Sie	BD 190, BD 440, 2N5194	data
619	Si-N	=BD 611: 80V	13h§	Sie	BD 441, 2N5192	data
620	Si-P	=BD 612: 80V	13h§	Sie	BD 442, 2N5195	data
633	Si-N	NF-L, 45V, 2A, 30W, >3MHz	17j§	Tix,Mic	BD 239, BD 241, BD 533, BD 933, ++	data
634	Si-P	NF-L, 45V, 2A, 30W, >3MHz	17j§	Tix,Mic	BD 240, BD 242, BD 534, BD 934, ++	data
635	Si-N	=BD 633: 60V	17j§	Tix,Mic	BD 239A, BD 241A, BD 535, BD 935, ++	data
636	Si-P	=BD 634: 60V	17j§	Tix,Mic	BD 240A, BD 242A, BD 536, BD 936, ++	data
637	Si-N	=BD 633: 100V	17j§	Tix,Mic	BD 239C, BD 241C, BD 937, 2SD712, ++	data
638	Si-P	=BD 634: 100V	17j§	Tix,Mic	BD 240C, BD 242C, BD 938, 2SB682, ++	data
643	Si-N-Darl+Di	NF-L, 60V, 8A, 62,5W, >10MHz, B>750	17j§	EUR	BD 895, BDW 73, BDX 33, BDX 53, ++	data
643 F	Si-N-Darl+Di	=BD 643: Iso, >20W	17c	Phi	BDT 63(A...C)F, 2SD1415...17, 2SD1826, ++	data
644	Si-P-Darl+Di	NF-L, 45V, 8A, 62,5W, >10MHz, B>750	17j§	EUR	BD 896, BDW 74, BDX 34, BDX 54, ++	data
644 F	Si-P-Darl+Di	=BD 644: Iso, >20W	17c	Phi	BDT 62(A...C)F, 2SB1020...22, 2SB1224, ++	data
645	Si-N-Darl+Di	=BD 643: 80V	17j§	EUR	BD 897, BDW 73A, BDX 33A, BDX 53A, ++	data
645 F	Si-N-Darl+Di	=BD 645: Iso, >20W	17c	Phi	BDT 63AF...CF, 2SD1415...16, 2SD1791, ++	data
646	Si-P-Darl+Di	=BD 644: 60V	17j§	EUR	BD 898, BDW 74A, BDX 34A, BDX 54A, ++	data
646 F	Si-P-Darl+Di	=BD 646: Iso, >20W	17c	Phi	BDT 62(A...C)F, 2SB1020...22, 2SB1224, ++	data
647	Si-N-Darl+Di	=BD 643: 100V	17j§	EUR	BD 899, BDW 73B, BDX 33B, BDX 53B, ++	data
647 F	Si-N-Darl+Di	=BD 647: Iso, >20W	17c	Phi	BDT 63BF...CF, 2SD1415, 2SD1791, ++	data
648	Si-P-Darl+Di	=BD 644: 80V	17j§	EUR	BD 900, BDW 74B, BDX 34B, BDX 54B, ++	data
648 F	Si-P-Darl+Di	=BD 648: Iso, >20W	17c	Phi	BDT 62AF...CF, 2SB1020...21, 2SB1283, ++	data
649	Si-N-Darl+Di	=BD 643: 120V	17j§	EUR	BD 901, BDW 73C, BDX 33C, BDX 53C, ++	data
649 F	Si-N-Darl+Di	=BD 649: Iso, >20W	17c	Phi	BDT 63CF, 2SD1590, 2SD1792, 2SD1590, ++	data
650	Si-P-Darl+Di	=BD 644: 100V	17j§	EUR	BD 902, BDW 74C, BDX 34C, BDX 54C, ++	data
650 F	Si-P-Darl+Di	=BD 650: Iso, >20W	17c	Phi	BDT 62AF...CF, 2SB1020, 2SB1283, ++	data
651	Si-N-Darl+Di	=BD 643: 140V	17j§	EUR	BDW 73D, BDX 33D, BDX 53E, BDT 21, ++	data
651 F	Si-N-Darl+Di	=BD 651: Iso, >20W	17c	Phi	2SD1590, 2SD1792	data
652	Si-P-Darl+Di	=BD 644: 120V	17j§	EUR	BDW 74D, BDX 34D, BDX 54E, BDT 20, ++	data
652 F	Si-P-Darl+Di	=BD 652: Iso, >20W	17c	Phi	BDT 62CF, 2SB1193	data
661	Si-N	NF-L, 32V, 4A, 36W	≈14j§	Sgs	(BD 243, BD 533, BD 539, 945, ++)[4]	data
661 K	Si-N	=BD 661: 50W	17j§		BD 243, BD 533, BD 539, BD 945, ++	data
662	Si-P	NF-L, 32V ,4A, 32W	≈14j§	Sgs	(BD 244, BD 534, BD 540, BD 946, ++)[4]	data
662 K	Si-P	=BD 662: 50W	17j§		BD 244, BD 534, BD 540, BD 946, ++	data
663(A,B)	Si-N	NF-L, 45V, 10A, 75W, >3MHz	17j§	Sgs	BD 705, BD 743, BD 805, BD 905	data
664(A,B)	Si-P	NF-L, 45V, 10A, 75W, >3MHz	17j§	Sgs	BD 706, BD 744, BD 806, BD 906	data
675(A)	Si-N-Darl+Di	NF-L, 45V, 4A, 40W, >10MHz, B>750	14h§	EUR	BD 775; 2N6038	data
676(A)	Si-P-Darl+Di	NF-L, 45V, 4A, 40W, >10MHz, B>750	14h§	EUR	BD 776, 2N6035	data
677(A)	Si-N-Darl+Di	=BD 675: 60V	14h§	EUR	BD 777, 2N6038	data
678(A)	Si-P-Darl+Di	=BD 676: 60V	14h§	EUR	BD 778, 2N6035	data
679(A)	Si-N-Darl+Di	=BD 675: 80V	14h§	EUR	BD 779, 2N6039	data
680(A)	Si-P-Darl+Di	=BD 676: 80V	14h§	EUR	BD 780, 2N6036	data
681	Si-N-Darl+Di	=BD 675: 100V	14h§	EUR	-	data
682	Si-P-Darl+Di	=BD 676: 100V	14h§	EUR	-	data
683	Si-N-Darl+Di	=BD 675: 140V	14h§	Phi	-	data
684	Si-P-Darl+Di	=BD 676: 120V	14h§	Phi	-	data
675H...680H	Si-N/P-Darl	=BD 675...680:	≈14j§	Sgs	(→BD 675...680)[5]	data
695(A)	Si-N-Darl	NF-L, 45V, 8A, 70W, >1MHz, B>750	15j§	Aeg,Mot	BD 643, BD 895, BDW 73, BDX 53, ++	data
696(A)	Si-P-Darl	NF-L, 45V, 8A, 70W, >1MHz, B>750	15j§	Aeg,Mot	BD 644, BD 896, BDW 74, BDX 54, ++	data
697(A)	Si-N-Darl	=BD 695: 60V	15j§	Aeg,Mot	BD 645, BD 897, BDW 73A, BDX 53A, ++	data
698(A)	Si-P-Darl	=BD 696: 60V	15j§	Aeg,Mot	BD 646, BD 898, BDW 74A, BDX 54A, ++	data
699(A)	Si-N-Darl	=BD 695: 80V	15j§	Aeg,Mot	BD 647, BD 899, BDW 73B, BDX 53B, ++	data
700(A)	Si-P-Darl	=BD 696: 80V	15j§	Aeg,Mot	BD 648, BD 900, BDW 74B, BDX 54B, ++	data
701	Si-N-Darl	=BD 695: 100V	15j§	Aeg,Mot	BD 649, BD 901, BDW 73C, BDX 53C, ++	data
702	Si-P-Darl	=BD 696: 100V	15j§	Aeg,Mot	BD 650, BD 902, BDW 74C, BDX 54C, ++	data
705	Si-N	NF-L, 45V, 12A, 75W, >3MHz	17j§	Tho	BD 743, BD 905	data
706	Si-P	NF-L, 45V, 12A, 75W, >3MHz	17j§	Tho	BD 744, BD 906	data
707	Si-N	=BD 705: 60V	17j§	Tho	BD 743A, BD 907	data
708	Si-P	=BD 706: 60V	17j§	Tho	BD 744A, BD 908	data
709	Si-N	=BD 705: 80V	17j§	Tho	BD 743B, BD 909	data
710	Si-P	=BD 706: 80V	17j§	Tho	BD 744B, BD 910	data
711	Si-N	=BD 705: 100V	17j§	Tho	BD 743C, BD 911	data
712	Si-P	=BD 706: 100V	17j§	Tho	BD 744C, BD 912	data
713	Si-N-Darl	NF-L, 45V, 4A, 36W, >1MHz, B>750	17j§	Sie	BD 643, BD 895, BDW 23, BDW 53, ++	data
714	Si-P-Darl	NF-L, 45V, 4A, 36W, >1MHz, B>750	17j§	Sie	BD 644, BD 896, BDW 24, BDW 54, ++	data
715	Si-N-Darl	=BD 713: 60V	17j§	Sie	BD 645, BD 897, BDW 23A, BDW 53A, ++	data
716	Si-P-Darl	=BD 714: 60V	17j§	Sie	BD 646, BD 898, BDW 24A, BDW 54A, ++	data
717	Si-N-Darl	=BD 713: 80V	17j§	Sie	BD 647, BD 899, BDW 23B, BDW 53B, ++	data

Type	Device	Short Description	Fig.	Manu	Comparision Types	More at
BD 718	Si-P-Darl	=BD 713: 80V	17j§	Sie	BD 648, BD 900, BDW 24B, BDW 54B, ++	data
BD 719	Si-N	NF/S-L, 60V, 4A, 36W, >3MHz	14h§	Phi	BD 189, BD 439, BD 441, 2N5191...92	data
BD 720	Si-P	NF/S-L, 60V, 4A, 36W, >3MHz	14h§	Phi	BD 190, BD 440, BD 442, 2N5194...95	data
BD 721	Si-N	=BD 719: 80V	14h§	Phi	BD 441, 2N5192	data
BD 722	Si-P	=BD 720: 80V	14h§	Phi	BD 442, 2N5195	data
BD 723	Si-N	=BD 719: 100V	14h§	Phi	-	data
BD 724	Si-P	=BD 720: 100V	14h§	Phi	-	data
BD 725	Si-N	=BD 719: 120V	14h§	Phi	-	data
BD 726	Si-P	=BD 720: 120V	14h§	Phi	-	data
BD 733	Si-N	NF-L, 25V, 4A, 40W, >3MHz	17j§	Nsc,Tix	BD 243, BD 533, BD 539, BD 943, ++	data
BD 734	Si-P	NF-L, 25V, 4A, 40W, >3MHz	17j§	Nsc,Tix	BD 244, BD 534, BD 540, BD 944, ++	data
BD 735	Si-N	=BD 733: 35V	17j§	Nsc,Tix	BD 243, BD 533, BD 539, BD 947, ++	data
BD 736	Si-P	=BD 734: 35V	17j§	Nsc,Tix	BD 244, BD 534, BD 540, BD 948, ++	data
BD 737	Si-N	=BD 733: 45V	17j§	Nsc,Tix	BD 243, BD 533, BD 539A, BD 947, ++	data
BD 738	Si-P	=BD 734: 45V	17j§	Nsc,Tix	BD 244, BD 534, BD 540A, BD 948, ++	data
BD 743	Si-N	NF-L, 50V, 15A, 90W, >5MHz	17j§	Tix	BD 907	data
BD 743 A	Si-N	=BD 743: 70V	17j§		BD 909	data
BD 743 B	Si-N	=BD 743: 90V	17j§		BD 911	data
BD 743 C	Si-N	=BD 743: 110V	17j§		-	data
BD 743 D	Si-N	=BD 743: 130V	17j§		-	data
BD 743 E	Si-N	=BD 743: 150V	17j§		-	data
BD 743 F	Si-N	=BD 743: 170V	17j§		-	data
BD 744	Si-P	NF-L, 50V, 15A, 90W, >5MHz	17j§	Tix	BD 908	data
BD 744 A	Si-P	=BD 744: 70V	17j§		BD 910	data
BD 744 B	Si-P	=BD 744: 90V	17j§		BD 912	data
BD 744 C	Si-P	=BD 744: 110V	17j§		-	data
BD 744 D	Si-P	=BD 744: 130V	17j§		-	data
BD 744 E	Si-P	=BD 744: 150V	17j§		-	data
BD 744 F	Si-P	=BD 744: 170V	17j§		-	data
BD 745	Si-N	NF-L, 50V, 20A, 115W, >5MHz	18j§	Tix	BD 249	data
BD 745 A	Si-N	=BD 745: 70V	18j§		BD 249A	data
BD 745 B	Si-N	=BD 745: 90V	18j§		BD 249B	data
BD 745 C	Si-N	=BD 745: 110V	18j§		BD 249C	data
BD 745 D	Si-N	=BD 745: 130V	18j§		BD 249D	data
BD 745 E	Si-N	=BD 745: 150V	18j§		BD 249D	data
BD 745 F	Si-N	=BD 745: 170V	18j§		• BD 249E	data
BD 746	Si-P	NF-L, 50V, 20A, 115W, >5MHz	18j§	Tix	BD 250	data
BD 746 A	Si-P	=BD 746: 70V	18j§		BD 250A	data
BD 746 B	Si-P	=BD 746: 90V	18j§		BD 250B	data
BD 746 C	Si-P	=BD 746: 110V	18j§		BD 250C	data
BD 746 D	Si-P	=BD 746: 130V	18j§		BD 250D	data
BD 746 E	Si-P	=BD 746: 150V	18j§		BD 250D	data
BD 746 F	Si-P	=BD 746: 170V	18j§		BD 250E	data
BD 750	Si-P	NF-L, 100V, 20A(ss), 200W, >4MHz	23a§	Rca	BD 318, 2N6029...6031, 2SA1117	data
BD 750 A	Si-P	=BD 750: 130V	23a§		2N6030...6031, MJ 15016, 2SA1117	data
BD 750 B	Si-P	=BD 750: 110V, 250W	23a§		-	data
BD 750 C	Si-P	=BD 750: 140V, 250W	23a§		-	data
BD 751	Si-N	NF-L, 100V, 20A(ss), 200W, >4MHz	23a§	Rca	BD 317, 2N5629...5631, 2SC2608	data
BD 751 A	Si-N	=BD 751: 130V	23a§		MJ 15015, 2N5630...5631, 2SC2608	data
BD 751 B	Si-N	=BD 751: 110V, 250W	23a§		-	data
BD 751 C	Si-N	=BD 751: 140V, 250W	23a§		-	data
BD 775	Si-N-Darl+Di	NF-L, 45V, 4A, 15W, >20MHz, B>750	14h§	Mot	BD 675, 2N6038	data
BD 776	Si-P-Darl+Di	NF-L, 45V, 4A, 15W, >20MHz, B>750	14h§	Mot	BD 676, 2N6035	data
BD 777	Si-N-Darl+Di	=BD 775: 60V	14h§	Mot	BD 677, 2N6038	data
BD 778	Si-P-Darl+Di	=BD 776: 60V	14h§	Mot	BD 678, 2N6035	data
BD 779	Si-N-Darl+Di	=BD 775: 80V	14h§	Mot	BD 679, 2N6039	data
BD 780	Si-P-Darl+Di	=BD 776: 80V	14h§	Mot	BD 680, 2N6036	data
BD 785	Si-N	NF-L, 60V, 4A, 15W, >50MHz	14h§	Mot	BDX 35...37, MJE 240...244	data
BD 786	Si-P	NF-L, 60V, 4A, 15W, >40MHz	14h§	Mot	MJE 250...254	data
BD 787	Si-N	=BD 785: 80V	14h§	Mot	BDX 35...37, MJE 240...244	data
BD 788	Si-P	=BD 786: 80V	14h§	Mot	MJE 250...254	data
BD 789	Si-N	=BD 785: 80V	14h§	Mot	BDX 35...37, MJE 240:.244	data
BD 790	Si-P	=BD 786: 80V	14h§	Mot	MJE 250...254	data
BD 791	Si-N	=BD 785: 100V	14h§	Mot	BDX 35...37, MJE 240...244	data
BD 792	Si-P	=BD 786: 100V	14h§	Mot	MJE 250...254	data
BD 795	Si-N	NF-L, 45V, 8A, 65W, >3MHz	17j§	Mot,Rca	BD 543A, BD 705, BD 805	data
BD 796	Si-P	NF-L, 45V, 8A, 65W, >3MHz	17j§	Mot,Rca	BD 544A, BD 706, BD 806	data
BD 797	Si-N	=BD 795: 60V	17j§	Mot,Rca	BD 543A, BD 707, BD 807	data
BD 798	Si-P	=BD 796: 60V	17j§	Mot,Rca	BD 544A, BD 708, BD 808	data
BD 799	Si-N	=BD 795: 80V	17j§	Mot,Rca	BD 543B, BD 709, BD 809	data
BD 800	Si-P	=BD 796: 80V	17j§	Mot,Rca	BD 544B, BD 710, BD 810	data
BD 801	Si-N	=BD 795: 100V	17j§	Mot,Rca	BD 543C, BD 711	data
BD 802	Si-P	=BD 796: 100V	17j§	Mot,Rca	BD 544C, BD 712	data
BD 805	Si-N	NF-L, 55V, 10A, 90W, >1,5MHz	17j§	Mot	BD 707, BD 743A, BD 907	data
BD 806	Si-P	NF-L, 55V, 10A, 90W, >1,5MHz	17j§	Mot	BD 708, .BD 744A, BD 908	data
BD 807	Si-N	=BD 805: 70V	17j§	Mot	BD 709, BD 743A, BD 909	data
BD 808	Si-P	=BD 806: 70V	17j§	Mot	BD 710, BD 744A, BD 910	data
BD 809	Si-N	=BD 805: 80V	17j§	Mot	BD 709, BD 743B, BD 909	data
BD 810	Si-P	=BD 806: 80V	17j§	Mot	BD 710, BD 744B, BD 910	data
BD 813(A)	Si-N	NF-L, 45V, 2A, 12,5W, >3MHz	13h§	Aeg,Phi	BD 175, BD 233, BD 615	data
BD 814(A)	Si-P	NF-L, 45V, 2A, 12,5W, >3MHz	13h§	Aeg,Phi	BD 176, BD 234, BD 616	data

Type	Device	Short Description	Fig.	Manu	Comparision Types	More at
BD 815(A)	Si-N	=BD 813: 60V	13h§	Aeg,Phi	BD 177, BD 235, BD 617	data
BD 816(A)	Si-P	=BD 814: 60V	13h§	Aeg,Phi	BD 178, BD 236, BD 618	data
BD 817(A)	Si-N	=BD 813: 100V	13h§	Aeg,Phi	BD 237, BD 443	data
BD 818(A)	Si-P	=BD 814: 100V	13h§	Aeg,Phi	BD 238	data
BD 825(A,B)	Si-N	NF-L, 45V, 1A, 8W, 250MHz	13h§	Aeg,Phi,++	BD 135, BD 226, BD 385, BD 839, ++	data
BD 826(A,B)	Si-P	NF-L, 45V, 1A, 8W, 75MHz	13h§	Aeg,Phi,++	BD 136, BD 227, BD 386, BD 840, ++	data
BD 827(A,B)	Si-N	=BD 825: 60V	13h§	Aeg,Phi,++	BD 137, BD 228, BD 385, BD 841, ++	data
BD 828(A,B)	Si-P	=BD 826: 60V	13h§	Aeg,Phi,++	BD 138, BD 229, BD 386, BD 842, ++	data
BD 829(A,B)	Si-N	=BD 825: 100V	13h§	Aeg,Phi,++	BD 139, BD 230, BD 389, BD 843, ++	data
BD 830(A,B)	Si-P	=BD 826: 100V	13h§	Aeg,Phi,++	BD 140, BD 231, BD 390, BD 844, ++	data
BD 833	Si-N	NF-L, 45V, 3A, 15W, 50MHz	13j§	Gen	2SC3252...3253	data
BD 834	Si-P	NF-L, 45V, 3A, 15W, 40MHz	13j§	Gen	2SA1288...1289	data
BD 835	Si-N	=BD 833: 60V	13j§	Gen	2SC3252...3253	data
BD 836	Si-P	=BD 834: 60V	13j§	Gen	2SA1288...1289	data
BD 837	Si-N	=BD 833: 100V	13j§	Gen	-	data
BD 838	Si-P	=BD 834: 100V	13j§	Gen	-	data
BD 839	Si-N	NF-L, 45V, 1,5A, 10W, 125MHz	13h§	Phi	BD 135, BD 226, BD 375	data
BD 840	Si-P	NF-L, 45V, 1,5A, 10W, 50MHz	13h§	Phi	BD 136, BD 227, BD 376	data
BD 841	Si-N	=BD 839: 60V	13h§	Phi	BD 137, BD 228, BD 377	data
BD 842	Si-P	=BD 840: 60V	13h§	Phi	BD 138, BD 229, BD 378	data
BD 843	Si-N	=BD 839: 100V	13h§	Phi	BD 139, BD 230, BD 379	data
BD 844	Si-P	=BD 840: 100V	13h§	Phi	BD 140, BD 231, BD 380	data
BD 845	Si-N	NF-L, 100V, 1,5A, 10W, >150MHz	13h§	Phi	BD 139, BD 379, BD 843, 2SC3117, 2SD669	data
BD 846	Si-P	NF-L, 100V, 1,5A, 10W, >150MHz	13h§	Phi	BD 140, BD 380, BD 844, 2SA1249, 2SB649	data
BD 847	Si-N	=BD 845: 120V	13h§	Phi	2SC3117, 2SD669	data
BD 848	Si-P	=BD 846: 120V	13h§	Phi	2SA1249, 2SB649	data
BD 849	Si-N	=BD 845: 140V	13h§	Phi	2SC3117, 2SD669	data
BD 850	Si-P	=BD 845: 140V	13h§	Phi	2SA1249, 2SB649	data
BD 861	Si-N-Darl	NF-L, 45V, 4A, 15W, 7MHz, B>750	13h§	Sie	BD 675, BD 775, 2N6038	data
BD 862	Si-P-Darl	NF-L, 45V, 4A, 15W, 7MHz, B>750	13h§	Sie	BD 676, BD 776, 2N6035	data
BD 863	Si-N-Darl	=BD 861: 60V	13h§a	Sie	BD 677, BD 777, 2N6038	data
BD 864	Si-P-Darl	=BD 862: 60V	13h§	Sie	BD 678, BD 778, 2N6035	data
BD 865	Si-N-Darl	=BD 861: 80V	13h§	Sie	BD 679, BD 779, 2N6039	data
BD 866	Si-P-Darl	=BD 862: 80V	13h§	Sie	BD 680, BD 780, 2N6036	data
BD 875	Si-N-Darl	NF-L, 60V, 1A, 9W(Tc=60°), 200MHz	14h§	Sie	BDX 42...44	data
BD 876	Si-P-Darl	NF-L, 60V, 1A, 9W(Tc=60°), 200MHz	14h§	Sie	BDX 45...47	data
BD 877	Si-N-Darl	=BD 875: 80V	14h§	Sie	BDX 43...44	data
BD 878	Si-P-Darl	=BD 876: 80V	14h§	Sie	BDX 46...47	data
BD 879	Si-N-Darl	=BD 875: 100V	14h§	Sie	BDX 44	data
BD 880	Si-P-Darl	=BD 876: 100V	14h§	Sie	BDX 47	data
BD 887	Si-P	NF-L, 30V, 20A, 62,5W, >50MHz	17j§	Sie	(BD 250, BD 746)[4]	data
BD 888	Si-P	=BD 887: 45V	17j§	Sie	(BD 250, BD 746)[4]	data
BD 895(A)	Si-N-Darl+Di	NF-L, 45V, 8A, 70W, >1MHz, B>750	17j§	Mot,Tix,++	BD 643, BDW 73, BDX 33, BDX 53, ++	data
BD 896(A)	Si-P-Darl+Di	NF-L, 45V, 8A, 70W, >1MHz, B>750	17j§	Mot,Tix,++	BD 644, BDW 74, BDX 34, BDX 54, ++	data
BD 897(A)	Si-N-Darl+Di	=BD 895: 60V	17j§	Mot,Tix,++	BD 645, BDW 73A, BDX 33A, BDX 53A, ++	data
BD 898(A)	Si-P-Darl+Di	=BD 896: 60V	17j§	Mot,Tix,++	BD 646, BDW 74A, BDX 34A, BDX 54A, ++	data
BD 899(A)	Si-N-Darl+Di	=BD 895: 80V	17j§	Mot,Tix,++	BD 647, BDW 73B, BDX 33B, BDX 53B, ++	data
BD 900(A)	Si-P-Darl+Di	=BD 896: 80V	17j§	Mot,Tix,++	BD 648, BDW 74B, BDX 34B, BDX 54B, ++	data
BD 901	Si-N-Darl+Di	=BD 895: 100V	17j§	Mot,Tix,++	BD 649, BDW 73C, BDX 33C, BDX 53C, ++	data
BD 902	Si-P-Darl+Di	=BD 896: 100V	17j§	Mot,Tix,++	BD 650, BDW 74C, BDX 34C, BDX 54C, ++	data
BD 905	Si-N	NF-L, 45V, 15A, 90W, >3MHz	17j§	Tho	BD 743	data
BD 905 Fl	Si-N	=BD 905: Iso; 35W	17c	Tho	BDT 81F	data
BD 906	Si-P	NF-L, 45V, 15A, 90W, >3MHz	17j§	Tho	BD 744	data
BD 907	Si-N	=BD 905: 60V	17j§	Tho	BD 743A	data
BD 908	Si-P	=BD 906: 60V	17j§	Tho	BD 744A	data
BD 908 Fl	Si-P	=BD 906: Iso, 60, 35W	17c	Sgs	BDT 82F, 2SA1444, 2SA1744	data
BD 908 Fl	Si-P	=BD 908: Iso, 35W	17c	Tho	BDT 84F	data
BD 909	Si-N	=BD 905: 80V	17j§	Tho	BD 743B	data
BD 910	Si-P	=BD 906: 80V	17j§	Tho	BD 744B	data
BD 911	Si-N	=BD 905: 100V	17j§	Tho	BD 743C	data
BD 912	Si-P	=BD 906: 100V	17j§	Tho	BD 744C	data
BD 933	Si-N	NF-L, 45V, 3A, 30W, >3MHz	17j§	Phi	BD 241, BD 533, BD 539A, BD 947, ++	data
BD 933 F	Si-N	=BD 933: Iso, >14W	17c		BD 947F, BDT 31F, 2SD1985, 2SD2012, ++	data
BD 934	Si-P	NF-L, 45V, 3A, 30W, >3MHz	17j§	Phi	BD 242, BD 534, BD 540A, BD 948, ++	data
BD 934 F	Si-P	=BD 934: Iso, >14W	17c		BD 948F, BDT 32F, 2SB1015, 2SB1094, ++	data
BD 935	Si-N	=BD 933: 60V	17j§	Phi	BD 241A, BD 535, BD 539A, BD 949, ++	data
BD 935 F	Si-N	=BD 935: Iso, >14W	17c		BD 949F, BDT 31F, 2SD1985, 2SD2012, ++	data
BD 936	Si-P	=BD 934: 60V	17j§	Phi	BD 242A, BD 536, BD 540A, BD 950, ++	data
BD 936 F	Si-P	=BD 936: Iso, >14W	17c		BD 950F, BDT 32F, 2SB1015, 2SB1094, ++	data
BD 937	Si-N	=BD 933: 100V	17j§	Phi	BD 241C, BD 539C, BD 953, 2SD712, ++	data
BD 937 F	Si-N	=BD 937: Iso, >14W	17c		BD 953F, BDT 31AF, 2SD1407, 2SD1586, ++	data
BD 938	Si-P	=BD 934: 100V	17j§	Phi	BD 242C, BD 540C, BD 954, 2SB682, ++	data
BD 938 F	Si-P	=BD 938: Iso, >14W	17c		BD 954F, BDT 32AF, 2SB1095, 2SB1294, ++	data
BD 939	Si-N	=BD 933: 120V	17j§	Phi	BD 241C, BD 539D, BD 955, 2SD959...961,++	data
BD 939 F	Si-N	=BD 939: Iso, >14W	17c		BD 955F, BDT 31BF, 2SC3566, 2SC4334, ++	data
BD 940	Si-P	=BD 934: 120V	17j§	Phi	BD 242C, BD 540D, BD 956, 2SB867...869,++	data
BD 940 F	Si-P	=BD 940: Iso, >14W	17c		BD 956F, BDT 32BF, 2SA1650	data
BD 941	Si-N	=BD 933: 140V	17j§	Phi	BD 241D, BD 243D, 2SC2516	data
BD 941 F	Si-N	=BD 941: Iso, >14W	17c		BDT 31CF, 2SC3566, 2SC4334	data
BD 942	Si-P	=BD 934: 140V	17j§	Phi	BD 242D, BD 244D	data
BD 942 F	Si-P	=BD 942: Iso, >14W	17c		BDT 32CF, 2SA1650	data

Type	Device	Short Description	Fig.	Manu	Comparision Types	More at
BD 943	Si-N	NF-L, 22V, 5A, 40W, >3MHz	17j§	Phi	BD 243, BD 539, BD 795, BD 943, ++	data
BD 943 F	Si-N	=BD 943: Iso, >15W	17c		BD 949F, BDT 41F, BDT 91F, 2SD1667, ++	data
BD 944	Si-P	NF-L, 22V, 5A, 40W, >3MHz	17j§	Phi	BD 244, BD 540, BD 796, BD 944, ++	data
BD 944 F	Si-P	=BD 944: Iso, >15W	17c		BD 950F, BDT 42F, BDT 92F, 2SB1134, ++	data
BD 945	Si-N	=BD 943: 32V	17j§	Phi	BD 243, BD 539, BD 795, BD 945, ++	data
BD 945 F	Si-N	=BD 945: Iso, >15W	17c		BD 949F, BDT 41F, BDT 91F, 2SD1667, ++	data
BD 946	Si-P	=BD 944: 32V	17j§	Phi	BD 244, BD 540, BD 796, BD 946, ++	data
BD 946 F	Si-P	=BD 946: Iso, >15W	17c		BD 950F, BDT 42F, BDT 92F, 2SB1134, ++	data
BD 947	Si-N	=BD 943: 45V	17j§	Phi	BD 243A, BD 539A, BD 975, BD 947, ++	data
BD 947 F	Si-N	=BD 947: Iso, >15W	17c		BD 949F, BDT 41F, BDT 91F, 2SD1667, ++	data
BD 948	Si-P	=BD 944: 45V	17j§	Phi	BD 244A, BD 540A, BD 796, BD 948, ++	data
BD 948 F	Si-P	=BD 948: Iso, >15W	17c		BD 950F, BDT 42F, BDT 92F, 2SB1134, ++	data
BD 949	Si-N	NF-L, 60V, 5A, 40W, >3MHz	17j§	Phi	BD 243A, BD 539A, BD 543A, BD 797, ++	data
BD 949 F	Si-N	=BD 949: Iso, >15W	17c		2DT 41F, BDT 91F, BDX 77F, 2SD1667, ++	data
BD 950	Si-P	NF-L, 60V, 5A, 40W, >3MHz	17j§	Phi	BD 244A, BD 540A, BD 544A, BD 798, ++	data
BD 950 F	Si-P	=BD 950: Iso, >15W	17c		BDT 42F, BDT 92F, BDX 78F, 2SB1134, ++	data
BD 951	Si-N	=BD 949: 80V	17j§	Phi	BD 243B, BD 539B, BD 543B, BD 799, ++	data
BD 951 F	Si-N	=BD 951: Iso, >15W	17c		BDT 41F, BDT 93F, BDX 77F, 2SD1940, ++	data
BD 952	Si-P	=BD 950: 80V	17j§	Phi	BD 244B, BD 540B, BD 544B, BD 800, ++	data
BD 952 F	Si-P	=BD 952: Iso, >15W	17c		BDT 42F, BDT 94F, BDX 78F, 2SB1294, ++	data
BD 953	Si-N	=BD 949: 100V	17j§	Phi	BD 243C, BD 539C, BD 543C, BD 801, ++	data
BD 953 F	Si-N	=BD 953: Iso, >15W	17c		BDT 41AF, BDT 95F, BDX 77F, 2SD1940, ++	data
BD 954	Si-P	=BD 950: 100V	17j§	Phi	BD 244C, BD 540C, BD 544C, BD 802, ++	data
BD 954 F	Si-P	=BD 954: Iso, >15W	17c		BDT 42AF, BDT 96F, BDX 78F, 2SB1294, ++	data
BD 955	Si-N	=BD 949: 120V	17j§	Phi	BD 543D, 2SD866	data
BD 955 F	Si-N	=BD 955: Iso, >15W	17c		BDT 41BF, 2SC4335	data
BD 956	Si-P	=BD 950: 120V	17j§	Phi	BD 544D, 2SB870	data
BD 956 F	Si-P	=BD 956: Iso, >15W	17c		BDT 42BF, 2SA1650	data
BD 975	Si-N-Darl	NF-L, 60V, 1A, 3,6W(Tc=60°), 200MHz	13h§	Sie	BDX 42...44	data
BD 976	Si-P-Darl	NF-L, 60V, 1A, 3,6W(Tc=60°), 200MHz	13h§	Sie	BDX 45...47	data
BD 977	Si-N-Darl	=BD 975: 80V	13h§	Sie	BDX 43..44	data
BD 978	Si-P-Darl	=BD 976: 80V	13h§	Sie	BDX 46..47	data
BD 979	Si-N-Darl	=BD 975: 100V	13h§	Sie	BDX 44	data
BD 980	Si-P-Darl	=BD 976: 100V	13h§	Sie	BDX 47	data
BD 1540	Si-N-Darl	S-L, 400/400V, 15A, 100W, 7MHz	23a§	Ssc	BU 930...932, 2SD572..573, 2SD711, ++	data
BD 1550	Si-N-Darl	=BD 1540: 500V	23a§	Ssc	BU 932, 2SD572...573, 2SD683, 2SD711, ++	data
BD 1560	Si-N-Darl	=BD 1540: 600/600V	23a§	Ssc	2SD573, 2SD606, 2SD683, 2SD711A, ++	data
BD 2530	Si-N-Darl	S-L, 300/300V, 25A, 100W, 7MHz	23a§	Ssc	BUT 13	data
BD 2540	Si-N-Darl	=BD 2530: 400/400V	23a§	Ssc	BUT 13	data
BD 2550	Si-N-Darl	=BD 2530: 500/500V	23a§	Ssc	BUT 13	data
BD9060	LDR	610nm Ro:0.15MOhm Vb:2V		Hei		data pinout
BD 9766 FV	LIN-IC	DC-AC Inverter Control IC, 12V	TSOP	Rhm	-	pdf pinout
BD 9882 F	IC	DC-AC Inverter Control IC	MDIP	Rhm		pdf pinout
BD 9882 FV	IC	DC-AC Inverter Control IC	TSOP	Rhm		pdf pinout
BD 9883 AF	IC	DC-AC Inverter Control AC	MDIP	Rhm		pdf pinout
BD 9883 FV	IC	DC-AC Inverter Control AC	TSOP	Rhm		pdf pinout

BDB...BDS

Type	Device	Short Description	Fig.	Manu	Comparision Types	More at
BDB 01A	Si-N	NF-Tr/E, 45V, 0,5A, 1W, >50MHz	7e(9mm)	Mot	2SC2235, 2SD667(A), 2SD863, 2SD1292,++	data
BDB 01B	Si-N	=BDB 01A: 60V	7e(9mm)		2SC2235, 2SD667(A), 2SD863, 2SD1292,++	data
BDB 01C	Si-N	=BDB 01A: 80V	7e(9mm)		2SC2235, 2SD667(A), 2SD1292, 2SD1812,++	data
BDB 01D	Si-N	=BDB 01A: 100V	7e(9mm)		2SC2235, 2SD667(A), 2SD1292, 2SD1812,++	data
BDB 02A	Si-P	NF-Tr/E, 45V, 0,5A, 1W, >50MHz	7e(9mm)	Mot	2SA965, 2SB647(A), 2SB764, 2SB1041,++	data
BDB 02B	Si-P	=BDB 02A: 60V	7e(9mm)		2SA965, 2SB647(A), 2SB764, 2SB1041,++	data
BDB 02C	Si-P	=BDB 02A: 80V	7e(9mm)		2SA965, 2SB647(A), 2SB1041, 2SB1212,++	data
BDB 02D	Si-P	=BDB 02A: 100V	7e(9mm)		2SA965, 2SA1275, 2SB647(A), 2SB1212,++	data
BDB 03	Si-N	NF-Tr/E, 60V, 1A, 1W, >150MHz	7e(9mm)	Mot	2SC2235, 2SD667(A), 2SD863, 2SD1292,++	data
BDB 04	Si-P	NF-Tr/E, 60V, 1A, 1W, >150MHz	7e(9mm)	Mot	2SA965, 2SB647(A), 2SB764, 2SB1041,++	data
BDB 05	Si-N	NF-Tr/E, 120V, 1A, 1W, >100MHz	7e(9mm)	Mot	2SC3228, 2SD667(A), 2SD1292, 2SD1812,++	data
BDB 06	Si-P	NF-Tr/E, 80V, 1A, >150MHz	7e(9mm)	Mot	2SA965, 2SA1275, 2SB647(A), 2SB1212,++	data
BDC 01(A...D)	Si-N	=BDB 01(A...D):	7c(9mm)	Mot	→BDB 01(A...D)	data
BDC 02(A...D)	Si-P	=BDB 02(A...D):	7c(9mm)	Mot	→BDB 02(A...D)	data
BDC 03	Si-N	NF-Tr/E, 25V, 1A, 1W, >50MHz	7c(9mm)	Mot	2SC2236, 2SD863, 2SD1146, 2SD1331,++	data
BDC 04	Si-P	NF-Tr/E, 25V, 1A, 1W, >50MHz	7c(9mm)	Mot	2SA966, 2SB764, 2SB978, 2SB892,++	data
BDC 05	Si-N	Vid, 300V, 0,5A, 1W, >60MHz	7c(9mm)	Mot	(BF 382, BF 461, BF 758, MPSU 10)[5]	data
BDC 06	Si-P	Vid, 300V, 0,5A, 1W, >60MHz	7c(9mm)	Mot	(BF 464, BF 761, MPSU 60)[5]	data
BDC 07	Si-N	=BDC 05: 250V	7c(9mm)	Mot	(BF 381, BF 460, BF 757, MPSU 10)[5]	data
BDC 08	Si-P	=BDC 06: 250V	7c(9mm)	Mot	(BF 463, BF 760, MPSU 60)[5]	data
BDD	Z-Di	=SM 15T 6V8C(Typ-Code/Stempel/marking)	71(8x5mm)	Tho	→SM 15T...	data
BDE	Z-Di	=SM 15T6V8CA(Typ-Code/Stempel/marking)	71(8x5mm)	Tho	→SM 15T...	data
BDF	Z-Di	=SM 15T 7V5C(Typ-Code/Stempel/marking)	71(8x5mm)	Tho	→SM 15T...	data
BDG	Z-Di	=SM 15T7V5CA(Typ-Code/Stempel/marking)	71(8x5mm)	Tho	→SM 15T...	data
BDL 31	Si-N	Uni, 40V, 5A, 1,35W, >100MHz	48zh§	Phi	-	data
BDL 32	Si-P	Uni, 15V, 5A, 1,35W, >100MHz	48zh§	Phi	-	data
BDM	Si-P	=2SA1733K-M (Typ-Code/Stempel/marking)	35	Rhm	→2SA1733K	data
BDM	Si-P	=2SA1808-M (Typ-Code/Stempel/marking)	35(2mm)	Rhm	→2SA1808	data
BDM	Si-P	=2SA1821-M (Typ-Code/Stempel/marking)	35(1,6mm)	Rhm	→2SA1821	data
BDN	Si-P	=2SA1733K-N (Typ-Code/Stempel/marking)	35	Rhm	→2SA1733K	data

Type	Device	Short Description	Fig.	Manu	Comparision Types	More at
BDN	Si-P	=2SA1808-N (Typ-Code/Stempel/marking)	35(2mm)	Rhm	→2SA1808	data
BDN	Si-P	=2SA1821-N (Typ-Code/Stempel/marking)	35(1,6mm)	Rhm	→2SA1821	data
BDN	Z-Di	=SM 15T 10C (Typ-Code/Stempel/marking)	71(8x5mm)	Tho	→SM 15T...	data
BDP	Si-P	=2SA1733K-P (Typ-Code/Stempel/marking)	35	Rhm	→2SA1733K	data
BDP	Si-P	=2SA1808-P (Typ-Code/Stempel/marking)	35(2mm)	Rhm	→2SA1808	data
BDP	Si-P	=2SA1821-P (Typ-Code/Stempel/marking)	35(1,6mm)	Rhm	→2SA1821	data
BDP	Si-P	=2SB1189-P (Typ-Code/Stempel/marking)	39	Rhm	→2SB1189	data
BDp	Si-P	=BCW 61D (Typ-Code/Stempel/marking)	35	Phi	→BCW 61D	data
BDP	Z-Di	=SM 15T 10CA(Typ-Code/Stempel/marking)	71(8x5mm)	Tho	→SM 15T...	data
BDP 31	Si-N	SMD, Uni, 70V, 3A, 1,35W, >60MHz	48j§	Phi	-	data
BDP 32	Si-P	SMD, Uni, 45V, 3A, 1,35W, >60MHz	48j§	Phi	-	data
BDP 947	Si-N	SMD, NF-Tr/E, lo-sat, 45V, 3A, 100MHz	48j§	Sie	BDP 951, BDP 953, BDP 955	data
BDP 948	Si-P	SMD, NF-Tr/E, lo-sat, 45V, 3A, 100MHz	48j§	Sie	BDP 952, BDP 954, BDP 956	data
BDP 949	Si-N	=BDP 947: 60V	48j§	Sie	BDP 951, BDP 953, BDP 955	data
BDP 950	Si-P	=BDP 948: 60V	48j§	Sie	BDP 952, BDP 954, BDP 956	data
BDP 951	Si-N	SMD, NF-Tr/E, lo-sat, 100V, 3A, 100MHz	48j§	Sie	-	data
BDP 952	Si-P	SMD, NF-Tr/E, lo-sat, 100V, 3A, 100MHz	48j§	Sie	-	data
BDP 953	Si-N	=BDP 951: 120V	48j§	Sie	-	data
BDP 954	Si-P	=BDP 952: 120V	48j§	Sie	-	data
BDP 955	Si-N	=BDP 951: 140V	48j§	Sie	-	data
BDP 956	Si-P	=BDP 952: 140V	48j§	Sie	-	data
BDP 4148	Si-Di	Di-Array, 8x 1N4148	16-DIP	Say	-	data
BDQ	Si-P	=2SB1189-Q (Typ-Code/Stempel/marking)	39	Rhm	→2SB1189	data
BDR	Si-P	=2SB1189-R (Typ-Code/Stempel/marking)	39	Rhm	→2SB1189	data
BDs	Si-P	=BCW 61D (Typ-Code/Stempel/marking)	35	Sie	→BCW 61D	data
BDS	Z-Di	=SM 15T 12C (Typ-Code/Stempel/marking)	71(8x5mm)	Tho	→SM 15T...	data
BDS 60 (A...C)	Si-P-Darl+Di	=BDT 60...: 3A, 8W	48j	Phi	-	data
BDS 61 (A...C)	Si-N-Darl+Di	=BDT 61...: 3A, 8W	48j	Phi	-	data
BDS 77	Si-N	=BDX 77: 3A, 8W	48j	Phi	-	data
BDS 78	Si-P	=BDX 78: 3A, 8W	48j	Phi	-	data
BDS 201...BDS 204	Si-N/P	=BD 201...204: 3A, 8W	48j	Phi	-	data
BDS 643...BDS 652	Si-N/P-Darl	=BD 643...652: 3A, 8W	48j	Phi	-	data
BDS 933...BDS 942	Si-N/P	=BD 933...942: 3A, 8W	48j	Phi	-	data
BDS 943...BDS 948	Si-N/P	=BD 943...948: 3A, 8W	48j	Phi	-	data
BDS 949...BDS 956	Si-N/P	=BD 949...956: 3A, 8W	48j	Phi	-	data

BDT

Type	Device	Short Description	Fig.	Manu	Comparision Types	More at
BDt	Si-P	=BCW 61D (Typ-Code/Stempel/marking)	35	Phi	→BCW 61D	data
BDT	Z-Di	=SM 15T 12CA(Typ-Code/Stempel/marking)	71(8x5mm)	Tho	→SM 15T...	data
BDT 20	Si-P-Darl+Di	NF-L, 130V, 8A, 62,5W, B>750	17j§	Phi	BDX 54E	data
BDT 21	Si-N-Darl	NF-L, 130V, 8A, 62,5W, B>750	17j§	Phi	BDX 53, 2SD1386, 2SD1025, 2SD1500	data
BDT 29	Si-N	NF-L, 80V, 1A, 30W, >3MHz	17j§a	Phi	BD 239B, BD 241B, BD 537, BD 937, ++	data
BDT 29 A	Si-N	=BDT 29: 100V	17j§		BD 239C, BD 241C, BD 937, 2SD1138, ++	data
BDT 29 AF	Si-N	=BDT 29: Iso, 100V, >15W	17c		BD 937F, BDT 31AF, 2SC3298, 2SD1586, ++	data
BDT 29 B	Si-N	=BDT 29: 120V	17j§		BD 239C, BD 241C, BD 939, 2SD1138, ++	data
BDT 29 BF	Si-N	=BDT 29: Iso, 120V, >15W	17c		BD 939F, BDT 31BF, 2SC3298, 2SD1587, ++	data
BDT 29 C	Si-N	=BDT 29: 140V	17j§		BD 239D, BD 241D, 2SD1138, 2SD1459, ++	data
BDT 29 CF	Si-N	=BDT 29: Iso, 140V, >15W	17c		BD 941F, BDT 31CF, 2SC3298, 2SD1587, ++	data
BDT 29 DF	Si-N	=BDT 29: Iso, 160V, >15W	17c		BDT 31DF, 2SC3298(A...B), 2SD1587, ++	data
BDT 29 F	Si-N	=BDT 29: Iso, >15W	17c		BD 937F, BDT 31F, 2SC3851, 2SD1408, ++	data
BDT 30	Si-P	NF-L, 80V, 1A, 30W, >3MHz	17j§	Phi	BD 240B, BD 242B, BD 538, BD 938, ++	data
BDT 30 A	Si-P	=BDT 30: 100V	17j§		BD 240C, BD 242C, BD 938, 2SB861, ++	data
BDT 30 AF	Si-P	=BDT 30: Iso, 100V, >15W	17c		BD 938F, BDT 32AF, 2SA1306, 2SB1095, ++	data
BDT 30 B	Si-P	=BDT 30: 120V	17j§		BD 240C, BD 242C, BD 940, 2SB861, ++	data
BDT 30 BF	Si-P	=BDT 30: Iso, 120V, >15W	17c		BD 940F, BDT 32BF, 2SA1306, 2SB1096, ++	data
BDT 30 C	Si-P	=BDT 30: 140V	17j§		BD 240D, BD 242D, 2SB861, 2SB1037, ++	data
BDT 30 CF	Si-P	=BDT 30: Iso, 140V, >15W	17c		BD 942F, BDT 32CF, 2SA1306, 2SB1096, ++	data
BDT 30 DF	Si-P	=BDT 30: Iso, 160V, >14W	17c		BDT 32DF, 2SA1306(A...B), 2SB1096, ++	data
BDT 30 F	Si-P	=BDT 30: Iso, >15W	17c		BD 938F, BDT 32F, 2SA1635, 2SB1017, ++	data
BDT 31	Si-N	NF-L, 80V, 3A, 40W, >3MHz	17j§	Phi	BD 241B, BD 537, BD 539B, BD 937, ++	data
BDT 31 A	Si-N	=BDT 31: 100V	17j§		BD 241C, BD 243C, BD 539C, BD 937, ++	data
BDT 31 AF	Si-N	=BDT 31: Iso, 100V, >15W	17c		BD 937F, BDT 953F, BDT 41AF, 2SD1586, ++	data
BDT 31 B	Si-N	=BDT 31: 120V	17j§		BD 241C, BD 243C, BD 539D, BD 939, ++	data
BDT 31 BF	Si-N	=BDT 31: Iso, 120V, >15W	17c		BD 939F, BD 955F, BDT 41BF, 2SC3566, ++	data
BDT 31 C	Si-N	=BDT 31: 140V	17j§		BD 241D, BD 243D, BD 941	data
BDT 31 CF	Si-N	=BDT 31: Iso, 140V, >15W	17c		BD 941F, BDT 41CF, 2SC3566, 2SC4334	data
BDT 31 DF	Si-N	=BDT 31: Iso, 160V, >15W	17c		2SC4153	data
BDT 31 F	Si-N	=BDT 31: Iso, >15W	17c		BD 937F, BDT 951F, BDT 41F, 2SD1586, ++	data
BDT 32	Si-P	NF-L, 80/40V, 3A, 40W, >3MHz	17j§	Phi	BD 242B, BD 538, BD 540B, BD 938, ++	data
BDT 32 A	Si-P	=BDT 32: 60/60V	17j§		BD 242B, BD 244B, BD 540B, BD 938, ++	data
BDT 32 AF	Si-P	=BDT 32: Iso, 60/60V, >15W	17c		BD 938F, BD 954F, BDT 42AF, 2SB1095, ++	data
BDT 32 B	Si-P	=BDT 32: 80/80V	17j§		BD 242C, BD 244C, BD 540D, BD 940, ++	data
BDT 32 BF	Si-P	=BDT 32: Iso, 80/80V, >15W	17c		BD 940F, BD 956F, BDT 42BF, 2SA1650, ++	data
BDT 32 C	Si-P	=BDT 32: 100/100V	17j§		BD 242C, BD 244C, BD 540D, BD 940	data
BDT 32 CF	Si-P	=BDT 32: Iso, 100/100V, >15W	17c		BD 942F, BDT 42CF, 2SA1650	data
BDT 32 DF	Si-P	=BDT 32: Iso, 160/120V, >15W	17c		-	data
BDT 32 F	Si-P	=BDT 32: Iso, >15W	17c		BD 938F, BD 952F, BDT 42F, 2SB1095, ++	data
BDT 41	Si-N	NF-L, 80V, 6A, 65W, >3MHz	17j§	Phi	BD 243B, BD 543B, BD 799, BD 809, ++	data

Type	Device	Short Description	Fig.	Manu	Comparision Types	More at
BDT 41 A	Si-N	=BDT 41: 100V	17j§		BD 243C, BD 543C, BD 801, 2SD866, ++	data
BDT 41 AF	Si-N	=BDT 41: Iso, 100V, >20W	17c		BDT 95F, BDX 77F, 2SD1411, 2SD1588, ++	data
BDT 41 B	Si-N	=BDT 41: 120V	17j§		BD 243C, BD 543D, 2SD866	data
BDT 41 BF	Si-N	=BDT 41: Iso, 120V, >20W	17c		2SC4335	data
BDT 41 C	Si-N	=BDT 41: 140V	17j§		BD 243D, 2SC4329	data
BDT 41 CF	Si-N	=BDT 41: Iso, 140V, >20W	17c		2SC4335	data
BDT 41 F	Si-N	=BDT 41: Iso, >20W	17c		BDT 93F, BDX 77F, 2SD1411, 2SD1588, ++	data
BDT 42	Si-P	NF-L, 80V, 6A, 65W, >3MHz	17j§	Phi	BD 244B, BD 544B, BD 800, BD 810, ++	data
BDT 42 A	Si-P	=BDT 42: 100V	17j§		BD 244C, BD 544C, BD 802, 2SB870, ++	data
BDT 42 AF	Si-P	=BDT 42: Iso, 100V, >20W	17c		BDT 96F, BDX 78F, 2SB1018, 2SB1290, ++	data
BDT 42 B	Si-P	=BDT 42: 120V	17j§		BD 244C, BD 544D, 2SB870	data
BDT 42 BF	Si-P	=BDT 42: Iso, 120V, >20W	17c		2SA1651	data
BDT 42 C	Si-P	=BDT 42: 140V	17j§		BD 244D	data
BDT 42 CF	Si-P	=BDT 42: Iso, 140V, >20W	17c		2SA1651	data
BDT 42 F	Si-P	=BDT 42: Iso, >20W	17c		BDT 94F, BDX 78F, 2SB1018, 2SB1290, ++	data
BDT 51	Si-N	NF/S-L, 60V, 15A, 90W	17j§	Phi	BD 907, BDT 81	data
BDT 52	Si-P	NF/S-L, 60V, 15A, 90W	17j§	Phi	BD 908, BDT 82	data
BDT 53	Si-N	=BDT 51: 80V	17j§	Phi	BD 909, BDT 83	data
BDT 54	Si-P	=BDT 52: 80V	17j§	Phi	BD 910, BDT 84	data
BDT 55	Si-N	=BDT 51: 100V	17j§	Phi	BD 911, BDT 85	data
BDT 56	Si-P	=BDT 52: 100V	17j§	Phi	BD 912, BDT 86	data
BDT 57	Si-N	=BDT 51: 120V	17j§	Phi	BDT 87	data
BDT 58	Si-P	=BDT 52: 120V	17j§	Phi	BDT 88	data
BDT 60	Si-P-Darl+Di	NF-L, 60V, 4A, 50W, >10MHz, B>750	17j§	Phi, Tix	BD 716, BDW 24A, BDW 54A, BDW 64A	data
BDT 60 A	Si-P-Darl+Di	=BDT 60: 80V	17j§		BD 718, BDW 24B, BDW 54B, BDW 64B	data
BDT 60 AF	Si-P-Darl+Di	=BDT 60: Iso, 80V, >17W	17c		BD 648F, 2SB1342, 2SB1282, 2SB1024, ++	data
BDT 60 B	Si-P-Darl+Di	=BDT 60: 100V	17j§		BDW 24C, BDW 54C, BDW 64C	data
BDT 60 BF	Si-P-Darl+Di	=BDT 60: Iso, 100V, >17W	17c		BD 650F, 2SB1024, 2SB1282, 2SB1098, ++	data
BDT 60 C	Si-P-Darl+Di	=BDT 60: 120V	17j§		BDW 54D, BDW 64D	data
BDT 60 CF	Si-P-Darl+Di	=BDT 60: Iso, 120V, 17W	17c		BD 652F, 2SB1193, 2SB1340	data
BDT 60 F	Si-P-Darl+Di	=BDT 60: Iso, >17W	17c		BD 646F, 2SB1257, 2SB1223, 2SB1342, ++	data
BDT 60 L	Si-P-Darl+Di	=BDT 60: 45V	17j§		BD 714, BDW 24, BDW 54, BDW 64	data
BDT 61	Si-N-Darl+Di	NF-L, 60V, 4A, 50W, >10MHz, B>750	17j§	Phi, Tix	BD 715, BDW 23A, BDW 53A, BDW 63A	data
BDT 61 A	Si-N-Darl+Di	=BDT 61: 80V	17j§		BD 717, BDW 23B, BDW 53B, BDW 63B	data
BDT 61 AF	Si-N-Darl+Di	=BDT 61: Iso, 80V, >17W	17c		BD 645F, 2SD1933, 2SD2014, 2SD1788, ++	data
BDT 61 B	Si-N-Darl+Di	=BDT 61: 100V	17j§		BDW 23C, BDW 53C, BDW 63C	data
BDT 61 BF	Si-N-Darl+Di	=BDT 61: Iso, 100V, >17W	17c		BD 647F, 2SD1414, 2SD1788, 2SD1589, ++	data
BDT 61 C	Si-N-Darl+Di	=BDT 61: 120V	17j§		BDW 53D, BDW 63D, 2SD1147	data
BDT 61 CF	Si-N-Darl+Di	=BDT 61: Iso, 120V, >17W	17c		BD 649F, 2SD1785, 2SD1590	data
BDT 61 F	Si-N-Darl+Di	=BDT 61: Iso, >17W	17c		BD 643F, 2SD1796, 2SD1825, 2SD1933, ++	data
BDT 61 L	Si-N-Darl+Di	=BDT 61: 45V	17j§		BD 713, BDW 23, BDW 53, BDW 63	data
BDT 62	Si-P-Darl+Di	NF-L, 60V, 10A, 90W, >10MHz, B>1000	17j§	Phi	BDT 64, BDW 94A, BDX 34A	data
BDT 62 A	Si-P-Darl+Di	=BDT 62: 80V	17j§		BDT 64A, BDW 94B, BDX 34B	data
BDT 62 AF	Si-P-Darl+Di	=BDT 62: Iso, 80V, >17W	17c		BDT 64AF, 2SB1100, 2SB1284, 2SB1259	data
BDT 62 B	Si-P-Darl+Di	=BDT 62: 100V	17j§		BDT 64B, BDW 94C, BDX 34C	data
BDT 62 BF	Si-P-Darl+Di	=BDT 62: Iso, 100V, >17W	17c		BDT 64BF, 2SB1100, 2SB1284, 2SB1259	data
BDT 62 C	Si-P-Darl+Di	=BDT 62: 120V	17j§		BDT 64C, BDX 34D	data
BDT 62 CF	Si-P-Darl+Di	=BDT 62: Iso, 120V, >17W	17c		BDT 64CF	data
BDT 62 F	Si-P-Darl+Di	=BDT 62: Iso, >17W	17c		BDT 64F, 2SB1225, 2SB1100, 2SB1284, ++	data
BDT 63	Si-N-Darl+Di	NF-L, 60V, 10A, 90W, >10MHz, B>1000	17j§	Phi	BDT 65, BDW 93A, BDX 33A	data
BDT 63 A	Si-N-Darl+Di	=BDT 63: 80V	17j§		BDT 65A, BDW 93B, BDX 33B	data
BDT 63 AF	Si-N-Darl+Di	=BDT 63: Iso, 80V, >17W	17c		BDT 65AF, 2SD1793, 2SD1591	data
BDT 63 B	Si-N-Darl+Di	=BDT 63: 100V	17j§		BDT 65B, BDW 93C, BDX 33C	data
BDT 63 BF	Si-N-Darl+Di	=BDT 63: Iso, 100V, >17W	17c		BDT 65BF, 2SD1793, 2SD1591	data
BDT 63 C	Si-N-Darl+Di	=BDT 63: 120V	17j§		BDT 65C, BDX 33D	data
BDT 63 CF	Si-N-Darl+Di	=BDT 63: Iso, 120V, >17W	17c		BDT 65CF, 2SD1591	data
BDT 63 F	Si-N-Darl+Di	=BDT 63: Iso, >17W	17c		BDT 65F, 2SD1827, 2SD1793, 2SD1591	data
BDT 64	Si-P-Darl+Di	NF-L, 60V, 12A, 125W, >10MHz, B>1000	17j§	Phi	BDW 45, BDW 94A	data
BDT 64 A	Si-P-Darl+Di	=BDT 64: 80V	17j§		BDW 46, BDW 94B	data
BDT 64 AF	Si-P-Darl+Di	=BDT 64: Iso, 80V, >22W	17c		-	data
BDT 64 B	Si-P-Darl+Di	=BDT 64: 100V	17j§		BDW 47, BDW 94C	data
BDT 64 BF	Si-P-Darl+Di	=BDT 64: Iso, 100V, >22W	17c		-	data
BDT 64 C	Si-P-Darl+Di	=BDT 64: 120V	17j§			data
BDT 64 CF	Si-P-Darl+Di	=BDT 64: Iso, 120V, >22W	17c		-	data
BDT 64 F	Si-P-Darl+Di	=BDT 64: Iso, >22W	17c		2SB1351	data
BDT 65	Si-N-Darl+Di	NF-L, 60V, 12A, 125W, >10MHz, B>1000	17j§	Phi	BDW 40, BDW 93A	data
BDT 65 A	Si-N-Darl+Di	=BDT 65: 80V	17j§		BDW 41, BDW 93B	data
BDT 65 AF	Si-N-Darl+Di	=BDT 65: Iso, 80V, >22W	17c		(BDW 93B,C)[3]	data
BDT 65 B	Si-N-Darl+Di	=BDT 65: 100V	17j§		BDW 42, BDW 93C	data
BDT 65 BF	Si-N-Darl+Di	=BDT 65: Iso, 100V, >22W	17c		(BDW 93C)[3]	data
BDT 65 C	Si-N-Darl+Di	=BDT 65: 120V	17j§			data
BDT 65 CF	Si-N-Darl+Di	=BDT 64: Iso, 120V, >22W	17c			data
BDT 65 F	Si-N-Darl+Di	=BDT 65: Iso, >22W	17c		(BDW 93A...C)[3]	data
BDT 81	Si-N	NF/S-L, 60V, 15A, 125W, 10MHz	17j§	Phi	BD 907	data
BDT 81 F	Si-N	=BDT 81: Iso, >21W	17c		2SD1964	data
BDT 82	Si-P	NF/S-L, 60V, 15A, 125W, 20MHz	17j§	Phi	BD 908	data
BDT 82 F	Si-P	=BDT 82: Iso, >21W	17c		-	data
BDT 83	Si-N	=BDT 81: 80V	17j§	Phi	BD 909	data
BDT 83 F	Si-N	=BDT 83: Iso, >21W	17c		2SD1964	data
BDT 84	Si-P	=BDT 82: 80V	17j§	Phi	BD 910	data
BDT 84 F	Si-P	=BDT 84: Iso, >21W	17c		-	data
BDT 85	Si-N	=BDT 81: 100V	17j§	Phi	BD 911	data
BDT 85 F	Si-N	=BDT 85: Iso, >21W	17c		2SD1964	data

Type	Device	Short Description	Fig.	Manu	Comparision Types	More at
BDT 86	Si-P	=BDT 82: 100V	17j§	Phi	BD 912	data
BDT 86 F	Si-P	=BDT 86: Iso, >21W	17c		-	data
BDT 87	Si-N	=BDT 81: 120V	17j§	Phi	-	data
BDT 87 F	Si-N	=BDT 87: Iso, >21W	17c		2SD1964	data
BDT 88	Si-P	=BDT 82: 120V	17j§	Phi	-	data
BDT 88 F	Si-P	=BDT 88: Iso, >21W	17c		-	data
BDT 91	Si-N	NF/S-L, 60V, 10A, 90W, >4MHz	17j§	Phi	BD 743A, BD 807, BD 907	data
BDT 91 F	Si-N	=BDT 91: Iso, >20W	17c		BDT 81F, 2SD1964	data
BDT 92	Si-P	NF/S-L, 60V, 10A, 90W, >4MHz	17j§	Phi	BD 744A, BD 808, BD 908	data
BDT 92 F	Si-P	=BDT 92: Iso, >20W	17c		BDT 82F	data
BDT 93	Si-N	=BDT 91: 80V	17j§	Phi	BD 743B, BD 809, BD 909	data
BDT 93 F	Si-N	=BDT 93: Iso, >20W	17c		BDT 83F, 2SD1964	data
BDT 94	Si-P	=BDT 92: 80V	17j§	Phi	BD 744B, BD 810, BD 910	data
BDT 94 F	Si-P	=BDT 93: Iso, >20W	17c		BDT 84F	data
BDT 95	Si-N	=BDT 91: 100V	17j§	Phi	BD 743C, BD 911	data
BDT 95 F	Si-N	=BDT 95: Iso, >20W	17c		BDT 85F, 2SD1964	data
BDT 96	Si-P	=BDT 92: 100V	17j§	Phi	BD 744C, BD 912	data
BDT 96 F	Si-P	=BDT 96: Iso, >20W	17c		BDT 86F	data

BDV

Type	Device	Short Description	Fig.	Manu	Comparision Types	More at
BDV 10	Si-N	NF/S-L, 120V, 5A, 26W(Tc=45°), >30MHz	17j§	Sie	MJE 15028, 2SC2527, 2SD772	data
BDV 11	Si-N	=BDV 10: 140V	17j§	Sie	MJE 15030, 2SD772	data
BDV 12	Si-N	=BDV 10	17j§	Sie	MJE 15028, 2SC2527, 2SD772	data
BDV 13	Si-N-Darl	NF-L, 45V, 4A, 36W, >3MHz, B>750	17j§	Sie	BD 713, BDW 23, BDW 53, BDW 63	data
BDV 14	Si-P-Darl	NF-L, 45V, 4A, 36W, >3MHz, B>750	17j§	Sie	BD 714, BDW 24, BDW 54, BDW 64	data
BDV 15	Si-N-Darl	=BDV 13: 60V	17j§	Sie	BD 715, BDW 23A, BDW 53A, BDW 63A	data
BDV 16	Si-P-Darl	=BDV 14: 60V	17j§	Sie	BD 716, BDW 24A, BDW 54A, BDW 64A	data
BDV 17	Si-N-Darl	=BDV 13: 80V	17j§	Sie	BD 717, BDW 23B, BDW 53B, BDW 63B	data
BDV 18	Si-P-Darl	=BDV 14: 80V	17j§	Sie	BD 718, BDW 24B, BDW 54B, BDW 64B	data
BDV 33	Si-N	NF-L, 45V, 5A, 40W, >3MHz	17j§	Sie	BD 243, BD 539A, BD 543A, BD 947, ++	data
BDV 34	Si-P	NF-L, 45V, 5A, 40W, >3MHz	17j§	Sie	BD 244, BD 540A, BD 544A, BD 948, ++	data
BDV 35	Si-N	=BDV 33: 60V	17j§	Sie	BD 243A, BD 539A, BD 543A, BD 949, ++	data
BDV 36	Si-P	=BDV 34: 60V	17j§	Sie	BD 244A, BD 540A, BD 544A, BD 950, ++	data
BDV 37	Si-N	=BDV 33: 80V	17j§	Sie	BD 243B, BD 539B, BD 543B, BD 951, ++	data
BDV 38	Si-P	=BDV 34: 80V	17j§	Sie	BD 244B, BD 540B, BD 544B, BD 952, ++	data
BDV 45	Si-N-Darl	NF-L, 60V, 8A, 62,5W, >3MHz, B>750	17j§	Sie	BD 645, BD 897, BDW 73A, BDX 53A, ++	data
BDV 46	Si-P-Darl	NF-L, 60V, 8A, 62,5W, >3MHz, B>750	17j§	Sie	BD 646, BD 898, BDW 74A, BDX 54A, ++	data
BDV 47	Si-N-Darl	=BDV 45: 80V	17j§	Sie	BD 647, BD 899, BDW 73B, BDX 53B, ++	data
BDV 48	Si-P-Darl	=BDV 46: 80V	17j§	Sie	BD 648, BD 900, BDW 74B, BDX 54B, ++	data
BDV 49	Si-N-Darl	=BDV 45: 100V	17j§	Sie	BD 649, BD 901, BDW 73C, BDX 53C, ++	data
BDV 50	Si-P-Darl	=BDV 46: 100V	17j§	Sie	BD 650, BD 902, BDW 74C, BDX 54C, ++	data
BDV 64	Si-P-Darl+Di	NF-L, 60V, 12A, 125W, B>1000	18j§	Mot,Phi,Tho	BDV 66, BDW 84A	data
BDV 64 A	Si-P-Darl+Di	=BDV 64: 80V	18j§	Tix	BDV 66A, BDW 84B	data
BDV 64 AF	Si-P-Darl+Di	=BDV 64A: Iso, >31W	16c		2SB1382, 2SB1448	data
BDV 64 B	Si-P-Darl+Di	=BDV 64: 100V	18j§		BDV 66B, BDW 84C	data
BDV 64 BF	Si-P-Darl+Di	=BDV 64B: Iso, >31W	16c		2SB1382, 2SB1448	data
BDV 64 C	Si-P-Darl+Di	=BDV 64: 120V	18j§		BDV 66C, BDW 84D	data
BDV 64 CF	Si-P-Darl+Di	=BDV 64C: Iso, >31W	16c		2SB1382	data
BDV 64 F	Si-P-Darl+Di	=BDV 64: Iso, >31W	16c		2SB1352, 2SB1382, 2SB1448	data
BDV 65	Si-N-Darl+Di	NF-L, 60V, 12A, 125W, B>1000	18j§	Mot,Phi,Tho	BDV 67, BDW 83A	data
BDV 65 A	Si-N-Darl+Di	=BDV 65: 80V	18j§	Tix	BDV 67A, BDW 83B	data
BDV 65 AF	Si-N-Darl+Di	=BDV 65A: Iso, >31W	16c		2SD2082	data
BDV 65 B	Si-N-Darl+Di	=BDV 65: 100V	18j§		BDV 67B, BDW 83C	data
BDV 65 BF	Si-N-Darl+Di	=BDV 65B: Iso, >31W	16c		2SD2082	data
BDV 65 C	Si-N-Darl+Di	=BDV 65: 120V	18j§		BDV 67C, BDW 83D	data
BDV 65 CF	Si-N-Darl+Di	=BDV 65C: Iso, >31W	16c		2SD2082	data
BDV 65 F	Si-N-Darl+Di	=BDV 65: Iso, >31W	16c		2SD2082	data
BDV 66	Si-P-Darl+Di	NF-L, 60V, 16A, 200W, B>1000	18j§	Aeg,Phi,Tho	BDW 84A, 2SB1079	data
BDV 66 A	Si-P-Darl+Di	=BDV 66: 100V	18j§		BDW 84C, 2SB1079	data
BDV 66 AF	Si-P-Darl+Di	=BDV 66A: Iso, >35W	16c		2SB1382, 2SB1448	data
BDV 66 B	Si-P-Darl+Di	=BDV 66: 120V	18j§		BDW 84D	data
BDV 66 BF	Si-P-Darl+Di	=BDV 66B: Iso, >35W	16c		2SB1382	data
BDV 66 C	Si-P-Darl+Di	=BDV 66: 140V	18j§		-	data
BDV 66 CF	Si-P-Darl+Di	=BDV 66C: Iso, >35W	16c		-	data
BDV 66 D	Si-P-Darl+Di	=BDV 66: 160V	18j§		-	data
BDV 66 DF	Si-P-Darl+Di	=BDV 66D: Iso, >35W	16c		-	data
BDV 67	Si-N-Darl+Di	NF-L, 60V, 16A, 200W, B>1000	18j§	Aeg,Phi,Tho	BDW 83A, 2SD1559	data
BDV 67 A	Si-N-Darl+Di	=BDV 67: 100V	18j§		BDW 83C, 2SD1559	data
BDV 67 AF	Si-N-Darl+Di	=BDV 67A: Iso, >35W	16c		2SD2082	data
BDV 67 B	Si-N-Darl+Di	=BDV 67: 120V	18j§		BDW 83D	data
BDV 67 BF	Si-N-Darl+Di	=BDV 67B: Iso, >35W	16c		2SD2082	data
BDV 67 C	Si-N-Darl+Di	=BDV 67: 140V	18j§		-	data
BDV 67 CF	Si-N-Darl+Di	=BDV 67C: Iso, >35W	16c		-	data
BDV 67 D	Si-N-Darl+Di	=BDV 67: 160V	18j§		-	data
BDV 67 DF	Si-N-Darl+Di	=BDV 67D: Iso, >35W	16c		-	data
BDV 91	Si-N	NF/S-L, 60V, 10A, 100W, >3MHz	18j§	Phi	BD 245A, BD 745A	data
BDV 92	Si-P	NF/S-L, 60V, 10A, 100W, >4MHz	18j§	Phi	BD 246A, BD 746A	data
BDV 93	Si-N	=BDV 91: 80V	18j§	Phi	BD 245B, BD 745B	data
BDV 94	Si-P	=BDV 92: 80V	18j§	Phi	BD 246B, BD 746B	data
BDV 95	Si-N	=BDV 91: 100V	18j§	Phi	BD 245C, BD 745C	data
BDV 96	Si-P	=BDV 92: 100V	18j§	Phi	BD 246C, BD 746C	data

Type	Device	Short Description	Fig.	Manu	Comparision Types	More at
BDW						
BDW	Z-Di	=SM 15T 15C (Typ-Code/Stempel/marking)	71(8x5mm)	Tho	→SM 15T...	data
BDW 10(A)	Si-N	NF/S-L, 140V, 15A, 180W, >1MHz	23a§	Mot	BDW 30, MJ 15015, 2SC1585, 2SC2607...08,+	data
BDW 12(A)	Si-N	=BDW 10: 160V	23a§	Mot	BDW 34, MJ 15015, 2SC1585, 2SC2607...08,+	data
BDW 14(A)	Si-N	=BDW 10: 180V	23a§	Mot	BDW 34, MJ 15015, 2SC1585, 2SC2607...08,+	data
BDW 16(A)	Si-N	=BDW 10: 200V	23a§	Mot	BDW 36, MJ 15015, 2SC1585, 2SC2607...08,+	data
BDW 21	Si-N	NF/S-L, 45V, 10A, 90W, >3MHz	23a§	Sgs	BD 245, BD 311, BD 315, BDW 51, ++	data
BDW 21 A	Si-N	=BDW 21: 60V	23a§		BD 245A, BD 311, BD 315, BDW 51A, ++	data
BDW 21 B	Si-N	=BDW 21: 80V	23a§		BD 245B, BD 313, BD 315, BDW 51B, ++	data
BDW 21 C	Si-N	=BDW 21: 100V	23a§		BD 245C, BD 317, BDW 51C, ++	data
BDW 22	Si-P	NF/S-L, 45V, 10A, 90W, >3MHz	23a§	Sgs	BD 246, BD 312, BD 316, BDW 52, ++	data
BDW 22 A	Si-P	=BDW 22: 60V	23a§		BD 246A, BD 312, BD 316, BDW 52A, ++	data
BDW 22 B	Si-P	=BDW 22: 80V	23a§		BD 246B, BD 314, BD 316, BDW 52B, ++	data
BDW 22 C	Si-P	=BDW 22: 100V	23a§		BD 246C, BD 316, BDW 52C, ++	data
BDW 23	Si-N-Darl+Di	NF-L, 45V, 6A, 50W, B>750	17j§	Sgs, Tix	BD 643, BD 895, BDW 63, BDX 53, ++	data
BDW 23 A	Si-N-Darl+Di	=BDW 23: 60V	17j§		BD 645, BD 897, BDW 63A, BDX 53A, ++	data
BDW 23 B	Si-N-Darl+Di	=BDW 23: 80V	17j§		BD 647, BD 899, BDW 63B, BDX 53B, ++	data
BDW 23 C	Si-N-Darl+Di	=BDW 23: 100V	17j§		BD 649, BD 901, BDW 63C, BDX 53C, ++	data
BDW 24	Si-P-Darl+Di	NF-L, 45V, 6A, 50W, B>750	17j§	Sgs, Tix	BD 644, BD 896, BDW 64, BDX 54, ++	data
BDW 24 A	Si-P-Darl+Di	=BDW 24: 60V	17j§		BD 646, BD 898, BDW 64A, BDX 54A, ++	data
BDW 24 B	Si-P-Darl+Di	=BDW 24: 80V	17j§		BD 648, BD 900, BDW 64B, BDX 54B, ++	data
BDW 24 C	Si-P-Darl+Di	=BDW 24: 100V	17j§		BD 650, BD 902, BDW 64C, BDX 54C, ++	data
BDW 25	Si-N	NF/S-L, 130V, 5A, 26W(Tc=45°), 30MHz	22a§	Sie	BDX 25, BDV 11, MJE 15030, 2SD772	data
BDW 26	Diac	Ub=22...30V, Ib<0,8mA, Itsm=1A	31	Tra		data
BDW 30	Si-N	NF/S-L, 140V, 30A, 250W, >1MHz	23a§	Mot	2N6276...6277	data
BDW 32	Si-N	=BDW 30: 160V	23a§	Mot	2N6276...6277	data
BDW 32	Diac	=BDW 26: Ub=28...36V	31	Tra	A9903, D32	data
BDW 34	Si-N	=BDW 30: 180V	23a§	Mot	2N6277	data
BDW 36	Si-N	=BDW 30: 200V	23a§	Mot	-	data
BDW 38	Diac	=BDW 26: Ub=34...42V	31	Tra	-	data
BDW 39	Si-N-Darl+Di	NF-L, 45V, 15A, 85W, >4MHz, B>1000	17j§	Mot	-	data
BDW 40	Si-N-Darl+Di	=BDW 39: 60V	17j§	Mot	-	data
BDW 41	Si-N-Darl+Di	=BDW 39: 80V	17j§	Mot	-	data
BDW 42	Si-N-Darl+Di	=BDW 39: 100V	17j§	Mot	-	data
BDW 44	Si-P-Darl+Di	NF-L, 45V, 15A, 85W, >4MHz, B>1000	17j§	Mot	-	data
BDW 45	Si-P-Darl+Di	=BDW 44: 60V	17j§	Mot	-	data
BDW 46	Si-P-Darl+Di	=BDW 44: 80V	17j§	Mot	-	data
BDW 47	Si-P-Darl+Di	=BDW 44: 100V	17j§	Mot	-	data
BDW 51	Si-N	NF/S-L, 45V, 15A, 125W, >3MHz	23a§	Sgs	BD 249, BD 315, BDY 55, 2N5629...31, ++	data
BDW 51 A	Si-N	=BDW 51: 60V	23a§		BD 249A, BD 315, BDY 55, 2N5629...31, ++	data
BDW 51 B	Si-N	=BDW 51: 80V	23a§		BD 249B, BD 315, BDY 55, 2N5629...31, ++	data
BDW 51 C	Si-N	=BDW 51: 100V	23a§		BD 249C, BD 317, BDY 55, 2N5629...31, ++	data
BDW 52	Si-P	NF/S-L, 45V, 15A, 125W, >3MHz	23a§	Sgs	BD 250, BD 316, 2N6029...6031, ++	data
BDW 52 A	Si-P	=BDW 52: 60V	23a§		BD 250A, BD 316, 2N6029...6031, ++	data
BDW 52 B	Si-P	=BDW 52: 80V	23a§		BD 250B, BD 316, 2N6029...6031, ++	data
BDW 52 C	Si-P	=BDW 52: 100V	23a§		BD 250C, BD 318, 2N6029...6031, ++	data
BDW 53	Si-N-Darl+Di	NF-L, 45V, 4A, 40W, >1MHz, B>750	17j§	Tix	BD 713, BDW 23, BDW 63, ++	data
BDW 53 A	Si-N-Darl+Di	=BDW 53: 60V	17j§		BD 715, BDW 23A, BDW 63A, ++	data
BDW 53 B	Si-N-Darl+Di	=BDW 53: 80V	17j§		BD 717, BDW 23B, BDW 63B, ++	data
BDW 53 C	Si-N-Darl+Di	=BDW 53: 100V	17j§		BDW 23C, BDW 63C	data
BDW 53 D	Si-N-Darl+Di	=BDW 53: 120V	17j§		BDW 63D, 2SD1147	data
BDW 54	Si-P-Darl+Di	NF-L, 45V, 4A, 40W, >1MHz, B>750	17j§	Tix	BD 714, BDW 24, BDW 64, ++	data
BDW 54 A	Si-P-Darl+Di	=BDW 54: 60V	17j§		BD 716, BDW 24A, BDW 64A, ++	data
BDW 54 B	Si-P-Darl+Di	=BDW 54: 80V	17j§		BD 718, BDW 24B, BDW 64B, ++	data
BDW 54 C	Si-P-Darl+Di	=BDW 54: 100V	17j§		BDW 24C, BDW 64C	data
BDW 54 D	Si-P-Darl+Di	=BDW 54: 120V	17j§		BDW 64D	data
BDW 55	Si-N	=BD 135: hi-rel Version	14h§	Phi	→BD 135	data
BDW 56	Si-P	=BD 136: hi-rel Version	14h§	Phi	→BD 136	data
BDW 57	Si-N	=BD 137: hi-rel Version	14h§	Phi	→BD 137	data
BDW 58	Si-P	=BD 138: hi-rel Version	14h§	Phi	→BD 138	data
BDW 59	Si-N	=BD 139: hi-rel Version	14h§	Phi	→BD 139	data
BDW 60	Si-P	=BD 140: hi-rel Version	14h§	Phi	→BD 140	data
BDW 63	Si-N-Darl+Di	NF-L, 45V, 6A, 60W, >1MHz, B>750	17j§	Tix	BD 643, BD 895, BDW 23, BDX 53, ++	data
BDW 63 A	Si-N-Darl+Di	=BDW 63: 60V	17j§	Tix	BD 645, BD 897, BDW 23A, BDX 53A, ++	data
BDW 63 B	Si-N-Darl+Di	=BDW 63: 80V	17j§		BD 647, BD 899, BDW 23B, BDX 53B, ++	data
BDW 63 C	Si-N-Darl+Di	=BDW 63: 100V	17j§		BD 649, BD 901, BDW 23C, BDX 53C, ++	data
BDW 63 D	Si-N-Darl+Di	=BDW 63: 120V	17j§		BD 651, BDT 21, BDX 53E	data
BDW 64	Si-P-Darl+Di	NF-L, 45V, 6A, 60W, >1MHz, B>750	17j§	Tix	BD 644, BD 896, BDW 24, BDX 54, ++	data
BDW 64 A	Si-P-Darl+Di	=BDW 64: 60V	17j§		BD 646, BD 898, BDW 24A, BDX 54A, ++	data
BDW 64 B	Si-P-Darl+Di	=BDW 64: 80V	17j§		BD 648, BD 900, BDW 24B, BDX 54B, ++	data
BDW 64 C	Si-P-Darl+Di	=BDW 64: 100V	17j§		BD 650, BD 902, BDW 24C, BDX 54C, ++	data
BDW 64 D	Si-P-Darl+Di	=BDW 64: 120V	17j§		BD 652, BDT 20, BDX 54E	data
BDW 69	MOS-N-FET-e	V-MOS, NF-L, 40V, 2A, 12,5W	13e	Tho	-	data
BDW 70	MOS-N-FET-e	=BDW 69: 60V	13e	Tho	-	data
BDW 71	MOS-N-FET-e	=BDW 69: 90V	13e	Tho	-	data
BDW 73	Si-N-Darl+Di	NF-L, 45V, 8A, 80W, >1MHz, B>750	17j§	Tix	BD 643, BD 895, BDX 33, BDX 53	data
BDW 73 A	Si-N-Darl+Di	=BDW 73: 60V	17j§		BD 645, BD 897, BDX 33A, BDX 53A	data
BDW 73 B	Si-N-Darl+Di	=BDW 73: 80V	17j§		BD 647, BD 899, BDX 33B, BDX 53B	data

Type	Device	Short Description	Fig.	Manu	Comparision Types	More at .
BDW 73 C	Si-N-Darl+Di	=BDW 73: 100V	17j§		BD 649, BD 901, BDX 33C, BDX 53C	data
BDW 73 D	Si-N-Darl+Di	=BDW 73: 120V	17j§		BD 651, BDX 33D, BDX 53E	data
BDW 74	Si-P-Darl+Di	NF-L, 45V, 8A, 80W, >1MHz, B>750	17j§	Tix	BD 644, BD 896, BDX 33, BDX 53	data
BDW 74 A	Si-P-Darl+Di	=BDW 74: 60V	17j§		BD 646, BD 898, BDX 33A, BDX 53A	data
BDW 74 B	Si-P-Darl+Di	=BDW 74: 80V	17j§		BD 648, BD 900, BDX 33B, BDX 53B	data
BDW 74 C	Si-P-Darl+Di	=BDW 74: 100V	17j§		BD 650, BD 902, BDX 33C, BDX 53C	data
BDW 74 D	Si-P-Darl+Di	=BDW 74: 120V	17j§		BD 652, BDX 33D, BDX 53E	data
BDW 83	Si-N-Darl+Di	NF-L, 45V, 15A, 150W, >1MHz, B>750	18j§	Sgs, Tix	BDV 67	data
BDW 83 A	Si-N-Darl+Di	=BDW 83: 60V	18j§		BDV 67	data
BDW 83 B	Si-N-Darl+Di	=BDW 83: 80V	18j§		BDV 67A	data
BDW 83 C	Si-N-Darl+Di	=BDW 83: 100V	18j§		BDV 67A	data
BDW 83 D	Si-N-Darl+Di	=BDW 83: 120V	18j§		BDV 67B	data
BDW 84	Si-P-Darl+Di	NF-L, 45V, 15A, 150W, >1MHz, B>750	18j§	Sgs, Tix	BDV 66	data
BDW 84 A	Si-P-Darl+Di	=BDW 84: 60V	18j§		BDV 66	data
BDW 84 B	Si-P-Darl+Di	=BDW 84: 80V	18j§		BDV 66A	data
BDW 84 C	Si-P-Darl+Di	=BDW 84: 100V	18j§		BDV 66A	data
BDW 84 D	Si-P-Darl+Di	=BDW 84: 120V	18j§		BDV 66B	data
BDW 91	Si-N-Darl+Di	NF/S, 180V, 4A, 10W(Tc=25°), 20MHz	2a§	Sgs	(2SC3186, 2SD1121)[4]	data
BDW 92	Si-P-Darl+Di	NF/S, 180V, 4A, 10W(Tc=25°), 20MHz	2a§	Sgs		data
BDW 93	Si-N-Darl+Di	NF-L, 45V, 12A, 80W, >20MHz, B>750	17j§	Sgs, Tix	BDT 65, BDW 39	data
BDW 93 A	Si-N-Darl+Di	=BDW 93: 60V	17j§		BDT 65, BDW 40	data
BDW 93 B	Si-N-Darl+Di	=BDW 93: 80V	17j§		BDT 65A, BDW 41	data
BDW 93 C	Si-N-Darl+Di	=BDW 93: 100V	17j§		BDT 65B, BDW 42	data
BDW 93 CFI	Si-N-Darl+Di	=BDW 93C: Iso, 40W	17c		BDT 65BF...CF	data
BDW 93 CFP	Si-N-Darl+Di	=BDW 93C: Iso, 33W	17c		BDT 65BF...CF	data
BDW 94	Si-P-Darl+Di	NF-L, 45V, 12A, 80W, >20MHz, B>750	17j§	Sgs, Tix	BDT 64, BDW 44	data
BDW 94 A	Si-P-Darl+Di	=BDW 94: 60V	17j§		BDT 64, BDW 45	data
BDW 94 B	Si-P-Darl+Di	=BDW 94: 80V	17j§		BDT 64A, BDW 46	data
BDW 94 C	Si-P-Darl+Di	=BDW 94: 100V	17j§		BDT 64B, BDW 47	data
BDW 94 CFI	Si-P-Darl+Di	=BDW 94C: Iso, 40W	17c		BDT 64BF...CF	data
BDW 94 CFP	Si-P-Darl+Di	=BDW 94C: Iso, 33W	17c		BDT 64BF...CF	data

BDX

Type	Device	Short Description	Fig.	Manu	Comparision Types	More at .
BDX	Z-Di	=SM 15T 15CA(Typ-Code/Stempel/marking)	71(8x5mm)	Tho	→SM 15T...	data
BDX 10(C,H)	Si-N	NF/S-L, 100V, 15A, 117W, >0,8MHz	23a§	Sgs	BD 317, BD 745C, BDW 51C, 2N3055, ++	data
BDX 11	Si-N	NF/S-L, 160V, 10A, 117W, >0,8MHz	23a§	Sgs	BDW 12, 2N3442, 2SD733, 2SD1047, ++	data
BDX 12	Si-N	NF/S-L, 140V, 10A, 100W, >0,8MHz	23a§	Sgs	BDW 10, 2N5631, 2SC1584, 2SD1047, ++	data
BDX 13	Si-N	NF/S-L, 50V, 15A, 117W, >0,8MHz	23a§	Sgs	BD 315, BD 745, BDW 51A, 2N5881, ++	data
BDX 14	Si-P	NF/S-L, 90V, 4A, 29W, >4MHz	22a§	Tho,Tix	BD 244B, BD 954, 2N5954, 2SB550, ++	data
BDX 15	Si-P	NF/S-L, -70V, 10A, 117W, >0,8MHz	23a§	Tix	BD 246B, BD 314, BDW 22A, 2N5875, ++	data
BDX 16	Si-P	NF/S-L, 160V, 3A, 25W, 4MHz	22a§	Tho	BD 242D, BUX 66, MJE 5850, 2SB720	data
BDX 18	Si-P	NF/S-L, 100V, 15A, 117W, 4MHz	23a§	Rca,Tho	BD 318, BD 746C, BDW 52C, MJ 2955, ++	data
BDX 18 N	Si-P	=BDX 18: 70V	23a§		BD 316, BD 746B, BDW 52B, MJ 2955, ++	data
BDX 20	Si-P	NF/S-L, 160V, 10A, 117W, >4MHz	23a§	Tho	BD 245D, 2SA1147, 2SB600, 2SB697, 2SB817	data
BDX 22	Si-N	NF/S-L, 160V, 10A, 37,5W, >0,8MHz	22a§	Sgs	BD 743F, BUV 27, 2SC2867	data
BDX 23	Si-N	NF/S-L, 95V, 15A, 117W, >0,8MHz	23a§	Rca,Sgs	BD 317, BD 745C, BDW 51C, 2N3055, ++	data
BDX 24	Si-N	NF/S-L, 50V, 4A, 29W, >0,8MHz	22a§	Rca,Sgs	BD 243, BD 535, 2N4231A, 2N6374, ++	data
BDX 25	Si-N	NF/S-L, 130V, 5A, 34W(Tc=45°), 30MHz	22a§	Sie	BDW 25, BDV 11, MJE 15030, 2SD772	data
BDX 26	Si-P	NF/S-L, 40V, 5A, 40W, >30MHz	23a§	Sie	BD 312, BD 246, 2SA1185, 2SA1292, ++	data
BDX 27	Si-P	NF/S-L, 40V, 5A, 50W(Tc=45°), 50MHz	22a§	Sie	BD 244, BD 948, 2SA1012, 2SA1289, ++	data
BDX 28	Si-P	=BDX 27: 60V	22a§	Sie	BD 244A, BD 950, 2SA1012, 2SA1289, ++	data
BDX 29	Si-P	=BDX 27: 80V	22a§	Sie	BD 244B, BD 952, MJE 15029, 2SA1289, ++	data
BDX 30	Si-P	=BDX 27: 125V	22a§	Sie	MJE 15031, 2SB869	data
BDX 31	Si-N	TV-HA, 2200V, 4A, 40W	23a§	Tix	2SD621, 2SD838	data
BDX 32	Si-N	=BDX 31: 1700V	23a§	Tix	2SD784...785	data
BDX 33	Si-N-Darl+Di	NF-L, 45V, 10A, 70W, >20MHz, B>750	17j§	Rca,Tho,++	BDT 63, BDT 65, BDW 93	data
BDX 33 A	Si-N-Darl+Di	=BDX 33: 60V	17j§		BDT 63, BDT 65, BDW 93A	data
BDX 33 B	Si-N-Darl+Di	=BDX 33: 80V	17j§		BDT 63A, BDT 65A, BDW 93B	data
BDX 33 C	Si-N-Darl+Di	=BDX 33: 100V	17j§		BDT 63B, BDT 65B, BDW 93C	data
BDX 33 D	Si-N-Darl+Di	=BDX 33: 120V	17j§		BDT 63C, BDT 65C, 2SD1126, 2SD 1607	data
BDX 33 E	Si-N-Darl+Di	=BDX 33: 140V	17j§		2SD1500	data
BDX 34	Si-P-Darl+Di	NF-L, 45V, 10A,.70W, >20MHz, B>750	17j§	Rca,Tho,++	BDT 62, BDT 64, BDW 94	data
BDX 34 A	Si-P-Darl+Di	=BDX 34: 60V	17j§		BDT 62, BDT 64, BDW 94A	data
BDX 34 B	Si-P-Darl+Di	=BDX 34: 80V	17j§		BDT 62A, BDT 64A, BDW 94B	data
BDX 34 C	Si-P-Darl+Di	=BDX 34: 100V	17j§		BDT 62B, BDT 64B, BDW 94C	data
BDX 34 D	Si-P-Darl+Di	=BDX 34: 120V	17j§		BDT 62C, BDT 64C	data
BDX 35	Si-N	NF/S-L, 100V, 5A, 15W(Tc=75°), 100MHz	14h§	Phi	-	data
BDX 36	Si-N	=BDX 35: 120/60V	14h§	Phi	-	data
BDX 37	Si-N	=BDX 35: 120/80V	14h§		-	data
BDX 40	Si-N	NF/S-L, 100V, 20A, 150W, >0,8MHz	23a§	Sgs	BDY 29, MJ 802, 2N3772, 2SD797	data
BDX 41	Si-N	NF/S-L, 50V, 30A, 150W, >0,8MHz	23a§	Sgs	BDY 29, MJ 802, 2N3771, 2SD630	data
BDX 42	Si-N-Darl+Di	NF-L, 60V, 1A, 5W(Tc=100°), 200MHz	14h§	Phi	-	data
BDX 43	Si-N-Darl+Di	=BDX 42: 80V	14h§	Phi	-	data
BDX 44	Si-N-Darl+Di	=BDX 42: 100V	14h§	Phi	-	data

Type	Device	Short Description	Fig.	Manu	Comparision Types	More at
BDX 45	Si-P-Darl+Di	NF-L, 60V, 1A, 5W(Tc=100°), 200MHz	14h§	Phi	-	data
BDX 46	Si-P-Darl+Di	=BDX 45: 80V	14h§	Phi	-	data
BDX 47	Si-P-Darl+Di	=BDX 45: 100V	14h§		-	data
BDX 50	Si-N	NF/S-L, 160V, 16A, 150W, >0,8MHz	23a§	Sgs	BD 249D, BD 745E, 2N3773, 2SC2608	data
BDX 51	Si-N	NF/S-L, 140V, 10A, 120W, >0,8MHz	23a§	Sgs	BD745E, BDW 10, 2N4348, 2N5634, 2SD425++	data
BDX 53	Si-N-Darl+Di	NF-L, 45V, 8A, 60W, 20MHz, B>750	17j§	Rca,Tho,++	BD 643, BD 895, BDW 73, BDX 33	data
BDX 53 A	Si-N-Darl+Di	=BDX 53: 60V	17j§		BD 645, BD 897, BDW 73A, BDX 33A	data
BDX 53 B	Si-N-Darl+Di	=BDX 53: 80V	17j§		BD 647, BD 899, BDW 73B, BDX 33B	data
BDX 53 BFI	Si-N-Darl+Di	=BDX 53B: Iso, 30W	17c	Tho	BD 645F, BDT 63AF, 2SD1416, 2SD2162, ++	data
BDX 53 BFP	Si-N-Darl+Di	=BDX 53B: Iso, 29W	17c	Tho	BD 645F, BDT 63AF, 2SD1416, 2SD2162, ++	data
BDX 53 C	Si-N-Darl+Di	=BDX 53: 100V	17j§		BD 649, BD 901, BDW 73C, BDX 33C	data
BDX 53 E	Si-N-Darl+Di	=BDX 53: 140V	17j§		BDX 33E, 2SD1386, 2SD1500	data
BDX 53 F	Si-N-Darl+Di	=BDX 53: 16CV	17j§		2SD1025	data
BDX 53 H	Si-N-Darl+Di	=BDX 53: 60V	17j§		BD 645, BD 897, BDW 73A, BDX 33A	data
BDX 53 S	Si-N-Darl+Di	NF/S, 150V, 6A, 15W(Tc=25°), 20MHz	2a§	Sgs		data
BDX 54	Si-P-Darl+Di	NF-L, 45V, 8A, 60W, 20MHz, B>750	17j§	Rca,Tho,++	BD 644, BD 896, BDW 74, BDX 34	data
BDX 54 A	Si-P-Darl+Di	=BDX 54: 60V	17j§		BD 646, BD 898, BDW 74A, BDX 34A	data
BDX 54 B	Si-P-Darl+Di	=BDX 54: 80V	17j§		BD 648, BD 900, BDW 74B, BDX 34B	data
BDX 54 BFI	Si-P-Darl+Di	=BDX 54B: Iso, 30W	17c	Tho	BD 648F, BDT 62AF, 2SB1021, 2SB1344, ++	data
BDX 54 C	Si-P-Darl+Di	=BDX 54: 100V	17j§		BD 650, BD 902, BDW 74C, BDX 34C	data
BDX 54 E	Si-P-Darl+Di	=BDX 54: 140V	17j§		-	data
BDX 54 F	Si-P-Darl+Di	=BDX 54: 160V	17j§		-	data
BDX 54 H	Si-P-Darl+Di	=BDX 54: 60V	17j§		BD 646, BD 898, BDW 74A, BDX 34A	data
BDX 54 S	Si-P-Darl+Di	NF/S, 150V, 6A, 15W(Tc=25°), 20MHz	2a§	Sgs		data
BDX 55	Si-N	NF/S, 100V, 7A, 10W(Tc=25°), >4MHz	2a§	Mot	BU 125, BUY 47...48, BUY 68, BUY 81	data
BDX 56	Si-N	=BDX 55: 120V	2a§	Mot	BU 125, BUY 47...48, BUY 68, BUY 81	data
BDX 57	Si-N	=BDX 55: 140V	2a§	Mot	BUY 47...48, BUY 68, BUY 81	data
BDX 60	Si-N	NF/S-L, 100V, 15A, 150W, >0,8MHz	23a§	Sgs	BD 317, BDW 51C, 2N3772, 2N5629...31, ++	data
BDX 61	Si-N	=BDX 60: 80V	23a§	Sgs	BD 315, BDW 51B, 2N3772, 2N5629...31, ++	data
BDX 62	Si-P-Darl+Di	NF/S-L, 60/60V, 8A, 90W, 7MHz, B>1000	23a§	Mot,Phi,Tho	BDX 84A, BDX 86A, MJ 900, MJ 2500, ++	data
BDX 62 A	Si-P-Darl+Di	=BDX 62: 80/80V	23a§		BDX 84B, BDX 86B, MJ 901, MJ 2501, ++	data
BDX 62 B	Si-P-Darl+Di	=BDX 62: 100/100V	23a§		BDX 84C, BDX 86C, 2SB638...639, ++	data
BDX 62 C	Si-P-Darl+Di	=BDX 62: 120/120V	23a§		BDX 64C	data
BDX 62 L	Si-P-Darl+Di	=BDX 62: 45/45V	23a§		BDX 84A, BDX 86, MJ 900, MJ 2500, ++	data
BDX 63	Si-N-Darl+Di	NF/S-L, 80/60V, 8A, 90W, 7MHz, B>1000	23a§	Mot,Phi,Tho	BDX 83A, BDX 85A, MJ 1000, MJ 3000, ++	data
BDX 63 A	Si-N-Darl+Di	=BDX 63: 100/80V	23a§		BDX 83B, BDX 85B, MJ 1001, MJ 3001, ++	data
BDX 63 B	Si-N-Darl+Di	=BDX 63: 120/100V	23a§		BDX 83C, BDX 85C, 2SD628...629, ++	data
BDX 63 C	Si-N-Darl+Di	=BDX 63: 140/120V	23a§		BDX 65C	data
BDX 63 L	Si-N-Darl+Di	=BDX 63: 60/45V	23a§		BDX 83A, BDX 85, MJ 1000, MJ 3000, ++	data
BDX 64	Si-P-Darl+Di	NF/S-L, 60/60V, 12A, 117W, 7MHz,B>1000	23a§	Mot,Phi,Tho	BDX 66, BDX 88A, MJ 4030, 2N6050	data
BDX 64 A	Si-P-Darl+Di	=BDX 64: 80/80V	23a§		BDX 66A, BDX 88B, MJ 4031, 2N6051	data
BDX 64 B	Si-P-Darl+Di	=BDX 64: 100/100V	23a§		BDX 66B, BDX 88C, MJ 4032, 2N6052	data
BDX 64 C	Si-P-Darl+Di	=BDX 64: 120/120V	23a§		BDX 66C	data
BDX 64 L	Si-P-Darl+Di	=BDX 64: 45/45V	23a§		BDX 66L, BDX 88, MJ 4030, 2N6050	data
BDX 65	Si-N-Darl+Di	NF/S-L, 80/60V, 12A, 117W, 7MHz,B>1000	23a§	Mot,Phi,Tho	BDX 67, BDX 87A, MJ 4033, 2N6057	data
BDX 65 A	Si-N-Darl+Di	=BDX 65: 100/80V	23a§		BDX 67A, BDX 87B, MJ 4034, 2N6058	data
BDX 65 B	Si-N-Darl+Di	=BDX 65: 120/100V	23a§		BDX 67B, BDX 87C, MJ 4035, 2N6059	data
BDX 65 C	Si-N-Darl+Di	=BDX 65: 140/120V	23a§		BDX 67C	data
BDX 65 L	Si-N-Darl+Di	=BDX 65: 60/45V	23a§		BDX 67L, BDX 87, MJ 4033, 2N6057	data
BDX 66	Si-P-Darl+Di	NF/S-L, 60/60V, 16A, 150W, 7MHz,B>1000	23a§	Mot,Phi,Tho	BDW 84A...D, MJ 4030, 2N6285	data
BDX 66 A	Si-P-Darl+Di	=BDX 66: 80/80V	23a§		BDW 84B...D, MJ 4031, 2N6286	data
BDX 66 B	Si-P-Darl+Di	=BDX 66: 100/100V	23a§		BDW 84C...D, MJ 4032, 2N6287	data
BDX 66 C	Si-P-Darl+Di	=BDX 66: 120/120V	23a§			data
BDX 66 L	Si-P-Darl+Di	=BDX 66: 45/45V	23a§		BDW 84(A...D), MJ 4030, 2N6285	data
BDX 67	Si-N-Darl+Di	NF/S-L, 80/60V, 16A, 150W, 7MHz,B>1000	23a§	Mot,Phi,Tho	BDW 83A...D, MJ 4033, 2N6282	data
BDX 67 A	Si-N-Darl+Di	=BDX 67: 100/80V	23a§		BDW 83C...D, MJ 4034, 2N6283	data
BDX 67 B	Si-N-Darl+Di	=BDX 67: 120/100V	23a§		MJ 4035, 2N6284	data
BDX 67 C	Si-N-Darl+Di	=BDX 67: 140/120V	23a§		-	data
BDX 67 L	Si-N-Darl+Di	=BDX 67: 60/45V	23a§		BDW 83(A...D), MJ 4033, 2N6282	data
BDX 68	Si-P-Darl+Di	NF/S-L, 60/60V, 25A, 200W, B>1000	23a§	Phi	MJ 11011, 2SB694	data
BDX 68 A	Si-P-Darl+Di	=BDX 68: 80/80V	23a§		MJ 11013, 2SB694	data
BDX 68 B	Si-P-Darl+Di	=BDX 68: 100/100V	23a§		MJ 11015, 2SB694	data
BDX 68 C	Si-P-Darl+Di	=BDX 68: 120/120V	23a§		MJ 11015	data
BDX 69	Si-N-Darl+Di	NF/S-L, 80/60V, 25A, 200W, B>1000	23a§	Phi	MJ 11012, 2SD730	data
BDX 69 A	Si-N-Darl+Di	=BDX 69: 100/80V	23a§		MJ 11014, 2SD730	data
BDX 69 B	Si-N-Darl+Di	=BDX 69: 120/100V	23a§		MJ 11016, 2SD730	data
BDX 69 C	Si-N-Darl+Di	=BDX 69: 140/120V	23a§		MJ 11016	data
BDX 70	Si-N	NF/S-L, 70V, 10A, 75W, >0,8MHz	17j§	Rca,Sgs	BD 709, BD 743A, BD 807, BD 909	data
BDX 71	Si-N	=BDX 70	17j§	Rca,Sgs	BD 709, BD 743A, BD 807, BD 909	data
BDX 72	Si-N	=BDX 70: 80V	17j§	Rca,Sgs	BD 709, BD 743B, BD 809, BD 909	data
BDX 73	Si-N	=BDX 72	17j§	Rca,Sgs	BD 709, BD 743B, BD 809, BD 909	data
BDX 74	Si-N	=BDX 70: 45V, 16A	17j§	Rca,Sgs	BD 743, BD 905	data
BDX 75	Si-N	=BDX 74	17j§	Rca,Sgs	BD 743, BD 905	data
BDX 77	Si-N	NF/S-L, 100V, 8A, 60W, >7MHz	17j§	Phi	BD 543C, BD 711, BD 801	data
BDX 77 F	Si-N	=BDX 77: Iso, >20W	17c		BDT 95F, 2SD1411, 2SD1588	data
BDX 78	Si-P	NF/S-L, 80V, 8A, 60W, >7MHz	17j§	Phi	BD 544B, BD 710, BD 800, BD 810	data
BDX 78 F	Si-P	=BDX 78: Iso, >20W	17c		BDT 96F, 2SB1018, 2SB1290, 2SA1396	data
BDX 83	Si-N-Darl+Di	NF/S-L, 40V, 10A, 125W, >20MHz, B>1000	23a§	Rca	BDX 85, BDX 87, MJ 3000, 2N6057	data
BDX 83 A	Si-N-Darl+Di	=BDX 83: 60V	23a§		BDX 85A, BDX 87A, MJ 3000, 2N6057	data
BDX 83 B	Si-N-Darl+Di	=BDX 83: 80V	23a§		BDX 85B, BDX 87B, MJ 3001, 2N6058	data

Type	Device	Short Description	Fig.	Manu	Comparision Types	More at
BDX 83 C	Si-N-Darl+Di	=BDX 83: 100V	23a§		BDX 85C, BDX 87C, 2N6059, 2SD628..629	data
BDX 84	Si-P-Darl+Di	NF/S-L, 40V, 10A, 125W, >20MHz, B>1000	23a§	Rca	BDX 86, BDX 88, MJ 2500, 2N6050	data
BDX 84 A	Si-P-Darl+Di	=BDX 84: 60V	23a§		BDX 86A, BDX 88A, MJ 2500, 2N6050	data
BDX 84 B	Si-P-Darl+Di	=BDX 84: 80V	23a§		BDX 86B, BDX 88B, MJ 2501, 2N6051	data
BDX 84 C	Si-P-Darl+Di	=BDX 84: 100V	23a§		BDX 86C, BDX 88C, 2N6052, 2SB638...639	data
BDX 85	Si-N-Darl+Di	NF/S-L, 45V, 10A, 100W, 10MHz, B>1000	23a§	Sgs	BDX 83A, BDX 87, MJ 3000, 2N6057	data
BDX 85 A	Si-N-Darl+Di	=BDX 85: 60V	23a§		BDX 83A, BDX 87A, MJ 3000, 2N6057	data
BDX 85 B	Si-N-Darl+Di	=BDX 85: 80V	23a§		BDX 83B, BDX 87B, MJ 3001, 2N6058	data
BDX 85 C	Si-N-Darl+Di	=BDX 85: 100V	23a§		BDX 83C, BDX 87C, 2N6059, 2SD628...629	data
BDX 86	Si-P-Darl+Di	NF/S-L, 45V, 10A, 100W, 10MHz, B>1000	23a§	Sgs	BDX 84A, BDX 88, MJ 2500, 2N6050	data
BDX 86 A	Si-P-Darl+Di	=BDX 86: 60V	23a§		BDX 84A, BDX 88A, MJ 2500, 2N6050	data
BDX 86 B	Si-P-Darl+Di	=BDX 86: 80V	23a§		BDX 84B, BDX 88B, MJ 2501, 2N6051	data
BDX 86 C	Si-P-Darl+Di	=BDX 86: 100V	23a§		BDX 84C, BDX 88C, 2N6052, 2SB638...639	data
BDX 87	Si-N-Darl+Di	NF/S-L, 45V, 12A, 120W, 25MHz, B>1000	23a§	Sgs	BDX 67, 2N6057, MJ 4033	data
BDX 87 A	Si-N-Darl+Di	=BDX 87: 60V	23a§		BDX 67, 2N6057, MJ 4033	data
BDX 87 B	Si-N-Darl+Di	=BDX 87: 80V	23a§		BDX 67A, 2N6058, MJ 4034	data
BDX 87 C	Si-N-Darl+Di	=BDX 87: 100V	23a§		BDX 67B, 2N6059, MJ 4035	data
BDX 88	Si-P-Darl+Di	NF/S-L, 45V, 12A, 120W, 35MHz, B>1000	23a§	Sgs	BDX 66, 2N6050, MJ 4030	data
BDX 88 A	Si-P-Darl+Di	=BDX 88: 60V	23a§		BDX 66, 2N6050, MJ 4030	data
BDX 88 B	Si-P-Darl+Di	=BDX 88: 80V	23a§		BDX 66A, 2N6051, MJ 4031	data
BDX 88 C	Si-P-Darl+Di	=BDX 88: 100V	23a§		BDX 66B, 2N6052, MJ 4032	data
BDX 91	Si-N	NF/S-L, 60V, 8A, 90W, >4MHz	23a§	Phi	BD 245A, BD 311, BDV 91, 2N5873, ++	data
BDX 92	Si-P	NF/S-L, 60V, 8A, 90W, >4MHz	23a§	Phi	BD 246A, BD 312, BDV 92, 2N5871	data
BDX 93	Si-N	=BDX 91: 80V	23a§	Phi	BD 245B, BD 313, BDV 93, 2N5874, ++	data
BDX 94	Si-P	=BDX 92: 80V	23a§	Phi	BD 246B, BD 314, BDV 94, 2N5872, ++	data
BDX 95	Si-N	=BDX 91: 100V	23a§	Phi	BD 245C, BDV 95, 2N5632...5633, ++	data
BDX 96	Si-P	=BDX 92: 100V	23a§	Phi	BD 246C, BDV 96, 2N6229...6230, ++	data

BDY

Type	Device	Short Description	Fig.	Manu	Comparision Types	More at
BDY 10	Si-N	NF/S-L, 50V, 2A, 150W, >1MHz	23a§	Phi	BD 311, BDW 21A, 2N3055, 2N4914, ++	data
BDY 11	Si-N	=BDY 10: 100V	23a§	Phi	BD 317, BDW 21C, 2N3055, 2N5632, ++	data
BDY 12	Si-N	NF/S-L, 60V, 3A, 26W(Tc=45°), 70MHz	22a§	Sie	BDW 25, BDX 25, BDV 10...12, 2SC3252	data
BDY 13	Si-N	=BDY 12: 80V	22a§	Sie	BDW 25, BDX 25, BDV 10...12, 2SC3252	data
BDY 15	Si-N	NF/S-L, 36V, 2,5A, 11,5W, 100MHz	22a§	Itt	BDW 25, BDX 25, BDV 10...12, 2SC3252	data
BDY 16	Si-N	=BDY 15: 64V	22a§	Itt	BDW 25, BDX 25, BDV 10...12, 2SC3252	data
BDY 17	Si-N	NF/S-L, 80V, 10A, 115W, 1MHz	23a§	Phi	BD 745B, BD 313, BDW 21B, 2N3055, ++	data
BDY 18	Si-N	=BDY 17: 120V	23a§	Phi	BD745D, BDW10, 2N5633, 2SC2706, 2SD1047+	data
BDY 19	Si-N	=BDY 17: 150V	23a§	Phi	BD745E, BDW12, 2N3442, 2SC2706, 2SD1047+	data
BDY 20	Si-N	NF/S-L, 100V, 15A, 117W, 1MHz	23a§	Phi	BD 317, BD 745C, BDW 51C, 2N3055, ++	data
BDY 21	Si-N	NF/S-L, 80V, 3A, 25W(Tc=45°), >30MHz	22a§/3Pin	Sie	BDW 25, BDX 25, BDV 10...12, 2SC3252	data
BDY.22	Si-N	=BDY 21: 100V	22a§/3Pin	Sie	BDW 25, BDX 25, BDV 10...12, 2SD772	data
BDY 23	Si-N	NF/S-L, 60V, 6A, 87,5W, >10MHz	23a§	Tho	BD 245A, BD 311, BDW 21A, 2N3055, ++	data
BDY 24	Si-N	=BDY 23: 100V	23a§	Tho.	BD 245C, BD 317, BDW 21C, 2N3055, ++	data
BDY 25	Si-N	=BDY 23: 200V	23a§	Tho	BD245F, BDW16, MJ15015, 2SC1585, 2SC2608	data
BDY 26	Si-N	=BDY 23: 300V	23a§	Tho	BUX 18, BUY 18, BUY 35, BUY 67, ++	data
BDY 27	Si-N	=BDY 23: 400V	23a§	Tho	BUX 18B, BUY 18S, BUY 67, BUY 77, ++	data
BDY 28	Si-N	=BDY 23: 500V	23a§	Tho	BUW 34, BUX 15, BUX 45, BUY 69C...70C, ++	data
BDY 29	Si-N	NF/S-L, 100V, 30A, 220W, >0,2MHz	23a§	Rca	BDW 30, BDY 57, MJ 802, 2SD797	data
BDY 34	Si-N	NF/S-L, 45V, 3A, 21W, >80MHz	14h§	Aeg	BD 785, MJE 240..242, 2SD1348	data
BDY 34	Si-N	NF/S-L, 45V, 3A, 13W(Tc=45°), >80MHz	22a§	Aeg	BDV 10...12, BDW 25, BDX 25, 2SC3252, ++	data
BDY 37	Si-N	NF/S-L, 160V, 16A, 150W, >0,2MHz	23a§	Rca	BDW 32, BDW 34, BDW 36, BDY 58, 2N3773	data
BDY 37 A	Si-N	=BDY 37: 250W	23a§		BDW 32, BDW 34, BDW 36, 2SC2608	data
BDY 38	Si-N	NF/S-L, 50V, 6A, 117W, 1MHz	23a§	Phi	BD 311, BD 745A, BDW 21A, 2N3055, ++	data
BDY 39	Si-N	NF/S-L, 100V, 15A, 117W, 1,1MHz	23a§	Sgs, Sie	BD 317, BD 745C, BDW 51C, 2N3055, ++	data
BDY 42	Si-N	S-L, 400V, 5A, 60W, 12MHz	23a§	Aeg	BUW 71, BUX 18B, BUY 18S, BUY 67, ++	data
BDY 43	Si-N	=BDY 42: 600V	23a§	Aeg	BUS 11, BUW 25, BUX 82, BUY 78...79, ++	data
BDY 44	Si-N	=BDY 42: 750V	23a§	Aeg	BUS 11, BUW 26, BUX 82, BUY 79, ++	data
BDY 45	Si-N	S-L, 400V, 15A, 95W, 13MHz	23a§	Aeg	BUW 44, BUX 13, BUX 25, BUY 50, ++	data
BDY 46	Si-N	=BDY 45: 600V	23a§	Aeg	BUS 13, BUW 45...46, BUX 48, 2SD641	data
BDY 47	Si-N	=BDY 45: 750V	23a§	Aeg	BUS 13, BUW 45...46, BUX 48, 2SC3094	data
BDY 48/200	Si-N	S-L, 200V, 3,5A, 100W, 5MHz	23a§	Aeg	BUX 16...18, BUY 18, BUY 35, BUY 67, ++	data
BDY 48/300	Si-N	=BDY 48/200: 300V	23a§	Aeg	BUX 16A...18A, BUY 18, BUY 35, BUY 67, ++	data
BDY 48/400	Si-N	=BDY 48/200: 400V	23a§	Aeg	BUX 16C...18C, BUY 18S, BUY 67, ++	data
BDY 49	Si-N	S-L, -/100V, 30A, 150W	23a§	Aeg	BDW 30, BDY 29, BDY 57, MJ 802, 2SD797	data
BDY 53	Si-N	NF/S-L, 100V, 12A, 60W, >20MHz	23a§	Tho	BD 317, BD 745C, BDW 51C, 2N3055, ++	data
BDY 54	Si-N	=BDY 53: 180V	23a§	Tho	BDW 14, BDW 16, BDY 54, BDY 56, ++	data
BDY 55	Si-N	NF/S-L, 100V, 15A, 115W, >10MHz	23a§	Rca,Tho	BD 317, BD 745C, BDW 51C, 2N3055, ++	data
BDY 56	Si-N	=BDY 55: 180V	23a§	Rca,Tho	BDW 14, BDW 16, BDY 56, 2SC2607, ++	data
BDY 57(A)	Si-N	NF/S-L, 120V, 25A, 175W, >7MHz	23a§	Rca,Tho	BDW 30, 2N6032, 2N6274...6275	data
BDY 58	Si-N	=BDY 57: 160V	23a§	Rca,Tho	BDW 32, BDW 34, 2N6276...6277	data
BDY 58 R	Si-N	=BDY 57: 250V	23a§	Rca	BUV 11...12, BUX 11...12, BUW 58, BUW 73	data
BDY 60	Si-N	NF/S-L, 120V, 5A, 15W(Tc=100°), 100MHz	23a§	Phi	BDY 90, 2SC2681, 2SC2706, 2SC2837	data
BDY 61	Si-N	=BDY 60: 100V	23a§	Phi	BDY 90...91, 2SC2681, 2SC2706, 2SC2837	data
BDY 62	Si-N	=BDY 60: 30V	23a§	Phi	BDY 90...92, 2SC2681, 2SC2706, 2SC2837	data
BDY 63	Si-N	S-L, 100V, 10A, 50W(Tc=100°), >40MHz	49a§	Tix	2N5542	data

Type	Device	Short Description	Fig.	Manu	Comparision Types	More at
BDY 64	Si-N	S-L, 100V, 30A, 100W(Tc=100°), >30MHz	49a§	Tix	2N2824..2825	data
BDY 65	Si-N	NF/S-L, 150V, 1A, 15W(Tc=100°), >30MHz	2a§	Tix	BUY 41	data
BDY 66	Si-N	=BDY 65:	49a§	Tix	2N5002, 2N5504, 2N5284...5285	data
BDY 67	Si-P	S-L, 100V, 5A, 30W(Tc=100°), >30MHz	50g	Tix	2N5002, 2N5504, 2N5284...5285	data
BDY 68	Si-P	=BDY 67:	49a§	Tix	2N5002, 2N5504, 2N5284...5285	data
BDY 69	Si-P	S-L, 100V, 12A, 50W(Tc=100°), >30MHz	49a§	Tix	2N5542	data
BDY 70	Si-P	S-L, 100V, 2A, 15W(Tc=100°), >30MHz	2a§	Tix	BUY 41	data
BDY 71	Si-N	NF/S-L, 90V, 4A, 29W, >0,8MHz	22a§	Rca,Tho	BD 243B, BD 953, BDY 78, 2N3054, ++	data
BDY 72	Si-N	NF/S-L, 150V, 3A, 25W, >0,8MHz	22a§	Tho	BD241D, BDY79, 2N3441, 2SC2516	data
BDY 73	Si-N	NF/S-L, 100V, 15A, 115W, >0,8MHz	23a§	Tho	BD 317, BD 745C, BDW 51C, 2N3055, ++	data
BDY 74	Si-N	=BDY 73: 150V	23a§	Tho	BD745E, BDW12, 2N3773, 2SC1584, 2SD1047+	data
BDY 75	Si-N	NF/S-L, 50V, 30A, 150W, >0,8MHz	23a§	Tho	BDY 29, MJ 802, 2N3771, 2SD630	data
BDY 76	Si-N	=BDY 75: 100V, 20A	23a§	Tho	BD 249C, BDY 29, MJ 802, 2N3772, 2SD797	data
BDY 77	Si-N	=BDY 75: 150V, 16A	23a§	Tho	BD 249D, BDW 32, BDY 58, 2N3773, 2SC2608	data
BDY 78	Si-N	NF/S-L, 90V, 4A, 25W, >0,8MHz	22a§	Tho	BD 243B, BD 953, BDY 71, 2N3054, ++	data
BDY 79	Si-N	=BDY 78: 150V	22a§	Tho	BD 243D, 2SC2516	data
BDY 80	Si-N	NF/S-L, 40V, 4A, 36W, 3MHz	17j§	Tho,Tos	BD 243, BD 533, BD 539, BD 947, ++	data
BDY 81	Si-N	=BDY 80: 60V	17j§	Tho,Tos	BD 243A, BD 535, BD 539A, BD 949, ++	data
BDY 82	Si-P	NF/S-L, 35V, 4A, 36W, 3MHz	17j§	Tho,Tos	BD 244, BD 534, BD 540, BD 948, ++	data
BDY 83	Si-P	=BDY 82: 50V	17j§	Tho,Tos	BD 244A, BD 536, BD 540A, BD 950, ++	data
BDY 87	Si-N-Darl	NF/S-L, 20V, 8A, 35W(Tc=45°), B=2500	(22)	Sie	-	data
BDY 88	Si-N-Darl	=BDY 87: 40V	(22)	Sie	-	data
BDY 89	Si-N-Darl	=BDY 87: 60V	(22)	Sie	-	data
BDY 90	Si-N	NF/S-L, 120V, 10A, 60W, 70MHz	23a§	Phi,Sgs,++	2SC2681, 2SC2706, 2SC2837	data
BDY 90 A	Si-N	=BDY 90: 12A	23a§	Phi,Sgs	(2SC2681, 2SC2706, 2SC2837)	data
BDY 91	Si-N	=BDY 90: 100V	23a§	Phi,Sgs	2SC2681, 2SC2706, 2SC2837	data
BDY 92	Si-N	=BDY 90: 80V	23a§	Phi,Sgs	2SC2681, 2SC2706, 2SC2837, 2SC3256	data
BDY 93	Si-N	S-L, 750/350V, 4A, 30W(Tc=75°), 10MHz	23a§	Phi,Tho	BUS 11, BUX 46...47, BUX 82...83, ++	data
BDY 94	Si-N	=BDY 93: 750/300V	23a§	Phi,Tho	BUS 11, BUX 46...47, BUX 82...83, ++	data
BDY 95	Si-N	=BDY 93: 400/250V	23a§	Phi,Tho	BUW 71, BUX 16C, BUX 45, BUY 67; ++	data
BDY 96	Si-N	S-L, 750/350V, 10A, 40W(Tc=90°), 10MHz	23a§	Phi,Tho	BUS 12, BUW 26, BUW 35...36, BUX 80, ++	data
BDY 97	Si-N	=BDY 96: 750/300V	23a§	Phi,Tho	BUS 12, BUW 26, BUW 35...36, BUX 80, ++	data
BDY 98	Si-N	=BDY 96: 400/250V	23a§	Phi,Tho	BUW 24, BUW 72, BUX 17B, BUX 43, ++	data
BDY 99	Si-N	=BDY 96: 750/250V	23a§	Phi	BUS 12, BUW 26, BUW 35...36, BUX 80, ++	data

BE

Type	Device	Short Description	Fig.	Manu	Comparision Types	More at
BE	Si-Di	=BAS 70L... (Typ-Code/Stempel/marking)	35	Ons	→BAS 70L	
BE	Si-N	=BCP 55 (Typ-Code/Stempel/marking)	48	Phi	→BCP 55	data
BE	Si-P	=BCW 61E (Typ-Code/Stempel/marking)	35	Sie	→BCW 61E	data
BE	Si-P	=BCX 53 (Typ-Code/Stempel/marking)	39	Rhm	→BCX 53	data
BE	Si-N	=BCX 55 (Typ-Code/Stempel/marking)	39	Sie,Val,Mot	→BCX 55	data
BE	Z-Di	=BZX 399-C6V8(Typ-Code/Stempel/marking)	71(1,7mm)	Phi	→BZX 399-...	
BE	MOS-P-FET-e	=NTTD 2P02 (Typ-Code/Stempel/marking)	8-SSMDIP	Ons	→NTTD 2P02	
BE	Z-Di	=P4 SMA-18A (Typ-Code/Stempel/marking)	71(5x2,5)	Fag	→P4 SMA-18A	data
BE	Z-Di	=SMBJ 12CA (Typ-Code/Stempel/marking)	71(5x3,5)	Mop	→SMBJ ...	data
BE	Si-Di	=1SV172 (Typ-Code/Stempel/marking)	35	Tos	→1SV172	data
BE	Si-Di	=1SV252 (Typ-Code/Stempel/marking)	35 (2mm)	Tos	→1SV252	data
BE	Si-P	=2SB1122 (Typ-Code/Stempel/marking)	39	Say	→2SB1122	data
BE	Si-N	=2SC3440-E (Typ-Code/Stempel/marking)	35	Mit	→2SC3440	data
BE	Si-N	=2SC3443-E (Typ-Code/Stempel/marking)	39	Mit	→2SC3443	data
BE 565	LIN-IC	=LM 565			→LM 565	
BED	Z-Di	=SM 15T 18C (Typ-Code/Stempel/marking)	71(8x5mm)	Tho	→SM 15T...	data
BEE	Z-Di	=SM 15T 18CA(Typ-Code/Stempel/marking)	71(8x5mm)	Tho	→SM 15T...	data
BEH	Z-Di	=SM 15T 22C (Typ-Code/Stempel/marking)	71(8x5mm)	Tho	→SM 15T...	data
BEK	Z-Di	=SM 15T 22CA(Typ-Code/Stempel/marking)	71(8x5mm)	Tho	→SM 15T...	data
BEL	Z-Di	=SM 15T 24C (Typ-Code/Stempel/marking)	71(8x5mm)	Tho	→SM 15T...	data
BEM	Z-Di	=SM 15T 24CA(Typ-Code/Stempel/marking)	71(8x5mm)	Tho	→SM 15T...	data
BEN	Z-Di	=SM 15T 27C (Typ-Code/Stempel/marking)	71(8x5mm)	Tho	→SM 15T...	data
BEN 139	Ge-P	UHF, 700MHz		Ben	AF 239(S)	data
BEP	Si-P	=2SB1260-P (Typ-Code/Stempel/marking)	39	Rhm	→2SB1260	data
BEP	Z-Di	=SM 15T 27CA(Typ-Code/Stempel/marking)	71(8x5mm)	Tho	→SM 15T...	data
BEQ	Si-P	=2SB1260-Q (Typ-Code/Stempel/marking)	39	Rhm	→2SB1260	data
BEQ	Z-Di	=SM 15T 30C (Typ-Code/Stempel/marking)	71(8x5mm)	Tho	→SM 15T...	data
BER	Si-P	=2SB1260-R (Typ-Code/Stempel/marking)	39	Rhm	→2SB1260	data
BER	Z-Di	=SM 15T 30CA(Typ-Code/Stempel/marking)	71(8x5mm)	Tho	→SM 15T...	data
BEs	Si-N	=BCP 55M (Typ-Code/Stempel/marking)	≈45	Sie	→BCP 55M	data
BES	Z-Di	=SM 15T 33C (Typ-Code/Stempel/marking)	71(8x5mm)	Tho	→SM 15T...	data
BET	Z-Di	=SM 15T 33CA(Typ-Code/Stempel/marking)	71(8x5mm)	Tho	→SM 15T...	data
BEU	Z-Di	=SM 15T 36C (Typ-Code/Stempel/marking)	71(8x5mm)	Tho	→SM 15T...	data
BEV	Z-Di	=SM 15T 36CA(Typ-Code/Stempel/marking)	71(8x5mm)	Tho	→SM 15T...	data
BEW	Z-Di	=SM 15T 39C (Typ-Code/Stempel/marking)	71(8x5mm)	Tho	→SM 15T...	data
BEX	Z-Di	=SM 15T 39CA(Typ-Code/Stempel/marking)	71(8x5mm)	Tho	→SM 15T...	data

Type	Device	Short Description	Fig.	Manu	Comparision Types	More at
BF						
BF	Si-Di	=1SS268 (Typ-Code/Stempel/marking)	35	Tos	→1SS268	data
BF	Si-Di	=1SS312 (Typ-Code/Stempel/marking)	35(2mm)	Tos	→1SS312	data
BF	Si-Di	=1SS364 (Typ-Code/Stempel/marking)	35(1,6mm)	Tos	→1SS364	data
BF	Si-P	=2SB1123 (Typ-Code/Stempel/marking)	39	Say	→2SB1123	data
BF	Si-N	=2SC3440-F (Typ-Code/Stempel/marking)	35	Mit	→2SC3440	data
BF	Si-N	=2SC3443-F (Typ-Code/Stempel/marking)	39	Mit	→2SC3443	data
BF	Si-P	=BCW 61FF (Typ-Code/Stempel/marking)	35	Sie	→BCW 61FF	data
BF	Si-N	=BCX 55-6 (Typ-Code/Stempel/marking)	39	Sie,Val,Mot	→BCX 55-6	data
BF	Z-Di	=BZX 399-C7V5(Typ-Code/Stempel/marking	71(1,7mm)	Phi	→BZX 399-...	data
BF	Si-Di	=KDS 112 (Typ-Code/Stempel/marking)	35(2mm)	Kec	→KDS 112	data
BF	C-Di	=KDS 112E (Typ-Code/Stempel/marking)	35(1,6m)	Kec	→KDS 112E	data
BF	Z-Di	=P4 SMA-20 (Typ-Code/Stempel/marking)	71(5x2,5)	Fag	→P4 SMA-20	data
BF	Z-Di	=SMBJ 13C (Typ-Code/Stempel/marking)	71(5x3,5)	Mop	→SMBJ ...	data
BF 1	Si-P-Darl	=RXTA-64 (Typ-Code/Stempel/marking)	39	Rhm	→RXTA-64	data
BF 2	Si-P-Darl	=RXTA-76 (Typ-Code/Stempel/marking)	39	Rhm	→RXTA-76	data
BF 108	Si-N	Vid, 140V, 0,15A, 0,86W, 180MHz	2a§	Tho	BF 257...259, BF 657...659, 2N5058...5059	data
BF 109	Si-N	Vid, 135V, 0,05A, 0,52W, >80MHz	2a§	Phi	BF 257...259, BF 657...659, 2N5058...5059	data
BF 110	Si-N	Vid, 160V, 0,04A, 0,75W, 150MHz	2a§	Aeg,Phi,Sie	BF 257...259, BF 657...659, 2N5058...5059	data
BF 111	Si-N	Vid, 200V, 0,08A, 0,8W, 120MHz	2a§	Sie	BF 258...259, BF 658...659, 2N5058...5059	data
BF 114	Si-N	Vid, 135V, 0,04A, 1,2W(Tc=60°), >80MHz	2a§	Aeg,Phi,Sie	BF 257...259, BF 657...659, 2N5058...5059	data
BF 115	Si-N	AM/FM-V/M/O/ZF, 230MHz	5k	Phi,Sie,++	BF 184...185, BF 240...241, BF 254...255,++	data
BF 117	Si-N	Vid, 140V, 0,1A, 0,68W, 80MHz	2a§	Itt	BF 257...259, BF 657...659, 2N5058...5059	data
BF 118	Si-N	Vid, 250V, 0,1A, 0,8W, 110MHz	2a§	Itt	BF 258...258, BF 658...659, 2N5058...5059	data
BF 119	Si-N	Vid, 160V, 0,1A, 0,8W, 110MHz	2a§	Itt	BF 257...259, BF 657...659, 2N5058...5059	data
BF 120	Si-N	Vid, TV-HA-O, 220V, 0,05A, 0,3W	2a§	Itt	BF 298...299, BF 422, (BF 258...259,++)[6]	data
BF 121	Si-N	AM/FM-V-re, 350MHz	6k	Itt	BF 198, BF 225, BF 310, BF 367, BF 596	data
BF 123	Si-N	TV-ZF, 550MHz	6k	Itt	BF 199, BF 224, BF 311, BF 373, BF 597	data
BF 125	Si-N	AM/FM-M/O/ZF, 450MHz	6k	Itt	BF 199, BF 224, BF 311, BF 373, BF 597	data
BF 127	Si-N	TV-ZF-re, 350MHz	6k	Itt	BF 198, BF 225, BF 310, BF 367, BF 596	data
BF 130	Si-N	Uni, 45V, 0,1A, 150MHz	8a	Itt	BC 167, BC 182, BC 237, BC 547, ++	data
BF 131	Si-N	=BF 130:	2a§	Itt	BC 167, BC 182, BC 237, BC 547, ++	data
BF 132	Si-N	Uni, 25V, 0,1A, 270MHz	2a§	Itt	BC 168, BC 183, BC 238, BC 548, ++	data
BF 133	Si-N	=BF 132:	8a	Itt	BC 168, BC 183, BC 238, BC 548, ++	data
BF 1:4	Si-N	FM-M/O, >600MHz	8a	Itt	BF 240, BF 254, BF 310, BF 314, ++	data
BF 136	Si-N	FM-V-re, 600MHz	8a	Itt	BF 240, BF 255, BF 495, BF 595, ++	data
BF 137	Si-N	Vid, 160V, 0,1A, 0,68W, 95MHz	2a§	Itt	BF 257...259, BF 657...659, 2N5058...5059	data
BF 138	Si-N	FM-V-re, 600MHz	8a	Itt	BF 240, BF 255, BF 310, BF 314, ++	data
BF 140	Si-N	Vid, 135V, 0,05A, 0,8W, 100MHz	2a§	Tho	BF 257...259, BF 657...659, 2N5058...5059	data
BF 140 A	Si-N	=BF 140: 150V	2a§		BF 257...259, BF 657...659, 2N5058...5059	
BF 140 D	Si-N	=BF 140: 180V	2a§		BF 258...259, BF 658...659, 2N5058...5059	
BF 140 R	Si-N	=BF 140: 0,03A, 0,3W	2a§		BF 297...299, BF 422, (BF 257...259, ++)[6]	
BF 140 S	Si-N	=BF 140: 0,03A	2a§		BF 257...259, BF 657...659, 2N5058...5059	
BF 152	Si-N	VHF-M/O, 800MHz	8a	Fch,Sgs	BF 224, BF 314, BF 496, BF 502...503, ++	data
BF 153	Si-N	AM-ZF, 400MHz	8a	Nsc,Sgs	BF 241, BF 254, BF 494, BF 594, ++	data
BF 154	Si-N	Vid-Tr, 30V, 0,05A, 0,3W, 400MHz	8a	Sgs	BF 199, BF 224, BF 311, BF 373, BF 597	data
BF 155	Si-N	UHF-V/M/O, 600MHz	5g	Sgs	BF 180...183, BF 689, BF 763, 2N2857, ++	data
BF 155 R	Si-N	Vid, 155V, 0,05A, 0,3W, >40MHz	2a§	-	BF 297...299, BF 422, (BF 257...259, ++)[6]	data
BF 155 S	Si-N	=BF 155 R: 0,8W	2a§	-	BF 257...259, BF 657...659, 2N5058...59	data
BF 156	Si-N	Vid, 120V, 0,1A, 0,8W, 60MHz	2a§	Sgs	BF 257...259, BF 657...659, 2N5058...5059	data
BF 157	Si-N	=BF 156: 150V	2a§	Sgs	BF 257...259, BF 657...659, 2N5058...5059	data
BF 157 B	Si-N	=BF 156: 175V	2a§		BF 257...259, BF 657...659, 2N5058...5059	data
BF 158	Si-N	TV-ZF, 800MHz	8a	Sgs	BF 199, BF 224, BF 311, BF 373, BF 597	data
BF 159	Si-N	TV-ZF, 800MHz	8a	Sgs	BF 199, BF 224, BF 311, BF 373, BF 597	data
BF 160	Si-N	AM/FM-ZF, 600MHz	8a	Nsc,Sgs	BF 241, BF 254, BF 494, BF 594, ++	data
BF 161	Si-N	UHF-V/M/O, 550MHz	5g	Sgs	BF 180...183, BF 689, BF 763, 2N2857, ++	data
BF 162	Si-N	VHF-V, re, 600MHz	8a	Sgs	BF 225, BF 314, BF 496, BF 502...503, ++	data
BF 163	Si-N	TV-ZF-re, 600MHz	8a	Sgs	BF 198, BF 225, BF 310, BF 367, BF 596	data
BF 164	Si-N	TV-ZF-re, 600MHz	8a	Sgs	BF 198, BF 225, BF 310, BF 367, BF 596	data
BF 165	Si-N	AM/FM-ZF-re, 300MHz	8a	Sgs	BF 240...241, BF 254...255, BF 494...495,++	data
BF 166	Si-N	VHF-V/M/O-re, 500MHz	5g	Sgs	BF 225, BF 314, BF 496, BF 502...503, ++	data
BF 167	Si-N	TV-ZF-re, 350MHz	5k	EUR	BF 198, BF 225, BF 310, BF 367, BF 596	data
BF 168	Si-N	TV-ZF, 550MHz	5k	Phi	BF 199, BF 224, BF 311, BF 373, BF 597	data
BF 169	Si-N	Vid-Tr, 50V, 0,05A, 0,3W, >200MHz	2a§	Tho	BC 167, BC 182, BC 237, BC 547, ++	data
BF 169 R	Si-N	=BF 169:	8a		BC 167, BC 182, BC 237, BC 547, ++	data
BF 170	Si-N	Vid, 160V, 0,05A, 0,8W, 100MHz	2a§		BF 257...259, BF 657...659, 2N5058...5059	data
BF 173	Si-N	TV-ZF, 550MHz	5k	EUR	BF 199, BF 224, BF 311, BF 373, BF 597	data
BF 174	Si-N	Vid, 150V, 0,1A, 0,8W, 80MHz	2a§	Sgs	BF 257...259, BF 657...659, 2N5058...5059	data
BF 175	Si-N	TV-ZF-re, 500MHz	5g	Sgs	BF 198, BF 225, BF 310, BF 367 BF 596	data
BF 176	Si-N	TV-ZF, 450MHz	8a	Sgs	BF 199, BF 224, BF 311, BF 373, BF 597	data

Type	Device	Short Description	Fig.	Manu	Comparision Types	More at
BF 177	Si-N	Vid, 100V, 0,04A, 0,7W, 120MHz	2a§	EUR	BF 257...259, BF 657...659, 2N5058...5059	data
BF 178	Si-N	Vid, 185V, 0,05A, 0,7W, 120MHz	2a§	EUR	BF 258...259, BF 658...659, 2N5058...5059	data
BF 179(A...C)	Si-N	=BF 178: 185...250V	2a§		BF 258...259, BF 658...659, 2N5058...5059	data
BF 180	Si-N	VHF/UHF-V, 675MHz	5g	Phi,Tho	BF 689, BF 763, 2N918, 2N2857, ++	data
BF 181	Si-N	VHF/UHF-M, 600MHz	5g	Phi,Tho	BF 689, BF 763, 2N918, 2N2857, ++	data
BF 182	Si-N	VHF/UHF-M, 650MHz	5g	Phi,Tho	BF 689, BF 763, 2N918, 2N2857, ++	data
BF 183	Si-N	VHF/UHF-O, 800MHz	5g	Phi,Tho	BF 689, BF 763, 2N918, 2N2857, ++	data
BF 184	Si-N	AM/FM-V/M/O/ZF, 300V	5k	EUR	BF 240, BF 254, BF 494, BF 594, ++	data
BF 185	Si-N	FM-V/M/O, 220MHz	5k	EUR	BF 241, BF 255, BF 495, BF 595, ++	data
BF 186	Si-N	Vid, 190V, 0,06A, 0,8W, 120MHz	2a§	Phi	BF 258...259, BF 658...659, 2N5058...5059	data
BF 187	Si-N	HF/ZF, 500MHz	2a§	Tho	BF 199, BF 224, BF 311, BF 373, BF 597	data
BF 188	Si-N	VHF-M/O, 600MHz	5g	Tho	BF 224, BF 314, BF 496, BF 502...503,++	data
BF 189	Si-N	AM/FM-ZF, 270MHz	5k		BF 240, BF 254, BF 494, BF 594, ++	data
BF 194	Si-N	AM/FM-V/M/O/ZF, 260MHz	11d	Aeg,Phi,Tho	BF 240, BF 254, BF 494, BF 594, ++	data
BF 195	Si-N	FM-V/M/O, 200MHz	11d	Aeg,Phi,Tho	BF 241, BF 255, BF 495, BF 595, ++	data
BF 196	Si-N	TV-ZF-re, 400MHz	11d, 12d	Aeg,Phi,Tho	BF 198, BF 225, BF 310, BF 367, BF 596	data
BF 197	Si-N	TV-ZF, 550MHz	11d, 12d	Aeg,Phi,Tho	BF 199, BF 224, BF 311, BF 373, BF 597	data
BF 198	Si-N	TV-ZF-re, 400MHz	7d	EUR	BF 225, BF 310, BF 367, BF 596, ++	data
BF 199	Si-N	TV-ZF, 550MHz	7d	EUR	BF 224, BF 311, BF 373, BF 597, ++	data
BF 200	Si-N	FM/VHF-V, 650MHz	5g	Mot,Phi,Tho	BF 225, BF 314, BF 496, BF 502...503, ++	data
BF 202	Si-N	VHF/UHF, 650MHz	2a§	Aeg	BF 689, BF 763, 2N918, 2N2857, ++	data
BF 203	Si-N	VHF/UHF, 900MHz	2a§	Aeg	BF 689, BF 763, 2N918, 2N2857, ++	data
BF 206	Si-N	VHF, 500MHz	5g	Tho	BF 225, BF 314, BF 496, BF 502...503, ++	data
BF 207	Si-N	VHF/ZF, 400MHz	5k	Tho	BF 198, BF 225, BF 310, BF 367, BF 596	data
BF 208	Si-N	VHF/ZF, 600MHz	5k	Tho	BF 199, BF 224, BF 311, BF 373, BF 597	data
BF 209	Si-N	VHF, 500MHz	5g	Tho	BF 225, BF 314, BF 496, BF 502...503, ++	data
BF 212	Si-N	UHF-O, 700MHz	5g	Tho	BF 180...183, BF 689, BF 763, 2N2857, ++	data
BF 213	Si-N	UHF-M, 600MHz	5g	Tho	BF 180...182, BF 689, BF 763, 2N2857, ++	data
BF 214	Si-N	AM/FM-V/M/O/ZF, 250MHz	5k	Tho	BF 240, BF 254, BF 494, BF 594, ++	data
BF 215	Si-N	FM-V/M/O, 250MHz	5k	Tho	BF 241, BF 255, BF 495, BF 595, ++	data
BF 216	Si-N	FM-V, 220MHz	7c	Aei	BF 241, BF 255, BF 495, BF 595, ++	data
BF 217	Si-N	FM-M, 240MHz	7c	Aei	BF 241, BF 255, BF 495, BF 595, ++	data
BF 218	Si-N	AM/FM-ZF, 220MHz	7c	Aei	BF 240, BF 254, BF 494, BF 594, ++	data
BF 219	Si-N	AM-V/M/O, 260MHz	7c	Aei	BF 240, BF 254, BF 494, BF 594, ++	data
BF 220	Si-N	AM/FM-O, 260MHz	7c	Aei	BF 240, BF 254, BF 494, BF 594, ++	data
BF 221 [Riz]	Si-N	HF/ZF, 135MHz	2a	Riz	BF 240...241, BF 254...255, BF 494...495,++	data
BF 222 [Riz]	Si-N	HF/ZF, 135MHz	2a	Riz	BF 240...241, BF 254...255, BF 494...495,++	data
BF 222 [SGS]	Si-N	FM-V-re, 400MHz	5g	Sgs	BF 225, BF 314, BF 496, BF 502...503, ++	data
BF 223 [AEG]	Si-N	TV-ZF, 750MHz	11d	Aeg	BF 199, BF 224, BF 311, BF 373, BF 597	data
BF 223 [Riz]	Si-N	HF/ZF, 135MHz	2a	Riz	BF 240...241, BF 254...255, BF 494...495,++	data
BF 224	Si-N	TV-ZF, VHF-M/O, 45V, 700MHz	7d	Tix	BF 199, BF 311, BF 373, BF 597, ++	data
BF 225	Si-N	TV-ZF-re, VHF-V, 700MHz	7d	Tix	BF 198, BF 310, BF 367, BF 596, ++	data
BF 226	Si-N	FM-M/O, 250MHz	5k	Tho	BF 240...241, BF 254...255, BF 494...495,++	data
BF 227	Si-N	Min, TV-ZF, 600MHz	36d	Aeg	(BF 199, BF 224, BF 311, BF 373, BF 597)[6]	data
BF 228	Si-N	Min, Nix, 100V, 50mA, 0,05W, >50MHz	36c	Aeg	BF 622, (BF 297...299, BF 422)[6]	data
BF 229	Si-N	Min, AM-V/M/O, AM/FM-ZF, 260MHz	36d	Aeg	(BF 240, BF 254, BF 494, BF 594, ++)[6]	data
BF 230	Si-N	Min, FM-V/M/O, 200MHz	36d	Aeg	(BF 241, BF 255, BF 495, BF 595, ++)[6]	data
BF 231 [Riz]	Si-N	AM/FM/VHF, 800MHz	2a§	Riz	BF 199, BF 224, BF 311, BF 373, BF 597	data
BF 232 [Riz]	Si-N	AM/FM/VHF, 300MHz	2a§	Riz	BF 199, BF 224, BF 311, BF 373, BF 597	data
BF 232 [Siemens]	Si-N	TV-ZF, 600MHz	5k	Sie	BF 199, BF 224, BF 311, BF 373, BF 597	data
BF 233 [CSF,SGS]	Si-N	AM/FM-ZF, 500MHz	8d	Nsc,Sgs	BF 240, BF 254, BF 494, BF 594, ++	data
BF 233 [Riz]	Si-N	AM/FM/VHF, 800MHz	2a§	Riz	BF 199, BF 224, BF 311, BF 373, BF 597	data
BF 234 [CSF,SGS]	Si-N	AM-V/M/O/ZF, 500MHz	8d	Sgs	BF 240...241, BF 254...255, BF 494...495,++	data
BF 234 [Riz]	Si-N	AM/FM/VHF, 300MHz	2a§	Riz	BF 199, BF 224, BF 311, BF 373, BF 597	data
BF 235 [CSF,SGS]	Si-N	FM-V/M, 500MHz	8d	Sgs	BF 241, BF 255, BF 495, BF 595, ++	data
BF 235 [Riz]	Si-N	AM/FM/VHF, 400MHz	2a§	Riz	BF 199, BF 224, BF 311, BF 373, BF 597	data
BF 236	Si-N	FM/VHF-O, 250MHz	8d	Tho	BF 240...241, BF 254...255, BF 494...495,++	data
BF 237	Si-N	FM-V/M/O/ZF	7d	Tix	BF 241, BF 255, BF 495, BF 595, ++	data
BF 238	Si-N	AM-V/M/O/ZF	7d	Tix	BF 240, BF 254, BF 494, BF 594, ++	data
BF 240(B)	Si-N	AM/FM-ZF-re, 400MHz	7d	EUR	BF 254...255, BF 494...495, BF 594...595,++	data
BF 240 B	Si-N	=BF 240:	10d	Pih	→BF 240	data
BF 240 [Riz]	Si-N	Uni, 15V, 0,1A, 0,3W, >90MHz	2a§	Riz	BC 168, BC 183, BC 238, BC 548, ++	data
BF 241(C,D)	Si-N	AM/FM-ZF, 400MHz	7d	EUR	BF 254...255, BF 494...495, BF 594...595,++	data
BF 241 [Riz]	Si-N	Uni, 30V, 0,1A, 0,3W, 125MHz	2a§	Riz	BC 168, BC 183, BC 238, BC 548, ++	data
BF 241 A [Riz]	Si-N	=BF 241[Riz]: 60V	2a§		BC 167, BC 182, BC 237, BC 546, ++	data
BF 241 C,D [Piher]	Si-N	=BF 241:	10d	Pih	→BF 241	data
BF 242 [Riz]	Si-N	Uni, 30V, 0,1A, 0,3W, 140MHz	2a§	Riz	BC 168, BC 183, BC 238, BC 548, ++	data

Type	Device	Short Description	Fig.	Manu	Comparision Types	More at
BF 242 A [Riz]	Si-N	=BF 242[Riz]: 60V	2a§		BC 167, BC 182, BC 237, BC 546, ++	data
BF 243 [Riz]	Si-N	Uni, 30V, 0,1A, 0,3W, 170MHz	2a§	Riz	BC 168, BC 183, BC 238, BC 548, ++	data
BF 243 [Texas]	Si-P	AM-V/M/O/ZF, >80MHz	7d	Tix	BF 324, BF 440...441, BF 450...451, ++	data
BF 244 [Riz]	Si-N	Uni, 30V, 0,1A, 0,3W, 170MHz	2a§	Riz	BC 168, BC 183, BC 238, BC 548, ++	data
BF 244	N-FET	NF/HF...VHF, 30V, Idss>2mA, Up<8V	7e	Phi,Sie,++	BFW 61, BFS 72, 2N3819, 2N3823, ++	data
BF 245	N-FET	=BF 244:	7f	Phi,Sie,++	→BF 244	data
BF 246	N-FET	FM/VHF, 25V, Idss>10mA, Up<14,5V	7e	Phi,Sie,++	-	data
BF 247	N-FET	=BF 246:	7f	Phi,Sie,++	-	data
BF 248	Si-N	NF-Tr, 30V, 0,6A, 0,4W, 250MHz	2a§	Tix	BC 337...338, BC 635, BC 637, BC 639, ++	data
BF 249	Si-P	NF-Tr, 30V, 0,6A, 0,4W, 250MHz	2a§	Tix	BC 327...328, BC 636, BC 638, BC 640, ++	data
BF 250	Si-N	NF-Tr, 15V, 0,6A, 0,4W, 20MHz	2a§	Tix	BC 337-338, BC 635, BC 637, BC 639, ++	data
BF 251	Si-N	TV-ZF-re, 600MHz	5k	Sgs	BF 167, BF 198, BF 225, BF 310, BF 367	data
BF 252	Si-N	TV-ZF, 400MHz	5k	Sgs	BF 167, BF 198, BF 225, BF 310, BF 367	data
BF 253	Si-N	AM-V/M/O, >150MHz	7d	Tho	BF 240, BF 254, BF 494, BF 594, ++	data
BF 254	Si-N	AM-V/M/O/ZF, 260MHz	7d	EUR	BF 240, BF 241, BF 494, BF 594, ++	data
BF 255	Si-N	FM-V/M/O, 200MHz	7d	EUR	BF 240, BF 241, BF 495, BF 595, ++	data
BF 255 [Riz]	Si-N	HF/ZF, 600MHz	5k	Riz	BF 199, BF 224, BF 311, BF 373, BF 597	data
BF 256	N-FET	VHF/UHF, 30V, Idss>3mA, Up<7,5V, 1GHz	7f	Mot,Phi,++	2N5397...5398, 2N5486	data
BF 256 L	N-FET	=BF 256:	7e	Tix	→BF 256	data
BF 257	Si-N	Vid, 160V, 0,1A, 0,8W, 90MHz	2a§	EUR	BF 657...659, 2N5058...5059	data
BF 257 A	Si-N	=BF 257: 180V	2a§	Tix	BF 658...659, 2N5058...5059	data
BF 257 B	Si-N	=BF 257: 220V	2a§	Tix	BF 658...659, 2N5058...5059	data
BF 257 C	Si-N	=BF 257: 220V	2a§	Tix	BF 658...659, 2N5058...5059	data
BF 257 D	Si-N	=BF 257: 160V	2a§	Tix	BF 657...659, 2N5058...5059	data
BF 257 G	Si-N	=BF 257: 200V	2a§	Tix	BF 658...659, 2N5058...5059	data
BF 257 N	Si-N	=BF 257: 180V	2a§	Tix	BF 658...659, 2N5058...5059	data
BF 257 S	Si-N	=BF 257: 140V	2a§	Tix	BF 657...659, 2N5058...5059	data
BF 258	Si-N	=BF 257: 250V	2a§	EUR	BF 658...659, 2N5058...5059	data
BF 258 A	Si-N	=BF 257: 280V	2a§	Tix	BF 659, BFS 89, 2N5058	data
BF 259(A,G)	Si-N	=BF 257: 300V	2a§	EUR	BF 659, BFS 89, 2N5058	data
BF 260	Si-N	VHF-V, re, 800MHz	5k	Sgs	BF 225, BF 314, BF 496, BF 502...503, ++	data
BF 261 [ATES,CSF]	Si-N	TV-ZF-re, 730MHz	5k	Sgs	BF 198, BF 225, BF 310, BF 367, BF 596	data
BF 261 [Riz]	Si-N	AM/FM/VHF	5g	Riz	BF 199, BF 225, BF 311, BF 373, BF 597	data
BF 262	Si-N	UHF-V, 650MHz	24a	Phi	BF 362, (BF 377...378, BF 763)[6]	data
BF 263	Si-N	UHF-M, >525MHz	24a	Phi	BF 363, (BF 377...378, BF 763)[6]	data
BF 264	Si-N	UHF-M, >400MHz	24a	Phi,Tho	BF 363, (BF 377...378, BF 763)[6]	data
BF 265	Si-N	UHF-V/M/O, 600MHz	5k	Riz	BF 180...183, BF 689, BF 763, 2N2857, ++	data
BF 266	Si-N	VHF-V-re, 400MHz	5g	Riz	BF 225, BF 314, BF 496, BF 502...503, ++	data
BF 267	Si-N	TV-ZF-re, 350MHz	5k	Riz	BF 198, BF 225, BF 310, BF 367, BF 596	data
BF 268	Si-N	VHF/UHF, 600MHz		Sgs	BF 180...183, BF 689, BF 763, 2N2857, ++	data
BF 269	Si-N			Sgs	-	data
BF 270	Si-N	TV-ZF, re, 600MHz	5k	Sgs	BF 198, BF 225, BF 310, BF 367, BF 596	data
BF 271	Si-N	TV-ZF, 900MHz	5k	Sgs	BF 199, BF 224, BF 311, BF 373, BF 597	data
BF 272(A,S)	Si-P	UHF-V-re, 850MHz	5g	Sgs	BF 316, BF 516, BF 606, BFR 38	data
BF 272 [Riz]	Si-N	TV-ZF, 775MHz	5k	Riz	BF 198, BF 225, BF 310, BF 367, BF 596	data
BF 273 [Riz]	Si-N	TV-ZF, 550MHz	5k	Riz	BF 199, BF 224, BF 311, BF 373, BF 597	data
BF 273 [SGS]	Si-N	AM-M/O/ZF, 700MHz	8d	Sgs	BF 240...241, BF 254...255, BF 494...495,++	data
BF 274 [SGS]	Si-N	AM/FM-ZF-re, 700MHz	8d	Sgs	BF 240...241, BF 254...255, BF 494...495,++	data
BF 275	Si-N	VHF, 500MHz	5g	Riz	BF 225, BF 314, BF 496, BF 502...502, ++	data
BF 277	Si-N	HF/ZF, 350MHz	6k	Tho	BF 225, BF 314, BF 496, BF 502...503, ++	data
BF 278	Si-N	HF/ZF, 550MHz	6k	Tho	BF 225, BF 314, BF 496, BF 502...503, ++	data
BF 279	Si-N	HF/ZF, 500MHz	6k	Tho	BF 225, BF 314, BF 496, BF 502...503, ++	data
BF 280	Si-N	HF/ZF, 500MHz	6k	Tho	BF 225, BF 314, BF 496, BF 502...503, ++	data
BF 281	Si-N	HF/ZF, 700MHz	6k	Tho	BF 225, BF 314, BF 496, BF 502...503, ++	data
BF 282	Si-N	HF/ZF, 600MHz	6k	Tho	BF 225, BF 314, BF 496, BF 502...503, ++	data
BF 283	Si-N	HF/ZF, 250MHz	6k	Tho	BF 225, BF 314, BF 496, BF 502...503, ++	data
BF 284	Si-N	HF/ZF, 250MHz	6k	Tho	BFW 225, BF 314, BF 496, BF 502...503, ++	data
BF 285	Si-N	HF/ZF, 250MHz	6k	Tho	BF 225, BF 314, BF 496, BF 502...503, ++	data
BF 286	Si-N	HF/ZF, 250MHz	6k	Tho	BF 225, BF 314, BF 496, BF 502...503, ++	data
BF 287	Si-N	AM-M/O/ZF, 600MHz	5k	Sgs	BF 241, BF 255, BF 495, BF 595, ++	data
BF 288	Si-N	AM/FM-ZF, re, 500MHz	5k	Sgs	BF 240, BF 254, BF 494, BF 594, ++	data
BF 290 [SGS]	Si-N	VHF-M/O, 1000MHz	5k	Sgs	BF 180...183, BF 689, BF 763, 2N2857, ++	data
BF 290 [Riz]	Si-N	Vid, 120V, 0,03A, 0,8W, 80MHz	2a§	Riz	BF 257...259, BF 657...659, 2N5058...5059	data
BF 291 [Riz]	Si-N	Vid, 150V, 0,03A, 0,8W, 80MHz	2a§	Riz	BF 257...259, BF 657...659, 2N5058...5059	data
BF 291 A [Riz]	Si-N	=BF 291[Riz]: 160V	2a§		BF 257...259, BF 657...659, 2N5058...5059	data
BF 291(A,B) [SGS]	Si-N	Uni, 50V, 0,1A, 0,36W, 260MHz	2a§	Sgs	BC 167, BC 182, BC 237, BC 547, ++	data
BF 292 A	Si-N	Vid, 150V, 0,3A, 0,8W, 66MHz	2a§	Sgs	BF 257...259, BFR 57...59, 2N5058...5059	data
BF 292 B	Si-N	=BF 292A: 190V	2a§		BF 258...259, BFR 58...59, 2N5058...5059	data
BF 292 C	Si-N	=BF 292A: 220V	2a§		BF 258...259, BFR 58...59, 2N5058...5059	data
BF 293(A,D)	Si-N	Uni, 50V, 0,1A, 0,36W, 250...380MHz	2a§	Sgs	BC 167, BC 182, BC 237, BC 547, ++	data
BF 294	Si-N	Vid, 160V, 0,1A, 0,8W, >40MHz	2a§	Sgs	BF 257...259, BF 657...659, 2N5058...5059	data
BF 297	Si-N	Vid, 160V, 0,1A, 0,625W, 95MHz	7a	Tix	BFR 87...89, BFT 57...59, (BF 257...259,++)[6]	data

Type	Device	Short Description	Fig.	Manu	Comparision Types	More at
BF 297 P	Si-N	=BF 297: 1W	30a	Tho	2SC1758, (BF 457...459, BF 857...859,++)[5]	data
BF 297 P2	Si-N	=BF 297: 1W	30b	Tho	BF 457...459, BF 615, BF 857...859, ++	data
BF 298	Si-N	=BF 297: 250V	7a	Tix	BFR 88...89, BFT 58...59, (BF 258...259,++)[6]	data
BF 298 P	Si-N	=BF 298: 1W	30a	Tho	2SC1758, (BF 458...459, BF 858...859,++)[5]	data
BF 298 P2	Si-N	=BF 298: 1W	30b	Tho	BF 458...459, BF 615, BF 858...859, ++	data
BF 299	Si-N	=BF 297: 300V	7a	Tix	BFR 89, BFT 59, (BF 259 ++)[6]	data
BF 299 P	Si-N	=BF 299: 1W	30a	Tho	2SC1758, (BF 459, BF 859,++)[5]	data
BF 299 P2	Si-N	=BF 299: 1W	30b	Tho	BF 459, BF 617, BF 859, ++	data
BF 302	Si-N	FM-ZF, 650MHz	5k	Sgs	BF 240...241, BF 254...255, BF 494...495,++	data
BF 303	Si-N	AM-V/M/O/ZF, 500MHz	5k	Sgs	BF 240...241, BF 254...255, BF 494...495,++	data
BF 304	Si-N	TV-V, 500MHz	5k	Sgs	BF 199, BF 224, BF 311, BF 373, BF 597	data
BF 305	Si-N	Vid, 185V, 0,1A, 0,6W, 100MHz	2a§	Sgs	BF 258...259, BF 658...659, 2N5058...5059	data
BF 306	Si-N	TV-ZF, 1000MHz	5k	Sgs	BF 199, BF 224, BF 311, BF 373, BF 597	data
BF 307	Si-N	TV-ZF	5k		BF 199, BF 224, BF 311, BF 373, BF 597	data
BF 308	Si-N	TV-ZF, >800MHz	5k	Sgs	BF 199, BF 224, BF 311, BF 373, BF 597	data
BF 309	Si-N	TV-ZF, >800MHz	5k	Sgs	BF 198, BF 225, BF 310, BF 367, BF 596	data
BF 310	Si-N	FM-V, TV-ZF, 580MHz	7a	Aeg,Sie	BF 198, BF 225, BF 314, BF 367, BF 596	data
BF 311	Si-N	TV-ZF, 750MHz	7d	Aeg,Phi	BF 199, BF 224, BF 373, BF 597	data
BF 314	Si-N	FM/VHF-V, 450MHz	7a	Aeg,Sie	BF 225, BF 496, BF 502...503, ++	data
BF 315	Si-P	AM/FM, 500MHz	2a§	Sgs	BF 324, BF 440...441, BF 450...451, ++	data
BF 316(A)	Si-P	UHF-M/O, 550...600MHz	5g	Sgs	BF 272, BF 516, BF 606, BFR 38	data
BF 317	Si-P	=BF 315:	8a	Sgs	→BF 315	data
BF 320	P-FET	NF/HF, 15V, Idss>0,3mA, Up<8V	7e	Tix	2N3820, 2N3909, 2N5800, 2N5462, ++	data
BF 321(A...F)	Si-N	Uni, 30V, 0,03A, 0,3W	8a	Tho	BC 168, BC 183, BC 238, BC 548, ++	data
BF 322	Si-N	NF/S, 30V, 0,6A, 0,4W, 250MHz	2a§	Tix	BC 140...141, BC 337...338, BC 635, ++	data
BF 323	Si-P	NF/S, 30V, 0,6A, 0,4W, 250MHz	2a§	Tix	BC 160...161, BC 327...328, BC 636, ++	data
BF 324	Si-P	FM/VHF-V, 450MHz	7a	Phi,Sie,++	BF 414, BF 506, BF 914, BF 936, BF 939	data
BF 325	Si-N	TV-ZF, 700MHz	7a	Tix	BF 198, BF 225, BF 310, BF 367, BF 596	data
BF 327	MOS-N-FET-d	Dual-Gate, VHF, 20V, Idss>20mA, Up<3,8V	25u	Phi		data
BF 329	Si-N	TV-ZF, re, 730MHz	11d	Sgs	BF 198, BF 225, BF 310, BF 367, BF 596	data
BF 330	Si-N	TV-ZF, 1000MHz	11d	Sgs	BF 199, BF 224, BF 311, BF 373, BF 597	data
BF 332	Si-N	AM/FM-M/O/ZF, 600MHz	11d	Sgs	BF 240...241, BF 254...255, BF 494...495,++	data
BF 333	Si-N	AM/FM-M/O/ZF, 400MHz	11d	Sgs	BF 240...241, BF 254...255, BF 494...495,++	data
BF 334	Si-N	AM-V/M/O/ZF, 430MHz	11d	Phi	BF 240...241, BF 254...255, BF 494...495,++	data
BF 335	Si-N	FM-V/M/O/ZF, 370MHz	11d	Phi	BF 240...241, BF 254...255, BF 494...495,++	data
BF 336	Si-N	Vid, 185V, 0,1A, 0,8W, 130MHz	2a§	Mot,Phi,Tho	BF 258...259, BF 658...659, 2N5058...5059	data
BF 336 D	Si-N	=BF 336: 0,5A	2a§	Tix	2SD413, 2SD576, 2SD624	data
BF 337	Si-N	=BF 336: 250V	2a§	Mot,Phi,Tho	BF 258...259, BF 658...659, 2N5058...5059	data
BF 337 D	Si-N	=BF 337: 0,5A	2a§	Tix.	2SD576, 2SD593	data
BF 338	Si-N	=BF 336: 300V	2a§	Mot,Phi,Tho	BF 259, BF 659, BFS 89, 2N5058	data
BF 338 D	Si-N	=BF 338: 0,5A	2a§	Tix	2SD593	data
BF 339	Si-P	VHF-V/M/O, 500MHz	7d	Tix	BF 324, BF 414, BF 506, BF 914, BF 936	data
BF 340	Si-P	AM-V/M/O/ZF, >80MHz, B>30	7d	Tix	BF 324, BF 440...441, BF 450...451, ++	data
BF 341	Si-P	=BF 430: B=45...150	7d	Tix	BF 324, BF 440...441, BF 450...451, ++	data
BF 342	Si-P	=BF 340: B=60...150	7d	Tix	BF 324, BF 440...441, BF 450...451, ++	data
BF 343	Si-P	=BF 340: B>30	7d	Tix	BF 324, BF 440...441, BF 450...451, ++	data
BF 344	Si-N	AM-V/M/O/ZF, 500MHz	2d	Sgs	BF 240...241, BF 254...255, BF 494...495,++	data
BF 345	Si-N	AM-V/M/O/ZF, 500MHz	2d	Sgs	BF 240...241, BF 254...255, BF 494...495,++	data
BF 346	N-FET	VHF-ra, 15V, Idss>2mA, Up<5,5V, 500MHz	7f	Tix	BF 256, BFS 71, 2N3822, 2N4416, ++	data
BF 347	N-FET	NF/HF, 30V, Idss>0,5mA, Up<0,5V	7e	Tix	BFW 13, 2N4338, 2SK68, 2SK106, 2SK118,++	data
BF 348	N-FET	FM/VHF-V/M, 40V, Idss>10mA, Up<6V	7f	Tix	BF 256, BFT 10, 2N5397, 2N5486	data
BF 350	MOS-N-FET-d	Dual-Gate, VHF, 15V, Idss>3mA, Up<5V	5t*	Tix	BF 900, 3SK40	data
BF 351	MOS-N-FET-d	Dual-Gate, VHF, 24V, Idss>5mA, Up<5V	5t*	Tix	3N201...206, 3N209...210, 3N211...213	data
BF 352	MOS-N-FET-d	Dual-Gate, VHF, 24V, Idss>5mA, Up<2V	5t*	Tix	BF 963, 3SK51, 3SK74, 3SK81	data
BF 353	MOS-N-FET-d	Dual-Gate, VHF, 24V, Idss>5mA, Up<3V	5t*	Tix	BFR 963, 3SK51, 3SK74, 3SK81	data
BF 354	MOS-N-FET-d	Dual-Gate, VHF, 24V, Idss>7mA, Up>3V	5t*	Tix	BF 963, 3SK41, 3SK74, 3SK81	data
BF 355	Si-N	Vid, 300V, 0,1A, 0,8W, >80MHz	2a§	Phi,Tix	BF 259, BF 659, 2N5058	data
BF 356	Si-N	Vid, 220V, 0,2A, 0,8W, 100MHz	7c	Tix	BF 298...299, BFT 58...59 (BF 258...259,++)[6]	data
BF 357(K,S)	Si-N	UHF-A, 30V, 0,05A, 1,6GHz	7a	Tix	BF 377...378, BF 689, BF 763, 2N2857	data
BF 359	Si-P	Min, UHF-V-ra, 850MHz	(40b)	Sgs	(BF 606, BF 679...680, BF 967...970, ++)[6]	data
BF 360	Si-P	Min, UHF-M/O, 750MHz	(40b)	Sgs	(BF 606, BF 679...680, BF 967...970, ++)[6]	data
BF 362	Si-N	UHF-V, re, 800MHz	24a	Aeg,Phi,++	(BF 689K, BF 763)[6]	data
BF 363	Si-N	UHF-M, 700MHz	24a	Aeg,Phi,++	(BF 689K, BF 763)[6]	data
BF 364	Si-N	AM-V/M/O/ZF, 260MHz	8d	Phi,Tho	BF 240, BF 254, BF 494, BF 594, ++	data
BF 365	Si-N	FM-V/M/O/ZF, 200MHz	8d	Phi,Tho	BF 241, BF 255, BF 495, BF 595, ++	data
BF 366	Si-N	FM/VHF, re, >400MHz	7e, 7f	Mot	BF 225, BF 314, BF 496, BF 502...503, ++	data
BF 367	Si-N	TV-ZF, re, 440MHz	7f	Mot	BF 198, BF 225, BF 310, BF 596, ++	data
BF 368	Si-N	AM/FM-M/O/ZF, >250MHz	7e	Mot,Mic	BF 240...241, BF 254...255, BF 494...495,++	data
BF 368 K	Si-N	=BF 368:	7f		→BF 368	data
BF 369	Si-N	AM/FM-M/O/ZF, >400MHz	7e	Mot,Mic	BF 240...241, BF 254...255, BF 494...495,++	data

Type	Device	Short Description	Fig.	Manu	Comparision Types	More at
BF 369 K	Si-N	=BF 369:	7f		→BF 369	data
BF 370	Si-N	TV-ZF-V (OFW/SAW-Filter), >500MHz	7a	Phi	BF 920...921S, BF 959	data
BF 370 R	Si-N		7d			data
BF 371	Si-N	TV-ZF, 720MHz	7f	Mot	BF 199, BF 224, BF 311, BF 373, BF 597	data
BF 372	Si-P	VHF-V-re, 850MHz	5g	Mot,Sgs	BF 506, BF 509, BF 914, BF 939	data
BF 373	Si-N	TV-ZF, 720MHz	7f	Mot	BF 199, BF 224, BF 311, BF 597, ++	data
BF 374	Si-N	FM/VHF, 30V, 100mA, >400MHz	7f	Mot	BF 225, BF 314, BF 496, BF 502...503, ++	data
BF 375	Si-N	FM/VHF, 30V, 400MHz	7f	Mot	BF 225, BF 314, BF 496, BF 502...503, ++	data
BF 377	Si-N	VHF/UHF, >1300MHz	7a	Aeg	BF 689, BF 763, 2N2857	data
BF 378	Si-N	=BF 377:	7d	Aeg	→BF 377	data
BF 379	Si-P	AM/FM, 350MHz	7a	Aeg,Mot	BF 324, BF 440...441, BF 450...451, ++	data
BF 380	Si-N	Vid-L, 180V, 0,5A, 10W, 90MHz	13g§	Mot	MPSU 10, BF 460...462, (BF 757...759)[5]	data
BF 381	Si-N	=BF 380: 250V	13g§	Mot	MPSU 10, BF 460...462, (BF 757...759)[5]	data
BF 382	Si-N	=BF 380: 300V	13g§	Mot	MPSU 10, BF 461...462, (BF 758...759)[5]	data
BF 384	Si-N	AM/FM-V/M/O/ZF, 800MHz	7d	Tix	BF 240...241, BF 254...255, BF 494...495,++	data
BF 385	Si-N	AM/FM-V/M/O/ZF, 800MHz	7d	Tix	BF 240...241, BF 254...255, BF 494...495,++	data
BF 387	Si-N	Vid, 100V, 0,05A, 0,8W, 120MHz	2a§	Sgs	BF 257...259, BF 657...659, 2N5058...5059	data
BF 388	Si-N	=BF 387: 160V	2a§	Sgs	BF 257...259, BF 657...659, 2N5058...5059	data
BF 389 B	Si-N	=BF 387: 220V	2a§		BF 258...259, BF 658...659, 2N5058...5059	data
BF 389 C	Si-N	=BF 387: 250V	2a§		BF 258...259, BF 658...659, 2N5058...5059	data
BF 390	Si-N	Vid, 310V, 0,1A, 0,6W, 120MHz	2a§	Sgs	BF 259, BF 659, BFS 89, 2N5058	data
BF 391(K,L,M)	Si-N	Vid, 200V, 0,5A, 0,625W, 70MHz	7e, 40	Mot,Sgs	BFP 22, MPSA 43	data
BF 391 P1	Si-N	=BF 391: 0,2A, 1W	30e	Tho	(BF 460...461, BF 757...759, MPSU 10, ++)[5]	data
BF 391 P2	Si-N	=BF 391: 0,2A, 1W	30b	Tho	BF 757...759, (BF 460...461, MPSU 10, ++)[5]	data
BF 392(K,L,M)	Si-N	=BF 391: 250V	7e, 40	Mot,Sgs	BFP 25, MPSA 42	data
BF 392 P1	Si-N	=BF 392: 0,2A, 1W	30e	Tho	(BF 460...461, BF 757...759, MPSU 10, ++)[5]	data
BF 392 P2	Si-N	=BF 392: 0,2A, 1W	30b	Tho	BF 757...759, (BF 460...461, MPSU 10, ++)[5]	data
BF 393(K,L,M)	Si-N	=BF 391: 300V	7e, 40	Mot,Sgs	BFP 25, MPSA 42	data
BF 393 P1	Si-N	=BF 393: 0,2A, 1W	30e	Tho	(BF 461, BF 758...759, MPSU 10, ++)[5]	data
BF 393 P2	Si-N	=BF 393: 0,2A, 1W	30b	Tho	BF 758...759, (BF 461, MPSU 10, ++)[5]	data
BF 394	Si-N	AM/FM-V/M/O/ZF, >80MHz	7f	Mot	BF 240...241, BF 254...255, BF 494...495,++	data
BF 395	Si-N	AM/FM-V/M/O/ZF, >80MHz	7f	Mot	BF 240...241, BF 254...255, BF 494...495,++	data
BF 397	Si-P	Nix, Vid, 90V, 0,1A, 0,625W	7a	Tix	BF 435...437, BSS 68, 2SA1370...1371	data
BF 398	Si-P	=BF 397: 150V	7a	Tix	BF 435...437, 2SC1370...1371	data
BF 400	Si-P	VHF, >700MHz	8d	Sgs	BF 509, BF 606, BF 914, BF 939, ++	data
BF 402	Si-P	VHF, >400MHz	8a	Sgs	BF 324, BF 414, BF 506, BF 914, BF 936++	data
BF 403	Si-N	SMD, NF, 30V 0,05A, 300MHz	35d(2mm)	Fer	-	data
BF 404	Si-P	SMD, NF, 30V 0,05A, 150MHz	35d(2mm)	Fer	-	data
BF 405	Si-N	SMD, NF, 45V 0,5A, >150MHz	35d(2mm)	Fer	-	data
BF 406	Si-P	SMD, NF, 45V 0,5A, >150MHz	35d(2mm)	Fer	-	data
BF 410	N-FET	VHF-ra, 20V, Idss>0,3mA, Up=1,5V	7f	Phi,Sie	2N5484, 2SK152, 2SK192	data
BF 411	Si-N	Nix, 110V, 0,05A, 0,3W, 120MHz	7a	Aeg	BF 422, BF 297...299, BF 483, BSS 38	data
BF 412	Si-N	=BF 411: 150V	7a	Aeg	BF 422, BF 297...299, BF 483	data
BF 413	Si-N	=BF 411: 200V	7a	Aeg	BF 422, BF 298...299, BF 483	data
BF 414	Si-P	FM/VHF-V, 40V, 25mA, 400MHz	7a	Aeg,Sgs,Sie	BF 324, BF 506, BF 914, BF 936, BF 939++	data
BF 415	Si-N	Vid-L, 250V, 0,2A, 6W, >50MHz	14h§	Tho,Tix	BF 615, BF 757...759, (BF 460...461, ++)[5]	data
BF 416	Si-P	Vid-L, 250V, 0,2A, 6W, >50MHz	14h§	Tho,Tix	BF 616, BF 760...762, (BF 464...465, ++)[5]	data
BF 417(G)	Si-N	=BF 415: 300V	14h§	Tho,Tix	BF 617, BF 758...759, (BF 461, ++)[5]	data
BF 418(G)	Si-P	=BF 416: 300V	14h§	Tho,Tix	BF 618, BF 761...762, (BF 465, ++)[5]	data
BF 419	Si-N	Vid-L,TV-HA-Tr, 300V, 0,1A, 6W(Tc=90°)	14h§	Phi	BF 417, BF 459, 2SC3417	data
BF 420(S)	Si-N	Vid, 300V, 25...50mA, 0,83W, >60MHz	7c	EUR	BF 299, BF 483, BFR 89, 2SC3468, ++	data
BF 420 A	Si-N	=BF 420: 0,5A, 0,8W	7c	Sie	BF 393, BFP 25, MPSA 42	data
BF 420 L	Si-N	=BF 420: 0,5A, 0,625W	7e	Sie	→BF 420A	data
		Philips: 50mA	7c	Phi		
BF 420 P	Si-N	=BF 420: 0,9W	30a	Tho	BF 461...462, MPSU 10	data
BF 420 P3	Si-N	=BF 420: 0,9W	30c	Tho	(BF 461...462, BF 758...759, MPSU 10)[5]	data
BF 421(S)	Si-P	Vid, 300V, 25...50mA, 0,83W, >60MHz	7c	EUR	BF 437, 2SA1371...1372, 2SA1251, ++	data
BF 421 A	Si-P	=BF 421: 0,5A, 0,8W	7c	Sie	BF 493, BFP 26, MPSA 92	data
BF 421 L	Si-P	=BF 421: 0,5A, 0,625W	7e	Sie	→BF 421A	data
		Philips: 50mA	7c	Phi		
BF 421 P	Si-P	=BF 421: 0,9W	30a	Tho	BF 464...465, MPSU 60	data
BF 421 P3	Si-P	=BF 421: 0,9W	30c	Tho	(BF 464...465, BF 761...762, MPSU 60)[5]	data
BF 422(S)	Si-N	=BF 420: 250V	7c	EUR	BF 298...299, BF 483, BFR 88...99, ++	data
BF 422 A	Si-N	=BF 422: 0,5A, 0,8W	7c	Sie	BF 392...393, BFP 25, MPSA 42	data
BF 422 L	Si-N	=BF 422: 0,5A, 0,625W	7e	Sie	→BF 422A	data
		Philips: 50mA	7c	Phi		
BF 422 P	Si-N	=BF 422: 0,9W	30a	Tho	BF 460...462, MPSU 10	data
BF 422 P3	Si-N	=BF 422: 0,9W	30c	Tho	(BF 460...462, BF 757...759, MPSU 10)[5]	data
BF 423(S)	Si-P	=BF 421: 250V	7c	EUR	BF 436...437, 2SA1371...1372, 2SA1251, ++	data
BF 423 A	Si-P	=BF 423: 0,5A, 0,8W	7c	Sie	BF 492...493, BFP 26, MPSA 92	data
BF 423 L	Si-P	=BF 423: 0,5A, 0,625W	7e	Sie	→BF 423A	data
		Philips: 50mA	7c	Phi		
BF 423 P	Si-P	=BF 423: 0,9W	30a	Tho	BF 463...465, MPSU 60	data
BF 423 P3	Si-P	=BF 423: 0,9W	30c	Tho	(BF 463...465, BF 760...762, MPSU 60)[5]	data

Type	Device	Short Description	Fig.	Manu	Comparision Types	More at
BF 424	Si-P	HF, 300MHz	7a	Itt	BF 324, BF 440...441, BF 450...451, ++	data
BF 430(L)	Si-N	SMD, UHF, 20V, 70mA, 7500MHz	44s	Mot	BFG 520, BFP 193	data
BF 431(L)	Si-N	SMD, UHF, 25V, 30mA, 3800MHz	44s	Mot	BFG 93A	data
BF 432(L)	Si-N	SMD, UHF, 15V, 1mA, 5000MHz	44s	Mot	=MRF 9331(L), BFG 92A	data
BF 433	Si-N	SMD, VHF/UHF, 30V, 200mA, 5500MHz	8-MDIP	Mot	=MRF 5812	data
BF 435	Si-P	Vid, 160V, 0,2A, 0,625W, >80MHz	7c	Tix	BF 491...493, BFP 23, BFP 26, MPSA 92...93	data
BF 436	Si-P	=BF 435: 250V	7c	Tix	BF 492...493, BFP 26, MPSA 92, 2SA1251	data
BF 437	Si-P	=BF 435: 300V	7c	Tix	BF 439, BFP 26, MPSA 92, 2SA1251	data
BF 439	Si-P	UHF-M/O, >900MHz	5g	Mot	BF 316, BF 516, BF 606, BFR 38	data
BF 440	Si-P	AM/FM, re, 250MHz	7d	Aeg	BF 324, BF 450, BF 540...542, ++	data
BF 441	Si-P	AM/FM, 250MHz	7d	Aeg	BF 324, BF 451, BF 540...542, ++	data
BF 450	Si-P	AM/FM, re, 375MHz	7d	Phi,Sie,++	BF 324, BF 440, BF 540...542, ++	data
BF 451	Si-P	AM/FM, 325MHz	7d	Phi,Sie,++	BF 324, BF 441, BF 540...542, ++	data
BF 454	Si-N	AM/FM-ZF, 400MHz	8d	Sgs	BF 240, BF 254, BF 494, BF 594, ++	data
BF 455	Si-N	AM/FM-V/M/O, 400MHz	8d	Sgs	BF 241, BF 255, BF 495, BF 595, ++	data
BF 456	Si-N	Vid-L, 160V, 0,1A, 7W, 90MHz	14h§	Sie,Tix	BF 457...459, BF 415, BF 417, 2SC3417	data
BF 457	Si-N	Vid-L, 160V, 0,1A, 10W(Tc=45°), 90MHz	14h§	EUR	BF 415, BF 417, 2SC3417	data
BF 458	Si-N	=BF 457: 250V	14h§	EUR	BF 415, BF 417, 2SC3417	data
BF 459	Si-N	=BF 457: 300V	14h§	EUR	BF 850, 2SC3417...3418	data
BF 460	Si-N	Vid-L, 250V, 0,5A, 10W, >45MHz	13g§	Mot,Tho	MPSU 10, (BF 758...759)[5]	data
BF 461	Si-N	=BF 460: 300V	13g§	Mot,Tho	MPSU 10, (BF 758...759)[5]	data
BF 462	Si-N	=BF 460: 350V	13g§	Mot,Tho	(BF 759)[5]	data
BF 463	Si-P	Vid-L, 250V, 0,5A, 10W, >20MHz	13g§	Mot,Tho	MPSU 60, (BF 761...762)[5]	data
BF 464	Si-P	=BF 463: 300V	13g§	Mot,Tho	MPSU 60, (BF 761...762)[5]	data
BF 465	Si-P	=BF 463: 350V	13g§	Mot,Tho	(BF 762)[5]	data
BF 460BA...465BA	Si-N/P	=BF 460...465	13g§	Tho	→BF 460...465	data
BF 460EA...465EA	Si-N/P	=BF 460...465:	30g§	Tho	→BF 460...465	data
BF 466	Si-N	NF/S/Vid-L, 200V, 1A, 10W, >100MHz	13g§	Mot,Tho	(BD 410, BF 666)[5]	data
BF 467	Si-N	=BF 466: 250V	13g§	Mot,Tho	(BD 410, BF 667)[5]	data
BF 468	Si-N	=BF 466: 300V	13g§		(BD 410, BF 668)[5]	data
BF 469(S)	Si-N	Vid-L, 250V, 0,03A, 2W(Tc=110°),>60MHz	14h§	EUR	BF 458...459, BF 415, BF 417, 2SC3424	data
BF 470(S)	Si-N	Vid-L, 250V, 0,03A, 2W(Tc=110°),>60MHz	14h§	EUR	BF 416, BF 418, 2SA1361, 2SA1353	data
BF 471(S)	Si-N	=BF 469: 300V	14h§	EUR	BF 459, BF 417, 2SC3417...3418	data
BF 472(S)	Si-N	=BF 470: 300V	14h§	EUR	BF 418, 2SA1353...1354	data
BF 479	Si-P	VHF/UHF-V, 2000MHz	24a	Aeg,Sgs	BF 679, BF 967...968, BF 979	data
BF 479 S	Si-P	=BF 479: 1300MHz	24a	Sgs	→BF 479	data
BF 479 T	Si-P	=BF 479: 1850MHz	24a	Aeg	→BF 479	data
BF 480	Si-N	UHF-V, 1500MHz	24e	Phi,Tho	BF 362, 2SC2464...66, 2SC2726, 2SC2728,++	data
BF 481	Si-N	UHF-V, 1200MHz	24e	Phi	BF 362, 2SC2464...66, 2SC2726, 2SC2728,++	data
BF 483	Si-N	Vid, 300V, 0,05A, 0,83W, >70MHz	7c	Phi	BF 420(A), 2SC3468...3469, 2SC4218, ++	data
BF 484	Si-P	Vid, 250V, 0,1A, 0,83W, >70MHz	7c	Phi	BF 421A, BF 436...437, 2SA1371, 2SA1624++	data
BF 485	Si-N	=BF 483: 350/300V	7c	Phi	2SC2267, 2SC3469, 2SC4166, 2SD1385	data
BF 486	Si-P	=BF 484: 300/300V	7c	Phi	BF 421A, BF 437, 2SA1371, 2SA1624, ++	data
BF 487	Si-N	=BF 483: 400/350V	7c	Phi	2SC2267, 2SC3469, 2SC4166, 2SD1385	data
BF 488	Si-P	=BF 484: 350/350V	7c	Phi	BF 493S, 2SA1251, 2SA1372, 2SB1209	data
BF 491(K,L,M)	Si-P	Vid, 200V, 0,5A, 0,625W, >50MHz	7e, 40	Fer,Mot,Tho	BFP 23, MPSA 93	data
BF 491 P1	Si-P	=BF 491: 0,2A, 1W	30e	Tho	(BF 463...464, BF 760...762, MPSU 60, ++)[5]	data
BF 491 P2	Si-P	=BF 491: 0,2A, 1W	30b	Tho	BF 760...762, (BF 463...464, MPSU 60, ++)[5]	data
BF 492(K,L,M)	Si-P	=BF 491: 250V	7e, 40	Fer,Mot,Tho	BFP 26, MPSA 92	data
BF 492 P1	Si-P	=BF 492: 0,2A, 1W	30e	Tho	(BF 463...464, BF 760...762, MPSU 60, ++)[5]	data
BF 492 P2	Si-P	=BF 492: 0,2A, 1W	30b	Tho	BF 760...762, (BF 463...464, MPSU 60, ++)[5]	data
BF 493(K,L,M)	Si-P	=BF 491: 300V	7e, 40	Fer,Mot,Tho	BFP 26, MPSA 92	data
BF 493 P1	Si-P	=BF 493: 0,2A, 1W	30e	Tho	(BF 464, BF 761...762, MPSU 60, ++)[5]	data
BF 493 P2	Si-P	=BF 493: 0,2A, 1W	30b	Tho	BF 761...762, (BF 464, MPSU 60, ++)[5]	data
BF 493 S	Si-P	=BF 491: 350V	7e(9mm)	Mot	2SB1074, (BF 465, BF 762, 2SA1156)[4]	data
BF 494(B)	Si-N	AM/FM, 30V, 30mA, 0,3W, 260MHz	7d	Phi,Nsc	BF 240...241, BF 254...255, BF 594...595,++	data
BF 495(C,D)	Si-N	AM/FM, 30V, 30mA, 0,3W, 200MHz	7d	Phi,Nsc	BF 240...241, BF 254...255, BF 594...595,++	data
BF 494...495 [Mot]	Si-N	=BF 494...495: 0,1A 0,35W	7e	Mot	BF 240...241, BF 254...255, BF 594...595,++	data
BF 496	Si-N	VHF-V, 550MHz	7a	Phi	BF 225, BF 314, BF 502...503, BF 505, ++	data
BF 497	Si-N	TV-ZF, 1000MHz	8a	Sgs	BF 199, BF 224, BF 311, BF 373, BF 597	data
BF 500(A)	Si-P	VHF-V/M/O, 400MHz	8a	Sgs	BF 324, BF 414, BF 506, BF 509, BF 914++	data
BF 501	Si-P	VHF-ra, 300MHz	8a	Sgs	BF 324, BF 414, BF 509, BF 914, BF 939++	data
BF 502	Si-N	VHF-M/O, 700MHz	7d	Sie	BF 225, BF 314, BF 496, BF 505, BF 507++	data
BF 502 S	Si-N	=BF 502: 900MHz	7d	Tho	→BF 502	data
BF 503	Si-N	VHF-M/O, 750MHz	7d	Sie	BF 225, BF 314, BF 496, BF 505, BF 507++	data
BF 504	Si-N	NF, 30V, 0,05A, 0,25W, 85MHz	2a§	Ucp	BC 168, BC 183, BC 238, BC 548, ++	data
BF 505	Si-N	VHF, >750MHz	7d	Sie	BF 502...503, BF 507, ++	data
BF 506	Si-P	VHF-V, 40V, 30mA, 550MHz	7a	Aeg,Mot,Sie	BF 324, BF 414, BF 509, BF 914, BF 936++	data
BF 506 A	Si-P	=BF 506: 25V, 25mA, 350MHz	7a	Sgs	BF 324, BF 414, BF 509, BF 914, BF 936++	data
BF 506	Si-N	NF, 45V, 0,05A, 0,25W, 95MHz	2a§	Ucp	BC 167, BC 182, BC 237, BC 547, ++	data
BF 507	Si-N	VHF-V/M, >750MHz	7d	Sie	BF 502...503, BF 505, ++	data
BF 509(S,T)	Si-P	VHF-V, re, 700...800MHz	7a	Aeg,Mot,Tho	BF 506, BF 914, BF 936, BF 939	data
BF 510...511 [Ucp]	Si-N	NF, 30V, 0,05A, 0,25W, 80MHz	2a§	Ucp	BC 168, BC 183, BC 238, BC 548, ++	data

Type	Device	Short Description	Fig.	Manu	Comparision Types	More at
BF 510	N-FET	SMD, VHF-ra	35e	Phi	2SK212	data
BF 511	N-FET	SMD, VHF-ra	35e	Phi	2SK212	data
BF 512	N-FET	SMD, VHF-ra	35e	Phi	-	data
BF 513	N-FET	SMD, VHF-ra	35e	Phi	-	data
BF 516	Si-P	VHF/UHF-M/O, 850MHz	5g	Sgs	BF 316, BF 439, BF 606, BFR 38	data
BF 517	Si-N	SMD, UHF-A/O, 20V, 25mA, 2GHz	35a	Sie	BFR 35, BFT 25, 2SC3098...3099	data
BF 519	Si-N	NF/HF, 70V, 0,05A, 0,3W, 150MHz	2a§	Ucp	BC 174, BC 182, BC 190, BC 546	data
BF 520	Si-N	NF/HF, 50V, 0,05A, 0,3W, 150MHz	2a§	Ucp	BC 167, BC 182, BC 237, BC 547, ++	data
BF 521	Si-N	NF/HF, 30V, 0,05A, 0,3W, 150MHz	2a§	Ucp	BC 168, BC 183, BC 238, BC 548, ++	data
BF 523	Si-N	TV-ZF, 500MHz	7d	Tix	BF 199, BF 224, BF 311, BF 373, BF 597	data
BF 536	Si-P	SMD, VHF-M/O, 350MHz	35a	Phi,Nsc	BF 568...569, BF 660	data
BF 540	Si-P	AM/FM, 130MHz	7d	Tix	BF 324, BF 440...441, BF 450...451, ++	data
BF 541	Si-P	AM/FM, 130MHz	7d	Tix	BF 324, BF 440...441, BF 450...451, ++	data
BF 542	Si-P	AM/FM, 130MHz	7d	Tix	BF 324, BF 440...441, BF 450...451, ++	data
BF 543	MOS-N-FET-d	=BF 544: SMD	35e	Sie	-	data
BF 544	MOS-N-FET-d	HF, 20V, 30mA, Idss>1,5mA, Up<1,5V	7d	Sie	-	data
BF 545	N-FET	=BF 245: SMD	35f	Phi	-	data
BF 547	Si-N	SMD, VHF/UHF-M/O, 30V, 0,03A, 120MHz	35a	Phi	BFS 17	data
BF 547 W	Si-N	=BF 547:	35a(2mm)	Phi	-	data
BF 550	Si-P	SMD, 40V, 25mA, Hr-/ZF, 325MHz	35a	Aeg,Phi,Sie	BF 536, BF 568...569, BF 660	data
BF 550 R	Si-P	=BF 550:	35d		BF 569R	data
BF 554	Si-N	SMD, AM...VHF, 30V, 30mA, 260MHz	35a	Sie,Tho	BF 599, BF 799, BFS 18..20, ++	data
BF 556	N-FET	=BF 256: SMD	35f	Phi	-	data
BF 559	Si-P	Min, VHF-re, 850MHz	(40b)	Sgs	(BF 606, BF 679...680, BF 967...970, ++)[6]	data
BF 560	Si-P	Min, VHF-M/O, 850MHz	(40b)	Sgs	(BF 606, BF 679...680, BF 967...970, ++)[6]	data
BF 562	Si-N	VHF-V, re, 600MHz	7a	Sie	BF 225, BF 314, BF 496, BF 502...503, ++	data
BF 568	Si-P	SMD, VHF-V, ra, 1100MHz	35a	Sie	BF 569, BF 579, BF 767	data
BF 569	Si-P	SMD, UHF-M/O, 40V, 30mA, 850MHz	35a	Aeg,Phi,Sie	BF 579, BF 767	data
BF 569 R	Si-P	=BF 569:	35d		BF 579R	data
BF 570	Si-N	TV-ZF-V, SAW-Filter, 40V, 0,1A,>500MHz	35a	Phi	BF 799	data
BF 576	Si-P	VHF/UHF, >800MHz	7c	Tix	BF 316, BF 516, BF 606, BFR 38	data
BF 579	Si-P	SMD, VHF/UHF, 20V, 30mA, 1600MHz	35a	Aeg,Phi,Sie	-	data
BF 579 R	Si-P	=BF 579:	35d		-	data
BF 583	Si-N	Vid-L, 300V, 0,1A, 5W, >70MHz	13h§	Phi	BF 600, BF 617, BF 758, BF 859, BF 880++	data
BF 584	Si-P	Vid-L, 250V, 0,1A, 5W, >70MHz	13h§	Phi	BF 616, BF 760, BF 848, BF 872, BF 890++	data
BF 585	Si-N	=BF 583: 350V	13h§	Phi	BF 600, BF 759, BF 880...81	data
BF 586	Si-P	=BF 584: 300V	13h§	Phi	BF 618, BF 761, BF 849, BF 890...91,++	data
BF 587	Si-N	=BF 583: 400V	13h§	Phi	BF 600, BF 881	data
BF 588	Si-P	=BF 584: 350V	13h§	Phi	BF 762, BF 890...91	data
BF 591	Si-N	Telefon, 210V, 0,15A, 1,3W(Ta=55°)	13h§	Phi	BF 615, BF 757...759, (BF 460...462)[5]	data
BF 593	Si-N	=BF 591: 250V	13h§	Phi	BF 615, BF 757...759, (BF 460...462)[5]	data
BF 594	Si-N	AM/FM, 35V, 30mA, 260MHz	7d	Tix	BF 240...241, BF 254...255, BF 494...495,++	data
BF 595	Si-N	AM/FM, 35V, 30mA, 260MHz	7d	Tix	BF 240...241, BF 254...255, BF 494...495,++	data
BF 596	Si-N	TV-ZF-re, 40V, 25mA, 400MHz	7d	Tix	BF 198, BF 225, BF 310, BF 367, ++	data
BF 597(A,B)	Si-N	TV-ZF, 40V, 25mA, 550MHz	7d	Tix	BF 199, BF 224, BF 311, BF 373, ++	data
BF 599	Si-N	SMD, VHF, ZF, 40V, 25mA, 550MHz	35a	Aeg,Sie,Tho	BF 799, BFS 20, 2SC3015, 2SC3374	data
BF 600	Si-N	S/Vid-L, 400V, 0,03A, >60MHz	13h§	Sie	BF 587, BF 881	data
BF 606(A,B)	Si-P	VHF/UHF, 370-1200MHz	7d	Phi,Sie,Sgs	BF 316, BF 439, BF 516, BFR 38	data
BF 615(BA)	Si-N	Vid-L, 250V, 0,2A, 10W, 70MHz	13h§	Tho,Tix	BF 757...759, (BF 460...462, MPSU 10)[5]	data
BF 616(BA)	Si-P	Vid-L, 250V, 0,2A, 10W, 70MHz	13h§	Tho,Tix	BF 760...762, (BF 463...465, MPSU 60)[5]	data
BF 617(BA,G)	Si-N	=BF 615: 300V	13h§	Tho,Tix	BF 758...759, (BF 461...462, MPSU 10)[5]	data
BF 618(BA,G)	Si-P	=BF 616: 300V	13h§	Tho,Tix	BF 761...762, (BF 464...465, MPSU 60)[5]	data
BF 615EA..618EA	Si-N/P	=BF 615..618:	30h§	Tho	→BF 615..618	data
BF 620	Si-N	SMD, Vid, 300V, 0,05A, >60MHz	39b	Phi, Ztx	BFN 18, BFN 20, 2SC3554, 2SC4189, ++	data
BF 621	Si-P	SMD, Vid, 300V, 0,05A, >60MHz	39b	Phi, Ztx	BFN 19, BFN 21, 2SA1384	data
BF 622	Si-N	=BF 620: 250V	39b	Phi,Sie,Ztx	BFN 16, BFN 18, BFN 20, 2SC3515, 2SC4189	data
BF 623	Si-P	=BF 621: 250V	39b	Phi,Sie,Ztx	BFN 17, BFN 19, BFN 21, 2SA1384	data
BF 630	Si-N	UHF, 2000MHz	7a	Sie	BF 377...378, BF 689K, BF 763	data
BF 639	Si-P	VHF/UHF, >700MHz	2a§	Tix	BF 316, BF 516, BF 606, BFR 38	data
BF 640	Si-P	VHF/UHF, 650MHz	2a§	Tix	BF 316, BF 516, BF 606, BFR 38	data
BF 642	Si-N	Vid, 300V, 0,5A, 0,625W, >50MHz	7a	Mot,Tho	BF 393, BF 420A, BFP 25, MPSA 42	data
BF 642 P	Si-N	=BF 642: 0,9W	30a	Tho	BF 393P1, MPSU 10, (BF 461, BF 758...759)[5]	data
BF 642 P2	Si-N	=BF 642: 0,9W	30b	Tho	BF 393P2, BF 758...759, (BF 461, MPSU 10)[5]	data
BF 643	Si-N	=BF 642: 200V	7a	Mot,Tho	BF391...393, BF422A, BFP 22, MPSA 42...43	data
BF 643 P	Si-N	=BF 643: 0,9W	30a	Tho	BF 391P1, MPSU 10, (BF 460, BF 757...759)[5]	data
BF 643 P2	Si-N	=BF 643: 0,9W	30b	Tho	BF 391P2, BF 757...759, (BF 460, MPSU 10)[5]	data
BF 657	Si-N	Vid, 160V, 0,1A, 1W, 90MHz	2a§	Sgs	BF 257...259, 2N5058...5059	data
BF 658	Si-N	=BF 657: 250V	2a§	Sgs	BF 258...259, 2N5058...5059	data
BF 659	Si-N	=BF 657: 300V	2a§	Sgs	BF 259, BFS 89, 2N5058	data
BF 660	Si-P	SMD, VHF-O, 40V, 25mA, 650MHz	35a	Phi,Sie	BF 536, BF 568...569, BF 579, BF 767	data

Type	Device	Short Description	Fig.	Manu	Comparision Types	More at
BF 660 R	Si-P	=BF 660:	35d		BF 569R, BF 579R	data
BF 660 W	Si-P	=BF 660:	35a(2mm)			data
BF 666	Si-N	Vid-L, 200V, 1A, 10W, >100MHz	13h§	Mot	BD 410, (BF 466...468)[5]	data
BF 667	Si-N	=BF 666: 250V	13h§	Mot	BD 410, (BF 467...468)[5]	data
BF 668	Si-N	=BF 666: 300V	13h§	Mot	BD 410, (BF 468)[5]	data
BF 679(M,S,T)	Si-P	UHF-V, re, 930...1000MHz	24e	Aeg,Sgs	BF 479, BF 779, BF 967...968, BF 979	data
BF 680(A,H)	Si-P	UHF-M/O, 650...750MHz	24e	Aeg,Sgs	BF 780, BF 967, BF 969...970, BF 979	data
BF 681	Si-P	UHF, 950MHz	24e	Aeg	BF 479, BF 779...780, BF 967...970, BF 979	data
BF 689	Si-N	UHF, 1000MHz	5g	Sie	BF 377...378, BF 763, 2N2857	data
BF 689 K	Si-N	=BF 689: 1800MHz	7a	Phi,Sie	BF 377...378, BF 763, 2N2857	data
BF 692	Si-P	Vid, 300V, 0,5A, 0,625W, >50	7a	Mot,Tho	BF 493, BF 421A, BFP 26, MPSA 92	data
BF 692 P	Si-P	=BF 692: 0,9W	30a	Tho	BF 493P1, MPSU 60, (BF 464, BF 761...762)[3]	data
BF 692 P2	Si-P	=BF 692: 0,9W	30b	Tho	BF 493P2, BF 761...762, (BF 464, MPSU 60)[5]	data
EF 693	Si-P	=BF 692: 200V	7a	Mot,Tho	BF491...493, BF423A, BFP23, MPSA 92...93	data
5F 693 P	Si-P	=BF 693: 0,9W	30a	Tho	BF 491P1, MPSU 60, (BF 463, BF 760...762)[5]	data
BF 693 P2	Si-P	=BF 693: 0,9W	30b	Tho	BF 491P2, BF 760...762, (BF 463, MPSU 60)[5]	data
BF 694(A,B)	Si-N	TV-ZF, 300MHz	7d	Tix	BF 199, BF 224, BF 311, BF 373, BF 597	data
BF 706	Si-P	AM/FM, >200MHz	7e	Mot	BF 324, BF 440...411, BF 450...451, ++	data
BF 709	Si-P	FM-V-re, >350MHz	7e	Mot	BF 324, BF 414, BF 506, BF 914, BF 936++	data
BF 715(BA)	Si-N	Vid-L, 250V, 0,03A, 6,25W, >60MHz	13h§	Tho,Tix	BF 858...859, BF 869, BF 861, BF 880	data
BF 716(BA)	Si-P	Vid-L, 250V, 0,03A, 6,25W, >60MHz	13h§	Tho,Tix	BF 848...849, BF 870, BF 872, BF 890	data
BF 717(BA)	Si-N	=BF 715: 300V	13h§	Tho,Tix	BF 859, BF 871, BF 880	data
BF 718(BA)	Si-P	=BF 716: 300V	13h§	Tho,Tix	BF 849, BF 872, BF 890	data
BF 715EA...718EA	Si-N/P	=BF 715...718:	30h§	Tho	→BF 715...718	data
BF 720	Si-N	=BF 620: 0,05A, 1,5W	48j§	Phi, Sie,++	-	data
BF 721	Si-P	=BF 621: 0,05A, 1,5W	48j§	Phi, Sie,++	-	data
BF 722	Si-N	=BF 622: 0,05A, 1,5W	48j§	Phi, Sie,++	-	data
BF 723	Si-P	=BF 623: 0,05A, 1,5W	48j§	Phi, Sie,++	-	data
BF 739	Si-P	VHF, >600MHz	7e°	Mot	BF 316, BF 516, BF 606, BF 939, BFR 38	data
BF 740	Si-P	VHF/UHF-re, >600MHz	7e°	Mot	BF 316, BF 516, BF 606, BFR 38	data
BF 747 [Phi]	Si-N	SMD, VHF/UHF-Tuner, M/O, 30V, 1200MHz	35a	Phi	BF 775	data
BF 747 [Eiy]	Si-P	20V, 0,1A 0,2W, 600MHz		Eiy	-	data
BF 749	Si-N	SMD, UHF, 20V, 25mA, 5000MHz	44ze	Phi	-	data
BF 750	Si-N	SMD, UHF, 15V, 35mA, 7000MHz	44ze	Phi	-	data
BF 752	Si-N	SMD, UHF, 15V, 35mA, 7000MHz	44s	Phi	-	data
BF 753	Si-N	SMD, UHF, 15V, 35mA, 5000MHz	35a	Phi	-	data
BF 757	Si-N	Vid-L, 250V, 0,5A, 10W, >45MHz	13h§	Mot,Nsc,Tho	(BF 460..462, MPSU 10)[5]	data
BF 758	Si-N	=BF 757: 300V	13h§	Mot,Nsc,Tho	(BF 461...462, MPSU 10)[5]	data
BF 759	Si-N	=BF 757: 350V	13h§	Mot,Nsc,Tho	(BF 462)[5]	data
BF 760	Si-P	Vid-L, 250V, 0,5A, 10W, >20MHz	13h§	Mot,Nsc,Tho	2SA1156, (BF 463...465, MPSU 60)[5]	data
BF 761	Si-P	=BF 760: 300V	13h§	Mot,Nsc,Tho	2SA1156, (BF 464...465, MPSU 60)[5]	data
BF 762	Si-P	=BF 760: 350V	13h§	Mot,Nsc,Tho	2SA1156, (BF 465)[5]	data
BF 757BA...762BA	Si-N/P	=BF 757..762: Flachmontage/flat mount.	13h§	Tho	→BF 757...762	data
BF 757EA...762EA	Si-N/P	=BF 757...762:	30h§	Tho	→BF 757...762	data
BF 763	Si-N	UHF-V/M/O, 25V, 25mA, 1800MHz	7f	Phi,Sie	BF 377...378, BF 689K, 2SC2570(A)	data
BF 767	Si-P	SMD, VHF/UHF-ra, 950MHz	35a	Phi,Sie	BF 568...569, BF 579	data
BF 770 A	Si-N	SMD, SATV-ZF, 20v, 0,05A, 6GHz	35a	Sie	2SC3513	data
BF 771	Si-N	SMD, SATV/VC-Tuner, 20V, 80mA, 8GHz	35a	Sie	BFR 193, BFR 520, 2SC3445	data
BF 771 W	Si-N	=BF 771:	35a(2mm)	Sie	2SC5085	data
BF 772	Si-N	=BF 771:	44s	Sie	BFG 520, BFP 193	data
BF 775	Si-N	SMD, SATV-Tuner, 20V, 0,03A, 5GHz	35a	Sie	BF 747, BFQ 29, BFT 75, 2SC3110, ++	data
BF 775 A	Si-N	=BF 775: 25V, 5800MHz	35a		BFQ 81, BFR 93, 2SC3110, 2SC3513	data
BF 775 W	Si-N	=BF 775:	35a(2mm)		2SC4903, 2SC5107	data
BF 777	Si-N	SMD, VHF/UHF-O, 30V, 50mA, 2200MHz	35a	Sie	BF 747, BF 775, BFR 53, 2SC3014, 2SC3773	data
BF 779	Si-P	UHF-V, 800MHz	24e	Tix	BF 479, BF 679, BF 967...968, BF 979	data
BF 780	Si-P	UHF-M/O, 700MHz	24e	Tix	BF 680, BF 967, BF 969...970	data
BF 787	Si-N	Vid-L, 250V, 0,1A, 10W, >60MHz	13h§	Mot	BF 615, BF 858..859, BF 880...881, ++	data
BF 788	Si-N	=BF 787: 300V	13h§	Mot	BF 617, BF 859, BF 880...881	data
BF 789	Si-N	=BF 787: 350V	13h§	Mot	BF 880...881	data
BF 790	Si-P	Vid-L, 250V, 0,1A, 10W, >60MHz	13h§	Mot	BF 616, BF 848...849, BF 890...891, ++	data
BF 791	Si-P	=BF 790: 300V	13h§	Mot	BF 618, BF 849, BF 890...891	data
BF 792	Si-P	=BF 790: 350V	13h§	Mot	BF 890...891	data
BF 799	Si-N	SMD, TV-ZF-E(SAW-Filter), 800...1100MHz	35a	Sie	BF 570	data
BF 799 W	Si-N	=BF 799:	35a(2mm)	Sie		data
BF 800	N-FET	NF/HF, 25V, Idss>0,3mA, Up<6V	5h+	Tix	-	data
BF 801	N-FET	NF/HF, 25V, Idss>0,3mA, Up<6V	5h+	Tix	-	data
BF 802	N-FET	NF/HF, 25V, Idss>0,3mA, Up<6V	5h+	Tix	-	data
BF 803	N-FET	NF/HF, 30V, Idss<0,8mA, Up<3V	5h+	Tix	-	data
BF 804	N-FET	NF/HF, 30V, Idss<1,2mA, Up<5V	5h+	Tix	-	data
BF 805	N-FET	NF/HF, 30V, Idss>3mA, Up<6V	5h+	Tix	-	data
BF 806	N-FET	NF/HF, 30V, Idss>3mA, Up<6V	5h+	Tix	-	data
BF 808	N-FET	NF/HF, 20V, Idss>1mA, Up<5V	5h+	Tix	-	data
BF 810(A)	N-FET	NF/HF, 30V, Idss>5mA, Up<6V	5h+	Tix	-	data
BF 811(A)	N-FET	NF/HF, 30V, Idss>5mA, Up<6V	5h+	Tix	-	data

Type	Device	Short Description	Fig.	Manu	Comparision Types	More at
BF 815	N-FET	NF/HF, 30V, Idss>15mA, Up<6V	5h+	Tix	-	data
BF 816	N-FET	NF/HF, 30V, Idss>15mA, Up<6V	5h+	Tix	-	data
BF 817	N-FET	NF/HF, 25V, Idss>10mA, Up<5V	5h+	Tix	-	data
BF 818	N-FET	NF/HF, 25V, Idss>10mA, Up<5V	5h+	Tix	-	data
BF 819	Si-N	Vid-L, TV-HA-Tr,300V, 0,1A, 6W(Tc=75°)	13h§	Phi	BF 617, BF 859, BF 880...881, MJE 340	data
BF 819 A	Si-N	=BF 819:	13g§		2SC1758, (BF 617, BF 859, BF 880...881)[5]	data
BF 820(S)	Si-N	SMD, Vid, 300/300V, 25...50mA, >60MHz	35a	Aeg, Phi	BFN 26, 2SC4061, 2SC4412, 2SC4497	data
BF 821(S)	Si-P	SMD, Vid, 300/300V, 25...50mA, >60MHz	35a	Aeg, Phi	BFN 27, 2SA1682, 2SA1721	data
BF 822(S)	Si-N	=BF 820: 250/250V	35a	Aeg, Phi	BFN 22, BFN 24, BFN 26, 2SC4412, ++	data
BF 823(S)	Si-P	=BF 821: 250/250V	35a	Aeg, Phi	BFN 23, BFN 25, BFN 27, 2SC1682, ++	data
BF 820W...823W	Si-N/P	=BF 820...823:	35a(2mm)		-	data
BF 824	Si-P	SMD, FM-V, 30V, 25mA, 450MHz	35a	Phi	BF 568	data
BF 824W	Si-P	=BF 824:	35a(2mm)		-	data
BF 840	Si-N	=BF 240: SMD	35a	Phi,Sie,++	BF 554, BΓS 18...19	data
BF 841	Si-N	=BF 241: SMD	35a	Phi,Sie,++	BF 554, BFS 18...19	data
BF 844	Si-N	Vid, 450V, 0,3A, 0,625W, >50MHz	7e	Mot	MPSA 44	data
BF 845	Si-N	=BF 844: 400V	7e	Mot	MPSA 44...45	data
BF 847	Si-P	Vid-L, 160V, 0,1A, 2,5W(Tc=100°),90MHz	13h§	Sie	BF 616, BF 760...762, BF 890...891	data
BF 848	Si-P	=BF 847: 270V	13h§	Sie	BF 618, BF 761...762, BF 890...891	data
BF 849	Si-P	=BF 847: 300V	13h§	Sie	BF 761...762, BF 890...891	data
BF 850	Si-N	Vid-L, 400V, 0,1A, >90MHz	14h§	Sie	2SC3418	data
BF 851	N-FET	AM-V, 25V, Idss=2...25mA	7d	Phi	-	data
BF 857	Si-N	Vid-L, 160V, 0,2A, 6W(Tc=75°), 90MHz	13h§	Phi,Sie,Tho	BF 615, BF 757...759, BF 880...881	data
BF 858	Si-N	=BF 857: 250V	13h§	Phi,Sie,Tho	BF 617, BF 758...759, BF 880...881	data
BF 859	Si-N	=BF 857: 300V	13h§	Phi,Sie,Tho	BF 758...759, BF 880...881	data
BF 857A...859A	Si-N	=BF 857...859:	13g§	Phi	BF 461...462	data
BF 857BA...859BA	Si-N	=BF 857...859: Flachmontage/flat mount.	13h§	Tho	→BF 857...859	data
BF 857EA...859EA	Si-N	=BF 857...859:	30h§	Tho	→BF 857...859	data
BF 860	Si-N	Vid-L, 400V, 0,1A, >90MHz		Sie	(BF 850, BF 881)	data
BF 861	N-FET	=BF 851: SMD, >1999: Fig.35f	35a,f	Phi	-	data
BF 862	N-FET	SMD, AM, 20V, Idss>10mA, Up=0,7V	35f	Phi	-	data
BF 869(A,S,SA)	Si-N	Vid-L, 250V, 0,05A, 5W, >60MHz	13h§	Aeg,Phi,++	BF 583, BF 615, BF 858...859, BF 880...881	data
BF 870(A,S,SA)	Si-P	Vid-L, 250V, 0,05A, 5W, >60MHz	13h§	Aeg,Phi,++	BF 616, BF 848...849, BF 890...891	data
BF 871(A,S,SA)	Si-N	=BF 869: 300V	13h§	Aeg,Phi,++	BF 583, BF 617, BF 859, BF 880...881	data
BF 872(A,S,SA)	Si-P	=BF 870: 300V	13h§	Aeg,Phi,++	BF 618, BF 849, BF 890...891	data
BF 869BA...872BA	Si-N/P	=BF 869...872: Flachmontage/flat mount.	13h§	Tho	→BF 869...872	data
BF 869EA...872EA	Si-N/P	=BF 869...872:	30h§	Tho	→BF 869...872	data
BF 876	Si-P		13h			
BF 879	Si-P	VHF/UHF	24e	Sie	BF 479, BF 679...680, BF 967...970, ++	data
BF 880	Si-N	Vid-L, 350V, 0,1A, >60MHz	13h§	Sie	BF 850, 2SC3418	data
BF 881	Si-N	Vid-L, 400V, 0,1A, >60MHz	13h§	Sie	BF 850, 2SC3418	data
BF 883(S)	Si-N	Vid-L, 300/275V, 0,05A, 7W, >60MHz	13h§	Aeg	BF 759, BF 859, BF 880...881	data
BF 885(S)	Si-N	Vid-L, 300V, 0,05A, 7W, >60MHz	13h§	Aeg	BF 759, BF 859, BF 880...881	data
BF 890	Si-P	Vid-L, 350V, 0,1A, >60MHz	13h§	Sie	BF 762, 2SA1156, 2SA1354, 2SB1011	data
BF 891	Si-P	Vid-L, 300V, 0,03A, >60MHz	13h§	Sie	2SA1156, 2SA1354, 2SB1011	data
BF 900	MOS-N-FET-d*	Dual-Gate, VHF/UHF, Idss>3mA, Up<5V	25u	Mot, Tix	BF 960, BF 965...966	data
BF 901	MOS-N-FET-e*	Dual-Gate, VHF/UHF, 12V, 30mA	44u	Phi	-	data
BF 901 R	MOS-N-FET-e*	=BF 901:	44v		-	data
BF 904(A)	MOS-N-FET-e*	Dual-Gate, VHF/UHF, 7V, 30mA	44u	Phi	BF 909	data
BF 904 R,AR	MOS-N-FET-e*	=BF 904:	44v		BF 909R	data
BF 904 WR,AWR	MOS-N-FET-e*	=BF 904:	44v(2mm)		BF 909WR	data
BF 905	MOS-N-FET-d*	Dual-Gate, UHF, 20V, Idss>2mA, Up<5V	25u	Tix	BF 900, BF 960, BF 965...966	data
BF 906	Si-P	AM/FM, 30/25V, 50mA, >200MHz	7a	Mot, Tix	BF 324, BF 440...441, BF 450...451, ++	data
BF 907	MOS-N-FET-d*	Dual-Gate, UHF, 20V, Idss>5mA, Up<3,5V	25u	Tix	BF 900, BF 960, BF 965...966	data
BF 908	MOS-N-FET-d*	Dual-Gate, VHF/UHF, 12V, 40mA	44u	Phi	BF 990	data
BF 908 R	MOS-N-FET-d*	=BF 908:	44v		-	data
BF 908 WR	MOS-N-FET-d*	=BF 908:	44v(2mm)		-	data
BF 909(A)	MOS-N-FET-e*	Dual-Gate, VHF/UHF, 7V, 40mA	44u	Phi	BF 904	data
BF 909 R,AR	MOS-N-FET-e*	=BF 909:	44v		BF 904R	data
BF 909 WR,AWR	MOS-N-FET-e*	=BF 909:	44v(2mm)		BF 904WR	data
BF 910	MOS-N-FET-d*	Dual-Gate, VHF/UHF, 20V, Idss>6mA	25u	Aeg,Tix	BF 900, BF 960, BF 965...966	data
BF 914	Si-P	FM/VHF-V, 850MHz	7a	Aeg	BF 324, BF 414, BF 506, BF 509, BF 939++	data
BF 915	MOS-N-FET-d*	Dual-Gate, VHF/UHF, Idss>15mA, Up<4V	25u	Tix	(BF 900, BF 960, BF 965...966)	data
BF 920	Si-N	VHF/UHF,TV-ZF(OFW/SAW-Filter), 1800MHz	24e	Tix	BF 921S, BF 959	data
BF 920 TS...923 TS	Si-N/P	=BF 420...423: 0,46W	7e	Aeg	→BF 420...423	data
BF 921(S)	Si-N	VHF/UHF,TV-ZF(OFW/SAW-Filter), 1800MHz	7d	Sgs,Tix	BF 920, BF 959	data
BF 926	Si-P	FM/VHF-V/M/O, 500MHz	7d	Phi,Sie	BF 324, BF 414, BF 506, BF 914, BF 936++	data
BF 930	MOS-N-FET-d	SMD, Dual-Gate, 12V, 40mA	44u	Sie	-	data
BF 936	Si-P	FM/VHF, 350MHz	7a	Mot,Nsc,Phi	BF 324, BF 414, BF 506, BF 509, BF 914++	data

Type	Device	Short Description	Fig.	Manu	Comparision Types	More at
BF 939	Si-P	FM/VHF-V-re, 750MHz	7a	Phi,Sie	BF 324, BF 414, BF 506, BF 509, BF 914++	data
BF 959	Si-N	VHF, TV-ZF (OFW/SAW-Filter), 1100MHz	7d	Aeg,Mot,Sie	BF 920, BF 921S	data
BF 960(S)	MOS-N-FET-d*	Dual-Gate, UHF, 20V, Idss>2mA, Up<2,7V	25u	EUR	BF 900, BF 965...966	data
BF 961	MOS-N-FET-d*	Dual-Gate, FM/VHF, Idss>2mA, Up<4V	25u	EUR	BF 963..964, BF 981...982, ++	data
BF 962	MOS-N-FET-d*	Dual-Gate, VHF-CATV, 20V	25u	Sie	BF 961, BF 963..964, BF 981..982, ++	data
BF 963	MOS-N-FET-d*	Dual-Gate, FM/VHF, Idss>6mA, Up<3,5V	25u	Aeg	BF 961, BF 964, BF 981..982, ++	data
BF 964(S)	MOS-N-FET-d*	Dual-Gate, FM/VHF, Idss>2mA, Up<2,5V	25u	Aeg,Phi,Sie	BF 961, BF 963, BF 981..982, ++	data
BF 965	MOS-N-FET-d*	HF, 10V, Ids=30mA, Up<4V	7d	Sie	-	data
BF 965	MOS-N-FET-d*	Dual-Gate, CATV-Tuner, 20V, 30mA	25u	Aeg,Phi,Sie	BF 960, BF 966	data
BF 966(S)	MOS-N-FET-d*	Dual-Gate, UHF, 20V, Idss>2mA, Up<2,5V	25u	Aeg,Phi,Sie	BF 900, BF 960, BF 965	data
BF 967	Si-P	UHF-V/M, re, 900MHz	24e	Phi,Sie	BF 479, BF 679...680, BF 779, BF 979	data
BF 968	Si-P	UHF-V, re, 1100MHz	24e	Sie	BF 479, BF 679, BF 779, BF 979	data
BF 969(S)	Si-P	UHF-M/O, 850MHz	24e	Sie	BF 479, BF 679...680, BF 779, BF 979	data
BF 970(A)	Si-P	UHF-M/O, 40V, 30mA, 900MHz	24e	Aeg,Phi,++	BF 479, BF 679...680, BF 779, BF 979	data
BF 979(S)	Si-P	UHF-V, 20V, 30mA, 1350...1600MHz	24e	Aeg,Phi,++	BF 479	data
BF 980(A)	MOS-N-FET-d*	Dual-Gate, UHF, 18V, Up<1,3V	25u	Phi	BF 960, BF 965...966	data
BF 981	MOS-N-FET-d*	Dual-Gate, FM/VHF, Idss>4mA, Up<2,5V	25u	Phi,Tho	BF 961, BF 963..964, BF 982, ++	data
BF 982(T)	MOS-N-FET-d*	Dual-Gate, FM/VHF, 20V, Up<1,3V	25u	Aeg,Phi	BF 961, BF 963..964, BF 981, ++	data
BF 987	MOS-N-FET-d	FM/VHF, 20V, Idss=5...18mA	7d	Aeg,Sie	-	data
BF 988	MOS-N-FET-d*	Dual-Gate, UHF, 12V, Idss>2mA, Up<2,5V	25u	Aeg,Phi,Sie	BF 960, BF 965...966	data
BF 989(S)	MOS-N-FET-d*	=BF 960: SMD	44u	Aeg,Phi,Sie	BF 996	data
BF 990(A)	MOS-N-FET-d*	=BF 980(A): SMD	44u	Phi	BF 989, BF 996	data
BF 990 AR	MOS-N-FET-d*	=BF 980(A): SMD	44v			data
BF 991	MOS-N-FET-d*	=BF 981: SMD	44u	Phi	BF 993, BF 994	data
BF 992(T)	MOS-N-FET-d*	=BF 982(T): SMD	44u	Phi	BF 993, BF 994	data
BF 992 R	MOS-N-FET-d*	=BF 982: SMD	44v			data
BF 993	MOS-N-FET-d*	=BF 963: SMD	44u	Aeg,Sie	BF 991, BF 994	data
BF 994(S)	MOS-N-FET-d*	=BF 964: SMD	44u	Aeg,Phi,Sie	BF 991, BF 993	data
BF 994 SR	MOS-N-FET-d*	=BF 964: SMD	44s			data
BF 995	MOS-N-FET-d*	=BF 965: SMD	44u	Aeg,Sie	BF 991, BF 993, BF 994	data
BF 996(S)	MOS-N-FET-d*	=BF 966: SMD	44u	Aeg,Phi,Sie	BF 989, BF 990	data
BF 996 SR	MOS-N-FET-d*	=BF 966: SMD	44s			data
BF 997	MOS-N-FET-d*	=BF 965: SMD	44u	Aeg,Phi,Sie	BF 989, BF 996	data
BF 998	MOS-N-FET-d*	=BF 988: SMD	44u	Aeg,Phi,Sie	-	data
BF 998 R	MOS-N-FET-d*	=BF 988: SMD	44v	Aeg, Sie	-	data
BF 998 W	MOS-N-FET-d*	=BF 988: SMD	44u(2mm)	Sie, Phi	-	data
BF 998 WR	MOS-N-FET-d*	=BF 988: SMD	44v(2mm)	Aeg, Phi	-	data
BF 999	MOS-N-FET-d	=BF 987: SMD	35d	Aeg,Sie	-	data
BF 1005(S)	MOS-N-FET-d	SMD, Dual-Gate, UHF, 8V, 25mA	44u	Sie	-	data
BF 1009(S)	MOS-N-FET-d	SMD, Dual-Gate, UHF, 12V, 25mA	44u	Sie	-	data
BF 1009 SW	MOS-N-FET-d	=BF 1009S:	44u(2mm)			data
BF 1012(S)	MOS-N-FET-d	SMD, Dual-Gate, UHF, 16V, 30mA	44u	Sie	-	data
BF 1100	MOS-N-FET-e*	Dual-Gate, VHF/UHF, 14V, 30mA	44u	Phi	-	data
BF 1100 R	MOS-N-FET-e*	=BF 1100:	44v		-	data
BF 1100 WR	MOS-N-FET-e*	=BF 1100:	44v(2mm)		-	data
BF 1101	MOS-N-FET-e*	Dual-Gate, VHF/UHF, 7V, 30mA	44u	Phi	-	data
BF 1101 R	MOS-N-FET-e*	=BF 1101:	44v		-	data
BF 1101 WR	MOS-N-FET-e*	=BF 1101:	44v(2mm)		-	data
BF 1102	MOS-N-FET-e*	2x Dual-Gate, VHF/UHF, 7V, 40mA	46	Phi	-	data
BF 1105	MOS-N-FET-e*	Dual-Gate, VHF/UHF, 7V, 30mA	44u	Phi	-	data
BF 1105 R	MOS-N-FET-e*	=BF 1105:	44v		-	data
BF 1105 WR	MOS-N-FET-e*	=BF 1105:	44v(2mm)		-	data
BF 1107	MOS-N-FET-d*	S, VC Tuner, 7V, 10mA, Up<4,5V	35b	Phi	-	data
BF 1107 W	MOS-N-FET-d*	=BF 1107:	35b(2mm)		-	data
BF 1108	MOS-N-FET-d*	FET + Diode(G→Pin3), HF-S, 7V, 10mA	44	Phi	-	data
BF 1108 R	MOS-N-FET-d*	=BF 1108:	44		-	data
BF 1109	MOS-N-FET-e*	Dual-Gate, VHF/UHF, 11V, 30mA	44u	Phi	-	data
BF 1109 R	MOS-N-FET-e*	=BF 1109:	44v		-	data
BF 1109 WR	MOS-N-FET-e*	=BF 1109:	44v(2mm)		-	data
BF 2000	MOS-N-FET-e	SMD, Dual-Gate, UHF, 12V, 30mA	44u	Sie	-	data
BF 2000 W	MOS-N-FET-e	=BF 2000:	44v(2mm)		-	data
BF 2030	MOS-N-FET-d	SMD, Dual-Gate, UHF, 14V, 40mA	44u	Sie	-	data
BF 2030 R	MOS-N-FET-d	=BF 2030:	44v		-	data
BF 2030 W	MOS-N-FET-d	=BF 2030:	44v(2mm)		-	data
BF 2040	MOS-N-FET-d	SMD, Dual-Gate, UHF, 14V, 40mA	44u	Sie	-	data
BF 2040 W	MOS-N-FET-d	=BF 2040:	44v(2mm)		-	data
BFC 520	Si-N	40V, 0,2A, 0,24W, 150MHz		Ucp	-	data
BFE 182	Si-N	=BFR 182:	44zk	Sie	-	data
BFE 183	Si-N	=BFR 183:	44zk	Sie	-	data
BFE 193	Si-N	=BFR 193:	44zk	Sie	-	data
BFE 196	Si-N	=BFP 196:	44zk	Sie	-	data

Type	Device	Short Description	Fig.	Manu	Comparision Types	More at
BFG						
BFG 16 A	Si-N	UHF-A, 40V, 0,15A, 1600MHz	48s§	Phi	-	data
BFG 17 A	Si-N	SMD, UHF-A, 25V, 25mA, 2800MHz	44zk	Phi	-	data
BFG 19 S	Si-N	UHF-A, 20V, 0,1A, 5300MHz	48s§	Sie	-	data
BFG 21 W	Si-N	SMD, UHF, 15V, 0,5A, Gp>10dB(1,9GHz)	44s(2mm)	Phi	-	data
BFG 23	Si-P	VHF/UHF-A, 15V, 35mA, 5GHz	24za	Phi	BFQ 75...76	data
BFG 25(A,X)	Si-N	SMD, UHF-A, 8V, 6,5mA, 5000MHz	44s	Phi	-	data
BFG 31	Si-P	UHF-A, 20V, 0,1A, 5000MHz	48s§	Phi	BFG 55	data
BFG 32	Si-P	VHF/UHF-A, 20V, 75mA, 4,5GHz	24za	Phi	BFQ 194	data
BFG 33	Si-N	SMD, UHF, 9V, 20mA, 12GHz	44zk	Phi	-	data
BFG 33 X	Si-N	=BFG 33:	44s		-	data
BFG 34	Si-N	VHF/UHF-A, 25V, 150mA, 3,7GHz	24za	Phi	BFG 96	data
BFG 35	Si-N	VHF/UHF-A, 25V, 0,15A, 4GHz	48s§	Phi	BFG 135	data
BFG 51	Si-P	VHF/UHF-A, 20V, 25mA, 5GHz	24za	Phi	BFQ 75...76	data
BFG 55	Si-P	UHF-A, 25V, 0,15A, 4GHz	48s§	Phi	BFG 31	data
BFG 65(T)	Si-N	UHF-A, 20V, 50mA, 7,5GHz	24za	Aeg,Phi	BFQ 65...66, 2SC3062, 2SC3584	data
BFG 67	Si-N	SMD, VHF/UHF-A, 20V, 0,05A, 7,5GHz	44zk	Aeg,Phi	BFG 197	data
BFG 67 R	Si-N	=BFG 67:	44(CBEE)		-	data
BFG 67 X	Si-N	=BFG 67:	44s		-	data
BFG 67 XR	Si-N	=BFG 67:	44ze		-	data
BFG 90 A	Si-N	UHF-A, 20V, 25mA, 5GHz	24za	Phi	BFQ 28, BFR 49, BFR 90	data
BFG 91 A	Si-N	UHF-A, 15V, 35mA, 6GHz	24za	Phi	BFQ 57, BFQ 58, BFQ 74	data
BFG 92 A	Si-N	SMD, VHF/UHF-A, 20V, 25mA, 5GHz	44zk	Aeg,Phi	BFG 67, BFG 93A	data
BFG 92 AX	Si-N	=BFG 92:	44s		-	data
BFG 92(A)XR	Si-N	=BFG 92:	44ze		-	data
BFG 93 A	Si-N	SMD, VHF/UHF-A, 20V, 35mA, 6GHz	44zk	Aeg,Phi	BFG 197	data
BFG 93 AX	Si-N	=BFG 93:	44s		-	data
BFG 93(A)XR	Si-N	=BFG 93:	44ze		-	data
BFG 94	Si-N	UHF-A, 15V, 60mA, 6GHz	48s§	Phi	BFG 35	data
BFG 96	Si-N	UHF-A, 20V, 150mA, 5GHz	24za	Phi	-	data
BFG 97	Si-N	VHF/UHF-A, 20V, 0,1A, 5,5GHz	48s§	Phi	BFG 135	data
BFG 135(A)	Si-N	VHF/UHF-A, 25V, 0,15A, 7GHz	48s§	Phi,Sie	-	data
BFG 193	Si-N	UHF-A, 20V, 0,08A, 8GHz	48s§	Sie	-	data
BFG 194	Si-P	UHF-A, ...1,5GHz, 20V, 0,1A, 5000MHz	48s§	Sie	BFG 55	data
BFG 195	Si-N	VHF/UHF-A, 20V, 0,1A, 0,5W, 7,5GHz	24za	Phi	2SC3358, 2SC3603	data
BFG 196	Si-N	UHF-A, ...1,5GHz, 20V, 0,1A, 7500MHz	48s§	Sie	BFG 135	data
BFG 197	Si-N	SMD, UHF, 20V, 0,1A, 7500MHz	44zk	Phi	-	data
BFG 197 X	Si-N	=BFG 197:	44s		-	data
BFG 197 XR	Si-N	=BFG 197:	44ze		-	data
BFG 198	Si-N	VHF/UHF-A, 20V, 0,1A, 8GHz	48s§	Phi	-	data
BFG 235	Si-N	UHF-A, ...2GHz, 25V, 0,3A, 5,5GHz	48s§	Sie	-	data
BFG 505	Si-N	SMD, UHF, 20V, 18mA, 9000MHz	44zk	Phi	-	data
BFG 505 X	Si-N	=BFG 505:	44s		-	data
BFG 505 XR	Si-N	=BFG 505:	44ze		2SC4095	data
BFG 520	Si-N	SMD, UHF, 20V, 70mA, 9000MHz	44zk	Phi	-	data
BFG 520 X	Si-N	=BFG 520:	44s		-	data
BFG 520 XR	Si-N	=BFG 520:	44ze		-	data
BFG 540	Si-N	SMD, UHF, 20V, 120mA, 9000MHz	44zk	Phi	-	data
BFG 540 X	Si-N	=BFG 540:	44s		-	data
BFG 540 XR	Si-N	=BFG 540:	44ze		-	data
BFG 541	Si-N	SMD, UHF, 20V, 120mA, 9000MHz	48s§	Phi	-	data
BFJ						
BFJ 17	Si-N	VHF-Tr, 60V, 1A, 400MHz	2a§	Riz	BFS 23, BFV 90, BFW 47, BFX 17	data
BFJ 18	Si-N	VHF-A, 30V, 550MHz	5g	Riz	BFS 62, BFW 41, BFX 18...19, BFX 31	data
BFJ 19	Si-N	VHF-A, 30V, 550MHz	5g	Riz	BFS 62, BFW 41, BFX 18...19, BFX 31	data
BFJ 21	Si-N	VHF, 400MHz		Riz	-	data
BFJ 22	Si-P	VHF/UHF, 700MHz		Riz	-	data
BFJ 22	Si-P	VHF/UHF, 700MHz		Riz	-	data
BFJ 45	Si-N	HF/S, 80V, 1A, 0,8W, 120MHz	2a§	Riz	BSW 65...68, BSX 45...47, 2N3107...3110, ++	data
BFJ 46	Si-N	HF/S, 80V, 1A, 0,8W, 120MHz	2a§	Riz	BSW 65...68, BSX 45...47, 2N3107...3110, ++	data
BFJ 47	Si-N	HF/S, 120V, 1A, 0,8W, 120MHz	2a§	Riz	BSS 42...43, BSW 67...68, BSX 47, ++	data
BFJ 48	Si-N	HF/S, 120V, 1A, 0,8W, 120MHz	2a§	Riz	BSS 42...43, BSW 67...68, BSX 47, ++	data
BFJ 49	Si-N	HF/S, 120V, 1A, 0,8W, 150MHz	2a§	Riz	BSS 42...43, BSW 67...68, BSX 47, ++	data
BFJ 50	Si-N	HF/S, 120V, 1A, 0,8W, 150MHz	2a§	Riz	BSS 42...43, BSW 67...68, BSX 47, ++	data
BFJ 51	Si-P	NF/HF, 35V, 0,8W, 60MHz	2a§	Riz	-	data
BFJ 52	Si-P	NF/HF, 35V, 0,8W, 75MHz	2a§	Riz	-	data

Type	Device	Short Description	Fig.	Manu	Comparision Types	More at
BFJ 53	Si-P	NF/HF, 64V, 0,8W, 120MHz	2a§	Riz	-	data
BFJ 54	Si-P	NF/HF, 64V, 0,8W, 120MHz	2a§	Riz	-	data
BFJ 57	Si-N	NF/HF, 125V, 0,8W, 40MHz	2a§	Riz	BF 257..259, BF 657...659, 2N5058...5059	data
BFJ 64	Si-P	HF/S, 40V, 0,5A, 0,7W, <50/-ns	2a§	Riz	BSW 23, 2N3072...3073, 2N3467...3468, ++	data
BFJ 67	Si-N	40V, 0,36W	2a§	Riz	-	data
BFJ 68	Si-N	40V, 0,36W	2a§	Riz	-	data
BFJ 70	Si-N	HF-V/M/O/ZF, 550MHz	5g	Riz	BF 199, BF 224, BF 311, BF 373, BF 597	data
BFJ 72	Si-N	Uni, 45V, 0,1A, 0,5W, >50MHz, B>30	2a§	Riz	BC 167, BC 182, BC 237, BC 547, ++	data
BFJ 73	Si-N	=BFJ 72: B>76	2a§	Riz	BC 167, BC 182, BC 237, BC 547, ++	data
BFJ 74	Si-N	NF/HF, 70V, 0,3W, >250MHz	2a§	Riz	BC 174, BC 182, BC 190, BC 546, ++	data
BFJ 75	Si-N	NF/HF, 40V, 0,3W, >300MHz	2a§	Riz	BC 167, BC 183, BC 237, BC 547, ++	data
BFJ 77	Si-N	VHF/UHF, >800MHz	5g	Riz	BF 689, BF 763, 2N918, 2N2857, ++	data
BFJ 78	Si-N	VHF/UHF, >600MHz	5g	Riz	BF 689, BF 763, 2N918, 2N2857, ++	data
BFJ 79	Si-N	VHF/UHF, >600MHz	5g	Riz	BF 689, BF 763, 2N918, 2N2857, ++	data
BFJ 92	Si-N	NF-ra, 50V, 0,1A, 0,3W, 45MHz, B>40	2a§	Riz	BC 184, BC 413..414, BC 550	data
BFJ 93	Si-N	=BFJ 92: B>100	2a§	Riz	BC 184, BC 413..414, BC 550	data
BFJ 98	Si-N	Vid, 150V, 0,1A, 0,8W, 90MHz	2a§	Riz	BF 257..259, BF 657...659, 2N5058...5059	data

BFN

Type	Device	Short Description	Fig.	Manu	Comparision Types	More at
BFN	Z-Di	=SM 15T 68C (Typ-Code/Stempel/marking)	71(8x5mm)	Tho	→SM 15T...	data
BFN 16	Si-N	SMD, Vid, 250V, 0,2A, >60MHz	39b	Sie, Ztx	2SC3554	data
BFN 17	Si-P	SMD, Vid, 250V, 0,2A, >60MHz	39b	Sie, Ztx	-	data
BFN 18	Si-N	=BFN 16: 300V	39b	Sie, Ztx	2SC3554	data
BFN 19	Si-P	=BFN 17: 300V	39b	Sie, Ztx	-	data
BFN 20	Si-N	SMD, Vid, 300V, 20...50mA, >60MHz	39b	Sie	BFN 18, 2SC3554	data
BFN 21	Si-P	SMD, Vid, 300V, 20...50mA, >60MHz	39b	Sie	BFN 19	data
BFN 22	Si-N	SMD, Vid, 250V, 20...50mA, >60MHz	35a	Sgs,Sie	BF 820, BF 822, 2SC4412	data
BFN 22 R	Si-N	=BFN 22:	35d		-	data
BFN 23	Si-P	SMD, Vid, 250V, 20...50mA, >60MHz	35a	Sgs,Sie	BF 821, BF 823, 2SA1682, 2SA1721	data
BFN 23 R	Si-N	=BFN 23:	35d		-	data
BFN 24	Si-N	=BFN 16:	35a	Sie	-	data
BFN 25	Si-P	=BFN 17:	35a	Sie	-	data
BFN 26	Si-N	=BFN 18:	35a	Sie	-	data
BFN 27	Si-P	=BFN 19:	35a	Sie	-	data
BFN 36	Si-N	=BFN 16: 1,5W	48j§	Sie, Ztx	-	data
BFN 37	Si-P	=BFN 17: 1,5W	48j§	Sie, Ztx	-	data
BFN 38	Si-N	=BFN 18: 1,5W	48j§	Sie, Ztx	-	data
BFN 39	Si-P	=BFN 19: 1,5W	48j§	Sie, Ztx	-	data

BFP

Type	Device	Short Description	Fig.	Manu	Comparision Types	More at
BFP	Si-P	=2SB1308-P (Typ-Code/Stempel/marking)	39	Rhm	→2SB1308	data
BFP	Z-Di	=SM 15T 68CA(Typ-Code/Stempel/marking)	71(8x5mm)	Tho	→SM 15T...	data
BFP 10	Si-N	UHF-A, 25V, 100mA, 0,35W, 4GHz	51r	Tho	BFG 96, BFP 96, BFR 96, BFQ 73	data
BFP 11	Si-N	S/Vid, 150V, 0,5A, 1,25W, 60MHz	7c	Tix	BF 391...393, BFP 22, BFP 25, MPSA 42...43	data
BFP 12	Si-N	=BFP 11: 250V	7c	Tix	BF 392...393, BFP 25, MPSA 42	data
BFP 13	Si-N	=BFP 11: 350V	7c	Tix	MPSA 44...45, 2SD1350	data
BFP 14	Si-N	=BFP 11: 450V	7c	Tix	MPSA 44, 2SD1350A	data
BFP 17	Si-N	=BFS 17P	44s	Sie	-	data
BFP 22	Si-N	Vid, 200V, 0,2/0,5A, 0,625W, >50MHz	7e	Sie	BF 391...393, MPSA 42...43	data
BFP 23	Si-P	Vid, 200V, 0,2/0,5A, 0,625W, >50MHz	7e	Sie	BF 491...493, MPSA 92...93, 2SB918	data
BFP 24				Sie	-	data
BFP 25	Si-N	=BFP 22: 300V	7e	Sie	BF 393, MPSA 42	data
BFP 26	Si-P	=BFP 23: 300V	7e	Sie	BF 493, MPSA 92	data
BFP 29	Si-N	=BFQ 29P:	44s	Sie	BFG 93A	data
BFP 35 A	Si-N	=BFR 34AP:	44s	Sie	BFG 93A	data
BFP 67	Si-N	=BFG 67:	44s	Aeg	-	data
BFP 67 R	Si-N	=BFG 67:	44ze	Aeg	-	data
BFP 67 W	Si-N	=BFG 67:	44s(2mm)	Aeg	-	data
BFP 81	Si-N	=BFQ 81:	44s	Aeg, Sie	-	data
BFP 90(A)	Si-N	UHF-A, 20V, 30mA, 0,25W, 5GHz	51s	Phi,Tho	BFQ 69, BFQ 85, BFR 14, BFR 91, ++	data
BFP 91(A)	Si-N	UHF-A, 15V, 50mA, 0,35W, 6GHz	51s	Phi,Tho	BFG 65, BFQ 65...66, 2SC3511, 2SC3062	data
BFP 92	Si-N	UHF-A, 22V, 0,35W, 3,5GHz	51s	Tho	BFQ 59...60, BFQ 70...71, BFT 97, ++	data
BFP 92 A	Si-N	=BFG 92A:	44s	Aeg	-	data
BFP 92 AW	Si-N	=BFG 92A:	44s(2mm)	Aeg	-	data
BFP 93 A	Si-N	=BFG 93A:	44s	Aeg, Sie	-	data
BFP 93 AW	Si-N	=BFG 93A:	44s(2mm)	Aeg	-	data
BFP 96	Si-N	UHF-A, 20V, 100mA, 0,5W, 4,5GHz	51s	Phi,Tho	BFG 96, BFQ 73, BFR 96	data
BFP 136 W	Si-N	SMD, UHF, DECT, 20V, 0,15A, 5,5GHz	44ze(2mm)	Sie	-	data
BFP 167	Si-N	ZF, 40V, 25mA, 250MHz	5g	Ucp	BF 167, BF 198, BF 225, BF 310, BF 596	data
BFP 173	Si-N	ZF, 40V, 25mA, 350MHz	5g	Ucp	BF 173, BF 199, BF 224, BF 311, BF 597	data
BFP 177	Si-N	Vid, 100V, 0,05A, 0,6W, 75MHz	2a§	Ucp	BF 257..259, BF 657...659, 2N5058...5059	data
BFP 178	Si-N	Vid, 160V, 0,05A, 0,6W, 75MHz	2a§	Ucp	BF 258..259, BF 658...659, 2N5058...5059	data
BFP 179	Si-N	Vid, 160V, 0,05A, 0,6W, 75MHz	2a§	Ucp	BF 258..259, BF 658...659, 2N5058...5059	data

Type	Device	Short Description	Fig.	Manu	Comparision Types	More at
BFP 180	Si-N	=BFR 180:	44s	Sie	-	data
BFP 180 W	Si-N	=BFR 180:	44s(2mm)	Sie		data
BFP 181(T)	Si-N	=BFR 181:	44s	Aeg, Sie	-	data
BFP 181 R,TR	Si-N	=BFR 181:	44ze	Sie	-	data
BFP 181 TRW	Si-N	=BFR 181:	44ze(2mm)	Aeg		
BFP 181 TW	Si-N	=BFR 181:	44s(2mm)	Aeg		data
BFP 181 W	Si-N	=BFR 181:	44s(2mm)	Sie		data
BFP 182(T)	Si-N	=BFR 182:	44s	Aeg, Sie		data
BFP 182 R	Si-N	=BFR 192:	44ze	Sie		data
BFP 182 TRW	Si-N	=BFR 182:	44ze(2mm)	Aeg		data
BFP 182 TW	Si-N	=BFR 182:	44s(2mm)	Aeg		data
BFP 182 W	Si-N	=BFR 182:	44s(2mm)	Sie		data
BFP 183(T)	Si-N	=BFR 183:	44s	Aeg, Sie		data
BFP 183 R,TR	Si-N	=BFR 183:	44ze	Sie		data
BFP 183 TRW	Si-N	=BFR 183:	44ze(2mm)	Aeg		
BFP 183 TW	Si-N	=BFR 183:	44s(2mm)	Aeg		data
BFP 183 W	Si-N	=BFR 183:	44s(2mm)	Sie		data
BFP 193(T)	Si-N	=BFR 193:	44s	Aeg, Sie	-	data
BFP 193 TRW	Si-N	=BFR 193:	44ze(2mm)	Aeg		
BFP 193 TW	Si-N	=BFR 193:	44s(2mm)	Aeg		data
BFP 193 W	Si-N	=BFR 193:	44s(2mm)	Sie		data
BFP 194	Si-P	=BFR 194:	44s	Sie		data
BFP 196	Si-N	SMD, UHF-A, 20V, 0,1A, 8GHz	44s	Sie		data
BFP 196 W	Si-N	=BFP 196:	44s(2mm)	Sie		data
BFP 280(T)	Si-N	=BFR 280:	44s	Aeg, Sie	-	data
BFP 280 TRW	Si-N	=BFR 280:	44ze(2mm)	Aeg		
BFP 280 TW	Si-N	=BFR 280:	44s(2mm)	Aeg		data
BFP 280 W	Si-N	=BFR 280:	44s(2mm)	Aeg		data
BFP 405	Si-N	SMD, UHF, 15V, 12mA, >20GHz	44s(2mm)	Sie		data
BFP 420	Si-N	SMD, UHF, 15V, 35mA, >20GHz	44s(2mm)	Sie		data
BFP 450	Si-N	SMD, UHF, 15V, 0,1A, 24GHz	44s(2mm)	Sie		data
BFP 490	Si-N	SMD, UHF, 15V, 0,6A, 17,5GHz	≈45(BECCE)	Sie		data
BFP 520	Si-N	SMD, UHF, 12V, 40mA, 45GHz	44s(2mm)	Sie	-	data
BFP 540	Si-N	SMD, UHF, 15V, 80mA, 29GHz	44s(2mm)	Sie		data
BFP 620	Si-N	SMD, UHF, 8V, 80mA, 70GHz	44s(2mm)	Sie	-	

BFQ

Type	Device	Short Description	Fig.	Manu	Comparision Types	More at
BFQ	Si-P	=2SB1308-Q (Typ-Code/Stempel/marking)	39	Rhm	→2SB1308	data
BFQ 10...16	N-FET	Dual, Ugs1-Ugs2 <5...50mV	81bx	Phi	2N3921...22, 2N3954...58, 2N5045...47	data
BFQ 17(P)	Si-N	SMD, VHF/UHF-A, 40V, 150mA, 1200MHz	39b	Mot,Phi,Sie	BFQ 64	data
BFQ 18(A)	Si-N	SMD, UHF-A, 30V, 150mA, 3,5GHz	39b	Mot,Phi,Sie	BFQ 64	data
BFQ 19(P,S)	Si-N	SMD, UHF-A, 20V, 75mA, 5GHz	39b	Mot,Phi,Sie	2SC3268	data
BFQ 20..21	N-FET	Dual, Ugs1-Ugs2 <25...50mV	81bx	Tix	2N3954...58, 2N4082...85	data
BFQ 22(S)	Si-N	UHF-A, 15V, 35mA, 0,15W, 5GHz	5g	Phi,Tho	BFQ 69, BFQ 85, BFR 14, BFR 91, ++	data
BFQ 23	Si-P	UHF-A, 15V, 35mA, 0,18W, 5GHz	24f	Phi,Sie	BFQ 56, BFQ 75...76	data
BFQ 23 C	Si-P	=BFQ 23:	51s	Phi	BFQ 56, BFQ 75...76	data
BFQ 24	Si-P	UHF-A, 15V, 35mA, 0,15W, 5GHz	5g	Phi	BFQ 24, BFQ 56, BFQ 75...76	data
BFQ 25..26	N-FET	Dual, 25V, Idss>10mA, Up<5V	81bx	Tix	2N5564...5566	data
BFQ 27	Si-N	UHF-ra, 15V, 25mA, 0,2W	52	Tho	BFQ 69, BFQ 85, BFR 14, BFR 91, ++	data
BFQ 28	Si-N	UHF-ra, 20V, 15mA, 0,2W, 5GHz	52r	Sie	BFG 90, BFQ 85, BFR 49, BFR 90...91, ++	data
BFQ 29(P)	Si-N	SMD, UHF-ra, 20V, 30mA, 4GHz	35a	Sie	BFR 92...93, BFT 75	data
BFQ 29 R	Si-N	=BFQ 29:	35d		BFR 92R...93R, BFT 75R	data
BFQ 30	Si-N	UHF-ra, 10V, 100mA, 0,25W, 2GHz	5g	Phi	BFG 34, BFT 12, BFX 59	data
BFQ 31(A)	Si-N	SMD, VHF, 30V, 100mA, 0,2W, >600MHz	35a	Fer,Tho	BFR 106	data
BFQ 31(A)R	Si-N	=BFQ 31:	35d		-	data
BFQ 32	Si-P	UHF-A, 20V, 75mA, 0,5W, 4,2GHz	24f	Phi,Sie	BFQ 79, BFT 96	data
BFQ 32 C	Si-P	=BFQ 32: 100mA, 4,5GHz	51s	Phi	BFQ 194	data
BFQ 32 S	Si-P	=BFQ 32: 100mA, 0,7W, 4,5GHz	24f	Phi	BFQ 194	data
BFQ 33	Si-N	UHF-A-ra, 9V, 20mA, 0,14W, 12GHz	52r	Phi	-	data
BFQ 33 C	Si-N	=BFQ 33:	51r		-	data
BFQ 34	Si-N	UHF-A, 25V, 150mA, 4GHz	55r	Phi	BFQ 68	data
BFQ 34 T	Si-N	=BFQ 34: 3,7GHz	24f		BFG 34, BFG 96	data
BFQ 35	Si-P	Vid, 160V, 0,2A, 0,8W, >80MHz	2a§	Tix	BFT 44...45, 2SB606, 2SB622	data
BFQ 36	Si-P	=BFQ 35: 250V	2a§	Tix	BFT 44...45, 2SB606, 2SB622	data
BFQ 37	Si-P	=BFQ 35: 300V	2a§	Tix	BFT 44, 2SB622	data
BFQ 38	Si-N	S/Vid, 300/250V, 1A, 0,6W, >20MHz	2a§	Phi	BUX 55, BUY 59...60	data
BFQ 39	Si-N	=BFQ 38: 300/300V	2a§	Phi	BUX 55, BUY 59...60	data
BFQ 40	Si-N	=BFQ 38: 450/350V	2a§	Phi	BUX 55, BUY 59...60	data
BFQ 41	Si-N	UHF-Tr/E, 50V, 350mA, 1GHz	55r	Aeg	BFR 65, MRF 626, 2N5944	data
BFQ 42	Si-N	VHF-Tr/E, 36V, 600mA, PQ=2W(175MHz)	2a§	Phi	BFW 46, BLY 33, MRF 227, 2N3924, ++	data
BFQ 43(S)	Si-N	VHF-Tr/E, 36V, 1,25A, PQ=4W(175MHz)	2e*	Phi	(2N6255)[5]	data
BFQ 44...45	N-FET	Dual, Ugs1-Ugs2 <25...50mV	81bx	Tix	2N5564...5566	data
BFQ 46...48	N-FET	Dual, Ugs1-Ugs2 <10...50mV	81bx	Tix	2N3954...58, 2N4082...85	data

Type	Device	Short Description	Fig.	Manu	Comparision Types	More at
BFQ 49(A...C)	N-FET	Dual, Ugs1-Ugs2 <10...50mV	81bx	Tix	2N3921..22, 2N3954...58, 2N5045...47	data
BFQ 51	Si-P	UHF-A, 20V, 25mA, 0,18W, 5GHz	24f	Phi,Sie	BFT 95, BFQ 23, BFQ 56, BFQ 75...76	data
BFQ 51 C	Si-P	=BFQ 51: 30mA, 0,25W	51s		BFT 95, BFQ 23, BFQ 56, BFQ 75...76	data
BFQ 52	Si-P	UHF-A, 20V, 25mA, 0,15W, 5GHz	5g	Phi	BFT 95, BFQ 23..24, BFQ 56, BFQ 75...76	data
BFQ 53	Si-N	UHF-A, 20V, 25mA, 0,15W, 5GHz	5g	Phi	BFR 14, BFR 90...92, BFQ 22, BFQ 85	data
BFQ 54 T	Si-P	UHF-A, 25V, 150mA, 1W, 4,5GHz	24f	Phi	-	data
BFQ 56	Si-P	UHF-A-ra, 20V, 30mA, 5GHz	24f	Sie	BFQ 23, BFQ 75...76	data
BFQ 57	Si-N	UHF-A, 25V, 35mA, 0,45W, 6,5GHz	52r	Sie	BFG 65, BFQ 65...66, BFQ 74	data
BFQ 58	Si-N	UHF-A, ra, 25V, 30mA, 0,45W, 6,5GHz	52r	Sie	BFG 65, BFQ 65...66, BFQ 74	data
BFQ 59	Si-N	UHF, ra, 27V, 35mA, 0,7W, 4GHz	51r	Sie	BFT 97, BFQ 60, BFQ 70...71	data
BFQ 60	Si-N	UHF, ra, 27V, 35mA, 0,7W, 4GHz	52r	Sie	BFT 97, BFQ 59, BFQ 70...71	data
BFQ 61	Si-P	SMD, UHF, ra, 15V, 30mA, 5GHz	35a	Sie	BFT 93	data
BFQ 62	Si-P	SMD, UHF, ra, 15V, 90mA, 5GHz	39b	Sie	-	data
BFQ 63	Si-N	VHF/UHF-A, ra, 20V, 75mA, 4,5GHz	5g	Phi,Tho	BFG 96, BFP 96, BFR 96, BFQ 73, ++	data
BFQ 64	Si-N	SMD, UHF-A, 30V, 200mA, 3GHz	39b	Sie	-	data
BFQ 65	Si-N	UHF-A, ra, 20V, 50mA, 0,3W, 7,5GHz	24f	Aeg,Phi	BFG 65, BFQ 66, 2SC3062	data
BFQ 66	Si-N	UHF-A, ra, 20V, 50mA, 0,3W, 7,5GHz	51s	Phi	BFG 65, BFQ 65, 2SC3062	data
BFQ 67	Si-N	SMD, UHF-A, ra, 20V, 0,05A, 8GHz	35a	Aeg, Phi	3SC3356, 2SC3445, 2SC3583, 2SC4592	data
BFQ 67 R	Si-N	=BFQ 67:	35d			
BFQ 67 W	Si-N	=BFQ 67:	35a(2mm)		3SC4593	data
BFQ 68	Si-N	UHF-A, 25V, 300mA, 4,5W, 4GHz	55r	Phi	-	data
BFQ 69	Si-N	UHF-A, ra, 25V, 30mA, 0,2W, 5,8GHz	24f	Sie	BFG 91, BFP 91, BFQ 57...58, BFQ 74, ++	data
BFQ 70	Si-N	UHF-A, ra, 20V, 30mA, 5GHz	51s	Sie	BFR 14, BFT 97, BFQ 59...60, BFQ 71, ++	data
BFQ 71	Si-N	UHF-A, ra, 20V, 30mA, 5,2GHz	51s	Sie	BFR 14, BFR 91, BFQ 69, BFQ 85, ++	data
BFQ 72	Si-N	UHF-A, ra, 20V, 50mA, 5,1GHz	51s	Sie	BFP 91, BFP 96, BFT 65, BFR 96	data
BFQ 73(S)	Si-N	UHF-A, ra, 20V, 90mA, 0,29W, 5GHz	51s	Sie	BFG 96, BFP 96, BFR 96	data
BFQ 74	Si-N	UHF-A, ra, 25V, 35mA, 6GHz	51s	Sie	BFG 65, BFQ 57...58, BFQ 65...66	data
BFQ 75	Si-P	UHF-A, ra, 15V, 50mA, 5GHz	51s	Sie	BFQ 23, BFQ 56, BFQ 76	data
BFQ 76	Si-P	UHF-A, ra, 20V, 30mA, 5GHz	51s	Sie	BFQ 23, BFQ 56, BFQ 75	data
BFQ 77	Si-N	UHF-A, ra, -/15V, 20mA, 0,25W, 7GHz	51s	Sie	BFG 65, BFQ 65...66, 2SC3586...87	data
BFQ 78	Si-P	UHF-A, ra, 25V, 50mA, 0,7W, 3,3GHz	51r	Sie	BFQ 32, 2SA1057...1058, 2SA1223	data
BFQ 79	Si-P	UHF-A, ra, -/15V, 75mA, 0,29W, 4,2GHz	51s	Sie	BFT 96, BFQ 32(S), 2SA1057, 2SA1223	data
BFQ 80	Si-N	UHF-A, ra, -/15V, 20mA, 0,29W, 7GHz	52r	Sie	BFG 65, BFQ 65...66, BFQ 77	data
BFQ 81	Si-N	SMD, UHF-A, ra, 25V, 0,03A, 4,2GHz	35a	Aeg,Sie	BFQ 29, BFR 93, BFT 75, 2SC3110	data
BFQ 82	Si-N	UHF-A, 20V, 0,08A, 8GHz	51s	Sie	BFQ 196, 2SC3603	data
BFQ 85	Si-N	UHF-A, ra, 20V, 40mA, 0,2W, 5GHz	25za	Sgs	BFG 91, BFR 14, BFR 91, BFQ 69	data
BFQ 88	Si-N	UHF-A, ra, 20V, 40...80mA, 5GHz	51r	Sgs	BFG 91, BFP 91, 2SC3511	data
BFQ 88 A	Si-N	=BFQ 88: 80mA, 0,35W	51r		BFG 96, BFR 96, BFQ 73	data
BFQ 89	Si-N	UHF-A, ra, 20V, 20mA, 0,2W, 6GHz	52r	Sgs	BFG 91, BFQ 57...58, BFQ 74, BFQ 77	data
BFQ 98	Si-P	UHF-A, ra, 20V, 40mA, 0,25W, 5GHz	52r	Sgs	BFT 96, BFQ 23, BFQ 56, BFQ 75...76	data
BFQ 131	Si-N	Monitor, hi-res, 25V, 0,15A, 4GHz	7c	Phi	-	data
BFQ 136	Si-N	UHF-A, 25V, 0,6A, 9W(Tc=110°), 4GHz	55r	Phi	-	data
BFQ 149	Si-P	=BFQ 32: SMD	39b	Phi	-	data
BFQ 151	Si-N	Tr f. Video Module, 20V, 0,1A, 3,5GHz	7c	Phi	-	data
BFQ 161	Si-N	Monitor, hi-res, 20V, 0,5A, 1GHz	7c	Phi	-	data
BFQ 162	Si-N	Monitor, hi-res, 20V, 0,5A, >1GHz	14h§	Phi	-	data
BFQ 166	Si-N	Monitor,hi-res, 20V, 0,5A, 2W, >1GHz	48s§	Phi	-	data
BFQ 181	Si-N	UHF-A, 20V, 0,02A, 8GHz	51s	Sie	BFQ 77	data
BFQ 182	Si-N	UHF-A, 20V, 35mA, 8,3GHz	51s	Sie	BFQ 66, BFQ 645	data
BFQ 193	Si-N	SMD, UHF-A, 20V, 0,08A, 7,5GHz	39b	Sie	2SC3301	data
BFQ 194	Si-P	=BFR 194:	51s	Sie		data
BFQ 196	Si-N	=BFR 196:	51s	Sie		data
BFQ 221	Si-N	Monitor, hi-res, 100V, 0,1A, 1GHz	7c	Phi	BFQ 231(A)	data
BFQ 222	Si-N	=BFQ 221: 5W	14h§	Phi	BFQ 232(A)	data
BFQ 225	Si-N	=BFQ 221: 3,75W	13h§	Phi	BFQ 235(A)	data
BFQ 226	Si-N	=BFQ 221: 3W(Tc=60°)	48s§	Phi	BFQ 236(A)	data
BFQ 231	Si-N	Monitor, hi-res, 100V, 0,3A, >1GHz	7c	Phi	-	data
BFQ 231 A	Si-N	=BFQ 231: 115V, >800MHz	7c		-	data
BFQ 232(A)	Si-N	=BFQ 231(A): 3W(Tc=115°)	14h§	Phi	BFQ 262(A)	data
BFQ 235(A)	Si-N	=BFQ 231(A): 3W(Tc=100°)	13h§	Phi	BFQ 265(A)	data
BFQ 236(A)	Si-N	=BFQ 231(A): 3W(Tc=115°)	48s§	Phi	-	data
BFQ 241	Si-P	Monitor, hi-res, 100V, 0,1A, 1GHz	7c	Phi	BFQ 251(A)	data
BFQ 242	Si-P	=BFQ 241: 5W	14h§	Phi	BFQ 252(A)	data
BFQ 245	Si-P	=BFQ 241: 3,75W	13h§	Phi	BFQ 255(A)	data
BFQ 246	Si-P	=BFQ 241: 3W(Tc=60°)	48s§	Phi	BFQ 256(A)	data
BFQ 251	Si-P	Monitor, hi-res, 100V, 0,3A, >1GHz	7c	Phi	-	data
BFQ 251 A	Si-P	=BFQ 231: 115V, >800MHz	7c		-	data
BFQ 252(A)	Si-P	=BFQ 251(A): 3W(Tc=115°)	14h§	Phi	-	data
BFQ 255(A)	Si-P	=BFQ 251(A): 3W(Tc=100°)	13h§	Phi	-	data
BFQ 256(A)	Si-P	=BFQ 251(A): 2W(Tc=115°)	48s§	Phi	-	data
BFQ 262	Si-N	Monitor, hi-res, 100V, 0,4A, 5W, >1GHz	14h§	Phi	-	data

Type	Device	Short Description	Fig.	Manu	Comparision Types	More at
BFQ 262 A	Si-N	=BFQ 262: 115V, >800MHz	14h§		-	data
BFQ 265(A)	Si-N	=BFQ 262(A): 5W(Tc=65°)	13h§	Phi	-	data
BFQ 645	Si-N	UHF-A, 25V, 0,04A, 9GHz	51s	Sie	-	data
BFR						
BFR	Si-P	=2SB1308-R (Typ-Code/Stempel/marking)	39	Rhm	→2SB1308	data
BFR	Si-N	=2SC4642K-R (Typ-Code/Stempel/marking)	35	Rhm	→2SC4642K	data
BFR	Si-N	=2SC4723-R (Typ-Code/Stempel/marking)	35(2mm)	Rhm	→2SC4723	data
BFR 10	Si-N	HF/S, 75V, 0,5A, 0,8W, 14/80ns	2a§	Sgs	BFX 96A...97A, BSS 27, 2N2218A...19A, ++	data
BFR 11	Si-N	=BFR 10: 0,4W	2a§	Sgs	BFX 94A...95A, BSW 63, 2N2221A...22A, ++	data
BFR 12	Si-N	VHF/UHF-Tr, 55V, 300mA, >480MHz	2a§	Aeg	BLW 10, (BFX 55, 2N4428, 2SC2852, ++)[6]	data
BFR 14	Si-N	UHF-A, ra, 20V, 30mA, 0,25W, 3,6GHz	51s	Sie	BFT 97, BFQ 59...60, BFQ 70...71	data
BFR 14 A	Si-N	UHF-A, ra, 20V, 30mA, 0,25W, 5GHz	52r*	Sie	BFG 91, BFR 91, BFQ 69, BFQ 85	data
BFR 14 B	Si-N	=BFR 14A: 6GHz	52r*	Sie	BFG 91, BFQ 57...58, BFQ 74	data
BFR 14 C	Si-N	=BFR 14A: 27V, 35mA, 0,7W, 4,3GHz	51r	Sie	BFR 91, BFQ 69, BFQ 71, BFQ 85	data
BFR 15(A)	Si-N	UHF-ra, 20V, 30mA, 0,2W, 3,3..4,5GHz	5k	Sie	BFS 55, BFT 66...67, BFQ 22, BFQ 63, ++	data
BFR 16	Si-N	NF, ra, 60V, 0,05A, 100MHz, B>150	2a§	Sgs	BC 382, BC 414, BC 550, 2N2483...84, ++	data
BFR 17	Si-N	=BFR 16: B>450	2a§	Sgs	BC 382, BC 414, BC 550, 2N2483...84, ++	data
BFR 18	Si-N	NF-Tr, 85V, 0,5A, 0,5W, 90MHz	2a§	Sgs	BC 639, BCX 24, 2N3700...01, 2SD667,++	data
BFR 19	Si-N	NF-Tr, 75V, 1A, 0,8W, 100MHz	2a§	Sgs	BC 140...141, 2N1990, 2N2102, 2N2405, ++	data
BFR 20	Si-N	NF-Tr/S, 75V, 1A, 0,8W, 130/450ns	2a§	Sgs	BC 140...141, BSX 59...61, 2N3444, ++	data
BFR 21	Si-N	NF-Tr/S, 120V, 1A, 0,8W, 130/450ns	2a§	Sgs	BSS 42...43, BSW 67...68, BSX 47	data
BFR 22	Si-N	NF/S, 120V, 1A, 1W, >120MHz	2a§	Sgs	BSS 42...43, BSW 67...68, BSX 47, 2N2102	data
BFR 23	Si-P	NF/S, 90V, 1A, 1W, <110/700ns	2a§	Sgs	BSW 40, 2N4036	data
BFR 24	Si-P	NF/S, 60V, 1A, 1W, <110/700ns	2a§	Sgs	BSW 40, 2N4030...4033, 2N4036...4037	data
BFR 25	Si-N	Nix, S, 120V, 0,1A, 0,375W, >50MHz	2a	Sgs	BF 279...299, BSS 38, BSX 21, ++	data
BFR 26	Si-N	VHF, 30V, 0,12A, 0,3W, >350MHz	2a§	Tho	BFX 59, (BFW 16...17)[6]	data
BFR 27	Si-N	Vid, 160V, 0,4A, 0,6W	2a§	Tho	2SD413, 2SD576, 2SD624	data
BFR 28	Si-N	VHF/UHF, 30V, 50mA, 0,2W, 1GHz	≈36e	Sie	BF 357, BF 377...378, BF 763, 2SC3037, ++	data
BFR 29	MOS-N-FET-d	NF/HF, 30V, Idss>10mA, Up<4V	5m+	Phi	3N142...143, 3N152, 3N154, 3N192...193	data
BFR 30	N-FET	SMD, Uni, 25V, Idss>4mA, Up<5V	35b	Mot,Phi	BF 512	data
BFR 31	N-FET	SMD, Uni, 25V, Idss>1mA, Ur <2,5V	35b	Mot,Phi	BF 511, 2SK425	data
BFR 34(A)	Si-N	UHF-A/V-ra; 20V, 30mA, 3,3..4,5GHz	24f	Aeg,Sie	BFT 97, BFQ 59...60, BFQ 70...71, 2N6620	data
BFR 35(A,AP)	Si-N	=BFR 34: SMD	35a	Aeg,Sie	BFR 92...93, BFT 75, BFQ 29, 2N6619	data
BFR 35(A)R	Si-N	=BFR 34: SMD	35d		BFR 92R...93R, BFT 75R, BFQ 29R	data
BFR 36(A)	Si-N	VHF/UHF-A/Tr, 40V, 200mA, 0,8W, 1,3GHz	2a§	Sgs,Tix	BFW 16...17, 2N4428, 2N5160	data
BFR 37	Si-N	VHF/UHF-A, 30V, 50mA, 0,25W, 1,4GHz	5k	Sgs	BFW 30, BFX 59, BFX 73	data
BFR 38	Si-P	VHF/UHF-ra, 40V, 20mA, 0,2W, 1GHz	5g	Sgs	BF 272, BF 316, BF 516, BF 606	data
BFR 39	Si-N	NF/S, 90V, 1A, 0,8W, >100MHz, <55/-ns	7c	Tix	BC 639, BFR 50, BFT 29, BFT 53	data
BFR 40	Si-N	=BFR 39: 70V	7c	Tix	BC 639, BFR 50, BSS 26, BSS 40...41	data
BFR 41	Si-N	=BFR 39: 60V	7c	Tix	BC 637, BFR 50, BSS 26, BSS 40...41, BFR 51	data
BFR 44(A...C)	Si-N	Dual, VHF, 30V, 0,05A, >600MHz	81bs	Tho	2N3423...3424	data
BFR 45	N-FET	NF/HF, 30V, Idss>2mA, Up<8V	5h+	Tix	BFW 10...11, BFS 72, 2N3823	data
BFR 46	MOS-P-FET-e*	NF-ra, 15V, Idss<0,01mA, Up=2V	2b	Fer	-	data
BFR 47	MOS-P-FET-e	=BFR 46: o.Gateschutz-D./wo. gate prot	2b	Fer	-	data
BFR 48	Si-N	VHF, 50V, 100mA, 0,45W, >600MHz	2a§	Tho	BFR 36, BFW 16...17, BFX 55	data
BFR 49	Si-N	UHF, ra, 20V, 25mA, 0,3W, 5GHz	52r	Phi	BFG 90, BFQ 28, BFQ 85, BFR 90...91, ++	data
BFR 50	Si-N	NF/S, 80V, 1A, 0,8W, >50MHz, <55/-ns	7c	Tix	BC 639, BFR 39, BFT 29, BFT 53	data
BFR 51	Si-N	=BFR 50: 60V	7c	Tix	BC 637, BFR 40, BSS 26, BSS 40...41	data
BFR 52	Si-N	=BFR 50: 40V	7c	Tix	BC 637, BFR 41, BSS 26, BSS 40...41, ++	data
BFR 53	Si-N	SMD, UHF-A, 18V, 50mA, 2GHz	35a	Phi,Tho	BFR 93, BFT 75, 2SC3014	data
BFR 53 R	Si-N	=BFR 53:	35d		BFR 93R, BFT 75R	data
BFR 54	Si-N	VHF/UHF, 40V, 100mA, >500MHz	7a	Phi	2SC2851, (BFX 55, 2SC2852)[6]	data
BFR 56	Si-N	NF/S, 60V, 1,2A, 1W	2a§	Tix	BC 140...141, BSW 65...68, BSX 45...47	data
BFR 57	Si-N	Vid, 160V, 0,2A, 1W, 100MHz	2a§	Tix	BF 257...259, BF 657...659, 2N5058...5059	data
BFR 58	Si-N	=BFR 57: 250V	2a§	Tix	BF 258...259, BF 658...659, 2N5058...5059	data
BFR 59	Si-N	=BFR 57: 300V	2a§	Tix	BF 259, BF 659, 2N5058, BFS 89	data
BFR 60	Si-P	NF/S, 80V, 1A, 0,8W, >50MHz, <55/-ns	7c	Tix	BC 640, BFR 79, BFT 20, BFT 69	data
BFR 61	Si-P	=BFR 60: 60V	7c	Tix	BC 638, BFR 80, 2N4026...4029, ++	data
BFR 62	Si-P	=BFR 60: 40V	7c	Tix	BC 638, BFR 81, 2N4026...4029, ++	data
BFR 63	Si-N	VHF-A/Tr, 40V, 200mA, PQ=0,15W(200MHz)	55r	Phi	BFR 64...65, BFT 91, BLX 96	data
BFR 64	Si-N	UHF-A/Tr, 40V, 200mA, PQ=0,09W(800MHz)	55r	Phi	BFT 91, BLX 96	data
BFR 65	Si-N	UHF-A/Tr, 40V, 400mA, PQ=0,45W(200MHz)	55r	Phi	BFT 91, BLX 96	data
BFR 67(A,B)	Si-N	Min, NF, 50V, 0,15A	(40b)	Sgs	(BC 167, BC 182, BC 237, BC 547, ++)[6]	data
BFR 68(A...C)	Si-N	=BFR 67: ra, 30V	(40b)	Sgs	(BC 169, BC 184, BC 239, BC 549, ++)[6]	data
BFR 69(VI,A,B)	Si-P	Min, NF, 50V, 0,15A	(40b)	Sgs	(BC 212, BC 257, BC 307, BC 557, ++)[6]	data
BFR 70(VI,A,B)	Si-P	=BFR 69: ra, 30V	(40b)	Sgs	(BC 214, BC 259, BC 309, BC 559, ++)[6]	data
BFR 71	Si-N	Min, NF, 85V, 0,2A, B>40	(40b)	Sgs	(BC 546, 2SC3245)[6]	data
BFR 72	Si-N	=BFR 71: B>100	(40b)	Sgs	(BC 546, 2SC3245)[6]	data
BFR 73	Si-P	Min, NF, 60V, 0,2A, B>40	(40b)	Sgs	(BC 556, 2SA1285)[6]	data
BFR 74	Si-P	=BFR 73: B>100	(40b)	Sgs	(BC 556, 2SA1285)[6]	data

167

Type	Device	Short Description	Fig.	Manu	Comparision Types	More at
BFR 75	Si-N	Min, VHF, 700MHz	(40b)	Sgs	(BF 225, BF 314, BF 496, BF 502...503,++)[6]	data
BFR 76	Si-P	Min, VHF, 450MHz	(40b)	Sgs	(BF 324, BF 414, BF 506, BF 914, ++)[6]	data
BFR 77	Si-N	NF/S, 120/80V, 1A, 0,8W, >50MHz	2a§	Sgs	BSW 67...68, BSX 47, 2N3019...3020, ++	data
BFR 78	Si-N	=BFR 77: 120/100V	2a§	Sgs	BSW 67...68, BSX 47, 2N3019...3020, ++	data
BFR 79	Si-P	NF/S, 90V, 1A, 0,8W, >100MHz, <55/-ns	7c	Tix	BC 640, BFR 60, BFT 20, BFT 69	data
BFR 80	Si-P	=BFR 79: 70V	7c	Tix	BC 640, BFR 60, 2N4026...4029, ++	data
BFR 81	Si-P	=BFR 79: 60V	7c	Tix	BC 638, BFR 61, 2N4026...4029, ++	data
BFR 83	Si-N	UHF, 40V, 0,2A, 1400MHz	55r		BFR 64...65	data
BFR 84	MOS-N-FET-d*	Dual-Gate, VHF, 20V, Idss>20mA,Up<3,8V	5t*	Phi	-	data
BFR 86(A,B)	Si-N	Vid, 120V, 0,2A, 0,8W, 130MHz	7c	Tix	BFT 57...59, BF 391...393, BF 422A, ++	data
BFR 87(A,B)	Si-N	=BFR 86: 160V	7c	Tix	BFT 57...59, BF 391...393, BF 422A, ++	data
BFR 88(A,B)	Si-N	=BFR 86: 250V	7c	Tix	BFT 58...59, BF 392...393, BF 422A, ++	data
BFR 89(A,B)	Si-N	=BFR 86: 300V	7c	Tix	BFT 59, BF 393, BF 420A	data
BFR 90(A,B,H)	Si-N	UHF-A, 20V, 25mA, 5GHz	24f	EUR	BFG 90, BFQ 28, BFR 14, BFR 49, ++	data
BFR 91(A,H)	Si-N	UHF-A, 15V, 35mA, 5...6GHz	24f	EUR	BFG 91, BFQ 69, BFQ 85, BFR 14	data
BFR 92(A,P)	Si-N	=BFR 90: SMD	35a	EUR	BFT 75, 2SC3011, 2SC3513, 2SC3161	data
BFR 92(A)R	Si-N	=BFR 90: SMD	35d		BFT 75R	data
BFR 92(A)W	Si-N	=BFR 90: SMD	35a(2mm)			data
BFR 93(A,P)	Si-N	=BFR 91: SMD	35a	EUR	BFT 75, 2SC3011, 2SC3513, 2SC3161	data
BFR 93(A)R	Si-N	=BFR 91: SMD	35d		BFT 75R	data
BFR 93(A)W	Si-N	=BFR 91: SMD	35a(2mm)			data
BFR 94	Si-N	UHF-A, 30V, 0,15A, 3,5GHz	55r	Phi	BFQ 34, BFQ 68, BFT 98	data
BFR 95	Si-N	VHF/UHF-A, 30V, 0,15A, 3,5GHz	2a§	Phi	(BFG 34, BFG 96)	data
BFR 96(H,S,T,TS)	Si-N	UHF-A/Tr, 20V, 75...100mA, 5GHz	24f	EUR	BFG 96, BFP 96, BFQ 73	data
BFR 97	Si-N	VHF/UHF-Tr/E, 55V, 0,5A, PQ>1W(400MHz)	2a§	Sgs	2N3866, 2SC2852	data
BFR 98	Si-N	VHF-Tr/E, 40V, 0,5A, PQ>1W(175MHz)	2a§	Sgs	BFQ 42, BFS 51, BLW 16, 2N4427, ++	data
BFR 99(A)	Si-P	VHF/UHF-V-ra, 25V, 50mA, 2..2,3GHz	5g	Mot,Sgs	BFT 96, BFQ 32, BFQ 78, 2SA1228	data
BFR 101 A	N-FET*	SMD, Uni, sym, 30V, Idss>0,2mA, Up<1V integr. Diode (Pin3)	44o	Phi	-	data
BFR 101 B	N-FET*	=3FR 101A: Idss>1mA, Up<2,5V	44o			data
BFR 106	Si-N	SMD, VHF/UHF-A, 20V, 0,1A, 5GHz	35a	Phi, Sie	2SC3356, 2SC3775	data
BFR 134	Si-N	UHF-A, 25V, 0,15A, 7GHz	24f	Phi	-	data
BFR 180	Si-N	SMD, UHF-A, 15V, 4mA, 4,4GHz	35a	Sie	-	data
BFR 180 W	Si-N	=BFR 180:	35a(2mm)			data
BFR 181(T)	Si-N	SMD, UHF-A, 20V, 0,02A, 8GHz	35a	Aeg, Sie	BFR 505, 2SC3585	data
BFR 181 W,TW	Si-N	=BFR 181:	35a(2mm)		BFS 505	data
BFR 182(T)	Si-N	SMD, UHF-A, 20V, 35mA, 8GHz	35a	Aeg, Sie	BFQ 67, 2SC3585	data
BFR 182 W,TW	Si-N	=BFR 182:	35a(2mm)			data
BFR 183(T)	Si-N	SMD, UHF-A, 20V, 65mA, 8GHz	35a	Aeg, Sie	BFR 193, BFR 520, 2SC3445	data
BFR 183 T,TW	Si-N	=BFR 183:	35a(2mm)		BFS 520, 2SC4593	data
BFR 193(T)	Si-N	SMD, UHF-A, ...2GHz, 20V, 0,08A, 8GHz	35a	Aeg, Sie	BFR 520, 2SC3356, 2SC3445	data
BFR 193 W,TW	Si-N	=BFR 193:	35a(2mm)		BFS 520	data
BFR 194	Si-P	SMD, UHF-A, ...1,5GHz, 20V, 0,1A, 5GHz	35a	Sie	-	data
BFR 194 T	Si-P	SMD, UHF-A, -/8V, 25mA, 3,5GHz	35a	Aeg	-	data
BFR 200	N-FET	SMD,asym.,lo Igss,30V, 20mA, Idss>0,2m	44o	Phi	-	data
BFR 280(T)	Si-N	SMD, UHF-A, 15V, 0,01A, >5GHz	35a	Aeg, Sie	BFR 92	data
BFR 280 W,TW	Si-N	=BFR 280:	35a(2mm)		BFR 92W	data
BFR 505	Si-N	SMD, SATV, 20V, 18mA, 9GHz	35a	Phi	-	data
BFR 520	Si-N	SMD, SATV, 20V, 70mA, 9GHz	35a	Phi	-	data
BFR 541	Si-N	=BFG 541:	25za	Phi	-	data
BFR 949 T	Si-N	SMD, UHF, 20V, 35mA, 8,5GHz	35a(1,6mm)	Sie		data
BFRC 96	Si-N	=BFR 96:	Chip			data

BFS

Type	Device	Short Description	Fig.	Manu	Comparision Types	More at
BFS	Si-N	=2SC4642K-S (Typ-Code/Stempel/-marking)	35	Rhm	→2SC4642K	data
BFS	Si-N	=2SC4723-S (Typ-Code/Stempel/-marking)	35(2mm)	Rhm	→2SC4723	data
BFs	Si-P	=BCW 61FF (Typ-Code/Stempel/-marking)	35	Sie	→BCW 61FF	data
BFS 10	Si-N	VHF-A/Tr, 55V, 400mA, 0,5W, >500MHz	2a§	Sgs	BFW 47, BFX 55, 2N3553	data
BFS 11	Si-N	VHF-V, ra, 800MHz	5k	Sgs	BF 225, BF 314, BF 496, BF 502, ++	data
BFS 12	Si-P	NF-Tr, 40V, 1A, 0,8W, >100MHz	2a§	Sgs	BC 160...161, BCX 60, 2N4234...4236, ++	data
BFS 13 E	Si-N	Min, NF-ra, 40V, 0,05A, 90MHz	36c	Sgs	BC 123	data
BFS 13 F	Si-N	=BFS 13E:	36f		BC 123	data
BFS 13 G	Si-N	=BFS 13E:	36e		BC 123	data
BFS 14 E	Si-P	Min, NF-ra, 40V, 0,05A, 40MHz	36c	Sgs	BC 203	data
BFS 14 F	Si-P	=BFS 14E:	36f		BC 203	data
BFS 14 G	Si-P	=BFS 14E:	36e		BC 203	data
BFS 15 E	Si-N	Min, NF/S, 40V, 0,05A, 400MHz	36c	Sgs	(BC 123)	data
BFS 15 F	Si-N	=BFS 15E:	36f		(BC 123)	data
BFS 15 G	Si-N	=BFS 15E:	36e		(BC 123)	data
BFS 16 E	Si-P	Min, NF/S, 40V, 0,05A, 210MHz	36c	Sgs	(BC 203)	data
BFS 16 F	Si-P	=BFS 16E:	36f		(BC 203)	data
BFS 16 G	Si-P	=BFS 16E:	36e		(BC 203)	data

Type	Device	Short Description	Fig.	Manu	Comparision Types	More at
BFS 17(A,P)	Si-N	SMD, VHF/UHF, 25V, 25mA, 1..2,8GHz	35a	EUR	2SC3005, 2SC3016, 2SC3161	data
BFS 17 R(AR)	Si-N	=BFS 17:	35d		-	data
BFS 17 S	Si-N	=BFS 17: Dual	46bh5(2mm)			data
BFS 17 W,AW	Si-N	=BFS 17:	35a(2mm)			data
BFS 18	Si-N	SMD, HF, 30V, 30mA, 200MHz	35a	EUR	BF 554, BF 599, BF 799, BFS 20	data
BFS 18 R	Si-N	=BFS 18:	35d		BFS 20R	data
BFS 19	Si-N	SMD, HF, 30V, 30mA, 260MHz	35a	EUR	BF 554, BF 599, BF 799, BFS 20	data
BFS 19 R	Si-N	=BFS 19:	35d		BFS 20R	data
BFS 20	Si-N	SMD, HF, 30V 25mA, 450MHz	35a	EUR	BF 599, BF 799, BFS 17, 2SC3015, 2SC3374	data
BFS 20 R	Si-N	=BFS 20:	35d		BFS 17R	data
BFS 20 W	Si-N	=BFS 20:	35a(2mm)		-	data
BFS 21(A)	N-FET	Dual, Ugs1-Ugs2 <10...20mV	2x 5h+	Phi	-	data
BFS 22(A,R)	Si-N	VHF-Tr/E, 36V, 0,75A, PQ=4W(175MHz)	2a§	Phi	BFW 46, BFQ 43, MRF 237, 2N3924	data
BFS 22 Q	Si-N	=BFS 22:	55r		BLY 85, 2N5589	data
BFS 23(A,R)	Si-N	VHF-Tr/E, 65V, 0,5A, PQ=4W(175MHz)	2a§	Phi	BFW 47, BLY 34, 2N3553	data
BFS 25 A	Si-N	SMD, UHF, 8V, 6,5mA, 5000MHz	35a(2mm)	Phi	BFS 505	data
BFS 26 E	Si-N	Min, HF/S, 20V, 0,05A, 550MHz	36c	Sgs	-	data
BFS 26 F	Si-N	=BFS 26E:	36f		-	data
BFS 26 G	Si-N	=BFS 26E:	36e		-	data
BFS 27 E	Si-P	Min, HF/S, 20V, 0,05A, 400MHz	36c	Sgs	-	data
BFS 27 F	Si-P	=BFS 27:	36f		-	data
BFS 27 G	Si-P	=BFS 27:	36e		-	data
BFS 28(R)	MOS-N-FET-d	Dual-Gate, NF/HF, 20V, Up<5V	5t*	Phi	3N201..206, 3N209...213	data
BFS 29(P)	Si-N	Min, NF/S, 45V, 0,2A, 150MHz		Tix	-	data
BFS 30(P)	Si-N	Min, NF/S, 45V, 0,2A, 150MHz		Tix	-	data
BFS 31(P)	Si-N	Min, NF/S, 45V, 0,2A, 150MHz		Tix	-	data
BFS 32(P)	Si-P	Min, NF/S, 45V, 0,2A, 200MHz		Tix	-	data
BFS 33(P)	Si-P	Min, NF/S, 45V, 0,2A, 200MHz		Tix	-	data
BFS 34(P)	Si-P	Min, NF/S, 45V, 0,2A, 200MHz		Tix	-	data
BFS 36	Si-N	SMD, NF, ra, 45V, 0,5A, >30MHz	35d(2mm)	Fer	(BC 817R, BCW 65R...66R, BCX 19R,++)[6]	data
BFS 36 A	Si-N	=BFS 36: 30V	35d(2mm)		(BC 818R, BCW 65R...66R, BCX 20R,++)[6]	data
BFS 37	Si-P	SMD, NF, ra, 45V, 0,5A, >30MHz	35d(2mm)	Fer	(BC 807R, BCW 67R...68R, BCX 17R,++)[6]	data
BFS 37 A	Si-P	=BFS 37: 30V	35d(2mm)		(BC 808R, BCW 67R...68R, BCX 18R,++)[6]	data
BFS 38	Si-N	SMD, NF/HF, 45V, 0,5A, >150MHz	35d(2mm)	Fer	(BC 817R, BCW 65R...66R, BCX 19R,++)[6]	data
BFS 38 A	Si-N	=BFS 38: 25V	35d(2mm)		(BC 818R, BCW 65R...66R, BCX 20R,++)[6]	data
BFS 39	Si-N	=BFS 38: 60V	35d(2mm)		(BCW 65R...66R, BCX 41R)[6]	data
BFS 40	Si-P	SMD, NF/HF, 45V, 0,5A, >150MHz	35d(2mm)	Fer	(BC 807R, BCW 67R...68R, BCX 17R,++)[6]	data
BFS 40 A	Si-P	=BFS 40: 25V	35d(2mm)		(BC 808R, BCW 67R...68R, BCX 18R,++)[6]	data
BFS 41	Si-P	=BFS 40: 60V	35d(2mm)		(BCW 68R, BCX 42R)[6]	data
BFS 42	Si-N	SMD, NF/HF, 60/30V, 1A, >60MHz	35d(2mm)	Fer	(BCW 65R...66R, BCX 41R)[6]	data
BFS 43	Si-N	=BFS 42: 60/60V	35d(2mm)	Fer	(BCW 65R...66R, BCX 41R)[6]	data
BFS 44	Si-P	SMD, NF/HF, 60/30V, 1A, >60MHz	35d(2mm)	Fer	(BCW 68R, BCX 42R)[6]	data
BFS 45	Si-P	=BFS 44: 60/60V	35d(2mm)	Fer	(BCW 68R, BCX 42R)[6]	data
BFS 46(A)	Si-N	SMD, VHF, 30V, 0,5A, >600MHz	35d(2mm)	Fer	-	data
BFS 50	Si-N	VHF/UHF-Tr/E, 36V, 0,4A, PQ>1W(400MHz)	2a§	Aeg	MRF 629, 2N3948	data
BFS 51	Si-N	VHF-O/Tr, 40V, 0,75A, PQ=1W(175MHz)	2a§	Aeg	BFQ 42, BFR 98, BLW 16, BLY 61, ++	data
BFS 55	Si-N	UHF-A, 20/12V, 50mA, 0,325W, 3,3GHz	5k	Sie	BFQ 63, BFQ 72, BFT 65	data
BFS 55 A	Si-N	=BFS 55: 20/15V, 4,5GHz	5k		BFQ 63, BFQ 72, BFT 65	data
BFS 57(P)	Si-N	Min, UHF, 25V, 50mA, 1700MHz		Tix	-	data
BFS 58(P)	Si-N	Min, UHF, 25V, 50mA, 2400MHz		Tix	-	data
BFS 59(K,L,M)	Si-N	NF/HF, 60V, 1A(ss), 0,5W, >150MHz,B>40	40e	Fer	BC 637, BSS 26, BSS 40...41, 2N2221...22++	data
BFS 60(K,L,M)	Si-N	=BFS 59: B>100	40e	Fer	BC 637, BSS 26, BSS 40...41, 2N2221...22++	data
BFS 61(K,L,M)	Si-N	=BFS 59: 80V	40e	Fer	BC 639, 2N2221A...2222A	data
BFS 62	Si-N	VHF, 40V, 25mA, >580MHz	5k	Aeg	BF 225, BF 314, BF 496, BF 502...503, ++	data
BFS 64	Si-N	VHF/UHF-A, 30V, 50mA, 0,4W	51r	Tho	BFW 93	data
BFS 65	Si-N	VHF/UHF-A/Tr, 40V, 200mA, 0,4W	51r	Tho	-	data
BFS 67(P)	N-FET	SMD, 50V, Idss>0,5mA, Up<6V	Chip	Tix	-	data
BFS 68(P)	N-FET	SMD, 30V, Idss>4mA, Up<8V	Chip	Tix	-	data
BFS 69	Si-P	Min, NF/HF, 30V, 0,1A, >50MHz	36c	Aeg	BC 202...203	data
BFS 70	N-FET	Uni, VHF, 50V, Idss>0,5mA, Up<4	5h+	Tix	2N3821	data
BFS 71	N-FET	Uni, VHF, 50V, Idss>2mA, Up<6V	5h+	Tix	2N3822	data
BFS 72	N-FET	VHF-ra, 30V, Idss>4mA, Up<8V	5h+	Tix	2N3823	data
BFS 73	N-FET	Chopper, sym, 50V, on<250Ω	5h+	Tix	2N3824	data
BFS 74	N-FET	Chopper, sym, 40V, Idss>50mA, Up<10V	2b&	Tix	BSV 78, 2N4856	data
BFS 75	N-FET	Chopper, sym, 40V, Idss>20mA, Up<6V	2b&	Tix	BSV 79, 2N4857	data
BFS 76	N-FET	Chopper, sym, 40V, Idss>8mA, Up<4V	2b&	Tix	BSV 80, 2N4858	data
BFS 77	N-FET	=BFS 74: 30V	2b&	Tix	BSV 78, 2N4859	data
BFS 78	N-FET	=BFS 75: 30V	2b&	Tix	BSV 79, 2N4860	data
BFS 79	N-FET	=BFS 76: 30V	2b&	Tix	BSV 80, 2N4861	data
BFS 80	N-FET	VHF-ra, 30V, Idss>5mA, Up<6V	5h+	Tix	BFW 11, 2N4416, 2N5245	data
BFS 85	Si-N	SMD, UHF-ra, 25V, 25mA, >1GHz	35d(2mm)	Fer	(BFS 17R)[6]	data
BFS 86	Si-N	VHF/UHF-A/Tr, 50V, 0,3A, >1GHz		Aeg	-	data
BFS 87	Si-N	VHF/UHF-A/Tr, 50V, 0,3A, 1,1GHz		Aeg	-	data
BFS 88	Si-N	SMD, UHF-ra, 30V, 25mA, >1GHz	35d(2mm)	Fer	(BFS 17R)[6]	data
BFS 89	Si-N	Vid, 300V, 0,15A, 0,58W, 90MHz	2a§	Sgs,Tix	BF 259, BF 659, BFR 59, 2N5058	data

Type	Device	Short Description	Fig.	Manu	Comparision Types	More at
BFS 90(A,B)	Si-P	NF/S, 140V, 0,1A, 0,8W	2a§	Tix	BFW 44, 2N3634...3637	data
BFS 91(A,B)	Si-P	=BFS 90: 80V	2a§	Tix	BSV 17, 2N3634...3637	data
BFS 92	Si-P	NF/S, 100V, 1A, 0,8W, >40MHz, <55/-ns	2a§	Phi,Tix	BCX 60, BSS 17, BSW 40, 2N3634...3637, ++	data
BFS 93	Si-P	=BFS 92	2a§	Phi,Tix	BCX 60, BSS 17, BSW 40, 2N3634...3637, ++	data
BFS 94	Si-P	=BFS 92: 80V	2a§	Phi,Tix	BCX 60, BSS 17, BSW 40, 2N3634...3637, ++	data
BFS 95	Si-P	=BFS 92: 40V	2a§	Phi,Tix	BCX 60, BSS 18, BSW 40, 2N3634...3637, ++	data
BFS 96(K,L,M)	Si-P	NF/HF, 60V, 1A(ss), 0,5W, >150MHz,B>40	40e	Fer	BC 638, BSW 24, 2N2906(A)...2907(A), ++	data
BFS 97(K,L,M)	Si-P	=BFS 96: B>100	40e	Fer	BC 638, BSW 24, 2N2906(A)...2907(A), ++	data
BFS 98(K,L,M)	Si-P	=BFS 96: 80V	40e	Fer	BC 640, 2N4027...4029	data
BFS 99	Si-N	Nix, 120V, 0,05A, 0,3W	2a§	Sgs	BF 297...299, BSS 38, BSX 21, ++	data
BFS 480	Si-N	SMD, Dual, UHF, 10V, 10mA, >5GHz	46bh5(2mm)	Sie	-	data
BFS 481	Si-N	SMD, Dual, UHF, 20V, 20mA, >6GHz	46bh5(2mm)	Sie	-	data
BFS 482	Si-N	SMD, Dual, UHF, 20V, 35mA, >6GHz	46bh5(2mm)	Sie	-	data
BFS 483	Si-N	SMD, Dual, UHF, 20V, 65mA, >6GHz	46bh5(2mm)	Sie	-	data
BFS 505	Si-N	=BFG 505:	35a(2mm)	Phi	-	data
BFS 520	Si-N	=BFG 520:	35a(2mm)	Phi	-	data
BFS 540	Si-N	=BFG 540:	35a(2mm)	Phi	-	data

BFT...BFU

Type	Device	Short Description	Fig.	Manu	Comparision Types	More at
BFT 10(A...C)	N-FET	S, VHF, 40V, Idss>10mA, Up<7V	7f	Tix	BF 256, BF 348, 2N5397, 2N5486	data
BFT 11	P-FET	S, VHF, 25V, Idss>10mA, Up<9,5V	7f	Tix	2N4343	data
BFT 12	Si-N	UHF-A/O, 25V, 150mA, 2GHz	24f	Sie	BFG 34, (BFW 16, BFR 95)[6]	data
BFT 13	Si-N	UHF-A, 25V, 20mA, 0,3W, 4GHz	24f	Tho	BFG 90, BFQ 28, BFR 49, BFR 90	data
BFT 13 A,B	Si-N	=BFT 13: 0,15W	52r		BFG 90, BFQ 28, BFR 49, BFR 90	data
BFT 14	Si-N	UHF-A/Tr, 25V, 60mA, 0,7W, 4GHz	24f	Tho	BFP 96, BFQ 72...73, BFR 96, BFT 65	data
BFT 14 A,B	Si-N	=BFT 14: 0,3W	52r		BFP 96, BFQ 72...73, BFR 96, BFT 65	data
BFT 15	Si-N	UHF-A/Tr, 25V, 150mA, 0,8W, 3GHz	24f	Tho	BFG 34, BFG 96	data
BFT 16	Si-N	UHF-A/Tr, 25V, 200mA, 0,7W, 3GHz	24f	Tho	(BFG 34, BFG 96)	data
BFT 17	Si-N	UHF, 30V, 50mA, 0,2W, >1,8GHz	5g	Tho	BFQ 63, BFS 55	data
BFT 18	Si-N	UHF-A/Tr, 25V, 50mA, 0,7W, 4GHz	24f	Tho	BFQ 72...73, BFT 65	data
BFT 18 A	Si-N	=BFT 18: 0,3W	52r		BFQ 72...73, BFT 65	data
BFT 19	Si-P	S/Vid, 200V, 1A, 1W, >25MHz	2a§	Rca	BFT 28A...C, MJ 5415...5416, 2N5415...5416	data
BFT 19 A	Si-P	=BFT 19: 300V	2a§		BFT 28C, MJ 5415...5416, 2N5416	data
BFT 19 B	Si-P	=BFT 19: 400V	2a§		MJ 5416	data
BFT 20	Si-P	NF/S, 80V, 1A, 0,36W, >60MHz, <55/-ns	2a§	Tix	BC 640, BFT 69, 2N4027, 2N4029, ++	data
BFT 21	Si-P	=BFT 20: 60V	2a§	Tix	BC 638, BFT 70, 2N4026...4029, ++	data
BFT 22	Si-P	=BFT 20: 40V	2a§	Tix	BC 638, BFT 71, 2N4026...4029, ++	data
BFT 23	Si-N	VHF, 25V, >600MHz		Aeg	-	data
BFT 24	Si-N	UHF-A, 8V, 2,5mA, 0,03W, 2,3GHz	24f	Phi	BFQ 59...60, BFQ 70, BFR 34, BFT 97	data
BFT 25	Si-N	SMD, UHF-A, 8V, 6,5mA, 0,03W, 2,3GHz	35a	Phi	BFQ 29, BFR 35, BFR 53, 2SC3099	data
BFT 25 A	Si-N	=BFT 25: 5000MHz	35a		BFR 92...93, 2SC3110	data
BFT 25 R	Si-N	=BFT 25:	35d		BFQ 29R, BFR 35R, BFR 53R, 2SC3089	data
BFT 26	Si-N	=4x BFW 93		Phi	-	data
BFT 27	Si-N	SMD, NF, 60V, 0,5A, >30MHz	35d(2mm)	Fer	(BCW 65R...66R, BCX 41R)[6]	data
BFT 28	Si-P	S/Vid, 150V, 1A, 1W, >25MHz	2a§	Rca	BFT 19(A,B), 2N5415...5416, MJ 5415...5416	data
BFT 28 A	Si-P	=BFT 28: 200V	2a§		BFT 19(A,B), 2N5415...5416, MJ 5415...5416	data
BFT 28 B	Si-P	=BFT 28: 250V	2a§		BFT 19A...B, 2N5416, MJ 5415...5416	data
BFT 28 C	Si-P	=BFT 28: 300V	2a§		BFT 19A...B, 2N5416, MJ 5415...5416	data
BFT 29	Si-N	NF/S, 90V, 1A, 0,36W, >100MHz,	2a§	Tix	BC 639, BFR 39, BFR 50, BFT 53	data
BFT 30	Si-N	=BFT 29: 70V	2a§	Tix	BC 639, BFR 50, BSS 26, BSS 40...41	data
BFT 31	Si-N	=BFT 29: 60V	2a§	Tix	BC 639, BFR 51, BSS 26, BSS 40...41, ++	data
BFT 32	Si-N	NF-Tr/E, 80V, 5A, 1W, >100MHz	2a§	Tix	BSS 45, BUX 34, BUY 80, 2N5336...5339	data
BFT 33	Si-N	=BFT 32: 100V	2a§	Tix	BUX 34, BUY 80, 2N5338...5339	data
BFT 34	Si-N	=BFT 32: 120V	2a§	Tix	BUX 34, BUY 80, 2N4895...4897	data
BFT 35	Si-P	NF-Tr/E, 80V, 5A, 1W, >100MHz	2a§	Tix	BSS 46, BUY 90, 2N6190...6191	data
BFT 36	Si-P	=BFT 35: 100V	2a§	Tix	BUY 90, 2N6192...6193	data
BFT 37	Si-P	=BFT 35: 120V	2a§	Tix	BUY 90	data
BFT 38	Si-P	UHF, 12V, 35mA, 4800MHz	5g§	Sie	BFQ 24	data
BFT 39	Si-N	NF/S, 90V, 1A, 0,8W, >100MHz	2a§	Tix	BC 141, BSW 39, BSS 42...43, BSX 46...47++	data
BFT 40	Si-N	=BFT 39: 70V	2a§	Tix	BC 140...141, BSW 39, BSX 45...47, ++	data
BFT 41	Si-N	=BFT 39: 60V	2a§	Tix	BC 140...141, BSW 39, BSX 45...47, ++	data
BFT 42	Si-N	NF-Tr, 125/110V, 1A, 0,8W, >50MHz	2a§	Tix	BSS 42...43, BSW 67...68, BSX 47, ++	data
BFT 43	Si-N	NF-Tr, 125/100V, 1A, 0,8W, >50MHz	2a§	Tix	BSS 42...43, BSW 67...68, BSX 47, ++	data
BFT 44	Si-P	S/Vid, 300V, 0,5A, 0,75W, 70MHz	2a§	Phi	BFT 19A...B, MJ 5415...5416, 2N5416, ++	data
BFT 45	Si-P	=BFT 44: 250V	2a§	Phi	BFT 19A...B, MJ 5415...5416, 2N5416, ++	data
BFT 46	N-FET	SMD, NF/HF, 25V, Idss>0,2mA, Up<1,2V	35b	Phi	-	data
BFT 47	Si-N	Vid, 160V, 0,2A, 0,8W, 110MHz	2a§	Tho	BF 257...259, BFR 57...59, 2N5058...5059	data
BFT 48	Si-N	=BFT 47: 250V	2a§	Tho	BF 258...259, BFR 58...59, 2N5058...5059	data
BFT 49	Si-N	=BFT 47: 300V	2a§	Tho	BF 259, BFR 59, BFS 89, 2N5058	data
BFT 50	Si-N	UHF-ra, 22V, 0,25W, 3,5GHz	5g	Tho	BFR 15, BFS 55, BFT 66...67, BFW 99, ++	data
BFT 51	Si-N	UHF-A/Tr, 35V, 0,4A, >3GHz	2a§	Phi	-	data
BFT 53	Si-N	NF/S, 80V, 1A, 0,36W, >50MHz, <55/-ns	2a§	Tix	BC 639, BFR 39, BFR 50, BFT 29	data
BFT 54	Si-N	=BFT 53: 60V	2a§	Tix	BC 637, BFT 30, BSS 26, BSS 40...41, ++	data

Type	Device	Short Description	Fig.	Manu	Comparision Types	More at
BFT 55	Si-N	=BFT 53: 40V	2a§	Tix	BC 637, BFT 30, BSS 26, BSS 40..41, ++	data
BFT 57	Si-N	Vid, 160V, 0,2A, 0,36W, 110MHz	2a§	Tix	BF 391...393, BFR 87...89, BF 422A, ++	data
BFT 58	Si-N	=BFT 57: 250V	2a§	Tix	BF 392...393, BFR 88...89, BF 422A, ++	data
BFT 59	Si-N	=BFT 57: 300V	2a§	Tix	BF 393, BFR 89, BF 420A	data
BFT 60	Si-P	NF/S, 80V, 1A, 0,8W, >60MHz, <55/-ns	2a§	Tix	BCX 60, BSW 40, 2N4031, 2N4033, ++	data
BFT 61	Si-P	=BFT 60: 60V	2a§	Tix	BC 161, BSW 40, 2N4030...4033, ++	data
BFT 62	Si-P	=BFT 60: 40V	2a§	Tix	BC 160...161, BSW 40, 2N4030...4033, ++	data
BFT 65	Si-N	UHF-A, ra, 20V, 50mA, 0,25W, 5GHz	24f	Sie	BFP 96, BFQ 72...73, BFR 96	data
BFT 66	Si-N	UHF-A, ra, 20V, 30mA, 0,2W, 4GHz	5k	Sie	BFQ 22, BFQ 63	data
BFT 66 S(E)	Si-N	=BFT 66: 25V	5g	Sgs	BFQ 22, BFQ 63	data
BFT 67	Si-N	UHF-A-ra, 20V, 30mA, 0,2W, 4GHz	5k	Sie	BFQ 22, BFQ 63	data
BFT 69	Si-P	NF/S, 90V, 1A, 0,36W, >100MHz	2a§	Tix	BC 640, BFR 60, BFR 79, BFT 20	data
BFT 70	Si-P	=BFT 69: 70V	2a§	Tix	BC 640, BFT 20, 2N4027, 2N4029, ++	data
BFT 71	Si-P	=BFT 69: 60V	2a§	Tix	BC 638, BFT 21, 2N4026...4029, ++	data
BFT 72	Si-N	Vid-L, 160V, 0,1A, 60MHz	14h§	Tho	BF 457...459, BF 415, BF 417	data
BFT 73	Si-N	=BFT 72: 250V	14h§	Tho	BF 458...459, BF 415, BF 417	data
BFT 74	Si-N	=BFT 72: 300V	14h§	Tho	BF 459, BF 417, 2SC3417	data
BFT 75	Si-N	SMD, UHF-A, ra, 20V, 50mA, 4,6GHz	35a	Sie	BFR 93, 2SC3161, 2SC3513	data
BFT 75 R	Si-N	=BFT 75:	35d		BFR 93R	data
BFT 79	Si-P	NF/S, 90V, 1A, 0,8W, >100MHz	2a§	Tix	BCX 60, BSW 40, 2N4036	data
BFT 80	Si-P	=BFT 79: 70V	2a§	Tix	BCX 60, BSW 40, 2N4031, 2N4033, 2N4036	data
BFT 81	Si-P	=BFT 79: 60V	2a§	Tix	BC 161, BSW 40, 2N4030...33, 2N4036...37++	data
BFT 82	Si-N	NF/S, 90V, 2A, 0,8W, >100MHz	7c	Tix	(MPS 651, 2SC3328)	data
BFT 83	Si-N	=BFT 82: 70V	7c	Tix	MPS 651, 2SC3328	data
BFT 84	Si-N	=BFT 82: 60V	7c	Tix	MPS 650...651, 2SC3328, 2SD1207	data
BFT 85	Si-P	NF/S, 90V, 2A, 0,8W, >100MHz	7c	Tix	(MPS 751, 2SA1315)	data
BFT 86	Si-P	=BFT 85: 70V	7c	Tix	MPS 751, 2SA1315	data
BFT 87	Si-P	=BFT 85: 60V	7c	Tix	MPS 750...751, 2SA1315	data
BFT 91	Si-N	UHF-Tr/E, 60V, 350mA, PQ>1,5W(470MHz)	55r	Aeg	BLW 89, BLW 92, BLY 76	data
BFT 92	Si-P	SMD, UHF-A, 20V, 25mA, 5GHz	35a	Phi,Sie	-	data
BFT 92 R	Si-P	=BFT 92:	35d		-	data
BFT 92 W	Si-P	=BFT 92:	35a(2mm)			data
BFT 93	Si-P	SMD, UHF-A, 15V, 35mA, 5GHz	35a	Phi,Sie	2SA1963	data
BFT 93 R	Si-P	=BFT 93:	35d		-	data
BFT 93 W	Si-P	=BFT 93:	35a(2mm)		-	data
BFT 95(A,B,H)	Si-P	UHF-A, 15V, 25mA, 3,6...5GHz	24f	Aeg,Sgs	BFQ 23, BFQ 51, BFQ 56, BFQ 75...76	data
BFT 96(A)	Si-P	UHF-A, 20V, 75mA, 3,6-4,5GHz	24f	Aeg,Sgs	BFQ 32S	data
BFT 97	Si-N	UHF-V, ra, 20V, 30mA, 0,2W, 5GHz	24f	Sie	BFQ 59...60, BFQ 70...71, BFR 14, BFR 91++	data
BFT 98	Si-N	UHF-A, 30V, 200mA, 3,3GHz	55r	Sie	BFQ 68	data
BFT 98 B	Si-N	=BFT 98:	51r			data
BFT 98 T	Si-N	=BFT 98: 150mA, 3,2GHz	24f		BFG 34	data
BFT 99	Si-N	UHF-A, 30V, 350mA, 3,3GHz	55r	Sie	(BFQ 68)	data
BFT 99 A	Si-N	=BFT 99:	51r	Sie		data
BFU 308	N-FET	Sym, VHF/UHF, 25V, Idss>12mA, Up<6,5V	2b	Phi	BF 256C, 2N5397, 2N5486	data
BFU 309	N-FET	=BFU 308: Idss>12mA, Up<4V	2b	Phi	-	data
BFU 310	N-FET	=BFU 308: Idss>24mA, Up<6,5V	2b	Phi	-	data

BFV

Type	Device	Short Description	Fig.	Manu	Comparision Types	More at
BFV 10	Si-N	SMD, HF/S, 50V, 0,8A, >200MHz	Chip	Tix	BFV 50	data
BFV 11	Si-N	SMD, HF/S, 50V, 0,8A, >200MHz	Chip	Tix	BFV 50	data
BFV 12	Si-N	SMD, HF/S, 60V, 0,8A, >250MHz	Chip	Tix	BFV 51	data
BFV 13	Si-P	SMD, Dual, 60V, 0,05A	Chip	Tix	BFV 15	data
BFV 14	Si-N	SMD, NF-Tr, 60V, 1A, >50MHz	Chip	Tix	BFV 16	data
BFV 15	Si-P	SMD, Dual, 60V, 0,05A	Chip	Tix	BFV 13	data
BFV 16	Si-N	SMD, NF-Tr, 100V, 1A	Chip	Tix	-	data
BFV 17	Si-N	SMD, NF, 80V, >60MHz	Chip	Tix	-	data
BFV 18	Si-N	SMD, NF, 80V, >150MHz	Chip	Tix	-	data
BFV 19	Si-N	SMD, NF/HF, 60V, 0,03A	Chip	Tix	BFV 62	data
BFV 20	Si-N	SMD, NF/S, 40V, 0,6A, >150MHz, B>40	Chip	Tix	-	data
BFV 21	Si-P	=BFV 20: B>100	Chip	Tix	-	data
BFV 22	Si-P	=BFV 20: 50V, B>100	Chip	Tix	-	data
BFV 23	Si-P	S, 12V, 0,2A, 0,36W, >400MHz, <60/-ns	2a§	Tix	BSV 21, BSW 25, BSW 37, BSX 29, ++	data
BFV 24	Si-P	S, 12V, 0,2A, 0,36W, >400MHz, <60/-ns	2a§	Tix	BSV 21, BSW 25, BSW 37, BSX 29, ++	data
BFV 25	Si-P	SMD, NF-ra, 60V, 0,03A, >30MHz, B>30	Chip	Tix	-	data
BFV 26	Si-P	=BFV 25: B>100	Chip	Tix	-	data
BFV 27	Si-N	SMD, SS, 15V, 0,05A, >500MHz, <15/-ns	Chip	Tix	BFV 39, BFV 42, BFV 47...48	data
BFV 28	Si-N	SMD, SS, 15V, 0,05A, >500MHz, <12/-ns	Chip	Tix	BFV 39, BFV 42, BFV 47...48	data
BFV 29	Si-P	SMD, S, 20V, 0,2A, >400MHz, <35/-ns	Chip	Tix	-	data
BFV 30	Si-P	SMD, S, 20V, 0,1A, >140MHz, <60/-ns	Chip	Tix	-	data
BFV 31	Si-P	SMD, S, 12V, 0,2A, >350MHz, <60/-ns	Chip	Tix	BFV 29	data
BFV 32	Si-P	SMD, S, 12V, 0,2A, >350MHz, <60/-ns	Chip	Tix	BFV 29	data
BFV 33	Si-P	SMD, NF/HF, 25V, 0,05A, >140MHz	Chip	Tix	BFV 20..22	data
BFV 34	Si-P	SMD, Chopper, sym, 15V, 0,1A, B>80	Chip	Tix	-	data
BFV 35	Si-P	=BFV 34: 25V, B>40	Chip	Tix	-	data
BFV 36	Si-P	=BFV 34: 40V, B>30	Chip	Tix	-	data
BFV 37	Si-N	SMD, Chopper, 30V, 0,1A	Chip	Tix	-	data

Type	Device	Short Description	Fig.	Manu	Comparision Types	More at
BFV 38	Si-N	=BFV 37: 45V	Chip	Tix	-	data
BFV 39	Si-N	SMD, SS, 40V, 0,2A, >500MHz, <12/-ns	Chip	Tix	BFV 42...44, BFV 48	data
BFV 40	Si-N	SMD, S, 25V, 0,2A, >200MHz, <40/-ns	Chip	Tix	BFV 49	data
BFV 41	Si-N	SMD, S, 20V, 0,2A, >250MHz, <45/-ns	Chip	Tix	BFV 39, BFV 42...44, BFV 47...48	data
BFV 42	Si-N	SMD, SS, 35V, 0,2A, >400MHz, <12/-ns	Chip	Tix	BFV 39, BFV 43...44, BFV 47...48	data
BFV 43	Si-N	SMD, SS, 30V, 0,2A, >300MHz, <20/-ns	Chip	Tix	BFV 39, BFV 42, BFV 47...48	data
BFV 44	Si-N	SMD, SS, 30V, 0,2A, >300MHz, <20/-ns	Chip	Tix	BFV 39, BFV 42, BFV 47...48	data
BFV 45	Si-N	SMD, S, 35V, >250MHz	Chip	Tix	-	data
BFV 46	Si-N	SMD, S, 35V, >300MHz	Chip	Tix	-	data
BFV 47	Si-N	SMD, SS, 30V, 0,2A, >400MHz, <15/-ns	Chip	Tix	BFV 39, BFV 42...44	data
BFV 48	Si-N	SMD, SS, 30V, 0,2A, >400MHz, <12/-ns	Chip	Tix	BFV 39, BFV 42...44	data
BFV 49	Si-N	SMD, S, 25V, 0,2A, >200MHz, <40/-ns	Chip	Tix	BFV 40	data
BFV 50	Si-N	SMD, S, 50V, 0,8A, >175MHz, <65/-ns	Chip	Tix	BFV 51...55	data
BFV 51	Si-N	SMD, S, 60V, 0,8A, <40/-ns	Chip	Tix	BFV 53...55	data
BFV 52	Si-N	SMD, S, 50V, 1A, >175MHz, <45/-ns	Chip	Tix	BFV 50...51, BFV 53...55	data
BFV 53	Si-N	SMD, S, 60V, 0,8A, <40/-ns	Chip	Tix	BFV 51, BFV 54, BFV 55	data
BFV 54	Si-N	SMD, S, 60V, >250MHz, <40/-ns	Chip	Tix	BFV 51, BFV 53, BFV 55	data
BFV 55	Si-N	SMD, S, 75V, 0,5A, >175MHz, <40/-ns	Chip	Tix	-	data
BFV 56	Si-N	HF/S, 60V, 1A, 0,5W, >200MHz, <40/-ns	2a§	Tix	BSS 26, BSS 40...41, 2N4014	data
BFV 56 A	Si-N	=BFV 56: 75V, >175MHz	2a§		2N4014	data
BFV 57	Si-N	HF/S, 50V, 0,5A, 0,8W, >300MHz,<35/-ns	2a§	Tix	BSX 48...49, BSV 59, 2N4013...4014, ++	data
BFV 57 A	Si-N	=BFV 57: 80V	2a§		2N4013...4014	data
BFV 58	Si-N	SMD, VHF, 60V, 500mA, >150MHz	Chip	Tix	BFV 12	data
BFV 59	Si-N	SMD, VHF/UHF, 25V, 50mA, >600MHz	Chip	Tix	-	data
BFV 60	Si-N	SMD, NF-ra, 30V, 0,03A, >30MHz, B>20	Chip	Tix	-	data
BFV 61	Si-N	=BFV 60: B>80	Chip	Tix	-	data
BFV 62	Si-N	SMD, NF-ra, 60V, 0,05A, >45MHz	Chip	Tix	-	data
BFV 63	Si-N	HF/S, 60V, 0,8A, 0,5W, 25/200ns	2a§	Tix	BSS 26, BSS 40...41, 2N2221...2222	data
BFV 63 A	Si-N	=BFV 63: 75V	2a§		BSW 63...64, 2N2221A..2222A	data
BFV 63 B	Si-N	=BFV 63: 25/175ns	2a§		BSS 26, BSS 40...41, 2N2221...2222	data
BFV 64(A,B)	Si-P	HF/S, 60V, 0,6A, 0,4W, 26/70ns	2a§	Tix	BSW 24, 2N2906...2907, 2N4026...4029	data
BFV 65	Si-N	HF/S, 40V, 0,3A, 0,36W, <40/40ns	2a§	Tix	BSS 10, BSX 26, BSX 39, 2N3261	data
BFV 65 A,B	Si-N	=BFV 65: <12/15ns	2a§		BSS 10, BSX 26, BSX 39, 2N3261	data
BFV 66(A)	Si-N	HF/S, 60V, 0,8A, 0,5W, >250MHz,<40/-ns	2a§	Tix	BSS 26, BSS 40...41, 2N2221...2222	data
BFV 67	Si-N	SS, 15V, 0,05A, 0,3W, >600MHz, <15/-ns	2a§	Tix	BSV 89...92, BSX 27, BSX 44, 2N2475, ++	data
BFV 68	Si-N	NF-ra, 45V, 0,05A, 0,36W, >30MHz	2a§	Tix	BC 382...384, BC 413...414, BC 550	data
BFV 68 A	Si-N	=BFV 68: 60V	2a§		2N2483...2484, 2N3117, 2SC2390, ++	data
BFV 69	Si-N	VHF/UHF, 30V, 50mA, 0,2W, >600MHz	2a§	Tix	BF 180...183, BF 689, BF 763, 2N2857, ++	data
BFV 69 A	Si-N	=BFV 69: 25V	2a§		BF 180...183, BF 689, BF 763, 2N2857, ++	data
BFV 70	Si-N/P	SMD, 2x NPN + 2x PNP, 60V, >200MHz	14-FLP	Tix	-	data
BFV 70 N	Si-N/P	=BFV 70:	14-DIP		-	data
BFV 71	Si-N/P	SMD, 2x NPN + 2x PNP, 40V, >200MHz	14-FLP	Tix	-	data
BFV 71 N	Si-N/P	=BFV 71:	14-DIP		-	data
BFV 72	Si-N	SMD, 3x NPN, 40V, 0,5A, <15/-ns	14-FLP	Tix	-	data
BFV 72 N	Si-N	=BFV 72:	14-DIP		-	data
BFV 73	Si-N/P	2x NPN + 2x PNP, 50V, 0,8A, >250MHz	14-FLP	Tix	-	data
BFV 73 N	Si-N/P	=BFV 73:	14-DIP		-	data
BFV 75	Si-N	SMD, 4x NPN, Chopper, 30V, 0,1A	10-FLP	Tix	-	data
BFV 75 N	Si-N	=BFV 75:	10-DIP		-	data
BFV 76	Si-N	SMD,4x NPN, Chopper, 30V, 0,1A, >20MHz	10-FLP	Tix	-	data
BFV 76 N	Si-N	=BFV 76:	10-DIP		-	data
BFV 77	Si-P	SMD, 4x PNP, 40V, 0,2A, >350MHz	10-FLP	Tix	-	data
BFV 77 N	Si-P	=BFV 77:	10-DIP		-	data
BFV 78	Si-N	SMD, 4x NPN, 40V, 0,2A, >350MHz	10-FLP	Tix	-	data
BFV 78 N	Si-N	=BFV 78:	10-DIP		-	data
BFV 79	Si-N	SMD, 4x NPN, 40V, 0,2A, >350MHz	14-FLP	Tix	-	data
BFV 79 N	Si-N	=BFV 79:	14-DIP		-	data
BFV 80	Si-N	SMD, VHF/UHF, 25V, 50mA, >500MHz	Chip	Tix	BFV 59	data
BFV 81	Si-P	HF/S, 12V, 0,2A, 0,3W, >400MHz,<60/-ns	24b	Mot,Tix	(BSV 21, BSW 25, BSW 37, BSX 29, ++)[6]	data
BFV 81 A,B	Si-P	=BFV 81: 20V	24b		(2N3209, 2N3905...3906, 2N4125...4126)[6]	data
BFV 82(A...C)	Si-P	NF/S, 25V, 50...100mA, 0,3W, >140MHz	24b	Mot,Tix	(2N3905...3906, 2N4125...4126)[6]	data
BFV 83(A...C)	Si-N	S, 40V, 0,2A, 0,3W, >300MHz, <40/-ns	24b	Mot,Tix	(BSW 41, BSY 63, 2N708, 2N4123)[6]	data
BFV 84(A,B)	Si-N	VHF/UHF, 25...30V, 50mA, >600MHz	24b	Mot,Tix	BF 362...363, (BF 377...378, BF 763)[6]	data
BFV 85(B)	Si-N	HF/S, 60V, 0,8A, 0,36W, >250MHz	24b	Mot,Tix	(BSS 26, BSS 40...41, 2N2221...2222)[6]	data
BFV 85 A,C	Si-N	=BFV 85(B): 75V	24b		(BSW 63...64, 2N2221A...2222A)[6]	data
BFV 85 D,E	Si-N	NF-ra, 45V, 0,03A, 0,3W, >30MHz	24b		(BC 382...384, BC 413...414, BC 550)[6]	data
BFV 85 F,G	Si-N	NF-ra, 60V, 0,05A, 0,3W, >30MHz	24b		(2N2483...2484, 2N3117, 2SC2390, ++)[6]	data
BFV 86(A...C)	Si-P	NF/S, 60V, 0,6A, 0,36W, <45/-ns	24b	Mot,Tix	(BSW 24, 2N2906...2907, 2N4026...4029)[6]	data
BFV 87(A,B)	Si-N	SS, 40V, 0,2A, 0,3W, <12/-ns	24b	Mot,Tix	(BSS 11, BSX 19...20, 2N2369)[6]	data
BFV 88(A...D)	Si-N	HF/S, 60V, 0,8A, 0,36W, >250MHz	24b	Mot,Tix	(BSS 26, BSS 40...41, 2N2221...2222)[6]	data
BFV 89	Si-N	Chopper, 30V, 0,1A, 0,3W, >20MHz	24b	Mot,Tix	-	data
BFV 89 A	Si-N	=BFV 89: 45V	24b		-	data
BFV 90(A,B)	Si-N	HF-Tr, 70V, 0,8A, 0,8W, >300MHz	2a§	Tix	BFX 17, BFX 96...97	data
BFV 91	Si-P	4x PNP, 12V, 0,2A, >400MHz, <60/-ns	14-FLP	Tix	-	data
BFV 91 N	Si-P	=BFV 91:	14-DIP		-	data
BFV 92	Si-N	4x NPN, 40V, 0,2A, >350MHz, <20/-ns	14-FLP	Tix	-	data
BFV 92 N	Si-N	=BFV 92:	14-DIP		-	data
BFV 93	Si-N	SMD, 4x NPN, 50V, 0,8A, >250MHz	10-FLP	Tix	-	data

Type	Device	Short Description	Fig.	Manu	Comparision Types	More at
BFV 93 A	Si-P	=BFV 93: PNP	10-FLP		-	data
BFV 93 AN	Si-P	=BFV 93N: PNP	14-DIP		-	data
BFV 93 N	Si-N	=BFV 93:	14-DIP			data
BFV 94	Si-N	SMD, 4x NPN, 50V, 0,8A, >250MHz	14-FLP	Tix	-	data
BFV 94 N	Si-N	=BFV 94:	14-DIP		-	data
BFV 95	Si-P	SMD, 4x PNP, 50V, 0,8A, >250MHz	14-FLP	Tix	-	data
BFV 95 N	Si-P	=BFV 95:	14-DIP		-	data
BFV 96	Si-N	SMD, 4x NPN, 40V, 0,8A, >175MHz	14-FLP	Tix	-	data
BFV 96 N	Si-N	=BFV 96:	14-DIP		-	data
BFV 97	Si-N	SMD, 4x NPN, 30V, 0,05A, >600MHz	14-FLP	Tix	-	data
BFV 97 N	Si-N	=BFV 97:	14-DIP		-	data
BFV 98	Si-N	SMD, 4x NPN, 45V, 0,03A, >30MHz	14-FLP	Tix	-	data
BFV 98 N	Si-N	=BFV 98:	14-DIP		-	data
BFV 99	Si-N	NF/S, 75V, 1A, 0,5W	2a§	Tix	BC 639, BSW 63...64, 2N2221A...2222A, ++	data
BFV 420	Si-N	Moni, Vid, 140V, 0,1A, 0,83W, >150MHz	7c	Phi	BF 420, BF 422, BFR 87...89, 2SC3467, ++	data
BFV 421	Si-P	Moni, Vid, 140V, 0,1A, 0,83W, >150MHz	7c	Phi	BF 421, BF 423, BF 435...37, 2SA1370, ++	data
BFV 469	Si-N	Moni, Buffer, 140V, 0,1A, 2W, >150MHz	14h§	Phi	BF 457...459, 2SD1609, 2SD2491, ++	data
BFVP 20	Si-N	SMD, 40V, 0,2A(ss), 150 MHz	Chip	Ucp	-	data

BFW

Type	Device	Short Description	Fig.	Manu	Comparision Types	More at
BFW	Z-Di	=SM 15T 100C(Typ-Code/Stempel/marking)	71(8x5mm)	Tho	→SM 15T...	data
BFW 10	N-FET	VHF-A, sym, 30V, Idss>8mA, Up<8V	5h+	Mot,Phi,Tix	BFS 72, 2N3823	data
BFW 11	N-FET	VHF-A, sym, 30V, Idss>4mA, Up<6V	5h+	Mot,Phi,Tix	BFS 71, 2N3822	data
BFW 12	N-FET	NF/HF-ra, 30V, Idss>1mA, Up<2,5V	5h+	Mot,Phi	(2N4340)	data
BFW 13	N-FET	NF/HF-ra, 30V, Idss>0,2mA, U<1,2V	5h+	Mot,Phi	(2N4338)	data
BFW 16(A)	Si-N	VHF/UHF-A/Tr, 40V, 150mA, 1200MHz	2a§	Phi,Sie,++	BFR 36, BLW 11	data
BFW 17(A)	Si-N	VHF-A/Tr, 40V, 150mA, 1100MHz	2a§	Phi,Sie,++	BFR 36, BLW 11	data
BFW 19	Si-N	VHF-Tr, 40V, 0,3A, >500MHz	2a§	Sgs	BFS 50, BFX 33, BFX 55, 2N3137T	data
BFW 20	Si-P	NF-ra, 60V, 0,2A, 0,36W, B>100	2a§	Mot,Sgs,Tix	2N3962, 2SA1137	data
BFW 21	Si-P	=BFW 20: 80V	2a§	Mot,Sgs,Tix	2N3963, 2SA970, 2SA1049, 2SA1136	data
BFW 22	Si-P	=BFW 20: 45V, B>250	2a§	Mot,Sgs,Tix	BC 214, BC 415...416, BC 560, 2N3964, ++	data
BFW 23	Si-P	=BFW 20: B>250	2a§	Mot,Sgs,Tix	2N3965, 2SA1137	data
BFW 24	Si-N	NF/S, 100V, 1A, 0,8W, >60MHz	2a§	Mot,Sgs	BC 141, BSX 46...47, 2N3107...3108, ++	data
BFW 25	Si-N	=BFW 24: 80V, >70MHz	2a§	Mot,Sgs	BC 140...141, BSX 45...47, 2N3107...3110,++	data
BFW 26	Si-N	=BFW 24: 80V	2a§	Mot,Sgs	BC 140...141, BSX 45...47, 2N3107...3110,++	data
BFW 27	MOS-P-FET-e	S, 30V, Up<6V	5j+	Aeg	-	data
BFW 29	Si-N	NF/S, 50V, 0,4A, 0,6W, >40MHz	2a§	Tho,Tix	BC 140...141, BC 302, 2N1613, 2N1711, ++	data
BFW 30	Si-N	VHF/UHF-A, 20V, 50mA, 1,6GHz	5g	Phi,Sie,++	BFR 37, BFX 59, BFX 73	data
BFW 31	Si-P	HF/S, 50V, 0,7A, 0,5W, 200MHz	2a§	Mot,Tix	BC 327, BC 638, 2N2906...2907, ++	data
BFW 32	Si-N	HF/S, 50V, 0,7A, 0,5W, 200MHz	2a§	Mot,Tix	BC 337, BC 637, 2N2221...2222, ++	data
BFW 33	Si-N	NF/S, 120V, 1A, 0,8W, >50MHz	2a§	Mot,Sgs	BSS 42...43, BSV 84, BSX 47, 2N1893, ++	data
BFW 34	Si-N	NF/S, 50V, 0,2A, 0,6W, >70MHz, B>40	2a§	Tho	BC 140...141, BSX 45...47, 2N4432, ++	data
BFW 35	Si-N	=BFW 34: B>80	2a§	Tho	BC 140...141, BSX 45...47, 2N5203, ++	data
BFW 36	Si-N	Vid, 180V, 0,4A, 0,6W, 120MHz	2a§	Tho	2SD413, 2SD624, 2SD576, 2N5073	data
BFW 37	Si-N	Vid, 130V, 0,2A, 0,6W, 100MHz	2a§	Tho	BFR 57...59, 2N5058...5059, BFS 89	data
BFW 38	Si-N	Vid, 180V, 0,4A, 0,6W, >40MHz	2a§	Tho	2SD413, 2SD624, 2SD576	data
BFW 39(A)	Si-N	Dual, 50V, 0,03A, >60MHz	81bs	Sgs,Tix	2N2915(A)	data
BFW 40(A)	Si-N	Dual, 50V, 0,03A, >60MHz	81bs	Sgs,Tix	2N2916(A)	data
BFW 41	Si-N	VHF/UHF, 30V, 30mA, 0,2W, >600MHz	5g	Tho	BFS 62, BFW 97, BFX 60, BFX 73, BFX 89++	data
BFW 42	Si-N	VHF/UHF-Tr, 40V, 100mA, 0,6W, >600MHz	2a§	Tho	BFR 36, BFW 16...17, BLW 11	data
BFW 43	Si-P	Vid, 150V, 0,1A, 0,4W, 50MHz	2a§	Mot, Sgs	BF 398, BF 435...437, 2SA1370...1372, ++	data
BFW 44	Si-P	=BFW 43: 0,7W	2a§	Sgs	BFQ 35...37, BFT 44...45, 2SB606, ++	data
BFW 45	Si-N	Vid, 165V, 0,05A, 0,8W, 120MHz	2a§	Phi	BF 257...259, BF 657...659, 2N5058...5059	data
BFW 46	Si-N	VHF-Tr/E, 36V, 500mA, PQ>4W(175MHz)	2a§	Mot,Phi,Tix	BFS 22, MRF 237, 2N3924	data
BFW 47	Si-N	VHF-Tr/E, 65V, 350mA, PQ>2,5W(175MHz)	2a§	Mot,Phi,Tix	BFS 23, BLY 34, 2N3553	data
BFW 51(A)	Si-N	Dual-ra, 50V, >60MHz, B>150	81bq	Mot,Sgs	2N2974, 2N2976, 2N2978	data
BFW 52(A)	Si-N	=BFW 51: B>300	81bq	Mot,Sgs	2N2975, 2N2977, 2N2979	data
BFW 54	N-FET	50V, Idss>2mA, Up<6V	2b&	Tix	-	data
BFW 55	N-FET	50V, Idss>2mA, Up<6V	2b&	Tix	-	data
BFW 56	N-FET	50V, Idss>2mA, Up<6V	2b&	Tix	-	data
BFW 57	Si-N	NF/S, 80V, 0,5A, 0,3W, >80MHz, B>80	11a	Phi	BC 639, BSX 33, 2N2221A...2222A	data
BFW 58	Si-N	=BSW 57: B>50	11a	Phi	BC 939, BSX 33, 2N2221A...2222A	data
BFW 59	Si-N	=BFW 57: 40V	11a	Phi	BC 635, BSV 59, BSX 48...49, 2N2221..22++	data
BFW 60	Si-N	=BFW 57: 40V, B>50	11a	Phi	BC 635, BSV 59, BSX 48...49, 2N2221..22++	data
BFW 61	N-FET	NF/HF, 25V, Idss>2mA, Up<8V	5h+	Phi	BF 244...245, BFS 72, 2N3819, 2N3823, ++	data
BFW 63	Si-N	VHF-re, 600MHz	5k	Sgs	BF 225, BF 314, BF 496, BF 502...503, ++	data
BFW 64	Si-N	=BFW 63: 650MHz	5k	Sgs	BF 225, BF 314, BF 496, BF 502...503, ++	data
BFW 66	Si-N	HF/S, 60V, 1A, 0,8W, 400MHz	2a§	Sgs	BC 140...141, BSW 27...29, BSX 59...61, ++	data
BFW 67	Si-N	Vid, 300V, 0,4A, 0,8W, 60MHz	2a§	Sgs	BFQ 38...40, (BF 259, BFR 59, BFS 89,++)[7]	data
BFW 68	Si-N	HF/S, 50V, 0,1A 0,36W, 30/240ns	2a§	Sgs	BC 547, BSV 59, BSX 48...49, 2N3903...04++	data

Type	Device	Short Description	Fig.	Manu	Comparision Types	More at
BFW 69	Si-N	VHF-Tr/E, 65V, 1A, >400MHz	2a§	Sgs	BFS 23, BLY 34	data
BFW 70	Si-N	VHF/UHF/ZF, 900MHz	5k	Sgs	BF 180...183, BF 689, BF 763, 2N2857, ++	data
BFW 71	Si-N	HF/S, 60V, 1A, 0.5W, 400MHz	2a§	Sgs	BSS 26, BSS 40...41, 2N4013...4014	data
BFW 72	Si-P			Sgs		data
BFW 73(A)	Si-N	UHF-O/Tr, 30V, 250mA, >950MHz, B>20	2a§	Sgs	(BFR 36, BFW 16...17, BLW 11)[6]	data
BFW 74	Si-N	=BFW 73: B>40	2a§	Sgs	(BFR 36, BFW 16...17, BLW 11)[6]	data
BFW 75	Si-N	UHF, 30V, 250mA, >950MHz	Koax	Sgs	-	data
BFW 76(A)	Si-N	UHF-Tr, 30V, 80mA, 1,3...1,5GHz	2a§	Sgs	BFR 37, BFW 30, BFX 59, BFX 73	data
BFW 77(A)	Si-N	=BFW 76:	5g	Sgs	BFR 37, BFW 30, BFX 59, BFX 73	data
BFW 78	Si-N	Min, UHF, 30V, 80mA, 1,5GHz		Sgs	-	data
BFW 79	Si-N	UHF, 30V, 80mA, 1,5GHz	Koax	Sgs	-	data
BFW 80	Si-N	NF/HF, 50V, 0,2A, 0,6W, >70MHz	2a§	Tho	BC 140...141, BC 301...302, 2N1613, ++	data
BFW 87	Si-P	NF/S, 60V, 0,5A, 0,3W, >100MHz, B>80	11a	Phi	BC 327, BC 638, BSW 24, 2N2906...2907, ++	data
BFW 88	Si-P	=BFW 87: B>40	11a	Phi	BC 327, BC 638, BSW 24, 2N2906...2907, ++	data
BFW 89	Si-P	=BFW 87: 40V	11a	Phi	BC 328, BC 636, BSW 24, 2N2906...2907, ++	data
BFW 90	Si-P	=BFW 87: 40V, B>40	11a	Phi	BC 328, BC 636, BSW 24, 2N2906...2907, ++	data
BFW 91	Si-P	=BFW 87: 20V, B>40	11a	Phi	BC 328, BC 636, BSW 24, 2N2906...2907, ++	data
BFW 92	Si-N	UHF-A, 25V, 25mA, 0,3W, 2,4GHz	24f	EUR	BFR 34, BFT 24, BFT 97	data
BFW 92 A	Si-N	UHF-A, 25V, 25mA, 0,3W, 3,5GHz	24f	Aeg	BFR 34, BFQ 70, BFT 97	data
BFW 93	Si-N	UHF-A, 18V, 50mA, 0,19W, 1,8GHz	24f	Phi,Sie,Tho	BFQ 59...60, BFQ 70, BFR 34, BFT 97	data
BFW 94	Si-N	UHF-A, 20V, 200mA, 3GHz	51r	Sgs	BFG 34	data
BFW 96	MOS-N-FET-d	NF/HF, S, 30V, Idss=1mA, Up<6,5V	5h+	Phi	BFX 63, BSV 22, BSV 81, BSX 83	data
BFW 97(K,L,M)	Si-N	VHF/UHF, 30V, 50mA, 0,25W, >600MHz	40e	Fer	BF 357, BFR 37, BFW 30, BFX 73	data
BFW 98	Si-N	UHF-Tr/E, 36V, 400mA, 1GHz	55r	Phi	BLW 42, BLW 79, BLX 67, 2N5944...5945	data
BFW 98 G	Si-N	=BFW 98:	51r		BLW 12, BLX 36	data
BFW 99(S)	Si-N	UHF-A, 12V, 20mA, 0,2W, 3GHz	5k	Sie	BFR 15, BFS 55, BFT 66...67	data
BFWP 21	MOS-N-FET	30V, 15mA, Up=5V	5h+	Ucp		data

BFX

Type	Device	Short Description	Fig.	Manu	Comparision Types	More at
BFX	Z-Di	=SM 15T100CA(Typ-Code/Stempel/marking)	71(8x5mm)	Tho	→SM 15T...	data
BFX 10	Si-N	Dual, 60V, >250MHz	81bs	Sgs	2N3409...3411	data
BFX 11	Si-P	Dual-ra, 45V, 0,5A, >130MHz	81bs	Mot,Sgs,Tix	2N3726...3727, 2N4015...4016	data
BFX 12	Si-P	Uni, 20V, 0,1A, 0,3W, 210MHz, B>20	2a§	Phi,Tix	BC 213, BC 258, BC 308, BC 558, ++	data
BFX 13	Si-P	=BFX 12: B>50	2a§	Phi,Tix	BC 213, BC 258, BC 308, BC 558, ++	data
BFX 14	Si-N	VHF/UHF-O, 25V, 300mA, 0,8W, 530MHz	2a§	Sgs	BFS 22...23, BFS 50, BFX 55	data
BFX 15	Si-N	Dual, 80V, 0,2A, >50MHz	81bs	Mot,Sgs,Tix	2N2060, 2N2223, 2N2480	data
BFX 16	Si-N	3x NPN, 45V, 50mA, >60MHz	82cd	Sgs,Tix	-	data
BFX 17	Si-N	VHF-Tr/E, 60V, 1A, PQ=1,8W(150MHz)	2a§	Sgs,Tix	BFS 23, BFW 47, BLY 34, 2N3553	data
BFX 18	Si-N	HF/ZF, 30V, 550MHz	5g	Sgs	BF 167, BF 198, BFX 31, BFX 60, BFX 73++	data
BFX 19	Si-N	HF/ZF, 30V, 550MHz	5g	Sgs	BF 167, BF 198, BFX 31, BFX 60, BFX 73++	data
BFX 20	Si-N	HF/ZF, 30V, 550MHz	5g	Sgs	BF 167, BF 198, BFX 31, BFX 60, BFX 73++	data
BFX 21	Si-N	HF/ZF, 30V, 550MHz	5g	Sgs	BF 167, BF 198, BFX 31, BFX 60, BFX 73++	data
BFX 29	Si-P	HF/S, 60V, 0,6A, 0,6W, <60/150ns	2a§	Nsc,Phi,Tix	BC 161, BC 303, 2N2904...2905, ++	data
BFX 30	Si-P	S, 65V, 0,6A, 0,6W, <50/290ns	2a§	Phi,Tix	BC 161, BC 303, 2N2904...2905, ++	data
BFX 31	Si-N	VHF, ra, re, 500MHz	5g	Sgs	BF 225, BF 314, BFX 60, BFX 73, ++	data
BFX 32	Si-N	VHF/UHF, 850MHz	9f	Aeg	BF 377...378, BF689, BF 763, 2N2857, ++	data
BFX 33	Si-N	VHF-A/Tr, 55V, 400mA, 600MHz	2a§	Aeg	BFS 23, BFW 47, BFX 55, 2N3866	data
BFX 34	Si-N	NF/S-Tr, 120V, 2A,5A(ss),0,87W, 100MHz	2a§	Aeg,Phi,++	BFT 34, 2N4895...4896	data
BFX 35	Si-P	NF/S, 40V, 0,6A, 0,4W, >200MHz	2a§	Mot,Sgs,Tix	BC 327, BC 636, BSW 24, 2N2906...07, ++	data
BFX 36	Si-P	Dual-ra, 60V, 0,1A, 110MHz	81bs	Mot,Sgs,Tix	2N3806...3811, 2N4016...4016	data
BFX 37	Si-P	NF-ra, 90V, 0,1A, 0,36W, 70MHz	2a§	Phi,Sgs,++	2N3963, 2SA970, 2SA1049, 2SA1136, ++	data
BFX 38	Si-P	NF/S, 55V, 1A, 0,8W, 33/160ns, B>85	2a§	Sgs,Tix,++	BC 161, BSV 82, BSW 40, 2N4030...4033	data
BFX 39	Si-P	=BFX 38: B>40	2a§	Sgs,Tix,++	BC 161, BSV 82, BSW 40, 2N4030...4033	data
BFX 40	Si-P	=BFX 38: 75V	2a§	Sgs,Tix,++	BC 461, BSV 82, BSW 40, 2N4031, 2N4033	data
BFX 41	Si-P	=BFX 38: 75V, B>40	2a§	Sgs,Tix,++	BC 461, BSV 82, BSW 40, 2N4031, 2N4033	data
BFX 42	Si-N	UHF, RadH, 1400MHz	2a§	Sgs	-	data
BFX 43	Si-N	VHF/S, 30V, 0,2A, >500MHz	2a§	Phi,Tix	BSS 10...12, BSX 19...20, 2N2368, ++	data
BFX 44	Si-N	=BFX 43: 40V	2a§	Phi,Tix	BSS 10...11, BSX 19...20, 2N2369, ++	data
BFX 45	Si-N	NF/S, 30V, 0,1A, 0,125W, <200/400ns	12a	Phi	BC 548, BSW 41, BSX 87...88, BSY 63, ++	data
BFX 47	Si-N	UHF-A, 30V, 20mA, >1GHz	5g	Phi	BFR 37, BFW 30, BFX 73	data
BFX 48	Si-P	HF/S, 30V, 0,1A, 0,36W, 20/95ns	2a§	Mot,Sgs,Tix	BSW 24, 2N3905...3906, 2N4034...4035	data
BFX 49	Si-N	UHF-A/Tr, 65V, 250mA, 1,3GHz	55r	Phi	BFR 64...65, BFT 91, BLX 91	data
BFX 49 G	Si-N	=BFX 49:	51r			data
BFX 50	Si-N	NF/S, 80V, 1A, 0,6W, >60MHz	2a§	Phi,Tix	BC 639, BSS 59, BSX 23, 2N4014	data
BFX 51	Si-N	NF/S, 60V, 1A, 0,6W, >50MHz	2a§	Phi,Tix	BC 637, BSS 26, BSS 40...41, 2N4014	data
BFX 52	Si-N	NF/S, 40V, 1A, 0,6W, >60MHz	2a§	Phi,Tix	BC 635, BSS 23, BSX 23, 2N4013...14, ++	data

Type	Device	Short Description	Fig.	Manu	Comparision Types	More at
BFX 53	Si-N	UHF, 20V, 25mA, >1,3GHz	24c	Aeg	BF 362...363, BFR 34, BFT 24	data
BFX 55	Si-N	VHF-A/Tr, 60V, 400mA, 700MHz	2a§	Sie	BFS 23, BFW 47, BFX 33, 2N3553, 2N3866	data
BFX 56(I...III)	Si-N	VHF-A/Tr, 60V, 500mA, >350MHz	2a§	Sie	BFS 23, BFW 47, BFX 55, 2N3553, 2N3866	data
BFX 57(I...III)	Si-N	VHF-A, 30V, 100mA, >600MHz	5g	Sie	BFR 36, BFW 16...17	data
BFX 58(I...III)	Si-N	VHF-A/Tr, 60V, 400mA, >600MHz	5g	Sie	BFS 23, BFW 47, BFX 55, 2N3553, 2N3866	data
BFX 58 D	Si-N	=BFX 58(I...III):	2a§		BFS 23, BFW 47, BFX 55, 2N3553, 2N3866	data
BFX 59(F,R)	Si-N	VHF/UHF-A/Tr, 30V, 100mA, 1GHz	5g	Sie	(BFR 37, BFW 30, BFX 73)[7]	data
BFX 60	Si-N	VHF, 40V, 25mA, 550MHz	5k	Sie	BF 225, BF 314, BF 496, BF 502...503, ++	data
BFX 61	Si-N	NF/S-Tr, 80V, 1A, 180MHz	2a§	Tho	BC 141, BSS 42, BSX 46...47, 2N3107...08++	data
BFX 62	Si-N	UHF-V/M/O, 675MHz	5g	Sie,Tho	BF 180...183, BF 689, BF 763, 2N2857, ++	data
BFX 63	MOS-N-FET-d	=BFW 96:	5m+	Phi	→BFW 96	data
BFX 65	Si-P	NF-ra, 45V, 0,05A, 0,36W, >40MHz	2a§	Aeg,Phi,++	BC 214, BC 415...416, BC 560	data
BFX 66	Si-N-Darl	100V, 0,2A, 0,5W, B>1600	5r	Mot,Sgs	2N998	data
BFX 67	Si-N-Darl	60V, 0,5A, 0,5W, B>7000	5r	Mot,Sgs	2N999	data
BFX 68	Si-N	NF/S, 75V, 1A, 0,7W, >70MHz	2a§	Mot,Sgs,Tix	BC 141, BSX 45...47, BSY 53...54, 2N1711++	data
BFX 68 A	Si-N	=BFX 68: 80V, 0,8W	2a§		BC 141, BSX 45...47, 2N3109...3110, ++	data
BFX 69	Si-N	NF/S, 75V, 1A, 0,8W, >60MHz	2a§	Mot,Sgs,Tix	BC 141, BSX 45...47, BSY 53...54, 2N1613++	data
BFX 69 A	Si-N	=BFX 69: 80V	2a§		BC 141, BSX 45...47, 2N3109...3110, ++	data
BFX 70	Si-N	Dual, 100V, 0,5A, 100MHz	81bs	Mot,Sgs,Tix	2N2060, 2N2223	data
BFX 71	Si-N	Dual, 100V, 0,5A, 100MHz	81bs	Mot,Sgs,Tix	2N2060, 2N2223	data
BFX 72	Si-N	Dual, 100V, 0,5A, 100MHz	81bs	Mot,Sgs,Tix	2N2060, 2N2223	data
BFX 73	Si-N	VHF/UHF, 30V, 50mA, 900MHz	5g	Mot,Sgs,Tix	BF 357, BFR 37, BFW 30, BFX 59	data
BFX 74	Si-P	NF-Tr, 50V, 0,5A, 0,6W, 90MHz	2a§	Mot,Sgs,Tix	BC 161, BC 303...304, 2N2904..2905, ++	data
BFX 74 A	Si-P	=BFX 74: 60V, 0,8W, 150MHz	2a§		BC 161, BC 303...304, 2N2904..2905, ++	data
BFX 75	Si-N	SMD, HF, 30V, 30mA, >200MHz, B>70	Chip	Phi	-	data
BFX 76	Si-N	=BFX 75: B>33	Cnip	Phi	-	data
BFX 77	Si-N	HF, 50V, 300MHz	5k	Tho	BF 167, BF 198, BF 225, BF 310, ++	data
BFX 78	MOS-N-FET-d	FM/VHF-ra, 15V, Idss=16mA	5n+	Sgs		data
BFX 79	Si-N/P	1x NPN + 1x PNP, 80V, 0,6A, 100MHz	81bs	Sgs	2N4854...4855	data
BFX 80	Si-N/P	1x NPN + 1x PNP, 60V, 0,2A, >40MHz	81bs	Sgs	2N4854...4855	data
BFX 81	Si-N/P	1x NPN + 1x PNP, 25V, 0,2A, >350MHz	81bs	Sgs	2N4854...4855	data
BFX 82	P-FET	NF/HF-ra, 25V, Idss=17mA, Up=5V	2a&	Sgs	-	data
BFX 83	P-FET	NF/HF-ra, 25V, Idss=6,8mA, Up=9V	2a&	Sgs	-	data
BFX 84	Si-N	NF/S, 100V, 1A, 0,8W, >50MHz, B>30	2a§	Phi,Tix,++	BC 141, BSX 46...47, 2N3107...3108, ++	data
BFX 85	Si-N	=BFX 84: B>70	2a§	Phi,Tix,++	BC 141, BSX 46...47, 2N3107...3108, ++	data
BFX 86	Si-N	=BFX 84: 40V, B>70	2a§	Phi,Tix,++	BC 140...141, BSX 45...47, 2N3107...3110,++	data
BFX 87	Si-P	S, 50V, 0,6A, 0,53W, <60/150ns	2a§	Phi,Tix,++	BSW 40, 2N2904...2905, 2N3468, ++	data
BFX 88	Si-P	=BFX 87: 40V	2a§	Phi,Tix,++	BSW 40, 2N2904...2905, 2N3468, ++	data
BFX 89	Si-N	UHF-A, 30V, 25mA, 1,3GHz	5g	EUR	BFR 37, BFW 30, BFW 92...93, BFX 73	data
BFX 90	Si-P	NF-ra, 180V, 0,05A, 0,4W, 60MHz	2a§	Mot,Sgs	2N3930, (BF 421, BF 423, BF 436...437)	data
BFX 91	Si-P	=BFX 90: 0,7W	2a§	Mot,Sgs	2N3931, (BFQ 36)	data
BFX 92	Si-N	NF-ra, 50V, 0,03A, 0,3W, 45MHz, B>40	2a§	Mot,Sgs,Tix	BC 382, BC 414, BC 550, 2N929, 2N3117	data
BFX 92 A	Si-N	NF-ra, 60V, 0,05A, 0,36W, >60MHz	2a§		2N2483, 2N3117, 2SC2390	data
BFX 93	Si-N	=BFX 92: B>100	2a§	Mot,Sgs,Tix	BC,382, BC 414, BC 550, 2N930, 2N3117	data
BFX 93 A	Si-N	=BFX 92A: B>100	2a§		2N2484, 2N3117, 2SC2390	data
BFX 94(A)	Si-N	=2N2221(A)	2a§	Mot,Sgs,Tix	→2N2221(A)	data
BFX 95(A)	Si-N	=2N2222(A)	2a§	Mot,Sgs,Tix	→2N2222(A)	data
BFX 96(A)	Si-N	=2N2218(A)	2a§	Mot,Sgs,Tix	→2N2218(A)	data
BFX 97(A)	Si-N	=2N2219(A)	2a§	Mot,Sgs,Tix	→2N2219(A)	data
BFX 98	Si-N	Vid, 150V, 0,1A, 0,8W, 90MHz	2a§	Sgs,Tix	BF 257...259, BF 657...659, 2N5058...5059	data
BFX 99	Si-N	Dual, 100V, 0,5A, 100MHz	81bs	Mot,Sgs,Tix	2N2060, 2N2223	data
BFX 152	Si-N	Vid, hi-res,100V, 0,3A, 0,83W, >500MHz	7e	Aeg	-	data
BFX 154	Si-N	Vid,hi-res,100V,0,3A,2W(Tc=110°),>500M	14h§	Aeg	2SC3599	data
BFX 155	Si-P	Vid,hi-res,100V,0,3A,2W(Tc=110°),>500M	14h§	Aeg	2SA1405	data
BFX 156	Si-N	=BFX 154: 5W(Tc=25°)	13h§	Aeg	(2SC3599)[4]	data
BFX 157	Si-P	=BFX 155: 5W(Tc=25°)	13h§	Aeg	(2SA1405)[4]	data

BFY..BFZ

Type	Device	Short Description	Fig.	Manu	Comparision Types	More at
BFY 10	Si-N	HF/S, 45V, 0,05A, 0,3W, 120MHz, B>25	2a	Phi	BC 107, BC 167, BC 182, BC 237, BC 546++	data
BFY 11	Si-N	=BFY 10: B>40	2a	Phi	BC 107, BC 167, BC 182, BC 237, BC 546++	data
BFY 12(B,C,D)	Si-N	HF/S, 60V, 0,5A, >180MHz, <40/720ns	2a§	Sie	BSW 39, BSW 51...52, 2N2218...2219, ++	data
BFY 13(B,C,D)	Si-N	=BFY 12: 80V, 0,35A	2a§	Sie	BSW 39, BSW 53...54, 2N2218A...2219A, ++	data
BFY 14(B,C,D)	Si-N	=BFY 12: 100V, 0,25A	2a§	Sie	BSW 39, 2N1893, 2N3723	data
BFY 15	Si-N	HF/S, 40V, 0,5A, 0,6W, 100MHz	2a§	Itt,Sgs	BC 140...141, BC 300...302, 2N2218...19, ++	data
BFY 16	Si-N	=BFY 15: 150MHz	2a§	Itt	BC 140...141, BC 300...302, 2N2218...19, ++	data
BFY 17	Si-N	HF/S, 40V, 0,1A, 0,6W, 245MHz	2a§	Itt	BC 140...141, BC 300...302, 2N2218...19, ++	data
BFY 18	Si-N	=BFY 17: 0,3W	2a§	Itt	BC 107, BSW 41, BSY 63, 2N2221...22, ++	data
BFY 19	Si-N	HF/S, 30V, 0,1A, 0,3W, 400MHz	2a§	Itt	BC 108, BSW 41, BSY 63, 2N2221...22, ++	data
BFY 20	Si-N	Dual, 40V, 0,2A, 245MHz	81bs	Itt	2N3409...3411	data
BFY 21	Si-N	Dual, 40V, 0,2A, >200MHz	81bs	Itt	2N3409...3411	data

Type	Device	Short Description	Fig.	Manu	Comparision Types	More at
BFY 22	Si-N	Min, NF, 5V, 0,05A, 20MHz, ß>30	36c	Itt	BC 121...123, BC 146	data
BFY 23(a)	Si-N	=BFY 22: ß>70	36c	Itt	BC 121...123, BC 146	data
BFY 24	Si-N	=BFY 22: ß>45	36c	Itt	BC 121...123, BC 146	data
BFY 25	Si-N	S, 60V, 0,2A, 0,6W, >200MHz, 22/-ns	2a§	Itt	BSW 27...28, BSW 51...52, 2N2218...19, ++	data
BFY 26	Si-N	=BFY 25: 0,36W	2a§	Itt	BSV 59, BSX 49, 2N2221...2222, ++	data
BFY 27	Si-N	=2N915	2a§	Aeg	→2N915	data
BFY 28	Si-N	NF/HF, 60V, 0,1A, 0,3W, 400MHz	2a§	Itt	BC 182, BC 190, BC 546, 2N2221...22, ++	data
BFY 29	Si-N	Min, NF, 45V, 0,05A, 20MHz, ß>30	36c	Itt	BC 123	data
BFY 30	Si-N	=BFY 29: ß>70	36c	Itt	BC 123	data
BFY 31	Si-N	NF/S, 75V, 0,5A, >60MHz	2a§	Sie	BC 140...141, BC 300...301, 2N2218A...19A++	data
BFY 33	Si-N	NF/HF/S, 50V, 0,5A, 0,75W, 80MHz	2a§	Sie,Tsm	BC 140...141, BC 300...301, 2N2218...19, ++	data
BFY 34	Si-N	=BFY 33: 75V	2a§	Mot,Sie,++	BC 140...141, BC 300...301, 2N2218A...19A++	data
BFY 37(i)	Si-N	NF/S, 25V, 0,1A, 0,3W, 270MHz, i: Iso	2a(§)	Itt	BC 108, BC 168, BC 183, BC 238, BC 548++	data
BFY 39...(i)	Si-N	NF/S, 45V, 0,1A, 0,3W, 150MHz, i: Iso	2a(§)	Itt	BC 107, BC 167, BC 183, BC 237, BC 547++	data
BFY 40	Si-N	NF/S, 60V, 0,8A, 0,8W, 60MHz	2a§	Itt,Tix	BC 140...141, BC 300...302, 2N2218...19, ++	data
BFY 41	Si-N	NF/S, 120V, 0,6A, 0,8W, 60MHz	2a§	Itt,Tix	BC 300, BSX 47, BSY 55...56, 2N1893, ++	data
BFY 42	Si-N	≈BC 107			→BC 107	data
BFY 43	Si-N	Vid, 140V, 0,1A, 0,8W, 60MHz	2a§	Itt	BF 257...259, BF 657...659, 2N5058...5059	data
BFY 44	Si-N	VHF-A/Tr/E, 80V, 1A, PQ=2,1W(180MHz)	2a§	Phi,Tix	BFS 23, BFW 47, BLY 34, 2N3553	data
BFY 45	Si-N	Nix, 140V, 0,03A, 0,7W, 130MHz	2a§	Sie,Tix	BF 257...259, BF 657...659, 2N5058...5059	data
BFY 46	Si-N	=2N1711	2a§	Mot,Sie,++	→2N1711	data
BFY 47	Si-N	Min, NF, 5V, 0,05A, 0,075W, 50MHz	36c	Sie	BC 121...123, BC 146	data
BFY 48	Si-N	=BFY 47: 30V	36c	Sie	BC 122...123	data
BFY 49	Si-N	=BFY 47: 45V	36c	Sie	BC 123	data
BFY 50(E)	Si-N	HF/S, 80V, 1A, 0,8W, 100MHz	2a§	EUR	BC 140...141, BSW 39, 2N3444	data
BFY 51	Si-N	=BFY 50: 60V	2a§	EUR	BC 140...141, BSW 27...28, 2N3252...53, ++	data
BFY 52	Si-N	=BFY 50: 40V	2a§	EUR	BC 140...141, BSW 27...29, 2N3252...53, ++	data
BFY 53	Si-N	=BFY 50: 40V	2a§	Phi	BC 140...141, BSW 27...29, 2N3252...53, ++	data
BFY 55	Si-N	HF/S, 80V, 1A, 0,7W, >60MHz	2a§	Mot,Phi,Tix	BC 140...141, BSX 45...47, 2N3707...10, ++	data
BFY 56(A,B)	Si-N	HF/S, 60...80V, 1A, 0,8W, 150/350ns	2a§	Aeg,Tix,++	BC 140...141 BSW 39, 2N3444	data
BFY 57	Si-N	NF/Vid, 125V, 0,1A, 0,8W, >40MHz	2a§	Sgs,Tix,++	BF 257...259, BF 657...659, 2N5058...5059	data
BFY 63	Si-N	VHF-A/Tr, 30V, 100mA, 750MHz	2a§	Sgs,Tix	BFR 36, BFW 16...17	data
BFY 64	Si-P	NF/S, 40V, 0,6A, 0,7W, 35/70ns	2a§	Sgs,Tix,++	BSV 82, BSW 23, BSW 40, 2N3467...68, ++	data
BFY 65	Si-N	Nix, 100V, 0,1A, 0,6W, >50MHz	2a§	Aeg,Tix	BF 257...259, BF 657...659, 2N5058...5059	data
BFY 66	Si-N	=2N918	5g	Aeg,Mot	→2N918	data
BFY 67(A,C)	Si-N	=2N1613	2a§	Mot,Phi,Tix	→2N1613	data
BFY 68(A)	Si-N	=2N1711	2a§	Mot,Phi,Tix	→2N1711	data
BFY 69(A,B)	Si-N	Min, NF, 25V, 0,1A, 0,1W, >50MHz	36c	Aeg	BC 122...123	data
BFY 70	Si-N	VHF-Tr/E, 60V, 1A, PQ=1,5W(180MHz)	2a§	Phi,Tix	BFS 23, BFW 47, BLY 34, 2N3553	data
BFY 72	Si-N	HF/S, 50V, 0,5A, 0,8W, 14/80ns	2a§	Nsc,Sgs,Tix	BSS 13, BSS 27...29, 2N3722...3725, ++	data
BFY 73	Si-N	HF/S, 60V, 0,8A, 0,8W, <60/-ns	2a§	Sgs,Tix	BSS 13, BSS 27...29, 2N3722...3725, ++	data
BFY 74	Si-N	HF/S, 60V, 0,1A, 0,36W, 360MHz, B>40	2a§	Mot,Sgs,Tix	BC 190, BC 546, 2N2221...2222, ++	data
BFY 75	Si-N	=BFY 74: B>65	2a§	Mot,Sgs,Tix	BC 190, BC 546, 2N2221...2222, ++	data
BFY 76	Si-N	NF-ra, 45V, 0,05A, 0,36W, 100MHz, B>30	2a§	Sgs,Tix,++	BC 184, BC 413...414, BC 550, ++	data
BFY 77	Si-N	=BFY 76: B>80	2a§	Sgs,Tix	BC 184, BC 413...414, BC 550, ++	data
BFY 78	Si-N	VHF, 25V, 50mA, 0,3W, 900MHz	2a§	Mot,Sgs,Tix	BFR 37, BFW 30, BFX 59, BFX 73	data
BFY 79	Si-N	HF/ZF-re, 30V, >400MHz	5g	Sgs	BF 167, BF 198, BF 225, BF 310, ++	data
BFY 80	Si-N	Nix, 100V, 0,1A, 0,26W, >50MHz	2a§	Aeg,Tix	BF 297...299, BFR 86...89, BSS 38, BSX 21	data
BFY 81	Si-N	Dual, 45V, 0,05A, >60MHz	81bs	Mot,Sgs,Tix	2N2639...2644, 2N2913...2920	data
BFY 82	Si-N	Dual, 60V, 0,1A, >250MHz	81bs	Sgs,Tix	-	data
BFY 83	Si-N	Dual, 100V, 0,2A, >50MHz	81bs	Sgs,Tix	2N2060, 2N2223, 2N2652	data
BFY 84	Si-N	Dual, 30V, 0,2A, >600MHz	81bs	Mot,Sgs,Tix	2N3423...3424	data
BFY 85(A,B)	Si-N	Dual, 45V, 0,1A, >50MHz	81bs	Aeg,Tix	2N2639...2644, 2N2913...2920, 2N2453	data
BFY 86(A,B)	Si-N	Dual, 45V, 0,1A, >60MHz	81bs	Aeg,Tix	2N2639...2644, 2N2913...2920, 2N2453	data
BFY 87(A)	Si-N	Min, NF, 25V, 0,1A, 0,05W, >50MHz	36c	Aeg	BC 122...123	data
BFY 88	Si-N	UHF-V/M/O, 850MHz	5k	Aeg	BF 180...183, BF 689, BF 763, 2N2857, ++	data
BFY 90(B)	Si-N	VHF/UHF-A/Tr, 30V, 25mA, 1,1GHz	5g	EUR	BFR 37, BFW 30, BFX 59, BFX 73	data
BFY 91	Si-N	Dual, 45V, 0,03A, 60MHz	81bs	Mot,Tho,Tix	2N2639...2644, 2N2913...2920	data
BFY 92	Si-N	Dual, 45V, 0,03A, 60MHz	81bs	Mot,Tho,Tix	2N2639...2644, 2N2913...2920	data
BFY 94	Si-P	NF/S, 50V, 0,6A, 0,8W, >100MHz	2a§	Mot,Tix	BC 161, BC 303...304, 2N2904...2905, ++	data
BFY 95	Si-P	NF/S, 30V, 0,6A, 0,36W, >100MHz	2a§	Tix	BC 327...328, BC 636, 2N2906...2907, ++	data
BFY 99	Si-N	VHF-A/Tr, 65V, 1A, PQ>2,5W(260MHz)	2a§	Sie	BFS 23, BFW 47, BLY 34, 2N3553	data
BFY 180(ES,P,H,S)	Si-N	UHF, hi-rel, 15V, 4mA, 6,5GHz	51r	Sie	-	
BFY 181(ES,P,H,S)	Si-N	UHF, hi-rel, 20V, 20mA, 8GHz	51r	Sie		
BFY 182(ES,P,H,S)	Si-N	UHF, hi-rel, 20V, 35mA, 8GHz	51r	Sie		
BFY 183(ES,P,H,S)	Si-N	UHF, hi-rel, 20V, 65mA, 8GHz	51r	Sie		
BFY 193(ES,P,H,S)	Si-N	UHF, hi-rel, 20V, 80mA, 8GHz	51r	Sie	-	
BFY 196(P,H,S)	Si-N	UHF, hi-rel, 20V, 100mA, 6,5GHz	51r	Sie	-	

Type	Device	Short Description	Fig.	Manu	Comparision Types	More at
BFY 280(ES,P,H,S)	Si-N	UHF, hi-rel, 15V, 10mA, 7,2GHz	51r	Sie	-	
BFY 405(ES,P,H,S)	Si-N	UHF, hi-rel, 15V, 12mA, 22GHz	51r	Sie	-	
BFY 420(ES,P,H,S)	Si-N	UHF, hi-rel, 15V, 35mA, 22GHz	51r	Sie	-	
BFY 450(ES,P,H,S)	Si-N	UHF, hi-rel, 15V, 100mA, 22GHz	51r	Sie	-	
BFZ 10	Si-P	NF/S, 15V, 10mA, 0,05W, 3,5MHz	1a	Phi	BC 213, BC 258, BC 308, BC 558, ++	data

BG...BL

Type	Device	Short Description	Fig.	Manu	Comparision Types	More at
BG	Si-Di	=1SS269 (Typ-Code/Stempel/marking)	35	Tos	→1SS269	data
BG	Si-Di	=1SS313 (Typ-Code/Stempel/marking)	35(2mm)	Tos	→1SS313	data
BG	Si-P	=2SB1124 (Typ-Code/Stempel/marking)	39	Say	→2SB1124	data
BG	Si-N	=2SC3440-G (Typ-Code/Stempel/marking)	35	Mit	→2SC3440	data
BG	Si-N	=2SC3443-G (Typ-Code/Stempel/marking)	39	Mit	→2SC3443	data
BG	Si-N	=BCP 55-10 (Typ-Code/Stempel/marking)	48	Phi	→BCP 55-10	data
BG	Si-N	=BCX 55-10 (Typ-Code/Stempel/marking)	39	Sie,Val,Mot	→BCX 55-10	data
BG	Si-P	=BCX 71G (Typ-Code/Stempel/marking)	35	Fer,Phi,Sie	→BCX 71G	data
BG	Z-Di	=BZX 399-C8V2(Typ-Code/Stempel/marking)	71(1,7mm)	Phi	→BZX 399-...	data
BG	Z-Di	=P4 SMA-20A (Typ-Code/Stempel/marking)	71(5x2,5)	Fag	→P4 SMA-20A	data
BG	Z-Di	=SMAJ 13A (Typ-Code/Stempel/marking)	71(5x2,5)	Tho	→SMAJ ...	data
BG	Z-Di	=SMBJ 13CA (Typ-Code/Stempel/marking)	71(5x3,5)	Mop	→SMBJ ...	data
BG 34 S	Si-Di	=SPB-G34S (Typ-Code/Stempel/marking)	30		→SPB-G34S	data
BG 2011 SM	GaAs-FET-IC	Breitb.-Vst./wide band amp., ~1GHz	44	Rhm	-	
BG 3011 F,CF	GaAs-FET-IC	Frequ.-Teiler/prescaler, 1:128/129	8-MDIP	Rhm	-	
BG 3012 F,CF	GaAs-FET-IC	Frequ.-Teiler/prescaler, 1:128/129	8-MDIP	Rhm	-	
BG 3013 F,CF	GaAs-FET-IC	Frequ.-Teiler/prescaler, 1:128/129	8-MDIP	Rhm	-	
BGA 310	LIN-IC	MMIC-Amplifier, UHF, 0..2,4GHz	44	Sie	-	data pdf
BGA 312	LIN-IC	MMIC-Amplifier, UHF, 0...2GHz	44	Sie	-	data pdf
BGA 318	LIN-IC	MMIC-Amplifier, UHF, 0...1,2GHz	44	Sie	-	data pdf
BGA 420	LIN-IC	MMIC-Amplifier, UHF, 1,8GHz	44(2mm)	Sie	-	data pdf
BGA 425	LIN-IC	MMIC-Amplifier, UHF, 1,8GHz	44(2mm)	Sie	-	data pdf
BGA 427	LIN-IC	MMIC-Amplifier, UHF, 1,8GHz	44(2mm)	Sie	-	data pdf
BGA 2709	LIN/OP-IC	Mmic Wideband Amp., 3.6GHz at 3dB	46	Phi	-	pdf pinout
BGD ...	Si-N/P	Module, Spezialtypen/special types				data
BGF 100	Dig. Audio	Microphone Filter and ESD Protection	QFP	Inf	-	pdf pinout
BGF 104	IC	HSMMC Interface Filter,ESD Protection	QFP	Inf	-	pdf pinout
BGF 105	IC	SIM Card Interface Filter,ESD Protect.	QFP	Inf	-	pdf pinout
BGF 109	IC	10-CH LCD Filter Array,ESD Protection	QFP	Inf	-	pdf pinout
BGF 110	IC	SD Card Interface ESD Protection	QFP	Inf	-	pdf pinout
BGF 200	Dig. Audio	Microphone Filter and ESD Protection	QFP	Inf	-	pdf pinout
BGH	Z-Di	=SM 15T 150C(Typ-Code/Stempel/marking)	71(8x5mm)	Tho	→SM 15T...	data
BGJ	Si-Di	=MBRS 1540 (Typ-Code/Stempel/marking)	71(5x3,5)	Ons	→MBRS 1540	
BGK	Z-Di	=SM 15T150CA(Typ-Code/Stempel/marking)	71(8x5mm)	Tho	→SM 15T...	data
BGp	Si-P	=BCX 71G (Typ-Code/Stempel/marking)	35	Phi	→BCX 71G	data
BGs	Si-P	=BCX 71G (Typ-Code/Stempel/marking)	35	Sie	→BCX 71G	data
BGt	Si-P	=BCX 71G (Typ-Code/Stempel/marking)	35	Phi	→BCX 71G	data
BGU	Z-Di	=SM 15T 200C(Typ-Code/Stempel/marking)	71(8x5mm)	Tho	→SM 15T...	data
BGV	Z-Di	=SM 15T200CA(Typ-Code/Stempel/marking)	71(8x5mm)	Tho	→SM 15T...	data
BGX 11/.....17/...	Thy	Thy Module/Modules		Phi	-	data
BGX 50(A)	Si-Br	SMD, HF Br-Gl, 50V, 0,1A, ..50MHz	44	Sie	-	data
BGX 88(/01)	Si	VHF/UHF-Modul		Phi		data
BGX 400	Si-Di	SMD, Dual, 400V, 0,25A, 1µs	35o1	Sie	-	
BGX 885	Si	VHF/UHF-Modul		Phi		data
BGY ...	Si-N/P, Di	Module, Spezialtypen/special types			-	data
BGY 887	Hybrid-IC	860 MHz, 21.5 dB gain push-pull amp.	SIP	Phi	-	pdf pinout
BH	Si-Di	=1SS295 (Typ-Code/Stempel/marking)	35	Tos	→1SS295	data
BH	Si-Di	=1SS350 (Typ-Code/Stempel/marking)	35	Say	→1SS350	data
BH	Si-P	=2SB1001-BH (Typ-Code/Stempel/marking)	39	Hit	→2SB1001	data
BH	Si-P	=2SB1125 (Typ-Code/Stempel/marking)	39	Say	→2SB1125	data
AB	Si-N	=BCP 56 (Typ-Code/Stempel/marking)	48	Ons	→BCP 56	
BH	Si-N	=BCX 56 (Typ-Code/Stempel/marking)	39	Mot,Sie,Val	→BCX 56	data
BH	Si-P	=BCX 71H (Typ-Code/Stempel/marking)	35	Mot,Phi,Sie Fer	→BCX 71H	data
BH	Z-Di	=BZX 399-C9V1(Typ-Code/Stempel/marking	71(1,7mm)	Phi	→BZX 399-...	data
BH	Z-Di	=P4 SMA-22 (Typ-Code/Stempel/marking)	71(5x2,5)	Fag	→P4 SMA-22	data
BH	Si-Di	=SC 017-2 (Typ-Code/Stempel/marking)	71(5x2,5)	Fjd	→SC 017-...	data
BH	Z-Di	=SMAJ 13CA (Typ-Code/Stempel/marking)	71(5x2,5)	Tho	→SMAJ ...	data
BH	Z-Di	=SMBJ 14C (Typ-Code/Stempel/marking)	71(5x3,5)	Mop	→SMBJ ...	data
BH-10	Si-N	=BCP 56-10 (Typ-Code/Stempel/marking)	48	Ons	→BCP 56	
BH-16	Si-N	=BCP 56-16 (Typ-Code/Stempel/marking)	48	Ons	→BCP 56	
BH 3532 FS	LIN-IC	SMD, Audio, Digital Volume Ctrl.	20-SMDIP	Rhm	-	pdf
BH 3561 F	LIN-IC	SMD, Cd, Endverst./post Amp., Filter	22-MDIP	Rhm	-	
BH 3562 F	LIN-IC	SMD, Cd, Endverst./post Amp., Filter	22-MDIP	Rhm	-	pdf
BH 3810 FS	LIN-IC	SMD, Cd,TV++, Vocal Fader	32-SMDIP		-	pdf
BH 3852 FS	LIN-IC	=BH 3852S: SMD	24-SMDIP		-	pdf

Type	Device	Short Description	Fig.	Manu	Comparision Types	More at
BH 3852 S	LIN-IC	Cd++, Dual Potentiometer (sound Ctrl.)	24-SDIP	Rhm	-	pdf
BH 3854 AFS	LIN-IC	=BH 3854AS: SMD	32-SMDIP		-	pdf
BH 3854 AS	LIN-IC	Cd++, Dual Potentiometer (sound Ctrl.)	32-SDIP	Rhm	-	pdf
BH 3856 FS	LIN-IC	=BH 3856S: SMD	32-MDIP		-	pdf
BH 3856 S	LIN-IC	Signal Processing (sound Ctrl.)	30-SDIP	Rhm	-	pdf
BH 3857 FV,AFV	LIN-IC	SMD, CD++, Audio Sound Controller	40-SSMDIP	Rhm	-	
BH 3857 FV	LIN-IC	SMD, Cd++, Audio Sound Controller	40-SSMDIP	Rhm	-	
BH 3864 F	LIN-IC	SMD, Cd++, Audio Sound Controller	24-MDIP	Rhm	-	
BH 4100 FV	LIN-IC	SMD, Telecom, Ir-fb Empf./receiver	14-SSMDIP	Rhm	-	pdf
BH 4126 FV	LIN-IC	SMD, Telecom, Wideband ZF-Dem. f. Phs	16-SSMDIP	Rhm	-	
BH 4128 FV	LIN-IC	SMD, 2. Mixer Zf f. Mobile Telephones	16-SSMDIP	Rhm	-	pdf
BH 6111 FV	LIN-IC	SMD, Telecom, Power Uni IC f. Pager	20-SSMDIP	Rhm	-	pdf
BH 6113 FV	LIN-IC*	SMD, Telecom, Power Uni IC f. Pager	16-SSMDIP	Rhm	-	pdf
BH 6620 K	LIN-IC	SMD, Fdd, Read/write Amplifier, Ucc=5V	32-MP	Rhm	-	
BH 6625 FS	BiCMOS-IC	SMD, Fdd, Read/write Amplifier, Ucc=5V	24-SMDIP	Rhm	-	pdf
BH 6626 FS	BiCMOS-IC	SMD, Fdd, Read/write Amplifier, Ucc=5V	24-SMDIP	Rhm	-	pdf
BH 6627 FS	BiCMOS-IC	SMD, Fdd, Read/write Amplifier, Ucc=5V	24-SMDIP	Rhm	-	pdf
BH 6628 AFS	BiCMOS-IC	SMD, Fdd, Read/write Amplifier, Ucc=5V	24-SMDIP	Rhm	-	pdf
BH 6629 BFS	BiCMOS-IC	SMD, FDD, Read/Write Amplifier, Ucc=5V	24-SMDIP	Rhm	-	
BH 6629 AFS	BiCMOS-IC	SMD, Fdd, Read/write Amplifier, Ucc=5V	24-SMDIP	Rhm	-	pdf
BH 7236 AF	LIN-IC	TV Peripherals, NTSC/PAL Color Encoder	24-SDIP	Rhm	-	pdf
BH 7332 FS	MOS-IC	Crt Display, I²c-bus Control Video Amp	32-SDIP	Rhm	-	
BH 7370 FS	BiCMOS-IC	SMD, TV,Vc, Zf Signal-proz.(japan,USA)	24-SMDIP	Rhm	-	
BH 7502 K1	LIN-IC	SMD, Vc, VHS-NTSC Video-Signal-proz.	80-MP	Rhm	-	
BH 7507 K1,AK1	LIN-IC	SMD,Vc, VHS-PAL Video-Signal-prozessor	80-MP	Rhm	-	
BH 7513 AKV	LIN-IC	SMD, Vc, Vhs-ntsc Video-signal-proz.	80-MP	Rhm	-	
BH 7517 K1	LIN-IC	SMD,Vc, Vhs-pal Video-signal-prozessor	80-MP	Rhm	-	
BH 7518 AKV	LIN-IC	SMD,Vc, Vhs-pal Video-signal-prozessor	80-MP	Rhm	-	
BH 7634 AS	LIN-IC	Vc, Audio/video Switch f. Scrambled TV (secam/pal) f. Canal Plus Decoder	24-SDIP	Rhm	-	pdf
BH 7733 S	LIN-IC	Audio, Eingangswähler/input Selector	24-SDIP	Rhm	-	
BH 7773 KS	BiCMOS-IC	SMD, Vc, HiFi Audio Signal-prozessor	100-MP	Rhm	-	
BH 7775 K	BiCMOS-IC	SMD, Vc, Audio Signal-prozessor	64-MP	Rhm	-	
BH 7776 K	BiCMOS-IC	SMD, Vc, HiFi Audio Signal-prozessor	64-MP	Rhm	-	pdf
BH 7777 K,AK	BiCMOS-IC	SMD, Vc, HiFi Audio Signal-prozessor	64-MP	Rhm	-	
BH 7778 AK	BiCMOS-IC	SMD, Vc, HiFi Audio Signal-prozessor	64-MP	Rhm	-	
BH 7779 K	BiCMOS-IC	SMD, Vc, Audio Signal-prozessor	64-MP	Rhm	-	
BH 7800 K	BiCMOS-IC	SMD, Vc, HiFi Audio Signal-prozessor	44-MP	Rhm	-	
BH 7801 BK	BiCMOS-IC	SMD, Vc, HiFi Audio Signal-prozessor	44-MP	Rhm	-	pdf
BH 7802 K	BiCMOS-IC	SMD, Vc, HiFi Audio Signal-prozessor	44-MP	Rhm	-	pdf
BH 8220 KS	MOS-IC	SMD, Telecom,Fax, Analog Signal Proz.	56-MP	Rhm	-	pdf
BH 9590 FP-Y	BiCMOS-IC	µComp, Scsi Bus, Active Terminator	25-SMDIP+b	Rhm	-	pdf
BH 9595 FP-Y	MOS-IC	µcomp, Scsi Bus, Active Terminator	25-SMDIP+b	Rhm	-	
BH 9596 FP-Y	MOS-IC	µcomp, Scsi Bus, Active Terminator	25-SMDIP+b	Rhm	-	pdf
BH 9610 K	BiCMOS-IC	SMD, Optical Disk, Laser Power Control	44-MP	Rhm	-	pdf
BHM	Si-N	=2SC4699K-M (Typ-Code/Stempel/marking)	35	Rhm	→2SC4699K	data
BHM	Si-N	=2SC4700-M (Typ-Code/Stempel/marking)	35(2mm)	Rhm	→2SC4700	data
BHN	Si-N	=2SC4699K-N (Typ-Code/Stempel/marking)	35	Rhm	→2SC4699K	data
BHN	Si-N	=2SC4700-N (Typ-Code/Stempel/marking)	35(2mm)	Rhm	→2SC4700	data
BHP	Si-P	=2SB1386-P (Typ-Code/Stempel/marking)	39	Rhm	→2SB1386	data
BHP	Si-N	=2SC4699K-P (Typ-Code/Stempel/marking)	35	Rhm	→2SC4699K	data
BHP	Si-N	=2SC4700-P (Typ-Code/Stempel/marking)	35(2mm)	Rhm	→2SC4700	data
BHp	Si-P	=BCX 71H (Typ-Code/Stempel/marking)	35	Phi	→BCX 71H	data
BHQ	Si-P	=2SB1386-Q (Typ-Code/Stempel/marking)	39	Rhm	→2SB1386	data
BHR	Si-P	=2SB1386-R (Typ-Code/Stempel/marking)	39	Rhm	→2SB1386	data
BHs	Si-N	=BCP 56M (Typ-Code/Stempel/marking)	≥45	Sie	→BCP 56M	data
BHs	Si-P	=BCX 71H (Typ-Code/Stempel/marking)	35	Sie	→BCX 71H	data
BHt	Si-P	=BCX 71H (Typ-Code/Stempel/marking)	35	Phi	→BCX 71H	data
BI	Si-P	=2SB1126 (Typ-Code/Stempel/marking)	39	Say	→2SB1126	data
BI	Si-Di	=SC 017-4 (Typ-Code/Stempel/marking)	71(5x2,5)	Fjd	→SC 017-...	data
BIQ	N-FET	=2SK541-BIQ (Typ-Code/Stempel/marking)	35	Hit	→2SK541	data
BIR	N-FET	=2SK541-BIR (Typ-Code/Stempel/marking)	35	Hit	→2SK541	data
BIS	N-FET	=2SK541-BIS (Typ-Code/Stempel/marking)	35	Hit	→2SK541	data
BIT 3105-SSO	DIG/LIN-IC	High Efficiency ZVS Ccfl Controller	TSOP		-	pdf pinout
BIT 3193-DP	IC	High Performance PWM Controller	DIP	Bit		pdf pinout
BIT 3193-SO	IC	High Performance PWM Controller	MDIP	Bit		pdf pinout
BIT 3193-SS	IC	High Performance PWM Controller	TSOP	Bit	-	pdf pinout
BJ	Si-P	=2SB1001-BJ (Typ-Code/Stempel/marking)	39	Hit	→2SB1001	data
BJ	Si-P	=2SB1302 (Typ-Code/Stempel/marking)	39	Say	→2SB1302	data
BJ	Si-N	=BCP 56 (Typ-Code/Stempel/marking)	48	Phi	→BCP 56	data
BJ	Si-N	=BCX 56-6 (Typ-Code/Stempel/marking)	39	Mot,Sie,Val	→BCX 56	data
BJ	Si-P	=BCX 71J (Typ-Code/Stempel/marking)	35	Mot,Phi,Sie	→BCX 71J	data
BJ	Z-Di	=BZX 399-C10 (Typ-Code/Stempel/marking	71(1,7mm)	Fer Phi	→BZX 399-...	data
BJE	Si-P	=2SB1427-E (Typ-Code/Stempel/marking)	39	Rhm	→2SB1427	data
BJp	Si-P	=BCX 71J (Typ-Code/Stempel/marking)	35	Phi	→BCX 71J	data
BJS	Si-P	=2SB1427-S (Typ-Code/Stempel/marking)	39	Rhm	→2SB1427	data
BJs	Si-P	=BCX 71J (Typ-Code/Stempel/marking)	35	Sie	→BCX 71J	data
BJU	Si-P	=2SB1427-U (Typ-Code/Stempel/marking)	39	Rhm	→2SB1427	data
BJU	Si-N	=2SD2226K-U (Typ-Code/Stempel/marking)	35	Rhm	→2SD2226K	data
BJU	Si-N	=2SD2351-U (Typ-Code/Stempel/marking)	35(2mm)	Rhm	→2SD2351	data
BJV	Si-N	=2SD2226K-V (Typ-Code/Stempel/marking)	35	Rhm	→2SD2226K	data
BJV	Si-N	=2SD2351-V (Typ-Code/Stempel/marking)	35(2mm)	Rhm	→2SD2351	data
BJW	Si-N	=2SD2226K-W (Typ-Code/Stempel/marking)	35	Rhm	→2SD2226K	data
BJW	Si-N	=2SD2351-W (Typ-Code/Stempel/marking)	35(2mm)	Rhm	→2SD2351	data

Type	Device	Short Description	Fig.	Manu	Comparision Types	More at
BK	Si-P	=2SB1323 (Typ-Code/Stempel/marking)	39	Say	→2SB1323	data
BK	Si-P	=2SB1590K (Typ-Code/Stempel/marking)	35	Rhm	→2SB1590K	data
BK	Si-N	=BCP 56-10 (Typ-Code/Stempel/marking)	48	Phi	→BCP 56	data
BK	Si-N	=BCX 56-10 (Typ-Code/Stempel/marking)	39	Mot,Sie,Val	→BCX 56	data
BK	Si-P	=BCX 71K (Typ-Code/Stempel/marking)	35	Fer,Phi,Sie	→BCX 71K	data
BK	Z-Di	=BZX 399-C11 (Typ-Code/Stempel/marking)	71(1,7mm)	Phi	→BZX 399-...	data
BK	Z-Di	=P4 SMA-22A (Typ-Code/Stempel/marking)	71(5x2,5)	Fag	→P4 SMA-22A	data
BK	Z-Di	=SMBJ 14CA (Typ-Code/Stempel/marking)	71(5x3,5)	Mop	→SMBJ ...	data
BK	Si-P+R	=XN 4114 (Typ-Code/Stempel/marking)	46	Mat	→XN 4114	data
BK	Si-P+R	=XP 4114 (Typ-Code/Stempel/marking)	46(2mm)	Mat	→XP 4114	data
BKJL	Si-Di	=MBRS 2040 (Typ-Code/Stempel/marking)	71(5x3,5)	Ons	→MBRS 2040	data
BKp	Si-P	=BCX 71K (Typ-Code/Stempel/marking)	35	Phi	→BCX 71K	data
BKs	Si-P	=BCX 71K (Typ-Code/Stempel/marking)	35	Sie	→BCX 71K	data
BL-3000	LED	rd+bl+gr+wht ø5mm l:567lm Vf:10.5V		Roi		data pdf pinout
BL	Si-P	=2SA1341 (Typ-Code/Stempel/marking)	35	Say	→2SA1341	data
BL	Si-P	=2SA1676 (Typ-Code/Stempel/marking)	35(2mm)	Say	→2SA1676	data
BL	Si-P	=2SB1324 (Typ-Code/Stempel/marking)	39	Say	→2SB1324	data
BL	Si-N	=BCP 56-16 (Typ-Code/Stempel/marking)	48	Phi	→BCP 56	data
BL	Si-N	=BCX 56-16 (Typ-Code/Stempel/marking)	39	Phi	→BCX 56	data
BL	Si-P	=BCX 71L (Typ-Code/Stempel/marking)	35	Sie	→BCX 71L	data
BL	Z-Di	=BZX 399-C12 (Typ-Code/Stempel/marking)	71(1,7mm)	Phi	→BZX 399-...	data
BL	Si-Di	=MBD 54DW (Typ-Code/Stempel/marking)	46(2mm)	Ons	→MBD 54DW	data
BL	Z-Di	=P4 SMA-24 (Typ-Code/Stempel/marking)	71(5x2,5)	Fag	→P4 SMA-24	data
BL	Z-Di	=SMBJ 15C (Typ-Code/Stempel/marking)	71(5x3,5)	Mop	→SMBJ ...	data

BLF

Type	Device	Short Description	Fig.	Manu	Comparision Types	More at
BLF 145	MOS-N-FET-e	V-MOS, SSB-L, 65V, 3A, PEP=30W(28MHz)	59z	Phi	-	data
BLF 146	MOS-N-FET-e	V-MOS, SSB-L, 65V, 7A, Vp>18dB(28MHz)	59z	Phi	-	data
BLF 147	MOS-N-FET-e	V-MOS, SSB-L, 65V, 13A,PEP=150W(28MHz)	59z	Phi	-	data
BLF 175	MOS-N-FET-e	V-MOS, SSB-L, 110V, 1,5A,PEP=8W(28MHz)	59z	Phi	-	data
BLF 177	MOS-N-FET-e	V-MOS, SSB-L, 110V, 7A,PEP=150W(28MHz)	59z	Phi	-	data
BLF 241	MOS-N-FET-e	V-MOS, VHF, 65V, 0,5A, PQ=3W(175MHz)	2a§	Phi	-	data
BLF 242	MOS-N-FET-e	V-MOS, VHF-L,65V, 0,5A,Vp=16dB(175MHz)	59z	Phi	-	data
BLF 244	MOS-N-FET-e	V-MOS, VHF-L,65V, 1,5A,Vp=17dB(175MHz)	59z	Phi	-	data
BLF 245	MOS-N-FET-e	V-MOS, VHF-L, 65V, 3A,Vp=16dB(175MHz)	59z	Phi	-	data
BLF 246	MOS-N-FET-e	V-MOS, FM/VHF, 65V, 7A, PQ=80W(108MHz)	59z	Phi	-	data
BLF 1046	MOS-N-FET-e	UHF, 65V, 4,5A, PQ=45W(960MHz, 26V)	62d*	Phi	-	data
BLF 1047	MOS-N-FET-e	UHF, 65V, 9A, PQ=70W(960MHz, 26V)	≈62d*	Phi	-	data
BLF 1048	MOS-N-FET-e	UHF, 65V, 9A, PQ=90W(960MHz, 26V)	≈62d*	Phi	-	data
BLF 2045	MOS-N-FET-e	UHF, 65V, 4,5A, PEP=30W(2GHz, 26V)	62d*	Phi	-	data
BLF 2047	MOS-N-FET-e	UHF, 65V, 9A, PEP=65W(2GHz, 26V)	≈62d*	Phi	-	data
BLL	Si-N	=2SC4771K-L (Typ-Code/Stempel/marking)	35	Rhm	→2SC4771K	data
BLL	Si-N	=2SC4772-L (Typ-Code/Stempel/marking)	35(2mm)	Rhm	→2SC4772	data
BLM	Si-N	=2SC4771K-M (Typ-Code/Stempel/marking)	35	Rhm	→2SC4771K	data
BLM	Si-N	=2SC4772-M (Typ-Code/Stempel/marking)	35(2mm)	Rhm	→2SC4772	data
BLN	Si-N	=2SC4771K-N (Typ-Code/Stempel/marking)	35	Rhm	→2SC4771K	data
BLN	Si-N	=2SC4772-N (Typ-Code/Stempel/marking)	35(2mm)	Rhm	→2SC4772	data
BLP	Si-P	=2SB1561-P (Typ-Code/Stempel/marking)	39	Rhm	→2SB1561	data
BLP	Si-N	=2SC4771K-P (Typ-Code/Stempel/marking)	35	Rhm	→2SC4771K	data
BLP	Si-N	=2SC4772-P (Typ-Code/Stempel/marking)	35(2mm)	Rhm	→2SC4772	data
BLQ	Si-P	=2SB1561-Q (Typ-Code/Stempel/marking)	39	Rhm	→2SB1561	data
BLS	LIN-IC	=BGA 310 (Typ-Code/Stempel/marking)	44	Sie	→BGA 310	data
BLS	LIN-IC	=BGA 420 (Typ-Code/Stempel/marking)	44(2mm)	Sie	→BGA 420	data
BLS	LIN-IC	=BGA 420 (Typ-Code/Stempel/marking)	44(2mm)	Sie	→BGA 420	data

BLT

Type	Device	Short Description	Fig.	Manu	Comparision Types	More at
BLT 50	Si-N	UHF-E, 20V, 0,5A, 2W, PQ=1,2W(470MHz)	48s§	Phi	-	data
BLT 80	Si-N	UHF-E, 20V, 0,25A, 2W, PQ=0,8W(900MHz)	48s§	Phi	-	data
BLT 81	Si-N	UHF-E, 20V, 0,5A, 2W, PQ=1,2W(900MHz)	48s§	Phi	-	data
BLT 90/SL	Si-N	UHF-Tr/E, 20V, 0,25A, PQ=0,75W(900MHz)	51r	Phi	-	data
BLT 91/SL	Si-N	UHF-Tr/E, 20V, 0,5A, PQ=1,5W(900MHz)	51r	Phi	-	data
BLT 92/SL	Si-N	UHF-L, 20V, 1,2A, PQ=3W(900MHz)	51r	Phi	-	data
BLT 93/SL	Si-N	UHF-L, 20V, 1,2A, PQ=6W(900MHz)	51r	Phi	-	data

BLU

Type	Device	Short Description	Fig.	Manu	Comparision Types	More at
BLU 10/12	Si-N	UHF-L, 36V, 2A, PQ=10W(470MHz)	57s	Phi	MRF 641	data
BLU 20/12	Si-N	UHF-L, 36V, 4A, PQ=20W(470MHz)	57s	Phi	MRF 641	data
BLU 30/12	Si-N	UHF-L, 36V, 6A, PQ=30W(470MHz)	57s	Phi	MRF 644	data
BLU 45/12	Si-N	UHF-L, 36V, 9A, PQ=45W(470MHz)	57s	Phi	MRF 648	data
BLU 50	Si-N	Dual, UHF-L, 60V, 1,8A, PQ=30W(400MHz)		Phi	-	data
BLU 51	Si-N	Dual, UHF-L, 60V, 2,5A, PQ=45W(400MHz)		Phi	-	data
BLU 52	Si-N	Dual, UHF-L, 60V, 4A, PQ=60W(400MHz)		Phi	-	data
BLU 53	Si-N	Dual, UHF-L, 60V, 5A, PQ=100W(400MHz)	≈61/8Pin	Phi	-	data
BLU 56	Si-N	UHF-E, 36V, 0,2A, 2W, PQ=1W(470MHz)	48s§	Phi	-	data
BLU 60/12	Si-N	UHF-L, 36V, 12A, PQ=60W(470MHz)	57s	Phi	MRF 648	data
BLU 86	Si-N	UHF-E, 32V, 0,2A, 2W, PQ=1W(900MHz)	48s§	Phi	-	data

Type	Device	Short Description	Fig.	Manu	Comparision Types	More at
BLU 97	Si-N	UHF-Tr/E, 36V, 1,2A, PQ=7W(470MHz)	55r	Phi	BLX 68, BLY 53	data
BLU 98	Si-N	UHF-Tr/E, 36V, 0,15A, PQ=0,5W(900MHz)	25za	Phi	-	data
BLU 99	Si-N	UHF-Tr/E, 36V, 0,8A, PQ=4W(900MHz)	55r	Phi	-	data

BLV

Type	Device	Short Description	Fig.	Manu	Comparision Types	More at
BLV 10	Si-N	VHF-L, 36V, 1,5A, PQ=8W(175MHz)	59r	Phi	MRF 221	data
BLV 11	Si-N	VHF-L, 36V, 3A, PQ=15W(175MHz)	59r	Phi	MRF 221..223	data
BLV 15/12	Si-N	VHF-L, 36V, 3A, PQ=15W(175MHz)	57s	Phi	MRF 215	data
BLV 20	Si-N	VHF-L, 65V, 0,9A, PQ=8W(175MHz)	59r	Phi	-	data
BLV 21	Si-N	VHF-L, 65V, 1,75A, PQ=15W(175MHz)	59r	Phi	-	data
BLV 25	Si-N	VHF-L, 65V, 17,5A, PQ=175W(108MHz)	57s	Phi	-	data
BLV 30	Si-N	VHF-Tr/E, 60V, 1,5A, PQ>1,5W(224MHz)	55r	Phi	-	data
BLV 30/12	Si-N	VHF-L, 36V, 6A, PQ=30W(175MHz)	57s	Phi	MRF 216	data
BLV 31	Si-N	VHF-Tr/E, 60V, 3A, PQ>5W(224MHz)	55r	Phi	BLW 75	data
BLV 32 F	Si-N	VHF-L, 60V, 4A, PQ>10W(224MHz)	57s	Phi	-	data
BLV 33	Si-N	VHF-L, 65V, 12,5A, PQ>19W(224MHz)	55r	Phi	-	data
BLV 33 F	Si-N	VHF-L, 65V, 12,5A, PQ>16W(224MHz)	57s	Phi	-	data
BLV 34 F	Si-N	VHF-L, 75V, 15A, PQ>130W(225MHz)	57s	Phi	-	data
BLV 36	Si-N	Dual, VHF-L, 65V, 10A, PQ=120W(224MHz)	≈61/8Pin	Phi	-	data
BLV 37	Si-N	Dual, VHF-L, 70V, 10A, PQ=250W(108MHz)	≈61/5Pin	Phi	-	data
BLV 38	Si-N	Dual, VHF-L, 70V, 10A, PQ=225W(224MHz)	≈61/5Pin	Phi	-	data
BLV 45/12	Si-N	VHF-L, 36V, 9A, PQ=45W(175MHz)	57s	Phi	MRF 243	data
BLV 57	Si-N	Dual, UHF-L, 50V, 2A, PQ>6W(860MHz)	≈61/8Pin	Phi	-	data
BLV 59	Si-N	UHF-L, 50V, 3A, PQ=30W(860MHz)	61s	Phi	-	data
BLV 75/12	Si-N	VHF-L, 36V, 15A, PQ=75W(175MHz)	57s	Phi	MRF 243, MRF 245	data
BLV 80/28	Si-N	VHF-L, 65V, 8,5A, PQ=80W(175MHz)	59r	Phi	BLW 78	data
BLV 90	Si-N	UHF-Tr/E, 36V, 0,2A, PQ=1W(900MHz)	55r	Phi	MRF 838A	data
BLV 90/SL	Si-N	=BLV 90:	51r		MRF 838	data
BLV 91	Si-N	UHF-Tr/E, 36V, 0,4A, PQ=2W(900MHz)	55r	Phi	MRF 817	data
BLV 91/SL	Si-N	=BLV 91:	51r		-	data
BLV 92	Si-N	UHF-Tr/E, 36V, 0,8A, PQ=4W(900MHz)	61s	Phi	MRF 840	data
BLV 93	Si-N	UHF-Tr/E, 36V, 1,6A, PQ=8W(900MHz)	61s	Phi	MRF 840	data
BLV 94	Si-N	UHF-L, 36V, 3A, PQ=12,5W(900MHz)	61v	Phi	MRF 842	data
BLV 95	Si-N	UHF-L, 36V, 5A, PQ=25W(900MHz)	61v	Phi	MRF 844	data
BLV 97	Si-N	UHF-L, 50V, 3A, PQ=30W(900MHz)	61v	Phi	-	data
BLV 98	Si-N	UHF-L, 50V, 1,5A, PQ=14W(900MHz)	61v	Phi	-	data
BLV 99	Si-N	UHF-Tr/E, 50V, 0,2A, PQ=2W(900MHz)	55r	Phi	BLX 91	data

BLW

Type	Device	Short Description	Fig.	Manu	Comparision Types	More at
BLW 10	Si-N	VHF/UHF-A/Tr, 55V, 0,4A, >500MHz	7a	Tix	(BFS 23, 2N3866)[6]	data
BLW 11	Si-N	VHF/UHF-A/Tr, 40V, 0,4A, >1,2GHz	2a§	Tix	BFR 36, BFW 16	data
BLW 12	Si-N	UHF-Tr/E, 36V, 0,4A, PQ>0,75W(470MHz)	51r	Tix	BLW 39, BLX 66	data
BLW 13	Si-N	UHF-L, 36V, 2A, PQ>3,75W(470MHz)	55r	Tix	BLW 43, BLX 68, BLY 53	data
BLW 14	Si-N	UHF-L, 36V, 2A, PQ>7W(470MHz)	55r	Tix	BLW 44, BLX 68, BLY 53B	data
BLW 15	Si-N	UHF-L, 36V, 3A, PQ>12W(470MHz)	55r	Tix	BLX 69	data
BLW 16	Si-N	VHF-L, 36V, 0,5A, PQ>1,4W(175MHz)	2a§	Tix	BFS 22, BFS 50..51, BLY 61, 2N4427, ++	data
BLW 17	Si-N	VHF-Tr/E, 36V, 0,5A, PQ>2W(175MHz)	51r	Fer,Tix	BLW 39, BLX 66	data
BLW 18	Si-N	VHF-L, 36V, 2A, PQ>5W(175MHz)	55r	Tix	BLW 37, BLY 62, BLY 83, 2N5590, ++	data
BLW 19	Si-N	VHF-L, 36V, 2A, PQ>8W(175MHz)	55r	Tix	BLW 37, BLY 83, MRF 212, 2N5590, ++	data
BLW 20	Si-N	VHF-L, 36V, 5A, PQ>25W(175MHz)	55r	Tix	BLW 31, 2N6084	data
BLW 21	Si-N	HF-L, 36V, 7A, 35MHz	55r	Tix	2N5708	data
BLW 22	Si-N	VHF-A/Tr, 40V, 0,4A, 1000MHz	55r	Tix	BFR 65, BFT 91, BLW 42, BLX 96	data
BLW 23	Si-N	VHF-L, 55V, 2A, PQ>5W(175MHz)	55r	Tix	BLV 31, BLW 19, BLY 83	data
BLW 24	Si-N	VHF-L, 60V, 2A, PQ>17W(175MHz)	55r	Tix	BLY 93, 2N5642	data
BLW 25	Si-N	VHF-L, 65V, 5A, PQ>40W(175MHz)	55r	Tix	BLY 94, 2N5643	data
BLW 26	Si-N	HF-L (SSB), 36V, 8A, >50MHz	59r	Tix	MRF 460, 2N5706, 2N5709, 2N5941	data
BLW 27	Si-N	HF-L (SSB), 36V, 10A, >50MHz	59r	Tix	MRF 460, 2N5709, 2N5942	data
BLW 29	Si-N	VHF-L, 36V, 2,75A, PQ=15W(175MHz)	55r	Phi	BLY 84, BLY 88, MRF 226, 2N6081	data
BLW 31	Si-N	VHF-L, 36V, 6A, PQ=28W(175MHz)	55r	Phi	BLW 20, 2N6084	data
BLW 32	Si-N	UHF-L, 50V, 0,65A, PQ=0,58W(860MHz)	55r	Phi,Tix	BLX 97	data
BLW 33	Si-N	UHF-L, 50V, 1,25A, PQ=1,07W(860MHz)	55r	Phi	BLW 98, BLX 98	data
BLW 34	Si-N	UHF-L, 50V, 2,25A, PQ=1,9W(860MHz)	55r	Phi	BLW 98, BLX 98	data
BLW 35	Si-N	VHF-L, 39V, 1,7A, PQ>7,5W(175MHz)	49a*	Aeg	BLY 36, BLY 58, BLY 79, 2N3927	data
BLW 36	Si-N	VHF-L, 39V, 3,5A, PQ>15,5W(175MHz)	49a*	Aeg	BLY 36	data
BLW 37	Si-N	=BLW 35:	55r	Aeg	BLW 19, BLY 62, BLY 83, MRF 212, 2N5590	data
BLW 38	Si-N	=BLW 36:	55r	Aeg	BLW 20, BLY 63, BLY 84, MRF 238, 2N6083	data
BLW 39	Si-N	UHF-Tr/E, 50V, 0,5A, PQ>2W(175MHz)	51r	Fer	BLW 17, BLX 66	data
BLW 42	Si-N	UHF-L, 50V, 0,7A, PQ>1,1W(470MHz)	55r	Aeg	BLX 92, 2N5944	data
BLW 43	Si-N	UHF-L, 50V, 1A, PQ>3,5W(470MHz)	55r	Aeg	BLU 97, BLW 80, BLX 68, BLX 93, BLY 53	data
BLW 44	Si-N	UHF-L, 50V, 2A, PQ>8W(470MHz)	55r	Aeg	BLW 14, BLW 81, BLX 94, 2N5946	data
BLW 45	Si-N	UHF-Tr/E, 25V, 0,15A, PQ=0,2W(1GHz)	55r	Tho	BLV 90, MRF 838A	data
BLW 46	Si-N	=BLW 45:	51r	Tho	MRF 816, MRF 838	data
BLW 47	Si-N	UHF-Tr/E, 25V, 0,2A, PQ=0,5W(1GHz)	55r	Tho	B'LV 90, MRF 838A	data

Type	Device	Short Description	Fig.	Manu	Comparision Types	More at
LW 48	Si-N	=BLW 47:	51r	Tho	MRF 816, MRF 838	data
LW 50 F	Si-N	AM-SSB-L, 110V, 2,5A, PEP=17W(28MHz)	59r	Phi	-	data
LW 60	Si-N	AM...VHF-L, 36V, 8A, PQ=45W(175MHz)	55r	Phi	BLY 90	data
LW 60 C	Si-N	=BLY 60: 9/22A	55r	Phi	-	data
LW 64	Si-N	VHF-L, 60V, 4A, PQ>15W(224MHz)	55r	Phi	BLW 75	data
LW 65	Si-N	NF/HF-Tr, 40V, 5A(ss), 80MHz	2a§	Sol	BUX 34, BUX 49, BUY 41, BUY 80, ++	data
LW 66	Si-N	=BLW 65: 60V	2a§	Sol	BUX 34, BUX 49, BUY 41, BUY 80, .. +	data
LW 67	Si-N	=BLW 65: 80V	2a§	Sol	BUX 34, BUX 49, BUY 41, BUY 80, ++	data
LW 68	Si-N	=BLW 65: 100V	2a§	Sol	BUX 34, BUX 49, BUY 41, BUY 80, ++	data
LW 69	Si-N	=BLW 65: 120V	2a§	Sol	BUX 34, BUX 49, BUY 41, BUY 80, ++	data
LW 70	Si-N	=BLW 65: 140V	2a§	Sol	BU 125S, BUX 50, BUY 49, BUY 80, ++	data
LW 71	Si-N	=BLW 65: 160V	2a§	Sol	BU 125S, BUX 50, BUY 49	data
LW 72	Si-N	=BLW 65: 180V	2a§	Sol	BU 125S, BUX 50, BUY 49	data
LW 73	Si-N	=BLW 65: 200V	2a§	Sol	BU 125S, BUX 50, BUY 49	data
LW 75	Si-N	VHF-L, 60V, 4A, PQ>14W(225MHz)	55r	Phi	BLW 64	data
LW 76	Si-N	AM/FM-L, 70V, 8A, PQ=80W(108MHz)	59r	Phi	-	data
LW 77	Si-N	AM/FM-L, 70V, 12A, PQ=130W(87,5MHz)	59r	Phi	BLW 96	data
LW 78	Si-N	AM...VHF-L, 70V, 10A, PQ=100W(170MHz)	59r	Phi	-	data
LW 79	Si-N	UHF-L, 36V, 0,5A, PQ=2W(470MHz)	55r	Phi	BLX 67, BLY 38, 2N5945	data
LW 80	Si-N	UHF-L, 36V, 1A, PQ=4W(470MHz)	55r	Phi	BLU 97, BLW 43, BLX 68, BLY 53	data
LW 81	Si-N	UHF-L, 36V, 2,5A, PQ=10W(470MHz)	55r	Phi	BLW 15, BLW 44, BLX 69, 2N5946	data
LW 82	Si-N	UHF-L, 36V, 7A, PQ=30W(470MHz)	57s	Phi	BLU 30/12, MRF 646	data
LW 83	Si-N	AM-SSB-L, 65V, 3A, PEP=30W(28MHz)	59r	Phi	2N5941	data
LW 84	Si-N	VHF-L, 65V, 3A, PQ=25W(175MHz)	59r	Phi	BLW 86, MRF 314	data
LW 85	Si-N	AM...VHF-L, 36V, 9A, PQ=45W(175MHz)	59r	Phi	-	data
LW 86	Si-N	VHF-L, 65V, 4A, PQ=45W(175MHz)	59r	Phi	MRF 315	data
LW 87	Si-N	VHF-L, 36V, 6A, PQ=25W(175MHz)	59r	Phi	MRF 224	data
LW 89	Si-N	UHF-L, 60V, 0,32A, PQ=2W(470MHz)	55r	Phi	BFT 91, BLW 92, BLY 76, MRF 5174	data
LW 90	Si-N	UHF-L, 60V, 0,62A, PQ=4W(470MHz)	55r	Phi	BLW 93, BLY 37	data
LW 91	Si-N	UHF-L, 60V, 1,5A, PQ=10W(470MHz)	55r	Phi	BLW 94, BLX 94, MRF 323	data
LW 92	Si-N	UHF-L, 60V, 0,7A, PQ>1,5W(470MHz)	55r	Aeg	BLW 90, BLX 92	data
LW 93	Si-N	UHF-L, 60V, 1A, PQ>4,5W(470MHz)	55r	Aeg	BLW 91, BLX 93, MRF 321	data
LW 94	Si-N	UHF-L, 60V, 2A, PQ>15W(470MHz)	55r	Phi	BLX 94, MRF 323	data
LW 95	Si-N	AM-SSB-L, 110V, 8A, PEP=160W(28MHz)	59r	Phi	BLW 96	data
LW 96	Si-N	AM/FM-L, 110V, 12A, PEP=200W(108MHz)	59r	Phi	-	data
LW 97	Si-N	AM-SSB-L, 65V, 15A, PQ=175W(28MHz)	59r	Phi	-	data
LW 98	Si-N	UHF-L, 50V, 2A, PQ=4,4W(860MHz)	55r	Phi	BLX 98, 2N4431	data
LW 99	Si-N	AM-SSB-L, 36V, 18A, PEP=80W(28MHz)	59r	Phi	MRF 421, MRF 454	data

BLX

Type	Device	Short Description	Fig.	Manu	Comparision Types	More at
BLX 10	Si-N	NF-L, 125V, 2A, 11W(Tc=50°), >10MHz	2a§	Tra	BU 125S, BUX 49...50, BUY 41	data
BLX 11	Si-N	=BLX 10: 145V	2a§	Tra	BU 125S, BUX 49...50, BUY 49	data
BLX 12	Si-N	=BLX 10: 170V	2a§	Tra	BU 125S, BUX 50...51, BUY 49	data
BLX 13(C)	Si-N	AM/FM-L, 65V, 3A, PQ=25W(70MHz)	55r	Phi	-	data
BLX 14	Si-N	AM/FM-L, 85V, 4A, PQ=50W(70MHz)	≈55r	Phi	-	data
BLX 15	Si-N	AM/FM-L, 110V, 6,5A, PQ=150W(108MHz)	≈55r	Phi	-	data
BLX 16	Si-N	NF-L, 125V, 5A, 15W(Tc=50°), >10MHz	2a§	Tra	BU 125, BUX 34, BUY 47...48, BUY 80...81	data
BLX 17	Si-N	=BLX 16: 145V	2a§	Tra	BUY 47...48, BUY 68, BUY 80...81	data
BLX 18	Si-N	=BLX 16: 170V	2a§	Tra	BUY 48	data
BLX 19	Si-N	NF-L, 125V, 5A, 75W(Tc=50°), >10MHz	49a§	Tra	2N1724, 2N5387...5389	data
BLX 20	Si-N	=BLX 19: 145V	49a§	Tra	2N1724, 2N5387...5389	data
BLX 21	Si-N	=BLX 19: 170V	49a§	Tra	2N1724A, 2N5387	data
BLX 22	Si-N	NF-L, 125V, 10A, 60W(Tc=50°), >10MHz	49a§	Tra	(2N5542)[6]	data
BLX 23	Si-N	=BLX 22: 145V	49a§	Tra	(2N5542)[6]	data
BLX 24	Si-N	=BLX 22: 170V	49a§	Tra	(2N5542)[6]	data
BLX 25	Si-N	NF-L, 125V, 30A, 150W(Tc=50°), >10MHz	49a§	Tra	2N2825, 2N6324...6325	data
BLX 26	Si-N	=BLX 25: 145V	49a§	Tra	2N2825, 2N6324...6325	data
BLX 27	Si-N	=BLX 25: 170V	49a§	Tra	2N6324...6325	data
BLX 28	Si-N	NF-L, 125V, 40A, 187W(Tc=50°), >10MHz	49a§	Tra	2N2825, 2N6324...6325	data
BLX 29	Si-N	=BLX 28: 145V	49a§	Tra	2N2825, 2N6324...6325	data
BLX 30	Si-N	=BLX 28: 170V	49a§	Tra	2N6324...6325	data
BLX 31	Si-N	NF-L, 125V, 60A, 300W(Tc=50°), >10MHz	49a§	Tra	2N6278...6281	data
BLX 32	Si-N	=BLX 31: 145V	49a§	Tra	2N6278...6281	data
BLX 33	Si-N	=BLX 31: 170V	49a§	Tra	-	data
BLX 34	Si-N	NF-L, 125V, 80A, 300W(Tc=50°), >10MHz	49a§	Tra	2N6309...6311	data
BLX 35	Si-N	=BLX 34: 145V	49a§	Tra	2N6310...6311	data
BLX 36	Si-N	=BLX 34: 170V	49a§	Tra	2N6311	data
BLX 37	Si-N	UHF, 35V, 0,4A, 3GHz		Phi	-	data
BLX 38	Si-N	UHF, 30V, 0,8A, 3GHz		Phi	-	data
BLX 39	Si-N	VHF-L, 65V, 3A, PQ=45W(170MHz)	55r	Phi	BLW 25, MRF 315A, 2N5643	data
BLX 40	Si-P	NF-L, 80V, 2A, 11W(Tc=50°), >20MHz	2a§	Tra	BUY 90, 2N6190...6193	data
BLX 41	Si-P	=BLX 40: 100V	2a§	Tra	BUY 90, 2N6192...6193	data
BLX 42	Si-P	=BLX 40: 120V	2a§	Tra	BUY 90	data
BLX 46	Si-P	NF-L, 80V, 5A, 15W(Tc=50°), >20MHz	2a§	Tra	BUY 90, 2N6190...6193	data
BLX 47	Si-P	=BLX 46: 100V	2a§	Tra	BUY 90, 2N6192...6193	data

Type	Device	Short Description	Fig.	Manu	Comparision Types	More at
BLX 48	Si-P	=BLX 46: 120V	2a§	Tra	BUY 90	data
BLX 49	Si-P	NF-L, 80V, 5A, 45W(Tc=50°), >20MHz	49a§	Tra	2N5003, 2N5005, 2N5286...87, 2N6182...85	data
BLX 50	Si-P	=BLX 49: 100V	49a§	Tra	2N5003, 2N5005, 2N5286...87, 2N6184...85	data
BLX 51	Si-P	=BLX 49: 120V	49a§	Tra	2N5286...5287	data
BLX 52	Si-P	NF-L, 80V, 10A, 60W(Tc=50°), >20MHz	49a§	Tra	2N6182...6185, 2N6186...6189	data
BLX 53	Si-P	=BLX 52: 100V	49a§	Tra	2N6184...6185, 2N6188...6189	data
BLX 54	Si-P	=BLX 52: 120V	49a§	Tra	(2N5290...5291)[6]	data
BLX 55	Si-P	NF-L, 80V, 30A, 150W(Tc=50°), >20MHz	49a§	Tra	(2N6380...6382)[6]	data
BLX 56	Si-P	=BLX 55: 100V	49a§	Tra	(2N6380...6382)[6]	data
BLX 57	Si-P	=BLX 55: 120V	49a§	Tra	(2N6381...6382)[6]	data
BLX 58	Si-P	NF-L, 80V, 40A, 187W(Tc=50°), >20MHz	49a§	Tra	2N6380...6382	data
BLX 59	Si-P	=BLX 58: 100V	49a§	Tra	2N6380...6382	data
BLX 60	Si-P	=BLX 58: 120V	49a§	Tra	2N6381...6382	data
BLX 61	Si-P	NF-L, 80V, 60A, 300W(Tc=50°), >20MHz	49a§	Tra	2N6061, 2N6063, 2N6380...6382	data
BLX 62	Si-P	=BLX 61: 100V	49a§	Tra	2N6061, 2N6063, 2N6380...6382	data
BLX 63	Si-P	=BLX 61: 120V	49a§	Tra	2N6381...6382	data
BLX 65	Si-N	UHF-Tr/E, 36V, 0,7A, PQ=2W(470MHz)	2a§	Phi,Tix	2N5913	data
BLX 65 E,ES	Si-N	=BLX 65:	2e*		-	data
BLX 66	Si-N	UHF-L, 36V, 0,7A, PQ=2,5W(470MHz)	51r	Phi,Tix	-	data
BLX 67	Si-N	UHF-L, 36V, 0,7A, PQ=3W(470MHz)	55r	Phi,Tix	BLW 43, BLW 80, 2N5945	data
BLX 68	Si-N	UHF-L, 36V, 1A, PQ=7,8W(470MHz)	55r	Phi,Tix	BLU 97, BLY 53	data
BLX 69(A)	Si-N	UHF-L, 36V, 3,5A, PQ=17W(470MHz)	55r	Phi	BLW 15	data
BLX 70	Si-N	NF-L, -/225V, 20A(ss), 100W(Tc=45°)	49a§	Tra	2N6325	data
BLX 71	Si-N	=BLX 70: -/250V	49a§	Tra	2N6325	data
BLX 72	Si-N	=BLX 70: -/300V	49a§	Tra	2N6325	data
BLX 73	Si-N	=BLX 70: -/375V	49a§	Tra	-	data
BLX 74	Si-N	NF-L, -/225V, 10A(ss), 50W(Tc=45°)	49a§	Tra	2N5388...5389, 2N5540	data
BLX 75	Si-N	=BLX 74: -/250V	49a§	Tra	2N5388...5389, 2N5540	data
BLX 76	Si-N	=BLX 74: -/300V	49a§	Tra	2N5389, 2N5540	data
BLX 77	Si-N	=BLX 74: -/375V	49a§	Tra	-	data
BLX 78	Si-N	NF-L, -/225V, 5A(ss), 30W(Tc=45°)	49a§	Tra	-	data
BLX 79	Si-N	=BLX 79: -/250V	49a§	Tra	-	data
BLX 80	Si-N	=BLX 79: -/300V	49a§	Tra	-	data
BLX 81	Si-N	=BLX 79: -/375V	49a§	Tra	-	data
BLX 82	Si-P	NF-L, 60V, 20A, 150W(Tc=50°), >20MHz	23a§	Tra	BD 250A, BD 367, BD 746A, 2N5883...5884	data
BLX 83	Si-P	=BLX 82: 80V	23a§	Tra	BD 250B, BD 369, BD 746B, 2N5884	data
BLX 84	Si-P	=BLX 82: 100V	23a§	Tra	BD 250C, BD 746C, MJ 4502	data
BLX 85	Si-N	NF-L, 60V, 20A, 150W(Tc=50°), >10MHz	23a§	Tra	BD 249A, BD 366, BD 745A, 2N5885...5886	data
BLX 86	Si-N	=BLX 85: 80V	23a§	Tra	BD 249B, BD 368, BD 745B, 2N5886	data
BLX 87	Si-N	=BLX 85: 100V	23a§	Tra	BD 249C, BD 745C, MJ 802	data
BLX 88	Si-N	VHF-A/Tr, 50V, 0,1A, >600MHz	2a§	Tho	BFS 23, BFW 16...17, BFX 33, 2N3866	data
BLX 89	Si-N	VHF-A/Tr, 50V, 0,5A, 900MHz	2a§	Fer	BFS 23, BFX 33, BFX 55, 2N3866	data
BLX 91(A)	Si-N	UHF-Tr/E, 65V, 0,4A, PQ=1W(470MHz)	55r	Phi	BFT 91, BLW 89, BLY 76	data
BLX 91 CB	Si-N	=BLX 91:	55v		-	data
BLX 92(A)	Si-N	UHF-Tr/E, 65V, 0,7A, PQ=2,5W(470MHz)	55r	Phi	BLW 90, BLW 92, BLY 37, MRF 5174	data
BLX 93(A)	Si-N	UHF-Tr/E, 65V, 1A, PQ=7W(470MHz)	55r	Phi	BLW 91, MRF 321	data
BLX 94(A,C)	Si-N	UHF-L, 65V, 2A, PQ=20W(470MHz)	55r	Phi	BLW 94, MRF 323	data
BLX 95(A)	Si-N	UHF-L, 65V, 3A, PQ=40W(470MHz)	55r	Phi	2N6104...6105	data
BLX 96	Si-N	UHF-Tr/E, 40V, 0,4A, PQ=0,6W(860MHz)	55r	Phi	BLW 32, BLX 91, 2N4429	data
BLX 97	Si-N	UHF-Tr/E, 40V, 0,8A, PQ=1,1W(860MHz)	55r	Phi	BLW 33, 2N4430	data
BLX 98	Si-N	UHF-L, 40V, 2A, PQ=4W(860MHz)	55r	Phi	BLW 98, 2N4431	data

BLY

Type	Device	Short Description	Fig.	Manu	Comparision Types	More at
BLY 10	Si-N	HF-L, 40V, 2A, 10W(Tc=45°), >50MHz	23a§	Itt	-	data
BLY 11	Si-N	=BLY 10: >100MHz	23a§	Itt	-	data
BLY 12	Si-N	HF-L, 60V, 1,5A, 25W, >60MHz	23a§	Itt	-	data
BLY 14	Si-N	VHF-Tr/E, 80V, 1A, PQ=3,6W(180MHz)	49a	Itt	BLY 60	data
BLY 15	Si-N	HF-L, 90V, 2A, 15W, 200MHz	23a§	Itt	-	data
BLY 15 A	Si-N	HF-L, 64V, 2A, 11,5W, 180MHz	22a§	Itt	-	data
BLY 16	Si-N	HF-L, 64V, 1,5A, 11W, 250MHz	22a§	Itt	-	data
BLY 17(A,C)	Si-N	HF-L, 100V, 10A, PQ=40W(30MHz)	38a§	Phi	-	data
BLY 20	Si-N	VHF-L, 45V, 1A, PQ=6W(180MHz)	49a	Phi	BLY 57, BLY 60, 2N3632, 2N3926	data
BLY 21	Si-N	VHF-L, 70V, 1A, PQ=12W(180MHz)	49a	Phi	BLY 60, 2N3632	data
BLY 22	Si-N	VHF-L, 65V, 1,5A, PQ>7,5W(175MHz)	49a	Sie	BLY 60, 2N3632	data
BLY 23	Si-N	VHF-L, 65V, 3A, PQ>13,5W(175MHz)	49a	Sie	BLY 60, 2N3632	data
BLY 25	Si-N	VHF-L, 120V, 5A, 195MHz	49a	Sgs	2N4116, 2N5002, 2N5004	data
BLY 26	Si-N	VHF-L, 100V, 5A, 135MHz	49a	Sgs	2N4115, 2N5002, 2N5004	data
BLY 27	Si-N	VHF-L, 80V, 1A, PQ>4W(250MHz)	49a*	Tho	BLY 59...60, 2N3375, 2N3632	data
BLY 28	Si-N	VHF-L, 80V, 1A, PQ>4W(125MHz)	49a*	Tho	BLY 59...60, 2N3375, 2N3632	data
BLY 29	Si-N	HF/S-L, 100V, 2A, 46MHz	49a	Sgs	2N2892...93, 2N4075...76, 2N4998, 2N5000	data
BLY 30	Si-N	=BLY 29: 50MHz	49a	Sgs	2N2892...93, 2N4075...76, 2N4998, 2N5000	data
BLY 33	Si-N	VHF-Tr/E, 66V, 0,5A, PQ=3W(175MHz)	2a§	Phi,Tix	BFS 23, BFW 47, 2N3553	data
BLY 34	Si-N	VHF-Tr/E, 40V, 0,5A, PQ=3W(175MHz)	2a§	Phi,Tix	BFS 22, BFW 46, 2N3924	data
BLY 35	Si-N	VHF-L, 66V, 2,5A, PQ=13W(175MHz)	49a*	Phi,Tix	BLY 60, 2N3632	data
BLY 36	Si-N	VHF-L, 40V, 2,5A, PQ=13W(175MHz)	49a*	Phi,Tix	BLY 58, 2N3927	data
BLY 37	Si-N	UHF-L, 65V, 0,75A, PQ=6W(470MHz)	55r	Phi,Tix	BLW 93, BLX 93, MRF 321	data
BLY 38	Si-N	UHF-L, 36V, 0,5A, PQ=2W(470MHz)	55r	Phi,Tix	BLW 43, BLW 79, BLX 67, 2N5945	data
BLY 39	Si-N	UHF-L, 60V, 2A, PQ>7W(470MHz)	55r	Fer	BLW 91, BLX 93, MRF 321	data

Type	Device	Short Description	Fig.	Manu	Comparision Types	More at
BLY 40	Si-N	HF-L, 100V, 10A, 125W, >40MHz		Tho	2N5072	data
BLY 47	Si-N	S-L, 100V, 3A, 40W, 300/2000ns, B>30	23a§	Tix	BDY 24	data
BLY 47 A	Si-N	=BLY 47:	22a§		2N6077...6079, 2N6233...6235	data
BLY 48	Si-N	=BLY 47: B>60	23a§	Tix	BDY 24	data
BLY 48 A	Si-N	=BLY 48:	22a§		2N6077...6079, 2N6233...6235	data
BLY 49	Si-N	=BLY 47: 250V	23a§	Tix	BDY 26	data
BLY 49 A	Si-N	=BLY 49:	22a§		2N6077...6079, 2N6233...6235	data
BLY 50	Si-N	=BLY 47: 250V, B>60	23a§	Tix	BDY 26	data
BLY 50 A	Si-N	=BLY 50:	22a§		2N6077...6079, 2N6233...6235	data
BLY 53	Si-N	UHF-L, 36V, 1,3A, PQ=6W(470MHz)	55r	Phi	BLU 97, BLW 14, BLX 68	data
BLY 53 A	Si-N	UHF-L, 36V, 1A, PQ=7,8W(470MHz)	55r	Phi	BLU 97, BLW 14, BLX 68	data
BLY 53 B	Si-N	UHF-L, 36V, 2A, PQ>8,3W(470MHz)	55r	Fer	BLW 14, BLW 44, 2N5946	data
BLY 55	Si-N	VHF-L, 40V, 1A, PQ=4W(175MHz)	49a*	Phi	BLY 57, BLY 78, 2N3926	data
BLY 57	Si-N	VHF-L, 36V, 1A, PQ>7W(175MHz)	49a*	Mot,Phi,Tix	BLY 78, 2N3926	data
BLY 58	Si-N	VHF-L, 36V, 1,5A, PQ>12W(175MHz)	49a*	Mot,Phi,Tix	BLY 79, 2N3927	data
BLY 59	Si-N	VHF-L, 65V, 0,5A, PQ>7,5W(100MHz)	49a	Mot,Phi,Tix	2N3375	data
BLY 60	Si-N	VHF-L, 65V, 1A, PQ>13,5W(175MHz)	49a	Mot,Phi,Tix	2N3632	data
BLY 61	Si-N	VHF-Tr/E, 36V, 0,5A, PQ>1W(175MHz)	2a§	Tix	BFR 98, BFS 22, BFS 51, BLW 16, 2N4427	data
BLY 62	Si-N	VHF-L, 36V, 2A, PQ>5W(175MHz)	55r	Tix	BLW 18...19, BLW 37, MRF 212, 2N5590	data
BLY 63	Si-N	VHF-L, 36V, 5A, PQ>15W(175MHz)	55r	Tix	BLW 20, BLW 31, BLW 38, 2N6084	data
BLY 64	Si-N	HF-L, 80V, 5A, 50W, >60MHz	49a§	Sgs	2N5002, 2N5004	data
BLY 65	Si-P	HF-L, 80V, 5A, 50W, >60MHz	49a§	Sgs	2N5003, 2N5005	data
BLY 66(A,B)	Si-N	HF-L, 100V, 2A, 30W, >50MHz	49a	Sgs	2N4998, 2N5000	data
BLY 67(A,B)	Si-P	HF-L, 100V, 2A, 30W, >50MHz	49a	Sgs	2N4999, 2N5001	data
BLY 68	Si-N	NF/HF-L, 100V, 3A, 25W, 100MHz	23a§	Sgs	BDY 90...91, 2SC2681, 2SC2706, 2SC2837	data
BLY 70	Si-N	NF/HF-L, 100V, 5A, 33W, 70MHz	23a§	Sgs	BDY 90...91, 2SC2681, 2SC2706, 2SC2837	data
BLY 72	Si-N	NF/HF-L, 80V, 10A, 100W, >30MHz	49a	Sgs	2N5006, 2N5008, 2N5288...5289	data
BLY 74	Si-N	VHF-L, 65V, 0,5A, PQ>3W(400MHz)	49a	Sgs,Tix	-	data
BLY 76	Si-N	UHF-Tr/E, 65V, 0,3A, PQ=2W(470MHz)	55r	Phi,Tix	BFT 91, BLW 89, BLW 92, BLX 92	data
BLY 78	Si-N	VHF-L, 40V, 1A, PQ>4,7W(175MHz)	49a*	Aeg,Tix	BLY 57, 2N3926	data
BLY 79	Si-N	VHF-L, 40V, 2A, PQ>11W(175MHz)	49a*	Aeg,Tix	BLY 58, 2N3927	data
BLY 80	Si-N	VHF-L, 40V, 1A, PQ>4W(175MHz)		Aeg	-	data
BLY 81	Si-N	VHF-L, 40V, 2A, PQ>11W(175MHz)		Aeg	-	data
BLY 82	Si-N	HF-L, 80V, 10A, 125W, >40MHz		Tho	BLY 40, 2N5072	data
BLY 83	Si-N	VHF-L, 66V, 2,5A, PQ>7W(175MHz)	55r	Phi,Tix	BLW 19, MRF 212, 2N5590	data
BLY 84	Si-N	VHF-L, 40V, 2,5A, PQ>13W(175MHz)	55r	Phi,Tix	BLW 29, BLY 88, MRF 226, 2N6081	data
BLY 85	Si-N	VHF-L, 40V, 1A, PQ>4W(175MHz)	55r	Phi,Tix	BLY 87, 2N6080	data
BLY 86	Si-N	TV-HA, 300/300V, 0,4A, 10W, 60MHz	22a§	Sgs	2N3739, 2SC1929, 2SC2022, 2SC2354,	data
BLY 87(A,C)	Si-N	VHF-L, 36V, 1,25A, PQ=8W(175MHz)	55r	Phi,Tix	BLW 19, BLW 37, MRF 212, 2N5590	data
BLY 88(A,C,T)	Si-N	VHF-L, 36V, 2,5A, PQ=15W(175MHz)	55r	Phi,Tix	BLW 29, BLY 84, MRF 226, 2N6081	data
BLY 89(A,C)	Si-N	VHF-L, 36V, 3,5A, PQ=25W(175MHz)	55r	Phi,Tix	MRF 209, MRF 238, 2N5591, 2N6082...83	data
BLY 90	Si-N	VHF-L, 36V, 8A, PQ=50W(175MHz)	≈55r	Phi	-	data
BLY 91(A,C)	Si-N	VHF-L, 65V, 0,75A, PQ=8W(175MHz)	55r	Phi,Tix	BLY 98, 2N5641	data
BLY 92(A,C)	Si-N	VHF-L, 65V, 1,5A, PQ=15W(175MHz)	55r	Phi,Tix	BLW 24, BLY 93	data
BLY 93(A,C)	Si-N	VHF-L, 65V, 2A, PQ=25W(175MHz)	55r	Phi,Tix	BLW 24	data
BLY 94	Si-N	VHF-L, 65V, 6A, PQ=50W(175MHz)	≈55r	Phi	BLW 25, 2N5643	data
BLY 95	Si-N	VHF-L, 55V, 1A, >400MHz		Aeg	-	data
BLY 96	Si-N	VHF-L, 55V, 2A, >500MHz		Aeg	-	data
BLY 97	Si-N	VHF-L, 60V, 1A, PQ>4W(175MHz)	55r	Phi,Tix	BLY 91, BLY 98, 2N5641	data
BLY 98	Si-N	VHF-L, 60V, 1A, PQ>7W(175MHz)	55r	Phi	BLY 91, BLY 97, 2N5641	data
BLY 99	Si-N	VHF-L, 30V, 0,5A, PQ>1W(470MHz)	2a§	Phi	BFS 50, BLX 65, MRF 629, 2N3948	data

BM...BO

Type	Device	Short Description	Fig.	Manu	Comparision Types	More at
BM	Si-P	=2SB1325 (Typ-Code/Stempel/marking)	39	Say	→2SB1325	data
BM	Si-N	=BCP 55-16 (Typ-Code/Stempel/marking)	48	Phi	→BCP 55	data
BM	Si-N	=BCX 55-16 (Typ-Code/Stempel/marking)	39	Phi	→BCX 55	data
BM	Si-P	=BSS 63 (Typ-Code/Stempel/marking)	35	Mot,Phi,Sie	→BSS 63	data
BM	Z-Di	=BZX 399-C13 (Typ-Code/Stempel/marking)	71(1,7mm)	Phi	→BZX 399-...	data
BM	Si-N/P+R	=KRX 102U (Typ-Code/Stempel/marking)	45(2mm)	Kec	→KRX 102U	data
BM	Z-Di	=P4 SMA-24A (Typ-Code/Stempel/marking)	71(5x2,5)	Fag	→P4 SMA-24A	data
BM	Z-Di	=SMAJ 15A (Typ-Code/Stempel/marking)	71(5x2,5)	Tho	→SMAJ ...	data
BM	Z-Di	=SMBJ 15CA (Typ-Code/Stempel/marking)	71(5x3,5)	Mop	→SMBJ ...	data
BM 339	KOP-IC	=LM 339			→LM 339	
BMp	Si-P	=BSS 63 (Typ-Code/Stempel/marking)	35	Phi	→BSS 63	data
BMQ	Si-N	=2SC4713K-Q (Typ-Code/Stempel/marking)	35	Rhm	→2SC4713K	data
BMQ	Si-N	=2SC4774-Q (Typ-Code/Stempel/marking)	35(2mm)	Rhm	→2SC4774	data
BMR	Si-N	=2SC4713K-R (Typ-Code/Stempel/marking)	35	Rhm	→2SC4713K	data
BMR	Si-N	=2SC4774-R (Typ-Code/Stempel/marking)	35(2mm)	Rhm	→2SC4774	data
BMS	Si-N	=2SC4713K-S (Typ-Code/Stempel/marking)	35	Rhm	→2SC4713K	data
BMS	Si-N	=2SC4774-S (Typ-Code/Stempel/marking)	35(2mm)	Rhm	→2SC4774	data
BMS	LIN-IC	=BGA 312 (Typ-Code/Stempel/marking)	44	Sie	→BGA 312	data
BMS	LIN-IC	=BGA 425 (Typ-Code/Stempel/marking)	44(2mm)	Sie	→BGA 425	data
BMS	LIN-IC	=BGA 425 (Typ-Code/Stempel/marking)	44(2mm)	Sie	→BGA 425	data
BMS	LIN-IC	=BGA 427 (Typ-Code/Stempel/marking)	44(2mm)	Sie	→BGA 427	data
BMS	LIN-iC	=BGA 427 (Typ-Code/Stempel/marking)	44(2mm)	Sie	→BGA 427	data
BMs	Si-P	=BSS 63 (Typ-Code/Stempel/marking)	35	Sie	→BSS 63	data
BMT	Si-P	=BSS 63 (Typ-Code/Stempel/marking)	35	Phi	→BSS 63	data

Type	Device	Short Description	Fig.	Manu	Comparision Types	More at
BN	Si-P	=2SB1394 (Typ-Code/Stempel/marking)	39	Say	→2SB1394	data
BN	Si-N+Di+R	=2SD2324 (Typ-Code/Stempel/marking)	35	Say	→2SD2324	data
BN	MOS-N-FET-e*	=2SK3290 (Typ-Code/Stempel/marking)	35	Hit	→2SK3290	
BN	Si-P	=BCW 61FN (Typ-Code/Stempel/marking)	35	Sie	→BCW 61FN	data
BN	Z-Di	=BZX 399-C15 (Typ-Code/Stempel/marking)	71(1,7mm)	Phi	→BZX 399-...	data
BN	Si-N/P+R	=KRX 104U (Typ-Code/Stempel/marking)	45(2mm)	Kec	→KRX 104U	data
BN	Z-Di	=P4 SMA-27 (Typ-Code/Stempel/marking)	71(5x2,5)	Fag	→P4 SMA-27	data
BN	Z-Di	=SMBJ 16C (Typ-Code/Stempel/marking)	71(5x3,5)	Mop	→SMBJ ...	data
BN 1 A3Q...L4Z	Si-R+R	=AN 1A3Q...L4Z: Fig.→	40c	Nec	-	data
BN 2 A3Q	Si-P+R	hi-beta, Rb=1k, Rbe=10kΩ, 25V, 0,1A	40c	Nec	-	data
BN 2 A4M	Si-P+R	=BN 2A3Q: Rb=10k, Rbe=10kΩ	40c	Nec	-	data
BN 2 A4P	Si-P+R	=BN 2A3Q: Rb=10k, Rbe=47kΩ	40c	Nec	-	data
BN 2 A4Z	Si-P+R	=BN 2A3Q: Rb=10k, Rbe=-	40c	Nec	-	data
BN 2 F4M	Si-P+R	=BN 2A3Q: Rb=22k, Rbe=22kΩ	40c	Nec	-	data
BN 2 F4N	Si-P+R	=BN 2A3Q: Rb=22k, Rbe=47kΩ	40c	Nec	-	data
BN 2 F4Z	Si-P+R	=BN 2A3Q: Rb=22k, Rbe=-	40c	Nec	-	data
BN 2 L3M	Si-P+R	=BN 2A3Q: Rb=Rbe=4,7kΩ	40c	Nec	-	data
BN 2 L3N	Si-P+R	=BN 2A3Q: Rb=4,7k, Rbe=10kΩ	40c	Nec	-	data
BN 2 L3Z	Si-P+R	=BN 2A3Q: Rb=4,7k, Rbe=-	40c	Nec	-	data
BN 2 L4L	Si-P+R	=BN 2A3Q: Rb=47k, Rbe=22kΩ	40c	Nec	-	data
BN 2 L4M	Si-P+R	=BN 2A3Q: Rb=47k, Rbe=47kΩ	40c	Nec	-	data
BN 2 L4Z	Si-P+R	=BN 2A3Q: Rb=47k, Rbe=-	40c	Nec	-	data
BN 3 L4Z	Si-P+R	Rb=47kΩ, int. Emitter-Di, 20V, 20mA	40c	Nec	-	data
BNs	Si-P	=BCW 61FN (Typ-Code/Stempel/marking)	35	Sie	→BCW 61FN	data
BNS	LIN-IC	=BGA 318 (Typ-Code/Stempel/marking)	44	Sie	→BGA 318	data
BO	Si-P	=2SA1200-O (Typ-Code/Stempel/marking)	39	Tos	→2SA1200	data
BO	Si-P	=2SB1396 (Typ-Code/Stempel/marking)	39	Say	→2SB1396	data
BO	Si-P	=BCW 61RA (Typ-Code/Stempel/marking)	35	Sie,Val	→BCW 61RA	data
BO	Si-P	=KTA1660-O (Typ-Code/Stempel/marking)	39	Kec	→KTA 1660	data
BO 12	Si-Di	Uni, 100V, 0,25A	31a	Tho	BA 157...159, BY 204/4, BY 208/600, ++	data
BO 22	Si-Di	=BO 12: 200V	31a	Tho	BA 157...159, BY 204/4, BY 208/600, ++	data
BO 42	Si-Di	=BO 12: 400V	31a	Tho	BA 157...159, BY 204/4, BY 208/6, ++	data
BO 62	Si-Di	=BO 12: 600V	31a	Tho	BA 158...159, BY 204/8, BY 208/600, ++	data
BO 82	Si-Di	=BO 12: 800V	31a	Tho	BA 159, BY 204/8, BY 208/800, ++	data
BO 102	Si-Di	=BO 12: 1000V	31a	Tho	BA 159, BY 204/10, BY 208/1000, ++	data
BOD 1-04...40(L,R)	Trigger-Di	Ub=400...4000V,Ib=1mA,Itsm=200A	26b	Bbc		data

BP..BQ

Type	Device	Short Description	Fig.	Manu	Comparision Types	More at
BP	Si-P	=2SB1397 (Typ-Code/Stempel/marking)	39	Say	→2SB1397	data
BP	MOS-P-FET-e*	=2SJ574 (Typ-Code/Stempel/marking)	35	Hit	→2SJ574	data
BP	SI-P	=BCW 61RB (Typ-Code/Stempel/marking)	35	Sie,Val	→BCW 61RB	data
BP	Z-Di	=BZX 399-C16 (Typ-Code/Stempel/marking)	71(1,7mm)	Phi	→BZX 399-...	data
BP	Z-Di	=P4 SMA-27A (Typ-Code/Stempel/marking)	71(5x2,5)	Fag	→P4 SMA-27A	data
BP	Z-Di	=SMBJ 16CA (Typ-Code/Stempel/marking)	71(5x3,5)	Mop	→SMBJ ...	data
BP 1 A3M...L3N	Si-P+R	=AP 1A3M...L3N: Fig.→	40c	Nec	-	data
BP 50 M05	SMPS-IC	DC-DC Converter, 5V, 0,5A	≈9-SIP	Rhm	-	
BP 50 M12	SMPS-IC	DC-DC Converter, 12V, 0,5A	≈9-SIP	Rhm	-	
BP 51 L05	SMPS-IC	DC-DC Converter, -5V, 0,1A	≈9-SIP	Rhm	-	pdf
BP 51 L12	SMPS-IC	DC-DC Converter, -12V, 0,1A	≈9-SIP	Rhm	-	pdf
BP 100	Si-Opto	Foto-Element, >230mV(1000Lux),50nA/Lux	Chip	Sie		
BP100	PC	850nm S:0.55mA/mW P:25nA/Lx		Sie		data pinout
BP100P	PC	850nm S:0.55mA/mW P:25nA/Lx		Sie		data pinout
BP101	PT	780nm P:>0.5nA/Lx	TO18	Sie		data pinout
BP103	PT	850(950)nm P:>1µA/mW/cm²	TO18	Sie		data pinout
BP103B	PT	850nm P:>3.2µA/mW/cm²	T1¾	Sie		data pdf pinout
BP103BF	PT	880(950)nm P:>3.2µA/mW/cm²	T1¾	Sie		data pdf pinout
BP104	PD	850nm S:0.62mA/mW P:55nA/Lx		Sie		data pdf pinout
BP104FS	PD	950nm S:0.7mA/mW P:34µA/mW/cm²		Sie		data pdf pinout
BP 3002	Hybrid-IC	Telecom, Telefonapparat/telephone unit		Rhm		
BP 3003	Hybrid-IC	Telecom, Telefonapparat/telephone unit		Rhm		
BP 3004	Hybrid-IC	Telecom, Telefonapparat/telephone unit		Rhm		
BP 3005	Hybrid-IC	Telecom, Telefonapparat/telephone unit		Rhm		
BP 3008	Hybrid-IC	Telecom, Telefon Set f. Fax		Rhm		
BP 3009	Hybrid-IC	Telecom, Telefon Set f. Fax		Rhm		
BP 3303(-7)	Hybrid-IC	Telecom, Telefon Set f. Fax, multifct.	48Pin	Rhm		
BP 3304(-7)	Hybrid-IC	Telecom, Telefon Set f. Fax, multifct.	48Pin	Rhm		
BP 3323	Hybrid-IC	Telecom, Telefon Set f. Fax, multifct.	46Pin	Rhm		
BP 3324(-7)	Hybrid-IC	Telecom, Telefon Set f. Fax, multifct.	46Pin	Rhm		
BP 3501	Hybrid-IC	Telecom, Lautspr./speakerphone Amp.	≈20-SIP	Rhm		
BP 3507(Z-1,Z-2)	Hybrid-IC	µcomp, Scsi Bus, Active Terminator	52Pin	Rhm		
BP 3510	Hybrid-IC	µcomp, Scsi Bus, Active Terminator	22-SIP	Rhm		
BP 3511	Hybrid-IC	Telecom, Lautspr./speakerphone Amp.	18-SIP	Rhm		
BP 5005	SMPS-IC	DC-DC Converter, 5V, 1A	≈9-SIP	Rhm		
BP 5020(X)	SMPS-IC	DC-DC Converter, 5V, 1A	≈9-SIP	Rhm		
BP 5021	SMPS-IC	DC-DC Converter, 5V, 0,5A	≈9-SIP	Rhm		
BP 5022	SMPS-IC	DC-DC Converter, 12V, 0,5A	≈9-SIP	Rhm		
BP 5030	Hybrid-IC	AC-DC Converter, 80...120V~, 5V, 100mA	≈10-SIP	Rhm		
BP 5032	Hybrid-IC	AC-DC Converter, 80...120V~, 24V, 50mA	≈10-SIP	Rhm		

Type	Device	Short Description	Fig.	Manu	Comparision Types	More at
BP 5034	Hybrid-IC	AC-DC Converter, 80...120V~, 12V, 100mA	≈10-SIP	Rhm	-	pdf
BP 5035	Hybrid-IC	AC-DC Converter, 80...120V~,-12V, 200mA	≈10-SIP	Rhm	-	pdf
BP 5040	Hybrid-IC	AC-DC Converter, 160..253V~, 5V, 100mA	≈8-SIP	Rhm	-	pdf
BP 5041	Hybrid-IC	AC-DC Converter, 160..253V~,12V, 100mA	≈8-SIP	Rhm	-	pdf
BP 5201	SMPS-IC	DC-DC Converter, 5V, 0,2A	≈7-SIP	Rhm	-	
BP 5302(F)	SMPS-IC	DC-DC Converter, -24V, 0,03A	≈9-SIP	Rhm	-	pdf
BP 5307	SMPS-IC	DC-DC Converter, 32V/40mA, 19V/40mA	Board	Rhm	-	pdf
BP 5310	SMPS-IC	DC-DC Converter, 12V, 0,12A	≈9-SIP	Rhm	-	pdf
BP 5311(X)	Hybrid-IC.	DC-DC Converter, 29,5V, 25mA	≈9-SIP	Rhm	-	
BP 5311	SMPS-IC	DC-DC Converter, 29,5V, 25mA	≈9-SIP	Rhm	-	pdf
BP 5313	SMPS-IC	DC-DC Converter, 40V, 60mA	≈11-SIP	Rhm	-	pdf
BP 5315	SMPS-IC	DC-DC Converter, 12V, 0,12A	Board	Rhm	-	
BP 5401	SMPS-IC	DC-DC Conv.,+5V/1A, +12V/1A, -12V/0,1A		Rhm	-	
BPF24	PD	850nm S:0.55mA/mW	TO48	Phi		data pinout
BPV10	PD	950nm S:0.5mA/mW P:70µA/mW/cm²		Tel		data pinout
BPV10NF	PD	940(870)nm S:0.5mA/mW P:60µA/mW/cm²		Tel,Tem,Vis		data pdf pinout
BPV11	PT	850(950)nm P:10µA/mW/cm²	T1¾	Tem,Vis		data pdf pinout
BPV11F	PT	920(950)nm P:45µA/mW/cm²	T1	Tem,Vis		data pdf pinout
BPV20F	PD	950nm S:0.6mA/mW P:60µA/mW/cm²	TO92	Tem,Vis		data pdf pinout
BPV20NFL	PD	940nm S:0.6mA/mW P:60µA/mW/cm²	TO92	Tem		data pinout
BPV21F	PD	950nm S:0.6mA/mW P:38µA/mW/cm²	TO92	Vis		data pdf pinout
BPV22F	PD	950nm S:0.6mA/mW P:80µA/mW/cm²		Tem,Vis		data pdf pinout
BPV22NF	PD	940nm S:0.6mA/mW P:85µA/mW/cm²		Vis		data pdf pinout
BPV23F	PD	950nm S:0.6mA/mW P:63µA/mW/cm²		Vis		data pdf pinout
BPV23NF	PD	940nm S:0.6mA/mW P:65µA/mW/cm²		Tem,Vis		data pdf pinout
BPV23NFL	PD	940nm S:0.6mA/mW P:65µA/mW/cm²		Tem,Vis		data pdf pinout
BPW ...	Opto					
BPW13	PT	780nm P:1nA/Lx	TO18	Tel		data pdf pinout
BPW14	PT	780nm P:10nA/Lx	TO18	Aeg		data pdf pinout
BPW16N	PT	780nm P:0.4nA/Lx		Tel,Vis		data pdf pinout
BPW17N	PT	780nm P:3nA/Lx		Tel,Vis		data pdf pinout
BPW20	PD	920nm P:60nA/Lx	TO5	Tem,Vis		data pdf pinout
BPW20RF	PD	920nm P:60nA/Lx	TO5	Vis		data pdf pinout
BPW21	PD	565nm S:0.34mA/mW P:7nA/Lx	TO5	Tel		data pinout
BPW22A	PT	800(950)nm P:>5µA/mW/cm²	SOD53F	Phi,Val		data pinout
BPW24	PD	900(950)nm S:0.55mA/mW	TO18	Tem,Vis		data pdf pinout
BPW28	PD	900(950)nm	TO18	Tel		data pinout
BPW32	PD	800(850)nm S:0.5mA/mW P:10nA/Lx		Sie		data pinout
BPW33	PD	800(850)nm S:0.59mA/mW P:75nA/Lx		Sie		data pdf pinout
BPW33S	PD	800(850)nm S:0.59mA/mW P:75nA/Lx		Sie		data pinout
BPW34	PD	880(830)nm S:0.65mA/mW P:50µA/mW/cm²		Sie,Osr		data pdf pinout
BPW34FAS	PD	880(830)nm S:0.65mA/mW P:50µA/mW/cm²		Sie,Osr		data pinout
BPW34S	PD	850nm S:0.62mA/mW P:80nA/Lx		Sie,Tel,Tem		data pdf pinout
BPW35	PD	750nm P:300µA/mW/cm²		Tel		data pinout
BPW36	PT	850nm P:>0.3µA/mW/cm²	TO18	Har,Qtc,Rca		data pdf pinout
BPW38	PT	850nm P:>15µA/mW/cm²	TO18	Har,Qtc,Rca		data pdf pinout
BPW39	PT	780nm P:2nA/Lx	TO92	Tel		data pinout
BPW40	PT	780nm P:6nA/Lx	T1¾	Tel		data pdf pinout
BPW41N	PD	950nm P:45µA/mW/cm²		Tel,Vis		data pdf pinout
BPW42	PT	780nm P:3nA/Lx	T1	Tel		data pinout
BPW43	PD	900(950)nm P:8µA/mW/cm²	T1¾	Tel,Vis		data pdf pinout
BPW46	PD	900(950)nm P:50µA/mW/cm²		Vis		data pdf pinout
BPW47	PT	800(950)nm P:1µA/mW/cm²	TO18	Tel		data pinout
BPW48	PD	930(870)nm S:0.6mA/mW P:4.5µA/mW/cm²	TO18	Tel		data pdf pinout
BPW50	PD	930nm P:45µA/mW/cm²	SOD67	Phi,Val		data pinout
BPW71	PT	800nm P:37µA/mW/cm²	SOT71A	Phi		data pinout
BPW75	PD	950nm P:48µA/mW/cm²		Tel		data pinout
BPW76	PT	800nm P:2nA/Lx	TO18	Tem,Vis		data pdf pinout
BPW77	PT	800nm P:30nA/Lx	TO18	Tel		data pinout
BPW77N	PT	850(950)nm P:20nA/Lx	TO18	Tem,Vis		data pdf pinout
BPW78	PT	780nm P:12nA/Lx		Tel,Vis		data pdf pinout
BPW79	PD	700nm	TO18	Tel		data pinout
BPW80	PD	950nm Vb:4.5...5.5V Ib:<35mA	TO18	Tel		data pdf pinout
BPW82	PD	920(870)nm P:45µA/mW/cm²		Tel,Vis		data pdf pinout
BPW83	PD	920(870)nm P:45µA/mW/cm²		Tel		data pinout
BPW85	PT	830(950)nm P:>3µA/mW/cm²	T1	Tel,Vis		data pdf pinout
BPW86	PD	810(870)nm S:0.5mA/mW P:4.5µA/mW/cm²	TO18	Tel		data pdf pinout
BPW96	PT	830(950)nm P:4.5µA/mW/cm²	T1¾	Tel,Vis		data pdf pinout
BPW97	PD	810(950)nm S:0.35mA/mW	TO18	Tem,Vis		data pdf pinout
BPX ...	Opto					
BPX25	PT	800nm P:10nA/Lx	SOT29/1	Phi		data pinout
BPX29	PT	800nm P:0.6nA/Lx	SOT29/2	Phi		data pinout
BPX 38	Si-Opto-N	Foto-Trans., 25V, Ip=1,5mA(1000Lux)	2a	Sie		
BPX38	PT	880(950)nm P:>1µA/mW/cm²	TO18	Sie,Tem		data pdf pinout
BPX40	PD	800nm P:14nA/Lx		Phi		data pinout
BPX41	PD	800nm P:40nA/Lx		Phi		data pinout
BPX42	PD	800nm P:140nA/Lx		Phi		data pinout
BPX 43	Si-Opto-N	Foto-Trans., 25V, Ip=7>2mA(1000Lux)	2a	Sie		
BPX43	PT	880(950)nm P:>2µA/mW/cm²	TO18	Tem,Vis		data pdf pinout
BPX 48	Si-Opto-Di	Dual Diff. Foto-Di, 10V, >15nA/Lux	common K	Sie		
BPX48	PD	920(950)nm S:0.55mA/mW P:7.5nA/Lx		Osr,Sie		data pdf pinout
BPX 48	Si-Opto-Di	Idark=100nA, <500ns, 0,85µm				
BPX60	PD	850(400)nm S:0.2mA/mW P:70nA/Lx	TO5	Sie		data pdf pinout
BPX61	PD	850nm P:35µA/mW/cm²	TO5	Phi,Sie		data pinout

Type	Device	Short Description	Fig.	Manu	Comparision Types	More at
BPX63	PD	800(850)nm S:0.5mA/mW P:10nA/Lx	TO18	Sie		data pdf pinout
BPX65	PD	850nm S:0.55mA/mW P:;0nA/Lx	TO18	Sie		data pdf pinout
BPX71	PT	800nm P:0.2µA/mW/cm²	SOT71A	Phi		data pinout
BPX72	PT	800nm P:>2.4nA/Lx	SOT70A	Phi,Sie		data pinout
BPX79	PC	800(400)nm S:0.19mA/mW P:170nA/Lx		Sie		data pdf pinout
BPX81	PT	850(950)nm P:>1.25µA/mW/cm²		Sie		data pdf pinout
BPX90	PD	950nm S:0.48mA/mW P:13nA/Lx		Sie		data pdf pinout
BPX91	PD	850(400)nm S:0.2mA/mW P:65nA/Lx		Sie		data pinout
BPX92	PD	830(850)nm S:0.5mA/mW P:9.5nA/Lx		Sie		data pinout
BPX99	PT	800(950)nm P:>4µA/mW/cm²	TO18	Tem		data pinout
BPY ...	Opto					
BPY 11(I,II,III)	Si-Opto	Foto-Element, >300mV(1000Lux),80nA/Lux	Chip	Sie	-	
BPY11P	PC	850nm S:0.55mA/mW P:>28nA/Lx		Sie		data pdf pinout
BPY 12	Si-Opto-Di	Foto-Di, 20V, >100nA/Lux, Idark<1mA	Chip	Sie		
BPY12	PD	920(850)nm S:0.6mA/mW P:180nA/Lx		Sie		data pdf pinout
BPY 43	Si-Opto	Foto-Element, >270mV(1000Lux),20nA/Lux	(13x2mmø)	Sie	-	
BPY 44	Si-Opto	Foto-Element, >330mV(1000Lux),27nA/Lux	(13x2mmø)	Sie	-	
BPY 45	Si-Opto	Foto-Element, >280mV(1000Lux) >1,45µA/Lux, 0,85µm	Chip	Sie	-	
BPY 46	Si-Opto	Foto-Element, >280mV(1000Lux) >1,3µA/Lux, 0,85µm	Chip	Sie	-	
BPY 47	Si-Opto	Foto-Element, >280mV(1000Lux) >1,3µA/Lux, 0,85µm	Chip	Sie	-	
BPY47P	PC	850nm S:0.51mA/mW P:1400nA/Lx		Sie		data pdf pinout
BPY 48	Si-Opto	Foto-Element, >280mV(1000Lux) >0,43µA/Lux, 0,85µm	Chip	Sie	-	
BPY48P	PC	850nm S:0.55mA/mW P:500nA/Lx		Sie		data pdf pinout
BPY 61(I,II,III)	Si-Opto-N	Foto-Trans., 32V, 0,45...1µm I: Ip=0,25...1, II: 0,9..2,6mA(1000Lux) III: Ip>2mA(1000Lux)	(13x2mmø) ohne/wo. Basis/Base	Sie		
BPY 62(I,II,III)	Si-Opto-N	Foto-Trans., 32V, 0,45...1µm	2a	Sie		
BPY62	PT	850(950)nm P:>4µA/mW/cm²	TO18	Sie		
BPY 62(I,II,III)	Si-Opto-N	I: Ip=1..2,5, II: 2..4, III: >3mA (1000Lux)				data pdf pinout
BPY 63	Si-Opto	Foto-Element, >280mV(1000Lux) >0,65µA/Lux, 0,85µm	Chip	Sie		
BPY63P	PC	830nm S:0.5mA/mW P:650nA/Lx		Sie		
BPY 64	Si-Opto	Foto-Element, >280mV(1000Lux) >0,23µA/Lux, 0,85µm	Chip	Sie		
BPY64P	PC	850nm S:0.5mA/mW P:250nA/Lx		Sie,Vac		data pdf pinout
BPY 73	Si-Opto	Solarzelle, 535mV, Ik>137mA, 0,7µm	Chip	Sie	-	
BPY 74	Si-Opto	Solarzelle, 585mV, Ik>130mA, 0,7µm	Chip	Sie	-	
BQ	Si-P	=2PB709AQ (Typ-Code/Stempel/marking)	35	Phi	→2PB709A	data
BQ	Si-P	=2SB1218A-Q (Typ-Code/Stempel/marking)	35(2mm)	Mat	→2SB1218A	data
BQ	Si-P	=2SB1627-Q (Typ-Code/Stempel/marking)	≈35	Mat	→2SB1627	data
BQ	Si-P	=2SB709A-Q (Typ-Code/Stempel/marking)	35	Mat	→2SB709A	data
BQ	Si-P	=2SB766A-Q (Typ-Code/Stempel/marking)	39	Mat	→2SB766A	data
BQ	Si-N	=2SC2412-BQ (Typ-Code/Stempel/marking)	≈35	Rhm	→2SC2412	data
BQ	Si-N	=2SC2412K-Q (Typ-Code/Stempel/marking)	35	Rhm	→2SC2412K	data
BQ	Si-N	=2SC4081-Q (Typ-Code/Stempel/marking)	35(2mm)	Rhm	→2SC4081	data
BQ	Z-Di	=BZX 399-C18 (Typ-Code/Stempel/marking)	71(1,7mm)	Phi	→BZX 399-...	data
BQ	Z-Di	=P4 SMA-30 (Typ-Code/Stempel/marking)	71(5x2,5)	Fag	→P4 SMA-30	data
BQ	Z-Di	=SMBJ 17C (Typ-Code/Stempel/marking)	71(5x3,5)	Mop	→SMBJ ...	data
BQ 20 Z 75 DBT	DIG-IC	Gas Gauge and Protection-enabled IC	QFN	Tix	-	pdf pinout
bq 2083-V1P2	LIN-IC	SMD, Sbs-compliant Gas Gauge IC	38-VSMDIP	Tix	-	pdf pinout
bq 2084	LIN-IC	SMD, SBS 1.1 Compliant Gas Gauge IC	38-VSMDIP	Tix	-	pdf
bq 2085	LIN-IC	SMD, SBS-compliant Gas Gauge IC	38-VSMDIP	Tix	-	pdf
BQ 20 Z 90 DBT	DIG-IC	SBS 1.1-compliant Gas Gauge enabled	QFN	Tix	-	pdf pinout
BQ 20 Z 90 DBTR	DIG-IC	SBS 1.1-compliant Gas Gauge enabled	QFN	Tix	-	pdf pinout
BQ 24230	LIN-IC	Li-ion Battery Charger, Power-path IC	TSOP	Tix	-	pdf pinout
BQ 24232	LIN-IC	Li-ion Battery Charger, Power-path IC	TSOP	Tix	-	pdf pinout
BQ 24314 DSG	DIG-IC	Overvolt. and Overcurr. Protection IC	QFN	Tix	-	pdf pinout
BQ 24314 DSJ	DIG-IC	Overvolt. and Overcurr. Protection IC	QFN	Tix	-	pdf pinout
BQ 24316 DSG	DIG-IC	Overvolt. and Overcurr. Protection IC	QFN	Tix	-	pdf pinout
BQ 24316 DSJ	DIG-IC	Overvolt. and Overcurr. Protection IC	QFN	Tix	-	pdf pinout
bq 29311	LIN-IC	SMD, 3- or 4-Cell Li-Ion/Li-Polymer Battery Protection IC, I²C	24-SSMDIP	Tix	-	pdf
bq 29312	LIN-IC	SMD, 2-,3- or 4-Cell Li-Ion/Li-Polymer Battery Protection IC, I²C	24-SSMDIP	Tix	-	pdf
bq 29400	LIN-IC	SMD, Voltage Prot. for 2-,3- or 4-Cell Li-Ion Batteries (2nd Protection)	8-SSMDIP	Tix	-	pdf
bq 29401	LIN-IC	SMD, Voltage Prot. for 2-,3- or 4-Cell Li-Ion Batteries (2nd Protection)	8-SSMDIP	Tix	-	pdf
BQ 4010 LYMA-70 N	IC	8k x 8 nonvolatile Sram (5 V, 3.3 V)	DIP	Tix	-	pdf pinout
BQ 4010 MA-70	IC	8k x 8 nonvolatile Sram (5 V, 3.3 V)	DIP	Tix	-	pdf pinout
BQ 4010 YMA-70	IC	8k x 8 nonvolatile Sram (5 V, 3.3 V)	DIP	Tix	-	pdf pinout
BQ 4010 YMA-70 N	IC	8k x 8 nonvolatile Sram (5 V, 3.3 V)	DIP	Tix	-	pdf pinout
BQ 4010 MA-85	IC	8k x 8 nonvolatile Sram (5 V, 3.3 V)	DIP	Tix	-	pdf pinout
BQ 4010 YMA-85	IC	8k x 8 nonvolatile Sram (5 V, 3.3 V)	DIP	Tix	-	pdf pinout
BQ 4010 YMA-85 N	IC	8k x 8 nonvolatile Sram (5 V, 3.3 V)	DIP	Tix	-	pdf pinout
BQ 4010 MA-150	IC	8k x 8 nonvolatile Sram (5 V, 3.3 V)	DIP	Tix	-	pdf pinout
BQ 4010 YMA-150	IC	8k x 8 nonvolatile Sram (5 V, 3.3 V)	DIP	Tix	-	pdf pinout

Type	Device	Short Description	Fig.	Manu	Comparision Types	More at
BQ 4010 YMA-150 N	IC	8k x 8 nonvolatile Sram (5 V, 3.3 V)	DIP	Tix	-	pdf pinout
BQ 4010 MA-200	IC	8k x 8 nonvolatile Sram (5 V, 3.3 V)	DIP	Tix	-	pdf pinout
BQ 4010 YMA-200	IC	8k x 8 nonvolatile Sram (5 V, 3.3 V)	DIP	Tix	-	pdf pinout

BR

Type	Device	Short Description	Fig.	Manu	Comparision Types	More at
BR	Si-P	=2PB709AR (Typ-Code/Stempel/marking)	35	Phi	→2PB709A	data
BR	Si-P	=2SB1218A-R (Typ-Code/Stempel/marking)	35(2mm)	Mat	→2SB1218A	data
BR	Si-P	=2SB1627-R (Typ-Code/Stempel/marking)	≈35	Mat	→2SB1627	data
BR	Si-P	=2SB709A-R (Typ-Code/Stempel/marking)	35	Mat	→2SB709A	data
BR	Si-P	=2SB766A-R (Typ-Code/Stempel/marking)	39	Mat	→2SB766A	data
BR	Si-N	=2SC2412-BR (Typ-Code/Stempel/marking)	≈35	Rhm	→2SC2412	data
BR	Si-N	=2SC2412K-R (Typ-Code/Stempel/marking)	35	Rhm	→2SC2412K	data
BR	Si-N	=2SC4081-R (Typ-Code/Stempel/marking)	35(2mm)	Rhm	→2SC4081	data
BR	Si-P	=BCW 61RC (Typ-Code/Stempel/marking)	35	Sie,Val	→BCW 61RC	data
BR	Z-Di	=BZX 399-C20 (Typ-Code/Stempel/marking	71(1,7mm)	Phi	→BZX 399-...	data
BR	Z-Di	=P4 SMA-30A (Typ-Code/Stempel/marking)	71(5x2,5)	Fag	→P4 SMA-30A	data
BR	Z-Di	=SMBJ 17CA (Typ-Code/Stempel/marking)	71(5x3,5)	Mop	→SMBJ ...	data
BR	Si-N+R	=XN 4214 (Typ-Code/Stempel/marking)	46	Mat	→XN 4214	data
BR	Si-N+R	=XP 4214 (Typ-Code/Stempel/marking)	46(2mm)	Mat	→XP 4214	data
BR 1	Si-P	=BSP 30 (Typ-Code/Stempel/marking)	48	Phi	→BSP 30	data
BR 1	Si-P	=BSR 30 (Typ-Code/Stempel/marking)	39	Val	→BSR 30	data
BR 2	Si-P	=BSP 31 (Typ-Code/Stempel/marking)	48	Phi	→BSP 31	data
BR 2	Si-P	=BSR 31 (Typ-Code/Stempel/marking)	39	Val	→BSR 31	data
BR 3	Si-P	=BSP 32 (Typ-Code/Stempel/marking)	48	Phi	→BSP 32	data
BR 3	Si-P	=BSR 32 (Typ-Code/Stempel/marking)	39	Val	→BSR 32	data
BR 4	Si-P	=BSP 33 (Typ-Code/Stempel/marking)	48	Phi	→BSP 33	data
BR 4	Si-P	=BSR 33 (Typ-Code/Stempel/marking)	39	Val	→BSR 33	data
BR 24 C01A	EEPROM-IC	ser, 1024 (128 x 8)Bit, 2,7...5,5V, I²C	8-DIP	Rhm	...2401...	pdf
BR 24 C02	EEPROM-IC	ser, 2048 (256 x 8)Bit, 2,7...5,5V, I²C	8-DIP	Rhm	...2402...	pdf
BR 24 C04	EEPROM-IC	ser, 4096 (512 x 8)Bit, 2,7...5,5V, I²C	8-DIP	Rhm	...2404...	pdf
BR 24 C01F...04F	EEPROM-IC	=BR 24C01...04: SMD	8-MDIP			
BR 28 C16 A-150	EEPROM-IC	CMOS, 2048 x 8 Bit, Ucc=4,5...5,5V	24-DIP	Rhm	...28C16...	
BR 28 C64	EEPROM-IC	CMOS, 8k x 8 Bit, Ucc=4,5...5,5V	28-DIP	Rhm	...28C64...	
BR 46 C15	EEPROM-IC	CMOS, 2048 x 8 Bit	24-DIP	Rhm		
BR 93 C46	EEPROM-IC	CMOS, ser, 1024 (64 x 16)Bit,4,5...5,5V	8-DIP	Rhm		
BR 93 C56A	EEPROM-IC	CMOS,ser, 2048(128W x 16)Bit,4,5...5,5V	8-DIP	Rhm		
BR 93 C56AF	EEPROM-IC	=BR 93C56A: SMD	8-MDIP	Rhm		
BR 93 CS46	EEPROM-IC	CMOS, ser, 1024 (64 x 16)Bit, 5V	8-DIP	Rhm		
BR 93 CS46F	EEPROM-IC	=BR 93CS46: SMD	8-MDIP	Rhm		
BR 93 LC46	EEPROM-IC	CMOS, ser, 1024(64 x 16)Bit, 2,7...5,5V	8-DIP	Rhm		
BR 93 LC46A	EEPROM-IC	=BR 93LC46: Ucc=4,5...5,5V	8-DIP	Rhm		
BR 93 LC46AF	EEPROM-IC	=BR 93LC46A: SMD	8-MDIP	Rhm		
BR 93 LC46F	EEPROM-IC	=BR 93LC46: SMD	8-MDIP	Rhm		
BR 93 LC56	EEPROM-IC	CMOS, ser, 2048(128x16)Bit, 2,7...5,5V	8-DIP	Rhm		pdf
BR 93 LC56A	EEPROM-IC	=BR 93LC56: Ucc=4,5...5,5V	8-DIP	Rhm		
BR 93 LC56AF	EEPROM-IC	=BR 93LC56A: SMD	8-MDIP	Rhm		
BR 93 LC56F	EEPROM-IC	=BR 93LC56: SMD	8-MDIP	Rhm		pdf
BR 93 LC66	EEPROM-IC	CMOS, ser. 4096(256x16)Bit, 2,7...5,5V	8-DIP	Rhm		pdf
BR 93 LC66A	EEPROM-IC	=BR 93LC66: Ucc=4,5...5,5V	8-DIP	Rhm		
BR 93 LC66ARF	EEPROM-IC	=BR 93LC66A: SMD	8-MDIP	Rhm		
BR 93 LC66RF	EEPROM-IC	=BR 93LC66: SMD	8-MDIP	Rhm		pdf
BR 93 LL46	EEPROM-IC	CMOS, ser, 1024 (64 x 16)Bit, 1,8...4V	8-DIP	Rhm	-	
BR 93 LL46F	EEPROM-IC	=BR 93LL46: SMD	8-MDIP	Rhm	-	pdf
BR 100(/03,DB3)	Diac	Ub=28...36V, Ib<0,1mA, Itsm=2A	31a	Phi	1N5761, N 413M, D 3202Y, D 0201YR	data
BR 101	Tetrode	Tetrode, 50V, 0,175A	5a§	Phi	BRY 39, BRY 62	data
BR 103	F-Thy	30V, 0,8A, Igt/Ih<0,2/<2mA, <10µs	7a	Sie,Tag	BRX 44...49, TAG 60A...Y, TAG 62A...Y, ++	data
BR 103 X	F-Thy	=BR 203: <6µs	2a	Tag	TAG 64A...Y, TAG 103X	data
BR 203	F-Thy	30V, 1A, Igt/Ih<0,2/<2mA, <10µs	2a	Tag	TAG 64A...Y, TAG 103X	data
BR 210-100...280	Si-Di	Breakover-Di(Diac,TAZ), 6A, 400W(1ms)	17(A1-A2)	Phi	-	data
BR 211-100...280	Si-Di	Breakover-Di(Diac,TAZ), 50W(1ms)	31(A1A2)	Phi	-	data
BR 211SM-100...280	Si-Di	=BR 211-...: SMD	71 (5x2,5)		-	data
BR 213-100...280	Si-Di	=BR 210-...: Dual	17(A1A2A3)	Phi	-	data
BR 216	Si-Di	Brakover-Di(Diac,TAZ), Dual, 110W(1ms)	17(A1A2A3)	Phi	-	data
BR 220-100...280	Si-Di	=BR 210-...: Dual	17(A1A2A3)	Phi	-	data
BR 303	F-Thy	30V, 1A, Igt/Ih<2/<5mA, <13µs	14h§	Sie,Tag	BR 403, TAG 66A...Y	data
BR 403	F-Thy	=BR 303:	13h§	Sie	BR 303, TAG 66A...Y	data
BR 2804 A	EEPROM-IC	512 x 8 Bit, 5V	24-DIP	Rhm	...2804...	
BR 2816 A	EEPROM-IC	2048 x 8 Bit, 5V	24-DIP	Rhm	...2816...	
BR 2864 A	EEPROM-IC	8192 x 8 Bit, 5V	28-DIP	Rhm	...2864...	
BR 2865 A	EEPROM-IC	8192 x 8 Bit, 5V	28-DIP	Rhm	...2865...	
BR 6116	CMOS-sRAM-IC	hi-speed, 2048 x 8 Bit	24-DIP	Rhm	...6116...	
BR 6216 A	CMOS-sRAM-IC	16k(2048 x 8)Bit, <150ns, 0...+70°	24-DIP	Rhm	-	
BR 6216 B-10LL	CMOS-sRAM-IC	16k(2048 x 8)Bit, <100ns, 0...+70°	24-SDIP	Rhm	-	
BR 6264 A	CMOS-sRAM-IC	hi-speed, 8192 x 8 Bit	28-DIP	Rhm	...6164...	
BR 6264 P	CMOS-sRAM-IC	hi-speed, 8192 x 8 Bit	28-DIP	Rhm	...6164...	
BR 6265	CMOS-sRAM-IC	64k(8k x 8)Bit; <120ns, 0...+70°	28-DIP	Rhm	-	pdf pinout
BR 6265 A-10LL	CMOS-sRAM-IC	64k(8k x 8)Bit, <100ns, 0...+70°	28-DIP	Rhm	-	pdf pinout
BR 6265 AF-10LL	CMOS-sRAM-IC	=BR 6265A-10LL: SMD	28-MDIP	Rhm	-	pdf pinout
BR 9021 A	EEPROM-IC	ser, 2048 (128 x 16)Bit, Ucc=4,5...5,5V	8-DIP	Rhm	-	
BR 9021 AF	EEPROM-IC	=BR 9021A: SMD	8-MDIP	Rhm	-	
BR 9021 B	EEPROM-IC	ser, 2048 (128 x 16)Bit, Ucc=2,7...5,5V	8-DIP	Rhm	-	

Type	Device	Short Description	Fig.	Manu	Comparision Types	More at
BR 9021 BF	EEPROM-IC	=BR 9021B: SMD	8-MDIP	Rhm	-	
BR 9040	EEPROM-IC	ser, 4096 (256 x 16)Bit, 2,7...5,5V	8-DIP	Rhm	-	pdf
BR 9040 F	EEPROM-IC	=BR 9040: SMD	8-MDIP		-	pdf
BR 9041 A	EEPROM-IC	ser, 4096 (256 x 16)Bit, 4,5...5,5V	8-DIP	Rhm	-	
BR 9041 ARF	EEPROM-IC	=BR 9041: SMD	8-MDIP		-	
BR 62256	sRAM-IC	CMOS, lo-power, 32k x 8 Bit	28-DIP	Rhm	-	
BRA 114 ECM	Si-P+R	SMD, Inverter, Driver, S	35a	Hit		data
BRA 114 EMP	Si-P+R	SMD, Inverter, Driver, S	≈35a	Hit		data
BRA 114 ETP	Si-P+R	Inverter, Driver, S	7c	Hit		data
BRA 123 ECM	Si-P+R	SMD, Inverter, Driver, S	35a	Hit		data
BRA 123 EMP	Si-P+R	SMD, Inverter, Driver, S	≈35a	Hit		data
BRA 123 ETP	Si-P+R	Inverter, Driver, S	7c	Hit		data
BRA 124 ECM	Si-P+R	SMD, Inverter, Driver, S	35a	Hit		data
BRA 124 EMP	Si-P+R	SMD, Inverter, Driver, S	≈35a	Hit		data
BRA 124 ETP	Si-P+R	Inverter, Driver, S	7c	Hit		data
BRA 143 ECM	Si-P+R	SMD, Inverter, Driver, S	35a	Hit		data
BRA 143 EMP	Si-P+R	SMD, Inverter, Driver, S	≈35a	Hit		data
BRA 143 ETP	Si-P+R	Inverter, Driver, S	7c	Hit		data
BRA 144 ECM	Si-P+R	SMD, Inverter, Driver, S	35a	Hit		data
BRA 144 EMP	Si-P+R	SMD, Inverter, Driver, S	≈35a	Hit		data
BRA 144 ETP	Si-P+R	Inverter, Driver, S	7c	Hit		data
BRC 114 ECM	Si-N+R	SMD, Inverter, Driver, S	35a	Hit		data
BRC 114 EMP	Si-N+R	SMD, Inverter, Driver, S	≈35a	Hit		data
BRC 114 ETP	Si-N+R	Inverter, Driver, S	7c	Hit		data
BRC 123 ECM	Si-N+R	SMD, Inverter, Driver, S	35a	Hit		data
BRC 123 EMP	Si-N+R	SMD, Inverter, Driver, S	≈35a	Hit		data
BRC 123 ETP	Si-N+R	Inverter, Driver, S	7c	Hit		data
BRC 124 ECM	Si-N+R	SMD, Inverter, Driver, S	35a	Hit		data
BRC 124 EMP	Si-N+R	SMD, Inverter, Driver, S	≈35a	Hit		data
BRC 124 ETP	Si-N+R	Inverter, Driver, S	7c	Hit		data
BRC 143 ECM	Si-N+R	SMD, Inverter, Driver, S	35a	Hit		data
BRC 143 EMP	Si-N+R	SMD, Inverter, Driver, S	≈35a	Hit		data
BRC 143 ETP	Si-N+R	Inverter, Driver, S	7c	Hit		data
BRC 144 ECM	Si-N+R	SMD, Inverter, Driver, S	35a	Hit		data
BRC 144 EMP	Si-N+R	SMD, Inverter, Driver, S	≈35a	Hit		data
BRC 144 ETP	Si-N+R	Inverter, Driver, S	7c	Hit		data
BRM 10	MOS-FET	SMD, hi-voltage Level Shifter	6-MDIP	Sie	-	data
BRT11	OC	LED/Triac Viso:5300V Vbs:1.1V		Sie		data pinout
BRT 11	LIN-IC	SITAC AC-Switch (FET+Thy), 400V, 0,3A~	6-DIP	Sie	-	pdf pinout
BRT12	OC	LED/Triac Viso:5300V Vbs:1.1V		Sie		data pinout
BRT 12	LIN-IC	=BRT 11: 600V	6-DIP	Sie	-	pdf pinout
BRT 13	LIN-IC	=BRT 11: 800V	6-DIP	Sie	-	pdf pinout
BRT21	OC	LED/Triac Viso:5300V Vbs:1.1V		Sie		data pdf pinout
BRT 21	LIN-IC	SITAC AC-Switch (FET+Thy), 400V, 0,3A~	6-DIP	Sie	-	pdf pinout
BRT 22	LIN-IC	=BRT 21: 600V	6-DIP	Sie	-	pinout
BRT 23	LIN-IC	=BRT 21: 800V	6-DIP	Sie	-	pinout

BRX

Type	Device	Short Description	Fig.	Manu	Comparision Types	More at
BRX 44	F-Thy	30V, 0,8A, Igt/Ih<0,2/<5mA, 8µs	7a	Mot,Phi,++	BRX 50...56, BRY 55S/30..800	data
BRX 45	F-Thy	=BRX 44: 60V	7a	Mot,Phi,++	BRX 51...56, BRY 55S/60...800	data
BRX 46	F-Thy	=BRX 44: 100V	7a	Mot,Phi,++	BRX 51...56, BRY 55S/100...800	data
BRX 47	F-Thy	=BRX 44: 200V	7a	Mot,Phi,++	BRX 52...56, BRY 55S/200...800	data
BRX 48	F-Thy	=BRX 44: 300V	7a	Mot,Phi,++	BRX 53...56, BRY 55S/300...800	data
BRX 49	F-Thy	=BRX 44: 400V	7a	Mot,Phi,++	BRY 54...56, BRY 55S/400...800S	data
BRX 50	F-Thy	50V, 0,6A(Tc=55°), Igt/Ih<0,2/<5mA	7a	Tag	BRX 45...49, BRY 55S/60...800	data
BRX 51	F-Thy	=BRX 50: 100V	7a	Tag	BRX 46...49, BRY 55S/100...800	data
BRX 52	F-Thy	=BRX 50: 200V	7a	Tag	BRX 47...49, BRY 55S/200...800	data
BRX 53	F-Thy	=BRX 50: 300V	7a	Tag	BRX 48...49, BRY 55S/300...800	data
BRX 54	F-Thy	=BRX 50: 400V	7a	Tag	BRX ..., BRY 55S/400...800	data
BRX 55	F-Thy	=BRX 50: 500V	7a	Tag	BRY 55S/500...800	data
BRX 56	F-Thy	=BRX 50: 600V	7a	Tag	BRY 55S/600...800	data
BRX 60	F-Thy	50V, 2,6A(Tc=100°), Igt/Ih<0,2/<5mA	14h§	Tag	TAG 66A..F, BR 303	data
BRX 61	F-Thy	=BRX 60: 100V	14h§	Tag	TAG 66A	data
BRX 62	F-Thy	=BRX 60: 200V	14h§	Tag	-	data
BRX 63	F-Thy	=BRX 60: 300V	14h§	Tag	-	data
BRX 64	F-Thy	=BRX 60: 400V	14h§	Tag	-	data
BRX 65	F-Thy	=BRX 60: 500V	14h§	Tag	-	data
BRX 66	F-Thy	=BRX 60: 600V	14h§	Tag	-	data
BRX 70-03...-50	50Hz-Thy	30..500V, 0,5A(Ta=55°), Igt<0,2mA	7a	Tho	BRX 45...49, BRX 50...56, BRY 55/...,++	data
BRX 71-03...-50	50Hz-Thy	30..500V, 0,5A(Ta=55°), Igt<0,2mA	7b	Tho	BRX 45...49, BRX 50...56, BRY 55/...,++	data
BRX 72-03...-50	50Hz-Thy	30..500V, 0,6A(Ta=55°), Igt<0,2mA	≈30b	Tho	(BRX 45...49, BRX 50...56, BRY 55/...,++)	
BRX 73-03...-50	50Hz-Thy	30..500V, 0,6A(Ta=55°), Igt<0,2mA	≈30a	Tho	(BRX 45...49, BRX 50...56, BRY 55/...,++)	
BRX 78-03...-50	Triac	30..500V, Igt<10mA	7b	Tho	MAC 94-..., MAC 95-..., Z 106-...	
BRX 79-03...-50	Triac	30..500V, Igt<10mA	≈30	Tho	(MAC 94-..., MAC 95-..., Z 106-...)	

Type	Device	Short Description	Fig.	Manu	Comparision Types	More at
BS 10-01A...07A	Triac	100...700V, 10A, Igt<50mA	21b§	Bbc	TW10N...C, TW12N...C, TW18N..., TW25N...	data
BS 10-01B...07B	Triac	=BS 10-..A:	29b§		TW10N...H, TW12N...H, T6401...	data
BS100	PD	880nm P:10.5nA/Lx		Sha		data pinout
BS 107	MOS-N-FET-e	V-MOS, LogL, 200V, 0,13A, 1W, <26Ω	7a	Phi,Sie,++	BS 108, BS 189, BSS 89, BSS 101, ++	data
BS 107 A	MOS-N-FET-e	=BS 107: <6,4Ω	7a	Mot,Phi,++	→BS 107	data
BS 107 P,PT	MOS-N-FET-e	=BS 107: 0,5W	40e	Fer, Ztx	→BS 107	data
BS 108	MOS-N-FET-e	V-MOS,LogL, 200V, 0,25A, 1W, <8Ω(0,1A)	7a	Itt,Phi,++	BSN 254, BSS 89, BSS 297	data
BS 109	MOS-N-FET-e	V-MOS, 300V, 0,15A, 0,83W	7a	Itt	BSS 124	data
BS 110	Si-N	SMD, S, 20V, 0,2A, <12/18ns	35d(2mm)	Fer	(BSV 52, 2SC3578)[6]	data
BS 112	MOS-N-FET-e	V-MOS, LogL, 170V, 0,2A, 0,83W, <10Ω	7a	Itt	BS 107...108, BSS 101, BS 189, BSS 89, ++	data
BS112	PD	700nm P:70nA/Lx		Sha		data pinout
BS115	PD	750(560)nm	TO18	Sha		data pinout
BS120	PD	560nm		Sha		data pdf pinout
BS142	PD	700nm P:220nA/Lx		Sha		data pdf pinout
BS 170 F	MOS-N-FET-e	V-MOS, LogL, 60V, 0,3A, 0,63W, <5Ω	7a	Phi,Sie,++	BSS 296, BST 70, 2SK422...423	data
BS 170 F	MOS-N-FET-e	=BS 170P: SMD, 0,15A	35a	Ztx	BSS 145, BST 82, 2SK1848	data
BS 170 P	MOS-N-FET-e	=BS 170: 0,27A, 0,625W	40e	Fer, Ztx	→BS 170	data
BS 189	MOS-N-FET-e	V-MOS, 200V, 0,2A, 0,83W	7c	Itt	BS 108, BSS 89, BST 74, BSS 88, ++	data
BS 192	MOS-P-FET-e	V-MOS, 200V, 0,18A, 0,83W	7c	Itt	BS 208, BSS 92	data
BS 208	MOS-P-FET-e	V-MOS, 200V, 0,2A, 0,83W, <14Ω, 5/20ns	7a	Itt,Phi,Six	BS 192, BS 209, BSS 92	data
BS 209	MOS-P-FET-e	V-MOS, 300V, 0,12A, 0,83W	7a	Itt	-	data
BS 212	MOS-P-FET-e	V-MOS, 170V, 0,2A, 0,83W	7a	Itt	BS 192, BS 208	data
BS 250	MOS-P-FET-e	V-MOS, 45V, 0,25A, 0,83W, <14Ω, 5/25ns	7a	Itt,Phi,++	BST 100	data
BS 250 F	MOS-P-FET-e	=BS 250P: SMD, 90mA, 0,33W	35a	Ztx	BSH 201, 2SJ210..211	data
BS 250 P	MOS-P-FET-e	=BS 250: 0,23A, 0,7W	40e	Ztx	→BS 250	data
BS 270	MOS-N-FET-e	V-MOS, 60V, 0,4A, 0,63W, <2Ω, <10/i0ns	7a	Nsc	BSS 296, BST 70, 2SK422...423	data
BS500	PD	560nm P:>5.5nA/Lx		Sha		data pinout
BS520	PD	560nm P:7.3nA/Lx		Sha		data pinout
BS530	PD	800nm P:500nA/Lx		Sha		data pinout
BS 850	MOS-P-FET-e	SMD, V-MOS, 60V, 0,25A, <5Ω, 5/25ns	35a	Gsi	BSH 201, 2SJ210	data
BS 870	MOS-N-FET-e	SMD, V-MOS, 60V, 0,25A, <5Ω, 5/25ns	35a	Gsi	BSS 145, 2SK1590	data

BSC...BSH

Type	Device	Short Description	Fig.	Manu	Comparision Types	More at
BSC 52	Si-N	SMD, 40V, 0,1A, 300MHz	Chip	Ucp	-	data
BSD 10	MOS-N-FET-d	SS, Chopper, 15V, 50mA, 1/5ns	5h+	Phi	-	data
BSD 12	MOS-N-FET-d	=BSD 10: 25V	5h+	Phi	-	data
BSD 20	MOS-N-FET-d	=BSD 10: SMD	44j	Phi	-	data
BSD 22	MOS-N-FET-d	=BSD 12: SMD	44j	Phi	-	data
BSD 212	MOS-N-FET-e	S, Chopper, sym, 40V, 50mA, 1/5ns	5h+	Phi	-	data
BSD 213	MOS-N-FET-e*	=BSD 212: + Gateschutz-Di/gate prot.	5h+	Phi	-	data
BSD 214	MOS-N-FET-e	S, Chopoer, sym, 40V, 50mA, 1/5ns	5h+	Phi	SD 210	data
BSD 215	MOS-N-FET-e*	=BSD 214: + Gateschutz-Di/gate prot.	5h+	Phi	SD 211	data
BSD 254	MOS-N-FET-d	V-MOS, 250V, 0,2A, 0,85W,<12Ω,<10/30ns	7c	Phi	-	data
BSD 254 A	MOS-N-FET-d	=BSD 254:	7a	Phi	-	data
BSD 254 AR	MOS-N-FET-d	=BSD 254:	7e	Phi	-	data
BSH 101	MOS-N-FET-e	SMD, V-MOS, LogL, 60V, 0,7A, <0,9Ω	35a	Phi	-	data
BSH 102	MOS-N-FET-e	SMD, V-MOS, LogL, 30V, 0,85A, <0,6Ω	35a	Phi	2SK2910	data
BSH 103	MOS-N-FET-e	SMD, V-MOS, LogL, 30V, 0,85A, <0,5Ω	35a	Phi	2SK2910	data
BSH 105	MOS-N-FET-e	SMD, V-MOS, LogL, 20V, 1,21A, <0,25Ω	35a	Phi, Gsi	IRLML 2402	data
BSH 106	MOS-N-FET-e	SMD, V-MOS, LogL, 20V, 1,43A, <0,25Ω	46bn4(2mm)	Phi	-	data
BSH 107	MOS-N-FET-e	SMD, V-MOS, LogL, 20V, 2,5A, <0,09Ω	46bn4	Phi	-	data
BSH 201	MOS-P-FET-e	SMD, V-MOS, LogL, 60V, 0,34A, <3,75Ω	35§	Phi	-	data
BSH 202	MOS-P-FET-e	SMD, V-MOS, LogL, 30V, 0,57A, <1,35Ω	35a	Phi	2SJ502	data
BSH 203	MOS-P-FET-e	SMD, V-MOS, LogL, 30V, 0,57A, <1,1Ω	35a	Phi	2SJ502	data
BSH 205	MOS-P-FET-e	SMD, V-MOS, LogL, 12V, 0,86A, <0,5Ω	35a	Phi	IRLML 6302	data
BSH 206	MOS-P-FET-e	SMD, V-MOS, LogL, 12V, 1A, <0,5Ω	46bn4(2mm)	Phi	-	data
BSH 207	MOS-P-FET-e	SMD, V-MOS, LogL, 12V, 2A, <0,15Ω	46bn4	Phi	-	data
BSH 299	MOS-P-FET-e	SMD, V-MOS, LogL, 50V, 0,2A, <10Ω	46bn4(2mm)	Phi	-	data

BSJ

Type	Device	Short Description	Fig.	Manu	Comparision Types	More at
BSJ 30	Si-N	S, -/30V, 0,8W, 250MHz	2a	Riz	BC 140...141, 2N1613, 2N1711, ++	data
BSJ 32	Si-N	S, -/40V, 0,8W, 300MHz	2a	Riz	BC 140...141, 2N1613, 2N1711, ++	data
BSJ 36	Si-P	S, 40V, 0,5A, 0,36W, <40/-ns	2a§	Riz	BSW 24, BSX 36, 2N2906..2907	data
BSJ 61	Si-N	S, 25V, 0,36W, <40/-ns	2a§	Riz	BSW 41, BSY 62...63, 2N706A, 2N2221..22++	data
BSJ 62	Si-N	S, 25V, 0,36W, <40/-ns	2a§	Riz	BSW 41, BSY 62...63, 2N706A, 2N2221..22++	data
BSJ 63	Si-N	S, 40V, 0,2A, 0,36W, <40/-ns	2a§	Riz	BSW 41, BSY 63, 2N708, 2N2221..22, ++	data
BSJ 65	Si-N	S, 25V, 0,36W, <40/-ns	2a§	Riz	BSW 41, BSY 62...63, 2N706A, 2N2221..22++	data
BSJ 66	Si-N	S, 40V, 0,2A, 0,36W, <35/-ns	2a§	Riz	BSW 41, BSY 63, 2N708, 2N2221..22, ++	data
BSJ 67	Si-N	S, 40V, 0,5A, 0,36W, <25/-ns	2a§	Riz	BSS 10, BSX 26, BSX 39, 2N3261	data
BSJ 68	Si-N	S, 40V, 0,5A, 0,36W, <9/-ns	2a§	Riz	BSS 10, BSX 26, BSX 39, 2N3261	data
BSJ 79	Si-N	Nix, Vid, -/120V, 0,03A, 0,3W, 120MHz	2a§	Riz	BF 297...299, BF 422, BF 483, ++	data
BSJ 108	N-FET	Sym, S, 25V, Idss>80mA, Up<10V, 10/6ns	2b	Phi	J 108, 2N5433	data
BSJ 109	N-FET	=BSJ 108: Idss>40mA, Up<6V	2b	Phi	J 109	data

Type	Device	Short Description	Fig.	Manu	Comparision Types	More at
BRY						
BRY 20	Tetrode	40V, 0,5A	5b§	Sie	-	data
BRY 21	Tetrode	=BRY 20: 80V	5b§	Sie	-	data
BRY 23	F-Thy	100V, 2,2A, Igt/Ih<0,2/<2,5mA, <22µs	2a§	Sgs	-	data
BRY 24	F-Thy	=BRY 23: 200V	2a§	Sgs	-	data
BRY 25	F-Thy	=BRY 23: 300V	2a§	Sgs	-	data
BRY 26	F-Thy	=BRY 23: 400V	2a§	Sgs	-	data
BRY 28	F-Thy	100V, 2,2A(Tc=85°), Igt/Ih<15/<25mA	2a§	Sgs	TAG 2,5S-...	data
BRY 29	F-Thy	=BRY 28: 200V	2a§	Sgs	TAG 2,5S-...	data
BRY 3C	F-Thy	=BRY 28: 300V	2a§	Sgs	TAG 2,5S-...	data
BRY 31	F-Thy	=BRY 28: 400V	2a§	Sgs	TAG 2,5S-...	data
BRY 32	F-Thy	50V, 1A(Tc=25°),Igt/Ih<0,5/<2mA, <15µs	2a§	Sgs	BRX 50...56	data
BRY 33	F-Thy	=BRY 32: 100V	2a§	Sgs	BRX 51...56	data
BRY 34	F-Thy	=BRY 32: 200V	2a§	Sgs	BRX 52...56	data
BRY 35	F-Thy	100V, 1A, Igt/Ih<0,1/<1mA, <22µs	2a§	Sgs	-	data
BRY 36	F-Thy	=BRY 35: 200V	2a§	Sgs	-	data
BRY 37	F-Thy	=BRY 35: 300V	2a§	Sgs	-	data
BRY 39(P,S,T)	Tetrode	70V, 0,25A	5a§	Mot,Phi	(BRY 21)[6]	data
BRY 40	Si-Di	PNPN-Diode, 300V	31a	Fer	-	data
BRY 41-50...-600	Triac	50...600V, 1A(Tc=75°), Igt/Ih<25/25mA	2a§	Ssc	T 2303A...M, T 2306A...F	data
BRY 42	50Hz-Thy	250V, 3A(Tc=90°),Igt/Ih<25/<25mA,<20µs		Itt	-	data
BRY 43	50Hz-Thy	=BRY 42: 400V		Itt	-	data
BRY 44	50Hz-Thy	=BRY 42: 500V		Itt	-	data
BRY 45/50...600	Triac	50...600V, 3A(Tc=75°), Igt/Ih<50/<25mA	2a§	Ssc	-	data
BRY 46	Tetrode	15/20V, 0,05A	6c	Itt	BR 101, BRY 39, MAS 32, MAS 39	data
BRY 49	Tetrode	30V, 0,3A	5d§	Aeg	BRY 39, (BRY 20...21)[6]	data
BRY 50	Tetrode	=BRY 49: 70V	5d§	Aeg	BRY 39, (BRY 21)[6]	data
BRY 51	Tetrode	=BRY 49: 120V	5d§	Aeg	-	data
BRY 52-50...-600	Triac	50...600V, 6A(Tc=75°), Igt/Ih<50/<50mA	2a§	Ssc	(BT 158/..., T 2806B...M, T 2856A...M,++)[4]	data
BRY 53	Si-Di	PNPN-Di, 90...125V, 0,04A, Itsm=5A	31a	Fer,	-	data
BRY 54/30...600(T)	50Hz-Thy	30...600V, 2,5A, Igt/Ih<20/25mA	2a§	Ssc	TAG 2,5-..., TAG 611-...	data
BRY 55/30...800(M)	Thy	30...800V, 0,8A, Igt/Ih<0,2/<5mA, 20µs	7a	Mot,Sie,Tho	BRY 58/..., TAG 70D...S, TAG 72D...S, ++	data
BRY 55/30S..800S	F-Thy	=BRY 55/...(M): <6µs	7a	Tho	-	data
BRY 56 A	PUT	70V, Ip=0,22µA, Iv=2µA	7c	Phi,Sie	-	data
BRY 56 B	PUT	70V, Ip=1µA, Iv=10µA	7c		-	data
BRY 56 C	PUT	70V, Ip=5µA, Iv=50µA	7c		-	data
BRY 57	Si-Di	PNPN-Di, 90...125V, 0,04A, Itsm=6A	31a	Fer	-	data
BRY 58/30...800	Thy	=BRY 55/...:	2a§	Tho	→BRY 55/...	data
BRY 59 A..B	Thy	30...60V, 0,43A(Ta=25°), Igt<0,1mA	7e	Tix	TAG 06A...YY, TIC 44...47, TIC 60...64, ++	data
BRY 61	PUT	70V, Ip=5µA, Iv=30µA	35b	Phi	-	data
BRY 62	Tetrode	Tetrode, SMD, 70V, 0,175A	44e	Phi	-	data
BRY 70/200...800	Thy	200...800V, 0,5A, Igt<10mA	7	Sie	-	data
BRY 71/200...800	Thy	200...800V, 0,5A, Igt<0,2mA	7	Sie	BRY 55/..., TAG 70..., TAG 72...,++	data
BS						
BS	Si-P	=2PB709AS (Typ-Code/Stempel/marking)	35	Phi	→2PB709A	data
BS	Si-P	=2SA1669 (Typ-Code/Stempel/marking)	35	Say	→2SA1669	data
BS	Si-P	=2SB1218A-S (Typ-Code/Stempel/marking)	35(2mm)	Mat	→2SB1218A	data
BS	Si-P	=2SB1627-S (Typ-Code/Stempel/marking)	≈35	Mat	→2SB1627	data
BS	Si-P	=2SB709A-S (Typ-Code/Stempel/marking)	35	Mat	→2SB709A	data
BS	Si-P	=2SB766A-S (Typ-Code/Stempel/marking)	39	Mat	→2SB766A	data
BS	Si-N	=2SC2412-BS (Typ-Code/Stempel/marking)	≈35	Rhm	→2SC2412	data
BS	Si-N	=2SC2412K-S (Typ-Code/Stempel/marking)	35	Rhm	→2SC2412K	data
BS	Si-N	=2SC4081-S (Typ-Code/Stempel/marking)	35(2mm)	Rhm	→2SC4081	data
BS	Si-N	=2SD2444K (Typ-Code/Stempel/marking)	35	Rhm	→2SD2444K	data
BS	Si-P	=BCW 61RD (Typ-Code/Stempel/marking)	35	Sie	→BCW 61RD	data
BS	Z-Di	=BZX 399-C22 (Typ-Code/Stempel/marking)	71(1,7mm)	Phi	→BZX 399-...	data
BS	Z-Di	=P4 SMA-33 (Typ-Code/Stempel/marking)	71(5x2,5)	Fag	→P4 SMA-33	data
BS	Z-Di	=SMBJ 18C (Typ-Code/Stempel/marking)	71(5x3,5)	Mop	→SMBJ ...	data
BS 1	Si-P-Darl+Di	=BSP 60 (Typ-Code/Stempel/marking)	48	Phi, Sie	→BSP 60	data
BS 1	Si-P-Darl+Di	=BST 60 (Typ-Code/Stempel/marking)	39	Phi, Val	→BST 60	data
BS 2	Si-P-Darl+Di	=BSP 61 (Typ-Code/Stempel/marking)	48	Phi, Sie	→BSP 61	data
BS 2	Si-P-Darl+Di	=BST 61 (Typ-Code/Stempel/marking)	39	Phi, Val	→BST 61	data
BS 3	Si-P-Darl+Di	=BSP 62 (Typ-Code/Stempel/marking)	48	Phi, Sie	→BSP 62	data
BS 3	Si-P-Darl+Di	=BST 62 (Typ-Code/Stempel/marking)	39	Phi, Val	→BST 62	data
BS 6-01A...07A	Triac	100...700V, 6A, Igt<50mA	21b§	Bbc	TW6N...C, BS8...A, TW8N...C, BS10...A, ++	data
BS 6-01B...07B	Triac	=BS 6-...A:	29b§	Bbc	TW6N...H, BS8...B, TW8N...H, BS10...B, ++	data
BS 7-02A...06A	Triac	200...600V, 6A, Igt<60mA	17§	Bbc	MAC216..., SC141..., T 2801..., TXC10K...M, ++	data
BS 08 A	SBS	175mA, Ub=7...9V, Is<0,2mA, Ih<1,5mA	7e(A2GA1)	Mit	2N4992	
BS 8-01A...07A	Triac	100...700V, 8A, Igt<50mA	21b§	Bbc	TW8N...C, BS10...A, TW10M...C, TW12N...C, ++	data
BS 8-01B...07B	Triac	=BS 8-...A:	29b§	Bbc	TW8N...H, BS10...B, TW10M...H, TW12N...H, ++	data
BS 9-02A...06A	Triac	200...600V, 8A, Igt<60mA	17h§	Bbc	T 2802...,MAC222..., TIC226..., TXD10K...,++	data

Type	Device	Short Description	Fig.	Manu	Comparision Types	More at
BSJ 110	N-FET	=BSJ 108: Idss>10mA, Up<4V	2b	Phi	BFS 76, BFS 79, J 110	data
BSJ 111	N-FET	Chopper, sym, 40V, Idss>20mA, Up<10V	7d	Phi	-	data
BSJ 112	N-FET	=BSJ 111: Idss>5mA, Up<5V	7d	Phi	-	data
BSJ 113	N-FET	=BSJ 111: Idss>2mA, Up<3V	7d	Phi	-	data
BSJ 174	P-FET	S, sym, 30V, Idss>20mA, Up<10V, 7/15ns	7a	Phi	-	data
BSJ 175	P-FET	S, sym, 30V, Idss>7mA, Up<6V, 15/30ns	7a	Phi	-	data
BSJ 176	P-FET	S, sym, 30V, Idss>2mA, Up<4V, 35/55ns	7a	Phi	-	data
BSJ 177	P-FET	S,sym,30V,Idss>1,5mA, Up<2,3V, 45/45ns	7a	Phi	-	data

BSN...BSO

Type	Device	Short Description	Fig.	Manu	Comparision Types	More at
BSN 10	MOS-N-FET-e	V-MOS, LogL, 50V, 0,175A, 0,83W, <15Ω	7c	Phi	BS 170, BSS 98, BST 72	data
BSN 10 A	MOS-N-FET-e	=BSN 10:	7a		BS 170, BSS 98, BST 72	data
BSN 12	MOS-N-FET-e	V-MOS, LogL, 50V, 0,150A, 0,83W, <20Ω	7c	Phi	BS 170, BSS 98, BST 72	data
BSN 12 A	MOS-N-FET-e	=BSN 12:	7a		BS 170, BSS 98, BST 72	data
BSN 20	MOS-N-FET-e	=BSN 10: SMD, 0,1A	35a	Phi, Gsi	BSS 123, BSS 138, BSS 145, 2SK1590	data
BSN 20 W	MOS-N-FET-e	=BSN 10: SMD, 0,1A	35a(2mm)	Phi	-	data
BSN 22	MOS-N-FET-e	=BSN 12: SMD, 0,1A	35a	Phi	BSS 123, BSS 138, BSS 145, 2SK1590	data
BSN 65 F	MOS-N-FET-e	V-MOS, LogL, 30V, 1,5A, 10W, <3Ω	13g§	Phi	-	data
BSN 204	MOS-N-FET-e	V-MOS,LogL, 200V, 0,25A, 1W, <8Ω(0,1A)	7c	Phi	BSS 88..89, BSN 254, BST 74	data
BSN 204 A	MOS-N-FET-e	=BSN 204:	7a		BSS 88..89, BSN 254, BST 74	data
BSN 205	MOS-N-FET-e	V-MOS, 200V, 0,3A, 1W, <6Ω, 5/15ns	7c	Phi	BSS 88..89, BSN 254, BST 74	data
BSN 254	MOS-N-FET-e	V-MOS, LogL, 250V, 0,3A, 1W, <7Ω(0,3A)	7c	Phi	BSN 274	data
BSN 254 A	MOS-N-FET-e	=BSN 254:	7a		BSN 274	data
BSN 274	MOS-N-FET-e	V-MOS,LogL, 270V, 0,25A, 1W,<8Ω(0,25A)	7c	Phi	BSN 304	data
BSN 274 A	MOS-N-FET-e	=BSN 274:	7a		BSN 304	data
BSN 304	MOS-N-FET-e	V-MOS,LogL, 300V, 0,25A, 1W,<8Ω(0,25A)	7c	Phi	-	data
BSN 304 A	MOS-N-FET-e	=BSN 304:	7a		-	data
BSO 215 C	MOS-NP-FET-e	SMD, V-MOS, N+P, LogL, 20V, 3,7A, <0,15Ω(3A)	8-MDIP/ca3	Sie	-	
BSO 220 N	MOS-N-FET-e	SMD, V-MOS, Dual, LogL, 20V, 3,2A, <0,2Ω(2,6A), <73/44ns	8-MDIP/ca3	Sie	-	data
BSO 302 SN	MOS-N-FET-e	SMD, V-MOS, LogL, 30V, 9,8A, <17mΩ <210/165ns	8-MDIP/ca1	Sie	-	data
BSO 304 SN	MOS-N-FET-e	SMD, V-MOS, LogL, 30V, 6,4A, <42mΩ <93/54ns	8-MDIP/ca1	Sie	-	data
BSO 305 N	MOS-N-FET-e	SMD, V-MOS, Dual, LogL, 30V, 6A,	8-MDIP/ca3	Sie	-	data
BSO 307 N	MOS-N-FET-e	SMD, V-MOS, Dual, LogL, 30V, 5A,	8-MDIP/ca3	Sie	-	data
BSO 315 C	MOS-NP-FET-e	SMD, V-MOS, N+P, LogL, 30V, N: 3,4A <0,15Ω(2,9A), P: 2,3A, <0,4Ω(1,8A)	8-MDIP/ca3	Sie	-	
BSO 604 NS2	MOS-N-FET-e	SMD, V-MOS, LogL, 55V, 5A, <48mΩ(2,6A)	8-MDIP/ca1	Sie	-	
BSO 612 CV	MOS-NP-FET-e	SMD, V-MOS, N+P, LogL, 60V, N: 3A <0,12Ω(3A), P: 2A, <0,3Ω(2A)	8-MDIP/ca3	Sie	-	
BSO 613 SPV	MOS-P-FET-e	SMD, V-MOS, 60V, 3,44A, <0,13Ω(3,44A) <31,5/66ns	8-MDIP/ca1	Sie	-	
BSO 615 C	MOS-NP-FET-e	SMD, V-MOS, N+P, LogL, 60V, N: 3,A <0,15Ω(2,7A), P: 2A, <0,45Ω(1,7A)	8-MDIP/ca3	Sie	-	
BSO 615 N	MOS-N-FET-e	SMD, V-MOS, Dual, LogL, 60V, 2,6A, <0,15Ω(2,6A), <45/55ns	8-MDIP/ca3	Sie	-	data
BSO 615 NV	MOS-N-FET-e	SMD, V-MOS, Dual, LogL, 60V, 3,1A, <0,12Ω(3,1A), <57/95ns	8-MDIP/ca3	Sie	-	data

BSP

Type	Device	Short Description	Fig.	Manu	Comparision Types	More at
BSP 15	Si-P	=BST 15: 1,5W	48j§	Phi, Ztx	-	data
BSP 16	Si-P	=BST 16: 1,5W	48j§	Phi,Mot,Ztx	-	data
BSP 17	MOS-N-FET-e	SMD, V-MOS, 50V, 3,2A, <0,1Ω, 60/95ns	48c§	Sie	BSP 319, BUK 483-60A	data
BSP 19	Si-N	=BST 39: 1,5W	48j§	Phi,Mot,Ztx	-	data
BSP 19 A(T1,T3)	Si-N	=BST 39: 400/350V, 1,5W	48j§	Ons		
BSP19 AT1	Si-N	SMD	48c§	Ons		data
BSP 20	Si-N	=BST 40: 1,5W	48j§	Phi,Mot,Ztx	-	data
BSP 30	Si-P	=BSR 30: 1,5W	48j§	Phi, Tho		data
BSP 030	MOS-N-FET-e	SMD, V-MOS, 30V, 10A, <30mΩ, <60/150ns	48c§	Phi	BUK 9830-30	data
BSP 31	Si-P	=BSR 31: 1,5W	48j§	Phi,Tho,Ztx	-	data
BSP 32	Si-P	=BSR 32: 1,5W	48j§	Phi, Tho	-	data
BSP 33	Si-P	=BSR 33: 1,5W	48j§	Phi,Tho,Ztx	-	data
BSP 40	Si-N	=BSR 40: 1,5W	48j§	Phi,Tho,Ztx	-	data
BSP 41	Si-N	=BSR 41: 1,5W	48j§	Phi,Tho,Ztx	-	data
BSP 42	Si-N	=BSR 42: 1,5W	48j§	Phi,Tho,Ztx	-	data
BSP 43	Si-N	=BSR 43: 1,5W	48j§	Phi,Tho,Ztx	-	data
BSP 50	Si-N-Darl+Di	=BST 50: 1,5W	48j§	Phi,Sie	-	data
BSP 51	Si-N-Darl+Di	=BST 51: 1,5W	48j§	Phi,Sie	-	data
BSP 52	Si-N-Darl+Di	=BST 52: 1,5W	48j§	Phi,Sie,Mot	-	data
BSP 60	Si-P-Darl+Di	=BST 60: 1,5W	48j§	Phi,Sie	-	data
BSP 61	Si-P-Darl+Di	=BST 61: 1,5W	48j§	Phi,Sie	-	data

Type	Device	Short Description	Fig.	Manu	Comparision Types	More at
BSP 62	Si-P-Darl+Di	=BST 62: 1,5W	48j§	Phi,Sie,Mot	-	data
BSP 75(A)	MOS-N-FET	SMD, V-MOS, LogL, 55V, >1A, <0,55Ω Low-side Switch	48c§	Sie	-	data
BSP 76	MOS-N-FET	SMD, V-MOS, LogL, 42V, 1,4A, <0,24Ω Low-side Switch	48c§	Sie	-	
BSP 77	MOS-N-FET	SMD, V-MOS, LogL, 42V, 2,17A, <0,12Ω Low-side Switch	48c§	Sie	-	
BSP 78	MOS-N-FET	SMD, V-MOS, LogL, 42V, 3A, <0,06Ω Low-side Switch	48c§	Sie	-	data
BSP 88	MOS-N-FET-e	SMD, V-MOS, LogL, 240V, 0,32A, <8Ω	48c§	Sie	BSP 126, BSP 127, BSP 130	data
BSP 89	MOS-N-FET-e	SMD, V-MOS, LogL, 240V, 0,36A, <6Ω	48c§	Phi, Sie	BSP 126, BSP 127, BSP 130	data
BSP 090	MOS-P-FET-e	SMD,V-MOS, 30V, 5,7A, <90mΩ, <25/190ns	48c§	Phi	-	data
BSP 92	MOS-P-FET-e	SMD,V-MOS,LogL, 240V, 0,2A, <20Ω(0,2A)	48c§	Phi, Sie	BSP 225	data
BSP 100	MOS-N-FET-e	SMD, V-MOS, 30V, 3,5A, <0,1Ω(2,2A)	48c§	Phi	BSP 17, BSP 319, IRFL 4105	data
BSP 103	MOS-N-FET-e	SMD, V-MOS, LogL, 35V, 0,7A, <1,8Ω(1A)	48c§	Phi	BSP 295	data
BSP 105	MOS-N-FET-e	SMD, V-MOS, LogL, 60V, 0,5A, <3Ω(1A)	48c§	Phi	BUK 581-100A	data
BSP 106	MOS-N-FET-e	SMD, V-MOS, 60V, 0,425A, <4Ω(0,2A)	48c§	Phi	BSP 105, BSP 108, BSP 110	data
BSP 107	MOS-N-FET-e	SMD, V-MOS, LogL, 200V, 0,2A, <28Ω	48c§	Phi	BSP 120, BSP 121, BSP 126, BSP 128, ++	data
BSP 108	MOS-N-FET-e	SMD, V-MOS, LogL, 80V, 0,5A, <3Ω, 2/10ns	48c§	Phi	BUK 581-100A	data
BSP 109	MOS-N-FET-e	SMD, V-MOS, LogL, 90V, 0,45A, <4Ω	48c§	Phi	BUK 581-100A	data
BSP 110	MOS-N-FET-e	SMD, V-MOS, LogL, 80V, 0,325A, <10Ω	48c§	Phi	BSP 109, BSP 123, BUK 581-100A	data
BSP 120	MOS-N-FET-e	SMD, V-MOS, LogL, 200V, 0,25A, <12Ω	48c§	Phi	BSP 88...89, BSP 126, BSP 128	data
BSP 121	MOS-N-FET-e	SMD, V-MOS, LogL, 200V, 0,35A, <6Ω	48c§	Phi	BSP 88...89, BSP 126, BSP 128	data
BSP 122	MOS-N-FET-e	SMD, V-MOS, LogL, 200V, 0,55A, <2,5Ω	48c§	Phi	BSP 297	data
BSP 123	MOS-N-FET-e	SMD, V-MOS, LogL, 100V, 0,38A, <6Ω	48c§	Sie	BUK 581-100A	data
BSP 124	MOS-N-FET-d	SMD,V-MOS, 250V, 0,25A, <12Ω, <10/30ns	48c§	Phi	BSP 129	data
BSP 125	MOS-N-FET-e	SMD, V-MOS, 600V, 0,12A, <45Ω(0,12A)	48c§	Sie	BSP 300	data
BSP 126	MOS-N-FET-e	SMD, V-MOS, LogL, 250V, 0,35A, <7Ω	48c§	Phi	BSP 88...89, BSP 127, BSP 130	data
BSP 127	MOS-N-FET-e	SMD, V-MOS, LogL, 270V, 0,35A, <8Ω	48c§	Phi	BSP 130, BSP 298	data
BSP 128	MOS-N-FET-e	SMD, V-MOS, LogL, 200V, 0,35A, <8Ω	48c§	Phi	BSP 88...89, BSP 126, BSP 130	data
BSP 129	MOS-N-FET-d	SMD, V-MOS, 240V, 0,2A, <20Ω(14mA)	48c§	Sie	(BSP 124)[7]	data
BSP 130	MOS-N-FET-e	SMD, V-MOS, LogL, 300V, 0,3A, <8Ω	48c§	Phi	BSP 298	data
BSP 135	MOS-N-FET-d	SMD, V-MOS, 600V, 0,1A, <60Ω(10mA)	48c§	Sie	-	data
BSP 145	MOS-N-FET-e	SMD, V-MOS, 450V, 0,25A, <14Ω(0,1A)	48c§	Phi	BSP 299	data
BSP 149	MOS-N-FET-d	SMD, V-MOS, 200V, 0,48A, <3,5Ω(30mA)	48c§	Sie	-	data
BSP 152	MOS-N-FET-e	SMD, V-MOS, 200V, 0,55A, <2,5Ω(0,75A)	48c§	Phi	BSP 122, BSP 297	data
BSP 170 P	MOS-P-FET-e	SMD, V-MOS, 60V, 1,9A, <0,3Ω(1,9A)	48c§	Sie	-	data
BSP 171	MOS-P-FET-e	SMD, V-MOS, LogL, 60V, 1,6A, <0,35Ω	48c§	Sie	-	data
BSP 171 P	MOS-P-FET-e	SMD, V-MOS, LogL, 60V, 1,9A, <0,3Ω	48c§	Sie	-	data
BSP 204	MOS-P-FET-e	V-MOS, 200V, 0,25A, 1W, <15Ω, 5/20ns	7c	Phi	(BS 192, BS 208, BSS 92, BSS 192)[7]	data
BSP 204A	MOS-P-FET-e	=BSP 204:	7a	Phi	→BSP 204	data
BSP 205	MOS-P-FET-e	SMD, V-MOS, 60V, 0,275A, <10Ω, 3/10ns	48c§	Phi	BSP 206	data
BSP 206	MOS-P-FET-e	SMD, V-MOS, 60V, 0,35A, <6Ω, 4/15ns	48c§	Phi	BSP 316	data
BSP 220	MOS-P-FET-e	SMD, V-MOS, LogL, 200V, 0,225A, <12Ω	48c§	Phi	BSP 317	data
BSP 225	MOS-P-FET-e	SMD, V-MOS, LogL, 250V, 0,225A, <15Ω	48c§	Phi	BSP 230	data
BSP 230	MOS-P-FET-e	SMD, V-MOS, LogL, 300V, 0,215A, <17Ω	48c§	Phi	BSP 255	data
BSP 250	MOS-P-FET-e	SMD, V-MOS, LogL, 30V, 3A, <0,4Ω(0,5A)	48c§	Phi	-	data
BSP 254	MOS-P-FET-e	V-MOS, LogL, 250V, 0,2A, <15Ω, 5/20ns	7c	Phi	BS 209, BSS 92	data
BSP 254 A	MOS-P-FET-e	=BSP 254:	7a	Phi	BSS 92	data
BSP 255	Si-P-FET-e	SMD, V-MOS, LogL, 300V, 0,325A, <17Ω	48c§	Phi	-	data
BSP 280	MOS-N-IGBT	SMD, 1000V, 0,5A, 35/140ns	48id	Sie	-	data
BSP 295	MOS-N-FET-e	SMD, V-MOS, LogL, 50V, 1,8A, <0,3Ω	48c§	Sie	BUK 581-100A	data
BSP 296	MOS-N-FET-e	SMD, V-MOS, LogL, 100V, 1A, <0,8Ω(1A)	48c§	Sie	BUK 582-100A	data
BSP 297	MOS-N-FET-e	SMD, V-MOS, LogL, 200V, 0,65A, <2Ω	48c§	Sie	BSP 122	data
BSP 298	MOS-N-FET-e	SMD, V-MOS, 400V, 0,5A, <3Ω(0,5A)	48c§	Sie	2SK3333	data
BSP 299	MOS-N-FET-e	SMD, V-MOS, 500V, 0,4A, <4Ω(0,4A)	48c§	Sie	-	data
BSP 300	MOS-N-FET-e	SMD, V-MOS, 800V, 0,19A, <20Ω(0,19A)	48c§	Sie	STN 1NB80	data
BSP 304	MOS-P-FET-e	V-MOS, LogL, 300V, 0,17A, <17Ω, 5/15ns	7c	Phi	-	data
BSP 304 A	MOS-P-FET-e	=BSP 304:	7a		-	data
BSP 308	MOS-N-FET-e	SMD, V-MOS, LogL, 30V, 4,7A, <0,05Ω	48c§	Sie	IRLL 024N, IRLL 2703	data
BSP 315	MOS-P-FET-e	SMD, V-MOS, LogL, 50V, 1,1A, <0,8Ω	48c§	Sie	BSP 171	data
BSP 315 P	MOS-P-FET-e	SMD, V-MOS, LogL, 60V, 1,1A, <0,8Ω	48c§	Sie	BSP 171	data
BSP 316	MOS-P-FET-e	SMD, V-MOS, LogL, 100V, 0,65A, <2,2Ω	48c§	Sie	-	data
BSP 317	MOS-P-FET-e	SMD, V-MOS,LogL, 200V, 0,37A, <6Ω	48c§	Sie	BSP 220	data
BSP 318	MOS-N-FET-e	SMD,V-MOS, 60V, 2,6A, <0,15Ω, 42/160ns	48c§	Sie	BUK 483-60A, BUK 583-60A	data
BSP 318 S	MOS-N-FET-e	=BSP 318: LogL, 27/35ns	48c§	Sie	BUK 583-60A	data
BSP 319	MOS-N-FET-e	SMD, V-MOS, LogL, 50V, 3,8A, <0,07Ω	48c§	Sie	BUK 583-60A	data
BSP 319 S	MOS-N-FET-e	SMD, V-MOS, 60V, 2,9A, <0,12Ω(2,9A)	48c§	Sie	BUK 483-60A, BUK 583-60A	data
BSP 324	MOS-N-FET-e	SMD, V-MOS, 400V, 0,17A, <25Ω(0,17A)	48c§	Sie	BSP 145, BSP 299	data
BSP 350	IC	High-side Switch, Ubb=4,9...45V	≈39°	Sie	-	

Type	Device	Short Description	Fig.	Manu	Comparision Types	More at
? 372	MOS-N-FET-e	SMD, V-MOS,LogL,100V,1,7A, <0,31Ω	48c§	Sie		data
? 373	MOS-N-FET-e	SMD,V-MOS, 100V, 1,7A, <0,3Ω, 40/145ns	48c§	Sie	IRFL 4310	data
? 450	MOS-IC	High-side Switch, Ubb=12...40V	≈39°	Sie	-	
? 452	MOS-IC	High-side Switch, Ubb=6...34V	≈39°	Sie		
? 550	MOS-IC	High-side Switch, Ubb=12...40V	≈39°	Sie		
? 603 S2L	MOS-N-FET-e	SMD, V-MOS, LogL, 55V, 5,2A, <40mΩ	48c§	Sie		
? 613 P	MOS-P-FET-e	SMD, V-MOS, 60V, 2,9A, <0,13Ω(2,9A)	48c§	Sie		
SR						
R 12	Si-P	SMD, S, 15V, 0,1A, >1,5GHz, <20/30ns	35a	Phi	2SA1764	data
R 12 R	Si-P	=BSR 12:	35d			data
R 13	Si-N	SMD, HF/S, 60V, 0,8A, <35/285ns	35a	Phi,Nsc,Tho	BSR 14, BSS 79, BSS 81	data
R 13 R	Si-N	=BSR 13:	35d		BSR 14R	data
R 14	Si-N	SMD, HF/S, 75V, 0,8A, <35/285ns	35a	Phi,Nsc,Tho	BSS 79, BSS 81	data
R 14 R	Si-N	=BSR 14:	35d			data
R 15	Si-P	SMD, HF/S, 60/40V, 0,6A, <50/110ns	35a	Phi,Nsc,Tho	BSS 80, BSS 82	data
R 15 R	Si-P	=BSR 15:	35d			data
R 16	Si-P	=BSR 15: 60/60V	35a	Phi,Nsc,Tho	BSS 80, BSS 82	data
R 16 R	Si-P	=BSR 16:	35d			data
SR 17(A)	Si-N	SMD, HF/S, 60V, 0,2A, <70/250ns	35a	Phi,Nsc	BSR 13...14, BSS 79, BSS 81	data
SR 17(A)R	Si-N	=BSR 17:	35d		BSR 13R...14R	data
SR 17A	Si-N	SMD, HF/S	35a	Phi, Val,Ns	2SC3739, BSR14, BSR19,	data
SR 18(A)	Si-P	SMD, HF/S, 40V, 0,2A, <70/300ns	35a	Phi,Nsc	BSR 15...16, BSS 80, BSS 82	data
SR 18(A)R	Si-P	=BSR 18:	35d		BSR 15R...16R	data
SR 18A	Si-P	SMD, HF/S	35a	Phi,Val,Nsc	2SA1464, 2SB1222, BSR15, BSR20,	data
SR 19	Si-N	SMD, HF/S, 160V, 0,6A, >100MHz	35a	Phi,Nsc		data
SR 19 A	Si-N	=BSR 19: 180V	35a		-	data
SR 20	Si-P	SMD, HF/S, 130V, 0,6A, >100MHz	35a	Phi,Nsc		data
SR 20 A	Si-P	=BSR 20: 160V	35a		-	data
SR 30	Si-P	SMD, NF/S, 70V, 1A, <500/650ns, B>40	39b	Phi, Ztx	-	data
SR 31	Si-P	=BSR 30: B>100	39b	Phi, Ztx	-	data
SR 32	Si-P	=BSR 30: 90V	39b	Phi	-	data
SR 33	Si-P	=BSR 30: 90V, B>100	39b	Phi, Ztx	-	data
SR 40	Si-N	SMD, NF/S, 70V, 1A, <250/1000ns, B>40	39b	Phi, Ztx	-	data
SR 41	Si-N	=BSR 40: B>100	39b	Phi, Ztx	-	data
SR 42	Si-N	=BSR 40: 90V	39b	Phi, Ztx	-	data
SR 43	Si-N	=BSR 40: 90V, B>100	39b	Phi, Ztx	-	data
SR 50	Si-N-Darl+Di	S, 60V, 1A, 0,8W, 350MHz, B>2000	7c	Phi	BC 618, BC 875, BC 877, BC 879	data
SR 51	Si-N-Darl+Di	=BSR 50: 80V	7c	Phi	BC 618, BC 877, BC 879	data
SR 52	Si-N-Darl+Di	=BSR 50: 100V	7c	Phi	BC 879	data
SR 55	Si-N-Darl	100V, 1A, >100MHz, B>1500	2a§	Sie	2SC1879, 2SD406, 2SD688	data
SR 56	N-FET	SMD, S, 40V, Idss>50mA, Up<10V	35b	Nsc,Phi,Tho	-	data
SR 57	N-FET	=BSR 56: Idss>20mA, Up<6V	35b	Nsc,Phi,Tho	-	data
SR 58	N-FET	=BSR 56: Idss>8mA, Up<4V	35b	Nsc,Phi,Tho	-	data
SR 59	Si-N-Darl	S, 15V, 5A, >100MHz, B>1500	2a§	Tix	2N5425...5426	data
SR 60	Si-P-Darl+Di	S, 60V, 1A, 0,8W, 350MHz, B>2000	7c	Phi	BC 876, BC 878, BC 880	data
SR 61	Si-P-Darl+Di	=BSR 60: 80V	7c	Phi	BC 878, BC 880	data
SR 62	Si-P-Darl+Di	=BSR 60: 100V	7c	Phi	BC 880	data
BSR 111	N-FET	=BSJ 111: SMD	35b	Phi	PMBFJ 111	data
BSR 112	N-FET	=BSJ 112: SMD	35b	Phi	PMBFJ 112	data
BSR 113	N-FET	=BSJ 113: SMD	35b	Phi	PMBFJ 113	data
BSR 174	P-FET	=BSJ 174: SMD	35b	Phi	PMBFJ 174	data
BSR 175	P-FET	=BSJ 175: SMD	35b	Phi	PMBFJ 175	data
BSR 176	P-FET	=BSJ 176: SMD	35b	Phi	PMBFJ 176	data
BSR 177	P-FET	=BSJ 177: SMD	35b	Phi	PMBFJ 177	data
BSS						
BSS 10	Si-N	SS, 40V, 0,5A, 0,3W, <13/16ns	2a§	Sgs	BSX 26, BSX 39, 2N3261	data
BSS 11	Si-N	SS, 40V, 0,2A, 0,36W, <12/18ns	2a§	Sgs	BSX 19...20, 2N2368, 2N2369(A), ++	data
BSS 12	Si-N	SS, 30V, 0,2A, 0,36W, <15/20ns	2a§	Sgs	BSS 11, BSW 38, 2N2368...69, 2N3011, ++	data
BSS 13	Si-N	S/Tr, 60V, 1A, 1W, 18/35ns	2a§	Sgs	BSS 27, BSV 77, 2N5189, 2SC1386, ++	data
BSS 14	Si-N	S/Tr, 75V, 2A, 1W, 18/35ns	2a§	Sgs	2N3506...3507, 2N5262	data
BSS 15	Si-N	HF/S, 100V, 2A, 1W, <80/800ns	2a§	Sgs	2N5320, 2SD854	data
BSS 16	Si-N	=BSS 15: 75V	2a§	Sgs	2N5320...5321, 2SD854	data
BSS 17	Si-P	HF/S, 100V, 2A, 1W, <100/1000ns	2a§	Sgs	2N5322	data
BSS 18	Si-P	=BSS 17: 75V	2a§	Sgs	BSV 82, 2N5322...5323	data
BSS 19	Si-N	Nix, 120V, 0,05A, 0,225W, >50MHz	9c	Tix	BF 297...299, BF 422, BSS 38, BSX 21	data
BSS 20	Si-N	=BSS 19: 160V	9c	Tix	BF 297...299, BF 422	data
BSS 21	Si-N	SS, 30V, 0,2A, 0,25W, <15/23ns	7a	Tix	BSS 11...12, BSW 38, 2N2368...2369(A), ++	data
BSS 22	Si-P	S, 12V, 0,2A, 0,25W, <60/75ns	7a	Tix	BSV 21, BSW 25, BSW 37, BSX 29, 2N3012++	data
BSS 23	Si-N	S/Tr, 45V, 1A, 0,5W, 25/40ns	2a§	Aeg,Tix	BSS 26, BSW 26, BSS 40...41, 2N4013...14	data
BSS 24	Si-N	4x NPN, 60V, 1A, <35ns	82ce	Sgs	-	data
BSS 25	Si-P	SMD, S, -/25V, 0,4A, <45/100ns	35a	Sie	BSR 15...16, BSS.80, BSS 82	data

Type	Device	Short Description	Fig.	Manu	Comparision Types	More at
BSS 26	Si-N	S/Tr, 60V, 1A, 0,36W, 15/40ns	2a§	Sgs	BSS 23, BSW 26, BSS 40...41, 2N4013...14	data
BSS 27	Si-N	S/Tr, 70V, 1A, 0,8W, <25/40ns	2a§	Phi,Tix	BSS 14, BSV 95, 2N3735, 2SC1386	data
BSS 28	Si-N	=BSS 27: 50V, <25/45ns	2a§	Phi,Tix	BSS 13, BSV 77, BSV 95, 2N3724A, ++	data
BSS 29	Si-N	=BSS 27: 50V, <30/50ns	2a§	Phi,Tix	BSS 13, BSV 77, BSV 95, 2N3724A, ++	data
BSS 30	Si-N	NF/S/Tr, 100V, 1A, 0,8W, 80MHz	2a§	Sgs	BC 141, BSW 66...68, BSX 46...47, ++	data
BSS 31	Si-N	=BSS 30: 100MHz	2a§	Sgs	BC 141, BSW 66...68, BSX 46...47, ++	data
BSS 32	Si-N	=BSS 30: 120V, 70MHz	2a§	Sgs	BSS 42...43, BSW 67...68, BSX 47, ++	data
BSS 33	Si-N	S/Vid, 200V, 0,4A, >40MHz	2a§	Tho,Tix	BFR 57...59, 2SD413, 2SD576, 2SD624	data
BSS 34	Si-N	HF/Nix, 100V, 0,2A, 0,625W, >90MHz	7c	Tix	BFR 86...89, BSS 38, BSX 21, 2SC3245	data
BSS 35	Si-N	=BSS 34: 120V	7c	Tix	BFR 86...89, BSS 38, BSX 21, 2SC3245	data
BSS 36	Si-N-Darl	0,6W, B>400	5r	Tho		data
BSS 37	Si-P	Nix, 110V, 0,1A, 0,2W, 95MHz	12d	Phi	BF 398, BF 435...437, BSS 68, BSV 68	data
BSS 38	Si-N	Nix, 120V, 0,1A, 0,3W, >60MHz	7a	Phi,Sie,Nsc	BF 297...299, BFR 86...89, BSX 21, ++	data
BSS 39	Si-P	Nix, 120V, 0,1A, 0,225W	9b	Tix	BF 398, BF 435...437, BSS 68, BSV 68	data
BSS 40	Si-N	S/Tr, 60/40V, 1A, 0,36W, <35/45ns	2a§	Phi,Tix	2N4014	data
BSS 41	Si-N	=BSS 40: 60/30V	2a§	Phi,Tix	2N4014	data
BSS 42	Si-N	S, 120V, 1,5A, 1W, 40/700ns	2a§	Aeg	BSV 84, BSW 67...68, BSX 47	data
BSS 43	Si-N	=BSS 42: 150V	2a§	Aeg	BSW 68, 2SC1860	data
BSS 44	Si-P	S/Tr, 65V, 5A, 0,87W, >70MHz	2a§	Aeg,Phi,Sgs	BSS 46, 2N6190...6193	data
BSS 45	Si-N	S/Tr, 85V, 5A, 0,87W, <300/1000ns	2a§	Aeg	2N5338...5339, 2N4895...4897	data
BSS 46	Si-P	S/Tr, 85V, 5A, 0,87W, <300/1000ns	2a§	Aeg	2N6190...6193	data
BSS 47	Si-N	SMD, Nix, 120V, 0,1A	35d(2mm)	Fer		data
BSS 48	Si-N	S/Vid, 300V, 1A, 1W, 120/500ns	2a§	Aeg	BFQ 38...40, 2N3439...40, 2SC1861...62	data
BSS 49	Si-N	=BSS 48: 400V	2a§	Aeg	BFQ 40, 2N3440, 2SC1862	data
BSS 50	Si-N-Darl+Di	S, 60V, 1A, 0,8W, 350MHz, B>2000	2a§	Mot,Phi	BCX 21	data
BSS 51	Si-N-Darl+Di	=BSS 50: 80V	2a§	Mot,Phi	-	data
BSS 52	Si-N-Darl+Di	=BSS 50: 100V	2a§	Mot,Phi	-	data
BSS 53(A,B)	Si-P	Vid, -/160V, 0,1A, 0,3W, 70MHz	2a§	Itt	BF 435...437, BF 491...493, BFP 23, BFP 26	data
BSS 54(A,B)	Si-P	=BSS 53: -/250V	2a§	Itt	BF 436...437, BF 492...493, BFP 26	data
BSS 55(A,B)	Si-P	=BSS 53: -/300V	2a§	Itt	BF 437, BF 493, BFP 26	data
BSS 56	Si-N	SMD, S/Tr, 100V, 0,5A, >80MHz	35d(2mm)	Fer	(BCX 41R)[6]	data
BSS 58 A	PUT	=BSV 58 A:	7e	Aeg	→BSV 58 A	data
BSS 58 B	PUT	=BSV 58 B:	7e	Aeg	→BSV 58 B	data
BSS 59	Si-N	HF/S, 140V, 1A, 0,5W, <200/750ns	2a§	Aeg	(BSS 43, BSW 68)[6]	data
BSS 60	Si-P-Darl+Di	S, 60V, 1A, 0,8W, 350MHz, B>2000	2a§	Mot,Phi	-	data
BSS 61	Si-P-Darl+Di	=BSS 60: 80V	2a§	Mot,Phi	-	data
BSS 62	Si-P-Darl+Di	=BSS 60: 100V	2a§	Mot,Phi	-	data
BSS 63	Si-P	SMD, Uni, 110V, 0,1A, 85MHz	35a	Phi,Sie,++	BCX 42, 2SA1257, 2SA1325, 2SA1330	data
BSS63 LT1	Si-P	SMD, Uni Typ-C.: BM(p,s)	35a	Ons	2SA1330, BSR20,	data
BSS 63 R	Si-P	=BSS 63:	35d		BCX 42R	data
BSS 64	Si-N	SMD, Uni, 120V, 0,1A, 100MHz	35a	Phi,Sie,++	BCX 41, 2SC3143, 2SC3340, 2SC3360	data
BSS 64 R	Si-N	=BSS 64:	35d		BCX 41R	data
BSS 65	Si-P	SMD, S, 12V, 0,1A, <60/90ns	35a	Fer	BSR 15...16, BSR 18	data
BSS 65 R	Si-P	=BSS 65:	35d		BSR 15R...16R, BSR 18R	data
BSS 66	Si-N	SMD, S, 60V, 0,1A, <70/250ns, B>50	35a	Fer	BSR 17	data
BSS 66 R	Si-N	=BSS 66:	35d		BSR 17R	data
BSS 67	Si-N	=BSS 66: B>100	35a	Fer	BSR 17	data
BSS 67 R	Si-N	=BSS 67:	35d		BSR 17R	data
BSS 68	Si-P	Nix, 110V, 0,1A, 0,3W, >50MHz	7a	Phi,Sie	BF 398, BF 435...437, BSV 68	data
BSS 69	Si-P	SMD, S, 40V, 0,1A, <75/225ns, B>50	35a	Fer	BSR 15...16, BSR 18	data
BSS 69 R	Si-P	=BSS 69:	35d		BSR 15R...16R, BSR 18R	data
BSS 70	Si-P	=BSS 69: B>100	35a	Fer	BSR 15...16, BSR 18	data
BSS 70 R	Si-P	=BSS 70:	35d		BSR 15R...16R, BSR 18R	data
BSS 71(S)	Si-N	S/Vid, 200V, 0,5A, 0,5W, 70MHz	2a§	Mot	BF 391...393, BFP 22, MPSA 42...43	data
BSS 72(S)	Si-N	=BSS 71: 250V	2a§	Mot	BF 392...393, BFP 25, MPSA 42	data
BSS 73(S)	Si-N	=BSS 71: 300V	2a§	Mot	BF 393, BFP 25, MPSA 42	data
BSS 74(S)	Si-P	S/Vid, 200V, 0,5A, 0,5W, 80MHz	2a§	Mot	BF 491...493, BFP 23, MPSA 92...93	data
BSS 75(S)	Si-P	=BSS 74: 250V	2a§	Mot	BF 492...493, BFP 26, MPSA 92	data
BSS 76(S)	Si-P	=BSS 74: 300V	2a§	Mot	BF 493, BFP 26, MPSA 92	data
BSS 77	Si-N	=BSS 48: 0,8W	2a§	Mot	BSS 48...49, 2N3440, 2SD413, 2SD624	data
BSS 78	Si-N	=BSS 72: 0,8W	2a§	Mot	BSS 48...49, 2N3440, 2SD576	data
BSS 79(B,C)	Si-N	SMD, NF/S, 75/40V, 0,8A, <35/310ns	35a	Mot,Sie,++	BSR 14, (BCX 41)	data
BSS 80(B,C)	Si-P	SMD, NF/S, 60/40V, 0,8A, <50/110ns	35a	Mot,Sie,++	(BCX 42)	data
BSS 81(B,C)	Si-N	=BSS 79: 75/35V	35a	Mot,Sie,++	BSR 14, (BCX 41)	data
BSS 82(B,C)	Si-P	=BSS 80: 60/60V	35a	Mot,Sie,++	(BCX 42)	data
BSS 83	MOS-N-FET-e*	SMD, HF, S, 15V, Up<2V, <45Ω, 1/5ns	44j	Phi,Tho		data
BSS 83 P	MOS-P-FET-e	SMD, V-MOS, LogL, 60V, 0,33A, <3Ω	35a	Sie	BSH 201	data
BSS 84	MOS-P-FET-e	SMD, V-MOS, LogL, 50V, 0,13A, 19/30ns	35a	Phi,Sie,++	BSS 284, 2SJ185, 2SJ210	data
BSS 84 P	MOS-P-FET-e	SMD, V-MOS, LogL, 60V, 0,17A, 20/50ns	35a	Sie	BSH 201, 2SJ210	data
BSS 87	MOS-N-FET-e	SMD, V-MOS, LogL, 240V, 0,29A, 16/55ns	39b	Phi,Sie	2SK848	data
BSS 88	MOS-N-FET-e	V-MOS, LogL, 240, 0,25A, 1W, <8Ω	7c	Phi,Sie	BSN 254	data

Type	Device	Short Description	Fig.	Manu	Comparision Types	More at
BSS 89	MOS-N-FET-e	V-MOS, LogL, 240V, 0,29A, 1W, <6Ω	7c	Phi,Sie,Six	BSN 254	data
BSS 91	MOS-N-FET-e	V-MOS,LogL, 200V, 0,35A, 1,5W, <6Ω	2a	Phi, Sie	BSS 89, BSN 254	data
BSS 92	MOS-P-FET-e	V-MOS, LogL, 240V, 0,15A, 1W, <20Ω	7c	Phi,Sie,Six	BSP 254, BSP 304	data
BSS 93	MOS-N-FET-e	=BSS 87: 0,5A, 2,5W	2a§	Sie	BSS 95	data
BSS 95	MOS-N-FET-e	V-MOS,LogL, ∟40V, 0,8A, 8,3W,<6Ω(0,5A)	13c	Sie		data
BSS 97	MOS-N-FET-e	V-MOS, 200V, 1,5A, 10W, <2Ω, 20/150ns	13c	Sie	(2SK296, 2SK375)[4]	data
BSS 98	MOS-N-FET-e	V-MOS, LogL, 50V, 0,3A, 0,63W, <3,5Ω	7a	Sie	BS 170, 2SK1336	data
BSS 99	Si-N-Darl	80V, 4A, >2MHz, B>750	2a	Sgs	2SC2208, (BD 679, BD 779, BD 865)[6]	data
BSS 100	MOS-N-FET-e	V-MOS, LogL, 100V, 0,22A, 0,63W, <6Ω	7a	Phi,Sie,Nsc	BS 189, BSS 89, BSS 91, BST 74, BST 76++	data
BSS 101	MOS-N-FET-e	V-MOS, LogL, 240V, 0,13A, 0,63W, <16Ω	7a	Sie,Nsc	BSS 88...89, BSN 254	data
BSS 110	MOS-P-FET-e	V-MOS, LogL, 50V, 0,17A, 0,63W, <10Ω	7a	Sie	BSS 92, 2SJ184, 2SJ460	data
BSS 119	MOS-N-FET-e	V-MOS, SMD, 100V, 0,17A, <6Ω, 9/24ns	35a	Sie	BSS 123, 2SK1589, 2SK1591	data
BSS 123(A)	MOS-N-FET-e	SMD, V-MOS, LogL, 100V, 0,17A, <6Ω 10/22ns, BSS 123(Ztx)≈BSS 123	35a	Mot,Phi,Sie Nsc, Ztx	BSS 119, 2SK1589, 2SK1591	data
BSS 124	MOS-N-FET-e	V-MOS, 400V, 0,12A, 1W, <28Ω, 15/33ns	7a, 7c	Sie	-	data
BSS 125	MOS-N-FET-e	V-MOS, 600V, 0,1A, 1W, <45Ω, 15/31ns	7c	Sie	-	data
BSS 129	MOS-N-FET-d	V-MOS, 240V, 0,15A, 1W, <20Ω, 14/40ns	7c	Sie, Six	DN 2530N3, ND 2410L	data
BSS 131	MOS-N-FET-e	SMD, V-MOS, 240V, 0,1A, <16Ω(0,1A)	35a	Phi, Sie	-	data
BSS 135	MOS-N-FET-d	V-MOS, 600V, 0,08A, 1W, <60Ω, 14/35ns	7c	Sie	-	data
BSS 138	MOS-N-FET-e	SMD, V-MOS, LogL, 50V, 0,22A, 3,5Ω 11/27ns	35a	Phi,Sie,Nsc Gsi,Ztx	BSH 101, 2SK1848	data
BSS 139	MOS-N-FET-d	SMD,V-MOS, 250V, 0,04A, <100Ω, 14/25ns	35a	Sie	-	data
BSS 145	MOS-N-FET-e	SMD, V-MOS, 65V, 0,22A, <3,5Ω, 11/27ns	35a	Sie	PMBF 170, BST 82, 2SK1848	data
BSS 149	MOS-N-FET-d	V-MOS, 200V, 0,35A, 1W,<3,5Ω, 27/110ns	7c	Sie	-	data
BSS 159	MOS-N-FET-d	SMD, V-MOS, 50V, 0,16A, <8Ω(70mA)	35a	Sie	BSS 169	data
BSS 169	MOS-N-FET-e	SMD, V-MOS, 100V, 0,12A, <12Ω(50mA)	35a	Sie	-	data
BSS 192	MOS-P-FET-e	=BSS 92: SMD	39b	Phi,Sie	2SK848	data
BSS 229	MOS-N-FET-d	=BSS 139: 0,07A	7c	Sie	-	data
BSS 284	MOS-P-FET-e	SMD, V-MOS, LogL, 50V, 0,13A, 19/30ns	35a	Sie	BSS 84, 2SJ185	data
BSS 295	MOS-N-FET-e	V-MOS, LogL, 50V, 1,4A, 1W, <0,3Ω	7c	Sie	2SK1272, 2SK1274, 2SK1733, 2SK1736, ++	data
BSS 296	MOS-N-FET-e	V-MOS, LogL, 100V, 0,8A, 1W, <0,8Ω	7c	Sie	2SK941, 2SK1484, 2SK1729	data
BSS 297	MOS-N-FET-e	V-MOS, LogL, 200V, 0,48A, 1W, <2Ω	7c	Sie	-	data
BSS 395	MOS-N-FET-e	V-MOS, LogL, 50V, 4,4A, 10W, <0,3Ω	13c	Sie	-	data
BSS 396	MOS-N-FET-e	V-MOS, LogL, 100V, 2A, 10W, <0,8Ω	13c	Sie	-	data
BSS 397	MOS-N-FET-e	V-MOS, LogL, 200V, 1,5A, 10W, <2Ω	13c	Sie	-	data
BSS 7728	MOS-N-FET-e	SMD, V-MOS, 60V, 0,15A, 11/27ns	35a	Sie	BSS 123, 2SK1590	data

BST

Type	Device	Short Description	Fig.	Manu	Comparision Types	More at
BST 15	Si-P	SMD, NF/S/Vid, 200V, 1A, >15MHz	39b	Phi, Ztx	-	data
BST 16	Si-P	=BST 15: 350V	39b	Phi, Ztx	-	data
BST 39	Si-N	SMD, NF/S/Vid, 450V, 1A, >70MHz	39b	Phi, Ztx	-	data
BST 40	Si-N	=BST 39: 300V	39b	Phi, Ztx	-	data
BST 50	Si-N-Darl+Di	SMD, 60V, 0,5/1,5A, 350MHz, B>2000	39b	Phi	2SD1470, 2SD1472, 2SD1511	data
BST 51	Si-N-Darl	=BST 50: 80V	39b	Phi, Ztx	2SD1472	data
BST 52	Si-N-Darl	=BST 50: 100V	39b	Phi, Ztx	2SD1472	data
BST 60	Si-P-Darl+Di	SMD, 60V, 0,5/1,5A, 350MHz, B>2000	39b	Phi	2SB1048	data
BST 61	Si-P-Darl	=BST 60: 80V	39b	Phi, Ztx	-	data
BST 62	Si-P-Darl	=BST 60: 100V	39b	Phi	-	data
BST 62-20	Si-P-Darl	=BST 60: 85V	39b	Ztx	-	data
BST 70	MOS-N-FET-e	V-MOS, 80V, 0,5A, 1W, <3Ω, <10/15ns	7c	Phi	BSS 296, 2SK423, 2SK940...41	data
BST 70 A	MOS-N-FET-e	=BST 70:	7a		→BST 70	data
BST 72	MOS-N-FET-e	V-MOS, 80V, 0,3A, 0,83W,<10Ω, <10/10ns	7c	Phi	BSS 88...89, BSS 91, 2SK1337	data
BST 72 A	MOS-N-FET-e	=BST 72:	7a		→BST 72	data
BST 74	MOS-N-FET-e	V-MOS, 200V, 0,25A, 1W, <12Ω, <10/25ns	7c	Phi	BSS 88...89, BSS 91, BSN 205	data
BST 74 A	MOS-N-FET-e	=BST 74:	7a		→BST 74	data
BST 76	MOS-N-FET-e	V-MOS, 180V, 0,3A, 1W, <10Ω, <10/15ns	7c	Phi	BSS 89, BSS 91, BSS 297	data
BST 76 A	MOS-N-FET-e	=BST 76:	7a		→BST 76	data
BST 78	MOS-N-FET-e	V-MOS,450V, 0,75A, 15W,<14Ω, <10/100ns	14b§	Phi	-	data
BST 80	MOS-N-FET-e	=BST 70: SMD	39b	Phi	2SK601, 2SK1078...1079	data
BST 82	MOS-N-FET-e	=BST 72: SMD	35a	Phi	-	data
BST 84	MOS-N-FET-e	=BST 74: SMD	39b	Phi	BSS 87	data
BST 86	MOS-N-FET-e	=BST 76: SMD	39b	Phi	BSS 87	data
BST 90	MOS-N-FET-e	=BST 70: 2,5W	2a§	Phi	-	data
BST 95	MOS-N-FET-e	V-MOS, S, 200V, 2A, <2Ω, <35/50ns	2a§	Phi	-	data
BST 97	MOS-N-FET-e	=BST 76: 0,4W	2a	Phi	-	data
BST 100	MOS-P-FET-e	V-MOS, 60V, 0,3A, 1W, <6Ω, 4/20ns	7a	Phi	2SJ198, 2SJ228, 2SJ231	data
BST 110	MOS-P-FET-e	V-MOS, 60V, 0,25A, 0,83W, <10Ω, 4/10ns	7a	Phi	BSP 204, 2SJ198, 2SJ228, 2SJ231	data
BST 120	MOS-P-FET-e	=BST 100: SMD	39b	Phi	2SJ199, 2SJ2i2...213	data
BST 122	MOS-P-FET-e	=BST 110: SMD	39b	Phi	2SJ199, 2SJ212..213	data
BST 124	MOS-N-FET-d	V-MOS, 250V, 0,45A, <12Ω, <10/30ns	14b	Phi		data

Type	Device	Short Description	Fig.	Manu	Comparision Types	More at
BSt A30 26	50Hz-Thy	400V, 0,6A(Ta=45°C), Igt/Ih<10/<40mA	27b	Sie	CS0,8-04, BStA3026M, CS1,2-04[9]	data
BSt A30 26 M	50Hz-Thy	400V, 0,8A(Ta=45°C), Igt/Ih<10/<40mA	27b	Sie	CS08-04, CS1,2-04,(TAG612-400,TAG606-400)[11]	data
BSt A30 33	50Hz-Thy	=BStA3026: 500V	27b		CS0,8-05, BStA3033M, CS1,2-05[9]	data
BSt A30 33 M	50Hz-Thy	=BStA3026M: 500V	27b		CS08-05,CS1,2-05,(TAG612-600,TAG606-600)[11]	data
BSt A30 40	50Hz-Thy	=BStA3026: 600V	27b		CS0,8-06, BStA3040M, CS1,2-06[9]	data
BSt A30 40 M	50Hz-Thy	=BStA3026M: 600V	27b		CS08-06,CS1,2-06,(TAG612-600,TAG606-600)[11]	data
BSt A30 46	50Hz-Thy	=BStA3026: 700V	27b		CS0,8-07, BStA3046M, CS1,2-07[9]	data
BSt A30 46 M	50Hz-Thy	=BStA3026M: 700V	27b		CS08-07,CS1,2-07,(TAG612-700,TAG606-700)[11]	data
BSt A30 53	50Hz-Thy	=BStA3026: 800V	27b		BStA3053M, (TAG613-800, TAG606-800)[11]	data
BSt A30 53 M	50Hz-Thy	=BStA3026M: 800V	27b		(TAG612-800, TAG606-800)[11]	data
BSt B01 06	50Hz-Thy	100V, 0,8A(Ta=45°C), Igt/Ih<10/<60mA	27b	Sie	CS0,8-02, BStA3026M, CS1,2-02[9]	data
BSt B01 13	50Hz-Thy	=BStB0106: 200V	27b		CS0,8-02, CS1,2-02, BStA3026M[9]	data
BSt B01 26	50Hz-Thy	=BStB0106: 400V	27b		CS0,8-04, BStA3026M, CS1,2-04[9]	data
BSt B01 33	50Hz-Thy	=BStB0106: 500V	27b		CS0,8-05, BStA3033M, CS1,2-05[9]	data
BSt B01 40	50Hz-Thy	=BStB0106: 600V	27b		CS0,8-06, BStA3040M, CS1,2-06[9]	data
BSt B01 46	50Hz-Thy	=BStB0106: 700V	27b		CS0,8-07, BStA3046M, CS1,2-07[9]	data
BSt B02 06	50Hz-Thy	100V, 3A(Tc=64°C), Igt/Ih<10/<60mA	28a	Sie	(TAG621-100, TAG626/100, T3,5N400,++)[11]	data
BSt B02 13	50Hz-Thy	=BStB0206: 200V	28a		(TAG621-200, TAG626-200, T3,5N400,++)[11]	data
BSt B02 26	50Hz-Thy	=BStB0206: 400V	28a		(TAG621-400, TAG626-400, T3,5N400,++)[11]	data
BSt B02 33	50Hz-Thy	=BStB0206: 500V	28a		(TAG621-500, TAG626-500, T3,5N600,++)[11]	data
BSt B02 40	50Hz-Thy	=BStB0206: 600V	28a		(TAG621-600, TAG626-600, T3,5N600,++)[11]	data
BSt B02 46	50Hz-Thy	=BStB0206: 700V	28a		(TAG621-700, TAG626-700, T3,5N800,++)[11]	data
BSt C01 06	50Hz-Thy	100V, 5A(Tc=82°C), Igt/Ih<35/<90mA	21b§	Sie	BStC0313, BStD0313, CS5-02, CS8-02[9]	data
BSt C01 13	50Hz-Thy	=BStC0106: 200V	21b§		BStC0313, BStD0313, CS5-02, CS8-02[9]	data
BSt C01 26	50Hz-Thy	=BStC0106: 400V	21b§		BStC0326, BStD0326, CS5-04, CS8-04[9]	data
BSt C01 33	50Hz-Thy	=BStC0106: 500V	21b§		BStC0340, BStD0340, CS5-06, CS8-06[9]	data
BSt C01 40	50Hz-Thy	=BStC0106: 600V	21b§		BStC0340, BStD0340, CS5-06, CS8-06[9]	data
BSt C01 46	50Hz-Thy	=BStC0106: 700V	21b§		BStC0353, BStD0353, CS5-08, CS8-08[9]	data
BSt C02 06	50Hz-Thy	100V, 3,5A(Tc=67°C), Igt/Ih<20/<60mA	28a	Sie	(TAG620-100, BStC1026, TAG625-100,++)[11]	data
BSt C02 13	50Hz-Thy	=BStC0206: 200V	28a		(TAG620-200, BStC1026, TAG625-200,++)[11]	data
BSt C02 26	50Hz-Thy	=BStC0206: 400V	28a		(TAG620-400, BStC1026, TAG625-400,++)[11]	data
BSt C02 33	50Hz-Thy	=BStC0206: 500V	28a		(TAG620-500, BStC1033, TAG625-500,++)[11]	data
BSt C02 40	50Hz-Thy	=BStC0206: 600V	28a		(TAG620-600, BStC1040, TAG625-600,++)[11]	data
BSt C02 46	50Hz-Thy	=BStC0206: 700V	28a		(TAG620-700, BStC1046, TAG625-700,++)[11]	data
BSt C03 13	50Hz-Thy	200V, 11,4A(Tc=85°C), Igt/Ih<40/<30mA	21b§	Sie	BStD0313, CS5-02, CS8-02[9]	data
BSt C03 26	50Hz-Thy	=BStC0313: 400V	21b§		BStD0326, CS5-04, CS8-04[9]	data
BSt C03 40	50Hz-Thy	=BStC0313: 600V	21b§		BStD0340, CS5-06, CS8-06[9]	data
BSt C03 53	50Hz-Thy	=BStC0313: 800V	21b§		BStD0353, CS5-08, CS8-08[9]	data
BSt C03 66	50Hz-Thy	=BStC0313: 1000V	21b§		BStD0366, CS5-10, CS8-10[9]	data
BSt C03 80	50Hz-Thy	=BStC0313: 1200V	21b§		BStD0380, CS5-12, CS8-12[9]	data
BSt C05 06	50Hz-Thy	100V, 5A(Tc=76°C), Igt/Ih<20/<60mA	22a§	Sie	TAG671-100, TAG676-100[9]	data
BSt C05 13	50Hz-Thy	=BStC0506: 200V	22a§		TAG671-200, TAG676-200[9]	data
BSt C05 26	50Hz-Thy	=BStC0506: 400V	22a§		TAG671-400, TAG676-400[9]	data
BSt C05 33	50Hz-Thy	=BStC0506: 500V	22a§		TAG671-500, TAG676-500[9]	data
BSt C05 40	50Hz-Thy	=BStC0506: 600V	22a§		TAG671-600, TAG676-600[9]	data
BSt C05 46	50Hz-Thy	=BStC0506: 700V	22a§		TAG671-700, TAG676-700[9]	data
BSt C06 06	F-Thy	100V, 3,2A(Tc=65°C), Igt/Ih<50/<100mA	22a§	Sie	S3700B, T3S400, TAG670S-100, TAG675S-100[9]	data
BSt C06 13	F-Thy	=BStC0606: 200V	22a§		S3700B, T3S400, TAG670S-200, TAG675S-200[9]	data
BSt C06 26	F-Thy	=BStC0606: 400V	22a§		S3700D, T3S400, TAG670S-400, TAG675S-400[9]	data
BSt C06 33	F-Thy	=BStC0606: 500V	22a§		S3700M, T3S500, TAG670S-500, TAG675S-500[9]	data
BSt C06 40	F-Thy	=BStC0606: 600V	22a§		S3700M, T3S600, TAG670S-600, TAG675S-600[9]	data
BSt C06 43	F-Thy	=BStC0606: 650V	22a§		T3S700, TAG670S-700, TAG675S-700[9]	data
BSt C06 46	F-Thy	=BStC0606: 700V	22a§		T3S700, TAG670S-700, TAG675S-700[9]	data
BSt C06 50	F-Thy	=BStC0606: 750V	22a§		TAG670S-800, TAG675S-800[9]	data
BSt C06 53	F-Thy	=BStC0606: 800V	22a§		TAG670S-800, TAG675S-800[9]	data
BSt C07 06	50Hz-Thy	100V, 0,9A(Tc=45°C), Igt/Ih<20/<60mA	27b	Sie	BStC3026, (TAG2,5-100, TAG611-100,++)[11]	data
BSt C07 13	50Hz-Thy	=BStC0706: 200V	27b		BStC3026, (TAG2,5-200, TAG611-200,++)[11]	data
BSt C07 26	50Hz-Thy	=BStC0706: 400V	27b		BStC3026, (TAG2,5-400, TAG611-400,++)[11]	data
BSt C07 33	50Hz-Thy	=BStC0706: 500V	27b		BStC3033, (TAG2,5-500, TAG611-500,++)[11]	data
BSt C07 40	50Hz-Thy	=BStC0706: 600V	27b		BStC3040, (TAG2,5-600, TAG611-600,++)[11]	data
BSt C07 46	50Hz-Thy	=BStC0706: 700V	27b		BStC3046, (TAG2,5-700, TAG611-700,++)[11]	data
BSt C09 26T92	F-Thy	400V, Igt<50mA	22a§	Sie	-	data
BSt C09 30T92	F-Thy	450V, Igt<50mA	22a§		-	data
BSt C10 26	50Hz-Thy	400V, 4A(Tc=85°C), Igt/Ih<25/<80mA	17h§	Sie	TAG625-400, C122D, TIC116D, CS6-04, ++[9]	data
BSt C10 26 M	50Hz-Thy	400V, 6A(Tc=85°C), Igt/Ih<25/<80mA	17h§	Sie	CS6-04, TAG660-400, S2800D, CS3,5-04, ++[9]	data
BSt C10 33	50Hz-Thy	=BStC1026: 500V	17h§		TAG625-500, C122E, TIC116E, CS6-06, ++[9]	data
BSt C10 33 M	50Hz-Thy	=BStC1026M: 500V	17h§		CS6-06, TAG660-500, S2800E, CS3,5-06, ++[9]	data
BSt C10 40	50Hz-Thy	=BStC1026: 600V	17h§		TAG625-600, C122M, TIC116M, CS6-06, ++[9]	data
BSt C10 40 M	50Hz-Thy	=BStC1026M: 600V	17h§		CS6-06, TAG660-600, S2800M, CS3,5-06, ++[9]	data
BSt C10 46	50Hz-Thy	=BStC1026: 700V	17h§		TAG625-700, TIC116S, CS6-07, BStC1046M++[9]	data
BSt C10 46 M	50Hz-Thy	=BStC1026M: 700V	17h§		CS6-07, TAG660-700, S2800S, CS3,5-07, ++[9]	data
BSt C10 53	50Hz-Thy	=BStC1026: 800V	17h§	TO-220	TAG625-800, TIC116N, CS6-08, BStC1053M++[9]	data
BSt C10 53 M	50Hz-Thy	=BStC1026M: 800V	17h§		CS6-08, TAG660-800, S2800N, 2N6399, ++[9]	data
BSt C12 26	F-Thy	400V, 2,5A(Tc=80°C), Igt/Ih<50/<100mA	17h§	Sie	S5800D, BT153, CSF11-04, CSF0,7-04	data
					(BStCC0233H, S3900E, 17088)[15]	data
BSt C12 33	F-Thy	=BStC1226: 500V	17h§		S5800E, BT153, CSF11-06	data
					(BStCC0233H, S3900E, 17088)[15]	data
BSt C12 40	F-Thy	=BStC1226: 600V	17h§		S5800M, CSF11-06	data
					(BStCC0240H, S3900MF, 17088)[15]	data

Type	Device	Short Description	Fig.	Manu	Comparision Types	More at
BSt C12 46	F-Thy	=BStC1226: 700V	17.i§		S5800S, CSF11-08 (BStCC0246H, S3900S, 17088)[15]	data data
BSt C12 50	F-Thy	=BStC1226: 750V	17h§		S5800N, CSF11-08 (BStCC0253H, S3900SF, 17088)[15]	data data
BSt C12 53	F-Thy	=BStC1226: 800V	17h§		S5800N, CSF11-08 (BStCC0253H, 17088)[15]	data data
BSt C30 26	50Hz-Thy	400V, 1A(Ta=45°C), Igt/Ih<25/<80mA	27b	Sie	(TAG2,5-400, TAG611-400, TAG605-400)[11]	data
BSt C30 33	50Hz-Thy	=BStC3026: 500V	27b		(TAG2,5-500, TAG611-500, TAG605-500)[11]	data
BSt C30 40	50Hz-Thy	=BStC3026: 600V	27b		(TAG2,5-600, TAG611-600, TAG605-600)[11]	data
BSt C30 46	50Hz-Thy	=BStC3026: 700V	27b		(TAG2,5-700, TAG611-700, TAG605-700)[11]	data
BSt C30 53	50Hz-Thy	=BStC3026: 800V	27b		(TAG2,5-800, TAG611-800, TAG605-800)[11]	data
BSt C31 26	50Hz-Thy	400V, 1,5A(Ta=85°C), Igt/Ih<25/<80mA	28a	Sie	BStC3126M, (TAG630-400, TAG620-400,++)[11]	data
BSt C31 26 M	50Hz-Thy	400V, 2,5A(Tc=85°C),Igt/Ih<25/<80mA	28a	Sie	(TAG630-400, TAG620-400, BStC1026,++)[11]	data
BSt C31 33	50Hz-Thy	=BStC3126: 500V	28a		BStC3133M, (TAG630-500, TAG620-500,++)[11]	data
BSt C31 33 M	50Hz-Thy	=BStC3126M: 500V	28a		(TAG630-500, TAG620-500, BStC1033,++)[11]	data
BSt C31 40	50Hz-Thy	=BStC3126: 600V	28a		BStC3140M, (TAG630-600, TAG620-600,++)[11]	data
BSt C31 40 M	50Hz-Thy	=BStC3126M: 600V	28a		(TAG630-600, TAG620-600, BStC1040,++)[11]	data
BSt C31 46	50Hz-Thy	=BStC3126: 700V	28a		BStC3146M, (TAG630-700, TAG620-700,++)[11]	data
BSt C31 46 M	50Hz-Thy	=BStC3126M: 700V	28a		(TAG630-700, TAG620-700, BStC1046,++)[11]	data
BSt C31 53	50Hz-Thy	=BStC3126: 800V	28a		BStC3153M, (TAG630-800, TAG620-800,++)[11]	data
BSt C31 53 M	50Hz-Thy	=BStC3126M: 800V	28a		(TAG630-800, TAG620-800, BStC1053,++)[11]	data
BSt CC01 26H	F-Thy+Di	TV-HA,400V,3,2A,Igt/Ih<50/<100mA,<3µs	22a§	Sie	S6087B, TD3F400H, TD4F400H[9]	data
BSt CC01 26R	F-Thy+Di	TV-HA,400V,3,2A,Igt/Ih<50/<100mA,<5µs	22a§	Sie	S6087A, TD3F400R, TD4F400R[9]	data
BSt CC01 33H	F-Thy+Di	=BStCC0126H: 500V	22a§		S6087B, TD3F500H, TD4F500H[9]	data
BSt CC01 33R	F-Thy+Di	=BStCC0126R: 500V	22a§		S6087A, TD3F500R, TD4F500R[9]	data
BSt CC01 40H	F-Thy+Di	=BStCC0126H: 600V	22a§		S6087B, TD3F600H, TD4F600H[9]	data
BSt CC01 40R	F-Thy+Di	=BStCC0126R: 600V	22a§		S6087A, TD3F600R, TD4F600R[9]	data
BSt CC01 46H	F-Thy+Di	=BStCC0126H: 700V	22a§		S6087B, TD3F700H, TD4F700H[9]	data
BSt CC01 46R	F-Thy+Di	=BStCC0126R: 700V	22a§		S6087A, TD3F700R, TD4F700R[9]	data
BSt CC01 53H	F-Thy+Di	=BStCC0126H: 800V	22a§		TD3F800H, TD4F800H[9]	data
BSt CC01 53R	F-Thy+Di	=BStCC0126R: 800V	22a§		TD3F800R, TD4F800R[9]	data
BSt CC02 33H	F-Thy+Di	TV-HA, 500V, 5A, Igt/Ih<50/<100mA, <3µs	17j§	Sie	S3900E, 17088	data
BSt CC02 33R	F-Thy+Di	TV-HA, 500V, 5A, Igt/Ih<50/<100mA, 5µs	17j§	Sie	S3901E, 17089	data
BSt CC02 40H	F-Thy+Di	=BStCC0233H: 600V	17j§		S3900MF, 17088	data
BSt CC02 40R	F-Thy+Di	=BStCC0233R: 600V	17j§		S3901MF, 17089	data
BSt CC02 46H	F-Thy+Di	=BStCC0233H: 700V	17j§		S3900S, 17088	data
BSt CC02 46R	F-Thy+Di	=BStCC0233R: 700V	17j§		S3901S, 17089	data
BSt CC02 53H	F-Thy+Di	=BStCC0233H: 800V	17j§		17088	data
BSt CC02 53R	F-Thy+Di	=BStCC0233R: 800V	17j§		17089	data
BSt CC02 60H	F-Thy+Di	=BStCC0233H: 900V	17j§		17088	data
BSt CC02 60R	F-Thy+Di	=BStCC0233R: 900V	17j§		17089	data
BSt D02 20	50Hz-Thy	300V, 8,5A(Tc=82°C), Igt/Ih<50/120mA	21b§	Sie	BTW42/600R	data
BSt D02 40	50Hz-Thy	=BStD0220: 600V	21b§		BTW42/600R	data
BSt D02 60	50Hz-Thy	=BStD0220: 900V	21b§		BTW42/1000R	data
BSt D03 13	50Hz-Thy	200V, 16A(Tc=85°C), Igt/Ih<30/<80mA	21b§	Sie	CS8-02[9]	data
BSt D03 26	50Hz-Thy	=BStD0313: 400V	21b§		CS8-04[9]	data
BSt D03 40	50Hz-Thy	=BStD0313: 600V	21b§		CS8-06[9]	data
BSt D03 53	50Hz-Thy	=BStD0313: 800V	21b§		CS8-08[9]	data
BSt D03 66	50Hz-Thy	=BStD0313: 1000V	21b§		CS8-10[9]	data
BSt D03 80	50Hz-Thy	=BStD0313: 1200V	21b§		CS8-12[9]	data
BSt D10 26	50Hz-Thy	400V, 8A(Tc=85°C), Igt/Ih<25/<80mA	17h§	Sie	BStD1026M, TIC126D, T7,5N400, TAG680-400[9]	data
BSt D10 26 M	50Hz-Thy	400V, 10A(Tc=85°C),Igt/Ih<25/<80mA	17h§	Sie	T9,5N400, CS15-04[9]	data
BSt D10 33	50Hz-Thy	=BStD1026: 500V	17h§		BStD1033M, TIC126E, T7,5N600, TAG680-500[9]	data
BSt D10 33 M	50Hz-Thy	=BStD1026M: 500V	17h§		T9,5N500, CS15-06[9]	data
BSt D10 40	50Hz-Thy	=BStD1026: 600V	17h§		BStD1040M, TIC126M, T7,5N600, TAG680-600[9]	data
BSt D10 40 M	50Hz-Thy	=BStD1026M: 600V	17h§		T9,5N600, CS15-06[9]	data
BSt D10 46	50Hz-Thy	=BStD1026: 700V	17h§		BStD1046M, TIC126S, T7,5N700, TAG680-700[9]	data
BSt D10 46 M	50Hz-Thy	=BStD1026M: 700V	17h§		T9,5N700, CS15-07[9]	data
BSt D10 53	50Hz-Thy	=BStD1026: 800V	17h§		BStD1052M, TIC126N, T7,5N800, TAG680-800[9]	data
BSt D10 53 M	50Hz-Thy	=BStD1026M: 800V	17h§		T9,5N800, CS15-08[9]	data
BSt D16 66 M	50Hz-Thy	1000V, 7,5A(Tc=85°C), Igt/Ih<10/<50mA	17h§	Sie	-	data
BSt D16 66 N	50Hz-Thy	=BStD1666M: Igt/Ih<20/<80mA	17h§	Sie	-	data
BSt D16 66 P	50Hz-Thy	=BStD1666M: Igt/Ih<50/<150mA	17h§		-	data
BSt D16 80 M	50Hz-Thy	=BStD1666M: 1200V	17h§		-	data
BSt D16 80 N	50Hz-Thy	=BStD1680M: Igt/Ih<20/<80mA	17h§		-	data
BSt D16 80 P	50Hz-Thy	=BStD1680M: Igt/Ih<50/<150mA	17h§		-	data
BSt D36 66 M	50Hz-Thy	1000V, 1A(Ta=45°C), Igt/Ih<10/<50mA	27b	Sie	(BStD1666M)[4]	data
BSt D36 66 N	50Hz-Thy	=BStD3666M: Igt/Ih<20/<80mA	27b		(BStD1666N)[4]	data
BSt D36 66 P	50Hz-Thy	=BStD3666M: Igt/Ih<50/<150mA	27b		(BStD1666P)[4]	data
BSt D36 80 M	50Hz-Thy	=BStD3666M: 1200V	27b		(BStD1680M)[4]	data
BSt D36 80 N	50Hz-Thy	=BStD3680M: Igt/Ih<20/<80mA	27b		(BStD1680N)[4]	data
BSt D36 80 P	50Hz-Thy	=BStD3680M: Igt/Ih<50/<150mA	27b		(BStD1680P)[4]	data
BSt D40 26	50Hz-Thy	400V, 12A, Igt<10mA	29b	Sie	T10N400H, S6200D, BStD4026M, C232D[9]	data
BSt D40 26 M	50Hz-Thy	400V, 15A, Igt<10mA	29b	Sie	C 232D[9]	data
BSt D40 33	50Hz-Thy	=BStD4026: 500V	29b		T10N500H, S6200M, BStD4033M, C232E[9]	data
BSt D40 33 M	50Hz-Thy	=BStD4026M: 500V	29b		C 232M[9]	data
BSt D40 40	50Hz-Thy	=BStD4026: 600V	29b		T10N600H, S6200M, BStD4040M, C232M[9]	data
BSt D40 40 M	50Hz-Thy	=BStD4026M: 600V	29b		C 232M[9]	data
BSt D40 46	50Hz-Thy	=BStD4026: 700V	29b		T10N700H, BStD4046M[9]	data
BSt D40 46 M	50Hz-Thy	=BStD4026M: 700V	29b		[9]	data
BSt D40 53	50Hz-Thy	=BStD4026: 800V	29b		T10N800H, BStD4053M[9]	data

Type	Device	Short Description	Fig.	Manu	Comparision Types	More at
BSt D40 53 M	50Hz-Thy	=BStD4026M: 800V	29b		-	data
BSt D41 26	50Hz-Thy	=BStD4026:	21b§		T10N400C, S6210D, BStD4126M, BStE4126,++[9]	data
BSt D41 26 M	50Hz-Thy	=BStD4026M:	21b§		C230D, BStE4126, BStE4126M[9]	data
BSt D41 33	50Hz-Thy	=BStD4033:	21b§		T10N500C, S6210M, BStD4133M, BStE4133,++[9]	data
BSt D41 33 M	50Hz-Thy	=BStD4033M:	21b§		C230E, BStE4133, BStE4133M[9]	data
BSt D41 40	50Hz-Thy	=BStD4040:	21b§		T10N600C, S6210M, BStD4140M, BStE4140,++[9]	data
BSt D41 40 M	50Hz-Thy	=BStD4040M:	21b§		C230M, BStE4140, BStE4140M[9]	data
BSt D41 46	50Hz-Thy	=BStD4046:	21b§		T10N700C, BStD4146M, BStE4146, BStE4146M[9]	data
BSt D41 46 M	50Hz-Thy	=BStD4046M:	21b§		BStE4146, BStE4146M[9]	data
BSt D41 53	50Hz-Thy	=BStD4053:	21b§		T10N800C, BStD4153M, BStE4153, BStE4153M[9]	data
BSt D41 53 M	50Hz-Thy	=BStD4053M:	21b§		BStE4153, BStE4153M[9]	data
BSt E03 26	F-Thy	400V, 0,6A(Ta=45°C),Igt/Ih<35/<50mA	27b	Sie	(S5800D, BT153, TAG650S-400,TAG655S-400)[11]	data
BSt E03 33	F-Thy	=BStE0326: 500V	27b		(S5800E, BT153, TAG650S-500,TAG655S-500)[11]	data
BSt E03 40	F-Thy	=BStE0326: 600V	27b		(S5800M, TAG650S-600, TAG655S-600)[11]	data
BSt E03 46	F-Thy	=BStE0326: 700V	27b		(S5800S, CSF11-08)[11]	data
BSt E04 26	F-Thy	400V, 4A(Tc=77°C),Igt/Ih<50/<100mA	22a§	Sie	T3S400, TAG670S-400, TAG675S-400[9]	data
BSt E04 33	F-Thy	=BStE0426: 500V	22a§		T3S500, TAG670S-500, TAG675S-500[9]	data
BSt E04 40	F-Thy	=BStE0426: 600V	22a§		T3S600, TAG670S-600, TAG675S-600[9]	data
BSt E04 46	F-Thy	=BStE0426: 700V	22a§		T3S700, TAG670S-700, TAG675S-700[9]	data
BSt E40 26	50Hz-Thy	400V, 18A, Igt<50mA	29b	Sie	BStE4026M[9]	data
BSt E40 26 M	50Hz-Thy	400V, 22A, Igt<50mA	29b	Sie	-	data
BSt E40 33	50Hz-Thy	=BStE4026: 500V	29b		BStE4033M[9]	data
BSt E40 33 M	50Hz-Thy	=BStE4026M: 500V	29b		-	data
BSt E40 40	50Hz-Thy	=BStE4026: 600V	29b		BStE4040M[9]	data
BSt E40 40 M	50Hz-Thy	=BStE4026M: 600V	29b		-	data
BSt E40 46	50Hz-Thy	=BStE4026: 700V	29b		BStE4046M[9]	data
BSt E40 46 M	50Hz-Thy	=BStE4026M: 700V	29b		-	data
BSt E40 53	50Hz-Thy	=BStE4026: 800V	29b		BStE4053M[9]	data
BSt E40 53 M	50Hz-Thy	=BStE4026M: 800V	29b		[9]	data
BSt E41 26	50Hz-Thy	=BStE4026:	21b§		CS13-04, T15,1N400, TAG16N400, BTW40/400[9]	data
BSt E41 26 M	50Hz-Thy	=BStE4026M:	21b§		T17N400, T24N400, TAG24N400, CS23-04, ++[9]	data
BSt E41 33	50Hz-Thy	=BStE4033:	21b§		CS13-06, T15,1N600, TAG16N600, BTW40/600[9]	data
BSt E41 33 M	50Hz-Thy	=BStE4033M:	21b§		T17N600, T24N600, TAG24N600, CS23-06, ++[9]	data
BSt E41 40	50Hz-Thy	=BStE4040:	21b§		CS13-06, T15,1N600, TAG16N600, BTW40/600[9]	data
BSt E41 40 M	50Hz-Thy	=BStE4040M:	21b§		T17N600, T24N600, TAG24N600, CS23-06, ++[9]	data
BSt E41 46	50Hz-Thy	=BStE4046:	21b§		CS13-08, T15,1N800, TAG16N800, BTW40/800[9]	data
BSt E41 46 M	50Hz-Thy	=BStE4046M:	21b§		T17N800, T24N800, TAG24N800, CS23-08, ++[9]	data
BSt E41 53	50Hz-Thy	=BStE4053:	21b§		CS13-08, T15,1N800, TAG16N800, BTW40/800[9]	data
BSt E41 53 M	50Hz-Thy	=BStE4053M:	21b§		T17N800, T24N800, TAG24N800, CS23-08, ++[9]	data

BSV

Type	Device	Short Description	Fig.	Manu	Comparision Types	More at
BSV 10	Si-P	NF/S, 40V, 1A, >50MHz, <500/-ns	2a§	Sie	BC 160...161, BSV 15...17, BSW 40, ++	data
BSV 11	Si-P	=BSV 10: 60V	2a§	Sie	BC 161, BSV 16...17, BSW 40, ++	data
BSV 12	Si-P	=BSV 10: 90V	2a§	Sie	BC 161, BCX 60, BSV 17, BSW 40, ++	data
BSV 15	Si-P	NF/S, 40V, 1A, >50MHz, <500/650ns	2a§	EUR	BC 160...161, BCX 60, BSV 82, BSW 40, ++	data
BSV 16	Si-P	=BSV 15: 60V	2a§	EUR	BC 161, BCX 60, BSV 82, BSW 40, ++	data
BSV 17	Si-P	=BSV 15: 90V	2a§	EUR	BCX 60, BSS 17, BSV 82, BSW 40, ++	data
BSV 19	Si-P	60V, 1A, 1W	2a§	Sie	-	data
BSV 20(A)	MOS-P-FET-e	S, Chopper, 30V, 0,2A	5n+	Sgs	-	data
BSV 21	Si-P	S, 12V, 0,2A, 0,36W, <60/90ns	2a§	Tix	BSW 25, BSW 37, BSX 29, 2N2894A, ++	data
BSV 22	MOS-N-FET-d	S, Chopper, 30V, 50mA	5h+	Phi	BFW 96, BSV 81	data
BSV 23(K,L,M)	Si-N	S, 25V, 0,2A, 0,3W, >200MHz	40e	Fer	BSS 11...12, BSW 38, BSX 19...20, 2N3011++	data
BSV 24(K,L,M)	Si-N	=BSV 23: 20V	40e	Fer	BSS 11...12, BSW 38, BSX 19...20, 2N3011++	data
BSV 25(K,L,M)	Si-N	SS, 30V, 0,5A, 0,3W, <15/20ns	40e	Fer	BSS 11...12, BSW 38, BSX 19...20, 2N3011++	data
BSV 26(K,L,M)	Si-N	SS, 40V, 0,5A, 0,3W, <12/18ns	40e	Fer	BSS 11, BSX 19...20, 2N2368...69(A), ++	data
BSV 27(K,L,M)	Si-N	SS, 30V, 0,5A, 0,3W, <15/20ns	40e	Fer	BSS 11...12, BSW 38, BSX 19...20, 2N3011++	data
BSV 28(K,L,M)	Si-N	Nix, 100V, 0,1A, 0,3W	40e	Fer	BF 297...299, BFR 86, BSS 38, BSX 21	data
BSV 29(K,L,M)	Si-N	=BSV 28: 120V	40e	Fer	BF 297...299, BFR 86, BSS 38, BSX 21	data
BSV 33(K,L,M)	Si-P	S, 12V, 0,2A, 0,3W, <60/60ns	40e	Fer	BSV 21, BSW 25, BSW 37, BSX 29, 2N2894++	data
BSV 34(A)	MOS-P-FET-e	Dual, 30V, 0,2A, on<500Ω	81bz+	Sgs	-	data
BSV 35	Si-N	SMD, SS, 40V, 0,5A, <12/18ns	35d(2mm)	Fer	-	data
BSV 35 A	Si-N	=BSV 35: 25V, <40/75ns	35d(2mm)		-	data
BSV 36	Si-N	SMD, S, 15V, 0,5A, <20/15ns	35d(2mm)	Fer	-	data
BSV 37	Si-P	SMD, S, 12V, 0,5A, <60/90ns	35d(2mm)	Fer	-	data
BSV 38(P)	N-FET	Min, Chopper, 25V, Idss>50mA, Up<10V		Tix	-	data
BSV 39(P)	N-FET	Min, Chopper, 25V, Idss>8mA, Up<6V		Tix	-	data
BSV 40	Si-N	NF/S, 40V, 0,1A, 0,36W, >300MHz, B>40	2a§	Itt	BSW 41, BSY 63, 2N708, 2N4123, ++	data
BSV 41	Si-N	=BSV 40: B>100	2a§	Itt	BSW 41, BSY 63, 2N708, 2N4123, ++	data
BSV 42	Si-P	NF/S, 70/70V, 0,5A, 0,6W, 200MHz	2a§	Itt	BC 161, BC 303, BSV 17, BSW 40, ++	data
BSV 43(A,B)	Si-P	=BSV 42: 60/60V	2a§	Itt	BC 161, BC 303...304, BSV 16...17, ++	data
BSV 44(A,B)	Si-P	=BSV 42: 60/40V	2a§	Itt	BC 161, BC 303...304, BSV 16...17, ++	data
BSV 45(A,B)	Si-P	=BSV 42: 30/30V	2a§	Itt	BC 160...161, BC 303...304, BSV 15...17, ++	data
BSV 46	Si-P	NF/S, 70/70V, 0,5A, 0,4W, 200MHz	2a§	Itt	BC 640, BCY 77, 2N2906A...2907A, ++	data
BSV 47(A,B)	Si-P	=BSV 46: 60/60V	2a§	Itt	BC 638, BCY 77, 2N2906A...2907A, ++	data
BSV 48(A,B)	Si-P	=BSV 46: 60/40V	2a§	Itt	BC 638, BCY 77, 2N2906(A)...2907(A), ++	data
BSV 49(A,B)	Si-P	=BSV 46: 30/30V	2a§		BC 636, BC 327, BCY 78...79, 2N2906...07++	data
BSV 50 E	Si-P	Min, S, 12V, 0,1A, 800MHz, <20/-ns	36c	Sgs	-	data

Type	Device	Short Description	Fig.	Manu	Comparision Types	More at
BSV 50 F	Si-P	=BSV 50E:	36f		-	data
BSV 50 G	Si-P	=BSV 50E:	36e		-	data
BSV 51	Si-N	Nix, 100V, 0,2A, 0,25W, >50MHz	9e	Aeg	BF 297...299, BFR 86, BSS 38, BSX 21	data
BSV 52	Si-N	SMD, S, 20V, 0,1A, <12/18ns	35a	Mot,Phi,++	BSV 65	data
BSV 52 R	Si-N	=BSV 52:	35d		BSV 65R	data
BSV 53(P)	Si-N	Min, S, 40V, 0,2A, 400MHz, <12/-ns		Tix	-	data
BSV 54(P)	Si-N	=BSV 53: 20V		Tix	-	data
BSV 55(P)	Si-P	Min, S, 20V, 0,2A, 400MHz, <30/-ns		Tix	-	data
BSV 55A(P)	Si-P	=BSV 55: 12V, <60/-ns		Tix	-	data
BSV 56 A	UJT-P	35V, Ip<6µA, Iv>4mA, η=0,56...0,75	5m§	Aeg	2N2646, 2N4871	data
BSV 56 B	UJT-P	=BSV 56 A: η=0,68...0,82	5m§		-	data
BSV 56 C	UJT-P	=BSV 56 A: Ip<25µA, η=0,47...0,82	5m§		-	data
BSV 57 A,B,C	UJT-P	=BSV 56 A,B,C:	7n	Aeg	→BSV 56 A,B,C	data
BSV 58 A	PUT	Ip<1µA, Iv>25µA	5o	Aeg	2N6028, 2N6120	data
BSV 58 B	PUT	=BSV 58 A: Ip<5µA, Iv>70µA	5o		2N6027, 2N6116, 2N6119, MPU131, MPU231	data
BSV 59	Si-N	S/Tr, 60V, 0,5A, 0,36W, 18/25ns	2a§	Sgs	BSX 49, 2N3301...3302, 2N4014	data
BSV 60	Si-N	S/Tr, 45V, 3A, 0,8W, <500/1000ns	2a§	Aeg,Fer	BSS 45, BSX 62...64, 2N4237...4239	data
BSV 61	Si-N	SMD, 20/9V, 0,05A, <12/22ns	Chip	Phi	-	data
BSV 62	Si-N	=BSV 61: 20/12V		Phi	-	data
BSV 63	Si-N	=BSV 61: 20/12V		Phi	-	data
BSV 64	Si-N	S/Tr, 100V, 2A, 100MHz, <600/1200ns	2a§	Fer,Phi	BCX 40, BSS 15, 2N4239, 2N5320	data
BSV 65(A,B)	Si-N	SMD, S, 20V, 0,15A, <20/40ns	35a	Sie	BSV 52, 2SC3578	data
BSV 65R(RA,RB)	Si-N	=BSV 52:	35d		BSV 52R	data
BSV 68	Si-P	Nix, 110V, 0,1A, 0,25W, 95MHz	2a§	Phi	BF 398, BF 435...437, BSS 68	data
BSV 69	Si-N	S/Tr, 45V, 1A, 0,8W, 25/40ns	2a§	Aeg	BSS 13, BSS 27...29, BSV 77, 2N5189, ++	data
BSV 71	Si-N	SMD, S, 15V, 0,05A, >600MHz, <20/-ns	Chip	Phi	-	data
BSV 74	Si-N	Ubr=110V, Ip=1,5A, Ih<400mA, <4/-ns	2a§	Phi	-	data
BSV 75	Si-N	Ubr=60V, Ip=1A, Ih<400mA, <4/-ns	2a§	Phi	-	data
BSV 76	Si-N	Ubr=90V, Ip=2A, Ih<400mA, <5/-ns	2a§	Phi	-	data
BSV 77	Si-N	S/Tr, 60V, 1A, 0,8W, 15/40ns	2a§	Sgs	BSS 13, BSS 27, BSV 95, 2N5189	data
BSV 78	N-FET	Chopper, sym, 40V, Idss>50mA, Up<11V	2b&	Phi	BFS 74, 2N4856	data
BSV 79	N-FET	Chopper, sym, 40V, Idss>20mA, Up<7V	2b&	Phi	BFS 75, 2N4857	data
BSV 80	N-FET	Chopper, sym, 40V, Idss>10mA, Up<5V	2b&	Phi	BFS 76, 2N4858	data
BSV 81	MOS-N-FET-d	Chopper, sym, 30V, 50mA	5m+	Phi	BFW 96, BSV 22	data
BSV 82	Si-P	NF/S, 80V, 2A, 1W, 50/250ns	2a§	Sgs	2N6303	data
BSV 83	Si-P	NF/S, 90V, 1A, 0,8W, <35/-ns	2a§	Sgs	BSW 40, 2N4036	data
BSV 84	Si-N	NF/S, 120V, 2A, 1W, <250/700ns	2a§	Sgs	(BSS 42...43, BSW 67...68, BSX 47)[7]	data
BSV 85	Si-N	NF/S, 50V, 1A, 0,36W, 250MHz	2a§	Sgs	BC 637, BSS 26, BSS 40...41, BSW 26, ++	data
BSV 86	Si-N	NF/S, 75V, 0,4A, 50/210ns, B>100	12a	Phi	BC 639, BSW 63...64, 2N2221A..2222A, ++	data
BSV 87	Si-N	=BSV 86: B>40	12a	Phi	BC 639, BSW 63...64, 2N2221A..2222A, ++	data
BSV 88	Si-N	=BSV 86: 60V	12a	Phi	BC 637, BSW 61...62, 2N2221..2222, ++	data
BSV 89	Si-N	SS, 25V, 0,1A, 0,36W, <12/18ns	2a§	Sgs	BSS 11...12, BSX 92...93, 2N2368...69, ++	data
BSV 90	Si-N	=BSV 89: 30V	2a§	Sgs	BSS 11...12, BSX 92...93, 2N2369...69, ++	data
BSV 91	Si-N	=BSV 89: 40V	2a§		BSS 11...12, BSX 92...93, 2N2368...69	data
BSV 92	Si-N	=BSV 89: 40V	2a§	Sgs	BSS 11...12, BSX 92...93, 2N2368...69, ++	data
BSV 95	Si-N	S/Tr, 80V, 1A, 0,8W, 15/40ns	2a§	Sgs	2N3735	data
BSV 96	Si-P	NF/S, 30V, 0,3A, 0,22W, 75MHz, B>100	12a	Phi	BC 328, BC 636, BCY 77...79, 2N2906...07++	data
BSV 97	Si-P	=BSV 96: B>40	12a	Phi	BC 328, BC 636, BCY 77...79, 2N2906...07++	data
BSV 98	Si-P	=BSV 96: B>30	12a	Phi	BC 328, BC 636, BCY 77...79, 2N2906...07++	data
BSV 99	Si-N	SS, 30V, 0,2A, 0,25W, <12/-ns	7a	Tix	BSS 11...12, BSX 92...93, 2N2368...69, ++	data
BSVP 20	Si-N	SMD, 40V, 0,2A(ss), 150MHz	Chip	Ucp	-	data
BSVP 30	Si-N	SMD, 40V, 0,1A, 300MHz	Chip	Ucp	-	data

BSW

Type	Device	Short Description	Fig.	Manu	Comparision Types	More at
BSW 10	Si-N	NF/S, 90V, 0,8A, 0,6W, 100/350ns	2a§	Aeg	BSS 42...43, BSW 66...68, BSX 46...47, ++	data
BSW 11	Si-N	Min, S, 25V, 0,03A, <25/50ns	36c	Aeg	-	data
BSW 12	Si-N	Min, S, 40V, 0,2A, <40/80ns	36c	Aeg	-	data
BSW 13	Si-N	Min, S, 20V, 0,05A, <20/40ns	(41c)	Sie	-	data
BSW 16	Si-N	8x NPN, 50V, 0,8A, <50/90ns	14-DIP	Sgs	-	data
BSW 17	Si-N	5x NPN, 120V, 0,1A, >40MHz	14-DIP	Sgs	-	data
BSW 18	Si-N	=BSW 17: SMD	14-FLP	Sgs	-	data
BSW 19	Si-P	S, 35V, 0,1A, 0,3W, <150/800ns	2a§	Aeg	BCY 77...79, 2N4034...35, 2N3905...06, ++	data
BSW 20	Si-P	=BSW 19:	7a	Aeg	BCY 77...79, 2N4034...35, 2N3905...06, ++	data
BSW 21	Si-P	S, 25V, 0,2A, 0,3W, 200/-ns, B>75	2a§	Mot,Tho,Tix	BCY 77...79, 2N2906...07, 2N3905...06, ++	data
BSW 21 A	Si-P	=BSW 21: 50V	2a§		BCY 77...79, 2N2906...07, 2N3250..51, ++	data
BSW 22	Si-P	=BSW 21: B>180	2a§	Mot,Tho,Tix	BCY 77...79, 2N2906...07, 2N3905...06, ++	data
BSW 22 A	Si-P	=BSW 22: 50V	2a§		BCY 77, 2N2906...07, 2N3250..51, ++	data
BSW 23	Si-P	=2N2904	2a§	Mot,Sgs,Tix	→2N2904	data
BSW 24	Si-P	=2N2906	2a§	Mot,Sgs,Tix	→2N2906	data
BSW 25	Si-P	=2N2894A	2a§	Mot,Sgs,Tix	→2N2894A	data
BSW 26	Si-N	S/Tr, 50V, 1A, 0,5W, <40/85ns	2a§	Tix	BSS 26, BSS 40...41	data

Type	Device	Short Description	Fig.	Manu	Comparision Types	More at
BSW 27	Si-N	S/Tr, 60V, 1A, 0,8W, <40/85ns	2a§	Tix	BSS 14, BSS 27, BSV 95, 2N3735, 2SC1386	data
BSW 28	Si-N	=BSW 27: <50/85ns	2a§	Tix	BSS 14, BSS 27, BSV 95, 2N3735, 2SC1386	data
BSW 29	Si-N	=BSW 27: 40V	2a§	Tix	BSS 14, BSS 27...29, BSV 95, 2N3735, ++	data
BSW 30	MOS-P-FET-e	HF/S, 30V, 0,5A	5n+	Sgs	-	data
BSW 31	MOS-P-FET-e	HF/S, 30V, 0,5A	5n+	Sgs	-	data
BSW 32	Si-N	Nix, 100V, 0,03A, 0,25W	7c	Tix	BF 297...299, BSS 38, BSX 21	data
BSW 33	Si-N	NF/S, 40V, 0,1A, 0,125W, <200/-ns	12a	Phi	BC 167, BC 182, BC 237, BC 547, ++	data
BSW 34	Si-N	=BSW 33: 50V	12a	Phi	BC 167, BC 182, BC 237, BC 547, ++	data
BSW 35	Si-N	=BSW 33: 60V	12a	Phi	BC 174, BC 182, BC 190, BC 546, ++	data
BSW 36	Si-P	NF/S, 32V, 0,5A, 0,8W, <60/150ns	2a§	Tix	BSW 23, 2N2904(A)...05(A), 2N3072...73	data
BSW 37	Si-P	S, 12V, 0,2A, 0,36W, <80/90ns	2a§	Tix	BSV 21, BSW 25, BSX 29, 2N2894A, ++	data
BSW 38	Si-N	SS, 30V, 0,3A, 0,36W, <15/20ns	2a§	Tix	BSS 10, BSX 19...20, BSX 28, 2N3261, ++	data
BSW 39	Si-N	NF/S, 100V, 1A, 0,79W, 50/300ns	2a§	Aeg	BSS 42, BSW 66...68, BSX 46...47, 2SD854	data
BSW 40	Si-P	NF/S, 100V, 1A, 0,79W, 50/300ns	2a§	Aeg	BSS 17, 2N5322	data
BSW 41(A)	Si-N	S, 40V, 0,3A, 0,32W, <60/60ns	2a§	Phi	BSX 48...49, BSY 63, 2N708, 2N4123, ++	data
BSW 42(A,B)	Si-N	S, 25...60V, 0,2A, 0,3W, 70/250ns, B>75	8a	Tho	BSV 59, BSX 49, 2N3301...02, 2N3903...04++	data
BSW 43(A,B)	Si-N	=BSW 42: B>180	8a	Tho	BSV 59, BSX 49, 2N3301...02, 2N3903...04++	data
BSW 44(A,B)	Si-P	S, 25...60V, 0,2A, 0,3W, 215/105ns, B>75	8a	Tho	BCY 77, BSW 24, 2N3250A...3251A, ++	data
BSW 45(A,B)	Si-P	=BSW 44: B>180	8a	Tho	BCY 77, BSW 24, 2N3250A...3251A, ++	data
BSW 42...45(A)K	Si-N/P	=BSW 42...45(A,B):	7e		→BSW 42...45(A,B)	data
BSW 49	Si-N	NF/S, 40V, 1A, 0,6W, <-/50ns	2a§	Tho,Tix	BSW 27...29, BSX 32, BSX 59...61, 2N3252++	data
BSW 50	Si-N	NF/S, 65V, 0,8A, 0,8W, 250MHz	2a§	Phi	BSW 27...28, BSX 32, BSX 59...61, 2N3252++	data
BSW 51	Si-N	=2N2218	2a§	Mot,Phi,Tix	→2N2218	data
BSW 52	Si-N	=2N2219	2a§	Mot,Phi,Tix	→2N2219	data
BSW 53	Si-N	=2N2218A	2a§	Mot,Phi,Tix	→2N2218A	data
BSW 54	Si-N	=2N2219A	2a§	Mot,Phi,Tix	→2N2219A	data
BSW 58	Si-N	SS, 40V, 0,5A, 0,125W, <7/18ns	12a	Phi	BSS 10, BSX 26, BSX 39, 2N3261	data
BSW 59	Si-N	SS, 30V, 0,5A, 0,125W, <7/21ns	12a	Phi	BSS 10, BSX 26, BSX28, BSX 39, 2N3261	data
BSW 60	Si-N	NF/S, 60V, 0,8A, 0,5W, 250MHz	2a§	Phi	BC 337, BC 637, BC 639, 2N2221...22, ++	data
BSW 61	Si-N	=2N2221	2a§	Mot,Phi,Tix	→2N2221	data
BSW 62	Si-N	=2N2222	2a§	Mot,Phi,Tix	→2N2222	data
BSW 63	Si-N	=2N2221A	2a§	Mot,Phi,Tix	→2N2221A	data
BSW 64	Si-N	=2N2222A	2a§	Mot,Phi,Tix	→2N2222A	data
BSW 65	Si-N	S/Tr, 80V, 1A, 0,7W, 500/1000ns	2a§	Phi	BC 141, BSS 15, BSS 42...43, BSX 45...47++	data
BSW 66(A)	Si-N	=BSW 65: 100V	2a§	Phi,Mot	BC 141, BSS 15, BSS 42...43, BSX 46...47++	data
BSW 67(A)	Si-N	=BSW 65: 120V	2a§	Phi,Mot,Sgs	BSS 42...43, BSV 84, BSX 47	data
BSW 68(A)	Si-N	=BSW 65: 150V	2a§	Phi,Mot,Sgs	BSS 43, 2SC1860	data
BSW 69	Si-N	Nix, 150V, 0,05A, 0,125W, 130MHz	12d	Phi	BF 297...299, BFR 87...89, BFT 57...59, ++	data
BSW 70	Si-N	Nix, 100V, 0,05A, 0,25W		Phi	BF 297...299, BFR 86, BSS 38, BSX 21, ++	data
BSW 72	Si-P	Uni, 40V, 0,5A(ss), 0,4W, >150Mhz,B>40	2a§	Itt	BC 327, BC 636, BCY 79, 2N2906...07, ++	data
BSW 73	Si-P	=BSW 72: B>100	2a§	Itt	BC 327, BC 636, BCY 79, 2N2906...07, ++	data
BSW 74	Si-P	=BSW 72: 75V	2a§	Itt	BC 640, 2SA1013, 2SB647	data
BSW 75	Si-P	=BSW 72: 75V, B>100	2a§	Itt	BC 640, 2SA1013, 2SB647	data
BSW 78	Si-N	=2N2368	7a	Itt	→2N2368	data
BSW 79	Si-N	=2N2369:	7a	Itt	→2N2369	data
BSW 80	Si-N	=2N2369A:	7a	Itt	→2N2369A	data
BSW 81	Si-P	=2N3012:	7a	Itt	→2N3012	data
BSW 82	Si-N	Uni, 40V, 0,5A(ss), 0,5W, >200MHz,B>40	2a§	Itt,Tix	BC 337, BC 635, BCY 59, 2N2221...22, ++	data
BSW 83	Si-N	=BSW 82: B>100	2a§	Itt,Tix	BC 337, BC 635, BCY 59, 2N2221...22, ++	data
BSW 84	Si-N	=BSW 82: 75V	2a§	Itt,Tix	BC 639, 2N3700...01, 2SD667	data
BSW 85	Si-N	=BSW 82: 75V, B>100	2a§	Itt,Tix	BC 639, 2N3700...01, 2SD667	data
BSW 88(A,B)	Si-N	S, 35V, 0,1A, 0,3W, <150/800ns	7c	Aeg	BSW 41, BSY 63, 2N708, 2N4123, ++	data
BSW 89(A,B)	Si-N	=BSW 88:	7a	Aeg	BSW 41, BSY 63, 2N708, 2N4123, ++	data
BSW 92	Si-N	NF/S, 18V, 0,2A, 0,3W, >150MHz	8a	Tho	BC 168, BC 183, BC 238, BC 548, ++	data
BSW 93	Si-P	HF/S, 30V, 1A, 1W, 25/65ns	2a§	Sgs	2N3467...68, 2SA717	data
BSW 95(A)	MOS-P-FET-e	S, Chopper, 30V, 50mA	5n+	Sgs		data
BSWP 30	MOS-P-FET-e	S, 25V, 0,5A	5g	Ucp		data

BSX

Type	Device	Short Description	Fig.	Manu	Comparision Types	More at
BSX 12(A)	Si-N	SS/Tr, 25V, 1A, 0,6W, 10/15ns	2a§	Fer,Phi,Sgs	BSS 27...29, BSV 69, BSV 77, 2N3426	data
BSX 19	Si-N	SS, 40V, 0,5A(ss), 0,36W, <7/18ns	2a§	Mot,Phi,++	BSS 10, BSX 26, BSX 39, 2N2368...69(A)	data
BSX 20	Si-N	SS, 40V, 0,5A(ss), 0,36W, <7/21ns	2a§	Mot,Phi,++	BSS 10, BSX 26, BSX 39, 2N2368...69(A)	data
BSX 21	Si-N	Nix, 120V, 0,1A, 0,3W, 120MHz	2a§	Phi,Tix	BF 297...299, BFR 86...89, BSS 38	data
BSX 22	Si-N	NF/S, 40V, 1,5A, 0,8W, 100MHz	2a§	Itt,Tix	BSX 65...68, BSX 45...47, 2N3107...10, ++	data
BSX 23	Si-N	=BSX 22: 90V	2a§	Itt,Tix	BSW 66...68, BSX 46...47, 2N3107...08, ++	data
BSX 24	Si-N	NF/S, 32V, 0,1A, 0,3W, 25/400ns	2a§	Itt,Tho,Tix	BC 548, BSW 41, BSY 63, 2N708, ++	data
BSX 25	Si-N	NF/S, 40V, 0,3A, 0,36W, >50MHz	2a§	Aeg,Tix	BC 337, BC 635, BCY 59, 2N2221...22, ++	data
BSX 26	Si-N	SS, 40V, 0,5A, 0,36W, 9/15ns	2a§	Sgs,Tix	BSS 10, BSX 39, 2N3261	data
BSX 27	Si-N	SS, 15V, 0,15A, 0,3W, <12/12ns	2a§	Sgs,Tix	BSS 10...12, BSX 92...93, 2N3011, ++	data
BSX 28	Si-N	SS, 30V, 0,5A, 0,36W, 9/13ns	2a§	Sgs,Tix	BSS 10, BSX 26, BSX 39, 2N3261	data
BSX 29	Si-P	HF/S, 12V, 0,2A, 0,36W, 25/35ns	2a§	Mot,Sgs,Tix	BSV 21, BSW 25, BSW 37, BSX 36, 2N2894A	data

Type	Device	Short Description	Fig.	Manu	Comparision Types	More at
BSX 30	Si-N	S/Tr, 60V, 0,8A, 0,8W, 22/22ns	2a§	Sgs,Tix	BSS 13, BSS 27, BSV 77, 2N5189	data
BSX 31	Si-N	Chopper, 15V	5z	Sgs	-	data
BSX 32	Si-N	S/Tr, 65V, 1A, 0,8W, 35/40ns	2a§	Mot,Sgs,++	BSS 13, BSS 27, BSV 77, 2N5189	data
BSX 33	Si-N	S/Tr, 85V, 1A, 0,5W, 120/350ns	2a§	Sgs,Tix	BC 639, 2N3700...3701, 2SD667	data
BSX 34	MOS-P-FET-e	Dual, 30V, 0,2A	81bz+	Sgs	-	data
BSX 35	Si-P	SS, 6V, 0,2A, 0,3W, <25/25ns	2a§	Sgs,Tix	BSX 29, BSX 36, 2N2894A	data
BSX 36	Si-P	S, 40V, 0,5A, 0,36W, 17/18ns	2a§	Mot,Sgs,Tix	BSX 29, BSX 35, 2N2894A	data
BSX 38(A,B)	Si-N	NF/S, 35V, 0,1A, 0,3W, <150/800ns	2a§	Aeg	BC 548, BCY 59, BSW 41, BSY 63, 2N708,++	data
BSX 39	Si-N	S/Tr, 45V, 0,5A, 0,36W, 9/15ns	2a§	Sgs,Tix,++	BSS 10, BSX 26, 2N3261	data
BSX 39	Si-N	SMD, S/Tr, 45V, 0,2A, <12/18ns	35d	Mot	-	data
BSX 40	Si-P	NF/S, 30V, 0,5A, 0,6W, >100MHz	2a§	Itt,Tix	BC 160...161, BC 303...304, BSV 15...17, ++	data
BSX 41	Si-P	=BSX 40: >150MHz	2a§	Itt,Tix	BC 160...161, BC 303...304, BSV 15...17, ++	data
BSX 44	Si-N	SS, 15V, 0,2A(ss), <20/15ns	2a§	Phi,Tix	BSV 89...92, BSX 27, BSX 92...93, 2N2475	data
BSX 45	Si-N	NF/S, 80V, 1A, >50MHz, <200/850ns	2a§	EUR	BSS 42...43, BSW 65...68, 2N3107...10, ++	data
BSX 46	Si-N	=BSX 45: 100V	2a§	EUR	BSS 42...43, BSW 66...68, 2N3107...08, ++	data
BSX 47	Si-N	=BSX 45: 120V	2a§	EUR	BSS 42...43, BSW 67...68, 2N3019...20, ++	data
BSX 48	Si-N	S/Tr, 50V, 0,6A, 35/60ns	2a§	Mot,Phi,Sie	BSV 59, 2N2221...2222, 2N3301...3302	data
BSX 49	Si-N	=BSX 48: 60V, 30/50ns	2a§	Mot,Phi,Sie	BSV 59, 2N2221...2222, 2N3301...3302	data
BSX 50	Si-N	NF/S, 120V, 1A, >50MHz, <200/850ns	2a§	Sie	BSS 42...43, BSW 67...68, 2N3019...20, ++	data
BSX 51	Si-N	NF/S, 25V, 0,2A, 0,3W, 70/145ns, B>75	2a§	Mot,Tho,Tix	BSW 41, BSY 62...63, 2N706A, 2N708, ++	data
BSX 51 A	Si-N	=BSX 51: 50V	2a§		BSV 59, 2N3301...3302, 2N3903...3904, ++	data
BSX 51 B	Si-N	=BSX 51: 60V	2a§		BSV 59, 2N3301...3302, 2N3903...3904, ++	data
BSX 52	Si-N	=BSX 51: B>180	2a§	Mot,Tho,Tix	BSW 41, BSY 62...63, 2N706A, 2N708, ++	data
BSX 52 A	Si-N	=BSX 51: 50V, B>180	2a§		BSV 59, 2N3301...3302, 2N3903...3904, ++	data
BSX 52 B	Si-N	=BSX 51: 60V, B>180	2a		BSV 59, 2N3301...3302, 2N3903...3904, ++	data
BSX 53(A,B)	Si-N	NF/S, 35V, 0,1A, 0,13W, <150/800ns	2a	Aeg	BC 548, BCY 59, BSW 41, BSY 62...63, ++	data
BSX 54(A,B)	Si-N	=BSX 53: 50V	2a	Aeg	BC 548, BCY 59, BSW 41, BSY 62...63, ++	data
BSX 59	Si-N	S/Tr, 70/45V, 1A, 0,8W, 17/45ns	2a§	Mot,Phi,Tix	BSS 27, BSV 95, 2N3735, 2SC1386	data
BSX 60	Si-N	=BSX 59: 70/30V, 17/58ns	2a§	Mot,Phi,Tix	BSS 27, BSV 95, 2N3735, 2SC1386	data
BSX 61	Si-N	=BSX 59: 18/70ns	2a§	Mot,Phi,Tix	BSS 27, BSV 95, 2N3735, 2SC1386	data
BSX 62	Si-N	NF/S, 60V, 3A, 0,8W, <300/1500ns	2a§	Phi,Sie	BSS 45, 2N4238...39, 2N5336...39, 2SC2214	data
BSX 63	Si-N	=BSX 62: 80V	2a§	Phi,Sie	BSS 45, 2N4238...39, 2N5336...39, 2SC2214	data
BSX 64	Si-N	=BSX 62: 100V	2a§	Phi,Sie	2N4239, 2N5338...39, 2SC2214	data
BSX 66	Si-N	NF/S, 30V, 0,1A, <200/400ns, B>40	2a§	Phi	BC 548, BCY 58...59, BSY 62...63, 2N708,++	data
BSX 67	Si-N	=BSX 66: B>60	2a§	Phi	BC 548, BCY 58...59, BSY 62...63, 2N708,++	data
BSX 68	Si-N	NF/S, 30/15V, 0,1A, <200/400ns, B>30	12a	Aeg,Phi	BC 548, BCY 58...59, BSY 62...63, 2N708,++	data
BSX 69	Si-N	=BSX 68: 30/20V, B>60	12a	Aeg,Phi	BC 548, BCY 58...59, BSY 62...63, 2N708,++	data
BSX 70	Si-N	NF/S, 75V, 0,5A, 0,5W, <70/250ns, B>40	2a§	Phi,Tix	BC 639, BSW 63...64, 2N2221A...22A, ++	data
BSX 71	Si-N	=BSX 70: B>100	2a§	Phi,Tix	BC 639, BSW 63...64, 2N2221A...22A, ++	data
BSX 72	Si-N	NF/S, 40V, 1A, 0,7W, 25/150ns	2a§	Aeg,Tix	BC 140...141, BSW 27...29, 2N3252...53, ++	data
BSX 73	Si-N	HF/S, 60V, 0,8A, 0,8W, 25/-ns, B>40	2a§	Aeg	BSS 13, BSS 27, BSV 77, 2N5189, 2SC1386	data
BSX 74	Si-N	=BSX 73: B>100	2a§	Aeg	BSS 13, BSS 27, BSV. 77, 2N5189, 2SC1386	data
BSX 75	Si-N	NF/S, 40V, 0,8A, 0,43W, 25/150ns	2a§	Aeg,Tix	BC 635, BSW 61...64, 2N2221(A)...22(A), ++	data
BSX 76	Si-N	NF/S, 20V, 0,1A, 0,3W, <30/-ns, B>35	2a§	Phi,Tix	BC 548, BSW 41, BSY 62...63, 2N706A, ++	data
BSX 77	Si-N	=BSX 76: 40V, B>40	2a§	Phi,Tix	BC 547, BSW 41, BSY 63, 2N708, 2N4123,++	data
BSX 78	Si-N	=BSX 76: 40V, B>80	2a§	Phi,Tix	BC 547, BSW 41, BSY 63, 2N708, 2N4123,++	data
BSX 79(A,B)	Si-N	NF/S, 50V, 0,1A, 0,375W, <150/800ns	2a§	Aeg,Tix	BC 547, BCY 59, 2N3903...3904, ++	data
BSX 80	Si-N	S, 35V, 0,23A, 0,15W, <40/80ns	9e	Aeg	BSW 41, BSY 63, 2N708, 2N4123, ++	data
BSX 81(A,B)	Si-N	NF/S, 30V, 0,1A, 0,3W, <150/800ns	9e	Aeg	BC 548, BSW 41, BSY 63, 2N708, 2N4123,++	data
BSX 82	MOS-N-FET-d	Chopper, 30V, 50mA	5m+	Phi	BFX 63	data
BSX 83	MOS-P-FET-e	Chopper, 30V, <250/-ns	5n+	Sgs	-	data
BSX 84	MOS-P-FET-e	Chopper, 30V, <250/-ns	5n+	Sgs	-	data
BSX 85	MOS-P-FET-e	Dual, Chopper, 30V, 0,2A, <50/-ns	81bz+	Sgs	MFE 3020...3021	data
BSX 86	MOS-P-FET-e	Dual, Chopper, 30V, 0,2A, <50/-ns.	81bz+	Sgs	MFE 3020...3021	data
BSX 87	Si-N	S, 40V, 0,5A, 0,36W, 25/25ns	2a§	Sgs,Tix	BSS 10, BSX 26, BSX 39, 2N3261	data
BSX 87 A	Si-N	=BSX 87: 9/15ns	2a§		BSS 10, BSX 26, BSX 39, 2N3261	data
BSX 88	Si-N	S, 40V, 0,5A, 0,36W, <40/75ns	2a§	Sgs,Tix	BSS 10, BSX 26, BSX 39, 2N3261	data
BSX 88 A	Si-N	=BSX 88: <30/70ns	2a§		BSS 10, BSX 26, BSX 39, 2N3261	data
BSX 89	Si-N	S, 25V, 0,5A, 0,3W, <40/75ns	2a§	Mot,Sgs,Tix	BSS 10, BSX 26, BSX 39, 2N914, 2N3261,++	data
BSX 90	Si-N	SS, 20V, 0,2A, 0,3W, <12/45ns, B>20	2a§	Mot,Sgs,Tix	BSS 10...12, BSX 19...20, 2N2368...69, ++	data
BSX 91	Si-N	=BSX 90: B>40	2a§	Mot,Sgs,Tix	BSS 10...12, BSX 19...20, 2N2368...69, ++	data
BSX 92	Si-N	SS, 40V, 0,15A, 0,36W, <12/15ns, B>20	2a§	Mot,Sgs,Tix	BSS 10...12, BSX 19...20, 2N2368...69, ++	data
BSX 93	Si-N	=BSX 92: B>40	2a§	Mot,Sgs,Tix	BSS 10...12, BSX 19...20, 2N2368...69, ++	data
BSX 94	Si-P	SS, 6V, 0,3W, <15/-ns	2a§	Sgs	BSX 36	data
BSX 94 A	Si-N	S, 60V, 0,4W, 400MHz	2a§	Sgs	BSV 59, BSX 49, 2N3301...3302, ++	data
BSX 95	Si-N	NF/S, 75V, 0,5A, 0,7W, 50/210ns, B>40	2a§	Phi,Tix	BSW 53...54, BSX 59...61, 2N2218A...19A, ++	data
BSX 96	Si-N	=BSX 95: B>100	2a§	Phi,Tix	BSW 53...54, BSX 59...61, 2N2218A...19A, ++	data
BSX 97	Si-N	HF/S, 40V, 0,5A, 0,4W, <30/-ns	2a§	Tho	BSV 59, BSX 48...49, BSX 87...88, 2N914,++	data
BSX 98	Si-N	HF/S, 25V, 0,2A, 0,25W, <40/-ns	11a	Sie	BSW 41, BSY 62...63, 2N706A, 2N4124, ++	data
BSX 99	Si-N	HF/S, 25V, 0,2A, 0,5W, <50/-ns	11a	Sie	BSW 41, BSY 62...63, 2N706A, 2N4124, ++	data

Type	Device	Short Description	Fig.	Manu	Comparision Types	More at
BSY						
BSY 10	Si-N	NF/S, 60V, 0,05A, 0,3W, 17,7/110ns	2a§	Phi,Tix	BSW 51...54, 2N2218...19, 2N3299...3300, ++	data
BSY 11	Si-N	=BSY 10: 45V	2a§	Phi,Tix	BSW 51...54, 2N2218...19, 2N3299...3300, ++	data
BSY 17	Si-N	SS, 20V, 0,2A, 0,35W, 7/25ns, B>20	2a§	Mot,Sie,Tix	BSS 10...12, BSX 19...20, 2N2368...69, ++	data
BSY 18	Si-N	=BSY 17: B>40	2a§	Mot,Sie,Tix	BSS 10...12, BSX 19...20, 2N2368...69, ++	data
BSY 19	Si-N	S, 40V, 0,2A, 0,32W, <40/70ns	2a§	Mot,Phi,Tix	BSW 41, BSY 63, 2N708, 2N4123, ++	data
BSY 20	Si-N	S, 25V, 0,05A, 0,3W, <40/75ns	2a§	Itt,Mot	BSW 41, BSY 62...63, 2N706B, 2N4123, ++	data
BSY 21	Si-N	=2N914	2a§	Aeg,Itt,Mot	→2N914	data
BSY 22	Si-N	=2N916	2a§	Itt,Mot	→2N916	data
BSY 23	Si-N	=2N834	2a§	Itt,Mot	→2N834	data
BSY 24	Si-N	NF/S, 40V, 0,5A, 0,6W, 19/375ns, B>15	2a§	Itt,Tix	BSW 51...54, 2N3299...3300, 2N2218...2219	data
BSY 25	Si-N	=BSY 24: 16/325ns, B>40	2a§	Itt,Tix	BSW 51...54, 2N3299...3300, 2N2218...2219	data
BSY 26	Si-N	SS, 20V, 0,1A, 0,3W, 20/31ns, B>15	2a§	Itt,Tix	BSV 89...92, BSX 92...93, 2N3011, ++	data
BSY 27	Si-N	=BSY 26: 19/35ns, B>25	2a§	Itt,Tix	BSV 89...92, BSX 92...93, 2N3011, ++	data
BSY 28	Si-N	SS, 15V, 0,1A, 0,3W, 14/22ns, B>20	2a§	Itt,Tix	BSV 89...92, BSX 44, BSX 92...93, 2N3011++	data
BSY 29	Si-N	=BSX 28: 12/20ns, B>40	2a§	Itt,Tix	BSV 89...92, BSX 44, BSX 92...93, 2N3011++	data
BSY 30	Si-N	S, 15V, 0,2A, 0,3W, <120/-ns, B>20	2a§	Sie	BSW 41, BSY 62...63, 2N706A, 2N4123, ++	data
BSY 31	Si-N	=BSY 30: <180/-ns, B>30	2a§	Sie	BSW 41, BSY 62...63, 2N706A, 2N4123, ++	data
BSY 32	Si-N	S, 20V, 0,1A, 0,1W, <27/130ns, B=32		Itt	BSW 41, BSY 62...63, 2N706A, 2N4123, ++	data
BSY 33	Si-N	=BSY 32: B=55		Itt	BSW 41, BSY 62...63, 2N706A, 2N4123, ++	data
BSY 34	Si-N	S/Tr, 60V, 0,6A, 0,8W, 30/50ns	2a§	Phi,Sie	BSS 27, BSV 77, BSV 95, 2SC1385...86, ++	data
BSY 35	Si-P	SMD, S, 75V, 0,05A	Chip	Itt	-	data
BSY 36	Si-N	S, 15V, 0,1A, 0,1W, <20/30ns		Itt	BSV 89...92, BSX 44, BSX 92...93, 2N3011++	data
BSY 38(A)	Si-N	=2N743	2a§	Mot,Phi,Tix	→2N743	data
BSY 39(A)	Si-N	=2N744	2a§	Mot,Phi,Tix	→2N744	data
BSY 40	Si-P	S, 25V, 0,1A, 0,3W, <25/100ns, B>20	2a§	Mot,Phi,Tix	2N3250...51, 2N3905...06, 2N4125...26	data
BSY 41	Si-P	=BSY 40: B>50	2a§	Mot,Phi,Tix	2N3250...51, 2N3905...06, 2N4125...26	data
BSY 42	Si-N	Dual, 20V, 0,2A, >200MHz	81	Itt	-	data
BSY 43	Si-N	Dual, 15V, 0,2A, >300MHz	81	Itt	-	data
BSY 44	Si-N	=2N1613	2a§	Aeg,Mot,Tix	→2N1613	data
BSY 45	Si-N	=2N1893	2a§	Aeg,Mot,Tix	→2N1893	data
BSY 46	Si-N	=2N2193	2a§	Aeg,Mot,Tix	→2N2193	data
BSY 47	Si-N	S, 20V, 0,1A, 0,1W, <27/130ns		Itt	BSW 41, BSY 62...63, 2N706A, 2N4124, ++	data
BSY 48	Si-N	S, 50V, 0,6A, 0,4W, <65/-ns	2a§	Sie	BSV 59, BSX 48, 2N2221...22, 2N3301...02++	data
BSY 49	Si-N	S, 60V, 0,6A, 0,4W, <50/-ns	2a§	Sie	BSV 59, BSX 49, 2N2221...22, 2N3301...02++	data
BSY 50	Si-N	S, 15V, 0,1A, 0,1W, <20/30ns		Itt	BSV 89...92, BSX 44, BSX 92...93, 2N3011++	data
BSY 51	Si-N	NF/S, 60V, 0,5A, 0,8W, 100MHz, B>40	2a§	EUR	BC 140...141, BC 300...302, 2N697, ++	data
BSY 52	Si-N	=BSY 51: B>100	2a§	EUR	BC 140...141, BC 300...302, 2N1420, ++	data
BSY 53	Si-N	NF/S, 75V, 0,75A, 0,8W, 100MHz, B>40	2a§	EUR	BC 140...141, BC 300...301, 2N1613, ++	data
BSY 54	Si-N	=BSY 53: B>100	2a§	EUR	BC 140...141, BC 300...301, 2N1711, ++	data
BSY 55	Si-N	NF/S, 120V, 0,5A, 0,8W, 100MHz, B>40	2a§	EUR	BC 300, BSW 67...68, BSX 47, 2N1893, ++	data
BSY 56	Si-N	=BSY 55: B>100	2a§	EUR	BC 300, BSW 67...68, BSX 47, 2N1893, ++	data
BSY 58	Si-N	S/Tr, 50V, 0,6A, 0,8W, 35/65ns	2a§	Phi,Sie	BSS 27...29, BSV 77, 2N3724A, 2SC1386, ++	data
BSY 59	Si-P	S, 30V, 0,8A, 0,28W, <500/850ns	11a	Sie	BC 327...328, BC 636, 2N2906...2907, ++	data
BSY 61(Y)	Si-N	S, 25V, 0,05...0,2A, 0,2W, <40/75ns	7c	Sie	BSW 41, BSY 62...63, 2N706A, 2N4124, ++	data
BSY 62(A,B)	Si-N	S, 25V, 0,2A, 0,35W, <40/75ns	2a§	Sie,Tix	BSW 41, 2N706A, 2N708, 2N4123...24, ++	data
BSY 63	Si-N	S, 40V, 0,2A, 0,35W, <40/75ns	2a§	Phi,Sie,Tix	BSW 41, 2N708, 2N3903...04, 2N4123, ++	data
BSY 65	Si-N	NF/S, 15V, 0,1A, 0,3W, 200MHz	2a§	Itt	BC 548, BCY 58...59, BSW 41, 2N706A, ++	data
BSY 66	Si-N	10V, 0,05W		Itt	-	data
BSY 67	Si-N	10V, 0,05W		Itt	-	data
BSY 68	Si-N	Nix, 120V, 0,05A, 0,6W, >20MHz	2a§	Phi,Tix	BF 257...259, BF 657...659, 2N5058...5059	data
BSY 70	Si-N	=2N706	2a§	Aeg,Mot,Tix	→2N706	data
BSY 71	Si-N	=2N1711	2a§	Aeg,Mot,Tix	→2N1711	data
BSY 72	Si-N	NF/S, 25V, 0,03A, 0,3W, 170MHz	2a§	Itt,Tix	BC 169, BC 184, BC 239, BC 549, ++	data
BSY 73	Si-N	NF/S, 25V, 0,1A, 0,3W, 145MHz	2a§	Itt,Tix	BC 168, BC 183, BC 238, BC 548, ++	data
BSY 74	Si-N	NF/S, 25V, 0,1A, 0,3W, 170MHz	2a§	Itt,Tix	BC 168, BC 183, BC 238, BC 548, ++	data
BSY 75	Si-N	NF/S, 40V, 0,25A, 0,3W, 145MHz	2a§	Itt,Tix	BC 167, BC 182, BC 237, BC 547, ++	data
BSY 76	Si-N	=BSY 75: 170MHz	2a§	Itt,Tix	BC 167, BC 182, BC 237, BC 547, ++	data
BSY 77	Si-N	=BSY 75: 80V	2a§	Itt,Tix	BC 546, 2SC2240, 2SC2459, 2SC3445	data
BSY 78	Si-N	=BSY 75: 80V, 170MHz	2a§	Itt,Tix	BC 546, 2SC2240, 2SC2459, 2SC3445	data
BSY 79	Si-N	Nix, 120V, 0,03A, 0,3W, 100MHz	2a§	Itt,Tix	BF 297...299, BF 422, BSS 38, BSX 21, ++	data
BSY 80	Si-N	NF/S, 25V, 0,1A, 0,3W, 210MHz	2a§	Itt,Tix	BC 168, BC 183, BC 238, BC 548, ++	data
BSY 81	Si-N	NF/S, 40V, 1A, 0,9W, 100MHz, B>40	2a§	Itt,Tix	BC 140...141, BSW 65...68, BSX 45...47, ++	data
BSY 82	Si-N	=BSY 81: 120MHz, B>100	2a§	Itt,Tix	BC 140...141, BSW 65...68, BSX 45...47, ++	data
BSY 83	Si-N	=BSY 81: 80V	2a§	Itt,Tix	BC 140...141, BSW 65...68, BSX 45...47, ++	data
BSY 84	Si-N	=BSY 81: 80V, 120MHz, B>100	2a§	Itt,Tix	BC 140...141, BSW 65...68, BSX 45...47, ++	data
BSY 85	Si-N	=BSY 81: 120V, 110MHz	2a§	Itt,Tix	BSS 42...43, BSW 67...68, BSX 47, 2N2102++	data
BSY 86	Si-N	=BSY 81: 120V, 130MHz, B>100	2a§	Itt,Tix	BSS 42...43, BSW 67...68, BSX 47, 2N2102++	data
BSY 87	Si-N	NF/S, 100V, 0,5A, 0,8W, 100MHz, B>40	2a§	Itt,Tix	BC 141, BC 300, BSW 66...68, BSX 46...47++	data
BSY 88	Si-N	=BSY 87: 145MHz, B>100	2a§	Itt,Tix	BC 141, BC 300, BSW 66...68, BSX 46...47++	data
BSY 89	Si-N	NF/S, Chopper, 25V, 0,1A, 0,3W	2a§	Itt,Tix		data

Type	Device	Short Description	Fig.	Manu	Comparision Types	More at
BSY 90	Si-N	NF/S, 60V, 0,5A, 0,8W, 170MHz	2a§	Itt,Tix	BC 140...141, BC 300...302, 2N2218...19, ++	data
BSY 91	Si-N	NF/S, 40V, 0,3A, 0,8W, >50MHz	2a§	Aeg,Itt,Tix	BC 140...141, BC 300...302, 2N2218...19, ++	data
BSY 92	Si-N	=BSY 91: 60V	2a§	Aeg,Tix	BC 140...141, BC 300...302, 2N2218...19, ++	data
BSY 93	Si-N	=BSY 91: 60V, 0,36W	2a§	Aeg,Tix	BC 337, BC 637, BSV 59, 2N2221...22, ++	data
BSY 95(A)	Si-N	HF/S, 20V, 0,1A, 0,3W, >200MHz	2a§	Phi,Tix,++	BSW 42, BSY 62...63, 2N706A, 2N4124, ++	data
BSYP 62	Si-N	NF/S, 25V, 0,2A, 0,36W, 200MHz	2a	Ucp	-	data
BSYP 63	Si-N	NF/S, 40V, 0,2A, 0,36W, 300MHz	2a	Ucp	-	data

BT

Type	Device	Short Description	Fig.	Manu	Comparision Types	More at
BT	Si-P	=2SB1048 (Typ-Code/Stempel/marking)	39	Hit	→2SB1048	data
BT	Si-N	=2SC4069 (Typ-Code/Stempel/marking)	35	Say	→2SC4069	data
BT	Si-P	=BCW 61RE (Typ-Code/Stempel/marking)	35	Sie	→BCW 61RE	data
BT	Z-Di	=BZX 399-C24 (Typ-Code/Stempel/marking)	71(1,7mm)	Phi	→BZX 399-...	data
BT	Z-Di	=P4 SMA-33A (Typ-Code/Stempel/marking)	71(5x2,5)	Fag	→P4 SMA-33A	data
BT	Z-Di	=SMBJ 18CA (Typ-Code/Stempel/marking)	71(5x3,5)	Mop	→SMBJ ...	data
BT 1	Si-P	=BSP 15 (Typ-Code/Stempel/marking)	48	Phi	→BSP 15	data
BT 1	Si-P	=BST 15 (Typ-Code/Stempel/marking)	39	Phi, Val	→BST 15	data
BT 2	Si-P	=BSP 16 (Typ-Code/Stempel/marking)	48	Phi, Ons	→BSP 16	data
BT 2	Si-P	=BST 16 (Typ-Code/Stempel/marking)	39	Phi, Val	→BST 16	data
Bt 4	Si-P+R	=PUMB 4 (Typ-Code/Stempel/marking)	46	Phi	→PUMB 4	data
BT 100 A/...R	50Hz-Thy	300...500V, 4,5A, Igt/Ih<10/<15mA	13n	Phi	(TAG 631..., TAG 632..., TAG 621...,++)[4]	data
Bt 101/300...500 R	50Hz-Thy	300...500V, 15A, Igt<10	21b§	Phi	-	data
Bt 101 BC,KC	D/A-IC	8 Bit, Video, 50, 30Msps, TTL Level	40-DIP		-	
BT 102/...R	50Hz-Thy	=BT 101/...R: Igt<50mA	21b§	Phi	BTW 42/...	data
Bt 102 BC	D/A-IC	8 Bit, Video, 75Msps, TTL Level	24-DIP		-	
BT 103	GTO-Thy	1100V, 2,6A(Tc=85°C), Igt<100mA, <1μs	22a§	Itt	-	data
BT 104	GTO-Thy	1200V, 3A(Tc=85°C), Igt<100mA, <1μs	22a§	Itt	-	data
Bt 104 KC	D/A-IC	12 Bit, Video, 30Msps, TTL Level	24-DIP		-	
BT 105	GTO-Thy	600V, 2,7A(Tc=85°C), Igt<100mA, <1μs	22a§	Itt	-	data
Bt 105 KC	D/A-IC	12 Bit, Video, 30Msps, TTL Level	24-DIP		-	
BT 106	F-Thy	700V, 1A(Tc=90°C), Igt/Ih<20/<25mA	21b§	Phi	-	data
BT 106 A...D	50Hz-Thy	100...400V, 4A, Igt/Ih<0,2/<3mA A=100V, B=200V, C=300V, D=400V	14h§	Phi	C 106 A, 2N6236...6241, (TIC 106...)[4]	data
Bt 106 BC,KC	D/A-IC	8 Bit, Video, 50, 30Msps, TTL Level	20-DIP		-	
BT 107	Thy	500V, 6,5A(Tc=60°C), Igt<10mA	21b§	Phi	-	data
BT 108	Thy	500V, 6,5A(Tc=60°C), Igt<50mA	21b§	Phi	BTW 42/600	data
Bt 108 BF300,C200	D/A-IC	8 Bit, Video, 300, 200Msps, Ecl Level	24-DIP		-	
BT 109	Thy	500V, 6,5A(Tc=75°C), Igt/Ih<10/<50mA	21b§	Phi	-	data
Bt 109 KC	D/A-IC	8 Bit, 3 Ch, Video, 250Msps, Ecl Level	40-DIP		-	
BT 110	Thy	400V, 6A(Tc=75°C), Igt/Ih<35/<20mA		Phi	-	data
BT 112/750R	F-Thy	TV-HA, 750V, 5A, Igt<50mA, <2,4μs	22a§	Tho	S 3703 SF, TU 3F 800H, (TD 3F 800 H)[1]	data
BT 113/700R	F-Thy	TV-HA, 700V, 5A, Igt<50mA, <4,2μs	22a§	Tho	S 3702 S, S 6080 C, (TD 3F 800 R)[1]	data
BT 114	F-Thy	TV-HA, 750V, 3,2A, Igt<50mA, <2,4μs	22a§	Ssc	S 3703 SF, TU 3F 800H, (TD 3F 800 H)[1]	data
BT 114	F-Thy	TV-HA, 700V, 3,2A, Igt<50mA, <4,2μs	22a§	Ssc	S 3702 S, S 6080 C, (TD 3F 800 R)[1]	data
BT 116	Thy	750V, 1A(Tc=100°C), Igt<20mA		Phi	-	data
BT 117	F-Thy	TV-HA, 750V, 3,2A(Tc=60°C), <2,4μs	17§	Ssc	BStC12 50, (17088)[1]	data
BT 118	F-Thy	TV-HA, 750V, 3,2A(Tc=60°C), <4,2μs	17§	Ssc	BT 154, (BStCC02 53R, 17089)[1]	data
BT 119	F-Thy	TV-HA, 750V, 3A, <2,4μs	22a§	Itt	S 3703, TU 3F 800 H, (TD 3F 800 H)[1]	data
BT 120	F-Thy	TV-HA, 700V, Igt/Ih<40/<100mA, <4,5μs	22a§	Itt	(TD 3F 800 R, TD 4F 800 R,)[1]	data
BT 121	F-Thy	TV-HA, 500V, 2,8A, <2,4μs	22a§	Itt	S 3705, S 3703, TU 3F500H, (TD 3F500H)[1]	data
BT 122	F-Thy	500V, 2,8A, Igt/Ih<40/<100mA, <2,4μs	22a§	Itt	BSt C09 30T92, BStE04...T, (TD 3F 500 H)[1]	data
BT 123	Triac	500V, 15A(Tc=70°C), Igt/Ih<35/<50mA	16§	Phi	(TAG 257-500)[4]	data
BT 124	F-Thy	500V, 4,5A(Tc=85°C), Igt/Ih<15/<25mA	16h§	Phi	-	data
BT 125/700 R	F-Thy	TV-HA, 700V, 5A, Igt<40mA, <4,5μs	22a§	Phi	S 3702, S 6080 C, (TD 3F 700 R)[1]	data
BT 126/750 R	F-Thy	TV-HA, 700V, 5A, Igt<40mA, <2,4μs	22a§	Phi	S 3703, TU 3F 800 H, (TD 3F 800 H)[1]	data
BT 127/...(R)	F-Thy	350...750V, 5A, Igt/Ih<40/<50mA, <10μs	22a§	Phi	TAG 670S-..., TAG 675S-...	data
BT 128/700 R	F-Thy+Di	TV-HA, 700V, 5A, Igt<40mA, <4,5μs	22a§	Phi	S 6087 A, BSt CC01 46R, TD 3F 800 R	data
BT 129/...R	F-Thy+Di	TV-HA, 600...750V, 5A, <2,4μs	22a§	Phi	BStCC0140...0153H, S 6087B, TD 3F..H	data
BT 131-...	Triac	500...800V, 1A≈, Igt/Ih<3...7/<5	7e	Phi	MAC 97-...	data
BT 132-...	Triac	500...600V, 1A≈, Igt/Ih<2...10/<10mA	7e	Phi	MAC 97-...	data
BT 132	Thy	-700V, +400V, 1A(Tc=85°C), Igt<50mA	21b§	Phi	(BTW 42/...)	data
BT 133	Thy	-100V, +400V, 1A(Tc=85°C), Igt<50mA	21b§	Phi	(BTW 42/...)	data
BT 134/...	Triac	500...800V, 4A(Tc=102°C), Igt/Ih<70/<15	≈14h§	Phi	-	data
BT 134/...B	Triac	=BT 134/...: Igt/Ih<10/<10mA	≈14h§		-	data
BT 134/...E	Triac	=BT 134/...: Igt/Ih<25/<15mA	≈14h§		-	data
BT 134/...F	Triac	=BT 134/...: Igt/Ih<70/<15mA	≈14h§		-	data
BT 134/...G	Triac	=BT 134/...: Igt/Ih<100/<30mA	≈14h§		-	data
BT 134W/...	Triac	500...800V, 1A(Tc=77°C), Igt/Ih<35/<15	48h§		-	data
BT 136/...	Triac	500...800V, 4A, Igt/Ih<70/<30mA	17h§	Phi	TXC10H..., TAG 230-..., TAG 231-...,++	data
BT 136/...D	Triac	500...800V, 4A, Igt/Ih<10/<15mA	17h§		TXC18G..., TAG 232-..., TAG 221-...,++	data
BT 136/...E	Triac	500...800V, 4A, Igt/Ih<25/<20mA	17h§		TAG 231-..., TAG 220-..., TAG 225-..., ++	data
BT 136/...F	Triac	500...800V, 4A, Igt/Ih<70/<30mA	17h§		TXC10H..., TAG 230-..., TAG 231-..., ++	data
BT 136/...G	Triac	500...800V, 4A, Igt/Ih<100/<45mA	17h§		TXC10H..., TAG 230-..., TAG 224-..., ++	data

Type	Device	Short Description	Fig.	Manu	Comparision Types	More at
BT 137/...	Triac	500...800V, 8A, Igt/Ih<70/<45mA	17h§	Phi	TAG 224-..., MAC 222A-..., TXD10H.., ++	data
BT 137/...D	Triac	500...800V, 8A, Igt/Ih<10/<20mA	17h§		TAG 226-..., MAC 228A-...	data
BT 137/...E	Triac	500...800V, 8A, Igt/Ih<25/<35mA	17h§		TAG 225-..., TAG 252-..., TAG 255-..., ++	data
BT 137/...F	Triac	500...800V, 8A, Igt/Ih<70/<45mA	17h§		TAG 224-..., MAC 222A-..., TXD10H.., ++	data
BT 137/...G	Triac	500...800V, 8A, Igt/Ih<100/<60mA	17h§		TAG 224-..., MAC 222A-..., TXD10H.., ++	data
BT 138/...	Triac	500...800V, 12A, Igt/Ih<70/<60mA	17h§	Phi	TAG 255-..., TAG 256-..., TXD10H..P, ++	data
BT 138/...E	Triac	500...800V, 12A, Igt/Ih<25/<40mA	17h§		TAG 257-...	data
BT 138/...F	Triac	500...800V, 12A, Igt/Ih<60/<60mA	17h§		TAG 255-..., TAG 256-..., TXD10H..P, ++	data
BT 138/...G	Triac	500...800V, 12A, Igt/Ih<100/<90mA	17h§		TAG 255-..., TAG 256-..., TXD10H..P, ++	data
BT 139/...	Triac	500...800V, 16A, Igt/Ih<70/<60mA	17h§	Phi	MAC 15A-..., TAG 280-..., TAG 281-..., ++	data
BT 139/...E	Triac	500...800V, 16A, Igt/Ih<25/<40mA	17h§			data
BT 139/...F	Triac	500...800V, 16A, Igt/Ih<70/<60mA	17h§		MAC 15A-..., TAG 280-..., TAG 281-..., ++	data
BT 139/...G	Triac	500...800V, 16A, Igt/Ih<100/<90mA	17h§		MAC 15A-..., TAG 280-..., TAG 281-..., ++	data
BT 143/400 R	F-Thy+Di	400V, 3,2A(Tc=85°C), Igt<40mA	22a§	Phi	BStCC 0126T91...0153T91, BT 127/750 R	data
BT 145/...R	Thy	500...800V, 16A, Igt/Ih<35/<60mA	17h§	Phi	2N6400...6405, BStD10..M, T9,5N..,CS 15...	data
BT 146	Thy	500V, 15A, Igt<35mA	22a§	Phi	2N6404, 2N6405, T9,5N.., BT 152/.., ++	data
BT 148	Thy	500V, 6,5A(Tc=60°C), Igt<35mA	22a§	Phi	2N6404, 2N6405, TIC 122D...E	data
BT 148W/...	Thy	400...600V, 0,6A, Igt/Ih<0,2/<6	48h§	Phi	-	data
BT 149 A...M	50Hz-Thy	50...600V, 1A, Igt/Ih<0,2/<5mA, 50µs	7a	Phi	TAG 59..., MCR 606-2...8, 2N6681...6685, ++	data
BT 150	Thy	A=100,B=200,D=400,E=500,F=50,M=600V 500V, 2,5A(Tc=98°C), Igt/Ih<0,2/<6mA	17h§	Phi	TIC106E, TAG623-500, TAG628-500, S4060E	data
BT 151/... R	F-Thy	500...800V, 12A, Igt/Ih<15/<20mA, 2µs	17h§	Phi	-	data
BT 151F/...	F-Thy	=BT 151/...R: Iso, 9A	17h		-	data
BT 152/... R	50Hz-Thy	400...800V, 20A, Igt/Ih<32/<60mA, 35µs	17h§	Phi	2N6504...6509	data
BT 153	F-Thy	TV-VA, 500V, 6A, Igt/Ih<40/<100mA	17h§	Phi	(BT 154)	data
BT 154	F-Thy	TV-HA, 750V, 8A, Igt<40mA, <2,4µs	17h§	Phi		data
BT 155/...RK	F-Thy	600...800V, 15A, Igt<100mA, <6µs	17h§	Phi	-	data
BT 155/...RN	F-Thy	600...800V, 15A, Igt<100mA, <9µs	17h§		-	data
BT 155/...RP	F-Thy	600...800V, 15A, Igt<100mA, <12µs	17h§		-	data
BT 157/...R	GTO-Thy	+1300...+1500V, 2,2A(Tc=80°C),Igt<200mA	17h§	Phi	BTW 58/1300...1500	data
BT 158/...	Triac	400...600V, 8A, Igt/Ih<40/<30mA	17h§	Mot	T 2806.., T 2856.., BT 162/.., ++	data
BT 162/...	Triac	400...600V, 12A, Igt/Ih<40/<30mA	17h§	Mot	T 6006.., TAG 257-...	data
BT 169 B...G	F-Thy	200...600V, 0,8A, Igt/Ih<0,2/<5mA	7e	Phi	BRX 47...49, BRY 55S/200...800	data
Bt 208 KC	A/D-IC	B=200V, D=400V, E=500V, G=600V Video 8Bit A/D-Conv. 20Msps, TTL	28-PLCC		-	
Bt 208 KC	A/D-IC	Video 8Bit A/D-Conv. 20Msps, TTL	24-DIP		-	
Bt 451-...	CMOS-D/A-IC	4 Bit, Video, CMOS/TTL Level -80: 20, -110: 27,5, -125: 31,25MHz	84-PGA		-	
Bt 453	D/A-IC	8 Bit, Video, 40Msps, TTL Level	40-DIP			pdf
Bt 457 KG125	D/A-IC	8 Bit, Video, 125Msps, TTL Level	PGA		-	
Bt 458 KG125	CMOS-D/A-IC	8 Bit, Video, 31,25MHz, CMOS/TTL Level	84-PGA		-	
Bt 471 KPJ40,80	D/A-IC	6 Bit, Video, 40, 80Msps, TTL Level	44-MP		-	
Bt 474 KPJ75,85	D/A-IC	8 Bit, Video, 75, 85Msps, CMOS/TTL L.	84-PLCC		-	
Bt 475 KPJ ...	D/A-IC	6 Bit, 3 Channel, Video, TTL Level KPJ35, 50, 66, 80: 35, 50, 66, 80Msps	44-PLCC		-	
Bt 476 KN50,KN66	D/A-IC	6 Bit, Video, 50, 66Msps, TTL Level	28-DIP		-	
Bt 477 KPJ ...	D/A-IC	8 Bit, 3Ch., Video, CMOS/TTL Level KPJ35, 50, 66, 80: 35, 50, 66, 80Msps	44-PLCC		-	
Bt 478 KPJ40,80	D/A-IC	8 Bit, Video, 40, 80Msps, TTL Level	44-MP		-	
Bt 492 ECL	CMOS-D/A-IC	8 Bit, Video, 360MHz, Ecl Level	68-PGA		-	
Bt 492 TTL	CMOS-D/A-IC	8 Bit, Video, 360MHz, TTL Level	68-PGA		-	
BT 500	Si-N	Vorspg.-Stabi f. Sender/RF Bias Source	49a	Mot		data
BT 500 A	Si-N	=BT 500:	59r			data
BT 1511	Z-Di, Z-IC	=THBT 150-11(Typ-Code/Stempel/marking)	8-MDIP	Tho	→THBT 150	data
BT 1512	Z-Di, Z-IC	=THBT 150-12(Typ-Code/Stempel/marking)	8-DIP	Tho	→THBT 150	data
BT 2011	Z-Di, Z-IC	=THBT 200-11(Typ-Code/Stempel/marking)	8-MDIP	Tho	→THBT 200	data
BT 2012	Z-Di, Z-IC	=THBT 200-12(Typ-Code/Stempel/marking)	8-DIP	Tho	→THBT 200	data
BT 2711	Z-Di, Z-IC	=THBT 270-11(Typ-Code/Stempel/marking)	8-MDIP	Tho	→THBT 270	data
BT 2712	Z-Di, Z-IC	=THBT 270-12(Typ-Code/Stempel/marking)	8-DIP	Tho	→THBT 270	data
BT 8028 C (...)	CMOS-IC	simple Melody Generator	30	Iks	-	pdf pinout
BT 8028 C (...) N	CMOS-IC	simple Melody Generator	30	Iks	-	pdf
BT 8031(...)	CMOS-IC	Cmos Lsi chip, prearranged melodies	30	Iks	-	pdf pinout
BT 8031(...)/CHIP	CMOS-IC	Cmos Lsi chip, prearranged melodies	30	Iks	-	pdf pinout
BT 8040 D	CMOS-IC	Melody Generator with Accompaniment	QFN	Iks	-	pdf pinout
BT 8040 N	CMOS-IC	Melody Generator with Accompaniment	QFN	Iks	-	pdf pinout

BTA

Type	Device	Short Description	Fig.	Manu	Comparision Types	More at
BTA 04-...A	Triac	200...800V, 4A(Tc=75°), Igt/Ih<25/<25mA	17h	Tho	TAG 421..., (TAG 231..., TAG 220...)[3]	data
BTA 04-...D	Triac	200...800V, 4A(Tc=75°), Igt/Ih<10/<15mA	17h		(TXC 18H..., TIC 216..., TAG 232...)[3]	data
BTA 04-...GP	Triac	200...800V, 4A(Tc=75°), Igt/Ih<5/<13mA	17h		(TXC 18H..., TIC 216..., TAG 232...)[3]	data
BTA 04-...S	Triac	200...800V, 4A(Tc=75°), Igt/Ih<10/<25mA	17h		(TAG 232..., TAG 231..., TXC 18G...)[3]	data
BTA 04-...T	Triac	200...800V, 4A(Tc=75°), Igt/Ih<5/<15mA	17h		(TAG 233..., TXC 18D..., TAG 222-..,)[3]	data
BTA 06-...A	Triac	200...800V, 6A(Tc=75°), Igt/Ih<25/<25mA	17h	Tho	TAG 421..., (TAG 220..., TAG 225...)[3]	data
BTA 06-...B	Triac	200...800V, 6A(Tc=75°), Igt/Ih<100/<50mA	17h		TAG 420..., (TAG 224..., TXC 10H...M)[3]	data
BTA 06-...C	Triac	200...800V, 6A(Tc=75°), Igt/Ih<50/<25mA	17h		TAG 420..., (TAG 224..., TXC 10H...M)[3]	data
BTA 06-...D	Triac	200...800V, 6A(Tc=75°), Igt/Ih<10/<15mA	17h		(TAG 221..., TXC 18G..M, TAG 226...)[3]	data

ype	Device	Short Description	Fig.	Manu	Comparision Types	More at
A 06-...GP	Triac	200...600V, 6A(Tc=75°), Igt/Ih<75/<13mA	17h		(MAC 216.., T 2801.., BS 7-..A, SC 141.)[3]	data
A 06-...S	Triac	200...800V, 6A(Tc=75°), Igt/Ih<10/<25mA	17h		(TAG 221.., TXC 18G..M, TAG 226..)[3]	data
A 06-...T	Triac	200...800V, 6A(Tc=75°), Igt/Ih<5/<15mA	17h		(TAG 222.., TXC 18D..M, TAG 227..)[3]	data
A 06-...AW,BW,CW	Triac	=BTA 06-..A,B,C: snubberless di/dt	17h		-	data
A 06-...SW,TW	Triac	=BTA 06-..S,T: LogL	17h			data
A 08-...A	Triac	200...800V, 8A(Tc=75°), Igt/Ih<25/<25mA	17h	Tho	TAG 426.., TAG 452.., TAG 457...	data
A 08-...B	Triac	200...800V, 8A(Tc=75°), Igt/Ih<100/<50mA	17h		TAG 425.., TAG 451.., TAG 456...	data
A 08-...C	Triac	200...800V, 8A(Tc=75°), Igt/Ih<50/<25mA	17h		TAG 425.., TAG 451.., TAG 456..	data
A 08-...S	Triac	200...800V, 8A(Tc=75°), Igt/Ih<10/<25mA	17h		(TAG 226.., MAC 228A-..)[3]	data
A 08-...AW,BW,CW	Triac	=BTA 08-..A,B,C: snubberless di/dt	17h		-	data
A 08-...SW,TW	Triac	=BTA 08-..S: LogL, TW: Igt/Ih<5/<15mA	17h			data
A 10-...B	Triac	200...800V, 10A(Tc=75°), Igt/Ih<100/<50mA	17h	Tho	TAG 456.., TAG 480.., TAG 481...	data
A 10-...C	Triac	200...800V, 10A(Tc=75°), Igt/Ih<50/<25mA	17h		TAG 456.., TAG 480.., TAG 481...	data
A 10-...AW,BW,CW	Triac	=BTA 10-..: snubberless di/dt	17h			data
A 12-...B	Triac	200...800V, 12A(Tc=75°), Igt/Ih<100/<50mA	17h	Tho	TAG 480.., TAG 481...	data
A 12-...C	Triac	200...800V, 12A(Tc=75°), Igt/Ih<50/<25mA	17h		TAG 480.., TAG 481...	data
A 12-...AW,BW,CW	Triac	=BTA 12-..: snubberless di/dt	17h		-	data
A 13-...B	Triac	200...800V, 12A, Igt/Ih<50/<50mA	17h		TAG 480.., TAG 481...	data
A 16-...B	Triac	200...800V, 16A(Tc=75°), Igt/Ih<100/<50mA	17h	Tho	(MAC 15A.., BT 139/.., TAG 280..)[3]	data
A 16-...C	Triac	200...800V, 16A(Tc=75°), Igt/Ih<50/<25mA	17h		(MAC 15A.., BT 139/.., TAG 280..)[3]	data
A 16-...AW,BW,CW	Triac	=BTA 16-..: snubberless di/dt	17h		-	data
A 20-...AW	Triac	Snubberless,400...800V, 20A, Igt<75mA	17h	Tho	-	
A 20-...BW	Triac	=BTA 20-..AW: Igt/Ih<50/<75mA	17h			
A 20-...CW	Triac	=BTA 20-..AW: Igt/Ih<35/<50mA	17h		-	
A 20 C	Triac	300V, 6A(Tc=80°), Igt/Ih<80/<100mA	17h§	Rca	TXC 10L40M, TW7N400, TIC 226C, ++	data
A 20 D	Triac	=BTA 20C: 400V	17h§		TXC 10L40M, TW7N400, TIC226D, ++	data
A 20 E	Triac	=BTA 20C: 500V	17h§		TXC 10L50M, TW7N600, TIC226E, ++	data
A 20 M	Triac	=BTA 20C: 600V	17h§		TXC 10L60M, TW7N600, TIC226M, ++	data
A 20 N	Triac	=BTA 20C: 800V	17h§		TXC 10L80M, TW7N800, TIC226N, ++	data
A 21 C	Triac	300V, 8A(Tc=80°), Igt/Ih<35/<100mA	17h§	Rca	TAG 225-.., TAG 257-...	data
A 21 D	Triac	=BTA 21C: 400V	17h§		TAG 225-.., TAG 257-...	data
A 21 E	Triac	=BTA 21C: 500V	17h§		TAG 225-.., TAG 257-...	data
A 21 M	Triac	=BTA 21C: 600V	17h§		TAG 225-.., TAG 257-...	data
A 21 N	Triac	=BTA 21C: 800V	17h§		TAG 225-.., TAG 257-...	data
A 22 B	Triac	200V, 10A(Tc=75°), Igt/Ih<60/<30mA	17h§	Rca	TAG 250-.., TAG 251-.., TXD 10H40M, ++	data
A 22 C	Triac	=BTA 22B: 300V	17h§		TAG 250-.., TAG 251-.., TXD 10H40M, ++	data
A 22 D	Triac	=BTA 22B: 400V	17h§		TAG 250-.., TAG 251-.., TXD 10H40M, ++	data
A 22 E	Triac	=BTA 22B: 500V	17h§		TAG 250-.., TAG 251-.., TXD 10H50M, ++	data
A 22 M	Triac	=BTA 22B: 600V	17h§		TAG 250-.., TAG 251-.., TXD 10H60M, ++	data
A 22 N	Triac	=BTA 22B: 800V	17h§		TAG 250-.., TAG 251-.., TXD 10H80M, ++	data
A 23 B	Triac	200V, 12A(Tc=70°), Igt/Ih<60/<30mA	17h§	Rca	TAG 255-.., TAG 256-.., BT 138/.., ++	data
A 23 C	Triac	=BTA 23B: 300V	17h§		TAG 255-.., TAG 256-.., BT 138/.., ++	data
A 23 D	Triac	=BTA 23B: 400V	17h§		TAG 255-.., TAG 256-.., BT 138/.., ++	data
A 23 E	Triac	=BTA 23B: 500V	17h§		TAG 255-.., TAG 256-.., BT 138/.., ++	data
A 23 M	Triac	=BTA 23B: 600V	17h§		TAG 255-.., TAG 256-.., BT 138/.., ++	data
A 23 N	Triac	=BTA 23B: 800V	17h§		TAG 255-.., TAG 256-.., BT 138/.., ++	data
A 25-...A	Triac	200...800V, 25A, Igt/Ih<150/<100mA	64h	Tho	(BTW 41/..)[3]	data
A 25-...B	Triac	200...700V, 25A(Tc=65°), Igt/Ih<80/<80mA	64h		(MAC 525A..,TAG 725-.., TAG 726-..)[3]	data
A 26-...A	Triac	200...800V, 25A, Igt/Ih<150/<100mA	18h	Tho	BTA 41-...	data
A 26-...B	Triac	200...800V, 25A, Igt/Ih<100/<80mA	18h		BTA 41-...	data
A 26-...AW,BW,CW	Triac	=BTA 26-..: snubberless di/dt	18h			
A 40-...A	Triac	200...800V, 40A, Igt/Ih<150/<100mA	64h	Tho	(BTW 41/..)[3]	data
A 40-...B	Triac	200...800V, 40A, Igt/Ih<80/<80mA	64h		(MAC 50A-..)[3]	data
A 41-...A	Triac	200...800V, 40A, Igt/Ih<150/<100mA	18h	Tho	-	data
A 41-...B	Triac	200...800V, 40A, Igt/Ih<100/<80mA	18h		-	data
A 140/...	Triac	500...800V, 25A(Tc=89°), Igt/Ih<70/<30mA	17h§	Phi	TXE 10.., MAC 223-.., MAC 223A-...	
B						
B 04-...A	Triac	=BTA 04-..A:	17h§	Tho	T 2506.., T 2806.., TAG 220...	data
B 04-...D	Triac	=BTA 04-..D:	17h§		TXC 18H.., TIC 216.., TXC 18H..M	data
B 04-...S	Triac	=BTA 04-..S:	17h§		TXC 18G.., TAG 232.., TAG 221.., ++	data
B 04-...T	Triac	=BTA 04-..T:	17h§		TXC 18D.., TAG 233.., TAG 222.., ++	data
B 06-...A	Triac	=BTA 06-..A:	17h§	Tho	T 2506.., T 2806.., T 6006...	data
B 06-...B	Triac	=BTA 06-..B:	17h§		TXC 10K/L..M, TW7N.., MAC 222-.., ++	data
B 06-...C	Triac	=BTA 06-..C:	17h§		TXC 10K/L..M, TW7N.., MAC 222-.., ++	data
B 06-...D	Triac	=BTA 06-..D:	17h§		TIC 216.., TXC 18H..M	data
B 06-...GP	Triac	=BTA 06-..GP:	17h§		TXC 10K/L..M, TW7N.., MAC 222-.., ++	data
B 06-...S	Triac	=BTA 06-..S:	17h§		TXC 18G..M, TAG 221.., TAG 226..	data
B 06-...T	Triac	=BTA 06-..T:	17h§		TAG 222.., TXC 18D..M, TAG 227...	data
B 06-...AW,BW,CW	Triac	=BTA 06-..AW,BW,CW	17h§		-	data
B 06-...SW,TW	Triac	=BTA 06-..SW,TW:	17h§			data
B 08-...A	Triac	=BTA 08-..A:	17h§	Tho	T 2806.., T 6006...	data
B 08-...B	Triac	=BTA 08-..B:	17h§		MAC 222-.., TIC 226.., T 2802.., ++	data
B 08-...C	Triac	=BTA 08-..C:	17h§		MAC 222-.., TIC 226.., T 2802.., ++	data
B 08-...GP	Triac	=BTA 08-..GP:	17h§		TAG 224-.., MAC 222A-.., BT 137/.., ++	data
B 08-...S	Triac	=BTA 08-..S:	17h§		TAG 226.., MAC 228A-...	data
B 08-...AW,BW,CW	Triac	=BTA 08-..AW,BW,CW:	17h§		-	data

Type	Device	Short Description	Fig.	Manu	Comparision Types	More at
BTB 08-..SW,TW	Triac	=BTA 08-..SW,TW:	17h§		-	data
BTB 10-..B	Triac	=BTA 10-..B:	17h§	Tho	TW9N.., TXD 10K/L..M, TIC 236.., ++	data
BTB 10-..C	Triac	=BTA 10-..C:	17h§		TAG 250-.., TAG 251-.., TXD 10H..M, ++	data
BTB 10-..AW,BW,CW	Triac	=BTA 10-..AW,BW,CW:	17h§		-	
BTB 12-..B	Triac	=BTA 12-..B:	17h§	Tho	TIC 236.., TW11N.., 2N6342..6345A, ++	data
ВТВ 12-..C	Triac	=BTA 12-..C:	17h§		TIC 236.., TW11N.., 2N6342..6345A, ++	data
BTB 12-..AW,BW,CW	Triac	=BTA 12-..AW,BW,CW:	17h§		-	
BTB 13-..B	Triac	=BTA 13-..B:	17h§	Tho	TIC 236.., TW11N.., 2N6342..6345A, ++	
BTB 15-..B	Triac	200..800V, 15A(Tc=75°),Igt/Ih<75/<50mA	17h§	Tho	MAC 15-.., TIC 246.., T 6001.., ++	data
BTB 16-..B	Triac	=BTA 16-..B:	17h§	Tho	MAC 15-.., TIC 246.., T 6001.., ++	data
BTB 16-..C	Triac	=BTA 16-..C:	17h§		MAC 15-.., TIC 246.., T 6001.., ++	data
BTB 16-..AW,BW,CW	Triac	=BTA 16-..AW,BW,CW:	17h§			
BTB 19-..B	Triac	200..700V,19A(Tc=75°),Igt/Ih<100/<50mA	17h§	Tho	TXE 10.., MAC 223-.., MAC 223A-...	data
BTB 20-..AW,BW,CW	Triac	=BTA 20-..AW,BW,CW:	17h§	Tho		
BTB 24-..B	Triac	=BTB 19-..B: 25A(Tc=75°)	17h§	Tho	TXE 10.., MAC 223-.., MAC 223A-...	data
BTB 25-..A	Triac	=BTA 25-..A:	64h		(BTW 41/...)[4]	data
BTB 25-..B	Triac	=BTA 25-..B:	64h		(MAC 525-.., MAC 25-.., MAC 50-...)[4]	data
BTB 26-..A	Triac	=BTA 26-..A:	18h§	Tho		data
BTB 26-..B	Triac	=BTA 26-..B:	18h§		·BTB 41-...	data
BTB 40-..A	Triac	=BTA 40-..A:	64h	Tho	(BTW 41/...)[4]	data
BTB 40-..B	Triac	=BTA 40-..B:	64h		(MAC 50-.., TAG 740-.., TAG 741-...)[4]	data
BTB 41-..A	Triac	=BTA 41-..A:	18h§	Tho	-	data
BTB 41-..B	Triac	=BTA 41-..B:	18h§		-	data

BTD

Type	Device	Short Description	Fig.	Manu	Comparision Types	More at
BTD 4 L..N	Diac	Ub=26..40V, Ib<0,05mA, Itsm=2A L=26..32V, M=29..37V, N=34..40V	31	Say	1N5761..62, D3202U,Y	
BTD 0105	Triac	50V, 1A(Tc=80°), Igt/Ih<25/<25mA	2d	Tra	T 2306.., TAG 200-.., T 2303...	data
BTD 0110	Triac	100V, 1A(Tc=80°), Igt/Ih<25/<25mA	2d		T 2306.., TAG 200-.., T 2303...	data
BTD 0120	Triac	200V, 1A(Tc=80°), Igt/Ih<25/<25mA	2d		T 2306.., TAG 200-.., T 2303...	data
BTD 0140	Triac	400V, 1A(Tc=80°), Igt/Ih<25/<25mA	2d		T 2306.., TAG 200-.., T 2303...	data
BTD 0160	Triac	600V, 1A(Tc=80°), Igt/Ih<25/<25mA	2d		T 2306.., TAG 200-.., T 2303...	data
BTD 0205-1	Triac	50V, 1,6A(Tc=80°), Igt/Ih=10/<10mA	2d	Tra	TAG 201.., T 2302.., TAG 202-.., ++	data
BTD 0210-1	Triac	100V, 1,6A(Tc=80°), Igt/Ih=10/<10mA	2d		TAG 201.., T 2302.., TAG 202-.., ++	data
BTD 0220-1	Triac	200V, 1,6A(Tc=80°), Igt/Ih=10/<10mA	2d		TAG 201.., T 2302.., TAG 202-.., ++	data
BTD 0240-1	Triac	400V, 1,6A(Tc=80°), Igt/Ih=10/<10mA	2d		TAG 201.., T 2302.., TAG 202-.., ++	data
BTD 02..-2	Triac	=BTD 02..-1: Igt/Ih=6/<10mA	2d		T 2301.., TAG 204-.., TAG 204A-...	data
BTD 02..-3	Triac	=BTD 02..-1: Igt/Ih=3/<10mA	2d		TAG 206-.., TAG 207-...	data
BTD 0305	Triac	50V, 3A(Tc=80°), Igt/Ih<50/<50mA	2d	Tra	-	data
BTD 0310	Triac	100V, 3A(Tc=80°), Igt/Ih<50/<50mA	2d			data
BTD 0320	Triac	200V, 3A(Tc=80°), Igt/Ih<50/<50mA	2d			data
BTD 0340	Triac	400V, 3A(Tc=80°), Igt/Ih<50/<50mA	2d			data
BTD 0360	Triac	600V, 3A(Tc=80°), Igt/Ih<50/<50mA	2d			data
BTD 03..-1	Triac	=BTD 03..: Igt/Ih<10/<15mA	2d		TAG 202-.., TAG 203-.., TAG 209-...	data
BTD 03..-2	Triac	=BTD 03..: Igt/Ih<4/<10mA	2d		TAG 204-.., TAG 205-...	data
BTD 03..-3	Triac	=BTD 03..: Igt/Ih<3/<5mA	2d		TAG 206-.., TAG 207-...	data
BTD 0605	Triac	50V, 6A(Tc=80°), Igt/Ih<50/<50mA	2d	Tra	-	data
BTD 0610	Triac	100V, 6A(Tc=80°), Igt/Ih<50/<50mA	2d		-	data
BTD 0620	Triac	200V, 6A(Tc=80°), Igt/Ih<50/<50mA	2d		-	data
BTD 0640	Triac	400V, 6A(Tc=80°), Igt/Ih<50/<50mA	2d		-	data
BTD 0660	Triac	600V, 6A(Tc=80°), Igt/Ih<50/<50mA	2d		-	data
BTE 5 M	Diac	Ub=29..37V, Ib<0,05mA, Itsm=2A	31	Say	1N5762, D3202Y	

BTL

Type	Device	Short Description	Fig.	Manu	Comparision Types	More at
BTL 0405	Triac	50V, 4A(Tc=80°), Igt/Ih<25/<25mA	17h§	Tra	T 2506.., T 2806.., TAG 220-.., ++	data
BTL 0410	Triac	100V, 4A(Tc=80°), Igt/Ih<25/<25mA	17h§		T 2506.., T 2806.., TAG 220-.., ++	data
BTL 0420	Triac	200V, 4A(Tc=80°), Igt/Ih<25/<25mA	17h§		T 2506.., T 2806.., TAG 220-.., ++	data
BTL 0440	Triac	400V, 4A(Tc=80°), Igt/Ih<25/<25mA	17h§		T 2506.., T 2806.., TAG 220-.., ++	data
BTL 0460	Triac	600V, 4A(Tc=80°), Igt/Ih<25/<25mA	17h§		T 2506.., T 2806.., TAG 220-.., ++	data
BTL 0605	Triac	50V, 6A(Tc=80°), Igt/Ih<50/<50mA	17h§	Tra	TW7N.., MAC 222-.., TIC 226.., T 2802...	data
BTL 0610	Triac	100V, 6A(Tc=80°), Igt/Ih<50/<50mA	17h§		TW7N.., MAC 222-.., TIC 226.., T 2802...	data
BTL 0620	Triac	200V, 6A(Tc=80°), Igt/Ih<50/<50mA	17h§		TW7N.., MAC 222-.., TIC 226.., T 2802...	data
BTL 0640	Triac	400V, 6A(Tc=80°), Igt/Ih<50/<50mA	17h§		TW7N.., MAC 222-.., TIC 226.., T 2802...	data
BTL 0660	Triac	600V, 6A(Tc=80°), Igt/Ih<50/<50mA	17h§		TW7N.., MAC 222-.., TIC 226.., T 2802...	data
BTL 0805	Triac	50V, 8A(Tc=80°), Igt/Ih<50/<50mA	17h§	Tra	TIC 226.., T 2802.., TXD 10K/L.., ++	data
BTL 0810	Triac	100V, 8A(Tc=80°), Igt/Ih<50/<50mA	17h§		TIC 226.., T 2802.., TXD 10K/L.., ++	data
BTL 0820	Triac	200V, 8A(Tc=80°), Igt/Ih<50/<50mA	17h§		TIC 226.., T 2802.., TXD 10K/L.., ++	data
BTL 0840	Triac	400V, 8A(Tc=80°), Igt/Ih<50/<50mA	17h§		TIC 226.., T 2802.., TXD 10K/L.., ++	data
BTL 0860	Triac	600V, 8A(Tc=80°), Igt/Ih<50/<50mA	17h§		TIC 226.., T·2802.., TXD 10K/L.., ++	data
BTL 1005	Triac	50V, 10A(Tc=80°), Igt/Ih<100/<100mA	17h§	Tra	TW9N.., TXD 10K/L.., TIC 236.., ++	data
BTL 1010	Triac	100V, 10A(Tc=80°), Igt/Ih<100/<100mA	17h§		TW9N.., TXD 10K/L.., TIC 236.., ++	data
BTL 1020	Triac	200V, 10A(Tc=80°), Igt/Ih<100/<100mA	17h§		TW9N.., TXD 10K/L.., TIC 236.., ++	data
BTL 1040	Triac	400V, 10A(Tc=80°), Igt/Ih<100/<100mA	17h§		TW9N.., TXD 10K/L.., TIC 236.., ++	data
BTL 1060	Triac	600V, 10A(Tc=80°), Igt/Ih<100/<100mA	17h§		TW9N.., TXD 10K/L.., TIC 236.., ++	data

Type	Device	Short Description	Fig.	Manu	Comparision Types	More at
BTL 1605	Triac	50V, 16A(Tc=80°), Igt/Ih<100/<100mA	17h§		SC 151..., MAC 15-..., TIC 246..., ++	data
BTL 1610	Triac	100V, 16A(Tc=80°), Igt/Ih<100/<100mA	17h§		SC 151..., MAC 15-..., TIC 246..., ++	data
BTL 1620	Triac	200V, 16A(Tc=80°), Igt/Ih<100/<100mA	17h§		SC 151..., MAC 15-..., TIC 246..., ++	data
BTL 1640	Triac	400V, 16A(Tc=80°), Igt/Ih<100/<100mA	17h§		SC 151..., MAC 15-..., TIC 246..., ++	data
BTL 1660	Triac	600V, 16A(Tc=80°), Igt/Ih<100/<100mA	17h§		SC 151..., MAC 15-..., TIC 246..., ++	data

BTM

Type	Device	Short Description	Fig.	Manu	Comparision Types	More at
BTM 0405...0460	Triac	=BTL 0405...0460:	17h§	Tra	→BTL 0405...0460	data
BTM 0605...0660	Triac	=BTL 0605...0660:	17h§	Tra	→BTL 0605...0660	data
BTM 1005...1060	Triac	=BTL 1005...1060:	17h§	Tra	→BTL 1005...1060	data
BTM 1605...1660	Triac	=BTL 1605...1660:	17h§	Tra	→BTL 1605...1660	data

BTR

Type	Device	Short Description	Fig.	Manu	Comparision Types	More at
BTR 59/..R	GTO-Thy	800...1300V, 10A(Tc=85°C), Igt<0,5mA	18h§	Phi	BTS 59/..R	data
BTR 0205	Triac	50V, 3A(Tc=75°), Igt/Ih<50/<25mA	22a§	Tra	TAG 260-..., TAG 265-..., T 4700...	data
BTR 0210	Triac	100V, 3A(Tc=75°), Igt/Ih<50/<25mA	22a§		TAG 260-..., TAG 265-..., T 4700...	data
BTR 0220	Triac	200V, 3A(Tc=75°), Igt/Ih<50/<25mA	22a§		TAG 260-..., TAG 265-..., T 4700...	data
BTR 0240	Triac	400V, 3A(Tc=75°), Igt/Ih<50/<25mA	22a§		TAG 260-..., TAG 265-..., T 4700...	data
BTR 0260	Triac	600V, 3A(Tc=75°), Igt/Ih<50/<25mA	22a§		TAG 260-..., TAG 265-..., T 4700...	data
BTR 0305	Triac	50V, 6A(Tc=75°), Igt/Ih<50/<50mA	22a§	Tra	TAG 260-..., TAG 265-..., T 4700...	data
BTR 0310	Triac	100V, 6A(Tc=75°), Igt/Ih<50/<50mA	22a§		TAG 260-..., TAG 265-..., T 4700...	data
BTR 0320	Triac	200V, 6A(Tc=75°), Igt/Ih<50/<50mA	22a§		TAG 260-..., TAG 265-..., T 4700...	data
BTR 0340	Triac	400V, 6A(Tc=75°), Igt/Ih<50/<50mA	22a§		TAG 260-..., TAG 265-..., T 4700...	data
BTR 0360	Triac	600V, 6A(Tc=75°), Igt/Ih<50/<50mA	22a§		TAG 260-..., TAG 265-..., T 4700...	data
BTR 0405	Triac	50V, 10A(Tc=75°), Igt/Ih<100/<50mA	22a§	Tra	TAG 265-..., T 4700...	data
BTR 0410	Triac	100V, 10A(Tc=75°), Igt/Ih<100/<50mA	22a§		TAG 265-..., T 4700...	data
BTR 0420	Triac	200V, 10A(Tc=75°), Igt/Ih<100/<50mA	22a§		TAG 265-..., T 4700...	data
BTR 0440	Triac	400V, 10A(Tc=75°), Igt/Ih<100/<50mA	22a§		TAG 265-..., T 4700...	data
BTR 0460	Triac	600V, 10A(Tc=75°), Igt/Ih<100/<50mA	22a§		TAG 265-..., T 4700...	data
BTR 0505	Triac	50V, 15A, Igt/Ih<80/<50mA	22a§	Tra	T 4700...	data
BTR 0510	Triac	100V, 15A, Igt/Ih<80/<50mA	22a§		T 4700...	data
BTR 0520	Triac	200V, 15A, Igt/Ih<80/<50mA	22a§		T 4700...	data
BTR 0530	Triac	300V, 15A, Igt/Ih<80/<50mA	22a§		T 4700...	data
BTR 0540	Triac	400V, 15A, Igt/Ih<80/<50mA	22a§		T 4700...	data
BTR 0550	Triac	500V, 15A, Igt/Ih<80/<50mA	22a§		T 4700...	data
BTR 0560	Triac	600V, 15A, Igt/Ih<80/<50mA	22a§		T 4700...	data
BTR 0605...0660	Triac	=BTL 0605...0660:	22a§	Tra	TAG 260-..., TAG 265-..., T 4700...	data
BTR 1005...1060	Triac	=BTL 1005...1060:	22a§	Tra	TAG 265-..., T 4700...	data
BTR 1605...1660	Triac	=BTL 1605...1660:	22a§	Tra	T 4700...	data

BTS

Type	Device	Short Description	Fig.	Manu	Comparision Types	More at
BTS 59/850R	GTO-Thy	850V, 15A(Tc=85°), Igt<300mA	18h§	Phi	-	data
BTS 59/1000R	GTO-Thy	=BTS 59/850R: 1000V	18h§		-	data
BTS 59/1200R	GTO-Thy	=BTS 59/850R: 1200V	18h§		-	data
BTS 100	MOS-P-FET-e	V-MOS, TEMPFET, 50V, 8A, 40W, <0,3Ω	17c§	Sie	-	data
BTS 110	MOS-N-FET-e	V-MOS, TEMPFET, 100V, 10A, 40W, <0,2Ω	17c§	Sie	-	data
BTS 112 A	MOS-N-FET-e	V-MOS, TEMPFET, 60V, 12A, 40W, <0,15Ω	17c§	Sie	-	data
BTS 113 A	MOS-N-FET-e	V-MOS,TEMPFET, 60V, 11,5A, 40W, <0,17Ω LogL, 70/85ns	17c§	Sie	-	data
BTS 114	MOS-N-FET-e	V-MOS, TEMPFET, 50V, 14A, 40W, <0,1Ω	17c§	Sie	-	data
BTS 114 A	MOS-N-FET-e	V-MOS,TEMPFET, 50V, 17A, 50W, <0,1Ω	17c§	Sie	-	data
BTS 115	MOS-N-FET-e	V-MOS, TEMPFET, 50V, 12,5A, 40W,<125mΩ	17c§	Sie	-	data
BTS 115 A	MOS-N-FET-e	V-MOS,TEMPFET, 50V, 15,5A, 50W, <0,12Ω LogL, 85/120ns	17c§	Sie	-	data
BTS 116	MOS-N-FET-e	V-MOS, TEMPFET, 50V, 14A, 40W, <0,1Ω	≈30c	Sie	-	data
BTS 117	MOS-N-FET-e	V-MOS, HITFET, LogL, Low-side Switch 60V, 3,5A, 50W, <0,1Ω(3,5A), 40/70ns	17 (InpDS)	Sie	-	data
BTS 117 E3045(A)	MOS-N-FET-e	=BTS 117:	30			data
BTS 118 D	MOS-N-FET-e	V-MOS, HITFET, LogL, Low-side Switch 42V, 2,4A, 21W, <0,12Ω(2,2A), 40/70µs	30c§ (InpDS)	Sie	-	
BTS 120	MOS-N-FET-e	V-MOS, TEMPFET, 100V, 19A, 75W, <0,1Ω	17c§	Sie	-	data
BTS 121 A	MOS-N-FET-e	V-MOS, TEMPFET, 100V, 22A, 75W, <0,1Ω	17c§	Sie	-	data
BTS 129	MOS-N-FET-e	V-MOS, TEMPFET, 60V, 27A, 75W, <0,05Ω	17c§	Sie	-	data
BTS 130	MOS-N-FET-e	V-MOS, TEMPFET, 50V, 27A, 75W, <0,05Ω	17c§	Sie	-	data
BTS 131	MOS-N-FET-e	V-MOS, TEMPFET, 50V, 25A, 75W, <0,06Ω	17c§	Sie	-	data
BTS 132	MOS-N-FET-e	V-MOS, TEMPFET, 60V, 24A, 75W, <0,065Ω	17c§	Sie	-	data
BTS 133	MOS-N-FET-e	V-MOS, HITFET, LogL, Low-side Switch 60V, 7A, 90W, <0,05Ω(7A), 40/70ns	17 (InpDS)	Sie	-	data
BTS 133 E3045(A)	MOS-N-FET-e	=BTS 133:	30		-	data
BTS 134 D	MOS-N-FET-e	V-MOS, HITFET, LogL, Low-side Switch 42V, 3,5A, 43W, <0,06Ω(3A), 60/60µs	30c§ (InpDS)	Sie	-	
BTS 136	MOS-N-FET-e	V-MOS, TEMPFET, 50V, 27A, 75W, <0,05Ω	≈30c§	Sie	-	data
BTS 140 A	MOS-N-FET-e	V-MOS, TEMPFET, 50V, 42A, 125W	17c§	Sie	-	data

Type	Device	Short Description	Fig.	Manu	Comparision Types	More at
BTS 141	MOS-N-FET-e	V-MOS, HITFET, LogL, Low-side Switch 60V, 12A, 149W, <28mΩ(12A), 40/70ns	17 (InpDS)	Sie	-	data
BTS 141 E3045(A)	MOS-N-FET-e	=BTS 141:	30			data
BTS 142 D	MOS-N-FET-e	V-MOS, HITFET, LogL, Low-side Switch 42V, 4,6A, 59W, <34mΩ(4,6A), 60/60µs	30c§ (InpDS)	Sie	-	data
BTS 149	MOS-N-FET-e	V-MOS, HITFET, LogL, Low-side Switch 60V, 19A, 240W, <18mΩ(16A), 40/70ns	17 (InpDS)	Sie	-	data
BTS 149 E3045(A)	MOS-N-FET-e	=BTS 149:	30		-	data
BTS 240 A	MOS-N-FET-e	V-MOS, TEMPFET, 50V, 58A, 170W, <18mΩ	18c§	Sie	-	data
BTS 244 Z	MOS-N-FET-e	V-MOS, TEMPFET, LogL, 55V, 35A, 170W <18mΩ(19A), <130/100ns	86/5Pin 30/5Pin	Sie	-	
BTS 247 Z	MOS-N-FET-e	V-MOS, TEMPFET, LogL, 55V, 33A, 120W <28mΩ(12A), <70/75ns	86/5Pin 30/5Pin	Sie	-	
BTS 282 Z	MOS-N-FET-e	V-MOS, TEMPFET, LogL, 49V, 36A, 300W <9,5mΩ(36A), <101/160ns	86/5Pin 30/5Pin	Sie	-	
BTS 307	MOS-N-FET	V-MOS, PROFET, 65V, 1,7A, 50W, <0,25Ω <80/70µs	86/5Pin	Sie	-	data
BTS 307 E3062(A)	MOS-N-FET	=BTS 307:	30/5Pin		-	data
BTS 308	MOS-N-FET	V-MOS, PROFET, 60V, 1,3A, 50W, <0,23Ω <50/55µs	86/5Pin	Sie	-	data
BTS 308 E3062(A)	MOS-N-FET	=BTS 308:	30/5Pin		-	data
BTS 409 L1	MOS-N-FET	V-MOS, PROFET, 43V, 2,3A, 18W, <0,2Ω <400/400µs	86/5Pin	Sie	-	data
BTS 409 L1E3062(A)	MOS-N-FET	=BTS 409L1:	30/5Pin		-	data
BTS 410	MOS-N-FET	V-MOS, PROFET, 50V, 75W, <0,22Ω	86/5Pin	Sie	-	data
BTS 410 E3062(A)	MOS-N-FET	=BTS 410:	30/5Pin		-	data
BTS 412 A	MOS-N-FET	V-MOS, PROFET, 45V, 11A, 75W, <0,4Ω	86/5Pin	Sie	-	data
BTS 412 B	MOS-N-FET	V-MOS, PROFET, 56V, 14A, 75W, <0,25Ω	86/5Pin	Sie	-	data
BTS 412... E3062(A)	MOS-N-FET	=BTS 412:	30/5Pin		-	data
BTS 413 A	MOS-N-FET	V-MOS, PROFET, 45V, 11A, 75W, <0,4Ω	86/5Pin	Sie	-	data
BTS 425 L1	MOS-N-FET	V-MOS, PROFET, 43V, 7A, 75W, <0,06Ω <400/450µs	86/5Pin	Sie	-	data
BTS 425 L1E3062(A)	MOS-N-FET	=BTS 425L1:	30/5Pin		-	data
BTS 426 L1	MOS-N-FET	V-MOS, PROFET, 43V, 7A, 75W, <0,06Ω <400/450µs	86/5Pin	Sie	-	data
BTS 426 L1E3062(A)	MOS-N-FET	=BTS 426L1:	30/5Pin		-	data
BTS 428 L2	MOS-N-FET	V-MOS, PROFET, 43V, 7A, 75W, <0,06Ω <200/200µs	30/5Pin	Sie	-	
BTS 430 K2	MOS-N-FET	V-MOS, PROFET, 50V, 11A, 125W, <38mΩ <260/60µs	86/5Pin	Sie	-	data
BTS 432	MOS-N-FET	V-MOS, PROFET, 50V, 125W, <38mΩ	86/5Pin	Sie	-	data
BTS 432 E2	MOS-N-FET	V-MOS, PROFET, 63V, 44A, 125W, <38mΩ	86/5Pin 30/5Pin	Sie	-	
BTS 436 L2	MOS-N-FET	V-MOS, PROFET, 43V, 9,8A, 75W, <38mΩ	86/5Pin 30/5Pin	Sie	-	
BTS 441 T	MOS-N-FET	V-MOS, PROFET, 43V, 21A, 125W, <20mΩ	86/5Pin 30/5Pin	Sie	-	
BTS 442 D2,E2	MOS-N-FET	V-MOS, PROFET, 63V, 21A, 167W, <18mΩ <350/130µs	86/5Pin	Sie	-	data
BTS 442... E3062(A)	MOS-N-FET	=BTS 442...:	30/5Pin		-	data
BTS 452 T	MOS-N-FET	V-MOS, PROFET, 40V, 2,2A, 41,6W, <0,2Ω 80/80µs	30/5Pin	Sie	-	
BTS 462 T	MOS-N-FET	V-MOS, PROFET, 40V, 4,4A, 41,6W, <0,1Ω 90/90µs	30/5Pin	Sie	-	
BTS 542	MOS-N-FET	V-MOS, PROFET, 50V, 170W, <20mΩ	86/5Pin	Sie	-	data
BTS 550 P	MOS-N-FET	V-MOS, PROFET, 63V, 80A, 230W, <5mΩ <400/200µs	18/5Pin	Sie	-	data
BTS 555 P	MOS-N-FET	V-MOS, PROFET, 63V, 108A, 310W, <3,6mΩ <550/240µs	18/5Pin	Sie	-	data
BTS 610 L1	MOS-N-FET	V-MOS, PROFET, 2-Channel 43V, 4,4A, 36W, <0,2Ω(1,8A),<400/400µs	86/7Pin	Sie	-	data
BTS 611 L1	MOS-N-FET	V-MOS, PROFET, 2-Channel 43V, 4,4A, 36W, <0,2Ω(1,8A),<400/400µs	86/7Pin	Sie	- •	data
BTS 611 L1E3128(A)	MOS-N-FET	=BTS 611L1:	30/7Pin		-	data
BTS 612 N1	MOS-N-FET	V-MOS, PROFET, 2-Channel 43V, 4,4A, 36W, <0,2Ω(1,8A),<400/400µs	86/7Pin	Sie	-	data
BTS 612 N1E3128(A)	MOS-N-FET	=BTS 612N1:	30/7Pin		-	data
BTS 620 L1	MOS-N-FET	V-MOS, PROFET, 2-Channel 43V, 8,5A, 75W, <0,1Ω(2A), <400/400µs	86/7Pin	Sie	-	data
BTS 621 L1	MOS-N-FET	V-MOS, PROFET, 2-Channel 43V, 8,5A, 75W, <0,1Ω(2A), <400/400µs	86/7Pin	Sie	-	data
BTS 621 L1E3128(A)	MOS-N-FET	=BTS 621L1:	30/7Pin		-	data
BTS 629	MOS-FET	Dimmer, 60V, 35W, <0,2Ω	86/7Pin	Sie	-	data
BTS 630	MOS-FET	Dimmer, 60V, 20A, 75W, <0,07Ω(3A)	86/7Pin	Sie	-	data

Type	Device	Short Description	Fig.	Manu	Comparision Types	More at
BTS 640 S2	MOS-N-FET	V-MOS, PROFET, 41V, 12,6A, 85W, <30mΩ <150/200µs	86/7Pin	Sie	-	data
BTS 640 S2E3128(A)	MOS-N-FET	=BTS 640S2:	30/7Pin			data
BTS 650 P	MOS-N-FET	V-MOS, PROFET, 63V, 46A, 170W, <8mΩ <320/180µs	86/7Pin	Sie	-	data
BTS 650 PE3128(A)	MOS-N-FET	=BTS 650S2:	30/7Pin			data
BTS 707	MOS-N-FET	SMD, V-MOS, PROFET, 2-Channel 65V, 2,8A, 2x<0,25Ω(2A), <80/70µs	20-MDIP	Sie	-	data
BTS 710 L1	MOS-N-FET	SMD, V-MOS, PROFET, 4-Channel 43V, 4,4A, 4x<0,2Ω(1,8A), <400/400µs	20-MDIP	Sie	-	data
BTS 711 L1	MOS-N-FET	SMD, V-MOS, PROFET, 4-Channel 43V, 4,4A, 4x<0,2Ω(2A), <400/400µs	20-MDIP	Sie		data
BTS 712 N1	MOS-N-FET	SMD, V-MOS, PROFET, 4-Channel 43V, 4,4A, 4x<0,2Ω(2A), <400/400µs	20-MDIP			data
BTS 720 L1	MOS-N-FET	SMD, V-MOS, PROFET, 4-Channel 43V, 6,3A, 4x<0,1Ω(2A), <400/400µs	20-MDIP	Sie		data
BTS 721 L1	MOS-N-FET	SMD, V-MOS, PROFET, 4-Channel 43V, 6,3A, 4x<0,1Ω(2A), <400/400µs	20-MDIP			data
BTS 725 L1	MOS-N-FET	SMD, V-MOS, PROFET, 2-Channel 43V, 6A, 2x<60mΩ(2A), <400/450µs	20-MDIP	Sie		data
BTS 726	MOS-N-FET	SMD, V-MOS, PROFET, 2-Channel 43V, 6A, 2x<60mΩ(2A), <400/450µs	20-MDIP			data
BTS 728 L2	MOS-N-FET	SMD, V-MOS, PROFET, 2-Channel 43V, 6A, 2x<60mΩ(2A), <200/200µs	20-MDIP			
BTS 730	MOS-N-FET	SMD, V-MOS, PWM Power Unit, 40V, <70mΩ	20-MDIP	Sie	-	data
BTS 733 L1	MOS-N-FET	SMD, V-MOS, PROFET, 2-Channel 43V, 7,3A, 2x<40mΩ(2A), <350/450µs	20-MDIP	Sie		data
BTS 734 L1	MOS-N-FET	SMD, V-MOS, PROFET, 2-Channel 43V, 7,3A, 2x<40mΩ(2A), <350/450µs	20-MDIP	Sie		data
BTS 736 L2	MOS-N-FET	SMD, V-MOS, PROFET, 2-Channel 43V, 7,3A, 2x<40mΩ(2A), <200/250µs	20-MDIP	Sie		
BTS 740 S2	MOS-N-FET	SMD, V-MOS, PROFET, 2-Channel 43V, 8,5A, 2x<30mΩ(5A), <150/200µs	20-MDIP	Sie		
BTS 770	MOS-IC	SMD, Trilith-IC, 4x Motor-Tr., 40V/8A	28-MDIP	Sie	-	data pdf pinout
BTS 771	MOS-IC	SMD, Trilith-IC, 4x Motor-Tr., 40V/10A	28-MDIP	Sie	-	data pdf pinout
BTS 780	LIN-IC	Trilith IC	SIP	Sie	-	pdf pinout
BTS 840 S2	MOS-N-FET	SMD, V-MOS, PROFET, 2-Channel 43V, 24A, 2x<30mΩ(5A), <150/200µs	20-MDIP	Sie	-	
BTS 903	MOS-P-FET-e	V-MOS, TEMPFET, 200V, 3,6A, 50W, <1,5Ω	86/5Pin	Sie	-	data
BTS 917	MOS-N-FET-e	V-MOS, HITFET, LogL, Low-side Switch 60V, 3,5A, 50W, <0,1Ω(3,5A), 40/70ns	86/5Pin	Sie	-	data
BTS 917 E3062(A)	MOS-N-FET-e	=BTS 917:	30/5Pin			data
BTS 933	MOS-N-FET-e	V-MOS, HITFET, LogL, Low-side Switch 60V, 7A, 90W, <60mΩ(7A), 40/70ns	86/5Pin	Sie	-	data
BTS 933 E3062(A)	MOS-N-FET-e	=BTS 933:	30/5Pin			data
BTS 941	MOS-N-FET-e	V-MOS, HITFET, LogL, Low-side Switch 60V, 12A, 149W, <28mΩ(12A), 40/70ns	86/5Pin	Sie	-	data
BTS 941 E3062(A)	MOS-N-FET-e	=BTS 941:	30/5Pin			data
BTS 949	MOS-N-FET-e	V-MOS, HITFET, LogL, Low-side Switch 60V, 19A, 240W, <22mΩ(19A), 40/70ns	86/5Pin	Sie	-	data
BTS 949 E3062(A)	MOS-N-FET-e	=BTS 949:	30/5Pin			data
BTS 950	MOS-N-FET-e	V-MOS, TEMPFET, 500V, 9A, 125W, <0,8Ω	18/5Pin	Sie	-	data
BTS 6510	MOS-N-FET	V-MOS, PROFET, 40V, 70A, 170W, <6mΩ 230/130µs	30/7Pin	Sie	-	
BTS 7700G	IC	Trilith IC, DC Motor Drv., double high side & dual low-side switch, Vs<42V	MDIP	Inf	-	pdf pinout
BTS 7710G	IC	Trilith IC, DC Motor Drv., double high side & dual low-side switch, Vs<42V	MDIP	Inf	-	pdf pinout
BTS 7710GP	IC	Trilith IC, DC Motor Drv., double high side & dual low-side switch, Vs<42V	SIL	Inf	-	pdf pinout
BTV						
BTV 24/...R	50Hz-Thy	600...1400V, 70A, Igt/Ih<100/<200mA	53b§	Phi	(T 35N..., T 45N...)[4]	data
BTV 34/...G	Triac	=BTW 34/...G	53b§	Phi	-	data
BTV 34/...H	Triac	=BTW 34/...H	53b§	Phi	-	data
BTV 58/...R	GTO-Thy	+600...+1000V, 10A, Igt<200mA	17h§	Phi	-	data
BTV 59/...R	GTO-Thy	+600...+1000V, 15A(Tc=60°), Igt<300mA	66h	Phi	BTV 70/...R	data
BTV 59D/...R	GTO-Thy+Di	=BTV 59/...R: integr. Diode	66h	Phi	BTV 70D/...R	data
BTV 60/...R	GTO-Thy	+850...+1200V, 25A(Tc=70°), Igt<500mA	66h	Phi	-	data
BTV 60D/...R	GTO-Thy+Di	=BTV 60/...R: integr. Diode	66h	Phi	-	data
BTV 70/...R	GTO-Thy	+850...+1200V; 15A(Tc=60°), Igt<300mA	66h	Phi	-	data
BTV 70D/...R	GTO-Thy+Di	=BTV 70/...R: integr. Diode	66h	Phi	-	data
BTV 159/...R	GTO-Thy	+850...+1200V, 15A, Igt<300mA	80	Phi	-	data
BTV 160/...R	GTO-Thy	+850...+1200V, 25A, Igt<300mA	80	Phi	-	data

209

Type	Device	Short Description	Fig.	Manu	Comparision Types	More at
BTW						
BTW 10/...	Triac	50...600V, 3A(Tc=75°), Igt/Ih<50/<25mA	22a§	Tra	TXC01.., T 4706.., BS7-...A, MAC 216-...++	data
BTW 11/...	Triac	50...600V, 6A(Tc=75°), Igt/Ih<50/<50mA	22a§	Tra	TXC 01.., TAG 262-.., TAG 265-.., ++	data
BTW 12/...	Triac	=BTW 11/...	29b§	Tra	BS 6-...B, TW 6N...H, BS 8-...B, TW 10N...H	data
BTW 13/...	Triac	=BTW 11/...	21b§	Tra	BS 6-...A, TW 6N...C, BS 8-...A, TW 8N...C++	data
BTW 14/...	Triac	50...600V, 10A(Tc=75°), Igt/Ih<100/<50mA	22a§	Tra	T 4706.., TAG 265-.., T 4700.., ++	data
BTW 15/...	Triac	=BTW 14/...	29b§	Tra	SC 246.., BS 10-...B, TW 10N...H, ++	data
BTW 16/...	Triac	=BTW 14/...	21b§	Tra	SC 245.., BS 10-...A, TW 10N...C, ++	data
BTW 17/...	Triac	=BTW 14/...	54h	Tra	SC 245.., BS 10-...A, TW 10N...C, ++	data
BTW 18/...	Triac	50...600V, 15A(Tc=75°), Igt/Ih<100/<50mA	29b§	Tra	SC 251.., SC 261.., 2N6157...6159, ++	data
BTW 19/...	Triac	=BTW 18/...	21b§	Tra	SC 250.., TXD 98.., TXE 99.., ++	data
BTW 20/...	Triac	50...600V, 25A(Tc=75°), Igt/Ih<100/<100m	21b§	Tra	TXE 99.., SC 260.., T 6410.., ++	data
BTW 21/...	Triac	=BTW 20/...	29b§	Tra	SC 261.., T 6407.., T 6405.., ++	data
BTW 23/...(RM,RU)	50Hz-Thy	600...1600V, 140A, Igt/Ih<150/<200mA	53b§	Phi	T 51 N...B, T 71 N...B, T 99 N...B, ++	data
BTW 24/...(RM,RU)	50Hz-Thy	600...1600V, 55A, Igt/Ih<150/<200mA	53b§	Phi	T 35 N.., T 45 N...	data
BTW 26/...	Triac	150...600V, 14A(Tc=50°), Igt/Ih<35/<20mA	66	Phi	MAC 525A-.., MAC 25A-.., TAG 725-.., ++	data
BTW 27/...R	50Hz-Thy	100...600V, 10A, Igt/Ih<50/<50mA	22a§	Tho	TAG 675-.., TAG 676-.., BStD 10.., ++	data
BTW 27 S,SA/...	F-Thy	+100...+600V, 7A, Igt/Ih<50/25mA, <6µs	22a§	Phi	(TAG 670S-.., T3S-.., TAG 675S-.., ++)	data
BTW 28(A)/...R	F-Thy	-300...+800V, 35A, Igt/Ih<150/45mA	21b§	Tho	BTW 63/.., CS 15,9-.., TAG 35S-...	data
BTW 30/...(RM,RU)	F-Thy	300...1200V, 24A, Igt/Ih<0,2/0,2A, <6µs	21b§	Phi,Tho	BTW 31/...	data
BTW 30/...RS	F-Thy	=BTW 30/...(RM,RU): <15µS	21b§		BTW 31/...	data
BTW 31/...RM,RU,RW	F-Thy	300...1200V, 31A, Igt/Ih<0,2/0,2A,<12µs	21b§	Phi		data
BTW 32/...(RM,RU)	F-Thy	300...1200V, 55A, Igt/Ih<0,15/0,2A,<25µs	53b§	Phi	T 31F-.., CSF 34-...	data
BTW 33/...(RM,RU)	F-Thy	300...1200V,110A,Igt/Ih<0,15/0,2A,<25µs	53b§	Phi	CS 38-.., CS 78-.., CS 79-.., T 71F...B	data
BTW 34/...(M,G,H,U)	Triac	600...1600V, 55A, Igt/Ih<0,2/<0,2A	53b§	Phi	-	data
BTW 35	50Hz-Thy	500V, 15A, Igt<10mA, <50µs	21b§	Phi	-	data
BTW 36/...RM	50Hz-Thy	200...600V, 31A(Tc=65°C), Igt<60mA		Phi	-	data
BTW 37(V)/...	Triac	400...1200V, 12A, Igt/Ih<0,1/<0,1A	21b§	Phi	BTW 43/...	data
BTW 38/...R	50Hz-Thy	600...1200V, 16A, Igt/Ih<50/<75mA	21b§	Phi	BTW 42/.., BStC03.., CS 5-.., BStD03	data
BTW 39/...	50Hz-Thy	50...1200V, 25A, Igt/Ih<80/20mA	21b§	Tho	BTW45/.., BTW 47/.., CS13-.., T15.1N..++	data
BTW 40/...R	50Hz-Thy	200...800V, 32A, Igt/Ih<75/<75mA	21b§	Phi	BTW 47/..,CS 13-..,T15.1N..,TAG 16N..,++	data
BTW 41/...(G,H)	Triac	400...800V, 40A, Igt/Ih<150/50mA		Phi	-	data
BTW 42/...R	50Hz-Thy	600...1200V, 16A, Igt/Ih<50/<75mA	21b§	Phi	-	data
BTW 43/...(G,H)	Triac	600...1200V, 15A, Igt/Ih<200/<100mA	21b§	Phi	-	data
BTW 44/...(M,U)	Triac	100...600V, 23A, Igt/Ih<0,2/<0,2A	53b§	Phi	-	data
BTW 45/...R	50Hz-Thy	200...1200V, 25A, Igt/Ih<75/<75mA	21b§	Phi	BTW 47/.., CS 13-.., T 15.1N.., CS16..++	data
BTW 46/...R	50Hz-Thy	200...600V, 25A, Igt/Ih<60/<40mA		Phi	-	data
BTW 47/...(RM,RU)	50Hz-Thy	200...1600V, 25A, Igt/Ih<100/<200mA	21b§	Phi	BTW 39/.., BTW 47/..,CS 13..,TAG 16N..++	data
BTW 48/...[Tho]	50Hz-Thy	200...1200V, 50A, Igt/Ih<60/30mA, 50µs	21b§	Tho	CS 23-...	data
BTW 48/...[Phi]	Triac	→≈BTX 94/...	21b§	Phi	→BTX 94/...	data
BTW 49/...	F-Thy	50..800V, 15A, Igt/Ih<200/75mA, <20µs	21b§	Tho	BTW 30/.., BTW 31/.., CS 15,9-.., ++	data
BTW 50/...	50Hz-Thy	100...1200V, 63A, Igt/Ih<150/50mA	21b§	Tho	-	data
BTW 52	Thy	60V, 8A(Tc=65°), Igt/Ih<20/25mA, 17µs	22a§	Itt	BStC0506, TAG 671-100, TAG 676-100, ++	data
BTW 53	Thy	=BTW 52: 120V	22a§	Itt	BStC0513, TAG 671-200, TAG 676-200, ++	data
BTW 54	Thy	=BTW 52: 240V	22a§	Itt	BStC0526, TAG 671-300, TAG 676-300, ++	data
BTW 55	Thy	=BTW 52: 480V	22a§	Itt	BStC0533, TAG 671-500, TAG 676-500, ++	data
BTW 56	Thy	=BTW 52: 600V	22a§	Itt	BStC0540, TAG 671-600, TAG 676-600, ++	data
BTW 58/...	GTO-Thy	+1000...+1500V, 7,5A, Igt<200mA	17h§	Phi,Sie	-	data
BTW 59/...R	GTO-Thy	+1300...+1500V, 12A, Igt<250mA	66	Phi	-	data
BTW 62/...RK	F-Thy	600...1000V, 28A,Igt/Ih<0,25/<0,4A,<4µs	66h	Phi	-	data
BTW 62/...RN	F-Thy	=BTW 62/...RK: <9µs	66h		-	data
BTW 62D/...	F-Thy+Di	=BTW 62/...RK: integr. Diode	66h		-	data
BTW 63/...RK	F-Thy	600...800V, 25A, Igt/Ih<0,25/<0,4A, 4µs	21b§	Phi	BTW 28/.., CS 15,9-.., TAG 35 S-...	data
BTW 63/...RN	F-Thy	=BTW 63/...RK: 6µs	21b§		BTW 28/.., CS 15,9-.., TAG 35 S-...	data
BTW 63/...RP	F-Thy	=BTW 63/...RK: 8µs	21b§		BTW 28/.., CS 15,9-.., TAG 35 S-...	data
BTW 66/...	50Hz-Thy	200...1200V, 25A, Igt/Ih<50/<20mA	64h	Tho	BTW 67/...	data
BTW 66/...N	50Hz-Thy	=BTW 66/..:	64h§		BTW 67/...N	data
BTW 67/...	50Hz-Thy	200...1200V, 40A, Igt/Ih<80/<20mA	64h	Tho	-	data
BTW 67/...N	50Hz-Thy	=BTW 67/..:	64h§		-	data
BTW 68/...	50Hz-Thy	200...1200V, 25A, Igt/Ih<50/<75mA	18h	Tho	BTW 69/...	data
BTW 68/...N	50Hz-Thy	=BTW 68/..:	18h§		BTW 69/...N	data
BTW 69/...	50Hz-Thy	200...1200V, 40A, Igt/Ih<80/<150mA	18h	Tho	-	data
BTW 69/...N	50Hz-Thy	=BTW 69/..:	18h§		-	data
BTW 92/...(R,RM,RU)	50Hz-Thy	200...1600V, 31A, Igt/Ih<100/<200mA	21b§	Phi	CS 16-.., CS 23-...	data
BTX						
BTX 12/...R	50Hz-Thy	100...700V, 20A, Igt/Ih<50/<40mA	53b§	Phi	BTX 13/...R, CS 42-.., CS 50-...	data
BTX 13/...R	50Hz-Thy	100...700V, 48A, Igt/Ih<50/40mA	53b§	Phi	CS 42-.., CS 50-...	data
BTX 18/...	50Hz-Thy	100...500V, 1,6A, Igt/Ih<5/5mA	2a§	Phi	BTX 30/.., T1N.., S 2600.., TAG 612..++	data
BTX 20	50Hz-Thy	200V, 78A(Tc=70°C), Igt/Ih<100/<100mA	53b§	Phi	-	data
BTX 21	50Hz-Thy	=BTX 20: 300V	53b§	Phi	-	data
BTX 22	50Hz-Thy	=BTX 20: 400V	53b§	Phi	-	data

Type	Device	Short Description	Fig.	Manu	Comparision Types	More at
BTX 23	50Hz-Thy	=BTX 20: 500V	53b§	Phi	-	data
BTX 24	50Hz-Thy	=BTX 20: 600V	53b§	Phi	-	data
BTX 25	50Hz-Thy	=BTX 20: 700V	53b§	Phi	-	data
BTX 26	50Hz-Thy	=BTX 20: 800V	53b§	Phi	-	data
BTX 29/..R	50Hz-Thy	200..1800V, 78A(Tc=110°C), Igt<100mA	53b§	Phi	CS 70-...	data
BTX 30/...	50Hz-Thy	50..600V, 1,6A, Igt/Ih<10/<25mA	2a§	Tag	T1N..., S2600..., TAG612-..., TAG613-...,++	data
BTX 31/..R	50Hz-Thy	50..1000V,7A(Tc=95°C), Igt/Ih15/<15mA	21b§	Tag	T 10N...C	data
BTX 31S/..R	F-Thy	=BTX 31/..R: <12µs			TAG 15S-..., T 12F...	data
BTX 32/..R	50Hz-Thy	50..1000V,10A(Tc=125°C)Igt/Ih<25/<25mA	21b§	Tag	T 10N...C, T 12N..., 2N5204..5207, ++	data
BTX 32/..S	F-Thy	=BTX 32/..R: <12µs	21b§		TAG 15S..., T 12F...	data
BTX 33/..R	50Hz-Thy	50..1000V,20A(Tc=125°C)Igt/Ih<50/<50mA	21b§	Phi	TAG15-..., TAG14N.., BTW39/.., CS13-..,++	data
BTX 33/..S	F-Thy	=BTX 33/..R: <12µs	21b§		T 12F-...	data
BTX 35/..R	50Hz-Thy	500..800V,8,2A(Tc=85°C),Igt/Ih<60/10mA	21b§	Phi	TAG 14N-..., BTW 39/..., BStE41..., ++	data
BTX 36/..R	50Hz-Thy	500..800V,11A(=85°C),Igt/Ih<40/10mA	21b§	Phi	2N681..692, T 12N..., 2N5204..5207, ++	data
BTX 37/..R	50Hz-Thy	500..800V,32A(Tc=85°C),Igt/Ih<80/10mA	53b§	Phi	CS 42-..., CS 50-..., CS 70-...	data
BTX 38/..R	50Hz-Thy	500..800V,62A(Tc=85°C),Igt/Ih<70/10mA	53b§	Phi	CS 42-..., CS 50-..., CS 70-...	data
BTX 41/..R	50Hz-Thy	200..1800V, 175A, Igt/Ih<300/<300mA	53b§	Phi	BStN 35/..., BStP 36/..., BStP 35/...	data
BTX 44/..R	50Hz-Thy	200..1800V, 55A, Igt/Ih<100/<100mA	53b§	Phi	CS 50-..., CS 70-...	data
BTX 45/..R	50Hz-Thy	200..1800V, 110A, Igt/Ih<100/<100mA	53b§	Phi	CS 50-..., CS 70-...	data
BTX 46/..R	50Hz-Thy	200..1800V, 150A, Igt/Ih<300/<300mA	53b§	Phi	BStN 35/..., BStP 36/..., BStP 35/...	data
BTX 47/..R	50Hz-Thy	1000..1400V,11A(Tc=85°),Igt/Ih<65/10mA	21b§	Phi	BTW47/..., TAG16N.., BTW92/.., CS16-..,++	data
BTX 48/..R	50Hz-Thy	1000..1400V,16A(Tc=85°),Igt/Ih<65/10mA	21b§	Phi	BTW47/..., TAG16N.., BTW92/.., CS16-..,++	data
BTX 49/..R	50Hz-Thy	600..1400V,60A(Tc=85°),Igt<100mA, 50µs	53b§	Phi	CS 42-..., CS 50-..., CS 70-...	data
BTX 50/..R	50Hz-Thy	600..1400V, 70A, Igt/Ih<100/<150mA	53b§	Phi	CS 42-..., CS 50-..., CS 70-...	data
BTX 51/..R	50Hz-Thy	500..800V, 70A, Igt/Ih<70/10mA, 20µs	53b§	Phi	-	data
BTX 52	50Hz-Thy	100V, 8A(Tc=65°C), Igt/Ih<0,2/<2,5mA	23a§	Sgs	-	data
BTX 53	50Hz-Thy	=BTX 52: 200V	23a§	Sgs	-	data
BTX 54	50Hz-Thy	=BTX 52: 300V	23a§	Sgs	-	data
BTX 55	50Hz-Thy	=BTX 52: 400V	23a§	Sgs	-	data
BTX 57	50Hz-Thy	100V, 8A(Tc=85°C), Igt/Ih<15/<25mA	23a§	Sgs	-	data
BTX 58	50Hz-Thy	=BTX 57: 200V	23a§	Sgs	-	data
BTX 59	50Hz-Thy	=BTX 57: 300V	23a§	Sgs	-	data
BTX 60	50Hz-Thy	=BTX 57: 400V	23a§	Sgs	-	data
BTX 64/..R	F-Thy	100..600V,8,5A(Tc=85°C),Igt/Ih<65/10mA	21b§	Phi	C 234-..., MCR 1718-..., TAG 12F..., ++	data
BTX 65/..R	F-Thy	100..600V, 12A(Tc=85°C),Igt/Ih<65/10mA	21b§	Phi	C 234-..., MCR 1718-..., TAG 12F..., ++	data
BTX 66/..R	F-Thy	100..600V, 32A(Tc=85°C),Igt/Ih<80/10mA	53b§	Phi	BStH 37.., BStH 34.., BStL 37.., ++	data
BTX 67/..R	F-Thy	100..600V, 62A(Tc=85°C),Igt/Ih<80/10mA	53b§	Phi	BStL 34.., BStH 34.., BStL 37.., ++	data
BTX 68/..R	50Hz-Thy	500..1000V,6,4A(Tc=85°C)Igt/Ih<30/10mA	21b§	Phi	BStC 03.., BStD 03.., CS 5-.., CS 8-...	data
BTX 70/..R	50Hz-Thy	50..1000V, 15A, Igt/Ih<50/<50mA	21b§	Tag	TAG9N-..., TAG15-..., BTW39/.., BTW45/...	data
BTX 70/..S	F-Thy	=BTX 70/..R: <12µs	21b§		TAG 15S..., T 12F...	data
BTX 71/..R	50Hz-Thy	50..1000V, 7A, Igt/Ih<15/<15mA	21b§	Tag	T 10N...C, T 12N...	data
BTX 71/..S	F-Thy	=BTX 71/..: <12µs	21b§		TAG 15S..., T 12F...	data
BTX 72/..R	50Hz-Thy	50..1000V,10A(Tc=95°C),Igt/Ih<25/<25mA	21b§	Tag	T 10N...C, T 12N...	data
BTX 72/..S	F-Thy	=BTX 72/..: <12µs	21b§		T 12F.., TAG 15S...	data
BTX 73/..(R,A,B)	50Hz-Thy	50..1000V,25A(Tc=60°C),Igt/Ih<50/<50mA	21b§	Tag	T 17N.., BTW 48/.., CS 23-...	data
BTX 73/..S	F-Thy	=BTX 73/..: <12µs	21b§		T 12F-...	data
BTX 74/..R	50Hz-Thy	50..1000V, 15A, Igt/Ih<50/<50mA	21b§	Tag	TAG 9N-.., TAG 15-.., BTW 39/.., ++	data
BTX 74/..S	F-Thy	50..1000V, 15A, Igt/Ih<50/<50mA <12µs	21b§		TAG 15S-.., T 12F...	data
BTX 75/..R	50Hz-Thy	100..400V,8,5A(Tc=110°C)Igt/Ih<65/10mA	21b§	Phi	TAG 9N-.., 2N1842..1850, TAG 15-.., ++	data
BTX 76/..R	50Hz-Thy	100..400V,12A(Tc=110°C),Igt/Ih<40/10mA	21b§	Phi	MCR 3918-.., 2N5168..5171, T 12N.., ++	data
BTX 81/..R(RM,RU)	50Hz-Thy	100..800V,20A(Tc=85°C),Igt/Ih<80/100mA	21b§	Phi	BTW 40/.., BStE41..M, T 17N.., CS16..,++	data
BTX 82/..R(RM,RU)	50Hz-Thy	100..800V,26A(Tc=85°C),Igt/Ih<80/100mA	21b§	Phi	BTW 48/.., CS 23-...	data
BTX 92/..R	50Hz-Thy	600.1600V, 16A, Igt/Ih<150/<200mA	21b§	Phi	BTW 47/.., CS 23-.., BTW 92/...	data
BTX 94/..(H,J)	Triac	100..1200V, 25A(Tc=85°), Igt<200mA	21b§	Phi		data
BTX 95/..R	50Hz-Thy	500..800V,10A(Tc=60°C),Igt/Ih<50/30mA	21b§	Phi,Tho	TAG 15-.., BTW 39/.., T 15.1N.., ++	data
BTX 0605	Triac	50V, 6A(Tc=80°C), Igt/Ih<50/<50mA	54b	Tra	SC 245A3, T 4120F, T 4121F, SC 250A3, ++	data
BTX 0610	Triac	=BTX 0605: 100V	54b		SC 245B3, T 4120E, T 4121E, SC 250B3, ++	data
BTX 0620	Triac	=BTX 0605: 200V	54b		SC 245A3, T 4120E, T 4221E, SC 250A3, ++	data
BTX 0640	Triac	=BTX 0605: 400V	54b		SC 245D3, T 4120E, T 4121E, SC 250D3, ++	data
BTX 0660	Triac	=BTX 0605: 600V	54b		SC 245M3, T 4120M, T 4121M, SC 250M3, ++	data
BTX 1005	Triac	50V, 10A(Tc=80°), Igt/Ih<100/<100mA	54b	Tra	SC 245B3, T 4120F, T 4121F, SC 250B3, ++	data
BTX 1010	Triac	=BTX 1005: 100V	54b		SC 245B3, T 4120E, T 4121E, SC 250B3, ++	data
BTX 1020	Triac	=BTX 1005: 200V	54b		SC 245B3, T 4120E, T 4121E, SC 250B3, ++	data
BTX 1040	Triac	=BTX 1005: 400V	54b		SC 245D3, T 4120E, T 4121E, SC 250D3, ++	data
BTX 1060	Triac	=BTX 1005: 600V	54b		SC 245M3, T 4120M, T 4121M, SC 250M3, ++	data
BTX 1605	Triac	50V, 16A(Tc=80°), Igt/Ih<100/<100mA	54b	Tra	SC 260B3, T 6421F, 2N6163..6165	data
BTX 1610	Triac	=BTX 1605: 100V	54b		SC 260B3, T 6421B, 2N6163..6165	data
BTX 1620	Triac	=BTX 1605: 200V	54b		SC 260B3, T 6421B, 2N6163..6165	data
BTX 1640	Triac	=BTX 1605: 400V	54b		SC 260D3, T 6421D, 2N6164..6165	data
BTX 1660	Triac	=BTX 1605: 600V	54b		SC 260M3, T 6421M, 2N6165	data
BTX 2505	Triac	50V, 25A(Tc=80°), Igt/Ih<100/<100mA	54b	Tra	SC 260B3, T 6421F, 2N6163..6165	data
BTX 2510	Triac	=BTX 2505: 100V	54b		SC 260B3, T 6421B, 2N6163..6165	data
BTX 2520	Triac	=BTX 2505: 200V	54b		SC 260B3, T 6421B, 2N6163..6165	data
BTX 2540	Triac	=BTX 2505: 400V	54b		SC 260D3, T 6421D, 2N6164..6165	data
BTX 2560	Triac	=BTX 2505: 600V	54b		SC 260M3, T 6421M, 2N6165	data

Type	Device	Short Description	Fig.	Manu	Comparision Types	More at
BTY						
BTY 10	50Hz-Thy	50V, 200A(Tc=125°C), Igt/Ih<200/<200mA		Aeg	-	data
BTY 11	50Hz-Thy	=BTY 10: 100V		Aeg	-	data
BTY 13	50Hz-Thy	=BTY 10: 400V		Aeg	-	data
BTY 15	50Hz-Thy	=BTY 10: 500V		Aeg	-	data
BTY 16	50Hz-Thy	=BTY 10: 600V		Aeg	-	data
BTY 20	50Hz-Thy	50V, 90A(Tc=125°C), Igt/Ih<80/<100mA	76b	Aeg	T 99N400E, T 159N400E	data
BTY 21	50Hz-Thy	=BTY 20: 100V	76b	Aeg	T 99N400E, T 159N400E	data
BTY 22	50Hz-Thy	=BTY 20: 200V	76b	Aeg	T 99N400E, T 159N400E	data
BTY 23	50Hz-Thy	=BTY 20: 400V	76b	Aeg	T 99N400E, T 159N400E	data
BTY 24	50Hz-Thy	=BTY 20: 500V	76b	Aeg	T 99N600E, T 159N600E	data
BTY 25	50Hz-Thy	=BTY 20: 600V	76b	Aeg	T 99N600E, T 159N600E	data
BTY 28	50Hz-Thy	100V, 6A(Tc=60°C), Igt/Ih<15/<4mA	21b§	Phi	T 6N100C, T 8N100C, BStD4126, T 10N100C	data
BTY 29	50Hz-Thy	=BTY 28: 150V	21b§	Phi	T 6N200C, T 8N200C, BStD4126, T 10N200C	data
BTY 30	50Hz-Thy	=BTY 28: 250V	21b§	Phi	T 6N300C, T 8N300C, BStD4126, T 10N300C	data
BTY 31	50Hz-Thy	=BTY 28: 300V	21b§	Phi	T 6N300C, T 8N300C, BStD4126, T 10N300C	data
BTY 34/...R	50Hz-Thy	100...400V, 6,4A(Tc=110°C), Igt<30mA	21b§	Phi	2N5168...5171, MCR 3918-..., 2N681...692	data
BTY 34	50Hz-Thy	100V, 6A(Tc=60°C), Igt/Ih<15/<2mA	21b§	Phi	T 6N100C, T 8N100C, BStD4126, T 10N100C	data
BTY 35	50Hz-Thy	=BTY 34: 150V	21b§	Phi	T 6N200C, T 8N200C, BStD4126, T 10N200C	data
BTY 36	50Hz-Thy	=BTY 34: 200V	21b§	Phi	T 6N200C, T 8N200C, BStD4126, T 10N200C	data
BTY 37	50Hz-Thy	=BTY 34: 250V	21b§	Phi	T 6N300C, T 8N300C, BStD4126, T 10N300C	data
BTY 38	50Hz-Thy	=BTY 34: 300V	21b§	Phi	T 6N300C, T 8N300C, BStD4126, T 10N300C	data
BTY 39	50Hz-Thy	=BTY 34: 400V	21b§	Phi	T 6N400C, T 8N400C, BStD4126, T 10N400C	data
BTY 41	50Hz-Thy	50V, 9,6A(Tc=60°C), Igt/Ih<80/<20mA	21b§	Phi	2N1843...1850, TAG 9N400, TAG 15-100	data
BTY 43	50Hz-Thy	=BTY 41: 150V	21b§	Phi	2N1845...1850, TAG 9N400, TAG 15-200	data
BTY 44	50Hz-Thy	=BTY 41: 200V	21b§	Phi	2N1846...1850, TAG 9N400, TAG 15-200	data
BTY 46	50Hz-Thy	=BTY 41: 300V	21b§	Phi	2N1848...1850, TAG 9N400, TAG 15-300	data
BTY 47	50Hz-Thy	=BTY 41: 400V	21b§	Phi	2N1850, TAG 9N400, TAG 15-400	data
BTY 50	50Hz-Thy	50V, 16A(Tc=60°C), Igt/Ih<40/<10mA	21b§	Phi	2N682...692, T 12N100, C 228A, C 35A, ++	data
BTY 51	50Hz-Thy	=BTY 50: 100V	21b§	Phi	2N683...692, T 12N100, C 228A, C 35A, ++	data
BTY 52	50Hz-Thy	=BTY 50: 150V	21b§	Phi	2N684...692, T 12N200, C 228B, C 35B, ++	data
BTY 53	50Hz-Thy	=BTY 50: 250V	21b§	Phi	2N686...692, T 12N300, C 228C, C 35C, ++	data
BTY 54	50Hz-Thy	=BTY 50: 300V	21b§	Phi	2N687...692, T 12N300, C 228C, C 35C, ++	data
BTY 57	50Hz-Thy	50V, 16A(Tc=60°C), Igt/Ih<40/<10mA	21b§	Phi	2N682...692, T 12N100, C 228A, C 35A, ++	data
BTY 58	50Hz-Thy	=BTY 57: 100V	21b§	Phi	2N683...692, T 12N100, C 228A, C 35A, ++	data
BTY 59	50Hz-Thy	=BTY 57: 150V	21b§	Phi	2N684...692, T 12N200, C 228B, C 35B, ++	data
BTY 60	50Hz-Thy	=BTY 57: 200V	21b§	Phi	2N685...692, T 12N200, C 228B, C 35B, ++	data
BTY 61	50Hz-Thy	=BTY 57: 250V	21b§	Phi	2N686...692, T 12N300, C 228C, C 35C, ++	data
BTY 62	50Hz-Thy	=BTY 57: 300V	21b§	Phi	2N687...692, T 12N300, C 228C, C 35C, ++	data
BTY 64	50Hz-Thy	50V, 110A(Tc=60°C), Igt/Ih<70/<20mA	21b§	Phi	-	data
BTY 65	50Hz-Thy	=BTY 64: 100V	21b§	Phi	-	data
BTY 66	50Hz-Thy	=BTY 64: 150V	21b§	Phi	-	data
BTY 67	50Hz-Thy	=BTY 64: 200V	21b§	Phi	-	data
BTY 68	50Hz-Thy	=BTY 64: 250V	21b§	Phi	-	data
BTY 69	50Hz-Thy	=BTY 64: 300V	21b§	Phi	-	data
BTY 70	50Hz-Thy	=BTY 64: 400V	21b§	Phi	-	data
BTY 79/...R	50Hz-Thy	100...1000V,10A(Tc=85°C)Igt/Ih<30/<75mA	21b§	Phi	BTW 42/..., BStC03..., CS 5-..., BStD03...	data
BTY 79/...R/05	50Hz-Thy	=	21b§	Phi	BTW 42/..., BStC03..., CS 5-..., BStD03...	data
BTY 79A/05	50Hz-Thy	50V, 10A(Tc=65°C), Igt/Ih<20/<25mA	21b§	Itt	BStD4126, S 6210A, T 10N100C	data
BTY 79A/10	50Hz-Thy	=BTY 79A/05: 100V	21b§		BStD4126, S 6210A, T 10N100C	data
BTY 79A/20	50Hz-Thy	=BTY 79A/05: 200V	21b§		BStD4126, S 6210B, T 10N200C	data
BTY 79A/30	50Hz-Thy	=BTY 79A/05: 300V	21b§		BStD4126, S 6210D, T 10N300C	data
BTY 79A/40	50Hz-Thy	=BTY 79A/05: 400V	21b§		BStD4126, S 6210D, T 10N400C	data
BTY 79A/50	50Hz-Thy	=BTY 79A/05: 500V	21b§		BStD4133, S 6210M, T 10N500C	data
BTY 80	50Hz-Thy	=BTY 79/200R	21b§	Phi	→BTY 79/...R	data
BTY 81	50Hz-Thy	=BTY 79/400R	21b§	Phi	→BTY 79/...R	data
BTY 84	50Hz-Thy	=BTY 79/100R	21b§	Phi	→BTY 79/...R	data
BTY 85	50Hz-Thy	=BTY 79/200R	21b§	Phi	→BTY 79/...R	data
BTY 86	50Hz-Thy	=BTY 79/300R	21b§	Phi	→BTY 79/...R	data
BTY 87/...R(RU,RM)	50Hz-Thy	100...800V, 10A(Tc=85°C),Igt/Ih<65/10mA	21b§	Phi	TAG15-..., BTW39/..., CS13-..., TAG16N...,++	data
BTY 88	50Hz-Thy	=BTY 91/100R	21b§	Phi	→BTY 91/...R	data
BTY 89	50Hz-Thy	=BTY 91/200R	21b§	Phi	→BTY 91/...R	data
BTY 90	50Hz-Thy	=BTY 91/300R	21b§	Phi	→BTY 91/...R	data
BTY 91/...R(RU,RM)	50Hz-Thy	100...1200V,14A(Tc=85°C),Igt/Ih<40/25mA	21b§	Phi	T 12N.., 2N5204...5207	data
BTY 92	50Hz-Thy	=BTY 95/100R	53b§	Phi	→BTY 95/...R	data
BTY 93	50Hz-Thy	=BTY 95/200R	53b§	Phi	→BTY 95/...R	data
BTY 94	50Hz-Thy	=BTY 95/300R	53b§	Phi	→BTY 95/...R	data
BTY 95/...R	50Hz-Thy	100...800V, 32A(Tc=85°C),Igt/Ih<80/10mA	53b§	Phi	CS 42-..., CS 50-..., CS 70-...	data
BTY 96	50Hz-Thy	=BTY 99/100R	53b§	Phi	→BTY 99/...R	data
BTY 97	50Hz-Thy	=BTY 99/200R	53b§	Phi	→BTY 99/...R	data
BTY 98	50Hz-Thy	=BTY 99/300R	53b§	Phi	→BTY 99/...R	data
BTY 99/...R	50Hz-Thy	100...800V, 62A(Tc=85°C),Igt/Ih<70/10mA	53b§	Phi	CS 70-...	data

Type	Device	Short Description	Fig.	Manu	Comparision Types	More at
BTZ						
BTZ 10	50Hz-Thy	50V, 24A(Tc=125°C), Igt/Ih<50/<30mA	76	Aeg	-	data
BTZ 11	50Hz-Thy	=BTZ 10: 100V	76	Aeg	-	data
BTZ 12	50Hz-Thy	=BTZ 10: 200V	76	Aeg	-	data
BTZ 13	50Hz-Thy	=BTZ 10: 400V	76	Aeg	-	data
BTZ 15	50Hz-Thy	=BTZ 10: 500V	76	Aeg	-	data
BTZ 16	50Hz-Thy	=BTZ 10: 600V	76	Aeg	-	data
BTZ 18	50Hz-Thy	200V, 6A(Tc=60°C), Igt/Ih<15/4mA	21b§	Aeg	T 6N200C, T 8N200C, BStD4126	data
BTZ 19	50Hz-Thy	=BTZ 18: 400V	21b§	Aeg	T 6N400C, T 8N400C, BStD4126	data
BTZ 21	50Hz-Thy	400V, 16A(Tc=60°C), Igt/Ih<40/10mA	21b§	Aeg	2N688..692, T 12N400, C 35D, MCR 64-6	data
BTZ 35	50Hz-Thy	50V, 3,5A, Igt/Ih<40/10mA	21b§	Aeg	2N1843...1850, TAG 9N400, TAG 15-100, ++	data
BTZ 36	50Hz-Thy	=BTZ 35: 100V	21b§	Aeg	2N1844...1850, TAG 9N400, TAG 15-100, ++	data
BTZ 37	50Hz-Thy	=BTZ 35: 200V	21b§	Aeg	2N1846...1850, TAG 9N400, TAG 15-200, ++	data
BTZ 38	50Hz-Thy	=BTZ 35: 350V	21b§	Aeg	2N1850, TAG 9N400, TAG 15-400,TAG 16N400	data
BTZ 39	50Hz-Thy	=BTZ 35: 450V	21b§	Aeg	2N1850, TAG 9N600, TAG 15-500,TAG 16N600	data
BU						
BU	Si-N	=2SD1005-BU (Typ-Code/Stempel/marking)	39	Nec	→2SD1005	data
BU	Si-Di	=BAT 54T1 (Typ-Code/Stempel/marking)	71(2,7mm)	Ons	→BAT 54T1	
BU	Si-P	=BCX 71RG (Typ-Code/Stempel/marking)	35	Sie,Val	→BCX 71RG	data
BU	Z-Di	=BZX 399-C27 (Typ-Code/Stempel/marking)	71(1,7mm)	Phi	→BZX 399-...	data
BU	Z-Di	=P4 SMA-36 (Typ-Code/Stempel/marking)	71(5x2,5)	Fag	→P4 SMA-36	data
BU	Z-Di	=SMBJ 20C (Typ-Code/Stempel/marking)	71(5x3,5)	Mop	→SMBJ ...	data
BU 4 S01...BU 4Sxx	CMOS-Logic	Standard CMOS-Logic 4000-Family	MDIP	Rhm	-	pdf
BU 100	Si-N	TV-HA(90°), 150/60V, 10A, 15W(Tc=75°)	23a§	Sgs	BU 109, BUY 69C, BUY 70C, (BU 606...608)[7]	data
BU 100 A	Si-N	=BU 100: 150/100V, 25W(Tc=100°)	23a§		BU 109, BUY 69C, BUY 70C, (BU 606...608)[7]	data
BU 102	Si-N	TV-HA(110°), 400/150V, 7A, 50W(Tc=50°)	23a§	Sgs	BU 104, BU 606, BU 608	data
BU 103	Si-N	S-L, TV-VA, 120/100V, 1A, 15W(Tc=25°)	2a§	Tho	BU 125(S), BUY 41, 2SC1860	data
BU 103 A	Si-N	TV-VA(110°), 120/100V, 1A, 30W	22a§		BUX 67, 2SC782...783, 2SD610	data
BU 104	Si-N	TV-HA(110°), 400/150V, 7A, 85W	23a§	Tho	BU 606, BU 608	data
BU 104 D	Si-N+Di	=BU 104: + integr. Damper-Diode	23a§		BU 606D, BU 608D	data
BU 104 DP	Si-N+Di	=BU 104 P: integr. Damper-Diode	17j§		BU 406D, BU 408D, 2SC3176	data
BU 104 P	Si-N	=BU 104: 50W	17j§		BU 406, BU 408, 2SC3175, 2SC3591	data
BU 105	Si-N	TV-HA, 1500V, 2,5A, 10W(Tc=90°)	23a§	Aeg,Phi,++	BU 205, BU 208(A), 2SC2928, 2SD350(A),++	data
BU 106	Si-N	TV-HA, 325/140V, 10A, 50W	23a§	Rca,Sgs,Tix	BU 109, BUY 69C, BUY 70C, (BU 606...608)[7]	data
BU 107	Si-N	TV-HA, 300/120V, 10A, 50W	23a§	Rca,Sgs,Tix	BU 109, BUY 69C, BUY 70C, (BU 606...608)[7]	data
BU 108	Si-N	CTV-HA, 1500/750V,5A(ss),12,5W(Tc=95°)	23a§	Aeg,Phi,++	BU 208(A), 2SC2928, 2SD350(A), 2SD820,++	data
BU 109	Si-N	TV-HA(110°), 330/120V, 10A, 85W	23a§	Tho	BUY 69C, BUY 70C, (BU 606...608)[7]	data
BU 109 D	Si-N+Di	=BU 109: integr. Damper-Diode	23a§		(BU 104D, BU 606D...608D)[7]	data
BU 109 DP	Si-N+Di	=BU 109P: integr. Damper-Diode	17j§		(BU 406D...408D, 2SC3176)[7]	data
BU 109 NP	Si-N	≈BU 109	23a§		(→BU 109)	
BU 109 P	Si-N	=BU 109: 50W	17j§		(BU 406...408, 2SC3175, 2SC3591)[7]	data
BU 110	Si-N	TV-HA, 330/150V, 10A, 60W(Tc=75°)	23a§	Sie	BU 109, BUY 69C, BUY 70C, (BU 606...608)[7]	data
BU 111	Si-N	S-L, TV-SN, 500/300V, 6A, 50W(Tc=75°)	23a§	Sie	BU 326(A), BU 426(A), BU 526	data
BU 112	Si-N	CTV-HA(90°), 550V, 10A, 85W(Tc=30°)	23a§	Tho	BU 526, BU 626A, BUY 69C	data
BU 113(S)	Si-N	TV-HA(110°), 700V, 10A, 85W(Tc=30°)	23a§	Tho	BU 526, BU 626A, BUY 69B	data
BU 114	Si-N	S-L, TV-SN, 350/225V, 6A, 50W(Tc=75°)	23a§	Sie	BU 104, BU 109, BU 526, BU 606...608	data
BU 115	Si-N	TV-HA, 800/600V, 15A, 50W(Tc=75°)	23a§	Sgs	(BU 526, BU 626A)[7]	data
BU 116	Si-N	=BU 115: 400/300V	23a§	Sgs	(BU 526, BU 626A)[7]	data
BU 117	Si-N	=BU 115: 250/200V	23a§	Sgs	(BU 526, BU 626A)[7]	data
BU 118	Si-N+Di	TV-HA, 400/200V, 7A, 60W	17j§	Phi	BU 104DP, BU 109DP, BU 406D, BU 408D, ++	data
BU 120	Si-N	S-L, TV-HA, 400/200V, 10A, 50W(Tc=75°)	23a§	Rca,Sgs	BUY 69C, BUY 70C, (BU 606, BU 608)[7]	data
BU 121	Si-N	S-L, TV-HA, 400/200V, 10A, 50W(Tc=75°)	23a§	Sgs	BUY 69C, BUY 70C, (BU 606, BU 608)[7]	data
BU 122	Si-N	TV-VA, 250/150V, 5A, 67W(Tc=75°)	23a§	Sgs	BU 104, BU 109, BU 606...608	data
BU 123	Si-N	=BU 122: 180/120V, 50W(Tc=75°)	23a§	Sgs	BU 104, BU 109, BU 606...608	data
BU 124	Si-N	TV-HA, 350/150V, 10A, 50W	18j§	Tix	(BU 426, BU 926)[7]	data
BU 124 A	Si-N	=BU 124: 400/150V	18j§		(BU 426, BU 926)[7]	data
BU 125	Si-N	S-L, TV-HA, 130/60V, 7A, 10W(Tc=50°)	2a§	Sgs	BUY 47...48, BUY 68, BUY 81,(BU 406..408)[6]	data
BU 125 S	Si-N	=BU 125: 250/150V, 3A	2a§		BUY 41, BUX 49, (BU 406...408)[6]	data
BU 126(A,S,T)	Si-N	S-L, TV-SN, 750V, 3A, 50W	23a§	EUR,Tos	BU 326, BU 426, BU 526, BU 926	data
BU 127	Si-N	S-L, 200/120V, 10A, 62W, 70MHz	23a§	Sgs	BU 109, BUY 18, BUY 69C, BUY 70C	data
BU 128	Si-N	=BU 127: 300/200V, 80MHz	23a§	Sgs	BU 109, BUY 18, BUY 69C, BUY 70C	data
BU 129	Si-N	TV-HA(110°), 400V, 5A, 25W(Tc=100°)	23a§	Tho	BU 104, BU 606, BU 608	data
BU 130	Si-N	TV-HA, 330/150V, 10A, 15W(Tc=100°)	23a§	Phi	BU 109, BUY 69C, BUY 70C, (BU 606...608)[7]	data
BU 131	Si-N	S-L, TV-HA, 750/300V, 10A, 40W(Tc=60°)	23a§	Phi	BU 526, BU 626A, BUY 69B, BUY 70B	data
BU 132	Si-N	TV-VA, 800/600V, 1A, 15W(Tc=97°)	23a§	Phi	BU 126, 2SC1101, 2SC1167	data
BU 133	Si-N	S-L, 750/250V, 3A, 30W(Tc=50°), 8MHz	23a§	Mot,Phi,Rca	BU 326, BUS 11, BUX 82, 2SC3091, 2SC3155	data
BU 134	Si-N	S-L, TV-SN, 500/350V, 4A, 85W	23a§	Tho	BU 326, BU 426, BU 526, BU 926	data
BU 135	Si-N	S-L, 500/250V, 3A, 30W(Tc=50°), 8MHz	23a§	Mot	BU 126, BU 326, BU 426, BUX 45	data
BU 136	Si-N	=BU 135: 600/250V, 7A	23a§	Mot	BU 326, BU 426, BU 526, BU 926	data
BU 137	Si-N	S-L, 1000/1000V, 12A, 70W	23a§	Tix	BU 157, BU 626A, BUS 13A, 2SD1094	data

Type	Device	Short Description	Fig.	Manu	Comparision Types	More at
BU 137 A	Si-N	=BU 137: 1200/1000V, 10A	23a§		BU 157, MJ 8504, 2SC3061, 2SD1279	data
BU 138	Si-N+Di	TV-HA, 160/125V, 10A, 60W(Tc=95°)	23a§	Tho	BU 109D, (BU 606D...608D)[7]	data
BU 139	Si-N+Di	=BU 138: 200/150V	23a§	Tho	BU 109D, (BU 606D...608D)[7]	data
BU 140	Si-N+Di	=BU 138: 280/175V	23a§	Tho	BU 109D, (BU 606D...608D)[7]	data
BU 141	Si-N+Di	=BU 138: 330/175V	23a§	Tho	BU 109D, (BU 606D...608D)[7]	data
BU 142	Si-N	S-L, 900/350V, 12A, 70W	23a§	Tho	BUS 13A, BUW 46, (BU 626A, 2SD1094)[7]	data
BU 143	Si-N	=BU 142: 800/350V	23a§	Tho	BUS 13, BUW 45...46, (BU 626A, 2SD1094)[7]	data
BU 144	Si-N	=BU 142: 700/350V	23a§	Tho	BUS 13, BUW 45...46, (BU 626A, 2SD1094)[7]	data
BU 157	Si-N	TV-HA, 1500/650V, 12A, 70W	23a§	Tix	BU 2525A, (BU 508(A), U 908)[7]	data
BU 180	Si-N-Darl+Di	TV-HA, 320V, 10A, 50W, B>200	18j§	Tix	(BU 284, BU 289, BU 826)[7]	data
BU 180 A	Si-N-Darl+Di	=BU 180: 400V	18j§		(BU 284, BU 826)[7]	data
BU 181	Si-N-Darl	TV-HA/SN, 600V, 10A, 65W, B>200	18j§	Tix	-	data
BU 181 A	Si-N-Darl	=BU 181: 800V	18j§		-	data
BU 184	Si-N-Darl+Di	TV-HA/SN, 400/200V, 8A, 60W	17j§	Tho	BU 806	data
BU 189	Si-N-Darl+Di	=BU 184: 330/150V	17j§	Tho	BU 806...807	data
BU 204(A)	Si-N	CTV-HA, 1300/600V, 2,5A, 10W(Tc=90°)	23a§	EUR,Tos	BU 207(A)...208(A), 2SC2928, 2SD350(A),++	data
BU 205(A)	Si-N	=BU 204: 1500/700V	23a§	EUR,Tos	BU 208(A), 2SC2928, 2SD350(A), 2SD820,++	data
BU 206	Si-N	=BU 204: 1700/800V	23a§	EUR,Tos	BU 209(A), 2SD784...785	data
BU 207(A)	Si-N	CTV-HA, 1300/600V, 5A, 12,5W(Tc=95°)	23a§	EUR,Hit,Tos	BU 208(A), 2SC2928, 2SD350(A), 2SD820,++	data
BU 208(A)	Si-N	=BU 207: 1500/700V	23a§	EUR,Hit,Tos	BU 508(A), 2SC2928, 2SD350(A), 2SD820,++	data
BU 208A	Si-N	TV, Horizontal Deflection Transistor	23a§	Ons		data
BU 208 D	Si-N+Di	=BU 208: integr. Damper-Diode	23a§		BU 508D, BU 800, 2SD1171...1175	data
BU 209(A)	Si-N	=BU 207: 1700/800V, 4A	23a§	EUR,Hit,Tos	2SD784...785	data
BU 210	Si-N	S-L, TV-HA, 400/250V, 12A, 85W(Tc=45°)	23a§	Sie	BUW 74, 2SC2123, (BU 626A, BUY 69C)[7]	data
BU 211	Si-N	=BU 210: 600/300V	23a§	Sie	BUW 75, 2SC2123, (BU 626A, BUY 69B)[7]	data
BU 212	Si-N	=BU 210: 750/350V	23a§	Sie	BUW 76...77, 2SC2123, (BU 626A, BUY 69B)[7]	data
BU 213	Si-N	S-L, 150/60V, 7,5A, 38W(Tc=45°)	22a§	Fer	BUW 64A...B, 2SC1863...1864, 2SC2334	data
BU 214	Si-N	S-L, 150/60V, 10A, 50W(Tc=45°)	22a§	Fer	BUV 27(A)...28, 2SC2867	data
BU 215	Si-N	S-L, 150/60V, 20A, 60W(Tc=45°)	22a§	Fer	BUV 26	data
BU 216	Si-N	S-L, 150/60V, 7,5A, 50W(Tc=45°)	23a§	Fer	BUW 86...87, 2SC2769	data
BU 217	Si-N	S-L, 150/60V, 10A, 60W(Tc=45°)	23a§	Fer	BUW 70, BUY 55, 2SC2769	data
BU 218	Si-N	S-L, 150/60V, 20A, 115W(Tc=45°)	23a§	Fer	BUV 10, BUW 57...58, BUX 10, BUX 70, ++	data
BU 221	Si-N	S-L, TV-HA, 800, 15A, 100W	23a§	Sgs	BUS 13(A), BUW 45...46, 2SC2123	data
BU 222	Si-N	S-L, 450/350V, 6A, 75W, >10MHz	23a§	Mot	BUX 15, BUX 18C, BUX 44, 2SC2656, ++	data
BU 222 A	Si-N	=BU 222: 525/475V	23a§		BUX 15, 2SC2536, 2SC3041, 2SD640, ++	data
BU 223	Si-N	S-L, 450/400V, 10A, 125W, 7,5MHz	23a§	Mot	BUW 24, BUW 72, BUX 14, 2SC2625, ++	data
BU 223 A	Si-N	=BU 223: 525/475V	23a§		BUW 34, BUY 69, BUW 25, 2SC2441, ++	data
BU 225	Si-N	TV-HA, 2200/800V, 2A, 10W(Tc=80°)	23a§	Aeg	BUY 71, 2SD621, 2SD838	data
BU 226	Si-N	TV-HA, 2000/800V, 1,5A, 32W	23a§	Aeg	BU 225, BUY 71, 2SD621, 2SD838	data
BU 284	Si-N-Darl+Di	=BU 184: 90W	18j§	Tho	BU 180A	data
BU 287	Si-N	TV-SN, 1300/600V, 8A	23a§	Aeg	BU 908	data
BU 289	Si-N-Darl+Di	=BU 189: 90W	18j§	Tho	BU 180	data
BU 306 F	Si-N	S-L, HA, 600/300V, 8A, 20W	17c	Phi	BUT 12(A)F, 2SC4056, 2SC4441	data
BU 307 F	Si-N	=BU 306F: 700/400V	17c	Phi	BUT 12(A)F, BUT 22CF	data
BU 308	Si-N	TV-HA, 1500/750V, 5A, 12,5W(Tc=95°)	23a§	Tix	BU 208(A), 2SC2928, 2SD350(A), 2SD820,++	data
BU 310	Si-N	TV-HA, 160/100V, 6A, 25W(Tc=100°)	23a§	Sie	BU 104, BU 109, BU 606...608	data
BU 311	Si-N	=BU 310: 200/125V	23a§	Sie	BU 104, BU 109, BU 606...608	data
BU 312	Si-N	=BU 310: 280/150V	23a§	Sie	BU 104, BU 109, BU 606...608	data
BU 322	Si-N-Darl	S-L, 450/400V, 7A, 100W	23a§	Mot	2SD520...521, 2SD528, 2SD705, 2SD707, ++	data
BU 322 A	Si-N-Darl	=BU 322: 525/475V	23a§		2SD520...521, 2SD528, 2SD705, 2SD707, ++	data
BU 323	Si-N-Darl	S-L, 500/350V, 10A, 175W, B>150	23a§	Mot,Rca	MJ 10013...10014, 2SD683	data
BU 323 A	Si-N-Darl	=BU 323: 600/400V	23a§		MJ 10013...10014, 2SD683	data
BU 323(A)P	Si-N-Darl	=BU 323: 125W	18j§		BUT 51P	data
BU 323 Z	Si-N-Darl+Di	=BU 323: 360V Clamp C., -/350V, 150W B=500...3400	18j§	Mot	-	data
BU 325	Si-N	S-L, 200/200V, 3A, 25W	14h§	Sgs	-	data
BU 326	Si-N	TV-SN, 800/375V, 6A, 60W(Tc=50°)	23a§	EUR	BU 426(A), BU 526, BU 926	data
BU 326 A,R	Si-N	=BU 326: 900/400V	23a§		BU 426A, BU 526, BU 926	data
BU 326(A)P	Si-N	=BU 326(A): 110W	18j§		BU 426(A), BU 926	data
BU 326 S	Si-N	=BU 326: 800/400V	23a§		BU 426(A), BU 526, BU 926	data
BU 361	Si-N	S-L, 800/800V, 12A, 70W	23a§	Tix	BU 157, BU 626A, BUW 77, 2SC2123	data
BU 406(H)	Si-N	TV-HA, 400/200V, 7A, 60W	17j§	EUR,Say,Tos	BU 104P, BU 408, 2SC3175, 2SC3591	data
BU 406	Si-N	TV-HA	17j§	Mot,Tho,Sgs		data
BU 406 D	Si-N+Di	=BU 406: integr. Damper-Diode	17j§		BU 104DP, BU 408D, 2SC3176	data
BU 406 F	Si-N	=BU 406: Iso, 18W	17c	Phi	BU 306...307F	data
BU 406 FI	Si-N	=BU 406: Iso, 25W	17c	Tho	BU 306...307F	data
BU 407(H)	Si-N	TV-HA, 330/150V, 7A, 60W	17j§	EUR,Say,Tos	BU 109P, BU 406, BU 408, 2SC3173, ++	data
BU 407	Si-N		17j§	Mot,Tho,Sgs	=BU406	data
BU 407 D	Si-N+Di	=BU 407: integr. Damper-Diode	17j§		BU 109DP, BU 406D, BU 408D, 2SC3174	data
BU 407 F	Si-N	=BU 407: Iso, 18W	17c	Phi	BU 306...307F	data
BU 408	Si-N	TV-HA, 400/200V, 7A, 60W	17j§		BU 104P, BU 406, 2SC3175, 2SC3591	data
BU 408 D	Si-N+Di	=BU 408: integr. Damper-Diode	17j§		BU 104DP, BU 406D, 2SC3176	data
BU 409	Si-N	S-L, 250/150V, 7A, 60W	17j§	Sgs	BU 104P, BU 406...408, BUT 54, BUT 56(A)	data
BU 410	Si-N+Di	TV-HA, 160/125V, 8A, 50W(Tc=50°)	23a§	Sie,Tos	BU 104D, BU 109D, BU 606D...608D	data
BU 411	Si-N+Di	=BU 410: 220/150V	23a§	Sie, Tos	BU 104D, BU 109D, BU 606D...608D	data

Type	Device	Short Description	Fig.	Manu	Comparision Types	More at
BU 412	Si-N+Di	=BU 410: 280/175V	23a§	Sie,Tos	BU 104D, BU 109D, BU 606D...608D	data
BU 413	Si-N+Di	=BU 410: 330/175V, 10A, 60W(Tc=50°)	23a§	Sie	BU 109D, (BU 104D, BU 606D...608D)[7]	data
BU 414	Si-N+Di	S-L, TV-HA, 800/400V, 8A, 60W(Tc=75°)	23a§	Sie	(BU 426(A), BU 526, BU 626A, BU 926)[2]	data
BU 414 B	Si-N+Di	=BU 414: 900/400V	23a§		(BU 426A, BU 526, BU 626A, BU 926)[2]	data
BU 415	Si-N+Di	S-L, TV-HA, 800/400V, 12A, 120W	23a§	Sie,Tos	(BU 626A, 2SC2123)[2]	data
BU 415 B	Si-N+Di	=BU 415: 900/400V	23a§		(BU 626A, 2SC2123)[2]	data
BU 426	Si-N	S-L, TV-SN, 800/375V, 6A, 70W(Tc=73°)	18j§	EUR	BU 926	data
BU 426 A	Si-N	=BU 426: 900/400V	18j§		BU 926	data
BU 426 F,AF	Si-N	=BU 426(A): Iso	18c	Aeg	-	data
BU 426 FI,AFI	Si-N	=BU 426(A): Iso	18c	Tho	-	data
BU 433	Si-N	=BU 426: 800/375V	18j§	Phi	BU 426(A), BU 926	data
BU 500	Si-N	CTV-HA, 1500/700V, 6A, 75W(Tc=30°)	23a§	Aeg,Mot,Tix	BU 508A, 2SC1308, 2SD649, 2SD821,++	data
BU 505	Si-N	TV-HA, 1500/700V, 2,5A, 75W	17j§	Phi,Sgs	BU 506	data
BU 505 D	Si-N+Di	=BU 505: integr. Damper-Diode	17j§		BU 506D	data
BU 505 DF	Si-N+Di	=BU 505D: Iso, 20W	17c		BU 506DF, BU 1506DX	data
BU 505 F	Si-N	=BU 505: Iso, 20W	17c		BU 506F, 2SD1575	data
BU 506	Si-N	TV-HA, SMPS, 1500/700V, 5A, 100W	17j§	Phi	-	data
BU 506 D	Si-N+Di	=BU 506: integr. Damper-Diode	17j§		-	data
BU 506 DF	Si-N+Di	=BU 506D: Iso, 20W	17c		BU 1506DX	data
BU 506 F	Si-N	=BU 506: Iso, 20W	17c		-	data
BU 508(A)	Si-N	CTV-HA, 1500/700V, 8A, 125W	18j§	EUR,Say	BU 908, BU 2508A, 2SC3687	data
BU 508 AF	Si-N	=BU 508A: Iso, 34W	16c	Phi	BU 2508AF, 2SC3886A, 2SC3896, 2SD1548	data
BU 508 AT	Si-N	=BU 508A: 75W	17j§	Tho, Dsi		
BU 508 AW	Si-N	=BU 508A:	16j§	Phi	→BU 508(A)	data
BU 508 AX	Si-N	=BU 508A: Iso, 45W	18c	Phi	BU 2508AF, 2SC3886A, 2SC3896, 2SD1548	data
BU 508 AXI	Si-N	=BU 508A: Iso, 40W	17c	Tho	BU 1508AX	data
BU 508 D,DW	Si-N+Di	=BU 508A: integr. Damper-Diode	18j§		BU 2508D, S 2055A, 2SC3683	data
BU 508 DF	Si-N+Di	=BU 508D: Iso, 34W	16c	Phi	BU 2508DF, S 2055AF, 2SC3893A, 2SC4124	data
BU 508 DFI	Si-N+Di	=BU 508D: Iso, 60W	18c	Tho	BU 2508DF, S 2055AF, 2SC3893A, 2SC4124	data
BU 508 DR	Si-N+Di	=BU 508D: integr. Rbe=25Ω	18j§	Aeg	-	data
BU 508 DRF	Si-N+Di	=BU 508DR: Iso, 34W	18c	Aeg	-	data
BU 508 DW	Si-N+Di	=BU 508A: integr. Damper-Diode	16j§	Phi	BU 2508D, S 2055A, 2SC3683	data
BU 508 DX	Si-N+Di	=BU 508D: Iso, 45W	18c	Phi	BU 2508DF, S 2055AF, 2SC3893A, 2SC4124	data
BU 508 DXI	Si-N+Di	=BU 508D: Iso, 40W	17c	Tho	BU 1508DX	data
BU 508 FI,AFI	Si-N	=BU 508(A): Iso, 50W	18c	Tho	BU 2508AF, 2SC3886A, 2SC3896, 2SD1548	data
BU 508 L	Si-N	≈BU 508A	18j§		→BU 2508A	data
BU 508 V	Si-N	≈BU 508A	18j§	Phi	→BU 2508A	data
BU 522	Si-N-Darl+Di	S-L, 400/375V, 7A, 75W, B>250	17j§	Mot	2SD1533, (BU 806, BU 810)	data
BU 522 A	Si-N-Darl	=BU 522: 450/425V	17j§		2SD1533, (BU 810)[1]	data
BU 522 B	Si-N-Darl	=BU 522: 475/450V	17j§		2SD1533, (BU 810)[1]	data
BU 526	Si-N	TV-SN, 900/400V, 8A, 86W	23a§	Aeg	BU 626A, BU 926, 2SD1094	data
BU 526 A	Si-N	=BU 526: 900/460V	23a§		BU 626A, 2SD1094	data
BU 536	Si-N	TV-SN, 1100/480V, 8A, 62W	23a§	Aeg	BU 902, BU 908	data
BU 546	Si-N	TV-SN, 1300/550V, 6A, 100W	23a§	Aeg	BU 903	data
BU 603	Si-N	TV-HA, SMPS, 1350/550V, 5A, 100W	17j§	Phi	-	data
BU 606	Si-N	TV-HA, 400/200V, 7A, 90W	23a§	Sgs	BU 104, BU 608, BUY 69C	data
BU 606 D	Si-N+Di	=BU 606: integr. Damper-Diode	23a§		BU 104D, BU 608D	data
BU 607	Si-N	TV-HA, 330/200V, 7A, 90W	23a§	Sgs	BU 104, BU 606, BU 608, BUY 69C	data
BU 607 D	Si-N+Di	=BU 607: integr. Damper-Diode	23a§		BU 104D, BU 606D, BU 608D	data
BU 608	Si-N	TV-HA, 400/200V, 7A, 90W	23a§	Sgs	BU 104, BU 606, BUY 69C	data
BU 608 D	Si-N+Di	=BU 608: integr. Damper-Diode	23a§		BU 104D, BU 606D	data
BU 626(A)	Si-N	TV-SN, 1000/400V, 10A, 100W	23a§	Sie	S 2530A, 2SD1094	data
BU 705	Si-N	TV-HA, 1500/700V, 2,5A, 75W	18j§	Aeg,Phi	BU 706, 2SC3483, 2SD1493...1494, ++	data
BU 705 D	Si-N+Di	=BU 705: integr. Damper-Diode	18j§		BU 706D, 2SC3479, 2SD1290..91,++	data
BU 705 DF	Si-N+Di	=BU 705D: Iso, 29W	16c		BU 706DF, 2SD1553, 2SD1649, 2SD1876,++	data
BU 705 F	Si-N	=BU 705: Iso, 29W	16c		BU 706F, 2SD1543, 2SD1653, 2SD1882,++	data
BU 706	Si-N	TV-HA/SN, 1500/700V, 5A, 100W	18j§	Phi,Tho	BU 508(A), 2SC3485...86, 2SD1496...97, ++	data
BU 706 D	Si-N+Di	=BU 706: integr. Damper-Diode	18j§		BU 508A, 2SC3481...82, 2SC3681, 2SD1878++	data
BU 706 DF	Si-N+Di	=BU 706D: Iso, 32W	16c		BU 508DF, 2SD1555...56, 2SD1651...52,++	data
BU 706 F	Si-N	=BU 706: Iso, 32W	16c		BU 508F, 2SC4142...43, 2SD1545...46, ++	data
BU 724	Si-N-Darl+Di	S-L, SMPS, 650/375V, 2A, 25W	≈14j§	Phi	(BUD 46(A), 2SC3259, 2SC3579)[4]	data
BU 724 A	Si-N-Darl+Di	=BU 724: 850/400V	≈14j§		-	data
BU 726	Si-N	TV, -/400V, 7A, 60W(Tc=50°)	23a§	Tho	BU 426(A), BU 526, BU 536, BU 926	data
BU 800(A,S)	Si-N+Di	CTV-HA, 1500/700V, 5A, 12,5W(Tc=95°)	23a§	Mot,Tho	BU 208D, BU 508D, 2SD1171...1175	data
BU 801	Si-N-Darl+Di	S-L, 600/400V, 3A, 40W, B>100	14h§		-	data
BU 806(/01)	Si-N-Darl+Di	TV-HA, 400/200V, 8A, 60W 350/750ns	17j§	Phi,Sgs,Tix Mot,Sam	BU 184	data
BU 806 AF	Si-N-Darl+Di	=BU 806: Iso	17c		-	
BU 806 FI	Si-N-Darl+Di	=BU 806: Iso, 30W	17c	Tho	-	data
BU 807	Si-N-Darl+Di	=BU 806: 330/150V	17j§	Phi,Sgs,Tix	BU 184, BU 189	data
BU 807 FI	Si-N-Darl+Di	=BU 807: Iso, 30W	17c	Tho	-	data
BU 808 DFI [SGS]	Si-N-Darl+Di	=BU 808FI: integr. Damper-Diode	18c	Sgs	-	data
BU 808 DXI [SGS]	Si-N-Darl+Di	=BU 808FI: Damper-Diode, Iso, 35W	17c	Sgs	-	data
BU 808 FI [SGS]	Si-N-Darl	S-L, 1400/700V, 5A, 50W, B>25	18c	Sgs	-	data
BU 808 [Philips]	Si-N	3Ph.-Motor-E, 1500V, 12A, 160W	23a§	Phi	BUX 88	data

Type	Device	Short Description	Fig.	Manu	Comparision Types	More at
BU 810	Si-N-Darl+Di	S-L,Speedup Di(E-B), 600/400V, 7A, 75W	17j§	Sgs	2SD798...799	data
BU 824	Si-N-Darl+Di	S-L, 650/375V, 0,5A, 12,5W, B>325	13h§	Phi	-	data
BU 826	Si-N-Darl+Di	S-L, 800/375V, 6A, 125W	18j§	Phi	-	data
BU 826 A	Si-N-Darl+Di	=BU 826: 1000/400V	18j§			data
BU 900(DT)	Si-N-Darl	3 Trans, 650/400V, 8A, 70W, B>7000	17j§	Tho	-	data
BU 902	Si-N	S-L, TV-SN, 1100/480V, 8A, 100W	18j§	Aeg	BU 508(A), BU 908	data
BU 902 F	Si-N	=BU 902: Iso	18c		BU 508AF, 2SC3886, 2SC4144	data
BU 903	Si-N	S-L, TV-SN, 1350/550V, 6A, 125W	18j§	Aeg,Phi	BU 508(A), BU 908	data
BU 903 F	Si-N	=BU 903: Iso	18c		BU 508AF, 2SC3885, 2SC3895, 2SC4143,++	data
BU 908	Si-N	S-L, TV-SN/HA, 1500/700V, 8A, 125W	18j§	Aeg	BU 508(A), BU 2508A	data
BU 908 AF	Si-N	=BU 908: Iso	18c		BU 508AF, BU 2508AF	data
BU 910	Si-N-Darl+Di	S-L, 400/350V, 6A, 60W, B>20	17j§	Sgs	BU 806, 2SD1114, 2SD1245	data
BU 911	Si-N-Darl+Di	=BU 910: 450/400V	17j§	Sgs	BU 806, 2SD1245	data
BU 912	Si-N-Darl+Di	=BU 910: 500/450V	17j§		2SD1245	data
BU 920	Si-N-Darl+Di	S-L, 400/350V, 10A, 120W, B>300	23a§	Sgs	BU 930...932, BUT 51P, BUW 66	data
BU 920 P	Si-N-Darl+Di	=BU 920: 105W	18j§		BU 930P...932P, BUT 51P	data
BU 920 PFI	Si-N-Darl+Di	=BU 920: 55W	18c		-	data
BU 920 T	Si-N-Darl+Di	=BU 920: 105W	17j§		2SD1162	data
BU 921	Si-N-Darl+Di	=BU 920: 450/400V	23a§	Sgs	BU 931...932, BUT 51P	data
BU 921 P	Si-N-Darl+Di	=BU 921: 105W	18j§		BU 931P...932P, BUT 51P	data
BU 921 PFI	Si-N-Darl+Di	=BU 921: 55W	18c		-	data
BU 921 T	Si-N-Darl+Di	=BU 921: 105W	17j§		2SD1162	data
BU 921 ZP	Si-N-Darl+Di	=BU 921: integr. Z-Diode, 350V, 125W	18j§			data
BU 921 ZPFI	Si-N-Darl+Di	=BU 921: integr. Z-Diode, 350V, 60W·	18c			data
BU 921 ZT	Si-N-Darl+Di	=BU 921: integr. Z-Diode, 350V, 100W	17j§			data
BU 921 ZTFI	Si-N-Darl+Di	=BU 921: integr. Z-Diode, 350V, 40W	17c		-	data
BU 922	Si-N-Darl+Di	=BU 920: 500/450V	23a§	Sgs	BU 932, BUT 51P	data
BU 922 P	Si-N-Darl+Di	=BU 922: 105W	18j§		BU 932P, BUT 51P	data
BU 922 PFI	Si-N-Darl+Di	=BU 922: 55W	18c		-	data
BU 922 T	Si-N-Darl+Di	=BU 922: 105W	17j§		2SD1162	data
BU 926	Si-N	S-L, TV-SN, 850V, 8A, 90W	18j§	Tho	BUV 47(A), BUV 70...71, BUV 89	data
BU 930	Si-N-Darl+Di	S-L, 400/350V, 15A, 150W, B>300 Kfz-Zündung/Electronic Ignition	23a§	Sgs	BU 941, BUT 13...15	data
BU 930 P	Si-N-Darl+Di	=BU 930: 105W	18j§		BU 941P, BUT 51P, 2SD1466	data
BU 930 Z	Si-N-Darl+Di	=BU 930: integr. Z-Diode, 175W	23a§		-	data
BU 930 ZP	Si-N-Darl+Di	=BU 930: integr. Z-Diode	18j§		-	data
BU 931	Si-N-Darl+Di	S-L, 500/400V, 15A, 175W, B>300 Kfz-Zündung/Electronic Ignition	23a§	Sgs	BU 941, BUT 13...15	data
BU 931 P	Si-N-Darl+Di	=BU 931: 135W	18j§		BU 941P, BUT 51P, 2SD1466	data
BU 931 PFI	Si-N-Darl+Di	=BU 931: Iso, 60W	18c		BU 941PFI	data
BU 931 R	Si-N-Darl+Di	=BU 931: 175W	23a§		BUT 13...15	data
BU 931 RP	Si-N-Darl+Di	=BU 931: 125W	18j§		BUT 51P, 2SD1466	data
BU 931 RPFI	Si-N-Darl+Di	=BU 931: Iso, 60W	18c		-	data
BU 931 SM	Si-N-Darl+Di	=BU 931: SMD, 125W	10-MDIP		BU 941SM	data
BU 931 T	Si-N-Darl+Di	=BU 931: 10A, 125W	17j§		BU 941T	data
BU 931 TFI	Si-N-Darl+Di	=BU 931: Iso, 45W	17c		BU 941TFI	data
BU 931 Z	Si-N-Darl+Di	=BU 931: integr. Z-Diode, 350V, 175W	23a§		BU 941Z	data
BU 931 ZP	Si-N-Darl+Di	=BU 931: integr. Z-Diode, 350V, 125W	18j§		BU 941ZP	data
BU 931 ZPFI	Si-N-Darl+Di	=BU 931: integr. Z-Diode, 350V, 60W	18c		BU 941ZPFI	data
BU 932	Si-N-Darl+Di	=BU 930: 500/450V	23a§	Sgs	BUT 13...15	data
BU 932 P	Si-N-Darl+Di	=BU 932: 105W	18j§		BUT 51P, 2SD1466	data
BU 932 R	Si-N-Darl+Di	=BU 932: 175W	23a§		BUT 13...15	data
BU 932 RP	Si-N-Darl+Di	=BU 932: 125W	18j§		BUT 51P, 2SD1466	data
BU 932 RPFI	Si-N-Darl+Di	=BU 932: 60W	18c		-	data
BU 941	Si-N-Darl+Di	S-L, 500/400V, 15A, 180W, B>300 Kfz-Zündung/Electronic Ignition	23a§	Tho	BUT 13...15	data
BU 941 P	Si-N-Darl+Di	=BU 941: 155W	18j§		-	data
BU 941 PFI	Si-N-Darl+Di	=BU 941: Iso, 65W	18c		-	data
BU 941 SM	Si-N-Darl+Di	=BU 941: SMD, 125W	10-MDIP		-	data
BU 941 T	Si-N-Darl+Di	=BU 941: 150W	17j§		-	data
BU 941 TFI	Si-N-Darl+Di	=BU 941: Iso, 55W	17c		-	data
BU 941 Z...	Si-N-Darl+Di	=BU 941...: 350/350V			-	data
BU 999	Si-N	NF/S-L, 160/140V, 25A, 106W	18j§	Sgs	BUX 69	data
BU 1085	Si-N	≈BU 108	23a§	Tix	→BU 108	
BU 1200...1206(L)	CMOS-IC	Gate Arrays		Rhm	-	
BU 1301 F	CMOS-IC	Ir-fb Encoder	20-MDIP	Rhm	-	
BU 1414 AK	MOS-IC	SMD, Video Cd, Digital NTSC Encoder	64-MP	Rhm	-	
BU 1417 AK	MOS-IC	SMD, Video Cd, Dig. NTSC/PAL Encoder	64-MP	Rhm	-	
BU 1418 K	MOS-IC	SMD, Video Cd, Digital NTSC Encoder	64-MP	Rhm		
BU 1419 K	MOS-IC	SMD, Video Cd, Dig. NTSC/PAL Encoder	64-MP	Rhm		
BU 1506 DX	Si-N+Di	CTV-HA, 1500V, 5A, 32W	17c	Phi		data
BU 1508 AX	Si-N	CTV-HA, 1500/700V, 8A, 35W	17c	Phi		data
BU 1508 DX	Si-N+Di	=BU 1508AX: integr. Damper-Diode	17c	Phi		data
BU 1706 A	Si-N	S-L, 1750/850V, 5A, 100W	17j§	Phi	BUL 310, BUL 416	data
BU 1706 AX	Si-N	=BU 1706A: Iso, 32W	17c	Phi	BUL 310PI	data
BU 1708 AX	Si-N	S-L, 1750/850V, 8A, 35W	17c	Phi	-	data
BU 1920 F	CMOS-IC	SMD, RDS/RBDS-decoder f. Europa/USA	16-MDIP	Rhm	-	pdf
BU 1922	CMOS-IC	Rds/rbds-decoder f. Europa/usa	16-DIP	Rhm	-	

Type	Device	Short Description	Fig.	Manu	Comparision Types	More at
BU 1922 F	CMOS-IC	=BU 1922: SMD	16-MDIP		-	
BU 2007 F	CMOS-Logic	Inverter, hi-speed, (=2xBU74HCU04)	8-MDIP	Rhm	-	
BU 2029	CMOS-IC	12 Bit Serial In Parallel Out Driver	16-DIP	Rhm	-	
BU 2029 F	CMOS-IC	=BU 2029: SMD	16-MDIP	Rhm	-	
BU 2040	CMOS-IC	12 Bit Serial In Parallel Out Driver	16-DIP	Rhm	-	
BU 2040 F	CMOS-IC	=BU 2040: SMD	16-MDIP	Rhm	-	
BU 2090(A)	CMOS-IC	12 Bit Serial In Parallel Out Driver	16-DIP	Rhm	-	
BU 2090 F,AF	CMOS-IC	SMD,12 Bit Serial In Parallel Out Driv	16-MDIP		-	
BU 2090 FS,AFS	CMOS-IC	SMD,12 Bit Serial In Parallel Out Driv	16-SMDIP		-	
BU 2092	CMOS-IC	12 Bit Serial In Parallel Out Driver	18-DIP	Rhm	-	pdf
BU 2092 F	CMOS-IC	=BU 2092: SMD	18-MDIP	Rhm	-	pdf
BU 2105 CH-2W	CMOS-IC	64 Bit Serial In Parallel Out Driver	Wafer	Rhm	-	
BU 2112 F	CMOS-IC	Ladestrg./battery Charge Controller	18-MDIP	Rhm	-	
BU 2114	CMOS-IC	8 Bit Shift Register, Latch Driver	18-DIP	Rhm	-	pdf
BU 2114 F	CMOS-IC	=BU 2114: SMD	18-MDIP	Rhm	-	pdf
BU 2120 CH-3W	CMOS-IC	64 Bit Thermal Driver	Chip	Rhm	-	
BU 2130 K1	MOS-IC	SMD, Fax, Halbton-/halftone Proz.	80-MP	Rhm	-	
BU 2133 KS	MOS-IC	SMD, Fax, Halbton-/halftone Proz.	100-MP	Rhm	-	
BU 2134 K,AK	MOS-IC	SMD, Fax, 32-Halbton-/Halftone Proz.	64-MP	Rhm	-	
BU 2135 K	MOS-IC	SMD, Fax, 64-Halbton-/Halftone Proz.	64-MP	Rhm	-	
BU 2140 KS	MOS-IC	SMD, Fax, Halbton-/halftone Proz.	100-MP	Rhm	-	
BU 2173 F	MOS-IC	SMD, Video Cd, Clock Generator	18-MDIP	Rhm	-	
BU 2191 F	MOS-IC	SMD, Hdd, Clock Generator	8-MDIP	Rhm	-	
BU 2302	CMOS-IC	CR Timer, Ucc=1,8...6V	8-DIP	Rhm	-	
BU 2302 F	CMOS-IC	=BU 2302: SMD	8-MDIP	Rhm	-	
BU 2305	CMOS-IC	CR Timer, Ucc=1,8...6V	8-DIP	Rhm	-	
BU 2305 F	CMOS-IC	=BU 2305: SMD	8-MDIP	Rhm	-	
BU 2400(L)	CMOS-µC-IC	4 Bit, 64x4 Bit RAM, 1024x8 Bit ROM	40-DIP	Rhm	-	
BU 2403(L)	CMOS-µC-IC	4 Bit, 16x4 Bit RAM, 640x8 Bit ROM	18-MDIP	Rhm	-	
BU 2404(L)	CMOS-µC-IC	4 Bit, 16x4 Bit RAM, 640x8 Bit ROM	18-DIP	Rhm	-	
BU 2405(L)	CMOS-µC-IC	4 Bit, 16x4 Bit RAM, 512x8 Bit ROM	16-DIP	Rhm	-	
BU 2406(L)	CMOS-µC-IC	4 Bit, 16x4 Bit RAM, 512x8 Bit ROM	18-MDIP	Rhm	-	
BU 2407(L)	CMOS-µC-IC	4 Bit, 16x4 Bit RAM, 512x8 Bit ROM	18-SQP	Rhm	-	
BU 2408 A,AL	CMOS-µC-IC	4 Bit, 16x4 Bit RAM, 512x8 Bit ROM	18-MDIP	Rhm	-	
BU 2409 A,AL	CMOS-µC-IC	4 Bit, 16x4 Bit RAM, 512x8 Bit ROM	18-DIP	Rhm	-	
BU 2411(L)	CMOS-µC-IC	4 Bit, 16x4 Bit RAM, 1024x8 Bit ROM	30-SDIP	Rhm	-	
BU 2412(L)	CMOS-µC-IC	4 Bit, 16x4 Bit RAM, 1024x8 Bit ROM	28-MDIP	Rhm	-	
BU 2413(L)	CMOS-µC-IC	4 Bit, 16x4 Bit RAM, 1024x8 Bit ROM	32-SDIP	Rhm	-	
BU 2414(L)	CMOS-µC-IC	4 Bit, 16x4 Bit RAM, 1024x8 Bit ROM	42-SDIP	Rhm	-	
BU 2415(L)	CMOS-µC-IC	4 Bit, 16x4 Bit RAM, 1024x8 Bit ROM	40-MDIP	Rhm	-	
BU 2416(L)	CMOS-µC-IC	SMD, 4 Bit, 16x4 Bit RAM, 1024x8 Bit R	44-MP	Rhm	-	
BU 2417(L)	CMOS-µC-IC	SMD, 4 Bit,16x4 Bit RAM,1024x8 Bit ROM	32-MP	Rhm	-	
BU 2418	CMOS-µC-IC	4Bit, 3V, 32x4 Bit RAM, 1024x8 Bit ROM	22-MDIP	Rhm	-	
BU 2418 L	CMOS-µC-IC	4Bit, 3V, 32x4 Bit RAM, 1024 Bit ROM	22-MDIP	Rhm	-	
BU 2419(L)	CMOS-µC-IC	4Bit, 3V, 32x4 Bit RAM, 1024x8 Bit ROM	22-DIP	Rhm	-	
BU 2421	µC-IC	4 Bit, 64x4 Bit RAM, 1024x8 Bit ROM	32-SDIP	Rhm	-	
BU 2422(L)	CMOS-µC-IC	4Bit, 3V, 32x4 Bit RAM, 1024x8 Bit ROM	22-SDIP	Rhm	-	
BU 2424(L)	CMOS-µC-IC	SMD,4 Bit, 64x4 Bit RAM,1024x8 Bit ROM	40-MP	Rhm	-	
BU 2425	CMOS-µC-IC	SMD,4 Bit, 64x4 Bit RAM,1024x8 Bit ROM	32-MP	Rhm	-	
BU 2430(L)	CMOS-µC-IC	4Bit, 3V, 32x4 Bit RAM, 1024x8 Bit ROM	20-MDIP	Rhm	-	
BU 2431	CMOS-µC-IC	4Bit, 3V, 32x4 Bit RAM, 1024x8 Bit ROM	20-DIP	Rhm	-	
BU 2456	CMOS-µC-IC	4Bit, 3V, 32x4 Bit RAM, 1024x8 Bit ROM	24-SMDIP	Rhm	-	
BU 2458	CMOS-µC-IC	4Bit, 3V, 32x4 Bit RAM, 1024x8 Bit ROM	22-MDIP	Rhm	-	
BU 2459	CMOS-µC-IC	4Bit, 3V, 32x4 Bit RAM, 1024x8 Bit ROM	22-DIP	Rhm	-	
BU 2460	CMOS-µC-IC	4Bit, 3V, 32x4 Bit RAM, 1024x8 Bit ROM	22-SDIP	Rhm	-	
BU 2461	CMOS-µC-IC	4Bit, 3V, 32x4 Bit RAM, 1024x8 Bit ROM	20-MDIP	Rhm	-	
BU 2462	CMOS-µC-IC	4Bit, 3V, 32x4 Bit RAM, 1024x8 Bit ROM	20-DIP	Rhm	-.	
BU 2463	CMOS-µC-IC	4Bit, 3V, 16x4 Bit RAM, 640x8 Bit ROM	18-MDIP	Rhm	-	
BU 2464	CMOS-µC-IC	4Bit, 3V, 16x4 Bit RAM, 640x8 Bit ROM	18-DIP	Rhm	-	
BU 2465	CMOS-µC-IC	4Bit, 3V, 16x4 Bit RAM, 640x8 Bit ROM	18-SDIP	Rhm	-	
BU 2466	CMOS-µC-IC	4Bit, 3V, 16x4 Bit RAM, 640x8 Bit ROM	16-MDIP	Rhm	-	
BU 2492	CMOS-µC-IC	SMD,4 Bit, 64x4 Bit RAM,external EPROM	100-MP	Rhm	-	
BU 2493	CMOS-µC-IC	SMD,4 Bit, 64x4 Bit RAM,external EPROM	100-MP	Rhm	-	
BU 2494	CMOS-µC-IC	SMD,4 Bit,32x4 Bit RAM, external EPROM	80-MP	Rhm	-	
BU 2495	CMOS-µC-IC	SMD,4 Bit,16x4 Bit RAM, external EPROM	64-MP	Rhm	-	
BU 2496	CMOS-µC-IC	4 Bit, 16x4 Bit RAM, external EPROM	42-SDIP	Rhm	-	
BU 2497	CMOS-µC-IC	4 Bit, 16x4 Bit RAM, external EPROM	42-SDIP	Rhm	-	
BU 2498	CMOS-µC-IC	SMD,4 Bit,64x4 Bit RAM, external EPROM	64-MP	Rhm	-	
BU 2506 DF	Si-N+Di	CTV-HA, 1500/700V, 5A, 45W	16c	Phi	BU 2508DF, 2SC3892A, 2SC4765, 2SC4916	data
BU 2506 DX	Si-N+Di	=BU 2506DF:	18c		→BU 2506DF	data
BU 2507 AF	Si-N	CTV-HA, 1500/700V, 8A, 45W	16c	Phi	BU 2508AF, 2SC3896, 2SC4542, 2SC4758,++	data
BU 2508 A,AW	Si-N	CTV-HA, 1500/700V, 8A, 125W	18j§	Phi	2SC3687...88	data
BU 2508 AF	Si-N	=BU 2508A: Iso, 45W	16c		2SC3886A, 2SC3896, 2SC4542, 2SC4758,++	data
BU 2508 AW	Si-N	=BU 2508A:	16j§		→BU 2508A	data
BU 2508 AX	Si-N	=BU 2508A: Iso, 45W	18c		→BU 2508AF	data
BU 2508 D	Si-N+Di	=BU 2508A: integr. Damper-Diode	18j§		2SC3683...84	data
BU 2508 DF	Si-N+Di	=BU 2508D: Iso, 45W	16c		2SC3893A	data
BU 2508 DW	Si-N+Di	=BU 2508D:	16j§		→BU 2508D	data
BU 2508 DX	Si-N+Di	=BU 2508D: Iso, 45W	18c		→BU 2508DF	data
BU 2515 AF	Si-N	PC Monitor HA, 1500/800V, 9A, 45W	16c	Phi	BU 2520AF, BUH 715, 2SC4542, 2SC4759	data
BU 2515 AX	Si-N	=BU 2515AF:	18c		BU 2520AF, BUH 715, 2SC4542, 2SC4759	data
BU 2515 DF	Si-N+Di	=BU 2515AF: integr. Damper-Diode	16c		BU 2520DF, 2SC4125, 2SC4531, 2SC4878	data
BU 2515 DX	Si-N+Di	=BU 2515AF: integr. Damper-Diode	18c		BU 2520DF, 2SC4125, 2SC4531, 2SC4878	data
BU 2520 A	Si-N	CTV-HA, hi-res, 1500/800V, 10A, 125W	18j§	Phi	BU 2522A, 2SC3688	data

Type	Device	Short Description	Fig.	Manu	Comparision Types	More at
BU 2520 AF	Si-N	=BU 2520A: Iso, 45W	16c		BU 2522AF, 2SC4542, 2SC4759	data
BU 2520 AW	Si-N	=BU 2520A:	16j§		→BU 2520A	data
BU 2520 AX	Si-N	=BU 2520A: Iso, 45W	18c		→BU 2520AF	data
BU 2520 D	Si-N+Di	=BU 2520A: integr. Damper-Diode	18j§		BU 2522D, 2SC3684	data
BU 2520 DF	Si-N+Di	=BU 2520D: Iso, 45W	16c		BU 2522DF, 2SC4125	data
BU 2520 DW	Si-N+Di	=BU 2520D:	16j§		→BU 2520D	data
BU 2520 DX	Si-N+Di	=BU 2520D: Iso, 45W	18c		→BU 2520DF	data
BU 2522 A	Si-N	CRT-HA, hi-res, 1500/800V, 10A, 125W	18j§	Phi	BU 2527A	data
BU 2522 AF	Si-N	=BU 2522A: Iso, 45W	16c		BU 2527AF, 2SC4542	data
BU 2522 AW	Si-N	=BU 2522A:	16j§		→BU 2522A	data
BU 2522 AX	Si-N	=BU 2522A: Iso, 45W	18c		→BU 2522AF	data
BU 2522 DF	Si-N+Di	=BU 2522A: Iso, 45W, int. Damper-Diode	16c		BU 2527DF, 2SC4125	data
BU 2522 DX	Si-N+Di	=BU 2522A: Iso, 45W, int. Damper-Diode	18c		→BU 2522DF	data
BU 2523 AF	Si-N	Moni/HDTV-HA, 1500/800V, 11A, 45W	16c	Phi	BU 2527AF, 2SC4891, 2SC5048, 2SC5251	data
BU 2523 AX	Si-N	=BU 2523AF:	18c		→BU 2523AF	data
BU 2523 DF	Si-N+Di	=BU 2523AF: integr. Damper-Diode	16c		BU 2527DF	data
BU 2523 DX	Si-N+Di	=BU 2523AF: integr. Damper-Diode	18c		BU 2527DF	data
BU 2525 A	Si-N	CTV-HA, hi-res, 1500/800V, 12A, 125W	18j§	Phi	BU 2527A, BUH 1015	data
BU 2525 AF	Si-N	=BU 2525A: Iso, 45W	16c		BU 2527AF, 2SC4890, 2SC5048, 2SC5251	data
BU 2525 AW	Si-N	=BU 2525A:	16j§	Phi	→BU 2525A	data
BU 2525 AX	Si-N	=BU 2525A: Iso, 45W	18c		→BU 2525AF	data
BU 2525 DF	Si-N+Di	=BU 2525A: Iso, 45W, int. Damper-Diode	16c		BU 2527DF,DX	data
BU 2525 DW	Si-N+Di	=BU 2525A: integr. Damper-Diode	18j§			data
BU 2525 DX	Si-N+Di	=BU 2525A: Iso, 45W, int. Damper-Diode	18c		BU 2527DF,DX	data
BU 2527 A	Si-N	CRT-HA, hi-res, 1500/800V, 12A, 125W	18j§	Phi	BUH 1015	data
BU 2527 AF	Si-N	=BU 2527A: Iso, 45W	16c		2SC4890...91, 2SC5048, 2SC5251	data
BU 2527 AW	Si-N	=BU 2527A:	16j§		BUH 1015	data
BU 2527 AX	Si-N	=BU 2527A: Iso, 45W	18c		→BU 2527AF	data
BU 2527 DF	Si-N+Di	=BU 2527A: Iso, 45W, int. Damper-Diode	16c			data
BU 2527 DX	Si-N+Di	=BU 2527A: Iso, 45W, int. Damper-Diode	18c			data
BU 2530 AL	Si-N	CTV-HA, large scr.,1500/800V, 16A, 125W	77j§	Phi	2SC3996, 2SC4289A	data
BU 2530 AW	Si-N	=BU 2530AL:	16j§		BUH 1015, BUH 1215	data
BU 2532 AL	Si-N	Moni HA, hi-res, 1500/800V, 16A, 125W	77j§	Phi	-	data
BU 2532 AW	Si-N	=BU 2532AL:	16j§		-	data
BU 2611(A)	MOS-IC	Am/fm-tuner PLL Frequ.-synthesizer	16-DIP	Rhm	-	pdf
BU 2611 AFS	MOS-IC	=BU 2611: SMD	20-SMDIP		-	pdf
BU 2611 AF,F	MOS-IC	=BU 2611: SMD	20-MDIP		-	pdf
BU 2614	MOS-IC	Am/fm-tuner PLL Frequ.-synthesizer	16-DIP	Rhm	-	pdf
BU 2614 F	MOS-IC	=BU 2614: SMD	16-MDIP		-	pdf
BU 2614 FS	MOS-IC	=BU 2614: SMD	16-SMDIP		-	pdf
BU 2615	MOS-IC	Am/fm-tuner PLL Frequ.-synthesizer	20-DIP	Rhm	-	pdf
BU 2615 F	MOS-IC	=BU 2615: SMD	20-MDIP		-	
BU 2615 FS	MOS-IC	=BU 2615: SMD	20-SMDIP		-	pdf
BU 2615 S	MOS-IC	=BU 2615:	22-SDIP		-	pdf
BU 2616	MOS-IC	Am/fm-tuner PLL Frequ.-synthesizer	18-DIP	Rhm	-	pdf
BU 2616 F	MOS-IC	=BU 2616: SMD	18-MDIP		-	pdf
BU 2618 FV	MOS-IC	SMD, Fm-tuner PLL Frequ.-synthesizer	16-SSMDIP	Rhm	-	pdf
BU 2619	MOS-IC	Am/fm-tuner PLL Frequ.-synthesizer	20-DIP	Rhm	-	
BU 2619 F	MOS-IC	=BU 2619: SMD	20-MDIP	Rhm	-	
BU 2620	MOS-IC	Am/fm-tuner PLL Frequ.-synthesizer	20-DIP	Rhm	-	
BU 2620 F	MOS-IC	=BU 2620: SMD	20-MDIP	Rhm	-	
BU 2621 F	MOS-IC	SMD,Am/fm-tuner PLL Frequ.-synthesizer	20-MDIP	Rhm	-	
BU 2622 S	MOS-IC	Am/fm-tuner PLL Frequ.-synthesizer	22-SDIP	Rhm	-	
BU 2624 AF	MOS-IC	SMD,Am/fm-tuner PLL Frequ.-synthesizer	20-MDIP	Rhm	-	
BU 2630 F	CMOS-IC	SMD, Telecom, Dual PLL Frequ. Synth.	16-MDIP	Rhm	-	pdf
BU 2630 FV	CMOS-IC	=BU 2630F:	16-SSMDIP		-	pdf
BU 2670 K	MOS-IC	SMD,LCD-controller f.spectrum Analyzer	64-MP	Rhm	-	pdf
BU 2701 F	CMOS-IC	Camera, PAL/SECAM Synchr.-signal-gen.	28-MDIP	Rhm	-	
BU 2702	CMOS-IC	VC++, Zeitgeber/time base	7-SIP	Rhm	-	
BU 2705 F	CMOS-IC	Camera, Pal/secal Synchr.-signal-gen.	28-MDIP	Rhm	-	
BU 2708 AF	Si-N	CTV-HA, 1700/825V, 8A, 45W	16c	Phi	BUH 517, 2SC4797	data
BU 2708 AX	Si-N	=BU 2708AF:	18c		→BU 2708AF	data
BU 2708 DF	Si-N+Di	=BU 2708AF: integr. Damper-Diode	16c		BUH 517D, 2SC4963	data
BU 2708 DX	Si-N+Di	=BU 2708AF: integr. Damper-Diode	18c		→BU 2708DF	data
BU 2710 S	CMOS-IC	Vc, Digital Servo	42-SDIP	Rhm	-	
BU 2720 AF	Si-N	CTV-HA, 1700/825V, 10A, 45W	16c	Phi	BU 2722AF, 2SC4879, 2SC5150	data
BU 2720 AX	Si-N	=BU 2720AF:	18c		→BU 2720AF	data
BU 2720 DF	Si-N+Di	=BU 2720AF: integr. Damper-Diode	16c		BU 2725DF, 2SC5143	data
BU 2720 DX	Si-N+Di	=BU 2720AF: integr. Damper-Diode	18c		→BU 2720DF	data
BU 2722 AF	Si-N	CRT-HA, hi-res, 1700/825V, 10A, 45W	16c	Phi	BU 2727AF,AX	data
BU 2722 AX	Si-N	=BU 2722AF:	18c		→BU 2722AF	data
BU 2725 AF	Si-N	CTV-HA, 1700/825V, 12A, 45W	16c	Phi	BU 2727AF,AX	data
BU 2725 AX	Si-N	=BU 2725AF:	18c		→BU 2725AF	data
BU 2725 DF	Si-N+Di	=BU 2725AF: integr. Damper-Diode	16c		-	data
BU 2725 DX	Si-N+Di	=BU 2725AF: integr. Damper-Diode	18c		-	data
BU 2727 A	Si-N	CRT-HA, hi-res, 1700/825V, 12A, 125W	18j§	Phi	-	data
BU 2727 AF	Si-N	=BU 2727A: Iso, 45W	16c		-	data
BU 2727 AW	Si-N	=BU 2727A:	16j§		-	data
BU 2727 AX	Si-N	=BU 2727A: Iso, 45W	18c		-	data
BU 2728 K	CMOS-IC	SMD, Vc, Title Memory Controller	44-MP	Rhm	-	

Type	Device	Short Description	Fig.	Manu	Comparision Types	More at
BU 2730 S	CMOS-IC	Vc, Digital Servo	30-SDIP	Rhm	-	
BU 2735 AS	IC	Power Supply Controller	DIP	Roh	-	pinout
BU 2735 S	IC	Power Supply Controller	DIP	Roh	-	pinout
BU 2762 AL	CMOS-IC	Vc, Color Signals & Test Patterns	18-SQP	Rhm	-	
BU 2763 F	CMOS-IC	=BU 2763S: SMD	18-MDIP	Rhm	-	
BU 2763 S	CMOS-IC	Vc, Color Signal-prozessor	18-SDIP	Rhm	-	
BU 2767 S	CMOS-IC	Vc, Servo, Special Playback(fine Slow)	32-SDIP	Rhm	-	
BU 2770 S	CMOS-IC	Vc, Digital Servo	42-SDIP	Rhm	-	
BU 2780 S	CMOS-IC	Vc, Digital Servo	30-SDIP	Rhm	-	pdf
BU 2790 S	CMOS-IC	Vc, Digital Servo	42-SDIP	Rhm	-	pdf
BU 2841 AFS	CMOS-IC	SMD, Vc, Color Signals & Test Patterns	20-SMDIP	Rhm	-	
BU 2842 FS	CMOS-ROM-IC	SMD, 128-Character OSD	20-SMDIP	Rhm	-	
BU 2846 AKV	CMOS-IC	SMD, Camcorder, Title Memory Control	48-MP	Rhm	-	
BU 2848 FS	CMOS-ROM-IC	SMD, 128-Character OSD, 12x18 Dot	20-SMDIP	Rhm	-	
BU 2857	CMOS-ROM-IC	Vc, 255-Character OSD, 12x18 Dot	28-DIP	Rhm	-	
BU 2857 F	CMOS-ROM-IC	=BU 2857: SMD	28-MDIP	Rhm	-	
BU 2858 FV	CMOS-ROM-IC	SMD, 127-Character OSD, 12x24 Dot	20-SSMDIP	Rhm	-	
BU 2870 FS	CMOS-ROM-IC	SMD, 256-Character OSD, 12x18 Dot	20-SMDIP	Rhm	-	
BU 2871 FS	CMOS-ROM-IC	=BU 2871S: SMD	20-SMDIP		-	
BU 2871 S	CMOS-ROM-IC	Vc, 128-Character OSD, 10x24 Dot	24-SDIP	Rhm	-	
BU 2873 FS	CMOS-ROM-IC	SMD, 64-Character OSD, 12x18 Dot	20-SMDIP	Rhm	-	
BU 2874 FV,AFV	CMOS-ROM-IC	SMD, 255-Character OSD, 12x24 Dot	20-SSMDIP	Rhm	-	
BU 2875	CMOS-ROM-IC	Vc, 127-Character OSD, 12x18 Dot	28-DIP	Rhm	-	
BU 2875 F	CMOS-ROM-IC	=BU 2875: SMD	28-MDIP		-	
BU 2878 FS	CMOS-ROM-IC	SMD, Vc, 128-Character OSD, 10x24 Dot	20-SMDIP	Rhm	-	
BU 2880	CMOS-IC.	Vc, Digital Servo, Digital Filter	24-SDIP	Rhm	-	
BU 2890	CMOS-IC	Vc, Digital Servo Controller	42-SDIP	Rhm	-	
BU 2902 F	CMOS-IC	SMD, Telecom, Ton-/hold Tone Gener.	18-MDIP	Rhm	-	
BU 2906 F	CMOS-IC	SMD, Telecom, Ton-/hold Tone Gener.	18-MDIP	Rhm	-	
BU 2907 F	CMOS-IC	SMD, Telecom, Ton-/hold Tone Gener.	18-MDIP	Rhm	-	
BU 2908 F	CMOS-IC	SMD, Telecom, Ton-/hold Tone Gener.	18-MDIP	Rhm	-	
BU 2911	CMOS-IC	Türgong/door Chime, Multi-melody	18-DIP	Rhm	-	
BU 3434	CMOS-µC-IC	4Bit, 5V, 96x4 Bit RAM, 2048x8 Bit ROM	24-MDIP	Rhm	-	
BU 3437	CMOS-µC-IC	4Bit, 5V, 64x4 Bit RAM, 1024x8 Bit ROM	32-SDIP	Rhm	-	
BU 3447	CMOS-µC-IC	4Bit, 5V, 64x4 Bit RAM, 1024x8 Bit ROM	32-MP	Rhm	-	
BU 3451	CMOS-µC-IC	4Bit, 3V, 16x4 Bit RAM, 640x8 Bit ROM	24-MDIP	Rhm	-	
BU 3456	CMOS-µC-IC	4Bit, 5V, 32x4 Bit RAM, 1024x8 Bit ROM	24-SMDIP	Rhm	-	
BU 3458	CMOS-µC-IC	4Bit, 5V, 32x4 Bit RAM, 1024x8 Bit ROM	22-MDIP	Rhm	-	
BU 3459	CMOS-µC-IC	4Bit, 5V, 32x4 Bit RAM, 1024x8 Bit ROM	22-DIP	Rhm	-	
BU 3460	CMOS-µC-IC	4Bit, 5V, 32x4 Bit RAM, 1024x8 Bit ROM	22-SDIP	Rhm	-	
BU 3461	CMOS-µC-IC	4Bit, 5V, 32x4 Bit RAM, 1024x8 Bit ROM	20-MDIP	Rhm	-	
BU 3462	CMOS-µC-IC	4Bit, 5V, 32x4 Bit RAM, 1024x8 Bit ROM	20-DIP	Rhm	-	
BU 3463	CMOS-µC-IC	4Bit, 5V, 16x4 Bit RAM, 640x8 Bit ROM	18-MDIP	Rhm	-	
BU 3464	CMOS-µC-IC	4Bit, 5V, 16x4 Bit RAM, 640x8 Bit ROM	18-DIP	Rhm	-	
BU 3465	CMOS-µC-IC	4Bit, 5V, 16x4 Bit RAM, 640x8 Bit ROM	18-SDIP	Rhm	-	
BU 3466	CMOS-µC-IC	4Bit, 5V, 16x4 Bit RAM, 640x8 Bit ROM	16-MDIP	Rhm	-	
BU 3616 K	CMOS-D/A-IC	SMD, 8 Bit, 3 Channel, Video	44-MP	Rhm	-	pdf
BU 3892 FV	CMOS-IC	SMD, Hdd, Cd-rom, Shock Sensor Ctrl.	16-SSMDIP	Rhm	-	
BU4051	MULT8	Vs:3...18V Ron:aOhm Ton:150ns at 5/0V	16DS	Roh		data pdf pinout
BU4052	MULT4	Vs:3...18V Ron:aOhm Ton:150ns at 5/0V	16DS	Roh		data pdf pinout
BU4053	MULT2	Vs:3...18V Ron:aOhm Ton:150ns at 5/0V	16DS	Roh		data pdf pinout
BU4066	SPST	Vs:3...18V Ron:Ohm Ton:40ns at 5/0V	14DS	Roh		data pdf pinout
BU 4506 AF	Si-N	CTV-HA, ...16kHz, 1500/800V, 5A, 45W	16c	Phi	2SC5042	data
BU 4506 AX	Si-N	=BU 4506AF:	18c		→BU 4506AF	data
BU 4506 AZ	Si-N	=BU 4506AF: 32W	17c		-	data
BU 4506 DF	Si-N+Di	=BU 4506AF: integr. Damper-Diode	16c		2SC4122, 2SC4293...4294	data
BU 4506 DX	Si-N+Di	=BU 4506AF: integr. Damper-Diode	18c		→BU 4506DF	data
BU 4506 DZ	Si-N+Di	=BU 4506AF: 32W, integr. Damper-Diode	17c		-	data
BU 4507 AF	Si-N	CTV/Monitor-HA, 1500/800V, 8A, 45W ..56kHz	16c	Phi	2SC4758	data
BU 4507 AX	Si-N	=BU 4507AF:	18c		→BU 4507AF	data
BU 4507 AZ	Si-N	=BU 4507AF: 32W	17c		-	pinout
BU 4507 DF	Si-N+Di	=BU 4507AF: integr. Damper-Diode	16c		2SC4124, 2SC5043	data
BU 4507 DX	Si-N+Di	=BU 4507AF: integr. Damper-Diode	18c		→BU 4507DF	data
BU 4507 DZ	Si-N+Di	=BU 4507AF: 32W, integr. Damper-Diode	17c		-	data
BU 4508 AF	Si-N	CTV/Monitor-HA, 1500/800V, 8A, 45W ..64kHz	16c	Phi	2SC4758	data
BU 4508 AX	Si-N	=BU 4508AF:	18c		→BU 4508AF	data
BU 4508 AZ	Si-N	=BU 4508AF: 32W	17c		-	data
BU 4508 DF	Si-N+Di	=BU 4508AF: integr. Damper-Diode	16c		2SC4124, 2SC5043	data
BU 4508 DX	Si-N+Di	=BU 4508AF: integr. Damper-Diode	18c		→BU 4508DF	data
BU 4508 DZ	Si-N+Di	=BU 4508AF: 32W, integr. Damper-Diode	17c		-	data
BU 4515 AF	Si-N	CTV/Monitor-HA, 1500/800V, 9A, 45W ..64kHz	16c	Phi	2SC4759, 2SC5044	data
BU 4515 AX	Si-N	=BU 4515AF:	18c		→BU 4515AF	data
BU 4515 DF	Si-N+Di	=BU 4515AF: integr. Damper-Diode	16c		2SC4125	data
BU 4515 DX	Si-N+Di	=BU 4515AF: integr. Damper-Diode	18c		→BU 4515DF	data
BU 4522 AF	Si-N	CTV/Monitor-HA, 1500/800V, 10A, 45W ..64kHz	16c	Phi	2SC4759, 2SC5044	data
BU 4522 AX	Si-N	=BU 4522AF:	18c		→BU 4522AF	data
BU 4522 DF	Si-N+Di	=BU 4522AF: integr. Damper-Diode	16c		2SC4125	data
BU 4522 DX	Si-N+Di	=BU 4522AF: integr. Damper-Diode	18c		→BU 4522DF	data

Type	Device	Short Description	Fig.	Manu	Comparision Types	More at
BU 4523 AF	Si-N	CTV/Monitor-HA, 1500/800V, 11A, 45W ...70kHz	16c	Phi	2SC4890, 2SC5251	data
BU 4523 AW	Si-N	=BU 4523AF: 125W	16j§		-	data
BU 4523 AX	Si-N	=BU 4523AF:	18c		→BU 4523AF	data
BU 4523 DF	Si-N	=BU 4523AF: integr. Damper-Diode	16c		-	data
BU 4523 DW	Si-N	=BU 4523AF: 125W, integr. Damper-Diode	18c		-	data
BU 4523 DX	Si-N	=BU 4523AF: integr. Damper-Diode	18c		-	data
BU 4525 AF	Si-N	CTV/Monitor-HA, 1500/800V, 12A, 45W ...70kHz	16c	Phi	2SC4890, 2SC5251	data
BU 4525 AL	Si-N	=BU 4525AF: 125W	77j§		2SC3995...3996	data
BU 4525 AW	Si-N	=BU 4525AF: 125W	16j§			data
BU 4525 AX	Si-N	=BU 4525AF:	18c		→BU 4525AF	data
BU 4525 DF	Si-N+Di	=BU 4525AF: integr. Damper-Diode	16c		2SC4125	data
BU 4525 DL	Si-N+Di	=BU 4525AF: 125W, integr. Damper-Diode	77j§			data
BU 4525 DW	Si-N+Di	=BU 4525AF: 125W, integr. Damper-Diode	16j§			data
BU 4525 DX	Si-N+Di	=BU 4525AF: integr. Damper-Diode	18c		→BU 4525DF	data
BU 4530 AL	Si-N	CTV/Monitor-HA, 1500/800V, 16A, 125W ...90kHz	77j§	Phi	-	data
BU 4530 AW	Si-N	=BU 4530AL:	16j§		-	data
BU 4540 AL	Si-N	CTV/Monitor-HA, 1500/800V, 25A, 125W ...110kHz	77j§	Phi	-	data
BU 4540 AW	Si-N	=BU 4540AL:	16j§		-	data
BU 4550 AL	Si-N	CTV/Monitor-HA, 1500/800V, 25A, 125W ...110kHz	77j§	Phi	-	data
BU4551	MULT8	Vs:3...18V Ron:aOhm Ton:360ns at 5/0V	16DS	Roh	-	data pdf pinout
BU 6550 KS	MOS-IC	SMD, Jpeg Proc. f. Dig. Still Camera	160-MP	Rhm	-	
BU 6836	MOS-IC	µcomp, Notebook Card Bus Controller	256-PGA	Rhm	-	
BU 8241 F	MOS-IC	SMD, Telecom, Cross Point Mixer	24-MDIP	Rhm	-	pdf
BU 8241 FS	MOS-IC	=BU 8241F:	24-SMDIP	Rhm	-	pdf
BU 8242 F	MOS-IC	SMD, Telecom, Cross Point Mixer	20-MDIP	Rhm	-	pdf
BU 8244 F	MOS-IC	SMD, Telecom, Cross Point Mixer	16-MDIP	Rhm	-	pdf
BU 8302 A	MOS-IC	Telecom, Wahl/tone & pulse dialer	24-DIP	Rhm	-	
BU 8304	MOS-IC	Telecom, Wahl/tone & pulse dialer	24-DIP	Rhm	-	
BU 8304 F	MOS-IC	=BU 8304: SMD	28-MDIP	Rhm	-	
BU 8307 F,CF	MOS-IC	=BU 8307S: SMD	24-MDIP	Rhm	-	pdf
BU 8307 S,CS	MOS-IC	Telecom, Wahl/tone & pulse dialer	22-SDIP	Rhm	-	pdf
BU 8309 AK	MOS-IC	=BU 8309AS: SMD	32-MP		-	
BU 8309 AS	MOS-IC	Telecom, Wahl/tone & pulse dialer	32-SDIP	Rhm	-	
BU 8310 AK	MOS-IC	SMD, Telecom, LED Panel Interface	64-MP	Rhm	-	pdf
BU 8311 KS	MOS-IC	SMD, Telecom, LED Panel Interface	56-MP	Rhm	-	pdf
BU 8313 K	MOS-IC	SMD, Telecom, LED Panel Interface	44-MP	Rhm	-	pdf
BU 8315 F	MOS-IC	=BU 8315S: SMD	24-MDIP	Rhm	-	pdf
BU 8315 S	MOS-IC	Telecom, LED Panel Interface	24-SDIP	Rhm	-	pdf
BU 8317 KV	MOS-IC	Telecom, LED Panel Interface	80-MP	Rhm	-	
BU 8320 A	MOS-IC	Telecom, Wahl/tone & pulse dialer	28-DIP	Rhm	-	
BU 8320 AF	MOS-IC	=BU 8320A: SMD	28-MDIP	Rhm	-	
BU 8321	MOS-IC	Telecom, Wahl/tone & pulse dialer	28-DIP	Rhm	-	
BU 8321 F	MOS-IC	=BU 8321: SMD	28-MDIP	Rhm	-	
BU 8322	MOS-IC	Telecom, Wahl/tone & pulse dialer	28-DIP	Rhm	-	
BU 8322 F	MOS-IC	=BU 8322: SMD	28-MDIP	Rhm	-	
BU 8323	MOS-IC	Telecom, Wahl/tone & pulse dialer	28-DIP	Rhm	-	
BU 8323 F	MOS-IC	=BU 8323: SMD	28-MDIP	Rhm	-	
BU 8325 K	MOS-IC	=BU 8325S: SMD	32-MP		-	
BU 8325 S	MOS-IC	Telecom, Wahl/tone & pulse dialer	32-SDIP	Rhm	-	
BU 8326	MOS-IC	Telecom, Wahl/tone & pulse dialer	22-DIP	Rhm	-	
BU 8329	MOS-IC	Telecom, Wahl/tone & pulse rep. dialer	22-DIP	Rhm	-	
BU 8329 F	MOS-IC	=BU 8329: SMD	24-MDIP		-	
BU 8330	MOS-IC	Telecom, Wahl/tone & pulse rep. dialer	22-DIP	Rhm	-	
BU 8331	MOS-IC	Telecom, Wahl/tone & pulse rep. dialer	22-DIP	Rhm	-	
BU 8710 AKS	MOS-IC	Telecom, Cordless T. ADPCM Transcoder	80-MP	Rhm	-	pdf
BU 8871	MOS-IC	Telecom, DTMF-empfänger/receiver	8-DIP	Rhm	-	
BU 8871 F	MOS-IC	=BU 8871: SMD	18-MDIP		-	pdf
BU 8872	MOS-IC	Telecom, Dtmf-empfänger/receiver	8-DIP	Rhm	-	pdf
BU 8872 FS	MOS-IC	=BU 8872: SMD	16-SMDIP		-	pdf
BU 8874	MOS-IC	Telecom, Dtmf-empfänger/receiver	8-DIP	Rhm	-	pdf
BU 8874 F	MOS-IC	=BU 8874: SMD	18-MDIP		-	pdf
BU 9206	MOS-IC	µcomp, PS/2 Mouse Controller	16-DIP	Rhm	-	pdf
BU 9250 F	CMOS-IC	=BU 9250S: SMD	18-MDIP	Rhm	-	
BU 9250 S	CMOS-IC	Audio Digital Delay f. Karaoke Echo	18-SDIP	Rhm	-	
BU 9251 F	CMOS-IC	=BU 9251S: SMD	18-MDIP	Rhm	-	
BU 9251 S	CMOS-IC	Audio Digital Delay f. Karaoke Echo	18-SDIP	Rhm	-	
BU 9252 F	CMOS-IC	=BU 9252S: SMD	18-MDIP	Rhm	-	pdf
BU 9252 S	CMOS-IC	Audio Digital Delay f. Karaoke Echo	18-SDIP	Rhm	-	pdf
BU 9253 AS	CMOS-IC	Karaoke Echo, A/D+D/A-Conv., Sram	18-SDIP	Rhm	-	pdf
BU 9253 FS	CMOS-IC	=BU 9253AS: SMD	16-SMDIP		-	pdf
BU 9255 FS	CMOS-IC	SMD, Audio Dig. Delay f. Karaoke Echo	16-SMDIP	Rhm	-	pdf
BU 9260 FS	CMOS-IC	SMD, Audio Dig. Key f. Karaoke Echo	24-SMDIP	Rhm	-	
BU 9262 FS	CMOS-IC	SMD, Audio Dig. Echo + Surround Sound	32-SMDIP	Rhm	-	
BU 9480 F	CMOS-D/A-IC	SMD, 16 Bit, Stereo Audio, Udd=3.. 5,5V	8-MDIP	Rhm	-	pdf
BU 9500 K	DIG-IC	SMD, Floppy-disk, Controller	64-MP	Rhm	-	
BU 9706 KS	MOS-IC	SMD, LCD Segment Treiber/driver	56-MP	Rhm	-	pdf
BU 9732 KS	MOS-IC	SMD, LCD Segment Treiber/driver	100-MP	Rhm	-	
BU 9742 KS	MOS-IC	SMD, LCD Segment Treiber/driver	100-MP	Rhm	-	

Type	Device	Short Description	Fig.	Manu	Comparision Types	More at
BU 9745 KS	MOS-IC	SMD, LCD Segment Treiber/driver		Rhm	-	
BU9761	2IN/2OUT	Vs:4...6V Ron:Ohm Ton:200ns at 5/0V	20S	Roh		data pdf pinout
BU 9785 KS	MOS-I/O-IC	ASIC, V.34 Modem f. PCs, D/a-converter	100-MP	Rhm	-	
BU 12001...12007	CMOS-IC	Gate Arrays		Rhm	-	
BU 12101...12113	CMOS-IC	Gate Arrays		Rhm	-	
BU 12206...12211	CMOS-IC	Gate Arrays		Rhm		
BU 12307...12311	CMOS-IC	Gate Arrays		Rhm		
BU 24201	CMOS-µC-IC	4 Bit, 128x4 Bit RAM, 2048x8 Bit ROM	32-SDIP	Rhm		
BU 24400(L)	CMOS-µC-IC	4 Bit, 256x4 Bit RAM, 4096x8 Bit ROM	64-MP	Rhm		
BU 24401 L	CMOS-µC-IC	4 Bit, 128x4 Bit RAM, 4096x8 Bit ROM	80-MP	Rhm		
BU 24403	CMOS-µC-IC	4 Bit, 256x4 Bit RAM, 4096x8 Bit ROM	80-MP	Rhm		
BU 24404	CMOS-µC-IC	4 Bit, 256x4 Bit RAM, 4096x8 Bit ROM	64-MP	Rhm		
BU 24407 L	CMOS-µC-IC	4 Bit, 128x4 Bit RAM, 4096x8 Bit ROM	64-MP	Rhm		
BU 24410 L	CMOS-µC-IC	4 Bit, 256x4 Bit RAM, 4096x8 Bit ROM	80-MP	Rhm		
BU 24421	CMOS-µC-IC	4Bit, 3V,128x4 Bit RAM, 4096x8 Bit ROM	64-MP	Rhm		
BU 24426	CMOS-µC-IC	4Bit, 3V,128x4 Bit RAM, 4096x8 Bit ROM	80-MP	Rhm		
BU 24805(L)	CMOS-µC-IC	4Bit, 3V,256x4 Bit RAM, 8192x8 Bit ROM	80-MP	Rhm		
BU 24807	CMOS-µC-IC	4Bit, 3V,128x4 Bit RAM, 8192x8 Bit ROM	64-MP	Rhm		
BU 24821	CMOS-µC-IC	4Bit, 3V,128x4 Bit RAM, 8192x8 Bit ROM	64-MP	Rhm		
BU 24822	CMOS-µC-IC	4Bit, 3V,256x4 Bit RAM, 8192x8 Bit ROM	80-MP	Rhm		
BU 34204	CMOS-µC-IC	4Bit, 5V,128x4 Bit RAM, 2048x8 Bit ROM	64-MP	Rhm		
BU 34424	CMOS-µC-IC	4Bit, 5V,256x4 Bit RAM, 4096x8 Bit ROM	64-MP	Rhm		
BU 38101	CMOS-µC-IC	8Bit, 5V, 96x8 Bit RAM, 2048x8 Bit ROM	32-MP	Rhm		
BU 38603	CMOS-µC-IC	8Bit, 5V,512x8 Bit RAM,16384x8 Bi. ROM	80-MP	Rhm	-	pdf
BU 38703	CMOS-µC-IC	8Bit, 5V,512x8 Bit RAM,24576x8 Bit ROM	80-MP	Rhm	-	pdf
BU 38803	CMOS-µC-IC	8Bit, 5V,512x8 Bit RAM,32768x8 Bit ROM	80-MP	Rhm	-	pdf
BU 38905	CMOS-µC-IC	8Bit, 896x8 Bit RAM, 48152x8 Bit ROM	80-MP	Rhm		
BUA	Z-Di	=SMBJ 6.0A (Typ-Code/Stempel/marking	71(5x3,5)	Tho	→SMBJ ...	data
BUB	Z-Di	=SMBJ 6.5A (Typ-Code/Stempel/marking	71(5x3,5)	Tho	→SMBJ ...	data
BUB 931 T	Si-N-Darl+Di	=BU 931: 10A, 125W	30j§	Tho	-	data
BUB 941 T,ZT	Si-N-Darl+Di	=BU 941: 150W	30j§	Tho	-	data
BUC	Z-Di	=SMBJ 8.5A (Typ-Code/Stempel/marking	71(5x3,5)	Tho	→SMBJ ...	data

BUD

Type	Device	Short Description	Fig.	Manu	Comparision Types	More at
BUD	Z-Di	=SMBJ 10A (Typ-Code/Stempel/marking	71(5x3,5)	Tho	→SMBJ ...	data
BUD 43 B	Si-N	SMPS,lo-drive,650/350V, 2A, 25W, 13MHz	30j§	Mot	-	data
BUD 44 D2	Si-N+Di	Ballast, lo-drive, 700/400V, 2A, 25W	30j§	Mot	-	data
BUD 46	Si-N-Darl	S-L, 850/400V, 4A, 70W	17j§	Tho	-	data
BUD 46 A	Si-N-Darl	=BUD 46: 1000/450V	17j§		-	data
BUD 47	Si-N-Darl	S-L, 850/400V, 8A, 100W	17j§	Tho	-	data
BUD 47 A	Si-N-Darl	=BUD 47: 1000/450V	17j§		-	data
BUD 48	Si-N-Darl	S-L, 850/400V, 16A, 150W	18j§	Tho	BUT 51P	data
BUD 48 A	Si-N-Darl	=BUD 48: 1000/450V	18j§		-	data
BUD 48 DI	Si-N-Darl+Di	=BUD 48: Iso, Damper-Di., 600V, 90W	18c		-	data
BUD 48 I, AI	Si-N-Darl	=BUD 48(A): Iso, 90W	18c		-	data
BUD 86 (-SMD)	Si-N	S-L, 800/400V, 0,5A, 20W(Tc=60°),20MHz	30j§	Aeg	2SC3405, 2SC3489, 2SC3555	data
BUD 87 (-SMD)	Si-N	=BUD 86: 1000/450V	30j§	Aeg	2SD1945	data
BUD 98	Si-N-Darl	S-L, 850/400V, 32A, 250W	18j§	Tho	-	data
BUD 98 I	Si-N-Darl	=BUD 98: Iso, 110W	18c		-	data
BUD 600 (-SMD)	Si-N	S-L,lo-sat, 600/250V, 2A, 12W, >4MHz	30j§	Aeg	BUD 620	data
BUD 616 A (-SMD)	Si-N	L,lo-sat, 1000/450V, 1,6A, 20W(Tc=50°)	30h§ (!)	Aeg	-	data
BUD 620 (-SMD)	Si-N	S-L, lo-sat, 700/400V, 4A, 25W, 8MHz	30j§	Aeg	-	data
BUD 630 (-SMD)	Si-N	S-L, lo-sat, 700/400V, 6A, 40W, >4MHz	30j§	Aeg	-	data
BUD 636 (-SMD)	Si-N	S-L, lo-sat, 1000/450V, 5A, 40W, >4MHz	30j§	Aeg	-	data
BUE	Z-Di	=SMBJ 12A (Typ-Code/Stempel/marking	71(5x3,5)	Tho	→SMBJ ...	data

BUF...BUJ

Type	Device	Short Description	Fig.	Manu	Comparision Types	More at
BUF	Z-Di	=SMBJ 13A (Typ-Code/Stempel/marking	71(5x3,5)	Tho	→SMBJ ...	data
BUF03	FOLLOWER	Vs:±18V Vi0:2mV f:63Mc	8Ad	Pmi		data pinout
BUF-03(A,B,E,F)J	LIN-IC	Buffer,Voltage Follower, ±18V, Iq=70mA	81	Pmi		data
BUF04	FAST-BUFF	Vs:±18V Vo:±13.5V Vi0:0.3mV f:110Mc	8DdS	And		data pinout
BUF 298 F,AF,V,AV	Si-N	=BUV 298...	80	Tho	-	data
BUF 405	Si-N	SMPS, lo-drive, 850/450V, 7,5A, 80W	17j§	Tho	BUF 646, MJE 18008	data
BUF 405 A	Si-N	=BUF 405: 1000/450V	17j§		MJE 18008	data
BUF 405 FI,AFI	Si-N	=BUF 405(A): Iso, 40W	17c		MJF 18008	data
BUF 405 XI,AXI	Si-N	=BUF 405(A): Iso, 35W	17c		MJF 18008	data
BUF 410	Si-N	SMPS, lo-drive, 850/450V, 15A, 125W	18j§	Tho	-	data
BUF 410 A	Si-N	=BUF 410: 1000/450V	18j§		-	data
BUF 410 AFI,FI	Si-N	=BUF 410(A): Iso, 60W	18c		-	data
BUF 410 AI	Si-N	=BUF 410A: Iso, 85W	18c		-	data
BUF 410 I	Si-N	=BUF 410: Iso, 85W	18c		-	data
BUF 420	Si-N	SMPS, lo-drive, 850/450V, 30A, 200W	18j§	Tho	-	data
BUF 420 A	Si-N	=BUF 420: 1000/450V	18j§		-	data
BUF 420 AI	Si-N	=BUF 420A: Iso, 115W	18c		-	data

Type	Device	Short Description	Fig.	Manu	Comparision Types	More at
BUF 420 AM	Si-N	=BUF 420A: 200W	23a§		-	data
BUF 420 I	Si-N	=BUF 420: Iso, 115W	18c		-	data
BUF 420 M	Si-N	=BUF 420: 200W	23a§		-	data
BUF 460(F)	Si-N	S-L, 850/450V, 80A, 270W	80zk	Tho	-	data
BUF 460 A,AF	Si-N	=BUF 460(F): 1000/450V	80zk			data
BUF 460 V,AV	Si-N	=BUF 460(F,AF)	80zk			data
BUF600	BUFF	Vs:±6V Vo:±3.3V Vi0:4mV f:320Mkc	d,8DS	Bub		data pinout
BUF601	BUFF	Vs:±6V Vo:±3V Vi0:1.5mV f:320Mkc	d,8DS	Bub		data pinout
BUF 620	Si-N	S-L, lo-sat, 700/400V, 4A, 40W, 8MHz	17j§	Aeg	-	data
BUF 630	Si-N	S-L, lo-sat, 700/400V, 6A, 50W, >4MHz	17j§	Aeg	-	data
BUF634	BUFF	Vs:±18V Vo:±1.7V Vi0:±30mV f:160Mkc	d,5S,8DPS	Bub		data pinout
BUF 636 A	Si-N	S-L, lo-sat, 1000/450V, 5A, 50W, >4MHz	17j§	Aeg		data
BUF 640	Si-N	S-L, 850/400V, 6A, 70W, <0,7/2,2µs	17j§	Aeg	BUT 11(A), BUT 18(A), BUV 46(A), ++	data
BUF 640 A	Si-N	=BUF 640: 1000/450V	17j§	·	BUT 11A, BUT 12A, BUT 18A, BUV 46A, ++	data
BUF 642	Si-N	S-L, 900/400V, 6A, 70W, <0,5/2,2µs	17j§	Aeg	BUT 11A, BUT 12A, BUT 18A, BUV 46A, ++	data
BUF 644	Si-N	S-L,lo-sat,700/400V, 8A, 70W, <1,2/2µs	17j§	Aeg		data
BUF 646	Si-N	S-L, 850/400V, 7A, 70W, -/3,25µs	17j§	Aeg	BUT 12(A), BUT 56(A), BUF 405(A)	data
BUF 646 A	Si-N	=BUF 646: 1000/450V	17j§		BUT 12A, BUT 56A, BUF 405A	data
BUF 650	Si-N	lo-sat, 700/400V, 10A, 70W,<0,6/1,65µs	17j§	Aeg		data
BUF 653	Si-N	lo-sat, 700/400V, 11A, 70W, >4MHz	17j§	Aeg	BUF 654	data
BUF 654	Si-N	lo-sat, 700/400V, 12A, 80W,<0,7/1,65µs	17j§	Aeg		data
BUF 656 B	Si-N	S-L, 1100/400V, 7A, 70W, 180ns	17j§	Aeg	(BUT 12A, BUT 56A, BUF 405A)[7]	data
BUF 660	Si-N	lo-sat, 700/400V, 14A, 80W, >4MHz	17j§	Aeg	BUF 654	data
BUF 672	Si-N	lo-sat, 900/450V, 11A, 80W, >4MHz	17j§	Aeg	(BUV 66A)[17]	data
BUF 725 D	Si-N+Di	lo-sat, 700/400V, 5A, 40W, tf=400ns	17j§	Aeg	-	data
BUF 742	Si-N	lo-sat, 900/450V, 5A, 50W, >4MHz	17j§	Aeg	BUF 636A	data
BUF 744	Si-N	lo-sat, 700/400V, 8A, 50W, <1,2/2µs	17j§	Aeg	BUF 644, BUF 650	data
BUG	Z-Di	=SMBJ 15A (Typ-Code/Stempel/marking	71(5x3,5)	Tho	→SMBJ ...	data
BUH	Z-Di	=SMBJ 18A (Typ-Code/Stempel/marking	71(5x3,5)	Tho	→SMBJ ...	data
BUH 2 M20P	Si-N	TV, Dyn Focus, 2000/900V, 30mA, 40W	17j§	Tho	2SC4579, 2SC4913	data
BUH 2 M20AP	Si-N	=BUH 2M20P: 2000/1200V	17j§		2SC4913	data
BUH 50	Si-N	SMPS, lo-drive, 800/500V, 4A, 50W	17j§	Mot	BUF 405(A)	data
BUH 51	Si-N	SMPS, lo-drive, 800/500V, 3A, 50W	14j§	Mot	-	data
BUH 100	Si-N	S-L, lo-drive, 700/400V, 10A, 10W	17j§	Mot	-	data
BUH 150	Si-N	S-L, lo-drive, 700/400V, 15A, 150W	17j§	Mot	-	data
BUH 313	Si-N	CRT-HA, hi-res, 1300/600V, 5A, 50W	18c	Tho	BU 2508AF, 2SC3884...85, 2SC3894...95,++	data
BUH 313 D	Si-N+Di	=BUH 313: integr. Damper-Diode	18c		BU 2508DF, 2SC3892(A)	data
BUH 315	Si-N	CRT-HA, hi-res, 1500/700V, 6A, 44W	18c	Tho	BU 2508AF, 2SC3885A, 2SC3895, 2SC4830	data
BUH 315 D	Si-N+Di	=BUH 315: integr. Damper-Diode	18c		BU 2508DF, 2SC3892A, 2SC4294, 2SD1652	data
BUH 315 DXI	Si-N+Di	=BUH 315D: Iso, 35W	17c		BU 1506DX, BU 1508DX	data
BUH 315 XI	Si-N	=BUH 315: Iso, 35W	17c		BU 1508AX	data
BUH 414 DXI	Si-N+Di	TV/Monitor-HA, 1400/700V, 5A, 35W	17c	Tho	BU 1506DX, BU 1508DX	data
BUH 415 DXI	Si-N+Di	TV/Monitor-HA, 1500/700V, 5A, 35W	17c	Tho	BU 1506DX, BU 1508DX	data
BUH 417	Si-N	CRT-HA, 1700/700V, 7A, 55W	18c	Tho	BUH 517	data
BUH 513	Si-N	CRT-HA, hi-res, 1300/700V, 8A, 50W	18c	Tho	BU 2508AF, 2SC3886(A), 2SC3896	data
BUH 515	Si-N	CRT-HA, hi-res, 1500/700V, 8A, 50W	18c	Tho	BU 2508AF, 2SC3886A, 2SC3896	data
BUH 515 D	Si-N+Di	=BUH 515: integr. Damper-Diode	18c		BU 2515DF,DX, BU 4508DF,DX, 2SC4124	data
BUH 515 DXI	Si-N+Di	=BUH 515: integr. Damper-Diode, 40W	17c		BU 1508DX	data
BUH 515 FP	Si-N	=BUH 515: 38W	17c		BU 1508AX	data
BUH 515 XI	Si-N	=BUH 515: 40W	17c		BU 1508AX	data
BUH 517	Si-N	CRT-HA, hi-res, 1700/700V, 8A, 60W	18c	Tho	2SC5150	data
BUH 517 D	Si-N+Di	=BUH 517: integr. Damper-Diode	18c		2SC5143	data
BUH 615	Si-N	CRT-HA, hi-res, 1500/700V, 8A, 54W	18c	Tho	BU 2508AF, 2SC3886A, 2SC3896, 2SC5148	data
BUH 615 D	Si-N+Di	=BUH 615: integr. Damper-Di, 55W	18c		BU 2508DF, 2SC3893A, 2SC4763, 2SC5149	data
BUH 713	Si-N	CTV-HA, hi-res, 1300/700V, 10A, 57W	18c	Tho	BU 2520AF, BU 2525AF, 2SC4924, 2SC4542	data
BUH 715	Si-N	CTV-HA, hi-res, 1500/700V, 10A, 57W	18c	Tho	BU 2520AF, BU 2525AF, 2SC4924, 2SC4542	data
BUH 1015(T)	Si-N	CRT-HA, hi-res, 1500/700V, 14A, 160W T: 16A	18j§	Tho	BUH 1215	data
BUH 1015 HI	Si-N	=BUH 1015: Iso, 70W	18c		2SC4891, 2SC5252	data
BUH 1215(T)	Si-N	CRT-HA, hi-res, 1500/700V, 16A, 200W T: 19A	18j§	Tho	-	data
BUI	Z-Di	=SMBJ 20A (Typ-Code/Stempel/marking	71(5x3,5)	Tho	→SMBJ ...	data
BUJ	Z-Di	=SMBJ 24A (Typ-Code/Stempel/marking	71(5x3,5)	Tho	→SMBJ ...	data
BUJ 101	Si-N	S, 700/400V, 1A, 2W, <880/1538ns HF Electronic Light Ballast	7b	Phi	-	data
BUJ 101 A	Si-N	S-L, 700/400V, 0,5A, 42W, <500/3300ns HF Electronic Light Ballast	17j§	Phi	BUL 44	data
BUJ 101 AX	Si-N	=BUJ 101A: Iso, 26W	17c		BUL 44F	data
BUJ 103 A	Si-N	S-L, 700/400V, 4A, 80W, <600/3630ns HF Electronic Light Ballast	17j§	Phi	BUH 50, BUL 45, BUL 128	data
BUJ 103 AX	Si-N	=BUJ 103A: Iso, 26W	17c		BUL 44F, BUL 128FP	data
BUJ 105 A	Si-N	S-L, 700/400V, 6A, 80W, <1000/4800ns HF Electronic Light Ballast	17j§	Phi	BUL 45, BUL 138, BUL 146	data
BUJ 105 AX	Si-N	=BUJ 105A: Iso, 26W	17c		BUL 146F	data

Type	Device	Short Description	Fig.	Manu	Comparision Types	More at
JJ 106 A	Si-N	S-L, 700/400V, 8A, 80W, <1000/4800ns HF Electronic Light Ballast	17j§	Phi	BUH 100, BUL 147, MJE 16106	data
JJ 106 AX	Si-N	=BUJ 106A: Iso, 26W	17c		BUL 147F, BUL 57PI, MJF 13007	data
JJ 202 A	Si-N	S-L, 850/450V, 2A, 40W, <500/4900ns HF Electronic Light Ballast	17j§	Phi	MJE 18002	data
JJ 202 AX	Si-N	=BUJ 202A: Iso, 18W	17c		MJF 18002	data
JJ 204 A	Si-N	S-L, 850/450V, 6A, 100W, <800/4470ns HF Electronic Light Ballast	17j§	Phi	BUF 405(A), MJE 18006	data
UJ 204 AX	Si-N	=BUJ 204A: Iso, 32W	17c		BUF 405(A)FI, MJF 18006	data
JJ 205 A	Si-N	S-L, 850/450V, 8A, 125W, <1000/4800ns HF Electronic Light Ballast	17j§	Phi	BUF 405(A), MJE 18008	data
JJ 205 AX	Si-N	=BUJ 205A: Iso, 32W	17c		BUF 405(A)FI, MJF 18008	data
JJ 301 A	Si-N	S-L, 1000/500V, 0,5A, 42W, <1000/4800 HF Electronic Light Ballast	17j§	Phi	MJE 18002	data
JJ 301 AX	Si-N	=BUJ 301A: Iso, 32W	17c		MJF 18002	data
JJ 303 A	Si-N	S-L, 1000/500V, 5A, 100W, <700/4450 HF Electronic Light Ballast	17j§	Phi	BUF 405A, MJE 18006	data
JJ 303 AX	Si-N	=BUJ 303A: Iso, 32W	17c		BUF 405 AFI, MJF 18006	data
JJ 304 A	Si-N	S-L, 1000/500V, 6A, 100W, <1000/4800 HF Electronic Light Ballast	17j§	Phi	BUF 405A, MJE 18006	data
JJ 304 AX	Si-N	=BUJ 304A: Iso, 35W	17c		BUF 405 AFI, MJF 18008	data
JJ 403 A	Si-N	S-L, 1200/550V, 6A, 100W, <500/3300 HF Electronic Light Ballast	17j§	Phi	MJE 18206	data
UJ 403 AX	Si-N	=BUJ 403A: Iso, 35W	17c		MJF 18206	data

BUK

Type	Device	Short Description	Fig.	Manu	Comparision Types	More at
UK	Z-Di	=SMBJ 26A (Typ-Code/Stempel/marking	71(5x3,5)	Tho	→SMBJ ...	data
UK 100-50 DL	MOS-N-FET-e	V-MOS, LogL, Temp./Overload Prot., 50V 13,5A, 40W, <125mΩ(7,5A), 48/75µs	17c§	Phi	-	data
UK 100-50 GL	MOS-N-FET-e	=BUK 100-50DL: 9,5/10,5µs	17c§		-	data
UK 100-50 GS	MOS-N-FET-e	=BUK 100-50DL: 15A, 5/15µs	17c§		-	data
UK 101-50 DL	MOS-N-FET-e	V-MOS, LogL, Temp./Overload Prot., 50V 26A, 75W, <60mΩ(13A), 92/130µs	17c§	Phi	-	data
UK 101-50 GL	MOS-N-FET-e	=BUK 101-50DL: 17,5/17µs	17c§		-	data
UK 101-50 GS	MOS-N-FET-e	=BUK 101-50DL: 29A, 7,5/27µs	17c§		-	data
UK 102-50 DL	MOS-N-FET-e	V-MOS, LogL, Temp./Overload Prot., 50V 45A, 125W, <35mΩ(25A), 180/240µs	17c§	Phi	-	data
UK 102-50 GL	MOS-N-FET-e	=BUK 102-50DL: 10/16µs	17c§		-	data
UK 102-50 GS	MOS-N-FET-e	=BUK 102-50DL: 50A, 7/22µs	17c§		-	data
UK 104-50 L,S(P)	MOS-N-FET-e	V-MOS, LogL, Temp./Overload Prot., 50V 15A, 40W, <125mΩ(7,5A), 21/145µs	86/5Pin	Phi		data
UK 105-50 L,S(P)	MOS-N-FET-e	V-MOS, LogL, Temp./Overload Prot., 50V 29A, 75W, <60mΩ(13A)• 33/225µs	86/5Pin	Phi		data
UK 106-50 L,S(P)	MOS-N-FET-e	V-MOS, LogL, Temp./Overload Prot., 50V 50A, 125W, <35mΩ(25A), 45/400µs	86/5Pin	Phi		data
UK 107-50 DL	MOS-N-FET-e	V-MOS, LogL, Temp./Overload Prot., 50V 0,5A, <0,2Ω(0,1A), 38/9µs	48c§	Phi	-	data
UK 107-50 DS	MOS-N-FET-e	=BUK 107-50DL: <175mΩ(0,1A), 20/9µs			-	data
UK 107-50 GL	MOS-N-FET-e	=BUK 107-50DL: 4,4/11,8µs			-	data
UK 108-50 DL	MOS-N-FET-e	V-MOS, LogL, Temp./Overload Prot., 50V 13,5A, 40W, <125mΩ(7,5A), 48/75µs	30c§	Phi	-	data
UK 108-50 GL	MOS-N-FET-e	=BUK 108-50DL: 9,5/10,5µs			-	data
UK 109-50 DL	MOS-N-FET-e	V-MOS, LogL, Temp./Overload Prot., 50V 16A, 75W, <60mΩ(13A), 92/130µs	30c§	Phi	-	data
BUK 109-50 GL	MOS-N-FET-e	=BUK 109-50DL: 17,5/17µs				data
BUK 110-50 DL	MOS-N-FET-e	V-MOS, LogL, Temp./Overload Prot., 50V 45A, 125W, <35mΩ(25A), 180/240µs,	30c§	Phi		data
BUK 110-50 GL	MOS-N-FET-e	=BUK 110-50DL: 10/16µs				data
BUK 114-50 ...	MOS-N-FET-e	=BUK 104-50...: SMD	30c§	Phi	-	data
BUK 116-50 ...	MOS-N-FET-e	=BUK 106-50...: SMD	30c§	Phi	-	data
BUK 200-50 Y	MOS-N-FET-e	V-MOS, High Side S, Temp./Overl. Prot. 50V, 10A, 62,5W, <0,1Ω(5A), 40/50µs	86/5Pin	Phi		data
BUK 201-50 Y	MOS-N-FET-e	V-MOS, High Side S, Temp./Overl. Prot. 50V, 15A, 83,3W, <60mΩ(7,5A), 40/50µs	86/5Pin	Phi		data
BUK 202-50 X	MOS-N-FET-e	V-MOS, High Side S, Temp./Overl. Prot. 50V, 20A, 125W, <38mΩ(10A), 140/70µs	86/5Pin	Phi	-,	data
BUK 202-50 Y	MOS-N-FET-e	=BUK 202-50X: int. ground resistor	86/5Pin			data
BUK 203-50 X	MOS-N-FET-e	V-MOS, High Side S, Temp./Overl. Prot. 50V, 4A, 50W, <220mΩ(2A), 40/35µs	86/5Pin	Phi	-	data
BUK 203-50 Y	MOS-N-FET-e	=BUK 202-50X: int. ground resistor	86/5Pin		-	data
BUK 204-50 ...	MOS-N-FET-e	=BUK 200-50...: SMD	30/5Pin	Phi	-	data
BUK 205-50 ...	MOS-N-FET-e	=BUK 201-50...: SMD	30/5Pin	Phi	-	data
BUK 206-50 ...	MOS-N-FET-e	=BUK 202-50...: SMD	30/5Pin	Phi	-	data
BUK 207-50 ...	MOS-N-FET-e	=BUK 203-50...: SMD	30/5Pin	Phi	-	data
BUK 416-100AE,BE	MOS-N-FET-e	V-MOS, 100V, 100...110A, 310W, <16mΩ	80	Phi	-	data
BUK 416-200AE,BE	MOS-N-FET-e	V-MOS, 200V, 55...63A, 310W, <45mΩ	80	Phi	-	data
BUK 416-1000AE,BE	MOS-N-FET-e	V-MOS, 1000V, 11...12A, 310W, <1Ω	80			data

Type	Device	Short Description	Fig.	Manu	Comparision Types	More at
BUK 417-500AE,BE	MOS-N-FET-e	V-MOS, 500V, 28..32A, 310W, 0,16Ω	80	Phi	-	data
BUK 426-50A,B	MOS-N-FET-e	V-MOS, 50V, 30A, 45W, <30mΩ	16c	Phi	2SK1223, 2SK1298, 2SK1424, 2SK1666	data
BUK 426-60A,B	MOS-N-FET-e	V-MOS, 60V, 30A, 45W, <30mΩ	16c	Phi	2SK1223, 2SK1298, 2SK1424, 2SK1666	data
BUK 426-100A,B	MOS-N-FET-e	V-MOS, 100V, 19..20A, 45W, <65mΩ	16c	Phi	2SK2008	data
BUK 426-200A,B	MOS-N-FET-e	V-MOS, 200V, 10..11A, 45W, <0,2Ω	16c	Phi	2SK1206, 2SK1225, 2SK1328	data
BUK 426-800A,B	MOS-N-FET-e	V-MOS, 800V, 2,1..2,4A, 45W, <4Ω	16c	Phi	-	data
BUK 426-1000A,B	MOS-N-FET-e	V-MOS, 1000V, 1,9..2,1A, 45W, <5Ω	16c	Phi	BUK 428-1000	data
BUK 427-400A,B	MOS-N-FET-e	V-MOS, 400V, 6,2..6,9A, 45W, <0,5Ω	16c	Phi	2SK1206, 2SK1225, 2SK1328	data
BUK 427-450B	MOS-N-FET-e	V-MOS, 450V, 5,6A, 45W, <0,6Ω	16c	Phi	2SK1206, 2SK1225, 2SK1328	data
BUK 427-500A,B	MOS-N-FET-e	V-MOS, 500V, 4,8..5,6A, 45W, <0,8Ω	16c	Phi	2SK1206, 2SK1523	data
BUK 427-600A,B	MOS-N-FET-e	V-MOS, 600V, 3,9..4,3A, 45W, <1,2Ω	16c	Phi	BUK 727-600, 2SK1463	data
BUK 428-500A,B	MOS-N-FET-e	V-MOS, 500V, 6,8..6,1A, 45W, <0,5Ω	16c	Phi	2SK1206, 2SK1523	data
BUK 428-800A,B	MOS-N-FET-e	V-MOS, 800V, 3..3,4A, 45W, <2Ω	16c	Phi	2SK809A, 2SK1463	data
BUK 428-1000A,B	MOS-N-FET-e	V-MOS, 1000V, 2,6..2,9A, 45W, <2,6Ω	16c	Phi	BUK 426-1000	data
BUK 436-50A,B	MOS-N-FET-e	V-MOS, 50V, 46..50A, 125W, <30mΩ	18c§	Phi	BUZ 346, 2SK1379	data
BUK 436-60A,B	MOS-N-FET-e	V-MOS, 60V, 46..50A, 125W, <30mΩ	18c§	Phi	2SK1379	data
BUK 436-100A,B	MOS-N-FET-e	V-MOS, 100V, 31..33A, 125W, <65mΩ	18c§	Phi	BUZ 345, 2SK850, 2SK906, 2SK1429, ++	data
BUK 436-200A,B	MOS-N-FET-e	V-MOS, 200V, 17..19A, 125W, <0,2Ω	18c§	Phi	BUZ 350, 2SK851, 2SK623, 2SK901, 2SK1491	data
BUK 436-800A,B	MOS-N-FET-e	V-MOS, 800V, 3,5..4A, 125W, <4Ω	18c§	Phi	BUZ 355...356, 2SK534, 2SK604, 2SK695	data
BUK 436-1000A,B	MOS-N-FET-e	V-MOS, 1000V, 3,1...3,5A, 125W, <5Ω	18c§	Phi	2SK696	data
BUK 436-...W	MOS-N-FET-e	=BUK 436-...	16c§	Phi		data
BUK 437-400A,B	MOS-N-FET-e	V-MOS, 400V, 12...14A, 180W, <0,5Ω	18c§	Phi	-	data
BUK 437-450B	MOS-N-FET-e	V-MOS, 450V, 11A, 180W, <0,6Ω	18c§	Phi	-	data
BUK 437-500A,B	MOS-N-FET-e	V-MOS, 500V, 10...11A, 180W, <0,8Ω	18c§	Phi	-	data
BUK 437-600A,B	MOS-N-FET-e	V-MOS, 600V, 7,8...9A, 180W, <1,2Ω	18c§	Phi	-	data
BUK 438-500A,B	MOS-N-FET-e	V-MOS, 500V, 13,5...15A, 220W, <0,5Ω	18c§	Phi	2SK1678	data
BUK 438-800A,B	MOS-N-FET-e	V-MOS, 800V, 6,6...7,6A, 220W, <2Ω	18c§	Phi	-	data
BUK 438-1000A,B	MOS-N-FET-e	V-MOS, 1000V, 5,7...6,5A, 220W, <2,6Ω	18c§	Phi	-	data
BUK 438-...W	MOS-N-FET-e	=BUK 438-...	16c§	Phi		data
BUK 439-60A	MOS-N-FET-e	V-MOS, 60V, 50A, 230W, <0,13Ω	18c§	Phi	BUK 539-60	data
BUK 441-60A,B	MOS-N-FET-e	V-MOS, 60V, 4,8..5A, 20W, <0,5Ω	17c	Phi	BUK 541-60, 2SK1260	data
BUK 441-100A,B	MOS-N-FET-e	V-MOS, 100V, 3A, 20W, <1,1Ω	17c	Phi	BUK 541-60, 2SK1264	data
BUK 442-50A,B	MOS-N-FET-e	V-MOS, 50V, 9,2...10A, 22W, <0,15Ω	17c	Phi	BUK 543-50, BUZ71(A)F, 2SK1093, 2SK1974+	data
BUK 442-60A,B	MOS-N-FET-e	V-MOS, 60V, 9,2...10A, 22W, <0,15Ω	17c	Phi	BUK 545-100, 2SK1093, 2SK1344, 2SK1974++	data
BUK 442-100A,B	MOS-N-FET-e	V-MOS, 100V, 6,1...6,6A, 22W, <0,3Ω	17c	Phi	BUK 443-100, BUZ 72(A)F, 2SK1261, ++	data
BUK 443-50A,B	MOS-N-FET-e	V-MOS, 50V, 12...13A, 25W, <0,1Ω	17c	Phi	2SK1094, 2SK1419	data
BUK 443-60A,B	MOS-N-FET-e	V-MOS, 60V, 12...13A, 25W, <0,1Ω	17c	Phi	2SK1094, 2SK1419	data
BUK 443-100A,B	MOS-N-FET-e	V-MOS, 100V, 8...9A, 25W, <0,2Ω	17c	Phi	BUK 543-100, BUZ 72(A)F, 2SK1261	data
BUK 444-200A,B	MOS-N-FET-e	V-MOS, 200V, 4,7...5,3A, 25W, <0,5Ω	17c	Phi	BUK 445-200, 2SK1478, 2SK1568...69, ++	data
BUK 444-400A,B	MOS-N-FET-e	V-MOS, 400V, 2,4...2,7A, 25W, <1,8Ω	17c	Phi	BUK 445-450, 2SK503, 2SK1833	data
BUK 444-450B	MOS-N-FET-e	V-MOS, 450V, 2,1A, 25W, <2,3Ω	17c	Phi	BUK 445-450, 2SK1833	data
BUK 444-500A,B	MOS-N-FET-e	V-MOS, 500V, 1,9...2,1A, 25W, <2,3Ω	17c	Phi	BUK 445-500, 2SK1758, 2SK1833...34	data
BUK 444-600A,B	MOS-N-FET-e	V-MOS, 600V, 1,5...1,6A, 25W, <4,5Ω	17c	Phi	BUK 445-600, 2SK1611, 2SK1758, 2SK1834++	data
BUK 444-800A,B	MOS-N-FET-e	V-MOS, 800V, 1,2...1,4A, 25W, <8Ω	17c	Phi	BUK 446-1000, 2SK808(A)	data
BUK 445-50A,B	MOS-N-FET-e	V-MOS, 50V, 20...21A, 30W, <45mΩ	17c	Phi	BUK 545-50, BUZ 11F, 2SK1214, 2SK1420,++	data
BUK 445-60A,B,H	MOS-N-FET-e	V-MOS, 60V, 20...22,5A, 30W, <45mΩ	17c	Phi	2SK943, 2SK1095, 2SK1214, 2SK1420, ++	data
BUK 445-100A,B	MOS-N-FET-e	V-MOS, 100V, 12...14A, 30W, <0,1Ω	17c	Phi	2SK1350, 2SK1431	data
BUK 445-200A,B	MOS-N-FET-e	V-MOS, 200V, 7...7,6A, 30W, <0,28Ω	17c	Phi	BUK 545-200, 2SK1478, 2SK1568...69	data
BUK 445-400A,B	MOS-N-FET-e	V-MOS, 400V, 3,6...4A, 30W, <1Ω	17c	Phi	2SK1231, 2SK1351, 2SK1377	data
BUK 445-450B	MOS-N-FET-e	V-MOS, 450V, 3,1A, 30W, <1,3Ω	17c	Phi	2SK1231, 2SK1351, 2SK1377	data
BUK 445-500A,B	MOS-N-FET-e	V-MOS, 500V, 2,9...3,1A, 30W, <1,5Ω	17c	Phi	2SK1572, 2SK1767	data
BUK 445-600A,B	MOS-N-FET-e	V-MOS, 600V, 2,2...2,5A, 30W, <2,5Ω	17c	Phi	BUK 446-800, 2SK1611, 2SK1758, 2SK1834	data
BUK 445-800A,B	MOS-N-FET-e	V-MOS, 800V, 1,7...2A, 30W, <4Ω	17c	Phi	2SK1275, 2SK1459, 2SK1611, 2SK1834	data
BUK 446-1000A,B	MOS-N-FET-e	V-MOS, 1000V, 1,5...1,7A, 30W, <5Ω	17c	Phi	-	data
BUK 451-60A,B	MOS-N-FET-e	V-MOS, 60V, 5A, 40W, <0,5Ω	17c§	Phi	IRF 511, IRF 513, 2SK346, 2SK463, ++	data
BUK 451-100A,B	MOS-N-FET-e	V-MOS, 100V, 3A, 40W, <1,1Ω	17c§	Phi	IRF 612, IRF 613, 2SK923	data
BUK 452-50A,B	MOS-N-FET-e	V-MOS, 50V, 14...15A, 60W, <0,15Ω	17c§	Phi	IRF 531, 2SK673, 2SK971, 2SK1416	data
BUK 452-60A,B	MOS-N-FET-e	V-MOS, 60V, 14...15A, 60W, <0,15Ω	17c§	Phi	IRF 531, 2SK673, 2SK971, 2SK1416	data
BUK 452-100A,B	MOS-N-FET-e	V-MOS, 100V, 10...11A, 60W, <0,3Ω	17c§	Phi	BUZ 20, BUZ 72, 2SK918, 2SK921	data
BUK 453-50A,B	MOS-N-FET-e	V-MOS, 50V, 20...22A, 75W, <0,1Ω	17c§	Phi	BUZ 10, BUZ 21, 2SK600, 2SK674, 2SK972++	data
BUK 453-60A,B	MOS-N-FET-e	V-MOS, 60V, 20...22A, 75W, <0,1Ω	17c§	Phi	BUZ 21, 2SK600, 2SK674, 2SK972, 2SK1417+	data
BUK 453-100A,B	MOS-N-FET-e	V-MOS, 100V, 13...14A, 75W, <0,2Ω	17c§	Phi	BUZ 21...22, 2SK919, 2SK929, 2SK1301, ++	data
BUK 453-500A,B	MOS-N-FET-e	V-MOS, 500V, 1,6...1,7A, 75W, <7Ω	17c§	Phi	BUZ 74, IRF 820, IRF 822, 2SK822, 2SK892	data
BUK 454-200A,B	MOS-N-FET-e	V-MOS, 200V, 8,2...9,2A, 75W, <0,5Ω	17c§	Phi	BUZ 31...32, 2SK459, 2SK477, 2SK1319, ++	data
BUK 454-400A,B	MOS-N-FET-e	V-MOS, 400V, 4,2...4,6A, 75W, <1,8Ω	17c§	Phi	BUZ 41...42, BUZ 60, 2SK552...53, 2SK1751+	data
BUK 454-450B	MOS-N-FET-e	V-MOS, 450V, 3,7A, 75W, <2,3Ω	17c§	Phi	BUZ 41...42, 2SK552...53, 2SK893, 2SK1751+	data
BUK 454-500A,B	MOS-N-FET-e	V-MOS, 500V, 3,3...3,7A, 75W, <2,8Ω	17c§	Phi	BUZ 41...42, 2SK553, 2SK893, 2SK1751, ++	data
BUK 454-600A,B	MOS-N-FET-e	V-MOS, 600V, 2,6...2,8A, 75W, <4,5Ω	17c§	Phi	2SK513, 2SK791...792, 2SK1600...1601, ++	data
BUK 454-800A,B	MOS-N-FET-e	V-MOS, 800V, 2...2,4A, 75W, <8Ω	17c§	Phi	BUZ 50, BUZ 80, 2SK1199, 2SK1323, ++	data
BUK 455-50A,B	MOS-N-FET-e	V-MOS, 50V, 38...41A, 125W, <45mΩ	17c§	Phi	-	data
BUK 455-60A,B,H	MOS-N-FET-e	V-MOS, 60V, 38...43A, 125W, <45mΩ A=41A, B=38A, H=43A	17c§	Phi	-	data
BUK 455-100A,B	MOS-N-FET-e	V-MOS, 100V, 23...26A, 125W, <0,1Ω	17c§	Phi		data
BUK 455-200A,B	MOS-N-FET-e	V-MOS, 200V, 13...14A, 125W, <0,28Ω	17c§	Phi		data
BUK 455-400A,B	MOS-N-FET-e	V-MOS, 400V, 6,5...7,3A, 100W, <1Ω	17c§	Phi		data
BUK 455-450B	MOS-N-FET-e	V-MOS, 450V, 5,7A, 100W, <1,3Ω	17c§	Phi		data
BUK 455-500A,B	MOS-N-FET-e	V-MOS, 500V, 5,3...5,7A, 100W, <1,5Ω	17c§	Phi		data
BUK 455-600A,B	MOS-N-FET-e	V-MOS, 600V, 4,5...4A, 100W, <2,5Ω	17c§	Phi		data

Type	Device	Short Description	Fig.	Manu	Comparision Types	More at
BUK 456-50A,B	MOS-N-FET-e	V-MOS, 50V, 51...52A, 150W, <0,03Ω	17c§	Phi	-	data
BUK 456-60A,B,H	MOS-N-FET-e	V-MOS, 60V, 51...60A, 150W, <0,03Ω	17c§	Phi	-	data
BUK 456-100A,B	MOS-N-FET-e	V-MOS, 100V, 32...34A, 150W, <65mΩ	17c§	Phi	-	data
BUK 456-200A,B	MOS-N-FET-e	V-MOS, 200V, 17...19A, 150W, <0,2Ω	17c§	Phi	-	data
BUK 456-800A,B	MOS-N-FET-e	V-MOS, 800V, 3,5...4A, 125W, <4Ω	17c§	Phi	-	data
BUK 456-1000A,B	MOS-N-FET-e	V-MOS, 1000V, 3,1...3,5A, 125W, <5Ω	17c§	Phi	-	data
BUK 457-400A,B	MOS-N-FET-e	V-MOS, 400V, 11...13A, 150W, <0,5Ω	17c§	Phi	-	data
BUK 457-450B	MOS-N-FET-e	V-MOS, 450V, 10A, 150W, <0,6Ω	17c§	Phi	-	data
BUK 457-500A,B	MOS-N-FET-e	V-MOS, 500V, 9...10A, 150W, <0,8Ω	17c§	Phi	-	data
BUK 457-600A,B	MOS-N-FET-e	V-MOS, 600V, 7,1...8A, 150W, <1,2Ω	17c§	Phi	-	data
BUK 462-100A	MOS-N-FET-e	V-MOS, 100V, 11A, 60W, <0,25Ω(5,5A)	30c§	Phi	2SK1620, 2SK1741, 2SK1907	data
BUK 463-100A	MOS-N-FET-e	V-MOS, 100V, 14A, 75W, <0,16Ω(5A)	30c§	Phi	IRFW 540A, 2SK1928, 2SK2107, 2SK2136-Z	data
BUK 464-200A	MOS-N-FET-e	V-MOS, 200V, 9,2A, 90W, <0,4Ω(3,5A)	30c§	Phi	IRFW 630A, 2SK1621, 2SK2252	data
BUK 465-100A	MOS-N-FET-e	V-MOS, 100V, 26A, 125W, <80mΩ(13A)	30c§	Phi	MTB 33N10E, IRF 3415S, 2SK2789	data
BUK 465-200A	MOS-N-FET-e	V-MOS, 200V, 14A, 125W, <230mΩ(7A)	30c§	Phi	IRFW 640A, MTB 16N25E, 2SK2134-Z	data
BUK 466-100A	MOS-N-FET-e	V-MOS, 100V, 34A, 150W, <57mΩ(15A)	30c§	Phi	MTB 33N10E, IRF 3415S, 2SK2789	data
BUK 466-200A	MOS-N-FET-e	V-MOS, 200V, 19A, 150W, <160mΩ(10A)	30c§	Phi	MTB 20N20E, 2SK2063, 2SK2107, 2SK2136-Z	data
BUK 471-...	MOS-N-FET-e	=BUK 441-...	17c	Phi	→BUK 441-...	data
BUK 472-...	MOS-N-FET-e	=BUK 442-...	17c	Phi	→BUK 442-...	data
BUK 473-...	MOS-N-FET-e	=BUK 443-...	17c	Phi	→BUK 443-...	data
BUK 474-...	MOS-N-FET-e	=BUK 444-...	17c	Phi	→BUK 444-...	data
BUK 475-...	MOS-N-FET-e	=BUK 445-...	17c	Phi	→BUK 445-...	data
BUK 476-...	MOS-N-FET-e	=BUK 446-...	17c	Phi	→BUK 446-...	data
BUK 481-60 A	MOS-N-FET-e	SMD, V-MOS, 60V, 1,6A, <0,35Ω(1,6A)	48c§	Phi	BSP 318	data
BUK 481-100 A	MOS-N-FET-e	=BUK 481-60A: 100V, 1A, <0,8Ω(1A)	48c§		BSP 296, IRFL 110	data
BUK 482-60 A	MOS-N-FET-e	SMD, V-MOS, 60V, 2,7A, <0,13Ω(2,7A)	48c§	Phi	BSP 17, IRFL 4105	data
BUK 482-100 A	MOS-N-FET-e	=BUK 482-60A: 100V, 1,8A, <0,28Ω(1,8A)	48c§		IRFL 4310	data
BUK 482-200 A	MOS-N-FET-e	=BUK 482-60A: 200V, 2A, <0,9Ω(2A)	48c§		-	data
BUK 483-60 A	MOS-N-FET-e	SMD, V-MOS, 60V, 3,2A, <0,1Ω(3,2A)	48c§	Phi	BUK 583-60A	data
BUK 539-60A	MOS-N-FET-e	V-MOS, LogL, 60V, 50A, 230W, <15mΩ	18c§	Phi	-	data
BUK 541-60A,B	MOS-N-FET-e	V-MOS, LogL, 60V, 5A, 20W, <0,5Ω	17c	Phi	BUK 554-200	data
BUK 541-100A,B	MOS-N-FET-e	V-MOS, LogL, 100V, 3A, 20W, <1,1Ω	17c	Phi	BUK 542-200	data
BUK 542-50A,B	MOS-N-FET-e	V-MOS,LogL, 50V, 8,4...9,2A, 22W,<0,18Ω	17c	Phi	BUK 543-50, 2SK1033, 2SK1256, 2SK1344,++	data
BUK 542-60A,B	MOS-N-FET-e	V-MOS,LogL, 60V, 8,4...9,2A, 22W,<0,18Ω	17c	Phi	BUK 543-60, 2SK1033, 2SK1256, 2SK1344,++	data
BUK 542-100A,B	MOS-N-FET-e	V-MOS,LogL, 100V, 5,6...6,3A,22W,<0,35Ω	17c	Phi	BUK 543-100, 2SK1261, 2SK1305, 2SK1556++	data
BUK 543-50A,B	MOS-N-FET-e	V-MOS, LogL, 50V, 12...13A, 25W, <0,1Ω	17c	Phi	BUK 545-100, 2SK1033, 2SK1256, 2SK1344++	data
BUK 543-60A,B	MOS-N-FET-e	V-MOS, LogL, 60V, 12...13A, 25W, <0,1Ω	17c	Phi	BUK 545-100, 2SK1033, 2SK1256, 2SK1344	data
BUK 543-100A,B	MOS-N-FET-e	V-MOS,LogL, 100V, 7,5...8,3A,25W,<0,22Ω	17c	Phi	BUK 545-100, 2SK1035, 2SK1261, 2SK1305++	data
BUK 545-50A,B	MOS-N-FET-e	V-MOS, LogL, 50V, 18...21A, 30W, <55mΩ	17c	Phi	2SK943, 2SK1095, 2SK1345...46, 2SK1951,++	data
BUK 545-60A,B	MOS-N-FET-e	V-MOS, LogL, 60V, 18...20A, 30W, <55mΩ H: 21A, <38mΩ(20A)	17c	Phi	2SK943, 2SK1095, 2SK1345...46, 2SK1951,++	data
BUK 545-100A,B	MOS-N-FET-e	V-MOS, LogL, 100V, 12...13A, 30W,<0,11Ω	17c	Phi	2SK1034...1035, 2SK1558, 2SK1306	data
BUK 545-200A,B	MOS-N-FET-e	V-MOS, LogL, 200V, 7...7,6A, 30W,<0,28Ω	17c	Phi	2SK2160...61	data
BUK 551-100A,B	MOS-N-FET-e	V-MOS, LogL, 100V, 3A, 40W, <1,1Ω	17c§	Phi	-	data
BUK 552-50A,B	MOS-N-FET-e	V-MOS, LogL, 50V, 13...14A, 60W, <0,18Ω	17c§	Phi	2SK971, 2SK1555	data
BUK 552-60A,B	MOS-N-FET-e	V-MOS, LogL, 60V, 14/56A, 60W, <0,18Ω	17c§	Phi	2SK971, 2SK1300...01, 2SK1559, 2SK1561,++	data
BUK 552-100A,B	MOS-N-FET-e	V-MOS,LogL, 100V, 8,5...10A, 60W,<0,35Ω	17c§	Phi	2SK1300...01, 2SK1559, 2SK1561	data
BUK 553-48 C	MOS-N-FET-e*	=BUK 553-50A: 30V, Gate-Prot.	17c§		2SK972, 2SK1910, 2SK1287, 2SK1302, ++	data
BUK 553-50A,B	MOS-N-FET-e	V-MOS, LogL, 50V, 20...21A, 75W, <0,1Ω	17c§	Phi	2SK972, 2SK1115...16, 2SK1302, 2SK1347,++	data
BUK 553-60A,B	MOS-N-FET-e	V-MOS, LogL, 60V, 20...21A, 75W, <0,1Ω	17c§	Phi	2SK972, 2SK1115...16, 2SK1302, 2SK1347,++	data
BUK 553-100A,B	MOS-N-FET-e	V-MOS,LogL, 100V, 12...13A, 75W,<0,22Ω	17c§	Phi	2SK1301, 2SK1559, 2SK1561	data
BUK 554-200A,B	MOS-N-FET-e	V-MOS, LogL, 200V, 8,2...9,2A,90W,<0,5Ω	17c§	Phi	BUK 555-200, IRL 640A	data
BUK 555-50A,B	MOS-N-FET-e	V-MOS, LogL, 50V, 35...39A, 125W, <55mΩ	17c§	Phi	MTP 35N06ZL	data
BUK 555-60A,B,H	MOS-N-FET-e	V-MOS, LogL, 60V, 35...41A, 125W, <55mΩ	17c§	Phi	-	data
BUK 555-100A,B	MOS-N-FET-e	V-MOS,LogL, 100V, 22...25A, 125W,<0,11Ω	17c§	Phi	-	data
BUK 555-200A,B	MOS-N-FET-e	V-MOS,LogL, 200V, 13...14A, 125W,<0,28Ω	17c§	Phi	-	data
BUK 556-60 A,C	MOS-N-FET-e	V-MOS, LogL, 60V, 50...60A, 150W, <26mΩ	17c§	Phi	2SK1542	data
BUK 562-100A	MOS-N-FET-e	V-MOS, LogL, 100V, 10A, 60W, <0,28Ω	30c§	Phi	2SK1620, 2SK1907	data
BUK 563-100A	MOS-N-FET-e	V-MOS, LogL, 100V, 13A, 75W, <0,18Ω	30c§	Phi	IRLW 530A, 2SK1559S, 2SK1908	data
BUK 564-200A	MOS-N-FET-e	V-MOS, LogL, 200V, 9,2A, 90W, <0,4Ω	30c§	Phi	IRLW 630A	data
BUK 565-100A	MOS-N-FET-e	V-MOS, LogL, 100V, 25A, 125W, <85mΩ	30c§	Phi	IRLW 540A	data
BUK 565-200A	MOS-N-FET-e	V-MOS, LogL, 200V, 14A, 125W, <0,23Ω	30c§	Phi	IRLW 640A	data
BUK 571-...	MOS-N-FET-e	=BUK 541-...	17c	Phi	→BUK 541-...	data
BUK 572-...	MOS-N-FET-e	=BUK 542-...	17c	Phi	→BUK 542-...	data
BUK 573-...	MOS-N-FET-e	=BUK 543-...	17c	Phi	→BUK 543-...	data
BUK 575-...	MOS-N-FET-e	=BUK 545-...	17c	Phi	→BUK 545-...	data
BUK 581-60 A	MOS-N-FET-e	SMD, V-MOS, LogL, 60V, 1,5A, <0,04Ω	48c§	Phi	IRLL 110	data
BUK 581-100 A	MOS-N-FET-e	=BUK 581-60A: 100V, 0,9A, <0,9Ω(0,9A)	48c§		BSP 296, IRLL 110	data
BUK 582-60 A	MOS-N-FET-e	SMD, V-MOS, LogL, 60V, 2,5A, <0,15Ω	48c§	Phi	IRLL 014(N)	data
BUK 582-100 A	MOS-N-FET-e	=BUK 582-60A: 100V, 1,7A, <0,31Ω(1,7A)	48c§		-	data
BUK 583-60 A	MOS-N-FET-e	SMD, V-MOS, LogL, 60V, 3,2A, <0,1Ω	48c§	Phi	-	data
BUK 617-500AE,BE	MOS-N-FET-e	FREDFET, 500V, 27...29A, 310W, <0,18Ω	80	Phi	-	data
BUK 627-400A,B	MOS-N-FET-e	FREDFET, 400V, 6,2...6,9A, 45W, <0,6Ω	16c	Phi	-	data
BUK 627-450B	MOS-N-FET-e	FREDFET, 450V, 5,6A, 45W, <0,65Ω	16c	Phi	-	data

Type	Device	Short Description	Fig.	Manu	Comparision Types	More at
BUK 627-500A,B,C	MOS-N-FET-e	FREDFET, 500V, 4,5...5,6A, 45W, 0,9Ω	16c	Phi	-	data
BUK 627-600A,B,C	MOS-N-FET-e	FREDFET, 600V, 3,5...4,3A, 45W, <1,4Ω	16c	Phi		data
BUK 637-400A,B	MOS-N-FET-e	FREDFET, 400V, 12...14A, 180W, <0,6Ω	18c§	Phi	BUK 638-500	data
BUK 637-450B	MOS-N-FET-e	FREDFET, 450V, 11A, 180W, <0,65Ω	18c§	Phi	BUK 638-500	data
BUK 637-500A,B,C	MOS-N-FET-e	FREDFET, 500V, 9,5...11A, 180W, <0,9Ω	18c§	Phi	BUK 638-500	data
BUK 637-600A,B,C	MOS-N-FET-e	FREDFET, 600V, 7...9A, 180W, <1,4Ω	18c§	Phi	BUK 638-800	data
BUK 638-500A,B	MOS-N-FET-e	FREDFET, 500V, 13...14,6A, 220W, <0,6Ω	18c§	Phi	-	data
BUK 638-800A,B	MOS-N-FET-e	FREDFET, 800V, 6,3...7,3A, 220W, <2,4Ω	18c§	Phi		data
BUK 638-1000A,B	MOS-N-FET-e	FREDFET, 1000V, 5,6...6,2A, 220W, <3Ω	18c§	Phi		data
BUK 655-450B	MOS-N-FET-e	FREDFET, 450V, 5,7A, 100W, <1,3Ω	17c§	Phi	-	data
BUK 655-500A,B,C	MOS-N-FET-e	FREDFET, 500V, 5...5,7A, 100W, <1,7Ω	17c§	Phi		data
BUK 657-400A,B	MOS-N-FET-e	FREDFET, 400V, 11...13A, 150W, <0,6Ω	17c§	Phi	-	data
BUK 657-450B	MOS-N-FET-e	FREDFET, 450V, 10A, 150W, <0,65Ω	17c§	Phi		data
BUK 657-500A,B,C	MOS-N-FET-e	FREDFET, 500V, 8,5...10A, 150W, <0,9Ω	17c§	Phi	-	data
BUK 657-600A,B,C	MOS-N-FET-e	FREDFET, 600V, 6,5...8A, 150W, <1,4Ω	17c§	Phi		data
BUK 793-60A	MOS-N-FET-e	Sensor-FET, 60V, 20A, 75W, <0,1Ω	86/5Pin	Phi	-	data
BUK 795-60A	MOS-N-FET-e	Sensor-FET, 60V, 38A, 125W, <45mΩ	86/5Pin	Phi	-	data
BUK 854-500 IS	MOS-N-IGBT	L, 500V, 15A, 85W	17id	Phi	GN 6015A, IGT 4 E10...11	data
BUK 854-800 A	MOS-N-IGBT	L, 800V, 12A, 85W, 50/370ns	17id	Phi	BUP 202	data
BUK 856-400 IZ	MOS-N-IGBT*	L, LogL, 350...500V, 20A, 125W	17id	Phi	MGP 20N40CL	data
BUK 856-450 IX	MOS-N-IGBT*	L, 400...500V, 15A, 125W	17id	Phi	IGT 4E10...11, GN 6015A	data
BUK 856-800 A	MOS-N-IGBT	L, 800V, 24A, 125W, 70/430ns	17id	Phi	BUP 203, BUP 213	data
BUK 866-400 IZ	MOS-N-IGBT*	L, LogL, 350...500V, 20A, 125W	30id	Phi	-	data
BUK 993-60A	MOS-N-FET-e	Sensor-FET,LogL, 60V, 18A, 75W, <0,12Ω	86/5Pin	Phi	-	data
BUK 995-60A	MOS-N-FET-e	Sensor-FET, LogL, 60V, 34A, 125W,<55mΩ	86/5Pin	Sie	-	data
BUK 7506-30	MOS-N-FET-e*	V-MOS, 30V, 75A, 188W, <6mΩ(25A)	17c§	Phi	IRF 1104	data
BUK 7508-55	MOS-N-FET-e*	V-MOS, 55V, 75A, 187W, <8mΩ(25A)	17c§	Phi	IRF 1010E	data
BUK 7510-30	MOS-N-FET-e*	V-MOS, 30V, 75A, 142W, <10,5mΩ(25A)	17c§	Phi	IRF 1104	data
BUK 7514-30	MOS-N-FET-e*	V-MOS, 30V, 69A, 125W, <14mΩ(25A)	17c§	Phi	IRF 1104	data
BUK 7514-55	MOS-N-FET-e*	V-MOS, 55V, 68A, 142W, <14mΩ(25A)	17c§	Phi	IRF 1010E	data
BUK 7518-30	MOS-N-FET-e*	V-MOS, 30V, 55A, 103W, <18mΩ(25A)	17c§	Phi	BUZ 100, MTP 52N06V, PRFZ 40	data
BUK 7518-55	MOS-N-FET-e*	V-MOS, 55V, 57A, 125W, <18mΩ(25A)	17c§	Phi	MTP 52N06V, MTP 60N06HD	data
BUK 7520-55	MOS-N-FET-e*	V-MOS, 55V, 52A, 116W, <20mΩ(25A)	17c§	Phi	MTP 52N06V, MTP 60N06HD	data
BUK 7524-55	MOS-N-FET-e*	V-MOS, 55V, 45A, 103W, <24mΩ(25A)	17c§	Phi	MTP 52N06V, PRFZ 42	data
BUK 7528-55	MOS-N-FET-e*	V-MOS, 55V, 40A, 96W, <28mΩ(20A)	17c§	Phi	MTP 52N06V, PRFZ 42	data
BUK 7535-55	MOS-N-FET-e*	V-MOS, 55V, 34A, 85W, <35mΩ(17A)	17c§	Phi	BUZ 22, 2SK1911, 2SK2513, 2SK2930	data
BUK 7575-55	MOS-N-FET-e*	V-MOS, 55V, 19,7A, 61W, <75mΩ(10A)	17c§	Phi	BUZ 21, 2SK1287, 2SK1910, 2SK2929	data
BUK 7606-30	MOS-N-FET-e*	V-MOS, 30V, 75A, 188W, <6mΩ(25A)	30c§	Phi	IRF 1010..., MTB 75N05HD	data
BUK 7608-30	MOS-N-FET-e*	V-MOS, 55V, 75A, 187W, <8mΩ(25A)	30c§	Phi	-	data
BUK 7610-30	MOS-N-FET-e*	V-MOS, 30V, 75A, 142W, <10,5mΩ(25A)	30c§	Phi	IRF 1010..., MTB 75N05HD	data
BUK 7614-30	MOS-N-FET-e*	V-MOS, 30V, 69A, 125W, <14mΩ(25A)	30c§	Phi	IRF 1010..., MTB 75N05HD	data
BUK 7614-55	MOS-N-FET-e*	V-MOS, 55V, 68A, 142W, <14mΩ(25A)	30c§	Phi	-	data
BUK 7618-30	MOS-N-FET-e*	V-MOS, 30V, 55A, 103W, <18mΩ(25A)	30c§	Phi	MTB 55N06Z, MTB 60N06HD	data
BUK 7618-55	MOS-N-FET-e*	V-MOS, 55V, 57A, 125W, <18mΩ(25A)	30c§	Phi	MTB 55N06Z, MTB 60N06HD	data
BUK 7620-55	MOS-N-FET-e*	V-MOS, 55V, 52A, 116W, <20mΩ(25A)	30c§	Phi	MTB 55N06Z, MTB 60N06HD	data
BUK 7624-55	MOS-N-FET-e*	V-MOS, 55V, 45A, 103W, <24mΩ(25A)	30c§	Phi	MTB 52N06V, MTB 55N06Z	data
BUK 7628-55	MOS-N-FET-e*	V-MOS, 55V, 40A, 96W, <28mΩ(20A)	30c§	Phi	2SK1792, 2SK1919, 2SK2376, 2SK2513-Z, ++	data
BUK 7635-55	MOS-N-FET-e*	V-MOS, 55V, 34A, 85W, <35mΩ(17A)	30c§	Phi	MTB 36N06V, 2SK1919, 2SK2266, 2SK2290,++	data
BUK 7675-55	MOS-N-FET-e*	V-MOS, 55V, 19,7A, 61W, <75mΩ(10A)	30c§	Phi	2SK1622, 2SK1918, 2SK2286, 2SK2311, ++	data
BUK 7830-30	MOS-N-FET-e*	SMD, V-MOS, 30V, 4,1A, <30mΩ(3,2A)	48c§	Phi	BSP 308, IRFL 024N, IRLL 2703	data
BUK 7840-55	MOS-N-FET-e*	SMD, V-MOS, 55V, 3,1A, <40mΩ(5A)	48c§	Phi	BSP 17, BUK 483-60A, IRFL 4105	data
BUK 7880-55	MOS-N-FET-e*	SMD, V-MOS, 55V, 3,5A, <80mΩ(5A)	48c§	Phi	BSP 17, BUK 483-60A, IRFL 4105	data
BUK 9120-48TC	MOS-N-FET-e*	V-MOS, Voltage Clamped, Temp. Sens. Di LogL, 40V, 52A, 116W, <20mΩ(20A)	30c§	Phi		data
BUK 9506-30	MOS-N-FET-e*	V-MOS, LogL, 30V, 75A, 187W, <6mΩ(25A)	17c§	Phi	IRL 1104, IRL 2505, IRL 3705N	data
BUK 9508-55	MOS-N-FET-e*	V-MOS, LogL, 55V, 75A, 187W, <8mΩ(25A)	17c§	Phi	IRL 3705N	data
BUK 9510-30	MOS-N-FET-e*	V-MOS, LogL, 30V, 75A, 142W, <10mΩ(25A)	17c§	Phi	IRL 1104, IRL 2505, IRL 3705N	data
BUK 9514-30	MOS-N-FET-e*	V-MOS, LogL, 30V, 69A, 125W,<14mΩ(25A)	17c§	Phi	IRL 1104, IRL 2505, IRL 3705N	data
BUK 9514-55	MOS-N-FET-e*	V-MOS, LogL, 55V, 68A, 142W,<14mΩ(25A)	17c§	Phi	IRL 3705N	data
BUK 9518-30	MOS-N-FET-e*	V-MOS, LogL, 30V, 55A, 103W,<18mΩ(25A)	17c§	Phi	BUK 556-60, BUZ 100L	data
BUK 9518-55	MOS-N-FET-e*	V-MOS, LogL, 55V, 57A, 125W,<18mΩ(25A)	17c§	Phi	BUK 556-60, MTP 52N06VL	data
BUK 9520-55	MOS-N-FET-e*	V-MOS, LogL, 55V, 52A, 116W,<20mΩ(25A)	17c§	Phi	BUK 556-60, MTP 52N06VL	data
BUK 9524-55	MOS-N-FET-e*	V-MOS, LogL, 55V, 45A, 103W,<24mΩ(25A)	17c§	Phi	BUK 556-60, MTP 52N06VL	data
BUK 9528-55	MOS-N-FET-e*	V-MOS, LogL, 55V, 40A, 96W, <28mΩ(20A)	17c§	Phi	IRL 2310, IRL 2910, 2SK1911, 2SK2513, ++	data
BUK 9535-55	MOS-N-FET-e*	V-MOS, LogL, 55V, 34A, 85W, <35mΩ(17A)	17c§	Phi	BUK 555-60, MTP 35N06ZL, 2SK1296, ++	data
BUK 9575-55	MOS-N-FET-e*	V-MOS, LogL, 55V, 20A, 61W, <75mΩ(10A)	17c§	Phi	BUK 555-100, 2SK1287, 2SK1910, 2SK2929	data
BUK 9606-30	MOS-N-FET-e*	V-MOS, LogL, 30V, 75A, 187W, <6mΩ(25A)	30c§	Phi	IRL 1004..., IRL 1104...	data
BUK 9608-55	MOS-N-FET-e*	V-MOS, LogL, 55V, 75A, 187W, <8mΩ(25A)	30c§	Phi	IRL 2505..., IRL 3705...	data
BUK 9610-30	MOS-N-FET-e*	V-MOS, LogL, 30V, 75A, 142W,<10mΩ(25A)	30c§	Phi	IRL 1004..., IRL 1104...	data

Type	Device	Short Description	Fig.	Manu	Comparision Types	More at
BUK 9614-30	MOS-N-FET-e*	V-MOS, LogL, 30V, 69A, 125W,<14mΩ(25A)	30c§	Phi	IRL 1004..., IRL 1104...	data
BUK 9614-55	MOS-N-FET-e*	V-MOS, LogL, 55V, 68A, 142W,<14mΩ(25A)	30c§	Phi	IRL 2505..., IRL 3705...	data
BUK 9618-30	MOS-N-FET-e*	V-MOS, LogL, 30V, 55A, 103W,<18mΩ(25A)	30c§	Phi	MTB 52N06VL	data
BUK 9618-55	MOS-N-FET-e*	V-MOS, LogL, 55V, 57A, 125W,<18mΩ(25A)	30c§	Phi	MTB 52N06VL	data
BUK 9620-55	MOS-N-FET-e*	V-MOS, LogL, 55V, 52A, 116W,<20mΩ(25A)	30c§	Phi	MTB 52N06VL	data
BUK 9624-55	MOS-N-FET-e*	V-MOS, LogL, 55V, 45A, 103W,<24mΩ(25A)	30c§	Phi	MTB 52N06VL	data
BUK 9628-55	MOS-N-FET-e*	V-MOS, LogL, 55V, 40A, 96W, <28mΩ(20A)	30c§	Phi	IRL 2910, 2SK2266, 2SK2290, 2SK2376, ++	data
BUK 9635-55	MOS-N-FET-e*	V-MOS, LogL, 55V, 34A, 85W, <35mΩ(17A)	30c§	Phi	MTB 35N06ZL, 2SK2288-Z, 2SK2411-Z, ++	data
BUK 9675-55	MOS-N-FET-e*	V-MOS, LogL, 55V, 20A, 61W, <75mΩ(10A)	30c§	Phi	2SK1918, 2SK2286, 2SK2311, 2SK2938, ++	data
BUK 9775-55	MOS-N-FET-e*	V-MOS, LogL, 55V, 11,7A, 19W,<75mΩ(7A)	30c§	Phi	IRLSZ 24A, 2SK1896, 2SK2285, 2SK2933, ++	data
BUK 9830-30	MOS-N-FET-e*	SMD, V-MOS, LogL, 30V, 12,8A, <30mΩ	48c§	Phi	-	data
BUK 9840-55	MOS-N-FET-e*	SMD, V-MOS, LogL, 55V, 10,7A, <40mΩ	48c§	Phi		data
BUK 9880-55	MOS-N-FET-e*	SMD, V-MOS, LogL, 55V, 7,5A, <80mΩ	48c§	Phi	BUK 9840-55	data
BUK 78150-55	MOS-N-FET-e*	SMD, V-MOS, 55V, 5,5A, <150mΩ(5A)	48c§	Phi	BUK 98150-55, IRLL 2705	data
BUK 98150-55	MOS-N-FET-e*	SMD, V-MOS, LogL, 55V, 5,5A, <150mΩ	48c§	Phi	BUK 9840-55, IRLL 2705	data

BUL...BUO

Type	Device	Short Description	Fig.	Manu	Comparision Types	More at
BUL	Z-Di	=SMBJ 28A (Typ-Code/Stempel/marking	71(5x3,5)	Tho	→SMBJ ...	data
BUL 26	Si-N	S-L, 600/300V, 4A, 60W, <-/1,63µs SMPS, Electronic Light Ballast	17j§	Tho	BUT 11(A), BUV 46(A), MJE 13070, 2SC3830	data
BUL 38 D	Si-N+Di	S-L, 800/400V, 5A, 80W, <-/1,9µs SMPS, Electronic Light Ballast	17j§	Tho	BUL 381D, (BUT 46, BUV 46(A), 2SC3047)[2]	data
BUL 39 D	Si-N+Di	S-L, 850/450V, 4A, 70W, <-/1,6µs SMPS, Electronic Light Ballast	17j§	Tho	(BUT 11(A), BUT 46, BUV 46(A), 2SC3047)[2]	data
BUL 43 B	Si-N	SMPS,lo-drive,650/350V, 2A, 40W, 13MHz	17j§	Mot	BUL 44	data
BUL 44	Si-N	SMPS,lo-drive,700/400V, 2A, 50W, 13MHz	17j§	Mot	BUH 50	data
BUL 44 D2	Si-N+Di	=BUL 44: int. Damper-Di	17j§		-	data
BUL 44 F	Si-N	=BUL 44: Iso. 25W	17c			data
BUL 45	Si-N	SMPS,lo-drive,700/400V, 5A, 75W, 12MHz	17j§	Mot	BUF 405(A)	data
BUL 45 D2	Si-N+Di	=BUL 45: int. Damper-Di	17j§		-	data
BUL 45 F	Si-N	=BUL 45: Iso, 35W	17c			data
BUL 48	Si-N	S-L, SN, 800/400V, 7A, 75W, <-/2,1µs	17j§	Tho	BUT 12(A), BUT 54, BUV 56(A)	data
BUL 49 D	Si-N+Di	S-L, 850/450V, 5A, 80W, <-/1,4µs DC-DC Converter, Halogen Lamps	17j§	Tho	(BUT 11(A), BUT 46, BUV 46(A), 2SC3047)[2]	data
BUL 57	Si-N	S-L, 700/400V, 8A, 85W, <-/1,71µs SMPS, Electronic Light Ballast	17j§	Tho	BUF 405(A), BUT 12(A), BUT 56(A)	data
BUL 57 PI	Si-N	=BUL 57: Iso, 35W	17c		BU 307F, BUT 12(A), BUT 22BF,CF	data
BUL 58 D	Si-N+Di	S-L, 800/450V, 8A, 85W, <-/1,98µs SMPS, Electronic Light Ballast	17j§	Tho	BUF 405(A), BUT 12(A), BUT 56(A)	data
BUL 59	Si-N	S-L, 850/400V, 8A, 90W, <-/950ns SMPS, Electronic Light Ballast	17j§	Tho	BUF 405(A), BUT 12(A), BUT 56(A)	data
BUL 67	Si-N	S-L, 700/400V, 10A, 100W, <-/3,38µs SMPS, Electronic Light Ballast	17j§	Tho	BUT 76(A), BUV 56(A)	data
BUL 87	Si-N	S-L, 700/400V, 12A, 110W, <-/3,56µs SMPS, Electronic Light Ballast	17j§	Tho	BUF 654	data
BUL 118	Si-N	S-L, 700/400V, 3A, 60W, <-/960ns Electronic Light Ballast	17j§	Tho	BUF 620, BUH 50	data
BUL 128	Si-N	S-L, 700/400V, 4A, 70W, <-/1,2µs SMPS, Electronic Light Ballast	17j§	Tho	BUF 620, BUH 50	data
BUL 128 D	Si-N+Di	=BUL 128: integr. Damper-Di, -/700ns	17j§		BUL 39D	data
BUL 128 FP	Si-N+Di	=BUL 128: Iso, 31W	17c		BUL 45F	data
BUL 138	Si-N	S-L, 800/400V, 5A, 80W, <-/1,5µs SMPS, Electronic Light Ballast	17j§	Tho	BUF 636A, BUT 18, BUT 46, BUT 56(A)	data
BUL 146	Si-N	SMPS, lo-drive, 700/400V, 6A, 100W	17j§	Mot	BUF 405(A)	data
BUL 146 F	Si-N	=BUL 146: Iso, 40W	17c		BUF 405(A)FI	data
BUL 147	Si-N	SMPS, lo-drive, 700/400V, 8A, 125W	17j§	Mot	BUF 405(A)	data
BUL 147 F	Si-N	=BUL 147: Iso, 45W	17c		BUF 405(A)FI	data
BUL 213	Si-N	S-L, 1300/600V, 3A, 60W, <-/6,42µs SMPS, Electronic Light Ballast	17j§	Tho	BU 505, 2SC4021	data
BUL 216	Si-N	S-L, 1600/800V, 4A, 90W, <-/4,02µs SMPS, Electronic Light Ballast	17j§	Tho		data
BUL 310	Si-N	S-L, 1600/500V, 5A, 75W, <-/2,06µs SMPS, Electronic Light Ballast	17j§	Tho	BU 1706A	data
BUL 310 FP	Si-N	=BUL 310: Iso, 36W	17c		BUX 1706AX	data
BUL 310 PI	Si-N	=BUL 310: Iso, 35W	17c		BUX 1706AX	data
BUL 381	Si-N	S-L, 800/400V, 5A, 70W, <-/3µs SMPS, Electronic Light Ballast	17j§	Tho	BUL 138, BUT 11(A), BUT 46, BUV 46(A)	data
BUL 381 D	Si-N+Di	=BUL 381: integr. Damper-Di	17j§		BUL 38D	data
BUL 382	Si-N	S-L, 800/400V, 5A, 70W, <-/2,72µs SMPS, Electronic Light Ballast	17j§	Tho	BUL 138, BUT 11(A), BUT 46, BUV 46(A)	data
BUL 382 D	Si-N+Di	=BUL 382: integr. Damper-Di	17j§		BUL 38D	data
BUL 410	Si-N	S-L, SN, 1000/450V, 7A, 75W, <-/2,11µs	17j§	Tho	BUT 12A, BUT 56A, BUV 56A, BUV 66A	data

Type	Device	Short Description	Fig.	Manu	Comparision Types	More at
BUL 416	Si-N	S-L, 1600/800V, 6A, 110W, <-/2,95µs SMPS, Electronic Light Ballast	17j§	Tho	-	data
BUL 510	Si-N	S-L, 1000/450V, 10A, 100W, <-/3,55µs SMPS, Electronic Light Ballast	17j§	Tho	MJE 18009	data
BUL 770	Si-N	S-L, 700/400V, 2,5A, 50W, <-/3,65µs HF Electronic Light Ballast	17j§	Pwi	BUL 118, BUL 128	
BUL 791	Si-N	S-L, 700/400V, 4A, 75W, <-/3,25µs HF Electronic Light Ballast	17j§	Pwi	BUL 45, BUL 138	
BUL 810	Si-N	S-L, 1000/450V, 15A, 125W, <-/2,41 SMPS, Electronic Light Ballast	18j§	Tho	BUV 48A, BUW 13A, BUW 133A, 2SC3552	data
BUL 903 ED	Si-N+Di	S-L, 900/400V, 5A, 70W, <1,2/1,05µs SMPS, Electronic Light Ballast	17j§	Tho	(BUT 11, BUT 46, BUV 46, 2SC3047)[2]	data
BULD 25 D,DR	Si-N+Di	SMD, S, 600/400V, 2A, <-/5,35µs HF Electronic Light Ballast	8-MDIP	Pwi	-	
BULD 25 SL	Si-N+Di	=BULD 25D,DR:	9c		-	
BULD 26(-1)	Si-N	S-L, 600/300V, 4A, 20W, <-/1,63µs	30j§	Tho	-	data
BULD 38(-1)	Si-N	SMPS, Electronic Light Ballast S-L, 800/400V, 5A, 30W, <-/1,3µs SMPS, Electronic Light Ballast	30j§	Tho	-	data
BULD 50 KC	Si-N+Di	S-L, 600/400V, 3,5A, 50W, <-/4,75µs HF Electronic Light Ballast	17j§	Pwi	BUL 39D, BUL 128D	
BULD 50 SL	Si-N+Di	=BULD 50KC:	9c		-	
BULD 85	Si-N+Di	S-L, 600V, 8A, 85W	17c	Tix	-	data
BULD 85 KC	Si-N+Di	S-L, 600/400V, 6A, 70W, <-/5,2µs HF Electronic Light Ballast	17j§	Pwi	BUL 58D	
BULD 118-1	Si-N	S-L, 700/400V, 2A, 20W, <-/4,9µs SMPS, Electronic Light Ballast	30j§	Tho	-	data
BULD 118-D1	Si-N+Di	=BULD 118-1: integr. Damper-Di	30j§	Tho		data
BULD 125	Si-N+Di	S-L, 600V, 12A, 125W	17c	Tix	-	data
BULD 125 KC	Si-N+Di	S-L, 600/400V, 8A, 85W, <-/5,25µs HF Electronic Light Ballast	17j§	Pwi	-	
BULD 128-D1	Si-N+Di	S-L, 700/400V, 4A, 35W, <-/3,2µs SMPS, Electronic Light Ballast	30j§	Tho	-	
BULD 138(-1)	Si-N	S-L, 800/400V, 5A, 30W, <-/1,5µs SMPS, Electronic Light Ballast	30j§	Tho	-	
BULK 26	Si-N	=BUL 26: 50W	≈14j§	Tho	-	data
BULK 26 D	Si-N+Di	=BUL 26: 50W, integr. Damper-Di	≈14j§	Tho	-	data
BULK 38 D	Si-N+Di	=BUL 38D: 60W	≈14j§	Tho	-	data
BULK 128 D	Si-N+Di	=BUL 128D: 55W	≈14j§	Tho	-	data
BULK 380 D	Si-N+Di	S-L, 750/400V, 5A, 60W SMPS, Electronic Light Ballast	≈14j§	Tho	-	data
BULK 381	Si-N	=BUL 381: 60W	≈14	Tho	-	data
BULK 382	Si-N	=BUL 382: 60W	≈14	Tho	-	data
BULT 118(D)	Si-N	=BULD 118-1(D-1): 45W	14h§	Tho	2SC3924	data
BUM	Z-Di	=SMBJ 30A (Typ-Code/Stempel/marking)	71(5x3,5)	Tho	→SMBJ ...	data
BUN	Z-Di	=SMBJ 33A (Typ-Code/Stempel/marking)	71(5x3,5)	Tho	→SMBJ ...	data
BUO	Z-Di	=SMBJ 58A (Typ-Code/Stempel/marking)	71(5x3,5)	Tho	→SMBJ ...	data

BUP

Type	Device	Short Description	Fig.	Manu	Comparision Types	More at
BUP 22	Si-N	S-L, 550/300V, 8A, 125W	18j§	Phi	BUV 47(A), BUW 12(A), 2SC3089, 2SC3449++	data
BUP 22 A	Si-N	=BUP 22: 650/350V	18j§		BUV 47(A), BUW 12(A), 2SC3089, 2SC3449++	data
BUP 22 B	Si-N	=BUP 22: 750/400V	18j§		BUV 47(A), BUW 12(A), 2SC3089, 2SC3449++	data
BUP 22 BF	Si-N	=BUP 22B: Iso	18c		2SC4429	data
BUP 22 C	Si-N	=BUP 22: 850/450V	18j§		BUV 47(A), BUW 12(A), 2SC3536	data
BUP 22 CF	Si-N	=BUP 22C: Iso	18c		2SC4429	data
BUP 23	Si-N	S-L, 550/300, 15A, 175W	18j§	Phi	BUX 98AP, 2SC3988	data
BUP 23 A	Si-N	=BUP 23: 650/350V	18j§		BUX 98AP, 2SC3988	data
BUP 23 B	Si-N	=BUP 23: 750/400V	18j§		BUX 98AP, 2SC3988	data
BUP 23 BF	Si-N	=BUP 23B: Iso	18c		BUX,98API	data
BUP 23 C	Si-N	=BUP 23: 850/450V	18j§		BUX 98AP	data
BUP 23 CF	Si-N	=BUP 23C: Iso	18c		BUX 98API	data
BUP 101	MOS-N-FET-e	Ring Emitter Trans., 1000V, 15A, 90W	18c	Sie	-	data
BUP 200	MOS-N-IGBT	L, 1200V, 3,6A, 50W, 50/185ns	17id	Sie	-	data
BUP 200 D	MOS-N-IGBT	=BUP 200: Freilauf/Free-wheel Di.		Sie	-	data
BUP 202	MOS-N-IGBT	L, 1000V, 12A, 100W, 50/195ns	17id	Sie	-	data
BUP 203	MOS-N-IGBT	L, 1000V, 23A, 165W, 70/220ns	17id	Sie	-	data
BUP 212	MOS-N-IGBT	L, 1200V, 22A, 125W, 210/690ns	17id	Sie	BUP 213	
BUP 213	MOS-N-IGBT	L, 1000V, 32A, 200W, 115/470ns	17id	Sie	-	data
BUP 300	MOS-N-IGBT	=BUP 200: 40W	18id	Sie	GT 8Q101	data
BUP 302	MOS-N-IGBT	=BUP 202: 125W	18id	Sie	GN 12015C, GT 15N101, GT 15Q101	data
BUP 303	MOS-N-IGBT	=BUP 203: 200W	18id	Sie	-	data
BUP 304	MOS-N-IGBT	L, 1000V, 35A, 310W, 52/250ns	18id	Sie	-	data

Type	Device	Short Description	Fig.	Manu	Comparision Types	More at
BUP 305 D	MOS-N-IGBT	=BUP 202: Freilauf/Free-wheel Di.	18id		-	data
BUP 306 D	MOS-N-IGBT	=BUP 203: Freilauf/Free-wheel Di.	18id		-	data
BUP 307	MOS-N-IGBT	=BUP 304: 1200V	18id	Sie		data
BUP 307 D	MOS-N-IGBT	=BUP 307: Freilauf/Free-wheel Di.	18id		BUP 314D	data
BUP 309	MOS-N-IGBT	L, 1700V, 25A, 310W, -/200ns	18id	Sie		data
BUP 313	MOS-N-IGBT	L, 1200V, 32A, 200W, 170/625ns	18id	Sie	BUP 307	data
BUP 313 D	MOS-N-IGBT	=BUP 313: Freilauf/Free-wheel Di.	18id		BUP 307D	data
BUP 314	MOS-N-IGBT	L, 1200V, 52A, 300W, 210/620ns	18id	Sie	-	data
BUP 314 D	MOS-N-IGBT	=BUP 314: 42A, Freilauf/Free-wheel Di.	18id		-	data
BUP 400	MOS-N-IGBT	L, 600V, 22A, 100W, 105/750ns	17id	Sie	BUK 856-800A, BUP 203	data
BUP 400 D	MOS-N-IGBT	=BUP 400: Freilauf/Free-wheel Di.	17id			data
BUP 401	MOS-N-IGBT	L, 600V, 29A, 125W, 100/750ns	17id	Sie		data
BUP 402	MOS-N-IGBT	L, 600V, 36A, 150W, 110/750ns	17id	Sie	BUP 402	data
BUP 403	MOS-N-IGBT	L, 600V, 42A, 150W, 130/750ns	17id	Sie	BUP 403	data
BUP 410 D	MOS-N-IGBT	L, 600V, 13A, 50W, 80/335ns int. Freilauf/Free-wheel Di.	17id	Sie	-	data
BUP 602 D	MOS-N-IGBT	=BUP 402: Freilauf/Free-wheel Di.	18id	Sie	BUP 603D	data
BUP 603 D	MOS-N-IGBT	=BUP 403: Freilauf/Free-wheel Di.	18id	Sie	-	data
BUP 604	MOS-N-IGBT	L, 600V, 80A, 300W, 140/870ns	18id	Sie		data
BUPD 1520	Si-N+Di	S-L, 1500/700V, 2A, 50W, sat<3V(0,5A) <700/1200ns, Inverter	17j§	Pwi	-	data
BUQ	Z-Di	=SMBJ 85A (Typ-Code/Stempel/marking	71(5x3,5)	Tho	→SMBJ ...	data

BUR

Type	Device	Short Description	Fig.	Manu	Comparision Types	More at
BUR 10	Si-N	NF/S-L, 100/80V, 5A, 30W(Tc=100°)	22a§	Sgs	-	data
BUR 11	Si-N	S-L, 300/200V, 20A, 175W	49a§	Sgs	2N6324...6325	data
BUR 12	Si-N	S-L, 200/120V, 10A, 40W	49a§	Sgs	2N5540	data
BUR 13	Si-N	S-L, 200/125V, 70A, 250W	49a§	Sgs		data
BUR 14	Si-N	NF/S, 150/120V, 3A, 7W(Tc=25°)	2a§	Sgs	BU 125S, BUX 49...53, 2SD625	data
BUR 15	Si-P	NF/S, 150/120V, 3A, 7W(Tc=25°)	2a§	Sgs	BUY 90	data
BUR 20	Si-N	S-L, 200/125V, 50A, 250W, 24MHz	23a§	Sgs	BUT 90, BUT 100, BUV 60	data
BUR 21	Si-N	S-L, 300/200V, 40A, 250W, 20MHz	23a§	Sgs	BUV 22, BUX 22, BUV 61	data
BUR 22	Si-N	S-L, 350/250V, 40A, 250W, 20MHz	23a§	Sgs	BUV 62	data
BUR 23	Si-N	S-L, 400/325V, 30A, 250W, 20MHz	23a§	Sgs	BUV 23, BUX 23, 2N6323	data
BUR 24	Si-N	=BUR 23: 450/400V	23a§	Sgs	2SC2761, 2SC2930	data
BUR 30	Si-P	S-L, -/125V, 250W	23a§	Sgs	-	data
BUR 31	Si-P	S-L, -/200V, 250W	23a§	Sgs	-	data
BUR 32	Si-P	S-L, -/250V, 250W	23a§	Sgs	-	data
BUR 33	Si-P	S-L, -/325V, 250W	23a§	Sgs	-	data
BUR 34	Si-P	S-L, -/400V, 250W	23a§	Sgs	-	data
BUR 50(S)	Si-N	S-L, 200/125V, 70A, 350W, 500/920ns	23a§	Sgs	BUS 50	data
BUR 51	Si-N	S-L, 300/200V, 60A, 350W, 350/1140ns	23a§	Sgs	-	data
BUR 52	Si-N	S-L, 350/250V, 60A, 350W, 300/1400ns	23a§	Sgs	-	data
BUR 53	Si-N	S-L, -/325V, 350W	23a§	Sgs	-	data
BUR 54	Si-N	S-L, -/400V, 350W	23a§	Sgs	-	data
BUR 55	Si-N	S-L, -/600V, 250W	23a§	Sgs	-	data
BUR 56	Si-N	S-L, -/700V, 250W	23a§	Sgs	-	data
BUR 60	Si-P	S-L, -/125V, 350W	23a§	Sgs	-	data
BUR 61	Si-P	S-L, -/200V, 350W	23a§	Sgs	-	data
BUR 62	Si-P	S-L, -/250V, 350W	23a§	Sgs	-	data

BUS

Type	Device	Short Description	Fig.	Manu	Comparision Types	More at
BUS	Z-Di	=SMBJ 130A (Typ-Code/Stempel/marking	71(5x3,5)	Tho	→SMBJ ...	data
BUS 11	Si-N	S-L, 850/400V, 5A, 100W	23a§	Phi	BUX 47, BUX 83, 2SC3048, 2SD802, ++	data
BUS 11 A	Si-N	=BUS 11: 1000/450V	23a§		BUX 83, MJ 8502, 2SC3060, 2SC3214, ++	data
BUS 12	Si-N	S-L, 850/400V, 8A, 125W	23a§	Phi	BUW 36, BUX 81, MJ 8504, 2SC3061, ++	data
BUS 12 A	Si-N	=BUS 12: 1000/450V	23a§		BUX 81, MJ 8504, 2SC3061, 2SC3215,++	data
BUS 13	Si-N	S-L, 850/400V, 15A, 175W	23a§	Phi	BUW 46, BUX 48, 2SC3593	data
BUS 13 A	Si-N	=BUS 13: 1000/450V	23a§		2SC3216	data
BUS 14	Si-N	S-L, 850/400V, 30A, 250W	23a§	Phi	BUS 98(A)	data
BUS 14 A	Si-N	=BUS 14: 1000/450V	23a§		BUS 98A	data
BUS 21	Si-N	S-L, 550/300V, 5A, 100W	23a§	Phi	BUS 11(A), BUX 82...83, 2SC3048, ++	data
BUS 21 A	Si-N	=BUS 21: 650/350V	23a§		BUS 11(A), BUX 82...83, 2SC3048, ++	data
BUS 21 B	Si-N	=BUS 21: 750/400V	23a§		BUS 11(A), BUX 82...83, 2SC3048, ++	data
BUS 21 C	Si-N	=BUS 21: 850/450V	23a§		BUS 11(A), BUX 83, 2SC3048, 2SC3156, ++	data
BUS 22	Si-N	S-L, 550/300V, 8A, 125W	23a§	Phi	BUS 12(A), BUW 36, BUX 47(A), 2SD811, ++	data
BUS 22 A	Si-N	=BUS 22: 650/350V	23a§		BUS 12(A), BUW 36, BUX 47(A), 2SD811, ++	data
BUS 22 B	Si-N	=BUS 22: 750/400V	23a§		BUS 12(A), BUW 36, BUX 47(A), 2SD811, ++	data
BUS 22 C	Si-N	=BUS 22: 850/450V	23a§		BUS 12(A), BUW 36, BUX 47(A), 2SD811, ++	data
BUS 23	Si-N	S-L, 550/300V, 15A, 175W	23a§	Phi	BUS 13(A), BUS 97(A), BUW 45...46, ++	data
BUS 23 A	Si-N	=BUS 23: 650/350V	23a§		BUS 13(A), BUS 97(A), BUW 45...46, ++	data
BUS 23 B	Si-N	=BUS 23: 750/400V	23a§		BUS 13(A), BUS 97(A), BUW 45...46, ++	data
BUS 23 C	Si-N	=BUS 23: 850/450V	23a§		BUS 13(A), BUS 97(A), BUW 46, BUX 48, ++	data

Type	Device	Short Description	Fig.	Manu	Comparision Types	More at
BUS 24 B	Si-N	S-L, 750/400V, 30A, 250W	23a§	Phi	BUS 14(A), BUS 98(A), BUX 348(A)	data
BUS 24 C	Si-N	=BUS 24B: 850/450V	23a§		BUS 14(A), BUS 98(A), BUX 348(A)	data
BUS 36	Si-N	S-L, 250/120V, 12A, 107W, >30MHz	17j§	Mot	BUV 27(A), BUV 28(A)	data
BUS 37	Si-N	=BUS 36: 300/150V	17j§	Mot	BUV 27A, BUV 28(A)	data
BUS 45 P	Si-N	S-L, 850/450V, 3A, 75W	17j§	Mot	BUT 11(A), BUV 46(A), 2SC3490...3491, ++	data
BUS 46 P	Si-N	S-L, 850/450V, 5A, 80W	17j§	Mot	BUT 11(A), BUV 46(A), 2SC3047	data
BUS 47	Si-N	S-L, 850/450V, 9A, 150W	23a§	Mot	BUS 12(A), BUW 36, BUX.81, MJ 8504	data
BUS 47 A	Si-N	=BUS 47: 1000/450V	23a§		BUS 12A, BUX 81, MJ 8504	data
BUS 47 AP	Si-N	=BUS 47: 1000/450V, 107W	18j§		BUV 47A, BUW 12A	data
BUS 47 P	Si-N	=BUS 47: 107W	18j§		BUV 47(A), BUW 12(A)	data
BUS 48	Si-N	S-L, 850/450V, 15A, 175W	23a§	Mot	BUS 13(A), BUW 46, BUX 48	data
BUS 48 A	Si-N	=BUS 48: 1000/450V	23a§		BUS 13A, 2SC3216	data
BUS 48 AP	Si-N	=BUS 48: 1000/450V, 125W	18j§		BUV 48A...C, BUW 13A, 2SC3552	data
BUS 48 P	Si-N	=BUS 48: 125W	18j§		BUV 48(A...C), BUW 13(A), 2SC3552	data
BUS 50	Si-N	S-L, 200/125V, 70A, 350W	23a§	Mot	BUR 50	data
BUS 51	Si-N	S-L, 300/200V, 50A, 350W	23a§	Mot	BUR 51	data
BUS 52	Si-N	=BUS 51: 350/250V, 40A	23a§	Mot	BUR 52	data
BUS 97	Si-N	S-L, 850/400V, 18A, 175W	23a§	Mot	BUS 13(A), BUX 48(A...C), 2SC3593, ++	data
BUS 97 A	Si-N	=BUS 97: 1000/450V	23a§		BUS 13A, BUX 48A...C, 2SC3593	data
BUS 98	Si-N	S-L, 850/450V, 30A, 250W, <900/2700ns	23a§	Mot	BUS 14(A), BUS 24C, BUX 348(A)	data
BUS 98 A	Si-N	=BUS 98: 1000/450V	23a§		BUS 14A, BUX 348A	data
BUS 131(H)	Si-N	S-L, 850/450V, 5A, 125W	23a§	Phi	BUS 11(A), BUX 83, 2SC3156, 2SD802, ++	data
BUS 131 A	Si-N	=BUS 131: 1000/500V	23a§		BUS 11A, BUX 83, 2SC3060, 2SC3214, ++	data
BUS 132(H)	Si-N	S-L, 850/450V, 8A, 150W	23a§	Phi	BUS 12(A), BUW 36, BUX 81, 2SC3061, ++	data
BUS 132 A	Si-N	=BUS 132: 1000/500V	23a§		BUS 12A, BUX 81, MJ 8504, 2SC3061, ++	data
BUS 133(H)	Si-N	S-L, 850/450V, 15A, 175W	23a§	Phi	BUS 13(A), BUS 97(A), BUW 46, BUX 48, ++	data
BUS 133 A	Si-N	=BUS 133: 1000/500V	23a§		BUS 13A, BUS 97A, BUX 48A...C	data

BUT

Type	Device	Short Description	Fig.	Manu	Comparision Types	More at
BUT	Z-Di	=SMBJ 154A (Typ-Code/Stempel/marking	71(5x3,5)	Tho	→SMBJ ...	data
BUT 11	Si-N	S-L, 850/400, 5A, 100W, <1/4,8µs	17j§	Phi,Tho,Tix	BUS 46P, BUV 46(A), 2SC3047	data
BUT 11 A	Si-N	=BUT 11: 1000/450V	17j§		BUV 46A, MJE 8502, 2SC3050	data
BUT 11 F,AF	Si-N	=BUT 11(A): Iso, 20W	17c	Phi	BUT 18(A)F, BUV 46FI,AFI	data
BUT 11 AI	Si-N	=BUT 11: 1000/450V, Electr. Ballast	17j§		BUF 405A, MJE 18006	data
BUT 11 APX	Si-N	=BUT 11: Iso, CTV-HA, 32W	17c		(BUT 11AF, BUT 18AF, BUV 46AF)	data
BUT 11 FI,AFI	Si-N	=BUT 11(A): Iso, 40W	17c	Tho	BUT 18(A)F, BUV 46FI,AFI, 2SC4898, ++	data
BUT 11 X,AX	Si-N	=BUT 11A: Iso, 32W	17c	Phi	→BUT 11AF	data
BUT 12	Si-N	S-L, 850/400, 8A, 125W, <1/4,8µs	17j§	Phi	BUT 76, BUV 56(A), BUV 66(A)	data
BUT 12 A	Si-N	=BUT 12: 1000/450V	17j§		BUT 76A, BUV 56A, BUV 66A	data
BUT 12 AI	Si-N	=BUT 12: 1000/450V, Electr. Ballast	17j§		BUF 405A, MJE 18008	data
BUT 12 F,AF	Si-N	=BUT 12(A): Iso, 23W	17c	Phi	BUT 22CF	data
BUT 12 FI,AFI	Si-N	=BUT 12(A): Iso	17c	Tho	BUT 22CF	data
BUT 13	Si-N-Darl+Di	S-L, 600/400V, 28A, 175W	23a§	Mot,Tho	MJ 10023	data
BUT 13 P	Si-N-Darl+Di	=BUT 13: 150W	18j§	Tho	BUF 420, BUT 98, BUX 98P	data
BUT 13 PFI	Si-N-Darl+Di	=BUT 13P: Iso, 60W	18c	Tho	BUF 420I, BUX 98PI	data
BUT 14	Si-N-Darl+Di	S-L, 850/500V, 25A, 175W	23a§	Mot	BUT 35...36	data
BUT 15	Si-N-Darl+Di	S-L, 1000/700V, 20A, 175W	23a§	Mot	MJ 10024...10025	data
BUT 16	Si-N-Darl+Di	S-L, 1400/1000V, 12A, 150W	23a§	Mot	-	data
BUT 18	Si-N	S-L, 850/400V, 6A, 110W, <1/1,8µs	17j§	Phi	BUT 12(A), BUT 56(A), BUT 76(A)	data
BUT 18 A	Si-N	=BUT 18: 1000/450V	17j§		BUT 12A, BUT 56A, BUT 76A	data
BUT 18(A)F	Si-N	=BUT 18(A): Iso, 33W	17c		BUF 405(A)FI, BUT 12(A)F	data
BUT 21 B	Si-N	S-L, 750/400V, 5A, 100W	17j§	Phi	BUT 12(A), BUT 56(A)	data
BUT 21 BF	Si-N	=BUT 21B: Iso, 20W	17c		BUT 11(A)F, BUT 12(A)F, BUT 18(A)F	data
BUT 21 C	Si-N	=BUT 21B: 850/450V	17j§		BUT 12(A), BUT 56(A)	data
BUT 21 CF	Si-N	=BUT 21C: Iso, 20W	17c		BUT 11(A)F, BUT 12(A)F, BUT 18(A)F	data
BUT 22 B	Si-N	S-L, 750/400V, 8A, 125W	17j§	Phi	BUT 12(A), BUT 76(A), BUV 56(A)	data
BUT 22 BF	Si-N	=BUT 22B: Iso, 23W	17c		BUT 12(A)F	data
BUT 22 C	Si-N	=BUT 22B: 850/450V	17j§		BUT 12(A), BUT 76(A), BUV 56(A)	data
BUT 22 CF	Si-N	=BUT 22C: Iso, 23W	17c		BUT 12(A)F	data
BUT 30(F,V)	Si-N	S-L, 200/125V, 100A, 250W	80zn	Tho	-	data
BUT 32(V)	Si-N	S-L, 400/300V, 80A, 250W	80zn	Tho	-	data
BUT 33	Si-N-Darl+Di	S-L, 600/400V, 56A, 250W, <-/4,9µs	23a§	Mot	MJ10015...10016	data
BUT 34	Si-N-Darl+Di	S-L, 850/500V, 50A, 250W, <-/4,5µs	23a§	Mot	-	data
BUT 35	Si-N-Darl+Di	S-L, 1000/700V, 40A, 250W, <1,2/4,5µs	23a§	Mot	-	data
BUT 36	Si-N-Darl+Di	S-L, 1400/1000V, 24A, 250W, <-/8,5µs	23a§	Mot	-	data
BUT 44 D	Si-N+Di	S-L, 700/400V, 4A, 63W, 160/-ns	17j§	Aeg		data
BUT 46	Si-N	S-L, SMPS, 850/400V, 5A, 75W	17j§	Aeg	BUT 11(A), BUT 18(A), BUV 46(A), ++	data
BUT 46 A	Si-N	=BUT 46: 1000/450V	17j§		BUT 11A, BUT 18A, BUV 46A, 2SC3050, ++	data
BUT 50 P	Si-N-Darl+Di	S-L, 850/500V, 8A, 100W	18j§	Mot	2SC3030, 2SC3032	data
BUT 51 P	Si-N-Darl+Di	S-L, 850/500V, 8A, 125W	18j§	Mot	BUD 48	data
BUT 54	Si-N	S-L, SN, 800/430V, 8A, 100W, 10MHz	17j§	Aeg	BUT 56(A), BUV 56	data
BUT 55	Si-N-Darl+Di	S-L, 400/400V, 12A, 105W, B>100	17j§	Tho	-	data
BUT 56	Si-N	S-L, SN, 800/400V, 8A, 100W, 10MHz	17j§	Aeg	BUT 12(A), BUT 54, BUV 56(A)	data
BUT 56 A	Si-N	=BUT 56: 1000/450V	17j§		BUT 12A, BUT 76A, BUV 56A	data
BUT 56 A/668	Si-N	≈BUT 56A	17j§		→BUT 56A	data
BUT 56 AF	Si-N	=BUT 56: Iso	17c		BUT 12AF	

Type	Device	Short Description	Fig.	Manu	Comparision Types	More at
BUT 56 H,PH	Si-N	≈BUT 56A	17j§		(→BUT 56A)	
BUT 56 T	Si-N	≈BUT 56A	17j§		→BUT 56A	data
BUT 57	Si-N-Darl+Di	S-L, 400/400V, 15A, 110W, B>200	18j§	Tho	BUT 51P	data
BUT 57 I	Si-N-Darl+Di	=BUT 57: Iso, 80W	18c		-	data
BUT 60	Si-N	S-L, 200/125V, 15A, 125W	17j§	Tho	BUS 36..37, BUV 27(A)	data
BUT 62	Si-N	=BUT 60: 400/300V	17j§	Tho	BUV 28(A)	data
BUT 70	Si-N	S-L, SMPS, lo-sat, 200/125V, 40A, 200W	18j§	Tho	BUW 60...61	data
BUT 70 I	Si-N	=BUT 70: Iso, 115W	18c		-	data
BUT 71	Si-N	S-L, SMPS, 300/200V, 40A, 175W	18j§	Tho	-	data
BUT 71 I	Si-N	=BUT 71: Iso, 100W	18c		-	data
BUT 72	Si-N	S-L, SMPS, 400/300V, 40A, 200W	18j§	Tho	-	data
BUT 72 I	Si-N	=BUT 72: Iso, 115W	18c		-	data
BUT 76	Si-N	S-L, SMPS, 850/400V, 10A, 100W	17j§	Aeg	BUV 56(A), BUV 66(A)	data
BUT 76 A	Si-N	=BUT 76: 1000/450V	17j§		BUV 56A, BUV 66A	data
BUT 76 AF	Si-N	=BUT 76: Iso	17c		BUT 12AF	
BUT 90	Si-N	S-L, lo-sat, 200/125V, 50A, 250W	23a§	Tho	BUT 100, (BUR 20)	data
BUT 91	Si-N	=BUT 90: 300/200V	23a§	Tho	BUT 102, (BUR 51..52)[17]	data
BUT 92	Si-N	S-L, 400/300V, 50A, 250W, <-/2,3µs	23a§	Tho	BUT 102, (BUR 52)[17]	data
BUT 92 A	Si-N	=BUT 90: 400/300V	23a§		BUT 102	data
BUT 93	Si-N	S-L, SN, 600/350V, 4A, 55W, 9MHz	17j§	Aeg	BUT 11(A), BUV 46(A), 2SC3086, 2SD841,++	data
BUT 93 D	Si-N+Di	integr. Damper-Diode	17j§		-	
BUT 98	Si-N	S-L, 850/450V, 30A, 200W	18j§	Aeg	-	data
BUT 98 A	Si-N	=BUT 98: 1000/450V	18j§		-	data
BUT 100	Si-N	S-L, lo-sat, 200/125V, 50A, 300W	23a§	Tho	BUT 90, (BUR 20, BUV 60...61)[17]	data
BUT 102	Si-N	S-L, lo-sat, 400/300V, 50A, 300W	23a§	Tho	BUT 92A	data
BUT 131(H)	Si-N	S-L, 850/450V, 5A, 80W	17j§	Phi	BUT 11(A), BUT 18(A), BUV 46(A), ++	data
BUT 131 A	Si-N	=BUT 131(H): 1000/450V	17j§		BUT 11A, BUT 18A, BUV 46A, 2SC3050, ++	data
BUT 211	Si-N	S-L, 850/400, 5A, 100W, <-/2,8µs Electric Lighting Ballast	17j§	Phi	BUT 18(A), BUS 46P, BUV 46(A), 2SC3047++	data
BUT 211 X	Si-N	=BUT 211: Iso	17c		BUT 11(A)F, BUT 21CF, 2SC3795A, ++	data
BUT 230(F,V)	Si-N	S-L, 200/125V, 200A, 300W	80zk	Tho	-	data
BUT 232(F,V)	Si-N	S-L, 400/300V, 140A, 300W, <-/5,4µs	80zk	Tho	-	data
BUTW 92	Si-N	S-L, 500/250V, 60A, 180W, <-/1,7µs	16j§	Tho	-	data
BUU	Z-Di	=SMBJ 170A (Typ-Code/Stempel/marking)	71(5x3,5)	Tho	→SMBJ ...	data

BUV

Type	Device	Short Description	Fig.	Manu	Comparision Types	More at
BUV	Z-Di	=SMBJ 188A (Typ-Code/Stempel/marking)	71(5x3,5)	Tho	→SMBJ ...	data
BUV 10	Si-N	NF/S-L, 160/125V, 25A, 150W, >8MHz	23a§	Mot	BUW 39, BUX 10, BUX 40, 2N6322	data
BUV 10 N	Si-N	=BUV 10: 175W	23a§		2N6322	data
BUV 11	Si-N	NF/S-L, 250/220V, 20A, 150W, >8MHz	23a§	Mot	BUW 58, BUW 73, BUX 11...12, 2N6322	data
BUV 11 N	Si-N	=BUV 11: 220/160V	23a§		BUW 58, BUW 73, BUX 11...12,, 2N6322	data
BUV 12	Si-N	S-L, 300/250V, 20A, 150W, >8MHz	23a§	Mot	BUW 73, BUX 12, 2N6322	data
BUV 18	Si-N	S-L, 120/60V, 50A, 250W, >8MHz	23a§	Tho	BUR 20, BUV 19..20, BUX 20	data
BUV 18 A	Si-N	=BUV 18: 120/100V	23a§		BUR 20, BUV 19..20, BUX 20	data
BUV 19	Si-N	S-L, 160/80V, 50A, 250W, >8MHz	23a§	Tho	BUR 20, BUV 20, BUX 20	data
BUV 20	Si-N	S-L, 160/125V, 50A, 250W, >8MHz	23a§	Mot, Tho	BUR 20, BUR 50, BUX 20	data
BUV 21	Si-N	S-L, 250/200V, 40A, 250W, >8MHz	23a§	Mot, Tho	BUR 21...22, BUV 22, BUV 61, BUX 21...22	data
BUV 21 N	Si-N	=BUV 21: 220/160V	23a§		BUR 21...22, BUV 22, BUV 61, BUX 21...22	data
BUV 22	Si-N	S-L, 300/250V, 40A, 250W, >8MHz	23a§	Mot, Tho	BUR 21...22, BUV 61...62, BUX 22	data
BUV 23	Si-N	S-L, 400/325V, 30A, 250W, >8MHz	23a§	Mot, Tho	BUX 23, 2N6323	data
BUV 24	Si-N	S-L, 450/400V, 20A, 250W, >8MHz	23a§	Mot, Tho	BUX 24	data
BUV 25	Si-N	S-L, 500/500V, 15A, 250W, >8MHz	23a§	Mot, Tho	BUX 25	data
BUV 26	Si-N	S-L, 180/90V, 20A, 85W	17j§	Mot,Phi,Tho	-	data
BUV 26 A	Si-N	=BUV 26: 200/100V	17j§		-	data
BUV 26 F,AF	Si-N	=BUV 26(A): Iso, 18W	17c	Phi	-	data
BUV 27	Si-N	S-L, 240/120V, 15A, 85W	17j§	Mot,Phi,Tho	BUS 36...37	data
BUV 27 A	Si-N	=BUV 27: 300/150V	17j§		BUS 37	data
BUV 27 F,AF	Si-N	=BUV 27(A): Iso, 18W	17c	Phi	-	data
BUV 28	Si-N	S-L, 400/200V, 12A, 85W	17j§	Mot,Phi,Tho	MJE 13008...13009	data
BUV 28 A	Si-N	=BUV 28: 450/225V	17j§		MJE 13008...13009	data
BUV 28 F,AF	Si-N	=BUV 28(A): Iso, 18W	17c	Phi	2SC4163	data
BUV 30	Si-N-Darl+Di	S-L, -/400V, 8A, 83W, B>250	17j§	Aeg	-	data
BUV 36	Si-N	S-L, 850/400V, 2A, 50W	17j§	Tho	BUT 11(A), BUS 45, BUV 46(A), BUX 85, ++	data
BUV 36 A	Si-N	=BUV 36: 1000/450V	17j§		BUT 11A, BUV 46A, BUX 85, 2SC3491, ++	data
BUV 37	Si-N-Darl	S-L, 450/400V, 15A, 125W, B>20	1dj§	Tho	2SD1466	data
BUV 39	Si-N	S-L, 160/90V, 25A, 120W	23a§	Tho	BUV 10, BUX 10, BUX 40, 2N6322	data
BUV 40	Si-N	S-L, 250/125V, 20A, 120W	23a§	Tho	BUV 11, BUW 58, BUX 11, 2N6322	data
BUV 41	Si-N	S-L, 300/200V, 15A, 120W	23a§	Tho	BUV 12, BUW 73, BUX 12, 2N6322	data
BUV 42	Si-N	S-L, 350/250V, 12A, 120W	23a§	Tho	BUW 74, BUX 13, BUY 50, MJ 15024	data
BUV 42 A	Si-N	=BUV 42: 400/300V	23a§		BUW 74, BUX 13, BUY 50, MJ 15024	data
BUV 46	Si-N	S-L, 850/400V, 6A, 85W, 12MHz	17j§	Phi,Tho,Tix	BUT 11(A), BUT 56(A), 2SC3047, ++	data
BUV 46 A	Si-N	=BUV 46: 1000/450V	17j§		BUT 11A, BUT 12A, BUT 56A, 2SC3050, ++	data
BUV 46 FI,AFI	Si-N	=BUV 46(A): Iso, 30W	17c	Tho	BUT 11(A)F, BUT 12(A)F, BUT 18(A)F	data
BUV 47(B)	Si-N	S-L, 850/400V, 9A, 120W, 7MHz	18j§	Aeg,Phi,Tho	BUV 48(A...C), BUW 12(A)...13(A), 2SC3552	data

Type	Device	Short Description	Fig.	Manu	Comparision Types	More at
BUV 47 A	Si-N	=BUV 47(B): 1000/450V	18j§		BUV 48A...C, BUW 12A.:13A, 2SC3552	data
BUV 47 AF	Si-N	=BUV 47A: Iso	18c	Aeg	2SC4429	data
BUV 47 F	Si-N	=BUV 47: Iso	18c	Aeg	BUP 22CF, 2SC4429	data
BUV 47 FI,AFI	Si-N	=BUV 47(A): Iso, 55W	18c	Tho	BUP 22CF, 2SC4429	data
BUV 48	Si-N	SMPS, 850/400V, 15A, 150W, 5MHz	18j§	Aeg,Phi,Tho Mot,Tix	BUW 13(A), 2SC3552	data
BUV 48 A	Si-N	=BUV 48: 1000/450V	18j§		BUW 13A, 2SC3552	data
BUV 48 AF	Si-N	=BUV 48A: Iso	18c	Aeg	BUX 98API	data
BUV 48 B	Si-N	=BUV 48: 1200/600V	18j§		2SC3644	data
BUV 48 C	Si-N	=BUV 48: 1200/700V	18j§		2SC3644	data
BUV 48 F	Si-N	=BUV 48: Iso	18c	Aeg	BUP 23CF, BUX 98(A)PI	data
BUV 48 FI,AFI	Si-N	=BUV 48(A): Iso, 65W	18c	Tho	BUP 23CF, BUX 98(A)PI	data
BUV 48 T	Si-N	=BUV 48: Io-sat	18j§		-	data
BUV 50	Si-N	S-L, 250/125V, 25A, 150W	23a§	Tho	BUV 11, BUX 11, 2N6322	data
BUV 51	Si-N	S-L, 300/200V, 20A, 150W	23a§	Tho	BUV 12, BUW 73, BUX 12, 2N6322	data
BUV 52	Si-N	S-L, 350/250V, 20A, 150W	23a§	Tho	BUV 23, BUX 23, 2N6323	data
BUV 52 A	Si-N	=BUV 52: 400/300V	23a§		BUV 23, BUX 23, 2N6323	data
BUV 54	Si-N-Darl	S-L, -/400V, 18A, 150W, B>20	23a§	Tho	BUT 15, MJ 10024...10025	data
BUV 54 A	Si-N-Darl	=BUV 54: -/600V	23a§		MJ 10025	data
BUV 56	Si-N	S-L, 850/400V, 9A, 70W	17j§	Tho	BUV 66(A)	data
BUV 56 A	Si-N	=BUV 56: 1000/450V	17j§		BUV 66A	data
BUV 60	Si-N	S-L, 250/125V, 50A, 250W	23a§	Mot, Tho	BUR 51...52	data
BUV 61	Si-N	S-L, 300/200V, 50A, 250W, <-/1,5µs	23a§	Tho	BUR 51...52, BUS 51, BUT 91	data
BUV 62	Si-N	S-L, 350/250V, 40A, 250W	23a§	Tho	BUR 22, BUR 52	data
BUV 62 A	Si-N	=BUV 62: 400/300V	23a§		-	data
BUV 63	Si-N	S-L, 600/350V, 1,5A, 10W	13h§	Aeg	BUV 93...95	data
BUV 66	Si-N	S-L, 850/400V, 15A, 100W	17j§	Tho	-	data
BUV 66 A	Si-N	=BUV 66: 1000/450V	17j§		-	data
BUV 70	Si-N	S-L, 1300/550V, 10A, 140W, 9MHz,	18j§	Aeg	2SC4023	data
BUV 70 F	Si-N	=BUV 70: Iso	18c		2SC4199(A)	data
BUV 71	Si-N	S-L, 1500/800V, 9A, 140W, 9MHz	18j§	Aeg	2SC4023	data
BUV 71 F	Si-N	=BUV 71: Iso	18c		2SC4199A	data
BUV 74	Si-N-Darl	S-L, 600/400V, 36A, 250W, B>20	23a§	Tho	MJ 10023	data
BUV 74 A	Si-N-Darl	=BUV 74: 1000/600V	23a§		BUT 35	data
BUV 82	Si-N	S-L, 850/400V, 6A, 100W, 6MHz	18j§	Phi	BUW 11(A), BUW 12(A), 2SC3153, 2SC3232++	data
BUV 83	Si-N	=BUV 82: 1000/450V	18j§	Phi	BUW 11A, BUW 12A, BUW 131A, 2SC3535, ++	data
BUV 89	Si-N	S-L, SN, 1200/800V, 8A, 125W, 7MHz	18j§	Phi	BU 908, BUV 70...71, 2SC3466	data
BUV 90(A)	Si-N-Darl+Di	S-L, 650/400V, 12A, 125W	18j§	Pni	BUT 51P	data
BUV 90 F	Si-N-Darl+Di	=BUV 90(A): Iso, 34W	18c		-	data
BUV 93	Si-N	S-L, SN, 600/350V, 2A, 15W, 12MHz	13h§	Aeg	-	data
BUV 94	Si-N	=BUV 93: 800/400V	13h§	Aeg	-	data
BUV 95	Si-N	=BUV 93: 1000/450V	13h§	Aeg	-	data
BUV 98(F,V)	Si-N	S-L, 850/400V, 30A, 150W, <1/3,8µs	80zn	Aeg,Phi,Tho	-	data
BUV 98 A(F,V)	Si-N	=BUV 98: 1000/450V	80zn		-	data
BUV 98 C(F,V)	Si-N	=BUV 98: 1200/700V	80zn		-	data
BUV 298(V)	Si-N	S-L, 850/400V, 60A, 250W, <-/4,9µs	80zn	Aeg,Phi,Tho	-	data
BUV 298A(V)	Si-N	=BUV 298: 1000/450V	80zn		-	data
BUV 298C(V)	Si-N	=BUV 298: 1200/700V	80zn		-	data

BUW

Type	Device	Short Description	Fig.	Manu	Comparision Types	More at
BUW	Z-Di	=SMBJ 48A (Typ-Code/Stempel/marking	71(5x3,5)	Tho	→SMBJ ...	data
BUW 11	Si-N	=BUS 11:	18j§	Phi,Sgs	BUV 82, BUW 131(A), 2SC3232, 2SC3387, ++	data
BUW 11 A	Si-N	=BUS 11A:	18j§		BUV 83, BUW 131A, 2SC3387, 2SC3535...36,+	data
BUW 11 AF	Si-N	=BUW 11A: Iso, 32W	16c	Phi	BUV 47AFI, 2SC4427	data
BUW 11 F	Si-N	=BUW 11: Iso, 32W	16c	Phi	BUV 47(A)FI, 2SC4427	data
BUW 12	Si-N	=BUS 12:	18j§	Phi,Tho	BUV 47(A), BUV 48(A...C), BUW 13(A)	data
BUW 12 A	Si-N	=BUS 12A:	18j§		BUV 47A, BUV 48A...C, BUW 13A	data
BUW 12 AF	Si-N	=BUW 12A: Iso, 34W	16c	Phi	BUV 47AFI, 2SC4429	data
BUW 12 F	Si-N	=BUW 12: Iso, 34W	16c	Phi	BUV 47(A)FI, 2SC4429	data
BUW 13	Si-N	=BUS 13:	18j§	Phi,Tho	BUV 48(A...C), 2SC3552	data
BUW 13 A	Si-N	=BUS 13A:	18j§		BUV 48A...C, 2SC3552	data
BUW 13 AF	Si-N	=BUW 13A: Iso, 37W	16c	Phi	BUV 48AFI, BUX 98API	data
BUW 13 F	Si-N	=BUW 13: Iso, 37W	16c	Phi	BUV 48(A)FI, BUX 98(A)PI	data
BUW 14	Si-N	S-L, 1000/450V, 0,5A, 20W, 20MHz	≈14h§	Phi	(BUX 87, 2SC3456, 2SD1527)[4]	data
BUW 16	Si-N	NF-L, 450/400V, 10A, 100W	23a§	Sie	BU 223, BUW 24, BUW 72, BUX 14, ++	data
BUW 17	Si-N	NF-L, 450/400V, 15A, 100W	23a§	Sie	BUV 25, BUW 44, BUX 25, 2SD641, ++	data
BUW 18	Si-N	NF-L, 250/200V, 30A, 200W	68a§	Sie	2SC1301, 2SC2249, 2SC2445	data
BUW 19	Si-N	=BUW 18: 450/400V	68a§	Sie	2SC1470, 2SC2250	data
BUW 22	Si-P	S-L, 400/350V, 6A, 100W	23a§	Sgs	BUW 32, BUW 42, MJ 15025	data
BUW 22 A	Si-P	=BUW 22: 450/400V	23a§		BUW 32A, BUW 42A	data
BUW 22 AP	Si-P	=BUW 22: 450/400V, 60W	17j§		2SA1500, MJE 5851...5852	data
BUW 22 P	Si-P	=BUW 22: 60W	17j§		MJ 5852	data
BUW 23	Si-P	S-L, 450/400V, 10A, 125W	23a§	Sgs	BUW 32, BUW 42	data
BUW 24	Si-N	S-L, 450/350V, 10A, 100W, 20MHz	23a§	Mot,Sgs	BU 223, BUW 72, BUX 14, BUX 17C	data
BUW 25	Si-N	S-L, 600/400V, 10A, 125W, 20MHz	23a§	Mot,Sgs	BUS 12(A), BUW 35...36, 2SC3046, ++	data
BUW 26	Si-N	S-L, 800/450V, 10A, 125W, 20MHz	23a§	Mot,Sgs	BUS 12(A), BUW 35...36, BUX 80...81, ++	data

Type	Device	Short Description	Fig.	Manu	Comparision Types	More at
BUW 28	Si-N-Darl	S-L, 350/350V, 10A, 100W, B>70	23a§	Sie	BU 323(A), BU 920...922, BUW 66, ++	data
BUW 29	Si-N-Darl	=BUW 28: 400V, B>45	23a§	Sie	BU 323(A), BU 920...922, BUW 66, ++	data
BUW 32	Si-P	S-L, 400/350V, 10A, 125W	23a§	Sgs	BUW 23, BUW 42(A)	data
BUW 32 A	Si-P	=BUW 32: 450/400V	23a§		BUW 23, BUW 42A	data
BUW 32 AP	Si-P	=BUW 32: 450/400V, 105W	18j§		BUW 42AP	data
BUW 32 P	Si-P	=BUW 32: 105W	18j§		BUW 42(A)P	data
BUW 32 PFI,APFI	Si-P	=BUW 32(A): Iso, 55W	18c		BUW 42(A)PFI	data
BUW 34	Si-N	S-L, 500/400V, 10A, 125W	23a§	Sgs,Tix	BUS 12(A), BUW 25, BUY 69C, 2SC3046, ++	data
BUW 35	Si-N	=BUW 34: 800/400V	23a§	Sgs,Tix	BUS 12(A), BUW 26, BUX 80...81, ++	data
BUW 36	Si-N	=BUW 34: 900/450V	23a§	Sgs,Tix	BUS 12A, BUX 81, BUY 69A, MJ 8504, ++	data
BUW 37	Si-N	S, 300V, 0,7A, 10W(Tc=25°), >25MHz	2a§	Tho	BSS 48...49, BUX 55, BUY 59...60, 2N3440	data
BUW 38	Si-N	S-L, 120/60V, 30A, 150W, >8MHz	23a§	Tho	BUV 21, BUW 39, BUX 21, BUX 39	data
BUW 39	Si-N	S-L, 160/80V, 30A, 150W, >8MHz	23a§	Tho	BUV 21, BUX 21	data
BUW 40	Si-N	S-L, 450/300V, 1A, 40W, >10MHz	17j§	Fer,Rca	TIP 49...50, 2SB859B	data
BUW 40 A	Si-N	=BUW 40: 550/350V	17j§		BUV 36(A), BUX 84...85	data
BUW 40 B	Si-N	=BUW 40: 650/400V	17j§		BUV 36(A), BUX 84...85	data
BUW 41	Si-N	S-L, 450/300V, 8A, 100W, >15MHz	17j§	Fer,Rca	BUT 56(A), BUV 56(A), 2SC2427, 2SC2739	data
BUW 41 A	Si-N	=BUW 41: 550/350V	17j§		BUT 56(A), BUV 56(A)	data
BUW 41 B	Si-N	=BUW 41: 650/400V	17j§		BUT 56(A), BUV 56(A)	data
BUW 42	Si-P	S-L, 400/350V, 15A, 150W	23a§	Sgs	MJ 15025	data
BUW 42 A	Si-P	=BUW 42: 450/400V	23a§		-	data
BUW 42 AP	Si-P	=BUW 42: 450/400V, 105W	18j§		-	data
BUW 42 P	Si-P	=BUW 42: 105W	18j§		-	data
BUW 42 PFI,APFI	Si-P	=BUW 42(A): Iso, 65W	18c		-	data
BUW 44	Si-N	S-L, 500/400V, 15A, 175W	23a§	Sgs,Tix	BUS 13(A), BUV 25, BUX 25, BUX 48, ++	data
BUW 45	Si-N	=BUW 44: 800/400V	23a§	Sgs,Tix	BUS 13(A), BUX 48(A...C)	data
BUW 46	Si-N	=BUW 44: 900/450V	23a§	Sgs,Tix	BUS 13A, BUX 48A...C	data
BUW 48	Si-N	=BUW 38:	18j§	Tho	BUX 69	data
BUW 49	Si-N	=BUW 39:	18j§	Tho	BUX 69	data
BUW 50	Si-N	=BUV 50:	18j§	Tho	-	data
BUW 51	Si-N	=BUV 51:	18j§	Tho	-	data
BUW 52	Si-N	=BUV 52:	18j§	Tho	-	data
BUW 52 I	Si-N	=BUW 52: Iso, 90W	18c		-	data
BUW 57	Si-N	S-L, SN, 150/125V, 20A, 120W, 15MHz	23a§	Sie	BUV 10...12, BUX 10...12, BUX 40	data
BUW 58	Si-N	=BUW 57: 250/160V	23a§	Sie	BUV 11...12, BUW 73, BUX 11...12	data
BUW 60	Si-N	S-L, 250/125V, 40A, 175W	18j§	Tho	-	data
BUW 60 I	Si-N	=BUW 60: Iso, 100W	18c		-	data
BUW 61	Si-N	S-L, 300/200V, 40A, 175W	18j§	Tho	-	data
BUW 61 I	Si-N	=BUW 61: Iso, 100W	18c		-	data
BUW 62	Si-N	S-L, 400/300V, 40A, 175W	18j§	Tho	-	data
BUW 62 I	Si-N	=BUW 62: Iso, 100W	18c		-	data
BUW 64 A	Si-N	S-L, 140/90V, 7A, 50W, >50MHz	17j§	Rca	BU 409, TIP 150...152, 2SC2867	data
BUW 64 B	Si-N	=BUW 64A: 160/110V	17j§		BU 409, TIP 150...152, 2SC2867	data
BUW 64 C	Si-N	=BUW 64A: 180/130V	17j§		BU 409, TIP 150...152, 2SC2867	data
BUW 66	Si-N-Darl+Di	S-L, 400/200V, 10A, 90W	23a§	Sgs	BU 323(A), BU 920...922, BUW 29, 2SD565	data
BUW 67	Si-N-Darl+Di	=BUW 66: 330/200V	23a§	Sgs	BU 323(A), BU 920...922, BUW 29, 2SD565	data
BUW 70	Si-N	S-L, SN, 150/100V, 10A, 80W	23a§	Sie	BU 223(A), BUW 24, BUX 17(A...C), ++	data
BUW 71	Si-N	S-L, SN, 450/400V, 5A, 100W	23a§	Sie	BU 223(A), BUW 24...25, BUX 17C, ++	data
BUW 72	Si-N	S-L, SN, 450/400V, 10A, 100W	23a§	Sie	BU 223(A), BUW 24...25, BUX 17C, ++	data
BUW 73	Si-N	S-L, SN, 300/200V, 20A, 120W	23a§	Sie	BUV 12, BUX 12, 2N6322	data
BUW 74	Si-N	S-L, SN, 400/250V, 12A, 120W, 20MHz	23a§	Mot,Sie	BUX 13, BUY 50, MJ 15024, 2SC3043	data
BUW 75	Si-N	=BUW 74: 600/300V	23a§	Mot,Sie	BUS 13(A), BUW 45...46, BUX 48, ++	data
BUW 76	Si-N	=BUW 74: 750/350V	23a§	Mot,Sie	BUS 13(A), BUW 45...46, BUX 48, ++	data
BUW 77	Si-N	=BUW 74: 800/400V	23a§	Mot,Sie	BUS 13(A), BUW 45...46, BUX 48, ++	data
BUW 81	Si-N-Darl	S-L, SN, 600/600V, 10A, 80W	23a§	Tix	MJ 10013...10014, 2SD685	data
BUW 81 A	Si-N-Darl	=BUW 81: 800/800V	23a§		BUT 16	data
BUW 84	Si-N	S-L, 800/400V, 2A, 50W, 20MHz	≈14j§	Phi	(BUV 36(A), BUX 84...85, 2SC3531)[6]	data
BUW 85	Si-N	=BUW 84: 1000/450V	≈14j§	Phi	(BUV 36A, BUX 85, 2SC3531)[6]	data
BUW 86	Si-N	S-L, 240/120V, 10A, 62,5W, 50MHz	23a§	Phi	BUX 17(A...C), BUW 24, BUW 72, BUW 74, ++	data
BUW 87	Si-N	=BUW 86: 300/150V	23a§	Phi	BUX 17A...C, BUW 24, BUW 72, BUW 74, ++	data
BUW 87 A	Si-N	=BUW 86: 400/200V	23a§		BUX 17B...C, BUW 24, BUW 72, BUW 74, ++	data
BUW 89	Si-N	=BUV 39:	18j§	Tho	BUW 49...50, BUX 69...70	data
BUW 90	Si-N	=BUV 40:	18j§	Tho	BUW 50...52, BUX 70	data
BUW 91	Si-N	=BUV 41:	18j§	Tho	BUW 51...52	data
BUW 92	Si-N	=BUV 42:	18j§	Tho	BUW 52	data
BUW 92 I	Si-N	=BUW 92: Iso, 75W	18c		2SC4297, 2SC4423	data
BUW 96	Si-P	S-L, 200/150V, 15A, 150W	18j§	Tho	BUW 42(A)P	data
BUW 131(H)	Si-N	S-L, 850/450V, 5A, 80W	18j§	Phi	BUV 82...83, BUW 11(A), 2SC3535...36, ++	data
BUW 131 A	Si-N	=BUW 131: 1000/450V	18j§		BUV 83, BUW 11A, 2SC3535...36	data
BUW 132(H)	Si-N	S-L, 850/450V, 8A, 125W	18j§	Phi	BUV 47(A), BUW 12(A), 2SC3637	data
BUW 132 A	Si-N	=BUW 132: 1000/450V	18j§		BUV 47A, BUW 12A	data
BUW 133(H)	Si-N	S-L, 850/450V, 15A, 135W	18j§	Phi	BUV 48(A), BUW 13(A), 2SC3638	data
BUW 133 A	Si-N	=BUW 133: 1000/450V	18j§		BUV 48A, BUW 13A, 2SC3552, 2SC3644	data

Type	Device	Short Description	Fig.	Manu	Comparision Types	More at
BUX						
BUX 10	Si-N	S-L, 160/125V, 25A, 150W, >8MHz	23a§	Mot,Sgs	BUV 10, BUW 39, 2N6322	data
BUX 10 A	Si-N	=BUX 10: 170/125V, >50MHz	23a§	Rca	BUV 10, BUW 39, 2N6322	data
BUX 10 P	Si-N	=BUX 10: 106W	18j§	Sgs	BUW 49...50, BUX 69...70	data
BUX 11	Si-N	S-L, 250/200V, 20A, 150W, >8MHz	23a§	Mot,Rca,Sgs	BUV 11...12, BUW 58, BUW 73, 2N6322	data
BUX 11 A	Si-N	=BUX 11: 200W	23a§	Rca	2N6322	data
BUX 11 N	Si-N	=BUX 11: 220/160V	23a§		BUV 11...12, BUW 58, BUW 73, 2N6322	data
BUX 12	Si-N	S-L, 300/250V, 20A, 150W, >8MHz	23a§	Mot,Rca,Sgs	BUV 12, BUW 73, 2N6322	data
BUX 13	Si-N	S-L, 400/325V, 15A, 150W, >8MHz	23a§	Mot,Rca,Sgs	BUV 25, BUW 44, BUX 25, MJ 15024	data
BUX 14	Si-N	S-L, 450/400V, 10A, 150W, >8MHz	23a§	Mot,Rca,Sgs	BUW 34, BUW 75...77, BUX 17C	data
BUX 15	Si-N	S-L, 500/500V, 8A, 150W, >8MHz	23a§	Mot,Rca,Sgs	BUW 34, BUW 75...77	data
BUX 16	Si-N	S-L, 250/200V, 5A, 100W, >5MHz	23a§	Rca	BU 222(A), BUW 71, BUX 44...45, ++	data
BUX 16 A	Si-N	=BUX 16: 325/250V	23a§		BU 222(A), BUW 71, BUX 44...45, ++	data
BUX 16 B	Si-N	=BUX 16: 375/300V	23a§		BU 222(A), BUW 71, BUX 44...45, ++	data
BUX 16 C	Si-N	=BUX 16: 425/350V	23a§		BU 222(A), BUW 71, BUX 44...45, ++	data
BUX 17	Si-N	S-L, 250/150V, 10A, 150W, >2,5MHz	23a§	Rca	BUW 34, BUW 44, BUW 75, BUX 14, ++	data
BUX 17 A	Si-N	=BUX 17: 350/250V	23a§		BUW 34, BUW 44, BUW 75, BUX 14, ++	data
BUX 17 B	Si-N	=BUX 17: 400/300V	23a§		BUW 34, BUW 44, BUW 75, BUX 14, ++	data
BUX 17 C	Si-N	=BUX 17: 450/350V	23a§		BUW 34, BUW 44, BUW 75, BUX 14, ++	data
BUX 18	Si-N	S-L, 250/200V, 8A, 120W	23a§	Rca	BUW 25, BUW 34, BUW 72, BUX 14, ++	data
BUX 18 A	Si-N	=BUX 18: 350/325V	23a§		BUW 25, BUW 34, BUW 72, BUX 14, ++	data
BUX 18 B	Si-N	=BUX 18: 400/375V	23a§		BUW 25, BUW 34, BUW 72, BUX 14, ++	data
BUX 18 C	Si-N	=BUX 18: 475/425V	23a§		BUW 25, BUW 34, BUW 72, BUX 14, ++	data
BUX 20	Si-N	S-L, 160/125V, 50A, 350W, >8MHz	23a§	Sgs	BUR 50...52	data
BUX 20 A	Si-N	=BUX 20: 40A, 140W, >50MHz	23a§		BUR 21...22, BUV 21...22, BUV 60...62	data
BUX 21	Si-N	S-L, 250/200V, 40A, 350W, >8MHz	23a§	Rca,Sgs	BUR 51...52, 2N6323	data
BUX 22	Si-N	S-L, 300/250V, 40A, 350W, >10MHz	23a§	Sgs	BUR 51...52, 2N6323	data
BUX 23	Si-N	S-L, 400/325V, 30A, 350W, >8MHz	23a§	Sgs	2N6323	data
BUX 24	Si-N	S-L, 450/400V, 20A, 350W, >8MHz	23a§	Sgs	-	data
BUX 25	Si-N	S-L, 500/500V, 15A, 350W, >8MHz	23a§	Sgs	-	data
BUX 26	Si-N	S-L, 750/350V, 6A, 60W(Tc=75°)	23a§	Sie	BUS 12(A), BUW 35...36, BUX 47, ++	data
BUX 27	Si-N	=BUX 26: 800/400V	23a§	Sie	BUS 12(A), BUW 35...36, BUX 47, ++	data
BUX 28	Si-N-Darl	S-L, 350/350V, 8A, 80W(Tc=55°), B>50	23a§	Sie	BU 323(A), BUW 29, BUW 81(A)	data
BUX 28 V	Si-N-Darl	=BUX 28: -/400V	23a§		BU 323(A), BUW 29, BUW 81(A)	data
BUX 29	Si-N-Darl	=BUX 28: 400/400V, B>30	23a§	Sie	BU 323(A), BUW 29, BUW 81(A)	data
BUX 30	Si-N-Darl+Di	S-L, -/400V, 10A, 90W, B>150	23a§	Aeg	BU 920...922, BUW 66	data
BUX 30 AV	Si-N-Darl+Di	=BUX 30: -/350V, B>250	23a§		BU 920...922, BUW 66	data
BUX 30 AVA	Si-N-Darl+Di	=BUX 30: -/250V, B>250	23a§		BU 920...922, BUW 66	data
BUX 30 AVB	Si-N-Darl+Di	=BUX 30: -/300V, B>250	23a§		BU 920...922, BUW 66	data
BUX 30 AVC	Si-N-Darl+Di	=BUX 30: -/350V, B>250	23a§		BU 920...922, BUW 66	data
BUX 31	Si-N	S-L, 800/400V, 8A, 150W, >15MHz	23a§	Fer,Rca	BUS 12(A), BUW 35...36, MJ 8504, 2SC3061	data
BUX 31 A	Si-N	=BUX 31: 900/450V	23a§		BUS 12A, BUW 36, MJ 8504, 2SC3061	data
BUX 31 B	Si-N	=BUX 31: 1000/500V	23a§		BUS 12A, MJ 8504, 2SC3061, 2SC3215	data
BUX 32	Si-N	S-L, 800/400V, 8A, 150W, >15MHz	23a§	Fer,Rca	BUS 12(A), BUW 35...36, MJ 8504, 2SC3061	data
BUX 32 A	Si-N	=BUX 32: 900/450V	23a§		BUS 12A, BUW 36, MJ 8504, 2SC3061	data
BUX 32 B	Si-N	=BUX 32: 1000/500V	23a§		BUS 12A, MJ 8504, 2SC3061, 2SC3215	data
BUX 33	Si-N	S-L, 800/400V, 12A, 150W, >15MHz	23a§	Rca	BUS 13(A), BUW 45...46, BUX 48, 2SC3593	data
BUX 33 A	Si-N	=BUX 33: 900/450V	23a§		BUS 13A, BUW 46, 2SC3593, 2SC3216	data
BUX 33 B	Si-N	=BUX 33: 1000/500V	23a§		BUS 13A, 2SC3216	data
BUX 34	Si-N	S-L, 120/60V, 5A, 20W(Tc=25°), >70MHz	2a§	Fer	BU 125, BUY 47...48, BUY 68, BUY 80	data
BUX 35	Si-N	S-L, 250/160V, 15A, 140W, >3MHz	23a§	Sgs	BUV 11...12, BUW 73, BUX 11...12, BUX 41	data
BUX 36	Si-N	=BUX 35: 250/200V	23a§	Sgs	BUV 11...12, BUW 73, BUX 11...12, BUX 41	data
BUX 37	Si-N-Darl-Di	S-L, -/400V, 15A, 35W(Tc=100°), B>100	23a§	Aeg,Rca,Tho	BU 930...932, BUT 51P	data
BUX 38	Si-N	S-L, 500/400V, 40A, 250W	68a§	Tho	2SC2366, 2SD642	data
BUX 39	Si-N	S-L, 120/90V, 30A, 120W, >8MHz	23a§	Mot,Phi,++	BUV 21...22, BUW 39, BUX 21...22, 2N6322	data
BUX 40(A)	Si-N	S-L 160/125V, 20A, 120W, >8MHz	23a§	Mot,Phi,++	BUV 10...12, BUW 58, BUX 10...12, 2N6322++	data
BUX 41	Si-N	S-L, 250/200V, 15A, 120W, >8MHz	23a§	Mot,Phi,++	BUV 11...12, BUW 58, BUW 73, BUX 11...12++	data
BUX 41 N	Si-N	=BUX 41: 220/160V, 18A	23a§		BUV 11...12, BUW 58, BUW 73, BUX 11...12++	data
BUX 42	Si-N	S-L, 300/250V, 12A, 120W, >8MHz	23a§	Mot,Rca,Sgs	BUW 44, BUW 74...77, BUX 13, BUY 50, ++	data
BUX 43	Si-N	S-L, 400/325V, 10A, 120W, >8MHz	23a§	Mot,Rca,Sgs	BUW 24...25, BUW 34, BUX 14, BUX 17C, ++	data
BUX 44	Si-N	S-L, 450/400V, 8A, 120W, >8MHz	23a§	Mot,Rca,Sgs	BUW 25, BUW 34, BUX 15, BUX 18C, ++	data
BUX 45	Si-N	S-L, 500/500V, 5A, 120W, >8MHz	23a§	Mot,Rca,Sgs	BUS 11(A), BUX 47, BUX 82...83, BUX 97,++	data
BUX 46	Si-N	=BUV 46:	23a§	Mot,Phi,Sgs	BUS 11(A), BUX 47, BUX 83, MJ 8502, ++	data
BUX 46 A	Si-N	=BUV 46A:	23a§		BUS 11A, BUX 83, MJ 8502, 2SC3060, ++	data
BUX 47(B)	Si-N	=BUV 47(B): 125W	23a§	Mot,Phi,++	BUS 12(A), BUW 36, BUX 81, MJ 8504, ++	data
BUX 47 A	Si-N	=BUV 47A: 125W	23a§		BUS 12A, BUX 81, MJ 8504, 2SC3061, ++	data
BUX 48(A...C)	Si-N	=BUV 48...: 175W	23a§	Mot,Phi,++	BUS 13(A), BUW 46, 2SC3216	data
BUX 48 A	Si-N	=BUV 48A: 175W	23a§		BUS 13A, 2SC3216	data
BUX 48 B	Si-N	=BUV 48B: 175W	23a§		2SC3216	data
BUX 48 C	Si-N	=BUV 48C: 175W	23a§		2SC3216	data
BUX 49	Si-N	S, 150/90V, 3,5A, 10W(Tc=25°), >8MHz	2a§	Tho	BU 125(S), BUX 50...53, BUY 49, 2SD625	data
BUX 50	Si-N	S, 200/125V, 3,5A, 10W(Tc=25°), >8MHz	2a§	Tho	BU 125(S), BUX 51...53, BUY 49, 2SD625	data
BUX 51	Si-N	S, 300/200V, 3,5A, 10W(Tc=25°), >8MHz	2a§	Tho	BUX 52...53, BUY 61...62	data
BUX 51 N	Si-N	=BUX 51: 250/160V	2a§		BUX 52...53, BUY 61...62	data
BUX 52	Si-N	S, 350/250V, 3,5A, 10W(Tc=25°), >8MHz	2a§	Tho	BUX 53, BUY 61...62	data
BUX 53	Si-N	S, 425/325V, 3A, 10W(Tc=25°), >8MHz	2a§	Tho	BUY 61...62	data
BUX 54	Si-N	S, 450/400V, 2A, 10W(Tc=25°), >8MHz	2a§	Tho	BUY 61...62, 2SC1862	data
BUX 55	Si-N	S, 500/500V, 1A, 10W(Tc=25°), >8MHz	2a§	Tho	BUY 59...60	data

Type	Device	Short Description	Fig.	Manu	Comparision Types	More at
BUX 56	Si-N-Darl	8A, 40W, B>200	17j§	Sie	-	data
BUX 57	Si-N-Darl	8A, 40W, B>100	17j§	Sie	-	data
BUX 59	Si-N	S-L, 120/90V, 8A, 70W, >8MHz	22a§	Tho	BUS 36...37, BUV 27(A), 2SC2867	data
BUX 60	Si-N	S-L, 160/125V, 8A, 70W, >8MHz	22a§	Tho	BUS 36...37, BUV 27(A), 2SC2867	data
BUX 61	Si-N	S-L, 250/200V, 8A, 70W, >8MHz	22a§	Tho	BUS 36...37, BUV 27(A), 2SC2867	data
BUX 62	Si-N	S-L, 300/250V, 7A, 70W, >8MHz	22a§	Tho	BUS 36...37, BUV 27(A), 2SC2867	data
BUX 63	Si-N	S-L, 400/325V, 5A, 70W, >8MHz	22a§	Tho	BUT 56(A), MJE 52T...53T, 2SC2440, ++	data
BUX 64	Si-N	S-L, 450/400V, 4A, 70W, >8MHz	22a§	Tho	BUT 93, BUT 56(A), MJE 53T, 2SC2440, ++	data
BUX 65	Si-N	S-L, 500/500V, 3A, 70W, >8MHz	22a§	Tho	BUT 93, BUT 56(A), TIP 75C, 2SC2826, ++	data
BUX 66	Si-P	S-L, 200/150V, 2A, 35W, >20MHz	22a§	Rca	2N6211...6214, 2SA1009(A), 2SB630	data
BUX 66 A	Si-P	=BUX 66: 300/250V	22a§		2N6212...14, 2SA1009(A), 2SA1236	data
BUX 66 B	Si-P	=BUX 66: 350/300V	22a§		2N6212...14, 2SA1009(A), 2SA1236	data
BUX 66 C	Si-P	=BUX 66: 400/350V	22a§		2N6213...14, 2SA1009A, 2SA1236	data
BUX 67	Si-N	S-L, 200/150V, 2A. 35W, >10MHz	22a§	Rca	2N3583...85, 2SC2023, 2SC3055, 2SD610, ++	data
BUX 67 A	Si-N	=BUX 67: 300/250V	22a§		2N3584...85, 2SC2023, 2SC2333, 2SC3055,++	data
BUX 67 B	Si-N	=BUX 67: 350/300V	22a§		2N3584...85, 2SC2534, 2SC2333, 2SC3055,++	data
BUX 67 C	Si-N	=BUX 67: 400/350V	22a§		2N3585, 2SC2534, 2SC2333, 2SC3055, ++	data
BUX 69	Si-N	S-L, 180/90V, 30A, 125W, 8MHz	18j§	Tho	-	data
BUX 70	Si-N	S-L, 250/150V, 20A, 125W, 8MHz	18j§	Tho	BUW 50...52	data
BUX 71	Si-N	S-L, 600/600V, 20A, 200W, 10MHz	38a§	Sie	-	data
BUX 72	Si-N	S-L, 500/500V, 40A, 200W, 10MHz	38a§	Sie	-	data
BUX 73	Si-N	S-L, 400/400V, 60A, 200W, 10MHz	38a§	Sie	-	data
BUX 74	Si-N	S-L, 300/300V, 100A, 200W, 10MHz	38a§	Sie	-	data
BUX 75	Si-N	S-L, 220/200V, 150A, 200W, 10MHz	38a§	Sie	-	data
BUX 76	Si-N	S-L, 120/100V, 200A, 200W, 10MHz	38a§	Sie	-	data
BUX 77	Si-N	S-L, 100/80V, 5A, 40W, >2,5MHz	22a§	Sgs	BU 409, 2N6233...35, 2SC2767, 2SC3035,++	data
BUX 78	Si-P	S-L, 100/80V, 5A, 40W, >2,5MHz	22a§	Sgs	MJE 5850...5852	data
BUX 80	Si-N	S-L, 800/400V, 10A, 100W(Tc=40°), 6MHz	23a§	Phi,Sie,++	BUS 12(A), BUW 26, BUW 35...36, MJ 8504++	data
BUX 81	Si-N	=BUX 80: 1000/450V	23a§	Phi,Sie,++	BUS 12A, MJ 8504, 2SC3061, 2SC3215, ++	data
BUX 82	Si-N	S-L, 800/400V, 6A, 60W(Tc=50°), 6MHz	23a§	Phi,Sie,++	BUS 11(A), BUX 47(A), MJ 8502, 2SC3060++	data
BUX 83	Si-N	=BUX 82: 1000/450V	23a§	Phi,Sie	BUS 11A, BUX 47A, MJ 8502, 2SC3060, ++	data
BUX 84(A)	Si-N	S-L, 800/400V, 2A, 40W, 20MHz	17j§a	Phi,Sie,++	BUV 36(A), MJE 8500, 2SC3178, 2SC3531,++	data
BUX 84 F	Si-N	=BUX 84: Iso, 18W	17c	Phi	BUT 11(A), 2SC3352, 2SC3794, 2SC4304,++	data
BUX 85	Si-N	=BUX 84: 1000/450V	17j§	Phi,Sie,++	BUV 36A, MJE 8501, 2SC3178, 2SC3531, ++	data
BUX 85 F	Si-N	=BUX 85: Iso, 18W	17c	Phi	BUT 11AF, 2SC3752, 2SC3978A, 2SC4234,++	data
BUX 86	Si-N	S-L,.800/400V, 0,5A, 20W(Tc=60°),20MHz	14h§	Aeg,Phi,Sie	BUV 94...95	data
BUX 87	Si-N	=BUX 86: 1000/450V	14h§	Aeg,Phi,Sie	BUV 95	data
BUX 86P...87P	Si-N	=BUX 86...87: 42W	≈14j§	Phi	-	data
BUX 88	Si-N	S-L, SN, 1500/800V, 12A, 160W, 7MHz	23a§	Phi	BU 808[Philips]	data
BUX 90	Si-N-Darl+Di	S-L, 650/400V, 12A, 125W	23a§	Phi	BUT 51P, (BUW 81A)[2]	data
BUX 91	Si-N	=BUX 71: 300W	49a§	Sie	-	data
BUX 92	Si-N	=BUX 72: 300W	49a§	Sie	-	data
BUX 93	Si-N	=BUX 73: 300W	49a§	Sie	-	data
BUX 94	Si-N	=BUX 74: 300W	49a§	Sie	-	data
BUX 95	Si-N	=BUX 75: 300W	49a§	Sie	-	data
BUX 96	Si-N	=BUX 76: 300W	49a§	Sie	-	data
BUX 97	Si-N	S-L, 750/350V, 6A, 60W(Tc=75°), 20MHz	23a§	Rca,Sgs,Tix	BUS 11(A), BUX 82...83, MJ8502, 2SC3048++	data
BUX 97 A	Si-N	=BUX 97: 800/400V	23a§		BUS 11(A), BUX 82...83, MJ8502, 2SC3048++	data
BUX 97 B	Si-N	=BUX 97: 800/450V	23a§		BUS 11(A), BUX 82...83, MJ8502, 2SC3048++	data
BUX 98	Si-N	S-L, 850/400V, 30A, 250W, 5MHz	23a§	Phi,Sgs,Tix	BUS 14(A), BUS 24C, BUS 98(A)	data
BUX 98 A	Si-N	=BUX 98: 1000/450V	23a§		BUS 14A, BUS 98A	data
BUX 98 AP	Si-N	=BUX 98A: 24A, 200W	18j§		-	data
BUX 98 API	Si-N	=BUX 98A: Iso, 30A, 100W	18c		-	data
BUX 98 B	Si-N	=BUX 98: 1000/600V	23a§		BUS 14A, BUS 98A	data
BUX 98 C	Si-N	=BUX 98: 1200/700V	23a§		-	data
BUX 98 P	Si-N	=BUX 98: 200W	18j§		-	data
BUX 98 PI	Si-N	=BUX 98: Iso, 100W	18c		-	data
BUX 99	Si-N	S-L, 730/300V, 1,5A, 28W, 4MHz	14h§	Phi	(BUV 94)[4]	data
BUX 100	Si-N	S-L, 600/300V, 2A, 60W, <650/6500ns	≈14j§	Phi	(BUT 93, BUX 84)[4]	data
BUX 127	Si-N-Darl	S-L, -/400V, 15A, 125W, B>200	18j§	Aeg	BU 931P...932P, BUV 37, BUT 51P	data
BUX 348	Si-N	S-L, 850/400V, 45A, 300W, <-/4,9µs	23a§	Phi,Tho	-	data
BUX 348 A	Si-N	=BUX 348: 1000/450V, 35A	23a§		BUS 14A, BUS 98A	data
BUXD 87-1(T4)	Si-N	S-L, 1000/450V, 0,5A, 20W, 20MHz	30j§	Tho	BUD 87, 2SD1945	data

BUY

Type	Device	Short Description	Fig.	Manu	Comparision Types	More at
BUY 10	Si-N	S-L, 40/20V, 0,8A, 10W(Tc=100°), 90MHz	23a§	Itt	(BDY 90...92)	data
BUY 11	Si-N	=BUY 10: 140MHz	23a§	Itt	(BDY 90...92)	data
BUY 12	Si-N	S-L, 210/80V, 10A, 70W(Tc=45°), 11MHz	23a§/3Pin	Sie	BUW 24, BUX 17, BUX 42...43, BUY 18, ++	data
BUY 12 S	Si-N	=BUY 12: 210/90V, 85W(Tc=45°)	23a§	Tho	BUW 24, BUX 17, BUX 42...43, BUY 18, ++	data
BUY 12 T	Si-N	=BUY 12: 200/-V, 50W(Tc=80°)	23a§	Tsm	BUW 24, BUX 17, BUX 42...43, BUY 18, ++	data
BUY 13	Si-N	=BUY 12: 120/70V	23a§/3Pin	Sie	BUW 70, BUX 17, BUX 42...43, BUY 18, ++	data
BUY 13 S	Si-N	=BUY 12: 120/70V, 85W(Tc=45°)	23a§	Tho	BUW 70, BUX 17, BUX 42...43, BUY 18, ++	data
BUY 14	Si-N	S-L, 60/60V, 8A, 35W(Tc=35°), 11MHz	22a§	Sie	BU 409, BUW 64A...C, 2SC2334	data
BUY 16	Si-N	S-L, 150/80V, 10A, 15W(Tc=100°),100MHz	49a§	Sgs	(2N5542)[6]	data
BUY 17	Si-N	=BUY 16: 120/60V	49a§	Sgs	(2N5288...5289)[6]	data

Type	Device	Short Description	Fig.	Manu	Comparision Types	More at
BUY 18	Si-N	S-L, 300/150V, 10A, 25W(Tc=100°),50MHz	23a§	Tho	BUW 24, BUW 34, BUW 72, BUX 17A...C, ++	data
BUY 18 S	Si-N	S-L, 400/200V, 7A, 50W(Tc=75°), 30MHz	23a§		BUW 24, BUW 34, BUW 72, BUX 17B...C, ++	data
BUY 19	Si-N	S-L, 80/40V, 10A, 20W(Tc=100°), 100MHz	49a§	Sgs	(2N5288...5289)[6]	data
BUY 20	Si-N	S-L, 200/120V, 10A, 85W, 25MHz	23a§	Tix	BUW 24, BUW 34, BUX 17(A...C), BUX 43, ++	data
BUY 21	Si-N	=BUY 20: 300/180V	23a§	Tix	BUW 24, BUW 34, BUX 17A...C, BUX 43, ++	data
BUY 21 A	Si-N	=BUY 20: 400/230V	23a§		BUW 24, BUW 34, BUX 17B...C, BUX 43, ++	data
BUY 22	Si-N	=BUY 20: 450/230V	23a§	Tix	BUW 24, BUW 34, BUX 17C, BUX 14, ++	data
BUY 23	Si-N	S-L, 600/250V, 10A, 85W, 25MHz	23a§	Tix	BUS 12(A), BUW 26, BUW 35...36, BUX 80,++	data
BUY 23 A	Si-N	=BUY 23: 700/300V	23a§		BUS 12(A), BUW 26, BUW 35...36, BUX 80,++	data
BUY 23 B	Si-N	=BUY 23: 700/400V	23a§		BUS 12(A), BUW 26, BUW 35...36, BUX 80,++	data
BUY 24	Si-N	S-L, 120/60V, 5A, 15W(Tc=75°), 100MHz	23a§	Sgs	BDY 90, BUW 86...87, BUX 16(A...C), ++	data
BUY 26	Si-N	S-L, 200/150V, 10A, 100W(Tc=45°)	53b§	Sie	-	data
BUY 27	Si-N	=BUY 26: 350/250V	53b§	Sie	-	data
BUY 28	Si-N	=BUY 26: 420/300V	53b§	Sie	-	data
BUY 29	Si-N	S-L, -/200V, 8A, 125W, >50MHz	23a§	Mot	BUX 18(A...C), BUX 43...44, BUY 18S, ++	data
BUY 30	Si-N	=BUY 29: -/250V	23a§	Mot	BUX 18(A...C), BUX 43...44, BUY 18S, ++	data
BUY 32/40	Si-N	S-L, -/40V, 6A, 60W, 0,8MHz	23a§	Aeg	BD 245(A...C), BUW 86...87, 2N3055, ++	data
BUY 32/70	Si-N	=BUY 32/40: -/70V	23a§		BD 245A...C, BUW 86...87, 2N3055, ++	data
BUY 32/100	Si-N	=BUY 32/40: -/100V	23a§		BD 245C, BUW 86...87, 2N3055, ++	data
BUY 33/40	Si-N	S-L, -/40V, 10A, 90W, 0,8MHz	23a§	Aeg	BD 245(A...C), BUW 86...87, 2N3055, ++	data
BUY 33/70	Si-N	=BUY 33/40: -/70V	23a§		BD 245A...C, BUW 86...87, 2N3055, ++	data
BUY 33/100	Si-N	=BUY 33/40: -/100V	23a§		BD 245C, BUW 86...87, 2N3055, ++	data
BUY 35	Si-N	S-L, 350/300V, 6A, 50W(Tc=50°), 20MHz	23a§	Sie	BUW 71, BUX 16B...C, BUX 18A...C, ++	data
BUY 38	Si-N	NF/S-L, 90/55V, 4A, 25W, >0,8MHz	23a§	Sgs	BD 243B...C, BD 543B...D, 2N3054, ++	data
BUY 39	Si-N	S-L, 100/80V, 5A, 30W(Tc=100°), >40MHz	50g	Tix	2N3996...3997	data
BUY 40	Si-N	=BUY 39:	≈50a§	Tix	2N3998...3999	data
BUY 41	Si-N	S-L, 125/80V, 3A, 15W(Tc=100°), >40MHz	2a§	Tix	BU 125(S), BUX 49...53, BUY 41, BUY 49	data
BUY 43	Si-N	NF/S-L, 50/40V, 4A, 31W(Tc=45°), 1MHz	22a§	Sie	BD 243(A...C), BD 543(A...D), BDW 25, ++	data
BUY 44	Si-N	S-L, 330/150V, 7A, 30W(Tc=125°), 15MHz	23a§	Sie	BUX 18A...C, BUX 43...44, BUY 18S, ++	data
BUY 46	Si-N	NF/S-L, 90/60V, 4A, 31W(Tc=45°)	22a§	Mot,Sgs,Sie	BD 243B...C, BD 543B...D, BDW 25, ++	data
BUY 47	Si-N	S-L, 150/120V, 7A, 10W(Tc=50°), 90MHz	2a§	Sgs	BUY 68, BUY 81	data
BUY 48	Si-N	=BUY 47: 200/170V	2a§	Sgs	-	data
BUY 49	Si-N	=BUY 47: 250/200V	2a§	Sgs	-	data
BUY 49 P	Si-N	=BUY 49: 3A, 15W	14h§	Mot,Sgs	-	data
BUY 49 S	Si-N	=BUY 49: 3A	2a§	Mot,Sgs	BU 125S, BUX 51...52, 2SD625	data
BUY 50	Si-N	S-L, 400/250V, 15A, 95W(Tc=45°), 13MHz	23a§	Aeg	BUV 25, BUW 44, BUX 13, MJ 15024, ++	data
BUY 51	Si-N	S-L, 60/60V, 30A, 150W, >10MHz	49a§	Tix	(2N2823...2825)[6]	data
BUY 51 A	Si-N	=BUY 51:	23a§		BUW 38...39, BUX 39, 2N6322	data
BUY 52	Si-N	=BUY 51	49a§	Tix	(2N2823...2825)[6]	data
BUY 52 A	Si-N	=BUY 51:	23a§		BUW 38...39, BUX 39, 2N6322	data
BUY 53	Si-N	=BUY 51: 100/100V	49a§	Tix	(2N2824...2825)[6]	data
BUY 53 A	Si-N	=BUY 51: 100/100V	23a§		BUW 38...39, BUX 39, 2N6322	data
BUY 54	Si-N	=BUY 51: 100/100V	49a§	Tix	(2N2824...2825)[6]	data
BUY 54 A	Si-N	=BUY 51: 100/100V	23a§		BUW 38...39, BUX 39, 2N6322	data
BUY 55	Si-N	S-L, 150/125V, 10A, 60W(Tc=75°), 20MHz	23a§	Sie	BUW 70, BUX 17(A...C), BUX 42, BUY 18, ++	data
BUY 56	Si-N	=BUY 55: 250/160V	23a§	Sie	BUW 74, BUX 17A...C, BUX 42, BUY 18, ++	data
BUY 57	Si-N	S-L, 150/125V, 15A, 117W, 20MHz	23a§	Sie	BUW 58, BUW 73, BUX 13, BUX 41, ++	data
BUY 58	Si-N	=BUW 57: 250/160V	23a§	Sie	BUW 58, BUW 73, BUX 13, BUX 41, ++	data
BUY 59	Si-N	NF/S, 500/325V, 1A, 10W(Tc=25°)	2a§	Tix	BUX 55, 2N5095, 2N5097...5099	data
BUY 60	Si-N	=BUY 59: 600/400V	2a§	Tix	2N5097...5099	data
BUY 61	Si-N	NF/S, 500/325V, 3A, 10W(Tc=25°)	2a§	Tix	(BUT 93, 2SC2826, 2SC3038)[6]	data
BUY 62	Si-N	=BUY 61: 600/400V	2a§	Tix	(BUT 93)[6]	data
BUY 63	Si-N	NF/S-L, 500/325V, 3A, 20W, >2,5MHz	22a§	Tix	BUT 93, MJ 4380...4381, 2SC1467, 2SC2826	data
BUY 64	Si-N	=BUY 63: 600/400V	22a§	Tix	BUT 93, MJ 4380...4381	data
BUY 65	Si-N	NF/S-L, 600/400V, 10A, 30W(Tc=100°)	22a§	Tix	BUV 56(A), BUV 66(A), MJE 13008...13009	data
BUY 66	Si-N	NF/S-L, 400/325V, 12,5A, 100W, >10MHz	22a§	Tix	BUW 44, BUW 74...77, BUX 13, BUY 50, ++	data
BUY 67	Si-N	NF/S-L, 400/350V, 5A, 75W(Tc=100°)	23a§	Tix	BUW 71, BUX 15, BUX 16C, BUX 44...45, ++	data
BUY 68	Si-N	NF/S, 100/60V, 7A, 10W(Tc=50°), >50MHz	2a§	Sgs	BU 125, BUY 47...49, BUY 81, 2N5338...39	data
BUY 69 A	Si-N	S-L, TV-HA, 1000/400V, 10A, 100W	23a§	Mot,Tix,++	BU 626A, 2SC2123, 2SD1094	data
BUY 69 B	Si-N	=BUY 69A: 800/325V	23a§		BU 626A, 2SC2123, 2SD1094	data
BUY 69 C	Si-N	=BUY 69A: 500/200V	23a§		BU 626A, 2SC2123, 2SD1094	data
BUY 70 A	Si-N	S-L, TV-HA, 1000/400V, 10A, 75W	23a§	Mot,Tix,Tos	BU 626A, 2SC2123, 2SD1094	data
BUY 70 B	Si-N	=BUY 70A: 800/325V	23a§		BU 626A, 2SC2123, 2SD1094	data
BUY 70 C	Si-N	=BUY 70A: 500/200V	23a§		BU 626A, 2SC2123, 2SD1094	data
BUY 71	Si-N	TV-HA, 2200/800V, 2A, 10W(Tc=80°)	23a§	Hit,Tix,Tos	BU 225, 2SD621, 2SD838	data
BUY 72	Si-N	S-L, 280/200V, 10A, 60W(Tc=75°)	23a§	Sie	BUW 74, BUX 17A...C, BUX 43, BUY 18, ++	data
BUY 73	Si-N	S-L, 280/200V, 15A, 117W, 20MHz	23a§	Sie	BUV 12, BUW 73, BUX 12...13, BUY 50, ++	data
BUY 74	Si-N	S-L, 400/250V, 12A, 110W, 15MHz	23a§	Sie	BUW 44...46, BUW 74...77, BUX 13, BUY 50++	data
BUY 75	Si-N	=BUY 74: 600/300V	23a§	Sie	BUS 13, BUW 45...46, BUW 75...77, BUX 48++	data
BUY 76	Si-N	=BUY 74: 750/350V	23a§	Sie	BUS 13, BUW 45...46, BUW 76...77, BUX 48++	data
BUY 77	Si-N	S-L, 400/250V, 8A, 60W(Tc=75°), 15MHz	23a§	Sie	BUW 24...25, BUW 34...36, BUX 15, BUX 44++	data
BUY 78	Si-N	=BUY 77: 600/300V	23a§	Sie	BUS 12, BUW 25, BUW 35...36, BUX 47, ++	data
BUY 79	Si-N	=BUY 77: 750/350V	23a§	Sie	BUS 12, BUW 26, BUW 35...36, BUX 47, ++	data

Type	Device	Short Description	Fig.	Manu	Comparision Types	More at
BUY 80	Si-N	S-L, 150V, 5A(ss), 10W(Tc=100°), 60MHz	2a§	Fer	-	data
BUY 81	Si-N	S-L, 150V, 7,5A(ss),12W(Tc=100°),60MHz	2a§	Fer	-	data
BUY 82	Si-N	S-L, 150V, 10A(ss), 15W(Tc=100°),60MHz	2a§	Fer	-	data
BUY 83	Si-N	S-L, 160/140V, 3A, 25W, 10MHz	22§	Aeg	TIP 75, 2N3441, 2SD422..423	data
BUY 84	Si-N	S-L, 800/300V, 15A, 100W	23a§	Sgs	BUS 13(A), BUW 45...46, BUX 48(A..C), ++	data
BUY 85	Si-N	=BUY 84: 600/250V	23a§	Sgs	BUS 13(A), BUW 45...46, BUX 48(A..C), ++	data
BUY 86	Si-N	S-L, 200/100V, 7A, 50W(Tc=50°), 100MHz	23a§	Phi	BUW 87, BUX 17...18(A..C), BUX 43...44, ++	data
BUY 87	Si-N	=BUY 86: 300/150V	23a§	Phi	BUW 72, BUX 17...18A..C, BUX 43...44, ++	data
BUY 88	Si-N	=BUY 86: 350/150V	23a§	Phi	BUW 72, BUX 17...18A..C, BUX 43...44, ++	data
BUY 89	Si-N	S-L, SN, 1500/800V, 6A, 80W(Tc=60°)	23a§	Phi	BU 908, BUX 88, 2SD649, 2SD821	data
BUY 90	Si-P	S-L, 150V, 5A(ss), 10W(Tc=100°), 60MHz	2a§	Fer	-	data
BUY 91	Si-P	S-L, 150V, 7,5A(ss),12W(Tc=100°), 60MHz	2a§	Fer	-	data
BUY 92	Si-P	S-L, 150V, 10A(ss), 15W(Tc=100°),60MHz	2a§	Fer	-	data
BUY 94	Si-N	S-L, 750/300V, 15A(ss), 100W	23a§	Sgs	BUS 12(A), BUW 26, BUW 35...36, BUX 80,++	data
BUY 95	Si-N	=BUY 94: 600/250V	23a§	Sgs	BUS 12(A), BUW 25, BUW 35...36, BUX 80,++	data
BUY 96	Si-N	=BUY 94: 450/275V	23a§	Sgs	BUS 12(A), BUW 24, BUW 34...36, BUX 80,++	data
BUYP 52	Si-N	S-L, 120V, 5A, 50W, 10MHz	23a§	Ucp	BD 245C, BUX 16, 2SC2908, 2SD2140,++	data
BUYP 53	Si-N	S-L, 80V, 5A, 50W, 10MHz	23a§	Ucp	BD 245B..F, BUX 16, 2SC2908, 2SD2140,++	data
BUYP 54	Si-N	S-L, 40V, 5A, 50W, 10MHz	23a§	Ucp	BD 245(B..F), BD 311, 2N5068...69, 2SD844	data

BUZ

Type	Device	Short Description	Fig.	Manu	Comparision Types	More at
BUZ	Z-Di	=SMBJ 5.0A (Typ-Code/Stempel/marking	71(5x3,5)	Tho	→SMBJ ...	data
BUZ 10	MOS-N-FET-e	V-MOS, 50V, 23A, 75W, <0,07Ω(16A)	17c§	Phi,Sie,Tho	BUZ 11, 2SK600, 2SK674, 2SK1417	data
BUZ 10 A...	MOS-N-FET-e	=BUZ 10: <0,12Ω(6A)	17c§		BUZ 11, 2SK600, 2SK674, 2SK1417	data
BUZ 10(A)L	MOS-N-FET-e	=BUZ 10: LogL, <0,07Ω(11,5A)	17c§	Sie	2SK942, 2SK972, 2SK1115, 2SK1910	data
BUZ 10 S2	MOS-N-FET-e	=BUZ 10: 60V, 24A	17c§	Sie	BUZ 11, 2SK600, 2SK674, 2SK1417	data
BUZ 11	MOS-N-FET-e	V-MOS, 50V, 30A, 75W, <0,04Ω(19A)	17c§	Phi,Sie,++	BUK 456-100, BUK 555-50, BUZ 12	data
BUZ 11 A...	MOS-N-FET-e	=BUZ 11: 26A, <0,055Ω(16A)	17c§		BUK 456-100, BUZ 12	data
BUZ 11 FI,P,S2FI	MOS-N-FET-e	=BUZ 11: Iso, 21A, 35W	17c	Sie,Tho	BUK 445-50, BUK 545-50, 2SK1214, 2SK1420	data
BUZ 11 AL	MOS-N-FET-e	=BUZ 11: LogL, 20A, 35W	17c§	Sie	2SK942, 2SK972, 2SK1115, 2SK1910,++	data
BUZ 11 S2	MOS-N-FET-e	=BUZ 11: 60V	17c§	Sie,Tho	BUK 456-100, 2SK856, 2SK1418	data
BUZ 12	MOS-N-FET-e	V-MOS, 50V, 42A, 125W, <28mΩ(32A)	17c§	Sie	PRFZ 42, 2SK856, 2SK1418	data
BUZ 12 A	MOS-N-FET-e	=BUZ 12:<35mΩ(32A)	17c§		PRFZ 42, 2SK856, 2SK1418	data
BUZ 12 AL	MOS-N-FET-e	=BUZ 12: LogL, <35mΩ(21A)	17c§		2SK1542, 2SK1911	data
BUZ 14	MOS-N-FET-e	V-MOS, 50V, 39A, 125W, <40mΩ(22A)	23a§	Phi,Sie,Tho	BUZ 15	data
BUZ 15	MOS-N-FET-e	V-MOS, 50V, 45A, 125W, <0,03Ω(29A)	23a§	Phi,Sie,Tho	BUK 439-60, BUZ 347, 2SK1258, 2SK1379	data
BUZ 15 S2	MOS-N-FET-e	=BUZ 15: 60V	23a§		BUK 439-60, BUZ 347, 2SK1258, 2SK1379	data
BUZ 16	MOS-N-FET-e	V-MOS, 50V, 48A, 125W, <18mΩ(40A)	23a§	Sie	BUK 439-60, BUZ 346, 2SK1258, 2SK1379	data
BUZ 17	MOS-N-FET-e	V-MOS, 50V, 32A, 83,3W, <40mΩ(22A)	66b	Sie	BUZ 18	data
BUZ 18	MOS-N-FET-e	V-MOS, 50V, 37A, 83,3W, <40mΩ(22A)	66b	Sie	-	data
BUZ 20	MOS-N-FET-e	V-MOS, 100V, 13,5A, 75W, <0,2Ω(8,5A)	17c§	Phi,Sie,++	IRF530, 2SK919, 2SK922, 2SK1301, 2SK1559	data
BUZ 21	MOS-N-FET-e	V-MOS, 100V, 21A, 75W, <85mΩ(13A)	17c§	Phi,Sie,Tho	BUZ 22, IRF 540, IRF 542, 2SK1428,++	data
BUZ 21 L	MOS-N-FET-e	=BUZ21: LogL, <85mΩ(10,5A)	17c§	Sie	BUK 555-100, 2SK1116, 2SK1302, 2SK1347++	data
BUZ 22	MOS-N-FET-e	V-MOS, 100V, 34A, 125W, <55mΩ(22A)	17c§	Sie	BUK 456-100	data
BUZ 23	MOS-N-FET-e	V-MOS, 100V, 10A, 78W, <0,2Ω(6A)	23a§	Phi,Sie	BUZ 35, 2SK398, 2SK401	data
BUZ 24	MOS-N-FET-e	V-MOS, 100V, 32A, 125W, <60mΩ(20A)	23a§	Phi,Sie,Tho	BUZ 349, 2SK561, 2SK1429, 2SK1433	data
BUZ 25	MOS-N-FET-e	V-MOS, 100V, 19A, 78W, <0,1Ω(9A)	23a§	Phi,Sie,Tho	BUZ 24, BUZ 36	data
BUZ 27	MOS-N-FET-e	V-MOS, 100V, 26A, 83,3W, <0,06Ω(20A)	66b	Sie	-	data
BUZ 28	MOS-N-FET-e	V-MOS, 100V, 18A, 70W, <0,1Ω(9A)	66b	Sie	BUZ 27	data
BUZ 30	MOS-N-FET-e	V-MOS, 200V, 7A, 75W, <0,75Ω(4,5A)	17c§	Phi,Sie	BUZ 73, 2SK477, 2SK741, 2SK1319, 2SK1667	data
BUZ 30 A	MOS-N-FET-e	=BUZ 30: 21A, 125W, <0,13Ω(13,5A)	17c§	Sie	-	data
BUZ 31	MOS-N-FET-e	V-MOS, 200V, 14,5A, 95W, <0,2Ω(9A)	17c§	Phi,Sie	BUK 546-200, IRF 640, IRF 642, 2SK891,++	data
BUZ 31 L	MOS-N-FET-e	=BUZ 31: LogL, 13,5A, 75W	17c§	Sie	BUK 555-200	data
BUZ 32	MOS-N-FET-e	V-MOS, 200V, 9,5A, 75W, <0,4Ω(6A)	17c§	Phi,Sie,Tho	BUZ 31, 2SK459, 2SK890, 2SK925, 2SK1221+	data
BUZ 33	MOS-N-FET-e	=BUZ 30: 7,2A, 78W	23a§	Phi,Sie	BUZ 34...35, 2SK293	data
BUZ 34	MOS-N-FET-e	=BUZ 31: 14A, 78W	23a§	Phi,Sie,Tho	BUZ 36	data
BUZ 35	MOS-N-FET-e	=BUZ 32: 9,9A, 78W	23a§	Phi,Sie	BUZ 34, 2SK401	data
BUZ 36	MOS-N-FET-e	V-MOS, 200V, 22A, 125W, <0,12(14A)	23a§	Phi,Sie	BUZ 341, BUZ 350, 2SK1491, 2SK1641,++	data
BUZ 37	MOS-N-FET-e	V-MOS, 200V, 13A, 70W, <0,2Ω(7A)	66b	Sie	BUZ 38	data
BUZ 38	MOS-N-FET-e	V-MOS, 200V, 18A, 83,3W, <12Ω(14A)	66b	Sie	-	data
BUZ 40	MOS-N-FET-e	V-MOS, 500V, 2,5A, 75W, <4,5Ω(2,5A)	17c§	Phi,Sie	BUZ 74, IRF 820, IRF 822, 2SK892	data
BUZ 40 B	MOS-N-FET-e	=BUZ 40: 8,5A, 0,8Ω(5,5A)	17c§		IRF 840, 2SK894, 2SK1496, 2SK1574,++	data
BUZ 41	MOS-N-FET-e	V-MOS, 500V, 5A, 62,5W, <1,1Ω(2,5A)	17c§	Phi,Sie,++	IRF840, IRF842, 2SK553, 2SK893, 2SK1246+	data
BUZ 41 A	MOS-N-FET-e	=BUZ 41: 4,5A, 75W, <1,5Ω(3A)	17c§	Sie	IRF830, 2SK553, 2SK893, 2SK1246, 2SK1751	data
BUZ 42	MOS-N-FET-e	V-MOS, 500V, 4A, 75W, <2Ω(2,6A)	17c§	Phi,Sie,Tho	IRF830, IRF832, 2SK553, 2SK893, 2SK1246+	data
BUZ 43	MOS-N-FET-e	V-MOS, 500V, 2,8A, 78W, <4,5Ω(2,5A)	23a§	Phi,Sie	BUZ 44(A), BUZ 46	data
BUZ 44(A)	MOS-N-FET-e	V-MOS, 500V, 4,8...5,6A, 78W, <1,1Ω	23a§	Phi,Sie	BUZ 46	data
BUZ 45	MOS-N-FET-e	V-MOS, 500V, 9,6A, 125W, <0,6Ω(5A)	23a§	Phi,Sie,++	BUZ339, BUZ384, 2SK512, 2SK724, 2SK1753+	data
BUZ 45 A	MOS-N-FET-e	=BUZ 45: 8,3A, <0,8Ω(5A)	23a§		→BUZ 45	data
BUZ 45 B	MOS-N-FET-e	=BUZ 45: 10A, <0,5Ω(5A)	23a§		→BUZ 45	data
BUZ 45 C	MOS-N-FET-e	=BUZ 45: 450V, 10A, <0,5Ω(5A)	23a§		→BUZ 45	data
BUZ 46	MOS-N-FET-e	V-MOS, 500V, 4,2A, 78W, <1,1Ω(2,5A)	23a§	Phi,Sie	BUZ 44	data
BUZ 47(A)	MOS-N-FET-e	V-MOS, 500V, 3,9...4,5A, 50...70W	66b	Sie	BUZ 48	data
BUZ 48(A)	MOS-N-FET-e	V-MOS, 500V, 6,8...7,8A, 83,3W, <0,8Ω	66b	Sie	-	data

Type	Device	Short Description	Fig.	Manu	Comparision Types	More at
BUZ 50	MOS-N-FET-e	V-MOS, 1000V, 2,8A, 62,5W, <3,5Ω(1,4A)	17c§	Phi,Sie	BUK 456-1000, BUZ 51	data
BUZ 50 A	MOS-N-FET-e	=BUZ 50: 2,5A, 75W, <5Ω(1,5A)	17c§		→BUZ 50	data
BUZ 50 B	MOS-N-FET-e	=BUZ 50: 2A, 75W, <8Ω(1,5A)	17c§		→BUZ 50	data
BUZ 50 C	MOS-N-FET-e	=BUZ 50: 2,3A, 75W, <6Ω(1,5A)	17c§		→BUZ 50	data
BUZ 51	MOS-N-FET-e	V-MOS, 1000V, 3,4A, 125W, <4Ω(2,2A)	17c§	Sie	BUK 456-1000	data
BUZ 53(A,C)	MOS-N-FET-e	V-MOS, L, 1000V, 2,6...3A, 78W	23c§	Phi,Sie	BUZ 54(A)	data
BUZ 54	MOS-N-FET-e	V-MOS, 1000V, 5,1A, 125W, <2Ω(2,5A)	23a§	Phi,Sie	BUZ 357...358, 2SK1205, 2SK1359, 2SK1773+	data
BUZ 54 A	MOS-N-FET-e	=BUZ 54: 4,5A, <2,6Ω(3,2A)	23a§		→BUZ 54	data
BUZ 57(A)	MOS-N-FET-e	=BUZ 50(A): 70W	66b	Sie	BUZ 58	data
BUZ 58(A)	MOS-N-FET-e	=BUZ 54(A): 83,3W	66b	Sie	-	data
BUZ 60	MOS-N-FET-e	V-MOS, 400V, 5,5A, 75W, <1Ω(3,5A)	17c§	Phi,Sie,++	BUZ41, IRF830, 2SK553, 2SK1246, 2SK1751+	data
BUZ 60 B	MOS-N-FET-e	=BUZ 60: 4,5A, <1,5Ω(2,5A)	17c§		→BUZ 60	data
BUZ 61	MOS-N-FET-e	V-MOS, 400V, 12,5A, 150W, <0,4Ω(8A)	17c§	Sie	BUZ 64, 2SK1378	data
BUZ 61 A	MOS-N-FET-e	=BUZ 61: 11A, <0,5Ω(8A)	17c§		→BUZ 61	data
BUZ 63(B)	MOS-N-FET-e	=BUZ 60(B): 78W	23a§	Phi,Sie	BUZ 44(A)	data
BUZ 64	MOS-N-FET-e	V-MOS, 400V, 11,5A, 125W, <0,4Ω	23a§	Phi,Sie	BUZ45, 2SK312, 2SK724, 2SK1488, 2SK1752+	data
BUZ 67	MOS-N-FET-e	=BUZ 64: 83,3W	66b	Sie	-	data
BUZ 70	MOS-N-FET-e	V-MOS, 60V, 12A, 40W, <0,15Ω(7,5A)	17c§	Sie	BUZ 20, BUZ 72, 2SK428, 2SK442, 2SK672++	data
BUZ 70 L	MOS-N-FET-e	=BUZ 70: LogL, <0,15Ω(6A)	17c§		BUK552-60, 2SK970...71, 2SK1114, 2SK1416+	data
BUZ 71	MOS-N-FET-e	V-MOS, 50V, 14A, 40W, <0,1Ω(9A)	17c§	Phi,Sie,++	BUZ 10, IRF 530, 2SK888, 2SK1416,++	data
BUZ 71 A...	MOS-N-FET-e	=BUZ 70: 13A, <0,12Ω(7A)	17c§		→BUZ 71	data
BUZ 71(A)F,FI,P	MOS-N-FET-e	=BUZ 71(A): Iso, 25...30W	17c		BUK 442-60, BUK 443-50, 2SK1093...91,++	data
BUZ 71(A)L	MOS-N-FET-e	=BUZ 70: LogL, <0,1Ω(7A)	17c§		BUZ 10(A)L, BUK 552-50, 2SK971, 2SK1555+	data
BUZ 71 S2	MOS-N-FET-e	=BUZ 71: 60V	17c§		BUZ 10S2, IRF 530, 2SK888, 2SK1416,++	data
BUZ 72	MOS-N-FET-e	V-MOS, 100V, 10A, 40W, <0,2Ω(6A)	17c§	Phi,Sie,++	BUZ20, IRF532, 2SK918, 2SK921, 2SK1427++	data
BUZ 72 A...	MOS-N-FET-e	=BUZ 72: 9A, <0,25Ω(6A)	17c§		→BUZ 72	data
BUZ 72(A)F	MOS-N-FET-e	=BUZ 72: Iso, 7...8A, 25W	17c		BUK443-100, BUK543-100, 2SK1261, 2SK1556	data
BUZ 72(A)L	MOS-N-FET-e	=BUZ 72: LogL	17c§		2SK1300...01, 2SK1559, 2SK1561	data
BUZ 73	MOS-N-FET-e	V-MOS, 200V, 7A, 40W, <0,4Ω(4,5A)	17c§	Phi,Sie++	BUZ 30...32, 2SK477, 2SK741, 2SK1319	data
BUZ 73 A	MOS-N-FET-e	=BUZ 73: 5,5A, <0,6Ω(4,5A)	17c§		→BUZ 73	data
BUZ 73(A)F	MOS-N-FET-e	=BUZ 73(A): Iso, ≤5A, 25W	17c		BUK 444-200	data
BUZ 73(A)L	MOS-N-FET-e	=BUZ 73: LogL	17c§		BUK 554-200	data
BUZ 74	MOS-N-FET-e	V-MOS, 500V, 2,4A, 40W, <3Ω(1,5A)	17c§	Phi,Sie,Tho	BUZ 40, IRF 820, 2SK382, 2SK892	data
BUZ 74 A	MOS-N-FET-e	=BUZ 74: 2,1A, <4Ω(1,5A)	17c§		→BUZ 74	data
BUZ 76	MOS-N-FET-e	V-MOS, 400V, 3A, 40W, <1,8Ω(2A)	17c§	Phi,Sie++	BUZ 60, IRF 720, 2SK1244, 2SK1493,++	data
BUZ 76 A	MOS-N-FET-e	=BUZ 76: 2,7A, <2,5Ω(2A)	17c§		→BUZ 76	data
BUZ 77 A	MOS-N-FET-e	V-MOS, 600V, 2,7A, 75W, <4Ω(1,7A)	17c§	Sie	BUZ 50, BUZ 80, 2SK513, 2SK858	data
BUZ 77 B	MOS-N-FET-e	=BUZ 77A: 2,9A, <3,5Ω(1,7A)	17c§	Sie	→BUZ 77A	data
BUZ 78	MOS-N-FET-e	V-MOS, 800V, 1,5A, 40W, <8Ω(1A)	17c§	Phi,Sie	BUZ 50, BUZ 80, 2SK602, 2SK1199, 2SK1324	data
BUZ 80	MOS-N-FET-e	V-MOS, 800V, 6,1A, 100W, <4Ω(2A)	17c§	Phi,Sie++	BUZ 81, 2SK1457, 2SK1643, 2SK1807, ++	data
BUZ 80 A	MOS-N-FET-e	=BUZ 80: 3A, 75W, <3Ω(1,5A)	17c§		BUZ 51, 2SK513, 2SK792, 2SK1793	data
BUZ 80 FI,AFI	MOS-N-FET-e	=BUZ 80: Iso, 2,1A, 40W	17c		2SK1356, 2SK1460	data
BUZ 81	MOS-N-FET-e	V-MOS, 800V, 4A, 125W, <2,5Ω(2,8A)	17c§	Sie	2SK1501, 2SK1639, 2SK1643, 2SK1807	data
BUZ 83(A)	MOS-N-FET-e	=BUZ 80: ≤3,2A, 78W	23a§		BUZ 54(A), BUZ 84(A), BUZ 307, 2SK415,++	data
BUZ 84	MOS-N-FET-e	V-MOS, 800V, 5,3A, 125W, <2Ω(3A)	23a§	Phi,Sie,++	BUZ 355...356, 2SK727, 2SK793, 2SK1760,++	data
BUZ 84 A	MOS-N-FET-e	=BUZ 84: 6A, <1,5Ω(3A)	23a§		2SK684, 2SK1032, 2SK1614, 2SK1968,++	data
BUZ 88(A)	MOS-N-FET-e	=BUZ 84(A): 83,3W	66b	Sie	-	data
BUZ 90	MOS-N-FET-e	V-MOS, 600V, 4,5A, 75W, <1,6Ω(2,8A)	17c§	Phi,Sie++	BUK 455/600, 2SK1117, 2SK1402, 2SK1809++	data
BUZ 90 A	MOS-N-FET-e	=BUZ 90: 4A, <2Ω(2,8A)	17c§		→BUZ 90	data
BUZ 90 AF	MOS-N-FET-e	=BUZ 90: Iso	17c			data
BUZ 91	MOS-N-FET-e	V-MOS, 600V, 8,5A, 150W, <0,8Ω(5A)	17c§	Sie	BUK 657-600	data
BUZ 91 A	MOS-N-FET-e	=BUZ 91: 8A, 0,9Ω(5A)	17c§		→BUZ 91	data
BUZ 92	MOS-N-FET-e	V-MOS, 600V, 3,3A, 80W, <3Ω(2A)	17c§	Sie	BUK 456-800, 2SK791...792, 2SK1600...01,++	data
BUZ 93	MOS-N-FET-e	=BUZ 92: 3,6A, <2,5Ω(2A)	17c§	Sie	BUK 456-800, 2SK791...792, 2SK1600...01,++	data
BUZ 94	MOS-N-FET-e	V-MOS, 600V, 7,8A, 125W, <0,9Ω(5A)	23a§	Sie	2SK684, 2SK1032	data
BUZ 100	MOS-N-FET-e	V-MOS, 50V, 60A, 250W, <18mΩ(60A)	17c§	Sie		data
BUZ 100 L	MOS-N-FET-e	=BUZ 100: LogL, <18mΩ(30A)	17c§		-	data
BUZ 100 S	MOS-N-FET-e	V-MOS, 55V, 77A, 170W, <15mΩ(55A)	17c§		IRF 1010E	data
BUZ 100 S E3045(A)	MOS-N-FET-e	=BUZ 100S:	30c§			data
BUZ 100 SL	MOS-N-FET-e	V-MOS, LogL, 55V, 70A, 170W,<18mΩ(50A)	17c§		IRL 3705N	data
BUZ 100SL E3045(A)	MOS-N-FET-e	=BUZ 100SL:	30c§			data
BUZ 101	MOS-N-FET-e	V-MOS, 55V, 29A, 100W, <0,06Ω(21A)	17c§	Sie	BUK 555-50, 2SK1291, 2SK1296, 2SK2411	data
BUZ 101 L	MOS-N-FET-e	=BUZ 101: LogL, <0,06Ω(14,5A)	17c§		BUK 555-50, 2SK1291, 2SK1296, 2SK2411	data
BUZ 101 S	MOS-N-FET-e	V-MOS, 55V, 22A, 55W, <0,05Ω(16A)	17c§		BUK 555-100, IRF 540...543, 2SK1417, ++	data
BUZ 101 S E3045(A)	MOS-N-FET-e	=BUZ 101S:	30c§		2SK1622, 2SK1918, 2SK1967, 2SK2286, ++	data
BUZ 101 SL	MOS-N-FET-e	V-MOS, LogL, 55V, 20A, 55W, <70mΩ(14A)	17c§		BUK 555-100, 2SK1115, 2SK1287, 2SK1910++	data
BUZ 101SL E3045(A)	MOS-N-FET-e	=BUZ 101SL:	30c§		2SK1622, 2SK1918, 2SK1967, 2SK2286, ++	data
BUZ 102	MOS-N-FET-e	V-MOS, 50V, 42A, 200W, <23mΩ(42A)	17c§	Sie	-	data
BUZ 102 AL	MOS-N-FET-e	=BUZ 102: LogL, <28mΩ(21A)	17c§			data
BUZ 102 S	MOS-N-FET-e	V-MOS, 55V, 52A, 120W, <18mΩ(37A)	17c§		BUK 556-60, MTP 52N06V	data
BUZ 102 S E3045(A)	MOS-N-FET-e	=BUZ 102S:	30c§		MTB 52N06V	data
BUZ 102 SL	MOS-N-FET-e	V-MOS, LogL, 55V, 47A, 120W,<24mΩ(33A)	17c§		BUK 556-60, MTP 52N06VL	data
BUZ 102SL E3045(A)	MOS-N-FET-e	=BUZ 102SL:	30c§		MTB 52N06VL	data
BUZ 103	MOS-N-FET-e	V-MOS, 50V, 40A, 120W, <40mΩ(28A)	17c§	Sie	BUZ 102, 2SK1418, 2SK1542, 2SK1911	data
BUZ 103 AL	MOS-N-FET-e	=BUZ 103: LogL, 35A, <0,05(17,5A)	17c§		BUZ 102L, 2SK1542, 2SK1911	data
BUZ 103 S	MOS-N-FET-e	V-MOS, 55V, 31A, 75W, <36mΩ(22A)	17c§		BUK 555-60, 2SK1291, 2SK1296, 2SK1411,++	data
BUZ 103 S E3045(A)	MOS-N-FET-e	=BUZ 103S:	30c§		MTB 36N06V, 2SK1900, 2SK2288-Z, ++	data
BUZ 103 SL	MOS-N-FET-e	V-MOS, LogL, 55V, 28A, 75W, <44mΩ(20A)	17c§		BUK 555-60, 2SK1291, 2SK1296, 2SK1411,++	data

Type	Device	Short Description	Fig.	Manu	Comparision Types	More at
BUZ 103SL E3045(A)	MOS-N-FET-e	=BUZ 103SL:	30c§		MTB 36N06VL, 2SK1900, 2SK2288-Z, ++	data
BUZ 104	MOS-N-FET-e	V-MOS, 50V, 17,5A, 60W, <0,1Ω(12,5A)	17c§	Sie	2SK1115, 2SK1287, 2SK1417, 2SK1910	data
BUZ 104 L	MOS-N-FET-e	=BUZ 104: LogL, <0,1Ω(8,5A)	17c§		2SK942, 2SK1115, 2SK1287, 2SK1910	data
BUZ 104 S	MOS-N-FET-e	V-MOS, 55V, 13,5A, 35W, <80mΩ(9,6A)	17c§		BUK 552-60, MTP 15N06V, 2SK1416, ++	data
BUZ 104 S E3045(A)	MOS-N-FET-e	=BUZ 104S:	30c§		2SK1648, 2SK1898, 2SK2284, 2SK2322, ++	data
BUZ 104 SL	MOS-N-FET-e	V-MOS, LogL, 55V, 12,5A, 35W, <110mΩ	17c§		BUK 552-60, MTP 15N06VL, 2SK2175, ++	data
BUZ 104SL E3045(A)	MOS-N-FET-e	=BUZ 104SL:	30c§		2SK1648, 2SK1898, 2SK2284, 2SK2322, ++	data
BUZ 110 S	MOS-N-FET-e	V-MOS, 55V, 80A, 200W, <10mΩ(66A)	17c§	Sie	IRF 1010E	data
BUZ 110 S E3045(A)	MOS-N-FET-e	=BUZ 110S:	30c§		-	data
BUZ 110 SL	MOS-N-FET-e	V-MOS,LogL, 55V, 80A, 200W, <15mΩ(59A)	17c§		IRL 3705N	data
BUZ 110SL E3045(A)	MOS-N-FET-e	=BUZ 110SL:	30c§		-	data
BUZ 111 S	MOS-N-FET-e	V-MOS, 55V, 80A, 300W, <8mΩ(80A)	17c§	Sie	-	data
BUZ 111 S E3045(A)	MOS-N-FET-e	=BUZ 111S:	30c§		-	data
BUZ 111 SL	MOS-N-FET-e	V-MOS,LogL, 55V, 80A, 300W, <10mΩ(80A)	17c§		-	data
BUZ 111SL E3045(A)	MOS-N-FET-e	=BUZ 111SL:	30c§		-	data
BUZ 171	MOS-P-FET-e	V-MOS, 50V, 8A, 40W, <0,3Ω(5A)	17c§	Sie, Six	2SJ122...123, 2SJ171	data
BUZ 172	MOS-P-FET-e	V-MOS, 100V, 5,5A, 40W, <0,6Ω(3,7A)	17c§	Sie	IRF 9520, IRF 9522, 2SJ134	data
BUZ 173	MOS-P-FET-e	V-MOS, 200V, 3,6A, 40W, <1,5Ω(2,3A)	17c§	Sie	IRF 9620, IRF 9630, MTP 5P25	data
BUZ 201	MOS-N-FET-e	FREDFET, 400V, 12,5A, 125W, 130/440ns	23a§	Sie	BUK 637-400	data
BUZ 202	MOS-N-FET-e	=BUZ 201: 11,5A	23a§	Sie	-	data
BUZ 205	MOS-N-FET-e	FREDFET, 400V, 6A, 75W, <1Ω(4A)	17c§	Sie	BUK 655-500, BUZ 215	data
BUZ 206	MOS-N-FET-e	=BUZ 205: 5A, <1,5Ω(4A)	17c§	Sie	BU 655-500, BUZ 215...216	data
BUZ 210	MOS-N-FET-e	FREDFET, 500V, 10,5A ,125W,<0,6Ω(6,5A)	23a§	Sie	BUK 737-500, BUZ 384, 2SK1516	data
BUZ 211	MOS-N-FET-e	=BUZ 210: 9A, <0,8Ω(6,5A)	23a§	Sie	BUZ 384...385, BUK 637-500, 2SK1516	data
BUZ 213	MOS-N-FET-e	=BUZ 210: 8,5A, 83,3W	66b	Sie	BUZ 384	data
BUZ 214	MOS-N-FET-e	=BUZ 210: 7A, 83,3W	66b	Sie	BUZ 385	data
BUZ 215	MOS-N-FET-e	FREDFET, 500V, 5A, 75W, <1,5Ω(3,2A)	17c§	Sie	BUK 655-500	data
BUZ 216	MOS-N-FET-e	=BUZ 215: 4,4A, <2Ω(3,2A)	17c§	Sie	-	data
BUZ 220	MOS-N-FET-e	FREDFET, 800V, 6,5A, 125W, 150/440ns	23a§	Sie	BUZ 380	data
BUZ 221	MOS-N-FET-e	=BUZ 220: 5,5A	23a§	Sie	BUZ 381	data
BUZ 230	MOS-N-FET-e	FREDFET, 1000V, 5,5A, 125W, <2Ω(3,5A)	23a§	Sie	BUZ 380	data
BUZ 231	MOS-N-FET-e	=BUZ 230: 4,9A, <2,6Ω(3,5A)	23a§	Sie	BUZ 381	data
BUZ 255	MOS-N-FET-e	V-MOS, 250V, 13A, 95W, <0,24Ω(8,5A)	17c§	Sie	2SK2133	data
BUZ 271	MOS-P-FET-e	V-MOS, 50V, 22A, 125W, <0,15Ω(14A)	17c§	Sie	2SJ174, 2SJ291	data
BUZ 272	MOS-P-FET-e	V-MOS, 100V, 15A, 125W, <0,3Ω(9,5A)	17c§	Sie	IRF 9540, IRF 9542	data
BUZ 305	MOS-N-FET-e	V-MOS, 800V, 7,5A, 150W, <1Ω(5A)	18c§	Sie	BUK 638-800, 2SK1032, 2SK1358, 2SK1502++	data
BUZ 307	MOS-N-FET-e	V-MOS, 800V, 3A, 75W, <3Ω(1,5A)	18c§	Phi,Sie	2SK726, 2SK792, 2SK954, 2SK1339	data
BUZ 308	MOS-N-FET-e	=BUZ 307: 2,6A, <4Ω(1,2A)	18c§	Phi,Sie	2SK726, 2SK792, 2SK954, 2SK1339	data
BUZ 309	MOS-N-FET-e	V-MOS, 1000V, 2,8A, 125W, <2Ω	18c§	Sie	2SK696	data
BUZ 310	MOS-N-FET-e	V-MOS, 1000V, 2,5A, 75W, <5Ω(1,6A)	18c§	Phi,Sie	2SK696	data
BUZ 311	MOS-N-FET-e	=BUZ 310: 2,3A, <6Ω(1,6A)	18c§	Phi,Sie	2SK696	data
BUZ 312	MOS-N-FET-e	V-MOS, 1000V, 6A, 150W, <1,5Ω(4A)	18c§	Sie	2SK1120, 2SK1934	data
BUZ 323	MOS-N-FET-e	V-MOS, 400V, 15A, 170W, <0,3Ω(9,5A)	18c§	Sie	BUZ 338, 2SK899, 2SK1610, 2SK1745,++	data
BUZ 325	MOS-N-FET-e	V-MOS, 400V, 12,5A, 125W, <0,35Ω(8A)	18c§	Sie	BUZ 338, 2SK899, 2SK1610, 2SK1745,++	data
BUZ 326	MOS-N-FET-e	=BUZ 325: 10,5A, <0,5Ω(6,5A)	18c§	Sie	BUZ 338...39, 2SK899, 2SK1610, 2SK1745,++	data
BUZ 330	MOS-N-FET-e	V-MOS, 500V, 9,5A, 125W, <0,6Ω(6A)	18c§	Phi,Sie,Mot	BUZ 338...339, 2SK724, 2SK1488, 2SK1753++	data
BUZ 331	MOS-N-FET-e	=BUZ 330: 8A, <0,8Ω(5,5A)	18c§	Phi,Sie	BUZ 338...339, 2SK724, 2SK1488, 2SK1753++	data
BUZ 332	MOS-N-FET-e	V-MOS, 600V, 8,5, 150W, <0,8Ω(5A)	18c§	Sie	2SK1032, 2SK1358, 2SK1614, 2SK1968,++	data
BUZ 332 A	MOS-N-FET-e	=BUZ 332: 8A, <0,9Ω(5A)	18c§	Sie	2SK1032, 2SK1358, 2SK1614, 2SK1968,++	data
BUZ 334	MOS-N-FET-e	V-MOS, 600V, 12A, 180W, <0,5Ω(7,5A)	18c§	Sie	2SK1573, 2SK1723	data
BUZ 338	MOS-N-FET-e	V-MOS, 500V, 13,5A, 180W, <0,4Ω(8,5A)	18c§	Sie	BUK 638-500, 2SK1678	data
BUZ 339	MOS-N-FET-e	V-MOS, 500V, 11,5A, 170W, <0,5Ω(7,5A)	18c§	Sie	BUK 638-500, 2SK1678	data
BUZ 341	MOS-N-FET-e	V-MOS, 200V, 33A, 170W, <0,07Ω(21A)	18c§	Sie	IRFP 250, MTW 32N20E, 2SK1675	data
BUZ 342	MOS-N-FET-e	V-MOS, 50V, 60A, 400W, <10mΩ(60A)	18c§	Sie	-	data
BUZ 344	MOS-N-FET-e	V-MOS, 100V, 50A, 170W, <35mΩ(32A)	18c§	Sie	IRFP 2410, 2SK1381, 2SK1434	data
BUZ 345	MOS-N-FET-e	V-MOS, 100V, 41A, 150W, <45mΩ(26A)	18c§	Sie	IRFP 150, MTW 45N10E, 2SK1381, 2SK1434	data
BUZ 346	MOS-N-FET-e	V-MOS, 50V, 58A, 170W, 18mΩ(47A)	18c§	Sie	BUK 439-60, BUK 539-60, 2SK1379, 2SK1423	data
BUZ 346 S2	MOS-N-FET-e	=BUZ 346: 60V	18c§		→BUZ 346	data
BUZ 347	MOS-N-FET-e	V-MOS, 50V, 45A, 125W, <30mΩ(29A)	18c§	Phi,Sie	BUK 439-60, 2SK857, 2SK1124, 2SK1665,++	data
BUZ 348	MOS-N-FET-e	=BUZ 347: 39A, <30mΩ(26A)	18c§	Phi,Sie	BUK 439-60, 2SK857, 2SK1124, 2SK1665,++	data
BUZ 349	MOS-N-FET-e	V-MOS, 100V, 32A, 125W, <0,06Ω(21A)	18c§	Phi,Sie	IRFP 140, 2SK850...51, 2SK906, 2SK1433,++	data
BUZ 350	MOS-N-FET-e	V-MOS, 200V, 22A, 125W, <120mΩ(14A)	18c§	Phi, Sie	BUZ 341, IRFP 240, IRFP 252, 2SK901, ++	data
BUZ 351	MOS-N-FET-e	V-MOS, 400V, 11,5A, 125W, <0,4Ω(5,5A)	18c§	Phi,Rca;Sie	BUZ 325...326, BUZ 338...339, 2SK1616,++	data
BUZ 353	MOS-N-FET-e	V-MOS, 500V, 9,5A, 125W, <0,6Ω(5,5A)	18c§	Sie,Tho	BUZ 330, BUZ 338...339, 2SK724, 2SK1785++	data
BUZ 354	MOS-N-FET-e	=BUZ 353: 8A	18c§	Sie,Sgs	BUZ 331, BUZ 338...339, 2SK724, 2SK1785++	data
BUZ 355	MOS-N-FET-e	V-MOS, 800V, 6A, 125W, <1,5Ω(3,9A)	18c§	Mot,Phi,Sie	BUZ 305, 2SK1032, 2SK1649, 2SK1794,++	data
BUZ 356	MOS-N-FET-e	=BUZ 355: 5,3A, <2Ω(3,9A)	18c§	Phi,Sie	2SK727, 2SK1649, 2SK1760, 2SK1794,++	data
BUZ 357	MOS-N-FET-e	V-MOS, 1000V, 5,1, 125W, <2Ω(3,2A)	18c§	Phi,Sie	BUZ 312, 2SK1205, 2SK1359, 2SK1773,++	data
BUZ 358	MOS-N-FET-e	=BUZ 357: 4,5A, <2,6Ω(3,2A)	18c§	Phi,Sie	BUZ 312, 2SK1205, 2SK1359, 2SK1773,++	data
BUZ 360	MOS-N-FET-e	FREDFET, 800V, 3,6A, 75W, 80/160ns	18c§	Sie	BUZ 380...381	data

Type	Device	Short Description	Fig.	Manu	Comparision Types	More at
BUZ 361	MOS-N-FET-e	=BUZ 360: 2,9A	18c§	Sie	BUZ 380...381	data
BUZ 376	MOS-N-FET-e	V-MOS, 800V, 6,5A, 125W, <1,5Ω	18c§	Sie	BUZ 355, 2SK684, 2SK1032, 2SK1614,++	data
BUZ 377	MOS-N-FET-e	V-MOS, 800V, 5,5A, 125W, <2Ω	18c§	Sie	BUZ 356, 2SK684, 2SK1032, 2SK1614,++	data
BUZ 380	MOS-N-FET-e	FREDFET, 1000V, 5,5A, 125W, <2Ω(3,5A)	18c§	Sie	BUK 638-1000	data
BUZ 381	MOS-N-FET-e	=BUZ 380: 4,9A, <2,6Ω(3,5A)	18c§	Sie	BUK 638-1000	data
BUZ 382	MOS-N-FET-e	FREDFET, 400V, 12,5A, 125W, <0,4Ω(8A)	18c§	Sie	BUK 637-400, BUK 638-500, 2SK1515...1516	data
BUZ 383	MOS-N-FET-e	=BUZ 382: 11,5A, <0,5Ω(7,5A)	18c§	Sie	BUK 637-400, BUK 638-500, 2SK1515...1516	data
BUZ 384	MOS-N-FET-e	FREDFET, 500V, 10,5A, 125W,<0,6Ω(6,5A)	18c§	Phi,Sie	BUK 637-500, BUK 638-500, 2SK1516	data
BUZ 385	MOS-N-FET-e	=BUZ 384: 9A, <0,8Ω(6,5A)	18c§	Phi,Sie	BUK 637-500, BUK 638-500, 2SK1516	data

BV...BX

Type	Device	Short Description	Fig.	Manu	Comparision Types	More at
BV	PIN-Di	=1SV233 (Typ-Code/Stempel/marking)	35	Say	→1SV233	data
BV	PIN-Di	=1SV247 (Typ-Code/Stempel/marking)	35	Say	→1SV247	data
BV	PIN-Di	=1SV247 (Typ-Code/Stempel/marking)	35 (2mm)	Say	→1SV247	data
BV	Si-N	=2SC5274 (Typ-Code/Stempel/marking)	35(1,6mm)	Rhm	→2SC5274	data
BV	Si-N	=2SD1005-BV (Typ-Code/Stempel/marking)	39	Nec	→2SD1005	data
BV	Z-Di	=BZX 399-C30 (Typ-Code/Stempel/marking)	71(1,7mm)	Phi	→BZX 399-...	data
BV	Z-Di	=P4 SMA-36A (Typ-Code/Stempel/marking)	71(5x2,5)	Fag	→P4 SMA-36A	data
BV	Z-Di	=SMBJ 20CA (Typ-Code/Stempel/marking)	71(5x3,5)	Mop	→SMBJ ...	data
BV 1	Si-P	=2SB624-BV1 (Typ-Code/Stempel/marking)	35	Nec	→2SB624	data
BV 2	Si-P	=2SB624-BV2 (Typ-Code/Stempel/marking)	35	Nec	→2SB624	data
BV 3	Si-P	=2SB624-BV3 (Typ-Code/Stempel/marking)	35	Nec	→2SB624	data
BV 4	Si-P	=2SB624-BV4 (Typ-Code/Stempel/marking)	35	Nec	→2SB624	data
BV 4	Si-Di	Uni, kV-Gl, 3kV, 0,05A	31a	Die	BYX 120G, 1N5181	data
BV 5	Si-P	=2SB624-BV5 (Typ-Code/Stempel/marking)	35	Nec	→2SB624	data
BV 6	Si-Di	Uni, kV-Gl, 6kV, 0,05A	31a	Die, Itt	1N5183	data
BV 8	Si-Di	kV-Gl, 7kV, 0,35A	31a	Die	BYX 110GP	data
BV 12	Si-Di	kV-Gl, 11kV, 0,35A	31a	Die	-	data
BV 16	Si-Di	kV-Gl, 15kV, 0,35A	31a	Die	-	data
BVA	Z-Di	=SMBJ 22A (Typ-Code/Stempel/marking	71(5x3,5)	Tho	→SMBJ ...	data
BW	Si-P	=2SA2018 (Typ-Code/Stempel/marking)	35(1,6mm)	Rhm	→2SA2018	
BW	Si-P	=2SA2018H (Typ-Code/Stempel/marking)	35(1,6mm)	Rhm	→2SA2018	
BW	Si-N	=2SD1005-BW (Typ-Code/Stempel/marking)	39	Nec	→2SD1005	data
BW	Si-P	=BCX 71RH (Typ-Code/Stempel/marking)	35	Sie,Val	→BCX 71RH	data
BW	Z-Di	=BZX 399-C33 (Typ-Code/Stempel/marking)	71(1,7mm)	Phi	→BZX 399-...	data
BW	Z-Di	=P4 SMA-39 (Typ-Code/Stempel/marking)	71(5x2,5)	Fag	→P4 SMA-39	data
BW	Z-Di	=SMBJ 22C (Typ-Code/Stempel/marking)	71(5x3,5)	Mop	→SMBJ ...	data
BW 1	Si-P	=2SB736-BW1 (Typ-Code/Stempel/marking)	35	Nec	→2SB736	data
BW 2	Si-P	=2SB736-BW2 (Typ-Code/Stempel/marking)	35	Nec	→2SB736	data
BW 3	Si-P	=2SB736-BW3 (Typ-Code/Stempel/marking)	35	Nec	→2SB736	data
BW 4	Si-P	=2SB736-BW4 (Typ-Code/Stempel/marking)	35	Nec	→2SB736	data
BW 5	Si-P	=2SB736-BW5 (Typ-Code/Stempel/marking)	35	Nec	→2SB736	data
BX	Si-N	=2SC5585 (Typ-Code/Stempel/marking)	35(1,6mm)	Rhm	→2SC5585	
BX	Si-N	=2SC5585H (Typ-Code/Stempel/marking)	35(1,6mm)	Rhm	→2SC5585	
BX	Si-N	=2SC5663 (Typ-Code/Stempel/marking)	35(1,2mm)	Rhm	→2SC5663	
BX	Si-P	=BCX 71RJ (Typ-Code/Stempel/marking)	35	Sie,Val	→BCX 71RJ	
BX	Z-Di	=BZX 399-C36 (Typ-Code/Stempel/marking)	71(1,7mm)	Phi	→BZX 399-...	data
BX	Z-Di	=P4 SMA-39A (Typ-Code/Stempel/marking)	71(5x2,5)	Fag	→P4 SMA-39A	data
BX	Z-Di	=SMBJ 22CA (Typ-Code/Stempel/marking)	71(5x3,5)	Mop	→SMBJ ...	data
BX 7667 W	Hybrid-IC	Vc(vhs), HiFi Audio-signal-prozessor	42P. Modul	Rhm	-	
BX 7854 W	Hybrid-IC	Vc(vhs), HiFi Audio-signal-prozessor	42P. Modul	Rhm	-	
BX 8059 W	Hybrid-IC	Vc(vhs), HiFi Audio-signal-prozessor	42P. Modul	Rhm	-	
BXY ...	Si-Di	Mikrowellendioden/micro wave diodes				data

BY

Type	Device	Short Description	Fig.	Manu	Comparision Types	More at
BY	Si-P	=2SA1200-Y (Typ-Code/Stempel/marking)	39	Tos	→2SA1200	data
BY	Si-N	=2SC3395 (Typ-Code/Stempel/marking)	35	Say	→2SC3395	data
BY	Si-N	=2SC4396 (Typ-Code/Stempel/marking)	35(2mm)	Say	→2SC4396	data
BY	MOS-N-FET-e	=2SK1334 (Typ-Code/Stempel/marking)	39	Hit	→2SK1334	data
BY	Si-P	=BCX 71RK (Typ-Code/Stempel/marking)	35	Sie,Val	→BCX 71RK	data
BY	Z-Di	=BZX 399-C39 (Typ-Code/Stempel/marking	71(1,7mm)	Phi	→BZX 399-...	data
BY	Si-P	=KTA1660-Y (Typ-Code/Stempel/marking)	39	Kec	→KTA 1660	data
BY	Z-Di	=P4 SMA-43 (Typ-Code/Stempel/marking)	71(5x2,5)	Fag	→P4 SMA-43	data
BY	Z-Di	=SMBJ 24C (Typ-Code/Stempel/marking)	71(5x3,5)	Mop	→SMBJ ...	data
BY 100(S)	Si-Di	Gl, 1250V, 0,4A	34a§	Aeg,Phi	BY 127, BY 133, BY 227, G 1M, 1N4007,++	data
BY 101	Si-Di	Gl, 650V, 0,4A	34a§	Aei	BY 126, BY 134, BY 226, 1N4005...4007, ++	data
BY 102	Si-Di	Gl, 750V, 0,6A	32a§	Itt	BY 127, BY 133, BY 227, 1N4006...4007, ++	data
BY 103	Si-Di	Gl, 1300V, 1A	34a§	Itt	BY 127, BY 133, BY 227, BYX 87, EM 513	data
BY 104	Si-Di	Gl, 1250V, 0,5A	34a§	Itt	BY 127, BY 133, BY 227, G 1M, 1N4007,++	data
BY 105	Si-Di	Gl, 1250V, 0,325A	34a§	Aei	BY 127, BY 133, BY 227, G 1M, 1N4007,++	data
BY 112	Si-Di	Gl, 600V, 0,7A	12b	Aeg	BY 126, BY 134, BY 226, 1N4004...4007,++	data
BY 113	Si-Di	=BY 112: 1200V	12b	Aeg	BY 127, BY 133, BY 227, G 1M, 1N4007,++	data
BY 114	Si-Di	Gl, 650V, 0,4A	34a§	Aei,Phi	BY 126, BY 133, BY 226, 1N4005...4007,++	data
BY 114 [Eiy]	Si-Di	Gl, Uni, 400V, 0,1A	2a	Eiy	BA 158...159, BA 199/450, BAY 89	data
BY 115	Si-Di	Gl, 600V, 0,5A	34a§	Itt	BY 126, BY 134, BY 226, 1N4005...4007, ++	data

Type	Device	Short Description	Fig.	Manu	Comparision Types	More at
BY 116	Si-Di	GI, 400V, 0,45A	34b&	Sie	BY 126, BY 134, BY 226, 1N4005...4007, ++	data
BY 118	Si-Di	TV-Booster-Diode, 300V, 5A	22g§	Phi	BY 229/..., BY 223/..., BYT 08/...	data
BY 120	Si-Di	GI, 400V, 0,45A	34b&	Sie	BY 126, BY 134, BY 226, 1N4005...4007, ++	data
BY 121	Si-Di	≈BY 120	34b&	Sie	BY 127, BY 133, BY 227, 1N4006...4007, ++	data
BY 122	Si-Br	GI-Br, 60/120V, 0,8A	42	Phi	B60C800, etc.	data
BY 123	Si-Br	=BY 122: 400/800V, 0,7A	42	Phi	B400C800, etc	data
BY 124	Si-Di	GI, 75V, 0,425A	12b	Aei	BY 126, BY 135, BY 226, 1N4001...4007, ++	data
BY 125 [AEI]	Si-Di	GI, 200V, 0,425A	12b	Aei	BY 126, BY 134, BY 226, 1N4003...4007, ++	data
BY 125 [Philips]	Si-Di	GI, 150V, 1A	31a	Phi	BY 126, BY 135, BY 226, 1N4003...4007, ++	data
BY 126(M,GP,MGP)	Si-Di	GI, 650V, 1A	31a	Phi,Mot,++	BY 127, BY 134, BY 226, 1N4005...4007, ++	data
BY 127(M,GP)	Si-Di	=BY 126: 1250V	31a	Phi,Mot,++	BY 133, BY 227, BYX 87, EM 513, 1N4007++	data
BY 128 [Riz]	Si-Di	GI, Uni, 800V, 0,2A	2c		BY 127, BY 133, BY 227, 1N4006...4007, ++	data
BY 130	Si-Di	GI, 650V, 0,55A	31a	Aei	BY 126, BY 134, BY 226, 1N4005...4007, ++	data
BY 133(GP)	Si-Di	GI, 1300V, 1A	31a	Itt,Mot,Gie	BY 127, BY 227, BYX 87, BYX 95, EM 513++	data
BY 134(GP)	Si-Di	=BY 133: 600V	31a	Itt,Mot,Gie	BY 126...127, BY 226...227, 1N4006...07, ++	data
BY 135(GP)	Si-Di	=BY 133: 150V	31a	Itt,Mot,Gie	BY 126...127, BY 226...227, 1N4003...07, ++	data
BY 137/...	Si-Di	GI, Uni, 900...1300V, 1A	31a	Tho	BY 127, BY 133, BY 227, 1N4007, EM 513++	data
BY 138	Si-Di	GI, contr.av., 800V, 1A	31a	Phi	BYW 55...56, 1N4248...4249, 1N5062,++	data
BY 139	Si-Di	GI, Uni, 450V, 0,165A	12b	Aei	BA 158...159, BAY 89...90, 1N4005...07, ++	data
BY 140(A)	Si-Di	TV-kV-GI, 12...15kV, 2,5mA	31a	Phi	BY 476, BY 609...610, BY 710...713	data
BY 141	Si-Di	GI, Uni, 50V, 0,125A	12b	Aei	BA 157...159, BAY 18...20, 1N5606...09, ++	data
BY 142	Si-Di	GI, 1000V, 1,1A	31a	Sie	BY 127, BY 133, BY 227, 1N4007, ++	data
BY 143	Si-Di	=BY 142: 600V	31a	Sie	BY 126, BY 134, BY 226, 1N4006...4007, ++	data
BY 144	Si-Di	TV-kV-GI, 12,5kV, 2mA	31a	Itt	BY 209, BY 409, BY 476, BY 710...713, ++	data
BY 145	Si-Di	TV-kV-GI, 24kV, 2mA	31a	Itt	BY 713...714, BY 723...724	data
BY 147	Si-Di	TV-Booster-Diode, 6,5kV, 0,25A	31a	Itt	BY 167, GA 5005	data
BY 151 N	Si-Di	GI. 400V, 1A	31a	Tho	BY 126, BY 134, BY 226, 1N4004...4007, ++	data
BY 152 N	Si-Di	=BY 151N: 800V	31a	Tho	BY 127, BY 133, BY 227, 1N4006...4007, ++	data
BY 154	Si-Di		31a			data
BY 156	Si-Di	GI, contr.av., 800V, 0,65A	31a	Gie,Tho	BYW 55...56, 1N4248...4249, 1N5062	data
BY 157[Gen.Instr.]	Si-Di	GI, contr.av., 800V, 0,3A	31a	Gie,Tho	BYW 55...56, 1N4248...4249, 1N5062	data
BY 157 [SSC]	Si-Di	GI, S, 400V, 0,4A, <300ns		Ssc	BY 208/..., BY 231/..., BY 245/..., RGP10...	data
BY 157/...A	Si-Di	GI, 200/320...1000/1600V, 0,3A, <300ns	31a	Tho	BY 208/..., BY 231/..., BY 245/..., RGP10...	data
BY 157/...B	Si-Di	GI, 200/320...1000/1600V, 0,4A, <300ns	31a	Tho	BY 208/..., BY 231/..., BY 245/..., RGP10...	data
BY 157/...C	Si-Di	GI, 200/320...1000/1600V, 0,6A, <300ns	31a	Tho	BY 208/..., BY 231/..., BY 245/..., RGP10...	data
BY 158[Gen.Instr.]	Si-Di	GI, contr.av., 400V, 0,65A	31a	Gen,Tho	BYW 54...56, 1N4246...4249, 1N5060...5062	data
BY 158 [SSC]	Si-Di	GI, S, 600V, 0,4A, <300ns		Ssc	BY 208/..., BY 231/..., BY 245/..., RGP10...	data
BY 159/...	Si-Br	GI-Br, 50...400V, 0,8A	8	Gie	B50C800...B400C800, etc	data
BY 164	Si-Br	GI-Br, 160V, 1,5A	33x1	Phi, Die	B150 C1500	data
BY 165(T)	Si-Di	TV-Booster-Diode, 6kV, 0,3A	(31a)	Gie,Tho	BY 167, GA 5005	data
BY 166	Si-Di	TV-Booster-Diode, 5kV, 0,3A	(31a)	Gie	BY 167, GA 5005	data
BY 167	Si-Di	TV-Booster-Diode, 7,5kV, 0,25A	31a	Aeg	GA 5005	data
BY 172	Si-Di	GI, Uni, 800V, 1,4A	31a	Gie	BY 227, BY 254...255, 1N5398...5399, ++	data
BY 173	Si-Di	=BY 172: 600V	31a	Gie	BY 226...227, BY 253...255, 1N5397...99, ++	data
BY 174	Si-Di	=BY 172: 400V	31a	Gie	BY 226...227, BY 252...255, 1N5395...99, ++	data
BY 176	Si-Di	TV-kV-GI, 15kV, 2,5mA	31a	Phi	BY 476, BY 609...610, BY 710...713	data
BY 177	Si-Di	GI, Uni, 400V, 1,4A	31a	Tho	BY 226...227, BY 252...255, 1N5395...99, ++	data
BY 178	Si-Di	=BY 177: 800V	31a	Tho	BY 227, BY 254...255, 1N5398...5399, ++	data
BY 179	Si-Br	GI-Br, 500V, 1,5A	33x1	Phi, Die	B400 C1000	data
BY 182	Si-Di	TV-kV-GI, 12kV, 2,5mA	31a	Phi	BY 209, BY 409, BY 476, BY 710...713	data
BY 183/...	Si-Di	GI, Uni, 50...60CV, 0,2A	31a	Tho	BA 157...159, BAY 88...90, BY 204/..., ++	data
BY 184	Si-Di	TV-GI, 1500/1800V, 5mA	31a	Phi	BY 203/16...203/20, SHG 1,5..2	data
BY 185	Si-Di	kV-GI, 35kV, 2,5mA		Phi	-	data
BY 186	Si-Di	TV-Booster-Diode, 150V, 5A	2e	Sgs	BY 500-..., MR 822...826, RGP 50C...D	data
BY 187(-01)	Si-Di	TV-KV-GI, 10...12kV, 2,5mA	31a	Phi	BY 209, BY 409, BY 476, BY 710...713	data
BY 188 A	Si-Di	TV-GI, 50V, 1,2A	31a	Phi	BY 218/..., BY 258/..., RGP 15, RGP 30, ++	data
BY 188 B	Si-Di	TV-GI, 50V, 1,2A	31a	Phi	BY 218/..., BY 258/..., RGP 15, RGP 30, ++	data
BY 189	Si-Di	TV-Damper-Diode, 900V, 4A	32b&	Itt	(BY 229/800, BY 277/750, BYW 19/800,++)[4]	data
BY 190	Si-Di	=BY 189: 700V	32b&	Itt	(BY 229/600, BY 277/600, BYW 19/800,++)[4]	data
BY 191/...(P)	Si-Di	GI, S, 250...400V, 4A, <500ns	32a§	Tho	(BY 229/..., BY 277/..., BYW 19/...,++)[4]	data
BY 192	Si-Di	TV-GI, S, 100V, 4A, <500ns	32b&	Itt	(BY 229/200, BY 277/600, BYW 19/800,++)[4]	data
BY 193	Si-Di	=BY 192: 200V	32b&	Itt	(BY 229/200, BY 277/600, BYW 19/800,++)[4]	data
BY 194	Si-Di	=BY 192: 400V	32b&.	Itt	(BY 229/400, BY 277/600, BYW 19/800,++)[4]	data
BY 195	Si-Di	=BY 192: 800V	32b&	Itt	(BY 229/800, BY 277/750, BYW 19/800,++)[4]	data
BY 196	Si-Di	TV-GI, S, 100V, 1,2A, <500ns	31a	Itt,Mot,Tho	BY 201/2, BYX 58/100, MR 811, RGP 10B,++	data

Type	Device	Short Description	Fig.	Manu	Comparision Types	More at
BY 197	Si-Di	=BY 196: 200V	31a	Itt,Mot,Tho	BY 201/2, BYX 58/200, MR 812, RGP 10D,++	data
BY 198	Si-Di	=BY 196: 400V	31a	Itt,Mot,Tho	BY 201/4, BYX 58/400, MR 814, RGP 10G,++	data
BY 199	Si-Di	=BY 196: 800V	31a	Itt,Mot,Tho	BY 231/800, BY 245/800, MR 817, RGP 10K	data
BY 200	Si-Di	=BY 196: 1200V	31a	Itt,Mot,Tho	BY 231/1200, BY 245/1200, BY 400	data
BY 201/2../6	Si-Di	TV-GI, 250...650V, 1A, <200ns	31a	Aeg,Tho	BYV 13...16, RGP 10D...J, RGP 15D...J, ++	data
BY 202/...	Si-Di	TV-GI, 250...650V, 1,5A, <350ns	31a°	Aeg	BY 218/..., BYV 95A...E, RGP 15D...J, ++	data
BY 203/12../25(S)	Si-Di	TV-GI, 1200...2500V, 0,25A, <300ns	31a	Aeg	SHG 1...2,5	data
BY 204/...	Si-Di	TV-GI, 400...1000V, 0,4A <550ns	31a	Aeg	BA 157...159, BY 208/:.., RGP 10G...M, ++	data
BY 205/...	Si-Di	TV-GI-L, 100...1000V, 3A, <850ns	17t§	Tix	BY 229/..., BYT 71/..., RGP 80B...M	data
BY 206(GP)	Si-Di	TV-GI, 350V, 0,4A, <300ns	31a	Phi,Gie,Mot	BA 157...159, BY 204/4, BY 406...407, ++	data
BY 207(GP)	Si-Di	=BY 206: 600V	31a	Phi,Mot,Gie	BA 158...159, BY 204/8, BY 407, ++	data
BY 208/...(GP)	Si-Di	TV-GI, 600...1000V, 0,75A, <350ns	31a	Phi,Mot	BY 245/..., BY 268...269, RGP 10J...M, ++	data
BY 209	Si-Di	TV-kV-GI, 12,5kV, 2,5mA	31a	Phi	BY 409, BY 476, BY 509, BY 710...713, ++	data
BY 210/...	Si-Di	TV-GI, 400...800V, 1A, <300ns	31a	Phi,Mot	BYV 13...16, BYV 95B...E, RGP 10G...M, ++	data
BY 211/2../6	Si-Di	TV-GI, 250...650V, 2A, <350ns	12a	Aeg	BY 218/..., BY 296...299, BYW 32...36, ++	data
BY 212/...R	Si-Di	TV-Damper-Diode, 500...750V, 4A, <300ns	32b&	Tho	(BY 229/800, BY 277/750, BYW 19/800,++)[4]	data
BY 213/...R	Si-Di	=BY 212: 600...700V	32b&	Tho	(BY 229/800, BY 277/750, BYW 19/800,++)[4]	data
BY 214/...	Si-Di	GI, 50...1000V, 6A	31a	Tho	MR 750...756	data
BY 215	Si-Di	2x 9kV, 0,4A		Phi	-	data
BY 216	Si-Di	2x 9kV, 0,4A		Phi	-	data
BY 217/...	Si-Di	GI, 50...400V, 1A	31a	Ssc	BY 126, BY 134, BY 226, 1N4001...07, ++	data
BY 218/...	Si-Di	GI, S, 100...800V, 2A, <200ns	31a	Ssc,Tho	BYW 33...36, BYT 77...78, RGP 30B...M, ++	data
BY 223	Si-Di	CTV-Damper-Diode, 1500V, 5A(ss)	26g§	Phi	(BY 359/1500)[5]	data
BY 224/...	Si-Br	GI-Br, 400...850V, 3,2A	≈33x1	Phi	B220C3200...B280C3200	data
BY 225/...	Si-Br	=BY 224: 100...200V	≈33x1	Phi	B80C3200...B150C3200	data
BY 226(GP,MGP)	Si-Di	GI, 650V, 1,5A	31a	Phi,Gie	BY 253..255, BY 259/600, 1N5397...99, ++	data
BY 227(GP,MGP)	Si-Di	=BY 226: 1250V	31a	Phi,Gie	BY 255, BY 259/1000, BY 350/1300, ++	data
BY 228/...	Si-Di	TV-Damper-Diode, 1000...1500V, 3A	31a	Aeg,Phi,++	BY 328, BY 558, BY 578	data
BY 229-...	Si-Di	TV-Damper-Di, 200...1000V, 7A, <135ns	17t§	Phi, Gsi	BY 359-..., RGP 80D...M	data
BY 229B-...	Si-Di	=BY 229-...:	30t§	Gsi	-	data
BY 229F-...	Si-Di	=BY 229-...: Iso	17d		BYR 29F-..., BY 359F-...	data
BY 229/...R	Si-Di	=BY 229-...	17t1&		(BY 359-..., RGP 80D...M)[5]	data
BY 229X-...	Si-Di	=BY 229-...: Iso	17d		BYR 29F-..., BY 359F-...	data
BY 230	Si-Di	750V, 15A(ss)	22	Sgs	-	data
BY 231/...	Si-Di	TV-GI, 800...1500V, 1,25A	31a	Sie	BY 228/..., BY 448	data
BY 233/...(A)	Si-Di	GL/S-L, 200...600V, 10A, <150ns	17t§	Tho	BY 329/..., RGP 80D...M	data
BY 238	Si-Di	GI, 1500V, 0,8A	12a	Tsm	BY 350/1500, DM 513, EM 516, GP 10W...Y	data
BY 239/...	Si-Di	GI-L, 200...1250V, 10A	17t1&	Tho	- .	data
BY 239/...A,L	Si-Di	=BY 239:	17t§		BYT 12P/...	data
BY 239/...R	Si-Di	=BY 239:	17t1&		-	data
BY 242	Si-Di	GI, 800V, 0,45A	12	Sie	BY 127, BY 133, BY 227, 1N4006...07, ++	data
BY 245/...	Si-Di	GI, 800...1200V, 1,3A	31a	Sie	BY 227, BY 255, BY 350/1300, BYX 87, ++	data
BY 246/...	Si-Di	GI, 600...1200V, 2,5A	31a	Sie	BY 255, BY 259/..., BYW 17/...	data
BY 249/...	Si-Di	GI-L, 300...600V, 7A(Tc=131°)	17t§	Phi	BY 229/..., BY 239/..., GP 80G...M	data
BY 249F/...	Si-Di	=BY 249/...: Iso	17d		BY 229F/..., (BY 239/..., GP 80G...M)[3]	data
BY 249/...R	Si-Di	=BY 249/...:	17t1&		BY 229/...R, BY 239/...R, (GP 80G...M)[5]	data
BY 250	Si-Di	GI, 1000V, 1,25A	31a	Sie	BY 127, BY 133, BY 227, 1N4007, ++	data
BY 251(G,GP)	Si-Di	GI, 200V, 3A	31a	Itt,Mot,Tho	BYW 17/200, (R)GP 30D...M, 1N5402...08, ++	data
BY 252(G,GP)	Si-Di	=BY 251: 400V	31a	Itt,Mot,Tho	BYW 17/400, (R)GP 30G...M, 1N5404...08, ++	data
BY 253(G,GP)	Si-Di	=BY 251: 600V	31a	Itt,Mot,Tho	BYW 17/600, (R)GP 30J...M, 1N5406...08, ++	data
BY 254(G,GP)	Si-Di	=BY 251: 800V	31a	Itt,Mot,Tho	BYW 17/800, (R)GP 30K...M, 1N5407...08, ++	data
BY 255(G,GP)	Si-Di	=BY 251: 1300V	31a	Itt,Mot,Tho	BY 228, BY 448, BY 458, BYW 17/1200	data
BY 256	Si-Br	GI-Br, 200V, 1,5A	33x1	Phi	B80C1500	data
BY 257	Si-Br	=BY 256: 600V	33x1	Phi	B280C1500	data
BY 258/...	Si-Di	TV-GI, 100...800V, 1,7A, 150ns	31a	Sie	BYV 12...16, BY 218/..., RGP 15B...M, ++	data
BY 259/...	Si-Di	GI, 150...1000V, 2,4A	31a	Sie	BY 251...255, BYW 17/..., (R)GP 30D...M, ++	data
BY 260/...	Si-Br	GI-Br, 200...600V, 12A	69	Phi	KBPC 15-02...06	data
BY 261/...	Si-Br	GI-Br, 200...600V, 25A	≈70	Phi	KBPC 25-02...06	data
BY 268	Si-Di	GI, S, 1400V, 0,8A, <400ns	31a	Aeg	BY 228, BY 231/1500, BY 448	data
BY 269	Si-Di	=BY 268: 1600V	31a	Aeg	(BY 228, BY 231/1500, BY 448)[7]	data
BY 277/...R	Si-Di	TV-Damper-Diode, 600...750V, 3A	26g&	Phi	BYW 19/...R, BYX 71/...R, (BY 229/...R)[6]	data
BY 278	Si-Di	TV-Damper, ...16kHz, 1700V, 5A, <1μs	31a	Phi		data
BY 288/...	Si-Di	TV-GI, 150...1000V, 0,32A, 300ns	27c	Sie	BY 208/..., BY 231/..., RGP 10B...M, ++	data
BY 289/...	Si-Di	TV-GI, 150...1000V, 0,52A, 300ns	27c	Sie	BY 208/..., BY 231/..., RGP 10B...M, ++	data
BY 290/...	Si-Di	TV-GI, 150...600V, 150ns	27c	Sie	BY 208/..., BY 231/..., RGP 10B...M, ++	data
BY 291/...	Si-Di	TV-GI, 75...600V, 1,1A, 150ns	27c	Sie	BY 258/..., BYX 55/..., RGP 15B...M, ++	data
BY 292/...	Si-Di	TV-GI, 75...300V, 1,3A, 150ns	27c	Sie	BY 258/..., BYX 55/..., RGP 15B...M, ++	data
BY 293/...	Si-Di	TV-GI, 75...300V, 3A, 150ns	≈28d	Sie	(BY 205/..., BY 229/..., BYV 87/...,++)[4]	data
BY 294/...	Si-Di	TV-GI, 75...600V, 2,5A, 150ns	≈28d	Sie	(BY 205/..., BY 229/..., BYV 87/...,++)[4]	data

Type	Device	Short Description	Fig.	Manu	Comparision Types	More at
295/...	Si-Di	TV-GI, 150...600V, 1,05A, 150ns	31a	Sie	BY 258/..., BYX 55/..., RGP 15D...M, ++	data
296(G,GP)	Si-Di	TV-GI, 200V, 2A, <500ns	31a	Itt,Mot,++	BY 218/200, BY 246/600, RGP 30D...M, ++	data
297(G,GP)	Si-Di	=BY 296: 300V	31a	Itt,Mot,++	BY 218/400, BY 246/600, RGP 30G...M, ++	data
298(G,GP)	Si-Di	=BY 296: 500V	31a	Itt,Mot,++	BY 218/600, BY 246/600, RGP 30J...M, ++	data
299(G,GP)	Si-Di	=BY 296: 1000V	31a	Itt,Mot,++	BY 228, BY 246/1000, BY 400, RGP 30M, ++	data
300/...	Si-Di	TV-GI, 400...700V, 3A, 150ns		Sie	(BY 205/..., BY 229/..., RGP 80G...M,++)[4]	data
302/...	Si-Di	TV-GI, 75...300V, 2,5A, 250ns	≈28d	Sie	(BY 205/..., BY 229/..., RGP 80B...M,++)[4]	data
312/...	Si-Di	GI, S, 75...300V, 1,7A, <250ns	31a	Sie	BYV 12...16, BY 218/..., RGP 15B...M, ++	data
Y 318/...	Si-Di	GI, S, 100...600V, 3A, <200ns	31a	Sie	BYW 16/..., BYW 72...76, RGP 30B...J, ++	data
Y 328	Si-Di	TV-Damper, 1500V, 3A, <500ns, ...32kHz	31a	Phi	-	data
Y 329-...	Si-Di	TV-Damper-Diode, 800...1200V, 7A,<135ns	17t§	Phi	BY 359-..., RGP 80M	data
Y 329-1500(S)	Si-Di	=BY 329-...: 1500V, 6A, <230(S: <160)ns	17t§		BY 359-1500	data
Y 329-1700 S	Si-Di	=BY 329-...: 1700V, 6A, <170ns	17t§		-	data
329F-...	Si-Di	=BY 329-...: Iso	17d		BY 229F-..., BY 359F-...	data
Y 329X-...	Si-Di	=BY 329-...: Iso	17d		BY 229X-..., BY 359X-...	data
Y 330	Si-Di	GI, S, 50V, 1A, <750ns	31a	Gie,Mot	BY 201/2, BYX 55/350, RGP 10A...M, ++	data
Y 331	Si-Di	=BY 330: 100V	31a	Gie,Mot	BY 201/2, BYX 55/350, RGP 10B...M, ++	data
Y 332	Si-Di	=BY 330: 200V	31a	Gie,Mot	BY 201/2, BYX 55/350, RGP 10D...M, ++	data
Y 333	Si-Di	=BY 330: 300V	31a	Gie,Mot	BY 201/3, BYX 55/350, RGP 10G...M, ++	data
Y 334	Si-Di	=BY 330: 400V	31a	Gie,Mot	BY 201/4, BYX 55/600, RGP 10G...M, ++	data
Y 335	Si-Di	=BY 330:	31a	Gie,Mot	BY 201/5, BYX 55/600, RGP 10J...M, ++	
Y 336	Si-Di	=BY 330: 500V	31a	Gie,Mot	BY 201/5, BYX 55/600, RGP 10J...M, ++	data
Y 337	Si-Di	=BY 330: 800V	31a	Gie,Mot	BY 231/800, BY 245/800, RGP 10K...M, ++	data
Y 338	Si-Di	=BY 330: 1000V	31a	Gie,Mot	BY 231/1000, BY 245/1200, RGP 10M, ++	data
Y 339	Si-Di	=BY 330: 1500V	31a	Gie,Mot	BY 231/1500, BY 228, BY 448	data
Y 350/...	Si-Di	GI, S, 1300...1500V, 1,5A, <4µs	31a	Sie	BY 228, BY 448	data
Y 359-...	Si-Di	TV-GI-L, 1000...1500V, 6,5A, <600ns	17t§	Phi	-	data
Y 359F-...	Si-Di	=BY 359-...: Iso	17d		-	data
Y 359..-1500 S	Si-Di	=BY 359-1500...: <350ns			-	data
Y 359X-...	Si-Di	=BY 359-...: Iso	17d		-	data
Y 360/600	Si-Di	GI, S, 600V, 1A, <400ns	31a	Sie	BY 201/4, BYX 55/600, RGP 10J...M, ++	data
Y 396(GP,P)	Si-Di	TV-GI, 100V, 3A, <500ns	31a	Itt,Mot,++	BY 318/100, BYW 14...16/100, RGP 30B...M++	data
Y 397(GP,P)	Si-Di	=BY 396: 200V	31a	Itt,Mot,++	BY 318/200, BYW 14...16/200, RGP 30D...M++	data
Y 398(GP,P)	Si-Di	=BY 396: 400V	31a	Itt,Mot,++	BY 318/400, BYW 14...16/400, RGP 30G...M++	data
Y 399(GP,P)	Si-Di	=BY 396: 800V	31a	Itt,Mot,++	BY 438, BYW 14...16/800, RGP 30K...M, ++	data
Y 399 S	Si-Di	=BY 396: 1000V	31a	Gie	BY 228, BY 438, BYW 96E, RGP 30M	data
BY 400	Si-Di	TV-GI, 1300V, 2A, <500ns	31a	Mot	BY 228, BY 448	data
BY 401	Si-Di	GI, 50V, 0,5A	31a	Mot,Tix	BY 126, BY 135, BY 226, 1N4001...4007, ++	data
BY 402	Si-Di	=BY 401: 100V	31a	Mot,Tix	BY 126, BY 134, BY 226, 1N4002...4007, ++	data
BY 403	Si-Di	=BY 401: 200V	31a	Mot,Tix	BY 126, BY 134, BY 226, 1N4003...4007, ++	data
BY 404	Si-Di	=BY 401: 400V	31a	Mot,Tix	BY 126, BY 134, BY 226, 1N4004...4007, ++	data
BY 405	Si-Di	=BY 401: 600V	31a	Mot,Tix	BY 126, BY 134, BY 226, 1N4005...4007, ++	data
BY 406(A)	Si-Di	=BY 206: 0,8A	31a	Phi,Mot	BY 201/4, BYX 55/350, RGP 10G...M, ++	data
BY 407(A)	Si-Di	=BY 207: 0,8A	31a	Phi,Mot	BY 201/6, BYX 55/600, RGP 10J...M, ++	data
BY 409(A)	Si-Di	TV-kV-GI, 12,5kV, 2,5mA	31a	Phi	BY 209, BY 509, BY 476, BY 710...713, ++	data
BY 410	Si-Di	GI, S, Uni, 100V, 1A, <100ns	31a	Itt	EGP 10B...D, FE 1A...D, BYX 92/...	data
BY 428	Si-Di	TV-Damper, hi-def, 1500V, 4A, <250ns	31a	Phi	-	data
BY 430F/1000	Si-Di	GI, S, Iso, 1000V, 5A, <110ns	17d	Phi	-	data
BY 431F/1000	Si-Di	GI, S, Iso, 1000V, 5A, <85ns	17d	Phi	-	data
BY 438	Si-Di	TV-Damper-Di, 1200V, 5A(ss)	31a	Phi	BY 228, BY 328, BY 428	data
BY 448	Si-Di	TV-Damper-Di, 1650V, 4A, <1000ns	31a	Aeg,Phi,Gie	BY 278	data
BY 458	Si-Di	=BY 448: 1200V	31a	Aeg,Phi,Gie	BY 228, BY 328, BY 438	data
BY 459-1500	Si-Di	CRT-GI-L, 1500V, 12A, <350ns (f. Multi-sync Monitor, ...82kHz)	17t§	Phi	MUR 10150E	data
BY 459F-1500	Si-Di	=BY 459-1500: Iso	17d		BY 479X-1700	data
BY 459..-1500 S	Si-Di	=BY 459-1500: 10A, <220ns				data
BY 459X-1500	Si-Di	=BY 459-1500: Iso	17d		BY 479X-1700	data
BY 476(A)	Si-Di	TV-kV-GI, 18kV, 2,5mA	31a	Phi	BY 711...713	data
BY 477	Si-Di	TV-kV-GI, 23kV, 2mA	31a	Phi	BY 713	data
BY 478	Si-Di	=BY 477: 27,5kV	31a	Phi	BY 713	data
BY 479X-1700	Si-Di	CRT-GI-L, 1500V, 10A, <350ns	17d	Phi	-	data
BY 500-...	Si-Di	TV-GI, 50...1000V, 5A, <200ns	31a	Gie,Mot,Die	MR 821...828	data
BY 505	Si-Di	GI, S, 2000/2200V, 0,05A, 200ns	31a	Phi	BY 203/20, SHG 2..2,5	data
BY 509	Si-Di	TV-kV-GI, 12,5kV, 4mA	31a	Phi	BY 609...610	data
BY 510	Si-Di	=BY 509: 17kV	31a	Phi	BY 610	data
BY 520-10...-20	Si-Di	GI, S, 1000...2000V, 0,5A, 500ns	31a	Gie	RGP 15-...	data
BY 527	Si-Di	GI, contr.av., 1250V, 0,8A	31a	Phi,Mot	BYW 18/1000, BYW 56, BYW 86, 1N4249	data
BY 530-...	Si-Di	GI, 50...1000V, 3A	31a	Gie	BY 251...255, BYW 17/..., 1N5400...5408, ++	data
BY 550-...	Si-Di	GI, 50...1000V, 5A	31a	Fag, Die	BY 214/..., BY 500-..., MR 750...760	data

Type	Device	Short Description	Fig.	Manu	Comparision Types	More at
BY 558	Si-Di	CRT-GI, 1500V, 2,5A, <250ns (f. Multi-sync Monitor)	31a	Phi	BY 328	data
BY 559	Si-Di	CRT-GI, 1500V, 10A, <1000ns	17t§	Phi	BY 459-1500, MUR 10150E	data
BY 559-1500U	Si-Di	=BY 559-1500: <120ns	17t§		MUR 10150E	data
BY 559X-1500U	Si-Di	=BY 559-1500: Iso, <120ns	17d		-	data
BY 578	Si-Di	CRT-GI, 1700V, 2,5A, <250ns (f. Multi-sync Monitor)	31a	Phi	-	data
BY 584	Si-Di	TV-GI, 1800V, 0,05A, 200ns	31a	Phi	BY 203/20, SHG 1,5..2	data
BY 588	Si-Di	TV-GI, Basis-Emitter-Di, 25V, 1,5A	31a	Phi	BYW 52, BYX 82, GP 15A...M, 1N5391...96,++	data
BY 601	Si-Di	GI, 50V, 1,5A	31a	Mot	BY 226..227, BY 251...255, 1N5391...99, ++	data
BY 602	Si-Di	=BY 601: 100V	31a	Mot	BY 226...227, BY 251...255, 1N5392...99, ++	data
BY 603	Si-Di	=BY 601: 200V	31a	Mot	BY 226...227, BY 251...255, 1N5393...99, ++	data
BY 604	6i-Di	=BY 601: 400V	31a	Mot	BY 226...227, BY 252...255, 1N5395...99, ++	data
BY 605	Si-Di	=BY 601: 600V	31a	Mot	BY 226...227, BY 253...255, 1N5397...99, ++	data
BY 606	Si-Di	=BY 601: 800V	31a	Mot	BY 227, BY 254...255, 1N5398...5399, ++	data
BY 607	Si-Di	=BY 601: 1000V	31a	Mot	BY 227, BY 255, BY 350/1300, 1N5399	data
BY 608	Si-Di	=BY 601: 1250V	31a	Mot	BY 227, BY 255, BY 350/1300	data
BY 609	Si-Di	TV-kV-GI, 15kV, 4mA, 200ns	31a(9mm)	Phi	BY 619, BY 710...713	data
BY 610	Si-Di	=BY 609: 17kV	31a(9mm)	Phi	BY 620, BY 710...713	data
BY 614	Si-Di	GI, S, 2200V, 0,05A, <300ns	31a	Phi	(BY 203/20, SHG 2,5)	data
BY 617	Si-Di	TV-kV-GI, 9kV, 4mA, 100ns	31a	Phi	BY 708...709	data
BY 619	Si-Di	TV-kV-GI, 15kV, 4mA, 100ns	31a	Phi	BY 609, BY 710...713	data
BY 620	Si-Di	=BY 619: 17kV	31a	Phi	BY 610, BY 710...713	data
BY 627	Si-Di	Impatt-Di, contr.av., 1250V, 2A	31a	Phi	-	data
BY 705	Si-Di	kV-GI, 5kV, 20mA, 200ns	31a(5mm)	Phi	BY 715, CY 5, HS 6	data
BY 706	Si-Di	=BY 705: 6kV	31a(5mm)	Phi	BY 716, CY 6, HS 6	data
BY 707	Si-Di	TV-kV-GI, 10kV, 4mA, 200ns	31a(9mm)	Phi	BY 609...610, BY 619...620, BY 717...719	data
BY 708	Si-Di	TV-kV-GI, 12kV, 4mA, 200ns	31a(9mm)	Phi	BY 609...610, BY 619...620, BY 718...719	data
BY 709	Si-Di	TV-kV-GI, 14kV, 4mA, 200ns	31a(9mm)	Phi	BY 609...610, BY 619...620, BY 719	data
BY 710	Si-Di	TV-kV-GI, 17kV, 3mA, 200ns	31a(11mm)	Phi	BY 610, BY 620, BY 711...714, BY 720...724	data
BY 711	Si-Di	TV-kV-GI, 19kV, 3mA, 200ns	31a(11mm)	Phi	BY 712...713, BY 721...724	data
BY 712	Si-Di	TV-kV-GI, 22kV, 3mA, 200ns	31a(12mm)	Phi	BY 713...714, BY 722...724	data
BY 713	Si-Di	TV-kV-GI, 24kV, 3mA, 200ns	31a(12mm)	Phi	BY 714, BY 723...724	data
BY 714	Si-Di	TV-kV-GI, 30kV, 3mA, 200ns	31a(12mm)	Phi	BY 724	data
BY 715	Si-Di	kV-GI, 5kV, 20mA, 100ns	31a(5mm)	Phi	BY 705, CY 5, HS 6	data
BY 716	Si-Di	=BY 705: 6kV	31a(5mm)	Phi	BY 706, CY 6, HS 6	data
BY 717	Si-Di	TV-kV-GI, 10kV, 4mA, 100ns	31a(9mm)	Phi	BY 609...610, BY 619...620, BY 707...709	data
BY 718	Si-Di	TV-kV-GI, 12kV, 4mA, 100ns	31a(9mm)	Phi	BY 609...610, BY 619...620, BY 708...709	data
BY 719	Si-Di	TV-kV-GI, 14kV, 4mA, 100ns	31a(9mm)	Phi	BY 609...610, BY 619...620, BY 709	data
BY 720	Si-Di	TV-kV-GI, 17kV, 3mA, 100ns	31a(11mm)	Phi	BY 610, BY 620, BY 710...714	data
BY 721	Si-Di	TV-kV-GI, 19kV, 3mA, 100ns	31a(11mm)	Phi	BY 711...714	data
BY 722	Si-Di	TV-kV-GI, 22kV, 3mA, 100ns	31a(12mm)	Phi	BY 712...714	data
BY 723	Si-Di	TV-kV-GI, 24kV, 3mA, 100ns	31a(12mm)	Phi	BY 713...714	data
BY 724	Si-Di	TV-kV-GI, 30kV, 3mA, 100ns	31a(12mm)	Phi	BY 714	data
BY 1600	Si-Di	kV-GI, 1,6kV, 3A	31a	Die	-	data
BY 1800	Si-Di	kV-GI, 1,8kV, 3A	31a	Die	-	data
BY 2000	Si-Di	kV-GI, 2kV, 3A	31a	Die	-	data
BY 4000	Si-Di	kV-GI, 3kV, 1,5A	31a	Die	-	data
BY 6000	Si-Di	kV-GI, 5kV, 1A	31a	Die	-	data
BY 8000	Si-Di	kV-GI, 7kV, 0,7A	31a	Die	-	data
BY 8004	Si-Di	kV-GI, contr.av., 4/5kV, 20mA, <100ns	31a	Phi	BY 8104, BY 8404	data
BY 8006	Si-Di	=BY 8004: 6/8kV, 10mA	31a	Phi	BY 8106, BY 8406	data
BY 8008	Si-Di	=BY 8004: 8/10kV, 5mA	31a	Phi	BY 8108, BY 8408	data
BY 8010	Si-Di	=BY 8004: 10/12kV, 5mA	31a	Phi	BY 8110, BY 8410	data
BY 8012	Si-Di	=BY 8004: 12/14kV, 5mA	31a	Phi	BY 8112, BY 8412	data
BY 8014	Si-Di	=BY 8004: 14/17kV, 5mA	31a	Phi	BY 8114, BY 8414	data
BY 8016	Si-Di	=BY 8004: 16/19kV, 3mA	31a	Phi	BY 8116, BY 8416	data
BY 8104	Si-Di	kV-GI, contr.av., 4/5kV, 20mA, <60ns	31a	Phi	-	data
BY 8106	Si-Di	=BY 8104: 6/8kV, 10mA	31a	Phi	-	data
BY 8108	Si-Di	=BY 8104: 8/10kV, 5mA	31a	Phi	-	data
BY 8110	Si-Di	=BY 8104: 10/12kV, 5mA	31a	Phi	-	data
BY 8112	Si-Di	=BY 8104: 12/14kV, 5mA	31a	Phi	-	data
BY 8114	Si-Di	=BY 8104: 14/17kV, 5mA	31a	Phi	-	data
BY 8116	Si-Di	=BY 8104: 16/19kV, 5mA	31a	Phi	-	data
BY 8206	Si-Di	kV-GI, contr.av., 6kV, 10mA, <45ns	31a	Phi	BY 9206	data
BY 8208	Si-Di	=BY 8206: 8kV, 5mA	31a	Phi	BY 9208	data
BY 8210	Si-Di	=BY 8206: 10kV, 5mA	31a	Phi	BY 9210	data
BY 8212	Si-Di	=BY 8206: 12kV, 5mA	31a	Phi	BY 9212	data
BY 8404	Si-Di	kV-GI, contr.av., 4/5kV, 20mA, <100ns	31a	Phi	BY 8004, BY 8104	data
BY 8406	Si-Di	=BY 8404: 6/8kV, 10mA	31a	Phi	BY 8006, BY 8106	data
BY 8408	Si-Di	=BY 8404: 8/10kV, 5mA	31a	Phi	BY 8008, BY 8108	data
BY 8410	Si-Di	=BY 8404: 10/12kV, 5mA	31a	Phi	BY 8010, BY 8110	data
BY 8412	Si-Di	=BY 8404: 12/14kV, 5mA	31a	Phi	BY 8012, BY 8112	data
BY 8414	Si-Di	=BY 8404: 14/17kV, 5mA	31a	Phi	BY 8014, BY 8114	data
BY 8416	Si-Di	=BY 8404: 16/19kV, 3mA	31a	Phi	BY 8016, BY 8116	data
BY 8418	Si-Di	=BY 8404: 18/22kV, 3mA	31a	Phi	-	data
BY 8420	Si-Di	=BY 8404: 20/24kV, 3mA	31a	Phi	-	data
BY 8424	Si-Di	=BY 8404: 24/30kV, 3mA	31a	Phi	-	data

Type	Device	Short Description	Fig.	Manu	Comparision Types	More at.
BY 9206	Si-Di	kV-Gl, contr.av., 6kV, 10mA, <35ns	31a	Phi	-	data
BY 9208	Si-Di	=BY 9206: 8kV, 5mA	31a	Phi		data
BY 9210	Si-Di	=BY 9206: 10kV, 5mA	31a	Phi		data
BY 9212	Si-Di	=BY 9206: 12kV, 5mA	31a	Phi		data
BY 9304	Si-Di	kV-Gl, contr.av., 4kV, 10mA, <100ns	31a	Phi	-	data
BY 9306	Si-Di	=BY 9304: 6kV, 10mA	31a	Phi		data
BY 9308	Si-Di	=BY 9304: 8kV, 5mA	31a	Phi		data
BY 9310	Si-Di	=BY 9304: 10kV, 5mA	31a	Phi		data
BY 9312	Si-Di	=BY 9304: 12kV, 5mA	31a	Phi		data
BY 9314	Si-Di	=BY 9304: 14kV, 5mA	31a	Phi		data
BY 9316	Si-Di	=BY 9304: 16kV, 3mA	31a	Phi		data
BY 9318	Si-Di	=BY 9304: 18kV, 3mA	31a	Phi		data
BY 9410	Si-Di	kV-Gl, contr.av., 10kV, 5mA, <100ns	31a	Phi	-	data
BY 9412	Si-Di	=BY 9410: 12kV, 5mA	31a	Phi		data
BY 9414	Si-Di	=BY 9410: 14kV, 5mA	31a	Phi		data
BY 9416	Si-Di	=BY 9410: 18kV, 3mA	31a	Phi		data
BYC 5-600	Si-Di	S-L, 600V, 5A(Tc=110°), <30ns SMPS, Lighting Ballast	17t§	Phi	FE 8J	data
BYC 5-600B	Si-Di	=BYC 5-600	30f§		-	data
BYC 8-600	Si-Di	S-L, 600V, 8A(Tc=110°), <40ns SMPS, Lighting Ballast	17t§	Phi	FE 8J	data
BYC 8-600B	Si-Di	=BYC 8-600:	30f§		-	data
BYC 10-600	Si-Di	S-L, 600V, 10A(Tc=114°), <40ns SMPS, Lighting Ballast	17t§	Phi	-	data
BYC 10-600B	Si-Di	=BYC 10-600:	30f§		-	data

BYD...BYR

Type	Device	Short Description	Fig.	Manu	Comparision Types	More at.
BYD 11 D..M	Si-Di	Gl, contr.av., 200...1000V, 0,58A D=200, G=400, J=600, K=800, M=1000V	31a	Phi	BYD 33D..M, BYV 26B..E, 1N4245..49	data
BYD 12 D..M	Si-Di	Gl, contr.av., 200...1000V, 0,82A D=200, G=400, J=600, K=800, M=1000V	31a	Phi	BYD 33D..M, BYV 26B..E, 1N4245..49	data
BYD 13 D..M	Si-Di	Gl, contr.av., 200...1000V, 1,4A D=200, G=400, J=600, K=800, M=1000V	31a	Phi	BYD 33D..M, BYW 52..56, 1N5060..5062	data
BYD 14 D..M	Si-Di	Gl, contr.av., 200...1000V, 2A, 2,5µs D=200, G=400, J=600, K=800, M=1000V	31a	Phi	BYW 52..56, 1N5060..5062	data
BYD 17 D..M	Si-Di	=BYD 13D..M: SMD	72a(3,4mm)	Phi	BYD 37D..M	data
BYD 31 D..M	Si-Di	S, contr.av., 200...1000V, 0,4A, <300ns D=200, G=400, J=600, K=800, M=1000V	31a	Phi	BYD 33D..N, BYV 26B..E, BYV 95...96/...	data
BYD 32 D..J	Si-Di	S, contr.av., 200...600V, 0,76A, <250ns D=200V, G=400V, J=600V	31a	Phi	BYD 33D..N, BYV 26B..E, BYV 95...96/...	data
BYD 33 D..M	Si-Di	S, contr.av., 200...1000V, 1,3A, <300ns D=200, G=400, J=600, K=800, M=1000V	31a	Phi	BYM 26A..E, BYV 36A..E, BYV 95...96/...	data
BYD 33 U...V	Si-Di	=BYD 33D..M: <500ns, U=1200V, V=1400V	31a	Phi	RGP 15-12...-14	data
BYD 34 D..M	Si-Di	S, contr.av., 200...1000V, 1,8A, <300ns D=200, G=400, J=600, K=800, M=1000V	31a	Phi	BYM 26A..E, BYW 95...96/...	data
BYD 37 D..M	Si-Di	=BYD 33D..M: SMD	72a(3,4mm)	Phi	-	data
BYD 43-16...-20	Si-Di	kV-Gl, FLT, 1,6..2kV, 0,68A, <300ns -16=1,6kV -18=1,8kV, -20=2kV	31a	Phi	-	data
BYD 43 U...V	Si-Di	kV-Gl, FLT, 1,2..1,4kV, 1,2A, <250ns U=1,2kV, V=1,4kV	31a	Phi	-	data
BYD 47-16...-20	Si-Di	SMD,kV-Gl, FLT, 1,6..2kV, 0,8A, <300ns -16=1,6kV -18=1,8kV, -20=2kV	72a(3,4mm)	Phi	-	data
BYD 52 D..J	Si-Di	S, contr.av., 200...600V, 0,47A, <30ns D=200V, G=400V, J=600V	31a	Phi	BYD 71D..G, BYV 26B..C	data
BYD 53 D..M	Si-Di	S, contr.av., 200...1000V, 0,75A, <30ns D=200V, G=400V, J=600V, K=800V, <75ns M=1000V, <75ns, U=1200V, <150ns V=1400V, <150ns	31a	Phi	BYV 26B..E	data
BYD 57 D...V	Si-Di	=BYD 53D...V: SMD, 1A	72a(3,4mm)	Phi	-	data
BYD 63	Si-Di	Ripple Blocking, 300V, 0,85A, <150ns	31a	Phi	BYV 26B..E, EGP 10F..G, FE 1F..H	data
BYD 67	Si-Di	=BYD 63: SMD	72a(3,4mm)	Phi	BYD 57G..M	data
BYD 71 A..G	Si-Di	S, contr.av., 50...400V, 0,5A, <50ns A=50, B=100, C=150, D=200, E=250V, F=300, G=400V	31a	Phi	EGP 10A..G, FE 1A..H	data
BYD 72 A..G	Si-Di	S, contr.av., 50...400V, 1A, <25ns A=50, B=100, C=150, D=200, E=250V, <50ns, F=300V, <50ns, G=400V, <50ns	31a	Phi	BYV 26B..E, EGP 10A..G, FE 1A..H	data
BYD 73 A..G	Si-Di	S, contr.av., 50...400V, 1,75A, <50ns A=50, B=100, C=150, D=200, E=250V, F=300, G=400V	31a	Phi	BYV 27/..., EGP 20A..G, FE 2A..H	data
BYD 74 A..G	Si-Di	S, contr.av., 50...400V, 2,4A, <50ns A=50, B=100, C=150, D=200, E=250V, F=300, G=400V	31a	Phi	BYV 27/..., EGP 20A..G, FE 2A..H	data
BYD 77 A..G	Si-Di	=BYD 73A..G: SMD	72a(3,4mm)	Phi	-	data
BYD 123	Si-Di	Gl, S, 200V, 1A, <25ns	31a	Phi	BYV 26B..E, EGP 10A..G, FE 1A..H	data
BYD 127	Si-Di	=BYD 123: SMD	72a(3,4mm)	Phi	BYD 57D..J	data
BYD 143	Si-Di	Gl, S, 400V, 1A, <50ns	31a	Phi	BYV 26B..E, EGP 10G, FE 1H	data
BYD 147	Si-Di	=BYD 143: SMD	72a(3,4mm)	Phi	BYD 57G..J	data
BYD 163	Si-Di	Gl, S, 600V, 1A, <50ns	31a	Phi	BYV 26C..E	data
BYD 167	Si-Di	=BYD 163: SMD	72a(3,4mm)	Phi	BYD 57J	data

245

Type	Device	Short Description	Fig.	Manu	Comparision Types	More at
BYD 1100	Si-Di	SMD, SS, 100V, 2,7A, <10ns	72a(3,4mm)	Phi	-	data
BYG 10 D..M	Si-Di	SMD, GI, Uni, 200...1000V, 1,5A D=200, G=400, J=600, K=800, M=1000V	71a(5x2,5)	Aeg	FS 2D..M	data
BYG 20 D..J	Si-Di	SMD, GI, S, 200...600V, 1,5A, <75ns D=200V, G=400V, J=600V	71a(5x2,5)	Aeg	-	data
BYG 21 K..M	Si-Di	SMD, GI, S, 800...1000V, 1,5A, <120ns K=800V, M=1000V	71a(5x2,5)	Aeg	-	data
BYG 22 A..D	Si-Di	SMD, GI, S, 50...200V, 2A, <25ns A=50V, B=100V, D=200V	71a(5x2,5)	Aeg	-	data
BYG 23 M	Si-Di	SMD, GI, S, 1000V, 1,5A, <75ns	71a(5x2,5)	Aeg	-	data
BYG 50 D..M	Si-Di	SMD, GI, contr.av, 200...1000V, 0,7A D=200, G=400, J=600, K=800, M=1000V	71a(5x2,5)	Phi	BYG 10D..M, BYG 20D..M, FR 2D..M	data
BYG 60 D..M	Si-Di	SMD, S, contr.av, 200...1000V, 0,65A <250ns, D=200, G=400, J=600V, K=800V, <300ns, M=1000V, <300ns	71a(5x2,5)	Phi	BYG 20D..M, BYG 21K..M, FR 2D..M	data
BYG 70 D..J	Si-Di	SMD, S, contr.av, 200...600V, 0,53A <30ns, D=200, G=400, J=600V	71a(5x2,5)	Phi	BYG 80D..J	data
BYG 80 A..J	Si-Di	SMD, S, contr.av, 50...600V, 1A <50ns, A=50, B=100, C=150, D=200V F=300,<50ns,G=400,<50ns, J=600V,<50ns	71a(5x2,5)	Phi	BYG 70D..J	data
BYG 85 B	Si-Di	SMD, CS, 100V, 2,5A, <12,5ns	71a(5x2,5)	Phi	-	data
BYG 90-20...-90	Si-Di	SMD, Schottky, 20...90V, 1A, Uf<0,55V	71a(5x2,5)	Phi	-	data
BYM 05-...	Si-Di	SMD, GI, 50...1000V, 0,5A	72a(3,4mm)	Gie, Tho	BYD 37D..M	data
BYM 06-...	Si-Di	SMD, S, 50...1000V, 0,5A, <500ns	72a(3,4mm)	Gie, Tho	BYD 37D..M	data
BYM 07-...	Si-Di	SMD, S, 50...400V, 0,5A, <50ns	72a(3,4mm)	Gie	BYD 77A..G	data
BYM 10-...	Si-Di	SMD, GI, 50...1600V, 1A	72a(3,4mm)	Gie, Tho	BYM 11-...	data
BYM 11-...	Si-Di	SMD, GI, 50...1000V, 1A, <500ns	72a(5mm)	Gie	-	data
BYM 12-...	Si-Di	SMD, S, 50...400V, 1A, <50ns	72a(5mm)	Gie	SFE 1A..H	data
BYW 13-..	Si-Di	SMD, Schottky, 20...60V, 1A, Uf<0,7(1A)	72a(5mm)	Gsi	SGL 41-..., TMBYV 10-...	data
BYM 26 A..E	Si-Di	S, contr.av., 200...1000V, 2,3A, <75ns A=200, B=400, C=600, D=800, E=1000V	31a	Phi, Fag	BYM 36A..E	data
BYM 26 F...G	Si-Di	=BYM 26A..E: 1200...1400V, 2,3A, <150ns F=1200, G=1400V	31a		-	data
BYM 30-...	Si-Di	SMD, GI, 50...1000V, 3A	72a(7,5mm)	Gie	-	data
BYM 31-...	Si-Di	SMD, S, 50...1000V, 3A, <500ns	72a(7,5mm)	Gie	-	data
BYM 32-...	Si-Di	SMD, S, 50...400V, 3A, <50ns	72a(7,5mm)	Gie	-	data
BYM 36 A..E	Si-Di	S, contr.av., 200...1000V, 3A, <150ns A=200, B=400, C=600, D=800, E=1000V	31a	Aeg,Phi,Fag	BYW 95A..C, BYW 96D..E	data
BYM 36 F...G	Si-Di	=BYM 36A..E: 1200...1400V, 2,9A, <250ns F=1200, G=1400V	31a		-	data
BYM 56 A..E	Si-Di	GI, contr.av., 200...1000V, 3,5A	31a	Phi	BYW 18/..., BYW 82..86	data
BYM 63	Si-Di	Ripple Blocking, 300V, 1A, <150ns	31a	Phi	BYV 26B..E, EGP 10F..G, FE 1F..H	data
BYM 99	Si-Di	SMD, SS, 600V, 0,8A, <15ns	72a	Phi	-	data
BYM 300 A...	Si-Di	Power Module		Sie		
BYM 600 A...	Si-Di	Power Module		Sie		
BYP 20-...	Si-Di	Dual, SMPS, 50...150V, 10A, <30ns	17p	Phi	BYQ 28/..., BYT 28/..., BYV 32/..., FE 16...	data
BYP 21-...	Si-Di	S-L, 50..200V, 8A, <25ns	17t§	Phi	BYV 29/..., BYW 80/..., FE 8A..J	data
BYP 22-...	Si-Di	Dual, S-L, 50..200V, 20A, <25ns	17p§	Phi	BYV 44/..., BYW 51/...	data
BYP 53/75../800	Si-Di	=BYY 53/..: Plastic Cover	75b&	Ztx	DS 25-01A...-06A, 1N3492R..3495R	data
BYP 54/75../800	Si-Di	=BYY 53/..: Plastic Cover	75a§	Ztx	1N3492..3495	data
BYP 57/75../800	Si-Di	=BYY 57/..: Plastic Cover	75b&	Ztx	SSi E1210...1240, 1N3660R..3665R	data
BYP 58/75../800	Si-Di	=BYY 57/..: Plastic Cover	75a§	Ztx	SSi E1110...1140, 1N3660..3665	data
BYP 59-...M,U	Si-Di	S-L, 300..400V, 60A, <60ns	32a	Phi	BYT 60/...	data
BYP 100	Si-Di	S-L (FRED), 1000V, 5A(Tc=90°), 55ns	18t	Sie	-	data
BYP 101	Si-Di	S-L (FRED), 1000V, 15A(Tc=90°), 80ns	18t	Sie	-	data
BYP 102	Si-Di	S-L (FRED), 1000V, 28A(Tc=90°), 130ns	18t	Sie	-	data
BYP 103	Si-Di	S-L (FRED), 1000V, 45A(Tc=90°), 140ns	18t	Sie	-	data
BYP 300	Si-Di	S-L (FRED), 1200V, 4A(Tc=90°), 50ns	18t	Sie	-	data
BYP 301	Si-Di	S-L (FRED), 1200V, 12A(Tc=90°), 80ns	18t	Sie	-	data
BYP 302	Si-Di	S-L (FRED), 1200V, 25A(Tc=90°), 130ns	18t	Sie	-	data
BYP 303	Si-Di	S-L (FRED), 1200V, 40A(Tc=90°), 140ns	18t	Sie	-	data
BYQ 27/...	Si-Di	Dual, SMPS, 50..200V, 10A, <20ns	≈14p§	Phi	-	data
BYQ 28-...	Si-Di	Dual, SMPS, 50..200V, 10A, <20ns	17p§	Phi, Gsi	BYP 20-..., BYV 32-..., FE 16A..J	data
BYQ 28 E-...	Si-Di	=BYQ 28-...: ESD protected	17p§		BYV 32E-...	data
BYQ 28 EB-...	Si-Di	=BYQ 28-...:	30p§		BYQ 30EB-...	data
BYQ 28 ED-...	Si-Di	=BYQ 28-...:	30p§		BYQ 30ED-...	data
BYQ 28 F-...	Si-Di	=BYQ 28-...: Iso	17p		BYV 32F-...	data
BYQ 28(E)X-...	Si-Di	=BYQ 28-...: Iso	17p		BYV 32F-...	data
BYQ 30-...	Si-Di	Dual, SMPS, 150...200V, 16A, <25ns	17p§	Phi	BYV 32-..., FE 16A..J	data
BYQ 30 EB-...	Si-Di	=BYQ 30-...:	30p§		-	data
BYQ 30 ED-...	Si-Di	=BYQ 30-...:	30p§		-	data
BYQ 30 EX-...	Si-Di	=BYQ 30-...: Iso	17p		-	data
BYQ 40 EW-...	Si-Di	Dual, SMPS, 150...200V, 40A, <40ns	16p§	Phi	BYV 52-..., BYT 60P-...	data
BYQ 60 EW-...	Si-Di	Dual, SMPS, 150...200V, 60A, <60ns	16p§	Phi	BYV 52-..., BYT 60P-...	data

Type	Device	Short Description	Fig.	Manu	Comparision Types	More at
BYQ 63	Si-Di	Ripple Blocking, 300V, 0,68A, <150ns	31a	Phi	BYV 26B..E, EGP 10F..G, FE 1F..H	data
BYR 28/...	Si-Di	Dual, SMPS, 500...800V, 10A, <80ns	17p§	Phi	-	data
BYR 28	Si-Di	=PBYR 280CT (Typ-Code/Stempel/marking)	48	Phi	→PBYR 280CT	data
BYR 29	Si-Di	=PBYR 290CT (Typ-Code/Stempel/marking)	48	Phi	→PBYR 290CT	data
BYR 29-...	Si-Di	SMPS, 500...800V, 8A, <75ns	17t§	Phi,Sie	MUR 850..8100	data
BYR 29 F-...	Si-Di	=BYR 29-...: Iso	17d		-	data
BYR 30-...(U)	Si-Di	SMPS, 500...700V, 14A, <100ns	32a§	Phi,Sie	BYV 31/...	data
BYR 34-...	Si-Di	Dual, SMPS, 500...800V, 20A, <80ns	17p§	Phi,Sie	BYV 44/...	data
BYR 79-...	Si-Di	SMPS, 500...800V, 14A, <100ns	17t§	Phi,Sie	BYT 87/...	data
BYR 210	Si-Di	=PBYR 2100CT(Typ-Code/Stempel/marking)	48	Phi	→PBYR 2100CT	data
BYR 215	Si-Di	=PBYR 2150CT(Typ-Code/Stempel/marking)	48	Phi	→PBYR 2150CT	data
BYR 220	Si-Di	=PBYR 220CT(Typ-Code/Stempel/marking)	48	Phi	→PBYR 220CT	data
BYR 225	Si-Di	=PBYR 225CT(Typ-Code/Stempel/marking)	48	Phi	→PBYR 225CT	data
BYR 240	Si-Di	=PBYR 240CT(Typ-Code/Stempel/marking)	48	Phi	→PBYR 240CT	data
BYR 245	Si-Di	=PBYR 245CT(Typ-Code/Stempel/marking)	48	Phi	→PBYR 245CT	data
BYR 280	Si-Di	=PBYR 280CT(Typ-Code/Stempel/marking)	48	Phi	→PBYR 280CT	data
BYR 290	Si-Di	=PBYR 290CT(Typ-Code/Stempel/marking)	48	Phi	→PBYR 290CT	data

BYS

Type	Device	Short Description	Fig.	Manu	Comparision Types	More at
BYS 05-...	Si-Di	Schottky-Gl, 20...40V, 15A	34a	Mot	(BYS 15)[4]	data
BYS 08-...	Si-Di	Schottky-Gl, 20...50V, 8A	34a	Mot	(BYS 15)[4]	data
BYS 10-...	Si-Di	SMD, Schottky-Gl, 25...45V, 1,5A	71a(5x2,5)	Aeg	BYS 11-90, BYS 12-90, SFPB-76	data
BYS 11-90	Si-Di	SMD, Schottky-Gl, 90V, 1,5A	71a(5x2,5)	Aeg	BYS 12-90	data
BYS 12-90	Si-Di	SMD, Schottky-Gl, 90V, 1,5A	71a(5x2,5)	Aeg	BYS 11-90	data
BYS 13-90	Si-Di	SMD, Schottky-Gl, ESD(20kV), 90V, 1,5A	71a(5x2,5)	Aeg		data
BYS 1	Si-Di	Schottky-Gl, 45V, 15A	32a§	Sie	BYS 31-50, BYS 32	data
BYS 16-...	Si-Di	Schottky-Gl, 20...40V, 15A	32a§	Mot	MBR 1520...1540, 1N5826...28	data
BYS 21-...	Si-Di	Schottky-Gl, 45...90V, 1A	31a	Sie	BYS 22-..., HRP 22, HRP 32	data
BYS 22-...	Si-Di	Schottky-Gl, 45...90V, 2A	31a	Sie	BYS 26-...	data
BYS 24-...	Si-Di	Dual, Schottky-Gl, 45...90V, 2x5A	17p	Sie	MBR 2060CT..20100CT	data
BYS 25-...	Si-Di	Schottky-Gl, 20...40V, 25A	32a§	Mot	BYV 121-..., MBR 2520...2540, 1N5829...31	data
BYS 26-...	Si-Di	Schottky-Gl, 45...90V, 3A	31a	Sie	BYS 27/..., SB 350...380	data
BYS 27-45	Si-Di	Schottky-Gl, 45V, 5A	31a	Sie	SB 550...580	data
BYS 28-...	Si-Di	Dual, Schottky-Gl, 45...90V, 2x15A	18p	Sie	BYV 73/..., MBR 3045PT...3060PT	data
BYS 30	Si-Di	Schottky-Gl, 45V, 30A	32a§	Sie	BYS 31/..., BYS 32, BYS 35/...	data
BYS 31-...	Si-Di	Schottky-Gl, 40...50V, 30A	32a§	Sie	BYS 32, BYS 35/..., MBR 3545	data
BYS 32	Si-Di	=BYS 31: 50V	32a§	Sie	BYS 31/50, BYS 35/50	data
BYS 35-...	Si-Di	Schottky-Gl, 20...50V, 35A	32a§	Mot	BYV 22/..., MBR 3520...3545	data
BYS 40-...	Si-Di	Schottky-Gl, 20...40V, 40A	32a§	Mot	MBR 4020...4040, 1N5832...34	data
BYS 41(-35)	Si-Di	Schottky-Gl, 35V, 30A	32a§	Sie	BYS 30...32, BYS 35-45, MBR 3535...3545	data
BYS 42(-45)	Si-Di	Dual, Schottky-Gl, 45V, 60A	23l§	Sie	-	data
BYS 50	Si-Di	Schottky-Gl, 45V, 60A	32a§	Sie	BYS 60-45, BYV 23/45, MBR 6045	data
BYS 51(-35)	Si-Di	Schottky-Gl, 35V, 60A	32a§	Sie	BYS 50, BYS 60-45, BYV 23/45, MBR 6035	data
BYS 60-...	Si-Di	Schottky-Gl, 20...50V, 60A	32a§	Mot	BYS 50..51, BYV 23/..., MBR 6020...6045	data
BYS 71-...	Si-Di	Schottky-Gl, 40...50V, 80A	32a§	Sie	BYS 72, BYS 75/..., MBR 8045	data
BYS 72	Si-Di	Schottky-Gl, 50V, 75A	32a§	Sie	BYS 75-50, BYS 76-45, BYV 23/45, MBR7545	data
BYS 75-...	Si-Di	Schottky-Gl, 20...50V, 75A	32a§	Mot	BYS 71/..., BYS 72, BYS 76-45, BYV 23/...	data
BYS 76(-45)	Si-Di	Schottky-Gl, 45V, 75A	32a§	Sie	BYS 71/..., BYS 72, BYS 75-45, BYV 23/45	data
BYS 79-...	Si-Di	Dual, Schottky-Gl, 40...50V, 30A	23l§	Sie	BYS 42-45, MBR 3045...3060	data
BYS 90...99	Si-Di	Schottky-Gl-Module		Sie		data

BYT

Type	Device	Short Description	Fig.	Manu	Comparision Types	More at
BYT 01-...	Si-Di	S, 200...400V, 1A, <50ns	31a	Tho	BYV 26B..E	data
BYT 03-...	Si-Di	S, 200...400V, 3A, <50ns	31a	Tho	(BYT 56G..M)	data
BYT 08-...	Si-Di	Gl-L, 200...400V, 8A, <35ns	17t§	Tho	BYV 29/..., ESM 980/..., FE 8D...G	data
BYT 08P-...A	Si-Di	contr.av. 200...1000V, 8A, <65ns	17t§	Tho		data
BYT 08PI-...	Si-Di	Iso	17d	Tho	BYQ 28F-...	data
BYT 11-...	Si-Di	S, 600...1000V, 1A, <100ns	31a	Tho	BYV 26C..E	data
BYT 12-...(M)	Si-Di	Gl-L, 200...1000V, 12A, <65ns	32a§	Tho	BYT 61/..., 1N3891...3893	data
BYT 12P..A	Si-Di	=BYT 12-...(M): contr.av.	17t§		(BYR 79-..., BYT 87/...)	data
BYT 12PI..A	Si-Di	=BYT 12-...(M): Iso	17d		-	data
BYT 12-...R	Si-Di	=BYT 12-...(M):	32b&		BYT 61/...R, 1N3891R...3893R	data
BYT 13-...	Si-Di	Gl/S, 600...1000V, 3A, <150ns	31a	Tho	BYT 56J..M	data
BYT 16P-...A	Si-Di	Dual, S-L, 200...400V, 16A, <35ns	17p§	Tho	BYV 34/..., FE 16D...J	data
BYT 25/...	Si-Br	Gl-Br, 50...1000V, 25A	70	Mot	KBPC 25-005...-10	data
BYT 28-...	Si-Di	Dual, S-L, 300...500V, 10A, <60ns	17p§	Phi	BYR 28-..., BYV 34-..., FE 16F..J	data
BYT 30-...(M)	Si-Di	Gl-L, 200...1000V, 30A, <70ns	32a§	Tho	BYT 65/..., BYV 92/...	data
BYT 30G-400	Si-Di	=BYT 30-...(M): 30A(Tc=100°)	30j§			data

Type	Device	Short Description	Fig.	Manu	Comparision Types	More at
BYT 30M-400	Si-Di	=BYT 30-...(M): SMD, 30A(Tc=100°)	10-MDIP		-	data
BYT 30P-...	Si-Di	=BYT 30-...(M):	18t§		-	data
BYT 30PI-...	Si-Di	=BYT 30P-...: Iso	18d		-	data
BYT 30-...R(RM)	Si-Di	=BYT 30/...(M):	32b&		BYT 65/...R, BYV 92/...R	data
BYT 40 Y	Si-Di	Gl, Uni, 1600V, 1A, <3µs	31a	Aeg	DM 513, EM 516, GP 10Y	data
BYT 41 A...M	Si-Di	Gl, Uni, 50...1000V, 1,25A, <2µs	31a	Aeg	BY 226...227, BYW 52...56, 1N5391...99, ++	data
		A=50, B=100, D=200, G=400, J=600				
BYT 42 A...M	Si-Di	Gl, S, 50...1000V, 1,25A, <150...200ns	31a	Aeg	BYM 26A...E, BYV 36A...E, MUR 105...1100	data
BYT 43 A...M	Si-Di	Gl, S, 50...1000V, 1A, <50...75ns	31a	Aeg	BYM 26A...E, BYV 36A...E, MUR 105...1100	data
BYT 44 A...D	Si-Di	Gl, S, 50...200V, 1,5A, <25ns	31a	Aeg	BYM 26A...E, BYV 27/..., FE 2A...H	data
		A=50, B=100, C=150, D=200V				
BYT 51 A...M	Si-Di	Gl, 50...1000V, 1,5A, <4µs	31a	Aeg	BY 127, BY 133, BYW 37...43, 1N4001...07++	data
		A=50, B=100, D=200, G=400, J=600V.				
BYT 52 A...M	Si-Di	Gl/S, 50...1000V, 1,4A, <200ns	31a	Aeg	BY 218/..., BYV 13...16, BYV 36A...E	data
		A=50, B=100, D=200, G=400, J=600V,				
BYT 53 A...G	Si-Di	Gl/S, 50...400V, 1,5A, <50ns	31a	Aeg	BYM 27..., BYV 27/..., EGP 20..., FE2...	data
		A=50, B=100, C=150, D=200,F=300,G=400V				
BYT 54 A...M	Si-Di	Gl/S, 50...1000V, 1,25A, <100ns	31a	Aeg	BYM 26A...E, BYV 36A...E	data
		A=50, B=100, D=200, G=400, J=600V,				
BYT 56 A...M	Si-Di	Gl/S, 50...1000V, 3A, <100ns	31a	Aeg	BYT 13/..., BYM 36A...E, (BYW 16/...)	data
		A=50, B=100, D=200, G=400, J=600V,				
BYT 60-...(M)	Si-Di	Gl-L, 200...1000V, 60A, <70ns	32a§	Tho	-	data
BYT 60P-...	Si-Di	=BYT 60-...:	18t§		-	data
BYT 60-...R,RM	Si-Di	=BYT 60-...:	32b&		-	data
BYT 61(B)-...(M)	Si-Di	Gl-L, 600...1000V, 12A, <200ns	32a§	Tho	BYT 12-..., BYX 62/..., BYX 66/...	data
BYT 61-...R,RM	Si-Di	=BYT 61(B)-...:	32b&		BYT 12-...R, BYX 62/...R, BYX 66/...R	data
BYT 62	Si-Di	Gl, contr.av., 2400V, 0,35A	31a	Aeg	-	data
BYT 65(B)-...(M)	Si-Di	Gl-L, 600...1000V, 30A, <200ns	32a§	Tho	BYT 30-..., BYX 64/..., BYX 67/...	data
BYT 65-...R,RM	Si-Di	=BYT 65(B)-...:	32b&		BYT 30-...R, BYX 64/...R, BYX 67/...R	data
BYT 67-...	Si-Di	S-L, 800...1000V, 50A, <100ns	32a§	Tho	-	data
BYT 71-...	Si-Di	Gl/S-L, 100...800V, 6A, <300ns	17t§	Tho	BY 233/..., BYT 86/..., RGP 80B...M	data
BYT 71-...A	Si-Di	=BYT 71-...: contr. av.	17t§		-	data
BYT 71F-...	Si-Di	=BYT 71-...: Iso	17d		BY 229F/...	data
BYT 71-...R	Si-Di	=BYT 71-...:	17t1&		BY 229/...R	data
BYT 75-...K	Si-Di	Gl-L, 200...600V, 120A	73a§	Tho	DS 85-02C...06C, 1N4588...4596	data
BYT 77	Si-Di	Gl, 800V, 3A, <250ns	31a	Aeg	BYT 13/..., BYT 56K...M, BYW 16/800	data
BYT 78	Si-Di	=BYT 77: 1000V	31a	Aeg	BYT 13/1000, BYT 56M	data
BYT 79-...	Si-Di	Gl/S-L, 300...500V, 12,5A, <60ns	17t§	Phi,Sie	BYT 12-..., BYT 87-..., BYR 79-...	data
BYT 85/...	Si-Di	Gl/S-L, 600...1000V, 4A, <80ns	17t§	Aeg	BY 329/..., BYR 29/...	data
BYT 86/...	Si-Di	Gl/S-L, 600...1000V, 8A, <80ns	17t§	Aeg	BY 329/..., BYR 29/...	data
BYT 86/1300	Si-Di	=BYT 86/...: 1300V, 5A, <150ns	17t§		BY 359/1300, BYT 106/1300	data
BYT 87/...	Si-Di	Gl/S-L, 600...1000V, 15A, <80ns	17t§	Aeg	BYR 79-...	data
BYT 106/1300	Si-Di	Gl/S-L, 1300V, 5A, <120ns	17t§	Aeg	RY 359/1300, BYT 86/1300	data
BYT 108/...	Si-Di	S-L, 200...400V, 8A, <50ns	17t§	Aeg	BYV 29/..., FE 8D...J, RGP 80D...M	data
BYT 115/...	Si-Di	S-L, 200...400V, 15A, <50ns	17t§	Aeg	BYV 79/200, MUR 1520...1560	data
BYT 130/...	Si-Di	Gl/S-L, 600...1000V, 30A, 90ns	17t§		-	data
BYT 200PIV-400	Si-Di	Dual, Gl/S-L, 400V, 100A, <100ns	80s3	Tho	-	data
BYT 230PI(V)200...400	Si-Di	Dual, Gl/S-L, 200...400V, 2x30A, <100ns	80s3	Phi,Tho	-	data
BYT 230PI(V)600...1200	Si-Di	Dual,Gl/S-L, 600...1200V, 2x60A, <100ns	80s3	Phi,Tho	-	data
BYT 231PI(V)...	Si-Di	=BYT 230...:	80s5	Tho	-	data
BYT 254(V)...	Si-Di	=BYT 54...:	80s5	Tho	-	data
BYT 260PI(V)...	Si-Di	=BYT 261...:	80s3	Tho	-	data
BYT 261PI(V)...	Si-Di	Dual, Gl/S-L, 200...1000V, 2x60A,<170ns	80s5	Tho	-	data
BYT 3400B(-TR)	Si-Di	Gl/S-L, 400V, 3A, <25ns	30f§	Tho	-	data

BYV

Type	Device	Short Description	Fig.	Manu	Comparision Types	More at
BYV 10-...(A)	Si-Di	Schottky-Gl, 20...60V, 1A, <30ns	31a	Phi,Tho	(BYS 21/..., SB 120...180, 1N5817...5819)	data
BYV 12	Si-Di	TV-Gl, 100...1000V, 1,5A, <300ns	31a	Aeg,Gie,Mot	BY 218/100, BYW 32...36, RGP 15B...M ,++	data
BYV 13	Si-Di	=BYV 12: 400V	31a	Gie,Mot	BY 218/400, BYW 34...36, RGP 15G...M ,++	data
BYV 14	Si-Di	=BYV 12: 600V	31a	Gie,Mot	BY 218/600, BYW 36, RGP 15J...M, ++	data
BYV 15	Si-Di	=BYV 12: 800V	31a	Gie,Mot	BY 218/800, BYT 77...78, RGP 15K...M, ++	data
BYV 16	Si-Di	=BYV 12: 1000V	31a	Gie,Mot	BYT 78, BYV 96E, RGP 15M	data
BYV 18-...	Si-Di	Dual, Schottky-Gl-L, 30...45V, 8,8A	17p§	Phi,Gie	BYS 24-..., BYV 118-..., MBR 2060CT	data
BYV 19/...	Si-Di	Schottky-Gl-L, 30...45V, 9A	17t§	Phi	BYV 39/..., MBR 1035...10100	data
BYV 20/...	Si-Di	Schottky-Gl-L, 30...45V, 12,5A	32a§	Phi	BYS 15, MBR 1535...1540, 1N5827...5828	data
BYV 21/...	Si-Di	Schottky-Gl-L, 30...45V, 27A	32a§	Phi	BYS 30...32, BYV 121/..., MBR 2535...2540	data
BYV 22/...	Si-Di	Schottky-Gl-L, 30...45V, 50A	32a§	Phi	BYS 50...51, BYS 60/..., MBR 6030...6045	data
BYV 23/...	Si-Di	Schottky-Gl-L, 30...45V, 70A	32a§	Phi	BYS 71...72, BYS 75/..., BYS 76,BYV 123/...	data
BYV 24/...	Si-Di	Gl/S-L, 800...1000V, 12A, <450ns	32a§	Phi	BYT 61/..., BYX 66/...	data
BYV 24/...R	Si-Di	=BYV 24:	32b&		BYT 61/...R, BYX 66/...R	data
BYV 25/...	Si-Br	Gl-Br, 50...1000V, 25A	70	Mot	KBPC 25-005...10	data

Type	Device	Short Description	Fig.	Manu	Comparision Types	More at
..V 26 A..E	Si-Di	S, contr.av., 200...1000V, 1A, <30ns A=200, B=400, C=600, D=800V, <75ns E=1000V, <75ns	31a	Aeg,Phi,Fag Gsi	EGP 10D...G, FE 1D..H	data
..V 26 F..G	Si-Di	=BYV 26A..E: 1200...1400V, <150ns F=1200, G=1400V	31a			data
..V 27-...	Si-Di	S, contr.av., 50...600V, 2...1,6A, <50ns 50...400V: 2A, 500...600V: 1,6A 50...200V: <25ns, 300...600V: <50ns	31a	Aeg,Gie,Phi	BYV 28-..., EGP 20A...G, FE 2A..H	data
..V 28-...	Si-Di	S, contr.av., 50...600V, 3,5A, <50ns 50...400V: 3,5A, 500...600V: 3,1A 50...200V: <25ns, 300...600V: <50ns	31a	Aeg,Gie,Phi	EGP 30A...G, FE 3A..H	data
..V 29-...	Si-Di	GI/S-L, 300...500V, 8A, <60ns	17t§	Phi,Sie	BYT 08-..., BYR 29-..., FE 8F..J	data
..V 29 F-...	Si-Di	=BYV 29-..: Iso	17d		BYR 29F-...	data
..V 30-...(M,U)	Si-Di	GI/S-L, 200...500V, 12,5A, <50ns	32a§	Phi,Sie,Tho	BYT 12-..., BYX 61/...	data
..V 30-...R,RM	Si-Di	=BYV 30-..(M,U): Iso	32b&		BYT 12-..R, BYX 61/..R	data
..V 31-...(U)	Si-Di	GI/S-L, 300...500V, 25A, <50ns	32a§	Phi,Sie	-	data
..V 32-...	Si-Di	Dual, SMPS, 50...200V, 18A, <25ns	17p§	Phi,Sie,++	BYT 16P-...A, BYW 51/..., FE 16A...G	data
..V 32 D-...	Si-Di	=BYV 32-..:	17n1	Gie	-	data
..V 32 E-...	Si-Di	=BYV 32-..: ESD protected	17p§		-	data
..V 32 EB-...	Si-Di	=BYV 32-..:	30p§		-	data
..V 32 EX-...	Si-Di	=BYV 32-..: Iso	17p		-	data
..V 32 F-...	Si-Di	=BYV 32-..: Iso	17p		-	data
..V 32 N-...	Si-Di	=BYV 32-..:	17q&	Gie	-	data
..V 33-...	Si-Di	Dual, Schottky-GI, 30...45V, 18A	17p§	Phi	BYV 43-..., BYV 133-..., MBR 1550...1560	data
..V 33F-...	Si-Di	=BYV 33-..: Iso	17p		BYV 43F-..., BYV 133F-...	data
..V 34-...	Si-Di	Dual, GI/S-L, 300...500V, 2x9A, <60ns	17p§	Phi	BYR 34-..., BYT 16P-..A, FE 16F..J	data
..V 36 A..E	Si-Di	S, contr.av., 200...1000V, 1,6A, <100ns A=200, B=400, C=600, D=800, <150ns E=1000V, <150ns, F=1200V, <250ns G=1400V, <250ns	31a	Phi	BYD 34D..M, BYM 26A...G	data
..V 37	Si-Di	GI, S, 800V, 2A, <300ns	31a	Aeg	BY 218/800, BYT 77...78, BYW 96D...E, ++	data
..V 38	Si-Di	=BYV 37: 1000V	31a	Aeg	BYT 78, BYW 96E	data
..V 39-...	Si-Di	Schottky-GI-L, 30...45V, 12,5A	17k	Phi	MBR 1635...1645	data
..V 40-...	Si-Di	Dual, GI/S, 100...200V, 1,35A, <25ns	48p§	Phi	-	data
..V 40 E-...	Si-Di	=BYV 40-...: ESD protected	48p§	Phi	-	data
..V 42-...	Si-Di	Dual, GI/S-L, 50...200V, 30A, <25ns	17p§	Phi	-	data
..V 42 E-...	Si-Di	=BYV 42-..: ESD protected	17p§		-	data
..V 42 EB-...	Si-Di	=BYV 42-..: ESD protected	30p§		-	data
..V 42 EX-...	Si-Di	=BYV 42-..: Iso, ESD protected	17p		-	data
..V 42 F-...	Si-Di	=BYV 42-..: Iso	17p		-	data
..V 43-...	Si-Di	Dual, Schottky-GI, 30...45V, 30A	17p§	Phi	BYV 143-..., MBR 2535CT..2545CT	data
..V 43-...	Si-Di	=BYV 43-..: Iso, 26A	17p		-	data
..V 44-...	Si-Di	Dual, GI/S-L, 300...500V, 27A, <60ns	17p§	Phi	BYR 34-...	data
..V 52-...	Si-Di	Dual, S-L, 50...200V, 2x30A, <50ns	18p§	Phi,Tho	-	data
..V 52PI-...	Si-Di	=BYV 52-..: Iso	18p		-	data
..V 54(V)-...	Si-Di	Dual, S-L, 50...200V, 2x50A, <60ns	80s3	Phi,Tho	-	data
..V 60-...	Si-Di	Dual, S-L, 850...1200V, 15A, <600ns	66a	Phi	-	data
..V 61	Si-Di	GI, S, 50V, 6A, <30ns	31a	Aeg	FE 6A..D	data
..V 62	Si-Di	=BYV 61: 100V	31a	Aeg	FE 6B..D	data
..V 63	Si-Di	=BYV 61: 150V	31a	Aeg	FE 6C..D	data
..V 71-...	Si-Di	=BYT 71-..:	17t1&	Tho	BY 229-..R	data
..V 71-..R	Si-Di	=BYT 71-..R:	17			data
..V 72-...	Si-Di	Dual, GI/S-L, 50...200V, 27A, <28ns	18p§	Phi	BYW 99P-..., MUR 3005PT...3060PT	data
..V 72 E-...	Si-Di	=BYV 72-..: ESD protected	18p§		-	data
..V 72 EW-...	Si-Di	=BYV 72-..: ESD protected	16p§		BYW 99P-..., MUR 3005PT...3060PT	data
..V 72(E)F-...	Si-Di	=BYV 72-..: Iso	16p		BYT 77PI-..., BYW 99PI-...	data
..V.73-...	Si-Di	Dual, Schottky-GI, 30...45V, 30A	18p§	Phi	BYS 79-..., MBR 4035PT...4060PT	data
..V 74-...	Si-Di	Dual, GI/S-L, 300...500V, 27A, <60ns	18p§	Phi	MUR 3030PT...3060PT	data
..V 74 F-...	Si-Di	=BYV 74-..: Iso, 20A	16p		BYT 30PI-...	data
..V 74 W-...	Si-Di	=BYV 74-..:	16p§		MUR 3030PT...3060PT	data
..V 79-...	Si-Di	SMPS, 50...200V, 12,7A, <35ns	17t§	Phi,Sie	BYR 79-..., BYT 79-...	data
..V 79 E-...	Si-Di	=BYV 79-..: ESD protected	17t§		-	data
..V 79 EB-...	Si-Di	=BYV 79-..: ESD protected	30f§		-	data
..V 87-..R	Si-Di	GI/S-L, 300...800V, 4A, <300ns	14t1	Tho	ESM 181/...	data
..V 88-...	Si-Di	GI/S, 200...1000V, 1A, <150ns	31a	Phi,Tho	BY 218/..., BYV 37...38, RGP 10D...M, ++	data
..V 92-...(M)	Si-Di	GI/S-L, 200...500V, 35A, <50ns	32a§	Phi,Gie	MR 862...866	data
..V 92-...R,RM	Si-Di	=BYV 92/..:	32b&		MR 862R...866R	data
..V 95 A...C	Si-Di	GI, contr.av., 200...600V, 1,5A, <250ns A=200V, B=400V, C=600V	31a	Phi, Gie	BYW 95A...C, BYW 96D,E	data
..V 96 D...E	Si-Di	=BYV 95: 800...1000V, <300ns D=800V, E=1000V	31a	Phi, Gie	BYW 96D,E	data

Type	Device	Short Description	Fig.	Manu	Comparision Types	More at
BYV 97 F...G	Si-Di	=BYV 95: 1200...1400V, 1,6A, <500ns F=1200V, G=1400V	31a	Phi	BYW 97F,G	data
BYV 98	Si-Di	kV-Gl, 2,1kV, 1A, <300ns	31a	Phi	-	data
BYV 98-50...-200	Si-Di	S, 50...200V, 4A, <35ns	31a	Aeg	EGP 50A...G, FE 5A...D, MUR 405...420	data
BYV 99	Si-Di	SMD, SS, 600V, 1A, <15ns	31a	Phi	-	data
BYV 116-...	Si-Di	Dual, Schottky-Gl, 20...35V, 10A	17p§	Phi	BYS 24-45, BYV 33-..., BYV 133-...	data
BYV 116 B-...	Si-Di	=BYV 116-...	30p§		U 10FWJ2C48M, 10FWJ2C48M	data
BYV 118-...	Si-Di	Dual, Schottky-Gl, 35...45V, 9A	17p§	Phi	BYS 24-45, BYV 33-..., BYV 133-...	data
BYV 118 B-...	Si-Di	=BYV 118-...	30p§		D 10SC4M	data
BYV 118 F-...	Si-Di	=BYV 118-...: Iso	17p		BYV 33F-..., BYV 133F-...	data
BYV 118 X-...	Si-Di	=BYV 118-...: Iso	17p		BYV 33F-..., BYV 133F-...	data
BYV 120-...(M)	Si-Di	Schottky-Gl, 35...45V, 13,5A	32a§	Phi	BYS 15, MBR 1535...1540, 1N2528	data
BYV 121-...(M)	Si-Di	Schottky-Gl, 35...45V, 27A	32a§	Phi	BYS 31-..., BYS 32, MBR 2535...40, 1N5831+	data
BYV 123-45	Si-Di	Schottky-Gl, 45V, 70A	32a§	Aeg,Sie	BYS 71-50, BYS 72, BYS 76-45, MBR 7545++	data
BYV 133-...	Si-Di	Dual, Schottky-Gl, 35...45V, 18A	17p§	Phi	BYV 33-..., MBR 2035...2045CT+	data
BYV 133 F-...	Si-Di	=BYV 133-...: Iso	17p		BYV 33F-..., BYV 43F-..., BYV 143F-...	data
BYV 133 X-...	Si-Di	=BYV 133-...: Iso	17p		BYV 33F-..., BYV 43F-..., BYV 143F-...	data
BYV 143-...	Si-Di	Dual, Schottky-Gl, 35...45V, 2x15A	17p§	Aeg,Phi,Sie	BYV 43-..., MBR 2535CT...2545CT	data
BYV 143 F-...	Si-Di	=BYV 143-...: Iso, 2x10A	17p		BYV 43F-...	data
BYV 143 X-...	Si-Di	=BYV 143-...: Iso, 2x10A	17p		BYV 43F-...	data
BYV 255(V)/...	Si-Di	Dual, S-L, 50..200V, 2x100A, <80ns	80s5	Tho	-	data
BYV 541(V)/...	Si-Di	=BYV 54(V)/...:	80s5	Tho	-	data
BYV 1100	Si-Di	Gl, SS, 100V, 1,7A, <10ns	31a	Phi	BYV 2100	data
BYV 2100	Si-Di	Gl, SS, 100V, 2A, <12,5ns	31a	Phi	BYV 4100	data
BYV 4100	Si-Di	Gl, SS, 100V, 4A, <15ns	31a	Phi	-	data
BYVB 32-...	Si-Di	Dual, S-L, 50..200V, 18A, <25ns	30p§	Gsi	-	data

BYW

Type	Device	Short Description	Fig.	Manu	Comparision Types	More at
BYW 07/...(M)	Si-Di	Gl/S-L, 50...200V, 70A, <50ns	32a§	Tho	BYT 60/..., BYW 08/..., BYW 94/...	data
BYW 07/...A	Si-Di	=BYW 07/...(M,R,RM): contr.av.	32a§,b&		-	data
BYW 07/...R,RM	Si-Di	=BYW 07/...(M):	32b&		BYT 60/...R	data
BYW 08/...(M)	Si-Di	Gl/S-L, 50...200V, 80A, <60ns	32a§	Tho		data
BYW 10/...	Si-Di	Gl, S, 50...1000V, 1,5A, <400ns	34a§	Tix	BY 218/..., BY 258/..., RGP 15A...M, ++	data
BYW 10/...R	Si-Di	=BYW 10/...:	34b&		BY 218/..., BY 258/..., RGP 15A...M, ++	data
BYW 11/...	Si-Di	Gl/S-L, 50...1000V, 6A, <400ns	32a§	Tix	BYT 61/..., BYX 50/..., BYX 66/...	data
BYW 11/...R	Si-Di	=BYW 11/...:	32b&		BYT 61/...R, BYX 50/...R, BYX 66/...R	data
BYW 12/...	Si-Di	Gl/S-L, 50...1000V, 15A, <400ns	32a§	Tix	BYX 30/...	data
BYW 12/...R	Si-Di	=BYW 12/...:	32b&		BYX 30/...R	data
BYW 13/...	Si-Di	Schottky-Gl, 25...30V, 40A	32a§	Phi	BYS 50...51, BYS 60/..., MBR 4030...4040	data
BYW 14/...	Si-Di	Gl, S, 100...800V, 3A, <750ns	31a	Tho	BYT 77...78, BYW 72...76, RGP 30B...M	data
BYW 15/...	Si-Di	=BYW 14/...: <500ns	31a	Tho	BYT 77...78, BYW 72...76, RGP 30B...M	data
BYW 16/...	Si-Di	=BYW 14/...: <200ns	31a	Tho	BYT 77...78, BYW 72...76, BYW 96D...E	data
BYW 17/...	Si-Di	Gl, 100...1200V, 3A	31a	Tho	BY 251...255, 1N5402...5408, GP 30B...M, ++	data
BYW 18/...	Si-Di	Gl, contr.av., 400...1000V, 3A	31a	Tho	BYW 84...86, BYW 95B,C, BYW 96D,E	data
BYW 19/...	Si-Di	TV-Gl-L, 800...1000V, 7A, <450ns	26h§	Phi	(BY 229/..., RGP 80K...M)[4]	data
BYW 19/...R	Si-Di	=BYW 19/...:	26j&		(BY 229/...R)[4]	data
BYW 20	Si-Bi	Gl-Br, 50V, 15A	70x	Mot,Sie	KBPC 15-005	data
BYW 21	Si-Br	=BYW 20: 100V	70x	Mot,Sie	KBPC 15-02	data
BYW 22	Si-Br	=BYW 20: 200V	70x	Mot,Sie	KBPC 15-02	data
BYW 23	Si-Br	=BYW 20: 300V	70x	Mot,Sie	KBPC 15-04	data
BYW 24	Si-Br	=BYW 20: 400V	70x	Mot,Sie	KBPC 15-04	data
BYW 25/...	Si-Di	Gl/S-L, 800...1000V, 40A, <450ns	32a§	Phi	BYT 67/...	data
BYW 25/...R	Si-Di	=BYW 25/...:	32b&		-	data
BYW 26	Si-Br	=BYW 20: 600V	70x	Mot,Sie	KBPC 15-06	data
BYW 27/...(GP)	Si-Di	Gl, 50...1000V, 1A	31a	Gie,Tho,++	BY 126...127, BY 133...135, 1N4001...07, ++	data
BYW 28	Si-Br	=BYW 20: 800V	70x	Mot,Sie	KBPC 15-08	data
BYW 28-...	Si-Di	Gl, S, 500...600V, 4A, <50ns	31a	Mot	MUR 450, MUR 460, MUR 480, MUR 4100	data
BYW 29-...	Si-Di	S-L, SMPS, 50...200V, 7,6A, <25ns	17t§	Phi,Sie,++	BYW 80-..., FE 8A...G	data
BYW 29-...A	Si-Di	=BYW 29-...: contr.av.	17t§		BYW 80-...A	data
BYW 29 E-...	Si-Di	=BYW 29-...: ESD protected	17t§		-	data
BYW 29 EB-...	Si-Di	=BYW 29-...: ESD protected	30f§		-	data
BYW 29 ED-...	Si-Di	=BYW 29-...: ESD protected	30f§		-	data
BYW 29 F-...	Si-Di	=BYW 29-...: Iso	17d		-	data
BYW 29 G-200	Si-Di	=BYW 29-...: 8A(Tc=120°)	30f§	Tho	-	data
BYW 29 M-200	Si-Di	=BYW 29-...: SMD	10-MDIP	Tho	-	data
BYW 29 X-...	Si-Di	=BYW 29-...: Iso	17d		-	data
BYW 30/...(U)	Si-Di	Gl/S-L, 50...200V, 10A, <35ns	32a§	Gie,Mot,Phi	BYT12/..., BYV30/..., BYW81/..., BYX61/...	data
BYW 31/...(U)	Si-Di	Gl/S-L, 50...200V, 23A, <50ns	32a§	Mot,Phi,Sie	MUR 2505...2520	data
BYW 32	Si-Di	TV-Gl, 200V, 2A, <200ns	31a	Aeg, Gie	BY 218/200, BYW 72...76, BYW 95A	data
BYW 33	Si-Di	=BYW 32: 300V	31a	Aeg, Gie	BY 218/400, BYW 73...76, BYW 95B, ++	data
BYW 34	Si-Di	=BYW 32: 400V	31a	Aeg, Gie	BY 218/400, BYW 74...76, BYW 95B, ++	data

Type	Device	Short Description	Fig.	Manu	Comparision Types	More at
YW 35	Si-Di	=BYW 32: 500V	31a	Aeg, Gie	BY 218/600, BYW 76, BYW 95C, ++	data
YW 36	Si-Di	=BYW 32: 600V	31a	Aeg, Gie	BY 218/600, BYW 76, BYW 95C	data
YW 37	Si-Di	Gl, 50V, 1A	31a	Gen,Gie,Sie	BY 126...127, BY 133...135, 1N4001...07, ++	data
YW 38	Si-Di	=BYW 37: 100V	31a	Gen,Gie,Sie	BY 126...127, BY 133...135, 1N4002...07, ++	data
YW 39	Si-Di	=BYW 37: 200V	31a	Gen,Gie,Sie	BY 126...127, BY 133...134, 1N4003...07, ++	data
YW 40	Si-Di	=BYW 37: 400V	31a	Gen,Gie,Sie	BY 126...127, BY 133...134, 1N4004...07, ++	data
YW 41	Si-Di	=BYW 37: 600V	31a	Gen,Gie,Sie	BY 126...127, BY 133...134, 1N4005...07, ++	data
YW 42	Si-Di	=BYW 37: 800V	31a	Gen,Gie,Sie	BY 127, BY 133, BY 227, 1N4006...07, ++	data
YW 43	Si-Di	=BYW 37: 1000V	31a	Gen,Gie,Sie	BY 127, BY 133, BY 227, 1N4007, ++	data
YW 44/...	Si-Br	Gl-Br, 200...800V, 4A		Phi	B150C5000...B600C5000, etc.	data
YW 45/...	Si-Br	Gl-Br, 200...800V, 6A		Phi	KBPC 6-02...08	data
YW 46/...	Si-Br	Gl-Br, 200...800V, 8A		Phi	KBPC 8-02...08	data
YW 47/...	Si-Br	Gl-Br, 200...800V, 12,5A		Phi	KBPC 15-02...08	data
YW 48/...	Si-Di	Gl/S-L, 100...300V, 12A, <200ns	32a§	Phi	BYT 12/..., BYV 30/..., 1N3890...3893, ++	data
YW 48/...R	Si-Di	=BYW 48/...:	32b&		BYT 12/..., BYV 30/...R, 1N3890R...3893R,++	data
YW 51-...	Si-Di	Dual, S-L, 50...200V, 2x10A, <35ns	17p§	Mot, Tho	BYP 22/..., BYV 42/..., BYV 44/...	data
YW 51F-...	Si-Di	=BYW 51-...: Iso	17p		-	data
YW 51G-200	Si-Di	=BYW 51-...: 2x10A(Tc=120°)	30p§		DF 20LC20U, U 20DL2C48A, 20DL2C48A	data
YW 51M-200	Si-Di	=BYW 51-...: SMD	10-MDIP		-	data
YW 52	Si-Di	Gl, contr.av., 200V, 2A, <4µs	31a	Aeg	BYW 18/400, BYW 82...86, 1N5060...5062	data
YW 53	Si-Di	=BYW 52: 400V	31a	Aeg	BYW 18/400, BYW 83...86, 1N5060...5062	data
YW 54	Si-Di	=BYW 52: 600V	31a	Aeg	BYW 18/600, BYW 84...86, 1N5061...5062	data
YW 55	Si-Di	=BYW 52: 800V	31a	Aeg	BYW 18/800, BYW 85...86, 1N5062	data
YW 56	Si-Di	=BYW 52: 1000V	31a	Aeg	BYW 18/1000, BYW 86	data
YW 58/...	Si-Di	Gl, S, 50...600V, 1A, <200ns	31a	Gen,Gie	BY 201/..., BYV 12...16, RGP 10A...M, ++	data
YW 59/...	Si-Di	Gl, S, 50...600V, 3A, <200ns	31a	Gen,Gie	BYW 16/..., BYW 72...76, BYW 95A...C, ++	data
YW 60	Si-Br	Gl-Br, 50V, 35A	70x	Mot,Sie	KBPC 35-005	data
YW 61	Si-Br	=BYW 60: 100V	70x	Mot,Sie	KBPC 35-02	data
YW 62	Si-Br	=BYW 60: 200V	70x	Mot,Sie	KBPC 35-02	data
YW 63	Si-Br	=BYW 60: 300V	70x	Mot,Sie	KBPC 35-04	data
YW 64	Si-Br	=BYW 60: 400V	70x	Mot,Sie	KBPC 35-04	data
YW 65/...	Si-Di	Gl, 200...1000V, 1A	31a	Gen,Gie	BY 126...127, BY 133...134, 1N4003...07, ++	data
YW 66	Si-Br	=BYW 60: 600V	70x	Mot,Sie	KBPC 35-06	data
YW 67/...	Si-Di	Gl, 200...800V, 3A	31a	Gen,Gie	BY 251...255, 1N5402...08, RGP 30D...M, ++	data
YW 68	Si-Br	=BYW 60: 800V	70x	Mot,Sie	KBPC 35-08	data
YW 69(A,B)	Si-Di	Schottky-Gl, 20...60V, 15A	32a§	Tho	BYS 15, BYV 120/...	data
YW 70(A,B)	Si-Di	Schottky-Gl, 20...60V, 25A	32a§	Tho	BYS 31...32, BYS 35/..., BYS 41	data
YW 71(A,B)	Si-Di	Schottky-Gl, 20...60V, 10A	32a§	Tho	BYS 15, BYV 120/..., MBR 1520...1540	data
YW 72	Si-Di	Gl, S, 200V, 3A, <200ns	31a	Aeg, Gie	BYM 36A, BYT 56D, BYV 28-200	data
YW 73	Si-Di	=BYW 72: 300V	31a	Aeg, Gie	BYM 36B, BYT 56G, BYV 28-300	data
YW 74	Si-Di	=BYW 72: 400V	31a	Aeg, Gie	BYM 36B, BYT 56G, BYV 28-400	data
YW 75	Si-Di	=BYW 72: 500V	31a	Aeg, Gie	BYM 36C, BYT 56J, BYV 28-500	data
YW 76	Si-Di	=BYW 72: 600V	31a	Aeg, Gie	BYM 36C, BYT 56J, BYV 28-600	data
YW 77-...(M)	Si-Di	Gl/S-L, 50...200V, 25A, <50ns	32a§	Mot, Tho,++	BYV 31/..., BYW 31/..., MUR 2505...2520	data
YW 77-...A...	Si-Di	=BYW 77-...(M,R,RM): contr.av.	32a§,b&		-	data
YW 77G-200	Si-Di	=BYW 51-...:	30j§		-	data
YW 77M-200	Si-Di	=BYW 77-...: SMD	10-MDIP		-	data
YW 77P-...	Si-Di	=BYW 77-...:	18t§		BYT 30P-...	data
YW 77PI-...	Si-Di	=BYW 77P-...: Iso	18d		BYT 30PI-...	data
YW 77-...R(RM)	Si-Di	=BYW 77-:	32b&		-	data
YW 78-...(M)	Si-Di	Gl/S-L, 50...200V, 50A, <60ns	32a§	Mot,Tho	BYW 07/...(M), BYW 93/..., BYW 94/...	data
YW 78-...A...	Si-Di	=BYW 78-...(M,R,RM): contr.av.	32a§,b&		BYW 07/...A...	data
YW 78-...R(RM)	Si-Di	=BYW 78-...:	32b&		BYW 07/...R(RM	data
BYW 79	Si-Br	Gl-Br, 1000V, 15A	70x	Mot	KBPC 15-10	data
BYW 80-...	Si-Di	Gl/S-L, 50...200V, 10A, <35ns	17t§	Mot,Tho	BYP 21-..., BYW 29/..., FE 8A...G	data
BYW 80-...A	Si-Di	=BYW 80/...: contr.av.	17k		-	data
BYW 80P-...	Si-Di	=BYW 80P-...: Iso	17d		BYW 29/...F	data
BYW 80PI-...	Si-Di	=BYW 80P-...: Iso	17d		BYW 29/...F	data
BYW 81-...(M)	Si-Di	Gl/S-L, 50...200V, 15A, <35ns	32a§	Mot,Tho	BYV 31/..., BYW 31/..., BYW 77/...(M)	data
BYW 81-...A...	Si-Di	=BYW 81-...(M,R,RM): contr.av.	32a§,b&		BYW 77/...A...	data
BYW 81M-200	Si-Di	=BYW 81-...: SMD	10-MDIP		-	data
BYW 81P-...	Si-Di	=BYW 81P-...(M):	17t§		BYV 79/...	data
BYW 81PI-...	Si-Di	BYW 81P-...: Iso	17d		-	data
BYW 81-...R,RM	Si-Di	=BYW 81-...:	32b&		BYW 77/...R(RM)	data
BYW 82	Si-Di	Gl, contr.av. 200V, 3A	31a	Aeg	BYW 18/400, BYW 95A	data
BYW 83	Si-Di	=BYW 82: 400V	31a	Aeg	BYW 18/400, BYW 95B	data
BYW 84	Si-Di	=BYW 82: 600V	31a	Aeg	BYW 18/600, BYW 95C	data
BYW 85	Si-Di	=BYW 82: 800V	31a	Aeg	BYW 18/800, BYW 96D	data
BYW 86	Si-Di	=BYW 82: 1000V	31a	Aeg	BYW 18/1000, BYW 96E	data
BYW 88-...(M)	Si-Di	Gl-L, 50...1200V, 12A	32a§	Tho	BYX 75...81, BYX 99/..., 1N4506...11, ++	data
BYW 88-...R,RM	Si-Di	=BYW 88-...:	32b&		BYX 75...81R, BYX 99/...R	data
BYW 89	Si-Br	Gl-Br, 1000V, 15A	70x	Mot	KBPC 15-10	data
BYW 90/...(R)	Si-Di	Gl-L, 50...1000V, 24A	(17)	Mot	-	data
BYW 91/...	Si-Di	=BYW 90/...: <200ns	(17)	Mot	-	data
BYW 92/...(U)	Si-Di	Gl/S-L, 50...200V, 23A, <50ns	32a§	Phi,Sie,++	BYX 65/...	data
BYW 92/...R	Si-Di	=BYW 92/...:	32b&		BYX 65/...R	data

Type	Device	Short Description	Fig.	Manu	Comparision Types	More at
BYW 93/...(U)	Si-Di	Gl/S-L, 50...200V, 50A, <60ns	32a§	Mot,Phi	BYW 07/..., BYW 78/..., BYW 94/...	data
BYW 94/...	Si-Di	Gl/S-L, 50...200V, 70A, <60ns	32a§	Mot,Phi	BYW 07/..., BYW 08/...	data
BYW 95 A..C	Si-Di	Gl, S, contr.av. 200...600V, 3A, <250ns A=200V, B=400V, C=600V	31a	Gie,Phi	BYM 36A...E	data
BYW 96 D..E	Si-Di	=BYW 95: 800...1000V, <300ns D=800V, E=1000V	31a	Gie,Phi	BYM 36A...E	data
BYW 97 F...G	Si-Di	Gl, S, contr.av. 1200...1400V, 3,3A <500ns, F=1200V, G=1400V	31a	Phi	(BY 228, BY 328)	data
BYW 98/...	Si-Di	Gl, S, 50...200V, 3A, <50ns	31a	Tho	BYV 28/.., EGP 30A...D, FE 3A...D	data
BYW 99-...	Si-Di	Dual, S-L, 50...200V, 2x15A, <50ns	23l§	Tho	-	data
BYW 99P-...	Si-Di	=BYW 99-..:	18p§		BYV 72/..., MUR 3005PT...3060PT	data
BYW 99PI-...	Si-Di	=BYW 99P-..: Iso	18p		BYV 52/...PI	data
BYW 99W-...	Si-Di	=BYW 99-..:	16p§		BYV 72/..., MUR 3005PT...3060PT	data
BYW 100/...	Si-Di	Gl/S, 50...200V, 1,5A, <35ns	31a	Tho	BYD 74A...G, BYV 27/..., EGP 20A...D	data
BYW 172 D...G	Si-Di	Gl, S, 200...400V, 3A, <100ns D=200V, F=300V, G=400V	31a	Aeg	BYT 56.., EGP 30D...G	data
BYW 178	Si-Di	Gl, S, 800V, 3A, <60ns	31a	Aeg	BYT 56K...M	data
BYW 4200B(-TR)	Si-Di	Gl/S-L, 200V, 4A(Tc=130°), <35ns	30f§	Tho	-	data
BYWB 29-...	Si-Di	Gl/S-L, 50...200V, 8A(Tc=125°), <25ns	30t§	Gsi	-	data

BYX

Type	Device	Short Description	Fig.	Manu	Comparision Types	More at
BYX 10(G,GP)	Si-Di	Gl, 800/1600V, 0,36A	31a	Mot,Phi	BY 231/1500, BY 269, EM 516, GP 10Y	data
BYX 11	Si-Di	Gl, 2000V, 0,01A	31a	Phi	BAY 91, BY 203/20, SHG 2..2,5	data
BYX 13/...	Si-Di	Gl-L, 400...1600V, 20A	73a§	Phi	-	data
BYX 13/..R	Si-Di	=BYX 13/..:	73b&		D21S/..., D24/...	data
BYX 14/...	Si-Di	Gl-L, 400...1200V, 150A	74a§	Phi	-	data
BYX 14/..R	Si-Di	=BYX 14/..:	74b&		-	data
BYX 15	Si-Di	Gl-L, 1200V, 40A	73a§	Phi	-	data
BYX 16	Si-Di	=BYX 15:	73b&	Phi	D30S/..., D34/...	data
BYX 20/200	Si-Di	Gl-L, 200V, 25A	75a§	Phi	1N3493...3495	data
BYX 20/200R	Si-Di	=BYX 20/200:	75b&		DS 25-02A...06A, 1N3493R..3495R	data
BYX 21(L)/...	Si-Di	Gl-L, 50...200V, 25A	75a§	Phi	1N3491...3495	data
BYX 21(L)/..R	Si-Di	=BYX 21(L)/..:	75b&		DS 25-01A...06A, 1N3491R..3495R	data
BYX 22/...	Si-Di	Gl, 200...1200V, 1,4A	34§	Phi	BY 226...227, GP 15D...M, 1N5393...99, ++	data
BYX 23/...	Si-Di	Gl-L, 400...1000V, 100A	73a§	Phi	(1N3291...32979)	data
BYX 24	Si-Di	Dual-Gl, 800V, 0,8A	42v	Phi	-	data
BYX 25/...	Si-Di	Gl-L, 600...1400, 20A	32a§	Phi	-	data
BYX 25/..R	Si-Di	=BYX 25/..:	32b&		DS 17-07A...16A, SSi E2040...E20100	data
BYX 26/...	Si-Di	Gl, 60...150V, 0,25A	31a	Phi	BA 157...159, BY 401...405, 1N4001...07, ++	data
BYX 27/...	Si-Di	Gl-L, contr.av., 400...1000V, 250A	74a§	Phi	-	data
BYX 28/...	Si-Di	Gl-L, 200...400V, 25A	75a§	Phi	1N3493...3495	data
BYX 28/..R	Si-Di	=BYX 28/..:	75b&		DS 25-02A...06A, 1N3493R..3495R	data
BYX 29/...	Si-Di	kV-Gl-L, contr.av., 75...150kV, 0,05A		Phi	-	data
BYX 30/...	Si-Di	Gl-L, contr.av., 200...600V, 14A,<350ns	32a§	Phi	BYX 46/...	data
BYX 30/..R	Si-Di	=BYX 30/..:	32b&		BYX 46/..R	data
BYX 32/...	Si-Di	Gl-L, 200...1600V, 100A	73a§	Phi	1N3289...3297	data
BYX 32/..R	Si-Di	=BYX 32/..:	73b&		DS 80-...A, 1N3289R..3297R	data
BYX 33/...(R)	Si-Di	Gl-L, 200...1600V, 250A	74a§,b&	Phi	-	data
BYX 34/...	Si-Di	Gl-L, contr.av., 200...500V, 60A,<600ns	73a§	Phi	BYT 60/...	data
BYX 34/..R	Si-Di	=BYX 34/..:	73b&		BYT 60/..R	data
BYX 35	Si-Di	kV-Gl, 37kV, 0,05A	31a	Phi	-	data
BYX 36/...	Si-Di	Gl, 150...600V, 1A	31a	Phi	BYW 27/..., BYX 55/..., 1N4003...4007, ++	data
BYX 38/...	Si-Di	Gl-L, 300...1200V, 6A	32a§	Mot,Phi	BYX 39/...	data
BYX 38/..R	Si-Di	=BYX 38/..:	32b&		BYX 39/..R	data
BYX 39/...	Si-Di	Gl-L, contr.av., 600...1400V, 6A	32a§	Phi	-	data
BYX 39/..R	Si-Di	=BYX 39/..:	32b&		-	data
BYX 40/...	Si-Di	Gl-L, contr.av., 600...1000V, 12A	32a§	Phi	1N4508...4511	data
BYX 40/..R	Si-Di	=BYX 40/..:	32b&		-	data
BYX 42/...	Si-Di	Gl-L, 300...1200V, 10A	32a§	Phi	BYW 88/..., BYX 98/..., BYX 99/..., ++	data
BYX 42/..R	Si-Di	=BYX 42/..:	32b&		BYW 88/..R, BYX 98/..R, BYX 99/..R, ++	data
BYX 42/..T	Si-Di	=BYX 42/..:	32b&	Tsm	BYW 88/..R, BYX 98/..R, BYX 99/..R, ++	data
BYX 45/..R	Si-Di	Gl, contr.av., 200...1400V, 1,5A	34b&	Phi	BYW 52...56, SSiB9860A...9890A	data
BYX 46/...	Si-Di	Gl-L, contr.av., 200...600V, 15A,<200ns	32a§	Phi	BYX 30/...	data
BYX 46/..R	Si-Di	=BYX 30/..:	32b&		BYX 30/..R	data
BYX 47	Si-Di	Gl, 2000V, 6A	31a	Phi	-	data
BYX 48/...	Si-Di	Gl-L, 300...1200V, 6A	32a§	Phi	BYX 38/..., BYX 39/..., BYX 42/..., ++	data
BYX 48/..R	Si-Di	=BYX 48/..:	32b&		BYX 38/..R, BYX 39/..R, BYX 42/..R, ++	data
BYX 49/...	Si-Di	Gl-L, 300...1200V, 6A	26h§	Phi	-	data
BYX 49/..R	Si-Di	=BYX 49/..:	26g&		-	data
BYX 50/...	Si-Di	Gl-L, contr.av., 200...600V, 7A, <500ns	32a§	Mot,Phi	-	data
BYX 50/..R	Si-Di	=BYX 50/..:	32b&		-	data

Type	Device	Short Description	Fig.	Manu	Comparision Types	More at
YX 51/...(R)	Si-Di	Gl-L, 1200...2000V, 400A	74a§,b&	Phi	-	data
YX 52/...	Si-Di	Gl-L, 300...1200V, 40A	32a§	Phi	BYX 56.., BYX 97/...	data
YX 52/...R	Si-Di	=BYX 52/...:	32b&		BYX 56...R, BYX 97/...R	data
YX 53	Si-Di	Gl, 2000V, 0,05A	31a	Ssc	BY 203/20, SHG 2..2,5, 1N5181	data
YX 54	Si-Di	=BYX 53: 3000V	31a	Ssc	1N5181	data
YX 55/...(GP,P)	Si-Di	Gl, S, 350...600V, 1,2A, <750ns	31a	Mot,Phi,Gie	BY 231/..., RGP 15G...M, MR 814...818, ++	data
YX 56/...	Si-Di	Gl-L, contr.av., 600...1400V, 40A	32a§	Phi	-	data
YX 56/...R	Si-Di	=BYX 56/...:	32b&		-	data
YX 57/...	Si-Di	Gl, contr.av., 500...600V, 0,4A, <200ns	31a	Tho	BYT 11/..., BYV 26C...E, BYV 95C	data
YX 58/...	Si-Di	Gl, S, 50...400V, 1A, <250ns	34a§	Tho	BYV 12...16, BYX 92/..., RGP 10A...M, ++	data
YX 59/...	Si-Di	Gl-L, 200...500V, 40A, <600ns	73a§,b	Phi	(BYX 56/.., BYX 97/...)	data
YX 60/...	Si-Di	Gl, 50...1000V, 0,4A	31a	Tho	BA 157...159, BYW 27/..., 1N4001...07, ++	data
YX 61/...(M)	Si-Di	Gl-L, 50...400V, 12A, <100ns	32a§	Mot,Tho	BYT 12/..., BYT 61/..., BYV 30/...	data
YX 61/...R,RM	Si-Di	=BYX 61/...:	32b&		BYT 12/...R, BYT 61/..., BYV 30/...R	data
YX 62/...(M)	Si-Di	Gl-L, 600...800V, 12A, <200ns	32a§	Tho	BYT 12/..., BYT 61/..., BYV 30/...	data
YX 62/...R,RM	Si-Di	=BYX 62/...:	32b&		BYT 12/...R, BYT 61/...R, BYV 30/...	data
YX 63/600(M)	Si-Di	Gl-L, 600V, 20A, <200ns	32a§	Tho	BYT 65/..., BYX 64/...	data
YX 63/600R,RM	Si-Di	=BYX 63/600(M):	32b&		BYT 65/...R, BYX 64/...R	data
YX 64/600(M)	Si-Di	Gl-L, 600V, 30A, <200ns	32a§	Tho	BYT 65/..., MR 866	data
YX 64/600R,RM	Si-Di	=BYX 64/...:	32b&		BYT 65/...R, MR 866R	data
YX 65/...(M)	Si-Di	Gl-L, 50...400V, 30A, <100ns	32a§	Mot,Tho	BYT 30/..., BYV 92/...	data
YX 65/...R,RM	Si-Di	=BYX 65/...:	32b&		BYT 30/...R, BYV 92/...R	data
YX 66/...(M)	Si-Di	Gl-L, 400...1000V, 12A, <500ns	32a§	Mot,Tho	BYT 61/...	data
YX 66/...R,RM	Si-Di	=BYX 66/...:	32b&		BYT 61/...R	data
YX 67/...	Si-Di	Gl-L, 400...1000V, 30A, <500ns	32a§	Tho	BYT 65/...	data
YX 67/...R	Si-Di	=BYX 67/...:	32b&		BYT 65/...R	data
YX 68	Si-Di	Gl, 1000V, 1A	34a§	Fer	BY 127, BY 133, BY 227, 1N4007, ++	data
YX 69	Si-Di	=BYX 68:	31a	Fer	BY 127, BY 133, BY 227, 1N4007, ++	data
YX 70/...	Si-Di	Gl, S, 100...600V, 1A, <200ns	31a	Phi	BY 201/..., BYX 92/..., RGP 10B...M	data
YX 71/...	Si-Di	TV-Gl, 350...600V, 7A, <450ns	26h§	Phi	BYW 19/..., (BY 229/...)[4]	data
YX 71/...R	Si-Di	=BYX 71/...:	26g&		BY 277/...R, BYW 19/...R, (BY 229/...R)[4]	data
YX 72/...	Si-Di	Gl-L, 150...500V, 10A	26h§	Phi	(BY 239/...)[4]	data
YX 72/...R	Si-Di	=BYX 72/...:	26g&		(BY 239/...R)[4]	data
YX 73(A,B)	Si-Di	Schottky-Gl, 20...60V, 5A	32a§	Tho	BYS 15, BYV 120/..., MBR 1520...1540	data
YX 74/...	Si-Br	Gl-Br, 100...800V, 1,5A	33x1	Aei	B80C1500...B600C1500	data
YX 75	Si-Di	Gl-L, 50V, 12A	32a§	Mot	BYW 88/50, BYX 99/300, 1N4506...4511	data
YX 75 R	Si-Di	=BYX 75:	32b&		BYW 88/50R, BYX 99/300R	data
YX 76	Si-Di	Gl-L, 100V, 12A	32a§	Mot	BYW 88/100, BYX 99/300, 1N4506...4511	data
YX 76 R	Si-Di	=BYX 76:	32b&		BYW 88/100R, BYX 99/300R	data
YX 77	Si-Di	Gl-L, 200V, 12A	32a§	Mot	BYW 88/200, BYX 99/300, 1N4506...4511	data
YX 77 R	Si-Di	=BYX 77:	32b&		BYW 88/200R, BYX 99/300R	data
YX 78	Si-Di	Gl-L, 400V, 12A	32a§	Mot	BYW 88/400, BYX 99/600, 1N4507...4511	data
YX 78 R	Si-Di	=BYX 78:	32b&		BYW 88/400R, BYX 99/600R	data
YX 79	Si-Di	Gl-L, 600V, 12A	32a§	Mot	BYW 88/600, BYX 99/600, 1N4508...4511	data
YX 79 R	Si-Di	=BYX 79:	32b&		BYW 88/600R, BYX 99/600R	data
YX 80	Si-Di	Gl-L, 800V, 12A	32a§	Mot	BYW 88/800, BYX 99/900, 1N4509...4511	data
YX 80 R	Si-Di	=BYX 80:	32b&		BYW 88/800R, BYX 99/900R	data
YX 81	Si-Di	Gl-L, 1000V, 12A	32a§	Mot	BYW 88/1000, BYX 99/1200, 1N4510...4511	data
YX 81 R	Si-Di	=BYX 81:	32b&		BYW 88/1000R, BYX 99/1200R	data
YX 82	Si-Di	Gl, 200V, 1,5A	31a	Aeg,Gie	BY 226..227, (R)GP 15D...M, 1N5393...99,++	data
YX 83	Si-Di	=BYX 82: 400V	31a	Aeg,Gie	BY 226..227, (R)GP 15G...M, 1N5395...99,++	data
YX 84	Si-Di	=BYX 82: 600V	31a	Aeg,Gie	BY 226..227, (R)GP 15J...M, 1N5397...99,++	data
YX 85	Si-Di	=BYX 82: 800V	31a	Aeg,Gie	BY 227, (R)GP 15K...M, 1N5398...5399, ++	data
YX 86	Si-Di	=BYX 82: 1000V	31a	Aeg,Gie	BY 227, BY 350/1300, (R)GP 15M, 1N5399	data
YX 87	Si-Di	=BYX 82: 1200V	31a	Aeg,Gie	BY 227, BY 255, BY 350/1300	data
YX 88	Si-Di	Schottky-Gl, 20V, 30A	75a§	Tho	-	data
YX 89(A,B)	Si-Di	Schottky-Gl, 20...60V, 50A	32a§	Tho	BYS 50...51, BYS 60/..., BYV 23/...	data
YX 90	Si-Di	kV-Gl, 7,5kV, 0,2A	31a	Phi	-	data
YX 90 G	Si-Di	=BYX 90: 8kV, 0,5A	31a		-	data
YX 91/...	Si-Di	kV-Gl, 115...225kV, 0,2A		Phi	-	data
YX 92/...	Si-Di	Gl, S, 50...400V, 1A, <100ns	34a§	Tho	BY 218/..., FE 1A...D, RGP 10A...G	data
YX 93	Si-Di	Schottky-Gl, 200V, 60A	32a§	Fer	-	data
YX 93 R	Si-Di	=BYX 93:	32b&		-	data
YX 94	Si-Di	Gl, 1250V, 1A	31a	Phi	BY 133, BY 255, EM 513, GP 10V...Y	data
YX 95	Si-Di	Gl, 1300V, 1A	31a	Phi	BY 133, BY 255, EM 513, GP 10V...Y	data
YX 96/...(U)	Si-Di	Gl-L, 300...1600V, 30A	32a§	Phi	-	data
YX 96/...R	Si-Di	=BYX 96/...:	32b&		-	data
YX 97/...	Si-Di	Gl-L, 300...1600V, 40A	32a§	Phi	BYX 56/...	data
YX 97/...R	Si-Di	=BYX 97/...:	32b&		BYX 56/...R	data
YX 98/...	Si-Di	Gl-L, 300...1200V, 10A	32a§	Mot,Phi	BYW 88/..., BYX 42/..., 1N4507...4511, ++	data
YX 98/...R	Si-Di	=BYX 98/...:	32b&		BYW 88/...R, BYX 42/...R	data
YX 99/...	Si-Di	Gl-L, 300...1200V, 15A	32a§	Mot,Phi	BYX 25/..., BYX 96/...	data
YX 99/...R	Si-Di	=BYX 99/...:	32b&		BYX 25/...R, BYX 96/...R	data

Type	Device	Short Description	Fig.	Manu	Comparision Types	More at
BYX 101 G	Si-Di	kV-Gl, contr.av., 10kV, 0,28A, <600ns	31a	Phi	-	data
BYX 102 G	Si-Di	kV-Gl, contr.av., 10kV, 0,25A, <600ns	31a	Phi		data
BYX 103 G	Si-Di	kV-Gl, contr.av., 10kV, 0,22A, <175ns	31a	Phi		data
BYX 104 G	Si-Di	kV-Gl, contr.av., 10kV, 0,16A, <50ns	31a	Phi		data
BYX 105 G	Si-Di	kV-Gl, contr.av., 5kV, 0,46A, <600ns	31a	Phi		data
BYX 106 G	Si-Di	kV-Gl, contr.av., 5kV, 0,4A, <350ns	31a	Phi		data
BYX 107 G	Si-Di	kV-Gl, contr.av., 5kV, 0,34A, <175ns	31a	Phi		data
BYX 108 G	Si-Di	kV-Gl, contr.av., 5kV, 0,24A, <50ns	31a	Phi		data
BYX 110 GP	Si-Di	kV-Gl, contr.av., ..400Hz, 9kV, 0,35A	31a	Phi	-	data
BYX 120 G	Si-Di	kV-Gl, Kfz-Zündg./ignition, 3kV, 0,1A	31a	Phi	-	data
BYX 132 G	Si-Di	kV-Gl, Kfz-Zündg./ignition, 2kV, 50mA	31a	Phi		data
BYX 133 G	Si-Di	kV-Gl, Kfz-Zündg./ignition, -3kV, 50mA	31a	Phi		data
BYX 134 G,GP	Si-Di	kV-Gl, Kfz-Zündg./ignition, 4kV, 50mA	31a	Phi		data
BYX 135 G	Si-Di	kV-Gl, Kfz-Zündg./ignition, 5kV, 50mA	31a	Phi		data

BYY

Type	Device	Short Description	Fig.	Manu	Comparision Types	More at
BYY 10	Si-Di	Gl, 800V, 0,5A		Phi	(BY 127, BY 133, BY 227, 1N4006...07,++)[4]	data
BYY 15	Si-Di	Gl-L, 800V, 36A	73a§	Phi	-	data
BYY 16	Si-Di	=BYY 15:	73b&	Phi	D30S/...	data
BYY 19	Si-Di	Gl-L, 1500V, 4A	32a§	Itt	(BY 359/1500)[4]	data
BYY 20(200)	Si-Di	Gl-L, 200V, 18A	75a§	Phi	1N3493..3495	data
BYY 21(200)	Si-Di	Gl-L, 200V, 18A	75b&	Phi	DS25-02A...06A, 1N3493R..3495R	data
BYY 22	Si-Di	Gl-L, 400V, 12A	73a§	Phi	-	data
BYY 23	Si-Di	=BYY 22:	73b&	Phi	D24/400B...1800B, D21S/1000...1400	data
BYY 24	Si-Di	Gl-L, 800V, 12A	73a§	Phi	-	data
BYY 25	Si-Di	=BYY 24:	73b&	Phi	D24/800B...1800B, D21S/1000...1400	data
BYY 27	Si-Di	Gl-L, 300V, 220A	73a§	Phi	1N3737..3744	data
BYY 28	Si-Di	=BYY 27: 500V.	73a§	Phi	1N3739..3744	data
BYY 29	Si-Di	=BYY 27: 650V	73a§	Phi	1N3740..3744	data
BYY 30	Si-Di	=BYY 27: 800V	73a§	Phi	1N3741..3744	data
BYY 31	Si-Di	Gl, 150V, 1A	34a§	Itt	BY 126, BY 135, BY 226, 1N4003...07, ++	data
BYY 32	Si-Di	=BYY 31: 300V	34a§	Itt	BY 126, BY 134, BY 226, 1N4004...07, ++	data
BYY 33	Si-Di	=BYY 31: 450V	34a§	Itt	BY 126, BY 134, BY 226, 1N4005...07, ++	data
BYY 34	Si-Di	=BYY 31: 600V	34a§	Itt	BY 126, BY 134, BY 226, 1N4005...07, ++	data
BYY 35	Si-Di	=BYY 31: 750V	34a§	Itt	BY 127, BY 133, BY 227, 1N4006...07, ++	data
BYY 36	Si-Di	=BYY 31: 900V	34a§	Itt	BY 127, BY 133, BY 227, 1N4007, ++	data
BYY 37	Si-Di	=BYY 31: 1050V	34a§	Itt	BY 127, BY 133, BY 227, 1N4007, ++	data
BYY 38	Si-Di	=BYY 27: 1000V	73a§	Phi	1N3742..3744	data
BYY 39	Si-Di	=BYY 27: 1200V	73a§	Phi	1N3743..3744	data
BYY 39/...	Si-Di	Gl-L, 200..2400V, 220A	73a§	Phi	1N3736..3744	data
BYY 39/...R	Si-Di	=BYY 39/...:	73b&		1N3736R..3744R	data
BYY 53/75../1500	Si-Di	Gl-L, 75...1500V, 25A(Tc=79°)	75b&	Ztx	DS 25-01A..-06A, 1N3492R..3495R	data
BYY 54/75../1500	Si-Di	=BYY 53/...:	75a§	Ztx	1N3492..3495	data
BYY 56	Si-Di	Gl, 1200V, 0,56A	12a	Aeg	BY 208/1000, BY 245/1200, (R)GP 10M, ++	data
BYY 57/75../1500	Si-Di	Gl-L, 75...1500V, 35A(Tc=110°)	75b&	Aeg, Ztx	SSi E1205...1240, 1N3660R..3665R	data
					SSi E4360..4383	
BYY 57A/75../800	Si-Di	=BYY 57/...: 50A	75b&	Ztx	-	data
BYY 58/75../1500	Si-Di	=BYY 57/...:	75a§	Aeg	SSi E1105...1140, 1N3660..3665	data
					SSi E4460..4483	
BYY 58A/75../800	Si-Di	=BYY 57/...: 50A	75a§	Ztx	MR 5005..5040	data
BYY 59	Si-Di	Gl, 300V, 1,2A	12a	Aeg	BY 226, (R)GP 15G...M, 1N5394..5399, ++	data
BYY 60	Si-Di	=BYY 59: 600V	12a	Aeg	BY 226, (R)GP 15J...M, 1N5397..5399, ++	data
BYY 61	Si-Di	=BYY 59: 900V	12a	Aeg	BY 227, BY 350/1300, (R)GP 15M, 1N5399	data
BYY 62	Si-Di	=BYY 59: 1200V	12a	Aeg	BY 227, BY 350/1300, GH 3E...F	data
BYY 67	Si-Di	Gl-L, 600V, 10A	73a§	Phi	-	data
BYY 68	Si-Di	=BYY 67:	73b&	Phi	D24/800B...1800B, D21S/1000...1400	data
BYY 69	Si-Di	Gl-L, 1000V, 10A	73a§	Phi	-	data
BYY 70	Si-Di	=BYY 69:	73b&	Phi	D24/1200B...1800B, D21S/1000...1400	data
BYY 71	Si-Di	Gl-L, 1200V, 10A	73a§	Phi	-	data
BYY 72	Si-Di	=BYY 71:	73b&	Phi	D24/1200B...1800B, D21S/1000...1400	data
BYY 73	Si-Di	Gl-L, 600V, 40A	73a§	Phi	-	data
BYY 74	Si-Di	=BYY 73:	73b&	Phi	D60/800...1400, DS42-07A...16A	data
BYY 75	Si-Di	Gl-L, 1000V, 40A	73a§	Phi	-	data
BYY 76	Si-Di	=BYY 75:	73b&	Phi	D60/1200...1400, DS42-11A...16A	data
BYY 77	Si-Di	Gl-L, 1200V, 36A	73a§	Phi	-	data
BYY 78	Si-Di	=BYY 77:	73b&	Phi	D30S/1200...1400, D34/1200B...1800B	data
BYY 88	Si-Di	Gl-L, 150V, 1...4A	32a§	Itt	BYX 38/300...1200, (BY 205/200...1000)[4]	data
BYY 89	Si-Di	=BYY 88: 300V	32a§	Itt	BYX 38/300...1200, (BY 205/400...1000)[4]	data
BYY 90	Si-Di	=BYY 88: 600V	32a§	Itt	BYX 38/600...1200, (BY 205/600...1000)[4]	data
BYY 91	Si-Di	=BYY 88: 1200V	32a§	Itt	BYX 38/1200, (BY 359/1300)[4]	data
BYY 92	Si-Di	=BYY 88: 1600V	32a§	Itt	(BY 359/1500)[4]	data
BYY 93	Si-Di	Gl-L, 800V, 150A	32a§	Phi	-	data
BYY 94	Si-Di	=BYY 93:	32b&	Phi	-	data
BYY 95	Si-Di	Gl-L, 1200V, 150A	32a§	Phi	-	data

Type	Device	Short Description	Fig.	Manu	Comparision Types	More at
Y 96	Si-Di	=BYY 95:	32b&	Phi	-	data

YZ

Type	Device	Short Description	Fig.	Manu	Comparision Types	More at
Z 10	Si-Di	Gl-L, 1200V, 6A	32a§	Phi	BYX 38/1200, (BY 359/1300)[4]	data
Z 11	Si-Di	=BYZ 10: 900V	32a§	Phi	BYX 38/900...1200V, (BY 359/1000,++)[4]	data
Z 12	Si-Di	=BYZ 10: 600V	32a§	Phi	BYX 38/600...1200, (BY 205/600...1000,++)[4]	data
Z 13	Si-Di	=BYZ 10: 300V	32a§	Phi	BYX 38/300...1200, (BY 205/400...1000,++)[4]	data
Z 14	Si-Di	Gl-L, 400V, 36A	73a§	Phi	-	data
Z 15	Si-Di	=BYZ 14:	73b&	Phi	D30S/1000...1400, D34/400B...1800B	data
Z 16	Si-Di	=BYZ 10:	32b&	Phi	BYX 38/1200R, (BY 359/1300)[4]	data
Z 17	Si-Di	=BYZ 11:	32b&	Phi	BYX 38/900R...1200R, (BY 359/1000,++)[4]	data
Z 18	Si-Di	=BYZ 12:	32b&	Phi	BYX 38/600R...1200R,(BY 205/600...1000,++)[4]	data
Z 19	Si-Di	=BYZ 13:	32b&	Phi	BYX 38/300R...1200R,(BY 205/400...1000,++)[4]	data

ZC

Type	Device	Short Description	Fig.	Manu	Comparision Types	More at
	Si-P	=BCX 71RL (Typ-Code/Stempel/marking)	35	Sie	→BCX 71RL	data
	Z-Di	=BZX 399-C43 (Typ-Code/Stempel/marking)	71(1,7 m)	Phi	→BZX 399-...	data
	Z-Di	=P4 SMA-43A (Typ-Code/Stempel/marking)	71(5x2,5)	Fag	→P4 SMA-43A	data
	Z-Di	=SMBJ 24CA (Typ-Code/Stempel/marking)	71(5x3,5)	Mop	→SMBJ ...	data
	Si-N/P+R	=XP 04313 (Typ-Code/Stempel/marking)	46(2mm)	Mat	→XP 04313	data
-050...-350	Z-Di	5..35V, 5%, 1W	31a	Njr	BZW22/.., BZX61/..., ZPY.., 1N4733...53,++	data
100	Z-Di	3,9V, 10%, 0,3W	31a	Phi	BZX55/C3V9, BZX79/C3V9, ZPD3,9, 1N4730++	data
102/0V7...3V4	Si-St	0,7..3,4V, 0,25W	31a	Aeg	BZX 75/...	data
A 100	Z-Di, IC	SMD, 18x TAZ, 6,8V, Common Anode	20-MDIP	Phi		
A 109	Z-Di, IC	SMD, 9x TAZ, 6,8V, Common Anode	20-MDIP	Phi		
A 109 TS	Z-Di, IC	=BZA 109:	20-SSMDIP			
A 408 B	Z-Di	SMD, 4x TAZ, -5..+5V	46	Phi		
A 420 A	Z-Di	SMD, 4x TAZ, 20V, Common Anode	46	Phi		
A 456 A	Z-Di	SMD, 4x TAZ, 5,6V, Common Anode	46	Phi		
A 462 A	Z-Di	SMD, 4x TAZ, 6,2V, Common Anode	46	Phi		

ZD...BZT

Type	Device	Short Description	Fig.	Manu	Comparision Types	More at
D 10/C3V3...200	Z-Di	3,3..200V, 5%, 1,3W	34a§	Sie	BZW22/.., BZX61/.., ZPY.., 1N5313...56,++	data
D 23/C3V9...270	Z-Di	Z, TAZ, 3,9..270V, 5%, 2,5W	31a	Phi	BZW 04/...	data
D 27/C3V6...510	Z-Di	SMD, Z,TAZ, 3,9..510V, 5%, 300W(100µs)	72a(3,4mm)	Phi	HZK .., RLZ 5221B...	data
G 03-10...-270	Z-Di	SMD, Z, TAZ, 10..270V, 5%, 600W(100µs)	71a(5x2,5)	Aeg, Phi	BZG 04/.., HZF .., MA 1Z...	data
G 04/8V2...430	Z-Di	TAZ, 8,2..430V, 5%, 300W(10/1000µs)	71a(5x2,5)	Aeg	-	data
G 05/3V3...100	Z-Di	Z, 3,3...100V, 5%, 1,25W, 60W(100µs)	71a(5x2,5)	Aeg	BZG 03/.., PTZ.., RD..FM	data
M 55/C2,4...75	Z-Di	=BZX 55/B,C...	72a(1,9mm)	Aeg	-	data
M 85/C2V7...200	Z-Di	=BZX 85/C...: SMD	72a(5mm)	Tho	-	data
T 03/C,D6V2...270	Z-Di	TAZ, 6,2..270V, 1,3W, 600W(100µs)	31a	Aeg, Phi	BZW 04/.., BZW 06/...	data
T 55/C2,4...75	Z-Di	=BZX 55/C...: 0,5W	72a(3,5mm)	Aeg, Phi	BZD 27/C...	data

ZV

Type	Device	Short Description	Fig.	Manu	Comparision Types	More at
V 10	Ref-Di	6,5V, 5%, 0,4W, ±0,01%/°C	31a	Phi	BZV 27, BZX 90, 1N4575, 1N4580,++	data
V 11	Ref-Di	=BZV 10: ±0,005%/°C	31a	Phi	BZV 28, BZX 91, 1N4576, 1N4581,++	data
V 12	Ref-Di	=BZV 10: ±0,002%/°C	31a	Phi	BZV 29, BZX 92, 1N4577, 1N4582,++	data
V 13	Ref-Di	=BZV 10: ±0,001%/°C	31a	Phi	BZV 30, BZX 93, 1N4578, 1N4583,++	data
V 14	Ref-Di	=BZV 10: ±0,0005%/°C	31a	Phi	BZV 31, BZX 94, 1N4579, 1N4584,++	data
V 15/C7V5...75	Z-Di	7,5..75V, 5%, 15W(Tc=82°)	26h§	Phi	-	data
V 15/C7V5R...75R	Z-Di	7,5..75V, 5%, 15W(Tc=82°)	26g&	Phi	-	data
V 16/C3V3...100	Z-Di	3,3..100V, 5%, 3W	34a§	Tho	BZT03/.., BZV40/.., BZV48/.., 1N5333...78	data
V 17/C5V6...56	Z-Di	5,6..56V, 5%, 0,25W, ra	31a	Tho	BZV39/.., 1N4099...4127	data
V 18/C5V6...12	Z-Di	5,6..12V, 5%, 1W	31a	Tho	BZW22/.., BZX61/.., ZPY..., 1N4734...42,++	data
V 19M/C4V7...47	Z-Di	4,7..47V, 5%, 0,4W	40d	Fer	BZX55/.., BZX79/.., ZPD.., 1N5728...52,++	data
V 19N/C4V7...47	Z-Di	=BZV 19M/...	31a	Phi	→BZV 19M/...	data
V 19P/C4V7...47	Z-Di	=BZV 19M/...	31a	Phi	→BZV 19M/...	data
V 20	Z-Di	120...280V	12	Tho	-	data
V 21 A	Ref-Di	8,5V, 0,4W, ±0,001%/°C	31a	Phi	1N3157, 1N4778, 1N4783	data
V 21 B	Ref-Di	=BZV 21A: ±0,002%/°C	31a		1N3156, 1N4777, 1N4782	data
V 21 C	Ref-Di	=BZV 21A: ±0,005%/°C	31a		1N3155, 1N4776, 1N4781	data
V 21 D	Ref-Di	=BZV 21A: ±0,01%/°C	31a		1N3154, 1N4775, 1N4780	data
V 22(A...D)	Ref-Di	=BZV 21(A...D): 12,5V	31a	Phi	1N4907...04, 1N4911...08, 1N4915..12,++	data
V 23(A...D)	Ref-Di	=BZV 21(A...D): 18,5V	31a	Phi		data
V 24(A...D)	Ref-Di	=BZV 21(A...D): 8,7V	31a	Phi	→BZV 21(A...D)	data
V 25(A...D)	Ref-Di	=BZV 21(A...D): 13V	31a	Phi		data
V 26(A...D)	Ref-Di	=BZV 21(A...D): 19V	31a	Phi	2N4924...22, 1N4928...25, 1N4932...29,++	data
V 27(A)	Ref-Di	6,2V, 0,4W, ±0,01%/°C	31a	Tho	BZV 10, BZX 90, 1N4575, 1N4580,++	data
V 28(A)	Ref-Di	=BZV 27(A): ±0,005%/°C	31a	Phi	BZV 11, BZX 91, 1N4576, 1N4581,++	data
V 29(A)	Ref-Di	=BZV 27(A): ±0,002%/°C	31a	Phi	BZV 12, BZX 92, 1N4577, 2N4582,++	data
V 30(A)	Ref-Di	=BZV 27(A): ±0,001%/°C	31a	Phi	BZV 13, BZX 93, 1N4578, 1N4583,++	data
V 31(A)	Ref-Di	=BZV 27(A): ±0,0005%/°C	31a	Phi	BZV 14, BZX 94, 1N4579, 1N4584,++	data

Type	Device	Short Description	Fig.	Manu	Comparision Types	More at
BZV 32(A...B)	Ref-Di	9V, 0,5W, ±0,01%/°C	31a	Tho	1N935, 1N4765	data
BZV 33(A...B)	Ref-Di	=BZV 32(A...B): ±0,005%/°C	31a	Tho	1N936, 1N4766, 1N4771	data
BZV 34(A...B)	Ref-Di	=BZV 32(A...B): ±0,002%/°C	31a	Tho	1N937, 1N4767, 1N4772	data
BZV 35(A...B)	Ref-Di	=BZV 32(A...B): ±0,001%/°C	31a	Tho	1N938, 1N4768, 1N4773	data
BZV 36(A...B)	Ref-Di	=BZV 32(A...B): ±0,0005%/°C	31a	Tho	1N939, 1N4769, 1N4774	data
BZV 37	Ref-Di	Ref, sym, TAZ, 6,5V, 5%, 0,4W	31a	Phi,Tho	-	data
BZV 38	Ref-Di	6,4V, 5%	31a	Phi,Sie	BZV 10...14, BZX 90...94, 1N4565...84,++	data
BZV 39/C0V8	Si-St	ra, 0,8V, 0,5W	31a	Sie,Tho	(BZX 55/C0V8)	data
BZV 39/C2V4...75	Z-Di	ra, 2,4...75V, 5%, 0,5W	31a	Sie,Tho	1N4099...4131	data
BZV 40/C3V3...200	Z-Di	3,3...200V, 5%, 5W	31a	Sie	BZV 48/..., 2N5338B...5388B	data
BZV 41/C2V7...47	Z-Di	2,7...47V, 5%, 0,75W	40d	Fer	BZV85/..., BZX61/..., ZPY..., 1N4728...56,++	data
BZV 42/C2V7...47	Z-Di	2,7...47V, 5%, 0,75W	40d	Fer	BZV85/..., BZX61/..., ZPY..., 1N4728...56,++	data
BZV 43(A...C)	Ref-Di	6,2V, 1...5%, 0,25W	31a	Tho	BZV 10...14, BZX 90...94, 1N4565...4584,++	data
BZV 44(A...C)	Ref-Di	6,2V, 1...5%, 0,25W	31a	Tho	BZV 10...14, BZX 90...94, 1N4565...4584,++	data
BZV 45(A...C)	Ref-Di	6,2V, 1...5%, 0,25W	31a	Tho	BZV 10...14, BZX 90...94, 1N4565...4584,++	data
BZV 46/C1V5	Si-St	1,35...1,55V(5mA), 0,25W	31a	Phi	BZ 102/1V4, BZX 75/C1V4, ZTE 1,5	data
BZV 46/C2V0	Si-St	2...2,3V, 0,25W	31a	Phi	BZ 102/2V1, BZX 75/C2V1, ZTE 2	data
BZV 47/C3V3...200	Z-Di	3,3...200V, 5%, 2W	31a	Tho	BZD 23/..., BZT 03/..., ZY ...	data
BZV 48/C3V3...200	Z-Di	3,3...200V, 5%, 5W	31a	Tho	BZV 40/..., 1N5333B...5388B	data
BZV 49/C0V8	Z-Di	SMD, 0,73...0,83V, 1W(Tc=25°)	39f	Sie	-	data
BZV 49/C2V4...75	Z-Di	SMD, 2,4...75V, 5%, 1W(Tc=25°)	39f, 39j	Phi,Sie,Ztx	BZX 78/...	data
		Sie: Fig. 39f, Phi, Ztx: Fig. 39j				
BZV 53A...B	Ref-Di	SMD, 5,89...6,51V(7,5mA)	35e	Tho	-	data
		A: ±0,01%/°C, B= 0,005%/°C				
BZV 54A...B	Ref-Di	SMD, 6,08...6,72V(0,5mA)	35e	Tho	-	data
		A: ±0,01%/°C, B= 0,005%/°C				
BZV 55/...0V8...200	Z-Di	=BZX 55/...: SMD	72a(3,4mm)	Phi,Tho	BZD 27/..., HZK.., RLZ 5221B...	data
BZV 58/3V3...200	Z-Di	3,3...200V, 5%, 1,43W	31a	Tho,Die,Fag	BZV 40/..., BZV 48/..., 1N5333B...5388B	data
BZV 60/C...	Z-Di	=BZV 49/C...: 0,5W	31a	Phi	BZW22/..., BZX55/..., ZPD..., 1N5221...67,++	data
BZV 61/20.../39	Z-Di	Z-L, TAZ, 20...39V, Izmax=42,5A	75/b&	Ztx	-	data
BZV 62/20.../39	Z-Di	=BZV 61/...:	75/a&	Ztx	-	data
BZV 80	Ref-Di	=BZV 53A: SMD	72a(3,4mm)	Phi	-	data
BZV 81	Ref-Di	=BZV 53B: SMD	72a(3,4mm)	Phi	-	data
BZV 85/C3V6...75	Z-Di	3,6...75V, 5%, 1W	31a	Phi	BZW22/..., BZX61.., ZPY..., 1N4729...61A,++	data
BZV 86/1V4...3V2	Si-St	1,4...3,2V(5mA), 0,33W	31a	Phi	BZ 102/..., BZX 75/...	data
BZV 87/1V4...3V2	Si-St	SMD, 1,4...3,2V(5mA), 0,35W	72a(3,4mm)	Phi	-	data
BZV 90/C2V4...C75	Z-Di	=BZV 49/C...: 1,5W, Pbr=40W(100µs)	48j§	Phi	-	data

BZW

Type	Device	Short Description	Fig.	Manu	Comparision Types	More at
BZW 1,5-5V8...376	Z-Di	TAZ, 5,8...376V, 5%, 1,5kW(1ms)	31a		1N5629...5665, 1N6267...6303	data
BZW 03/...7V5...510	Z-Di	TAZ, 7,5...510V, 5...10%, 1kW(0,1ms)	31a	Aeg, Phi	BZW07/..., BZW70/..., BZX70/..., BZY93/...	data
BZW 04/5V5...376(P)	Z-Di	TAZ, 5,5...376V, 5%, 400W(1ms)	31a	Tho, Fag	BZW 06/..., 1N6102...6137	data
BZW 04/...B	Z-Di	=BZW 04/...: bidirektional	31		BZW 06/...B, 1N6102...6137	data
BZW 06/5V8...376(P)	Z-Di	TAZ, 5,8...376V, 5%, 600W(1ms)	31a	Tho, Fag	BZW70/..., BZX70/..., BZY93/...	data
BZW 06/...B	Z-Di	=BZW 06/...: bidirektional	31		-	data
BZW 07/10...110	Z-Di	TAZ, 10...110V, 700W(1ms)	31a	Tho	BZW 70/..., BZX 70/..., BZY 93/...	data
BZW 07/10B...110B	Z-Di	=BZW 07/...: bidirektional	31		-	data
BZW 10/12...15	Z-Di-Br	TAZ-Br, 12...15V	33x1	Phi	-	data
BZW 11/7V0...171	Z-Di	TAZ, 7,5...171V, 1kW(1ms)	31a	Tho	1N5630...5663	data
BZW 11/7V0B...171B	Z-Di	=BZW 11/...: bidirektional	31a		1N6036...6069, 1N6139...6172	data
BZW 12/11...390	Z-Di	TAZ, 11...390V, 420W(10ms)	32a§	Tho	-	data
BZW 13/11...390	Z-Di	TAZ, 11...390V, 1,2kW(10ms)	32a§	Tho	-	data
BZW 14	Z-Di	TAZ, 12V	31a	Phi	-	data
BZW 20	Si-N+Di+R	Regelbaustein/voltage ctrl. device	35a	Sie	-	data
BZW 22/C1	Si-St	0,65...0,75V(5mA), 1,3W	31a	Sie	ZPY 1, ZY 1	data
BZW 22/C2V7...200	Z-Di	2,7...200V, 5%, 1,3W	31a	Aeg,Sie	BZX61/..., BZX85/..., BZY97/..., 2N5913...56	data
BZW 25/12...120	Z-Di	TAZ, 12...120V, 2,5kW(1ms)	17t§	Tho	-	data
BZW 30/12...324	Z-Di	TAZ, 12...324V, 3kW(1ms)	31a	Tho	-	data
BZW 50/8V2...180	Z-Di	TAZ, 8,2...180V, 5kW(1ms)	31a	Tho	-	data
BZW 50/8V2B...180B	Z-Di	=BZW 50/...: bidirektional	31		-	data
BZW 70/5V6...62	Z-Di	TAZ, 5,6...62V, 700W(1ms)	31a	Phi	BZX 70/..., BZY 93/...	data
BZW 86/7V5...62	Z-Di	TAZ, 7,5...62V, 13kW(1ms)	73a§	Phi	-	data
BZW 86/7V5R...62R	Z-Di	=BZW 86/...:	73b&		-	data
BZW 91/5V6...62	Z-Di	TAZ, 5,6...62V, 9,5kW(1ms)	32a§	Phi	BZY 91/...	data
BZW 91/5V6R...62R	Z-Di	=BZW 91/...:	32b&		BZY 91/...R	data
BZW 93/5V6...62	Z-Di	TAZ, 5,6...62V, 400W(1ms)	32a§	Phi	BZY 93/...	data
BZW 93/5V6R...62R	Z-Di	=BZW 93/...:	32b&		BZY 93/...R	data
BZW 95/5V6...62	Z-Di	TAZ, 5,6...62V, 400W(1ms)	34a§	Phi	BZW 04/...	data

Type	Device	Short Description	Fig.	Manu	Comparision Types	More at
BZW 96/3V9...7V5	Z-Di	TAZ, 3,9...7,5V, 110W(1ms)	34a§		BZW 04/...	data
BZW 100/20...24	Z-Di	TAZ, 20...24V, 1,8kW(15ms)	31a	Tho	-	data

BZX

Type	Device	Short Description	Fig.	Manu	Comparision Types	More at
BZX 10	Z-Di	6,2V, 5%, 0,4W	31a	Fch,Sgs	BZX55/..., BZX79/..., ZPD..., 1N5234...57,++	data
BZX 11	Z-Di	=BZX 10: 6,8V	31a	Fch,Sgs	→BZX 10	data
BZX 12	Z-Di	=BZX 10: 7,5V	31a	Fch,Sgs	→BZX 10	data
BZX 13	Z-Di	=BZX 10: 8,2V	31a	Fch,Sgs	→BZX 10	data
BZX 14	Z-Di	=BZX 10: 9,1V	31a	Fch,Sgs	→BZX 10	data
BZX 15	Z-Di	=BZX 10: 10V	31a	Fch,Sgs	→BZX 10	data
BZX 16	Z-Di	=BZX 10: 11V	31a	Fch,Sgs	→BZX 10	data
BZX 17	Z-Di	=BZX 10: 12V	31a	Fch,Sgs	→BZX 10	data
BZX 18	Z-Di	=BZX 10: 13V	31a	Fch,Sgs	→BZX 10	data
BZX 19	Z-Di	=BZX 10: 15V	31a	Fch,Sgs	→BZX 10	data
BZX 20	Z-Di	=BZX 10: 16V	31a	Fch,Sgs	→BZX 10	data
BZX 21	Z-Di	=BZX 10: 18V	31a	Fch,Sgs	→BZX 10	data
BZX 22	Z-Di	=BZX 10: 20V	31a	Fch,Sgs	→BZX 10	data
BZX 23	Z-Di	=BZX 10: 22V	31a	Fch,Sgs	→BZX 10	data
BZX 24	Z-Di	=BZX 10: 24V	31a	Fch,Sgs	→BZX 10	data
BZX 25	Z-Di	=BZX 10: 27V	31a	Fch,Sgs	→BZX 10	data
BZX 26	Z-Di	=BZX 10: 30V	31a	Fch,Sgs	→BZX 10	data
BZX 27	Z-Di	=BZX 10: 33V	31a	Fch,Sgs	→BZX 10	data
BZX 29/C3V3...100	Z-Di	3,3...100V, 5%, 1,5W	31a	Phi	BZV47/..., BZY97/..., ZY..., 1N5913...49,++	data
BZX 30/C3V3...27	Z-Di	3,3...27V, 5%, 0,25W	31a	Tho	BZX55/..., BZX79/..., ZPD..., 1N5226...54,++	data
BZX 30/D3V3...27	Z-Di	=BZX 30/C...: 10%	31a		→BZX 30/C...	data
BZX 31/C3V6...9V1	Z-Di	3,6...9,1V, 5%, 0,25W	31a	Tho	BZX55/..., BZX79/..., ZPD..., 1N5227...39,++	data
BZX 32/C3V6...27	Z-Di	3,6...27V, 5%, 5W(Tc=25°)	32a§	Tho	BZX 98/.., 1N2970...88	data
BZX 32/D3V6...27	Z-Di	=BZX 32/C...: 10%	32a§		→BZX 32/C...	data
BZX 33	Ref-Di	8,6V, 0,2W, ±0,5mV/°C	31a	Tho	1N3154...3157, 1N4775...4784	data
BZX 34	Ref-Di	8,6V, 0,2W, ±0,1mV/°C	31a	Tho	1N3154...3157, 1N4775...4784	data
BZX 35	Ref-Di	10V, 0,2W, ±1,5mV/°C	31a	Tho	-	data
BZX 36	Ref-Di	6,2V, 0,2W, -0,5...+1mV/°C	31a	Tho	BZV 10...14, BZX 90...94, 1N4575...4584,++	data
BZX 43	Ref-Di	6,7V, ±0,001%/°C	2d&	Fch,Sgs	BZV 13, BZX 48, BZX 93, 1N4578, 1N4890++	data
BZX 44	Ref-Di	6,7V, ±0,002%/°C	2d&	Fch,Sgs	BZV 12, BZX 49, BZX 92, 1N4577, 1N4582++	data
BZX 45	Ref-Di	6,7V, ±0,003%/°C	2d&	Fch,Sgs	BZV 12, BZX 49, BZX 92, 1N4577, 1N4582++	data
BZX 46/C2V7...200	Z-Di	2,7...200V, 5%, 0,5W	31a	Itt,Phi	BZX55/..., BZX79/..., ZPD..., 1N5223...81,++	data
BZX 47	Ref-Di	6,5V, 10mA, ±0,0005%/°C	2c§	Phi	BZV 14, BZX 94, 1N4574, 1N4579, 1N4584++	data
BZX 48	Ref-Di	6,5V, ±0,001%/°C	2c§	Phi	BZV 13, BZX 43, BZX 93, 1N4578, 1N4583++	data
BZX 49	Ref-Di	6,5V, ±0,002%/°C	2c§	Phi	BZV 12, BZX 44, BZX 92, 1N4577, 1N4582++	data
BZX 50	Ref-Di	6,5V, ±0,005%/°C	2c§	Phi	BZV 11, BZX 45, BZX 91, 1N4576, 1N4581++	data
BZX 51	Ref-Di	8,4V, 25mA, 0,25W, ±0,01%/°C	31a	Aeg	1N3154, 1N4775, 1N4780	data
BZX 52	Ref-Di	8,4V, 25mA, 0,25W, ±0,005%/°C	31a	Aeg	1N3155, 1N4776, 1N4781	data
BZX 53	Ref-Di	8,4V, 25mA, 0,25W, ±0,002%/°C	31a	Aeg	1N3156, 1N4777, 1N4782	data
BZX 54	Ref-Di	8,4V, 25mA, 0,25W, ±0,001%/°C	31a	Aeg	1N3157, 1N4778, 1N4783	data
BZX 55/A2V4...200	Z-Di	=BZX 55/C...: 1%	31a	Aeg	→BZX 55/C...	data
BZX 55/B2V4...200	Z-Di	=BZX 55/C...: 2%	31a	Aeg	→BZX 55/C...	data
BZX 55/C0V8	Si-St	0,73...0,83V(5mA), 0,5W	31a	Itt,Sie,Tho	BZ 102/0V7, BZX 83/C0V8, ZPD 1	data
BZX 55/C2V4...200	Z-Di	2,4...200V, 5%, 0,5W	31a	Phi,Sie,++	BZW 22/..., BZX79/..., ZPD..., 1N5221...81++	data
BZX 55/D2V4...200	Z-Di	=BZX 55/D...: 10%	31a	Sie	→BZX 55/C...	data
BZX 57	Z-Di	7,5V, 5%, 0,25W	31a	Tho	BZX55/C7V5, BZX79/C7V5, ZPD7,5, 1N5236++	data
BZX 58/C6V8...10	Z-Di	6,8...10V, 5%, 0,25W	31a	Tho	BZX55/..., BZX79/..., ZPD..., 1N5235...40,++	data
BZX 59/C11...27	Z-Di	11...27V, 5%, 0,25W	31a	Tho	BZX55/..., BZX79/..., ZPD..., 1N5241...54,++	data
BZX 60/C30...56	Z-Di	30...56V, 5%, 0,25W	31a	Tho	BZX55/..., BZX79/..., ZPD..., 1N5256...63,++	data
BZX 61/C3V6...200	Z-Di	3,6...200V, 5%, 1,3W	31a	Phi,Tho	BZW22/..., BZX85/..., ZPY..., 1N5914...56,++	data
BZX 62	Si-St	0,65...0,75V(5mA), 0,25W	31a	Phi,Tho	BA 220, BZ 102/0V8, BZX 55/0V8, ZPD 1	data
BZX 63/C6V8...10	Z-Di	6,8...10V, 5%, 0,25W	31a	Tho	→BZX 58/...	data
BZX 64/C11...27	Z-Di	11...27V, 5%, 0,25W	31a	Tho	→BZX 59/...	data
BZX 65/C30...56	Z-Di	30...56V, 5%, 0,25W	31a	Tho	→BZX 60/...	data
BZX 66	Ref-Di	6,8V, 5%, 0,25W, -0,005...+0,08%/°C	2a§	Tho	BZV 14, BZX 43...45, BZX 47...50, 1N4895++	data
BZX 67/C12...200	Z-Di	12...200V, 5%, 10,7W(Tc=45°)	≈32a§	Aeg	BZX 98/..., 1N2976...3015	data
BZX 68 A	Z-Di	58...61V, 10,7W(Tc=45°)	≈32a§	Aeg	(BZX 98/C62, 1N3000)	data
BZX 68 B	Z-Di	=BZX 68A: 61...63V	≈32a§		BZX 98/C62, 1N3000	data
BZX 68 C	Z-Di	=BZX 68A: 63...66V	≈32a§		(BZX 98/C62, 1N3000)	data
BZX 69/C7V5...12	Z-Di	7,5...12V, 5%, 0,25W	31a	Tho	BZX55/..., BZX79/..., ZPD..., 1N5236...42,++	data
BZX 70/C7V5...75	Z-Di	Z, TAZ, 7,5...75V, 5%, 2,5W	31a	Phi	BZD 23/..., BZT 03/..., BZW 70/...	data
BZX 71/B5V1...24	Z-Di	5,1...24V, 2%, 0,4W	31a	Aeg	BZX55/..., BZX79/..., ZPD..., 1N5231...52,++	data
BZX 71/C5V1...24	Z-Di	=BZX 71/B...: 5%	31a		→BZX 71/B...	data
BZX 72(A...C)	Ref-Di	9V(5mA), 5%, 0,0567W A: ±0,002, B: ±0,002, C: ±0,004%/°C	31a	Fer	1N938...936, 1N4768...66, 1N4773...71	data
BZX 74/C5V6...12	Z-Di	5,6...12V, 5%, 0,4W	31a	Ssc	BZX55/..., BZX79/..., ZPD..., 1N5232...42,++	data
BZX 75/C1V4...3V6	Si-St	1,4...3,6V(10mA), 5%, 0,4W	31a	Fer,Phi	BZ 102/..., BZV 86/...	data
BZX 76	Z-Di	13V, 5%, 0,4W	31a	Tho	BZX55/C13, BZX79/C13, ZPD13, 1N5243,++	data

Type	Device	Short Description	Fig.	Manu	Comparision Types	More at
BZX 77/D5V6...9V1	Z-Di	SMD, 5,6...9,1V, 10%	Chip	Tix	-	data
BZX 78/C5V1...75	Z-Di	SMD, 5,1...75V, 5%	39j	Phi	BZV 49/...	data
BZX 79/A2V4...200	Z-Di	=BZX 79/B...: 1%	31a		→BZX 79/B...	data
BZX 79/B2V4...200	Z-Di	2,4...200V, 2%, 0,4W	31a	Mot,Phi,Tho	BZX55/..., BZX85/..., ZPD..., 1N5221...81,++	data
BZX 79/C2V4...200	Z-Di	=BZX 79/B...: 5%	31a		→BZX 79/B...	data
BZX 79/F2V4...200	Z-Di	=BZX 79/B...: 3%	31a		→BZX 79/B...	data
BZX 80/C6V8...10	Z-Di	6,8...10V, 5%, 0,8W	2d	Tho	→BZX 58/...	data
BZX 81/C11...27	Z-Di	11...27V, 5%, 0,8W	2d	Tho	→BZX 59/...	data
BZX 82/C30...56	Z-Di	30...56V, 5%, 0,8W	2d	Tho	→BZX 60/...	data
BZX 83/C0V8	Si-St	0,73...0,83V(5mA), 0,4A, 0,5W	31a	Aeg,Sie,++	BZW 22/C0V8, ZPY 1	data
BZX 83/C2V4...75	Z-Di	2,4...75V, 5%, 0,5W	31a	Aeg,Sie,++	BZX55/..., BZX79/..., ZPD..., 1N5221...67,++	data
BZX 84/A2V4...75	Z-Di	=BZX 84/C...: 1%	35e	Sie	-	data
BZX 84/B2V4...75	Z-Di	=BZX 84/C...: 2%	35e	Sie	-	data
BZX 84/C0V8	Si-St	SMD, 0,73...0,83V(5mA)	35e	Sie	-	data
BZX 84/C2V4...75	Z-Di	SMD, 2,4...75V, 5%	35e	Phi,Sie,++	HZM ..., RD ...M	data
BZX 85 B2V7...39	Z-Di	=BZX 85(/)C2V7...39: 2%	31a	Aeg	BZW22/..., BZX61/..., ZPY..., 1N5913...56,++	data
BZX 85(/)C2V7...200	Z-Di	2,7...200V, 5%, 1,3W	31a	Aeg,Sie,++	BZW22/..., BZX61/..., ZPY..., 1N5913...56,++	data
BZX 86-E36	Z-Di	35...45V	73a§	Phi	-	data
BZX 87/C4V7...75	Z-Di	4,7...75V, 5%, 1,3W	31a	Phi	BZW22/..., BZX61/..., ZPY..., 1N5917...46,++	data
BZX 88/C2V7...47	Z-Di	SMD, 2,7...47V, 5%	35e(2mm)	Fer	-	data
BZX 89/C7V5...12	Z-Di	7,5...12V, 5%, 0,4W	2	Tho	BZX55/..., BZX79/..., ZPD..., 1N5236...42,++	data
BZX 90	Ref-Di	6,5V(7,5mA), 5%, 0,4W, ±0,01%/°C	31a	Phi	BZV 10, 1N4565, 1N4570, 1N4575, 1N4580++	data
BZX 91	Ref-Di	6,5V(7,5mA), 5%, 0,4W, ±0,005%/°C	31a	Phi	BZV 11, 1N4566, 1N4571, 1N4576, 1N4581++	data
BZX 92	Ref-Di	6,5V(7,5mA), 5%, 0,4W, ±0,002%/°C	31a	Phi	BZV 12, 1N4567, 1N4572, 1N4577, 1N4582++	data
BZX 93	Ref-Di	6,5V(7,5mA), 5%, 0,4W, ±0,001%/°C	31a	Phi	BZV 13, 1N4568, 1N4573, 1N4578, 1N4583++	data
BZX 94	Ref-Di	6,5V(7,5mA), 5%, 0,4W, ±0,0005%/°C	31a	Phi	BZV 14, 1N4569, 1N4574, 1N4579, 1N4584++	data
BZX 95/C4V7...36	Z-Di	4,7...36V, 5%, 0,4W	31a	Hit	BZX55/..., BZX79/..., ZPD..., 1N5230...58,++	data
BZX 96/C2V4...33	Z-Di	2,4...33V, 5%, 0,4W	31a	Tho	BZX55/..., BZX79/..., ZPD..., 1N5221...57,++	data
BZX 97/C0V8	Si-St	0,73...0,83V(5mA), 0,5W	31a	Itt,Sie	BA 315, BZX 62, BZ 102/0V7, BZX 55/C0V8	data
BZX 97/C2V4...62	Z-Di	2,4...62V, 5%, 0,5W	31a	Itt,Sie	BZX55/..., BZX79/..., ZPD..., 1N5221...65,++	data
BZX 98/C3V9...200	Z-Di	3,9...200V, 5%, 13W(Tc=85°)	32a§	Sie	BZY 93/...	data
BZX 99-C2V4...15	Z-Di	SMD, lo-noise, 2,4...15V, ±5%	35e	Phi	BZX 84/..., HZM ..., RD ...M	data
BZX 284-...2V4...C75	Z-Di	SMD, 2,4...75V, 0,4W, -B: ±2%, -C: ±5%	71a(2x1,2)	Phi	DTZ ..., HZU..., RD... S, UDZ...	data
BZX 399	Z-Di	SMD, 1,8...43V, ±5%, 0,3W	71a(1,7mm)	Phi	DTZ ..., HZU ..., RD ...S	data

BZY

Type	Device	Short Description	Fig.	Manu	Comparision Types	More at
BZY 1	Si-St	0,7...0,9V, 0,1A, 1,25W(Ta=25°)	32a§	Eiy		data
BZY 5	Z-Di	5,5V, 10%, 1,25W(Ta=25°)	32a§	Eiy	BZX 98/...	data
BZY 6	Z-Di	=BZY 5: 6,5V	32a§	Eiy	→BZY 5	data
BZY 7	Z-Di	=BZY 5: 7,5V	32a§	Eiy	→BZY 5	data
BZY 8	Z-Di	=BZY 5: 8,5V	32a§	Eiy	→BZY 5	data
BZY 10	Z-Di	=BZY 5: 10V	32a§	Eiy	→BZY 5	data
BZY 12	Z-Di	=BZY 5: 12V	32a§	Eiy	→BZY 5	data
BZY 14	Z-Di	5,6V, 10%, 3,5W(Tc=45°)	32a§	Aeg	BZY 98/...	data
BZY 15	Z-Di	=BZY 14: 6,8V	32a§	Aeg	→BZY 14	data
BZY 16	Z-Di	=BZY 14: 8,2V	32a§	Aeg	→BZY 14	data
BZY 16/C...	Z-Di	=BZY 16/C...		Ssc	→BZV 16/C...	data
BZY 17	Z-Di	=BZY 14: 10V	32a§	Aeg	→BZY 14	data
BZY 17/C...	Z-Di	=BZY 17/C...		Ssc	→BZV 17/C...	data
BZY 18	Z-Di	=BZY 14: 12V	32a§	Aeg	→BZY 14	data
BZY 18/C...	Z-Di	=BZY 18/C...		Ssc	→BZV 18/C...	data
BZY 19	Z-Di	=BZY 14: 15V	32a§	Aeg	→BZY 14	data
BZY 20	Z-Di	=BZY 14: 18V	32a§	Aeg	→BZY 14	data
BZY 21	Z-Di	=BZY 14: 22V	32a§	Aeg	→BZY 14	data
BZY 22	Ref-Di	8,4V(5mA), 5%, 0,2W, ±0,01%/°C	33z	Itt	-	data
BZY 23	Ref-Di	8,4V(5mA), 5%, 0,2W, ±0,005%/°C	33z	Itt	-	data
BZY 24	Ref-Di	8,4V(5mA), 5%, 0,2W, ±0,002%/°C	33z	Itt	-	data
BZY 25	Ref-Di	8,4V(5mA), 5%, 0,2W, ±0,001%/°C	33z	Itt	-	data
BZY 56	Z-Di	4,7V, 5%, 0,28W	31a	Phi	BZX55/..., BZX79/..., ZPD..., 1N5230...,++	data
BZY 57	Z-Di	=BZY 56: 5,1V	31a	Phi	→BZY 56	data
BZY 58	Z-Di	=BZY 56: 5,6V	31a	Phi	→BZY 56	data
BZY 59	Z-Di	=BZY 56: 6,2V	31a	Phi	→BZY 56	data
BZY 60	Z-Di	=BZY 56: 6,8V	31a	Phi	→BZY 56	data
BZY 61	Z-Di	=BZY 56: 7,5V	31a	Phi	→BZY 56	data
BZY 62	Z-Di	=BZY 56: 8,2V	31a	Phi	→BZY 56	data
BZY 63	Z-Di	=BZY 56: 9,1V	31a	Phi	→BZY 56	data
BZY 64	Z-Di	=BZY 56: 4,3V, 15%	31a	Phi	→BZY 56	data
BZY 65	Z-Di	=BZY 56: 5,1V, 15%	31a	Phi	→BZY 56	data
BZY 66	Z-Di	=BZY 56: 6,2V, 15%	31a	Phi	→BZY 56	data
BZY 67	Z-Di	=BZY 56: 7,5V, 15%	31a	Phi	→BZY 56	data

Type	Device	Short Description	Fig.	Manu	Comparision Types	More at
BZY 68	Z-Di	=BZY 56: 9,1V, 15%	31a	Phi	→BZY 56	data
BZY 69	Z-Di	=BZY 56: 12V, 15%	31a	Phi	→BZY 56	data
BZY 70	Ref-Di	8,1V(100mA), 5%, 0,002%/°C		Itt	-	data
BZY 71	Ref-Di	=BZY 70: 2,5%, 0,001%/°C			-	data
BZY 74	Z-Di	6,2V, 15%	32a§	Phi	BZX 98/C6V2	data
BZY 75	Z-Di	=BZY 74: 7,5V	32a§	Phi	BZX 98/C7V5	data
BZY 76	Z-Di	=BZY 74: 9,2V	32a§	Phi	BZX 98/C9V2	data
BZY 78(P)	Z-Di	5,3V, 5%, 0,28W	31a	Phi	BZX55/C5V1, BZX79/C5V1, ZPD5,1, 1N5231++	data
BZY 83/C4V7...24V5	Z-Di	4,7..24,5V, 5%, 0,25W	2c	Aeg,Sie	BZX55/..., BZX79/..., ZPD..., 1N5230...53,++	data
BZY 83/D1	Si-St	0,62...0,78V(5mA), 0,2A, 0,25W	2c	Aeg,Sie	BZ 102/0V8, BZX 55/C0V8, ZPD 1	data
BZY 83/D4V7...22	Z-Di	=BZY 83/C..: 10%	2c	Aeg,Sie	→BZY 83/C...	data
BZY 84/D1	Si-St	0,7...0,9V(100mA), 14W(Tc=45°)	≈32a§	Sie	-	data
BZY 84/D5V6...12	Z-Di	5,6...12V, 10%, 14W(Tc=45°)	≈32a§	Sie	BZX 98/...	data
BZY 85/B2V7...33	Z-Di	2,7..33V, 2%, 0,4W	31a	Aeg,Sie	BZX55/..., BZX79/..., ZPD..., 1N5223...57,++	data
BZY 85/C2V7...33	Z-Di	=BZY 85/B..: 5%	31a		→DZY 85/B...	data
BZY 85/D1	Si-St	0,62...0,78V(5mA), 0,2A, 0,4W	31a	Aeg,Sie	BZ 102/0V7, BZX 55/C0V8, ZPD 1	data
BZY 85/D4V7...22	Z-Di	=BZY 85/B..: 10%	31a		→BZY 85/B...	data
BZY 87/0V7...3V4	Si-St	0,7..3,4V(5mA), 0,25W	31a	Aeg	BZ 102/...	data
BZY 88/C0V7	Si-St	0,71...0,8V(5mA), 0,25A, 0,4W	31a	Mot,Phi	BZ 102/0V7, BZX 55/C0V8, ZPD 1	data
BZY 88/C2V7...36	Z-Di	2,7..36V, 5%, 0,4W	31a	Mot,Phi	BZX55/..., BZX79/..., ZPD..., 1N5223...58,++	data
BZY 91/C7V5...75	Z-Di	Z, TAZ, 7,5...75V, 5%, 75W(Tc=65°)	32a§	Phi	BZW 91/...	data
BZY 91/C7V5R...75R	Z-Di	=BZY 91/C..:	32b&		BZW 91/...R	data
BZY 92/C3V9...36	Z-Di	3,9..36V, 5%, 1,1W	34a	Aeg	BZW22/..., BZX61/..., ZY..., 1N5915...38	data
BZY 93/C6V8...75	Z-Di	Z, TAZ, 6,8...75V, 5%, 20W(Tc=75°)	32a§	Phi	BZW 93/...	data
BZY 93/C6V8R...75R	Z-Di	=BZY 93/C..:	32b&		BZW 93/...R	data
BZY 94/C10...75	Z-Di	10...75V, 5%, 0,4W	31a	Phi	BZX55/..., BZX79/..., ZPD..., 1N5240...67,++	data
BZY 95/C10...75	Z-Di	10...75V, 5%, 1,5W	34a§	Aeg,Phi	BZW22/..., BZX61/..., ZY..., 2N5925...46	data
BZY 96/C4V7...10	Z-Di	4,7...10V, 5%, 1,5W	34a§	Phi	BZW22/..., BZX61/..., ZY..., 1N5917...25,++	data
BZY 97/C3V3...200	Z-Di	3,3..200V, 5%, 1,5W	31a	Sie,Tho,++	BZW22/..., BZX61/..., ZY..., 1N5913...56,++	data

BZZ

Type	Device	Short Description	Fig.	Manu	Comparision Types	More at
BZZ 10	Z-Di	6,15V, 10%, 0,28W	31a	Phi	BZX55/..., BZX79/..., ZPD..., 1N5233...,++	data
BZZ 11	Z-Di	=BZZ 10: 6,55V	31a	Phi	→BZZ 10	data
BZZ 12	Z-Di	=BZZ 10: 7,25V	31a	Phi	→BZZ 10	data
BZZ 13	Z-Di	=BZZ 10: 8,05V	31a	Phi	→BZZ 10	data
BZZ 14	Z-Di	5,6V, 5%, 10W(Tc=50°)	32a§	Phi	BZX 98/C5V6	data
BZZ 15	Z-Di	=BZZ 14: 6,2V	32a§	Phi	BZX 98/C6V2	data
BZZ 16	Z-Di	=BZZ 14: 6,8V	32a§	Phi	BZX 98/C6V8	data
BZZ 17	Z-Di	=BZZ 14: 7,5V	32a§	Phi	BZX 98/C7V5	data
BZZ 18	Z-Di	=BZZ 14: 8,2V	32a§	Phi	BZX 98/C8V2	data
BZZ 19	Z-Di	=BZZ 14: 9,1V	32a§	Phi	BZX 98/C9V1	data
BZZ 20	Z-Di	=BZZ 14: 10V	32a§	Phi	BZX 98/C10	data
BZZ 21	Z-Di	=BZZ 14: 11V	32a§	Phi	BZX 98/C11	data
BZZ 22	Z-Di	=BZZ 14: 12V	32a§	Phi	BZX 98/C12	data
BZZ 23	Z-Di	=BZZ 14: 13V	32a§	Phi	BZX 98/C13	data
BZZ 24	Z-Di	=BZZ 14: 15V	32a§	Phi	BZX 98/C15	data
BZZ 25	Z-Di	=BZZ 14: 16V	32a§	Phi	BZX 98/C16	data
BZZ 26	Z-Di	=BZZ 14: 18V	32a§	Phi	BZX 98/C18	data
BZZ 27	Z-Di	=BZZ 14: 20V	32a§	Phi	BZX 98/C20	data
BZZ 28	Z-Di	=BZZ 14: 22V	32a§	Phi	BZX 98/C22	data
BZZ 29	Z-Di	=BZZ 14: 24V	32a§	Phi	BZX 98/C24	data

C

Type	Device	Short Description	Fig.	Manu	Comparision Types	More at
C...	...-N	→2SC.., z.B./e.g. "C1398"=2SC1398→	Japantypen	JAP		
C...	...-N	→KSC.., z.B./e.g. "C1009"=KSC1009→	Samsung	Sam		
C...	...-N	→KTC.., z.B./e.g. "C 200"=KTC 200→	KEC	Kec		
C...	IC	→µPC... (NEC !)		Nec		
C	Si-P	=2SB1219 (Typ-Code/Stempel/marking)	35(2mm)	Mat	→2SB1219	data
C	Si-P	=2SB710 (Typ-Code/Stempel/marking)	35	Mat	→2SB710	data
C	Si-P	=2SB767 (Typ-Code/Stempel/marking)	39	Mat	→2SB767	data
C	N-FET	=2SK1375 (Typ-Code/Stempel/marking)	35(2mm)	Say	→2SK1375	data
C	GaAs-N-FET	=2SK649 (Typ-Code/Stempel/marking)	51	Mat	→2SK649	data
C	MOS-N-FET-d	=3SK303 (Typ-Code/Stempel/marking)	44	Mat	→3SK303	data
C	MOS-N-FET-d	=3SK307 (Typ-Code/Stempel/marking)	44(2mm)	Mat	→3SK307	data
C	C-Di	=BB 179B (Typ-Code/Stempel/marking)	71(1,3mm)	Phi	→BB 179B	data
C	C-Di	=BB 833 (Typ-Code/Stempel/marking)	71(1,7mm)	Sie	→BB 833	data
C	Si-Di	=DSE 010 (Typ-Code/Stempel/marking)	71(1,7mm)	Say	→DSE 010	data
C	C-Di	=HVR 17 (Typ-Code/Stempel/marking)	71(2,7mm)	Hit	→HVR 17	data
C	C-Di	=HVU 89 (Typ-Code/Stempel/marking)	71(1,7mm)	Hit	→HVU 89	data
C	Si-Di	=MA 2S784 (Typ-Code/Stempel/marking)	71(1,7mm)	Mat	→MA 784	data
C	Diode	=RB 521S-30 (Typ-Code/Stempel/marking)	71(1,3mm)	Rhm	→RB 521S-30	data
C	Si-Di	=SB005-09CP (Typ-Code/Stempel/marking)	35	Say	→SB 005-09CP	data
C 1	Si-Di	=1SS352 (Typ-Code/Stempel/marking)	71(1,7mm)	Tos	→1SS352	data
C 1	Si-Di	=1SS368 (Typ-Code/Stempel/marking)	71(1,3mm)	Tos	→1SS368	data
C 1	Si-Di	=1SS387 (Typ-Code/Stempel/marking)	71(1,3mm)	Tos	→1SS387	data
C 1	MOS-P-FET-e*	=2SJ559 (Typ-Code/Stempel/marking)	35(1,6mm)	Nec	→2SJ559	data

Type	Device	Short Description	Fig.	Manu	Comparision Types	More at
C 1	Si-P	=BCW 29 (Typ-Code/Stempel/marking)	35	Fer,Mot,Phi	→BCW 29	data
C 1	Si-N	=BFS 38A (Typ-Code/Stempel/marking)	35(2mm)	Fer	→BFS 38A	data
C 1	Si-N/P+R	=FMC 1A (Typ-Code/Stempel/marking)	45	Rhm	→FMC 1A	data
C 1	Si-Di	=HSM 88(A)S (Typ-Code/Stempel/marking)	35	Hit	→HSM 88(A)S	data
C 1	Si-Di	=HSM 88(A)SR(Typ-Code/Stempel/marking)	35	Hit	→HSM 88(A)SR	data
C 1	Z-Di	=MAZF 043 (Typ-Code/Stempel/marking)	71(1,7mm)	Mat	→MAZF 043	data
C 1	Z-Di	=MAZN 043 (Typ-Code/Stempel/marking)	71(1,3mm)	Mat	→MAZN 043	data
C 1	Z-Di	=MMSZ 5221B (Typ-Code/marking)	71(2,7mm)	Ons	→MMSZ 5221B	
C 1	N-FET	=SST 111 (Typ-Code/Stempel/marking)	35	Six	→SST 111	
C 1	Si-N/P+R	=UMC 1N (Typ-Code/Stempel/marking)	45(2mm)	Rhm	→UMC 1N..5N	data
C 1 D	LIN-IC	=µPC 2708T (Typ-Code/Stempel/marking)	46	Nec	→µPC 2708T	data
C 1 E	LIN-IC	=µPC 2709TB (Typ-Code/Stempel/marking)	46(2mm)	Nec	→µPC 2709TB	
C 1 E	LIN-IC	=µPC 2709TB (Typ-Code/Stempel/marking)	46(2mm)	Nec	→µPC 2709TB	
C 1 E	LIN-IC	=µPC 2709T (Typ-Code/Stempel/marking)	46	Nec	→µPC 2709T	
C 1 F	LIN-IC	=µPC 2710T (Typ-Code/Stempel/marking)	46	Nec	→µPC 2710T	
C 1 G	LIN-IC	=µPC 2711T (Typ-Code/Stempel/marking)	46	Nec	→µPC 2711T	
C 1 G	LIN-IC	=µPC 2711TB (Typ-Code/Stempel/marking)	46(2mm)	Nec	→µPC 2711TB	
C 1 G	LIN-IC	=µPC 2711TB (Typ-Code/Stempel/marking)	46(2mm)	Nec	→µPC 2711TB	
C 1 G	Si-N	=HN1C01F-GR(Typ-Code/Stempel/marking)	46	Tos	→HN 1C01F	data
C 1 G	Si-N	=HN1C01FU-GR(Typ-Code/Stempel/marking)	46(2mm)	Tos	→HN 1C01FU	data
C 1 G	LIN-IC	=µPC 2711T (Typ-Code/Stempel/marking)	46	Nec	→µPC 2711T	
C 1 H	LIN-IC	=µPC 2712TB (Typ-Code/Stempel/marking)	46(2mm)	Nec	→µPC 2712TB	
C 1 H	LIN-IC	=µPC 2712TB (Typ-Code/Stempel/marking)	46(2mm)	Nec	→µPC 2712TB	
C 1 H	LIN-IC	=µPC 2712T (Typ-Code/Stempel/marking)	46	Nec	→µPC 2712T	
C 1 J	LIN-IC	=µPC 2713T (Typ-Code/Stempel/marking)	46	Nec	→µPC 2713T	
C 1 K	LIN-IC	=µPC 2714T (Typ-Code/Stempel/marking)	46	Nec	→µPC 2714T	
C 1 L	LIN-IC	=µPC 2715T (Typ-Code/Stempel/marking)	46	Nec	→µPC 2715T	
C 1 L	LIN-IC	=µPC 2715T (Typ-Code/Stempel/marking)	46	Nec	→µPC 2715T	
C 1 M	LIN-IC	=µPC 2723T (Typ-Code/Stempel/marking)	46	Nec	→µPC 2723T	
C 1 O	Si-N	=KSC 1632-O (Typ-Code/Stempel/marking)	35	Sam	→KSC 1632	data
C 1 p	Si-P	=BCW 29 (Typ-Code/Stempel/marking)	35	Phi	→BCW 29	data
C 1 P	LIN-IC	=µPC 2726T (Typ-Code/Stempel/marking)	46	Nec	→µPC 2726T	
C 1 Q	LIN-IC	=µPC 2745TB (Typ-Code/Stempel/marking)	46(2mm)	Nec	→µPC 2745TB	
C 1 Q	LIN-IC	=µPC 2745TB (Typ-Code/Stempel/marking)	46(2mm)	Nec	→µPC 2745TB	
C 1 Q	LIN-IC	=µPC 2745T (Typ-Code/Stempel/marking)	46	Nec	→µPC 2745T	
C 1 R	LIN-IC	=µPC 2746TB (Typ-Code/Stempel/marking)	46(2mm)	Nec	→µPC 2746TB	
C 1 R	LIN-IC	=µPC 2746TB (Typ-Code/Stempel/marking)	46(2mm)	Nec	→µPC 2746TB	
C 1 R	LIN-IC	=µPC 2746T (Typ-Code/Stempel/marking)	46	Nec	→µPC 2746T	
C 1 S	LIN-IC	=µPC 2747T (Typ-Code/Stempel/marking)	46	Nec	→µPC 2747T	
C 1 t	Si-P	=BCW 29 (Typ-Code/Stempel/marking)	35	Phi	→BCW 29	data
C 1 T	LIN-IC	=µPC 2748T (Typ-Code/Stempel/marking)	46	Nec	→µPC 2748T	
C 1 U	LIN-IC	=µPC 2749T (Typ-Code/Stempel/marking)	46	Nec	→µPC 2749T	
C 1 W	LIN-IC	=µPC 2756T (Typ-Code/Stempel/marking)	46	Nec	→µPC 2756T	
C1X19X	LED	cy ca. ø4mm I:40lm Vf:3.5V If:350mA		Seo		data pdf pinout
C1X29X	LED	cy ca. ø4mm I:78lm Vf:3.5V If:700mA		Seo		data pdf pinout
C1X49X	LED	cy ca. ø4mm I:150lm Vf:3.5V		Seo		data pdf pinout
C 1 X	LIN-IC	=µPC 2757T (Typ-Code/Stempel/marking)	46	Nec	→µPC 2757T	
C 1 Y	LIN-IC	=µPC 2758T (Typ-Code/Stempel/marking)	46	Nec	→µPC 2758T	
C 1 Y	Si-N	=HN1C01F-Y (Typ-Code/Stempel/marking)	46	Tos	→HN 1C01F	
C 1 Y	Si-N	=HN1C01FU-Y (Typ-Code/Stempel/marking)	46(2mm)	Tos	→HN 1C01FU	data
C 1 Y	LIN-IC	=µPC 2758T (Typ-Code/Stempel/marking)	46	Nec	→µPC 2758T	data
C 1 Z	LIN-IC	=µPC 2762TB (Typ-Code/Stempel/marking)	46(2mm)	Nec	→µPC 2762TB	
C 1 Z	LIN-IC	=µPC 2762TB (Typ-Code/Stempel/marking)	46(2mm)	Nec	→µPC 2762TB	
C 1 Z	LIN-IC	=µPC 2762T (Typ-Code/Stempel/marking)	46	Nec	→µPC 2762T	
C 2	Si-P	=BCW 30 (Typ-Code/Stempel/marking)	35	Fer,Mot,Phi	→BCW 30	data
C 2	Si-P	=BFQ 32C (Typ-Code/Stempel/marking)	51	Val	→BFQ 32C	data
C 2	Si-N	=BFS 38 (Typ-Code/Stempel/marking)	35(2mm)	Fer	→BFS 38	data
C 2	Z-Di	=EDZ 5.6B (Typ-Code/Stempel/marking)	71(1,3mm)	Rhm	→EDZ 5.6B	data
C 2	Si-N/P+R	=FMC 2A (Typ-Code/Stempel/marking)	45	Rhm	→FMC 2A	data
C 2	Si-Di	=HSC 276 (Typ-Code/Stempel/marking)	71(1,3mm)	Hit	→HSC 276	data
C 2	Si-Di	=HSM 276S (Typ-Ccde/Stempel/marking)	35	Hit	→HSM 276S	data
C 2	Z-Di	=MAZF 047 (Typ-Code/Stempel/marking)	71(1,7mm)	Mat	→MAZF 047	data
C 2	Z-Di	=MAZN 047 (Typ-Code/Stempel/marking)	71(1,3mm)	Mat	→MAZN 047	data
C 2	Z-Di	=MMSZ 5222B (Typ-Code/marking)	71(2,7mm)	Ons	→MMSZ 5222B	
C 2	Si-Di	=SMBYT03-200(Typ-Code/Stempel/marking)	71(8mm)	Tho	→SMBYT 03-...	data
C 2	N-FET	=SST 112 (Typ-Code/Stempel/marking)	35	Six	→SST 112	data
C 2	Z-Di	=UDZ 5.6B (Typ-Code/Stempel/marking)	71(1,7mm)	Rhm	→UDZ 2.0B...36B	data
C 2	Z-Di	=UDZS 5.6B (Typ-Code/Stempel/marking)	71(1,7mm)	Rhm	→UDZS 5.1B...10B	data
C 2	Si-N/P+R	=UMC 2N (Typ-Code/Stempel/marking)	45(2mm)	Rhm	→UMC 1N...5N	data
C 2 A	LIN-IC	=µPC 2763TB (Typ-Code/Stempel/marking)	46(2mm)	Nec	→µPC 2763TB	
C 2 A	LIN-IC	=µPC 2763TB (Typ-Code/Stempel/marking)	46(2mm)	Nec	→µPC 2763TB	
C 2 A	LIN-IC	=µPC 2763T (Typ-Code/Stempel/marking)	46	Nec	→µPC 2763T	
C 2 B	LIN-IC	=µPC 8102T (Typ-Code/Stempel/marking)	46	Nec	→µPC 8102T	
C 2 C	LIN-IC	=µPC 8103T (Typ-Code/Stempel/marking)	46	Nec	→µPC 8103T	
C 2 D	LIN-IC	=µPC 8106T (Typ-Code/Stempel/marking)	46	Nec	→µPC 8106T	
C 2 F	LIN-IC	=µPC 8108T (Typ-Code/Stempel/marking)	46	Nec	→µPC 8108T	
C 2 G	LIN-IC	=µPC 8109T (Typ-Code/Stempel/marking)	46	Nec	→µPC 8109T	
C 2 H	LIN-IC	=µPC 2771TB (Typ-Code/Stempel/marking)	46(2mm)	Nec	→µPC 2771TB	
C 2 H	LIN-IC	=µPC 2771TB (Typ-Code/Stempel/marking)	46(2mm)	Nec	→µPC 2771TB	
C 2 H	LIN-IC	=µPC 2771T (Typ-Code/Stempel/marking)	46	Nec	→µPC 2771T	
C 2 K	LIN-IC	=µPC 8112T (Typ-Code/Stempel/marking)	46	Nec	→µPC 8112T	
C 2 L	LIN-IC	=µPC 2776TB (Typ-Code/Stempel/marking)	46(2mm)	Nec	→µPC 2776TB	
C 2 L	LIN-IC	=µPC 2776TB (Typ-Code/Stempel/marking)	46(2mm)	Nec	→µPC 2776TB	
C 2 L	LIN-IC	=µPC 2776T (Typ-Code/Stempel/marking)	46	Nec	→µPC 2776T	
C 2 M	LIN-IC	=µPC 8119T (Typ-Code/Stempel/marking)	46	Nec	→µPC 8119T	

Type	Device	Short Description	Fig.	Manu	Comparision Types	More at
C 2 N	LIN-IC	=µPC 8120T (Typ-Code/Stempel/marking)	46	Nec	→µPC 8120T	
C 2 p	Si-P	=BCW 30 (Typ-Code/Stempel/marking)	35	Phi	→BCW 30	data
C 2 S	LIN-IC	=µPC 2791TB (Typ-Code/Stempel/marking)	46(2mm)	Nec	→µPC 2791TB	
C 2 t	Si-P	=BCW 30 (Typ-Code/Stempel/marking)	35	Phi	→BCW 30	data
C 2 T	LIN-IC	=µPC 2792TB (Typ-Code/Stempel/marking)	46(2mm)	Nec	→µPC 2792TB	
C2V4 PH	Z-Di	=BZX 55-C2V4(Typ-Code/Stempel/marking)	31	Phi	→BZX 55-C...	data
C2V4 PH	Z-Di	=BZX 79-C2V4(Typ-Code/Stempel/marking)	31	Phi	→BZX 79-C...	data
C2V7 PH	Z-Di	=BZX 55-C2V7(Typ-Code/Stempel/marking)	31	Phi	→BZX 55-C...	data
C2V7 PH	Z-Di	=BZX 79-C2V7(Typ-Code/Stempel/marking)	31	Phi	→BZX 79-C...	data
C 3	Si-N	=BFS 39 (Typ-Code/Stempel/marking)	35(2mm)	Fer	→BFS 39	data
C 3	Si-N/P+R	=FMC 3A (Typ-Code/Stempel/marking)	45	Rhm	→FMC 3A	data
C 3	Si-Di	=HSM 88(S)R (Typ-Code/Stempel/marking)	35	Hit	→HSM 88(S)R	data
C 3	Si-Di	=KDS 122 (Typ-Code/Stempel/marking)	35(2mm)	Kec	→KDS 122	data
C 3	Si-Di	=KDS 226 (Typ-Code/Stempel/marking)	35	Kec	→KDS 226	data
C 3	Si-P	=KST 4126 (Typ-Code/Stempel/marking)	35	Sam	→KST 4126	data
C 3	Si-P	=MMBT 4126 (Typ-Code/Stempel/marking)	35	Mot,Sam	→MMBT 4126	data
C 3	Z-Di	=MMSZ 5223B (Typ-Code/marking)	71(2,7mm)	Ons	→MMSZ 5223B	
C 3	Si-Di	=SMBYT03-300(Typ-Code/Stempel/marking)	71(8mm)	Tho	→SMBYT 03-...	
C 3	N-FET	=SST 113 (Typ-Code/Stempel/marking)	35	Six	→SST 113	data
C 3	Si-N/P+R	=UMC 3N (Typ-Code/Stempel/marking)	45(2mm)	Rhm	→UMC 1N...5N	data
C 3	Si-Di	=1PS302 (Typ-Code/Stempel/marking)	35(2mm)	Phi	→1PS302	data
C 3	Si-Di	=1SS226 (Typ-Code/Stempel/marking)	35	Tos	→1SS226	data
C 3	Si-Di	=1SS302 (Typ-Code/Stempel/marking)	35(2mm)	Tos	→1SS302	data
C 3	Si-Di	=1SS362 (Typ-Code/Stempel/marking)	35(1,6mm)	Tos	→1SS362	data
C 3 A	Si-N	=HN 1C03F-A (Typ-Code/Stempel/marking)	46	Tos	→HN 1C03F	data
C 3 B	Si-N	=HN 1C03F-B (Typ-Code/Stempel/marking)	46	Tos	→HN 1C03F	data
C 3 T	Si-Di	=1PS226 (Typ-Code/Stempel/marking)	35	Phi	→1PS226	data
C3V0 PH	Z-Di	=BZX 55-C3V0(Typ-Code/Stempel/marking)	31	Phi	→BZX 55-C...	data
C3V0 PH	Z-Di	=BZX 79-C3V0(Typ-Code/Stempel/marking)	31	Phi	→BZX 79-C...	data
C3V3 PH	Z-Di	=BZX 55-C3V3(Typ-Code/Stempel/marking)	31	Phi	→BZX 55-C...	data
C3V3 PH	Z-Di	=BZX 79-C3V3(Typ-Code/Stempel/marking)	31	Phi	→BZX 79-C...	data
C3V6 PH	Z-Di	=BZX 55-C3V6(Typ-Code/Stempel/marking)	31	Phi	→BZX 55-C...	data
C3V6 PH	Z-Di	=BZX 79-C3V6(Typ-Code/Stempel/marking)	31	Phi	→BZX 79-C...	data
C3V6 PH	Z-Di	=BZV 85-C3V6(Typ-Code/Stempel/marking)	31	Phi	→BZV 85-C...	data
C3V9 PH	Z-Di	=BZX 55-C3V9(Typ-Code/Stempel/marking)	31	Phi	→BZX 55-C...	data
C3V9 PH	Z-Di	=BZX 79-C3V9(Typ-Code/Stempel/marking)	31	Phi	→BZX 79-C...	data
C3V9 PH	Z-Di	=BZV 85-C3V9(Typ-Code/Stempel/marking)	31	Phi	→BZV 85-C...	data
C3X18X	LED	cy ca. ø4mm I:40lm Vf:3.5V If:350mA		Seo		data pdf pinout
C3X28X	LED	cy ca. ø4mm I:71lm Vf:4V If:700mA		Seo		data pdf pinout
C 4	Si-P	=2SA811 (Typ-Code/Stempel/marking)	35	Nec	→2SA811	data
C 4	Si-P	=BCW 29R (Typ-Code/Stempel/marking)	35	Val,Fer	→BCW 29R	data
C 4	Si-P	=BFS 40A (Typ-Code/Stempel/marking)	35(2mm)	Fer	→BFS 40A	data
C 4	Si-N/P+R	=FMC 4A (Typ-Code/Stempel/marking)	45	Rhm	→FMC 4A	data
C 4	Si-Di	=HSM 88WK (Typ-Code/Stempel/marking)	35	Hit	→HSM 88WK	data
C 4	Si-Di	=MC 804 (Typ-Code/Stempel/marking)	35	Mit	→MC 804	data
C 4	Z-Di	=MMSZ 5224B (Typ-Code/marking)	71(2,7mm)	Ons	→MMSZ 5224B	
C 4	Si-Di	=SMBYT03-400(Typ-Code/Stempel/marking)	71(8mm)	Tho	→SMBYT 03-...	data
C 4	Si-N/P+R	=UMC 4N (Typ-Code/Stempel/marking)	45(2mm)	Rhm	→UMC 1N...5N	data
C04 02	Si-Di	=ERC 04-02 (Typ-Code/Stempel/marking)	31a	Fjd	→ERC 04-...(F)	data
C04 04	Si-Di	=ERC 04-04 (Typ-Code/Stempel/marking)	31a	Fjd	→ERC 04-...(F)	data
C04 06	Si-Di	=ERC 04-06 (Typ-Code/Stempel/marking)	31a	Fjd	→ERC 04-...	data
C04 10	Si-Di	=ERC 04-10 (Typ-Code/Stempel/marking)	31a	Fjd	→ERC 04-...	data
C4V3 PH	Z-Di	=BZX 55-C4V3(Typ-Code/Stempel/marking)	31	Phi	→BZX 55-C...	data
C4V3 PH	Z-Di	=BZX 79-C4V3(Typ-Code/Stempel/marking)	31	Phi	→BZX 79-C...	data
C4V3 PH	Z-Di	=BZV 85-C4V3(Typ-Code/Stempel/marking)	31	Phi	→BZV 85-C...	data
C4V7 PH	Z-Di	=BZX 55-C4V7(Typ-Code/Stempel/marking)	31	Phi	→BZX 55-C...	data
C4V7 PH	Z-Di	=BZX 79-C4V7(Typ-Code/Stempel/marking)	31	Phi	→BZX 79-C...	data
C4V7 PH	Z-Di	=BZV 85-C4V7(Typ-Code/Stempel/marking)	31	Phi	→BZV 85-C...	data
C 5	Si-P	=2SA811-C5 (Typ-Code/Stempel/marking)	35	Nec	→2SA811	data
C 5	Si-P	=BCW 30R (Typ-Code/Stempel/marking)	35	Fer,Val	→BCW 30R	data
C 5	Si-P	=BFS 40 (Typ-Code/Stempel/marking)	35(2mm)	Fer	→BFS 40	data
C 5	GaAs-N-FET-d	=CFY 25-17 (Typ-Code/Stempel/marking)	51	Sie	→CFY 25	data
C 5	Si-N/P+R	=FMC 5A (Typ-Code/Stempel/marking)	45	Rhm	→FMC 5A	data
C 5	Si-Di	=HSM 107S (Typ-Code/Stempel/marking)	35	Hit	→HSM 107S	data
C 5	Si-P	=MMBA 811C5 (Typ-Code/Stempel/marking)	35	Sam	→MMBA 811	data
C 5	Z-Di	=MMSZ 5225B (Typ-Code/marking)	71(2,7mm)	Ons	→MMSZ 5225B	data
C 5	MOS-P-FET-e	=Si 2315DS (Typ-Code/Stempel/marking)	35	Six	→Si 2315DS	data
C 5	Z-Di	=UDZ 30B (Typ-Code/Stempel/marking)	71(1,7mm)	Rhm	→UDZ 2.0B...36B	data
C 5	Si-N/P+R	=UMC 5N (Typ-Code/Stempel/marking)	45(2mm)	Rhm	→UMC 1N...5N	data
C05 06	Si-Di	=ERC 05-06 (Typ-Code/Stempel/marking)	31a	Fjd	→ERC 05-...	data
C05 08	Si-Di	=ERC 05-08 (Typ-Code/Stempel/marking)	31a	Fjd	→ERC 05-...	data
C5V1 PH	Z-Di	=BZX 55-C5V1(Typ-Code/Stempel/marking)	31	Phi	→BZX 55-C...	data
C5V1 PH	Z-Di	=BZX 79-C5V1(Typ-Code/Stempel/marking)	31	Phi	→BZX 79-C...	data
C5V1 PH	Z-Di	=BZV 85-C5V1(Typ-Code/Stempel/marking)	31	Phi	→BZV 85-C...	data
C5V6 PH	Z-Di	=BZX 55-C5V6(Typ-Code/Stempel/marking)	31	Phi	→BZX 55-C...	data
C5V6 PH	Z-Di	=BZX 79-C5V6(Typ-Code/Stempel/marking)	31	Phi	→BZX 79-C...	data
C5V6 PH	Z-Di	=BZV 85-C5V6(Typ-Code/Stempel/marking)	31	Phi	→BZV 85-C...	data
C 5 R	GaAs-FET	=CF 930R (Typ-Code/Stempel/marking)	44	Aeg	→CF 930R	data
C 6	Si-P	=2SA811-C6 (Typ-Code/Stempel/marking)	35	Nec	→2SA811	data
C 6	Si-P	=BFS 41 (Typ-Code/Stempel/marking)	35(2mm)	Fer	→BFS 41	data
C 6	GaAs-N-FET-d	=CFY 25-20 (Typ-Code/Stempel/marking)	51	Sie	→CFY 25	data
C 6	Si-N/P+R	=FMC 6A (Typ-Code/Stempel/marking)	45	Rhm	→FMC 6A	data
C 6	Si-Di	=HSM 198S (Typ-Code/Stempel/marking)	35	Hit	→HSM 198S	data
C 6	Si-P	=MMBA 811C6 (Typ-Code/Stempel/marking)	35	Sam	→MMBA 811	data
C06-13	Si-Di	=ERC 06-13 (Typ-Code/Stempel/marking)	31a	Fjd	→ERC 06-...	data

Type	Device	Short Description	Fig.	Manu	Comparision Types	More at
C06-15	Si-Di	=ERC 06-15 (Typ-Code/Stempel/marking)	31a	Fjd	→ERC 06-...	data
C6V2 PH	Z-Di	=BZX 55-C6V2(Typ-Code/Stempel/marking)	31	Phi	→BZX 55-C...	data
C6V2 PH	Z-Di	=BZX 79-C6V2(Typ-Code/Stempel/marking)	31	Phi	→BZX 79-C...	data
C6V2 PH	Z-Di	=BZV 85-C6V2(Typ-Code/Stempel/marking)	31	Phi	→BZV 85-C...	data
C6V8 PH	Z-Di	=BZX 55-C6V8(Typ-Code/Stempel/marking)	31	Phi	→BZX 55-C...	data
C6V8 PH	Z-Di	=BZX 79-C6V8(Typ-Code/Stempel/marking)	31	Phi	→BZX 79-C...	data
C6V8 PH	Z-Di	=BZV 85-C6V8(Typ-Code/Stempel/marking)	31	Phi	→BZV 85-C...	data
C 7	Si-P	=2SA811-C7 (Typ-Code/Stempel/marking)	35	Nec	→2SA811	data
C 7	Si-N	=2SC5231-C7 (Typ-Code/Stempel/marking)	35(1,6mm)	Say	→2SC5231	data
C 7	Si-P	=BCF 29 (Typ-Code/Stempel/marking)	35	Phi	→BCF 29	data
C 7	GaAs-N-FET-d	=CFY 25-23 (Typ-Code/Stempel/marking)	51	Sie	→CFY 25	data
C 7	Si-N/P+R	=FMC 7A (Typ-Code/Stempel/marking)	45	Rhm	→FMC 7A	data
C 7	Si-Di	=HSM 88WA (Typ-Code/Stempel/marking)	35	Hit	→HSM 88WA	data
C 7	Si-P	=MMBA 811C7 (Typ-Code/Stempel/marking)	35	Sam	→MMBA 811	data
C 7 p	Si-P	=BCF 29 (Typ-Code/Stempel/marking)	35	Phi	→BCF 29	data
C7V5 PH	Z-Di	=BZX 55-C7V5(Typ-Code/Stempel/marking)	31	Phi	→BZX 55-C...	data
C7V5 PH	Z-Di	=BZX 79-C7V5(Typ-Code/Stempel/marking)	31	Phi	→BZX 79-C...	data
C7V5 PH	Z-Di	=BZV 85-C7V5(Typ-Code/Stempel/marking)	31	Phi	→BZV 85-C...	data
C 8	Si-P	=2SA811-C8 (Typ-Code/Stempel/marking)	35	Nec	→2SA811	data
C 8	Si-N	=2SC5231-C8 (Typ-Code/Stempel/marking)	35(1,6mm)	Say	→2SC5231	data
C 8	Si-Di	=BAV 21WS (Typ-Code/Stempel/marking)	71(1,7mm)	Gsi	→BAV 21WS	data
C 8	Si-P	=BCF 30 (Typ-Code/Stempel/marking)	35	Phi	→BCF 30	data
C 8	Si-Di	=HSM 198SR (Typ-Code/Stempel/marking)	35	Hit	→HSM 198SR	data
C 8	Si-P	=MMBA 811C8 (Typ-Code/Stempel/marking)	35	Sam	→MMBA 811	data
C 8 p	Si-P	=BCF 30 (Typ-Code/Stempel/marking)	35	Phi	→BCF 30	data
C8V2 PH	Z-Di	=BZX 55-C8V2(Typ-Code/Stempel/marking)	31	Phi	→BZX 55-C...	data
C8V2 PH	Z-Di	=BZX 79-C8V2(Typ-Code/Stempel/marking)	31	Phi	→BZX 79-C...	data
C8V2 PH	Z-Di	=BZV 85-C8V2(Typ-Code/Stempel/marking)	31	Phi	→BZV 85-C...	data
C 9	Si-Di	=1SS307 (Typ-Code/Stempel/marking)	35	Tos	→1SS307	data
C 9	Si-N	=2SC5231-C9 (Typ-Code/Stempel/marking)	35(1,6mm)	Say	→2SC5231	data
C 9	Si-P	=BCF 30R (Typ-Code/Stempel/marking)	35	Val	→BCF 30R	data
C 9	Si-Di	=HSM 276SR (Typ-Code/Stempel/marking)	35	Hit	→HSM 276SR	data
C9V1 PH	Z-Di	=BZX 55-C9V1(Typ-Code/Stempel/marking)	31	Phi	→BZX 55-C...	data
C9V1 PH	Z-Di	=BZX 79-C9V1(Typ-Code/Stempel/marking)	31	Phi	→BZX 79-C...	data
C9V1 PH	Z-Di	=BZV 85-C9V1(Typ-Code/Stempel/marking)	31	Phi	→BZV 85-C...	data
C 10	C-Di	≈BA 101	31a	old	→BA 101	
C10A29	IC	7 stage Divider	DIP		AY-1-1007,C10130,PL4C07C	pinout
C10 PH	Z-Di	=BZX 55-C10 (Typ-Code/Stempel/marking)	31	Phi	→BZX 55-C...	data
C10 PH	Z-Di	=BZX 79-C10 (Typ-Code/Stempel/marking)	31	Phi	→BZX 79-C...	data
C10 PH	Z-Di	=BZV 85-C10 (Typ-Code/Stempel/marking)	31	Phi	→BZV 85-C...	data
C11A1-HX	LED	bl+gr+rd+wht+yl ca. ø4mm l:10lm		Roi		data pdf pinout
C11 PH	Z-Di	=BZX 55-C11 (Typ-Code/Stempel/marking)	31	Phi	→BZX 55-C...	data
C11 PH	Z-Di	=BZX 79-C11 (Typ-Code/Stempel/marking)	31	Phi	→BZX 79-C...	data
C11 PH	Z-Di	=BZV 85-C11 (Typ-Code/Stempel/marking)	31	Phi	→BZV 85-C...	data
C12 06	Si-Di	=ERC 12-06 (Typ-Code/Stempel/marking)	31a	Fjd	→ERC 12-...	data
C12 08	Si-Di	=ERC 12-08 (Typ-Code/Stempel/marking)	31a	Fjd	→ERC 12-...	data
C12 PH	Z-Di	=BZX 55-C12 (Typ-Code/Stempel/marking)	31	Phi	→BZX 55-C...	data
C12 PH	Z-Di	=BZX 79-C12 (Typ-Code/Stempel/marking)	31	Phi	→BZX 79-C...	data
C12 PH	Z-Di	=BZV 85-C12 (Typ-Code/Stempel/marking)	31	Phi	→BZV 85-C...	data
C13 06	Si-Di	=ERC 13-06 (Typ-Code/Stempel/marking)	31a	Fjd	→ERC 13-...	data
C13 08	Si-Di	=ERC 13-08 (Typ-Code/Stempel/marking)	31a	Fjd	→ERC 13-...	data
C13 PH	Z-Di	=BZX 55-C13 (Typ-Code/Stempel/marking)	31	Phi	→BZX 55-C...	data
C13 PH	Z-Di	=BZX 79-C13 (Typ-Code/Stempel/marking)	31	Phi	→BZX 79-C...	data
C13 PH	Z-Di	=BZV 85-C13 (Typ-Code/Stempel/marking)	31	Phi	→BZV 85-C...	data
C 15	C-Di	≈BA 102	31a	old	→BA 102	
C 15	Si-P	=2SA1612-C15(Typ-Code/Stempel/marking)	35(2mm)	Nec	→2SA1612	data
C 15	Si-P	=2SA811A-C15(Typ-Code/Stempel/marking)	35	Nec	→2SA811A	data
C15 PH	Z-Di	=BZX 55-C15 (Typ-Code/Stempel/marking)	31	Phi	→BZX 55-C...	data
C15 PH	Z-Di	=BZX 79-C15 (Typ-Code/Stempel/marking)	31	Phi	→BZX 79-C...	data
C15 PH	Z-Di	=BZV 85-C15 (Typ-Code/Stempel/marking)	31	Phi	→BZV 85-C...	data
C 16	Si-P	=2SA1612-C16(Typ-Code/Stempel/marking)	35(2mm)	Nec	→2SA1612	data
C 16	Si-P	=2SA811A-C16(Typ-Code/Stempel/marking)	35	Nec	→2SA811A	data
C16 PH	Z-Di	=BZX 55-C16 (Typ-Code/Stempel/marking)	31	Phi	→BZX 55-C...	data
C16 PH	Z-Di	=BZX 79-C16 (Typ-Code/Stempel/marking)	31	Phi	→BZX 79-C...	data
C16 PH	Z-Di	=BZV 85-C16 (Typ-Code/Stempel/marking)	31	Phi	→BZV 85-C...	data
C 17	Si-P	=2SA1612-C17(Typ-Code/Stempel/marking)	35(2mm)	Nec	→2SA1612	data
C 17	Si-P	=2SA811A-C17(Typ-Code/Stempel/marking)	35	Nec	→2SA811A	data
C 18	Si-P	=2SA1612-C18(Typ-Code/Stempel/marking)	35(2mm)	Nec	→2SA1612	data
C 18	Si-P	=2SA811A-C18(Typ-Code/Stempel/marking)	35	Nec	→2SA811A	data
C18 PH	Z-Di	=BZX 55-C18 (Typ-Code/Stempel/marking)	31	Phi	→BZX 55-C...	data
C18 PH	Z-Di	=BZX 79-C18 (Typ-Code/Stempel/marking)	31	Phi	→BZX 79-C...	data
C18 PH	Z-Di	=BZV 85-C18 (Typ-Code/Stempel/marking)	31	Phi	→BZV 85-C...	data
C 20	C-Di	≈BA 102	31a	old	→BA 102	
C 20	N-FET	=2SK595-20 (Typ-Code/Stempel/marking)	35	Say	→2SK595	data
C20A1-HX	LED	bl+gr+rd+wht+yl ca. ø4mm l:10lm		Roi		data pdf pinout
C20C3-HXXX	LED	rd/gr/bl+yl/gr/bl+yl/rd/bl ca. ø4mm		Roi		data pdf pinout
C20 PH	Z-Di	=BZX 55-C20 (Typ-Code/Stempel/marking)	31	Phi	→BZX 55-C...	data
C20 PH	Z-Di	=BZX 79-C20 (Typ-Code/Stempel/marking)	31	Phi	→BZX 79-C...	data
C20 PH	Z-Di	=BZV 85-C20 (Typ-Code/Stempel/marking)	31	Phi	→BZV 85-C...	data
C 21	N-FET	=2SK595-21 (Typ-Code/Stempel/marking)	35	Say	→2SK595	data
C 22	N-FET	=2SK595-22 (Typ-Code/Stempel/marking)	35	Say	→2SK595	data
C22 PH	Z-Di	=BZX 55-C22 (Typ-Code/Stempel/marking)	31	Phi	→BZX 55-C...	data
C22 PH	Z-Di	=BZX 79-C22 (Typ-Code/Stempel/marking)	31	Phi	→BZX 79-C...	data

Type	Device	Short Description	Fig.	Manu	Comparision Types	More at
C22 PH	Z-Di	=BZV 85-C22 (Typ-Code/Stempel/marking)	31	Phi	→BZV 85-C...	data
C 23	N-FET	=2SK595-23 (Typ-Code/Stempel/marking)	35	Say	→2SK595	data
C 24	N-FET	=2SK595-24 (Typ-Code/Stempel/marking)	35	Say	→2SK595	data
C24 PH	Z-Di	=BZX 55-C24 (Typ-Code/Stempel/marking)	31	Phi	→BZX 55-C...	data
C24 PH	Z-Di	=BZX 79-C24 (Typ-Code/Stempel/marking)	31	Phi	→BZX 79-C...	data
C24 PH	Z-Di	=BZV 85-C24 (Typ-Code/Stempel/marking)	31	Phi	→BZV 85-C...	data
C 25	Si-P	=2SA1247-C25(Typ-Code/Stempel/marking)	35	Nec	→2SA1247	data
C25-04	Si-Di	=ERC 25-04 (Typ-Code/Stempel/marking)	31a	Fjd	→ERC 25-...	data
C25-06	Si-Di	=ERC 25-06 (Typ-Code/Stempel/marking)	31a	Fjd	→ERC 25-...	data
C 26	Si-P	=2SA1247-C26(Typ-Code/Stempel/marking)	35	Nec	→2SA1247	data
C 27	Si-P	=2SA1247-C27(Typ-Code/Stempel/marking)	35	Nec	→2SA1247	data
C27 PH	Z-Di	=BZX 55-C27 (Typ-Code/Stempel/marking)	31	Phi	→BZX 55-C...	data
C27 PH	Z-Di	=BZX 79-C27 (Typ-Code/Stempel/marking)	31	Phi	→BZX 79-C...	data
C27 PH	Z-Di	=BZV 85-C27 (Typ-Code/Stempel/marking)	31	Phi	→BZV 85-C...	data
C 28	Si-P	=2SA1247-C28(Typ-Code/Stempel/marking)	35	Nec	→2SA1247	data
C30 01	Si-Di	=ERC 30-01 (Typ-Code/Stempel/marking)	31a	Fjd	→ERC 30-...	data
C30 02	Si-Di	=ERC 30-02 (Typ-Code/Stempel/marking)	31a	Fjd	→ERC 30-...	data
C30 PH	Z-Di	=BZX 55-C30 (Typ-Code/Stempel/marking)	31	Phi	→BZX 55-C...	data
C30 PH	Z-Di	=BZX 79-C30 (Typ-Code/Stempel/marking)	31	Phi	→BZX 79-C...	data
C30 PH	Z-Di	=BZV 85-C30 (Typ-Code/Stempel/marking)	31	Phi	→BZV 85-C...	data
C33 PH	Z-Di	=BZX 55-C33 (Typ-Code/Stempel/marking)	31	Phi	→BZX 55-C...	data
C33 PH	Z-Di	=BZX 79-C33 (Typ-Code/Stempel/marking)	31	Phi	→BZX 79-C...	data
C33 PH	Z-Di	=BZV 85-C33 (Typ-Code/Stempel/marking)	31	Phi	→BZV 85-C...	data
C35-02	Si-Di	=ERC 35-02 (Typ-Code/Stempel/marking)	31a	Fjd	→ERC 35-...	data
C36 PH	Z-Di	=BZX 55-C36 (Typ-Code/Stempel/marking)	31	Phi	→BZX 55-C...	data
C36 PH	Z-Di	=BZX 79-C36 (Typ-Code/Stempel/marking)	31	Phi	→BZX 79-C...	data
C36 PH	Z-Di	=BZV 85-C36 (Typ-Code/Stempel/marking)	31	Phi	→BZV 85-C...	data
C38 04	Si-Di	=ERC 38-04 (Typ-Code/Stempel/marking)	31a	Fjd	→ERC 38-...	data
C38 05	Si-Di	=ERC 38-05 (Typ-Code/Stempel/marking)	31a	Fjd	→ERC 38-...	data
C38 06	Si-Di	=ERC 38-06 (Typ-Code/Stempel/marking)	31a	Fjd	→ERC 38-...	data
C38A8-HX	LED	bl+gr+rd+wht+yl ca. ø4mm I:80lm		Roi		data pdf pinout
C39 PH	Z-Di	=BZX 55-C39 (Typ-Code/Stempel/marking)	31	Phi	→BZX 55-C...	data
C39 PH	Z-Di	=BZX 79-C39 (Typ-Code/Stempel/marking)	31	Phi	→BZX 79-C...	data
C39 PH	Z-Di	=BZV 85-C39 (Typ-Code/Stempel/marking)	31	Phi	→BZV 85-C...	data
C43 PH	Z-Di	=BZX 55-C43 (Typ-Code/Stempel/marking)	31	Phi	→BZX 55-C...	data
C43 PH	Z-Di	=BZX 79-C43 (Typ-Code/Stempel/marking)	31	Phi	→BZX 79-C...	data
C43 PH	Z-Di	=BZV 85-C43 (Typ-Code/Stempel/marking)	31	Phi	→BZV 85-C...	data
C47 PH	Z-Di	=BZX 55-C47 (Typ-Code/Stempel/marking)	31	Phi	→BZX 55-C...	data
C47 PH	Z-Di	=BZX 79-C47 (Typ-Code/Stempel/marking)	31	Phi	→BZX 79-C...	data
C47 PH	Z-Di	=BZV 85-C47 (Typ-Code/Stempel/marking)	31	Phi	→BZV 85-C...	data
C51 PH	Z-Di	=BZX 55-C51 (Typ-Code/Stempel/marking)	31	Phi	→BZX 55-C...	data
C51 PH	Z-Di	=BZX 79-C51 (Typ-Code/Stempel/marking)	31	Phi	→BZX 79-C...	data
C51 PH	Z-Di	=BZV 85-C51 (Typ-Code/Stempel/marking)	31	Phi	→BZV 85-C...	data
C56 PH	Z-Di	=BZX 55-C56 (Typ-Code/Stempel/marking)	31	Phi	→BZX 55-C...	data
C56 PH	Z-Di	=BZX 79-C56 (Typ-Code/Stempel/marking)	31	Phi	→BZX 79-C...	data
C56 PH	Z-Di	=BZV 85-C56 (Typ-Code/Stempel/marking)	31	Phi	→BZV 85-C...	data
C62 PH	Z-Di	=BZX 55-C62 (Typ-Code/Stempel/marking)	31	Phi	→BZX 55-C...	data
C62 PH	Z-Di	=BZX 79-C62 (Typ-Code/Stempel/marking)	31	Phi	→BZX 79-C...	data
C62 PH	Z-Di	=BZV 85-C62 (Typ-Code/Stempel/marking)	31	Phi	→BZV 85-C...	data
C68 PH	Z-Di	=BZX 55-C68 (Typ-Code/Stempel/marking)	31	Phi	→BZX 55-C...	data
C68 PH	Z-Di	=BZX 79-C68 (Typ-Code/Stempel/marking)	31	Phi	→BZX 79-C...	data
C68 PH	Z-Di	=BZV 85-C68 (Typ-Code/Stempel/marking)	31	Phi	→BZV 85-C...	data
C 71	Si-P	=BCW 61A (Typ-Code/Stempel/marking)	35	Val	→BCW 61A	data
C 72	Si-P	=BCW 61B (Typ-Code/Stempel/marking)	35	Val	→BCW 61B	data
C 73	Si-P	=BCW 61C (Typ-Code/Stempel/marking)	35	Val	→BCW 61C	data
C 74	Si-P	=BCW 61D (Typ-Code/Stempel/marking)	35	Val	→BCW 61D	data
C75 PH	Z-Di	=BZX 55-C75 (Typ-Code/Stempel/marking)	31	Phi	→BZX 55-C...	data
C75 PH	Z-Di	=BZX 79-C75 (Typ-Code/Stempel/marking)	31	Phi	→BZX 79-C...	data
C75 PH	Z-Di	=BZV 85-C75 (Typ-Code/Stempel/marking)	31	Phi	→BZV 85-C...	data
C 77	Si-P	=BCF 29R (Typ-Code/Stempel/marking)	35	Val	→BCF 29R	data
C 78 L03 BZ,CZ	CMOS-Z-IC	lo-drop, 3V (BZ=5%, CZ=10%), 0,1A	7	Tho	-	
C 78 L03 CD	CMOS-Z-IC	=C78 L03BZ,CZ: SMD	8-MDIP	Tho	-	
C 78 L05 BZ,CZ	CMOS-Z-IC	lo-drop, 5V (BZ=5%, CZ=10%), 0,1A	7	Tho	-	
C 78 L05 CD	CMOS-Z-IC	=C78 L05BZ,CZ: SMD	8-MDIP	Tho	-	
C81-004	Si-Di	=ERC 81-004 (Typ-Code/Stempel/marking)	31a	Fjd	→ERC 81-...	data
C81-006	Si-Di	=ERC 81-006 (Typ-Code/Stempel/marking)	31a	Fjd	→ERC 81-...	data
C84-009	Si-Di	=ERC 84-009 (Typ-Code/Stempel/marking)	31a	Fjd	→ERC 84-...	data
C 90 M	Si-Di	=ERC90M... (Typ-Code/Stempel/marking)	15d	Fui	→ERC 90M...	
C 91	Si-P	=BCV 62 (Typ-Code/Stempel/marking)	44	Val	→BCV 62	data
C91-02	Si-Di	=ERC 91-02 (Typ-Code/Stempel/marking)	31a	Fjd	→ERC 91-...	data
C 92	Si-P	=BCV 62A (Typ-Code/Stempel/marking)	44	Val	→BCV 62A	data
C 93	Si-P	=BCV 62B (Typ-Code/Stempel/marking)	44	Val	→BCV 62B	data
C 94	Si-P	=BCV 62C (Typ-Code/Stempel/marking)	44	Val	→BCV 62C	data
C 95	Si-P	=BCV 64 (Typ-Code/Stempel/marking)	44	Phi	→BCV 64	data
C 96	Si-P	=BCV 64B (Typ-Code/Stempel/marking)	44	Phi	→BCV 64B	data
C 101	Si-N+R	=RC 101S (Typ-Code/Stempel/marking)	40	Say	→RC 101S	data
C 104	Si-N+R	=RC 104S (Typ-Code/Stempel/marking)	40	Say	→RC 104S	data
C 106 A	50Hz-Thy	100V, 2,2A(Tc=45°C), Igt/Ih<0,2/<3mA Mot: Fig. 14h§	13h§	Gen,Mot,Rca	TAG 106A, X 0403B, TAG 108A, C 108A,++	data
C 106 B	50Hz-Thy	=C 106A: 200V	13h§		TAG 106B, X 0403B, TAG 108B, C.108B, ++	data
C 106 C	50Hz-Thy	=C 106A: 300V	13h§		TAG 106C, X 0403D, TAG 108C, C 108C, ++	data
C 106 D	50Hz-Thy	=C 106A: 400V	13h§		TAG 106D, X 0403D, TAG 108D, C 108D, ++	data
C 106 E	50Hz-Thy	=C 106A: 500V	13h§		TAG 106E, X 0403M, TAG 108E, C 108E, ++	data
C 106 F	50Hz-Thy	=C 106A: 50V	13h§		TAG 106F, X 0403B, TAG 108F, C 108F, ++	data
C 106 G	50Hz-Thy	=C 106A: 150V	13h§		TAG 106G, X 0403B, TAG 108G, C 108G, ++	data

Type	Device	Short Description	Fig.	Manu	Comparision Types	More at
C 106 M	50Hz-Thy	=C 106A: 600V	13h§		TAG 106M, X 0403M, TAG 108M, C 108M, ++	data
C 106 Q	50Hz-Thy	=C 106A: 15V	13h§		TAG 106Q, X 0403B, TAG 108Q, C 108Q, ++	data
C 106 Y	50Hz-Thy	=C 106A: 30V	13h§		TAG 106Y, X 0403B, TAG 108Y, C 108Y, ++	data
C 107 A	50Hz-Thy	100V, 2A(Tc=45°C), Igt/Ih<0,5/<3mA	13h§	Gen,Rca	TAG 107A, X 0403B, C 108A, TAG 108A, ++	data
C 107 B	50Hz-Thy	=C 107A: 200V	13h§		TAG 107B, X 0403B, C 108B, TAG 108B, ++	data
C 107 C	50Hz-Thy	=C 107A: 300V	13h§		TAG 107C, X 0403D, C 108C, TAG 108C, ++	data
C 107 D	50Hz-Thy	=C 107A: 400V	13h§		TAG 107D, X 0403D, C 108D, TAG 108D, ++	data
C 107 E	50Hz-Thy	=C 107A: 500V	13h§		TAG 107E, X 0403M, C 108E, TAG 108E, ++	data
C 107 F	50Hz-Thy	=C 107A: 50V	13h§		TAG 107F, X 0403B, C 108F, TAG 108F, ++	data
C 107 G	50Hz-Thy	=C 107A: 150V	13h§		TAG 107G, X 0403B, C 108G, TAG 108G, ++	data
C 107 M	50Hz-Thy	=C 107A: 600V	13h§		TAG 107M, X 0403M, C 108M, TAG 108M, ++	data
C 107 Q	50Hz-Thy	=C 107A: 15V	13h§		TAG 107Q, X 0403B, C 108Q, TAG 108Q, ++	data
C 107 Y	50Hz-Thy	=C 107A: 30V	13h§		TAG 107Y, X 0403B, C 108Y, TAG 108Y, ++	data
C 108 A	50Hz-Thy	100V, 3,3A(Tc=45°C), Igt/Ih<0,2/<3mA	13h§	Gen,Rca	TAG 108A, (TIC 106A, TAG 623-100)[4]	data
C 108 B	50Hz-Thy	=C 108A: 200V	13h§		TAG 108B, (TIC 106B, TAG 623-200)[4]	data
C 108 C	50Hz-Thy	=C 108A: 300V	13h§		TAG 108C, (TIC 106C, TAG 623-300)[4]	data
C 108 D	50Hz-Thy	=C 108A: 400V	13h§		TAG 108D, (TIC 106D, TAG 623-400)[4]	data
C 108 E	50Hz-Thy	=C 108A: 500V	13h§		TAG 108E, (TIC 106E, TAG 623-500)[4]	data
C 108 F	50Hz-Thy	=C 108A: 50V	13h§		TAG 108F, (TIC 106F, TAG 623-100)[4]	data
C 108 G	50Hz-Thy	=C 108A: 150V	13h§		TAG 108G, (TIC 106B, TAG 623-200)[4]	data
C 108 M	50Hz-Thy	=C 108A: 600V	13h§		TAG 108M, (TIC 106M, TAG 623-600)[4]	data
C 108 Q	50Hz-Thy	=C 108A: 15V	13h§		TAG 108Q, (TIC 106Y, TAG 623-100)[4]	data
C 108 Y	50Hz-Thy	=C 108A: 30V	13h§		TAG 108Y, (TIC 106Y, TAG 623-100)[4]	data
C 116 A	50Hz-Thy	100V, 8A, Igt/Ih<25/<30mA	13h§	Gen	(TAG625-100, C122A, TIC116A, CS6-02,++)[4]	data
C 116 B	50Hz-Thy	=C 116A: 200V	13h§		(TAG625-200, C 122B, TIC116B, CS6-02,++)[4]	data
C 116 C	50Hz-Thy	=C 116A: 300V	13h§		(TAG625-300, C 122C, TIC116C, CS6-04,++)[4]	data
C 116 D	50Hz-Thy	=C 116A: 400V	13h§		(TAG625-400, C 122D, TIC116D, CS6-04,++)[4]	data
C 116 E	50Hz-Thy	=C 116A: 500V	13h§		(TAG625-500, C 122E, TIC116E, CS6-06,++)[4]	data
C 116 F	50Hz-Thy	=C 116A: 50V	13h§		(TAG625-100, C 122F, TIC116F, CS6-02,++)[4]	data
C 116 M	50Hz-Thy	=C 116A: 600V	13h§		(TAG625-600, C 122M, TIC116M, CS6-06,++)[4]	data
C118	FET	Vs:±18V Vo:88dB Vo:±10V Vi0:1mV	5Y	B&h,Cpd		data pinout
C 118 A	50Hz-Thy	100V; 8A, Igt/Ih<0,2/<6mA	13h§	Gen	(TAG628-100, S 4060A)[4]	data
C 118 B	50Hz-Thy	=C 118A: 200V	13h§		(TAG 628-200, S 4060B, MCR 72-4)[4]	data
C 118 C	50Hz-Thy	=C 118A: 300V	13h§		(TAG 628-300, S 4060C, MCR 72-5)[4]	data
C 118 D	50Hz-Thy	=C 118A: 400V	13h§		(TAG 628-400, S 4060D, MCR 72-6)[4]	data
C 118 E	50Hz-Thy	=C 118A: 500V	13h§		(TAG 628-500, S 4060E, MCR 72-7)[4]	data
C 118 F	50Hz-Thy	=C 118A: 50V	13h§		(TAG 628-100, S 4060F, MCR 72-2)[4]	data
C 118 M	50Hz-Thy	=C 118A: 600V	13h§		(TAG 628-600, S 4060M, MCR 72-8)[4]	data
C 122 A	50Hz-Thy	100V, 8A(Tc=75°C), Igt/Ih<25/<30mA	17h§	Mot,Rca	TIC116A, BStC1026M CS6-02, TAG660-100++	data
C 122 B	50Hz-Thy	=C 122A: 200V	17h§		TIC116B, BStC1026M, CS6-02, TAG660-200++	data
C 122 C	50Hz-Thy	=C 122A: 300V	17h§		TIC116C, BStC1026M, CS6-04, TAG660-400++	data
C 122 D	50Hz-Thy	=C 122A: 400V	17h§		TIC116D, BStC1026M, CS6-04, TAG660-400++	data
C 122 E	50Hz-Thy	=C 122A: 500V	17h§		TIC116E, BStC1033M, CS6-06, TAG660-500++	data
C 122 F	50Hz-Thy	=C 122A: 50V	17h§		TIC116F, BStC1026M, CS6-02, TAG660-100++	data
C 122 M	50Hz-Thy	=C 122A: 600V	17h§		TIC116M, BStC1040M, CS6-02, TAG660-600++	data
C 203 A	F-Thy	100V, 0,8A(Tc=25°C), Igt/Ih<0,2/<5mA	7e	Gen	BRX 46...49, BRX 51...56, BRY55/...S	data
C 203 B	F-Thy	=C 203A: 200V	7e		BRX 47...49, BRX 52...56, BRY55/...S	data
C 203 C	F-Thy	=C 203A: 300V	7e		BRX 48...49, BRX 53...56, BRY55/...S	data
C 203 D	F-Thy	=C 203A: 400V	7e		BRX 49, BRX 54...56, BRY55/...S	data
C 203 Y	F-Thy	=C 203A: 30V	7e		BRX 44...49, BRX 50...56, BRY55/...S	data
C 203 YY	F-Thy	=C 203A: 60V	7e		BRX 45...49, BRX 51...56, BRY55/...S	data
C 205 A	F-Thy	100V, 1,2A, Igt/Ih<0,2/<5mA	7e	Gen,Mot	(BRX 61...66)[4]	data
C 205 B	F-Thy	=C 205A: 200V	7e		(BRX 62...66)[4]	data
C 205 C	F-Thy	=C 205A: 300V	7e		(BRX 63...66)[4]	data
C 205 D	F-Thy	=C 205A: 400V	7e		(BRX 64...66)[4]	data
C 205 Y	F-Thy	=C 205A: 30V	7e		(BRX 60...66)[4]	data
C 205 YY	F-Thy	=C 205A: 60V	7e		(BRX 61...66)[4]	data
C218	FET	Vs:±18V Vu:88dB Vo:±10V Vi0:1mV	12A	B&h,Cpd		data pinout
C228	FET	Vs:±18V Vu:88dB Vo:±10V Vi0:1mV	12A	B&h,Cpd		data pinout
C 228 A	50Hz-Thy	100V, 35A, Igt/Ih<40/<75mA	21b§	Gen,Mot	2N3896...3899, MCR3935-3, C35A, TAG35-100	data
C 228 B	50Hz-Thy	=C 228A: 200V	21b§		2N3897...3899, MCR3935-4, C35B, TAG35-200	data
C 228 C	50Hz-Thy	=C 228A: 300V	21b§		2N3898...3899, MCR3935-5, C35C, TAG35-300	data
C 228 D	50Hz-Thy	=C 228A: 400V	21b§		2N3898...3899, MCR3935-6, C35D, TAG35-400	data
C 228 E	50Hz-Thy	=C 228A: 500V	21b§		2N3899, MCR3935-7, C35E, TAG35-500	data
C 228 F	50Hz-Thy	=C 228A: 50V	21b§		2N3896...3899, MCR3935-2, C35F, TAG35-100	data
C 228 M	50Hz-Thy	=C 228A: 600V	21b§		2N3899, MCR3935-8, C35M, TAG35-600	data
C 228 U	50Hz-Thy	=C 228A: 25V	21b§		2N3896...3899, MCR3935-1, C35U, TAG35-100	data
C 228 A...U 3	50Hz-Thy	=C228 A...U:	54b			data
C 229 A...U	50Hz-Thy	=C228 A...U:	29b§	Gen,Mot	2N3870...3873, MCR3835..., MCR63...	data
C 230 A	50Hz-Thy	100V, 25A, Igt/Ih<25/<50mA	21b§	Gen,Mot	BStE4126N, BStE4126MN	data
C 230 B	50Hz-Thy	=C 230A: 200V	21b§		BStE4126N, BStE4126MN	data
C 230 C	50Hz-Thy	=C 230A: 300V	21b§		BStE4126N, BStE4126MN	data
C 230 D	50Hz-Thy	=C 230A: 400V	21b§		BStE4126N, BStE4126MN	data
C 230 E	50Hz-Thy	=C 230A: 500V	21b§		BStE4133N, BStE4133MN	data
C 230 F	50Hz-Thy	=C 230A: 50V	21b§		BStE4126N, BStE4126MN	data
C 230 M	50Hz-Thy	=C 230A: 600V	21b§		BStE4140N, BStE4140MN	data
C 230 U	50Hz-Thy	=C 230A: 25V	21b§		BStE4126N, BStE4126MN	data
C 230 A...U 3	50Hz-Thy	=C230 A...U:	54b			data
C 231 A...U	50Hz-Thy	=C230 A...U: Igt/Ih<9/<50mA	21b§	Gen,Mot	-	data
C 231 A...U 3	50Hz-Thy	=C230 A...U: Igt/Ih<9/<50mA	54b			data
C 232 A...U	50Hz-Thy	=C230 A...U:	29b§	Gen,Mot	BStE40...N, BStE40...MN	data
C 233 A...U	50Hz-Thy	=C231 A...U:	29b§	Gen,Mot		data

Type	Device	Short Description	Fig.	Manu	Comparision Types	More at
C238	FET	Vs:±15V Vo:±10V Vi0:1mV f:10Mc	12A	Cpd		data pinout
C238C	FET	Vs:±15V Vo:±10V Vi0:2mV f:10Mc	12A	Cpd		data pinout
C 310	IC	Wideband Crt Distortion Correct.device	DIC	Itr	-	pdf pinout
C 311	IC	Wideband Crt Distortion Correct.device	DIC	Itr	-	pdf pinout
C 312	IC	Wideband Crt Distortion Correct.device	DIC	Itr	-	pdf pinout
C 410	IC	Wideband Crt Distortion Correct.device	DIC	Itr	-	pdf pinout
C 411	IC	Wideband Crt Distortion Correct.device	DIC	Itr	-	pdf pinout
C438	HI-SPEED	Vs:±22V Vu:100dB Vo:±10V Vi0:1mV	14D	B&h,Cpd		data pinout
C438C	HI-SPEED	Vs:±22V Vu:100dB Vo:±10V Vi0:2mV	14D	B&h,Cpd		data pinout
C 500 D	LIN-IC	Analogprozessor f. A/D-Conv., 14 Bit	18-DIP	Hfo	TL 500C	
C 501 D	LIN-IC	Analogprozessor f. A/D-Conv., 11 Bit	18-DIP	Hfo	TL 501C	
C 502 D	LIN-IC	Digitalprozessor, 4,5-Digit-Displ.-Tr.	20-DIP	Hfo	TL 502C	
C 504 D	LIN-IC	Digitalprozessor, 3,5/4,5-Digit-Displ.	28-DIP	Hfo		
C 515	CMOS-IC	8-Bit Cmos Microcontroller	MP	Sie	-	pdf pinout
C 515-LN	CMOS-IC	8-Bit Cmos Microcontroller	MP	Sie	-	pdf pinout
C 520 D	A/D-IC	3-Digit A/D-Converter, Dual-Slope	16-DIP	Hfo	AD 2020	
C 560 D	D/A-IC	8 Bit D/a-converter	16-DIC	Hfo	(AD 558 JD)	
C 565 D	D/A-IC	12 Bit D/A-Converter, ~C 5650D	24-DIC	Hfo	AD 565 JN, µA 565...	
C 570 D	A/D-IC	8 Bit A/D-Converter	18-DIC	Hfo	AD 570	
C 571 D	A/D-IC	10 Bit A/D-Converter	18-DIC	Hfo	AD 571, µA 571...	
C 574 C	A/D-IC	12 Bit A/D-Converter, µP-Interface	28-DIP	Hfo		
C 670 C,CGn	A/D-IC	8 Bit A/D-Converter	18-DIC	Hfo	-	
C 1398	Si-N	≈BC 237			→BC 237	
C 1406(G)	Si-Br	GI-Br, 100V, 3,2..4,8A	33x		B40C3700/2200	data
(SSi) C 1710	Si-Di	→SSi C 1710	27c	Sie		data
(SSi) C 1720	Si-Di	→SSi C 1720	27c	Sie		data
(SSi) C 1740	Si-Di	→SSi C 1740	27c	Sie		data
(SSi) C 1760	Si-Di	→SSi C 1760	27c	Sie		data
(SSi) C 1780	Si-Di	→SSi C 1780	27c	Sie		data
(SSi) C 2620	Si-Di	→SSi C 2620	31a	Sie		
(SSi) C 2630	Si-Di	→SSi C 2630	31a	Sie		
C 4160	Si-N	≈BC 237	7e		→BC 237	
(SSi) C 4610	Si-Di	→SSi C 4610		Sie		
(SSi) C 4620	Si-Di	→SSi C 4620		Sie		
C 5122 M	IC	Cb Transceiver PLL IC	DIP	Npc	-	pdf pinout
C 5122 P	IC	Cb Transceiver PLL IC	DIP	Npc	-	pdf pinout
C 515-1RN	CMOS-IC	8-Bit Cmos Microcontroller	MP	Sie	-	pdf pinout
C 5650 D	D/A-IC	10 Bit D/A-Converter, ~C 565D	24-DIC	Hfo		
C 5658 D	D/A-IC	8 Bit D/a-converter	24-DIP	Hfo		
C 7136 D	A/D-IC	3,5-Digit A/D-Converter	40-DIP	Hfo	ICL 7136	
C9960	LDR	610nm Ro:0.27MOhm Vb:2V		Hei		data pinout
C10130	IC	7-stage Divider	DIP		AY-1-1007,C10A29,PL4C07C	pinout
82 C 205	DIG-IC	LCD Monitor Controller	MP	Opi	-	pdf pinout
82 C 861	DIG-IC	PCI-to-USB Bridge	MP	Opi	-	pdf pinout
82 C 862	DIG-IC	Dual Controller PCI-USB Host Bridge	MP	Opi	-	pdf pinout
82 C 863	DIG-IC	Dual Controller Pci-usb Host Bridge	MP	Opi	-	pdf pinout

CA

Type	Device	Short Description	Fig.	Manu	Comparision Types	More at
CA	Si-P	=2SA1866 (Typ-Code/Stempel/marking)	35(1,6mm)	Say	→2SA1866	data
CA	Si-P	=2SA1885 (Typ-Code/Stempel/marking)	35(1,6mm)	Rhm	→2SA1885	data
CA	Si-P	=2SA1886 (Typ-Code/Stempel/marking)	35(2mm)	Rhm	→2SA1886	data
CA	Si-N	=2SC3645 (Typ-Code/Stempel/marking)	39	Say	→2SC3645	data
CA	Si-N	=2SD1368-CA (Typ-Code/Stempel/marking)	39	Hit	→2SD1368	data
CA	C-Di	=BB 510 (Typ-Code/Stempel/marking)	35	Itt	→BB 510	
CA	Si-N	=BCP 68 (Typ-Code/Stempel/marking)	48	Phi, Ons	→BCP 68	data
CA	Si-P	=BCW 61RA (Typ-Code/Stempel/marking)	35	Fer	→BCW 61RA	data
CA	Si-N	=BCX 68 (Typ-Code/Stempel/marking)	39	Sie,Val,Mot	→BCX 68	data
CA	Si-N	=BFS 18 (Typ-Code/Stempel/marking)	35	Sie	→BFS 18	data
CA	Si-N	=CPH 3201 (Typ-Code/Stempel/marking)	35	Say	→CPH 3201	data
CA	Si-P	=KTX 301U-Y (Typ-Code/Stempel/marking)	45(2mm)	Kec	→KTX 301U	data
CA	Z-Di	=MMSZ 4691 (Typ-Code/marking)	71(2,7mm)	Ons	→MMSZ 4691	
CA	Si-N+R	=RN 1444A (Typ-Code/Stempel/marking)	35	Tos	→RN 1444	
CA	N-FET	=SST 4391 (Typ-Code/Stempel/marking)	35	Six	→SST 4391	data
CA	Si-N/P+R	=XN 4314 (Typ-Code/Stempel/marking)	46	Mat	→XN 4314	data
CA	Si-N/P+R	=XP 4314 (Typ-Code/Stempel/marking)	46(2mm)	Mat	→XP 4314	data
CA	C-Di	=ZMV 829B (Typ-Code/Stempel/marking)	71(1,7mm)	Ztx	→ZMV 829	data
CA	MOS-P-FET-e	=µPA503T (Typ-Code/Stempel/marking)	45	Nec	→µPA503T	data
CA15A8	2xOP	Vs:±16V Vu:100dB Vi0:1mV	8A	Rca		data pinout
CA080	BIMOS	Vs:±18V Vu:106dB Vo:±13.5V Vi0:2mV	8ADd	Rca		data pinout
CA080A	BIMOS	Vs:±18V Vu:106dB Vo:±13.5V Vi0:3mV	8D	Rca		data pinout
CA 080(A,B)E	BiMOS-OP-IC	=CA 080...S,T: Fig.→	8-DIP	Rca	→Serie 080	data pinout
CA 080(A,B,C)S,T	BiMOS-OP-IC	MOS Inp./Outp., ±18V	81	Rca	→Serie 080	data pinout
CA081	BIMOS	Vs:±18V Vu:106dB Vo:±13.5V Vi0:3mV	8DAd	Rca		data pinout
CA 081(A,B)E	BiMOS-OP-IC	=CA 081...S,T: Fig.→	8-DIP	Rca	→Serie 080	data pinout
CA 081(A,B,C)S,T	BiMOS-OP-IC	MOS Inp./Outp., ±18V	81	Rca	→Serie 080	data pinout
CA082	2xBIMOS	Vs:±18V Vu:106dB Vo:±13.5V Vi0:2mV	8ADd	Rca		data pinout
CA082A	2xBIMOS	Vs:±18V Vu:106dB Vo:±13.5V Vi0:3mV	8D	Rca		data pinout
CA 082(A,B)E	BiMOS-OP-IC	=CA 082...S,T: Fig.→	8-DIP	Rca	→Serie 080	data pinout
CA 082(A,B,C)S,T	BiMOS-OP-IC	Dual, MOS Inp./Outp., ±18V	81	Rca	→Serie 080	data pinout
CA083	2xBIMOS	Vs:±18V Vu:106dB Vo:±13.5V Vi0:3mV	14Dd	Rca		data pinout
CA083B	2xBIMOS	Vs:±18V Vu:106dB Vo:±13.5V Vi0:2mV	14D	Rca		data pinout

Type	Device	Short Description	Fig.	Manu	Comparision Types	More at
CA 083(A,B)E	BiMOS-OP-IC	Dual, MOS Inp./Outp., ±18V	14-DIP	Rca	→Serie 080	data pinout
CA084	4xBIMOS	Vs:18V Vu:106dB Vo:±13.5V Vi0:5mV	14Dd	Rca		data pinout
CA084A	4xBIMOS	Vs:±18V Vu:106dB Vo:±13.5V Vi0:3mV	14D	Rca		data pinout
CA084B	4xBIMOS	Vs:±18V Vu:106dB Vo:±13.5V Vi0:2mV	14D	Rca		data pinout
CA 084(A,B)E	BiMOS-OP-IC	Quad, MOS Inp./Outp., ±18V	14-DIP	Rca	→Serie 080	data pinout
CA084T	BIMOS	Vs:±18V Vu:200V/mV Vo:24/0V f:5Mc	8A	Rca		data pinout
CA101	UNI	Vs:±22V Vu:104dB Vo:±14V Vi0:1mV	8DA	Rca		data pinout
CA101A	UNI	Vs:±22V Vu:104dB Vo:±14V Vi0:0.7mV	8DA	Rca		data pinout
CA 101 (A)E,G	OP-IC	Uni, Serie 101, ±22V, -55…+125°	8-DIP,DIC	Rca	→Serie 101	data pinout
CA 101 (A)S,T	OP-IC	=CA 101E,G: Fig.→	81	Rca	→Serie 101	data pinout
CA107	UNI	Vs:±22V Vu:106dB Vo:±14V Vi0:0.6mV	8DA	Rca		data pinout
CA 107 E,G	OP-IC	Uni, Serie 107, ±22V, -55…+125°	8-DIP,DIC	Rca	→Serie 107	data pinout
CA 107 S,T	OP-IC	=CA 107…: Fig.→	81	Rca	→Serie 107	data pinout
CA108	LO-POWER	Vs:±20V Vu:110dB Vo:±14V Vi0:0.7mV	8A	Rca		data pinout
CA108A	LO-POWER	Vs:±20V Vu:110dB Vo:±14V Vi0:0.3mV	8A	Rca		data pinout
CA 108(A)S,T	OP-IC	Uni, Serie 108, ±20V, -55…+125°	81	Rca	→Serie 108	data pinout
CA111	COMP	Vs:±18V Vu:106dB Vi0:0.7mV	8DA	Rca		data pinout
CA 111 E,G	KOP-IC	Uni, Serie 111, ±18V, -55…+125°	8-DIP,DIC	Rca	→Serie 111	data pinout
CA 111 S,T	KOP-IC	=CA 111E,G: Fig.→	81	Rca	→Serie 111	data pinout
CA124	4xOP	Vs:±16V Vu:100dB Vi0:2mV	14DdS	Rca		data pinout
CA 124 E,G	OP-IC	Quad, Serie 124, ±16V, -55…+125°	14-DIP,DIC	Rca	→Serie 124	data pdf pinout
CA139	4xCOMP	Vs:36V Vu:200V/mV Vo:TTL Vi0:2mV	14DdS	Har		data pdf pinout
CA139A	4xCOMP	Vs:36V Vi0:1mV	14DdS	Har		data pinout
CA 139(A)E,G	KOP-IC	Quad, Serie 139, ±18V, -55…+125°	14-DIP,DIC	Rca	→Serie 139	data pdf pinout
CA158	2xOP	Vs:±16V Vu:100dB Vi0:2mV	8DSA	Rca		data pdf pinout
CA158A	2xOP	Vs:±16V Vu:100dB Vi0:1mV	8DSA	Rca		data pinout
CA 158(A)E,G	OP-IC	Dual, Serie 158, ±16V, -55…+125°	8-DIP,DIC	Rca	→Serie 158	data pdf pinout
CA 158(A)S,T	OP-IC	=CA 158E,G: -25…+85°	81	Rca	→Serie 158	data pdf pinout
CA201	UNI	Vs:±22V Vu:104dB Vo:±14V Vi0:1mV	8DA	Rca		data pinout
CA201A	UNI	Vs:±22V Vu:104dB Vo:±14V Vi0:0.7mV	8DA	Rca		data pinout
CA 201 (A)E,G	OP-IC	=CA 101…: -25…+85°	8-DIP,DIC	Rca	→Serie 101	data pinout
CA 201 (A)S,T	OP-IC	=CA 101…: -25…+85°	81	Rca	→Serie 101	data pinout
CA207	UNI	Vs:±22V Vu:106dB Vo:±14V Vi0:0.6mV	8DA	Rca		data pinout
CA 207 E,G	OP-IC	=CA 107…: -25…+85°	8-DIP,DIC	Rca	→Serie 107	data pinout
CA 207 S,T	OP-IC	=CA 107…: -25…+85°	81	Rca	→Serie 107	data pinout
CA208	LO-POWER	Vs:±20V Vu:110dB Vo:±14V Vi0:0.7mV	8A	Rca		data pinout
CA208A	LO-POWER	Vs:±20V Vu:110dB Vo:±14V Vi0:0.3mV	8A	Rca		data pinout
CA 208 (A)S,T	OP-IC	=CA 108…: -25…+85°	81	Rca	→Serie 208	data pinout
CA211	COMP	Vs:±18V Vu:106dB Vi0:0.7mV	8DA	Rca		data pinout
CA 211 E,G	KOP-IC	=CA 111…: -25…+85°	8-DIP,DIC	Rca	→Serie 111	data pinout
CA 211 S,T	KOP-IC	=CA 111…: -25…+85°	81	Rca	→Serie 111	data pinout
CA224	4xOP	Vs:±16V Vu:100dB Vi0:2mV	14DdS	Rca		data pinout
CA 224 E,G	OP-IC	=CA 124…: -25…+85°	14-DIP,DIC	Rca	→Serie 124	data pdf pinout
CA239	4xCOMP	Vs:36V Vu:200V/mV Vo:TTL Vi0:2mV	14DdS	Har		data pdf pinout
CA239A	4xCOMP	Vs:36V Vu:200V/mV Vo:TTL Vi0:1mV	14DdS	Har		data pinout
CA 239(A)E,G	KOP-IC	=CA 139…: -25…+85°	14-DIP	Rca	→Serie 139	data pdf pinout
CA258	2xOP	Vs:±16V Vu:100dB Vi0:2mV	8DSA	Rca		data pdf pinout
CA258A	2xOP	Vs:±16V Vu:100dB Vi0:1mV f:1Mc	8DSA	Rca		data pinout
CA 258 E,G	OP-IC	=CA 158…: -25…+85°	8-DIP,DIC	Rca	→Serie 158	data pdf pinout
CA 258 S,T	OP-IC	=CA 158…: -25…+85°	81	Rca	→Serie 158	data pdf pinout
CA 270	LIN-IC	TV, Synchron-dem., Video-Verst./Amp.	16-QIP	Rca	TCA 270 S	data pdf pinout
CA300	COMP	Vs:±6V Vi0:5mV	12Ad	Rca		data pinout
CA301A	UNI	Vs:±22V Vu:104dB Vo:±14V Vi0:2mV	8DdA	Rca		data pinout
CA 301 AS,AT	OP-IC	=CA 101…: 0…+70°	81	Rca	→Serie 101	data pinout
CA 301 (A)E,G	OP-IC	=CA 101…: 0…+70°	8-DIP,DIC	Rca	→Serie 101	data pinout
CA307	UNI	Vs:±22V Vu:104dB Vo:±13V Vi0:2mV	8DdA	Rca		data pinout
CA 307 E,G	OP-IC	=CA 107…: 0…+70°	8-DIP,DIC	Rca	→Serie 107	data pinout
CA 307 S,T	OP-IC	=CA 107…: 0…+70°	81	Rca	→Serie 107	data pinout
CA308	LO-POWER	Vs:±20V Vu:110dB Vo:±14V Vi0:2mV	8DdA	Rca		data pinout
CA308	ER+	Io=12mA Vo:1.8…26V Vin:30V P:630mW	8DA	Rca		data pinout
CA308A	LO-POWER	Vs:±20V Vu:110dB Vo:±14V Vi0:0.3mV	8A	Rca		data pinout
CA 308 E	OP-IC	=CA 108…: 0…+70°	8-DIP	Rca	→Serie 108	data pinout
CA 308(A)S,T	OP-IC	=CA 108…: 0…+70°	81	Rca	→Serie 108	data pinout
CA309A4	PROG-POWER	Vs:±18V Vu:100dB Vi0:0.4mV f:30Mc	8D	Har		data pdf pinout
CA311	COMP	Vs:±18V Vu:106dB Vi0:2mV	8DdA	Rca		data pinout
CA 311 E,G	KOP-IC	=CA 111…: 0…+70°	8-DIP,DIC	Rca	→Serie 111	data pinout
CA 311 S,T	KOP-IC	=CA 111…: 0…+70°	81	Rca	→Serie 111	data pinout
CA324	4xOP	Vs:±16V Vu:100dB Vi0:2mV	14DdS	Rca		data pinout
CA 324 E,G	OP-IC	=CA 124…: 0…+70°	14-DIP,DIC	Rca	→Serie 124	data pdf pinout
CA339	4xCOMP	Vs:36V Vu:200V/mV Vo:TTL Vi0:2mV	14DdS	Har		data pdf pinout
CA339A	4xCOMP	Vs:36V Vu:200V/mV Vo:TTL Vi0:1mV	14DdS	Har		data pinout
CA 339(A)E,G	KOP-IC	=CA 139…: 0…+70°	14-DIP	Rca	→Serie 139	data pdf pinout
CA358	2xOP	Vs:±16V Vu:100dB Vo:±13.5V Vi0:2mV	8DdSA	Rca		data pdf pinout
CA358A	2xOP	Vs:±16V Vu:100dB Vo:±13.5V Vi0:2mV	8DSA	Rca		data pinout
CA 358 E,G	OP-IC	=CA 158…: 0…+70°	8-DIP,DIC	Rca	→Serie 158	data pdf pinout
CA 358 S,T	OP-IC	=CA 158…: 0…+70°	81	Rca	→Serie 158	data pdf pinout
CA526A0	2xBIMOS	Vs:±8V Vu:110dB Vo:13.3/0V Vi0:2mV	8S	Har		data pdf pinout
CA 555(C)E	LIN-IC	Timer, Ucc=4,5…18V, Iout=0,2A	8-DIP	Rca	→NE 555N	pinout
CA 555 S,T	LIN-IC	=CA 555(C)E: Fig.→	81	Rca	→NE 555N	pinout
CA580	HI-GAIN	Vs:±15V Vu:174dB Vo:±10V Vi0:0.02mV	8A	Itr		data pinout
CA723	ER+	Io=150mA Vo:2…37V P:800mW	14D,10A	Rca		data pdf pinout
CA 723(C)E	Z-IC	+2…37V, 0,15A, -55…+125°	14-DIP	Rca	…723…	data pinout
CA 723(C)T	Z-IC	=CA 723(C)E: Fig.→	82	Rca	…723…	data pinout
CA741	UNI	Vs:44V Vu:200kV/mV Vo:±13V Vi0:1mV	8D	Har		data pinout
CA 741 E,F,G,M	OP-IC	Uni, Serie 741, ±22V, -55…+125°	8-DIP,DIC	Rca	→Serie 741	data pinout

Type	Device	Short Description	Fig.	Manu	Comparision Types	More at
CA 741 S,T	OP-IC	=CA 741E,F,G,#: Fig.→	81	Rca	→Serie 741	data pinout
CA 741 CE,CF,CG,CM	OP-IC	=CA 741E,F,G,M: ±18V, 0...+70°	8-DIP,DIC	Rca	→Serie 741	data pinout
CA 741 CS,CT	OP-IC	=CA 741E,F,G,M: ±18V, 0...+70°	81	Rca	→Serie 741	data pinout
CA741D	UNI	Vs:±18V Vu:106dB Vo:±13V Vi0:2mV	d	Rca		data pinout
CA741L	UNI	Vs:±22V Vu:106dB Vo:±14V Vi0:1mV	d	Rca		data pinout
CA741S	UNI	Vs:±18V Vu:106dB Vo:±13V Vi0:2mV	8A	Rca		data pinout
CA741T	UNI	Vs:36V Vu:200kV/mV Vo:±13V Vi0:2mV	8A	Har		data pinout
CA747	2xOP	Vs:±22V Vu:106dB Vo:±13V Vi0:2mV	14Dd,10A	Rca		data pinout
CA 747 E,F,G	OP-IC	Dual, Serie 747, ±22V, -55...+125°	14-DIP,DIC	Rca		data pinout
CA 747 T	OP-IC	=CA 747E,F,G: Fig.→	82	Rca	→Serie 747	data pinout
CA747C	2xOP	Vs:±18V Vu:106dB Vo:±13V Vi0:2mV	14Dd,10A	Rca		data pinout
CA 747 CE,CF,CG	OP-IC	=CA 747E,F,G: ±18V, 0...+70°	14-DIP,DIC	Rca	→Serie 747	data pinout
CA 747 CT	OP-IC	=CA 747E,F,G: ±18V, 0...+70°	82	Rca	→Serie 747	data pinout
CA 748	UNI	Vs:±22V Vu:106dB Vo:±13V Vi0:1mV	8DA	Rca		data pinout
CA 748 E,F,G	OP-IC	Uni, Serie 748, ±22V, -55...+125°	8-DIP,DIC	Rca	→Serie 748	data pinout
CA 748 S,T	OP-IC	=CA 748E,F,G: Fig.→	81	Rca	→Serie 748	data pinout
CA748C	UNI	Vs:±18V Vu:106dB Vo:±13V Vi0:2mV	8DdA	Rca		data pinout
CA 748 CE,CF,CG	OP-IC	=CA 748E,F,G: ±18V, 0...+70°	8-DIP,DIC	Rca	→Serie 748	data pinout
CA 748 CT	OP-IC	=CA 748E,F,G: ±18V, 0...+70°	81	Rca	→Serie 748	data pinout
CA748M	UNI	Vs:±22V Vu:106dB Vo:±14V Vi0:1mV	8D	Rca		data pinout
CA 758(E)	LIN-IC	PLL FM MPX Stereo-Decoder, Ucc=10...16V	16-DIP	Rca	LM 1800, MC 1311, ULX 2244, µA 758 C	pdf pinout
CA 920(A)E	LIN-IC	TV, HA-Synchr., Osc.	16-DIP	Rca	TBA 920 S	
CA 1190(Q)	LIN-IC	TV, Ton-ZF, Nf-e	12-QIP+b	Rca	TDA 1190, TDA 3190	
CA 1191(E)	LIN-IC	TV, Ton-ZF, Nf-e	16-DIP	Rca	TDA 1190, TDA 3190	
CA 1310(A)E	LIN-IC	PLL FM MPX Stereo-Decoder, Ucc=8...16V	14-DIP	Rca	A 290D, LM 1310, MC 1310, SN 76115	
CA 1352	LIN-IC	TV, Video-ZF, AGC(neg.)	14-DIP	Rca	MC 1352	
CA 1391 E	LIN-IC	TV, Hor. Synchr.-proc., pos. Signal	8-DIP	Rca	MC 1391	pdf pinout
CA 1394 E	LIN-IC	TV, Hor. Synchr.-proc., neg. Signal Input	8-DIP	Rca	MC 1394	pdf pinout
CA 1398 E	LIN-IC	Ctv, Chroma-prozessor	14-DIP	Rca		pdf
CA1458	2xOP	Vs:36V Vu:106dB Vo:±13V Vi0:2mV	8DdA	Har		data pdf pinout
CA 1458 E,G,M	OP-IC	Dual, Serie 158, ±18V, 0...+70°	8-DIP,DIC	Rca	→Serie 158	data pinout
CA 1458 S,T	OP-IC	=CA 1458E,G,M: Fig.→	81	Rca	→Serie 158	data pinout
CA 1524(E)	SMPS-IC	SMPS/PWM Controller, 100mA, -25...+125°	16-DIP	Rca	→SG 1524	pdf pinout
CA1558	2xOP	Vs:44V Vu:200kV/mV Vo:±13V Vi0:1mV	8DdA	Har		data pdf pinout
CA 1558 E,G,M	OP-IC	Dual, Serie 158, ±22V, -55...+125°	8-DIP,DIC	Rca	→Serie 158	data pinout
CA 1558 S,T	OP-IC	=CA 1558E,G,M: Fig.→	81	Rca	→Serie 158	data pinout
CA 1724	Si-N	4x NPN-Trans., 40V, 1A, 38/185ns	14-DIP	Rca	DH 3724, FPQ 3724, MPQ 3724, SP 3724	
CA 1725	Si-N	=CA 1724: 50V	14-DIP	Rca	DH 3725, FPQ 3725, MPQ 3725, SP 3725	
CA 2002	LIN-IC	NF-E, 28V, 3,5A, 8W(14V/2Ω)	86/5Pin	Rca	TDA 2002 V	
CA 2002 M	LIN-IC	=CA 2002	86/5Pin	Rca	TDA 2002 H	
CA 2004	LIN-IC	NF-E, 28V, 3,5A, 12W(24V/4Ω)	86/5Pin	Rca		
CA 2004 M	LIN-IC	=CA 2004	86/5Pin	Rca		
CA 2111(A)E	LIN-IC	Tv-ton-zf	14-DIP	Rca	LM 2111, MC 1357(P), ULN 2111(A)	pdf
CA 2111(A)Q	LIN-IC	=CA 2111(A)E:	14-QIP	Rca	LM 2111, MC 1357(P), ULN 2111(A)	pdf pinout
CA 2136 A	LIN-IC	FM/TV-Ton-ZF	14-DIP	Rca	LM 1841, ULN 2136A	
CA 2524(E)	SMPS-IC	=CA 1524: -40...+85°	16-DIP	Rca	→SG 2524	pdf pinout
CA 2902 ...	OP-IC	=LM 2902...		Rca	... 124.., ... 224..., ... 2902...	
CA2904	2xOP	Vs:±13V Vu:100dB Vo:±3.5V Vi0:2mV	8DS	Rca		data pdf pinout
CA 2904 E,G	OP-IC	Dual, Serie 158, ±13V, -55...+125°	8-DIP	Rca	→Serie 158	data pdf pinout
CA3000	COMP	Vs:±6V Vu:28dB Vi0:8mV	10Ad	Rca		data pinout
CA 3000	LIN-IC	Diff.-Verst./DC Amp., ±6V, -55...+125°	82	Rca	-	data pdf
CA3001	COMP	Vs:±6V Vu:28dB Vi0:1.5mV	12Ad	Rca		data pinout
CA 3001	LIN-IC	Video-,Breitb./wideb./verst./amplifier	83	Rca	HA 1110	data
CA 3002	LIN-IC	ZF-Verst./IF Amp., ±8V, -55...+125°	82	Rca	-	
CA 3004	LIN-IC	HF-Verst./RF Amp., ±6V, -55...+125°	83	Rca	-	data
CA 3005	LIN-IC	HF-Verst./RF Amp., ±6V, -55...+125°	83	Rca	-	data
CA3006	COMP	Vs:±6V Vi0:1mV	12A	Rca		data pinout
CA 3006	LIN-IC	HF-Verst./RF Amp., ±6V, -55...+125°	83	Rca	-	data
CA3007	COMP	Vs:±6V Vi0:5mV	12A	Rca		data pinout
CA 3007	LIN-IC	NF-Verst./AF Amp., ±6V, -55...+125°	83	Rca	-	data
CA3008	UNI	Vs:±10V Vu:60dB Vo:±6V Vi0:1.08mV	14F	Rca		data pinout
CA 3008(A)	OP-IC	=CA 3029(A): -55...+125°	14-FLP	Rca		data pinout
CA3008A	UNI	Vs:±10V Vu:60dB Vo:±6V Vi0:0.9mV	14F	Rca		data pinout
CA3010	UNI	Vs:±10V Vu:60dB Vo:±6V Vi0:1.08mV	12A	Rca		data pinout
CA 3010(A)	CP-IC	=CA 3029(A): -55...+125°	83	Rca		data pdf pinout
CA3010A	UNI	Vs:±10V Vu:60dB Vo:±6V Vi0:0.9mV	12A	Rca		data pdf pinout
CA 3011	LIN-IC	TV/CATV, Breitb./Wideb. Amp., ...20MHz	82	Rca	LM 3011	pdf
CA 3012	LIN-IC	TV/CATV, Breitb./Wideb. Amp., ...20MHz	82	Rca	-	pdf
CA 3013	LIN-IC	TV/CATV, Breitb./Wideb. Amp., ...20MHz	82	Rca	-	pdf
CA 3014	LIN-IC	TV/CATV, Breitb./Wideb. Amp., ...20MHz	82	Rca	-	pdf
CA3015	UNI	Vs:±20V Vu:70dB Vo:±7V Vi0:1.37mV	d,12A	Rca		data pinout
CA 3015(A)	OP-IC	=CA 3030(A): -55...+125°	83	Rca		data pdf pinout
CA3016	UNI	Vs:±20V Vu:70dB Vo:±7V Vi0:1.37mV	14F	Rca		data pinout
CA 3016(A)	OP-IC	=CA 3030(A): -55...+125°	14-FLP	Rca		data pinout
CA3016A	UNI	Vs:±20V Vu:70dB Vo:±7V Vi0:1mV	12A,14F	Rca		data pdf pinout
CA 3018	Si-N	2xNPN+1xNPN-Darl.-Trans., 20V, 0,05A	83	Rca		
CA 3018 A	Si-N	=CA 3018: 30V	83	Rca		
CA 3019	LIN-IC	6-Di-Array(2xDi + 1xDi-Br), S, 25mA	82	Rca		pdf
CA3020	WIDEBAND	Vs:9V f:8Mc	12A	Har		data pinout
CA 3020	LIN-IC	Breitb./Wideb. Amp., PQ=0,5W(9V)	83	Rca		pdf
CA3020A	WIDEBAND	Vs:12V Vu:75dB f:8Mc	12A	Har		data pinout
CA 3020 A	LIN-IC	=CA 3020: PQ=1W(12V)	83	Rca		pdf
CA 3021	OP-IC	Breitb./Wideb., 2,4MHz,+18V,-55...+125°	83	Rca	CA 3022, CA 2023	
CA 3022	OP-IC	Breitb./Wideb., 7,5MHz,+18V,-55...+125°	83	Rca	CA 3023	

267

Type	Device	Short Description	Fig.	Manu	Comparision Types	More at
CA 3023	OP-IC	Breitb./Wideb., 16MHz, +18V,-55...+125°	83	Rca	-	
CA 3026	LIN-IC	=CA 3054: Fig.→	83	Rca		pdf
CA 3028 A	LIN-IC	Diff. Amplifier, ...120MHz, -55...125°C	81	Rca		pdf pinout
CA 3028 AE	LIN-IC	Diff. Amplifier, ...120MHz, -55...125°C	8-DIP	Rca		pdf pinout
CA 3028 AF	LIN-IC	Diff. Amplifier, ...120MHz, -55...125°C	8-DIC	Rca		pdf
CA 3028 AM	LIN-IC	SMD, Diff. Amp., ...120MHz, -55...125°C	8-MDIP	Rca		pdf pinout
CA 3028 B	LIN-IC	Diff. Amplifier, ...120MHz, -55...125°C	81	Rca		pdf pinout
CA 3028 BE	LIN-IC	Diff. Amplifier, ...120MHz, -55...125°C	8-DIP	Rca		pdf pinout
CA 3028 BF	LIN-IC	Diff. Amplifier, ...120MHz, -55...125°C	8-DIC	Rca		pdf
CA 3028 BM	LIN-IC	SMD, Diff. Amp., ...120MHz, -55...125°C	8-MDIP	Rca		pdf pinout
CA 3028 S,T	LIN-IC	=CA 3028AF,BF: Fig.→	81	Rca	-	
CA3029	UNI	Vs:±10V Vu:60dB Vo:±6V Vi0:0.9mV	14D	Rca		data pdf pinout
CA 3029(A)	OP-IC	Uni, ±10V, 0...+70°	14-DIP	Rca	CA 3037	data pdf pinout
CA 3030	UNI	Vs:±20V Vu:70dB Vo:±7V Vi0:1mV	14D	Rca		data pdf pinout
CA 3030(A)	OP-IC	Uni, ±20V, 0...+70°	14-DIP	Rca	CA 3038	data pdf pinout
CA3031	UNI	Vs:-7/14V Vu:71dB Vo:5V Vi0:0.5mV	8A	Rca		data pinout
CA 3031	OP-IC	Uni, Serie 712, +14/-7V, -55...+125°	81	Rca	→Serie 712	data pinout
CA3032	UNI	Vs:-7/14V Vu:71dB Vo:5V Vi0:1.5mV	8A	Rca		data pinout
CA 3032	OP-IC	=CA 3031: 0...+70°	81	Rca	→Serie 712	data pinout
CA3033	UNI	Vs:±13V Vu:93dB Vi0:2.9mV	14Dd	Rca		data pinout
CA 3033(A)	OP-IC	Uni, ±13V, -55...+125°	14-DIP	Rca	-	data pinout
CA 3034(V1)	LIN-IC	TV/CTV, Breitb./Wideb. Amp., Dem.f.afc	82	Rca		
CA 3035(V1)	LIN-IC	Breitb./Wideb. Amp., 129dB(40kHz)	82	Rca		pdf
CA 3036	Si-N	2x Darl.-Trans., 30V, 0,05A, >150MHz	82	Rca		
CA3037	UNI	Vs:±10V Vu:60dB Vo:±6V Vi0:1.08mV	14D	Rca		data pinout
CA 3037(A)	OP-IC	=CA 3029(A): -55...+125°	14-DIC	Rca	-	data pinout
CA3038	UNI	Vs:±20V Vu:70dB Vo:±7V Vi0:1.37mV	14D	Rca		data pinout
CA 3038(A)	OP-IC	=CA 3030(A): -55...+125°	14-DIC	Rca		data pinout
CA3038A	UNI	Vs:±10V Vu:60dB Vo:±6V Vi0:0.9mV	14D	Rca		data pinout
CA 3039	LIN-IC	6-Di-Array, S, 5V, 25mA, 1ns, 0,65pF	83	Rca		pdf
CA 3040	LIN-IC	Video-,Breitb./wideb./verst./amplifier	83	Rca		pdf
CA 3041	LIN-IC	TV, Ton-ZF, Dem., NF-Tr. f. Roehren-E	14-QIP	Rca	-	pdf pinout
CA 3042	LIN-IC	TV, Ton-ZF, Dem., Nf-tr. f. Trans.-e	14-QIP	Rca	-	pdf pinout
CA 3043	LIN-IC	FM-ZF, Dem., NF-V/Tr	83	Rca	-	pdf
CA 3044(V1)	LIN-IC	Tv/ctv, Breitb./Wideb. Amp., Afc	82	Rca		pdf
CA 3045(F)	LIN-IC	5x NPN-Trans., 20V, 0,05A, -55...+125°	14-DIC	Rca	HA 1127, LM 4045	pdf
CA 3046	LIN-IC	=CA 3045: -40...+85°	14-DIP	Rca	LM 3046, TBA 331, µA 3046	pdf
CA3047	UNI	Vs:±13V Vu:90dB Vi0:2.6mV	14D	Rca		data pinout
CA 3047(A)	OP-IC	Uni, ±13V, 0...+70°	14-DIP	Rca	-	data pinout
CA3047A	UNI	Vs:±20V Vu:70dB Vo:±7V Vi0:1mV	14D	Rca		data pinout
CA3048	4xOP	Vs:16V Vu:58dB f:kc	16D	Rca		data pinout
CA 3048	LIN-IC	Quad, Uni AC Verst./Amp., 16V	16-DIP	Rca		
CA 3049 T	LIN-IC	=CA 3102E: Fig.→	83	Rca		
CA 3050	LIN-IC	Dual, Diff.-Verst./Amp., -55...+125°	14-DIC	Rca		
CA 3051	LIN-IC	=CA 3050: -40...+85°	14-DIP	Rca		
CA3052	4xOP	Vs:16V Vu:58dB f:300kc	16D	Rca		data pinout
CA 3052	LIN-IC	Quad, AC Verst./Amp., NF-V, +16V	16-DIP	Rca		
CA 3053	LIN-IC	Diff. Amplifier, ...120MHz, -55...125°C	81	Rca		pdf pinout
CA 3053 E	LIN-IC	Diff. Amplifier, ...120MHz, -55...125°C	8-DIP	Rca		pdf pinout
CA 3053 F	LIN-IC	Diff.-Verst./Amp., ...120MHz, -55...125°	8-DIC	Rca		
CA 3053 S	LIN-IC	Diff. Amplifier, ...120MHz, -55...125°C	81	Rca		
CA 3053 T	LIN-IC	Diff. Amplifier, ...120MHz, -55...125°C	81	Rca		
CA 3054	LIN-IC	Dual, Breitb./Wideband Amp., ...120MHz	14-DIP	Rca	TBA 341	pdf
CA3055	ER+	Io=150mA Vo:1.6...36V P:630mW	8A	Rca		data pinout
CA 3058	LIN-IC	Nullspg.-s/zero-voltage Switch f. Thy	14-DIC	Rca	-	data
CA 3059	LIN-IC	Nullspg.-s/zero-voltage Switch f. Thy	14-DIP	Rca	-	data pdf
CA3060	3xOTA	Vs:±18V Vo:13/14V Vi0:1mV f:0.11Mc	16D	Rca		data pinout
CA 3060 AD,BD	OP-IC	=CA 3060D: ±18V	14-DIC	Rca		data pdf pinout
CA3060D	3xOTA	Vs:±7V Vo:5/6V Vi0:1mV f:0.11Mc	16D	Rca		data pinout
CA 3060 D	OP-IC	3xTransconductance OP, ±7V, -55...+125°	14-DIC	Rca		data pdf pinout
CA3060E	3xOTA	Vs:36V Vi0:1mV f:20kkc	16D	Har		data pinout
CA 3060 E	OP-IC	=CA 3060D: ±18V, -40...+85°	14-DIP	Rca	-	data pdf pinout
CA3060H	3xOTA	Vs:±18V Vo:13/14V Vi0:1mV f:0.11Mc	d	Har		data pinout
CA 3062	LIN-IC	Foto-detektor, Verst./power Amp.	83	Rca	-	
CA 3064	LIN-IC	TV, Automatic Fine Tuning, Afc	82	Rca		pdf pinout
CA 3064 E	LIN-IC	TV, Automatic Fine Tuning, Afc	14-DIP	Rca	NTE783	pdf pinout
CA 3065(E)	LIN-IC	TV-Ton-ZF, FM-ZF, Dem, NF-Tr	14-QIP	Rca	AN 241, HA 1125, KA 2101, LA 1365, LM 30	pdf pinout
CA 3066	LIN-IC	Ctv, Chroma-signal-prozessor	16-DIP	Rca		
CA 3067	LIN-IC	Ctv, Chroma-demodulator	16-DIP	Rca		
CA 3068	LIN-IC	TV, Video-ZF, Ton-ZF, Agc	20-QIP	Rca		
CA 3070	LIN-IC	Ctv, Chroma-system, VCO, Hue Control	16-DIP	Rca		pdf
CA 3071	LIN-IC	Ctv, Chroma-system, Chroma Verst./amp.	14-DIP	Rca		pdf
CA 3072	LIN-IC	Ctv, Chroma-system, Color Matrix	14-DIP	Rca		pdf
CA 3075	LIN-IC	HiFi, FM-ZF-Limiter, Dem., NF-V	14-QIP	Rca	LM 3075	
CA 3076	LIN-IC	Breitb./Wideband FM-ZF-Limiter	81	Rca		pdf
CA3078	LO-POWER	Vs:±7V Vu:92dB Vo:±5.3V Vi0:1.3mV	8DdSA	Har		data pdf pinout
CA3078A	LO-POWER	Vs:±18V Vu:100dB Vo:±5.3V Vi0:0.7mV	8DdSA	Har		data pdf pinout
CA 3078 A(E,S,T)	OP-IC	=CA 3078E: ±18V, -55...+125°		Rca		data pdf pinout
CA 3078 E	OP-IC	Uni, lo-power, ±7V, 0...+70°	8-DIP	Rca		data pdf pinout
CA 3078 S,T	OP-IC	=CA 3078(A)E: Fig.→	81	Rca		data pdf pinout
CA 3079	LIN-IC	Nullspg.-s/zero-voltage Switch f. Thy	14-DIP	Rca		data pdf
CA3080	OTA	Vs:±18V Vi0:0.3mV f:2Mc	8DSAd	Har		data pinout
CA 3080(S)	OP-IC	=CA 3080E: Fig.→	81	Rca		data pdf pinout
CA 3080 A, AS	OP-IC	=CA 3080E: -55...+125°		Rca		data pdf pinout
CA 3080 E	OP-IC	Transconductance OP, ±18V, 0...+70°	8-DIP	Rca		data pdf pinout

Type	Device	Short Description	Fig.	Manu	Comparision Types	More at
CA 3081(F)	LIN-IC	7x NPN Trans.-Array, 20V, 0,1A, 0,75W	16-DIP,DIC	Rca	-	pdf
CA 3082(F)	LIN-IC	7x NPN Trans.-Array, 20V, 0,1A, 0,75W	16-DIP,DIC	Rca	-	pdf
CA 3083	LIN-IC	5x NPN-Trans.-Array, 20V, 0,1A, 450MHz	16-DIP	Rca	MB 5331	pdf
CA 3084(E)	LIN-IC	PNP-Trans.-Array, 40V, 10mA	14-DIP	Rca	-	data pinout
CA3085	ER+	Io=100mA Vo:1.7...46V Vin:30...50V	8AD	Rca	-	data pdf pinout
CA 3085(S)	Z-IC	1,8...26V, 12mA, -55...+125°	81	Rca	-	data pdf pinout
CA 3085 A(S)	Z-IC	=CA 3085(S): 1,7...36V, 100mA	81	Rca	-	data pdf pinout
CA 3085 B(S)	Z-IC	=CA 3085(S): 1,7...46V, 100mA	81	Rca	-	data pdf pinout
CA 3085(A,B)E,F	Z-IC	=CA 3085(A,B)(S): Fig.→	8-DIP,DIC	Rca	-	data pdf pinout
CA 3086(F)	LIN-IC	5xNPN-Trans.-Array, 20V, 0,05A, 550MHz	14-DIP,DIC	Rca	LM 3086, TBA 331, µA 3086	pdf
CA 3088 E	LIN-IC	Am Radio, Ucc=6...16V	16-DIP	Rca	-	pdf
CA 3089(E,N)	LIN-IC	TV-Ton-ZF, FM-ZF, Agc, Afc	16-DIP	Rca	LM 3089, TCA 3089, TDA 1200(A)	pdf
CA 3090(A)	LIN-IC	FM MPX Stereo-Decoder, Ucc=10...16V	16-DIP	Rca	-	
CA 3090 (A)Q	LIN-IC	=CA 3090(A)	16-QIP	Rca	-	
CA 3091 D	LIN-IC	4-Quadrant Multiplier, 18V, -55...+125°	14-DIC	Rca	-	
CA 3093 E	LIN-IC	3x NPN-Trans., 2x Z-Diode, 1x Diode	16-DIP	Rca	-	
CA3094	PROG-POWER	Vs:±12V Vu:100dB Vi0:0.4mV f:30Mc	8DdSA	Har	-	data pdf pinout
CA3094A	PROG-POWER	Vs:±18V Vu:100dB Vo:12/15V Vi0:0.4mV	8DSA	Rca	-	data pdf pinout
CA 3094 AE	OP-IC	=CA 3094E: ±18V	8-DIP	Rca	-	data pdf pinout
CA3094B	PROG-POWER	Vs:±22V Vu:100dB Vi0:0.4mV f:30Mc	8DSA	Har	-	data pinout
CA 3094 BE	OP-IC	=CA 3094E: ±22V	8-DIP	Rca	-	data pdf pinout
CA 3094 E	OP-IC	S, hi-power, progr., ±12V, -55...+125°	8-DIP	Rca	-	data pdf pinout
CA 3094 S,T	OP-IC	=CA 3094(A,B)E: Fig.→	81	Rca	-	
CA 3095 E	LIN-IC	Diff.-Verst./Amp., 3x NPN-Trans.	16-DIP	Rca	-	
CA 3096 AE,E	LIN-IC	3x NPN(45/35V)-/2x PNP(40/40V)-Trans.	16-DIP	Rca	-	pdf
CA 3096 CE	LIN-IC	=CA 3096AE,E: 24/24V	16-DIP	Rca	-	pdf
CA 3097 E	LIN-IC	Thyristor/PUT/NPN-/PNP-Trans.-array	16-DIP	Rca	-	
CA3098	PROG-COMP	Vs:±8V Vi0:±10mV	8D	Har	-	data pdf pinout
CA 3098 E	KOP-IC	Dual Inp. Schmitt T., 16V, -55...+125°	8-DIP	Rca	-	data pdf pinout
CA 3098 S,T	KOP-IC	=CA 3098E: Fig.→	81	Rca	-	data pdf pinout
CA 3099 E	KOP-IC	progr., Memory, 16V, -55...+125°	14-DIP	Rca	-	data
CA3100	WIDEBAND	Vs:±18V Vu:61dB Vo:±11V Vi0:±1mV	8DdSA	Har	-	data pdf pinout
CA 3100 E	OP-IC	Breitb./Wideb., 38MHz, ±18V, -40...+85°	8-DIP	Rca	-	data pdf pinout
CA 3100 S,T	OP-IC	=CA 3100E: -55...+125°	81	Rca	-	data pdf pinout
CA 3102 E	LIN-IC	Dual, Breitb./Wideband Amp., ...500MHz	14-DIP	Rca	-	pdf
CA3105	POWER-OP	Vs:28V Vo:±8.5V Vi0:1.3mV f:5Mc	5P	Rca	-	data pinout
CA 3105(M)	OP-IC	Power-OP, 28V, 3,5A, 15W, 0...+125°	86/5Pin	Rca	-	data pinout
CA 3118 AT	LIN-IC	=CA 3118T: 50/40V	83	Rca	-	
CA 3118 T	LIN-IC	4x NPN-Trans.-Array, 40/30V, 0,05A	83	Rca	-	
CA 3120 E	LIN-IC	TV, Signal-prozessor	16-DIP	Rca	-	
CA 3121 E	LIN-IC	Ctv, Chroma-verst./amp., Chroma Dem.	16-DIP	Rca	-	
CA 3123 E	LIN-IC	Am Radio, Ucc=9V	14-DIP	Rca	-	
CA 3125 E	LIN-IC	Ctv, Chroma-demodulator	16-DIP	Rca	-	
CA 3126 Q	LIN-IC	Ctv, Chroma-prozessor	16-QIP	Rca	-	pdf
CA 3127 E	LIN-IC	5x NPN-Trans.-Array, 20V, 20mA, >1GHz	14-DIP	Rca	-	pdf
CA 3128 Q	LIN-IC	Ctv, Pal Chroma-prozessor	16-QIP	Rca	-	
CA3130	BIMOS	Vs:16V Vu:110dB Vo:13.3/0V Vi0:8mV	8DdSA	Rca	-	data pdf pinout
CA3130A	BIMOS	Vs:16V Vu:110dB Vo:13.3/0V Vi0:2mV	8DSA	Rca	-	data pdf pinout
CA3130B	BIMOS	Vs:16V Vu:110dB Vo:13.3/0V Vi0:0.8mV	8A	Rca	-	data pinout
CA 3130(A,B)E	BiMOS-OP-IC	MOS Inp./Outp., +16(±8)V, -55...+125°	8-DIP	Rca	CA 5130...	data pdf pinout
CA 3130(A,B)S,T	BiMOS-OP-IC	=CA 3130(A,B)E: Fig.→	81	Rca	CA 5130...	data pdf pinout
CA 3134 EM	LIN-IC	TV-Ton-ZF, Dem, NF-E, 5W(30V/16Ω)	16-DIP+d	Rca	-	
CA 3134 QM	LIN-IC	=CA 3134EM:	16-QIP+d	Rca	-	
CA 3135 E	LIN-IC	Ctv, Luminanz-prozessor	16-DIP	Rca	-	
CA 3136 P	LIN-IC	Ctv, Video PLL Synchron Demodulator	16-DIP+c	Rca	-	
CA 3137 E	LIN-IC	Ctv, Chroma-demodulator	16-DIP	Rca	-	
CA 3138 AE	Si-N	=CA 3138: 25V	14-DIP	Rca	-	
CA 3138 E	Si-N	4x NPN-Trans., 20V, 1A, 106/857ns	14-DIP	Rca	-	
CA 3139 E	LIN-IC	TV, Autom. Feinabstimmung/fine Tuning	14-DIP	Rca	-	
CA 3139 Q	LIN-IC	=CA 3139E: Fig.→	14-QIP	Rca	-	
CA3140	BIMOS	Vs:±18V Vu:100dB Vo:13/14V Vi0:5mV	8Dd	Rca	-	data pdf pinout
CA3140A	BIMOS	Vs:±18V Vu:100dB Vi0:2mV f:4.5Mc	8DSA	Har	-	data pdf pinout
CA3140B	BIMOS	Vs:±22V Vu:100dB Vo:13/14V Vi0:0.8mV	8A	Har	-	data pinout
CA3140E	BIMOS	Vs:±18V Vu:100dB Vi0:5mV f:4.5Mc	8D	Har	-	data pdf pinout
CA 3140(A,B)E	BiMOS-OP-IC	MOS Inp., ±18(B=±22V)V, -55...+125°	8-DIP	Rca	-	data pdf pinout
CA3140M	BIMOS	Vs:±18V Vu:100dB Vi0:5mV f:4.5Mc	8S	Har	-	data pinout
CA3140S	BIMOS	Vs:±18V Vu:100dB Vo:13/14V Vi0:5mV	8A	Har,Rca	-	data pinout
CA 3140(A,B)S,T	BiMOS-OP-IC	=CA 3140(A,B)E: Fig.→	81	Rca	-	data pdf pinout
CA3140T	BIMOS	Vs:±18V Vu:100dB Vi0:5mV f:4.5Mc	8A	Har	-	data pdf pinout
CA 3141 E	LIN-IC	10-Di-Array, 30V, 25mA, 50ns	16-DIP	Rca	-	pdf
CA 3142 E	LIN-IC	TV, Signal-prozessor	16-DIP	Rca	-	
CA 3143 E	LIN-IC	Ctv, Luminanz-prozessor	14-DIP	Rca	-	
CA 3144 E	LIN-IC	Ctv, Luminanz-prozessor	16-DIP	Rca	-	
CA 3145 E	LIN-IC	Ctv, Chroma-verst./amp., Chroma Dem.	16-DIP	Rca	-	
CA 3146 AE	LIN-IC	5x NPN-Trans.-Array, 50/40V, 0,05A	14-DIP	Rca	-	pdf
CA 3146 E	LIN-IC	5x NPN-Trans.-Array, 40/30V, 0,05A	14-DIP	Rca	-	pdf
CA 3151 E	LIN-IC	Ctv, Chroma-prozessor, Demodulator	24-DIP	Rca	-	
CA 3152 E	BiMOS-OP-IC	MOS Input, TV Tuning Interf., 18V	18-DIP	Rca	-	
CA 3153 E	LIN-IC	TV, Video ZF, Demodulator, Agc	16-DIP	Rca	-	
CA 3154 E	LIN-IC	TV, HA Synchr.-prozessor, Agc	16-DIP	Rca	-	pdf
CA 3156 E	LIN-IC	Ctv, Video-, Chroma-prozessor	16-DIP	Rca	-	
CA 3157 E	LIN-IC	TV, Ha/va Digital Synchr.-system	14-DIP	Rca	-	
CA 3158 E	LIN-IC	Ctv, Chroma-system, VCO, Hue Control	16-DIP	Rca	-	
CA 3159 E	LIN-IC	TV, Ha Prozessor, Agc Detector	16-DIP	Rca	-	
CA3160	BIMOS	Vs:16V Vu:110dB Vo:13.3/0V Vi0:6mV	8DdSA	Rca	-	data pdf pinout

Type	Device	Short Description	Fig.	Manu	Comparision Types	More at
CA3160A	BIMOS	Vs:16V Vu:110dB Vo:13.3/0V Vi0:2mV	8DSA	Rca		data pdf pinout
CA3160B	BIMOS	Vs:16V Vu:110dB Vo:13.3/0V Vi0:0.8mV	8A	Rca		data pinout
CA 3160(A,B)E	BiMOS-OP-IC	MOS Inp./Outp., +16(±8)V, -55...+125°	8-DIP	Rca	CA 5160...	data pdf pinout
CA 3160(A,B)S,T	BiMOS-OP-IC	=CA 3160(A,B)E: Fig.→	81	Rca	CA 5160...	data pdf pinout
CA 3161 E	LIN-IC	BCD→7-Segment Decoder, +7V	16-DIP	Rca		pdf
CA 3162 E	A/D-IC	Dual Slope, 3-Digit MPX BCD Output, 7V	16-DIP	Rca		pdf
CA 3163 E	LIN-IC	HF-Teiler/Prescaler, 90...1000MHz	14-DIP	Rca		
CA 3164 E	BiMOS-IC	Detector, Alarm System, Ucc=7...11V	14-DIP	Rca		
CA 3165 E1	LIN-IC	=CA 3165E: 5 Outputs	14-DIP	Rca		
CA 3165 E	LIN-IC	Näherungssch./Proximity Det., 2 Outp.	8-DIP	Rca		pdf
CA 3166 E	BiMOS-OP-IC	MOS Input, TV Tuning Interf., 35V	18-DIP	Rca		pdf
CA 3168 E	LIN-IC	2-Digit BCD→7-Segment Decoder, +6V	24-DIP	Rca		
CA 3169(M)	LIN-IC	Solenoid/Motor Driver, 17V, 2,5/3A	86/5Pin	Rca		
CA 3170 E	LIN-IC	Ctv, Chroma-system, VCO, Hue Control	16-DIP	Rca		
CA 3172 E	LIN-IC	Ctv, Chroma-demodulator	14-DIP	Rca		
CA 3179 E	ECL-IC	HF-Teiler/Prescaler, 1,25GHz	14-DIP	Rca		
CA 3183 AE	LIN-IC	5x NPN-Trans.-Array, 50/40V, 0,075A	16-DIP	Rca		pdf
CA 3183 E	LIN-IC	5x NPN-Trans.-Array, 40/30V, 0,075A	16-DIP	Rca		pdf
CA 3189 E	LIN-IC	HiFi, FM-ZF, Dem., Afc, Agc(progr.)	16-DIP	Rca	TCA 3189, LM 3189N	pdf
CA 3190 E	LIN-IC	TV, Ha/va Digital Synchr.-system	14-DIP	Rca		
CA 3191 E	LIN-IC	TV, Video ZF, Synchr., Agc(pos./neg.)	16-DIP	Rca		
CA 3192 E	LIN-IC	TV, Video ZF, Dem., Video-verst./amp.	16-DIP	Rca		
CA3193	BIMOS	Vs:±18V Vu:115dB Vo:±13.5V f:1.2Mc	8DSA	Rca		data pinout
CA3193A	BIMOS	Vs:±18V Vu:115dB Vo:±13.5V f:1.2Mc	8DSA	Har		data pdf pinout
CA 3193 AE,AS,AT	BiMOS-OP-IC	=CA 3193E: ±18V, -25...+85°		Rca		data pdf pinout
CA3193B	BIMOS	Vs:±22V Vu:115dB Vo:±13.5V f:1.2Mc	8DA	Rca		data pinout
CA 3193 BE,BS,BT	BiMOS-OP-IC	=CA 3193E: ±22V, -55...+125°		Rca		data pdf pinout
CA3193E	BIMOS	Vs:±18V Vu:110dB Vo:±13.5V f:1.2Mc	8D	Har		data pinout
CA 3193 E	BiMOS-OP-IC	Bipolar Inp/Outp,hi-prec,±18V, 0...+70°	8-DIP	Rca		data pdf pinout
CA 3193 S,T	BiMOS-OP-IC	=CA 3193E: Fig.→	81	Rca		data pdf pinout
CA3193T	BIMOS	Vs:±18V Vu:110dB Vo:±13.5V f:1.2Mc	8A	Har		data pinout
CA 3194 E	LIN-IC	Ctv, Pal Luminanz/chroma-prozessor	24-DIP	Rca		pdf
CA 3195	LIN-IC	PLL FM MPX Stereo-Decoder, Ucc=10...16V	16-DIP	Rca		
CA 3199 E	ECL-IC	HF-Teiler/Prescaler, 1,3GHz, 1:4	8-DIP	Rca		
CA 3201 E	LIN-IC	Ctv, Chroma-prozessor, Demodulator	24-DIP	Rca		
CA 3202 E	LIN-IC	TV, Ha/va Digital Synchr.-system	14-DIP	Rca		
CA 3207 E	BiMOS-IC	Sequencer Driver f. FLT, Ucc=55V	22-DIP	Rca		
CA 3208 E	BiMOS-IC	Segment Latch-Driver f. FLT, Ucc=55V	22-DIP	Rca		
CA 3209 E	LIN-IC	HiFi, FM-ZF, Dem., Agc(progr.)	16-DIP	Rca		
CA 3210 E	LIN-IC	TV, Ha/va Digital Synchr.-system	24-DIP	Rca		
CA 3211 E	LIN-IC	HF-Teiler/Prescaler, 1GHz, 1:256	18-DIP	Rca		
CA 3215 E	LIN-IC	FM-ZF, Detector Limiter	16-DIP	Rca		
CA 3216 E	LIN-IC	Ctv, Chroma-prozessor	24-DIP	Rca		
CA 3217 E	LIN-IC	Ctv, Chroma-, Luminanz-prozessor	28-DIP	Rca		
CA 3219 E	LIN-IC	Quad Power NAND Driver	16-DIP	Rca		
CA 3221 E	LIN-IC	Ctv, Chroma-verst./amp., Demodulator	16-DIP	Rca		
CA 3223 E	LIN-IC	TV, Ha/va Digital Synchr.-system	24-DIP	Rca		
CA 3227 E	LIN-IC	5x NPN-Trans.-Array, 12V, 20mA, >3GHz	16-DIP	Rca		pdf
CA 3228(E)	LIN-IC	Speed-control System	24-DIP	Rca		pdf
CA3240	BIMOS	Vs:22V Vu:100V/mV Vo:±9.7V Vi0:0.8mV	8ADd	Rca		data pinout
CA3240A	2xBIMOS	Vs:±18V Vu:96dB Vi0:2mV f:4.5Mc	8D	Rca		data pdf pinout
CA3240AE1	2xBIMOS	Vs:±18V Vi0:2mV f:4.5Mc	14D	Har		data pdf pinout
CA3240E1	2xBIMOS	Vs:±18V Vi0:5mV f:4.5Mc	14D	Har		data pinout
CA 3240(A)E1	BiMOS-OP-IC	=CA 3140(A)E: Dual	14-DIP	Rca		data pdf pinout
CA 3240(A)E	BiMOS-OP-IC	=CA 3140(A)E: Dual	8-DIP	Rca		data pdf pinout
CA 3246 E	LIN-IC	5x NPN-Trans.-Array, 12V, 20mA, >3GHz	14-DIP	Rca		
CA3260	2xBIMOS	Vs:16V Vu:100dB Vo:6mV f:4Mc	8DdA	Rca		data pdf pinout
CA3260A	2xBIMOS	Vs:16V Vu:100dB Vo:2mV f:4Mc	8DA	Rca		data pdf pinout
CA3260B	2xBIMOS	Vs:16V Vu:100dB Vi0:1mV f:4Mc	8A	Rca		data pinout
CA 3260(A,B)E	BiMOS-OP-IC	=CA 3160(A,B)E: Dual	8-DIP	Rca	CA 5260...	data pdf pinout
CA 3260(A,B)S,T	BiMOS-OP-IC	=CA 3160(A,B)E: Dual	81	Rca	CA 5260...	data pdf pinout
CA 3275	LIN-IC	2x Mosfet Bridge Drv, ±0,15A, 30kHz	14-DIP	Har		pdf
CA3280	2xOP	Vs:36V Vu:100dB Vo:>±12V Vi0:0.7mV	16Dd	Har		data pinout
CA 3280	OP-IC	Dual, Variable, ±18V, 0...+70°	16-DIP	Rca		data pdf pinout
CA3280A	2xOP	Vs:36V Vu:100dB Vo:>±12V Vi0:0.25mV	16D	Har		data pinout
CA 3280 A	OP-IC	=CA 3280: -55...+125°	16-DIP	Rca		data pdf pinout
CA3280AF3	2xOTA	Vs:±18V Vu:100dB Vo:>±12.5V f:9Mc	16D	Har		data pdf pinout
CA3290	2xBIMOS	Vs:±18V Vu:118dB Vi0:3mV	8ADd	Rca		data pinout
CA3290A	2xBIMOS	Vs:±18V Vu:800V/mV Vo:TTL Vi0:4mV	8DdA	Har		data pinout
CA3290AE1	2xBIMOS	Vs:±18V Vu:800V/mV Vo:TTL Vi0:4mV	14D	Har		data pinout
CA3290B	2xBIMOS	Vs:±18V Vu:118dB Vi0:3mV	8A	Rca		data pinout
CA3290E1	2xBIMOS	Vs:±18V Vu:800V/mV Vi0:7.5mV	14D	Har		data pinout
CA 3290(A)E1	BiMOS-OP-IC	=CA 3290(A,B)E: Fig.→	14-DIP	Rca		data pinout
CA 3290(A,B)E	BiMOS-KOP-IC	MOS Inp., ±18(B=±22)V, -55...+125°	8-DIP	Rca		data pdf pinout
CA 3290(A,B)S,T	BiMOS-OP-IC	=CA 3290(A,B)E: Fig.→	81	Rca		data pdf pinout
CA 3300(D)	CMOS-A/D-IC	6 Bit, hi-speed, Video, 15MHz	18-DIC	Rca		data pdf pinout
CA 3302...	KOP-IC	=LM 3302...	14-DIP	Rca	→LM 3302...	
CA 3306 E,AE	CMOS-A/D-IC	6 Bit, video, >15 Msps, CMOS/TTL Level	18-DIP	Rca		
CA 3308(D)	CMOS-A/D-IC	8 Bit, hi-speed, Video, 15MHz	24-DIC	Rca		pdf
CA 3318 C	CMOS-A/D-IC	8 Bit, Video, >15 Msps, Cmos Level	24-DIP	Rca		
CA3401	4xOP	Vs:18V Vu:66dB Vo:14.2/0V f:5Mc	14Dd	Rca		
CA 3401 E,G	OP-IC	Quad, +18V, 5MHz, -55...+125°	14-DIP,DIC	Rca	MC 3401...	data pinout
CA3410AE1	4xBIMOS	Vs:36V Vu:100V/mV Vi0:3mV f:5.4Mc	14D	Rca		data pinout
CA3410E1	4xBIMOS	Vs:36V Vu:100V/mV Vi0:8mV f:5.4Mc	14D	Rca		data pinout
CA 3410(A)E	BiMOS-OP-IC	Quad, MOS Inp., ±18V, -55...+125°	8-DIP	Rca		data pinout

Type	Device	Short Description	Fig.	Manu	Comparision Types	More at
CA3420	BIMOS	Vs:22V Vu:100dB Vi0:5mV f:0.5Mc	8DdA	Har		data pdf pinout
CA3420A	BIMOS	Vs:22V Vu:100dB Vi0:2mV f:0.5Mc	8DdA	Har		data pinout
CA 3420 E,AE,BE	BiMOS-OP-IC	PMOS Input, 22V, -55...+125°	8-DIP	Rca	CA 5420...	data pdf pinout
CA 3420(A,B)S,T	BiMOS-OP-IC	=CA 3420E...: Fig.→	81	Rca	CA 5420...	data pdf pinout
CA3440	BIMOS	Vs:25V Vu:100dB Vo:±3.2V Vi0:5mV	d,8SA	Har		data pinout
CA3440A	BIMOS	Vs:25V Vu:100dB Vo:±3.2V Vi0:2mV	8DdSA	Har		data pinout
CA3440B	BIMOS	Vs:±12.5V Vu:100dB Vi0:0.8mV f:63kkc	8DdA	Rca		data pinout
CA3440E	BIMOS	Vs:25V Vu:100dB Vo:±3.2V Vi0:5mV	8D	Har		data pinout
CA 3440 E,AE,BE	BiMOS-OP-IC	PMOS Inp., lo-power,±12,5V, -55...+125°	8-DIP	Rca	-	data pdf pinout
CA 3440(A,B)S,T	BiMOS-OP-IC	=CA 3440E...: Fig.→	81	Rca	-	data pdf pinout
CA3450	HI-SPEED	Vs:±7.25V Vu:70dB Vo:±4.1V Vi0:8mV	16Dd	Har		data pinout
CA 3450 E	OP-IC	Video Line Driver, ±7,25V, -40...+85°	16-DIP	Rca	-	data pdf pinout
CA 3458 T	OP-IC	Dual, Serie 158, ±15V, 0...+70°,0,5V/µs	81	Rca	→Serie 158	data
CA3493	BIMOS	Vs:±18V Vu:110dB Vo:±13.5V Vi0:0.3mV	8DA	Rca		data pinout
CA3493A	BIMOS	Vs:±18V Vu:115dB Vo:±13.5V f:1.2Mc	8DA	Rca		data pinout
CA 3493 AE,AS,AT	BiMOS-OP-IC	=CA 3493E: ±18V, -25...+85°		Rca	-	data pinout
CA3493B	BIMOS	Vs:±18V Vu:125dB Vo:±13.5V Vi0:40µµV	8DA	Rca		data pinout
CA 3493 BE,BS,BT	BiMOS-OP-IC	=CA 3493E: ±22V, -55...+125°		Rca	-	data pinout
CA 3493 E	BiMOS-OP-IC	Bipolar Inp/Outp,hi-prec,±18V, 0...+70°	8-DIP	Rca	-	data pinout
CA 3493 S,T	BiMOS-OP-IC	=CA 3493E: Fig.→	81	Rca	-	data pinout
CA 3524(E)	SMPS-IC	=CA 1524: 0...+70°	16-DIP	Rca	→SG 3524	pdf
CA 3541 D	LIN-IC	Kernspeicher/core Memory, Verst./amp.	16-DIC	Rca	-	
CA 3558 T	OP-IC	=CA 3458T: ±22V, -55...+125°	81	Rca	→Serie 158	data
CA 3600 E	MOS-IC	6x N-FET/3x P-FET, 15V, 10mA	14-DIP	Rca		
CA3620	2xBIMOS	Vs:±8V Vu:110dB Vo:13.3/0V Vi0:2mV	8AD	Rca		data pinout
CA3620B	2xBIMOS	Vs:±8V Vu:110dB Vo:13.3/0V Vi0:0.8mV	8A	Rca		data pinout
CA 3741 CT	OP-IC	=CA 3741T: 0...+70°C	81		→Serie 741	data
CA 3741 T	OP-IC	Uni, Serie 741, ±15, -55...+125°	81	Rca	→Serie 741	data
CA 3747 CE	OP-IC	=CA 3747T: 0...+70°	14-DIP		→Serie 747	data
CA 3747 CT	OP-IC	=CA 3747T: 0...+70°	82		→Serie 747	data
CA 3747 T	OP-IC	Dual, Serie 747, ±15V, -55...+125°	82	Rca	→Serie 747	data
CA5130	BIMOS	Vs:±8V Vu:110dB Vo:15/0V Vi0:8mV	8A	Har		data pinout
CA5130A	BIMOS	Vs:±8V Vu:110dB Vo:13.3/0V Vi0:2mV	8DSAd	Har		data pinout
CA 5130(A)E	BiMOS-OP-IC	MOS Inp./Outp., +16(±8)V, -55...+125°	8-DIP	Rca	CA 3130...	data pinout
CA 5130(A)M	BiMOS-OP-IC	=CA 5130(A)E: SMD	8-MDIP		-	data pinout
CA 5130(A)S,T	BiMOS-OP-IC	=CA 5130(A)E: Fig.→	81		CA 3130...	data pinout
CA5160	BIMOS	Vs:±8V Vu:110dB Vo:13.3/0V Vi0:6mV	8DdSA	Har		data pdf pinout
CA5160A	BIMOS	Vs:±8V Vu:110dB Vo:13.3/0V Vi0:2mV	8DdSA	Har		data pinout
CA 5160(A)E	BiMOS-OP-IC	MOS Inp./Outp., +16(±8)V, -55...+125°	8-DIP	Rca	CA 3160...	data pinout
CA 5160(A)S,T	BiMOS-OP-IC	=CA 5160(A)E: Fig.→	81		CA 3160...	data pinout
CA5260	2xBIMOS	Vs:±8V Vu:110dB Vo:13.3/0V Vi0:6mV	8DdSA	Har		data pdf pinout
CA5260A	2xBIMOS	Vs:±8V Vu:110dB Vo:13.3/0V Vi0:2mV	8DdSA	Har		data pdf pinout
CA 5260(A)E	BiMOS-OP-IC	=CA 5160...: Dual	8-DIP		CA 3260...	data pinout
CA 5260(A)M	BiMOS-OP-IC	=CA 5160...: SMD, Dual	8-MDIP		-	data pinout
CA 5260(A)S,T	BiMOS-OP-IC	=CA 5160...: Dual	81		CA 3260...	data pinout
CA5420	BIMOS	Vs:±11V Vu:85dB Vo:4.6/0V Vi0:1.5mV	8DdSA	Har		data pinout
CA5420A	BIMOS	Vs:±11V Vu:85dB Vo:4.6/0V Vi0:1mV	8DdSA	Har		data pinout
CA 5420(A)E	BiMOS-OP-IC	PMOS Input, 22V, -55...+125°	8-DIP	Rca	CA 3420...	data pinout
CA 5420(A)S,T	BiMOS-OP-IC	=CA 5420E...: Fig.→	81		CA 3420...	data pinout
CA5470	4xBIMOS	Vs:±8V Vu:90dB Vo:4.4/0V Vi0:6mV	14DdS	Har		data pdf pinout
CA 5470 E	BiMOS-OP-IC	Quad, MOS Inp., ±16(±8)V, -55...+125°	8-DIP	Rca	-	data pinout
CA 5470 M	BiMOS-OP-IC	=CA 5470E: SMD	8-MDIP		-	data pinout
CA6078A	LO-POWER	Vs:±18V Vu:100dB Vo:±14V Vi0:1.4mV	d,8A	Rca		data pinout
CA 6078 AS,AT	OP-IC	lo-power, lo-volt, ±18V, -55...+125°	81	Rca	-	data pinout
CA6741	LO-NOISE	Vs:±22V Vu:106dB Vo:±13V Vi0:1mV	8A	Rca		data pinout
CA6741C	LO-NOISE	Vs:±18V Vu:100dB Vo:±14V Vi0:1mV	8A	Rca		data pinout
CA 6741 CS,CT	OP-IC	=CA 6741S,T: ±18V, 0...+70°	81	Rca	→Serie 741	data pinout
CA 6741 S,T	OP-IC	Serie 741, lo-noise, ±22V, -55...+125°	81	Rca	→Serie 741	data pinout
CA 7607 E	LIN-IC	TV, Video-zf f. SAW-filter, AGC(f.FET)	16-DIP	Rca		
CA 7611 E	LIN-IC	TV, Video-filter, AGC(f.npn)	16-DIP	Rca		
CA30394A	PROG-POWER	Vs:±18V Vu:100dB Vi0:0.4mV f:30Mc	8A	Har		data pdf pinout
CAC	Si-N	=BC 868 (Typ-Code/Stempel/marking)	39	Val	→BC 868	data
CAT 823	IC	System Supervisory Voltage Reset, Watchdog	45	Cat	-	pdf pinout
CAT 824	IC	System Supervisory Voltage Reset, Watchdog	45	Cat	-	pdf pinout
CAT 825	IC	System Supervisory Voltage Reset, Watchdog	45	Cat	-	pdf pinout
CAT 28 F 010 /N	CMOS-RAM-IC	1 Megabit Cmos Flash Memory	DIP	Cat		pdf pinout
CAT 28 F 010 /P	CMOS-RAM-IC	1 Megabit Cmos Flash Memory	DIP	Cat		pdf pinout
CAT 28 F 010 /T	CMOS-RAM-IC	1 Megabit Cmos Flash Memory	DIP	Cat		pdf pinout
CAT 28 F 020 /N	CMOS-RAM-IC	2 Megabit Cmos Flash Memory	DIP	Cat		pdf pinout
CAT 28 F 020 /P	CMOS-RAM-IC	2 Megabit Cmos Flash Memory	DIP	Cat		pdf pinout
CAT 28 F 020 /T	CMOS-RAM-IC	2 Megabit Cmos Flash Memory	DIP	Cat		pdf pinout
CAT 4016 W-T 1	IC	16-Channel Constant Current LED Driver	45	Cat		pdf pinout
CAT 4016 VSR-T 2	IC	16-Channel Constant Current LED Driver	45	Cat		pdf pinout
CAT 4016 VS-T 2	IC	16-Channel Constant Current LED Driver	45	Cat		pdf pinout
CAT 4016 Y-T 2	IC	16-Channel Constant Current LED Driver	45	Cat		pdf pinout
CAT 4016 HV 6-T 2	IC	16-Channel Constant Current LED Driver	45	Cat		pdf pinout

Type	Device	Short Description	Fig.	Manu	Comparision Types	More at
CB...CC						
CB	Si-N	=2SC3646 (Typ-Code/Stempel/marking)	39	Say	→2SC3646	data
CB	Si-N	=2SC4997 (Typ-Code/Stempel/marking)	35(1,6mm)	Rhm	→2SC4997	data
CB	Si-N	=2SC4998 (Typ-Code/Stempel/marking)	35(2mm)	Rhm	→2SC4998	data
CB	Si-N	=2SD1368-CB (Typ-Code/Stempel/marking)	39	Hit	→2SD1368	data
CB	Si-P	=BCW 61RB (Typ-Code/Stempel/marking)	35	Fer	→BCW 61RB	data
CB	Si-N	=BCX 68-10 (Typ-Code/Stempel/marking)	39	Sie,Val,Mot	→BCX 68-10	data
CB	Si-N	=BFS 19 (Typ-Code/Stempel/marking)	35	Sie	→BFS 19	data
CB	Si-P	=KTX 301U-GR(Typ-Code/Stempel/marking)	45(2mm)	Kec	→KTX 301U	data
CB	Si-N+R	=RN 1444B (Typ-Code/Stempel/marking)	35	Tos	→RN 1444	
CB	Si-N	=RXT 2222A (Typ-Code/Stempel/marking)	39	Rhm	→RXT 2222A	data
CB	N-FET	=SST 4392 (Typ-Code/Stempel/marking)	35	Six	→SST 4392	data
CB	Si-N/P+R	=XN 4315 (Typ-Code/Stempel/marking)	46	Mat	→XN 4315	data
CB	Si-N/P+R	=XP 4315 (Typ-Code/Stempel/marking)	46(2mm)	Mat	→XP 4315	data
CB	C-Di	=ZMV 830B (Typ-Code/Stempel/marking)	71(1,7mm)	Ztx	→ZMV 830	data
CB	MOS-P-FET-e	=µPA573T (Typ-Code/Stempel/marking)	45(2mm)	Nec	→µPA573T	data
CBC	Si-N	=BC 868-... (Typ-Code/Stempel/marking)	39	Phi	→BC 868	data
CBG	Si-N	=2SC3324-GR (Typ-Code/Stempel/marking)	35	Tos	→2SC3324	data
CBH	Z-Di	=SMBJ 22CA (Typ-Code/Stempel/marking)	71(5x3,5)	Tho	→SMBJ ...	data
CBJ	Z-Di	=SMBJ 40CA (Typ-Code/Stempel/marking)	71(5x3,5)	Tho	→SMBJ ...	data
CBL	Si-N	=2SC3324-BL (Typ-Code/Stempel/marking)	35	Tos	→2SC3324	data
CBM	Z-Di	=SMBJ 70CA (Typ-Code/Stempel/marking)	71(5x3,5)	Tho	→SMBJ ...	data
CBN	Si-N	=2SC4132-N (Typ-Code/Stempel/marking)	39	Rhm	→2SC4132	data
CBP	Si-N	=2SC4132-P (Typ-Code/Stempel/marking)	39	Rhm	→2SC4132	data
CBQ	Si-N	=2SC4132-Q (Typ-Code/Stempel/marking)	39	Rhm	→2SC4132	data
CBQ	Z-Di	=SMBJ 100CA (Typ-Code/Stempel/marking)	71(5x3,5)	Tho	→SMBJ ...	data
CBR	Si-N	=2SC4132-R (Typ-Code/Stempel/marking)	39	Rhm	→2SC4132	data
CBT 3125 D	Logic-IC	Quadruple Fet bus switch	MDIP	Phi	-	pdf pinout
CBT 3125 DB,PW	Logic-IC	Quadruple Fet bus switch	TSOP	Phi	-	pdf pinout
CBT 3125 DS	Logic-IC	Quadruple Fet bus switch	TSOP	Phi	-	pdf pinout
CBT 3126 D	IC	Quadruple Fet bus switch	MDIP	Phi	-	pdf pinout
CBT 3126 DB,PW	IC	Quadruple Fet bus switch	TSOP	Phi	-	pdf pinout
CBT 3126 DS	IC	Quadruple Fet bus switch	TSOP	Phi	-	pdf pinout
CBT 3244 A	Logic-IC	SMD, Octal bus switch, 5V APW,ADS,ADB: 20-SSMDIP, AD:20-MDIP	20-SSMDIP	Phi	-	pdf pinout
CBT 3244 D	IC	Octal bus switch, quad output enables	MDIP	Phi	-	pdf pinout
CBT 3244 DS,DB	IC	Octal bus switch, quad output enables	SDIP	Phi	-	pdf pinout
CBT 3244 PW	IC	Octal bus switch, quad output enables	TSOP	Phi	-	pdf pinout
CBT 3245 ABQ	IC	Octal bus switch	QFN	Phi	-	pdf pinout
CBT 3245 AD	IC	Octal bus switch	MDIP	Phi	-	pdf pinout
CBT 3245 ADS,DB	IC	Octal bus switch	MDIP	Phi	-	pdf pinout
CBT 3245 APW	IC	Octal bus switch	TSOP	Phi	-	pdf pinout
CBT 3245 D	IC	Octal bus switch	MDIP	Phi	-	pdf pinout
CBT 3245 DB,PW	IC	Octal bus switch	TSOP	Phi	-	pdf pinout
CBT 3251 D	IC	1-of-8 Fet multiplexer/demultiplexer	MDIP	Phi	-	pdf pinout
CBT 3251 DB,DS	IC	1-of-8 Fet multiplexer/demultiplexer	TSOP	Phi	-	pdf pinout
CBT 3251 PW	IC	1-of-8 Fet multiplexer/demultiplexer	TSOP	Phi	-	pdf pinout
CBT 3253 D	IC	2x 1-of-4 Fet multiplexer/ demultiplexer	MDIP	Phi	-	pdf pinout
CBT 3253 DB,DS	IC	2x 1-of-4 Fet multiplexer/ demultiplexer	TSOP	Phi	-	pdf pinout
CBT 3253 PW	IC	2x 1-of-4 Fet multiplexer/ demultiplexer	TSOP	Phi	-	pdf pinout
CBT 3257 D	IC	4x 1-of-2 multiplexer/demultiplexer	MDIP	Phi	-	pdf pinout
CBT 3257 DB,DS	IC	4x 1-of-2 multiplexer/demultiplexer	TSOP	Phi	-	pdf pinout
CBT 3257 PW	IC	4x 1-of-2 multiplexer/demultiplexer	TSOP	Phi	-	pdf pinout
CBT 3306 D	Logic-IC	Dual bus switch	MDIP	Phi	-	pdf pinout
CBT 3306 PW	Logic-IC	Dual bus switch	TSOP	Phi	-	pdf pinout
CBT 3384 D	Logic-IC	10-bit bus switch, 5-bit output enables	MDIP	Phi	-	pdf pinout
CBT 3384 DB,DK	Logic-IC	10-bit bus switch, 5-bit output enables	TSOP	Phi	-	pdf pinout
CBT 3384 PW	Logic-IC	10-bit bus switch, 5-bit output enables	TSOP	Phi	-	pdf pinout
CBT 3857 PW	Logic-IC	10-bit bus switch, 10 kΩ pull-down termination resistor	TSOP	Phi	-	pdf pinout
CC	Si-P	=2SA1122-C (Typ-Code/Stempel/marking)	35	Hit	→2SA1122	data
CC	Si-P	=2SA1364-C (Typ-Code/Stempel/marking)	39	Mit	→2SA1364	data
CC	Si-N	=2SC3647 (Typ-Code/Stempel/marking)	39	Say	→2SC3647	data
CC	Si-N	=2SD1368-CC (Typ-Code/Stempel/marking)	39	Hit	→2SD1368	data
CC	Si-P	=BCW 61RC (Typ-Code/Stempel/marking)	35	Fer	→BCW 61RC	data
CC	Si-N	=BCX 68-16 (Typ-Code/Stempel/marking)	39	Mot,Sie,Val	→BCX 68	data
CC	Si-N	=BF 544 (Typ-Code/Stempel/marking)	35	Sie	→BF 544	data
CC	Z-Di	=MMSZ 4678 (Typ-Code/marking)	71(2,7mm)	Ons	→MMSZ 4678	data
CC	N-FET	=SST 4393 (Typ-Code/Stempel/marking)	35	Six	→SST 4393	data
CC	C-Di	=ZMV 831B (Typ-Code/Stempel/marking)	71(1,7mm)	Ztx	→ZMV 831	data
CC 6168 F	Si-N	≈BF 254			→BF 254	
CC 6168 G	Si-N	≈BC 239			→BC 239	
CC 6225(F)	Si-N	≈BF 255			→BF 255	
CC 6227(F)	Si-N	≈BF 254			→BF 254	
CC 2510 F 8 RSP	µP-IC	Low-power Soc, MCU, Memory, Transceiv.	QFN	Tix	-	pdf pinout
CC 2510 F 8 RSPR	µP-IC	Low-power Soc, Mcu, Memory, Transceiv.	QFN	Tix	-	pdf pinout
CC 2511 F 8 RSP	µP-IC	Low-power Soc, Mcu, Memory, Transceiv.	QFN		-	pdf pinout
CC 2511 F 8 RSPR	µP-IC	Low-power Soc, Mcu, Memory, Transceiv.	QFN		-	pdf pinout
CC 62266	Si-N	≈BF 254			→BF 254	
CC 62276	Si-N	≈BF 254			→BF 254	
CC 2510 F 16 RSP	µP-IC	Low-power Soc, Mcu, Memory, Transceiv.	QFN		-	pdf pinout
CC 2510 F 16 RSPR	µP-IC	Low-power Soc, Mcu, Memory, Transceiv.	QFN		-	pdf pinout
CC 2510 F 32 RSP	µP-IC	Low-power Soc, Mcu, Memory, Transceiv.	QFN		-	pdf pinout
CC 2510 F 32 RSPR	µP-IC	Low-power Soc, Mcu, Memory, Transceiv.	QFN		-	pdf pinout

ype	Device	Short Description	Fig.	Manu	Comparision Types	More at
C 2511 F 16 RSP	µP-IC	Low-power Soc, Mcu, Memory, Transceiv.	QFN			pdf pinout
C 2511 F 16 RSPR	µP-IC	Low-power Soc, Mcu, Memory, Transceiv.	QFN			pdf pinout
C 2511 F 32 RSP	µP-IC	Low-power Soc, Mcu, Memory, Transceiv.	QFN			pdf pinout
C 2511 F 32 RSPR	µP-IC	Low-power Soc, Mcu, Memory, Transceiv.	QFN			pdf pinout
C A	Si-N	=2SC3326A (Typ-Code/Stempel/marking)	35	Tos	→2SC3326A	data
C B	Si-N	=2SC3326B (Typ-Code/Stempel/marking)	35	Tos	→2SC3326B	data
CC	Si-N	=BC 868-... (Typ-Code/Stempel/marking)	39	Phi	→BC 868	data
CS 2001(G)	Si-N	≈BF 254			→BF 254	
CS 2004(B,D)	Si-N	≈BC 238			→BC 238	
CS 2006(G)	Si-N	≈BF 255			→BF 255	
CS 2008(G,GF)	Si-N	≈BF 254			→BF 254	
CS 2053(EFG)	Si-N	≈BC 337			→BC 338	
CU-06A	MOS-IC	FB	40-DIP			
CU 2000...200x	HMOS-µC-IC	TV, 8 Bit, 96x8 Bit RAM, 3,875kB ROM	40-DIP	Itt		
CU 2030...203x	HMOS-µC-IC	TV, 8 Bit, 120x8 Bit RAM, 6,5kB ROM	40-DIP	Itt		
CU 2050...205x	HMOS-µC-IC	TV, 8 Bit, 256x8 Bit RAM, 8kB ROM	40-DIP	Itt		
CU 2070...207x ...	HMOS-µC-IC	TV, 8 Bit, 256x8 Bit RAM, 16kB ROM	40-DIP	Itt		
CU 3000...	CMOS-IC	TV, Zentral-prozessor, ROM-less	PLCC	Itt		pdf
CU 3001	CMOS-IC	TV, Zentral-prozessor with ROM	PLCC	Itt		pdf
CU-N6 10CR525	MOS-IC	FB	40-DIP			

CD

	Device	Short Description	Fig.	Manu	Comparision Types	More at
CD	Si-P	=2SA1122-D (Typ-Code/Stempel/marking)	35	Hit	→2SA1122	data
CD	Si-P	=2SA1364-D (Typ-Code/Stempel/marking)	39	Mit	→2SA1364	data
CD	Si-P	=2SA1366-D (Typ-Code/Stempel/marking)	35	Mit	→2SA1366	data
CD	Si-N	=2SC3648 (Typ-Code/Stempel/marking)	39	Say	→2SC3648	data
CD	Si-N	=2SC4673 (Typ-Code/Stempel/marking)	39	Say	→2SC4673	data
CD	Si-P	=BCW 61RD (Typ-Code/Stempel/marking)	35	Fer	→BCW 61RD	data
CD	Si-N	=BCX 68-25 (Typ-Code/Stempel/marking)	39	Sie,Val,Mot	→BCX 68-25	data
CD	Si-N	=BSS 81B (Typ-Code/Stempel/marking)	35	Sie	→BSS 81B	data
CD	Si-N	≈CPH 3204 (Typ-Code/Stempel/marking)	35	Say	→CPH 3204	data
CD	Si-N	=KTX 401U-Y (Typ-Code/Stempel/marking)	45(2mm)	Kec	→KTX 401U	data
CD	Z-Di	=MMSZ 4679 (Typ-Code/marking)	71(2,7mm)	Ons	→MMSZ 4679	
CD	Z-Di	=P4 SMA-47 (Typ-Code/Stempel/marking)	71(5x2,5)	Fag	→P4 SMA-47	data
CD	Si-N	=RXT 3904 (Typ-Code/Stempel/marking)	39	Rhm	→RXT 3904	data
CD	Z-Di	=SMBJ 26C (Typ-Code/Stempel/marking)	71(5x3,5)	Mop	→SMBJ ...	data
CD	C-Di	=ZMV 832B (Typ-Code/Stempel/marking)	71(1,7mm)	Ztx	→ZMV 832	data
CD 0000	Ge-Di	≈1N34	31a		→1N34	
CD 0014	Si-Di	≈BAV 20			→BAV 20	
CD 0014 N(NA,NG)	Si-N	≈BC 337			→BC 337	
CD 021	Si-Di	≈BA 127	31a		→BA 127	
CD22M3494	ARRAY4x8	Vs:4...15V Ron:Ohm Ton:50ns at 12/0V	40D,44C	Isi	MT8816	data pdf pinout
CD 82C55A-5	µP-Periph	CMOS, Ppi, 3x8 Programmable I/O Pins, TTL Compatible, 5MHz, 0..70°C	DIC	Har	...82C55...	pdf pinout
CD 82C55A	µP-Periph	CMOS, Ppi, 3x8 Programmable I/O Pins, TTL Compatible, 8MHz, 0...70°C	DIC	Har	...82C55...	pdf pinout
CD 0089	Si-Di	≈BA 100	31a		→BA 100	
CD 0099	Si-Di	≈BA 100	31a		→BA 100	
CD 0140	Z-Di	7,5V	31a		BZW22/..., BZX61/..., ZYP..., 1N5236,++	
CD 624	Si-Di	≈RGP 30	31a		→RGP 30 G...M	
CD 951	Si-P	≈BC 309	7e		→BC 309	
CD 1602	Si-P	≈BC 309			→BC 309	
CD 3226 E	CMOS-IC	µComp.→video Disk Interface Buffer	14-DIP	Rca		
CD 5000 D	Si-N	≈BF 255	2a		→BF 255	
CD 9000	Si-P	≈BC 307	2a		→BC 307	
CD 15000 C,D	Si-N	≈BC 237	2a		→BC 237	
CD 15000 E	Si-N	≈BF 254	2a		→BF 254	
CDC	Si-N	=2SC4673-C (Typ-Code/Stempel/marking)	39	Say	→2SC4673	data
CDC	Si-N	=BC 868-... (Typ-Code/Stempel/marking)	39	Phi	→BC 868	data
CDC 8002	Si-N	≈BC 639			→BC 639	
CDC 9002	Si-P	≈BC 640			→BC 640	
CDCE 421 RGER	IC	Wide- Range, Low- Jitter, Crystal-Oscillator Clock Generator	DIP	Tix		pdf pinout
CDCE 421 RGET	IC	Wide- Range, Low- Jitter, Crystal-Oscillator Clock Generator	DIP	Tix		pdf pinout
CDCE 937	IC	Program. 3-PLL VCXO Clock Synthesizer	QFP	Tix		pdf pinout
CDCE 421 RGERG 4	IC	Wide- Range, Low- Jitter, Crystal-Oscillator Clock Generator	DIP	Tix		pdf pinout
CDCE 421 RGETG 4	IC	Wide- Range, Low- Jitter, Crystal-Oscillator Clock Generator	DIP	Tix		
CDCEL 937	IC	Program. 3-PLL VCXO Clock Synthesizer	QFP	Tix		pdf pinout
CDD	Si-N	=2SC4673-D (Typ-Code/Stempel/marking)	39	Say	→2SC4673	data
CDD 5000	Ge-Di	≈AA 119	31a		→AA 119	
CDE	Si-N	=2SC4673-E (Typ-Code/Stempel/marking)	39	Say	→2SC4673	data
CDG 00	Si-Di	≈BA 216	31a		→BA 216	
CDG 20	Si-Di	≈1N4148	31a		→1N4148	
CDG 21	Ge-Di	≈AA 119	31a		→AA 119	

Type	Device	Short Description	Fig.	Manu	Comparision Types	More at
CDG 22	Ge-Di	≈AA 119	31a		→AA 119	
CDG 23	Si-Di	≈BA 127	31a		→BA 127	
CDG 24	Si-Di	≈BA 216	31a		→BA 216	
CDG 25	Si-Di	≈BA 127	31a		→BA 127	
CDG 26	Si-Di	≈BA 127	31a		→BA 127	
CDG 27	Si-Di	≈BA 216	31a		→BA 216	
CDs	Si-N	=BSS 81B (Typ-Code/Stempel/marking)	35	Sie	→BSS 81B	data
CDT 1310	Ge-P	NF/S-L, 40V, 5A, 45W	23a§	Itt	AL 102, AUY 28, 2N1539...43, 2N1544...48	data
CDT 1311	Ge-P	=CDT 1310: 60V	23a§	Itt	AL 102, AUY 28, 2N1540..43, 2N1545..48	data
CDT 1312	Ge-P	=CDT 1310: 80V	23a§	Itt	AL 102, AUY 28, 2N1541..43, 2N1546..48	data
CDT 1313	Ge-P	=CDT 1310: 100V	23a§	Itt	AL 102, AUY 37, 2N1542..43, 2N1547...48	data
CDT 1315	Ge-P	NF/S-L, 100V, 8A, 45W	23a§	Stc	AL 100, AUY 37, 2N2290, 2N2293	data
CDT 1319	Ge-P	NF/S-L, 40V, 5A, 45W	23a§	Stc	AL 102, AUY 28, 2N1539...43, 2N1544...48	data
CDT 1320	Ge-P	=CDT 1319: 50V	23a§	Stc	AL 102, AUY 28, 2N1540..43, 2N1545..48	data
CDT 1321	Ge-P	=CDT 1319: 80V	23a§	Stc	AL 102, AUY 28, 2N1541..43, 2N1546..48	data
CDT 1322	Ge-P	=CDT 1319: 100V	23a§	Stc	AL 102, AUY 37, 2N1542..43, 2N1547..48	data
CDT 1349(A)	Ge-P	≈2N2063..2064	23a§	Clv	→2N2063..2064	data
CDT 1350(A)	Ge-P	≈2N2065...2066	23a§	Clv	→2N2065...2066	data

CE...CK

Type	Device	Short Description	Fig.	Manu	Comparision Types	More at
CE	Si-P	=2SA1122-E (Typ-Code/Stempel/marking)	35	Hit	→2SA1122	data
CE	Si-P	=2SA1364-E (Typ-Code/Stempel/marking)	39	Mit	→2SA1364	data
CE	Si-P	=2SA1366-E (Typ-Code/Stempel/marking)	35	Mit	→2SA1366	data
CE	Si-N	=2SC3649 (Typ-Code/Stempel/marking)	39	Say	→2SC3649	data
CE	Si-P	=BCP 69 (Typ-Code/Stempel/marking)	48	Phi, Ons	→BCP 69	data
CE	Si-P	=BCX 69 (Typ-Code/Stempel/marking)	39	Sie,Val,Mot	→BCX 69	data
CE	Si-N	=BSS 79B (Typ-Code/Stempel/marking)	35	Fer,Mot,Sie	→BSS 79B	data
CE	Si-N	=CPH 3205 (Typ-Code/Stempel/marking)	35	Say	→CPH 3205	data
CE	Si-N	=KTX 401U-GR(Typ-Code/Stempel/marking)	45(2mm)	Kec	→KTX 401U	data
CE	Z-Di	=MMSZ 4680 (Typ-Code/marking)	71(2,7mm)	Ons	→MMSZ 4680	
CE	Z-Di	=P4 SMA-47A (Typ-Code/Stempel/marking)	71(5x2,5)	Fag	→P4 SMA-47A	data
CE	Z-Di	=SMBJ 26CA (Typ-Code/Stempel/marking)	71(5x3,5)	Mop	→SMBJ ...	data
CE	C-Di	=ZMV 833B (Typ-Code/Stempel/marking)	71(1,7mm)	Ztx	→ZMV 833	data
CE 1 A3Q	Si-N+Di+R	hi-beta,Rb=1k,Rbe=10kΩ, Z-Di, 60V, ±2A	9b	Nec	-	data
CE 1 F3P	Si-N+Di+R	=CE 1A3Q: Rb=2,2k, Rbe=10kΩ	9b	Nec	-	
CE 1 N2R	Si-N+Di+R	=CE 1A3Q: Rb=0,68k, Rbe=10kΩ	9b	Nec	-	data
CE 2 A3Q...N2R	Si-N+Di+R	=CE 1A3Q...N2R: ohne/wo. Z-Di	9b	Nec	-	data
CEC	Si-P	=BC 869 (Typ-Code/Stempel/marking)	39	Val	→BC 869	data
CEN	Si-N	=2SC4505-N (Typ-Code/Stempel/marking)	39	Rhm	→2SC4505	data
CEO	Si-N	=2SC3325-O (Typ-Code/Stempel/marking)	35	Tos	→2SC3325	data
CEP	Si-N	=2SC4505-P (Typ-Code/Stempel/marking)	39	Rhm	→2SC4505	data
CEQ	Si-N	=2SC4505-Q (Typ-Code/Stempel/marking)	39	Rhm	→2SC4505	data
CEs	Si-N	=BSS 79B (Typ-Code/Stempel/marking)	35	Sie	→BSS 79B	data
CEY	Si-N	=2SC3325-Y (Typ-Code/Stempel/marking)	35	Tos	→2SC3325	data
CF	Si-P	=2SA1366-F (Typ-Code/Stempel/marking)	35	Mit	→2SA1366	data
CF	Si-N	=2SC3650 (Typ-Code/Stempel/marking)	39	Say	→2SC3650	data
CF	Si-P	=BCX 69-10 (Typ-Code/Stempel/marking)	39	Sie,Val,Mot	→BCX 69-10	data
CF	Si-N	=BSS 79C (Typ-Code/Stempel/marking)	35	Fer,Mot,Sie	→BSS 79C	data
CF	Z-Di	=MMSZ 4681 (Typ-Code/marking)	71(2,7mm)	Ons	→MMSZ 4681	
CF	Z-Di	=P4 SMA-51 (Typ-Code/Stempel/marking)	71(5x2,5)	Fag	→P4 SMA-51	data
CF	Z-Di	=SMBJ 28C (Typ-Code/Stempel/marking)	71(5x3,5)	Mop	→SMBJ ...	data
CF	C-Di	=ZMV 834B (Typ-Code/Stempel/marking)	71(1,7mm)	Ztx	→ZMV 834	data
CF 1	GaAs-FET	=CF 910 (Typ-Code/Stempel/marking)	44	Aeg	→CF 910	data
CF 3	GaAs-FET	=CF 912 (Typ-Code/Stempel/marking)	44	Aeg	→CF 912	data
CF 4	GaAs-FET	=CF 922 (Typ-Code/Stempel/marking)	44	Aeg	→CF 922	data
CF 5	GaAs-FET	=CF 930 (Typ-Code/Stempel/marking)	44	Aeg	→CF 930	data
CF 100	GaAs-N-FET-d	Dual-Gate, UHF, ...2GHz, 10V, 80mA	25u	Aeg	-	data
CF 121	GaAs-N-FET-d	=CF 100: integr. Gate-Schutz/protect.	25u	Aeg	-	data
CF 221	GaAs-N-FET-d	Dual-Gate, UHF, ...2GHz, 10V, 80mA	25r	Aeg	-	data
CF 300	GaAs-N-FET-d	Dual-Gate, UHF, ...2GHz, 10V, 80mA	25u	Aeg	-	data
CF 400	GaAs-N-FET-d	Dual-Gate, UHF, ...2GHz, 10V, 80mA	25r	Aeg	-	data
CF 739	GaAs-N-FET-d	SMD, Dual-Gate, UHF, 10V, 80mA	44u	Sie	-	data
CF 750	GaAs-N-FET	SMD, Dual-Gate, 0,4...3GHz, 8V, 80mA	44(SGDGnd)	Sie	-	data
CF 910	GaAs-N-FET-d	=CF 100: SMD	44u	Aeg	-	data
CF 912	GaAs-N-FET-d	=CF 121: SMD	44u	Aeg	-	data
CF 922	GaAs-N-FET-d	=CF 221: SMD	44u	Aeg	-	data
CF 930	GaAs-N-FET-d	=CF 300: SMD	44u	Aeg	-	data
CF 930R	GaAs-N-FET-d	=CF 300: SMD	44-?	Aeg	-	data
CF 940	GaAs-N-FET-d	=CF 400: SMD	44u	Aeg	-	data
CF 1155	IC	Melody Lsis	MP	Npc	-	pdf pinout
CF 1156	IC	Melody Lsis	MP	Npc	-	pdf pinout
CF 2386	N-FET	Uni, 25V, 0,5W, Idss<9mA, Up<-8V	2a	Tdy	BF 244...245, BFW 10, 2N3819, 2N4222, ++	data
CF 5005 BLA	IC	High-freq. Crystal Oscill. Module Ics	MP	Npc	-	pdf pinout
CF 5005 BLB	IC	High-freq. Crystal Oscill. Module Ics	MP	Npc	-	pdf pinout
CF 5017 ALA	IC	3rd Overtone Crystal Oscil. Module Ics	MP	Npc	-	pdf pinout
CF 5017 ALB	IC	3rd Overtone Crystal Oscil. Module Ics	MP	Npc	-	pdf pinout

ype	Device	Short Description	Fig.	Manu	Comparision Types	More at
F 5017 ALC	IC	3rd Overtone Crystal Oscil. Module Ics	MP	Npc	-	pdf pinout
F 5017 ALD	IC	3rd Overtone Crystal Oscil. Module Ics	MP	Npc	-	pdf pinout
F 5018 ALA	IC	2.5V 3rd Crystal Oscillator Module Ics	MP	Npc	-	pdf pinout
F 5018 ALB	IC	2.5V 3rd Crystal Oscillator Module Ics	MP	Npc	-	pdf pinout
F 5018 ALC	IC	2.5V 3rd Crystal Oscillator Module Ics	MP	Npc	-	pdf pinout
F 5018 ALD	IC	2.5V 3rd Crystal Oscillator Module Ics	MP	Npc	-	pdf pinout
F 5018 ALE	IC	2.5V 3rd Crystal Oscillator Module Ics	MP	Npc	-	pdf pinout
5020 ALA	IC	3rd Overtone Crystal Oscil. Module Ics	MP	Npc	-	pdf pinout
5020 ALB	IC	3rd Overtone Crystal Oscil. Module Ics	MP	Npc	-	pdf pinout
5020 ALC	IC	3rd Overtone Crystal Oscil. Module Ics	MP	Npc	-	pdf pinout
F 5020 ALD	IC	3rd Overtone Crystal Oscil. Module Ics	MP	Npc	-	pdf pinout
F 5026 MLA	IC	Crystal Oscillator Module Ics	MP	Npc	-	pdf pinout
5034 AA	IC	350MHz PECL-output Oscillator Ics	MP	Npc	-	pdf pinout
5034 AB	IC	350MHz PECL-output Oscillator Ics	MP	Npc	-	pdf pinout
F 5034 BA	IC	350MHz PECL-output Oscillator Ics	MP	Npc	-	pdf pinout
F 5034 BB	IC	350MHz PECL-output Oscillator Ics	MP	Npc	-	pdf pinout
F 5034 DA	IC	350MHz PECL-output Oscillator Ics	MP	Npc	-	pdf pinout
F 5034 DB	IC	350MHz PECL-output Oscillator Ics	MP	Npc	-	pdf pinout
F 5034 LA	IC	350MHz PECL-output Oscillator Ics	MP	Npc	-	pdf pinout
F 5034 LB	IC	350MHz PECL-output Oscillator Ics	MP	Npc	-	pdf pinout
F 5034 MA	IC	350MHz PECL-output Oscillator Ics	MP	Npc	-	pdf pinout
F 5034 MB	IC	350MHz PECL-output Oscillator Ics	MP	Npc	-	pdf pinout
F 5035 ALA	IC	1.8V, Crystal Oscillator Module Ics	MP	Npc	-	pdf pinout
F 5035 ALB	IC	1.8V, Crystal Oscillator Module Ics	MP	Npc	-	pdf pinout
F 5035 ALC	IC	1.8V, Crystal Oscillator Module Ics	MP	Npc	-	pdf pinout
F 5035 ALD	IC	1.8V, Crystal Oscillator Module Ics	MP	Npc	-	pdf pinout
F 5072 AA	IC	155MHz VCXO Module Ics	MP	Npc	-	pdf pinout
F 5072 BA	IC	155MHz VCXO Module Ics	MP	Npc	-	pdf pinout
F 5705 AA	CMOS-IC	Analog Watch Stepping Motor Driver	LLP	Npc	-	pdf pinout
F 5732 EA	CMOS-IC	Analog clock Cmos IC	LLP	Npc	-	pdf pinout
F 5732 FA	CMOS-IC	Analog clock Cmos IC	LLP	Npc	-	pdf pinout
F 5732 GA	CMOS-IC	Analog clock Cmos IC	LLP	Npc	-	pdf pinout
F 5732 HA	CMOS-IC	Analog clock Cmos IC	LLP	Npc	-	pdf pinout
F 5732 JA	CMOS-IC	Analog clock Cmos IC	LLP	Npc	-	pdf pinout
F 5732 KA	CMOS-IC	Analog clock Cmos IC	LLP	Npc	-	pdf pinout
F 5734 BA	CMOS-IC	Analog clock Cmos IC	LLP	Npc	-	pdf pinout
F 5741 AA	CMOS-IC	Analog clock Cmos IC	LLP	Npc	-	pdf pinout
F 5741 AB	CMOS-IC	Analog clock Cmos IC	LLP	Npc	-	pdf pinout
F 5741 BA	CMOS-IC	Analog clock Cmos IC	LLP	Npc	-	pdf pinout
F 5741 BB	CMOS-IC	Analog clock Cmos IC	LLP	Npc	-	pdf pinout
F 5746 AAB	CMOS-IC	Analog clock Cmos IC	LLP	Npc	-	pdf pinout
F 5746 ABA	CMOS-IC	Analog clock Cmos IC	LLP	Npc	-	pdf pinout
F 5746 ADC	CMOS-IC	Analog clock Cmos IC	LLP	Npc	-	pdf pinout
F 5746 AEA	CMOS-IC	Analog clock Cmos IC	LLP	Npc	-	pdf pinout
F 5746 AFA	CMOS-IC	Analog clock Cmos IC	LLP	Npc	-	pdf pinout
F 5746 AGA	CMOS-IC	Analog clock Cmos IC	LLP	Npc	-	pdf pinout
F 5746 BCA	CMOS-IC	Analog clock Cmos IC	LLP	Npc	-	pdf pinout
F 5750 A	CMOS-IC	Analog clock Cmos IC	LLP	Npc	-	pdf pinout
F 5750 C	CMOS-IC	Analog clock Cmos IC	LLP	Npc	-	pdf pinout
F 5750 D	CMOS-IC	Analog clock Cmos IC	LLP	Npc	-	pdf pinout
F 5760 AA	IC	Cmos Analog Clock IC	LLP	Npc	-	pdf pinout
F 5760 BA	IC	Cmos Analog Clock IC	LLP	Npc	-	pdf pinout
F 5760 CC	IC	Cmos Analog Clock IC	LLP	Npc	-	pdf pinout
F 5760 DA	IC	Cmos Analog Clock IC	LLP	Npc	-	pdf pinout
F 5760 EA	IC	Cmos Analog Clock IC	LLP	Npc	-	pdf pinout
F 5760 FC	IC	Cmos Analog Clock IC	LLP	Npc	-	pdf pinout
F 5760 GB	IC	Cmos Analog Clock IC	LLP	Npc	-	pdf pinout
F 5760 HA	IC	Cmos Analog Clock IC	LLP	Npc	-	pdf pinout
F 5760 JA	IC	Cmos Analog Clock IC	LLP	Npc	-	pdf pinout
F 5761 CA	IC	Cmos Analog Clock IC, Alarm Function	LLP	Npc	-	pdf pinout
F 5761 EA	IC	Cmos Analog Clock IC, Alarm Function	LLP	Npc	-	pdf pinout
F 5761 HA	IC	Cmos Analog Clock IC, Alarm Function	LLP	Npc	-	pdf pinout
F 5761 LB	IC	Cmos Analog Clock IC, Alarm Function	LLP	Npc	-	pdf pinout
F 5761 MB	IC	Cmos Analog Clock IC, Alarm Function	LLP	Npc	-	pdf pinout
F 5761 NB	IC	Cmos Analog Clock IC, Alarm Function	LLP	Npc	-	pdf pinout
F 8140 A	IC	EL Driver IC	TSOP	Npc	-	pdf pinout
F 8141 A	IC	EL Sheet Driver	TSOP	Npc	-	pdf pinout
F 8141 B	IC	EL Sheet Driver	TSOP	Npc	-	pdf pinout
F 8223 A	IC	Fsk Decoder and Dtmf Receiver IC	LLP	Npc	-	pdf pinout
F 9501 A	BiCMOS-IC	Radio Controlled Clock Receiver IC	TSOP	Npc	-	pdf pinout
F 9502 A	BiCMOS-IC	Radio Controlled Clock Receiver IC	TSOP	Npc	-	pdf pinout
F 9503 A	BiCMOS-IC	Radio Controlled Clock Receiver IC	TSOP	Npc	-	pdf pinout
CF 5008 A1	IC	VCXO Module IC	MP	Npc	-	pdf pinout
CF 5008 A2	IC	VCXO Module IC	MP	Npc	-	pdf pinout
CF 5008 A3	IC	VCXO Module IC	MP	Npc	-	pdf pinout
CF 5014 AL 1	IC	Crystal Oscillator Module Ics	MP	Npc	-	pdf pinout
CF 5014 AL 2	IC	Crystal Oscillator Module Ics	MP	Npc	-	pdf pinout
CF 5014 AL 3	IC	Crystal Oscillator Module Ics	MP	Npc	-	pdf pinout
CF 5014 AL 4	IC	Crystal Oscillator Module Ics	MP	Npc	-	pdf pinout
CF 5014 AL 5	IC	Crystal Oscillator Module Ics	MP	Npc	-	pdf pinout
CF 5015 AL 1	IC	Crystal Oscillator Module Ics	MP	Npc	-	pdf pinout
CF 5015 BL 1	IC	Crystal Oscillator Module Ics	MP	Npc	-	pdf pinout
CF 5015 AL 2	IC	Crystal Oscillator Module Ics	MP	Npc	-	pdf pinout
CF 5015 BL 2	IC	Crystal Oscillator Module Ics	MP	Npc	-	pdf pinout
CF 5015 AL 3	IC	Crystal Oscillator Module Ics	MP	Npc	-	pdf pinout

Type	Device	Short Description	Fig.	Manu	Comparision Types	More at
CF 5015 BL 3	IC	Crystal Oscillator Module Ics	MP	Npc	-	pdf pinout
CF 5015 AL 4	IC	Crystal Oscillator Module Ics	MP	Npc		pdf pinout
CF 5015 BL 4	iC	Crystal Oscillator Module Ics	MP	Npc		pdf pinout
CF 5015 AL 5	IC	Crystal Oscillator Module Ics	MP	Npc		pdf pinout
CF 5015 BL 5	IC	Crystal Oscillator Module Ics	MP	Npc		pdf pinout
CF 5016 AL 1	IC	Crystal Oscillator Module Ics	MP	Npc		pdf pinout
CF 5016 AL 2	IC	Crystal Oscillator Module Ics	MP	Npc		pdf pinout
CF 5016 AL 3	IC	Crystal Oscillator Module Ics	MP	Npc		pdf pinout
CF 5016 AL 4	IC	Crystal Oscillator Module Ics	MP	Npc		pdf pinout
CF 5016 AL 5	IC	Crystal Oscillator Module Ics	MP	Npc		pdf pinout
CF 5026 AL 1	IC	Crystal Oscillator Module Ics	MP	Npc		pdf pinout
CF 5026 BL 1	IC	Crystal Oscillator Module Ics	MP	Npc		pdf pinout
CF 5026 AL 2	IC	Crystal Oscillator Module Ics	MP	Npc		pdf pinout
CF 5026 AL 3	IC	Crystal Oscillator Module Ics	MP	Npc		pdf pinout
CF 5026 AL 4	IC	Crystal Oscillator Module Ics	MP	Npc		pdf pinout
CF 5026 AL 5	IC	Crystal Oscillator Module Ics	MP	Npc		pdf pinout
CF 5026 AL 6	IC	Crystal Oscillator Module Ics	MP	Npc		pdf pinout
CF 5029 A–2	IC	Crystal Oscillator Module IC	MP	Npc		pdf pinout
CF 5036 A 1	IC	2.5V LVPECL Output Oscillator Ics	MP	Npc		pdf pinout
CF 5036 B 1	IC	2.5V LVPECL Output Oscillator Ics	MP	Npc		pdf pinout
CF 5036 C 1	IC	2.5V LVPECL Output Oscillator Ics	MP	Npc		pdf pinout
CF 5036 D 1	IC	2.5V LVPECL Output Oscillator Ics	MP	Npc		pdf pinout
CF 5036 E 1	IC	2.5V LVPECL Output Oscillator Ics	MP	Npc		pdf pinout
CF 5036 F 1	IC	2.5V LVPECL Output Oscillator Ics	MP	Npc		pdf pinout
CF 5036 G 1	IC	2.5V LVPECL Output Oscillator Ics	MP	Npc		pdf pinout
CF 5036 V 1	IC	2.5V LVPECL Output Oscillator Ics	MP	Npc		pdf pinout
CF 5036 W 1	IC	2.5V LVPECL Output Oscillator Ics	MP	Npc		pdf pinout
CF 5036 A 2	IC	2.5V LVPECL Output Oscillator Ics	MP	Npc		pdf pinout
CF 5036 B 2	IC	2.5V LVPECL Output Oscillator Ics	MP	Npc		pdf pinout
CF 5036 C 2	IC	2.5V LVPECL Output Oscillator Ics	MP	Npc		pdf pinout
CF 5036 D 2	IC	2.5V LVPECL Output Oscillator Ics	MP	Npc		pdf pinout
CF 5036 E 2	IC	2.5V LVPECL Output Oscillator Ics	MP	Npc		pdf pinout
CF 5036 F 2	IC	2.5V LVPECL Output Oscillator Ics	MP	Npc		pdf pinout
CF 5036 G 2	IC	2.5V LVPECL Output Oscillator Ics	MP	Npc		pdf pinout
CF 5036 V 2	IC	2.5V LVPECL Output Oscillator Ics	MP	Npc		pdf pinout
CF 5036 W 2	IC	2.5V LVPECL Output Oscillator Ics	MP	Npc		pdf pinout
CF 5037 A 1	IC	2.5V LVDS Output Oscillator Ics	MP	Npc		pdf pinout
CF 5037 B 1	IC	2.5V LVDS Output Oscillator Ics	MP	Npc		pdf pinout
CF 5037 C 1	IC	2.5V LVDS Output Oscillator Ics	MP	Npc		pdf pinout
CF 5037 D 1	IC	2.5V LVDS Output Oscillator Ics	MP	Npc		pdf pinout
CF 5037 E 1	IC	2.5V LVDS Output Oscillator Ics	MP	Npc		pdf pinout
CF 5037 F 1	IC	2.5V LVDS Output Oscillator Ics	MP	Npc		pdf pinout
CF 5037 V 1	IC	2.5V LVDS Output Oscillator Ics	MP	Npc		pdf pinout
CF 5037 B 2	IC	2.5V LVDS Output Oscillator Ics	MP	Npc		pdf pinout
CF 5037 C 2	IC	2.5V LVDS Output Oscillator Ics	MP	Npc		pdf pinout
CF 5037 D 2	IC	2.5V LVDS Output Oscillator Ics	MP	Npc		pdf pinout
CF 5037 E 2	IC	2.5V LVDS Output Oscillator Ics	MP	Npc		pdf pinout
CF 5037 F 2	IC	2.5V LVDS Output Oscillator Ics	MP	Npc		pdf pinout
CF 5037 V 2	IC	2.5V LVDS Output Oscillator Ics	MP	Npc		pdf pinout
CF 5074 A-1	IC	VCXO Module IC with Built-in Varicap	MP	Npc		pdf pinout
CF 5074 B-1	IC	80MHz VCXO IC	LLP	Npc		pdf pinout
CF 5074 A-3	IC	VCXO Module IC with Built-in Varicap	MP	Npc		pdf pinout
CF 5762 AB 0	IC	Analog Chime Clock Cmos IC	MP	Npc		pdf pinout
CFD470	PD	880nm S:0.5mA/mW	TO18	Cla		data pdf pinout
CFE320	IRED	850nm ø3.2mm I:71lm Vf:1.8V If:100mA	TO46	Cla		data pdf pinout
CFE370	IRED	850nm ø3.2mm I:71lm Vf:1.7V If:100mA	TO18	Cla		data pdf pinout
CFK 10	GaAs-N-FET-d	Dual-Gate, UHF, ..2GHz, 10V, 50mA	51zc	Aeg		data
CFK 12	GaAs-N-FET-d	=CFK 10: integr. Gate-Schutz/protect.	51zc	Aeg		data
CFK 22	GaAs-N-FET-d	=CFK 10: integr. Gate-Schutz/protect.	51zd	Aeg		data
CFK 30	GaAs-N-FET-d	Dual-Gate, UHF, ..2GHz, 10V, 80mA	51zc	Aeg		data
CFK 40	GaAs-N-FET-d	Dual-Gate, UHF, ..2GHz, 10V, 80mA	51zd	Aeg		data
CFM 13026	N-FET	Uni, 40V, Idss<50mA, Up<-7V	2a	Tsc	BF 348, BFT 10, 2N4091...4092	data
CFQ	Si-N	=2SD2150-Q (Typ-Code/Stempel/marking)	39	Rhm	→2SD2150	data
CFR	Si-N	=2SD2150-R (Typ-Code/Stempel/marking)	39	Rhm	→2SD2150	data
CFS	Si-N	=2SD2150-S (Typ-Code/Stempel/marking)	39	Rhm	→2SD2150	data
CFs	Si-N	=BSS 79C (Typ-Code/Stempel/marking)	35	Sie	→BSS 79C	data
CFX 13	GaAs-FET-N	Mikro-W., ≈12GHz, 5V, 100mA	52y	Phi	-	data
CFX 14	GaAs-FET	=CFX 13: ≈16GHz	52y	Phi		data
CFX 21	GaAs-FET-N	Mikro-W., ≈11GHz, 8V, 110mA	52y	Phi		data
CFX 30	GaAs-FET-N	Mikro-W., ≈11GHz, 15V, 130mA	62f*	Phi	-	data
CFX 31	GaAs-FET-N	Mikro-W., ≈11GHz, 15V, 250mA	62f*	Phi		data
CFX 32	GaAs-FET-N	Mikro-W., ≈8,5GHz, 15V, 500mA	62f*	Phi		data
CFX 33	GaAs-FET-N	Mikro-W., ≈8,5GHz, 15V, 600mA	62f*	Phi		data
CFY 10	GaAs-N-FET-d	Mikro-W., ≈6GHz, 5V, 100mA	52z	Sie		data
CFY 11	GaAs-FET	Mikro-W., ≈6GHz, 5V, 100mA	52z	Sie		data
CFY 12	GaAs-FET	Mikro-W., ≈6GHz, 5V, 100mA	52z	Sie		data
CFY 13	GaAs-FET	Mikro-W., ≈6GHz, 5V, 100mA	52z	Sie		data
CFY 14	GaAs-FET	Mikro-W., ≈6GHz, 5V, 100mA	51z	Sie		data
CFY 15	GaAs-FET	Mikro-W., ≈6GHz, 5V, 100mA	52z	Sie		data
CFY 16	GaAs-FET	Mikro-W., ≈12GHz, 5V, 100mA	52z	Sie		data
CFY 17	GaAs-FET	Mikro-W., ≈6GHz, 5V, 100mA	51z	Sie		data
CFY 18	GaAs-FET	Mikro-W., ≈6GHz, 5V, 100mA	51z	Sie		data
CFY 19	GaAs-N-FET-d	Mikro-W., ≈6GHz, 5V, 100mA	51z	Sie		data

Type	Device	Short Description	Fig.	Manu	Comparision Types	More at
CFY 20	GaAs-FET	Dual-Gate	52	Sie	-	data
CFY25(-17,-20,-25)	GaAs-N-FET-d	Mikro-W., 5V, 80mA	51z	Sie	-	data
CFY 30	GaAs-N-FET-d	SMD, Mikro-W., ≈4...6GHz, 7V, 80mA	44za	Sie	-	data
CFY 35(-20,-23)	GaAs-N-FET-d	SMD, Mikro-W., 5V, 60mA	≈44za	Sie	-	data
CFY 65(-12,-14)	GaAs-N-FET-d	Min, Mikro-W.(HEMT-FET), 4V, 70mA	51z	Sie	-	data
CFY 66	GaAs-N-FET	Min, Mikro-W., 3,5V, 60mA	51z	Sie	-	data
CFY 75(-13,-15)	GaAs-N-FET-d	SMD, UHF-V, ...20GHz, 5V, 70mA	≈44za	Sie	-	data
CFY 76-08	GaAs-N-FET	SMD, Mikro-W.(HEMT-FET), 3,5V, 60mA	≈44za	Sie	-	data
CFY 77(-08,-10)	GaAs-N-FET-d	SMD, Mikro-W.(HEMT-FET), 3,5V, 60mA	≈44	Sie	-	data
CG	Si-P	=2SA1163-GR (Typ-Code/Stempel/marking)	35	Tos	→2SA1163	data
CG	Si-P	=2SA1587-GR (Typ-Code/Stempel/marking)	35	Tos	→2SA1587	data
CG	Si-N	=2SC3651 (Typ-Code/Stempel/marking)	39	Say	→2SC3651	data
CG	Si-P	=BCX 69-16 (Typ-Code/Stempel/marking)	39	Sie,Val,Mot	→BCX 69-16	data
CG	Si-P	=BCX 71RG (Typ-Code/Stempel/marking)	35	Fer	→BCX 71RG	data
CG	Si-N	=BSS 81C (Typ-Code/Stempel/marking)	35	Sie	→BSS 81C	data
CG	Si-N	=CPH 3207 (Typ-Code/Stempel/marking)	35	Say	→CPH 3207	data
CG	Si-P	=KTA2017-GR (Typ-Code/Stempel/marking)	35(2mm)	Kec	→KTA 2017	data
CG	Z-Di	=P4 SMA-51A (Typ-Code/Stempel/marking)	71(5x2,5)	Fag	→P4 SMA-51A	data
CG	Z-Di	=SMAJ 28A (Typ-Code/Stempel/marking)	71(5x2,5)	Tho	→SMAJ ...	data
CG	Z-Di	=SMBJ 28CA (Typ-Code/Stempel/marking)	71(5x3,5)	Mop	→SMBJ ...	data
CG	C-Di	=ZMV 835B (Typ-Code/Stempel/marking)	71(1,7mm)	Ztx	→ZMV 835	data
CG 1	Si-Di	TV-Damper-Di, 1400V, 1,5A	31a	Gie	BY 228, BY 328, BY 448	data
CG 2	Si-Di	=CG 1: 2A	31a	Gie	BY 228, BY 328, BY 448	data
CG 3	Si-Di	=CG 1: 3A	31a	Gie	BY 228, BY 328, BY 448	data
CG 61 H	Ge-Di	≈AA 133	31a		→AA 133	data
CG 64 H	Ge-Di	≈1N34			→1N34	data
CGC	Si-P	=BC 869-... (Typ-Code/Stempel/marking)	39	Phi	→BC 869	data
CGG	Si-N	=2SC3426-GR (Typ-Code/Stempel/marking)	35	Tos	→2SC3426	data
CGL	Si-N	=2SC3426-BL (Typ-Code/Stempel/marking)	35	Tos	→2SC3426	data
CGP	Si-N	=2SC5053-P (Typ-Code/Stempel/marking)	39	Rhm	→2SC5053	data
CGQ	-Si-N	=2SC5053-Q (Typ-Code/Stempel/marking)	39	Rhm	→2SC5053	data
CGR	Si-N	=2SC5053-R (Typ-Code/Stempel/marking)	39	Rhm	→2SC5053	data
CGs	Si-N	=BSS 81C (Typ-Code/Stempel/marking)	35	Sie	→BSS 81C	data
CXG 1012N	IC	SMD, High-freq. SPDT Antenna Switch	SSMDIP	Son	-	pdf pinout
CGY	Si-N	=2SC3426-Y (Typ-Code/Stempel/marking)	35	Tos	→2SC3426	data
CGY 10...14	(GaAs)-Di	Gunn Di., UHF		Aeg	-	data
CGY 20...0918	GaAs-FET-IC	Mmic, Mikro-w./broadband Amplifier		Sie	-	data
CH	Si-Di	=1SS351 (Typ-Code/Stempel/marking)	35	Say	→1SS351	data
CH	Si-P	=2SB1002-CH (Typ-Code/Stempel/marking)	39	Hit	→2SB1002	data
CH	Si-N	=2SC4272 (Typ-Code/Stempel/marking)	39	Say	→2SC4272	data
CH	Si-P	=BCX 69-25 (Typ-Code/Stempel/marking)	39	Mot,Sie,Val	→BCX 69	data
CH	Si-P	=BSS 80B (Typ-Code/Stempel/marking)	35	Mot, Sie	→BSS 80B	data
CH	Z-Di	=MMSZ 4682 (Typ-Code/marking)	71(2,7mm)	Ons	→MMSZ 4682	
CH	Z-Di	=P4 SMA-56 (Typ-Code/Stempel/marking)	71(5x2,5)	Fag	→P4 SMA-56	data
CH	Z-Di	=SMAJ 28CA (Typ-Code/Stempel/marking)	71(5x2,5)	Tho	→SMAJ ...	data
CH	Z-Di	=SMBJ 30C (Typ-Code/Stempel/marking)	71(5x3,5)	Mop	→SMBJ ...	data
CHC	Si-P	=BC 869-... (Typ-Code/Stempel/marking)	39	Phi	→BC 869	data
CHO	Si-N	=2SC3437-O (Typ-Code/Stempel/marking)	35	Tos	→2SC3437	data
CHO	Si-N	=2SC4667-O (Typ-Code/Stempel/marking)	35(2mm)	Tos	→2SC4667	data
CHR	Si-N	=2SC3437-R (Typ-Code/Stempel/marking)	35	Tos	→2SC3437	data
CHR	Si-N	=2SC4667-R (Typ-Code/Stempel/marking)	35(2mm)	Tos	→2SC4667	data
CHs	Si-P	=BSS 80B (Typ-Code/Stempel/marking)	35	Sie	→BSS 80B	data
CHY	Si-N	=2SC3437-Y (Typ-Code/Stempel/marking)	35	Tos	→2SC3437	data
CHY	Si-N	=2SC4667-Y (Typ-Code/Stempel/marking)	35(2mm)	Tos	→2SC4667	data
CI	Si-N	=2SC4080 (Typ-Code/Stempel/marking)	39	Say	→2SC4080	data
CI	Si-N	=2SD1777 (Typ-Code/Stempel/marking)	44	Hit	→2SD1777	data
CIC 7641	LIN-IC	→TA 7641BP	16-DIP		→TA 7641BP	data
CIC 9185	LIN-IC	→KA 2412A	14-DIP		KA 2412A, LS 285A	data
CIP 3250A	DIG-IC	Component Interface Processor	PLCC	Mas	-	pdf pinout
CJ	Si-P	=2SB1002-CJ (Typ-Code/Stempel/marking)	39	Hit	→2SB1002	data
CJ	Si-N	=2SC4390 (Typ-Code/Stempel/marking)	39	Say	→2SC4390	data
CJ	MOS-N-FET-d	=2SK1067 (Typ-Code/Stempel/marking)	35(2mm)	Say	→2SK1067	data
CJ	MOS-N-FET-d	=2SK543 (Typ-Code/Stempel/marking)	35	Say	→2SK543	data
CJ	Si-P	=BSS 80C (Typ-Code/Stempel/marking)	35	Mot, Sie	→BSS 80C	data
CJ	Si-N	=CPH 3209 (Typ-Code/Stempel/marking)	35	Say	→CPH 3209	data
CJ	Z-Di	=MMSZ 4683 (Typ-Code/marking)	71(2,7mm)	Ons	→MMSZ 4683	
CJs	Si-P	=BSS 80C (Typ-Code/Stempel/marking)	35	Sie	→BSS 80C	data
CK	Si-N	=2SC4520 (Typ-Code/Stempel/marking)	39	Say	→2SC4520	data
CK	Si-N	=2SD999-CK (Typ-Code/Stempel/marking)	39	Nec	→2SD999	data
CK	Si-P	=BCX 71RK (Typ-Code/Stempel/marking)	35	Fer	→BCX 71RK	data
CK	Si-N	=CPH 3210 (Typ-Code/Stempel/marking)	35	Say	→CPH 3210	data
CK	Z-Di	=MMSZ 4684 (Typ-Code/marking)	71(2,7mm)	Ons	→MMSZ 4684	
CK	Z-Di	=P4 SMA-56A (Typ-Code/Stempel/marking)	71(5x2,5)	Fag	→P4 SMA-56A	data
CK	Z-Di	=SMAJ 30A (Typ-Code/Stempel/marking)	71(5x2,5)	Tho	→SMAJ ...	data
CK	Z-Di	=SMBJ 30CA (Typ-Code/Stempel/marking)	71(5x3,5)	Mop	→SMBJ ...	data
CK	Si-P+R	=XN 6114 (Typ-Code/Stempel/marking)	46	Mat	→XN 6114	data
CK	Si-P+R	=XP 6114 (Typ-Code/Stempel/marking)	46(2mm)	Mat	→XP 6114	data
CK 300	IC	Digital-uhr/digital clock	28-DIP			

Type	Device	Short Description	Fig.	Manu	Comparision Types	More at
CL...CR						
CL	Si-P	=2SA1163-BL (Typ-Code/Stempel/marking)	35	Tos	→2SA1163	data
CL	Si-P	=2SA1342 (Typ-Code/Stempel/marking)	35	Say	→2SA1342	data
CL	Si-P	=2SA1587-BL (Typ-Code/Stempel/marking)	35(2mm)	Tos	→2SA1587	data
CL	Si-P	=2SA1677 (Typ-Code/Stempel/marking)	35(2mm)	Say	→2SA1677	data
CL	Si-N	=2SC4521 (Typ-Code/Stempel/marking)	39	Say	→2SC4521	data
CL	Si-N	=2SD999-CL (Typ-Code/Stempel/marking)	39	Nec	→2SD999	data
CL	Si-P	=BSS 82B (Typ-Code/Stempel/marking)	35	Fer,Mot,Sie	→BSS 82B	data
CL	Si-P	=KTA2017-BL (Typ-Code/Stempel/marking)	35(2mm)	Kec	→KTA 2017	data
CL	Z-Di	=P4 SMA-62 (Typ-Code/Stempel/marking)	71(5x2,5)	Fag	→P4 SMA-62	data
CL	Z-Di	=SMAJ 30CA (Typ-Code/Stempel/marking)	71(5x2,5)	Tho	→SMAJ ...	data
CL	Z-Di	=SMBJ 33C (Typ-Code/Stempel/marking)	71(5x3,5)	Mop	→SMBJ ...	data
CL3P	LDR	615nm Ro:0.08MOhm Vb:4V		Cla		data pinout
CL5M	LDR	550nm Ro:7.3MOhm Vb:10V	TO8	Cla		data pinout
CL5P	LDR	550nm Ro:6MOhm Vb:10V		Cla		data pinout
CL7P	LDR	550nm Ro:18.7MOhm Vb:10V		Cla		data pinout
CL9P	LDR	550nm Ro:10MOhm Vb:10V		Cla		data pinout
CL9P5	LDR	550nm Ro:0.67MOhm Vb:10V		Cla		data pdf pinout
CL 055(A...D)	Si-P	NF-Tr, 25V, 1A, 0,625W, 120MHz	7e	Mic	BC 327...328, BC 636, BC 638, BC 640, ++	data
CL 055...P	Si-P	=CL 055...: 0,75W	30e		2SB968, 2SB1181...82, (BD 506, BD 508),++[6]	data
CL 066(A...D)	Si-N	NF-Tr, 25V, 1A, 0,625W, 120MHz	7e	Mic	BC 337...338, BC 635, BC 637, BC 639, ++	data
CL 066...P	Si-N	=CL 066...: 0,75W	30e		2SD1078, 2SD1295, (BD 505, BD 507), ++[6]	data
CL 116 P	Si-P	≈BC 307	7a		→BC 307	
CL 151-3(A...C)	Si-N	Min, Uni, -/10V, 0,1A, 0,1W	≈36b	Mic	BC 122...123, BC 146	data
CL 151-4(A...C)	Si-N	Min, Uni, -/10V, 0,1A, 0,1W	7c		BC 122...123, BC 146	data
CL 152-3(A...C)	Si-P	Min, Uni, -/10V, 0,1A, 0,1W	≈36b	Mic	BC 200, BC 202...203	data
CL 152-4(A...C)	Si-P	Min, Uni, -/10V, 0,1A, 0,1W	7c		BC 200, BC 202..203	data
CL 155(A...D)	Si-P	NF-Tr, 30V, 1,5A, 0,625W, 120MHz	7e	Mic	MPS 650...51, 2SC2236, 2SD1146, 2SD1207	data
CL 155...P	Si-P	=CL 155..: 0,75W	30e		2SB968, 2SD1181...82, (BD 506, BD 508),++[6]	data
CL 166(A...D)	Si-N	NF-Tr, 30V, 1,5A, 0,625W, 120MHz	7e	Mic	MPS 750...51, 2SA966, 2SA1315, 2SA1382	data
CL 166...P	Si-N	=CL 166...: 0,75W	30e		2SD1078, 2SD1295, (BD 505, BD 507), ++[6]	data
CL 168	Si-N	NF-Tr/E, -/7V, 3A, 0,625W, 120MHz	7c	Mic	2SD879, 2SD1347, 2SD1507, 2SD1617, ++	data
CL 169	Si-N	NF-Tr/E, -/9V, 3A, 0,625W, 100MHz	7c	Mic	2SD879, 2SD1347, 2SD1507, 2SD1617, ++	data
CL 266	Si-N	NF-Tr/E, -/60V, 2A, 0,625W	7c	Mic	MPS 651, 2SC3328, 2SC3669, 2SC4145	data
CL 266 P	Si-N	=CL 166...: 0,75W	30e		2SD1079, 2SD1281...82, (BD 519, BD 527)++[6]	data
CL600	LDR	615nm Ro:2.3MOhm Vb:8V		Cla		data pinout
CL700	LDR	550nm Ro:18.7MOhm Vb:10V	TO5	Cla		data pinout
CL 855(A...C)	Si-P	NF-Tr, 70V, 1A, 0,625W, 150MHz	7e	Mic	BC 640, 2SB647, 2SA1013	data
CL 866(A...C)	Si-N	NF-Tr, 70V, 1A, 0,625W, 150MHz	7e	Mic	BC 639, 2SD667, 2N3700...3701, ++	data
CL900	LDR	550nm Ro:67MOhm Vb:10V	TO18	Cla		data pinout
CLA7	OC	LED/NPN Viso:7000V CTR:>20% Ibs:10mA		Cla		data pinout
CLA7AA	OC	LED/NPN Viso:7000V CTR:>10% Ibs:10mA		Cla		data pinout
CLA7D	OC	LED/Dar Viso:7000V CTR:>400%		Cla		data pinout
CLA7DA	OC	LED/Dar Viso:7000V CTR:>200%		Cla		data pinout
CLA60	OC	LED/NPN Viso:10000V CTR:>40% Ibe:4mA		Cla		data pinout
CLA60AA	OC	LED/NPN Viso:10000V CTR:>20% Ibe:2mA		Cla		data pinout
CLA60AB	OC	LED/NPN Viso:10000V CTR:>10% Ibe:1mA		Cla		data pinout
CLA65	OC	LED/Dar Viso:10000V CTR:>400%		Cla		data pinout
CLA65AA	OC	LED/Dar Viso:10000V CTR:>200%		Cla		data pinout
CLA90A	OC	LED/PD+IC Viso:10000V Vbs:<1.5V		Cla		data pinout
CLA90AA	OC	LED/PD+IC Viso:10000V Vbs:<1.5V		Cla		data pinout
CLA101	PT	860nm P:>150µA/mW/cm²		Cla		data pdf pinout
CLB 2711	KOP-IC	Dual, +14/-7V, 50mA, 0...+70°C	82	Idr		
CLB 2711	KOP-IC	Dual, +14/-7V, 50mA, 0...+70°C	14-DIP	Idr		
CLC 011	CMOS-IC	Serial Digital Video Decoder	PLCC	Nsc		pdf pinout
CLC103A	OP	Vs:±16V Vo:>±11V Vi0:10mV f:155Mc	24D	Com		data pdf pinout
CLC104A	OP	Vs:±16V Vu:14.2dB Vo:<±3.5V f:1100Mc	14Y	Com		data pinout
CLC109A	BUFF	Vs:±7V Vo:3.8/2.V Vi0:1mV f:270Mc	8DSd	Nsc		data pinout
CLC110A	BUFF	Vs:±7V Vo:±4V Vi0:2mV f:730Mc	8DSd	Nsc		data pinout
CLC111A	BUFF	Vs:±7V Vo:3.5/0V Vi0:2mV f:800Mc	8DSd	Nsc		data pdf pinout
CLC114A	4xVIDEO-BUFF	Vs:±7V Vo:±4V Vi0:±0.5mV f:200Mc	14DSd	Nsc		data pdf pinout
CLC115A	4xBUFF	Vs:±7V Vo:±3.7V Vi0:±2mV f:700Mc	14DSd	Nsc		data pinout
CLC200A	OP	Vs:±17V Vo:±12V Vi0:10mV f:100Mc	12A	Com		data pdf pinout
CLC201A	HI-REL	Vs:±20V Vo:±12V Vi0:0.5mV f:95Mc	12A	Com		data pinout
CLC203A	HI-REL	Vs:±20V Vo:±11V Vi0:0.5mV f:160Mc	24D	Com		data pinout
CLC210A	OP	Vs:±40V Vo:±33V Vi0:5mV f:50Mc	12A	Com		data pinout
CLC220A	OP	Vs:±17V Vo:±12V Vi0:10mV f:200Mc	12A	Com		data pinout
CLC231A	OP	Vs:±20V Vo:±12V Vi0:1mV f:65Mc	12A	Com		data pdf pinout
CLC300A	OP	Vs:±16V Vo:10/0V Vi0:10mV f:85Mc	24D	Com		data pinout
CLC 400(...)JP	OP-IC	±5<7V, ±70mA, 200MHz, 700V/µs, 12ns	8-DIP	Com		data
CLC400A	FAST-SET	Vs:±7V Vo:±3.5V Vi0:2mV f:200Mc	8DS	Nsc		data pdf pinout
CLC 400(...)JE	OP-IC	=CLC 400JP: SMD	8-MDIP			
CLC 401(...)JP	OP-IC	±5<7V, ±70mA, 150MHz, 1200V/µs, 10ns	8-DIP	Com		
CLC401A	FAST-SET	Vs:±50V Vo:2.8/0V Vi0:3mV f:150Mc	8DS	Nsc		data pdf pinout
CLC 401(...)JE	OP-IC	=CLC 401JP: SMD	8-MDIP			
CLC 402(...)JP	OP-IC	±5<7V, ±55mA, 175MHz, 800V/µs, 25ns	8-DIP	Com		
CLC 402(...)JE	OP-IC	=CLC 402JP: SMD	8-MDIP			
CLC 404(...)JP	OP-IC	±5<7V, ±70mA, 175MHz, 2600V/µs, 10ns	8-DIP	Com		data
CLC404A	OP	Vs:±7V Vo:±3.3V Vi0:2mV f:175Mc	8DSd,5S	Com		data pinout
CLC 404(...)JE	OP-IC	=CLC 404JP: SMD	8-MDIP			
CLC405A	LO-POWER	Vs:±7V Vo:3.5/2.V Vi0:1mV f:110Mc	8DSd	Nsc		data pinout

Type	Device	Short Description	Fig.	Manu	Comparision Types	More at
CLC 406(...)JP	OP-IC	±5<7V, ±70mA, 160MHz, 1500V/µs, 12ns	8-DIP	Com	-	
CLC406A	WIDEBAND	Vs:±7V Vo:3/2.7V Vi0:2mV f:160Mc	8S,5SD	Nsc		data pdf pinout
CLC 406(...)JE	OP-IC	=CLC 406JP: SMD	8-MDIP		-	
CLC407A	PROG-BUFF	Vs:±7V Vo:4/3.3V Vi0:1mV f:110Mc	8DSd	Nsc		data pinout
CLC 409(...)JP	OP-IC	±5<7V, ±70mA, 350MHz, 1200V/µs, 8ns	8-DIP	Com	-	
CLC409A	WIDEBAND	Vs:±7V Vo:±3.5V Vi0:0.5mV f:350Mc	8S,5SDd	Nsc		data pdf pinout
CLC 409(...)JE	OP-IC	=CLC 409JP: SMD	8-MDIP		-	
CLC 410(...)JP	OP-IC	±5<7V, ±70mA, 200MHz, 700V/µs, 12ns	8-DIP	Com	-	data
CLC410A	VIDEO-OP	Vs:±7V Vo:±3.5V Vi0:2mV f:200Mc	8DSd	Nsc		data pinout
CLC 410(...)JE	OP-IC	=CLC 410JP: SMD	8-MDIP		-	
CLC411A	VIDEO-OP	Vs:±18V Vo:±4.5V Vi0:2mV f:200Mc	8DSd	Nsc		data pdf pinout
CLC412A	2xVIDEO-OP	Vs:±7V Vo:3/2.9V Vi0:±2mV f:250Mc	8D,20CSd	Nsc		data pdf pinout
CLC 414(...)JP	OP-IC	Quad,±5V, ±70mA, 90MHz, 1000V/µs, 16ns	8-DIP	Com	-	
CLC414A	4xLO-POWER	Vs:±7V Vo:±2.8V Vi0:2mV f:90Mc	14DSd	Nsc		data pdf pinout
CLC 414(...)JE	OP-IC	=CLC 414JP: SMD	14-MDIP		-	
CLC 415(...)JP	OP-IC	Quad,±5V, ±70mA, 160MHz, 1500V/µs,12ns	8-DIP	Com	-	data
CLC415A	4xWIDEBAND	Vs:±7V Vo:±2.6V Vi0:2mV f:160Mc	14SD	Nsc		data pdf pinout
CLC415A8	4xOP	Vs:±7V Vo:±2.6V Vi0:2mV f:160Mc	14D	Com		data pinout
CLC 415(...)JE	OP-IC	=CLC 415JP: SMD	14-MDIP		-	
CLC416A	2xLO-POWER	Vs:±7V Vo:3.5/2.V Vi0:1mV f:120Mc	8SD	Nsc		data pinout
CLC417A	2xPROG-BUFF	Vs:±7V Vo:4/3.4V Vi0:1mV f:120Mc	8SD	Nsc		data pdf pinout
CLC420A	FB-OP	Vs:±7V Vo:±2.9V Vi0:1mV f:300Mc	8DSd	Nsc		data pinout
CLC420B	FB-OP	Vs:±7V Vo:±2.9V Vi0:0.5mV f:300Mc	8DS	Nsc		data pinout
CLC425A	WIDEBAND	Vs:±7V Vu:96dB Vo:±3.4V Vi0:±0.1mV	8DS,5Sd	Nsc		data pdf pinout
CLC426A	FB-OP	Vs:±7V Vu:64dB Vo:±3.5V Vi0:1mV	8DSd	Nsc		data pinout
CLC428A	2xFB-OP	Vs:±7V Vu:60dB Vo:±3.5V Vi0:1mV	8DSd	Nsc		data pdf pinout
CLC430A	WIDEBAND	Vs:±16.5V Vo:±14V Vi0:1mV f:100Mc	8DS	Nsc		data pinout
CLC431	2xWIDEBAND	Vs:±16.5V Vo:±14V Vi0:3mV f:62Mc	14DSd	Nsc		data pinout
CLC432	2xWIDEBAND	Vs:±16.5V Vo:±14V Vi0:3mV f:62Mc	8DSd	Nsc ·		data pdf pinout
CLC440A	FB-OP	Vs:±6V Vo:±2.5V Vi0:1mV f:260Mc	8DS	Nsc		data pinout
CLC446A	FB-OP	Vs:±6V Vo:±3.1V Vi0:2mV f:400Mc	8DSd	Nsc		data pinout
CLC449A	WIDEBAND	Vs:±6V Vo:2.9/0V Vi0:3mV f:1100Mc	8SD	Nsc		data pinout
CLC450A	FB-AMP	Vs:14V Vo:±3.8V Vi0:2mV f:135Mc	8S,5SDd	Nsc		data pinout
CLC451A	PROG-BUFF	Vs:14V Vo:±3.8V Vi0:3mV f:100Mc	8S,5SDd	Nsc		data pinout
CLC452A	FB-AMP	Vs:14V Vo:±3.8V Vi0:1mV f:160Mc	8DS,5Sd	Nsc		data pinout
CLC453A	PROG-BUFF	Vs:14V Vo:±3.8V Vi0:7mV f:130Mc	8S,5SDd	Nsc		data pinout
CLC 500(...)JP	OP-IC	±5<7V, ±70mA, 150MHz, 800V/µs, 18ns	14-DIP	Com	-	
CLC 500(...)JE	OP-IC	=CLC 500JP: SMD	14-MDIP		-	
CLC 501(...)JP	OP-IC	±5<7V, ±70mA, 80MHz, 1200V/µs, 12ns	8-DIP	Com	-	
CLC501A	CLAMP-OP	Vs:±7V Vo:±3.5V Vi0:1.5mV f:75Mc	8SD	Nsc		data pdf pinout
CLC 501(...)JE	OP-IC	=CLC 501JP: SMD	8-MDIP		-	
CLC502	CLAMP-OP	Vs:±7V Vo:±3.5V Vi0:0.5mV f:150Mc	8DS	Com		data pdf pinout
CLC 502(...)JP	OP-IC	±5<7V, ±70mA, 150MHz, 800V/µs, 18ns	8-DIP	Com	-	data
CLC502A	LO-GAIN	Vs:±7V Vo:±3.5V Vi0:0.5mV f:150Mc	8SD	Nsc		
CLC 502(...)JE	OP-IC	=CLC 502JP: SMD	8-MDIP		-	
CLC 505(...)JP	OP-IC	±5<7V, ±45mA, 150MHz, 1700V/µs, 12ns	8-DIP	Com	-	
CLC505A	HI-SPEED	Vs:±7V Vo:±3.3V Vi0:2mV f:150Mc	8SdD	Nsc		data pdf pinout
CLC 505(...)JE	OP-IC	=CLC 505JP: SMD	8-MDIP		-	
CLC520A	WIDEBAND	Vs:±7V Vo:±3.2V Vi0:±120mV f:100Mkc	14DSd	Nsc		data pinout
CLC522A	WIDEBAND	Vs:±7V Vi0:±85mV	20C,14DSd	Nsc		data pinout
CLC532	MUX-OP	Vs:±7V Vi0:1mV	14SDd	Nsc		data pdf pinout
CLC532A	MUX-OP	Vs:±7V Vi0:±3.5mV	20C,14D	Nsc		data pinout
CLC532A8	MUX-OP	Vs:±7V Vi0:1mV	14D,20C	Nsc		data pdf pinout
CLC533	MUX-OP	Vs:±7V Vi0:1mV	16D,20CSd	Nsc		data pdf pinout
CLC 912 ACP-1...-3	D/A-IC	12 Bit, Video, >20 MHz, TTL Level	24-DIP	Com	-	
CLC 920 ACD	A/D-IC	10 Bit, Video, >20 Msps, TTL Level	64-SDIP	Com	-	
CLC 925 A1	A/D-IC	12 Bit, Video, >10 Msps, TTL Level	40-SDIP	Com	-	
CLC 926 A1	A/D-IC	12 Bit, Video, >10 Msps, TTL Level	40-SDIP	Com	-	
CLC5506	PROG-GAIN	Vs:6V Vu:25.75dB f:600Mc	14Sd	Nsc		data pdf pinout
CLC5523I	WIDEBAND	Vs:±7V Vo:±3V Vi0:50mV	8SD	Nsc		data pdf pinout
CLC5526	PROG-GAIN	Vs:6V Vu:30dB	20SD	Nsc		data pdf pinout
CLC5602I	2xVIDEO-AMP	Vs:14V Vo:±3.8V Vi0:2mV f:135Mc	8SD	Nsc		data pdf pinout
CLC5612I	2xPROG-BUFF	Vs:14V Vo:±3.8V Vi0:3mV f:90Mc	8SD	Nsc		data pdf pinout
CLC5622I	2xVIDEO-AMP	Vs:14V Vo:±3.8V Vi0:1mV f:160Mc	8SD	Nsc		data pdf pinout
CLC5623I	3xVIDEO-AMP	Vs:14V Vo:±3.8V Vi0:1mV f:148Mc	14SD	Nsc		data pdf pinout
CLC5632I	2xPROG-BUFF	Vs:14V Vo:±3.8V Vi0:7mV f:130Mc	8SD	Nsc		data pdf pinout
CLC5633I	3xPROG-BUFF	Vs:14V Vo:±3.8V Vi0:7mV f:130Mc	14SD	Nsc		data pdf pinout
CLC5644I	4xFB-AMP	Vs:14V Vo:±2.8V Vi0:2.5mV f:25Mc	14SD	Nsc		data pdf pinout
CLC5654I	4xFB-AMP	Vs:14V Vo:±2.6V Vi0:2.5mV f:40Mc	14SD	Nsc		data pdf pinout
CLC5665I	WIDEBAND	Vs:±16V Vo:±14V Vi0:1mV f:20Mc	8SD	Nsc		data pdf pinout
CLC 5956 IMTD	A/D-IC	SMD, 12 Bit A/D-Converter, 65MSPS, T&H	48-VSMDIP	Nsc	-	data pdf pinout
CLC 5957 MTD	A/D-IC	SMD, 12 Bit A/D-Converter, 70MSPS, T&H	48-VSMDIP	Nsc	-	data pdf pinout
CLC 5958 SLB	A/D-IC	SMD, 14 Bit A/D-Converter, 52MSPS, T&H	48-CSP	Nsc	-	data pdf pinout
CLD31	PD	860nm		Cla,Tel		data pdf pinout
CLD41	PD	860nm		Cla,Tel		data pdf pinout
CLD42	PD	860nm		Cla,Tel		data pdf pinout
CLD56	PD	860nm		Cla,Tel		data pdf pinout
CLD71	PD	860nm		Cla,Tel		data pinout
CLD72	PD	860nm		Cla,Tel		data pinout
CLD141	PD	860nm	TO46	Cla,Tel		data pdf pinout
CLD142	PD	860nm	TO46	Cla,Tel		data pdf pinout
CLD156	PD	880nm	TO5	Cla,Tel		data pdf pinout
CLD160	PD	860nm	TO5	Cla,Tel		data pdf pinout
CLD171	PD	860nm		Cla		data pdf pinout
CLD185	PD	860nm	TO5	Cla,Tel		data pdf pinout

Type	Device	Short Description	Fìg.	Manu	Comparision Types	More at
CLD240	PD	850nm	TO46	Cla		data pdf pinout
CLD340	PD	880nm	TO46	Cla,Tel		data pdf pinout
CLD370F	PD	850nm	T1¾	Cla,Tel		data pdf pinout
CLE100	IRED	850nm ø3.2mm I:71lm Vf:1.4V If:20mA	PLCC-2	Cla		data pdf pinout
CLE100F	IRED	940nm ø3.2mm I:71lm Vf:<1.5V If:20mA	PLCC-2	Cla		data pdf pinout
CLE110	IRED	940nm ø3.2mm I:71lm Vf:<1.5V If:20mA	PLCC-2	Cla		data pdf pinout
CLE110F	IRED	940nm ø3.2mm I:71lm Vf:<1.5V If:20mA	PLCC-2	Cla		data pdf pinout
CLE130	IRED	940nm ø4.8mm I:>1.5mlm Vf:<1.8V	TO46	Cla		data pdf pinout
CLE130E	IRED	940nm ø4.8mm I:>0.3mlm Vf:1.5V	TO46	Cla		data pdf pinout
CLE130W	IRED	940nm ø4.8mm I:>1.5mlm Vf:<1.8V	TO46	Cla		data pdf pinout
CLE135	IRED	940nm ø4.8mm I:>1.8mlm Vf:1.5V	TO46	Cla		data pdf pinout
CLE230	IRED	880nm ø4.8mm I:>>3.6mlm Vf:<1.8V	TO46	Cla		data pdf pinout
CLE230W	IRED	890nm ø4.8mm I:>5mlm Vf:<2V If:100mA	TO46	Cla		data pdf pinout
CLE234	IRED	880nm ø4.5mm I:>5mlm Vf:1.65V	TO46	Cla		data pdf pinout
CLE234E	IRED	880nm ø4.2mm I:>5mlm Vf:<2V If:100mA	TO46	Cla		data pdf pinout
CLE235	IRED	880nm ø4.8mm I:>0.5mlm Vf:1.4V	TO46	Cla		data pdf pinout
CLE250	IRED	880nm ø1.5mm I:6mW/Sr Vf:<1.8V	TO46	Cla		data pdf pinout
CLE300	IRED	880nm ø1.5mm I:6mW/Sr Vf:1.4V	PLCC-2	Cla		data pdf pinout
CLE320	IRED	810nm ø4.5mm I:6mW/Sr Vf:<2.5V	TO46	Cla		data pdf pinout
CLE330	IRED	850nm ø4.2mm I:0.5mlm Vf:1.7V	TO46	Cla		data pdf pinout
CLE331	IRED	850nm ø4.8mm I:0.5mlm Vf:3V If:100mA	TO46	Cla		data pdf pinout
CLE332	IRED	850nm ø4.5mm I:0.5mlm Vf:1.8V	TO46	Cla		data pdf pinout
CLE333	IRED	850nm ø4.5mm I:0.5mlm Vf:1.8V	TO46	Cla		data pdf pinout
CLE334	IRED	850nm ø4.1mm I:0.5mlm Vf:1V If:100mA	TO46	Cla		data pdf pinout
CLE335	IRED	850nm ø4.8mm I:3.5mlm Vf:1.7V	TO46	Cla		data pdf pinout
CLE336	IRED	850nm ø4.5mm I:3.5mlm Vf:1.7V	TO46	Cla		data pdf pinout
CLE351	IRED	850nm ø4.8mm I:0.3mlm Vf:<2.2V	PLCC-2	Cla		data pdf pinout
CLE355	IRED	850nm ø4.8mm I:0.3mlm Vf:<2.2V	PLCC-2	Cla		data pdf pinout
CLE367	IRED	850nm ø5mm I:0.3mlm Vf:1.45V	T1¾	Cla		data pdf pinout
CLE400	IRED	850nm ø1.4mm I:0.3mlm Vf:<1.4V		Cla		data pdf pinout
CLE400F	IRED	880nm ø1.4mm I:0.3mlm Vf:1.4V		Cla		data pdf pinout
CLE435	LED	rd ø4.5mm I:0.3mlm Vf:1.8V If:20mA	TO46	Cla		data pdf pinout
CLE436	LED	rd ø4.8mm I:72mlm Vf:<2.2V If:20mA	TO46	Cla		data pdf pinout
CLE445	LED	rd ø1.4mm I:95mlm Vf:<2.4V If:20mA		Cla		data pdf pinout
CLE501	LED	uv ø4.5mm I:95mlm Vf:3V If:20mA	TO46	Cla		data pdf pinout
CLE509	LED	uv ø4.5mm I:95mlm Vf:3V If:180mA	TO46	Cla		data pdf pinout
CLE535	LED	gr ø4.5mm I:1.2mlm Vf:3.2V If:20mA	TO46	Cla		data pdf pinout
CLE536	LED	bl ø4.5mm I:887.5mcd Vf:3.2V If:20mA	TO46	Cla		data pdf pinout
CLE539	LED	bl ø4.5mm I:635mlm Vf:3.2V If:20mA	TO46	Cla		data pdf pinout
CLE545	LED	gr ø1.4mm I:1.5lm Vf:3.2V If:20mA		Cla		data pdf pinout
CLE546	LED	bl ø1.4mm I:113mlm Vf:3.2V If:20mA		Cla		data pdf pinout
CLE549	LED	bl ø1.4mm I:820mlm Vf:3.2V If:20mA		Cla		data pdf pinout
CLED1	IRED	940nm ø3.7mm I:820mlm Vf:<1.8V	TO46	Cla		data pinout
CLED5	IRED	940nm ø1.3mm I:820mlm Vf:<1.8V	TO18	Cla		data pinout
CLED155	IRED	880nm ø3.7mm I:820mlm Vf:<1.8V	TO18	Cla		data pinout
CLED305	IRED	880nm ø0.25mm I:820mlm Vf:<1.5V	PILL	Cla		data pinout
CLED400	IRED	880nm ø1mm I:820mlm Vf:<1.7V If:50mA		Cla		data pinout
CLI2	OC	LED/NPN Viso:2500V CTR:30-100%		Cla		data pinout
CLI3	OC	LED/NPN Viso:2500V CTR:30-100%		Cla		data pinout
CLI4	OC	LED/NPN Viso:2500V CTR:>150% Ibe:2mA		Cla		data pinout
CLI5	OC	LED/NPN Viso:2500V CTR:>20% Ibs:50mA		Cla	OPI2252	data pinout
CLI6	OC	LED/NPN Viso:2500V CTR:>50% Ibs:50mA		Cla	OPI2253	data pinout
CLI7	OC	LED/NPN Viso:2500V CTR:>50% Ibs:16mA		Cla	OPI2251	data pinout
CLI8	OC	LED/NPN Viso:2500V CTR:>20% Ibs:50mA		Cla	OPI2252	data pinout
CLI9	OC	LED/NPN Viso:2500V CTR:6-30% Ibe:2mA		Cla		data pinout
CLI10	OC	LED/Dar Viso:2500V CTR:>800%		Cla		data pinout
CLI11	OC	LED/Dar Viso:2500V CTR:>500%		Cla	OPI3250	data pinout
CLI12	OC	LED/Dar Viso:2500V CTR:>300%		Cla	OPI3250	data pinout
CLI13	OC	LED/Dar Viso:2500V CTR:>100%		Cla	OPI3250	data pinout
CLI14	OC	LED/Dar Viso:2500V CTR:>100%		Cla	OPI3250	data pinout
CLI25	OC	bid. LED/NPN Viso:2500V CTR:>20%		Cla		data pinout
CLI26	OC	bid. LED/NPN Viso:2500V CTR:>40%		Cla		data pinout
CLI27	OC	bid. LED/Dar Viso:2500V CTR:>400%		Cla		data pinout
CLI28	OC	bid. LED/Dar Viso:2500V CTR:>600%		Cla		data pinout
CLI200	PI	LED/Dar CTR:10% Vbs:<1.5V Ibs:16mA		Cla		data pdf pinout
CLI250	PI	LED/NPN CTR:>2.5% Vbs:1.2V Ibs:20mA		Cla		data pdf pinout
CLI300	PI	LED VR/NPN CTR:>5% Vbs:<1.5V		Cla		data pinout
CLI385	PS	LED/PD+IC Vbs:<1.5V Ibs:10mA		Cla		data pdf pinout
CLI400	PI	LED/NPN CTR:>2.5% Vbs:1.2V Ibs:20mA		Cla		data pdf pinout
CLI500	PI	LED/NPN CTR:>2.5% Vbs:<1.6V Ibs:20mA		Cla		data pdf pinout
CLI506	OC	LED/NPN Viso:2500V CTR:>6% Ibs:50mA		Cla	OPI2251	data pdf pinout
CLI506B	OC	LED/NPN Viso:2500V CTR:>6% Ibs:50mA		Cla		data pdf pinout
CLI600	PI	LED/NPN CTR:>2% Vbs:<1.8V Ibs:16mA		Cla		data pdf pinout
CLI700	PI	LED/2 CTR:>0.2% Vbs:<1.8V Ibs:16mA		Cla		data pinout
CLI710	PS	LED/NPN CTR:>1% Vbs:1.3V Ibs:20mA		Cla		data pdf pinout
CLI800	PI	LED/NPN CTR:>2% Vbs:<1.5V Ibs:16mA		Cla		data pdf pinout
CLI800W	PI	LED/NPN CTR:>2% Vbs:<1.5V Ibs:16mA		Cla		data pinout
CLI811	PI	LED/NPN CTR:>2% Vbs:<1.5V Ibs:16mA		Cla		data pinout
CLI840W	PI	LED/Dar CTR:>10% Vbs:<1.5V Ibs:16mA		Cla		data pinout
CLI841W	PI	LED/Dar CTR:>10% Vbs:<1.5V Ibs:16mA		Cla		data pinout
CLI890	PI	LED/PD+IC Vbs:<1.8V Ibs:16mA		Cla		data pinout
CLI890A	PI	LED/PD+IC Vbs:<1.8V Ibs:16mA		Cla		data pinout
CLK 5010	CMOS-IC	Kfz-quarzuhr/car quartz clock	16-DIP	Itt	-	
CLK 5011	CMOS-IC	Kfz-quarzuhr/car quartz clock	16-DIP	Itt		
CLL130	PD	860nm Vb:4.5...18V Ib:12mA	TO18	Cla		data pdf pinout

Type	Device	Short Description .	Fig.	Manu	Comparision Types	More at
CLL4040	PD	850nm Vb:5V Ib:<10mA		Cla		data pinout
CLM5H10A	OC	10V La/LDR Viso:2500Vrms Vbs:8V		Cla		data pinout
CLM7H16A	OC	16V La/LDR Viso:2500Vrms Vbs:13.2V		Cla		data pinout
CLM50	OC	LED/LDR Viso:2500V Vbs:2V Ibs:16mA		Cla		data pinout
CLM51	OC	LED/LDR Viso:2500V Vbs:2V Ibs:≤40mA		Cla		data pinout
CLM60	OC	bid. LED/LDR Viso:2500V Vbs:2V		Cla		data pinout
CLM61	OC	bid. LED/LDR Viso:2500V Vbs:2V		Cla		data pinout
CLM3006A	OC	6V La/LDR Viso:2500Vrms Vbs:5V		Cla		data pinout
CLM3012A	OC	12V La/LDR Viso:2500Vrms Vbs:10V		Cla		data pinout
CLM3120A	OC	Ne La/LDR Viso:2500Vrms Ibs:2mA		Cla		data pinout
CLM3500	OC	Ne La/LDR Viso:1600V Ibs:2.5mA		Cla		data pdf pinout
CLM3500/2	OC	Ne La/2 LDR Viso:1600V Ibs:1.2mA		Cla		data pinout
CLM3700	OC	Ne La/LDR Viso:1600V Ibs:0.3mA		Cla		data pinout
CLM4006A	OC	6V La/LDR Viso:2500Vrms Vbs:5V		Cla		data pinout
CLM4012A	OC	12V La/LDR Viso:2500Vrms Vbs:10V		Cla		data pinout
CLM4120A	OC	Ne La/LDR Viso:2500Vrms Ibs:2mA		Cla		data pdf pinout
CLM6000	OC	LED/LDR Viso:2000V Vbs:<2V Ibs:20mA		Cla		data pdf pinout
CLM6200	OC	LED/LDR Viso:2000V Vbs:<2V Ibs:16mA		Cla		data pinout
CLM6500	OC	LED/LDR Viso:2000V Vbs:<2.5V		Cla		data pinout
CLM7000	OC	LED/LDR Viso:750V Vbs:<2V Ibs:20mA		Cla		data pinout
CLM7100	OC	LED/LDR Viso:750V Vbs:<2V Ibs:20mA		Cla		data pinout
CLM7200	OC	LED/LDR Viso:750V Vbs:<2V Ibs:20mA		Cla		data pinout
CLM8000	OC	LED/LDR Viso:2500V Vbs:<2.5V		Cla		data pdf pinout
CLM8200	OC	LED/2 LDR Viso:2500V Vbs:<2V		Cla		data pinout
CLM8500	OC	LED/2 LDR Viso:2000V Vbs:<2.8V		Cla		data pinout
CLM8500HV	OC	LED/LDR Viso:2000V Vbs:<2.5V		Cla		data pinout
CLM8600	OC	LED/LDR Viso:3000V Vbs:<2.5V		Cla		data pdf pinout
CLM9000	OC	LED/LDR Viso:7000V Vbs:<2.8V		Cla		data pinout
CLR130	PT	940nm P:>66µA/mW/cm²	TO18	Cla		data pinout
CLR130W	PT	940nm P:>23µA/mW/cm²	TO18	Cla		data pdf pinout
CLR335	PT	850nm P:5µA/mW/cm²	TO18	Cla		data pinout
CLR2050	PT	780nm P:>7µA/mW/cm²	TO18	Cla		data pinout
CLR2170	PT	780nm P:>30µA/mW/cm²	TO18	Cla		data pinout
CLR4180	PT	780nm P:>6µA/mW/cm²		Cla		data pinout
CLR5101	PT	780nm P:>50µA/mW/cm²	TO18	Cla		data pinout
CLs	Si-P	=BSS 82B (Typ-Code/Stempel/marking)	35	Sie	→BSS 82B	data
CLT130	PT	780nm P:2.7µA/mW/cm²	TO18	Cla		data pdf pinout
CLT130W	PT	780nm P:0.5µA/mW/cm²	TO18	Cla		data pdf pinout
CLT135	PT	940nm P:12µA/mW/cm²	TO18	Cla		data pdf pinout
CLT250	PT	820nm P:0.35µA/mW/cm²		Cla		data pdf pinout
CLT435	PT	820nm P:2µA/mW/cm²	TO18	Cla		data pdf pinout
CLT2010	PT	820nm P:>0.2µA/mW/cm²	TO18	Cla		data pinout
CLT2130	PT	820nm P:>0.67µA/mW/cm²	TO18	Cla		da pinout
CLT2164	PT	820nm P:>3µA/mW/cm²	TO18	Cla		data pinout
CLT4140	PT	820nm P:>0.4µA/mW/cm²		Cla		data pinout
CLT4200	PT	820nm P:>0.1µA/mW/cm²		Cla		data pinout
CLT5160	PT	820nm P:>0.28µA/mW/cm²	TO18	Cla		data pinout
CLY 2	GaAs-N-FET-d	SMD, UHF, ...3GHz, 12V, 600mA	≈44(GSDDSGSie		-	data
CLY 5	GaAs-N-FET-d	Idss=300..600mA, Up=-3,8...-1,8V SMD, UHF, 0,4.±2,5GHz, 12V, 1,2A	48za§	Sie	-	data
CLY 10	GaAs-N-FET-d	Idss=0,6..1,2A, Up=-3,8...-1,8V SMD, UHF, 0,4..2,5GHz, 12V, 2,1A	48za§	Sie	-	data
CLY 15	GaAs-N-FET-d	Idss=1,2..2,4A, Up=-3,8...-1,8V SMD, UHF, 0,4..2,5GHz, 12V, 5A	48za§	Sie	-	data
		Idss=2,4..4,8A, Up=-3,8...-1,8V				
CM	Si-N	=2SC4504 (Typ-Code/Stempel/marking)	39	Say	→2SC4504	data
CM	Si-N	≈2SD999-CM (Typ-Code/Stempel/marking)	39	Nec	→2SD999	data
CM	Si-P	=BSS 82C (Typ-Code/Stempel/marking)	35	Fer,Mot,Sie	→BSS 82C	data
CM	Si-N	=CPH 3212 (Typ-Code/Stempel/marking)	35	Say	→CPH 3212	data
CM	Z-Di	=MMSZ 4685 (Typ-Code/marking)	71(2,7mm)	Ons	→MMSZ 4685	
CM	Z-Di	=P4 SMA-62A (Typ-Code/Stempel/marking)	71(5x2,5)	Fag	→P4 SMA-62A	data
CM	Z-Di	=SMAJ 33A (Typ-Code/Stempel/marking)	71(5x2,5)	Tho	→SMAJ ...	data
CM	Z-Di	=SMBJ 33CA (Typ-Code/Stempel/marking)	71(5x3,5)	Mop	→SMBJ ...	data
CM 4-482 B	Opto					
CM 101 S	IC	USB 2CH Audio Controller	TSOP	Cme	-	
CM 101 S	IC	USB 2CH Audio Controller	TSOP	Cme	-	pdf pinout
CM 102 A	IC	USB 2CH Audio Controller	DIC	Cme	-	pdf pinout
CM 102 S	IC	USB 2CH Audio Controller	TSOP	Cme	-	pdf pinout
CM 103	IC	16-Bit Stereo USB Audio Cntrl	MP	Cme	-	pdf pinout
CM 106 -F+	IC	USB Audio I/O Cntrl.	MP	Cme	-	pdf pinout
CM 106 -F	IC	USB Audio I/O Cntrl.	MP	Cme	-	pdf pinout
CM 106 -L+	IC	USB Audio I/O Cntrl.	MP	Cme	-	pdf pinout
CM 106 -L	IC	USB Audio I/O Cntrl.	MP	Cme	-	pdf pinout
CM 108	DIG-IC	USB Audio I/O Controller	QFP	Cme	-	pdf pinout
CM 109	DIG-IC	USB Audio I/O Controller	QFP	Cme	-	pdf pinout
CM 112	IC	USB Audio I/O Cntrl.	MP	Cme	-	pdf pinout
CMI-8738	Dig. Audio	PCI-Based HRTF 3D Extension Positional Audio Chip	QFP	Cme	-	pdf pinout
CMI 8768+	Dig. Audio	PCI 8CH Integrated Sound Chip	QFP	Cme	-	pdf pinout
CMI 8768	Dig. Audio	PCI 8CH Integrated Sound Chip	QFP	Cme	-	pdf pinout
CMI 8787	Dig. Audio	High Performance PCI Audio Processor	TSOP	Cme	-	pdf pinout

Type	Device	Short Description	Fig.	Manu	Comparision Types	More at
CMI 8788	IC	High Performance PCI Audio Processor	TSOP	Cme	-	pdf pinout
CMI 8738-037-SX	Dig. Audio	C3DX PCI- Based HRTF 3D Extension Positional Audio Chip	QFP	Cme		pdf pinout
CMP-01(E,C)P	KOP-IC	Fast, 36V, Iq=75mA, <180ns, -55...+125° C,E: 0...+70°	8-DIP	Pmi		data pinout
CMP01C	FAST-COMP	Vs:36V Vu:500V/mV Vi0:0.4mV	8AD	Pmi		data pinout
CMP01E	FAST-COMP	Vs:36V Vu:500V/mV Vi0:0.3mV	8AD	Pmi		data pinout
CMP01GR	FAST-COMP	Vs:36V Vi0:2.8mV	d	Pmi		data pinout
CMP01J	FAST-COMP	Vs:36V Vu:500V/mV Vi0:0.3mV	8A	Pmi		data pinout
CMP-01(E,C)J	KOP-IC	=CMP-01...P:	81			data pinout
CMP01N	FAST-COMP	Vs:36V Vi0:0.8mV	d	Pmi		data pinout
CMP01Z/883	FAST-COMP	Vs:36V Vu:500V/mV Vi0:0.3mV	8D	Pmi		data pinout
CMP-01(E,C)Z	KOP-IC	=CMP-01...P:	8-DIC			data pinout
CMP-02(E,C)P	KOP-IC	36V, Iq=75mA, C: -40...+85°, E: 0...+70°	8-DIP	Pmi		data pinout
CMP02C	COMP	Vs:36V Vu:500V/mV Vi0:0.4mV	8ADS	Pmi		data pinout
CMP02EP	COMP	Vs:36V Vu:500V/mV Vi0:0.3mV	8D	Pmi		data pinout
CMP02GR	COMP	Vs:36V Vi0:2.8mV	d	Pmi		data pinout
CMP-02(E,C)J	KOP-IC	=CMP-02...P:	81			data pinout
CMP02N	COMP	Vs:36V Vi0:0.8mV	d	Pmi		data pinout
CMP-02(E,C)S	KOP-IC	=CMP-02...P: SMD	8-MDIP			data pinout
CMP04	4xCOMP	Vs:±18V Vu:200V/mV Vi0:0.4mV	14DS	And		data pinout
CMP-04(F)P	KOP-IC	Quad, ±18V, Iin=50mA, -40...+85°	14-DIP	Pmi	→Serie 139	data pinout
CMP04G	4xCOMP	Vs:±18V Vi0:2mV	d	Pmi		data pinout
CMP04N	4xCOMP	Vs:±18V Vi0:1mV	d	Pmi		data pinout
CMP-04(F)S	KOP-IC	=CMP-04...P: SMD	14-MDIP			data pinout
CMP-04(B,F)Y	KOP-IC	=CMP-04...P: B:-55...+125°, F:-25...+85°	14-DIC			data pinout
CMP-05(G)P	KOP-IC	hi-speed, +6/-18, 38ns, -40...+85°	8-DIP	Pmi		data pinout
CMP05B	COMP	Vs:-6/18V Vu:16V/mV Vi0:0.15mV	8AD	Pmi		data pinout
CMP05C	COMP	Vs:-6/18V Vu:14V/mV Vi0:0.4mV	8AD	Pmi		data pinout
CMP05F	COMP	Vs:-6/18V Vu:16V/mV Vi0:0.15mV	8AD	Pmi		data pinout
CMP05G	COMP	Vs:-6/18V Vi0:1mV	d,8DS	Pmi		data pinout
CMP-05(B,C,F,G)J	KOP-IC	=CMP-05...P: B,C: -55...+125°, F,G: -40...+85°	81			data pinout
CMP-05(G)S	KOP-IC	=CMP-05...P: SMD	8-MDIP			data pinout
CMP-05(B,C,F,G)Z	KOP-IC	=CMP-05...P: B,C: -55...+125°, F,G: -40...+85°	8-DIC			data pinout
CMP-08(B,F)Z	KOP-IC	hi- speed, +6, 5/ - 6, Iq=30mA B: -55... +125°, F: - 40... +85°	8-DIC	Pmi		data pinout
CMP08B	COMP	Vs:-6/6.5V Vi0:±3mV	8D	Pmi		data pinout
CMP08F	COMP	Vs:-6/6.5V Vi0:±2.5mV	8SD	Pmi		data pinout
CMP08N	COMP	Vs:-6/6.5V Vi0:±2.2mV	d	Pmi		data pinout
CMP-08(F)S	KOP-IC	=CMP-08...Z: SMD	8-MDIP		-	data pinout
CMP401	4xCOMP	Vs:16V Vu:10V/mV Vi0:<3mV	16DS	And		data pdf pinout
CMP402	4xCOMP	Vs:16V Vu:10V/mV Vi0:<3mV	16DS	And		data pdf pinout
CMP404	4xCOMP	Vs:±18V Vu:400V/mV Vi0:<1mV	14D	Pmi		data pinout
CMP404FY	4xCOMP	Vs:±18V Vu:400V/mV Vi0:<2mV	14D	Pmi		data pinout
CMP404G	4xCOMP	Vs:±18V Vi0:<2mV	d	Pmi		data pinout
CMP-404(A,E,F)Y	KOP-IC	Quad, ±18V, Iin=20mA, A: -55...+125° E,F: -40...+85°	14-DIC	Pmi	-	data pinout
CMs	Si-P	=BSS 82C (Typ-Code/Stempel/marking)	35	Sie	→BSS 82C	data
CMS 01	Si-Di	SMD, Schottky, 30V, 3A, Uf<0,37V(3A)	71a(4x2,4)	Tos	(D 1FH3, SFPA-73, SFPJ-73, U 3FWJ44N)[6]	
CMS 02	Si-Di	SMD, Schottky, 30V, 3A, Uf<0,4V(3A)	71a(4x2,4)	Tos	(D 1FH3, SFPA-73, SFPJ-73, U 3FWJ44N)[6]	
CMS 03	Si-Di	SMD, Schottky, 30V, 3A, Uf<0,45V(3A)	71a(4x2,4)	Tos	(D 1FH3, SFPA-73, SFPJ-73, U 3FWJ44N)[6]	
CMS 04	Si-Di	SMD, Schottky, 30V, 5A, Uf<0,37V(5A)	71a(4x2,4)	Tos		
CMS 05	Si-Di	SMD, Schottky, 30V, 5A, Uf<0,45V(5A)	71a(4x2,4)	Tos		
CMS 06	Si-Di	SMD, Schottky, 30V, 2A, Uf<0,37V(2A)	71a(4x2,4)	Tos	(D 1FP3, SFPA-63, SFPJ-63, U 2FWJ44N)[6]	
CMS 07	Si-Di	SMD, Schottky, 30V, 2A, Uf<0,45V(2A)	71a(4x2,4)	Tos	(D 1FP3, SFPA-63, SFPJ-63, U 2FWJ44N)[6]	
CMS 08	Si-Di	SMD, Schottky, 30V, 1A, Uf<0,37V(1A)	71a(4x2,4)	Tos	(BYG 90-30, D 1FS4, MA 735, U 1FWJ44N)[6]	
CMS 09	Si-Di	SMD, Schottky, 30V, 1A, Uf<0,45V(1A)	71a(4x2,4)	Tos	(BYG 90-30, D 1FS4, MA 735, U 1FWJ44N)[6]	
CMS 10	Si-Di	SMD, Schottky, 40V, 1A, Uf<0,55V(1A)	71a(4x2,4)	Tos	(BYG 90-40, D 1FS4, MA 736, U 1GWJ44N)[6]	
CMS 11	Si-Di	SMD, Schottky, 40V, 2A, Uf<0,55V(2A)	71a(4x2,4)	Tos	(RB 060L-40, SFPB-74)[6]	
CMY 90..211	Ga~s-FET-IC	Mmic, SMD, Telecom Amp./conv./mixer		Sie		data
CN	Si-N	=2SC4548 (Typ-Code/Stempel/marking)	39	Say	→2SC4548	data
CN	Si-N	=2SC4853 (Typ-Code/Stempel/marking)	35(2mm)	Say	→2SC4853	data
CN	Si-N	=2SC4854 (Typ-Code/Stempel/marking)	35	Say	→2SC4854	data
CN	Si-N	=2SC4855 (Typ-Code/Stempel/marking)	44	Say	→2SC4855	data
CN	Si-N	=2SC5537 (Typ-Code/Stempel/marking)	35(1,4mm)	Say	→2SC5537	data
CN	MOS-N-FET-e*	=2SK3348 (Typ-Code/Stempel/marking)	35(2mm)	Hit	→2SK3348	data
CN	Z-Di	=MMSZ 4686 (Typ-Code/marking)	71(2,7mm)	Ons	→MMSZ 4686	
CN	Z-Di	=P4 SMA-68 (Typ-Code/Stempel/marking)	71(5x2,5)	Fag	→P4 SMA-68	data
CN	Z-Di	=SMAJ 33CA (Typ-Code/Stempel/marking	71(5x2,5)	Tho	→SMAJ ...	data
CN	Z-Di	=SMBJ 36C (Typ-Code/Stempel/marking)	71(5x3,5)	Mop	→SMBJ ...	data
CN 3	Si-N	=2SC4853-3 (Typ-Code/Stempel/marking)	35(2mm)	Say	→2SC4853	data
CN 3	Si-N	=2SC4854-3 (Typ-Code/Stempel/marking)	35	Say	→2SC4854	data
CN 3	Si-N	=2SC4855-3 (Typ-Code/Stempel/marking)	44	Say	→2SC4855	data
CN 4	Si-N	=2SC4853-4 (Typ-Code/Stempel/marking)	35(2mm)	Say	→2SC4853	data
CN 4	Si-N	=2SC4854-4 (Typ-Code/Stempel/marking)	35	Say	→2SC4854	data
CN 4	Si-N	=2SC4855-4 (Typ-Code/Stempel/marking)	44	Say	→2SC4855	data
CN 5	Si-N	=2SC4853-5 (Typ-Code/Stempel/marking)	35(2mm)	Say	→2SC4853	data
CN 5	Si-N	=2SC4854-5 (Typ-Code/Stempel/marking)	35	Say	→2SC4854	data
CN 5	Si-N	=2SC4855-5 (Typ-Code/Stempel/marking)	44	Say	→2SC4855	data

Type	Device	Short Description	Fig.	Manu	Comparision Types	More at
CN 515 T	OP-IC	=µA 709: -55...+125°		Fer	→µA 709	
CNG35	OC	LED/NPN Viso:4400V CTR:40-160%		Phi,Val		data pdf pinout
CNG36	OC	LED/NPN Viso:4400V CTR:80-200%		Phi,Val		data pdf pinout
CNG82	OC	LED/NPN Viso:5300V CTR:40-160%		Phi		data pinout
CNR21	OC	LED/Triac Viso:15000V Vbs:1.25V		Aeg,Tel		data pinout
CNR36	OC	LED/PD+NPN Viso:4400V CTR:40%		Phi,Val		data pinout
CNR50	OC	LED/PD+IC Viso:7070V Vbs:1.5V		Phi		data pinout
CNW11AV	OC	LED/NPN Viso:5656V CTR:>20% Ibs:10mA		Phi		data pinout
CNW82	OC	LED/NPN Viso:8340V CTR:40-150%		Phi		data pdf pinout
CNW83	OC	LED/NPN Viso:8340V CTR:40-150%		Phi		data pdf pinout
CNW84	OC	LED/NPN Viso:8340V CTR:63-300%		Phi		data pdf pinout
CNW135	OC	LED/PD+NPN Viso:7070V CTR:40%		Hpa,Phi		data pinout
CNW138	OC	LED/PD+Dar Viso:7070V CTR:3000%		Hpa,Phi		data pinout
CNW4562	OC	LED/PD+NPN Viso:5000Vrms CTR:52%		Hpa		data pinout
CNX21	OC	LED/NPN Viso:10000V CTR:50% Ibs:10mA		Phi		data pinout
CNX35	OC	LED/NPN Viso:4400V CTR:40-160%		Phi,Rca,Val	CNY1	data pdf pinout
CNX36	OC	LED/NPN Viso:4400V CTR:80-200%		Phi,Rca,Val	CNY17	data pdf pinout
CNX38	OC	LED/NPN Viso:4400V CTR:60-210%		Phi,Rca,Val	CNY17	data pdf pinout
CNX39	OC	LED/NPN Viso:4400V CTR:60-100%		Phi,Val		data pinout
CNX44	OC	LED/NPN Viso:1000V CTR:100% Ibs:10mA		Phi		data pinout
CNX48	OC	LED/Dar Viso:4400V CTR:>500% Ibe:5mA		Phi,Val		data pinout
CNX62	OC	LED/NPN Viso:5300V CTR:150% Ibs:10mA		Phi		data pdf pinout
CNX71	OC	LED/NPN Viso:5300V CTR:40-160%		Phi		data pdf pinout
CNX72	OC	LED/NPN Viso:5300V CTR:40-160%		Phi,Val		data pinout
CNX82	OC	LED/NPN Viso:5300V CTR:10-250%		Phi,Val		data pdf pinout
CNX83	OC	LED/NPN Viso:5300V CTR:10-250%		Phi,Val		data pdf pinout
CNX91	OC	LED/NPN Viso:800V CTR:30-200%		Phi,Val		data pinout
CNX92	OC	LED/NPN Viso:800V CTR:30-200%		Phi,Val		data pinout
CNY...	Opto					
CNY17	OC	LED/NPN Viso:5300V CTR:160-320%		Phi,Rca,Sie	MCT273,H11A5100,MCT210,MCT2210	data pdf pinout
CNY17F-1	OC	LED/NPN Viso:4400V CTR:160-320%		Phi,Sie,Vis	MOC8111	data pdf pinout
CNY17F-2	OC	LED/NPN Viso:4400V CTR:63-125%		Qtc,Vis	MOC8111	data pdf pinout
CNY17F-3	OC	LED/NPN Viso:4400V CTR:100-200%		Qtc,Vis	MOC8111	data pdf pinout
CNY17GF-1	OC	LED/NPN Viso:5300V CTR:40-80%		Phi,Sie		data pinout
CNY17GF-2	OC	LED/NPN Viso:5300V CTR:63-125%		Sie		data pinout
CNY17GF-3	OC	LED/NPN Viso:5300V CTR:100-200%		Sie		data pinout
CNY17GF-4	OC	LED/NPN Viso:5300V CTR:160-320%		Sie		data pinout
CNY18	OC	LED/NPN Viso:500V CTR:100-200%		Aeg,Tel		data pinout
CNY21	OC	LED/NPN Viso:10000V CTR:60% Ibs:10mA		Aeg,Tel	CNX21T	data pdf pinout
CNY22	OC	LED/NPN Viso:2800V CTR:<25% Ibs:8mA		Val		data pinout
CNY23	OC	LED/NPN Viso:2000V CTR:<50% Ibs:8mA		Val		data pinout
CNY28	PI	LED/NPN CTR:>2% Vbs:<1.7V Ibs:10mA		Har,Qtc		data pdf pinout
CNY29	PI	LED/Dar CTR:>40% Vbs:<1.7V Ibs:10mA		Har,Qtc,Rca		data pdf pinout
CNY30	OC	LED/SCR Viso:2250Vrms Vbs:1.5V		Har,Iso,Rca	IL400,PS3001	data pdf pinout
CNY31	OC	LED/Dar Viso:3750Vrms CTR:>400%		Har,Iso,Rca		data pinout
CNY32	OC	LED/NPN Viso:3750Vrms CTR:>20%		Har,Iso,Rca		data pdf pinout
CNY33	OC	LED/NPN Viso:3750Vrms CTR:>20%		Har,Iso,Rca		data pdf pinout
CNY34	OC	LED/SCR Viso:2250Vrms Vbs:1.5V		Har,Iso,Rca	IL400	data pdf pinout
CNY35	OC	bid. LED/NPN Viso:3750Vrms CTR:>10%		Har,Iso,Rca	IL250,CNY71	data pdf pinout
CNY36	PI	LED/NPN CTR:2-8% Vbs:<1.7V Ibs:10mA		Aeg,Tel		data pinout
CNY42	OC	LED/NPN Viso:2800V CTR:<25% Ibs:8mA		Val		data pinout
CNY43	OC	LED/NPN Viso:2000V CTR:<50% Ibs:8mA		Val		data pinout
CNY44	OC	LED/NPN Viso:1000V CTR:<30% Ibs:10mA		Val		data pinout
CNY47	OC	LED/NPN Viso:2250Vrms CTR:20-60%		Har,Rca,Thr	SFH600	data pdf pinout
CNY47A	OC	LED/NPN Viso:2250Vrms CTR:>40%		Har,Rca,Thr	SFH600	data pdf pinout
CNY48	OC	LED/Dar Viso:2250Vrms CTR:>600%		Har,Rca	4N32	data pdf pinout
CNY50	OC	LED/NPN Viso:1000V CTR:80% Vbs:1.15V		Phi		data pinout
CNY51	OC	LED/NPN Viso:5000V CTR:>100%		Har,Rca	SFH601,CNY75B	data pdf pinout
CNY57	OC	LED/NPN Viso:4400V CTR:50% Vbs:1.15V		Phi	K102P	data pinout
CNY57A	OC	LED/NPN Viso:4400V CTR:100% Ibs:10mA		Phi,Val	CQY80N	data pinout
CNY62	OC	LED/NPN Viso:5300V CTR:50% Vbs:1.2V		Phi,Val		data pinout
CNY64	OC	LED/NPN Viso:8200V CTR:100-200%		Aeg,Tel,Vis		data pdf pinout
CNY65	OC	LED/NPN Viso:11600V CTR:50-300%		Aeg,Tel,Vis		data pdf pinout
CNY66	OC	LED/NPN Viso:15000V CTR:50-300%		Aeg,Tel,Vis		data pdf pinout
CNY70	PS	LED/NPN CTR:1.5-2.5% Vbs:1.25V		Aeg,Vis		data pdf pinout
CNY71	OC	bid. LED/NPN Viso:5300V CTR:>20%		Aeg,Tel	CNY35	data pinout
CNY74-2	DOC	LED/NPN Viso:2500V CTR:50-300%		Aeg,Tel,Tem	FDC880	data pdf pinout
CNY74-2H	DOC	LED/NPN Viso:5000V CTR:1000% Ibe:5mA		Vis		data pinout
CNY74-4	QOC	LED/NPN Viso:2500V CTR:50-600%		Aeg,Tel	FDC880	data pinout
CNY74-4H	QOC	LED/NPN Viso:5000V CTR:1000% Ibe:5mA		Vis		data pinout
CNY75A	OC	LED/NPN Viso:5300V CTR:63-125%		Aeg,Tel,Vis	SFH601,CNY17,GFH600	data pdf pinout
CNY75B	OC	LED/NPN Viso:5300V CTR:100-200%		Aeg,Tel,Vis	SFH601,CNY17,GFH600	data pdf pinout
CNY75C	OC	LED/NPN Viso:5300V CTR:160-320%		Aeg,Tel,Vis	SFH601,CNY17,GFH600	data pdf pinout
CNY75GA	OC	LED/NPN Viso:5300V CTR:63-125%		Aeg,Tel,Vis	SFH601G	data pdf pinout
CNY75GB	OC	LED/NPN Viso:5300V CTR:100-200%		Aeg,Tel,Vis	SFH601G	data pdf pinout
CNY75GC	OC	LED/NPN Viso:5300V CTR:160-320%		Aeg,Tel,Vis	SFH601G	data pdf pinout
CO	Si-N	=2SC2881-O (Typ-Code/Stempel/marking)	39	Tos	→2SC2881	data
CO	Si-N	=2SC4209-O (Typ-Code/Stempel/marking)	35	Tos	→2SC4209	data
CO	Si-N	=2SC5087-O (Typ-Code/Stempel/marking)	44	Tos	→2SC5087	data
CO	Si-N	=KTC4373-O (Typ-Code/Stempel/marking)	39	Kec	→KTC 4373	data
COP 620 C	IC	8-Bit Cmos ROM Based Microcontrollers	DIP	Nsc	-	pdf pinout
COP 622 C	IC	8-Bit Cmos ROM Based Microcontrollers	DIP	Nsc	-	pdf pinout
COP 640 C	IC	8-Bit Cmos ROM Based Microcontrollers	DIP	Nsc	-	pdf pinout
COP 642 C	IC	8-Bit Cmos ROM Based Microcontrollers	DIP	Nsc	-	pdf pinout
COP 820 C	IC	8-Bit Cmos ROM Based Microcontrollers	DIP	Nsc	-	pdf pinout

Type	Device	Short Description	Fig.	Manu	Comparision Types	More at
COP 822 C	IC	8-Bit Cmos ROM Based Microcontrollers	DIP	Nsc	-	pdf pinout
COP 840 C	IC	8-Bit Cmos ROM Based Microcontrollers	DIP	Nsc	-	pdf pinout
COP 842 C	IC	8-Bit Cmos ROM Based Microcontrollers	DIP	Nsc	-	pdf pinout
COP 920 C	IC	8-Bit Cmos ROM Based Microcontrollers	DIP	Nsc	-	pdf pinout
COP 922 C	IC	8-Bit Cmos ROM Based Microcontrollers	DIP	Nsc	-	pdf pinout
COP 940 C	IC	8-Bit Cmos ROM Based Microcontrollers	DIP	Nsc	-	pdf pinout
COP 942 C	IC	8-Bit Cmos ROM Based Microcontrollers	DIP	Nsc	-	pdf pinout
CP	Si-P	=2SB767-P (Typ-Code/Stempel/marking)	39	Mat	→2SB767	data
CP	Si-N	=2SC2411-CP (Typ-Code/Stempel/marking)	≈35	Rhm	→2SC2411	data
CP	Si-N	=2SC2411K-P (Typ-Code/Stempel/marking)	35	Rhm	→2SC2411K	data
CP	Si-N	=2SC4097-P (Typ-Code/Stempel/marking)	35(2mm)	Rhm	→2SC4097	data
CP	Si-N	=2SC4705 (Typ-Code/Stempel/marking)	39	Say	→2SC4705	data
CP	MOS-P-FET-e*	=2SJ586 (Typ-Code/Stempel/marking)	35(2mm)	Hit	→2SJ586	data
CP	Si-N	=HC 1L2N (Typ-Code/Stempel/marking)	39	Nec	→HC 1...	data
CP	Z-Di	=MMSZ 4687 (Typ-Code/marking)	71(2,7mm)	Ons	→MMSZ 4687	
CP	Z-Di	=P4 SMA-68A (Typ-Code/Stempel/marking)	71(5x2,5)	Fag	→P4 SMA-68A	data
CP	Z-Di	=SMBJ 36CA (Typ-Code/Stempel/marking)	71(5x3,5)	Mop	→SMBJ ...	data
CP 8 C...G	50Hz-Thy	200...600V, 30A(Tc=85°), Igt/Ih<50/60mA C=200, D=300, E=400, F=500, G=600V	21b§	Hit	BTW48/..., CS23-..., MCR64-..., TAG35-...	
CP 9 C...G	50Hz-Thy	=CP 8 C...G:	≈21b§	Hit	-	
CP 82C55A-5	µP-Periph	CMOS, Ppi, 3x8 Programmable I/O Pins, TTL Compatible, 5MHz, 0...70°C	DIP	Har	..82C55...	pdf pinout
CP 82C55A	µP-Periph	CMOS, Ppi, 3x8 Programmable I/O Pins, TTL Compatible, 8MHz, 0...70°C	DIP	Har	..82C55...	pdf pinout
CP431	EZ+	Io=200mA Vo:2.5...36V P:800mW	3SN	Cer		data pdf pinout
CP432	EZ+	Io=250mA Vo:1.24..20V P:780mW	3SN,4S	Cer		data pdf pinout
CP 1600	µC-IC	µComp., 8x16 Bit RAM	40-DIC	Itt		data pdf pinout
CPC1810	PC	820nm	TO39	Cpc		data pdf pinout
CPC1822	PC	820nm		Cpc		data pdf pinout
CPC1824	PC	820nm	SOIC-16	Cpc		data pdf pinout
CPC1831	PC	820nm	SOIC-8	Cpc		data pdf pinout
CPC1832	PC	820nm	SOIC-16	Cpc		data pdf pinout
CPH 3101	Si-P	SMD, S, lo-sat, 30V, 2A, 60/375ns	35a	Say	-	data
CPH 3104	Si-P	SMD, S, lo-sat, 15V, 1,5A, 300MHz	35a	Say	-	data
CPH 3105	Si-P	SMD, S, lo-sat, 80V, 3A, 30/245ns	35a	Say	-	data
CPH 3106	Si-P	SMD, S, lo-sat, 15V, 3A, 30/100ns	35a	Say	-	data
CPH 3107	Si-P	SMD, S, lo-sat, 15V, 6A, 30/134ns	35a	Say	-	data
CPH 3109	Si-P	SMD, S, lo-sat, 30V, 3A, 50/295ns	35a	Say	-	data
CPH 3110	Si-P	SMD, S, lo-sat, 30V, 5A, 30/207ns	35a	Say	-	data
CPH 3112	Si-P	SMD, S, lo-sat, 50V, 5A, 39/250ns	35a	Say	-	data
CPH 3114	Si-P	SMD, S, lo-sat, 15V, 1,5A, 350MHz	35a	Say	-	data
CPH 3115	Si-P	SMD, S, lo-sat, 30V, 1,5A, 450MHz	35a	Say	-	data
CPH 3116	Si-P	SMD, S, lo-sat, 50V, 1A, 420MHz	35a	Say	-	data
CPH 3120	Si-P	SMD, lo-sat R(BE)=1.6k	35d	Say	-	data
CPH 3201	Si-N	SMD, S, lo-sat, 30V, 2A, 60/525ns	35a	Say	-	data
CPH 3204	Si-N	SMD, S, lo-sat, 15V, 1,5A, 200MHz	35a	Say	-	data
CPH 3205	Si-N	SMD, S, lo-sat, 80V, 3A, 35/322ns	35a	Say	-	data
CPH 3206	Si-N	SMD, S, lo-sat, 15V, 3A, 30/221ns	35a	Say	-	data
CPH 3207	Si-N	SMD, S, lo-sat, 15V, 6A, 32/260ns	35a	Say	-	data
CPH 3209	Si-N	SMD, S, lo-sat, 40V, 3A, 30/315ns	35a	Say	-	data
CPH 3210	Si-N	SMD, S, lo-sat, 40V, 5A, 30/334ns	35a	Say	-	data
CPH 3212	Si-N	SMD, S, lo-sat, 60V, 5A, 32/448ns	35a	Say	-	data
CPH 3214	Si-N	SMD, S, lo-sat, 15V, 1,5A, 450MHz	35a	Say	-	data
CPH 3215	Si-N	SMD, S, lo-sat, 40V, 1,5A, 500MHz	35a	Say	-	data
CPH 3216	Si-N	SMD, S, lo-sat, 40V, 1A, 420MHz	35a	Say	-	data
CPH 3217	Si-N	SMD, DC/DC-Converter, Driver		Say		data
CPH 3303	MOS-P-FET-e*	SMD, V-MOS, LogL, 20V, 1,6A, <0,48Ω	35a	Say	2SJ557	data
CPH 3304	MOS-P-FET-e*	SMD, V-MOS, LogL, 30V, 1,6A, <0,55Ω	35a	Say	2SJ557	data
CPH 3305	MOS-P-FET-e*	SMD, V-MOS, LogL, 60V, 0,8A, <1,8Ω	35a	Say	-	data
CPH 3306	MOS-P-FET-e*	SMD, V-MOS, LogL, 60V, 1A, <1,15Ω	35a	Say	-	data
CPH 3403	MOS-N-FET-e*	SMD, V-MOS, LogL, 20V, 2,2A, <0,22Ω	35a	Say	Si 2302DS	data
CPH 3404	MOS-N-FET-e*	SMD, V-MOS, LogL, 30V, 2,2A, <0,15Ω	35a	Say	2SK3105	data
CPH 3405	MOS-N-FET-e*	SMD, V-MOS, LogL, 60V, 1,2A, <0,68Ω	35a	Say	-	data
CPH 3406	MOS-N-FET-e*	SMD, V-MOS, LogL, 60V, 1,6A, <0,44Ω	35a	Say	-	data
CPH 5501	Si-P	SMD, DC/DC Converter, lo-sat,[Tr.1 60V,(=2SA1338)] 50V,	45ba5	Say		data
CPH 5503	Si-N	SMD, DC/DC Converter, lo-sat	45ba5	Say	=cph3209	data
CPH 5504	Si-N	SMD, DC/DC Converter, lo-sat	45ba5	Say	=cph3205	data
CPH 5505	Si-P	SMD, DC/DC Converter, lo-sat	45ba5	Say	=cph3109	data
CPH 5506	Si-N+P	SMD, DC/DC Converter, lo-sat	45ba5	Say	=cph3115+cph3215	data
CPH 5601	MOS-P-FET-e*	SMD,V-MOS, Dual, LogL, 20V, 1A, <0,89Ω	45bb1	Say		data
CPH 5602	MOS-N-FET-e*	SMD,V-MOS,Dual,LogL, 20V, 1,4A, <0,36Ω	45bb1	Say		data
CPH 5603	MOS-P-FET-e*	SMD,V-MOS, Dual, LogL, 30V, 1A, <0,97Ω	45bb1	Say		data
CPH 5604	MOS-N-FET-e*	SMD,V-MOS,Dual,LogL, 30V, 1,4A, <0,49Ω	45bb1	Say		data
CPH 5605	MOS-NP-FET-e	=CPH 5602 + CPH 5601	45bb1	Say		data
CPH 5606	MOS-NP-FET-e	=CPH 5604 + CPH 5603	45bb1	Say		data
CPH 5608	MOS-N-FET-e*	SMD,V-MOS,Dual,LogL, 100V, 0,4A, <3,8Ω	45bb1	Say		data
CPH 5701	Si-P+Di	SMD, S, lo-sat, 15V, 3A, 30/100ns int. Schottky-Di, 15V, 1A (SBS 004)	45bc1	Say	-	data
CPH 5702	Si-N+Di	SMD, S, lo-sat, 40V, 3A, 30/315ns int. Schottky-Di, 30V, 0,7A(SB07-03C)	45bc1	Say	-	data

Type	Device	Short Description	Fig.	Manu	Comparision Types	More at
CPH 5703	Si-N+Di	SMD, S, lo-sat, 60V, 3A, 35/322ns int. Schottky-Di, 60V, 0,5A(SB05-05CP)	45bc1	Say	-	data
CPH 5704	Si-N+Di	SMD, S, lo-sat, 15V, 3A, 30/221ns int. Schottky-Di, 15V, 1A (SBS 004)	45bc1	Say		data
CPH 5705	Si-P+Di	SMD, DC/DC Converter, lo-sat (=CPH 3106) 30V,	45bc1	Say	-	data data
CPH 5801	MOS-N-FET-e*	SMD, V-MOS, LogL, 20V, 1,4A, <0,36Ω int. Schottky-Di, 30V, 1A (SBS 005)	45bb2	Say		data
CPH 5902	Si-N+N-FET	SMD, 2 Transistors includet	45bc3	Say		data
CPH 6001	Si-N	SMD, UHF ra, 20V, 100mA, 6,7GHz	46	Say	-	data
CPH 6002	Si-N	SMD, VHF/UHF/CATV, 40V, 300mA, 4GHz	46	Say		data
CPH 6101	Si-P	SMD, lo-sat, DC-DC, 30V, 2A, 150MHz	46	Say	-	data
CPH 6102	Si-P	SMD, lo-sat, DC-DC, 60V, 1A, 150MHz	46	Say	-	data
CPH 6103	Si-P	SMD, lo-sat, DC-DC, 60V, 2A, 150MHz	46	Say	-	data
CPH 6104	Si-P	SMD, lo-sat, DC-DC, 15V, 1,5A, 300MHz	46	Say	-	data
CPH 6201	Si-N	SMD, lo-sat, DC-DC, 30V, 2A, 150MHz	46	Say	-	data
CPH 6202	Si-N	SMD, lo-sat, DC-DC, 60V, 1A, 150MHz	46	Say	-	data
CPH 6203	Si-N	SMD, lo-sat, DC-DC, 60V, 2A, 150MHz	46	Say	-	data
CPH 6204	Si-N	SMD, lo-sat, DC-DC, 15V, 1,5A, 200MHz	46	Say	-	data
CPH 6301	MOS-P-FET-e*	SMD, V-MOS, LogL, 20V, 3A, <0,22Ω	46bn4	Say	-	data
CPH 6302	MOS-P-FET-e*	SMD, V-MOS, LogL, 30V, 3A, <0,27Ω	46bn4	Say	-	data
CPH 6351	MOS-P-FET-e*	SMD, V-MOS, LogL, 20V, 3A, <0,2Ω	46	Say	-	data
CPH 6352	MOS-P-FET-e*	SMD, V-MOS, LogL, 30V, 3A, <0,24Ω(0,5A	46	Say	-	data
CPH 6401	MOS-N-FET-e*	SMD, V-MOS, LogL, 20V, 4A, <0,075Ω	46	Say	-	data
CPH 6402	MOS-N-FET-e*	SMD, V-MOS, LogL, 30V, 4A, <0,155Ω	46bn4	Say	-	data
CPH 6403	MOS-N-FET-e*	SMD, V-MOS, LogL, 20V, 6A, <0,038Ω	46	Say	-	data
CPH 6404	MOS-N-FET-e*	SMD, V-MOS, LogL, 30V, 6A, <68mΩ	46bn4	Say	-	data
CPH 6405	MOS-N-FET-e*	SMD, V-MOS, LogL, 30V, 6A, <0,061Ω	46	Say	-	data
CPH 6406	MOS-N-FET-e*	SMD, V-MOS, LogL, 60V, 3A, <0,21Ω	46bn4	Say	-	data
CPH 6501	Si-N	SMD, DC/DC-Converter, Driver	46bh1	Say	öCPH 3215	data
CPH 6701	Si-P+Di	SMD, S, lo-sat, 15V, 3A, 30/100ns int. Schottky-Di, 11V, 0,5A (SBS 001)	46	Say	-	data
CPH 6702	Si-P+Di	SMD, DC/DC Converter, (=CPH 3114) 15V,	46bn5	Say	-	data data
CPV 362 M4F	MOS-N-IGBT	6xIGBT+Damper-Di, 1-10kHz,600V, 6x8,8A	13-HYB	Inr	-	data
CPV 362 M4K	MOS-N-IGBT	6xIGBT+Damper-Di, >5kHz, 600V, 6x5,7A	13-HYB	Inr	-	data
CPV 362 M4U	MOS-N-IGBT	6xIGBT+Damper-Di, >5kHz, 600V, 6x7,2A	13-HYB	Inr	-	data
CPV 363 M4F	MOS-N-IGBT	6xIGBT+Damper-Di, 1-10kHz, 600V, 6x16A	13-HYB	Inr	-	data
CPV 363 M4K	MOS-N-IGBT	6xIGBT+Damper-Di, >5kHz, 600V, 6x11A	13-HYB	Inr	-	data
CPV 363 M4U	MOS-N-IGBT	6xIGBT+Damper-Di, >5kHz, 600V, 6x13A	13-HYB	Inr	-	data
CPV 364 M4F	MOS-N-IGBT	6xIGBT+Damper-Di, 1-10kHz, 600V, 6x27A	13-HYB	Inr	-	data
CPV 364 M4K	MOS-N-IGBT	6xIGBT+Damper-Di, >5kHz, 600V, 6x24A	13-HYB	Inr	-	data
CPV 364 M4U	MOS-N-IGBT	6xIGBT+Damper-Di, >5kHz, 600V, 6x20A	13-HYB	Inr	-	data
CQ	Si-P	=2PB1219Q (Typ-Code/Stempel/marking)	35(2mm)	Phi	→2PB1219	data
CQ	Si-P	=2PB710Q (Typ-Code/Stempel/marking)	35	Phi	→2PB710	data
CQ	Si-P	=2SB1219-Q (Typ-Code/Stempel/marking)	35(2mm)	Mat	→2SB1219	data
CQ	Si-P	=2SB1582-Q (Typ-Code/Stempel/marking)	≈35	Mat	→2SB1582	data
CQ	Si-P	=2SB710-Q (Typ-Code/Stempel/marking)	35	Mat	→2SB710	data
CQ	Si-P	=2SB767-Q (Typ-Code/Stempel/marking)	39	Mat	→2SB767	data
CQ	Si-N	=2SC2411-CQ (Typ-Code/Stempel/marking)	≈35	Rhm	→2SC2411	data
CQ	Si-N	=2SC2411K-Q (Typ-Code/Stempel/marking)	35	Rhm	→2SC2411K	data
CQ	Si-N	=2SC4097-Q (Typ-Code/Stempel/marking)	35(2mm)	Rhm	→2SC4097	data
CQ	Si-N	=HC 1A3M (Typ-Code/Stempel/marking)	39	Nec	→HC 1...	data
CQ	Z-Di	=P4 SMA-75 (Typ-Code/Stempel/marking)	71(5x2,5)	Fag	→P4 SMA-75	data
CQ	Z-Di	=SMBJ 40C (Typ-Code/Stempel/marking)	71(5x3,5)	Mop	→SMBJ ...	data
CQ 223-2 M	-IC	Epoxy Molded Silicon Triac	FLP	Cen	-	pdf pinout
CQ 223-2 N	-IC	Epoxy Molded Silicon Triac	FLP	Cen	-	pdf pinout
CQF24	IRED	850nm ø1.5mm I:5mW/Sr Vf:1.9V	TO46	Phi,Val		data pinout
CQL10A	IR LD+MD	820nm ø1.5mm I:5mW/Sr Vf:2.5V	SOT148	Phi,Val		data pinout
CQL13A	IR LD+MD	820nm ø5.4mm I:5mW/Sr If:175mA		Phi,Son		data pinout
CQL16A	IR LD+MD	820nm 5.4x3mm I:5mW/Sr If:100mA		Phi,Son		data pinout
CQL20	RED LD+MD	rd 780nm 5.4x3mm I:5mW/Sr Vf:1.7V	SOT184A	Phi,Son		data pinout
CQS51	LED	rd ø4.9mm I>0.7mcd Vf:2V If:10mA	T1¾	Phi,Val		data pinout
CQS54	LED	rd ø3.1mm I>0.7mcd Vf:2V If:10mA	T1	Phi,Val		data pinout
CQS82AL	LED	rd ø4.9mm I>10mcd Vf:1.75V If:10mA	T1¾	Phi,Val		data pinout
CQS82L	LED	rd ø4.9mm I>0.7mcd Vf:2V If:20mA	T1¾	Phi,Val		data pinout
CQS84L	LED	gr ø4.9mm I>1mcd Vf:2.1V If:10mA	T1¾	Phi,Val		data pinout
CQS86L	LED	yl ø4.9mm I>0.7mcd Vf:2.1V If:10mA	T1¾	Phi,Val		data pinout
CQS93	LED	rd ø3mm I>0.7mcd Vf:2.2V If:20mA	T1	Phi,Val		data pinout
CQS95	LED	gr ø3mm I>1.6mcd Vf:2.2V If:10mA	T1	Phi,Val		data pinout
CQS97	LED	yl ø3mm I>1.6mcd Vf:2.2V If:10mA	T1	Phi,Val		data pinout
CQT10B	LED	rd/gr 5x2mm I:1.5mcd Vf:2.2V If:20mA	SOD76A1	Phi,Val		data pinout
CQT10BL	LED	rd/gr 5x2mm I:1.5mcd Vf:2.2V If:20mA	SOD76L	Phi,Val		data pinout
CQT24	LED	rd/gr ø4.9mm I>3mcd Vf:2.2V If:20mA	T1¾	Phi,Val		data pinout
CQT60	LED	rd/gr 1.2x5mm I:1.5mcd Vf:2.2V	SOD75L	Phi,Val		data pinout
CQT70	LED	rd/gr 3.2x5.4mm I:1.5mcd Vf:2.2V	SOD77L	Phi,Val		data pinout
CQT80	LED	rd/gr 5x5mm I:2mcd Vf:2.2V If:20mA	SOD74L	Phi,Val		data pinout
CQV10	LED	rd+yl+gr ø3mm I>4mcd Vf:2.4V If:20mA	T1	Sie		data pinout

Type	Device	Short Description	Fig.	Manu	Comparision Types	More at
CQV16	LED	rd+yl+gr 3.2x4mm I>1.6mcd Vf:2.4V		Sie		data pinout
CQV20	LED	rd+yl+gr ø5mm I>4mcd Vf:2.4V If:20mA	T1¾	Sie		data pinout
CQV26	LED	rd+yl+gr 5x5.8x5.8mm I>1.6mcd		Sie		data pinout
CQV30	LED	rd+yl+gr ø2.8mm I>6.3mcd Vf:2.4V	T1	Sie		data pinout
CQV36	LED	rd+yl+gr 2.4x4.8mm I>2.5mcd Vf:2.4V		Sie		data pinout
CQV41	LED	rd+yl+gr ø2.8mm I>2.5mcd Vf:2.4V	T1	Sie		data pinout
CQV51	LED	rd+yl+gr ø5.1mm I>63mcd Vf:2.4V	T1¾	Sie		data pinout
CQV70	LED	srd 5.4x3.2mm I>0.7mcd Vf:2.1V	SOD77A2	Phi,Val		data pinout
CQV70A	LED	rd 5.4x3.2mm I>0.7mcd Vf:1.75V	SOD77L	Phi,Val		data pinout
CQV70L	LED	rd 5.4x3.2mm I>0.7mcd Vf:2.1V	SOD77L	Phi,Val		data pinout
CQV70U	LED	rd 5.4x3.2mm I>0.7mcd Vf:2V If:10mA	SOD77L	Phi,Val		data pinout
CQV71A	LED	gr 5.4x3.2mm I>0.7mcd Vf:2.1V	SOD77L	Phi,Val		data pinout
CQV72	LED	yl 5.4x3.2mm I>0.7mcd Vf:2.1V	SOD77L	Phi,Val		data pinout
CQV80AL	LED	rd 5x5mm I>0.7mcd Vf:1.75V If:10mA	SOD74L	Phi,Val		data pinout
CQV80L	LED	rd 5x5mm I>0.7mcd Vf:2.1V If:10mA	SOD74L	Phi,Val		data pinout
CQV80U	LED	rd 5x5mm I>0.7mcd Vf:2V If:10mA	SOD74L	Phi,Val		data pinout
CQV81L	LED	gr 5x5mm I>0.7mcd Vf:2.1V If:10mA	SOD74L	Phi,Val		data pinout
CQV82L	LED	yl 5x5mm I>0.7mcd Vf:2.1V If:10mA	SOD74L	Phi,Val		data pinout
CQW10A	LED	rd 5x2.5mm I>0.7mcd Vf:1.75V If:10mA	SOD76L	Phi,Val		data pinout
CQW10B	LED	rd 5x2.5mm I>0.7mcd Vf:2.1V If:10mA	SOD76L	Phi,Val		data pinout
CQW10U	LED	rd 5x2.5mm I>0.7mcd Vf:2V If:10mA	SOD76L	Phi,Val		data pinout
CQW11B	LED	gr 5x2.5mm I>0.7mcd Vf:2.1V If:10mA	SOD76L	Phi,Val		data pinout
CQW12B	LED	yl 5x2.5mm I>0.7mcd Vf:2.1V If:10mA	SOD76L	Phi,Val		data pinout
CQW13	IRED	950nm ø5mm I:170mW/Sr Vf:2.4V	T1¾	Aeg		data pinout
CQW14	IRED	950nm ø5mm I:120mW/Sr Vf:2.7V	T1¾	Aeg		data pinout
CQW15	IRED	950nm ø5mm I:300mW/Sr Vf:5V	T1¾	Tel		data pinout
CQW20	LED	rd ø2mm I>2.5mcd Vf:1.75V If:10mA	SOD79	Phi		data pinout
CQW21	LED	gr ø2mm I>1.5mcd Vf:2.1V If:10mA	SOD79	Phi		data pinout
CQW22	LED	yl ø2mm I>1.5mcd Vf:2.1V If:20mA	SOD79	Phi		data pinout
CQW24	LED	rd ø4.9mm I>4mcd Vf:1.75V If:10mA	T1¾	Phi,Val		data pinout
CQW54	LED	rd ø3.1mm I>3mcd Vf:1.75V If:10mA	T1	Phi,Val		data pinout
CQW60	LED	rd 5x1mm I>0.7mcd Vf:1.75V If:10mA	SOD75B	Phi,Val		data pinout
CQW60A	LED	rd 5x1mm I>0.7mcd Vf:1.75V If:10mA	SOD75L	Phi,Val		data pinout
CQW60L	LED	rd 5x1mm I>0.7mcd Vf:1.75V If:10mA	SOD75L	Phi,Val		data pinout
CQW60U	LED	rd 5x1mm I>0.7mcd Vf:2V If:10mA	SOD75L	Phi,Val		data pinout
CQW61	LED	gr 5x1mm I>0.7mcd Vf:2.1V If:10mA	SOD75L	Phi,Val		data pinout
CQW62	LED	yl 5x1mm I>0.7mcd Vf:2.1V If:10mA	SOD75L	Phi,Val		data pinout
CQW89A	IRED	830nm ø4.9mm I:>9mW/Sr Vf:1.7V	T1¾	Phi,Val		data pinout
CQW93	LED	rd ø3.1mm I>3mcd Vf:1.75V If:10mA	T1	Phi,Val		data pinout
CQW95	LED	gr+yl ø3.1mm I>3mcd Vf:2.1V If:10mA	T1	Phi,Val		data pinout
CQX...	Opto					
CQX10	LED	rd+gr+yl 2.5x5mm I:4.6mcd Vf:2.4V		Aeg		data pinout
CQX13	LED	gr ø5.1mm I>2.8mcd Vf:2.4V If:20mA	T1¾	Sie		data pinout
CQX14	IRED	940nm ø4.7mm I>2.8mcd Vf:1.7V	TO46	Qtc,Rca		data pdf pinout
CQX18	IRED	950nm ø4.7mm I>0.25mW/Sr Vf:1.2V	TO92	Tel		data pinout
CQX19	IRED	950nm ø8.15mm I:40mW/Sr Vf:1.2V	TO39	Tel,Tem		data pinout
CQX20	IR LD	820nm ø8.15mm I:40mW/Sr Vf:2V		Aeg		data pinout
CQX21	LED	rd+ord+gr+yl ø5mm I:3mcd Vf:5V	T1¾	Aeg		data pinout
CQX22	LED	rd ø5mm I:1.6mcd Vf:5V If:20mA	T1¾	Aeg		data pinout
CQX23	LED	rd ø5.1mm I>2.8mcd Vf:2.4V If:20mA	T1¾	Sie		data pinout
CQX24	LED	rd ø4.9mm I>16mcd Vf:2.2V If:10mA	T1¾	Phi,Val		data pinout
CQX25	LED	rd+gr+yl ø3mm I:5mcd Vf:2.4V If:20mA	T1	Aeg		data pinout
CQX28	LED	rd ø4.75mm I:0.8mcd Vf:1.6V If:20mA	TO18	Tem		data pinout
CQX29	LED	gr ø4.75mm I:1mcd Vf:2.4V If:20mA	TO18	Tem		data pinout
CQX30	LED	yl ø4.75mm I:1mcd Vf:2.4V If:20mA	TO18	Tem		data pinout
CQX31	LED	rd/gr ø4.75mm I:0.8mcd Vf:1.6V	TO18	Tem		data pinout
CQX32	LED	rd/yl ø4.75mm I:0.8mcd Vf:2.4V	TO18	Tem		data pinout
CQX33	LED	yl ø5.1mm I>2.8mcd Vf:2.4V If:20mA	T1¾	Sie		data pinout
CQX35	LED	rd+gr+yl+ord ø5mm I:40mcd Vf:2.2V	T1¾	Aeg		data pinout
CQX40	LED	ord 2.5x5mm I:5mcd Vf:2.2V If:20mA		Aeg		data pinout
CQX42	LED	ord ø3mm I:15mcd Vf:2.2V If:20mA	T1	Aeg		data pinout
CQX46	IRED	950nm ø3mm I:10mW/Sr Vf:1.4V	T1	Aeg		data pinout
CQX47	IRED	950nm ø5mm I:300mW/Sr Vf:5V		Tel		data pinout
CQX48	IRED	950nm ø1.5mm I:>1mW/Sr Vf:1.25V		Tel,Tem		data pdf pinout
CQX48B	IRED	950nm ø1.5mm I:>2mW/Sr Vf:1.25V		Tel,Tem,Vis		data pdf pinout
CQX51	LED	rd ø4.9mm I>1.6mcd Vf:2.1V If:10mA	T1¾	Phi,Val		data pinout
CQX51L	LED	rd ø4.9mm I>1.6mcd Vf:2.1V If:10mA	T1¾	Phi,Val		data pinout
CQX54	LED	rd ø4.9mm I>10mcd Vf:2.1V If:10mA	T1¾	Phi,Val		data pinout
CQX54D	LED	rd ø4.9mm I>3mcd Vf:2.1V If:10mA	T1¾	Phi,Val		data pinout
CQX54L	LED	rd ø4.9mm I>10mcd Vf:2.1V If:10mA	T1¾	Phi,Val		data pinout
CQX64	LED	gr ø4.9mm I>10mcd Vf:2.1V If:10mA	T1¾	Phi,Val		data pinout
CQX64D	LED	gr ø4.9mm I>3mcd Vf:2.1V If:10mA	T1¾	Phi,Val		data pinout
CQX64L	LED	gr ø4.9mm I>10mcd Vf:2.1V If:10mA	T1¾	Phi,Val		data pinout
CQX74	LED	yl ø4.9mm I>10mcd Vf:2.1V If:10mA	T1¾	Phi,Val		data pinout
CQX74D	LED	yl ø4.9mm I>3mcd Vf:2.1V If:10mA	T1¾	Phi,Val		data pinout
CQX74L	LED	yl ø4.9mm I>10mcd Vf:2.1V If:10mA	T1¾	Phi,Val		data pinout
CQX95	LED	crd/gr ø5mm I:6mcd Vf:2.4V If:20mA	T1¾	Aeg		data pinout
CQX96	LED	gr ø5mm I:70mcd Vf:2.4V If:20mA	T1¾	Aeg		data pinout
CQY...	Opio					
CQY11	IRED	880nm ø4.8mm I:>1.25mW/Sr Vf:1.3V	TO18	Phi		data pinout
CQY17	IRED	950nm ø4.6mm I:>15mW/Sr Vf:1.35V	TO18	Sie		data pinout
CQY24B	LED	rd ø4.9mm I>0.7mcd Vf:1.7V If:10mA	T1¾	Phi,Val		data pinout
CQY31	IRED	900nm ø3.9mm I:1mW/Sr Vf:1.25V	TO18	Aeg		data pinout
CQY32	IRED	900nm ø3.9mm I:10mW/Sr Vf:1.25V	TO18	Aeg		data pinout
CQY33N	IRED	950nm ø3.9mm I:7mW/Sr Vf:1.4V	TO18	Aeg		data pinout

Type	Device	Short Description	Fig.	Manu	Comparision Types	More at
CQY34N	IRED	950nm ø3.9mm I:18mW/Sr Vf:1.4V	TO18	Aeg		data pinout
CQY35N	IRED	950nm ø3.9mm I:36mW/Sr Vf:1.4V	TO18	Aeg		data pinout
CQY36N	IRED	950nm ø1.8mm I:1.6mW/Sr Vf:1.3V		Tel,Tem,Vis		data pdf pinout
CQY37N	IRED	950nm ø1.8mm I:4.5mW/Sr Vf:1.3V		Tel,Tem,Vis		data pdf pinout
CQY38	LED	ord ø5mm I:12mcd Vf:2.2V If:20mA	T1¾	Aeg		data pinout
CQY40	LED	rd ø5mm I:1.6mcd Vf:1.6V If:20mA	T1¾	Aeg		data pinout
CQY41	LED	ord ø5mm I:1.6mcd Vf:1.6V If:20mA		Aeg		data pinout
CQY41N	LED	rd+ord ø1.8mm I:5mcd Vf:2.2V If:20mA		Aeg		data pinout
CQY49	IRED	930nm ø4.8mm I:5mW/Sr Vf:1.3V	TO18	Phi		data pinout
CQY50	IRED	930nm ø1.5mm I:>0.45mW/Sr Vf:1.3V	DO31	Phi		data pinout
CQY53S	LED	rd ø4.2mm I>0.4mcd Vf:1.4V If:10mA	FO81	Phi		data pinout
CQY54A	LED	rd ø3.1mm I>0.7mcd Vf:1.7V If:20mA	T1	Phi,Val		data pinout
CQY58A	IRED	930nm ø2.6mm I:>1mW/Sr Vf:1.2V	SOD53F	Phi,Val		data pinout
CQY72	LED	gr ø5mm I:2mcd Vf:2.7V If:20mA		Aeg		data pinout
CQY73	LED	gr ø1.8mm I:2mcd Vf:2.7V If:20mA		Aeg		data pinout
CQY74	LED	yl ø5mm I:3mcd Vf:2.4V If:20mA	T1¾	Aeg		data pinout
CQY75	LED	yl ø1.8mm I:3mcd Vf:2.4V If:20mA		Aeg		data pinout
CQY77	IRED	950nm ø4.6mm I:>20mW/Sr Vf:1.35V	TO18	Sie		data pinout
CQY78	IRED	950nm ø4.6mm I:>2.5mW/Sr Vf:1.35V	TO18	Sie		data pinout
CQY80	OC	LED/NPN Viso:4000V CTR:>50% Ibs:50mA		Aeg,Har,Rca		data pdf pinout
CQY80N	OC	LED/NPN Viso:5300V CTR:90% Vbs:1.25V		Aeg,Tel,Vis	CNY17,CNY57A,FCD810C,FCD825	data pdf pinout
CQY80NG	OC	LED/NPN Viso:5300V CTR:90% Vbs:1.25V		Aeg,Tel,Vis		data pdf pinout
CQY85NA	LED	rd+gr+yl ø3mm I:5mcd Vf:2.4V If:20mA	T1	Aeg		data pinout
CQY89A	IRED	930nm ø4.8mm I:>9mW/Sr Vf:1.4V	T1¾	Phi,Rtc,Val		data pinout
CQY94B	LED	gr ø4.9mm I:>0.7mW/Sr Vf:2.1V	T1¾	Phi,Val		data pinout
CQY95	LED	gr ø3.1mm I:>0.7mW/Sr Vf:2.1V	T1	Phi,Val		data pinout
CQY96	LED	yl ø4.9mm I:>0.7mW/Sr Vf:2.1V	T1¾	Phi,Val		data pinout
CQY97A	LED	yl ø3.1mm I:>0.7mW/Sr Vf:2.1V	T1	Phi,Val		data pinout
CQY98	IRED	950nm ø5mm I:85mW/Sr Vf:2.7V	T1¾	Aeg		data pdf pinout
CQY99	IRED	950nm ø5mm I:60mW/Sr Vf:2.7V	T1¾	Aeg		data pdf pinout
CR	Si-P	=2PB1219R (Typ-Code/Stempel/marking)	35(2mm)	Phi	→2PB1219	data
CR	Si-P	=2PB710R (Typ-Code/Stempel/marking)	35	Phi	→2PB710	data
CR	Si-P	=2SB1219-R (Typ-Code/Stempel/marking)	35(2mm)	Mat	→2SB1219	data
CR	Si-P	=2SB1582-R (Typ-Code/Stempel/marking)	≈35	Mat	→2SB1582	data
CR	Si-P	=2SB710-R (Typ-Code/Stempel/marking)	35	Mat	→2SB710	data
CR	Si-P	=2SB767-R (Typ-Code/Stempel/marking)	39	Mat	→2SB767	data
CR	Si-N	=2SC2411-CR (Typ-Code/Stempel/marking)	≈35	Rhm	→2SC2411	data
CR	Si-N	=2SC2411K-R (Typ-Code/Stempel/marking)	35	Rhm	→2SC2411K	data
CR	Si-N	=2SC4097-R (Typ-Code/Stempel/marking)	35(2mm)	Rhm	→2SC4097	data
CR	Si-N	=2SC4422 (Typ-Code/Stempel/marking)	39	Hit	→2SC4422	data
CR	Si-N	=HC 1F3M (Typ-Code/Stempel/marking)	39	Nec	→HC 1...	data
CR	Z-Di	=P4 SMA-75A (Typ-Code/Stempel/marking)	71(5x2,5)	Fag	→P4 SMA-75A	data
CR	Z-Di	=SMBJ 40CA (Typ-Code/Stempel/marking)	71(5x3,5)	Mop	→SMBJ ...	data
CR	Si-N	=XP 6210 (Typ-Code/Stempel/marking)	46(2mm)	Mat	→XP 6210	data
CR 02 AM-...	Thy	≈BRX 49	7b		→BRX 49	
CR 08 B	Thy	=MCR 08B (Typ-Code/Stempel/marking)	48	Ons	→MCR 08B	
CR 08 M	Thy	=MCR 08M (Typ-Code/Stempel/marking)	48	Ons	→MCR 08M	
CR 126	Si-Di	Gl, 1400V, 1,5A	31a	Mic	BY 228; BY 448, BY 350/1500	data
CR 1296	Hybrid-IC	Monitor, hi-res, Video-Tr, 10V, 140MHz	14-SIL	Phi	-	pdf
CR 2424...8727	Hybrid-IC	Monitor, hi-res, Video-Tr	-SIL	Phi	-	
CR 2424 H,S	Hybrid-IC	Monitor, hi-res, Video-Tr, 70V,>130MHz		Mot, Phi	-	
CR 2425	Hybrid-IC	Monitor, hi-res, Video-Tr, 70V, >130MHz		Mot	-	
CR 2427	Hybrid-IC	Monitor, hi-res, Video-Tr, 70V,>130MHz	5-SIL	Phi	-	
CR 3424 (S)	Hybrid-IC	Monitor, hi-res, Video-Tr, 90V,>130MHz		Phi	-	
CR 3427	Hybrid-IC	Monitor, hi-res, Video-Tr, 90V,>130MHz	5-SIL	Phi	-	
CR 5427	Hybrid-IC	Monitor, hi-res, 3x Video-Tr, 80V	11-SIL	Phi	-	pdf
CR 5527 S	Hybrid-IC	Monitor, hi-res, 3x Video-Tr, 90V	12-SIL	Phi	-	pdf
CR 5627	Hybrid-IC	Monitor, hi-res, 3x Video-Tr, 90V	12-SIL	Phi	-	pdf
CR 6627	Hybrid-IC	Monitor, hi-res, 3x Video-Tr, 90V	12-SIL	Phi	-	pdf
CR 6727(A)	Hybrid-IC	Monitor, hi-res, 3x Video-Tr, 70V	12-SIL	Phi	-	pdf
CR 6927(A)	Hybrid-IC	Monitor, hi-res, 3x Video-Tr, 90V	12-SIL	Phi	-	pdf
CR 8727	Hybrid-IC	Monitor, hi-res, 3x Video-Tr, 100/10V	12-SIL	Phi	-	
CRS 01	Si-Di	SMD, Schottky, 30V, 1A, Uf<0,37V(0,7A)	71a(2,6mm)	Tos	(M 1FH3, M 1FM3)[6]	
CRS 02	Si-Di	SMD, Schottky, 30V, 1A, Uf<0,4V(0,7A)	71a(2,6mm)	Tos	(M 1FH3, M 1FM3)[6]	
CRS 03	Si-Di	SMD, Schottky, 30V, 1A, Uf<0,45V(0,7A)	71a(2,6mm)	Tos	(M 1FH3, M 1FM3)[6]	
CRS 04	Si-Di	SMD, Schottky, 40V, 1A, Uf<0,49V(0,7A)	71a(2,6mm)	Tos	(M 1FS4)[6]	
CRS 05	Si-Di	SMD, Schottky, 30V, 1A, Uf<0,45V(0,7A)	71a(2,6mm)	Tos	(M 1FH3, M 1FM3)[6]	
CRS 06	Si-Di	SMD, Schottky, 20V, 1A, Uf<0,36V(1A)	71a(2,6mm)	Tos	(M 1FH3, M 1FM3)[6]	
CRS 07	Si-Di	SMD, Schottky, 20V, 1A, Uf<0,33V(1A)	71a(2,6mm)	Tos	(M 1FH3, M 1FM3)[6]	
CRS 08	Si-Di	SMD, Schottky, 30V, 1,5A, Uf<0,36V(1,5A)	71a(2,6mm)	Tos	(M 1FH3, M 1FM3)[6]	
CRS 09	Si-Di	SMD, Schottky, 30V, 1,5A, Uf<0,46V(1,5A)	71a(2,6mm)	Tos	(M 1FH3, M 1FM3)[6]	
CRT 1544	Ge-P	NF/S-L, 60V, 25A, 90W	23a§	Itt	2N1651...1653, 2N2285...2287	data
CRT 1545	Ge-P	NF/S-L, 80V, 25A, 90W	23a§	Itt	2N1652...1653, 2N2286...2287	data
CRT 1552	Ge-P	NF/S-L, 40V, 25A, 90W	23a§	Itt	2N1651...1653, 2N2285...2287	data
CRT 1553	Ge-P	NF/S-L, 100V, 25A, 90W	23a§	Itt	2N1652...1653, 2N2286...2287	data
CRT 1592	Ge-P	NF/S-L, 80V, 35A	23a§	Itt	2N5693...5696	data

Type	Device	Short Description	Fig.	Manu	Comparision Types	More at
CS						
CS	Si-N	=HC 1F3P (Typ-Code/Stempel/marking)	39	Nec	→HC 1...	data
CS	Z-Di	=P4 SMA-82 (Typ-Code/Stempel/marking)	71(5x2,5)	Fag	→P4 SMA-82	data
CS	Z-Di	=SMBJ 43C (Typ-Code/Stempel/marking)	71(5x3,5)	Mop	→SMBJ ...	data
CS	Si-P	=2PB1219S (Typ-Code/Stempel/marking)	35(2mm)	Phi	→2PB1219	data
CS	Si-P	=2PB710S (Typ-Code/Stempel/marking)	35	Phi	→2PB710	data
CS	Si-P	=2SA1682 (Typ-Code/Stempel/marking)	35	Say	→2SA1682	data
CS	Si-P	=2SB1219-S (Typ-Code/Stempel/marking)	35(2mm)	Mat	→2SB1219	data
CS	Si-P	=2SB1582-S (Typ-Code/Stempel/marking)	≈35	Mat	→2SB1582	data
CS	Si-P	=2SB710-S (Typ-Code/Stempel/marking)	35	Mat	→2SB710	data
CS	Si-P	=2SB767-S (Typ-Code/Stempel/marking)	39	Mat	→2SB767	data
CS	Si-N	=2SD1870 (Typ-Code/Stempel/marking)	39	Hit	→2SD1870	data
CS 0,6-02...-12	50Hz-Thy	200...1200V, 1,5A(Tc=85°C), Igt<10mA	27e	Bbc	BStA30...M, (TAG 632-..., TAG 621-...)[4]	data
CS 0,8-02...-07	50Hz-Thy	200...700V,0,8A(Ta=45°C),Igt/Ih<10/20mA	30b	Bbc	BStD36...	data
CS 1-02...-12	50Hz-Thy	200...1200V,2A(Tc=85°C),Igt/Ih<15/40mA	27e	Bbc	BStC07..., (TAG 630-..., TAG 620-...)[4]	data
CS 1,2-02...-07	50Hz-Thy	200...700V,0,3A(Ta=45°C),Igt/Ih<15/25mA	30b	Bbc	TAG 620-..., BStC10..., TIC 116... ++	data
CS 2,5-04...-06	50Hz-Thy	200...600V,3,2A(Tc=85°C),Igt/Ih<15/25mA	17h§	Bbc	T 3,5N..., TAG 661-..., TAG 662-... ++	data
CS 3-02...-07	50Hz-Thy	200...700V, 5A(Tc=85°C), Igt/Ih<10/20mA	17h§	Bbc	S 2800..., TAG 665-..., TIC 126... ++	data
CS 3,5-02...-07	50Hz-Thy	200...700V, 6A(Tc=85°C),I gt/Ih<15/25mA	17h§	Bbc	S 2800..., TAG 665-..., TIC 126... ++	data
CS 6-02...08	50Hz-Thy	200...800V, 6A(Tc=85°C),Igt/Ih<15/25mA	17h§	Bbc	T 9,5N..., BStD10..M	data
CS 10-02...08	50Hz-Thy	200...800V,10A(Tc=85°C),Igt/Ih<25/100mA	17h§	Bbc		data
CS 10 D,S	Si-Br	Schottky-Br, 20V, 10V≈, 1A	4-DIP	Die		data
CS 0014	Si-Di	≈BA 100	31a		→BA 100	data
CS 15-02...08	50Hz-Thy	200...800V,15A(Tc=85°C),Igt/Ih<25/100mA	17h§	Bbc		data
CS23	HI-SPEED	Vs:±15V Vu:92dB Vo:±10V f:100Mc	7Y	Ilc		data pinout
CS24	HI-SPEED	Vs:±15V Vu:80dB Vo:±10V f:25Mc	8Y	Ilc		data pinout
CS 0080	Si-Di	≈BA 100	31a		→BA 100	
CS 82C55A-5	µP-Periph	CMOS, Ppi, 3x8 Programmable I/O Pins, TTL Compatible, 5MHz, 0...70°C	PLCC	Har	..82C55...	pdf pinout
CS 82C55A	µP-Periph	CMOS, Ppi, 3x8 Programmable I/O Pins, TTL Compatible, 8MHz, 0...70°C	PLCC	Har	..82C55...	pdf pinout
CS 387	LIN/Z-IC	Alternator Voltage Reg. Darlington Drv	GRID	Ons		pdf pinout
CS 1245(F...T)	Si-N	≈BC 337			→BC 337	
CS 1250(E,F)	Si-N	≈BC 639			→BC 639	
CS 1251(E,F)	Si-P	≈BC 640			→BC 640	
CS 1303	Si-P	≈BC 640			→BC 640	
CS 1312 F,G	Si-P	≈BF 450	2a		→BF 450	
CS 1312 H	Si-P	≈BC 309	2a		→BC 309	
CS 1312 I	Si-P	≈BC 307	2a		→BC 307	
CS 1506 F,G	Si-P	≈BC 307	2a		→BC 307	
CS 1508 E,G	Si-N	≈BF 255	2a		→BF 255	
CS 1509 E,F	Si-N	≈BF 254	2a		→BF 254	
CS 1655	Si-N	≈BF 494	7c		→BF 494	
CS 1659	Si-N	≈BC 338	7c		→BC 338	
CS 1660	Si-P	≈BC 328	7c		→BC 328	
CS 1702	Si-N	≈BC 337			→BC 337	
CS 1774	Si-N	≈BC 548	7c		→BC 548	
CS 1909	Si-N	≈BC 337	2a		→BC 337	
CS 1910	Si-P	≈BC 327	2a		→BC 327	
CS 1914(H)	Si-P	≈BC 307	2a		→BC 307	
CS 1978(A)	Si-N	≈BC 337	7e		→BC 337	
CS 3341	LIN/Z-IC	Alternator Voltage Reg. Darlington Drv	MDIP	Ons		pdf pinout
CS 3351	LIN/Z-IC	Alternator Voltage Reg. Darlington Drv	MDIP	Ons		pdf pinout
CS 3361	LIN-IC	Alternator Voltage Reg. Fet Drv	MDIP	Ons		pdf pinout
CS 4121 EDW	IC	Prec. Air-core Tach/speedo Drv.	MDIP	Ons		pdf pinout
CS 4121 ENF	IC	Prec. Air-core Tach/speedo Drv.	DIC	Ons		pdf pinout
CS 42 L 55	Dig. Audio	Stereo codec w/class H Headphone Amp	PLCC	Cir		pdf pinout
CS 4350	D/A-IC	192 kHz Stereo DAC with Integrated PLL	TSOP	Cir		pdf pinout
CS 44 L 10-KZ	Dig. Audio	Digital PWM Headphone Monitor	TSOP	Cir		pdf pinout
CS 4525	Dig. Audio	30 W Digital TV Amp, Integrated ADC	TSOP	Cir		pdf pinout
CS5005	OC	LED/MOSFET Viso:4000Vrms Vbs:1.2V		Inr		data pinout
CS5010	OC	LED/MOSFET Viso:4000Vrms Vbs:1.2V		Inr		data pinout
CS 5609	Si-N	≈BC 337			→BC 337	
CS 5610	Si-P	≈BC 327			→BC 327	
CS6005	OC	LED/MOSFET Viso:4000Vrms Vbs:1.2V		Inr		data pinout
CS6010	OC	LED/MOSFET Viso:4000Vrms Vbs:1.2V		Inr		data pinout
CS 6203 H,I	Si-P	≈BC 309	7e		→BC 309	
CS 6208	Si-N	≈BC 337	2a		→BC 337	
CS 6209	Si-P	≈BC 327	2a		→BC 327	
CS 6305(A)	Si-P	≈BC 307	2a		→BC 307	
CS 8190 EDW	IC	Prec. Air-core Tach/speedo Drv.	MDIP	Ons		pdf pinout
CS 8190 ENF	IC	Prec. Air-core Tach/speedo Drv.	DIC	Ons		pdf pinout
CS 8204	LIN-IC	→KA 2410	8-DIP	Chy	KA 2410, ML 8204, TA 31001	
CS 8205	LIN-IC	→KA 2411	8-DIP	Chy	KA 2411, ML 8205, TA 31002	
CS 8422	IC	24-bit,192 kHz,Async Sample Rate Conv.	QFN	Cir		pdf pinout
CS 9003	Si-N	≈BC 547	7		→BC 547	
CS 9010	Si-N	≈BC 547	7e		→BC 547	data
CS 9011(D...I)	Si-N	=SS 9011	7e	Mic, Nsc	→SS 9011	data
CS 9012(D...I)	Si-P	=SS 9012	7e	Mic, Nsc	→SS 9012	data
CS 9013(D...I)	Si-N	=SS 9013	7e	Mic, Nsc	→SS 9013	data
CS 9014(A...C)	Si-N	=SS 9014	7e	Mic, Nsc	→SS 9014	data
CS 9015(A...C)	Si-P	=SS 9015	7e	Mic, Nsc	→SS 9015	data

Type	Device	Short Description	Fig.	Manu	Comparision Types	More at
S 9016(D..H)	Si-N	=SS 9016	7e	Mic, Nsc	→SS 9016	data
S 9017	Si-N	AM/FM, -/18V	7e		BF 225, BF 255, BF 314, BF 495, BF 595++	data
S 9018(D..H)	Si-N	=SS 9018	7e	Mic, Nsc	→SS 9018	data
S 9020(G,H)	Si-P	≈BF 450	7e		→BF 450	data
S 9021	Si-N	≈BF 495	7		→BF 495	data
S 9022	Si-N	≈BC 547	7		→BC 547	data
S 9102(B)	Si-P	≈BC 640			→BC 640	
S 9103(B,C)	Si-N	≈BC 639			→BC 639	
S 9126	Si-N	≈BC 548	7e		→BC 548	data
S 9127	Si-P	≈BC 558	7e		→BC 558	data
S 42324-CQZ	Dig. Audio	10-In, 6-Out, 2 Vrms Audio codec	QFP	Cir	-	pdf pinout
S 42324-DQZ	Dig. Audio	10-In, 6-Out, 2 Vrms Audio codec	QFP	Cir	-	pdf pinout
S 493005	Dig. Audio	Multi-standard Audio Decoder Family	PLCC	Cir	-	pdf pinout
S 493105	Dig. Audio	Multi-standard Audio Decoder Family	PLCC	Cir	-	pdf pinout
S 493115	Dig. Audio	Multi-standard Audio Decoder Family	PLCC	Cir	-	pdf pinout
S 493122	Dig. Audio	Multi-standard Audio Decoder Family	PLCC	Cir	-	pdf pinout
S 493253	Dig. Audio	Multi-standard Audio Decoder Family	PLCC	Cir	-	pdf pinout
S 493254	Dig. Audio	Multi-standard Audio Decoder Family	PLCC	Cir	-	pdf pinout
S 493263	Dig. Audio	Multi-standard Audio Decoder Family	PLCC	Cir	-	pdf pinout
S 493264	Dig. Audio	Multi-standard Audio Decoder Family	PLCC	Cir	-	pdf pinout
S 493292	Dig. Audio	Multi-standard Audio Decoder Family	PLCC	Cir	-	pdf pinout
S 493295	Dig. Audio	Multi-standard Audio Decoder Family	PLCC	Cir	-	pdf pinout
S 493302	Dig. Audio	Multi-standard Audio Decoder Family	PLCC	Cir	-	pdf pinout

CSF...CW

Type	Device	Short Description	Fig.	Manu	Comparision Types	More at
CSF 0,7-04	F-Thy	400V, 3A(Tc=85°C), Igt/Ih<50/<200mA	17h§	Bbc	BT153, TAG650S-400, TAG655S-400	data
CSF 02 AM 1	F-Thy	50V, 0,26A(Ta=45°C),Igt/Ih<0,25/<2,3mA	7b	Bbc	BRX45...46, BRX 50...51, TAG60F, TAG62F	data
CSF 02 AM 2	F-Thy	=CSF02AM1: 100V	7b		BRX46...47, BRX 50...51, TAG60A, TAG 62A	data
CSF 02 AM 4	F-Thy	=CSF02AM1: 200V	7b		BRX47...48, BRX 52...54, BRY55S/200	data
CSF 02 AM 8	F-Thy	=CSF02AM1: 400V	7b		BRX49, BRX 54...56, BRY55S/400	data
CSF 02 AM 10	F-Thy	=CSF02AM1: 500V	7b		BRX 55...56, BRY55S/500	data
CSF 7,9-04...08	F-Thy	400...800V, 14A, Igt/Ih<100/<100mA	21b§	Bbc	-	data
CSF 11-02...08	F-Thy	200...800V, 10A(Tc=85°C),Igt/Ih<35/80mA	17h§	Bbc	-	data
CSM 2B2	50Hz-Thy	200V, 2A(Tc=71°), Igt/Ih=1/1,5mA	13b	Hit	C 107..., TAG 107..., C 106..., TAG 106...	data
CSM 2B4	50Hz-Thy	=CSM2B2: 400V	13b		C 107..., TAG 107..., C 106..., TAG 106...	data
CSM 3B2	50Hz-Thy	200V, 3A(Tc=63°), Igt/Ih=1/1,5mA	13b	Hit	C 107..., TAG 107..., C 106..., TAG 106...	data
CSM 3B4	50Hz-Thy	=CSM3B2: 400V	13b		C 107..., TAG 107..., C 106..., TAG 106...	data
CSM 5B2	50Hz-Thy	200V, 5A(Tc=101°), Igt/Ih=30/25mA	17b	Hit	TAG 651-..., TAG 656-..., (CS 2,5-...,++)[3]	data
CSM 5B4	50Hz-Thy	=CSM5B2: 400V	17b		TAG 651-..., TAG 656-..., (CS 2,5-...,++)[3]	data
CSY 240	GaAs-FET-IC	SMD, Mmic, Telecom, 0,5...3GHz	≈46	Sie	-	data
CT	Si-N	=2SC4984 (Typ-Code/Stempel/marking)	39	Say	→2SC4984	data
CT	Si-N-Darl+Di	=2SD1472 (Typ-Code/Stempel/marking)	39	Hit	→2SD1472	data
CT	Si-N	=2SD1935 (Typ-Code/Stempel/marking)	35	Say	→2SD1935	data
CT	Si-N	=HC 1L2Q (Typ-Code/Stempel/marking)	39	Nec	→HC 1...	data
CT	Z-Di	=MMSZ 4688 (Typ-Code/marking)	71(2,7mm)	Ons	→MMSZ 4688	
CT	Z-Di	=P4 SMA-82A (Typ-Code/Stempel/marking)	71(5x2,5)	Fag	→P4 SMA-82A	data
CT	Z-Di	=SMBJ 43CA (Typ-Code/Stempel/marking)	71(5x3,5)	Mop	→SMBJ ...	data
CT 005...06	Si-Br	50..600V, 2A	33x1	Mic	B..C2200	data
CT 15 SM-24	MOS-N-IGBT	L, 1200V, 15A, 250W, 200/400ns	18id	Mit	-	data
CT 20 AS-8	MOS-N-IGBT	L, Strobo Flasher, 400V, 20A, Uth<7V	30id	Mit	-	data
CT 20 ASJ-8	MOS-N-IGBT	L, Strobo Flasher, 400V, 20A, Uth<1,5V	30id	Mit	-	data
CT 20 ASL-8	MOS-N-IGBT	L, Strobo Flasher, 400V, 20A, Uth<4V	30id	Mit	-	data
CT 20 TM-8	MOS-N-IGBT	=CT 20AS-8:	17ic	Mit	-	data
CT 20 VM-8	MOS-N-IGBT	=CT 20AS-8:	30id	Mit	-	data
CT 20 VML-8	MOS-N-IGBT	=CT 20ASL-8:	30id	Mit	-	data
CT 20 VS-8	MOS-N-IGBT	=CT 20AS-8:	30id	Mit	-	data
CT 20 VSL-8	MOS-N-IGBT	=CT 20ASL-8:	30id	Mit	-	data
CT 25 AS-8	MOS-N-IGBT	L, Strobo Flasher, 400V, 25A, Uth<7V	30id	Mit	-	data
CT 25 ASJ-8	MOS-N-IGBT	L, Strobo Flasher, 400V, 25A, Uth<1,5V	30id	Mit	-	data
CT 30 SM-12	MOS-N-IGBT	L, 600V, 30A, 250W, 165/385ns	18id	Mit	-	data
CT 30 TM-8	MOS-N-IGBT	L, Strobo Flasher, 400V, 30A	17ic	Mit	-	data
CT 30 VM-8	MOS-N-IGBT	=CT 30TM-8:	30id	Mit	-	data
CT 30 VS-8	MOS-N-IGBT	=CT 30TM-8:	30id	Mit	-	data
CT 35 SM-8	MOS-N-IGBT	L, Strobo Flasher, 400V, 35A	18id	Mit	-	data
CT 40 MH-8	MOS-N-IGBT	L, Strobo Flasher, 400V, 40A	17ic	Mit	-	data
CT 60 AM-18B	MOS-N-IGBT	L, int. FW-Di(E→C), 1000V, 60A, 200W	77id	Mit	-	data
CT 60 AM-20	MOS-N-IGBT	L, int. FW-Di(E→C), 1000V, 60A, 250W	77id	Mit	-	data
CT 75 AM-12	MOS-N-IGBT	L, 600V, 75A, 300W, 170/440ns	77id	Mit	GN 6075E, GT 80J101	data
CT 1010	DIG-IC	TV, Frequ.-Teiler/prescaler, :64		Pls	-	
CT 1011	DIG-IC	TV, Frequ.-synthesizer		Pls	-	
CT 1117	DIG-IC	TV, Synthesizer Tuning Interface		Pls	-	
CT 1133	DIG-IC	TV, Synthesizer Tuning, 100-Ch. ROM		Pls	-	
CT 1134	DIG-IC	TV, Control Logic f. CT 1133		Pls	-	
CT 1650	DIG-IC	TV, µP-synthesizer Control		Pls	-	
CT 2010	DIG-IC	TV, Frequ.-T./prescaler, 1GHz:380/400	8-DIP	Pls	-	
CT 2012	MOS-IC	TV, PLL Frequ.-synthesizer	24-DIP	Pls	-	
CT 2014	DIG-IC	TV, Synthesizer Tuning Control	40-DIP	Pls	-	
CT 2015	DIG-IC	TV, Synthesizer Tuning Control	24-DIP	Pls	-	

Type	Device	Short Description	Fig.	Manu	Comparision Types	More at
CT 2017	DIG-IC	TV, Synthesizer Tuning Interface		Pls	-	
CT 2030	ROM-IC	TV, ROM f. 100Kanäle/Channel (pal)	16-DIP	Pls	-	
CT 2031	ROM-IC	=CT 2030: f. Frankr./French Standard	16-DIP	Pls	-	
CT 2032	ROM-IC	=CT 2030: f. Usa	16-DIP	Pls	-	
CT 2033	ROM-IC	=CT 2030: f. England/British Isles	16-DIP	Pls	-	
CT 2200	NMOS-IC	5 Bit→13-Segment Decoder	24-DIP	Pls	-	
CT 3244 A	Logic-IC	SMD, Octal bus switch, 5V A, ADS: 20-SSMDIP, AD:20- MDIP	20-SSMDIP	Phi	-	pdf pinout pdf pinout
CT 7004	IC	Digital-uhr/digital clock			-	
CT-2512-001-1	DIG-IC	Dual Redundant Remote Terminal	CSP	Aer	-	pdf pinout
CT-2512-001-2	DIG-IC	Dual Redundant Remote Terminal	CSP	Aer	-	pdf pinout
CT-2512-201-1	DIG-IC	Dual Redundant Remote Terminal	CSP	Aer	-	pdf pinout
CT-2512-201-2	DIG-IC	Dual Redundant Remote Terminal	CSP	Aer	-	pdf pinout
CTA2P1N	\N	\N	\N	\N	\N	
CTB-23	Si-Di	Dual, Schottky-Gl, 30V, 4A, 200ns	17p§	Sak	-	data
CTB-23 L	Si-Di	Dual, Schottky-Gl, 30V, 10A, 100ns	17p§	Sak	-	data
CTB-24	Si-Di	Dual, Schottky-Gl, 40V, 4A, 200ns	17p§	Sak	-	data
CTB-24 L	Si-Di	Dual, Schottky-Gl, 40V, 10A, 100ns	17p§	Sak	-	data
CTB-33	Si-Di	Dual, Schottky-Gl, 30V, 15A, 100ns	18p§	Sak	-	data
CTB-33 M	Si-Di	Dual, Schottky-Gl, 30V, 30A, 100ns	18p§	Sak	-	data
CTB-33 S	Si-Di	Dual, Schottky-Gl, 30V, 12A, 100ns	18p§	Sak	-	data
CTB-34	Si-Di	Dual, Schottky-Gl, 40V, 15A, 100ns	18p§	Sak	-	data
CTB-34 M	Si-Di	Dual, Schottky-Gl, 40V, 30A, 100ns	18p§	Sak	-	data
CTB-34 S	Si-Di	Dual, Schottky-Gl, 40V, 12A, 100ns	18p§	Sak	-	data
.CTB-G 14	Si-Di	Schottky-Gl, 40V, 3A(Tc=115°)	17t§	Sak	BYV 19/40, MBR 745	data
CTB-G 14L	Si-Di	Schottky-Gl, 40V, 6A(Tc=110°)	17t§	Sak	BYV 19/40, MBR 745	data
CTG-11 R	Si-Di	Dual, S-L, 100V, 5A, 100ns	17q&	Sak	-	data
CTG-12 R	Si-Di	=CTG-11R: 200V	17q&	Sak	-	data
CTG-14 R	Si-Di	=CTG-11R: 400V	17q&	Sak	-	data
CTG-... S	Si-Di	=CTG-...R:	17p§		-	data
CTG-21 R	Si-Di	Dual, S-L, 100V, 10A, 100ns	17q&	Sak	-	data
CTG-22 R	Si-Di	=CTG-21R: 200V	17q&	Sak	-	data
CTG-23 R	Si-Di	=CTG-21R: 300V	17q&	Sak	-	data
CTG-24 R	Si-Di	=CTG-21R: 400V, 8A	17q&	Sak	-	data
CTG-... S	Si-Di	=CTG-...R:	17p§		BYR 28/..., BYT 28/...	data
CTG-31 R	Si-Di	Dual, S-L 100V, 20A, 100ns	18q&	Sak	-	data
CTG-32 R	Si-Di	=CTG-31R: 200V	18q&	Sak	-	data
CTG-33 R	Si-Di	=CTG-31R: 300V	18q&	Sak	-	data
CTG-34 R	Si-Di	=CTG-31R: 400V, 16A	18q&	Sak	-	data
CTG-... S	Si-Di	=CTG-...R:	18p§		BYV 74/..., MUR 3010..3040	data
CTG-G 12	Si-Di	Gl/S-L, 200V, 5A, 100ns	17	Sak	BY 229/200, BY 233/200, RGP 80D..M, ++	data
CTG-G 12 S	Si-Di	=CTG-G12: 4A(104°)	17t§		BY 229/200, BY 233/200, RGP 80D..M, ++	data
CTL-11 S	Si-Di	Dual, S-L, 100V, 5A, 40ns	17p§	Sak	MUR 610	data
CTL-12 S	Si-Di	=CTL-11S: 200V	17p§	Sak	MUR 620	data
CTL-21 S	Si-Di	Dual, S-L, 100V, 10A, 40ns	17p§	Sak	BYR 28/..., BYT 28/,,	data
CTL-22 S	Si-Di	=CTL-21S: 200V	17p§	Sak	BYR 28/..., BYT 28/...	data
CTL-... R	Si-Di	=CTL-...S:	17q&	Sak	-	data
CTL-31 S	Si-Di	Dual, S-L, 100V, 20A, 40ns	18p§	Sak	BYV 74/...	data
CTL-32 S	Si-Di	=CTL-31S: 200V	18p§	Sak	BYV 74/...	data
CTL-G 11 S	Si-Di	Gl/S-L, 100V, 6A(Tc=124), <50ns	17t§	Sak	BYT 08/..., BYW 29/..., FE 8B..J	data
CTL-G 12 S	Si-Di	=CTL-G11S: 200V	17t§	Sak	BYT 08/..., BYW 29/..., FE 8D..J	data
CTL-G 21 S	Si-Di	Gl/S-L, 100V, 10A(Tc=108), <50ns	17t§	Sak	BYT 79/..., BYV 79/..., MUR 1510..1560	data
CTL-G 22 S	Si-Di	=CTL-G 21S: 200V	17t§	Sak	BYT 79/..., BYV 79/..., MUR 1520..1560	data
CTM-20 R	Si-Di	=CTM-21R: 50V	17q&	Sak	-	data
CTM-21 R	Si-Di	Dual, Gl-L, 100V, 8A	17q&	Sak	-	data
CTM-22 R	Si-Di	=CTM-21R: 200V	17q&	Sak	-	data
CTM-24 R	Si-Di	=CTM-21R: 400V	17q&	Sak	-	data
CTM-26 R	Si-Di	=CTM-21R: 600V	17q&	Sak	-	data
CTM-... S	Si-Di	=CTM-...R:	17p§		BY 239/..., GP 80A..M	data
CTM-... U	Si-Di	=CTM-...R: Volt. Doubler	17		-	data
CTM-31 R	Si-Di	Dual, Gl-L, 100V, 15A	18q&	Sak	-	data
CTM-32 R	Si-Di	=CTM-31R: 200V	18q&	Sak	-	data
CTM-34 R	Si-Di	=CTM-31R: 400V	18q&	Sak	-	data
CTM-... S	Si-Di	=CTM-...R:	18p§		BYV 74/..., MUR 3010PT..3060PT	data
CTP 1104	Ge-P	NF-L, 30V, 3A, 40W	23a§	Itt	AD 149, AL 102, AUY 19..20, 2SB449	data
CTP 1108	Ge-P	≈2N2061	23a§	Itt	→2N2061	data
CTP 1109	Ge-P	≈2N2062	23a§	Itt	→2N2062	data
CTP 1111	Ge-P	NF-L, 80V, 3A, 45W	23a§	Itt	AL 102, AUY 20, 2N1541..43, 2N1546..48	data
CTP 1500	Ge-P	NF/S-L, 100V, 15A, 50W(Tc=45°)	23a§	Itt	AUY 37, 2N1552, 2N1556, 2N1560	data
CTP 1503	Ge-P	=CTP 1500: 80V	23a§	Itt	2N1551..52, 2N1555..56, 2N1559..60	data
CTP 1504	Ge-P	=CTP 1500: 60V	23a§	Itt	2N1550..52, 2N1554..56, 2N1558..60	data
CTP 1508	Ge-P	NF/S-L, 40V, 15A, 50W(Tc=45°)	23a§	Itt	2N1549..52, 2N1553..56, 2N1557..60	data
CTP 1544	Ge-P	NF/S-L, 60V, 25A, 50W(Tc=45°)	23a§	Itt	2N1651..1653, 2N2285..2287	data
CTP 1545	Ge-P	=CTP 1544: 80V	23a§	Itt	2N1652..1653, 2N2286..2287	data
CTP 1552	Ge-P	NF/S-L, 40V, 25A, 50W(Tc=45°)	23a§	It	2N1651..1653, 2N2285..2287	data
CTP 1553	Ge-P	=CTP 1552: 100V	23a§	Itt	2N1652..1653, 2N2286..2287	data
CTS	Si-N	=2SC4984-S (Typ-Code/Stempel/marking)	39	Say	→2SC4984	data
CTT	Si-N	=2SC4984-T (Typ-Code/Stempel/marking)	39	Say	→2SC4984	data
CTU	Si-N	=2SC4984-U (Typ-Code/Stempel/marking)	39	Say	→2SC4984	data

Type	Device	Short Description	Fig.	Manu	Comparision Types	More at
TU-11 R	Si-Di	Dual, S-L, 100V, 6A, 400ns	17q&	Sak	-	data
TU-12 R	Si-Di	=CTU-11R: 200V	17q&	Sak	-	data
TU-14 R	Si-Di	=CTU-11R: 400V	17q&	Sak	-	data
TU-16 R	Si-Di	=CTU-11R: 600V	17q&	Sak	-	data
TU-... S	Si-Di	=CTU-11R:	17p§	Sak	BYR 28/..., BYT 28/...	data
TU-20 R	Si-Di	=CTU-21R: 50V	17q&	Sak	-	data
TU-21 R	Si-Di	Dual, S-L, 100V, 8A, 400ns	17q&	Sak	-	data
TU-22 R	Si-Di	=CTU-21R: 200V	17q&	Sak	-	data
TU-24 R	Si-Di	=CTU-21R: 400V	17q&	Sak	-	data
TU-26 R	Si-Di	=CTU-21R: 600V	17q&	Sak	-	data
TU-... S	Si-Di	=CTU-...R:	17p§		BYR 28/..., BYT 28/...	data
TU-... U	Si-Di	=CTU-...R: Volt. Doubler	17		-	data
TU-31 R	Si-Di	Dual, S-L, 100V, 12A, 400ns	18q&	Sak	-	data
TU-32 R	Si-Di	=CTU-31R: 200V	18q&	Sak	-	data
TU-34 R	Si-Di	=CTU-31R: 400V	18q&	Sak	-	data
TU-36 R	Si-Di	=CTU-31R: 600V	18q&	Sak	-	data
TU-38 R	Si-Di	=CTU-31R: 800V	18q&	Sak	-	data
TU-... S	Si-Di	=CTU-31R:	18p§		BYV 74/..., MUR 3010PT...3060PT	data
TU-... U	Si-Di	=CTU-31R: Volt. Doubler	18		-	data
TU-G 20 R	Si-Di	Gl, S-L, 1300V, 4A, 400ns	17d	Sak	BYT 86/1300, BYT 106/1300	data
TU-G 30 R	Si-Di	Gl, S-L, 1300V, 6A, 400ns	18d	Sak	-	data
TU-S 16 R	Si-Di	Gl, S-L, 600V, 6A, <400ns	17t1&	Sak	BYV 71/600	data
CU	Z-Di	=DF 3A6.8LFE(Typ-Code/Stempel/marking)	35(1,6mm)	Tos	→DF 3A6.8LFE	
CU	Z-Di	=DF 3A6.8LFU(Typ-Code/Stempel/marking)	35(2mm)	Tos	→DF 3A6.8LFU	
CU	Z-Di	=DF 5A6.8LF (Typ-Code/Stempel/marking)	45	Tos	→DF 5A6.8LF	
CU	Z-Di	=DF 5A6.8LFU(Typ-Code/Stempel/marking)	45(2mm)	Tos	→DF 5A6.8LFU	
CU	Si-N	=HC 1F2Q (Typ-Code/Stempel/marking)	39	Nec	→HC 1...	data
CU	Z-Di	=MMSZ 4689 (Typ-Code/marking)	71(2,7mm)	Ons	→MMSZ 4689	
CU	Z-Di	=P4 SMA-91 (Typ-Code/Stempel/marking)	71(5x2,5)	Fag	→P4 SMA-91	data
CU	Z-Di	=SMBJ 45C (Typ-Code/Stempel/marking)	71(5x3,5)	Mop	→SMBJ ...	data
CUJ	Z-Di	=SMBJ 40A (Typ-Code/Stempel/marking)	71(5x3,5)	Tho	→SMBJ ...	data
CUM	Z-Di	=SMBJ 70A (Typ-Code/Stempel/marking)	71(5x3,5)	Tho	→SMBJ ...	data
CUQ	Z-Di	=SMBJ 100A (Typ-Code/Stempel/marking)	71(5x3,5)	Tho	→SMBJ ...	data
CV	PIN-Di	=1SV234 (Typ-Code/Stempel/marking)	35	Say	→1SV234	data
CV	PIN-Di	=1SV246 (Typ-Code/Stempel/marking)	35(2mm)	Say	→1SV246	data
CV	Z-Di	=MMSZ 4690 (Typ-Code/marking)	71(2,7mm)	Ons	→MMSZ 4690	
CV	Z-Di	=P4 SMA-91A (Typ-Code/Stempel/marking)	71(5x2,5)	Fag	→P4 SMA-91A	data
CV	Z-Di	=SMBJ 45CA (Typ-Code/Stempel/marking)	71(5x3,5)	Mop	→SMBJ ...	data
CV	Si-P	=XN 4404 (Typ-Code/Stempel/marking)	46	Mat	→XN 4404	data
CVPU 2210	NMOS-IC	Ctv, NTSC Kamm-/comb Filter Video-Proz	40-DIP	Itt	-	
CVPU 2233	NMOS-IC	Ctv, NTSC Kamm-/comb Filter Video-proz	40-DIP	Itt	-	
CVPU 2235	NMOS-IC	Ctv, NTSC Kamm-/comb Filter Video-proz	40-DIP	Itt	-	
CVPU 2270	NMOS-IC	Ctv, NTSC Kamm-/comb Filter Video-proz	40-DIP	Itt	-	
CW	Z-Di	=P4 SMA-100 (Typ-Code/Stempel/marking)	71(5x2,5)	Fag	→P4 SMA-100	data
CW	Z-Di	=SMBJ 48C (Typ-Code/Stempel/marking)	71(5x3,5)	Mop	→SMBJ ...	data
CW	Si-N/P+R	=XN 4381 (Typ-Code/Stempel/marking)	46	Mat	→XN 4381	data
CW 01 B	Thy	100V, 0,2A(Tc=57°), Igt/Ih 1/5mA	9b	Hit	MCR 1906-..., 2N1595...1599, TAG 2-...	data
CW 01 C	Thy	=CW01B: 200V	9b		MCR 1906-..., 2N1595...1599, TAG 2-...	data
CW 12 B	50Hz-Thy	100V, 0,2A(Tc=57°), Igt/Ih 1/5mA	7b	Hit	MCR 1906-..., 2N1595...1599, TAG 2-...	data
CW 12 C	50Hz-Thy	=CW12C: 200V	7b		MCR 1906-..., 2N1595...1599, TAG 2-...	data

CX...CZ

Type	Device	Short Description	Fig.	Manu	Comparision Types	More at
CX	MOS-N-FET-d	=3SK227 (Typ-Code/Stempel/marking)	44	Mat	→2SK227	data
CX	MOS-N-FET-d	=3SK271 (Typ-Code/Stempel/marking)	44(2mm)	Mat	→3SK271	data
CX	Si-N	=HC 1A4A (Typ-Code/Stempel/marking)	39	Nec	→HC 1...	data
CX	Z-Di	=P4 SMA-100A(Typ-Code/Stempel/marking)	71(5x2,5)	Fag	→P4 SMA-100A	data
CX	Z-Di	=MMSZ 4692 (Typ-Code/marking)	71(2,7mm)	Ons	→MMSZ 4692	
CX	Z-Di	=SMAJ 48A (Typ-Code/Stempel/marking)	71(5x2,5)	Tho	→SMAJ ...	data
CX	Z-Di	=SMBJ 48CA (Typ-Code/Stempel/marking)	71(5x3,5)	Mop	→SMBJ ...	data
CX 701	Si-N	TV-VA, 150/120V, 2A, 25W	17j§	Mic	BD 239D, 2SD578, 2SD608(A), 2SD1138	data
CX 701 A	Si-N	=CX 701: 180/150V	17j§		BD 239E, 2SD578, 2SD608A, 2SD1138	data
CX 702	Si-N	TV-HA, 160/80V, 5A, 40W	17j§	Mic	BU 104P, BU 406...408, 2SD823	data
CX 702 A	Si-N	=CX 702: 200/100V	17j§		BU 104P, BU 406...408, 2SD823	data
CX 703	Si-N	Vid, 160V, 0,1A, 0,625W, >50MHz	7e	Mic	BFR 87...89, BFT 57...59, (BF 257...259,++)[6]	data
CX 703 A	Si-N	=CX 703: 200V	7e		BFR 88...89, BFT 58...59, (BF 258...259,++)[6]	data
CX 703 B	Si-N	=CX 703: 250V	7e		BFR 88, BFT 58, (BF 259, ++)[6]	data
CX 704(A...C)	Si-N	NF/S-L, 60V, 4A, 30W, >3MHz	17j§	Mic	BD 243A, BD 535, BD 539A, BD 949	data
CX 705	Si-N	NF/S-L, 55V, 7A, 75W, >0,5MHz	23a§	Mic	BD 245A, BD 311, BDV 91, 2N5873, ++	data
CX 705 A	Si-N	=CX 705: 70V	23a§		BD 245A, BD 313, BDV 93, 2N5874, ++	data
CX 754(A...C)	Si-P	NF/S-L, 60V, 4A, 30W, >3MHz	17j§	Mic	BD 244A, BD 536, BD 540A, BD 950	data
CX 894	LIN-IC	TV, Vc, Video Switch, Ucc=9(8...10)V	9-SIP	Son	-	
CX 901	Si-N	NF, Z-Diode	7e	Mic	-	data
CX 904(B...E)	Si-N	Uni, ra, 45V, 0,1A, 0,3W, 200MHz	7e	Mic	BC 184, BC 413...414, BC 550	data
CX 906(A...D)	Si-N	NF-Tr, 45V, 0,5A, 0,5W, 200MHz	7e	Mic	BC 337, BC 635, BC 637, BC 639, ++	data
CX 908(B...D)	Si-N	NF-Tr, 45V, 1A, 0,625W, 150MHz	7e	Mic	BC 337, BC 635, BC 635, BC 639, ++	data
CX 917	Si-N	AM-V, FM-ZF, 330MHz	7e	Mic	BF 240...241, BF 254, BF 494, BF 594, ++	data
CX 918	Si-N	FM/VHF, TV-ZF, 620MHz	7e	Mic	BF 198, BF 225, BF 310, BF 367, BF 596++	data

Type	Device	Short Description	Fig.	Manu	Comparision Types	More at
CX 954(B...E)	Si-P	Uni, ra, 45V, 0,1A, 0,3W, 200MHz	7e	Mic	BC 214, BC 415...416, BC 560	data
CX 956(A...D)	Si-P	NF-Tr, 45V, 0,5A, 0,5W, 200MHz	7e	Mic	BC 327, BC 636, BC 638, BC 640, ++	data
CX 958(B...D)	Si-P	NF-Tr, 45V, 1A, 0,625W, 150MHz	7e	Mic	BC 327, BC 636, BC 638, BC 640, ++	data
CX 20026	LIN-IC	TV, Stereo-mpx-decoder f. Japan	28-DIP	Son	-	
CX 20034	IC	SMD, VTR Record/Playback Amp, 5V	QFP	Son		pdf pinout
CX 20036	IC	SMD, brushless (capstan) motor drive, 3-phase, Ucc=4...9V Imot<850mA	MDIP	Son		pdf
CX 20037A	IC	SMD, VTR audio signal processor, FM (de-)modulation, noise reduction	QFP	Son		pdf pinout
CX 20051 A	D/A-IC	10 Bit, Video, >30Msps, Ecl Level	28-DIP	Son		
CX 20052 A	A/D-IC	8 Bit, Video, >20Msps, Ecl Level	28-DIP	Son		
CX 20060	LIN-IC	Vc, Video-Umschalter/Switch	6-DIP	Son		pdf
CX 20061	LIN-IC	=CX 20060: Fig.→	8-SIP	Son		pdf pinout
CX 20095 A	LIN-IC	SMD, Video-Leitungstreiber/line driver	14-MDIP	Son		
CX 20099	IC	SMD, 8mm VTR Pcm audio noise reduction	MDIP	Son		pdf pinout
CX 20100	LIN-IC	Ctv, RGB Interface	28-DIP	Son		
CX 20102	IC	VTR 8mm, Pcm Audio HF/RF proc., +5V	MDIP	Son		pdf pinout
CX 20106 A	LIN-IC	IR-FB, Empf.-vorverst./receiver preamp	8-SIP	Son	KA 2184	pinout
CX 20125	LIN-IC	Video, Dynamic Picture	8-SIP	Son		
CX 20126	LIN-IC	TV, Stereo-mpx-decoder f. Japan	28-DIP	Son		
CX 20148	IC	8mm VTR Pcm audio noise reduction	DIP	Son		pdf pinout
CX 20158	LIN-IC	SMD, Video-umschalter/switch	14-MDIP	Son		
CX 20159	LIN-IC	SMD, TV, Stromversorgung/power supply	16-MDIP	Son		
CX 20183	LIN-IC	SMD, TV, Video-ZF, Ton-zf	24-MDIP	Son		
CX 20186	LIN-IC	Video-leitungstreiber/line driver	14-DIP	Son		
CX20197	2xOP	Vs:±15V Vu:120dB Vo:±12.5V Vi0:0.3mV	8D	Son		data pinout
CX20198	2xOP	Vs:±15V Vu:120dB Vo:±12.5V Vi0:0.3mV	10I	Son		data pinout
CX 20201 A-1	D/A-IC	10 Bit, Video, >160Msps, Ecl Level	28-DIP	Son	DAC-330	
CX 20201 A-2	D/A-IC	9 Bit, Video, >160Msps, Ecl Level	28-DIP	Son		
CX 20201 A-3	D/A-IC	8 Bit, Video, >160Msps, Ecl Level	28-DIP	Son		
CX 20202 A-1	D/A-IC	10 Bit, Video, >160Msps, Ecl Level	28-DIP	Son	DAC-330	
CX 20202 A-2	D/A-IC	9 Bit, Video, >160Msps, Ecl Level	28-DIP	Son		
CX 20202 A-3	D/A-IC	8 Bit, Video, >160Msps, Ecl Level	28-DIP	Son		
CX 20206	D/A-IC	8 Bit, RGB 3x ADC, 35Msps, TTL Level	42-SDIP	Son		pdf pinout
CX 20220 A-1	A/D-IC	10 Bit, Video, >20Msps, Ecl Level	28-DIP	Son		
CX 20220 A-2	A/D-IC	9 Bit, Video, >20Msps, Ecl Level	28-DIP	Son		
CX 23038	CMOS-IC	Digital Video, progr. Shift Register	28-DIP	Son		pdf pinout
CX 23880	CMOS-IC	PCI Audio/video Broadcast Decoder	MP	Son		pdf pinout
CX 23881	CMOS-IC	PCI Audio/video Broadcast Decoder	MP	Son	-	pdf pinout
CXA 1011 M	LIN-IC	=CXA 1011P: SMD	16-MDIP			
CXA 1011 P	LIN-IC	dbx-TV, Noise suppr.,	16-DIP	Son	KA 2270	
CXA 1012 AS	LIN-IC	Ctv, Y/chroma/Synchr.-signal (japan)	48-SDIP	Son		
CXA 1013 AS	LIN-IC	Ctv, Y/chroma/synchr.-signal (USA)	48-SDIP	Son		
CXA 1016 K,P	A/D-IC	8 Bit, Video, >30Msps, Ecl Level	28-DIP	Son		
CXA 1019	LIN-IC	AM-Radio, NF-V+E, Ucc=2...8,5V	28-SDIP		KA 22426	pdf pinout
CXA 1024 S	LIN-IC	Ctv, RGB Interface, Weiss-/white Bal.	48-SDIP	Son	-	
CXA 1034 P	LIN-IC	Recorder, 2x Nf-v+e			AN 7108, KA 22132	
CXA 1044 BP	LIN-IC	hi-res Display, RGB Predriver	18-DIP+g	Son	-	
CXA 1047M	IC	SMD, 8mm VCR, Recording Y/C signal processing, amplifier	MDIP	Son		pdf pinout
CXA 1056 K,P	A/D-IC	8 Bit, Video, >50Msps, Ecl Level	28-DIC,DIP	Son		
CXA 1066 K	A/D-IC	8 Bit, Video, >100Msps, Ecl Level	68-LCC	Son	-	pdf
CXA 1076 AK	A/D-IC	8 Bit, Video ADC, >200Msps, Ecl Level	68-LCC	Son	CXA1166K	
CXA 1077 M	LIN-IC	SMD, 2x Diff. Video-Amplifier, 180MHz	14-MDIP	Son	-	pdf pinout
CXA 1081(Q)	LIN-IC	=KA 9201		Son	KA 9201	
CXA 1082BQ	MOS-IC	Servo Signal Processor for CD Player	48-MP	Son	KA 8309	pdf pinout
CXA 1082BS	MOS-IC	Servo Signal Processor for CD Player	MDIP	Son	KA 8309	pdf pinout
CXA 1096 P	A/D-IC	8 Bit, Video, >20Msps, TTL Level	28-DIP	Son		
CXA 1100 P	LIN-IC	Dolby B Type Noise Reduction System	DIP	Son	KA 2271	pdf pinout
CXA 1101 M	LIN-IC	SMD, Dolby B Type Noise Reduction	MDIP	Son	KA 2271	pdf pinout
CXA 1101 P	LIN-IC	Dolby B Type Noise Reduction System	DIP	Son	KA 2271	pdf pinout
CXA 1102	LIN-IC			Son	KA 22712	pdf
CXA 1102 M	LIN-IC	SMD, Dolby B Type Noise Reduction	MDIP	Son	KA 2271	pdf pinout
CXA 1102 P	LIN-IC	Dolby B Type Noise Reduction System	DIP	Son	KA 2271	pdf pinout
CXA 1103AM	LIN-IC	transimpedance, low noise amp for photo detectors, 40MHz, Vcc=4.75...9.5V	MDIP	Son	-	pdf pinout
CXA 1106 P	D/A-IC	8 Bit, Video, >35Msps, TTL Level	24-DIP	Son		pdf
CXA 1106 M	D/A-IC	SMD, 8 Bit, Video, >35Msps, TTL Level	-MDIP	Son		pdf
CXA 1110 BS	LIN-IC	Ctv/vc, Video-ZF, Ton-zf	30-SDIP	Son		
CXA 1113 AS	LIN-IC	TV,Stereo-mpx-decoder, NF-ctrl., Japan	30-SDIP	Son		
CXA 1114 M	LIN-IC	SMD, Audio/video-umsch./switch, I²C	28-MDIP	Son		pdf pinout
CXA 1114 P	LIN-IC	Audio-umschalter/switch, I²c-bus	DIP	Son		pdf pinout
CXA 1124 BQ	LIN-IC	SMD, US Audio Multiplexing IC, dbx-tv, Mts-decoder, Usa Standard	48-MP	Son	-	pdf pinout
CXA 1124 BS	LIN-IC	Us Audio Multiplexing IC, dbx-tv, Mts-decoder, Usa Standard	42-SDIP	Son	-	pdf pinout
CXA 1125 P	LIN-IC	TV, Catv-tuner	14-DIP	Son		
CXA 1126 S	LIN-IC	TV, Stereo-mpx-decoder f. Japan	28-SDIP	Son		
CXA 1138 AM	LIN-IC	=CXA 1138AS: SMD	28-MDIP	Son	CXA 1438M	
CXA 1138 AS	LIN-IC	Tv/vc, Stereo-mpx-decoder f. EIAJ	22-SDIP	Son	CXA 1438S	
CXA 1145 M	LIN-IC	=CXA 1145P: SMD	24-MDIP	Son	-	pdf
CXA 1145 P	LIN-IC	Video, Rgb-encoder (ntsc/pal)	24-DIP	Son	-	pdf
CXA 1156 Q	D/A-IC	SMD, 8Bit, 3 Ch, Video, >360Msps, Ecl	44-MP	Son	-	
CXA 1157 M	LIN-IC	SMD, Sat Amp, Video Switch, Ucc=5V	16-MDIP	Son	-	
CXA 1158 P	LIN-IC	Display, Synchr. Generator	28-DIP	Son	-	

Type	Device	Short Description	Fig.	Manu	Comparision Types	More at
CXA 1163M	LIN-IC	SMD, Dolby B Type Noise Reduction	MDIP	Son	KA 2271	pdf pinout
CXA 1163P	LIN-IC	Dolby B Type Noise Reduction System	DIP	Son	KA 2271	pdf pinout
CXA 1165 AL	LIN-IC	TV, Catv-tuner	16-SQP	Son	-	
CXA 1165 M	LIN-IC	=CXA 1165P: SMD	14-MDIP	Son	-	pdf
CXA 1165 P	LIN-IC	TV, Catv-tuner	14-DIP	Son	-	pdf
CXA 1166K	A/D-IC	8Bit high-speed Flash ADC, 250MSPS, EIL ±0.5LSB, EDL ±0.5LSB, -5.2V	LCC	Son	-	pdf pinout
CXA 1176 AK	A/D-IC	8 Bit, Video ADC, >300Msps, Ecl Level	68-LCC	Son	CXA†166K	
CXA 1198AP	IC	Stereo Cassette Deck, Recording Equalizer, Vcc=8... 15V	DIP	Son	-	pdf pinout
CXA 1200BQ	IC	8mm VCR, Brightness/Color signal proc.	QFP	Son	-	pdf pinout
CXA 1201M	IC	VCR, Image I/O signal processing, 5V	MDIP	Son	-	pdf pinout
CXA 1201Q	IC	SMD, VCR, Image I/O signal proc., 5V	QFP	Son	-	pdf pinout
CXA 1202Q-Z	IC	VCR, Record/Playback amplifier, +5V	QFP	Son	-	pdf pinout
CXA 1202R	IC	VCR, Record/Playback amplifier, +5V	QFP	Son	-	pdf pinout
CXA 1203M	IC	SMD, 8mm VCR, Pal Jog, 5V	MDIP	Son	-	pdf pinout
CXA 1203N	IC	SMD, 8mm VCR, Pal Jog, 5V	SSMDIP	Son	-	pdf pinout
CXA 1204Q	IC	8mm VCR, Atf (Auto. Track Finding), 5V	QFP	Son	-	pdf pinout
CXA 1207AQ	IC	SMD, 8mm VCR, Luminance signal proc.	QFP	Son	-	pdf pinout
CXA 1207AR	IC	SMD, 8mm VCR, Luminance signal proc.	QFP	Son	-	pdf pinout
CXA 1208Q	IC	SMD, 8mm VCR, color signal processor	QFP	Son	-	pdf pinout
CXA 1208R	IC	SMD, 8mm VCR, color signal processor	QFP	Son	-	pdf pinout
CXA 1209 P	LIN-IC	hi-res Display, RGB Predriver	18-DIP+g	Son	-	pdf pinout
CXA 1211 M	LIN-IC	SMD, 2x Breitb.-Verst., wideband VCA	8-MDIP	Son	-	pdf
CXA 1213 AS,BS	LIN-IC	Ctv, Y/chroma/synchr.-signal Proz.	48-SDIP	Son	-	
CXA 1214 P	LIN-IC	Ctv, SECAM-Decoder	24-DIP	Son	-	pdf
CXA 1215 P	LIN-IC	Ctv, RGB Interface (idtv)	28-DIP	Son	-	
CXA 1216 P	LIN-IC	Ctv, RGB Interface (idtv)	28-DIP	Son	-	
CXA 1218 S	LIN-IC	Tv++, NTSC/PAL-Decoder	28-SDIP	Son	-	pdf pinout
CXA 1219 M	LIN-IC	=CXA 1219P: SMD	24-MDIP	Son	-	pdf pinout
CXA 1219 P	LIN-IC	NTSC/PAL-Encoder, Audio Buffer	24-DIP	Son	-	pdf pinout
CXA 1227 Q	LIN-IC	SMD, Tv++, Secam-decoder	32-MP	Son	-	pdf pinout
CXA 1228 S	LIN-IC	Tv++, Ntsc/pal-decoder	28-SDIP	Son	-	pdf pinout
CXA 1229 M	LIN-IC	=CXA 1229P: SMD	24-MDIP	Son	-	pdf pinout
CXA 1229 P	LIN-IC	TV,Vc, Ntsc/pal-encoder, Audio Buffer	24-DIP	Son	-	pdf pinout
CXA 1234AR	IC	SMD, VTR Record/Playback Amp, 5V	QFP	Son	-	pdf pinout
CXA 1236 Q	D/A-IC	SMD, 8 Bit, Video, >500Msps, Ecl Level	44-MP	Son	-	
CXA 1237AR	IC	SMD, VTR audio signal processor, FM (de-)modulation, noise reduction	QFP	Son	-	pdf pinout
CXA 1238M	IC	SMD, AM/FM-Radio, stereo, 2...9V	MDIP	Son	-	pdf pinout
CXA 1238S	IC	AM/FM-Radio, stereo, 2...9V	SDIP	Son	-	pdf pinout
CXA 1260Q-Z	D/A-IC	SMD, 3x(RGB) 8Bit DAC, 35Msps, TTL in	44-QFP	Son	-	pdf
CXA 1261 M	LIN-IC	IR-FB, Empf.-vorverst./receiver preamp	8-MDIP	Son	-	
CXA 1262N	IC	Tape Deck, Pre&Power Amp 0.22W (8Ohm, 2.8V), Agc, Record Mute, Vcc=1.8...4.2V	SSMDIP	Son	-	
CXA 1264 AS	LIN-IC	dbx-tv, Zenith Audio-decoder, Us Std.	42-SDIP	Son	-	
CXA 1268 P	LIN-IC	Display, Synchr. Generator	28-DIP	Son	-	
CXA 1271	LIN-IC	=KA 9201Q	32-MP	Son	KA 9201Q	
CXA 1272	MOS-IC	=KA 9221	48-MP	Son	KA 9221	
CXA 1276K	A/D-D/A-IC	8Bit 500MSPS Flash ADC, Vcc=-5.2V	LCC	Son	-	pdf pinout
CXA 1279 AS	LIN-IC	TV, Tonsteuerung/sound processor	22-SDIP	Son	-	pdf pinout
CXA 1280N	IC	SMD, AM/FM-Radio, 0.5W PAmp, 2...7.5V	SSMDIP	Son	-	pdf pinout
CXA 1296 P	A/D-IC	8 Bit, Video, >20Msps, TTL Level	28-DIP	Son	-	
CXA 1299 P	LIN-IC	hi-res Display, RGB Predriver	18-DIP+g	Son	-	
CXA 1314 P	LIN-IC	Video-Umschalter/switch, I²c-bus	16-DIP	Son	-	pdf pinout
CXA 1315 M	D/A-IC	=CXA 1315P: SMD	16-MDIP	Son	-	pdf
CXA 1315 P	D/A-IC	TV, 8 Bit, 5 Channel, I²c-bus	16-DIP	Son	-	pdf
CXA 1330S	IC	Dolby B/C Noise Reduction, Vcc<16V	SDIP	Son	-	pdf pinout
CXA 1331M	IC	SMD, Dolby B/C Noise Reduct., Vcc<16V	MDIP	Son	-	pdf pinout
CXA 1331S	IC	Dolby B/C Noise Reduction, Vcc<16V	SDIP	Son	-	pdf pinout
CXA 1332M	IC	SMD, Dolby B/C Noise Reduct., Vcc<16V	MDIP	Son	-	pdf pinout
CXA 1332S	IC	Dolby B/C Noise Reduction, Vcc<16V	SDIP	Son	-	pdf pinout
CXA 1352AS	IC	Stereo Graphic Equalizer, 100/400/ 1k/4k/10k Hz, Volume, Balance, 4...10V	SDIP	Son	-	pdf pinout
CXA 1355 L	LIN-IC	TV, VHF/UHF/CATV-Tuner, ZF/IF amp, 9V	16-SQP	Son	-	pdf pinout
CXA 1360Q	IC	Fdd, read/write amplifier, +5V	QFP	Son	-	pdf pinout
CXA 1362Q	IC	Fdd, read/write amplifier, +5V	QFP	Son	-	pdf pinout
CXA 1365 S	LIN-IC	Display, Crt Synchron Discriminator	28-SDIP	Son	-	pdf pinout
CXA 1366 S	LIN-IC	Display, Ha/va Correction	28-SDIP	Son	-	pdf pinout
CXA 1372AQ	IC	SMD, Rf signal processing servo amplifier for CD Player, 3.6... 11V	QFP	Son	-	pdf pinout
CXA 1372AS	IC	Rf signal processing servo amplifier for CD Player, 3.6...11V	SDIP	Son	-	pdf pinout
CXA 1372BQ	IC	SMD, Rf signal processing servo amplifier for CD Player, 3.6... 11V	QFP	Son	-	pdf pinout
CXA 1372BS	IC	Rf signal processing servo amplifier for CD Player, 3.6...11V	SDIP	Son	-	pdf pinout
CXA 1372Q	IC	SMD, Rf signal processing servo amplifier for CD Player, 3.6... 11V	QFP	Son	-	pdf pinout
CXA 1372S	IC	Rf signal processing servo amplifier for CD Player, 3.6...11V	SDIP	Son	-	pdf pinout
CXA 1385 Q	LIN-IC	LCD Ctv, Y/chroma/synchr.-signal Proz.	32-MP	Son	-	pdf
CXA 1387 S	LIN-IC	CTV/Monitor, Aperture Compensation	30-SDIP	Son	-	
CXA 1398M	IC	SMD, Double Cassette Deck, Recording Equalizer, Vcc=9...16V	MDIP	Son	-	pdf pinout

Type	Device	Short Description	Fig.	Manu	Comparision Types	More at
CXA 1398P	IC	Double Cassette Deck, Recording Equalizer, Vcc=9... 16V	DIP	Son	-	pdf pinout
CXA 1401 M	LIN-IC	SMD, DBS-tuner, SAT FM-demodulation	24-MDIP	Son	-	
CXA 1403 AM	LIN-IC	SMD, Dbs-tuner, 4-Phase PSK Demod.	20-MD!P	Son	-	
CXA 1409 AQ	LIN-IC	SMD, VC++, Video-Treiber/driver	32-MP	Son	-	
CXA 1410 M	LIN-IC	SMD, Vc, Video-umschalter/switch	24-MDIP	Son	-	pdf
CXA 1413 L	LIN-IC	Tv/vc, Vertical Correlator	8-SIP	Son	-	
CXA 1414 P	LIN-IC	Video-umschalter/switch, I²c-bus	16-DIP	Son	-	pdf pinout
CXA 1420 P	LIN-IC	Ctv, Aperture Compens., Dyn. Picture	24-DIP	Son	-	
CXA 1434 P	LIN-IC	Audio/video-umschalter/switch, I²c-bus	28-DIP	Son	-	pdf pinout
CXA 1438 M	LIN-IC	=CXA 1438S: SMD	28-MDIP	Son	-	
CXA 1438 S	LIN-IC	Tv/vc, Stereo-mpx-decoder f. Eiaj	22-SDIP	Son	-	
CXA 1446 S	LIN-IC	Ctv/vc, PLL Split System, Vif Video-ZF, Sif Ton-ZF, Japan/usa Standard	30-SDIP	Son	-	pdf pinout
CXA 1450 M	LIN-IC	SMD, Vc, Video-umschalter/switch	8-MDIP	Son	-	
CXA 1451 M	LIN-IC	SMD, Vc, Videoschalter/switch & drive	16-MDIP	Son	-	pdf
CXA 1464 AS	LIN-IC	Ctv, Y/chroma/synchr.-signal (japan)	48-SDIP	Son	-	
CXA 1465 AS	LIN-IC	Ctv, Y/chroma/synchr.-signal (usa)	48-SDIP	Son	-	
CXA 1470 AM	LIN-IC	=CXA 1470AS: SMD	28-MDIP	Son	-	
CXA 1470 AS	LIN-IC	Display, HA/VA-signal-generator	28-SDIP	Son	-	
CXA 1477 AS	BiMOS-IC	Ctv, Y/chroma/synchr.-signal Proz., I²C	48-SDIP	Son	-	
CXA 1485 Q	LIN-IC	SMD, LCD Ctv, RGB-Treiber/driver	32-MP	Son	-	
CXA 1488R	IC	SMD, VTR 8mm, Recording/playback FM Audio signal processing, PAL/NTSC	QFP	Son	-	pdf pinout
CXA 1490 M	LIN-IC	=CXA 1490S: SMD	28-MDIP	Son	-	
CXA 1490 S	LIN-IC	Tv/vc, Stereo-mpx-decoder f. Eiaj	22-SDIP	Son	-	
CXA 1498M	LIN-IC	SMD, Stereo Tape, Rec/play Equal. Amp	MDIP	Son	-	pdf pinout
CXA 1498S	LIN-IC	Stereo Tapedeck, Rec/play Equal. Amp	SDIP	Son	-	pdf pinout
CXA 1511 L	LIN-IC	TV, Vc, FB IR-vorverst./preamp.	8-SIP	Son	-	pdf
CXA 1511 M	LIN-IC	=CXA 1511L: SMD	8-MDIP		-	pdf
CXA 1518 Q	LIN-IC	=CXA 1518S: SMD	32-MP		-	pdf
CXA 1518 S	LIN-IC	TV, Stereo-mpx-decoder, Eiaj Standard	22-SDIP	Son	-	pdf
CXA 1519 M	LIN-IC	=CXA 1519S: SMD	28-MDIP		-	
CXA 1519 S	LIN-IC	=CXA 1438S: + D-FAX Filter	22-SDIP	Son	-	
CXA 1520 M	LIN-IC	=CXA 1520S: SMD	28-MDIP		-	
CXA 1520 S	LIN-IC	=CXA 1490S: + D-FAX Filter	22-SDIP	Son	-	
CXA 1521 M	LIN-IC	SMD, Video-Verstärker/gain control amp.	8-MDIP	Son	-	
CXA 1526 P	LIN-IC	Ctv, Dynamic Convergence, I²c-bus	16-DIP	Son	-	
CXA 1534 Q	LIN-IC	SMD, Us Audio Multiplexing IC, dbx-tv, Mts-decoder, Usa Standard	48-MP	Son	-	pdf pinout
CXA 1534 S	LIN-IC	Us Audio Multiplexing IC, dbx-tv, Mts-decoder, Usa Standard	42-SDIP	Son	-	pdf pinout
CXA 1536Q	IC	SMD, Vcr 8mm, Stereo FM Audio Matrix	QFP	Son	-	pdf pinout
CXA 1538	IC	(SMD), FM Stereo/AM Radio, Vcc=2...9V	SDIP	Son	-	pdf pinout
CXA 1545 AS	LIN-IC	Audio/video-umschalter/switch, I²c-bus	48-SDIP	Son	-	pinout
CXA 1558 L	LIN-IC	Vc, Video-umschalter/switch & driver	16-SQP	Son	-	
CXA 1568M	IC	SMD, 4 stereo chan. selector switch, Micro in Mute, Audio Mute, Agc, +12V	MDIP	Son	-	pdf pinout
CXA 1568S	IC	4 stereo chan. selector switch, Micro in Mute, Audio Mute, Agc, Vcc=12V	SDIP	Son	-	pdf pinout
CXA 1578M	IC	SMD, Stereo Cassette Deck, Recording Equalizer Amplifier, Rec Mute, 9...16V	MDIP	Son	-	pdf pinout
CXA 1578P	IC	Stereo Cassette Deck, Recording Equalizer Amplifier, Rec Mute, 9... 16V	DIP	Son	-	pdf pinout
CXA 1579M	IC	SMD, Stereo Cassette Deck, Recording Equalizer, Vcc=9...16V	MDIP	Son	-	pdf pinout
CXA 1579P	IC	Stereo Cassette Deck, Recording Equalizer, Vcc=9... 16V	DIP	Son	-	pdf pinout
CXA 1580Q	IC	SMD, Dolby B & Playback Equalizer, Vcc=6.5...11V	QFP	Son	-	pdf pinout
CXA 1585 Q	LIN-IC	SMD, Tv++, RGB-Decoder NTSC/PAL, +5V	32-MP	Son	-	pdf pinout
CXA 1587 S	LIN-IC	Ctv, Y/chroma/RGB-signal Prozessor	48-SDIP	Son	-	
CXA 1597M	LIN-IC	SMD, Stereo Recording Equalizer Amp	MDIP	Son	-	pdf pinout
CXA 1597P	LIN-IC	Stereo Recording Equalizer Amplifier	DIP	Son	-	pdf pinout
CXA 1599Q	IC	SMD, Cassette Deck, All-in-one except Dolby Noise Reduction	QFP	Son	-	pdf pinout
CXA 1600M	IC	SMD, Am Radio & PAmp, Vcc=1.8...4.5V	MDIP	Son	-	pdf pinout
CXA 1600P	IC	Am Radio & Power Amp, Vcc=1.8...4.5V	DIP	Son	-	pdf pinout
CXA 1619M	IC	SMD, FM/AM Radio & PAmp, Vcc=2...7.5V	MDIP	Son	-	pdf pinout
CXA 1619S	IC	FM/AM Radio & Power Amp, Vcc=2...8.5V	SDIP	Son	-	pdf pinout
CXA 1644P	IC	Echo Effect Generator, Vcc=4...6V	DIP	Son	-	pdf pinout
CXA 1645 M	LIN-IC	SMD, TV, Vc, Rgb-encoder (ntsc/pal)	24-MDIP	Son	-	pdf pinout
CXA 1645 P	LIN-IC	TV, Vc, Rgb-encoder (ntsc/pal)	24-DIP	Son	-	pdf pinout
CXA 1665AM-S	IC	SMD, All Band TV Tuner, VHF/CATV/UHF	MDIP	Son	-	pdf pinout
CXA 1673M	IC	SMD, 2 Chan. Surround, Bass Boost, 1.8...10V	MDIP	Son	-	pdf pinout
CXA 1673P	IC	2 Chan. Surround, Bass Boost, 1.8...10V	DIP	Son	-	pdf pinout
CXA 1686 A	LIN-IC	SMD, Digital TV, Clock Signal Generator	30-MDIP	Son	-	
CXA 1691M	IC	SMD, FM/AM Radio & PAmp, Vcc=2...7.5V	MDIP	Son	-	pdf pinout
CXA 1691S	IC	FM/AM Radio & Power Amp, Vcc=2...8.5V	SDIP	Son	-	pdf pinout
CXA 1695L	IC	All Band TV Tuner, VHF/CATV/UHF, 9V	SQP	Son	-	pdf pinout
CXA 1700AQ	IC	SMD, 8mm VCR, Luminance & Color Signal Processing, 5V	QFP	Son	-	pdf
CXA 1700AR	IC	SMD, 8mm VCR, Luminance & Color Signal Processing, 5V	QFP	Son	-	pdf

Type	Device	Short Description	Fig.	Manu	Comparision Types	More at
CXA 1709 P	LIN-IC	Display, RGB Pre-driver, 250MHz	16-DIP	Son	-	pdf
CXA 1726 AM	LIN-IC	=CXA 1726AS: SMD	30-MDIP			pdf
CXA 1726 AS	LIN-IC	Display, Dynamic Focus/convergence	30-SDIP	Son	-	pdf
CXA 1727 Q	LIN-IC	Tv/camera, TV ID Signal Adding/detect.	32-MP	Son	-	
CXA 1734 S	LIN-IC	dbx-tv, MTS-decoder, Usa Standard, I²C	30-SDIP	Son	-	pdf pinout
CXA 1742Q	IC	SMD, analog cellular(amps) FM IF, IF filter, RSSI, Vcc=2.7...3.6V	QFP	Son	-	pdf pinout
CXA 1744AR	IC	SMD, dig. cordless telephone Europe CT-2, IF amp, Mixer, RSSI, Detector	QFP	Son	-	pdf pinout
CXA 1779 P	LIN-IC	Display, RGB Pre-driver, 150MHz	28-DIP	Son	-	pdf
CXA 1784 AS	LIN-IC	Zenith TV, Us Audio Multiplexing Decoder, dbx noise reduction, I²C	42-SDIP	Son	-	pdf pinout
CXA 1785AR	IC	SMD, RGB Decoder/Driver for LCD panels, NTSC/PAL compatible	QFP	Son	-	pdf
CXA 1790 Q,R	BiMOS-IC	TV, NTSC/PAL 8 Bit A/D-Decoder, I²C	64-MP	Son	-	
CXA 1792S	IC	2 Channel, 5-Element Graphic Equalizer, Vcc=4...10V	SDIP	Son	-	
CXA 1799 P	LIN-IC	Display, RGB Pre-driver, 150MHz	28-DIP	Son	-	
CXA 1814N	IC	SMD, Vcr 8mm, Automa:c Track Finding	SSMDIP	Son	-	pdf pinout
CXA 1829N	IC	SMD, Hard Disk Drive, Thin Film Heads, 8 Channel Read/write Amplifier, 5V	SSMDIP	Son	-	pdf pinout
CXA 1845 Q	LIN-IC	SMD, Audio/video Switch, I²c-bus	64-MP	Son	-	pdf pinout
CXA 1846M	IC	SMD, Electr. Vol. Control, Vcc=6...12V	MDIP	Son	-	pdf pinout
CXA 1846N	IC	SMD, Electr. Vol. Control, Vcc=6...12V	SSMDIP	Son	-	pdf pinout
CXA 1851N	IC	SMD, Mobile Communications, Up/Down Converter 900MHz Band, Vcc=2.7...4.5 V	SSMDIP	Son	-	pdf pinout
CXA 1852N	IC	SMD, Mobile Communications, Quadrature Modulator 900MHz Band, Vcc=2.7...5.0V	SSMDIP	Son	-	pdf pinout
CXA 1853Q	IC	SMD, RGB driver for LCD panels, 5V	QFP	Son	-	pdf pinout
CXA 1855 Q	LIN-IC	SMD, Audio/video Switch, +9V, I²c-bus	48-MP	Son	-	pdf pinout
CXA 1855 S	LIN-IC	Audio/video Switch, +9V, I²c-bus	48-SDIP	Son	-	pdf pinout
CXA 1870 S	LIN-IC	Ctv,Y/chroma/Synchr.-signal(NTSC), I²C	42-SDIP	Son	-	pdf
CXA 1871 S	LIN-IC	Ctv, Y/chroma/Synchr., (pal/ntsc), I²C	48-SDIP	Son	-	pdf
CXA 1875 AM	D/A-IC	=CXA 1875AP: SMD	16-MDIP		-	pdf
CXA 1875 AP	D/A-IC	TV, 8 Bit, 5 Channel, I²c-bus	16-DIP	Son	-	pdf
CXA 1910Q	IC	SMD, Dolby B with Playback Equalizer	QFP	Son	-	pdf pinout
CXA 1911Q	IC	SMD, Dolby B/C, Playback Equalizer	QFP	Son	-	pdf pinout
CXA 1940N	IC	SMD, Hard Disk, Thin Film Head, 4 Channel Read/write Amplifier, +5V	SSMDIP	Son	-	pdf pinout
CXA 1950 Q	LIN-IC	SMD, Tv++, RGB-Decoder (ntsc/pal)	32-MP	Son	-	
CXA 1951 AQ	BiCMOS-IC	Gps Down Converter	MP	Son	-	pdf pinout
CXA 1951Q	IC	SMD, Gps Down Converter, Vcc=2.7...5.5V	QFP	Son	-	pdf pinout
CXA 1992R	IC	SMD, RF/HF Signal Processing Servo Amplifier, Anti-shock	QFP	Son	-	pdf
CXA 2000 Q	LIN-IC	SMD,Ctv, Y/chroma/RGB, (pal/ntsc), I²C	64-MP	Son	-	pdf
CXA 2016 S	LIN-IC	Display, Synchr.-signal Processing	22-SDIP	Son	-	pdf
CXA 2020 M	LIN-IC	=CXA 2020S: SMD	28-MDIP		-	pdf
CXA 2020 S	LIN-IC	TV, Stereo-mpx-decoder, Eiaj Standard	22-SDIP	Son	-	pdf
CXA 2021 S	LIN-IC	TV, Sound-Prozessor, I²C-Bus, Ucc=12V	22-SDIP	Son	-	pdf
CXA 2022 S	LIN-IC	TV, Sound-Prozessor, I²C-Bus, Ucc=12V	30-SDIP	Son	-	pdf
CXA 2025 AS	LIN-IC	Ctv, Y/chroma/rgb-signal (ntsc), I²C	48-SDIP	Son	-	pdf
CXA 2040 Q	LIN-IC	Video-umschalter/switch, I²c-bus	32-MP	Son	-	pdf pinout
CXA 2050 S	LIN-IC	Ctv, Y/chroma/rgb, (pal/ntsc), I²C	64-SDIP	Son	-	pdf
CXA 2096 N	LIN-IC	Digital Ccd Camera Head Amplifier	QFP	Son	-	pdf pinout
CXA 2134Q	IC	Us Audio Multiplex. Decoder, I2C	SMDIP	Son	-	pdf pinout
CXA 2164Q	IC	SMD, Us Audio Multiplex. Decoder, I2C	QFP	Son	-	pdf pinout
CXA 2174S	IC	Us Audio Multiplex. Decoder, I2C	SMDIP	Son	-	pdf pinout
CXA 2500N	LIN-IC	SMD, low volt. cassette tape recorder, dual recording/playback pre-amp	SSMDIP	Son	-	pdf pinout
CXA 2555Q	IC	Rf Amplifier for CD Player/cd-rom	QFP	Son	-	pdf pinout
CXA 2647 N	DIG-IC	Rf Signal Processor for CD Players	TSOP	Son	-	pdf pinout
CXA 2765 ER	CMOS-IC	5-channel 3-LD Driver for Optical Disk Drive	QFN	Son	-	pdf pinout
CXA 3025N	IC	SMD, All Band TV Tuner, Vhf/catv/uhf	HSMDIP	Son	-	pdf pinout
CXA 3197 R	D/A-IC	10-bit 125MSPS D/A Converter	QFP	Son	-	pdf pinout
CXA 3221 AN	Si-	RX Gain Control Amplifier	TSOP	Son	-	pdf pinout
CXA 3222 AN	Si-	Tx Gain Control Amplifier	TSOP	Son	-	pdf pinout
CXA 3238TN	IC	SMD, 6 Channel Read/write Amp. for Gmr-ind. Heads in Hard Disk Drives	VSMDIP	Son	-	
CXA 3239TN	IC	SMD, 4 Channel Read/write Amp. for Gmr-ind. Heads in Hard Disk Drives	VSMDIP	Son	-	pdf pinout
CXA 3266 Q	CMOS-IC	PLL IC for LCD Monitor/projector	QFP	Son	-	pdf pinout
CXA 3268 AR	CMOS-IC	Driver/timing Generator for Color LCD Panels	QFP	Son	-	pdf pinout
CXA 3286 R	Si-	8-bit 160MSPS Flash A/D Converter	TSOP	Son	-	pdf pinout
CXA 3299 TN	BiMOS-IC	Wideband Pre-amplifier	TSOP	Son	-	pdf pinout
CXA 3355 AER	BiCMOS-IC	Gps Down Converter IC	QFN	Son	-	pdf pinout
CXA 3355 ER	BiCMOS-IC	Gps Down Converter IC	QFN	Son	-	pdf pinout
CXA 3355 TQ	BiCMOS-IC	Gps Down Converter IC	QFN	Son	-	pdf pinout
CXA 3510N	IC	SMD, 4 Channel Read/write Amp. for Thin Film Head or Hard Disk Drive, 5V	SSMDIP	Son	-	pdf pinout
CXA 3668 N	BiCMOS-IC	Reception Analog Signal Processor IC f. Infrared Space Dig. Audio Communication	MDIP	Son	-	
CXA 3685 ER	CMOS-IC	Direct Conversion IC for Digital Satellite Broadcast Tuner	QFN	Son	-	pdf pinout

Type	Device	Short Description	Fig.	Manu	Comparision Types	More at
CXA 3691 AEN	BiCMOS-IC	High-speed Buffer Amplifier for Ccd Image Sensor	46	Son	-	pdf pinout
CXA 3691 EN	BiCMOS-IC	High-speed Buffer Amplifier for Ccd Imager	46	Son	-	pdf pinout
CXA 3741 AUR	BiMOS-IC	High-speed Buffer Amplifier for Ccd Image Sensor	QFN	Son	-	pdf pinout
CXA 3741 UR	BiCMOS-IC	High-speed Buffer Amplifier for Ccd Imager	QFN	Son	-	pdf pinout
CXA4558	2xHI-GAIN	Vs:±18V Vu:>15V/mV Vo:>±10V Vi0:2mV	8AD	Son	-	data pinout
CXA4559	2xOP	Vs:±18V Vu:>15V/mV Vo:>±10V Vi0:2mV	8AD,9I	Son	-	data pinout
CXA 7002 R	BiCMOS-IC	I2C Bus Compatible Audio Video (AV) Switch & Electronic Volume Control	QFP	Son	-	pdf pinout
CXA 8038	SMPS-IC	Resonance Mode Switching Regulator	DIP		-	pdf pinout
CXA 20116	A/D-IC	8 Bit, Video, >100Msps, Ecl Level	42-DIP	Son	-	
CXB 1010 G	LIN-IC	Digital Video, 16x16 Bit Multiplier	85-PGA	Son	-	pdf pinout
CXB 1549 Q	IC	Laser Diode Driver	QFP	Son	-	pdf pinout
CXB 1573 R	LIN-IC	Post-amp,Optical Fiber Communication Receiver	QFP	Son	-	pdf
CXB 1583Q	IC	SMD, 266Mbaud Fibre Chan. Transceiver, built-in PLL, ANSI X3T11, 3.3V	QFP	Son	-	pdf pinout
CXB 1589R	IC	SMD, Gigabit Ethernet, 10Bit 1.25 GBaud Transceiver, TTL/ECL, 3.3V	QFP	Son	-	pdf pinout
CXB 1810 FN	IC	Post Amplifier, Opt. Fiber Receiver	QFP	Son	-	pdf pinout
CXB 1828 ER	IC	2.5Gbps Laser Diode Driver	TSOP	Son	-	pdf pinout
CXD 1030 M	CMOS-IC	SMD, Camera, Signal-prozessor	28-MDIP	Son	-	
CXD 1050 A	CMOS-IC	Crt, Sreen Display, 128Zeichen/charac.	24-DIP	Son	-	
CXD 1158 M	CMOS-IC	SMD, Camera, Signal-prozessor	30-MDIP	Son	-	
CXD 1159 Q	CMOS-IC	SMD, Camera, Signal-prozessor	32-MP	Son	-	pdf
CXD 1161	CMOS-D/A-IC	=KDA 0316	20-DIP	Son	KDA 0316	
CXD 1167(Q)	MOS-IC	=KS 9210: SMD	80-MP	Son	KS 9210	
CXD 1170 M	CMOS-D/A-IC	SMD, 6 Bit, Video, >40Msps, CMOS/TTL L	24-MDIP	Son	-	pdf
CXD 1171 M	CMOS-D/A-IC	SMD, 8 Bit, Video, >40Msps, CMOS/TTL L	24-MDIP	Son	-	pdf
CXD 1172 AP	CMOS-A/D-IC	6 Bit, Video, >20Msps, TTL/CMOS Level	16-DIP	Son	-	pdf
CXD 1172 M	CMOS-A/D-IC	6 Bit, Video, >20Msps, TTL/CMOS Level	16-MDIP	Son	-	
CXD 1175 AP	CMOS-A/D-IC	8 Bit, Video, >20Msps, TTL Level	24-DIP	Son	-	pdf
CXD 1175 AM	CMOS-A/D-IC	SMD, 8 Bit, Video, >20Msps, TTL Level	24-MDIP	Son	-	pdf
CXD 1199AQ	IC	SMD, CDROM Decoder, ADPCM decoder, 5V	QFP	Son	-	pdf
CXD 1217 M	CMOS-IC	SMD, Camera, Signal-prozessor	28-MDIP	Son	-	pdf
CXD 1229 Q	CMOS-IC	SMD,Digital Video, Synchr. Signal, Afc	48-MP	Son	-	pdf pinout
CXD 1253AQ	IC	SMD, Timing-Generator for ICX026/027CK, ICX044/045AK	QFP	Son	-	pdf pinout
CXD 1253AR	IC	SMD, Timing-Generator for ICX026/027CK, ICX044/045AK	QFP	Son	-	
CXD 1803AQ	IC	SMD, CD-ROM Decoder, ADPCM decoder, Compat. with Cd-rom, Cd-i, CD-ROM XA	QFP	Son	-	pdf
CXD 1803AR	IC	SMD, CD-ROM Decoder, ADPCM decoder, Compat. with Cd-rom, Cd-i, CD-ROM Xa	QFP	Son	-	pdf
CXD 1804AR	IC	SMD, CD-ROM decoder, Cd-rom, Cd-i, Cd-rom, Xa, Fast Scsi controller	QFP	Son	-	pdf
CXD 1807Q	IC	SMD, Cd-g Decoder (karaoke), NTSC/PAL	QFP	Son	-	pdf
CXD 1812Q	IC	SMD, CD-ROM Decoder, Atapi Interface, Cd-rom, Cd-i, CD-ROM Xa compatible	QFP	Son	-	pdf
CXD 1812R	IC	SMD, CD-ROM Decoder, Atapi Interface, Cd-rom, Cd-i, CD-ROM Xa compatible	QFP	Son	-	pdf
CXD 1910 AQ	MOS-IC	SMD, Vc, NTSC/PAL Digital Encoder	64-MP	Son	-	pdf
CXD 1968 AR	CMOS-IC	Dvb-t Demodulator	QFP	Son	-	pdf pinout
CXD 2011 Q	CMOS-IC	SMD, Digital Tv/vc, Kamm-/comb Filter	80-MP	Son	-	
CXD 2012 Q	CMOS-IC	SMD, Dbs-tuner, Audio Pcm Demodulator	80-MP	Son	-	
CXD 2018 Q	CMOS-IC	SMD,Ctv, VA-proz., Multi-standard, I²C	48-MP	Son	-	
CXD 2020 Q	CMOS-IC	SMD, Signal-proz. f. Muse-ntsc Conver.	100-MP	Son	-	
CXD 2021 Q	CMOS-IC	SMD, Time Conversion f. Muse-ntsc Conv	100-MP	Son	-	
CXD 2023 Q	CMOS-IC	SMD, Digital Tv/vc, Kamm-/comb Filter	80-MP	Son	-	
CXD 2024 Q	CMOS-IC	SMD, Digital Tv/vc, Kamm-/comb Filter	80-MP	Son	-	
CXD 2027Q	IC	SMD, Dbs Audio Signal Processor, 5V	QFP	Son	-	pdf pinout
CXD 2027R	IC	SMD, Dbs Audio Signal Processor, 5V	QFP	Son	-	pdf pinout
CXD 2043 Q	CMOS-IC	SMD, Digital Tv/vc, Kamm-/comb Filter.	80-MP	Son	-	pdf pinout
CXD 2044 Q	CMOS-IC	SMD, Digital Tv/vc, Kamm-/comb Filter	80-MP	Son	-	pdf pinout
CXD 2053 AM	CMOS-IC	=CXD 2053AS: SMD	28-MDIP	Son	-	pdf
CXD 2053 AS	CMOS-IC	Wide TV, EDTV-II Id Processor	28-SDIP	Son	-	pdf
CXD 2105 AQ	CMOS-IC	SMD,Digital Vc, dig. Kamm-/comb Filter	80-MP	Son	-	
CXD 2122 AQ	CMOS-IC	SMD, Tv/camera, TV Id Signal Codec	32-MP	Son	-	
CXD 2131 Q	CMOS-IC	SMD, TV,Vc, NTSC Decoder/encoder	32-MP	Son	-	pdf
CXD 2200 G	CMOS-IC	Digital Video Processor	180-PGA	Son	-	
CXD 2300 Q	CMOS-IC	8-bit 18MSPS Video A/D Converter	QFP	Son	-	pdf pinout
CXD 2301Q	IC	SMD, 8Bit 30MSPS VideoADC & Amp/Clamp	QFP	Son	-	pdf pinout
CXD 2302 Q	CMOS-IC	8-bit 50MSPS Video A/D Converter	QFP	Son	-	pdf pinout
CXD 2303 AQ	CMOS-IC	8-bit 3-channel 50 MSPS Video A/D Converter with clamp function	QFP	Son	-	pdf pinout
CXD 2305 Q	CMOS-D/A-IC	SMD, Crt, 10 Bit, 50 Msps, RGB 3 Chan.	100-MP	Son	-	
CXD 2311 AR	CMOS-IC	10-bit 20MSPS Video A/D Converter	QFP	Son	-	pdf pinout
CXD 2315 Q	CMOS-IC	10-bit 80MSPS 1ch D/A Converter	QFP	Son	-	pdf pinout
CXD 2343 S	CMOS-IC	Display, 10 Bit Binary Counter, TTL	30-SDIP	Son	-	
CXD 2412AQ	IC	SMD, LCD Panel Timing Generator, 5V	QFP	Son	-	pdf pinout
CXD 2422R	IC	SMD, Ccd Camera Timing Generator, 5V	QFP	Son	-	pdf pinout
CXD 2436Q	IC	SMD, timing signal generator for the VGA	QFP	Son	-	pdf pinout

Type	Device	Short Description	Fig.	Manu	Comparision Types	More at
		LCD panel LCX012 driver				
CXD 2442Q	IC	SMD, Timing Generator for LCD Panel LCX016(SVGA)/LCX012BL(VGA) driver, 5V	QFP	Son	-	pdf pinout
CXD 2450 R	CMOS-IC	Timing Generator for Progressive Scan Ccd Image Sensor	QFP	Son	-	pdf pinout
CXD 2460 R	CMOS-IC	Timing Generator for Progressive Scan Ccd Image Sensor	QFP	Son	-	pdf pinout
CXD-2508AQ	IC	SMD, CD Digital Signal Processor	QFP	Son	-	pdf
CXD 2508AR	IC	SMD, CD Digital Signal Processor	QFP	Son	-	pdf
CXD 2917Q	IC	SMD, Digital Audio Data Modulation & Transmission, +5V	QFP	Son	-	pdf pinout
CXD 2931R	IC	SMD, Gps (global Positioning System)	QFP	Son	-	pdf pinout
CXD 2951 GA-2	CMOS-IC	Single Chip Gps Lsi	PGA	Son	-	pdf pinout
CXD 2951 GA-4	CMOS-IC	Single Chip Gps Lsi	LGA	Son	-	pdf
CXD 2951 GL-4	CMOS-IC	Single Chip Gps Lsi	PGA	Son	-	pdf pinout
CXD 2956 AGL-1	CMOS-IC	Gps Baseband Lsi	PGA	Son	-	pdf pinout
CXD 3057 R	CMOS-IC	CD Dig.signal Processor, Rf Amp, Dig.servo+high & Bass Boost	QFP	Son	-	pdf pinout
CXD 3059 AR	CMOS-IC	CD Digit. Signal Proces. w. Built-in Rf Amplifier, Digital Servo	QFP	Son	-	pdf pinout
CXD 3607 R	CMOS-IC	Timing Generator for Progressive Scan Ccd Image Sensor	QFN	Son	-	pdf pinout
CXD 4016 R	CMOS-IC	Transmission Digital Signal Processor IC for IR Spatial Digit. Audio Com.	QFP	Son	-	pdf pinout
CXD 4017 R	CMOS-IC	Reception Digital Signal Processor IC f. IR Spatial Digit. Audio Com.	QFP	Son	-	pdf pinout
CXD 5091 AGG	CMOS-IC	Ultra low power Audio Decoder Lsi	LGA	Son	-	pdf
CXD 7500 M		SMD, Regelwdst./vari. resistor volt. control, ±15mA	8-MDIP	Son	-	
CXD 7501 M	MOS-P	=CXD 7500M:	45	Son	-	
CXG 1009 TN	IC	High Isolation SPDT Switch	TSOP	Son	-	pdf pinout
CXG 1010N	IC	SMD, Phs (personal Handyphone System), Amplifier, 21.5dbm	SSMDIP	Son	-	pdf pinout
CXG 1039 TN	IC	High Isolation Absorptive SPDT Switch MMIC	TSOP	Son	-	pdf pinout
CXG 1090 EN	CMOS-Logic	High Power 2 x 4 Antenna Switch Mmic with Integrated Control Logic	QFN	Son	-	pdf pinout
CXG 1091TN	MIC	SMD, Dualband GSM 900/1800, Antenna Switch, SP4T (Ant→TX1,TX2,RX1,RX2)	VSMDIP	Son	-	pdf pinout
CXG 1106 EN	IC	High Power 2 x 4 Antenna Switch Mmic	TSOP	Son	-	pdf pinout
CXG 1144 AEN	CMOS-IC	High Power DPDT Switch with Logic Control	CSP	Son	-	pdf pinout
CXG 1166 AER	CMOS-IC	High Power 3 × 5 Antenna Switch Mmic with Integrated Control Logic	QFN	Son	-	pdf pinout
CXG 1172 UF	LIN-IC	Jphemt DPDT Switch, Logic Control	MLP	Son	-	pdf pinout
CXG 1173 UF	CMOS-IC	High Power SPDT Switch with Logic Control	QFN	Son	-	pdf pinout
CXG 1174 UR	CMOS-IC	High Power DP3T Switch, Logic Control	QFN	Son	-	pdf pinout
CXG 1175 UR	GaAs-IC	High Power DPDT Switch with Logic Control	QFN	Son	-	pdf pinout
CXG 1176 UR	CMOS-IC	High Power DPDT Switch with Logic Control	QFN	Son	-	pdf pinout
CXG 1177 UR	CMOS-IC	High Power DP3T Switch with Logic Control	QFN	Son	-	pdf pinout
CXG 1180 EQ	GaAs-IC	GSM Quad-band Antenna Switch	QFN	Son	-	pdf pinout
CXG 1188 UR	CMOS-IC	High Power DPDT Switch with Logic Control	QFN	Son	-	pdf pinout
CXG 1189 AXR	GaAs-IC	High Power SPDT Switch	QFN	Son	-	pdf pinout
CXG 1190 AEQ	GaAs-IC	GSM/UMTS Dual Mode Antenna Switch	QFN	Son	-	pdf pinout
CXG 1192 UR	CMOS-IC	SP3T × 2 Antenna Switch for GSM Quad-Band	QFN	Son	-	pdf pinout
CXG 1194 UR	GaAs	SP4T Antenna Switch for GSM	QFN	Son	-	pdf pinout
CXG 1194 XR	GaAs-IC	SP4T Antenna Switch for Gsm/umts	QFN	Son	-	pdf pinout
CXG 1195 XR	GaAs-IC	High Power SP5T Antenna Switch Mmic for Gsm/umts Dual Mode	QFN	Son	-	pdf pinout
CXG 1198 AEQ	GaAs-IC	SP8T Gsm/umts Dual Mode Antenna Switch	QFN	Son	-	pdf pinout
CXG 1198 BEQ	GaAs-IC	SP8T Gsm/umts Dual Mode Antenna Switch	QFN	Son	-	pdf pinout
CXG 1199 UR	GaAs-IC	High-frequency SPDT Switch	QFN	Son	-	pdf pinout
CXG 1210 UR	GaAs-IC	High Power 3P3T Switch with Logic Control	QFN	Son	-	pdf pinout
CXG 1212 UR	GaAs-IC	High Power 3P3T Switch with Logic Control	QFN	Son	-	pdf pinout
CXG 1213 XR	GaAs-IC	High Power DPDT Switch with Logic Control	QFN	Son	-	pdf pinout
CXG 1214 UR	GaAs-IC	High Power 3P3T Switch with Logic Control	QFN		-	pdf pinout
CXG 1215 UR	GaAs-IC	High Power DP3T Switch with Logic Control	QFN	Son	-	pdf pinout
CXG 1216 UR	GaAs-IC	High Power DP4T Switch with Logic Control	QFN	Son	-	pdf pinout
CXG 1224 XR	GaAs-IC	SP4T+SP4T Antenna Switch for Gsm/umts	QFN	Son	-	pdf pinout
CXG 1228 XR	GaAs-IC	SP8T Antenna Switch for Gsm/umts	QFN	Son	-	pdf pinout
CXG 1230 EQ	GaAs-IC	SP9T Gsm/umts Dual Mode Antenna Switch	QFN	Son	-	pdf pinout
CXG 7001 FN	GaAs-IC	Power Amplifier/antenna Switch + Low Noise Down Conversion Mixer for Phs	QFN	Son	-	pdf pinout

Type	Device	Short Description	Fig.	Manu	Comparision Types	More at
CXG 7003 FN	LIN-IC	Power Amplifier/antenna Switch	MLP	Son	-	pdf pinout
CXK 5B16120J	sRAM-IC	SMD, 64k x 16Bit BiCMOS SRAM, 3.3V	MDIP	Son		pdf pinout
CXK 5B16120TM	sRAM-IC	SMD, 64k x 16Bit BiCMOS SRAM, 3.3V	HSMDIP	Son		pdf pinout
CXK 5B18120TM	sRAM-IC	SMD, 64k x 18Bit BiCMOS SRAM, 3.3V	HSMDIP	Son		pdf pinout
CXK 5B41020TM	sRAM-IC	SMD, 256k x 4Bit BiCMOS SRAM, 3.3V	MDIP	Son		pdf pinout
CXK 5B81020J	sRAM-IC	SMD, 128k x 8Bit BiCMOS SRAM, 3.3V	MDIP	Son		pdf pinout
CXK 5B81020TM	sRAM-IC	SMD, 128k x 8Bit BiCMOS SRAM, 3.3V	MDIP	Son		pdf pinout
CXK 5T16100TM	sRAM-IC	SMD, 64k x 16Bit Cmos Static RAM, 3.3V	HSMDIP	Son		pdf pinout
CXK 5V16100TM	sRAM-IC	SMD, 64k x 16Bit Cmos Static RAM, 3.3V	HSMDIP	Son		pdf pinout
CXK 5V8512TM	sRAM-IC	SMD, 64k x 8Bit Cmos Static RAM, 3.3V	VSMDIP	Son		pdf pinout
CXK 1202 Q	CMOS-IC	SMD, Digital Tv/vc, dig. Delay Line	32-MP	Son		pdf pinout
CXK 1202 S	CMOS-IC	=CXK 1202Q: Fig.→	28-SDIP	Son		pdf pinout
CXK 1203 Q,R	CMOS-IC	SMD, Digital Tv/vc, dig. Line Memory	48-MP	Son		pdf pinout
CXK 1206 AM,ATM	CMOS-VRAM-IC	SMD, Dig. Video Signal Field Memory, 293760 x 4Bit, 30ns	38-SMDIP	Son		
CXD 1206 ATM	CMOS-VRAM-IC	=CXK 1206AM:	44-SSDIP			
CXK 1207 M	CMOS-VRAM-IC	SMD, Digital Video Signal Field Memory	38-SMDIP	Son		
CXK 5864 AP/BP	sRAM-IC	MOS, 8k x 8Bit				
CXK 48324 Q	CMOS-VRAM-IC	SMD, Digital Video Signal Field Memory	64-MP	Son		
CXK 48324 R	CMOS-VRAM-IC	=CXK 48324Q: Fig.→	80-MP			
CXK 58512M	sRAM-IC	SMD, 64k x 8Bit Cmos Static RAM	MDIP	Son		pdf pinout
CXK 58512TM	sRAM-IC	SMD, 64k x 8Bit Cmos Static RAM	VSMDIP	Son		pdf pinout
CXK 581000A	sRAM-IC	(SMD,) 128k x 8Bit Cmos Static RAM	DIP	Son		pdf pinout
CXK 582000	sRAM-IC	SMD, 256k x 8Bit Cmos Static RAM	MDIP	Son		pdf pinout
CXL 1506 M	CMOS-CCD-IC	SMD, Video, Ccd Delay Line (pal)	16-MDIP	Son		pdf
CXL 1510M	IC	SMD, Ccd Delay Line for Multi System	MDIP	Son		pdf pinout
CXL 1511M	IC	SMD, Ccd Delay Line for Pal	MDIP	Son		pdf pinout
CXL1512M	IC	SMD, Ccd Delay Line for NTSC, +5V	MDIP	Son		pdf pinout
CXL 5001 M	CMOS-CCD-IC	SMD, Video, Ccd Delay Line (ntsc)	8-MDIP			pdf
CXL 5001 P	CMOS-CCD-IC	Video, Ccd Delay Line (ntsc)	8-DIP	Son		pdf
CXL 5002 M	CMOS-CCD-IC	SMD, Video, Ccd Delay Line (ntsc)	8-MDIP			pdf
CXL 5002 P	CMOS-CCD-IC	Video, Ccd Delay Line (ntsc)	8-DIP	Son		pdf
CXL 5005 M	CMOS-CCD-IC	SMD, Video, Ccd Delay Line (ntsc)	14-MDIP			pdf
CXL 5005 P	CMOS-CCD-IC	Video, Ccd Delay Line (ntsc)	14-DIP	Son		pdf
CXL 5504 M	CMOS-CCD-IC	SMD, Video, Ccd Delay Line (ntsc)	8-MDIP			pdf
CXL 5504 P	CMOS-CCD-IC	Video, Ccd Delay Line (ntsc)	8-DIP	Son		pdf
CXL 5509 M	CMOS-CCD-IC	=CXL 5509P: SMD	16-MDIP			pdf
CXL 5509 P	CMOS-CCD-IC	Video, Ccd Delay Line (ntsc)	16-DIP	Son		pdf
CXL 5512M	IC	SMD, Ccd 1H Delay Line for NTSC, +5V	MDIP	Son		pdf pinout
CXL 5512P	IC	Ccd 1H Delay Line for NTSC, +5V	DIP	Son		pdf pinout
CXL 5513M	IC	SMD, Cmos Ccd 1H Delay Line, NTSC	MDIP	Son		pdf pinout
CXL 5513P	IC	Cmos Ccd 1H Delay Line, NTSC	DIP	Son		pdf pinout
CXL 5514M	IC	SMD, Cmos Ccd 1H Delay Line, Pal	MDIP	Son		pdf pinout
CXL 5514P	IC	Cmos Ccd 1H Delay Line, Pal	DIP	Son		pdf pinout
CXL 5515M	IC	SMD, Cmos Ccd 1H Delay Line, Pal	MDIP	Son		pdf pinout
CXL 5515P	IC	Cmos Ccd 1H Delay Line, Pal	DIP	Son		pdf pinout
CXL 5520 M	CMOS-CCD-IC	=CXL 5520P: SMD	16-MDIP			pdf
CXL 5520 P	CMOS-CCD-IC	Video, Ccd Delay Line (ntsc/pal/secam)	16-DIP	Son		pdf
CXM 3512 EQ	GaAs-IC	SP9T Gsm/umts Dual Mode Antenna Switch	QFN	Son		pdf pinout
CXM 3513 ER	GaAs-IC	High Power DP3T Switch with Logic Control	QFN	Son		pdf pinout
CXP 50P116	µC-IC	Cmos 4Bit Single Chip Microcomputer	QFP	Son		pdf
CXP 80P624A	µC-IC	SMD, Cmos 8Bit Single Chip Microcomp.	QFP	Son		pdf
CXP 401	µC-IC	Cmos 4Bit Single Chip Microcomputer	QFP	Son		pdf
CXP 508L4	µC-IC	SMD, Cmos 4Bit Single Chip Microcomp.	QFP	Son		pdf
CXP 508L4	µC-IC	Cmos 4Bit Single Chip Microcomputer	SDIP	Son		pdf
CXP 508L6	µC-IC	Cmos 4Bit Single Chip Microcomputer	SDIP	Son		pdf
CXP 508L6	µC-IC	SMD, Cmos 4Bit Single Chip Microcomp.	QFP	Son		pdf
CXP 819P60	µC-IC	Cmos 8Bit Single Chip Microcomputer	QFP	Son		pdf
CXP 856P40	µC-IC	8Bit µC, ADC, serial interface, timer counter, Prom version of CXP85640	SDIP	Son		pdf pinout
CXP 856P40	µC-IC	SMD, 8Bit µC, ADC, serial interface, timer/counter, Prom vers. of CXP85640	QFP	Son		pdf pinout
CXP 1021Q	µC-IC	SMD, 4Bit single chip µC (SPC500 series), CD Player System Controller	QFP	Son		pdf pinout
CXP 1042Q	µC-IC	SMD, 4Bit single chip µC (SPC500 series), CD Player System Controller	QFP	Son		pdf pinout
CXP 5084	µC-IC	SMD, Cmos 4Bit Single Chip Microcomp.	QFP	Son		pdf
CXP 5084	µC-IC	Cmos 4Bit Single Chip Microcomputer	SDIP	Son		pdf
CXP 5086	µC-IC	SMD, Cmos 4Bit Single Chip Microcomp.	QFP	Son		pdf
CXP 5086	µC-IC	Cmos 4Bit Single Chip Microcomputer	SDIP	Son		pdf
CXP 50112	µC-IC	Cmos 4Bit Single Chip Microcomputer	QFP	Son	CXP50116	pdf
CXP 50116	µC-IC	Cmos 4Bit Single Chip Microcomputer	QFP	Son	CXP50P116	pdf
CXP 80116	µC-IC	Cmos 8Bit Single Chip Microcomputer	QFP	Son	-	pdf
CXP 80316	µC-IC	Cmos 8Bit Single Chip Microcomputer	SDIP	Son		pdf
CXP 80620	µC-IC	SMD, Cmos 8Bit Single Chip Microcomp.	QFP	Son		pdf
CXP 80620A	µC-IC	SMD, Cmos 8Bit Single Chip Microcomp.	QFP	Son		pdf
CXP 80624	µC-IC	SMD, Cmos 8Bit Single Chip Microcomp.	QFP	Son		pdf
CXP 80624A	µC-IC	SMD, Cmos 8Bit Single Chip Microcomp.	QFP	Son		pdf
CXP 80712A	µC-IC	SMD, Cmos 8Bit Single Chip Microcomp.	QFP	Son		pdf
CXP 80716A	µC-IC	SMD, Cmos 8Bit Single Chip Microcomp.	QFP	Son		pdf
CXP 80720	µC-IC	SMD, Cmos 8Bit Single Chip Microcomp.	QFP	Son		pdf
CXP 80720A	µC-IC	SMD, Cmos 8Bit Single Chip Microcomp.	QFP	Son		pdf
CXP 80724	µC-IC	SMD, Cmos 8Bit Single Chip Microcomp.	QFP	Son		pdf
CXP 80724A	µC-IC	SMD, Cmos 8Bit Single Chip Microcomp.	QFP	Son		pdf

Type	Device	Short Description	Fig.	Manu	Comparision Types	More at
CXP 80732	µC-IC	SMD, Cmos 8Bit Single Chip Microcomp.	QFP	Son		pdf
CXP 80740	µC-IC	SMD, Cmos 8Bit Single Chip Microcomp.	QFP	Son		pdf
CXP 81952	µC-IC	Cmos 8Bit Single Chip Microcomputer	QFP	Son	-	pdf
CXP 81960	µC-IC	Cmos 8Bit Single Chip Microcomputer	QFP	Son		pdf
CXP 82316	µC-IC	SMD, 8Bit Single Chip µC, ADC, Serial Interface, Fluorescent Display Ctrl.	QFP	Son		pdf pinout
CXP 82320	µC-IC	SMD, 8Bit Single Chip µC, ADC, Serial Interface, Fluorescent Display Ctrl.	QFP	Son		pdf pinout
CXP 82324	µC-IC	SMD, 8Bit Single Chip µC, ADC, Serial Interface, Fluorescent Display Ctrl.	QFP	Son	-	pdf pinout
CXP 82832	CMOS-IC	Cmos 8-bit Single Chip Microcomputer	TSOP	Son	-	pdf pinout
CXP 82840	CMOS-IC	Cmos 8-bit Single Chip Microcomputer	TSOP	Son	-	pdf pinout
CXP 82852	CMOS-IC	Cmos 8-bit Single Chip Microcomputer	TSOP	Son		pdf pinout
CXP 82860	CMOS-IC	Cmos 8-bit Single Chip Microcomputer	TSOP	Son		pdf pinout
CXP 85632	µC-IC	SMD, Cmos 8-bit Single Chip µC, ROM 32kBytes, ADC, Serial interface ...	QFP	Son	-	pdf pinout
CXP 85632	µC-IC	Cmos 8-bit Single Chip Microcomputer, ROM 32kBytes, ADC, Serial interface ...	SDIP	Son	-	pdf pinout
CXP 85640	µC-IC	SMD, Cmos 8-bit Single Chip µC, ROM 40kBytes, ADC, Serial interface ...	QFP	Son		pdf pinout
CXP 85640	µC-IC	Cmos 8-bit Single Chip Microcomputer, ROM 40kBytes, ADC, Serial interface ...	SDIP	Son		pdf pinout
CXP 88852	µC-IC	SMD, Cmos 8-bit Single Chip µC, ROM 52kBytes, ADC, Serial interface ...	QFP	Son		pdf pinout
CXP 88860	µC-IC	SMD, Cmos 8-bit Single Chip µC, ROM 60kBytes, ADC, Serial interface ...	QFP	Son		pdf pinout
CY	Si-N	=2SC2881-Y (Typ-Code/Stempel/marking)	39	Tos	→2SC2881	data
CY	Si-N	=2SC3396 (Typ-Code/Stempel/marking)	35	Say	→2SC3396	data
CY	Si-N	=2SC4209-Y (Typ-Code/Stempel/marking)	35	Tos	→2SC4209	data
CY	Si-N	=2SC4397 (Typ-Code/Stempel/marking)	35(2mm)	Say	→2SC4397	data
CY	Si-N	=2SC5087-Y (Typ-Code/Stempel/marking)	44	Tos	→2SC5087	data
CY	Si-N	=2SC5229 (Typ-Code/Stempel/marking)	39	Say	→2SC5229	data
CY	MOS-P-FET-e	=2SJ186 (Typ-Code/Stempel/marking)	39	Hit	→2SJ186	data
CY	Si-N	=BFS 18R (Typ-Code/Stempel/marking)	35	Sie	→BFS 18R	data
CY	Si-N	=KTC4373-Y (Typ-Code/Stempel/marking)	39	Kec	→KTC 4373	data
CY	Z-Di	=MMSZ 4693 (Typ-Code/marking)	71(2,7mm)	Ons	→MMSZ 4693	
CY	Z-Di	=P4 SMA-110 (Typ-Code/Stempel/marking	71(5x2,5)	Fag	→P4 SMA-110	data
CY	Z-Di	=SMAJ 48CA (Typ-Code/Stempel/marking)	71(5x2,5)	Tho	→SMAJ ...	data
CY	Z-Di	=SMBJ 51C (Typ-Code/Stempel/marking)	71(5x3,5)	Mop	→SMBJ ...	data
CYQ	Si-N	=2SC4661-Q (Typ-Code/Stempel/marking)	35(2mm)	Mat	→2SC4661	data
CYR	Si-N	=2SC4661-R (Typ-Code/Stempel/marking)	35(2mm)	Mat	→2SC4661	data
CYS	Si-N	=2SC4661-S (Typ-Code/Stempel/marking)	35(2mm)	Mat	→2SC4661	data
CZ	Si-N	=BFS 19R (Typ-Code/Stempel/marking)	35	Sie	→BFS 19R	data
CZ	Si-N	=2SC5347 (Typ-Code/Stempel/marking)	39	Say	→2SC5347	
CZ	Z-Di	=MMSZ 4694 (Typ-Code/marking)	71(2,7mm)	Ons	→MMSZ 4694	
CZ	Z-Di	=P4 SMA-110A(Typ-Code/Stempel/marking	71(5x2,5)	Fag	→P4 SMA-110A	data
CZ	Z-Di	=SMBJ 51CA (Typ-Code/Stempel/marking)	71(5x3,5)	Mop	→SMBJ ...	data
CZD	Si-N	=2SC5347-D (Typ-Code/Stempel/marking)	39	Say	→2SC5347	data
CZE	Si-N	=2SC5347-E (Typ-Code/Stempel/marking)	39	Say	→2SC5347	data
CZF	Si-N	=2SC5347-F (Typ-Code/Stempel/marking)	39	Say	→2SC5347	data

D						
D...	...-N	→2SD..., z.B./e.g. "D731" = 2SD731→	Japantypen	JAP		
D...	...-N	→KSD..., z.B./e.g. "D1616"=KSD1616→	Samsung	Sam		
D...	...-N	→KTD..., z.B./e.g. "D 998"=KTD 998→	KEC	Kec		
D...	IC	→µPD... (NEC !)		Nec		
D	Diode	=1SS376 (Typ-Code/Stempel/marking)	71(1,7mm)	Rhm	→1SS376	data
D	Si-P	=2SA1687 (Typ-Code/Stempel/marking)	35(2mm)	Say	→2SA1687	data
D	Si-P	=2SB1219A (Typ-Code/Stempel/marking)	35(2mm)	Mat	→2SB1219A	data
D	Si-P	=2SB710A (Typ-Code/Stempel/marking)	35	Mat	→2SB710A	data
D	Si-P	=2SB789 (Typ-Code/Stempel/marking)	39	Mat	→2SB789	data
D	N-FET	=2SK2219 (Typ-Code/Stempel/marking)	35(2mm)	Say	→2SK2219	data
D	MOS-N-FET-d	=3SK304 (Typ-Code/Stempel/marking)	44	Mat	→3SK304	data
D	MOS-N-FET-d	=3SK308 (Typ-Code/Stempel/marking)	44(2mm)	Mat	→3SK308	data
D	C-Di	=BB 659 (Typ-Code/Stempel/marking)	71(1,3mm)	Sie	→BB 659	data
D	PIN-Di	=HVU 187 (Typ-Code/Stempel/marking)	71(1,7mm)	Hit	→HVU 187	data
D	Si-Di	=RB 551V-30 (Typ-Code/Stempel/marking)	71(1,7mm)	Rhm	→RB 551V-30	data
D	Si-Di	=SB 02-09CP (Typ-Code/Stempel/marking)	35	Say	→SB 02-09CP	data
D 0	MOS-N-FET-e	=Si 2320DS (Typ-Code/Stempel/marking)	35	Six	→Si 2320DS	data
D 1	Se-Di	Blitzschutz/lightning protection	12	Hfo		data
D 1	Si-N	=2SC5277-D1 (Typ-Code/Stempel/marking)	35(1,6mm)	Say	→2SC5277	data
D 1	MOS-N-FET-e*	=2SK3107 (Typ-Code/Stempel/marking)	35(1,6mm)	Nec	→2SK3107	data
D 1	Si-Di	=BAW 63 (Typ-Code/Stempel/marking)	35(2mm)	Fer	→BAW 63	data
D 1	Si-N	=BCW 31 (Typ-Code/Stempel/marking)	35	Fer, Phi	→BCW 31	data
D 1	Z-Di	=DL 04-18 (Typ-Code/Stempel/marking)	71(5x2,5)	Shi	→DL 04-18	data
D 1	Si-N/P+R	=IMD 1A (Typ-Code/Stempel/marking)	46	Rhm	→IMD 1A	data
D 1	Z-Di	=MAZF 051 (Typ-Code/Stempel/marking)	71(1,7mm)	Mat	→MAZF 051	data
D 1	Z-Di	=MAZN 051 (Typ-Code/Stempel/marking)	71(1,3mm)	Mat	→MAZN 051	data
D 1	Z-Di	=MMSZ 5226B (Typ-Code/marking)	71(2,7mm)	Ons	→MMSZ 5226B	
D 01	Si-Di	=SD 914 (Typ-Code/Stempel/marking)	35	Tho	→SD 914	data
D 1	MOS-N-FET-e	=SST 211 (Typ-Code/Stempel/marking)	35	Six	→SST 211	data
D 1 CA20	Si-Di	4 Di Array, 200V, 1A	5-SIP	Shi	Pin-Code: KAAAA	data

Type	Device	Short Description	Fig.	Manu	Comparision Types	More at
D 1 CA40	Si-Di	=D 1CA20: 400V	5-SIP		Pin-Code: KAAAA	data
D 1 CA60	Si-Di	=D 1CA20: 600V	5-SIP		Pin-Code: KAAAA	data
D 1 CA...R	Si-Di	=D 1CA...:	5-SIP		Pin-Code: AKKKK	data
D 1 CAK20	Si-Di	4 Di Array, 200V, 0,7A, <300ns	5-SIP	Shi	Pin-Code: KAAAA	data
D 1 CAK20R	Si-Di	=D 1 CAK20:	5-SIP		Pin-Code: AKKKK	data
D 1 CS20	Si-Di	3 Di Array, 200V, 1A	4-SIP	Shi	Pin-Code: KAAA	data
D 1 CS20R	Si-Di	=D 1CS20:	4-SIP		Pin-Code: AKKK	data
D 1 CSK20	Si-Di	3 Di Array, 200V, 0,7A, <300ns	4-SIP	Shi	Pin-Code: KAAA	data
D 1 F10	Si-Di	SMD, GI, Uni, 100V, Ifsm=25A, 1A	71a(5x2,5)	Shi	BYG 10D...M, U 1BC44...JC44	data
D 1 F20	Si-Di	=D 1 F10: 200V	71a(5x2,5)		BYG 10D...M, U 1DC44...JC44	data
D 1 F40	Si-Di	=D 1 F10: 400V	71a(5x2,5)		BYG 10G...M, U 1GC44...JC44	data
D 1 F60	Si-Di	=D 1 F10: 600V	71a(5x2,5)		BYG 10J...M, U 1JC44	data
D 1 F60A	Si-Di	=D 1 F10: 600V, Ifsm=45A	71a(5x2,5)			data
D 1 FH3	Si-Di	SMD, Schottky, 30V, 3A	71a(5x2,5)	Shi	SFPA-73, SFPJ-73, U 3FWJ44N	data
D 1 FL20	Si-Di	SMD, GI, S, 200V, 1,1A, <50ns	71a(5x2,5)	Shi	ER 2D...M, MURS 120, MURS 160	data
D 1 FL20U	Si-Di	=D 1FL20: <35ns	71a(5x2,5)		ER 2D...M, MURS 120, MURS 160	data
D 1 FL40	Si-Di	=D 1FL20: 400V, 0,8A	71a(5x2,5)		ER 2G...M, MURS 160	data
D 1 FM3	Si-Di	SMD, Schottky, 30V, 3A	71a(5x2,5)		SFPA-73, SFPJ-73, U 3FWJ44N	data
D 1 FP3	Si-Di	SMD, Schottky, 30V, 2A	71a(5x2,5)		SFPA-63, SFPJ-63, U 2FWJ44N	data
D 1 FS4	Si-Di	SMD, Schottky-GI, 40V, 1,1A	71a(5x2,5)	Shi	HRF 22, MBRS 140, SB 11-04HP, SFPB-54,++	data
D 1 FS4A	Si-Di	=D1 FS4: 1,5A	71a(5x2,5)		BYS 10-45, SFPB-64	data
D 1 FS6	Si-Di	=D1 FS4: 60V	71a(5x2,5)		HRF 32, MBRS 1100	data
D 1 G	Si-P	=KSA 812-G (Typ-Code/Stempel/marking)	35	Sam	→KSA 812	data
D 1 G	Si-P	=HN 1A01F-GR(Typ-Code/Stempel/marking)	46	Tos	→HN 1A01F	data
D 1 G	Si-P	=HN1A01FU-GR(Typ-Code/Stempel/marking)	46(2mm)	Tos	→HN 1A01FU	data
D 1 JA20	Si-Di	4 Di Array, 200V, 1A	42/8Pin	Shi	Pin-Code: AAAAKKKK	data
D 1 JA40	Si-Di	=D 1JA20: 400V	42/8Pin		Pin-Code: AAAAKKKK	data
D 1 JA60	Si-Di	=D 1JA20: 600V	42/8Pin		Pin-Code: AAAAKKKK	data
D 1 JAK20	Si-Di	4 Di Array, 200V, 0,7A, <300ns	42/8Pin	Shi	Pin-Code: AAAAKKKK	data
D 1 JS20	Si-Di	3 Di Array, 200V, 1A	42/6Pin	Shi	Pin-Code: AAAKKK	data
D 1 L	Si-P	=KSA 812-L (Typ-Code/Stempel/marking)	35	Sam	→KSA 812	data
D 1 N20	Si-Di	GI, Uni, 200V, 1A(Ta=25°)	31a	Shi	BY 126...127, BY 133...135, 1N4003...07, ++	data
D 1 N40	Si-Di	=D 1N20: 400V	31a		BY 126...127, BY 133...134, 1N4004...07, ++	data
D 1 N60	Si-Di	=D 1N20: 600V	31a		BY 126...127, BY 133...134, 1N4005...07, ++	data
D 1 NF60	Si-Di	GI, S, 600V, 0,8A, <400ns	31a	Shi	BYV 26C	data
D 1 NL20U	Si-Di	GI, S, 200V, 1A, <35ns	31a	Shi	BYV 26B...C, EGP 10D...G, FE 1D...H	data
D 1 NL40	Si-Di	GI, S, 400V, 0,9A, <50ns	31a	Shi	BYV 26B...C, EGP 10G, FE 1H	data
D 1 NS4	Si-Di	Schottky, 40V, 1A, Uf<0,55V(1A)	31a	Shi	BYV 10-40, SB 150, 1N5819	data
D 1 NS6	Si-Di	Schottky, 60V, 1A, Uf<0,58V(1A)	31a	Shi	BYV 10-60, SB 160, MBR 160	data
D 1 O	Si-P	=KSA 812-O (Typ-Code/Stempel/marking)	35	Sam	→KSA 812	data
D 1 p	Si-N	=BCW 31 (Typ-Code/Stempel/marking)	35	Phi	→BCW 31	data
D 1 1	Si-N	=BCW 31 (Typ-Code/Stempel/marking)	35	Phi	→BCW 31	data
D 1 U40	Si-Di	Dual, 400V, 0,4A	4-DIP	Shi	-	data
D 1 UK20	Si-Di	Dual, 200V, 0,35A	4-DIP	Shi	-	data
D1X19X	LED	bl ø5mm I:60mW/Sr Vf:3.5V If:350mA		Seo		data pdf pinout
D 1 Y	Si-P	=KSA 812-Y (Typ-Code/Stempel/marking)	35	Sam	→KSA 812	data
D 1 Y	Si-P	=HN 1A01F-Y (Typ-Code/Stempel/marking)	46	Tos	→HN 1A01F	data
D 1 Y	Si-P	=HN 1A01FU-Y(Typ-Code/Stempel/marking)	46(2mm)	Tos	→HN 1A01FU	data
D2	UNI	Vs:±15V Vu:85dB Vo:±11V Vi0:6mV	7Y	Ilc		data pinout
D 2	Si-N	=2SC5277-D2 (Typ-Code/Stempel/marking)	35(1,6mm)	Say	→2SC5277	data
D 2	Si-Di	=BAW 63A (Typ-Code/Stempel/marking)	35(2mm)	Fer	→BAW 63A	data
D 2	Si-N	=BCW 32 (Typ-Code/Stempel/marking)	35	Fer,Mot,Phi	→BCW 32	data
D 2	Z-Di	=DL 04-28 (Typ-Code/Stempel/marking)	71(5x2,5)	Shi	→DL 04-28	data
D 2	Si-N/P+R	=IMD 2A (Typ-Code/Stempel/marking)	46	Rhm	→IMD 2A	data
D 2	Z-Di	=MAZF 056 (Typ-Code/Stempel/marking)	71(1,7mm)	Mat	→MAZF 056	data
D 2	Z-Di	=MAZN 056 (Typ-Code/Stempel/marking)	71(1,3mm)	Mat	→MAZN 056	data
D 2	Z-Di	=MMSZ 5227B (Typ-Code/Stempel/marking)	71(2,7mm)	Ons	→MMSZ 5227B	data
D 2	Si-P	=MSB 92AW (Typ-Code/Stempel/marking)	35(2mm)	Ons	→MSB 92W	
D 2	Si-N/P+R	=UMD 2N (Typ-Code/Stempel/marking)	46(2mm)	Rhm	→IMD 2A...6A	
D 2 C01	MOS-NP-FET-e	=MMDF 2C01HD(Typ-Code/Stempel/marking)	8-MDIP	Mot	→MMDF 2C01HD	data
D 2 C02	MOS-NP-FET-e	=MMDF 2C02HD(Typ-Code/Stempel/marking)	8-MDIP	Mot	→MMDF 2C02HD	data
D 2 C03	MOS-NP-FET-e	=MMDF 2C03HD(Typ-Code/Stempel/marking)	8-MDIP	Mot	→MMDF 2C03HD	data
D 2 E	Si-Di	=RB 491D (Typ-Code/Stempel/marking)	35	Rhm	→RB 491D	data
D 2 F10	Si-Di	SMD, GI, Uni, 100V, 1,4A	71a(8mm)	Shi	-	data
D 2 F20	Si-Di	=D 2F10: 200V	71a(8mm)		-	data
D 2 F40	Si-Di	=D 2F10: 400V	71a(8mm)		-	data
D 2 F60	Si-Di	=D 2F10: 600V	71a(8mm)		-	data
D 2 FL20U	Si-Di	SMD, GI, S, 200V, 1,5A, <35ns	71a(8mm)	Shi	-	data
D 2 FL40	Si-Di	SMD, GI, S, 400V, 1,3A, <50ns	71a(8mm)	Shi	-	data
D 2 FS4	Si-Di	SMD, Schottky-Di, 40V, 1,6A	71a(7,6x4)	Shi	-	data
D 2 FS6	Si-Di	SMD, Schottky-Di, 60V, 1,5A	71a(7,6x4)	Shi	-	data
D 2 L	Si-Di	=D 2L20U (Typ-Code/Stempel/marking)	31	Shi	→D 2L20U	data
D 2 L	Si-Di	=D 2L40 (Typ-Code/Stempel/marking)	31	Shi	→D 2L40	data
D 2 L20U	Si-Di	GI, S, 200V, 1,5A, <35ns	31a	Shi	BYD 73D...G, EGP 20D...G, FE 2D...H	data
D 2 L40	Si-Di	GI, S, 400V, 1,5A, <50ns	31a	Shi	BYD 73G, EGP 20G, FE 2H	data

Type	Device	Short Description	Fig.	Manu	Comparision Types	More at
? p	Si-N	=BCW 32 (Typ-Code/Stempel/marking)	35	Phi	→BCW 32	data
? P01	MOS-P-FET-e	=MMDF 2P01HD(Typ-Code/Stempel/marking)	8-MDIP	Mot	→MMDF 2P01HD	data
? P02	MOS-P-FET-e	=MMDF 2P02HD(Typ-Code/Stempel/marking)	8-MDIP	Mot	→MMDF 2P02HD	data
? P03	MOS-P-FET-e	=MMDF 2P03HD(Typ-Code/Stempel/marking)	8-MDIP	Mot	→MMDF 2P03HD	data
? S4M	Si-Di	Schottky, 40V, 2A, Uf<0,55V(2A)	31a	Shi	BYS 22-45, BYS 26-45, 1N5822	data
? S6M	Si-Di	Schottky, 60V, 2A, Uf<0,58V(2A)	31a	Shi	BYS 22-90, HRP 34, MBR 360, SB 360	data
? SB20	Si-Br	Gl-Br, 200V, 1,5A(Ta=25°)	33x1	Shi	B140 C1500	data
? SB40	Si-Br	=D 2SB20: 400V	33x1		B250 C1500	data
? SB60	Si-Br	=D 2SB20: 600V	33x1		B380 C1500	data
? SB60L	Si-Br	=D 2SB20: 600V	33x1			data
? SB80	Si-Br	=D 2SB20: 800V	33x1		B500 C1500	data
? SBA20	Si-Br	Gl-Br, 200V, 1,5A(Ta=25°)	33x1	Shi	B140 C1500	data
? SBA40	Si-Br	=D 2SBA20: 400V	33x1		B250 C1500	data
? SBA60	Si-Br	=D 2SBA20: 600V	33x1		B380 C1500	data
? t	Si-N	=BCW 32 (Typ-Code/Stempel/marking)	35	Phi	→BCW 32	data
T 918	Si-N	Dual, 30V, 0,05A, >600MHz	81bs	Tix	2N3423...3424	data
T 2218	Si-N	Dual, 60V, 0,8A, >250MHz, B>40	81bs	Tix	2N3409...3411	data
T 2218A	Si-N	=D2T 2218: 75V, >300MHz	81bs		(2N3409...3411)[7]	data
T 2219	Si-N	=D2T 2218: B>100	81bs	Tix	2N3409...3411	data
T 2219A	Si-N	=D2T 2218A: B>100	81bs		(2N3409...3411)[7]	data
T 2904(A)	Si-P	Dual, 60V, 0,6A, >200MHz, B>40	81bs	Tix	2N4015...4016	data
T 2905(A)	Si-P	=D2T2904: B>100	81bs	Tix	2N4015...4016	data
	UNI	Vs:±15V Vu:100dB Vo:±11V f:1Mc	9Y	Ilc		data pinout
3	Si-Di	=BAW 63B (Typ-Code/Stempel/marking)	35(2mm)	Fer	→BAW 63B	data
3	Si-N	=BCW 33 (Typ-Code/Stempel/marking)	35	Fer, Phi	→BCW 33	data
3	Si-N/P+R	=IMD 3A (Typ-Code/Stempel/marking)	46	Rhm	→IMD 3A	data
3	Si-Di	=KDS 187 (Typ-Code/Stempel/marking)	35	Kec	→KDS 187	data
3	Z-Di	=MMSZ 5228B (Typ-Code/marking)	71,2,7mm)	Ons	→MMSZ 5228B	data
3	MOS-N-FET-e	=SST 213 (Typ-Code/Stempel/marking)	35	Six	→SST 213	data
3	Si-N/P+R	=UMD 3N (Typ-Code/Stempel/marking)	46(2mm)	Rhm	→IMD 2A...6A	data
3	Si-Di	=1SS 187 (Typ-Code/Stempel/marking)	35	Tos	→2SS 187	data
3 -	Si-N	=2SC5277-D3 (Typ-Code/Stempel/marking)	35(1,6mm)	Say	→2SC5277	data
3 A	Si-Di	=RB 400D (Typ-Code/Stempel/marking)	35	Rhm	→RB 400D	data
3 B	Si-Di	=RB 420D (Typ-Code/Stempel/marking)	35	Rhm	→RB 420D	data
3 C	Si-Di	=RB 421D (Typ-Code/Stempel/marking)	35	Rhm	→RB 421D	data
3 E	Si-Di	=RB 411D (Typ-Code/Stempel/marking)	35	Rbm	→RB 411D	data
3 F60	Si-Di	SMD, Gl, Uni, 600V, 3A	71a(8mm)	Shi		data
3 FP3	Si-Di	SMD, Schottky, 30V, 3A	71a(7,6x4)	Shi	.	data
3 FS4A	Si-Di	SMD, Schottky, 40V, 2,6A	71a(7,6x4)	Shi	.	data
3 FS6	Si-Di	SMD, Schottky, 60V, 3A	71a(7,6x4)	Shi	.	data
3 G	Si-Di	=RB 471E (Typ-Code/Stempel/marking)	45	Rhm	→RB 471E	data
3 H	Si-Di	=RB 705D (Typ-Code/Stempel/marking)	35	Rhm	→RB 705D	data
3 J	Si-Di	=RB 706D-40 (Typ-Code/Stempel/marking)	35	Rhm	→RB 706D-40	data
3 L	Si-Di	=RB 425D (Typ-Code/Stempel/marking)	35	Rhm	→RB 425D	data
3 L20U	Si-Di	Gl, S-L, 200V, 2,5A(Tc=136°), <35ns	15d	Shi	MA 690	data
3 L60	Si-Di	Gl, S-L, 600V, 3A(Tc=127°), <50ns	15d	Shi	BY 431F-1000	data
3 N02	MOS-N-FET-e	=MMDF 3N02HD(Typ-Code/Stempel/marking)	8-MDIP	Mot	→MMDF 3N02HD	data
3 N03	MOS-N-FET-e	=MMDF 3N03HD(Typ-Code/Stempel/marking)	8-MDIP	Mot	→MMDF 3N03HD	data
3 N04H	MOS-N-FET-e	=MMDF 3N04HD (Typ-Code/Stempel/marking)	8-MDIP	Ons	→MMDF 3N04HD	
3 N06	MOS-N-FET-e	=MMDF 3N06HD (Typ-Code/Stempel/marking)	8-MDIP	Ons	→MMDF 3N06HD	
3 N06V	MOS-N-FET-e	=MMDF 3N06VL (Typ-Code/Stempel/marking)	8-MDIP	Ons	→MMDF 3N06VL	
3 P	Si-Di	=RB 731U (Typ-Code/Stempel/marking)	46	Rhm	→RB 731U	data
3 p	Si-N	=BCW 33 (Typ-Code/Stempel/marking)	35	Phi	→BCW 33	data
3 Q	Si-Di	=RB 495D (Typ-Code/Stempel/marking)	35	Rhm	→BB 495D	data
3 S3M	Si-Di	Schottky, 30V, 3A, Uf<0,55V(3A)	31a	Shi	BYS 26-45, MBR 330, SB 330, 1N5821...22	data
3 S4M	Si-Di	Schottky, 40V, 3A, Uf<0,55V(3A)	31a	Shi	BYS 26-45, MBR 340, SB 340, 1N5822	data
3 S6M	Si-Di	Schottky, 60V, 3A, Uf<0,58V(3A)	31a	Shi	HRP 34, MBR 360, SB 360	data
3 SB10	Si-Br	Gl-Br, 100V, 4A(Tc=108°)	33x1	Shi	B80 C4000	data
3 SB20	Si-Br	=D 3SB10: 200V	33x1		B140 C4000	data
3 SB40	Si-Br	=D 3SB10: 400V	33x1		B250 C4000	data
3 SB60	Si-Br	=D 3SB10: 600V	33x1		B380 C4000	data
3 SB80	Si-Br	=D 3SB10: 800V	33x1		B500 C4000	data
3 SB80Z	Si-Br	=D 3SB10: contr.av., 800V	33x1		-	data
3 SBA20	Si-Br	Gl-Br, 200V, 4A(Tc=108°)	33x1	Shi	B140 C4000	data
3 SBA60	Si-Br	=D 3SBA10: 600V	33x1		B380 C4000	data
3 t	Si-N	=BCW 33 (Typ-Code/Stempel/marking)	35	Phi	→BCW 33	data
3 V	Si-Di	=RB 415D (Typ-Code/Stempel/marking)	35	Rhm	→RB 415D	data
X18X	LED	bl ø5mm I:60mW/Sr Vf:3.5V If:350mA		Seo		data pdf pinout
X28X	LED	bl ø5mm I:60mW/Sr Vf:4V If:700mA		Seo		data pdf pinout
	UNI	Vs:±15V Vu:90dB Vo:±10V f:3Mc		Ilc		data pinout
4	Si-P	=2SA1252-4 (Typ-Code/Stempel/marking)	35	Say	→2SA1252	data
4	Si-Di	=BAW 63B (Typ-Code/Stempel/marking)	35(2mm)	Fer	→BAW 64	data
4	Si-N	=BCW 31R (Typ-Code/Stempel/marking)	35	Val,Fer	→BCW 31R	data
4	Si-Di	=MMBD 4148SE(Typ-Code/Stempel/marking)	35	Nsc	→MMBD 4148SE	data
4	Z-Di	=MMSZ 5229B (Typ-Code/marking)	71(2,7mm)	Ons	→MMSZ 5229B	data
C/5E22F	LO-NOISE	Vs:±18V Vu:140dB Vo:±10V Vi0:0.03mV	7Y	Acc.		data pinout
4 L20U,LA20	Si-Di	Gl, S-L, 200V, 4A(Tc=130°), <35ns	15d	Shi	MA 690	data
4 L40	Si-Di	Gl, S-L, 400V, 4A(Tc=125°), <50ns	15d	Shi	BY 431F-1000	data

Type	Device	Short Description	Fig.	Manu	Comparision Types	More at
D 4 N01	MOS-N-FET-e	=MMDF 4N01HD(Typ-Code/Stempel/marking)	8-MDIP	Mot	→MMDF 4N01HD	data
D 4 N01Z	MOS-N-FET-e	=MMDF 4N01Z (Typ-Code/Stempel/marking)	8-MDIP	Mot	→MMDF 4N01Z	data
D 4 SB 60L	Si-Br	Gl-Br, 600V, 4A(Tc=111°), <10µs	33x1	Shi	B380 C4000	data
D 4 SB 80	Si-Br	Gl-Br, 800V, 4A(Tc=108°)	33x1	Shi	B500 C4000	data
D 4 SBL 20U	Si-Br	Gl-Br, 200V, 4A(Tc=108°), <35ns	33x1	Shi	-	data
D 4 SBL 40	Si-Br	Gl-Br, 400V, 4A(Tc=91°), <50ns	33x1	Shi	-	data
D 4 SBS4	Si-Br	Schottky-Br, 40V, 4A(Tc=116°)	33x1	Shi	-	data
D 4 SBS6	Si-Br	Schottky-Br, 60V, 4A(Tc=89°)	33x1	Shi	-	data
D 4 SC4M	Si-Di	Schottky, Dual, 40V, 4A(Tc=113°)	15p	Shi	MA 749(A)	data
D 4 SC6M	Si-Di	Schottky, Dual, 60V, 4A(Tc=138°)	15p	Shi	HRW 34F, MA 755	data
D5	UNI	Vs:±15V Vu:96dB Vo:±10V f:2Mc	7Y	Ilc		data pinout
D 5	Z-Di	=DL 03-58 (Typ-Code/Stempel/marking)	71(5x2,5)	Shi	→DL 03-58	data
D 5	Si-Di	=MMBD 4148CC(Typ-Code/Stempel/marking)	35	Nsc	→MMBD 4148CC	data
D 5	Z-Di	=MMSZ 5230B (Typ-Code/marking)	71(2,7mm)	Ons	→MMSZ 5230B	data
D 5	Si-Di	=SMBYW04-50(Typ-Code/Stempel/marking)	71(8mm)	Tho	→SMBYW 04-...	data
D 5	MOS-N-FET-e	=SST 215 (Typ-Code/Stempel/marking)	35	Six	→SST 215	data
D 5	Si-P	=2SA1252-5 (Typ-Code/Stempel/marking)	35	Say	→2SA1252	data
D 5	Si-N	=2SC1622-D5 (Typ-Code/Stempel/marking)	35	Nec	→2SC1622	data
D 5	Si-Di	=BAW 65 (Typ-Code/Stempel/marking)	35(2mm)	Fer	→BAW 65	data
D 5	Si-N	=BCW 32R (Typ-Code/Stempel/marking)	35	Val,Fer	→BCW 32R	data
D5C	UNI	Vs:±15V Vu:96dB Vo:±10V f:2Mc	7Y	Ilc		data pinout
D 5 E29	UJT-P	Iv>25mA, Ip<25µA	5m§	Gen	-	data
D 5 E35	UJT-P	Iv>10mA	5m§	Gen	-	data
D 5 E36	UJT-P	Iv>10mA	5m§	Gen	-	data
D 5 E37	UJT-P	Iv>4mA, Ip<25µA	5m§	Gen	-	data
D 5 E43	UJT-P	Iv>6mA, Ip<2µA	5m§	Gen	-	data
D 5 E44	UJT-P	Iv>4mA, Ip<5µA	5m§	Gen	-	data
D 5 E45	UJT-P	Iv>8mA, Ip<2µA	5m§	Gen	2N2647	data
D 5 F	PIN-Di	=RN 739D (Typ-Code/Stempel/marking)	35	Rhm	→RN 739D	data
D 5 K1	UJT-N	Iv>1mA, Ip<5µA	5m	Gen	2N6114	data
D 5 K2	UJT-N	Iv>1mA, Ip<15µA	5m	Gen	2N6115	data
D 5 L60	Si-Di	S-L, 600V, 5A(Tc=123°), <50ns	15d	Shi	BY 431F-1000	data
D 5 LC20U	Si-Di	Dual, S-L, 200V, 5A(Tc=132°), <35ns	15p	Shi	MA 649, MA 653	data
D 5 LC20UR	Si-Di	=D 5LC20U:	15q	Shi	-	data
D 5 LC40	Si-Di	Dual, S-L, 400V, 5A(Tc=125°), <50ns	15p	Shi	MA 693	data
D 5 LCA20	Si-Di	Dual, S-L, 200V, 5A(Tc=132°), <50ns	15p	Shi	MA 649, MA 653	data
D 5 LD20U	Si-Di	=D 5LC20U:	15o1	Shi	-	data
D 5 S3M	Si-Di	Schottky, 30V, 5A(Tc=110°)	15d	Shi	MA 2D749, MBRF 745	data
D 5 S4M	Si-Di	Schottky, 40V, 5A(Tc=131°)	15d	Shi	MA 2D749, MBRF 745	data
D 5 S6M	Si-Di	Schottky, 60V, 5A(Tc=130°)	15d	Shi	MA 2D755	data
D 5 S9M	Si-Di	Schottky, 90V, 5A(Tc=101°)	15d	Shi	MA 2D760(A)	data
D 5 SB10	Si-Br	Gl-Br, 100V, 6A(Tc=110°)	33x1	Shi	B80 C6000	data
D 5 SB20	Si-Br	=D 5SB10: 200V	33x1		B140 C6000	data
D 5 SB40	Si-Br	=D 5SB10: 400V	33x1		B250 C6000	data
D 5 SB60	Si-Br	=D 5SB10: 600V	33x1		B380 C6000	data
D 5 SB80	Si-Br	=D 5SB10: 800V	33x1		B500 C6000	data
D 5 SBA20	Si-Br	Gl-Br, 200V, 6A(Tc=110°)	33x1	Shi	B140 C6000	data
D 5 SBA60	Si-Br	=D 5SBA20: 600V	33x1		B380 C6000	data
D 5 SC4M	Si-Di	Schottky, Dual, 40V, 5A(Tc=113°)	15p	Shi	BYV 118F-..., MA 749(A)	data
D 5 SC4MR	Si-Di	=S 5SC4M:	15q		-	data
D6	UNI	Vs:±15V Vu:96dB Vo:±10V f:1.5Mc	7Y	Ilc		data pinout
D 6	Si-Di	=BAW 66 (Typ-Code/Stempel/marking)	35(2mm)	Fer	→BAW 66	data
D 6	Si-N	=BCW 33R (Typ-Code/Stempel/marking)	35	Val,Fer	→BCW 33R	data
D 6	Si-N/P+R	=IMD 6A (Typ-Code/Stempel/marking)	46	Rhm	→IMD 6A	data
D 6	Si-Di	=MMBD 4148CA(Typ-Code/Stempel/marking)	35	Nsc	→MMBD 4148CA	data
D 6	Si-N/P+R	=UMD 6N (Typ-Code/Stempel/marking)	46(2mm)	Rhm	→IMD 2A..6A	data
D 6	Si-P	=2SA1252-6 (Typ-Code/Stempel/marking)	35	Say	→2SA1252	data
D 6	Si-N	=2SC1622-D6 (Typ-Code/Stempel/marking)	35	Nec	→2SC1622	data
D 6 E	Z-Di	=FTZ 5.6E (Typ-Code/Stempel/marking)	45	Rhm	→FTZ 5.6E	data
D 6 F	Z-Di	=FTZ 6.8E (Typ-Code/Stempel/marking)	45	Rhm	→FTZ 6.8E	data
D 6 F	Z-Di	=UMZ 6.8E (Typ-Code/Stempel/marking)	45(2mm)	Rhm	→UMZ 6.8E	data
D 6 L20U	Si-Di	Gl, S-L, 200V, 6A(Tc=118°), <35ns	15d	Shi	MA 690	data
D 6 N03	MOS-N-FET-e	=MMDF 6N03HD (Typ-Code/Stempel/marking)	8-MDIP	Ons	→MMDF 6N03HD	
D 6 SB 60L	Si-Br	Gl-Br, 600V, 6A(Tc=112°), <10µs	33x1	Shi	B380 C6000	data
D 6 SB 80	Si-Br	Gl-Br, 800V, 6A(Tc=110°)	33x1	Shi	B500 C6000	data
D7	UNI	Vs:±15V Vu:100dB Vo:±11V f:1Mc	9Y	Ilc		data pinout
D 7	Si-P	=2SA1252-7 (Typ-Code/Stempel/marking)	35	Say	→2SA1252	data
D 7	Si-N	=2SC1662-D7 (Typ-Code/Stempel/marking)	35	Nec	→2SC1662	data
D 7	Si-Di	=BAW 67 (Typ-Code/Stempel/marking)	35(2mm)	Fer	→BAW 67	data
D 7	Si-N	=BCF 32 (Typ-Code/Stempel/marking)	35	Phi	→BCF 32	data
D07-13	Si-Di	=ERD 07-13 (Typ-Code/Stempel/marking)	31a	Fjd	→ERD 07-...	data
D07-15	Si-Di	=ERD 07-15 (Typ-Code/Stempel/marking)	31a	Fjd	→ERD 07-...	data
D 7 p	Si-N	=BCF 32 (Typ-Code/Stempel/marking)	35	Phi	→BCF 32	data
D 7 t	Si-N	=BCF 32 (Typ-Code/Stempel/marking)	35	Phi	→BCF 32	data
D8	UNI	Vs:±15V Vo:95dB Vo:±11V f:5Mc	7Y	Ilc		data pinout
D 8	Si-Di	=BAW 68 (Typ-Code/Stempel/marking)	35(2mm)	Fer	→BAW 68	data
D 8	Si-N	=BCF 33 (Typ-Code/Stempel/marking)	35	Phi	→BCF 33	data
D 8	Si-N/P+R	=IMD 8A (Typ-Code/Stempel/marking)	46	Rhm	→IMD 8A	data

Type	Device	Short Description	Fig.	Manu	Comparision Types	More at
) 8	Si-Di	=ZUMD 54 (Typ-Code/Stempel/marking)	35(2mm)	Ztx	→ZUMD 54	data
) 8	Si-N	=2SC1662-D8 (Typ-Code/Stempel/marking)	35	Nec	→2SC1662	data
) 8 C	Si-Di	=ZUMD 54C (Typ-Code/Stempel/marking)	35(2mm)	Ztx	→ZUMD 54C	data
) 8 L60	Si-Di	GI, S-L, 600V, 8A(Tc=106°), <70ns	15d	Shi	BYR 29F-600	data
) 8 LC20U	Si-Di	Dual, S-L, 200V, 8A(Tc=122°), <35ns	15p	Shi	BYQ 28F-200, MA 650	data
) 8 LC20UR	Si-Di	=D 8LC20U:	15q			data
) 8 LC40	Si-Di	Dual, S-L, 400V, 8A(Tc=114°), <50ns	15p	Shi	MA 694	data
) 8 LCA20	Si-Di	Dual, S-L, 200V, 8A(Tc=122°), <50ns	15p	Shi	BYQ 28F-200, MA 650	data
) 8 LD20U	Si-Di	=D 8LC20U:	15o1		-	data
) 8 LD40U	Si-Di	=D 8LC40U:	15o1		-	data
D9	UNI	Vs:±15V Vu:99dB Vo:±10V f:10Mc	7Y	Ilc		data pinout
D 9	Si-N/P+R	=IMD 9A (Typ-Code/Stempel/marking)	46	Rhm	→IMD 9A	data
D 9	Si-Di	=1SS401 (Typ-Code/Stempel/marking)	35(2mm)	Tos	→1SS401	
D10	UNI	Vs:±15V Vu:92dB Vo:±11V f:1Mc	7Y	Ilc		data pdf pinout
D 10	Si-N/P+R	=IMD 10A (Typ-Code/Stempel/marking)	46	Rhm	→IMD 10A	data
D 10	Si-Di	=SMBYW04-100(Typ-Code/Stempel/marking)	71(8mm)	Tho	→SMBYW 04-...	data
D 10 L20U	Si-Di	GI, S-L, 200V, 10ATc=111°), <35ns	15d	Shi	BYT 12PI-600	data
D 10 LC20U	Si-Di	Dual, S-L, 200V, 10A(Tc=114°), <35ns	15p	Shi	BYQ 28F-200, MA 650	data
D 10 LC20UR	Si-Di	=D 10LC20U:	15q			data
D 10 LC40U	Si-Di	Dual, S-L, 400V, 10A(Tc=100°), <50ns	15p	Shi	MA 694	data
D 10 LCA20	Si-Di	Dual, S-L, 200V, 10A(Tc=114°), <50ns	15p	Shi	BYQ 28F-200, MA 650	data
D 10 N05	MOS-N-FET-e	=RFD 10N05 (Typ-Code/Stempel/marking)	30	Rca	→RFD 10N05	data
D 10 SBS4	Si-Br	Schottky-Br, 40V, 10A(Tc=67°)	33x1	Shi	-	data
D 10 SC3M	Si-Di	Schottky, Dual, 30V, 10A(Tc=102°)	15p	Shi	BYV 33F-..., BYV 133F-..., MA 750(A)	data
D 10 SC4M	Si-Di	Schottky, Dual, 40V, 10A(Tc=102°)	15p	Shi	BYV 33F-..., BYV 133F-..., MA 750(A)	data
D 10 SC6M	Si-Di	Schottky, Dual, 60V, 10A(Tc=92°)	15p	Shi	MA 756	data
D 10 SC9M	Si-Di	Schottky, Dual, 90V, 10A(Tc=111°)	15p	Shi	HRW 36F, MA 761	data
D 10 SC..MR	Si-Di	=D 10SC..M:	15q		-	data
D 10 SD6M	Si-Di	=D 10SC6M:	15o1		-	data
D 10 XB20	Si-Br	GI-Br, 200V, 10A(Tc=100°)	33x1	Shi	B140 C10000	data
D 10 XB20H	Si-Br	=D 10XB20: 10A(Tc=112°)	33x1		B140 C10000	data
D 10 XB40	Si-Br	=D 10XB20: 400V	33x1		B250 C10000	data
D 10 XB60	Si-Br	=D 10XB20: 600V	33x1		B380 C10000	data
D 10 XB60H	Si-Br	=D 10XB20: 600V, 10A(Tc=112°)	33x1		B380 C10000	data
D 10 XB80	Si-Br	=D 10XB20: 800V	33x1		B500 C10000	data
D 10 XB80H	Si-Br	=D 10XB20: 800V, 10A(Tc=112°)	33x1		B500 C10000	data
D11	UNI	Vs:±15V Vu:100dB Vo:±10V f:0.8Mc	15Y,7Y	Ilc		data pinout
D12	UNI	Vs:±15V Vu:100dB Vo:±11V f:1Mc	7Y	Ilc		data pinout
D 12	Si-N/P+R	=UMD 12N (Typ-Code/Stempel/marking)	46(2mm)	Rhm	→UMD 12N	data
D 12 E026	Si-N	Dual, 45V, 0,03A, >30MHz	81bs	Gen	BFW 39...40, 2N2913..2918	data
D 12 E109	Si-N	Dual, 60V, 0,05A	81bs	Gen	(2N2919...2920)[6]	data
D 12 E126	Si-N	=D12E026:	81bs	Gen	(BFW 39...40, 2N2913..2918)[6]	data
D 13 H1	SAS	Ub:+7...+9V,-14...-18V, Itsm:+1A,-0,5A	7e	Gen	-	data
D 13 H2	SAS	=D 13 H1	7e			data
D 13 K1	PUT	Iv=70µA, Ip=5µA	2a	Gen	2N6027, 2N6116, 2N6119, MPU 131, MPU 231	data
D 13 K2	PUT	Iv=50µA, Ip=2µA	2a	Gen	2N6117, MPU 132, MPU 232	data
D 13 K3	PUT	Iv=50µA, Ip=1µA	2a	Gen	2N6118, MPU 133, MPU 233	data
D 13 T1	PUT	Iv=70µA, Ip=5µA	7a	Gen	2N6027, 2N6116, 2N6119, MPU 131, MPU 231	data
D 13 T2	PUT	Iv=25µA, Ip=1µA	7a	Gen	2N6028, 2N6120	data
D13V	ER+	Io=40mA Vo:10...40V P:400mW	3A	Gie		data pinout
D 13 V	Z-IC	+10...40V, 40mA	2	Gen		data pinout
D14	HI-SPEED	Vs:±15V Vu:86dB Vo:±10V f:20Mc	7Y	Ilc		data pinout
D 14	Si-N/P+R	=IMD 14 (Typ-Code/Stempel/marking)	46	Rhm	→IMD 14	data
D 14 N05	MOS-N-FET-e	=RFD 14N05 (Typ-Code/Stempel/marking)	30	Rca	→RFD 14N05	data
D15	HI-OHM	Vs:±15V Vu:100dB Vo:±10V Vi0:0.75mV	7Y	Ilc		data pdf pinout
D 15	Si-N	=2SC1622A-D15(Typ-Code/Stempel/mark.)	35	Nec	→2SC1622A	data
D 15	Si-N	=2SC4180-D15(Typ-Code/Stempel/marking)	35(2mm)	Nec	→2SC4180	data
D 15	Si-Di	=SMBYW04-150(Typ-Code/Stempel/marking)	71(8mm)	Tho	→SMBYW 04-...	data
D 15 LC20U	Si-Di	Dual, S-L, 200V, 15A(Tc=104°), <35ns	15p	Shi	MA 652	data
D 15 SCA4M	Si-Di	Schottky, Dual, 40V, 15A(Tc=117°)	15p	Shi	BYV 33F-..., BYV 133F-..., MA 752(A)	data
D 15 VD40	Si-Di	400V, 15A(Tc=94°)		Shi	-	data
D 15 XB20	Si-Br	GI-Br, 200V, 15A(Tc=100°)	33x1	Shi	B140 C15000	data
D 15 XB20H	Si-Br	GI-Br, 200V, 15A(Tc=100°)	33x1	Shi	B140 C15000	data
D 15 XB40	Si-Br	=D 15XB20: 400V	33x1		B250 C15000	data
D 15 XB60	Si-Br	=D 15XB20: 600V	33x1		B380 C15000	data
D 15 XB60H	Si-Br	=D 15XB20H: 600V	33x1		B380 C15000	data
D 15 XB80	Si-Br	=D 15XB20: 800V	33x1		B500 C15000	data
D16	UNI	Vs:±15V Vu:100dB Vo:±11V f:1Mc	7Y	Ilc		data pinout
D 16	Si-N/P+R	=IMD 16A (Typ-Code/Stempel/marking)	46	Rhm	→IMD 16A	data
D 16	Si-N	=2SC1662A-D16 (Typ-Code/Stempel/mark.)	35	Nec	→2SC1662A	data
D 16	Si-N	=2SC4180-D16(Typ-Code/Stempel/marking)	35(2mm)	Nec	→2SC4180	data
D 16 E7	Si-N	25V, 0,2W, 135MHz	7	Gen	-	data
D 16 E9	Si-N	30V, 0,2W, 135MHz	7	Gen	-	data
D 16 G6	Si-N	UHF-O, >500MHz	7c	Gen	BF 377...378, BF 763, 2N918, 2N2857	data
D 16 K1...K4	Si-N	VHF, 580...650MHz	7c	Gen	BF 225, BF 314, BF 496, BF 502...503, ++	data
D 16 N05	MOS-N-FET-e	=RFD 16N05 (Typ-Code/Stempel/marking)	30	Rca	→RFD 16N05	data

Type	Device	Short Description	Fig.	Manu	Comparision Types	More at
D 16 P1..P4	Si-N-Darl	NF, 18..20V, 0,2A, 0,32W, >60MHz	7c	Gen	BC 517, BC 875, BC 877, BC 879, MPSA 25	data
D17	HI-OHM	Vs:±15V Vu:100dB Vo:±10V Vi0:0.5mV	7Y	Ilc		data pinout
D 17	Si-N	=2SC1662A-D17 (Typ-Code/Stempel/mark.)	35	Nec	→2SC1662A	data
D 17	Si-N	=2SC4180-D17(Typ-Code/Stempel/marking)	35(2mm)	Nec	→2SC4180	data
D18	UNI	Vs:±15V Vu:92dB Vo:±11V f:1.5Mc	7Y	Ilc		data pinout
D 18	Si-N	=2SC1662A-D18 (Typ-Code/Stempel/mark.)	35	Nec	→2SC1662A	data
D 18	Si-N	=2SC4180-D18(Typ-Code/Stempel/marking)	35(2mm)	Nec	→2SC4180	data
D19	HI-OHM	Vs:±15V Vu:110dB Vo:±10V Vi0:0.5mV	7Y	Ilc		data pinout
D 20	Si-Di	=SMBYW04-200(Typ-Code/Stempel/marking)	71(8mm)	Tho	→SMBYW 04-...	data
D20B	HI-VOLT	Vs:±26V Vu:100dB Vo:±20V Vi0:0.5mV	15Y	Ilc		data pinout
D 20 LC20U	Si-Di	Dual, S-L, 200V, 20A(Tc=112°), <35ns	18p	Shi	BYV 72F-200, MA 651	data
D 20 LC40	Si-Di	Dual, S-L, 400V, 20A(Tc=102°), <50ns	18p	Shi	BYV 74F-400, MA 695	data
D 20 SC9M	Si-Di	Schottky, Dual, 90V, 20A(Tc=111°)	18p	Shi		data
D 20 XB20	Si-Br	Gl-Br, 200V, 20A(Tc=87°)	33x1	Shi	B140 C20000	data
D 20 XB40	Si-Br	=D 20XB20: 400V	33x1		B250 C20000	data
D 20 XB60	Si-Br	=D 20XB20: 600V	33x1		B380 C20000	data
D 20 XB80	Si-Br	=D 20XB20: 800V	33x1		B500 C20000	data
D 21	N-FET	=2SK2219-21 (Typ-Code/Stempel/marking)	35(2mm)	Say	→2SK2219	data
D 21	N-FET	=2SK931-21 (Typ-Code/Stempel/marking)	35	Say	→2SK931	data
D21B	HI-OHM	Vs:±15V Vu:110dB Vo:±10V Vi0:0.5mV	15Y	Ilc		data pinout
D22	UNI	Vs:±15V Vu:100dB Vo:±10V f:1Mc	15Y	Ilc		data pinout
D 22	N-FET	=2SK2219-22 (Typ-Code/Stempel/marking)	35(2mm)	Say	→2SK2219	data
D 22	N-FET	=2SK931-22 (Typ-Code/Stempel/marking)	35	Say	→2SK931	data
D23	HI-OHM	Vs:±24V Vu:88dB Vo:±20V Vi0:1mV	7Y	Ilc		data pinout
D 23	N-FET	=2SK2219-23 (Typ-Code/Stempel/marking)	35(2mm)	Say	→2SK2219	data
D 23	N-FET	=2SK931-23 (Typ-Code/Stempel/marking)	35	Say	→2SK931	data
D24	HI-SPEED	Vs:±15V Vu:100dB Vo:±10V f:100Mc	7Y	Ilc		data pinout
D 24	N-FET	=2SK931-24 (Typ-Code/Stempel/marking)	35	Say	→2SK931	data
D 24 A3391(A)	Si-N	Min, 25V, 0,1A, 120MHz	≈24c	Gen	BC 122...123	data
D 24 A3392	Si-N	Min, 25V, 0,1A, 140MHz	≈24c	Gen	BC 122...123	data
D 24 A3393	Si-N	Min, 25V, 0,1A, 140MHz	≈24c	Gen	BC 122...123	data
D 24 A3394	Si-N	Min, 25V, 0,1A, 140MHz	≈24c	Gen	BC 122...123	data
D 24 A3900(A)	Si-N	Min, 18V, 0,1A, 160MHz	≈24c	Gen	BC 122...123	data
D25	HI-OHM	Vs:±15V Vu:100dB Vo:±10V Vi0:0.75mV	7Y	Ilc		data pinout
D 25	Si-N	=2SC3115-D25(Typ-Code/Stempel/marking)	35	Nec	→2SC3115	data
D 25	N-FET	=2SK931-25 (Typ-Code/Stempel/marking)	35	Say	→2SK931	data
D 25 SC6M	Si-Di	Schottky, Dual, 60V, 25A(Tc=117°)	18p	Shi	-	data
D 25 SC6MR	Si-Di	=D 25SC6M:	18q			data
D 25 XB20	Si-Br	Gl-Br, 200V, 25A(Tc=98°)	33x1	Shi	B140 C25000	data
D 25 XB40	Si-Br	=D 25XB20: 400V	33x1		B250 C25000	data
D 25 XB60	Si-Br	=D 25XB20: 600V	33x1		B380 C25000	data
D 25 XB80	Si-Br	=D 25XB20: 800V	33x1		B500 C25000	data
D26	HI-VOLT	Vs:±28V Vu:98dB Vo:±20V f:1Mc	7Y	Ilc		data pinout
D 26	Si-N	=2SC3115-D26(Typ-Code/Stempel/marking)	35	Nec	→2SC3115	data
D 26 B1..B2	Si-N	Min, S, 40V, <12/18ns	36b	Gen	-	data
D 26 C1..C5	Si-N	Min, Uni, 25V, 0,09W	36b	Gen		data
D 26 E1..E7	Si-N	Min, NF, 18..45V, 0,09W	36b	Gen	BC 122...123	data
D 26 G1	Si-N	Min, VHF/UHF, >600MHz	36b	Gen	BC 123	data
D 26 P1..P3	Si-N-Darl	Min, 18..25V, 0,09W	36b	Gen	-	data
D27	HI-OHM	Vs:±15V Vu:100dB Vo:±11V Vi0:1mV	7Y	Ilc		data pinout
D 27	Si-N	=2SC3115-D27(Typ-Code/Stempel/marking)	35	Nec	→2SC3115	data
D 27 C1	Si-N	→D42C5		Gen		data
D 27 C2	Si-N	→D42C4		Gen		data
D 27 C3	Si-N	→D42C2		Gen		data
D 27 C4	Si-N	→D42C1		Gen		data
D 27 D1	Si-P	→D43C5		Gen		data
D 27 D2	Si-P	→D43C4		Gen		data
D 27 D3	Si-P	→D43C2		Gen		data
D 27 D4	Si-P	→D43C1		Gen		data
D28	UNI	Vs:±15V Vu:110dB Vo:±10V Vi0:2mV	7Y	Ilc		data pinout
D 28	Si-N	=2SC3115-D28(Typ-Code/Stempel/marking)	35	Nec	→2SC3115	data
D 28	Si-Di	=BAS 70W (Typ-Code/Stempel/marking)	35(2mm)	Tho	→BAS 70W	data
D28 04	Si-Di	=ERD 28-04 (Typ-Code/Stempel/marking)	31a	Fjd	→ERD 28-...	data
D28 06	Si-Di	=ERD 28-06 (Typ-Code/Stempel/marking)	31a	Fjd	→ERD 28-...	data
D28 08	Si-Di	=ERD 28-08 (Typ-Code/Stempel/marking)	31a	Fjd	→ERD 28-...	data
D 28 A5..A13	Si-N	NF-Tr/E, 35..50V, 0,5A	13	Gen	BD 385, BD 415, BD 827, BD 841	data
D 28 B	Si-N	S/Vid, 150V, 0,1A	13	Gen	BF 615, BF 617, BF 857...859	data
D 28 C1..C8	Si-N	→D40C1...C8		Gen		data
D 28 D1..D10	Si-N	→D40D1...D10		Gen		data
D 28 E	Si-N	→D40N3		Gen		data
D 29	Si-Di	=BAS 70-06W (Typ-Code/Stempel/marking)	35(2mm)	Tho	→BAS 70-06W	data
D29-02	Si-Di	=ERD 29-02 (Typ-Code/Stempel/marking)	31a	Fjd	→ERD 29-...	data
D29-04	Si-Di	=ERD 29-04 (Typ-Code/Stempel/marking)	31a	Fjd	→ERD 29-...	data
D29-06	Si-Di	=ERD 29-06 (Typ-Code/Stempel/marking)	31a	Fjd	→ERD 29-...	data
D 29 A4..A12	Si-P	Uni, 35..60V, 0,5A, 0,33W	7c	Gen	BC 327, BC 638, BC 640	data
D 29 E1..E10(J1)	Si-P	NF-Tr, 35..70V, 0,75A, 0,5W	7c	Gen	BC 327, BC 638, BC 640	data
D 29 F1..F7	Si-P	Uni, 40..60V, 0,1A, 0,36W, >90MHz	7c	Gen	BC 212, BC 257, BC 307, BC 556, ++	data

Type	Device	Short Description	Fig.	Manu	Comparision Types	More at
D 30	Si-Di	=BAS 70-05W (Typ-Code/Stempel/marking)	35(2mm)	Tho	→BAS 70-05W	data
D 30 A1...A5	Si-P	Min, Uni, 25V, 0,09W	36b	Gen	BC 202..203	data
D30B	UNI	Vs:±24V Vu:84dB Vo:±18.5V Vi0:0.5mV	15Y	Ilc		data pinout
D 30 SC4M	Si-Di	Schottky, Dual, 40V, 30A(Tc=112°)	18p	Shi		data
D 30 VC60	Si-Di	Dual, 600V, 30A(Tc=124°)	70/3Pin	Shi	-	data
D 30 VTA160	Si-Br	3-Phase Br, 1600V, 30A(Tc=105°)		Shi		data
D31	UNI	Vs:±24V Vu:86dB Vo:±17.5V Vi0:5mV	7Y	Ilc		data pinout
D 31	Si-Di	=BAS 70-04W (Typ-Code/Stempel/marking)	35(2mm)	Tho	→BAS 70-04W	data
D 31 B	Si-P	→D40D		Gen		data
D 32	Diac	Ub=28...36V, Ib<0,3mA, Itsm=1A	31	Tag	A 9903	data
D32	UNI	Vs:±15V Vu:88dB Vo:±11V Vi0:5mV	7Y	Ilc		data pinout
D32-01	Si-Di	=ERD 32-01 (Typ-Code/Stempel/marking)	31a	Fjd	→ERD 32-...	data
D32-02	Si-Di	=ERD 32-02 (Typ-Code/Stempel/marking)	31a	Fjd	→ERD 32-...	data
D 32 H1...H9	Si-N	Uni, 60...100V, 0,5A, 0,5W	7a	Gen	BC 639, 2N3700...3701, 2SD667	data
D 32 K1...K2	Si-N	NF/S-Tr, 30...50V, 0,75A, 0,5W	7a	Gen	BC 337, BC 637, BC 639, 2N3700...3701, ++	data
D 32 L1...L6	Si-N-Darl	25...40V, 0,5A, >80MHz	7a	Gen	BC 517, BC 875, BC 877, BC 879, MPSA 25	data
D 32 P1...P4	Si-N	AM/FM-ZF	7a	Gen	BF 240...241, BF 254, BF 494, BF 594, ++	data
D 32 S1...S10	Si-N	NF, 30...60V, 0,1A, 0,4W	7a	Gen	BC 174, BC 182, BC 190, BC 546, ++	data
D 32 W7...W14	Si-N	Uni, 100...120V, 0,1A, 0,4W, >75MHz	7a	Gen	2SC2240, 2SC2459, 2SC3245	data
D33	UNI	Vs:±15V Vu:88dB Vo:±11V Vi0:5mV	7Y	Ilc		data pinout
D 33 D1...D6	Si-N	Uni, 40...50V, 0,5A, 0,4W	7c	Gen	BC 337, BC 637, BC 639, 2N3700...01, ++	data
D 33 D21...D30(J1)	Si-N	NF-Tr, 35...70V, 0,75A, 0,5W	7c	Gen	BC 337, BC 637, BC 639, 2N3700...01, ++	data
D 33 K1...K3	Si-N	Uni, 50...80V, 1A, 0,33W	7c	Gen	BC 637, BC 639, 2N3700...01, 2SD667, ++	data
D34	POWER-OP	Vs:±24V Vu:86dB Vo:±17.5V Vi0:5mV	8Y	Ilc		data pinout
D 34 C1...C6	Si-P-Darl	25...40V, 0,5A, >80MHz	7a	Gen	BC 516, BC 876, BC 878, BC 880, MPSA 75	data
D 34 J1...J9	Si-P	Uni, 60...100V, 0,5A, 0,5W, >60MHz	7a	Gen	BC 640, 2SA965, 2SB647	data
D35	POWER-OP	Vs:±28V Vu:86dB Vo:±17.5V Vi0:5mV	9Y	Ilc		data pinout
D38-04	Si-Di	=ERD 38-04 (Typ-Code/Stempel/marking)	31a	Fjd	→ERD 38-...	data
D38-05	Si-Di	=ERD 38-05 (Typ-Code/Stempel/marking)	31a	Fjd	→ERD 38-...	data
D38-06	Si-Di	=ERD 38-06 (Typ-Code/Stempel/marking)	31a	Fjd	→ERD 38-...	data
D 38 H1...H9	Si-N	NF-Tr, 60...100V, 0,5A, 0,5A, >80MHz	7e	Gen	BC 637, BC 639, 2N3700...01, 2SD667, ++	data
D 38 L1...L6	Si-N-Darl	25...40V, 0,5A, >80MHz	7e	Gen	BC 517, BC 875, BC 877, BC 879, MPSA 25	data
D 38 S1...S10	Si-N	Uni, 30...60V, 0,1A, 0,4W, 100MHz	7e	Gen	BC 174, BC 182, BC 190, BC 546, ++	data
D 38 V1...V3	Si-N	S, Vid, 200...300V, 0,1A, 0,5W, >50MHz	7e	Gen	BF 298...299, BFR 88...89, BFT 58...59, ++	data
D 38 W7...W14	Si-N	NF/S, 100...120V, 0,1A, 0,4W, >75MHz	7e	Gen	2SC2240, 2SC2459, 2SC3245	data
D 39 C1...C6	Si-P-Darl	25...40V, 0,5A, >80MHz	7e	Gen	BC 516, BC 876, BC 878, BC 880, MPSA 75	data
D 39 J1...J9	Si-P	NF-Tr, 60...100V, 0,5A, 0,5W, >60MHz	7e	Gen	BC 638, BC 640, 2SA965, 2SB647	data
D 40 C1...C8(U)	Si-N-Darl	NF/S-L, 30...50V, 0,5A, 6,25W, 75MHz	13g§	Gen	MPSU 45	data
D 40 D1...D14(U)	Si-N	NF/S-L, 45...90V, 1A, 6,25W, 200MHz	13g§	Gen	BD 419, BD 519, BD 529	data
D 40 E1...E7	Si-N	NF/S-L, 45...90V, 2A, 8W, 230MHz	13g§	Gen	BD 525, BD 527, BD 529	data
D 40 K1...K4	Si-N-Darl	NF/S-L, 30...50V, 2A, 10W, 75MHz, B>10k	13g§	Gen	MPSU 45	data
D 40 N1...N5(U)	Si-N	S/Vid, 250...300V, 0,1A, 6,25W, 80MHz	13g§	Gen	BF 382, BF 461...462, MPSU 10, 2SC1758,++	data
D 40 P1...P5(U)	Si-N	Vid, 200...300V, 0,5A, 6,25W, >50MHz	13g§	Gen	BF 382, BF 461...462, MPSU 10	data
D 40 V1...V6	Si-N	S/Vid, 300...400V, 0,1A, >50MHz	13g§	Gen	BF 462, (BF 881)[5]	data
D41	HI-VOLT	Vs:±150V Vu:100dB Vo:±125V Vi0:5mV	6Y	Ilc		data pinout
D 41 D1...D14(U)	Si-P	NF/S-L, 45...90V, 1A, 6,25W, 150MHz	13g§	Gen	BD 420, BD 520, BD 530	data
D 41 E1...E7	Si-P	NF/S-L, 45...90V, 2A, 8W, 175MHz	13g§	Gen	BD 526, BD 528, BD 530	data
D 41 K1...K4	Si-P-Darl	NF/S-L, 30...50V, 2A, 10W, 100MHz,B>10k	13g§	Gen,Mot,Nsc	MPSU 95	data
D42	POWER-OP	Vs:±24V Vu:86dB Vo:±18V Vi0:5mV		Ilc		data pinout
D 42		=2SD2228-D42(Typ-Code/Stempel/marking)	35(2mm)	Nec	→2SD2228	data
D 42 C1...C12(J,U)	Si-N	NF/S-L, 40...90V, 3A, 12,5W, 50MHz	13j§	Gen	(BDV 10...12)[6]	data
D 42 D1...D6	Si-N-Darl	NF/S-L, 50...90V, 4A, 12W, 50MHz	13g	Gen	(BD 679, BD 681, BD 779, 2N6039)[5]	data
D 42 R1...R2	Si-N	Vid-L, 250...300V, 1A, 55MHz	13j§	Gen	(BF 468, BF 668)[5]	data
D 42 T1...T8	Si-N	S/Vid-L, 300...400V, 2A, 15W, 45MHz	13g§, 13j§	Gen	(BUV 93...95)[5]	data
D 43	Si-N	=2SD2228-D43(Typ-Code/Stempel/marking)	35(2mm)	Nec	→2SD2228	data
D 43 C1...C12(J,U)	Si-P	NF/S-L, 40...90V, 3A, 12,5W, 40MHz	13j§	Gen	(BD 242(A...F), 2SA1293)[4]	data
D 43 D1...D6	Si-P-Darl	NF/S-L, 50...90V, 4A, 12W, 50MHz, B=10k	13g	Gen	(BD 680, BD 682, BD 780, 2N6036)[5]	data
D 44	Si-N	=2SD2228-D44(Typ-Code/Stempel/marking)	35(2mm)	Nec	→2SD2228	data
D 44 C1...C12	Si-N	NF/S-L, 40...90V, 4A, 30W, 50MHz	17j§	Gen	BD 243(A...C), BD 539(A...C), BD 953, ++	data
D 44 D1...D6	Si-N-Darl+Di	NF/S-L, 50...90V, 6A, 30W, 50MHz, B>2k	17j§	Gen	BDT 63(A...C), BDX 33(A...D)	data
D 44 E1...E3	Si-N-Darl	NF/S-L, 40...80V, 10A, 50W, B>1000	17j§	Gen	BDT 65(A...C), BDW 93(A...C)	data
D 44 H1...H11	Si-N	NF/S-L, 30...80V, 10A, 50W, 50MHz	17j§	Gen,Mot,++	BD 711, BD 809, 2SC2527	data
D 44 Q1...Q5	Si-N	S-L, 200...300V, 4A, 31W, 50MHz	17j§	Gen	BU 406...409, MJE 51T...53T, TIP 75, ++	data
D 44 R1...R8	Si-N	NF/S-L, 400...500V, 1A, 31W, 40MHz	17j§	Gen	BUW 40A...B, BUX 84...85, TIP 50	data
D 44 T1...T8	Si-N	S-L, 300...400V, 2A, 45MHz	17j§	Gen	BUW 40(A...B), BUX 84...85, TIP 49...59, ++	data
D 44 TD3...TD5	Si-N	S-L, lo-sat, 400...600V, 2A, 50W	17j§	Gen	2SC3148	data
D 44 TE3...TE5	Si-N	NF/S-L, lo-sat, 400...600V, 4A, 75W	17j§	Gen	2SC3148	data
D 44 TQ1...TQ2	Si-N	S-L, SMPS, 650...750V, 12A, 100W	17j§	Gen	BUV 66(A)	data
D 44 VH1...VH10	Si-N	S-L, lo-sat, 50...100V, 15A, 83W, 50MHz	17j§	Gen		data
D 44 VM1...VM10	Si-N	S-L, lo-sat, 50...100V, 8A, 50W, 50MHz	17j§	Gen		data
D 45	Si-Di	≈BA 170	31a		→BA 170	data
D 45	Si-N	=2SD2228-D45(Typ-Code/Stempel/marking)	35(2mm)	Nec	→2SD2228	data
D 45 C1...C12	Si-P	NF/S-L, 40...90V, 4A, 30W, 40MHz	17j§	Gen	BD 244(A...C), BD 540(A...C), BD 954, ++	data
D 45 D1...D6	Si-P-Darl+Di	NF/S-L, 50...90V, 6A, 30W, 50MHz, B>2k	17j§	Gen	BDT 62(A...C), BDX 34(A...D)	data

Type	Device	Short Description	Fig.	Manu	Comparision Types	More at
D 45 E1...E3	Si-P-Darl	NF/S-L, 40...80V, 10A, 50W, B>1000	17j§	Gen	BDT 64(A...C), BDW 94(A...C)	data
D 45 H1...H12	Si-P	NF/S-L, 30...80V, 10A, 50W, 40MHz	17j§	Gen	BD 712, BD 810, 2SA1077,++	data
D 45 VH1...VH10	Si-P	S-L, lo-sat, 40...100V, 15A, 83W, 50MHz	17j§	Gen	-	data
D 45 VM1...VM10	Si-P	S-L, lo-sat, 50...100V, 8A, 50W, 50MHz	17j§	Gen	-	data
D 46	Si-Di	=BAT 46W (Typ-Code/Stempel/marking)	35(2mm)	Tho	→BAT 46W	data
D 46 TQ1...TQ2	Si-N	S-L, 650...750V, 12A, 110W	18j§	Gen	BUV 48(A...C), BUW 13(A), BUW 133(A), ++	data
D 47	Z-Di	=BZX 84/C3V9(Typ-Code/Stempel/marking)	35	Tho	→BZX 84	data
D 48	Z-Di	=BZX 84/C4V3(Typ-Code/Stempel/marking)	35	Tho	→BZX 84	data
D 49	Si-Di	=BAY 84 (Typ-Code/Stempel/marking)	35	Tho	→BAY 84	data
D 51	Si-N	=2SD780A-D51(Typ-Code/Stempel/marking)	35	Nec	→2SD780A	data
D 52	Si-Di	≈BY 133	31a		→BY 133	
D 52	Si-N	=2SD780A-D52(Typ-Code/Stempel/marking)	35	Nec	→2SD780A	data
D 52	Thy	=SoBRY55/30 (Typ-Code/Stempel/marking)	35	Tho	→SoBRY55/30	data
D 53	Si-N	=2SD780A-D53(Typ-Code/Stempel/marking)	35	Nec	→2SD780A	data
D 53	Si-Di	=BAY 85 (Typ-Code/Stempel/marking)	35	Tho	→BAY 85	data
D 54	Z-Di	=BZX 84/C3V3(Typ-Code/Stempel/marking)	35	Tho	→BZX 84	data
D 54	Si-N	=2SD780A-D54(Typ-Code/Stempel/marking)	35	Nec	→2SD780A	data
D 54 A7D	Si-N-Darl+Di	Hammer-Tr, 100/100V, 7A, 30W, B>2k	17c	Gen	BD 647F, 2SD1415, 2SD1791, 2SD1830, ++	data
D 54 D6D	Si-N-Darl+Di	S-L, 600/400V, 6A, 25W, B>600	17c	Gen	2SD1409	data
D 54 FY7D	Si-N-Darl+Di	Hammer-Tr, 80/80V, 7A, 30W, B>2000	17c	Gen	BD 645F, 2SD1415, 2SD1791, 2SD1830, ++	data
D 54 H6D	Si-N-Darl	S-L, 300/250V, 6A, 25W, B>2000	17c	Gen	2SD1410, 2SD2017	data
D 55	Si-N	=2SD780A-D55(Typ-Code/Stempel/marking)	35	Nec	→2SD780A	data
D 55 A7D	Si-N-Darl+Di	Hammer-Tr, 100/100V, 7A, 30W, B>2000	17c	Gen	BD 647F, 2SD1415, 2SD1791, 2SD1830, ++	data
D 55 FY7D	Si-N-Dan+Di	Hammer-Tr, 80/80V, 7A, 30W, B>2000	17c	Gen	BD 645F, 2SD1415, 2SD1791, 2SD1830, ++	data
D 56 W1...W2	Si-N	S-L, 1400V, 5A, 78W	23a§	Gen	BU 208(A), BU 508(A), 2SC2928, 2SD820,++	data
D 58	Si-Di	=FLLD 258 (Typ-Code/Stempel/marking)	35	Ztx	→FLLD 258	data
D 62	Si-Di	=BAT 53 (Typ-Code/Stempel/marking)	35	Tho	→BAT 53	data
D 63	Si-Di	=FLLD 263 (Typ-Code/Stempel/marking)	35	Ztx	→FLLD 263	data
D 64	Z-Di	=BZX 84/C3V6(Typ-Code/Stempel/marking)	35	Tho	→BZX 84	data
D 64 DS5...DS7	Si-N-Darl+Di	S-L, 500...700V, 20A, 125W, B>100	23a§	Gen	BUT 15, MJ 10009, MJ 10024...25	data
D 64 DV5...DV7	Si-N-Darl+Di	S-L, 500...700V, 50A, 185W, B>100	23a§	Gen	BUT 34, MJ 10016	data
D 64 ES5...ES7	Si-N-Darl+Di	=D64DS: integr. Speedup-Diode	23a§	Gen	-	data
D 64 EV5...EV7	Si-N-Darl+Di	=D64DV: integr. Speedup-Diode	23a§	Gen	-	data
D 64 TS3...TS5	Si-N	S-L, 195W	23a§	Gen	-	data
D 64 VS3...VS5	Si-N	S-L, 450...550V, 15A, 195W	23a§	Gen	BUS 13(A), BUS 23(A...B), BUW 45...46	data
D 65	Thy	=SoBRY55/30R(Typ-Code/Stempel/marking)	35	Tho	→SoBRY55/30R	data
D 66 DS5...DS7	Si-N-Darl+Di	=D64DS5...7: Iso, 62,5W	66b	Gen	-	data
D 66 DV5...DV7	Si-N-Darl+Di	=D64DV5...7: Iso, 125W	66b	Gen	-	data
D 66 DW1...DW3	Si-N-Darl+Di	S-L, 800...900V, 50A, 167W	66b	Gen	-	data
D 66 ES5...ES7	Si-N-Darl+Di	=D64ES5...7: Iso, 62,5W	66b	Gen	-	data
D 66 EV5...EV7	Si-N-Darl+Di	=D64EV5...7: Iso, 125W	66b	Gen	-	data
D 66 EW1...EW3	Si-N-Darl+Di	=D66DW1...3: integr. Speedup-Diode	66b	Gen	-	data
D 66 GV5...GV7	Si-N-Darl+Di	S-L, 600...700V, 50A, 125W, Speedup-Di.	66b	Gen	-	data
D 66	Thy	=SoBRY55/60 (Typ-Code/Stempel/marking)	35	Tho	→SoBRY55/60	data
D 67	Thy	=SoBRY55/60R(Typ-Code/Stempel/marking)	35	Tho	→SoBRY55/60R	data
D 67 DE5...DE7	Si-N-Darl+Di	S-L, 500...700V, 100A, 312W		Gen	-	data
D 67 FP5...FP7	Si-N-Darl+Di	S-L, 500...700V, 100A, 312W		Gen	-	data
D 70 F2T1	Si-N	SMD, S, lo-sat, 50V, 2A, 100/1100ns	39b	Gen	2SD1623, 2SC4409	data
D 70 G.05T1	Si-N	SMD, S, 200/150V, 0,05A	39b	Gen	BF 622, 2SC2880, 2SC4372, 2SC4080	data
D 70 Y.8T1	Si-N	SMD, NF, 35V, 0,8A	39b	Gen	BCX 54...56, 2SC3444, 2SC4539, 2SD1623,++	data
D 70 Y1.5T1	Si-N	SMD, NF, 30V, 1,5A	39b	Gen	2SC3439, 2SC4377, 2SD1766	data
D 71	Si-N	=BCW 60A (Typ-Code/Stempel/marking)	35	Val	→BCW 60A	data
D 71 F2T1	Si-P	SMD, S, lo-sat, 50V, 2A, 100/1100ns	39b	Gen	2SA1681, 2SB1123	data
D 71 G.05T1	Si-P	SMD, S, 150/150V, 0,05A	39b	Gen	BF 623, 2SA1200, 2SA1660, 2SB1046...47	data
D 71 Y.8T1	Si-P	SMD, NF, 35V, 0,8A	39b	Gen	BCX 51...53, 2SA1364, 2SB1122, 2SB1260,++	data
D 71 Y1.5T1	Si-P	SMD, NF, 30V, 1,5A	39b	Gen	2SA1213, 2SA1681, 2SB1313, 2SB1123,++	data
D 72	Si-N	=BCW 60B (Typ-Code/Stempel/marking)	35	Val	→BCW 60B	data
D 72 F5T1..2	Si-N	S-L, lo-sat, 60V, 5A, 20W	30j§	Gen	2SC3074, 2SD1803	data
D 72 FY4D1..2	Si-N-Darl+Di	S-L, 100/80V, 4A, 15W, B>2000	30j§	Gen	2SC4345, 2SD1223, 2SD1520	data
D 72 K3D1..2	Si-N-Darl+Di	S-L, 60/40V, 3A, 15W, B>2000	30j§	Gen	2SD1222...23, 2SD1520, 2SD1817	data
D 72 Y1.5D1..2	Si-N-Darl	Uni, 30V, 1,5A, 10W, B>4000	30j§	Gen	2SD1759, 2SD1980	data
D 73	Si-Di	=BAT 54W (Typ-Code/Stempel/marking)	35(2mm)	Tho	→BAT 54W	data
D 73	PIN-Di	=BA 579A (Typ-Code/Stempel/marking)	35	Tho	→BA 579A	data
D 73	Si-N	=BCW 60C (Typ-Code/Stempel/marking)	35	Val	→BCW 60C	data
D 73 F5T1..2	Si-P	S-L, lo-sat, 60/50V, 5A, 20W	30j§	Gen	2SA1244, 2SB1203...04	data
D 73 FY4D1..2	Si-P-Darl+Di	S-L, 100/80V, 4A, 15W, B>2000	30j§	Gen	2SB908, 2SB1072	data
D 73 K3D1..2	Si-P-Darl+Di	S-L, 60/40V, 3A, 15W, B>2000	30j§	Gen	2SB907, 2SB1214, 2SB1303	data
D 74	PIN-Di	=BA 579C (Typ-Code/Stempel/marking)	35	Tho	→BA 579C	data
D 74	Si-Di	=BAT 54AW (Typ-Code/Stempel/marking)	35(2mm)	Tho	→BAT 54AW	data
D 74	Si-N	=BCW 60D (Typ-Code/Stempel/marking)	35	Val	→BCW 60D	data
D 74......D 78...	Si-N/P	General Electric: Transistor-Arrays		Gen		data
D 75	PIN-Di	=BA 579S (Typ-Code/Stempel/marking)	35	Tho	→BA 579S	data
D 76	Si-Di	=BAR 18 (Typ-Code/Stempel/marking)	35	Tho	→BAR 18	data

Type	Device	Short Description	Fig.	Manu	Comparision Types	More at
77	Si-Di	=BAT 54CW (Typ-Code/Stempel/marking)	35(2mm)	Tho	→BAT 54CW	data
77	Si-N	=BCF 32R (Typ-Code/Stempel/marking)	35	Val	→BCF 32R	data
78	Si-Di	≈1N4148	31a		→1N4148	
78	Si-Di	=BAT 54SW (Typ-Code/Stempel/marking)	35(2mm)	Tho	→BAT 54SW	data
81	Si-Di	=BAR 43A (Typ-Code/Stempel/marking)	35	Tho	→BAR 43A	data
81	Si-N	=BCF 33R (Typ-Code/Stempel/marking)	35	Val	→BCF 33R	data
82	Si-Di	=BAR 43C (Typ-Code/Stempel/marking)	35	Tho	→BAR 43C	data
83	Si-Di	=BB 503DK (Typ-Code/Stempel/marking)	35	Sie,Tho	→BB 503DK	data
84	Si-Di	=BAT 54A (Typ-Code/Stempel/marking)	35	Tho	→BAT 54A	data
84	Si-Di	=BAT 18DK (Typ-Code/Stempel/marking)	35	Tho	→BAT 18DK	data
85	Si-Di	=BAT 17DS (Typ-Code/Stempel/marking)	35	Tho	→BAT 17DS	data
86	Si-Di	=BAT 54 (Typ-Code/Stempel/marking)	35	Tho	→BAT 54	data
87	Si-Di	=BAT 54C (Typ-Code/Stempel/marking)	35	Tho	→BAT 54C	data
88	Si-Di	=BAT 54S (Typ-Code/Stempel/marking)	35	Tho	→BAT 54S	data
89	Si-Di	=BAT 74 (Typ-Code/Stempel/marking)	44	Tho	→BAT 74	data
91	Si-N	=BCV 61 (Typ-Code/Stempel/marking)	44	Val	→BCV 61	data
92	Si-N	=BCV 61A (Typ-Code/Stempel/marking)	44	Val	→BCV 61	data
93	Si-N	=BCV 61B (Typ-Code/Stempel/marking)	44	Val	→BCV 61	data
94	Si-Di	=BAR 42 (Typ-Code/Stempel/marking)	35	Tho	→BAR 42	data
94	Si-N	=BCV 61C (Typ-Code/Stempel/marking)	44	Val	→BCV 61	data
94	Si-Di	=ZUMD 70-04 (Typ-Code/Stempel/marking)	35(2mm)	Ztx	→ZUMD 70-04	data
95	Si-Di	=BAR 43 (Typ-Code/Stempel/marking)	35	Tho	→BAR 43	data
95	Si-N	=BCV 63 (Typ-Code/Stempel/marking)	44	Phi	→BCV 63	data
95	Si-Di	=ZUMD 70-05 (Typ-Code/Stempel/marking)	35(2mm)	Ztx	→ZUMD 70-05	data
96	Si-N	=BCV 63B (Typ-Code/Stempel/marking)	44	Phi	→BCV 63	data
96	Si-Di	=BAS 70-04 (Typ-Code/Stempel/marking)	35	Tho	→BAS 70-04	data
97	Si-Di	=BAS 70-05 (Typ-Code/Stempel/marking)	35	Tho	→BAS 70-05	data
98	Si-Di	=BAS 70-06 (Typ-Code/Stempel/marking)	35	Tho	→BAS 70-06	data
99	Si-Di	=BAS 70-07 (Typ-Code/Stempel/marking)	44	Tho	→BAS 70-07	data
D 100 C,D	TTL-Logic	=... 7400 (TTL)	14-DIC,DIP	Hfo	... 7400... (TTL)	data
D 103 C,D	TTL-Logic	=... 7403 (TTL)	14-DIC,DIP	Hfo	... 7403... (TTL)	data
D 104 C,D	TTL-Logic	=... 7404 (TTL)	14-DIC,DIP	Hfo	... 7404... (TTL)	data
D 108 C,D	TTL-Logic	=... 7408 (TTL)	14-DIC,DIP	Hfo	... 7408... (TTL)	data
D 110 C,D	TTL-Logic	=... 7410 (TTL)	14-DIC,DIP	Hfo	... 7410... (TTL)	data
D 118	Si-Di	≈1N4148	31a		→1N4148	
D 120 C,D	TTL-Logic	=... 7420 (TTL)	14-DIC,DIP	Hfo	... 7420... (TTL)	data
D 120 LC40(B)	Si-Di	S-L, Dual, 400V, 120A(Tc=60°), <100ns B: 120A(Tc=95°)	(Module)	Shi	-	data
D 120 SC3M	Si-Di	Schottky, Dual, 30V, 120A(Tc=99°)	(Module)	Shi	-	data
D 120 SC4M	Si-Di	Schottky, Dual, 40V, 120A(Tc=90°)	(Module)	Shi	-	data
D 120 SC6M	Si-Di	Schottky, Dual, 60V, 120A(Tc=85°)	(Module)	Shi	-	data
D 120 SC7M	Si-Di	Schottky, Dual, 70V, 120A(Tc=84°)	(Module)	Shi	-	data
D 121 C,D	TTL-Logic	=... 74121 (TTL)	14-DIC,DIP	Hfo	... 74121... (TTL)	data
D 122 D	LIN-IC	2-Kanal-Leseverst./2-channel read amp.	16-DIP	Hfo	SN 7522N	
D 123 D	LIN-IC	2-Kanal-Leseverst./2-channel read amp.	16-DIP	Hfo	SN 7523N	
D 126 C,D	TTL-Logic	=... 7426 (TTL)	14-DIC,DIP	Hfo	... 7426... (TTL)	data
D 129	Si-Di	≈BA 127	31a		→BA 127	
D 130 C,D	TTL-Logic	=... 7430 (TTL)	14-DIC,DIP	Hfo	... 7430... (TTL)	data
D 140 C,D	TTL-Logic	=... 7440 (TTL)	14-DIC,DIP	Hfo	... 7440... (TTL)	data
D 146 C,D	TTL-Logic	=... 7446 (TTL)	16-DIC,DIP	Hfo	... 7446... (TTL)	data
D 147	C-Di	≈BA 136	31a		→BA 136	
D 147 C,D	TTL-Logic	=... 7447 (TTL)	16-DIC,DIP	Hfo	... 7447... (TTL)	data
D 150 C,D	TTL-Logic	=... 7450 (TTL)	14-DIC,DIP	Hfo	... 7450... (TTL)	data
D 151 C,D	TTL-Logic	=... 7451 (TTL)	14-DIC,DIP	Hfo	... 7451... (TTL)	data
D 153 C,D	TTL-Logic	=... 7453 (TTL)	14-DIC,DIP	Hfo	.. 7453... (TTL)	data
D 154 C,D	TTL-Logic	=... 7454 (TTL)	14-DIC,DIP	Hfo	... 7454... (TTL)	data
D 160 C,D	TTL-Logic	=... 7460 (TTL)	14-DIC,DIP	Hfo	... 7460... (TTL)	data
D 172 C,D	TTL-Logic,	=... 7472 (TTL)	14-DIC,DIP	Hfo	... 7472... (TTL)	data
D 174 C,D	TTL-Logic	=... 7474 (TTL)	14-DIC,DIP	Hfo	... 7474... (TTL)	data
D 175 C,D	TTL-Logic	=... 7475 (TTL)	16-DIC,DIP	Hfo	... 7475... (TTL)	data
D 180 SC3M	Si-Di	Schottky, Triple, 30V, 180A(Tc=94°)	(Module)	Shi	-	data
D 180 SC4M	Si-Di	Schottky, Triple, 40V, 180A(Tc=83°)	(Module)	Shi	-	data
D 180 SC6M	Si-Di	Schottky, Triple, 60V, 180A(Tc=78°)	(Module)	Shi	-	data
D 180 SC7M	Si-Di	Schottky, Triple, 70V, 180A(Tc=77°)	(Module)	Shi	-	data
D 181 C,D	TTL-Logic	=... 7481 (TTL)	14-DIC,DIP	Hfo	... 7481... (TTL)	data
D 191 C,D	TTL-Logic	=... 7491 (TTL)	14-DIC,DIP	Hfo	... 7491... (TTL)	data
D 192 C,D	TTL-Logic	=... 74192 (TTL)	16-DIC,DIP	Hfo	... 74192... (TTL)	data
D 193 C,D	TTL-Logic	=... 74193 (TTL)	16-DIC,DIP	Hfo	... 74193... (TTL)	data
D 195 C,D	TTL-Logic	=... 7495 (TTL)	14-DIC,DIP	Hfo	... 7495... (TTL)	data
D 200 C,D	TTL-Logic	=... 74H00 (TTL)	14-DIC,DIP	Hfo	... 74H00... (TTL)	data
D 200 LC40B	Si-Di	S-L, Dual, 400V, 200A(Tc=52°), <150ns	(Module)	Shi	-	data
D 201 C,D	TTL-Logic	=... 74H01 (TTL)	14-DIC,DIP	Hfo	... 74H01... (TTL)	data
D 204 C,D	TTL-Logic	=... 74H04 (TTL)	14-DIC,DIP	Hfo	... 74H04... (TTL)	data
D 210 C,D	TTL-Logic	=... 74H10 (TTL)	14-DIC,DIP	Hfo	... 74H10... (TTL)	data
D 220 C,D	TTL-Logic	=... 74H20 (TTL)	14-DIC,DIP	Hfo	... 74H20... (TTL)	data
D 228	Si-Di	≈BA 159	31a		→BA 159	

Type	Device	Short Description	Fig.	Manu	Comparision Types	More at
D 230 C,D	TTL-Logic	=... 74H30 (TTL)	14-DIC,DIP	Hfo	... 74H30... (TTL)	data
D 232	Si-Di	≈BA 159	31a		→BA 159	
D 240 C,D	TTL-Logic	=... 74H40 (TTL)	14-DIC,DIP	Hfo	... 74H40... (TTL)	data
D 240 LC40	Si-Di	S-L, Dual, 400V, 240A(Tc=77°), <150ns	(Module)	Shi	-	data
D 240 SC3M,MH	Si-Di	Schottky, Dual, 30V, 240A(Tc=90°)	(Module)	Shi	-	data
D 240 SC4M,MH	Si-Di	Schottky, Dual, 40V, 240A(Tc=77°)	(Module)	Shi	-	data
D 240 SC6M,MH	Si-Di	Schottky, Dual, 60V, 240A(Tc=71°)	(Module)	Shi	-	data
D 240 SC7M	Si-Di	Schottky, Dual, 70V, 240A(Tc=70°)	(Module)	Shi	-	data
D 251 C,D	TTL-Logic	=.. 74H51 (TTL)	14-DIC,DIP	Hfo	... 74H51... (TTL)	data
D 254 C,D	TTL-Logic	=.. 74H54 (TTL)	14-DIC,DIP	Hfo	74H54... (TTL)	data
D 274 C,D	TTL-Logic	=.. 74H74 (TTL)	14-DIC,DIP	Hfo	74H74... (TTL)	data
D 335	Ge-Di	≈AA 119	31a		→AA 119	
D 336	Si-Di	≈1N4148	31a		→1N4148	
D 345 D	TTL-Logic	BCD→7-Segm.-Decoder, 0...+70°	16-DIP	Hfo	-	
D 346 D	TTL-Logic	BCD→7-Segm.-Decoder, 0...+70°	16-DIP	Hfo	-	
D 347 D	TTL-Logic	BCD→7-Segm.-Decoder, 0...+70°	16-DIP	Hfo	-	
D 348 D	TTL-Logic	BCD→7-Segm.-Decoder, 0...+70°	16-DIP	Hfo	74LS247 (TTL)	
D 351 D	TTL-Logic	Teiler/divider, 0...+75°	14-DIP	Hfo	74LS247 (TTL)	
D 352	Si-Di	≈BA 282	31a		→BA 282	
D 355 D	TTL-Logic	Timer	18-DIP	Hfo	-	
D 356 D	TTL-Logic	Timer	18-DIP	Hfo	-	
D 360 SC3M	Si-Di	Schottky, Triple, 30V, 360A(Tc=81°)	(Module)	Shi	-	data
D 360 SC4M	Si-Di	Schottky, Triple, 40V, 360A(Tc=64°)	(Module)	Shi	-	data
D 360 SC5M	Si-Di	Schottky, Triple, 50V, 360A(Tc=63°)	(Module)	Shi	-	data
D 360 SC6M	Si-Di	Schottky, Triple, 60V, 360A(Tc=58°)	(Module)	Shi	-	data
D 360 SC7M	Si-Di	Schottky, Triple, 70V, 360A(Tc=57°)	(Module)	Shi	-	data
D 377	Si-Di	≈FDH 600	31a		→FDH 600	
D 380	Se-Di	Dual-GI			M40C4	
D 394 D	LIN-IC	Schrittmotorstrg./stepper motor ctrl.	18-DIP	Hfo	-	
D 395 D	TTL-Logic	Schrittmotorstrg./stepper motor ctrl.	18-DIP	Hfo	-	
D 410 D	TTL-Logic	3x Treiber/driver (AND-Gates)	16-DIP	Hfo	SAA 1029	
D 461 D	TTL-Logic	Dual-treiber/driver f. Mos-memory	14-DIP	Hfo	SN 75361	
D 469 ADJ	LIN-IC	4x Power Driver, hi-current	14-DIP	Six	-	pdf pinout
D 469 ADN	LIN-IC	=D 469ADJ:	20-PLCC		-	pdf pinout
D 473	Si-Di	≈BA 127	31a		→BA 127	
D 474	Si-Di	≈1N4148	31a		→1N4148	
D 475	Si-Di	≈1N4148	31a		→1N4148	
D 491 D	TTL-Logic	4-Segm.-Treiber/driver	14-DIP	Hfo		
D 492 D	TTL-Logic	6x Digit-Treiber/driver	14-DIP	Hfo	SN 75492	
D610P	LED	rd+rd/gr 2.5x5mm I>0.15mcd Vf:<18V		Tem		data pinout
D 716 X	LIN-IC	Thermo-printer-control		Hfo	-	
D 718 D	LIN-IC	16 Bit Ser./Par.-Converter		Hfo	(UAA 2022)	
D 764 A	Si-Di	≈BAV 45	31a		→BAV 45	
D 780C-1	NMOS-µC-IC	Z80-Family, 8 Bit, 5V, 4MHz	40-DIP	Nec	-	pdf pinout
D 797	Si-Di	≈1N4148	31a		→1N4148	
D 837(A)	Si-Di	≈1N4148	31a		→1N4148	
D 838	Si-Di	≈BA 159	31a		→BA 159	
D1160	LDR	610nm Ro:2.4MOhm Vb:2V		Hei		data pinout
D 1201 A	Si-Di	=1N4002	31a	Rca	→1N4002	data
D 1201 B	Si-Di	=1N4003	31a	Rca	→1N4003	data
D 1201 D	Si-Di	=1N4004	31a	Rca	→1N4004	data
D 1201 F	Si-Di	=1N4001	31a	Rca	→1N4001	data
D 1201 M	Si-Di	=1N4005	31a	Rca	→1N4005	data
D 1201 N	Si-Di	=1N4006	31a	Rca	→1N4006	data
D 1201 P	Si-Di	=1N4007	31a	Rca	→1N4007	data
D 1300 A...D	Si-Di	Uni, 150...525V, 0,25A A=100/150V, B=200/300V, D=400/525V	2c	Rca	BA 157...159, BA 199/.., BY 204/..., ++	data
D 2005	MOS-N-FET-e*	=FTD 2005 (Typ-Code/Stempel/marking)	8-SSMDIP	Say	→FTD 2005	data
D 2007	MOS-N-FET-e*	=FTD 2007 (Typ-Code/Stempel/marking)	8-SSMDIP	Say	→FTD 2007	data
D 2011	MOS-N-FET-e*	=FTD 2011 (Typ-Code/Stempel/marking)	8-SSMDIP	Say	→FTD 2011	data
D 2013	MOS-N-FET-e*	=FTD 2013 (Typ-Code/Stempel/marking)	8-SSMDIP	Say	→FTD 2013	data
D 2014	MOS-N-FET-e*	=FTD 2014 (Typ-Code/Stempel/marking)	8-SSMDIP	Say	→FTD 2014	data
D 2015	MOS-N-FET-e*	=FTD 2015 (Typ-Code/Stempel/marking)	8-SSMDIP	Say	→FTD 2015	data
D 2017	MOS-N-FET-e*	=FTD 2017 (Typ-Code/Stempel/marking)	8-SSMDIP	Say	→FTD 2017	data
D 2019	MOS-N-FET-e*	=FTD 2019 (Typ-Code/Stempel/marking)	8-SSMDIP	Say	→FTD 2019	data
D 2022	MOS-N-FET-e*	=FTD 2022 (Typ-Code/Stempel/marking)	8-SSMDIP	Say	→FTD 2022	data
D 2101(S)	Si-Di	TV-Clamp-Diode, 600V, 1A, <700ns	34a§	Rca	BYX 55/600, MR 816..818, RGP 10J...M, ++	data
D 2103 S,SF	Si-Di	TV-Damper-Diode, 750V, 3A, <500ns	34b&	Rca	BY 399, RGP 30K..M, (BY 229/800...1000)[6]	data
D 2201 A...N	Si-Di	GI, S, 50...800V, 1A, <500ns A=100, B=200, D=400, F=50,M=600,N=800V	31a	Rca	BY 218/..., BY 258/.., RGP 15A..M, ++	data
D-2240 A	Hybrid-IC	NF-E, ±25V, >20W(±22V/8Ω)				
D 2406 A...M	Si-Di	GL/S-L, 100...600V 6A, <350ns A=100, B=200, C=300, D=400,F=50,M=600V	32a§	Rca	BYX 50/...	data
D 2406 A-R..M-R	Si-Di	=D 2406A...M:	32b&		BYX 50/...R	data
D 2412 A...M	Si-Di	GI/S-L, 100...600V, 12A, <350ns	32a§	Rca	BYT 61/..., BYX 62/..., BYV 24/.., ++	data
D 2412 A-R..M-R	Si-Di	=D 2412A..M:	32b&		BYT 61/...R, BYX 62/..R, BYV 24/..R, ++	data
D 2520 A...M	Si-Di	GI/S-L, 100...600V, 20A, <350ns	32a§	Rca	BYT 65/..., BYX 63/..., BYX 64/..., ++	data

Type	Device	Short Description	Fig.	Manu	Comparision Types	More at
D 2520 A-R...M-R	Si-Di	=D 2520A...M:	32b&		BYT 65/...R, BYX 63/...R, BYX 64/...R, ++	data
D 2540 A...M	Si-Di	Gl/S-L, 100...600, 40A, <350ns	32a§	Rca	BYW 25/..., MR 861...866, MR 871...876	data
D 2540 A-R...M-R	Si-Di	=D 2540A...M:	32b&		BYW 25/...R, MR 861R...866R, MR 871R...876R	data
D 2600 EF,M	Si-Di	TV-Clamp-Di, 550...600V, 0,5A, <700ns EF=550V, M=600V	31a	Rca	BY 407, BY 208/600, BY 201/6, RGP 10J,++	data
D 2601 A..N	Si-Di	TV-Damper-Di, 100...800V, 1A, <500ns A=100, B=200, D=400, DF=450, E=500V, EF=550. M=600, N=800V	31a	Rca	BY 258/..., BYV 12...16, RGP 15B...M, ++	data
D-2950 A	Hybrid-IC	NF-E, 50V, >20W(44V/8Ω)			-	
D 3202 U	Diac	Ub=25...40V, Ib<25µA, Itsm=2A	31	Rca	1N5760, N 413L	data
D 3202 Y	Diac	=D 3202 U: Ub=29...35V	31		1N5761, N 413M, BR 100, DO 201YR	data
D 3702	LIN-IC	Telecom, 5x Relay Driver, 65V Output	-DIC	Pls	-	
D 4042	NMOS-IC	Telecom, Dual 1008 Bit Shift Register	14-DIC	Pls	-	
D 4803 DC	LIN-IC	8 Bit Treiber/driver	16-DIP	Hfo	-	
D-4902	Hybrid-Z-IC	Z-IC, +5V, 1,5A			-	
D-4903	Hybrid-Z-IC	Z-IC, +12V, 1,8A			-	
D-4904	Hybrid-Z-IC	Z-IC, +24V, 1,5A			-	
D 6221 VC	LIN-IC	4x Darl.-Schalter/switch, 50V, 1,8A	15-SQL	Hfo	L 6221N	
D9960	LDR	610nm Ro:4.2MOhm Vb:2V		Hei		data pinout
D-Q	Si-P	=2PB1219AQ (Typ-Code/Stempel/marking)	35(2mm)	Phi	→2PB1219A	data
D-R	Si-P	=2PB1219AR (Typ-Code/Stempel/marking)	35(2mm)	Phi	→2PB1219A	data
D-S	Si-P	=2PB1219AS (Typ-Code/Stempel/marking)	35(2mm)	Phi	→2PB1219A	data

DA...DM

Type	Device	Short Description	Fig.	Manu	Comparision Types	More at
DA	Si-N	=2SC4919 (Typ-Code/Stempel/marking)	35(1,6mm)	Say	→2SC4919	data
DA	Si-N	=2SD1418-DA (Typ-Code/Stempel/marking)	39	Hit	→2SD1418	data
DA	Si-N	=2SD1618 (Typ-Code/Stempel/marking)	39	Say	→2SD1618	data
DA	Si-N	=2SD1664 (Typ-Code/Stempel/marking)	39	Rhm	→2SD1664	data
DA	Si-P	=BCW 67A (Typ-Code/Stempel/marking)	35	Fer, Sie	→BCW 67A	data
DA	Si-N	=BF 622 (Typ-Code/Stempel/marking)	39	Phi,Sie,Val	→BF 622	data
DA	Si-N	=BF 722 (Typ-Code/Stempel/marking)	48	Phi	→BF 722	data
DA	Z-Di	=MMSZ 4708 (Typ-Code/marking)	71(2,7mm)	Ons	→MMSZ 4708	data
DA	MOS-N-FET-e*	=SSM 3K03FE (Typ-Code/Stempel/marking	35(1,6mm)	Tos	→SSM 3K03FE	data
DA	C-Di	=ZMV 829 (Typ-Code/Stempel/marking)	71(1,7mm)	Ztx	→ZMV 829	data
DA	MOS-N-FET-e	=µPA502T (Typ-Code/Stempel/marking)	45	Nec	→µPA502T	data
DA 2	Si-Di	=SD BAX 12 (Typ-Code/Stempel/marking)	35	Tho	→SD BAX 12	data
DA 3	Diac	Ub=28...36V, Ib<0,3mA, Itsm=1A	31	Tho	A 9903, D 32	data
DA 3 N	Si-Di	3 Di Array, 35V, 0,1A, <4ns	4-SIP	Org	Pin-Code: AKKK	data
DA 3 P	Si-Di	=DA 3N:	4-SIP	Org	Pin-Code: KAAA	data
DA 4	Diac	Ub=35...45V, Ib<0,15mA, Itsm=2A	31	Tho	1N5762, N 413N, D 3202U	data
DA 4 N	Si-Di	4 Di Array, 35V, 0,1A, <4ns	5-SIP	Org	Pin-Code: AKKKK	data
DA 4 P	Si-Di	=DA 4N:	5-SIP	Org	Pin-Code: KAAAA	data
DA 5	Si-Di	=BAR 43S (Typ-Code/Stempel/marking)	35	Tho	→BAR 43S	data
DA 6	Z-Di	=BZV 53A (Typ-Code/Stempel/marking)	35	Tho	→BZV 53A	data
DA 6 NC	Si-Di	6 Di Array, 35V, 0,1A, <4ns	7-SIP	Org	Pin-Code: KKKAKKK	data
DA 6 PC	Si-Di	=DA 6NC:	7-SIP	Org	Pin-Code: AAAKAAA	data
DA 7	Z-Di	=BZV 53B (Typ-Code/Stempel/marking)	35	Tho	→BZV 53B	data
DA 8	Z-Di	=BZV 54A (Typ-Code/Stempel/marking)	35	Tho	→BZV 54A	data
DA 8 N	Si-Di	8 Di Array, 35V, 0,1A, <4ns	9-SIP	Org	Pin-Code: AKKKKKKKK	data
DA 8 P	Si-Di	=DA 8N:	9-SIP	Org	Pin-Code: KAAAAAAAA	data
DA 9	Z-Di	=BZV 54B (Typ-Code/Stempel/marking)	35	Tho	→BZV 54B	data
DA15	HI-OHM	Vs:±15V Vu:100dB Vo:±10V Vi0:0.75mV	7Y	Ilc		data pinout
DA17	HI-OHM	Vs:±15V Vu:100dB Vo:±10V Vi0:0.5mV	7Y	Ilc		data pinout
DA18	UNI	Vs:±15V Vu:92dB Vo:±11V f:1.5Mc	7Y	Ilc		data pinout
DA21B	HI-OHM	Vs:±15V Vu:110dB Vo:±10V Vi0:0.5mV	15Y	Ilc		data pinout
DA28	UNI	Vs:±15V Vu:110dB Vo:±10V Vi0:2mV	7Y	Ilc		data pinout
DA 103	Si-Di	SMD, 10V, 15mA, Uf=1,5...1,9V(1mA)	35c	Rhm	-	data
DA 106	Si-Di	SMD, 8V, 15mA, Uf=1,3...1,7V(1mA)	35c	Rhm	-	data
DA 106 K	Si-Di	=DA 106:	35c		-	data
DA 106 U	Si-Di	=DA 106:	35c(2mm)		-	data
DA 112	Si-Di	SMD, SS, 80V, 0,1A, <6ns	35f(2mm)	Rhm	-	data
DA 113	Si-Di	=DA 112:	35d(2mm)	Rhm	BAL 99W	data
DA 114	Si-Di	=DA 112:	35e(2mm)	Rhm	BAS 16W	data
DA 115	Si-Di	=DA 112:	35c(2mm)	Rhm	BAL 74W	data
DA 116	Si-Di	SMD, SS, 80V, 0,1A, <6ns	35f	Rhm	-	data
DA 118	Si-Di	=DA 116:	35d	Rhm	BAL 99	data
DA 119	Si-Di	=DA 116:	35c	Rhm	-	data
DA 120	Si-Di	SMD, SS, 80V, 0,1A, <6ns	35f(1,6mm)	Rhm	-	data
DA 121(TT1)	Si-Di	=DA 120:	35e(1,6mm)	Rhm, Ons	-	data
DA 122	Si-Di	=DA 120:	35c(1,6mm)	Rhm	-	data
DA 123	Si-Di	=DA 120:	35d(1,6mm)	Rhm	-	data
DA 146	Si-Di	S, TV, 100V, 0,4A, 300ns	31a	Fag	BY 204/4, BY 206, BY 406, BY 208/600, ++	data
DA 148	Si-Di	S, TV, 300V, 0,4A, 300ns	31a	Fag	BY 204/4, BY 206, BY 406, BY 208/600, ++	data
DA 148 V	Si-Di	S, TV, 300V, 0,35A, 320ns	31a	Fag	BY 204/4, BY 206, BY 406, BY 208/600, ++	data
DA 151	Si-Di	S, TV, 600V, 0,4A, 300ns	31a	Fag	BY 204/8, BY 207, BY 407, BY 208/600, ++	data
DA 203	Si-Di	Dual, Uni, 20V, 0,1A	9n1	Rhm	MA 206, MA 207	data
DA 204	Si-Di	SMD, Dual, Uni, 20V, 0,1A	35n	Rhm	-	data

Type	Device	Short Description	Fig.	Manu	Comparision Types	More a
DA 204 K	Si-Di	=DA 204:	35o		-	data
DA 204 U	Si-Di	=DA 204:	35o(2mm)			data
DA 210(S)	Si-Di	Dual, Uni, 20V, 0,1A	41n1	Rhm	MA 206, MA 207	data
DA 216	Si-Di	Dual, Uni, 20V, 0,1A	9n1	Rhm	MA 206, MA 207	data
DA 217	Si-Di	SMD, Dual, Uni, 80V, 0,1A	35o	Rhm	-	data
DA 218 S	Si-Di	Dual, Uni, 80V, 0,1A	41o1	Rhm	MC 931	data
DA 221	Si-Di	SMD, Dual, Uni, 20V, 0,1A	35o(1,6mm)	Rhm	-	data
DA 223	Si-Di	SMD, Dual, Uni, 40V, 0,1A	35o(1,6mm)	Rhm	-	data
DA 223 K	Si-Di	=DA 223:	35o		-	data
DA 226 U	Si-Di	SMD, Dual, Uni, 40V, 0,1A	35o(2mm)	Rhm	-	data
DA 227	Si-Di	SMD, Dual, Uni, 20V, 0,1A	44(2mm)	Rhm		data
DA 228 K	Si-Di	SMD, Dual, Uni, 20V, 0,1A	35o	Rhm	-	data
DA 246	Si-Di	S, TV, 200V, 1,4A, 750ns	31a	Fag	BY 296...299, BYV 36A...E, RGP 15D...M, ++	data
DA 248	Si-Di	S, TV, 300V, 1,4A, 750ns	31a	Fag	BY 297...299, BYV 36B...E, RGP 15G...M, ++	data
DA 251	Si-Di	S, TV, 500V, 1,4A, 750ns	31a	Fag	BY 298...299, BYV 36C...E, RGP 15J..M, ++	data
DA 0601	LIN-IC	6x Si-Dioden-Array, gem./comm. Cath.	7-SIP	Hit	-	
DA 0602	LIN-IC	6x Si-Dioden-Array, gem./comm. Anode	7-SIP	Hit	-	
DA 811 A	Si-Di	8 Di Array, 100V, 0,6A	9-SIP	Die	Pin-Code: AKKKKKKKK	data
DA 811 K	Si-Di	=DA 811A:	9-SIP		Pin-Code: KAAAAAAAA	data
DA 814 A	Si-Di	8 Di Array, 100V, 0,6A	9-SIP	Die	Pin-Code: AKKKKKKKK	data
DA 814 K	Si-Di	=DA 814A:	9-SIP		Pin-Code: KAAAAAAAA	data
DA 4148 A	Si-Di	8 Di Array, SS, 75V, 0,15A	9-SIP	Die	Pin-Code: AKKKKKKKK	data
DA 4148 K	Si-Di	=DA 4148A:	9-SIP		Pin-Code: KAAAAAAAA	data
DA 56-11 EWA	Logic	Dual Digit Numeric Display	DIP	Kin	-	pdf pinout
DA 8110 A	Si-Di	8 Di Array, 1000V, 0,6A	9-SIP	Die	Pin-Code: AKKKKKKKK	data
DA 8110 K	Si-Di	=DA 8110A:	9-SIP		Pin-Code: KAAAAAAAA	data
DAC-01(...)Y	D/A-IC	6 Bit, ±12...18V, Settling <3μs	14-DIC	Pmi	-	
DAC-02(...)X	D/A-IC	10 Bit, ±18V, Settling 2μs	18-DIC	Pmi		
DAC-03(...)X	D/A-IC	10 Bit, ±18V, Settling 2μs	18-DIC	Pmi		
DAC-08...F,N,DC,PC	D/A-IC	8 Bit, multiplying, hi-speed	16-DIC,DIP	Ray	NE 5007...5008, SE 5008, DAC 0800...0802	
DAC-10 BD,CD	D/A-IC	=DAC-10(F,G)D: -55...+125°C BD: ±0,05%, CD: ±0,01%	18-DIC			
DAC-10 FD,GD	D/A-IC	10 Bit, multiplying, hi-speed, 0...+70° FD: ±0,05%, GD: ±0,01%	18-DIC	Ray		
DAC-20(C)P,Q	D/A-IC	2 Digit BCD, multiplying, ±18V, 85ns	16-DIP,DIC	Pmi		
DAC-86(...)X	D/A-IC	COMDAC Companding(μ-255 Law), ±18V	18-DIC	Pmi		
DAC-88(...)X	D/A-IC	COMDAC Companding(μ-255 Law), ±18V	18-DIC	Pmi		
DAC-89(...)X	D/A-IC	COMDAC Companding(A Law), ±18V	18-DIC	Pmi		
DAC-100(...)Q	D/A-IC	10 Bit, ±6...18V, Settling <225ns	16-DIC	Pmi		
DAC-108	D/A-IC	8 Bit, ±4,5...18V, Settling 70ns	16-DIC	Pmi		
DAC-210(...)X	D/A-IC	10 Bit, ±4,5...18V, Settling 1,5μs	18-DIC	Pmi		
DAC-312(...)P,R	D/A-IC	12 Bit, multipl, ±18V, Settling 250ns	20-DIP,DIC	Pmi		
DAC-312(...)S	D/A-IC	SMD,12 Bit, multipl, ±18V,Settl. 250ns	20-MDIP			
DAC-330	D/A-IC	10 Bit, Video, >160Msps, Ecl Level	28-DIP		CX 20201A-1, CX 20202A-1	
DAC-400	GaAs-D/A-IC	12 Bit, hi-speed, Settling 5ns		Pmi		
DAC 0800 ...	CMOS-A/D-IC	8 Bit, Settling 100ns, ±18V	16-DIC,DIP	Nsc		
DAC 0801 ...	CMOS-A/D-IC	8 Bit, Settling 100ns, ±18V	16-DIC,DIP	Nsc		
DAC 0802 ...	CMOS-A/D-IC	8 Bit, Settling 100ns, ±18V	16-DIC	Nsc		
DAC 0806 D	D/A-IC	SMD, 8Bit, multiplying, hi-speed, 0...+75°	16-MDIP	Sgs		pdf pinout
DAC 0806 LCJ	D/A-IC	8 Bit, multiplying, hi-speed, 0...+75°	16-DIC	Sgs	MC 1408L6	pdf pinout
DAC 0806 LCM	D/A-IC	SMD, 8Bit, multiplying, hi-speed, 0...+75°	16-MDIP	Nsc		pdf pinout
DAC 0806 LCN	D/A-IC	8 Bit, multiplying, hi-speed, 0...+75°	16-DIP	Sgs	MC 1408P6	pdf pinout
DAC 0807 D	D/A-IC	SMD, 8 Bit, multiplying, hi-speed, 0...+75°	16-MDIP	Sgs		pdf pinout
DAC 0807 LCJ	D/A-IC	8 Bit, multiplying, hi-speed, 0...+75°	16-DIC	Sgs	MC 1408L7	pdf pinout
DAC 0807 LCM	D/A-IC	SMD, 8Bit, multiplying, hi-speed, 0...+75°	16-MDIP	Nsc		pdf pinout
DAC 0807 LCN	D/A-IC	8 Bit, multiplying, hi-speed, 0...+75°	16-DIP	Sgs	MC 1408P7	pdf pinout
DAC 0808 D	D/A-IC	SMD, 8 Bit, multiplying, hi-speed, 0...+75°	16-MDIP	Sgs	-	pdf pinout
DAC 0808 LCJ	D/A-IC	8 Bit, multiplying, hi-speed, 0...+75°	16-DIC	Sgs	MC 1408L8	
DAC 0808 LCM	D/A-IC	SMD, 8Bit, multiplying, hi-speed, 0...+75°	16-MDIP	Nsc		pdf pinout
DAC 0808 LCN	D/A-IC	8 Bit, multiplying, hi-speed, 0...+75°	16-DIP	Sgs	MC 1408P8	pdf pinout
DAC 0808 LD	D/A-IC	SMD, 8 Bit, multiplying, hi-speed, -55...+125°	16-MDIP	Sgs	MC 1508L8	pdf pinout
DAC 0808 LJ	D/A-IC	8 Bit, multiplying, hi-speed, -55...+125°	16-DIC	Sgs	MC 1508L8	pdf pinout
DAC 0830 LCD	CMOS-D/A-IC	8 Bit, 0,05% Linearity, -40...+85°	20-DIC	Nsc		
DAC 0831 LCD	CMOS-D/A-IC	8 Bit, 0,1% Linearity, -40...+85°	20-DIC	Nsc		
DAC 0832 LCD	CMOS-D/A-IC	8 Bit, 0,2% Linearity, -40...+85°	20-DIC	Nsc		
DAC 0830...832 LCN	CMOS-D/A-IC	=DAC 0830LCD...0832LCD: 0...+70°	20-DIP	Nsc		
DAC 0830...832 LD	CMOS-D/A-IC	=DAC 0830LCD...0832LCD: -55...+125°	20-DIC	Nsc		
DAC-888(...)X	D/A-IC	8 Bit, multiplying, Settling 300ns	18-DIC	Pmi		
DAC 1000 LCD	CMOS-D/A-IC	10 Bit, 0,05% Linearity, -40...+85°	24-DIC	Nsc		
DAC 1001 LCD	CMOS-D/A-IC	10 Bit, 0,1% Linearity, -40...+85°	24-DIC	Nsc		
DAC 1002 LCD	CMOS-D/A-IC	10 Bit, 0,2% Linearity, -40...+85°	24-DIC	Nsc		
DAC 1000...1002 LCN	CMOS-D/A-IC	=DAC 1000LCD...1002LCD: 0...+70°	24-DIP	Nsc		
DAC 1000...1002 LD	CMOS-D/A-IC	=DAC 1000LCD...1002LCD: -55...+125°	24-DIC	Nsc		
DAC 1006 LCD	CMOS-D/A-IC	10 Bit, 0,05% Linearity, -40...+85°	20-DIC	Nsc		
DAC 1007 LCD	CMOS-D/A-IC	10 Bit, 0,1% Linearity, -40...+85°	20-DIC	Nsc		

Type	Device	Short Description	Fig.	Manu	Comparision Types	More at
DAC 1008 LCD	CMOS-D/A-IC	10 Bit, 0,2% Linearity, -40...+85°	20-DIC	Nsc	-	
DAC 1006...1008 LCN	CMOS-D/A-IC	=DAC 1006LCD...1008LCD: 0...+70°	20-DIP	Nsc	-	
DAC 1006...1008 LD	CMOS-D/A-IC	=DAC 1006LCD...1008LCD: -55...+125°	20-DIC	Nsc	-	
DAC 1020 LCD	D/A-IC	10 Bit, 0,05% Linearity, -40...+85°	16-DIC	Nsc	-	
DAC 1021 LCD	D/A-IC	10 Bit, 0,1% Linearity, -40...+85°	16-DIC	Nsc	-	
DAC 1022 LCD	D/A-IC	10 Bit, 0,2% Linearity, -40...+85°	16-DIC	Nsc	-	
DAC 1020...1022 LCN	D/A-IC	=DAC 1020LCD...1022LCD: 0...+70°	16-DIP	Nsc	-	
DAC 1020...1022 LD	D/A-IC	=DAC 1020LCD...1022LCD: -55...+125°	16-DIC	Nsc	-	
DAC 1200 HCD	D/A-IC	12 Bit, 0,012% Linearity, -40...+85°	24-DIC	Nsc	-	
DAC 1201 HCD	D/A-IC	12 Bit, 0,049% Linearity, -40...+85°	24-DIC	Nsc	-	
DAC 1200...1201 HD	D/A-IC	=DAC 1200HCD...1201HCD: -55...+125°	24-DIC	Nsc	-	
DAC 1208 LCD	D/A-IC	12 Bit, 0,012% Linearity, -40...+85°	24-DIC	Nsc	-	
DAC 1209 LCD	D/A-IC	12 Bit, 0,024% Linearity, -40...+85°	24-DIC	Nsc	-	
DAC 1210 LCD	D/A-IC	12 Bit, 0,05% Linearity, -40...+85°	24-DIC	Nsc	-	
DAC 1218 LCD	D/A-IC	12 Bit, 0,012% Linearity, -40...+85°	18-DIC	Nsc	-	
DAC 1219 LCD	D/A-IC	12 Bit, 0,024% Linearity, -40...+85°	18-DIC	Nsc	-	
DAC 1218...1219 LD	D/A-IC	=DAC 1218LCD...1219LCD: -55...+125°	18-DIC	Nsc	-	
DAC 1220 LCD	D/A-IC	12 Bit, 0,05% Linearity, -40...+85°	18-DIC	Nsc	-	
DAC 1221 LCD	D/A-IC	12 Bit, 0,1% Linearity, -40...+85°	18-DIC	Nsc	-	
DAC 1222 LCD	D/A-IC	12 Bit, 0,2% Linearity, -40...+85°	18-DIC	Nsc	-	
DAC 1220...1222 LCN	D/A-IC	=DAC 1220LCD...1222LCD: 0...+70°	18-DIP	Nsc	-	
DAC 1220...1222 LD	D/A-IC	=DAC 1220LCD...1222LCD: -55...+125°	18-DIC	Nsc	-	
DAC 1230 LCD	D/A-IC	12 Bit, 0,012% Linearity, -40...+85°	20-DIC	Nsc	-	
DAC 1231 LCD	D/A-IC	12 Bit, 0,024% Linearity, -40...+85°	20-DIC	Nsc	-	
DAC 1232 LCD	D/A-IC	12 Bit, 0,05% Linearity, -40...+85°	20-DIC	Nsc	-	
DAC 1280 ACD	D/A-IC	12 Bit, 0,012% Linearity, 0...+70°	24-DIC	Nsc	-	
DAC 1280 CD	D/A-IC	12 Bit, 0,024% Linearity, 0...+70°	24-DIC	Nsc	-	
DAC 1280...J	D/A-IC	=DAC 1280(A): -25...+85°	24-DIC	Nsc	-	
DAC 1285 ACD,HCD	D/A-IC	12 Bit, 0,012% Linearity, -25...+85°	24-DIC	Nsc	-	
DAC 1285 AD	D/A-IC	=DAC 1285ADC,HDC: -55...+125°	24-DIC	Nsc	-	
DAC1408(A...C)DC,PC	D/A-IC	8 Bit, multiplying, hi-speed, 0...+70°	16-DIC,DIP	Fch	AM 1408..., MC 1408...	
DAC 1508 DM	D/A-IC	=DAC 1408...: -55...+125°	16-DIC	Fch	AM 1508..., MC 1508...	
DAC 2932	D/A-IC	SMD, Dual, 12 Bit DAC, +3V, 40MSPS Demultiplexer 2:1, 4x 12 Bit ser.DACs	48-QFP	Bub	-	pdf pinout
DAC-4881 BS	D/A-IC	=DAC-4881DS,FS: -55...+125°C	28-DIC		-	
DAC-4881 DS,FS	D/A-IC	12 Bit, hi-performance, 0...+70°	28-DIC	Ray	-	
DAC-4888 BD	D/A-IC	=DAC-4888DD,FD: -55...+125°C	28-DIC		-	
DAC-4888 DD,FD	D/A-IC	8 Bit, µP Interface Latches, 0...+70°	24-DIC	Ray	-	
DAC-6012(A)CN	D/A-IC	12 Bit, multiplying, hi-speed, 0...+70°	20-DIP	Ray	-	
DAC-6012(A)MD	D/A-IC	=DAC-6012(A)CN: -55...+125°	20-DIC		-	
DAC-8012(...)P,R	CMOS-D/A-IC	12 Bit, multiplying, Memory	20-DIP,DIC	Pmi	-	
DAC-8012(...)PC	CMOS-D/A-IC	12 Bit, multiplying, Memory	20-PLCC		-	
DAC-8043(...)P,Z	CMOS-D/A-IC	12 Bit, multiplying, Serial Input	8-DIP,DIC	Pmi	-	
DAC-8143(...)P,Q	CMOS-D/A-IC	12 Bit, Serial Daisy Chain	16-DIP,DIC	Pmi	-	
DAC-8143(...)S	CMOS-D/A-IC	SMD, 12 Bit, Serial Daisy Chain	16-MDIP		-	
DAC-8212(...)P,V	CMOS-D/A-IC	Dual, 12 Bit, multiplying, Buffer	24-DIP,DIC	Pmi	-	
DAC-8212(...)PC	CMOS-D/A-IC	Dual, 12 Bit, multiplying, Buffer	28-PLCC		-	
DAC 8221 (A,E,F)W	D/A-IC	Dual,12Bit buff. multiplying D/A-Conv. Single Supply, +5...15V	24-DIC	And	-	pdf pinout
DAC 8221 (F-H)P	D/A-IC	Dual,12Bit buff. multiplying D/A-Conv. Single Supply, +5...15V	24-DIP	And	-	pdf pinout
DAC 8221 HS	D/A-IC	SMD, Dual,12Bit buff. multiplying DAC Single Supply, 0...+70°C, +5...15V	24-MDIP	And	-	pdf pinout
DAC-8221(...)P,W	CMOS-D/A-IC	Dual, 12 Bit, multiplying, Buffer	24-DIP,DIC	Pmi	-	
DAC-8221(...)S	CMOS-D/A-IC	SMD, Dual, 12 Bit, multiplying, Buffer	24-MDIP		-	
DAC-8222(...)P,W	CMOS-D/A-IC	Dual, 12 Bit, multiplying, 2x Buffer	24-DIP,DIC	Pmi	-	
DAC-8222(...)TC	CMOS-D/A-IC	Dual, 12 Bit, multiplying, 2x Buffer	28-LCC		-	
DAC-8228(...)P,R	CMOS-D/A-IC	Dual, 8 Bit, +12...+15V	20-DIP,DIC	Pmi	-	
DAC-8228(...)S	CMOS-D/A-IC	SMD, Dual, 8 Bit, +12...+15V	20-MDIP		-	
DAC-8229(...)P,R	CMOS-D/A-IC	Dual, 8 Bit, multiplying	20-DIP,DIC	Pmi	-	
DAC-8229(...)S	CMOS-D/A-IC	SMD, Dual, 8 Bit, multiplying	20-MDIP		-	
DAC-8248(...)P,W	CMOS-D/A-IC	Dual, 12 Bit, Double Input Buffer	24-DIP,DIC	Pmi	-	
DAC-8248(...)S	CMOS-D/A-IC	SMD, Dual, 12 Bit, Double Input Buffer	24-MDIP		-	
DAC-8408(...)P,T	CMOS-D/A-IC	Quad, 8 Bit, multiplying, Memory	28-DIP,DIC	Pmi	-	
DAC-8408(...)PC	CMOS-D/A-IC	Quad, 8 Bit, multiplying, Memory	28-PLCC		-	
DAC-8408(...)S	CMOS-D/A-IC	SMD, Quad, 8 Bit, multiplying, Memory	28-MDIP		-	
DAC-8412	CMOS-D/A-IC	Quad, 12 Bit, Double Input Buffer	28-DIP,DIC	Pmi	-	pdf
DAC-8426(...)P,R	CMOS-D/A-IC	Quad, 8 Bit, int. 10V Reference	20-DIP,DIC	Pmi	PM-7226	
DAC-8426(...)S	CMOS-D/A-IC	SMD, Quad, 8 Bit, int. 10V Ref.	20-MDIP		-	
DAC-8565 DS,JS	D/A-IC	12 Bit, hi-speed, 0...+70°	24-SDIP	Ray	-	
DAC-8565 SS	D/A-IC	=DAC-8565DS,JS: -55...+125°	24-SDIP		-	
DAC 8574 IPW	IC	quad16-BIT,digital-to-analog Conv.	TSOP	Tix	-	pdf pinout
DAC 8574 IPWR	IC	quad16-BIT,digital-to-analog Conv.	TSOP	Tix	-	pdf pinout
DAC-8800(...)P,R	CMOS-D/A-IC	Octal, 8 Bit, Serial Data Input	20-DIP,DIC	Pmi	-	
DAC-8800(...)S	CMOS-D/A-IC	SMD, Octal, 8 Bit, Serial Data Input	20-MDIP		-	
DAC-8840(...)	CMOS-D/A-IC	Octal, 8 Bit, multiplying, 4 Quadrant	24-DIP,DIC	Pmi	-	
DAC-8840(...)S	CMOS-D/A-IC	SMD, Octal, 8 Bit,Multiplying,4 Quadr.	24-MDIP		-	
DAF 811 A	Si-Di	8 Di Array, 100V, 0,6A, <350ns	9-SIP	Die	Pin-Code: AKKKKKKKK	data
DAF 811 K	Si-Di	=DAF 811A:	9-SIP		Pin-Code: KAAAAAAAA	data
DAF 814 A	Si-Di	8 Di Array, 400V, 0,6A, <350ns	9-SIP	Die	Pin-Code: AKKKKKKKK	data
DAF 814 K	Si-Di	=DAF 814A:	9-SIP		Pin-Code: KAAAAAAAA	data
DAN 201	Si-Di	Dual, SS, 80V, 0,1A, <4ns	9p	Rhm	MA 155WK, MA 176WK, 1SS201	data
DAN 202	Si-Di	SMD, Dual, SS, 80V, 0,1A, <4ns	35p	Rhm	-	data
DAN 202 K,C	Si-Di	=DAN 202:	35l		-	data
DAN 202 U	Si-Di	=DAN 202:	35l(2mm)		-	data

Type	Device	Short Description	Fig.	Manu	Comparision Types	More at
DAN 207-100...-400	Si-Di	Dual, Gl, 150...500V, 0,2A -100: 150V, -200: 300V, -400: 500V	9p	Rhm		data
DAN 208-100, -200	Si-Di	Dual, Gl, 150, 300V, 1A	9p	Rhm	DAN 213-...	data
DAN 209	Si-Di	Dual, SS, 80V, 0,1A, <4ns	7p	Rhm	MA 155WK, MA 176WK, 1SS201	data
DAN 209 S	Si-Di	=DAN 209:	41p	Rhm	MA 155WK, MA 176WK, 1SS201	data
DAN 212 K,C	Si-Di	SMD, SS, 80V, 0,1A, <4ns	35e	Rhm	BAS 16	data
DAN 213-100, -200	Si-Di	Dual, Gl, 150, 300V, 1A	9p	Rhm	DAN 208-...	data
DAN 215	Si-Di	Dual, SS, 80V, 0,1A, <4ns	9p	Rhm	MA 155WK, MA 176WK, 1SS201	data
DAN 217(C)	Si-Di	SMD, Dual, Uni, 80V, 0,1A	35o	Rhm	-	data
DAN 222	Si-Di	SMD, Dual, SS, 80V, 0,1A, <4ns	35l(1,6mm)	Rhm, Ons		data
DAN 235 E	Si-Di	=DAN 235K:	35l(1,6mm)		-	data
DAN 235 K	Si-Di	SMD, HF-Band-S, 35V, <1,2pF(6V)	35l	Rhm	-	data
DAN 235 U	Si-Di	=DAN 235K:	35l(2mm)		-	data
DAN 401	Si-Di	4 Di Array, 80V, 25mA, <4ns	5-SIP	Rhm	Pin-Code: KAAAA	data
DAN 403	Si-Di	4 Di Array, 80V, 0,1A	9-SIP	Rhm	Pin-Code: KAKAKAKA-	data
DAN 601	Si-Di	=DAN 401: 6 Di	7-SIP	Rhm	Pin-Code: AAAKAAA	data
DAN 801	Si-Di	=DAN 401: 8 Di	9-SIP	Rhm	Pin-Code: AAAAKAAAA	data
DAN 803	Si-Di	=DAN 401: 8 Di	9-SIP	Rhm	Pin-Code: KAAAAAAAA	data
DAP	Si-N	=2SD1664-P (Typ-Code/Stempel/marking)	39	Rhm	→2SD1664	data
DAP 201	Si-Di	Dual, SS, 80V, 0,1A, <4ns	9q	Rhm	MA 155WA, MA 176WA, 1SS200	data
DAP 202	Si-Di	SMD, Dual, SS, 80V, 0,1A, <4ns	35q	Rhm	-	data
DAP 202 K,C	Si-Di	=DAP 202:	35m		-	data
DAP 202 U	Si-Di	=DAP 202:	35m(2mm)		-	data
DAP 207-100...-400	Si-Di	Dual, Gl, 150...500V, 0,2A -100: 150V, -200: 300V, -400: 500V	9q	Rhm	-	data
DAP 208-100, -200	Si-Di	Dual, Gl, 150, 300V, 1A	9q	Rhm	DAP 213-...	data
DAP 209	Si-Di	Dual, SS, 80V, 0,1A, <4ns	7q	Rhm	MA 155WA, MA 176WA, 1SS200	data
DAP 209 S	Si-Di	=DAP 209:	41q		MA 155WA, MA 176WA, 1SS200	data
DAP 213-100, -200	Si-Di	Dual, Gl, 150, 300V, 1A	9q	Rhm	DAP 208-...	data
DAP 215	Si-Di	Dual, SS, 80V, 0,1A, <4ns	9q	Rhm	MA 155WA, MA 176WA, 1SS200	data
DAP 222	Si-Di	SMD, Dual, SS, 80V, 0,1A, <4ns	35m(1,6mm)	Rhm, Ons	-	data
DAP 236 K	Si-Di	SMD, HF-Band-S, 35V, <1,2pF(6V)	35m	Rhm	-	data
DAP 236 U	Si-Di	=DAP 236K:	35m(2mm)		-	data
DAP 401	Si-Di	=DAN 401:	5-SIP	Rhm	Pin-Code: AKKKK	data
DAP 601	Si-Di	=DAN 401: 6 Di	7-SIP	Rhm	Pin-Code: KKKAKKK	data
DAP 801	Si-Di	=DAN 401: 8 Di	9-SIP	Rhm	Pin-Code: KKKKAKKKK	data
DAP 803	Si-Di	=DAN 401: 8 Di	9-SIP	Rhm	Pin-Code: AKKKKKKKK	data
DAQ	Si-N	=2SD1664-Q (Typ-Code/Stempel/marking)	39	Rhm	→2SD1664	data
DAR	Si-N	=2SD1664-R (Typ-Code/Stempel/marking)	39	Rhm	→2SD1664	data
DAs	Si-P	=BCW 67A (Typ-Code/Stempel/marking)	35	Sie	→BCW 67A	data
DB	Si-N	=2SD1418-DB (Typ-Code/Stempel/marking)	39	Rhm	→2SD1418	data
DB	Si-N	=2SD1619 (Typ-Code/Stempel/marking)	39	Say	→2SD1619	data
DB	Si-N	=2SD1766 (Typ-Code/Stempel/marking)	39	Rhm	→2SD1766	data
DB	Si-P	=BCW 67B (Typ-Code/Stempel/marking)	35	Fer,Mot,Sie	→BCW 67B	data
DB	Si-P	=BF 623 (Typ-Code/Stempel/marking)	39	Phi,Sie,Val	→BF 623	data
DB	Si-P	=BF 723 (Typ-Code/Stempel/marking)	48	Phi	→BF 723	data
DB	C-Di	=ZMV 830 (Typ-Code/Stempel/marking)	71(1,7mm)	Ztx	→ZMV 830	data
DB	MOS-N-FET-e	=µPA572T (Typ-Code/Stempel/marking)	45(2mm)	Nec	→µPA572T	data
DB 1	Si-Di	=BAR 43A (Typ-Code/Stempel/marking)	35		→BAR 43A	data
DB 1 A..P	Si-Br	Gl-Br, 50...1000V, 1A	4-DIP	Gen,Rca	A 0503...0580, DF 01...10	data
DB 2	Si-Di	=BAR 43B (Typ-Code/Stempel/marking)	35	Tho	→BAR 43B	data
DB 3	Diac	Ub=28...36V, Ib<0,3mA	31	Tho	-	data
DB 3-l-35	Diac	Ub=28...36V, Ib<0,05mA	31		-	data
DB 3 TG	Diac	Ub=30...34V, Ib<15µA	31		-	data
DB 4	Diac	=DB 3: Ub=35...45V	31	Tho	-	data
DB 6	Diac	=DB 3: Ub=56...70V	31	Tho	-	data
DB 6	Si-Di	=BAT 46AW (Typ-Code/Stempel/marking)	35(2mm)	Tho	→BAT 46AW	data
DB 6	Si-Di	=GSD 2004S (Typ-Code/Stempel/marking)	35	Gsi	→GSD 2004S	data
DB 6	Si-Di	=GSD 2004SW (Typ-Code/Stempel/marking)	71(2,7mm)	Gsi	→GSD 2004SW	data
DB 6	Si-Di	=BAY 85S (Typ-Code/Stempel/marking)	35	Tho	→BAY 85S	data
DB15	HI-OHM	Vs:±15V Vu:100dB Vo:±10V Vi0:0.75mV	7Y	Ilc		data pinout
DB17	HI-OHM	Vs:±15V Vu:100dB Vo:±10V Vi0:0.5mV	7Y	Ilc		data pinout
DB18	UNI	Vs:±15V Vu:92dB Vo:±11V f:1.5Mc	7Y	Ilc		data pinout
DB19	HI-OHM	Vs:±15V Vu:110dB Vo:±10V Vi0:0.5mV	7Y	Ilc		data pinout
DB21	HI-OHM	Vs:±15V Vu:110dB Vo:±10V Vi0:0.5mV	15Y	Ilc		data pinout
DB25	HI-OHM	Vs:±15V Vu:100dB Vo:±10V Vi0:0.75mV	7Y	Ilc		data pinout
DB28	UNI	Vs:±15V Vu:110dB Vo:±10V Vi0:2mV	7Y	Ilc		data pinout
DBA 30 B...G	Si-Br	Gl-Br, 100...600V, 3A B=100V, C=200V, E=400V, G=600V	79x3	Say	B250 C3300	data
DBB	Z-Di	=SMAJ 6.0CA (Typ-Code/Stempel/marking)	71(5x2,5)	Tho	→SMAJ ...	data

ype	Device	Short Description	Fig.	Manu	Comparision Types	More at
B 04 B...G	Si-Br	Gl-Br, 100..600V, 0,4A	4-DIP	Say	B... C400	data
		B=100V, C=200V, E=400V, G=600V				
B 08 B...G(LT)	Si-Br	Gl-Br, 100..600V, 0,8A	4-DIP	Say	B... C800	data
		B=100V, C=200V, E=400V, G=600V				
B 10 B...G	Si-Br	Gl-Br, 100..600V, 1A	4-DIP	Say	B... C1000	data
		B=100V, C=200V, E=400V, G=600V				
C	Z-Di	=SMAJ 6.5CA (Typ-Code/Stempel/marking)	71(5x2,5)	Tho	→SMAJ ...	data
C 10 C...G	Si-Br	Gl-Br, 200..600V, 1A	4-DIP	Say	B... C1000	data
		C=200V, E=400V, G=600V				
C 146-4(A...C)	Si-N	=BC 146:	36b		→BC 146	data
C 201	Si-P	=BC 201:	36b		→BC 201	data
C 202	Si-P	=BC 202:	36b		→BC 202	data
C 203	Si-P	=BC 203:	36b		→BC 203	data
D 10 C,G	Si-Br	Gl-Br, 200..600V, 1A		Say	B... C1000	data
		C=200V, G=600V				
F 10 B...G	Si-Br	Gl-Br, 100..600V, 1A	33x1	Say	B... C1000	data
		B=100V, C=200V, E=400V, G=600V				
F 20(T)B...G	Si-Br	Gl-Br, 100..600V, 2A	33x1	Say	B... C2000	data
		B=100V, C=200V, E=400V, G=600V				
F 40(T)B...G	Si-Br	Gl-Br, 100..600V, 4A	33x1	Say	B... C4000	data
		B=100V, C=200V, E=400V, G=600V				
F 60(T)B...G	Si-Br	Gl-Br, 100..600V, 6A	33x1	Say	B... C6000	data
		B=100V, C=200V, E=400V, G=600V				
F 100 C,G	Si-Br	Gl-Br, 200..600V, 10A	33	Say	B... C10000	data
		C=200V, G=600V				
F 150 C,G	Si-Br	Gl-Br, 200..600V, 15A(Tc=100°)	33x1	Say	B... C15000	data
		C=200V, G=600V				
F 200 C,G	Si-Br	Gl-Br, 200..600V, 20A	33	Say	B... C20000	data
		C=200V, G=600V				
F 250 C,G	Si-Br	Gl-Br, 200..600V, 25A(Tc=98°)	33x1	Say	B... C25000	data
		C=200V, G=600V				
H	Z-Di	=SMAJ 8.5CA (Typ-Code/Stempel/marking)	71(5x2,5)	Tho	→SMAJ ...	data
K	Z-Di	=SMAJ 12CA (Typ-Code/Stempel/marking)	71(5x2,5)	Tho	→SMAJ ...	data
L 339	LIN-IC	Quad comparator, 2...36V or ±1..±18V	DIP		..339..	pdf pinout
P	Si-N	=2SD1766-P (Typ-Code/Stempel/marking)	39	Rhm	→2SD1766	data
Q	Si-N	=2SD1766-Q (Typ-Code/Stempel/marking)	39	Rhm	→2SD1766	data
Q	Z-Di	=SMAJ 18CA (Typ-Code/Stempel/marking)	71(5x2,5)	Tho	→SMAJ ...	data
R	Si-N	=2SD1766-R (Typ-Code/Stempel/marking)	39	Rhm	→2SD1766	data
R	Z-Di	=SMAJ 20CA (Typ-Code/Stempel/marking)	71(5x2,5)	Tho	→SMAJ ...	data
S	Si-P	=BCW 67B (Typ-Code/Stempel/marking)	35	Sie	→BCW 67B	data
S	Z-Di	=SMAJ 22CA (Typ-Code/Stempel/marking)	71(5x2,5)	Tho	→SMAJ ...	data
T	Z-Di	=SMAJ 24CA (Typ-Code/Stempel/marking)	71(5x2,5)	Tho	→SMAJ ...	data
U	Z-Di	=SMAJ 26CA (Typ-Code/Stempel/marking)	71(5x2,5)	Tho	→SMAJ ...	data
Z	Z-Di	=SMAJ 40CA (Typ-Code/Stempel/marking)	71(5x2,5)	Tho	→SMAJ ...	data
	Si-N	=2SC3444-C (Typ-Code/Stempel/marking)	39	Mit	→2SC3444	data
	Si-N	=2SD1418-DC (Typ-Code/Stempel/marking)	39	Rhm	→2SD1418	data
	Si-N	=2SD1620 (Typ-Code/Stempel/marking)	39	Say	→2SD1620	data
	Si-N	=2SD1767 (Typ-Code/Stempel/marking)	39	Rhm	→2SD1767	data
	Si-P	=BCW 67C (Typ-Code/Stempel/marking)	35	Fer, Sie	→BCW 67C	data
	Si-N	=BF 620 (Typ-Code/Stempel/marking)	39	Phi, Val	→BF 620	data
	Si-N	=BF 720 (Typ-Code/Stempel/marking)	48	Phi, Ons	→BF 720	data
	Si-N	=BFN 20 (Typ-Code/Stempel/marking)	39	Sie	→BFN 20	data
	Z-Di	=MMSZ 4695 (Typ-Code/marking)	71(2,7mm)	Ons	→MMSZ 4695	
	MOS-N-FET-e*	=SSM 3K04FE (Typ-Code/Stempel/marking)	35(1,6mm)	Tos	→SSM 3K04FE	data
	MOS-N-FET-e*	=SSM 3K04FS (Typ-Code/Stempel/marking)	35(1,6mm)	Tos	→SSM 3K04FS	data
	MOS-N-FET-e*	=SSM 3K04FU (Typ-Code/Stempel/marking)	35(2mm)	Tos	→SSM 3K04FU	data
	MOS-N-FET-e*	=SSM 6N04FU (Typ-Code/Stempel/marking)	46(2mm)	Tos	→SSM 6N04FU	data
	C-Di	=ZMV 831 (Typ-Code/Stempel/marking)	71(1,7mm)	Ztx	→ZMV 831	data
8	UNI	Vs:±15V Vu:95dB Vo:±11V f:5Mc	7Y	Ilc		data pinout
15	HI-OHM	Vs:±15V Vu:98dB Vo:±10V Vi0:1.5mV	7Y	Ilc		data pinout
16	UNI	Vs:±15V Vu:100dB Vo:±11V f:1Mc	7Y	Ilc		data pinout
17	HI-OHM	Vs:±15V Vu:100dB Vo:±10V Vi0:0.5mV	7Y	Ilc		data pinout
018	I/O-IC	Ser.-par./par.-ser. Converter	40-DIP		DS 8609DC	
19	HI-OHM	Vs:±15V Vu:110dB Vo:±10V Vi0:0.5mV	7Y	Ilc		data pinout
021	DIG-IC	8 Bit Bustreiber/bus driver, bidirect.	20-DIP		DS 8638DC	
25	HI-OHM	Vs:±15V Vu:100dB Vo:±10V Vi0:0.75mV	7Y	Ilc		data pinout
34	Diac	Ub=30...38V, Ib<50µA	31	Tho	-	data
38	Diac	Ub=34...42V, Ib<50µA	31	Tho	-	data
42	Diac	Ub=39...45V, Ib<50µA	31	Tho	-	data
A 010	Si-Di	SMD, Dual, SS, 85V, 0,1A, <4ns	35m	Say	BAW 56, 1SS181	data
A 015	Si-Di	SMD, Dual, SS, 75V, 0,1A, <8ns	35m	Say	BAW 56, 1SS181	data
B 010	Si-Di	SMD, Dual, SS, 85V, 0,1A, <4ns	35l	Say	BAV 70, 1SS184	data
B 015	Si-Di	SMD, Dual, SS, 75V, 0,1A, <8ns	35l	Say	BAV 70, 1SS184	data
C 010	Si-Di	SMD, Dual, SS, 85V, 0,1A, <4ns	35o	Say	BAV 99, HSM 123, HSM 124S	data
D 010	Si-Di	SMD, Dual, Uni, 20V, 0,1A	35o	Say	BAS 31, BAV 99, HSM 124S	data
D 015	Si-Di	Dual, SS, 75V, 0,1A, <15ns	40q	Say	MA 155WA, MC 911, 1SS236	data
E 015	Si-Di	Dual, SS, 75V, 0,1A, <12ns	40p	Say	MA 155WK, MC 921, 1SS234	data
F 010	Si-Di	SMD, Dual, SS, 85V, 0,1A, <4ns	35m(2mm)	Say	-	data
F 015	Si-Di	SMD, Dual, SS, 75V, 0,1A, <6ns	35m(2mm)	Say	-	data
G 010	Si-Di	SMD, Dual, Uni, 85V, 0,15A	35l (2mm)	Say	-	data

Type	Device	Short Description	Fig.	Manu	Comparision Types	More at
DCG 10	Si-Di	SMD, Dual, S, 200V, 1,5A	39p	Say	-	data
DCG 015	Si-Di	SMD, Dual, SS, 75V, 0,1A, <6ns	35l(2mm)	Say	-	data
DCH 10	Si-Di	SMD, Dual, S, 200V, 1,5A	39q	Say	-	data
DCJ 010	Si-Di	SMD, Dual, 20V, 0,1A	35o(2mm)	Say	-	data
DCP	Si-N	=2SD1767-P (Typ-Code/Stempel/marking)	39	Rhm	→2SD1767	data
DCQ	Si-N	=2SD1767-Q (Typ-Code/Stempel/marking)	39	Rhm	→2SD1767	data
DCR	Si-N	=2SD1767-R (Typ-Code/Stempel/marking)	39	Rhm	→2SD1767	data
DCs	Si-P	=BCW 67C (Typ-Code/Stempel/marking)	35	Sie	→BCW 67C	data
DD	Si-N	=2SC2463-D (Typ-Code/Stempel/marking)	35	Hit	→2SC2463	data
DD	Si-N	=2SC3441-D (Typ-Code/Stempel/marking)	35	Mit	→2SC3441	data
DD	Si-N	=2SC3444-D (Typ-Code/Stempel/marking)	39	Mit	→2SC3444	data
DD	Si-N	=2SC4933-D (Typ-Code/Stempel/marking)	35(2mm)	Hit	→2SC4933	data
DD	Si-N	=2SD1419-DD (Typ-Code/Stempel/marking)	39	Hit	→2SD1419	data
DD	Si-N	=2SD1621 (Typ-Code/Stempel/marking)	39	Say	→2SD1621	data
DD	Si-N	=BFN 16 (Typ-Code/Stempel/marking)	39	Sie	→BFN 16	data
DD	Z-Di	=MMSZ 4696 (Typ-Code/marking)	71(2,7mm)	Ons	→MMSZ 4696	data
DD	Z-Di	=P4 SMA-120 (Typ-Code/Stempel/marking)	71(5x2,5)	Fag	→P4 SMA-120	data
DD	Z-Di	=SM 6T 6V8 (Typ-Code/Stempel/marking)	71(6x4mm)	Tho	→SM 6T...	data
DD	Z-Di	=SMBJ 54C (Typ-Code/Stempel/marking)	71(5x3,5)	Mop	→SMBJ ...	data
DD	MOS-P-FET-e*	=SSM 3J02F (Typ-Code/Stempel/marking)	35	Tos	→SSM 3J02F	data
DD	C-Di	=ZMV 832 (Typ-Code/Stempel/marking)	71(1,7mm)	Ztx	→ZMV 832	data
DD 20 R	Si-Di	CRT, Damper, hi-def, 1500V, 2A, <1μs	17f	Say	-	data
DD 50 R	Si-Di	CRT, Damper, hi-def, 1500V, 5A, <1μs	18f	Say	-	data
DD 52 RC	Si-Di	CRT, Damper, hi-def, 1500V, 5A, <1,5μs	18d	Say	5THZ 52, 5TUZ 52	data
DD 54 RC	Si-Di	CRT, Damper, hi-def, 1500V, 5A, <1,5μs	17d	Say	DTV 16F, DTV 32F	data
DD 54 SC	Si-Di	CRT, Damper, hi-def, 1600V, 5A, <1,5μs	17d	Say	-	data
DD 82 RC	Si-Di	CRT, Damper, hi-def, 1500V, 8A, <1,5μs	18d	Say	FMV-G5FS	data
DD 82 SC	Si-Di	CRT, Damper, hi-def, 1600V, 8A, <1,5μs	18d	Say	FMR-G5HS	data
DD 84 RC	Si-Di	CRT, Damper, hi-def, 1500V, 8A, <1,5μs	17d	Say	DTV 16F, DTV 32F	data
DD 84 SC	Si-Di	CRT, Damper, hi-def, 1600V, 8A, <1,5μs	17d	Say	-	data
DD 300	Si-Di	kV-Gl, 3kV, 20mA, <55ns	31a	Die	-	data
DD 600	Si-Di	=DD 300: 6kV	31a	Die	-	data
DD 1000	Si-Di	=DD 300: 10kV	31a	Die	-	data
DD 1200	Si-Di	=DD 300: 12kV	31a	Die	-	data
DD 1300	Si-Di	=DD 300: 13kV	31a	Die	-	data
DD 1600	Si-Di	=DD 300: 16kV	31a	Die	-	data
DD 1800	Si-Di	=DD 300: 18kV	31a	Die	-	data
DD 7661	LIN-IC	TV-Tuning-prozessor	40-DIP	-		
DD 7662	LIN-IC	Tv-tuning-prozessor	16-DIP	-		
DDX-1050	Dig. Audio	All-digital High Efficiency Power Amp.	TSOP	Apg	-	pdf pinout
DDX-1060	Dig. Audio	All-digital High Efficiency Power Amp.	TSOP	Apg	-	pdf pinout
DDX-1080	Dig. Audio	All-digital High Efficiency Power Amp.	TSOP	Apg	-	pdf pinout
DDX-2050	Dig. Audio	All-digital High Efficiency Power Amp.	TSOP	Apg	-	pdf pinout
DDX-2052	Dig. Audio	All-digital High Efficiency Power Amp.	TSOP	Apg	-	pdf pinout
DDX-2060	Dig. Audio	All-digital High Efficiency Power Amp.	TSOP	Apg	-	pdf pinout
DDX-2062	Dig. Audio	All-digital High Efficiency Power Amp.	TSOP	Apg	-	pdf pinout
DDX-2100	Dig. Audio	All-digital High Efficiency Power Amp.	TSOP	Apg	-	pdf pinout
DDX-2102	Dig. Audio	All-digital High Efficiency Power Amp.	TSOP	Apg	-	pdf pinout
DDX-2120	Dig. Audio	All-digital High Efficiency Power Amp.	TSOP	Apg	-	pdf pinout
DDX-2160	Dig. Audio	All-digital High Efficiency Power Amp.	TSOP	Apg	-	pdf pinout
DDX-2200	Dig. Audio	All-digital High Efficiency Power Amp.	TSOP	Apg	-	pdf pinout
DDX-2240	Dig. Audio	All-digital High Efficiency Power Amp.	TSOP	Apg	-	pdf pinout
DDXI-2051	Dig. Audio	2.0/ 2.1- Ch Integrated Audio Controller, Power Output	TSOP	Apg	-	pdf pinout
DDXI-2101	Dig. Audio	2/2.1 Ch Integrated Audio Controller, Power Output	TSOP	Apg	-	pdf pinout
DDXI-2161	Dig. Audio	2/2.1 Ch Integrated Audio Controller, Power Output	TSOP	Apg	-	pdf pinout
DE	Si-P	=2SA1037KLN-E (Typ-Code/Stempel/mark.)	35	Rhm	→2SA1037KLN	data
DE	Si-N	=2SC2463-E (Typ-Code/Stempel/marking)	35	Hit	→2SC2463	data
DE	Si-N	=2SC3441-E (Typ-Code/Stempel/marking)	35	Mit	→2SC3441	data
DE	Si-N	=2SC3444-E (Typ-Code/Stempel/marking)	39	Mit	→2SC3444	data
DE	Si-N	=2SC4933-E (Typ-Code/Stempel/marking)	35(2mm)	Hit	→2SC4933	data
DE	Si-N	=2SD1419-DE (Typ-Code/Stempel/marking)	39	Hit	→2SD1419	data
DE	Si-N	=2SD1622 (Typ-Code/Stempel/marking)	39	Say	→2SD1622	data
DE	Si-N-Darl	=2SD1834 (Typ-Code/Stempel/marking)	39	Rhm	→2SD1834	data
DE	Si-N	=BFN 18 (Typ-Code/Stempel/marking)	39	Sie	→BFN 18	data
DE	Z-Di	=MMSZ 4697 (Typ-Code/marking)	71(2,7mm)	Ons	→MMSZ 4697	data
DE	Z-Di	=P4 SMA-120A(Typ-Code/Stempel/marking)	71(5x2,5)	Fag	→P4 SMA-120A	data
DE	Z-Di	=SM 6T 6V8A (Typ-Code/Stempel/marking)	71(6x4mm)	Tho	→SM 6T...	data
DE	Z-Di	=SMBJ 54CA (Typ-Code/Stempel/marking)	71(5x3,5)	Mop	→SMBJ ...	data
DE	MOS-P-FET-e*	=SSM 3J01F (Typ-Code/Stempel/marking)	35	Tos	→SSM 3J01F	data
DE	MOS-P-FET-e*	=SSM 3J01T (Typ-Code/Stempel/marking)	35	Tos	→SSM 3J01T	data
DE	C-Di	=ZMV 833 (Typ-Code/Stempel/marking)	71(1,7mm)	Ztx	→ZMV 833	data
DE 3 L20U	Si-Di	Gl, S, 200V, 3A(Tc=113°), <35ns	30d1§	Shi	-	data
DE 3 L40	Si-Di	Gl, S, 400V, 3A(Tc=99°), <50ns	30d1§	Shi	-	data
DE 3 S4M	Si-Di	Schottky-Gl, 40V, 3A(Tc=100°)	30d1§	Shi	-	data
DE 3 S6M	Si-Di	Schottky-Gl, 60V, 3A(Tc=117°)	30d1§	Shi	-	data
DE 5 L60	Si-Di	Gl, S, 600V, 5A(Tc=57°), <50ns	30t§	Shi	-	data
DE 5 LC20U	Si-Di	Dual, Gl, S, 200V, 5A(Tc=81°), <35ns	30p§	Shi	-	data
DE 5 LC40	Si-Di	Dual, Gl, S, 400V, 3A(Tc=61°), <50ns	30p§	Shi	-	data

314

pe	Device	Short Description	Fig.	Manu	Comparision Types	More at
PC3(M)	Si-Di	Dual, Schottky, 30V, 5A(Tc=90°)	30p§	Shi	DE 5SC3ML, DE 5SC4M	data
S4M	Si-Di	Schottky-Gl, 40V, 5A(Tc=101°)	30d1§	Shi	-	data
S500	R+	Io=500mA	5Y	Npp		data pinout
S6M	Si-Di	Schottky-Gl, 60V, 5A(Tc=96°)	30d1§	Shi		data
SC3ML	Si-Di	Dual, Schottky-Gl, 30V, 5A(Tc=110°)	30p§	Shi	DE 5PC3	data
SC4M	Si-Di	Dual, Schottky-Gl, 40V, 5A(Tc=101°)	30p§	Shi		data
SC6M	Si-Di	Dual, Schottky-Gl, 60V, 5A(Tc=92°)	30p§	Shi	-	data
10 P3	Si-Di	Schottky-Gl, 30V, 10A(Tc=95°)	30§	Shi		data
10 PC3	Si-Di	Dual, Schottky-Gl, 30V, 10A(Tc=97°)	30p§	Shi	DE 10SC3L, PBYR 1040CTD	data
10 S3L	Si-Di	Schottky-Gl, 30V, 10A(Tc=124°)	30t§	Shi		data
10 SC3L	Si-Di	Dual, Schottky-Gl, 30V, 10A(Tc=124°)	30p§	Shi	DE 10PC3	data
10 SC4	Si-Di	Dual, Schottky-Gl, 40V, 10A(Tc=132°)	30p§	Shi	PBYR 1040CTD	data
2D100	R+	Io=100mA	7Y	Npp		data pinout
5	HI-OHM	Vs:±15V Vu:100dB Vo:±10V Vi0:0.75mV	7Y	Ilc		data pinout
5D100	R+	Io=100mA	7Y	Npp		data pinout
7	HI-OHM	Vs:±15V Vu:100dB Vo:±10V Vi0:0.5mV	7Y	Ilc		data pinout
9	HI-OHM	Vs:±15V Vu:110dB Vo:±10V Vi0:0.5mV	7Y	Ilc		data pinout
1	HI-OHM	Vs:±15V Vu:110dB Vo:±10V Vi0:0.5mV		Ilc		data pinout
4D100	R+	Io=100mA	7Y	Npp		data pinout
5	HI-OHM	Vs:±15V Vu:100dB Vo:±10V Vi0:0.75mV	7Y	Ilc		data pinout
	Si-N	=2SC2463-F (Typ-Code/Stempel/marking)	35	Hit	→2SC2463	data
	Si-N	=2SC3441-F (Typ-Code/Stempel/marking)	35	Mit	→2SC3441	data
	Si-N	=2SC4933-F (Typ-Code/Stempel/marking)	35(2mm)	Hit	→2SC4933	data
	Si-N	=2SD1623 (Typ-Code/Stempel/marking)	39	Say	→2SD1623	data
	Si-P	=BCW 68F (Typ-Code/Stempel/marking)	35	Fer, Sie	→BCW 68F	data
	Si-N	=BCX 56 (Typ-Code/Stempel/marking)	39	Rhm	→BCX 56	data
	Si-P	=BF 621 (Typ-Code/Stempel/marking)	39	Phi, Val	→BF 621	data
	Si-P	=BF 721 (Typ-Code/Stempel/marking)	48	Phi, Ons	→BF 721	data
	Si-P	=BFN 21 (Typ-Code/Stempel/marking)	39	Sie	→BFN 21	data
	Z-Di	=MMSZ 4698 (Typ-Code/marking)	71(2,7mm)	Ons	→MMSZ 4698	
	Z-Di	=P4 SMA-130 (Typ-Code/Stempel/marking)	71(5x2,5)	Fag	→P4 SMA-130	data
	Z-Di	=SM 6T 7V5 (Typ-Code/Stempel/marking)	71(6x4mm)	Tho	→SM 6T...	data
	Z-Di	=SMBJ 58C (Typ-Code/Stempel/marking)	71(5x3,5)	Mop	→SMBJ ...	data
	MOS-N-FET-e*	=SSM 3K05FU (Typ-Code/Stempel/marking)	35(2mm)	Tos	→SSM 3K05FU	
	MOS-N-FET-e*	=SSM 5N05FU (Typ-Code/Stempel/marking)	45(2mm)	Tos	→SSM 5N05FU	
	MOS-N-FET-e*	=SSM 6N05FU (Typ-Code/Stempel/marking)	46(2mm)	Tos	→SSM 6N05FU	
	C-Di	=ZMV 834 (Typ-Code/Stempel/marking)	71(1,7mm)	Ztx	→ZMV 834	data
01	C-Di	AFC	31a			
01(M,S)	Si-Br	Gl-Br, 100V, 1A	4-DIP	Gie,Die,Fag	A 0526, DB 1B	data
02(M,S)	Si-Br	=DF 01(M), 200V	4-DIP	Gie	A 0553, DB 1B	data
2 S12FU	Z-Di	SMD, ESD-Prot., 12(11,4...12,6)V	71a(1,7mm)	Tos	-	
3 A2.2..8.2FE,	Z-Di	SMD, ESD-Prot., Dual, 2,2..8,2V ±5% LFE: High-Speed	35m(1,6mm)	Tos	-	
3 A2.2..8.2FU,	Z-Di	SMD, ESD-Prot., Dual, 2,2..8,2V ±5% LFU: High-Speed	35m(2mm)	Tos	-	
04(M,S)	Si-Br	=DF 01(M): 400V	4-DIP	Gie	A 0580, DB 1D	data
005(M,S)	Si-Br	=DF 01(M): 50V	4-DIP	Gie	A 0506, DB 1A	data
5 A2.2..8.2F,	Z-Di	SMD, ESD-Prot., Quad, 2,2..8,2V ±5% LF: High-Speed	45bb	Tos	-	
5 A2.2..8.2FU,	Z-Di	SMD, ESD-Prot., Quad, 2,2..8,2V ±5% LJE: High-Speed	45bb(2mm)	Tos	-	
5 A2.2..8.2JE,	Z-Di	SMD, ESD-Prot., Quad, 2,2..8,2V ±5% LJE: High-Speed	45bb(1,6mm)	Tos	-	
5 VD60	Si-Di	Gl, Uni, 600V, 5A(Tc=140°)	30o1	Shi	-	data
06(M,S)	Si-Br	=DF 01(M): 600V	4-DIP	Gie	DB 1M	data
6 A6.8FU(T1)	Z-Di	SMD, Quad, TAZ, 6,8V, 75W(8x20µs)	46	Ons	-	
08(M,S)	Si-Br	=DF 01(M): 800V	4-DIP	Gie	DB 1N	data
8 A6.8FK	Z-Di	SMD, ESD-Prot., 7 Z-Di, 6,8V ±5%	8-MDIP/ck	Tos	-	
10(M,S)	Si-Br	=DF 01(M): 1000V	4-DIP	Gie	DB 1P	data
10 L60	Si-Di	Gl, S, 600V, 10A(Tc=105°), <50ns	30d1§	Shi	-	data
10 LC20U	Si-Di	Dual, Gl, S, 200V, 10A(Tc=127°), <35ns	30p§	Shi	BYQ 28EB-200, (U)10DL2C48A	data
10 SC4M	Si-Di	Dual, Schottky-Gl, 40V, 10A(Tc=125°)	30p§	Shi	DF 15SC4M	data
15 SC4M	Si-Di	Dual, Schottky-Gl, 40V, 15A(Tc=129°)	30p§	Shi	MBRB 1545CT, PBYR 1540CTB	data
15 VD60	Si-Di	Gl, Uni, 600V, 15A(Tc=103°)	30o1	Shi	-	data
20 L60	Si-Di	Gl, S, 600V, 20A(Tc=84°), <70ns	30d1§	Shi	-	data
20 LC20U	Si-Di	Dual, Gl, S, 200V, 20A(Tc=114°), <35ns	30p§	Shi	BYW 51G-200, (U)20DL2C48A	data
20 LC30	Si-Di	Dual, Gl, S, 300V, 20A(Tc=105°), <30ns	30p§	Shi	U 20FL2C48A, 20FL2C48A	data
20 PC3M	Si-Di	Dual, Schottky-Gl, 30V, 20A(Tc=105°)	30p§	Shi	PBYR 2035CTB, (U)20FWJ2C48M	data
20 SC4M	Si-Di	Dual, Schottky-Gl, 40V, 20A(Tc=122°)	30p§	Shi	PBYR 2040CTB, PBYR 2045CTB	data
20 SC9M	Si-Di	Dual, Schottky-Gl, 90V, 20A(Tc=111°)	30p§	Shi	MBRB 20100CT	data
25 SC6M	Si-Di	Dual, Schottky-Gl, 60V, 25A(Tc=115°)	30p§	Shi		data
30 PC3M	Si-Di	Dual, Schottky-Gl, 30V, 30A(Tc=97°)	30p§	Shi	MBRB 3030CT, (U)30FWJ2C48M	data
30 SC3ML	Si-Di	Dual, Schottky-Gl, 30V, 30A(Tc=119°)	30p§	Shi	MBRB 3030CT, (U)30FWJ2M48M	data
30 SC4M	Si-Di	Dual, Schottky-Gl, 40V, 30A(Tc=112°)	30p§	Shi	MBRB 2545CT, PBYR 2545CTB, (U)30GWJ2C48C	data
40 SC3L	Si-Di	Dual, Schottky-Gl, 30V, 40A(Tc=112°)	30p§	Shi		data

Type	Device	Short Description	Fig.	Manu	Comparision Types	More at
DF 320(DJ,DP)	CMOS-IC	Telecom, Tastwahl/digital dialing	18-DIC,DIP	Itt	MV 4320	
DF 321	CMOS-IC	Telecom, Tastwahl/digital dialing	28-DIC	Itt	-	
DF 322(DJ,DP)	CMOS-IC	Telecom, Tastwahl/digital dialing	18-DIC,DIP	Itt	-	
DF 323(DJ,DP)	CMOS-IC	Telecom, Tastwahl/digital dialing	18-DIC,DIP	Itt	-	
DF 820(DJ,DK)	CMOS-IC	Telecom, Tastwahl/digital dialing	18-DIP,DIC	Itt	-	
DF 821(DJ,DK)	CMOS-IC	Telecom, Tastwahl/digital dialing	18-DIP,DIC	Itt	-	
DF 822(DJ,DK)	CMOS-IC	Telecom, Tastwahl/digital dialing	18-DIP,DIC	Itt	-	
DF 823(DJ,DK)	CMOS-IC	Telecom, Tastwahl/digital dialing	18-DIP,DIC	Itt	-	
DF 824(DJ,DK)	CMOS-IC	Telecom, Tastwahl/digital dialing	18-DIP,DIC	Itt	-	
DFA 007	Ge-Di	≈AA 143	31a		→AA 143	
DFA 08 C...E	Si-Di	SMD, Gl, S, 200...400V, 0,8A, <300ns C=200V, E=400V	71a(5x2,5)	Say	BYG 20D...M	data
DFA 12 C...E	Si-Di	SMD, Gl, S, 200...400V, 1,2A, <300ns C=200V, E=400V	71a(7,6x4)	Say	BYG 20D...M	data
DFB 03 R,T,W	Si-Di	Gl, 1500...2000V, 0,3A, <1µs R=1500V, T=1700V, W=2000V	31a	Say	BY 203/.., BYD 43-20	data
DFB 20(T)B...L	Si-Di	S, 100...1000V, 2A, <150...300ns B=100,C=200,E=400,G=600,J=800,L=1000V	31a	Say	BYM 26A...E, BYM 36A...E, BYT 56G...M	data
DFC 03 G...W	Si-Di	Gl, 600...2000V, 0,3A, <2...2,5µs G=600, J=800, L=1000, N=1200, R=1500, T=1700, W=2000V	31a	Say	BA 157...159, BY 203/.., BYD 43-20	data
DFC 05 J...R	Si-Di	Gl, 800...1500V, 0,5A, <1µs J=800V, L=1000V, N=1200V, R=1500V	31a	Say	DM513, DM516, EM513, EM516, GP 10Q...Y	data
DFC 15(T)B...R	Si-Di	Gl, S, 100...1500V, 1,5A, <150...300ns B=100, C=200, E=400, G=600, J=800V, L=1000V, N=1200V, R=1500V	31a	Say	BYV 12...16, BYV 36A...E, RGP 15-10...-20	data
DFD 05(T)B...T	Si-Di	Gl, S, 100...1700V, 0,5A, <150...300ns B=100, C=200, E=400, G=600, J=800V, L=1000, N=1200, R=1500, T=1700	31a	Say	BYT 11/.., BYT 52B...M, RGP 15-10...-20	data
DFD 15 C...G	Si-Di	Gl, S, 200...600V, 1,5A, <100ns C=200V, E=400V, G=600V	31a	Say	BYM 26A...C, BYV 27-..., BYV 36A...C	data
DFD 30(T)B...L	Si-Di	Gl, S, 100...1000V, 3A, <150...300ns B=100,C=200,E=400,G=600,J=800,L=1000V	31a	Say	BYM 36A...E, BYT 56B...M	data
DFE 30 C...G	Si-Di	Gl, S, 200...600V, 3A, <100ns	31a	Say	BYM 36A...E, BYT 56D...M	data
DFH 10(T)B...R	Si-Di	Gl, S, 100...1500V, 1A, <150...300ns B=100, C=200, E=400, G=600, J=800V, L=1000; N=1200, R=1500V	31a	Say	BYT 11/.., BYT 52B...E, RGP 15-10...-20	data
DFJ 10 C...G	Si-Di	Uni, S, 200...600V, 1A, <100ns C=200V, E=400V, G=600V	31a	Say	BYT 43D...J, BYV 26A...E	data
DFP	Si-N	=2SD1898-P (Typ-Code/Stempel/marking)	39	Rhm	→2SD1898	data
DFQ	Si-N	=2SD1898-Q (Typ-Code/Stempel/marking)	39	Rhm	→2SD1898	data
DFR	Si-N	=2SD1898-R (Typ-Code/Stempel/marking)	39	Rhm	→2SD1898	data
DFs	Si-P	=BCW 68F (Typ-Code/Stempel/marking)	35	Sie	→BCW 68F	data
DG	Si-N	=2SC2713-GR (Typ-Code/Stempel/marking)	35	Tos	→2SC2713	data
DG	Si-N	=2SC4117-GR (Typ-Code/Stempel/marking)	35(2mm)	Tos	→2SC4117	data
DG	Si-N	=2SD1624 (Typ-Code/Stempel/marking)	39	Say	→2SD1624	data
DG	Si-P	=BCW 68G (Typ-Code/Stempel/marking)	35	Fer, Sie	→BCW 68G	data
DG	Si-P	=BFN 17 (Typ-Code/Stempel/marking)	39	Sie	→BFN 17	data
DG	Si-N	=KTC4077-GR (Typ-Code/Stempel/marking)	35(2mm)	Kec	→KTC 4077	data
DG	Z-Di	=P4 SMA-130A(Typ-Code/Stempel/marking)	71(5x2,5)	Fag	→P4 SMA-130A	data
DG	Si-N-Darl	=RXTA-14 (Typ-Code/Stempel/marking)	39	Rhm	→RXTA-14	data
DG	Z-Di	=SM 6T 7V5A (Typ-Code/Stempel/marking)	71(6x4mm)	Tho	→SM 6T...	data
DG	Z-Di	=SMBJ 58CA (Typ-Code/Stempel/marking)	71(5x3,5)	Mop	→SMBJ ...	data
DG	C-Di	=ZMV 835 (Typ-Code/Stempel/marking)	71(1,7mm)	Ztx	→ZMV 835	data
DG 1	Si-Di	TV-Damper-Di, 1500V, 1,5A, <20µs	31a	Gie	BY 228, BY 328, BY 448	data
DG 2	Si-Di	=DG 1: 2A	31a	Gie	BY 228, BY 328, BY 448	data
DG 3	Si-Di	=DG 1: 3A	31a	Gie	BY 228, BY 328, BY 448	data
DG 13	Ge-Di	TV Damper-Di, 120V, 3A	23f&	Say	AY 102, AY 106	data
DG 14	Ge-Di	=DG 13: 250V	23f&	Say	AY 102	data
DG 123 A,B	PMOS-IC	Analog Switch(SPST), 36V, 30mA A: -55...+125°, B: -25...+85°	14-DIC,DIP	Six	-	
DG 125 A,B	PMOS-IC	Analog Switch(SPST), 36V, 30mA A: -55...+125°, B: -25...+85°	14-DIC,DIP	Six	-	
DG 126 A,B	BiMOS-IC	Analog Switch, JFET Input, 36V, 30mA A: -55...+125°, B: -25...+85°	14-DIC	Six	-	
DG 129	BiMOS-IC	Analog Switch, JFET Input, 36V, <30Ω		Six	-	
DG 130	Ge-Di	TV Damper-Di, 120V, 3A	23d1&	Say	AY 102, AY 106	data
DG 133 A,B	BiMOS-IC	Analog Switch, JFET Input, 36V, 30mA A: -55...+125°, B: -25...+85°	14-DIC	Six	-	
DG 134 A,B	BiMOS-IC	Analog Switch, JFET Input, 36V, 30mA A: -55...+125°, B: -25...+85°	14-DIC	Six	-	
DG 140	Ge-Di	=DG 130: 250V	23d1&	Say	AY 102	data
DG 140 A,B	BiMOS-IC	Analog Switch, JFET Input, 36V, 30mA A: -55...+125°, B: -25...+85°	14-DIC	Six	-	
DG 141 A,B	BiMOS-IC	Analog Switch, JFET Input, 36V, 30mA A: -55...+125°, B: -25...+85°	14-DIC	Six	-	
DG 172 A,B	PMOS-IC	Analog Switch(SPST), 36V, 20mA A: -55...+125°, B: -25...+85°	14-DIC	Six	-	

Type	Device	Short Description	Fig.	Manu	Comparision Types	More at
DG180	SPST	Ron:Ohm Ton:240ns at 15/-15V	14D,10AF	Vis	-	data pdf pinout
DG 180 AA	BiMOS-IC	Analog Switch, JFET Input, -55...+125° 36V, 200mA, on<20Ω	82	Six	-	
DG 180 B...	BiMOS-IC	=DG 180A..: -25...+85°, on<25Ω			-	
DG181	SPST	Ron:Ohm Ton:85ns at 15/-15V	10A,14DF	Vis		data pdf pinout
DG 181 AA	BiMOS-IC	Analog Switch, JFET Input, -55...+125° 36V, 30mA, on<60Ω	82	Six	-	pdf
DG 181 B...	BiMOS-IC	=DG 181A..: -25...+85°, on<75Ω			-	
DG182	SPST	Ron:Ohm Ton:120ns at 15/-15V	10A,14DF	Vis		data pdf pinout
DG 182 AA	BiMOS-IC	Analog Switch, JFET Input, -55...+125° 36V, 30mA, on<150Ω	82	Six	TL 182...	pdf
DG 182 B...	BiMOS-IC	=DG 182A..: -25...+85°, on<150Ω			TL 182...	
DG 180...182...P	BiMOS-IC	=DG 180...182(A,B)A:	14-DIP		-	
DG 183 AP	BiMOS-IC	Analog Switch, JFET Input, -55...+125° 36V, 200mA, on<20Ω	16-DIP	Six	-	
DG 183 BP	BiMOS-IC	=DG 183AP: -25...+85°, on<25Ω	16-DIP			pdf
DG184	SPST	Ron:Ohm Ton:85ns at 15/-15V	16D,14F	Vis		data pdf pinout
DG 184 AP	BiMOS-IC	Analog Switch, JFET Input, -55...+125° 36V, 30mA, on<60Ω	16-DIP	Six	-	
DG 184 BP	BiMOS-IC	=DG 184AP: -25...+85°, on<75Ω	16-DIP			pdf
DG185	SPST	Ron:Ohm Ton:120ns at 15/-15V	16D,14F	Vis		data pdf pinout
DG 185 AP	BiMOS-IC	Analog Switch, JFET Input, -55...+125° 36V, 30mA, on<150Ω	16-DIP	Six	TL 185...	pdf
DG 185 BP	BiMOS-IC	=DG 185AP: -25...+85°, on<150Ω	16-DIP		TL 185...	pdf
DG186	SPST	Ron:Ohm Ton:240ns at 15/-15V	10A,14D	Vis	-	data pdf pinout
DG 186 AA	BiMOS-IC	Analog Switch, JFET Input, -55...+125° 36V, 200mA, on<20Ω	82	Six	-	
DG 186 B...	BiMOS-IC	=DG 186A..: -25...+85°, on<25Ω			-	
DG187	SPST	Ron:Ohm Ton:85ns at 15/-15V	10A,14DF	Vis		data pdf pinout
DG 187 AA	BiMOS-IC	Analog Switch, JFET Input, -55...+125° 36V, 30mA, on<60Ω	82	Six	-	
DG 187 B...	BiMOS-IC	=DG 187A..: -25...+85°, on<75Ω			-	
DG188	SPST	Ron:Ohm Ton:120ns at 15/-15V	10A,14DF	Vis		data pdf pinout
DG 188 AA	BiMOS-IC	Analog Switch, JFET Input, -55...+125° 36V, 30mA, on<150Ω	82	Six	TL 188...	pdf
DG 188 B...	BiMOS-IC	=DG 188A..: -25...+85°, on<150Ω			TL 188...	
DG 186...188...P	BiMOS-IC	=DG 186...188(A,B)A:	14-DIP		-	
DG189	SPDT	Ron:Ohm Ton:240ns at 15/-15V	16D	Vis		data pdf pinout
DG 189 AP	BiMOS-IC	Analog Switch, JFET Input, -55...+125° 36V, 200mA, on<20Ω	16-DIP	Six	-	
DG 189 BP	BiMOS-IC	=DG 189AP: -25...+85°, on<25Ω	16-DIP			
DG190	SPDT	Ron:Ohm Ton:85ns at 15/-15V	16D,14F	Vis		data pdf pinout
DG 190 AP	BiMOS-IC	Analog Switch, JFET Input, -55...+125° 36V, 30mA, on<60Ω	16-DIP	Six	-	pdf
DG 190 BP	BiMOS-IC	=DG 190AP: -25...+85°, on<75Ω	16-DIP			pdf
DG191	SPDT	Ron:Ohm Ton:120ns at 15/-15V	16D,14F	Vis		data pdf pinout
DG 191 AP	BiMOS-IC	Analog Switch, JFET Input, -55...+125° 36V, 30mA, on<150Ω	16-DIP	Six	TL 191...	pdf
DG 191 BP	BiMOS-IC	=DG 191AP: -25...+85°, on<150Ω	16-DIP		TL 191...	pdf
DG200	SPST	Ron:Ohm Ton:440ns at 15/-15V	10A,14D	Vis	-	data pdf pinout
DG 200 AAA	CMOS-IC	2x Analog Switch, 44V, 30mA,-55...+125° on<100Ω, <1000/425ns	82	Six	-	
DG 200 AB...	CMOS-IC	=DG 200AA..: -25...+85°			-	
DG 200 ACJ	CMOS-IC	=DG 200AA..: 0...+70°	14-DIP		-	
DG200B	SPST	Ron:Ohm Ton:300ns at 15/-15V	14D	Vis		data pdf pinout
DG 200 A ... K	CMOS-IC	=DG 200A ... A:	14-DIC		-	
DG201	SPST	Ron:Ohm Ton:480ns at 15/-15V	16D	Vis		data pdf pinout
DG 201 AAK	CMOS-IC	4x Analog Switch, 44V, 20mA,-55...+125° on<250Ω, <600/450ns, Logic 0 = On	16-DIC	Six	-	pdf
DG 201 ABK	CMOS-IC	=DG 201AAK: -25...+85°	16-DIC		-	
DG 201 ACJ,ACK	CMOS-IC	=DG 201AAK: 0...+70°	16-DIP,DIC		-	
DG 201 ACSE	CMOS-IC	SMD,4x Analog Switch, 44V, 20mA	16-MDIP			pdf
DG 201 ADY	CMOS-IC	=DG 201AAK: SMD, -40...+85°	16-MDIP		-	
DG201HS	SPST	Ron:Ohm Ton:48ns at 15/-15V	16D,20CS	Vis		data pdf pinout
DG202	SPST	Vs:±4.5...±18V Ron:Ohm: at 15/-15V	16DSd	Max		data pdf pinout
DG202A	SPST	Vs:±4.5...±18V Ron:Ohm at 15/-15V	16DSC	Max		data pdf pinout
DG 202 CJ	CMOS-IC	=DG 201A... ..: Logic 0 = OFF	16-DIP	Six	-	pdf
DG211	SPST	Vs:±4.5...±22V Ron:Ohm at 15/-15V	16DS	Isi		data pdf pinout
DG 211 CJ	CMOS-IC	4x Analog Switch, 44V, 20mA, 0...+70° on<175Ω, <1000/500ns, Logic 0 = On	16-DIP	Six	-	pdf
DG 211 CSE	CMOS-IC	SMD,4x Analog Switch, 44V, 20mA	16-MDIP		-	pdf
DG 211 DY	CMOS-IC	=DG 211CJ: SMD, -40...+85°	16-MDIP			
DG212	SPST	Vs:±4.5...±18V Ron:Ohm at 15/-15V	16DSdC	Max		data pdf pinout
DG 212 CJ	CMOS-IC	=DG 211CJ: Logic 0 = OFF	16-DIP	Six	-	pdf
DG 212 CSE	CMOS-IC	SMD,=DG 211CJ: Logic 0 = OFF	16-MDIP		-	pdf
DG 212 DY	CMOS-IC	=DG 212CJ: SMD, -40...+85°	16-MDIP			
DG213	SPST	Vs:3..40V Ron:Ohm at 15/-15V	16CS	Vis		data pdf pinout
DG 221	CMOS-IC	4x Analog Switch, 44V, <90Ω, <550ns		Six		
DG221B	SPST	Ron:Ohm at 15/-15V	16DS	Vis		data pdf pinout
DG 243 A,B	CMOS-IC	Dual, Analog Switch(SPDT), 44V, 30mA A: -55...+125°, B: -25...+85°	14-DIC	Six	-	pdf
DG 271 AK	CMOS-IC	4x Analog Switch, 44V, 20mA,-55...+125° on<75Ω, <65/80ns	16-DIC	Six	-	pdf
DG271B	SPST	Ron:Ohm Ton:55ns at 15/-15V	16DS	Vis		data pdf pinout
DG 271 CJ	CMOS-IC	=DG 271AK: 0...+70°	16-DIP			

Type	Device	Short Description	Fig.	Manu	Comparision Types	More at
DG 271 DY	CMOS-IC	=DG 271AK: SMD, -40...+85°	16-MDIP		-	
DG300	SPST	Vs:±5..±15V Ron:Ohm Ton:70ns at 30/0V	10A,14D	Max		data pdf pinout
DG300A	SPST	Vs:±5..±15V Ron:Ohm at 15/-15V	10A,14D	Vis		data pdf pinout
DG 300 AAA	CMOS-IC	2x1 Analog Switch,44V, 30mA, -55...+125° on<75Ω, <300/250ns	82	Six	-	
DG 300 AB...	CMOS-IC	=DG 300AA...: -25...+85°		Six		
DG 300 ACJ	CMOS-IC	=DG 300AAA: 0...+70°	14-DIP	Six		pdf
DG300B	SPST	Vs:±5..±15V Ron:Ohm at 15/-15V	14D	Vis		data pdf pinout
DG 300 A ... K	CMOS-IC	=DG 300A ... A:	14-DIC	Six		
DG301	SPDT	Vs:±5..±15V Ron:Ohm Ton:70ns at 30/0V	10A,14D	Max		
DG301A	SPDT	Vs:±5..±15V Ron:Ohm at 15/-15V	10A,14D	Vis		data pdf pinout
DG 301 AAA	CMOS-IC	1x2 Analog Switch,44V, 30mA, -55...+125° on<75Ω, <300/250ns	82	Six	-	data pdf pinout
DG 301 AB...	CMOS-IC	=DG 301AA..A: -25...+85°		Six		
DG 301 ABA	CMOS-IC	=DG 301AAA: -25...+85°	82	Six		pdf
DG 301 ACJ	CMOS-IC	=DG 301AAA: 0...+70°	14-DIP	Six		pdf
DG301B	SPDT	Vs:±5..±15V Ron:Ohm at 15/-15V	14D	Vis		data pdf pinout
DG 301 A ... K	CMOS-IC	=DG 301A ... A:	14-DIC	Six		
DG302	DPST	Vs:±5..±15V Ron:Ohm Ton:70ns at 30/0V	14D,16Sd	Max		
DG302A	DPST	Vs:±5..±15V Ron:Ohm at 15/-15V	14D,16Sd	Vis		data pdf pinout
DG 302 AAK	CMOS-IC	2x2 Analog Switch,44V, 30mA, -55...+125° on<75Ω, <300/250ns	14-DIC	Six		data pdf pinout
DG 302 ACJ	CMOS-IC	=DG 302AAK: 0...+70°	14-DIP	Six		
DG302B	DPST	Vs:±5..±15V Ron:Ohm at 15/-15V	14D	Vis		data pdf pinout
DG303	SPDT	Vs:±5..±15V Ron:Ohm Ton:70ns at 30/0V	14D,16Sd	Max		data pdf pinout
DG303A	SPDT	Vs:±5..±15V Ron:Ohm at 15/-15V	14D,16Sd	Vis		data pdf pinout
DG 303 AAK	CMOS-IC	2x2 Analog Switch,44V, 30mA, -55...+125° on<75Ω, <300/250ns	14-DIC	Six		pdf
DG 303 ACJ	CMOS-IC	=DG 303AAK: 0...+70°	14-DIP	Six		pdf
DG303B	SPDT	Vs:±5..±15V Ron:Ohm at 15/-15V	14DS	Vis		data pdf pinout
DG304	SPST	Vs:±5..±15V Ron:Ohm at 30/0V	10A,14D	Max		data pdf pinout
DG304A	SPST	Vs:±5..±15V Ron:Ohm at 15/-15V	14D,10A	Vis		data pdf pinout
DG 304 AAK	CMOS-IC	2x1 Analog Switch,44V, 30mA, -55...+125° on<75Ω, <250/150ns	14-DIC	Six		
DG 304 ACJ	CMOS-IC	=DG 304AAK: 0...+70°	14-DIP			pdf
DG304B	SPST	Vs:±5..±15V Ron:Ohm at 15/-15V	14D	Vis		data pdf pinout
DG305	SPDT	Vs:±5..±15V Ron:Ohm at 30/0V	10A,14D	Max		data pdf pinout
DG305A	SPDT	Vs:±5..±15V Ron:Ohm at 15/-15V	10A,14D	Vis		data pdf pinout
DG 305 AAA	CMOS-IC	1x2 Analog Switch,44V, 30mA, -55...+125° on<75Ω, <250/150ns	82	Six		
DG306	DPST	Vs:±5..±15V Ron:Ohm at 30/0V	14D,16Sd	Max		data pdf pinout
DG306A	DPST	Vs:±5..±15V Ron:Ohm at 15/-15V	14D,16Sd	Vis		data pdf pinout
DG 306 ACJ	CMOS-IC	2x2 Analog Switch,44V, 30mA, -55...+125° on<75Ω, <250/150ns	14-DIP	Six		pdf
DG306B	SPST	Vs:±5..±15V Ron:Ohm at 15/-15V	14D	Vis		data pdf pinout
DG307	SPDT	Vs:±5..±15V Ron:Ohm at 30/0V	14D,16Sd	Max		data pdf pinout
DG307A	SPDT	Vs:±5..±15V Ron:Ohm at 15/-15V	14D,16Sd	Vis		data pdf pinout
DG 307 AAK	CMOS-IC	2x2 Analog Switch,44V, 30mA, -55...+125° on<75Ω, <250/150ns	14-DIC	Six	-	
DG 307 ABK	CMOS-IC	=DG 307AAK: -25...+85°	14-DIC			
DG 307 ACJ	CMOS-IC	=DG 307AAK: 0...+70°	14-DIP		-	
DG307B	SPDT	Vs:±5..±15V Ron:Ohm at 15/-15V	14D	Vis		pdf
DG308A	SPST	Vs:±5..±20V Ron:Ohm at 15/-15V	16DdS	Max,Vis		data pdf pinout
DG 308 AAK	CMOS-IC	4x Analog Switch, 44V, 20mA, -55...+125° on<150Ω, <200/150ns, Logic 0 = OFF	16-DIC	Six		data pdf pinout pdf
DG 308 ABK	CMOS-IC	=DG 308AAK: -25...+85°, on<125Ω	16-DIC			
DG 308 ACJ	CMOS-IC	=DG 308AAK: 0...+70°, on<125Ω	16-DIP			
DG 308 ACK	CMOS-IC	=DG 308AAK: 0...+70°, on<125Ω	16-DIC			pdf
DG 308 ADY	CMOS-IC	=DG 308AAK: SMD, -40...+85°, on<125Ω	16-MDIP			pdf
DG308B	SPST	Vs:±4..±22V Ron:Ohm at 15/-15V	16DS	Vis		pdf
DG309	SPST	Vs:±5..±20V Ron:Ohm at 15/-15V	16DdS	Max,Vis		data pdf pinout
DG 309 AK	CMOS-IC	=DG 308AAK: Logic 0 = On	16-DIC	Six		data pdf pinout
DG309B	SPST	Vs:±4..±22V Ron:Ohm at 15/-15V	16DS	Vis		pdf
DG 309 CJ	CMOS-IC	=DG 309AK: 0...+70°, on<125Ω	16-DIP			data pdf pinout
DG 309 DY	CMOS-IC	=DG 309AK: SMD, -40...+85°, on<125Ω	16-MDIP			pdf
DG333	SPDT	Vs:±4..±22V Ron:Ohm at 15/-15V	20DS	Vis		pdf
DG381	SPST	Vs:±5..±15V Ron:Ohm Ton:70ns at 30/0V	10A,14D	Max		data pdf pinout
DG381A	SPST	Vs:±5..±15V Ron:Ohm Ton:70ns at 30/0V	14D,10A	Max		data pdf pinout
DG 381 AAK	CMOS-IC	2x1 Analog Switch,44V, 30mA, -55...+125° on<75Ω, <300/250ns	14-DIC	Six		data pdf pinout
DG 381 ABK	CMOS-IC	=DG 381AAK: -25...+85°	14-DIC			
DG 381 ACJ	CMOS-IC	=DG 381AAK: 0...+70°	14-DIP		-	pdf
DG381B	SPST	Vs:±5..±15V Ron:Ohm at 15/-15V	14D	Vis		pdf
DG384	DPST	Vs:±5..±15V Ron:Ohm Ton:70ns at 30/0V	16DSd	Max		data pdf pinout
DG384A	DPST	Vs:±5..±15V Ron:Ohm at 15/-15V	16DSd	Vis		data pdf pinout
DG 384 AAK	CMOS-IC	2x2 Analog Switch,44V, 30mA, -55...+125° on<75Ω, <300/250ns	16-DIC	Six		data pdf pinout
DG 384 ACJ	CMOS-IC	=DG 384AAK: 0...+70°	16-DIP			
DG384B	DPST	Vs:±5..±15V Ron:Ohm at 15/-15V	14D	Vis		pdf
DG387	SPDT	Vs:±5..±15V Ron:Ohm Ton:70ns at 30/0V	10A,14D	Max		data pdf pinout
DG387A	SPDT	Vs:±5..±15V Ron:Ohm Ton:70ns at 30/0V	10A,14D	Max		data pdf pinout
DG 387 AAA	CMOS-IC	1x2 Analog Switch,44V, 30mA, -55...+125° on<75Ω, <300/250ns	82	Six	-	data pdf pinout
DG 387 AAK	CMOS-IC	=DG 387AAA:	14-DIC			
DG 387 ACJ	CMOS-IC	=DG 387AAA: 0...+70°	14-DIP		-	pdf

Type	Device	Short Description	Fig.	Manu	Comparision Types	More at
G387B	SPDT	Vs:±5..±15V Ron:Ohm at 15/-15V	14D	Vis	-	data pdf pinout
G390	SPDT	Vs:±5..±15V Ron:Ohm Ton:70ns at 30/0V	16DSd	Max		data pdf pinout
G390A	SPDT	Vs:±5..±15V Ron:Ohm Ton:70ns at 30/0V	16DSd	Max		data pdf pinout
G 390 AAK	CMOS-IC	2x2 Analog Switch,44V, 30mA,-55...+125° on<75Ω, <300/250ns	16-DIC	Six	-	
G-390 ABK	CMOS-IC	=DG 390AAK: -25...+85°	16-DIC			pdf
G 390 ACJ	CMOS-IC	=DG 390AAK: 0...+70°	16-DIP			pdf
G 390 ACK	CMOS-IC	=DG 390AAK: 0...+70°	16-DIC			pdf
G390B	SPDT	Vs:±5..±15V Ron:Ohm at 15/-15V	16D	Vis		data pdf pinout
G401	SPST	Ron:Ohm Ton:100ns at 15/-15V	20C,16DS	Vis		data pdf pinout
G 401 AK	CMOS-IC	2x1 Analog Switch,44V, 30mA,-55...+125° on<45Ω, <150/100ns	16-DIC	Six	-	pdf
G 401 DJ	CMOS-IC	=DG 401AK: -40...+85°, on<55Ω	16-DIP			pdf
G403	SPDT	Ron:Ohm Ton:100ns at 15/-15V	16D,20CS	Vis		data pdf pinout
G 403 AK	CMOS-IC	2x2 Analog Switch,44V, 30mA,-55...+125° on<45Ω, <150/100ns	16-DIC	Six	-	pdf
G 403 DJ	CMOS-IC	=DG 403AK: -40...+85°, on<55Ω	16-DIP			pdf
G 403 DY	CMOS-IC	=DG 403AK: SMD, -40...+85°, on<55Ω	16-MDIP			pdf
G405	DPST	Ron:Ohm Ton:100ns at 15/-15V	20C,16DS	Vis		data pdf pinout
G 405 AK	CMOS-IC	2x2 Analog Switch,44V, 30mA,-55...+125° on<45Ω, <150/100ns	16-DIC	Six	-	pdf
G 405 DJ	CMOS-IC	=DG 405AK: -40...+85°, on<55Ω	16-DIP			pdf
G 405 DY	CMOS-IC	=DG 405AK: SMD, -40...+85°, on<55Ω	16-MDIP			pdf
G406	MULT16	Ron:Ohm Ton:150ns at 15/-15V	28DCS	Isi		data pdf pinout
G 406 AK	CMOS-IC	16-Ch. Analog Multiplexer, -55...+125° 44V, 20mA, on<125Ω, <200/150ns	28-DIC	Six	-	pdf
G406B	MULT16	Vs:±5..±20V Ron:Ohm at 15/-15V	28DCS	Vis		data pdf pinout
G 406 DJ	CMOS-IC	=DG 406AK: -40...+85°	28-DIP			pdf
G 406 DN	CMOS-IC	=DG 406AK: -40...+85°	28-PLCC			
G407	MULT16	Ron:Ohm Ton:150ns at 15/-15V	28DCS	Isi		data pdf pinout
G 407 AK	CMOS-IC	2x8-Ch. Analog Multiplexer, -55...+125° 44V, 20mA, on<125Ω, <200/150ns	28-DIC	Six	-	pdf
G407B	MULT8	Vs:±5..±20V Ron:Ohm at 15/-15V	28DCS	Vis		data pdf pinout
G 407 DJ	CMOS-IC	=DG 407AK: -40...+85°	28-DIP			pdf
G408	MULT8	Ron:Ohm Ton:115ns at 15/-15V	20C,16DS	Vis		data pdf pinout
G 408 AK	CMOS-IC	8-Ch. Analog Multiplexer, -55...+125° 44V, 20mA, on<125Ω, <225/150ns	16-DIC	Six	-	pdf
G 408 DJ	CMOS-IC	=DG 408AK: -40...+85°	16-DIP			pdf
G 408 DY	CMOS-IC	=DG 408AK: SMD, -40...+85°	16-MDIP			pdf
G408L	MULT8	Ron:Ohm Ton:43ns at 5/0V	16D,20CS	Vis		data pdf pinout
G409	MULT4	Ron:Ohm Ton:115ns at 15/-15V	20C,16DS	Vis		data pdf pinout
G 409 AK	CMOS-IC	2x4-Ch. Analog Multiplexer, -55...+125° 44V, 20mA, on<125Ω, <225/150ns	16-DIC	Six	-	pdf
G 409 DJ	CMOS-IC	=DG 409AK: -40...+85°	16-DIP			pdf
G 409 DY	CMOS-IC	=DG 409AK: SMD, -40...+85°	16-MDIP			pdf
G409L	MULT4	Ron:Ohm Ton:43ns at 5/0V	16D,20CS	Vis		data pdf pinout
G411	SPST	Ron:Ohm Ton:110ns at 15/-15V	20C,16DS	Vis		data pdf pinout
G 411 AK	CMOS-IC	4x Analog Switch, 44V, 30mA,-55...+125° on<45Ω, <220/160ns, Logic 0 = On	16-DIC	Six	-	pdf
G 411 DJ	CMOS-IC	=DG 411AK: -40...+85°	16-DIP			pdf
G 411 DY	CMOS-IC	=DG 411AK: SMD, -40...+85°	16-MDIP			pdf
G411HS	SPST	Ron:Ohm Ton:68ns at 15/-15V	16D,20CHS	Vis		data pdf pinout
G411L	SPST	Vs:3...12V Ron:Ohm Ton:27ns at 5/0V	16S	Vis		data pdf pinout
G412	SPST	Ron:Ohm Ton:110ns at 15/-15V	20C,16DS	Vis		data pdf pinout
G 412 AK	CMOS-IC	4x Analog Switch, 44V, 30mA,-55...+125° on<45Ω, <220/160ns, Logic 0 = On	16-DIC	Six	-	pdf
G 412 DJ	CMOS-IC	=DG 412AK: -40...+85°	16-DIP			pdf
G 412 DY	CMOS-IC	=DG 412AK: SMD, -40...+85°	16-MDIP			pdf
G412HS	SPST	Ron:Ohm Ton:68ns at 15/-15V	16DHS	Vis		data pdf pinout
G412L	SPST	Vs:3...12V Ron:Ohm Ton:27ns at 5/0V	16S	Vis		data pdf pinout
G413	SPST	Ron:Ohm Ton:110ns at 15/-15V	20C,16DS	Vis		data pdf pinout
G 413 AK	CMOS-IC	4x Analog Switch, 44V, 30mA,-55...+125° on<45Ω, <220/160ns Sw1+4: Logic 0=OFF, Sw2+3: Logic 0=ON	16-DIC	Six	-	pdf
G 413 DJ	CMOS-IC	=DG 413AK: -40...+85°	16-DIP			pdf
G 413 DY	CMOS-IC	=DG 413AK: SMD, -40...+85°	16-MDIP			pdf
G413HS	SPST	Ron:Ohm Ton:68ns at 15/-15V	16DHS	Vis		data pdf pinout
G413L	SPST	Vs:3...12V Ron:Ohm Ton:27ns at 5/0V	16S	Vis		data pdf pinout
G417	SPDT	Vs:±4.5..20V Ron:Ohm at 15/-15V	8DS	Max		data pdf pinout
G 417 AK	CMOS-IC	Analog Switch, 44V, 30mA, -55...+125° 44V, 30mA, on<45Ω, <250/210ns, Log.0=ON	8-DIC	Six	-	pdf
G417B	SPST	Ron:Ohm Ton:62ns at 15/-15V	8DS	Vis		data pdf pinout
G 417 DJ	CMOS-IC	=DG 417AK: -40...+85°	8-DIP			pdf
G 417 DY	CMOS-IC	=DG 417AK: SMD, -40...+85°	8-MDIP			pdf
G417L	SPST	Vs:3...12V Ron:Ohm Ton:37ns at 5/0V	8S	Vis		data pdf pinout
G418	SPDT	Vs:9...36V Ron:Ohm Ton:150ns at 12/0V	8DS	Max		data pdf pinout
G 418 AK	CMOS-IC	Analog Switch, 44V, 30mA, -55...+125° 44V, 30mA, <45Ω, <250/210ns, Log.0=OFF	8-DIC	Six	-	pdf
G418B	SPST	Ron:Ohm Ton:62ns at 15/-15V	8DS	Vis		data pdf pinout
DG 418 DJ	CMOS-IC	=DG 418AK: -40...+85°	8-DIP			pdf
DG 418 DY	CMOS-IC	=DG 418AK: SMD, -40...+85°	8-MDIP			pdf
DG418L	SPST	Vs:3...12V Ron:Ohm Ton:37ns at 5/0V	8S	Vis		data pdf pinout
DG419	SPDT	Vs:±4.5..20V Ron:Ohm at 15/-15V	8DS	Max		data pdf pinout
DG 419 AK	CMOS-IC	Analog Switch, 44V, 30mA, -55...+125°	8-DIC	Six	-	pdf

Type	Device	Short Description	Fig.	Manu	Comparision Types	More at
		44V, 30mA, on<45Ω, <250/210ns Sw1: Logic 0=ON, Sw2: Logic 0=OFF				
DG419B	SPST	Ron:Ohm Ton:62ns at 15/-15V	8DS	Vis		data pdf pinout
DG 419 DJ	CMOS-IC	=DG 419AK: -40...+85°	8-DIP		·	pdf
DG 419 DY	CMOS-IC	=DG 419AK: SMD, -40...+85°	8-MDIP			pdf
DG419L	SPDT	Vs:3...12V Ron:Ohm Ton:37ns at 5/0V	8S	Vis		data pdf pinout
DG 421 AK	CMOS-IC	2x1 Analog Switch,44V, 40mA,-55...+125° on<45Ω, <300/200ns	16-DIC	Six	·	pdf
DG 421 DJ	CMOS-IC	=DG 421AK: -40...+85°	16-DIP			pdf
DG 423 AK	CMOS-IC	2x2 Analog Switch,44V, 40mA,-55...+125° on<45Ω, <300/200ns	16-DIC	Six		pdf
DG 423 DJ	CMOS-IC	=DG 423AK: -40...+85°	16-DIP			pdf
DG 423 DN	CMOS-IC	=DG 423AK: -40...+85°	20-PLCC			pdf
DG 425 AK	CMOS-IC	2x2 Analog Switch, 44V, 40mA,-55...+125° on<45Ω, <300/200ns	16-DIC	Six		pdf
DG 425 DJ	CMOS-IC	=DG 425AK: -40...+85°	16-DIP			pdf
DG428	MULT8	Ron:Ohm Ton:90ns at 15/-15V	18D,20C	Vis		data pdf pinout
DG 428	CMOS-IC	8-Ch. Analog Multiplexer, -15...+15V		Six		pdf
DG429	MULT8	Ron:Ohm Ton:90ns at 15/-15V	18D,20CS	Vis		data pdf pinout
DG 429	CMOS-IC	2x4-Ch. Analog Multiplexer, -15...+15V		Six		pdf
DG441	SPST	Ron:Ohm Ton:150ns at 15/-15V	16DS	Isi		data pdf pinout
DG 441 AK	CMOS-IC	4x Analog Switch, 44V, 30mA,-55...+125° on<100Ω, <250/170ns, Logic 0 = On	16-DIC	Six	·	pdf
DG441B	SPST	Ron:Ohm Ton:120ns at 15/-15V	16DHS	Vis		data pdf pinout
DG 441 DJ	CMOS-IC	=DG 441AK: -40...+85°	16-DIP		·	pdf
DG 441 DY	CMOS-IC	=DG 441AK: SMD, -40...+85°	16-MDIP			pdf
DG441L	SPST	Vs:3...12V Ron:Ohm Ton:27ns at 5/0V	16S	Vis		data pdf pinout
DG442	SPST	Ron:Ohm Ton:150ns at 15/-15V	16DS	Vis		data pdf pinout
DG 442 AK	CMOS-IC	4x Analog Switch, 44V, 30mA,-55...+125° on<100Ω, <250/170ns, Logic 0 = OFF	16-DIC	Six		pdf
DG442B	SPST	Ron:Ohm Ton:120ns at 15/-15V	16DHS	Vis		data pdf pinout
DG 442 DJ	CMOS-IC	=DG 442AK: -40...+85°	16-DIP			pdf
DG 442 DY	CMOS-IC	=DG 442AK: SMD, -40...+85°	16-MDIP			pdf
DG442L	SPST	Vs:3...12V Ron:Ohm Ton:27ns at 5/0V	16S	Vis	·	data pdf pinout
DG444	SPST	Ron:Ohm Ton:120ns at 15/-15V	16DS	Vis		data pdf pinout
DG444B	SPST	Ron:Ohm Ton:120ns at 15/-15V	16DHS	Vis		data pdf pinout
DG 444 DJ	CMOS-IC	4x Analog Switch, 44V, 30mA, -40...+85° on<100Ω, <250/170ns, Logic 0 = On	16-DIP	Six		pdf
DG 444 DY	CMOS-IC	=DG 444AK: SMD	16-MDIP			pdf
DG445	SPST	Ron:Ohm Ton:120ns at 15/-15V	16DS	Vis		data pdf pinout
DG445B	SPST	Ron:Ohm Ton:120ns at 15/-15V	16DHS	Vis		data pdf pinout
DG 445 DJ	CMOS-IC	4x Analog Switch, 44V, 30mA, -40...+85° on<100Ω, <250/170ns, Logic 0 = OFF	16-DIP	Six		pdf
DG 445 DY	CMOS-IC	=DG 445AK: SMD	16-MDIP		·	pdf
DG458	MULT8	Vs:±4.5...±18V Ron:Ohm at 15/-15V	16D,20C	Vis		data pdf pinout
DG 458 AK	CMOS-IC	8-Ch. Analog Multiplexer, -55...+125° on<500Ω, <250/500ns	16-DIC	Six		pdf
DG 458 DJ	CMOS-IC	=DG 458AK: -40...+85°	16-DIP			pdf
DG459	MULT4	Vs:±4.5...±18V Ron:Ohm at 15/-15V	16D,20C	Vis		data pdf pinout
DG 459 AK	CMOS-IC	2x4-Ch. Analog Multiplexer, -55...+125° on<500Ω, <250/500ns	16-DIC	Six		
DG 459 DJ	CMOS-IC	=DG 459AK: -40...+85°	16-DIP			pdf
DG485	ARRAY8	Ron:Ohm Ton:170ns at 15/0V	20C,18D	Vis		data pdf pinout
DG 485 AK	CMOS-IC	8x Analog Switch, 44V, 30mA,-55...+125° on<125Ω, <200/275ns	18-DIC	Six	·	
DG 485 DJ	CMOS-IC	=DG 485AK: -40...+85°	18-DIP		·	pdf
DG 485 DN	CMOS-IC	=DG 485AK: -40...+85°	20-PLCC		·	pdf
DG506	MULT16	Vs:±4.5...±18V Ron:230Ohm at 15/-15V	28DdS	Max	·	data pdf pinout
DG 506 AAK	CMOS-IC	16-Ch. Analog Multiplexer, -55...+125° 44V, 20mA, on<500Ω	28-DIC	Six	·	pdf
DG 506 ABK,ABR	CMOS-IC	=DG 506AAK: -25...+85°, on<550Ω	28-DIC,DIP			
DG 506 ACJ,ACK	CMOS-IC	=DG 506AAK: 0...+70°, on<550Ω	28-DIP,DIC			
DG 506 ACWI	CMOS-IC	SMD,16-Ch. Analog Multiplexer, 0...+70°	28-MDIP			pdf
DG 506 ADN	CMOS-IC	=DG 506AAK: -40...+85°, on<550Ω	28-PLCC			
DG507	MULT8	Vs:±4.5...±18V Ron:230Ohm at 15/-15V	28DdS	Max		data pdf pinout
DG 507 AAK	CMOS-IC	2x8-Ch. Analog Multiplexer, -55...+125° 44V, 20mA, on<500Ω	28-DIC	Six	·	pdf
DG 507 ABK,ABR	CMOS-IC	=DG 507AAK: -25...+85°, on<550Ω	28-DIC,DIP			
DG 507 ACJ,ACK	CMOS-IC	=DG 507AAK: 0...+70°, on<550Ω	28-DIP,DIC			
DG 507 ACWI	CMOS-IC	SMD,2x8-Ch. Analog Multiplexer,0...+70°	28-MDIP			pdf
DG508A	MULT8	Ron:270Ohm 1μsat 15/-15V	16D	Isi		data pdf pinout
DG 508 AAK	CMOS-IC	8-Ch. Analog Multiplexer, -55...+125° 44V, 20mA, on<500Ω	16-DIC	Six	DG 458...	pdf
DG 508 ABK	CMOS-IC	=DG 508AAK: -25...+85°, on<550Ω	16-DIC		·	pdf
DG 508/ACJ,ACK	CMOS-IC	=DG 508AAK: 0...+70°, on<550Ω	16-DIP,DIC			
DG 508 ACWE	CMOS-IC	SMD,8-Ch. Analog Multiplexer, 0...+70°	16-MDIP			pdf
DG 508 ADY	CMOS-IC	=DG 508AAK: SMD, -40...+85°, on<550Ω	16-MDIP			pdf
DG 509 AAK	CMOS-IC	2x4-Ch. Analog Multiplexer, -55...+125° 44V, 20mA, on<500Ω	16-DIC	Six	DG 459...	pdf
DG 509 ABK	CMOS-IC	=DG 509AAK: -25...+85°, on<550Ω	16-DIC		·	pdf
DG 509 ACJ,ACK	CMOS-IC	=DG 509AAK: 0...+70°, on<550Ω	16-DIP,DIC			
DG 509 ACWE	CMOS-IC	SMD,2x4-Ch. Analog Multiplexer,0...+70°	16-MDIP			pdf
DG 509 ADY	CMOS-IC	=DG 509AAK: SMD, -40...+85°, on<550Ω	16-MDIP			pdf
DG528	MULT8	Ron:Ohm 1μsat 15/-15V	18D,20CdS	Vis		data pdf pinout
DG 528 AK	CMOS-IC	8-Ch. Analog Multiplexer, -55...+125°	18-DIC	Six	·	pdf

Type	Device	Short Description	Fig.	Manu	Comparision Types	More at
		44V, 20mA, on<500Ω			-	pdf
DG 528 BK	CMOS-IC	=DG 528AK: -25...+85°, on<550Ω	18-DIC		-	pdf
DG 528 CJ,CK	CMOS-IC	=DG 528AK: 0...+70°, on<550Ω	18-DIP,DIC			
DG 528 DN	CMOS-IC	=DG 528AK: 0...+70°, on<550Ω	20-PLCC		-	pdf
DG529	MULT4	Vs:±4,5...±20V Ron:Ohm at 15/-15V	18D,20CdS	Max		data pdf pinout
DG 529 AK	CMOS-IC	2x4-Ch. Analog Multiplexer, -55...+125°	18-DIC	Six		pdf
		44V, 20mA, on<500Ω				
DG 529 BK	CMOS-IC	=DG 529AK: -25...+85°, on<550Ω	18-DIC		-	pdf
DG 529 CJ	CMOS-IC	=DG 529AK: 0...+70°, on<550Ω	18-DIP		-	pdf
DG534A	MULT4/2	Ron:Ohm at 15/-3V	20DC	Vis		data pdf pinout
DG 534 AP	CMOS-IC	4-Ch. Wideband Multiplexer, -55...+125°	20-DIP	Six	-	
		21V, 20mA, on<120Ω, <500/300ns				
DG 534 DJ	CMOS-IC	=DG 534AP: -40...+85°	20-DIP		-	
DG 534 DN	CMOS-IC	=DG 534AP: -40...+85°	20-PLCC		-	
DG535	MULT16	Ron:Ohm at 15/0V	28D	Vis		data pdf pinout
DG 535 AP	CMOS-IC	16-Ch. Wideband Multiplexer,-55...+125°	28-DIP	Six	-	
		18V, 20mA, on<120Ω, <300/150ns				
DG 535 DJ	CMOS-IC	=DG 535AP: -40...+85°	28-DIP		-	
DG536	MULT16	Ron:Ohm at 15/0V	44C	Vis		data pdf pinout
DG 536 AM	CMOS-IC	16-Ch. Wideband Multiplexer,-55...+125°	44-PLCC	Six	-	
		18V, 20mA, on<120Ω, <300/150ns				
DG 536 DN	CMOS-IC	=DG 536AM: -40...+85°	44-PLCC		-	
DG538A	MULT8/4	Ron:Ohm at 15/-3V	28DC	Vis		data pdf pinout
DG 538 AP	CMOS-IC	8-Ch. Wideband Multiplexer, -55...+125°	28-DIP	Six	-	
		21V, 20mA, on<120Ω, <500/300ns				
DG 538 DJ	CMOS-IC	=DG 538AP: -40...+85°	28-DIP		-	
DG 538 DN	CMOS-IC	=DG 538AP: -40...+85°	28-PLCC		-	
DG540	SPST	Ron:Ohm Ton:45ns at 15/-3V	20DC	Vis		data pdf pinout
DG 540 AP	CMOS-IC	Wideband/Video 'T' Switch, -55...+125°	20-DIP	Six	-	pdf
		21V, 20mA, <100Ω, <130/85ns, Log.0=OFF				
DG 540 DJ	CMOS-IC	=DG 540AP: -40...+85°, on<75Ω	20-DIP		-	pdf
DG 540 DN	CMOS-IC	=DG 540AP: -40...+85°, on<75Ω	20-PLCC		-	pdf
DG541	SPST	Ron:Ohm Ton:45ns at 15/-3V	16DS	Vis		data pdf pinout
DG 541 AP	CMOS-IC	Wideband/Video 'T' Switch, -55...+125°	16-DIP	Six	-	pdf
		21V, 20mA, <100Ω, <130/85ns, Log.0=OFF				
DG 541 DJ	CMOS-IC	=DG 541AP: -40...+85°, on<75Ω	16-DIP		-	pdf
DG 541 DY	CMOS-IC	=DG 541AP: SMD, -40...+85°, on<75Ω	16-MDIP		-	pdf
DG542	SPST	Ron:Ohm Ton:55ns at 15/-3V	16DS	Vis		data pdf pinout
DG 542 AP	CMOS-IC	Wideband/Video 'T' Switch, -55...+125°	16-DIP	Six	-	pdf
		21V, 20mA, on<100Ω, <160/85ns Sw1+2:				
		Logic 0=OFF, Sw3+4: Logic 0=ON				
DG 542 DJ	CMOS-IC	=DG 542AP: -40...+85°, on<75Ω	16-DIP		-	pdf
DG 542 DY	CMOS-IC	=DG 542AP: SMD, -40...+85°, on<75Ω	16-MDIP		-	
DG 601 AK	CMOS-IC	4x Analog Switch, 22V, 30mA,-55...+125°	16-DIC	Six	-	
		on<50Ω, 45/30ns				
DG 601 DJ	CMOS-IC	=DG 601AK: -40...+85°	16-DIP		-	
DG 601 DY	CMOS-IC	=DG 601AK: SMD, -40...+85°	16-MDIP		-	
DG611	SPST	Ron:Ohm Ton:12ns at 15/-3V	20C,16DS	Vis		data pdf pinout
DG 611	CMOS-IC	4x Analog Switch, 21V, <45Ω, <35ns		Six	-	pdf
DG612	SPST	Ron:Ohm Ton:12ns at 15/-3V	20C,16DS	Vis		data pdf pinout
DG 612	CMOS-IC	4x Analog Switch, 21V, <45Ω, <35ns		Six		pdf
DG613	SPST	Ron:Ohm Ton:12ns at 15/-3V	20C,16DS	Vis		data pdf pinout
DG 613	CMOS-IC	4x Analog Switch, 21V, <45Ω, <35ns		Six	-	pdf
DG641	SPST	Ron:Ohm Ton:50ns at 15/-5V	16DS	Vis		data pdf pinout
DG 641	CMOS-IC	4x Analog Switch, 21V, <15Ω, <70ns		Six	-	pdf
DG642	SPDT	Ron:Ohm Ton:50ns at 15/-5V	8DS	Vis		data pdf pinout
DG 642	CMOS-IC	1x Analog Switch, 21V, <100Ω, <8ns		Six		pdf
DG643	SPST	Ron:Ohm Ton:50ns at 15/-5V	16DS	Vis		data pdf pinout
DG 643	CMOS-IC	1x Analog Switch, 21V, <70Ω, <15ns		Six	-	pdf
DG884	ARRAY4x8	Ron:Ohm at 15/-3V	44C	Vis		data pdf pinout
DG 884 DN	CMOS-IC	8x4 Wideband Crosspoint, -40...+85° 21V,	44-PLCC	Six		
		20mA, on<120Ω, <500/300ns				
DG 894 DN	CMOS-IC	SCART/S-VHS Video Switch, -40...+85° 19V,	28-DIP	Six		
		20mA, on<150Ω, <200/180ns				
DG2001	SPDT	Vs:1.8...2.2V Ron:Ohm Ton:20ns at 5/0V	6S	Vis		data pdf pinout
DG2002	SPDT	Vs:1.8...5.5V Ron:Ohm Ton:8ns at 5/0V	6S	Vis		data pdf pinout
DG2003	SPST	Vs:1.8...5.5V Ron:Ohm Ton:13ns at 5/0V	8S	Vis		data pdf pinout
DG2004	SPST	Vs:1.8...5.5V Ron:Ohm Ton:13ns at 5/0V	8S	Vis		data pdf pinout
DG2005	SPST	Vs:1.8...5.5V Ron:Ohm Ton:13ns at 5/0V	8S	Vis		data pdf pinout
DG2011	SPDT	Vs:1.8...5.5V Ron:Ohm Ton:75ns at 2/0V	16S	Vis		data pdf pinout
DG2012	SPDT	Vs:1.8...5.5V Ron:Ohm Ton:17ns at 5/0V	6S	Vis		data pdf pinout
DG2015	SPDT	Vs:2.7...3.3V Ron:Ohm Ton:40ns at 3/0V	16H	Vis		data pdf pinout
DG2016	SPDT	Vs:1.8...5.5V Ron:Ohm Ton:23ns at 5/0V	10S	Vis		data pdf pinout
DG2017	DPDT	Vs:2...5.5V Ron:Ohm Ton:46ns at 3/0V	16H	Vis		data pdf pinout
DG2020	SPDT	Vs:2.7...5.5V Ron:Ohm 3µsat 5/0V	6S	Vis		data pdf pinout
DG2026	SPDT	Vs:1.8...5.5V Ron:Ohm Ton:23ns at 5/0V	10S	Vis		data pdf pinout
DG2031	SPDT	Vs:1.8...5.5V Ron:Ohm Ton:34ns at 3/0V	10S	Vis		data pdf pinout
DG2032	SPDT	Vs:1.8...5.5V Ron:Ohm Ton:28ns at 3/0V	12H	Vis		data pdf pinout
DG2034	MULT4	Vs:1.8...5.5V Ron:Ohm Ton:18ns at 5/0V	10S,12H	Vis		data pdf pinout
DG2035	SPDT	Vs:1.8...5.5V Ron:Ohm Ton:34ns at 3/0V	10S	Vis		data pdf pinout
DG2037	SPST	Vs:1.8...5.5V Ron:Ohm Ton:19ns at 5/0V	8S	Vis		data pdf pinout
DG2038	SPST	Vs:1.8...5.5V Ron:Ohm Ton:19ns at 5/0V	8S	Vis		data pdf pinout
DG2039	SPST	Vs:1.8...5.5V Ron:Ohm Ton:19ns at 5/0V	8S	Vis		data pdf pinout
DG2041	SPST	Vs:1.8...5.5V Ron:Ohm Ton:13ns at 5/0V	16HS	Vis		data pdf pinout
DG2042	SPST	Vs:1.8...5.5V Ron:Ohm Ton:13ns at 5/0V	16HS	Vis		data pdf pinout

Type	Device	Short Description	Fig.	Manu	Comparision Types	More at
DG2043	SPST	Vs:1.8...5.5V Ron:Ohm Ton:13ns at 5/0V	16HS	Vis		data pdf pinout
DG2131	SPDT	Ron:Ohm Ton:39ns at 3/0V	8S,10S	Vis		data pdf pinout
DG2301	SPST	Vs:1.8...5.5V Ron:aOhm at 5/0V	5S	Vis		data pdf pinout
DG2302	SPST	Vs:4.5...5.5V Ron:aOhm Ton:5ns at 5/0V	5S	Vis		data pdf pinout
DG2303	SPST	Vs:1.65...5.5V at 5/0V	5S	Vis		data pdf pinout
DG2531	SPDT	Vs:1.8...5.5V Ron:Ohm Ton:40ns at 3/0V	10S	Vis		data pdf pinout
DG2532	SPDT	Vs:1.8...5.5V Ron:Ohm Ton:40ns at 3/0V	10S	Vis		data pdf pinout
DG2535	SPDT	Ron:Ohm Ton:52ns at 3/0V	10S	Vis		data pdf pinout
DG2536	SPDT	Ron:Ohm Ton:52ns at 3/0V	10S	Vis		data pdf pinout
DG2714	DPDT	Vs:1.6...3.6V Ron:Ohm Ton:28ns at 3/0V	6S	Vis		data pdf pinout
DG2718	DPDT	Vs:1.65...3.6V Ron:Ohm at 3/0V	16H	Vis		data pdf pinout
DG2741	SPST	Vs:1.6...3.6V Ron:Ohm Ton:20ns at 3/0V	8S	Vis		data pdf pinout
DG2742	SPST	Vs:1.6...3.6V Ron:Ohm Ton:20ns at 3/0V	8S	Vis		data pdf pinout
DG2743	SPST	Vs:1.6...3.6V Ron:Ohm Ton:20ns at 3/0V	8S	Vis		data pdf pinout
DG3000	SPDT	Vs:1.8...5.5V Ron:Ohm Ton:24ns at 5/0V	6G	Vis		data pdf pinout
DG3001	SPST	Vs:1.8...5.5V Ron:Ohm Ton:47ns at 3/0V	6G	Vis		data pdf pinout
DG3002	SPST	Vs:1.8...5.5V Ron:Ohm Ton:47ns at 3/0V	6G	Vis		data pdf pinout
DG3003	SPDT	Vs:1.8...5.5V Ron:Ohm Ton:47ns at 3/0V	6G	Vis		data pdf pinout
DG3157	SPDT	Vs:1.65...5.5V Ron:6Ohm at 5/0V	6S	Vis	NC7SB3157,NLASB3157,STG3157	data pdf pinout
DG3408	MULT8	Vs:2.7...12V Ron:Ohm Ton:42ns at 12/0V	16G	Vis		data pdf pinout
DG3409	MULT4	Vs:2.7...12V Ron:Ohm Ton:74ns at 5/0V	16G	Vis		data pdf pinout
DG3535	SPDT	Ron:Ohm Ton:52ns at 3/0V	10G	Vis		data pdf pinout
DG3536	SPDT	Ron:Ohm Ton:52ns at 3/0V	10G	Vis		data pdf pinout
DG4599	SPDT	Vs:2.25...5.5V Ron:Ohm Ton:8ns at 5/0V	6S	Vis		data pdf pinout
DG5043	SPDT	Ron:aOhm at 15/-15V	16D	Vis		data pdf pinout
DG 5043 CJ	CMOS-IC	2x2 Analog Switch, 22V, 30mA, 0...+70° 44V, 30mA, on<75Ω, <1200/700ns	16-DIP	Six		
DG 5143	SPDT	Ron:aOhm at 15/-15V	16D	Vis		data pdf pinout
DG 5143	CMOS-IC	2x2 Analog Switch, 36V, <50Ω, <175ns		Six		
DG9232	SPST	Vs:2.7...12V Ron:Ohm Ton:50ns at 3/0V	8S	Vis		data pdf pinout
DG9233	SPST	Vs:2.7...12V Ron:Ohm Ton:35ns at 5/0V	8S	Vis		data pdf pinout
DG9262	SPST	Vs:2.7...12V Ron:Ohm Ton:50ns at 5/0V	8S	Vis		data pdf pinout
DG9263	SPST	Vs:2.7...12V Ron:Ohm Ton:50ns at 5/0V	8S	Vis		data pdf pinout
DG9408	MULT4	Vs:2.7...12V Ron:Ohm Ton:42ns at 12/0V	16H	Vis		data pdf pinout
DG9409	MULT4	Vs:2.7...12V Ron:Ohm Ton:74ns at 5/0V	16H	Vis		data pdf pinout
DG9411	SPDT	Ron:Ohm Ton:9ns at 5/0V	6S	Vis		data pdf pinout
DG9414	MULT4	Vs:3...12V Ron:Ohm Ton:56ns at 5/0V	10S	Vis		data pdf pinout
DG9415	MULT2	Vs:3...12V Ron:Ohm Ton:56ns at 5/0V	10S	Vis		data pdf pinout
DG9421	SPST	Ron:Ohm Ton:38ns at 5/-5V	6S	Vis		data pdf pinout
DG9422	SPST	Ron:Ohm Ton:38ns at 5/-5V	6S	Vis		data pdf pinout
DG9424	SPST	Vs:2.7...12V Ron:Ohm Ton:71ns at 5/0V	16S	Vis		data pdf pinout
DG9425	SPST	Vs:2.7...12V Ron:Ohm Ton:71ns at 5/0V	16S	Vis		data pdf pinout
DG9426	SPST	Vs:2.7...12V Ron:Ohm Ton:71ns at 5/0V	16S	Vis		data pdf pinout
DG9432	SPST	Vs:2.7...12V Ron:Ohm Ton:33ns at 5/0V	8S	Vis		data pdf pinout
DG9433	SPST	Vs:2.7...12V Ron:Ohm Ton:33ns at 5/0V	8S	Vis		data pdf pinout
DG9434	SPST	Vs:2.7...12V Ron:Ohm Ton:33ns at 5/0V	8S	Vis		data pdf pinout
DG9461	SPDT	Vs:2.7...12V Ron:Ohm Ton:35ns at 5/0V	6S,8S	Vis		data pdf pinout
DGQ	Si-N	=2SD1963-Q (Typ-Code/Stempel/marking)	39	Rhm	→2SD1963	data
DGR	Si-N	=2SD1963-R (Typ-Code/Stempel/marking)	39	Rhm	→2SD1963	data
DGS	Si-N	=2SD1963-S (Typ-Code/Stempel/marking)	39	Rhm	→2SD1963	data
DGs	Si-P	=BCW 68G (Typ-Code/Stempel/marking)	35	Sie	→BCW 68G	data
DH	Si-P	=2SB1025-DH (Typ-Code/Stempel/marking)	39	Hit	→2SB1025	data
DH	Si-N	=2SD1625 (Typ-Code/Stempel/marking)	39	Say	→2SD1625	data
DH	Si-P	=BCW 68H (Typ-Code/Stempel/marking)	35	Fer, Sie	→BCW 68H	data
DH	Si-P	=BFN 19 (Typ-Code/Stempel/marking)	39	Sie	→BFN 19	data
DH	Z-Di	=MMSZ 4699 (Typ-Code/marking)	71(2,7mm)	Ons	→MMSZ 4699	
DH	Z-Di	=P4 SMA-150 (Typ-Code/Stempel/marking)	71(5x2,5)	Fag	→P4 SMA-150	data
DH	Si-N-Darl	=RXTA-28 (Typ-Code/Stempel/marking)	39	Rhm	→RXTA-28	data
DH	Z-Di	=SMBJ 60C (Typ-Code/Stempel/marking)	71(5x3,5)	Mop	→SMBJ ...	data
DH	MOS-P-FET-e*	=SSM 3J05FU (Typ-Code/Stempel/marking)	35	Tos	→SSM 3J05FU	data
DH	MOS-P-FET-e*	=SSM 5P05FU (Typ-Code/Stempel/marking)	45(2mm)	Tos	→SSM 5P05FU	
DH	MOS-P-FET-e*	=SSM 6P05FU (Typ-Code/Stempel/marking)	46(2mm)	Tos	→SSM 6P05FU	
DHs	Si-P	=BCW 68H (Typ-Code/Stempel/marking)	35	Sie	→BCW 68H	data
DI	Si-N	=2SD1626 (Typ-Code/Stempel/marking)	39	Say	→2SD1626	data
DI	Si-N+R	=XN 421L (Typ-Code/Stempel/marking)	46	Mat	→XN 421L	data
DI-	Si-N	=2SC3867 (Typ-Code/Stempel/marking)	35	Hit	→2SC3867	data
DIR 1701 E	Dig. Audio	Digital Audio Interface Receiver	TSOP	Bub		pdf pinout
DIR 1703 E	Dig. Audio	Digital Audio Interface Receiver	TSOP	Bub		pdf pinout
DJ	Si-P	=2SB1025-DJ (Typ-Code/Stempel/marking)	39	Hit	→2SB1025	data
DJ	Si-N	=2SD1627 (Typ-Code/Stempel/marking)	39	Say	→2SD1627	data
DJ	Z-Di	=MMSZ 4700 (Typ-Code/marking)	71(2,7mm)	Ons	→MMSZ 4700	
DJ	MOS-N-FET-e*	=SSM 3K09FU (Typ-Code/Stempel/marking)	35(2mm)	Tos	→SSM 3K09FU	
DJ	MOS-N-FET-e*	=SSM 6N09FU (Typ-Code/Stempel/marking)	46(2mm)	Tos	→SSM 6N09FU	
DJ 4	N-FET	=3SK180-4 (Typ-Code/Stempel/marking)	44	Say	→3SK180	data
DJ 5	N-FET	=3SK180-5 (Typ-Code/Stempel/marking)	44	Say	→3SK180	data
DJ 6	N-FET	=3SK180-6 (Typ-Code/Stempel/marking)	44	Say	→3SK180	data
DJQ	Si-N	=2SD2098-Q (Typ-Code/Stempel/marking)	39	Rhm	→2SD2098	data
DJR	Si-N	=2SD2098-R (Typ-Code/Stempel/marking)	39	Rhm	→2SD2098	data
DJS	Si-N	=2SD2098-S (Typ-Code/Stempel/marking)	39	Rhm	→2SD2098	data
DK	Si-P	=2SB1025-DK (Typ-Code/Stempel/marking)	39	Hit	→2SB1025	data
DK	Si-P	=2SB798-DK (Typ-Code/Stempel/marking)	39	Nec	→2SB798	data
DK	Si-N	=2SD1628 (Typ-Code/Stempel/marking)	39	Say	→2SD1628	data
DK	MOS-N-FET-e*	=2SK2909 Typ-Code/Stempel/marking)	35	Say	→2SK2909	data
DK	Si-P	=BCX 42 (Typ-Code/Stempel/marking)	35	Sie, Tho	→BCX 42	data
DK	Z-Di	=MMSZ 4701 (Typ-Code/marking)	71(2,7mm)	Ons	→MMSZ 4701	

Type	Device	Short Description	Fig.	Manu	Comparision Types	More at
DK	Z-Di	=SMBJ 60CA (Typ-Code/Stempel/marking)	71(5x3,5)	Mop	→SMBJ ...	data
DK	MOS-P-FET-e*	=SSM 3J09FU (Typ-Code/Stempel/marking)	35(2mm)	Tos	→SSM 3J09FU	data
DK	MOS-P-FET-e*	=SSM 6P09FU (Typ-Code/Stempel/marking)	46(2mm)	Tos	→SSM 6P09FU	data
DK15	HI-OHM	Vs:±15V Vu:100dB Vo:±10V Vi0:0.75mV	7Y	Ilc		data pinout
DK17	HI-OHM	Vs:±15V Vu:100dB Vo:±10V Vi0:0.5mV	7Y	Ilc		data pinout
DK19	HI-OHM	Vs:±15V Vu:110dB Vo:±10V Vi0:0.5mV	7Y	Ilc		data pinout
DK21	HI-OHM	Vs:±15V Vu:110dB Vo:±10V Vi0:0.5mV	15Y	Ilc		data pinout
DK25	HI-OHM	Vs:±15V Vu:100dB Vo:±10V Vi0:0.75mV	7Y	Ilc		data pinout
DKP	Si-N	=2SC4672-P (Typ-Code/Stempel/marking)	39	Rhm	→2SC4672	data
DKQ	Si-N	=2SC4672-Q (Typ-Code/Stempel/marking)	39	Rhm	→2SC4672	data
DKs	Si-P	=BCX 42 (Typ-Code/Stempel/marking)	35	Sie	→BCX 42	data
DKS 21	Si-N-Darl	NF, 30V, 0,5A, 0,6W, 200MHz, B>2000	7e	Sam	BC 517, BC 617, BC 875, MPSA 25..29, ++	data
DKS 22	Si-N-Darl	=DKS 21: B>10000	7e	Sam	BC 517, BC 617, MPSA 25..29, ++	data
DKS 23	Si-P-Darl	NF, 30V, 0,5A, 0,6W, 175MHz, B>2000	7e	Sam	BC 516, BC 876, MPSA 63..64, MPSA 75...77	data
DKS 24	Si-P-Darl	=DKS 23: B>1000	7e	Sam	BC 516, MPSA 63..64, MPSA 75...77	data
DL	Si-N	=KTC4077-BL (Typ-Code/Stempel/marking)	35(2mm)	Kec	→KTC 4077	data
DL	Z-Di	=P4 SMA-150A(Typ-Code/Stempel/marking)	71(5x2,5)	Fag	→P4 SMA-150A	data
DL	Z-Di	=SMBJ 64C (Typ-Code/Stempel/marking)	71(5x3,5)	Mop	→SMBJ ...	data
DL	Si-Di	=U 1DL44A (Typ-Code/Stempel/marking)	71(5x2,5)	Tos	→U 1DL44A	data
DL	Si-P	=2SA1343 (Typ-Code/Stempel/marking)	35	Say	→2SA1343	data
DL	Si-P	=2SB1026-DL (Typ-Code/Stempel/marking)	39	Hit	→2SB1025	data
DL	Si-P	=2SB798-DL (Typ-Code/Stempel/marking)	39	Nec	→2SB798	data
DL	Si-N	=2SC2713-BL (Typ-Code/Stempel/marking)	35	Tos	→2SC2713	data
DL	Si-N	=2SC4117-BL (Typ-Code/Stempel/marking)	35(2mm)	Tos	→2SC4117	data
DL	Si-N	=2SD2099 (Typ-Code/Stempel/marking)	39	Say	→2SD2099	data
DL 000 D,DG	TTL-Logic	=... 74LS00	14-DIP	Hfo	... 74LS00... (TTL)	data
DL 000 SC	TTL-Logic	=DL 000D,DG: SMD	14-MDIP	Hfo		data
DL 002 D,DG	TTL-Logic	=... 74LS02	14-DIP	Hfo	... 74LS02... (TTL)	data
DL 002 SC	TTL-Logic	=DL 002D,DG: SMD	14-MDIP	Hfo		data
DL 03-58	Z-Di	SMD, TAZ, 58±6V, 300W(10/1000µs)	71a(5x2,5)	Shi	-	data
DL 003 D,DG	TTL-Logic	=... 74LS03	14-DIP	Hfo	... 74LS03... (TTL)	data
DL 003 S,SC	TTL-Logic	=DL 003D: SMD	14-MDIP	Hfo		data
DL 04-18	Z-Di	SMD, TAZ, 18±1,5V, 400W(10/1000µs)	71a(5x2,5)	Shi	-	data
DL 04-28	Z-Di	SMD, TAZ, 28±4V, 400W(10/1000µs)	71a(5x2,5)	Shi	-	data
DL 004 D,DG	TTL-Logic	=... 74LS04	14-DIP	Hfo	... 74LS04... (TTL)	data
DL 005 DC	TTL-Logic	=... 74LS05	14-DIP	Hfo	... 74LS05... (TTL)	data
DL 008 D,DG	TTL-Logic	=... 74LS08	14-DIP	Hfo	... 74LS08... (TTL)	data
DL 008 SC	TTL-Logic	=DL 008D,DG: SMD	14-MDIP	Hfo		data
DL 010 D,DG	TTL-Logic	=... 74LS10	14-DIP	Hfo	... 74LS10... (TTL)	data
DL 010 SC	TTL-Logic	=DL 010D,DG: SMD	14-MDIP	Hfo		data
DL 011 D	TTL-Logic	=... 74LS11	14-DIP	Hfo	... 74LS11... (TTL)	data
DL 011 SC	TTL-Logic	=DL 011D,DG: SMD	14-MDIP	Hfo		data
DL 014 D	TTL-Logic	=... 74LS14	14-DIP	Hfo	... 74LS14... (TTL)	data
DL15	HI-OHM	Vs:±15V Vu:98dB Vo:±10V Vi0:1.5mV	7Y	Ilc		data pinout
DL 016 DC	TTL-Logic	=... 74LS16	14-DIP	Hfo	... 74LS16... (TTL)	data
DL17	HI-OHM	Vs:±15V Vu:98dB Vo:±10V Vi0:1mV	7Y	Ilc		data pinout
DL19	HI-OHM	Vs:±15V Vu:108dB Vo:±10V Vi0:1mV	7Y	Ilc		data pinout
DL 020 D,DG	TTL-Logic	=... 74LS20	14-DIP	Hfo	... 74LS20... (TTL)	data
DL 020 SC	TTL-Logic	=DL 020D,DG: SMD	14-MDIP	Hfo		data
DL21	HI-OHM	Vs:±15V Vu:108dB Vo:±10V Vi0:1mV	15Y	Ilc		data pinout
DL 021 D	TTL-Logic	=... 74LS21	14-DIP	Hfo	... 74LS21... (TTL)	data
DL 021 SC	TTL-Logic	=DL 021D,DG: SMD	14-MDIP	Hfo		data
DL25	HI-OHM	Vs:±15V Vu:98dB Vo:±10V Vi0:1.5mV	7Y	Ilc		data pinout
DL 026 DC	TTL-Logic	=... 74LS26	14-DIP	Hfo	... 74LS26... (TTL)	data
DL 030 D,DG	TTL-Logic	=... 74LS30	14-DIP	Hfo	... 74LS30... (TTL)	data
DL 030 SC	TTL-Logic	=DL 030D,DG: SMD	14-MDIP	Hfo		data
DL 032 D	TTL-Logic	=... 74LS32	14-DIP	Hfo	... 74LS32... (TTL)	data
DL 032 SC	TTL-Logic	=DL 032D,DG: SMD	14-MDIP	Hfo		data
DL 037 D	TTL-Logic	=... 74LS37	14-DIP	Hfo	... 74LS37... (TTL)	data
DL 038 D	TTL-Logic	=... 74LS38	14-DIP	Hfo	... 74LS38... (TTL)	data
DL 040 D	TTL-Logic	=... 74LS40	14-DIP	Hfo	... 74LS40... (TTL)	data
DL 051 D	TTL-Logic	=... 74LS51	14-DIP	Hfo	... 74LS51... (TTL)	data
DL 074 D	TTL-Logic	=... 74LS74	14-DIP	Hfo	... 74LS74... (TTL)	data
DL 083 D	TTL-Logic	=... 74LS83	16-DIP	Hfo	... 74LS83... (TTL)	data
DL 086 D	TTL-Logic	=... 74LS86	14-DIP	Hfo	... 74LS86... (TTL)	data
DL 090 D	TTL-Logic	=... 74LS90	14-DIP	Hfo	... 74LS90... (TTL)	data
DL 093 D	TTL-Logic	=... 74LS93	14-DIP	Hfo	... 74LS93... (TTL)	data
DL 112 D	TTL-Logic	=... 74LS112	16-DIP	Hfo	... 74LS112... (TTL)	data
DL 123 D	TTL-Logic	=... 74LS123	16-DIP	Hfo	... 74LS123... (TTL)	data
DL 132 D	TTL-Logic	=... 74LS132	14-DIP	Hfo	... 74LS132... (TTL)	data
DL 155 D	TTL-Logic	=... 74LS155	16-DIP	Hfo	... 74LS155... (TTL)	data
DL 164 D	TTL-Logic	=... 74LS164	14-DIP	Hfo	... 74LS164... (TTL)	data
DL 175 D	TTL-Logic	=... 74LS175	16-DIP	Hfo	... 74LS175... (TTL)	data
DL 192 D	TTL-Logic	=... 74LS192	16-DIP	Hfo	... 74LS192... (TTL)	data
DL 193 D	TTL-Logic	=... 74LS193	16-DIP	Hfo	... 74LS193... (TTL)	data
DL 194 D	TTL-Logic	=... 74LS194	16-DIP	Hfo	... 74LS194... (TTL)	data
DL 251 D	TTL-Logic	=... 74LS251	16-DIP	Hfo	... 74LS251... (TTL)	data
DL 253 D	TTL-Logic	=... 74LS253	16-DIP	Hfo	... 74LS253... (TTL)	data
DL 257 D.	TTL-Logic	=... 74LS257	16-DIP	Hfo	... 74LS257... (TTL)	data
DL 259 D	TTL-Logic	=... 74LS259	16-DIP	Hfo	... 74LS259... (TTL)	data
DL 295 D	TTL-Logic	=... 74LS295	14-DIP	Hfo	... 74LS295... (TTL)	data

Type	Device	Short Description	Fig.	Manu	Comparision Types	More at
DL 299 D	TTL-Logic	=... 74LS299	20-DIP	Hfo	... 74LS299... (TTL)	data
DL 374 D	TTL-Logic	=... 74LS374	20-DIP	Hfo	... 74LS374... (TTL)	data
DL 540 D	TTL-Logic	=... 74LS540	20-DIP	Hfo	... 74LS540... (TTL)	data
DL 541 D	TTL-Logic	=... 74LS541	20-DIP	Hfo	... 74LS541... (TTL)	data
DL 2631 D	I/O-IC	Leitungs-Tr./line driver, CCITT V.11.	16-DIP	Hfo	AM 26LS31	
DL 2632 D	LIN-IC	Empfänger/receiver f. DL 2631D	16-DIP	Hfo	AM 26LS32	
DL 8121 D	LIN-IC	8 Bit Komparator	20-DIP	Hfo	AmZ 8121	
DL 8127 D	TTL-IC	System Clock f. 16 Bit µComp.	24-DIP	Hfo	AmZ 8127	
DL 8640 DC	TTL-IC	Bus-empfänger/bus receiver	14-DIP	Hfo	DS 8640N	
DL 8641 DC	TTL-IC	Bus-empfänger+treiber,bus receiver+drv	16-DIP	Hfo	DS 8641N	
DL 60278	LIN-IC	=TDA 7270 S	16-DIP	Sgs	→TDA 7270S	
DL 75113 DC	TTL-IC	Leitungssender/line driver	16-DIP	Hfo	SN 75113N	
DLA/C	LED	gr+srd+yl ø20mm l>6mcd Vf:2.1V		Kin		data pdf pinout
DLA 5 G	Si-Di	GI, S, 600V, 5A, <50ns	17	Say	BY 430F-1000, BY 431-1000	data
DLA 8 G	Si-Di	GI, S, 600V, 8A, <70ns	17	Say	BYR 29F-600	data
DLA 11 C	Si-Di	SMD, GI, S, 200V, 1,1A, <50ns	71a(5x2,5)	Say	BYG 80D...J, 2D...M, MURS 120, MURS 160	data
DLA 15 C	Si-Di	SMD, GI, S, 200V, 1,5A, <35ns	71a(7,6x4)	Say		data
DLC/6	LED	gr+srd+yl ø20mm l>6mcd Vf:2.1V		Kin		data pdf pinout
DLC 20 C,E	Si-Di	GI, S, 200...400V, 2A, <35ns C=200V, E=400V	31a	Say	BYV 27/..., FE 2D...H	data
DLCA 5 C,E	Si-Di	GI, S, Dual, 5A, C=200V, <35ns E=400V, <50ns	17p	Say		data
DLCA 10 C,E	Si-Di	GI, S, Dual, 10A, C=200V, <35ns E=400V, <50ns	17p	Say	FE 16D...J	data
DLE 30 B...E	Si-Di	GI, S, 100...400V, 3A, <35ns B=100V, C=200V, E=400V	31a	Say	BYV 28/..., FE 3B...H	data
DLF 30 C...E	Si-Di	GI, S, 200...400V, 3A, <35ns C=200V, E=400V	31a	Say	BYV 28/..., FE 3B...H	data
DLM 10 B...E	Si-Di	GI, S, 100...400V, 1A, <35ns B=100V, C=200V, E=400V	31a	Say	BYV 26A...B, FE 1B...H	data
DLN	Si-N	=2SD2167-N (Typ-Code/Stempel/marking)	39	Rhm	→2SD2167	data
DLN 10 C,E	Si-Di	GI, S, 200...400V, 1A, <50ns C=200V, E=400V	31a	Say	BYM 26A...E, BYV 36A...E, MUR 120...1100	data
DLP	Si-N	=2SD2167-P (Typ-Code/Stempel/marking)	39	Rhm	→2SD2167	data
DLP 05 LC-7-F	LIN-IC	low Capacitance Unidirectional TVS	45	Dii	-	pdf pinout
DLPA 006	\N	\N	\N	\N	\N	\N
DLQ	Si-N	=2SD2167-Q (Typ-Code/Stempel/marking)	39	Rhm	→2SD2167	data
DM	Si-P	=2SB1026-DM (Typ-Code/Stempel/marking)	39	Hit	→2SB1025	data
DM	Si-P	=2SB798-DM (Typ-Code/Stempel/marking)	39	Nec	→2SB798	data
DM	Si-N	=2SD1998 (Typ-Code/Stempel/marking)	39	Say	→2SD1998	data
DM	Si-N-Darl	=2SD2170 (Typ-Code/Stempel/marking)	39	Rhm	→2SD2170	data
DM	Z-Di	=MMSZ 4702 (Typ-Code/marking)	71(2,7mm)	Ons	→MMSZ 4702	data
DM	Z-Di	=P4 SMA-160 (Typ-Code/Stempel/marking)	71(5x2,5)	Fag	→P4 SMA-160	data
DM	Z-Di	=SMBJ 64CA (Typ-Code/Stempel/marking)	71(5x3,5)	Mop	→SMBJ ...	data
DM-106 B	Si	Magneto-Resitance Element	40	Son	-	data
DM-111	Si	Magneto-Resitance Element	12	Son	-	data
DM-112	Si	Magneto-Resitance Element	12	Son	-	data
DM 133	Si-Di	≈2x 1N4007			→1N4007	data
DM-211	Si	Magneto-Resitance Element	12	Son	-	data
DM-230	Si	Magneto-Resitance Element	42	Son	-	data
DM-231	Si	Magneto-Resitance Element	42	Son	-	data
DM-232	Si	Magneto-Resitance Element	42	Son	-	data
DM-233	Si	Magneto-Resitance Element	42	Son	-	data
DM 513	Si-Di	kV-GI, 1600V, 1A	31a	Tho	BY 228, EM 516, GP 10Y	data
DM 516	Si-Di	=DM 513: 1800V	31a	Tho	RGP 15-18...-20	data
DM 2502(C)J,N	TTL-IC	8Bit Successive Approx. Register f.A/D	16-DIC,DIP	Nsc	-	
DM 2503(C)J,N	TTL-IC	8Bit Successive Approx. Register f.A/D	16-DIC,DIP	Nsc	-	
DM 2504(C)J,N	TTL-IC	12Bit Success. Approx. Register f.A/D	16-DIC,DIP	Nsc	-	
DMP 6101 A	MOS-I/O-IC	Microprocessor Interface Modules	MSIP	Cry	-	pdf pinout
DMP 6201 A	MOS-I/O-IC	Microprocessor Interface Modules	MSIP	Cry	-	pdf pinout
DMP 6202 A	MOS-I/O-IC	Microprocessor Interface Modules	MSIP	Cry	-	pdf pinout
DMP 6301 A	MOS-I/O-IC	Microprocessor Interface Modules	MSIP	Cry	-	pdf pinout
DMP 6402 A	MOS-I/O-IC	Microprocessor Interface Modules	MSIP	Cry	-	pdf pinout
DM 9000 AE	DIG-IC	Ethernet Controller	MP	Dav	-	pdf pinout
DM 9000 AEP	DIG-IC	Ethernet Controller	MP	Dav	-	pdf pinout
DMA 2270	MOS-IC	Ctv, D2-MAC-Decoder	68-PLCC	Itt	-	
DMA 2275	CMOS-IC	Ctv, MAC Descrambling-prozessor	68-PLCC	Itt	-	pdf
DMV 16	Si-Di	Dual, CRT HA, D1: Damper, 1500V, 5A, <300ns, D2: Modulation,600V, 3A, <95ns	17o1	Tho	-	data
DMV 32	Si-Di	Dual, CRT HA, D1: Damper, 1500V, 6A, <185ns, D2: Modulation,600V, 6A, <50ns	17o1	Tho	-	data
DMV 56	Si-Di	Dual, CRT HA, D1: Damper, 1500V, 6A, <135ns, D2: Modulation,600V, 6A, <50ns	17o1	Tho	-	data

Type	Device	Short Description	Fig.	Manu	Comparision Types	More at
DN...DR						
DN	Si-N	=2SD1999 (Typ-Code/Stempel/marking)	39	Say	→2SD1999	data
DN	MOS-N-FET-e*	=2SK3349 (Typ-Code/Stempel/marking)	35(1,6mm)	Hit	→2SK3349	
DN	Z-Di	=MMSZ 4703 (Typ-Code/marking)	71(2,7mm)	Ons	→MMSZ 4703	
DN	Z-Di	=P4 SMA-160A(Typ-Code/Stempel/marking)	71(5x2,5)	Fag	→P4 SMA-160A	data
DN	Z-Di	=SM 6T 10 (Typ-Code/Stempel/marking)	71(6x4mm)	Tho	→SM 6T...	data
DN	Z-Di	=SMBJ 70C (Typ-Code/Stempel/marking)	71(5x3,5)	Mop	→SMBJ ...	data
DN 74 LSxx...	TTL-Logic	Standard TTL-Logic 74LS-Serie		Mat	siehe http://www.ecadata.de→	data
DN 811	LIN-IC	12/16-Zähler/counter	14-DIC	Mat	-	
DN 819	LIN-IC	i2L-Frequ.-Teiler/divider	4-SIP	Mat	-	
DN 834	LIN-IC	Hall-element, Schalter/switch	4-SIP	Mat	-	
DN 835	LIN-IC	Hall-element, linear	4-SIP	Mat	-	
DN 837	LIN-IC	Hall-element, Schalter/switch	4-SIP	Mat	-	
DN 838	LIN-IC	Hall-element, Schalter/switch	4-SIP	Mat	-	
DN 839	LIN-IC	Hall-element, Schalter/switch	4-SIP	Mat	-	
DN 850	LIN-IC	Multivibrator, monostab.	14-DIC	Ma*	-	
DN 851	LIN-IC	4 Bit Zähler/counter, reversibel	14-DIC	Mat	-	
DN 852	LIN-IC	Bin.→oct. Decoder	16-DIC	Mat	-	
DN 2530 N3	MOS-N-FET-d	V-MOS, 300V, 175mA, 0,74W, <12Ω(0,15A)	7e	Stx	-	data
DN 2530 N8	MOS-N-FET-d	=DN 2530N3: SMD, 0,2A	39b		-	data
DN 2535 N3	MOS-N-FET-d	V-MOS, 350V, 120mA, 1W, <25Ω(0,12A)	7e	Stx	-	data
DN 2535 N5	MOS-N-FET-d	=DN 2535N3: 0,5A	17c§		-	data
DN 2540 N3	MOS-N-FET-d	=DN 2535N3: 400V	7e	Stx	-	data
DN 2540 N5	MOS-N-FET-d	=DN 2535N3: 400V, 0,5A	17c§		-	data
DN 2540 N8	MOS-N-FET-d	=DN 2535N3: SMD, 400V, 0,17A	39b		-	data
DN 2634 N3	MOS-N-FET-d	V-MOS, 240V, 0,3A, 1W, <4Ω(0,2A)	7e	Stx	-	data
DN 26:0 N3	MOS-N-FET-d	V-MOS, 400V, 0,25A, 1W, <6Ω(0,15A)	7e	Stx	-	data
DN 6835	LIN-IC	Hall-element, linear	3-SIP	Mat	-	
DN 6836	LIN-IC	Hall-element, linear	3-SIP	Mat	-	
DN 6837	LIN-IC	Hall-element, Schalter/switch	3-SIP	Mat	-	
DN 6838	LIN-IC	Hall-element, Schalter/switch	3-SIP	Mat	-	
DN 6839	LIN-IC	Hall-element, Schalter/switch	3-SIP	Mat	-	
DN 6844 S	LIN-IC	Hall-Verst./Amplifier, Ucc=3,6...16V	≤4-MDIP	Mat	-	
DN 6845 S	LIN-IC	Hall-Verst./Amplifier, Ucc=3,6...16V	≤4-MDIP	Mat	-	
DN 6846 S	LIN-IC	Hall-Verst./Amplifier, Ucc=3,6...16V	≤4-MDIP	Hit	-	
DN 6847(S,SE,TE)	LIN-IC	Hall-Verst./Amplifier, Ucc=4,5...16V	Diverse	Mat	-	pdf
DN 6848(S,SE,TE)	LIN-IC	Hall-Verst./Amplifier, Ucc=4,5...16V	Diverse	Mat	-	pdf
DN 6849(S,SE,TE)	LIN-IC	Hall-Verst./Amplifier, Ucc=4,5...16V	Diverse	Mat	-	pdf
DN 6851	LIN-IC	Hall-Verst./Amplifier, Ucc=3,6...16V	3-SIP	Mat	-	pdf
DN 6852	LIN-IC	Hall-Verst./Amplifier, Ucc=3,6...16V	3-SIP	Mat	-	pdf
DN 6853	LIN-IC	Hall-Verst./Amplifier, Ucc=3,6...16V	3-SIP	Mat	-	pdf
DN 8640 S	MOS-IC	3 Bit Shift Register Latch Driver	36-MDIP	Mat	-	
DN 8643 S	MOS-IC	24 Bit Shift Register Latch Driver	36-MDIP	Mat	-	pdf
DN 8646 FBP	MOS-IC	SMD, 4x8 Bit Shift Register Latch Drv.	44-MP	Mat	-	
DN 8648 FBP	MOS-IC	SMD, 32 Bit Shift Register Latch Driver	44-MP	Mat	-	pdf
DN 8650	MOS-IC	Darlington Driver Array, 5V, 0,5A	16-DIP	Mat	-	
DN 8661	MOS-IC	Darlington Driver Array, 50V, 0,5A	16-DIP	Mat	-	
DN 8667 NS	MOS-IC	SMD, LED Panel Driver	20-MDIP	Mat	-	pdf
DN 8680	MOS-IC	Darlington Driver Array, 50V, 1,5A	16-DIP	Mat	-	
DN 8690	MOS-IC	Darlington Driver Array, 60V, 1,5A	16-DIP	Mat	-	pdf
DN 8695	MOS-IC	Darlington Driver Array, 50V, 1,5A	23-SQL	Mat	-	pdf
DN 8897(S,SE,TE)	LIN-IC	Hall-Verst./Amplifier, Ucc=4,5...16V	Diverse	Mat	-	pdf
DN 8898 SE,TE	LIN-IC	Hall-Verst./Amplifier, Ucc=4,5...16V	3-SIP	Mat	-	
DN 8899(S,SE,TE)	LIN-IC	Hall-Verst./Amplifier, Ucc=4,5...16V	Diverse	Mat	-	pdf
DNE	Si-N	=2SD2153-E (Typ-Code/Stempel/marking)	39	Rhm	→2SD2153	data
DNU	Si-N	=2SD2153-U (Typ-Code/Stempel/marking)	39	Rhm	→2SD2153	data
DNV	Si-N	=2SD2153-V (Typ-Code/Stempel/marking)	39	Rhm	→2SD2153	data
DNW	Si-N	=2SD2153-W (Typ-Code/Stempel/marking)	39	Rhm	→2SD2153	data
DO	Si-P	=2SA1201-O (Typ-Code/Stempel/marking)	39	Tos	→2SA1201	data
DO	Si-P	=2SA1620-O (Typ-Code/Stempel/marking)	35	Tos	→2SA1620	data
DO	Si-N	=2SC5092-O (Typ-Code/Stempel/marking)	44	Tos	→2SC5092	data
DO	Si-N	=2SD1997 (Typ-Code/Stempel/marking)	39	Say	→2SD1997	data
DO	Si-P	=KTA1661-O (Typ-Code/Stempel/marking)	39	Kec	→KTA 1661	data
DO 201 YR	Diac	Ub=26...32V, Ib<50µA, Itsm=2A	31	Tag	1N5761, N 413M, BR 100, D 3202Y	data
DP	Si-P	=2SB789-P (Typ-Code/Stempel/marking)	39	Mat	→2SB789	data
DP	Si-N	=2SD2100 (Typ-Code/Stempel/marking)	39	Say	→2SD2100	data
DP	Si-N-Darl	=2SD2195 (Typ-Code/Stempel/marking)	39	Rhm	→2SD2195	data
DP	MOS-P-FET-e*	=2SJ587 (Typ-Code/Stempel/marking)	35(1,6mm)	Hit	→2SJ587	
DP	Si-P	=HQ 1L2N (Typ-Code/Stempel/marking)	39	Nec	→HQ 1...	data
DP	Z-Di	=MMSZ 4704 (Typ-Code/marking)	71(2,7mm)	Ons	→MMSZ 4704	
DP	Z-Di	=P4 SMA-170 (Typ-Code/Stempel/marking)	71(5x2,5)	Fag	→P4 SMA-170	data
DP	Z-Di	=SM 6T 10A (Typ-Code/Stempel/marking)	71(6x4mm)	Tho	→SM 6T...	data
DP	Z-Di	=SMBJ 70CA (Typ-Code/Stempel/marking)	71(5x3,5)	Mop	→SMBJ ...	data
DP1110	OC	LED/MOSFET Viso:3750Vrms Vbs:1.2V		Inr		data pinout
DP1210	OC	LED/MOSFET Viso:3750Vrms Vbs:1.2V		Inr		data pinout
DP1610	OC	LED VR/MOSFET Viso:3750Vrms Vbs:1.2V		Inr		data pinout
DP2110	OC	LED/MOSFET Viso:3750Vrms Vbs:1.2V		Inr		data pinout
DP2210	OC	LED/MOSFET Viso:3750Vrms Vbs:1.2V		Inr		data pinout

Type	Device	Short Description	Fig.	Manu	Comparision Types	More a
DP2610	OC	LED/MOSFET Viso:3750Vrms Vbs:1.2V		Inr		data pinou
DP6110	OC	LED/MOSFET Viso:3750Vrms Vbs:1.2V		Inr		data pinou
DP6210	OC	LED/MOSFET Viso:3750Vrms Vbs:1.2V		Inr		data pinou
DP6610	OC	LED/MOSFET Viso:3750Vrms Vbs:1.2V		Inr		data pinou
DPA 422 G	LIN-IC	Highly Integrated DC-DC Converter IC, max Out 10W	47	Pwi	-	pdf pinout
DPA 422 P	LIN-IC	Highly Integrated DC-DC Converter IC, max Out 10W	DIC	Pwi	-	pdf pinout
DPA 423 G	LIN-IC	Highly Integrated DC-DC Converter IC, max Out 18W	47	Pwi	-	pdf pinout
DPA 423 P	LIN-IC	Highly Integrated DC-DC Converter IC, max Out 18W	DIC	Pwi	-	pdf pinout
DPA 423 R	LIN-IC	Highly Integrated DC-DC Converter IC, max Out 18W	87	Pwi	-	pdf pinout
DPA 424 G	LIN-IC	Highly Integrated DC-DC Converter IC, max Out 35W	47	Pwi	-	pdf pinout
DPA 424 P	LIN-IC	Highly Integrated DC-DC Converter IC, max Out 35W	DIC	Pwi	-	pdf pinout
DPA 424 R	LIN-IC	Highly Integrated DC-DC Converter IC, max Out 35W	87	Pwi	-	pdf pinout
DPA 425 G	LIN-IC	Highly Integrated DC-DC Converter IC, max Out 70W	47	Pwi	-	pdf pinout
DPA 425 P	LIN-IC	Highly Integrated DC-DC Converter IC, max Out 70W	DIC	Pwi	-	pdf pinout
DPA 425 R	LIN-IC	Highly Integrated DC-DC Converter IC, max Out 70W	87	Pwi	-	pdf pinout
DPA 426 P	LIN-IC	Highly Integrated DC-DC Converter IC, max Out 100W	DIC	Pwi	-	pdf pinout
DPA 426 R	LIN-IC	Highly Integrated DC-DC Converter IC, max Out 100W	87	Pwi	-	pdf pinout
DPA1110	OC	LED/MOSFET Viso:4000Vrms Vbs:1.2V		Inr		data pinout
DPAD 1	Si-Di	=PAD 1: Dual, 0,5W	81/5Pin	Six		data
DPAD 5	Si-Di	=PAD 5: Dual, 0,5W	81/5Pin	Six		data
DPAD 50	Si-Di	=PAD 50: Dual, 0,5W	81/5Pin	Six		data
DPU 2500	NMOS-IC	Ctv, Ablenk-/deflection-prozessor	40-DIP	Itt	-	
DPU 2540	NMOS-IC	Ctv, Ablenk-/deflection-prozessor	40-DIP	Itt	-	
DPU 2543	NMOS-IC	Ctv, Ablenk-/deflection-prozessor	40-DIP	Itt	-	
DPU 2544	NMOS-IC	Ctv, Ablenk-/deflection-prozessor	40-DIP	Itt	-	
DPU 2545	NMOS-IC	Ctv, Ablenk-/deflection-prozessor	40-DIP	Itt	-	
DPU 2553	NMOS-IC	Ctv, Ablenk-/defl.-proz., normal-scan	40-DIP	Itt		pinout
DPU 2554	NMOS-IC	Ctv, Ablenk-/defl.-proz., double-scan	40-DIP	Itt		pinout
DPU 2555	NMOS-IC	Ctv, Ablenk-/defl.-proz., normal-scan	40-DIP	Itt		pinout
DQ	Si-P	=2PB1219AQ (Typ-Code/Stempel/marking)	35(2mm)	Phi	→2PB1219A	data
DQ	Si-P	=2PB710AQ (Typ-Code/Stempel/marking)	35	Phi	→2PB710A	data
DQ	Si-P	=2SB1219A-Q (Typ-Code/Stempel/marking)	35(2mm)	Mat	→2SB1219A	data
DQ	Si-P	=2SB710A-Q (Typ-Code/Stempel/marking)	35	Mat	→2SB710A	data
DQ	Si-P	=2SB789-Q (Typ-Code/Stempel/marking)	39	Mat	→2SB789	data
DQ	Si-N-Darl+Di	=2SD2176 (Typ-Code/Stempel/marking)	39	Say	→2SD2176	data
DQ	Si-P	=HQ 1A3M (Typ-Code/Stempel/marking)	39	Nec	→HQ 1...	data
DQ	Z-Di	=P4 SMA-170A(Typ-Code/Stempel/marking)	71(5x2,5)	Fag	→P4 SMA-170A	data
DQ	Z-Di	=SMBJ 75C (Typ-Code/Stempel/marking)	71(5x3,5)	Mop	→SMBJ ...	data
DQN	Si-N	=2SD2211-N (Typ-Code/Stempel/marking)	39	Rhm	→2SD2211	data
DQP	Si-N	=2SD2211-P (Typ-Code/Stempel/marking)	39	Rhm	→2SD2211	data
DQQ	Si-N	=2SD2211-Q (Typ-Code/Stempel/marking)	39	Rhm	→2SD2211	data
DR	Si-P	=2PB1219AR (Typ-Code/Stempel/marking)	35(2mm)	Phi	→2PB1219A	data
DR	Si-P	=2PB710AR (Typ-Code/Stempel/marking)	35	Phi	→2PB710A	data
DR	Si-P	=2SA1037KLN-R (Typ-Code/Stempel/mark.)	35	Rhm	→2SA1037KLN	data
DR	Si-P	=2SB1219A-R (Typ-Code/Stempel/marking)	35(2mm)	Mat	→2SB1219A	data
DR	Si-P	=2SB710A-R (Typ-Code/Stempel/marking)	35	Mat	→2SB710A	data
DR	Si-P	=2SB789-R (Typ-Code/Stempel/marking)	39	Mat	→2SB789	data
DR	Si-N	=2SC4643 (Typ-Code/Stempel/marking)	39	Hit	→2SC4643	data
DR	Si-N	=2SC5092-R (Typ-Code/Stempel/marking)	44	Tos	→2SC5092	data
DR	Si-N-Darl	=2SD2212 (Typ-Code/Stempel/marking)	39	Rhm	→2SD2212	data
DR	Si-P	=HQ 1F3M (Typ-Code/Stempel/marking)	39	Nec	→HQ 1...	data
DR	Z-Di	=P4 SMA-180 (Typ-Code/Stempel/marking)	71(5x2,5)	Fag	→P4 SMA-180	data
DR	Z-Di	=SMBJ 75CA (Typ-Code/Stempel/marking)	71(5x3,5)	Mop	→SMBJ ...	data
DRA 01 B...E	50Hz-Thy	100...400V, 0,1A, Igt/Ih<0,2/=3mA B=100V, C=200V, E=400V	7b	Say	BRX51...56, BRY55/..., MCR100-..., TAG06-...	data
DRA 2 TB...TG	50Hz-Thy	100...600V, 2A(Tc=45°), Igt/Ih<0,2/=3mA TB=100V, TC=200V, TE=400V, TG=600V	13h§	Say	C 106-..., TAG 106-...	data
DRA 3 B...G	50Hz-Thy	100...600V, 3A(Tc=87°) B=100V, C=200V, E=400V, G=600V	13h§	Say	C 108-..., TAG 108-...	data
DRA 03 TB...TG	50Hz-Thy	100...600V, 0,3A, Igt/Ih<0,2/=4mA TB=100V, TC=200V, TE=400V, TG=600V	7b	Say	BRX51...56, BRY55/..., MCR100-..., TAG06-...	data
DRA 5 B...G	50Hz-Thy	100...600V, 5A(Tc=91°), Igt/Ih<40/<60mA B=100V, C=200V, E=400V, G=600V	17h§	Say	MCR 218-..., TAG 625-..., TIC 116-..., ++	data
DRA 8 B...G	50Hz-Thy	100...600V, 8A(Tc=83°), Igt/Ih<40/<60mA B=100V, C=200V, E=400V, G=600V	17h§	Say	TAG 680-..., TIC 122-..., TIC 126-..., ++	data
DRB 2 B...E	F-Thy	100...400V, 2A(Tc=35°),Igt/Ih<1,5/=12mA B=100V, C=200V, E=400V	13h§	Say	S 3060...	
DRB 3 B...E	F-Thy	100...400V, 3A(Tc=87°) B=100V, C=200V, E=400V	13h§	Say	CSF 0,7..., S5800...5802...	
DRC 2 E	50Hz-Thy	400V, 2A(Tc=30°), Igt/Ih<50/<120mA	30b	Say		
DRC 3 E	50Hz-Thy	400V, 3A(Tc=65°), Igt/Ih<50/<120mA	17h§	Say	BStC10..., TAG 620-...	

...ype	Device	Short Description	Fig.	Manu	Comparision Types	More at
..O 3 B...G	50Hz-Thy	100...600V, 3A, Igt/Ih<15/=60mA	17h§	Say	TAG 620-...	data
...ONB 21 D	IC	Complex Array For Dual Relay Driver	46	Dii	-	pdf pinout
...E 3 B...G	50Hz-Thy	100...600V, 3A(Tc=90°), Igt/Ih<0,2/=4mA	13h§	Say	C 108-..., TAG 108-...	data
...R 114	Si-Di	≈BA 159			→BA 159	
...V101	DRIVER	Vs:60V	7SP	Bub		data pdf pinout
...V 600 RTJR	IC	Directpath Stereo Line Driver	QFN	Tix	-	pdf pinout
...V 600 RTJT	IC	Directpath Stereo Line Driver	QFN	Tix	-	pdf pinout
...V 601	IC	DirectpathÖ Stereo Line Driver, Adjustable Gain	QFN	Tix	-	pdf pinout
...V 603 PW	LIN-IC	3VRMS Line Driver with Adjustable Gain	TSOP	Tix		pdf pinout
...V1100	DIFF-DRV	Vs:6V Vo:<4.875V Vi0:5mV	8DS	Bub		data pdf pinout
...V 1100P	LIN-IC	High Power diff. Drv, xDSL, +5V, 230mA	8-DIP	Bub		diff pinout
...V 1100U	LIN-IC	SMD, High Power diff. Drv, xDSL, +5V	8-MDIP	Bub		pdf pinout
...V 1101U	LIN-IC	SMD, High Power diff. Drv, xDSL, +5V	8-MDIP	Bub		pdf pinout
...V 600 RTJRG 4	IC	Directpath Stereo Line Driver	QFN	Tix	-	pdf pinout
...V 600 RTJTG 4	IC	Directpath Stereo Line Driver	QFN	Tix	-	pdf pinout
...S						
	Si-P	=2PB1219AS (Typ-Code/Stempel/marking)	35(2mm)	Phi	→2PB1219A	data
	Si-P	=2PB710AS (Typ-Code/Stempel/marking)	35	Phi	→2PB710A	data
	Si-P	=2SA1037KLN-S (Typ-Code/Stempel/mark.)	35	Rhm	→2SA1037KLN	data
	Si-P	=2SA1728 (Typ-Code/Stempel/marking)	35	Say	→2SA1728	data
	Si-P	=2SB1219A-S (Typ-Code/Stempel/marking)	35(2mm)	Mat	→2SB1219A	data
	Si-P	=2SB710A-S (Typ-Code/Stempel/marking)	35	Mat	→2SB710A	data
	Si-P	=2SB789-S (Typ-Code/Stempel/marking)	39	Mat	→2SB789	data
	Si-N	=2SD1946 (Typ-Code/Stempel/marking)	39	Hit	→2SD1946	data
	Si-P	=BCX 42R (Typ-Code/Stempel/marking)	35	Sie,Tho	→BCX 42R	data
	Si-P	=HQ 1F3P (Typ-Code/Stempel/marking)	39	Nec	→HQ 1...	data
	Z-Di	=P4 SMA-180A(Typ-Code/Stempel/marking)	71(5x2,5)	Fag	→P4 SMA-180A	data
	Z-Di	=SMBJ 78C (Typ-Code/Stempel/marking)	71(5x3,5)	Mop	→SMBJ ...	data
...S 0,8-04B...16B	Si-Di	GI, Uni, 400...1600V, 1A		Bbc	BY 133...134, BY 126...127, GP 10G...Y, ++	data
...S 0,9-04A...16A	Si-Di	GI, Uni, 400...1600V, 1,2A	12c	Bbc	BY 226...228, BY 252...255, BY 350/..., ++	data
...S 1-04B...16B	Si-Di	GI, Uni, 400...1600V, 1,2A		Bbc	BY 226...228, BY 252...255, BY 350/..., ++	data
...S 1 H..P	Si-Di	GI, Uni, 300...1800V, 1,1A H=1800, K=1500, M=1000, N=500, P=300V	34a	Fjd	BY 226...228, BY 251...255, GP 15B...M, ++	data
...S 1,2-04A...16A	Si-Di	GI, Uni, 400...1600V, 1,7A	31a	Bbc	BY 228, BY 252...255, BY 448, GP 20G...M++	data
...S 1,8-04A...16A	Si-Di	GI, Uni, 400...1600V, 1,7A	12c	Bbc	BY 228, BY 252...255, BY 448, GP 20G...M++	data
...S 2-04A...16A	Si-Di	GI, Uni, 400...1600V, 1,8A	34b	Bbc	BY 228, BY 252...255, BY 448, GP 30G...M++	data
...S 2 H..P	Si-Di	=DS 1H..P: 1,65A	34a	Fjd	BY 228, BY 251...255, BYW 52...56, ++	data
...S 2 S	Si-Di	≈BA 100	31a		→BA 100	
...S 5 S(R)	Si-Di	Dual, GI			2xBY126...127, 2xBY133...135, 2x1N4002...07	
...S 6-04A...16A	Si-Di	GI-L, 400...1600V, 10A	32b&	Bbc	BYX 42/...R, BYX 98/...R, SSi D04..., ++	data
...S 9-04A...16A	Si-Di	GI-L, 400...1600V, 11A	32b&	Bbc	BYW 88/...R, BYX 99/...R, SSi E20..., ++	data
...S 10-01B...05B	Si-Di	GI-L, 100...500V, 12A	32b&	Bbc	BYW 88/...R, BYX 99/...R, SSi E20..., ++	data
...S 13 A,B	Si-Di	GI, 300...500V, 1,5A	31a	Say	BY 226...227, BY 252...255, GP 20G...M, ++	data
...S 13 C...J	Si-Di	GI-L, 200...800V, 20A C=200V, E=400V, G=600V, J=800V	32a§	Hit	1N3765...3768, 1N4525...4530	data
...S 14 C...J	Si-Di	=DS 13 C...J:	32b&	Hit	D 24/...C, 1N3765...3768R	data
...S 14 C88	CMOS-IC	→KS 5788	14-DIP	Nsc	KS 5788	pdf
...S 14 C89(A)	CMOS-IC	→KS 5789	14-DIP	Nsc	KS 5789	pdf
...S 15 A,B	Si-Di	GI, S, A=600, B=800V, 0,3A, <1,5µs	31a	Say	BA 158...159, BY 126...127, BY 204/..., ++	data
...S 15 C,E,G	Si-Di	GI, Uni, C=300, E=500, G=800V, 0,2A	12	Say	BA 157...159, BY 126...127, BY 204/..., ++	data
...S 16(N)A...(N)E	Si-Di	GI, Uni, 50...800V, 0,5A	31a	Say	BY 126...127, BY 133...135, 1N4001...07, ++	data
...S 17	Si-Di	Dual, Uni, 100V, 2x0,5A	12p	Say	2xBY126...127, 2xBY133...135, 2x1N4002...07	data
...S 18	Si-Di	=DS 17:	12q	Say	2xBY126...127, 2xBY133...135, 2x1N4002...07	data
...S 19	Si-Di	=DS 17:	12n1	Say	2xBY126...127, 2xBY133...135, 2x1N4002...07	data
...S 20	Si-Di	≈BA 127			→BA 127	
...S 23	Si-St	≈BZX 75/C2V1			→BZX 75/C2V1	
...S 26 LS 31	I/O-IC	→AM 26LS31		Nsc	→AM 26LS31	pdf
...S 26 LS 32	I/O-IC	→AM 26LS32		Nsc	→AM 26LS32	pdf
...S 26 LS 31...32...	I/O-IC	4x Line Driver, RS422, 0...+70°		Nsc	→AM 26LS31...32...	
...S 38	Si-Di	Uni, 70V, 0,15A	31a	Say	BA 188...190, BAY 19...21, 1N4148, ++	
...S 113 A..C	Si-Di	TV Damper-Di, 1000...1500V, 1,5A A=1000V, B=1300V, C=1500V	34b&	Say	BY 228, BY 448	data
...S 118 A..E	Si-Di	GI, Uni, 200...1000V, 1,5A A=200, B=400, C=600, D=800, E=1000V	34	Say	BY 226...227, BY 251...255, GP 20B...M, ++	data
...S 130 A..E	Si-Di	GI, Uni, 100...800V, 1A A=800, B=600, C=400, D=200, E=1000V	31a	Say	BY 126...127, BY 133...135, 1N4002...07, ++	data
...S 130 NA..NE	Si-Di	=DS 130A...E	31a	Say	→DS 130A...E	data
...S 130 TA..TE	Si-Di	=DS 130A...E	31a	Say	→DS 130A...E	data
...S 130 YA..YE	Si-Di	=DS 130A...E	31a	Say	→DS 130A...E	data
...S 131 A,B	Si-Di	Dual, GI, 100...200V, 2x0,9A	12e	Say	2xBY251...255, 2xBY259/..., 2xGP20B..M, ++	data
...S 132 A,B	Si-Di	=DS 131A,B	12q	Say	→DS 131A,B	data
...S 133 A,B	Si-Di	=DS 131A,B:	12n1	Say	→DS 131A,B	data

Type	Device	Short Description	Fig.	Manu	Comparision Types	More
DS 135 C...F	Si-Di	Gl, Uni, 50..400V, 1A C=400V, D=200V, E=100V, F=50V	31a	Say	BY 126...127, BY 133...135, 1N4001...07, ++	data
DS 140	Si-Di	TV Damper-Di, 750V, 13A(ss)	34b&	Say	BY 399, BYW 14...16/..., BYW 96/...	data
DS 140 DC	TTL-Logic	Schottky-Interface, Leist.-/power-tr	14-DIP	Hfo	SN 74S140	
DS 150 A...C	Si-Di	Gl, Uni, 200...600V, 1,5A A=200, B=400, C=600V	31a	Say	BY 226...227, BY 251...255, GP 20D...M, ++	data
DS 157 DC	TTL-Logic	Schottky-Interface, Multiplexer	16-DIP	Hfo	SN 74S157	
DS 160 A,B	Si-Di	kV-Gl, A=9kV, B=12kV, 0,55A		Say	-	data
DS 185 D..F	Si-Br	Gl-Br, 50..200V, 1,3A D=200V, E=100V, F=50V	33x	Say	B30C1500...B150C1500	data
DS 230 A...D	Si-Di	Gl, S, 300...1000V, 0,3A, <300ns A=300V, B=600V, C=800V, D=1000V	31a	Say	BA 157...159, BY 204/.., BY 208/..., ++	data
DS 330 A...D	Si-Di	Gl, S, 400...1000V, 0,13A, <300ns A=400V, B=600V, C=800V, D=1000V	31a	Say	BA 157...159, BY 204/.., BY 208/..., ++	data
DS 410	Si-Di	≈BA 127	31a		→BA 127	
DS 430	Si-St	0,55..0,8V(1,5mA)	31a	Say	BA 216, BA 314...315	data
DS 441	Si-Di	Uni, 35V, 0,1A	31a	Say	BA 127, BA 147/50, BA 187...190, 1N4148++	data
DS 442	Si-Di	SS, 35V, 0,12A, <4ns	31a	Say	BAW 62, BAW 76, BAX 95, 1N4148...49, ++	data
DS 442 X	Si-Di	=DS 442: 55V, 0,13A, <2ns	31a	Say	BAW 62, BAW 76, BAX 95, 1N4148...49, ++	data
DS 446	Si-Di	SS, 105V, 0,2A, 2<4ns	31a	Say	BAV 14	data
DS 448	Si-Di	SS, 35V, 0,12A, <4ns	31a	Say	BAW 62, BAW 76, BAX 95, 1N4148...49, ++	data
DS 452	Si-Di	Uni, 220V, 0,1A	31a	Say	BAY 20, BAY 46, BAY 88, BA 157...159	data
DS 454	Si-Di	=DS 452: 320V	31a	Say	BA 157...159, BAY 46, BAY 88...89, ++	data
DS 462	Si-Di	Uni, S, 250V, 0,2A, <60ns	31a	Say	BAY 21, BAY 46, BAY 88, BA 157...159	data
DS 464	Si-Di	=DS 462: 300V	31a	Say	BAY 21, BAY 46, BAY 88, BA 157...159	data
DS 1080L	IC	SMD, Spread-spectrum Clock Synthesiser from 16MHz to 134MHz, Vcc=3.0...3.6V	SSMDIP	Max	-	pdf pinou
DS 1086LU	IC	SMD, Spread-spectrum Clock Synthesiser from 130kHz to 66.6MHz, Vcc=2.7...3.6V	SSMDIP	Max	-	pdf pinou
DS 1086U	IC	SMD, Spread-spectrum Clock Synthesiser from 260kHz to 130MHz, Vcc=5V	SSMDIP	Max	-	pdf pinou
DS 1086Z	IC	SMD, Spread-spectrum Clock Synthesiser from 260kHz to 130MHz, Vcc=5V	MDIP	Max	-	pdf pinou
DS 1181 L	LIN-IC	20MHz - 134MHz Spread-Spectrumm Clock Modulator f. LCD Panels	TSOP	Max	-	pdf pinou
DS 1488	I/O-IC	4x Leitungstr./line driver f. RS232	14-DIP	Nsc	→µA 1488	pdf
DS 1489(A)	I/O-IC	4x Leitungsempf./line receiver f.RS232	14-DIP	Nsc	→µA 1489	pdf
DS 1631 (A)U	LIN-IC	High-precis. Digit. Thermometer & Thermostat	TSOP	Max	-	pdf pinou
DS 1631	LIN-IC	High-precis. Digit. Thermometer & Thermostat	DIC	Max	-	pdf pinou
DS 1631 S/Z	LIN-IC	High-precis. Digit. Thermometer & Thermostat	MDIP	Max	-	pdf pinou
DS 1722 S	LIN-IC	Digital Thermometer w. SPI/3-Wire Interface	TSOP	Max	-	pdf pinou
DS 1722 U	LIN-IC	Digital Thermometer w. SPI/3-Wire Interface	DIC	Max	-	pdf pinou
DS 1731 U	LIN-IC	High-precis. Digit. Thermometer & Thermostat	TSOP	Max	-	pdf pinou
DS 18 B 20	IC	Programmable Resolution 1-Wire Digital Thermometer	DFN	Dal	-	pdf pinou
DS 18 B 20 U	IC	Programmable Resolution 1-Wire Digital Thermometer	DFN	Dal	-	pdf pinou
DS 18 B 20 Z	IC	Programmable Resolution 1-Wire Digital Thermometer	DFN	Dal	-	pdf pinou
DS 1841	LIN-IC	Temperature-Controlled, NV, I2C, Log. Resistor	MLP	Max	-	pdf pinou
DS 2001 CJ	TTL-IC	High Cur/volt Darlington Drives	DIP	Nsc	-	pdf pinou
DS 2001 CM	TTL-IC	High Cur/volt Darlington Drives	DIP	Nsc	-	pdf pinou
DS 2001 CN	TTL-IC	High Cur/volt Darlington Drives	DIP	Nsc	-	pdf pinou
DS 2001 MJ	TTL-IC	High Cur/volt Darlington Drives	DIP	Nsc	-	pdf pinou
DS 2001 TJ	TTL-IC	High Cur/volt Darlington Drives	DIP	Nsc	-	pdf pinou
DS 2001 TM	TTL-IC	High Cur/volt Darlington Drives	DIP	Nsc	-	pdf pinou
DS 2001 TN	TTL-IC	High Cur/volt Darlington Drives	DIP	Nsc	-	pdf pinou
DS 2002 CJ	TTL-IC	High Cur/volt Darlington Drives	DIP	Nsc	-	pdf pinou
DS 2002 CM	TTL-IC	High Cur/volt Darlington Drives	DIP	Nsc	-	pdf pinou
DS 2002 CN	TTL-IC	High Cur/volt Darlington Drives	DIP	Nsc	-	pdf pinou
DS 2002 MJ	TTL-IC	High Cur/volt Darlington Drives	DIP	Nsc	-	pdf pinou
DS 2002 TJ	TTL-IC	High Cur/volt Darlington Drives	DIP	Nsc	-	pdf pinou
DS 2002 TM	TTL-IC	High Cur/volt Darlington Drives	DIP	Nsc	-	pdf pinou
DS 2002 TN	TTL-IC	High Cur/volt Darlington Drives	DIP	Nsc	-	pdf pinou
DS 2003 CJ	IC	High Cur/volt Darlington Drivers	DIP	Nsc	-	pdf pinou
DS 2003 CM	IC	High Cur/volt Darlington Drivers	DIP	Nsc	-	pdf pinou
DS 2003 CN	IC	High Cur/volt Darlington Drivers	DIP	Nsc	-	pdf pinou
DS 2003 MJ	IC	High Cur/volt Darlington Drivers	DIP	Nsc	-	pdf pinou
DS 2003 TJ	IC	High Cur/volt Darlington Drivers	DIP	Nsc	-	pdf pinou
DS 2003 TM	IC	High Cur/volt Darlington Drivers	DIP	Nsc	-	pdf pinou
DS 2003 TN	IC	High Cur/volt Darlington Drivers	DIP	Nsc	-	pdf pinou
DS 2004 CJ	IC	High Cur/volt Darlington Drivers	DIP	Nsc	-	pdf pinou
DS 2004 CM	IC	High Cur/volt Darlington Drivers	DIP	Nsc	-	pdf pinout

e	Device	Short Description	Fig.	Manu	Comparision Types	More at
4 CN	IC	High Cur/volt Darlington Drivers	DIP	Nsc	-	pdf pinout
04 MJ	IC	High Cur/volt Darlington Drivers	DIP	Nsc	-	pdf pinout
4 TJ	IC	High Cur/volt Darlington Drivers	DIP	Nsc	-	pdf pinout
4 TM	IC	High Cur/volt Darlington Drivers	DIP	Nsc	-	pdf pinout
04 TN	IC	High Cur/volt Darlington Drivers	DIP	Nsc	-	pdf pinout
3	IC	1-Wire Dual Channel Addressable Switch	TSOP	Max	-	pdf pinout
0 DC	TTL-IC	Schottky-Interface, 4 Bit Shifter	16-DIP	Hfo	Am 25S10N	
0 DC	TTL-IC	Schottky-Interface, 4x Bus-I/O	16-DIP	Hfo	Am 26S10N	
03U+	IC	SMD, SHA-1 Battery Pack Authentication	SSMDIP	Max	-	pdf pinout
03U	IC	SMD, SHA-1 Battery Pack Authentication	SSMDIP	Max	-	pdf pinout
04G+	IC	SMD, 1280Bit EEPROM with SHA- 1 Authentication	VSMDIP	Max	-	pdf pinout
04G	IC	SMD, 1280Bit EEPROM with SHA- 1 Authentication	VSMDIP	Max	-	pdf pinout
05U+	IC	SMD, SHA-1 Authentication Master	SSMDIP	Max	-	pdf pinout
14E+	IC	SMD, Quad Loose Cell NiMH Charger, 1 to 4 NiMH (NiCd)Cells, VDD=4...5.5V	SSMDIP	Max	-	pdf pinout
38	IC	Stand-alone Fuel-gauge IC with LED Display Drivers	TSOP	Max	-	pdf pinout
02	CMOS-TTL-IC	Stratum 3 Timing Card IC w. Synchronous Ethernet Support	PGA	Max	-	pdf pinout
36	I/O-IC	=MC 3486L,P	16-DIC,DIP	Mot	→MC 3486	pdf
37	I/O-IC	=MC 3487L,P	16-DIC,DIP	Mot	→MC 3487	pdf
24	LIN-IC	Low-jitter, Precision Clock Generator w. Four Out.	MLP	Max	-	pdf pinout
12 U	D/A-IC	Dual-Channel, I2C Sink/Source Current DAC	GRID	Max	-	pdf pinout
05 D	TTL-IC	Schottky, 1/8-Binärdecoder/binary dec.	16-DIP	Hfo	i 8205	
12 D	TTL-IC	Schottky, 8 Bit Bustreiber/bus driver	24-DIP	Hfo	i 8212	
16 D	TTL-IC	Schottky, 4 Bit Bustreiber/bus driver	16-DIP	Hfo	i 8216	
82 D	TTL-IC	Schottky, 8 Bit Bustreiber/bus driver	20-DIP	Hfo	i 8282	
33 D	TTL-IC	Schottky, 8 Bit Bustreiber/bus driver	20-DIP	Hfo	i 8283	
86 D	TTL-IC	Schottky, 8 Bit Bustreiber/bus driver	20-DIP	Hfo	i 8286	
87 D	TTL-IC	Schottky, 8 Bit Bustreiber/bus driver	20-DIP	Hfo	i 8287	
01 DC	I/O-IC	Dma-logic		Hfo		
09 DC	I/O-IC	Ser.-par./par.-ser. Converter	40-DIP	Hfo	DC 018	
38 DC	DIG-IC	8 Bit Bustreiber/bus driver, bidirect.	20-DIP	Hfo	DC 021	
40 N	TTL-IC	Bus-empfänger, bus receiver	14-DIP	Nsc	DL 8640DC	pdf
41 N	TTL-IC	Bus-empfänger+treiber,bus receiver+drv	16-DIP	Nsc	DL 8641DC	pdf
65 CJ	TTL-IC	High Cur/volt Darlington Drives	DIP	Nsc	-	pdf pinout
65 CN	TTL-IC	High Cur/volt Darlington Drives	DIP	Nsc	-	pdf pinout
65 MJ	TTL-IC	High Cur/volt Darlington Drives	DIP	Nsc	-	pdf pinout
65 TJ	TTL-IC	High Cur/volt Darlington Drives	DIP	Nsc	-	pdf pinout
65 TN	TTL-IC	High Cur/volt Darlington Drives	DIP	Nsc	-	pdf pinout
66 CJ	TTL-IC	High Cur/volt Darlington Drives	DIP	Nsc	-	pdf pinout
66 CN	TTL-IC	High Cur/volt Darlington Drives	DIP	Nsc	-	pdf pinout
66 MJ	TTL-IC	High Cur/volt Darlington Drives	DIP	Nsc	-	pdf pinout
66 TJ	TTL-IC	High Cur/volt Darlington Drives	DIP	Nsc	-	pdf pinout
66 TN	TTL-IC	High Cur/volt Darlington Drives	DIP	Nsc	-	pdf pinout
67 CJ	IC	High Cur/volt Darlington Drivers	DIP	Nsc	-	pdf pinout
67 CN	IC	High Cur/volt Darlington Drivers	DIP	Nsc	-	pdf pinout
67 MJ	IC	High Cur/volt Darlington Drivers	DIP	Nsc	-	pdf pinout
67 TJ	IC	High Cur/volt Darlington Drivers	DIP	Nsc	-	pdf pinout
67 TN	IC	High Cur/volt Darlington Drivers	DIP	Nsc	-	pdf pinout
68 CJ	TTL-IC	High Cur/volt Darlington Drives	DIP	Nsc	-	pdf pinout
68 CN	TTL-IC	High Cur/volt Darlington Drives	DIP	Nsc	-	pdf pinout
68 MJ	TTL-IC	High Cur/volt Darlington Drives	DIP	Nsc	-	pdf pinout
68 TJ	TTL-IC	High Cur/volt Darlington Drives	DIP	Nsc	-	pdf pinout
68 TN	TTL-IC	High Cur/volt Darlington Drives	DIP	Nsc	-	pdf pinout
431-A1	IC	1024-Bit,1-Wire Eeprom,Automotive Apps	TSOP	Max	-	pdf pinout
4 T 101 GN	DIG-IC	Single Tdm-over-packet Chip	GRID	Max	-	pdf pinout
4 T 102 GN	DIG-IC	Dual Tdm-over-packet Chip	GRID	Max	-	pdf pinout
4 T 104 GN	DIG-IC	Quad Tdm-over-packet Chip	GRID	Max	-	pdf pinout
4 T 108 GN	DIG-IC	Octal Tdm-over-packet Chip	GRID	Max	-	pdf pinout
5107...	I/O-IC	→µA 75107...		Nsc	→µA 75107...	
5108...	I/O-IC	→µA 75108...		Nsc	→µA 75108...	
5150...	I/O-IC	→µA 75150...		Nsc	→µA 75150...	
5154...	I/O-IC	→µA 75154...		Nsc	→µA 75154...	
5450...	I/O-IC	→µA 75450...		Nsc	→µA 75450...	
5491...	I/O-IC	→µA 75491...		Nsc	→µA 75491...	
0612 DC	DIG-IC	Taktgenerator/clock generator	18-DIP	Hfo		
0CF365 MTD	DIG/LIN-IC	LVDS transmitter, 18, 24-Bit Panel display	TSOP	Nsc	-	pdf pinout
0C385 MTD	DIG/LIN-IC	LVDS transmitter, 18, 24-Bit Panel display	TSOP	Nsc	-	pdf pinout
0C385 SLC	DIG/LIN-IC	LVDS transmitter, 18, 24-Bit Panel display	LGA	Nsc	-	pdf
0C387 A	DIG/LIN-IC	LVDS display interface	FLP	Nsc	-	pdf pinout
0C388 A	DIG/LIN-IC	LVDS display interface	FLP	Nsc	-	pdf pinout
0 CF 388 A	DIG/LIN-IC	LVDS display interface	FLP	Nsc	-	pdf pinout
010	Si-Di	SMD, SS, 85V, 0,1A, <4ns	35f	Say	-	data
10 G,J,L	Si-Di	Gl, Uni, 600...1000V, 1A G=600V, J=800V, L=1000V	31a	Say	BY 126...127, BY 133...134, 1N4005...4007++	data

Type	Device	Short Description	Fig.	Manu	Comparision Types	More at
DSA 12(T)B...L	Si-Di	Gl, Uni, 100...1000V, 1,2A B=100V, C=200V, E=400V, G=600V, J=800V L=1000V	31a	Say	BY 133...135, BY 227...227, GP 15B..M, ++	data
DSA 14 C,G	Si-Di	SMD, Uni, 200...600V, 1,1A C=200V, G=600V	71a(7,6x4)	Say	-	data
DSA 015	Si-Di	SMD, SS, 75V, 0,1A, <8ns	35f	Say	1SS222	data
DSA 17 C,E,G	Si-Di	Gl, Uni, 200...600V, 1,7A C=200V, E=400V, G=600V	31a	Say	BY 251...255, MR 502...510, 1N5402...08, ++	data
DSA 20(T)B...L	Si-Di	Gl, Uni, 100...1000V, 2A B=100V, C=200V, E=400V, G=600V, J=800V L=1000V	31a	Say	BY 251...255, MR 502...510, 1N5402...08, ++	data
DSA 26 B,C,E,G	Si-Di	Gl, Uni, 100...600V, 2,6A B=100V, C=200V, E=400V, G=600V	31a	Say	BY 251...255, MR 501...510, 1N5401...08, ++	data
DSB 010	Si-Di	SMD, SS, 85V, 0,1A, <4ns	35e	Say	1SS221	data
DSB 015	Si-Di	SMD, SS, 75V, 0,1A, <8ns	35e	Say	1SS220	data
DSB 15(T)B...L	Si-Di	Gl, Uni, 100...1000V, 1,5A B=100V, C=200V, E=400V, G=600V, J=800V L=1000V	31a	Say	BY 227...227, GP 15B..M, 1N5392...99, ++	data
DSC 010	Si-Di	SMD, SS, 85V, 0,1A, <4ns	35d	Say	BAL 99	data
DSC 015	Si-Di	SMD, Uni, S, 250V, 0,15A, 30ns	35e	Say	BAS 21, 1SS250	data
DSC 30(T)B...L	Si-Di	Gl, Uni, 100...1000V, 3A B=100V, C=200V, E=400V, G=600V, J=800V L=1000V	31a	Say	BY 251...255, MR 501...510, 1N5401...08, ++	data
DSD 010	Si-Di	SMD, SS, 85V, 0,1A, <4ns	35c	Say	-	data
DSD 015	Si-Di	SMD, Uni, S, 350V, 0,15A, 30ns	35e	Say	-	data
DSE 010	Si-Di	SMD, SS, 90V, 0,1A, <4ns	71a(1,7mm)	Say	HSU 119, MA 111, 1SS352, 1SS353	data
DSE 015	Si-Di	SMD, SS, 75V, 0,1A, <8ns	35f(2mm)	Say	-	data
DSF 10(T) B...L	Si-Di	Gl, Uni, 100...1000V, 1A B=100V, C=200V, E=400V, G=600V, J=800V L=1000V	31a	Say	BY 126...127, BY 133...134, 1N4002...4007++	data
DSG 10 G,J,L	Si-Di	Gl, Uni, 600...1000V, 1A G=600V, J=800V, L=1000V	31a	Say	BY 126...127, BY 133...134, 1N4005...4007++	data
DSH 015	Si-Di	SMD, SS, 75V, 0,1A, <8ns	35e(2mm)	Say	-	data
DSK 10 B..L	Si-Di	Gl, Uni, 100...1000V, 1A B=100V, C=200V, E=400V, G=600V, J=800V L=1000V	31a	Say	BY 126...127, BY 133...134, 1N4002...4007++	data
DSL 10 B,C,E	Si-Di	Gl, Uni, 100...400V, 1A B=100V, C=200V, E=400V	31a	Say	BY 126...127, BY 133...134, 1N4002...4007++	data
DSM 10 C,E,G	Si-Di	SMD, Gl, Uni, 200...600V, 0,75A C=200V, E=400V, G=600V	71a(5x2,5)	Say	D 1F20...F60, U 1DC44...JC44	data
DSP 10	Si-Di	SMD, Uni, 600V, 0,8A	71a(2,8mm)	Say	-	data
DSR 10 C,E	Si-Di	Gl, Uni, 200...400V, 1A C=200V, E=400V	31a	Say	BY 126...127, BY 133...134, 1N4003...4007++	data

DT...DV

Type	Device	Short Description	Fig.	Manu	Comparision Types	More at
DT	Si-N	=XN 6A554 (Typ-Code/Stempel/marking)	46	Mat	→XN 6A554	data
DT	Si-N	=2SC4112 (Typ-Code/Stempel/marking)	35	Say	→2SC4112	data
DT	Si-N-Darl	=2SD1471-DT (Typ-Code/Stempel/marking)	39	Hit	→2SD1471	data
DT	Si-P	=BCW 67RA (Typ-Code/Stempel/marking)	35	Sie	→BCW 67RA	data
DT	Si-P	=HQ 1L2Q (Typ-Code/Stempel/marking)	39	Nec	→HQ 1...	data
DT	Z-Di	=MMSZ 4705 (Typ-Code/marking)	71(2,7mm)	Ons	→MMSZ 4705	
DT	Z-Di	=P4 SMA-200 (Typ-Code/Stempel/marking)	71(5x2,5)	Fag	→P4 SMA-200	data
Dt 0	Si-N/P+R	=PUMD 10 Typ-Code/Stempel/marking)	46	Phi	→PUMD 10	data
DT	Z-Di	=SMBJ 78CA (Typ-Code/Stempel/marking)	71(5x3,5)	Mop	→SMBJ ...	data
Dt 1	Si-N/P+R	=PUMD 12 Typ-Code/Stempel/marking)	46	Phi	→PUMD 12	data
DT 1-02...12	50Hz-Thy	200...1200V, 1,3A(Tc=135°C), Igt=15mA	27e	Tho	BStA30...M, BStC07...	data
Dt 2	Si-N/P+R	=PUMD 2 (Typ-Code/Stempel/marking)	46	Phi	→PUMD 2	data
DT 3 C144W(A)	Si-N+R	3xNPN, Rb=47k, Rbe=22kΩ, 50V, 30mA	9-SQP	Rhm	-	data
Dt 3	Si-N/P+R	=PUMD 3 (Typ-Code/Stempel/marking)	46	Phi	→PUMD 3	data
DT 5 A113Z(A)	Si-P+R	5xPNP, Rb=1k, Rbe=10kΩ, 50V, 100mA	12-SQP	Rhm	-	data
DT 5 A114E(A)	Si-P+R	5xPNP, Rb=Rbe=10kΩ, 50V, 50mA	12-SQP	Rhm	-	data
DT 5 A124E(A)	Si-P+R	5xPNP, Rb=Rbe=22kΩ, 50V, 30mA	12-SQP	Rhm	-	data
DT 5 A143E(A)	Si-P+R	5xPNP, Rb=Rbe=4,7kΩ, 50V, 100mA	12-SQP	Rhm	-	data
DT 5 A143T(A)	Si-P+R	5xPNP, Rb=4,7kΩ, Rbe=-, 50V, 100mA	12-SQP	Rhm	-	data
DT 5 A143X(A)	Si-P+R	5xPNP, Rb=4,7k, Rbe=10kΩ, 50V, 100mA	12-SQP	Rhm	-	data
DT 5 A144E(A)	Si-P+R	5xPNP, Rb=Rbe=47kΩ, 50V, 30mA	12-SQP	Rhm	-	data
DT 5 B123E	Si-P+R	5xPNP, Rb=Rbe=2,2kΩ, 50V, 500mA	12-SQP	Rhm	-	data
DT 5 C113Z(A)	Si-N+R	5xNPN, Rb=1k, Rbe=10kΩ, 50V, 100mA	12-SQP	Rhm	-	data
DT 5 C114E(A)	Si-N+R	5xNPN, Rb=Rbe=10kΩ, 50V, 50mA	12-SQP	Rhm	-	data
DT 5 C124E(A)	Si-N+R	5xNPN, Rb=Rbe=22kΩ, 50V, 30mA	12-SQP	Rhm	-	data
DT 5 C143E(A)	Si-N+R	5xNPN, Rb=Rbe=4,7kΩ, 50V, 100mA	12-SQP	Rhm	-	data
DT 5 C143T(A)	Si-N+R	5xNPN, Rb=4,7kΩ, Rbe=-, 50V, 100mA	12-SQP	Rhm	-	data
DT 5 C143X(A)	Si-N+R	5xNPN, Rb=4,7k, Rbe=10kΩ, 50V, 100mA	12-SQP	Rhm	-	data

ype	Device	Short Description	Fig.	Manu	Comparision Types	More at
5 C144E(A)	Si-N+R	5xNPN, Rb=Rbe=47kΩ, 50V, 30mA	12-SQP	Rhm	-	data
5 D123D(A)	Si-N+R	5xNPN, Rb=Rbe=2,2kΩ, 50V, 500mA	12-SQP	Rhm	-	data
5	Si-N/P+R	=PUMD 6 (Typ-Code/Stempel/marking)	46	Phi	→PUMD 6	data
06-02...10	50Hz-Thy	200...1000V, 0,8A(Tc=125°C), Igt=10mA	2a§	Tho	TAG 612-..., S 2600..., TAG 613-..., ++	data
9	Si-N/P+R	=PUMD 9 (Typ-Code/Stempel/marking)	46	Phi	→PUMD 9	data
5111	Z-Di, Z-IC	=THDT 51-11 (Typ-Code/Stempel/marking)	8-MDIP	Tho	→THDT 51	data
5112	Z-Di, Z-IC	=THDT 51-12 (Typ-Code/Stempel/marking)	8-DIP	Tho	→THDT 51	data
6511	Z-Di, Z-IC	=THDT 65-11 (Typ-Code/Stempel/marking)	8-MDIP	Tho	→THDT 65	data
6512	Z-Di, Z-IC	=THDT 65-12 (Typ-Code/Stempel/marking)	8-DIP	Tho	→THDT 65	data
A 1 C...E	Triac	200...400V, 1A≈(Tc=74°),Igt/Ih=10/<10mA C=200V, E=400V	7e	Say	Z 0101..., Z 0104..., Z 0106	data
A1D3R(A,F,L,S,V)	Si-P+R	S, Rb=2,7k, Rbe=1kΩ, 50V, 70mA, 0,3W	9c, 41c	Rhm	-	data
A 1 D3RE	Si-P+R	=DTA 1D3R...: SMD	35a(1,6mm)		-	data
A 1 D3RK(A)	Si-P+R	=DTA 1D3R...: SMD	35a(2,9mm)		-	data
A 1 D3RU(A)	Si-P+R	=DTA 1D3R...: SMD	35a(2mm)		-	data
A 2 B...E	Triac	100...400V, 2A≈(Tc=70°),Igt/Ih<15/<25mA B=100V, C=200V, E=400V	13h§	Say	TAG 137-..., TAG 138-...	data
A 3 C...G	Triac	200...600V, 3A≈(Tc=77°),Igt/Ih<45/=25mA C=200V, E=400V, F=500V, G=600V	13h§	Say	Z 0409..., Z 0410...	
A 05 B...E	Triac	100...400V, 0,5A≈, Igt/Ih<15/<25mA B=100V, C=200V, E=400V	7b	Say	MAC 94(A)-..., MAC 95(A)-...	data
A 6 C...G(-N)	Triac	200...600V,6A≈(Tc=104°),Igt/Ih<50/<50mA C=200V, E=400V, G=600V	17h§	Say	MAC 216A-..., T 2500..., TAG 220-...	data
A 08 E	Triac	400V, 0,8A≈(Tc=60°), Igt/Ih<10/<10mA	7e	Say	MAC 94(A)-..., MAC 95(A)-..., MAC 96(A)-...	data
A 8 C...G	Triac	200...600V, 8A≈(Tc=97°),Igt/Ih<50/<50mA C=200V, E=400V, F=500V, G=600V	17h§	Say	MAC 222-..., MAC 228-..., TIC 225..226...,+	
A 10 C...G	Triac	200...600V, 10A≈(91°),Igt/Ih<80/=30mA C=200V, E=400V, F=500V, G=600V	17h§	Say	BTA 22-..., TAG 250-..., TAG 251-...	
A 12 C...G	Triac	200...600V, 12A≈(Tc=92°) C=200V, E=400V, F=500V, G=600V	17h§	Say	BT 162/.., SC 149..., TIC 236..., ++	
A 16 C...G	Triac	200...600V, 16A≈(Tc=80°) C=200V, E=400V, F=500V, G=600V	17h§	Say	SC 151..., MAC 15-..., TIC 246..., ++	
A 25 C...G	Triac	200...600V, 25A≈(73°), Igt/Ih<50/<70mA C=200V, E=400V, G=600V	65b	Say	MAC 525(A)-...	
A113T(A,F,L,S,V)	Si-P+R	S, Rb=1kΩ, Rbe=-, 50V, 100mA, 0,3W	9c, 41c	Rhm	-	data
A 113 TE	Si-P+R	=DTA 113T...: SMD	35a(1,6mm)		-	data
A 113 TC,TK(A)	Si-P+R	=DTA 113T...: SMD	35a(2,9mm)			data
A 113 TKA	Si-P+R	SMD, Digital Transistor, RB=1k, 2SA1037AK 50V, 100mA,	35a	Rhm	DTAJ14GKA, DTA114TKA, DTA115GKA,	data data
A 113 TU(A)	Si-P+R	=DTA 113T...: SMD	35a(2mm)			data
A113Z(A,F,L,S,V)	Si-P+R	S, Rb=1k, Rbe=10kΩ, 50V, 100mA, 0,3W	9c, 41c	Rhm	AN 1A3Q, UN 4119	data
A 113 ZE	Si-P+R	=DTA 113Z...: SMD	35a(1,6mm)		UN 9119	data
A 113 ZC,ZK(A)	Si-P+R	=DTA 113Z...: SMD	35a(2,9mm)		FN 1A3Q, UN 2119	data
A 113 ZKA	Si-P+R	SMD, Digital Transistor, Inverter, RB=1k RBE=10k 2SA1037AK 100mA,	35bd3	Rhm		data
A 113 ZSA	Si-P+R	Digital Transistor, Inverter, RB=1k RBE=10k 2SA1037AK 100mA,	78bd4	Rhm		data
A 113 ZU(A)	Si-P+R	=DTA 113Z...: SMD	35a(2mm)		UN 5119	data
A 113 ZUA	Si-P+R	SMD, Digital Transistor, Inverter, RB=1k RBE=10k 2SA1037AK 100mA,	35bd3	Rhm		data data
A114E(A,F,L,S,V)	Si-P+R	S, Rb=Rbe=10kΩ, 50V, 50mA, 0,3W	9c, 41c	Rhm	AN 1A4M, RN 2002, UN 4111, 2SA1348,++	data
A 114 EE	Si-P+R	=DTA 114E...: SMD	35a(1,6mm)		RN 2102, UN 9111	data
A 114 EC,EK(A)	Si-P+R	=DTA 114E...: SMD	35a(2,9mm)		FN 1A4M, RN 2402, UN2111, 2SA1344,++	data
A 114 EKA	Si-P+R	SMD, Digital Transistor, Inverter, RB=10k RBE=10k 2SA1037AK 100mA,	35bd3	Rhm		data
A 114 ESA	Si-P+R	Digital Transistor, Inverter, RB=10k RBE=10k 2SA1037AK 100mA,	78bd4	Rhm		data
A 114 EU(A)	Si-P+R	=DTA 114E...: SMD	35a(2mm)		RN 2302, UN 5111, 2SA1678	data
A 114 EUA	Si-P+R	SMD, Digital Transistor, Inverter, RB=10k RBE=10k 2SA1037AK 100mA,	35bd3	Rhm		data data
A114G(A,F,L,S,V)	Si-P+R	S, Rb=-, Rbe=10kΩ, 50V, 100mA, 0,3W	9c, 41c	Rhm	-	data
A 114 GE	Si-P+R	=DTA 114G...: SMD	35a(1,6mm)		-	data
A 114 GC,GK(A)	Si-P+R	=DTA 114G...: SMD	35a(2,9mm)		-	data
A 114 GKA	Si-P+R	SMD, Digital Transistor, Inverter, RBE=10k 2SA1037AK 50V, 100mA,	35a	Rhm	DTA113TKA, DTA114TKA, DTA115GKA,	data
A 114 GSA	Si-P+R	Digital Transistor, Inverter, RBE=10k 2SA1037AK 50V, 100mA,	78b	Rhm	DTA114TSA, DTA115TSA, DTA124GSA,	data
A 114 GU(A)	Si-P+R	=DTA 114G...: SMD	35a(2mm)		-	data
A 114 GUA	Si-P+R	SMD, Digital Transistor, Inverter, RBE=10k 2SA1037AK 50V, 100mA,	35a	Rhm	DTA114TUA, DTA115GUA, DTA115TUA,	data
A114T(A,F,L,S,V)	Si-P+R	S, Rb=10kΩ, Rbe=-, 50V, 100mA, 0,3W	9c, 41c	Rhm	AN 1A4Z, RN 2011, UN 4115, 2SA1497,++	data
TA 114 TE	Si-P+R	=DTA 114T...: SMD	35a(1,6mm)		RN 2111, UN 9115	data
TA 114 TC,TK(A)	Si-P+R	=DTA 114T...: SMD	35a(2,9mm)		FN 1A4Z, RN 2411, UN 2115, 2SA1496,++	data
TA 114 TKA	Si-P+R	SMD, Digital Transistor, Inverter, RB=10k 2SA1037AK 50V, 100mA,	35a	Rhm	DTA113TKA, DTA114GKA, DTA115GKA,	data data
TA 114 TSA	Si-P+R	Digital Transistor, Inverter, RB=10k 2SA1037AK 50V, 100mA,	78b	Rhm	DTA114GSA, DTA115TSA, DTA124GSA,	data
TA 114 TU(A)	Si-P+R	=DTA 114T...: SMD	35a(2mm)		RN 2311, UN 5115	data
TA 114 TUA	Si-P+R	SMD, Digital Transistor, Inverter, RB=10k 2SA1037AK 50V, 100mA,	35a	Rhm	DTA114GUA, DTA115GUA, DTA115TUA,	data data
TA114W(A,F,L,S,V)	Si-P+R	S, Rb=10k, Rbe=4,7kΩ, 50V, 100mA, 0,3W	9c, 41c	Rhm	UN 411K	data

Type	Device	Short Description	Fig.	Manu	Comparision Types	More at
DTA 114 WE	Si-P+R	=DTA 114W..: SMD	35a(1,6mm)		UN 911K	data
DTA 114 WC,WK(A)	Si-P+R	=DTA 114W..: SMD	35a(2,9mm)		UN 211K	data
DTA 114 WK4	Si-P+R	SMD, Digital Transistor, Inverter, RB=10k RBE=47k 2SA1037AK 100mA,	35bd3	Rhm		data
DTA 114 WSA	Si-P+R	Digital Transistor, Inverter, RB=10k RBE=47k 2SA1037AK 100mA,	78bd4	Rhm		data
DTA 114 WU(A)	Si-P+R	=DTA 114W..: SMD	35a(2mm)		UN 511K	data
DTA 114 WUA	Si-P+R	SMD, Digital Transistor, Inverter, RB=10k RBE=47k 2SA1037AK 100mA,	35bd3	Rhm		data
DTA114Y(A,F,L,S,V)	Si-P+R	S, Rb=10k, Rbe=47kΩ, 50V, 100mA, 0,3W	9c, 41c	Rhm	AN 1A4P, RN 2007, UN 4114, 2SA1564,++	data
DTA 114 YE	Si-P+R	=DTA 114Y..: SMD	35a(1,6mm)		RN 2107, UN 9114	data
DTA 114 YC,YK(A)	Si-P+R	=DTA 114Y..: SMD	35a(2,9mm)		FN 1A4P, RN 2407, UN 2114, 2SA1563,++	data
DTA 114 YKA	Si-P+R	SMD, Digital Transistor, Inverter, RB=10k RBE=47k 2SA1037AK 100mA,	35bd3	Rhm		data
DTA 114 YSA	Si-P+R	Digital Transistor, Inverter, RB=10k RBE=47k 2SA1037AK 100mA,	78bd4	Rhm		data
DTA 114 YU(A)	Si-P+R	=DTA 114Y..: SMD	35a(2mm)		RN 2307, UN 5114	data
DTA 114 YUA	Si-P+R	SMD, Digital Transistor, Inverter, RB=10k RBE=47k 2SA1037AK 100mA,	35bd3	Rhm		data
DTA115E(A,F,L,S,V)	Si-P+R	S, Rb=Rbe=100kΩ, 50V, 20mA, 0,3W	9c, 41c	Rhm	-	data
DTA 115 EE	Si-P+R	=DTA 115E..: SMD	35a(1,6mm)			data
DTA 115 EC,EK(A)	Si-P+R	=DTA 115E..: SMD	35a(2,9mm)			data
DTA 115 EKA	Si-P+R	SMD, Digital Transistor, Inverter, RB=100k RBE=100k 2SA1037AK 100mA,	35bd3	Rhm		data
DTA 115 ESA	Si-P+R	Digital Transistor, Inverter, RB=100k RBE=100k 2SA1037AK 100mA,	78bd4	Rhm		data
DTA 115 EU(A)	Si-P+R	=DTA 115E..: SMD	35a(2mm)			data
DTA 115 EUA	Si-P+R	SMD, Digital Transistor, Inverter, RB=100k RBE=100k 2SA1037AK 100mA,	35bd3	Rhm		data
DTA115G(A,F,L,S,V)	Si-P+R	S, Rb=-, Rbe=100kΩ, 50V, 100mA, 0,3W	9c, 41c	Rhm	-	data
DTA 115 GE	Si-P+R	=DTA 115G..: SMD	35a(1,6mm)		-	data
DTA 115 GC,GK(A)	Si-P+R	=DTA 115G..: SMD	35a(2,9mm)		-	data
DTA 115 GKA	Si-P+R	SMD, Digital Transistor, Inverter, RBE=100k 2SA1037AK 50V, 100mA,	35a	Rhm	DTA113TKA, DTA114GKA, DTA114TKA,	data
DTA 115 GU(A)	Si-P+R	=DTA 115G..: SMD	35a(2mm)		-	data
DTA 115 GUA	Si-P+R	SMD, Digital Transistor, Inverter, RBE=100k 2SA1037AK 50V, 100mA,	35a	Rhm	DTA114GUA, DTA114TUA, DTA115TUA,	data
DTA115T(A,F,L,S,V)	Si-P+R	S, Rb=100kΩ, Rbe=-, 50V, 100mA, 0,3W	9c, 41c	Rhm	-	data
DTA 115 TE	Si-P+R	=DTA 115T..: SMD	35a(1,6mm)		-	data
DTA 115 TH	Si-P+R	SMD, Digital Transistor, Inverter, RB=100k 2SA1037AK 50V, 100mA,	35a	Rhm	DTA114GE, DTA114TE, DTA115TE, DTA124TE,	data
DTA 115 TC,TK(A)	Si-P+R	=DTA 115T..: SMD	35a(2,9mm)			data
DTA 115 TKA	Si-P+R	SMD, Digital Transistor, Inverter, RB=100k 2SA1037AK 50V, 100mA,	35a	Rhm	DTA113TKA, DTA114GKA, DTA114TKA,	data
DTA 115 TSA	Si-P+R	Digital Transistor, Inverter, RB=100k 2SA1037AK 50V, 100mA,	78b	Rhm	DTA114GSA, DTA114TSA, DTA124GSA,	data
DTA 115 TU(A)	Si-P+R	=DTA 115T..: SMD	35a(2mm)		-	data
DTA 115 TUA	Si-P+R	SMD, Digital Transistor, Inverter, RB=100k 2SA1037AK 50V, 100mA,	35a	Rhm	DTA114GUA, DTA114TUA, DTA115GUA,	data
DTA115U(A,F,L,S,V)	Si-P+R	S, Rb=100k, Rbe=10kΩ, 50V, 100mA, 0,3W	9c, 41c	Rhm	-	data
DTA 115 UE	Si-P+R	=DTA 115U..: SMD	35a(1,6mm)		-	data
DTA 115 UC,UK(A)	Si-P+R	=DTA 115U..: SMD	35a(2,9mm)		-	data
DTA 115 UU(A)	Si-P+R	=DTA 115U..: SMD	35a(2mm)		-	data
DTA123E(A,F,L,S,V)	Si-P+R	S, Rb=Rbe=2,2kΩ, 50V, 100mA, 0,3W	9c, 41c	Rhm	-	data
DTA 123 EE	Si-P+R	=DTA 123E..: SMD	35a(1,6mm)		-	data
DTA 123 EC,EK(A)	Si-P+R	=DTA 123E..: SMD	35a(2,9mm)		-	data
DTA 123 EKA	Si-P+R	SMD, Digital Transistor, Inverter, RB=2.2k RBE=2.2k 2SA1037AK 100mA,	35bd3	Rhm		data
DTA 123 ESA	Si-P+R	Digital Transistor, Inverter, RB=2.2k RBE=2.2k 2SA1037AK 100mA,	78bd4	Rhm		data
DTA 123 EU(A)	Si-P+R	=DTA 123E..: SMD	35a(2mm)			data
DTA 123 EUA	Si-P+R	SMD, Digital Transistor, Inverter, RB=2.2k RBE=2.2k 2SA1037AK 100mA,	35bd3	Rhm		data
DTA123J(A,F,L,S,V)	Si-P+R	S, Rb=2,2k, Rbe=47kΩ, 50V, 100mA, 0,3W	9c, 41c	Rhm	KSR 2013, RN 2005	data
DTA 123 JE	Si-P+R	=DTA 123J..: SMD	35a(1,6mm)		RN 2105	data
DTA 123 JC,JK(A)	Si-P+R	=DTA 123J..: SMD	35a(2,9mm)		KSR 2113, RN 2405, UN 211M	data
DTA 123 JKA	Si-P+R	SMD, Digital Transistor, Inverter, RB=2.2k RBE=47k 2SA1037AK 100mA,	35bd3	Rhm		data
DTA 123 JSA	Si-P+R	Digital Transistor, Inverter, RB=2.2k RBE=47k 2SA1037AK 100mA,	78bd4	Rhm		data
DTA 123 JU(A)	Si-P+R	=DTA 123J..: SMD	35a(2mm)		RN 2305	data
DTA 123 JUA	Si-P+R	SMD, Digital Transistor, Inverter, RB=2.2k RBE=47k 2SA1037AK 100mA,	35bd3	Rhm		data
DTA123Y(A,F,L,S,V)	Si-P+R	S, Rb=2,2k, Rbe=10kΩ, 50V, 100mA, 0,3W	9c, 41c	Rhm	UN 411H, 2SA1503	data
DTA 123 YE	Si-P+R	=DTA 123Y..: SMD	35a(1,6mm)		UN 911H	data
DTA 123 YC,YK(A)	Si-P+R	=DTA 123Y..: SMD	35a(2,9mm)		UN 211H, 2SA1502	data
DTA 123 YKA	Si-P+R	SMD, Digital Transistor, Inverter, RB=2.2k RBE=10k 2SA1037AK 100mA,	35bd3	Rhm		data
DTA 123 YSA	Si-P+R	Digital Transistor, Inverter, RB=2.2k RBE=10k 2SA1037AK 100mA,	78bd4	Rhm		data
DTA 123 YU(A)	Si-P+R	=DTA 123Y..: SMD	35a(2mm)		UN 511H, 2SA1722	data
DTA 123 YUA	Si-P+R	SMD, Digital Transistor, Inverter, RB=2.2k RBE=10k 2SA1037AK 100mA,	35bd3	Rhm		data
DTA124E(A,F,L,S,V)	Si-P+R	S, Rb=22k, Rbe=22kΩ, 50V, 100mA, 0,3W	9c, 41c	Rhm	AN 1F4M, RN 2003, UN 4112, 2SA1346,++	data

Type	Device	Short Description	Fig.	Manu	Comparision Types	More at
DTA 124 ECA	Si-P+R	SMD, Digital Transistor, Inverter, RB=22k RBE=22k 2SA1037AK 100mA,	35bd3	Rhm		data data
DTA 124 EE	Si-P+R	=DTA 124E..: SMD	35a(1,6mm)		RN 2103, UN 9112	data
DTA 124 EC,EK(A)	Si-P+R	=DTA 124E..: SMD	35a(2,9mm)		FN 1F4M, RN 2403, UN 2112, 2SA1342,++	data
DTA 124 EKA	Si-P+R	SMD, Digital Transistor, Inverter, RB=22k RBE=22k 2SA1037AK 100mA,	35bd3	Rhm		data data
DTA 124 ESA	Si-P+R	Digital Transistor, Inverter, RB=22k RBE=22k 2SA1037AK 100mA,	78bd4	Rhm		data data
DTA 124 EU(A)	Si-P+R	=DTA 124E..: SMD	35a(2mm)		RN 2303, UN 5112, 2SA1677	data
DTA 124 EUA	Si-P+R	SMD, Digital Transistor, Inverter, RB=22k RBE=22k 2SA1037AK 100mA,	35bd3	Rhm		data data
DTA124G(A,F,L,S,V)	Si-P+R	S, Rb=-, Rbe=22kΩ, 50V, 100mA, 0,3W	9c, 41c	Rhm	2SA1574	data
DTA 124 GE	Si-P+R	=DTA 124G..: SMD	35a(1,6mm)			data
DTA 124 GC,GK(A)	Si-P+R	=DTA 124G..: SMD	35a(2,9mm)		2SA1573	data
DTA 124 GKA	Si-P+R	SMD, Digital Transistor, Inverter, RBE=22k 2SA1037AK 50V, 100mA,	35a	Rhm	DTA113TKA, DTA114GKA, DTA114TKA,	data data
DTA 124 GSA	Si-P+R	Digital Transistor, Inverter, RBE=22k 2SA1037AK 50V, 100mA,	78b	Rhm	DTA114GSA, DTA114TSA, DTA115TSA,	data data
DTA 124 GU(A)	Si-P+R	=DTA 124G..: SMD	35a(2mm)		-	data
DTA124T(A,F,L,S,V)	Si-P+R	S, Rb=22kΩ, Rbe=-, 50V, 100mA, 0,3W	9c, 41c	Rhm	AN 1F4Z, KSR 2011, UN 4117, 2SA1590	data
DTA 124 TE	Si-P+R	=DTA 124T..: SMD	35a(1,6mm)		RN 2112, UN 9117	data
DTA 124 TH	Si-P+R	SMD, Digital Transistor, Inverter, RB=22k 2SA1037AK 50V, 100mA,	35a	Rhm	DTA114GE, DTA114TE, DTA115TE, DTA115TH,	data data
DTA 124 TC,TK(A)	Si-P+R	=DTA 124T..: SMD	35a(2,9mm)		FN 1F4Z, KSR 2111, UN 2117, 2SA1598	data
DTA 124 TKA	Si-P+R	SMD, Digital Transistor, Inverter, RB=22k 2SA1037AK 50V, 100mA,	35a	Rhm	DTA113TKA, DTA114GKA, DTA114TKA,	data data
DTA 124 TSA	Si-P+R	Digital Transistor, Inverter, RB=22k 2SA1037AK 50V, 100mA,	78b	Rhm	DTA114GSA, DTA114TSA, DTA115TSA,	data data
DTA 124 TU(A)	Si-P+R	=DTA 124T..: SMD	35a(2mm)		RN 2312, UN 5117	data
DTA 124 TUA	Si-P+R	SMD, Digital Transistor, Inverter, RB=22k 2SA1037AK 50V, 100mA,	35a	Rhm	DTA114GUA, DTA114TUA, DTA115GUA,	data data
DTA124X(A,F,L,S,V)	Si-P+R	S, Rb=22k, Rbe=47kΩ, 50V, 100mA, 0,3W	9c, 41c	Rhm	AN 1F4N, KSR 2007, RN 2008	data
DTA 124 XE	Si-P+R	=DTA 124X..: SMD	35a(1,6mm)		RN 2108	data
DTA 124 XC,XK(A)	Si-P+R	=DTA 124X..: SMD	35a(2,9mm)		FN 1F4N, KSR 2107, RN 2408	data
DTA 124 XKA	Si-P+R	SMD, Digital Transistor, Inverter, RB=22k RBE=47k 2SA1037AK 100mA,	35bd3	Rhm		data data
DTA 124 XSA	Si-P+R	Digital Transistor, Inverter, RB=22k RBE=47k 2SA1037AK 100mA,	78bd4	Rhm		data data
DTA 124 XU(A)	Si-P+R	=DTA 124X..: SMD	35a(2mm)		RN 2308	data
DTA 124 XUA	Si-P+R	SMD, Digital Transistor, Inverter, RB=22k RBE=47k 2SA1037AK 100mA,	35bd3	Rhm		data data
DTA125T(A,F,L,S,V)	Si-P+R	S, Rb=200kΩ, Rbe=-, 50V, 100mA, 0,3W	9c, 41c	Rhm	-	data
DTA 125 TC,TK	Si-P+R	=DTA 125T..: SMD	35a(2,9mm)			data
DTA 125 TKA	Si-P+R	SMD, Digital Transistor, Inverter, RB=200k 2SA1037AK 50V, 100mA,	35a	Rhm	DTA113TKA, DTA114GKA, DTA114TKA,	data data
DTA 125 TSA	Si-P+R	Digital Transistor, Inverter, RB=200k 2SA1037AK 50V, 100mA,	78b	Rhm	DTA114GSA, DTA114TSA, DTA115TSA,	data data
DTA 125 TU	Si-P+R	=DTA 125T..: SMD	35a(2mm)		-	data
DTA 125 TUA	Si-P+R	SMD, Digital Transistor, Inverter, RB=200k 2SA1037AK 50V, 100mA,	35a	Rhm	DTA114GUA, DTA114TUA, DTA115GUA,	data data
DTA143E(A,F,L,S,V)	Si-P+R	S, Rb=Rbe=4,7kΩ, 50V, 100mA, 0,3W	9c, 41c	Rhm	AN 1L3M, RN 2001, UN 411L, 2SA1656,++	data
DTA 143 ECA	Si-P+R	SMD, Digital Transistor, Inverter, RB=4.7k RBE=4.7k 2SA1037AK 100mA,	35bd3	Rhm		data data
DTA 143 EE	Si-P+R	=DTA 143E..: SMD	35a(1,6mm)		RN 2101, UN 911L	data
DTA 143 EC,EK(A)	Si-P+R	=DTA 143E..: SMD	35a(2,9mm)		FN 1L3M, RN 2401, UN 211L, 2SA1655,++	data
DTA 143 EKA	Si-P+R	SMD, Digital Transistor, Inverter, RB=4.7k RBE=4.7k 2SA1037AK 100mA,	35bd3	Rhm		data data
DTA 143 ESA	Si-P+R	Digital Transistor, Inverter, RB=4.7k RBE=4.7k 2SA1037AK 100mA,	78bd4	Rhm		data data
DTA 143 EU(A)	Si-P+R	=DTA 143E..: SMD	35a(2mm)		RN 2301, UN511L	data
DTA 143 EUA	Si-P+R	SMD, Digital Transistor, Inverter, RB=4.7k RBE=4.7k 2SA1037AK 100mA,	35bd3	Rhm		data data
DTA143T(A,F,L,S,V)	Si-P+R	S, Rb=4,7kΩ, Rbe=-, 50V, 100mA, 0,3W	9c, 41c	Rhm	AN 1L3Z, RN 2010, UN 4116, 2SA1511,++	data
DTA 143 TE	Si-P+R	=DTA 143T..: SMD	35a(1,6mm)		RN 2110, UN 9116	data
DTA 143 TC,TK(A)	Si-P+R	=DTA 143T..: SMD	35a(2,9mm)		FN 1L3Z, RN 2410, UN 2116, 2SA1510,++	data
DTA 143 TKA	Si-P+R	SMD, Digital Transistor, Inverter, RB=4.7k 2SA1037AK 50V, 100mA,	35a	Rhm	DTA113TKA, DTA114GKA, DTA114TKA,	data data
DTA 143 TSA	Si-P+R	Digital Transistor, Inverter, RB=4.7k 2SA1037AK 50V, 100mA,	78b	Rhm	DTA114GSA, DTA114TSA, DTA115TSA,	data data
DTA 143 TU(A)	Si-P+R	=DTA 143T..: SMD	35a(2mm)		RN 2310, UN 5116	data
DTA 143 TUA	Si-P+R	SMD, Digital Transistor, Inverter, RB=4.7k 2SA1037AK 50V, 100mA,	35a	Rhm	DTA114GUA, DTA114TUA, DTA115GUA,	data data
DTA143X(A,F,L,S,V)	Si-P+R	S, Rb=4,7k, Rbe=10kΩ, 50V, 100mA, 0,3W	9c, 41c	Rhm	AN 1L3N, KSR 2005, UN 411F, 2SA1654	data
DTA 143 XE	Si-P+R	=DTA 143X..: SMD	35a(1,6mm)		UN 911F	data
DTA 143 XC,XK(A)	Si-P+R	=DTA 143X..: SMD	35a(2,9mm)		FN 1L3N, KSR 2105, UN 211F, 2SA1653	data
DTA 143 XKA	Si-P+R	SMD, Digital Transistor, Inverter, RB=4.7k RBE=10k 2SA1037AK 100mA,	35bd3	Rhm		data data
DTA 143 XSA	Si-P+R	Digital Transistor, Inverter, RB=4.7k RBE=10k 2SA1037AK 100mA,	78bd4	Rhm		data data
DTA 143 XU(A)	Si-P+R	=DTA 143X..: SMD	35a(2mm)		UN 511F	data
DTA 143 XUA	Si-P+R	SMD, Digital Transistor, Inverter, RB=4.7k RBE=10k 2SA1037AK 100mA,	35bd3	Rhm		data data
DTA143Y(A,F,L,S,V)	Si-P+R	S, Rb=4,7k, Rbe=22kΩ, 50V, 100mA, 0,3W	9c, 41c	Rhm	-	data
DTA 143 YE	Si-P+R	=DTA 143Y..: SMD	35a(1,6mm)		-	data
DTA 143 YC,YK(A)	Si-P+R	=DTA 143Y..: SMD	35a(2,9mm)			data

Type	Device	Short Description	Fig.	Manu	Comparision Types	More at
DTA 143 YU(A)	Si-P+R	=DTA 143Y..: SMD	35a(2mm)		-	data
DTA143Z(A,F,L,S,V)	Si-P+R	S, Rb=4,7k, Rbe=47kΩ, 50V, 100mA, 0,3W	9c, 41c	Rhm	RN 2006, 2SA1591, 2SA1616	data
DTA 143 ZE	Si-P+R	=DTA 143Z..: SMD	35a(1,6mm)		RN 2106	data
DTA 143 ZC,ZK(A)	Si-P+R	=DTA 143Z..: SMD	35a(2,9mm)		RN 2406, 2SA1597	data
DTA 143 ZKA	Si-P+R	SMD, Digital Transistor, Inverter, RB=4.7k RBE=47k 2SA1037AK 100mA,	35bd3	Rhm		data data
DTA 143 ZSA	Si-P+R	Digital Transistor, Inverter, RB=4.7k RBE=47k 2SA1037AK 100mA,	78bd4	Rhm		data
DTA 143 ZU(A)	Si-P+R	=DTA 143Z..: SMD	35a(2mm)		RN 2306	data
DTA 143 ZUA	Si-P+R	SMD, Digital Transistor, Inverter, RB=4.7k RBE=47k 2SA1037AK 100mA,	35bd3	Rhm		data data
DTA144E(A,F,L,S,V)	Si-P+R	S, Rb=47kΩ, 50V, 30mA, 0,3W	9c, 41c	Rhm	AN 1L4M, RN 2004, UN 4113, 2SA1345,++	data
DTA 144 EE	Si-P+R	=DTA 144E..: SMD	35a(1,6mm)		RN 2104, UN 9113	data
DTA 144 EC,EK(A)	Si-P+R	=DTA 144E..: SMD	35a(2,9mm)		FN 1L4M, RN 2404, UN 2113, 2SA1341,++	data
DTA 144 EKA	Si-P+R	SMD, Digital Transistor, Inverter, RB=47k RBE=47k 2SA1037AK 100mA,	35bd3	Rhm		data data
DTA 144 ESA	Si-P+R	Digital Transistor, Inverter, RB=47k RBE=47k 2SA1037AK 100mA,	78bd4	Rhm		data
DTA 144 EU(A)	Si-P+R	=DTA 144E..: SMD	35a(2mm)		RN 2304, UN 5113, 2SA1676	data
DTA 144 EUA	Si-P+R	SMD, Digital Transistor, Inverter, RB=47k RBE=47k 2SA1037AK 100mA,	35bd3	Rhm		data
DTA144G(A,F,L,S,V)	Si-P+R	S, Rb=-, Rbe=47kΩ, 50V, 100mA, 0,3W	9c, 41c	Rhm	2SA1572	data
DTA 144 GE	Si-P+R	=DTA 144G..: SMD	35a(1,6mm)			data
DTA 144 GC,GK(A)	Si-P+R	=DTA 144G..: SMD	35a(2,9mm)		2SA1571	data
DTA 144 GKA	Si-P+R	SMD, Digital Transistor, Inverter, RBE=47k 2SA1037AK 50V, 100mA,	35a	Rhm	DTA113TKA, DTA114GKA, DTA114TKA,	data data
DTA 144 GU(A)	Si-P+R	=DTA 144G..: SMD	35a(2mm)		-	data
DTA 144 GUA	Si-P+R	SMD, Digital Transistor, Inverter, RBE=47k 2SA1037AK 50V, 100mA,	35a		DTA114GUA, DTA114TUA, DTA115GUA,	data data
DTA144T(A,F,L,S,V)	Si-P+R	S, Rb=47kΩ, Rbe=-, 50V, 100mA, 0,3W	9c, 41c	Rhm	AN 1L4Z, KSR 2012, UN 4110, 2SA1509	data
DTA 144 TCA	Si-P+R	SMD, Digital Transistor, Inverter, RB=47k 2SA1037AK 50V, 100mA,	35a	Rhm		data data
DTA 144 TE	Si-P+R	=DTA 144T..: SMD	35a(1,6mm)		RN 2113, UN 9110	data
DTA 144 TC,TK(A)	Si-P+R	=DTA 144T..: SMD	35a(2,9mm)		FN 1L4Z, KSR 2112, UN 2110, 2SA1508	data
DTA 144 TKA	Si-P+R	SMD, Digital Transistor, Inverter, RB=47k 2SA1037AK 50V, 100mA,	35a	Rhm	DTA113TKA, DTA114GKA, DTA114TKA,	data
DTA 144 TSA	Si-P+R	Digital Transistor, Inverter, RB=47k 2SA1037AK 50V, 100mA,	78b	Rhm	DTA114GSA, DTA114TSA, DTA115TSA,	data data
DTA 144 TU(A)	Si-P+R	=DTA 144T..: SMD	35a(2mm)		RN 2313, UN 5110	data
DTA 144 TUA	Si-P+R	SMD, Digital Transistor, Inverter, RB=47k 2SA1037AK 50V, 100mA,	35a	Rhm	DTA114GUA, DTA114TUA, DTA115GUA,	data
DTA124V(A,F,L,S,V)	Si-P+R	S, Rb=47k, Rbe=10kΩ, 50V, 30mA, 0,3W	9c, 41c	Rhm	UN 411D	data
DTA 144 VE	Si-P+R	=DTA 144V..: SMD	35a(1,6mm)		UN 911D	data
DTA 144 VC,VK(A)	Si-P+R	=DTA 144V..: SMD	35a(2,9mm)		UN 211D	data
DTA 144 VKA	Si-P+R	SMD, Digital Transistor, Inverter, RB=47k RBE=10k 2SA1037AK 100mA,	35bd3	Rhm		data data
DTA 144 VSA	Si-P+R	Digital Transistor, Inverter, RB=47k RBE=10k 2SA1037AK 100mA,	78bd4	Rhm		data
DTA 144 VU(A)	Si-P+R	=DTA 144V..: SMD	35a(2mm)		UN 511D	data
DTA 144 VUA	Si-P+R	SMD, Digital Transistor, Inverter, RB=47k RBE=10k 2SA1037AK 100mA,	35bd3	Rhm		data data
DTA144W(A,F,L,S,V)	Si-P+R	S, Rb=47k, Rbe=22kΩ, 50V, 100mA, 0,3W	9c, 41c	Rhm	AN 1L4L, RN 2009, UN 411E, 2SA1347,++	data
DTA 144 WE	Si-P+R	=DTA 144W..: SMD	35a(1,6mm)		RN 2109, UN 911E	data
DTA 144 WC,WK(A)	Si-P+R	=DTA 144W..: SMD	35a(2,9mm)		FN 1L4L, RN 2409, UN 211E, 2SA1343,++	data
DTA 144 WKA	Si-P+R	SMD, Digital Transistor, Inverter, RB=47k RBE=22k 2SA1037AK 100mA,	35bd3	Rhm		data data
DTA 144 WSA	Si-P+R	Digital Transistor, Inverter, RB=47k RBE=22k 2SA1037AK 100mA,	78bd4	Rhm		data
DTA 144 WU(A)	Si-P+R	=DTA 144W..: SMD	35a(2mm)		RN 2309, UN 511E	data
DTA 144 WUA	Si-P+R	SMD, Digital Transistor, Inverter, RB=47k RBE=22k 2SA1037AK 100mA,	35bd3	Rhm		data data
DTA214Y(A,F,L,S,V)	Si-P+R	S, Rb=10k, Rbe=47kΩ, 50V, 100mA, 0,3W	9c, 41c	Rhm	AN 1A4P, RN 2007, UN 4114, 2SA1564,++	data
DTA 214 YE	Si-P+R	=DTA 214Y..: SMD	35a(1,6mm)		RN 2107, UN 9114	data
DTA 214 YC,YK(A)	Si-P+R	=DTA 214Y..: SMD	35a(2,9mm)		FN 1A4P, RN 2407, UN 2114, 2SA1563,++	data
DTA 214 YU(A)	Si-P+R	=DTA 214Y..: SMD	35a(2mm)		RN 2307, UN 5114	data
DTB 005...10	Si-Br	50...1000V, 1A	4-DIP	Mic	B...C1000	data
DTB 3 B...G	Triac	100...600V, 3A≈(Tc=71°),Igt/Ih<30/<25mA B=100V, C=200V, E=400V, G=600V	13h§	Say	TAG 136-..., TAG 137-...	data
DTB 8 C...G	Triac	200...600V, 8A≈(Tc=86°) C=200V, E=400V, F=500V, G=600V	17h§	Say	T 2800..., T 2850..., TAC 224...227..., ++	
DTB 10 C...G	Triac	200...600V, 10A≈(Tc=86°) C=200V, E=400V, F=500V, G=600V	17h§	Say	BTA 22..., TAG 250-...TAG 252-..., ++	
DTB 12 C...G	Triac	200...600V, 12A≈(Tc=75°) C=200V, E=400V, F=500V, G=600V	17h§	Say	BTA 23..., TAG 255-...TAG 257-..., ++	
DTB 16 C...G	Triac	200...600V, 16A≈(Tc=75°) C=200V, E=400V, F=500V, G=600V	23	Say	(TIC 253...)[4]	
DTB113E(A,F,L,S,V)	Si-P+R	S, Rb=Rbe=1kΩ, 50V, 500mA, 0,6W	9c, 41c	Rhm	RN 2221	data
DTB 113 EC,EK	Si-P+R	=DTB 113E..: SMD	35a(2,9mm)			data
DTB 113 EK	Si-P+R	SMD, Digital Transistor, Inverter, RB=1k RBE=1k 2SA1036K 500mA,	35bd3	Rhm		data data
DTB 113 ES	Si-P+R	Digital Transistor, Inverter, RB=1k RBE=1k 2SA1036K 500mA,	78bd4	Rhm		data data
DTB113Z(A,F,L,S,V)	Si-P+R	S, Rb=1kΩ, Rbe=10kΩ, 50V, 500mA, 0,6W	9c, 41c	Rhm	RN 2226	data

Type	Device	Short Description	Fig.	Manu	Comparision Types	More at
DTB 113 ZC,ZK	Si-P+R	=DTB 113Z..: SMD	35a(2,9mm)		-	data
DTB 113 ZK	Si-P+R	SMD, Digital Transistor, Inverter, RB=1k RBE=10k 2SA1036K 500mA,	35bd3	Rhm		data / data
DTB 113 ZS	Si-P+R	Digital Transistor, Inverter, RB=1k RBE=10k 2SA1036K 500mA,	78bd4	Rhm		data / data
DTB114E(A,F,L,S,V)	Si-P+R	S, Rb=Rbe=10kΩ, 50V, 500mA, 0,6W	9c, 41c	Rhm	RN 2224, UN 4123, 2SA1522, 2SA1526	data
DTB 114 EC,EK	Si-P+R	=DTB 114E..: SMD	35a(2,9mm)		UN 2123, 2SA1518	data
DTB 114 EK	Si-P+R	SMD, Digital Transistor, Inverter, RB=10k RBE=10k 2SA1036K 500mA,	35bd3	Rhm		data / data
DTB 114 ES	Si-P+R	Digital Transistor, Inverter, RB=10k RBE=10k 2SA1036K 500mA,	78bd4	Rhm		data / data
DTB114G(A,F,L,S,V)	Si-P+R	S, Rb=-, Rbe=10kΩ, 50V, 500mA, 0,6W	9c, 41c	Rhm	-	data
DTB 114 GC,GK	Si-P+R	=DTB 114G..: SMD	35a(2,9mm)		-	data
DTB 114 GK	Si-P+R	SMD, Digital Transistor, Inverter, RBE=10k 2SA1036K 50V, 500mA,	35a	Rhm		data / data
DTB114T(A,F,L,S,V)	Si-P+R	S, Rb=10kΩ, Rbe=-, 50V, 500mA, 0,6W	9c, 41c	Rhm	-	data
DTB 114 TC,TK	Si-P+R	=DTB 114T..: SMD	35a(2,9mm)			data
DTB 114 TK	Si-P+R	SMD, Digital Transistor, RB=10k 40V, 500mA,	35a	Rhm	DTB114GK, DTB123TK, DTB143TK, DTB143TS,	data / data
DTB122J(A,F,L,S,V)	Si-P+R	S, Rb=220Ω,Rbe=4,7kΩ, 50V, 500mA, 0,6W	9c, 41c	Rhm	-	data
DTB 122 JC,JK	Si-P+R	=DTB 122J..: SMD	35a(2,9mm)		-	data
DTB 122 JK	Si-P+R	SMD, Digital Transistor, Inverter, RB=0.22k RBE=4.7k 2SA1036K 500mA,	35bd3	Rhm		data / data
DTB123E(A,F,L,S,V)	Si-P+R	S, Rb=Rbe=2,2kΩ, 50V, 500mA, 0,6W	9c, 41c	Rhm	RN 2222, UN 4121, 2SA1525, 2SA1529	data
DTB 123 EC,EK	Si-P+R	=DTB 123E..: SMD	35a(2,9mm)		UN 2121, 2SA1521	data
DTB 123 EK	Si-P+R	SMD, Digital Transistor, Inverter, RB=2.2k RBE=2.2k 2SA1036K 500mA,	35bd3	Rhm		data / data
DTB 123 ES	Si-P+R	SMD, Digital Transistor, Inverter, RB=2.2k RBE=2.2k 2SA1036K 500mA,	78bd4	Rhm		data / data
DTB123T(A,F,L,S,V)	Si-P+R	S, Rb=2,2kΩ, Rbe=-, 50V, 500mA, 0,6W	9c, 41c	Rhm	-	data
DTB 123 TC,TK	Si-P+R	=DTB 123T..: SMD	35a(2,9mm)		-	data
DTB 123 TK	Si-P+R	SMD, Digital Transistor, Inverter, RB=2,2k 2SA1036AK 40V, 500mA,	35a	Rhm	DTB114GK, DTB114TK, DTB143TK, DTB143TS,	data / data
DTB123Y(A,F,L,S,V)	Si-P+R	S, Rb=2,2k, Rbe=10kΩ, 50V, 500mA, 0,6W	9c, 41c	Rhm	RN 2227, UN 4124, 2SA1524, 2SA1528	data
DTB 123 YC	Si-P+R	SMD, Digital Transistor, Inverter, RB=2.2k Rbe=10k 2SA1036K 500mA,	35bd3	Rhm		data / data
DTB 123 YC,YK	Si-P+R	=DTB 123Y..: SMD	35a(2,9mm)		UN 2124, 2SA1520	data
DTB 123 YK	Si-P+R	SMD, Digital Transistor, Inverter, RB=2.2k RBE=10k 2SA1036K 500mA,	35bd3	Rhm		data / data
DTB 123 YS	Si-P+R	Digital Transistor, Inverter, RB=2.2k RBE=10k 2SA1036K 500mA,	78bd4	Rhm		data / data
DTB133H(A,F,L,S,V)	Si-P+R	S, Rb=3,3k, Rbe=10kΩ, 50V, 500mA, 0,6W	9c, 41c	Rhm	-	data
DTB 133 HC,HK	Si-P+R	=DTB 133Y..: SMD	35a(2,9mm)		-	data
DTB 133 HK	Si-P+R	SMD, Digital Transistor, Inverter, RB=3.3k RBE=10k 2SA1036K 500mA,	35bd3	Rhm		data / data
DTB 133 HS	Si-P+R	SMD, Digital Transistor, Inverter, RB=3.3k RBE=10k 2SA1036K 500mA,	78bd4	Rhm		data / data
DTB143E(A,F,L,S,V)	Si-P+R	S, Rb=Rbe=4,7kΩ, 50V, 500mA, 0,6W	9c, 41c	Rhm	RN 1223, UN 4122, 2SA1523, 2SA1527	data
DTB 143 EC	Si-P+R	SMD, Digital Transistor, Inverter, RB=4.7k RBE=4.7k 2SA1036K 500mA,	35bd3	Rhm		data / data
DTB 143 EC,EK	Si-P+R	=DTB 143E..: SMD	35a(2,9mm)		UN 2122, 2SA1519	data
DTB 143 EK	Si-P+R	SMD, Digital Transistor, Inverter, RB=4.7k RBE=4.7k 2SA1036K 500mA,	35bd3	Rhm		data / data
DTB 143 ES	Si-P+R	Digital Transistor, Inverter, RB=4.7k RBE=4.7k 2SA1036K 500mA,	78bd4	Rhm		data / data
DTB143T(A,F,L,S,V)	Si-P+R	S, Rb=4,7kΩ, Rbe=-, 50V, 500mA, 0,6W	9c, 41c	Rhm	-	data
DTB 143 TC,TK	Si-P+R	=DTB 143T..: SMD	35a(2,9mm)			data
DTB 143 TK	Si-P+R	SMD, Digital Transistor, Inverter, RB=4,7k 2SA1036AK 40V, 500mA,	35a	Rhm	DTB114GK, DTB114TK, DTB123TK, DTB143TS,	data / data
DTB 143 TS	Si-P+R	Digital Transistor, Inverter, RB=4,7k 2SA1036AK 40V, 500mA,	35a	Rhm	DTB114GK, DTB114TK, DTB123TK, DTB143TK,	data / data
DTB163T(A,F,L,S,V)	Si-P+R	S, Rb=6,8kΩ, Rbe=-, 50V, 500mA, 0,6W	9c, 41c	Rhm	-	data
DTB 163 TC,TK	Si-P+R	=DTB 163T..: SMD	35a(2,9mm)		-	data
DTC1D3R(A,F,L,S,V)	Si-N+R	S, Rb=2,7k, Rbe=1kΩ, 50V, 70mA, 0,3W	9c, 41c	Rhm	-	data
DTC 1 D3RE	Si-N+R	=DTC 1D3R..: SMD	35a(1,6mm)		-	data
DTC 1 D3 RC,RK(A)	Si-N+R	=DTC 1D3R..: SMD	35a(2,9mm)		-	data
DTC 1 D3RU(A)	Si-N+R	=DTC 1D3R..: SMD	35a(2mm)		-	data
DTC 8 C...G(-N)	Triac	200...600V,8A≈(Tc=105°),Igt/Ih<50/<50mA C=200V, E=400V, G=600V	17h§	Say	T 2800..., T 2850..., TAG 224-...	
DTC 10 C-N...G-N	Triac	200...600V, 10A≈(98°), Igt/Ih<50/<50mA	17h§	Say	BTA 22-..., TAG 250-..., TAG 251-...	data
DTC 12 C...G(-N)	Triac	200...600V, 12A≈(98°), Igt/Ih<50/<50mA C=200V, E=400V, G=600V	17h§	Say	BTA23-..., BT138-..., TAG255-..., TAG256-...	data
DTC 16 C...G	Triac	200...600V, 16A≈(Tc=75°)	23	Say	(TIC 253...)[4]	
DTC 113... 144		C=200V, E=400V, F=500V, G=600V Suffix f. 100mA Typen/Series	#			data
DTC113Z(A,F,L,S,V)	Si-N+R	S, Rb=1k, Rbe=10kΩ, 50V, 100mA, 0,3W	9c, 41c	Rhm	AA 1A3Q, UN 4219	data
DTC 113 ZE	Si-N+R	=DTC 113Z..: SMD	35a(1,6mm)		UN 9219	data
DTC 113 ZC,ZK(A)	Si-N+R	=DTC 113Z..: SMD	35a(2,9mm)		FA 1A3Q, UN 2219	data
DTC 113 ZKA	Si-N+R	SMD, Digital Transistor, Inverter, RB=1k RBE=10k 2SC2412K 100mA,	35bd3	Rhm		data / data
DTC 113 ZSA	Si-N+R	Digital Transistor, Inverter, RB=1k	78bd4	Rhm		data

Type	Device	Short Description	Fig.	Manu	Comparision Types	More at
DTC 113 ZU(A)	Si-N+R	RBE=10k 2SC2412K 100mA, =DTC 113Z..: SMD	35a(2mm)		UN 5219	data data
DTC 113 ZUA	Si-N+R	SMD, Digital Transistor, Inverter, RB=1k RBE=10k 2SC2412K 100mA,	35bd3	Rhm		data data
DTC114E(A,F,L,S,V)	Si-N+R	S, Rb=Rbe=10kΩ, 50V, 50mA, 0,3W	9c, 41c	Rhm	AA 1A4M, RN 1002, UN 4211, 2SC3402,++	data
DTC 114 ECA	Si-N+R	SMD, Digital Transistor, Inverter, RB=10k RBE=10k 2SC2412K 100mA,	35bd3	Rhm		data data
DTC 114 EE	Si-N+R	=DTC 114E..: SMD	-35a(1,6mm)		RN 1102, UN 9211	data
DTC 114 EC,EK(A)	Si-N+R	=DTC 114E..: SMD	35a(2,9mm)		FA 1A4M, RN 1402, UN 2211, 2SC3398,++	data
DTC 114 EKA	Si-N+R	SMD, Digital Transistor, Inverter, RB=10k RBE=10k 2SC2412K 100mA,	35bd3	Rhm		data data
DTC 114 ESA	Si-N+R	Digital Transistor, Inverter, RB=10k RBE=10k 2SC2412K 100mA,	78bd4	Rhm		data data
DTC 114 EU(A)	Si-N+R	=DTC 114E..: SMD	35a(2mm)		RN 1302, UN 5211, 2SC4398	data
DTC 114 EUA	Si-N+R	SMD, Digital Transistor, Inverter, RB=10k RBE=10k 2SC2412K 100mA,	35bd3	Rhm		data data
DTC114G(A,F,L,S,V)	Si-N+R	S, Rb=-, Rbe=10kΩ, 50V, 100mA, 0,3W	9c, 41c	Rhm	-	data
DTC 114 GE	Si-N+R	=DTC 114G..: SMD	35a(1,6mm)		-	data
DTC 114 GC,GK(A)	Si-N+R	=DTC 114G..: SMD	35a(2,9mm)		-	data
DTC 114 GKA	Si-N+R	SMD, Digital Transistor, Inverter, RBE=10k 2SC2412K 50V, 100mA,	35a	Rhm	DTC114TKA, DTC115GKA, DTC115TKA,	data
DTC 114 GSA	Si-N+R	Digital Transistor, Inverter, RBE=10k 2SC2412K 50V, 100mA,	78b	Rhm	DTC114TSA, DTC115TSA, DTC124GSA,	data
DTC 114 GU(A)	Si-N+R	=DTC 114G..: SMD	35a(2mm)			data
DTC 114 GUA	Si-N+R	SMD, Digital Transistor, Inverter, RBE=10k 2SC2412K 50V, 100mA,	35a	Rhm	DTC114TUA, DTC115GUA, DTC115TUA,	data
DTC114T(A,F,L,S,V)	Si-N+R	S, Rb=10kΩ, Rbe=-, 50V, 100mA, 0,3W	9c, 41c	Rhm	AA 1A4Z, RN 1011, UN 4215, 2SC3860,++	data
DTC 114 TCA	Si-N+R	SMD, Digital Transistor, Inverter, RB=10k 2SC2412K 50V, 100mA,	35a	Rhm		data data
DTC 114 TE	Si-N+R	=DTC 114T..: SMD	35a(1,6mm)		RN 1111, UN 9215	data
DTC 114 TC,TK(A)	Si-N+R	=DTC 114T..: SMD	35a(2,9mm)		FA 1A4Z, RN 1411, UN 2215, 2SC3859,++	data
DTC 114 TKA	Si-N+R	SMD, Digital Transistor, Inverter, RB=10k 2SC2412K 50V, 100mA,	35a	Rhm	DTC114GKA, DTC115GKA, DTC115TKA,	data
DTC 114 TSA	Si-N+R	Digital Transistor, Inverter, RB=10k 2SC2412K 50V, 100mA,	78b	Rhm	DTC114GSA, DTC115TSA, DTC124GSA,	data
DTC 114 TU(A)	Si-N+R	=DTC 114T..: SMD	35a(2mm)		RN 1311, UN 5215	data
DTC 114 TUA	Si-N+R	SMD, Digital Transistor, Inverter, RB=10k 2SC2412K 50V, 100mA,	35a	Rhm	DTC114GUA, DTC115GUA, DTC115TUA,	data
DTC114W(A,F,L,S,V)	Si-N+R	S, Rb=10k, Rbe=4,7kΩ, 50V, 100mA, 0,3W	9c, 41c	Rhm	UN 421K	data
DTC 114 WE	Si-N+R	=DTC 114W..: SMD	35a(1,6mm)		UN 921K	data
DTC 114 WC,WK(A)	Si-N+R	=DTC 114W..: SMD	35a(2,9mm)		UN 221K	data
DTC 114 WKA	Si-N+R	SMD, Digital Transistor, Inverter, RB=10k RBE=47k 2SC2412K 100mA,	35bd3	Rhm		data data
DTC 114 WSA	Si-N+R	Digital Transistor, Inverter, RB=10k RBE=47k 2SC2412K 100mA,	78bd4	Rhm		data data
DTC 114 WU(A)	Si-N+R	=DTC 114W..: SMD	35a(2mm)		UN 521K	data
DTC 114 WUA	Si-N+R	SMD, Digital Transistor, Inverter, RB=10k RBE=47k 2SC2412K 100mA,	35bd3	Rhm		data data
DTC114Y(A,F,L,S,V)	Si-N+R	S, Rb=10k, Rbe=47kΩ, 50V, 70mA, 0,3W	9c, 41c	Rhm	AA 1A4P, RN 1007, UN 4214, 2SC4048,++	data
DTC 114 YE	Si-N+R	=DTC 114Y..: SMD	35a(1,6mm)		RN 1107, UN 9214	data
DTC 114 YC,YK(A)	Si-N+R	=DTC 114Y..: SMD	35a(2,9mm)		FA 1A4P, RN 1407, UN 2214, 2SC4047,++	data
DTC 114 YKA	Si-N+R	SMD, Digital Transistor, Inverter, RB=10k RBE=47k 2SC2412K 100mA,	35bd3	Rhm		data data
DTC 114 YSA	Si-N+R	Digital Transistor, Inverter, RB=10k RBE=47k 2SC2412K 100mA,	78bd4	Rhm		data data
DTC 114 YU(A)	Si-N+R	=DTC 114Y..: SMD	35a(2mm)		RN 1307, UN 5214	data
DTC 114 YUA	Si-N+R	SMD, Digital Transistor, Inverter, RB=10k RBE=47k 2SC2412K 100mA,	35bd3	Rhm		data data
DTC115E(A,F,L,S,V)	Si-N+R	S, Rb=Rbe=100kΩ, 50V, 20mA, 0,3W	9c, 41c	Rhm	-	data
DTC 115 EE	Si-N+R	=DTC 115E..: SMD	35a(1,6mm)			data
DTC 115 EC,EK(A)	Si-N+R	=DTC 115E..: SMD	35a(2,9mm)			data
DTC 115 EKA	Si-N+R	SMD, Digital Transistor, Inverter, RB=100k RBE=100k 2SC2412K 100mA,	35bd3	Rhm		data data
DTC 115 ESA	Si-N+R	Digital Transistor, Inverter, RB=100k RBE=100k 2SC2412K 100mA,	78bd4	Rhm		data data
DTC 115 EU(A)	Si-N+R	=DTC 115E..: SMD	35a(2mm)			data
DTC 115 EUA	Si-N+R	SMD, Digital Transistor, Inverter, RB=100k RBE=100k 2SC2412K 100mA,	35bd3	Rhm		data data
DTC115G(A,F,L,S,V)	Si-N+R	S, Rb=-, Rbe=100kΩ, 50V, 100mA, 0,3W	9c, 41c	Rhm	-	data
DTC 115 GE	Si-N+R	=DTC 115G..: SMD	35a(1,6mm)		-	data
DTC 115 GC,GK(A)	Si-N+R	=DTC 115G..: SMD	35a(2,9mm)		-	data
DTC 115 GKA	Si-N+R	SMD, Digital Transistor, Inverter, RBE=100k 2SC2412K 50V, 100mA,	35a	Rhm	DTC114GKA, DTC114TKA, DTC115TKA,	data
DTC 115 GU(A)	Si-N+R	=DTC 115G..: SMD	35a(2mm)			data
DTC 115 GUA	Si-N+R	SMD, Digital Transistor, Inverter, RBE=100k 2SC2412K 50V, 100mA,	35a	Rhm	DTC114GUA, DTC114TUA, DTC115TUA,	data
DTC115T(A,F,L,S,V)	Si-N+R	S, Rb=100kΩ, Rbe=-, 50V, 100mA, 0,3W	9c, 41c	Rhm	-	data
DTC 115 TE	Si-N+R	=DTC 115T..: SMD	35a(1,6mm)		-	data
DTC 115 TH	Si-N+R	SMD, Digital Transistor, Inverter, RB=100k 2SC2412K 50V, 100mA,	35a	Rhm	DTC114TE, DTC124TE, DTC143TE, DTC144GE,	data data
DTC 115 TC,TK(A)	Si-N+R	=DTC 115T..: SMD	35a(2,9mm)			data
DTC 115 TKA	Si-N+R	SMD, Digital Transistor, Inverter, RB=100k 2SC2412K 50V, 100mA,	35a	Rhm	DTC114GKA, DTC114TKA, DTC115GKA,	data
DTC 115 TSA	Si-N+R	Digital Transistor, Inverter, RB=100k 2SC2412K 50V, 100mA,	78b	Rhm	DTC114GSA, DTC114TSA, DTC124GSA,	data data

Type	Device	Short Description	Fig.	Manu	Comparision Types	More at
DTC 115 TU(A)	Si-N+R	=DTC 115T...: SMD	35a(2mm)			data
DTC 115 TUA	Si-N+R	SMD, Digital Transistor, Inverter, RB=100k 2SC2412K 50V, 100mA,	35a	Rhm	DTC114GUA, DTC114TUA, DTC115GUA,	data data
DTC115U(A,F,L,S,V)	Si-N+R	S, Rb=100k, Rbe=10kΩ, 50V, 20mA, 0,3W	9c, 41c	Rhm		data
DTC 115 UC,UK	Si-N+R	=DTC 115U...: SMD	35a(2,9mm)			data
DTC 115 UU	Si-N+R	=DTC 115U...: SMD	35a(2mm)			data
DTC123E(A,F,L,S,V)	Si-N+R	S, Rb=Rbe=2,2kΩ, 50V, 100mA, 0,3W	9c, 41c	Rhm	-	data
DTC 123 EE	Si-N+R	=DTC 123E...: SMD	35a(1,6mm)		-	data
DTC 123 EC,EK(A)	Si-N+R	=DTC 123E...: SMD	35a(2,9mm)		-	data
DTC 123 EKA	Si-N+R	SMD, Digital Transistor, Inverter, RB=2.2k RBE=2.2k 2SC2412K 100mA,	35bd3	Rhm		data data
DTC 123 ESA	Si-N+R	Digital Transistor, Inverter, RB=2.2k RBE=2.2k 2SC2412K 100mA,	78bd4	Rhm		data data
DTC 123 EU(A)	Si-N+R	=DTC 123E...: SMD	35a(2mm)			data
DTC 123 EUA	Si-N+R	SMD, Digital Transistor, Inverter, RB=2.2k RBE=2.2k 2SC2412K 100mA,	35bd3	Rhm		data data
DTC123J(A,F,L,S,V)	Si-N+R	S, Rb=2.2k, Rbe=47kΩ, 50V, 100mA, 0,3W	9c, 41c	Rhm	RN 1005	data
DTC 123 JE	Si-N+R	=DTC 123J...: SMD	35a(1,6mm)		RN 1105	data
DTC 123 JC,JK(A)	Si-N+R	=DTC 123J...: SMD	35a(2,9mm)		KSR 1113, RN 1405	data
DTC 123 JKA	Si-N+R	SMD, Digital Transistor, Inverter, RB=2.2k RBE=47k 2SC2412K 100mA,	35bd3	Rhm		data data
DTC 123 JSA	Si-N+R	Digital Transistor, Inverter, RB=2.2k RBE=47k 2SC2412K 100mA,	78bd4	Rhm		data data
DTC 123 JUA	Si-N+R	SMD, Digital Transistor, Inverter, RB=2.2k RBE=47k 2SC2412K 100mA,	35bd3	Rhm		data data
DTC 123 JU(A)	Si-N+R	=DTC 123J...: SMD	35a(2mm)		RN 1305	data
DTC 123 TKA	Si-N+R	SMD, Digital Transistor, RB=2.2k, 2SC2412K 50V, 100mA,	35a	Rhm	DTC114GKA, DTC114TKA, DTC115GKA,	data data
DTC123Y(A,F,L,S,V)	Si-N+R	S, Rb=2.2k, Rbe=10kΩ, 50V, 100mA, 0,3W	9c, 41c	Rhm	UN 421H, 2SC3864	data
DTC 123 YE	Si-N+R	=DTC 123Y...: SMD	35a(1,6mm)		UN 921H	data
DTC 123 YC,YK(A)	Si-N+R	=DTC 123Y...: SMD	35a(2,9mm)		UN 921H, 2SC3863	data
DTC 123 YKA	Si-N+R	SMD, Digital Transistor, Inverter, RB=2.2k RBE=10k 2SC2412K 100mA,	35bd3	Rhm		data data
DTC 123 YSA	Si-N+R	Digital Transistor, Inverter, RB=2.2k RBE=10k 2SC2412K 100mA,	78bd4	Rhm		data data
DTC 123 YU(A)	Si-N+R	=DTC 123Y...: SMD	35a(2mm)		UN 521H, 2SC4498	data
DTC 123 YUA	Si-N+R	SMD, Digital Transistor, Inverter, RB=2.2k RBE=10k 2SC2412K 100mA,	35bd3	Rhm		data data
DTC124E(A,F,L,S,V)	Si-N+R	S, Rb=Rbe=22kΩ, 50V, 30mA, 0,3W	9c, 41c	Rhm	AA 1F4M, RN 1003, UN 4212, 2SC3400,++	data
DTC 124 ECA	Si-N+R	SMD, Digital Transistor, Inverter, RB=22k RBE=22k 2SC2412K 100mA,	35bd3	Rhm		data data
DTC 124 EE	Si-N+R	=DTC 124E...: SMD	35a(1,6mm)		RN 1103, UN 9212	data
DTC 124 EC,EK(A)	Si-N+R	=DTC 124E...: SMD	35a(2,9mm)		FA 1F4M, RN 1403, UN 2212, 2SC3396,++	data
DTC 124 EKA	Si-N+R	SMD, Digital Transistor, Inverter, RB=22k RBE=22k 2SC2412K 100mA,	35bd3	Rhm		data data
DTC 124 ESA	Si-N+R	Digital Transistor, Inverter, RB=22k RBE=22k 2SC2412K 100mA,	78bd4	Rhm		data data
DTC 124 EU(A)	Si-N+R	=DTC 124E...: SMD	35a(2mm)		RN 1303, UN 5212, 2SC4397	data
DTC 124 EUA	Si-N+R	SMD, Digital Transistor, Inverter, RB=22k RBE=22k 2SC2412K 100mA,	35bd3	Rhm		data data
DTC124G(A,F,L,S,V)	Si-N+R	S, Rb=-, Rbe=22kΩ, 50V, 100mA, 0,3W	9c, 41c	Rhm	2SC4070	data
DTC 124 GE	Si-N+R	=DTC 124G...: SMD	35a(1,6mm)			data
DTC 124 GC,GK(A)	Si-N+R	=DTC 124G...: SMD	35a(2,9mm)		2SC4069	data
DTC 124 GKA	Si-N+R	SMD, Digital Transistor, Inverter, RBE=22k 2SC2412K 50V, 100mA,	35a	Rhm	DTC114GKA, DTC114TKA, DTC115GKA,	data data
DTC 124 GSA	Si-N+R	Digital Transistor, Inverter, RBE=22k 2SC2412K 50V, 100mA,	78b	Rhm	DTC114GSA, DTC114TSA, DTC115TSA,	data data
DTC 124 GU(A)	Si-N+R	=DTC 124G...: SMD	35a(2mm)			data
DTC 124 GUA	Si-N+R	SMD, Digital Transistor, Inverter, RBE=22k 2SC2412K 50V, 100mA,	35a	Rhm	DTC114GUA, DTC114TUA, DTC115GUA,	data data
DTC124T(A,F,L,S,V)	Si-N+R	S, Rb=22kΩ, Rbe=-, 50V, 100mA, 0,3W	9c, 41c	Rhm	AA 1F4Z, KSR 1011, UN 4217, 2SC4121,++	data
DTC 124 TE	Si-N+R	=DTC 124T...: SMD	35a(1,6mm)		RN 1112, UN 9217	data
DTC 124 TC,TK(A)	Si-N+R	=DTC 124T...: SMD	35a(2,9mm)		FA 1F4Z, KSR 1111, UN 2217, 2SC4120	data
DTC 124 TKA	Si-N+R	SMD, Digital Transistor, Inverter, RB=22k 2SC2412K 50V, 100mA,	35a	Rhm	DTC114GKA, DTC114TKA, DTC115GKA,	data data
DTC 124 TSA	Si-N+R	Digital Transistor, Inverter, RB=22k 2SC2412K 50V, 100mA,	78b	Rhm	DTC114GSA, DTC114TSA, DTC115TSA,	data data
DTC 124 TU(A)	Si-N+R	=DTC 124T...: SMD	35a(2mm)		RN 1312, UN 5217	data
DTC 124 TUA	Si-N+R	SMD, Digital Transistor, Inverter, RB=22k 2SC2412K 50V, 100mA,	35a	Rhm	DTC114GUA, DTC114TUA, DTC115GUA,	data data
DTC124X(A,F,L,S,V)	Si-N+R	S, Rb=22k, Rbe=47kΩ, 50V, 100mA, 0,3W	9c, 41c	Rhm	AA 1F4N, KSR 1007, RN 1008	data
DTC 124 XE	Si-N+R	=DTC 124X...: SMD	35a(1,6mm)		RN 1108	data
DTC 124 XC,XK(A)	Si-N+R	=DTC 124X...: SMD	35a(2,9mm)		FA 1F4N, KSR 1107, RN 1408	data
DTC 124 XKA	Si-N+R	SMD, Digital Transistor, Inverter, RB=22k RBE=47k 2SC2412K 100mA,	35bd3	Rhm		data data
DTC 124 XSA	Si-N+R	Digital Transistor, Inverter, RB=22k RBE=47k 2SC2412K 100mA,	78bd4	Rhm		data data
DTC 124 XU(A)	Si-N+R	=DTC 124X...: SMD	35a(2mm)		RN 1308	data
DTC 124 XUA	Si-N+R	SMD, Digital Transistor, Inverter, RB=22k RBE=47k 2SC2412K 100mA,	35bd3	Rhm		data data
DTC125T(A,F,L,S,V)	Si-N+R	S, Rb=200kΩ, Rbe=-, 50V, 100mA, 0,3W	9c, 41c	Rhm	-	data
DTC 125 TC,TK	Si-N+R	=DTC 125T...: SMD	35a(2,9mm)			data
DTC 125 TKA	Si-N+R	SMD, Digital Transistor, Inverter, RB=200k 2SC2412K 50V, 100mA,	35a	Rhm	DTC114GKA, DTC114TKA, DTC115GKA,	data data

Type	Device	Short Description	Fig.	Manu	Comparision Types	More at
DTC 125 TSA	Si-N+R	Digital Transistor, Inverter, RB=200k 2SC2412K 50V, 100mA,	78b	Rhm	DTC114GSA, DTC114TSA, DTC115TSA,	data data
DTC 125 TU	Si-N+R	=DTC 125T...: SMD	35a(2mm)		-	data
DTC 125 TUA	Si-N+R	SMD, Digital Transistor, Inverter, RB=200k 2SC2412K 50V, 100mA,	35a	Rhm	DTC114GUA, DTC114TUA, DTC115GUA,	data data
DTC143E(A,F,L,S,V)	Si-N+R	S, Rb=4,7k,Rbe=4,7kΩ, 50V, 100mA, 0,3W	9c, 41c	Rhm	AA 1L3M, RN -1001, UN 421L, 2SC4363,++	data
DTC 143 EE	Si-N+R	=DTC 143E...: SMD	35a(1,6mm)		RN 1101, UN 921L	data
DTC 143 EC,EK(A)	Si-N+R	=DTC 143E...: SMD	35a(2,9mm)		FA 1L3M, RN 1401, UN 221L, 2SC5362,++	data
DTC 143 EKA	Si-N+R	SMD, Digital Transistor, Inverter, RB=4.7k RBE=4.7k 2SC2412K 100mA,	35bd3	Rhm		data data
DTC 143 ESA	Si-N+R	Digital Transistor, Inverter, RB=4.7k RBE=4.7k 2SC2412K 100mA,	78bd4	Rhm		data data
DTC 143 EU(A)	Si-N+R	=DTC 143E...: SMD	35a(2mm)		RN 1301, UN 521L	data
DTC 143 EUA	Si-N+R	SMD, Digital Transistor, Inverter, RB=4.7k RBE=4.7k 2SC2412K 100mA,	35bd3	Rhm		data data
DTC143T(A,F,L,S,V)	Si-N+R	S, Rb=4,7kΩ, Rbe=-, 50V, 100mA, 0,3W	9c, 41c	Rhm	AA 1L3Z, RN 1010, UN 4216, 2SC3901,++	data
DTC 143 TE	Si-N+R	=DTC 143T...: SMD	35a(1,6mm)		RN 1110, UN 9216	data
DTC 143 TC,TK(A)	Si-N+R	=DTC 143T...: SMD	35a(2,9mm)		FA 1L3Z, RN 1410, UN 2216, 2SC3900,++	data
DTC 143 TKA	Si-N+R	SMD, Digital Transistor, Inverter, RB=4.7k 2SC2412K 50V, 100mA,	35a	Rhm	DTC114GKA, DTC114TKA, DTC115GKA,	data data
DTC 143 TSA	Si-N+R	Digital Transistor, Inverter, RB=4.7k 2SC2412K 50V, 100mA,	78b	Rhm	DTC114GSA, DTC114TSA, DTC115TSA,	data data
DTC 143 TU(A)	Si-N+R	=DTC 143T...: SMD	35a(2mm)		RN 1310, UN 5216	data
DTC 143 TUA	Si-N+R	SMD, Digital Transistor, Inverter, RB=4.7k 2SC2412K 50V, 100mA,	35a	Rhm	DTC114GUA, DTC114TUA, DTC115GUA,	data data
DTC143X(A,F,L,S,V)	Si-N+R	S, Rb=4,7k, Rbe=10kΩ, 50V, 100mA, 0,3W	9c, 41c	Rhm	AA 1L3N, KSR 1005, UN 421F, 2SC4361	data
DTC 143 XE	Si-N+R	=DTC 143X...: SMD	35a(1,6mm)		UN 921F	data
DTC 143 XC,XK(A)	Si-N+R	=DTC 143X...: SMD	35a(2,9mm)		FA 1L3N, UN 221F, 2SC4360	data
DTC 143 XKA	Si-N+R	SMD, Digital Transistor, Inverter, RB=4.7k RBE=10k 2SC2412K 100mA,	35bd3	Rhm		data data
DTC 143 XSA	Si-N+R	Digital Transistor, Inverter, RB=4.7k RBE=10k 2SC2412K 100mA,	78bd4	Rhm		data data
DTC 143 XU(A)	Si-N+R	=DTC 143X...: SMD	35a(2mm)		UN 521F	data
DTC 143 XUA	Si-N+R	SMD, Digital Transistor, Inverter, RB=4.7k RBE=10k 2SC2412K 100mA,	35bd3	Rhm		data data
DTC143Z(A,F,L,S,V)	Si-N+R	S, Rb=4,7k, Rbe=47kΩ, 50V, 100mA, 0,3W	9c, 41c	Rhm	RN 1006, 2SC4133, 2SC4195	data
DTC 143 ZCA	Si-N+R	SMD, Digital Transistor, Inverter, RB=4.7k RBE=47k 2SC2412K 100mA,	35bd3	Rhm		data data
DTC 143 ZE	Si-N+R	=DTC 143Z...: SMD	35a(1,6mm)		RN 1106	data
DTC 143 ZC,ZK(A)	Si-N+R	=DTC 143Z...: SMD	35a(2,9mm)		RN 1406, 2SC4146	data
DTC 143 ZKA	Si-N+R	SMD, Digital Transistor, Inverter, RB=4.7k RBE=47k 2SC2412K 100mA,	35bd3	Rhm		data data
DTC 143 ZSA	Si-N+R	Digital Transistor, Inverter, RB=4.7k RBE=47k 2SC2412K 100mA,	78bd4	Rhm		data data
DTC 143 ZU(A)	Si-N+R	=DTC 143Z...: SMD	35a(2mm)		RN 1306	data
DTC 143 ZUA	Si-N+R	SMD, Digital Transistor, Inverter, RB=4.7k RBE=47k 2SC2412K 100mA,	35bd3	Rhm		data
DTC144E(A,F,L,S,V)	Si-N+R	S, Rb=Rbe=47kΩ, 50V, 30mA, 0,3W	9c, 41c	Rhm	AA 1L4M, RN 1004, UN 4213, 2SC3399,++	data
DTC 144 EE	Si-N+R	=DTC 144E...: SMD	35a(1,6mm)		RN 1104, UN 9213	data
DTC 144 EC,EK(A)	Si-N+R	=DTC 144E...: SMD	35a(2,9mm)		FA 1L4M, RN 1404, UN 2213, 2SC3395,++	data
DTC 144 EKA	Si-N+R	SMD, Digital Transistor, Inverter, RB=47k RBE=47k 2SC2412K 100mA,	35bd3	Rhm		data data
DTC 144 ESA	Si-N+R	Digital Transistor, Inverter, RB=47k RBE=47k 2SC2412K 100mA,	78bd4	Rhm		data data
DTC 144 EU(A)	Si-N+R	=DTC 144E...: SMD	35a(2mm)		RN 1304, UN 5213, 2SC4396	data
DTC 144 EUA	Si-N+R	SMD, Digital Transistor, Inverter, RB=47k RBE=47k 2SC2412K 100mA,	35bd3	Rhm		data data
DTC144G(A,F,L,S,V)	Si-N+R	S, Rb=-, Rbe=47kΩ, 50V, 100mA, 0,3W	9c, 41c	Rhm	2SC4067	data
DTC 144 GE	Si-N+R	=DTC 144G...: SMD	35a(1,6mm)			data
DTC 144 GC,GK(A)	Si-N+R	=DTC 144G...: SMD	35a(2,9mm)		2SC4066	data
DTC 144 GKA	Si-N+R	SMD, Digital Transistor, Inverter, RB=47k 2SC2412K 50V, 100mA,	35a	Rhm	DTC114GKA, DTC114TKA, DTC115GKA,	data data
DTC 144 GSA	Si-N+R	Digital Transistor, Inverter, RBE=47k 2SC2412K 50V, 100mA,	78b	Rhm	DTC114GSA, DTC114TSA, DTC115TSA,	data data
DTC 144 GU(A)	Si-N+R	=DTC 144G...: SMD	35a(2mm)		-	data
DTC 144 GUA	Si-N+R	SMD, Digital Transistor, Inverter, RB=47k 2SC2412K 50V, 100mA,	35a	Rhm	DTC114GUA, DTC114TUA, DTC115GUA,	data data
DTC144T(A,F,L,S,V)	Si-N+R	S, Rb=47kΩ, Rbe=-, 50V, 100mA, 0,3W	9c, 41c	Rhm	AA 1L4Z, KSR 1012, UN 4210, 2SC3899	data
DTC 144 TE	Si-N+R	=DTC 144T...: SMD	35a(1,6mm)		RN 1113, UN 9210	data
DTC 144 TC,TK(A)	Si-N+R	=DTC 144T...: SMD	35a(2,9mm)		FA 1L4Z, KSR 1112, UN 2210, 2SC3898	data
DTC 144 TKA	Si-N+R	SMD, Digital Transistor, Inverter, RB=47k 2SC2412K 50V, 100mA,	35a	Rhm	DTC114GKA, DTC114TKA, DTC115GKA,	data data
DTC 144 TSA	Si-N+R	Digital Transistor, Inverter, RB=47k 2SC2412K 50V, 100mA,	78b	Rhm	DTC114GSA, DTC114TSA, DTC115TSA,	data data
DTC 144 TU(A)	Si-N+R	=DTC 144T...: SMD	35a(2mm)		RN 1313, UN 5210	data
DTC 144 TUA	Si-N+R	SMD, Digital Transistor, Inverter, RB=47k 2SC2412K 50V, 100mA,	35a	Rhm	DTC114GUA, DTC114TUA, DTC115GUA,	data data
DTC144V(A,F,L,S,V)	Si-N+R	S, Rb=47k, Rbe=10kΩ, 50V, 30mA, 0,3W	9c, 41c	Rhm	UN 421D	data
DTC 144 VE	Si-N+R	=DTC 144V...: SMD	35a(1,6mm)		UN 921D	data
DTC 144 VC,VK(A)	Si-N+R	=DTC 144V...: SMD	35a(2,9mm)		UN 221D	data
DTC 144 VKA	Si-N+R	SMD, Digital Transistor, Inverter, RB=47k RBE=10k 2SC2412K 100mA,	35bd3	Rhm		data data
DTC 144 VSA	Si-N+R	Digital Transistor, Inverter, RB=47k RBE=10k 2SC2412K 100mA,	78bd4	Rhm		data data
DTC 144 VU(A)	Si-N+R	=DTC 144V...: SMD	35a(2mm)		UN 521D	data

Type	Device	Short Description	Fig.	Manu	Comparision Types	More at
DTC 144 VUA	Si-N+R	SMD, Digital Transistor, Inverter, RB=47k RBE=10k 2SC2412K 100mA,	35bd3	Rhm		data data
DTC144W(A,F,L,S,V)	Si-N+R	S, RB=47k, Rbe=22kΩ, 50V, 30mA, 0,3W	9c, 41c	Rhm	AA 1L4L, RN 1009, UN 421E, 2SC3401,++	data
DTC 144 WE	Si-N+R	=DTC 144W...: SMD	35a(1,6mm)	Rhm	RN 1109, UN 921E	data
DTC 144 WC,WK(A)	Si-N+R	=DTC 144W...: SMD	35a(2,9mm)		FA 1L4L, RN 1409, UN 221E, 2SC3397,++	data
DTC 144 WKA	Si-N+R	SMD, Digital Transistor, Inverter, RB=47k RBE=22k 2SC2412K 100mA,	35bd3	Rhm		data data
DTC 144 WSA	Si-N+R	Digital Transistor, Inverter, RB=47k RBE=22k 2SC2412K 100mA,	78bd4	Rhm		data data
DTC 144 WU(A)	Si-N+R	=DTC 144W...: SMD	35a(2mm)		RN 1309, UN 521E	data
DTC 144 WUA	Si-N+R	SMD, Digital Transistor, Inverter, RB=47k RBE=22k 2SC2412K 100mA,	35bd3	Rhm		data data
DTC314T(A,F,L,S,V)	Si-N+R	S, Rb=10kΩ, Rbe=-, 30V, 600mA, 0,3W	9c, 41c	Rhm	DTD 114T...	data
DTC 314 TC,TK	Si-N+R	=DTC 314T...: SMD	35a(2,9mm)		DTD 114TK	data
DTC 314 TK	Si-N+R	SMD, Digital Transistor, Inverter, RB=10k 2SD1757K 15V, 600mA,	35a	Rhm	DTC323TK, DTC343TK, DTC363TK,	data data
DTC 314 TS	Si-N+R	Digital Transistor, Inverter, RB=10k 2SD1757K 15V, 600mA,	78b	Rhm	DTC323TS, DTC343TS, DTC363TS,	data data
DTC 314 TU	Si-N+R	SMD, Digital Transistor, Inverter, RB=10k 2SD1757K 15V, 600mA,	35a	Rhm	DTC323TU,	data data
DTC323T(A,F,L,S,V)	Si-N+R	S, Rb=2,2kΩ, Rbe=-, 30V, 600mA, 0,3W	9c, 41c	Rhm	DTD 123T...	data
DTC 323 TC,TK	Si-N+R	=DTC 323T...: SMD	35a(2,9mm)		DTD 123TK	data
DTC 323 TK	Si-N+R	SMD, Digital Transistor, Inverter, RB=2.2k 2SD1757K 15V, 600mA,	35a	Rhm	DTC314TK, DTC343TK, DTC363TK,	data data
DTC 323 TS	Si-N+R	Digital Transistor, Inverter, RB=2.2k 2SD1757K 15V, 600mA,	78b	Rhm	DTC314TS, DTC343TS, DTC363TS,	data data
DTC 323 TU	Si-N+R	SMD, Digital Transistor, Inverter, RB=2.2k 2SD1757K 15V, 600mA,	35a	Rhm	DTC314TU,	data data
DTC343T(A,F,L,S,V)	Si-N+R	S, Rb=4,7kΩ, Rbe=-, 30V, 600mA, 0,3W	9c, 41c	Rhm	DTD 143T...	data
DTC 343 TC,TK	Si-N+R	=DTC 343T...: SMD	35a(2,9mm)		DTD 143TK	data
DTC 343 TK	Si-N+R	SMD, Digital Transistor, Inverter, RB=4.7k 2SD1757K 15V, 600mA,	35a	Rhm	DTC314TK, DTC323TK, DTC363TK,	data data
DTC 343 TS	Si-N+R	Digital Transistor, Inverter, RB=4.7k 2SD1757K 15V, 600mA,	78b	Rhm	DTC314TS, DTC323TS, DTC363TS,	data
DTC363E(A,F,L,S,V)	Si-N+R	S, Rb=Rbe=6,8kΩ, 20V, 600mA, 0,3W	9c, 41c	Rhm	2SD1676	data
DTC 363 EC,EK	Si-N+R	=DTC 363E...: SMD	35a(2,9mm)			data
DTC 363 EK	Si-N+R	SMD, Digital Transistor, Inverter, RB=6.8k RBE=6.8k 2SD1757K 600mA,	35bd3	Rhm		data
DTC 363 ES	Si-N+R	Digital Transistor, Inverter, RB=6.8k RBE=6.8k 2SD1757K 600mA,	78bd4	Rhm		data data
DTC 363 EU	Si-N+R	SMD, Digital Transistor, Inverter, RB=6.8k RBE=6.8k 2SD1757K 600mA,	35bd3	Rhm		data data
DTC363T(A,F,L,S,V)	Si-N+R	S, Rb=6,8kΩ, Rbe=-, 30V, 600mA, 0,3W	9c, 41c	Rhm	DTD 163T...	data
DTC 363 TC,TK	Si-N+R	=DTC 363T...: SMD	35a(2,9mm)		DTD 163TK	data
DTC 363 TK	Si-N+R	SMD, Digital Transistor, Inverter, RB=6.8k 2SD1757K 15V, 600mA,	35a	Rhm	DTC314TK, DTC323TK, DTC343TK,	data
DTC 363 TS	Si-N+R	Digital Transistor, Inverter, RB=6.8k 2SD1757K 15V, 600mA,	78b	Rhm	DTC314TS, DTC323TS, DTC343TS,	data data
DTD113E(A,F,L,S,V)	Si-N+R	S, Rb=Rbe=1kΩ, 50V, 500mA, 0,6W	9c, 41c	Rhm	RN 1221	data
DTD 113 EC,EK	Si-N+R	=DTD 113E...: SMD	35a(2,9mm)		-	data
DTD 113 EK	Si-N+R	SMD, Digital Transistor, Inverter, RB=1k RBE=1k 2SC2411K 500mA,	35bd3	Rhm		data
DTD 113 ES	Si-N+R	Digital Transistor, Inverter, RB=1k RBE=1k 2SC2411K 500mA,	78bd4	Rhm		data data
DTD113Z(A,F,L,S,V)	Si-N+R	S, Rb=1kΩ, Rbe=10kΩ, 50V, 500mA, 0,6W	9c, 41c	Rhm	RN 1226	data
DTD 113 ZC,ZK	Si-N+R	=DTD 113Z..: SMD	35a(2,9mm)		-	data
DTD 113 ZK	Si-N+R	SMD, Digital Transistor, Inverter, RB=1k RBE=10k 2SC2411K 500mA,	35bd3	Rhm		data data
DTD 113 ZS	Si-N+R	Digital Transistor, Inverter, RB=1k RBE=10k 2SC2411K 500mA,	78bd4	Rhm		data data
DTD 113 ZU	Si-N+R	SMD, Digital Transistor, Inverter, RB=1k RBE=10k 2SC2411K 500mA,	35bd3	Rhm		data data
DTD114E(A,F,L,S,V)	Si-N+R	S, Rb=Rbe=10kΩ, 50V, 500mA, 0,6W	9c, 41c	Rhm	RN 1224, UN 4223, 2SC3916, 2SC 3920	data
DTD 114 EC,EK	Si-N+R	=DTD 114E...: SMD	35a(2,9mm)		UN 2223, 2SC3912	data
DTD 114 EK	Si-N+R	SMD, Digital Transistor, Inverter, RB=10k RBE=10k 2SC2411K 500mA,	35bd3	Rhm		data
DTD 114 ES	Si-N+R	Digital Transistor, Inverter, RB=10k RBE=10k 2SC2411K 500mA,	78bd4	Rhm		data data
DTD114G(A,F,L,S,V)	Si-N+R	S, Rb=-, Rbe=10kΩ, 50V, 500mA, 0,6W	9c, 41c	Rhm		data
DTD 114 GC,GK	Si-N+R	=DTD 114G...: SMD	35a(2,9mm)			data
DTD 114 GK	Si-N+R	SMD, Digital Transistor, Inverter, RBE=10k 2SC2411K 50V, 500mA,	35a	Rhm		data data
DTD114T(A,F,L,S,V)	Si-N+R	S, Rb=10kΩ, Rbe=-, 50V, 500mA, 0,6W	9c, 41c	Rhm	DTC 314T...	data
DTD 114 TC,TK	Si-N+R	=DTD 114T...: SMD	35a(2,9mm)		DTC 314TK	data
DTD122J(A,F,L,S,V)	Si-N+R	S, Rb=220Ω,Rbe=4,7kΩ, 50V, 500mA, 0,6W	9c, 41c	Rhm	-	data
DTD 122 JC,JK	Si-N+R	=DTD 122J..: SMD	35a(2,9mm)		-	data
DTD 122 JK	Si-N+R	SMD, Digital Transistor, Inverter, RB=0.22k RBE=4.7k 2SC2411K 500mA,	35bd3	Rhm		data data
DTD123E(A,F,L,S,V)	Si-N+R	S, Rb=Rbe=2,2kΩ, 50V, 500mA, 0,6W	9c, 41c	Rhm	RN 1222, UN 4221, 2SC3919, 2SC 3923	data
DTD 123 EC,EK	Si-N+R	=DTD 123E...: SMD	35a(2,9mm)		UN 2221, 2SC3915	data
DTD 123 EK	Si-N+R	SMD, Digital Transistor, Inverter, RB=2.2k RBE=2.2k 2SC2411K 500mA,	35bd3	Rhm		data data

Type	Device	Short Description	Fig.	Manu	Comparision Types	More at
DTD 123 ES	Si-N+R	Digital Transistor, Inverter, RB=2.2k RBE=2.2k 2SC2411K 500mA,	78bd4	Rhm		data
						data
DTD123T(A,F,L,S,V)	Si-N+R	S, Rb=2,2kΩ, Rbe=-, 50V, 500mA, 0,6W	9c. 41c	Rhm	DTC 323T...	data
DTD 123 TC,TK	Si-N+R	=DTD 123T...: SMD	35a(2.9mm)	Rhm	DTC 323TK	data
DTD 123 TK	Si-N+R	SMD, Digital Transistor, Inverter, RB=2.2k 2SC2411K 40V, 500mA,	35a	Rhm	DTD114GK, DTD143TK,	data
						data
DTD 123 TS	Si-N+R	Digital Transistor, Inverter, RB=2,2k 2SC2411K 40V, 500mA,	78b	Rhm		data
						data
DTD123Y(A,F,L,S,V)	Si-N+R	S, Rb=2,2k, Rbe=10kΩ, 50V, 500mA, 0,6W	9c. 41c	Rhm	RN 1227, UN 4224, 2SC3918, 2SC3922	data
DTD 123 YC,YK	Si-N+R	=DTD 123Y...: SMD	35a(2.9mm)	Rhm	UN 2224, 2SC3914	data
DTD 123 YK	Si-N+R	SMD, Digital Transistor, Inverter, RB=2.2k RBE=10k 2SC2411K 500mA,	35bd3	Rhm		data
						data
DTD 123 YS	Si-N+R	Digital Transistor, Inverter, RB=2.2k RBE=10k 2SC2411K 500mA,	78bd4	Rhm		data
						data
DTD133H(A,F,L,S,V)	Si-N+R	S, Rb=3,3k, Rbe=10kΩ, 50V, 500mA, 0,6W	9c. 41c	Rhm	-	data
DTD 133 HC,HK	Si-N+R	=DTD 133H...: SMD	35a(2.9mm)	Rhm		data
DTD 133 HK	Si-N+R	SMD, Digital Transistor, Inverter, RB=3.3k RBE=10k 2SC2411K 500mA,	35bd3	Rhm		data
DTD 133 HS	Si-N+R	Digital Transistor, Inverter, RB=3.3k RBE=10k 2SC2411K 500mA,	78bd4	Rhm		data
						data
DTD143E(A,F,L,S,V)	Si-N+R	S, Rb=Rbe=4,7kΩ, 50V, 500mA, 0,6W	9c. 41c	Rhm	RN 1223, UN 4222, 2SC3917, 2SC 3921	data
DTD 143 EC	Si-N+R	Digital Transistor, Inverter, RB=4.7k RBE=4.7k 2SC2411K 500mA,	78bd4	Rhm		data
						data
DTD 143 EC,EK	Si-N+R	=DTD 143E...: SMD	35a(2,9mm)		UN 2222, 2SC3913	data
DTD 143 EK	Si-N+R	SMD, Digital Transistor, Inverter, RB=4.7k RBE=4.7k 2SC2411K 500mA,	35bd3	Rhm		data
						data
DTD 143 ES	Si-N+R	SMD, Digital Transistor, Inverter, RB=4.7k RBE=4.7k 2SC2411K 500mA,	35bd3	Rhm		data
						data
DTD143T(A,F,L,S,V)	Si-N+R	S, Rb=4,7kΩ, Rbe=-, 50V, 500mA, 0,6W	9c. 41c	Rhm	DTC 343T...	data
DTD 143 TC,TK	Si-N+R	=DTD 143T...: SMD	35a(2,9mm)	Rhm	DTC 343TK	data
DTD 143 TK	Si-N+R	SMD, Digital Transistor, Inverter, RB=4,7k 2SC2411K 40V, 500mA,	35a	Rhm	DTD114GK, DTD123TK,	data
						data
DTD163T(A,F,L,S,V)	Si-N+R	S, Rb=6,8kΩ, Rbe=-, 50V, 500mA, 0,6W	9c. 41c	Rhm	DTC 363T...	data
DTD 163 TC,TK	Si-N+R	=DTD 163T...: SMD	35a(2,9mm)	Rhm	DTC 363TK	data
DTDG 14 GP	Si-N+R	SMD, Rbe=10kΩ, Z-Di, hi-beta, 60V, 1A	39b	Rhm	-	data
DTDG 23 YP	Si-N+R	SMD, Rb=2,2k, Rbe=10kΩ, Z-Di, 60V, 1A hi-beta>300	39b	Rhm	-	data
DTDK 14 GP	Si N+R+Di	SMD, Rbe=10kΩ, -/10V, 2A	39b	Rhm		data
DTDM 12 ZP	Si-N+R+Di	SMD, Rb=100Ω, Rbe=1kΩ, 30V, 3A	39b	Rhm	-	data
DTE 16 E	Triac	400V, 16A≈(Tc=75°), Igt/Ih<80/=30mA	65b	Say	MAC 515A-..., MAC 525A-...	
DTF 16 C...G	Triac	200...600V, 16A≈(78°), Igt/Ih<50/<50mA	65b	Say	MAC 515(A)-..., MAC 525(A)-...	
DTI 2222	NMOS-IC	C=200V, E=400V, G=600V Ctv, Color-prozessor	40-DIP	Itt	-	
DTI 2223	NMOS-IC	Ctv, dig. Color-prozessor	40-DIP	Itt		pinout
DTM 6 C...G(-N)	Triac	200...600V, 6A≈(Tc=70°),Igt/Ih<50/<50mA	17b	Say	TAG 420-... (MAC 216-... T 2500...)³	data
DTM 8 C...G(-N)	Triac	C=200V, E=400V, G=600V 200...600V, 8A≈(Tc=83°),Igt/Ih<50/<50mA	17b	Say	TAG425-..., TAG451-... (T2800..., T2850...)³	data
DTM 10 C-N...G-N	Triac	C=200V, E=400V, G=600V 200...600V, 10A≈(83°), Igt/Ih<50/<50mA C-N=200V, E-N=400V, G-N=600V	17b	Say	TAG 456-..., TAG 457-... (BTA 22-...)³	data
DTM 12 C...G(-N)	Triac	200...600V, 12A≈(73°), Igt/Ih<50/<50mA C=200V, E=400V, G=600V	17b	Say	TAG480-..., TAG481-...,(BTA23-...,BT138-...)³	data
DTN 6 E...G(-N)	Triac	400...600V, 6A≈(Tc=90°),Igt/Ih<20/<50mA E=400V, G=600V	17b	Say	TAG 420-... (MAC 216-... T 2500...)³	
DTN 8 C...G	Triac	400...600V, 8A≈(Tc=83°),Igt/Ih<20/<50mA E=400V, G=600V	17b	Say	TAG425-..., TAG451-..., (T2800..., T2850...)³	
DTN 12 E...G	Triac	400...600V, 12A≈(72°), Igt/Ih<20/<50mA E=400V, G=600V	17b	Say	TAG480-..., TAG481-...,(BTA23-...,BT138-...)³	
DTP	Si-N	=2SD2391-P (Typ-Code/Stempel/marking)	39	Rhm	→2SD2391	data
DtQ	Si-P	=2PB1219AQ (Typ-Code/Stempel/marking)	35(2mm)	Phi	→2PB1219A	data
DTQ	Si-N	=2SD2391-Q (Typ-Code/Stempel/marking)	39	Rhm	→2SD2391	data
DtR	Si-P	=2PB1219AR (Typ-Code/Stempel/marking)	35(2mm)	Phi	→2PB1219A	data
DtS	Si-P	=2PB1219AS (Typ-Code/Stempel/marking)	35(2mm)	Phi	→2PB1219A	data
DTS 401	Si-N	TV-VA, -/400V, 2A, 75W(Tc=75°)	23a§	Del	2SC1101, 2SC3151, 2SC3533	data
DTS 402	Si-N	TV-HA, 700/400V, 3,5A, 100W(Tc=75°)	23a§	Del	BU 706, 2SC3484, 2SD1098, 2SD1495, ++	data
DTS 403	Si-N	S-L, 400/400V, 3,5A, 100W(Tc=75°)	23a§	Del	BUW,71, BUX 45, TIP 54, 2SC3083, ++	data
DTS 409	Si-N	S-L, 400/400V, 3,5A, 100W(Tc=75°)	23a§	Del	BUW 71, BUX 45, TIP 54, 2SC3083, ++	data
DTS 410	Si-N	S-L, 200/200V, 3,5A, 100W(Tc=75°)	23a§	Del	BUW 71, BUX 16(A...C), 2SC2908, 2SD1018++	data
DTS 411	Si-N	S-L, 300/300V, 3,5A, 100W(Tc=75°)	23a§	Del	BUW 71, BUX 16A...C, TIP 52, 2SC4799,++	data
DTS 413	Si-N	S-L, -/400V, 2A, 75W(Tc=75°)	23a§	Del	BUX 46(A), BUX 82, 2SC3151, 2SC3533,++	data
DTTi 5516	IC	Demodulator and decoder IC f. digit. video set-top box.	GRID	Stm		pdf
DTV 16 D	Si-Di	CRT-HA, Damper-Di, 1500V, 5A, <300ns	17t§	Tho	BY 459/1500, DTV 32D	data
DTV 16F	Si-Di	=DTV 16D: Iso	17d		BY 459F-1500, DTV 32F	data
DTV 32-1000A	Si-Di	CRT-HA, Damper-Diode, 1000V, 3A, <72ns	31a	Tho	BYT 56M, (BYW 96E)	data
DTV 32-1200A	Si-Di	CRT-HA,Damper-Diode, 1200V, 6A, <600ns	17t§	Tho	BY 329/1200, DTV 32D	data
DTV 32-1500A	Si-Di	=DTV 32-1200A: 1500V	17t§		BY 359/1500, DTV 32B	data
DTV 32(F)-...B	Si-Di	=DTV 32(F)-... <175ns	17t§		DTV 56D	data
DTV 32F-1200...1500	Si-Di	=DTV 32-...: Iso	17d		BY 459F-1500, DTV 56F	data
DTV 32 D	Si-Di	CRT-HA, Damper-Di, 1500V, 6A, <175ns	17t§	Tho	DTV 56D	data
DTV 32 F	Si-Di	=DTV 32D: Iso	17d		DTV 56F	data

Type	Device	Short Description	Fig.	Manu	Comparision Types	More at
DTV 56 D	Si-Di	CRT-HA, Damper-Di, 1500V, 6A, <135ns	17t§	Tho	DTV 64D	data
DTV 56 F	Si-Di	=DTV 56D: Iso	17d		DTV 64F	data
DTV 64-1200C	Si-Di	CRT-HA,Damper-Diode, 1200V, 6A, <100ns	17t§	Tho	DTV 82D	data
DTV 64F-1200C	Si-Di	=DTV 64-1200C: Iso	17d		DTV 82F	data
DTV 64 D	Si-Di	CRT-HA, Damper-Di, 1500V, 6A, <135ns	17t§	Tho	DTV 82D	data
DTV 64 F	Si-Di	=DTV 62D: Iso	17d		DTV 82F	data
DTV 82 D	Si-Di	CRT-HA, Damper-Di, 1500V, 6A, <125ns	17t§	Tho	DTV 110D	data
DTV 82 F	Si-Di	=DTV 82D: Iso	17d		DTV 110F	data
DTV 110 D	Si-Di	CRT-HA, Damper-Di, 1500V, 6A, <115ns	17t§	Tho	-	data
DTV 110 F	Si-Di	=DTV 110D: Iso	17d		-	data
DTZ 2.0...36...	Z-Di	SMD, 2,0...36V, 0,2W	71(1,7mm)	Rhm	HZU ..., RD ...S	data
DU	GaAs-N-FET-d	=3SK241 (Typ-Code/Stempel/marking)	44	Mat	→3SK241	data
DU	GaAs-N-FET-d	=3SK272 (Typ-Code/Stempel/marking)	44(2mm)	Mat	→3SK272	data
DU	Si-P	=BCW 67RB (Typ-Code/Stempel/marking)	35	Sie	→BCW 67RB	data
DU	Si-P	=HQ 1F2Q (Typ-Code/Stempel/marking)	39	Nec	→HQ 1...	data
DU	Z-Di	=MMSZ 4706 (Typ-Code/marking)	71(2,7mm)	Ons	→MMSZ 4706	data
DU	Z-Di	=P4 SMA-200A(Typ-Code/Stempel/marking)	71(5x2,5)	Fag	→P4 SMA-200A	data
DU	Z-Di	=SMBJ 85C (Typ-Code/Stempel/marking)	71(5x3,5)	Mop	→SMBJ ...	data
DU	Si-N/P+R	=XP 03311 (Typ-Code/Stempel/marking)	45(2mm)	Mat	→XP 03311	data
DUB	Z-Di	=SMAJ 6.0A (Typ-Code/Stempel/marking)	71(5x2,5)	Tho	→SMAJ ...	data
DUC	Z-Di	=SMAJ 6.5A (Typ-Code/Stempel/marking)	71(5x2,5)	Tho	→SMAJ ...	data
DUH	Z-Di	=SMAJ 8.5A (Typ-Code/Stempel/marking)	71(5x2,5)	Tho	→SMAJ ...	data
DUK	Z-Di	=SMAJ 12A (Typ-Code/Stempel/marking)	71(5x2,5)	Tho	→SMAJ ...	data
DUQ	Z-Di	=SMAJ 18A (Typ-Code/Stempel/marking)	71(5x2,5)	Tho	→SMAJ ...	data
DUR	Z-Di	=SMAJ 20A (Typ-Code/Stempel/marking)	71(5x2,5)	Tho	→SMAJ ...	data
DUS	Z-Di	=SMAJ 22A (Typ-Code/Stempel/marking)	71(5x2,5)	Tho	→SMAJ ...	data
DUT	Z-Di	=SMAJ 24A (Typ-Code/Stempel/marking)	71(5x2,5)	Tho	→SMAJ ...	data
DUU	Z-Di	=SMAJ 26A (Typ-Code/Stempel/marking)	71(5x2,5)	Tho	→SMAJ ...	data
DUZ	Z-Di	=SMAJ 40A (Typ-Code/Stempel/marking)	71(5x2,5)	Tho	→SMAJ ...	data
DV	Si-N	=2SC4755 (Typ-Code/Stempel/marking)	35(2mm)	Mat	→2SC4755	data
DV	Si-N	=2SC4782 (Typ-Code/Stempel/marking)	35	Mat	→2SC4782	data
DV	Z-Di	=MMSZ 4707 (Typ-Code/marking)	71(2,7mm)	Ons	→MMSZ 4707	data
DV	Z-Di	=P4 SMA-220 (Typ-Code/Stempel/marking)	71(5x2,5)	Fag	→P4 SMA-220	data
DV	Z-Di	=SMBJ 85CA (Typ-Code/Stempel/marking)	71(5x3,5)	Mop	→SMBJ ...	data
DV	Si-N/P+R	=XP 03383 (Typ-Code/Stempel/marking)	45(2mm)	Mat	→XP 03383	data
DV 1	Si-N	=2SD596-DV1 (Typ-Code/Stempel/marking)	35	Nec	→2SD596	data
DV 2	Si-N	=2SD596-DV2 (Typ-Code/Stempel/marking)	35	Nec	→2SD596	data
DV-2	Si-Varistor	VDR/Varistor, 0,2A, 1,2V(1mA)	31	Shi		data
DV 3	Si-N	=2SD596-DV3 (Typ-Code/Stempel/marking)	35	Nec	→2SD596	data
DV-3	Si-Varistor	VDR/Varistor, 0,15A, 1,8V(1mA)	31	Shi		data
DV 4	Si-N	=2SD596-DV4 (Typ-Code/Stempel/marking)	35	Nec	→2SD596	data
DV-4	Si-Varistor	VDR/Varistor, 0,1A, 2,35V(1mA)	31	Shi		data
DV 5	Si-N	=2SD596-DV5 (Typ-Code/Stempel/marking)	35	Nec	→2SD596	data
DV-5	Si-Varistor	VDR/Varistor, 0,08A, 3V(1mA)	31	Shi		data
DV 07	Hybrid-IC	Kabeltreiber/cable driver		Fui	-	
DV 08	Hybrid-IC	Kabeltreiber/cable driver		Fui	-	
DV 24	Hybrid-IC	Hammertreiber/hammer driver		Fui	-	
DV 1202 S	MOS-N-FET-e	V-MOS, VHF, 50V, 0,5A, PQ>2,5W(175MHz)	59z	Six	-	data
DV 1202 W	MOS-N-FET-e	=DV 1202S:	≈61z		-	data
DV 1205 S	MOS-N-FET-e	V-MOS, VHF, 50V, 1A, PQ>5W(175MHz)	59z	Six	-	data
DV 1205 W	MOS-N-FET-e	=DV 1205S:	≈61z		-	data
DV 1210 S	MOS-N-FET-e	V-MOS, VHF, 50V, 2A, PQ>9W(175MHz)	59z	Six	-	data
DV 1210 W	MOS-N-FET-e	=DV 1210S:	≈61z		-	data
DV 1220 S	MOS-N-FET-e	V-MOS, VHF, 50V, 4A, PQ>18W(175MHz)	59z	Six	-	data
DV 1220 W	MOS-N-FET-e	=DV 1220S:	≈61z		-	data
DV 1230 T	MOS-N-FET-e	V-MOS, VHF, 50V, 6A, PQ>27W(175MHz)	57z	Six	-	data
DV 1230 W	MOS-N-FET-e	=DV 1230T:	≈61z		-	data
DV 1240 T	MOS-N-FET-e	V-MOS, VHF, 50V, 8A, PQ>36W(175MHz)	57z	Six	-	data
DV 1240 W	MOS-N-FET-e	=DV 1240T:	≈61z		-	data
DV 2805 S	MOS-N-FET-e	V-MOS, VHF, 80V, 0,5A, PQ>5W(175MHz)	59z	Six	-	data
DV 2805 W	MOS-N-FET-e	=DV 2805S:	≈61z		-	data
DV 2810 S	MOS-N-FET-e	V-MOS, VHF, 80V, 1A, PQ>10W(175MHz)	59z	Six	-	data
DV 2810 W	MOS-N-FET-e	=DV 2810S:	≈61z		-	data
DV 2820 S	MOS-N-FET-e	V-MOS, VHF, 80V, 2A, PQ>20W(175MHz)	59z	Six	-	data
DV 2820 W	MOS-N-FET-e	=DV 2820S:	≈61z		-	data
DV 2820 Z	MOS-N-FET-e	=DV 2820S:	55z		-	data
DV 2840 S	MOS-N-FET-e	V-MOS, VHF, 80V, 4A, PQ=40W(175MHz)	59z	Six	-	data
DV 2840 T	MOS-N-FET-e	=DV 2840S:	57z		-	data
DV 2840 V	MOS-N-FET-e	=DV 2840S: Push-Pull, PQ>40W(175W)	≈61/8Pin		-	data
DV 2840 W	MOS-N-FET-e	=DV 2840S:	≈61z		-	data
DV 2880 T	MOS-N-FET-e	V-MOS, VHF, 80V, 8A, PQ=80W(175MHz)	57z	Six	-	data
DV 2880 U	MOS-N-FET-e	=DV 2880T:	59z		-	data
DV 2880 V	MOS-N-FET-e	=DV 2880T: Push-Pull, PQ>80W(175W)	≈61/8Pin		-	data
DV 2880 W	MOS-N-FET-e	=DV 2880S:	≈61z		-	data
DV 28120 T	MOS-N-FET-e	V-MOS, VHF, 80V, 12A, PQ=120W(175MHz)	57z	Six	-	data
DV 28120 U	MOS-N-FET-e	=DV 28120T:	59z		-	data
DV 28120 V	MOS-N-FET-e	=DV 28120T: Push-Pull, PQ>120W(175W)	≈61/8Pin		-	data
DVP	Si-N	=2SC4755-P (Typ-Code/Stempel/marking)	35(2mm)	Mat	→2SC4755	data
DVP	Si-N	=2SC4782-P (Typ-Code/Stempel/marking)	35	Mat	→2SC4782	data
DVQ	Si-N	=2SC4755-Q (Typ-Code/Stempel/marking)	35(2mm)	Mat	→2SC4755	data

Type	Device	Short Description	Fig.	Manu	Comparision Types	More at
DVQ	Si-N	=2SC4782-Q (Typ-Code/Stempel/marking)	35	Mat	→2SC4782	data
DVR-8	R+	Io=200mA P:2.4W	3P	Del		data pinout
DVR	Si-N	=2SC4755-R (Typ-Code/Stempel/marking)	35(2mm)	Mat	→2SC4755	data
DVR	Si-N	=2SC4782-R (Typ-Code/Stempel/marking)	35	Mat	→2SC4782	data
DVV	Si-N	=2SD2537-V (Typ-Code/Stempel/marking)	39	Rhm	→2SD2537	data
DVW	Si-N	=2SD2537-W (Typ-Code/Stempel/marking)	39	Rhm	→2SD2537	data

DW

Type	Device	Short Description	Fig.	Manu	Comparision Types	More at
DW	Si-P	=BCW 67RC (Typ-Code/Stempel/marking)	35	Sie	→BCW 67RC	data
DW	Z-Di	=SM 6T 15 (Typ-Code/Stempel/marking)	71(6x4mm)	Tho	→SM 6T...	data
DW	Z-Di	=SMBJ 90C (Typ-Code/Stempel/marking)	71(5x3,5)	Mop	→SMBJ ...	data
DW 1	Si-N	=2SD780-DW1 (Typ-Code/Stempel/marking)	35	Nec	→2SD780	data
DW 2	Si-N	=2SD780-DW2 (Typ-Code/Stempel/marking)	35	Nec	→2SD780	data
DW 3	Si-N	=2SD780-DW3 (Typ-Code/Stempel/marking)	35	Nec	→2SD780	data
DW 4	Si-N	=2SD780-DW4 (Typ-Code/Stempel/marking)	35	Nec	→2SD780	data
DW 5	Si-N	=2SD780-DW5 (Typ-Code/Stempel/marking)	35	Nec	→2SD780	data
DW 542/...	Si-Di	≈SKE 4F2/04			→SKE 4F2/04	
DW 6089	Si-P	≈BC 154	8a		→BC 154	
DW 6170	Si-N	≈BC 237	8a		→BC 237	
DW 6208	Si-N	≈BC 237	8a		→BC 237	
DW 6335	Si-N	≈BC 237	8a		→BC 237	
DW 6577	Si-N	≈BF 199	8a		→BF 199	
DW 6618	Si-N	≈BC 337	8a		→BC 337	
DW 6619	Si-P	≈BC 327	8a		→BC 327	
DW 6737	Si-N	≈BC 237	2a		→BC 237	
DW 6969	Si-P	≈BC 307	2a		→BC 307	
DW 7000	Si-N	≈BF 198	2a		→BF 198	
DW 7035	Si-N	≈BC 237	8a		→BC 237	
DW 7039	Si-N	≈BSX 24	2a		→BSX 24	
DW 7050	Si-N	≈BF 199	8a		→BF 199	
DW 7975	Si-P	≈BC 154	8a		→BC 154	

DX...DZ

Type	Device	Short Description	Fig.	Manu	Comparision Types	More at
DX	Si-P	=BCW 68RF (Typ-Code/Stempel/marking)	35	Sie	→BCW 68RF	data
DX	Si-P	=HQ 1A4A (Typ-Code/Stempel/marking)	39	Nec	→HQ 1...	data
DX	Z-Di	=MMSZ 4709 (Typ-Code/marking)	71(2,7mm)	Ons	→MMSZ 4709	
DX	Z-Di	=P4 SMA-220A(Typ-Code/Stempel/marking)	71(5x2,5)	Fag	→P4 SMA-220A	data
DX	Z-Di	=SM 6T 15A (Typ-Code/Stempel/marking)	71(6x4mm)	Tho	→SM 6T...	data
DX	Z-Di	=SMBJ 90CA (Typ-Code/Stempel/marking)	71(5x3,5)	Mop	→SMBJ ...	data
DX 0038 CE	Si-Di	≈RGP 30M			→RGP 30M	
DX 0048 CE	Si-Di	≈1N4148			→1N4148	
DX 0055 CE	Si-Di	≈RGP 30M			→RGP 30M	
DX 0073 CE	Si-Di	≈BA 159			→BA 159	
DX 0081 TA	Si-Di	≈RGP 30A...M			→RGP 30A...M	
DX 0086 TA	Si-Di	≈RGP 30G...M			→RGP 30G...M	
DX 0101 CE	Si-Di	≈BA 157			→BA 157...159	
DX 0107 TA	Si-Br	Gl-Br	33x1		B380 C2200	
DX 0113 TA	Si-Di	≈BA 176			→BA 176	
DX 0115 CE	Si-Di	≈RGP 30M			→RGP 30M	
DX 0117 TA	Si-Di	≈RGP 30M			→RGP 30M	
DX 0118 CE	Si-Br	Gl-Br	33x1		B40 C3300	
DX 0124 TA	Si-Di	≈BY 255			→BY 255	
DX 0125 CE	Si-Di	≈RGP 30M			→RGP 30M	
DX 0128 CE	Si-Di	≈RGP 30M	31a		→RGP 30M	
DY	Si-P	=2SA1201-Y (Typ-Code/Stempel/marking)	39	Tos	→2SA1201	data
DY	Si-P	=2SA1620-Y (Typ-Code/Stempel/marking)	35	Tos	→2SA1620	data
DY	Si-N	=2SC3397 (Typ-Code/Stempel/marking)	35	Say	→2SC3397	data
DY	MOS-N-FET-e*	=2SK1579 (Typ-Code/Stempel/marking)	39	Hit	→2SK1579	data
DY	Si-P	=BCW 68RG (Typ-Code/Stempel/marking)	35	Sie	→BCW 68RG	data
DY	Si-P	=KTA1661-Y (Typ-Code/Stempel/marking)	39	Kec	→KTA 1661	data
DY	Z-Di	=MMSZ 4710 (Typ-Code/marking)	71(2,7mm)	Ons	→MMSZ 4710	
DY	Z-Di	=SMBJ 100C (Typ-Code/Stempel/marking)	71(5x3,5)	Mop	→SMBJ ...	data
DZ...	Z-Di		31a			
DZ	Si-P	=BCW 68RH (Typ-Code/Stempel/marking)	35	Sie	→BCW 68RH	data
DZ	Z-Di	=SMBJ 100CA (Typ-Code/Stempel/marking)	71(5x3,5)	Mop	→SMBJ ...	data
DZ	Si-N	=XN 4505 (Typ-Code/Stempel/marking)	46	Mat	→XN 4505	data
DZ 23 C2V7...C51	Z-Di	SMD, Dual, 2,7...51V, 5%	35l	Aeg		data
DZB 6.2...30(C)	Z-Di	-6,2...30V, ±10%, C=±5%, 1W	31a	Say	BZW 22/.., BZX 61/.., BZX 85/.., ++	data
DZD 2.0...47(X,Y,Z)	Z-Di	SMD, 2,0...47V, ±5%, X,Y,Z≈ -5,±2,5,+5%	35e	Say	BZX 84/...	data
DZF 6,8..36	Z-Di	6,8...36V, 10%, 1W	31a	Say	BZW22/.., BZX61..., ZPY..., 1N4736..53, ++	data

Type	Device	Short Description	Fig.	Manu	Comparision Types	More at
E						
E..... [Motorola]	Si-N	→E...., z.B./e.g. "E13007" =MJE13007	17	Mot		
E...[SEC,Samsung]	Si-N	→E...., z.B./e.g. "E13007" =KSE13007	17	Sam		
E	Diode	=1SS380 (Typ-Code/Stempel/marking)	71(1,7mm)	Rhm	→1SS380	data
E	Si-P	=2SA1022 (Typ-Code/Stempel/marking)	35	Mat	→2SA1022	data
E	Si-P	=2SA1532 (Typ-Code/Stempel/marking)	35(2mm)	Mat	→2SA1532	data
E	Si-P	=2SA1688 (Typ-Code/Stempel/marking)	35	Say	→2SA1688	data
E	Si-P	=2SA1688 (Typ-Code/Stempel/marking)	35	Say	→2SA1688	data
E	Si-P	=2SA1790 (Typ-Code/Stempel/marking)	35(1,6mm)	Mat	→2SA1790	data
E	Si-P	=2SB789A (Typ-Code/Stempel/marking)	39	Mat	→2SB789A	data
E	C-Di	=BB 689 (Typ-Code/Stempel/marking)	71(1,3mm)	Sie	→BB 689	data
E	C-Di	=HVU 17 (Typ-Code/Stempel/marking)	71(1,7mm)	Hit	→HVU 17	data
E	Si-D	=SB002-15CP (Typ-Code/Stempel/marking)	35	Say	→SB 002-15CP	data
E 1	Si-P	=2SA1737 (Typ-Code/Stempel/marking)	39	Mat	→2SA1737	data
E 1	Si-N	=BFS 17 (Typ-Code/Stempel/marking)	35	Aeg,Mot,Phi Sie	→BFS 17	data
E 1	Si-N	=BFS 17W (Typ-Code/Stempel/marking)	35(2mm)	Phi,Aeg	→BFS 17W	data
E 01	Si-N+R	=DTDG 14GP (Typ-Code/Stempel/marking)	39	Rhm	→DTDG 14GP	data
E 1	Si-Di	=HRB 0103A (Typ-Code/Stempel/marking)	35(2mm)	Hit	→HRB 0103A	data
E 1	Z-Di	=MAZF 062 (Typ-Code/Stempel/marking)	71(1,7mm)	Mat	→MAZF 062	data
E 1	Z-Di	=MAZN 062 (Typ-Code/Stempel/marking)	71(1,3mm)	Mat	→MAZN 062	data
E 1	Z-Di	=MMSZ 5231B (Typ-Code/marking)	71(2,7mm)	Ons	→MMSZ 5231B	
E 1 H	Si-N	=BFS 17H (Typ-Code/Stempel/marking)	35	Ztx	→BFS 17H	data
E 1 L	Si-N	=BFS 17L (Typ-Code/Stempel/marking)	35	Ztx	→BFS 17L	data
E 1 p	Si-N	=BFS 17 (Typ-Code/Stempel/marking)	35	Phi	→BFS 17	data
E 2	Si-P	=2SA1226-2 (Typ-Code/Stempel/marking)	35	Say	→2SA1226	data
E 2	Si-Di	=BAL 99 (Typ-Code/Stempel/marking)	35	Fer	→BAL 99	data
E 2	Si-N	=BFS 17A (Typ-Code/Stempel/marking)	35	Aeg, Phi	→BFS 17A	data
E 02	Si-N+R	=DTDG 23YP (Typ-Code/Stempel/marking)	39	Rhm	→DTDG 23YP	data
E 2	Z-Di	=EDZ 6.2B (Typ-Code/Stempel/marking)	71(1,3mm)	Rhm	→EDZ 6.2B	data
E 2	Si-Di	=HRB 0103B (Typ-Code/Stempel/marking)	35(2mm)	Hit	→HRB 0103B	data
E 2	Z-Di	=MAZF 068 (Typ-Code/Stempel/marking)	71(1,7mm)	Mat	→MAZF 068	data
E 2	Z-Di	=MAZN 068 (Typ-Code/Stempel/marking)	71(1,3mm)	Mat	→MAZN 068	data
E 2	Z-Di	=MMSZ 5232B (Typ-Code/marking)	71(2,7mm)	Ons	→MMSZ 5232B	
E 2	Z-Di	=UDZ 6.2B (Typ-Code/Stempel/marking)	71(1,7mm)	Rhm	→UDZ 2.0B...36B	data
E 2	Z-Di	=UDZS 6.2B (Typ-Code/Stempel/marking)	71(1,7mm)	Rhm	→UDZS 5.1B...10B	data
E 2 p	Si-N	=BFS 17A (Typ-Code/Stempel/marking)	35	Phi	→BFS 17A	
E 2 P102	MOS-P-FET-e	=NTMSD2P102L(Typ-Code/Stempel/marking)	8-MDIP	Ons	→NTMSD 2P102L	
E 3	Si-Di	=1SS190 (Typ-Code/Stempel/marking)	35	Tos	→1SS190	data
E 3	Si-P	=2SA1226-3 (Typ-Code/Stempel/marking)	35	Say	→2SA1226	data
E 3	Si-P	=2SA1256-3 (Typ-Code/Stempel/marking)	35	Say	→2SA1256	data
E 3	Si-Di	=BAR 99 (Typ-Code/Stempel/marking)	35	Fer	→BAR 99	data
E 3	Si-Di	=KDS 190 (Typ-Code/Stempel/marking)	35	Kec	→KDS 190	data
E 3	Z-Di	=MMSZ 5233B (Typ-Code/marking)	71(2,7mm)	Ons	→MMSZ 5233B	
E 3 P03	MOS-P-FET-e	=NTMS 3P03 (Typ-Code/Stempel/marking)	8-MDIP	Ons	→NTMS 3P03	
E 3 P102	MOS-P-FET-e	=NTMSD3P102(Typ-Code/Stempel/marking)	8-MDIP	Ons	→NTMSD 3P102	
E 3 P303	MOS-P-FET-e	=NTMSD3P303(Typ-Code/Stempel/marking)	8-MDIP	Ons	→NTMSD 3P303	
E 4	Si-P	=2SA1226-4 (Typ-Code/Stempel/marking)	35	Say	→2SA1226	data
E 4	Si-P	=2SA1256-4 (Typ-Code/Stempel/marking)	35	Say	→2SA1256	data
E 4	Si-N	=BFS 17R (Typ-Code/Stempel/marking)	35	Val,Aeg	→BFS 17R	data
E 4	Z-Di	=MMSZ 5234B (Typ-Code/marking)	71(2,7mm)	Ons	→MMSZ 5234B	
E 4 N01	MOS-N-FET-e	=NTMS 4N01 (Typ-Code/Stempel/marking)	8-MDIP	Ons	→NTMS 4N01	
E 4 P01	MOS-P-FET-e	=NTMS 4P01 (Typ-Code/Stempel/marking)	8-MDIP	Ons	→NTMS 4P01	
E 5	Si-P	=2SA1256-5 (Typ-Code/Stempel/marking)	35	Say	→2SA1256	data
E 5	Si-N	=BFS 17AR (Typ-Code/Stempel/marking)	35	Val,Aeg	→BFS 17AR	data
E 5	Z-Di	=MMSZ 5235B (Typ-Code/marking)	71(2,7mm)	Ons	→MMSZ 5235B	
E 5	Z-Di	=UDZ 33B (Typ-Code/Stempel/marking)	71(1,7mm)	Rhm	→UDZ 2.0B...36B	data
E 5 P02	MOS-P-FET-e	=NTMS 5P02 (Typ-Code/Stempel/marking)	8-MDIP	Ons	→NTMS 5P02	
E 6	Si-Di	=ZC 2800E (Typ-Code/Stempel/marking)	35	Fer	→ZC 2800E	data
E 6 N02	MOS-N-FET-e	=NTMD 6N02 (Typ-Code/Stempel/marking)	8-MDIP	Ons	→NTMD 6N02	
E 6 P02	MOS-P-FET-e	=NTMD 6P02 (Typ-Code/Stempel/marking)	8-MDIP	Ons	→NTMD 6P02	
E 7	Si-Di	=ZC 2810E (Typ-Code/Stempel/marking)	35	Fer	→ZC 2810E	data
E 8	Si-Di	=ZC 2811E (Typ-Code/Stempel/marking)	35	Fer	→ZC 2811E	data
E 9	Si-Di	=ZC 5800E (Typ-Code/Stempel/marking)	35	Fer	→ZC 5800E	data
E 10 P02	MOS-P-FET-e	=NTMS 10P02 (Typ-Code/Stempel/marking)	8-MDIP	Ons	→NTMS 10P02	
E 15	Si-P	=BF 747 (Typ-Code/Stempel/marking)	35	Phi	→BF 747	data
E 16	Si-N	=BF 547W (Typ-Code/Stempel/marking)	35(2mm)	Phi	→BF 547W	data
E 16	Si-Di	=STPS 160U (Typ-Code/Stempel/marking)	71(5x3,5)	Tho	→STPS 160U	data
E 21	N-FET	=2SK932-21 (Typ-Code/Stempel/marking)	35	Say	→2SK932	data
E 22	N-FET	=2SK932-22 (Typ-Code/Stempel/marking)	35	Say	→2SK932	data
E 23	N-FET	=2SK932-23 (Typ-Code/Stempel/marking)	35	Say	→2SK932	data
E 24	N-FET	=2SK932-24 (Typ-Code/Stempel/marking)	35	Say	→2SK932	data
E28F020	EEPROM-IC	2048K (256K x 8Bit) CMOS, Flash Memory	TSOP	Int		pdf pinout
E41-15	Si-Di	=ERE 41-15 (Typ-Code/Stempel/marking)	31a	Fjd	→ERE 41-...	data
E 100 C,D	TTL-Logic	=... 8400 (TTL)	14-DIC,DIP	Hfo	... 8400... (TTL)	data
E 0100 AD	F-Thy	100V, 0,5A(Tc=40°C), Igt/Ih<0,02/=5mA	2a	Tag		data
E 0100 FD	F-Thy	=E 0100AD: 60V	2a			data
E 0100 YD	F-Thy	=E 0100AD: 30V	2a			data
E 0102 AA	F-Thy	100V, 0,5A(Tc=40°C), Igt/Ih=0,2/5mA	7e	Tag	BRX 46..47, BRX 51..52, TAG 62...	data
E 0102 FA	F-Thy	=E 0102AA: 60V	7e		BRX 46..47, BRX 51..52, TAG 62...	data
E 0102 YA	F-Thy	=E 0102AA: 30V	7e		BRX 46..47, BRX 51..52, TAG 62...	data
E 0102 AB	F-Thy	100V, 0,5A(Tc=40°C), Igt/Ih=0,2/5mA	7a	Tag	BRX 46..47, BRX 51..52, TAG 62...	data
E 0102 FB	F-Thy	=E 0102AB: 60V	7a		BRX 46..47, BRX 51..52, TAG 62...	data

Type	Device	Short Description	Fig.	Manu	Comparision Types	More at
E 0102 YB	F-Thy	=E 0102AB: 30V	7a		BRX 46...47, BRX 51...52, TAG 62...	data
E 0102 AD	F-Thy	100V, 0,5A(Tc=40°C), Igt/Ih=0,2/5mA	2a	Tag	BRX 46...47, BRX 51...52, TAG 62...	data
E 0102 FD	F-Thy	=E 0102AD: 60V	2a		BRX 46...47, BRX 51...52, TAG 62...	data
E 0102 YD	F-Thy	=E 0102AD: 30V	2a		BRX 46...47, BRX 51...52, TAG 62...	data
E 0102 AG	F-Thy	100V, 0,65A(Tc=85°C), Igt/Ih=0,2/5mA	2a	Tag	BRX 46...47, BRX 51...52, TAG 62...	data
E 0102 FG	F-Thy	=E 0102AG: 60V	2a		BRX 46...47, BRX 51...52, TAG 62...	data
E 0102 YG(9004)	F-Thy	=E 0102AG: 30V	2a		BRX 46...47, BRX 51...52, TAG 62...	data
E 0102 YB	F-Thy	=BR 103	7a	Tag	→BR 103	
E 103 C,D	TTL-Logic	=... 8403 (TTL)	14-DIC,DIP	Hfo	... 8403... (TTL)	data
E 104 C,D	TTL-Logic	=... 8404 (TTL)	14-DIC,DIP	Hfo	... 8404... (TTL)	data
E 108 C,D	TTL-Logic	=... 8408 (TTL)	14-DIC,DIP	Hfo	... 8408... (TTL)	data
E 110 C,D	TTL-Logic	=... 8410 (TTL)	14-DIC,DIP	Hfo	... 8410... (TTL)	data
E 120 C,D	TTL-Logic	=... 8420 (TTL)	14-DIC,DIP	Hfo	... 8420... (TTL)	data
E 121 C,D	TTL-Logic	=... 84121 (TTL)	14-DIC,DIP	Hfo	... 84121... (TTL)	data
E 126 C,D	TTL-Logic	=... 8426 (TTL)	14-DIC,DIP	Hfo	... 8426... (TTL)	data
E 130 C,D	TTL-Logic	=... 8430 (TTL)	14-DIC,DIP	Hfo	... 8430... (TTL)	data
E 140 C,D	TTL-Logic	=. 8440 (TTL)	14-DIC,DIP	Hfo	... 8440... (TTL)	data
E 146 C,D	TTL-Logic	=... 8446 (TTL)	16-DIC,DIP	Hfo	... 8446... (TTL)	data
E 147 C,D	TTL-Logic	=... 8447 (TTL)	16-DIC,DIP	Hfo	... 8447... (TTL)	data
E 150 C,D	TTL-Logic	=... 8450 (TTL)	14-DIC,DIP	Hfo	... 8450... (TTL)	data
E 151 C,D	TTL-Logic	=... 8451 (TTL)	14-DIC,DIP	Hfo	... 8451... (TTL)	data
E 153 C,D	TTL-Logic	=... 8453 (TTL)	14-DIC,DIP	Hfo	... 8453... (TTL)	data
E 154 C,D	TTL-Logic	=... 8454 (TTL)	14-DIC,DIP	Hfo	... 8454... (TTL)	data
E 160 C,D	TTL-Logic	=... 8460 (TTL)	14-DIC,DIP	Hfo	... 8460... (TTL)	data
E 172 C,D	TTL-Logic	=... 8472 (TTL)	14-DIC,DIP	Hfo	... 8472... (TTL)	data
E 174 C,D	TTL-Logic	=... 8474 (TTL)	14-DIC,DIP	Hfo	... 8474... (TTL)	data
E 175 C,D	TTL-Logic	=... 8475 (TTL)	16-DIC,DIP	Hfo	... 8475... (TTL)	data
E 181 C,D	TTL-Logic	=... 8481 (TTL)	14-DIC,DIP	Hfo	... 8481... (TTL)	data
E 191 C,D	TTL-Logic	=... 8491 (TTL)	14-DIC,DIP	Hfo	... 8491... (TTL)	data
E 192 C,D	TTL-Logic	=... 84192 (TTL)	16-DIC,DIP	Hfo	... 84192... (TTL)	data
E 193 C,D	TTL-Logic	=... 84193 (TTL)	16-DIC,DIP	Hfo	... 84193... (TTL)	data
E 195 C,D	TTL-Logic	=... 8495 (TTL)	14-DIC,DIP	Hfo	... 8495... (TTL)	data
E 204 C,D	TTL-Logic	=... 84H04 (TTL)	14-DIC,DIP	Hfo	... 84H04... (TTL)	data
E 274 D	TTL-Logic	=... 84H74 (TTL)	14-DIP	Hfo	... 84H74... (TTL)	data
E 300	N-FET	HF, 25V, Idss>6mA, Up<6V	8b	Nsc,Six	BF 256B...C, BF 348, BFT 10A, 2N5397	data
E 304	N-FET	VHF, 30V, Idss>5mA, Up<6V	8b	Nsc,Six	BFS 80, BFW 11, 2N4416, 2N5245	data
E 305	N-FET	=E 304: Idss>1mA, Up<3V	8b	Nsc,Six	2N5457, 2SK246, 2SK330	data
E 308	N-FET	VHF, 25V, Idss>12mA, Up<6,5V	8b	Nsc	BFT 10B...C, 2SK125	data
E 309	N-FET	=E 308: Up<4V	8b	Nsc		data
E 310	N-FET	=E 308: Idss>24mA	8b	Nsc	BFT 10C, 2SK125	data
E 310 D	LIN-IC	Kfz-blinkgeber/car blinking generator	16-DIP	Hfo	-	
E 311	N-FET	=E 308: Idss<30mA, Up<4V	8b	Nsc		data
E 312	N-FET	=E 308: Idss<60mA	8b	Nsc	BFT 10C, 2SK125	data,
E 345 D	TTL-Logic	BCD→7-Segm.-Decoder, -25...+85°	16-DIP	Hfo	-	
E 346 D	TTL-Logic	BCD→7-Segm.-Decoder, -25...+85°	16-DIP	Hfo	-	
E 347 D	TTL-Logic	BCD→7-Segm.-Decoder, -25...+85°	16-DIP	Hfo	84LS247 (TTL)	
E 348 D	TTL-Logic	BCD→7-Segm.-Decoder, -25...+85°	16-DIP	Hfo	84LS247 (TTL)	
E 351 D	TTL-Logic	Teiler/divider, -25...85°	14-DIP	Hfo	-	
E 355 D	TTL-Logic	Timer, -25...+85°	18-DIP	Hfo	-	
E 356 D	TTL-Logic	Timer, -25...+85°	18-DIP	Hfo	-	
E 412 D	LIN-IC	3x AND-Gate(Treiber/driver, Tri-State)	18-DIP	Hfo	-	
E 435 E	LIN-IC	Leistungstreiber/power driver		Hfo	(FZL 135S)	
E 1007	LIN-IC	=TEA 1007	8-DIP	Aeg	→TEA 1007	
E 1156/1201	MOS-IC	Analogarmbanduhr/analogic watch	Chip		U 117(X)	
E 1617	Si-N	≈BD 115	2a§		→BD 115	
E 1694	Si-P	≈BC 143	2a§		→BC 143	
E 5565	N-FET	≈MPF 111	7e		→MPF 111	
E 6008	Si-P	≈BC 154			→BC 154	
E 7133	Si-P	≈BF 506			→BF 506	
E 7134	Si-P	≈BF 969			→BF 969	
E 7140	MOS-N-FET-d	≈BF 960	25u		→BF 960	
E 7142	Si-P	≈BF 506			→BF 506	
E 7150	Si-P	≈BF 970			→BF 970	
E 7359	MOS-N-FET-d	≈BF 966	25u		→BF 966	
E 7606	Ge-Di	≈2x AA 119			→AA 119	
E-TDA 7478 AD	Dig. Audio	Single Chip Rds Demodulator	TSOP	Stm		pdf pinout

EA...EC

Type	Device	Short Description	Fig.	Manu	Comparision Types	More at
EA	Si-P/N	=µPA504T (Typ-Code/Stempel/marking)	45	Nec	→µPA504T	data
EA	Si-P/N	=µPA574T (Typ-Code/Stempel/marking)	45(2mm)	Nec	→µPA574T	data
EA	Si-P	=2SA1022-A (Typ-Code/Stempel/marking)	35	Mat	→2SA1022	data
EA	Si-P	=2SA1532-A (Typ-Code/Stempel/marking)	35(2mm)	Mat	→2SA1532	data
EA	Si-P	=2SA1790-A (Typ-Code/Stempel/marking)	35(1,6mm)	Mat	→2SA1790	data
EA	Si-N	=2SC4920 (Typ-Code/Stempel/marking)	35(1,6mm)	Say	→2SC4920	data
EA	Si-N	=2SC5415 (Typ-Code/Stempel/marking)	39	Say	→2SC5415	data
EA	Si-N	=2SD1420-EA (Typ-Code/Stempel/marking)	39	Hit	→2SD1420	data
EA	Si-N	=BCW 65A (Typ-Code/Stempel/marking)	35	Fer, Sie	→BCW 65A	data
EA	Si-N	=HC 2A4A (Typ-Code/Stempel/marking)	39	Nec	→HC 2...	data
EA	Z-Di	=MMSZ 4711 (Typ-Code/marking)	71(2,7mm)	Ons	→MMSZ 4711	
EA 03	Si-Di	Schottky, 30V, 1A	31a	Sak	BYV 10-30, MBR 150, SB130, 1N5818...5819	data

Type	Device	Short Description	Fig.	Manu	Comparision Types	More at
EA 961	Si-P	≈2SB681	23a§		→2SB681	
EA 7316	MOS-IC	Digitaluhr/digital clock		Tos	-	
EA 7317 B	MOS-IC	Digitaluhr/digital clock + Datum/date		Tos	-	
EAE	Si-N	=2SC5415-E (Typ-Code/Stempel/marking)	39	Say	→2SC5415	data
EAF	Si-N	=2SC5415-F (Typ-Code/Stempel/marking)	39	Say	→2SC5415	data
EAs	Si-N	=BCW 65A (Typ-Code/Stempel/marking)	35	Sie	→BCW 65A	data
EB	Si-P	=2SA1022-B (Typ-Code/Stempel/marking)	35	Mat	→2SA1022	data
EB	Si-P	=2SA1532-B (Typ-Code/Stempel/marking)	35(2mm)	Mat	→2SA1532	data
EB	Si-P	=2SA1790-B (Typ-Code/Stempel/marking)	35(1,6mm)	Mat	→2SA1790	data
EB	Si-P	=2SA1888-B (Typ-Code/Stempel/marking)	≈35	Mat	→2SA1888	data
EB	Si-N	=2SC5551 (Typ-Code/Stempel/marking)	39	Say	→2SC5551	data
EB	Si-N	=2SD1420-EB (Typ-Code/Stempel/marking)	39	Hit	→2SD1420	data
EB	N-FET	=2SK1842 (Typ-Code/Stempel/marking)	35	Mat	→2SK1842	data
EB	N-FET	=2SK2380 (Typ-Code/Stempel/marking)	35(1,6mm)	Mat	→2SK2380	data
EB	Si-N	=BCW 65B (Typ-Code/Stempel/marking)	35	Fer,Mot,Sie	→BCW 65B	data
EBA	Z-Di	=SMAJ 43CA (Typ-Code/Stempel/marking)	71(5x2,5)	Tho	→SMAJ ...	data
EBE	Si-N	=2SC5551-E (Typ-Code/Stempel/marking)	39	Say	→2SC5551	data
EBF	Si-N	=2SC5551-F (Typ-Code/Stempel/marking)	39	Say	→2SC5551	data
EBF	Z-Di	=SMAJ 58CA (Typ-Code/Stempel/marking)	71(5x2,5)	Tho	→SMAJ ...	data
EBI	Z-Di	=SMAJ 70CA (Typ-Code/Stempel/marking)	71(5x2,5)	Tho	→SMAJ ...	data
EBL	Z-Di	=SMAJ 85CA (Typ-Code/Stempel/marking)	71(5x2,5)	Tho	→SMAJ ...	data
EBN	Z-Di	=SMAJ 100CA (Typ-Code/Stempel/marking)	71(5x2,5)	Tho	→SMAJ ...	data
EB P	N-FET	=2SK1842-P: (Typ-Code/Stempel/marking)	35	Sam	→2SK1842	data
EB Q	N-FET	=2SK1842-Q: (Typ-Code/Stempel/marking)	35	Sam	→2SK1842	data
EB Q	N-FET	=2SK2380-Q: (Typ-Code/Stempel/marking)	35(1,6mm)	Mat	→2SK2380	data
EBQ	Z-Di	=SMAJ 130CA (Typ-Code/Stempel/marking)	71(5x2,5)	Tho	→SMAJ ...	data
EB R	N-FET	=2SK1842-R: (Typ-Code/Stempel/marking)	35	Sam	→2SK1842	data
EB R	N-FET	=2SK2380-R: (Typ-Code/Stempel/marking)	35(1,6mm)	Mat	→2SK2380	data
EB S	N-FET	=2SK1842-S: (Typ-Code/Stempel/marking)	35	Sam	→2SK1842	data
EB S	N-FET	=2SK2380-S: (Typ-Code/Stempel/marking)	35(1,6mm)	Mat	→2SK2380	data
EBs	Si-N	=BCW 65B (Typ-Code/Stempel/marking)	35	Sie	→BCW 65B	data
EBT	Z-Di	=SMAJ 154CA (Typ-Code/Stempel/marking)	71(5x2,5)	Tho	→SMAJ ...	data
EBV	Z-Di	=SMAJ 188CA (Typ-Code/Stempel/marking)	71(5x2,5)	Tho	→SMAJ ...	data
EC	Si-P	=2SA1022-C (Typ-Code/Stempel/marking)	35	Mat	→2SA1022	data
EC	Si-P	=2SA1368-C (Typ-Code/Stempel/marking)	39	Mit	→2SA1368	data
EC	Si-P	=2SA1532-C (Typ-Code/Stempel/marking)	35(2mm)	Mat	→2SA1532	data
EC	Si-P	=2SA1790-C (Typ-Code/Stempel/marking)	35(1,6mm)	Mat	→2SA1790	data
EC	Si-P	=2SA1888-C (Typ-Code/Stempel/marking)	≈35	Mat	→2SA1888	data
EC	Si-N	=2SC2732 (Typ-Code/Stempel/marking)	35	Hit	→2SC2732	data
EC	Si-N	=2SC4462 (Typ-Code/Stempel/marking)	35(2mm)	Hit	→2SC4422	data
EC	Si-N	=2SD1420-EC (Typ-Code/Stempel/marking)	39	Hit	→2SD1420	data
EC	MOS-N-FET-d	=3SK247 (Typ-Code/Stempel/marking)	44	Mat	→3SK247	data
EC	Si-N	=BCW 65C (Typ-Code/Stempel/marking)	35	Fer,Mot,Sie	→BCW 65C	data
EC	Z-Di	=MMSZ 4712 (Typ-Code/marking)	71(2,7mm)	Ons	→MMSZ 4712	data
EC 3 H01B	Si-N	SMD, VHF ra, 20V, 50mA, 5GHz	Chip	Say	-	data
EC 3 H02C	Si-N	SMD, VHF/UHF ra, 20V, 70mA, 7GHz	Chip	Say	-	data
EC 3 H03B	Si-N	SMD, VHF/UHF ra, 20V, 100mA, 7,5GHz	Chip	Say	-	data
EC 3 H04B	Si-N	SMD, VHF/UHF ra, 9V, 100mA, 8GHz	Chip	Say	-	data
EC 3 H04C	Si-N	SMD, VHF/UHF ra, 9V, 100mA, 8GHz	Chip	Say	-	data
EC 3 H05B	Si-N	SMD, VHF/UHF ra, 16V, 50mA, 9GHz	Chip	Say	-	data
EC 3 H07B	Si-N	SMD, UHF/SHF ra, 9V, 30mA, 12,5GHz	Chip	Say	-	data
EC 961	Si-N	≈2SD551	23a§		→2SD551	
ECG 1081 A	LIN-IC	4.8W Audio Power Amp	SIL	Ecg	-	pdf pinout
ECG 1321	LIN-IC	55W AF Power Amp	SIL	Ecg	-	pdf
ECG 1328	LIN-IC	50W AF Power Amp	HYB	Ecg	-	pdf
ECG 1883	Hybrid-Z-IC	VCR, +12,1V/0,8A, +12V/0,8A, +5,3V/1A	HYB(44x25)	Nte	NTE1883,STK5372,STK5372H	pdf
ECG 2043	IC	12 Note Top Octave Synthesizer	DIP		MO86B,M50242,S50242,50242,AY-1-0212	pinout
ECO 0100..9999	Si-Di, Z-Di	→0100...9999		Ntn		
ECs	Si-N	=BCW 65C (Typ-Code/Stempel/marking)	35	Sie	→BCW 65C	data
ED						
ED	Si-P	=BCV 28 (Typ-Code/Stempel/marking)	39	Phi, Sie	→BCV 28	data
ED	Z-Di	=MMSZ 4713 (Typ-Code/marking)	71(2,7mm)	Ons	→MMSZ 4713	
ED	Z-Di	=SM 6T 18 (Typ-Code/Stempel/marking)	71(6x4mm)	Tho	→SM 6T...	data
ED	Z-Di	=SMBJ 110C (Typ-Code/Stempel/marking)	71(5x3,5)	Mop	→SMBJ ...	data
ED	Si-N/P	=XP 4654 (Typ-Code/Stempel/marking)	46(2mm)	Mat	→XP 4654	data
ED	Z-Di	=Z1 SMA-7V5 (Typ-Code/Stempel/marking)	71(5x2,5)	Fag	→Z1 SMA-7V5	data
ED	Si-P	=2SA1368-D (Typ-Code/Stempel/marking)	39	Mit	→2SA1368	data
ED	Si-N	=2SD1421-ED (Typ-Code/Stempel/marking)	39	Hit	→2SD1421	data
ED 3 P03	MOS-P-FET-e	=NTMD 3P03 (Typ-Code/Stempel/marking)	8-MDIP	Ons	→NTMD 3P03	
ED 592	Si-N	≈BF 254	7d		→BF 254	
ED 1401(A...E)	Si-N	≈ED 1402	7e	Nsc	→ED 1402	data
ED 1402(A...E)	Si-N	Uni, 28V, 0,2A, 0,5W, >100MHz	7e	Nsc	BC 183, BC 238, BC 548, 2SC1815, 2SD 767	data
		Philips:	7a	Phi		
ED 1501(A...E)	Si-N	≈ED 1502	7e	Nsc	→ED 1502	data
ED 1502(A...E)	Si-N	AM/FM, 40V, 25mA, 0,5W, >361MHz	7e	Nsc	BF 240...241, BF 254...255, BF 594...595,++	data
		Philips:	7a	Phi		
ED 1601(A...E)	Si-P	≈ED 1602	7e	Nsc	→ED 1602	data
ED 1602(A...F)	Si-P	Uni, 25V, 100mA, 0,5W, >100MHz	7e	Nsc	BC 309, BC 214, BC 559, 2SA1137	data
		Philips:	7a	Phi		
ED 1701(K..N)	Si-N	≈ED 1702	7e	Nsc	→ED 1702	data

345

Type	Device	Short Description	Fig.	Manu	Comparision Types	More at
ED 1702(K..N)	Si-N	NF-Tr/E, 30V, 0,5A, 0,625W, >80MHz	7e	Nsc	BC 337...338, BC 635, 2SD1225...1226, ++	data
		Philips:	7a	Phi		
ED 1801(K..N)	Si-P	≈ED 1802	7e	Nsc	→ED 1802	data
ED 1802(K..N)	Si-P	NF-Tr/E, 30V, 0,5A, 0,625W, >80MHz	7e	Nsc	BC 327...328, BC 636, 2SA1426, 2SB909, ++	data
		Philips:	7a	Phi		
ED 2502	Si-N	≈BF 255	7a		→BF 255	
ED 8050	Si-N	=SS 8050	7e		→SS 8050	
ED 8550	Si-P	=SS 8550	7e		→SS 8550	
6 ED 003 L 06-F	LIN-IC	Integrated 3 Phase Gate Driver	16-MDIP	Inf	-	pdf pinout
EDD 1014	Si-Di	=4x BAW 56, common Anode	9-SIP		-	data
EDD 1015	Si-Di	=4x BAV 56, common Cathode	9-SIP		-	data
EDF 1 AM...DM	Si-Br	Gl-Br, S, 50..200V, 1A, <50ns	4-DIP	Gsi	-	data
		AM=50V, BM=100V, CM=150V, DM=200V				
EDF 1 AS...DS	Si-Br	Gl-Br, S, 50..200V, 1A, <50ns	4-DIP	Gsi		data
		AS=50V, BS=100V, CS=150V, DS=200V				
EDZ 4.7B..6.8B	Z-Di	SMD, 4.7..6.8V, 5%, 0,1W	71(1,3mm)	Rhm	-	data
EE	Si-P	=2SA1368-E (Typ-Code/Stempel/marking)	39	Mit	→2SA1368	data
EE	Si-N	=2SD1421-EE (Typ-Code/Stempel/marking)	39	Hit	→2SD1421	data
EE	Si-P	=BCV 48 (Typ-Code/Stempel/marking)	39	Phi, Sie	→BCV 48	data
EE	Z-Di	=MMSZ 4714 (Typ-Code/marking)	71(2,7mm)	Ons	→MMSZ 4714	
EE	Z-Di	=SM 6T 18A (Typ-Code/Stempel/marking)	71(6x4mm)	Tho	→SM 6T...	data
EE	Z-Di	=SMBJ 110CA (Typ-Code/Stempel/marking)	71(5x3,5)	Mop	→SMBJ ...	data
EE	Z-Di	=Z1 SMA-6V8 (Typ-Code/Stempel/marking)	71(5x2,5)	Fag	→Z1 SMA-6V8	data

EF

Type	Device	Short Description	Fig.	Manu	Comparision Types	More at
EF	Si-N	=BCV 29 (Typ-Code/Stempel/marking)	39	Phi, Sie	→BCV 29	data
EF	Si-N	=BCW 66F (Typ-Code/Stempel/marking)	35	Fer, Sie	→BCW 66F	data
EF	Z-Di	=MMSZ 4715 (Typ-Code/marking)	71(2,7mm)	Ons	→MMSZ 4715	
EF	Z-Di	=SMBJ 120C (Typ-Code/marking)	71(5x3,5)	Mop	→SMBJ ...	data
EF	Z-Di	=Z1 SMA-8V2 (Typ-Code/Stempel/marking)	71(5x2,5)	Fag	→Z1 SMA-8V2	data
EF 68HC...	µC-IC	8 Bit µcomp, Family 6800...		Tho	-	
EF 4440	NMOS-IC	µcomp, Peripherie Interface (arinc)	28-DIP	Tho	-	
EF 4442	NMOS-IC	µcomp, Peripherie(ARINC) f.6800,6802µC	28-DIP	Tho	-	pdf
EF 4443 DP	NMOS-IC	Drehzahlregler/motor speed control	16-DIP	Tho	-	
EF 6800..6854	µC-IC	8 Bit µcomp, Family 6800..., Peripherie		Tho	-	
EF 6843	I/O-IC	Floppy Disk Controller	40-DIP	Tho	-	
EF 6844	I/O-IC	DMA Controller, 1MByte/s, 4Kan./chann.	40-DIP	Tho	-	
EF 6845	I/O-IC	Crt Controller, 1MByte/s, 4Kan./chann.	40-DIP	Tho	-	
EF 6846(C1)	MOS-IC	Combo Chip, 2k ROM, Timer	40-DIP	Tho	-	
EF 6862	I/O-IC	Digital Modulator, ser., 1200/2400bps	24-DIP	Tho	-	
EF 7331	NMOS-IC	Telecom, Zeitmatrix, Register etc.	28-DIP	Tho	-	
EF 7332	NMOS-IC	Telecom, Taktgenerg. 2,048MHz	16-DIP	Tho	-	
EF 7333	NMOS-IC	Telecom, Kommunikation, 2,048MHz	24-DIP	Tho	-	
EF 7442	NMOS-IC	Telecom, Zeitbasis/time base, Multipl.	16-DIP	Tho	-	
EF 7445	NMOS-IC	Telecom, Encoder-decoder-strg./ctrl.	24-DIP	Tho	-	
EF 7910	NMOS-IC	Telecom, Modem, AMD 7910-kompatibel	28-DIP	Tho	-	
EF 8307	A/D-IC	Telecom, A/D-Converter, 7 Bit, 20MHz	24-DIP	Tho	-	
EF 8308	A/D-IC	Telecom, A/D-Converter, 8 Bit, 20MHz	24-DIP	Tho	-	
EF 8408	D/A-IC	Telecom, A/D-Converter, 8 Bit, 20MHz	16-DIP	Tho	-	
EF 9241	NMOS-IC	Demultiplexer (ceefax, Didon, Antiope)	40-DIP	Tho	-	
EF 9340	MOS-IC	Semigraphik Display-prozessor	40-DIP	Tho	-	
EF 9341	MOS-IC	Zeichengenerator/character generator	40-DIP	Tho	-	
EF 9345 FN	MOS-IC	=EF 9345P: Fig.→	44-PLCC	Tho	-	
EF 9345 P	HMOS-IC	Semigraphik Display-prozessor	40-DIP	Tho	-	
EF 9364 A	I/O-IC	Video-Controller, 50Hz/625Zeilen/lines	24-DIP	Tho	-	
EF 9364 B	I/O-IC	=EF 9364A: 60Hz/525Zeilen/lines	24-DIP	Tho	-	
EF 9365	MOS-IC	Graphik Display-prozessor	40-DIP	Tho	-	
EF 9366	MOS-IC	Graphik Display-prozessor	40-DIP	Tho	-	
EF 9367 P, P3	MOS-IC	Graphik Display-prozessor	40-DIP	Tho	-	
EF 9368	MOS-IC	Graphik Display-prozessor	64-FLP	Tho	-	
EF 9369 FN	MOS-IC	=EF 9369P: Fig.→	28-PLCC	Tho	-	
EF 9369 P	MOS-IC	Color-palette-decoder	28-DIP	Tho	-	
EF 68000...68901	µC-IC	µcomp, Family 6800..., Peripherie etc.		Tho	-	
EF 84108	D/A-IC	Telecom, A/D-Converter, 8 Bit, 5MHz	16-DIP	Tho	-	
EFB 4443	NMOS-IC	Drehzahlregler/motor speed control	16-DIP	Tho	-	
EFB 7189	CMOS-IC	Telecom, Frequ.-gener., µComp.-interf.	16-DIP	Tho	-	
EFB 7303	CMOS-IC	Telecom, ser./par. Converter, 8/8 Bit	16-DIP	Tho	-	
EFB 7310	CMOS-IC	Telecom, 4x4x1 Matrix f. analog Signal	16-DIP	Tho	-	
EFB 7334	CMOS-IC	Telecom, Demultipl. f. 8xEF7331,EF7333	24-DIP	Tho	-	
EFB 7335	CMOS-IC	Telecom, 1 Kanal/Channel Signalling C.	24-DIP	Tho	-	
EFB 7336	CMOS-IC	Telecom, 8 Kanal/Channel Controller	22-DIP	Tho	-	
EFB 7356	CMOS-IC	Telecom, 1 Kanal/Channel Codec	22-DIP	Tho	-	
EFB 7360	CMOS-IC	Telecom, 1 Kanal/Channel Codec, Filter	16/18-DIP	Tho	-	
EFB 7441	CMOS-IC	Telecom, Zeitbasis/time base, Multipl.	28-DIP	Tho	-	
EFB 7443	CMOS-IC	Telecom, Encoder, TMS 3863 kompat.	28-DIP	Tho	-	
EFB 7444	CMOS-IC	Telecom, Decoder, A-law	28-DIP	Tho	-	
EFB 7446	CMOS-IC	Telecom, Bin.→HDB3-Transcoder	16-DIP	Tho	-	
EFB 7447	CMOS-IC	Telecom, 1/16-Multiplexer	28-DIP	Tho	-	
EFB 7510	CMOS-IC	Telecom, FSK-Modem, 75/1200Baud, V.23	18-DIP	Tho	-	
EFB 7512	CMOS-IC	Telecom, FSK-Modem, 75/1200Baud, V.23	22-DIP	Tho	-	
EFB 7513	CMOS-IC	Telecom, FSK-Modem, 75/1200Baud, V.23	22-DIP	Tho	-	

Type	Device	Short Description	Fig.	Manu	Comparision Types	More at
EFB 7910	NMOS-IC	Telecom, FSK-Modem, Ccitt, Bell	28-DIP	Tho	-	
EFB 7912	CMOS-IC	Telecom, Filter f. 1Kanal/chann. Codec	16-DIP	Tho	-	
EFB 8305 A	CMOS-IC	Telecom, Programm. Timer	16-DIP	Tho	-	
EFB 9151	CMOS-IC	Telecom, Tastwahl/key dialing	18-DIP	Tho	-	
EFB 9158	CMOS-IC	Telecom, Tastwahl/key dialing	22-DIP	Tho	-	
EFD 108	Ge-Di	≈SFD 108	31a		→SFD 108	
EFD 7130	PMOS-IC	Telecom, Telefonnummernbeschränkung	28-DIP	Tho	-	
EFG ...0	CMOS-IC	Gate-arrays, kundenspez./customized	28-DIP	Tho	-	
EFG 850 XY	CMOS-IC	Telecom, Switched-capacitor-filter	8/16-DIP	Tho	-	
EFG 7189(PD)	CMOS-IC	Telecom, DTMF, par. Interface	14-DIP	Tho	-	pinout
EFG 7515	CMOS-IC	Telecom, DPSK-Modem, V.22, Bell 212A	28-DIP	Tho	-	
EFG 71891(PD)	CMOS-IC	Telecom, Dtmf, ser. Interface	8-DIP	Tho	-	pdf pinout
EFs	Si-N	=BCW 66F (Typ-Code/Stempel/marking)	35	Sie	→BCW 66F	data
EFV 1	Si-Di	≈BY 228	31a		→BY 228	
EFZ ...	ECL/TTL-IC	Gate-arrays, kundenspez./customized		Tho		
EG...EK						
EG	Si-N	=BCV 49 (Typ-Code/Stempel/marking)	39	Phi, Sie	→BCV 49	data
EG	Si-N	=BCW 66G (Typ-Code/Stempel/marking)	35	Fer, Sie	→BCW 66G	data
EG	Z-Di	=SMBJ 120CA (Typ-Code/Stempel/marking)	71(5x3,5)	Mop	→SMBJ ...	data
EG	Z-Di	=Z1 SMA-9V1 (Typ-Code/Stempel/marking)	71(5x2,5)	Fag	→Z1 SMA-9V1	data
EG 01	Si-Di	GI/S, 400V, 0,7A, 100ns	31a	Sak	BYV 26B...E, BYV 36B...E, RGP 10G...M,++	data
EG 01 A	Si-Di	GI/S, 600V, 0,5A, 100ns	31a	Sak	BYV 26C...E, BYV 36C...E, RGP 10J...M,++	data
EG 01 C	Si-Di	GI/S, 1000V, 0,5A, 100ns	31a	Sak	BYV 26E, BYV 36E, BYT 52M, RGP 10M,++	data
EG 01 Y	Si-Di	GI/S, 70V, 1A, 100ns	31a	Sak	BYV 26B...E, BYV 36A...E, RGP 10A...M,++	data
EG 01 Z	Si-Di	GI/S, 200V, 0,7A, 100ns	31a	Sak	BYV 26B...E, BYV 36A...E, RGP 10D...M,++	data
EG 1	Si-Di	GI/S, 400V, 0,8A, 100ns	31a	Sak	BYV 26B...E, BYV 36B...E, RGP 10G...M,++	data
EG 1 A	Si-Di	GI/S, 600V, 0,6A, 100ns	31a	Sak	BYV 26C...E, BYV 36C...E, RGP 10J...M,++	data
EG 1 Y	Si-Di	GI/S, 70V, 1,1A, 100ns	31a	Sak	BYV 26B...E, BYV 36A...E, RGP 10A...M,++	data
EG 1 Z	Si-Di	GI/S, 200V, 0,8A, 100ns	31a	Sak	BYV 26B...E, BYV 36A...E, RGP 10D...M,++	data
EGF 1 A...D	Si-Di	SMD, GI, S, 50...200V, 1A, <50ns A=50V, B=100V, C=150V, D=200V	71a(5x2,7)	Gsi	BYG 80A...D, D 1FL20	data
EGL 27 A...G	Si-Di	=BYM 32-...		Gie	→BYM 32-...	data
EGL 34 A...G	Si-Di	=BYM 07-..., A=50, B=100, C=150V D=200, F=300, G=400V		Gie	→BYM 07-...	data
EGL 41 A...G	Si-Di	=BYM 12-..., A=50, B=100, C=150V D=200, F=300, G=400V		Gie	→BYM 12-...	data
EGP 10 A...G	Si-Di	GI, S, 50...400V, 1A, <50ns A=50, B=100, C=150, D=200,F=300,G=400V	31a	Gie, Fag	BYV 26B...E, FE 1A...H	data
EGP 20 A...G	Si-Di	GI, S, 50...400V, 2A, <50ns A=50, B=100, C=150, D=200,F=300,G=400V	31a	Gie, Fag	BYD 74A...G, BYV 27-..., FE 2A...H	data
EGP 30 A...G	Si-Di	GI, S, 50...400V, 3A, <50ns A=50, B=100, C=150, D=200,F=300,G=400V	31a	Gie, Fag	BYV 28-..., FE 3A...H	data
EGP 50 A...G	Si-Di	GI, S, 50...400V, 5A, <50ns A=50, B=100, C=150, D=200,F=300,G=400V	31a	Gie, Fag	BYV 61...63, FE 5A...D, FE 6A...H	data
EGs	Si-N	=BCW 66G (Typ-Code/Stempel/marking)	35	Sie	→BCW 66G	data
EH	Si-P	=2SB1027-EH (Typ-Code/Stempel/marking)	39	Hit	→2SB1027	data
EH	Si-N	=BCW 66H (Typ-Code/Stempel/marking)	35	Fer, Sie	→BCW 66H	data
EH	Z-Di	=MMSZ 4716 (Typ-Code/marking)	71(2,7mm)	Ons	→MMSZ 4716	data
EH	Z-Di	=SM 6T 22 (Typ-Code/Stempel/marking)	71(6x4mm)	Tho	→SM 6T...	data
EH	Z-Di	=SMBJ 130C (Typ-Code/Stempel/marking)	71(5x3,5)	Mop	→SMBJ ...	data
EH	Z-Di	=Z1 SMA-10 (Typ-Code/marking)	71(5x2,5)	Fag	→Z1 SMA-10	data
EH 1(A,Z)	Si-Di	GI, S, 200...600V, 0,6A, 4µs EH 1=400V, A=600V, Z=200V	31a	Sak	BY 126...127, BY 133...134, 1N4003...07,++	data
EHA2505	OP	Vs:40V Vu:>10V/mV Vo:±12V Vi0:<10mV	8AD	Ela		data pinout
EHA2515	OP	Vs:40V Vu:>5V/mV Vo:±12V Vi0:<14mV	8D	Ela		data pinout
EHA2525	OP	Vs:40V Vu:>5V/mV Vo:±12V Vi0:<14mV	8AD	Ela		data pinout
EHA2605	HI-SPEED	Vs:45V Vu:>70V/mV Vo:±12V Vi0:<7mV	8AD	Ela		data pinout
EHA2625	HI-SPEED	Vs:45V Vu:>70V/mV Vo:±12V Vi0:<7mV	14D,8AD	Ela		data pinout
EHD-RD 3053 N,NA	Hybrid-Z-IC	-5V, 1A		Mat	-	
EHD-RD 3053 PA	Hybrid-Z-IC	+5V, 1A		Mat	-	
EHD-RD 3053 R,RA	Hybrid-Z-IC	+5V, 1A		Mat	-	
EHD-RD 3053 S	Hybrid-Z-IC	+5V, 1A		Mat	-	
EHD-RD 3053 V	Hybrid-Z-IC	+5V, 3A		Mat	-	
EHD-RD 3093 PA	Hybrid-Z-IC	+9V, 1A		Mat	-	
EHD-RD 3093 R,RA	Hybrid-Z-IC	+9V, 1A		Mat	-	
EHD-RD 3093 S	Hybrid-Z-IC	+9V, 1A		Mat	-	
EHD-RD 3123 N,NA	Hybrid-Z-IC	-12V, 1A		Mat	-	
EHD-RD 3123 PA	Hybrid-Z-IC	+12V, 1A		Mat	-	
EHD-RD 3123 R,RA	Hybrid-Z-IC	+12V, 1A		Mat	-	
EHD-RD 3123 S	Hybrid-Z-IC	+12V, 1A		Mat	-	
EHD-RD 3123 V	Hybrid-Z-IC	+12V, 3A		Mat	-	
EHD-RD 3153 N,NA	Hybrid-Z-IC	-15V, 1A		Mat	-	
EHD-RD 3153 PA	Hybrid-Z-IC	+15V, 1A		Mat	-	
EHD-RD 3153 R,RA	Hybrid-Z-IC	+15V, 1A		Mat	-	
EHD-RD 3153 S	Hybrid-Z-IC	+15V, 1A		Mat	-	
EHD-RD 3183 PA	Hybrid-Z-IC	+18V, 1A		Mat	-	
EHD-RD 3183 R,RA	Hybrid-Z-IC	+18V, 1A		Mat	-	
EHD-RD 3183 S	Hybrid-Z-IC	+18V, 1A		Mat	-	

Type	Device	Short Description	Fig.	Manu	Comparision Types	More at
EHD-RD 3243 N,NA	Hybrid-Z-IC	-24V, 1A		Mat	-	
EHD-RD 3243 PA	Hybrid-Z-IC	+24V, 1A		Mat	-	
EHD-RD 3243 R,RA	Hybrid-Z-IC	+24V, 1A		Mat	-	
EHD-RD 3243 S	Hybrid-Z-IC	+24V, 1A		Mat	-	
EHs	Si-N	=BCW 66H (Typ-Code/Stempel/marking)	35	Sie	→BCW 66H	data
EI	Si-P+R	=UN 211M (Typ-Code/Stempel/marking)	35	Mat	→UN 211M	data
EI	Si-P+R	=UN 511M (Typ-Code/Stempel/marking)	35(2mm)	Mat	→UN 511M	data
EI-	MOS-N-FET-d	=3SK182 (Typ-Code/Stempel/marking)	44	Hit	→3SK182	data
EJ	Si-P	=2SB1027-EJ (Typ-Code/Stempel/marking)	39	Hit	→2SB1027	data
EJ	Z-Di	=MMSZ 4717 (Typ-Code/marking)	71(2,7mm)	Ons	→MMSZ 4717	
EJ 4	N-FET	=3SK181-4 (Typ-Code/Stempel/marking)	44	Say	→3SK181	
EJ 5	N-FET	=3SK181-5 (Typ-Code/Stempel/marking)	44	Say	→3SK181	data
EJ 6	N-FET	=3SK181-6 (Typ-Code/Stempel/marking)	44	Say	→3SK181	data
EJ 5027	Si-N	≈2N3055	23a§		→2N3055	
EK	Si-P	=2SB1027-EK (Typ-Code/Stempel/marking)	39	Hit	→2SB1027	data
EK	Si-N	=2SD1001-EK (Typ-Code/Stempel/marking)	39	Nec	→2SD1001	data
EK	Si-N	=BCX 41 (Typ-Code/Stempel/marking)	35	Sie, Tho	→BCX 41	data
EK	Z-Di	=SM 6T 22A (Typ-Code/Stempel/marking)	71(6x4mm)	Tho	→SM 6T...	data
EK	Z-Di	=SMBJ 130CA (Typ-Code/Stempel/marking)	71(5x3,5)	Mop	→SMBJ ...	data
EK	Si-P+R	=XN 111M (Typ-Code/Stempel/marking)	45	Mat	→XN 111M	data
EK	Si-P+R	=XP 0111M (Typ-Code/Stempel/marking)	45(2mm)	Mat	→XP 0111M	data
EK	Z-Di	=Z1 SMA-11 (Typ-Code/Stempel/marking)	71(5x2,5)	Fag	→Z1 SMA-11	data
EK 02	Si-Di	Schottky, S, 20V, 1A, 100ns	31a	Sak	-	data
EK 03	Si-Di	Schottky, S, 30V, 1A, 200ns	31a	Sak	-	data
EK 04	Si-Di	Schottky, S, 40V, 1A, 200ns	31a	Sak	-	data
EK 06	Si-Di	Schottky, S, 60V, 0,7A, 100ns	31a	Sak	-	data
EK 09	Si-Di	Schottky, S, 90V, 0,7A, 100ns	31a	Sak	-	data
EK 12	Si-Di	Schottky, S, 20V, 2A, 100ns	31a	Sak	MBR 320...390, SB 320...390, 1N5820...5822	data
EK 13	Si-Di	Schottky, S, 30V, 1,5A, 200ns	31a	Sak	-	data
EK 14	Si-Di	Schottky, S, 40V, 1,5A, 200ns	31a	Sak	-	data
EK 16	Si-Di	Schottky, S, 60V, 1,5A, 100ns	31a	Sak	-	data
EK 19	Si-Di	Schottky, S, 90V, 1,5A, 100ns	31a	Sak	-	data
EKs	Si-N	=BCX 41 (Typ-Code/Stempel/marking)	35	Sie	→BCX 41	data

EL

Type	Device	Short Description	Fig.	Manu	Comparision Types	More at
EL	Si-P	=2SA1344 (Typ-Code/Stempel/marking)	35	Say	→2SA1344	data
EL	Si-P	=2SA1678 (Typ-Code/Stempel/marking)	35(2mm)	Say	→2SA1678	data
EL	Si-P	=2SB1028-EL (Typ-Code/Stempel/marking)	39	Hit	→2SB1028	data
EL	Si-N	=2SD1001-EL (Typ-Code/Stempel/marking)	39	Nec	→2SD1001	data
EL	Z-Di	=SM 6T 24 (Typ-Code/Stempel/marking)	71(6x4mm)	Tho	→SM 6T...	data
EL	Z-Di	=SMBJ 150C (Typ-Code/Stempel/marking)	71(5x3,5)	Mop	→SMBJ ...	data
EL	Si-N+R	=UN 221M (Typ-Code/Stempel/marking)	35	Mat	→UN 221M	data
EL	Si-N+R	=UN 521M (Typ-Code/Stempel/marking)	35(2mm)	Mat	→UN 521M	data
EL	Si-N+R	=UNR 921M (Typ-Code/Stempel/marking)	35(1,6mm)	Mat	→UNR 921M	data
EL	Z-Di	=Z1 SMA-12 (Typ-Code/Stempel/marking)	71(5x2,5)	Fag	→Z1 SMA-12	data
EL 1	Si-Di	GI/S, 350V, 1,5A, 50ns	31a	Sak	BYD 73G, EGP 20G, FE 2H	data
EL 1 Z	Si-Di	GI/S, 200V, 1,5A, 50ns	31a	Sak	BYD 73D...G, EGP 20D...G, FE 2D...H	data
EL 02 Z	Si-Di	GI/S, 200V, 1,5A, 40ns	31a	Sak	BYD 73D...G, EGP 20D...G, FE 2D...H	data
EL 133	N-FET	≈BF 245	7f		→BF 245	
EL 220/7128	Si-N	≈2N3856			→2N3856	
EL 400(C)...	OP-IC	±5<7V, ±70mA, 200MHz, 700V/µs, 12ns	8-DIP	Ela		
EL400C	FB-AMP	Vs:±7V Vo:3.4/0V Vio:2mV f:200Mc	8DS	Ela		
EL 692	Si-P	≈BC 393			→BC 393	data pdf pinout
EL2001	BUFF	Vs:±18V Vo:±11V Vio:2mV f:70Mc	8D,20S	Ela		data pdf pinout
EL2002A	BUFF	Vs:±18V Vo:±11V Vio:5mV f:180Mc	8D	Ela		data pdf pinout
EL2002C	BUFF	Vs:±18V Vo:±11V Vio:10mV f:180Mc	20S,8D	Ela		data pdf pinout
EL2003C	VIDEO-DRV	Vs:±18V Vo:±13.5V Vio:5mV f:100Mc	20S,8D	Ela		data pdf pinout
EL2005CG	FAST-BUFF	Vs:±20V Vo:12.5/0V Vio:3mV f:140Mc	12A	Max		data pdf pinout
EL 2005 CG	OP-)	→MAX 460IGC		Ela	-	data
EL2005G	FAST-BUFF	Vs:±20V Vo:12.5/0V Vio:2mV f:140Mc	12A	Max		data pdf pinout
EL 2005 G	OP-IC	→MAX 460MGC		Ela		data
EL2008CT	BUFF	Vs:±18V Vo:±11V Vio:10mV f:55Mc	5P	Ela		data pdf pinout
EL2009CT	BUFF	Vs:±18V Vo:±11V Vio:±60mV f:90Mc	5P	Ela		data pdf pinout
EL 2020(C)...	OP-IC	±5...18V, ±32,5mA, 50MHz, 500V/µs, 50ns	8-DIP	Ela		
EL2020C	FB-AMP	Vs:±18V Vu:80dB Vo:±13V Vio:3mV	20S,8D	Ela		data pdf pinout
EL 2022(C)...	OP-IC	±5...15V, ±100mA, 165MHz, 1900V/µs,22ns	83	Ela		
EL 2030(C)...	OP-IC	±5...18V, ±65mA, 120MHz, 2000V/µs, 40ns	8-DIP	Ela	-	
EL2030C	FB-AMP	Vs:±18V Vu:70dB Vo:±13V Vio:10mV	20S,8D	Ela		data pdf pinout
EL2033CN	VIDEO-DRV	Vs:±18V Vo:±13.5V Vio:10mV f:100Mc	8D	Ela		data pdf pinout
EL2038	HI-FREQ	Vs:35V Vo:±12V Vio:0.5mV f:1.1Gmc	14D,20C	Ela		data pdf pinout
EL2044C	OP	Vs:36V Vu:1.5V/mV Vo:±13.6V f:60Mc	8D	Ela		data pinout
EL2044CN	HI-SPEED	Vs:±18V Vu:1500V/mV Vo:±13.6V	8D	Ela		data pdf pinout
EL2044CS	HI-SPEED	Vs:±18V Vu:1500V/mV Vo:±13.6V	8S	Ela		data pdf pinout
EL2045C	HI-SPEED	Vs:±18V Vu:3000V/mV Vo:±13.6V	8DS	Ela		data pdf pinout
EL 2070(C)...	OP-IC	±5<7V, ±70mA, 200MHz, 700V/µs, 12ns	8-DIP	Ela		
EL2070C	FB-AMP	Vs:±7V Vo:3.4/0V Vio:2mV f:200Mc	8DS	Ela		data pdf pinout
EL 2071(C)...	OP-IC	±5<7V, ±70mA, 150MHz, 1200V/µs, 10ns	8-DIP	Ela	-	
EL2071C	FB-AMP	Vs:±7V Vo:3.4/0V Vio:3mV f:150Mc	8DS	Ela		data pinout
EL2072C	BUFF	Vs:±7V Vo:±4V Vio:2mV f:730Mc	8DS	Ela		data pdf pinout

Type	Device	Short Description	Fig.	Manu	Comparision Types	More at
L2073	OP	Vs:±7V Vu:1V/mV Vo:±4V Vi0:0.2mV	8D	Ela		data pdf pinout
L2073C	OP	Vs:±7V Vu:1V/mV Vo:±4V Vi0:0.2mV	8D	Ela		data pdf pinout
L2073CN	FB-AMP	Vs:±7V Vu:1000V/mV Vo:±3.6V f:200Mc	8D	Ela		data pdf pinout
L2073CS	FB-AMP	Vs:±7V Vu:1000V/mV Vo:±3.6V f:200Mc	8S	Ela		data pdf pinout
L2074C	FB-AMP	Vs:±7V Vu:1000V/mV Vo:±3.6V f:400Mc	8DS	Ela		data pdf pinout
L2075C	FB-AMP	Vs:±7V Vu:2800V/mV Vo:±3.6V f:2000Mc	8DS	Ela		data pdf pinout
L2099CT	VIDEO-AMP	Vs:±16.5V Vo:±11V Vi0:5mV f:50Mc	5P	Ela		data pdf pinout
L 2120(C)...	OP-IC	±5...15V, 100MHz, 750V/µs, 50ns	8-DIP	Ela		
L2120C	FB-AMP	Vs:±16.5V Vu:66dB Vo:±3.5V Vi0:4mV	8DS	Ela		data pdf pinout
L 2120 C(N,S)	OP-IC	Wideband Current Feedback Amp. 100MHz	8-DIP	Ela		pdf
L2130C	FB-AMP	Vs:±6V Vu:66dB Vo:3.5/0V Vi0:2mV	8DS	Ela		data pinout
L2150C	SS-OP	Vs:12.6V Vu:80dB Vo:10.8/0V Vi0:±2mV	8DS	Ela		
L2150CW	SS-OP	Vs:12.6V Vu:80dB Vo:10.8/0V Vi0:±3mV	5S	Ela		
L2157C	SS-OP	Vs:12.6V Vu:80dB Vo:10.8/0V Vi0:±2mV	8DS	Ela		data
L2160C	FB-AMP	Vs:±16.5V Vo:±13.5V Vi0:2mV f:130Mc	8DS	Ela		data
L2165C	FB-AMP	Vs:±16.5V Vo:±13V Vi0:1mV f:30Mc	8DS	Ela		data pinout
L2166C	FB-AMP	Vs:±16.5V Vo:±13.5V Vi0:2mV f:110Mc	8DS	Ela		data pinout
L2170	FB-AMP	Vs:±6V Vo:±4V Vi0:2.5mV f:70Mc	8DS,5S	Ela		data
L2171	FB-AMP	Vs:±7V Vo:3.4/0V Vi0:3mV f:150Mc	8DS	Ela		data pinout
L 2171(C)...	OP-IC	±5<7V, ±70mA, 150MHz, 1200V/µs, 10ns	8-DIP	Ela		
L2175	FB-AMP	Vs:±16.5V Vo:±13V Vi0:1mV f:120Mc	8DS	Ela		data pinout
L2176	FB-AMP	Vs:±6V Vo:±4V Vi0:2.5mV f:70Mc	8DS	Ela		data
L2180	FB-AMP	Vs:±6V Vo:±4V Vi0:2.5mV f:250Mc	8DS,5S	Ela		data
L2186	FB-AMP	Vs:±6V Vo:±4V Vi0:2.5mV f:250Mc	8DS	Ela		data
L2210C	2xVIDEO-OP	Vs:±9V Vu:250V/mV Vo:±2.5V Vi0:10mV	8DS	Ela		data
L2211C	2xVIDEO-OP	Vs:±9V Vu:380V/mV Vo:±2.5V Vi0:5mV	8DS	Ela		data
L 2232(C)...	OP-IC	Dual, ±4...18V, ±30mA,,60MHz, 400V/µs	8-DIP	Ela		
L2244C	2xHI-SPEED	Vs:±18V Vu:1500V/mV Vo:±13.6V f:60Mc	8DS	Ela		data pdf pinout
L2245C	2xHI-SPEED	Vs:±18V Vu:3000V/mV Vo:±13.6V	8DS	Ela		data pdf pinout
L2250C	2xSS-OP	Vs:12.6V Vu:80dB Vo:10.8/0V Vi0:±4mV	8DS	Ela		data pdf pinout
L2257C	2xSS-OP	Vs:12.6V Vu:80dB Vo:10.8/0V Vi0:±2mV	14DS	Ela		data
L2260	2xFB-AMP	Vs:±16.5V Vo:±13.5V Vi0:2mV f:130Mc	8DS	Ela		data
L2270	2xFB-AMP	Vs:±6V Vo:±4V Vi0:2.5mV f:70Mc	8DS	Ela		data
L2276	2xFB-AMP	Vs:±6V Vo:±4V Vi0:2.5mV f:70Mc	14DS	Ela		data
L2280	2xFB-AMP	Vs:±6V Vo:±4V Vi0:2.5mV f:250Mc	8DS	Ela		data
L2286	2xFB-AMP	Vs:±6V Vo:±4V Vi0:2.5mV f:250Mc	14DS	Ela		data
L2310C	3xVIDEO-OP	Vs:±9V Vu:250V/mV Vo:±2.5V Vi0:10mV	14DS	Ela		data
L2311C	3xVIDEO-OP	Vs:±9V Vu:380V/mV Vo:±2.5V Vi0:5mV	14DS	Ela		data
L2357C	3xSS-OP	Vs:12.6V Vu:80dB Vo:10.8/0V Vi0:±2mV	16DS	Ela		data
L2360C	3xFB-AMP	Vs:±16.5V Vo:±13.5V Vi0:2mV f:130Mc	16DS	Ela		data
L2386C	3xFB-AMP	Vs:±6V Vo:±4V Vi0:2.5mV f:250Mc	16DS	Ela		data
L2410C	4xVIDEO-OP	Vs:±9V Vu:250V/mV Vo:±2.5V Vi0:10mV	14DS	Ela		data
L2411C	4xVIDEO-OP	Vs:±9V Vu:380V/mV Vo:±2.5V Vi0:5mV	14DS	Ela		data
L2444C	4xHI-SPEED	Vs:±18V Vu:1500V/mV Vo:±13.6V f:60Mc	14DS	Ela		data
L2445C	4xHI-SPEED	Vs:±18V Vu:3000V/mV Vo:±13.6V	14DS	Ela		data
L2450C	4xSS-OP	Vs:12.6V Vu:80dB Vo:10.8/0V Vi0:±4mV	14DS	Ela		data
L2460C	4xFB-AMP	Vs:±16.5V Vo:±13.5V Vi0:2mV f:130Mc	14DS	Ela		data
L2470C	4xFB-AMP	Vs:±6V Vo:±4V Vi0:2.5mV f:70Mc	14DS	Ela		data
L2480C	2xFB-AMP	Vs:±6V Vo:±4V Vi0:2.5mV f:250Mc	14DS	Ela		data
L4393C	3xFB-AMP	Vs:±16.5V Vo:±13V Vi0:2mV f:80Mc	16DS	Ela		data
L4430C	VIDEO-AMP	Vs:±16.5V Vo:±12.8V Vi0:2mV f:20Mc	8DS	Ela		data
L4431C	VIDEO-AMP	Vs:±16.5V Vo:±13V Vi0:2mV f:14Mc	8DS	Ela		data
L4451C	WIDEBAND	Vs:±16.5V Vo:±12.8V Vi0:7mV f:10Mc	14DS	Ela		data
L4452C	WIDEBAND	Vs:±16.5V Vo:±12.8V Vi0:<10mV f:10Mc	14DS	Ela		data
ELA-660-22X60-3-6	LED	rd ø20mm I:4lm Vf:14V If:250mA		Roi		data
ELA-810-22X60-12-6	IRED	810nm ø20mm I:4lm Vf:9.5V If:250mA		Roi		data
ELA-810-22X60-2-6	LED	rd ø20mm I:4lm Vf:9.5V If:250mA		Roi		data
ELA-880-22X60-2-6	LED	rd ø20mm I:4lm Vf:9.5V If:250mA		Roi		data
ELA-920-22X60-12-6	LED	rd ø20mm I:4lm Vf:9.5V If:250mA		Roi		data
ELA-920-22X60-2-6	LED	rd ø20mm I:4lm Vf:9.5V If:250mA		Roi		data
ELJ-640-225	LED	rd ø14mm I:19lm Vf:19V If:100mA		Roi		data
ELJ-646-205B	LED	rd ø14mm I:3.3lm Vf:13V If:100mA		Roi		data
ELJ-660-225B	LED	rd ø14mm I:2.9lm Vf:13V If:100mA		Roi		data
ELJ-660-245B	LED	rd ø14mm I:3lm Vf:13V If:100mA		Roi		data
ELJ-810-228B	IRED	810nm ø14mm I:1.2W/SrmW/Sr Vf:13V		Roi		data
ELJ-880-228B	IRED	880nm ø14mm I:1.2W/SrmW/Sr Vf:10.5V		Roi		data
ELJ-920-228B	IRED	810nm ø14mm I:1.2W/SrmW/Sr Vf:9.5V		Roi		data
ELJ-950-228B	IRED	950nm ø14mm I:1.2W/SrmW/Sr Vf:8V		Roi		data

EM

Type	Device	Short Description	Fig.	Manu	Comparision Types	More at
EM	Si-P	=2SB1028-EM (Typ-Code/Stempel/marking)	39	Hit	→2SB1028	data
EM	Si-N	=2SD1001-EM (Typ-Code/Stempel/marking)	39	Nec	→2SD1001	data
EM	MOS-P-FET-e*	=2SJ501 (Typ-Code/Stempel/marking)	35	Say	→2SJ501	data
EM	Z-Di	=SM 6T 24A (Typ-Code/Stempel/marking)	71(6x4mm)	Tho	→SM 6T...	data
EM	Z-Di	=SMBJ 150CA (Typ-Code/Stempel/marking)	71(5x3,5)	Mop	→SMBJ ...	data
EM	Si-N+R	=XN 121M (Typ-Code/Stempel/marking)	45	Mat	→XN 121M	data
EM	Z-Di	=Z1 SMA-13 (Typ-Code/Stempel/marking)	71(5x2,5)	Fag	→Z1 SMA-13	data
EM 01(A,Y,Z)	Si-Di	Gl, Uni, 100...600V, 1A EM 01=400V, A=600V, Y=100V, Z=200V	31a	Sak	BY 126...127, BY 133...134, 1N4002...07,++	data
EM 1(A,B,C,Y,Z)	Si-Di	Gl, Uni, 100...1000V, 1A EM 1=400, A=600, B=800, C=1000, Y=100V Z=200V	31a	Sak	BY 126...127, BY 133...134, 1N4002...07,++	data

Type	Device	Short Description	Fig.	Manu	Comparision Types	More at
EM 2(A,B)	Si-Di	Gl, Uni, 400...800V, 1,2A EM 2=400V, A=600V, B=800V	31a	Sak	BY 226...227, BYW 53...56, GP 15G...M,++	data
EM 502	Si-Di	=1N4003	31a	Itt	→1N4003	data
EM 504	Si-Di	=1N4004	31a	Itt	→1N4004	data
EM 506	Si-Di	=1N4005	31a	Itt	→1N4005	data
EM 508	Si-Di	=1N4006	31a	Itt	→1N4006	data
EM 510	Si-Di	=1N4007	31a	Itt	→1N4007	data
EM 513	Si-Di	=1N4007: Itt: 1300V, Tho,Die: 1600V	31a	Itt,Tho,Die	BY 228, BY 231/1500, DM 513, GP 10Y	data
EM 516	Si-Di	=1N4007: Itt: 1600V, Tho,Die: 1800V	31a	Itt,Tho,Die	(BY 228, BY 231/1500, DM 516, GP 10Y)[7]	data
EM 518	Si-Di	=1N4007: 2000V	31a	Tho,Die	RGP 15-20	data
EM 6151	IC	Windowed Watchdog,Reset,Sleep Mode Funct.	TSOP	Emm	-	pdf pinout
EM 6323	IC	Reset Circuit, Manual Reset, Watchdog	TSOP	Emm	-	pdf pinout
EM 6324	µP-Periph	Reset Circuit, Manual Reset, Watchdog	45	Emm	-	pdf pinout
EM 6517	IC	Ultra Low Power Microcontr.,ADC,EEPROM	TSOP	Emm	-	pdf pinout
EM 6521	IC	Microcontroller, 4x20 LCD Driver	QFP	Emm	-	pdf pinout
EM 6522	IC	Microcontroller, 4x20 LCD Driver	QFP	Emm	-	pdf pinout
EM 6540	IC	4 bit Microcontroller	TSOP	Emm	-	pdf pinout
EM 6607	DIG-IC	microcontroller, 4 high drive outputs	TSOP	Emm	-	pdf pinout
EM 6625	DIG-IC	Microcontroller with 4x20 LCD Driver	QFP	Emm	-	pdf pinout
EM 6626	DIG-IC	Microcontroller with 4x20 LCD Driver	QFP	Emm	-	pdf pinout
EM 6635	DIG-IC	Microcontr.,RC,32kHz osci. 9 hi. drive outp.	QFP	Emm	-	pdf pinout
EM 6680 SO-8	IC	Ultra Low Power 8-pin Microcontroller	QFP	Emm	-	pdf pinout
EM 6682 SO-8	IC	Ultra Low Power 8-pin Microcontroller	QFP	Emm	-	pdf pinout
EM 6152 V 50	IC	5V Automotive Reg., Windowed Watchdog	TSOP	Emm	-	pdf pinout
EM 6152 V 50 P	IC	5V Automotive Reg., Windowed Watchdog	TSOP	Emm	-	pdf pinout
EM 6152 V 53	IC	5V Automotive Reg., Windowed Watchdog	TSOP	Emm	-	pdf pinout
EM 6152 V 53 P	IC	5V Automotive Reg., Windowed Watchdog	TSOP	Emm	-	pdf pinout
EM 6152 V 55	IC	5V Automotive Reg., Windowed Watchdog	TSOP	Emm	-	pdf pinout
EM 6607 /28	DIG-IC	microcontroller, 4 high drive outputs	TSOP	Emm	-	pdf pinout
EM 6617 SO 24	DIG-IC	Low Power Microcontroller, ADC, EEPROM	TSOP	Emm	-	pdf pinout
EM 6617 SO 28	DIG-IC	Low Power Microcontroller, ADC, EEPROM	TSOP	Emm	-	pdf pinout
EM 6617 TP 28	DIG-IC	Low Power Microcontroller, ADC, EEPROM	TSOP	Emm	-	pdf pinout
EM 6640 TP 16	DIG-IC	Microcontroller, Eeprom, RC Oscillator	QFP	Emm	-	pdf pinout
EM 6680 SO-14	IC	Ultra Low Power 8-pin Microcontroller	QFP	Emm	-	pdf pinout
EM 6682 SO-14	IC	Ultra Low Power 8-pin Microcontroller	QFP	Emm	-	pdf pinout
EM 6821 TQ 52	IC	Low Power µController, 4x20 LCD Driver	QFP	Emm	-	pdf pinout
EM 51256 C-10J	CMOS-IC	32K x 8 High Speed Sram	MDIP	Etr	-	pdf pinout
EM 51256 C-10P	CMOS-IC	32K x 8 High Speed Sram	MDIP	Etr	-	pdf pinout
EM 51256 C-12J	CMOS-IC	32K x 8 High Speed Sram	MDIP	Etr	-	pdf pinout
EM 51256 C-12P	CMOS-IC	32K x 8 High Speed Sram	MDIP	Etr	-	pdf pinout
EM 51256 C-15J	CMOS-IC	32K x 8 High Speed Sram	MDIP	Etr	-	pdf pinout
EM 51256 C-15P	CMOS-IC	32K x 8 High Speed Sram	MDIP	Etr	-	pdf pinout
EM 51256 C-15TS	CMOS-IC	32K x 8 High Speed Sram	MDIP	Etr	-	pdf pinout
EMA 2	Si-P+R	SMD, 2x DTA144E, Dual Digital Transistor, com. Emitter	45ba1	Rhm	=UMA2N	data data
EMA 3	Si-P+R	SMD, 2x DTA143T, Dual Digital Transistor, com. Emitter 50V, 100mA,	45ba1	Rhm	EMA4, EMA6,	data data
EMA 4	Si-P+R	SMD, 2x DTA114T, Dual Digital Transistor, com. Emitter 50V, 100mA,	45ba1	Rhm	EMA3, EMA6,	data data
EMA 5	Si-P+R	SMD, 2x DTA123J, Dual Digital Transistor, com. Emitter	45ba1	Rhm	=UMA5N	data data
EMA 6	Si-P+R	SMD, 2x DTA114T, Dual Digital Transistor, com. Emitter 50V, 100mA,	45ba1	Rhm	EMA3, EMA4,	data data
EMA 7	Si-P+R	SMD, 2x DTA143X, Dual Digital Transistor, com. Emitter	45ba1	Rhm	=UMA7N	data data
EMA 8	Si-P+R	SMD, 2x DTA114Y, Dual Digital Transistor, com. Emitter	45ba1	Rhm	=UMA8N	data data
EMA 11	Si-P+R	SMD, 2x DTA143Z, Dual Digital Transistor, com. Emitter 100mA,	45ba1	Rhm	=UMA11N	data data
EMB 2	Si-P+R	SMD, 2x DTA144E, Dual Digital Transistor 100mA,	46bh1	Rhm	=UMB2N	data data
EMB 3	Si-P+R	SMD, 2x DTA143T, Dual Digital Transistor 50V, 100mA,	46bh1	Rhm	KRA751U, KRA752U, KRA753U, KRA754U,	data data
EMB 4	Si-P+R	SMD, 2x DTA114T, Dual Digital Transistor 50V, 100mA,	46bh1	Rhm	=UMB4N	data data
EMB 6	Si-P+R	SMD, 2x DTA144E, Dual Digital Transistor	46bl1	Rhm	=UMB6N	data data
EMB 9	Si-P+R	SMD, 2x DTA144Y, Dual Digital Transistor 100mA,	46bh1	Rhm	=UMB9N	data data
EMB 10	Si-P+R	SMD, 2x DTA123J, Dual Digital Transistor 100mA,	46bh1	Rhm	=UMB10N	data data
EMB 11	Si-P+R	SMD, 2x DTA114E, Dual Digital Transistor 100mA,	46bh1	Rhm	=UMB11N	data data
EMC 2	Si-N/P+R	SMD, DTA124E + DTC124E, Dual Digital Transistor 100mA,	45ba4	Rhm	=UMC2N	data data
EMC 3	Si-N/P+R	SMD, DTA114E + DTC1124E, Dual Digital Transistor 100mA,	45ba4	Rhm	=UMC3N	data data
EMC 4	Si-N/P+R	SMD, DTA114Y + DTC144E, Dual Digital Transistor 100mA,	45ba4	Rhm	=UMC4N	data data
EMC 5	Si-N/P+R	SMD, DTA143X + DTC144E, Dual Digital Transistor 100mA,	45ba4	Rhm	=UMC5N	data data
EMD 2	Si-P/N+R	SMD, DTA124E + DTC124E, Dual Digital Transistor	46bh1	Rhm	=UMD2N	data data

Type	Device	Short Description	Fig.	Manu	Comparision Types	More at
EMD 3	Si-P/N+R	SMD, DTA114E + DTC114E, Dual Digital Transistor	46bh1	Rhm	=UMD3N	data data
EMD 6	Si-P/N+R	SMD, DTA143T + DTC143T, Dual Digital Transistor 50V, 100mA,	46bh1	Rhm	=UMD6N	data data
EMD 9	Si-P/N+R	SMD, DTA114Y + DTC114Y, Dual Digital Transistor	46bh1	Rhm	=UMD9N	data
EMD 12	Si-N/P+R	SMD, DTA144E + DTC144E, Dual Digital Transistor	46bh1	Rhm		data data
EMG 1	Si-N+R	SMD, 2x DTC124E, Dual Digital Transistor, com. Emitter	45ba1	Rhm	=UMG1N	data data
EMG 2	Si-N+R	SMD, 2x DTC144E, Dual Digital Transistor, com. Emitter 100mA,	45ba1	Rhm	=UMG2N	data data
EMG 3	Si-N+R	SMD, 2x DTC143T, Dual Digital Transistor, com. Emitter 50V, 100mA,	45ba1	Rhm	EMG4, EMG6,	data data
EMG 4	Si-N+R	SMD, 2x DTC114T, Dual Digital Transistor, com. Emitter 50V, 100mA,	45ba1	Rhm	EMG3, EMG6,	data
EMG 5	Si-N+R	SMD, 2x DTC114Y, Dual Digital Transistor, com. Emitter 100mA,	45ba1	Rhm	=UMG5N	data data
EMG 6	Si-N+R	SMD, 2x DTC114T, Dual Digital Transistor, com. Emitter 50V, 100mA,	45ba1	Rhm	EMG3, EMG4,	data data
EMG 8	Si-N+R	SMD, 2x DTC143Z, Dual Digital Transistor, com. Emitter 100mA,	45ba1	Rhm	=UMG8N	data
EMG 9	Si-N+R	SMD, 2x DTC114E, Dual Digital Transistor, com. Emitter 100mA,	45ba1	Rhm	=UMG9N	data data
EMG 11	Si-N+R	SMD, 2x DTC123J, Dual Digital Transistor, com. Emitter	45ba1	Rhm	=UMG11N	data data
EMH 1	Si-N+R	SMD, 2x DTC124E, Dual Digital Transistor	46bh1	Rhm	=UMH1N	data
EMH 2	Si-N+R	SMD, 2x DTC144E, Dual Digital Transistor 100mA,	46bh1	Rhm	=UMH2N	data data
EMH3	Si-N+R	SMD, 2x DTC143T, Dual Digital Transistor 50V, 100mA,	46bh1	Rhm	EMH4,	data data
EMH 4	Si-N+R	SMD, 2x DTC114T, Dual Digital Transistor 50V, 100mA,	46bh1	Rhm	EMH3,	data
EMH 6	Si-N+R	SMD, 2x DTC144E, Dual Digital Transistor	46bl1	Rhm	=UMH6N	data data
EMH 9	Si-N+R	SMD, 2x DTC114Y, Dual Digital Transistor 100mA,	46bh1	Rhm	=UMH9N	data data
EMH 10	Si-N+R	SMD, 2x DTC123J, Dual Digital Transistor 100mA,	46bh1	Rhm	=UMH10N	data
EMH 11	Si-N+R	SMD, 2x DTC114E, Dual Digital Transistor 100mA,	46bh1	Rhm	=UMH11N	data data
EMS 1	Si-P	SMD, 2x 2SA1037AK, com. Emitter 50V, 150mA,	45ba1	Rhm	=FMS1A	data data
EMS 2	Si-P	SMD, 2x 2SA1037AK, com. Base/Basis 50V, 150mA,	45ba3	Rhm	=FMS2A	data data
EMT 1	Si-P	SMD, 2x SA1037AK	46bh1	Rhm	=IMT1A	data
EMT 2	Si-P	SMD, 2x 2SA1037AK	46bl1	Rhm	=IMT2(A)	data
EMT 3	Si-P	SMD, 2x 2SA1037AK	46bh2	Rhm	=IMT3(A)	data
EMT 18	Si-P	SMD, 2x 2SA2018	46bh1	Rhm	=UMT18N	data
EMW 1	Si-N	SMD, 2x 2SC2412K, com. Emitter 50V, 150mA,	45ba1	Rhm	=FMW1	data data
EMW 2	Si-N	SMD, 2x 2SC2412AK, com. Base/Basis 50V, 150mA,	45ba3	Rhm	=UMW2N	data data
EMX 1	Si-N	SMD, 2x 2SC2412K, Uni	46bh1	Rhm	IMX1, IMX17,	data
EMX 2	Si-N	SMD, 2x 2SC2412K	46bl1	Rhm	=UMX2N	data
EMX 3	Si-N	SMD, 2x 2SC2412AK	46bh2	Rhm	=UMX3N	data
EMX 4	Si-N	SMD, 2x 2SC3837K, TV Tuner, Mixer, Osc. 18V, 50mA,	46bh2	Rhm	=UMX4N	data data
EMX 5	Si-N	SMD, 2x 2SC3838K, TV Tuner, Mixer, Osc. 11V, 50mA,	46bh2	Rhm	EMX4,	data
EMX 18	Si-N	SMD, 2x 2SC5585, Uni	46bh1	Rhm	=UMX18N	data
EMY 1	Si-PN	SMD, 2SA1037AK + 2SC2412K, com. Emitter 50V, 150mA,	45ba1	Rhm	=UMY1N	data data
EMZ 1	Si-PN	SMD, 2SA1037AK + 2SC2412ZK, Power Management 50V, 150mA,	46bh1	Rhm	=UMZ1N	data data
EMZ 2	Si-PN	SMD, 2SA1037AK + 2SC2412K, Power Management 50V, 150mA,	46bh2	Rhm	=UMZ2N	data data
EMZ 7	Si-N+P	SMD, 2SA2018 + 2SC5585, Uni 12V, 500mA,	46bh1	Rhm	=UMZ7N	data data
EMZ 8	Si-N+P	SMD, 2SA2018 + 2SC2412K, Uni 12V, 150mA,	46bh1	Rhm	EMZ7,	data data
EMZ 6.8N	Z-Di	SMD, Dual, 6,47...7,14V, <40Ω(5mA)	35m(1,6mm)	Rhm		data

EN

Type	Device	Short Description	Fig.	Manu	Comparision Types	More at
EN	Si-N	=2SC4860 (Typ-Code/Stempel/marking)	35(2mm)	Say	→2SC4860	data
EN	Si-N	=2SC4861 (Typ-Code/Stempel/marking)	35	Say	→2SC4861	data
EN	Si-N	=2SC4862 (Typ-Code/Stempel/marking)	44	Say	→2SC4862	data
EN	MOS-N-FET-e*	=2SK3288 (Typ-Code/Stempel/marking)	35	Hit	→2SK3288	
EN	MOS-N-FET-e*	=2SK3378 (Typ-Code/Stempel/marking)	35(2mm)	Hit	→2SK3378	
EN	Z-Di	=SM 6T 27 (Typ-Code/Stempel/marking)	71(6x4mm)	Tho	→SM 6T...	data
EN	Z-Di	=SMBJ 160C (Typ-Code/Stempel/marking)	71(5x3,5)	Mop	→SMBJ ...	data
EN	Si-N	=XN 4506 (Typ-Code/Stempel/marking)	46	Mat	→XN 4506	data
EN	Si-N	=XP 4506 (Typ-Code/Stempel/marking)	46(2mm)	Mat	→XP 4506	data

Type	Device	Short Description	Fig.	Manu	Comparision Types	More at
EN	Z-Di	=Z1 SMA-15 (Typ-Code/Stempel/marking)	71(5x2,5)	Fag	→Z1 SMA-15	data
EN 01 Z	Si-Di	GlS, 200V, 1A, 100ns	31a	Sak	BYD 72D...G, EGP 10D...G, FE 1D...H	data
EN 3	Si-N	=2SC4860-3 (Typ-Code/Stempel/marking)	35(2mm)	Say	→2SC4860	data
EN 3	Si-N	=2SC4861-3 (Typ-Code/Stempel/marking)	35	Say	→2SC4861	data
EN 3	Si-N	=2SC4862-3 (Typ-Code/Stempel/marking)	44	Say	→2SC4862	data
EN 4	Si-N	=2SC4860-4 (Typ-Code/Stempel/marking)	35(2mm)	Say	→2SC4860	data
EN 4	Si-N	=2SC4861-4 (Typ-Code/Stempel/marking)	35	Say	→2SC4861	data
EN 4	Si-N	=2SC4862-4 (Typ-Code/Stempel/marking)	44	Say	→2SC4862	data
EN 5	Si-N	=2SC4860-5 (Typ-Code/Stempel/marking)	35(2mm)	Say	→2SC4860	data
EN 5	Si-N	=2SC4861-5 (Typ-Code/Stempel/marking)	35	Say	→2SC4861	data
EN 5	Si-N	=2SC4862-5 (Typ-Code/Stempel/marking)	44	Say	→2SC4862	data
EN 697	Si-N	=2N697: 0,3W	8a	Fch,Nsc	→2N697	data
EN 706	Si-N	=2N706: 0,2W	8a	Fch	→2N706	data
EN 708	Si-N	=2N708: 0,2W	8a	Fch	→2N708	data
EN 718 A	Si-N	=2N718A: 0,22W	8a	Fch	→2N718 A	data
EN 722	Si-P	=2N722: 0,2W	8a	Fch,Nsc	→2N722	data
EN 744	Si-N	=2N744: 0,2W	8a	Fch	→2N744	data
EN 870	Si-N	=2N870: 0,22W	8a	Fch	→2N870	data
EN 871	Si-N	=2N871: 0,22W	8a	Fch	→2N871	data
EN 914	Si-N	=2N914: 0,2W	8a	Fch	→2N914	data
EN 915	Si-N	=2N915: 0,2W	8a	Fch	→2N915	data
EN 916	Si-N	=2N916: 0,2W	8a	Fch	→2N916	data
EN 918	Si-N	=2N918: 0,2W	8a	Fch,Nsc	→2N918	data
EN 930	Si-N	=2N930: 0,2W	8a	Fch,Nsc,Mic	→2N930	data
EN 956	Si-N	=2N956: 0,22W	8a	Fch,Nsc	→2N956	data
EN 1132	Si-P	=2N1132: 0,3W	8a	Fch,Nsc	→2N1132	data
EN 1613	Si-N	=2N1613: 0,3W	8a	Fch	→2N1613	data
EN 1711	Si-N	=2N1711: 0,3W	8a	Fch	→2N1711	data
EN 2218	Si-N	=2N2218: 0,35W	8a	Fch	→2N2218	data
EN 2219	Si-N	=2N2219: 0,35W	8a	Fch	→2N2219	data
EN 2221	Si-N	=2N2221: 0,2W	8a	Fch	→2N2221	data
EN 2222	Si-N	=2N2222: 0,2W	8a	Fch,Nsc	→2N2222	data
EN 2369 A	Si-N	=2N2369A: 0,2W	8a	Fch	→2N2369A	data
EN 2484	Si-N	=2N2484: 0,2W	8a	Fch,Nsc	→2N2484	data
EN 2894 A	Si-P	=2N2894A: 0,2W	8a	Fch	→2N2894A	data
EN 2904	Si-P	=2N2904: 0,3W	8a	Fch	→2N2904	data
EN 2905	Si-P	=2N2905: 0,3W	8a	Fch,Nsc	→2N2905	data
EN 2906	Si-P	=2N2906: 0,2W	8a	Fch	→2N2906	data
EN 2907	Si-P	=2N2907: 0,2W	8a	Fch,Nsc	→2N2907	data
EN 3009	Si-N	=2N3009: 0,2W	8a	Fch	→2N3009	data
EN 3011	Si-N	=2N3011: 0,2W	8a	Fch	→2N3011	data
EN 3013	Si-N	=2N3013: 0,2W	8a	Fch	→2N3013	data
EN 3014	Si-N	=2N3014: 0,2W	8a	Fch	→2N3014	data
EN 3250	Si-P	=2N3250: 0,2W	8a	Fch	→2N3250	data
EN 3502	Si-P	=2N3502: 0,3W	8a	Fch,Nsc	→2N3502	data
EN 3504	Si-P	=2N3504: 0,2W	8a	Fch,Nsc	→2N3504	data
EN 3903	Si-N	=2N3903: 0,31W	8a	Fch	→2N3903	data
EN 3904	Si-N	=2N3904: 0,31W	8a	Fch	→2N3904	data
EN 3905	Si-P	=2N3905: 0,31W	8a	Fch	→2N3905	data
EN 3906	Si-P	=2N3906: 0,31W	8a	Fch	→2N3906	data
EN 3962	Si-P	=2N3962: 0,2W	8a	Fch	→2N3962	data
EN 4123	Si-N	=2N4123: 0,2W	8a	Fch	→2N4123	data
EN 4124	Si-N	=2N4124: 0,2W	8a	Fch	→2N4124	data
EN 4125	Si-P	=2N4125: 0,2W	8a	Fch	→2N4125	data
EN 4126	Si-P	=2N4126: 0,2W	8a	Fch	→2N4126	data
EN 5172	Si-N	=2N5172: 0,2W	8a	Fch	→2N5172	data
EN 5310DC	SMPS-IC	SMD, 1A Voltage Mode Synchronous Buck PWM, DC/DC Converter, 5MHz, 0...70°C	DFN	Epi	-	pdf pinout
EN 5310DI	SMPS-IC	SMD, 1A Voltage Mode Synchronous Buck PWM, DC/DC Conv., 5MHz, -40...+85°C	DFN	Epi	-	pdf pinout
EN 5312QI	SMPS-IC	SMD, 1A Synchronous Buck Regulator, integrated Inductor, MOSFETs, PWM and compensation , 5MHz, -40...+85°C	QFN	Epi	-	pdf pinout
EN 5330DC	SMPS-IC	SMD, 3A Voltage Mode Synchronous Buck PWM DC/ DC Converter, integrated Inductor, 5MHz, 0... 70°C	DFN	Epi	-	pdf pinout
EN 5330DI	SMPS-IC	SMD, 3A Voltage Mode Synchronous Buck PWM DC/ DC Converter, integrated Inductor, - 40...+85°C	DFN	Epi	-	pdf pinout
EN 5335QI	SMPS-IC	SMD, 3A Voltage Mode Synchronous Buck PWM DC/DC Conv., integrated inductor, 3pin progr. output, 5MHz, -40...+85°C	QFN	Epi	-	pdf pinout
EN 5336QI	SMPS-IC	SMD, 3A Voltage Mode Synchronous Buck PWM DC/DC Conv., integrated inductor, resistor prog. out, 5MHz, -40...+85°C	QFN	Epi	-	pdf pinout
EN 5360DC	SMPS-IC	SMD, 6A Voltage Mode Synchronous Buck PWM DC/ DC Converter, integrated Inductor, 5MHz, 0... 70°C	DFN	Epi	-	pdf pinout
EN 5360DI	SMPS-IC	SMD, 6A Voltage Mode Synchronous Buck PWM DC/ DC Converter, integrated Inductor, 5MHz, - 40...+85°C	DFN	Epi	-	pdf pinout
EN 5365QC	SMPS-IC	SMD, 6A Voltage Mode Synchr. Buck PWM	QFN	Epi	-	pdf pinout

Type	Device	Short Description	Fig.	Manu	Comparision Types	More at
EN 5365QI	SMPS-IC	DC/DC Conv., integr. Inductor, 3pin progr. output, 5MHz, 0...70°C SMD, 6A Voltage Mode Synchr. Buck PWM	QFN	Epi	-	pdf pinout
EN 5366QI	SMPS-IC	DC/DC Conv., integr. Inductor, 3pin progr. output, 5MHz, -40...+85°C SMD, 6A Voltage Mode Synchronous Buck PWM DC/DC Conv., integrated inductor, resistor prog. out, 5MHz, -40...+85°C	QFN	Epi	-	pdf pinout
ENB 121...850 D-10A	Varistor	VDR/Varistors	36	Fjd		
ENB 121...881 D-14A	Varistor	VDR/Varistors	36	Fjd		
ENB 201...881 D-20A	Varistor	VDR/Varistors	36	Fjd		
EO	Si-N	=2SC2882-O (Typ-Code/Stempel/marking)	39	Tos	→2SC2882	data
EO	Si-N	=2SC3265-O (Typ-Code/Stempel/marking)	35	Tos	→2SC3265	data
EO	Si-N	=2SC5097-O (Typ-Code/Stempel/marking)	44	Tos	→2SC5097	data
EO	Si-N	=KTC 3265-O (Typ-Code/Stempel/marking)	35	Kec	→KTC 3265	data
EO	Si-N	=KTC4374-O (Typ-Code/Stempel/marking)	39	Kec	→KTC 4374	data
EO	Si-N	=XP 5555 (Typ-Code/Stempel/marking)	46(2mm)	Mat	→XP 5555	data
EP	Si-P	=2SB789A-P (Typ-Code/Stempel/marking)	39	Mat	→2SB789A	data
EP	Z-Di	=SM 6T 27A (Typ-Code/Stempel/marking)	71(6x4mm)	Tho	→SM 6T...	data
EP	Z-Di	=SMBJ 160CA (Typ-Code/Stempel/marking)	71(5x3,5)	Mop	→SMBJ ...	data
EP	Si-N	=XN 4556 (Typ-Code/Stempel/marking)	46	Mat	→XN 4556	data
EP	Z-Di	=Z1 SMA-16 (Typ-Code/Stempel/marking)	71(5x2,5)	Fag	→Z1 SMA-16	data
EP 01 C	Si-Di	GI, S, 1000V, 0,2A, 200ns	31a	Sak	BA 159, BYT 11/1000, BYT 52M, BYV 26E	data
EP20XX-150X1	LED	am+bl+cy+gr+or ø11.2mm I=3cd Vf:2.4V		Roi		data pdf pinout
EP 5352QI	SMPS-IC	SMD, 500mA Synchronous Buck Regulators, Integrated Inductor, 5MHz, 3pin output voltage select, -40...+85°C	QFN	Epi	-	pdf pinout
EP 5362QI	SMPS-IC	SMD, 600mA Synchronous Buck Regulators, Integrated Inductor, 5MHz, 3pin output voltage select, -40...+85°C	QFN	Epi	-	pdf pinout
EP 5382QI	SMPS-IC	SMD, 800mA Synchronous Buck Regulators, Integrated Inductor, 5MHz, 3pin output voltage select, -40...+85°C	QFN	Epi	-	pdf pinout

EQ

Type	Device	Short Description	Fig.	Manu	Comparision Types	More at
EQ	Si-P	=2SB789A-Q (Typ-Code/Stempel/marking)	39	Mat	→2SB789A	data
EQ	Z-Di	=SM 6T 30 (Typ-Code/Stempel/marking)	71(6x4mm)	Tho	→SM 6T...	data
EQ	Z-Di	=SMBJ 170C (Typ-Code/Stempel/marking)	71(5x3,5)	Mop	→SMBJ ...	data
EQ	N-FET	=XP 1D874 (Typ-Code/Stempel/marking)	45(2mm)	Mat	→XP 1D874	data
EQ	Z-Di	=Z1 SMA-18 (Typ-Code/Stempel/marking)	71(5x2,5)	Fag	→Z1 SMA-18	data
EQ 5352DI	SMPS-IC	SMD, 500mA Synchronous Buck Regulators, Integrated Inductor, 5MHz, 3pin output voltage select, -40...+85°C	DFN	Epi	-	pdf pinout
EQ 5362DI	SMPS-IC	SMD, 600mA Synchronous Buck Regulators, Integrated Inductor, 5MHz, 3pin output voltage select, -40...+85°C	DFN	Epi		pdf pinout
EQ 5382DI	SMPS-IC	SMD, 800mA Synchronous Buck Regulators, Integrated Inductor, 5MHz, 3pin output voltage select, -40...+85°C	DFN	Epi		pdf pinout
EQA 01-05...-35	Z-Di	5...35V, 0,5W	31a	Fjd	BZX55/..., BZX83/..., ZPD..., 1N5231...58,++	data
EQB 01-05...-35	Z-Di	5...35V, 1W	31a	Fjd	BZW22/..., BZX61/..., ZPY..., 1N5918...38,++	data

ER

Type	Device	Short Description	Fig.	Manu	Comparision Types	More at
ER 0082	EAROM-IC	128x1 Bit	18-DIP		MN 1213	
ER	Si-P	=2SB789A-R (Typ-Code/Stempel/marking)	39	Mat	→2SB789A	data
ER	Si-N	=2SC4807 (Typ-Code/Stempel/marking)	39	Hit	→2SC4807	data
ER	Si-N	=2SC5097-R (Typ-Code/Stempel/marking)	44	Tos	→2SC5097	data
ER	Z-Di	=SM 6T 30A (Typ-Code/Stempel/marking)	71(6x4mm)	Tho	→SM 6T...	data
ER	Z-Di	=SMBJ 170CA (Typ-Code/Stempel/marking)	71(5x3,5)	Mop	→SMBJ ...	data
ER	Si-N/P	=XN 4683 (Typ-Code/Stempel/marking)	46	Mat	→XN 4683	data
ER	Si-N/P	=XP 4683 (Typ-Code/Stempel/marking)	46(2mm)	Mat	→XP 4683	data
ER	Z-Di	=Z1 SMA-20 (Typ-Code/Stempel/marking)	71(5x2,5)	Fag	→Z1 SMA-20	data
ER 1 A...E	Si-Di	SMD, GI, S, 50...300V, 0,6A, <35ns A=50V, B=100V, C=150V, D=200V, E=300V	71a(5x3,5)	Die	D 1FL20, D 1FL40, MURS 120	data
ER 2	Si-Di	GI, 200V, 1A	31a	Ssc	BY 126...127, BY 133...134, 1N4003...07, ++	data
ER 2 A...E	Si-Di	SMD, GI, S, 50...300V, 1,5A, <35ns A=50V, B=100V, C=150V, D=200V, E=300V	71a(5x3,5)	Die	BYG 22A...D	data
ER 900	Diac	=BDW 32	31	Tra	A 9903, D 32	data
ER 1400	EAROM-IC	100x14 Bit	14-DIP	Gie	M5G1400, PCB 1400	
ERA 15-01...-10	Si-Di	GI, 100...1000V, 1A	31a	Fjd	BY 126...127, BY 133...135, 1N4002...07, ++	data
ERA 17-02...-04	Si-Di	GI, ESD-proof, 200...400V, 1A	31a	Fjd	BY 126...127, BY 133...135, 1N4004...07, ++	data
ERA 22-02...-10	Si-Di	GI, S, 200...1000V, 0,5A, <400ns	31a	Fjd	BY 208/600...1000, BYD 53D...M	data
ERA 32-01...-02	Si-Di	GI, S, 100...200V, 1A, <100ns	31a	Fjd	BYD 72B...G, EGP 10B...G, FE 1B...H	data
ERA 34-10	Si-Di	GI, S, 1000V, 0,1A, <150ns	31a	Fjd	BYD 53M	data
ERA 38-04...-06	Si-Di	GI, S, 400...600V, 0,5A, <50ns	31a	Fjd	BYD 53G...J, BYV 26B...C	data
ERA 81-004	Si-Di	Schottky-GI, 40V, 1A, Uf<0,55V(1A)	31a	Fjd	BYS 21-45, BYV 10-40, MBR 150, 1N5819,++	data

Type	Device	Short Description	Fig.	Manu	Comparision Types	More at
ERA 83-004	Si-Di	Schottky-Gl, 40V, 1A, Uf<0,55V(1A)	31a	Fjd	BYS 21-45, BYV 10-40, MBR 150, 1N5819,++	data
ERA 83-006	Si-Di	Schottky-Gl, 60V, 1A, Uf<0,58V(1A)	31a	Fjd	BYV 10-60, MBR 160, SB 160	data
ERA 84-009	Si-Di	Schottky-Gl, 90V, 1A, Uf<0,9V(1A)	31a	Fjd	BYS 21-90, MBR 190, SB 190	data
ERA 85-009	Si-Di	Schottky-Gl, 90V, 1A, Uf<0,82V(1A)	31a	Fjd	BYS 21-90, MBR 190, SB 190	data
ERA 91-02	Si-Di	Gl, S, 200V, 0,5A, <35ns	31a	Fjd	BYD 72D...G	data
ERA 92-02	Si-Di	Gl, S, 200V, 1A, <35ns	31a	Fjd	BYD 72D...G	data
ERB 06-13	Si-Di	CTV-Damper, 1300, 1A, <4µs	31a	Fjd,Aeg,Sie	BY 228, BY 328, BY 428, BY 558	data
ERB 06-15	Si-Di	CTV-Damper, 1500, 1A, <4µs	31a	Fjd	BY 228, BY 328, BY 428, BY 558	data
ERB 12-01...-10	Si-Di	Gl, 100...1000V, 1A	31a	Fjd	BY 126...127, BY 133...135, 1N4002...07, ++	data
ERB 24-04C...06C	Si-Di	Gl, S, 400...600V, 1A, <700ns	31a	Fjd	BYX 55/600, MR 814...818, RGP 10G...M, ++	
ERB 24-04D...06D	Si-Di	Gl, S, 400...600V, 0,7A, <1µs	31a	Fjd	BYX 55/600, MR 814...818, RGP 10G...M, ++	
ERB 26-20	Si-Di	TV-Gl, 2000V, 0,2A, <4µs	31a	Fjd	BY 203/20, SHG 2...2,5	data
ERB 28-04...06	Si-Di	Gl, S, 400...600V, 0,5A, <400ns	31a	Fjd	BYX 55/600, MR 814...818, RGP 10G...M, ++	
ERB 28-04D...06D	Si-Di	Gl, S, 400...600V, 0,5A, <1µs	31a	Fjd	BYX 55/600, MR 814...818, RGP 10G...M, ++	
ERB 29-02...04	Si-Di	Gl, S, 200...400V, 0,8A, <4µs	31a	Fjd	BYX 55/600, MR 812...818, RGP 10D...M, ++	
ERB 30-13	Si-Di	Gl, Uni, 1300V, 1A, <4µs	31a	Fjd	BY 228, BY 231/1400, EM 513, GP 10V	data
ERB 30-15	Si-Di	=ERB 30-13: 1500V	31a		BY 228, BY 231/1500, EM 516, GP 10W	data
ERB 32-01...-02	Si-Di	Gl, S, 100...200V, 1,2A, <100ns	31a	Fjd	BYD 72B...G, EGP 10B...H, FE 1B...H	data
ERB 37-08...-10	Si-Di	Gl, S, 800...1000V, 1A, <250ns	31a	Fjd	BYT 52K...M, BYV 26D...E, MUR 1100, ++	data
ERB 38-04...-06	Si-Di	Gl, S, 400...600V, 0,8A, <50ns	31a	Fjd	BYD 53G...J, BYV 26B...C	data
ERB 43-02...-08	Si-Di	Gl, S, 200...800V, 0,5A, <400ns	31a	Fjd	BY 208/600...1000, BYD 53D...M	data
ERB 44-02...-10	Si-Di	Gl, S, 200...1000V, 1A, <400ns	31a	Fjd	BYD 33D...M, BYT 52D...M, BYV 26A...E	data
ERB 81-004	Si-Di	Schottky-Gl, 40V, 1,7A, Uf<0,55V(2A)	31a	Fjd	BYS 22-45, MBR 340, SB 340, 1N5822, ++	data
ERB 83-004	Si-Di	Schottky-Gl, 40V, 1,7A, Uf<0,55V(2A)	31a	Fjd	BYS 22-45, MBR 340, SB 340, 1N5822, ++	data
ERB 83-006	Si-Di	Schottky-Gl, 60V, 2A, Uf<0,58V(2A)	31a	Fjd	BYS 26-90, MBR 360, SB 360	data
ERB 84-009	Si-Di	Schottky-Gl, 90V, 2A, Uf<0,9V(2A)	31a	Fjd	BYS 26-90, MBR 390, SB 390	data
ERB 91-02	Si-Di	Gl, S, 200V, 1A, <35ns	31a	Fjd	BYD 72D...G	data
ERB 93-02	Si-Di	Gl, S, 200V, 1,5, <35ns	31a	Fjd	BYD 73D...G, BYD 74D...G	data
ERC 01-02...-10	Si-Di	Gl, 200...1000V, 1,5A	31a	Fjd,Aeg,Sie	BYD 13D...M, BY 226...227, GP 15D...M, ++	data
ERC 01-02F...-04F	Si-Di	Gl, 200...400V, 1,8A	31a	Fjd	BYD 13D...M, BY 226...227, GP 15D...M, ++	data
ERC 04-02...-10	Si-Di	Gl, 200...1000V, 1,2A	31a	Fjd	BYD 13D...M, BY 226...227, GP 15D...M, ++	data
ERC 04-02F...-0,4F	Si-Di	Gl, 200...400V, 1,5A	31a	Fjd	BYD 13D...M, BY 226...227, GP 15D...M, ++	data
ERC 05-06...-08	Si-Di	Gl, 600...800V, 1,2A	31a	Fjd,Aeg,Sie	BYD 13J...M, BY 226...227, GP 15J...M, ++	data
ERC 06-13	Si-Di	S, CTV-Damper, 1300V, 1,5A, <4µs	31a	Fjd,Aeg,Sie	BY 228, BY 328, BY 428, BY 558	data
ERC 06-15	Si-Di	=ERC 06-13: 1500V	31a		BY 228, BY 328, BY 428, BY 558	data
ERC 12-06...-08	Si-Di	Gl, 600...800V, 1,2A	31a	Fjd	BYD 13J...M, BY 226...227, GP 15J...M, ++	data
ERC 13-06...-08	Si-Di	Gl, 600...800V, 1,2A	31a	Fjd	BYD 13J...M, BY 226...227, GP 15J...M, ++	data
ERC 20-02...-08	Si-Di	S-L, 200...800V, 5A(Tc=125°), <400ns	17t§	Fjd	BYR 29-500...-800, BYT 71-200...-800, ++	data
ERC 20-02M...-08M	Si-Di	=ERC 20-...: Iso	15d	Fjd	BY 229F-200...-800, BY 430F-1000, ++	data
ERC 24-04...06	Si-Di	Gl, S, 400...600V, 1A, <400ns	31a	Fjd	BYX 55/600, MR 814...818, RGP 10G...M, ++	data
ERC 25-04...06	Si-Di	Gl, S, 400...600V, 1,2A, <400ns	31a	Fjd	BYD 33G...M, BYV 95B...E, RGP 15G...M, ++	data
ERC 26-13	Si-Di	Gl, Uni, 1300V, 1,5A, <4µs	31a	Fjd	BY 228, BY 328, BY 428, BY 558	data
ERC 26-15	Si-Di	=ERC 26-13: 1500V	31a		BY 228, BY 328, BY 428, BY 558	data
ERC 27-13	Si-Di	Gl, Uni, 1300V, 1A, <4µs	31a	Fjd	BY 228, BY 328, BY 428, BY 558	data
ERC 27-15	Si-Di	=ERC 27-13: 1500V	31a		BY 228, BY 328, BY 428, BY 558	data
ERC 30-01...-02	Si-Di	Gl, S, 100...200V, 1,5A, <100ns	31a	Fjd	BYD 73B...G, EGP 20B...G, FE 2B...H	data
ERC 33-02	Si-Di	Gl, S, 200V, 0,8A, <100ns	31a	Fjd	BYV 26B...E, FE 1D...H	data
ERC 35-02	Si-Di	Gl, S, 200V, 2,5A, <100ns	31a	Fjd	BYM 36A, BYV 28-600, EGP 30D, FE 3D	data
ERC 38-04...-06	Si-Di	Gl, S, 400...600V, 1A, <50ns	31a	Fjd	BYV 26B...E, MUR 140...160	data
ERC 46-02...04	Si-Di	Gl, S, 200...400V, 1,5A, <500ns	31a	Fjd	BYV 95A...C, BYD 34D...M, RGP 15D...M, ++	
ERC 62-004	Si-Di	Schottky, S-L, 45V, 10A(Tc=110°)	17t§	Fjd	BYV 39-45, MBR 1045, PBYR 1045	data
ERC 62M-004	Si-Di	=ERC 62-...: Iso, 10A(Tc=101°)	15d	Fjd	MBRF 1045, PBYR 1045F,X	data
ERC 80-004	Si-Di	Schottky, S-L, 40V, 5A(Tc=102°)	17t§	Fjd	BYV 19-45, MBR 745, PBYR 745	
ERC 81-004	Si-Di	Schottky-Gl, 40V, 2,6A, Uf<0,55V(3A)	31a	Fjd	BYS 26-45, MBR 340, SB 340, 1N5822	data
ERC 81-006	Si-Di	Schottky-Gl, 60V, 3A, Uf<0,58V(3A)	31a	Fjd	BYS 26-90, MBR 360, SB 360	data
ERC 84-009	Si-Di	Schottky-Gl, 90V, 3A, Uf<0,8V(3A)	31a	Fjd	BYS 26-90, MBR 390, SB 390	data
ERC 90-02	Si-Di	Gl, S, 200V, 5A(Tc=125°), <35ns	17d	Fjd	-	
ERC 90 G-02	Si-Di	Gl, S, 200V, 5A(Tc=134°), <35ns	15d	Fjd	-	
ERC 90 M-02...-03	Si-Di	Gl, S, 200...300V, 5A(Tc=134°), <35ns	15d	Fjd	-	
ERC 91-02	Si-Di	Gl, S, 200V, 3A, <35ns	31a	Fjd	BYV 28-200...-600, EGP 30D...G, FE 3D...H	data
ERD 07-13	Si-Di	S, CTV-Damper, 1300V, 1,5A, <1,5µs	31a	Fjd	BY 228, BY 328, BY 428, BY 558	data
ERD 07-15	Si-Di	S, CTV-Damper, 1500V, 1,5A, <1,5µs	31a	Fjd	BY 228, BY 328, BY 428, BY 558	data
ERD 08M-15	Si-Di	S-L, CTV-Damper, 1500V, 5A, <1,5µs	18d	Fjd	5THZ52, 5TUZ52(C), 5VUZ52	data
ERD 28-04...08	Si-Di	Gl, S, 400...800V, 1,5A, <400ns	31a	Fjd	BYD 34G...M, BYV 95B...C, RGP 15G...M, ++	data
ERD 29-02...-06	Si-Di	TV-Gl, 200...600V, 2,5A, <400ns	31a	Aeg,Fjd,Sie	BYW 14/600, BYW 95C, RGP 30D...M, ++	
ERD 32-01...-02	Si-Di	Gl, S, 100...200V, 3A, <100ns	31a	Fjd	BYV 28/200, EGP 30D...G, FE 3D...H, ++	data
ERD 33-02	Si-Di	Gl, S, 200V, 2A, <100ns	31a	Fjd	BYV 27/200, EGP 20D...G, FE 2D...H, ++	data

354

Type	Device	Short Description	Fig.	Manu	Comparision Types	More at
ᴿD 38-04...-06	Si-Di	Gl, S, 400..600V, 1,5A, <50ns	31a	Fjd	BYM 26B..C, BYV 27-200..600	data
ᴿD 80-004	Si-Di	Schottky, S-L, 40V, 15A(Tc100°)	18t§	Fjd		data
ᴿD 2808	Si-Di	→ERD 28-08				
ᴿE 41-15	Si-Di	CTV-Damper, di-def, 1500V, 3A, <2,2µs	31a	Fjd	BY 328, BY 428	data

S

Type	Device	Short Description	Fig.	Manu	Comparision Types	More at
S	Si-P	=2SA1745 (Typ-Code/Stempel/marking)	35(2mm)	Say	→2SA1745	data
S	Si-P	=2SA1753 (Typ-Code/Stempel/marking)	35	Say	→2SA1753	data
S	Si-P	=2SB789A-S (Typ-Code/Stempel/marking)	39	Mat	→2SB789A	data
S	Si-N+Di	=2SD1974 (Typ-Code/Stempel/marking)	39	Hit	→2SD1974	data
S	Si-N	=BCX 41R (Typ-Code/Stempel/marking)	35	Sie,Tho	→BCX 41R	data
S	Z-Di	=SM 6T 33 (Typ-Code/Stempel/marking)	71(6x4mm)	Tho	→SM 6T...	data
S	Z-Di	=Z1 SMA-22 (Typ-Code/Stempel/marking)	71(5x2,5)	Fag	→Z1 SMA-22	data
S 1 A..D	Si-Di	SMD, Gl, S, 50..200V, 1A, <25ns A=50V, B=100V, C=150V, D=200V	71a(5x2,5)	Gsi	BYG 80A..D	data
S 1,01(A,Z)	Si-Di	Gl, S, 200..600V, 0,7A, 1,5µs ES 1,01=400V, A=600V, Z=200V	31a	Sak	BY 126...127, BY 133...134, 1N4003..07,++	data
S 1 F, ES 01 F	Si-Di	Gl, S, 1500V, 0,5A, <400ns	31a	Sak	BY 269, BY 231/1500, RGP 15-16	data
S 2 A..D	Si-Di	SMD, Gl, S, 50..200V, 2A, <30ns A=50V, B=100V, C=150C, D=200V	71a(5x3,5)	Gsi	FES 2A..D	data
S 2 F,G	Si-Di	SMD, Gl, S, F=300V, G=400V, 2A, <50ns	71a(5x3,5)		FES 2F...G	data
S 3 A..D	Si-Di	SMD, Gl, S, 50..200V, 3A, <30ns A=50V, B=100V, C=150C, D=200V	71a(8x6)	Gsi	-	data
S 3 F,G	Si-Di	SMD, Gl, S, F=300V, G=400V, 3A, <50ns	71a(8x65)			data
SAB 33(CS)	Si-Di	Dual, S-L, 200V, 5A(Tc=110°), <45ns	17p§	Fjd	BYQ 28-200, MUR 620	data
SAB 34M(C)	Si-Di	Dual, S-L, 200V, 5A(Tc=110°), <45ns	15p	Fjd	BYQ 28F-200, MA 649, MA 653	data
SAB 82-004	Si-Di	Dual, Schottky, S-L, 40V, 5A(Tc=103°)	17p§	Fjd	BYV 18-40, BYV 118-40, BYS 24-45	data
SAB 85-009	Si-Di	Dual, Schottky-Gl, 90V, 5A(Tc=95°)	17p§	Fjd	BYS 24-90, HRW 34, MBR 2090CT	data
SAB 92-02	Si-Di	Dual, S-L, 200V, 5A(Tc=120°), <35ns	17p§	Fjd	BYQ 28-200, MUR 620	data
ESAC 25-02C...-04C	Si-Di	Dual, 200..400V, 10A(Tc=106°), <400ns	17p§	Fjd	BYT 28/300.. 500	data
ESAC 25-...D	Si-Di	=ESAC 25-...C:	17n1		-	data
ESAC 25-...N	Si-Di	=ESAC 25-...C:	17q&		-	data
ESAC 33-02C	Si-Di	Dual, S-L, 200V, 8A(Tc=98°), <100ns	17p§	Fjd	BYQ 28-200, BYT 28-300...-500	data
ESAC 33-...D	Si-Di	=ESAC 33-...C:	17n1		-	data
ESAC 33-...N	Si-Di	=ESAC 33-...C:	17q&		-	data
ESAC 34M(C)	Si-Di	Dual, S-L, 200V, 10A(Tc=100°), <45ns	15p	Fjd	BYQ 28F-200, BYQ 28X-200	data
ESAC 39-04C...-06C	Si-Di	Dual, 400..600V, 5A(Tc=115°), <50ns	17p§	Fjd	BYT 16P400A, FE 16G...J	data
ESAC 39-...D	Si-Di	=ESAC 39-...C:	17n1		-	data
ESAC 39-...N	Si-Di	=ESAC 39-...C:	17q&		-	data
ESAC 61-004	Si-Di	Dual, Schottky, S-L, 40V, 12A(Tc100°)	18p§	Fjd	BYS 28-45, BYV 43-40, MBR 3045PT	data
ESAC 63-004	Si-Di	Dual, Schottky, S-L, 45V, 20A(Tc=92°)	17p§	Fjd	MBR 2045CT, PBYR 2045CT	data
ESAC 75-005...-02	Si-Di	Dual,S, 50..200V, 8A(Tc=110°), <300ns	22I§	Fjd	(BYQ 28/..., BYT 28/...)[4]	data
ESAC 81-004	Si-Di	Dual, Schottky-Gl, 40V, 8A(Tc=90°)	22I§	Fjd	(BYS 24/45, BYV 118/40)[4]	data
ESAC 82-004	Si-Di	Dual, Schottky, S-L, 40V, 10A(Tc=97°)	17p§	Fjd	BYS 24-45, BYV 33-40, BYV 133-40	data
ESAC 83-004	Si-Di	Dual, Schottky, S-L, 40V, 20A(Tc=95°)	18p§	Fjd	BYS 28-45, BYV 73-40, MBR 3045PT	data
ESAC 83M-006	Si-Di	Dual, Schottky, S-L, 60V, 20A(Tc=78°)	18p	Fjd	MA 751, PBYR 3040PTF	data
ESAC 85M-006	Si-Di	Dual, Schottky, S-L, 90V, 10A(Tc=87°)	15p	Fjd	HRW 36F, MA 761	data
ESAC 87-009	Si-Di	Dual, Schottky, S-L, 90V, 16A(Tc=90°)	18p§	Fjd	BYS 28-90, PBYR 30100PT	data
ESAC 87M-009	Si-Di	Dual, Schottky, S-L, 90V, 16A(Tc=90°)	18p	Fjd	MA 762	data
ESAC 93-02	Si-Di	Dual, S-L, 200V, 12A(Tc=115°), <35ns	18p§	Fjd	BYV 72-200, BYW 99-200	data
ESAC 93-03	Si-Di	Dual, S-L, 300V, 12A(Tc=116°), <35ns	18p§	Fjd	MUR 3030PT	data
ESAC 93M-02	Si-Di	Dual, S-L, 200V, 12A(Tc=116°), <35ns	18p	Fjd	BYV 72F-200, BYW 99PI-200	data
ESAD 25-02C...-04C	Si-Di	Dual, S-L, 200..400V, 15A(Tc=100°) <400ns	18p§	Fjd	BYV 74-300...500, MUR 3020PT...3040PT	data
ESAD 25-...D	Si-Di	=ESAD 25-...C:	18n1		-	data
ESAD 25M-...C	Si-Di	=ESAD 25-...C: Iso	18p		-	data
ESAD 25M-...D	Si-Di	=ESAD 25-...C: Iso	18n1		-	data
ESAD 25M-...N	Si-Di	=ESAD 25-...C: Iso	18q		-	data
ESAD 25-...N	Si-Di	=ESAD 25-...C:	18q&		-	data
ESAD 33-02C	Si-Di	Dual, S-L, 200V, 15A(Tc=88°), <100ns	18p§	Fjd	BYV 72/200, BYW 99P/200, MUR 3020PT	data
ESAD 33-...D	Si-Di	=ESAD 33-...C:	18n1		-	data
ESAD 33-...N	Si-Di	=ESAD 33-...C:	18q&		-	data
ESAD 39-04C...-06C	Si-Di	Dual, S-L, 400..600V, 10A(Tc=98°) <50ns	18p§	Fjd		data
ESAD 39-...D	Si-Di	=ESAD 39-...C:	18n1		-	data
ESAD 39M-...C	Si-Di	=ESAD 39-...C: Iso	18p		-	data
ESAD 39M-...D	Si-Di	=ESAD 39-...C:	18n1		-	data
ESAD 39M-...N	Si-Di	=ESAD 39-...C: Iso	18q		-	data
ESAD 39-...N	Si-Di	=ESAD 39-...C:	18q&		-	data
ESAD 75-005...-02	Si-Di	Dual,S, 50..200V, 16A(Tc=110°), <300ns	23I§	Fjd	(BYV 72/..., BYW 99P/..., MUR 3005...20PT)[4]	data
ESAD 81-004	Si-Di	Dual, Schottky-Gl, 40V, 15A(Tc=100°)	23I§	Fjd	MBR 3045CT	data

Type	Device	Short Description	Fig.	Manu	Comparision Types	More at
ESAD 83-004	Si-Di	Dual, Schottky, S-L, 40V, 30A(Tc=90°)	18p§	Fjd	BYS 28-45, BYV 73-40, MBR 3045PT	data
ESAD 83-006	Si-Di	Dual, Schottky, S-L, 60V, 30A(Tc=90°)	18p§	Fjd	MBR 3060PT, PBYR 3060PT	data
ESAD 83M-004	Si-Di	Dual, Schottky, S-L, 40V, 30A(Tc=79°)	18p	Fjd	PBYR 3040PT	data
ESAD 83M-006	Si-Di	Dual, Schottky, S-L, 60V, 30A(Tc=75°)	18p	Fjd	-	data
ESAD 85-009	Si-Di	Dual, Schottky, S-L, 90V, 25A(Tc=90°)	18p§	Fjd	BYS 28-90, PBYR 30100PT	data
ESAD 85M-009	Si-Di	Dual, Schottky, S-L, 90V, 25A(Tc=76°)	18p	Fjd	MA 762	data
ESAD 92-02	Si-Di	Dual, S-L, 200V, 20A(Tc=115°), <40ns	18p§	Fjd	BYV 72-200, BYW 99-200, MUR 30°0PT	data
ESAD 92-03	Si-Di	Dual, S-L, 300V, 20A(Tc=110°), <40ns	18p§	Fjd	-	data
ESAD 92M-02	Si-Di	=ESAD 92-02: Iso, 20A(Tc=108°)	18p	Fjd	BYV 72F-200, BYW 99PI-200	data
ESAD 92M-03	Si-Di	=ESAD 92-03: Iso, 20A(Tc=96°)	18p	Fjd		data
ESAE 83-004	Si-Di	Dual, Schottky, S-L, 40V, 60A(Tc=84°)	18p§	Fjd	MBR 6045PT, PBYR6040WT	data
ESAE 83-006	Si-Di	Dual, Schottky, S-L, 60V, 60A(Tc=71°)	18p§	Fjd		data
ESDA 6 V1M3	LIN-IC, Z-Di	SMD,18x TAZ(6,1V) f.Centronics Interf.	20-MDIP	Tho		
ESDA 6 V1S3	LIN-IC, Z-Di	SMD, 18x TAZ(6,1V) f. Data Line Prot.	20-MDIP	Tho	-	
ESDA 6 V1U1	LIN-IC, Z-Di	SMD, 6x TAZ(6,1V) f. RS 423 Interface	8-MDIP	Tho		
ESDA 25 B1	LIN-IC, Z-Di	SMD, 6x TAZ(25V) f. RS 232 Interface	8-MDIP	Tho		
ESDA 25 B5	LIN-IC, Z-Di	SMD, 8x TAZ(25V) f. RS 232 Interface	16-MDIP	Tho		

ESM

Type	Device	Short Description	Fig.	Manu	Comparision Types	More at
ESM 22-100...600(N)	Triac	100...600V, 2,5A, Igt/Ih<40/<30mA	2a§	Tho	T 2303..., TAG 208-...	data
ESM 23-100...600	Triac	100...600V, 6A, Igt<80mA	22a§	Tho	TAG 260-..., TAG 265-..., T 4700...	data
ESM 28	Si-N	NF-L, 30V, 4A, 25W, 3MHz	17j§	Tho	BD 243, BD 533, BD 539, BD 947, ++	data
ESM 29	Si-P	NF-L, 30V, 4A, 25W, 3MHz	17j§	Tho	BD 244, BD 534, BD 540, BD 948, ++	data
ESM 168	LIN-IC	Prellschutz-flipflop/chatter supress.	5	Tho		
ESM 188M/450...750	F-Thy	TV-HA, 450...750V, 5A, Igt<60mA, <2,7µs	22a§	Tho	S 3703..., (BStCC01...H, TD3F...H)[1]	data
ESM 189M/450...750	F-Thy	TV-HA, 450...750V, 5A, Igt<60mA, <5µs	22a§	Tho	S 6080C..., (BStCC01...R, TD3F...R)[1]	data
ESM 206 EV	CMOS-IC	Schrittmotorstg./stepper motor control	14-DIP	Tho		
ESM 217	Si-N-Darl	NF-L, 60V, 10A, 70W, >4MHz, B>1000	17j§	Tho	BDT 63(A...C), BDW 93A...C, BDX 33A...D, ++	data
ESM 218	Si-N-Darl	=ESM 217: 80V	17j§	Tho	BDT 63A...C, BDW 93B...C, BDX 33B...D, ++	data
ESM 222 R	LIN-IC	=TDA 1042: 'profess. Version	23/8Pin	Tho	-	
ESM 227(A)	LIN-IC	Motorregler/speed ctrl., 3,8...18V,1,8A	14-QIP	Tho		
ESM 227 N	LIN-IC	=ESM 227: Fig.→	14-QIP+d	Tho	UL 1901	
ESM 228M/450...750	F-Thy+Di	TV-HA, 450...750V, 5A, Igt<60mA, <2,7µs	22a§	Tho	BStCC01...H, TD3F...H, TD4F...H	data
ESM 229M/450...750	F-Thy+Di	TV-HA, 450...750V, 5A, Igt<60mA, <5µs	22a§	Tho	BStCC01...R, TD3F...R, TD4F...R	data
ESM 231(N)	LIN-IC	NF-E, 15W(18V/4Ω)	14-QIP+d	Tho	TBA 790, TCA 150, TDA 1042	
ESM 249R/500	F-Thy+Di	500V, 5A, Igt<60mA, <25µs	22a§	Tho	TD3F500H, TD4F500H, S 6087B	
ESM 261	Si-P-Darl	NF-L, 60V, 10A, 70W, >4MHz, B>1000	17j§	Tho	BDT 62(A...C), BDW 94A...C, BDX 34A...D, ++	data
ESM 262	Si-P-Darl	=ESM 261: 80V	17j§	Tho	BDT 62A...C, BDW 94B...C, BDX 34B...D, ++	data
ESM 273	LIN-IC	=TDA 1104(SP)	17-SQL	Tho	TDA 1104(SP)	
ESM 302 EV	CMOS-IC	Rhythmus-generator/rhythm generator	16-DIP	Tho		
ESM 303	LIN-IC	Motor Control		Tho		
ESM 310(BP)	LIN-IC	=TDA 1100(SP)	11-SIL	Tho	TDA 1100(SP)	
ESM 312	LIN-IC	Motorregler/speed control, 3...18V, 1A	8-DIP+b	Tho		
ESM 313/...R	F-Thy	100...600V, 40A, Igt/Ih<150/<200mA,<5µs	23a§	Tho	-	data
ESM 352	LIN-IC	=TEA 1000	14-DIP	Tho	TEA 1000	
ESM 374	Z-IC	+12V, 140mA	7c	Tho		
ESM 375...	LIN-IC	=TEA 1035...	...-DIP	Tho	TEA 1035...	
ESM 381	DIG-IC	High Immunity Logic, Phase Cell, ..60V	14-DIP	Tho	-	
ESM 382	DIG-IC	High Immunity Logic, Ctrl. Gate, ..60V	14-DIP	Tho	-	
ESM 383	DIG-IC	High Immunity Logic, Gate, 32...60V	14-DIP	Tho	-	
ESM 400(A)	Si-N	TV-VA, 170V, 1,5A, 20W, 1MHz	17j§	Tho	BD 239E, 2SD386, 2SD578, 2SD1138, ++	data
ESM 416	LIN-IC	=SAY 115: Fig.→	14-DIP	Tho	(SAY 115)	
ESM 427	LIN-IC	Ionisations-rauchmelder/smoke detector	14-DIP	Tho		
ESM 432	LIN-IC	=ESM 432C: Fig.→	14-QIP	Tho		
ESM 432 C	LIN-IC	=ESM 532 C: 30V, 3,5A, 20W(28V/4Ω)	11-SIL	Tho	ESM 532 C, TDA 1111	
ESM 432 N	LIN-IC	=ESM 432C: Fig.→	14-QIP+d	Tho		
ESM 463	SMPS-IC	=TEA 1001	17-SQL	Tho	TEA 1001	
ESM 504/...	50Hz-Thy	50...600V, 4A, Igt/Ih<20/<50mA, 25µs	17h§	Tho	TAG 630-..., TAG 620-..., TIC 116..., ++	data
ESM 508/...	50Hz-Thy	50...600V, 8A, Igt/Ih<40/<60mA	17h§	Tho	TAG 655-..., 2N6394...6399, BStD10..., ++	data
ESM 509/...	50Hz-Thy	50...600V, 10A, Igt/Ih<40/<60mA	17h§	Tho	2N6394...6399, BStD10..., BT 152/..., ++	data
ESM 532	LIN-IC	=ESM 532C: Fig.→	14-QIP	Tho	-	
ESM 532 C	LIN-IC	NF-E, 32V, 3,5A, 20W(28V/4Ω)	11-SIL	Tho	TDA 1111(SP)	
ESM 532 N	LIN-IC	=ESM 532C: Fig.→	14-QIP+d	Tho		
ESM 566	LIN-IC	=TEB 1013: ohne Basisvorwdst.	11-SIL	Tho	(L 702SP, TEB 1013)	
ESM 567	LIN-IC	=TEA 1034	16-DIP	Tho	TEA 1034	
ESM 568 A	LIN-IC	=TEA 2015A	13-SQL	Tho	TEA 2015A	
ESM 573 C	LIN-IC	=TEA 1020(SP)	17-SQL	Tho	TEA 1020(SP)	
ESM 585	IC	Teletext-demod.(antiope)		Tho	-	
ESM 586	IC	Teletext-demod.(antiope)		Tho	-	
ESM 620	LIN-IC	=TDA 1099	11-SIL	Tho	TDA 1099	
ESM 621	LIN-IC	=TDA 1098	14-DIP	Tho	TDA 1098	
ESM 631 CM	LIN-IC	Näherungssch./proximity det., 4...35V	81	Tho		
ESM 631 DP	LIN-IC	=ESM 631CM: Fig.→	8-DIP	Tho		
ESM 631 FP	LIN-IC	=ESM 631CM: SMD	8-MDIP	Tho		
ESM 632 C	LIN-IC	=ESM 532 C: 26V, 3,5A, 14W(24V/4Ω)	11-SIL	Tho	ESM 532 C, TDA 1111	
ESM700	R+	Io=220mA Vo:9.77...10.23V	3P	Nuc,Ses		data pinout
ESM 700	Z-IC	=TCA 700	14d	Tho	→TCA 700	data pinout
ESM 707	LIN-IC	Tacho-generator	8-DIP	Tho		

Type	Device	Short Description	Fig.	Manu	Comparision Types	More at
ESM 732 C	LIN-IC	=ESM 532 C: 18V, 3,5A, 8W(14V/4Ω)	11-SIL	Tho	ESM 532 C, TDA 1111	
ESM 740(G)	Thy	asym. -5...300V, 10A, Igt<50mA, <15μs	30b	Tho	-	
ESM 765/100...800A	Si-Di	Gl/S-L, 100...800V, 10A(Tc=100°),<300ns	17t§	Tho	BYT 12P-.., BYT 79-..., BYV 79-...	data
ESM 765/...PI	Si-Di	=ESM 765/...: Iso	17d	Tho	BYT 12PI-...	data
ESM 900	LIN-IC	=TCA 900: Fig.→	5	Tho	-	
ESM 901	LIN-IC	=TCA 900: Fig.→	4-DIP+b	Tho	-	
ESM 910	LIN-IC	=TCA 910: Fig.→	5	Tho	-	
ESM 911	LIN-IC	=TCA 910: Fig.→	4-DIP+b	Tho	-	
ESM 1231 C	LIN-IC	=TDA 1103(SP)	11-SIL	Tho	TDA 1103(SP)	
ESM 1350 P	LIN-IC	TV-ZF AM/FM-ZF	8-DIP	Tho	MC 1350	
ESM1406	R+	Io=1A Vo:5.7...6.3V P:1.2W	3P	Nuc,Ses		data pinout
ESM 1406	Z-IC	+6V, 0,78A	14b	Tho	... 78N06 (TO-126)	data pinout
ESM1410	R+	Io=900mA Vo:9...11V P:1.2W	3P	Nuc,Ses		data pinout
ESM 1410	Z-IC	+10V, 0,68A	14b	Tho	... 78N10 (TO-126)	data pinout
ESM 1532 C	LIN-IC	=TDA 1102(SP)	11-SIL	Tho	TDA 1102(SP), TDA 1111(SP)	
ESM 1600 B	KOP-IC	Quad, Pegelumsetzer/level shift	14-DIP	Tho	-	pdf
ESM 1600 BFP	KOP-IC	=ESM 1600B: SMB	14-MDIP	Tho	-	pdf
ESM 1601	LIN-IC	=TDE 1064, TDF 1064		Tho	TDE 1064, TDF 1064	
ESM 1602 B	KOP-IC	Quad, Pegelumsetzer/level shift	14-DIP	Tho	-	pdf
ESM 1607 G	LIN-IC	Relay & Lamp Driver, Ucc=36V	84	Tho	-	
ESM 1631 CM	LIN-IC	=ESM 631CM: 4...20V	81	Tho	ESM 631CM	
ESM 1631 DP	LIN-IC	=ESM 631DP: 20V	8-DIP	Tho	ESM 631DP	
ESM 1631 FP	LIN-IC	=ESM 631FP: 20V	8-MDIP	Tho	ESM 631FP	
ESM 1637	KOP-IC	50V, 1A	84	Tho	-	
ESM 2633	Si-N	≈2N3055	23a§	Tho	→2N3055	
ESM 2666	Si-N	S-L, 1500/600V, 6A, 75W(Tc=95°)	23a§	Tho	BU-508A, 2SD350A, 2SD649, 2SD821,++	data
ESM 2667	Si-N	=ESM2666: 1500/700V	23a§	Tho	BU 508A, 2SD350A, 2SD821, 2SD649,++	data
ESM 2668	Si-N	≈BU 526	23a§	Tho	→BU 526	
ESM 2725	Si-N	≈BF 459		Tho	→BF 459	
ESM 2731	Si-N	≈BU 526	23a§	Tho	→BU 526	
ESM 2808	Si-N+Di	=BU 800	23a§	Tho	→BU 800	
ESM 4629	Thy		17	Tho	-	
ESM 7040	MOS-IC	8x Latch+Treiber/driver, 50V, 0,5A	24-DIP	Tho	-	
ESM 7545 DF,V	Si-N-Darl+Di	S-L, 600/450V, 75A, 250W	80	Tho		data
ESMT 5070 DF,V	Si-N-Darl+Di	Triple, S-L, 1000/700V, 50A, 300W	80	Tho		data

ET...EZ

Type	Device	Short Description	Fig.	Manu	Comparision Types	More at
ET	Si-N	=2SC4120 (Typ-Code/Stempel/marking)	35	Say	→2SC4120	data
ET	Si-N-Darl	=2SD1471-ET (Typ-Code/Stempel/marking)	39	Hit	→2SD1471	data
ET	GaAs-N-FET-d	=3SK273 (Typ-Code/Stempel/marking)	44	Mat	→3SK273	data
ET	Si-N	=BCW 65RA (Typ-Code/Stempel/marking)	35	Sie	→BCW 65RA	data
ET	Z-Di	=SM 6T 33A (Typ-Code/Stempel/marking)	71(6x4mm)	Tho	→SM 6T...	data
ET	Z-Di	=Z1 SMA-24 (Typ-Code/Stempel/marking)	71(5x2,5)	Fag	→Z1 SMA-24	data
ET 020	Si-PNPN-Di	(Diac), 170V, 0,6A≈, Ubo=190..210V	31a	Sak	-	
ET 2128(J,N)...	sRAM-IC	2k x 8Bit, 150...200ns	24-DIP	Tho	...4016...	data
ET 2147 H...	sRAM-IC	4k x 1Bit, 35...70ns	18-DIP	Tho	...2147...	data
ET 2716 Q...	EPROM-IC	2k x 8Bit, 350...450ns, 525/132mW	24-DIP	Tho	...2716...	data
ET 2764 Q...	EPROM-IC	8k x 8Bit, 150...450ns, 525/132mW	28-DIP	Tho	...2764...	
ET 4116 J,N...	dRAM-IC	16k x 1Bit, 150...200ns	16-DIP	Tho	...4116...	
ET 4164 J,N...	dRAM-IC	64k x 1Bit, 150...200ns	16-DIP	Tho	...4164...	
ET 93...	μC-IC	8 Bit μComp.		Tho	-	
ET 94...	μC-IC	4 Bit μComp.		Tho	-	
ET 27128 Q...	EPROM-IC	16k x 8Bit, 150...450ns, 525/105mW	28-DIP	Tho	...27128...	
ET 90400	μC-IC	4 Bit μComp.		Tho	-	
ETC 2716 Q...	EPROM-IC	2k x 8Bit, 350...450ns	24-DIP	Tho	...2716...	data
ETC 2732 Q...	EPROM-IC	4k x 8Bit, 350...450ns	24-DIP	Tho	...2732...	data
ETC 5040	CMOS-IC	Telecom, Pcm Codec-filter	16-DIP	Tho	M 5912, KT 3040, TP 3040, μA 5912, 2912	
ETC 5051	CMOS-IC	Telecom, 1-Kan./chnl. μ-Law Par.-Combo	20-DIP	Tho	-	
ETC 5054	CMOS-IC	Telecom, μ-Law Combo Codec	16-DIP,DIC	Tho	KT 3030, TP 3054, μA 3054, 2916	pdf pinout
ETC 5054 FN(-X)	CMOS-IC	=ETC 5054: Fig.→	20-PLCC		-	pinout
ETC 5056	CMOS-IC	Telecom, 1-Kan./chnl. A-law Par.-Combo	20-DIP	Tho	-	pinout
ETC 5057(D)(-X)	CMOS-IC	Telecom, A-law Combo Codec	16-DIP,DIC	Tho	KT 3032, TP 3057, μA 3057, 2917	pinout
ETC 5057 FN(-X)	CMOS-IC	=ETC 5057: Fig.→	20-PLCC		-	pinout
ETC 5057 N(-X)	CMOS-IC	=ETC 5057: SMD	16-MDIP		-	
ETC 5064	CMOS-IC	Telecom, μ-Law Combo Codec	20-DIP,DIC	Tho	KT 3031, KT 3064, TP 3064	pdf pinout
ETC 5064 FN(-X)	CMOS-IC	=ETC 5064: Fig.→	20-PLCC		-	pinout
ETC 5067	CMOS-IC	Telecom, A-law Combo Codec	20-DIP,DIC	Tho	KT 3033, TP 3067	pdf pinout
ETC 5067 FN(-X)	CMOS-IC	=ETC 5067: Fig.→	20-PLCC		-	pinout
ETC 92...	μC-IC	16 Bit μComp.		Tho	-	
ETC 93...	μC-IC	8 Bit μComp.		Tho	-	
ETC 94...	μC-IC	4 Bit μComp.		Tho	-	
ETL 2128(J,N)...	sRAM-IC	2k x 8Bit, 150...200ns	24-DIP	Tho	...4016...	data
ETL 2147 H...	sRAM-IC	4k x 1Bit, 55...70ns	18-DIP	Tho	...2147...	data
ETL 93...	μC-IC	8 Bit μComp.		Tho	-	
ETL 94...	μC-IC	4 Bit μComp.		Tho	-	
ET P	GaAs-N-FET-d	=3SK273-P (Typ-Code/Stempel/marking)	44	Mat	→3SK273	data
ET Q	GaAs-N-FET-d	=3SK273-Q (Typ-Code/Stempel/marking)	44	Mat	→3SK273	data
ET R	GaAs-N-FET-d	=3SK273-R (Typ-Code/Stempel/marking)	44	Mat	→3SK273	data
ET S	GaAs-N-FET-d	=3SK273-S (Typ-Code/Stempel/marking)	44	Mat	→3SK273	data
EU	Si-N	=BCW 65RB (Typ-Code/Stempel/marking)	35	Sie	→BCW 65RB	data

Type	Device	Short Description	Fig.	Manu	Comparision Types	More at
EU	Z-Di	=SM 6T 36 (Typ-Code/Stempel/marking)	71(6x4mm)	Tho	→SM 6T...	data
EU	Si-N	=XP 1554 (Typ-Code/Stempel/marking)	45(2mm)	Mat	→XP 1554	data
EU	Z-Di	=Z1 SMA-27 (Typ-Code/Stempel/marking)	71(5x2,5)	Fag	→Z1 SMA-27	data
EU (0)1(A,Z)	Si-Di	GI, S, 200...600V, 0,25A, 400ns	31a	Sak	BA 157...159, BY 204/..., BYX 57/...	data
		EU(0)1=400V, A=600V, Z=200V				
EU (0)2(A,Z)	Si-Di	GI, S, 200...600V, 1A, 400ns	31a	Sak	BYD 33J..M, BYT 11/600, BYT 52J..M,++	data
		EU(0)2=400V, A=600V, Z=200V				
EU 2 YX	Si-Di	GI, S, 100V, 1,2A, 200ns	31a	Sak	BYD 33D..M, BYT 11/600, BYT 52B..M,++	data
EUA	Z-Di	=SMAJ 43A (Typ-Code/Stempel/marking)	71(5x2,5)	Tho	→SMAJ ...	data
EUF	Z-Di	=SMAJ 58A (Typ-Code/Stempel/marking)	71(5x2,5)	Tho	→SMAJ ...	data
EUI	Z-Di	=SMAJ 70A (Typ-Code/Stempel/marking)	71(5x2,5)	Tho	→SMAJ ...	data
EUL	Z-Di	=SMAJ 85A (Typ-Code/Stempel/marking)	71(5x2,5)	Tho	→SMAJ ...	data
EUN	Z-Di	=SMAJ 100A (Typ-Code/Stempel/marking)	71(5x2,5)	Tho	→SMAJ ...	data
EUQ	Z-Di	=SMAJ 130A (Typ-Code/Stempel/marking)	71(5x2,5)	Tho	→SMAJ ...	data
EUT	Z-Di	=SMAJ 154A (Typ-Code/Stempel/marking)	71(5x2,5)	Tho	→SMAJ ...	data
EUV	Z-Di	=SMAJ 188A (Typ-Code/Stempel/marking)	71(5x2,5)	Tho	→SMAJ ...	data
EV	PIN-Di	=1SV241 (Typ-Code/Stempel/marking)	45	Say	→1SV241	data
EV	Z-Di	=SM 6T 36A (Typ-Code/Stempel/marking)	71(6x4mm)	Tho	→SM 6T...	data
EV	Si-P+R	=UN 2154 (Typ-Code/Stempel/marking)	35	Mat	→UN 2154	data
EV	Z-Di	=Z1 SMA-3u (Typ-Code/Stempel/marking)	71(5x2,5)	Fag	→Z1 SMA-30	data
EW	Si-N	=BCW 65RC (Typ-Code/Stempel/marking)	35	Sie	→BCW 65RC	data
EW	Z-Di	=SM 6T 39 (Typ-Code/Stempel/marking)	71(6x4mm)	Tho	→SM 6T...	data
EW	Si-P, R	=UN 211N (Typ-Code/Stempel/marking)	35	Mat	→UN 211N	data
EW	Z-Di	=Z1 SMA-33 (Typ-Code/Stempel/marking)	71(5x2,5)	Fag	→Z1 SMA-33	data
EX	Si-N	=2SD2402-EX (Typ-Code/Stempel/marking)	39	Nec	→2SD2402	data
EX	Si-N	=BCW 66RF (Typ-Code/Stempel/marking)	35	Sie	→BCW 66RF	data
EX	Z-Di	=SM 6T 39A (Typ-Code/Stempel/marking)	71(6x4mm)	Tho	→SM 6T...	data
EX	Si-N+R	=UN 221N (Typ-Code/Stempel/marking)	35	Mat	→UN 221N	data
EX	Si-N+R	=UNR 921N (Typ-Code/Stempel/marking)	35(1,6mm)	Mat	→UNR 921N	data
EX	Z-Di	=Z1 SMA-36 (Typ-Code/Stempel/marking)	71(5x2,5)	Fag	→Z1 SMA-36	data
EX 0022 TA	Z-Di	11,5V	31a			
EX 0048 CE	Z-Di	6,2V	31a			
EX 0074 CE	Z-Di	115V	31a			
EXB 919	Hybrid-Z-IC	Z-IC, +100V, 0,2A		Fjd	-	
EY	Si-N	=2SC2882-Y (Typ-Code/Stempel/marking)	39	Tos	→2SC2882	data
EY	Si-N	=2SC3265-Y (Typ-Code/Stempel/marking)	35	Tos	→2SC3265	data
EY	Si-N	=2SC3398 (Typ-Code/Stempel/marking)	35	Say	→2SC3398	data
EY	Si-N	=2SC4398 (Typ-Code/Stempel/marking)	35(2mm)	Say	→2SC4398	data
EY	Si-N	=2SD2402-EY (Typ-Code/Stempel/marking)	39	Nec	→2SD2402	data
EY	MOS-N-FET-e*	=2SK1697 (Typ-Code/Stempel/marking)	39	Hit	→2SK1697	data
EY	Si-N	=BCW 66RG (Typ-Code/Stempel/marking)	35	Sie	→BCW 66RG	data
EY	Si-N	=KTC3265-Y (Typ-Code/Stempel/marking)	35	Kec	→KTC 3265	data
EY	Si-N	=KTC4374-Y (Typ-Code/Stempel/marking)	39	Kec	→KTC 4374	data
EY	Si-P+R	=UN 211T (Typ-Code/Stempel/marking)	35	Mat	→UN 211T	data
EY	Z-Di	=Z1 SMA-39 (Typ-Code/Stempel/marking)	71(5x2,5)	Fag	→Z1 SMA-39	data
EYV-320(D)	Si-Di	≈BA 127	31a		→BA 127	
EZ	Si-N	=2SD2402-EZ (Typ-Code/Stempel/marking)	39	Nec	→2SD2402	data
EZ	Si-N	=BCW 66RH (Typ-Code/Stempel/marking)	35	Sie	→BCW 66RH	data
EZ	Si-N+R	=UN 221T (Typ-Code/Stempel/marking)	35	Mat	→UN 221T	data
EZ	Si-N	=XP 05543 (Typ-Code/Stempel/marking)	46(2mm)	Mat	→XP 05543	data
EZ	Z-Di	=Z1 SMA-43 (Typ-Code/Stempel/marking)	71(5x2,5)	Fag	→Z1 SMA-43	data
EZ-055...-372	Z-Di	5...37V, 0,4W	31a	Njr	BZX55/.., BZX83/..., ZPD..., 1N5231...58,++	data
EZ 0150	Z-Di	Avalanche + int. Thyristor, 140...160V	31a	Sak	-	data
eZ 80190	µP-IC	8Bit CPU, Z80-compatible (64kB) mode or		Zil	-	pdf pinout
		full 24Bit (16MB) addressing mode				
F						
F.	Si-Di	=1PS79SB10 (Typ-Code/Stempel/marking)	71(1,3mm)	Phi	→1PS79SB10	data
F	C-Di	=HVU 352 (Typ-Code/Stempel/marking)	71(1,7mm)	Hit	→HVU 352	data
F	Si-Di	=MA 2S331 (Typ-Code/Stempel/marking)	71(1,3mm)	Mat	→MA 2S331	data
F	Z-Di	=MAZF 075 (Typ-Code/Stempel/marking)	71(1,7mm)	Mat	→MAZF 075	data
F	Z-Di	=MAZN 075 (Typ-Code/Stempel/marking)	71(1,3mm)	Mat	→MAZN 075	data
F	Z-Di	=MAZS 033 (Typ-Code/Stempel/marking)	71(1,3mm)	Mat	→MAZS 033	data
F	Si-Di	=SB 01-15CP (Typ-Code/Stempel/marking)	35	Say	→SB 01-15CP	data
F	Si-P	=2SA1034 (Typ-Code/Stempel/marking)	35	Mat	→2SA1034	data
F	Si-P	=2SA1531 (Typ-Code/Stempel/marking)	35(2mm)	Mat	→2SA1531	data
F	Si-N	=2SC4399 (Typ-Code/Stempel/marking)	35(2mm)	Say	→2SC3499	data
F 1	Si-N	=BFS 18 (Typ-Code/Stempel/marking)	35	Phi, Tho	→BFS 18	data
F 1	Si-Di	=FRS 1A (Typ-Code/Stempel/marking)	71(5x2,5)	Fag	→FRS 1A	data
F 1	PIN-Di	=JDP 2S01E (Typ-Code/Stempel/marking)	71(1,2mm)	Tos	→JDP 2S01E	
F.1	PIN-Di	=JDP 2S01T (Typ-Code/Stempel/marking)	71(1,2mm)	Tos	→JDP 2S01T	
F 1	PIN-Di	=JDP 2S01U (Typ-Code/Stempel/marking)	71(1,7mm)	Tos	→JDP 2S01U	
F 1	Si-N	=KST 1009F1 (Typ-Code/Stempel/marking)	35	Sam	→KST 1009	data
F 1	MOS-N-FET-e	=MMBF 1374 (Typ-Code/Stempel/marking)	35(2mm)	Ons	→MMBF 1374	
F 1	Z-Di	=MMSZ 5236B (Typ-Code/marking)	71(2,7mm)	Ons	→MMSZ 5236B	
F1-10W4DHCBB-H	LED	wht ø9.3mm I=25cd Vf:1.5V If:80mA		Roi		data pdf pinout
F1-10W4DHCVB-H	LED	wht ø9mm I=22cd Vf:3V If:80mA		Roi		data pdf pinout
F 1	Si-N	=2SC1009-F1 (Typ-Code/Stempel/marking)	35	Mot,Nec	→2SC1009	data
F 1 B1CA	Si-Di	Dual, S-L, 100V, 10A(Tc=106°), <400ns	17q&	Kec		data

Type	Device	Short Description	Fig.	Manu	Comparision Types	More at
F 1 B2CA	Si-Di	=F 1B1CA: 200V	17q&		-	data
F 1 B...CAI	Si-Di	=F 1B...CA: Iso	17q			data
F 1 B...CC	Si-Di	=F 1B...CA:	17p§		BYP 20-..., BYQ 28-..., BYT 28-...	data
F 1 B...CCI	Si-Di	=F 1B...CA: Iso	17p		-	data
F 1 C57...5120	Z-Di	TAZ, 7,5...340V	71a(5x2,5)	Org	BZG 04/...	data
F 1 E23	MOS-N-FET-e	=2SK1195		Shi		
F 1 E50	MOS-N-FET-e	=2SK1672		Shi		
F 1 E90	MOS-N-FET-e	=2SK1533		Shi		
F 1 F16	Si-Di	SMD, Uni, S, 1600V, 0,5A, 1500ns	71a(1,7mm)	Org	-	data
F 1 H2	Si-Di	SMD, GI, S, 200V, 1A, 400ns	71a(1,7mm)	Org	-	data
F 1 H4	Si-Di	=F 1H2: 400V	71a(1,7mm)	Org	-	data
F 1 H4	Si-Di	=F 1H2: 400V	71a(5x2,5)	Org	BYG 20G...M, BYG 21K..M, FR 2G...M	data
F 1 H6	Si-Di	=F 1H2: 600V	71a(5x2,5)	Org	BYG 20J...M, BYG 21K...M, FR 2J...M	data
F 1 J2	Si-Di	SMD, Schottky-Di, 20V, 1A, 30ns	71a(5x2,5)	Org	MA 735, MBRS 130	data
F 01 J3	Si-Di	SMD, Schottky-Di, 30V, 0,1A, 30ns	71a(1,7mm)	Org	-	data
F 1 J3U	Si-Di	SMD, Schottky-Di, 30V, 1A, 30ns	71a(5x2,5)	Org	-	data
F 1 J4	Si-Di	SMD, Schottky-Di, 40V, 1A, 30ns	71a(5x2,5)	Org	-	data
F 1 J6	Si-Di	SMD, Schottky-Di, 60V, 1A, 30ns	71a(5x2,5)	Org	-	data
F 1 J9	Si-Di	SMD, Schottky-Di, 90V, 0,7A, 30ns	71a(5x2,5)	Org	-	data
F 1 N05	MOS-N-FET-e	=MMDF 1N05E (Typ-Code/Stempel/marking)	8-MDIP	Mot	→MMDF 1N05E	data
F 1 N2	Si-Di	SMD, GI, 200V, 1A	71a(1,7mm)	Org	-	data
F 1 N4	Si-Di	=F 1N2: 400V	71a(1,7mm)	Org	-	data
F 1 N4	Si-Di	=F 1N2: 400V	71a(5x2,5)	Org	BYG 10G...M, D 1F40...60, U 1GC44...JC44	data
F 1 N6	Si-Di	=F 1N2: 600V	71a(5x2,5)	Org	BYG 10J...M, D 1F60, U 1DCJC44	data
F 1 O	Si-P	=KSA 1182-O (Typ-Code/Stempel/marking)	35	Sam	→KSA 1182	data
F 1 p	Si-N	=BFS 18 (Typ-Code/Stempel/marking)	35	Phi	→BFS 18	data
F 1 P2	Si-Di	SMD, GI,S, 200V, 1A, 50ns	71a(1,7mm)	Org	-	data
F 1 P2S	Si-Di	=F 1P2: 60ns	71a(1,7mm)	Org	-	data
F 1 S0P05	MOS-P-FET-e	=RF 1S0P05 (Typ-Code/Stempel/marking)	30	Rca	→RF 1S0P05	data
F 1 S0P06	MOS-P-FET-e	=RF 1S0P06 (Typ-Code/Stempel/marking)	30	Rca	→RF 1S0P06	data
F 1 S25N06	MOS-N-FET-e	=RF 1S25N06 (Typ-Code/Stempel/marking)	30	Rca	→RF 1S25N06	data
F 1 S45N06	MOS-N-FET-e	=RF 1S45N06 (Typ-Code/Stempel/marking)	30	Rca	→RF 1S45N06	data
F 1 S50N06	MOS-N-FET-e	=RF 1S50N06 (Typ-Code/Stempel/marking)	30	Rca	→RF 1S50N06	data
F 1 S60P03	MOS-N-FET-e	=RF 1S60P03 (Typ-Code/Stempel/marking)	30	Rca	→RF 1S60P03	data
F 1 S70N03	MOS-N-FET-e	=RF 1S70N03 (Typ-Code/Stempel/marking)	30	Rca	→RF 1S70N03	data
F 1 S70N06	MOS-N-FET-e	=RF 1S70N06 (Typ-Code/Stempel/marking)	30	Rca	→RF 1S70N06	data
F 1 V61B	Si-Varistor	SMD, VDR/Varistor, 0,15A, 2,3V(1mA)	71(5x2,5)	Org		data
F 1 V62B	Si-Varistor	SMD, VDR/Varistor, 0,15A, 2,3V(1mA)	71(5x2,5)	Org		data
F 1	Si-Varistor	=VR 61F1 (Typ-Code/Stempel/marking)	71(5x2,5)	Shi	→VR 61F1	data
F 1 Y	Si-P	=KSA 1182-Y (Typ-Code/Stempel/marking)	35	Sam	→KSA 1182	data
F 2	Si-N	=BFS 19 (Typ-Code/Stempel/marking)	35	Phi, Tho	→BFS 19	data
F 2	Z-Di	=EDZ 6.8B (Typ-Code/Stempel/marking)	71(1,3mm)	Rhm	→EDZ 6.8B	data
F 2	Si-Di	=FRS 1B (Typ-Code/Stempel/marking)	71(5x2,5)	Fag	→FRS 1B	data
F.2	PIN-Di	=JDP 2S02T (Typ-Code/Stempel/marking)	71(1,2mm)	Tos	→JDP 2S02T	
F 2	Si-N	=KST 1009F2 (Typ-Code/Stempel/marking)	35	Sam	→KST 1009	
F 2	Z-Di	=MMSZ 5237B (Typ-Code/marking)	71(2,7mm)	Ons	→MMSZ 5237B	
F 02	N-FET	=SO 4091 (Typ-Code/Stempel/marking)	35	Tho	→SO 4091	
F 2	Z-Di	=UDZ 6.8B (Typ-Code/Stempel/marking)	71(1,7mm)	Rhm	→UDZ 2.0B...36B	
F 2	Z-Di	=UDZS 6.8B (Typ-Code/Stempel/marking)	71(1,7mm)	Rhm	→UDZS 5.1B...10B	
F 2	Si-N	=2SC1009-F2 (Typ-Code/Stempel/marking)	35	Mot,Nec	→2SC1009	
F 2	Si-N	=2SC2814-F2 (Typ-Code/Stempel/marking)	35	Say	→2SC2814	
F 2 B1CA	Si-Di	Dual, GI/S-L, 100V, 20A, 400ns	18q	Kec	-	data
F 2 B2CA	Si-Di	=F 2B1CA: 200V	18q		-	data
F 2 B...CC	Si-Di	=F 2B...CA:	18p		BYV 72F-..., BYV 74F-..., MA 651	data
F 2 C02E	MOS-NP-FET-e	=MMDF 2C02E (Typ-Code/Stempel/marking)	8-MDIP	Mot	→MMDF 2C02E	data
F 02 J3	Si-Di	SMD, Schottky-Di, 30V, 0,2A, 30ns	71a(1,7mm)	Org	-	data
F 2 J3U	Si-Di	SMD, Schottky-Di, 30V, 3A, 50ns	71a(6,4mm)	Org	-	data
F 2 J4	Si-Di	SMD, Schottky-Di, 40V, 3A, 50ns	71a(6,4mm)	Org	-	data
F 2 J4S	Si-Di	SMD, Schottky-Di, 40V, 3A, 50ns	71a(6,4mm)	Org	-	data
F 2 J6	Si-Di	SMD, Schottky-Di, 60V, 3A, 50ns	71a(6,4mm)	Org	-	data
F 2 J6S	Si-Di	SMD, Schottky-Di, 60V, 3A, 50ns	71a(6,4mm)	Org	-	data
F 2 J9	Si-Di	SMD, Schottky-Di, 90V, 2A, 50ns	71a(6,4mm)	Org	-	data
F 2 N02E	MOS-N-FET-e	=MMDF 2N02E (Typ-Code/Stempel/marking)	8-MDIP	Mot	→MMDF 2N02E	data
F 2 N05Z	MOS-N-FET-e*	=MMDF 2N05Z (Typ-Code/Stempel/marking)	8-MDIP	Ons	→MMDF 2N05Z	data
F 2 p	Si-N	=BFS 19 (Typ-Code/Stempel/marking)	35	Phi	→BFS 19	
F 2 P02E	MOS-N-FET-e	=MMDF 2P02E (Typ-Code/Stempel/marking)	8-MDIP	Mot	→MMDF 2P02E	data
F 2 P2	Si-Di	SMD, GI, S, 200V, 2A, 50ns	71a(5x2,5)	Org	-	data
F 2 t	Si-N	=BFS 19 (Typ-Code/Stempel/marking)	35	Phi	→BFS 19	data
F2V4 PH	Z-Di	=BZX 79-F2V4 (Typ-Code/Stempel/marking)	31	Phi	→BZX 79-F...	data
F2V7 PH	Z-Di	=BZX 79-F2V7 (Typ-Code/Stempel/marking)	31	Phi	→BZX 79-F...	data
F 3	Si-Di	=1SS193 (Typ-Code/Stempel/marking)	35	Tos	→1SS193	data
F 3	Si-N	=2SC1009-F3 (Typ-Code/Stempel/marking)	35	Mot,Nec	→2SC1009	data
F 3	Si-N	=2SC2814-F3 (Typ-Code/Stempel/marking)	35	Say	→2SC2814	data
F 3	Si-N	=BF 840 (Typ-Code/Stempel/marking)	35	Val	→BF 840	data
F 3	Si-Di	=FRS 1D (Typ-Code/Stempel/marking)	71(5x2,5)	Fag	→FRS 1D	data
F 3	Si-Di	=KDS 193 (Typ-Code/Stempel/marking)	35	Kec	→KDS 193	data

Type	Device	Short Description	Fig.	Manu	Comparision Types	More at
F 3	Si-N	=KST 1009F3 (Typ-Code/Stempel/marking)	35	Sam	→KST 1009	data
F 3	Z-Di	=MMSZ 5238B (Typ-Code/marking)	71(2,7mm)	Ons	→MMSZ 5238B	
F 03	N-FET	=SO 4391 (Typ-Code/Stempel/marking)	35	Tho	→SO 4391	data
F3L60U	Si-Di	=SF 3L60U (Typ-Code/Stempel/marking)	17	Shi	→SF 3L60U	data
F 3 N08L	MOS-N-FET-e	=RFD 3N08L (Typ-Code/Stempel/marking)	30	Rca	→RFD 3N08L	data
F 3 T	Si-Di	=1PS193 (Typ-Code/Stempel/marking)	35	Phi	→1PS193	data
F3V0 PH	Z-Di	=BZX 79-F3V0 (Typ-Code/Stempel/marking)	31	Phi	→BZX 79-F...	data
F3V3 PH	Z-Di	=BZX 79-F3V3 (Typ-Code/Stempel/marking)	31	Phi	→BZX 79-F...	
F 3 V50	MOS-N-FET-e	=2SK1244		Shi		
F3V6 PH	Z-Di	=BZX 79-F3V6 (Typ-Code/Stempel/marking)	31	Phi	→BZX 79-F...	data
F 3 V90	MOS-N-FET-e	=2SK1534		Shi		
F3V9 PH	Z-Di	=BZX 79-F3V9 (Typ-Code/Stempel/marking)	31	Phi	→BZX 79-F...	data
F 3 W90	MOS-N-FET-e	=2SK1536		Shi		
F 4	Si-Di	=FRS 1G (Typ-Code/Stempel/marking)	71(5x2,5)	Fag	→FRS 1G	data
F 4	Z-Di	=MMSZ 5239B (Typ-Code/marking)	71(2,7mm)	Ons	→MMSZ 5239B	
F 4	Si-N	=2SC1009-F4 (Typ-Code/Stempel/marking)	35	Mot,Nec	→2SC1009	data
F 4	Si-N	=2SC2814-F4 (Typ-Code/Stempel/marking)	35	Say	→2SC2814	data
F 4	Si-N	=BFS 18R (Typ-Code/Stempel/marking)	35	Tho,Val	→BFS 18R	data
F 4	Si-N	=KST 1009F4 (Typ-Code/Stempel/marking)	35	Sam	→KST 1009	data
F4V3 PH	Z-Di	=BZX 79-F4V3 (Typ-Code/Stempel/marking)	31	Phi	→BZX 79-F...	
F4V7 PH	Z-Di	=BZX 79-F4V7 (Typ-Code/Stempel/marking)	31	Phi	→BZX 79-F...	
F 5	Si-N	=BFS 19R (Typ-Code/Stempel/marking)	35	Aeg,Tho,Val	→BFS 19R	
F 5	Si-Di	=FRS 1J (Typ-Code/Stempel/marking)	71(5x2,5)	Fag	→FRS 1J	data
F 5	Si-N	=KST 1009F5 (Typ-Code/Stempel/marking)	35	Sam	→KST 1009	data
F 5	Z-Di	=MMSZ 5240B (Typ-Code/marking)	71(2,7mm)	Ons	→MMSZ 5240B	
F 05	Si-N	=TSDF 1205 (Typ-Code/Stempel/marking)	44	Aeg	→TSDF 1205	data
F 5	Z-Di	=UDZ 36B (Typ-Code/Stempel/marking)	71(1,7mm)	Rhm	→UDZ 2.0B...36B	data
F 5	Si-Di	=1SS250 (Typ-Code/Stempel/marking)	35	Tos	→1SS250	data
F 5	Si-Di	=1SS370 (Typ-Code/Stempel/marking)	35(2mm)	Tos	→1SS370	data
F 5	Si-Di	=1SS403 (Typ-Code/Stempel/marking)	71(1,7mm)	Tos	→1SS403	data
F 5	Si-N	=2SC1009-F5 (Typ-Code/Stempel/marking)	35	Mot,Nec	→2SC1009	data
F 5	Si-N	=2SC2814-F5 (Typ-Code/Stempel/marking)	35	Say	→2SC2814	data
F 5 A1CA	Si-Di	Dual, GI/S-L, 100V, 5A, 200ns	17q	Kec	-	data
F 5 A2CA	Si-Di	=F 5A1CA: 200V	17q			data
F 5 A...CC	Si-Di	=F 5A..CA:	17p		BYP 20-..., BYQ 28-..., BYT 28-...	data
F5D1	IRED	880nm ⌀4.7mm I:>28mlm Vf:1.7V	TO18	Fcs,Har,Qtc		data pdf pinout
F5E1	IRED	880nm ⌀4.7mm I:>28mlm Vf:1.7V	TO18	Fcs,Har,Qtc		data pdf pinout
F 05 E23	MOS-N-FET-e	=2SK1194		Shi		
F5F1	IRED	945nm ⌀1.98mm I:0.28mW/Sr Vf:1.5V		Har,Qtc		data pdf pinout
F5G1	IRED	880nm ⌀1.98mm I:0.6mW/Sr Vf:1.5V		Har,Qtc		data pdf pinout
F5L60U	Si-Di	=SF 5L60U (Typ-Code/Stempel/marking)	17	Shi	→SF 5L60U	data
F5LC20U	Si-Di	=SF 5LC20U (Typ-Code/Stempel/marking)	17	Shi	→SF 5LC20U	data
F5LC30	Si-Di	=SF 5LC30 (Typ-Code/Stempel/marking)	17	Shi	→SF 5LC30	data
F5LC40	Si-Di	=SF 5LC40 (Typ-Code/Stempel/marking)	17	Shi	→SF 5LC40	data
F5S4	Si-Di	=SF 5S4 (Typ-Code/Stempel/marking)	17	Shi	→SF 5S4	data
F5S6	Si-Di	=SF 5S6 (Typ-Code/Stempel/marking)	17	Shi	→SF 5S6	data
F5SC3L	Si-Di	=SF 5SC3L (Typ-Code/Stempel/marking)	17	Shi	→SF 5SC3L	data
F5SC4	Si-Di	=SF 5SC4 (Typ-Code/Stempel/marking)	17	Shi	→SF 5SC4	data
F5V1 PH	Z-Di	=BZX 79-F5V1 (Typ-Code/Stempel/marking)	31	Phi	→BZX 79-F...	data
F 5 V50	MOS-N-FET-e	=2SK1246		Shi		
F5V6 PH	Z-Di	=BZX 79-F5V6 (Typ-Code/Stempel/marking)	31	Phi	→BZX 79-F...	data
F 5 W50	MOS-N-FET-e	=2SK1537		Shi		
F 6	Si-Di	=D 1NF60 (Typ-Code/Stempel/marking)	31	Shi	→D 1NF60	data
F 6	Si-Di	=FRS 1K (Typ-Code/Stempel/marking)	71(5x2,5)	Fag	→FRS 1K	data
F 6	Si-N	=2SC2223-F6 (Typ-Code/Stempel/marking)	35	Nec	→2SC2223	data
F 06C ..A	Si-Di	=F 06C..C:	17q&		-	data
F 06C 05C..60C	Si-Di	Dual, S, L, 50...600V, 6A, <150...250ns 05=50, 10=100, 15=150, 20=200V, <150ns 30=300, 40=400, 50=500, 60=600V,<250ns	17p§	Mop	BYR 28-..., BYT 28-...	
F 06C ..D	Si-Di	=F 06C..C:	17n1			data
F 6 F35VX2	MOS-N-FET-e	V-MOS, 350V, 6A, 30W, 0,78Ω, 55/110ns	17c	Shi	2SK1992, 2SK2357	data
F6L20U	Si-Di	=SF 6L20U (Typ-Code/Stempel/marking)	17	Shi	→SF 6L20U	data
F 6 V25	MOS-N-FET-e	=2SK1391		Shi		
F6V2 PH	Z-Di	=BZX 79-F6V2 (Typ-Code/Stempel/marking)	31	Phi	→BZX 79-F...	data
F6V8 PH	Z-Di	=BZX 79-F6V8 (Typ-Code/Stempel/marking)	31	Phi	→BZX 79-F...	data
F 7	Si-Di	=BAV 99RW (Typ-Code/Stempel/marking)	35(2mm)	Ons	→BAV 99RW	data
F 7	Si-Di	=FRS 1M (Typ-Code/Stempel/marking)	71(5x2,5)	Fag	→FRS 1M	data
F 7	Si-Di	=HSB 83J (Typ-Code/Stempel/marking)	35(2mm)	Hit	→HSB 83J	data
F 7	Si-Di	=HSM 83 (Typ-Code/Stempel/marking)	35	Hit	→HSM 83	data
F 07	N-FET	=SO 4392 (Typ-Code/Stempel/marking)	35	Tho	→SO 4392	data
F7V5 PH	Z-Di	=BZX 79-F7V5 (Typ-Code/Stempel/marking)	31	Phi	→BZX 79-F...	data
F 7 W90	MOS-N-FET-e	=2SK1538		Shi		
F 8	Si-P	=BF 824 (Typ-Code/Stempel/marking)	35	Phi	→BF 824	data
F 8	Si-Di	=HSM 122 (Typ-Code/Stempel/marking)	35	Hit	→HSM 122	data
F 8	Si-Di	=SM-1XF08 (Typ-Code/Stempel/marking)	31	Org	→SM-1XF08	data
F08	N-FET	=SO 4393 (Typ-Code/Stempel/marking)	35	Tho	→SO 4393	data

Type	Device	Short Description	Fig.	Manu	Comparision Types	More at
F 8 -	Si-P	=BF 824W (Typ-Code/Stempel/marking)	35(2mm)	Phi	→BF 824W	data
F 08A 05P...60P	Si-Di	S, L, 50...600V, 8A, <150...250ns 05=50, 10=100, 15=150, 20=200V, <150ns 30=300, 40=400, 50=500, 60=600V,<250ns	17t§	Mop	BYR 29-..., BYT 08-..., BYT 86-...	data
F 08A ...R	Si-Di	=F 08A..P:	17t1&			data
F 8 p	Si-P	=BF 824 (Typ-Code/Stempel/marking)	35	Phi	→BF 824	data
F 8 t	Si-P	=BF 824W (Typ-Code/Stempel/marking)	35(2mm)	Phi	→BF 824W	data
F 8 t	Si-P	=BF 824 (Typ-Code/Stempel/marking)	35	Phi	→BF 824	data
F8V2 PH	Z-Di	=BZX 79-F8V2 (Typ-Code/Stempel/marking	31	Phi	→BZX 79-F...	data
F 9	Si-Di	=1SS321 (Typ-Code/Stempel/marking)	35	Tos	→1SS321	data
F09	N-FET	=SO 3966 (Typ-Code/Stempel/marking)	35	Tho	→SO 3966	data
F9V1 PH	Z-Di	=BZX 79-F9V1 (Typ-Code/Stempel/marking	31	Phi	→BZX 79-F...	data
F10	N-FET	=SO 4092 (Typ-Code/Stempel/marking)	35	Tho	→SO 4092	data
F 10 F35VX2	MOS-N-FET-e	V-MOS, 350V, 10A, 40W, 0,36Ω, 90/190ns	17c	Shi	2SK2960	data
F10L60U	Si-Di	=SF 10L60U (Typ-Code/Stempel/marking)	17	Shi	→SF 10L60U	data
F10LC20U	Si-Di	=SF 10LC20U (Typ-Code/Stempel/marking)	17	Shi	→SF 10LC20U	data
F10LC40	Si-Di	=SF 10LC40 (Typ-Code/Stempel/marking)	17	Shi	→SF 10LC40	data
F10 PH	Z-Di	=BZX 79-F10 (Typ-Code/Stempel/marking)	31	Phi	→BZX 79-F...	data
F10SC3L	Si-Di	=SF 10SC3L (Typ-Code/Stempel/marking)	17	Shi	→SF 10SC3L	data
F10SC4(R)	Si-Di	=SF 10SC4(R)(Typ-Code/Stempel/marking)	17	Shi	→SF 10SC4(R)	data
F10SC6	Si-Di	=SF 10SC6 (Typ-Code/Stempel/marking)	17	Shi	→SF 10SC6	data
F10SC9	Si-Di	=SF 10SC9 (Typ-Code/Stempel/marking)	17	Shi	→SF 10SC9	data
F 10 V25	MOS-N-FET-e	=2SK1393		Shi		
F 10 W50	MOS-N-FET-e	=2SK1248		Shi		
F 10 W90	MOS-N-FET-e	=2SK1539		Shi		
F11 PH	Z-Di	=BZX 79-F11 (Typ-Code/Stempel/marking)	31	Phi	→BZX 79-F...	data
F 11	N-FET	=SO 4093 (Typ-Code/Stempel/marking)	35	Tho	→SO 4093	data
F 12	Si-N	=2SC2223-F12(Typ-Code/Stempel/marking)	35	Nec	→2SC2223	data
F 12	Si-N	=2SC4178-F12(Typ-Code/Stempel/marking)	35(2mm)	Nec	→2SC4178	data
F 12C ..A	Si-Di	=F 12C...C:	17q&		-	data
F 12C 05C...60C	Si-Di	Dual, S, L, 50...600V, 12A, <150...250ns 05=50, 10=100, 15=150, 20=200V, <150ns 30=300, 40=400, 50=500, 60=600V,<250ns	17p§	Mop	BYT 16P-..., BYV 34-..., FE 16A..J	data
F 12C ...D	Si-Di	=F 12C...C:	17n1			data
F12 PH	Z-Di	=BZX 79-F12 (Typ-Code/Stempel/marking)	31	Phi	→BZX 79-F...	data
F 12	N-FET	=SO 245B (Typ-Code/Stempel/marking)	35	Tho	→SO 245	data
F 13	Si-N	=2SC2223-F13(Typ-Code/Stempel/marking)	35	Nec	→2SC2223	data
F 13	Si-N	=2SC4178-F13(Typ-Code/Stempel/marking)	35(2mm)	Nec	→2SC4178	data
F 13	N-FET	=BFR 30R (Typ-Code/Stempel/marking)	35	Tho	→BFR 30R	data
F13 PH	Z-Di	=BZX 79-F13 (Typ-Code/Stempel/marking)	31	Phi	→BZX 79-F...	data
F 14	N-FET	=BFR 31R (Typ-Code/Stempel/marking)	35	Tho	→BFR 31R	data
F 14	Si-N	=2SC2223-F14(Typ-Code/Stempel/marking)	35	Nec	→2SC2223	data
F 14	Si-N	=2SC4178-F14(Typ-Code/Stempel/marking)	35(2mm)	Nec	→2SC4178	data
F 14 A..J	Si-Di	GI, Uni, 100...1000V, 1A A=100, B=200, C=300, D=400, E=500V, F=600, H=800, J=1000V	31a	Nec	BY 126...127, BY 133...135, 1N4002...07, ++	data
F 14 N05	MOS-N-FET-e	=RFD 14N05 (Typ-Code/Stempel/marking)	30	Rca	→RFD 14N05	data
F 14 N06	MOS-N-FET-e	=RFD 14N06 (Typ-Code/Stempel/marking)	30	Rca	→RFD 14N06	data
F 15 P05	MOS-P-FET-e	=RFD 15P05 (Typ-Code/Stempel/marking)	30	Rca	→RFD 15P05	data
F 15 P06	MOS-P-FET-e	=RFD 15P06 (Typ-Code/Stempel/marking)	30	Rca	→RFD 15P06	data
F15 PH	Z-Di	=BZX 79-F15 (Typ-Code/Stempel/marking)	31	Phi	→BZX 79-F...	data
F15	N-FET	=SO 5432 (Typ-Code/Stempel/marking)	35	Tho	→SO 5432	data
F 15 W50	MOS-N-FET-e	=2SK1249		Shi		
F 16C ..A	Si-Di	=F 16C...C:	17q&		-	data
F 16C 05C...60C	Si-Di	Dual, S, L, 50...600V, 16A, <150...250ns 05=50, 10=100, 15=150, 20=200V, <150ns 30=300, 40=400, 50=500, 60=600V,<250ns	17p§	Mop	BYT 16P-..., BYV 34-..., FE 16A..J	data
F 16C ...D	Si-Di	=F 16C...C:	17n1		-	data
F 16	Si-Di	=F 1F16 (Typ-Code/Stempel/marking)	71(1,7mm)	Org	→F 1F16	data
F 16 N05	MOS-N-FET-e	=RFD 16N05 (Typ-Code/Stempel/marking)	30	Rca	→RFD 15N05	data
F 16 N06	MOS-N-FET-e	=RFD 16N06 (Typ-Code/Stempel/marking)	30	Rca	→RFD 15N06	data
F16 PH	Z-Di	=BZX 79-F16 (Typ-Code/Stempel/marking)	31	Phi	→BZX 79-F...	data
F 16	Si-Di	=SM-1XF16 (Typ-Code/Stempel/marking)	31	Org	→SM-1XF16	data
F16	N-FET	=SO 5432R (Typ-Code/Stempel/marking)	35	Tho	→SO 5432R	data
F17	N-FET	=SO 5433 (Typ-Code/Stempel/marking)	35	Tho	→SO 5433	data
F 18 N10CS	MOS-FET-N-e	=RFB 18N10CS(Typ-Code/Stempel/marking)	17/5Pin	Rca	→RFB 18N10CS	data
F18 PH	Z-Di	=BZX 79-F18 (Typ-Code/Stempel/marking)	31	Phi	→BZX 79-F...	data
F18	N-FET	=SO 5433R (Typ-Code/Stempel/marking)	35	Tho	→SO 5433R	data
F19	N-FET	=SO 5434 (Typ-Code/Stempel/marking)	35	Tho	→SO 5433R	data
F 20	Si-Di	=SM-1XF20 (Typ-Code/Stempel/marking)	31	Org	→SM-1XF20	data
F 20	N-FET	=SO 245BR (Typ-Code/Stempel/marking)	35	Tho	→SO 245	data
F 20	Si-N	=TSDF 1220 (Typ-Code/Stempel/marking)	44	Aeg	→TSDF 1220	data
F20L60U	Si-Di	=SF 20L60U (Typ-Code/Stempel/marking)	17	Shi	→SF 20L60U	data
F20LC30	Si-Di	=SF 20LC30 (Typ-Code/Stempel/marking)	17	Shi	→SF 20LC30	data
F20 PH	Z-Di	=BZX 79-F20 (Typ-Code/Stempel/marking)	31	Phi	→BZX 79-F...	data
F20SC3L	Si-Di	=SF 20SC3L (Typ-Code/Stempel/marking)	17	Shi	→SF 20SC3L	data
F20SC4	Si-Di	=SF 20SC4 (Typ-Code/Stempel/marking)	17	Shi	→SF 20SC4	data
F 20 W25	MOS-N-FET-e	=2SK1395		Shi		
F 20 W50	MOS-N-FET-e	=2SK1250		Shi		

Type	Device	Short Description	Fig.	Manu	Comparision Types	More at
F 21	N-FET	=SO 245A (Typ-Code/Stempel/marking)	35	Tho	→SO 245	data
F22 PH	Z-Di	=BZX 79-F22 (Typ-Code/Stempel/marking)	31	Phi	→BZX 79-F...	data
F 22	N-FET	=SO 4091R (Typ-Code/Stempel/marking)	35	Tho	→SO 4091R	data
F 23 N06LE	MOS-N-FET-e	=RF1S23N06LE(Typ-Code/Stempel/marking)	30	Rca	→RF 1S23N06LE	data
F 23	N-FET	=SO 4391R (Typ-Code/Stempel/marking)	35	Tho	→SO 4391R	data
F24 PH	Z-Di	=BZX 79-F24 (Typ-Code/Stempel/marking)	31	Phi	→BZX 79-F...	data
F 24	N-FET	=SO 245AR (Typ-Code/Stempel/marking)	35	Tho	→SO 245AR	data
F 25	N-FET	=SO 245C (Typ-Code/Stempel/marking)	35	Tho	→SO 245C	data
F 26	N-FET	=SO 245CR (Typ-Code/Stempel/marking)	35	Tho	→SO 245CR	data
F27 PH	Z-Di	=BZX 79-F27 (Typ-Code/Stempel/marking)	31	Phi	→BZX 79-F...	data
F 27	N-FET	=SO 4392R (Typ-Code/Stempel/marking)	35	Tho	→SO 4392R	data
F 28	N-FET	=SO 4393R (Typ-Code/Stempel/marking)	35	Tho	→SO 4393R	data
F 29	N-FET	=SO 3966R (Typ-Code/Stempel/marking)	35	Tho	→SO 3966R	data
F 30	N-FET	=SO 4092R (Typ-Code/Stempel/marking)	35	Tho	→SO 4092R	data
F 30D ..A	Si-Di	=F 30D...C:	16q&			data
F 30D 05C..60C	Si-Di	Dual, S, L, 50...600V, 30A, <150...250ns 05=50, 10=100, 15=150, 20=200V, <150ns 30=300, 40=400, 50=500, 60=600V,<250ns	16p§	Mop	BYV 72-..., BYV 74-..., BYW 99P-...	data
F 30D ..D	Si-Di	=F 30D...C:	16n1		-	data
F 30 N06LE	MOS-N-FET-e	=RFP 30N06LE(Typ-Code/Stempel/marking)	17	Rca	→RFP 30N06LE	data
F30 PH	Z-Di	=BZX 79-F30 (Typ-Code/Stempel/marking)	31	Phi	→BZX 79-F...	data
F 30 S54(DC)	CMOS-IC	Telecom, µ-Law Codec, Filter	16-DIC	Fch	-	
F 30 S57(DC)	CMOS-IC	Telecom, A-law Codec, Filter	16-DIC	Fch	-	
F 30 S64(DC)	CMOS-IC	Telecom, µ-Law Codec, Filter	20-DIC	Fch	-	
F 30 S67(DC)	CMOS-IC	Telecom, A-law Codec, Filter	20-DIC	Fch	-	
F30SC3L	Si-Di	=SF 30SC3L (Typ-Code/Stempel/marking)	17	Shi	→SF 30SC3L	data
F30SC4	Si-Di	=SF 30SC4 (Typ-Code/Stempel/marking)	17	Shi	→SF 30SC4	data
F 30 W25	MOS-N-FET-e	=2SK1396		Shi		
F 31	Si-N	=BF 841 (Typ-Code/Stempel/marking)	35	Val	→BF 841	data
F 31	N-FET	=SO 4093R (Typ-Code/Stempel/marking)	35	Tho	→SO 4093R	data
F 32	N-FET	=SO 5434R (Typ-Code/Stempel/marking)	35	Tho	→SO 5434R	data
F33 PH	Z-Di	=BZX 79-F33 (Typ-Code/Stempel/marking)	31	Phi	→BZX 79-F...	data
F36 PH	Z-Di	=BZX 79-F36 (Typ-Code/Stempel/marking)	31	Phi	→BZX 79-F...	data
F39 PH	Z-Di	=BZX 79-F39 (Typ-Code/Stempel/marking)	31	Phi	→BZX 79-F...	data
F 40 N10LE	MOS-N-FET-e	=RF1S40N10LE(Typ-Code/Stempel/marking)	30	Rca	→RF 1S40N10LE	data
F 40 W25	MOS-N-FET-e	=2SK1397		Shi		
F43 PH	Z-Di	=BZX 79-F43 (Typ-Code/Stempel/marking)	31	Phi	→BZX 79-F...	data
F 44 H11	Si-N	=MJF 44H11 (Typ-Code/Stempel/marking)	17	Ons	→MJF 44H11	
F 45 H11	Si-N	=MJF 45H11 (Typ-Code/Stempel/marking)	17	Ons	→MJF 45H11	
F 45 N03L	MOS-N-FET-e	=RF 1S45N03L(Typ-Code/Stempel/marking)	30	Rca	→RF 1S45N03L	data
F47 PH	Z-Di	=BZX 79-F47 (Typ-Code/Stempel/marking)	31	Phi	→BZX 79-F...	data
F 50	Si-N	=TSDF 1250 (Typ-Code/Stempel/marking)	44	Aeg	→TSDF 1250	data
F51 PH	Z-Di	=BZX 79-F51 (Typ-Code/Stempel/marking)	31	Phi	→BZX 79-F...	data
F56 PH	Z-Di	=BZX 79-F56 (Typ-Code/Stempel/marking)	31	Phi	→BZX 79-F...	data
F62 PH	Z-Di	=BZX 79-F62 (Typ-Code/Stempel/marking)	31	Phi	→BZX 79-F...	data
F68 PH	Z-Di	=BZX 79-F68 (Typ-Code/Stempel/marking)	31	Phi	→BZX 79-F...	data
F75 PH	Z-Di	=BZX 79-F75 (Typ-Code/Stempel/marking)	31	Phi	→BZX 79-F...	data
F0094	IRED	950nm &0.3x0.3x0.18mm I=3cd Vf:1.35V	CHIP	Osr		data pdf pinout
F 099	Si-Di	≈BA 159	31a		→BA 159	data
F 100...117-L	µP-IC,MOS-IC	16 Bit µP System + Peripherie		Fer		pdf
F 114 B..F	Si-Di	GI, S, 200...600V, 0,8A, <200ns	31a	Nec	BY 201/..., BYV 13...16, RGP 10D..M, ++	data
F0118G	IRED	950nm &0.3x0.3x0.18mm I=3cd Vf:1.4V	CHIP	Osr		data pdf pinout
F0118J	IRED	950nm &0.3x0.3x0.24mm I=3cd Vf:1.3V	CHIP	Osr		data pdf pinout
F 133	Si-Di	≈BY 133	31a		→BY 133	
F0235D	IRED	950nm &0.25x0.25mm I=3cd Vf:1.4V	CHIP	Osr		data pdf pinout
F0235F	IRED	950nm &0.25x0.25mm I=3cd Vf:1.4V	CHIP	Osr		data pdf pinout
F0279E	LED	rd &0.2x0.2x0.22mm I=3cd Vf:<2.3V	CHIP	Osr		data pdf pinout
F0280E	LED	srd &0.2x0.2x0.22mm I:145mlm If:20mA	CHIP	Osr		data pdf pinout
F0281E	LED	or &0.2x0.2x0.1mm I:235mlm Vf:<2.3V	CHIP	Osr		data pdf pinout
F0282E	LED	yl &0.2x0.2x0.1mm I:190mlm Vf:<2.3V	CHIP	Osr		data pdf pinout
F0283C	LED	gr &0.17x0.17mm I:8mlm Vf:<2.2V	CHIP	Osr		data pdf pinout
F0283E	LED	gr &0.2x0.2mm I:8mlm Vf:<2.3V	CHIP	Osr		data pdf pinout
F0284B	LED	bl &0.26x0.26mm I:8mlm Vf:3.5V	CHIP	Osr		data pdf pinout
F0285B	LED	bl &0.26x0.26mm I:8mlm Vf:3.5V	CHIP	Osr		data pdf pinout
F0285 3	LED	gr &0.26x0.26mm I:8mlm Vf:3.5V	CHIP	Osr		data pdf pinout
F0292C	LED	gr &0.17x0.17mm I:3mlm Vf:2.2V	CHIP	Osr		data pdf pinout
F0292E	LED	gr &0.2x0.2mm I:>17mlm Vf:2.3V	CHIP	Osr		data pdf pinout
F0308A	LED	bl &0.26x0.26mm I:>17mlm Vf:3.8V	CHIP	Osr		data pdf pinout
F0347C	LED	yl &0.17x0.17mm I:12mlm Vf:2.2V	CHIP	Osr		data pdf pinout
F0348C	LED	or &0.17x0.17mm I:14mlm Vf:2.2V	CHIP	Osr		data pdf pinout
F0349C	LED	rd &0.17x0.17mm I:10mlm Vf:<2.2V	CHIP	Osr		data pdf pinout
F0372A	RCLED	rd &0.17x0.17mm I:0.7mW/Sr Vf:1.9V	CHIP	Osr		data pdf pinout
F418	LO-BIAS	Vs:±18V Vu:88dB Vo:±10V Vio:1mV	14D	B&h,Cpd		data pinout
F418C	LO-BIAS	Vs:±18V Vu:88dB Vo:±10V Vio:2mV	14D	B&h,Cpd		data pinout
F 420	Si-N	≈BF 393	7c		→BF 393	
F0460B	LED	bl &0.26x0.26mm I:0.7mW/Sr Vf:3.65V	CHIP	Osr		data pdf pinout
F0470B	LED	bl &0.26x0.26mm I:0.7mW/Sr Vf:3.65V	CHIP	Osr		data pdf pinout
F0496A	VCSEL	&0.27x0.22mm I:0.7mW/Sr Vf:2.1V	CHIP	Osr		data pdf pinout
F0527B	LED	gr &0.26x0.26mm I:>17mlm Vf:3.65V	CHIP	Osr		data pdf pinout

Type	Device	Short Description	Fig.	Manu	Comparision Types	More at
F0594A	LED	&0.3x0.3mm I:>17mlm Vf:1.4V If:100mA	CHIP	Osr		data pdf pinout
F 0810 BH	50Hz-Thy	200V, 5,1A(Tc=85°C), Igt/Ih <25/75mA	17h§	Tag	TIC 116.., BStC10...M, CS6-.., TAG 660...	data
F 0810 DH	50Hz-Thy	=F 0810BH: 400V	17h§		TIC 116.., BStC10...M, CS6-.., TAG 660...	data
F 0810 MH	50Hz-Thy	=F 0810BH: 600V	17h§		TIC 116.., BStC10...M, CS6-.., TAG 660...	data
F 0810 NH	50Hz-Thy	=F 0810BH: 800V	17h§		TIC 116.., BStC10...M, CS6-.., TAG 660...	data
F0950A	LED	&0.3x0.3mm I:>17mlm Vf:1.5V	CHIP	Osr		data pdf pinout
F0950B	LED	&0.3x0.3mm I:>17mlm Vf:1.5V	CHIP	Osr		data pdf pinout
F1047	IRED	880nm &0.3x0.3mm I=22cd Vf:<1.9V	CHIP	Osr		data pdf pinout
F1048B	IRED	880nm &0.4x0.4mm I=22cd Vf:<1.9V	CHIP	Osr		data pdf pinout
F1235	IRED	950nm &0.2x0.2mm I=22cd Vf:1.45V	CHIP	Osr		data pdf pinout
F1235A	IRED	950nm &0.2x0.2mm I=22cd Vf:1.5V	CHIP	Osr		data pdf pinout
F1372	LED	rd &0.22x0.22mm I=22cd Vf:<2.4V	CHIP	Osr		data pdf pinout
F 1612 BH	50Hz-Thy	200V, 10A(Tc=85°C), Igt/Ih 25/100mA	17h§	Tag	T 9,5N.., CS15-.., BStD10...M	data
F 1612 DH	50Hz-Thy	=F 1612BH: 400V	17h§	Tag	T 9,5N.., CS15-.., BStD10...M	data
F 1612 MH	50Hz-Thy	=F 1612BH: 600V	17h§		T 9,5N.., CS15-.., BStD10...M	data
F 1612 NH	50Hz-Thy	=F 1612BH: 800V	17h§		T 9,5N.., CS15-.., BStD10...M	data
F1998A	LED	srd &0.3x0.3mm I:>650mlm Vf:<2.5V	CHIP	Osr		data pdf pinout
F2000	LED	rd &0.7x0.7mm I:>8lm Vf:<2.5V	CHIP	Osr		data pdf pinout
F2001	LED	or &0.3x0.3mm I:>1150mlm Vf:<2.5V	CHIP	Osr		data pdf pinout
F2002	LED	yl &0.7x0.7mm I:>8lm Vf:<2.6V	CHIP	Osr		data pdf pinout
F2003	LED	gr &0.3x0.3mm I:>28mlm Vf:<2.5V	CHIP	Osr		data pdf pinout
F 2212 DC,PC	LIN-IC	Telecom, Full Duplex Modem, ..2400Bps	28-DIC,DIP	Fch		
F 2212 QC	LIN-IC	=F 2212DC,PC: SMD	28-MDIP			
F 2224 DC,PC	LIN-IC	Telecom, Full Duplex Modem, ..2400Bps	28-DIC,DIP	Fch		
F 2224 QC	LIN-IC	=F 2224DC,PC: SMD	28-MDIP			
F 3054(DC)	CMOS-IC	Telecom, μ-Law Codec, Filter	16-DIC	Fch		
F 3055 LE	MOS-N-FET-e	=RFD 3055LE (Typ-Code/Stempel/marking)	30	Rca	→RFD 3055LE	data
F 3057(DC)	CMOS-IC	Telecom, A-law Codec, Filter	16-DIC	Fch		
F 4116D	dRAM-IC	16384x1 Bit	16-DIC	Fch	... 4116...	pdf pinout
F 4116P	dRAM-IC	16384x1 Bit	16-DIP	Fch	... 4116...	pdf pinout
F 9010..9022	Si-N/P	→CS 9010...9022	7			
F9222L	SMPS-IC	PWM controller incl. 2x MOSFET, Vcc=16.5V	SQP	Fjd		pdf pinout
F10390	LED	rd/gr/bl ø9mm I=22cd Vf:3.5V		Seo		data pdf pinout
F10392	LED	rd/gr/bl ø9mm I=22cd Vf:3.5V		Seo		data pdf pinout
F 40098 BPC	CMOS-Logic	6x invert. Treiber/driver, Tri-State	16-DIP	Fch	U 40098, V 40098BPC	
F50380	LED	rd/gr/bl &0.3x0.3mm I:>28mlm Vf:3.5V		Seo		data pdf pinout
F50381	LED	rd/gr/bl &0.3x0.3mm I:>28mlm Vf:3.5V		Seo		data pdf pinout
FA	Si-N	=2SC2619-A (Typ-Code/Stempel/marking)	35	Hit	→2SC2619	data
FA	Si-N	=2SC4921 (Typ-Code/Stempel/marking)	35(1,6mm)	Say	→2SC4921	data
FA	Si-N	=2SC5376-A (Typ-Code/Stempel/marking)	35(1,6mm)	Tos	→2SC5376	
FA	Si-N	=2SC5564 (Typ-Code/Stempel/marking)	39	Say	→2SC5564	data
FA	Si-N	=BFP 81 (Typ-Code/Stempel/marking)	44	Aeg, Sie	→BFP 81	data
FA	Si-N	=BFQ 17 (Typ-Code/Stempel/marking)	39	Mot,Sie,Val	→BFQ 17	data
FA	Si-N	=BSV 65A (Typ-Code/Stempel/marking)	35	Sie	→BSV 65A	data
FA	MOS-P-FET-e*	=CPH 5601 (Typ-Code/Stempel/marking)	45	Say	→CPH 5601	data
FA	Si-P	=HQ 2A4A (Typ-Code/Stempel/marking)	39	Nec	→HQ 2...	data
FA	MOS-P/N-FETe	=μPA505T (Typ-Code/Stempel/marking)	45	Nec	→μPA505T	data
FA 1 A3Q	Si-N+R	=AA 1A3Q: SMD	35a	Nec	DTC 113ZK, UN 2219	data
FA 1 A4M	Si-N+R	=AA 1A4M: SMD	35a	Nec	DTC 113EK, RN 1402, UN 2211, 2SC3398,++	data
FA 1 A4P	Si-N+R	=AA 1A4P: SMD	35a	Nec	DTC 114YK, RN 1407, UN 2214, 2SC4047,++	data
FA 1 A4Z	Si-N+R	=AA 1A4Z: SMD	35a	Nec	DTC 114TK, RN 1411, UN 2215, 2SC3859,++	data
FA 1 F4M	Si-N+R	=AA 1F4M: SMD	35a	Nec	DTC 124EK, RN 1403, UN 2212, 2SC3396,++	data
FA 1 F4N	Si-N+R	=AA 1F4N: SMD	35a	Nec	DTC 124XK, KSR 1107, RN 1408	data
FA 1 F4Z	Si-N+R	=AA 1F4Z: SMD	35a	Nec	DTC 124TK, KSR 1111, UN 2217, 2SC4120	data
FA 1 L3M	Si-N+R	=AA 1L3M: SMD	35a	Nec	DTC 143EK, RN 1401, UN 221L, 2SC4362,++	data
FA 1 L3N	Si-N+R	=AA 1L3N: SMD	35a	Nec	DTC 143XK, KSR 1105, UN 221F, 2SC4360	data
FA 1 L3Z	Si-N+R	=AA 1L3Z: SMD	35a	Nec	DTC 143TK, RN 1410, UN 2216, 2SC3900,++	data
FA 1 L4L	Si-N+R	=AA 1L4L: SMD	35a	Nec	DTC 144WK, RN 1409, UN 221E, 2SC3397,++	data
FA 1 L4M	Si-N+R	=AA 1L4M: SMD	35a	Nec	DTC 144EK, RN 1404, UN 2213, 2SC3395,++	data
FA 1 L4Z	Si-N+R	=AA 1L4Z: SMD	35a	Nec	DTC 144TK, KSR 1112, UN 2210, 2SC3898	data
FA 3	Si-N	=2SC4179-FA3(Typ-Code/Stempel/marking)	35(2mm)	Nec	→2SC4179	data
FA 3 L4Z	Si-N+R	=BA 3L4Z: SMD	35a	Nec	-	data
FA 4	Si-N	=2SC4179-FA4(Typ-Code/Stempel/marking)	35(2mm)	Nec	→2SC4179	data
FA 38 SA50LC	MOS-N-FET-e	V-MOS, 500V, 38A, 500W, <0,13Ω(38A)	80ze	Inr	-	data
FA 57 SA50LC	MOS-N-FET-e	V-MOS, 500V, 57A, 625W, <0,08Ω(57A)	80ze	Inr	-	data
FA502	HI-SPEED	Vs:±15V Vo:±5V Vi0:5mV f:50Mc	8Y	Itr		data pinout
FA530	HI-SPEED	Vs:±15V Vu:94dB Vo:±10V Vi0:2mV	7Y	Itr		data pinout
FA531	HI-SPEED	Vs:±15V Vu:94dB Vo:±10V Vi0:2mV	7Y	Itr		data pinout
FA540	HI-SPEED	Vs:±15V Vu:106dB Vo:±10V Vi0:1mV	7Y	Itr		data pinout
FA541	HI-SPEED	Vs:±15V Vu:106dB Vo:±10V Vi0:1mV	7Y	Itr		data pinout
FA550	HI-SPEED	Vs:±15V Vu:83dB Vo:±10V Vi0:30mV	8A	Itr		data pinout
FA551	HI-SPEED	Vs:±15V Vu:103dB Vo:±10V Vi0:15mV	8A	Itr		data pinout
FA5202	4xOP	Vs:12V Vo:4.6/0V Vi0:±2mV	16D	Itr		data pinout
FA 5304AP	IC	Switching power supply control, pos. volt. detect.	DIP	Fjd		pdf pinout
FA 5304APS	IC	SMD, Switching power supply control, pos. volt. detect.	DIP	Fjd		pdf pinout
FA 5305AP	IC	Switching power supply control, neg. volt. detect.	DIP	Fjd		pdf pinout

Type	Device	Short Description	Fig.	Manu	Comparision Types	More at
FA 5305APS	IC	SMD, Switching power supply control, neg. volt. detect.	DIP	Fjd	-	pdf pinout
FA 5331...32P	LIN-IC	Pfc Control IC, +30V	16-DIP	Fjd	-	pdf pinout
FA 5331...32PM	LIN-IC	Pfc Control IC, +30V	16-MDIP	Fjd	-	pdf pinout
FA 5502 M	CMOS-IC	Pfc System	MDIP	Fjd	-	pdf pinout
FA 5502 P	CMOS-IC	Pfc System	DIC	Fjd	-	pdf pinout
FA 7610/12/17 CP	SMPS-IC	Pwm-type SMPS Control, Step-down Step-Up, Flyback, Soft-Start, 3.6..20V	8-DIP	Fcs	-	pdf pinout
FA 7610/12/17 CPN	SMPS-IC	SMD, PWM-type SMPS Control, Step-down Step-Up, Flyback, Soft-Start, 3.6..20V	8-MDIP	Fcs	-	pdf pinout
FA8025	R+	Io=3.5A Vo:14.2...14.8V P:3.5W		Hit		data pinout
FA 8025	Hybrid-Z-IC	Z-IC, +14,2...14,8V, 3,5A		Hit		
FAN 100	LIN-IC	Primary-side-control PWM Ctrl.	MDIP	Fch	-	pdf pinout
FAN 102	LIN-IC	Primary-side-control PWM Ctrl.	MDIP	Fch	-	pdf pinout
FAN 50 FC 3	IC	8-Bit Programmable, Buck Controller	QFN	Fch	-	pdf pinout
FAN1084	ER+	Io=5.5A Vo:1.5...2.6V Vin:4.75..5.25V	3P	Fcs		data pdf pinout
FAN 1084 D	IC	4.5A Adj. Linear Reg.	87	Fch	-	pdf pinout
FAN 1084 MC	IC	4.5A Adj. Linear Reg.	87	Fch	-	pdf pinout
FAN 1084 T	IC	4.5A Adj. Linear Reg.	17	Fch	-	pdf pinout
FAN1086	ER+	Io=2A Vo:1.5...5V Vin:1.5...5.75V	3P,4S	Fcs		data pdf pinout
FAN 1086 D 25	LIN-IC	1.5A, 2.5V Fix. Low Dropout Lin. Reg.	87	Fch	-	pdf pinout
FAN 1086 D 285	LIN-IC	1.5A, 2.85V Fix. Low Dropout Lin. Reg.	87	Fch	-	pdf pinout
FAN 1086 D 33	LIN-IC	1.5A, 3.3V Fix. Low Dropout Lin. Reg.	87	Fch	-	pdf pinout
FAN 1086 D 5	LIN-IC	1.5A, 5V Fix. Low Dropout Lin. Reg.	87	Fch	-	pdf pinout
FAN 1086 D	LIN-IC	1.5A Adj. Low Dropout Lin. Reg.	87	Fch	-	pdf pinout
FAN 1086 M 25	LIN-IC	1.5A, 2.5V Fix. Low Dropout Lin. Reg.	30	Fch	-	pdf pinout
FAN 1086 M 285	LIN-IC	1.5A, 2.85V Fix. Low Dropout Lin. Reg.	30	Fch	-	pdf pinout
FAN 1086 M 33	LIN-IC	1.5A, 3.3V Fix. Low Dropout Lin. Reg.	30	Fch	-	pdf pinout
FAN 1086 M 5	LIN-IC	1.5A, 5V Fix. Low Dropout Lin. Reg.	30	Fch	-	pdf pinout
FAN 1086 M	LIN-IC	1.5A Adj. Low Dropout Lin. Reg.	30	Fch	-	pdf pinout
FAN 1086 S 25	LIN-IC	1.5A, 2.5V Fix. Low Dropout Lin. Reg.	48	Fch	-	pdf pinout
FAN 1086 S 285	LIN-IC	1.5A, 2.85V Fix. Low Dropout Lin. Reg.	48	Fch	-	pdf pinout
FAN 1086 S 33	LIN-IC	1.5A, 3.3V Fix. Low Dropout Lin. Reg.	48	Fch	-	pdf pinout
FAN 1086 S 5	LIN-IC	1.5A, 5V Fix. Low Dropout Lin. Reg.	48	Fch	-	pdf pinout
FAN 1086 S	LIN-IC	1.5A Adj. Low Dropout Lin. Reg.	48	Fch	-	pdf pinout
FAN 112	R+	Io=1.5A Vo:1.14...1.26V Vin:7.2V	3P,4S	Fcs		data pdf pinout
FAN 1112 D	LIN-IC	1A 1.2V Low Dropout Lin. Reg.	30	Fch	-	pdf pinout
FAN 1112 S	LIN-IC	1A 1.2V Low Dropout Lin. Reg.	48	Fch	-	pdf pinout
FAN1117A	ER+	Io=1.5A Vin:1.5...7V	3P,4S	Fcs		data pdf pinout
FAN 1117 AD	LIN-IC	1A Adj. Low Dropout Linear Reg.	30	Fch	-	pdf pinout
FAN 1117 AD XX	LIN-IC	1A Fix. Low Dropout Linear Reg.	30	Fch	-	pdf pinout
FAN 1117 AS	LIN-IC	1A Adj. Low Dropout Linear Reg.	48	Fch	-	pdf pinout
FAN 1117 AS XX	LIN-IC	1A Fix. Low Dropout Linear Reg.	48	Fch	-	pdf pinout
FAN 1117 AT	LIN-IC	1A Adj. Low Dropout Linear Reg.	17	Fch	-	pdf pinout
FAN 1117 AT XX	LIN-IC	1A Fix. Low Dropout Linear Reg.	17	Fch	-	pdf pinout
FAN1539	R+	Io=1A Vo:3.234...3.336V	8H	Fcs		data pdf pinout
FAN1539B	R+	Io=1A Vo:3.234...3.336V	6H	Fcs		data pdf pinout
FAN 1539 B	LIN-IC	3.3V, 1A LDO w. Low Quiescent Current	MLP	Fch		pdf pinout
FAN1540	R+	Io=1.3A Vo:3.217...3.383V	3P,8H,6H	Fcs		data pdf pinout
FAN1540B	R+	Io=1.3A Vo:3.217...3.383V	6H	Fcs		data pdf pinout
FAN 1540 B	LIN-IC	3.3V 1.3A LDO w. Low Quiescent Current	MLP	Fch		pdf pinout
FAN1581	ER+	Vo:1.25...5.7V Vin:3...7V	5P	Fcs		data pdf pinout
FAN 1581 D 15	LIN-IC	5A 1.5V Ultra Low Dropout Linear Reg.	87	Fch	-	pdf pinout
FAN 1581 D 25	LIN-IC	5A 2.5V Ultra Low Dropout Linear Reg.	87	Fch	-	pdf pinout
FAN 1581 D	LIN-IC	5A Adj. Ultra Low Dropout Linear Reg.	87	Fch	-	pdf pinout
FAN 1581 M 12	LIN-IC	5A 1.2V Ultra Low Dropout Linear Reg.	87	Fch	-	pdf pinout
FAN 1581 M 25	LIN-IC	5A 2.5V Ultra Low Dropout Lin. Reg.	87	Fch	-	pdf pinout
FAN 1581 M	LIN-IC	5A Adj. Ultra Low Dropout Linear Reg.	87	Fch	-	pdf pinout
FAN 1581 T	LIN-IC	5A Adj. Ultra Low Dropout Linear Reg.	86	Fch	-	pdf pinout
FAN1582	ER+	Vo:1.25...5.7V Vin:2.05...5.5V	5P	Fcs		data pdf pinout
FAN 1582 M 15	LIN-IC	3A 1.5V Ultra Low Dropout Linear Reg.	87	Fch	-	pdf pinout
FAN 1582 M 25	LIN-IC	3A 2.5V Ultra Low Dropout Linear Reg.	87	Fch	-	pdf pinout
FAN 1582 M	LIN-IC	3A Adj. Ultra Low Dropout Linear Reg.	87	Fch	-	pdf pinout
FAN1585A	ER+	Vo:1.5...3.6V Vin:5V	3P	Fcs		data pdf pinout
FAN 1585 AD	LIN-IC	5A Adj. Low Dropout Linear Reg.	87	Fch	-	pdf pinout
FAN 1585 AM 15	LIN-IC	5A 1.5V Low Dropout Linear Reg.	30	Fch	-	pdf pinout
FAN 1585 AM 18	LIN-IC	5A 1.8V Low Dropout Linear Reg.	30	Fch	-	pdf pinout
FAN 1585 AM	LIN-IC	5A Adj. Low Dropout Linear Reg.	30	Fch	-	pdf pinout
FAN 1585 AT 15	LIN-IC	5A 1.5V Low Dropout Linear Reg.	17	Fch	-	pdf pinout
FAN 1585 AT 18	LIN-IC	5A 1.8V Low Dropout Linear Reg.	17	Fch	-	pdf pinout
FAN 1585 AT	LIN-IC	5A Adj. Low Dropout Linear Reg.	17	Fch	-	pdf pinout
FAN1587A	ER+	Vin:1.5...8.25V	3P	Fcs		data pdf pinout
FAN 1587 AD15	LIN-IC	3A 1.5V Low Dropout Linear Reg.	87	Fch	-	pdf pinout
FAN 1587 AD33	LIN-IC	3A 3.3V Low Dropout Linear Reg.	87	Fch	-	pdf pinout
FAN 1587 AD	LIN-IC	3A Adj. Low Dropout Linear Reg.	87	Fch	-	pdf pinout
FAN 1587 AM 15	LIN-IC	3A 1.5V Low Dropout Linear Reg.	30	Fch	-	pdf pinout
FAN 1587 AM 33	LIN-IC	3A 3.3V Low Dropout Linear Reg.	30	Fch	-	pdf pinout
FAN 1587 AM	LIN-IC	3A Adj. Low Dropout Linear Reg.	30	Fch	-	pdf pinout
FAN 1587 AT 15	LIN-IC	3A 1.5V Low Dropout Linear Reg.	17	Fch	-	pdf pinout
FAN 1587 AT 33	LIN-IC	3A 3.3V Low Dropout Linear Reg.	17	Fch	-	pdf pinout
FAN 1587 AT	LIN-IC	3A Adj. Low Dropout Linear Reg.	17	Fch	-	pdf pinout
FAN1589	R+	Vo:1.176...1.224V Vin:3.3...7V	3P	Fcs		data pdf pinout
FAN 1589 D	LIN-IC	2.7A, 1.2V Low Dropout Linear Reg.	87	Fch	-	pdf pinout
FAN 1589 M(C)	LIN-IC	2.7A, 1.2V Low Dropout Linear Reg.	30	Fch	-	pdf pinout
FAN 1589 T	LIN-IC	2.7A, 1.2V Low Dropout Linear Reg.	17	Fch	-	pdf pinout

ype	Device	Short Description	Fig.	Manu	Comparision Types	More at
1616A	ER+	Vin:1.5...5.75V	3P,4S	Fcs		data pdf pinout
1616 AD	LIN-IC	0.5A Adj. Low Dropout Linear Reg.	30	Fch	-	pdf pinout
1616 AD XX	LIN-IC	0.5A Fix. Low Dropout Linear Reg.	30	Fch	-	pdf pinout
1616 AS	LIN-IC	0.5A Adj. Low Dropout Linear Reg.	48	Fch	-	pdf pinout
1616 AS XX	LIN-IC	0.5A Fix. Low Dropout Linear Reg.	48	Fch	-	pdf pinout
1655	R+	Vo:1.31...1.39V Vin:2.5V P:1.4mW	8H,14S,16S	Fcs		data pdf pinout
1851 AMX	IC	Ground Fault Interrupter	QFN	Fch	-	pdf pinout
1851 AN	IC	Ground Fault Interrupter	QFN	Fch	-	pdf pinout
1950	R+	Vo:2.375...2.625V Vin:3.5...8V	3P	Fcs		data pdf pinout
2001	-IC	1A High-efficiency Step-down DC-DC Converter	MLP	Fch	-	pdf pinout
2002	-IC	1A High-efficiency Step-down DC-DC Converter	MLP	Fch	-	pdf pinout
2011 (I)	-IC	1.5A Low Voltage Current Mode Synchronous PWM Buck Regulator	MLP	Fch	-	pdf pinout
2012 (I)	-IC	1.5A Low Voltage Current Mode Synchronous PWM Buck Regulator	MLP	Fch	-	pdf pinout
2013	-IC	2A Low-Voltage Current-Mode Synchronous PWM Buck Regulator	MLP	Fch	-	pdf pinout
2103 (E)MPX	BiCMOS-IC	3A, 24V Input Integrated Synchronous Buck Regulator	MLP	Fcs	-	pdf pinout
2106 (E)MPX	BiCMOS-IC	6A, 24V Input Integrated Synchronous Buck Regulator	MLP	Fcs	-	pdf pinout
2500	ER+	Vo:1.32...6.4V Vin:6.5V	5S	Fcs		data pdf pinout
2501	R+	Vin:6.5V	5S	Fcs		data pdf pinout
2502	ER+	Vo:1.32...6.35V Vin:6.5V	5S	Fcs		data pdf pinout
2503	R+	Vin:6.5V	5S	Fcs		data pdf pinout
2504	ER+	Vo:1.32...6.33V Vin:6.5V	5S	Fcs		data pdf pinout
2508	ER+	Vo:1.32...6.45V Vin:6.5V	5S	Fcs		data pdf pinout
2509	R+	Vin:6.5V	5S	Fcs		data pdf pinout
2510	ER+	Vo:1.32...6.4V Vin:6.5V	5S	Fcs		data pdf pinout
2511	R+	Vin:6.5V	5S	Fcs		data pdf pinout
2512	ER+	Vo:1.32...6.35V Vin:6.5V	5S	Fcs		data pdf pinout
2513	R+	Vin:6.5V	5S	Fcs		data pdf pinout
2514	ER+	Vo:1.32...6.33V Vin:6.5V	5S	Fcs		data pdf pinout
2515	R+	Vin:6.5V	5S	Fcs		data pdf pinout
2518	ER+	Vo:1.32...6.45V Vin:6.5V	5S	Fcs		data pdf pinout
2519	R+	Vin:6.5V	5S	Fcs		data pdf pinout
2558	ER+	Io=180mA Vo:1.0...5.5V Vin:5.5V	6H,5SS	Fcs		data pdf pinout
2559	R+	Io=180mA Vin:2.7...5.5V	6HS	Fcs		data pdf pinout
2560	R+	Io=350mA Vin:1.6...5.5V	5G	Fcs		data pdf pinout
N 3100 CMPX	MOS-IC	Single 2A High-speed, Low-side Gate Driver, 58 - 120°C/W, Cmos	MLP	Fch	-	pdf pinout
N 3100 CSX	MOS-IC	Single 2A High-speed, Low-side Gate Driver, 150 - 248°C/W, Cmos	45	Fch	-	pdf pinout
N 3100 TMPX	MOS-IC	Single 2A High-speed, Low-side Gate Driver, 58 - 120°C/W, TTL	MLP	Fch	-	pdf pinout
N 3100 TSX	MOS-IC	Single 2A High-speed, Low-side Gate Driver, 150 - 248°C/W, TTL	45	Fch	-	pdf pinout
N 3800	BiCMOS-IC	M/ S Audio Amplifier, Microphone Pre-Amplifier, EMU Interface	MLP	Fch	-	pdf pinout
N4174	RR-OP	Vs:6V Vu:98dB Vo:2.69V Vi0:±6mV	5S	Fcs	KM4170	data pdf pinout
N4274	2xRR-OP	Vs:6V Vu:98dB Vo:2.69V Vi0:±6mV	5S	Fcs	KM4270	data pdf pinout
N 4800 IM(X)	BiCMOS-IC	Low Start-up Current Pfc/pwm Controller Combos	MDIP	Fch	-	pdf pinout
N 4800 IN	BiCMOS-IC	Low Start-up Current PFC/PWM Controller Combos	DIC	Fch	-	pdf pinout
N 4803 CP-1	BiCMOS-IC	Pfc and PWM Controller Combo, 67kHz / 67kHz	DIC	Fch	-	pdf pinout
N 4803 CP-2	BiCMOS-IC	Pfc and PWM Controller Combo, 67kHz / 134kHz	DIC	Fch	-	pdf pinout
N 4803 CS-1	BiCMOS-IC	Pfc and PWM Controller Combo, 67kHz / 67kHz	MDIP	Fch	-	pdf pinout
N 4803 CS-2	BiCMOS-IC	Pfc and PWM Controller Combo, 67kHz / 134kHz	MDIP	Fch	-	pdf pinout
N 4810 M(X)	BiCMOS-IC	Power Factor Correction Controller	TSOP	Fch	-	pdf pinout
N 4810 N	BiCMOS-IC	Power Factor Correction Controller	DIC	Fch	-	pdf pinout
N 4855	-IC	0, 5A Boost Regulator: Adj. Out., Shutdown, Low Battery Detect	TSOP	Fch	-	pdf pinout
N 5009 M	IC	Dual Bootstrapped 12V Mosfet Driver	QFN	Fch	-	pdf pinout
N 5009 MP	IC	Dual Bootstrapped 12V Mosfet Driver	QFN	Fch	-	pdf pinout
N 5018 B	IC	6-Bit VID Contr. 2-4 Phase VR10.X Contr.	QFN	Fch	-	pdf pinout
N 5026	IC	Dual DDR/Dual-output PWM Controller	QFN	Fch	-	pdf pinout
N 5029	IC	8-Bit Programmable Sy. Buck Controller	QFN	Fch	-	pdf pinout
N 5031	IC	8-Bit Programma., Sync Buck Controller	MLP	Fch	-	pdf pinout
N 5032	IC	8-Bit Programma., Sync Buck Controller	MLP	Fch	-	pdf pinout
N 5033	IC	8-Bit Programma., Sync Buck Controller	MLP	Fch	-	pdf pinout
N 5068	IC	DDR-1/DDR-2 plus Acpi Regulator Combo	MLP	Fcs	-	pdf pinout
N 5069	IC	PWM and LDO Controller Combo	MLP	Fcs	-	pdf pinout
N 5069 E	IC	PWM and LDO Controller Combo	MLP	Fcs	-	pdf pinout
N 5078	IC	DDR/ACPI Regulator Combo	MLP	Fcs	-	pdf pinout
N 5099	IC	Synchronous Buck PWM & LDO Controller	MLP	Fcs	-	pdf pinout
N 5109	MOS	Dual Bootstrapped 12V Mosfet Driver	MLP	Fcs	-	pdf pinout
N 5109 B	MOS	Dual Bootstrapped 12V Mosfet Driver	MLP	Fcs	-	pdf pinout

Type	Device	Short Description	Fig.	Manu	Comparision Types	More at
FAN 5182	IC	Adj. Output, Sync. Buck Controller	MLP	Fcs	-	pdf pinout
FAN 5232	IC	Adjustable PWM Buck Controller	TSOP	Fch	-	pdf pinout
FAN 5233	IC	System Electronics Regulator	TSOP	Fch	-	pdf pinout
FAN 5234	IC	Mobile-friendly Pwm Controller	TSOP	Fch	-	pdf pinout
FAN 5236	IC	Dual Mobile-friendly DDR / Dual-output PWM Controller	TSOP	Fch	-	pdf pinout
FAN 5240	IC	Multi-phase PWM Controller AMD Mobile CPU	TSOP	Fch	-	pdf pinout
FAN 5307 /MLP	IC	High- Efficiency Step- Down DC- DC Converter	45	Fcs	-	pdf pinout
FAN 5307 /SOT	IC	High- Efficiency Step- Down DC- DC Converter	45	Fcs	-	pdf pinout
FAN 5308	IC	800mA Step-down DC-DC Converter	45	Fcs	-	pdf pinout
FAN 5330	IC	Seri. LED Driver,30V Integrated Switch	45	Fcs	-	pdf pinout
FAN 5331	LIN-IC	High Efficiency Serial LED Driver and Oled Supply with 20V Integrated Switch	45	Fch	-	pdf pinout
FAN 5332 A	LIN-IC	LED Driver & Oled Supply, 30V Integrated Switch	45	Fch	-	pdf pinout
FAN 5333 (A,B)	LIN-IC	High Efficiency, High Current Serial LED Driver with 30V Integrated Switch	45	Fch	-	pdf pinout
FAN 5336	LIN-IC	1.5MHz Boost Regulator with 33V Integrated Fet Switch	MLP	Fcs	-	pdf pinout
FAN 5350 MPX	LIN-IC	3MHz, 600mA Step-down DC-DC Converter in Chip-scale and Mlp Packaging	MLP	Fcs	-	pdf pinout
FAN 5350 UCX	LIN-IC	3MHz, 600mA Step-down DC-DC Converter in Chip-scale and Mlp Packaging	GRID	Fcs	-	pdf pinout
FAN 5601	LIN-IC	Regulated Step-down Charge Pump DC/DC Converter, Output Voltage 1.3V/1.8V	MLP	Fcs	-	pdf pinout
FAN 5602 MP33X	LIN-IC	Step-up/step-down Charge Pump Regulated DC/DC Converter, Vout 3,3V	MLP	Fcs	-	pdf pinout
FAN 5602 MP45X	LIN-IC	Step-up/step-down Charge Pump Regulated DC/DC Converter, Vout 4,5V	MLP	Fcs	-	pdf pinout
FAN 5602 MP5X	LIN-IC	Step-up/step-down Charge Pump Regulated DC/DC Converter, Vout 5V	MLP	Fcs	-	pdf pinout
FAN 5607	IC	LED Driver, Adaptive Cha. Pump DC/DC Conv.	QFN	Fch	-	pdf pinout
FAN 5608	IC	FAN5608 Serial / Parallel LED Driver	QFN	Fch	-	pdf pinout
FAN 5609	IC	LED Driver, Adaptive Charge Pump DC/DC Conv.	QFN	Fch	-	pdf pinout
FAN 5609 /MLP	IC	LED Driver, Adaptive Charge Pump DC/DC Conv.	QFN	Fch	-	pdf pinout
FAN 5611	IC	Low-Dropout LED Drivers	QFN	Fch	-	pdf pinout
FAN 5612	IC	Low-Dropout LED Drivers	QFN	Fch	-	pdf pinout
FAN 5613	IC	Low-Dropout LED Drivers	QFN	Fch	-	pdf pinout
FAN 5614	IC	Low-Dropout LED Drivers	QFN	Fch	-	pdf pinout
FAN 5616	IC	Constant-current LED Driver	QFN	Fch	-	pdf pinout
FAN 5617	IC	Constant-current LED Driver	QFN	Fch	-	pdf pinout
FAN 5631	LIN-IC	Regulated Step-down Charge Pump DC/DC Converter	MLP	Fcs	-	pdf pinout
FAN 5632	LIN-IC	Regulated Step-down Charge Pump DC/DC Converter	MLP	Fcs	-	pdf pinout
FAN 5645	IC	Indicator LED Blinker, Single-wire Interf.	QFN	Fch	-	pdf pinout
FAN 5665	LIN-IC	High-Efficiency, Adaptive Charge Pump 5V Boost	GRID	Fcs	-	pdf pinout
FAN 6520 A	MOS-IC	Single Synchronous Buck PWM Controller	MDIP	Fch	-	pdf pinout
FAN 6520 B	MOS-IC	Single Synchronous Buck PWM Controller	MDIP	Fch	-	pdf pinout
FAN 7071	LIN-IC	Landing Correction IC	SIL	Fch	-	pdf pinout
FAN 7310	LIN-IC	LCD Backlight Inverter Drive Integrated Circuit	MDIP	Fch	-	pdf pinout
FAN 7311	LIN-IC	LCD Backlight Inverter Drive IC	MDIP	Fch	-	pdf pinout
FAN 7311 B	LIN-IC	LCD Backlight Inverter Drive IC	TSOP	Fch	-	pdf pinout
FAN 7313	LIN-IC	LCD Backlight Inverter Drive IC	TSOP	Fch	-	pdf pinout
FAN 7314	LIN-IC	LCD Backlight Inverter Drive IC	TSOP	Fch	-	pdf pinout
FAN 7315	LIN-IC	LCD Backlight Inverter Drive IC	TSOP	Fch	-	pdf pinout
FAN 7361	LIN-IC	High-Side Gate Driver, up to +600V	MDIP	Fch	-	pdf pinout
FAN 7362	LIN-IC	High-Side Gate Driver, up to +600V	MDIP	Fch	-	pdf pinout
FAN 7371	LIN-IC	High-current High-Side Gate Drive IC, up to +600V	TSOP	Fch	-	pdf pinout
FAN 7380	LIN-IC	Half-Bridge Gate Driver, up to +600V	TSOP	Fch	-	pdf pinout
FAN 7382 M1	LIN-IC	High- and Low-side Gate Driver, up to +600V	DIC	Fch	-	pdf pinout
FAN 7382 M 1 X	LIN-IC	High- and Low-side Gate Driver, up to +600V	DIC	Fch	-	pdf pinout
FAN 7382 M	LIN-IC	High- and Low-side Gate Driver, up to +600V	DIC	Fch	-	pdf pinout
FAN 7382 MX	LIN-IC	High- and Low-side Gate Driver, up to +600V	DIC	Fch	-	pdf pinout
FAN 7382 N	LIN-IC	High- and Low-side Gate Driver, up to +600V	DIC	Fch	-	pdf pinout
FAN 7383	LIN-IC	Half-Bridge Gate-drive IC	TSOP	Fcs	-	pdf pinout
FAN 7384	LIN-IC	Half-Bridge Gate-drive IC, up to +600V.	TSOP	Fcs	-	pdf pinout
FAN 7385	LIN-IC	Dual-channel High-Side Gate-drive IC, up to +600V	TSOP	Fcs	-	pdf pinout
FAN 7387 M	IC	High-Voltage Gate Driver	TSOP	Fch	-	pdf

Type	Device	Short Description	Fig.	Manu	Comparision Types	More at
FAN 7527 BM	LIN-IC	Power Factor Correction Controller	TSOP	Fcs	-	pdf pinout
FAN 7527 BN	LIN-IC	Power Factor Correction Controller	DIC	Fcs	-	pdf pinout
FAN 7528 M(X)	LIN-IC	Dual-output, Critical Conduction Mode Pfc Controller	TSOP	Fcs	-	pdf pinout
FAN 7528 N	LIN-IC	Dual-output, Critical Conduction Mode Pfc Controller	DIP	Fcs	-	pdf pinout
FAN 7529 M(X)	LIN-IC	Critical Conduction Mode Pfc Controller	TSOP	Fcs	-	pdf pinout
FAN 7529 N	LIN-IC	Critical Conduction Mode Pfc Controller	DIC	Fcs	-	pdf pinout
FAN 7530 M(X)	LIN-IC	Critical Conduction Mode Pfc Controller	TSOP	Fcs	-	pdf pinout
FAN 7530 N	LIN-IC	Critical Conduction Mode Pfc Controller	DIC	Fcs	-	pdf pinout
FAN 7532 M(X)	LIN-IC	Ballast Control IC	TSOP	Fcs	-	pdf pinout
FAN 7544 M(X)	LIN-IC	Simple Ballast Controller	TSOP	Fcs	-	pdf pinout
FAN 7544 N	LIN-IC	Simple Ballast Controller	DIC	Fcs	-	pdf pinout
FAN 7547 AM(X)	LIN-IC	LCD Backlight Inverter Drive IC	TSOP	Fcs	-	pdf pinout
FAN 7548	LIN-IC	Dual LCD Back Light Inverter Drive IC	TSOP	Fcs	-	pdf pinout
FAN 7554	LIN-IC	Versatile PWM Controller	DIC	Fcs	-	pdf pinout
FAN 7554 D	LIN-IC	Versatile PWM Controller	TSOP	Fcs	-	pdf pinout
FAN 7585	LIN-IC	Intelligent Voltage Mode PWM IC	DIC	Fcs	-	pdf pinout
FAN 7601 G	LIN-IC	Green Current Mode PWM Controller	TSOP	Fcs		pdf pinout
FAN 7601 M	LIN-IC	Green Current Mode PWM Controller	TSOP	Fcs		pdf pinout
FAN 7601 N	LIN-IC	Green Current Mode PWM Controller	DIC	Fcs		pdf pinout
FAN 7602 BM(X)	LIN-IC	Green Current-Mode PWM Controller	TSOP	Fcs		pdf pinout
FAN 7602 BN	LIN-IC	Green Current-Mode PWM Controller	DIC	Fcs		pdf pinout
FAN 7680 M	LIN-IC	PC Power Supply Outputs Monitoring IC	TSOP	Fcs		pdf pinout
FAN 7680 N	LIN-IC	PC Power Supply Outputs Monitoring IC	DIC	Fcs		pdf pinout
FAN 7685 M(X)	LIN-IC	PC Power Supply Output Monitoring IC	TSOP	Fcs		pdf pinout
FAN 7685 N	LIN-IC	PC Power Supply Output Monitoring IC	DIC	Fcs		pdf pinout
FAN 7686 M(X)	LIN-IC	PC Power Supply Output Monitoring IC	TSOP	Fcs		pdf pinout
FAN 7686 N	LIN-IC	PC Power Supply Output Monitoring IC	DIC	Fcs		pdf pinout
FAN 7687 (A)M(X)	LIN-IC	PC Power Supply Output Monitoring IC	TSOP	Fcs		pdf pinout
FAN 7687 (A)N	LIN-IC	PC Power Supply Output Monitoring IC	DIC	Fcs		pdf pinout
FAN 7710 N	LIN-IC	Ballast Control IC for Compact Fluorescent Lamps, up to +550V	DIC	Fcs	-	pdf pinout
FAN 7711 M(X)	LIN-IC	Ballast Control Integrated Circuit, up to +600V	TSOP	Fcs		pdf pinout
FAN 7711 N	LIN-IC	Ballast Control Integrated Circuit, up to +600V	DIC	Fcs		pdf pinout
FAN 7842 M(X)	LIN-IC	High and Low Side Gate Driver	TSOP	Fcs		pdf pinout
FAN 8040	LIN-IC	4-Ch. Motor Driver	TSOP	Fcs		pdf pinout
FAN 8048	IC	2 DC-DC Converter & 4 Ch PWM Motor IC	MP	Fch		pdf pinout
FAN 8434	LIN-IC	Camcorder 3 in 1 Motor Driver	MP	Fcs		pdf pinout
FAN 53418	LIN-IC	Synchronous DC-DC Mosfet Driver	MDIP	Fcs		pdf pinout
FAN 73832 M(X)	LIN-IC	Half-Bridge Gate-drive IC, up to +600V.	TSOP	Fcs		pdf pinout
FAN 73832 N	LIN-IC	Half-Bridge Gate-drive IC, up to +600V.	DIC	Fcs		pdf pinout
FAN 1084 D 15	IC	1.5V, 4.5A Fixed Linear Reg.	87	Fch		pdf pinout
FAN 1084 MC 15	IC	1.5V, 4.5A Fixed Linear Reg.	87	Fch		pdf pinout
FAN 1084 T 15	IC	1.5V, 4.5A Fixed Linear Reg.	87	Fch		pdf pinout
FAN 1084 D 33	IC	3.3V, 4.5A Fixed Linear Reg.	87	Fch		pdf pinout
FAN 1084 M 33	IC	3.3V, 4.5A Fixed Linear Reg.	87	Fch		pdf pinout
FAN 1084 MC 33	IC	3.3V, 4.5A Fixed Linear Reg.	87	Fch		pdf pinout
FAN 1084 T 33	IC	4.5A Adjustable/Fixed Linear Regulator	87	Fch		pdf pinout
FAP-450	MOS-N-FET-e	V-MOS, 500V, 14A, 190W, <0,38Ω(8A)	18c§	Fjd	IRFP 450, IRFP 462	data
FAs	Si-N	=BFP 81 (Typ-Code/Stempel/marking)	44	Aeg, Sie	→BFP 81	data
FAT-XXX	IRED	rd 870nm ca. ø16mm I:0.6mW/Sr		Roi		data pdf pinout
FAT-XXX-40	LED	rd ca. ø16mm I:0.6mW/Sr Vf:13V		Roi		data pdf pinout
FB	Si-N	=2SC2619-B (Typ-Code/Stempel/marking)	35	Hit	→2SC2619	data
FB	Si-N	=2SC3053-B (Typ-Code/Stempel/marking)	35	Mit	→2SC3053	data
FB	Si-N	=2SC4258-B (Typ-Code/Stempel/marking)	35(2mm)	Mit	→2SC4258	data
FB	Si-N	=2SC5016 (Typ-Code/Stempel/marking)	≈35	Mat	→2SC5016	data
FB	Si-P	=2SC5216 (Typ-Code/Stempel/marking)	35	Mat	→2SC5216	data
FB	Si-N	=2SC5376-B (Typ-Code/Stempel/marking)	35(1,6mm)	Tos	→2SC5376	
FB	Si-N	=2SC5565 (Typ-Code/Stempel/marking)	39	Say	→2SC5565	data
FB	Si-N	=BFP 17 (Typ-Code/Stempel/marking)	44	Sie	→BFP 17	data
FB	Si-N	=BFQ 19 (Typ-Code/Stempel/marking)	39	Sie, Val	→BFQ 19	data
FB	Si-N	=BSV 65B (Typ-Code/Stempel/marking)	35	Sie	→BSV 65B	data
FB	MOS-N-FET-e*	=CPH 5602 (Typ-Code/Stempel/marking)	45	Say	→CPH 5602	data
FB	MOS-N-FET-e*	=CPH 6406 (Typ-Code/Stempel/marking)	46	Say	→CPH 6406	data
FB	MOS-N-FET-e*	=MCH 6602 (Typ-Code/Stempel/marking)	46(2mm)	Say	→MCH 6602	data
FB 1 A3M...L3N	Si-N+R	=AB 1A3M...L3N: SMD	35a	Nec	-	data
FB 180 SA10	MOS-N-FET-e	V-MOS, 100V, 180A, 480W, <6,5mΩ(180A)	80ze	Inr		data
FB 2060A,B	Si-N	=2N2060A,B:	81bq	Fch	→2N2060A,B	data
FB 3423	Si-N	=2N3423:	81bq	Fch	→2N3423	data
FB 3424	Si-N	=2N3424:	81bq	Fch	→2N3424	data
FB 3726	Si-P	=2N3726:	81bq	Fch	→2N3726	data
FB 3727	Si-P	=2N3727:	81bq	Fch	→2N3727	data
FB 3728	Si-N	=2N3728:	81bq	Fch	→2N3728	data
FB 3729	Si-N	=2N3729:	81bq	Fch	→2N3729	data
FB 4015	Si-P	=2N4015:	81bq	Fch	→2N4015	data
FB 4016	Si-P	=2N4016:	81bq	Fch	→2N4016	data
FBA	Z-Di	=SMCJ 5.0CA (Typ-Code/Stempel/marking)	71(8x5mm)	Tho	→SMCJ ...	data

Type	Device	Short Description	Fig.	Manu	Comparision Types	More at
FBB	Z-Di	=SMCJ 6.0CA (Typ-Code/Stempel/marking)	71(8x5mm)	Tho	→SMCJ ...	data
FBC	Z-Di	=SMCJ 6.5CA (Typ-Code/Stempel/marking)	71(8x5mm)	Tho	→SMCJ ...	data
FBC 737	Si-N	≈BC 737			→BC 737	
FBC 738	Si-N	≈BC 738			→BC 738	
FBD	Z-Di	=SMCJ 8.5CA (Typ-Code/Stempel/marking)	71(8x5mm)	Tho	→SMCJ ...	data
FBF	Z-Di	=SMCJ 10CA (Typ-Code/Stempel/marking)	71(8x5mm)	Tho	→SMCJ ...	data
FBH	Z-Di	=SMCJ 12CA (Typ-Code/Stempel/marking)	71(8x5mm)	Tho	→SMCJ ...	data
FBI	Z-Di	=SMCJ 13CA (Typ-Code/Stempel/marking)	71(8x5mm)	Tho	→SMCJ ...	data
FBJ	Z-Di	=SMCJ 15CA (Typ-Code/Stempel/marking)	71(8x5mm)	Tho	→SMCJ ...	data
FBL	Z-Di	=SMCJ 18CA (Typ-Code/Stempel/marking)	71(8x5mm)	Tho	→SMCJ ...	data
FBM	Z-Di	=SMCJ 20CA (Typ-Code/Stempel/marking)	71(8x5mm)	Tho	→SMCJ ...	data
FBN	Z-Di	=SMCJ 22CA (Typ-Code/Stempel/marking)	71(8x5mm)	Tho	→SMCJ ...	data
FBO	Z-Di	=SMCJ 24CA (Typ-Code/Stempel/marking)	71(8x5mm)	Tho	→SMCJ ...	data
FBP	Z-Di	=SMCJ 26CA (Typ-Code/Stempel/marking)	71(8x5mm)	Tho	→SMCJ ...	data
FBQ	Z-Di	=SMCJ 28CA (Typ-Code/Stempel/marking)	71(8x5mm)	Tho	→SMCJ ...	data
FBR	Z-Di	=SMCJ 30CA (Typ-Code/Stempel/marking)	71(8x5mm)	Tho	→SMCJ ...	data
FBS	Z-Di	=SMCJ 33CA (Typ-Code/Stempel/marking)	71(8x5mm)	Tho	→SMCJ ...	data
FBU	Z-Di	=SMCJ 40CA (Typ-Code/Stempel/marking)	71(8x5mm)	Tho	→SMCJ ...	data
FBW	Z-Di	=SMCJ 48CA (Typ-Code/Stempel/marking)	71(8x5mm)	Tho	→SMCJ ...	data
FBZ	Z-Di	=SMCJ 58CA (Typ-Code/Stempel/marking)	71(8x5nm)	Tho	→SMCJ ...	data
FC	Si-N	=2SC2619-C (Typ-Code/Stempel/marking)	35	Hit	→2SC2619	data
FC	Si-N	=2SC3053-C (Typ-Code/Stempel/marking)	35	Mit	→2SC3053	data
FC	Si-N	=2SC3438-C (Typ-Code/Stempel/marking)	39	Mit	→2SC3438	data
FC	Si-N	=2SC4258-C (Typ-Code/Stempel/marking)	35(2mm)	Mit	→2SC4258	data
FC	Si-N	=BFP 29 (Typ-Code/Stempel/marking)	44	Sie	→BFP 29	data
FC	Si-N	=BFQ 64 (Typ-Code/Stempel/marking)	39	Sie	→BFQ 64	data
FC	MOS-P-FET-e*	=CPH 5603 (Typ-Code/Stempel/marking)	45	Say	→CPH 5603	data
FC	Si-P+R	=UN 211V (Typ-Code/Stempel/marking)	35	Mat	→UN 211V	data
FC 11	N-FET	SMD, Dual, =2x 2SK772	45ba2	Say	-	data
FC 12	Si-N+N-FET	SMD, Dual, =2SC4639 + 2SK2394	45	Say	-	data
FC 13	N-FET	SMD, Dual, =2x 2SK303	46bn	Say	-	data
FC 18	Si-N+N-FET	SMD, Dual, =2SC4639 + 2SK2394	45	Say	-	data
FC 21	Si-N+N-FET	SMD, Dual, =2SC2812 + 2SK931	45	Say	-	data
FC 101	Si-P	SMD, Dual, =2x 2SA1622	46bh1	Say	-	data
FC 102	Si-N	SMD, Dual, =2SC4211	46bh1	Say	-	data
FC 103	Si-P	SMD, Dual, =2x 2SA1622	45ba1	Say	-	data
FC 104	Si-N	SMD, Dual, =2SC4211	45ba1	Say	-	data
FC 105	Si-P+R	SMD, Dual, =2x 2SA1341	46bh1	Say	-	data
FC 106	Si-N+R	SMD, Dual, =2x 2SC3395	46bh1	Say	-	data
FC 107	Si-P+R	SMD, Dual, =2x 2SA1341	45ba1	Say	-	data
FC 108	Si-N+R	SMD, Dual, =2x 2SC3395	45ba1	Say	-	data
FC 109	Si-P+R	SMD, Dual, =2x 2SA1342	46bh1	Say	-	data
FC 110	Si-N+R	SMD, Dual, =2x 2SC3396	46bh1	Say	-	data
FC 111	Si-P+R	SMD, Dual, =2x 2SA1342	45ba1	Say	-	data
FC 112	Si-N+R	SMD, Dual, =2x 2SC3396	45ba1	Say	-	data
FC 113	Si-P+R	SMD, Dual, =2x 2SA1344	46bh1	Say	-	data
FC 114	Si-N+R	SMD, Dual, =2x 2SC3398	46bh1	Say	-	data
FC 115	Si-P+R	SMD, Dual, =2x 2SA1344	45ba1	Say	-	data
FC 116	Si-N+R	SMD, Dual, =2x 2SC3398	45ba1	Say	-	data
FC 117	Si-P	SMD, Dual, =2x 2SA1745	46bh1	Say	-	data
FC 118	Si-N	SMD, Dual, =2x 2SC4555	46bh1	Say	-	data
FC 119	Si-N	SMD, Dual, =2x 2SC2814	46bl1	Say	-	data
FC 120	Si-N	SMD, Dual, =2x 2SC3142	46bl1	Say	-	data
FC 121	Si-P+R	SMD, Dual, =2x 2SA1502	45ba1	Say	-	data
FC 123	Si-P+R	SMD, Dual, =2x 2SA1508	46bh1	Say	-	data
FC 124	Si-N+R	SMD, Dual, =2x 2SC3898	46bh1	Say	-	data
FC 125	Si-P+R	SMD, Dual, =2x 2SA1508	45ba1	Say	-	data
FC 126	Si-N+R	SMD, Dual, =2x 26C3898	45ba1	Say	-	data
FC 127	Si-P+R	SMD, Dual, =2x 2SA1496	46bh1	Say	-	data
FC 128	Si-N+R	SMD, Dual, =2x 2SC3859	46bh1	Say	-	data
FC 129	Si-P+R	SMD, Dual, =2x 2SA1496	45ba1	Say	-	data
FC 130	Si-N+R	SMD, Dual, =2x 2SC3859	45ba1	Say	-	data
FC 131	Si-P+R	SMD, Dual, =2x 2SA1563	46bh1	Say	-	data
FC 132	Si-N+R	SMD, Dual, =2x 2SC4047	46bh1	Say	-	data
FC 133	Si-P+R	SMD, Dual, =2x 2SA1563	45ba1	Say	-	data
FC 134	Si-N+R	SMD, Dual, =2x 2SC4047	45ba1	Say	-	data
FC 135	Si-P+R	SMD, Dual, =2x 2SA1510	46bh1	Say	-	data
FC 136	Si-N+R	SMD, Dual, =2x 2SC3900	46bh1	Say	-	data
FC 137	Si-P+R	SMD, Dual, =2x 2SA1510	45ba1	Say	-	data
FC 138	Si-N+R	SMD, Dual, =2x 2SC3900	45ba1	Say	-	data
FC 139	Si-N	SMD, Dual, =2x 2SC3689	46bh1	Say	-	data
FC 140	Si-N	SMD, Dual, =2x 2SC4452	46bh6	Say	-	data
FC 142	Si-P+R	SMD, Dual, =2x 2SA1653	45ba1	Say	-	data
FC 143	Si-N+R	SMD, Dual, =2x 2SC4360	45ba1	Say	-	data
FC 144	Si-N+R	SMD, Dual, =2x 2SC3863	45ba1	Say	-	data
FC 145	Si-P+R	SMD, Dual, =2x 2SA1597	45	Say	-	data
FC 146	Si-N+R	SMD, Dual, =2x 2SC4146	45	Say	-	data
FC 147	Si-P+R	SMD, Dual, =2x 2SA1573	46	Say	-	data

Type	Device	Short Description	Fig.	Manu	Comparision Types	More at
FC 148	Si-N+R	SMD, Dual, =2x 2SC4069	46	Say	-	data
FC 149	Si-P	SMD, Dual, =2x 2SA1813	46bh1	Say	-	data
FC 150	Si-P+N	SMD, Dual, =2SA1813 + 2SC4413	46bh1	Say	-	data
FC 151	Si-P	SMD, Dual, =2x 2SA1669	46bh7	Say	-	data
FC 152	Si-N	SMD, Dual, =2x 2SC4270	46bh2	Say	-	data
FC 154	Si-N+P	SMD, Dual, =2SC4270 + 2SA1699	46bh2	Say	-	data
FC 155	Si-P+R	SMD, Dual, T1: 20V, 0,5A, 150MHz Rb=4,7kΩ, Rbe=10kΩ, T2: 20V, 0,5A 400MHz, Rb/Rbe=-	46bh2	Say	-	data
FC 156	Si-N	SMD, Dual, =2x 2SC5226	46bh2	Say	-	data
FC 157	Si-N	SMD, Dual, =2x 2SC5245	46bh1	Say	-	data
FC 601	Si-P+Di	SMD, S, 50V, 0,1A, 200MHz, Rb=10kΩ Rbe=47kΩ, int. Schottky-Di, 30V, 70mA	45	Say	-	data
FC 801	Si-Di	SMD, Quad, 75V, 0,1A, <8ns	46bv	Say	-	data
FC 802	Si-Di	SMD, Quad, 75V, 0,1A, <5ns	46bu	Say	-	data
FC 803	Si-Di	SMD, Quad, Schottky, 35V, 0,07A, <10ns	46bu	Say	-	data
FC 804	Si-Di	SMD, Dual, Schottky, 35V, 0,2A, <10ns	45an	Say	-	data
FC 805	Si-Di	SMD, Dual, Schottky, 35V, 0,5A, <10ns	45an	Say	-	data
FC 806	Si-Di	SMD, Dual, Schottky, 55V, 0,1A, <10ns	45an	Say	-	data
FC 807	Si-Di	SMD, Quad, 80V, 0,1A	45	Say	-	data
FC 808	Si-Di	SMD, Quad, 80V, 0,1A	45	Say	-	data
FC 809	Si-Di	SMD, Dual, Schottky, 35V, 0,07A, <10ns	45an	Say	-	data
FC 810	Si-Di	SMD, Dual, Schottky, 15V, 0,7A, <10ns	45an	Say	-	data
FC 901	Si-Di	SMD, Dual, 80V, 0,1A	46	Say	-	data
FC 902	Si-Di	SMD, Dual, 80V, 0,1A	46	Say	-	data
FC 903	Si-Di	SMD, Triple, 85V, 0,1A, <4ns	46	Say	-	data
24 FC 1025 P	EEPROM-IC	1024K I2C Cmos Serial EEPROM	DIP	Mcp	-	pdf pinout
24 FC 1025 SM	EEPROM-IC	1024K I2C Cmos Serial EEPROM	DIP	Mcp	-	pdf pinout
FCB 61 C65(L,LL)P	CMOS-sRAM-IC	8k x 8Bit, 55ns, 0...70°	28-DIP	Phi	-	
FCB 61 C65(L,LL)T	CMOS-sRAM-IC	=FCB 61...P: SMD	28-MDIP	Phi	-	
FCD810	OC	LED/NPN Viso:6000Vac CTR:5% Vbs:1.2V		Fch	4N25,OPI2251,PS2010	data pinout
FCD810C	OC	LED/NPN Viso:5000Vac CTR:25%		Fch	CQY80N,PS2021	data pinout
FCD820	OC	LED/NPN Viso:6000Vac CTR:50% Ibe:5mA		Fch	4N25,OPI2252,PS2010	data pinout
FCD825	OC	LED/NPN Viso:6000Vac CTR:80% Ibe:8mA		Fch	CQY80N,OPI2153	data pinout
FCD830	OC	LED/NPN Viso:6000Vac CTR:50% Ibe:5mA		Fch	CQY80N,PS2021	data pinout
FCD831	OC	LED/NPN Viso:6000Vac CTR:15%		Fch	CQY80N,PS2021	data pinout
FCD836	OC	LED/NPN Viso:6000Vac CTR:10% Ibe:1mA		Fch	CQY80N,PS2021	data pinout
FCD850	OC	LED/Dar Viso:6000Vac CTR:>1500%		Fch		data pinout
FCD860	OC	LED/Dar Viso:6000Vac CTR:>200%		Fch	4N30	data pinout
FCD865	OC	LED/Dar Viso:6000Vac CTR:>400%		Fch	4N30	data pinout
FCD880	DOC	LED/NPN Viso:2500V CTR:20% Vbs:1.25V		Fch	CNY74-2	data pinout
FCD890	DOC	LED/Dar Viso:2500V CTR:>200% Ibe:2mA		Fch	CNY74-2	data pinout
FCF 61 C65(L,LL)T	CMOS-sRAM-IC	=FCB 61...: SMD, -40...+85°	28-MDIP	Phi	-	
FCM 7010	MOS-IC	Digital-uhr/digital clock	40-DIL		-	
FCS 6208...6209	Si-N/P	→CS 6208...6209	7			
FCS 9010...9022	Si-N/P	→CS 9010...9022	7			
FCW100	LED	wht &2.0x1.5x0.45mm I:6500mcd		Seo		data pdf pinout
FCW110	LED	wht &3.2x1.8x0.55mm I:6500mcd		Seo		data pdf pinout
FCX 458	Si-N	SMD, S, 400/400V, 225mA, >50MHz	39b	Ztx	2SC4548	data
FCX 491	Si-N	SMD, Tr/E, 80V, 1A, >150MHz	39b	Ztx	2SC4409, 2SD2533	data
FCX 491 A	Si-N	=FCX 491: 40V	39b		BCX 54..55, 2SC4539, 2SC5209, 2SD1615,++	data
FCX 493	Si-N	SMD, Tr/E, 120/100V, 1A, >150MHz	39b	Ztx	2SC3646, 2SD1418, 2SD1419, 2SD1615A	data
FCX 495	Si-N	SMD, Tr/E, 170/150V, 1A, >100MHz	39b	Ztx	2SC3649, 2SD1420, 2SD1421	data
FCX 558	Si-P	SMD, S, 400/400V, 200mA, >50MHz	39b	Ztx	2SA1740	data
FCX 589	Si-P	SMD, Tr/E, 50V, 1A, >100MHz	39b	Ztx	2SA1890, 2SB1115A, 2SB1260	data
FCX 591	Si-P	SMD, Tr/E, 80V, 1A, >150MHz	39b	Ztx	2SA1890, 2SB1115A, 2SB1260	data
FCX 591 A	Si-P	=FCX 591: 40V	39b		BCX 51..52, 2SA1364, 2SB1115, 2SB1122,++	data
FCX 593	Si-P	SMD, Tr/E, 120/100V, 1A, >50MHz	39b	Ztx	2SA1416, 2SB1025, 2SB1026	data
FCX 596	Si-P	SMD, Tr/E, 220/200V, 0,3A, >150MHz	39b	Ztx		data
FCX 617	Si-N	SMD, S, lo-sat, 15V, 3A, 120MHz	39b	Ztx	2SD1620	data
FCX 619	Si-N	SMD, S, lo-sat, 50V, 2,75A, 165MHz	39b	Ztx	2SD1624	data
FCX 688 B	Si-N	SMD, hi-beta, lo-sat, 12V, 3A, 150MHz	39b	Ztx		data
FCX 690 B	Si-N	SMD, hi-beta, lo-sat, 45V, 3A, 150MHz	39b	Ztx	2SC3439	data
FCX 717	Si-P	SMD, S, lo-sat, 12V, 3A, 110MHz	39b	Ztx	2SB1301, 2SB1518, 2SB1628	data
FCX 718	Si-P	SMD, S, lo-sat, 20V, 2,5A, 150MHz	39b	Ztx	2SB1301, 2SB1518, 2SB1628	data
FCX 789 A	Si-P	SMD, hi-beta, lo-sat, 25V, 3A, 100MHz	39b	Ztx	-	data
FCX 790 A	Si-P	SMD, hi-beta, lo-sat, 50V, 2A, 100MHz	39b	Ztx	-	data
FCX 1047 A	Si-N	SMD, hi-beta, lo-sat, 35V, 4A, 150MHz	39b	Ztx	-	data
FCX 1051 A	Si-N	SMD, hi-beta, lo-sat, 150V, 3A, 155MHz	39b	Ztx	-	data
FCX 1053 A	Si-N	SMD, hi-beta, lo-sat, 150V, 3A, 140MHz	39b	Ztx	-	data
FCX 1147 A	Si-P	SMD, lo-sat, 15V, 3A, 115MHz	39b	Ztx	2SB1301, 2SB1518, 2SB1628	data
FCX 1149 A	Si-P	SMD, lo-sat, 30V, 3A, 135MHz	39b	Ztx	2SB1518	data
FCX 1151 A	Si-P	SMD, lo-sat, 45V, 3A, 145MHz	39b	Ztx	2SA1736, 2SB1124	data

Type	Device	Short Description	Fig.	Manu	Comparision Types	More at
FD	Si-N	=2SC3053-D (Typ-Code/Stempel/marking)	35	Mit	→2SC3053	data
FD	Si-N	=2SC3438-D (Typ-Code/Stempel/marking)	39	Mit	→2SC3438	data
FD	Si-N	=2SC4258-D (Typ-Code/Stempel/marking)	35(2mm)	Mit	→2SC4258	data
FD	Si-N	=2SC5567 (Typ-Code/Stempel/marking)	39	Say	→2SC5567	data
FD	Si-P	=BCV 26 (Typ-Code/Stempel/marking)	35	Sie	→BCV 26	data
FD	Si-N	=BFP 35A (Typ-Code/Stempel/marking)	44	Sie	→BFP 35A	data
FD	Si-N	=BFQ 17P (Typ-Code/Stempel/marking)	39	Sie	→BFQ 17P	data
FD	MOS-N-FET-e*	=CPH 5604 (Typ-Code/Stempel/marking)	45	Say	→CPH 5604	data
FD	MOS-N-FET-e*	=MCH 6604 (Typ-Code/Stempel/marking)	46(2mm)	Say	→MCH 6604	data
FD	Si-N+R	=UN 221V (Typ-Code/Stempel/marking)	35	Mat	→UN 221V	data
FD	Z-Di	=Z1 SMA-47 (Typ-Code/Stempel/marking)	71(5x2,5)	Fag	→Z1 SMA-47	data
FD 100	Si-Di	SS, 75V, <4ns	31a	Fch	BAW 62, BAW 76, BAX 95, 1N4148...49, ++	data
FD 111	Si-Di	SS, 75V, <5ns	31a	Fch	BAW 62, BAW 76, BAX 95, 1N4148...49, ++	data
FD 200	Si-Di	S, 200V, <50ns	31a	Fch	BA 197...198, BAV 20...21, BAW 50	data
FD 222	Si-Di	S, 150V, <60ns	31a	Fch	BA 196...198, BAV 16, BAV 20...21, BAW 50	data
FD 300	Si-Di	Uni, 150, Ir<1nA(125V)	31a	Fch	BAS 45, FDH 300, FDH 333	data
FD 333	Si-Di	Uni, 150, Ir<3nA(125V)	31a	Fch	BAS 45, FDH 300, FDH 333	data
FD 600	Si-Di	SS, 75V, <4ns	31a	Fch	BAV 12, BAV 14, BAW 26...27, BAX 81	data
FD 700	Si-Di	SS, 20V, 0,05A, <0,7ns	31a	Fch	FDH 700	data
FD 777	Si-Di	SS, 8V, 0,05A, <0,75ns	31a	Fch	FDH 777	data
FD 6666	Si-Di	SS, 75V, <5ns	31a	Fch	BAW 26...27, BAV 12, BAV 14, BAX 81	data
FDA200	DOC	LED/PD Viso:3750Vrms Vbs:1.2V		Cpc,The		data pinout
FDB 6670 AL	MOS-N-FET-e	V-MOS, LogL, 30V, 80A, 75W, <8,5mΩ	30c§	Fch	IRL 1004S, IRL 1104S, 2SK3430-S,-ZJ	data
FDB 8030 L	MOS-N-FET-e	V-MOS, LogL, 30V, 80A, 187W, <4,5mΩ	30c§	Fch	IRL 1004S, IRL 1104S	data
FDC6323L	LOAD+	Vin:3..8V Il:1.5A	6S	Fcs		data pdf pinout
FDC6324L	LOAD+	Vin:3..20V Il:1.5A	6S	Fcs		data pdf pinout
FDC6325L	LOAD+	Vin:2.5..8V Il:1.8A Ron:115mOhm	6S	Fcs		data pdf pinout
FDC6326L	LOAD+	Vin:3..20V Il:1.8A Ron:140mOhm	6S	Fcs		data pdf pinout
FDC6329L	LOAD+	Vin:2.5..8V Il:2.5A Ron:47mOhm	6S	Fcs		data pdf pinout
FDC6330L	LOAD+	Vin:3..20V Il:2.3A Ron:81mOhm	6S	Fcs		data pdf pinout
FDC6331L	LOAD+	Vin:±8V Il:2.8A Ron:34mOhm	6S	Fcs		data pdf pinout
FDC6901L	LOAD+	Vin:-0.5..10V Il:3A Ron:105mOhm	6S	Fcs		data pdf pinout
FDG6331L	LOAD+	Vin:±8V Il:0.8A Ron:155mOhm	6S	Fcs		data pdf pinout
FDG6342L	LOAD+	Vin:±8V Il:1.5A Ron:125mOhm	6S	Fcs		data pdf pinout
FDH 300	Si-Di	Uni, 125V, 0,2A, Ir<1nA(125V)	31a	Fch, Nsc	BAS 45	data
FDH 333	Si-Di	Uni, 125V, 0,2A, Ir<3nA(125V)	31a	Fch, Nsc	BAS 45	data
FDH 400	Si-Di	Uni, S, 175V, 0,2A, <50ns	31a	Fch, Nsc	BA 197...198, BAV 20...21, BAW 50	data
FDH 444	Si-Di	Uni, S, 125V, 0,2A, <60ns	31a	Fch, Nsc	BA 196...198, BAV 15...16, BAV 19...21	data
FDH 600	Si-Di	SS, 50V, 0,2A, <4ns	31a	Fch, Nsc	BAW 62, BAW 76, BAX 95, 1N4148...49, ++	data
FDH 666	Si-Di	SS, 25V, 0,2A, <4ns	31a	Fch, Nsc	BAW 62, BAW 76, BAX 95, 1N4148...49, ++	data
FDH 700	Si-Di	SS, 20V, 0,05A, <0,7ns	31a	Fch	FD 700	data
FDH 777	Si-Di	SS, 8V, 0,05A, <0,75ns	31a	Fch	FD 777	data
FDH 900	Si-Di	SS, 40V, 0,2A, <4ns	31a	Fch, Nsc	BAV 10, BAW 24...27, BAX 81, 1N4150, ++	data
FDH 999	Si-Di	SS, 25V, 0,2A, <5ns	31a	Fch, Nsc	BAV 10, BAW 24...27, BAX 81, 1N4150, ++	data
FDH 1000	Si-Di	S, 50V, 0,2A	31a	Fch, Nsc	BAW 24...27	data
FDH 3595	Si-Di	Uni, S, 150V, 0,2A, <3µs, (≈1N3595)	31a	Nsc	BA 157...159, BAW 52, BAY 20...21, ++	data
FDLL 300	Si-Di	=FDH 300: SMD	72a(3,4mm)	Fch, Nsc	BAS 45L	data
FDLL 333	Si-Di	=FDH 333: SMD	72a(3,4mm)	Fch, Nsc	BAS 45L	data
FDLL 400	Si-Di	=FDH 400: SMD	72a(3,4mm)	Fch, Nsc	-	data
FDLL 444	Si-Di	=FDH 444: SMD	72a(3,4mm)	Fch, Nsc	-	data
FDLL 456(A)	Si-Di	=1N456(A): SMD	72a(3,4mm)	Fch, Nsc	BAS 32, LL 4148	data
FDLL 457(A)	Si-Di	=1N457(A): SMD	72a(3,4mm)	Fch, Nsc	BAS 32, LL 4148	data
FDLL 458(A)	Si-Di	=1N458(A): SMD	72a(3,4mm)	Fch, Nsc	HSK 83	data
FDLL 459(A)	Si-Di	=1N459(A): SMD	72a(3,4mm)	Fch, Nsc	HSK 83	data
FDLL 461 A	Si-Di	=1N461A: SMD	72a(3,4mm)	Fch, Nsc	BAS 32, LL 4148	data
FDLL 462 A	Si-Di	=1N462A: SMD	72a(3,4mm)	Fch, Nsc	BAS 32, LL 4148	data
FDLL 463 A	Si-Di	=1N463A: SMD	72a(3,4mm)	Fch, Nsc	HSK 83	data
FDLL 482 B	Si-Di	=1N482B: SMD	72a(3,4mm)	Fch, Nsc	BAS 32, LL 4148	data
FDLL 483 B	Si-Di	=1N483B: SMD	72a(3,4mm)	Fch, Nsc	BAS 32, LL 4148	data
FDLL 484 B	Si-Di	=1N484B: SMD	72a(3,4mm)	Fch, Nsc	HSK 83	data
FDLL 485 B	Si-Di	=1N485B: SMD	72a(3,4mm)	Fch, Nsc	HSK 83	data
FDLL 600	Si-Di	=FDH 600: SMD	72a(3,4mm)	Fch, Nsc	BAS 32, LL 4148	data
FDLL 625	Si-Di	=1N625: SMD	72a(3,4mm)	Fch, Nsc	BAS 32, LL 4148	data
FDLL 626	Si-Di	=1N626: SMD	72a(3,4mm)	Fch, Nsc	BAS 32, LL 4148	data
FDLL 627	Si-Di	=1N627: SMD	72a(3,4mm)	Fch, Nsc	LL 4148	data
FDLL 628	Si-Di	=1N628: SMD	72a(3,4mm)	Fch, Nsc	HSK 83	data
FDLL 629	Si-Di	=1N629: SMD	72a(3,4mm)	Fch, Nsc	HSK 83	data
FDLL 658	Si-Di	=1N658: SMD	72a(3,4mm)	Fch, Nsc	BAS 32, LL 4148	data
FDLL 659	Si-Di	=1N659: SMD	72a(3,4mm)	Fch, Nsc	BAS 32, LL 4148	data
FDLL 660	Si-Di	=1N660: SMD	72a(3,4mm)	Fch, Nsc	LL 4148	data
FDLL 661	Si-Di	=1N661: SMD	72a(3,4mm)	Fch, Nsc	HSK 83	data
FDLL 666	Si-Di	=FDH 666: SMD	72a(3,4mm)	Fch	BAS 32, LL 4148	data
FDLL 700	Si-Di	=FDH 700: SMD	72a(3,4mm)	Fch	-	data
FDLL 777	Si-Di	=FDH 777: SMD	72a(3,4mm)	Fch	-	data
FDLL 900	Si-Di	=FDH 900: SMD	72a(3,4mm)	Fch	BAS 32, LL 4148	data

Type	Device	Short Description	Fig.	Manu	Comparision Types	More at
DLL 914 ...	Si-Di	=1N914...: SMD	72a(3,4mm)	Fch, Nsc	BAS 32, LL 4148	data
DLL 916 ...	Si-Di	=1N916...: SMD	72a(3,4mm)	Fch, Nsc	BAS 32, LL 4148	data
DLL 920	Si-Di	=1S920: SMD	72a(3,4mm)	Fch, Nsc	BAS 45L, HSK 83	data
DLL 921	Si-Di	=1S921: SMD	72a(3,4mm)	Fch, Nsc	BAS 45L, HSK 83	data
DLL 922	Si-Di	=1S922: SMD	72a(3,4mm)	Fch, Nsc	BAS 45L, HSK 83	data
DLL 923	Si-Di	=1S923: SMD	72a(3,4mm)	Fch, Nsc	HSK 83	data
DLL 999	Si-Di	=FDH 999: SMD	72a(3,4mm)	Fch	BAS 32, LL 4148	data
DLL 1000	Si-Di	=FDH 1000: SMD	72a(3,4mm)	Fch	BAS 45L	data
DLL 3064	Si-Di	=1N3064: SMD	72a(3,4mm)	Fch, Nsc	BAS 32, LL 4148	data
DLL 3070	Si-Di	=1N3070: SMD	72a(3,4mm)	Fch, Nsc		data
DLL 3595	Si-Di	=1N3595: SMD	72a(3,4mm)	Fch, Nsc	HSK 83, HSK 122	data
DLL 3600	Si-Di	=1N3600: SMD	72a(3,4mm)	Fch, Nsc	BAS 32, LL 4148	data
DLL 4009	Si-Di	=1N4009: SMD	72a(3,4mm)	Fch, Nsc	BAS 32, LL 4148	data
DLL 4148...4154	Si-Di	=1N4148...4154: SMD	72a(3,4mm)	Fch, Nsc	BAS 32, LL 4148	data
DLL 4305	Si-Di	=1N4305: SMD	72a(3,4mm)	Fch, Nsc	BAS 32, LL 4148	data
DLL 4446...4450	Si-Di	=1N4446...4450: SMD	72a(3,4mm)	Fch, Nsc	BAS 32, LL 4148	data
DLL 4454	Si-Di	=1N4454: SMD	72a(3,4mm)	Fch, Nsc	BAS 32, LL 4148	data
DLL 4938	Si-Di	=1N4938: SMD	72a(3,4mm)	Fch	-	data
DLL 6099	Si-Di	=1N6099: SMD	72a(3,4mm)	Fch, Nsc	HSK 83, HSK 122	data
FDMF 6700	IC	Driver plus Fet Multi-chip Module	QFN	Fch	-	pdf pinout
FDMF 6704	MOS	High Frequency DrMOS Module	MLP	Fch	-	pdf pinout
FDMF 8700	IC	Driver plus Fet Multi-chip Module	QFN	Fch	-	pdf pinout
FDMF 8705	IC	Driver plus Fet Multi-chip Module	QFN	Fch	-	pdf pinout
FDMS 2380	IC	Dual Integrated Solenoid Driver	QFN	Fch	-	pdf pinout
FDN 600	Si-Di	=FDH 600	31a	Fch	→FDH 600	data
FDN 666	Si-Di	=FDH 666	31a	Fch	→FDH 666	data
FDp	Si-P	=BCV 26 (Typ-Code/Stempel/marking)	35	Phi	→BCV 26	data
FDP 6670 AL	MOS-N-FET-e	V-MOS, LogL, 30V, 80A, 75W, <8,5mΩ	17c§	Fch	IRL 1004, IRL 2505S, 2SK3430	
FDP 8030 L	MOS-N-FET-e	V-MOS, LogL, 30V, 80A, 187W, <4,5mΩ	17c§	Fch	IRL 1004, IRL 2505S	
FDR 300	Si-Di	S, RadH, 250V, <325ns	31a	Fch	-	data
FDR 600	Si-Di	SS, RadH, 75V, <4ns	31a	Fch	-	data
FDR 700	Si-Di	SS, RadH, 30V, <0,7ns	31a	Fch	-	data
FDt	Si-P	=BCV 26 (Typ-Code/Stempel/marking)	35	Phi	→BCV 26	data
FE	Si-N	=2SC3438-E (Typ-Code/Stempel/marking)	39	Mit	→2SC3438	data
FE	Si-N	=2SC5568 (Typ-Code/Stempel/marking)	39	Say	→2SC5568	data
FE	Si-P	=BCV 46 (Typ-Code/Stempel/marking)	35	Sie,Phi,Tho	→BCV 46	data
FE	Si-N	=BFP 93A (Typ-Code/Stempel/marking)	44	Aeg, Sie	→BFP 93A	data
FE	Si-N	=BFQ 19P (Typ-Code/Stempel/marking)	39	Sie	→BFQ 19P	data
FE	MOS-NP-FET-e	=CPH 5605 (Typ-Code/Stempel/marking)	45	Say	→CPH 5605	data
FE	Si-P+R	=UN 211Z (Typ-Code/Stempel/marking)	35	Mat	→UN 211Z	data
FE 1 A..H	Si-Di	GI, S, 50...400V, 1A, <30ns A=50, B=100, C=150, D=200, E=250V, F=300, G=350, H=400V	31a	Gie, Die	BYV 26A...C, EGP 10A...G	data
FE 2 A..H	Si-Di	GI, S, 50...400V, 2A, <35ns	31a	Gie, Die	BYV 27-..., EGP 20A...G	data
FE 3 A..H	Si-Di	GI, S, 50...400V, 3A, <35ns	31a	Gie, Die	BYV 28-..., EGP 30A...G	data
FE 5 A..D	Si-Di	GI, S, 50...200V, 5A, <35ns	31a	Gie	BYV 61...63, EGP 50A...G	data
FE 6 A..H	Si-Di	GI, S, 50...400V, 6A, <35ns	31a	Gie, Die	BYV 61...63	data
FE 8 A..J	Si-Di	GI/S-L, 50...600V, 8A, <35ns A=50, B=100, C=150, D=200, F=300V, G=400, H=500, J=600V	17t§	Gie, Tho	BYC 8-600, MUR 805...860	data
FE 16 A..J	Si-Di	Dual, GI/S-L, 50...600V, 16A, <35ns	17p§	Gie, Tho	BYV 34/...	data
FE 16 AD...JD	Si-Di	=FE 16A..J:	17n1		FED 16AT...JT	data
FE 16 AN...JN	Si-Di	=FE 16A..J:	17q&		FEN 16AT...JT	data
FE 30 A..J	Si-Di	Dual, GI/S-L, 50...600V, 30A, <35...50ns	23l§	Gie, Tho	BYV 72/..., BYV 74/..., MUR 3005...3060PT	data
FE 30 AD...JD	Si-Di	=FE 30A..J:	23n		-	data
FE 30 AN...JN	Si-Di	=FE 30A..J:	23m&		-	data
FE 0654A...C	N-FET	HF/S 25V, Idss=1...40mA, Up<8V	8b	Fch,Tsc	BF 244A...C, BF 245A...C, 2N5163	data
FE 0655A...C	N-FET	Uni, 30V, Idss>3mA, Up<10V	8b	Fch,Tsc	BFS 74...79, BSV 78...80, 2N4856...4861	data
FE 1718A...E	Si-P	SMD, Dual, 40V, >400MHz	10-MDIP	Fch	-	data
FE 2060A,B	Si-N	=2N2060A,B: SMD	10-MDIP	Fch	-	data
FE 2223(A)	Si-N	=2N2223(A): SMD	10-MDIP	Fch	-	data
FE 2913...2920(A)	Si-N	=2N2913...2920(A): SMD	10-MDIP	Fch	-	data
FE 3423	Si-N	=2N3423: SMD	10-MDIP	Fch	-	data
FE 3424	Si-N	=2N3424: SMD	10-MDIP	Fch	-	data
FE 3726	Si-P	=2N3726: SMD	10-MDIP	Fch	-	data
FE 3727	Si-P	=2N3727: SMD	10-MDIP	Fch	-	data
FE 3728	Si-N	=2N3728: SMD	10-MDIP	Fch	-	data
FE 3729	Si-N	=2N3729: SMD	10-MDIP	Fch	-	data
FE 3819	N-FET	=2N3819: 0,3W	8b	Fch,Tsc	→2N3819	data
FE 4015...4025	Si-P	=2N4015...4025: SMD	10-MDIP	Fch	-	data
FE 4302...4304	N-FET	=2N4302...4304: 0,3W	8b	Fch,Tsc	→2N4302...4303	data

Type	Device	Short Description	Fig.	Manu	Comparision Types	More at
FE 5245...5247	N-FET	=2N5245...5247: 0,36W	8f	Fch,Tsc	→2N5245...5247	data
FE 5257...5259	N-FET	=2N5257...5259: 0,31W	8b	Fch,Tsc	→2N5257...5259	data
FE 5284...5286	N-FET	=2N5284...5286: 0,31W	8f	Fch,Tsc	→2N5284...5286	data
FED 16 AT...JT	Si-Di	=FEP 16AT...JT:	17n1	Gie	FE 16AD...JD	data
FED 30 AP...JP	Si-Di	=FEP 30AP...JP:	18n1	Gie		data
FEN 16 AT...JT	Si-Di	=FEP 16AT...JT:	17q&	Gie	FE 16AN...JN	data
FEN 30 AP...JP	Si-Di	=FEP 30AP...JP:	18q&	Gie	-	data
FEp	Si-P	=BCV 46 (Typ-Code/Stempel/marking)	35	Phi	→BCV 46	data
FEP 6 AT...DT	Si-Di	Dual, Gl/S-L, 50...200V, 6A, <35ns A=50, B=100, C=150, D=200V	17p§	Gsi	BYP 21-..., BYT 08-...	data
FEP 16 AT...JT	Si-Di	Dual, Gl/S-L, 50...600V, 16A, <35...50ns A=50, B=100, C=150, D=200, F=300V, G=400, H=500, J=600V	17p§	Gie	BYV 32/..., FE 16A..J	data
FEP 30 AT...JT	Si-Di	Dual, Gl/S-L, 50...600V, 30A, <35...50ns A=50, B=100, C=150, D=200, F=300V, G=400, H=500, J=600V	16p§	Gie	BYV 72/..., MUR 3005PT...3060PT	data
FEPB 6 AT...DT	Si-Di	Dual, Gl/S-L, 50...200V, 6A, <35ns A=50, B=100, C=150, D=200V	30p§	Gsi	BYQ 28EB-...,	data
FEPB 16 AT...JT	Si-Di	Dual, Gl/S-L, 50...600V, 16A, <35...50ns A=50, B=100, C=150, D=200, F=300V, G=400, H=500, J=600V	30p§	Gsi	BYV 32EB-...	data
FER 8 AT...JT	Si-Di	=FES 8AT...JT:	17t1&	Gie	-	data
FER 16 AT...JT	Si-Di	=FES 16AT...JT:	17t1&	Gie	-	data
FEs	Si-P	=BCV 46 (Typ-Code/Stempel/marking)	35	Sie	→BCV 46	data
FEs	Si-N	=BFP 93A (Typ-Code/Stempel/marking)	44	Sie	→BFP 93A	data
FES 1 A...G	Si-Di	SMD, Gl, S, 50...400V, 1A, <50ns A=50V, B=100V, D=200, F=300V, G=400V	71a(5x2,5)	Die	D 1FL20, MURS 120, MURS 160	data
FES 2 A...G	Si-Di	SMD, Gl, S, 50...400V, 2A, <50ns A=50V, B=100V, D=200, F=300V, G=400V	71a(5x3,5)	Die	BYG 22A...D	data
FES 8 AT...JT	Si-Di	Gl, S-L, 50...600V, 8A, <35...50ns A=50, B=100, C=150, D=200, F=300V, G=400, H=500, J=600V	17t§	Gie	FE 8A..J, MUR 805...860	data
FES 16 AT...JT	Si-Di	Gl, S-L, 50...600V, 16A, <35...50ns A=50, B=1C0, C=150, D=200, F=300V, G=400, H=500, J=600V	17t§	Gie	BYT 16P-..., BYV 32-..., FE 16A..J	data
FESB 8 AT...JT	Si-Di	Gl, S-L, 50...600V, 8A, <35...50ns A=50, B=100, C=150, D=200, F=300V, G=400, H=500, J=600V	30t§	Gsi		data
FESB 16 AT...JT	Si-Di	Gl, S-L, 50...600V, 16A, <35...50ns A=50, B=100, C=150, D=200, F=300V, G=400, H=500, J=600V	30t§	Gsi	BYT 16P-..., BYV 32-..., FE 16A..J	data
FEt	Si-P	=BCV 46 (Typ-Code/Stempel/marking)	35	Phi	→BCV 46	data
FF	Si-N	=BCV 27 (Typ-Code/Stempel/marking)	35	Sie,Phi,Tho	→BCV 27	data
FF	Si-N	=BFQ 18A (Typ-Code/Stempel/marking)	39	Val	→BFQ 18A	data
FF	MOS-NP-FET-e	=CPH 5606 (Typ-Code/Stempel/marking)	45	Say	→CPH 5606	data
FF	MOS-N-FET-e*	=MCH 6606 (Typ-Code/Stempel/marking)	46(2mm)	Say	→MCH 6606	data
FF	Si-N+R	=UN 221Z (Typ-Code/Stempel/marking)	35	Mat	→UN 221Z	data
FF	Z-Di	=Z1 SMA-51 (Typ-Code/Stempel/marking)	71(5x2,5)	Fag	→Z1 SMA-51	data
FF 1001	Si-Di	Gl, S, 50V, 1A, <150ns	31a	Fag	BY 201/3, BYT 52A, BYV 12, RGP 10A, ++	data
FF 1002	Si-Di	=FF 1001: 100V	31a	Fag	BY 201/3, BYT 52B, BYV 12, RGP 10B, ++	data
FF 1003	Si-Di	=FF 1001: 200V	31a	Fag	BY 201/3, BYT 52D, BYV 13, RGP 10D, ++	data
FF 1004	Si-Di	=FF 1001: 400V	31a	Fag	BY 201/4, BYT 52G, BYV 13, RGP 10G, ++	data
FF 1005	Si-Di	=FF 1001: 500V	31a	Fag	BY 201/5, BYT 52J, BYV 14, RGP 10J, ++	data
FF 1501	Si-Di	Gl, S, 100V, 1,5A, <150...250ns	31a	Fag	BYD 73B, BYV 12, BYV 36A, RGP 15B, ++	data
FF 1502	Si-Di	=FF 1501: 200V	31a	Fag	BYD 73D, BYV 13, BYV 36A, RGP 15D, ++	data
FF 1504	Si-Di	=FF 1501: 400V	31a	Fag	BYD 73G, BYV 13, BYV 36B, RGP 15G, ++	data
FF 1506	Si-Di	=FF 1501: 600V, <250ns	31a	Fag	BYD 34J, BYV 14, BYV 36C, RGP 15J, ++	data
FF 1508	Si-Di	=FF 1501: 800V, <250ns	31a	Fag	BYD 34K, BYV 15, BYV 36D, RGP 15K, ++	data
FF 1510	Si-Di	=FF 1501: 1000V, <250ns	31a	Fag	BYD 34M, BYV 16, BYV 36E, RGP 15M, ++	data
FFp	Si-N	=BCV 27 (Typ-Code/Stempel/marking)	35	Phi	→BCV 27	data
FFs	Si-N	=BCV 27 (Typ-Code/Stempel/marking)	35	Phi	→BCV 27	data
FFt	Si-N	=BCV 27 (Typ-Code/Stempel/marking)	35	Phi	→BCV 27	data
FG	Si-N	=BCV 47 (Typ-Code/Stempel/marking)	35	Phi, Sie	→BCV 47	data
FG	Si-N	=BFQ 149 (Typ-Code/Stempel/marking)	39	Phi	→BFQ 149	data
FG	Z-Di	=Z1 SMA-56 (Typ-Code/Stempel/marking)	71(5x2,5)	Fag	→Z1 SMA-56	data
FG 1	Ge-P	≈2SB33			→2SB33	data
FG 2 N	Si-Di	TV-Damper-Di, 300V, 1A	31a	Fjd	BY 201/3, BYX 55/350, MR 813, RGP 10G,++	data
FG 40 N10L	MOS-N-FET-e	=RFG 40N10 (Typ-Code/Stempel/marking)	16	Rca	→RFG 40N10(LE)	data
FGp	Si-N	=BCV 47 (Typ-Code/Stempel/marking)	35	Phi	→BCV 47	data
FGs	Si-N	=BCV 47 (Typ-Code/Stempel/marking)	35	Sie	→BCV 47	data
FGs	Si-N	=BFQ 19S (Typ-Code/Stempel/marking)	39	Sie	→BFQ 19S	data
FGt	Si-N	=BCV 47 (Typ-Code/Stempel/marking)	35	Phi	→BCV 47	data

Type	Device	Short Description	Fig.	Manu	Comparision Types	More at
FH	Si-N	=BFN 24 (Typ-Code/Stempel/marking)	35	Sie	→BFN 24	data
FH	MOS-N-FET-e*	=CPH 5608 (Typ-Code/Stempel/marking)	45	Say	→CPH 5608	data
FH	Z-Di	=Z1 SMA-62 (Typ-Code/Stempel/marking)	71(5x2,5)	Fag	→Z1 SMA-62	data
FH 102	Si-N	SMD, Dual, =2x 2SC5226	46bh3(2mm)	Say	-	data
FH 103	Si-N	SMD, Dual, =2x 2SC4867	46bh3(2mm)	Say	-	data
FH 104	Si-N	SMD, Dual, =2x 2SC4853	46bh3(2mm)	Say	-	data
FH 105	Si-N	SMD, Dual, =2x 2SC5245	46bh8(2mm)	Say	-	data
FH 201	Si-N	SMD, Dual, =2SC4871 + 2SC4867	46bh3(2mm)	Say	-	data
FH 202	Si-N	SMD, Dual, =2SC5245 + TS 4162	46bh4(2mm)	Say	-	data
FH 203	Si-N	SMD, Dual, =2SC5245 + 2SC5415	46bh4(2mm)	Say	-	data
FHP	Si-N	=2SC5020-P (Typ-Code/Stempel/marking)	35(2mm)	Mat	→2SC5020	data
FHP 3194	IC	4:1 High-speed Multiplexer	QFN	Fch	-	pdf pinout
FHP 3392	IC	±5V, Triple 2:1, Video Multiplexer	TSOP	Fch	-	pdf pinout
FHQ	Si-N	=2SC5020-Q (Typ-Code/Stempel/marking)	35(2mm)	Mat	→2SC5020	data
FHR	Si-N	=2SC5020-R (Typ-Code/Stempel/marking)	35(2mm)	Mat	→2SC5020	data
FHs	Si-N	=BFN 24 (Typ-Code/Stempel/marking)	35	Sie	→BFN 24	data
FI	Si-N/P+R	=XN 431L (Typ-Code/Stempel/marking)	46	Mat	→XN 431L	data
FI-	MOS-N-FET-d	=3SK186 (Typ-Code/Stempel/marking)	44	Hit	→3SK186	data
FI-	MOS-N-FET-d	=3SK217 (Typ-Code/Stempel/marking)	44	Hit	→3SK217	data
FIN 12 AC /BGA	IC	12-Bit Bi-Direct. Serial-/Deserializer	TSOP	Fch	-	pdf pinout
FIN 12 AC /MLP	IC	12-Bit Bi-Direct. Serial-/Deserializer	TSOP	Fch	-	pdf pinout
FIN 24 /BGA	IC	22-Bit Bi-Direct. Serial-/Deserializer	TSOP	Fch	-	pdf pinout
FIN 24 C /BGA	IC	24-Bit Bi-Direct. Serial-/Deserializer	TSOP	Fch	-	pdf pinout
FIN 24 C /MLP	IC	24-Bit Bi-Direct. Serial-/Deserializer	TSOP	Fch	-	pdf pinout
FIN 24 /MLP	IC	22-Bit Bi-Direct. Serial-/Deserializer	TSOP	Fch	-	pdf pinout
FIN 212 AC /MLP	IC	12-Bit Serializer Deserializer	TSOP	Fch	-	pdf pinout
FIN 224 AC /BGA	IC	22-Bit Bi-Direct. Serial-/Deserializer	TSOP	Fch	-	pdf pinout
FIN 224 AC /MLP	IC	22-Bit Bi-Direct. Serial-/Deserializer	TSOP	Fch	-	pdf pinout
FIN 324 C /BGA	IC	24-Bit Serializer / Deserializer	TSOP	Fch	-	pdf pinout
FIN 324 C /MLP	IC	24-Bit Serializer / Deserializer	TSOP	Fch	-	pdf pinout
FIN 1001 M5(X)	IC	3.3V LVDS 1-Bit High Speed Dif. Driver	45	Fcs	-	pdf pinout
FIN 1002 M5(X)	IC	LVDS 1-Bit High Speed Differential Receiver	45	Fcs	-	pdf pinout
FIN 1017 K8X	IC	3.3V LVDS 1-Bit High Speed Differential Driver	TSOP	Fcs	-	pdf pinout
FIN 1017 M(X)	IC	3.3V LVDS 1-Bit High Speed Differential Driver	MDIP	Fcs	-	pdf pinout
FIN 1018 K8X	IC	3.3V LVDS 1-Bit High Speed Differential Receiver	TSOP	Fch	-	pdf pinout
FIN 1018 M(X)	IC	3.3V LVDS 1-Bit High Speed Differential Receiver	MDIP	Fch	-	pdf pinout
FIN 1019 M	IC	3.3V LVDS High Speed Differential Driver/Receiver	MDIP	Fcs	-	pdf pinout
FIN 1019 MTC	IC	3.3V LVDS High Speed Differential Driver/Receiver	TSOP	Fcs	-	pdf pinout
FIN 1022 M	IC	2 X 2 LVDS High Speed Crosspoint Switch	MDIP	Fch	-	pdf pinout
FIN 1022 MTC	IC	2 X 2 LVDS High Speed Crosspoint Switch	TSOP	Fch	-	pdf pinout
FIN 1027 AM(X)	IC	3.3V LVDS 2-Bit High Speed Differential Driver	MDIP	Fch	-	pdf pinout
FIN 1027 K8X	IC	3.3V LVDS 2-Bit High Speed Differential Driver	TSOP	Fch	-	pdf pinout
FIN 1027 M(X)	IC	3.3V LVDS 2-Bit High Speed Differential Driver	MDIP	Fch	-	pdf pinout
FIN 1027 MPX	IC	3.3V LVDS 2-Bit High Speed Differential Driver	MLP	Fch	-	pdf pinout
FIN 1028 K8X	IC	3.3V LVDS 2-Bit High Speed Differential Receiver	TSOP	Fch	-	pdf pinout
FIN 1028 M	IC	3.3V LVDS 2-Bit High Speed Differential Receiver	MDIP	Fch	-	pdf pinout
FIN 1028 MPX	IC	3.3V LVDS 2-Bit High Speed Differential Receiver	MLP	Fch	-	pdf pinout
FIN 1031 M	IC	3.3V LVDS 4-Bit High Speed Differential Driver	MDIP	Fcs	-	pdf pinout
FIN 1031 MTC	IC	3.3V LVDS 4-Bit High Speed Differential Driver	TSOP	Fcs	-	pdf pinout
FIN 1032 M	IC	3.3V LVDS 4-Bit Differential Receiver	TSOP	Fch	-	pdf pinout
FIN 1032 MTC	IC	3.3V LVDS 4-Bit Differential Receiver	TSOP	Fch	-	pdf pinout
FIN 1047 M	IC	3.3V LVDS 4-Bit Flow-Through Diff. Driver	TSOP	Fch	-	pdf pinout
FIN 1047 MTC	IC	3.3V LVDS 4-Bit Flow-Through Diff. Driver	TSOP	Fch	-	pdf pinout
FIN 1048 M	IC	3.3V LVDS 4-Bit Flow-Through,Diff. Receiver	TSOP	Fch	-	pdf pinout
FIN 1048 MTC	IC	3.3V LVDS 4-Bit Flow-Through,Diff. Receiver	TSOP	Fch	-	pdf pinout
FIN 1049 MTC	IC	LVDS Dual Line Driver,Dual Line Recei.	TSOP	Fch	-	pdf pinout
FIN 1101 M	IC	LVDS Single Port High Speed Repeater	TSOP	Fch	-	pdf pinout
FIN 1101 MX	IC	LVDS Single Port High Speed Repeater	TSOP	Fch	-	pdf pinout
FIN 1102 MTC	IC	LVDS 2 Port High Speed Repeater	TSOP	Fch	-	pdf pinout
FIN 1104 MTC	IC	LVDS 4 Port High Speed Repeater	TSOP	Fch	-	pdf pinout
FIN 1108 MTD	IC	LVDS 8-Port, High-speed Repeater	TSOP	Fch	-	pdf pinout
FIN 1108 MTDX	IC	LVDS 8-Port, High-speed Repeater	TSOP	Fch	-	pdf pinout
FIN 1215 MTDX	IC	LVDS 21-Bit Serializers / De-Serializers	TSOP	Fch	-	pdf pinout
FIN 1216 MTDX	IC	LVDS 21-Bit Serializers / De-Serializers	TSOP	Fch	-	pdf pinout
FIN 1217 MTD	IC	LVDS 21-Bit Serializers / De-Serializers	TSOP	Fch	-	pdf pinout

Type	Device	Short Description	Fig.	Manu	Comparision Types	More at
FIN 1217 MTDX	IC	LVDS 21-Bit Serializers / De-Serializers	TSOP	Fch	-	pdf pinout
FIN 1218 MTDX	IC	LVDS 21-Bit Serializers / De-Serializers	TSOP	Fch	-	pdf pinout
FIN 1531 M	IC	5V LVDS 4-Bit High Speed Differential Receiver	TSOP	Fch	-	pdf pinout
FIN 1531 MTC	IC	5V LVDS 4-Bit High Speed Differential Receiver	TSOP	Fch	-	pdf pinout
FIN 1532 M	IC	5V LVDS 4-Bit High Speed Differential Receiver	TSOP	Fch	-	pdf pinout
FIN 1532 MTC	IC	5V LVDS 4-Bit High Speed Differential Receiver	TSOP	Fch	-	pdf pinout
FIN 3383 MTD	TTL-I/O-IC	28- Bit Flat Panel Display Link Serializers/ Deserializers	TSOP	Fch	-	pdf pinout
FIN 3384 MTD	TTL-I/O-IC	28- Bit Flat Panel Display Link Serializers/ Deserializers	TSOP	Fch	-	pdf pinout
FIN 3385 MTD	TTL-I/O-IC	28- Bit Flat Panel Display Link Serializers/ Deserializers	TSOP	Fch	-	pdf pinout
FIN 3386 MTD	TTL-I/O-IC	28- Bit Flat Panel Display Link Serializers/ Deserializers	TSOP	Fch	-	pdf pinout
FIN 1101 K 8 X	IC	LVDS Single Port High Speed Repeater	TSOP	Fch	-	pdf pinout
FIN 212 AC /BGA 36	IC	12-Bit Serializer Deserializer	TSOP	Fch	-	pdf pinout
FIN 212 AC /BGA 42	IC	12-Bit Serializer Deserializer	TSOP	Fch	-	pdf pinout
FJ	N-FET	=2SK1069 (Typ-Code/Stempel/marking)	35(2mm)	Say	→2SK1069	data
FJ	Si-N	=BFN 26 (Typ-Code/Stempel/marking)	35	Sie	→BFN 26	data
FJ 3	N-FET	=2SK771-3 (Typ-Code/Stempel/marking)	35	Say	→2SK771	data
FJ 4	N-FET	=2SK771-4 (Typ-Code/Stempel/marking)	35	Say	→2SK771	data
FJ 5	N-FET	=2SK771-5 (Typ-Code/Stempel/marking)	35	Say	→2SK771	data
FJ 2501	Si-P-Darl	≈BDX 34A	17j§		→BDX 34(A..D)	
FJ 3001	Si-N-Darl	≈BDX 33A	17j§		→BDX 33(A..D)	
FJs	Si-N	=BFN 26 (Typ-Code/Stempel/marking)	35	Sie	→BFN 26	data
FK	Si-P	=2SB800-FK (Typ-Code/Stempel/marking)	39	Nec	→2SB800	data
FK	Si-P	=BFN 25 (Typ-Code/Stempel/marking)	35	Sie	→BFN 25	data
FK	Si-N+R	=XN 421N (Typ-Code/Stempel/marking)	46	Mat	→XN 421N	data
FK	Z-Di	=Z1 SMA-68 (Typ-Code/Stempel/marking)	71(5x2,5)	Fag	→Z1 SMA-68	data
FK 7 KM-12	MOS-N-FET-e*	=FK 7UM-12: Iso, 35W	17c	Mit		data
FK 7 SM-12	MOS-N-FET-e*	=FK 7UM-12:	18c§	Mit		data
FK 7 UM-12	MOS-N-FET-e*	FREDFET, 900V, 7A, 125W, <1,63Ω(3A)	17c§	Mit		data
FK 7 VS-12	MOS-N-FET-e*	=FK 7UM-12:	30c§	Mit		data
FK 10 KM-9	MOS-N-FET-e*	=FK 10UM-9: Iso, 35W	17c	Mit	-	data
FK 10 KM-10	MOS-N-FET-e*	=FK 10UM-10: Iso, 35W	17c	Mit	-	data
FK 10 KM-12	MOS-N-FET-e*	=FK 10UM-12: Iso, 40W	17c	Mit	-	data
FK 10 SM-9	MOS-N-FET-e*	=FK 10UM-9:	18c§	Mit	-	data
FK 10 SM-10	MOS-N-FET-e*	=FK 10UM-10:	18c§	Mit	-	data
FK 10 SM-12	MOS-N-FET-e*	=FK 10UM-12:	18c§	Mit	-	data
FK 10 UM-9	MOS-N-FET-e*	FREDFET, 450V, 10A, 125W, <0,92Ω(5A)	17c§	Mit	-	data
FK 10 UM-10	MOS-N-FET-e*	FREDFET, 500V, 10A, 125W, <1,13Ω(5A)	17c§	Mit	-	data
FK 10 UM-12	MOS-N-FET-e*	FREDFET, 600V, 10A, 125W, <1,18Ω(5A)	17c§	Mit	-	data
FK 10 VS-9	MOS-N-FET-e*	=FK 10UM-9:	30c§	Mit	-	data
FK 10 VS10	MOS-N-FET-e*	=FK 10UM-10:	30c§	Mit	-	data
FK 10 VS12	MOS-N-FET-e*	=FK 10UM-12:	30c§	Mit	-	data
FK 14 KM-9	MOS-N-FET-e*	=FK 14UM-9: Iso, 40W	17c	Mit	-	data
FK 14 KM-10	MOS-N-FET-e*	=FK 14UM-10: Iso, 40W	17c	Mit	-	data
FK 14 SM-9	MOS-N-FET-e*	=FK 14UM-9:	18c§	Mit	-	data
FK 14 SM-10	MOS-N-FET-e*	=FK 14UM-10:	18c§	Mit	-	data
FK 14 SM-12	MOS-N-FET-e*	FREDFET, 600V, 14A, 250W, <0,75Ω(7A)	18c§	Mat	-	data
FK 14 UM-9	MOS-N-FET-e*	FREDFET, 450V, 14A, 150W, <0,65Ω(7A)	17c§	Mit	-	data
FK 14 UM-10	MOS-N-FET-e*	FREDFET, 500V, 14A, 150W, <0,8Ω(7A)	17c§	Mit	-	data
FK 14 VS-9	MOS-N-FET-e*	=FK 14UM-9:	30c§	Mit	-	data
FK 14 VS-10	MOS-N-FET-e*	=FK 14UM-10:	30c§	Mit	-	data
FK 16 KM-5	MOS-N-FET-e*	=FK 16UM-5: Iso, 35W	17c	Mit	-	data
FK 16 KM-6	MOS-N-FET-e*	=FK 16UM-6: Iso, 35W	17c	Mit	-	data
FK 16 SM-5	MOS-N-FET-e*	=FK 16UM-5:	18c§	Mit	-	data
FK 16 SM-6	MOS-N-FET-e*	=FK 16UM-6:	18c§	Mit	-	data
FK 16 UM-5	MOS-N-FET-e*	FREDFET, 250V, 16A, 125W, <0,31Ω(8A)	17c§	Mit	-	data
FK 16 UM-6	MOS-N-FET-e*	FREDFET, 300V, 16A, 125W, <0,41Ω(8A)	17c§	Mit	-	data
FK 16 VS-5	MOS-N-FET-e*	=FK 16UM-5:	30c§	Mit	-	data
FK 16 VS-6	MOS-N-FET-e*	=FK 16UM-6:	30c§	Mit	-	data
FK 18 SM-9	MOS-N-FET-e*	FREDFET, 450V, 18A, 250W, <0,41Ω(9A)	18c§	Mit	-	data
FK 18 SM-10	MOS-N-FET-e*	FREDFET, 500V, 18A, 250W, <0,5Ω(9A)	18c§	Mit	-	data
FK 18 SM-12	MOS-N-FET-e*	FREDFET, 600V, 18A, 275W, <0,54Ω(9A)	18c§	Mit	-	data
FK 20 KM-5	MOS-N-FET-e*	=FK 20UM-5: Iso, 40W	17c	Mit	-	data
FK 20 KM-6	MOS-N-FET-e*	=FK 20UM-6: Iso, 40W	17c	Mit	-	data
FK 20 SM-5	MOS-N-FET-e*	=FK 20UM-5:	18c§	Mit	-	data
FK 20 SM-6	MOS-N-FET-e*	=FK 20UM-6:	18c§	Mit	-	data
FK 20 SM-9	MOS-N-FET-e*	FREDFET, 450V, 20A, 275W, <0,3Ω(10A)	18c§	Mit	-	data
FK 20 SM-10	MOS-N-FET-e*	FREDFET, 500V, 20A, 275W, <0,36Ω(10A)	18c§	Mit	-	data
FK 20 UM-5	MOS-N-FET-e*	FREDFET, 250V, 20A, 150W, <0,24Ω(10A)	17c§	Mit	-	data
FK 20 UM-6	MOS-N-FET-e*	FREDFET, 300V, 20A, 150W, <0,33Ω(10A)	17c§	Mit	-	data
FK 20 VS-5	MOS-N-FET-e*	=FK 20UM-5:	30c§	Mit	-	data
FK 20 VS-6	MOS-N-FET-e*	=FK 20UM-6:	30c§	Mit	-	data
FK 25 SM-5	MOS-N-FET-e*	FREDFET, 250V, 25A, 250W, <0,16Ω(12A)	18c§	Mit	-	data
FK 25 SM-6	MOS-N-FET-e*	FREDFET, 300V, 25A, 250W, <0,21Ω(12A)	18c§	Mit	-	data

Type	Device	Short Description	Fig.	Manu	Comparision Types	More at
FK 30 SM-5	MOS-N-FET-e*	FREDFET, 250V, 30A, 275W, <108mΩ(15A)	18c§	Mit	-	data
FK 30 SM-6	MOS-N-FET-e*	FREDFET, 300V, 30A, 275W, <143mΩ(15A)	18c§	Mit	-	data
FKE 4 F2/...	Si-Di	→SKE 4F2/...	33z	Skr		
FKs	Si-P	=BFN 25 (Typ-Code/Stempel/marking)	35	Sie	→BFN 25	data
FL	Si-P	=2SA1434 (Typ-Code/Stempel/marking)	35	Say	→2SA1434	data
FL	Si-P	=2SB800-FL (Typ-Code/Stempel/marking)	39	Nec	→2SB800	data
FL	Si-P	=BFN 27 (Typ-Code/Stempel/marking)	35	Sie	→BFN 27	data
FL	Z-Di	=Z1 SMA-75 (Typ-Code/Stempel/marking)	71(5x2,5)	Fag	→Z1 SMA-75	data
FL 005...08	Si-Br	50...800V, 5A	33x1	Mic	B...C5000	data
FLI 2300	DIG-IC	digital video format converter	QFP_a	Gns	-	pdf pinout
FLI 2301	DIG-IC	digital video format converter	QFP_a	Gns	-	pdf pinout
FLI 2310	DIG-IC	digital video format converter	QFP_a	Gns	-	pdf pinout
FLI 5921	DIG-IC	LCD Monitor Controllers (1290x1024)	QFP	Gns	-	pdf
FLI 5961	DIG-IC	LCD Monitor Controllers (1600x1200)	QFP	Gns	-	pdf
FLI 5962	DIG-IC	LCD Monitor Controllers (1920x1200)	QFP	Gns	-	pdf
FLI 5962 H	DIG-IC	LCD Monitor Controllers (1920x1200)	QFP	Stm	-	pdf
FLI 5968 H	DIG-IC	LCD Monitor Controllers (1920x1200)	QFP	Stm	-	pdf
FLI 8120	DIG-IC	Crt TV Controller	QFP_a	Gns	-	pdf
FLI 8122	DIG-IC	LCD TV Controller	QFP_a	Gns	-	pdf
FLI 8125	DIG-IC	Crt/lcd TV Controller	QFP_a	Gns	-	pdf
FLI 8532	DIG-IC	LCD TV Controller	QFP_a	Gns	-	pdf
FLI 8538	DIG-IC	LCD TV Controller	QFP_a	Gns	-	pdf
FLI 8548 H	DIG-IC	LCD TV Controller with HDMI	QFP_a	Gns	-	pdf
FLI 8638	DIG-IC	LCD TV Controller	QFP_a	Gns	-	pdf
FLI 8668	DIG-IC	Dual-channel LCD TV Controller	QFP_a	Gns	-	pdf
FLLD 258	Si-Di	SMD, Dual, S, Picoampere, 100V, 0,25A <400ns, Ir<3nA(50V)	35l	Ztx	-	data
FLLD 261	Si-Di	=FLLD 258: Ir<5nA(100V)	35o	Ztx	-	data
FLLD 263	Si-Di	=FLLD 258: Ir<5nA(100V)	35m	Ztx	-	data
FLs	Si-P	=BFN 27 (Typ-Code/Stempel/marking)	35	Sie	→BFN 27	data
FLV...	Opto					
FLV104A	LED	rd ø5mm I:6500mcd Vf:2V If:100mA	T1¾	Fch		data pinout
FLV110	LED	rd ø5mm I:6500mcd Vf:1.7V If:20mA	T1¾	Fch		data pinout
FLV140	LED	rd ø5mm I:6500mcd Vf:1.7V If:20mA	T1¾	Fch		data pinout
FLV150	LED	rd ø5mm I:6500mcd Vf:1.7V If:20mA	T1¾	Fch		data pinout
FLV160	LED	rd ø5mm I:6500mcd Vf:1.7V If:20mA	T1¾	Fch		data pinout
FLV251	LED	rd ø5mm I:5mcd Vf:2.1V If:10mA	T1¾	Fch		data pinout
FLV252	LED	rd ø5mm I:8mcd Vf:2.1V If:10mA	T1¾	Fch		data pinout
FLV310	LED	gr ø5mm I:3.2mcd Vf:2.3V If:20mA	T1¾	Fch		data pinout
FLV340	LED	gr ø5mm I:3.2mcd Vf:2.3V If:20mA	T1¾	Fch		data pinout
FLV350	LED	gr ø5mm I:3.2mcd Vf:2.3V If:20mA	T1¾	Fch		data pinout
FLV360	LED	gr ø5mm I:3.2mcd Vf:2.3V If:20mA	T1¾	Fch		data pinout
FLV410	LED	yl ø5mm I:3.2mcd Vf:2.3V If:20mA	T1¾	Fch		data pinout
FLV440	LED	yl ø5mm I:3.2mcd Vf:2.3V If:20mA	T1¾	Fch		data pinout
FLV450	LED	yl ø5mm I:3.2mcd Vf:2.3V If:20mA	T1¾	Fch		data pinout
FLV460	LED	yl ø5mm I:3.2mcd Vf:2.3V If:20mA	T1¾	Fch		data pinout
FLV510	LED	rd ø5mm I:3mcd Vf:2.1V If:10mA	T1¾	Fch		data pinout
FM	Si-P	=2SB800-FM (Typ-Code/Stempel/marking)	39	Nec	→2SB800	data
FM	Si-Di	=BB 804 (Typ-Code/Stempel/marking)	35	Aeg,Sie	→BB 804	data
FM	Z-Di	=Z1 SMA-82 (Typ-Code/Stempel/marking)	71(5x2,5)	Fag	→Z1 SMA-82	data
FM 24 C256 ...M8	EEPROM-IC	256K-Bit 2-Wire Bus Interface Serial EEPROM	MDIP	Fch	-	pdf pinout
FM 24 C256 ...N	EEPROM-IC	256K-Bit 2-Wire Bus Interface Serial EEPROM	DIC	Fch	-	pdf pinout
FM 35-120	Si-Di	TV-Damper-Diode, 1200V, 1,5A	31a	Mop	BY 228, BY 328, BY 438, BY 558	data
FM 35-150	Si-Di	TV-Damper-Diode, 1500V, 1,5A	31a	Mop	BY 228, BY 328, BY 428, BY 558	data
FMA 1(A)	Si-P+R	SMD, Dual, Rb=Rbe=22kΩ, 50V, 100mA	45ba1	Rhm	-	data
FMA 1A	Si-P+R	SMD, 2x DTA124E, Dual Digital Transistor, com. Emitter	45ba1	Rhm	=UMA1N	data
FMA 2(A)	Si-P+R	SMD, Dual, Rb=Rbe=47kΩ, 50V, 30mA	45ba1	Rhm	-	data
FMA 2A	Si-P+R	SMD, 2x DTA144E, Dual Digital Transistor, com. Emitter	45ba1	Rhm	=UMA2N	data
FMA 3(A)	Si-P+R	SMD, Dual, Rb=4,7kΩ, 50V, 100mA	45ba1	Rhm	-	data
FMA 3A	Si-P+R	SMD, 2x DTA143T, Dual Digital Transistor, com. Emitter 50V, 100mA,	45ba1	Rhm	FMA4A, FMA6A,	data
FMA 4(A)	Si-P+R	SMD, Dual, Rb=10kΩ, 50V, 100mA	45ba1	Rhm	-	data
FMA 4A	Si-P+R	SMD, 2x DTA114T, Dual Digital Transistor, com. Emitter 50V, 100mA,	45ba1	Rhm	FMA3A, FMA6A,	data
FMA 5(A)	Si-P+R	SMD,Dual,Rb=2,2k, Rbe=47kΩ, 50V, 100mA	45ba1	Rhm	-	data
FMA 5A	Si-P+R	SMD, 2x DTA123J, Dual Digital Transistor, com. Emitter	45ba1	Rhm	=UMA5N	data
FMA 6(A)	Si-P+R	SMD, Dual, Rb=47kΩ, 50V, 100mA	45ba1	Rhm	-	data
FMA 6A	Si-P+R	SMD, 2x DTA114T, Dual Digital Transistor, com. Emitter 50V, 100mA,	45ba1	Rhm	FMA3A, FMA4A,	data
FMA 7 A	Si-P+R	SMD,Dual,Rb=4,7k, Rbe=10kΩ, 50V, 100mA	45ba1	Rhm	-	data
FMA 8 A	Si-P+R	SMD,Dual, Rb=10k, Rbe=47kΩ, 50V, 100mA	45ba1	Rhm	-	data
FMA 9 A	Si-P+R	SMD, Dual, Rb=10k, Rbe=10kΩ, 50V, 50mA	45ba1	Rhm	-	data
FMA 10 A	Si-P+R	SMD, Dual, Rb=1k, Rbe=10kΩ, 50V, 100mA	45ba1	Rhm	-	data
FMA 11 A	Si-P+R	SMD,Dual,Rb=4,7k, Rbe=47kΩ, 50V, 100mA	45ba1	Rhm	-	data
FMB-22 H	Si-Di	Dual, Schottky, S, 20V, 15A, 100ns	17p	Sak	-	data
FMB-22 L	Si-Di	Dual, Schottky, S, 20V, 10A, 100ns	17p	Sak	-	data

Type	Device	Short Description	Fig.	Manu	Comparision Types	More at
FMB-23	Si-Di	Dual, Schottky, S, 30V, 4A, 100ns	17p	Sak	-	data
FMB-23 L	Si-Di	Dual, Schottky, S, 30V, 10A, 100ns	17p	Sak	-	data
FMB-24	Si-Di	Dual, Schottky, S, 40V, 4A, 100ns	17p	Sak	-	data
FMB-24 H	Si-Di	Dual, Schottky, S, 40V, 15A, 100ns	17p	Sak	-	data
FMB-24 L	Si-Di	Dual, Schottky, S, 40V, 10A, 100ns	17p	Sak	-	data
FMB-24 M	Si-Di	Dual, Schottky, S, 40V, 6A, 100ns	17p	Sak	-	data
FMB-26	Si-Di	Dual, Schottky, S, 60V, 4A, 100ns	17p	Sak	-	data
FMB-26 L	Si-Di	Dual, Schottky, S, 60V, 10A, 100ns	17p	Sak	-	data
FMB-29	Si-Di	Dual, Schottky, S, 90V, 4A, 100ns	17p	Sak	-	data
FMB-29 L	Si-Di	Dual, Schottky, S, 90V, 8A, 100ns	17p	Sak	-	data
FMB-32	Si-Di	Dual, Schottky, S, 20V, 20A, 100ns	18p	Sak		data
FMB-32 M	Si-Di	Dual, Schottky, S, 20V, 30A, 100ns	18p	Sak		data
FMB-34	Si-Di	Dual, Schottky, S, 40V, 15A, 100ns	18p	Sak		data
FMB-34 M	Si-Di	Dual, Schottky, S, 40V, 30A, 100ns	18p	Sak		data
FMB-34 S	Si-Di	Dual, Schottky, S, 40V, 12A, 100ns	18p	Sak		data
FMB-36	Si-Di	Dual, Schottky, S, 60V, 15A, 100ns	18p	Sak		data
FMB-36 M	Si-Di	Dual, Schottky, S, 60V, 30A, 100ns	18p	Sak		data
FMB-39	Si-Di	Dual, Schottky, S, 90V, 15A, 100ns	18p	Sak		data
FMB-39 M	Si-Di	Dual, Schottky, S, 90V, 20A, 100ns	13p	Sak		data
FMB-G 12 L	Si-Di	Schottky, S, 20V, 5A, 100ns	17d	Sak	-	data
FMB-G 14	Si-Di	Schottky, S, 40V, 3A, 100ns	17d	Sak	-	data
FMB-G 14 L	Si-Di	Schottky, S, 40V, 5A, 100ns	17d	Sak	-	data
FMB-G 16 L	Si-Di	Schottky, S, 60V, 5A, 100ns	17d	Sak	-	data
FMB-G 19 L	Si-Di	Schottky, S, 90V, 4A, 100ns	17d	Sak	-	data
FMB-G 22 H	Si-Di	Schottky, S, 20V, 10A, 100ns	17d	Sak	-	data
FMB-G 24 H	Si-Di	Schottky, S, 40V, 10A, 100ns	17d	Sak	-	data
FMC 1(A)	Si-N/P+R	SMD, N+P, Rb=4,7kΩ, 50V, 100mA	45	Rhm	-	data
FMC 1A	Si-P/N+R	SMD, DTA143T + DTC143T, Dual Digital	45ba4	Rhm	=UMC1N	data
		Transistor 50V, 100mA,				data
FMC 2(A)	Si-N/P+R	SMD, N+P, Rb=Rbe=22kΩ, 50V, 100mA	45	Rhm	-	data
FMC 2A	Si-N/P+R	SMD, DTA124E + DTC124E, Dual Digital	45ba4	Rhm	=UMC2N	data
		Transistor 100mA,				data
FMC 3(A)	Si-N/P+R	SMD, N+P, Rb=Rbe=10kΩ, 50V, 50mA	45	Rhm	-	data
FMC 3A	Si-N/P+R	SMD, DTA114E + DTC114E, Dual Digital	45ba4	Rhm	=UMC3N	data
		Transistor 100mA,				data
FMC 4 A	Si-N/P+R	SMD, N+P, Rb=Rbe=47kΩ, 50V, 100mA	45	Rhm	-	data
FMC 5(A)	Si-N/P	SMD, T1=N, T2=P Rb=Rbe=47k*	45	Rhm	FMC1(A), FMC2(A), FMC6(A),	data
		Rbe/Rb=0,8...1,2 50V, 100mA,				data
FMC 5 A	Si-N/P+R	SMD, N+P, Rb=Rbe=47kΩ, 50V, 100mA	45	Rhm	-	data
FMC 6(A)	Si-N/P	SMD, T1=N, T2=P. Rb=Rbe=100k*	45	Rhm	FMC1(A), FMC2(A), FMC5(A),	data
		Rbe/Rb=0,8...1,2 50V, 100mA,				data
FMC 6 A	Si-N/P+R	SMD, N+P, Rb=Rbe=100kΩ, 50V, 100mA	45	Rhm	-	data
FMC 7 A	Si-N/P+R	SMD, N+P, Rb=Rbe=47kΩ, 50V, 100mA	45	Rhm	-	data
FMC-26 U	Si-Di	Dual, S, 600V, 3A, 70ns	17n1	Sak	-	data
FMC-28 U	Si-Di	=FMC 26U: 800V	17n1	Sak	-	data
FMC-G 28 S	Si-Di	S, 800V, 3A, 70ns	17d	Sak	BY 430F/1000, BY 431F/1000	data
FME-24 H	Si-Di	Dual, Schottky-Gl, 40V, 15A	17p	Sak	BYV 33F-40, BYV 133F-40	data
FME-24 L	Si-Di	Dual, Schottky-Gl, 40V, 10A	17p	Sak	BYV 33F-40, BYV 133F-40	data
FMG 1(A)	Si-N+R	SMD, Dual, Rb=Rbe=22kΩ, 50V, 30mA	45ba1	Rhm	-	data
FMG 1A	Si-N+R	SMD, 2x DTC124E, Dual Digital	45ba1	Rhm	=UMG1N	data
		Transistor, com. Emitter				data
FMG 2(A)	Si-N+R	SMD, Dual, Rb=Rbe=47kΩ, 50V, 30mA	45ba1	Rhm	-	data
FMG 2A	Si-N+R	SMD, 2x DTC144E, Dual Digital	45ba1	Rhm	=UMG2N	data
		Transistor, com. Emitter 100mA,				data
FMG 3(A)	Si-N+R	SMD, Dual, Rb=4,7kΩ, 50V, 100mA	45ba1	Rhm	-	data
FMG 3A	Si-N+R	SMD, 2x DTC143T, Dual Digital	45ba1	Rhm	FMG4A, FMG6A, FMG13,	data
		Transistor, com. Emitter 50V, 100mA,				data
FMG 4(A)	Si-N+R	SMD, Dual, Rb=10kΩ, 50V, 100mA	45ba1	Rhm	-	data
FMG 4A	Si-N+R	SMD, 2x DTC114T, Dual Digital	45ba1	Rhm	FMG3A, FMG6A, FMG13,	data
		Transistor, com. Emitter 50V, 100mA,				data
FMG 5(A)	Si-N+R	SMD, Dual, Rb=10k,Rbe=47kΩ, 50V, 100mA	45ba1	Rhm	-	data
FMG 5A	Si-N+R	SMD, 2x DTC114Y, Dual Digital	45ba1	Rhm	=UMG5N	data
		Transistor, com. Emitter 100mA,				data
FMG 6(A)	Si-N+R	SMD, Dual, Rb=47kΩ, 50V, 100mA	45ba1	Rhm	-	data
FMG 6A	Si-N+R	SMD, 2x DTC114T, Dual Digital	45ba1	Rhm	FMG3A, FMG4A, FMG13,	data
		Transistor, com. Emitter 50V, 100mA,				data
FMG 7(A)	Si-N+R	SMD, Dual, Rb=10kΩ, 50V, 100mA	45	Rhm	-	data
FMG 7A	Si-N+R	SMD, 2x DTC114T, Dual Digital	45ba8	Rhm	=UMG7N	data
		Transistor 50V, 100mA,				data
FMG 8(A)	Si-N+R	SMD,Dual, Rb=4,7k,Rbe=47kΩ, 50V, 100mA	45ba1	Rhm	-	data
FMG 8A	Si-N+R	SMD, 2x DTC143Z, Dual Digital	45ba1	Rhm	=UMG8N	data
		Transistor, com. Emitter 100mA,				data
FMG 9 A	Si-N+R	SMD, Dual, Rb=Rbe=10kΩ, 50V, 50mA	45ba1	Rhm	-	data
FMG 10 A	Si-N+R	SMD, Dual, Rb=1k, Rbe=10kΩ, 50V, 100mA	45ba1	Rhm	-	data
FMG 11 A	Si-N+R	SMD,Dual,Rb=2,2k, Rbe=47kΩ, 50V, 100mA	45ba1	Rhm	-	data
FMG-11 R	Si-Di	Dual, S, 100V, 5A, 100ns	17q	Sak	-	data
FMG-11 S	Si-Di	=FMG 11R:	17p	Sak	MA 649, MA 653	data
FMG 12	Si-N+R	SMD, Dual, Rb=2,2kΩ, 30V, 600mA	45ba1	Rhm	-	data
FMG-12 R	Si-Di	=FMG 11R: 200V	17q	Sak	-	data
FMG-12 S	Si-Di	=FMG 11R: 200V	17p	Sak	MA 649, MA 653	data

Type	Device	Short Description	Fig.	Manu	Comparision Types	More at
G 13	Si-N+R	=IMH 3:	45ba1	Rhm	-	data
G-13 R	Si-Di	=FMG 11R: 300V	17q	Sak	-	data
G-13 S	Si-Di	=FMG 11R: 300V	17p		MA 653, MA 693	data
G-14 R	Si-Di	=FMG 11R: 400V	17q	Sak	-	data
G-14 S	Si-Di	=FMG 11R: 400V	17p		MA 693	data
G-21 R	Si-Di	Dual, S, 100V, 10A, 100ns	17q	Sak	-	data
G-21 S	Si-Di	=FMG 21R:	17p		BYQ 28F/100, MA 650	data
G-22 R	Si-Di	=FMG 21R: 200V	17q	Sak	-	data
G-22 S	Si-Di	=FMG 21R: 200V	17p		BYQ 28F/200, MA 650	data
G-23 R	Si-Di	=FMG 21R: 300V	17q	Sak	-	data
G-23 S	Si-Di	=FMG 21R: 300V	17p		MA 654, MA 694	data
G-24 R	Si-Di	=FMG 21R: 400V, 8A	17q	Sak	-	data
G-24 S	Si-Di	=FMG 21R: 400V, 8A	17p		MA 694	data
G-26 R	Si-Di	=FMG 21R: 600V, 6A	17q	Sak	-	data
G-26 S	Si-Di	=FMG 21R: 600V, 6A	17p		-	data
G-31 R	Si-Di	Dual, S, 100V, 20A, 100ns	18q	Sak	-	data
G-31 S	Si-Di	=FMG 31R:	18p		BYV 72F/100, MA 651	data
G-32 R	Si-Di	=FMG 31R: 200V	18q	Sak	-	data
G-32 S	Si-Di	=FMG 31R: 200V	18p		BYV 72F/200, MA 651	data
G-33 R	Si-Di	=FMG 31R: 300V	18q	Sak	-	data
G-33 S	Si-Di	=FMG 31R: 300V	18p		BYV 74F/300, MA 655	data
G-34 R	Si-Di	=FMG 31R: 400V, 16A	18q	Sak	-	data
G-34 S	Si-Di	=FMG 31R: 400V, 16A	18p		BYV 74F/400, MA 695	data
G-36 R	Si-Di	=FMG 31R: 600V, 15A	18q	Sak	-	data
G-36 S	Si-Di	=FMG 31R: 600V, 15A	18p		-	data
G-G2 CS	Si-Di	S, 1000V, 3A, 100ns	17d	Sak	BY 430F/1000, BY 431F/1000	data
G-G3 CS	Si-Di	S, 1000V, 5A, 150ns	18d	Sak	-	data
G-G26 S	Si-Di	S, 600V, 4A, 100ns	17d	Sak	BY 430F/1000, BY 431F/1000	data
G-G36 S	Si-Di	S, 600V, 8A, 100ns	18d	Sak	-	data
J 1 A	Si-P+R	SMD, Emit.-Di, Rb=Rbe=47kΩ, 50V, 100mA	45	Rhm	-	data
L-11 R	Si-Di	Dual, S, 100V, 5A, 40ns	17q	Sak	-	data
L-11 S	Si-Di	=FML 11R:	17p		MA 649	data
L-12 R	Si-Di	=FML 11R: 200V	17q	Sak	-	data
L-12 S	Si-Di	=FML 11R: 200V	17p		MA 649, MA 653	data
L-13 R	Si-Di	=FML 11R: 300V, 50ns	17q	Sak	-	data
L-13 S	Si-Di	=FML 11R: 300V, 50ns	17p		MA 653, MA 693	data
L-14 R	Si-Di	=FML 11R: 400V, 50ns	17q	Sak	-	data
L-14 S	Si-Di	=FML 11R: 400V, 50ns	17p		MA 693	data
L-21 R	Si-Di	Dual, S, 100V, 10A, 40ns	17q	Sak	-	data
L-21 S	Si-Di	=FML-21R:	17p		BYQ 28F/100, MA 650	data
L-22 R	Si-Di	=FML-21R: 200V	17q	Sak	-	data
L-22 S	Si-Di	=FML-21R: 200V	17p		BYQ 28F/200, MA 650, MA 654	data
L-23 R	Si-Di	=FML-21R: 300V, 50ns	17q	Sak	-	data
L-23 S	Si-Di	=FML-21R: 300V, 50ns	17p		MA 654, MA 694	data
L-24 R	Si-Di	=FML-21R: 400V, 50ns	17q	Sak	-	data
L-24 S	Si-Di	=FML-21R: 400V, 50ns	17p		MA 694	data
L-31 R	Si-Di	Dual, S, 100V, 20A, 40ns	18q	Sak	-	data
L-31 S	Si-Di	=FML-31R:	18p		BYV 72F/100, MA 651	data
L-32 R	Si-Di	=FML-31R: 200V	18q	Sak	-	data
L-32 S	Si-Di	=FML-31R: 200V	18p		BYV 72F/200, MA 651, MA 655	data
L-33 R	Si-Di	=FML-31R: 300V, 50ns	18q	Sak	-	data
L-33 S	Si-Di	=FML-31R: 300V, 50ns	18p		BYV 74F/300, MA 655, MA 695	data
L-34 R	Si-Di	=FML-31R: 400V, 50ns	18q	Sak	-	data
L-34 S	Si-Di	=FML-31R: 400V, 50ns	18p		BYV 74F/400, MA 695	data
L-36 R	Si-Di	=FML-31R: 600V, 65ns	18q	Sak	-	data
L-36 S	Si-Di	=FML-31R: 600V, 65ns	18p		-	data
L-G12 S	Si-Di	S, 200V, 5A, 40ns	17d	Sak	BY 430F/1000, BY 431F/1000	data
L-G13 S	Si-Di	=FML-12GS: 300V, 50ns	17d	Sak	BY 430F/1000, BY 431F/1000	data
L-G14 S	Si-Di	=FML-12GS: 400V, 50ns	17d	Sak	BY 430F/1000, BY 431F/1000	data
L-G16 S	Si-Di	=FML-12GS: 600V, 50ns	17d	Sak	BY 430F/1000, BY 431F/1000	data
L-G22 S	Si-Di	S, 200V, 10A, 40ns	17d	Sak	BYT-12PI/600	data
L-G26 S	Si-Di	S, 600V, 10A, 65ns	17d	Sak	BYT 12PI/600	data
M-22 R	Si-Di	Dual, 200V, 10A	17q	Sak	-	data
M-22 S	Si-Di	=FMM-22R:	17p		BYQ 28F/200, MA 650	data
M-24 R	Si-Di	=FMM-22R: 400V	17q	Sak	-	data
M-24 S	Si-Di	=FMM-22R: 400V	17p		MA 694	data
M-26 R	Si-Di	=FMM-22R: 600V	17q	Sak	-	data
M-26 S	Si-Di	=FMM-22R: 600V	17p		-	data
M-31 R	Si-Di	Dual, 100V, 20A	18q	Sak	-	data
M-31 S	Si-Di	=FMM-31R:	18p		BYV 72F/100, MA651	data
M-32 R	Si-Di	=FMM-31R: 200V	18q	Sak	-	data
M-32 S	Si-Di	=FMM-31R: 200V	18p		BYV 72F/200, MA651	data
M-34 R	Si-Di	=FMM-31R: 400V	18q	Sak	-	data
M-34 S	Si-Di	=FMM-31R: 400V	18p		BYV 72F/400, MA695	data
M-36 R	Si-Di	=FMM-31R: 600V	18q	Sak	-	data
M-36 S	Si-Di	=FMM-31R: 600V	18p		-	data
MD 914	Si-Di	=1N914: SMD	35e	Fer	MMBD 914, PMBD 914	data
MT 38 A	Si-N-Darl	SMD, Uni, 80V, 0,3A, B>1000	39b	Ztx	BCV 49, BST 51..52, 2SD1511	data

Type	Device	Short Description	Fig.	Manu	Comparision Types	More at
FMMT 38 B	Si-N-Darl	=FMMT 38A: B>4000	39b		BCV 49, BST 51...52, 2SD1511	data
FMMT 38 C	Si-N-Darl	=FMMT 38A: B>10000	39b		BCV 49, BST 51...52, 2SD1511	data
FMMT 413	Si-N	=ZTX 413: SMD, 0,1A	35a	Ztx	-	data
FMMT 415	Si-N	=ZTX 415: SMD	35a	Ztx		data
FMMT 417	Si-N	=ZTX 417: SMD	35a	Ztx	-	data
FMMT 449	Si-N	=ZTX 449: SMD	35a	Ztx		data
FMMT 451	Si-N	=ZTX 451: SMD	35a	Ztx		data
FMMT 455	Si-N	=ZTX 455: SMD	35a	Ztx		data
FMMT 458	Si-N	=ZTX 458: SMD	35a	Ztx		data
FMMT 489	Si-N	SMD, Tr/E, 50V, 1A, >150MHz	35a	Ztx	MMBT 489	data
FMMT 491(A)	Si-N	=FCX 491(A):	35a	Ztx	-	data
FMMT 493	Si-N	=FCX 493:	35a	Ztx		data
FMMT 494	Si-N	=FCX 493: 140/120V, >100MHz	35a	Ztx		data
FMMT 495	Si-N	=FCX 495:	35a	Ztx		data
FMMT 497	Si-N	SMD, S,Vid, 300/300V, 0,5A, >75MHz	35a	Ztx	MMBTA 42	data
FMMT 549(A)	Si-P	=ZTX 549(A): SMD	35a	Ztx	-	data
FMMT 551	Si-P	=ZTX 551: SMD	35a	Ztx		data
FMMT 555	Si-P	=ZTX 555: SMD	35a	Ztx		data
FMMT 558	Si-P	=ZTX 558: SMD, 0,15A	35a	Ztx		data
FMMT 576	Si-P	=ZTX 576: SMD	35a	Ztx		data
FMMT 589	Si-P	=FCX 589:	35a	Ztx	MMBT 589	data
FMMT 591(A)	Si-P	=FCX 591(A):	35a	Ztx	-	data
FMMT 593	Si-P	=FCX 593:	35a	Ztx		data
FMMT 596	Si-P	=FCX 596:	35a	Ztx	-	data
FMMT 597	Si-P	SMD, S,Vid, 300/300V, 0,2A, >75MHz	35a	Ztx	MMBTA 92	data
FMMT 614	Si-N-Darl	SMD, Uni, 120/100V, 0,5A, B>15000	35a	Ztx	-	data
FMMT 617	Si-N	SMD, S, lo-sat, 15V, 3A, 120/160ns	35a	Ztx	-	data
FMMT 618	Si-N	SMD, S, lo-sat, 20V, 2,5A, 170/400ns	35a	Ztx	-	data
FMMT 619	Si-N	SMD, S, lo-sat, 50V, 2A, 170/750ns	35a	Ztx	-	data
FMMT 624	Si-N	SMD, S, lo-sat, 125V, 1A, 60/1300ns	35a	Ztx	-	data
FMMT 625	Si-N	SMD, S, lo-sat, 155V, 1A, 160/1500ns	35a	Ztx	-	data
FMMT 717	Si-P	SMD, S, lo-sat, 12V, 2,5A, 70/130ns	35a	Ztx	-	data
FMMT 718	Si-P	SMD, S, lo-sat, 20V, 1,5A, 40/670ns	35a	Ztx	-	data
FMMT 720	Si-P	SMD, S, lo-sat, 40V, 1,5A, 40/435ns	35a	Ztx	-	data
FMMT 722	Si-P	SMD, S, lo-sat, 70V, 1,5A, 40/700ns	35a	Ztx	-	data
FMMT 723	Si-P	SMD, S, lo-sat, 100V, 1A, 50/760ns	35a	Ztx	-	data
FMMT 918	Si-N	=2N918: SMD	35a	Ztx	→MMBT 918	data
FMMT 2222(A)	Si-N	=2N2222(A): SMD	35a	Ztx	→MMBT 2222(A)	data
FMMT 2222(A)R	Si-N	=2N2222(A): SMD	35d	Ztx		data
FMMT 2369(A)	Si-N	=2N2369(A): SMD	35a	Ztx	→MMBT 2369(A)	data
FMMT 2484	Si-N	=2N2484: SMD	35a	Ztx	→MMBT 2484	data
FMMT 2907(A)	Si-P	=2N2907(A): SMD	35a	Ztx	→MMBT 2907(A)	data
FMMT 2907(A)R	Si-P	=2N2907(A): SMD	35d	Ztx		data
FMMT 3903...3904	Si-N	=2N3903...3904: SMD	35a	Ztx	→MMBT 3903...3904	data
FMMT 3905...3906	Si-P	=2N3905...3906: SMD	35a	Ztx	→MMBT 3905...3906	data
FMMT 4124	Si-N	=2N4124: SMD	35a	Ztx	→MMBT 4124	data
FMMT 4125	Si-P	=2N4125: SMD	35a	Ztx	→MMBT 4125	data
FMMT 5087	Si-P	=2N5087: SMD	35a	Ztx	→MMBT 5087	data
FMMT 5087 R	Si-P	=2N5087: SMD	35a	Ztx		data
FMMT 5179	Si-N	=2N5179: SMD	35a	Ztx	→MMBT 5179	data
FMMT 6517	Si-N	=2N6517: SMD	35a	Ztx	→MMBT 6517	data
FMMT 6520	Si-N	=2N6520: SMD	35a	Ztx	→MMBT 6520	data
FMMTA 05...06	Si-N	=MPSA 05...06: SMD	35a	Ztx	→MMBTA 05...06	data
FMMTA 05R...06R	Si-N	=MPSA 05...06: SMD	35d	Ztx		data
FMMTA 12...14	Si-N-Darl	=MPSA 12...14: SMD	35a	Ztx	→MMBTA 12...14	data
FMMTA 20	Si-N	=MPSA 20: SMD	35a	Ztx	→MMBTA 20	data
FMMTA 42...43	Si-N	=MPSA 42...43: SMD	35a	Ztx	→MMBTA 42...43	data
FMMTA 42R...43R	Si-N	=MPSA 42...43: SMD	35d	Ztx		data
FMMTA 55...56	Si-P	=MPSA 55...56: SMD	35a	Ztx	→MMBTA 55...56	data
FMMTA 55R...56R	Si-P	=MPSA 55...56: SMD	35d	Ztx		data
FMMTA 63...64	Si-P-Darl	=MPSA 63...64: SMD	35a	Ztx	→MMBTA 63...64	data
FMMTA 70	Si-P	=MPSA 70: SMD	35a	Ztx	→MMBTA 70	data
FMMTA 92...93	Si-P	=MPSA 92...93: SMD	35a	Ztx	→MMBTA 92...93	data
FMMTA 92R...93R	Si-P	=MPSA 92...93: SMD	35d	Ztx	-	data
FMMTH 10	Si-N	=MPSH 10: SMD	35a	Ztx	→MMBTH 10	data
FMMV 105 G	C-Di	SMD, Tuning, 30V, 1,8...2,8pF(25V)	35e	Ztx	-	data
FMMV 109	C-Di	SMD, Tuning, 30V, 26...32pF(25V)	35e	Ztx	-	data
FMMV 2101	C-Di	=MV 2101: SMD	35e	Ztx	-	data
FMMV 2103	C-Di	=MV 2103: SMD	35e	Ztx	-	data
FMMV 2104	C-Di	=MV 2104: SMD	35e	Ztx	-	data
FMMV 2105	C-Di	=MV 2105: SMD	35e	Ztx	-	data
FMMV 2107	C-Di	=MV 2107: SMD	35e	Ztx	-	data
FMMV 2108	C-Di	=MV 2108: SMD	35e	Ztx	-	data
FMMV 2109	C-Di	=MV 2109: SMD	35e	Ztx	-	data
FMMV 3102	C-Di	=MV 3102: SMD	35e	Ztx	-	data
FMN-G12 S	Si-Di	S, 200V, 5A, 100ns	17d	Sak	BY 229F/200, BYW 29/200F	data

Type	Device	Short Description	Fig.	Manu	Comparision Types	More at
FMP-2 FUR	Si-Di	Dual, Moni Damper-Di, D1: 1500V, 5A, 700ns, D2: 600V, 5A, 100ns	17o1	Sak	-	data
FMP-3 FU	Si-Di	Dual, Moni Damper-Di, D1: 1500V, 5A, 700ns, D2: 600V, 5A, 100ns	18n1	Sak	-	data
FMP 18 N05	MOS-N-FET-e	V-MOS, 50V, 18A, 75W, <0,1Ω(10A)	17c§	Fch, Nsc	BUZ 21..22, IRF 540...543, 2SK674	data
FMP 18 N06	MOS-N-FET-e	=FMP 18N05: 60V	17c§		BUZ 21..22, IRF 540...543, 2SK674	data
FMP 20 N05	MOS-N-FET-e	V-MOS, 50V, 20A, 75W, <85mΩ(10A)	17c§	Fch, Nsc	BUZ 21..22, IRF 540...543, 2SK674	data
FMP 20 N06	MOS-N-FET-e	=FMP 20N05: 60V	17c§		BUZ 21..22, IRF 540...543, 2SK674	data
FMP 30 N05	MOS-N-FET-e	V-MOS, 50V, 30A, 100W, 0,05Ω	17c§	Fch	BUZ 11...12, PRFZ 42, 2SK856	data
FMP 35 N05	MOS-N-FET-e	V-MOS, 50V, 35A, 100W, 0,04Ω	17c§	Fch	BUZ 11...12, PRFZ 42, 2SK856	data
FMP-G 2 FS	Si-Di	Moni Damper-Di, 1500V, 5A, 700ns	17d	Sak	DTV 32F, DTV 56F, DTV 64F	data
FMP-G 5 HS	Si-Di	Moni Damper-Di, 1800V, 8A, 1000ns	18d	Sak	-	data
FMP-G12 S	Si-Di	S, 200V, 5A, 150ns	17d	Sak	BY 229F/200, BYW 29/200F	data
FMPS-A 05	Si-N	=MPSA 05	7e	Fch	→MPSA 05	data
FMPS-A 06	Si-N	=MPSA 06	7e	Fch	→MPSA 06	data
FMPS-A 55	Si-P	=MPSA 55	7e	Fch	→MPSA 55	data
FMPS-A 56	Si-P	=MPSA 56	7e	Fch	→MPSA 56	data
FMQ 1 A	Si-N/P	SMD, N+P, Re=Rc=470Ω, 20V, 30mA	45	Rhm	-	data
FMQ 2	Si-N/P	SMD, N+P, 40V, N: 30mA, P: 500mA	45	Rhm	-	data
FMQ-3 GU	Si-Di	Dual, Moni Damper-Di, D1: 1700V, 5A, 700ns, D2: 800V, 5A, 70ns	18n1	Sak	-	data
FMQ-G 2 FLS	Si-Di	Moni Damper-Di, 1500V, 10A, 1200ns	17d	Sak	BY 459F-1500, BY 559X-1500U	data
FMQ-G 2 FS	Si-Di	Moni Damper-Di, 1500V, 10A, 500ns	17d	Sak	BY 459F-1500, BY 559X-1500U	data
FMQ-G 5 FMS	Si-Di	Moni Damper-Di, 1500V, 10A, 500ns	18d	Sak	-	data
FMQ-G 5 GS	Si-Di	Moni Damper-Di, 1700V, 10A, 500ns	18d	Sak	-	data
FMR-G 5 HS	Si-Di	TV Damper-Di, 1800V, 10A, <1800ns	18d	Sak	-	data
FMS 1(A)	Si-P	SMD, Dual, Uni, 50V, 100mA, 140MHz	45ba1	Rhm	-	data
FMS 1A	Si-P	SMD, 2x 2SA1037AK, com. Emitter 50V, 150mA,	45ba1	Rhm	=UMS1N	data data
FMS 2(A)	Si-P	SMD, Dual, Uni, 50V, 100mA, 140MHz	45ba3	Rhm	-	data
FMS 3	Si-P	SMD, Dual, Uni, 120V, 50mA, 140MHz	45ba1	Rhm	-	data
FMS 4	Si-P	SMD, Dual, Uni, 120V, 50mA, 140MHz	45ba3	Rhm	-	data
FMS 3110	IC	10 bit, Triple Video D/A Converters	MP	Fcs	-	pdf pinout
FMS 3115	IC	10 bit, Triple Video D/A Converters	MP	Fcs	-	pdf pinout
FMS 3810	IC	Triple Video D/A Converters 150 Ms/s	QFP	Fch	-	pdf pinout
FMS 3815	IC	Triple Video D/A Converters 150 Ms/s	QFP	Fch	-	pdf pinout
FMS 3818	IC	Triple Video D/A Converters, 180 Ms/s	QFP	Fch	-	pdf pinout
FMS 6141 CS	IC	Sing.-CH 4th-Order Video Filter Driver	TSOP	Fch		pdf pinout
FMS 6141 CSX	IC	Sing.-CH 4th-Order Video Filter Driver	TSOP	Fch		pdf pinout
FMS 6143	IC	3-CH 4th-Order Video Filter Driver	TSOP	Fch		pdf pinout
FMS 6145	IC	5-CH 4th-Order Video Filter Driver	TSOP	Fch		pdf pinout
FMS 6146	IC	6-CH 4th-Order Video Filter Driver	TSOP	Fch		pdf pinout
FMS 6151	IC	Ultra Portable Video Filter Driver	TSOP	Fch		pdf pinout
FMS 6203	IC	Ultra Portable Video Filter Driver	TSOP	Fch		pdf pinout
FMS 6243	IC	3-CH, SD Video Filter Drivers Ext. Delay Contr.	TSOP	Fch		pdf pinout
FMS 6246	IC	6-CH, 6th Order SD/PS Video Filter Driver	TSOP	Fch	-	pdf pinout
FMS 6346	CMOS-IC	Six Channel, 6th-Order SD/HD Video Filter Driver	TSOP	Fcs	-	pdf pinout
FMS 6363	CMOS-IC	Low Cost 3 Ch. 6th Order HD Video Filter Driver	MDIP	Fcs	-	pdf pinout
FMS 6366	CMOS-IC	Selectable YPbPr HD/SD 4:2:2 Video Filter Driver	MDIP	Fcs	-	pdf pinout
FMS 6400 -1	CMOS-IC	2 Ch. Video Drivers with Integrated Filters	MDIP	Fcs	-	pdf pinout
FMS 6400	CMOS-IC	2 Ch. Video Drivers with Integrated Filters	MDIP	Fcs	-	pdf pinout
FMS 6403	CMOS-IC	Triple Video Drivers, HD/PS/SD/Bypass Filters f. RGB, YPbPr Signals	TSOP	Fcs	-	pdf pinout
FMS 6406	CMOS-IC	Precision S- Video Filter, Summed Composite Out., Sound Trap	MDIP	Fcs	-	pdf pinout
FMS 6407	CMOS-IC	Triple Video Drivers, Select. HD/ Progressive/ SD/ Bypass Filters	TSOP	Fcs	-	pdf pinout
FMS 6408	CMOS-IC	Prec. Triple Video Filter Driver f. RGB, Yuv Signals	TSOP	Fcs	-	pdf pinout
FMS 6410 B	-IC	Dual-CH. Video Drivers, Integrated Filters, Composite Video Summer	MDIP	Fcs	-	pdf pinout
FMS 6418 B	-IC	Triple Video Driver w. Select. HD/SD Video Filt. f. RGB, YUV, YPbPr Signals	MDIP	Fcs	-	pdf pinout
FMS 6418 BM	-IC	Triple Video Driver w. Select. HD/SD Video Filt. f. RGB, Yuv, YPbPr Signals	TSOP	Fcs	-	pdf pinout
FMS 6419	-IC	Select. RGB (YUV) HD/SD Video Filter Driver w. Y, C, Composite Out.	MDIP	Fcs	-	pdf pinout
FMS 6501	CMOS-IC	12 In. / 9 Out. Video Switch Matrix w. Input Clamp, Input Bias Circuitry	MDIP	Fcs	-	pdf pinout
FMS 6502	CMOS-IC	8-Inp., 6-Out. Video Switch Matrix w. Output Drivers, Input Clamp, and Bias	MDIP	Fcs	-	pdf pinout
FMS 6690	CMOS-IC	6 Ch., 6th Order SD/PS/HD Video Filter Driver	TSOP	Fcs	-	pdf pinout

Type	Device	Short Description	Fig.	Manu	Comparision Types	More at
FMS 6141 C5X	IC	Sing.-CH 4th-Order Video Filter Driver	TSOP	Fch	-	pdf pinout
FMU-12 R	Si-Di	Dual, S, 200V, 5A, 400ns	17q	Sak	-	data
FMU-12 S	Si-Di	=FMU-12R:	17p		MA 649, MA 653	data
FMU-14 R	Si-Di	=FMU-12R: 400V	17q	Sak		data
FMU-14 S	Si-Di	=FMU-12R: 400V	17p		MA 693	data
FMU-16 R	Si-Di	=FMU-12R: 600V	17q	Sak		data
FMU-16 S	Si-Di	=FMU-12R: 600V	17p			data
FMU-21 R	Si-Di	Dual, S, 100V, 10A, 400ns	17q	Sak	-	data
FMU-21 S	Si-Di	=FMU-21R:	17p		BYQ 28F/100	data
FMU-22 R	Si-Di	=FMU-21R: 200V	17q	Sak		data
FMU-22 S	Si-Di	=FMU-21R: 200V	17p		BYQ 28F/200, MA 650	data
FMU-24 R	Si-Di	=FMU-21R: 400V	17q	Sak		data
FMU-24 S	Si-Di	=FMU-21R: 400V	17p		MA 694	data
FMU-26 R	Si-Di	=FMU-21R: 600V	17q	Sak	-	data
FMU-26 S	Si-Di	=FMU-21R: 600V	17p			data
FMU-32 R	Si-Di	Dual, S, 200V, 20A, 400ns	18q	Sak		data
FMU-32 S	Si-Di	=FMU-32R:	18p		BYV72F/200, MA 651	data
FMU-34 R	Si-Di	=FMU-32R: 400V	18q	Sak		data
FMU-34 S	Si-Di	=FMU-32R: 400V	18p		BYV 74F/400, MA 695	data
FMU-36 R	Si-Di	=FMU-32R: 600V	18q	Sak		data
FMU-36 S	Si-Di	=FMU-32R: 600V	18p			data
FMU-G2YXS	Si-Di	S, 100V, 10A, 200ns	17d	Sak		data
FMU-G16 S	Si-Di	S, 600V, 5A, 400ns	17d	Sak	BY 430F/1000, BY 431F/1000	data
FMU-G26 S	Si-Di	S, 600V, 10A, 400ns	17d	Sak	BYT 12PI/600	data
FMV-3 FU	Si-Di	Dual, TV Damper-Di, D1: 1500V, 5A, 4µs, D2: 600V, 5A, 400ns	18n1	Sak	-	data
FMV-3 GU	Si-Di	Dual, TV Damper-Di, D1: 1700V, 5A, 2µs, D2: 600V, 5A, 400ns	18n1	Sak	-	data
FMV-G 5 FS	Si-Di	TV Damper-Di, 1500V, 10A, 2000ns	18d	Sak		data
FMW 1	Si-N	SMD, Dual, Uni, 60V, 150mA, 180MHz	45ba1	Rhm		data
FMW 2	Si-N	SMD, Dual, Uni, 60V, 150mA, 180MHz	45ba3	Rhm		data
FMW 3	Si-N	SMD, Dual, Uni, 120V, 50mA, 140MHz	45ba1	Rhm		data
FMW 4	Si-N	SMD, Dual, Uni, 120V, 50mA, 140MHz	45ba3	Rhm		data
FMW 5	Si-N	SMD.-Dual, Uni, 50V, 100mA	45ba1	Rhm		data
FMW 6	Si-N	SMD, Dual, HF, 30V, 50mA, >600MHz	45	Rhm		data
FMW 7	Si-N	SMD, Dual, HF, 20V, 50mA, >1400MHz	45	Rhm		data
FMW 8	Si-N	SMD, Dual, HF, 20V, 50mA, >1400MHz	45ba1	Rhm		data
FMW 9	Si-N	SMD, Dual, HF, 20V, 50mA, >1400MHz	45ba3	Rhm		data
FMW 10	Si-N	SMD, Dual, HF, 30V, 50mA, >600MHz	45ba1	Rhm		data
FMW 11	Si-N	SMD, Dual, HF, 30V, 50mA, >600MHz	45ba3	Rhm		data
FMW 12	Si-N	SMD, Dual, HF, 50V, 100mA	45	Rhm		data
FMW 13	Si-N	SMD, Dual, HF, 30V, 50mA, >900MHz	45ba3	Rhm		data
FMX-12 S	Si-Di	Dual, S, 200V, 5A, 30ns	17p	Sak	MA 649, MA 653	data
FMX-22 S	Si-Di	Dual, S, 200V, 10A, 30ns	17p	Sak	BYQ 28F/200, MA 650, MA 654	data
FMX-22 SL	Si-Di	Dual, S, 200V, 15A, 30ns	17p	Sak	BYQ 30EX/200, MA 652	data
FMX 23 S	Si-Di	Dual, S-L, 300V, 2x10A(Tc=103°), 30ns	17p	Sak		data
FMX 32 S	Si-Di	Dual, S-L, 200V, 2x20A(Tc=115°), 30ns	18p	Sak	BYV 52/200PI	data
FMX 32 SL	Si-Di	Dual, S-L, 200V, 20A, 30ns	18p	Sak	BYW 99PI/200	data
FMX 33 S	Si-Di	Dual, S-L, 300V, 2x20A(Tc=108°), 30ns	18p	Sak		data
FMX-G12 S	Si-Di	S, 200V, 5A, 30ns	17d	Sak	BY 430F/1000, BY 431F/1000	data
FMX-G22 S	Si-Di	S, 200V, 10A, 30ns	17d	Sak	BYT 12PI/600	data
FMY 1(A)	Si-P/N	SMD, P+N, 50V, 100mA, 140+180MHz	45ba1	Rhm	-	data
FMY 1A	Si-PN	SMD, 2SA1037AK + 2SC2412K, com. Emitter 50V, 150mA,	45ba1	Rhm	=UMY1N	data
FMY 3(A)	Si-P/N	SMD, P+N, 50V, 100mA, 140+180MHz	45	Rhm		data
FMY 4(A)	Si-P/N	SMD, P+N, 50V, 100mA, 140+180MHz	45	Rhm	-	data
FMY 4A	Si-PN	SMD, 2SA1037AK + 2SC2412K, Power Management 50V, 150mA,	45ba7	Rhm	=UMY4N	data
FMY 5	Si-P/N	SMD, P+N, Uni, 120V, 50mA, 140MHz	45ba1	Rhm	-	data
FMY 6	Si-P/N	SMD, P+N, Uni, 40V, 500mA, 200+250MHz	45ba1	Rhm	-	data

FN...FS

Type	Device	Short Description	Fig.	Manu	Comparision Types	More at
FN	Si-N	=2SC4863 (Typ-Code/Stempel/marking)	35(2mm)	Say	→2SC4863	data
FN	Si-N	=2SC4864 (Typ-Code/Stempel/marking)	35	Say	→2SC4864	data
FN	Si-N	=2SC4865 (Typ-Code/Stempel/marking)	44	Say	→2SC4865	data
FN	Z-Di	=Z1 SMA-91 (Typ-Code/Stempel/marking)	71(5x2,5)	Fag	→Z1 SMA-91	data
FN 1 A3Q	Si-P+R	=AN 1A3Q: SMD	35a	Nec	DTA 113ZK, UN 2119	data
FN 1 A4M	Si-P+R	=AN 1A4M: SMD	35a	Nec	DTA 114EK, RN 2402, UN 2111, 2SA1344,++	data
FN 1 A4P	Si-P+R	=AN 1A4P: SMD	35a	Nec	DTA 114YK, RN 2407, UN 2114, 2SA1563,++	data
FN 1 A4Z	Si-P+R	=AN 1A4Z: SMD	35a	Nec	DTA 114TK, RN 2411, UN 2115, 2SA1496,++	data
FN 1 F4M	Si-P+R	=AN 1F4M: SMD	35a	Nec	DTA 124EK, RN 2403, UN 2112, 2SA1342,++	data
FN 1 F4N	Si-P+R	=AN 1F4N: SMD	35a	Nec	DTA 124XK, KSR 2107, RN 2408	data
FN 1 F4Z	Si-P+R	=AN 1F4Z: SMD	35a	Nec	DTA 124TK, KSR 2111, UN 2117, 2SA1589	data
FN 1 L3M	Si-P+R	=AN 1L3M: SMD	35a	Nec	DTA 143EK, RN 2401, UN 211L, 2SA1655,++	data
FN 1 L3N	Si-P+R	=AN 1L3N: SMD	35a	Nec	DTA 143XK, KSR 2105, UN 211F, 2SA1653	data
FN 1 L3Z	Si-P+R	=AN 1L3Z: SMD	35a	Nec	DTA 143TK, RN 2410, UN 2116, 2SA1510,++	data
FN 1 L4L	Si-P+R	=AN 1L4L: SMD	35a	Nec	DTA 144WK, RN 2409, UN 211E, 2SA1343,++	data
FN 1 L4M	Si-P+R	=AN 1L4M: SMD	35a	Nec	DTA 144EK, RN 2404, UN 2113, 2SA1341,++	data
FN 1 L4Z	Si-P+R	=AN 1L4Z: SMD	35a	Nec	DTA 144TK, KSR 2112, UN 2110, 2SA1508	data

Type	Device	Short Description	Fig.	Manu	Comparision Types	More at
3	Si-N	=2SC4863-3 (Typ-Code/Stempel/marking)	35(2mm)	Say	→2SC4863	data
3	Si-N	=2SC4864-3 (Typ-Code/Stempel/marking)	35	Say	→2SC4864	data
3	Si-N	=2SC4865-3 (Typ-Code/Stempel/marking)	44	Say	→2SC4865	data
4	Si-N	=2SC4863-4 (Typ-Code/Stempel/marking)	35(2mm)	Say	→2SC4863	data
4	Si-N	=2SC4864-4 (Typ-Code/Stempel/marking)	35	Say	→2SC4864	data
4	Si-N	=2SC4865-4 (Typ-Code/Stempel/marking)	44	Say	→2SC4865	data
5	Si-N	=2SC4863-5 (Typ-Code/Stempel/marking)	35(2mm)	Say	→2SC4863	data
5	Si-N	=2SC4864-5 (Typ-Code/Stempel/marking)	35	Say	→2SC4864	data
5	Si-N	=2SC4865-5 (Typ-Code/Stempel/marking)	44	Say	→2SC4865	data
D 500	Opto					
H 111	Logic	Dual 2 input NOR-Gate	FLP	Sie	-	pdf
H 121	Logic	Single 5 input OR/NOR-Gate	FLP	Sie	-	pdf
H 131	Logic	3 input OR/NOR-Gate, 2 Expansionpins	FLP	Sie	-	pdf
H 141	Logic	5 input Expander for FNH 131/161, FNJ101	FLP	Sie	-	pdf
H 151	Logic	Halbaddierer/half adder	FLP	Sie	-	pdf
H 161	Logic	3 input OR/NOR-Gate, 2 Expansionpins	FLP	Sie	-	pdf
H 171	Logic	Dual 2 input NOR-Gate	FLP	Sie	-	pdf
J 101	Logic	RS- FF with 2 Set- , 2 Reset- and 2 Expansion pins	FLP	Sie	-	pdf
Y 101	Logic	Bias driver, reference voltage for FN Logic Family	FLP	Sie		pdf
	Si-P	=2SA1202-O (Typ-Code/Stempel/marking)	39	Tos	→2SA1202	data
	Si-N	=2SC2716-O (Typ-Code/Stempel/marking)	35	Tos	→2SC2716	data
	Si-P	=KTA1662-O (Typ-Code/Stempel/marking)	39	Kec	→KTA 1662	data
	Si-N	=KTC3878-O (Typ-Code/Stempel/marking)	35	Kec	→KTC 3878(A)	data
	Si-N	=XP 5A554 (Typ-Code/Stempel/marking)	46(2mm)	Mat	→XP 5A554	data
D617	OC	LED/PT Viso:&5000V CTR:>56 *)%		Fcs		data pdf pinout
D814	OC	bid. LED/PT Viso:&5000V CTR:50. .150%		Fcs		data pdf pinout
D817	OC	LED/PT Viso:&5000V CTR:300...600*)%		Fcs		data pdf pinout
D3180	OC	LED/2 Viso:&5000V Vbs:1.43V Ibe:<5mA		Fcs		data pdf pinout
D3181	OC	LED/2 Viso:&5000V Vbs:1.5V Ibs:<18mA		Fcs		data pdf pinout
DB100	OC	LED/PT Viso:&>2500V CTR:150...300%		Fcs		data pdf pinout
DM121	OC	LED/PT Viso:&>3750V CTR:200...400*)%		Fcs		data pdf pinout
DM124	OC	LED/PT Viso:&>3750V CTR:>50 *)%		Fcs		data pdf pinout
DM2701	OC	LED/PT Viso:&>3750V CTR:80...160 *)%		Fcs		data pdf pinout
DM2705	OC	bid. LED/PT Viso:&>3750V Vbs:<1.3V		Fcs		data pdf pinout
DM3011	OC	LED/Triac Viso:>3750Vrms Vbs:1.2V		Fcs		data pdf pinout
DM3012	OC	LED/Triac Viso:>3750Vrms Vbs:1.2V		Fcs		data pdf pinout
DM3022	OC	LED/Triac Viso:>3750Vrms Vbs:1.2V		Fcs		data pdf pinout
DM3023	OC	LED/Triac Viso:>3750Vrms Vbs:1.2V		Fcs		data pdf pinout
DM3052	OC	LED/Triac Viso:>3750Vrms Vbs:1.2V		Fcs		data pdf pinout
DM3053	OC	LED/Triac Viso:>3750Vrms Vbs:1.2V		Fcs		data pdf pinout
I	N-FET	=SO 4416 (Typ-Code/Stempel/marking)	35	Tho	→SO 4416	data
	Z-Di	=SM 6T 68 (Typ-Code/Stempel/marking)	71(6x4mm)	Tho	→SM 6T...	data
	Z-Di	=Z1 SMA-100 (Typ-Code/Stempel/marking)	71(5x2,5)	Fag	→Z1 SMA-100	data
1 A3M...L3N	Si-P+R	=AP 1A3M...L3N: SMD	35a	Nec	-	data
3 V50	MOS-N-FET-e	=2SK1245		Shi		
3 V90	MOS-N-FET-e	=2SK1535		Shi		
5 V50	MOS-N-FET-e	=2SK1247		Shi		
08 C	Triac	200V, 30A(Tc=90°C), Igt/Ih=50/30mA	21b§	Hit	T 6417..., T 6410..., 2N5444...5446	data
08 D	Triac	=FP 08C: 300V	21b§		T 6417..., T 6410..., 2N5445...5446	data
08 E	Triac	=FP 08C: 400V	21b§		T 6417..., T 6410..., 2N5445...5446	data
08 F	Triac	=FP 08C: 500V	21b§		T 6417..., T 6410..., 2N5446	data
08 G	Triac	=FP 08C: 600V	21b§		T 6417..., T 6410..., 2N5446	data
10 V25	MOS-N-FET-e	=2SK1394		Shi		
10 W50	MOS-N-FET-e	=2SK1523		Shi		
14 N05L	MOS-N-FET-e	=RFP 14N05L (Typ-Code/Stempel/marking)	17	Rca	→RFP 14N05L	data
14 N06L	MOS-N-FET-e	=RFP 14N06L (Typ-Code/Stempel/marking)	17	Rca	→RFP 14N06L	data
15 W50	MOS-N-FET-e	=2SK1524		Shi		
23 N06L	MOS-N-FET-e	=RFP 23N06LE(Typ-Code/Stempel/marking)	17	Rca	→RFP 23N06LE	data
40 N10L	MOS-N-FET-e	=RFP 40N10LE(Typ-Code/Stempel/marking)	17	Rca	→RFP 40N10LE	data
45 N03L	MOS-N-FET-e	=RFP 45N03L (Typ-Code/Stempel/marking)	17	Rca	→RFP 45N03L	data
101	Si-P+Di	SMD, S, lo-sat, 30V, 2A, 60/375ns + Schottky-Di, 50V, 0,5A (SB 05-05CP)	45bc2(5mm)	Say	-	data
102	Si-P+Di	SMD, S, lo-sat, 15V, 3A, 25/210ns + Schottky-Di, 30V, 0,7A (SB 07-03C)	45bc2(5mm)	Say	-	data
103	Si-P+Di	SMD, S, lo-sat, 30V, 2A, 60/375ns + Schottky-Di, 30V, 0,7A (SB 07-03C)	45bc2(5mm)	Say	-	data
104	Si-P+Di	SMD, S, lo-sat, 50V, 1,5A, 50/270ns + Schottky-Di, 50V, 0,5A (SB 05-05CP)	45bc2(5mm)	Say	-	data
105	Si-P+Di	SMD, S, lo-sat, 60V, 2A, 60/480ns + Schottky-Di, 50V, 0,5A (SB 05-05CP)	45bc2(5mm)	Say	-	data
106	Si-P+Di	SMD, S, lo-sat, 15V, 3A, 30/220ns + Schottky-Di, 15V, 1A (SB 10-05C)	45bc2(5mm)	Say	-	data
107	Si-P+Di	SMD, S, lo-sat, 15V, 3A, 25/210ns + Schottky-Di, 11V, 0,5A (SBS 001)	45bc2(5mm)	Say	-	data
108	Si-P+Di	SMD, S, lo-sat, 30V, 2A, 60/375ns + Schottky-Di, 15V, 1A (SB 01-015CP)	45bc2(5mm)	Say	-	data
201	Si-N	SMD, Dual, HF, 30V, 0,3A, 2,2GHz	45ba1(5mm)	Say	-	data
202	Si-P+N	SMD, Dual, 60V, 0,5A, P:200, N:250MHz	45ba1(5mm)	Say	-	data

Type	Device	Short Description	Fig.	Manu	Comparision Types	More at
FP 203	Si-P+N	SMD, Dual, 60V, 1A, 150MHz	45ba1(5mm)	Say	-	data
FP 204	Si-P+N	SMD, Dual, 60V, 2A, 150MHz	45ba1(5mm)	Say	-	data
FP 205	Si-P+N	SMD, Dual, 120V, 1A, 120MHz	45ba1(5mm)	Say	-	data
FP 206	Si-P+N	SMD, Dual, 50V, 0,5A, 350MHz	45ba1(5mm)	Say	-	data
FP 207	Si-P+N	SMD, Dual, 60V, 1,5A	45ba1(5mm)	Say	-	data
FP 208	Si-P	SMD, Dual, Tr, 30V, 2A, 150MHz	45ba1(5mm)	Say	-	data
FP 209	Si-N	SMD, Dual, Tr, 30V, 2A, 150MHz	45ba1(5mm)	Say	-	data
FP 210	Si-P	SMD, Dual, Tr, 60V, 2A, 150MHz	45ba1(5mm)	Say	-	data
FP 211	Si-N	SMD, Dual, Tr, 60V, 2A, 150MHz	45ba1(5mm)	Say	-	data
FP 212	Si-P+N	SMD, Dual, 200V, 0,1A, 150MHz	45ba1(5mm)	Say	-	data
FP 213	Si-P+Di+R	SMD, Dual, 25V, 2A, 300MHz, Rbe=1,6kΩ	45ba1(5mm)	Say	-	data
FP 214	Si-N+Di+R	SMD, Dual, 25V, 2A, 200MHz, Rbe=1,6kΩ	45ba1(5mm)	Say	-	data
FP 215	Si-P	SMD, Dual, HF, 30V, 2A, 1,5GHz	45ba1(5mm)	Say	-	data
FP 216	Si-N	SMD, Dual, Tr, 120V, 1A, 120MHz	45ba1(5mm)	Say	-	data
FP 218	Si-P	SMD, Dual, Tr, 120V, 1A, 120MHz	45ba1(5mm)	Say	-	data
FP 301	Si-N+Di	SMD, S, lo-sat, 30V, 2A, 60/525ns + Schottky-Di, 30V, 0,7A (SB 07-03C)	45bc1(5mm)	Say	-	data
FP 302	Si-N+Di	SMD, S, lo-sat, 60V, 1,5A, 50/330ns + Schottky-Di, 50V, 0,5A (SB 05-05CP)	45bc1(5mm)	Say		data
FP 303	Si-N+Di	SMD, S, lo-sat, 60V, 2A, 60/580ns + Schottky-Di, 50V, 0,5A (SB 05-05CP)	45bc1(5mm)	Say		data
FP 304	Si-N+Di	SMD, S, lo-sat, 30V, 3A, 200MHz + Schottky-Di, 30V, 0,7A (SB 07-03C)	45bc1(5mm)	Say		data
FP 401	MOS-N-FET-e*	SMD,V-MOS,Dual, LogL, 250V, 0,4A, <12Ω	45bb1(5mm)	Say	-	data
FP 402	MOS-N-FET-e*	SMD,V-MOS, Dual, LogL, 20V, 1A, <0,75Ω	45bb1(5mm)	Say	-	data
FP 403	MOS-N-FET-e*	SMD,V-MOS, Dual,LogL, 20V, 0,5A, <1,4Ω	45bb1(5mm)	Say	-	data
FP 501	MOS-N-FET-e*	SMD, V-MOS, LogL, 30V, 0,5A, <0,4Ω int. Schottky-Di (S→D), 30V, 0,5A	45(5mm)	Say		data
FP 502	MOS-N-FET-e*	SMD, V-MOS, LogL, 11V, 2A, <0,32Ω int. Schottky-Di (S→D)	45(5mm)	Say		data
FP 601	MOS-N-FET-e*	SMD, V-MOS, LogL, 20V, 1A, <0,55Ω + Schottky-Di (S→D), 30V, 0,7A	45(5mm)	Say		data
FP 3055 LE	MOS-N-FET-e	=RFP 3055LE (Typ-Code/Stempel/marking)	17	Rca	→RFP 3055LE	data
FPA 6101 MTCX	IC	2.4W Stereo Audio Power Amp., Headph. Driver	TSOP	Fch		pdf pinout
FPA 6101 MXL	IC	2.4W Stereo Audio Power Amp., Headph. Driver	TSOP	Fch		pdf pinout
FPAB 30 BH 60	IC	Smart Power Module,Front-end Rectifier	DIP	Fch		pdf pinout
FPD 87208 AXA	CMOS-IC	XGA/WXGA TFT-LCD Timing Controller	MP	Nsc		pdf pinout
FPD 87346	CMOS-IC	(SVGA)XGA/WXGA TFT-LCD Timing Contro.	MP	Nsc		pdf pinout
FPD 87352 CXA	CMOS-IC	+3.3V TFT-LCD Timing Controller	MP	Nsc		pdf pinout
FPD 87392 BXB	CMOS-IC	+3.3V TFT-LCD Timing Controller	MP	Nsc		pdf pinout
FPF1003	LOAD+	Vin:-0.3...+6V Il:2A Ron:20mOhm	6G	Fcs		data pdf pinout
FPF 1003	IC	SMD, low Rds P-channel Mosfet load switch, ESD Protected, Vin=1.2...5.5V	GRID	Fcs		pdf pinout
FPF1004	LOAD+	Vin:-0.3...+6V Il:2A Ron:20mOhm	6G	Fcs		data pdf pinout
FPF1005	LOAD+	Vin:-0.3...+6V Il:1.5A Ron:50mOhm	7H	Fcs		data pdf pinout
FPF 1005	IC	SMD, low Rds P-channel Mosfet load switch, ESD Protected, Vin=1.2...5.5V	MLP	Fcs	-	pdf pinout
FPF1006	LOAD+	Vin:-0.3...+6V Il:1.5A Ron:50mOhm	7H	Fcs		data pdf pinout
FPF 1006	IC	SMD, low Rds P-channel Mosfet load switch, ESD Protected, Vin=1.2...5.5V	MLP	Fcs	-	pdf pinout
FPF2000	LOAD+	Vin:-0.3...+6V Ilim:75mA Il:1.5A	5S	Fcs		data pdf pinout
FPF2001	LOAD+	Vin:-0.3...+6V Ilim:75mA Il:1.5A	5S	Fcs		data pdf pinout
FPF2002	LOAD+	Vin:-0.3...+6V Ilim:75mA Il:1.5A	5S	Fcs		data pdf pinout
FPF2003	LOAD+	Vin:-0.3...+6V Ilim:75mA Il:1.5A	5S	Fcs		data pdf pinout
FPF2004	LOAD+	Vin:-0.3...+6V Ilim:150mA Il:1.5A	5S	Fcs		data pdf pinout
FPF2005	LOAD+	Vin:-0.3...+6V Ilim:150mA Il:1.5A	5S	Fcs		data pdf pinout
FPF2006	LOAD+	Vin:-0.3...+6V Ilim:150mA Il:1.5A	5S	Fcs		data pdf pinout
FPF2007	LOAD+	Vin:-0.3...+6V Ilim:150mA Il:1.5A	5S	Fcs		data pdf pinout
FPF2108	LOAD+	Vin:-0.3...+6V Ilim:600mA Il:1.5A	5S	Fcs		data pdf pinout
FPF2116	LOAD+	Vin:-0.3...+6V Ilim:200mA Il:1.5A	5S	Fcs		data pdf pinout
FPF2123	LOAD+	Vin:-0.3...+6V Ilim:0.15...1.5A Il:1.5A	5S	Fcs		data pdf pinout
FPF2140	LOAD+	Vin:-0.3...6V Ilim:300mA Il:1.5A	6H	Fcs		data pdf pinout
FPF2144	LOAD+	Vin:-0.3...6V Ilim:600mA Il:1.5A	6H	Fcs		data pdf pinout
FPF2148	LOAD+	Vin:-0.3...6V Ilim:300mA Il:1.5A	6H	Fcs		data pdf pinout
FPF2163	LOAD+	Vin:-0.3...6V Ilim:150...1500mA Il:1.5A	6H	Fcs		data pdf pinout
FPF2164	LOAD+	Vin:-0.3...6V Ilim:150...1500mA Il:1.5A	6H	Fcs		data pdf pinout
FPF2165	LOAD+	Vin:-0.3...6V Ilim:150...1500mA Il:1.5A	6H	Fcs		data pdf pinout
FPF2174	LOAD+	Vin:-0.3...6V Ilim:300mA Il:1.5A	8H	Fcs		data pdf pinout
FPF2180	LOAD+	Vin:-0.3...6V Ilim:300mA Il:1.5A	6G	Fcs		data pdf pinout
FPF2182	LOAD+	Vin:-0.3...6V Ilim:300mA Il:1.5A	6G	Fcs		data pdf pinout
FPF2184	LOAD+	Vin:-0.3...6V Ilim:600mA Il:1.5A	6G	Fcs		data pdf pinout
FPF2193	LOAD+	Vin:-0.3...6V Ilim:0.15...1.5A Il:1.5A	6G	Fcs		data pdf pinout
FPF2194	LOAD+	Vin:-0.3...6V Ilim:0.15...1.5A Il:1.5A	6G	Fcs		data pdf pinout
FPF2195	LOAD+	Vin:-0.3...6V Ilim:0.15...1.5A Il:1.5A	6G	Fcs		data pdf pinout
FPF2500	LOAD+	Vin:-0.3...+20V Ilim:0.5...2A Il:1.5A	5S	Fcs		data pdf pinout
FPF2503	LOAD+	Vin:-0.3...+20V Ilim:600mA Il:1.5A	5S	Fcs		data pdf pinout
FPF2505	LOAD+	Vin:-0.3...+20V Ilim:1.2A Il:1.5A	5S	Fcs		data pdf pinout
FPQ 3467	Si-P	4x PNP, 40V, 1A, >175MHz, <40/-ns	14-DIP	Fch		data
FPQ 3468	Si-P	4x PNP, 50V, 1A, >150MHz, <40/-ns	14-DIP	Fch		data

Type	Device	Short Description	Fig.	Manu	Comparision Types	More at
FPQ 3724	Si-N	4x NPN, 40V, 1A, >300MHz, <35/-ns	14-DIP	Fch	CA 1724, DH 3724, MPQ 3724, SP 3724	data
FPQ 3725	Si-N	4x NPN, 50V, 1A, >300MHz, <35/-ns	14-DIP	Fch, Rca	CA 1725, DH 3725, FPQ 3725, SP 3725	data
FPT100	PT	800nm P:>0.16µA/mW/cm²		Fch		data pdf pinout
FPT101	PT	760nm P:>0.18µA/mW/cm²		Fch		data pinout
FPT102	PD	830nm S:0.6mA/mW P:1µA/mW/cm²		Fch		data pinout
FPT120	PT	830nm P:>3.2µA/mW/cm²	T1¾	Fch		data pdf pinout
FPT130	PT	830nm P:>1.2µA/mW/cm²		Fch		data pdf pinout
FPT131	PT	830nm P:0.28µA/mW/cm²		Fch		data pinout
FPT132	PT	830nm P:0.9µA/mW/cm²		Fch		data pinout
FPT136	PT	830nm P:0.18µA/mW/cm²		Fch		data pinout
FPT400	PT	830nm P:5µA/mW/cm²		Fch		data pinout
FPT500	PT	830nm P:>8µA/mW/cm²	TO18	Fch		data pinout
FPT500A	PT	830nm P:>8µA/mW/cm²	TO18	Fch		data pinout
FPT570	PT	830nm P:18µA/mW/cm²	TO18	Fch		data pinout
FPT610	PT	830nm P:1µA/mW/cm²		Fch		data pinout
FPT700	PT	830nm P:0.2µA/mW/cm²	T1	Fch		data pinout
FPT720	PD	940nm P:10µA/mW/cm²		Fch		data pinout
FPT 6004	Si-P-Darl	≈BD 902	17j§		→BD 902	
FPT 6005	Si-N-Darl	≈BD 901	17j§		→BD 901	
FQ	Si-P	=2PA1576Q (Typ-Code/Stempel/marking)	35(2mm)	Phi	→2PA1576	data
FQ	Si-P	=2SA1037-FQ (Typ-Code/Stempel/marking)	≈35	Rhm	→2SA1037	data
FQ	Si-P	=2SA1037K-Q (Typ-Code/Stempel/marking)	35	Rhm	→2SA1037K	data
FQ	Si-P	=2SA1576-Q (Typ-Code/Stempel/marking)	35(2mm)	Rhm	→2SA1576	data
FQ	Si-P	=2SA1774-FQ (Typ-Code/Stempel/marking)	35(1,6mm)	Rhm	→2SA1774	data
FQ	Si-P	=2SB1610 (Typ-Code/Stempel/marking)	≈35	Mat	→2SB1610	data
FQ	Si-P	=2SB1618 (Typ-Code/Stempel/marking)	35(2mm)	Mat	→2SB1618	data
FQ	Z-Di	=SM 6T 68A (Typ-Code/Stempel/marking)	71(6x4mm)	Tho	→SM 6T...	data
FQ	Z-Di	=Z1 SMA-110 (Typ-Code/Stempel/marking)	71(5x2,5)	Fag	→Z1 SMA-110	data
FQ 08 C	Triac	200V, 20A(Tc=100°C), Igt/Ih 50/30mA	21b§	Hit	TXE 99..., SC 260..., T 6411..., T 6410...	data
FQ 08 D	Triac	=FQ 08C: 300V	21b§		TXE 99..., SC 260..., T 6411..., T 6410...	data
FQ 08 E	Triac	=FQ 08C: 400V	21b§		TXE 99..., SC 260..., T 6411..., T 6410...	data
FQ 08 F	Triac	=FQ 08C: 500V	21b§		TXE 99..., SC 260..., T 6411..., T 6410...	data
FQ 08 G	Triac	=FQ 08C: 600V	21b§		TXE 99..., SC 260..., T 6411..., T 6410...	data
FQ 3467	Si-P	=FPQ 3467: SMD	14-MDIP	Fch	-	data
FQ 3468	Si-P	=FPQ 3468: SMD	14-MDIP	Fch	-	data
FQ 3724	Si-N	=FPQ 3724: SMD	14-MDIP	Fch	-	data
FQ 3725	Si-N	=FPQ 3725: SMD	14-MDIP	Fch	-	data
FR	Si-P	=2PA1576R (Typ-Code/Stempel/marking)	35(2mm)	Phi	→2PA1576	data
FR	Si-P	=2SA1034-R (Typ-Code/Stempel/marking)	35	Mat	→2SA1034	data
FR	Si-P	=2SA1037-FR (Typ-Code/Stempel/marking)	≈35	Rhm	→2SA1037	data
FR	Si-P	=2SA1037K-R (Typ-Code/Stempel/marking)	35	Rhm	→2SA1037K	data
FR	Si-P	=2SA1531-R (Typ-Code/Stempel/marking)	35(2mm)	Mat	→2SA1531	data
FR	Si-P	=2SA1576-R (Typ-Code/Stempel/marking)	35(2mm)	Rhm	→2SA1576	data
FR	Si-P	=2SA1774-FR (Typ-Code/Stempel/marking)	35(1,6mm)	Rhm	→2SA1774	data
FR	Si-N	=2SC2716-R (Typ-Code/Stempel/marking)	35	Tos	→2SC2716	data
FR	Si-N	=2SC4988 (Typ-Code/Stempel/marking)	39	Hit	→2SC4988	data
FR	Si-N	=2SD2472 (Typ-Code/Stempel/marking)	≈35	Mat	→2SD2472	data
FR	Si-N	=2SD2482 (Typ-Code/Stempel/marking)	35(2mm)	Mat	→2SD2482	data
FR	Si-N	=KTC3878-R (Typ-Code/Stempel/marking)	35	Kec	→KTC 3878(S)	data
FR	Z-Di	=Z1 SMA-120 (Typ-Code/Stempel/marking)	71(5x2,5)	Fag	→Z1 SMA-120	data
FR 1/...	Si-Di	GI			BY 126...127, BY 133...135, 1N4002...07, ++	
FR 1 A..M	Si-Di	SMD, GI, S, 50...1000V, 0,6A, <500ns A=50, B=100, D=200, G=400, J=600V, K=800, M=1000V	71a(5x3,7)	Die	BYG 20G...M, FR 2A..M	data
FR 1L	Si-Di	=TFR 1L (Typ-Code/Stempel/marking)	31	Tos	→TFR 1L...T	data
FR 1N	Si-Di	=TFR 1N (Typ-Code/Stempel/marking)	31	Tos	→TFR 1L...T	data
FR 1Q	Si-Di	=TFR 1Q (Typ-Code/Stempel/marking)	31	Tos	→TFR 1L...T	data
FR 1T	Si-Di	=TFR 1T (Typ-Code/Stempel/marking)	31	Tos	→TFR 1L...T	data
FR 2/-02...-12	Si-Di	GI, Uni, 200...1200V, 1,1A	31a	Fjd	BY 126...127, BY 133...135, 1N4003...07, ++	data
FR 2 A..M	Si-Di	SMD, GI, S, 50...1000V, 1,5A, <500ns A=50, B=100, D=200, G=400, J=600V, K=800, M=1000V	71a(5x3,7)	Die	BYG 20G...M, BYG 21K...M	data
FR 2L	Si-Di	=TFR 2L (Typ-Code/Stempel/marking)	31	Tos	→TFR 2L...T	data
FR 2N	Si-Di	=TFR 2N (Typ-Code/Stempel/marking)	31	Tos	→TFR 2L...T	data
FR 2Q	Si-Di	=TFR 2Q (Typ-Code/Stempel/marking)	31	Tos	→TFR 2L...T	data
FR 2T	Si-Di	=TFR 2T (Typ-Code/Stempel/marking)	31	Tos	→TFR 2L...T	data
FR 3L	Si-Di	=TFR 3L (Typ-Code/Stempel/marking)	31	Tos	→TFR 3L...T	data
FR 3N	Si-Di	=TFR 3N (Typ-Code/Stempel/marking)	31	Tos	→TFR 3L...T	data
FR 3Q	Si-Di	=TFR 3Q (Typ-Code/Stempel/marking)	31	Tos	→TFR 3L...T	data
FR 3T	Si-Di	=TFR 3T (Typ-Code/Stempel/marking)	31	Tos	→TFR 3L...T	data
FR 4L	Si-Di	=TFR 4L (Typ-Code/Stempel/marking)	31	Tos	→TFR 4L...T	data
FR 4N	Si-Di	=TFR 4N (Typ-Code/Stempel/marking)	31	Tos	→TFR 4L...T	data
FR 4Q	Si-Di	=TFR 4Q (Typ-Code/Stempel/marking)	31	Tos	→TFR 4L...T	data
FR 4T	Si-Di	=TFR 4T (Typ-Code/Stempel/marking)	31	Tos	→TFR 4L...T	data
FR 34	Si-Di	GI, Uni, 100V, 0,2A, Uf<1V(0,2A)	31a	Ssc	BAW 52, BAY 19...21, BAY 45	data
FR 101	Si-Di	GI, 50V, 1A	31a	Mic	BY 126...127, BY 226...227, 1N4001...07,++	data
FR 102	Si-Di	=FR 101: 100V	31a	Mic	BY 126...127, BY 226...227, 1N4002...07,++	data
FR 103	Si-Di	=FR 101: 200V	31a	Mic	BY 126...127, BY 226...227, 1N4003...07,++	data
FR 104	Si-Di	=FR 101: 400V	31a	Mic	BY 126...127, BY 226...227, 1N4004...07,++	data
FR 105	Si-Di	=FR 101: 600V	31a	Mic	BY 126...127, BY 226...227, 1N4005...07,++	data

Type	Device	Short Description	Fig.	Manu	Comparision Types	More at
FR 106	Si-Di	=FR 101: 800V	31a	Mic	BY 127, BY 227, 1N4006..07,++	data
FR 107	Si-Di	=FR 101: 1000V	31a	Mic	BY 127, BY 227, 1N4007,++	data
FR 151	Si-Di	GI, 50V, 1,5A	31a	Mic	BY 226..227, BY 251..255, 1N5391..99,++	data
FR 152	Si-Di	=FR 151: 100V	31a	Mic	BY 226..227, BY 251..255, 1N5392..99,++	data
FR 153	Si-Di	=FR 151: 200V	31a	Mic	BY 226..227, BY 251..255, 1N5393..99,++	data
FR 154	Si-Di	=FR 151: 400V	31a	Mic	BY 226..227, BY 252..255, 1N5395..99,++	data
FR 155	Si-Di	=FR 151: 600V	31a	Mic	BY 226..227, BY 253..255, 1N5397..99,++	data
FR 156	Si-Di	=FR 151: 800V	31a	Mic	BY 227, BY 254..255, 1N5398..99,++	data
FR 157	Si-Di	=FR 151: 1000V	31a	Mic	BY 227, BY 255, 1N53999,++	data
FR 301	Si-Di	GI, 50V, 3A	31a	Mic	BY 251..255, BY 396..399, 1N5400..08,++	data
FR 302	Si-Di	=FR 301: 100V	31a	Mic	BY 251..255, BY 396..399, 1N5401..08,++	data
FR 303	Si-Di	=FR 301: 200V	31a	Mic	BY 251..255, BY 397..399, 1N5402..08,++	data
FR 304	Si-Di	=FR 301: 400V	31a	Mic	BY 252..255, BY 398..399, 1N5404..08,++	data
FR 305	Si-Di	=FR 301: 600V	31a	Mic	BY 253..255, BY 399, 1N5406..08,++	data
FR 306	Si-Di	=FR 301: 800V	31a	Mic	BY 254..255, BY 399, 1N5407..08,++	data
FR 307	Si-Di	=FR 301: 1000V	31a	Mic	BY 255, BY 399S, 1N5408,++	data
FR 601	Si-Di	GI, 50V, 6A	31a	Mic	BY 214/50../1000, MR 750...760	data
FR 602	Si-Di	=FR 601: 100V	31a	Mic	BY 214/100../1000, MR 751...760	data
FR 603	Si-Di	=FR 601: 200V	31a	Mic	BY 214/200../1000, MR 752...760	data
FR 604	Si-Di	=FR 601: 400V	31a	Mic	BY 214/400../1000, MR 754...760	data
FR 605	Si-Di	=FR 601: 600V	31a	Mic	BY 214/600../1000, MR 756...760	data
FR 606	Si-Di	=FR 601: 800V	31a	Mic	BY 214/800../1000, MR 758...760	data
FR 607	Si-Di	=FR 601: 1000V	31a	Mic	BY 214/1000, MR 760	data
FR 3205 CC	Si-Di	Dual, S-L, 50V, 32A, <50ns	16	Fch	BYT 30P/200, BYV 72/50, BYW 99P/50	data
FR 3210 CC	Si-Di	=FR 3205: 100V	16	Fch	BYT 30P/200, BYV 72/100, BYW 99P/100	data
FR 3215 CC	Si-Di	=FR 3205: 150V	16	Fch	BYT 30P/200, BYV 72/150, BYW 99P/150	data
FR 3220 CC	Si-Di	=FR 3205: 180V	16	Fch	BYT 30P/200, BYV 72/200, BYW 99P/200	data
FR 4001	Ge-Di	≈AY 103K	3		→AY 103K	
FR 4001 [Thomson]	Si-Di	=BYX 58/400		Tho	→BYX 58/...	data
FR-	Si-N	=2SC5772 (Typ-Code/Stempel/marking)	35	Hit	→2SC5772	
FRH 101	Si-Di	≈1N4001			→1N4001	
FRM 3205 CC	Si-Di	Dual, S-L, 50V, 32A, <50ns	23	Fch	BYT 30P/200, BYV 72/50, BYW 99P/50	data
FRM 3210 CC	Si-Di	=FRM 3205: 100V	23	Fch	BYT 30P/200, BYV 72/100, BYW 99P/100	data
FRM 3215 CC	Si-Di	=FRM 3205: 150V	16	Fch	BYT 30P/200, BYV 72/150, BYW 99P/150	data
FRM 3220 CC	Si-Di	=FRM 3205: 180V	16	Fch	BYT 30P/200, BYV 72/200, BYW 99P/200	data
FRP 805	Si-Di	S-L, 50V, 8A, <50ns	17t§	Fch	BYP 21-50, BYT 79/300, BYW 29/50, FE 8A	data
FRP 810	Si-Di	=FRP 805: 100V	17t§	Fch	BYP 21-100, BYT 79/300, BYW 29/100, FE8B	data
FRP 815	Si-Di	=FRP 805: 150V	17t§	Fch	BYP 21-150, BYT 79/300, BYW 29/150, FE8C	data
FRP 820	Si-Di	=FRP 805: 180V	17t§	Fch	BYP 21-200, BYT 79/300, BYW 29/200, FE8D	data
FRP 840	Si-Di	S-L, 400V, 8A, <75ns	17t§	Fch	BYT 79/400, BYV 29/400, FE 8G	data
FRP 850	Si-Di	=FRP 840: 500V	17t§	Fch	BYT 79/500, BYV 29/500, FE 8H	data
FRP 860	Si-Di	=FRP 840: 600V	17t§	Fch	BYT 12P/600, FE 8J	data
FRP 1005	Si-Di	S-L, 50V, 10A, <50ns	17t§	Fch	BYT 79/300, BYV 79/50	data
FRP 1010	Si-Di	=FRP 1005: 100V	17t§	Fch	BYT 79/300, BYV 79/100	data
FRP 1015	Si-Di	=FRP 1005: 150V	17t§	Fch	BYT 79/300, BYV 79/150	data
FRP 1020	Si-Di	=FRP 1005: 180V	17t§	Fch	BYT 79/300, BYV 79/200	data
FRP 1605	Si-Di	S-L, 50V, 16A, <50ns	17t§	Fch	BYV 79/50, MUR 1505	data
FRP 1610	Si-Di	=FRP 1605: 100V	17t§	Fch	BYV 79/100, MUR 1510	data
FRP 1615	Si-Di	=FRP 1605: 150V	17t§	Fch	BYV 79/150, MUR 1515	data
FRP 1620	Si-Di	=FRP 1605: 180V	17t§	Fch	BYV 79/200, MUR 1520	data
FRP 1605...1620 CC	Si-Di	=FRP 1605...1620: Dual	17p§		BYV 32/..., BYP 22/..., FE 16A...D	data
FRP 2005 CC	Si-Di	Dual, S-L, 50V, 20A, <50ns	17p§	Fch	BYW 51/50, BYP 22-50, BYV 44/300	data
FRP 2010 CC	Si-Di	=FRP 2005CC: 100V	17p§	Fch	BYW 51/100, BYP 22-100, BYV 44/300	data
FRP 2015 CC	Si-Di	=FRP 2005CC: 150V	17p§	Fch	BYW 51/150, BYP 22-150, BYV 44/300	data
FRP 2020 CC	Si-Di	=FRP 2005CC: 180V	17p§	Fch	BYP 22-200, BYV 44/300	data
FRS 1 A..M	Si-Di	SMD, GI, S, 50...1000V, 1A, <300ns A=50, B=100, D=200, G=400, J=600V, K=800, M=1000V	71a(5x2,5)	Die	BYG 20G..M, BYG 21K..M, FR 2A...M	data
FRS 2 A..M	Si-Di	SMD, GI, S, 50...1000V, 1,5A, <500ns A=50, B=100, D=200, G=400, J=600V, K=800, M=1000V	71a(5x3,5)	Die	BYG 20G..M, BYG 21K..M, FR 2A...M	data
FS	Si-P	=2PA1576S (Typ-Code/Stempel/marking)	35(2mm)	Phi	→2PA1576	data
FS	Si-P	=2SA1034-S (Typ-Code/Stempel/marking)	35	Mat	→2SA1034	data
FS	Si-P	=2SA1037-FS (Typ-Code/Stempel/marking)	≈35	Rhm	→2SA1037	data
FS	Si-P	=2SA1037K-S (Typ-Code/Stempel/marking)	35	Rhm	→2SA1037K	data
FS	Si-P	=2SA1531-S (Typ-Code/Stempel/marking)	35(2mm)	Mat	→2SA1531	data
FS	Si-P	=2SA1576-S (Typ-Code/Stempel/marking)	35(2mm)	Rhm	→2SA1576	data
FS	Si-P	=2SA1763 (Typ-Code/Stempel/marking)	35(2mm)	Say	→2SA1763	data
FS	Si-P	=2SA1764 (Typ-Code/Stempel/marking)	35	Say	→2SA1764	data
FS	Si-P	=2SA1774-FS (Typ-Code/Stempel/marking)	35(1,6mm)	Rhm	→2SA1774	data
FS	Si-P	=2SB1519 (Typ-Code/Stempel/marking)	39	Hit	→2SB1519	data
FS	Si-P	=2SB1611 (Typ-Code/Stempel/marking)	≈35	Mat	→2SB1611	data
FS	Si-P	=2SB1619 (Typ-Code/Stempel/marking)	35(2mm)	Mat	→2SB1619	data
FS	Z-Di	=Z1 SMA-130 (Typ-Code/Stempel/marking)	71(5x2,5)	Fag	→Z1 SMA-130	data
FS 1/...	Si-Di	≈RGP 10G..M			→RGP 10G..M	
FS 1 AS-16A	MOS-N-FET-e*	=FS 1 UM-16A: 55W	30c§	Mit	-	data
FS 1 AS-18A	MOS-N-FET-e*	=FS 1 UM-18A: 55W	30c§	Mit	-	data

Type	Device	Short Description	Fig.	Manu	Comparision Types	More at
FS 1 G..M	Si-Di	SMD, Gl, Uni, 400...1000V, 1A, 4µs G=400V, J=600V, K=800V, M=1000V	71a(5x2,5)	Fag	BYG 10G..M	data
FS 1 KM-16A	MOS-N-FET-e*	=FS 1 UM-16A: Iso, 25W	17c	Mit	BUK 444-800, 2SK808	data
FS 1 KM-18A	MOS-N-FET-e*	=FS 1 UM-18A: Iso, 25W	17c	Mit	BUK 446-1000, 2SK808A	data
FS 1 UM-16A	MOS-N-FET-e*	V-MOS, 800V, 1A, 65W, <12,3Ω(0,5A)	17c§	Mit	BUZ 78, 2SK2733	data
FS 1 UM-18A	MOS-N-FET-e*	V-MOS, 900V, 1A, 65W, <15Ω(0,5A)	17c§	Mit	BUZ 50, 2SK2733	data
FS 1 VS-16A	MOS-N-FET-e*	=FS 1 UM-16A:	30c§	Mit	2SK1647, 2SK1980	data
FS 1 VS-18A	MOS-N-FET-e*	=FS 1 UM-18A:	30c§	Mit	2SK1647	data
FS 2 A..M	Si-Di	SMD, Gl, Uni, 50...1000V, 1,5, 4µs A=50, B=100, D=200, G=400, J=600V, K=800, M=1000V	71a(5x2,5)	Fag	BYG 10G..M	data
FS 2 AS-14A	MOS-N-FET-e*	=FS 2 UM-14A: 55W	30c§	Mit	-	data
FS 2 KM-12	MOS-N-FET-e*	=FS 2 UM-12: Iso, 30W	17c	Mit	BUK 445-600, 2SK1758, 2SK1953, 2SK2043	data
FS 2 KM-14A	MOS-N-FET-e*	=FS 2 UM-14A: Iso, 25W	17c	Mit	BUK 446-800, 2SK1142, 2SK1611, 2SK1834	data
FS 2 KM-16A	MOS-N-FET-e*	=FS 2 UM-16A: Iso, 30W	17c	Mit	BUK 446-800, 2SK1142, 2SK1611, 2SK1834	data
FS 2 KM-18A	MOS-N-FET-e*	=FS 2 UM-18A: Iso, 30W	17c	Mit	2SK1174, 2SK1459, 2SK1994, 2SK2478, ++	data
FS 2 UM-12	MOS-N-FET-e*	V-MOS, 600V, 2A, 60W, <6,4Ω(1A)	17c§	Mit	BUZ 80, 2SK1922	data
FS 2 UM-14A	MOS-N-FET-e*	V-MOS, 700V, 2A, 65W, <9,75Ω(1A)	17c§	Mit	BUZ 80, 2SK1323	data
FS 2 UM-16A	MOS-N-FET-e*	V-MOS, 800V, 2A, 85W, <6Ω(1A)	17c§	Mit	BUZ 80, 2SK1323	data
FS 2 UM-18A	MOS-N-FET-e*	V-MOS, 900V, 2A, 85W, <7,3Ω(1A)	17c§	Mit	BUZ 50, 2SK1199, 2SK1324, 2SK1338	data
FS 2 VS-12	MOS-N-FET-e*	=FS 2 UM-12:	30c§	Mit	MTB 2N60E, 2SK1746	data
FS 2 VS-14A	MOS-N-FET-e*	=FS 2 UM-14A:	30c§	Mit	2SK1980	data
FS 2 VS-16A	MOS-N-FET-e*	=FS 2 UM-16A:	30c§	Mit	2SK1980	data
FS 2 VS-18A	MOS-N-FET-e*	=FS 2 UM-18A:	30c§	Mit	2SK1647	data
FS 3 KM-9	MOS-N-FET-e*	=FS 3 UM-9: Iso, 30W	17c	Mit	BUK 445-450, 2SK1444, 2SK1862, 2SK2431	data
FS 3 KM-10	MOS-N-FET-e*	=FS 3 UM-10: Iso, 30W	17c	Mit	BUK 445-500, 2SK1863, 2SK2144, 2SK2750++	data
FS 3 KM-14A	MOS-N-FET-e*	=FS 3 UM-14A: Iso, 30W	17c	Mit	2SK1356, 2SK1460, 2SK1995, 2SK2476, ++	data
FS 3 KM-16A	MOS-N-FET-e*	=FS 3 UM-16A: Iso, 30W	17c	Mit	2SK1356, 2SK1460, 2SK1995, 2SK2480, ++	data
FS 3 KM-18A	MOS-N-FET-e*	=FS 3 UM-18A: Iso, 30W	17c	Mit	2SK1356, 2SK1460, 2SK1995, 2SK2480, ++	data
FS 3 SM-14A	MOS-N-FET-e*	=FS 3 UM-14A:	18c§	Mit	BUZ 307, 2SK1339, 2SK2667, 2SK2719, ++	data
FS 3 SM-16A	MOS-N-FET-e*	=FS 3 UM-16A:	18c§	Mit	BUZ 307, 2SK1339, 2SK2667, 2SK2719, ++	data
FS 3 SM-18A	MOS-N-FET-e*	=FS 3 UM-18A:	18c§	Mit	2SK1339, 2SK2667, 2SK2719	data
FS 3 UM-9	MOS-N-FET-e*	V-MOS, 450V, 3A, 60W, <3,5Ω(1A)	17c§	Mit	2SK1154, 2SK1439, 2SK1493, 2SK1494	data
FS 3 UM-10	MOS-N-FET-e*	V-MOS, 500V, 3A, 60W, <4,4Ω(1A)	17c§	Mit	2SK1154, 2SK1244, 2SK1494	data
FS 3 UM-14A	MOS-N-FET-e*	V-MOS, 700V, 3A, 85W, <4,75Ω(1,5A)	17c§	Mit	BUK 456-800, 2SK1600, 2SK2603	data
FS 3 UM-16A	MOS-N-FET-e*	V-MOS, 800V, 3A, 100W, <3,3Ω(1,5A)	17c§	Mit	-	data
FS 3 UM-18A	MOS-N-FET-e*	V-MOS, 900V, 3A, 100W, <4Ω(1,5A)	17c§	Mit	-	data
FS 3 VS-9	MOS-N-FET-e*	=FS 3 UM-9:	30c§	Mit	2SK1493Z, 2SK1494Z, 2SK1690, 2SK1721	data
FS 3 VS-10	MOS-N-FET-e*	=FS 3 UM-10:	30c§	Mit	2SK1494Z, 2SK1721	data
FS 3 VS-14A	MOS-N-FET-e*	=FS 3 UM-14A:	30c§	Mit	2SK1846, 2SK1858, 2SK2883	data
FS 3 VS-16A	MOS-N-FET-e*	=FS 3 UM-16A:	30c§	Mit	-	data
FS 3 VS-18A	MOS-N-FET-e*	=FS 3 UM-18A:	30c§	Mit	-	data
FS 4 KM-12	MOS-N-FET-e*	=FS 4 UM-12: Iso, 35W	17c	Mit	2SK1118, 2SK2097, 2SK2118, 2SK2137, ++	data
FS 4 UM-12	MOS-N-FET-e*	V-MOS, 600V, 4A, 90W, <2,6Ω(2A)	17c§	Mit	BUK 455-600, IRFBC 40, 2SK1117	data
FS 4 VS-12	MOS-N-FET-e*	=FS 4 UM-12:	30c§	Mit	2SK1915	data
FS 5 KM-5	MOS-N-FET-e*	=FS 5 UM-5: Iso, 30W	17c	Mit	2SK2010, 2SK2108, 2SK2345	data
FS 5 KM-6	MOS-N-FET-e*	=FS 5 UM-6: Iso, 30W	17c	Mit	BUK 445-400, 2SK1377, 2SK2345	data
FS 5 KM-9	MOS-N-FET-e*	=FS 5 UM-9: Iso, 35W	17c	Mit	2SK1990, 2SK1992, 2SK2114, 2SK2353, ++	data
FS 5 KM-10	MOS-N-FET-e*	=FS 5 UM-10: Iso, 35W	17c	Mit	2SK1991, 2SK1993, 2SK2115, 2SK2354, ++	data
FS 5 KM-14A	MOS-N-FET-e*	=FS 5 UM-14A: Iso, 35W	17c	Mit	2SK2100, 2SK2101	data
FS 5 KM-16A	MOS-N-FET-e*	=FS 5 UM-16A: Iso, 35W	17c	Mit	2SK1808, 2SK1985, 2SK2100, 2SK2101	data
FS 5 KM-18A	MOS-N-FET-e*	=FS 5 UM-18A: Iso, 30W	17c	Mit	2SK1808, 2SK1985	data
FS 5 SM-14A	MOS-N-FET-e*	=FS 5 UM-14A:	18c§	Mit	2SK695, 2SK1341, 2SK1794, 2SK2610, ++	data
FS 5 SM-16A	MOS-N-FET-e*	=FS 5 UM-16A:	18c§	Mit	2SK695, 2SK1341, 2SK1794, 2SK2610, ++	data
FS 5 SM-18A	MOS-N-FET-e*	=FS 5 UM-18A:	18c§	Mit	2SK1341, 2SK1773, 2SK1794, 2SK2610, ++	data
FS 5 UM-5	MOS-N-FET-e*	V-MOS, 250V, 5A, 60W, <1,3Ω(2A)	17c§	Mit	IRF 624, 2SK924, 2SK1391, 2SK1921	data
FS 5 UM-6	MOS-N-FET-e*	V-MOS, 300V, 5A, 60W, <1,6Ω(2A)	17c§	Mit	BUZ 60, IRF 730, 2SK1440, 2SK2355	data
FS 5 UM-9	MOS-N-FET-e*	V-MOS, 450V, 5A, 90W, <1,4Ω(2A)	17c§	Mit		data
FS 5 UM-10	MOS-N-FET-e*	V-MOS, 500V, 5A, 90W, <1,8Ω(2A)	17c§	Mit		data
FS 5 UM-14A	MOS-N-FET-e*	V-MOS, 700V, 5A, 100W, <2,6Ω(2A)	17c§	Mit	-	data
FS 5 UM-16A	MOS-N-FET-e*	V-MOS, 800V, 5A, 125W, <2,3Ω(2A)	17c§	Mit	-	data
FS 5 UM-18A	MOS-N-FET-e*	V-MOS, 900V, 5A, 125W, <2,8Ω(2A)	17c§	Mit	-	data
FS 5 VS-5	MOS-N-FET-e*	=FS 5 UM-5:	30c§	Mit	2SK2790	data
FS 5 VS-6	MOS-N-FET-e*	=FS 5 UM-6:	30c§	Mit	2SK1308, 2SK2838	data
FS 5 VS-9	MOS-N-FET-e*	=FS 5 UM-9:	30c§	Mit	-	data
FS 5 VS-10	MOS-N-FET-e*	=FS 5 UM-10:	30c§	Mit	-	data
FS 5 VS-14A	MOS-N-FET-e*	=FS 5 UM-14A:	30c§	Mit	-	data
FS 5 VS-16A	MOS-N-FET-e*	=FS 5 UM-16A:	30c§	Mit	-	data
FS 5 VS-18A	MOS-N-FET-e*	=FS 5 UM-18A:	30c§	Mit	-	data
FS 6 S 1265 RETU	SMPS-IC	high voltage power Sensefet and current mode PWM contr., 650V 0.7Ohm	86	Fcs	-	pdf pinout
FS 6 S 1265 REYDTU	SMPS-IC	high voltage power Sensefet and current mode PWM contr., 650V 0.7Ohm	86	Fcs	-	pdf pinout
FS 7 KM-5	MOS-N-FET-e*	=FS 7 UM-5: Iso, 30W	17c	Mit	2SK1987, 2SK2341, 2SK2417, 2SK2425	data
FS 7 KM-12	MOS-N-FET-e*	=FS 7 UM-12: Iso, 35W	17c	Mit	2SK2564, 2SK2645, 2SK2708, 2SK2740	data
FS 7 KM-14A	MOS-N-FET-e*	=FS 7 UM-14A: Iso, 30W	17c	Mit	-	data
FS 7 KM-16A	MOS-N-FET-e*	=FS 7 UM-16A: Iso, 40W	17c	Mit	-	data
FS 7 KM-18A	MOS-N-FET-e*	=FS 7 UM-18A: Iso, 40W	17c	Mit	-	data
FS 7 SM-12	MOS-N-FET-e*	=FS 7 UM-12:	18c§	Mit	-	data
FS 7 SM-14A	MOS-N-FET-e*	=FS 7 UM-14A:	18c§	Mit	-	data
FS 7 SM-16A	MOS-N-FET-e*	=FS 7 UM-16A:	18c§	Mit	-	data
FS 7 SM-18A	MOS-N-FET-e*	=FS 7 UM-18A:	18c§	Mit	-	data

Type	Device	Short Description	Fig.	Manu	Comparision Types	More at
FS 7 UM-5	MOS-N-FET-e*	V-MOS, 250V, 7A, 75W, <0,8Ω(3A)	17c§	Mit	IRF 634, 2SK1319, 2SK1320	data
FS 7 UM-12	MOS-N-FET-e*	V-MOS, 600V, 7A, 125W, <1,3Ω(3A)	17c§	Mit	-	data
FS 7 UM-14A	MOS-N-FET-e*	V-MOS, 700V, 7A, 125W, <1,82Ω(3A)	17c§	Mit	-	data
FS 7 UM-16A	MOS-N-FET-e*	V-MOS, 800V, 7A, 150W, <1,64Ω(3A)	17c§	Mit	-	data
FS 7 UM-18A	MOS-N-FET-e*	V-MOS, 900V, 7A, 150W, <2Ω(3A)	17c§	Mit	-	data
FS 7 VS-5	MOS-N-FET-e*	=FS 7 UM-5:	30c§	Mit	MTB 9N25E	data
FS 7 VS-12	MOS-N-FET-e*	=FS 7 UM-12:	30c§	Mit	-	data
FS 7 VS-14A	MOS-N-FET-e*	=FS 7 UM-14A:	30c§	Mit	-	data
FS 7 VS-16A	MOS-N-FET-e*	=FS 7 UM-16A:	30c§	Mit	-	data
FS 7 VS-18A	MOS-N-FET-e*	=FS 7 UM-18A:	30c§	Mit	-	data
FS 10 KM-5	MOS-N-FET-e*	=FS 10 UM-5: Iso, 35W	17c	Mit	2SK1762 2SK2011, 2SK2426	data
FS 10 KM-6	MOS-N-FET-e*	=FS 10 UM-6: Iso, 35W	17c	Mit	-	data
FS 10 KM-9	MOS-N-FET-e*	=FS 10 UM-9: Iso, 35W	17c	Mit	2SK2475	data
FS 10 KM-10	MOS-N-FET-e*	=FS 10 UM-10: Iso, 35W	17c	Mit	2SK2475, 2SK2842...43	data
FS 10 KM-12	MOS-N-FET-e*	=FS 10 UM-12: Iso, 40W	17c	Mit	2SK2843	data
FS 10 KM-14A	MOS-N-FET-e*	=FS 10 UM-14A: Iso, 40W	17c	Mit	-	data
FS 10 SM-9	MOS-N-FET-e*	=FS 10 UM-9:	18c§	Mit	BUZ 339, IRFPC 50, 2SK1488, 2SK2699	data
FS 10 SM-10	MOS-N-FET-e*	=FS 10 UM-10:	18c§	Mit	BUZ 339, IRFPC 50, 2SK1488, 2SK2699	data
FS 10 SM-12	MOS-N-FET-e*	=FS 10 UM-12:	18c§	Mit	BUZ 334, IRFPC 50, 2SK2699	data
FS 10 SM-14A	MOS-N-FET-e*	=FS 10 UM-14A:	18c§	Mit	-	data
FS 10 SM-16A	MOS-N-FET-e*	V-MOS, 800V, 10A, 200W, <0,98Ω(5A)	18c§	Mit	-	data
FS 10 SM-18A	MOS-N-FET-e*	V-MOS, 900V, 10A, 200W, <1,2Ω(5A)	18c§	Mit	-	data
FS 10 UM-5	MOS-N-FET-e*	V-MOS, 250V, 10A, 90W, <0,52Ω(5A)	17c§	Mit	BUK 657-400, BUZ 64, IRF 740, 2SK1378	data
FS 10 UM-6	MOS-N-FET-e*	V-MOS, 300V, 10A, 90W, <0,68Ω(5A)	17c§	Mit	BUK 657-400, BUZ 64, IRF 740, 2SK1378	data
FS 10 UM-9	MOS-N-FET-e*	V-MOS, 450V, 10A, 125W, <0,73Ω(5A)	17c§	Mit	-	data
FS 10 UM-10	MOS-N-FET-e*	V-MOS, 500V, 10A, 125W, <0,9Ω(5A)	17c§	Mit	-	data
FS 10 UM-12	MOS-N-FET-e*	V-MOS, 600V, 10A, 150W, <0,94Ω(5A)	17c§	Mit	-	data
FS 10 UM-14A	MOS-N-FET-e*	V-MOS, 700V, 10A, 150W, <1,3Ω(5A)	17c§	Mit	-	data
FS 10 VS-5	MOS-N-FET-e*	=FS 10 UM-5:	30c§	Mit	MTB 10N40E	data
FS 10 VS-6	MOS-N-FET-e*	=FS 10 UM-6:	30c§	Mit	MTB 10N40E	data
FS 10 VS-9	MOS-N-FET-e*	=FS 10 UM-9:	30c§	Mit	-	data
FS 10 VS-10	MOS-N-FET-e*	=FS 10 UM-10:	30c§	Mit	-	data
FS 10 VS-12	MOS-N-FET-e*	=FS 10 UM-12:	30c§	Mit	-	data
FS 10 VS-14A	MOS-N-FET-e*	=FS 10 UM-14A:	30c§	Mit	-	data
FS 12 KM-5	MOS-N-FET-e*	=FS 12 UM-5: Iso, 35W	17c	Mit	2SK1036, 2SK1762, 2SK2011, 2SK2426	data
FS 12 UM-5	MOS-N-FET-e*	V-MOS, 250V, 12A, 100W, <0,4Ω(6A)	17c§	Mit	BUZ 255	data
FS 12 VS-5	MOS-N-FET-e*	=FS 12 UM-5:	30c§	Mit	MTB 16N25E	data
FS 14 KM-9	MOS-N-FET-e*	=FS 14 UM-9: Iso, 40W	17c	Mit	2SK2704	data
FS 14 KM-10	MOS-N-FET-e*	=FS 14 UM-10: Iso, 40W	17c	Mit	-	data
FS 14 SM-9	MOS-N-FET-e*	=FS 14 UM-9:	18c§	Mit	BUK 638-500, BUZ 338, 2SK1678, 2SK2698	data
FS 14 SM-10	MOS-N-FET-e*	=FS 14 UM-10:	18c§	Mit	BUK 638-500, BUZ 338, 2SK1678, 2SK2698	data
FS 14 SM-12	MOS-N-FET-e*	V-MOS, 600V, 14A, 250W, <0,6Ω(7A)	18c§	Mit	-	data
FS 14 SM-14A	MOS-N-FET-e*	V-MOS, 700V, 14A, 200W, <0,78Ω(7A)	18c§	Mit	-	data
FS 14 SM-16A	MOS-N-FET-e*	V-MOS, 800V, 14A, 275W, <0,7Ω(7A)	18c§	Mit	-	data
FS 14 SM-18A	MOS-N-FET-e*	V-MOS, 900V, 14A, 275W, <0,85Ω(7A)	18c§	Mit	-	data
FS 14 UM-9	MOS-N-FET-e*	V-MOS, 450V, 14A, 150W, <0,52Ω(7A)	17c§	Mit	-	data
FS 14 UM-10	MOS-N-FET-e*	V-MOS, 500V, 14A, 150W, <0,64Ω(7A)	17c§	Mit	-	data
FS 14 VS-9	MOS-N-FET-e*	=FS 14 UM-9:	30c§	Mit	-	data
FS 14 VS-10	MOS-N-FET-e*	=FS 14 UM-10:	30c§	Mit	-	data
FS 16 KM-5	MOS-N-FET-e*	=FS 16 UM-5: Iso, 35W	17c	Mit	2SK2012, 2SK2255, 2SK2711, 2SK2739	data
FS 16 KM-6	MOS-N-FET-e*	=FS 16 UM-6: Iso, 35W	17c	Mit	2SK2739	data
FS 16 KM-9	MOS-N-FET-e*	=FS 16 UM-9: Iso, 40W	17c	Mit	-	data
FS 16 KM-10	MOS-N-FET-e*	=FS 16 UM-10: Iso, 40W	17c	Mit	-	data
FS 16 SM-5	MOS-N-FET-e*	=FS 16 UM-5:	18c§	Mit	MTW 16N40E	data
FS 16 SM-6	MOS-N-FET-e*	=FS 16 UM-6:	18c§	Mit	MTW 16N40E	data
FS 16 SM-9	MOS-N-FET-e*	=FS 16 UM-9:	18c§	Mit	2SK1573, 2SK1678, 2SK1745, 2SK2698	data
FS 16 SM-10	MOS-N-FET-e*	=FS 16 UM-10:	18c§	Mit	2SK1573, 2SK1678, 2SK1745, 2SK2698	data
FS 16 UM-5	MOS-N-FET-e*	V-MOS, 250V, 16A, 125W, <0,25Ω(8A)	17c§	Mit	-	data
FS 16 UM-6	MOS-N-FET-e*	V-MOS, 300V, 16A, 125W, <0,33Ω(8A)	17c§	Mit	-	data
FS 16 UM-9	MOS-N-FET-e*	V-MOS, 450V, 16A, 150W, <0,45Ω(8A)	17c§	Mit	-	data
FS 16 UM-10	MOS-N-FET-e*	V-MOS, 500V, 16A, 150W, <0,56Ω(8A)	17c§	Mit	-	data
FS 16 VS-5	MOS-N-FET-e*	=FS 16 UM-5:	30c§	Mit	MTB 16N25E	data
FS 16 VS-6	MOS-N-FET-e*	=FS 16 UM-6:	30c§	Mit	-	data
FS 16 VS-9	MOS-N-FET-e*	=FS 16 UM-9:	30c§	Mit	-	data
FS 16 VS-10	MOS-N-FET-e*	=FS 16 UM-10:	30c§	Mit	-	data
FS 18 SM-9	MOS-N-FET-e*	V-MOS, 450V, 18A, 250W, <0,33Ω(9A)	18c§	Mit	MTW 20N50E, 2SK1409, 2SK1411	data
FS 18 SM-10	MOS-N-FET-e*	V-MOS, 500V, 18A, 250W, <0,4Ω(9A)	18c§	Mit	MTW 20N50E, 2SK1411	data
FS 18 SM-14A	MOS-N-FET-e*	V-MOS, 700V, 18A, 275W, <0,55Ω(9A)	18c§	Mit	-	data
FS 20 KM-5	MOS-N-FET-e*	=FS 20 UM-5: Iso, 40W	17c	Mit	2SK1818	data
FS 20 KM-6	MOS-N-FET-e*	=FS 20 UM-6: Iso, 40W	17c	Mit	-	data
FS 20 SM-5	MOS-N-FET-e*	=FS 20 UM-5:	18c§	Mit	2SK1641, 2SK1673...74	data
FS 20 SM-6	MOS-N-FET-e*	=FS 20 UM-6:	18c§	Mit	MTW 24N40E, 2SK1674	data
FS 20 SM-12	MOS-N-FET-e*	V-MOS, 600V, 20A, 275W, <0,43Ω(10A)	18c§	Mit	-	data
FS 20 UM-5	MOS-N-FET-e*	V-MOS, 250V, 20A, 150W, <0,19Ω(10A)	17c§	Mit	-	data
FS 20 UM-6	MOS-N-FET-e*	V-MOS, 300V, 20A, 150W, <0,26Ω(10A)	17c§	Mit	-	data
FS 20 VS-5	MOS-N-FET-e*	=FS 20 UM-5:	30c§	Mit	-	data
FS 20 VS-6	MOS-N-FET-e*	=FS 20 UM-6:	30c§	Mit	-	data
FS 22 SM-9	MOS-N-FET-e*	V-MOS, 450V, 22A, 275W, <0,24Ω(11A)	18c§	Mit	MTW 20N50E	data
FS 22 SM-10	MOS-N-FET-e*	V-MOS, 500V, 22A, 275W, <0,29Ω(11A)	18c§	Mit	MTW 20N50E	data
FS23	HI-SPEED	Vs:±15V Vu:106dB Vo:±10V f:20Mc	6Y	IIc		data pinout
FS24	HI-SPEED	Vs:±15V Vu:86dB Vo:±10V Vi0:10mV	9Y	IIc		data pinout
FS 30 SM-5	MOS-N-FET-e*	V-MOS, 250V, 30A, 250W, <0,13Ω(15A)	18c§	Mit	2SK1675...76	data

Type	Device	Short Description	Fig.	Manu	Comparision Types	More at
FS 30 SM-6	MOS-N-FET-e*	V-MOS, 300V, 30A, 250W, <0,17Ω(15A)	18c§	Mit	2SK1676	data
FS 40 SM-5	MOS-N-FET-e*	V-MOS, 250V, 40A, 275W, <86mΩ(20A)	18c§	Mit	2SK1397	data
FS 40 SM-6	MOS-N-FET-e*	V-MOS, 300V, 40A, 275W, <146mΩ(20A)	18c§	Mit	-	data
FS125	HI-SPEED	Vs:±15V Vu:100dB Vo:±2.3V Vi0:5mV	7Y	Cml		data pinout
FSDM 311	DMOS-IC	Green Mode Fairchild Power Switch	DIC	Fch	-	pdf pinout
FSDM 311 L	DMOS-IC	Green Mode Fairchild Power Switch	DIC	Fch	-	pdf pinout
FS 6 X 0420 RJ	IC	Fairchild Power Switch(fps)	85	Fch	-	pdf pinout
FS 6 X 0720 RJ	IC	Fairchild Power Switch(fps)	85	Fch	-	pdf pinout
FS 6 X 1220 RD	IC	Fairchild Power Switch(fps)	85	Fch	-	pdf pinout
FS 6 X 01220 RJ	IC	Fairchild Power Switch(fps)	85	Fch	-	pdf pinout
FS 6 X 1220 RTYDTU	IC	Fairchild Power Switch(fps)	85	Fch	-	pdf pinout
FS 6 S 1565 RB	IC	Fairchild Power Switch(fps)	85	Fch	-	pdf pinout
FS 7 M 0680 TU	IC	Fairchild Power Switch(fps)	85	Fch	-	pdf pinout
FS 7 M 0680 YDTU	IC	Fairchild Power Switch(fps)	85	Fch	-	pdf pinout
FS 7 M 0880 TU	IC	Fairchild Power Switch(fps)	85	Fch	-	pdf pinout
FS 7 M 0880 YDTU	IC	Fairchild Power Switch(fps)	85	Fch	-	pdf pinout
FS 8 S 0765 RCB	IC	Fairchild Power Switch(fps)	85	Fch	-	pdf pinout
FS 8 S 0965 RCB	IC	Fairchild Power Switch(fps)	85	Fch	-	pdf pinout
FS7M 0880 (YD)TU	LIN-IC	Power Switch, PWM Controller & Sensefet	86	Fch	-	pdf pinout
FSA66	SPST	Vs:1.65...5.5V Ron:3Ohm at 5/0V	6M,5S	Fcs		data pdf pinout
FSA201	DPDT	Vs:3.6/5.5V Ron:Ohm at 5/2.7/0V	10HS	Fcs		data pdf pinout
FSA 201 L	DIG-IC	USB2.0 FS & Audio Switches	MLP	Fcs		pdf pinout
FSA 201 M	DIG-IC	USB2.0 FS & Audio Switches	TSOP	Fcs		pdf pinout
FSA 223 L/U	DIG-IC	USB2.0 HS(480Mbps) and Audio Switches	MLP	Fcs		pdf pinout
FSA 223 M	DIG-IC	USB2.0 HS(480Mbps) and Audio Switches	TSOP	Fcs		pdf pinout
FSA266	SPST	Vs:1.65...5.5V Ron:Ohm Ton:2ns at 5/0V	8SM	Fcs		data pdf pinout
FS A 66 M 5 X	IC	UHS Single SPST Normally Open Analog Switch	85	Fch		pdf pinout
FSA 66 P 5 X	IC	Uhs Single SPST Normally Open Analog Switch	85	Fch	-	pdf pinout
FSA 66 L 6 X	IC	Uhs Single SPST Normally Open Analog Switch	85	Fch	-	pdf pinout
FSA859	SPDT	Vs:1.65...5.5V Ron:Ohm at 5.5/0V	8G	Fcs		data pdf pinout
FSA1156	SPST	Vs:1.65...5.5V Ron:Ohm at 5/0V	6MS	Fcs		data pdf pinout
FSA 1156 L	CMOS-IC	Low-RON SPST Analog SW, Vcc: 1.65-5.5V	MLP	Fcs		pdf pinout
FSA 1156 P	CMOS-IC	Low-RON SPST Analog SW, Vcc: 1.65-5.5V	46	Fcs		pdf pinout
FSA1157	SPST	Vs:1.65...5.5V Ron:Ohm at 5/0V	6MS	Fcs		data pdf pinout
FSA 1157 L	CMOS-IC	Low-RON SPST Analog SW, Vcc: 1.65-5.5V	MLP	Fcs		pdf pinout
FSA 1157 P	CMOS-IC	Low-RON SPST Analog SW, Vcc: 1.65-5.5V	46	Fcs		pdf pinout
FSA1256	SPST	Vs:1.65...5.5V Ron:Ohm at 5/0V	8M	Fcs		data pdf pinout
FSA 1256 (A)L	CMOS-IC	Low-RON Low-voltage, Dual SPDT Analog Switch, Low-ICCT "aö Option	MLP	Fcs		pdf pinout
FSA1256A	SPST	Vs:2.7...5.5V Ron:Ohm Ton:10ns at 5/0V	8M	Fcs		data pdf pinout
FSA1257	SPST	Vs:1.65...5.5V Ron:Ohm at 5/0V	8M	Fcs		data pdf pinout
FSA 1257 (A)L	CMOS-IC	Low-RON Low-voltage, Dual SPDT Analog Switch, Low-icct "aö Option	MLP	Fcs		pdf pinout
FSA1257A	SPST	Vs:2.7...5.5V Ron:Ohm Ton:10ns at 5/0V	8M	Fcs		data pdf pinout
FSA1258	SPST	Vs:1.65...5.5V Ron:Ohm at 5/0V	8M	Fcs		data pdf pinout
FSA 1258 (A)L	CMOS-IC	Low-RON Low-voltage, Dual SPDT Analog Switch, Low-icct "aö Option	MLP	Fcs		pdf pinout
FSA1258A	SPST	Vs:2.7...5.5V Ron:Ohm Ton:10ns at 5/0V	8M	Fcs		data pdf pinout
FSA1259	SPST	Vs:1.65...5.5V Ron:Ohm at 5.5/0V	8S	Fcs		data pdf pinout
FSA1259A	SPST	Vs:1.65...5.5V Ron:Ohm at 5.5/0V	8S	Fcs		data pdf pinout
FSA2156	SPST	Vs:1.65...4.6V Ron:Ohm at 4/0V	6MS	Fcs		data pdf pinout
FSA 2156 L	CMOS-IC	Low-Voltage SPST 0.4 Analog Switch	MLP	Fcs	-	pdf pinout
FSA 2156 M	CMOS-IC	Low-Voltage SPST 0.4 Analog Switch	46	Fcs	-	pdf pinout
FSA2257	SPDT	Vs:1.65...5.5V Ron:Ohm at 5/0V	10H,14SS	Fcs		data pdf pinout
FSA 2257 L	CMOS-IC	Low RON Low- Voltage Dual SPDT Bi-Directional Analog Switch	MLP	Fcs		pdf pinout
FSA 2257 M	CMOS-IC	Low Ron Low- Voltage Dual SPDT Bi-Directional Analog Switch	TSOP	Fcs	-	pdf pinout
FSA 2257 MT	CMOS-IC	Low Ron Low- Voltage Dual SPDT Bi-Directional Analog Switch	TSOP	Fcs	-	pdf pinout
FSA2267	SPDT	Vs:2.3...4.3V Ron:Ohm at 3.3/0V	10HS	Fcs		data pdf pinout
FSA2267A	SPDT	Vs:2.3...4.3V Ron:Ohm Ton:37ns at 4/0V	10HS	Fcs		data pdf pinout
FSA 2267 (A)L	CMOS-IC	0.35Ω Low-Voltage Dual-SPDT Analog Switch	MLP	Fcs	-	pdf pinout
FSA 2267 (A)M	CMOS-IC	0.35Ω Low-Voltage Dual-SPDT Analog Switch	TSOP	Fcs	-	pdf pinout
FSA 2268 (T)	CMOS-IC	Low-Voltage Dual-SPDT (0.4Ω) Analog Switch, 16kV ESD	MLP	Fcs	-	pdf
FSA 2269	CMOS-IC	Low-Voltage Dual-SPDT (0.4Ω) Analog Switch	MLP	Fcs	-	pdf pinout
FSA 2270 (T)	CMOS-IC	Low-Voltage Dual-SPDT (0.4Ω) Analog Switch	MLP	Fcs	-	pdf pinout
FSA 2271 (T)	CMOS-IC	Low-Voltage Dual-SPDT (0.4Ω) Analog Switch	MLP	Fcs	-	pdf pinout
FSA 2367 B	CMOS-IC	Low Ron (0.75Ω) Triple- SPDT, Negative-Swing Audio Source Switch	QFN	Fcs	-	pdf pinout
FSA 2367 M	CMOS-IC	Low Ron (0.75Ω) Triple- SPDT, Negative-Swing Audio Source Switch	TSOP	Fcs	-	pdf pinout
FSA 2380 M	CMOS-IC	Low Ron (0.75Ω) 3:1 Negative Swing Audio Source Switch	TSOP	Fcs	-	pdf
FSA 2380 Q	CMOS-IC	Low Ron (0.75Ω) 3:1 Negative Swing Audio Source Switch	QFN	Fcs	-	pdf

Type	Device	Short Description	Fig.	Manu	Comparision Types	More at
FSA2467	DPDT	Vs:1.65...4.3V Ron:Ohm at 4/0V	16H	Fcs		data pdf pino
FSA 2467	CMOS-IC	0.4Ω Low-Voltage Dual DPDT Analog Switch	MLP	Fcs	-	pdf pinout
FSA 266 K 8 X	IC	Dual SPST Norm. Open Analog-/2-Bit Bus-switch	46	Fch		pdf pinout
FSA 266 L 8 X	IC	Dual SPST Norm. Open Analog-/2-Bit Bus-switch	46	Fch		pdf pinout
FSA2859	DPDT	Vs:1.65...5.5V Ron:Ohm at 5.5/0V	12G	Fcs		data pdf pino
FSA3157	SPDT	Vs:1.65...5.5V Ron:3Ohm at 5/0V	6MS	Fcs		data pdf pino
FSA3259	SP3T	Vs:1.65...5.5V Ron:5Ohm at 5/0V	16H	Fcs		data pdf pino
FSA 3259	IC	Dual SP3T Analog Switch	85	Fch	-	pdf pino
FSA3357	SP3T	Vs:1.65...5.5V Ron:5Ohm at 5/0V	8SM	Fcs		pdf pinout
FSA4157	SPDT	Vs:1.65...5.5V Ron:Ohm at 5/0V	6MS	Fcs		data pdf pino
FSA 4157	IC	Low-voltage, 1Ω SPDT Analog Switch	85	Fch	-	data pdf pino
FSA4157A	SPDT	Vs:1.65...5.5V Ron:Ohm at 3.3/0V	6MS	Fcs		data pdf pinou
FSA 4157 AL	IC	Low-voltage, 1Ω SPDT Analog Switch	85	Fch	-	pdf pinout
FSA 4157 L	IC	Low-voltage, 1Ω SPDT Analog Switch	85	Fch		pdf pinout
FSA4159	SPDT	Vs:1.65...5.5V Ron:Ohm at 5/0V	6MS	Fcs		pdf pinout
FSA5157	SPDT	Vs:1.65...4.3V Ron:Ohm at 4/0V	6MS	Fcs		data pdf pinou
FSA 3157 L 6 X	IC	SPDT Analog Switch, 2:1Multi- / Demultiplexer Bus Switch	85	Fch		pdf pinout
FSA 3157 P 6 X	IC	SPDT Analog Switch, 2:1Multi- / Demultiplexer Bus Switch	85	Fch	-	pdf pinout
FS A 3357 K 8 X	IC	SP3T Analog Switch(3:1 Multi- / Demultiplexer)	85	Fch	-	pdf pinout
FS A 3357 L 8 X	IC	SP3T Analog Switch(3:1 Multi- / Demultiplexer)	85	Fch	-	pdf pinout
FSA 4159 L 6 X	IC	1Ω SPDT Analog Switch, Power- Off Isolation	85	Fch	-	pdf pinout
FSA 4159 P 6 X	IC	1Ω SPDT Analog Switch, Power- Off Isolation	85	Fch	-	pdf pinout
FSA 5157 L 6 X	IC	0.4Ω Low-Voltage SPDT Analog Switch	85	Fch		pdf pinout
FSA 5157 P 6 X	IC	0.4Ω Low-Voltage SPDT Analog Switch	85	Fch		pdf pinout
FSAL200	MULT2	Vs:3...5.5V Ron:Ohm Ton:10ns at 5/0V	16S	Fcs		data pdf pinou
FSAM 10 SH 60	LIN-IC	SPM (smart Power Module),600V-10A 3Ph.	DIC	Fch		pdf pinout
FSAM 10 SM 60 A	LIN-IC	SPM (smart Power Module),600V-10A 3Ph.	DIC	Fch		pdf pinout
FSAM 15 SH 60	LIN-IC	SPM, Smart Power Module, 600V-15A 3-Ph	DIP	Fch		pdf pinout
FSAM 15 SL 60	LIN-IC	SPM, Smart Power Module, 600V-15A 3-Ph	DIP	Fch		pdf pinout
FSAM 20 SL 60	LIN-IC	SPM, Smart Power Module, 600V-20A 3Ph.	DIC	Fch		pdf pinout
FSAM 15 SH 60 A	IC	SPM (smart Power Module)	85	Fch		pdf pinout
FSAM 20 SM 60 A	IC	SPM (smart Power Module)	85	Fch		pdf pinout
FSAT66	SPST	Vs:1.65...5.5V Ron:3Ohm at 5/0V	6M,5S	Fcs		data pdf pinout
FSAT 66	IC	Single SPST Normally Open Analog Switch	85	Fch		pdf pinout
FSAT 66 L	IC	Single SPST Normally Open Analog Switch	85	Fch	-	pdf pinout
FSAU3157	SPDT	Vs:1.65...5.5V Ron:3Ohm at 5/0V	6S	Fcs		data pdf pinout
FSAU 3157	IC	SPDT Analog Switch,-2V Undershoot Prote.	46	Fch		pdf pinout
FSAV330	SPDT	Vs:4...5.5V Ron:Ohm Ton:5.2ns at 5/0V	16S	Fcs		data pdf pinout
FSAV 330	IC	4 Channel 2:1 Video Switch	46	Fch	-	pdf pinout
FSAV331	MULT4	Vs:4.25...5.25V Ron:Ohm at 5/0V	16S	Fcs		data pdf pinout
FSAV 331	IC	Dual Channel 4:1 Video Switch	46	Fch	-	pdf pinout
FSAV 332	SPST	Vs:4.25...5.25V Ron:Ohm at 5/0V	14S,16S	Fcs	-	data pdf pinout
FSAV 332	IC	Quad Video Switch, Individual Enables	46	Fch	-	pdf pinout
FSAV 332 MTC	IC	Quad Video Switch, Individual Enables	46	Fch	-	pdf pinout
FSAV430	MULT2	Vs:3...3.6V Ron:Ohm at 3.3/0V	16HS	Fcs		data pdf pinout
FSAV 430 BQX	IC	1.1GHz 4 Channel 2:1 Video Switch	46	Fch	-	pdf pinout
FSAV 430 MTC	IC	1.1GHz 4 Channel 2:1 Video Switch	46	Fch	-	pdf pinout
FSAV 430 QSC	IC	1.1GHz 4 Channel 2:1 Video Switch	46	Fch	-	pdf pinout
FSAV433	MULT3	Vs:2.3...3.6V Ron:Ohm at 3/0V	20HS	Fcs		data pdf pinout
FSAV 433 BQX	IC	(550MHz) Three-CH 3:1 Video Switch	46	Fch	-	pdf pinout
FSAV 433 MTC	IC	(550MHz) Three-CH 3:1 Video Switch	46	Fch	-	pdf pinout
FSAV 433 MTCX	IC	(550MHz) Three-CH 3:1 Video Switch	46	Fch	-	pdf pinout
FSAV450	MULT2	Vs:4.5...5.5V Ron:Ohm at 5/0V	16HS	Fcs		data pdf pinout
FSAV 450 BQX	IC	800MHz 4 Channel 2:1 Video Switch	46	Fch	-	data pdf pinout
FSAV 450 MTC	IC	800MHz 4 Channel 2:1 Video Switch	46	Fch	-	pdf pinout
FSAV 450 QSC	IC	800MHz 4 Channel 2:1 Video Switch	46	Fch	-	pdf pinout
FSB 32560	IC	Smart Power Module	46	Fch		pdf pinout
FSB 50250	CMOS-IC	Smart Power Module (SPM®), 500V, 2.0A	MDIP	Fcs	-	pdf pinout
FSB 50325 T	CMOS-IC	Smart Power Module (SPM®), 250V, 3.0A	MDIP	Fcs	-	pdf pinout
FSB 50450 S	CMOS-IC	Smart Power Module (SPM®), 500V, 3.0A	MDIP	Fcs	-	pdf pinout
FSB 50450 T	CMOS-IC	Smart Power Module (SPM®), 500V, 3.0A	MDIP	Fcs	-	pdf pinout
FSB 50550 T	CMOS-IC	Smart Power Module (SPM®), 500V, 3.5A	MDIP	Fcs	-	pdf pinout
FSB 50825 TB	LIN-IC	Smart Power Module, 250V 8.0A 3-Ph.	DIP	Fch		pdf pinout
FSBB 15 CH 60	LIN-IC	Smart Power Module, 600V, 15A, 3-Ph.	DIC	Fch		pdf pinout
FSBB 15 CH60B(T)	CMOS-IC	Smart Power Module, 600V, 15A	MDIP	Fcs	-	pdf pinout
FSBB 20 CH60(BT/B)	CMOS-IC	Smart Power Module, 600V, 20A	MDIP	Fcs	-	pdf pinout
FSBB 30 CH60(B)	CMOS-IC	Smart Power Module, 600V, 30A	MDIP	Fcs	-	pdf pinout
FSBB 30 CH60C	LIN-IC	Smart Power Module, 600V, 30A, 3-Ph.	DIC	Fch		pdf pinout
FSBF 3 CH 60 B	IC	Smart Power Module	DIP	Fch		pdf pinout
FSBF 5 CH 60 B	IC	Smart Power Module	DIP	Fch		pdf pinout
FSBF 10 CH 60 B	IC	Smart Power Module	DIP	Fch		pdf pinout
FSBF 10 CH 60 BT	IC	Smart Power Module	DIP	Fch	-	pdf pinout
FSBF 15 CH 60 BT	IC	Smart Power Module	DIP	Fch		pdf pinout
FSBS 10 CH 60	IC	Smart Power Module	DIP	Fch		pdf pinout
FSBS 10 CH 60 F	IC	Smart Power Module	DIP	Fch		pdf pinout
FSBS 15 CH 60	IC	Smart Power Module	DIP	Fch		pdf pinout
FSBS 15 CH 60 F	IC	Smart Power Module	DIP	Fch		pdf pinout

Type	Device	Short Description	Fig.	Manu	Comparision Types	More at
FSCM 0465 RG	IC	Green Mode Fairchild Power SW	86	Fch	-	pdf pinout
FSCM 0465 RI	IC	Green Mode Fairchild Power SW	87	Fch	-	pdf pinout
FSCM 0465 RJ	IC	Green Mode Fairchild Power SW	87	Fch	-	pdf pinout
FSCM 0565 RG	SMPS-IC	Green Mode Fairchild Power SW	86	Fch	-	pdf pinout
FSCM 0565 RI	SMPS-IC	Green Mode Fairchild Power SW	87	Fch	-	pdf pinout
FSCM 0565 RJ	SMPS-IC	Green Mode Fairchild Power SW	87	Fch	-	pdf pinout
FSCM 0765 RG	SMPS-IC	Green Mode Fairchild Power Switch	86	Fch	-	pdf pinout
FSCM 0765 RI	SMPS-IC	Green Mode Fairchild Power Switch	87	Fch	-	pdf pinout
FSCM 0765 RJ	SMPS-IC	Green Mode Fairchild Power Switch	87	Fch	-	pdf pinout
FSCQ 0565 RT	IC	Green Mode Fairchild Power Switch	DIP	Fch	-	pdf pinout
FSCQ 0765 RT	IC	Green Mode Fairchild Power Switch	DIP	Fch	-	pdf pinout
FSCQ 0965 RT	IC	Green Mode Fairchild Power Switch	DIP	Fch	-	pdf pinout
FSCQ 1265 RT	IC	Green Mode Fairchild Power Switch	DIP	Fch	-	pdf pinout
FSCQ 1465 RT	IC	Green Mode Fairchild Power Switch	DIP	Fch	-	pdf pinout
FSCQ 1565 RP	IC	Green Mode Fairchild Power Switch	DIP	Fch	-	pdf pinout
FSCQ 1565 RT	IC	Green Mode Fairchild Power Switch	DIP	Fch	-	pdf pinout
FSDH 0165 D	IC	Power Switch	DIP	Fch	-	pdf pinout
FSDH 0265 RL	SMPS-IC	Fairchild Power Switch, PWM, 100kHz	MDIP	Fch	-	pdf pinout
FSDH 0265 RN	SMPS-IC	Fairchild Power Switch, PWM, 100kHz	DIP	Fch	-	pdf pinout
FSDH 321	SMPS-IC	Green Mode Fairchild Power Switch, PWM, 100kHz	DIP	Fch	-	pdf pinout
FSDH 321 L	SMPS-IC	Green Mode Fairchild Power Switch, PWM, 100kHz	MDIP	Fch	-	pdf pinout
FSDL 0165 RL	SMPS-IC	Fairchild Power Switch, PWM, 50kHz	MDIP	Fch	-	pdf pinout
FSDL 0165 RN	SMPS-IC	Fairchild Power Switch, PWM, 50kHz	DIP	Fch	-	pdf pinout
FSDL 321	SMPS-IC	Green Mode Fairchild Power Switch, PWM, 50kHz	DIP	Fch	-	pdf pinout
FSDL 321 L	IC	Green Mode Fairchild Power Switch, PWM, 100kHz	MDIP	Fch	-	pdf pinout
FSDL 0365 RL	SMPS-IC	Fairchild Power Switch, PWM, 50kHz	MDIP	Fch	-	pdf pinout
FSDL 0365 RN	SMPS-IC	Fairchild Power Switch, PWM, 50kHz	DIC	Fch	-	pdf pinout
FSDL 0365 RNB	SMPS-IC	Fairchild Power Switch, PWM, 50kHz	DIP	Fch	-	pdf pinout
FSDM 101	SMPS-IC	Green Mode Fairchild Power Switch	DIP	Fch	-	pdf pinout
FSDM 0165 RN	IC	Green Mode Fairchild Power Switch	DIP	Fch	-	pdf pinout
FSDM 0265 RL	IC	Fairchild Power Switch, PWM, 67KHz	MDIP	Fch	-	pdf pinout
FSDM 0265 RN	IC	Fairchild Power Switch, PWM, 67KHz	DIP	Fch	-	pdf pinout
FSDM 0265 RNB	SMPS-IC	Green Mode Fairchild Power Switch (fps)	DIP	Fch	-	pdf pinout
FSDM 311RN	IC	Green Mode Fairchild Power Switch	DIP	Fch	-	pdf pinout
FSDM 321RN	IC	Green Mode Fairchild Power Switch	DIP	Fch	-	pdf pinout
FSDM 0365 RL	IC	Fairchild Power Switch, PWM 67kHz	MDIP	Fch	-	pdf pinout
FSDM 0365 RN	IC	Fairchild Power Switch, PWM 67kHz	DIP	Fch	-	pdf pinout
FSDM 0365 RNB	SMPS-IC	Fairchild Power Switch, PWM, 67kHz	DIP	Fch	-	pdf pinout
FSDM 0465 RB	SMPS-IC	Green Mode Fairchild Power Switch (fps)	86	Fch	-	pdf pinout
FSDM 0465 RE	IC	Green Mode Fairchild Power Switch	DIP	Fch	-	pdf pinout
FSDM 0565 R	IC	Green Mode Fairchild Power Switch	DIP	Fch	-	pdf pinout
FSDM 0565 RB	IC	Green Mode Fairchild Power Switch	DIP	Fch	-	pdf pinout
FSDM 0565 RBI	IC	Green Mode Fairchild Power Switch	DIP	Fch	-	pdf pinout
FSDM 0565 RE	IC	Green Mode Fairchild Power Switch	DIP	Fch	-	pdf pinout
FSDM 0765 RE	IC	Green Mode Fairchild Power Switch	DIP	Fch	-	pdf pinout
FSDM 1265 RB	IC	Green Mode Fairchild Power Switch	DIP	Fch	-	pdf pinout
FSDM 07652 R	IC	Green Mode Fairchild Power Switch	DIP	Fch	-	pdf pinout
FSDM 07652 RB	IC	Green Mode Fairchild Power Switch	DIP	Fch	-	pdf pinout
FSDM 12652 RB	SMPS-IC	Green Mode Fairchild Power Switch (fps)	86	Fch	-	pdf pinout
FSES 0765 RG	SMPS-IC	Green Mode Fairchild Power Switch	86	Fch	-	pdf pinout
FSFR 1800	IC	Fairchild Power Switch (fps)	QILP	Fch	-	pdf pinout
FSFR 1900	IC	Fairchild Power Switch (fps)	QILP	Fch	-	pdf pinout
FSFR 2000	IC	Fairchild Power Switch (fps)	QILP	Fch	-	pdf pinout
FSFR 2100	IC	Fairchild Power Switch (fps)	QILP	Fch	-	pdf pinout
FSHDMI 04 BQX	IC	Wide-bandwidth Diff. Signaling Hdmi Switch	QILP	Fch	-	pdf pinout
FSHDMI 04 MTDX	IC	Wide-bandwidth Diff. Signaling Hdmi Switch	QILP	Fch	-	pdf pinout
FSHDMI 04 QSPX	IC	Wide-bandwidth Diff. Signaling Hdmi Switch	QILP	Fch	-	pdf pinout
FSHDMI 08 BQX	IC	Wide-bandwidth, Hdmi Switch,Ddc+cec Multiplexer	QILP	Fch	-	pdf pinout
FSHDMI 08 MTDX	IC	Wide-bandwidth, Hdmi Switch,DDC+CEC Multiplexer	QILP	Fch	-	pdf pinout
FSM 3 B2	Triac	200V, 3A(Tc=67°C), Igt/Ih=40/20mA	13h§	Hit	TAG 136..., ZO 410...	data
FSM 3 B4	Triac	=FSM 3B2: 400V	13h§		TAG 136..., ZO 410...	data
FSM 6 B2	Triac	200V, 6A(Tc=103°C), Igt/Ih=30/20mA	17b	Hit	T 2856..., TAG 426-..., TAG 452-...	data
FSM 6 B4	Triac	=FSM 6B2: 400V	17b		T 2856..., TAG 426-..., TAG 452-...	data
FSM 10 B4	Triac	400V, 10A(Tc=100°C), Igt/Ih=30/20mA	17b	Hit	TAG 457-..., (BT 162/..., TAG 257-...)[3]	data
FSM 16 C2	Triac	200V, 16A(Tc=76°C), Igt/Ih=50/30mA	65h§	Hit	MAC 515-..., MAC 525-..., MAC 515A-...	data
FSM 16 C4	Triac	=FSM 16C2: 400V	65h§		MAC 515-..., MAC 525-..., MAC 515A-...	data
FSM 16 C6	Triac	=FSM 16C2: 600V	65h§		MAC 515-..., MAC 525-..., MAC 515A-...	data
FSM 20 C2	Triac	=FSM 16C2: 20A(Tc=74°C)	65h§	Hit	MAC 525-..., MAC 525A-...	data
FSM 20 C4	Triac	=FSM 16C4: 20A(Tc=74°C)	65h§		MAC 525-..., MAC 525A-...	data
FSM 30 C2	Triac	=FSM 16C2: 30A(Tc=63°C)	65h§		(MAC 50-..., MAC 50A-..., TAG 740-...)[4]	data
FSM 30 C4	Triac	=FSM 16C4: 30A(Tc=63°C)	65h§		(MAC 50-..., MAC 50A-..., TAG 740-...)[4]	data
FSP 100 DC	CMOS-IC	Progr. Digital Filter	40-DIP	Fch		
FSP 100 LC	CMOS-IC	=FSP 100DC:	44-LCC	Fch	-	
FSQ 100	IC	Green Mode Fairchild Power Switch	QILP	Fch	-	pdf pinout
FSQ 0165 RL	IC	Green Mode Fairchild Power Switch	DIP	Fch	-	pdf pinout
FSQ 0165 RN	IC	Green Mode Fairchild Power Switch	DIP	Fch	-	pdf pinout

Type	Device	Short Description	Fig.	Manu	Comparision Types	More at
FSQ 0170 RNA	IC	Green Mode Fairchild Power Switch	DIP	Fch	-	pdf pinout
FSQ 211	IC	Green Mode Fairchild Power Switch	DIP	Fch	-	pdf pinout
FSQ 211 L	IC	Green Mode Fairchild Power Switch	DIP	Fch	-	pdf pinout
FSQ 0265 RL	IC	Green Mode Fairchild Power Switch	DIP	Fch	-	pdf pinout
FSQ 0265 RN	IC	Green Mode Fairchild Power Switch	DIP	Fch	-	pdf pinout
FSQ 0270 RNA	IC	Green Mode Fairchild Power Switch	DIP	Fch	-	pdf pinout
FSQ 311	IC	Green Mode Fairchild Power Switch	DIP	Fch	-	pdf pinout
FSQ 311 L	IC	Green Mode Fairchild Power Switch	DIP	Fch	-	pdf pinout
FSQ 321	IC	Green Mode Fairchild Power Switch	DIP	Fch	-	pdf pinout
FSQ 321 L	IC	Green Mode Fairchild Power Switch	DIP	Fch	-	pdf pinout
FSQ 0365 RL	IC	Green Mode Fairchild Power Switch	DIP	Fch	-	pdf pinout
FSQ 0365 RN	IC	Green Mode Fairchild Power Switch	DIP	Fch	-	pdf pinout
FSQ 0370 RNA	IC	Green Mode Fairchild Power Switch	DIP	Fch	-	pdf pinout
FSQ 510	IC	Green Mode Fairchild Power Switch	DIP	Fch	,	pdf pinout
FSQ 510 H	IC	Green Mode Fairchild Power Switch	DIP	Fch	-	pdf pinout
FSQ 0565 R	IC	Green Mode Fairchild Power Switch	DIP	Fch	-	pdf pinout
FSQ 0765 R	IC	Green Mode Fairchild Power Switch	DIP	Fch	-	pdf pinout
FSS 101	MOS-P-FET-e*	SMD, V-MOS, LogL, 20V, 5A, <58mΩ(5A)	8-MDIP	Say	-	data
FSS 102	MOS-P-FET-e*	SMD, V-MOS, LogL, 30V, 5A, <54mΩ(5A)	8-MDIP	Say	-	data
FSS 104	MOS-P-FET-e*	SMD, V-MOS, LogL, 30V, 6A, <38mΩ(6A)	8-MDIP	Say	-	data
FSS 106	MOS-P-FET-e*	SMD, V-MOS, LogL, 30V, 7A, <30mΩ(7A)	8-MDIP	Say	-	data
FSS 107	MOS-P-FET-e*	SMD, V-MOS, LogL, 20V, 8A, <23mΩ(8A)	8-MDIP	Say	-	data
FSS 108	MOS-P-FET-e*	SMD, V-MOS, LogL, 30V, 8A, <22mΩ(8A)	8-MDIP	Say	-	data
FSS 110	MOS-P-FET-e*	SMD, V-MOS, LogL, 60V, 2A, <0,65Ω(1A)	8-MDIP	Say	-	data
FSS 201	MOS-N-FET-e*	SMD, V-MOS, LogL, 20V, 7A, <46mΩ(2A)	8-MDIP/ca1	Say	-	data
FSS 202	MOS-N-FET-e*	SMD, V-MOS, LogL, 30V, 7A, <50mΩ(4A)	8-MDIP/ca1	Say	-	data
FSS 203	MOS-N-FET-e*	SMD, V-MOS, LogL, 20V, 8A, <37mΩ(2A)	8-MDIP	Say	-	data
FSS 204	MOS-N-FET-e*	SMD, V-MOS, LogL, 30V, 8A, <37mΩ(4A)	8-MDIP/ca1	Say	-	data
FSS 206	MOS-N-FET-e*	SMD, V-MOS, LogL, 30V, 9A, <26mΩ(4A)	8-MDIP/ca1	Say	-	data
FSS 207	MOS-N-FET-e*	SMD, V-MOS, LogL, 20V, 10A, <13mΩ(10A)	8-MDIP	Say	-	data
FSS 208	MOS-N-FET-e*	SMD, V-MOS, LogL, 30V, 10A, <13mΩ(10A)	8-MDIP	Say	-	data
FSS 210	MOS-N-FET-e*	SMD, V-MOS, LogL, 40V, 9A, <40mΩ(4A)	8-MDIP/ca1	Say	-	data
FSS 212	MOS-N-FET-e*	SMD, V-MOS, LogL, 30V, 8A, <58mΩ(4A)	8-MDIP/ca1	Say	-	data
FSS 214	MOS-N-FET-e*	SMD, V-MOS, LogL, 30V, 9A, <20mΩ(9A)	8-MDIP	Say	-	data
FSS 216	MOS-N-FET-e*	SMD, V-MOS, LogL, 30V, 10A, <28mΩ(4A)	8-MDIP/ca1	Say	-	data
FSS 232	MOS-N-FET-e*	SMD, V-MOS, LogL, 30V, 9A, <33mΩ(4A)	8-MDIP/ca1	Say	-	data
FSS 237	MOS-N-FET-e*	SMD, V-MOS, LogL, 20V, 14A, <12mΩ(7A)	8-MDIP/ca1	Say	-	data
FSS 238	MOS-N-FET-e*	SMD, V-MOS, LogL, 30V, 14A, <14mΩ(7A)	8-MDIP/ca1	Say	-	data
FSS 239	MOS-N-FET-e*	SMD, V-MOS, LogL, 20V, 7A, <42mΩ(2A)	8-MDIP/ca1	Say	-	data
FSS 242	MOS-N-FET-e*	SMD, V-MOS, LogL, 30V, 8A, <40mΩ(4A)	8-MDIP/ca1	Say	-	data
FSS 244	MOS-N-FET-e*	SMD, V-MOS, LogL, 30V, 10A, <28mΩ(4A)	8-MDIP	Say	-	data
FSS 250	MOS-N-FET-e*	SMD, V-MOS, LogL, 30V, 7A, <54mΩ(4A)	8-MDIP/ca1	Say	-	data
FS 6 S 0765 RCB-TU	IC	Fairchild Power Switch(fps)	85	Fch	-	pdf pinout
FS 6 S0765RCB-YDTU	IC	Fairchild Power Switch(fps)	85	Fch	-	pdf pinout
FST 1045	Si-Di	Schottky-Gl, Dual, 45V, 10A	17p§	Sie	BYS 24-45, HRW 36, MBR 2060CT	data
FST 1090	Si-Di	Schottky-Gl, Dual, 90V, 10A	17p§	Sie	BYS 24-90, HRW 36, MBR 2090CT	data
FST 2045	Si-Di	Schottky-Gl, Dual, 45V, 20A	17p§	Sie	BYV 43-45, BYV 143-45, MBR 2045CT	data
FST 2090	Si-Di	Schottky-Gl, Dual, 90V, 20A	17p§	Sie		data
FST 3045	Si-Di	Schottky-Gl, Dual, 45V, 30A	16p§	Sie	BYS 28-45, BYV 73/45, MBR 3045PT	data
FST 3060	Si-Di	Schottky-Gl, Dual, 60V, 30A	16p§	Sie	BYS 28-90, MBR 3060PT	data
FST 3090	Si-Di	Schottky-Gl, Dual, 90V, 30A	16p§	Sie	BYS 28-90	data
FST 5050	Si-Di	Schottky-Gl, Dual, 50V, 50A	16p§	Sie	MBR 6045PT	data
FST 5090	Si-Di	Schottky-Gl, Dual, 90V, 50A	16p§	Sie		data
FST 16232 MEA	DIG-IC	Synchronous 16- Bit to 32- Bit Multiplexer/ Demultiplexer Bus Switch	TSOP	Fch	-	pdf pinout
FST 16232 MTD	DIG-IC	Synchronous 16- Bit to 32- Bit Multiplexer/ Demultiplexer Bus Switch	TSOP	Fch	-	pdf pinout
FST 16245 MTD	DIG-IC	16-Bit Bus Switch	TSOP	Fcs		pdf pinout
FST 16292 MEA	DIG-IC	12-Bit-24-Bit Multi/Demultiplexer Bus Switch	TSOP	Fch	-	pdf pinout
FST 16292 MTD	DIG-IC	12-Bit-24-Bit Multi/Demultiplexer Bus Switch	TSOP	Fch	-	pdf pinout
FST 32211 G	DIG-IC	40/48-Bit Bus Switch	GRID	Fch		pdf pinout
FST 162861 MTD	DIG-IC	20-Bit Bus Switch,25Ohm Series Resistors in Outputs	TSOP	Fch	-	pdf pinout
FST 162861 MTDX_NL	DIG-IC	20-Bit Bus Switch,25Ohm Series Resistors in Outputs	TSOP	Fch		pdf pinout
FT						
FT	Si-P	=2SA1034-T (Typ-Code/Stempel/marking)	35	Mat	→2SA1034	data
FT	Si-P	=2SA1531-T (Typ-Code/Stempel/marking)	35(2mm)	Mat	→2SA1531	data
FT	Si-N	=2SC4146 (Typ-Code/Stempel/marking)	35	Say	→2SC4146	data
FT	Si-N	=2SD2473 (Typ-Code/Stempel/marking)	≈35	Mat	→2SD2473	data
FT	Si-N	=2SD2483 (Typ-Code/Stempel/marking)	35(2mm)	Mat	→2SD2483	data
FT	Z-Di	=Z1 SMA-150 (Typ-Code/Stempel/marking)	71(5x2,5)	Fag	→Z1 SMA-150	data
FT 1 M,N,P	Si-Di	Gl, S, 140...600V, 0,3A, <3µs M=600V, N=300V, P=140V	31a	Fjd	BA 157...159, BY 204/..., BY 208/..., ++	data
FT 06 B	Triac	100V, 6A(Tc=88°C), Igt <75mA	22a§	Hit	TAG 260-..., TAG 265-..., T 4700...	data
FT 06 C	Triac	=FT 06B: 200V	22a§		TAG 260-..., TAG 265-..., T 4700...	data
FT 06 D	Triac	=FT 06B: 300V	22a§		TAG 260-..., TAG 265-..., T 4700...	data

Type	Device	Short Description	Fig.	Manu	Comparision Types	More at
FT 06 E	Triac	=FT 06B: 400V	22a§		TAG 260-..., TAG 265-..., T 4700...	data
FT 07 C	Triac	200V, 6A(Tc=90°C), Igt/Ih=50/20mA	22a§	Hit	TAG 260-..., TAG 265-..., T 4700...	data
FT 07 D	Triac	=FT 07C: 300V	22a§		TAG 260-..., TAG 265-..., T 4700...	data
FT 07 E	Triac	=FT 07C: 400V	22a§		TAG 260-..., TAG 265-..., T 4700...	data
FT 07 F	Triac	=FT 07C: 500V	22a§		TAG 260-..., TAG 265-..., T 4700...	data
FT 07 G	Triac	=FT 07C: 600V	22a§		TAG 260-..., TAG 265-..., T 4700...	data
FT 08 C	Triac	200V, 6A(Tc=110°C), Igt/Ih=50/20mA	21b§	Hit	BS6-...A, TW6N...C, BS8-...A, TW8N...C, ++	data
FT 08 D	Triac	=FT 08C: 300V	21b§		BS6-...A, TW6N...C, BS8-...A, TW8N...C, ++	data
FT 08 E	Triac	=FT 08C: 400V	21b§		BS6-...A, TW6N...C, BS8-...A, TW8N...C, ++	data
FT 08 F	Triac	=FT 08C: 500V	21b§		BS6-...A, TW6N...C, BS8-...A, TW8N...C, ++	data
FT 08 G	Triac	=FT 08C: 600V	21b§		BS6-...A, TW6N...C, BS8-...A, TW8N...C, ++	data
FT 12 C	Triac	200V, 6A, Igt/Ih=75/20mA	17h§	Hit	MAC 216-..., T 2801..., SC 141..., BS7-...	data
FT 12 E	Triac	=FT 12C: 400V	17h§		MAC 216-..., T 2801..., SC 141..., BS7-...	data
FT 17	Si-N	UHF, >1GHz, PQ>35mW(1GHz)	2a§	Fch	BF 357, BF 378...379, BFW 30, BFX 59, ++	data
FT 034 A	Si-N	S-L, 150V, 10A, 15W(Tc=100°), >80MHz	49a§	Fch	2N5542	data
FT 034 B	Si-N	=FT 34A: 120V	49a§	Fch	2N5542	data
FT 034 C	Si-N	S, 150V, 5A, 0,8W, >80MHz	2a§	Fch	BU 125, BUY 48...49, BUY 80...81, ++	data
FT 034 D	Si-N	=FT 34C: 120V	2a§	Fch	BU 125, BUY 48...49, BUY 80...81, ++	data
FT 40	Si-N	UHF, 1,4GHz, RadH	2a§	Sgs	-	data
FT 45	Si-N	VHF, >425MHz	5g	Sgs	BF 173, BF 199, BF 224, BF 311, BF 373++	data
FT 107	MOS-N-FET-e	=MMFT 107 (Typ-Code/Stempel/marking)	48	Ons	→MMFT 107	
FT 107 A	Si-N	NF-ra, 30V, 0,05A, 0,26W, B>1200	2a§	Fch	2SC3112...13, 2SD1010...11, 2SD1512	data
FT 107 B	Si-N	=FT 107A: 45V, B>600	2a§	Fch	2SC3112...13, 2SD1010...11, 2SD1512	data
FT 107 C	Si-N	=FT 107A: 60V, B>150	2a§	Fch	2N2483...2484, 2N3117	data
FT 109	Si-N	HF/S, 15V, 0,2A, 0,3W, >600MHz	2a§	Fch	BSS 10...12, BSY 17...18, 2N2369(A), ++	data
FT 118	Si-N	VHF, 20V, 500MHz	5k	Fch	BF 173, BF 199, BF 224, BF 311, BF 373++	data
FT 123	Si-N	Vid, 300V, 0,1A, 1W, >60MHz	2a§	Fch	BF 259, BF 659, BFS 89, 2N5058	data
FT 232 BM	DIG-IC	USB Uart (USB - Serial) I.C.	MP	Ftd		pdf pinout
FT 317(A)	Si-N	≈BD 237			→BD 237	
FT 400A,B	Si-P	NF-L, 80V, 8A, 30W(Tc=100°), 120MHz	49a	Fch	2N6186...6189	data
FT 401	Si-N	S-L, -/400V, 2A, 100W, >2MHz	23a§	Fch	BUX 46, 2SC3091, 2SC3099, 2SC3155, ++	data
FT 402	Si-N	=FT 401: 3,5A	23a§	Fch	BUX 46, 2SC3091, 2SC3099, 2SC3155, ++	data
FT 410	Si-N	S-L, -/200V, 7,5A, 100W, 5MHz	23a§	Fch	BUX 47, 2SC3090, 2SC3092, 2SD811, ++	data
FT 411	Si-N	=FT 410: -/300V	23a§	Fch	BUX 47, 2SC3090, 2SC3092, 2SD811, ++	data
FT 413	Si-N	S-L, -/400V, 7,5A, 100W, 5MHz	23a§	Fch	BUX 47, 2SC3090, 2SC3092, 2SD811, ++	data
FT 417(A)	Si-P	≈BD 238			→BD 238	
FT 423	Si-N	S-L, -/400V, 7,5A, 100W, 5MHz	23a§	Fch	BUX 47, 2SC3090, 2SC3092, 2SD811, ++	data
FT 430	Si-N	S-L, -/400V, 10A, 100W, B=15...45	23a§	Fch	BUS 12, BUW 26, BUW 12, BUW 35...36, ++	data
FT 431	Si-N	=FT 430: B=15...35	23a§	Fch	BUS 12, BUW 26, BUW 12, BUW 35...36, ++	data
FT 0654(A...C)	N-FET	=FE 0654(A...C):	2b&	Fch	→FE 0654(A...C)	data
FT 0655(A...C)	N-FET	=FE 0655(A...C):	2b&	Fch	→FE 0655(A...C)	data
FT 701	MOS-P-FET-e	Dual, 30V, 0,2A, Up<0,6V, 30/-ns	81bz	Fch	-	data
FT 703	MOS-P-FET-e	S, 40V, 0,5A, Up<0,6V, 66/-ns	5n+	Fch	-	data
FT 704	MOS-P-FET-e	S, 25V, 20mA, Up<0,6V, 150/-ns	5n+	Fch	-	data
FT 709	Si-N	≈2N709	2a§	Fch	→2N709	data
FT 960	MOS-N-FET-e	=MMFT 960 (Typ-Code/Stempel/marking)	48	Ons	→MMFT 960	
FT 1210	Si-N	≈2N917	5k	Fch	→2N917	data
FT 1310	Si-N	SS, 5V, <15/-ns	2a§	Fch	BSS 10...12, BSY 17...18, 2N2368...69, ++	data
FT 1315	Si-N	SS, 30V, 0,2A, 0,36W, 12/18ns	2a§	Fch	BSS 10...12, BSY 17...18, 2N2368...69, ++	data
FT 1324(B,C)	Si-N	VHF, 25V, 0,5A, >640MHz	2a§	Fch	BFX 55, 2SC2852	data
FT 1341	Si-N	S, 25V, 0,1A, 0,36W, <65/-ns	2a§	Fch	BSW 41, BSY 62...63, 2N706A, 2N708, ++	data
FT 1702	Si-P	HF/S, 12V, 0,1A, 0,3W, <60/75ns	2a§	Fui	BSV 21, BSW 25, BSW 37, BSX 36, ++	data
FT 1718(A...E)	Si-P	Dual, 40V, 0,1A, >400MHz	81bs	Fch	BFX 11, 2N3726...3727, 2N4015...4016	data
FT 1724	Si-N	=2N1724	49a	Fch	→2N1724	data
FT 1725	Si-N	=2N1725	49a	Fch	→2N1725	data
FT 1746	Si-P	Uni, 35V, 0,36W, 150MHz	2a§	Fui	BC 213, BC 258, BC 308, BC 558, ++	data
FT 1869...1874	50Hz-Thy	=2N1869...1874	2a§	Fch	→2N1869...1874	data
FT 1881...1885	50Hz-Thy	=2N1881...1885	2a	Fch	→2N1881...1885	data
FT 2009...2013	50Hz-Thy	=2N2009...2013	2a§	Fch	→2N2009...2013	data
FT 2368	Si-N	→2N2368	2a§	Fch		data
FT 2369	Si-N	→2N2369	2a§	Fch		data
FT 2383	Si-N	→2N2383	2a§	Fch		data
FT 2384	Si-N	→2N2384	2a§	Fch		data
FT 2955	Si-P	≈BD 204	17j§		→BD 204	
FT 2974	Si-N	=2N2974	81bq	Fch	→2N2974	data
FT 2975	Si-N	=2N2975	81bq	Fch	→2N2975	data
FT 2978	Si-N	=2N2978	81bq	Fch	→2N2978	data
FT 2979	Si-N	=2N2979	81bq	Fch	→2N2979	data
FT 3055	Si-N	≈BD 203	17j§		→BD 203	

Type	Device	Short Description	Fig.	Manu	Comparision Types	More at
FT 3352	Si-N	100/±70/100V, 2A, 0,8W, hFE>50, hFC>20	2a§	Fui	-	data
FT 3553	Si-N	100/±70/100V, 3A, 40W, hFE>50, hFC>20	22a§	Fui	-	data
FT 3567	Si-N	=2N3567: 0,5W	8a	Fch	→2N3567	data
FT 3568	Si-N	=2N3568: 0,5W	8a	Fch	→2N3568	data
FT 3569	Si-N	=2N3569: 0,5W	8a	Fch	→2N3569	data
FT 3641	Si-N	=2N3641: 0,45W	8a	Fch	→2N3641	data
FT 3642	Si-N	=2N3642: 0,45W	8a	Fch	→2N3642	data
FT 3643	Si-N	=2N3643: 0,45W	8a	Fch	→2N3643	data
FT 3644	Si-P	=2N3644: 0,45W	8a	Fch	→2N3644	data
FT 3645	Si-P	=2N3645: 0,45W	8a	Fch	→2N3645	data
FT 3722	Si-N	=2N3722: 0,5W	8a	Fch	→2N3722	data
FT 3723	Si-N	=2N3723	2a§	Fch	→2N3723	data
FT 3820	P-FET	=2N3820: 0,2W	8b	Fch	→2N3820	data
FT 3883	Si-N	100/±70/100V, 15A,120W, hFE>50, hFC>20	23a§	Fui	-	data
FT 4017...4025	Si-P	=2N4017...4025: 0,5W	81bq	Fch	→(2N4017...4025)[5]	data
FT 4354...4356	Si-P	=2N4354...4356: 0,5W	8a	Fch	→2N4354...4356	data
FT 5040...5041	Si-P	=2N5040...5041: 0,5W	8a	Fch	→2N5040...5041	data
FT 5415	Si-P	=2N5415: 0,5W	8a	Fch	→2N5415	data
FT 5701 AM	Si-N	4x NPN, Memory Drv., 80V, 1A, <35/60ns	14-DIP	Fui	-	data
FT 5702 M	Si-N	4x NPN, Memory Drv., 60V, 1A, <40/60ns	14-DIP	Fui	-	data
FT 5709 M	Si-N	4x NPN, S, 20V, 0,2A, <12/18ns	14-DIP	Fui	-	data
FT 5712 M	Si-N	4x NPN, S, 30V, 0,2A, <-/300ns	14-DIP	Fui	-	data
FT 5713 M	Si-N	4x NPN, S, 50V, 0,5A, <-/950ns	14-DIP	Fui	-	data
FT 5714 M	Si-P	4x PNP, S, 50V, 0,2A, <-/350ns	14-DIP	Fui	-	data
FT 5715 M	Si-N/P	2x N+P, S, 30V, 0,2A, <-/350ns	14-DIP	Fui	-	data
FT 5723 M	Si-N	4x NPN, 80V, 1A, >100MHz, <-/1650ns	14-DIP	Fui	-	data
FT 5746 M	Si-N	5x NPN, S, 250/250V, 0,5A, 100MHz	14-DIP	Fui	-	data
FT 5747 M	Si-N	5x NPN, S, 250/250V, 0,5A, 100MHz	14-DIP	Fui	-	data
FT 5748 M	Si-P	4x PNP, S, 60V, 1,5A, >150MHz	14-DIP	Fui	-	data
FT 5753 M	Si-N-Darl+Di	2x2 NPN, 150/100V, ±1,5A, B>2000	12-SIL	Fui	-	data
FT 5754 M	Si-N-Darl+Di	2x2 NPN, 150/100V, ±3A, B>2000	12-SIL	Fui	-	data
FT 5755 M,ML	Si-N-Darl+Di	2x2 NPN, 150/100V, 5A, B>2000	12-SIL	Fui	-	data
FT 5756 M	Si-N-Darl+Di	4x NPN, 150/100V, ±1,5A, B>2000	12-SIL	Fui	-	data
FT 5757 M	Si-N-Darl+Di	4x NPN, 150/100V, ±3A, B>2000	12-SIL	Fui	-	data
FT 5758 M,ML	Si-N-Darl+Di	4x NPN, 150/100V, 5A, B>2000	12-SIL	Fui	-	data
FT 5759 M	Si-P-Darl+Di	4x PNP, 150/100V, ±1,5A, B>2000	12-SIL	Fui	-	data
FT 5760 M	Si-P-Darl+Di	4x PNP, 100/100V, ±3A, B>2000	12-SIL	Fui	-	data
FT 5761 M	Si-P-Darl+Di	4x PNP, 100/100V, ±5A, B>2000	12-SIL	Fui	-	data
FT 5763 M	Si-N-Darl+Di	2x2 NPN, 150/100V, ±1,5A, B>2000	12-SIP	Fui	-	data
FT 5764 M	Si-N-Darl+Di	2x2 NPN, 150/100V, ±3A, B>2000	12-SIP	Fui	-	data
FT 5766 M	Si-N-Darl+Di	4x NPN, 150/100V, ±1,5A, B>2000	12-SIP	Fui	-	data
FT 5767 M	Si-N-Darl+Di	4x NPN, 150/100V, ±3A, B>2000	12-SIP	Fui	-	data
FT 5769 M	Si-P-Darl+Di	4x NPN, 100/100V, ±1,5A, B>2000	12-SIP	Fui	-	data
FT 5770 M	Si-P-Darl+Di	4x NPN, 100/100V, ±3A, B>2000	12-SIP	Fui	-	data
FT 5776 M	Si-N/P-Darl	2xNPN+2xPNP, 100/100V, ±1,5A, B>2000	12-SIL	Fui	-	data
FT 5777 M	Si-N/P-Darl	2xNPN+2xPNP, 100/100V, ±3A, B>2000	12-SIL	Fui	-	data
FT 5778 M	Si-N/P-Darl	2xNPN+2xPNP, 100/100V, ±5A, B>2000	12-SIL	Fui	-	data
FT 5786 M	Si-N/P-Darl	2xNPN+2xPNP, 100/100V, ±1,5A, B>2000	12-SIP	Fui	-	data
FT 5787 M	Si-N/P-Darl	2xNPN+2xPNP, 100/100V, ±3A, B>2000	12-SIP	Fui	-	data
FT 6011(D)	MOS-N-FET-e	4x V-MOS, LogL, 60V, 3,5A, <0,4Ω(2A) D: 2x2 V-MOS, 2x2 Flyback-Di	12-SIL	Fui	-	data
FT 6012(D)	MOS-N-FET-e	4x V-MOS, LogL, 60V, 6,5A, <0,16Ω(4A)	12-SIL	Fui	-	data
FT 6015	MOS-N-FET-e	4x V-MOS, LogL, 30V, 10A, <65mΩ(5A)	12-SIL	Fui	-	data
FT 6021(D)	MOS-N-FET-e	4x V-MOS, LogL, 60V, 3A, <0,4Ω(2A)	12-SIP	Fui	-	data
FT 6022(D)	MOS-N-FET-e	4x V-MOS, LogL, 60V, 6A, <0,16Ω(4A)	12-SIP	Fui	-	data
FT 6025	MOS-N-FET-e	4x V-MOS, LogL, 30V, 9A, <65mΩ(5A)	12-SIP	Fui	-	data
FT 6045	MOS-N-FET-e	V-MOS, LogL, 30V, 17A, <65mΩ(10A)	17p	Fui	2SK1115, 2SK1287, 2SK1910, 2SK2334, ++	data
FT 6046	MOS-N-FET-e	V-MOS, LogL, 30V, 45A, <26mΩ(20A)	17p	Fui	2SK1542, 2SK1911	data
FT 6110(D)	MOS-N-FET-e	4x V-MOS, LogL, 120V, 2A, <1,3Ω(1A) D: 2x2 V-MOS, 2x2 Flyback-Di	12-SIL	Fui	-	data
FT 6111(D)	MOS-N-FET-e	4x V-MOS, LogL, 120V, 3A, <0,6Ω(1,5A)	12-SIL	Fui	-	data
FT 6112(D)	MOS-N-FET-e	4x V-MOS, LogL, 120V, 4,5A, <0,4Ω(3A)	12-SIL	Fui	-	data
FT 6120(D)	MOS-N-FET-e	4x V-MOS, LogL, 120V, 1,5A, <1,3Ω(1A)	12-SIP	Fui	-	data
FT 6121(D)	MOS-N-FET-e	4x V-MOS, LogL, 120V, 2,5A,<0,6Ω(1,5A)	12-SIP	Fui	-	data
FT 6122(D)	MOS-N-FET-e	4x V-MOS, LogL, 120V, 4A, <0,4Ω(3A)	12-SIP	Fui	-	data
FT 6211	MOS-N-FET-e	4x V-MOS, LogL, 200V, 3A, <0,8Ω(1,5A)	12-SIL	Fui	-	data
FT 6221	MOS-N-FET-e	4x V-MOS, LogL, 200V, 2,5A,<0,8Ω(1,5A)	12-SIP	Fui	-	data
FT 6240	MOS-N-FET-e	V-MOS, 200V, 5A, 45W, <0,8Ω(3A)	17c§	Fui	IRF 620, 2SK924, 2SK1391, 2SK1921, ++	data
FT 6241	MOS-N-FET-e	V-MOS, 200V, 8A, 65W, <45mΩ(5A)	17c§	Fui	BUZ 30, BUZ 73, IRF 630, 2SK1319, ++	data
FT 6540	MOS-N-FET-e	V-MOS, 500V, 4A, 70W, <2,2Ω(2A)	17c§	Fui	BUZ 41A...42, IRF 830, 2SK893, 2SK1246,++	data
FT 6840	MOS-N-FET-e	V-MOS, 800V, 2A, 70W, <9Ω(1A)	17c§	Fui	BUZ 81, 2SK1457, 2SK1501, 2SK1643, ++	data
FT 7207 A	Si-N	NF/S-L, 120V, 5A, 30W(Tc=100°), >70MHz	49a	Fch	2N5284...5285	data
FT 7207 B	Si-N	=FT 7207A: 100V	49a	Fch	2N5002, 2N5004, 2N5284...5285	data
FTD 1001	MOS-P-FET-e*	SMD, V-MOS, Dual,LogL, 20V, 2A, 0,115Ω	8-SSMDIP	Say	-	data
FTD 1002	MOS-P-FET-e*	SMD, V-MOS, Dual,LogL, 20V, 2A, 0,115Ω	8-SSMDIP	Say	-	data

Type	Device	Short Description ·	Fig.	Manu	Comparision Types	More at
TD 2002	MOS-N-FET-e*	SMD, V-MOS, Dual, LogL, 30V, 3A, 0,06Ω	8-SSMDIP/	Say	-	data
TD 2003	MOS-N-FET-e*	SMD, V-MOS, Dual, LogL, 20V, 2,2A, 0,18Ω	8-SSMDIP/	Say	- ·	data
TD 2005	MOS-N-FET-e*	SMD, V-MOS, Dual, LogL, 20V, 1A <0,36Ω(0,5A)	8-SSMDIP/ ca4	Say	-	data
TD 2007	MOS-N-FET-e*	SMD, V-MOS, Dual, LogL, 100V, 0,8A <0,95Ω(0,4A)	8-SSMDIP/ ca4	Say	-	data
TD 2011	MOS-N-FET-e*	SMD, V-MOS, Dual, LogL, 20V, 5A <40mΩ(2A)	8-SSMDIP/ ca4	Say	- ·	data
TD 2013	MOS-N-FET-e*	SMD, V-MOS, Dual, LogL, 30V, 4,5A <48mΩ(2A)	8-SSMDIP/ ca4	Say	-	data
TD 2014	MOS-N-FET-e*	SMD, V-MOS, Dual, LogL, 20V, 4A <59mΩ(2A)	8-SSMDIP/ ca4	Say	-	data
TD 2015	MOS-N-FET-e*	SMD, V-MOS, Dual, LogL, 30V, 4A <73mΩ(4A)	8-SSMDIP/ ca4	Say	-	data
TD 2016	MOS-N-FET-e*	SMD, V-MOS, Dual, LogL, 30V, 4A <66mΩ(2A)	8-SSMDIP	Say	-	data
TD 2017	MOS-N-FET-e*	SMD, V-MOS, Dual, LogL, 20V, 5A <29mΩ(2A)	8-SSMDIP/ ca4	Say	-	data
TD 2019	MOS-N-FET-e*	SMD, V-MOS, Dual, LogL, 30V, 5A <33mΩ(2A)	8-SSMDIP/ ca4	Say	-	data
TD 2022	MOS-N-FET-e*	SMD, V-MOS, Dual, LogL, 30V, 4,5A <55mΩ(4A)	8-SSMDIP/ ca4	Say	-	data
tF	Si-P	=PUMT 1 (Typ-Code/Stempel/marking)	46	Phi	→PUMT 1	data
tQ	Si-P	=2PA1576Q (Typ-Code/Stempel/marking)	35(2mm)	Phi	→2PA1576	data
tR	Si-P	=2PA1576R (Typ-Code/Stempel/marking)	35(2mm)	Phi	→2PA1576 *	data
tS	Si-P	=2PA1576S (Typ-Code/Stempel/marking)	35(2mm)	Phi ,	→2PA1576	data
TS 1001	MOS-P-FET-e*	SMD, V-MOS, LogL, 20V, 4A, <58mΩ(4A)	8-SSMDIP	Say	-	data
TS 1002	MOS-P-FET-e*	SMD, V-MOS, LogL, 30V, 4A, <0,12Ω(2A)	8-SSMDIP/ ca5	Say	- · .	data
TS 1003	MOS-P-FET-e*	SMD, V-MOS, LogL, 20V, 3A, <90mΩ(3A)	8-SSMDIP	Say	-	data
TS 1004	MOS-P-FET-e*	SMD, V-MOS, LogL, 30V, 3A, <0,19Ω(1A)	8-SSMDIP/ ca5	Say	- ·	data
TS 2001	MOS-N-FET-e*	SMD, V-MOS, LogL, 20V, 5A, <46mΩ(2A)	8-SSMDIP/ ca5	Say	-	data
TS 2002	MOS-N-FET-e*	SMD, V-MOS, LogL, 30V, 5A, <50mΩ(3A)	8-SSMDIP/ ca5	Say	-	data
TS 2003	MOS-N-FET-e*	SMD, V-MOS, LogL, 20V, 4A, <70mΩ(2A)	8-SSMDIP/ ca5	Say	-	data
TS 2004	MOS-N-FET-e*	SMD, V-MOS, LogL, 30V, 4A, <78mΩ(4A)	8-SSMDIP/ ca5	Say	-	data
TS 2005	MOS-N-FET-e*	SMD, V-MOS, LogL, 100V, 2A, <0,33Ω(2A)	8-SSMDIP/ ca5	Say	-	data
TS 2011	MOS-N-FET-e*	SMD, V-MOS, LogL, 20V, 8A, <26mΩ(2A)	8-SSMDIP/ ca5	Say	-	data
TS 2012	MOS-N-FET-e*	SMD, V-MOS, LogL, 30V, 8A, <31mΩ(4A)	8-SSMDIP/ ca5	Say	-	data
TS 2015	MOS-N-FET-e*	SMD, V-MOS, LogL, 30V, 7A, <31mΩ(2A)	8-SSMDIP	Say	- ·	data
TSO 706...6571	Si-N/P	=2N706..2N6571: SMD	35	Fch	-	data
TSOA 05...51	Si-N/P	=MPSA 05...MPSA 51: SMD	35	Fch	-	data
tZ	Si-N/P	=PUMZ 1 (Typ-Code/Stempel/marking)	46	Phi	→PUMZ 1	data
TZ 5.6 E	Z-Di	SMD, Quad, 5,31...5,92V, <60Ω(5mA)	45(KAKKK)	Rhm	-	data
TZ 6.8 E	Z-Di	SMD, Quad, 6,47...7,14V, <40Ω(5mA)	45(KAKKK)	Rhm	-	data

FU..FZ

Type	Device	Short Description ·	Fig.	Manu	Comparision Types	More at
FU	Z-Di	=DF 3A8.2LFE(Typ-Code/Stempel/marking)	35(1,6mm)	Tos	→DF 3A8.2LFE	data
FU	Z-Di	=DF 3A8.2LFU(Typ-Code/Stempel/marking)	35(2mm)	Tos	→DF 3A8.2LFU	data
FU	Z-Di	=DF 5A8.2LF (Typ-Code/Stempel/marking)	45	Tos	→DF 5A8.2LF	data
FU	Z-Di	=DF 5A8.2LFU(Typ-Code/Stempel/marking)	45(2mm)	Tos	→DF 5A8.2LFU	data
FU	Z-Di	=Z1 SMA-160 (Typ-Code/Stempel/marking)	71(5x2,5)	Fag	→Z1 SMA-160	data
FU 1 M,N,P	Si-Di	Gl, S, 140...600V, 0,2A, <400ns M=600V, N=300V, P=140V	31a	Fjd	BA 157...159, BY 204/..., BY 208/..., ++	data
FU 12 B	Triac	100V, 3A, Igt/Ih=30/20mA	17h§	Hit	TAG 230-..., TAG 231-..., TAG 220-..., ++	data
FU 12 C	Triac	=FU 12B: 200V	17h§		TAG 230-..., TAG 231-..., TAG 220-..., ++	data
FU 12 D	Triac	=FU 12B: 300V	17h§		TAG 230-..., TAG 231-..., TAG 220-..., ++	data
FU 12 E	Triac	=FU 12B: 400V	17h§		TAG 230-..., TAG 231-..., TAG 220-..., ++	data
FUA	Z-Di	=SMCJ 5.0A (Typ-Code/Stempel/marking)	71(8x5mm)	Tho	→SMCJ ...	data
FUB	Z-Di	=SMCJ 6.0A (Typ-Code/Stempel/marking)	71(8x5mm)	Tho	→SMCJ ...	data
FUC	Z-Di	=SMCJ 6.5A (Typ-Code/Stempel/marking)	71(8x5mm)	Tho	→SMCJ ...	data
FUD	Z-Di	=SMCJ 8.5A (Typ-Code/Stempel/marking)	71(8x5mm)	Tho	→SMCJ ...	data
FUF	Z-Di	=SMCJ 10A (Typ-Code/Stempel/marking)	71(8x5mm)	Tho	→SMCJ ...	data
FUF 5400	Si-Di	Gl, S, 50V, 3A, <50ns	31a	Fag	BYV 28/50, EGP 30A, FE 3A	data
FUF 5401	Si-Di	=FUF 5400: 100V	31a	Fag	BYV 28/100, EGP 30B, FE 3B	data
FUF 5402	Si-Di	=FUF 5400: 200V	31a	Fag	BYV 28/200, EGP 30D, FE 3D	data
FUF 5404	Si-Di	=FUF 5400: 400V	31a	Fag	EGP 30G, FE 3H	data
FUF 5406	Si-Di	=FUF 5400: 600V	31a	Fag	BYW 178	data
FUF 5407	Si-Di	=FUF 5400: 800V	31a	Fag	BYW 178	data
FUH	Z-Di	=SMCJ 12A (Typ-Code/Stempel/marking)	71(8x5mm)	Tho	→SMCJ ...	data

Type	Device	Short Description	Fig.	Manu	Comparision Types	More at
FUI	Z-Di	=SMCJ 13A (Typ-Code/Stempel/marking)	71(8x5mm)	Tho	→SMCJ …	data
FUJ	Z-Di	=SMCJ 15A (Typ-Code/Stempel/marking)	71(8x5mm)	Tho	→SMCJ …	data
FUL	Z-Di	=SMCJ 18A (Typ-Code/Stempel/marking)	71(8x5mm)	Tho	→SMCJ …	data
FUM	Z-Di	=SMCJ 20A (Typ-Code/Stempel/marking)	71(8x5mm)	Tho	→SMCJ …	data
FUN	Z-Di	=SMCJ 22A (Typ-Code/Stempel/marking)	71(8x5mm)	Tho	→SMCJ …	data
FUO	Z-Di	=SMCJ 24A (Typ-Code/Stempel/marking)	71(8x5mm)	Tho	→SMCJ …	data
FUP	Z-Di	=SMCJ 26A (Typ-Code/Stempel/marking)	71(8x5mm)	Tho	→SMCJ …	data
FUQ	Z-Di	=SMCJ 28A (Typ-Code/Stempel/marking)	71(8x5mm)	Tho	→SMCJ …	data
FUR	Z-Di	=SMCJ 30A (Typ-Code/Stempel/marking)	71(8x5mm)	Tho	→SMCJ …	data
FUS	Z-Di	=SMCJ 33A (Typ-Code/Stempel/marking)	71(8x5mm)	Tho	→SMCJ …	data
FUSION 878 A	DIG-IC	PCI Video Decoder	MP	Con	-	pdf pinout
FUU	Z-Di	=SMCJ 40A (Typ-Code/Stempel/marking)	71(8x5mm)	Tho	→SMCJ …	data
FUW	Z-Di	=SMCJ 48A (Typ-Code/Stempel/marking)	71(8x5mm)	Tho	→SMCJ …	data
FUZ	Z-Di	=SMCJ 58A (Typ-Code/Stempel/marking)	71(8x5mm)	Tho	→SMCJ …	data
FV	PIN-Di	=1SV248 (Typ-Code/Stempel/marking)	35 (2mm)	Say	→1SV248	data
FV	PIN-Di	=1SV250 (Typ-Code/Stempel/marking)	35	Say	→1SV250	data
FV	Z-Di	=Z1 SMA-180 (Typ-Code/Stempel/marking)	71(5x2,5)	Fag	→Z1 SMA-180	data
FV 1043	C-Di	≈BB 103			→BB 103	
FVP 12030 IM 3LEG1	IC	Energy Recovery	DIP	Fch	-	pdf pinout
FVP 18030 IM 3LSG1	IC	Sustain	DIP	Fch	-	pdf pinout
FW	Z-Di	=Z1 SMA-200 (Typ-Code/Stempel/marking)	71(5x2,5)	Fag	→Z1 SMA-200	data
FW 101	MOS-P-FET-e*	SMD, V-MOS, Dual, LogL, 12V, 3A <0,115Ω(3A)	8-MDIP	Say	-	data
FW 102	MOS-P-FET-e*	SMD, V-MOS, Dual, LogL, 20V, 3A, <0,1Ω	8-MDIP	Say	-	data
FW 103	MOS-P-FET-e*	SMD, V-MOS, Dual, LogL, 30V, 3A <205mΩ(3A), 130/500ns	8-MDIP/ca3	Say	-	data
FW 104	MOS-P-FET-e*	SMD, V-MOS, Dual, LogL, 12V, 2A <0,4Ω(2A)	8-MDIP	Say	-	data
FW 106	MOS-P-FET-e*	SMD, V-MOS, Dual, LogL, 12V, 2A <0,4Ω(1A)	8-MDIP	Say	-	data
FW 107	MOS-P-FET-e*	SMD, V-MOS, Dual, LogL, 60V, 2A <0,31Ω(2A)	8-MDIP	Say	-	data
FW 108	MOS-P-FET-e*	SMD, V-MOS, Dual, LogL, 60V, 2A <0,33Ω(2A)	8-MDIP	Say	-	data
FW 111	MOS-P-FET-e*	SMD, V-MOS, Dual, LogL, 20V, 5A <98mΩ(2A), 135/225ns	8-MDIP/ca3	Say	-	data
FW 113	MOS-P-FET-e*	SMD, V-MOS, Dual, LogL, 30V, 5A <120mΩ(2A), 165/175ns	8-MDIP/ca3	Say	-	data
FW 114	MOS-P-FET-e*	SMD, V-MOS, Dual, LogL, 20V, 3A <130mΩ(1A), 155/165ns	8-MDIP/ca3	Say	-	data
FW 115	MOS-P-FET-e*	SMD, V-MOS, Dual, LogL, 30V, 3A <190mΩ(1A), 40/105ns	8-MDIP/ca3	Say	-	data
FW 201	MOS-N-FET-e*	SMD, V-MOS, Dual, LogL, 20V, 5A <50mΩ(5A)	8-MDIP	Say	-	data
FW 202	MOS-N-FET-e*	SMD, V-MOS, Dual, LogL, 20V, 5A <65mΩ(5A), 115/310ns	8-MDIP/ca3	Say	-	data
FW 203	MOS-N-FET-e*	SMD, V-MOS, Dual, LogL, 30V, 5A <78mΩ(5A), 215/310ns	8-MDIP/ca3	Say	-	data
FW 204	MOS-N-FET-e*	SMD, V-MOS, Dual, LogL, 20V, 2A <320mΩ(0,5A), 35/85ns	8-MDIP/ca3	Say	-	data
FW 206	MOS-N-FET-e*	SMD, V-MOS, Dual, LogL, 30V, 5A <65mΩ(5A)	8-MDIP	Say	-	data
FW 208	MOS-N-FET-e*	SMD, V-MOS, Dual, LogL, 60V, 3A <195mΩ(2A), 43/160ns	8-MDIP/ca3	Say	-	data
FW 211	MOS-N-FET-e*	SMD, V-MOS, Dual, LogL, 20V, 6A <48mΩ(2A), 220/300ns	8-MDIP/ca3	Say	-	data
FW 213	MOS-N-FET-e*	SMD, V-MOS, Dual, LogL, 30V, 7A <50mΩ(5A), 195/170ns	8-MDIP/ca3	Say	-	data
FW 214	MOS-N-FET-e*	SMD, V-MOS, Dual, LogL, 20V, 5A <70mΩ(2A), 270/200ns	8-MDIP/ca3	Say	-	data
FW 215	MOS-N-FET-e*	SMD, V-MOS, Dual, LogL, 30V, 5A <78mΩ(5A), 130/150ns	8-MDIP/ca3	Say	-	data
FW 221	MOS-N-FET-e*	SMD, V-MOS, Dual, LogL, 20V, 7A <40mΩ(2A), 320/310ns	8-MDIP/ca3	Say	-	data
FW 223	MOS-N-FET-e*	SMD, V-MOS, Dual, LogL, 30V, 7A 58mΩ(4A)	8-MDIP	Say	-	data
FW 231	MOS-N-FET-e*	SMD, V-MOS, Dual, LogL, 20V, 9A <27mΩ(2A), 464/500ns	8-MDIP/ca3	Say	-	data
FW 232	MOS-N-FET-e*	SMD, V-MOS, Dual, LogL, 30V, 8A <32mΩ(2A)	8-MDIP	Say	-	data
FW 233	MOS-N-FET-e*	SMD, V-MOS, Dual, LogL, 30V, 8A <34mΩ(4A), 222/205ns	8-MDIP/ca3	Say	-	data
FW 236	MOS-N-FET-e*	SMD, V-MOS, Dual, LogL, 20V, 6A <59mΩ(2A)	8-MDIP	Say	-	data
FW 237	MOS-N-FET-e*	SMD, V-MOS, Dual, LogL, 30V, 5A <66mΩ(2A)	8-MDIP	Say	-	data
FW 238	MOS-N-FET-e*	SMD, V-MOS, Dual, LogL, 30V, 5A <73mΩ(4A)	8-MDIP	Say	-	data
FW 306	MOS-NP-FET-e	SMD, V-MOS, Dual, LogL, T1: N, 30V, 5A <65mΩ(5A), T2: 30V, 3A, <160mΩ(3A)	8-MDIP	Say	-	data
FW 307	MOS-NP-FET-e	SMD, V-MOS, Dual, LogL, T1: N, 250V, 1A <1,6Ω(1A), T2: P, 250V, 0,7A, <4Ω(0,7A)	8-MDIP	Say	-	data

Type	Device	Short Description	Fig.	Manu	Comparision Types	More at
FW 313	MOS-NP-FET-e	SMD, V-MOS, Dual, LogL, T1: N, 30V, 7A, <50mΩ, T2: P, 30V, 5A, <120mΩ	8-MDIP/ca3	Say	-	data
FWC9801	LDR	575nm Ro:1.5MOhm Vb:2V		Sie		data pinout
FX	Si-P	=2SB1571-FX (Typ-Code/Stempel/marking)	39	Nec	→2SB1571	data
FX	Z-Di	=SM 6T 100 (Typ-Code/Stempel/marking)	71(6x4mm)	Tho	→SM 6T...	data
FX 107(A...C)	Si-N	=FT 107(A...C):	≈24	Fch	→(FT 107(A...C))[6]	data
FX 203	MOS-P-FET-e*	SMD, V-MOS, LogL, 30V, 4A, <0,29Ω(2A)	47bn5	Say	-	data
FX 205	MOS-P-FET-e*	SMD, V-MOS, LogL, 60V, 2A, <0,65Ω(1A)	47bn5	Say	-	data
FX 207	MOS-P-FET-e*	SMD, V-MOS, LogL, 12V, 4A, <0,3Ω(1A)	47bn5	Say	-	data
FX 208	MOS-N-FET-e*	SMD, V-MOS, LogL, 20V, 4A, <125mΩ(1A)	47bn5	Say	-	data
FX 211	MOS-N-FET-e*	SMD, V-MOS, LogL, 250V, 1,5A, <4Ω	47	Say	-	data
FX 212	MOS-N-FET-e*	SMD, V-MOS, LogL, 250V, 2A, <1,6Ω	47	Say	-	data
FX 216	MOS-N-FET-e*	SMD, V-MOS, LogL, 150V, 3A, <0,7Ω	47	Say	-	data
FX 401	Si-P+Di	SMD, S, 30V, 2A, 320MHz, 60/375ns int. Schottky-Di, 30V, 3A (SB 30-03P)	47	Say	-	data
FX 501	Si-P	SMD, Dual, 25V, 5A, 350MHz, 40/210ns	47bh9	Say	-	data
FX 502	Si-N	SMD, Dual, 60V, 5A, 220MHz, 30/340ns	47bh9	Say	-	data
FX 503	Si-P	SMD, Dual, 60V, 3A, 150MHz, 70/485ns	47bh9	Say	-	data
FX 504	Si-N	SMD, Dual, 60V, 3A, 170MHz, 70/685ns	47bh9	Say	-	data
FX 505	Si-P	SMD, Dual, 60V, 5A, 150MHz, 50/470ns	47bh9	Say	-	data
FX 506	Si-N	SMD, Dual, 60V, 5A, 180MHz, 50/520ns	47bh9	Say	-	data
FX 507	Si-N	SMD, Dual, 120V, 2A, 120MHz, 80/1050ns	47bh9	Say	-	data
FX 508	Si-N	SMD, Dual, 120V, 3A, 180MHz, 100/950ns	47bh9	Say	-	data
FX 509	Si-P	SMD, Dual, 120V, 3A, 180MHz, 100/850ns	47bh9	Say	-	data
FX 510	Si-P	SMD, Dual, 180V, 1,5A, 150MHz, 60/750ns	47bh9	Say	-	data
FX 601	MOS-P-FET-e*	SMD, V-MOS, Dual, LogL, 12V, 1A, 0,63Ω	47bn6	Say	-	data
FX 602	MOS-N-FET-e*	SMD, V-MOS, Dual, LogL, 20V, 2A, 0,25Ω	47bn6	Say	-	data
FX 603	MOS-P-FET-e*	SMD, V-MOS, Dual, LogL, 30V, 1A, 1,1Ω	47bn6	Say	-	data
FX 604	MOS-N-FET-e*	SMD, V-MOS, Dual, LogL, 30V, 2A, 0,38Ω	47bn6	Say	-	data
FX 605	MOS-P-FET-e*	SMD, V-MOS, Dual, LogL, 60V, 1A, 1,6Ω	47bn6	Say	-	data
FX 606	MOS-N-FET-e*	SMD, V-MOS, Dual, LogL, 60V, 2A, 0,6Ω	47bn6	Say	-	data
FX 607	MOS-N-FET-e*	SMD, V-MOS, Dual, LogL, 150V, 1,2A, 2,2Ω	47bn6	Say	-	data
FX 609	MOS-N-FET-e*	SMD, V-MOS, LogL, 20V, 2A, <0,2Ω(1A)	47	Say	-	data
FX 611	MOS-N-FET-e*	SMD, V-MOS, LogL, 30V, 2A, <0,26Ω(1A)	47	Say	·	data
FX 802	Si-P+Di	SMD, S, 25V, 5A, 320MHz, 40/210ns + 2x Schottky-Di, 30V, 2A (SB 20W03P)	47	Say	-	data
FX 803	Si-N+Di	SMD, S, 60V, 5A, 220MHz, 30/340ns + 2x Schottky-Di, 30V, 2A (SB 20W03P)	47	Say		data
FX 851	MOS-P-FET-e*	SMD, V-MOS, LogL, 30V, 1A, <1,1Ω(0,5A) int. Schottky-Di, 30V, 0,7A (SB07-03P)	47	Say	-	data
FX 852	MOS-N-FET-e*	SMD, V-MOS, LogL, 30V, 2A, <0,38Ω(1A) int. Schottky-Di, 30V, 0,7A (SB07-03P)	47	Say	-	data
FX 853	MOS-N-FET-e*	SMD, V-MOS, LogL, 30V, 2A, <0,38Ω(1A) int. Schottky-Di, 50V, 0,5A (SB05-05P)	47	Say	-	data
FX 854	MOS-P-FET-e*	SMD, V-MOS, LogL, 60V, 1A, <1,6Ω(0,5A) int. Schottky-Di, 50V, 0,5A (SB05-05P)	47	Say	-	data
FX 855	MOS-N-FET-e*	SMD, V-MOS, LogL, 60V, 2A, <0,6Ω(1A) int. Schottky-Di, 90V, 0,5A (SB05-09P)	47	Say	-	data
FX 856	MOS-N-FET-e*	SMD, V-MOS, LogL, 30V, 2A, <0,48Ω(1A) int. Schottky-Di, 30V, 0,7A, <10ns	47	Say	-	data
FX 901	Si-P+MOS-FET	SMD, Trans. 15V, 3A, 400MHz, 25/210ns V-MOS-e*, LogL, 11V, 2A, <0,32Ω(0,5A) Schottky-Di (S→D)	47	Say	-	data
FX 914	Si-N	=2N914: 0,25W	≈24	Fch	→(2N914)[6]	data
FX 918	Si-N	=2N918: 0,25W	≈24	Fch	→(2N918)[6]	data
FX 2368..2369(A)	Si-N	=2N2368..2369(A): 0,25W	≈24	Fch	→(2N2368..2369(A))[6]	data
FX 2483..2484	Si-N	=2N2483..2484: 0,25W	≈24	Fch	→(2N2483..2484)[6]	data
FX 2894(A)	Si-P	=2N2894(A): 0,25W	≈24	Fch	→(2N2894(A))[6]	data
FX 3299..3300	Si-N	=2N3299..3300: 0,25W	≈24	Fch	→(2N3299..3300)[6]	data
FX 3502..3503	Si-P	=2N3502..3503: 0,25W	≈24	Fch	→(2N3502..3503)[6]	data
FX 3724..3725	Si-N	=2N3724..3725: 0,25W	≈24	Fch	→(2N3724..3725)[6]	data
FX 3962..3965	Si-P	=2N3962..3965: 0,25W	≈24	Fch	→(2N3962..3965)[6]	data
FX 4046..4047	Si-N	=2N4046..4047: 0,25W	≈24	Fch	→(2N4046..4047)[6]	data
FX 4207	Si-P	=2N4207: 0,25W	≈24	Fch	→(2N4207)[6]	data
FX 4960	Si-N	=2N4960: 0,25W	≈24	Fch	→(2N4960)[6]	data
FXT 38 C	Si-N	=BCX 38C:	40c	Ztx	→BCX 38C	data
FXT 449	Si-N	=ZTX 449:	40c	Ztx	→ZTX 449	data
FXT 450	Si-N	=ZTX 450:	40c	Ztx	→ZTX 450	data
FXT 451	Si-N	=ZTX 451:	40c	Ztx	→ZTX 451	data
FXT 453	Si-N	=ZTX 453:	40c	Ztx	→ZTX 453	data
FXT 455	Si-N	=ZTX 455:	40c	Ztx	→ZTX 455	data
FXT 458	Si-N	=ZTX 458:	40c	Ztx	→ZTX 458	data
FXT 549	Si-P	=ZTX 549:	40c	Ztx	→ZTX 549	data
FXT 550	Si-P	=ZTX 550:	40c	Ztx	→ZTX 550	data
FXT 551	Si-P	=ZTX 551:	40c	Ztx	→ZTX 551	data
FXT 553	Si-P	=ZTX 553:	40c	Ztx	→ZTX 553	data

Type	Device	Short Description	Fig.	Manu	Comparision Types	More at
FXT 555	Si-P	=ZTX 555:	40c	Ztx	→ZTX 555	data
FXT 557	Si-P	=ZTX 557:	40c	Ztx	→ZTX 557	data
FXT 601 B	Si-N-Darl	=ZTX 601B:	40c	Ztx	→ZTX 601B	data
FXT 603	Si-N-Darl	=ZTX 603:	40c	Ztx	→ZTX 603	data
FXT 605	Si-N-Darl	=ZTX 605:	40c	Ztx	→ZTX 605	data
FXT 614	Si-N-Darl	=ZTX 614:	40c	Ztx	→ZTX 614	data
FXT 649	Si-N	=ZTX 649:	40c	Ztx	→ZTX 649	data
FXT 651	Si-N	=ZTX 651:	40c	Ztx	→ZTX 651	data
FXT 653	Si-N	=ZTX 653:	40c	Ztx	→ZTX 653	data
FXT 655	Si-N	=ZTX 655:	40c	Ztx	→ZTX 655	data
FXT 657	Si-N	=ZTX 657:	40c	Ztx	→ZTX 657	data
FXT 688 B	Si-N	=ZTX 688B:	40c	Ztx	→ZTX 688B	data
FXT 690 B	Si-N	=ZTX 690B:	40c	Ztx	→ZTX 690B	data
FXT 704	Si-P-Darl	=ZTX 704:	40c	Ztx	→ZTX 704	data
FXT 705	Si-P-Darl	=ZTX 705:	40c	Ztx	→ZTX 705	data
FXT 749	Si-P	=ZTX 749:	40c	Ztx	→ZTX 749	data
FXT 751	Si-P	=ZTX 751:	40c	Ztx	→ZTX 751	data
FXT 753	Si-P	=ZTX 753:	40c	Ztx	→ZTX 753	data
FXT 755	Si-P	=ZTX 755:	40c	Ztx	→ZTX 755	data
FXT 757	Si-P	=ZTX 757:	40c	Ztx	→ZTX 757	data
FXT 788 B	Si-P	=ZTX 788B:	40c	Ztx	→ZTX 788B	data
FXT 790 A	Si-P	=ZTX 790A:	40c	Ztx	→ZTX 790A	data
FXT 2222 A	Si-N	=2N2222A:	40c	Ztx	→2N2222A	data
FXT 2907 A	Si-P	=2N2907A:	40c	Ztx	→2N2907A	data
FXT 3866	Si-N	=2N3866:	40c	Ztx	→2N3866	data
FXTA 42	Si-N	=MPSA 42:	40c	Ztx	→MPSA 42	data
FXTA 92	Si-P	=MPSA 92:	40c	Ztx	→MPSA 92	data
FY	Si-P	=2SA1202-Y (Typ-Code/Stempel/marking)	39	Tos	→2SA1202	data
FY	Si-P	=2SB1571-FY (Typ-Code/Stempel/marking)	39	Nec	→2SB1571	data
FY	Si-N	=2SC2761-Y (Typ-Code/Stempel/marking)	35	Tos	→2SC2761	data
FY	Si-N	=2SC3661 (Typ-Code/Stempel/marking)	35	Say	→2SC3661	data
FY	MOS-N-FET-e*	=2SK1698 (Typ-Code/Stempel/marking)	39	Hit	→2SK1698	data
FY	Si-N	=BSV 65RA (Typ-Code/Stempel/marking)	35	Sie	→BSV 65RA	data
FY	Si-P	=KTA1662-Y (Typ-Code/Stempel/marking)	39	Kec	→KTA 1662	data
FY	Si-N	=KTC3878-Y (Typ-Code/Stempel/marking)	35	Kec	→KTC 3878(S)	data
FY	Z-Di	=SM 6T 100A (Typ-Code/Stempel/marking)	71(6x4mm)	Tho	→SM 6T...	data
FZ	Si-P	=2SB1571-FZ (Typ-Code/Stempel/marking)	39	Nec	→2SB1571	data
FZ	Si-N	=BSV 65RB (Typ-Code/Stempel/marking)	35	Sie	→BSV 65RB	data
FZE 1658 G	IC	8 x Digital Sensor Interface	TSOP	Sie	-	pdf pinout
FZH 101..291	Logic-IC	LSL Logic (slow, hi-rel), 11,4...17V		Phi, Sie	-	pdf
FZH 101A	Logic	Quad 2 input NAND	DIP	Sie	-	pdf
FZH 105A	Logic	Quad 2 input NAND	DIP	Sie	-	pdf
FZH 111A	Logic	Quad 2 input NAND, N-Pin	DIP	Sie	-	pdf
FZH 115B	Logic	Quad 2 input NAND, N-Pin	DIP	Sie	-	pdf
FZH 121	Logic	Dual 5 input NAND	DIP	Sie	-	pdf
FZH 125	Logic	Dual 5 input NAND	DIP	Sie	-	pdf
FZH 131	Logic	Dual 5 input NAND	DIP	Sie	-	pdf
FZH 135	Logic	Dual 5 input NAND	DIP	Sie	-	pdf
FZH 141	Logic	Dual 5 input NAND	DIP	Sie	-	pdf
FZH 142	Logic	Dual 5 input NAND	DIP	Sie	-	pdf
FZH 145	Logic	Dual 5 input NAND	DIP	Sie	-	pdf
FZH 151	Logic	And/or combination	DIP	Sie	-	pdf
FZH 155	Logic	And/or combination	DIP	Sie	-	pdf
FZH 161	Logic	4x TTL→LSL Levelshifter	DIP	Sie	-	pdf
FZH 165	Logic	4x TTL→LSL Levelshifter	DIP	Sie	-	pdf
FZH 171	Logic	Dual 5 input NAND, Expansionpin	DIP	Sie	-	pdf
FZH 175	Logic	Dual 5 input NAND, Expansionpin	DIP	Sie	-	pdf
FZH 181	Logic	4x TTL→LSL Levelshifter	DIP	Sie	-	pdf
FZH 185	Logic	4x TTL→LSL Levelshifter	DIP	Sie	-	pdf
FZH 191	Logic	Triple 3 input NAND, N-Pin	DIP	Sie	-	pdf
FZH 195	Logic	Triple 3 input NAND, N-Pin	DIP	Sie	-	pdf
FZH 201	Logic	Hex Inverter with Strobe	DIP	Sie	-	pdf
FZH 211	Logic	Quad 2 input NAND, open collector	DIP	Sie	-	pdf
FZH 211 S	Logic	Quad 2 input NAND, open collector, Pegelwandler/level shift, 0...+70°	16-DIP	Sie	-	pdf
FZH 215B	Logic	Quad 2 input NAND, open collector	DIP	Sie	-	pdf
FZH 215 S	LIN-IC	=FZH 211S: -25...+85°	16-DIP	Sie	-	pdf
FZH 231	Logic	Dual 5 input NAND, open collector	DIP	Sie	-	pdf
FZH 235	Logic	Dual 5 input NAND, open collector	DIP	Sie	-	pdf
FZH 241	Logic	Dual 4 input NAND, Schmitt-Trigger	DIP	Sie	-	pdf
FZH 245B	Logic	Dual 4 input NAND, Schmitt-Trigger	DIP	Sie	-	pdf
FZH 251	Logic	Quad 2 input and	DIP	Sie	-	pdf
FZH 255B	Logic	Quad 2 input and	DIP	Sie	-	pdf
FZH 261	Logic	2x NAND, 4x Inverter	DIP	Sie	-	pdf
FZH 265B	Logic	2x NAND, 4x Inverter	DIP	Sie	-	pdf
FZH 271	Logic	Quad 2 input ExOR	DIP	Val	-	pdf
FZH 271	Logic	Quad 2 input ExOR	DIP	Sie	-	pdf
FZH 275B	Logic	Quad 2 input ExOR	DIP	Sie	-	pdf
FZH 281	Logic	Quad 2 input NOR	DIP	Sie	-	pdf
FZH 285B	Logic	Quad 2 input NOR	DIP	Sie	-	pdf
FZH 291	Logic	Quad 2 input or	DIP	Val	-	pdf
FZH 291	Logic	Quad 2 input or	DIP	Sie	-	pdf
FZH 295B	Logic	Quad 2 input or	DIP	Sie	-	pdf
FZH 301	Logic	Quad 2 input NOR, protected	DIP	Sie	-	pdf
FZH 305	Logic	Quad 2 input NOR, protected	DIP	Sie	-	pdf

Type	Device	Short Description	Fig.	Manu.	Comparision Types	More at
ZJ 101...161	Logic-IC	LSL Logic (slow, hi-rel), 11,4...17V		Phi, Sie	-	pdf
ZJ 101	Logic	Jk-master/slave Flip-flop	DIP	Sie	-	pdf
ZJ 105	Logic	Jk-master/slave Flip-flop	DIP	Sie	-	pdf
ZJ 111	Logic	Jk-master/slave Flip-flop	DIP	Sie	-	pdf
ZJ 115	Logic	Jk-master/slave Flip-flop	DIP	Sie	-	pdf
ZJ 121	Logic	Dual, Jk-master/slave Ff, Set/reset	DIP	Sie	-	pdf
ZJ 125	Logic	Dual, Jk-master/slave Ff, Set/reset	DIP	Sie	-	pdf
ZJ 131	Logic	Quad Data Flip-flop	DIP	Sie	-	pdf
ZJ 135	Logic	Quad Data Flip-flop	DIP	Sie	-	pdf
ZJ 141A	Logic	4 Bit Decimal Counter, Set (each Bit), common Reset	DIP	Sie	-	pdf
ZJ 145A	Logic	4 Bit Decimal Counter, Set (each Bit), common Reset	DIP	Sie		pdf
ZJ 151A	Logic	4 Bit Binary Counter, Set (each Bit), common Reset	DIP	Sie	-	pdf
ZJ 155A	Logic	4 Bit Binary Counter, Set (each Bit), common Reset	DIP	Sie		pdf
ZJ 161	Logic	4 Bit Shift Register, Serial<→Parallel	DIP	Sie	-	pdf
ZJ 165	Logic	4 Bit Shift Register, Serial<→Parallel	DIP	Sie	-	pdf
ZK 101	Logic-IC	LSL Logic (slow, hi-rel), 11,4...17V		Phi, Sie	-	
ZK 101	Logic	Monoflop	DIP	Sie	-	pdf
ZK 105	Logic	Monoflop	DIP	Sie	-	pdf
ZL 101...141	Logic-IC	LSL Logic (slow, hi-rel), 11,4...17V		Phi, Sie		
ZL 101	Logic	BCD to Indicator Tube Decoder (1 of 10 Decoder) with Driver	DIP	Sie	-	pdf
ZL 105	Logic	BCD to Indicator Tube Decoder (1 of 10 Decoder) with Driver	DIP	Sie	-	pdf
ZL 111	Logic	BCD to 7-Segment Decoder with Driver (Open Collector 16.5V 20mA)	DIP	Sie	-	pdf
ZL 121	Logic	Driver (short-circuit-proof), 3 Inputs NOR, 20V Imax=400mA	DIP	Sie		pdf
ZL 121S	Logic	Driver (short-circuit-proof), 3 Inputs NOR, 30V Imax=400mA	DIP	Sie	-	pdf
ZL 125	Logic	Driver (short-circuit-proof), 3 Inputs NOR, 20V Imax=400mA	DIP	Sie	-	pdf
ZL 125S	Logic	Driver (short-circuit-proof), 3 Inputs NOR, 30V Imax=400mA	DIP	Sie	-	pdf
ZL 131	Logic	Driver (short-circuit-proof), 4 Inputs OR, 20V Imax=400mA	DIP	Sie	-	pdf
ZL 131S	Logic	Driver (short-circuit-proof), 4 Inputs OR, 30V Imax=400mA	DIP	Sie	-	pdf
ZL 135	Logic	Driver (short-circuit-proof), 4 Inputs OR, 20V Imax=400mA	DIP	Sie	-	pdf
ZL 135 S	Logic	Driver (short-circuit-proof), 4 Inputs OR, 30V Imax=400mA	DIP	Sie	E 435E	pdf
ZL 141	Logic	Driver (short-circuit-proof) 20V	DIP	Sie	-	pdf
ZL 141S	Logic	Driver (short-circuit-proof) 30V	DIP	Sie	-	pdf
ZL 145	Logic	Driver (short-circuit-proof) 20V	DIP	Sie	-	pdf
ZL 145S	Logic	Driver (short-circuit-proof) 30V	DIP	Sie	-	pdf
ZL 4141B	Logic	4 Bit, Driver (short-circuit-proof)	DIP	Sie	-	pdf
ZL 4141 D	LIN-IC	4x Treiber/Drv., Ucc=4,5...35V, 0...+70°	18-DIP	Sie	-	pdf
ZL 4145B	Logic	4 Bit, Driver (short-circuit-proof)	DIP	Sie	-	pdf
ZL 4145 D	LIN-IC	=FZL 4141D: -25...+85°	18-DIP	Sie	-	pdf
ZL 4146 G	LIN-IC	4x Tr./Drv., Ucc=4,5...40V, -25...+125°	20-MDIP	Sie	-	pdf pinout
ZT 458	Si-N	=ZTX 458: SMD	48j§	Ztx	-	data
ZT 489	Si-N	=FMMT 489:	48j§	Ztx	-	data
ZT 491	Si-N	=FCX 491:	48j§	Ztx	-	data
ZT 493	Si-N	=FCX 493:	48j§	Ztx	-	data
ZT 549	Si-P	=ZTX 549: SMD	48j§	Ztx	-	data
ZT 559	Si-P	=ZTX 559: SMD	48j§	Ztx	-	data
ZT 591	Si-P	=FCX 591:	48j§	Ztx	-	data
ZT 593	Si-P	=FCX 593:	48j§	Ztx	-	data
ZT 600	Si-N-Darl	=ZTX 600: SMD, 2A	48j§	Ztx	-	data
ZT 603	Si-N-Darl	=ZTX 603: SMD, 2A	48j§	Ztx	-	data
ZT 604	Si-N-Darl	=ZTX 604: SMD, 1,5A	48j§	Ztx	-	data
ZT 605	Si-N-Darl	=ZTX 605: SMD, 1,5A	48j§	Ztx	-	data
ZT 649	Si-N	=ZTX 649: SMD, 3A	48j§	Ztx	-	data
ZT 651	Si-N	=ZTX 651: SMD, 3A	48j§	Ztx	-	data
ZT 653	Si-N	=ZTX 653: SMD	48j§	Ztx	-	data
ZT 655	Si-N	=ZTX 655: SMD	48j§	Ztx	-	data
ZT 657	Si-N	=ZTX 657: SMD	48j§	Ztx	-	data
ZT 658	Si-N	=ZTX 658: SMD	48j§	Ztx	-	data
ZT 688 B	Si-N	=ZTX 688B: SMD, 4A	48j§	Ztx	-	data
ZT 689 B	Si-N	=ZTX 689B: SMD	48j§	Ztx	-	data
ZT 690 B	Si-N	=ZTX 690B: SMD, 3A	48j§	Ztx	-	data
ZT 692 B	Si-N	=ZTX 692B: SMD, 2A	48j§	Ztx	-	data
ZT 694 B	Si-N	=ZTX 694B: SMD, 1A	48j§	Ztx	-	data
ZT 696 B	Si-N	=ZTX 696B: SMD	48j§	Ztx	-	data
ZT 704	Si-P-Darl	=ZTX 704: SMD, 1,5A	48j§	Ztx	-	data
ZT 705	Si-P-Darl	=ZTX 705: SMD, 2A	48j§	Ztx	-	data
ZT 749	Si-P	=ZTX 749: SMD, 3A	48j§	Ztx	-	data
ZT 751	Si-P	=ZTX 751: SMD, 3A	48j§	Ztx	-	data
ZT 753	Si-P	=ZTX 753: SMD	48j§	Ztx	-	data
ZT 755	Si-P	=ZTX 755: SMD	48j§	Ztx	-	data
ZT 757	Si-P	=ZTX 757: SMD	48j§	Ztx	-	data

Type	Device	Short Description	Fig.	Manu	Comparision Types	More at
FZT 758	Si-P	=ZTX 758: SMD	48j§	Ztx	-	data
FZT 788 B	Si-P	=ZTX 788B: SMD	48j§	Ztx	-	data
FZT 789 A	Si-P	=ZTX 789A: SMD, 3A	48j§	Ztx	-	data
FZT 790 A	Si-P	=ZTX 790A: SMD, 3A	48j§	Ztx	-	data
FZT 792 A	Si-P	=ZTX 792A: SMD, 2A	48j§	Ztx	-	data
FZT 795 A	Si-P	=ZTX 795A: SMD	48j§	Ztx	-	data
FZT 796 A	Si-P	=ZTX 796A: SMD	48j§	Ztx	-	data
FZT 849	Si-N	=ZTX 849: SMD, 7A	48j§	Ztx	-	data
FZT 851	Si-N	=ZTX 851: SMD, 6A	48j§	Ztx	-	data
FZT 853	Si-N	=ZTX 853: SMD, 6A	48j§	Ztx	-	data
FZT 855	Si-N	=ZTX 855: SMD, 5A	48j§	Ztx	-	data
FZT 857	Si-N	=ZTX 857: SMD, 3,5A	48j§	Ztx	-	data
FZT 869	Si-N	=ZTX 869: SMD, 7A	48j§	Ztx	-	data
FZT 948	Si-P	=ZTX 948: SMD, 6A	48j§	Ztx	-	data
FZT 949	Si-P	=ZTX 949: SMD, 5,5A	48j§	Ztx	-	data
FZT 951	Si-P	=ZTX 951: SMD, 5A	48j§	Ztx	-	data
FZT 953	Si-P	=ZTX 953: SMD, 5A	48j§	Ztx	-	data
FZT 955	Si-P	=ZTX 955: SMD, 4A	48j§	Ztx	-	data
FZT 956	Si-P	=ZTX 956: SMD, 2A	48j§	Ztx	-	data
FZT 957	Si-P	=ZTX 957: SMD	48j§	Ztx	-	data
FZT 958	Si-P	=ZTX 958: SMD	48j§	Ztx	-	data
FZT 968	Si-P	=ZTX 968: SMD, 6A	48j§	Ztx	-	data
FZT 1047 A	Si-N	SMD, hi-beta, lo-sat, 35V, 5A, 150MHz B=300...1200	48j§	Ztx	-	data
FZT 1048 A	Si-N	SMD, hi-beta, lo-sat, 50V, 5A, 150MHz B=300...1200	48j§	Ztx	-	data
FZT 1049 A	Si-N	SMD, hi-beta, lo-sat, 80V, 5A, 180MHz B=300...1200	48j§	Ztx	-	data
FZT 1051 A	Si-N	SMD, hi-beta, lo-sat, 150V, 5A, 155MHz B=270...1200	48j§	Ztx	-	data
FZT 1053 A	Si-N	SMD, hi-beta, lo-sat, 150V, 4A, 140MHz B=300...1200	48j§	Ztx	-	data
FZT 2222 A	Si-N	=2N2222A: SMD	48j§	Ztx	-	data
FZT 2907(A)	Si-P	=2N2907(A): SMD	48j§	Ztx	-	data
FZT 4403	Si-P	=2N4403: SMD	48j§	Ztx	-	data
FZTA 14	Si-N	=MPSA 14: SMD	48j§	Ztx	-	data
FZTA 42	Si-N	=MPSA 42: SMD	48j§	Ztx	-	data
FZTA 63	Si-P-Darl	=MPSA 63: SMD	48j§	Ztx	-	data
FZTA 64	Si-P-Darl	=MPSA 64: SMD	48j§	Ztx	-	data
FZTA 92	Si-P	=MPSA 92: SMD	48j§	Ztx	-	data
FZY 101	Logic	Dual Voltage Generator	DIP	Sie	-	pdf
FZY 105	Logic	Dual Voltage Generator	DIP	Sie	-	pdf
G						
G.	Si-Di	=1PS79SB70 (Typ-Code/Stempel/marking)	71(1,3mm)	Phi	→1PS79SB70	data
G	PIN-Di	=BAR 63-02W (Typ-Code/Stempel/marking)	71(1,3mm)	Sie	→BAR 63-02W	data
G	PIN-Di	=BAR 63-03W (Typ-Code/Stempel/marking)	71(1,7mm)	Sie	→BAR 63-03W	data
G	Z-Di	=MAZF 082 (Typ-Code/Stempel/marking)	71(1,7mm)	Mat	→MAZF 082	data
G	Z-Di	=MAZN 082 (Typ-Code/Stempel/marking)	71(1,3mm)	Mat	→MAZN 082	data
G	Si-Di	=S 5566G (Typ-Code/Stempel/marking)	31	Tos	→S 5566G	data
G	Si-Di	=SB 007-03Q (Typ-Code/Stempel/marking)	35(2mm)	Say	→SB 007-03Q	data
G	Si-Di	=SB007-03CP (Typ-Code/Stempel/marking)	35	Say	→SB 007-03CP	data
G 1	Si-N	=BFS 20 (Typ-Code/Stempel/marking)	35	Fer,Phi,Tho	→BFS 20	data
G 1	Si-N+R	=FMG 1A (Typ-Code/Stempel/marking)	45	Rhm	→FMG 1A	data
G 1	Si-Di	=FRS 2A (Typ-Code/Stempel/marking)	71(5x3,5)	Fag	→FRS 2A	data
G 1	Si-N	=MMBT 5551 (Typ-Code/Stempel/marking)	35	Mot	→MMBT 5551	data
G 1	Si-N+R	=UMG 1N (Typ-Code/Stempel/marking)	45(2mm)	Rhm	→UMG 1N...11N	data
G 1 A..M	Si-Di	Gl, Uni, 50...1000V, 1A A=50, B=100, D=200, G=400, J=600V, K=800, M=1000V	31a	Gie	BY 126...127, BY 133...135, 1N4001..07, ++	data
G 1	Si-N	=D70Y1.5T1 (Typ-Code/Stempel/marking)	39	Gen	→D70Y1.5T1	data
G 1 F	Si-N	=BC 847B (Typ-Code/Stempel/marking)	35	Rhm	→BC 847B	data
G 1 K	Si-N	=BC 848B (Typ-Code/Stempel/marking)	35	Rhm	→BC 848B	data
G 1 K	Si-N	=BC 848BW (Typ-Code/Stempel/marking)	35(2mm)	Rhm	→BC 848BW	data
G 1 L	Si-N	=BC 848C (Typ-Code/Stempel/marking)	35	Rhm	→BC 848C	data
G 1 p	Si-N	=BFS 20 (Typ-Code/Stempel/marking)	35	Phi	→BFS 20	data
- G 1	Si-N	=PMST 5551 (Typ-Code/Stempel/marking)	35(2mm)	Phi	→PMST 5551	data
G 1 1	Si-N	=BFS 20 (Typ-Code/Stempel/marking)	35	Phi	→BFS 20	data
G1X19X	LED	gr ø5mm I:18lm Vf:3.5V If:350mA		Seo		data pdf pinout
G1X29X	LED	gr ø5mm I:93lm Vf:3.5V If:700mA		Seo		data pdf pinout
G1X49X	LED	gr ø5mm I:189lm Vf:3.5V If:1.4AmA		Seo		data pdf pinout
G 2	Si-P	=BF 550 (Typ-Code/Stempel/marking)	35	Val,Aeg	→BF 550	data
G 2	Si-N	=BF 599 (Typ-Code/Stempel/marking)	35	Kec	→BF 599	data
G 2	GaAs-FET-IC	=CGY 50 (Typ-Code/Stempel/marking)	44	Sie		data
G 2	Si-N+R	=FMG 2A (Typ-Code/Stempel/marking)	45	Rhm	→FMG 2A	data
G 2	Si-Di	=FRS 2B (Typ-Code/Stempel/marking)	71(5x3,5)	Fag	→FRS 2B	data
G 2	Si-N+R	=UMG 2N (Typ-Code/Stempel/marking)	45(2mm)	Rhm	→UMG 1N...11N	data

Type	Device	Short Description	Fig.	Manu	Comparision Types	More at
G 2 A..M	Si-Di	GI, Uni, 50...1000V, 2A, 1,5µs A=50, B=100, D=200, G=400, J=600V, K=800, M=1000V	31a	Gie	BY 251..255, BY 259/..., GP 20A..M, ++	data
G 2 N 2955	Si-P	≈BD 318	23a§		→BD 318	
G 2 N 3055	Si-N	≈BD 317	23a§		→BD 317	
G 2 SBA 20, 60	Si-Br	GI-Br, 20=200V, 60=600V, 1,5A	33x1	Gsi	B140...420 C1500	
G 3	PIN-Di	=BAR 63 (Typ-Code/Stempel/marking)	35	Sie	→BAR 63	data
G 3	Si-P	=BF 536 (Typ-Code/Stempel/marking)	35	Val	→BF 536	data
G 3	Si-N+R	=FMG 3A (Typ-Code/Stempel/marking)	45	Rhm	→FMG 3A	data
G 3	Si-Di	=FRS 2D (Typ-Code/Stempel/marking)	71(5x3,5)	Fag	→FRS 2D	data
G 3	Si-Di	=KDS 196 (Typ-Code/Stempel/marking)	35	Kec	→KDS 196	data
G 3	Si-P	=MMBT 589 (Typ-Code/Stempel/marking)	35	Ons	→MMBT 589	
G 3	Si-P	=MMBT 6589 (Typ-Code/Stempel/marking)	42	Ons	→MMBT 6589	
- G 3	Si-N	=PMST 5551 (Typ-Code/Stempel/marking)	35(2mm)	Phi	→PMST 5551	data
G 3	Si-N+R	=UMG 3N (Typ-Code/Stempel/marking)	45(2mm)	Rhm	→UMG 1N...11N	data
G 3	Si-Di	=1SS196 (Typ-Code/Stempel/marking)	35	Tos	→1SS196	data
G 3	Si-P	=2SA1257-3 (Typ-Code/Stempel/marking)	35	Say	→2SA1257	data
G 3	Sj-N	=2SC2107-G3 (Typ-Code/Stempel/marking)	35	Nec	→2SC2107	data
G 3 A..M	Si-Di	GI, Uni, 50...1000V, 3A, 3µs A=50, B=100, D=200, G=400, J=600V, K=800, M=1000V	31a	Gie	BY 251..255, GP 30A...M, 1N5400...08, ++	data
G 3 F	Si-P	=BC 857B (Typ-Code/Stempel/marking)	35	Rhm	→BC 857B	data
G 3 K	Si-P	=BC 858B (Typ-Code/Stempel/marking)	35	Rhm	→BC 858B	data
G 3 K	Si-P	=BC 858BW (Typ-Code/Stempel/marking)	35(2mm)	Rhm	→BC 858BW	data
G 3 SBA 20, 60	Si-Br	GI-Br, 20=200V, 60=600V, 4A	33x1	Gsi	B140...420 C4000	
G3X18X	LED	gr ø5mm I:18lm Vf:3.5V If:350mA		Seo		data pdf pinout
G3X28X	LED	gr ø5mm I:84lm Vf:4V If:700mA		Seo		data pdf pinout
G 4	PIN-Di	=BAR 63-04 (Typ-Code/Stempel/marking)	35	Sie	→BAR 63-04	data
G 4	Si-N	=BFS 20R (Typ-Code/Stempel/marking)	35	Aeg,Fer,Tho Val	→BFS 20R	data
G 4	Si-N+R	=FMG 4A (Typ-Code/Stempel/marking)	45	Rhm	→FMG 4A	data
G 4	Si-Di	=FRS 2G (Typ-Code/Stempel/marking)	71(5x3,5)	Fag	→FRS 2G	data
G 4	Si-P	=MBT 35200M (Typ-Code/Stempel/marking)	46	Ons	→MBT 35200M	
G 4	Si-N+R	=UMG 4N (Typ-Code/Stempel/marking)	45(2mm)	Rhm	→UMG 1N...11N	data
G 4	Si-P	=2SA1257-4 (Typ-Code/Stempel/marking)	35	Say	→2SA1257	data
G 4	Si-N	=2SC2107-G4 (Typ-Code/Stempel/marking)	35	Nec	→2SC2107	data
G 4 A..M	Si-Di	GI, Uni, 50...1000V, 3A, 3µs A=50, B=100, D=200, G=400, J=600V, K=800, M=1000V	31a	Gie	BY 251..255, GP 30A...M, 1N5400...08, ++	data
G 4 s	PIN-Di	=BAR 63-04W (Typ-Code/Stempel/marking)	35(2mm)	Sie	→BAR 63-04W	data
G 5	PIN-Di	=BAR 63-05 (Typ-Code/Stempel/marking)	35	Sie	→BAR 63-05	data
G 5	Si-N+R	=FMG 5A (Typ-Code/Stempel/marking)	45	Rhm	→FMG 5A	data
G 5	Si-Di	=FRS 2J (Typ-Code/Stempel/marking)	71(5x3,5)	Fag	→FRS 2J	data
G 5	Si-N+R	=UMG 5N (Typ-Code/Stempel/marking)	45(2mm)	Rhm	→UMG 1N...11N	data
G 5	Si-P	=2SA1257-5 (Typ-Code/Stempel/marking)	35	Say	→2SA1257	data
G 5	Si-N	=2SC2107-G5 (Typ-Code/Stempel/marking)	35	Nec	→2SC2107	data
G 5	Si-P	=BF 550R (Typ-Code/Stempel/marking)	35	Val,Aeg	→BF 550R	data
G 5 s	PIN-Di	=BAR 63-05W (Typ-Code/Stempel/marking)	35(2mm)	Sie	→BAR 63-05W	data
G 5 SBA 20, 60	Si-Br	GI-Br, 20=200V, 60=600V, 6A	33x1	Gsi	B140...420 C6000	
G 5/2	Ge-Di	≈OA 70	31a		→OA 70	
G 5/61	Ge-Di	≈AA 119	31a		→AA 119	
G 6	PIN-Di	=BAR 63-06 (Typ-Code/Stempel/marking)	35	Sie	→BAR 63-06	data
G 6	Si-N+R	=FMG 6A (Typ-Code/Stempel/marking)	45	Rhm	→FMG 6A	data
G 6	Si-Di	=FRS 2K (Typ-Code/Stempel/marking)	71(5x3,5)	Fag	→FRS 2K	data
G 6	Si-N+R	=UMG 6N (Typ-Code/Stempel/marking)	45(2mm)	Rhm	→UMG 1N...11N	data
G 6	Si-N	=2SC2107-G6 (Typ-Code/Stempel/marking)	35	Nec	→2SC2107	data
G 6	Si-P	=BF 569 (Typ-Code/Stempel/marking)	35	Val	→BF 569	data
G 6 s	PIN-Di	=BAR 63-06W (Typ-Code/Stempel/marking)	35(2mm)	Sie	→BAR 63-06W	data
G 7	Si-P	=BF 579 (Typ-Code/Stempel/marking)	35	Val,Aeg	→BF 579	data
G 7	Si-N+R	=FMG 7A (Typ-Code/Stempel/marking)	45	Rhm	→FMG 7A	data
G 7	Si-Di	=FRS 2M (Typ-Code/Stempel/marking)	71(5x3,5)	Fag	→FRS 2M	data
G 7	Si-N+R	=UMG 7N (Typ-Code/Stempel/marking)	45(2mm)	Rhm	→UMG 1N...11N	data
G 8	Si-P	=BF 660 (Typ-Code/Stempel/marking)	35	Val	→BF 660	data
G 8	Si-N+R	=FMG 8A (Typ-Code/Stempel/marking)	45	Rhm	→FMG 8A	data
G 8	Si-N+R	=UMG 8N (Typ-Code/Stempel/marking)	45(2mm)	Rhm	→UMG 1N...11N	data
G 9	Si-P	=BF 767 (Typ-Code/Stempel/marking)	35	Val	→BF 767	data
G 9	Si-N+R	=FMG 9A (Typ-Code/Stempel/marking)	45	Rhm	→FMG 9A	data
G 9	Si-N+R	=UMG 9N (Typ-Code/Stempel/marking)	45(2mm)	Rhm	→UMG 1N...11N	data
G 10	Si-N+R	=FMG 10A (Typ-Code/Stempel/marking)	45	Rhm	→FMG 10A	data
G 10	Si-N+R	=UMG 10N (Typ-Code/Stempel/marking)	45(2mm)	Rhm	→UMG 1N...11N	data
G 11	Si-N+R	=FMG 11A (Typ-Code/Stempel/marking)	45	Rhm	→FMG 11A	data
G 11	Si-N+R	=UMG 11N (Typ-Code/Stempel/marking)	45(2mm)	Rhm	→UMG 1N...11N	data
G 11	MOS-N-FET-e	=2SK1133 (Typ-Code/Stempel/marking)	35	Nec	→2SK1133	data
G 11	Si-Di	=STPS 1H100U(Typ-Code/Stempel/marking)	71(5x3,5)	Tho	→STPS 1H100U	data
G 12	Si-N+R	=FMG 12A (Typ-Code/Stempel/marking)	45	Rhm	→FMG 12A	data
G 12	MOS-N-FET-e	=2SK1399 (Typ-Code/Stempel/marking)	35	Nec	→2SK1399	data
G 13	Si-N+R	=FMG 13A (Typ-Code/Stempel/marking)	45	Rhm	→FMG 13A	data
G 13	MOS-N-FET-e*	=2SK1580 (Typ-Code/Stempel/marking)	35(2mm)	Nec	→2SK1580	data
G 14	MOS-N-FET-e*	=2SK1581 (Typ-Code/Stempel/marking)	35	Nec	→2SK1581	data
G 15	MOS-N-FET-e*	=2SK1582 (Typ-Code/Stempel/marking)	35	Nec	→2SK1582	data

Type	Device	Short Description	Fig.	Manu	Comparision Types	More at
G 15 N35CL	MOS-N-IGBT*	=MGB 15N35CL(Typ-Code/Stempel/marking)	30	Ons	→MGB 15N35CL	
G 15 N35CL	MOS-N-IGBT*	=MGP 15N35CL(Typ-Code/Stempel/marking)	17	Ons	→MGP 15N35CL	
G 15 N40CL	MOS-N-IGBT*	=MGB 15N40CL(Typ-Code/Stempel/marking)	30	Ons	→MGB 15N40CL	
G 15 N40CL	MOS-N-IGBT*	=MGP 15N40CL(Typ-Code/Stempel/marking)	17	Ons	→MGP 15N40CL	
G 16	MOS-N-FET-e*	=2SK1590 (Typ-Code/Stempel/marking)	35	Nec	→2SK1590	data
G 17	MOS-N-FET-e*	=2SK1589 (Typ-Code/Stempel/marking)	35	Nec	→2SK1589	data
G 18	MOS-N-FET-e*	=2SK1591 (Typ-Code/Stempel/marking)	35	Nec	→2SK1591	data
G 19	MOS-N-FET-e*	=2SK1657 (Typ-Code/Stempel/marking)	35	Nec	→2SK1657	data
G 19 N35CL	MOS-N-IGBT*	=MGB 19N35CL(Typ-Code/Stempel/marking)	30	Ons	→MGB 19N35CL	
G 19 N35CL	MOS-N-IGBT*	=MGP 19N35CL(Typ-Code/Stempel/marking)	17	Ons	→MGP 19N35CL	
G 20	MOS-N-FET-e*	=2SK1658 (Typ-Code/Stempel/marking)	35(2mm)	Nec	→2SK1658	data
G 21	MOS-N-FET-e*	=2SK1958 (Typ-Code/Stempel/marking)	35(2mm)	Nec	→2SK1958	data
G 22	MOS-N-FET-e*	=2SK2090 (Typ-Code/Stempel/marking)	35(2mm)	Nec	→2SK2090	data
G 23	MOS-N-FET-e*	=2SK2158 (Typ-Code/Stempel/marking)	35	Nec	→2SK2158	data
G 23	Si-Di	=STPS 1L30A (Typ-Code/Stempel/marking)	71(5x2,5)	Tho	→STPS 1L30A	data
G 23	Si-Di	=STPS 2L25U (Typ-Code/Stempel/marking)	71(5x3,5)	Tho	→STPS 2L25U	data
G 24	MOS-N-FET-e*	=2SK2858 (Typ-Code/Stempel/marking)	35(2mm)	Nec	→2SK2858	data
G 25	MOS-N-FET-e*	=2SK3054 (Typ-Code/Stempel/marking)	35(2mm)	Nec	→2SK3054	data
G 30	Si-Di	=STPS 2L30A (Typ-Code/Stempel/marking)	71(5x2,5)	Tho	→STPS 2L30A	data
G 61	Si-P	=BF 569R (Typ-Code/Stempel/marking)	35	Fer	→BF 569R	data
G 81	Si-P	=BF 660R (Typ-Code/Stempel/marking)	35	Val	→BF 660R	data
G 088	Si-St	≈BZ 102/0V7	31a		→BZ 102/0V7	
G 150	LIN-IC	Telecom, Teilnehmerschltg. (SLIC)	28-DIP	Sie	-	
G 300 HHCK 12P2	CMOS/LIN-IC	Plastik Half-Bridge IGBT Module	80	Inr	-	pdf pinout
G 450 HHBK 06P2	CMOS/LIN-IC	Plastik Half-Bridge IGBT Module	80	Inr	-	pdf pinout
G 580	Ge-Di	Uni, 20V, 0,2A	31a	Itt	AA 139, 1N270	data
G1126	PD	610(560)nm S:0.17mA/mW	TO5	Ham		data pdf pinout
G1127	PD	610(560)nm S:0.17mA/mW	TO8	Ham		data pdf pinout
G1735	PD	710nm S:0.4mA/mW	TO18	Ham		data pdf pinout
G1736	PD	710nm S:0.4mA/mW	TO5	Ham		data pdf pinout
G1737	PD	710nm S:0.4mA/mW	TO8	Ham		data pdf pinout
G1738	PD	710nm S:0.4mA/mW		Ham		data pdf pinout
G1740	PD	710nm S:0.4mA/mW		Ham		data pdf pinout
G1746	PD	710nm S:0.22mA/mW	TO5	Ham		data pdf pinout
G1747	PD	710nm S:0.22mA/mW	TO8	Ham		data pdf pinout
G2119	PD	610(560)nm S:0.17mA/mW		Ham		data pdf pinout
G3297	PD	710nm S:0.4mA/mW	TO18	Ham		data pdf pinout
G4176	PD	850nm S:0.3mA/mW	TO5	Ham		data pdf pinout
G5645	PD	470nm S:0.05mA/mW	TO18	Ham		data pdf pinout
G5842	PD	370nm		Ham		data pdf pinout
G5851	PIN PD	1750nm S:1.1mA/mW	TO8	Ham		data pdf pinout
G5852	PIN PD	1950nm S:1.2mA/mW	TO8	Ham		data pdf pinout
G5853	PIN PD	2300nm S:1.1mA/mW	TO8	Ham		data pdf pinout
G 6004	Si-P-Darl	≈BD 700	15j§		→BD 700	
G 6005	Si-N-Darl	≈BD 699	15j§		→BD 699	
G6262	PD	470nm S:0.05mA/mW		Ham		data pdf pinout
G6742	PIN PD	1550nm S:0.95mA/mW		Ham		data pdf pinout
G 6903 SZ	IC	Green Mode Pfc/flyback-pwm Controller	TSOP	Sgc		pdf pinout
G7096	PD	1500(1300)nm S:0.4mA/mW	TO5	Ham		data pdf pinout
G7189	PD	470nm S:0.02mA/mW		Ham		data pdf pinout
G8198-01	PIN PD	1550nm S:0.9mA/mW		Ham		data pdf pinout
G8198-02	PIN PD	1550nm S:0.9mA/mW		Ham		data pdf pinout
G8371	PIN PD	1750nm S:1.1mA/mW	TO5	Ham		data pdf pinout
G8372	PIN PD	1950nm S:1.2mA/mW	TO5	Ham		data pdf pinout
G8373	PIN PD	2300nm S:1.1mA/mW	TO5	Ham		data pdf pinout
G8421	PIN PD	1750nm S:1.1mA/mW	TO18	Ham		data pdf pinout
G8422	PIN PD	1950nm S:1.2mA/mW	TO18	Ham		data pdf pinout
G8423	PIN PD	2300nm S:1.1mA/mW	TO18	Ham		data pdf pinout
G 8870	LIN-IC	→KT 3170			KT 3170, MT 8870	data pdf pinout

GA

Type	Device	Short Description	Fig.	Manu	Comparision Types	More at
GA	Si-N	=µPA600T (Typ-Code/Stempel/marking)	46	Nec	→µPA600T	data
GA	Si-N	=2SC4922 (Typ-Code/Stempel/marking)	35(1,6mm)	Say	→2SC4922	data
GA	Si-N	=2SD1463-GA (Typ-Code/Stempel/marking)	39	Hit	→2SD1463	data
GA	Si-Di	=BAW 78A (Typ-Code/Stempel/marking)	39	Sie	→BAW 78A	data
GA	Si-N	=BFR 35 (Typ-Code/Stempel/marking)	35	Sie	→BFR 35	data
GA	Si-Di	=GF 1A (Typ-Code/Stempel/marking)	71(5x2,7)	Gsi	→GF 1A	data
GA	Si-P	=S 416 T (Typ-Code/Stempel/marking)	35	Aeg	→S 416 T	data
GA 1	Si-P	=2SA1871-GA1(Typ-Code/Stempel/marking)	39	Nec	→2SA1871	data
GA 1 A3Q	Si-N+R	=AA 1A3Q: SMD	35a(2mm)	Nec	DTC 113ZU	data
GA 1 A4M	Si-N+R	=AA 1A4M: SMD	35a(2mm)	Nec	DTC 114EU, 2SC4398	data
GA 1 A4P	Si-N+R	=AA 1A4P: SMD	35a(2mm)	Nec	DTC 114YU	data
GA 1 A4Z	Si-N+R	=AA 1A4Z: SMD	35a(2mm)	Nec	DTC 114TU	data
GA 1 F4M	Si-N+R	=AA 1F4M: SMD	35a(2mm)	Nec	DTC 124EU, 2SC4397	data
GA 1 F4N	Si-N+R	=AA 1F4N: SMD	35a(2mm)	Nec	DTC 124XU	data
GA 1 F4Z	Si-N+R	=AA 1F4Z: SMD	35a(2mm)	Nec	DTC 124TU	data
GA 1 L3M	Si-N+R	=AA 1L3M: SMD	35a(2mm)	Nec	DTC 143EU	data
GA 1 L3N	Si-N+R	=AA 1L3N: SMD	35a(2mm)	Nec	DTC 143XU	data
GA 1 L3Z	Si-N+R	=AA 1L3Z: SMD	35a(2mm)	Nec	DTC 143TU	data
GA 1 L4L	Si-N+R	=AA 1L4L: SMD	35a(2mm)	Nec	DTC 144WU	data
GA 1 L4M	Si-N+R	=AA 1L4M: SMD	35a(2mm)	Nec	DTC 144EU, 2SC4396	data

Type	Device	Short Description	Fig.	Manu	Comparision Types	More at
GA 1 L4Z	Si-N+R	=AA 1L4Z: SMD	35a(2mm)	Nec	DTC 144TU	data
GA 2	Si-P	=2SA1871-GA2(Typ-Code/Stempel/marking)	39	Nec	→2SA1871	data
GA 3	Si-P	=2SA1871-GA3(Typ-Code/Stempel/marking)	39	Nec	→2SA1871	data
GA 6	Si-Di	=STPS 160A (Typ-Code/Stempel/marking)	71(5x2,5)	Tho	→STPS 160A	data
GA 75 ... [IR]	MOS-IGBT	Int. Rectifier Power Module		Inr	-	
GA 100	Ge-Di	Uni, 26V, 0,02A	31a	Hfo	AA 132...134, 1N34, 1N54, 1N60	data
GA 100 ... [IR]	MOS-IGBT	Int. Rectifier Power Module		Inr	-	
GA 101	Ge-Di	Uni, 50V, 0,015A	31a	Hfo	AA 132...134, 1N34, 1N54, 1N60	data
GA 102	Ge-Di	Uni, 60V, 0,012A	31a	Hfo	AA 132...134, 1N34, 1N54, 1N60	data
GA 103	Ge-Di	Uni, 90V, 0,01A	31a	Hfo	AA 117...118, AA 132...133	data
GA 104	Ge-Di	Uni, 115V, 0,01A	31a	Hfo	AA 117...118, AA 132...133	data
GA 105	Ge-Di	Uni, 26V, 0,02A	31a	Hfo	AA 132...134, 1N34, 1N54, 1N60	data
GA 106	Ge-Di	S, 35V, 0,03A(ss), 500ns	31a	Hfo	AAY 49, AAZ 15, AAZ 17, 1N276	data
GA 107	Ge-Di	S, 90V, 0,15A(ss), 500ns	31a	Hfo	AAZ 15	data
GA 108	Ge-Di	Uni, 90V, 0,02A	31a	Hfo	AA 117...118, AA 132...133	data
(2)GA 109	Ge-Di	FM-Dem, 50V, 0,015A	31a	Hfo	AA 113, 1N34, 1N54, 1N60	data
(2)GA 113	Ge-Di	Dem, Diskr, 35V, 0,03A(ss)	31a	Hfo	AA 113, AA 119, 1N34, 1N54, 1N60	data
(4)GA 114	Ge-Di	4x Ge-Di, Ring-Dem, 35V, 0,03A(ss)	31a	Hfo	→4x GA 113	data
GA 150 ... [IR]	MOS-IGBT	Int. Rectifier Power Module		Inr	-	
GA 200 ... [IR]	MOS-IGBT	Int. Rectifier Power Module		Inr	-	
GA 250 ... [IR]	MOS-IGBT	Int. Rectifier Power Module		Inr	-	
GA 300 ... [IR]	MOS-IGBT	Int. Rectifier Power Module		Inr	-	
GA 400 ... [IR]	MOS-IGBT	Int. Rectifier Power Module		Inr	-	
GA 500 ... [IR]	MOS-IGBT	Int. Rectifier Power Module		Inr	-	
GA 5005(A...T)	Si-Di	TV-Booster-Di, 6...7kV, 0,3...0,44A		Gie,Tho	-	data
GAL 16.....600...	DIG-IC	Logic Gate Arrays	DIP, PLCC	Tho		
GAY 60	Ge-Di	Uni, S, 20V, 0,075A	31a	Hfo	AA 132...134, 1N34, 1N54, 1N60	data
GAY 61	Ge-Di	Uni, S, 20V, 0,1A	31a	Hfo	AA 135...136, AA 139	data
GAY 62	Ge-Di	Uni, S, 20V, 0,1A	31a	Hfo	AA 135...136, AA 139	data
GAY 63	Ge-Di	Uni, S, 40V, 0,1A	31a	Hfo	AA 136, 1N270	data
GAY 64	Ge-Di	Uni, S, 80V, 0,075A	31a	Hfo	1N270	data
GAZ 14	Ge-Di	S, 25V, 0,02A, 500ns	31a	Hfo	AAY 49, AAZ 15, AAZ 17, 1N276	data
GAZ 15	Ge-Di	Uni, S, 25V, 0,02A	31a	Hfo	AA 132...134, 1N34, 1N54, 1N60	data
GAZ 16	Ge-Di	=GAZ 14	31a	Hfo	→GAZ 14	data
GAZ 17	Ge-Di	=GAZ 15	31a	Hfo	→GAZ 15	data

GB

Type	Device	Short Description	Fig.	Manu	Comparision Types	More at
GB	Si-N	=2SD1463-GB (Typ-Code/Stempel/marking)	39	Hit	→2SD1463	data
GB	Si-Di	=BAW 78B (Typ-Code/Stempel/marking)	39	Sie	→BAW 78B	data
GB	Si-N	=BFR 35A (Typ-Code/Stempel/marking)	35	Sie	→BFR 35A	data
GB	Si-Di	=GF 1B (Typ-Code/Stempel/marking)	71(5x2,7)	Gsi	→GF 1B	data
GB 1	Si-P	=2SB1578-GB1(Typ-Code/Stempel/marking)	39	Nec	→2SA1578	data
GB 2	Si-P	=2SB1578-GB2(Typ-Code/Stempel/marking)	39	Nec	→2SA1578	data
GB 3	Si-P	=2SB1578-GB3(Typ-Code/Stempel/marking)	39	Nec	→2SA1578	data
GB 3	Si-Di	=STPS 1L30U (Typ-Code/Stempel/marking)	71(5x3,5)	Tho	→STPS 1L30U	data
GB 4	Si-Di	=STPS 1L40A (Typ-Code/Stempel/marking)	71(5x2,5)	Tho	→STPS 1L40A	data
GBA	Z-Di	=SMCJ 60CA (Typ-Code/Stempel/marking)	71(8x5mm)	Tho	→SMCJ ...	data
GBB	Z-Di	=SMCJ 70CA (Typ-Code/Stempel/marking)	71(8x5mm)	Tho	→SMCJ ...	data
GBC 107	Si-N	=BC 237	7a		→BC 237	data
GBC 108	Si-N	=BC 238	7a		→BC 238	data
GBC 109	Si-N	=BC 239	7a		→BC 239	data
GBD 179	Si-N	≈BD 179	14h§		→BD 179	
GBD 189	Si-N	≈BD 189	14h§		→BD 189	
GBD 190	Si-P	≈BD 190	14h§		→BD 190	
GBD 266	Si-P-Darl	≈BD 700	15j§		→BD 700	
GBD 267	Si-N-Darl	≈BD 699	15j§		→BD 699	
GBD 645	Si-N-Darl	≈BD 699	15j§		→BD 699	
GBD 646	Si-P-Darl	≈BD 700	15j§		→BD 700	
GBE	Z-Di	=SMCJ 85CA (Typ-Code/Stempel/marking)	71(8x5mm)	Tho	→SMCJ ...	data
GBG	Z-Di	=SMCJ 100CA (Typ-Code/Stempel/marking)	71(8x5mm)	Tho	→SMCJ ...	data
GBI	Z-Di	=SMCJ 130CA (Typ-Code/Stempel/marking)	71(8x5mm)	Tho	→SMCJ ...	data
GBL	Z-Di	=SMCJ 154CA (Typ-Code/Stempel/marking)	71(8x5mm)	Tho	→SMCJ ...	data
GBL 01	Si-Br	GI-Br, 100V, 4A	33x1	Gsi	B70 C4000	
GBL 02	Si-Br	GI-Br, 200V, 4A	33x1	Gsi	B140 C4000	
GBL 04	Si-Br	GI-Br, 400V, 4A	33x1	Gsi	B280 C4000	
GBL 005	Si-Br	GI-Br, 50V, 4A	33x1	Gsi	B35 C4000	
GBL 06	Si-Br	GI-Br, 600V, 4A	33x1	Gsi	B420 C4000	
GBL 08	Si-Br	GI-Br, 800V, 4A	33x1	Gsi	B560 C4000	
GBL 10	Si-Br	GI-Br, 1000V, 4A	33x1	Gsi	B700 C4000	
GBM	Z-Di	=SMCJ 170CA (Typ-Code/Stempel/marking)	71(8x5mm)	Tho	→SMCJ ...	data
GBN	Z-Di	=SMCJ 188CA (Typ-Code/Stempel/marking)	71(8x5mm)	Tho	→SMCJ ...	data
GBPC 12(-W)-...	Si-Br	GI-Br, 50...1000V, 12A 01=100, 02=200, 04=400, 005=50V 06=600, 08=800, 10=1000V	79x3	Gsi	KBPC 12-...	
GBPC 15(-W)-...	Si-Br	GI-Br, 50...1000V, 15A 01=100, 02=200, 04=400, 005=50V 06=600, 08=800, 10=1000V	79x3	Gsi	KBPC 15-...	

Type	Device	Short Description	Fig.	Manu	Comparision Types	More at
GBPC 25(-W)-...	Si-Br	Gl-Br, 50...1000V, 25A 01=100, 02=200, 04=400, 005=50V 06=600, 08=800, 10=1000V	79x3	Gsi	KBPC 25-...	
GBPC 35(-W)-...	Si-Br	Gl-Br, 50...1000V, 35A 01=100, 02=200, 04=400, 005=50V 06=600, 08=800, 10=1000V	79x3	Gsi	KBPC 35-...	
GBPC 101	Si-Br	Gl-Br, 100V, 3A	79x3	Gsi	B70 C3300	
GBPC 102	Si-Br	Gl-Br, 200V, 3A	79x3	Gsi	B140 C3300	
GBPC 104	Si-Br	Gl-Br, 400V, 3A	79x3	Gsi	B280 C3300	
GBPC 106	Si-Br	Gl-Br, 600V, 3A	79x3	Gsi	B420 C3300	
GBPC 108	Si-Br	Gl-Br, 800V, 3A	79x3	Gsi	B560 C3300	
GBPC 110	Si-Br	Gl-Br, 1000V, 3A	79x3	Gsi	B800 C3300	
GBPC 601	Si-Br	Gl-Br, 100V, 6A	79x3	Gsi	B70 C6000	
GBPC 602	Si-Br	Gl-Br, 200V, 6A	79x3	Gsi	B140 C6000	
GBPC 604	Si-Br	Gl-Br, 400V, 6A	79x3	Gsi	B280 C6000	
GBPC 606	Si-Br	Gl-Br, 600V, 6A	79x3	Gsi	B420 C6000	
GBPC 608	Si-Br	Gl-Br, 800V, 6A	79x3	Gsi	B560 C6000	
GBPC 610	Si-Br	Gl-Br, 1000V, 6A	79x3	Gsi	B700 C6000	
GBPC 1005	Si-Br	Gl-Br, 50V, 3A	79x3	Gsi	B35 C3300	
GBPC 6005	Si-Br	Gl-Br, 50V, 6A	79x3	Gsi	B35 C6000	
GBU 4 A..M	Si-Br	Gl-Br, 50...1000V, 4A A=50, B=100, D=200, G=400, J=600V K=800, M=1000V	33x1	Gsi	B35...700 C4000	
GBU 6 A..M	Si-Br	Gl-Br, 50...1000V, 6A A=50, B=100, D=200, G=400, J=600V K=800, M=1000V	33x1	Gsi	B35...700 C6000	
GBU 8 A..M	Si-Br	Gl-Br, 50...1000V, 8A A=50, B=100, D=200, G=400, J=600V K=800, M=1000V	33x1	Gsi	B35...700 C8000	

GC

Type	Device	Short Description	Fig.	Manu	Comparision Types	More at
GC	Si-N	=2SC2734 (Typ-Code/Stempel/marking)	35	Hit	→2SC2734	data
GC	Si-N	=2SC4264 (Typ-Code/Stempel/marking)	35(2mm)	Hit	→2SC4264	data
GC	Si-Di	=BAW 78C (Typ-Code/Stempel/marking)	39	Sie	→BAW 78C	data
GC	Si-Di	=U 1GC44 (Typ-Code/Stempel/marking)	71(5x2,5)	Tos	→U 1GC44	data
GC 4	Si-Di	=STPS 1L40U (Typ-Code/Stempel/marking)	71(5x3,5)	Tho	→STPS 1L40U	data
GC 100	Ge-P	NF-V, 15V, 15mA, 0,03W	2a	Hfo	AC 122, AC 125...126, AC 151	data
GC 101	Ge-P	NF-V, 15V, 15mA, 0,03W	2a	Hfo	AC 122, AC 125...126, AC 151	data
GC 102	Ge-P	NF-V/Tr, 15V, 50mA	2a	Hfo	AC 122, AC 125...126, AC 151	data
GC 103	Ge-P	NF-V, 15V, 15mA	2a	Hfo	AC 122, AC 125...126, AC 151	data
GC 104	Ge-P	=GC 103: ra	2a	Hfo	AC 151r, ACY 32	data
GC 111	Ge-P	NF, 80V, 0,125A, 0,07W	2a	Hfo	ACY 39, 2N2042...2043	data
GC 112	Ge-P	NF, 80V, 0,125A, 0,07W	2a	Hfo	ACY 39, 2N2042...2043	data
GC 115	Ge-P	NF-V/Tr, 20V, 0,15A, 0,07W	2a	Hfo	AC 122, AC 125...126, AC 151	data
GC 116	Ge-P	NF-V/Tr, 20V, 0,15A, 0,07W	2a	Hfo	AC 122, AC 125...126, AC 151	data
GC 117	Ge-P	=GC 116: ra	2a	Hfo	AC 151r, ACY 32	data
GC 118	Ge-P	=GC 116: ra	2a	Hfo	AC 151r, ACY 32	data
GC 120	Ge-P	NF-Tr/E, 20V, 0,15A, 0,07W	2a	Hfo	AC 128, AC 152...153, AC 188	data
GC 121	Ge-P	NF-Tr/E, 20V, 0,15A, 0,07W	2a	Hfo	AC 128, AC 152...153, AC 188	data
GC 122	Ge-P	NF/S, 30V, 0,15A, 0,07W	2a	Hfo	AC 128, AC 152...153, AC 188	data
GC 123	Ge-P	=GC 122: 60V	2a	Hfo	ACY 24, ASY 48, ASY 77	data
GC 181(A)[Grundig]	Si-P	≈BC 181	7e		→BC 181	
GC 189 [Grundig]	Si-N	≈BC 238	7e		→BC 238	
GC 195 [Grundig]	Si-N-Darl	≈BC 517	7a, 7e		→BC 517	
GC 196 [Grundig]	Si-N	≈BC 337	7e		→BC 337	
GC 197 [Grundig]	Si-P	≈BC 327	7e		→BC 327	
GC 198 [Grundig]	Si-P	≈BC 161			→BC 161	
GC 214 [Grundig]	Si-P	≈BC 214	7a		→BC 214	
GC 216	Ge-P	NF-Tr/E, 20V, 0,1A, 0,075W	2a	Hfo	AC 122, AC 125...126, AC 151	data
GC 217	Ge-P	NF-Tr/E, 20V, 0,1A, 0,075W	2a	Hfo	AC 122, AC 125...126, AC 151	data
GC 221	Ge-P	NF-Tr/E, 20V, 0,1A, 0,075W	2a	Hfo	AC 122, AC 125...126, AC 151	data
GC 223	Ge-P	=GC 221: 60V	2a	Hfo	ACY 24, ASY 48, ASY 77	data
GC 223(A,B)[Grun.]	Si-N	≈BC 337	7a		→BC 337	
GC 237 [Grundig]	Si-N	≈BC 237	7a		→BC 237	
GC 238 [Grundig]	Si-N	≈BC 238	7a		→BC 238	
GC 239 [Grundig]	Si-N	≈BC 239	7a		→BC 239	
GC 269 [Grundig]	Si-P	≈BC 328	7a		→BC 328	
GC 300	Ge-P	NF-E, 20V, 0,5A, 0,6W	2a	Hfo	AC 128, AC 152...153, AC 188	data
GC 301	Ge-P	NF-Tr/E, 32V, 0,5A	2a	Hfo	AC 128, AC 153, AC 188	data
GC 307 [Grundig]	Si-P	≈BC 307	7a		→BC 307	
GC 308 [Grundig]	Si-P	≈BC 308	7a		→BC 308	
GC 309 [Grundig]	Si-P	≈BC 309	7a		→BC 309	
GC 371 [Grundig]	Si-N	≈BC 338	7a		→BC 338	
GC 372 [Grundig]	Si-P	≈BC 328	7a		→BC 328	
GC 373 [Fairch.]	Si-N	≈BC 368	7a		→BC 368	
GC 373 [Valvo]	Si-N	≈BC 368	7c		→BC 368	
GC 374 [Fairch.]	Si-P	≈BC 369	7a		→BC 369	
GC 374 [Valvo]	Si-P	≈BC 369	7c		→BC 369	

Type	Device	Short Description	Fig.	Manu	Comparision Types	More at
GC 500	Ge-P	NF-Tr/E, 24V, 0,3A, 0,55W	1a	Hfo	AC 128, AC 152...153, AC 188	data
GC 501	Ge-P	NF-Tr/E, 24V, 0,3A, 0,55W	1a	Hfo	AC 128, AC 152...153, AC 188	data
GC 502	Ge-P	NF-Tr/E, 32V, 0,3A, 0,55W	1a	Hfo	AC 128, AC 152...153, AC 188	data
GC 503	Ge-P	Min, NF, 7V, 5mA, 0,01W	37d	Hfo	OC 57...60	data
GC 504	Ge-P	Min, NF, 7V, 5mA, 0,01W	37d	Hfo	OC 57...60	data
GC 505	Ge-P	Min, NF, 7V, 5mA, 0,01W	37d	Hfo	OC 57...60	data
GC 506	Ge-P	Min, NF, 7V, 5mA, 0,01W	37d	Hfo	OC 57...60	data
GC 507	Ge-P	NF-Tr, 32V, 0,125A, 0,125W	1a	Hfo	AC 122, AC 125...126, AC 151	data
GC 508	Ge-P	NF-Tr, 32V, 0,125A, 0,125W	1a	Hfo	AC 122, AC 125...126, AC 151	data
GC 509	Ge-P	NF-Tr, 60V, 0,125A, 0,125W	1a	Hfo	ACY 24, ASY 48, ASY 77	data
GC 510	Ge-P	NF-E, 32V, 1A, 0,2W	1a	Hfo	AC 128, AC 153, AC 188	data
GC 510 K	Ge-P	=GC 510: 0,3W	3a	Hfo	AC 128K, AC 153K, AC 188K	data
GC 511	Ge-P	NF-E, 25V, 1A, 0,2W	1a	Hfo	AC 128, AC 153, AC 188	data
GC 511 K	Ge-P	=GC 511: 0,3W	3a		AC 128K, AC 153K, AC 188K	data
GC 512	Ge-P	NF-E, 25V, 1A, 0,2W	1a	Hfo	AC 128, AC 153, AC 188	data
GC 512 K	Ge-P	=GC 512: 0,3W	3a		AC 128K, AC 153K, AC 188K	data
GC 515	Ge-P	NF-Tr/E, 32V, 0,125A, 0,125W	1a	Hfo	AC 122, AC 125...126, AC 151	data
GC 516	Ge-P	NF-Tr/E, 32V, 0,125A, 0,125W	1a	Hfo	AC 122, AC 125...126, AC 151	data
GC 517	Ge-P	NF-Tr/E, 32V, 0,125A, 0,125W	1a	Hfo	AC 122, AC 125...126, AC 151	data
GC 518	Ge-P	NF-Tr/E, 32V, 0,125A, 0,125W	1a	Hfo	AC 122, AC 125...126, AC 151	data
GC 519	Ge-P	NF-Tr/E, 32V, 0,125A, 0,125W	1a	Hfo	AC 122, AC 125...126, AC 151	data
GC 520	Ge-N	NF-E, 32V, 1A, 0,2W	1a	Hfo	AC 176, AC 187	data
GC 520 K	Ge-N	=GC 520: 0,3W	3a		AC 176K, AC 187K	data
GC 521	Ge-N	NF-E, 25V, 1A, 0,2W	1a		AC 176, AC 187	data
GC 521 K	Ge-N	=GC 521: 0,3W	3a		AC 176K, AC 187K	data
GC 522	Ge-N	NF-E, 20V, 1A, 0,2W	1a	Hfo	AC 176, AC 187	data
GC 522 K	Ge-N	=GC 522: 0,3W	3a		AC 176K, AC 187K	data
GC 525	Ge-N	NF-Tr/E, 15V, 0,125A, 0,125W	1a	Hfo	AC 127, AC 176, AC 187	data
GC 526	Ge-N	NF-Tr/E, 32V, 0,125A, 0,125W	1a	Hfo	AC 127, AC 176, AC 187	data
GC 527	Ge-N	NF-Tr/E, 32V, 0,125A, 0,125W	1a	Hfo	AC 127, AC 176, AC 187	data
GCN 53	Ge-N	NF-Tr/E, 30V, 0,25A, 0,125W	1a	Hfo	AC 127, AC 176, AC 187	data
GCN 54	Ge-N	NF-Tr/E, 48V, 0,25A, 0,125W	1a	Hfo	-	data
GCN 55	Ge-P	NF-Tr/E, 32V, 0,125A, 0,125W	1a	Hfo	AC 128, AC 152...153, AC 188	data
GCN 56	Ge-P	NF-Tr/E, 60V, 0,125A, 0,125W	1a	Hfo	ACY 24, ASY 48, ASY 77	data

GD

Type	Device	Short Description	Fig.	Manu	Comparision Types	More at
GD	Si-Di	=BAW 78D (Typ-Code/Stempel/marking)	39	Sie	→BAW 78D	data
GD	Si-Di	=GF 1D (Typ-Code/Stempel/marking)	71(5x2,7)	Gsi	→GF 1D	data
GD s	Si-Di	=BAW 78M (Typ-Code/Stempel/marking)	≈45	Sie	→BAW 78M	data
GD	Z-Di	=Z2 SMB-7V5 (Typ-Code/Stempel/marking)	71(5x3,5)	Fag	→Z2 SMB-7V5	data
GD 100	Ge-P	NF-L, 20V, 1,3A, 2W(Tc=45°)	22a§	Hfo	AD 162	data
GD 110	Ge-P	NF-L, 20V, 1,3A, 2W(Tc=45°)	22a§	Hfo	AD 162	data
GD 114 [Grundig]	Si-P	≈BD 246B	16h§		→BD 246B	
GD 115 [Grundig]	Si-N	≈BD 245B	16h§		→BD 245B	
GD 120	Ge-P	NF-L, 33V, 1,3A, 2W(Tc=45°)	22a§	Hfo	AD 162	data
GD 125	Ge-P	NF-L, 66V, 1,3A, 2W(Tc=45°)	22a§	Hfo	-	data
GD 130	Ge-P	NF-L, 66V, 1,3A, 2W(Tc=45°)	22a§	Hfo	-	data
GD 133 [Grundig]	Si-P	≈BD 140	14h§		→BD 140	
GD 134 [Grundig]	Si-P	≈BD 246B			→BD 246B	
GD 135 [Grundig]	Si-N	≈BD 245B			→BD 245B	
GD 142 [Grundig]	Si-N	≈2N3055	23a§		→2N3055	
GD 150	Ge-P	NF-L, 20V, 3A, 5,3W	22a§	Hfo	(AD 162, 2SB474)[7]	data
GD 151 [Grundig]	Si-N	≈BD 433	14h§		→BD 433	
GD 152 [Grundig]	Si-P	≈BD 434	14h§		→BD 434	
GD 160(A...C)	Ge-P	NF-L, 20V, 3A, 5,3W	22a§	Hfo	(AD 162, 2SB474)[7]	data
GD 170(A...C)	Ge-P	=GD 160: 33V	22a§	Hfo	(AD 162, 2SB474)[7]	data
GD 175(A...C)	Ge-P	=GD 160: 50V	22a§	Hfo	-	data
GD 180(A...C)	Ge-P	=GD 160: 66V	22a§	Hfo	-	data
GD 183 [Grundig]	Si-P	≈BD 136	14h§		→BD 136	
GD 190	Ge-P	NF-L, 30V, 1,5A	22a§	Hfo	AD 162	data
GD 191	Ge-P	NF-L, 40V, 1,5A	22a§	Hfo	AD 162	data
GD 192	Ge-P	NF-L, 50V, 1,5A	22a§	Hfo	-	data
GD 200	Ge-P	NF-L, 30V, 6A, 12W(Tc=50°)	23a§	Hfo	AL 102...103, AUY 22(A), AUY 28	data
GD 203 [Grundig]	Si-N	≈BD 243B	17j§		→BD 243B	
GD 204 [Grundig]	Si-P	≈BD 244B	17j§		→BD 244B	
GD 207 [Grundig]	Si-N	≈BD 243B	17j§		→BD 243B	
GD 210	Ge-P	NF-L, 60V, 6A, 12W(Tc=50°)	23a§	Hfo	AL 102...103, AUY 22(A), AUY 28	data
GD 220	Ge-P	NF-L, 80V, 6A, 12W(Tc=50°)	23a§	Hfo	AL 102...103, AUY 22(A), AUY 28	data
GD 240	Ge-P	NF/S-L, 30V, 3A, 10W	22a§	Hfo	(AD 162, 2SB474)[7]	data
GD 241	Ge-P	NF/S-L, 40V, 3A, 10W	22a§	Hfo	(AD 162, 2SB474)[7]	data
GD 241(A,B)[Grun.]	Si-N	≈BD 241A,B	17j§		→BD 241A,B	
GD 242	Ge-P	NF/S-L, 50V, 3A, 10W	22a§	Hfo	-	data
GD 243	Ge-P	NF/S-L, 65V, 3A, 10W	22a§	Hfo	-	data
GD 243 [Grundig]	Si-N	≈BD 243	17j§		→BD 243	
GD 244	Ge-P	NF/S-L, 75V, 3A, 10W	22a§	Hfo	-	data
GD 340 [Grundig]	Si-P	≈BD 438	14h§		→BD 438	
GD 341 [Grundig]	Si-N	≈BD 437	14h§		→BD 437	

Type	Device	Short Description	Fig.	Manu.	Comparision Types	More at
GD 361 [Grundig]	Si-N	≈BD 433	14h§		→BD 433	
GD 362 [Grundig]	Si-P	≈BD 434	14h§		→BD 434	
GD 363 [Grundig]	Si-N	≈BD 433	14h§		→BD 433	
GD 364 [Grundig]	Si-P	≈BD 434	14h§		→BD 434	
GD 384 [Grundig]	Si-N	≈BD 525	13j§		→BD 525	
GD 607	Ge-N	NF-L, 32V, 1A, 4W(Tc=60°)	22a§	Hfo	AD 161	data
GD 608	Ge-N	NF-L, 25V, 1A, 4W(Tc=60°)	22a§	Hfo	AD 161	data
GD 609	Ge-N	NF-L, 20V, 1A, 4W(Tc=60°)	22a§	Hfo	AD 161	data
GD 617	Ge-P	NF-L, 32V, 1A, 4W(Tc=60°)	22a§	Hfo	AD 162	data
GD 618	Ge-P	NF-L, 25V, 1A, 4W(Tc=60°)	22a§	Hfo	AD 162	data
GD 619	Ge-P	NF-L, 20V, 1A, 4W(Tc=60°)	22a§	Hfo	AD 162	data
GDC 21 D 003	IC	VSB Receiver	QFP	Hun		pdf pinout
GDC 21 D 301 A	IC	Transport Decoder	QFP	Hun		pdf pinout
GDC 21 D 401 B	IC	Video Decoder	QFP	Hun		pdf pinout
GDC 21 D 601	IC	32-Bit RISC MCU	QFP	Hun		pdf pinout
GDC 21 D 701 C	IC	Video Display Processor	QFP	Lgp		pdf pinout
GDE	Z-Di	=1SMC 5.0A (Typ-Code/Stempel/marking)	71(7,9mm)	Ons	→1SMC 5.0A...78A	
GDG	Z-Di	=1SMC 6.0A (Typ-Code/Stempel/marking)	71(7,9mm)	Ons	→1SMC 5.0A...78A	
GDK	Z-Di	=1SMC 6.5A (Typ-Code/Stempel/marking)	71(7,9mm)	Ons	→1SMC 5.0A...78A	
GDM	Z-Di	=1SMC 7.0A (Typ-Code/Stempel/marking)	71(7,9mm)	Ons	→1SMC 5.0A...78A	
GDP	Z-Di	=1SMC 7.5A (Typ-Code/Stempel/marking)	71(7,9mm)	Ons	→1SMC 5.0A...78A	
GDR	Z-Di	=1SMC 8.0A (Typ-Code/Stempel/marking)	71(7,9mm)	Ons	→1SMC 5.0A...78A	
GDs	Si-Di	=BAW 78A (Typ-Code/Stempel/marking)	≈45	Sie	→BAW 78M	data
GDT	Z-Di	=1SMC 8.5A (Typ-Code/Stempel/marking)	71(7,9mm)	Ons	→1SMC 5.0A...78A	
GDV	Z-Di	=1SMC 9.0A (Typ-Code/Stempel/marking)	71(7,9mm)	Ons	→1SMC 5.0A...78A	
GDX	Z-Di	=1SMC 10A (Typ-Code/Stempel/marking)	71(7,9mm)	Ons	→1SMC 5.0A...78A	
GDZ	Z-Di	=1SMC 11A (Typ-Code/Stempel/marking)	71(7,9mm)	Ons	→1SMC 5.0A...78A	

GE

Type	Device	Short Description	Fig.	Manu.	Comparision Types	More at
GE	Si-P	=2SA1455K-E (Typ-Code/Stempel/marking)	35	Rhm	→2SA1455K	data
GE	Si-Di	=BAW 79A (Typ-Code/Stempel/marking)	39	Sie	→BAW 79A	data
GE	Si-N	=BFR 35AP (Typ-Code/Stempel/marking)	35	Sie	→BFR 35AP	data
GE	Z-Di	=Z2 SMB-6V8 (Typ-Code/Stempel/marking)	71(5x3,5)	Fag	→Z2 SMB-6V8	data
GE 1001	Si-Di	Gl, S, 50V, 1A, <35ns	31a	Gen,Rca	BYV 26B, EGP 10A, FE 1A	data
GE 1002	Si-Di	=GE 1001: 100V	31a	Gen,Rca	BYV 26B, EGP 10B, FE 1B	data
GE 1003	Si-Di	=GE 1001: 150V	31a	Gen,Rca	BYV 26B, EGP 10C, FE 1C	data
GE 1004	Si-Di	=GE 1001: 200V	31a	Gen,Rca	BYV 26B, EGP 10D, FE 1D	data
GE 1101	Si-Di	Gl, S, 50V, 2,5A, <35ns	31a	Gen,Rca	BYV 28/50, EGP 30A, FE 3A	data
GE 1102	Si-Di	=GE 1101: 100V	31a	Gen,Rca	BYV 28/100, EGP 30B, FE 3B	data
GE 1103	Si-Di	=GE 1101: 150V	31a	Gen,Rca	BYV 28/150, EGP 30C, FE 3C	data
GE 1104	Si-Di	=GE 1101: 200V	31a	Ge,Rca	BYV 28/200, EGP 30D, FE 3D	data
GE 1301	Si-Di	Gl, S, 50V, 6A, <35ns	31a	Gen,Rca	BYV 61, FE 6A	data
GE 1302	Si-Di	=GE1301: 100V	31a	Gen,Rca	BYV 62, FE 6B	data
GE 1303	Si-Di	=GE1301: 150V	31a	Gen,Rca	BYV 63, FE 6C	data
GE 1304	Si-Di	=GE1301: 200V	31a	Gen,Rca	FE 6D	data
GE3009	OC	LED/Triac Viso:7500Vrms Vbs:1.15V		Har,Rca	K3010P	data pdf pinout
GE3020	OC	LED/Triac Viso:3750Vrms Vbs:1.15V		Har,Rca	K3020P	data pdf pinout
GE 3055 P	Si-N	NF/S-L, 80V, 10A, 70W	18j§	Gen	BD 245B...F, BD 545B...D, 2SD1187, ++	data
GE 5060	Si-N-Darl+Di	S-L, 400/350V, 20A, 125W, B>100	23a§	Gen	BUT 13, MJ 10000...01, MJ 10004...05	data
GE 5061	Si-N-Darl+Di	=GE 5060: 450/400V	23a§	Gen	BUT 13, MJ 10000...01, MJ 10004...05	data
GE 5062	Si-N-Darl+Di	=GE 5060: 500/450V	23a§	Gen	BUT 13, MJ 10001, MJ 10005, MJ 10008...09	data
GE 6060	Si-N-Darl+Di	≈GE 5060: int. Speedup-Diode(E→B)	23a§	Gen	MJ 10000...5, MJ 10008...09	data
GE 6061	Si-N-Darl+Di	≈GE 5061: int. Speedup-Diode(E→B)	23a§	Gen	MJ 10004...5, MJ 10008...09	data
GE 6062	Si-N-Darl+Di	≈GE 5062: int. Speedup-Diode(E→B)	23a§	Gen	MJ 10005, MJ 10008...09	data
GE 6251	Si-N-Darl+Di	S-L, Speedup-Di., 450/400V, 10A, 125W	23a§	Gen	MJ 10006...07, MJ 10013...14	data
GE 6252	Si-N-Darl+Di	=GE 6251: 500/450V	23a§	Gen	MJ 10007, MJ 10013...14	data
GE 6253	Si-N-Darl+Di	=GE 6251: 550/500V	23a§	Gen	MJ 10013...14	data
GE 10000...10009	Si-N-Darl+Di	=MJ 10000...10009	23a§	Gen	→MJ 10000...10009	data
GE 10015...10016	Si-N-Darl+Di	=MJ 10015...10016	23a§	Gen	→MJ 10015...10016	data
GE 10020...10023	Si-N-Darl+Di	=MJ 10020...10023	23a§	Gen	→MJ 10020...10023	data
GE 13070P...13071P	Si-N	=MJ 13070...13071: 100W	18j§	Gen	BUW 11(A), BU 131(A), BUV 82...83, ++	data
GE 13080P...13081P	Si-N	=MJ 13080...13081: 110W	18j§	Gen	BUW 12(A), BU 132(A), BUP 22B...C, ++	data
GE 13080T...13081T	Si-N	=MJ 13080...13081: 90W	17j§	Gen	BUT 12(A), BUT 54(A), BUT 56(A), ++	data
GE 13100P...13101P	Si-N	=MJ 13100...13101: 125W	18j§	Gen	BUX 98(A)P, 2SC3988	data
GEE	Z-Di	=1SMC 12A (Typ-Code/Stempel/marking)	71(7,9mm)	Ons	→1SMC 5.0A...78A	
GEG	Z-Di	=1SMC 13A (Typ-Code/Stempel/marking)	71(7,9mm)	Ons	→1SMC 5.0A...78A	
GEK	Z-Di	=1SMC 14A (Typ-Code/Stempel/marking)	71(7,9mm)	Ons	→1SMC 5.0A...78A	
GEM	Z-Di	=1SMC 15A (Typ-Code/Stempel/marking)	71(7,9mm)	Ons	→1SMC 5.0A...78A	
GEP	Z-Di	=1SMC 16A (Typ-Code/Stempel/marking)	71(7,9mm)	Ons	→1SMC 5.0A...78A	
GEPS2001	OC	LED/NPN Viso:2500Vrms CTR:>30%		Har,Rca	CQY80N	data pinout
GER	Z-Di	=1SMC 17A (Typ-Code/Stempel/marking)	7i(7,9mm)	Ons	→1SMC 5.0A...78A	
GER 4001...4007	Si-Di	=1N4001...4007	31a	Gen	→1N4001...4007	data
GEs	Si-N	=BFR 35AP (Typ-Code/Stempel/marking)	35	Sie	→BFR 35AP	data
GES 92	Si-N	Uni, 40V, 0,4A, 0,625W, 100MHz	7e	Gen	BC 337, BC 635, 2SC3377, 2SC3939, ++	data
GES 93	Si-P	Uni, 40V, 0,4A, 0,625W, 100MHz	7e	Gen	BC 327, BC 636, 2SB910, 2SD1116, ++	data
GES 97	Si-N	Uni, 60V, 0,1A, 0,36W, 100MHz	7e	Gen	BC 174, BC 546, 2SC2240, 2SC2459, ++	data
GES 98	Si-N	Uni, 80V, 0,1A, 0,36W, 100MHz	7e	Gen	BC 546, 2SC2240, 2SC2459, 2SC3378, ++	data

Type	Device	Short Description	Fig.	Manu	Comparision Types	More at
GES 929..930	Si-N	=2N929..930: 0,36W	7e	Gen	→2N929..930	data
GES 2218..2219(A)	Si-N	=2N2218..2219(A): 0,625W	7e	Gen	→2N2218..2219(A)	data
GES 2221..2222(A)	Si-N	=2N2221..2222(A): 0,36W	7e	Gen	→2N2221..2222(A)	data
GES 2483	Si-N	=2N2483: 0,36W	7e	Gen	→2N2483	data
GES 2646	UJT-P	=2N2646	7o	Gen	→2N2646	data
GES 2647	UJT-P	=2N2647	7o	Gen	→2N2647	data
GES 2904..2905(A)	Si-P	=2N2904..2905(A): 0,6W	7e	Gen	→2N2904..2905(A)	data
GES 2906..2907(A)	Si-P	=2N2906..2907(A): 0,36W	7e	Gen	→2N2906..2907(A)	data
GES 3053	Si-N	=2N3053: 0,625W	7e	Gen	BC 337A, BC 637, BC 639, 2SD1616, ++	data
GES 3414..3417	Si-N	=2N3414..3417: 0,625W	7e	Gen	→2N3414..3417	data
GES 3565..3569	Si-N	=2N3565..3569: 0,625W	7e	Gen	→2N3565..3569	data
GES 4121..4122	Si-P	=2N4121..4122: 0,2W	7e	Gen	→2N4121..4122	data
GES 4248	Si-P	=2N4248: 0,4W	7e	Gen	→2N4248	data
GES 4891..4894	UJT-P	=2N4891..4894	7o	Gen	→2N4891..4894	data
GES 5305..5308(A)	Si-N	=2N5305..5308(A): 0,9W	7e	Gen	→2N5305..5308(A)	data
GES 5368..5375	Si-N/P	=2N5368..5375: 0,36W	7e	Gen	→2N5368..5375	data
GES 5401	Si-P	=2N5401	7e	Gen	→2N5401	data
GES 5447..5451	Si-N/P	=2N5447..5451: 0,36W	7e	Gen	→2N5447..5451	data
GES 5551	Si-N	=2N5551	7e	Gen	→2N5551	data
GES 5810..5823	Si-N/P	=2N5810..5823: 0,5W	7c	Gen	→2N5810..5823	data
GES 5824..5828(A)	Si-N	=2N5824..5828(A): 0,36W	7c	Gen	→2N5824..5828(A)	data
GES 6000..6007	Si-N/P	=2N6000..6007: 0,4W	7e	Gen	→2N6000..6007	data
GES 6010..6017	Si-N/P	=2N6010..6017: 0,5W	7e	Gen	→2N6010..6017	data
GES 6027	PUT	=2N6027	7e	Gen	→2N6027	data
GES 6028	PUT	=2N6028	7e	Gen	→2N6028	data
GES 6218..6221	Si-N	=2N6218..6221: 0,5W	7e	Gen	→2N6218..6221	data
GES 6222	Si-N	=2N6222: 0,36W	7e	Gen	→2N6222	data
GES 6224	Si-N	=2N6224: 0,36W	7e	Gen	→2N6224	data
GES 6426..6427	Si-N	=2N6426..6427: 0,625W	7e	Gen	→2N6426..6427	data
GES 6560..6563	Si-N/P	=2N6560..6563: 0,625W	7e	Gen	→2N6560..6563	data
GET	Z-Di	=1SMC 18A (Typ-Code/Stempel/marking)	71(7,9mm)	Ons	→1SMC 5.0A...78A	
GET 706	Si-N	=2N706: 0,36W	7a	Gen	→2N706	data
GET 708	Si-N	=2N708: 0,36W	7a	Gen	→2N708	data
GET 914	Si-N	=2N914: 0,36W	7a	Gen	→2N914	data
GET 929	Si-N	=2N929: 0,36W	7a	Gen	→2N929	data
GET 930	Si-N	=2N930: 0,36W	7a	Gen	→2N930	data
GET 2221(A)	Si-N	=2N2221(A): 0,36W	7a	Gen	→2N2221(A)	data
GET 2222(A)	Si-N	=2N2222(A): 0,36W	7a	Gen	→2N2222(A)	data
GET 2369	Si-N	=2N2369: 0,36W	7a	Gen	→2N2369	data
GET 2483	Si-N	=2N2483: 0,36W	7a	Gen	→2N2483	data
GET 2484	Si-N	=2N2484: 0,36W	7a	Gen	→2N2484	data
GET 2904..2907	Si-P	=2N2904..2907: 0,36W	7a	Gen	→2N2904..2907	data
GET 3013..3014	Si-N	=2N3013..3014: 0,36W	7a	Gen	→2N3013..3014	data
GET 3562..3563	Si-N	=2N3562..3563: 0,25W	7a	Gen	→2N3562..3563	data
GET 3638(A)	Si-P	=2N3638(A): 0,25W	7a	Gen	→2N3638(A)	data
GET 3646	Si-N	=2N3646: 0,36W	7a	Gen	→2N3646	data
GET 3903..3906	Si-N/P	=2N3903..3906: 0,31W	7a	Gen	→2N3903..3906	data
GET 4870	UJT-P	=2N4870:	7o	Gen	→2N4870	data
GET 4871	UJT-P	=2N4871:	7o	Gen	→2N4871	data
GET 5305..5308(A)	Si-N	=2N5305..5308(A): 0,4W	7a	Gen	→2N5305..5308(A)	data
GEV	Z-Di	=1SMC 20A (Typ-Code/Stempel/marking)	71(7,9mm)	Ons	→1SMC 5.0A...78A	
GEX	Z-Di	=1SMC 22A (Typ-Code/Stempel/marking)	71(7,9mm)	Ons	→1SMC 5.0A...78A	
GEZ	Z-Di	=1SMC 24A (Typ-Code/Stempel/marking)	71(7,9mm)	Ons	→1SMC 5.0A...78A	

GF

Type	Device	Short Description	Fig.	Manu	Comparision Types	More at
GF	Si-Di	=BAW 79B (Typ-Code/Stempel/marking)	39	Sie	→BAW 79B	data
GF	Si-N	=BFN 22 (Typ-Code/Stempel/marking)	35	Sie	→BFN 22	data
GF	Si-N	=BFR 92P (Typ-Code/Stempel/marking)	35	Sie	→BFR 92P	data
GF	Z-Di	=Z2 SMB-8V2 (Typ-Code/Stempel/marking)	71(5x3,5)	Fag	→Z2 SMB-8V2	data
GF 1 A..M	Si-Di	SMD, Gl, 50...1000V, 1A, 2µs A=50, B=100, D=200, G=400, J=600V K=800, M=1000V	71a(5x2,7)	Gsi		data
GF 100	Ge-P	AM-ZF	2a	Hfo	AF 127, AF 200, AF 139, AF 239(S)	data
GF 105	Ge-P	AM-V/M/O	2a	Hfo	AF 126, AF 200, AF 139, AF 239(S)	data
GF 108	Ge-P	AM-V/M/O	2a	Hfo	AF 126, AF 200, AF 139, AF 239(S)	data
GF 120	Ge-P	AM-V/M/ZF, 30MHz	5k	Hfo	AF 126, AF 200, AF 139, AF 239(S)	data
GF 121	Ge-P	AM-V/M, 50MHz	5k	Hfo	AF 126, AF 200, AF 139, AF 239(S)	data
GF 122	Ge-P	AM-V/M/O, 50MHz	5k	Hfo	AF 126, AF 200, AF 139, AF 239(S)	data
GF 125	Ge-P	FM-ZF, 60MHz	5k	Hfo	AF 125...126, AF 200, AF 139, AF 239(S)	data
GF 126	Ge-P	AM-ZF	5k	Hfo	AF 125...126, AF 200, AF 139, AF 239(S)	data
GF 127	Ge-P	AM-V, 75MHz	5k	Hfo	AF 125, AF 200, AF 139, AF 239(S)	data
GF 128	Ge-P	AM/FM, 100MHz	5k	Hfo	AF 124...125, AF 200, AF 139, AF 239(S)	data
GF 129	Ge-P	AM-V/M/O, 75MHz	5k	Hfo	AF 124...125, AF 200, AF 139, AF 239(S)	data
GF 130	Ge-P	FM-ZF	5k	Hfo	AF 125...126, AF 200, AF 139, AF 239(S)	data
GF 131	Ge-P	FM-M, 85MHz	5k	Hfo	AF 124...125, AF 200, AF 139, AF 239(S)	data
GF 132	Ge-P	FM-V, 85MHz	5k	Hfo	AF 124, AF 200, AF 139, AF 239(S)	data
GF 134	Ge-P	VHF, 180MHz	5k	Hfo	AF 106, AF 109R, AF 239(S), AF 306	data
GF 135	Ge-P	VHF, 150MHz	5g	Hfo	AF 106, AF 109R, AF 239(S), AF 306	data
GF 136	Ge-P	VHF, 150MHz	5g	Hfo	AF 106, AF 109R, AF 239(S), AF 306	data
GF 137	Ge-P	VHF, 180MHz	5g	Hfo	AF 106, AF 109R, AF 239(S), AF 306	data

Type	Device	Short Description	Fig.	Manu	Comparision Types	More at
GF 138	Ge-P	VHF, 180MHz	5g	Hfo	AF 106, AF 109R, AF 239(S), AF 306	data
GF 139	Ge-P	AM-V/M, FM-ZF	5k	Hfo	AF 106, AF 109R, AF 200, AF 239(S)	data
GF 140	Ge-P	VHF-V/M/O, 300MHz	2a	Hfo	AF 139, AF 239(S)	data
GF 141	Ge-P	VHF-V/M/O, 300MHz	2a	Hfo	AF 139, AF 239(S)	data
GF 142	Ge-P	VHF-V/M/O, 300MHz	2a	Hfo	AF 139, AF 239(S)	data
GF 143	Ge-P	VHF-V/M/O, 300MHz	2a	Hfo	AF 139, AF 239(S)	data
GF 145	Ge-P	UHF-V/M/O, >25CMHz	5g	Hfo	AF 139, AF 239(S)	data
GF 146	Ge-P	UHF-V/M/O, >600MHz	5g	Hfo	AF 139, AF 239(S)	data
GF 147	Ge-P	UHF, >600MHz	5g	Hfo	AF 139, AF 239(S)	data
GF 180	Ge-P	HF	5k	Hfo	AF 106, AF 109R, AF 200, AF 306	data
GF 181	Ge-P	HF, 100MHz	5k	Hfo	AF 106, AF 109R, AF 200, AF 306	data
GF 268	N-FET	≈BF 245	7f		→BF 245	
GF 501	Ge-P	HF, 300MHz	2a	Hfo	AF 139, AF 239(S)	data
GF 502	Ge-P	HF, 300MHz	2a	Hfo	AF 139, AF 239(S)	data
GF 503	Ge-P	HF, 300MHz	2a	Hfo	AF 139, AF 239(S)	data
GF 504	Ge-F	HF, 300MHz	2a	Hfo	AF 139, AF 239(S)	data
GF 505	Ge-P	HF, 170MHz	5g	Hfo	AF 106, AF 109R, AF 239(S), AF 306	data
GF 506	Ge-P	HF, 170MHz	5g	Hfo	AF 106, AF 109R, AF 239(S), AF 306	data
GF 507	Ge-P	HF, 250MHz	5g	Hfo	AF 106, AF 109R, AF 239(S), AF 306	data
GF 515	Ge-P	HF, 60MHz	5k	Hfo	AF 124...127, AF 200, AF 139, AF 239(S)	data
GF 516	Ge-P	HF, 60MHz	5k	Hfo	AF 124...127, AF 200, AF 139, AF 239(S)	data
GF 517	Ge-P	HF, 50MHz	5k	Hfo	AF 124...127, AF 200, AF 139, AF 239(S)	data
GF 522	N-FET	≈BF 245	7f		→BF 245	
GF 757	Si-N	≈BF 757			→BF 757	
GF 758	Si-N	≈BF 758			→BF 758	
GF 759	Si-N	≈BF 759			→BF 759	
GF 760	Si-P	≈BF 760			→BF 760	
GF 761	Si-P	≈BF 761			→BF 761	
GF 762	Si-P	≈BF 762			→BF 762	
GF 2208	MOS-N-FET-e	SMD, V-MOS, 30V, 13A, <11mΩ(11A)	8-MDIP/ca1	Gsi	-	data
GF 2301	MOS-P-FET-e	SMD, V-MOS, LogL, 20V, 2,3A, <0,19Ω <85/190ns	35a	Gsi	2SJ557	data
GF 2304	MOS-N-FET-e	SMD, V-MOS, LogL, 30V, 2,5A, <0,19Ω <50/55ns	35a	Gsi	Si 2304DS, 2SK3105	data
GF 2524	MOS-N-FET-e	SMD, Dual, V-MOS, 30V, T1: 5,8A, <55mΩ (4,7A), T2: 7,8A, <28mΩ(6,3A)	8-MDIP/ca3	Gsi	-	data
GF 2918	MOS-N-FET-e	SMD,Dual,V-MOS, 30V, 7,8A, <28mΩ(6,3A) <40/112ns	8-MDIP/ca3	Gsi	-	data
GF 3441	MOS-P-FET-e	SMD, V-MOS, LogL, 20V, 3,3A, <135mΩ (3,5A), <80/150ns	46bn4	Gsi	-	data
GF 4126	MOS-N-FET-e	SMD, V-MOS, LogL, 20V, 7,2A, <40mΩ 86,2A), <52/100ns	8-MDIP/ca1	Gsi	-	data
GF 4410	MOS-N-FET-e	SMD, V-MOS, LogL, 30V, 10A, <20mΩ(5A) <35/180ns	8-MDIP/ca1	Gsi	-	data
GF 4412	MOS-N-FET-e	SMD, V-MOS, 30V, 7A, <42mΩ(3,5A) <22/51ns	8-MDIP/ca1	Gsi	-	data
GF 4420	MOS-N-FET-e	SMD, V-MOS, 30V, 12,5A, <13mΩ(4,5A) <32/236ns	8-MDIP/ca1	Gsi	-	data
GF 4425	MOS-P-FET-e	SMD, V-MOS, LogL, 30V, 11A, <23mΩ (8,5A), <50/350ns	8-MDIP/ca1	Gsi	-	data
GF 4435	MOS-P-FET-e	SMD, V-MOS, 30V, 8A, <35mΩ, <60/200ns	8-MDIP/ca1	Gsi	-	data
GF 4450	MOS-N-FET-e	SMD,V-MOS, 60V, 7,5A, <30mΩ, <50/157ns	8-MDIP/ca1	Gsi	-	data
GF 4800	MOS-N-FET-e	SMD, V-MOS, 30V, 9A, <33mΩ, <25/55ns	8-MDIP/ca1	Gsi	-	data
GF 4810	MOS-N-FET-e	SMD, V-MOS, LogL, 30V, 10A, <20mΩ(5A) <45/130ns, int. Schottky-Di, 30V, 4A	8-MDIP	Gsi	-	data
GF 4824	MOS-N-FET-e	SMD, Dual, V-MOS, 30V, T1: 4,7A, <65mΩ (3,7A), T2: 9A, <27mΩ(7,3A)	8-MDIP/ca3	Gsi	-	data
GF 4936	MOS-N-FET-e	SMD,Dual,V-MOS, 30V, 5,8A, <55mΩ(4,7A) <32/75ns	8-MDIP/ca3	Gsi	-	data
GF 4953	MOS-P-FET-e	SMD,Dual,V-MOS, 30V, 4,9A, <95mΩ(3,6A) <35/100ns	8-MDIP/ca3	Gsi	-	data
GF 6968 A	MOS-N-FET-e	SMD, Dual, V-MOS, LogL, 20V, 6,2A <30mΩ(5,3A), <80/150ns	8-SSMDIP/ ca4	Gsi	-	data
GF 6968 E	MOS-N-FET-e*	SMD, Dual, V-MOS, LogL, 20V, 6,5A <30mΩ(5,5A), <1450/8200ns, ESD Prot.	8-SSMDIP/ ca4	Gsi	-	data
GF 9410	MOS-N-FET-e	SMD, V-MOS, 30V, 7A, <50mΩ, <90/290ns	8-MDIP/ca1	Gsi	-	data
GF 9926	MOS-N-FET-e	SMD, Dual, V-MOS, LogL, 20V, 6A, <40mΩ (5,2A), <200/200ns	8-MDIP/ca3	Gsi	-	data
GFB 50 N03	MOS-N-FET-e	V-MOS, 30V, 50A, 62,5W, <20mΩ(20A) <40/110ns	30c§	Gsi	MTB 52N06V, PHB 50N06LT, 2SK2499-Z, ++	data
GFB 60 N03	MOS-N-FET-e	V-MOS, 30V, 60A, 62,5W, <16mΩ(25A) <40/110ns	30c§	Gsi	MTB 60N06HD, PHB 60N06LT, 2SK2499-Z, ++	data

Type	Device	Short Description	Fig.	Manu	Comparision Types	More at
GFB 70 N03	MOS-N-FET-e	V-MOS, 30V, 70A, 62,5W, <11mΩ(30A) <28/229ns	30c§	Gsi	IRF 1010NL,NS, MTB 75N05HD, PHB 80N06LT	data
GFB 75 N03	MOS-N-FET-e	V-MOS, 30V, 80A, 69,4W, <9,5mΩ(31A) <55/189ns	30c§	Gsi	IRF 1010NL,NS, MTB 75N05HD, PHB 80N06LT	data
GFB 7030 BL	MOS-N-FET-e	V-MOS, LogL, 30V, 60A, 65W, <12mΩ(25A) <50/97ns	30c§	Gsi	MTB 52N06VL, PHB 60N06LT, 2SK2499-Z, ++	data
GFC 6303	MOS-P-FET-e	SMD, Dual, V-MOS, LogL, 20V, 1,9A <0,25Ω(1,7A), <30/34ns	46	Gsi	-	data
GFD 30 N03	MOS-N-FET-e	V-MOS, 30V, 43A, 44,5W, <21mΩ(17A)	30c§	Gsi	IRLR 2905	data
GFD 50 N03	MOS-N-FET-e	V-MOS, 30V, 65A, 62,5W, <12mΩ(13A)	30c§	Gsi	-	data
GFD 50 N03A	MOS-N-FET-e	V-MOS, 30V, 78A, 70W, <10mΩ(13A)	30c§	Gsi	-	data
GFD1300	PD	1300nm S:>0.75mA/mW Resp:>6V/mW		Hon		data pinout
GFE	Z-Di	=1SMC 26A (Typ-Code/Stempel/marking)	71(7,9mm)	Ons	→1SMC 5.0A...78A	
GFE1300	IRED	1300nm 9/125-62.5/125µmm I:84lm		Hon		data pinout
GFG	Z-Di	=1SMC 28A (Typ-Code/Stempel/marking)	71(7,9mm)	Ons	→1SMC 5.0A...78A	
GFH600	OC	LED/NPN Viso:4000Vac CTR:160-320%		Har,Rca	CNY75A	data pinout
GFH601	OC	LED/NPN Viso:3750Vrms CTR:160-320%		Har,Rca		data pinout
GFK	Z-Di	=1SMC 30A (Typ-Code/Stempel/marking)	71(7,9mm)	Ons	→1SMC 5.0A...78A	
GFM	Z-Di	=1SMC 33A (Typ-Code/Stempel/marking)	71(7,9mm)	Ons	→1SMC 5.0A...78A	
GFP	Z-Di	=1SMC 36A (Typ-Code/Stempel/marking)	71(7,9mm)	Ons	→1SMC 5.0A...78A	
GFP 50 N03	MOS-N-FET-e	V-MOS, 30V, 50A, 62,5W, <20mΩ(20A) <40/110ns	17c§	Gsi	BUK 456-50, BUZ 100, MTP 52N06V	data
GFP 60 N03	MOS-N-FET-e	V-MOS, 30V, 60A, 62,5W, <16mΩ(25A) <40/110ns	17c§	Gsi	BUK 456-50, BUZ 100, MTP 60N06HD	data
GFP 70 N03	MOS-N-FET-e	V-MOS, 30V, 70A, 62,5W, <11mΩ(30A) <28/229ns	17c§	Gsi	IRF 1104, MTP 75N05HD, PHP 80N06LT	data
GFP 75 N03	MOS-N-FET-e	V-MOS, 30V, 80A, 69,4W, <9,5mΩ(31A) <55/189ns	17c§	Gsi	IRF 1104, MTP 75N05HD, PHP 80N06LT	data
GFR	Z-Di	=1SMC 40A (Typ-Code/Stempel/marking)	71(7,9mm)	Ons	→1SMC 5.0A...78A	
GFs	Si-N	=BFR 92P (Typ-Code/Stempel/marking)	35	Sie	→BFR 92P	data
GFT	Z-Di	=1SMC 43A (Typ-Code/Stempel/marking)	71(7,9mm)	Ons	→1SMC 5.0A...78A	
GFT 20 A6	GTO-Thy	600V, 10A(Tc=60°C), Igt/Ih 100/200mA	22a§	Hit	-	data
GFT 20 B12	GTO-Thy	=GFT 20A6: 1200V, 7A(Tc=60°C)	22a§	Hit	-	data
GFT 50 A6	GTO-Thy	600V, 20A(Tc=60°C), Igt/Ih 300/600mA	23a§	Hit	-	data
GFT 50 B12	GTO-Thy	=GFT 50A6: 1200V, 18A(Tc=60°C)	23a§	Hit	-	data
GFV	Z-Di	=1SMC 45A (Typ-Code/Stempel/marking)	71(7,9mm)	Ons	→1SMC 5.0A...78A	
GFX	Z-Di	=1SMC 48A (Typ-Code/Stempel/marking)	71(7,9mm)	Ons	→1SMC 5.0A...78A	
GFZ	Z-Di	=1SMC 51A (Typ-Code/Stempel/marking)	71(7,9mm)	Ons	→1SMC 5.0A...78A	

GG...GL

Type	Device	Short Description	Fig.	Manu	Comparision Types	More at
GG	Si-P	=2SA1369-G (Typ-Code/Stempel/marking)	39	Mit	→2SA1369	data
GG	Si-Di	=BAW 79C (Typ-Code/Stempel/marking)	39	Sie	→BAW 79C	data
GG	Si-P	=BF 579R (Typ-Code/Stempel/marking)	35	Aeg	→BF 579R	data
GG	Si-N	=BFR 93P (Typ-Code/Stempel/marking)	35	Sie	→BFR 93P	data
GG	Si-Di	=GF 1G (Typ-Code/Stempel/marking)	71(5x2,7)	Gsi	→GF 1G	data
GG	Z-Di	=Z2 SMB-9V1 (Typ-Code/Stempel/marking)	71(5x3,5)	Fag	→Z2 SMB-9V1	data
GGE	Z-Di	=1SMC 54A (Typ-Code/Stempel/marking)	71(7,9mm)	Ons	→1SMC 5.0A...78A	
GGG	Z-Di	=1SMC 58A (Typ-Code/Stempel/marking)	71(7,9mm)	Ons	→1SMC 5.0A...78A	
GGK	Z-Di	=1SMC 60A (Typ-Code/Stempel/marking)	71(7,9mm)	Ons	→1SMC 5.0A...78A	
GGM	Z-Di	=1SMC 64A (Typ-Code/Stempel/marking)	71(7,9mm)	Ons	→1SMC 5.0A...78A	
GGP	Z-Di	=1SMC 70A (Typ-Code/Stempel/marking)	71(7,9mm)	Ons	→1SMC 5.0A...78A	
GGR	Z-Di	=1SMC 75A (Typ-Code/Stempel/marking)	71(7,9mm)	Ons	→1SMC 5.0A...78A	
GGT	Z-Di	=1SMC 78A (Typ-Code/Stempel/marking)	71(7,9mm)	Ons	→1SMC 5.0A...78A	
GH	Si-P	=2SA1369-H (Typ-Code/Stempel/marking)	39	Mit	→2SA1369	data
GH	Si-Di	=BAW 79D (Typ-Code/Stempel/marking)	39	Sie	→BAW 79D	data
GH	Si-Di	=U 05GH44 (Typ-Code/Stempel/marking)	71(5x2,5)	Tos	→U 05GH44	data
GH	Z-Di ●	=Z2 SMB-10 (Typ-Code/Stempel/marking)	71(5x3,5)	Fag	→Z2 SMB-10	data
GH 1 E,F	Si-Di	Gl, 1300...1500V, 1A, <6µs E=1300V, F=1500V	31a	Sak	BY 228, BY 448, DM 513, EM 516, GP 10W	data
GH 3 E,F	Si-Di	Gl, 1300...1500V, 1,5A, <10µs E=1300V, F=1500V	31a	Sak	BY 228, BY 448, BY 350/...	data
GI 1-1200	Si-Di	Gl, 1200V, 1A, <25µs	31a	Gie	BY133, BY350/1300, CH 3E, EM 513, GP 10Q	data
GI 1-1400	Si-Di	=GI 1-1200: 1400V	31a		BY 350/1500, CH 3E, EM 516, GP 10V	data
GI 1-1600	Si-Di	=GI 1-1200: 1600V	31a		DM 513, DM 516, EM 516, GP 10Y	data
GI 250-1	Si-Di	kV-Gl, 1000V, 0,25A	31a	Gie	BA 159, BY 203/12, BY 204/10, MR 250-1	data
GI 250-2	Si-Di	=GI 250-1: 2000V	31a		BY 203/20, GP 02-20, MR 250-2	data
GI 250-3	Si-Di	=GI 250-1: 3000V	31a		GP 02-30, MR 250-3	data
GI 250-4	Si-Di	=GI 250-1: 4000V	31a		GP 02-40, MR 250-4	data
GI 500	Si-Di	Gl, 50V, 3A, 2µs	31a	Gie	BY 251, G3A, GP 30A, MR 500, 1N5400, ++	data
GI 501	Si-Di	=GI 500: 100V	31a		BY 251, G3B, GP 30B, MR 501, 1N5401, ++	data
GI 502	Si-Di	=GI 500: 200V	31a		BY 251, G3D, GP 30D, MR 502, 1N5402, ++	data
GI 504	Si-Di	=GI 500: 400V	31a		BY 252, G3G, GP 30G, MR 504, 1N5404, ++	data
GI 506	Si-Di	=GI 500: 600V	31a		BY 253, G3J, GP 30J, MR 506, 1N5406, ++	data
GI 508	Si-Di	=GI 500: 800V	31a		BY 254, G3K, GP 30K, MR 508, 1N5407, ++	data
GI 510	Si-Di	=GI 500: 1000V	31a		BY 255, G3M, GP 30M, MR 510, 1N5408, ++	data
GI 750	Si-Di	Gl, 50V, 6A, 2,5µs	31a	Gie	BY 214/50, BY 500-100, MR 750	data

Type	Device	Short Description	Fig.	Manu	Comparision Types	More at
GI 751	Si-Di	=MR 750: 100V	31a	Gie	BY 214/100, BY 500-100, MR 751	data
GI 752	Si-Di	=MR 750: 200V	31a	Gie	BY 214/200, BY 500-200, MR 752	data
GI 754	Si-Di	=MR 750: 400V	31a	Gie	BY 214/400, BY 500-400, MR 754	data
GI 756	Si-Di	=MR 750: 600V	31a	Gie	BY 214/600, BY 500-600, MR 756	data
GI 758	Si-Di	=MR 750: 800V	31a	Gie	BY 214/800, BY 500-800, MR 758	data
GI 810	Si-Di	GI, S, 50V, 1A, <750ns	31a	Gie	BYD 33D, BYT 52A, RGP 10A, MR 810, ++	data
GI 811	Si-Di	=GI 810: 100V	31a	Gie	BYD 33D, BYT 52B, RGP 10B, MR 811, ++	data
GI 812	Si-Di	=GI 810: 200V	31a	Gie	BYD 33D, BYT 52D, RGP 10D, MR 812, ++	data
GI 814	Si-Di	=GI 810: 400V	31a	Gie	BYD 33G, BYT 52G, RGP 10G, MR 814, ++	data
GI 816	Si-Di	=GI 810: 600V	31a	Gie	BYD 33J, BYT 52J, RGP 10J, MR 816, ++	data
GI 817	Si-Di	=GI 810: 800V	31a	Gie	BYD 33K, BYT 52K, RGP 10K, MR 817, ++	data
GI 818	Si-Di	=GI 810: 1000V	31a	Gie	BYD 33M, BYT 52M, RGP 10M, MR 818, ++	data
GI 820	Si-Di	GI, S, TV, 50V, 5A, <200ns	31a	Gie	BY 500-100, EGP 50A, MR 820	data
GI 821	Si-Di	=GI 820: 100V	31a	Gie	BY 500-100, EGP 50B, MR 821	data
GI 822	Si-Di	=GI 820: 200V	31a	Gie	BY 500-200, EGP 50D, MR 822	data
GI 824	Si-Di	=GI 820: 400V	31a	Gie	BY 500-400, EGP 50G, MR 824[4]	data
GI 826	Si-Di	=GI 820: 600V	31a	Gie	BY 500-600, MR 826	data
GI 828	Si-Di	=GI 820: 800V	31a	Gie	BY 500-800, MR 828	data
GI 850	Si-Di	GI, S, 50V, 3A, <200ns	31a	Gie	BYM 36A, BYT 56A, BYW 72, MR 850, ++	data
GI 851	Si-Di	=GI 850: 100V	31a	Gie	BYM 36A, BYT 56B, BYW 72, MR 851, ++	data
GI 852	Si-Di	=GI 850: 200V	31a	Gie	BYM 36A, BYT 56D, BYW 72, MR 852, ++	data
GI 854	Si-Di	=GI 850: 400V	31a	Gie	BYM 36B, BYT 56G, BYW 74, MR 854, ++	data
GI 856	Si-Di	=GI 850: 600V	31a	Gie	BYM 36C, BYT 56J, BYW 76, MR 856, ++	data
GI 858	Si-Di	=GI 850: 800V	31a	Gie	BYM 36D, BYT 56K, BYT 13-800	data
GI 910	Si-Di	GI, S, 50V, 3A, <750ns	31a	Gie	BYW 95A, BYW 72, RGP 30A, MR 850	data
GI 911	Si-Di	=GI 910: 100V	31a	Gie	BYW 95A, BYW 72, RGP 30B, MR 851	data
GI 912	Si-Di	=GI 910: 200V	31a	Gie	BYW 95A, BYW 72, RGP 30D, MR 852	data
GI 914	Si-Di	=GI 910: 400V	31a	Gie	BYW 95B, BYW 74, RGP 30G, MR 854	data
GI 916	Si-Di	=GI 910: 600V	31a	Gie	BYW 95C, BYW 76, RGP 30J, MR 856	data
GI 917	Si-Di	=GI 910: 800V	31a	Gie	BY 399, BYW 96D, BYT 77, RGP 30K	data
GI 918	Si-Di	=GI 910: 1000V	31a	Gie	BY 399S, BYW 96E, BYT 78, RGP 30M	data
GI 1001	Si-Di	GI, S, 50V, 1A, <25ns	31a	Gie	BYV 26A, FE 1A	data
GI 1002	Si-Di	=GI 1001: 100V	31a	Gie	BYV 26A, FE 1B	data
GI 1003	Si-Di	=GI 1001: 150V	31a	Gie	BYV 26A, FE 1C	data
GI 1004	Si-Di	=GI 1001: 200V	31a	Gie	BYV 26A, FE 1D	data
GI 1101	Si-Di	GI, S, 50V, 2,5A, <25ns	31a	Gie	BYV 28-50, FE 3A	data
GI 1102	Si-Di	GI, S, 100V, 2,5A, <25ns	31a	Gie	BYV 28-100, FE 3B	data
GI 1103	Si-Di	GI, S, 150V, 2,5A, <25ns	31a	Gie	BYV 28-150, FE 3C	data
GI 1104	Si-Di	GI, S, 200V, 2A, <50ns	31a	Gie	BYV 28-200, FE 3D	data
GI 1301	Si-Di	GI, S, 50V, 6A, <30ns	31a	Gie	BYV 61, FE 6A, (BYT 08/200, RGP 80A,++)[4]	data
GI 1302	Si-Di	GI, S, 100V, 6A, <30ns	31a	Gie	BYV 62, FE 6B, (BYT 08/200, RGP 80B,++)[4]	data
GI 1303	Si-Di	GI, S, 150V, 6A, <30ns	31a	Gie	BYV 63, FE 6C, (BYT 08/200, RGP 80C,++)[4]	data
GI 1304	Si-Di	GI, S, 200V, 5A, <50ns	31a	Gie	FE 6D, (BYT 08/200, RGP 80D,++)[4]	data
GI 1401	Si-Di	GI, S, 50V, 8A, <35ns	17t§	Gie	BYP 21-50, BYW 29-50, BYW 80-50, FE 8A	data
GI 1402	Si-Di	=GI 1401: 100V	17t§	Gie	BYP 21-100, BYW 29-100, BYW80-100, FE 8B	data
GI 1403	Si-Di	=GI 1401: 150V	17t§	Gie	BYP 21-150, BYW 29-150, BYW80-150, FE 8C	data
GI 1404	Si-Di	=GI 1401: 200V	17t§	Gie	BYP 21-200, BYW 29-200, BYW80-200, FE 8D	data
GI 1401R...1404R	Si-Di	=GI 1401...1404:	17t1		-	data
GI 2401	Si-Di	Dual, GI, S, 50V, 16A, <35ns	17p§	Gie	BYV 32-50, FE 16A, MUR 1605CT	data
GI 2402	Si-Di	=GI 2401: 100V	17p§	Gie	BYV32-100, FE 16B, MUR 1610CT	data
GI 2403	Si-Di	=GI 2401: 150V	17p§	Gie	BYV 32-150, FE 16C, MUR 1615CT	data
GI 2404	Si-Di	=GI 2401: 200V	17p§	Gie	BYV32-200, FE 16D, MUR 1620CT	data
GI 2401R..2404R	Si-Di	=GI 2401...2404:	17q&		BYV 32N/..., FE 16..N	data
GI 5823	Si-Di	Schottky-GI, 20V, 5A	31a	Gie	SB 520, 1N5823	data
GI 5824	Si-Di	=GI 5823: 30V	31a	Gie	SB 530, 1N5824	data
GI 5825	Si-Di	=GI 5823: 40V	31a	Gie	SB 540, 1N5825	data
GIA	Si-N-Darl	=2SC3957-GIA(Typ-Code/Stempel/marking)	44	Hit	→2SC3957	data
GIB	Si-N-Darl	=2SC3957-GIB(Typ-Code/Stempel/marking)	44	Hit	→2SC3957	data
GIB 1401	Si-Di	GI, S, 50V, 8A(Tc=125°), <35ns	30t§	Gsi	-	data
GIB 1402	Si-Di	=GIB 1401: 100V	30t§	Gsi	-	data
GIB 1403	Si-Di	=GIB 1401: 150V	30t§	Gsi	-	data
GIB 1404	Si-Di	=GIB 1401: 200V	30t§	Gsi	-	data
GIB 2401	Si-Di	Dual, GI, S, 50V, 16A(Tc=125°), <35ns	30p§	Gsi	BYQ 30EB-150, MURB 1620CT	data
GIB 2402	Si-Di	=GIB 2401: 100V	30p§	Gsi	BYQ 30EB-150, MURB 1620CT	data
GIB 2403	Si-Di	=GIB 2401: 150V	30p§	Gsi	BYQ 30EB-150, MURB 1620CT	data
GIB 2404	Si-Di	=GIB 2401: 200V	30p§	Gsi	BYQ 30EB-200, MURB 1620CT	data
GJ	Si-Di	=GF 1J (Typ-Code/Stempel/marking)	71(5x2,7)	Gsi	→GF 1J	data
GJ 2	N-FET	=2SK968-2 (Typ-Code/Stempel/marking)	35	Say	→2SK968	data
GJ 3	N-FET	=2SK968-3 (Typ-Code/Stempel/marking)	35	Say	→2SK968	data
GJ 4	N-FET	=2SK968-4 (Typ-Code/Stempel/marking)	35	Say	→2SK968	data
GJ 5	N-FET	=2SK968-5 (Typ-Code/Stempel/marking)	35	Say	→2SK968	data
GK	Si-N	=2SD1615-GK (Typ-Code/Stempel/marking)	39	Nec	→2SD1615	data
GK	MOS-N-FET-e*	=2SK2969 Typ-Code/Stempel/marking)	35	Say	→2SK2969	data
GK	Si-Di	=GF 1K (Typ-Code/Stempel/marking)	71(5x2,7)	Gsi	→GF 1K	data
GK	Z-Di	=SM 6T 150 (Typ-Code/Stempel/marking)	71(6x4mm)	Tho	→SM 6T...	data
GK	Z-Di	=Z2 SMB-11 (Typ-Code/Stempel/marking)	71(5x3,5)	Fag	→Z2 SMB-11	data
GL-55	LED	gr 9/125-62.5/125µmm l:1.3mcd		Sie		data pinout
GL	Si-P	=2SA1496 (Typ-Code/Stempel/marking)	35	Say	→2SA1496	data
GL	Si-N	=2SD1615-GL (Typ-Code/Stempel/marking)	39	Nec	→2SD1615	data

Type	Device	Short Description	Fig.	Manu	Comparision Types	More at
GL	Z-Di	=SM 6T 150A (Typ-Code/Stempel/marking)	71(6x4mm)	Tho	→SM 6T...	data
GL	Z-Di	=Z2 SMB-12 (Typ-Code/Stempel/marking)	71(5x3,5)	Fag	→Z2 SMB-12	data
GL1XX1	LED	gr+srd+or+yl ø2mm I:2.6mcd Vf:1.9V		Sha		data pdf pinout
GL1XX3	LED	gr+rd ø1.6mm I:250mcd Vf:1.85V		Sha		data pinout
GL2XX6	LED	gr+srd+or+yl ø2mm I:2mcd Vf:1.9V		Sha		data pdf pinout
GL3BX44	LED	bl ø3mm I:4mcd Vf:3.1V If:20mA	T1	Sha		data pinout
GL3ED8	LED	gr/rd ø3.1mm I:250mcd Vf:2V If:20mA	T1	Sha		data pinout
GL3XX2	LED	gr+rd ø3mm I:300mcd Vf:1.85V If:20mA	T1	Sha		data pdf pinout
GL3XX4	LED	srd+or+yl+gr ø3mm I:110mcd Vf:1.75V	T1	Sha		data pinout
GL3XX8	LED	gr+srd+or+yl ø3mm I:60mcd Vf:1.75V	T1	Sha		data pinout
GL5BX4	LED	bl ø5mm I:8mcd Vf:3.1V If:20mA	T1¾	Sha		data pinout
GL5UR2K	LED	rd ø4.8mm I:2000mcd Vf:1.85V If:20mA	T1¾	Sha		data pinout
GL5UR3K	LED	rd ø4.8mm I:3000mcd Vf:1.85V If:20mA	T1¾	Sha		data pinout
GL5XX2	LED	gr/rd+yl/rd ø5mm I:1mcd Vf:2V		Sha		data pinout
GL5XX4	LED	gr+srd+or+yl ø5mm I:100mcd Vf:1.75V	T1¾	Sha		data pinout
GL5XX5	LED	gr+rd+yl/rd ø5mm I:9mcd Vf:2V	T1¾	Sha		data pinout
GL5XX6	LED	gr/rd+srd ø5mm I:11mcd Vf:1.75V	T1¾	Sha		data pinout
GL5XX7	LED	gr+srd+or+yl ø4.9mm I:110mcd If:20mA	T1¾	Sha		data pinout
GL5XX8	LED	gr+srd+or+yl ø5mm I:80mcd Vf:1.75V	T1¾	Sha		data pinout
GL6CU11	LED	gr/rd ø5mm I:150mcd Vf:1.8V If:20mA	T1¾	Sha		data pdf pinout
GL6XX11T	LED	gr+rd ø5mm I:300mcd Vf:1.85V If:20mA	T1¾	Sha		data pinout
GL6XX23T	LED	yl+rd ø5mm I:300mcd Vf:1.85V If:20mA	T1¾	Sha		data pinout
GL6XX3T	LED	rd ø5mm I:600mcd Vf:1.75V If:20mA	T1¾	Sha		data pinout
GL8XX2	LED	gr+srd+yl 5x2.5mm I:25mcd Vf:1.75V		Sha		data pinout
GL8XX21	LED	gr+srd+or+yl 2x5mm I:16mcd Vf:1.85V		Sha		data pdf pinout
GL8XX22	LED	gr+srd+or+yl 5x5mm I:5mcd Vf:1.75V		Sha		data pinout
GL8XX23	LED	gr+srd+yl 2x4.5mm I:5mcd Vf:1.75V		Sha		data pinout
GL8XX25	LED	gr+srd+or+yl 2x3.2mm I:1.8mcd If:5mA		Sha		data pdf pinout
GL8XX28	LED	gr+srd+yl 4x4.5x4.5mm I:0.9mcd		Sha		data pinout
GL8XX4	LED	gr+srd+yl 2x3.1mm I:150mcd Vf:1.85V		Sha		data pinout
GL8XX42	LED	gr+srd+yl 1.8x3.9mm I:4mcd Vf:1.75V		Sha		data pinout
GL8XX48	LED	gr+srd+yl 3x3mm I:10mcd Vf:2V		Sha		data pinout
GL8XX5	LED	gr/rd+srd+yl/rd 2x5mm I:0.8mcd Vf:2V		Sha		data pinout
GL 27 A..M	Si-Di	=BYM 30-...		Gie	→BYM 30-...	data
GL 34 A..J	Si-Di	=BYM 05-..., A=50, B=100, D=200V G=400, J=600V		Gie	→BYM 05-...	data
GL 41 A..Y	Si-Di	=BYM 10-..., A=50, B=100, D=200, G=400V J=600, K=800, M=1000, T=1300, Y=1600V		Gie	→BYM 10-...	data
GL105X11	LED	yl+rd/gr 4.5x2mm I:0.25mcd Vf:1.9V		Sha		data pinout
GL105X8	LED	yl+rd/gr 4.5x2mm I:0.9mcd Vf:1.9V		Sha		data pinout
GL107X12	LED	yl+rd/gr 4.5x2mm I:0.3mcd Vf:1.9V		Sha		data pinout
GL107X8	LED	yl+rd/gr 4.5x2mm I:0.9mcd Vf:1.9V		Sha		data pinout
GL112X13	LED	yl+rd/gr 4.5x2mm I:0.3mcd Vf:1.9V		Sha		data pinout
GL112X9	LED	yl+rd/gr 4.5x2mm I:0.3mcd Vf:1.9V		Sha		data pinout
GL 317	Z-IC	+1,7...37V, 0,01...1,5A, -20...+100°	17l		... 117..., ... 217..., ... 317...	
GL360	IRED	950nm ø3.2mm I:5mW/Sr Vf:1.3V	T1	Sha		data pdf pinout
GL371	IRED	950nm ø3mm I:5mW/Sr Vf:1.3V If:40mA	T1	Sha		data pdf pinout
GL380	IRED	950nm ø3mm I:20mW/Sr Vf:1.3V If:50mA	T1	Sha		data pdf pinout
GL390	IRED	950nm 2x3.1mm I:16mW/Sr Vf:1.3V		Sha		data pdf pinout
GL420	IRED	950nm ø2.5mm I:60mcd Vf:1.2V If:20mA		Sha		data pdf pinout
GL430	IRED	950nm ø1.8mm I:60mcd Vf:1.2V If:20mA		Sha		data pdf pinout
GL450	IRED	950nm ø1.5mm I:60mcd Vf:1.2V If:20mA		Sha		data pdf pinout
GL453	IRED	950nm ø1.5mm I:60mcd Vf:1.2V If:20mA		Sha		data pdf pinout
GL460	IRED	950nm ø2mm I:60mcd Vf:1.2V If:20mA		Sha		data pdf pinout
GL480	IRED	950nm ø1.5mm I:60mcd Vf:1.2V If:20mA		Sha		data pdf pinout
GL 494	LIN-IC	PWM Cntrl. IC	DIP	Hyn		pdf pinout
GL514	IRED	950nm ø4.7mm I:60mcd Vf:1.35V	TO18	Sha		data pdf pinout
GL518	IRED	950nm ø4.7mm I:60mcd Vf:1.3V If:50mA	TO18	Sha		data pdf pinout
GL527	IRED	940nm ø5mm I:23mW/Sr Vf:1.3V If:50mA	T1¾	Sha		data pdf pinout
GL527V	IRED	940nm ø5mm I:23mW/Sr Vf:1.2V If:50mA	T1¾	Sha		data pdf pinout
GL533	IRED	940nm ø4.2mm I:23mW/Sr Vf:1.3V		Sha		data pdf pinout
GL537	IRED	950nm ø5mm I:30mW/Sr Vf:1.3V If:50mA	T1¾	Sha		data pdf pinout
GL550	IRED	880nm ø5mm I:30mW/Sr Vf:1.5V If:50mA	T1¾	Sha		data pdf pinout
GL 3276 A	IC	Infrared Preamplifier IC	MDIP	Hyn	-	pdf pinout
GL 3277	IC	Preamp for Remote Cntrl.	MDIP	Hyn	-	pdf pinout
GL 3361	IC	Low Power Narrow Band FM IF	DIC	Hyn	-	pdf pinout
GL 3812	CMOS/LIN-IC	Audio/video Switch	DIC	Hyn	-	pdf pinout
GL 3820	IC	Video Switch	DIC	Hyn	-	pdf pinout
GL4100	IRED	950nm ø3mm I:60mcd Vf:1.2V If:20mA		Sha		data pdf pinout
GL4600	IRED	950nm ø2mm I:60mcd Vf:1.2V If:20mA		Sha		data pdf pinout
GL4800	IRED	950nm ø0.8mm I:60mcd Vf:1.2V If:20mA		Sha		data pdf pinout
GL4910	IRED	850nm ø3.5mm I:60mcd Vf:1.5V If:50mA		Sha		data pdf pinout
GL 6551	IC	Compander	DIC	Hyn	-	pdf pinout
GL 6552	IC	Low Voltage Compander	MDIP	Hyn	-	pdf pinout
GL 6840 A	IC	Electronic 2-Tone Ringer	47	Hyn	-	pdf pinout
GL 6840 B	IC	Electronic 2-Tone Ringer	47	Hyn	-	pdf pinout
GL 6850	IC	2-Tone Ringer	47	Hyn	-	pdf pinout
GL 6851	IC	2-Tone Ringer	47	Hyn	-	pdf pinout
GL 6962 (A)	IC	Low Voltage Universal Speech Network	DIC	Hyn	-	pdf pinout
GL 7101	IC	Earth Leakage Current Detector	DIP		-	pdf pinout
GL 7805	Z-IC	~LM 7805	17b		→LM 7805	
GL 7812	Z-IC	~LM 7812	17b		→LM 7812	
GLC 555	LIN-IC	=NE 555CH,CM: SMD	8-MDIP		→NE 555...	
GLC 556	LIN-IC	=NE 556CH,CM: SMD	8-MDIP		→NE 556...	
GLC 4558	OP-IC	SMD, Dual, Serie 158, ±18V, 0...+70°	8-MDIP		→Serie 158	data

Type	Device	Short Description	Fig.	Manu	Comparision Types	More at
GLC 4559	OP-IC	=GLC 4558: lo-noise	8-MDIP		→Serie 158	data
GLL 4735...4763	Z-Di	=1N4735...4763: SMD	72a(5mm)	Gie		data
GLT 12	Z-Di	=BZW 25/12:	32	Tho	→BZW 25/12	data
GLT 24	Z-Di	=BZW 25/24:	32	Tho	→BZW 25/24	data
GLT 47	Z-Di	=BZW 25/47:	32	Tho	→BZW 25/47	data
GLT 120	Z-Di	=BZW 25/120:	32	Tho	→BZW 25/120	data
GLXCL	LED	gr/rd ø5mm l:100mcd Vf:1.8V If:20mA	T1¾	Sha		data pinout

GM...GO

Type	Device	Short Description	Fig.	Manu	Comparision Types	More at
GM	Si-N	=2SD1615-GM (Typ-Code/Stempel/marking)	39	Nec	→2SD1615	data
GM	Si-Di	=GF 1M (Typ-Code/Stempel/marking)	71(5x2,7)	Gsi	→GF 1M	data
GM	Z-Di	=Z2 SMB-13 (Typ-Code/Stempel/marking)	71(5x3,5)	Fag	→Z2 SMB-13	data
GM-1 A...Z	Si-Di	GI, Uni, 200...1000V, 1,3A A=600V, B=800V, C=1000V, Z=200V	31a	Sak	BY 226...227, BY 251...255, 1N5391...99, ++	data
GM 3 A...Z	Si-Di	GI, Uni, 100...1000V, 2,7A A=600, B=800, C=1000, Y=100, Z=200V	31a	Sak	BY 251...255, GP 30B...M, 1N5401...08, ++	data
GM 30 A..M	Si-Di	Dual, GI-L, 50...1000V, 30A A=50, B=100, D=200, G=400, J=600V, K=800, M=1000V	23I§	Gie	BYV 72/.., BYW 99P/.., RP 30AP...MP	
GM 30 AD...MD	Si-Di	=GM 30A..M:	23n		-	data
GM 30 AN...MN	Si-Di	=GM 30A..M:	23m&		-	data
GM 0290(A)	Ge-P	UHF-V/M/O, 750MHz	5g	Tix	AF 139, AF 239(S)	data
GM 0378(A)	Ge-P	VHF, 600MHz	5g	Tix	AF 139, AF 239(S)	data
GM 0656(A)	Ge-P	UHF-O	5g	Tix	AF 139, AF 239(S)	data
GM 0760	Ge-P	VHF-V	5g	Tix	AF 106, AF 109R, AF 239(S), AF 306	data
GM 0761	Ge-P	VHF-M	5g	Tix	AF 106, AF 109R, AF 239(S), AF 306	data
GM 0936 TQ	IC	Voice-band Audio codec for Cdma	DIP		-	pdf pinout
GM 2221	DIG-IC	LCD Monitor Controller	MP	Gns	-	pdf pinout
GM 5221	DIG-IC	LCD Monitor Controller	MP	Gns	-	pdf pinout
GM 5621	IC	2x Input LCD Controller	MP	Stm	-	pdf
GM 5626	IC	2x Input LCD Controller	MP	Stm	-	pdf
GM 5766	IC	2x Input LCD Ctrl. f. UXGA/WSXGA+ App.	MP	Stm	-	pdf
GM 5862 H	IC	LCD Monitor Ctrl. w. RTC up to WUXGA	MP	Stm	-	pdf
GM 5868 H	IC	LCD Monitor Ctrl. w. Rtc up to Wuxga	MP	Stm	-	pdf
GM 6486	IC	33 Output LED Driver	DIP		-	pdf pinout
GM 6535	Logic	60 MHz Universal Programmable Dual PLL Freq. synth.	DIP		-	pdf pinout
GM 69010 H	IC	Displayport, HDMI, & Component In. Receiver	MP	Stm		pdf
GM 16 C 550	IC	asynchronous communications element with FIFOs	DIP		-	pdf pinout
GM 62093	CMOS-IC	→KS 093			KS 093, M 093, MT 8812	pdf
GM 62093 A	CMOS-IC	12x8 Crosspoint Switch with Control MEMORY	DIP		-	pdf pinout
GM 0936 TQ-44	IC	Voice-band Audio codec for Cdma	DIP		-	pdf pinout
GM 6486-44	IC	33 Output LED Driver	DIP		-	pdf pinout
GM 16 C 550-44	IC	asynchronous communications element with FIFOs	DIP		-	pdf pinout
GMA 01	Si-Di	SS, 60V, 0,12A, <4ns	31a	Say	BAW 62, BAW 76, BAX 95, 1N4148...49, ++	data
GMA 01 U	Si-Di	=GMA 01: 105V	31a	Say	BAV 14	data
GMA 02	Si-Di	S, 250V, 0,2A, <75ns	31a	Say	BA 198, MA 188, BAV 21, 1SS82...83	data
GMB 01(U)	Si-Di	=GMA 01	31a	Say	→GMA 01(U)	data
GMJ 2955	Si-P	≈BD 208	14h§		→BD 208	
GMJ 3055	Si-N	≈BD 207	14h§		→BD 207	
GMS 81 C 5016 /DIP	CMOS-IC	Cmos Single Chip 8-Bit Microcontroller	TSOP	Hun	-	pdf pinout
GMS 81 C 5016/PLCC	CMOS-IC	Cmos Single Chip 8-Bit Microcontroller	TSOP	Hun	-	pdf pinout
GMS 81 C 5016/QFP	CMOS-IC	Cmos Single Chip 8-Bit Microcontroller	TSOP	Hun	-	pdf pinout
GMS 81 C 5016 /SOP	CMOS-IC	Cmos Single Chip 8-Bit Microcontroller	TSOP	Hun	-	pdf pinout
GMS 81 C 5024 /DIP	CMOS-IC	Cmos Single Chip 8-Bit Microcontroller	TSOP	Hun	-	pdf pinout
GMS 81 C 5024/PLCC	CMOS-IC	Cmos Single Chip 8-Bit Microcontroller	TSOP	Hun	-	pdf pinout
GMS 81 C 5024/QFP	CMOS-IC	Cmos Single Chip 8-Bit Microcontroller	TSOP	Hun	-	pdf pinout
GMS 81 C 5024 /SOP	CMOS-IC	Cmos Single Chip 8-Bit Microcontroller	TSOP	Hun	-	pdf pinout
GMS 81 C 5032 /DIP	CMOS-IC	Cn s Single Chip 8-Bit Microcontroller	TSOP	Hun	-	pdf pinout
GMS 81 C 5032/PLCC	CMOS-IC	Cmos Single Chip 8-Bit Microcontroller	TSOP	Hun	-	pdf pinout
GMS 81 C 5032/QFP	CMOS-IC	Cmos Single Chip 8-Bit Microcontroller	TSOP	Hun	-	pdf pinout
GMS 81 C 5032 /SOP	CMOS-IC	Cmos Single Chip 8-Bit Microcontroller	TSOP	Hun	-	pdf pinout
GN	Si-N	=2SC4867 (Typ-Code/Stempel/marking)	35(2mm)	Say	→2SC4867	data
GN	Si-N	=2SC4868 (Typ-Code/Stempel/marking)	35	Say	→2SC4868	data
GN	Si-N	=2SC4869 (Typ-Code/Stempel/marking)	44	Say	→2SC4869	data
GN	Si-N	=2SC5489 (Typ-Code/Stempel/marking)	35(1,4mm)	Say	→2SC5489	data
GN	Si-N	=2SC5503 (Typ-Code/Stempel/marking)	44(2mm)	Say	→2SC5503	data
GN	Z-Di	=Z2 SMB-15 (Typ-Code/Stempel/marking)	71(5x3,5)	Fag	→Z2 SMB-15	data
GN 1 A3Q	Si-P+R	=AN 1A3Q: SMD	35a(2mm)	Nec	DTA 113ZU	data
GN 1 A4M	Si-P+R	=AN 1A4M: SMD	35a(2mm)	Nec	DTA 114EU, 2SA1678	data
GN 1 A4P	Si-P+R	=AN 1A4P: SMD	35a(2mm)	Nec	DTA 114YU	data
GN 1 A4Z	Si-P+R	=AN 1A4Z: SMD	35a(2mm)	Nec	DTA 114TU	data
GN 1 F4M	Si-P+R	=AN 1F4M: SMD	35a(2mm)	Nec	DTA 124EU, 2SA1677	data
GN 1 F4N	Si-P+R	=AN 1F4N: SMD	35a(2mm)	Nec	DTA 124XU	data
GN 1 F4Z	Si-P+R	=AN 1F4Z: SMD	35a(2mm)	Nec	DTA 124TU	data

Type	Device	Short Description	Fig.	Manu	Comparision Types	More at
GN 1 L3M	Si-P+R	=AN 1L3M: SMD	35a(2mm)	Nec	DTA 143EU	data
GN 1 L3N	Si-P+R	=AN 1L3N: SMD	35a(2mm)	Nec	DTA 143XU	data
GN 1 L3Z	Si-P+R	=AN 1L3Z: SMD	35a(2mm)	Nec	DTA 143TU	data
GN 1 L4L	Si-P+R	=AN 1L4L: SMD	35a(2mm)	Nec	DTA 143WU	data
GN 1 L4M	Si-P+R	=AN 1L4M: SMD	35a(2mm)	Nec	DTA 144EU, 2SA1676	data
GN 1 L4Z	Si-P+R	=AN 1L4Z: SMD	35a(2mm)	Nec	DTA 144TU	data
GN 3	Si-N	=2SC4867-3 (Typ-Code/Stempel/marking)	35(2mm)	Say	→2SC4867	data
GN 3	Si-N	=2SC4868-3 (Typ-Code/Stempel/marking)	35	Say	→2SC4868	data
GN 3	Si-N	=2SC4869-3 (Typ-Code/Stempel/marking)	44	Say	→2SC4869	data
GN 4	Si-N	=2SC4867-4 (Typ-Code/Stempel/marking)	35(2mm)	Say	→2SC4867	data
GN 4	Si-N	=2SC4868-4 (Typ-Code/Stempel/marking)	35	Say	→2SC4868	data
GN 4	Si-N	=2SC4869-4 (Typ-Code/Stempel/marking)	44	Say	→2SC4869	data
GN 4	Si-N	=2SC5503-4 (Typ-Code/Stempel/marking)	44(2mm)	Say	→2SC5503	data
GN 5	Si-N	=2SC4867-5 (Typ-Code/Stempel/marking)	35(2mm)	Say	→2SC4867	data
GN 5	Si-N	=2SC4868-5 (Typ-Code/Stempel/marking)	35	Say	→2SC4868	data
GN 5	Si-N	=2SC4869-5 (Typ-Code/Stempel/marking)	44	Say	→2SC4869	data
GN 5	Si-N	=2SC5503-5 (Typ-Code/Stempel/marking)	44(2mm)	Say	→2SC5503	data
GN 4530 C	MOS-N-IGBT*	L, 450V, 30A, 150W, <1400/1200ns	18id	Hit	GN 6030C, IGT 8E20..21	data
GN 6010 A	MOS-N-IGBT*	L, 600V, 10A, 50W, <350/500ns	17id	Hit	BUK 854-800A	data
GN 6015 A	MOS-N-IGBT*	L, 600V, 15A, 60W, <400/600ns	17id	Hit	BUP 400	data
GN 6020 C	MOS-N-IGBT*	L, 600V, 20A, 100W, <400/600ns	18id	Hit	GT 25J101	data
GN 6030 C	MOS-N-IGBT*	L, 600V, 30A, 150W, <600/800ns	18id	Hit	BUP 304	data
GN 6050 E	MOS-N-IGBT*	L, 600V, 50A, 200W, <1000/1200ns	77id	Hit	GT 50J101...J102	data
GN 6075 E	MOS-N-IGBT*	L, 600V, 75A, 250W, <1000/1200ns	77id	Hit	GT 80J101	data
GN 9060 E	MOS-N-IGBT*	L, 900V, 60A, 200W	77id	Hit	GT 60M102...M104	data
GN 12015 C	MOS-N-IGBT*	L, 1200V, 15A, 150W	18id	Hit	GT 15Q101	data
GN 12030 E	MOS-N-IGBT*	L, 1200V, 30A, 200W	77id	Hit	MGY 25N120	data
GN 12050 E	MOS-N-IGBT*	L, 1200V, 50A, 200W	77id	Hit		data
GO	Si-P	=2SA2883-O (Typ-Code/Stempel/marking)	39	Tos	→2SA2883	data
GO	Si-N	=2SC2996-O (Typ-Code/Stempel/marking)	35	Tos	→2SC2996	data
GO	Si-N	=KTC4375-O (Typ-Code/Stempel/marking)	39	Kec	→KTC 4375	data

GP...GR

Type	Device	Short Description	Fig.	Manu	Comparision Types	More at
GP	Si-N	=2SD1615A-GP(Typ-Code/Stempel/marking)	39	Nec	→2SD1615A	data
GP	Z-Di	=Z2 SMB-16 (Typ-Code/Stempel/marking)	71(5x3,5)	Fag	→Z2 SMB-16	data
GP1A	PI	LED VR/PD+IC Ibe:16.5mA		Sha		data pdf pinout
GP1L	PI	LED/Dar CTR:>30% Vbs:1.25V Ibs:20mA		Sha		data pdf pinout
GP1S	PI	LED/NPN CTR:2.5-30% Vbs:1.2V Ibe:2mA		Sha		data pinout
GP1U10X	PD	950nm Vb:4.7...5.3V Ib:<5mA		Sha		data pdf pinout
GP1U50X	PD	950nm Vb:4.7...5.3V Ib:<5mA		Sha		data pdf pinout
GP1U52X	PD	950nm Vb:4.7...5.3V Ib:<5mA		Sha		data pdf pinout
GP1U52Y	PD	950nm Vb:4.7...5.3V Ib:<5mA		Sha		data pinout
GP1U56R	PD	950nm Vb:4.7...5.3V Ib:<5mA		Sha		data pdf pinout
GP1U57X	PD	950nm Vb:4.7...5.3V Ib:<5mA		Sha		data pdf pinout
GP1U58X	PD	950nm Vb:4.7...5.3V Ib:<5mA		Sha		data pdf pinout
GP1U58Y	PD	950nm Vb:4.7...5.3V Ib:<5mA		Sha		data pdf pinout
GP1U59X	PD	950nm Vb:4.7...5.3V Ib:<5mA		Sha		data pinout
GP1U70R	PD	950nm Vb:4.7...5.3V Ib:<5mA		Sha		data pdf pinout
GP1U72Q	PD	950nm Vb:4.7...5.3V Ib:<5mA		Sha		data pdf pinout
GP1U72R	PD	950nm Vb:4.7...5.3V Ib:<5mA		Sha		data pinout
GP1U77R	PD	950nm Vb:4.7...5.3V Ib:<5mA		Sha		data pdf pinout
GP1U78Q	PD	950nm Vb:4.7...5.3V Ib:<5mA		Sha		data pdf pinout
GP1U78R	PD	950nm Vb:4.7...5.3V Ib:<5mA		Sha		data pinout
GP1U90X	PD	950nm Vb:4.7...5.3V Ib:<5mA		Sha		data pdf pinout
GP1UM28	PD	950nm Vb:4.7...5.5V Ib:<1.5mA		Sha		data pdf pinout
GP 02-20...-40	Si-Di	kV-Gl, 2...4kV, 0,25A 20=2, 25=2,5, 30=3, 35=3,5, 40=4kV	31a	Gie	BYX 108G, MR 250-2...-5	data
GP2A	PS	LED/PD+IC Ibe:<30mA		Sha		data pdf pinout
GP2L	PS	LED/Dar CTR:100-375% Vbs:1.2V		Sha		data pinout
GP2S	PS	LED/NPN CTR:2.5-15% Vbs:1.2V		Sha		data pdf pinout
GP2TD	PS	LED/2 CTR:1.3-48% Vbs:<1.5V Ibe:15mA		Sha		data pinout
GP 08 A..J	Si-Di	Gl, Uni, 50...600V, 0,8A, 2µs	31a	Gie	BY 126...127, BY 133...135, 1N4001...07, ++	data
GP 10 A..M	Si-Di	Gl, Uni, 50...1000V, 1A A=50, B=100, D=200, G=400, J=600V, K=800, M=1000V	31a	Gie,Fag,Mic	BY 126...127, BY 133...135, 1N4001...07, ++	data
GP 10 N...Y	Si-Di	=GP 10A..M: 1100...1600V N=1100, Q=1200, T=1300, V=1400V, W=1500, Y=1600V	31a		BY 350/1500, GH 3F, DM 513, EM 516, ++	data
GP 15 A..M	Si-Di	Gl, Uni, 50...1000V, 1,5A A=50, B=100, D=200, G=400, J=600V, K=800, M=1000V	31a	Gie, Fag	BY 226...227, BY 251...255, 1N5391...99, ++	data
GP 20 A..M	Si-Di	Gl, Uni, 50...1000V, 2A, 2,5µs	31a	Gie	BY 251...255, MR 500...510, 1N5400...08, ++	data
GP 30 A..M	Si-Di	Gl, Uni, 50...1000V, 3A, 3µs	31a	Gie, Fag	BY 251...255, MR 500...510, 1N5400...08, ++	data
GP 80 A..M	Si-Di	Gl-L, 50...1000V, 8A	17t§	Gie	BY 239/..., BY 359/..., RGP 80...	data
GP 140	Si-N-Darl	≈BDW 83C	18j§		→BDW 83C	
GP 145	Si-P-Darl	≈BDW 84C	18j§		→BDW 84C	
GPP 10 A...M	Si-Di	≈GP 10A...M	31a	Sym	→GP 10A...M	data
GPP 15 A...M	Si-Di	≈GP 15A...M	31a	Sym	→GP 15A...M	data

Type	Device	Short Description	Fig.	Manu	Comparision Types	More at
GPP 20 A...M	Si-Di	≈GP 20A...M	31a	Sym	→GP 20A...M	data
GPP 30 A...M	Si-Di	≈GP 30A...M	31a	Sym	→GP 30A...M	data
GPP 60 A...M	Si-Di	GI, 50...1000V, 6A	31a	Sym	BY 214/..., MR 750...760	data
		A=50, B=100, D=200, G=400, J=600V,				
		K=800, M=1000V				
GPS-...	Si-N/P	≈MPS ...			→MPS ...	
GPU 2243	NMOS-IC	Ctv, digit. SECAM-Chroma-Prozessor	40-DIP	Itt		
GQ	Si-N	=2SD1615A-GQ(Typ-Code/Stempel/marking)	39	Nec	→2SD1615A	data
GQ	Z-Di	=Z2 SMB-18 (Typ-Code/Stempel/marking)	71(5x3,5)	Fag	→Z2 SMB-18	data
GR	Si-P	=2SA1455K-R (Typ-Code/Stempel/marking)	35	Rhm	→2SA1455K	data
GR	Si-N	=2SC2996-R (Typ-Code/Stempel/marking)	35	Tos	→2SC2996	data
GR	Si-N	=2SC4990 (Typ-Code/Stempel/marking)	39	Hit	→2SC4990	data
GR	Z-Di	=Z2 SMB-20 (Typ-Code/Stempel/marking)	71(5x3,5)	Fag	→Z2 SMB-20	data

GS

Type	Device	Short Description	Fig.	Manu	Comparision Types	More at
GS	Si-P	=2SA1455K-S (Typ-Code/Stempel/marking)	35	Rhm	→2SA1455K	data
GS	Si-P	=2SB1520 (Typ-Code/Stempel/marking)	39	Hit	→2SB1520	data
GS	Z-Di	=Z2 SMB-22 (Typ-Code/Stempel/marking)	71(5x3,5)	Fag	→Z2 SMB-22	data
GS 100(B...D)	Ge-P	S, 25V, 0,05A, 0,03W, <1500/-ns	2a	Hfo	AC 122, AC 125...126, AC 151, ASY 26...27	data
GS 109(B...D)	Ge-P	S, 20V, 0,05A, 0,083W, <1500/-ns	2a	Hfo	AC 122, AC 125...126, AC 151, ASY 26...27	data
GS 110	Ge-P	S, 20V, 0,3A, 0,083W, <1500/-ns	2a	Hfo	AC 128, AC 152...153, AC 188, ASY 76...77	data
GS 111(B...E)	Ge-P	S, 20V, 0,2A, 0,083W, <1200/1500ns	2a	Hfo	AC 128, AC 152...153, AC 188, ASY 76...77	data
GS 112(B...E)	Ge-P	=GS 111: <900/1500ns	2a	Hfo	AC 128, AC 152...153, AC 188, ASY 76...77	data
GS 121(B...D)	Ge-P	S, 20V, 0,1A, 0,15W, <10µs	2a	Hfo	AC 128, AC 152...153, AC 188, ASY 76...77	data
GS 122	Ge-P	=GS 121: 30V	2a	Hfo	AC 128, AC 152...153, AC 188, ASY 76...77	data
GS 8050 T	Si-N	Uni, 40V, 0,8A, 0,625W, 100MHz	7e	Gsi	BC 635, BC 637, 2SD1225, 2SD1616(A), ++	data
GS 8550 T	Si-P	Uni, 40V, 0,8A, 0,625W, 100MHz	7e	Nsc	BC 636, BC 638, 2SB909, 2SB1116(A), ++	data
GS 9012(D...H)	Si-P	NF-E, 40V, 0,5A, 0,625W	7e	Gsi	BC 327, BC 636, BC 638, BC 640, ++	data
GS 9013(D...H)	Si-N	NF-E, 40V, 0,5A, 0,625W	7e	Gsi	BC 328, BC 635, BC 637, BC 639, ++	data
GS 9014(A...D)	Si-N	Uni, ra, 50V, 0,1A, 0,45W, 270MHz	7e	Gsi	BC 414, BC 550, 2SC2240, 2SC2675, ++	data
GS 9015(A...C)	Si-P	Uni, ra, 50V, 0,1A, 0,45W, 190MHz	7e	Gsi	BC 416, BC 560, 2SA970, 2SA1137, ++	data
GSA 15 B...G	Si-Di	GI, Uni, 100...600V, 1,5A	31a	Say	BY 226...227, BY 251...255, 1N5391...99, ++	data
GSA 17 B...E	Si-Di	GI, Uni, 100...400V, 1,7A	31a	Say	BY 251...255, MR 500...510, 1N5400...08, ++	data
GSA 26 B...E	Si-Di	GI, Uni, 100...400V, 2,6A	31a	Say	BY 251...255, MR 500...510, 1N5400...08, ++	data
GSA 30 B...J	Si-Di	GI, Uni, 100...800V, 3A	31a	Say	BY 251...255, MR 500...510, 1N5400...08, ++	data
GSC 215	GaAs-Di	Dual, Schottky, 150V, 5A, 7ns	17p§	Sak	-	data
GSC 218	GaAs-Di	Dual, Schottky, 180V, 5A, 7ns	17p§	Sak	-	data
GSC 235	GaAs-Di	Dual, Schottky, 350V, 2x2A, <5ns	17p§	Sak	-	data
GSC 315	GaAs-Di	Dual, Schottky, 150V, 14A, 10ns	18p§	Sak	-	data
GSC 318	GaAs-Di	Dual, Schottky, 180V, 14A, 10ns	18p§	Sak	-	data
GSD 2004 S	Si-Di	SMD, Dual, 300V, 225mA, <50ns	35o	Gsi	-	data
GSD 2004 SW	Si-Di	=GSD 2004S:	71a(2,7mm)		-	data
GSF 18 R	GaAs-Di	Schottky, 180V, 7A, 10ns	17j	Sak	-	data
GS-H9012	Si-P	=SS 9012			→SS 9012	
GS-H9032	Si-P	→CS 9012				
GS-H9033	Si-N	→CS 9013				
GSI 510	MOS-N-IGBT	=IGT 5E10CS			→IGT 5E10CS	data
GSI 525	MOS-N-IGBT	=IGT 7E20CS			→IGT 7E20CS	data
GSI 550	MOS-N-IGBT	=IGT 7E50CS			→IGT 7E50CS	data

GT...GZ

Type	Device	Short Description	Fig.	Manu	Comparision Types	More at
GT	Si-N	=2SC4168 (Typ-Code/Stempel/marking)	35	Say	→2SC4168	data
GT	Si-N	=2SC4443 (Typ-Code/Stempel/marking)	35(2mm)	Say	→2SC4443	data
GT	Si-N-Darl+Di	=2SD2423 (Typ-Code/Stempel/marking)	39	Hit	→2SD2423	data
GT	Z-Di	=SM 6T 200 (Typ-Code/Stempel/marking)	71(6x4mm)	Tho	→SM 6T...	data
GT	Z-Di	=Z2 SMB-24 (Typ-Code/Stempel/marking)	71(5x3,5)	Fag	→Z2 SMB-24	data
GT 1	Si-P	=BCX 17 (Typ-Code/Stempel/marking)	35	Rhm	→BCX 17	data
GT 5 G101	MOS-N-IGBT	L, 400V, 5A, 20W, 900/2000ns	≈30ic	Tos	-	data
GT 5 J301	MOS-N-IGBT	L, 600V, 5A, 28W, 400/300ns	17ic	Tos	-	data
		int. Freilauf-/Free-wheel Diode(E→C)				
GT 5 J311(SM)	MOS-N-IGBT	L, 600V, 5A, 45W, 400/500ns	30id	Tos		data
		int. Freilauf-/Free-wheel Diode(E→C)				
GT 8 G101	MOS-N-IGBT*	L, LogL, 400V, 8A, 20W, 1100/2000ns	≈30ic	Tos	-	data
GT 8 J101	MOS-N-IGBT	L, 600V, 8A, 30W, 400/500ns	17ic	Tos	-	data
GT 8 J102(SM)	MOS-N-IGBT	=GT 8J101: 50W	30id	Tos		data
GT 8 N101	MOS-N-IGBT	L, 1000V, 8A, 100W, 200/600ns	18id	Tos	BUP 302	data
GT 8 Q101	MOS-N-IGBT	L, 1200V, 8A, 100W, 400/800ns	18id	Tos	BUP 301	data
GT 8 Q102(SM)	MOS-N-IGBT*	=GT 8Q101:	30id	Tos		data
GT 10 G101	MOS-N-IGBT	L, 400V, 10A, 30W, 150/4500ns	17ic	Tos	STHI 10N50R	data
GT 10 G102	MOS-N-IGBT	L, 400V, 10A, 30W, 150/4500ns	17ic	Tos	STHI 10N50R	data
GT 10 J301	MOS-N-IGBT	L, 600V, 10A, 90W, 400/700ns	18id	Tos	BUP 305D	data
		int. Freilauf-/Free-wheel Diode(E→C)				
GT 10 J303	MOS-N-IGBT	L, 600V, 10A, 30W, 400/500ns	17ic	Tos		data
		int. Freilauf-/Free-wheel Diode(E→C)				
GT 10 J311	MOS-N-IGBT	L, 600V, 10A, 80W, 400/500ns	30id	Tos		data
		int. Freilauf-/Free-wheel Diode(E→C)				

Type	Device	Short Description	Fig.	Manu	Comparision Types	More at
GT 10 J312(SM)	MOS-N-IGBT	L, 600V, 10A, 60W, 400/400ns int. Freilauf-/Free-wheel Diode(E→C)	30id	Tos		data
GT 10 Q101	MOS-N-IGBT	L, 1200V, 10A, 140W, 300/500ns	18id	Tos	GN 12015C, MGW 12N120	
GT 10 Q301	MOS-N-IGBT	L, 1200V, 10A, 140W, 300/500ns int. Freilauf-/Free-wheel Diode(E→C)	18id	Tos	MGW 12N120D	data
GT 15 G101	MOS-N-IGBT	L, 400V, 15A, 40W, 150/4500ns	17ic	Tos	GT 15J102	data
GT 15 J101	MOS-N-IGBT	L, 600V, 15A, 100W, 400/500ns	18id	Tos	GN 6020C	data
GT 15 J102	MOS-N-IGBT	=GT 15J101: 35W	17ic	Tos	-	data
GT 15 J103(SM)	MOS-N-IGBT	=GT 15J101: 70W	30id	Tos	-	data
GT 15 J301	MOS-N-IGBT	L, 600V, 15A, 35W, 400/500ns	17ic	Tos	-	data
GT 15 J311(SM)	MOS-N-IGBT	L, 600V, 15A, 70W, 400/500ns int. Freilauf-/Free-wheel Diode(E→C)	30id	Tos	-	data
GT 15 M321	MOS-N-IGBT	L, 900V, 15A, 55W, 300/500ns int. Freilauf-/Free-wheel Diode(E→C)	16ic	Tos	-	data
GT 15 N101	MOS-N-IGBT	L, 1000V, 15A, 150W, 400/800ns	18id	Tos	BUP 303, GN 12015C	data
GT 15 Q101	MOS-N-IGBT	L, 1200V, 15A, 150W, 400/800ns	18id	Tos	GN 12015C, MGW 12N120	data
GT 15 Q102	MOS-N-IGBT	L, 1200V, 15A, 170W, 120/560ns	18id	Tos	GN 12015C, MGW 12N120	
GT 15 Q301	MOS-N-IGBT	L, 1200V, 15A, 170W, 120/560ns int. Freilauf-/Free-wheel Diode(E→C)	18id	Tos	-	data
GT 15 Q311	MOS-N-IGBT	L, 1200V, 15A, 160W, 120/560ns int. Freilauf-/Free-wheel Diode(E→C)	30id	Tos	-	data
GT 20 D101	MOS-N-IGBT	L, NF, 250V, 20A, 180W	77id	Tos	-	data
GT 20 D201	MOS-P-IGBT	L, NF, 250V, 20A, 180W	77id	Tos	-	data
GT 20 G101(SM)	MOS-N-IGBT	L, 400V, 20A, 60W, 150/4500ns	30id	Tos	-	data
GT 20 G102(SM)	MOS-N-IGBT	L, 400V, 20A, 60W, 150/4500ns	30id	Tos	-	data
GT 20 J101	MOS-N-IGBT	S-L, 600V, 20A, 130W, 400/500ns	18id	Tos	GN 6020C, GT 25J101	
GT 20 J301	MOS-N-IGBT	L, 600V, 20A, 130W, 400/700ns int. Freilauf-/Free-wheel Diode(E→C)	18id	Tos	-	
GT 20 J311	MOS-N-IGBT	L, 600V, 20A, 120W, 400/500ns int. Freilauf-/Free-wheel Diode(E→C)	30id	Tos	-	data
GT 25 G101(SM)	MOS-N-IGBT	L, 400V, 25A, 75W, 150/4500ns	30id	Tos	-	data
GT 25 G102(SM)	MOS-N-IGBT	L, 400V, 25A, 75W, 150/4500ns	30id	Tos	-	data
GT 25 J101	MOS-N-IGBT	L, 600V, 25A, 150W, 400/500ns	18id	Tos	GN 6030C	data
GT 25 Q101	MOS-N-IGBT	L, 1200V, 25A, 200W, 400/800ns	77id	Tos	GN 12030E	data
GT 25 Q301	MOS-N-IGBT	L, 1200V, 25A, 200W, 300/680ns int. Freilauf-/Free-wheel Diode(E→C)	77id	Tos	MGY 25N120D	data
GT 30 J101	MOS-N-IGBT	S-L, 600V, 30A, 155W, 400/700ns	18id	Tos	GN 6030C	
GT 30 J301	MOS-N-IGBT	L, 600V, 30A, 155W, 400/700ns int. Freilauf-/Free-wheel Diode(E→C)	18id	Tos	-	data
GT 30 J311	MOS-N-IGBT	L, 600V, 30A, 145W, 400/700ns int. Freilauf-/Free-wheel Diode(E→C)	30id	Tos	-	data
GT 30 J322	MOS-N-IGBT	L, 600V, 30A, 75W, 300/400ns int. Freilauf-/Free-wheel Diode(E→C)	16ic	Tos	-	data
GT 40 T101	MOS-N-IGBT	L, 1500V, 40A, 200W, 700/500ns	77id	Tos	SGL 40N150	data
GT 40 T301	MOS-N-IGBT	L, 1500V, 40A, 200W, 460/600ns int. Freilauf-/Free-wheel Diode(E→C)	77id	Tos	SGL 40N150D	
GT 50 G321	MOS-N-IGBT	L, 400V, 50A, 130W, 430/540ns int. Freilauf-/Free-wheel Diode(E→C)	77id	Tos	GT 50J301, MGY 30N60D	data
GT 50 J101	MOS-N-IGBT	L, 600V, 50A, 200W, 400/500ns	77id	Tos	GN 6050E	
GT 50 J102	MOS-N-IGBT	L, 600V, 50A, 200W, 400/500ns	77id	Tos	GN 6050E	
GT 50 J301	MOS-N-IGBT	=GT 50J101: int. FW Diode(E→C,<200ns)	77id	Tos	MGY 30N60D	
GT 50 J322	MOS-N-IGBT	L, 600V, 50A, 130W, 300/400ns int. Freilauf-/Free-wheel Diode(E→C)	77id	Tos	GT 50J301, MGY 30N60D	
GT 50 M101	MOS-N-IGBT	L, 900V, 50A, 180W, 350/800ns	77id	Tos	GN 9060E	data
GT 60 J101	MOS-N-IGBT	L, 600V, 60A, 200W, 400/500ns	77id	Tos	GN 6075E	data
GT 60 J321	MOS-N-IGBT	L, 600V, 60A, 350/800ns int. Freilauf-/Free-wheel Diode(E→C)	77id	Tos	MGY 40N60D	
GT 60 J322	MOS-N-IGBT	L, 600V, 60A, 200W, 350/1500ns int. Freilauf-/Free-wheel Diode(E→C)	77id	Tos	MGY 40N60D	
GT 60 M102	MOS-N-IGBT	L, 900V, 60A, 200W, 350/800ns	77id	Tos	GN 9060E	
GT 60 M103	MOS-N-IGBT	L, 900V, 60A, 200W, 350/800ns	77id	Tos	GN 9060E	
GT 60 M104	MOS-N-IGBT	L, 900V, 60A, 200W, 350/500ns	77id	Tos	GN 9060E	data
GT 60 M301	MOS-N-IGBT	=GT 50J101: int. FW Diode(E→C,<2,5µs)	77id	Tos	-	
GT 60 M303	MOS-N-IGBT	L, 900V, 60A, 170W, 460/600ns int. Freilauf-/Free-wheel Diode(E→C)	77id	Tos	GT 60M103	
GT 60 N321	MOS-N-IGBT	L, 1000V, 60A, 170W, 330/700ns int. Freilauf-/Free-wheel Diode(E→C)	77id	Tos	-	data
GT 80 J101	MOS-N-IGBT	L, 600V, 80A, 200W, 500/700ns	77id	Tos	-	data
GT 322	Ge-P	HF, >80MHz	5k	Gdc,GUS	AF 139, AF 200, AF 239(S)	data
GT 328(A,B)	Ge-P	VHF, >400MHz	5g	Gdc,GUS	AF 139, AF 239(S)	data
GT 346(A,B)	Ge-P	VHF/UHF, 600...700MHz	5g	Gdc,GUS	AF 139, AF 239(S)	data
GU	Z-Di	=SM 6T 200A (Typ-Code/Stempel/marking)	71(6x4mm)	Tho	→SM 6T...	data
GU	Si-Di	=U 1GU44 (Typ-Code/Stempel/marking)	71(5x2,5)	Tos	→U 1GU44	data
GU	Z-Di	=Z2 SMB-27 (Typ-Code/Stempel/marking)	71(5x3,5)	Fag	→Z2 SMB-27	data
GU 1	Si-N	=BCX 19 (Typ-Code/Stempel/marking)	35	Rhm	→BCX 19	data
GU-1(A,B,Z)	Si-Di	GI/S, 200...800V, 0,8A, <1µs GU-1: 400V, A=600V, B=800V, Z=200V	31a	Sak	BA157...159, BY204/..., BY208/..., BY268,++	data
GU-1 E	Si-Di	GI/S, 1300V, 0,3A, <2µs	31a	Sak	BY 228, BY 231/1400, DM 513, GP 10V...Y	data
GU-1 F	Si-Di	=GU-1E: 1500V	31a	Sak	BY 228, BY 231/1500, DM 516, GP 10Y	data

Type	Device	Short Description	Fig.	Manu	Comparision Types	More at
GU-3(A,B,Z)	Si-Di	GI, S, 200...800V, 1A, <500ns	31a	Sak	BYT 52A...M, BYV 12...16, RG 1A...M, ++	data
		GU-3: 400V, A=600V, B=800V, Z=200V				
GU-3 SY SZ	Si-Di	GI, S, 100...200V, 3A, <500ns	31a	Sak	BYW 95A...C, BYW 96D...E, RG 3A...M, ++	data
GUA	Z-Di	=SMCJ 60A (Typ-Code/Stempel/marking)	71(8x5mm)	Tho	→SMCJ ...	data
GUB	Z-Di	=SMCJ 70A (Typ-Code/Stempel/marking)	71(8x5mm)	Tho	→SMCJ ...	data
GUE	Z-Di	=SMCJ 85A (Typ-Code/Stempel/marking)	71(8x5mm)	Tho	→SMCJ ...	data
GUG	Z-Di	=SMCJ 100A (Typ-Code/Stempel/marking)	71(8x5mm)	Tho	→SMCJ ...	data
GUI	Z-Di	=SMCJ 130A (Typ-Code/Stempel/marking)	71(8x5mm)	Tho	→SMCJ ...	data
GUL	Z-Di	=SMCJ 154A (Typ-Code/Stempel/marking)	71(8x5mm)	Tho	→SMCJ ...	data
GUM	Z-Di	=SMCJ 170A (Typ-Code/Stempel/marking)	71(8x5mm)	Tho	→SMCJ ...	data
GUN	Z-Di	=SMCJ 188A (Typ-Code/Stempel/marking)	71(8x5mm)	Tho	→SMCJ ...	data
GUR 440	Si-Di	S, 400V, 4A, <45ns	31a	Gsi	BYW 28-500, EGP 50G, MUR 440	data
GUR 460	Si-Di	S, 600V, 4A, <45ns	31a	Gsi	BYW 28-600, MUR 460	data
GV	PIN-Di	=1SV249 (Typ-Code/Stempel/marking)	35 (2mm)	Say	→1SV249	data
GV	PIN-Di	=1SV251 (Typ-Code/Stempel/marking)	35	Say	→1SV251	data
GV	Z-Di	=SM 6T 220 (Typ-Code/Stempel/marking)	71(6x4mm)	Tho	→SM 6T...	data
GV	Z-Di	=Z2 SMB-30 (Typ-Code/Stempel/marking)	71(5x3,5)	Fag	→Z2 SMB-30	data
GW	Z-Di	=SM 6T 220A (Typ-Code/Stempel/marking)	71(6x4mm)	Tho	→SM 6T...	data
GW	Si-Di	=U 1GFWJ44 (Typ-Code/Stempel/marking)	71(5x2,5)	Tos	→U 1GWJ44	data
GW	Z-Di	=Z2 SMB-33 (Typ-Code/Stempel/marking)	71(5x3,5)	Fag	→Z2 SMB-33	data
GX	Si-N	=2SD2403-GX (Typ-Code/Stempel/marking)	39	Nec	→2SD2403	data
GX	Z-Di	=Z2 SMB-36 (Typ-Code/Stempel/marking)	71(5x3,5)	Fag	→Z2 SMB-36	data
GXB 5005	CMOS-IC	SMD Type Gps Receiver Module	CSP	Son	-	pdf pinout
GXB 5205	CMOS-IC	Antenna Embedded Gps Receiver Module		Son		pdf pinout
GXB 5210	-IC	Antenna Embedded Gps Receiver Module		Son		pdf pinout
GXB 10147 A	ECL-RAM-IC	128x1 Bit, <12ns, Ucc=-5,2V	16-DIC	Sie		
GXB 10148	ECL-PROM-IC	256x4 Bit, <20ns, Ucc=-5,2V	16-DIC	Sie		
GXB 100473	ECL-RAM-IC	64x4 Bit, <8ns, Ucc=-4,5V	24-FLP	Sie		
GXB 100474	ECL-RAM-IC	1024x4 Bit, <25ns, Ucc=-4,5V	24-FLP	Sie		
GXs	Si-Di	=BGX 400 (Typ-Code/Stempel/marking)	35	Sie	→BGX 400	
GY	Si-P	=2SA2883-Y (Typ-Code/Stempel/marking)	39	Tos	→2SA2883	data
GY	Si-N	=2SC2996-Y (Typ-Code/Stempel/marking)	35	Tos	→2SC2996	data
GY	Si-N	=2SC3689 (Typ-Code/Stempel/marking)	35	Say	→2SC3689	data
GY	Si-N	=2SC4413 (Typ-Code/Stempel/marking)	35(2mm)	Say	→2SC4413	data
GY	Si-N	=2SD2403-GY (Typ-Code/Stempel/marking)	39	Nec	→2SD2403	data
GY	Si-N	=KTC4375-Y (Typ-Code/Stempel/marking)	39	Kec	→KTC 4375	data
GY	Z-Di	=Z2 SMB-39 (Typ-Code/Stempel/marking)	71(5x3,5)	Fag	→Z2 SMB-39	data
GZ	Si-N	=2SD2403-GZ (Typ-Code/Stempel/marking)	39	Nec	→2SD2403	data
GZ	Si-N	=BFR 35AR (Typ-Code/Stempel/marking)	35	Sie	→BFR 35AR	data
GZ	Z-Di	=Z2 SMB-43 (Typ-Code/Stempel/marking)	71(5x3,5)	Fag	→Z2 SMB-43	data
GZA 2.0...51(X,Y,Z)	Z-Di	2,0...51V, 0,5W, X,Y,Z≈-6,±2,5,+6%	31a	Say	BZW 22/.., BZX 55/.., BZX 85/.., ZPD...,+	data
GZB 2.0...36(B,C)	Z-Di	2...36V, 1W, B≈6%, C≈0...+12%	31a	Say	BZV 85/.., BZW 22/.., BZX 61/.., ZPY...,+	data
GZB 1500 S1	LIN-IC	US-FB-Sender/transmitter	24-DIP	Phi	-	
GZB 1504	LIN-IC	Us-fb-sender/transmitter		Phi	-	
GZE 2.0...39	Z-Di	2,0...39V, 0,4W	31a	Say	BZW 22/.., BZX 55/.., BZX 85/.., ZPD...,+	data
GZS 2.0...39(R...Z)	Z-Di	2,0...39V, 0,4W, R,X,Y,Z≈-8...+3%	31a	Say	BZW 22/.., BZX 55/.., BZX 85/.., ZPD...,+	data

H

Type	Device	Short Description	Fig.	Manu	Comparision Types	More at
H	Si-P	=2SA1035 (Typ-Code/Stempel/marking)	35	Mat	→2SA1035	data
H	Si-P	=2SA1531A (Typ-Code/Stempel/marking)	35(2mm)	Mat	→2SA1531A	data
H	Si-P	=2SB956 (Typ-Code/Stempel/marking)	39	Mat	→2SB956	data
H	Si-N	=2SC4446 (Typ-Code/Stempel/marking)	35(2mm)	Say	→2SC4446	data
H	GaAs-N-FET	=2SK1100 (Typ-Code/Stempel/marking)	51	Mat	→2SK1100	data
H	GaAs-N-FET	=2SK1229 (Typ-Code/Stempel/marking)	51	Hit	→2SK1229	data
H	N-FET	=2SK2076 (Typ-Code/Stempel/marking)	35	Say	→2SK2076	data
H	C-Di	=BB 141 (Typ-Code/Stempel/marking)	71(1,3mm)	Phi	→BB 141	data
H	C-Di	=BB 659C (Typ-Code/Stempel/marking)	71(1,3mm)	Sie	→BB 659C	data
H	C-Di	=BBY 51-03W (Typ-Code/Stempel/marking)	71(1,3mm)	Sie	→BBY 51-03W	data
H	C-Di	=HVU 354 (Typ-Code/Stempel/marking)	71(1,7mm)	Hit	→HVU 354	data
H	Si-Di	=MA 2S376 (Typ-Code/Stempel/marking)	71(1,3mm)	Mat	→MA 2S376	data
H	Z-Di	=MAZF 091 (Typ-Code/Stempel/marking)	71(1,7mm)	Mat	→MAZF 091	data
H	Z-Di	=MAZN 091 (Typ-Code/Stempel/marking)	71(1,3mm)	Mat	→MAZN 091	data
H	Z-Di	=MAZS 036 (Typ-Code/Stempel/marking)	71(1,3mm)	Mat	→MAZS 036	data
H	Si-Di	=SB 02-03C (Typ-Code/Stempel/marking)	35	Say	→SB 02-03C	data
H	Si-Di	=SB 02-03Q (Typ-Code/Stempel/marking)	35(2mm)	Say	→SB 02-03Q	data
H-HP803CB	LED	cy ø6mm I:40lm Vf:3.8V If:700mA		Roi		data pdf pinout
H-HP803NB	LED	bl ø6mm I:16lm Vf:4V If:700mA		Roi		data pdf pinout
H-HP803NO	LED	am ø6mm I:36lm Vf:3V If:700mA		Roi		data pdf pinout
H-HP803NR	LED	rd ø6mm I:40lm Vf:3V If:700mA		Roi		data pdf pinout
H-HP803NW	LED	wht ø6mm I:55lm Vf:4V If:700mA		Roi		data pdf pinout
H-HP803PG	LED	gr ø6mm I:55lm Vf:3.8V If:700mA		Roi		data pdf pinout
H-HP803WW	LED	wht ø6mm I:45lm Vf:4V If:700mA		Roi		data pdf pinout
H 0	Si-N	=2SC4252 (Typ-Code/Stempel/marking)	35(2mm)	Tos	→2SC4252	data
H 0	Si-N	=2SC4255 (Typ-Code/Stempel/marking)	35	Tos	→2SC4255	data
H 1	Si-P	=BCW 69 (Typ-Code/Stempel/marking)	35	Fer, Phi	→BCW 69	data
H 1	Si-N	=BFS 46 (Typ-Code/Stempel/marking)	35	Fer	→BFS 42	data
H 1	Si-P	=D71Y1.5T1 (Typ-Code/Stempel/marking)	39	Gen	→D71Y1.5T1	data
H 1	Si-Di	=HSU 119 (Typ-Code/Stempel/marking)	71(1,7mm)	Hit	→HSU 119	data
H 1	PIN-Di	=HVM 187WK (Typ-Code/Stempel/marking)	35	Hit	→HVM 187WK	data
H 1	Si-N+R	=IMH 1A (Typ-Code/Stempel/marking)	46	Rhm	→IMH 1A	data
H 1	Z-Di	=MMSZ 5241B (Typ-Code/marking)	71(2,7mm)	Ons	→MMSZ 5241B	

Type	Device	Short Description	Fig.	Manu	Comparision Types	More at
1	Si-N	=MSD 42W (Typ-Code/Stempel/marking)	35(2mm)	Ons	→MSD 42W	
1	N-FET	=SST 4416 (Typ-Code/Stempel/marking)	35	Six	→SST 4416	data
1	Si-N+R	=UMH 1N (Typ-Code/Stempel/marking)	46(2mm)	Rhm	→UMH 1N...14N	data
1 O	Si-P	=KSA 2755-O (Typ-Code/Stempel/marking)	35	Sam	→KSA 2755	data
1 O	Si-P	=KSA 2859-O (Typ-Code/Stempel/marking)	35	Sam	→KSA 2859	data
1 p	Si-P	=BCW 69 (Typ-Code/Stempel/marking)	35	Phi	→BCW 69	data
1 R	Si-P	=KSA 2755-R (Typ-Code/Stempel/marking)	35	Sam	→KSA 2755	data
1 t	Si-P	=BCW 69 (Typ-Code/Stempel/marking)	35	Phi	→BCW 69	data
1 Y	Si-P	=KSA 2755-Y (Typ-Code/Stempel/marking)	35	Sam	→KSA 2755	data
1 Y	Si-P	=KSA 2859-Y (Typ-Code/Stempel/marking)	35	Sam	→KSA 2859	data
2	Si-P	=BCW 70 (Typ-Code/Stempel/marking)	35	Fer,Mot,Phi	→BCW 70	data
2	Si-N	=BFS 46A (Typ-Code/Stempel/marking)	35	Fer	→BFS 42A	data
2	Si-Di	=F 1H2 (Typ-Code/Stempel/marking)	71(1,7mm)	Org	→F 1H2	data
2	PIN-Di	=HVM 13 (Typ-Code/Stempel/marking)	35	Hit	→HVM 13	data
2	Si-N+R	=IMH 2A (Typ-Code/Stempel/marking)	46	Rhm	→IMH 2A	data
2	Z-Di	=MMSZ 5242B (Typ-Code/marking)	71(2,7mm)	Ons	→MMSZ 5242B	
2	Si-Di	=SM-1XH02 (Typ-Code/Stempel/marking)	31	Org	→SM-1XH02	data
2	Z-Di	=UDZ 7.5B (Typ-Code/Stempel/marking)	71(1,7mm)	Rhm	→UDZ 2.0B...36B	data
2	Z-Di	=UDZS 7.5B (Typ-Code/Stempel/marking)	71(1,7mm)	Rhm	→UDZS 5.1B...10B	data
2	Si-N+R	=UMH 2N (Typ-Code/Stempel/marking)	46(2mm)	Rhm	→UMH 1N...14N	data
2 O	Si-N	=KSC 2756-Y (Typ-Code/Stempel/marking)	35	Sam	→KSC 2756	data
2 p	Si-P	=BCW 70 (Typ-Code/Stempel/marking)	35	Fer,Mot,Phi	→BCW 70	data
2 R	Si-N	=KSC 2756-R (Typ-Code/Stempel/marking)	35	Sam	→KSC 2756	data
2 t	Si-P	=BCW 70 (Typ-Code/Stempel/marking)	35	Phi	→BCW 70	data
2 Y	Si-N	=KSC 2756-Y (Typ-Code/Stempel/marking)	35	Sam	→KSC 2756	data
3	Si-P	=2SA956-H3 (Typ-Code/Stempel/marking)	35	Nec	→2SA956	data
3	Si-P	=BCW 89 (Typ-Code/Stempel/marking)	35	Fer,Phi,Tho	→BCW 89	data
3	Si-Di	=D 1FH3 (Typ-Code/Stempel/marking)	71(5x2,5)	Shi	→D 1FH3	data
3	PIN-Di	=HVM 187S (Typ-Code/Stempel/marking)	35	Hit	→HVM 187S	data
3	Si-N+R	=IMH 3A (Typ-Code/Stempel/marking)	46	Rhm	→IMH 3A	data
3	C-Di	=KDV 153 (Typ-Code/Stempel/marking)	35	Kec	→KDV 153	data
3	Z-Di	=MMSZ 5243B (Typ-Code/marking)	71(2,7mm)	Ons	→MMSZ 5243B	
3	Si-N+R	=UMH 3N (Typ-Code/Stempel/marking)	46(2mm)	Rhm	→UMH 1N...14N	data
3 O	Si-N	=KSC 2757-O (Typ-Code/Stempel/marking)	35	Sam	→KSC 2757	data
3 p	Si-P	=BCW 89 (Typ-Code/Stempel/marking)	35	Phi	→BCW 89	data
3 R	Si-N	=KSC 2757-R (Typ-Code/Stempel/marking)	35	Sam	→KSC 2757	data
3 t	Si-P	=BCW 89 (Typ-Code/Stempel/marking)	35	Phi	→BCW 89	data
3 Y	Si-N	=KSC 2757-Y (Typ-Code/Stempel/marking)	35	Sam	→KSC 2757	data
4	Si-N+R	=IMH 4A (Typ-Code/Stempel/marking)	46	Rhm	→IMH 4A	data
4	Z-Di	=MMSZ 5244B (Typ-Code/marking)	71(2,7mm)	Ons	→MMSZ 5244B	
4	N-FET	=SST 5484 (Typ-Code/Stempel/marking)	35	Six	→SST 5484	data
4	Si-N+R	=UMH 4N (Typ-Code/Stempel/marking)	46(2mm)	Rhm	→UMH 1N...14N	data
4	Si-P	=2SA956-H4 (Typ-Code/Stempel/marking)	35	Nec	→2SA956	data
4	Si-N	=2SC3134-4 (Typ-Code/Stempel/marking)	35	Say	→2SC3134	data
4	Si-P	=BCW 69R (Typ-Code/Stempel/marking)	35	Fer,Val	→BCW 69R	data
4	Si-Di	=F 1H4 (Typ-Code/Stempel/marking)	71(1,7mm)	Org	→F 1H4	data
4	Si-Di	=F 1H4 (Typ-Code/Stempel/marking)	71(5x2,5)	Org	→F 1H4	data
4	PIN-Di	=HVM 121WK (Typ-Code/Stempel/marking)	35	Hit	→HVM 121WK	data
4	Si-Di	=SM-1XH04 (Typ-Code/Stempel/marking)	31	Org	→SM-1XH04	data
H4T/4416	OP	Vs:±16V Vu:94dB Vo:±14V f:0.5Mc		Acc		data pinout
4 Z	Si-N	=KSC 2758 (Typ-Code/Stempel/marking)	35	Sam	→KSC 2758	data
5	Si-N+R	=IMH 5A (Typ-Code/Stempel/marking)	46	Rhm	→IMH 5A	data
5	Si-Di	=MBD 770DW (Typ-Code/Stempel/marking)	46(2mm)	Ons	→MBD 770DW	data
5	Z-Di	=MMSZ 5245B (Typ-Code/marking)	71(2,7mm)	Ons	→MMSZ 5245B	
5	N-FET	=SST 5485 (Typ-Code/Stempel/marking)	35	Six	→SST 5485	data
5	Si-N+R	=UMH 5N (Typ-Code/Stempel/marking)	46(2mm)	Rhm	→UMH 1N...14N	data
5	Si-P	=2SA956-H5 (Typ-Code/Stempel/marking)	35	Nec	→2SA956	data
5	Si-N	=2SC3134-5 (Typ-Code/Stempel/marking)	35	Say	→2SC3134	data
5	Si-P	=BCW 70R (Typ-Code/Stempel/marking)	35	Fer,Val	→BCW 70R	data
5	PIN-Di	=HVM 14 (Typ-Code/Stempel/marking)	35	Hit	→HVM 14	data
H 5 M4C500L-...	RAM-IC	Video, Frame Memory, 91820 x 6Bit -5: 50ns, -6: 60ns, -100: 100ns	28-SQP	Mit		
H 5N 0301SM	MOS-N-FET-e*	SMD, V-MOS, LogL, 30V, 0,05A, <13Ω	35a(1,6mm)	Hit	2SK1824, 2SK3107	
H 5 O	Si-N	=KSC 2223-O (Typ-Code/Stempel/marking)	35	Sam	→KSC 2223	data
H 5 R	Si-N	=KSC 2223-R (Typ-Code/Stempel/marking)	35	Sam	→KSC 2223	data
H 5 Y	Si-N	=KSC 2223-Y (Typ-Code/Stempel/marking)	35	Sam	→KSC 2223	data
H 6	Si-P	=BCW 89R (Typ-Code/Stempel/marking)	35	Fer	→BCW 89R	data
H 6	Si-Di	=F 1H6 (Typ-Code/Stempel/marking)	71(5x2,5)	Org	→F 1H4	data
H 6	PIN-Di	=HVB 14S (Typ-Code/Stempel/marking)	35(2mm)	Hit	→HVB 14S	data
H 6	PIN-Di	=HVM 14S (Typ-Code/Stempel/marking)	35	Hit	→HVM 14S	data
H 6	Si-N+R	=IMH 6A (Typ-Code/Stempel/marking)	46	Rhm	→IMH 6A	data
H 6	Si-Di	=SM-1XH06 (Typ-Code/Stempel/marking)	31	Org	→SM-1XH06	data
H 6	N-FET	=SST 5486 (Typ-Code/Stempel/marking)	35	Six	→SST 5486	data
H 6	Si-N+R	=UMH 6N (Typ-Code/Stempel/marking)	46(2mm)	Rhm	→UMH 1N...14N	data
H 6	Si-P	=2SA956-H6 (Typ-Code/Stempel/marking)	35	Nec	→2SA956	data
H 6	Si-N	=2SC3134-6 (Typ-Code/Stempel/marking)	35	Say	→2SC3134	data
H6O	Si-N	=KSC 2759-O (Typ-Code/Stempel/marking)	35	Sam	→KSC 2759	data
H6R	Si-N	=KSC 2759-R (Typ-Code/Stempel/marking)	35	Sam	→KSC 2759	data
H6Y	Si-N	=KSC 2759-Y (Typ-Code/Stempel/marking)	35	Sam	→KSC 2759	data
H 7	Si-N+R	=IMH 7A (Typ-Code/Stempel/marking)	46	Rhm	→IMH 7A	data
H 7	Si-N+R	=UMH 7N (Typ-Code/Stempel/marking)	46(2mm)	Rhm	→UMH 1N...14N	data
H 7	Si-N	=2SC3134-7 (Typ-Code/Stempel/marking)	35	Say	→2SC3134	data
H 7	Si-P	=BCF 70 (Typ-Code/Stempel/marking)	35	Phi	→BCF 70	data
H 7	PIN-Di	=HVM 14SR (Typ-Code/Stempel/marking)	35	Hit	→HVM 14SR	data
H 8	Si-N+R	=IMH 8A (Typ-Code/Stempel/marking)	46	Rhm	→IMH 8A	data

Type	Device	Short Description	Fig.	Manu	Comparision Types	More at
H 8	Si-Di	=SM-1XH08 (Typ-Code/Stempel/marking)	31	Org	→SM-1XH08	data
H 8	Si-N+R	=UMH 8N (Typ-Code/Stempel/marking)	46(2mm)	Rhm	→UMH 1N...14N	data
H 08A 05P...60P	Si·Di	S L, 50...600V, 8A, <50...100ns 05=50, 10=100, 15=150V, <50ns 20=200, 30=300, 40=400, 50=500V, <75ns 60=600V, <100ns	17t§	Mop	FE 8A...J, MUR 805...860	data
H 08A 05R...60R	Si-Di	=H 08A...P:	17t1&			data
H 8 D1010	Hybrid-IC	Vorverstärker/ pre-amplifier			-	
H 8 D1011	Hybrid-IC	Vorverstärker/ pre-amplifier			-	
H 8 D1029	Hybrid-IC	Vorverstärker/ pre-amplifier			-	
H 8 D1044	Hybrid-IC	Vorverstärker/ pre-amplifier			-	
H 8 D1063	Hybrid-IC	Vorverstärker/ pre-amplifier			-	
H 8 D1153	Hybrid-IC	Vorverstärker/ pre-amplifier			-	
H 8 Z	Si-N	=KSC 2734: (Typ-Code/Stempel/marking)	35	Sam	→2SC2734	data
H 9	Si-N+R	=IMH 9A (Typ-Code/Stempel/marking)	46	Rhm	→IMH 9A	data
H 9	Si-N+R	=UMH 9N (Typ-Code/Stempel/marking)	46(2mm)	Rhm	→UMH 1N...14N	data
H 9	Si-Di	=1SS344 (Typ-Code/Stempel/marking)	35	Tos	→1SS344	data
H9F/5639	OP	Vs:±18V Vu:100dB Vo:±10V f:3Mc		Acc		data pinout
H9F/5645	FET	Vs:±18V Vu:100dB Vi0:±10mV		Acc		data pinout
H 9 Z	Si-N	=KSC 3120: (Typ-Code/Stempel/marking)	35	Sam	→2SC3120	data
H 10	Si-N+R	=IMH 10A (Typ-Code/Stempel/marking)	46	Rhm	→IMH 10A	data
H 10	Si-N+R	=UMH 10N (Typ-Code/Stempel/marking)	46(2mm)	Rhm	→UMH 1N...14N	data
H 11	Si-N+R	=IMH 11A (Typ-Code/Stempel/marking)	46	* Rhm	→IMH 11A	data
H 11	Si-N+R	=UMH 11N (Typ-Code/Stempel/marking)	46(2mm)	Rhm	→UMH 1N...14N	data
H 11	MOS-P-FET-e	=2SJ166 (Typ-Code/Stempel/marking)	35	Nec	→2SJ166	data
H11A	OC	LED/NPN Viso:7500Vrms CTR:90%		Mot	IL1,CQY80N,OPI2253,FCD825	data pinout
H11A1-HX	LED	bl+gr+rd+wht+yl ca. ø4mm l:10lm		Roi		data pdf pinout
H11A10	OC	LED/NPN Viso:2500V CTR:30% Vbs:1.15V		Har,Rca		data pdf pinout
H11A5100	OC	LED/NPN Viso:4000V CTR:>100%		Har,Rca,Thr	SFH601,CNY17,4N35	data pinout
H11A520	OC	LED/NPN Viso:4000V CTR:>20% Ibs:10mA		Har,Rca,Thr	SFH601,IL1,OPI2252,FCD810	data pinout
H11A550	OC	LED/NPN Viso:4000V CTR:>50% Ibs:20mA		Har,Rca,Thr	SFH601,IL1,OPI2253	data pinout
H11AA	OC	bid. LED/NPN Viso:7500Vrms CTR:150%		Qtc	TLP530,IL250,CNY71,OPI2500	data pinout
H11AA1	OC	bid. LED/NPN Viso:5300Vrms CTR:>20%		Vis	TLP530,IL250,CNY71,OPI2500	data pdf pinout
H11AG1	OC	LED/NPN Viso:5300Vrms CTR:>300%		Qtc		data pinout
H11AG2	OC	LED/NPN Viso:5300Vrms CTR:>200%		Qtc		data pinout
H11AG3	OC	LED/NPN Viso:5300Vrms CTR:>100%		Qtc		data pinout
H11AV	OC	LED/NPN Viso:5300Vrms CTR:70%		Qtc		data pinout
H11B	OC	LED/Dar Viso:7500Vrms CTR:>100%		Mot	IL30,4N32,OPI3150,PS2022	data pinout
H11B1	OC	LED/Dar Viso:4400V CTR:>100%		Fch,Gin,Har	IL30,OPI3150	data pdf pinout
H11B255	OC	LED/Dar Viso:2820V CTR:>100%		Har,Phi,Rca	IL30,OPI3152	data pdf pinout
H 11	Si-P	=BCX 71G (Typ-Code/Stempel/marking)	35	Val	→BCX 71G	data
H11BX522	OC	LED/Dar		Rca	OPI3250	data pinout
H11C	OC	LED/SCR Viso:2500V Vbs:1.2V Ibs:10mA		Mot	IL400,PS3001(1)	data pinout
H11D	OC	LED/NPN Viso:7500Vrms CTR:>10%		Mot	OPI6000	data pinout
H11F1	OC	LED/FET Viso:5300Vrms Vbs:1.1V		Qtc		data pinout
H11F2	OC	LED/FET Viso:5300Vrms Vbs:1.1V		Qtc		data pinout
H11F3	OC	LED/FET Viso:5300Vrms Vbs:1.1V		Qtc		data pinout
H11G	OC	LED/Dar Viso:2500V CTR:>200% Ibe:2mA		Mot	OPI6000	data pinout
H11G45	OC	LED/Dar Viso:3750Vrms CTR:>500%		Har,Rca		data pinout
H11J	OC	LED/Triac Viso:2500Vrms Vbs:1.2V		Har,Iso,Rca	K3010P,OPI3010	data pdf pinout
H11K	OC	LED/2 Viso:2250Vrms CTR:>500%		Har,Rca		data pdf pinout
H11L	OC	LED/PD+IC Viso:%7500V Vbs:1.2V		Fcs,Har,Mot		data pinout
H11M	OC	LED/SCR Viso:3750Vrms Vbs:1.3V		Har,Rca,Thr		data pdf pinout
H11N1	OC	LED/PD+IC Viso:5300Vrms Vbs:1.6V		Qtc		data pinout
H11N2	OC	LED/PD+IC Viso:5300Vrms Vbs:1.6V		Qtc		data pinout
H11N3	OC	LED/PD+IC Viso:5300Vrms Vbs:1.6V		Qtc		data pinout
H11V	OC	LED/PD+IC Viso:3750Vrms Vbs:1.5V		Har,Rca		data pinout
H11V2	OC	LED/PD+IC Viso:3750Vrms Vbs:1.5V		Har,Rca		data pinout
H11V3	OC	LED/PD+IC Viso:3750Vrms Vbs:1.5V		Har,Rca		data pinout
H 12	MOS-P-FET-e	=2SJ185 (Typ-Code/Stempel/marking)	35	Nec	→2SJ185	data
H 12	Si-P	=BCX 71H (Typ-Code/Stempel/marking)	35	Val	→BCX 71H	data
H 12	Si-Di	=SM-1XH12 (Typ-Code/Stempel/marking)	31	Org	→SM-1XH12	data
H 13	MOS-P-FET-e	=2SJ202 (Typ-Code/Stempel/marking)	35(2mm)	Nec	→2SJ202	data
H 13	Si-P	=BCX 71J (Typ-Code/Stempel/marking)	35	Val	→BCX 71J	data
H 14	Si-N+R	=IMH 14A (Typ-Code/Stempel/marking)	46	Rhm	→IMH 14A	data
H 14	Si-N+R	=UMH 14N (Typ-Code/Stempel/marking)	46(2mm)	Rhm	→UMH 1N...14N	data
H 14	MOS-P-FET-e	=2SJ203 (Typ-Code/Stempel/marking)	35	Nec	→2SJ203	data
H 14	N-FET	=2SK2076-14 (Typ-Code/Stempel/marking)	35	Say	→2SK2076	data
H 14	Si-P	=BCX 71K (Typ-Code/Stempel/marking)	35	Val	→BCX 71K	data
H 15	Si-N+R	=IMH 15A (Typ-Code/Stempel/marking)	46	Rhm	→IMH 15A	data
H 15	MOS-P-FET-e	=2SJ204 (Typ-Code/Stempel/marking)	35	Nec	→2SJ204	data
H 15	N-FET	=2SK2076-15 (Typ-Code/Stempel/marking)	35	Say	→2SK2076	data
H 16	MOS-P-FET-e	=2SJ210 (Typ-Code/Stempel/marking)	35	Nec	→2SJ210	data
H 16C 05A...60A	Si-Di	=H 16C...C:	17q&			data
H 16C 05C...60C	Si-Di	Dual, S-L, 50...600V, 16A, <50...100ns 05=50, 10=100, 15=150V, <50ns 20=200, 30=300, 40=400, 50=500V, <75ns 60=600V, <100ns	17p§	Mop	BYV 32-..., BYV 34-..., FE 16A...J	data
H 16C 05D...60D	Si-Di	=H 16C...C:	17n1			data
H 16	Si-Di	=SM-1XH16 (Typ-Code/Stempel/marking)	31	Org	→SM-1XH16	data
H 17	MOS-P-FET-e	=2SJ209 (Typ-Code/Stempel/marking)	35	Nec	→2SJ209	data
H 18	MOS-P-FET-e	=2SJ211 (Typ-Code/Stempel/marking)	35	Nec	→2SJ211	data
H 19	MOS-P-FET-e	=2SJ461 (Typ-Code/Stempel/marking)	35	Nec	→2SJ461	data
H 21	MOS-P-FET-e	=2SJ463A (Typ-Code/Stempel/marking)	35(2mm)	Nec	→2SJ463A	data

Type	Device	Short Description	Fig.	Manu	Comparision Types	More at
21AX	PI	LED/NPN CTR:20% Vbs:1.34V lbs:60mA		Har,Mot,Qtc		data pdf pinout
21BX	PI	LED/Dar CTR:250% Vbs:1.34V lbs:60mA		Har,Mot,Qtc	MSAXXXX	data pdf pinout
21L	PI	LED/PD+IC Vbs:<1.5V lbs:20mA		Fcs		data pdf pinout
21LX	PI	LED/PD+IC Vbs:<1.6V lbs:20mA Ibe:5mA		Har,Qtc,Rca		data pdf pinout
22AX	PI	LED/NPN CTR:20% Vbs:1.34V lbs:60mA		Har,Mot,Qtc		data pdf pinout
22BX	PI	LED/Dar CTR:250% Vbs:1.34V lbs:60mA		Har,Mot,Qtc	MSAXXXX	data pdf pinout
22L	PI	LED/PD+IC Vbs:<1.5V lbs:20mA		Fcs		data pdf pinout
22LX	PI	LED/PD+IC Vbs:<1.6V lbs:20mA Ibe:5mA		Har,Qtc,Rca		data pdf pinout
23A1	MP	LED/Dar CTR:>75% Vbs:<1.7V lbs:60mA		Har,Qtc,Rca		data pdf pinout
23L1	MP	LED/PD+IC Vbs:<1.5V lbs:20mA Ibe:5mA		Har,Qtc,Rca		data pinout
24A1	OC	LED/NPN Viso:5300Vrms CTR:>100%		Har,Qtc,Rca	SFH610,OPI7010	data pdf pinout
24A2	OC	LED/NPN Viso:5300Vrms CTR:>20%		Har,Qtc,Rca	SFH610,OPI7002	data pdf pinout
24B1	OC	LED/Dar Viso:5300Vrms CTR:>1000%		Har,Qtc,Rca	SFH610,OPI7010	data pdf pinout
24B2	OC	LED/Dar Viso:5300Vrms CTR:>400%		Har,Qtc,Rca	OPI7002	data pdf pinout
30D 05A...60A	Si-Di	=H 30D...C:	16q&			data
30D 05C...60C	Si-Di	Dual, S-L, 50...600V, 30A, <50...100ns 05=50, 10=100, 15=150V, <50ns 20=200, 30=300, 40=400, 50=500V, <75ns 60=600V, <100ns	16p§	Mop	BYV 72-..., BYV 74-..., BYW 99P-...	data
30D 05D...60D	Si-Di	=H 30D...C:	16n1		-	data
31	Si-P	=BCW 89R (Typ-Code/Stempel/marking)	35	Val	→BCW 89R	data
32	Si-Di	≈1N4148	31a		→1N4148	
50	HI-OHM	Vs:±15V Vu:86dB Vo:±10V Vi0:1mV	12A	Ilc		data pinout
60	HI-OHM	Vs:±15V Vu:92dB Vo:±10V f:10Mc	12A	Ilc		data pinout
71	Si-P	=BCF 70R (Typ-Code/Stempel/marking)	35	Val	→BCF 70R	data
74A1	OC	LED/NPN Viso:2500V lbs:12mA Ibe:1mA		Har,Rca		data pdf pinout
74C1	OC	LED/SCR Viso:2500Vrms Ibe:1mA		Har,Rca	IL400	data pdf pinout
74C2	OC	LED/SCR Viso:2500Vrms Ibe:1mA		Har,Rca	IL400	data pdf pinout
78 L05(A/B)A	LIN/Z-IC	5V, 0.1A, 3-Terminal positiv voltage Reg.	7	Hsm	-	pdf pinout
78 L05(A/B)M	LIN/Z-IC	5V, 0.1A, 3-Terminal positiv voltage Reg.	48	Hsm	-	pdf pinout
78 L06(A/B)A	LIN/Z-IC	6V, 0.1A, 3-Terminal positiv voltage Reg.	7	Hsm	-	pdf pinout
78 L06(A/B)M	LIN/Z-IC	6V, 0.1A, 3-Terminal positiv voltage Reg.	48	Hsm	-	pdf pinout
78 L08(A/B)A	LIN/Z-IC	8V, 0.1A, 3-Terminal positiv voltage Reg.	7	Hsm	-	pdf pinout
78 L08(A/B)M	LIN/Z-IC	8V, 0.1A, 3-Terminal positiv voltage Reg.	48	Hsm	-	pdf pinout
78 L09(A/B)A	LIN/Z-IC	9V, 0.1A, 3-Terminal positiv voltage Reg.	7	Hsm	-	pdf pinout
78 L09(A/B)M	LIN/Z-IC	9V, 0.1A, 3-Terminal positiv voltage Reg.	48	Hsm	-	pdf pinout
78 L12(A/B)A	LIN/Z-IC	12V, 0.1A, 3-Terminal positiv voltage Reg.	7	Hsm	-	pdf pinout
78 L12(A/B)M	LIN/Z-IC	12V, 0.1A, 3-Terminal positiv voltage Reg.	48	Hsm	-	pdf pinout
H 431 AA	IC	2.495 ±2%, Adj. shunt reg.	7	Hsm	-	pdf pinout
H 431 AM	IC	2.495 ±2%, Adj. shunt reg.	48	Hsm	-	pdf pinout
H 431 AN	IC	2.495 ±2%, Adj. shunt reg.	35	Hsm	-	pdf pinout
H 431 AS	IC	2.495 ±2%, Adj. shunt reg.	MDIP	Hsm	-	pdf pinout
H 431 BA	IC	2.495 ±1%, Adj. shunt reg.	7	Hsm	-	pdf pinout
H 431 BM	IC	2.495 ±1%, Adj. shunt reg.	48	Hsm	-	pdf pinout
H 431 BN	IC	2.495 ±1%, Adj. shunt reg.	35	Hsm	-	pdf pinout
H 431 BS	IC	2.495 ±1%, Adj. shunt reg.	MDIP	Hsm	-	pdf pinout
H 431 CA	IC	2.495 ±0.5%, Adj. shunt reg.	7	Hsm	-	pdf pinout
H 431 CM	IC	2.495 ±0.5%, Adj. shunt reg.	48	Hsm	-	pdf pinout
H 431 CN	IC	2.495 ±0.5%, Adj. shunt reg.	35	Hsm	-	pdf pinout
H 431 CS	IC	2.495 ±0.5%, Adj. shunt reg.	MDIP	Hsm	-	pdf pinout
H 432 AA	IC	1.24V ±2%, Adjustable Shunt Regulator	7	Hsm	-	pdf pinout
H 432 AM	IC	1.24V ±2%, Adjustable Shunt Regulator	48	Hsm	-	pdf pinout
H 432 AN	IC	1.24V ±2%, Adjustable Shunt Regulator	35	Hsm	-	pdf pinout
H 432 BA	IC	1.24V ±1%, Adjustable Shunt Regulator	7	Hsm	-	pdf pinout
H 432 BM	IC	1.24V ±1%, Adjustable Shunt Regulator	48	Hsm	-	pdf pinout
H 432 BN	IC	1.24V ±1%, Adjustable Shunt Regulator	35	Hsm	-	pdf pinout
H 432 CA	IC	1.24V ±0.5%, Adjustable Shunt Regulator	7	Hsm	-	pdf pinout
H 432 CM	IC	1.24V ±0.5%, Adjustable Shunt Regulator	48	Hsm	-	pdf pinout
H 432 CN	IC	1.24V ±0.5%, Adjustable Shunt Regulator	35	Hsm	-	pdf pinout
H 432 DA	IC	1.25V ±2%, Adjustable Shunt Regulator	7	Hsm	-	pdf pinout
H 432 DM	IC	1.25V ±2%, Adjustable Shunt Regulator	48	Hsm	-	pdf pinout
H 432 DN	IC	1.25V ±2%, Adjustable Shunt Regulator	35	Hsm	-	pdf pinout
H 432 EA	IC	1.25V ±1%, Adjustable Shunt Regulator	7	Hsm	-	pdf pinout
H 432 EM	IC	1.25V ±1%, Adjustable Shunt Regulator	48	Hsm	-	pdf pinout
H 432 EN	IC	1.25V ±1%, Adjustable Shunt Regulator	35	Hsm	-	pdf pinout
H 432 FA	IC	1.25V ±0.5%, Adjustable Shunt Regulator	7	Hsm	-	pdf pinout
H 432 FM	IC	1.25V ±0.5%%, Adjustable Shunt Regulator	48	Hsm	-	pdf pinout
H 432 FN	IC	1.25V ±0.5%, Adjustable Shunt Regulator	35	Hsm	-	pdf pinout
H 580	LIN-IC	→SAS 580	18-DIP	Sgs	SAS 580	
H 590	LIN-IC	→SAS 590	18-DIP	Sgs	SAS 590	
H 629	MOS-IC	Orgelgatter, gates f. el. organ, 1x12	14-DIP	Sgs	-	
H 770	LIN-IC	→SN 29770	16-DIP	Sgs	SN 29770	
H 771	LIN-IC	→SN 29771	16-DIP	Sgs	SN 29771	
H 772	LIN-IC	→SN 29772	16-DIP	Sgs	SN 29772	

Type	Device	Short Description	Fig.	Manu	Comparision Types	More at
H 773	LIN-IC	→SN 29773	16-DIP	Sgs	SN 29773	
H 1117 E-1.8	LIN/Z-IC	1.2A, 1.8V, Low Dropout Positive Voltage Reg.	17	Hsm	-	pdf pinout
H 1117 E-2.5	LIN/Z-IC	1.2A, 2.5V, Low Dropout Positive Voltage Reg.	17	Hsm	-	pdf pinout
H 1117 E-3.3	LIN/Z-IC	1.2A, 3.3V, Low Dropout Positive Voltage Reg.	17	Hsm	-	pdf pinout
H 1117 E-5	LIN/Z-IC	1.2A, 5V, Low Dropout Positive Voltage Reg.	17	Hsm	-	pdf pinout
H 1117 E-Adj	LIN/Z-IC	1.2A, 1-4V, Low Dropout Positive Voltage Reg.	17	Hsm		pdf pinout
H 1117 J-1.8	LIN/Z-IC	1.2A, 1.8V, Low Dropout Positive Voltage Reg.	30	Hsm	-	pdf pinout
H 1117 J-2.5	LIN/Z-IC	1.2A, 2.5V, Low Dropout Positive Voltage Reg.	30	Hsm	-	pdf pinout
H 1117 J-3.3	LIN/Z-IC	1.2A, 3.3V, Low Dropout Positive Voltage Reg.	30	Hsm	-	pdf pinout
H 1117 J-5	LIN/Z-IC	1.2A, 5V, Low Dropout Positive Voltage Reg.	30	Hsm	-	pdf pinout
H 1117 J-Adj	LIN/Z-IC	1.2A, 1-4V, Low Dropout Positive Voltage Reg.	30	Hsm		pdf pinout
H 1117 M/SJ-1.8	LIN/Z-IC	1.2A, 1.8V, Low Dropout Positive Voltage Reg.	48	Hsm		pdf pinout
H 1117 M/SJ-2.5	LIN/Z-IC	1.2A, 2.5V, Low Dropout Positive Voltage Reg.	48	Hsm		pdf pinout
H 1117 M/SJ-3.3	LIN/Z-IC	1.2A, 3.3V, Low Dropout Positive Voltage Reg.	48	Hsm		pdf pinout
H 1117 M/SJ-5	LIN/Z-IC	1.2A, 5V, Low Dropout Positive Voltage Reg.	48	Hsm		pdf pinout
H 1117 M/SJ-Adj	LIN/Z-IC	1.2A, 1-4V, Low Dropout Positive Voltage Reg.	48	Hsm		pdf pinout
H 2584	Si-P-Darl	Reg, 20V, 10A, 65W, B=2000..60000	17j§	Hsm	BDT 62..., BDW 94..., BDX 34..., 2SB1550	data
H 3842 P	IC	Current Mode PWM Cntrl.	DIC	Hsm	-	pdf pinout
H 3842 S	IC	Current Mode PWM Cntrl.	MDIP	Hsm	-	pdf pinout
H6010C	UNI	Vs:±15V Vu:86dB Vo:±10V Vi0:0.6mV	7Y	Ite		data pinout
H 6060	IC	Self Recovering Watchdog	DIC	Emm	-	pdf pinout
H 7805 E	LIN/Z-IC	5V, 1A, 3-terminal positiv voltage reg.	17	Hsm	-	pdf pinout
H 7805 J/I	LIN/Z-IC	5V, 1A, 3-terminal positiv voltage reg.	30	Hsm	-	pdf pinout
H 7805 NE	LIN/Z-IC	5V, 1A, 3-terminal positiv voltage reg.	14	Hsm	-	pdf pinout
H 7806 E	LIN/Z-IC	6V, 1A, 3-terminal positiv voltage reg.	17	Hsm	-	pdf pinout
H 7806 J/I	LIN/Z-IC	6V, 1A, 3-terminal positiv voltage reg.	30	Hsm	-	pdf pinout
H 7806 NE	LIN/Z-IC	6V, 1A, 3-terminal positiv voltage reg.	14	Hsm	-	pdf pinout
H 7808 E	LIN/Z-IC	8V, 1A, 3-terminal positiv voltage reg.	17	Hsm	-	pdf pinout
H 7808 J/I	LIN/Z-IC	8V, 1A, 3-terminal positiv voltage reg.	30	Hsm	-	pdf pinout
H 7808 NE	LIN/Z-IC	8V, 1A, 3-terminal positiv voltage reg.	14	Hsm	-	pdf pinout
H 7809 E	LIN/Z-IC	9V, 1A, 3-terminal positiv voltage reg.	17	Hsm	-	pdf pinout
H 7809 J/I	LIN/Z-IC	9V, 1A, 3-terminal positiv voltage reg.	30	Hsm	-	pdf pinout
H 7809 NE	LIN/Z-IC	9V, 1A, 3-terminal positiv voltage reg.	14	Hsm	-	pdf pinout
H 7812 E	LIN/Z-IC	12V, 1A, 3-terminal positiv voltage reg.	17	Hsm	-	pdf pinout
H 7812 J/I	LIN/Z-IC	12V, 1A, 3-terminal positiv voltage reg.	30	Hsm	-	pdf pinout
H 7812 NE	LIN/Z-IC	12V, 1A, 3-terminal positiv voltage reg.	14	Hsm	-	pdf pinout
H 31002 P	IC	Bipolar Tone Ringer IC	DIC	Hsm	-	pdf pinout
H 34063 AP	LIN-IC	DC-DC converter IC	DIC	Hsm	-	pdf pinout
H 34063 AS	LIN-IC	DC-DC converter IC	MDIP	Hsm	-	pdf pinout
HD 66724	-IC	Graphics LCD Controller/driver	CSP	Hit	-	pdf pinout
HD 66725	-IC	Graphics LCD Controller/driver	CSP	Hit	-	pdf pinout
HD 404304 F	DIG-IC	4-bit single-chip microcomputer	DIP	Hit	-	pdf pinout

HA

Type	Device	Short Description	Fig.	Manu	Comparision Types	More at
HA	Si-P	=µPA601T (Typ-Code/Stempel/marking)	46	Nec	→µPA601T	data
HA0-5020	WIDEBAND	Vs:±18V Vo:±12.7V Vi0:2mV f:12.7Mkc	d	Har		data pinout
HA-OP07C	OP	Vs:±22V Vu:400V/mV Vo:±12.6V f:0.6Mc	8AD	Har		data pinout
HA-OP07E	OP	Vs:±22V Vu:450V/mV Vo:±12.6V f:0.6Mc	8AD	Har		data pinout
HA-OP27E	LO-NOISE	Vs:±22V Vu:1500V/mV Vo:±13.6V f:8Mc	8AD	Har		data pinout
HA-OP27F	LO-NOISE	Vs:±22V Vu:1300V/mV Vo:±13.5V f:8Mc	8AD	Har		data pinout
HA	Si-P	=2SA1883 (Typ-Code/Stempel/marking)	35(1,6mm)	Say	→2SA1883	data
HA	Si-N	=2SC2804 (Typ-Code/Stempel/marking)	25	Tos	→2SC2804	data
HA	Si-N	=2SC3119 (Typ-Code/Stempel/marking)	35	Tos	→2SC3119	data
HA	Si-N	=2SD1464-HA (Typ-Code/Stempel/marking)	39	Hit	→2SD1464	data
HA	Si-P	=BSS 25 (Typ-Code/Stempel/marking)	35	Sie	→BSS 25	data
HA	GaAs-N-FET-d	=CFY 65-12 (Typ-Code/Stempel/marking)	51	Sie	→CFY 55	data
HA 0-xxxx	OP/KOP-IC	→HA-xxxx	Chip	Har		data
HA1-2400	4xPROG	Vs:45V Vu:150V/mV Vo:±12V Vi0:4mV	16D	Har		data pinout
HA1-2404	4xPROG	Vs:45V Vu:150V/mV Vo:±12V Vi0:4mV	16D	Har		data pinout
HA1-2405	4xPROG	Vs:45V Vu:150V/mV Vo:±12V Vi0:4mV	16D	Har		data pinout
HA1-2406	4xPROG	Vs:45V Vo:±12V Vi0:7mV f:30Mc	16D	Har		data pinout
HA1-2420	SAMP&HOLD	Vs:±20V Vo:>±10V Vi0:2mV f:2.5Mc	14D	Har		data pinout
HA1-2425	SAMP&HOLD	Vs:±20V Vo:>±10V Vi0:3mV f:2.5Mc	14D	Har		data pinout
HA1-2444	4xPROG	Vs:±17.5V Vu:±71dB Vo:±10V Vi0:±7mV	16D	Har		data pinout
HA1-2500	OP	Vs:40V Vu:>10V/mV Vo:±12V Vi0:<10mV	8D	Har		data pinout
HA1-2502	OP	Vs:40V Vu:>10V/mV Vo:±12V Vi0:<10mV	8D	Har		data pinout
HA1-2505	OP	Vs:40V Vu:>10V/mV Vo:±12V Vi0:<10mV	8D	Har		data pinout

Type	Device	Short Description	Fig.	Manu	Comparision Types	More at
HA1-2510	OP	Vs:40V Vu:>5V/mV Vo:±12V Vi0:<14mV	8D	Har		data pinout
HA1-2512	OP	Vs:40V Vu:>5V/mV Vo:±12V Vi0:<14mV	8D	Har		data pinout
HA1-2515	OP	Vs:40V Vu:>5V/mV Vo:±12V Vi0:<14mV	8D	Har		data pinout
HA1-2520	OP	Vs:40V Vu:>5V/mV Vo:±12V Vi0:<14mV	8D	Har		data pinout
HA1-2522	OP	Vs:40V Vu:>5V/mV Vo:±12V Vi0:<14mV	8D	Har		data pinout
HA1-2525	OP	Vs:40V Vu:>5V/mV Vo:±12V Vi0:<14mV	8D	Har		data pinout
HA1-2539	HI-SLEW	Vs:±17.5V Vo:>±10V Vi0:8mV f:600Mc	14D	Har		data pinout
HA1-2540	WIDEBAND	Vs:±17.5V Vo:>±10V Vi0:8mV f:400Mc	14D	Har		data pinout
HA1-2541	WIDEBAND	Vs:±17.5V Vo:±11V Vi0:0.8mV f:40Mc	14D	Har		data pinout
HA1-2542	WIDEBAND	Vs:±17.5V Vo:±11V Vi0:5mV f:70Mc	14D	Har		data pinout
HA1-2544	OP	Vs:33V Vo:±11V Vi0:<20mV f:40Mc	8D	Har		data pinout
HA1-2620	HI-SPEED	Vs:±22V Vu:150V/mV Vo:±12V Vi0:0.5mV	14D	Har,Isi		data pinout
HA1-2622	HI-SPEED	Vs:±22V Vu:150V/mV Vo:±12V Vi0:3mV	14D	Har,Isi		data pinout
HA1-2625	HI-SPEED	Vs:±22V Vu:150V/mV Vo:±12V Vi0:3mV	14D	Har,Isi		data pinout
HA1-2650	2xOP	Vs:±20V Vu:40V/mV Vo:±14V Vi0:1.5mV	14D	Har		data pinout
HA1-2655	2xOP	Vs:±20V Vu:40V/mV Vo:±14V Vi0:2mV	14D	Har		data pinout
HA1-2700	UNI	Vs:±15V Vu:112dB Vo:±12V Vi0:5mV	14D	Har		data pinout
HA1-2704	UNI	Vs:±15V Vu:112dB Vo:±12V Vi0:6mV	14D	Har		data pinout
HA1-2705	UNI	Vs:±15V Vu:106dB Vo:±12V Vi0:7mV	14D	Har		data pinout
HA1-2720	PROG	Vs:±22V Vu:120V/mV Vo:±12V Vi0:2mV	8D	Har		data pinout
HA1-2730	2xPROG	Vs:±22V Vu:100V/mV Vo:±13.5V Vi0:2mV	14D	Har		data pinout
HA1-2735	2xPROG	Vs:±22V Vu:100V/mV Vo:±13.5V Vi0:2mV	14D	Har		data pinout
HA1-2740	4xPROG	Vs:±22V Vu:100V/mV Vo:±14V Vi0:2mV	16D	Har		data pinout
HA1-2839	HI-SLEW	Vs:±17.5V Vo:>±10V Vi0:0.6mV f:600Mc	14D	Har		data pinout
HA1-2840	HI-SLEW	Vs:±17.5V Vo:>±10V Vi0:0.6mV f:600Mc	14D	Har	HA-2540/883,AD840	data pinout
HA1-2841	WIDEBAND	Vs:±17.5V Vo:±10V Vi0:±4mV f:50Mc	14D	Har		data pinout
HA1-2842	WIDEBAND	Vs:±17.5V Vo:±10V Vi0:±4mV f:80Mc	14D	Har		data pinout
HA1-2850	LO-POWER	Vs:±17.5V Vo:±11V Vi0:0.6mV f:470Mc	14D	Har	HA-2540/883,AD840	data pinout
HA1-4156	4xOP	Vs:±20V Vu:50V/mV Vo:±13.7V Vi0:1mV	14D	Har		data pinout
HA1-4600	4xOP	Vs:±20V Vu:250V/mV Vo:±13V Vi0:0.3mV	14D	Har		data pinout
HA1-4602	4xOP	Vs:±20V Vu:250V/mV Vo:±13V Vi0:3mV	14D	Har		data pinout
HA1-4605	4xOP	Vs:±20V Vu:250V/mV Vo:±13V Vi0:3mV	14D	Har		data pinout
HA1-4620	4xHI-SPEED	Vs:±20V Vu:250V/mV Vo:±13V Vi0:0.3mV	14D	Har		data pinout
HA1-4622	4xHI-SPEED	Vs:±20V Vu:250V/mV Vo:±13V Vi0:3mV	14D	Har		data pinout
HA1-4625	4xHI-SPEED	Vs:±20V Vu:250V/mV Vo:±13V Vi0:3mV	14D	Har		data pinout
HA1-4741	4xOP	Vs:±20V Vu:100V/mV Vo:±12.5V f:3.5Mc	14D	Har		data pinout
HA1-4900	4xCOMP	Vs:33V Vi0:2mV	16D	Har		data pinout
HA1-4902	4xCOMP	Vs:33V Vi0:2mV	16D	Har		data pinout
HA1-4905	4xCOMP	Vs:33V Vi0:4mV	16D	Har		data pinout
HA1-5004	WIDEBAND	Vs:±20V Vo:±11.5V Vi0:1mV f:50Mkc	14D	Har		data pinout
HA1-5064	4xFET	Vs:±20V Vu:25V/mV Vo:±12V Vi0:<6mV	14D	Har		data pinout
HA1-5084	4xFET	Vs:±40V Vu:88dB Vo:±10V Vi0:<5mV	14D	Har		data pinout
HA1-5104	4xLO-NOISE	Vs:±20V Vo:±12V Vi0:0.5mV f:60Mc	14D	Har		data pinout
HA1-5114	4xLO-NOISE	Vs:±20V Vo:±12V Vi0:0.5mV f:60Mc	14D	Har		data pinout
HA1-5134	4xHI-PREC	Vs:±20V Vo:13.5/0V Vi0:50μμV f:4Mc	14D	Har		data pinout
HA1-5144	4xμ-POWER	Vs:40V Vu:100dB Vo:>0...4V Vi0:0.5mV	14D	Har		data pinout
HA1-5192	FAST-SET	Vs:35V Vo:±8V Vi0:3mV f:150Mc	14D	Har		data pinout
HA1-5195	FAST-SET	Vs:35V Vo:±8V Vi0:3mV f:>150Mc	14D	Har		data pinout
HA1-5320	SAMP&HOLD	Vs:±20V Vi0:0.2mV f:2Mc f:600kc	14D	Har		data pinout
HA1-5330	SAMP&HOLD	Vs:±20V Vi0:0.2mV f:4.5Mc f:1400kc	14D	Har		data pinout
HA1-5340	SAMP&HOLD	Vs:±18V Vi0:<1.5mV f:10Mc f:900kc	14D	Har		data pinout
HA1-8023	3xLO-POWER	Vs:±22V Vu:80dB Vo:±5V Vi0:2mV	16D	Har		data pinout
HA 1-xxxx	OP/KOP-IC	→HA-xxxx	...-DIC			data
HA2-2050	HI-SPEED	Vs:±15V Vu:73dB Vo:±10V Vi0:30mV	8A	Har		data pinout
HA2-2055	HI-SPEED	Vs:±15V Vu:73dB Vo:±10V Vi0:65mV	8A	Har		data pinout
HA2-2060	HI-SPEED	Vs:±15V Vu:98dB Vo:±10V Vi0:30mV	8A	Har		data pinout
HA2-2065	HI-SPEED	Vs:±15V Vu:98dB Vo:±10V Vi0:65mV	8A	Har		data pinout
HA2-2500	HI-PREC	Vs:40V Vo:±12V Vi0:2mV f:12Mc	8A	Har		data pinout
HA2-2502	HI-PREC	Vs:40V Vo:±12V Vi0:4mV f:12Mc	8A	Har		data pinout
HA2-2505	HI-PREC	Vs:40V Vo:±12V Vi0:4mV f:12Mc	8A	Har		data pinout
HA2-2510	HI-SLEW	Vs:40V Vo:±12V Vi0:4mV f:12Mc f:1Mkc	8A	Har		data pinout
HA2-2512	HI-SLEW	Vs:40V Vo:±12V Vi0:5mV f:12Mc f:1Mkc	8A	Har		data pinout
HA2-2515	HI-SLEW	Vs:40V Vo:±12V Vi0:5mV f:12Mc f:1Mkc	8A	Har		data pinout
HA2-2520	HI-SLEW	Vs:40V Vo:±12V Vi0:4mV f:20Mc f:2Mkc	8A	Har		data pinout
HA2-2522	HI-SLEW	Vs:40V Vo:±12V Vi0:5mV f:20Mc f:2Mkc	8A	Har		data pinout
HA2-2525	HI-SLEW	Vs:40V Vo:±12V Vi0:5mV f:20Mc f:2Mkc	8A	Har		data pinout
HA2-2529	HI-SLEW	Vs:40V Vo:±12V Vi0:2mV f:20Mc	8A	Har		data pinout
HA2-2541	WIDEBAND	Vs:±17.5V Vo:±11V Vi0:0.8mV f:40Mc	12A	Har		data pinout
HA2-2542	WIDEBAND	Vs:±17.5V Vo:±11V Vi0:5mV f:70Mc	12A	Har		data pinout
HA2-2544	VIDEO-AMP	Vs:±17.5V Vo:±11V Vi0:6mV f:50Mc	8A	Har		data pinout
HA2-2548	HI-SLEW	Vs:40V Vu:130dB Vo:±12V Vi0:300μμV	8A	Har		data pinout
HA2-2600	HI-SPEED	Vs:45V Vo:±12V Vi0:0.5mV f:12Mc	8A	Har		data pinout
HA2-2602	HI-SPEED	Vs:45V Vo:±12V Vi0:3mV f:12Mc f:75kc	8A	Har		data pinout
HA2-2605	HI-SPEED	Vs:45V Vo:±12V Vi0:3mV f:12Mc f:75kc	8A	Har		data pinout
HA2-2620	WIDEBAND	Vs:45V Vo:±12V Vi0:0.5mV f:100Mc	8A	Har		data pinout
HA2-2622	WIDEBAND	Vs:45V Vo:±12V Vi0:3mV f:100Mc	8A	Har		data pinout
HA2-2625	WIDEBAND	Vs:45V Vo:±12V Vi0:3mV f:100Mc	8A	Har		data pinout
HA2-2640	HI-VOLT	Vs:±50V Vo:>±35V Vi0:2mV f:4Mc	8A	Har		data pinout
HA2-2645	HI-VOLT	Vs:±50V Vo:>±35V Vi0:2mV f:4Mc	8A	Har		data pinout
HA2-2650	2xOP	Vs:±20V Vu:40V/mV Vo:±14V Vi0:1.5mV	8A	Har		data pinout
HA2-2655	2xOP	Vs:±20V Vu:40V/mV Vo:±14V Vi0:2mV	8A	Har		data pinout
HA2-2700	UNI	Vs:±15V Vu:112dB Vo:±12V Vi0:5mV	8A	Har		data pinout
HA2-2704	UNI	Vs:±15V Vu:112dB Vo:±12V Vi0:6mV	8A	Har		data pinout
HA2-2705	LO-POWER	Vs:±15V Vu:106dB Vo:±12V Vi0:7mV	8A	Har		data pinout
HA2-2720	PROG	Vs:±22V Vu:120V/mV Vo:±12V Vi0:2mV	8A	Har		data pinout

Type	Device	Short Description	Fig.	Manu	Comparision Types	More at
HA2-2900	CHOPPER	Vs:±15V Vo:±10V Vi0:0.06mV f:3Mc	8A	Har		data pinout
HA2-2904	CHOPPER	Vs:±15V Vo:±10V Vi0:0.05mV f:3Mc	8A	Har		data pinout
HA2-2905-5	CHOPPER	Vs:±15V Vo:±10V Vi0:0.08mV f:3Mc	8A	Har		data pinout
HA2-5002	BUFF	Vs:±22V Vo:±13.5V Vi0:5mV f:110Mc	8A	Har		data pinout
HA2-5033	VIDEO-BUFF	Vs:±20V Vo:±10V Vi0:5mV f:250Mc	12A	Har		data pinout
HA2-5062	2xFET	Vs:±20V Vu:25V/mV Vo:±12V Vi0:3mV	8A	Har		data pinout
HA2-5082	2xFET	Vs:±20V Vu:106dB Vo:±12V Vi0:3mV	8A	Har		data pinout
HA2-5100	FET	Vs:±40V Vu:103.5dB Vo:±13V Vi0:0.5mV	8A	Har		data pinout
HA2-5101	LO-NOISE	Vs:±20V Vo:±13V Vi0:0.5mV f:100Mc	8A	Har		data pinout
HA2-5102	2xLO-NOISE	Vs:±20V Vo:±12V Vi0:0.5mV f:60Mc	8A	Har		data pinout
HA2-5105	FET	Vs:±40V Vu:100dB Vo:±12V Vi0:0.5mV	8A	Har		data pinout
HA2-5110	FET	Vs:±40V Vu:103.5dB Vo:±13V Vi0:0.5mV	8A	Har		data pinout
HA2-5111	LO-NOISE	Vs:±20V Vo:±13V Vi0:0.5mV f:100Mc	8A	Har		data pinout
HA2-5112	2xLO-NOISE	Vs:±20V Vo:±12V Vi0:0.5mV f:60Mc	8A	Har		data pinout
HA2-5115	FET	Vs:±40V Vu:100dB Vo:±12V Vi0:0.5mV	8A	Har		data pinout
HA2-5127	LO-NOISE	Vs:±22V Vo:±13.5V Vi0:30μμV f:8.5Mc	8A	Har		data pinout
HA2-5130	HI-PREC	Vs:±40V Vu:140dB Vo:±12V Vi0:10μμV	8A	Har		data pinout
HA2-5135	HI-PREC	Vs:±40V Vu:140dB Vo:±12V Vi0:10μμV	8A	Har		data pinout
HA2-5137	LO-NOISE	Vs:±22V Vo:±13.5V Vi0:30μμV f:80Mc	8A	Har		data pinout
HA2-5141	FET	Vs:40V Vu:100dB Vo:>0...3V Vi0:0.7mV	8A	Har		data pinout
HA2-5142	2xμ-POWER	Vs:40V Vu:100dB Vo:>0...4V Vi0:0.5mV	8A	Har		data pinout
HA2-5147	LO-NOISE	Vs:±22V Vo:±13.5V Vi0:30mV f:140Mc	8A	Har		data pinout
HA2-5160	JFET	Vs:±20V Vo:±11V Vi0:1mV f:100Mc	8A	Har		data pinout
HA2-5162	JFET	Vs:±20V Vo:±11V Vi0:3mV f:100Mc	8A	Har		data pinout
HA2-5170	JFET	Vs:44V Vo:±12V Vi0:0.1mV f:8Mc	8A	Har		data pinout
HA2-5177	LO-OFFSET	Vs:±22V Vu:150dB Vo:±13V Vi0:20μμV	8A	Har		data pinout
HA2-5180	FET	Vs:40V Vu:120dB Vo:±12V Vi0:1mV	8A	Har		data pinout
HA2-5190	FAST-SET	Vs:35V Vo:±8V Vi0:3mV f:150Mc	12A	Har		data pinout
HA2-5195	FAST-SET	Vs:35V Vo:±8V Vi0:3mV f:>150Mc	12A	Har		data pinout
HA2-5221	LO-NOISE	Vs:36V Vu:>±106dB Vo:±12V f:>100Mc	8A	Har		data pinout
HA2-7712	BIMOS	Vs:±9V Vu:115dB Vo:>±4.95V f:1Mc	8A	Har		data pinout
HA2-7713	BIMOS	Vs:±9V Vu:115dB Vo:>±4.95V ft:120kkc	8A	Har		data pinout
HA2-909	UNI	Vs:±15V Vu:87dB Vo:±12V Vi0:6mV	8A	Har		data pinout
HA2-911	UNI	Vs:±15V Vu:87dB Vo:±12V Vi0:6mV	8A	Har		data pinout
HA 2-xxxx	OP/KOP-IC	→HA-xxxx	81, 83	Har		data
HA3-2405	4xPROG	Vs:45V Vu:150V/mV Vo:±12V Vi0:4mV	16D	Har		data pinout
HA3-2406	4xPROG	Vs:45V Vo:±12V Vi0:7mV f:30Mc	16D	Har		data pinout
HA3-2425	SAMP&HOLD	Vs:±20V Vo:>±10V Vi0:3mV f:2.5Mc	14D	Har		data pinout
HA3-2444	4xPROG	Vs:±17.5V Vu:76dB Vo:±11V Vi0:4mV	16D	Har		data pinout
HA3-2500	HI-SPEED	Vs:±20V Vu:30V/mV Vo:±12V Vi0:2mV	8D	Har		data pinout
HA3-2502	HI-SPEED	Vs:±20V Vu:25V/mV Vo:±12V Vi0:4mV	8D	Har		data pinout
HA3-2505	HI-SPEED	Vs:40V Vo:±12V Vi0:4mV f:12Mc	8D	Har		data pinout
HA3-2510	HI-SLEW	Vs:±20V Vu:15V/mV Vo:±12V Vi0:4mV	8D	Har		data pinout
HA3-2512	HI-SLEW	Vs:±20V Vu:15V/mV Vo:±12V Vi0:5mV	8D	Har		data pinout
HA3-2515	HI-SLEW	Vs:40V Vo:±12V Vi0:5mV f:12Mc f:1Mkc	8D	Har		data pinout
HA3-2520	HI-SLEW	Vs:±20V Vu:15V/mV Vo:±12V Vi0:4mV	8D	Har		data pinout
HA3-2522	HI-SLEW	Vs:±20V Vu:15V/mV Vo:±12V Vi0:5mV	8D	Har		data pinout
HA3-2525	HI-SLEW	Vs:40V Vo:±12V Vi0:5mV f:20Mc f:2Mkc	8D	Har		data pinout
HA3-2529	HI-SLEW	Vs:40V Vo:±12V Vi0:2mV f:20Mc	8D	Har		data pinout
HA3-2539	HI-SLEW	Vs:±17.5V Vo:>±10V Vi0:8mV f:600Mc	14D	Har		data pinout
HA3-2540	WIDEBAND	Vs:17V Vo:>±10V Vi0:8mV f:400Mc	14D	Har		data pinout
HA3-2542	WIDEBAND	Vs:±17.5V Vo:±11V Vi0:5mV f:70Mc	14D	Har		data pinout
HA3-2544	HI-SLEW	Vs:±17.5V Vo:±11V Vi0:6mV f:50Mc	8D	Har		data pinout
HA3-2548	HI-SLEW	Vs:40V Vu:130dB Vo:±12V Vi0:300μμV	8D	Har		data pinout
HA3-2605	WIDEBAND	Vs:45V Vo:±12V Vi0:3mV f:12Mc f:75kc	8D	Har		data pinout
HA3-2625	WIDEBAND	Vs:45V Vo:±12V Vi0:3mV f:100Mc	8D	Har		data pinout
HA3-2655	2xOP	Vs:±20V Vu:40V/mV Vo:±14V Vi0:2mV	14D	Har		data pinout
HA3-2705	LO-POWER	Vs:±15V Vu:106dB Vo:±12V Vi0:7mV	8D	Har		data pinout
HA3-2740	4xPROG	Vs:±22V Vu:100V/mV Vo:±14V Vi0:2mV	16D	Har		data pinout
HA3-2839	HI-SLEW	Vs:±17.5V Vo:>±10V Vi0:0.6mV f:600Mc	14D	Har		data pinout
HA3-2840	HI-SLEW	Vs:±17.5V Vo:>±10V Vi0:0.6mV f:600Mc	8D	Har	EL2039	data pinout
HA3-2841	WIDEBAND	Vs:±17.5V Vo:±10.5V Vi0:1mV f:50Mc	8D	Har	AD841,EL2041	data pinout
HA3-2842	WIDEBAND	Vs:±17.5V Vo:±11V Vi0:1mV f:80Mc	8D	Har	AD842	data pinout
HA3-2850	LO-POWER	Vs:±17.5V Vo:±11V Vi0:0.6mV f:470Mc	8D	Har	AD840,EL2040	data pinout
HA3-4156	4xOP	Vs:±20V Vu:50V/mV Vo:±13.7V Vi0:1mV	14D	Har		data pinout
HA3-4600	4xOP	Vs:±20V Vu:250V/mV Vo:±13V Vi0:0.3mV	14D	Har		data pinout
HA3-4605	4xOP	Vs:±20V Vu:250V/mV Vo:±13V Vi0:3mV	14D	Har		data pinout
HA3-4625	4xHI-SPEED	Vs:±20V Vu:250V/mV Vo:±13V Vi0:3mV	14D	Har		data pinout
HA3-4741	4xOP	Vs:±20V Vu:50V/mV Vo:±13.7V f:3.5Mc	14D	Har		data pinout
HA3-4905	4xCOMP	Vs:33V Vi0:4mV	16D	Har		data pinout
HA3-5002	BUFF	Vs:±22V Vo:±13.5V Vi0:5mV f:110Mc	8D	Har		data pinout
HA3-5004	WIDEBAND	Vs:±20V Vo:±9.5V Vi0:1mV f:100Mkc	14D	Har		data pinout
HA3-5020	WIDEBAND	Vs:±18V Vo:±12.7V Vi0:2mV f:12.7Mkc	8D	Har		data pinout
HA3-5033	VIDEO-BUFF	Vs:±20V Vo:±10V Vi0:5mV f:250Mc	8D	Har		data pinout
HA3-5062	2xFET	Vs:±20V Vu:25V/mV Vo:±12V Vi0:3mV	8D	Har		data pinout
HA3-5064	4xFET	Vs:±20V Vu:25V/mV Vo:±12V Vi0:<15mV	14D	Har		data pinout
HA3-5082	2xFET	Vs:±20V Vu:106dB Vo:±12V Vi0:5mV	8D	Har		data pinout
HA3-5084	4xFET	Vs:±40V Vu:88dB Vo:±10V Vi0:<2mV	14D	Har		data pinout
HA3-5101	LO-NOISE	Vs:±20V Vo:±13V Vi0:0.5mV f:100Mc	8D	Har		data pinout
HA3-5102	2xLO-NOISE	Vs:±20V Vo:±12V Vi0:0.5mV f:60Mc	8D	Har		data pinout
HA3-5104	4xLO-NOISE	Vs:±20V Vo:±12V Vi0:0.5mV f:60Mc	14D	Har		data pinout
HA3-5111	LO-NOISE	Vs:±20V Vo:±13V Vi0:0.5mV f:100Mc	8D	Har		data pinout
HA3-5112	2xLO-NOISE	Vs:±20V Vo:±12V Vi0:0.5mV f:60Mc	8D	Har		data pinout
HA3-5114	4xLO-NOISE	Vs:±20V Vo:±12V Vi0:0.5mV f:60Mc	14D	Har		data pinout
HA3-5127	LO-NOISE	Vs:±22V Vo:±13.5V Vi0:30μμV f:8.5Mc	8D	Har		data pinout

Type	Device	Short Description	Fig.	Manu	Comparision Types	More at
HA3-513	HI-PREC	Vs:±40V Vu:140dB Vo:±12V Vi0:10µµV	8D	Har		data pinout
HA3-5137	LO-NOISE	Vs:±22V Vo:±13.5V Vi0:30µµV f:80Mc	8D	Har		data pinout
HA3-5141	FET	Vs:40V Vu:100dB Vo:>0...3V Vi0:0.7mV	8D	Har		data pinout
HA3-5142	2xµ-POWER	Vs:40V Vu:100dB Vo:>0...4V Vi0:0.5mV	8D	Har		data pinout
HA3-5144	4xµ-POWER	Vs:40V Vu:100dB Vo:>0...4V Vi0:0.5mV	14D	Har		data pinout
HA3-5147	LO-NOISE	Vs:±22V Vo:±13.5V Vi0:30mV f:140Mc	8D	Har		data pinout
HA3-5170	FET	Vs:±44V Vu:>98dB Vo:>±10V Vi0:0.1mV	8D	Har		data pinout
HA3-5177	LO-OFFSET	Vs:±22V Vu:150dB Vo:±13V Vi0:20µµV	8D	Har		data pinout
HA3-5180	FET	Vs:40V Vu:120dB Vo:±12V Vi0:1mV	8D	Har		data pinout
HA3-5221	LO-NOISE	Vs:35V Vu:128dB Vo:±12.5V Vi0:0.3mV	8D	Har		data pinout
HA3-5222	2xLO-NOISE	Vs:35V Vu:128dB Vo:±12.5V Vi0:0.3mV	16D	Har		data pinout
HA3-5232	2xHI-PREC	Vs:36V Vo:>12/0V Vi0:100µµV f:0.5Mc	8D	Har		data pinout
HA3-5234	4xHI-PREC	Vs:36V Vo:>12/0V Vi0:100µµV f:0.5Mc	14D	Har		data pinout
HA3-5320	SAMP&HOLD	Vs:±20V Vi0:0.5mV f:2Mc f:600kc	14D	Har		data pinout
HA3-5330	SAMP&HOLD	Vs:±20V Vi0:0.2mV f:4.5Mc f:1400kc	14D	Har		data pinout
HA3-5340	SAMP&HOLD	Vs:±18V Vi0:<1.5mV f:10Mc f:900kc	14D	Har		data pinout
HA3-7712	BIMOS	Vs:±9V Vu:115dB Vo:>±4.95V f:1Mc	8D	Har		data pinout
HA3-7713	BIMOS	Vs:±9V Vu:115dB Vo:>±4.95V ft:120kkc	8D	Har		data pinout
HA3-8023	3xLO-POWER	Vs:±22V Vu:80dB Vo:±5V Vi0:2mV	16D	Har		data pinout
HA3B2840	HI-SLEW	Vs:±17.5V Vo:>±10V Vi0:0.6mV f:600Mc	14D	Har	EL2039	data pinout
HA3B2841	WIDEBAND	Vs:±17.5V Vo:±10.5V Vi01:1mV f:50Mc	14D	Har	AD841,EL2041	data pinout
HA3B2842	WIDEBAND	Vs:±17.5V Vo:±11V Vi01:1mV f:80Mc	14D	Har	AD842	data pinout
HA3B2850	LO-POWER	Vs:±17.5V Vo:±11V Vi0:0.6mV f:470Mc	14D	Har	AD840,EL2040	data pinout
HA 3-xxxx	OP/KOP-IC	→HA-xxxx	...-DIP	Har		data
HA4-2522	HI-SLEW	Vs:40V Vo:±10V Vi0:±10mV f:>10Mc	20C	Har		data pinout
HA4-2539	HI-SPEED	Vs:±17V Vu:30V/mV Vo:>±10V Vi0:3mV	d	Har		data pinout
HA4-2544	HI-SLEW	Vs:±17.5V Vo:±10V Vi0:±15mV f:>45Mc	20C	Har		data pinout
HA4-2640	HI-VOLT	Vs:±50V Vo:±35V Vi0:±4mV f:>45kc	20C	Har		data pinout
HA4-4741	4xOP	Vs:±20V Vu:100V/mV Vo:±13.7V f:3.5Mc	d,20c	Har		data pinout
HA4-4900	4xOP	Vs:±16V Vu:400V/mV Vi0:2mV	d	Har		data pinout
HA4-4902	4xOP	Vs:±16V Vu:400V/mV Vi0:2mV	d	Har		data pinout
HA4-4905	4xOP	Vs:±16V Vu:400V/mV Vi0:4mV	d	Har		data pinout
HA4-5002	BUFF	Vs:±22V Vo:±10V Vi0:±20mV f:110Mc	20C	Har		data pinout
HA4-5020	WIDEBAND	Vs:±18V Vo:±12V Vi0:±8mV f:105Mc	20C	Har	EL2020/883	data pinout
HA4-5033	VIDEO-BUFF	Vs:±20V Vo:±10V Vi0:5mV f:250Mc	20C	Har		data pinout
HA4-5102	2xLO-NOISE	Vs:±20V Vo:±10V Vi0:±2mV f:8Mc	20C	Har		data pinout
HA4-5104	4xLO-NOISE	Vs:±20V Vo:±10V Vi0:±2.5mV f:8Mc	20C	Har		data pinout
HA4-5112	2xLO-NOISE	Vs:±20V Vo:±10V Vi0:±2mV f:>54Mc	20C	Har		data pinout
HA4-5114	4xLO-NOISE	Vs:±20V Vo:±10V Vi0:±2.5V f:>40Mc	20C	Har		data pinout
HA4-5127	LO-NOISE	Vs:±20V Vo:±11.5V Vi0:±0.1mV f:>5Mc	20C	Har		data pinout
HA4-5130	HI-PREC	Vs:±40V Vu:140dB Vo:±12V Vi0:10µµV	d	Har		data pinout
HA4-5134	4xHI-PREC	Vs:±20V Vo:±12V Vi0:±0.2mV f:>3Mc	20C	Har		data pinout
HA4-5135	HI-PREC	Vs:±40V Vu:140dB Vo:±12V Vi0:10µµV	d	Har		data pinout
HA4-5137	LO-NOISE	Vs:±22V Vo:±11.5V Vi0:±0.1mV f:>60Mc	20C	Har		data pinout
HA4-5142	2xLO-POWER	Vs:35V Vo:±10V Vi0:±6µµV f:>12.7kc	20C	Har		data pinout
HA4-5147	LO-NOISE	Vs:±20V Vo:±11.5V Vi0:±0.1mV	20C	Har		data pinout
HA4-5170	FET	Vs:±44V Vu:>100dB Vo:>±10V Vi0:0.1mV	d	Har		data pinout
HA4-5177	LO-OFFSET	Vs:±22V Vu:126dB Vo:±12V Vi0:±60µµV	20C	Har		data pinout
HA4-5221	LO-NOISE	Vs:36V Vu:>±106dB Vo:±12V f:>100Mc	20C	Har		data pinout
HA4-5320	SAMP&HOLD	Vs:±20V Vo:±10V Vi0:±1mV	20C	Har		data pinout
HA4-5340	SAMP&HOLD	Vs:±18V Vo:±10V Vi0:±1.5mV	20C	Har		data pinout
HA4P2405	4xPROG	Vs:±22V Vu:150V/mV Vo:±10V Vi0:4mV	20C	Har		data pinout
HA4P2425	SAMP&HOLD	Vs:±20V Vo:>±10V Vi0:3mV f:2.5Mc	20C	Har		data pinout
HA4P2505	HI-SPEED	Vs:±20V Vu:25V/mV Vo:±12V Vi0:4mV	20C	Har		data pinout
HA4P2525	HI-SLEW	Vs:40V Vo:±12V Vi0:5mV f:20Mc f:2Mkc	20C	Har		data pinout
HA4P2539	HI-SLEW	Vs:±17.5V Vo:>±10V Vi0:8mV f:600Mc	20C	Har		data pinout
HA4P2540	WIDEBAND	Vs:±17.5V Vo:>±10V Vi0:8mV f:400Mc	20C	Har		data pinout
HA4P2544	HI-SLEW	Vs:±17.5V Vo:±11V Vi0:6mV f:50Mc	20C	Har		data pinout
HA4P26	HI-SPEED	Vs:±22V Vu:150V/mV Vo:±12V Vi0:3mV	20C	Har		data pinout
HA4P4741	4xOP	Vs:±20V Vu:50V/mV Vo:±13.7V Vi0:1mV	20C	Har		data pinout
HA4P4905	4xCOMP	Vs:33V Vi0:4mV	20C	Har		data pinout
HA4P5002	BUFF	Vs:±22V Vo:±13.9V Vi0:5mV f:110Mc	20C	Har		data pinout
HA4P5101	LO-NOISE	Vs:±20V Vo:±13V Vi0:0.5mV f:100Mc	20C	Har		data pinout
HA4P5102	2xLO-NOISE	Vs:±20V Vo:±12V Vi0:0.5mV f:60Mc	20C	Har		data pinout
HA4P5104	4xLO-NOISE	Vs:±20V Vo:±12V Vi0:0.5mV f:60Mc	20C	Har		data pinout
HA4P5114	4xLO-NOISE	Vs:±20V Vo:±12V Vi0:0.5mV f:60Mc	20C	Har		data pinout
HA4P5144	4xLO-POWER	Vs:35V Vo:±13V Vi0:2mV f:0.4Mc	20C	Har		data pinout
HA 4P-xxxx	OP/KOP-IC	→HA-xxxx	...-PLCC	Har		data
HA7-2500	HI-SPEED	Vs:40V Vo:±12V Vi0:2mV f:12Mc	8D	Har		data pinout
HA7-2502	HI-SPEED	Vs:40V Vo:±12V Vi0:4mV f:12Mc	8D	Har		data pinout
HA7-2505	HI-SPEED	Vs:20V Vu:25V/mV Vo:±12V Vi0:4mV	8D	Har		data pinout
HA7-2510	HI-SLEW	Vs:40V Vo:±12V Vi0:4mV f:12Mc f:1Mkc	8D	Har		data pinout
HA7-2512	HI-SLEW	Vs:40V Vo:±12V Vi0:5mV f:12Mc f:1Mkc	8D	Har		data pinout
HA7-2515	HI-SLEW	Vs:40V Vo:±12V Vi0:5mV f:12Mc f:1Mkc	8D	Har		data pinout
HA7-2520	HI-SLEW	Vs:40V Vo:±12V Vi0:4mV f:20Mc f:2Mkc	8D	Har		data pinout
HA7-2522	HI-SLEW	Vs:40V Vo:±12V Vi0:5mV f:20Mc f:2Mkc	8D	Har		data pinout
HA7-2525	HI-SLEW	Vs:40V Vo:±12V Vi0:5mV f:20Mc f:2Mkc	8D	Har		data pinout
HA7-2529	HI-SLEW	Vs:40V Vo:±12V Vi0:2mV f:20Mc	8D	Har		data pinout
HA7-2544	HI-SLEW	Vs:±17.5V Vo:±11V Vi0:6mV f:50Mc	8D	Har		data pinout
HA7-2548	HI-SLEW	Vs:40V Vu:130dB Vo:±12V Vi0:300µµV	8D	Har		data pinout
HA7-2600	WIDEBAND	Vs:45V Vo:±12V Vi0:0.5mV f:12Mc	8D	Har		data pinout
HA7-2602	WIDEBAND	Vs:45V Vo:±12V Vi0:3mV f:12Mc f:75kc	8D	Har		data pinout
HA7-2605	WIDEBAND	Vs:45V Vo:±12V Vi0:3mV f:12Mc f:75kc	8D	Har		data pinout
HA7-2620	WIDEBAND	Vs:45V Vo:±12V Vi0:0.5mV f:100Mc	8D	Har		data pinout
HA7-2622	WIDEBAND	Vs:45V Vo:±12V Vi0:3mV f:100Mc	8D	Har		data pinout

Type	Device	Short Description	Fig.	Manu	Comparision Types	More at
HA7-2625	WIDEBAND	Vs:45V Vo:±12V Vi0:3mV f:100Mc	8D	Har		data pinout
HA7-2640	HI-VOLT	Vs:±50V Vo:>±35V Vi0:2mV f:4Mc	8D	Har		data pinout
HA7-2645	HI-VOLT	Vs:±50V Vo:>±35V Vi0:2mV f:4Mc	8D	Har		data pinout
HA7-2720	PROG	Vs:±22V Vu:120V/mV Vo:±12V Vi0:2mV	8D	Har		data pinout
HA7-2725	PROG	Vs:±22V Vu:120V/mV Vo:±12V Vi0:2mV	8D	Har		data pinout
HA7-2840	HI-SLEW	Vs:±17.5V Vo:>±10V Vi0:0.6mV f:600Mc	8D	Har	EL2039	data pinout
HA7-2841	WIDEBAND	Vs:±17.5V Vo:±10V Vi0:±4mV f:50Mc	8D	Har		data pinout
HA7-2842	WIDEBAND	Vs:±17.5V Vo:±10V Vi0:±4mV f:80Mc	8D	Har		data pinout
HA7-2850	LO-POWER	Vs:±17.5V Vo:±11V Vi0:0.6mV f:470Mc	8D	Har	HA-2540/883,AD840	data pinout
HA7-5002	BUFF	Vs:±22V Vo:±13.5V Vi0:5mV f:110Mc	8D	Har		data pinout
HA7-5020	WIDEBAND	Vs:±18V Vo:±12.7V Vi0:2mV f:12.7Mkc	8D	Har	EL2020/883	data pinout
HA7-5062	2xFET	Vs:±20V Vu:25V/mV Vo:±12V Vi0:3mV	8D	Har		data pinout
HA7-5082	2xFET	Vs:±20V Vu:106dB Vo:±12V Vi0:3mV	8D	Har		data pinout
HA7-5101	LO-NOISE	Vs:±20V Vo:±13V Vi0:0.5mV f:100Mc	8D	Har		data pinout
HA7-5102	2xLO-NOISE	Vs:±20V Vo:±12V Vi0:0.5mV f:60Mc	8D	Har		data pinout
HA7-5111	LO-NOISE	Vs:±20V Vo:±13V Vi0:0.5mV f:100Mc	8D	Har		data pinout
HA7-5112	2xLO-NOISE	Vs:±20V Vo:±12V Vi0:0.5mV f:60Mc	8D	Har		data pinout
HA7-5127	LO-NOISE	Vs:±22V Vo:±13.5V Vi0:30μμV f:8.5Mc	8D	Har		data pinout
HA7-5130	HI-PREC	Vs:±20V Vu:140dB Vo:±12V Vi0:10μμV	8D	Har		data pinout
HA7-5135	HI-PREC	Vs:±20V Vu:140dB Vo:±12V Vi0:10μμV	8D	Har		data pinout
HA7-5137	LO-NOISE	Vs:±22V Vo:±13.5V Vi0:30μμV f:80Mc	8D	Har		data pinout
HA7-5141	FET	Vs:40V Vu:100dB Vo:>0...3V Vi0:0.7mV	8D	Har		data pinout
HA7-5142	2xLO-POWER	Vs:40V Vu:100dB Vo:>0...4V Vi0:0.5mV	8D	Har		data pinout
HA7-5147	LO-NOISE	Vs:±22V Vo:±13.5V Vi0:30mV f:140Mc	8D	Har		data pinout
HA7-5170	JFET	Vs:44V Vo:±12V Vi0:0.1mV f:8Mc	8D	Har		data pinout
HA7-5177	LO-OFFSET	Vs:±22V Vu:150dB Vo:±13V Vi0:20μμV	8D	Har		data pinout
HA7-5180	FET	Vs:40V Vu:120dB Vo:±12V Vi0:1mV	8D	Har		data pinout
HA7-5221	LO-NOISE	Vs:36V Vu:>±106dB Vo:±12V f:>100Mc	8D	Har		data pinout
HA7-5222	2xLO-NOISE	Vs:36V Vu:>±106dB Vo:±12V f:>100Mc	8D	Har		data pinout
HA 7-xxxx	OP/KOP-IC	→HA-xxxx	8-DIP,DIC	Har		data
HA9-2500	HI-SPEED	Vs:±15V Vu:86dB Vo:±10V Vi0:8mV	14F	Har,Isi		data pinout
HA9-2502	HI-SPEED	Vs:±15V Vu:83dB Vo:±10V Vi0:10mV	14F	Har,Isi		data pinout
HA9-2505	HI-SPEED	Vs:±15V Vu:83dB Vo:±10V Vi0:10mV	14F	Har,Isi		data pinout
HA9-2510	HI-SPEED	Vs:±15V Vu:80dB Vo:±10V Vi0:11mV	14F	Har,Is.		data pinout
HA9-2512	HI-SPEED	Vs:±15V Vu:77dB Vo:±10V Vi0:14mV	14F	Har,Isi		data pinout
HA9-2515	HI-SPEED	Vs:±15V Vu:77dB Vo:±10V Vi0:14mV	14F	Har,Isi		data pinout
HA9-2520	HI-SPEED	Vs:±15V Vu:80dB Vo:±10V Vi0:11mV	14F	Har,Isi		data pinout
HA9-2522	HI-SPEED	Vs:±15V Vu:77dB Vo:±10V Vi0:14mV	14F	Har,Isi		data pinout
HA9-2525	HI-SPEED	Vs:±15V Vu:77dB Vo:±10V Vi0:14mV	14F	Har,Isi		data pinout
HA9-2600	HI-SPEED	Vs:±15V Vu:100dB Vo:±10V Vi0:6mV	10F	Har,Isi		data pinout
HA9-2602	HI-SPEED	Vs:±15V Vu:98dB Vo:±10V Vi0:7mV	10F	Har,Isi		data pinout
HA9-2605	HI-SPEED	Vs:±15V Vu:98dB Vo:±10V Vi0:7mV	10F	Har,Isi		data pinout
HA9-2620	HI-SPEED	Vs:±22V Vu:150V/mV Vo:±12V Vi0:0.5mV	14F	Isi		data pinout
HA9-2622	HI-SPEED	Vs:±22V Vu:150V/mV Vo:±12V Vi0:3mV	14F	Isi		data pinout
HA9-2625	HI-SPEED	Vs:±22V Vu:150V/mV Vo:±12V Vi0:3mV	14F	Isi		data pinout
HA9-5147	LO-NOISE	Vs:±22V Vo:±13.5V Vi0:30mV f:140Mc	8S	Har		data pinout
HA9-909	UNI	Vs:±15V Vu:87dB Vo:±12V Vi0:6mV	14F	Har		data pinout
HA9-911	UNI	Vs:±15V Vu:86dB Vo:±11V Vi0:7.5mV	14F	Har,Isi		data pinout
HA9P2406	4xPROG	Vs:45V Vo:±12V Vi0:7mV f:30Mc	16S	Har		data pinout
HA9P2425	SAMP&HOLD	Vs:±20V Vo:>±10V Vi0:3mV f:2.5Mc	14S	Har		data pinout
HA9P2444	4xPROG	Vs:±17.5V Vu:76dB Vo:±11V Vi0:4mV	16S	Har		data pinout
HA9P2525	HI-SLEW	Vs:40V Vo:±12V Vi0:5mV f:20Mc f:2Mkc	8S	Har		data pinout
HA9P2529	HI-SLEW	Vs:40V Vo:±12V Vi0:2mV f:20Mc	8S	Har		data pinout
HA9P2539	HI-SLEW	Vs:±17.5V Vo:>±10V Vi0:8mV f:600Mc	14S	Har		data pinout
HA9P2540	WIDEBAND	Vs:±17.5V Vo:>±10V Vi0:8mV f:400Mc	14S	Har		data pinout
HA9P2544	HI-SLEW	Vs:±17.5V Vo:±11V Vi0:6mV f:50Mc	8S	Har		data pinout
HA9P2548	HI-SLEW	Vs:±20V Vu:130dB Vo:±12V Vi0:100μμV	8S	Har		data pinout
HA9P2548-5	HI-SLEW	Vs:40V Vu:130dB Vo:±12V Vi0:300μμV	16S	Har		data pinout
HA9P2548-9	HI-SLEW	Vs:±20V Vu:130dB Vo:±12V Vi0:300μμV	8S	Har		data pinout
HA9P2605-5	HI-SLEW	Vs:45V Vo:±12V Vi0:3mV f:12Mc f:75kc	8S	Har		data pinout
HA9P2605-9	HI-SLEW	Vs:±22.5V Vo:±12V Vi0:5mV f:12Mc	8S	Har		data pinout
HA9P2625	WIDEBAND	Vs:45V Vo:±12V Vi0:3mV f:100Mc	8S	Har		data pinout
HA9P2705	LO-POWER	Vs:±15V Vu:106dB Vo:±12V Vi0:7mV	8S	Har		data pinout
HA9P2840	HI-SLEW	Vs:±17.5V Vo:>±10V Vi0:0.6mV f:600Mc	8S	Har	EL2039	data pinout
HA9P2841	WIDEBAND	Vs:±17.5V Vo:±10.5V Vi0:1mV f:50Mc	8S	Har	AD841,EL2041	data pinout
HA9P2842	WIDEBAND	Vs:±17.5V Vo:±11V Vi0:1mV f:80Mc	8S	Har	AD842	data pinout
HA9P2850	LO-POWER	Vs:±17.5V Vo:±11V Vi0;0.6mV f:470Mc	8S	Har	AD840,EL2040	data pinout
HA9P4741-2	4xOP	Vs:±22V Vo:±12.5V Vi0:0.5mV f:3.5Mc	16S	Har		data pinout
HA9P4741-5	4xOP	Vs:±22V Vo:±12.5V Vi0:1mV f:3.5Mc	16S	Har		data pinout
HA9P4741-9	4xOP	Vs:±20V Vo:±12.5V Vi0:<5mV f:3.5Mc	16S	Har		data pinout
HA9P4905	4xCOMP	Vs:33V Vi0:4mV	16S	Har		data pinout
HA9P5002	BUFF	Vs:±22V Vo:±10.7V Vi0:5mV f:110Mc	8S	Har		data pinout
HA9P5004	WIDEBAND	Vs:±20V Vo:±9.5V Vi0:1mV f:100Mkc	14S	Har		data pinout
HA9P5020	WIDEBAND	Vs:±18V Vo:±12.7V Vi0:2mV f:12.7Mkc	8S	Har		data pinout
HA9P5033-2	BUFF	Vs:±20V Vo:±10V Vi0:5mV f:250Mc	8S	Har		data pinout
HA9P5033-5	VIDEO-BUFF	Vs:±20V Vo:±10V Vi0:5mV f:250Mc	8S	Har		data pinout
HA9P5033-9	BUFF	Vs:±20V Vo:±10V Vi0:15mV f:250Mc	8S	Har		data pinout
HA9P5101	LO-NOISE	Vs:±20V Vo:±13V Vi0:0.5mV f:100Mc	8S	Har		data pinout
HA9P5102	2xLO-NOISE	Vs:±20V Vo:±12V Vi0:0.5mV f:60Mc	16S	Har		data pinout
HA9P5104	4xLO-NOISE	Vs:±20V Vo:±12V Vi0:0.5mV f:60Mc	16S	Har		data pinout
HA9P5111	LO-NOISE	Vs:±20V Vo:±13V Vi0:0.5mV f:100Mc	8S	Har		data pinout
HA9P5112	2xLO-NOISE	Vs:±20V Vo:±12V Vi0:0.5mV f:60Mc	16S	Har		data pinout
HA9P5114	4xLO-NOISE	Vs:±20V Vo:±12V Vi0:0.5mV f:60Mc	16S	Har		data pinout
HA9P5127	LO-NOISE	Vs:±22V Vo:±13.5V Vi0:30μμV f:8.5Mc	8S	Har		data pinout
HA9P5137	LO-NOISE	Vs:±22V Vo:±13.5V Vi0:30μμV f:80Mc	8S	Har		data pinout

Type	Device	Short Description	Fig.	Manu	Comparision Types	More at
A9P5142	2xLO-POWER	Vs:35V Vo:±13V Vi0:2mV f:0.4Mc	16S	Har		data pinout
A9P5144	4xLO-POWER	Vs:35V Vo:±13V Vi0:2mV f:0.4Mc	16S	Har		data pinout
A9P5177	LO-OFFSET	Vs:±22V Vu:150dB Vo:±13V Vi0:20µµV	8S	Har		data pinout
A9P5190	WIDEBAND	Vs:35V Vo:±8V Vi0:3mV f:150Mc	14S	Har		data pinout
A9P5195-5	FAST-SET	Vs:35V Vo:±8V Vi0:3mV f:>150Mc	14S	Har		data pinout
A9P5195-9	WIDEBAND	Vs:35V Vo:±8V Vi0:3mV f:150Mc	14S	Har		data pinout
A9P5221	LO-NOISE	Vs:35V Vu:128dB Vo:±12.5V Vi0:0.3mV	8S	Har		data pinout
A9P5222	2xLO-NOISE	Vs:35V Vu:128dB Vo:±12.5V Vi0:0.3mV	16S	Har		data pinout
A9P5232	2xHI-PREC	Vs:36V Vo:>12/0V Vi0:100µµV f:0.5Mc	8S	Har		data pinout
A9P5234	4xHI-PREC	Vs:36V Vo:>12/0V Vi0:100µµV f:0.5Mc	16S	Har		data pinout
A9P5320-5	SAMP&HOLD	Vs:±20V Vi0:0.5mV f:2Mc f:600kc	16S	Har		data pinout
A9P5320-9	SAMP&HOLD	Vs:±20V Vi0:0.2mV f:2Mc f:600kc	16S	Har		data pinout
A9P5340-5	SAMP&HOLD	Vs:±18V Vi0:<1.5mV f:10Mc f:900kc	16S	Har		data pinout
A9P7712	BIMOS	Vs:±9V Vu:115dB Vo:>±4.95V f:1Mc	8S	Har		data pinout
A9P7713	BIMOS	Vs:±9V Vu:115dB Vo:>±4.95V ft:120kkc	8S	Har		data pinout
A 9P-xxxx	OP/KOP-IC	→HA-xxxx	...-MDIP	Har		data
A 17 L431 (A)LP	IC	Shunt Regulator	45	Ren	-	pdf pinout
A 17 L431 (A)P	IC	Shunt Regulator	7	Ren	-	pdf pinout
A 17 L431 ALTP	IC	Shunt Regulator	35	Ren	-	pdf pinout
A 17 L431 UP	IC	Shunt Regulator	48	Ren	-	pdf pinout
A 17 L432 ALTP	IC	Shunt Regulator	35	Ren	-	pdf pinout
A 17 L432 UP	IC	Shunt Regulator	48	Ren	-	pdf pinout
A 178L02(P),A(PA)	Z-IC	+2V, 0,15A, ±8%, A=±5%	7b	Hit	... 78L02... (TO-92)	pdf
A 178L05(P),A(PA)	Z-IC	+5V, 0,15A, ±8%, A=±5%	7b	Hit	... 78L05... (TO-92)	pdf
A 178L06(P),A(PA)	Z-IC	+6V, 0,15A, ±8%, A=±5%	7b	Hit	... 78L06... (TO-92)	pdf
A 178L08(P),A(PA)	Z-IC	+8V, 0,15A, ±8%, A=±5%	7b	Hit	... 78L08... (TO-92)	pdf
A 178L09(P),A(PA)	Z-IC	+9V, 0,15A, ±8%, A=±5%	7b	Hit	... 78L09... (TO-92)	pdf
A 178L10(P),A(PA)	Z-IC	+10V, 0,15A, ±8%, A=±5%	7b	Hit	... 78L10... (TO-92)	pdf
A 178L12(P),A(PA)	Z-IC	+12V, 0,15A, ±8%, A=±5%	7b	Hit	... 78L12... (TO-92)	pdf
A 178L15(P),A(PA)	Z-IC	+15V, 0,15A, ±8%, A=±5%	7b	Hit	... 78L15... (TO-92)	pdf
A 178L56(P),A(PA)	Z-IC	+5,6V, 0,15A, ±8%, A=±5%	7b	Hit	... 78L56... (TO-92)	pdf
A 178 L02...L15UA	Z-IC	=HA 178L02...L15: SMD	39b	Hit	M 5278LxxM, TA 78LxxF, ...78Lxx...(SOT-89)	pdf
A 178 M05(P,PJ)...	Z-IC	+5V, 0,5A, ±4%	17b	Hit	... 78M05... (TO-220)	data pdf pinout
A178M05P	R+	Io=500mA Vo:4.8...5.2V Vin:10V P:7.5W	3P	Hit		data pdf pinout
A 178 M06(P,PJ)...	Z-IC	+6V, 0,5A, ±4%	17b	Hit	... 78M06... (TO-220)	data pdf pinout
A178M06P	R+	Io=500mA Vo:5.75...6.25V Vin:11V	3P	Hit		data pdf pinout
A 178 M07(P,PJ)...	Z-IC	+7V, 0,5A, ±4%	17b	Hit	... 78M07... (TO-220)	data pdf pinout
A178M07P	R+	Io=500mA Vo:6.72...7.28V Vin:12.5V	3P	Hit		data pdf pinout
A 178 M08(P,PJ)...	Z-IC	+8V, 0,5A, ±4%	17b	Hit	... 78M08... (TO-220)	data pdf pinout
A178M08P	R+	Io=500mA Vo:7.7...8.3V Vin:14V P:7.5W	3P	Hit		data pdf pinout
A 178 M09(P,PJ)...	Z-IC	+9V, 0,5A, ±4%	17b	Hit	... 78M09... (TO-220)	data pdf
A 178 M12(P,PJ)...	Z-IC	+12V, 0,5A, ±4%	17b	Hit	... 78M12... (TO-220)	data pdf
A178M12P	R+	Io=500mA Vo:11.5...12.5V Vin:19V	3P	Hit		data pdf pinout
A 178 M15(P,PJ)...	Z-IC	+15V, 0,5A, ±4%	17b	Hit	... 78M15... (TO-220)	data pdf pinout
A178M15P	R+	Io=500mA Vo:14.4...15.6V Vin:23V	3P	Hit		data pdf pinout
A 178 M18(P,PJ)...	Z-IC	+18V, 0,5A, ±4%	17b	Hit	... 78M18... (TO-220)	data pdf pinout
A178M18P	R+	Io=500mA Vo:17.3...18.7V Vin:27V	3P	Hit		data pdf pinout
A 178 M20(P,PJ)...	Z-IC	+20V, 0,5A, ±4%	17b	Hit	... 78M20... (TO-220)	data pdf pinout
A178M20P	R+	Io=500mA Vo:19.2...20.8V Vin:29V	3P	Hit		data pdf pinout
A 178 M24(P,PJ)...	Z-IC	+24V, 0,5A, ±4%	17b	Hit	... 78M24... (TO-220)	data pdf pinout
A 178 MxxFM,FMP	Z-IC	=HA 178 M05...M24(P,PJ): Iso	17b	Hit	... 78Mxx... (TO-220 Iso)	
A178M24P	R+	Io=500mA Vo:23...25V Vin:33V P:7.5W	3P	Hit		data pdf pinout
A 179L05(P),A(PA)	Z-IC	-5V, 0,15A, ±4%, A=±2%	7a	Hit	... 79L05... (TO-92)	pdf
A 179L06(P),A(PA)	Z-IC	-6V, 0,15A, ±4%, A=±2%	7a	Hit	... 79L06... (TO-92)	pdf
A 179L08(P),A(PA)	Z-IC	-8V, 0,15A, ±4%, A=±2%	7a	Hit	... 79L08... (TO-92)	pdf
A 179L09(P),A(PA)	Z-IC	-9V, 0,15A, ±4%, A=±2%	7a	Hit	... 79L09... (TO-92)	pdf
A 179L10(P),A(PA)	Z-IC	-10V, 0,15A, ±4%, A=±2%	7a	Hit	... 79L10... (TO-92)	pdf
A 179L12(P),A(PA)	Z-IC	-12V, 0,15A, ±4%, A=±2%	7a	Hit	... 79L12... (TO-92)	pdf
A 179L15(P),A(PA)	Z-IC	-15V, 0,15A, ±4%, A=±2%	7a	Hit	... 79L15... (TO-92)	pdf
A 179 L05...L15UA	Z-IC	=HA 179L05...L15: SMD	39a	Hit	... 79Lxx (SOT-89)	pdf
A 179 M05 FM,FMP	Z-IC	Iso, -5V, 0,5A	17c	Hit	... 79M05...(TO-220 Iso)	
A 179 M06 FM,FMP	Z-IC	Iso, -6V, 0,5A	17c	Hit	... 79M06... (TO-220 Iso)	
A 179 M06 FM,FMP	Z-IC	Iso, -12V, 0,5A	17c	Hit	... 79M12...(TO-220 Iso)	
A 179 M12 FM,FMP	Z-IC	Iso, -12V, 0,5A	17c	Hit	... 79M12...(TO-220 Iso)	
A 179 M15 FM,FMP	Z-IC	Iso, -15V, 0,5A	17c	Hit	... 79M15...(TO-220 Iso)	
A456	ARRAY8x8	Vs:±4.5...±5.5V Ton:185ns at 5/-5V	44C	Isi	GX4404	data pdf pinout
A457	ARRAY8x8	Vs:±4.5...±5.5V Ton:185ns at 5/-5V	44CF	Isi	GX4404	data pdf pinout
A 1107	LIN-IC	Vc, Ton-zf		Hit		
A 1108	LIN-IC	TV-AFT	14-DIP	Hit		
A 1110	LIN-IC	Video-verstärker/amplifier	83	Hit	CA 3001	
A 1115	LIN-IC	Stereo-decoder	14-DIP	Hit	HA 1120	
A 1120	LIN-IC	Stereo-decoder	14-DIP	Hit	HA 1115	
A 1124	LIN-IC	TV, Ton-ZF, Nf-tr	14-DIP	Hit	→HA 1125	
A 1125	LIN-IC	TV, Ton-ZF, Nf-tr	14-DIP	Hit	AN 241, CA 3065, KA 2101, LA 1365, LM 30	
A 1126	LIN-IC	TV-AFT	14-DIP	Hit	LA 1364, M 5135, TA 7070	
A 1127(P)	LIN-IC	TV, 5x Trans.-Array, 20V, 50mA, 460MHz	14-DIC,DIP	Hit	CA 3045	pdf
A 1127 FP	LIN-IC	=HA 1127(P): SMD	14-MDIP	Hit	-	pdf
A 1137(W)	LIN-IC	FM-ZF, NF-V, Afc	16-DIP	Hit	-	
A 1138	LIN-IC	AM-V/M/O/ZF + Dem	16-DIP	Hit	-	
A 1142	LIN-IC	Stereo-decoder	14-DIP	Hit	-	
A 1144	LIN-IC	TV, Video-zf	14-DIP	Hit	-	pinout
A 1148	LIN-IC	Ctv, VA-korrektur/correction	14-DIP	Hit	-	
A 1149	LIN-IC	FM-muting	14-DIP	Hit	-	
A 1150	LIN-IC	FM-ZF	16-DIP	Hit	-	
A 1151	LIN-IC	AM-HF + Zf	14-DIP	Hit	-	

Type	Device	Short Description	Fig.	Manu	Comparision Types	More at
HA 1152	LIN-IC	Ctv, Video-zf	14-DIP	Hit	-	
HA 1154	LIN-IC	Vc, Ton-zf		Hit	-	
HA 1156(W)	LIN-IC	Stereo-decoder	14-DIP	Hit	-	
HA 1160	LIN-IC	TV, HA-synchr. + O	7-SIP	Hit	-	
HA 1166(W)	LIN-IC			Hit	-	
HA 1167	LIN-IC	ZF-amplifier, noise eliminator	DIP	Hit	-	pinout
HA 1173	LIN-IC	FM-demodulator	14-DIP	Hit	-	
HA 1190	LIN-IC	Sensor f. 4 Tasten/keys	16-DIP	Hit	SAS 560S	
HA 1194	LIN-IC	Sensor f. 4 Tasten/keys	16-DIP	Hit	SAS 570S	
HA 1196	LIN-IC	Stereo-decoder	16-DIP	Hit	-	pdf
HA 1197	LIN-IC	Am-tuner + Zf	16-DIP	Hit	-	
HA 1199	LIN-IC	Am-tuner for Car Radio	16-DIP	Hit	-	pdf pinout
HA 1201	LIN-IC	FM-ZF	8-DIP	Hit	(=HA 1211)[10]	
HA 1202	LIN-IC	FM-ZF	8-DIP	Hit	-	
HA 1203	LIN-IC	FM-ZF	8-DIP	Hit	-	
HA 1211	LIN-IC	FM-ZF	8-SIP	Hit	(=HA 1201)[10]	
HA1303	OP	Vs:±18V Vu:95dB Vo:±13V Vi0:1mV		Hit	-	data pinout
HA 1303	OP-IC		82	Hit	-	data
HA 1306(W)	LIN-IC	NF-E, 18V, 2,25A, 3,5W(13V/4Ω)	10-QIP+f	Hit	-	
HA 1314	LIN-IC			Hit	-	
HA 1317(V,W,VU,WU)	LIN-IC	NF-E, 36V, 8W(26V/8Ω)	10-DIP+g	Hit	-	
HA 1319	LIN-IC	NF-V+E, 12V, 1W(6V/4Ω)	14-DIP	Hit	-	
HA 1322(W)	LIN-IC	NF-E, 18V, 2,25A, 5,5W(13V/4Ω)	10-DIP+f	Hit	-	
HA 1324	LIN-IC	NF-E, 18V, 2,25A, 4,5W(13V/4Ω)	10-DIP+g	Hit	-	
HA 1325	LIN-IC	NF-E, 20V, 1,25A, 2W(13V/8Ω)	12-DIP+a	Hit	-	
HA 1327	LIN-IC	4-Kanal-/channel-Decoder	16-DIP	Hit	-	
HA 1328	LIN-IC	Matrix f. 4-Kanal-/channel-Decoder	16-DIP	Hit	-	
HA 1329	LIN-IC	NF-E, 9V, 1,4A, 2,5W(6V/8Ω)	12-DIP+a	Hit	-	
HA 1333	LIN-IC	CD-4-Demodulator	16-DIP	Hit	-	
HA 1334	LIN-IC	CD-4-Demodulator	16-DIP	Hit	-	
HA 1338	LIN-IC	NF-E, 33V, 4,1A, 6W(24V/8Ω)	10-QIP+f	Hit	-	pdf pinout
HA 1339(A)	LIN-IC	NF-E, 18V, 4,5A, 5,5W(13V)	10-SILP+a	Hit	-	
HA 1339(A)R	LIN-IC	=HA1339: spiegelb.Pinbel./rev.pinning	10-SILP+a	Hit	-	
HA 1342(A)	LIN-IC	NF-E, 18V, 4,5A, 5,5W(13V)	10-SILP+a	Hit	-	
HA 1342(A)R	LIN-IC	=HA1342: spiegelb.Pinbel./rev.pinning	10-SILP+a	Hit	-	
HA 1345 V	LIN-IC	NF-E, 36V, 8W(26V/8Ω)	10-SILP	Hit	-	
HA 1350	LIN-IC	NF-E, ±30V, 7,5A, 18W(±25V/8Ω)	10-SILP	Hit	HA 1370	
HA 1361	LIN-IC	NF-E, 9V, 2,25A, 1W(6V/4Ω)	12-DIP+b	Hit	-	
HA 1364	LIN-IC	TV-Ton-ZF+E, 1,5W(24V/16Ω)	12-QIP+b	Hit	-	
HA 1366 W	LIN-IC	NF-E, 18V, 4,5A, 5,5W(13V/4Ω)	10-SILP+a	Hit	-	
HA 1366 WR	LIN-IC	=HA1366W:spiegelb.Pinbel./rev.pinning	10 SILP+a	Hit	-	
HA 1367(A)	LIN-IC	Rec., NF-V+E, 9V, 2,25A, 2,2W(6V/4Ω)	20-DIP+b	Hit	-	
HA 1368	LIN-IC	NF-E, 18V, 4,5A, 5,3W(13V/4Ω)	10-SILP+a	Hit	-	
HA 1368 R	LIN-IC	=HA1368:spiegelb.Pinbel./rev.pinning	10-SILP+a	Hit	-	
HA 1370	LIN-IC	NF-E, ±30V, 7,5A, 18W(±25V/4...8Ω)	10-SILP	Hit	-	
HA 1371	LIN-IC	NF-E, 15V, 3A, 7,3W(9V/4Ω)	12-QIP+b	Hit	-	
HA 1372	LIN-IC	NF-E, 18V, 4,5A, 5,5W(13V/4Ω)	10-DIP+f	Hit	-	
HA 1374	LIN-IC	2x NF-E, 22V, 2,8A, 2x3W(15V/8Ω)	10-SILP+a	Hit	-	
HA 1374 A	LIN-IC	=HA 1374: 25V, 3,2A, 2x4W(17V/8Ω)	10-SILP+a	Hit	-	
HA 1377(A)	LIN-IC	2x NF-E, 18V, 4,5A, 2x5,8W(13V/4Ω)	12-SIL	Hit	-	
HA 1384(A)	LIN-IC	NF-E, 18V, 4A, 20W(13V/4Ω, BTL)	12-SIL	Hit	-	
HA 1385	LIN-IC	Ctv, VA-E, Ucc=110V	17/5Pin	Hit	-	
HA 1388	LIN-IC	NF-E, 18V, 4A, 18W(13V/4Ω)	12-SIL	Hit	-	
HA 1389	LIN-IC	NF-E, 30V, 3,73A, 7W(22V/8Ω)	10-SILP+a	Hit	-	
HA 1389 R	LIN-IC	=HA1389: spiegelb.Pinbel./rev.pinning	10-SILP+a	Hit	-	
HA 1392	LIN-IC	2x NF-E, 4A, 20V, 2x4,3W(12V/4Ω)	12-SIL	Hit	-	
HA 1393	LIN-IC	NF-E, 18V, 4A, 19W(13V/4Ω, BTL)	12-SIL	Hit	-	
HA 1394	LIN-IC	2x NF-E, 35V, 4,5A, 2x8,2W(25V/8Ω)	12-SIL	Hit	-	
HA 1396	LIN-IC	NF-E, 18V, 4A, 20W(13V/4Ω)	12-SIL	Hit	-	
HA 1397	LIN-IC	NF-E, ±30V, 7,5A, 20W(±22V/8Ω)	12-SIL	Hit	-	pdf
HA 1398	LIN-IC	NF-E	12-SIL	Hit	-	
HA 1403	Z-IC	Tuner-Stab., 33V, 10mA	2d	Hit	→TAA 550	
HA 1406	LIN-IC	NF-V	8-SIP	Hit	-	
HA 1451	LIN-IC	2x NF-V	14-DIP	Hit	-	
HA 1452(W)	LIN-IC	2x NF-V	14-DIP	Hit	-	
HA 1457	LIN-IC	Audio Preamplifier low noise, Ucc=±25V	8-SIP	Hit	KAA1457,ECG1666	
HA 1457W	LIN-IC	Audio Preamplifier low noise, Ucc=±25V	8-SIP	Hit	KAA1457,ECG1666	
HA 1607	LIN-IC	Monostab. Multivibrator	8-DIP	Hit	-	
HA 1806 M	KOP-IC	Dual, 18V, -30...+80°	82	Hit	-	
HA1807	2xOP	Vs:18V Vu:100dB Vi0:1mV	14D	Hit	-	data pinout
HA 1807	KOP-IC	Dual, 18V, -30...+80°	14-DIP	Hit	-	
HA 1807 M	KOP-IC	=HA 1807: Fig.→	83	Hit	-	
HA1812	OP	Vs:20V Vu:100dB Vi0:3mV	8D	Hit	-	data pinout
HA 1812(GS,PS)	KOP-IC	20V, 200mA, -20...+75°	8-DIC,DIP	Hit	-	
HA1813	OP	Vs:18V Vu:100dB Vi0:1mV	8D	Hit	-	data pinout
HA 1813(PS)	KOP-IC	18V, -20...+75°	8-DIP	Hit	-	
HA 1835 P	Z-IC	Watch Dog Timer, Reset, 5V	14-DIP	Hit	-	pdf
HA 1848 P	Z-IC	Watch Dog Timer, Reset, 5V	14-DIP	Hit	-	pdf
HA 1902	LIN-IC	2x Sensor-Verstärker/amplifier	16-DIP	Hit	-	
HA(1)-2400(-2)	OP-IC	Quad, progr, +45(±22)V, -55...+125°	16-DIC	Har	-	data pinout
HA(1)-2404(-4)	OP-IC	=HA-2400...: -25...+85°		Har	-	data pinout
HA(1)-2405(-5)	OP-IC	=HA-2400...: 0...+75°		Har	-	data pinout
HA(1)-2406(-5,-9)	OP-IC	Quad, progr, +45(±22)V, 30MHz, 20V/μs -5: 0...+75°, -9: -40...+85°	16-DIC	Har	-	data pinout
HA(3)-2406...	OP-IC		16-DIP		-	pinout

pe	Device	Short Description	Fig.	Manu	Comparision Types	More at
9P)-2406...	OP-IC	SMD	16-MDIP		-	pinout
-2420-2	LIN-IC	S&H Amp, 40V, 2.5MHz, -55...+125°C	14-DIC	Har	-	data pdf pinout
-2425-5	LIN-IC	S&H Amp, 40V, 2.5MHz, 0...+75°C	14-DIC	Har	-	data pdf pinout
-2425-5	LIN-IC	S&H Amp, 40V, 2.5MHz, 0...+75°C	14-DIP	Har	-	data pdf pinout
4,4B)-2420...	LIN-IC		20-(P)LCC		-	
1,3)-2425(-2)	LIN-IC	Sample & Hold, +40V, 2,5MHz, 0...+75°	14-DIC,DIP	Har	-	data pinout
P-2425-5	LIN-IC	S&H Amp, 40V, 2.5MHz, 0...+75°C	20-PLCC	Har	-	data pdf pinout
4,4B)-2425...	LIN-IC		20-(P)LCC		-	pinout
2)-2500(-2)	OP-IC	+40V, 12MHz, 30V/µs, -55...+125°	81	Har	-	data pinout
7)-2500...	OP-IC		8-DIC		-	pinout
2)-2502(-2)	OP-IC	+40V, 12MHz, 30V/µs, -55...+125°	81	Har	-	data pinout
7)-2502...	OP-IC		8-DIC		-	pinout
2)-2505(-5,-9)	OP-IC	+40V, 12MHz, 30V/µs -5: 0...+75°, -9: -40...+85°	81	Har	-	data pinout
3)-2505...	OP-IC		8-DIP		-	pinout
7)-2505...	OP-IC		8-DIC		-	pinout
9P)-2505...	OP-IC	SMD	8-MDIP		-	pinout
2)-2510(-2)	OP-IC	+40V, 12MHz, 65V/µs, -55...+125°	81	Har	-	data pinout
7)-2510...	OP-IC		8-DIC		-	pinout
2)-2512(-2)	OP-IC	+40V, 12MHz, 60V/µs, -55...+125°	81	Har	-	data pinout
7)-2512...	OP-IC		8-DIC		-	pinout
2)-2515(-5,-9)	OP-IC	+40V, 12MHz, 60V/µs -5: 0...+75°, -9: -40...+85°	81	Har	-	data pinout
3)-2515...	OP-IC		8-DIP		-	pinout
7)-2515...	OP-IC		8-DIC		-	pinout
9P)-2515...	OP-IC	SMD	8-MDIP		-	pinout
2)-2520(-2)	OP-IC	+40V, 20MHz, 120V/µs, -55...+125°	81	Har	-	data pinout
7)-2520...	OP-IC		8-DIC		-	pinout
2)-2522(-2)	OP-IC	+40V, 20MHz, 120V/µs, -55...+125°	81	Har	-	data pinout
7)-2522...	OP-IC		8-DIC		-	pinout
2)-2525(-5,-9)	OP-IC	+40V, 20MHz, 120V/µs -5: 0...+75°, -9: -40...+85°	81	Har	-	data pinout
3)-2525...	OP-IC		8-DIP		-	pinout
7)-2525...	OP-IC		8-DIC		-	pinout
9P)-2525...	OP-IC	SMD	8-MDIP		-	pinout
2)-2529(-2,-5)	OP-IC	hi-slew, +40V, 20MHz, 150V/µs -2: -55...+125°, -5: 0...+75°	81	Har	-	data pinout
3)-2529...	OP-IC		8-DIP		-	pinout
7)-2529...	OP-IC		8-DIC		-	pinout
9P)-2529...	OP-IC	SMD	8-MDIP		-	pinout
2530	OP	Vs:40V Vo:±12V Vi0:0.8mV f:70Mc	8A	Har	-	data pinout
2535	OP	Vs:40V Vo:±12V Vi0:0.8mV f:70Mc	8A	Har	-	data pinout
1)-2539(C)(-...)	OP-IC	Wideband, +35(±17, 5), 600MHz, 600V/µs - 2: - 55... +125°, - 5: 0... +75°, - 9: - 40... +85°	14-DIC	Har	-	data pinout
3)-2539...	OP-IC		8-DIP		-	pinout
9P)-2539...	OP-IC	SMD	8-MDIP		-	pinout
1)-2540(C)(-...)	OP-IC	Wideband, +35(±17,5)V, 400MHz, 400V/µs - 2: - 55... +125°, - 5: 0... +75°, - 9: - 40... +85°	14-DIC	Har	-	data pinout
3)-2540...	OP-IC		8-DIP		-	pinout
9P)-2540...	OP-IC	SMD	8-MDIP		-	pinout
1)-2541(-2,-5)	OP-IC	Wideband, +35(±17,5)V, 40MHz, 250V/µs -2: -55...+125°, -5: 0...+75°	14-DIC	Har	-	data pinout
2)-2541...	OP-IC		83		-	pinout
1)-2542(-2,-5)	OP-IC	Wideband, +35(±17,5)V, 70MHz, 300V/µs -2: -55...+125°, -5: 0...+75°	14-DIC	Har	-	data pinout
2)-2542...	OP-IC		83		-	pinout
3)-2542...	OP-IC		14-DIP		-	pinout
(2)-2544(C)(-...)	OP-IC	Video, +33V, 50MHz, 150V/µs -2: -55...+125°, -5: 0...+75°, -9: -40...+85°	81	Har	-	data pinout
3)-2544...	OP-IC		8-DIP		-	pinout
7)-2544...	OP-IC		8-DIC		-	pinout
9P)-2544...	OP-IC	SMD	8-MDIP		-	pinout
1)-2546(-5,-9)	LIN-IC	Wideband Multiplier, 35V, 300V/µs -5: 0...+75°, -9: -40...+85°	16-DIC	Har	-	
1)-2547(-5,-9)	LIN-IC	Wideband Multiplier, 35V, 100MHz -5: 0...+75°, -9: -40...+85°	16-DIC	Har	-	
(2)-2548(A)(-...)	OP-IC	Wideband, +40(±20)V, 150MHz, 120V/µs -5: 0...+75°, -9: -40...+85°	81	Har	-	data pinout
7)-2548...	OP-IC		8-DIC		-	pinout
(2)-2600(-2)	OP-IC	+45(±22)V, 12MHz, 7V/µs, -55...+125°	81	Har	-	data pinout
(7)-2600...	OP-IC		8-DIC		-	pinout
(2)-2602(-2)	OP-IC	+45(±22)V, 12MHz, 7V/µs, -55...+125°	81	Har	-	data pinout
(7)-2602...	OP-IC		8-DIC		-	pinout
(2)-2605(-5,-9)	OP-IC	+45(±22)V, 12MHz, 7V/µs -5: 0...+75°, -9: -40...+85°	81	Har	-	data pinout
(3)-2605...	OP-IC		8-DIP		-	pinout
(7)-2605...	OP-IC		8-DIC		-	pinout
(9P)-2605...	OP-IC	SMD	8-MDIP		-	pinout
(2)-2620(-2)	OP-IC	+45(±22)V, 100MHz, 35V/µs, -55...+125°	81	Har	-	data pinout
(7)-2620...	OP-IC		8-DIC		-	pinout
(2)-2622(-2)	OP-IC	+45(±22)V, 100MHz, 35V/µs, -55...+125°	81	Har	- - -	data pinout
(7)-2622...	OP-IC		8-DIC		-	pinout
(2)-2625(-2)	OP-IC	+45(±22)V, 100MHz, 35V/µs -5: 0...+75°,	81	Har	-	data pinout

Type	Device	Short Description	Fig.	Manu	Comparision Types	More at
		-9: -40...+85°				
HA(3)-2625...	OP-IC		8-DIP		-	pinout
HA(7)-2625...	OP-IC		8-DIC		-	pinout
HA(9P)-2625...	OP-IC	SMD	8-MDIP		-	pinout
HA(2)-2640(-2)	OP-IC	+100(±50)V, 4MHz, 5V/µs, -55...+125°	81	Har	-	data pinout
HA(7)-2640...	OP-IC		8-DIC		-	pinout
HA(2)-2645(-5)	OP-IC	+100(±50)V, 4MHz, 5V/µs, 0...+70°	81	Har	-	data pinout
HA(7)-2645...	OP-IC		8-DIC		-	pinout
HA(1)-2650(-2)	OP-IC	Dual, +40(±20)V, 5V/µs, -55...+125°	14-DIC	Har	-	data pinout
HA(2)-2650...	OP-IC		81		-	pinout
HA(1)-2655(-5)	OP-IC	Dual, +40(±20)V, 5V/µs, 0...+70°	14-DIC	Har	-	data pinout
HA(2)-2655...	OP-IC		81		-	pinout
HA(1)-2720(-2)	OP-IC	progr, +45(±22)V, 10MHz, -55...+125°	8-DIC	Har	-	data pinout
HA(2)-2720...	OP-IC		81		-	pinout
HA(1)-2725(-5)	OP-IC	progr, +45(±22)V, 10MHz, 0...+70°	8-DIC	Har	-	data
HA(2)-2725...	OP-IC		81		-	pinout
HA-2820	IC	Phase Locked Loop	DIP	Har	-	pdf pinout
HA4201B	ARRAY4	Vs:±4.5...±5.5V Ron:Ohm at 5/-5V	8S	Isi	GX4404	data pdf pinout
HA4314B	ARRAY4	Vs:±4.5...±5.5V Ron:Ohm at 5/0V	16S,14SD	Isi		data pdf pinout
HA4344B	ARRAY4	Vs:±4.5...±5.5V Ron:Ohm at 5/-5V	16S	Isi	GX4404	data pdf pinout
HA4404B	ARRAY4	Vs:±4.5...±5.5V Ron:Ohm at 5/-5V	16S	Isi	GX4404	data pdf pinout
HA4600C	VIDEO-BUFF	Vs:±6V Vo:±2.8V f:250Mc	8SD	Har	GB4600	data pdf pinout
HA(1)-4741(-...)	OP-IC	Quad, +40(±20)V, 3,5MHz, 1,6V/µs -2:-55...+125°, -5:0...+75°,-9:-40...+85°	14-DIC	Har	→Serie 124	data pinout
HA(3)-4741...	OP-IC		14-DIP			pinout
HA(9P)-4741...	OP-IC	SMD	16-MDIP		-	pinout
HA(1)-4900(-2)	KOP-IC	+33(±16)V, <200ns, -55...+125°	16-DIC	Har	-	data pinout
HA(1)-4902(-2)	KOP-IC	+33(±16)V, <200ns, -55...+125°	16-DIC	Har	-	data pinout
HA(1)-4905(-2)	KOP-IC	+33(±16)V, <200ns, 0...+75°	16-DIC	Har	-	data pinout
HA(2)-5002(-...)	LIN-IC	Wideband Buffer, 44V, ±0,2A, 1300V/µs -2:-55...+125°, -5:0...+75°,-9:-40...+85°	81	Har	-	data pinout
HA(3)-5002...	LIN-IC		8-DIP		-	pinout
HA(7)-5002...	LIN-IC		8-DIC		-	pinout
HA(9P)-5002...	LIN-IC	SMD	8-MDIP		-	pinout
HA(1)-5004(-5,-9)	OP-IC	Current Feedback Amp., ±22V, 1200V/µs -5: 0...+75°, -9: -40...+85°	14-DIC	Har	-	data pinout
HA5013	3xVIDEO-AMP	Vs:±18V Vo:±2.5V Vi0:0.8mV f:28Mkc	d,14SD	Har		data pinout
HA5022	2xFB-AMP	Vs:±18V Vo:±3V Vi0:0.8mV f:125Mc	d,16SD	Har		data pdf pinout
HA5022MJ/883	2xFB-AMP	Vs:±18V Vo:±2.5V Vi0:±3mV f:125Mc	16D	Har		data pinout
HA5023	2xFB-AMP	Vs:±18V Vo:±3V Vi0:0.8mV f:125Mc	d,8SD	Har		data pdf pinout
HA5023MJ/883	2xFB-AMP	Vs:±18V Vo:±2.5V Vi0:±3mV f:125Mc	8D	Har		data pinout
HA5024	4xFB-AMP	Vs:±18V Vo:±3V Vi0:0.8mV f:125Mc	d,20SD	Har		data pinout
HA5025	4xFB-AMP	Vs:±18V Vo:±3V Vi0:0.8mV f:125Mc	d,14SD	Har		data pdf pinout
HA(2)-5033(-...)	LIN-IC	Video Buffer, 40V, 250MHz, 1300V/µs -2:-55...+125°, -5:0...+75°,-9:-40...+85°	83	Har	-	pinout
HA(3)-5033...	LIN-IC		8-DIP		-	pinout
HA(9P)-5033...	LIN-IC	SMD	8-MDIP			pinout
HA(2)-5101(-...)	OP-IC	lo-noise, +40(±20)V, 10MHz, 10V/µs -2:-55...+125°, -5:0...+75°,-9:-40...+85°	81	Har	-	data pinout
HA(3)-5101...	OP-IC		8-DIP		-	pinout
HA(7)-5101...	OP-IC		8-DIC		-	pinout
HA(9P)-5101...	OP-IC	SMD	8-MDIP		-	pinout
HA(2)-5102(-...)	OP-IC	lo-noise, +40(±20)V, 8MHz, 3V/µs -2:-55...+125°, -5:0...+75°,-9:-40...+85°	81	Har	-	data pinout
HA(3)-5102...	OP-IC		8-DIP		-	pinout
HA(7)-5102...	OP-IC		8-DIC		-	pinout
HA(9P)-5102...	OP-IC	SMD	16-MDIP		-	pinout
HA(1)-5104(-...)	OP-IC	=HA(2)-5102...: Quad	14-DIC	Har	-	data pinout
HA(3)-5104(-...)	OP-IC		14-DIP		-	pinout
HA(9P)-5104...	OP-IC	SMD	16-MDIP		-	pinout
HA(2)-5111(-...)	OP-IC	lo-noise, +40(±20)V, 100MHz, 50V/µs -2:-55...+125°, -5:0...+75°,-9:-40...+85°	81	Har	-	data pinout
HA(3)-5111...	OP-IC		8-DIP		-	pinout
HA(7)-5111...	OP-IC		8-DIC		-	pinout
HA(9P)-5111...	OP-IC	SMD	8-MDIP		-	pinout
HA(2)-5112(-...)	OP-IC	lo-noise, +40(±20)V, 60MHz, 20V/µs -2:-55...+125°, -5:0...+75°,-9:-40...+85°	81	Har	-	data pinout
HA(3)-5112...	OP-IC		8-DIP		-	pinout
HA(7)-5112...	OP-IC		8-DIC		-	pinout
HA(9P)-5112...	OP-IC	SMD	16-MDIP		-	pinout
HA(1)-5114(-...)	OP-IC	=HA(2)-5104...: Quad	14-DIC	Har	-	data pinout
HA(3)-5114	OP-IC		14-DIP		-	pinout
HA(9P)-5114...	OP-IC	SMD	16-MDIP		-	pinout
HA(2)-5127(A)(-...)	OP-IC	lo-noise, ±22V, 8,5MHz, 10V/µs -2: -55...+125° -5: 0...+75°	81	Har	-	data pinout
HA(7)-5127...	OP-IC		8-DIC		-	pinout
HA(2)-5130(-2,-5)	OP-IC	hi-prec, +40V, 2,5MHz, 0,8V/µs -2: -55...+125°, -5: 0...+75°	81	Har	-	data pinout
HA(7)-5130...	OP-IC		8-DIC		-	pinout
HA(1)-5134(A)(-...)	OP-IC	4x hi-prec, +40V, 4MHz, 1V/µs -2: -55...+125°, -5: 0...+75°	81	Har	-	data pinout
HA(2)-5135(-2,-5)	OP-IC	hi-prec, +40V, 2,5MHz, 0,8V/µs	81	Har	-	data pinout
HA(7)-5135...	OP-IC		8-DIC		-	pinout
HA(2)-5137(A)(-...)	OP-IC	lo-noise, ±22V, 63MHz, 20V/µs -2: -55...+125°, -5: 0...+75°	81	Har	-	data pinout

Type	Device	Short Description	Fig.	Manu	Comparision Types	More at
HA(7)-5137...	OP-IC		8-DIC		-	pinout
HA(1)-5141(-..)	OP-IC	lo-power, +35V, 0,4MHz, 1,5V/µs -2:-55...+125°, -5:0...+75°,-9:-40...+85°	8-DIC	Har	-	data
HA(2)-5141...	OP-IC		81		-	pinout
HA(3)-5141...	OP-IC		8-DIP		-	pinout
HA(2)-5142(-..)	OP-IC	=HA(1)5141...: Dual	81	Har	-	data pinout
HA(3)-5142...	OP-IC		8-DIP		-	pinout
HA(7)-5142...	OP-IC		8-DIC		-	pinout
HA(1)-5144(-..)	OP-IC	=HA(1)5141...: Quad	14-DIC	Har	-	data pinout
HA(3)-5144...	OP-IC		14-DIP		-	pinout
HA(2)-5147(A)(-..)	OP-IC	lo-noise, ±22V, 120MHz, 35V/µs -2:-55...+125°, -5: 0...+75°	81	Har	-	data pinout
HA(7)-5147...	OP-IC		8-DIC		-	pinout
HA(2)-5151(-..)	OP-IC	lo-power, +35V, 1,3MHz, 6V/µs	81	Har	-	
HA(3)-5151...	OP-IC		8-DIP		-	
HA(7)-5151...	OP-IC		8-DIC		-	
HA(2)-5152(-..)	OP-IC	=HA(2)5151...: Dual	81	Har	-	
HA(3)-5152...	OP-IC		8-DIP		-	
HA(7)-5152...	OP-IC		8-DIC		-	
HA(1)-5154(-..)	OP-IC	=HA(2)5151...: Quad	14-DIC	Har	-	
HA(3)-5154...	OP-IC		8-DIP		-	
HA(2)-5160(-2,-5)	OP-IC	JFET Inp., +40V, 100MHz, 120V/µs -2:-55...+125°, -5: 0...+75°	81	Har	-	data pinout
HA(2)-5162(-5)	OP-IC	JFET Inp., +40V, 100MHz, 70V/µs	81	Har	-	data pinout
HA(2)-5170(-2,-5)	OP-IC	JFET Inp., +44V, 8MHz, 8V/µs - 2: -55...+125°, - 5: 0...+75°	81	Har	-	data pinout
HA(7)-5170...	OP-IC		8-DIC		-	pinout
HA(2)-5177(A)(-..)	OP-IC	lo-offset, +44(±22)V, 2MHz, Offs<0,1mV -2:-55...+125°,-5: 0...+75°, A:Offs<60µV	81	Har	-	data pinout
HA(7)-5177...	OP-IC		8-DIC		-	pinout
HA(2)-5180(-2,-5)	OP-IC	lo-power, JFET Inp., +40V, 2MHz, 7V/µs -2: -55...+125°, -5: 0...+75°	81	Har	-	data pinout
HA(7)-5180...	OP-IC		8-DIC		-	pinout
HA(1)-5190(-2)	OP-IC	Wideband, +35V, 150MHz, -55...+125°	14-DIC	Har	-	data pinout
HA(2)-5190...	OP-IC		83		-	pinout
HA(1)-5195(-2)	OP-IC	=HA(1)-5190: 0...75°	14-DIC	Har	-	data pinout
HA(2)-5195...	OP-IC		83		-	pinout
HA(9P)-5195...	OP-IC	SMD	14-MDIP		-	pinout
HA(2)-5221(-5,-9)	OP-IC	lo-noise, +35V, 100MHz, 25V/µs -5: 0...+75°, -9: -40...+85°	81	Har	-	data pinout
HA(7)-5221...	OP-IC		8-DIC		-	pinout
HA(7)-5227(-5,-9)	OP-IC	=HA(2)-5221...: Dual	8-DIC	Har	-	data
HA5232	2x HI-PREC	Vs:±36V Vi0:<0.1mV f:<0.8Mc	16S,8D	Har	-	data pinout
HA5234	4xHI-PREC	Vs:±36V Vi0:<0.1mV f:<0.8Mc	16S,14D	Har	-	data pinout
HA(1)-5320(-2,-8)	LIN-IC	Sample & Hold, 40V, 2MHz, -55...+125°	14-DIC	Har	-	pinout
HA(4)-5320...	LIN-IC		20-(P)LCC		-	
HA(1)-5330(-..)	LIN-IC	Sample & Hold, ±20V, 1,4MHz -2:-55...+125°, -4:-25...+85°,-5:0...+70°	14-DIC	Har	-	pinout
HA(4)-5330...	LIN-IC		20-(P)LCC		-	
HA(1)-5340(-..)	LIN-IC	Sample & Hold, 36V, 10MHz -5: 0...+75°, -9: -40...+85°	14-DIC	Har	-	pinout
HA5351I	SAMP&HOLD	Vs:11V Vu:108dB Vi0:±2mV f:40Mc	8SD	Har	-	data pdf pinout
HA 11107	IC			Hit		
HA 11120	LIN-IC	Am-tuner + AM/FM-ZF	16/20-DIP	Hit		
HA 11122	LIN-IC	2x Recorder-A/W-Verst./amplifier	16-DIP	Hit		
HA 11123(W)	LIN-IC	AM/FM-ZF + Afc + Agc	16/20-DIP	Hit		
HA 11211	LIN-IC	Am-tuner + AM/FM-ZF	16/20-DIP	Hit		
HA 11215(A)	LIN-IC	Ctv, Zf + Video + Agc	24-DIP	Hit		
HA 11218	LIN-IC	TV, Ha/va-synchr. + Osc.	16/20-DIP	Hit		
HA 11219	LIN-IC	FM-noise-suppr.	16-DIP	Hit		
HA 11220	LIN-IC	Ctv, Video-zf	22-DIP	Hit		
HA 11221	LIN-IC	TV, Video-zf	16-DIP	Hit		
HA 11222	LIN-IC	Ctv, NTSC-Chroma	28-DIP	Hit		
HA 11223(W)	LIN-IC	Stereo-decoder	16-DIP	Hit		
HA 11225	LIN-IC	FM-ZF	16-DIP	Hit		
HA 11226	LIN-IC	2x Dolby B System	16/20-DIP	Hit		
HA 11226 MP	LIN-IC	=HA 11226: SMD	18-MP	Hit		
HA 11227	LIN-IC	Stereo-decoder	16-DIP	Hit	AN7410, BA1330, KA2261, LA3361, TA7604	
HA 11229	LIN-IC	Tv-ton-zf	14-DIP	Hit	-	
HA 11235	LIN-IC	Ctv, HA-/VA-synchr. + Osc.	16/20-DIP	Hit	-	pdf
HA 11236	LIN-IC			Hit		
HA 11238	LIN-IC	Ctv, Video-zf	22-DIP	Hit	-	
HA 11244	LIN-IC	TV, Synchr.	16-DIP	Hit	-	pdf
HA 11247	LIN-IC	Ctv, Ntsc-chroma	16/20-DIP	Hit	-	pdf
HA 11247 MP	LIN-IC	=HA 11247: SMD	18-MP	Hit	-	pdf
HA 11251	LIN-IC	FM/AM Tuner, Zf + Demod.	16-DIP	Hit	NTE1490	pdf
HA 11401	LIN-IC	TV, Video	16-DIP	Hit	-	
HA 11405	LIN-IC	Ctv, Video-zf	16-DIP	Hit	-	
HA 11408 A	LIN-IC	TV, Video	28-DIP	Hit	-	
HA 11409	LIN-IC	Ctv, NTSC-VIR	16-DIP	Hit	-	
HA 11410	LIN-IC	TV	28-DIP	Hit	-	
HA 11412 A	LIN-IC	Ctv, Ntsc-chroma	28-DIP	Hit	-	pdf
HA 11414	LIN-IC	TV, Synchr.	16-DIP	Hit	-	
HA 11417	LIN-IC	Ctv, Ntsc-chroma + Demod	24-DIP	Hit	-	
HA 11423	LIN-IC	Ctv, Synchr., Blanking	16/20-DIP	Hit	-	pdf

Type	Device	Short Description	Fig.	Manu	Comparision Types	More at
HA 11423 MP	LIN-IC	=HA 11423: SMD	18-MP	Hit	-	
HA 11431	LIN-IC	Ctv, Luminanz/chroma System (ntsc)	28-DIP	Hit	-	
HA 11431 NT	LIN-IC	=HA 11431: Fig.→	30-SDIP	Hit	-	
HA 11433	LIN-IC	TV-stereo	16-DIP	Hit	-	
HA 11436 A	LIN-IC	Ctv, Luminanz/chroma-system (ntsc)	28-DIP	Hit	-	
HA 11440(A)	LIN-IC	Ctv, Video-ZF-System, Iagc<0,5mA	16-DIP	Hit	-	
HA 11441	LIN-IC	TV, HA-/VA-Osc.+synchr., Blanking	16/20-DIP	Hit	-	
HA 11441 MP	LIN-IC	=HA 11441: SMD	18-MP	Hit	-	
HA 11442(A)	LIN-IC	=HA11440(A): Iagc<1,5mA	16-DIP	Hit	-	
HA 11443	LIN-IC	TV, Video-/Ton-ZF	28-DIP	Hit	-	
HA 11445	LIN-IC	TV, Video-/ton-zf	28-DIP	Hit	-	
HA 11465 A	LIN-IC	Ctv, Video-Verst./ampl., Synchr.	16/20-DIP	Hit	-	pdf
HA 11466 S	LIN-IC	HF-modulator f. Vc, µcomp	16-DIP	Hit	-	
HA 11476 NT	LIN-IC	Tv/vc, Video-/Ton-ZF, Video-signal	30-SDIP	Hit	-	
HA 11477 NT	LIN-IC	Video-/audio-umschalter/switch	30-SDIP	Hit	-	
HA 11485 BNT	LIN-IC	Tv/vc, Video-/Ton-ZF, FM-Dem.	30-SDIP	Hit	-	
HA 11498	LIN-IC	Digital Video-prozessor	42-DIP	Hit	-	
HA 11505	LIN-IC	Video-Verst./amplifier	24-SDIP+b	Hit	-	
HA 11508	LIN-IC	Ctv, Video-/audio-umschalter/switch	20-DIP	Hit	-	pdf pinout
HA 11510 NT	LIN-IC	Ctv/vc, FB-interface	30-SDIP	Hit	-	
HA 11511 CNT	LIN-IC	Ctv, Video-/chroma-prozessor (ntsc)	42-SDIP	Hit	-	
HA 11513	LIN-IC	5Kanal/Channel Video-Umschalter/Switch	16-DIP	Hit	-	pdf
HA 11517 BNT	LIN-IC	TV, Ha-/va-osc.+synchr., Blanking	30-SDIP	Hit	-	pdf
HA 11525 MP	LIN-IC	SMD, Vc, Digital Signal-proz. (ntsc)	28-MP	Hit	-	
HA 11530 MP	LIN-IC	SMD, Vc, Digital Signal-proz. (ntsc)	28-MP	Hit	-	
HA 11531 NT	LIN-IC	Vc, Video-/chroma-prozessor (ntsc)	42-SDIP	Hit	-	
HA 11532 MP	LIN-IC	SMD,Vc, Digital-chroma-prozessor(ntsc)	28-MP	Hit	-	
HA 11533 NT	LIN-IC	RGB-Video-Verst./amplifier	24-SDIP+b	Hit	-	pdf
HA 11535 MP	LIN-IC	SMD,Vc, Chroma-/ha-/va-prozessor (pal)	28-MP	Hit	-	
HA 11536 MP	LIN-IC	SMD, ~HA 11535: R-Y-, B-Y-Signal (pal)	28-MP	Hit	-	
HA 11539 NT	LIN-IC	Ctv/vc, Tuning-system Interface	22-SDIP	Hit	-	pdf
HA 11544	LIN-IC	Multiplex Video-Switch	16-DIP	Hit	-	
HA 11545 A	LIN-IC	FM, Video Hf-modulator	16-DIP	Hit	-	
HA 11556 NT	LIN-IC	Ctv/vc, Video-ZF-System	22-SDIP	Hit	-	
HA 11559 NT	LIN-IC	Ctv, Luminanz-signal Proz., Delay Line	30-SDIP	Hit	-	
HA 11560 FP	LIN-IC	SMD, Vc, Hf-modulator	16-MDIP	Hit	-	pdf
HA 11560 T	LIN-IC	SMD, Vc, RF/HF-Modulator	16-SSMDIP	Hit	-	pdf
HA 11561	LIN-IC	Ctv, Color-signal Matrix	16-DIP	Hit	-	
HA 11566 NT	LIN-IC	Ctv/vc, Video-/ton-zf f. SAW-filter	22-SDIP	Hit	-	
HA 11567 MP	LIN-IC	SMD,Vc, Chroma-/ha-/va-prozessor (pal)	28-MP	Hit	-	
HA 11569 FS	LIN-IC	SMD, Ctv/vc, PIP-prozessor (ntsc)	80-MP	Hit	-	
HA 11571 AF,BF	LIN-IC	SMD,Sat-TV, BS Tuner PLL, Agc, Afc, Zf	28-MP/4TAB	Hit	-	
HA 11575 F	LIN-IC	SMD,Sat-tv, BS Tuner PLL, Agc, Afc, Zf	28-MP/4TAB	Hit	-	
HA 11579	LIN-IC	SMD, Ctv, Vc, Pip-prozessor (ntsc)	56-MP	Hit	-	
HA 11580	LIN-IC	Ctv, Chroma-signal	24-DIP	Hit	-	
HA 11701	LIN-IC	Vc, Fm-signal	16/20-DIP	Hit	-	
HA 11702	LIN-IC	Vc, Kopfvorverst./head preampl.	16/20-DIP	Hit	-	
HA 11703	LIN-IC	Vc, Fm-signal	20-DIP	Hit	-	
HA 11704	LIN-IC	Vc, Color-signal	16/20-DIP	Hit	-	
HA 11705	LIN-IC	Vc, Color-afc	16/20-DIP	Hit	-	
HA 11706	LIN-IC	Vc, Color-apc	16/20-DIP	Hit	-	
HA 11707	LIN-IC	Vc, Servo	28-DIP	Hit	-	
HA 11710	LIN-IC	Vc, Chroma	16/20-DIP	Hit	-	
HA 11711	LIN-IC	Vc, Servo Controller	28-DIP	Hit	-	
HA 11712 A	LIN-IC	Vc, NTSC, Vir	16-DIP	Hit	-	
HA 11713	LIN-IC	Vc, Servo	16-DIP	Hit	-	
HA 11714	LIN-IC	Vc, Servo	16/20-DIP	Hit	-	
HA 11715	LIN-IC	Vc, Servo	12-QIP+a	Hit	-	
HA 11716	LIN-IC	Vc, Lumin.-signal	20-DIP	Hit	-	
HA 11717	LIN-IC	Vc, Color-afc	16/20-DIP	Hit	-	
HA 11718	LIN-IC	Vc, Lumin.-signal	16-DIP	Hit	-	
HA 11719	LIN-IC	Camera	16/20-DIP	Hit	-	
HA 11720	LIN-IC	Camera	16/20-DIP	Hit	-	
HA 11722	LIN-IC	VC-Servo	14-DIP	Hit	-	
HA 11724	LIN-IC	Vc, A/w-verst.	28-DIP	Hit	KA 2984	
HA 11725	LIN-IC	Vc, Video-signal	28-DIP	Hit	KA 2985	
HA 11726	LIN-IC	Vc, Chroma-signal	28-DIP	Hit	KA 2986	
HA 11727	LIN-IC	Vc, Chroma-signal	28-DIP	Hit	KA 2987	
HA 11732	LIN-IC	Camera	16-DIP	Hit	-	
HA 11738	LIN-IC			Hit	-	
HA 11741	LIN-IC	Vc, Servo-controller	28-DIP	Hit	KA 2988, µPC 1536C	
HA 11744	LIN-IC	Vc, Video-a/w-verst./amplifier	28-DIP	Hit	KA 2944, µPC 1534C	
HA 11744 NT	LIN-IC	=HA 11744: Fig.→	30-SDIP	Hit	-	
HA 11745	LIN-IC	Vc, Video-verst./amplifier	28-DIP	Hit	KA 2945, µPC 1524A	
HA 11745 NT	LIN-IC	=HA 11745: Fig.→	30-SDIP	Hit	-	
HA 11752	LIN-IC	=HA 11772: + FM-AGC	16-DIP	Hit	-	
HA 11770 AMP	LIN-IC	SMD, Camera, Pal/secam Prozessor	44-MP	Hit	-	
HA 11772	LIN-IC	Vc, 2-Kanal/Channel Vorverst./Pre-Amp.	16-DIP	Hit	HA 11752	
HA 11776 AMP	LIN-IC	SMD, Camera, Encoder	28-MP	Hit	-	
HA 11787	LIN-IC	Vc, Drop Out Compensator	16-DIP	Hit	-	
HA 11844 BMP	LIN-IC	SMD, Camera, Weiss/white-balance	28-MP	Hit	-	
HA 11856 ANT	LIN-IC	Vc, Color-signal-prozessor (ntsc)	42-SDIP	Hit	-	
HA 11864 MP	LIN-IC	SMD, Vc, VHS-C, FM-signal-prozessor	44-MP	Hit	-	
HA 11870 ANT	LIN-IC	Vc, 4-Kanal/Channel Vorverst./Pre-Amp.	30-SDIP	Hit	-	
HA 11876 MP	LIN-IC	=HA 11876NT: SMD	44-MP	Hit	-	

Type	Device	Short Description	Fig.	Manu	Comparision Types	More at
11876 NT	LIN-IC	Vc, Vhs-c,4K./Chan.-Vorverst./Pre-Amp.	42-SDIP	Hit	-	
11882 AMP	LIN-IC	SMD, Camera, 4-Kanal/Channel-Prozessor	44-MP	Hit	-	
11883 MP	LIN-IC	SMD, Camera, Encoder	18-MP	Hit	-	
12001 W	LIN-IC	Recorder-steuerung/mechan. control	22-DIP	Hit	-	
12002(W)	LIN-IC	Lautspr.-schutz/speaker Protection	8-SIP	Hit	-	
12003	IC			Hit		
12005	LIN-IC	NF-V-ra, Dolby	16-DIP	Hit	-	pdf
12006	LIN-IC	Recorder-A/W-Verst./amplifier	16/20-DIP	Hit	-	pdf
12009	LIN-IC	7-Segm.-Decoder f. Digital Tuning	42-DIP	Hit	-	
12010	LIN-IC	FLT-treiber/driver	16-DIP	Hit	-	
12016	LIN-IC	PLL FM MPX Stereo-Decoder, Ucc=13V	16-DIP	Hit	-	
12017	OP	Vs:±26.5V Vu:105dB Vo:14.7/0V	8I	Hit	-	data pdf pinout
12017	LIN-IC	NF-V, ra, Ucc=±24V	8-SIP	Hit	-	
12018	LIN-IC	PLL FM MPX Stereo-decoder	16-SQP	Hit		
12019	LIN-IC	Recorder, Pegel-anz./level meter, log.	16-DIP	Hit		
12020	LIN-IC	Recorder-Equalizer	28-DIP	Hit		
12022	LIN-IC	Dual Amp., Ucc=22,5V	16/20-DIP	Hit		
12024	LIN-IC	Recorder	16/20-DIP	Hit		
12026	LIN-IC	PLL FM MPX Stereo-decoder	16-DIP	Hit		
12027	LIN-IC	Recorder, Dolby B System	16/20-DIP	Hit		
12028	LIN-IC	Recorder, Verstärker/amplifier	22-DIP	Hit		
12029	LIN-IC	SMD, 8 Bit µComp-Interface	18-MP	Hit		
12030	LIN-IC	8 Bit µComp-Interface	16-DIP	Hit		
12031	LIN-IC	PLL FM MPX Stereo-decoder	28-DIP	Hit		
12032	LIN-IC	Anzeigetreiber/meter driver	18-DIP	Hit		
12038	IC	SMD, Dolby C Noise reduction		Hit		pdf pinout
12062AMP	IC	R-dat, Data Strobe IC, PLL clock gen., ASPC	MP	Hit	-	pdf pinout
12075 MP	LIN-IC	SMD, 8mm-VC, Fm-prozessor	44-MP	Hit	-	
12093 MP	LIN-IC	=HA 12093NT: SMD	44-MP	Hit		
12093 NT	LIN-IC	Vc, VHS-HiFi-FM	42-SDIP	Hit		
12095 NT	LIN-IC	Cd, Vorverst./preamp., Servo Amp.	42-SDIP	Hit		
12096 NT	D/A-IC	16 Bit, Digital Audio, Ucc=±5V	22-SDIP	Hit		
12108 MP	D/A-IC	=HA 12108NT: SMD	44-MP	Hit		
12108 NT	D/A-IC	16 Bit, Stereo, Digital Audio, Ucc=5V	22-SDIP	Hit		
12127 ANT	LIN-IC	Stereo FM Demodulation, Cx NR System	56-SDIP	Hit		
12132 MP	A/D-D/A-IC	SMD, Dual, 16 Bit, Dig. Audio (R-DAT)	44-MP	Hit		
12133 MP	LIN-IC	SMD, R-dat, Dual, Rec/play Amp, Equal.	44-MP	Hit		
12134 A	LIN-IC	Dual Dolby-B Nr System, Ucc=6,5...16V	16-DIP	Hit		pdf
12135 A	LIN-IC	Dual Dolby-B Nr System, Ucc=8...16V	16-DIP	Hit		pdf
12136 A	LIN-IC	Dual Dolby-B Nr System, Ucc=9,5...16V	16-DIP	Hit		pdf
12136 AF	LIN-IC	=HA 12134...12136: SMD	16-MDIP	Hit		pdf
A 12141 NT	LIN-IC	Dual Dolby-B/C Nr System, Ucc=7,5...16V	30-SDIP	Hit		pdf
A 12142 NT	LIN-IC	Dual Dolby-B/C Nr System, Ucc=9,5...16V	30-SDIP	Hit		pdf
A 12144 FP	LIN-IC	FM-M/O/ZF, Agc, f. Auto-/car Radio	16-MDIP	Hit		
A 12151 MA	LIN-IC	Dual Dolby-B/C Nr System, Ucc=7,5...16V	44-MP	Hit		
A 12153 MA	LIN-IC	Dual Dolby-B/C Nr System, Ucc=9,5...16V	44-MP	Hit		
A 12155 NT	LIN-IC	R-dat, Rec/play Amp, Equal., Ucc=5V	56-MP	Hit		
A 12155 NT	LIN-IC	Recorder, 2x Signal Prozessor, Dolby	64-SDIP	Hit		pdf
A 12156 MA	LIN-IC	SMD, Cd, Vorverst./preamp., Tracking	44-MP	Hit		
A 12157 NT	LIN-IC	Recorder, Audio Prozessor, Dolby-B/C	64-SDIP	Hit		pdf
A 12158	LIN-IC	=HA 12158NT: SMD	56-MP	Hit		
A 12158 NT	LIN-IC	Cd, Servo System, Ucc=5V	56-SDIP	Hit		
A 12160	LIN-IC	SMD,Recorder, Audio Prozessor, Dolby-B	56-MP	Hit		pdf
A 12161 FP	LIN-IC	Dual Dolby-B/C Nr System, Ucc=7,5...16V	28-MDIP	Hit		pdf
A 12162 FP	LIN-IC	Dual Dolby-B/C Nr System, Ucc=9,5...16V	28-MDIP	Hit		pdf
A 12163	LIN-IC	SMD,Recorder, Audio Prozessor, Dolby-B	56-MP	Hit		pdf
A 12164	LIN-IC	SMD, Recorder, Audio Proz., Dolby-B	56-MP	Hit		pdf
A 12165	LIN-IC	SMD,Recorder, Audio Prozessor, Dolby-B	56-MP	Hit		pdf
A 12166 F	LIN-IC	SMD,Recorder, Audio Prozessor, Dolby-B	48-MP	Hit		pdf
A 12167 FB	LIN-IC	SMD, Recorder, Audio Proz., Dolby-B/C	80-MP	Hit		pdf
A 12168 NT	LIN-IC	SMD, Cd, Servo System, Ucc=5V	56-MP	Hit		
A 12169 FB	LIN-IC	SMD, Recorder, Audio Proz., Dolby-B/C	80-MP	Hit		pdf
A 12170 NT	LIN-IC	Dual Dolby-B/C Nr System, Ucc=12...16V	30-SDIP	Hit		pdf
A 12171 NT	LIN-IC	Recorder, Audio Prozessor, Dolby-B	56-SDIP	Hit		
A 12172 NT	LIN-IC	Recorder, Audio Prozessor, Dolby-B	56-SDIP	Hit		
A 12173	LIN-IC	SMD, Recorder, Audio Proz., Dolby-B/C	56-MP	Hit		pdf
A 12174	LIN-IC	SMD, Recorder, Audio Proz., Dolby-B/C	56-MP	Hit		pdf
A 12175	LIN-IC	SMD, Recorder, Audio Proz., Dolby-B/C	56-MP	Hit		pdf
A 12177	LIN-IC	SMD, Recorder, Audio Proz., Dolby-B/C	56-MP	Hit		pdf
A 12178 F	LIN-IC	SMD, Cd, Servo System, Ucc=5V	56-MP	Hit		
A 12179 F	LIN-IC	SMD,Recorder, Audio Prozessor, Dolby-B	56-MP	Hit		pdf
A 12180 F	LIN-IC	SMD,FM-ZF(DYNAS System), Mute, Ucc<16V	56-MP	Hit		
A 12181 FP	LIN-IC	AM-stoerunterdr./noise Reduction	16-MDIP	Hit		pdf
A 12182 F	LIN-IC	=HA 12182NT: SMD	56-MP	Hit		
A 12182 NT	LIN-IC	Recorder, Audio Prozessor, Dolby-B	56-SDIP	Hit		
HA 12187FP	IC	Automotive Audio, bus driver/receiver IC	MDIP	Ren	-	pdf pinout
HA 12192 F	LIN-IC	SMD, Car Recorder, Audio Processor Dolby-B	28-MDIP	Hit		pdf
HA 12193 F	LIN-IC	SMD, Car Recorder, Audio Processor Dolby-B	28-MDIP	Hit		pdf
HA 12194 F	LIN-IC	SMD, Car Recorder, Audio Processor Dolby-B	28-MDIP	Hit		pdf
HA 12197 F	LIN-IC	SMD, Car Recorder, Audio Processor	28-MDIP	Hit		pdf
HA 12198 F	LIN-IC	SMD, Car Recorder, Audio Processor	28-MDIP	Hit		pdf

Type	Device	Short Description	Fig.	Manu	Comparision Types	More at
HA 12199 F	LIN-IC	SMD, Car Recorder, Audio Processor	28-MDIP	Hit	-	pdf
HA 12212 F	LIN-IC	SMD, Car Recorder, Audio Processor Dolby-B	40-MDIP	Hit	-	pdf
HA 12215 F	LIN-IC	SMD,Recorder, Audio Processor, Dolby-B	56-MP	Hit	-	pdf
HA 12216...18F	IC	Car Deck, audio signal proc., Dolby B NR, playback amp.	QFP	Ren	-	pdf
HA 12221..23F	IC	Car Deck, audio signal proc., playback amp.	QFP	Ren		pdf
HA 12232FP	LIN-IC	Car audio, audio signal & equalizer preamp	MDIP	Ren	-	pdf pinout
HA 12240FP	IC	Automotive Audio, bus driver/receiver IC	MDIP	Ren	-	pdf pinout
HA 12402	LIN-IC	AM-V, AM/FM-ZF+NF, 0,45W(6V/8Ω)	16-DIP	Hit	A 283D, KA 22424, TA 7613AP, TDA 1083, U	pinout
HA 12404	LIN-IC	AM/FM-Frequ.-Zähl./count.+7-Segm. Dec.	22-DIP	Hit		
HA 12405	LIN-IC	AM/FM-Frequ.-Zähl./count.+7-Segm. Dec.	22-DIP	Hit		
HA 12411	LIN-IC	FM-ZF, Agc, Afc	16-DIP	Hit		
HA 12412	LIN-IC	FM-ZF, Dem, Mute, Afc, Ucc=13V	16-DIP	Hit		
HA 12413	LIN-IC	AM/FM-ZF, Dem, Meter, Ucc=3...16V	16-DIP	Hit	BA 4220, KA 2243	pdf
HA 12417	LIN-IC	Am-tuner for Car Radio	16-SQP	Hit		pdf pinout
HA 12418	LIN-IC	FM-ZF	16-SQP	Hit		
HA 12419	LIN-IC	FM-impulszähler/pulse count	16-DIP	Hit		
HA 12427	LIN-IC	AM-Tuner, Zf	16-DIP	Hit		
HA 12428	LIN-IC	AM-Tuner, ZF, Agc, Ucc=7,5...16V	16+4-DIP	Hit		
HA 12428 V	LIN-IC	=HA 12428: Fig.→	21-SQP	Hit		
HA 12430	LIN-IC	AM-Tuner, AM/FM-ZF, Agc, Ucc=3...12V	20-DIP	Hit		
HA 12438 CFP	LIN-IC	FM-Tuner, FM-ZF, Agc, Ucc=6,5...10V	16-MDIP	Hit		
HA 13001	LIN-IC	2x NF-E, 18V, 4,5A, 2x5,5W(13V/4Ω)	12-SIL	Hit		
HA 13002	SMPS-IC	Schaltregler/switching regulator	16-DIP	Hit		
HA 13003	SMPS-IC	S-Reg., 300kHz, Ucc=±10V	14-DIP	Hit		
HA 13006	LIN-IC	Video-disk, Motorregler/motor control	20-DIP+b	Hit		
HA 13007	LIN-IC	4x Relais/Motor-Tr., 5V, 0,7A	16-DIP	Hit	-	pdf
HA 13102	LIN-IC	2x NF-E, 18V, 4A, 2x5,5W(13V/4Ω)	12-SIL	Hit		
HA 13108	LIN-IC	2x NF-E, 18V, 4A, 2x5,5W(13V/4Ω)	12-SIL	Hit		
HA 13115	LIN-IC	2x NF-E, 40V, 2x4A, 2x12W(32V/8Ω)	12-SIL	Hit		
HA 13116	LIN-IC	NF-E, 18V, 4A, 20W(13V/4Ω, BTL)	15-SQL	Hit		pdf
HA 13117	LIN-IC	NF-E, 18V, 4A, 14W(13V/4Ω, BTL)	15-SQL	Hit		pdf
HA 13118	LIN-IC	NF-E, 18V, 4A, 18W(13V/4Ω, BTL)	15-SQL	Hit		pdf
HA 13119	LIN-IC	2x NF-E, 18V, 4A, 2x5,5W(13V/4Ω)	15-SQL	Hit		pdf pinout
HA 13127	LIN-IC	2x NF-E,18V 4A, 2x17W(BTL,14V/4Ω),50dB	16-SQL	Hit	HA 13128	pdf
HA 13128	LIN-IC	2x NF-E,18V 4A, 2x22W(BTL,14V/4Ω),50dB	16-SQL	Hit		pdf
HA 13130	LIN-IC	=HA 13127: Voltage Gain=40dB	16-SQL	Hit	HA 13135	pdf
HA 13135	LIN-IC	=HA 13128: Voltage Gain=40dB	16-SQL	Hit		pdf
HA 13143	LIN-IC	Cd, 4x BTL Treiber/Driver, Ucc=7...10V	26-SMDIP+b	Hit		pdf
HA 13150	LIN-IC	21W 4- Chan. BTL Power IC, 18V, 4A, 4x21W(BTL, 14V/ 4Ohm)	23-SQL	Hit		pdf pinout
HA 13150 A	LIN-IC	4x NF-E, 18V, 4A, 4x18W(BTL,13V/4Ohm)	23-SQL	Hit	-	pdf
HA 13151	LIN-IC	14W 4- Chan. BTL Power IC, 18V, 3A, 4x14W(BTL, 13V/ 4Ohm)	23-SQL	Hit	-	pdf pinout
HA 13152	LIN-IC	14W 4- Chan. BTL Power IC, 18V, 3A, 4x14W(BTL, 13V/ 4Ohm)	23-SQL	Hit	HA 13151	
HA 13153A	LIN-IC	15W 4- Chan. BTL Power IC, 18V, 3A, 4x15W(BTL, 13V/ 4Ohm)	23-SQL	Hit	HA 13154A	pdf pinout
HA 13154A	LIN-IC	15W 4- Chan. BTL Power IC, 18V, 3A, 4x15W(BTL, 13V/ 4Ohm)	23-SQL	Hit	HA 13153A	pdf pinout
HA 13155	LIN-IC	33W 4- Chan. BTL Power IC 18V, 4A, 4x33W(BTL, 13V/ 4Ohm)	23-SQL	Hit		
HA 13156	LIN-IC	38W 4- Chan. BTL Power IC 18V, 4A, 4x38W(BTL, 13V/ 4Ohm)	23-SQL	Hit		pdf pinout
HA 13158A	LIN-IC	34W 4- Chan. BTL Power IC ·18V, 4A, 4x34W(BTL, 13V/ 4Ohm)	23-SQL	Hit		pdf pinout
HA 13159	LIN-IC	37W 4- Chan. BTL Power IC 18V, 4A, 4x37W(BTL, 13V/ 4Ohm)	23-SQL	Hit		pdf pinout
HA 13164A	LIN-IC	car audio, voltage reg., 5V, 5.7V, 8V, 9V, 10V	SQL	Ren	-	pdf pinout
HA 13403(V)	LIN-IC	Vc, 3-Ph. Brushless Motor-Tr., 1,5A	23-SQL	Hit	-	pdf
HA 13403 MP	LIN-IC	=HA 13403: SMD, 1A	28-MP	Hit	-	pdf
HA 13406 W	LIN-IC	5,25" Hdd, 3-Ph. Brushless Motor-tr.	23-SQL	Hit	-	pdf
HA 13408	LIN-IC	9x Leistungs-Tr/Power Driver, 5V, 1,5A	23-SQL	Hit	-	pdf
HA 13412	LIN-IC	3-Phase Brushless Motor-Tr., 1A	23-SQL	Hit	-	pdf
HA 13415	LIN-IC	4x Solenoid Driver, Ucc=5V, 0,6A	16-DIP	Hit		
HA 13421 A	LIN-IC	3..5,25" Fdd, 2-Ph. Stepping Motor-tr.	16-DIP	Hit		pdf
HA 13421 AMP	LIN-IC	=HA 13421A: SMD	18-MP	Hit		pdf
HA 13426	LIN-IC	5,25" Hdd, Spindle Motor-Tr, 12V, 3A	23-SQL	Hit		pdf
HA 13431	LIN-IC	5,25" Fdd, Spindle Motor-Tr, 12V, 1A	23-SQL	Hit		
HA 13432	LIN-IC	3...3,5" Fdd, Spindle Motor-Tr,12V,0,5A	24-SDIP+b	Hit		
HA 13432 MP	LIN-IC	=HA 13432: SMD	28-MP	Hit		pdf
HA 13439 AMP	LIN-IC	SMD, 2-Phase Brushless Fan Motor-tr.	18-MP	Hit		pdf
HA 13440 MP	LIN-IC	SMD, Fdd, 3-Ph. Brushless DC Motor-tr.	28-MP	Hit		
HA 13441(V)	LIN-IC	Hdd, 3-Ph. Brushless DC Motor-Tr., 2A	23-SQL	Hit		pdf
HA 13442(V)	LIN-IC	=HA 13441(V): 4A	23-SQL	Hit		pdf
HA 13444	LIN-IC	4x Solenoid Driver, Ucc=5V, 0,6A	16-DIP	Hit		
HA 13447	LIN-IC	Hdd, Voice Coil Motor-Tr., 12V, 2,8A	23-SQL	Hit		pdf
HA 13449 MP	SMPS-IC	SMD,S-Reg., 4 Outputs, 200kHz, Ucc=12V	28-MP	Hit		
HA 13455	LIN-IC	Vc, 3-Phase Brushless Motor-Tr., S-Reg	24-SDIP+b	Hit	-	
HA 13456 A	LIN-IC	Fdd, 3-Phase Brushless DC Motor-tr.	24-SDIP+b	Hit	-	pdf
HA 13456 AMP	LIN-IC	=HA 13456A: SMD	28-MP	Hit	-	pdf

Type	Device	Short Description	Fig.	Manu	Comparision Types	More at
HA 13457 NT	LIN-IC	Vc, 3-Phase Motor-Tr., 1A	24-SDIP+b	Hit	-	
HA 13460 FP	LIN-IC	SMD,Brushless DC Fan Motor-Tr., 2x1,5A	16-MDIP	Hit	-	
HA 13464 MP	LIN-IC	SMD, 3-Phase Brushless Motor-Tr., 0,4A	28-MP	Hit	-	
HA 13467 NT	LIN-IC	Vc, 3-Phase Brushless, Motor-Tr., 1A	24-SDIP+b	Hit	-	
HA 13468 MP	LIN-IC	SMD, 3-Phase Brushless Motor-Tr., 0,7A	28-MP	Hit	-	
HA 13470	LIN-IC	Hdd, Voice Coil Motor-Tr., 12V, 1,67A	15-SQL	Hit	-	
HA 13471(A)	LIN-IC	Hdd, 3-Ph. Brushless Motor-Tr., 12V,2A	23-SQL	Hit	-	pdf
HA 13472(A)	LIN-IC	=HA 13471(A): 4A	23-SQL	Hit	-	
HA 13473 MP	LIN-IC	=HA 13473NT: SMD	28-MP	Hit	-	
HA 13473 NT	LIN-IC	Fdd, 3-Ph.Brushl. Motor-Tr., 12V, 0,7A	24-SDIP+b	Hit	-	
HA 13475 FP	LIN-IC	=HA 13475MP: Fig.→	16-MDIP	Hit	-	pdf
HA 13475 MP	LIN-IC	SMD, 2-Ph. Stepping Motor-Tr., 0,33A	18-MP	Hit	-	pdf
HA 13476 S	LIN-IC	SMD, Hdd, 3-Phase Brushl.Motor-Tr., 1A	26-MDIP+b	Hit	-	
HA 13480 AS-02	LIN-IC	SMD, 3-Phase Motor-Tr., 24V, 0,7A	26-MDIP+b	Hit	-	
HA 13480 S	LIN-IC	SMD, 3-Phase Motor-Tr., 24V, 0,7A	26-MDIP+b	Hit	-	pdf
HA 13481 AFP	LIN-IC	SMD, Hdd, 3-Phase Motor-Tr., 12V, 2A	26-MDIP+b	Hit	-	
HA 13481 S	LIN-IC	SMD, Hdd, 3-Phase Motor-Tr., 12V, 2A	26-MDIP+b	Hit	-	pdf
HA 13482	LIN-IC	Hdd, 3-Phase Brushl.Motor-Tr., 12V, 4A	23-SQL	Hit	-	
HA 13483 ANT	LIN-IC	=HA 13483NT: 0,8A	24-SDIP+b	Hit	-	
HA 13483 (A)MP	LIN-IC	=HA 13483NT: SMD	28-MP	Hit	-	
HA 13483 NT	LIN-IC	Fdd, 3-Ph.Brushl. Motor-Tr., 12V, 0,7A	24-SDIP+b	Hit	-	
HA 13488	LIN-IC	Hdd, 3-Phase Brushl.Motor-Tr., 12V, 4A	23-SQL	Hit	-	
HA 13490	LIN-IC	Hdd, Voice Coil Motor-Tr., 12V, 0,8A	16-DIP	Hit	-	
HA 13490 FP	LIN-IC	=HA 13490: SMD	16-MDIP	Hit	-	
HA 13490 MP	LIN-IC	=HA 13490: SMD	18-MP	Hit	-	
HA 13491 S	LIN-IC	SMD, Hdd, 3-Phase Motor-Tr., 5/12V, 1A	26-MDIP+b	Hit	-	pdf
HA 13492	LIN-IC	4x Solenoid Driver, Ucc=7..25V, 0,8A	15-SIL	Hit	-	pdf
HA 13493 MP	LIN-IC	Fdd, 3-Ph.Brushl. Motor-Tr., 5V, 0,7A	28-MP	Hit	-	
HA 13499 AMP,MP	LIN-IC	Fdd, 3-Ph.Brushl. Motor-Tr., 5V, 0,7A	28-MP	Hit	-	
HA 13501 F	LIN-IC	=HA 13501S: SMD	56-MP	Hit	-	
HA 13501 S	LIN-IC	Hdd, 3-Ph.Brushl. Motor-Tr., 5V, 1,2A	26-MDIP+b	Hit	-	pdf
HA 13509 F	LIN-IC	SMD,Hdd,Voice Coil Motor-Tr., 5V, 0,4A	28-MP/4TAB	Hit	-	
HA 13511 F	LIN-IC	Fdd, 3-Ph.Brushl. Motor-Tr., 5V, 0,7A	28-MP/4TAB	Hit	-	
HA 13513 F,NF	LIN-IC	Fdd, 3-Ph.Brushl. Motor-Tr., 5V, 0,7A	28-MP/4TAB	Hit	-	
HA 13518 FP	LIN-IC	Hdd, 3-Ph.Brushl. Motor-Tr., 5V, 1,2A	26-MDIP+b	Hit	-	
HA 13520 F	LIN-IC	SMD,Hdd,Spindel + Voice Coil Motor-tr.	28-MP/4TAB	Hit	-	
HA 13524(S)	LIN-IC	Hdd, Voice Coil Motor-Tr., 12V, 2,8A	23-SQL/SIL	Hit	-	
HA 13524 FP	LIN-IC	=HA 13524(S): SMD	26-MDIP+b	Hit	-	
HA 13525FP,AFP,BFP	LIN-IC	Hdd, Spindel + Voice Coil Motor-tr.	26-SMDIP+b	Hit	-	
HA 13529 FP	LIN-IC	Hdd, Voice Coil Motor-Tr., 12V, 0,8A	26-MDIP+b	Hit	-	
HA 13532 NT	LIN-IC	3-Ph. Stepping Motor-Tr., Ucc=24V	24-SDIP+b	Hit	-	
HA 13534 FP	LIN-IC	3-Ph.Brushl. Motor-Tr., 24V, 0,7A	26-MDIP+b	Hit	-	pdf
HA 13535 FP	LIN-IC	3-Ph.Brushl. Motor-Tr., 24V, 0,7A	26-SMDIP+b	Hit	-	
HA 13536 F	LIN-IC	Fdd, 3-Ph.Brushl. Motor-Tr., 5V, 0,7A	28-QFP+a	Hit	-	data pdf pinout
HA 13537 F	LIN-IC	Fdd, 3-Ph.Brushl. Motor-Tr., 12V, 0,8A	28-MP/4TAB	Hit	-	
HA 13537 NT	LIN-IC	=HA 13537F: Fig.→	24-SDIP+b	Hit	-	
HA 13539 FP	LIN-IC	Hdd, Voice Coil Motor-Tr., 12V, 1A	26-MDIP+b	Hit	-	
HA 13540 F	LIN-IC	Hdd, Voice Coil Motor-Tr., 12V, 1,5A	80-MP	Hit	-	
HA 13542 FP	LIN-IC	Hdd, 3-Ph.Brushl. Motor-Tr., 12V, 2A	26-MDIP+b	Hit	-	
HA 13544 F	LIN-IC	Hdd, Spindel + Voice Coil Motor-tr.	28-MP/4TAB	Hit	-	
HA 13549 FP	LIN-IC	SMD,Hdd, Voice Coil Motor-Tr., 12V, 1A	26-MDIP+b	Hit	-	
HA 13552	LIN-IC	3-Ph. Brushl. Motor-Tr., 24V, 2A	23-SQL	Hit	-	
HA 13555 T	LIN-IC	SMD, Spindel + Voice Coil Motor-tr.	48-MP/4TAB	Hit	-	
HA 13557 FH,AFH	LIN-IC	SMD, Spindel + Voice Coil Motor-tr.	48-QFP+a	Hit	-	data pdf pinout
HA 13558 FH,AFH	LIN-IC	SMD, Cd-rom, Motor-tr. Combo	48-QFP+a	Hit	-	data pdf pinout
HA 13559 FP	LIN-IC	Hdd, Voice Coil Motor-Tr., 12V, 1,5A	26-SMDIP+b	Hit	-	
HA 13561 F	LIN-IC	SMD, Spindel + Voice Coil Motor-tr.	80-QFP	Hit	-	data pdf pinout
HA 13563(V)	LIN-IC	3-Ph. Brushl. Motor-Tr., 24V, 2A	23-SQL	Hit	-	data pdf pinout
HA 13565 F	LIN-IC	SMD, 3-Ph. Brushl. Motor-Tr., 6V, 0,7A	28-QFP+a	Hit	-	data pdf pinout
HA 13605	LIN-IC	3-Ph.Brushl. Motor-Tr., 24V, 4,5A	23-SQL	Hit	-	
HA 13606(S)	LIN-IC	Hdd, 3-Ph.Brushl. Motor-Tr., 12V, 4,5A	23-SQP/SIL	Hit	-	
HA 13609 ANT	LIN-IC	3-Ph.Brushl. Motor-tr.	23-SQL	Hit	-	pdf
HA 13615 FP	LIN-IC	SMD, Schritt-/Stepping Motor-Tr., 1,5A	26-SMDIP+b	Hit	-	
HA 13705(C)	LIN-IC	Hi-Side Solenoid Driver, Udd=7...25V	17/5Pin	Hit	-	pdf
HA 16103 FPJ	Z-IC	=HA16103PJ: SMD	20-MDIP	Hit	-	pdf
HA 16103 PJ	Z-IC	Watch Dog Timer, Reset, 5V	18-DIP	Hit	-	pdf
HA 16107 FP	LIN-IC	=HA 16107P: SMD	16-MDIP	Hit	-	pdf
HA 16107 P	LIN-IC	PWM Schaltregler/switching Regulator	16-DIP	Hit	-	pdf
HA 16108 FP	LIN-IC	=HA 16108P: SMD	16-MDIP	Hit	-	pdf
HA 16108 P	LIN-IC	PWM Schaltregler/switching Regulator	16-DIP	Hit	-	pdf
HA 16109 FP	LIN-IC	=HA 16109P: SMD	16-MDIP	Hit	-	
HA 16109 P	LIN-IC	PWM Schaltregler/switching Regulator	16-DIP	Hit	-	pdf
HA 16111 FP	LIN-IC	=HA 16111P: SMD	16-MDIP	Hit	-	
HA 16111 P	LIN-IC	PWM Schaltregler/switching Regulator	16-DIP	Hit	-	
HA 16112 F	CMOS-IC	Spg.-Teiler/Volt. Divider, 1,25..8,75V	8-MDIP	Hit	-	
HA 16113 FPJ	Z-IC	SMD, Dual, Watch Dog Timer, Reset, 5V	24-MDIP	Hit	-	pdf
HA 16114 FP,FPJ	LIN-IC	SMD,PWM Schaltregler/switching Reg.	16-MDIP	Hit	-	
HA 16114 P	LIN-IC	PWM Schaltregler/switching Regulator	16-DIP	Hit	-	pdf
HA 16115 F	CMOS-IC	Spg.-Teiler/Volt. Divider, 1,25..8,75V	8-MDIP	Hit	-	
HA 16116 FP,FPJ	LIN-IC	SMD, Dual, PWM Schaltreg./switch. Reg.	20-MDIP	Hit	-	
HA 16117 FA,FB,FC	CMOS-IC	SMD, Watch Dog Timer, Reset	8-MDIP	Hit	-	
HA16118	2xCMOS	Vs:15V Vu:80dB Vo:4.2/0V Vio:<10mV	8S	Hit	-	data pinout
HA 16118 FP	CMOS-OP-IC	SMD, Dual, 15V, Iout=±5mA, -20...+75°	8-MDIP	Hit	-	data pinout
HA 16118 FPJ	CMOS-OP-IC	=HA 16118FP: -40...+85°	8-MDIP	Hit	-	data pinout
HA16119	2xCMOS	Vs:15V Vu:80dB Vo:4.3/0V Vio:<10mV	8S	Hit	-	data pinout

431

Type	Device	Short Description	Fig.	Manu	Comparision Types	More at
HA 16119 FP	CMOS-OP-IC	SMD, Dual, 15V, Iout=±5mA, -20...+75°	8-MDIP	Hit	-	data pinout
HA 16120 FP,FPJ	LIN-IC	SMD, PWM Schaltregler/switching Reg.	16-MDIP	Hit	-	
HA 16121 FP,FPJ	LIN-IC	SMD, Dual, PWM Schaltreg./switch. Reg.	20-MDIP	Hit	-	
HA 16128 L,LP,LPJ	CMOS-OP-IC	SMD, 5<7V, Iout=±5mA, -20...+75°	45	Hit	-	
HA 16128 L,LP	CMOS-OP-IC	SMD, 5<7V, Iout=±5mA, -20...+75°	45	Hit	-	
HA 16128 LPJ	CMOS-OP-IC	=HA 16128L,LP: -40...+85°	45	Hit	-	
HA 16129 FPJ	Z-IC	SMD, Watch Dog Timer, 5V ± 0,075	20-MDIP	Hit	-	.pdf
HA 16141 FP	IC	Pfc and PWM Cntrl.	MDIP	Ren	-	pdf
HA 16141 P	IC	Pfc and PWM Cntrl.	DIC	Ren	-	pdf pinout
HA 16142 FP	IC	Pfc and PWM Cntrl.	MDIP	Ren	-	pdf pinout
HA 16142 P	IC	Pfc and PWM Cntrl.	DIC	Ren	-	pdf pinout
HA 16341 FP	IC	Redundant Secondary Switching Power Supply Cntrl.	QIP_k	Ren	-	pdf pinout
HA 16341 NT	IC	Redundant Secondary Switching Power Supply Cntrl.	QIP_k	Ren	-	pdf pinout
HA 16342 FP	IC	Redundant Secondary Switching Power Supply Cntrl.	QIP_k	Ren	-	pdf pinout
HA 16342 NT	IC	Redundant Secondary Switching Power Supply Cntrl.	QIP_k	Ren	-	pdf pinout
HA 16503(P)	LIN-IC	Motor-steuerung/motor control	14-DIP	Hit	-	
HA 16506	LIN-IC	Camera, Verstärker/amplifier	14-DIP	Hit	-	
HA 16564	LIN-IC	Camera, Verstärker/amplifier	14-DIP	Hit	-	
HA 16603 FP	LIN-IC	=HA 16603P: SMD	16-MDIP	Hit	-	
HA 16603 P	LIN-IC	Münz-sensor/coin Sensor, Contactl. Sw.	16-DIP	Hit	-	pdf
HA 16604	LIN-IC	Breaker	8-SIP	Hit	-	
HA 16605 W	LIN-IC	Burner Controller	20-DIP	Hit	-	
HA 16613 A	A/D-IC	8 Bit, Dual Slope, Ucc=+5V	28-DIP	Hit	-	
HA 16617 P	LIN-IC	FLT Display Treiber/driver	18-DIP	Hit	-	
HA 16619 P	LIN-IC	Flt Display Treiber/driver	18-DIP	Hit	-	
HA 16628 P	LIN+D/A-IC	5 Bit, Position Amp. (f. DC Motor)	16-DIP	Hit	-	
HA 16629 P	LIN-IC	Tachometer F/v-converter (f. DC Motor)	16-DIP	Hit	-	
HA 16631 MP	LIN-IC	=HA 16631P: SMD	18-MP	Hit	-	
HA 16631 P	LIN-IC	Fdd, Leseverst./Read Amp., Ucc=5/12V	18-DIP	Hit	KA 6201	
HA 16632 AP	LIN-IC	Fdd, VCO, Data Separate, Ucc=5V	28-DIP	Hit	-	
HA 16636 AP	LIN-IC	Fehlstrom-/ground Fault Interrupter	8-SIP	Hit	-	
HA 16640 NT	BiMOS-IC	Fdd, Schreib-/write + Mechan. Ctrl.	42-SDIP	Hit	-	
HA 16642 MP	LIN-IC	=HA 16642NT: SMD	44-MP	Hit	-	
HA 16642 NT	LIN-IC	Fdd, Schreib-lese/read-write Prozessor	42-SDIP	Hit	-	
HA 16643 MP	BiMOS-IC	SMD, Fdd, Mechanik-/mechanism Ctrl.	44-MP	Hit	-	
HA 16650 MP	BiMOS-IC	SMD, Fdd, Mechanik-/mechanism Ctrl.	44-MP	Hit	-	
HA 16651 MP	LIN-IC	SMD,Fdd, Schreib-lese/read-write Proz.	44-MP	Hit	-	
HA 16652 MP	LIN-IC	SMD,Hdd,-Schreib-lese/read-write Proz.	28-MP	Hit	-	
HA 16652 P4	LIN-IC	Hdd, Schreib-lese/read-write Prozessor	22-DIP	Hit	-	
HA 16652 P6	LIN-IC	Hdd, Schreib-lese/read-write Prozessor	28-DIP	Hit	-	
HA 16654 AFP	LIN-IC	=HA 16654APS: SMD	14-MDIP	Hit	-	pdf
HA 16654 APS	LIN-IC	PWM Schaltregler/switching Regulator	8-DIP	Hit	-	pdf
HA 16656 MP	LIN-IC	SMD, Hdd, Lese-/read Pulse Generator	44-MP	Hit	-	
HA 16658 MA,MP	LIN-IC	=HA 16658NT: SMD	44-MP	Hit	-	
HA 16658 NT	LIN-IC	Hdd, VCO, Read Data (mfm, RLL)	42-SDIP	Hit	-	
HA 16662 MP	LIN-IC	SMD, Hdd, Interface(ST-506), Ucc=5V	44-MP	Hit	-	
HA 16663 MP	LIN-IC	SMD, Hdd, Lese-/read Pulse Generator	18-MP	Hit	-	pdf
HA 16664 AFP	LIN-IC	=HA 16664APS: SMD	14-MDIP	Hit	-	pdf
HA 16664 APS	LIN-IC	PWM Schaltregler/switching Regulator	8-DIP	Hit	-	pdf
HA 16666 FP	LIN-IC	=HA 16666P: SMD	16-MDIP	Hit	-	pdf
HA 16666 P	LIN-IC	PWM Schaltregler/switching Regulator	16-DIP	Hit	-	pdf
HA 16670 MP	LIN-IC	SMD, Hdd, Position Signal Generator	44-MP	Hit	-	
HA 16671 MP	LIN-IC	SMD, Hdd, VCM Servo Controller	44-MP	Hit	-	
HA 16672 MP	LIN-IC	SMD, Hdd, Vcm Servo Controller	44-MP	Hit	-	pdf
HA 16676 MP	LIN-IC	SMD, Hdd, Lese-/read Pulse Generator	44-MP	Hit	-	pdf
HA 16681 MA,MP	LIN-IC	SMD, Fdd, Schreib-lese/read-write	44-MP	Hit	-	
HA 16682 MP	LIN-IC	SMD, Hdd, Interface(ST-506), Ucc=5V	44-MP	Hit	-	pdf
HA 16686 MA,MP	LIN-IC	SMD, Hdd, Lese-/read Pulse Generator	44-MP	Hit	-	
HA 16688 MP	LIN-IC	SMD, Hdd, Schreib-lese/read-write	44-MP	Hit	-	pdf
HA 16689 MP	LIN-IC	SMD, Hdd, Schreib-lese/read-write	44-MP	Hit	-	
HA 16697 MP	LIN-IC	SMD, Fdd, Schreib-lese/read-write	44-MP	Hit	-	
HA 16721 MP	LIN-IC	SMD, Flt Display-Tr., 32 Segm.	44-MP	Hit	-	
HA 16722 MP	LIN-IC	SMD, Flt Display-Tr., 32 Segm.	44-MP	Hit	-	
HA 16853 MP	IC	SMD, Optical Communication Receiver Amplifier, TTL, 10MBit/s, +5V	MP	Hit	-	pdf pinout
HA 17008 RFP	D/A-IC	=HA 17008RG,RP: SMD	16-MDIP	Hit	-	pdf
HA 17008 RG,RP	D/A-IC	8 Bit, Multiplying, Ucc=15V, 85ns	16-DIC,DIP	Hit	-	pdf
HA 17012 G,P(B,C)	D/A-IC	12 Bit, Multiplying, Ucc=±15V, 250ns	20-DIC,DIP	Hit	-	
HA17080	JFET	Vs:±18V Vu:106dB Vi0:5mV f:3Mc	8D	Hit	-	data pinout
HA17080A	JFET	Vs:±18V Vu:106dB Vi0:6mV f:3Mc	8D	Hit	-	data pinout
HA 17080(A)GS	OP-IC	=HA 17080(A)PS: -40...+85°	8-DIC	Hit	TL 080...	pdf
HA 17080(A)PS	OP-IC	JFET, Serie 080, ±18V, -20...+75°	8-DIP	Hit	→Serie 080	data pdf pinout
HA17082	2xJFET	Vs:±18V Vu:106dB Vi0:5mV f:3Mc	8D	Hit	-	data pinout
HA 17082(A)GS,PS	OP-IC	=HA 17080...: Dual	8-DIP,DIC	Hit	TL 082...	data pinout
HA17082A	2xJFET	Vs:±18V Vu:106dB Vi0:6mV f:3Mc	8D	Hit	-	data pinout
HA17083	2xJFET	Vs:±18V Vu:106dB Vi0:5mV f:3Mc	14D	Hit	-	data pinout
HA 17083(A)G,P	OP-IC	=HA 17080...: Dual	14-DIP,DIC	Hit	TL 083...	data pinout
HA17083A	2xJFET	Vs:±18V Vu:106dB Vi0:6mV f:3Mc	14D	Hit	-	data pinout
HA17084	4xJFET	Vs:±18V Vu:106dB Vi0:5mV f:3Mc	14D	Hit	-	data pinout
HA 17084(A)G,P	OP-IC	=HA 17080...: Quad	14-DIP,DIC	Hit	TL 084...	data pinout
HA17084A	4xJFET	Vs:±18V Vu:106dB Vi0:6mV f:3Mc	14D	Hit	-	data pinout

Type	Device	Short Description	Fig.	Manu	Comparision Types	More at
HA17301	4xOP	Vu:63dB Vo:14/0V f:2.6Mc	14D	Hit		data pinout
HA 17301 G,P	OP-IC	Quad, +28V, 50mA, 2,6MHz, -20...+75°	14-DIP	Hit	MC 3301	data pinout
HA 17301 G	OP-IC	=HA 17301P: -40...+85°	14-DIC	Hit		data pinout
HA17324	4xOP	Vs:32V Vu:90dB Vo:13.6/0V Vio:2mV	14DS	Hit		data pdf pinout
HA 17324 F,FP	OP-IC	=HA 17324(G,P): SMD	14-MDIP	Hit	... 124..., ... 224..., ... 324..., ... 2902...	data pinout
HA 17324(G,P)	OP-IC	Quad, Serie 124, ±16V, 50mA, -20...+75°	14-DIC,DIP	Hit	→Serie 124	data pinout
HA17339	4xCOMP	Vs:36V Vu:200V/mV Vi0:2mV	14DS	Hit		data pdf pinout
HA 17339	KOP-IC	Quad, Serie 139, ±18V, 20mA, -20...+75°	14-DIP	Hit	→Serie 139	data pdf pinout
HA 17339 F	KOP-IC	=HA 17339: SMD	14-MDIP	Hit	... 139..., ... 239..., ... 339..., ... 2901...	data pdf pinout
HA17358	2xOP	Vs:32V Vu:90dB Vo:13.6/0V Vi0:3mV	8DS	Hit		data pdf pinout
HA 17358	OP-IC	Dual, Serie 158, ±16V, 50mA, -20...+75°	8-DIP	Hit	→Serie 158	data pdf pinout
HA 17358 F	OP-IC	=HA 17358: SMD	8-MDIP	Hit	... 158..., ... 258..., ... 358..., ... 1458...	data pdf pinout
HA 17384 (S,H)	LIN-IC	PWM Schaltreg./sw. Reg., Current Mode	MDIP	Ren	-	pdf pinout
HA 17384 (S,H)PS	LIN-IC	PWM Schaltreg./sw. Reg., Current Mode	8-DIP	Ren	-	pdf pinout
HA 17385 HRP	LIN-IC	=HA 17384PS: SMD	MDIP	Hit		pdf pinout
HA 17385 HPS	LIN-IC	PWM Schaltreg./sw. Reg., Current Mode	8-DIP	Ren	-	pdf pinout
HA17393	2xCOMP	Vs:36V Vu:200V/mV Vi0:2mV	8DS	Hit		data pdf pinout
HA 17393	KOP-IC	Dual, Serie 193, ±18V, -20...+75°	8-DIP	Hit	→Serie 193	data pdf pinout
HA 17393 F	KOP-IC	=HA 17393: SMD	8-MDIP	Hit	... 193..., ... 293..., ... 393..., ... 2903...	data pdf pinout
HA 17408 G,P	D/A-IC	8 Bit, Multiplying, Ucc=5V, 250ns	16-DIC,DIP	Hit		
HA 17431 FP,FPA(J)	IC	=HA 17431P...: Fig.→	MDIP	Ren	µPC 1093G	pdf pinout
HA 17431 P,PA,PJ	IC	+2,5...40V, 1...100mA, Uref=2,495V	7	Ren	TA 76431S, µPC 1093J	pdf pinout
HA 17431PS,PSA,PSJ	Ref-Z-IC	=HA 17431P...: Fig.→	8-DIP	Hit		pdf
HA 17431 UA.UPA	IC	=HA 17431P...: Fig.→	48	Ren	TA 76431F, µPC 1093T	pdf pinout
HA 17432 UPA	IC	Shunt Regulator, Vref=2.495V	48	Ren	TA 76431F, µPC 1093T	pdf
HA 17451 AP	LIN-IC	=HA 17451P: o./wo. Latch Circuits	16-DIP	Hit		pdf
HA 17451 FP,AFP	LIN-IC	=HA 17451(A)P: SMD	16-MDIP	Hit	-	
HA 17451 P	LIN-IC	Dual, PWM Schaltreg./sw. Reg.,3,3...40V	16-DIP	Hit		
HA17458	2xOP	Vs:±18V Vu:100dB Vo:±14V Vi0:2mV	8DSA	Hit		data pdf pinout
HA 17458 F,FP,FPJ	OP-IC	=HA 17458GS,PS: SMD	8-MDIP	Hit	... 1458..., ... 1558..., ... 4558...	data pdf pinout
HA 17458 GS,PS	OP-IC	Dual, Serie 158, ±18V, -25...+75°	8-DIC,DIP	Hit	→Serie 158	data pdf pinout
HA 17458 M	OP-IC	=HA 17458GS,PS: Fig.→	81	Hit	... 1458..., ...1558..., ... 4558...	data pdf pinout
HA 17458 PSJ	OP-IC	=HA 17458GS,PS: -40...+85°	8-DIP	Hit	... 1558...	data pdf pinout
HA17474	4xOP	Vs:±20V Vu:94dB Vo:±12.5V Vi0:1mV	14DS	Hit		data pdf pinout
HA 17474(G,P,RP)	OP-IC	Quad, ±20V, 1,9V/µs, -20...+75°	14-DIC,DIP	Hit	(... 124..., ... 224..., ... 324..., ... 2902.)	data pinout
HA 17494 G,P	LIN-IC	PWM Controller (=TL 494)	16-DIC,DIP	Hit	→TL 494	
HA 17524 FP	LIN-IC	=HA 17524G,P: SMD	16-MDIP	Hit	SG 3524CD	pdf pinout
HA 17524 G,P	LIN-IC	PWM Schaltregler/switching Regulator	16-DIC,DIP	Hit	KA 3524, LM 3524, SG 3524, UC 3524A	pinout
HA 17555(GS,PS)(J)	LIN-IC	=NE 555, 0...+70°, GS,PS=-20...+75°	8-DIC,DIP	Hit	→NE 555	pdf
HA 17555 F,FP	LIN-IC	=NE 555: SMD	8-MDIP	Hit	→NE 555	pdf
HA17558	2xOP	Vs:±18V Vu:106dB Vo:±13V Vi0:1mV	8DS	Hit		data pdf pinout
HA 17558(GS,PS)	OP-IC	Dual, ±18V, 1V/µs, -20...+75°	8-DIC,DIP	Hit	(... 1458..., ... 1558..., ... 4558...)	data pdf pinout
HA 17558 F,FP,FPJ	OP-IC	=HA 17558...: SMD	8-MDIP	Hit	(... 1458..., ... 1558..., ... 4558...)	data pdf pinout
HA 17558 PSJ	OP-IC	=HA 17558(GS,PS): -40...+85°	8-DIP	Hit	(... 1558...)	data pdf pinout
HA 17592(G,P)	LIN-IC	Diff. Video-Verst./Amp., ...90MHz	14-DIC,DIP	Hit	-	
HA 17711	KOP-IC	Dual	14-DIP	Hit		
HA17715	HI-SPEED	Vs:±18V Vu:89.5dB Vo:±13V Vi0:2mV	10A	Hit		data pinout
HA 17715 G	OP-IC	=HA 17715M: Fig.→	14-DIC	Hit		data pinout
HA 17715 M	OP-IC	hi-speed, Serie 115, ±18V, -25...+75°	82	Hit	→Serie 115	data pinout
HA17723	ER+	Io=150mA Vo:2...37V P:800mW	14D,10A	Hit		data pinout
HA 17723 F	Z-IC	=HA 17723M: SMD, 0...+70°	14-MDIP	Hit	... 723...	pdf
HA 17723(G,P)	Z-IC	=HA 17723M: Fig.→	14-DIC,DIP	Hit	... 723...	data pdf pinout
HA 17723 M	Z-IC	+2...37V, 0,15A, -25...+70	82	Hit	... 723...	data pdf pinout
HA 17733 M	LIN-IC	Diff. Video-Verst./Ampl., ...120MHz	82	Hit	LM 733..., µA 733...	
HA 17733(G,P)	LIN-IC	=HA 17733M: Fig.→	14-DIP	Hit	KA 733..., LM 733..., µA 733...	pinout
HA17741	UNI	Vs:±18V Vu:106dB Vo:±13V Vi0:1mV	8D,14DA	Hit		data pdf pinout
HA 17741 G	OP-IC	Uni, Serie 741, ±18, 1V/µs, -20...+75°	14-DIC	Hit	→Serie 741	data pdf pinout
HA 17741(GS,PS)	OP-IC	=HA 17741G: Fig.→	8-DIC,DIP	Hit	→Serie 741	data pdf pinout
HA 17741 M	OP-IC	=HA 17741G: Fig.→	81	Hit	→Serie 741	data pdf pinout
HA 17741 PSJ	OP-IC	=HA 17741G: -40...+85°	8-DIP	Hit	→Serie 741	data pdf pinout
HA17747	2xOP	Vs:±18V Vu:106dB Vo:±13V Vi0:2mV	14D,10A	Hit		data pdf pinout
HA 17747(G,P)	OP-IC	Dual, Serie 747, ±18V, 1V/µs,-20...+75°	14-DIC,DIP	Hit	→Serie 747	data pinout
HA 17747 M	OP-IC	=HA 17747G: Fig.→	82	Hit	→Serie 747	data pinout
HA 17805P,V,VP,VPJ	Z-IC	+5V, 1A, V,VP:-20...+75°, VPJ:-40...+85°	17b	Hit	... 7805... (TO-220)	data
HA 17805P,V,VP,VPJ	Z-IC	+5V, 1A, V,VP:-20...+75°, VPJ:-40...+85°	17b	Hit	... 7805... (TO-220)	data pdf pinout
HA17805P	R+	Io=750mA Vo:4.8...5.2V P:15W	3P	Hit		data pinout
HA 17806P,V,VP,VPJ	Z-IC	+6V, 1A	17b	Hit	... 7806... (TO-220)	data
HA 17806P,V,VP,VPJ	Z-IC	+6V, 1A V,VP:-20...+75°, VPJ:-40...+85°	17b	Hit	... 7806... (TO-220)	data pdf pinout
HA17806P	R+	Io=550mA Vo:5.7...6.2V P:15W	3P	Hit		data pinout
HA 17807P,V,VP,VPJ	Z-IC	+7V, 1A	17b	Hit	... 7807... (TO-220)	data
HA 17807P,V,VP,VPJ	Z-IC	+7V, 1A V,VP:-20...+75°, VPJ:-40...+85°	17b	Hit	... 7807... (TO-220)	data pdf pinout
HA 17808P,V,VP,VPJ	Z-IC	+8V, 1A	17b	Hit	... 7808... (TO-220)	data
HA 17808P,V,VP,VPJ	Z-IC	+8V, 1A V,VP:-20...+75°, VPJ:-40...+85°	17b	Hit	... 7808... (TO-220)	data pinout
HA 17809P,V,VP,VPJ	Z-IC	+9V, 1A	17b	Hit	... 7809... (TO-220)	data
HA 17809P,V,VP,VPJ	Z-IC	+9V, 1A V,VP:-20...+75°, VPJ:-40...+85°	17b	Hit	... 7809... (TO-220)	data pdf pinout
HA 17810P,V,VP,VPJ	Z-IC	+10V, 1A	17b	Hit	... 7810... (TO-220)	data
HA 17810P,V,VP,VPJ	Z-IC	+10V, 1A V,VP:-20...+75°, VPJ:-40...+85°	17b	Hit	... 7810... (TO-220)	data pdf pinout
HA 17812P,V,VP,VPJ	Z-IC	+12V, 1A	17b	Hit	... 7812... (TO-220)	data
HA 17812P,V,VP,VPJ	Z-IC	+12V, 1A V,VP:-20...+75°, VPJ:-40...+85°	17b	Hit	... 7812... (TO-220)	data pinout
HA17812P	R+	Io=350mA Vo:11...12V	3P	Hit		data
HA 17815P,V,VP,VPJ	Z-IC	+15V, 1A	17b	Hit	... 7815... (TO-220)	data
HA 17815P,V,VP,VPJ	Z-IC	+15V, 1A V,VP:-20...+75°, VPJ:-40...+85°	17b	Hit	... 7815... (TO-220)	data pinout
HA17815P	R+	Io=1.5A Vo:14...15V	3P	Hit		data
HA 17818P,V,VP,VPJ	Z-IC	+18V, 1A	17b	Hit	... 7818... (TO-220)	data
HA 17818P,V,VP,VPJ	Z-IC	+18V, 1A V,VP:-20...+75°, VPJ:-40...+85°	17b	Hit	... 7818... (TO-220)	data pdf pinout

Type	Device	Short Description	Fig.	Manu	Comparision Types	More at
HA17818P	R+	Io=200mA Vo:17...18V	3P	Hit		data pinout
HA 17824P,V,VP,VPJ	Z-IC	+24V, 1A	17b	Hit	... 7824... (TO-220)	data
HA 17824P,V,VP,VPJ	Z-IC	+24V, 1A V,VP:-20...+75°, VPJ:-40...+85°	17b	Hit	... 7824... (TO-220)	data pdf pinout
HA17824P	R+	Io=150mA Vo:23...25V	3P	Hit		data pinout
HA 17885P,V,VP,VPJ	Z-IC	+8,5V, 1A	17b	Hit	... 7885... (TO-220)	
HA 178xx FM,FMP	Z-IC	=HA 17805...17885: Iso	17b	Hit	... 78xx... (TO-220 Iso)	
HA17901	4xCOMP	Vs:36V Vu:200V/mV Vi0:2mV	14SD	Hit		data pdf pinout
HA 17901(P)	KOP-IC	Quad, Serie 139, ±18V, 20mA, -20...+75°	14-DIP	Hit	→Serie 139	data pdf pinout
HA 17901 FP,FPJ,RP	KOP-IC	=HA 17901(P,PJ): SMD	14-MDIP	Hit	→Serie 139	data pdf pinout
HA 17901 G,PJ	KOP-IC	=HA 17901(P): -40...+85°	14-DIC,DIP	Hit	→Serie 139	data pdf pinout
HA17902	4xOP	Vs:28V Vu:90dB Vo:13.6/0V Vi0:3mV	14SD	Hit		data pdf pinout
HA 17902(P)	OP-IC	Quad, Serie 124, 28V, 50mA, -20...+75°	14-DIP	Hit	→Serie 124	data pdf pinout
HA 17902 FP,FPJ	OP-IC	=HA 17902(P,PJ): SMD	14-MDIP	Hit	→Serie 124	data pdf pinout
HA 17902 G,PJ	OP-IC	=HA 17902(P): -40...+85°	14-DIC,DIP	Hit	→Serie 124	data pdf pinout
HA17903	2xCOMP	Vs:36V Vu:200V/mV Vi0:2mV	8SD	Hit		data pdf pinout
HA 17903(PS)	KOP-IC	Dual, Serie 193, ±18V, -20...+75°	8-DIP	Hit	→Serie 193	data pdf pinout
HA 17903 FP,FPJ	KOP-IC	=HA 17903(PS,PSJ): SMD	8-MDIP	Hit	... 193..., ...293..., ... 393..., ... 2903...	data pdf pinout
HA 17903 G,PSJ	KOP-IC	=HA 17903(PS): -40...+85°	8-DIC,DIP	Hit	... 193..., ...293..., ... 2903...	data pdf pinout
HA17904	2xOP	Vs:32V Vu:90dB Vo:13.6/0V Vi0:3mV	8SD	Hit		data pdf pinout
HA 17904 FP,FPJ	OP-IC	=HA 17904PS,PSJ: SMD	8-MDIP	Hit	... 158..., ... 258..., ... 358..., ... 2904...	data pdf pinout
HA 17904 GS	OP-IC	=HA 17904PS,PSJ: -40...+85°	8-DIC	Hit	... 158..., ... 258..., ... 2904...	pdf
HA 17904(PS)	OP-IC	Dual, Serie 158, ±16V, 50mA, -20...+75°	8-DIP	Hit	→Serie 158	data pdf pinout
HA 17904 PSJ	OP-IC	=HA 17904PS: -40...+85°	8-DIP	Hit	... 158..., ... 258..., ... 2904...	data pdf pinout
HA 17905 FM,FMP	Z-IC	Iso, -5V, 1A	17c	Hit	... 7905... (TO-220 Iso)	
HA 17906 FM,FMP	Z-IC	Iso, -6V, 1A	17c	Hit	... 7906... (TO-220 Iso)	
HA 17908 FM,FMP	Z-IC	Iso, -8V, 1A	17c	Hit	... 7908... (TO-220 Iso)	
HA 17909 FM,FMP	Z-IC	Iso, -9V, 1A	17c	Hit	... 7909... (TO-220 Iso)	
HA 17910 FM,FMP	Z-IC	Iso, -10V, 1A	17c	Hit	... 7910... (TO-220 Iso)	
HA 17912 FM,FMP	Z-IC	Iso, -12V, 1A	17c	Hit	... 7912... (TO-220 Iso)	
HA 17915 FM,FMP	Z-IC	Iso, -15V, 1A	17c	Hit	... 7915... (TO-220 Iso)	
HA 17952 FM,FMP	Z-IC	Iso, -5,2V, 1A	17c	Hit	... 7952... (TO-220 Iso)	
HA 19202	A/D-IC	4 Bit, Ucc=5V, 10MHz	22-DIP	Hit	-	
HA 19203 MP	A/D-IC	SMD, 4 Bit, Ucc=5V, 10MHz	18-MP	Hit	-	
HA 19209 C,TP	A/D-IC	8 Bit Flash, Vc, hi-speed, lo-power	28-DIC,DIP	Hit	-	
HA 19209 MP	A/D-IC	=HA 19209C,TP: SMD	44-MP	Hit	-	
HA 19210 C,TP	A/D-IC	8 Bit Flash, Vc, hi-speed, lo-power	28-DIC,DIP	Hit	-	
HA 19210 MP	A/D-IC	=HA 19210C,TP: SMD	44-MP	Hit	-	
HA 19211(B)MP	A/D-IC	=HA 19211(B)P: SMD	44-MP	Hit	-	pdf
HA 19211(B)NT	A/D-IC	=HA 19211(B)P:	30-SDIP	Hit	-	pdf
HA 19211 P	A/D-IC	8 Bit Flash, Vc, hi-speed, lo-power	28-DIP	Hit	-	
HA 19212 MP	A/D-IC	=HA 19212P: SMD	44-MP	Hit	-	
HA 19212 NT	A/D-IC	=HA 19212P:	30-SDIP	Hit	-	
HA 19212 P	A/D-IC	8 Bit Flash, Vc, hi-speed, lo-power	28-DIP	Hit	-	
HA 19213 MP	A/D-IC	=HA 19213NT: SMD	28-MP	Hit	-	
HA 19213 NT	A/D-IC	7 Bit Flash, hi-speed, lo-power	30-SDIP	Hit	-	
HA 19214 NT	A/D-IC	10 Bit Flash, Vc,hi-speed, lo-power	42-SDIP	Hit	-	
HA 19216(C)	A/D-IC	6 Bit Flash, hi-speed, Ucc=+5V, 30MHz	18-DIP,DIC	Hit	-	
HA 19216 MP	A/D-IC	=HA 19216(C): SMD	28-MP	Hit	-	
HA 19217 MP	A/D-IC	8 Bit, Video, >30Msps, LSTTL Level	44-MP	Hit	-	
HA 19503 ANT	D/A-IC	6 Bit, Clock Generator, Ucc=+5V	30-SDIP	Hit	-	
HA 19505	D/A-IC	10 Bit, Ucc=+5V, 50MHz	20-DIP	Hit	-	
HA 19507 MP	D/A-IC	=HA 19507NT: SMD	28-MP	Hit	-	
HA 19507 NT	D/A-IC	6 Bit, Clock Generator, Ucc=+5V	30-SDIP	Hit	-	
HA 19508 A	D/A-IC	6 Bit, Ucc=+5V, 30MHz	16-DIP	Hit	-	
HA 19508 AMP	D/A-IC	=HA 19508A: SMD	18-MP	Hit	-	
HA 19510 A	D/A-IC	8 Bit Flash, hi-speed, lo-power, 50MHz CMOS/TTL Level	18-DIP	Hit	-	
HA 19510 AMP	D/A-IC	=HA 19510A: SMD	18-MP	Hit	-	
HA 21001 MS	GaAs-IC	SMD, VHF/UHF-Tuner, ZF, Ucc=8...10V	18-MP	Hit	-	
HA 21005	GaAs-IC	BS-Tuner, 0,95...1,75GHz, Ucc=8,5...9,5V	44	Hit	-	
HA 21006 MP	GaAs-IC	BS-Tuner, 0,95...1,75GHz, Ucc=8,5...9,5V	18-MP	Hit	-	
HA 21008	GaAs-IC	BS Tuner, 0,95...2,05GHz, Ucc=4,5...5,5V	44	Hit	-	
HA 21009 MS	GaAs-IC	BS Tuner, 0,95...2,05GHz, Ucc=4,5...5,5V	18-MP	Hit	-	
HA 21010 MS	GaAs-IC	BS Tuner, 0,95...1,35GHz, Ucc=4,5...5,5V	18-MP	Hit	-	
HA23080-5	OP	Vs:±14V Vo:±13.8V Vi0:0.3mV f:2Mkc	8AD	Har		data pinout
HA 118001 MP	LIN-IC	SMD, Vc, Fm-signal-prozessor	44-MP	Hit	-	
HA 118002 FP	LIN-IC	Camera, Vorverst./Pre-Amp.	16-MDIP	Hit	-	
HA 118003 MP	LIN-IC	=HA 118003NT: SMD	28-MP	Hit	-	
HA 118003 NT	LIN-IC	Camera, Matrix-verstärker/amplifier	22-SDIP	Hit	-	
HA 118010 MP	LIN-IC	SMD, Camera, 4-Kanal/Channel-Prozessor	44-MP	Hit	-	
HA 118019 NT	LIN-IC	Vc, Vorverst./pre-amplifier	30-SDIP	Hit	-	pdf pinout
HA 118041 NT	LIN-IC	Vc, Vorverst./pre-amplifier	22-SDIP	Hit	-	
HA 118058	LIN-IC	Vc, Digital-schalter/switch	16-DIP	Hit	-	
HA 118059	LIN-IC	VC(S-VHS), 3x Video-Schalter/Switch	16-DIP	Hit	-	
HA 118059 FP	LIN-IC	=HA 118059: SMD	16-MDIP	Hit	-	
HA 118070	LIN-IC	Vc, Video-Schalter/Switch	8-DIP	Hit	-	
HA 118070 FP	LIN-IC	=HA 118070: SMD	8-MDIP	Hit	-	
HA 118082 MA	LIN-IC	SMD, Camera, 4-Kanal/Channel-Prozessor	44-MP	Hit	-	
HA 118084	LIN-IC	VC(VHS, S-VHS), Signal-prozessor	18-DIP	Hit	-	
HA 118084 MP	LIN-IC	=HA 118084: SMD	18-MP	Hit	-	
HA 118088	LIN-IC	Vc, Pip Switch	22-SDIP	Hit	-	
HA 118088 MP	LIN-IC	=HA 118088: SMD	28-MP	Hit	-	
HA 118099	LIN-IC	VC(S-VHS), 3x Video-Schalter/Switch	16-DIP	Hit	-	
HA 118099 FP	LIN-IC	=HA 118099: SMD	16-MDIP	Hit	-	
HA 118104	LIN-IC	Vc, 3x Video-Schalter/Switch	16-DIP	Hit	-	pdf pinout

Type	Device	Short Description	Fig.	Manu	Comparision Types	More at
HA 118104 FP	LIN-IC	=HA 118104: SMD	16-MDIP	Hit	-	pdf pinout
HA 118105 MA	LIN-IC	=HA 118105NT: SMD	56-MP	Hit	-	pdf
HA 118105(D)NT	LIN-IC	Vc, 1-Chip VHS-Prozessor	56-SDIP	Hit	-	pdf
HA 118118 MA	LIN-IC	SMD, Camera, Weiss-/white Balance	44-MP	Hit	-	
HA 118120 AMA	LIN-IC	SMD, Camera, Signal-prozessor	44-MP	Hit	-	
HA 118121 FP,VFP	LIN-IC	SMD, Camera, EVF(electr. View Finder)	16-MDIP	Hit	-	
HA 118124 MA	LIN-IC	SMD, Vc(vhs, S-vhs), Signal-prozessor	56-MP	Hit	-	pdf pinout
HA 118129 MP	LIN-IC	SMD, FM-ZF, Agc, Afc, 430MHz	28-MP	Hit	-	
HA 118142 AMA	LIN-IC	SMD, Camera, 4-Kanal/Channel-Prozessor	44-MP	Hit	-	
HA 118144 AF	LIN-IC	SMD, Camera, Signal-prozessor	48-MP	Hit	-	pdf
HA 118162 NT	LIN-IC	VC(S-VHS), Vorverst./pre-amplifier	30-SDIP	Hit	-	pdf pinout
HA 118191 NT	LIN-IC	Vc(vhs), Vorverst./pre-amplifier	30-SDIP	Hit	-	pdf pinout
HA 118285 BF	LIN-IC	=HA 118285BNT: SMD	56-MP	Hit	-	
HA 118285 BNT	LIN-IC	Vc(vhs), Multistandard-prozessor	56-SDIP	Hit	-	
HA 166008 FP	LIN-IC	=HA 166008MP:	24-MDIP	Hit	-	
HA 166008 MP	LIN-IC	SMD, Hdd, Schreib-lese/read-write	28-MP	Hit	-	pdf
HA 166009 MP	LIN-IC	SMD, Hdd, Schreib-lese/read-write	28-MP	Hit	-	pdf
HA 166010 MP	LIN-IC	SMD, Hdd, Schreib-lese/read-write	44-MP	Hit	-	pdf
HA 166024 FP	LIN-IC	SMD, Hdd, Schreib-lese/read-write	16-MDIP	Hit	-	pdf
HA 166025 FP	LIN-IC	SMD, Hdd, Schreib-lese/read-write	20-MDIP	Hit	-	pdf
HADC 77200 A,B	CMOS-A/D-IC	8 Bit, Video, >125 Msps, Ecl Level	48-DIP			
HAF 1001	MOS-P-FET-e	V-MOS, TEMPFET, SENSEFET, LogL, 60V 15A, 50W, <130mΩ(7,5A), 43,5/61µs	17c§	Hit		data
HAF 1002(L,S)	MOS-P-FET-e	=HAF 1001:	30c§	Hit	-	data
HAF 2001	MOS-N-FET-e	V-MOS, TEMPFET, SENSEFET, LogL, 60V 20A, 50W, <65mΩ(10A), 36,5/60µs	17c§	Hit		data
HAF 2002	MOS-N-FET-e	=HAF 2001: 30W	17c	Hit	-	data
HAF 2005	MOS-N-FET-e	V-MOS, TEMPFET, SENSEFET, LogL, 60V 40A, 30W, <33mΩ(20A), 71,8/49µs	17c	Hit		data
HAF 2006 R	MOS-N-FET-e	V-MOS, Dual, TEMPFET, SENSEFET, LogL 60V, 3A, 2W, <0,24Ω(2A), 8,6/11,4µs	8-MDIP	Hit	-	data
HAF 2007(L,S)	MOS-N-FET-e	V-MOS, TEMPFET, SENSEFET, LogL, 60V 5A, 20W, <0,16Ω(2,5A)	30c§	Hit		data
HAF 2008	MOS-N-FET-e	V-MOS, TEMPFET, SENSEFET, LogL, 60V 20A, 30W, <60mΩ(10A)	17c§	Hit		data
HAO	Si-N	=KSC 3123 (Typ-Code/Stempel/marking)	35	Sam	→KSC 3123	data
HAT 1001 F	MOS-P-FET-e*	SMD, V-MOS, LogL, 20V, 3,5A, <70mΩ(2A)	8-MDIP/ca1	Hit	-	data
HAT 1002 F	MOS-P-FET-e*	SMD, V-MOS, LogL, 30V, 3,5A, <70mΩ(2A)	8-MDIP/ca1	Hit	-	data
HAT 1004 F	MOS-P-FET-e*	SMD,V-MOS,Dual,LogL, 20V, 2,5A, <0,12Ω	8-MDIP/ca3	Hit	-	data
HAT 1005 F	MOS-P-FET-e*	SMD, V-MOS, LogL, 30V, 3,5A, <90mΩ(2A)	8-MDIP/ca1	Hit	-	data
HAT 1006 F	MOS-P-FET-e*	SMD,V-MOS, LogL, 60V, 2,5A, <0,14Ω(2A)	8-MDIP/ca1	Hit	-	data
HAT 1007 F	MOS-P-FET-e*	SMD, V-MOS, LogL, 20V, 3,5A, <80mΩ(2A)	8-MDIP/ca1	Hit	-	data
HAT 1008 F	MOS-P-FET-e*	SMD, V-MOS, LogL, 30V, 2,5A, <0,12Ω(2A)	8-MDIP/ca2	Hit	-	data
HAT 1009 F	MOS-P-FET-e*	SMD,V-MOS,Dual,LogL, 30V, 2,5A, <0,16Ω	8-MDIP/ca3	Hit	-	data
HAT 1016 F	MOS-P-FET-e*	SMD,V-MOS,Dual,LogL, 30V, 4,5A, <0,18Ω	8-MDIP/ca3	Hit	-	data
HAT 1020 R	MOS-P-FET-e*	SMD, V-MOS, LogL, 20V, 5A, <70mΩ(3A)	8-MDIP/ca1	Hit	-	data
HAT 1021 R	MOS-P-FET-e*	SMD, V-MOS, LogL, 20V, 5,5A, <60mΩ(3A)	8-MDIP/ca1	Hit	-	data
HAT 1023 R	MOS-P-FET-e*	SMD, V-MOS, LogL, 20V, 7A, <40mΩ(4A)	8-MDIP/ca1	Hit	-	data
HAT 1024 R	MOS-P-FET-e*	SMD,V-MOS,Dual,LogL, 30V, 2,5A, <0,16Ω	8-MDIP/ca3	Hit	-	data
HAT 1025 R	MOS-P-FET-e*	SMD, V-MOS,Dual,LogL, 20V, 4,5A, <95mΩ	8-MDIP/ca3	Hit	-	data
HAT 1026 R	MOS-P-FET-e*	SMD, V-MOS, LogL, 30V, 7A, <37mΩ(4A)	8-MDIP/ca1	Hit	-	data
HAT 1029 R	MOS-P-FET-e*	SMD,V-MOS,Dual,LogL, 20V, 3,5A, <0,23Ω	8-MDIP/ca3	Hit	-	data
HAT 1031 T	MOS-P-FET-e*	SMD, V-MOS,Dual,LogL, 20V, 2,5A, <0,28Ω	8-SSMDIP/ca4	Hit	-	data
HAT 1033 T	MOS-P-FET-e*	SMD, V-MOS, LogL, 20V, 3,5A, <90mΩ	8-SSMDIP/ca5	Hit	-	data
HAT 1036 R	MOS-P-FET-e*	SMD, V-MOS, LogL, 30V, 12A, <34mΩ(6A)	8-MDIP/ca1	Hit	-	data
HAT 1043 M	MOS-P-FET-e	SMD, V-MOS, LogL, 20V, 4,4A,<110mΩ(3A)	46/bn4	Hit	-	data
HAT 1044 M	MOS-P-FET-e	SMD, V-MOS, LogL, 30V, 4,5A,<105mΩ(3A)	46/bn4	Hit	-	data
HAT 1083 R,RJ	MOS-P-FET-e*	SMD,V-MOS,Dual,LogL, 60V, 3,5A, <0,23Ω RJ: Avalanche Energy Rated(1,05mJ)	8-MDIP/ca3	Hit		data
HAT 2001 F	MOS-N-FET-e*	SMD, V-MOS, LogL, 30V, 5A, <45mΩ(3A)	8-MDIP/ca1	Hit	-	data
HAT 2002 F	MOS-N-FET-e*	SMD, V-MOS, LogL, 30V, 5A, <40mΩ(3A)	8-MDIP/ca1	Hit	-	data
HAT 2003 F	MOS-N-FET-e*	SMD,V-MOS,Dual,LogL, 30V, 2,5A, <0,15Ω	8-MDIP/ca3	Hit	-	data
HAT 2004 F	MOS-N-FET-e*	SMD,V-MOS,Dual,LogL, 15V, 3,5A, <70mΩ	8-MDIP/ca3	Hit	-	data
HAT 2005 F	MOS-N-FET-e*	SMD, V-MOS, LogL, 20V, 3,5A, <65mΩ(2A)	8-MDIP/ca2	Hit	-	data
HAT 2006 F	MOS-N-FET-e*	SMD, V-MOS, LogL, 60V, 4A, <60mΩ(2A)	8-MDIP/ca1	Hit	-	data
HAT 2007 F	MOS-N-FET-e*	SMD, V-MOS, LogL, 30V, 4A, <70mΩ(2A)	8-MDIP/ca2	Hit	-	data
HAT 2008 F	MOS-N-FET-e*	SMD,V-MOS,Dual,LogL, 20V, 3,5A, <75mΩ	8-MDIP/ca3	Hit	-	data
HAT 2009 F	MOS-N-FET-e*	SMD,V-MOS,Dual,LogL, 30V, 3,5A, <80mΩ	8-MDIP/ca3	Hit	-	data
HAT 2010 F	MOS-N-FET-e*	SMD,V-MOS,Dual,LogL, 30V, 3,5A, <75mΩ	8-MDIP/ca3	Hit	-	data
HAT 2016 R	MOS-N-FET-e*	SMD,V-MOS,Dual,LogL, 30V, 6,5A, <45mΩ	8-MDIP/ca3	Hit	-	data
HAT 2019 R	MOS-N-FET-e*	SMD, V-MOS, LogL, 30V, 8A, <27mΩ(4A)	8-MDIP/ca1	Hit	-	data
HAT 2020 R	MOS-N-FET-e*	SMD, V-MOS, LogL, 30V, 8A, <28mΩ(4A)	8-MDIP/ca1	Hit	-	data
HAT 2022 R	MOS-N-FET-e*	SMD, V-MOS, LogL, 30V, 11A, <15mΩ(6A)	8-MDIP/ca1	Hit	-	data
HAT 2024 R	MOS-N-FET-e*	SMD,V-MOS,Dual,LogL, 30V, 5,5A, <110mΩ	8-MDIP/ca3	Hit	-	data
HAT 2025 R	MOS-N-FET-e*	SMD, V-MOS, LogL, 30V, 8A, <50mΩ(4A)	8-MDIP/ca1	Hit	-	data
HAT 2026 R	MOS-N-FET-e*	SMD, V-MOS, LogL, 20V, 11A, <21mΩ(6A)	8-MDIP/ca1	Hit	-	data
HAT 2027 R	MOS-N-FET-e*	SMD, V-MOS, Dual, LogL, 20V, 7A, <38mΩ	8-MDIP/ca3	Hit	-	data

Type	Device	Short Description	Fig.	Manu	Comparision Types	More at
HAT 2028 R,RJ	MOS-N-FET-e*	SMD,V-MOS, Dual, LogL, 60V, 4A, <0,16Ω RJ: Avalanche Energy Rated(1,37mJ)	8-MDIP/ca3	Hit	-	data
HAT 2029 R	MOS-N-FET-e*	SMD,V-MOS,Dual,LogL, 28V, 7,5A, <43mΩ	8-MDIP/ca3	Hit	-	data
HAT 2031 T	MOS-P-FET-e*	SMD, V-MOS,Dual,LogL, 20V, 3,5A, <98mΩ	8-SSMDIP/ ca4	Hit	-	data
HAT 2033 R,RJ	MOS-N-FET-e*	SMD, V-MOS, LogL, 60V, 7A, <53mΩ(4A) RJ: Avalanche Energy Rated	8-MDIP/ca1	Hit	-	data
HAT 2036 R	MOS-N-FET-e	SMD, V-MOS, LogL, 30V, 12A, <30mΩ(6A)	8-MDIP/ca1	Hit	-	data
HAT 2037 T	MOS-P-FET-e*	SMD, V-MOS, LogL, 28V, 5,5A, <38mΩ(3A)	8-SSMDIP/ ca5	Hit	-	data
HAT 2040 T	MOS-P-FET-e	SMD, V-MOS, LogL, 30V, 15A, <13mΩ(8A)	8-MDIP/ca1	Hit	-	data
HAT 3001 F	MOS-NP-FETe*	SMD, N+P-V-MOS, LogL, 30V, 2,5A N: <0,15Ω(4A), P: <0,16Ω(4A)	8-MDIP/ca3	Hit	-	data
HAT 3004 R	MOS-NP-FETe*	SMD, N+P-V-MOS, LogL, 30V, N: 5,5A <65mΩ(3A), P: 3,5A, <0,16Ω(2A)	8-MDIP/ca3	Hit	-	data
HAT 3006 R	MOS-NP-FETe*	SMD, N+P-V-MOS, LogL, 30V, N: 6,5A <45mΩ(4A), P: 4,5A, <90mΩ(3A)	8-MDIP/ca3	Hit	-	data
HAT 3008 R	MOS-NP-FETe*	SMD, N+P-V-MOS, LogL, 60V, N: 5A <84mΩ(3A), P: 3,5A, <0,23Ω(2A)	8-MDIP/ca3	Hit	-	data

HB...HZ

Type	Device	Short Description	Fig.	Manu	Comparision Types	More at
HB	Si-N	=2SC3120 (Typ-Code/Stempel/marking)	35	Tos	→2SC3120	data
HB	Si-N	=2SC3137 (Typ-Code/Stempel/marking)	25	Say	→2SC3137	data
HB	Si-N	=2SC4245 (Typ-Code/Stempel/marking)	35(2mm)	Tos	→2SC4245	data
HB	Si-N	=2SD1464-HB (Typ-Code/Stempel/marking)	39	Hit	→2SD1464	data
HB	Si-N	=BFN 22 (Typ-Code/Stempel/marking)	35	Sie	→BFN 22	data
HB	GaAs-FET	=CFY 75-13 (Typ-Code/Stempel/marking)	44	Sie	-	data
HB	GaAs-N-FET-d	=CFY 65-14 (Typ-Code/Stempel/marking)	51	Sie	→CFY 55	data
HB5B164	LED	bl 5.0x2.0mm I:12mcd Vf:3.5V If:10mA		Hrv		data pdf pinout
HB50	HI-OHM	Vs:±15V Vu:86dB Vo:±10V f:4Mc	12A	Ilc		data pinout
HB2200	LED	srd 8.4x3.6mm I:6mcd Vf:1.8V If:10mA		Hrv		data pdf pinout
HB2255	LED	srd 8.9x8.9mm I:12mcd Vf:1.8V		Hrv		data pdf pinout
HB2285C	LED	srd 18.8x8.9mm I:12mcd Vf:1.8V		Hrv		data pdf pinout
HB2300	LED	rd 8.4x3.6mm I:4mcd Vf:2V If:10mA		Hrv		data pdf pinout
HB2400	LED	yl 8.4x3.6mm I:3mcd Vf:2.1V If:10mA		Hrv		data pdf pinout
HB2500	LED	gr 8.4x3.6mm I:3mcd Vf:2.1V If:10mA		Hrv		data pdf pinout
HB2655	LED	rd 8.9x8.9mm I:6mcd Vf:2V If:10mA		Hrv		data pdf pinout
HB2685C	LED	rd 18.8x8.9mm I:12mcd Vf:2.1V		Hrv		data pdf pinout
HB2755	LED	yl 8.9x8.9mm I:5mcd Vf:2.1V If:10mA		Hrv		data pdf pinout
HB2785C	LED	yl 18.8x8.9mm I:10mcd Vf:2.1V		Hrv		data pdf pinout
HB2855	LED	gr 8.9x8.9mm I:5mcd Vf:2.1V If:10mA		Hrv		data pdf pinout
HB2885C	LED	gr 18.8x8.9mm I:10mcd Vf:2.2V		Hrv		data pdf pinout
HB52164	LED	srd 5.0x2.0mm I:10mcd Vf:1.8V		Hrv		data pdf pinout
HB52173	LED	srd 12.9x6.5mm I:12mcd Vf:1.8V		Hrv		data pdf pinout
HB52184	LED	srd 5.0x2.0mm I:10mbd Vf:1.8V		Hrv		data pdf pinout
HB53164	LED	yl 5.0x2.0mm I:2mcd Vf:2.1V If:10mA		Hrv		data pdf pinout
HB53173	LED	yl 12.9x6.5mm I:5mcd Vf:2.1V If:10mA		Hrv		data pdf pinout
HB53184	LED	yl 5.0x2.0mm I:2mcd Vf:2.1V If:10mA		Hrv		data pdf pinout
HB54164	LED	gr 5.0x2.0mm I:2mcd Vf:2.2V If:10mA		Hrv		data pdf pinout
HB54173	LED	gr 12.9x6.5mm I:5mcd Vf:2.2V If:10mA		Hrv		data pdf pinout
HB54184	LED	gr 5.0x2.0mm I:2mcd Vf:2.2V If:10mA		Hrv		data pdf pinout
HB57164	LED	rd 5.0x2.0mm I:2mcd Vf:2.1V If:10mA		Hrv		data pdf pinout
HB57173	LED	rd 12.9x6.5mm I:6mcd Vf:2.1V If:10mA		Hrv		data pdf pinout
HB57184	LED	rd 5.0x2.0mm I:2mcd Vf:2.1V If:10mA		Hrv		data pdf pinout
HBs	Si-N	=BFN 22 (Typ-Code/Stempel/marking)	35	Sie	→BFN 22	data
HBTS901	LED	bl 5.0x2.0mm I:200mcd Vf:3.3V		Seo		data pdf pinout
HBTS902	LED	bl 5.0x2.0mm I:200mcd Vf:3.3V		Seo		data pdf pinout
HC	Si-N	=2SC2733 (Typ-Code/Stempel/marking)	35	Hit	→2SC2733	data
HC	Si-N	=2SC3121 (Typ-Code/Stempel/marking)	35	Tos	→2SC3121	data
HC	Si-N	=2SC4246 (Typ-Code/Stempel/marking)	35(2mm)	Tos	→2SC4246	data
HC	Si-N	=2SC4463 (Typ-Code/Stempel/marking)	35(2mm)	Hit	→2SC4463	data
HC	Si-P	=BFN 23 (Typ-Code/Stempel/marking)	35	Sie	→BFN 23	data
HC	GaAs-FET	=CFY 75-15 (Typ-Code/Stempel/marking)	44	Sie	-	data
HC	Si-N/P+R	=XP 0431N (Typ-Code/Stempel/marking)	46(2mm)	Mat	→XP 0431N	data
HC 1 A3M...L2Q	Si-N+R	=AC 1A3M...L2Q: SMD	39b	Nec	-	data
HC50	HI-OHM	Vs:±15V Vu:86dB Vo:±10V f:4Mc	12A	Ilc		data pinout
HC 2000 H	Hybrid-IC	Op-Amp., 75V, 7A, PQ=100W(eff)		Rca	-	
HC 2500	Hybrid-IC	Op-Amp., 75V, 7A, PQ=100W(eff)		Rca	-	
HC 5513 IMA02	LIN-IC	Telecom, Subscriber Line Interface	28-PLCC	Har	PBL 3764	
HC 5513 IPA02	LIN-IC	=HC 5513IMA02:	22-DIP		PBL 3764	
HC 16701	LIN-IC	Thermal Head Driver, Ucc=5V, 0,1A	Chip	Hit		
HCLP2611	OC	LED/PD+IC Viso:2500Vrms Vbs:1.55V		Fcs,Hpa		data pdf pinout
HCLP7601	OC	LED/PD+IC Viso:600Vrms Vbs:1.5V		Hpa		data pdf pinout
HCMP96850	FAST-COMP	Vs:±6V Vi0:±3mV	d,16D	Spt		data pinout
HCMP96870A	2xFAST-COMP	Vs:±6V Vi0:±5mV	20C,16Dd	Spt		data pinout
HCPL062	DOC	LED/PD+IC Viso:>2500Vrms Vbs:<1.8V		Fcs		data pinout
HCPL0201	OC	LED/PD+IC Viso:2500Vrms Vbs:1.5V		Hpa		data pdf pinout
HCPL0500	OC	LED/PD+NPN Viso:2500Vrms CTR:19%		Hpa		data pdf pinout
HCPL0501	OC	LED/PD+NPN Viso:2500Vrms CTR:25%		Hpa		data pdf pinout
HCPL0600	OC	LED/PD+IC Viso:2500Vrms CTR:600%		Fcs,Hpa		data pdf pinout
HCPL0637	DOC	LED/PD+IC Viso:>3750Vrms Vbs:1.8V		Fcs		data pdf pinout

Type	Device	Short Description	Fig.	Manu	Comparision Types	More at
HCPL0700	OC	LED/PD+Dar Viso:2500Vrms CTR:2000%		Agt,Hpa		data pinout
HCPL1930	DOC	LED/PD+IC Viso:1500V Ibs:10mA		Hpa		data pinout
HCPL1931	DOC	LED/PD+IC Viso:1500V Ibs:10mA		Hpa		data pinout
HCPL2200	OC	LED/PD+IC Viso:2500Vrms Vbs:1.5V		Hpa		data pinout
HCPL2201	OC	LED/PD+IC Viso:2500Vrms Vbs:1.5V		Hpa		data pinout
HCPL2231	DOC	LED/PD+IC Viso:2500Vrms Vbs:1.5V		Hpa		data pinout
HCPL2300	OC	LED/PD+IC Viso:3000V Vbs:1.3V		Hpa		data pdf pinout
HCPL2400	OC	LED/PD+IC Viso:2500Vrms Vbs:1.3V		Hpa		data pinout
HCPL2430	DOC	LED/PD+IC Viso:2500Vrms Vbs:1.3V		Hpa		data pinout
HCPL2502	OC	LED/PD+NPN Viso:3000V CTR:15-22%		Hpa,Tix	OPI2502	data pinout
HCPL2503	OC	LED/PD+NPN Viso:3000V CTR:>9%		Qtc		data pinout
HCPL2530	DOC	LED/PD+NPN Viso:2500V CTR:18%		Qtc		data pinout
HCPL2531	DOC	LED/PD+NPN Viso:1500V CTR:21%		Qtc		data pinout
HCPL2601	OC	LED/PD+IC Viso:3000V CTR:600%		Gin,Qtc		data pinout
HCPL2602	OC	LED/PD+IC Viso:2500Vrms Vbs:2V		Hpa		data pinout
HCPL2611	OC	LED/PD+IC Viso:3000V Ibs:5mA		Qtc		data pinout
HCPL2630	DOC	LED/PD+IC Viso:2500Vrms Vbs:1.55V		Qtc	OPI2630	data pinout
HCPL2631	DOC	LED/PD+IC Viso:2500Vrms Vbs:1.55V		Qtc		data pinout
HCPL2730	DOC	LED/PD+Dar Viso:3000V CTR:2000%		Qtc		data pinout
HCPL2731	DOC	LED/PD+Dar Viso:3000V CTR:2000%		Qtc		data pinout
HCPL3000	OC	LED/MOSFET Viso:2500Vrms Vbs:1.6V		Hpa		data pinout
HCPL3700	OC	IC+LED/PD+Dar Viso:2500Vrms Vbs:5V		Hpa		data pinout
HCPL4100	OC	IC+LED/PD+IC Viso:2500Vrms Ibe:<8mA		Hpa		data pdf pinout
HCPL4200	OC	IC+LED/PD+IC Viso:2500Vrms Ibe:<8mA		Hpa		data pdf pinout
HCPL4502	OC	LED/PD+NPN Viso:2500Vrms CTR:25%		Hpa		data pdf pinout
HCPL4562	OC	LED/PD+NPN Viso:2500Vrms CTR:45%		Hpa		data pinout
HCPL4661	DOC	LED/PD+IC Viso:2500Vrms Vbs:1.55V		Hpa		data pinout
HCPL5200	OC	LED/PD+IC Viso:2500Vrms Vbs:1.5V		Hpa		data pinout
HCPL5230	DOC	LED/PD+IC Viso:2500Vrms Vbs:1.5V		Hpa		data pinout
HCPL5400	OC	LED/PD+IC Viso:2500Vrms Vbs:1.3V		Hpa		data pinout
HCPL5430	DOC	LED/PD+IC Viso:2500Vrms Vbs:1.3V		Hpa		data pinout
HCPL5431	DOC	LED/PD+IC Viso:2500Vrms Vbs:1.3V		Hpa		data pinout
HCPL5530	DOC	LED/PD+NPN Viso:2500V CTR:18%		Hpa		data pinout
HCPL5531	DOC	LED/PD+NPN Viso:1500V CTR:21%		Hpa		data pinout
HCPL5700	OC	LED/PD+Dar Viso:2500Vrms CTR:2000%		Hpa		data pinout
HCPL5701	OC	LED/PD+Dar Viso:2500Vrms CTR:2000%		Hpa		data pinout
HCPL5730	DOC	LED/PD+Dar Viso:2500Vrms CTR:1600%		Hpa		data pinout
HCPL5731	DOC	LED/PD+Dar Viso:2500Vrms CTR:1600%		Hpa		data pinout
HCPL5760	OC	LED/PD+Dar Ibs:2.6mA		Hpa		data pinout
HCPL5761	OC	LED/PD+Dar Ibs:2.6mA		Hpa		data pinout
HCPL7100	OC	IC+LED/PD+IC Viso:960Vrms Vbs:4.5V		Hpa		data pinout
HCPL7800	OC	IC+LED/PD+IC Viso:960Vrms Ibe:<22mA		Hpa		data pdf pinout
HCPL 7800	OPTO-IC	Motor control, high Cmr Isolation Amp	DIP	Hpa	-	pdf pinout
HCPL 7800A	OPTO-IC	Motor control, high Cmr Isolation Amp	DIP	Hpa	-	pdf pinout
HCPL 7800B	OPTO-IC	Motor control, high Cmr Isolation Amp	DIP	Hpa	-	pdf pinout
HCPL45032	OC	LED/PD+NPN Viso:2500Vrms CTR:25%		Hpa		data pdf pinout
HCPL45340	DOC	LED/PD+NPN Viso:2500V CTR:18%		Hpa		data pinout
HCs	Si-P	=BFN 23 (Typ-Code/Stempel/marking)	35	Sie	→BFN 23	data
HD	Si-N	=2SC3122 (Typ-Code/Stempel/marking)	35	Tos	→2SC3122	data
HD	Si-N	=2SC4249 (Typ-Code/Stempel/marking)	35(2mm)	Tos	→2SC4249	data
HD	Z-Di	=Z2 SMB-47 (Typ-Code/Stempel/marking)	71(5x3,5)	Fag	→Z2 SMB-47	data
HD 1 A3M...L3N	Si-N+R	=AD 1A3M...L3N: SMD	39b	Nec	-	data
HD 2 A3M...L3N	Si-N+R	=AD 2A3M...L3N: SMD	39b	Nec	-	data
HD 26 C31	CMOS-I/O-IC	4x Leitungstr./Line Driver f. RS422A	16-DIP	Hit	Pincompatible: HD 26LS31	pdf
HD 26 C32A	CMOS-I/O-IC	4x Leit.-Empf./Line Receiv. f.RS422/3A	16-DIP	Hit	Pincompatible: HD 26LS32(A)	pdf
HD 26 LS31	I/O-IC	4x Leitungstr./Line Driver f. RS422A	16-DIP	Hit	Pincompatible: HD 26C31	pdf pinout
HD 26 LS32(A)	I/O-IC	4x Leit.-Empf./Line Receiv. f.RS422/3A	16-DIP	Hit	Pincompatible: HD 26C32A	pdf
HD 26 LS33A	I/O-IC	4x Leit.-Empf./Line Receiv. f.RS422/3A	16-DIP	Hit	-	
HD 29 C 3486	CMOS-I/O-IC	=HD 293486: CMOS-Techn.	16-DIP	Hit	-	
HD 29 C 3487	CMOS-I/O-IC	=HD 293487: CMOS-Techn.	16-DIP	Hit	-	
HD74ALVC1G66	SPST	Vs:1.2...3.6V Ron:6Ohm at 3.3/0V	5S	Ren		data pdf pinout
HD74ALVC2G66	SPST	Vs:1.2...3.6V Ron:6Ohm at 3.3/0V	8S	Ren		data pdf pinout
HD74HC1G66	SPST	Vs:2...6V Ron:Ohm at 6/0V	5S	Ren		data pdf pinout
HD74HCT1G66	SPST	Vs:4.5...5.5V Ron:Ohm Ton:10ns at 5/0V	5S	Ren		data pdf pinout
HD74LV1G66A	SPST	Vs:1.65...5.5V Ron:Ohm Ton:5ns at 5/0V	5S	Ren		data pdf pinout
HD74LV1GT66A	SPST	Vs:3...5.5V Ron:Ohm Ton:5ns at 5/0V	5S	Ren		data pdf pinout
HD74LV2G66A	SPST	Vs:1.65...5.5V Ron:Ohm Ton:5ns at 5/0V	8S	Ren		data pdf pinout
HD74LV2GT66A	SPST	Vs:3...5.5V Ron:Ohm Ton:5ns at 5/0V	8S	Ren		data pdf pinout
HD74LV4051A	MULT8	Vs:2...5.5V Ron:Ohm Ton:5.5ns at 5/0V	16S	Ren		data pdf pinout
HD74LV4053A	MULT2	Vs:2...5.5V Ron:Ohm Ton:5ns at 5/0V	16S	Ren		data pdf pinout
HD74LV4066A	SPST	Vs:2...5.5V Ron:Ohm Ton:5ns at 5/0V	14S	Ren		data pdf pinout
HD74LVC1G66	SPST	Vs:1.65...5.5V Ron:Ohm at 5/0V	5G	Ren		data pdf pinout
HD74LVC2G66	SPST	Vs:1.65...5.5V Ron:Ohm at 5/0V	8G	Ren		data pdf pinout
HD 74 S,LS,ALSxxx	TTL-Logic	Standard TTL-Logic 7:(A,L)S-Serie		Hit	siehe http://www.ecadata.de→	data
HD74UH4066	SPST	Vs:2...6V Ron:Ohm at 6/0V	5S	Ren		data pdf pinout
HD 82 A	Si-Di	8 Di Array, 200V, 0,6A, (≈1N4002)	9-SIP	Shi	Pin-Code: AKKKKKKKK	data
HD 82 K	Si-Di	=HD 82A:	9-SIP		Pin-Code: KAAAAAAAA	data
HD 84 A	Si-Di	8 Di Array, 400V, 0,6A, (≈1N4004)	9-SIP	Shi	Pin-Code: AKKKKKKKK	data
HD 84 K	Si-Di	=HD 84A:	9-SIP		Pin-Code: KAAAAAAAA	data
HD 87 A	Si-Di	8 Di Array, 1000V, 0,6A, (≈1N4007)	9-SIP	Shi	Pin-Code: AKKKKKKKK	data
HD 87 K	Si-Di	=HD 87A:	9-SIP		Pin-Code: KAAAAAAAA	data

Type	Device	Short Description	Fig.	Manu	Comparision Types	More at
HD 151 BF854	CMOS-IC	2.5V PLL Clock Buffer	TSOP	Ren	-	pdf pinout
HD 2904	I/O-IC	3x Leitungsstreiber/Line Driver	16-DIC	Hit	-	
HD 2905	I/O-IC	3x Leitungsempfänger/Line Receiver	16-DIC	Hit	-	
HD 2912	DIG-IC	4x TTL-MOS Clock Driver f. Memory Ics	16-DIC	Hit	-	
HD 2915	I/O-IC	3x Leitungsempfänger/Line Receiver	16-DIC	Hit	-	
HD 2916	DIG-IC	4x TTL-MOS Clock Driver f. Memory Ics	16-DIC	Hit	-	
HD 2919(P)	DIG-IC	Druckertreiber/printer Driver	16-DIP	Hit	-	
HD 2922	DIG-IC	4x TTL-MOS Clock Driver f. Memory Ics	16-DIC	Hit	-	
HD 2923	DIG-IC	4x ECL-MOS Driver f. Memory Ics	16-DIC	Hit	-	
HD 6350	I/O-IC	→KS 5824	24-DIP	Hit	KS 5824, MC 6850	pdf
HD 7400...	TTL-Logic	Standard TTL-Logic 74xx-Serie		Hit	siehe http://www.ecadata.de→	data
HD 10101...10231	ECL-Logic	ECL-Logic Family 10..,→MB 10101...		Hit	→MB 10101...	
HD 10551	ECL-Logic	FM-Frequ.-Teiler/divider, 1:10...1:44	8-SIP	Hit	-	pdf
HD 17903 GS/PS	KOP-IC	Dual	8-DIC,DIP	Hit	→LM 2903	
HD 29026 A(AP)	CCD-IC	2x CCD-Treiber/Driver, 12V, Iq=±1A	8-DIP	Hit	-	
HD 29026 AF(AFP)	CCD-IC	=HD 29026A: SMD	8-MDIP			
HD 29027(P)	CCD-IC	2x CCD-Treiber/Driver, 6V, Iq=±1A	8-DIP	Hit	-	pdf
HD 29027 F(FP)	CCD-IC	=HD 29027: SMD	8-MDIP			
HD 29028(P)	CCD-IC	2x CCD-Treiber/Driver, 12V, Iq=±1A	8-DIP	Hit	-	pdf
HD 29028 F(FP)	CCD-IC	=HD 29028: SMD	8-MDIP			
HD 29029	CCD-IC	2x CCD-Treiber/Driver, 12V, Iq=±0,5A	8-DIP	Hit	-	pdf
HD 29050	I/O-IC	2x Line Driver/Receiver f.RS422/3A	16-DIP	Hit	-	pdf
HD 29051	I/O-IC	2x Line Driver/Receiver f.RS422/3A	16-DIP	Hit	-	pdf
HD 29412	I/O-IC	2x Leitungsstr./Line Driver f. RS422A	14-DIP	Hit	-	
HD 29413	I/O-IC	4x Leit.-Empf./Line Receiv. f.RS422/3A	16-DIP	Hit	Pincompatible: HD 26C32A	
HD 29468	I/O-IC	3x Line Driver/Receiver f. IBM360/370	16-DIP	Hit	-	pdf
HD 38980 C	MOS-IC	Digital-alarm-uhr/clock, fluorescent display, 18V fmax=10kHz	40-DIP	Hit	-	
HD 38991 A	MOS-IC	Digital-alarm-uhr/clock, LED-Tr.	40-DIP	Hit	-	
HD 42851	MOS-IC	Cb-pll-synthesizer	24-DIP	Hit	-	
HD 42853	MOS-IC	Cb-pll-synthesizer	22-DIP	Hit	-	
HD 42854	MOS-IC	Cb-scanner f.kanalwahl/channel select	16/20-DIP	Hit	-	
HD 42855	MOS-IC	Cb-scanner f.kanalwahl/channel select	16/20-DIP	Hit	-	
HD 43880	MOS-IC	Analoguhr/analog clock (4MHz)	8-DIP	Hit	-	
HD 43890	CMOS-IC	Tischuhr/desk-top clock	8-DIP	Hit	-	
HD 44007(A)	CMOS-IC	Camera, Color-signal (ntsc/pal/secam)	28-DIP	Hit	-	pdf
HD 44015	LIN-IC	PLL-Frequ.-synthesizer f. Dig. Tuning	22-DIP	Hit	-	
HD 44231 P	CMOS-IC	Telecom, A-law Codec, Filter	16-DIP	Hit	MB 6026A	pdf
HD 44232 P	CMOS-IC	Telecom, µ-Law Codec, Filter	16-DIP	Hit	MB 6025A	pdf
HD 44233 P	CMOS-IC	Telecom, A-law Codec, Filter	16-DIP	Hit	MB 6026A	pdf
HD 44234 P	CMOS-IC	Telecom, µ-Law Codec, Filter	16-DIP	Hit	MB 6025A	pdf
HD 44235 C	CMOS-IC	Telecom, A-law Codec, Filter	16-DIP	Hit	MB 6056A	
HD 44236 C	CMOS-IC	Telecom, µ-Law Codec, Filter	16-DIP	Hit	MB 6055A	
HD 44273 P	CMOS-IC	Telecom, A-law Codec, Filter	16-DIP	Hit	MB 6022A	
HD 44274 P	CMOS-IC	Telecom, µ-Law Codec, Filter	16-DIP	Hit	MB 6021A	
HD 44277 P	CMOS-IC	Telecom, A-law Codec, Filter	16-DIP	Hit	MB 6052A	
HD 44278 P	CMOS-IC	Telecom, µ-Law Codec, Filter	16-DIP	Hit	MB 6051A	
HD 44752	LIN-IC	PLL-Frequ.-synth.-contr.(am/fm)	42-DIP	Hit	-	
HD 44840	LIN-IC			Hit	-	
HD 49201 A	MOS-IC	SMD, Cd, Signal Prozessor, Udd=5V	80-MP	Hit	-	
HD 49202 NT	CMOS-IC	Cd, Digital Audio-Filter, Udd=5V	30-SDIP	Hit	-	
HD 49211 BFS	CMOS-IC	SMD, R-dat, Signal Prozessor	100-MP	Hit	-	
HD 49212	CMOS-IC	R-dat, Digital Servo, µComp. Interface	80-MP	Hit	-	
HD 49215	MOS-IC	SMD, Cd, Signal Proz., µComp. Interfac	80-MP	Hit	-	
HD 49217 AFS	MOS-IC	Cd-rom, Signal Proz., Double Spee	100-MP	Hit	-	
HD 49224A	µP-Periph	Error detection, correction for Dat/data app's	MP	Hit	-	pdf pinout
HD 49226 AFS	CMOS-IC	SMD, R-dat, Signal Proz., µComp.interf	100-MP	Hit	-	
HD 49227FS	µP-Periph	RAM Contr. for Dat/data app's	FLP	Hit	-	pdf
HD 49228 FS	CMOS-IC	R-dat, Digital Servo, µComp. Interface	80-MP	Hit	-	
HD 49229	CMOS-IC	R-dat, Data Extractor, Clock Generator	56-MP	Hit	-	
HD 49232 FS	MOS-IC	Cd, Digital Signal Proz., µC Interf.	80-MP	Hit	-	
HD 49233 AFS	MOS-IC	SMD, Cd, Signal Proz., µComp. Interf.	80-MP	Hit	-	
HD 49303	CMOS-A/D-IC	8 Bit, Tv/vc, hi-speed, lo-power	30-DIP	Hit	-	
HD 49304(ANT)	CMOS-D/A-IC	TV, 8 Bit, 3 Ch., Udd=+5V, 50MHz	42-SDIP	Hit	-	
HD 49306 AF	CMOS-A/D-IC	SMD, TV, 9 Bit, Udd=+5V	48-MP	Hit	-	
HD 49307	CMOS-D/A-IC	SMD, TV, 8 Bit, 3 Ch., Udd=+5V, >30MHz	56-MP	Hit	-	
HD 49315 F,AF	CMOS-A/D-IC	SMD, VC/Camera, 10 Bit, Digital Video	48-MP	Hit	-	
HD 49409 FS	LIN-IC	SMD, Tv/vc, PIP-control, A/D-D/A-Conv.	100-MP	Hit	-	
HD 49410 FS	LIN-IC	SMD, Tv/vc, Pip-control, A/d-d/a-conv.	100-MP	Hit	-	
HD 49412 FS	LIN-IC	SMD, Tv/vc, Pip-control, A/d-d/a-conv.	100-MP	Hit	-	
HD 49417 AFS	LIN-IC	SMD, Tv/vc, Muse Decoder	100-MP	Hit	-	
HD 49420 FS	LIN-IC	SMD, Tv/vc, Pip-control, A/d-d/a-conv.	80-MP	Hit	-	
HD 49421 FS	LIN-IC	SMD, Tv/vc, Pip-control, A/d-d/a-conv.	80-MP	Hit	-	
HD 49704 FP	CMOS-IC	SMD, Vc, Vhs-c, Servo-controller	80-MP	Hit	-	
HD 49723	LIN-IC			Hit	KA 8307	
HD 49733 NT	CMOS-IC	Vc, Servo-Controller, Ucc=5V	56-SDIP	Hit	-	
HD 49740 NT	MOS-IC	Vc, On-screen Display	42-SDIP	Hit	-	
HD 49741 NT	CMOS-IC	Vc, Servo-Controller, Ucc=5V	56-SDIP	Hit	-	
HD 49748	CMOS-IC	→KA 8316	56-SDIP	Hit	KA 8316	
HD 49781 F	CMOS-IC	=HD 49781NT: SMD	56-MP	Hit	-	
HD 49781 NT	CMOS-IC	Vc, Servo-Controller, Ucc=5V	56-SDIP	Hit	-	
HD 49783 FP	CMOS-IC	=HD 49783NT: SMD	16-MDIP	Hit	-	
HD 49783 NT	CMOS-IC	Vc, Synchronized Serial Bus	16-DIP	Hit	-	
HD 49801 FB	CMOS-IC	SMD, Camera, Digital Signal-prozessor	100-MP	Hit	-	

Type	Device	Short Description	Fig.	Manu	Comparision Types	More at
D 61927	MOS-IC	Camera, Color-signal (ntsc)	22-DIP	Hit	-	
D 64951	BiMOS-I/O-IC	SMD, µComp., Scsi Interface	68-MP	Hit	-	
D 64961	CMOS-I/O-IC	µComp., Scsi Controller		Hit	-	
D 66773 R	-IC	Graphics Controller Driver for TFT LCD panels	CSP	Ren	-	pdf pinout
D 75107 AG,AP	I/O-IC	2x Leitungsempfänger/Line Receiver	14-DIC,DIP	Hit	→µA 75107...	
D 75108 AG,AP	I/O-IC	2x Leitungsempfänger/Line Receiver	14-DIC,DIP	Hit	→µA 75108...	
D 75109 G,P	I/O-IC	2x Leitungstreiber/Line Driver	14-DIC,DIP	Hit	SN 75109...	
D 75110 G,P	I/O-IC	2x Leitungstreiber/Line Driver	14-DIC,DIP	Hit	SN 75110...	
D 75153	I/O-IC	4x Leitungstr./Line Driver f. RS422A	16-DIP	Hit	-	
D 75154 G,P	I/O-IC	4x Leitungsempf./Line Receiver f.RS232	16-DIC,DIP	Hit	→µA 75154	
D 75159	I/O-IC	2x Leitungstr./Line Driver f. RS422A	14-DIP	Hit	-	
D 75160 A	I/O-IC	8x Uni Interface Bus Transceiver	20-DIP	Hit	-	
D 75161 A	I/O-IC	8x Uni Interface Bus Transceiver	20-DIP	Hit	-	pdf pinout
D 75173	I/O-IC	4x Leit.-Empf./Line Receiv. f.RS422/3A	16-DIP	Hit	-	pdf
D 75175	I/O-IC	4x Leit.-Empf./Line Receiv. f.RS422/3A	16-DIP	Hit	-	
D 75188 G,P	I/O-IC	4x Leitungstr./Line Driver f. RS232	14-DIC,DIP	Hit	→µA 1488	
D 75189 G,P	I/O-IC	4x Leitungempf./Line Receiver f. RS232	14-DIC,DIP	Hit	→µA 1489	
D 75450 AG,AP	I/O-IC	2x Interface Driver, positive NAND	14-DIC,DIP	Hit	→µA 75450...	
D 75451 AG,AP	I/O-IC	2x Interface Driver, positive and	8-DIC,DIP	Hit	→µA 75451...	
D 75452 G,P	I/O-IC	2x Interface Driver, positive NAND	8-DIC,DIP	Hit	→µA 75452...	
D 75453 G,P	I/O-IC	2x Interface Driver, positive or	8-DIC,DIP	Hit	→µA 75453...	
D 75454 G,P	I/O-IC	2x Interface Driver, positive NOR	8-DIC,DIP	Hit	-	
D 100101...100194	ECL-Logic	ECL-Logic Family 100...	...-DIC	Hit	-	
D100101F...100194F	ECL-Logic	SMD,ECL-Logic Family 100...	...-FLP	Hit	-	
D 151005	MOS-I/O-IC	8x Inverter Buffer/Driver	20-DIP	Hit	-	pdf
D 151011	DIG-IC	2x BCD Progr. Counter	20-DIP	Hit	-	pdf
D 151012	DIG-IC	8 Bit Binary Progr. Counter	16-DIP	Hit	-	pdf
D 151015	I/O-IC	9 Bit Level Shifter/Transceiver	24-DIP	Hit	-	pdf
D 151240	I/O-IC	8x Buffer/Line Driver	20-DIP	Hit	-	
D 151244	I/O-IC	8x Buffer/Line Driver	20-DIP	Hit	-	
D 151245	I/O-IC	8x Bus Transceiver	20-DIP	Hit	-	
D 153007	MOS-IC	SMD, Hdd, 2-7 Code Encoder/Decoder	44-MP	Hit	-	
D 153009	MOS-IC	SMD, Hdd, 2-7 Code Encoder/Decoder	44-MP	Hit	-	pdf
D 153011	MOS-IC	SMD, Hdd, 2-7 RLL Encoder/Decoder	44-MP	Hit	-	pdf pinout
D 153013	MOS-IC	SMD, Hdd, 2-7 Code Encoder/Decoder	44-MP	Hit	-	pdf
D 153129(-65)	CMOS-D/A-IC	6 Bit, Video, 50(65)Msps, TTL Level	28-DIP	Hit	-	
D 153510 F50,F135	D/A-IC	SMD, 8 Bit, 3 Kanal/Channel, 50/135MHz	56-MP	Hit	-	
D 293486	I/O-IC	4x Leit.-Empf./Line Receiv. f.RS422/3A	16-DIP	Hit	-	
D 293487	I/O-IC	4x Leitungstr./Line Driver f. RS422A	16-DIP	Hit	-	
D 404302 RF	DIG-IC	4-bit single-chip microcomputer	DIP	Hit	-	pdf pinout
D 404302 RP	DIG-IC	4-bit single-chip microcomputer	DIP	Hit	-	pdf pinout
D 404302 RS	DIG-IC	4-bit single-chip microcomputer	DIP	Hit	-	pdf pinout
D 404304	CMOS-IC	Microcomputer 4-Bit w. I/O RAM/ROM	DIP	Hit	-	pdf pinout
D 404304	CMOS-IC	Microcomputer 4-Bit w. I/O RAM/ROM	FLP	Hit	-	pdf pinout
D 404304 P	DIG-IC	4-bit single-chip microcomputer	DIP	Hit	-	pdf pinout
D 404304 S	DIG-IC	4-bit single-chip microcomputer	DIP	Hit	-	pdf pinout
D 431115	MOS-IC	Analoguhr/analog clock (4MHz)	8-DIP	Hit	-	
D 440072	CMOS-IC	Camera, Color-signal(ntsc/pal/secam)	28-DIP	Hit	-	pdf
D 614042	µC-IC	4-Bit µ-Cop, 4096x10-bit ROM,245x4-bit RAM	DIP	Hit	-	pdf pinout
D 614042 (PLCC)	µC-IC	4-Bit µ-Cop, 4096x10-bit ROM,245x4-bit RAM	PLCC	Hit	-	pdf pinout
D 614045	µC-IC	4-Bit µ-Cop, 4096x10-bit ROM,245x4-bit RAM	DIP	Hit	-	pdf pinout
D 614045 (PLCC)	µC-IC	4-Bit µ-Cop, 4096x10-bit ROM,245x4-bit RAM	PLCC	Hit	-	pdf pinout
D 614048	µC-IC	4-Bit µ-Cop, 4096x10-bit ROM,245x4-bit RAM	DIP	Hit	-	pdf pinout
D 614048 (PLCC)	µC-IC	4-Bit µ-Cop, 4096x10-bit ROM,245x4-bit RAM	PLCC	Hit	-	pdf pinout
D 4074308 C	DIG-IC	4-bit single-chip microcomputer	DIP	Hit	-	pdf pinout
D 4074308 F	DIG-IC	4-bit single-chip microcomputer	DIP	Hit	-	pdf pinout
D 4074308 P	DIG-IC	4-bit single-chip microcomputer	DIP	Hit	-	pdf pinout
D 4074308 S	DIG-IC	4-bit single-chip microcomputer	DIP	Hit	-	pdf pinout
DA-0505...-2415	SMPS-IC	DC-DC Converter	≈7-SIP	Shi	-	
DB-0505...-2415	SMPS-IC	DC-DC Converter	≈7-SIP	Shi	-	
DF 0505..4815	SMPS-IC	DC-DC Converter	≈7-SIP	Shi	-	
DF 82 A,K	Si-Di	=HD 82A,K: 0,5A, <300ns	9-SIP	Shi	-	data
DF 84 A,K	Si-Di	=HD 84A,K: 0,5A, <300ns	9-SIP	Shi	-	data
DF-0515...-4815	SMPS-IC	DC-DC Converter	≈7-SIP	Shi	-	
DNS-2000	CMOS-IC	Optical Mouse Sensor	DIP	Agt	-	pdf pinout
DSP-4820	LED	rd+srd+yl+gr 1.52x58mm I:1.9mcd		Hpa		data pinout
DSP-4832	LED	rd/yl/gr 25.27x58mm I:1.9mcd Vf:2.1V		Hpa		data pinout
HE	Si-N	=2SC3123 (Typ-Code/Stempel/marking)	35	Tos	→2SC3123	data
HE	Si-N	=2SC4250 (Typ-Code/Stempel/marking)	35(2mm)	Tos	→2SC4250	data
HE	Si-N	=XP 05554 (Typ-Code/Stempel/marking)	46(2mm)	Mat	→XP 05554	data
HE1301ML	IRED	1300nm ø1.4mm I:1.9mcd Vf:1.5V		Hit		data pinout
HE1301R	IRED	1300nm ø0.4mm I:1.9mcd Vf:1.5V		Hit		data pinout
HE1301SG	IRED	1300nm ø4mm I:1.9mcd Vf:1.5V		Hit		data pinout
HE1302ML	IRED	1300nm ø1.4mm I:1.9mcd Vf:1.5V		Hit		data pinout

Type	Device	Short Description	Fig.	Manu	Comparision Types	More
HE7601	IRED	770nm ø4mm I:1.9mcd Vf:2.5V If:200mA	TO46	Hit		data po
HE8402	IRED	800nm ø50µmm I:1.9mcd Vf:<2.5V		Hit		data pi
HE8403	IRED	840nm ø4.65mm I:1.9mcd Vf:<2.5V		Hit		data pi
HE8404	IRED	820nm ø4mm I:1.9mcd Vf:2.5V If:200mA	TO46	Hit		data pi
HE8801	IRED	880nm ø4mm I:1.9mcd Vt:1.7V If:150mA		Hit		data pi
HE8804	IRED	880nm ø4mm I:1.9mcd Vf:<2.3V	TO46	Hit		data pi
HE8805	IRED	880nm ø4mm I:1.9mcd Vf:1.7V If:150mA		Hit		data pi
HE8806	IRED	880nm ø4mm I:1.9mcd Vf:1.7V If:150mA		Hit		data pi
HE8807	IRED	880nm ø4.65mm I:1.9mcd Vf:1.7V		Hit		data pi
HE8811	IRED	820nm ø4mm I:1.9mcd Vf:<2.5V	TO46	Hit		data po
HE 97131	CCD-IC	Photosensor Array, 1024 Bit	≈24-DIC	Hit		
HE 97132	CCD-IC	Photosensor Array, 2048 Bit	≈24-DIC	Hit		
HEF4016B	SPST	Vs:3...15V Ron:140Ohm Ton:90ns at 5/0V	14DS	Phi		data pe
HEF4051B	MULT8	Vs:3...15V Ron:115Ohm at 5/0V	16DS	Phi		data pe
HEF4052B	MULT4	Vs:3...15V Ron:115Ohm at 5/0V	16DS	Phi		data pe
HEF4053B	MULT2	Vs:3...15V Ron:115Ohm at 5/0V	16DS	Phi		data pe
HEF4066B	SPST	Vs:3...15V Ron:115Ohm Ton:80ns at 5/0V	14DS	Phi		data pe
HEF4067B	MULT16	Vs:3...15V Ron:115Ohm at 5/0V	24DS	Phi		data pe
HEF 4750 VD	MOS-IC	VHF/UHF-Frequ.-Synthesizer, >1GHz	28-DIC	Phi	-	pdf
HEF 4751 VD	MOS-IC	=HEF 4751VP: Keramik	28-DIC	Phi	-	pdf
HEF 4751 VP	MOS-IC	Frequ.-Teiler/divider f. HEF 4750	28-DIP	Phi	-	pdf
HEF 4751 VT	MOS-IC	=HEF 4751VP: SMD	26-MDIP	Phi	-	pdf
HEPC601	ER+	Io=200mA Vo:3.6...32V P:1W	10A,8A	Mot		data p
HEPC6049G	ER+	Io=250mA Vo:2.5...32V P:3W	10A	Mot		data p
HEPC6049R	ER+	Io=600mA Vo:2.5...32V P:680mW	9O	Mot		data p
HEPC6051	2xOP	Vs:±15V Vu:83dB Vo:±12V Vi0:10mV	14D	Mot		data p
HEPC6052	UNI	Vs:±15V Vu:86dB Vo:±12V Vi0:7.5mV	8AD	Mot		data p
HEPC6053	UNI	Vs:±15V Vu:83dB Vi0:7.5mV	8A,14D	Mot		data p
HEPC6054	ER-	Io=200mA Vo:-3.8...-32V	10A,9O	Mot		data p
HER 101	Si-Di	GI, S, 50V, 1A, <50ns	31a	Mop	BYT 43A, BYV 26A, EGP 10A, FE 1A, ++	data
HER 102	Si-Di	GI, S, 100V, 1A, <50ns	31a	Mop	BYT 43B, BYV 26A, EGP 10B, FE 1B, ++	data
HER 103	Si-Di	GI, S, 200V, 1A, <50ns	31a	Mop	BYT 43D, BYV 26A, EGP 10D, FE 1D, ++	data
HER 104	Si-Di	GI, S, 300V, 1A, <75ns	31a	Mop	BYT 43G, BYV 26B, EGP 10F, FE 1F, ++	data
HER 105	Si-Di	GI, S, 400V, 1A, <75ns	31a	Mop	BYT 43G, BYV 26B, EGP 10G, FE 1H, ++	data
HER 106	Si-Di	GI, S, 600V, 1A, <100ns	31a	Mop	BYT 43J, BYV 26C, MUR 160	data
HER 107	Si-Di	GI, S, 800V, 1A, <100ns	31a	Mop	BYT 43K, BYV 26D, MUR 180	data
HER 108	Si-Di	GI, S, 1000V, 1A, <100ns	31a	Mop	BYT 43M, BYV 26E, MUR 1100	data
HER 301	Si-Di	GI, S, 50V, 3A, <50ns	31a	Mop	BYV 28-50, EGP 30A, FE 3A	data
HER 302	Si-Di	GI, S, 100V, 3A, <50ns	31a	Mop	BYV 28-100, EGP 30B, FE 3B	data
HER 303	Si-Di	GI, S, 200V, 3A, <50ns	31a	Mop	BYV 28-200, EGP 30D, FE 3D	data
HER 304	Si-Di	GI, S, 300V, 3A, <75ns	31a	Mop	BYV 28-300, EGP 30E, FE 3F	data
HER 305	Si-Di	GI, S, 400V, 3A, <75ns	31a	Mop	BYV 28-400, EGP 30G, FE 3H	data
HER 306	Si-Di	GI, S, 600V, 3A, <100ns	31a	Mop	BYT 56J, BYV 28-600, BYW 28-600	data
HER 307	Si-Di	GI, S, 800V, 3A, <100ns	31a	Mop	BYT 56K	data
HER 308	Si-Di	GI, S, 1000V, 3A, <100ns	31a	Mop	BYT 56M	data
HER 501	Si-Di	GI, S, 50V, 5A, <50ns	31a	Mop	EGP 50A, FE 5A	data
HER 502	Si-Di	GI, S, 100V, 5A, <50ns	31a	Mop	EGP 50B, FE 5B	data
HER 503	Si-Di	GI, S, 200V, 5A, <50ns	31a	Mop	EGP 50D, FE 5D	data
HER 504	Si-Di	GI, S, 300V, 5A, <75ns	31a	Mop	EGP 50G, FE 6F	data
HER 505	Si-Di	GI, S, 400V, 5A, <75ns	31a	Mop	EGP 50G, FE 6H	data
HER 506	Si-Di	GI, S, 600V, 5A, <100ns	31a	Mop	-	data
HER 507	Si-Di	GI, S, 800V, 5A, <100ns	31a	Mop	-	data
HER 508	Si-Di	GI, S, 1000V, 5A, <100ns	31a	Mop	-	data
HF	Si-N	=2SC2805 (Typ-Code/Stempel/marking)	25	Tos	→2SC2805	data
HF	Si-N	=2SC3124 (Typ-Code/Stempel/marking)	35	Tos	→2SC4251	data
HF	Si-N	=2SC4251 (Typ-Code/Stempel/marking)	35(2mm)	Tos	→2SC4251	data
HF	Si-P	=BFN 23 (Typ-Code/Stempel/marking)	35	Sie	→BFN 23	data
HF	Si-N	=KTC 3882 (Typ-Code/Stempel/marking)	35	Kec	→KTC 3882	data
HF	Si-N	=KTC 4082 (Typ-Code/Stempel/marking)	35(2mm)	Kec	→KTC 4082	data
HF	Z-Di	=Z2 SMB-51 (Typ-Code/Stempel/marking)	71(5x3,5)	Fag	→Z2 SMB-51	data
HF 7739	MOS-N-IGBT	=GN 4530C		Hit	→GN 4530C	data
HF 7743	MOS-N-IGBT	=GN 6030C		Hit	→GN 6030C	data
HF 7745	MOS-N-IGBT	=GN 6050E		Hit	→GN 6050E	data
HF 7747	MOS-N-IGBT	=GN 6075E		Hit	→GN 6075E	data
HF 7749	MOS-N-IGBT	=GN 12015C		Hit	→GN 12015C	data
HF 7751	MOS-N-IGBT	=GN 12030E		Hit	→GN 12030E	data
HF 7753	MOS-N-IGBT	=GN 12050E		Hit	→GN 12050E	data
HF 7755	MOS-N-IGBT	=GN 6020C		Hit	→GN 6020C	data
HF 7757	MOS-N-IGBT*	Iso-Gate Trans, TV-HA,1700V, 20A, 150W	77id	Hit	-	data
HF 7761	MOS-N-IGBT	=GN 9060E		Hit	→GN 9060E	data
HF 75045	MOS-N-IGBT	=GN 6015A		Hit	→GN 6015A	data
HF 75049	MOS-N-IGBT	=GN 6010A		Hit	→GN 6010A	data
HF 75051	MOS-N-IGBT*	Iso-Gate Trans, 450V, 30A, 30W Camera Strobes	30id	Hit	-	data
HFA1-0001	HI-SLEW	Vs:±6V Vo:±3.5V Vi0:6mV f:350Mc	14D	Har		data
HFA1-0003	HI-SPEED	Vs:-8/18V Vi0:1mV f:270Mc	16D	Har		data
HFA(1)-0001(-5,-9)	OP-IC	hi-slew, +12(±6)V, 350MHz, 1000V/µs -5: 0...+75°, -9: -40...+85°	14-DIC	Har	-	data
HFA(3)-0001...	OP-IC		8-DIP			
HFA(9P)-0001...	OP-IC	SMD	16-MDIP			

Device	Short Description	Fig.	Manu	Comparision Types	More at
02	LO-NOISE	Vs:±6V Vu:105V/mV Vo:±3.9V Vi0:0.6mV	8A	Har		data pinout
03	HI-SPEED	Vs:-8/18V Vi0:1mV f:270Mc	10A	Har		data pinout
05	HI-SLEW	Vs:±6V Vu:230V/mV Vo:±3.5V Vi0:6mV	8A	Har		data pinout
002(-5,-9)	OP-IC	lo-noise, +12(±6)V, 1GHz, 250V/µs -5: 0...+75°, -9: -40...+85°	81	Har	-	data pinout
002...	OP-IC		8-DIP		-	
0002...	OP-IC	SMD	8-MDIP		-	
01	HI-SLEW	Vs:±6V Vo:±3.5V Vi0:6mV f:350Mc	8D	Har		data pinout
02	LO-NOISE	Vs:±6V Vu:105V/mV Vo:±3.9V Vi0:0.6mV	8D	Har		data pinout
03	HI-SPEED	Vs:-8/18V Vi0:1mV f:270Mc	8D,16D	Har		data pinout
05	HI-SLEW	Vs:±6V Vu:230V/mV Vo:±3.5V Vi0:6mV	8D	Har		data pinout
005(-5,-9)	OP-IC	hi-slew, +12(±6)V, 300MHz, 420V/µs -5: 0...+75°, -9: -40...+85	81	Har		data pinout
005...	OP-IC		8-DIP		-	
0005...	OP-IC	SMD	8-MDIP		-	
02	LO-NOISE	Vs:±6V Vu:105V/mV Vo:±3.9V Vi0:0.6mV	8D	Har		data pinout
03	HI-SPEED	Vs:-8/18V Vi0:1mV f:270Mc	8D	Har		data pinout
05	HI-SLEW	Vs:±6V Vu:230V/mV Vo:±3.5V Vi0:6mV	8D	Har		data pinout
001	HI-SLEW	Vs:±6V Vo:±3.5V Vi0:6mV f:350Mc	16S	Har		data pinout
002	LO-NOISE	Vs:±6V Vu:105V/mV Vo:±3.9V Vi0:0.6mV	8S	Har		data pinout
003	HI-SPEED	Vs:-8/18V Vi0:1mV f:270Mc	8S,16S	Har		data pinout
005	HI-SLEW	Vs:±6V Vu:230V/mV Vo:±3.5V Vi0:6mV	8S	Har		data pinout
	HI-SPEED	Vs:12V Vi0:2mV f:300Mkc	d,8SD	Har		data pdf pinout
0MJ/883	HI-SPEED	Vs:12V Vo:±3V Vi0:±6mV	8D	Har		data pinout
0Y	HI-SPEED	Vs:12V Vi0:<2mV f:<670Mc	d	Har		data pinout
	FB-AMP	Vs:12V Vi0:2mV f:600Mc	d,8SD	Har		data pdf pinout
8I	VIDEO-OP	Vs:12V Vi0:8mV f:200Mc	8SD	Har		data pdf pinout
5	VIDEO-OP	Vs:±5.5V Vo:±3.4V Vi0:2mV f:330Mc	d,8SD	Har		data pdf pinout
5	VIDEO-OP	Vs:±5.5V Vo:±3.4V Vi0:2mV f:315Mc	d,8SD	Har		data pinout
9	VIDEO-OP	Vs:±6V Vo:±3.2V Vi0:1mV f:450Mc	d,8SD	Har		data pdf pinout
9	BUFF	Vs:12V Vo:±3.3V Vi0:8mV f:750Mc	d,8SD	Har		data pdf pinout
0MJ/883	BUFF	Vs:12V Vo:±3V Vi0:±25mV f:750Mc	8D	Har		data pinout
0Y	HI-SLEW	Vs:12V Vo:<±3.3V Vi0:<8mV f:<100Mc	d	Har		data pinout
2	PROG-GAIN	Vs:±6V Vi0:8mV f:850Mc f:300Mkc	d,8SD	Har		data pdf pinout
2MJ/883	PROG-GAIN	Vs:12V Vo:±3V Vi0:±25mV f:850Mc	8D	Har		data pinout
3	PROG-GAIN	Vs:12V Vi0:8mV f:850Mc f:300Mkc	d,8SD	Har		data pdf pinout
3M	PROG-GAIN	Vs:12V Vo:±3V Vi0:±25mV f:850Mc	8D,20C	Har		data pinout
4	PROG-GAIN	Vs:±6V Vi0:8mV f:850Mc f:220Mc	d,8SD	Har		data pdf pinout
5	PROG-GAIN	Vs:±6V Vo:±3.2V Vi0:2mV f:225Mc	d,8SD	Har		data pdf pinout
5MJ/883	PROG-GAIN	Vs:±6V Vo:±3V Vi0:±10mV f:225Mc	8D	Har		data pinout
0	HI-SPEED	Vs:12V Vi0:2mV f:300Mkc	d,8SD	Har		data pdf pinout
0MJ/883	FB-AMP	Vs:12V Vo:±3V Vi0:±6mV f:850Mc	8D	Har		data pinout
0Y	HI-SPEED	Vs:12V Vi0:<2mV f:<670Mc	d	Har		data pinout
0	FB-AMP	Vs:±6V Vi0:2mV f:850Mc f:300Mkc	d,8SD	Har		data pdf pinout
0M	FB-AMP	Vs:12V Vo:±3V Vi0:±6mV f:850Mc	8D,20C	Har		data pinout
0Y	HI-SPEED	Vs:12V Vo:<±3.3V Vi0:<2mV f:<300Mc	d	Har		data pinout
5	VIDEO-OP	Vs:±5.5V Vo:±3.4V Vi0:2mV f:360Mc	d,8SD	Har		data pdf pinout
5M	VIDEO-OP	Vs:12V Vo:±3V Vi0:±5mV f:360Mc	8D,20C	Har		data pinout
5	VIDEO-OP	Vs:±5.5V Vo:±3.4V Vi0:2mV f:330Mc	d,8SD	Har		data pinout
5MJ/883	VIDEO-OP	Vs:12V Vo:±3V Vi0:±5mV f:360Mc	8D	Har		data pinout
9	VIDEO-OP	Vs:±6V Vo:±3.2V Vi0:1mV f:450Mc	d,8SD	Har		data pdf pinout
5	2xVIDEO-OP	Vs:±5.5V Vo:±3.4V Vi0:2mV f:400Mc	d,8SD	Har		data pinout
2I	2xBUFF	Vs:±5.5V Vo:±3.2V Vi0:2mV f:350Mc	8SD	Har		data pdf pinout
2MJ/883	2xBUFF	Vs:12V Vo:±3V Vi0:±10mV f:340Mc	8D	Har		data pinout
5	2xVIDEO-OP	Vs:±5.5V Vo:±3.4V Vi0:2mV f:530Mc	d,14SD	Har		data pinout
5MJ/883	2xVIDEO-OP	Vs:12V Vo:±3V Vi0:±5mV f:530Mc	14D	Har		data pinout
5	4xVIDEO-OP	Vs:±5.5V Vo:±3.4V Vi0:2mV f:560Mc	d,14SD	Har		data pinout
2	4xPROG-GAIN	Vs:±5.5V Vo:±3.2V Vi0:2mV f:350Mc	d,14SD	Har		data pinout
2MJ/883	4xPROG-GAIN	Vs:12V Vo:±2.5V Vi0:±10mV f:340Mc	14D	Har		data pinout
0	PD	850nm Vb:4.5...5.5V Ib:15mA	TO46	Hon		data pinout
2	PD	850nm S:<0.6mA/mW	TO46	Hon		data pinout
3	PD	850nm S:<0.33mA/mW	TO46	Hon		data pinout
0	PD	850nm Vb:4.5...5.5V Ib:15mA	TO18	Hon		data pinout
2	PD	850nm S:<0.58mA/mW	TO18	Hon		data pinout
3	PD	850nm Vb:4.5...5.5V Ib:13mA	TO18	Hon		data pinout
6	PD	850nm Resp:5V/mW Vb:4.5...5.5V	TO18	Hon		data pinout
9	PD	850nm Vb:4.5...5.5V Ib:6mA	TO18	Hon		data pinout
1	PD	850nm Vb:4.5...5.5V Ib:6mA	TO18	Hon		data pinout
3	PD	850nm S:<0.33mA/mW	TO18	Hon		data pinout
8	PD	850nm Resp:8V/mW Vb:<6V Ib:11mA	TO18	Hon		data pinout
2	PD	850nm Vb:4.5...5.5V Ib:20mA		Hon		data pinout
3	PD	850nm Vb:4.5...5.5V Ib:15mA		Hon		data pinout
6	PD	850nm Resp:80µV/mW Vb:<6V Ib:11mA		Hon		data pinout
1	PD	850nm Resp:7.5V/mW Vb:<6V Ib:9mA		Hon		data pinout
2	PD	850nm Resp:7.5V/mW Vb:<6V Ib:9mA		Hon		data pinout
2	PD	850nm Vb:4.5...5.5V Ib:20mA		Hon		data pinout
3	PD	850nm Vb:4.5...5.5V Ib:15mA		Hon		data pinout
6	PD	850nm Vb:<6V Ib:15mA		Hon		data pinout
1	PD	850nm Vb:<6V Ib:9mA		Hon		data pinout
0	IRED	850nm 50/125-200/230µmm I:1.9mcd	TO46	Hon		data pinout
3	IRED	850nm 50/125-200/230µmm I:1.9mcd	TO46	Hon		data pinout
0	IRED	850nm 50/125-200/230µmm I:1.9mcd	TO18	Hon		data pinout
3	IRED	850nm 50/125-200/230µmm I:1.9mcd	TO18	Hon		data pinout
6	IRED	850nm 50/125-200/230µmm I:1.9mcd	TO18	Hon		data pinout
0	IRED	850nm 50/125-200/230µmm I:1.9mcd	TO46	Hon		data pinout

Type	Device	Short Description	Fig.	Manu	Comparision Types	More at
HFE4053	IRED	850nm 50/125-200/230µmm I:1.9mcd	TO46	Hon		data pinout
HFE4070	IRED	850nm 50/125-200/230µmm I:1.9mcd	TO18	Hon		data pinout
HFE4073	IRED	850nm 50/125-200/230µmm I:1.9mcd	TO18	Hon		data pinout
HFE4211	IRED	850nm 50/125-100/140µmm I:1.9mcd		Hon		data pinout
HFE4213	IRED	850nm 50/125-100/140µmm I:1.9mcd		Hon		data pinout
HFE4216	IRED	850nm 62.5/125µmm I:1.9mcd Vf:1.6V		Hon		data pinout
HFE4217	IRED	850nm 62.5/125µmm I:1.9mcd Vf:1.6V		Hon		data pinout
HFE4218	IRED	850nm 62.5/125µmm I:1.9mcd Vf:1.6V		Hon		data pinout
HFE4222	IRED	850nm 50/125µmm I:1.9mcd Vf:1.84V		Hon		data pinout
HFE4401	IRED	850nm 50/125-100/140µmm I:1.9mcd	FIBER-DIP	Hon		data pinout
HFE4403	IRED	850nm 50/125-100/140µmm I:1.9mcd	FIBER-DIP	Hon		data pinout
HFE4870	IRED	850nm 50/125-100/140µmm I:1.9mcd		Hon		data pinout
HFE4871	IRED	850nm 50/125-100/140µmm I:1.9mcd		Hon		data pinout
HFE4872	IRED	850nm 50/125-100/140µmm I:1.9mcd		Hon		data pinout
HFE4873	IRED	850nm 50/125-100/140µmm I:1.9mcd		Hon		data pinout
HFS23	HI-SPEED	Vs:±15V Vu:96dB Vo:±10V f:30Mc	12A	Ilc		data pinout
HFSD-1 A..Z	Si-Di	GI, S, 200...1000V, 0,3A, <800ns HFSD-1=400V, A=600V, B=800V, C=1000C, Z=200V	31a	Sak	BA 157...159, BY 204/.., BY 208/.., ++	data
HFX6015	LED	rd 1000µmm I:1.9mcd Vf:2.5V If:50mA		Hon		data pinout
HG	Si-N	=2SC2806 (Typ-Code/Stempel/marking)	25	Tos	→2SC2806	data
HG	Si-N	=2SC3439-G (Typ-Code/Stempel/marking)	39	Mit	→2SC3439	data
HG	Z-Di	=Z2 SMB-56 (Typ-Code/Stempel/marking)	71(5x3,5)	Fag	→Z2 SMB-56	data
HGE 122..24005	SMPS-IC	DC-DC Converter	Module	Shi	-	
HGF 051..12001	SMPS-IC	DC-DC Converter	Module	Shi	-	
HGG 122..24005	SMPS-IC	DC-DC Converter	Module	Shi	-	
HH	Si-N	=2SC4253 (Typ-Code/Stempel/marking)	35(2mm)	Tos	→2SC4253	data
HH	Si-N	=2SC3125 (Typ-Code/Stempel/marking)	35	Tos	→2SC3125	data
HH	Si-N	=2SC3439-H (Typ-Code/Stempel/marking)	39	Mit	→2SC3439	data
HH	Si-N	=KTC 3881 (Typ-Code/Stempel/marking)	35	Kec	→KTC 3881(S)	data
HH	Si-N	=KTC 4081 (Typ-Code/Stempel/marking)	35(2mm)	Kec	→KTC 4081	data
HH	Z-Di	=Z2 SMB-62 (Typ-Code/Stempel/marking)	71(5x3,5)	Fag	→Z2 SMB-62	data
HH 8090 B	SMPS-IC	Schaltregler/switching Regulator	≈10-SIP	Mat	-	
HH 8360	Hybrid-IC	8 Bit D/a-converier	≈13-SIL	Mat	-	
HHs	C-Di	=BBY 51-07 (Typ-Code/Stempel/marking)	44	Sie	→BBY 51-07	data
HI	Si-N	=2SC4247 (Typ-Code/Stempel/marking)	35(2mm)	Tos	→2SC4247	data
HI-0200	SPST	Ron:Ohm Ton:240ns at 15/-15V	14D,10A	Isi		data pdf pinout
HI-0201	SPST	Vs:±4.5..±18V Ron:Ohm at 15/-15V	d,16D,14D	Max		data pdf pinout
HI-0303	SPDT	Ron:Ohm Ton:210ns at 15/-15V	14DS	Isi		data pdf pinout
HI-0390	SPDT	Ron:Ohm Ton:210ns at 15/-15V	16D	Isi,Max		data pdf pinout
HI-05042	SPDT	Ron:Ohm Ton:370ns at 15/-15V	16D	Isi		data pdf pinout
HI-05043	SPDT	Ron:Ohm Ton:370ns at 15/-15V	16D	Isi		data pdf pinout
HI-05047	QPST	Ron:Ohm Ton:370ns at 15/-15V	16D	Isi		data pdf pinout
HI-05049	DPST	Ron:Ohm Ton:370ns at 15/-15V	16D	Isi		data pdf pinout
HI-05051	SPDT	Ron:Ohm Ton:370ns at 15/-15V	16D	Isi		data pdf pinout
HI-0506	MULT16	Ron:Ohm Ton:250ns at 15/-15V	28DCS	Isi		data pdf pinout
HI-0507	MULT8	Ron:Ohm Ton:250ns at 15/-15V	28D,16D	Isi		data pdf pinout
HI-0508	MULT8	Ron:Ohm Ton:250ns at 15/-15V	16DS	Isi,Max		data pdf pinout
HI-0509	MULT4	Ron:Ohm Ton:250ns at 15/-15V	16D,20CS	Isi,Max		data pdf pinout
HI-0516	MULT16/8	Ron:Ohm Ton:120ns at 15/-15V	28D	Isi		data pdf pinout
HI-0518	MULT8/4	Ron:Ohm Ton:120ns at 15/-15V	18D	Isi		data pdf pinout
HI-0524	MULT4	Ron:Ohm Ton:180ns at 15/-15V	18D	Isi		data pdf pinout
HI-0539	MULT4	Vs:±5..±18V Ron:650Ohm at 15/-15V	16D	Isi		data pdf pinout
HI-0546	MULT16	Ron:Ohm Ton:300ns at 15/-15V	28DCS	Isi		data pdf pinout
HI-0547	MULT8	Ron:Ohm Ton:300ns at 15/-15V	28DCS	Isi		data pdf pinout
HI-0548	MULT8	Ron:Ohm Ton:300ns at 15/-15V	16D,20CS	Isi		data pdf pinout
HI-0549	MULT4	Ron:Ohm Ton:300ns at 15/-15V	16D,20CS	Isi		data pdf pinout
HI-1818A	MULT8	Ron:Ohm Ton:300ns at 15/-15V	16D	Isi		data pdf pinout
HI-1828A	MULT4	Ron:Ohm Ton:300ns at 15/-15V	16D	Isi		data pdf pinout
HI-200	SPST	Vs:±15V Ron:Ohm at 15/-15V	14D	Isi		data pdf pinout
HI-303	SPDT	Vs:±15V Ron:aOhm at 15/-15V	14D	Isi		data pdf pinout
HI-5043	SPDT	Ron:Ohm Ton:370ns at 15/-15V	16S	Isi		data pdf pinout
HI-5051	SPDT	Ron:Ohm Ton:370ns at 15/-15V	16S	Isi		data pdf pinout
HI-508 . (-2,-5)	CMOS-IC	Fault Protected 8 Ch. Multiplexer -2: -55...+125°, -5: 0...+75°	16-DIP,DIC	Max	-	
HI-509 A(-2,-5)	CMOS-IC	Fault Protected 4 Ch. Multiplexer	16-DIP,DIC	Max	-	
HI 5703	A/D-IC	SMD, 10 Bit, 40Msps	28-MDIP	Har	-	pdf
HI 5714	A/D-IC	SMD, 8 Bit, 40, 60, 75Msps	28-MDIP	Har	-	pdf
HI 5731 BIB	D/A-IC	=HI 5731BIP: SMD	28-MDIP			pdf
HI 5731 BIP	D/A-IC	12 Bit, 100MHz	28-DIP	Har		pdf
HI 5735 KCB	D/A-IC	=HI 5735KCP: SMD	28-MDIP			pdf
HI 5735 KCP	D/A-IC	12 Bit, Video, 80MHz	28-DIP	Har		pdf
HI 5746	A/D-IC	SMD, 10 Bit, 40Msps	28-MDIP	Har	-	pdf
HI 5766	A/D-IC	SMD, 10 Bit, 60Msps	28-MDIP	Har	-	pdf
HI 5804	A/D-IC	SMD, 12 Bit, 5Msps	28-MDIP	Har	-	pdf
HI 5805	A/D-IC	SMD, 12 Bit, 5Msps	28-MDIP	Har	-	pdf
HI 7188	A/D-IC	SMD, 16 Bit, 8-Channel	44-MP	Har	-	pdf
HI-	MOS-N-FET-d	=3SK188 (Typ-Code/Stempel/marking)	44	Hit	→3SK188	data
HIC106	ER+	Io=1A Vo:2..37.5V P:2.7W	12A	Tix		data pinout
HIC107	ER-	Io=1A Vo:-2..-37V P:2.7W	12A	Tix		data pinout
HIN 200	I/O-IC	SMD, 5xLine Drv, 0xReceiver f. RS 232	20-MDIP	Har	-	
HIN 201	I/O-IC	SMD, 2xLine Drv, 2xReceiver f. RS 232	16-MDIP	Har	-	pdf
HIN 202	I/O-IC	(SMD),2xLine Drv, 2xReceiver f. RS 232	16-(M)DIP	Har	-	pdf

ype	Device	Short Description	Fig.	Manu	Comparision Types	More at
N 204	I/O-IC	SMD, 4xLine Drv, 0xReceiver f. RS 232	16-MDIP	Har	-	
N 206	I/O-IC	(SMD),4xLine Drv, 3xReceiver f. RS 232	24-(M)DIP	Har	-	pdf
N 207	I/O-IC	(SMD),5xLine Drv, 3xReceiver f. RS 232	24-(M)DIP	Har	-	pdf
N 208	I/O-IC	(SMD),4xLine Drv, 4xReceiver f. RS 232	24-(M)DIP	Har	-	pdf
N 209	I/O-IC	SMD, 3xLine Drv, 5xReceiver f. RS 232	24-MDIP	Har	-	
N 211	I/O-IC	SMD, 4xLine Drv, 5xReceiver f. RS 232	28-MDIP	Har	-	pdf
N 213	I/O-IC	SMD, 4xLine Drv, 5xReceiver f. RS 232	28-MDIP	Har	-	pdf
N 230	I/O-IC	SMD, 5xLine Drv, 0xReceiver f. RS 232	20-MDIP	Har	-	pdf
N 231	I/O-IC	SMD, 2xLine Drv, 2xReceiver f. RS 232	16-MDIP	Har	-	pdf
N 232	I/O-IC	(SMD),2xLine Drv, 2xReceiver f. RS 232	16-(M)DIP	Har	-	pdf
N 234	I/O-IC	SMD, 4xLine Drv, 0xReceiver f. RS 232	16-MDIP	Har	-	pdf
N 236	I/O-IC	(SMD),4xLine Drv, 3xReceiver f. RS 232	24-(M)DIP	Har	-	pdf
N 237	I/O-IC	(SMD),5xLine Drv, 3xReceiver f. RS 232	24-(M)DIP	Har	-	pdf
N 238	I/O-IC	(SMD),4xLine Drv, 4xReceiver f. RS 232	24-(M)DIP	Har	-	pdf
N 239	I/O-IC	(SMD),3xLine Drv, 5xReceiver f. RS 232	24-(M)DIP	Har	-	pdf
N 240	I/O-IC	SMD, 5xLine Drv, 5xReceiver f. RS 232	44-MP	Har	-	pdf
N 241	I/O-IC	SMD, 4xLine Drv, 5xReceiver f. RS 232	28-MDIP	Har	-	pdf
P 2030	LIN-IC	MCT/IGBT Gate Drv, 30V, 6A, 180kHz	28-PLCC	Har	-	
P 2100	LIN-IC	SMD, Mosfet Half Bridge Drv, 100V, 2A	8-MDIP	Har	-	pdf
P 2500	LIN-IC	Mosfet Half Bridge Drv, 500V, 2A	14-,16-DIP	Har	-	pdf
P 4020	LIN-IC	SMD, Mosfet Bridge Drv, 0,5A, 100kHz	20-MDIP	Har	-	pdf
P 4080(A)	LIN-IC	Mosfet Bridge Drv, 1..80V, 2,5A, 1MHz	20-DIP	Har	-	pdf
P 4081(A)	LIN-IC	Mosfet Bridge Drv, 1..80V, 2,5A, 1MHz	20-DIP	Har	-	pdf
P 4082	LIN-IC	Mosfet Bridge Drv, 1..80V, 1,25A	16-DIP	Har	-	pdf
P 5010	LIN-IC	Mosfet Half Bridge, Reg, 5V→3,3V/7A	≈7-SIP	Har	-	pdf
P 5011	LIN-IC	Mosfet Half Bridge, Reg, 5V→3,3V/7A	≈7-SIP	Har	-	pdf
P 5500	LIN-IC	MOSFETIGBT Half Bridge Drv, 500V, 2,3A	20-DIP	Har	-	
T 5609..5610	Si-N/P	→CS 5609..5610				
T 9010..9022	Si-N/P	→CS 9010..9022				
	Si-N	=2SC3439-J (Typ-Code/Stempel/marking)	39	Mit	→2SC3439	data
	Si-N	=2SC3602 (Typ-Code/Stempel/marking)	25	Tos	→2SC3602	data
	Si-N	=2SC3828 (Typ-Code/Stempel/marking)	44	Tos	→2SC3828	data
	Si-N	=2SD1006-HK (Typ-Code/Stempel/marking)	39	Nec	→2SD1006	data
	Z-Di	=Z2 SMB-68 (Typ-Code/Stempel/marking)	71(5x3,5)	Fag	→Z2 SMB-68	data
Z 101(S)	LIN-IC	Hall-IC, Magnet-gabelschranke/vane Sw.		Sie		
Z 121	LIN-IC	Hall-IC, Magnet-gabelschranke/vane Sw.		Sie		pdf
	Si-N	=2SD1006-HL (Typ-Code/Stempel/marking)	39	Nec	→2SD1006	data
	Si-P	=2SA1502 (Typ-Code/Stempel/marking)	35	Say	→2SA1502	data
	Si-P	=2SA1722 (Typ-Code/Stempel/marking)	35(2mm)	Say	→2SA1722	data
	Si-N	=2SC3862 (Typ-Code/Stempel/marking)	35	Tos	→2SC3862	data
	Si-N/P+R	=XN 04382 (Typ-Code/Stempel/marking)	46	Mat	→XN 04382	data
	Z-Di	=Z2 SMB-75 (Typ-Code/Stempel/marking)	71(5x3,5)	Fag	→Z2 SMB-75	data
1221A	IR LD+MA	1200nm ø2mm I:1.9mcd Vf:1.1V If:45mA		Hit		data pinout
1221AC	IR LD+MA	1200nm ø1mm I:1.9mcd Vf:1.1V If:45mA		Hit		data pinout
1221C	IR LD+MA	1200nm ø5mm I:1.9mcd Vf:1.2V If:40mA		Hit		data pinout
1321AC	IR LD+MA	1300nm ø2mm I:1.9mcd Vf:1.25V		Hit		data pinout
1321FG	IR LD+MD	1310nm ø2mm I:1.9mcd If:50mA		Hit		data pinout
1322A	IR LD+MA	1310nm ø2mm I:1.9mcd Vf:1.2V If:70mA		Hit		data pinout
1322AC	IR LD+MA	1310nm ø1mm I:1.9mcd Vf:1.2V If:70mA		Hit		data pinout
1326	FIB IR LD+MD	1310nm 125µmm I:1.9mcd If:18mA		Hit		data pdf pinout
1326GN	IR LD+MD	1310nm ø2.1mm I:1.9mcd If:20mA		Hit		data pdf pinout
1327	FIB IR LD+MD	1310nm 125µmm I:1.9mcd If:18mA		Hit		data pdf pinout
1328DJS	IR LD+MD	1310nm 125µmm I:1.9mcd Vf:1.6V		Hit		data pdf pinout
1336DJS	IR LD+MD	1310nm 125µmm I:1.9mcd Vf:1.6V		Hit		data pdf pinout
1341A	IR LD+MA	1310nm 125µmm I:1.9mcd If:45mA		Hit		data pinout
1341AC	IR LD+MA	1310nm ø1mm I:1.9mcd If:45mA		Hit		data pinout
1341FG	IR LD+MD	1310nm 125µmm I:1.9mcd If:45mA		Hit		data pinout
1352	FIB IR LD+MD	1310nm 125µmm I:1.9mcd If:20mA		Hit		data pinout
1362	FIB IR LD+MA	1310nm 125µmm I:1.9mcd		Hit		data pinout
1521A	IR LD+MA	1550nm 125µmm I:1.9mcd Vf:1.2V		Hit		data pinout
1521AC	IR LD+MA	1550nm ø1mm I:1.9mcd Vf:1.2V If:50mA		Hit		data pinout
1541	FIB IR LD+MD	1550nm 125µmm I:1.9mcd		Hit		data pinout
1541A	IR LD+MA	1550nm 125µmm I:1.9mcd If:50mA		Hit		data pinout
1541AC	IR LD+MA	1550nm ø1mm I:1.9mcd If:50mA		Hit		data pinout
1541FG	IR LD+MD	1550nm ø2mm I:1.9mcd If:50mA		Hit		data pinout
1551	FIB IR LD+MA	1550nm ø2mm I:1.9mcd		Hit		data pinout
1553	FIB IR LD+MO	1557nm ø2mm I:1.9mcd		Hit		data pdf pinout
1560A	FIB IR LD+MA	1550nm ø2mm I:1.9mcd		Hit		data pdf pinout
1566AF	FIB IR LD	1550nm ø2mm I:1.9mcd If:50mA		Hit		data pinout
6312	RED LD+MD	rd 635nm ø2.1mm I:1.9mcd Vf:1.8V		Hit		data pinout
6314	RED LD+MD	rd 635nm ø2.1mm I:1.9mcd Vf:1.8V		Hit		data pinout
6315	RED LD+MD	rd 635nm ø2.1mm I:1.9mcd Vf:1.8V		Hit		data pinout
6319G	RED LD+MD	rd 635nm ø2.1mm I:1.9mcd Vf:<2.7V		Hit		data pinout
6320G	RED LD+MD	rd 635nm ø2.1mm I:1.9mcd Vf:<2.7V		Hit		data pinout
6321G	RED LD+MD	rd 635nm ø2.1mm I:1.9mcd Vf:<2.7V		Hit		data pinout
6324MG	RED LD+MD	rd 635nm ø2.1mm I:1.9mcd Vf:<2.7V		Hit		data pinout
6325G	RED LD+MD	rd 635nm ø2.1mm I:1.9mcd Vf:2.2V		Hit		data pinout
6326G	RED LD+MD	rd 635nm ø2.1mm I:1.9mcd Vf:2.2V		Hit		data pinout
6327MG	RED LD+MD	rd 635nm ø2.1mm I:1.9mcd Vf:2.2V		Hit		data pinout
6328MG	RED LD+MD	rd 635nm ø2.1mm I:1.9mcd Vf:2.2V		Hit		data pinout
6331G	RED LD+MD	rd 635nm ø2.1mm I:1.9mcd Vf:2.2V		Hit		data pinout
6332G	RED LD+MD	rd 635nm ø2.1mm I:1.9mcd Vf:2.2V		Hit		data pinout
6333MG	RED LD+MD	rd 635nm ø2.1mm I:1.9mcd Vf:2.2V		Hit		data pdf pinout
6334MG	RED LD+MD	rd 635nm ø2.1mm I:1.9mcd Vf:2.2V		Hit		data pdf pinout

Type	Device	Short Description	Fig.	Manu	Comparision Types	More at
HL6501MG	RED LD+MD	rd 658nm ø2.1mm I:1.9mcd Vf:2.6V		Hit		data pdf pinout
HL6503MG	RED LD	rd 664nm ø2.1mm I:1.9mcd Vf:2.6V		Hit		data pdf pinout
HL6504FM	RED LD	rd 664nm ø2.1mm I:1.9mcd Vf:2.6V		Hit		data pdf pinout
HL6712	RED LD+MD	rd 670nm ø2.1mm I:1.9mcd		Hit		data pinout
HL6713	RED LD+MD	rd 670nm ø2.1mm I:1.9mcd Vf:2.3V		Hit		data pdf pinout
HL6714	RED LD+MD	rd 670nm ø2.1mm I:1.9mcd		Hit		data pinout
HL6720	RED LD+MD	rd 670nm ø2.1mm I:1.9mcd Vf:<2.7V		Hit		data pdf pinout
HL6722	RED LD+MD	rd 670nm ø2.1mm I:1.9mcd Vf:<2.7V		Hit		data pdf pinout
HL6724	RED LD+MD	rd 670nm ø2.1mm I:1.9mcd Vf:<2.7V		Hit		data pdf pinout
HL6726	RED LD+MD	rd 685nm ø2.1mm I:1.9mcd Vf:<2.8V		Hit		data pinout
HL6727	RED LD+MD	rd 690nm ø2.1mm I:1.9mcd Vf:<3V		Hit		data pdf pinout
HL6733FM	RED LD	rd 690nm ø2.1mm I:1.9mcd Vf:2.5V		Hit		data pdf pinout
HL6734FM	RED LD+MD	rd 690nm ø2.1mm I:1.9mcd Vf:2.6V		Hit		data pdf pinout
HL6738MG	RED LD+MD	rd 690nm ø2.1mm I:1.9mcd Vf:2.5V		Hit		data pdf pinout
HL7801	IR LD+MD	rd 780nm ø2.1mm I:1.9mcd Vf:1.8V		Hit		data pdf pinout
HL7802	IR LD+MD	785nm ø2mm I:1.9mcd Vf:1.8V If:90mA		Hit		data pdf pinout
HL7806	IR LD+MD	785nm ø2mm I:1.9mcd Vf:1.8V If:60mA		Hit		data pdf pinout
HL7831	IR LD+MD	785nm ø2mm I:1.9mcd Vf:1.8V If:65mA		Hit		data pdf pinout
HL7832	IR LD+MD	785nm ø2mm I:1.9mcd Vf:1.8V If:65mA		Hit		data pdf pinout
HL7836	IR LD+MD	785nm ø2.1mm I:1.9mcd Vf:<2.7V		Hit		data pdf pinout
HL7838	IR LD+MD	rd 780nm ø2mm I:1.9mcd		Hit		data pdf pinout
HL7843	IR LD+MD	785nm ø2.1mm I:1.9mcd Vf:<2.3V		Hit		data pdf pinout
HL7851	IR LD+MD	785nm ø2.1mm I:1.9mcd Vf:2.3V		Hit		data pdf pinout
HL7852	IR LD+MD	785nm ø2.1mm I:1.9mcd Vf:2.3V		Hit		data pdf pinout
HL7853	IR LD+MD	785nm ø2.1mm I:1.9mcd Vf:2.3V		Hit		data pdf pinout
HL7859	IR LD+MD	830nm ø2.1mm I:1.9mcd		Hit		data pdf pinout
HL8311G	IR LD+MD	830nm ø2mm I:1.9mcd Vf:1.9V If:90mA		Hit		data pdf pinout
HL8312E	IR LD+MD	830nm ø2mm I:1.9mcd Vf:1.9V If:90mA		Hit		data pdf pinout
HL8312G	IR LD+MD	830nm ø2mm I:1.9mcd Vf:1.9V If:90mA		Hit		data pdf pinout
HL8314E	IR LD+MD	830nm ø2mm I:1.9mcd Vf:1.9V If:120mA		Hit		data pdf pinout
HL8314G	IR LD+MD	830nm ø2mm I:1.9mcd Vf:1.9V If:120mA		Hit		data pdf pinout
HL8315E	IR LD+MD	830nm ø2mm I:1.9mcd Vf:1.8V If:75mA		Hit		data pdf pinout
HL8325	IR LD+MD	785nm ø2.1mm I:1.9mcd Vf:<2.5V		Hit		data pdf pinout
HLC1395	PS	LED/NPN CTR:>6% Vbs:<1.6V Ibs:20mA		Hon		data pdf pinout
HLC1397	PS	LED/Dar CTR:>35% Vbs:<1.6V Ibs:20mA		Hon		data pinout
HLC2701	PD	900(880)nm Vb:4.5...5.5V Ib:<7mA		Hon		data pdf pinout
HLC2705	PD	900(880)nm Vb:4.5...5.5V Ib:<12mA		Hon		data pdf pinout
HLCP	LED	rd 8.89x195mm I:30mcd Vf:1.8V If:3mA		Agt,Hpa		data pdf pinout
HLCP-J100	LED	rd 1.52x58mm I:1mcd Vf:1.6V If:1mA		Agt,Hpa		data pdf pinout
HLD 00006...05003	SMPS-IC	DC-DC Converter	Module	Shi	-	
HLE 00006...12003	SMPS-IC	DC-DC Converter	Module	Shi	-	
HLM1-BL00	LED	am ø13.3mm I:15000mcd Vf:1.9V	T4	Hpa		data pinout
HLM 358 P	LIN/OP-IC	Low Power Dual Op Amp	DIC	Hsm	-	pdf pinout
HLM 358 S	LIN/OP-IC	Low Power Dual Op Amp	MDIP	Hsm	-	pdf pinout
HLMA-CH00	LED	or+am ø5mm I:3000mcd Vf:1.9V If:20mA	T1¾	Hpa	-	data pdf pinout
HLMA-DH00	LED	or+am ø5mm I:1000mcd Vf:1.9V If:20mA	T1¾	Hpa		data pdf pinout
HLMA-KH00	LED	or+am ø3mm I:200mcd Vf:1.9V If:20mA	T1	Hpa		data pdf pinout
HLMA-PG00	LED	rd ø1.8mm I>25mcd Vf:1.9V If:20mA	T1	Agt		data pdf pinout
HLMA-PH00	LED	or ø1.8mm I>25mcd Vf:1.9V If:20mA	T1	Agt		data pdf pinout
HLMA-PJ00	LED	am ø1.8mm I>25mcd Vf:1.9V If:20mA	T1	Agt		data pdf pinout
HLMA-PL00	LED	am ø1.8mm I>63mcd Vf:1.9V If:20mA	T1	Agt		data pdf pinout
HLMA-QF00	LED	rd ø1.8mm I>160mcd Vf:1.9V If:20mA	T1	Agt		data pdf pinout
HLMA-QG00	LED	rd ø1.8mm I>125mcd Vf:1.9V If:20mA	T1	Agt		data pdf pinout
HLMA-QH00	LED	or+am ø1.8mm I:500mcd Vf:1.9V	T1	Agt,Hpa		data pdf pinout
HLMA-QJ00	LED	am ø1.8mm I:500mcd Vf:1.9V If:20mA	T1	Agt		data pdf pinout
HLMA-QL00	LED	am ø1.8mm I>400mcd Vf:1.9V If:20mA	T1	Agt		data pdf pinout
HLMP-0300	LED	srd+yl+gr 7.2x2.4mm I:5mcd Vf:2.3V		Hpa,Qtc		data pinout
HLMP-0800	LED	rd/gr 7.2x2.4mm I:1.4mcd Vf:2.2V	T1¾	Hpa		data pinout
HLMP-0805	LED	gr/yl 7.2x2.4mm I:1.4mcd Vf:2.6V	T1¾	Agt		data pinout
HLMP-1000	LED	rd ø3mm I:2.5mcd Vf:1.6V If:20mA	T1	Hpa,Qtc		data pinout
HLMP-1100	LED	rd ø3mm I:2.5mcd Vf:5V If:13mA	T1	Hpa		data pinout
HLMP-1120	LED	rd ø3mm I:2.5mcd Vf:5V If:13mA	T1	Hpa		data pinout
HLMP-1300	LED	srd+yl+gr ø3mm I:8.5mcd Vf:2.1V	T1	Hpa		data pinout
HLMP-1320	LED	srd+yl+gr ø3mm I:12mcd Vf:2.1V	T1	Hpa,Qtc		data pinout
HLMP-1340	LED	rd+srd+yl+gr ø3mm I:45mcd Vf:2.2V	T1	Agt,Hpa,Qtc		data pdf pinout
HLMP-1400	LED	yl ø3mm I:5mcd Vf:2V If:10mA	T1	Hpa		data pdf pinout
HLMP-1440	LED	yl ø3mm I>23.5mcd Vf:2.1V If:20mA	T1	Agt		data pdf pinout
HLMP-1540	LED	gr ø3mm I:60mcd Vf:2.2V If:20mA	T1	Agt		data pdf pinout
HLMP-1600	LED	srd ø3mm I:8mcd Vf:5V If:10mA	T1	Hpa		data pinout
HLMP-1601	LED	srd ø3mm I:8mcd Vf:12V If:13mA	T1	Hpa		data pinout
HLMP-1620	LED	yl ø3mm I:8mcd Vf:5V If:10mA	T1	Hpa		data pinout
HLMP-1621	LED	yl ø3mm I:8mcd Vf:12V If:13mA	T1	Hpa		data pinout
HLMP-1640	LED	gr ø3mm I:8mcd Vf:5V If:12mA	T1	Hpa		data pinout
HLMP-1641	LED	gr ø3mm I:8mcd Vf:12V If:13mA	T1	Hpa		data pinout
HLMP-1700	LED	srd+yl+gr ø3mm I:2.3mcd Vf:1.9V	T1	Agt,Qtc		data pdf pinout
HLMP-1719	LED	yl ø3mm I:2.1mcd Vf:1.8V If:2mA	T1	Agt		data pdf pinout
HLMP-1790	LED	gr ø3mm I:2.4mcd Vf:1.9V If:2mA	T1	Agt,Qtc		data pdf pinout
HLMP-2300	LED	srd+yl+gr 8.89x3.81mm I:25mcd Vf:2V		Agt,Hpa,Qtc		data pdf pinout
HLMP-2350	LED	srd+yl+gr 195x3.81mm I:50mcd Vf:2V		Agt,Hpa,Qtc		data pdf pinout
HLMP-2600	LED	srd+yl+gr 8.89x3.81mm I:25mcd		Agt,Hpa		data pdf pinout
HLMP-2620	LED	srd+yl+gr 8.89x3.81mm I:25mcd		Agt,Hpa		data pdf pinout
HLMP-2635	LED	srd+yl+gr 8.89x195mm I:50mcd Vf:2.2V		Agt,Hpa		data pdf pinout
HLMP-2655	LED	srd+yl+gr 8.89x8.89mm I:50mcd		Agt,Hpa,Qtc		data pdf pinout
HLMP-2670	LED	srd+yl+gr 8.89x8.89mm I:50mcd		Agt,Hpa,Qtc		data pdf pinout
HLMP-2685	LED	srd+yl+gr 8.89x195mm I:100mcd		Agt,Hpa,Qtc		data pdf pinout

Type	Device	Short Description	Fig.	Manu	Comparision Types	More at
LMP-2950	LED	rd/yl+rd/gr 8.89x8.89mm I:50mcd		Agt,Hpa		data pdf pinout
LMP-3000	LED	rd ø5mm I:2.5mcd Vf:1.6V If:20mA	T1¾	Hpa		data pinout
LMP-3105	LED	rd ø5mm I:3mcd Vf:5V If:13mA	T1¾	Hpa		data pinout
LMP-3112	LED	rd ø5mm I:3mcd Vf:12V If:13mA	T1¾	Hpa		data pinout
LMP-3200	LED	rd ø5mm I:2mcd Vf:1.6V If:20mA	T1¾	Hpa		data pinout
LMP-3201	LED	srd ø5mm I:4mcd Vf:1.6V If:20mA	T1¾	Hpa		data pinout
LMP-3300	LED	srd+yl+gr ø5mm I:5.2mcd Vf:2.1V	T1¾	Hpa		data pinout
LMP-3315	LED	srd ø5mm I:60mcd Vf:2.2V If:20mA	T1¾	Cml,Hpa,Qtc		data pdf pinout
LMP-3330	LED	yl+gr ø5mm I:70mcd Vf:2.1V If:10mA	T1¾	Hpa		data pinout
LMP-3350	LED	srd ø5mm I:3.5mcd Vf:1.9V If:10mA	T1¾	Hpa		data pinout
LMP-3351	LED	srd ø5mm I:7mcd Vf:1.9V If:10mA	T1¾	Hpa		data pinout
LMP-3365	LED	srd ø5mm I:10mcd Vf:1.9V If:10mA	T1¾	Hpa		data pinout
LMP-3366	LED	srd ø5mm I:18mcd Vf:1.9V If:10mA	T1¾	Hpa		data pinout
LMP-3390	LED	srd+yl+gr ø5mm I:55mcd Vf:2.2V	T1¾	Agt,Hpa		data pdf pinout
LMP-3450	LED	yl ø5mm I:4mcd Vf:2V If:10mA	T1¾	Hpa		data pinout
LMP-3451	LED	yl ø5mm I:10mcd Vf:2V If:10mA	T1¾	Hpa		data pinout
LMP-3465	LED	yl ø5mm I:12mcd Vf:2V If:10mA	T1¾	Hpa		data pinout
LMP-3466	LED	yl ø5mm I:18mcd Vf:2V If:10mA	T1¾	Hpa		data pinout
LMP-3490	LED	yl ø5mm I:55mcd Vf:2.1V If:20mA	T1¾	Agt		data pdf pinout
LMP-3553	LED	gr ø5mm I:3.2mcd Vf:2.1V If:10mA	T1¾	Hpa		data pinout
LMP-3554	LED	gr ø5mm I:10mcd Vf:2.1V If:10mA	T1¾	Hpa		data pinout
LMP-3567	LED	gr ø5mm I:7mcd Vf:2.1V If:10mA	T1¾	Hpa		data pinout
LMP-3568	LED	gr ø5mm I:15mcd Vf:2.1V If:10mA	T1¾	Hpa		data pinout
LMP-3590	LED	gr ø5mm I:55mcd Vf:2.2V If:20mA	T1¾	Agt		data pdf pinout
LMP-3600	LED	srd ø5mm I:8mcd Vf:5V If:10mA	T1¾	Hpa		data pinout
LMP-3601	LED	srd ø5mm I:8mcd Vf:12V If:13mA	T1¾	Hpa		data pinout
LMP-3650	LED	yl ø5mm I:8mcd Vf:5V If:10mA	T1¾	Hpa		data pinout
LMP-3661	LED	yl ø5mm I:8mcd Vf:12V If:13mA	T1¾	Hpa		data pinout
LMP-3680	LED	gr ø5mm I:8mcd Vf:5V If:12mA	T1¾	Hpa		data pinout
LMP-3681	LED	gr ø5mm I:8mcd Vf:12V If:13mA	T1¾	Hpa		data pinout
LMP-3707	LED	srd ø5mm I>90.2mcd Vf:1.9V If:20mA	T1¾	Agt		data pdf pinout
LMP-3750	LED	srd+yl+gr ø5mm I:265mcd Vf:2.2V	T1¾	Agt,Hpa,Qtc		data pdf pinout
LMP-3760	LED	srd+yl+gr ø5mm I:11mcd Vf:2.1V	T1¾	Hpa		data pinout
LMP-3807	LED	yl ø5mm I:140mcd Vf:2.1V If:20mA	T1¾	Agt		data pdf pinout
LMP-3850	LED	yl ø5mm I:150mcd Vf:2.1V If:20mA	T1¾	Agt		data pdf pinout
LMP-3907	LED	gr ø5mm I>111.7mcd Vf:2.2V If:20mA	T1¾	Agt		data pdf pinout
LMP-3950	LED	gr ø5mm I:265mcd Vf:2.2V If:20mA	T1¾	Agt		data pdf pinout
LMP-4000	LED	rd/gr ø5mm I:4.2mcd Vf:2.2V If:10mA	T1¾	Hpa		data pinout
LMP-4015	LED	gr/yl ø5mm I>20mcd Vf:2.6V If:20mA	T1¾	Agt		data pinout
LMP-4100	LED	rd ø4.7mm I:1000mcd Vf:1.8V If:20mA	T1¾	Hpa,Qtc		data pinout
LMP-4700	LED	srd+yl+gr ø5mm I:3mcd Vf:1.9V If:2mA	T1¾	Qtc		data pinout
LMP-4719	LED	yl ø5mm I:2.1mcd Vf:1.8V If:2mA	T1¾	Agt		data pdf pinout
LMP-4740	LED	gr ø5mm I:2.3mcd Vf:1.9V If:2mA	T1¾	Agt		data pdf pinout
LMP-6203	LED	rd+srd+yl+gr ø1.65mm I:3mcd Vf:2.1V		Hpa		data pinout
LMP-6204	LED	rd+srd+yl ø1.65mm I:3mcd Vf:2V		Hpa		data pinout
LMP-6205	LED	rd+srd+yl ø1.65mm I:3mcd Vf:2V		Hpa		data pinout
LMP-6206	LED	rd+srd+yl+gr ø1.65mm I:3mcd Vf:2.1V		Hpa		data pinout
LMP-6208	LED	rd+srd+yl+gr ø1.65mm I:3mcd Vf:2.1V		Hpa		data pinout
LMP-6600	LED	srd+yl+gr ø1.65mm I:2mcd Vf:5V		Hpa,Qtc		data pinout
LMP-6620	LED	srd+yl ø1.65mm I:2mcd Vf:5V If:3.5mA		Hpa,Qtc		data pinout
LMP-8100	LED	rd ø5mm I:400mcd Vf:1.85V If:20mA	T1¾	Hpa		data pinout
LMP-8109	LED	rd+srd+yl+or+gr ø4.7mm I:260mcd	T1¾	Hpa		data pinout
LMP-8115	LED	rd+srd+yl+gr ø4.7mm I:75mcd Vf:2.2V	T1¾	Hpa		data pinout
LMP-8150	LED	rd ø13.3mm I:15000mcd Vf:1.85V	T4	Hpa		data pinout
LMP-AB01	LED	bl ø5mm I>240mcd Vf:3.5V If:20mA	T1¾	Agt		data pdf pinout
LMP-AD06	LED	rd ø5mm I:3200mcd Vf:2.3V If:20mA	T1¾	Agt		data pdf pinout
LMP-BB01	LED	gr+bl ø5mm I>1150mcd Vf:3.5V If:20mA	T1¾	Agt		data pdf pinout
LMP-BD06	LED	rd+am ø5mm I:2500mcd Vf:2.3V If:20mA	T1¾	Agt		data pdf pinout
LMP-C008	LED	rd ø5mm I>2900mcd Vf:1.9V If:20mA	T1¾	Agt		data pdf pinout
LMP-C025	LED	rd ø5mm I>500mcd Vf:1.9V If:20mA	T1¾	Agt		data pdf pinout
LMP-C027	LED	rd ø5mm I>500mcd Vf:1.9V If:20mA	T1¾	Agt		data pdf pinout
LMP-C208	LED	am ø5mm I>2600mcd Vf:1.9V If:20mA	T1¾	Agt		data pdf pinout
LMP-C225	LED	am ø5mm I>450mcd Vf:1.9V If:20mA	T1¾	Agt		data pdf pinout
LMP-C608	LED	rd ø5mm I>1000mcd Vf:1.9V If:20mA	T1¾	Agt		data pdf pinout
LMP-C625	LED	rd ø5mm I>500mcd Vf:1.9V If:20mA	T1¾	Agt		data pdf pinout
LMP-C627	LED	rd ø5mm I>500mcd Vf:1.9V If:20mA	T1¾	Agt		data pdf pinout
LMP-CB15	LED	bl ø5mm I>1150mcd Vf:3.8V If:20mA	T1¾	Agt		data pdf pinout
LMP-CB18	LED	bl ø5mm I:3500mcd Vf:3.4V If:20mA	T1¾	Agt		data pdf pinout
LMP-CB28	LED	bl ø5mm I:2400mcd Vf:3.4V If:20mA	T1¾	Agt		data pdf pinout
LMP-CB30	LED	bl ø5mm I>680mcd Vf:3.8V If:20mA	T1¾	Agt		data pdf pinout
LMP-CB38	LED	bl ø5mm I>1500mcd Vf:3.4V If:20mA	T1¾	Agt		data pdf pinout
LMP-CE15	LED	gr ø5mm I>2500mcd Vf:3.8V If:20mA	T1¾	Agt		data pdf pinout
LMP-CE18	LED	cy ø5mm I>11800mcd Vf:3.3V If:20mA	T1¾	Agt		data pdf pinout
LMP-CE23	LED	gr ø5mm I>2500mcd Vf:3.8V If:20mA	T1¾	Agt		data pdf pinout
LMP-CE28	LED	cy ø5mm I>7400mcd Vf:3.3V If:20mA	T1¾	Agt		data pdf pinout
LMP-CE30	LED	gr ø5mm I>1500mcd Vf:3.8V If:20mA	T1¾	Agt		data pdf pinout
LMP-CE38	LED	cy ø5mm I>3900mcd Vf:3.3V If:20mA	T1¾	Agt		data pdf pinout
LMP-CM15	LED	gr ø5mm I>4200mcd Vf:3.8V If:20mA	T1¾	Agt		data pdf pinout
LMP-CM18	LED	gr ø5mm I:12000mcd Vf:3.2V If:20mA	T1¾	Agt		data pdf pinout
LMP-CM28	LED	gr ø5mm I:9300mcd Vf:3.2V If:20mA	T1¾	Agt		data pdf pinout
LMP-CM30	LED	gr ø5mm I>2500mcd Vf:3.8V If:20mA	T1¾	Agt		data pdf pinout
LMP-CM38	LED	gr ø5mm I>5000mcd Vf:3.2V If:20mA	T1¾	Agt		data pdf pinout
LMP-CW15	LED	wht ø5mm I>4200mcd Vf:3.6V If:20mA	T1¾	Agt		data pdf pinout
LMP-CW18	LED	wht ø5mm I:6400mcd Vf:3.4V If:20mA	T1¾	Agt		data pdf pinout
LMP-CW23	LED	wht ø5mm I>2500mcd Vf:3.6V If:20mA	T1¾	Agt		data pdf pinout

Type	Device	Short Description	Fig.	Manu	Comparision Types	More at
HLMP-CW28	LED	wht ø5mm I:6400mcd Vf:3.4V If:20mA	T1¾	Agt		data pdf pinout
HLMP-CW30	LED	wht ø5mm I>1800mcd Vf:3.6V If:20mA	T1¾	Agt		data pdf pinout
HLMP-CW38	LED	wht ø5mm I:6400mcd Vf:3.4V If:20mA	T1¾	Agt		data pdf pinout
HLMP-CW70	LED	wht ø5mm I>400mcd Vf:3.6V If:20mA	T1¾	Agt		data pdf pinout
HLMP-CW78	LED	wht ø5mm I:570mcd Vf:3.4V If:20mA	T1¾	Agt		data pdf pinout
HLMP-D101	LED	rd ø4.7mm I:240mcd Vf:1.8V If:20mA	T1¾	Hpa,Qtc		data pinout
HLMP-D105	LED	rd ø4.7mm I:240mcd Vf:1.8V If:20mA	T1¾	Qtc		data pinout
HLMP-D150	LED	rd ø3.2mm I:3mcd Vf:1.6V If:1mA	T1	Hpa,Qtc		data pinout
HLMP-D155	LED	rd ø4.7mm I:10mcd Vf:1.6V If:1mA	T1¾	Qtc		data pinout
HLMP-D400	LED	or+gr ø5mm I:3mcd Vf:2.1V If:10mA	T1¾	Hpa,Qtc		data pinout
HLMP-D640	LED	gr ø5mm I:21mcd Vf:2.2V If:20mA	T1¾	Agt,Hpa,Qtc		data pdf pinout
HLMP-DB00	LED	bl ø5mm I:12mcd Vf:3V If:20mA	T1¾	Hpa		data pinout
HLMP-DB25	LED	bl ø3mm I>30mcd Vf:4V If:20mA	T1	Agt		data pdf pinout
HLMP-ED16	LED	rd ø5mm I>3200mcd Vf:2V If:20mA	T1¾	Agt		data pdf pinout
HLMP-ED25	LED	rd ø5mm I>2500mcd Vf:2V If:20mA	T1¾	Agt		data pdf pinout
HLMP-ED31	LED	rd ø5mm I>1900mcd Vf:2V If:20mA	T1¾	Agt		data pdf pinout
HLMP-ED80	LED	rd ø5mm I:7.2mW/Sr Vf:2.35V If:20mA	T1¾	Agt		data pdf pinout
HLMP-EG08	LED	rd ø5mm I>8000mcd Vf:1.9V If:20mA	T1¾	Agt		data pdf pinout
HLMP-EG15	LED	rd ø5mm I>2750mcd Vf:1.9V If:20mA	T1¾	Agt		data pdf pinout
HLMP-EG24	LED	rd ø5mm I>1300mcd Vf:1.9V If:20mA	T1¾	Agt		data pdf pinout
HLMP-EG30	LED	rd ø5mm I>1000mcd Vf:1.9V If:20mA	T1¾	Agt		data pdf pinout
HLMP-EH08	LED	or ø5mm I>8000mcd Vf:1.94V If:20mA	T1¾	Agt		data pdf pinout
HLMP-EH15	LED	or ø5mm I>2750mcd Vf:1.94V If:20mA	T1¾	Agt		data pdf pinout
HLMP-EH16	LED	or ø5mm I>3200mcd Vf:2.08V If:20mA	T1¾	Agt		data pdf pinout
HLMP-EH24	LED	or ø5mm I>1650mcd Vf:1.94V If:20mA	T1¾	Agt		data pdf pinout
HLMP-EH25	LED	or ø5mm I>2500mcd Vf:2.08V If:20mA	T1¾	Agt		data pdf pinout
HLMP-EH30	LED	or ø5mm I>1300mcd Vf:1.94V If:20mA	T1¾	Agt		data pdf pinout
HLMP-EH31	LED	or ø5mm I>1900mcd Vf:2.08V If:20mA	T1¾	Agt		data pdf pinout
HLMP-EJ08	LED	or ø5mm I>8000mcd Vf:1.98V If:20mA	T1¾	Agt		data pdf pinout
HLMP-EJ15	LED	or ø5mm I>1650mcd Vf:1.98V If:20mA	T1¾	Agt		data pdf pinout
HLMP-EJ24	LED	or ø5mm I>1300mcd Vf:1.98V If:20mA	T1¾	Agt		data pdf pinout
HLMP-EJ30	LED	or ø5mm I>765mcd Vf:1.98V If:20mA	T1¾	Agt		data pdf pinout
HLMP-EL08	LED	am ø5mm I>6200mcd Vf:2.02V If:20mA	T1¾	Agt		data pdf pinout
HLMP-EL15	LED	am ø5mm I>3600mcd Vf:2.02V If:20mA	T1¾	Agt		data pdf pinout
HLMP-EL16	LED	or+am ø5mm I>4200mcd Vf:2.15V	T1¾	Agt		data pdf pinout
HLMP-EL24	LED	am ø5mm I>2170mcd Vf:2.02V If:20mA	T1¾	Agt		data pdf pinout
HLMP-EL25	LED	am ø5mm I>2500mcd Vf:2.15V If:20mA	T1¾	Agt		data pdf pinout
HLMP-EL30	LED	am ø5mm I>1650mcd Vf:2.02V If:20mA	T1¾	Agt		data pdf pinout
HLMP-EL31	LED	am ø5mm I>1900mcd Vf:2.15V If:20mA	T1¾	Agt		data pdf pinout
HLMP-FW00	LED	wht ø5mm I:240mcd Vf:3.6V If:20mA	T1¾	Agt		data pdf pinout
HLMP-FW79	LED	wht ø5mm I:570mcd Vf:3.4V If:20mA	T1¾	Agt		data pdf pinout
HLMP-HB54	LED	bl ø5.2mm I:96mcd Vf:3.8V If:20mA	T1¾	Agt		data pdf pinout
HLMP-HB59	LED	bl 5.2x3.8mm I:500mcd Vf:3.4V	T1¾	Agt		data pdf pinout
HLMP-HD57	LED	rd 5.2x3.8mm I:870mcd Vf:2.2V	T1¾	Agt		data pdf pinout
HLMP-HD59	LED	gr 5.2x3.8mm I:1700mcd Vf:3.2V	T1¾	Agt		data pdf pinout
HLMP-HM54	LED	gr ø5.2mm I>450mcd Vf:3.8V If:20mA	T1¾	Agt		data pdf pinout
HLMP-K101	LED	rd ø3.2mm I:65mcd Vf:1.8V If:20mA	T1	Hpa,Qtc		data pinout
HLMP-K400	LED	or ø3mm I:6.5mcd Vf:1.9V If:10mA	T1	Hpa,Qtc		data pinout
HLMP-K600	LED	gr ø3mm I:4.5mcd Vf:2.1V If:10mA	T1	Hpa,Qtc		data pinout
HLMP-K640	LED	gr ø3mm I:21mcd Vf:2.2V If:20mA	T1	Agt,Hpa,Qtc		data pinout
HLMP-LB11	LED	bl+gr ø3.7mm I>680mcd Vf:3.8V	T1	Hpa		data pdf pinout
HLMP-LD16	LED	rd ø4mm I>450mcd Vf:2V		Agt		data pdf pinout
HLMP-M200	LED	srd+yl+gr ø1.6mm I:7mcd Vf:2.3V		Qtc		data pinout
HLMP-M201	LED	srd+yl+gr ø1.6mm I:10mcd Vf:2.3V		Qtc		data pinout
HLMP-M250	LED	srd+yl+gr ø1.6mm I:10mcd Vf:2.3V		Qtc		data pinout
HLMP-M251	LED	srd+yl+gr ø1.6mm I:16mcd Vf:2.3V		Qtc		data pinout
HLMP-MD16	LED	rd ø4mm I>450mcd Vf:2V		Agt		data pdf pinout
HLMP-NS35	LED	bl ø3mm I:390mcd Vf:3.5V If:20mA	T1	Agt		data pdf pinout
HLMP-P002	LED	rd ø1.65mm I:2.5mcd Vf:1.6V If:10mA		Hpa		data pinout
HLMP-P005	LED	rd+srd+yl+or+gr ø1.65mm I:1.5mcd		Hpa		data pinout
HLMP-P102	LED	rd ø1.65mm I:30mcd Vf:1.8V If:10mA		Hpa		data pinout
HLMP-P202	LED	srd ø1.65mm I:5mcd Vf:1.8V If:10mA		Hpa		data pinout
HLMP-P402	LED	yl ø1.65mm I:4mcd Vf:2V If:10mA		Hpa		data pinout
HLMP-P502	LED	gr ø1.65mm I:6mcd Vf:2.1V If:10mA		Hpa		data pinout
HLMP-Q101	LED	rd+srd+yl+or ø1.65mm I:1.5mcd		Hpa		data pinout
HLMP-Q105	LED	srd+yl+gr ø1.65mm I:55mcd Vf:1.8V		Hpa,Qtc		data pinout
HLMP-Q150	LED	srd+yl+gr ø1.7mm I:1.8mcd Vf:1.6V	T1	Hpa,Qtc		data pinout
HLMP-Q155	LED	rd ø1.65mm I:4mcd Vf:1.6V If:10mA		Hpa		data pinout
HLMP-R100	LED	rd 7.2x2.4mm I:7.5mcd Vf:1.8V		Hpa		data pinout
HLMP-S100	LED	rd+srd+yl+or+gr 2x5mm I:3mcd Vf:2.2V		Hpa		data pinout
HLMP-SG10	LED	bl+rd+am+gr ø4mm I>345mcd Vf:3.5V	T1¾	Agt		data pdf pinout
HLMP-T200	LED	srd+yl+or+gr 5.72x3.18mm I:6mcd		Hpa		data pinout
HLMT-PG00	LED	rd ø1.8mm I>40mcd Vf:2V If:20mA	T1	Agt		data pdf pinout
HLMT-PH00	LED	or ø1.8mm I>40mcd Vf:2V If:20mA	T1	Agt		data pdf pinout
HLMT-PL00	LED	am ø1.8mm I>40mcd Vf:2V If:20mA	T1	Agt		data pdf pinout
HLMT-QG00	LED	rd ø1.8mm I>250mcd Vf:2V If:20mA	T1	Agt		data pdf pinout
HLMT-QH00	LED	or ø1.8mm I>1000mcd Vf:2V If:20mA	T1	Agt		data pdf pinout
HLMT-QL00	LED	am ø1.8mm I:1000mcd Vf:2V If:20mA	T1	Agt		data pdf pinout
HLP20R	IRED	760nm ø0.6mm I:1000mcd Vf:2V		Hit		data pinout
HLP20RG	IRED	760nm ø4.2mm I:1000mcd Vf:2V		Hit		data pinout
HLP30RA	IRED	760nm ø0.6mm I:1000mcd Vf:2V		Hit		data pinout
HLP30RB	IRED	800nm ø0.6mm I:1000mcd Vf:1.7V		Hit		data pinout
HLP30RC	IRED	840nm ø0.6mm I:1000mcd Vf:1.7V		Hit		data pinout
HLP30RD	IRED	880nm ø0.6mm I:1000mcd Vf:1.7V		Hit		data pinout
HLP30RGA	IRED	760nm ø4.2mm I:1000mcd Vf:2V		Hit		data pinout

Type	Device	Short Description	Fig.	Manu	Comparision Types	More at
HLP30RGB	IRED	800nm ø4.2mm I:1000mcd Vf:1.7V		Hit		data pinout
HLP30RGC	IRED	840nm ø4.2mm I:1000mcd Vf:1.7V		Hit		data pinout
HLP30RGD	IRED	880nm ø4.2mm I:1000mcd Vf:1.7V		Hit		data pinout
HLP40RA	IRED	760nm ø4.2mm I:1000mcd Vf:2V		Hit		data pinout
HLP40RB	IRED	800nm ø0.6mm I:1000mcd Vf:1.7V		Hit		data pinout
HLP40RC	IRED	840nm ø0.6mm I:1000mcd Vf:1.7V		Hit		data pinout
HLP40RD	IRED	880nm ø0.6mm I:1000mcd Vf:1.7V		Hit		data pinout
HLP40RGA	IRED	760nm ø4.2mm I:1000mcd vf:2V		Hit		data pinout
HLP40RGB	IRED	800nm ø4.2mm I:1000mcd Vf:1.7V		Hit		data pinout
HLP40RGC	IRED	840nm ø4.2mm I:1000mcd Vf:1.7V		Hit		data pinout
HLP40RGD	IRED	880nm ø4.2mm I:1000mcd Vf:1.7V		Hit		data pinout
HLP50RB	IRED	800nm ø0.6mm I:1000mcd Vf:1.7V		Hit		data pinout
HLP50RC	IRED	840nm ø0.6mm I:1000mcd Vf:1.7V		Hit		data pinout
HLP50RD	IRED	880nm ø0.6mm I:1000mcd Vf:1.7V		Hit		data pinout
HLP50RGB	IRED	800nm ø4.2mm I:1000mcd Vf:1.7V		Hit		data pinout
HLP50RGC	IRED	840nm ø4.2mm I:1000mcd Vf:1.7V		Hit		data pinout
HLP50RGD	IRED	880nm ø4.2mm I:1000mcd Vf:1.7V		Hit		data pinout
HLP60RB	IRED	800nm ø0.6mm I:1000mcd Vf:1.7V		Hit		data pinout
HLP60RC	IRED	840nm ø0.6mm I:1000mcd Vf:1.7V		Hit		data pinout
HLP60RD	IRED	880nm ø0.6mm I:1000mcd Vf:1.7V		Hit		data pinout
HLP60RGB	IRED	800nm ø4.2mm I:1000mcd Vf:1.7V		Hit		data pinout
HLP60RGC	IRED	840nm ø4.2mm I:1000mcd Vf:1.7V		Hit		data pinout
HLP60RGD	IRED	880nm ø4.2mm I:1000mcd Vf:1.7V		Hit		data pinout
HLP1400	IR LD	830nm ø2mm I:1000mcd Vf:1.8V If:30mA		Hit		data pinout
HLP1500	IR LD	830nm ø50µmm I:1000mcd Vf:1.9V		Hit		data pinout
HLP1600	IR LD	830nm ø0.9mm I:1000mcd Vf:1.9V		Hit		data pinout
HLP5400	IR LD+MA	1200nm ø5mm I:1000mcd Vf:1.1V		Hit		data pinout
HLP5600	IR LD+MA	1300nm ø5mm I:1000mcd Vf:1.2V		Hit		data pinout
HM	Si-N	=2SC3547B (Typ-Code/Stempel/marking)	35	Tos	→2SC3547B	data
HM	Si-N	=2SC4248 (Typ-Code/Stempel/marking)	35(2mm)	Tos	→2SC4248	data
HM	Si-N	=2SD1006-HM (Typ-Code/Stempel/marking)	39	Nec	→2SD1006	data
HM	Z-Di	=Z2 SMB-82 (Typ-Code/Stempel/marking)	71(5x3,5)	Fag	→Z2 SMB-82	data
HM 2105...100480	RAM-IC			Hit		data
HM 32-120	Si-Di	TV-Damper-Diode, 1200V, 1,5A	17	Mop	BY 329-1200, BY 359-1300	data
HM 32-150	Si-Di	TV-Damper-Diode, 1500V, 1,5A	17	Mop	BY 329-1500, BY 359-1500	data
HM 6514	sRAM-IC	1k x 4 Bit			US 224D20	
HM 50256AP	dRAM-IC	256k x 1Bit DRAM, +5V, 20mW Standby	DIP	Hit	-	pdf pinout
HM 50256CP	dRAM-IC	256k x 1Bit DRAM, +5V, 20mW Standby	PLCC	Hit	-	pdf pinout
HM 50256P	dRAM-IC	256k x 1Bit DRAM, +5V, 20mW Standby	DIP	Hit	-	pdf pinout
HM 50256ZP	dRAM-IC	256k x 1Bit DRAM, +5V, 20mW Standby	SQP	Hit	-	pdf
HM 53051 FP-...	CMOS-RAM-IC	=HM 53051P: SMD	28-MDIP	Hit	-	pdf
HM 53051 P-...	CMOS-RAM-IC	Tv/vc, 262144x4 Bit, Frame Memory	18-DIP	Hit	-	pdf
HM 53461 JP-...	CMOS-VRAM-IC	=HM 53461P: SMD	24-MDIP	Hit	-	pdf
HM 53461 P-...	CMOS-VRAM-IC	Tv/vc, 65536x4 Bit, Video RAM	24-DIP	Hit	-	
HM 53461 ZP-...	CMOS-VRAM-IC	=HM 53461P: Fig.→	24-SQP	Hit	-	
HM 53462 JP-...	CMOS-VRAM-IC	=HM 53462P: SMD	24-MDIP	Hit	-	
HM 53462 P-...	CMOS-VRAM-IC	Tv/vc, 65536x4 Bit, Video RAM	24-DIP	Hit	-	
HM 53462 ZP-...	CMOS-VRAM-IC	=HM 53462P: Fig.→	24-SQP	Hit	-	
HM 63021 FP-...	RAM-IC	=HM 63021P: SMD	28-MDIP	Hit	-	
HM 63021 P-...	RAM-IC	Tv/vc, 2048x8 Bit, Line Memory	28-DIP	Hit	-	
HM 530281 TT-...	CMOS-RAM-IC	Tv/vc, 331776x8 Bit, Frame Memory	44-SMDIP	Hit	-	
HM 530283 FP-...	CMOS-RAM-IC	Tv/vc, 331776x8 Bit, Frame Memory	32-MDIP	Hit	-	
HM 534251 JP-...	VRAM-IC	SMD, Video, 256kBx4 Bit, Video RAM -10:<100ns, -12:<120ns, -15:<150ns	28-MDIP	Hit	-	
HM 534251 ZP-...	VRAM-IC	=HM 5334251JP...:	28-SQP			
HM 534252 JP-...	VRAM-IC	SMD, Video, 256kBx4 Bit, Video RAM -10:<100ns, -12:<120ns, -15:<150ns	28-MDIP	Hit	-	
HM 534252 ZP-...	VRAM-IC	=HM 5334252JP...:	28-SQP			
HM 534253 JP-...	VRAM-IC	SMD, Video, 256kBx4 Bit, Video RAM -10:<100ns, -12:<120ns, -15:<150ns	28-MDIP	Hit	-	
HM 534253 ZP-...	VRAM-IC	=HM 5334253JP...:	28-SQP			
HM 538121 JP-...	VRAM-IC	SMD, Video, 128kBx8 Bit, Video RAM -10:<100ns, -12:<120ns, -15:<150ns	40-MDIP	Hit	-	
HM 538122 JP-...	VRAM-IC	SMD, Video, 128kBx8 Bit, Video RAM -10:<100ns, -12:<120ns, -15:<150ns	40-MDIP	Hit	-	
HM 538123 JP-...	VRAM-IC	SMD, Video, 128kBx8 Bit, Video RAM -10:<100ns, -12:<120ns, -15:<150ns	40-MDIP	Hit	-	
HMCS 404 AC (PLCC)	µC-IC	4-Bit µ-Cop, 4096x10-bit ROM,245x4-bit RAM	PLCC	Hit	-	pdf pinout
HMCS 404 AC	µC-IC	4-Bit µ-Cop, 4096x10-bit ROM,245x4-bit RAM	DIP	Hit	-	pdf pinout
HMCS 404C (PLCC)	µC-IC	4-Bit µ-Cop, 4096x10-bit ROM,245x4-bit RAM	PLCC	Hit	-	pdf pinout
HMCS 404C	µC-IC	4-Bit µ-Cop, 4096x10-bit ROM,245x4-bit RAM	DIP	Hit	-	pdf pinout
HMCS 404 CL (PLCC)	µC-IC	4-Bit µ-Cop, 4096x10-bit ROM,245x4-bit RAM	PLCC	Hit	-	pdf pinout
HMCS 404 CL	µC-IC	4-Bit µ-Cop, 4096x10-bit ROM,245x4-bit RAM	DIP	Hit	-	pdf pinout
HMHA281	OC	LED/PT Viso:>2500Vrms Vbs:<1.3V		Fcs		data pdf pinout
HMHA2801	OC	LED/PT Viso:>2500Vrms Vbs:<1.4V		Fcs		data pdf pinout
HMHAA280	OC	bid. LED/PT Viso:>2500Vrms Vbs:<1.4V		Fcs		data pdf pinout
HN	Si-N	=2SC4214 (Typ-Code/Stempel/marking)	44	Tos	→2SC4214	data
HN	Si-N	=2SC4244 (Typ-Code/Stempel/marking)	35(2mm)	Tos	→2SC4244	data
HN	Si-N	=2SC4871 (Typ-Code/Stempel/marking)	35(2mm)	Say	→2SC4871	data

Type	Device	Short Description	Fig.	Manu	Comparision Types	More at
HN	Si-N	=2SC5540 (Typ-Code/Stempel/marking)	35(1,4mm)	Say	→2SC5540	data
HN	Si-N+R	=XP 0121N (Typ-Code/Stempel/marking)	45(2mm)	Mat	→XP 0121N	data
HN	Z-Di	=Z2 SMB-91 (Typ-Code/Stempel/marking)	71(5x3,5)	Fag	→Z2 SMB-91	data
HN 25044...613128	ROM/PROM-IC			Hit	-	
HN 1 A01F	Si-P	SMD, Dual, 50V, 0,15A, >80MHz	46bh1	Tos	-	data
HN 1 A01FE	Si-N	=HN 1 A01F:	46(1,6mm)			
HN 1 A01FU	Si-P	=HN 1 A01F:	46bh1(2mm)			data
HN 1 B01F	Si-N/P	SMD, N+P, 50V, 0,15A, 120MHz	46bh1	Tos	-	data
HN1B01 FDW1 T1	Si-N/P	SMD, Dual B1/B2=0,95 T1=PNP, T2=NPN 50V, 200mA,	46bh1	Ons	-	data
						data
HN 1 B01FU	Si-N/P	=HN 1 B01F:	46bh1(2mm)			data
HN' 1 B04FE	Si-N/P	=HN 1B04FU:	46(1,6mm)	Tos		
HN 1 B04FU	Si-N/P	SMD, P+N, 50V, 0,15A, 120MHz	46bh1(2mm)	Tos	-	data
HN 1 C01F	Si-N	SMD, Dual, 60V, 0,15A, >80MHz	46bh1	Tos	-	data
HN 1 C01FE	Si-N	=HN 1 C01F:	46(1,6mm)			
HN 1 C01FU	Si-N	=HN 1 C01F:	46bh1(2mm)			data
HN 1 C03F	Si-N	SMD, Dual, hi-Ueb, 50V, 0,3A, 30MHz	46bh1	Tos		data
HN 1 C03FU	Si-N	=HN 1 C03F:	46bh1(2mm)	Tos		data
HN 1 D01F	Si-Di	SMD, 4 Di, SS, 85V, 0,1A, <4ns	46bj	Tos	-	data
HN 1 D01FU	Si-Di	=HN 1 D01F:	46bj(2mm)		-	data
HN 1 D02F	Si-Di	SMD, 4 Di, SS, 85V, 0,1A, <4ns	46bk	Tos	-	data
HN 1 D02FU	Si-Di	=HN 1 D02F:	46bk(2mm)		-	data
HN 1 D03F	Si-Di	SMD, 4 Di, SS, 85V, 0,1A, <4ns	46bh	Tos	Pin-Code: A1A2A3+4K3K4K1+2	data
HN 1 D03FU	Si-Di	=HN 1 D03F:	46bh(2mm)			data
HN 1 J02FU	MOS-P-FET-e*	SMD, Dual, LogL, 20V, 50mA, <40Ω(10mA)	46bn3(2mm)	Tos	-	data
HN 1 K02FU	MOS-N-FET-e*	SMD, Dual, LogL, 20V, 50mA, <40Ω(10mA)	46bn3(2mm)	Tos	-	data
HN 1 K03FU	MOS-N-FET-e*	SMD, Dual, LogL, 20V, 0,1A, <12Ω(10mA)	46bn3(2mm)	Tos	-	data
HN 1 K04FU	MOS-N-FET-e*	=2x SK1827: SMD	46bn3(2mm)	Tos		
HN 1 K05FU	MOS-N-FET-e*	=2x SK2824: SMD	46bn3(2mm)	Tos		
HN 1 K06FU	MOS-N-FET-e*	=2x SK2037: SMD	46bn3(2mm)	Tos		
HN 1 L02FU	MOS-NP-FET-e	SMD, N+P, LogL, 20V, 50mA, <40Ω(10mA)	46bn3(2mm)	Tos	-	data
HN 1 L03FU	MOS-NP-FET-e	=2SK1827 + 2SJ346: SMD	46(2mm)	Tos	-	
HN 1 V01H(-A,-B)	C-Di	SMD, 4 Di, AM-Tuning	8-MDIP	Tos	-	data
HN 1 V02H(-A,-B)	C-Di	=HN 1 V01H...: 2 Di	8-MDIP	Tos	-	data
HN 2 A01FE	Si-N	=HN 2 A01FU:	46(1,6mm)		-	
HN 2 A01FU	Si-P	SMD, Dual, 50V, 0,15A, >80MHz	46(2mm)			data
HN 2 C01FE	Si-N	=HN 2 C01FU:	46(1,6mm)			
HN 2 C01FU	Si-N	SMD, Dual, 60V, 0,15A, >80MHz	46(2mm)			data
HN 2 D01F	Si-Di	SMD, 3 Di, SS, 85V, 0,08A, <4ns	46bg	Tos	Pin-Code: K1K2K3A3A2A1	data
HN 2 D01FU	Si-Di	=HN 2 D01F:	46bg(2mm)		-	data
HN 2 D02FU	Si-Di	SMD, 3 Di, SS, 85V, 0,08A, <4ns	46(2mm)	Tos	Pin-Code: A1A2A3K3K2K1	data
HN 2 S01F	Si-Di	SMD, 3 Schottky Di, S, 15V, 0,1A	46bg	Tos	-	data
HN 2 S01FU	Si-Di	=HN 2S01F:	46bg(2mm)		-	data
HN 2 S03FU	Si-Di	SMD, 3 Schottky Di, S, 25V, 0,05A	46bg(2mm)	Tos	-	data
HN 2 V02H(-A,-B)	C-Di	=HN 1 V01H...: 3 Di	8-MDIP	Tos	-	data
HN 3	Si-N	=2SC4871-3 (Typ-Code/Stempel/marking)	35(2mm)	Say	→2SC4871	data
HN 3 BO1F	Si-P/N	SMD, P+N, 50V, 0,15A, 120MHz	46bh2	Tos	-	data
HN 3 CO1F	Si-N	SMD, Dual, VHF, 30V, 50mA, 1400MHz	46bh2	Tos	-	data
HN 3 CO2F	Si-N	SMD, Dual, UHF, 30V, 50mA, 2400MHz	46bh2	Tos	-	data
HN 3 CO3F	Si-N	SMD, Dual, UHF, 20V, 30mA, 4000MHz	46bh2	Tos	-	data
HN 3 CO6F	Si-N	SMD, Dual, UHF, 20V, 80mA, 7000MHz	46bh2	Tos	-	data
HN 3 CO7F	Si-N	SMD, Dual, UHF, 20V, 40mA, 10GHz	46bh2	Tos	-	data
HN 3 CO8F	Si-N	SMD, Dual, UHF, 20V, 15mA, 10GHz	46bh2	Tos	-	data
HN 3 G01J	N-FET+Si-N	SMD, N-FET(20V)+NPN-Trans(60V, 0,15A)	45	Tos	-	data
HN 4	Si-N	=2SC4871-4 (Typ-Code/Stempel/marking)	35(2mm)	Say	→2SC4871	data
HN 4 K03JU	MOS-N-FET-e*	SMD, Dual, LogL, 20V, 0,1A, <12Ω(10mA)	45ba2(2mm)	Tos	-	data
HN 5	Si-N	=2SC4871-5 (Typ-Code/Stempel/marking)	35(2mm)	Say	→2SC4871	data
HNLD 12...0500	SMPS-IC	DC-DC Converter	Module	Shi	-	
HO	Si-P	=2SA1203-O (Typ-Code/Stempel/marking)	39	Tos	→2SA1203	data
HO	Si-N	=2SC5257-O (Typ-Code/Stempel/marking)	44	Tos	→2SC5257	
HO	Si-P	=KTA1663-O (Typ-Code/Stempel/marking)	39	Kec	→KTA 1663	date
HO	Si-N/P+R	=XP 04286 (Typ-Code/Stempel/marking)	46(2mm)	Mat	→XP 04286	data
HOA086	PI	LED/NPN CTR:>9% Vbs:<1.6V Ibs:20mA		Hon		data pinout
HOA087	PI	LED/NPN CTR:>1% Vbs:<1.6V Ibs:20mA		Hon		data pinout
HOA088	PI	LED/NPN CTR:>9% Vbs:<1.6V Ibs:20mA		Hon		data pinout
HOA089	PI	LED/NPN CTR:>9% Vbs:<1.6V Ibs:20mA		Hon		data pinout
HOA0149	PS	LED/NPN CTR:>2.5% Vbs:<1.6V Ibs:20mA		Hon		data pinout
HOA696	PI	LED/PD+IC Vbs:<1.6V Ibs:20mA		Hon		data pinout
HOA697	PI	LED/PD+IC Vbs:<1.6V Ibs:20mA		Hon		data pinout
HOA698	PI	LED/PD+IC Vbs:<1.6V Ibs:20mA		Hon		data pinout
HOA699	PI	LED/PD+IC Vbs:<1.6V Ibs:20mA		Hon		data pinout
HOA0708	PS	LED/Dar CTR:>2.5% Vbs:<1.6V Ibs:20mA		Hon		data pinout
HOA0825	PI	LED/NPN CTR:>2.5% Vbs:<1.6V Ibs:20mA		Hon		data pdf pinout
HOA0901	PI	LED/2 Vbs:1.6V Ibs:20mA Ibe:<7mA		Hon		data pdf pinout
HOA0902	PI	LED/2 Vbs:1.6V Ibs:20mA Ibe:<12mA		Hon		data pdf pinout
HOA0960	PI	LED/PD+IC Vbs:<1.6V Ibs:20mA		Hon		data pinout
HOA0970	PI	LED/PD+IC Vbs:<1.6V Ibs:20mA		Hon		data pinout
HOA1160	PS	LED/Dar CTR:>16% Vbs:<1.6V Ibs:20mA		Hon		data pinout

Type	Device	Short Description	Fig.	Manu	Comparision Types	More at
HOA1180	PS	LED/Dar CTR:>6.6% Vbs:<1.6V Ibs:20mA		Hon		data pdf pinout
HOA1404	PS	LED/Dar CTR:>10% Vbs:<1.6V Ibs:20mA		Hon		data pdf pinout
HOA1405	PS	LED/NPN CTR:>4% Vbs:<1.6V Ibs:20mA		Hon		data pinout
HOA1406	PS	LED/Dar CTR:>5% Vbs:<1.6V Ibs:20mA		Hon		data pdf pinout
HOA1870	PI	LED/Dar CTR:>10% Vbs:<1.6V Ibs:20mA		Hon		data pdf pinout
HOA1872	PI	LED/Dar CTR:>20% Vbs:<1.6V Ibs:20mA		Hon		data pdf pinout
HOA1873	PI	LED/Dar CTR:>20% Vbs:<1.6V Ibs:20mA		Hon		data pdf pinout
HOA1874	PI	LED/Dar CTR:>20% Vbs:<1.6V Ibs:20mA		Hon		data pdf pinout
HOA1875	PI	LED/Dar CTR:>6% Vbs:<1.6V Ibs:20mA		Hon		data pdf pinout
HOA1876	PI	LED/Dar CTR:>6% Vbs:<1.6V Ibs:20mA		Hon		data pdf pinout
HOA1877	PI	LED/Dar CTR:>7.5% Vbs:<1.6V Ibs:20mA		Hon		data pdf pinout
HOA1879	PI	LED/NPN CTR:>2.5% Vbs:<1.6V Ibs:20mA		Hon		data pdf pinout
HOA1881	PI	LED/Dar CTR:>20% Vbs:<1.6V Ibs:20mA		Hon		data pdf pinout
HOA1882	PI	LED/Dar CTR:>20% Vbs:<1.6V Ibs:20mA		Hon		data pdf pinout
HOA1883	PI	LED/Dar CTR:>20% Vbs:<1.6V Ibs:20mA		Hon		data pdf pinout
HOA1884	PI	LED/Dar CTR:>20% Vbs:<1.6V Ibs:20mA		Hon		data pdf pinout
HOA1885	PI	LED/Dar CTR:>20% Vbs:<1.6V Ibs:20mA		Hon		data pdf pinout
HOA1886	PI	LED/Dar CTR:>20% Vbs:<1.6V Ibs:20mA		Hon		data pdf pinout
HOA2000	PI	LED/PD+IC Vbs:<1.6V Ibs:20mA		Hon		data pdf pinout
HOA2498	PS	LED/Dar CTR:>5% Vbs:<1.6V Ibs:50mA		Hon		data pdf pinout
HOA2762	PI	LED/Dar CTR:>20% Vbs:<1.6V Ibs:20mA		Hon		data pdf pinout
HOA7720	PI	LED/PD+IC Vbs:<1.6V Ibs:20mA		Hon		data pinout
HOS050	OP	Vs:±18V Vu:100dB Vo:±10V Vi0:45mV	12A	And		data pinout
HOS050A	OP	Vs:±18V Vu:>100dB Vo:±10V Vi0:15mV	12A	And		data pinout
HOS060SH/883	FET	Vs:±18V Vu:100dB Vo:±10V Vi0:±0.5mV	12A	And		data pinout
HOS100A	OP	Vs:40V Vo:±13V Vi0:10mV f:125Mc	12A	And		data pinout
HOS100S	OP	Vs:40V Vo:±13V Vi0:5mV f:125Mc	12A	And		data pinout
HOVL4A30	LED	am ø4mm I:300mcd Vf:2V If:20mA	T1	Hrv		data pdf pinout
HOVL4B30	LED	bl ø4mm I:400mcd Vf:3.6V If:20mA	T1	Hrv		data pdf pinout
HOVL4BG30	LED	gr ø4mm I:650mcd Vf:3.6V If:20mA	T1	Hrv		data pdf pinout
HOVL4CB30	LED	bl ø4mm I:90mcd Vf:3.8V If:20mA	T1	Hrv		data pdf pinout
HOVL4E30	LED	or ø4mm I:300mcd Vf:2V If:20mA	T1	Hrv		data pdf pinout
HOVL4R30	LED	rd ø4mm I:190mcd Vf:2V If:20mA	T1	Hrv		data pdf pinout
HOVL4SR30	LED	srd ø4mm I:320mcd Vf:2V If:20mA	T1	Hrv		data pdf pinout
HOVL4TG30	LED	gr ø4mm I:550mcd Vf:3.6V If:20mA	T1	Hrv		data pdf pinout
HOVL4Y30	LED	yl ø4mm I:280mcd Vf:2V If:20mA	T1	Hrv		data pdf pinout
HOVL5A40	LED	am ø4mm I:500mcd Vf:2V If:20mA	T1	Hrv		data pdf pinout
HOVL5B40	LED	bl ø4mm I:450mcd Vf:3.6V If:20mA	T1	Hrv		data pdf pinout
HOVL5BG40	LED	gr ø4mm I:1300mcd Vf:3.6V If:20mA	T1	Hrv		data pdf pinout
HOVL5E40	LED	or ø4mm I:700mcd Vf:2V If:20mA	T1	Hrv		data pdf pinout
HOVL5R40	LED	rd ø4mm I:800mcd Vf:2V If:20mA	T1	Hrv		data pdf pinout
HOVL5SR40	LED	srd ø4mm I:400mcd Vf:2V If:20mA	T1	Hrv		data pdf pinout
HOVL5TG40	LED	gr ø4mm I:950mcd Vf:3.6V If:20mA	T1	Hrv		data pdf pinout
HOVL5UR40	LED	rd ø4mm I:1500mcd Vf:2.2V If:20mA	T1	Hrv		data pdf pinout
HOVL5Y40	LED	yl ø4mm I:600mcd Vf:2V If:20mA	T1	Hrv		data pdf pinout
HP	Si-P	=2SA1036-HP (Typ-Code/Stempel/marking)	≈35	Rhm	→2SA1036	data
HP	Si-P	=2SA1036K-P (Typ-Code/Stempel/marking)	35	Rhm	→2SA1036K	data
HP	Si-P	=2SA1577-P (Typ-Code/Stempel/marking)	35(2mm)	Rhm	→2SA1577	data
HP	Si-N	=2SC4527 (Typ-Code/Stempel/marking)	35	Tos	→2SC4527	data
HP	Si-N	=2SD1007-HP (Typ-Code/Stempel/marking)	39	Nec	→2SD1007	data
HP	Z-Di	=Z2 SMB-100 (Typ-Code/Stempel/marking)	71(5x3,5)	Fag	→Z2 SMB-100	data
HP803CB	LED	cy ø6mm I:40lm Vf:3.8V If:700mA		Roi		data pdf pinout
HP803NB	LED	bl ø6mm I:16lm Vf:4V If:700mA		Roi		data pdf pinout
HP803NO	LED	am ø6mm I:36lm Vf:3V If:700mA		Roi		data pdf pinout
HP803NR	LED	rd ø6mm I:40lm Vf:3V If:700mA		Roi		data pdf pinout
HP803NW	LED	wht ø6mm I:55lm Vf:4V If:700mA		Roi		data pdf pinout
HP803PG	LED	gr ø6mm I:55lm Vf:3.8V If:700mA		Roi		data pdf pinout
HP803WW	LED	wht ø6mm I:45lm Vf:4V If:700mA		Roi		data pdf pinout
HP 5082-4657	Opto					
HP 7800	OPTO-IC	Motor control, high Cmr Isolation Amp	DIP	Hpa	HCPL7800, HCPL7800A, HCPL7800B	pdf pinout
HPA 72 R	Si-N+Di	HA, 1500/800, 7A, 60W	18c	Say	BU 508DF, 2SC3892A, 2SC4123...24, 2SD2251	data
HPA 100 R	Si-N+Di	HA, hi-def, 1500/800V, 10A, 150W	77j§	Say	(2SC3995, 2SC4288A)[2]	data
HPA 150 R	Si-N+Di	HA, hi-def, 1500/800V, 15A, 150W	77j§	Say	(2SC3996, 2SC4289A)[2]	data
HPA 201 R	Si-N+Di	HA, hi-def, 1500/800V, 20A, 100W	77j§	Say	(2SC3998, 2SC4290A, 2SC4897, 2SD2356)[2]	data
HPA 251 R	Si-N+Di	HA, hi-def, 1500/800V, 25A, 120W	77j§	Say	(2SC3998, 2SC4789, 2SC5047)[2]	data
HPG 8	Si-Di	kV-Gl, 8kV, 0,05A	31a	Gie	1N5184	data
HPG 10	Si-Di	=HPG 8: 10kV, 0,04A	31a	Gie	1N5184	data
HPG 15	Si-Di	=HPG 8: 15kV, 0,03A	31a	Gie	1N3054	data
HPG 20	Si-Di	=HPG 8: 20kV, 0,02A	31a	Gie	1N3056	data
HPRA320A	LED	am ø6mm I:1000mcd Vf:1.9V If:30mA		Hrv		data pdf pinout
HPRA320AE	LED	yl ø6mm I:1200mcd Vf:1.9V If:30mA		Hrv		data pdf pinout
HPRA320B	LED	bl ø6mm I:800mcd Vf:3.6V If:30mA		Hrv		data pdf pinout
HPRA320BG	LED	gr ø6mm I:900mcd Vf:3.6V If:30mA		Hrv		data pdf pinout
HPRA320E	LED	or ø6mm I:1300mcd Vf:1.9V If:30mA		Hrv		data pdf pinout
HPRA320R	LED	rd ø6mm I:1500mcd Vf:1.9V If:30mA		Hrv		data pdf pinout
HPRA320SR	LED	srd ø6mm I:1400mcd Vf:1.9V If:30mA		Hrv		data pdf pinout
HPRA320TG	LED	gr ø6mm I:900mcd Vf:3.6V If:30mA		Hrv		data pdf pinout
HPRA320Y	LED	yl ø6mm I:1800mcd Vf:1.9V If:30mA		Hrv		data pdf pinout
HPRA320YG	LED	gr ø6mm I:800mcd Vf:1.9V If:30mA		Hrv		data pdf pinout
HPRA321A	LED	am ø6mm I:1900mcd Vf:1.9V If:30mA		Hrv		data pdf pinout
HPRA321AE	LED	yl ø6mm I:1900mcd Vf:1.9V If:30mA		Hrv		data pdf pinout
HPRA321B	LED	bl ø6mm I:900mcd Vf:3.6V If:30mA		Hrv		data pdf pinout
HPRA321BG	LED	gr ø6mm I:1350mcd Vf:3.6V If:30mA		Hrv		data pdf pinout
HPRA321E	LED	or ø6mm I:2250mcd Vf:1.9V If:30mA		Hrv		data pdf pinout

Type	Device	Short Description	Fig.	Manu	Comparision Types	More at
HPRA321R	LED	rd ø6mm I:1600mcd Vf:1.9V If:30mA		Hrv		data pdf pinout
HPRA321SR	LED	srd ø6mm I:1900mcd Vf:1.9V If:30mA		Hrv		data pdf pinout
HPRA321TG	LED	gr ø6mm I:1350mcd Vf:3.6V If:30mA		Hrv		data pdf pinout
HPRA321UR	LED	rd ø6mm I:2250mcd Vf:2.2V If:30mA		Hrv		data pdf pinout
HPRA321Y	LED	yl ø6mm I:3000mcd Vf:1.9V If:30mA		Hrv		data pdf pinout
HPRA321YG	LED	gr ø6mm I:1350mcd Vf:1.9V If:30mA		Hrv		data pdf pinout
HPRA328A	LED	am ø6mm I:800mcd Vf:1.9V If:30mA		Hrv		data pdf pinout
HPRA328AE	LED	yl ø6mm I:1050mcd Vf:1.9V If:30mA		Hrv		data pdf pinout
HPRA328B	LED	bl ø6mm I:300mcd Vf:3.6V If:30mA		Hrv		data pdf pinout
HPRA328BG	LED	gr ø6mm I:1050mcd Vf:3.6V If:30mA		Hrv		data pdf pinout
HPRA328E	LED	or ø6mm I:800mcd Vf:1.9V If:30mA		Hrv		data pdf pinout
HPRA328R	LED	rd ø6mm I:1050mcd Vf:1.9V If:30mA		Hrv		data pdf pinout
HPRA328SR	LED	srd ø6mm I:600mcd Vf:1.9V If:30mA		Hrv		data pdf pinout
HPRA328TG	LED	gr ø6mm I:1200mcd Vf:3.6V If:30mA		Hrv		data pdf pinout
HPRA328UR	LED	rd ø6mm I:2100mcd Vf:2.2V If:30mA		Hrv		data pdf pinout
HPRA328Y	LED	yl ø6mm I:750mcd Vf:1.9V If:30mA		Hrv		data pdf pinout
HPRA328YG	LED	gr ø6mm I:380mcd Vf:1.9V If:30mA		Hrv		data pdf pinout
HPWA-DH00	LED	or ø3mm I:1.5..4.2lm Vf:2.2V If:70mA		Lux		data pdf pinout
HPWA-DH02	LED	or ø3mm I:1.5..4.2lm Vf:2.2V If:70mA		Lux		data pdf pinout
HPWA-MH00	LED	or ø3mm I:1.5..4.2lm Vf:2.2V If:70mA		Lux		data pdf pinout
HPWA-MH02	LED	or ø3mm I:1.5..4.2lm Vf:2.2V If:70mA		Lux		data pdf pinout
HPWT-BD00	LED	rd ø3mm I:3..7.3lm Vf:2.6V If:70mA		Lux		data pdf pinout
HPWT-BD02	LED	rd ø3mm I:3..7.3lm Vf:2.6V If:70mA		Lux		data pdf pinout
HPWT-BH00	LED	or ø3mm I:4..12lm Vf:2.6V If:70mA		Lux		data pdf pinout
HPWT-BH02	LED	or ø3mm I:4..12lm Vf:2.6V If:70mA		Lux		data pdf pinout
HPWT-BL00	LED	am ø3mm I:2..4.8lm Vf:2.6V If:70mA		Lux		data pdf pinout
HPWT-BL02	LED	am ø3mm I:2..4.8lm Vf:2.6V If:70mA		Lux		data pdf pinout
HPWT-DD00	LED	rd ø3mm I:3..7.3lm Vf:2.6V If:70mA		Lux		data pdf pinout
HPWT-DD02	LED	rd ø3mm I:3..7.3lm Vf:2.6V If:70mA		Lux		data pdf pinout
HPWT-DH00	LED	or ø3mm I:4..12lm Vf:2.6V If:70mA		Lux		data pdf pinout
HPWT-DH02	LED	or ø3mm I:4..12lm Vf:2.6V If:70mA		Lux		data pdf pinout
HPWT-DL00	LED	am ø3mm I:2..4.8lm Vf:2.6V If:70mA		Lux		data pdf pinout
HPWT-DL02	LED	am ø3mm I:2..4.8lm Vf:2.6V If:70mA		Lux		data pdf pinout
HPWT-MB00	LED	bl ø3mm I:>1lm Vf:3.8V If:50mA		Lux		data pdf pinout
HPWT-MC00	LED	cy ø3mm I:>3lm Vf:3.8V If:50mA		Lux		data pdf pinout
HPWT-MD00	LED	rd ø3mm I:3..7.3lm Vf:2.6V If:70mA		Lux		data pdf pinout
HPWT-MD02	LED	rd ø3mm I:3..7.3lm Vf:2.6V If:70mA		Lux		data pdf pinout
HPWT-MG00	LED	gr ø3mm I:>3lm Vf:3.9V If:50mA		Lux		data pdf pinout
HPWT-MH00	LED	or ø3mm I:4..12lm Vf:2.6V If:70mA		Lux		data pdf pinout
HPWT-MH02	LED	or ø3mm I:4..12lm Vf:2.6V If:70mA		Lux		data pdf pinout
HPWT-ML00	LED	am ø3mm I:2..4.8lm Vf:2.6V If:70mA		Lux		data pdf pinout
HPWT-ML02	LED	am ø3mm I:2..4.8lm Vf:2.6V If:70mA		Lux		data pdf pinout
HPWT-RD00	LED	rd ø3mm I:3..7.3lm Vf:2.6V If:70mA		Lux		data pdf pinout
HPWT-RD02	LED	rd ø3mm I:3..7.3lm Vf:2.6V If:70mA		Lux		data pdf pinout
HPWT-RH00	LED	or ø3mm I:4..12lm Vf:2.6V If:70mA		Lux		data pdf pinout
HPWT-RH02	LED	or ø3mm I:4..12lm Vf:2.6V If:70mA		Lux		data pdf pinout
HPWT-RL00	LED	am ø3mm I:2..4.8lm Vf:2.6V If:70mA		Lux		data pdf pinout
HPWT-RL02	LED	am ø3mm I:2..4.8lm Vf:2.6V If:70mA		Lux		data pdf pinout
HQ	Si-P	=2SA1036-HQ (Typ-Code/Stempel/marking)	≈35	Rhm	→2SA1036	data
HQ	Si-P	=2SA1036K-Q (Typ-Code/Stempel/marking)	35	Rhm	→2SA1036K	data
HQ	Si-P	=2SA1577-Q (Typ-Code/Stempel/marking)	35(2mm)	Rhm	→2SA1577	data
HQ	Si-P	=2SB956-Q (Typ-Code/Stempel/marking)	39	Mat	→2SB956	data
HQ	Si-N	=2SC3928-Q (Typ-Code/Stempel/marking)	35	Mit	→2SC3928	data
HQ	Si-N	=2SC4155-Q (Typ-Code/Stempel/marking)	35(2mm)	Mit	→2SC4155	data
HQ	Si-N	=2SD1007-HQ (Typ-Code/Stempel/marking)	39	Nec	→2SD1007	data
HQ	Z-Di	=Z2 SMB-110 (Typ-Code/Stempel/marking)	71(5x3,5)	Fag	→Z2 SMB-110	data
HQ 1 A3M..L3N	Si-P+R	=AQ 1A3M..L3N: SMD	39b	Nec	-	data
HR-AR57G62B66BC-C	LED	rgb ø3mm I:4.5lm Vf:2V If:150mA		Roi		data pdf pinout
HR-AR58G05B24BC-C/	LED	rgb ø3mm I:6lm Vf:3.4V If:350mA		Roi		data pdf pinout
HR	Si-P	=2SA1035-R (Typ-Code/Stempel/marking)	35	Mat	→2SA1035	data
HR	Si-P	=2SA1036-HR (Typ-Code/Stempel/marking)	≈35	Rhm	→2SA1036	data
HR	Si-P	=2SA1036K-R (Typ-Code/Stempel/marking)	35	Rhm	→2SA1036K	data
HR	Si-P	=2SA1531A-R (Typ-Code/Stempel/marking)	35(2mm)	Mat	→2SA1531A	data
HR	Si-P	=2SA1577-R (Typ-Code/Stempel/marking)	35(2mm)	Rhm	→2SA1577	data
HR	Si-P	=2SB956-R (Typ-Code/Stempel/marking)	39	Mat	→2SB956	data
HR	Si-N	=2SC3928-R (Typ-Code/Stempel/marking)	35	Mit	→2SC3928	data
HR	Si-N	=2SC4155-R (Typ-Code/Stempel/marking)	35(2mm)	Mit	→2SC4155	data
HR	Si-N	=2SC5257-R (Typ-Code/Stempel/marking)	44	Tos	→2SC5257	
HR	Si-N	=2SD1007-HR (Typ-Code/Stempel/marking)	39	Nec	→2SD1007	data
HR	Z-Di	=Z2 SMB-120 (Typ-Code/Stempel/marking)	71(5x3,5)	Fag	→Z2 SMB-120	data
HR 1 A3M..L3N	Si-P+R	=AR 1A3M..L3N: SMD	39b	Nec	-	data
HR 5 A,B	Si-Di	Gl, Uni, 100..200V, 0,4A	31a	Hit	BA 157...159, BY 204/..., BY 208/..., ++	data
HR 51...	Si-Di	≈1N4001..4007			→1N4001...4007	
HR 100(R)	Si-Di	Gl, Uni, 100V, 3A	31a	Tho	BY 251...255, GP 30B...M, 1N5401...08, ++	data
HR 200(R)	Si-Di	=HR 100: 200V	31a	Tho	BY 251...255, GP 30D...M, 1N5402...08, ++	data
HR 400(R)	Si-Di	=HR 100: 400V	31a	Tho	BY 252...255, GP 30G...M, 1N5404...08, ++	data
HR 400(R)	Si-Di	≈BYX 71/350	26		→BYX 71/350	
HR 600(R)	Si-Di	=HR 100: 600V	31a	Tho	BY 253...255, GP 30J...M, 1N5406...08, ++	data
HR 800(R)	Si-Di	=HR 100: 800V	31a	Tho	BY 254...255, GP 30K...M, 1N5407...08, ++	data
HR 1000(R)	Si-Di	=HR 100: 1000V	31a	Tho	BY 255, GP 30M, MR 510, 1N5408, ++	data
HR1101	PD	1300nm S:0.7mA/mW	TO18	Hit		data pinout
HR1102	PD	1300nm S:0.7mA/mW		Hit		data pinout

Type	Device	Short Description	Fig.	Manu	Comparision Types	More at
HR1103	PD	1300nm S:0.85mA/i ıW	TO18	Hit		data pinout
HR1104	PD	1300nm S:0.85mA/mW	TO18	Hit		data pinout
HR1105	PD	1300nm S:0.85mA/mW	TO18	Hit		data pinout
HR1107	PD	1300nm S:0.78mA/mW		Hit		. data pinout
HR1201	PD	1300nm S:0.8mA/mW		Hit		data pinout
HR8101	PD	830nm S:>0.4mA/mW	TO18	Hit		data pinout
HR8102	PD	830nm S:>0.4mA/mW	TO18	Hit		data pinout
HR8202	PD	830nm	TO18	Hit		data pinout
HRB 0103 A	Si-Di	=HRU 0103A:	35c(2mm)	Hit	-	data
HRB 0103 B	Si-Di	=HRU 0103A: Dual	35o(2mm)	Hit	-	data
HRC 0202 A	Si-Di	SMD, Dual, Schottky-Di, 20V, 0,2A	35l(2mm)	Hit	-	data
HRF 22	Si-Di	SMD, Schottky-Di, SS, 40V, 1A	71(5x2,5)	Hit	D 1FS4, MBRS 140, SB 11-04HP	data
HRF 32	Si-Di	SMD, Schottky-Di, SS, 90V, 1A	71(5x2,5)	Hit	MBRS 1100	data
HRF 0302 A	Si-Di	SMD, Schottky-Di, 20V, 3A	71(8mm)	Hit	-	data
HRF 0502 A	Si-Di	SMD, Schottky-Di, 20V, 5A	71(8mm)	Hit	-	data
HRF 0503 A	Si-Di	SMD, Schottky-Di, 35V, 5A	71(8mm)	Hit	-	data
HRP 22	Si-Di	Schottky-Di, SS, 50V, 1A	31a	Hit	BYV 10-60, HRP 100, MBR 150, SB 150, ++	data
HRP 24	Si-Di	Schottky-Di, 40V, 3A	31a	Hit	BYS 26-45, MBR 340, SB 340, 1N5822, ++	data
HRP 32	Si-Di	Schottky-Di, SS, 90V, 1A	31a	Hit	BYS 21-90, MBR 190, SB 190	data
HRP 34	Si-Di	Schottky-Di, 60V, 3A	31a	Hit	BYS 26-90, MBR 360, SB 360	data
HRP 100	Si-Di	Schottky-Di, SS, 50V, 1A	31a	Hit	BYV 10-60, HRP 22, MBR 150, SB 150, ++	data
HRU 0103 A	Si-Di	SMD, Schottky-Di, 30V, 0,1A	71a(1,7mm)	Hit	BAS 140W	data
HRU 0203 A	Si-Di	=HRW 0203A:	71a(1,7mm)	Hit	BAT54J, 1PS 76B10	data
HRU 0302 A	Si-Di	=HRW 0302A:	71a(1,7mm)	Hit	MA 2ZD02	data
HRW 26	Si-Di	Dual, Schottky-Gl, 40V, 2x5A(Tc=95°)	17p§	Hit	BYS 24-45, MBR 2060CT..20100CT	data
HRW 26 F	Si-Di	=HRW 26: Iso	17p	Hit	MA 750(A), MBRF 2060CT..20100CT	data
HRW 34	Si-Di	Dual, Schottky-Gl, 90V, 5A(Tc=95°)	17p§	Hit	BYS 24-90, MBR 2090CT..20100CT	data
HRW 34 F	Si-Di	=HRW 34: Iso	17p	Hit	MA 760, MBRF 20100CT	data
HRW 36	Si-Di	Dual, Schottky-Gl, 90V, 2x5A(Tc=95°)	17p§	Hit	BYS 24-90, MBR 2090CT..20100CT	data
HRW 36 F	Si-Di	=HRW 36: Iso	17p	Hit	MA 761, MBRF 20100CT	data
HRW 37 F	Si-Di	Dual, Schottky-Gl, 90V, 2x10A(Tc=95°)	17p	Hit		data
HRW 0202 A,B	Si-Di	SMD, Dual, Schottky-Di, 20V, 0,2A	35l	Hit	BAT 54C, BAT 64-05	data
HRW 0203 A	Si-Di	SMD, Schottky-Di, 30V, 0,2A	35e	Hit	BAT 54, BAT 64	data
HRW 0302 A	Si-Di	SMD, Schottky-Di, 20V, 0,3A	35e	Hit	1SS344	data
HRW 0502 A	Si-Di	SMD, Schottky-Di, 20V, 0,5A	35e	Hit	1SS344	data
HRW 0503 A	Si-Di	SMD, Schottky-Di, 30V, 0,5A	35e	Hit		data
HRW 0702 A	Si-Di	SMD, Schottky-Di, 20V, 0,7A	35e	Hit	1SS349	data
HRW 0703 A	Si-Di	SMD, Schottky-Di, 30V, 0,7A	35e	Hit		data
HRW 1002 A(L,S)	Si-Di	Dual, Schottky, 20V, 10A	30p	Hit	BYV 116B-20, BYV 116B-25	data
HRW 1002 B	Si-Di	=HRW 1002A:	17p	Hit	MA 750(A)	data
HRW 2502 A(L,S)	Si-Di	Dual, Schottky, 20V, 25A	30p	Hit	PBYL 2520CTB, PBYL 2525CTB	data
HRW 2502 B	Si-Di	=HRW 2502A:	17p	Hit	BYV 43F/30../45	data
HS	N-FET	=2SK2988 (Typ-Code/Stempel/marking)	35(1,6mm)	Mat	→2SK2988	data
HS-0546	MULT16	Ron:Ohm Ton:300ns at 15/-15V	28D	Isi		data pdf pinout
HS-0547	MULT8	Ron:Ohm Ton:300ns at 15/-15V	28D	Isi		data pdf pinout
HS-201	SPST	Ron:Ohm Ton:185ns at 15/-15V	16FD	Isi		data pinout
HS	Si-P	=2SA1035-S (Typ-Code/Stempel/marking)	35	Mat	→2SA1035	data
HS	Si-P	=2SA1531A-S (Typ-Code/Stempel/marking)	35(2mm)	Mat	→2SA1531A	data
HS	Si-P	=2SB956-S (Typ-Code/Stempel/marking)	39	Mat	→2SB956	data
HS	Si-N	=2SC3928-S (Typ-Code/Stempel/marking)	35	Mit	→2SC3928	data
HS	Si-N	=2SC4155-S (Typ-Code/Stempel/marking)	35(2mm)	Mit	→2SC4155	data
HS	Si-N	=2SD2540 (Typ-Code/Stempel/marking)	39	Hit	→2SC2540	data
HS	N-FET	=2SK2751: (Typ-Code/Stempel/marking)	35	Mat	→2SK2751	data
HS	Z-Di	=Z2 SMB-130 (Typ-Code/Stempel/marking)	71(5x3,5)	Fag	→Z2 SMB-130	data
HS 11	Si-Di	SMD, Gl, S, 50V, 1A, <50ns	71a(5x2,5)	Mop	BYG 80A, D 1FL20	data
HS 12	Si-Di	SMD, Gl, S, 100V, 1A, <50ns	71a(5x2,5)	Mop	BYG 80B, D 1FL20	data
HS 13	Si-Di	SMD, Gl, S, 200V, 1A, <50ns	71a(5x2,5)	Mop	BYG 80D, D 1FL20	data
HS 14	Si-Di	SMD, Gl, S, 300V, 1A, <75ns	71a(5x2,5)	Mop	BYG 80F, U 1GU44	data
HS 15	Si-Di	SMD, Gl, S, 400V, 1A, <75ns	71a(5x2,5)	Mop	BYG 80G, U 1GU44	data
HS 16	Si-Di	SMD, Gl, S, 600V, 1A, <100ns	71a(5x2,5)	Mop	BYG 80J, U 1GU44	data
HS 17	Si-Di	SMD, Gl, S, 800V, 1A, <100ns	71a(5x2,5)	Mop	BYG 20K	data
HS 18	Si-Di	SMD, Gl, S, 1000V, 1A, <100ns	71a(5x2,5)	Mop	BYG 20M	data
HS-82C55ARH	µP-Periph	CMOS, Ppi, 3x8 Programmable I/O Pins, TTL Compatible, Radiation Hardened	DIP	Har	..82C55...	pdf pinout
HS-26 C 32 RH	CMOS-IC	RS-422 Line Receiver	DIP	Isi		pdf pinout
HS-26 CT 32 RH	CMOS-IC	RS-422 Line Receiver	DIP	Isi		pdf pinout
HS2700	R+	Io=±10mA Vo:9.998...10.00V P:300mW	14D	Sip		data pinout
HS2701	R-	Io=±10mA Vo:-9.998...-10.00V P:300mW	14D	Sip		data pinout
HS2702	R±	Io=±10mA Vo:±9.998...±10.00V P:300mW	14D	Sip		data pinout
HS 5157	MOS-N-FET-e	SMD, V-MOS, 60V, 0,5A, <1,7Ω(0,25A)	39	Hit	BST 80, 2SK1592, 2SK1697, 2SK2109, ++	data
HS 5159	MOS-N-FET-e	SMD, V-MOS, 100V, 0,3A, <4,5Ω(0,15A)	39	Hit	2SK1698	data
HS 5161	MOS-N-FET-e	SMD, V-MOS, 30V, 1A, <0,7Ω(0,5A)	39	Hit	2SK1485, 2SK1726, 2SK1728	data
HS 5305...5308(A)	Si-N	=2N5305...5308(A): 0,9W	7c°	Gen	→2N5305..5308(A)	data
HS 5810...5823	Si-N/P	=2N5810...5823: 0,7W	7a°	Gen	→2N5810..5823	data
HS 6010...6017	Si-N/P	=2N6010...6017: 0,7W	7a°	Gen	→2N6010...6017	data
HS 6018...6021	Si-N	=2N6018...6021: 0,7W	7a°	Gen	→2N6018...6021	data
HSB 83 J	Si-Di	=HSM 83:	35e(2mm)	Hit	-	data
HSB 88 WS	Si-Di	SMD, 2x2 Schottky-Di, Tuner, 10V, 15mA	≈8-MDIP	Hit	-	data

Type	Device	Short Description	Fig.	Manu	Comparision Types	More at
HSB 123	Si-Di	=HSM 123:	35o(2mm)	Hit	-	data
HSB 124 SJ	Si-Di	=HSM 124S:	35o(2mm)	Hit	-	data
HSB 2836	Si-Di	=HSM 2836C:	35m(2mm)	Hit	BAW 66, BAW 68	data
HSB 2838	Si-Di	=HSM 2838C:	35l(2mm)	Hit	BAW 64, BAW 65, BAW 67	data
HSC 101 WK	Si-Di	SMD, Dual, Dem, SS, 30V, 35mA	35l(2mm)	Hit	-	data
HSC200	OC	LED/NPN Viso:1000V CTR:>100%		Hon		data pinout
HSC 276	Si-Di	=1SS276: SMD	71a(1,3mm)	Hit		data
HSC 277	C-Di	SMD, VHF, Band-S, 35V, <0,7Ω(100MHz)	71a(1,3mm)	Hit	-	data
HSC300	OC	LED/NPN Viso:3000V CTR:>50% Ibs:10mA		Hon		data pinout
HSC 2692	Si-Di	=HSM 2692	35e(2mm)	Hit	-	data
HSC 2693	Si-Di	=HSM 2693	35l(2mm)	Hit	-	data
HSC 2694	Si-Di	=HSM 2694	35m(2mm)	Hit	-	data
HSC 2836	Si-Di	=HSM 2836	35m(2mm)	Hit	-	data
HSC 2838	Si-Di	=HSM 2838	35l(2mm)	Hit	-	data
HSE 11	GaAs-Di	Schottky-Di, SHF-Converter, 4V, 50mA	Chip	Hit	-	data
HSE 11 S	GaAs-Di	=HSE 11: Dual	35o			data
HSE 50	GaAs-Di	Schottky-Di, SHF-Mixer, 4V, 50mA	Chip	Hit	-	data
HSG 3xxx...7xxx	DIG-IC	Gate Arrays		Tho	-	
HSG 3020...3250	CMOS-IC	HCMOS Gate Arrays		Sgs	-	
HSG 5080...5600	CMOS-IC	HCMOS Gate Arrays		Sgs	-	
HSK 23	Si-St	SMD, 0,58..0,65V(3mA)	72a(3,4mm)	Hit		data
HSK 83	Si-Di	SMD, 300V, 0,15A, <100ns	72a(3,4mm)	Hit	-	data
HSK 110	C-Di	SMD, Band-S, VHF-/UHF-Tuner	72a(3,4mm)	Hit	BA 682...683	data
HSK 120	Si-Di	SMD, Dem, SS, 70V, 0,15A, <3ns	72a(3,4mm)	Hit	-	data
HSK 122	Si-Di	SMD, S, 410V, 0,15A, <10µs	72a(3,4mm)	Hit	-	data
HSK 151	Si-Di	SMD, Schottky, TV-Tuner, UHF-M, 3V	72a(3,4mm)	Hit	TMMBAT 29	data
HSK 277	C-Di	SMD, Band-S, VHF-/UHF-Tuner	72a(3,4mm)	Hit	BA 683	data
HSM 83	Si-Di	SMD, 300V, 0,1A, <100ns	35e	Hit	-	data
HSM 88 S	Si-Di	SMD, Dual, Schottky, CATV-M, 10V, 15mA	35o	Hit	BAT 68-04, 1SS271	data
HSM 88 AS	Si-Di	=HSM 88S:	35o		BAT 68-04	data
HSM 88 SR	Si-Di	=HSM 88S:	35n		-	data
HSM 88 ASR	Si-Di	=HSM 88S:	35n		-	data
HSM 88 WA	Si-Di	=HSM 88S:	35m		BAT 68-06	data
HSM 88 WK	Si-Di	=HSM 88S:	35l		BAT 68-05	data
HSM 107 S	Si-Di	SMD, Dual, Schottky, 8V, Ifm=0,1A	35o	Hit	BAT 68-04	data
HSM 109 WK	Si-Di	=HSM 122WK: D1=10V, D2=10V	35l	Hit	-	data
HSM 112 WK	Si-Di	SMD, Dual, D1=8V, D2=10V, 30mA	35l	Hit	-	data
HSM 113 WK	Si-Di	SMD, Dual, 20V, D1=0,2A, D2=0,1A	35l	Hit	-	data
HSM 122	Si-Di	SMD, Uni, 420V, 0,1A, <10µs	35e	Hit	1SS311	data
HSM 123	Si-Di	SMD, Dual, SS, 85V, 0,1A, <3ns	35o	Hit	BAV 99, MA 157A	data
HSM 124 S	Si-Di	SMD, Dual, S, 85V, 0,1A, <100ns	35o	Hit	BAV 99	data
HSM 125 WK	Si-Di	SMD, Dual, 20V, D1=0,5A, D2=0,1A	35l	Hit	-	data
HSM 126 S	Si-Di	SMD, Schottky-Di, 20V, 0,2A	35o	Hit	BAT 54S, BAT 64-04	data
HSM 198 S	Si-Di	SMD, Dual, Schottky, 10V, 30mA, <1,5pF	35o	Hit	1SS366	data
HSM 198 SR	Si-Di	=HSM 198S:	35n	Hit	-	data
HSM 221 C	Si-Di	SMD, SS, 85V, 0,1A, <3ns	35e	Hit	MMBD 914, MMBD 4148	data
HSM 223 C	Si-Di	=HSM 221C:	35f	Hit	-	data
HSM 276 S	Si-Di	SMD, Dual, Schottky, 3V, 30mA, <0,9pF	35o	Hit	MA 730, 1SS295, 1SS366	data
HSM 276 SR	Si-Di	=HSM 276S:	35n	Hit	-	data
HSM 402 S	Si-Di	=HSS 402: SMD, Dual	35o	Hit	-	data
HSM 402 WK	Si-Di	=HSS 402: SMD, Dual	35l		-	data
HSM 2692	C-Di	SMD, Tuner-Band-S, <0,9Ω(100MHz)	35e	Hit	-	data
HSM 2693(A)	C-Di	=HSM 2692: Dual	35l	Hit	1SS268	data
HSM 2694(A)	C-Di	=HSM 2692: Dual	35m	Hit	1SS269	data
HSM 2836 C	Si-Di	SMD, Dual, S, 85V, 0,1A, <20ns	35m	Hit	BAW 56, KDS 181, 1SS181	data
HSM 2838 C	Si-Di	SMD, Dual, SS, 85V, 0,1A, <3ns	35l	Hit	BAV 70, KDS 184, 1SS184	data
HSMA-A100	LED	am ø2.2mm I:220mcd Vf:1.9V If:210mA	PLCC-2	Agt		data pdf pinout
HSMA-S670	LED	am 1x1.25x0.8mm I>16mcd Vf:1.9V		Agt		data pdf pinout
HSMA-S690	LED	am 1.6x0.8x0.7mm I>16mcd Vf:1.9V		Agt		data pdf pinout
HSMB-A100	LED	bl ø2.2mm I>50mcd Vf:3.4V If:20mA	PLCC-2	Agt		data pdf pinout
HSMC-A100	LED	rd ø2.2mm I>160mcd Vf:1.9V If:20mA	PLCC-2	Agt		data pdf pinout
HSMC-S660	LED	rd 3x2x1mm I>16mcd Vf:1.9V If:20mA		Agt		data pdf pinout
HSMC-S670	LED	rd 2x1.25x0.8mm I>16mcd Vf:1.9V		Agt		data pdf pinout
HSMC-S690	LED	rd 1.6x0.8x0.7mm I>16mcd Vf:1.9V		Agt		data pdf pinout
HSMD-A100	LED	or ø2.2mm I:15mcd Vf:2.2V If:20mA	PLCC-2	Agt		data pdf pinout
HSMD-S660	LED	or 3x2x1mm I>16mcd Vf:1.9V If:20mA		Agt		data pdf pinout
HSMD-S670	LED	or 2x1.25x0.8mm I>16mcd Vf:1.9V		Agt		data pdf pinout
HSMD-S690	LED	or 1.6x0.8x0.7mm I>16mcd Vf:1.9V		Agt		data pdf pinout
HSME-A100	LED	gr ø2.2mm I>30mcd Vf:1.9V If:20mA	PLCC-2	Agt		data pdf pinout
HSME-C110	LED	gr 3.2x1x1.5mm I>16mcd Vf:2.1V	PLCC-2	Agt		data pdf pinout
HSME-C120	LED	gr 1.6x0.6x1mm I>16mcd Vf:2.1V	PLCC-2	Agt		data pdf pinout
HSME-C150	LED	gr 3.2x1.6x1.1mm I>16mcd Vf:2.1V	PLCC-2	Agt		data pdf pinout

Type	Device	Short Description	Fig.	Manu	Comparision Types	More at
HSME-C170	LED	gr 2x1.25x0.8mm I>16mcd Vf:2 1V	PLCC-2	Agt		data pdf pinout
HSME-C177	LED	gr 2x1.25x0.4mm I>16mcd Vf:2.1V	PLCC-2	Agt		data pdf pinout
HSME-C190	LED	gr 1.6x0.8x0.8mm I>16mcd Vf:2.1V	PLCC-2	Agt		data pdf pinout
HSME-C191	LED	gr 1.6x0.8x0.6mm I>16mcd Vf:2.1V	PLCC-2	Agt		data pdf pinout
HSME-C197	LED	gr 1.6x0.8x0.4mm I>16mcd Vf:2.1V	PLCC-2	Agt		data pdf pinout
HSME-C265	LED	gr 3.4x1.25x1.1mm I>16mcd Vf:2.1V	PLCC-2	Agt		data pdf pinout
HSMF-A201	LED	rd/gr ø2.2mm I:20mcd Vf:2.2V If:20mA	PLCC-4	Agt		data pdf pinout
HSMF-A202	LED	rd/yl ø2.2mm I:12mcd Vf:2.2V If:20mA	PLCC-4	Agt		data pdf pinout
HSMF-A203	LED	rd/gr ø2.2mm I:8mcd Vf:2.2V If:20mA	PLCC-4	Agt		data pdf pinout
HSMF-A204	LED	or/gr ø2.2mm I:20mcd Vf:2.2V If:20mA	PLCC-4	Agt		data pdf pinout
HSMF-A205	LED	or/gr ø2.2mm I:8mcd Vf:2.2V If:20mA	PLCC-4	Agt		data pdf pinout
HSMF-A206	LED	yl/gr ø2.2mm I:20mcd Vf:2.2V If:20mA	PLCC-4	Agt		data pdf pinout
HSMF-A211	LED	rd/gr ø2.2mm I:20mcd Vf:2.2V If:20mA	PLCC-4	Agt		data pdf pinout
HSMF-A212	LED	rd/yl ø2.2mm I:12mcd Vf:2.2V If:20mA	PLCC-4	Agt		data pdf pinout
HSMF-A222	LED	rd/am ø2.2mm I:80mcd Vf:1.9V If:20mA	PLCC-4	Agt		data pdf pinout
HSMF-A226	LED	am/gr ø2.2mm I:60mcd Vf:1.9V If:20mA	PLCC-4	Agt		data pdf pinout
HSMF-A227	LED	rd/bl ø2.2mm I:10mcd Vf:3.9V If:20mA	PLCC-4	Agt		data pdf pinout
HSMF-A228	LED	am/bl ø2.2mm I:10mcd Vf:3.9V If:20mA	PLCC-4	Agt		data pdf pinout
HSMF-A301	LED	rd/gr/bl ø2.2mm I:10mcd Vf:3.9	PLCC-4	Agt		data pdf pinout
HSMF-A331	LED	rd/gr/bl ø2.2mm I:10mcd Vf:3.9V	PLCC-4	Agt		data pdf pinout
HSMF-A332	LED	rd/gr/bl ø2.2mm I:10mcd Vf:3.9V	PLCC-4	Agt		data pdf pinout
HSMF-A341	LED	rd/gr/bl ø2.2mm I:40mcd Vf:3.9V	PLCC-4	Agt		data pdf pinout
HSMF-A342	LED	rd/gr/bl ø2.2mm I:40mcd Vf:3.9V	PLCC-4	Agt		data pdf pinout
HSMF-C117	LED	rd/gr/bl 3.2x2.7x1.1mm I>10mcd	T1	Agt		data pdf pinout
HSMF-C655	LED	rd/gr ø1.1mm I:9mcd Vf:2.2V If:20mA		Hpa		data pinout
HSMG-A100	LED	gr ø2.2mm I>8mcd Vf:2.2V If:10mA	PLCC-2	Agt		data pdf pinout
HSMH-A100	LED	rd ø2.2mm I>40mcd Vf:1.9V If:20mA	PLCC-2	Agt		data pdf pinout
HSMH-C650	LED	or+gr+rd+srd+yl ø1.1mm I:5mcd		Hpa		data pinout
HSMH-C670	LED	or+gr+rd+srd+yl ø1.1mm I:5mcd		Hpa		data pinout
HSMH-T400	LED	or+gr+rd+srd ø2.2mm I:6mcd Vf:1.9V	PLCC-2	Hpa		data pinout
HSMH-T500	LED	or+gr+rd+srd+yl ø2.2mm I:5mcd Vf:2V	PLCC-2	Hpa		data pinout
HSMH-T600	LED	or+gr+rd+srd+yl ø2.2mm I:5mcd Vf:2V	PLCC-2	Hpa		data pinout
HSMH-T700	LED	or+gr+rd+yl ø2.2mm I:5mcd Vf:2V	PLCC-2	Hpa		data pinout
HSMJ-A100	LED	or ø2.2mm I:200mcd Vf:1.9V If:20mA	PLCC-2	Agt		data pdf pinout
HSMK-A100	LED	cy ø2.2mm I>100mcd Vf:3.4V If:20mA	PLCC-2	Agt		data pdf pinout
HSML-A100	LED	or ø2.2mm I:220mcd Vf:1.9V If:20mA	PLCC-2	Agt		data pdf pinout
HSMM-A100	LED	gr ø2.2mm I>200mcd Vf:3.4V If:20mA	PLCC-2	Agt		data pdf pinout
HSMS-A100	LED	rd ø2.2mm I:15mcd Vf:2.2V If:20mA	PLCC-2	Agt		data pdf pinout
HSMU-A100	LED	am ø2.2mm I>160mcd Vf:2.2V If:20mA	PLCC-2	Agt		data pdf pinout
HSMV-A100	LED	or ø2.2mm I:350mcd Vf:2.2V If:20mA	PLCC-2	Agt		data pdf pinout
HSMY-A100	LED	yl+am ø2.2mm I:12mcd Vf:2.2V If:20mA	PLCC-2	Agt		data pdf pinout
HSMZ-A100	LED	rd ø2.2mm I>250mcd Vf:2.2V If:20mA	PLCC-2	Agt		data pdf pinout
HSP 50110 LC-52	MOS-IC	SAT-receiver, Digital Signal Process.	-PLCC	Har	-	
HSP 50210 LC-52	MOS-IC	Sat-receiver, Digital Signal Process.	-PLCC	Har	-	
HSR 101	Si-Di	SMD, Schottky,Dem,S, 30V, 35mA, <1,5pF	71a(2,7mm)	Hit	-	data
HSR 276	Si-Di	=1SS276: SMD	71a(2,7mm)	Hit		data
HSR 277	C-Di	=1SS277: SMD	71a(2,7mm)	Hit		data
HSR312	OC	LED/2 Viso:4000Vrms Vbs:<1.6V		Fcs		data pdf pinout
HSR412	OC	LED/2 Viso:4000Vrms Vbs:<1.6V		Fcs		data pdf pinout
HSS 81	Si-Di	=1SS81:	31a	Hit	-	data
HSS 82	Si-Di	=1SS82:	31a	Hit	-	data
HSS 83	Si-Di	=1SS83:	31a	Hit	-	data
HSS 100	Si-Di	Schottky, Dem, SS, 60V, 35mA, <2,2pF	31a	Hit	BAT 83, 1N6263	data
HSS 101	Si-Di	Schottky, Dem, SS, 30V, 35mA, <1,5pF	31a	Hit	BAT 81, 1N6263	data
HSS 102	Si-Di	Schottky, Dem, SS, 70V, 35mA, <2,2pF	31a	Hit	BAR 28, 1N5711, 1SS285	data
HSS 104	Si-Di	Dem, SS, 40V, 0,11A, <3ns	31a	Hit	BAW 62, BAW 76, BAX 95, 1N4148, ++	data
HSS 271	Si-Di	Dem, SS, 60V, 0,12A, <4ns	31a	Hit	BAW 62, BAW 76, BAX 95, 1N4148, ++	data
HSS 400 J	Si-Di	Tiefpass/lowpass Diode, 100V, 0,1A	31a	Hit	-	data
HSS 401 J	Si-Di	Tiefpass/lowpass Diode, 100V, 0,1A	31a	Hit	-	data
HSS 402 J	Si-Di	Tiefpass/lowpass Diode, 100V, 0,1A	31a	Hit	-	data
HSU 83	Si-Di	=HSM 83:	71a(1,7mm)	Hit	1SS376	data
HSU 88	Si-Di	=1SS88: SMD	71a(1,7mm)	Hit	-	data
HSU 119	Si-Di	SMD, SS, 85V, 0,1A, <3ns	71a(1,7mm)	Hit	MA 111, 1SS352, 1SS353	data
HSU 276	Si-Di	=1SS276: SMD	71a(1,7mm)	Hit	HSU 88, 1SS315	data
HSU 277	C-Di	=1SS277: SMD	71a(1,7mm)	Hit	BA 592, MA 77, 1SS314, 1SS356	data
HSU 402 J	Si-Di	=HSS 402J: SMD	71a(1,7mm)	Hit		data
HT-110D	LED	am &3.2x1.5x1.0mm I:9mcd Vf:2.1V		Hrv		data pdf pinout
HT-110NB	LED	bl &3.2x1.5x1.0mm I:26mcd Vf:2.8V		Hrv		data pdf pinout
HT-110NG	LED	gr &3.2x1.5x1.0mm I:50mcd Vf:2.8V		Hrv		data pdf pinout
HT-110SD	LED	or &3.2x1.5x1.0mm I:9mcd Vf:2.1V		Hrv		data pdf pinout
HT-110TW	LED	wht &3.2x1.5x1.0mm I:70mcd Vf:2.8V		Hrv		data pdf pinout
HT-110UD	LED	am &3.2x1.5x1.0mm I:160mcd Vf:1.9V		Hrv		data pdf pinout
HT-110UR	LED	rd &3.2x1.5x1.0mm I:20mcd Vf:1.8V		Hrv		data pdf pinout
HT-110USD	LED	or &3.2x1.5x1.0mm I:120mcd Vf:1.9V		Hrv		data pdf pinout
HT-110UY	LED	yl &3.2x1.5x1.0mm I:120mcd Vf:1.9V		Hrv		data pdf pinout
HT-110UYG	LED	gr &3.2x1.5x1.0mm I:60mcd Vf:2V		Hrv		data pdf pinout
HT-110Y	LED	yl &3.2x1.5x1.0mm I:9mcd Vf:2.1V		Hrv		data pdf pinout
HT-110YG	LED	gr &3.2x1.5x1.0mm I:16mcd Vf:2.2V		Hrv		data pdf pinout
HT-121D	LED	am &1.6x1.0x0.6mm I:9mcd Vf:2.1V		Hrv		data pdf pinout

Type	Device	Short Description	Fig.	Manu	Comparision Types	More at
HT-121NB	LED	bl &1.6x1.0x0.6mm I:26mcd Vf:2.8V		Hrv		data pdf pinout
HT-121NG	LED	gr &1.6x1.0x0.6mm I:50mcd Vf:2.8V		Hrv		data pdf pinout
HT-121SD	LED	or &1.6x1.0x0.6mm I:9mcd Vf:2.1V		Hrv		data pdf pinout
HT-121TW	LED	wht &1.6x1.0x0.6mm I:70mcd Vf:2.8V		rHv		data pdf pinout
HT-121UD	LED	am &1.6x1.0x0.6mm I:120mcd Vf:1.9V		Hrv		data pdf pinout
HT-121UR	LED	rd &1.6x1.0x0.6mm I:20mcd Vf:1.8V		Hrv		data pdf pinout
HT-121USD	LED	or &1.6x1.0x0.6mm I:140mcd Vf:1.9V		Hrv		data pdf pinout
HT-121UY	LED	yl &1.6x1.0x0.6mm I:120mcd Vf:1.9V		Hrv		data pdf pinout
HT-121UYG	LED	gr &1.6x1.0x0.6mm I:90mcd Vf:2V		Hrv		data pdf pinout
HT-121Y	LED	yl &1.6x1.0x0.6mm I:9mcd Vf:2.1V		Hrv		data pdf pinout
HT-121YG	LED	gr &1.6x1.0x0.6mm I:16mcd Vf:2.2V		Hrv		data pdf pinout
HT-150D	LED	am %3.2x1.6x0.6mm I:9mcd Vf:2.1V		Hrv		data pdf pinout
HT-150NB	LED	bl %3.2x1.6x0.6mm I:25mcd Vf:2.8V		Hrv		data pdf pinout
HT-150NG	LED	gr %3.2x1.6x0.6mm I:45mcd Vf:2.8V		Hrv		data pdf pinout
HT-150SD	LED	or %3.2x1.6x0.6mm I:14mcd Vf:2.1V		Hrv		data pdf pinout
HT-150TW	LED	wht %3.2x1.6x0.6mm I:60mcd Vf:2.8V		Hrv		data pdf pinout
HT-150UD	LED	am %3.2x1.6x0.6mm I:120mcd Vf:1.9V		Hrv		data pdf pinout
HT-150UR	LED	rd %3.2x1.6x0.6mm I:21mcd Vf:1.8V		Hrv		data pdf pinout
HT-150USD	LED	or %3.2x1.6x0.6mm I:120mcd Vf:1.9V		Hrv		data pdf pinout
HT-150UY	LED	yl %3.2x1.6x0.6mm I:90mcd Vf:1.9V		Hrv		data pdf pinout
HT-150UYG	LED	gr %3.2x1.6x0.6mm I:60mcd Vf:2V		Hrv		data pdf pinout
HT-150Y	LED	yl %3.2x1.6x0.6mm I:9mcd Vf:2.1V		Hrv		data pdf pinout
HT-150YG	LED	gr %3.2x1.6x0.6mm I:20mcd Vf:2.2V		Hrv		data pdf pinout
HT-155D/YG	LED	am/gr &3.2x2.7x1.1mm I:20mcd Vf:2.2V		Hrv		data pdf pinout
HT-155SD/YG	LED	or/gr &3.2x2.7x1.1mm Vf:2.2V		Hrv		data pdf pinout
HT-155UD/NB	LED	am/bl &3.2x2.7x1.1mm I:56mcd Vf:3.3V		Hrv		data pdf pinout
HT-155UD/UYG	LED	am/gr &3.2x2.7x1.1mm I:90mcd Vf:2V		Hrv		data pdf pinout
HT-155USD/NB	LED	yl/bl &3.2x2.7x1.1mm I:56mcd Vf:3.3V		Hrv		data pdf pinout
HT-155USD/UYG	LED	or/gr &3.2x2.7x1.1mm I:90mcd Vf:2V		Hrv		data pdf pinout
HT-155UY/NB	LED	yl/bl &3.2x2.7x1.1mm I:90mcd Vf:1.9V		Hrv		data pdf pinout
HT-155UY/UYG	LED	yl/gr &3.2x2.7x1.1mm I:90mcd Vf:2V		Hrv		data pdf pinout
HT-155Y/YG	LED	yl/gr &3.2x2.7x1:1mm I:20mcd Vf:2.2V		Hrv		data pdf pinout
HT-170D	LED	am &2.0x1.3x0.8mm I:9mcd Vf:2.1V		Hrv		data pdf pinout
HT-170NB	LED	bl &2.0x1.3x0.8mm I:25mcd Vf:2.8V		Hrv		data pdf pinout
HT-170NG	LED	gr &2.0x1.3x0.8mm I:45mcd Vf:2.8V		Hrv		data pdf pinout
HT-170SD	LED	or &2.0x1.3x0.8mm I:14mcd Vf:2.1V		Hrv		data pdf pinout
HT-170TW	LED	wht &2.0x1.3x0.8mm I:60mcd Vf:2.8V		Hrv		data pdf pinout
HT-170UD	LED	am &2.0x1.3x0.8mm I:90mcd Vf:1.9V		Hrv		data pdf pinout
HT-170UR	LED	rd &2.0x1.3x0.8mm I:21mcd Vf:1.8V		Hrv		data pdf pinout
HT-170USD	LED	or &2.0x1.3x0.8mm I:60mcd Vf:1.9V		Hrv		data pdf pinout
HT-170UY	LED	yl &2.0x1.3x0.8mm I:90mcd Vf:1.9V		Hrv		data pdf pinout
HT-170UYG	LED	gr &2.0x1.3x0.8mm I:50mcd Vf:2V		Hrv		data pdf pinout
HT-170Y	LED	yl &2.0x1.3x0.8mm I:9mcd Vf:2.1V		Hrv		data pdf pinout
HT-170YG	LED	gr &2.0x1.3x0.8mm I:18mcd Vf:2.2V		Hrv		data pdf pinout
HT-210D/YG	LED	am/gr &3.2x1.5x1.0mm I:16mcd Vf:2.2V		Hrv		data pdf pinout
HT-210SD/YG	LED	or/gr &3.2x1.5x1.0mm I:16mcd Vf:2.2V		Hrv		data pdf pinout
HT-210UD/NB	LED	am/bl &3.2x1.5x1.0mm I:60mcd Vf:3.3V		Hrv		data pdf pinout
HT-210UD/UYG	LED	am/gr &3.2x1.5x1.0mm I:60mcd Vf:2V		Hrv		data pdf pinout
HT-210USD/NB	LED	yl/bl &3.2x1.5x1.0mm I:60mcd Vf:3.3V		Hrv		data pdf pinout
HT-210USD/UYG	LED	or/gr &3.2x1.5x1.0mm I:60mcd Vf:2V		Hrv		data pdf pinout
HT-210UY/NB	LED	yl/bl &3.2x1.5x1.0mm I:120mcd		Hrv		data pdf pinout
HT-210UY/UYG	LED	yl/gr &3.2x1.5x1.0mm I:60mcd Vf:2V		Hrv		data pdf pinout
HT-210Y/YG	LED	yl/gr &3.2x1.5x1.0mm I:16mcd Vf:2.2V		Hrv		data pdf pinout
HT-260D	LED	am &3.2x1.2x1.1mm I:9mcd Vf:2.1V		Hrv		data pdf pinout
HT-260NB	LED	bl &3.2x1.2x1.1mm I:25mcd Vf:2.8V		Hrv		data pdf pinout
HT-260NG	LED	gr &3.2x1.2x1.1mm I:45mcd Vf:2.8V		Hrv		data pdf pinout
HT-260SD	LED	or &3.2x1.2x1.1mm I:14mcd Vf:2.1V		Hrv		data pdf pinout
HT-260TW	LED	wht &3.2x1.2x1.1mm I:60mcd Vf:2.8V		Hrv		data pdf pinout
HT-260UD	LED	am &3.2x1.2x1.1mm I:90mcd Vf:1.9V		Hrv		data pdf pinout
HT-260UR	LED	rd &3.2x1.2x1.1mm I:14mcd Vf:1.8V		Hrv		data pdf pinout
HT-260USD	LED	or &3.2x1.2x1.1mm I:60mcd Vf:1.9V		Hrv		data pdf pinout
HT-260UY	LED	yl &3.2x1.2x1.1mm I:63mcd Vf:1.9V		Hrv		data pdf pinout
HT-260UYG	LED	gr &3.2x1.2x1.1mm I:60mcd Vf:2V		Hrv		data pdf pinout
HT-260Y	LED	yl &3.2x1.2x1.1mm I:9mcd Vf:2.1V		Hrv		data pdf pinout
HT-260YG	LED	gr &3.2x1.2x1.1mm I:14mcd Vf:2.2V		Hrv		data pdf pinout
HT-280D	LED	am &1.0x0.5x0.4mm I:9mcd Vf:2.1V		Hrv		data pdf pinout
HT-280SD	LED	or &1.0x0.5x0.4mm I:11mcd Vf:2.1V		Hrv		data pdf pinout
HT-280UD	LED	am &1.0x0.5x0.4mm I:90mcd Vf:1.9V		Hrv		data pdf pinout
HT-280USD	LED	or &1.0x0.5x0.4mm I:60mcd Vf:1.9V		Hrv		data pdf pinout
HT-280UY	LED	yl &1.0x0.5x0.4mm I:60mcd Vf:1.9V		Hrv		data pdf pinout
HT-280UYG	LED	gr &1.0x0.5x0.4mm I:50mcd Vf:2V		Hrv		data pdf pinout
HT-280Y	LED	yl &1.0x0.5x0.4mm I:9mcd Vf:2.1V		Hrv		data pdf pinout
HT-280YG	LED	gr &1.0x0.5x0.4mm I:18mcd Vf:2.2V		Hrv		data pdf pinout
HT-297AD/UYG	LED	am/gr &1.6x0.8x0.5mm I:90mcd Vf:2V		Hrv		data pdf pinout
HT-297USD/NB	LED	or/bl &1.6x0.8x0.5mm I:80mcd Vf:3.3V		Hrv		data pdf pinout
HT-297USD/NG	LED	or/gr &1.6x0.8x0.5mm I:200mcd		Hrv		data pdf pinout
HT-297USD/UYG	LED	or/gr &1.6x0.8x0.5mm I:90mcd Vf:2V		Hrv		data pdf pinout
HT-297UY/UYG	LED	yl/gr &1.6x0.8x0.5mm I:60mcd Vf:1.9V		Hrv		data pdf pinout
HT-311FCH	LED	or/gr/bl &3.2x1.5x1.0mm I:90mcd		Hrv		data pdf pinout
HT-311FDH	LED	am/gr/bl &3.2x1.5x1.0mm I:90mcd		Hrv		data pdf pinout
HT-358FCH	LED	or/gr/bl &3.2x2.7x1.1mm I:90mcd		Hrv		data pdf pinout
HT-358FDH	LED	am/gr/bl.&3.2x2.7x1.1mm I:90mcd		Hrv		data pdf pinout
HT-372FCH	LED	or/gr/bl &2.0x1.3x0.5mm I:90mcd		Hrv		data pdf pinout
HT-372FDH	LED	am/gr/bl &2.0x1.3x0.5mm I:90mcd		Hrv		data pdf pinout
HT-374FCH	LED	or/gr/bl &2.0x1.3x0.5mm I:90mcd		Hrv		data pdf pinout

Type	Device	Short Description	Fig.	Manu	Comparision Types	More at
HT-374FDH	LED	am/gr/bl &2.0x1.3x0.5mm l:90mcd		Hrv		data pdf pinout
HT-C365FCH	LED	or/gr/bl &6.0x5.0x0.8mm l:160mcd		Hrv		data pdf pinout
HT-C365FCHSP	LED	or/gr/bl &6.0x5.0x0.8mm l:80mcd		Hrv		data pdf pinout
HT-C365TW	LED	wht &6.0x5.0x0.8mm l:2500mcd Vf:3.3V		Hrv		data pdf pinout
HT-C365TWCT	LED	wht &6.0x5.0x0.8mm l:2500mcd Vf:3.3V		Hrv		data pdf pinout
HT-C365TWFC	LED	wht &6.0x5.0x0.8mm l:4000mcd Vf:3.3V		Hrv		data pdf pinout
HT-C365TWSP	LED	wht &6.0x5.0x0.8mm l:2500mcd Vf:3.3V		Hrv		data pdf pinout
HT-D466NB	LED	bl &6.0x5.0x0.8mm l:25lm Vf:6.4V		Hrv		data pdf pinout
HT-D466NG	LED	gr &6.0x5.0x0.8mm l:100lm Vf:6.4V		Hrv		data pdf pinout
HT-D466TW	LED	wht &6.0x5.0x0.8mm l:150lm Vf:6.4V		Hrv		data pdf pinout
HT-D468NB	LED	bl &6.0x5.0x0.8mm l:25lm Vf:6.4V		Hrv		data pdf pinout
HT-D468NG	LED	gr &6.0x5.0x0.8mm l:100lm Vf:6.4V		Hrv		data pdf pinout
HT-D468TW	LED	wht &6.0x5.0x0.8mm l:150lm Vf:6.4V		Hrv		data pdf pinout
HT-D476NB	LED	bl &6.0x5.0x0.8mm l:25lm Vf:6.4V		Hrv		data pdf pinout
HT-D476NG	LED	gr &6.0x5.0x0.8mm l:100lm Vf:6.4V		Hrv		data pdf pinout
HT-D476TW	LED	wht &6.0x5.0x0.8mm l:150lm Vf:6.4V		Hrv		data pdf pinout
HT-D478NB	LED	bl &6.0x5.0x0.8mm l:25lm Vf:6.4V		Hrv		data pdf pinout
HT-D478NG	LED	gr &6.0x5.0x0.8mm l:100lm Vf:6.4V		Hrv		data pdf pinout
HT-D478TW	LED	wht &6.0x5.0x0.8mm l:150lm Vf:6.4V		Hrv		data pdf pinout
HT-F118NB	LED	bl &3.2x1.0x0.8mm l:180mcd Vf:3.3V		Hrv		data pdf pinout
HT-F118NG	LED	gr &3.2x1.0x0.8mm l:250mcd Vf:3.3V		Hrv		data pdf pinout
HT-F118TW	LED	wht &3.2x1.0x0.8mm l:715mcd Vf:3.3V		Hrv		data pdf pinout
HT-F195	LED	bl+gr+wht &1.6x0.8x0.4mm l:90mcd		Hrv		data pdf pinout
HT-F195U	LED	am+or+yl+gr &1.6x0.8x0.4mm l:80mcd		Hrv		data pdf pinout
HT-LB1131XXX	LED	bl+gr+wht+rd+am &1.6x0.8x0.4mm		Hrv		data pdf pinout
HT-LB2141XXX	LED	bl+gr+wht+rd+am &1.6x0.8x0.4mm		Hrv		data pdf pinout
HT-LB2331XXX	LED	bl+gr+wht+rd+am &1.6x0.8x0.4mm		Hrv		data pdf pinout
HT-LB3233XXX	LED	bl+gr+wht+rd+am &1.6x0.8x0.4mm		Hrv		data pdf pinout
HT-LB3262XXX	LED	bl+gr+wht+rd+am &1.6x0.8x0.4mm		Hrv		data pdf pinout
HT-LB3432XXX	LED	bl+gr+wht+rd+am &1.6x0.8x0.4mm		Hrv		data pdf pinout
HT-LB4431XXX	LED	bl+gr+wht+rd+am &1.6x0.8x0.4mm		Hrv		data pdf pinout
HT-MR16XXX-A1	LED	bl+gr+wht+rd+am+pi &1.6x0.8x0.4mm		Hrv		data pdf pinout
HT-MR16XXX-B1	LED	bl+gr+wht+rd+am+pi &1.6x0.8x0.4mm		Hrv		data pdf pinout
HT-MR16XXX-C2	LED	bl+gr+wht+rd+am+pi &1.6x0.8x0.4mm		Hrv		data pdf pinout
HT-P166NB	LED	bl &1.6x0.8x0.4mm l:8lm Vf:3.4V		Hrv		data pdf pinout
HT-P166NG	LED	gr &1.6x0.8x0.4mm l:25lm Vf:3.4V		Hrv		data pdf pinout
HT-P166TW	LED	wht &1.6x0.8x0.4mm l:30lm Vf:3.4V		Hrv		data pdf pinout
HT-P166US	LED	rd &1.6x0.8x0.4mm l:18lm Vf:2.2V		Hrv		data pdf pinout
HT-P166USD	LED	rd &1.6x0.8x0.4mm l:22lm Vf:2.1V		Hrv		data pdf pinout
HT-P166UY	LED	am &1.6x0.8x0.4mm l:18lm Vf:2.2V		Hrv		data pdf pinout
HT-P168NB	LED	bl &1.6x0.8x0.4mm l:8lm Vf:3.4V		Hrv		data pdf pinout
HT-P168NG	LED	gr &1.6x0.8x0.4mm l:25lm Vf:3.4V		Hrv		data pdf pinout
HT-P168TW	LED	wht &1.6x0.8x0.4mm l:30lm Vf:3.4V		Hrv		data pdf pinout
HT-P168US	LED	rd &1.6x0.8x0.4mm l:18lm Vf:2.2V		Hrv		data pdf pinout
HT-P168USD	LED	rd &1.6x0.8x0.4mm l:17lm Vf:2.1V		Hrv		data pdf pinout
HT-P168UY	LED	am &1.6x0.8x0.4mm l:18lm Vf:2.2V		Hrv		data pdf pinout
HT-P176NB	LED	bl &1.6x0.8x0.4mm l:8lm Vf:3.4V		Hrv		data pdf pinout
HT-P176NG	LED	gr &1.6x0.8x0.4mm l:25lm Vf:3.4V		Hrv		data pdf pinout
HT-P176TW	LED	wht &1.6x0.8x0.4mm l:30lm Vf:3.4V		Hrv		data pdf pinout
HT-P176US	LED	rd &1.6x0.8x0.4mm l:18lm Vf:2.2V		Hrv		data pdf pinout
HT-P176USD	LED	rd &1.6x0.8x0.4mm l:22lm Vf:2.1V		Hrv		data pdf pinout
HT-P176UY	LED	am &1.6x0.8x0.4mm l:18lm Vf:2.2V		Hrv		data pdf pinout
HT-P178NB	LED	bl &1.6x0.8x0.4mm l:8lm Vf:3.4V		Hrv		data pdf pinout
HT-P178NG	LED	gr &1.6x0.8x0.4mm l:25lm Vf:3.4V		Hrv		data pdf pinout
HT-P178TW	LED	wht &1.6x0.8x0.4mm l:30lm Vf:3.4V		Hrv		data pdf pinout
HT-P178US	LED	rd &1.6x0.8x0.4mm l:18lm Vf:2.2V		Hrv		data pdf pinout
HT-P178USD	LED	rd &1.6x0.8x0.4mm l:22lm Vf:2.1V		Hrv		data pdf pinout
HT-P178UY	LED	am &1.6x0.8x0.4mm l:18lm Vf:2.2V		Hrv		data pdf pinout
HT-P266NB	LED	bl &1.6x0.8x0.4mm l:12lm Vf:3.4V		Hrv		data pdf pinout
HT-P266NG	LED	gr &1.6x0.8x0.4mm l:50lm Vf:3.4V		Hrv		data pdf pinout
HT-P266TW	LED	wht &1.6x0.8x0.4mm l:70lm Vf:3.4V		Hrv		data pdf pinout
HT-P266US	LED	rd &1.6x0.8x0.4mm l:35lm Vf:2.2V		Hrv		data pdf pinout
HT-P266UY	LED	am &1.6x0.8x0.4mm l:30lm Vf:2.2V		Hrv		data pdf pinout
HT-P268NB	LED	bl &1.6x0.8x0.4mm l:12lm Vf:3.4V		Hrv		data pdf pinout
HT-P268NG	LED	gr &1.6x0.8x0.4mm l:50lm Vf:3.4V		Hrv		data pdf pinout
HT-P268TW	LED	wht &1.6x0.8x0.4mm l:70lm Vf:3.4V		Hrv		data pdf pinout
HT-P268US	LED	rd &1.6x0.8x0.4mm l:35lm Vf:2.2V		Hrv		data pdf pinout
HT-P268UY	LED	am &1.6x0.8x0.4mm l:30lm Vf:2.2V		Hrv		data pdf pinout
HT-P276NB	LED	bl &1.6x0.8x0.4mm l:12lm Vf:3.4V		Hrv		data pdf pinout
HT-P276NG	LED	gr &1.6x0.8x0.4mm l:50lm Vf:3.4V		Hrv		data pdf pinout
HT-P276TW	LED	wht &1.6x0.8x0.4mm l:70lm Vf:3.4V		Hrv		data pdf pinout
HT-P276US	LED	rd &1.6x0.8x0.4mm l:35lm Vf:2.2V		Hrv		data pdf pinout
HT-P276UY	LED	am &1.6x0.8x0.4mm l:30lm Vf:2.2V		Hrv		data pdf pinout
HT-P278NB	LED	bl &1.6x0.8x0.4mm l:12lm Vf:3.4V		Hrv		data pdf pinout
HT-P278NG	LED	gr &1.6x0.8x0.4mm l:50lm Vf:3.4V		Hrv		data pdf pinout
HT-P278TW	LED	wht &1.6x0.8x0.4mm l:70lm Vf:3.4V		Hrv		data pdf pinout
HT-P278US	LED	rd &1.6x0.8x0.4mm l:35lm Vf:2.2V		Hrv		data pdf pinout
HT-P278UY	LED	am &1.6x0.8x0.4mm l:30lm Vf:2.2V		Hrv		data pdf pinout
HT-P366NB	LED	bl &1.6x0.8x0.4mm l:17lm Vf:3.4V		Hrv		data pdf pinout
HT-P366NG	LED	gr &1.6x0.8x0.4mm l:70lm Vf:3.4V		Hrv		data pdf pinout
HT-P366TW	LED	wht &1.6x0.8x0.4mm l:100lm Vf:3.4V		Hrv		data pdf pinout
HT-P366US	LED	rd &1.6x0.8x0.4mm l:50lm Vf:2.2V		Hrv		data pdf pinout
HT-P366UY	LED	am &1.6x0.8x0.4mm l:45lm Vf:2.2V		Hrv		data pdf pinout
HT-P368NB	LED	bl &1.6x0.8x0.4mm l:17lm Vf:3.4V		Hrv		data pdf pinout
HT-P368NG	LED	gr &1.6x0.8x0.4mm l:70lm Vf:3.4V		Hrv		data pdf pinout

Type	Device	Short Description	Fig.	Manu	Comparision Types	More at
HT-P368TW	LED	wht &1.6x0.8x0.4mm I:100lm Vf:3.4V		Hrv		data pdf pinou
HT-P368US	LED	rd &1.6x0.8x0.4mm I:50lm Vf:2.2V		Hrv		data pdf pinou
HT-P368UY	LED	am &1.6x0.8x0.4mm I:45lm Vf:2.2V		Hrv		data pdf pinou
HT-P376NB	LED	bl &1.6x0.8x0.4mm I:17lm Vf:3.4V		Hrv		data pdf pinou
HT-P376NG	LED	gr &1.6x0.8x0.4mm I:70lm Vf:3.4V		Hrv		data pdf pinou
HT-P376TW	LED	wht &1.6x0.8x0.4mm I:100lm Vf:3.4V		Hrv		data pdf pinou
HT-P376US	LED	rd &1.6x0.8x0.4mm I:50lm Vf:2.2V		Hrv		data pdf pinou
HT-P376UY	LED	am &1.6x0.8x0.4mm I:45lm Vf:2.2V		Hrv		data pdf pinou
HT-P378NB	LED	bl &1.6x0.8x0.4mm I:17lm Vf:3.4V		Hrv		data pdf pinou
HT-P378NG	LED	gr &1.6x0.8x0.4mm I:70lm Vf:3.4V		Hrv		data pdf pinou
HT-P378TW	LED	wht &1.6x0.8x0.4mm I:100lm Vf:3.4V		Hrv		data pdf pinou
HT-P378US	LED	rd &1.6x0.8x0.4mm I:50lm Vf:2.2V		Hrv		data pdf pinou
HT-P378UY	LED	am &1.6x0.8x0.4mm I:45lm Vf:2.2V		Hrv		data pdf pinou
HT-S91D	LED	am &1.6x0.8x0.6mm I:9mcd Vf:2.1V		Hrv		data pdf pinou
HT-S91NB	LED	bl &1.6x0.8x0.6mm I:25mcd Vf:2.8V		Hrv		data pdf pinou
HT-S91NG	LED	gr &1.6x0.8x0.6mm I:45mcd Vf:2.8V		Hrv		data pdf pinou
HT-S91SD	LED	or &1.6x0.8x0.6mm I:12mcd Vf:2.1V		Hrv		data pdf pinou
HT-S91TW	LED	wht &1.6x0.8x0.6mm I:60mcd Vf:2.8V		Hrv		data pdf pinou
HT-S91UD	LED	am &1.6x0.8x0.6mm I:110mcd Vf:1.9V		Hrv		data pdf pinou
HT-S91UR	LED	rd &1.5x0.8x0.6mm I:16mcd Vf:1.8V		Hrv		data pdf pinou
HT-S91USD	LED	or &1.6x0.8x0.6mm I:120mcd Vf:1.9V		Hrv		data pdf pinou
HT-S91UY	LED	yl &1.6x0.8x0.6mm I:140mcd Vf:1.9V		Hrv		data pdf pinou
HT-S91UYG	LED	gr &1.6x0.8x0.6mm I:60mcd Vf:2V		Hrv		data pdf pinou
HT-S91Y	LED	yl &1.6x0.8x0.6mm I:8mcd Vf:2.1V		Hrv		data pdf pinou
HT-S91YG	LED	gr &1.6x0.8x0.6mm I:20mcd Vf:2.2V		Hrv		data pdf pinou
HT-V116NB	LED	bl &3.2x0.95x0.6mm I:180mcd Vf:3.3V		Hrv		data pdf pinou
HT-V116NG	LED	gr &3.2x0.95x0.6mm I:250mcd Vf:3.3V		Hrv		data pdf pinou
HT-V116TW	LED	wht &3.2x0.95x0.6mm I:630mcd Vf:3.3V		Hrv		data pdf pinou
HT-V150NB	LED	bl &3.2x1.6x1.1mm I:25mcd Vf:2.8V		Hrv		data pdf pinou
HT-V150NG	LED	gr &3.2x1.6x1.1mm I:160mcd Vf:3.3V		Hrv		data pdf pinou
HT-V150TW	LED	wht &3.2x1.6x1.1mm I:60mcd Vf:2.8V		Hrv		data pdf pinou
HT-V150UD	LED	am &3.2x1.6x1.1mm I:120mcd Vf:1.9V		Hrv		data pdf pinou
HT-V150USD	LED	or &3.2x1.6x1.1mm I:120mcd Vf:1.9V		Hrv		data pdf pinou
HT-V150UY	LED	yl &3.2x1.6x1.1mm I:90mcd Vf:1.9V		Hrv		data pdf pinou
HT-V150UYG	LED	gr &3.2x1.6x1.1mm I:50mcd Vf:2V		Hrv		data pdf pinou
HT-V153TW	LED	wht &3.2x2.7x0.8mm I:2600mcd Vf:3.3V		Hrv		data pdf pinou
HT-V155NB	LED	bl &3.2x2.7x1.1mm I:25mcd Vf:2.8V		Hrv		data pdf pinou
HT-V155NG	LED	gr &3.2x2.7x1.1mm I:280mcd Vf:3.3V		Hrv		data pdf pinou
HT-V155TW	LED	wht &3.2x2.7x1.1mm I:60mcd Vf:2.8V		Hrv		data pdf pinou
HT-V155UD	LED	am &3.2x2.7x1.1mm I:180mcd Vf:1.9V		Hrv		data pdf pinou
HT-V155USD	LED	or &3.2x2.7x1.1mm I:180mcd Vf:1.9V		Hrv		data pdf pinou
HT-V155UY	LED	yl &3.2x2.7x1.1mm I:120mcd Vf:1.9V		Hrv		data pdf pinou
HT-V155UYG	LED	gr &3.2x2.7x1.1mm I:100mcd Vf:2V		Hrv		data pdf pinou
HT-V170NB	LED	bl &2.0x1.2x0.8mm I:25mcd Vf:2.8V		Hrv		data pdf pinou
HT-V170NG	LED	gr &2.0x1.2x0.8mm I:280mcd Vf:3.3V		Hrv		data pdf pinou
HT-V170TW	LED	wht &2.0x1.2x0.8mm I:60mcd Vf:2.8V		Hrv		data pdf pinou
HT-V170UD	LED	am &2.0x1.2x0.8mm I:90mcd Vf:1.9V		Hrv		data pdf pinou
HT-V170USD	LED	or &2.0x1.2x0.8mm I:140mcd Vf:1.9V		Hrv		data pdf pinou
HT-V170UY	LED	yl &2.0x1.2x0.8mm I:70mcd Vf:1.9V		Hrv		data pdf pinou
HT-V170UYG	LED	gr &2.0x1.2x0.8mm I:60mcd Vf:2V		Hrv		data pdf pinou
HT-V335NB	LED	bl &4.0x1.5x0.8mm I:180mcd Vf:3.3V		Hrv		data pdf pinou
HT-V335NG	LED	gr &4.0x1.5x0.8mm I:250mcd Vf:3.3V		Hrv		data pdf pinou
HT-V335TW	LED	wht &4.0x1.5x0.8mm I:715mcd Vf:3.3V		Hrv		data pdf pinou
HT-V350FCH	LED	or/gr/bl &3.2x2.7x1.1mm I:90mcd		Hrv		data pdf pinou
HT-V350FDH	LED	am/gr/bl &3.2x2.7x1.1mm I:90mcd		Hrv		data pdf pinou
HT	Si-P	=2SA1035-T (Typ-Code/Stempel/marking)	35	Mat	→2SA1035	data
HT	Si-P	=2SA1531A-T (Typ-Code/Stempel/marking)	35(2mm)	Mat	→2SA1531A	data
HT	Si-P	=2SB956-T (Typ-Code/Stempel/marking)	39	Mat	→2SB956	data
HT	Si-N	=2SC3928-T (Typ-Code/Stempel/marking)	35	Mit	→2SC3928	data
HT	Si-N	=2SC4155-T (Typ-Code/Stempel/marking)	35(2mm)	Mit	→2SC4155	data
HT	Si-N	=2SC5378 (Typ-Code/Stempel/marking)	35(2mm)	Mat	→2SC5378	data
HT	Si-N	=2SC5379 (Typ-Code/Stempel/marking)	35(1,6mm)	Mat	→2SC5379	data
HT	Z-Di	=Z2 SMB-150 (Typ-Code/Stempel/marking)	71(5x3,5)	Fag	→Z2 SMB-150	data
Ht 1	Si-N+R	=PUMH 11 (Typ-Code/Stempel/marking)	46	Phi	→PUMH 11	data
HT 2	Si-N	SMD, Uni, 90V, 0,1A, 0,3W, <500/2000ns	35a	Fer	2SC4050	data
Ht 2	Si-N+R	=PUMH 1 (Typ-Code/Stempel/marking)	46	Phi	→PUMH 1	data
HT 3	Si-P	SMD, Uni, 90V, 0,1A, 0,3W, <500/1000ns	35a	Fer	2SA1566	data
Ht 4	Si-N+R	=PUMH 4 (Typ-Code/Stempel/marking)	46	Phi	→PUMH 4	data
Ht 7	Si-N+R	=PUMH 7 (Typ-Code/Stempel/marking)	46	Phi	→PUMH 7	data
HT 12 A /DIP	LIN-IC	2^12 Series of Encoders	TSOP	Hol	-	pdf pinout
HT 12 A /SOP	LIN-IC	2^12 Series of Encoders	TSOP	Hol	-	pdf pinout
HT 12 D /DIP	CMOS-IC	2^12 Series of Decoders	DIP	Hol	-	pdf pinout
HT 12 D /SOP	CMOS-IC	2^12 Series of Decoders	DIP	Hol	-	pdf pinout
HT 12 E /DIP	LIN-IC	2^12 Series of Encoders	TSOP	Hol	-	pdf pinout
HT 12 E /SOP	LIN-IC	2^12 Series of Encoders	TSOP	Hol	-	pdf pinout
HT 12 F /DIP	CMOS-IC	2^12 Series of Decoders	DIP	Hol	-	pdf pinout
HT 12 F /SOP	CMOS-IC	2^12 Series of Decoders	DIP	Hol	-	pdf pinout
HT 12 M	LIN-IC	2^12 Decoder Doorbell	LLP	Hol	-	pdf pinout
HT 36 A 0	LIN-IC	8-Bit Music Synthesizer MCU	TSOP	Hol	-	pdf pinout
HT 36 A 4	LIN-IC	Music Synthesizer 8-Bit MCU	TSOP	Hol	-	pdf pinout
HT 36 B 4	LIN-IC	Music Synthesizer 8-Bit MCU	TSOP	Hol	-	pdf pinout
HT 36 M 4	LIN-IC	Music Synthesizer 8-Bit MCU	TSOP	Hol	-	pdf pinout
HT 38 A 5 /DIP	LIN-IC	12 Melody Music Generator	TSOP	Hol	-	pdf pinout
HT 38 A 5 /SOP	LIN-IC	12 Melody Music Generator	TSOP	Hol	-	pdf pinout

Type	Device	Short Description	Fig.	Manu	Comparision Types	More at
T 0440 LG	MOS-IC	SMD,Telecom, 2x Mosfet Drv, Uiso=±400V	8-MDIP	Stx	-	pdf
T 600	LIN-IC	3^18 Series of Encoders	TSOP	Hol	-	pdf pinout
T 604 L	LIN-IC	3^18 Series of Decoders	TSOP	Hol	-	pdf pinout
T 0610	DIG-IC	33x120 LCD Driver	LLP	Hol	-	pdf pinout
T 614	LIN-IC	3^18 Series of Decoders	TSOP	Hol	-	pdf pinout
T 6 P 20 A	LIN-IC	2^24 Otp Encoder	TSOP	Hol	-	pdf pinout
T 6 P 20 B	LIN-IC	2^24 Otp Encoder	TSOP	Hol	-	pdf pinout
T 6 P 20 D	LIN-IC	2^24 Otp Encoder	TSOP	Hol	-	pdf pinout
T 0638 LG	MOS-IC	SMD, Telecom, Electronic Line Switch	8-MDIP	Stx	-	pdf
T 672 A	LIN-IC	13.56MHz RFID Transponder	TSOP	Hol	-	pdf pinout
T 672 B	LIN-IC	13.56MHz RFID Transponder	TSOP	Hol	-	pdf pinout
T 680	LIN-IC	3^18 Series of Encoders	TSOP	Hol	-	pdf pinout
T 692	LIN-IC	3^18 Series of Decoders	TSOP	Hol	-	pdf pinout
T 0740 LG	MOS-IC	SMD, Telecom, Mosfet Drv, Uiso=±400V	8-MDIP	Stx	-	pdf
T1104	4xHI-TEMP	Vs:13V Vu:115dB Vo:±4.8V Vi0:2mV	14D	Hon		data pdf pinout
T 1117	LIN-IC	1A General Purpose LDO	LLP	Hol	-	pdf pinout
T 1380	LIN-IC	Serial Timekeeper Chip	LLP	Hol	-	pdf pinout
T 1381	LIN-IC	Serial Timekeeper Chip	LLP	Hol	-	pdf pinout
T 1602 L	CMOS-IC	40 Dot Matrix LCD Segment Driver	DIP	Hol	-	pdf pinout
T 1608 L	CMOS-IC	2x20 Channel LCD Driver	DIP	Hol	-	pdf pinout
T 1609 L	CMOS-IC	2x40 Channel LCD Driver	DIP	Hol	-	pdf pinout
T 1611 C	DIG-IC	Timer with Dialer Interface	LLP	Hol	-	pdf pinout
T 1613 C	CMOS-IC	Timer with Dialer Interface	LLP	Hol	-	pdf pinout
T 1616 C	CMOS-IC	Timer with Dialer Interface	LLP	Hol	-	pdf pinout
T 1620	LIN-IC	RAM Mapping 32x4 LCD Controller,I/O MCU	DIP	Hol	-	pdf pinout
T 1621	LIN-IC	RAM Mapping 32x4 LCD Controller,I/O MCU	DIP	Hol	-	pdf pinout
T 1621 B /SSOP	LIN-IC	RAM Mapping 32x4 LCD Controller,I/O MCU	DIP	Hol	-	pdf pinout
T 1621 B /LQFP	LIN-IC	RAM Mapping 32x4 LCD Controller,I/O MCU	DIP	Hol	-	pdf pinout
T 1621 D	LIN-IC	RAM Mapping 32x4 LCD Controller,I/O MCU	DIP	Hol	-	pdf pinout
T 1622 /LQFP	LIN-IC	RAM Mapping 32x8 LCD Controller,I/O MCU	DIP	Hol	-	pdf pinout
T 1623	LIN-IC	RAM Mapping 48x8 LCD Controller,I/O MCU	DIP	Hol	-	pdf pinout
T 1625	LIN-IC	RAM Mapping 64x8 LCD Controller,I/O MCU	DIP	Hol	-	pdf pinout
T 1626	LIN-IC	RAM Mapping 48x16 LCD Controller,I/O MCU	DIP	Hol	-	pdf pinout
T 1632	LIN-IC	32x8 & 24x16 LED Driver	DIP	Hol	-	pdf pinout
T 1632 C	LIN-IC	32x8 & 24x16 LED Driver	DIP	Hol	-	pdf pinout
T 1647	LIN-IC	4- Level Gray Scale 64x16 LCD Controller, I/ O MCU	DIP	Hol	-	pdf pinout
T 1647 A	LIN-IC	64x16 LCD Controller for I/O MCU	DIP	Hol	-	pdf pinout
T 1650	LIN-IC	64x32 LCD Controller for I/O MCU	DIP	Hol	-	pdf pinout
T 1660	LIN-IC	96x32 LCD Controller for I/O MCU	DIP	Hol	-	pdf pinout
T 1670	LIN-IC	128x32 LCD Controller for I/O MCU	DIP	Hol	-	pdf pinout
T-2050	CMOS-IC	Five Lamp/led Flash Driver	MDIP	Hol	-	pdf pinout
T 23 B 60	LIN-IC	60x11 Pixel Data Bank 8-Bit Mask MCU	TSOP	Hol	-	pdf pinout
T 36 FA /28SOP	LIN-IC	Music Synthesizer 8-Bit MCU	TSOP	Hol	-	pdf pinout
T 36 FA /32SOP	LIN-IC	Music Synthesizer 8-Bit MCU	TSOP	Hol	-	pdf pinout
T 48 RA 0-3	LIN-IC	Remote Type 8-Bit MCU	DIP	Hol	-	pdf pinout
T 6012 /DIP	LIN-IC	3^12 Series of Encoders	TSOP	Hol	-	pdf pinout
T 6012 /SOP	LIN-IC	3^12 Series of Encoders	TSOP	Hol	-	pdf pinout
T 6014 /DIP	LIN-IC	3^12 Series of Encoders	TSOP	Hol	-	pdf pinout
T 6014 /SOP	LIN-IC	3^12 Series of Encoders	TSOP	Hol	-	pdf pinout
T 6026	LIN-IC	Remote Control Encoder	TSOP	Hol	-	pdf pinout
T 6030 /DIP	LIN-IC	3^12 Series of Decoders	TSOP	Hol	-	pdf pinout
T 6030 /SOP	LIN-IC	3^12 Series of Decoders	TSOP	Hol	-	pdf pinout
T 6032 /DIP	LIN-IC	3^12 Series of Decoders	TSOP	Hol	-	pdf pinout
T 6032 /SOP	LIN-IC	3^12 Series of Decoders	TSOP	Hol	-	pdf pinout
T 6034 /DIP	LIN-IC	3^12 Series of Decoders	TSOP	Hol	-	pdf pinout
T 6034 /SOP	LIN-IC	3^12 Series of Decoders	TSOP	Hol	-	pdf pinout
T 6207	LIN-IC	3^18 Series of Encoders	TSOP	Hol	-	pdf pinout
T 6221	LIN-IC	Multi-purpose Encoders	TSOP	Hol	-	pdf pinout
T 6222	LIN-IC	Multi-purpose Encoders	TSOP	Hol	-	pdf pinout
T 6230	LIN-IC	Infrared Remote Encoder	TSOP	Hol	-	pdf pinout
T 6720	LIN-IC	13.56MHz RFID Transponder	TSOP	Hol	-	pdf pinout
T 6740	LIN-IC	13.56MHz RFID Transponder	TSOP	Hol	-	pdf pinout
T 6751 A	LIN-IC	Camera Motor Driver (1.5 Channel)	TSOP	Hol	-	pdf pinout
T 6751 B	LIN-IC	Camera Motor Driver (1.5 Channel)	TSOP	Hol	-	pdf pinout
T 7218	LIN/Z-IC	1.8V, 0.3A TinyPowerTM LDO	48	Hol	-	pdf pinout
T 7225	LIN/Z-IC	2.5V, 0.3A TinyPowerTM LDO	48	Hol	-	pdf pinout
T 7227	LIN/Z-IC	2.7V, 0.3A TinyPowerTM LDO	48	Hol	-	pdf pinout
T 7230	LIN/Z-IC	3V, 0.3A TinyPowerTM LDO	48	Hol	-	pdf pinout
T 7233	LIN/Z-IC	3.3V, 0.3A TinyPowerTM LDO	48	Hol	-	pdf pinout
T 7250	LIN/Z-IC	5V, 0.3A TinyPowerTM LDO	48	Hol	-	pdf pinout
T 7318 A	CMOS-Z-IC	1.8V, 0.25A, Low Power Consumption LDO	48	Hol	-	pdf pinout
T 7325 A	CMOS-Z-IC	2.5V, 0.25A, Low Power Consumption LDO	48	Hol	-	pdf pinout
T 7327 A	CMOS-Z-IC	2.7V, 0.25A, Low Power Consumption LDO	48	Hol	-	pdf pinout
T 7330 A	CMOS-Z-IC	3V, 0.25A, Low Power Consumption LDO	48	Hol	-	pdf pinout
T 7333 A	CMOS-Z-IC	3.3V, 0.25A, Low Power Consumption LDO	48	Hol	-	pdf pinout
T 7335 A	CMOS-Z-IC	3.5V, 0.25A, Low Power Consumption LDO	48	Hol	-	pdf pinout
T 7350 A	CMOS-Z-IC	5V, 0.25A, Low Power Consumption LDO	48	Hol	-	pdf pinout
T 7430	CMOS-Z-IC	-3.0V Negative Voltage Reg.	48	Hol	-	pdf pinout
T 7500	CMOS-IC	Clinical Thermometer	QFN	Hol	-	pdf pinout
T 7501	CMOS-IC	Clinical Thermometer	QFN	Hol	-	pdf pinout
T 7610 A	CMOS-IC	General Purpose PIR Cntrl	DIC	Hol	-	pdf pinout
T 7610 B	CMOS-IC	General Purpose Pir Cntrl	DIC	Hol	-	pdf pinout
T 7611 A	CMOS-IC	General Purpose Pir Cntrl	DIC	Hol	-	pdf pinout
T 7611 B	CMOS-IC	General Purpose Pir Cntrl	DIC	Hol	-	pdf pinout

Type	Device	Short Description	Fig.	Manu	Comparision Types	More at
HT 7660	CMOS-Z-IC	Switched-capacitor Voltage Converter	MDIP	Hol	-	pdf pinout
HT 7718	CMOS-IC	1.8V, 0.1A, Pfm Step-up DC/DC Converter	48	Hol	-	pdf pinout
HT 7727	CMOS-IC	2.7V, 0.1A, Pfm Step-up DC/DC Converter	48	Hol	-	pdf pinout
HT 7727 A	CMOS-IC	2.7V, 0.2A, Pfm Step-up DC/DC Converter	48	Hol	-	pdf pinout
HT 7730	CMOS-IC	3V, 0.1A, Pfm Step-up DC/DC Converter	48	Hol	-	pdf pinout
HT 7730 A	CMOS-IC	3V, 0.2A, Pfm Step-up DC/DC Converter	48	Hol	-	pdf pinout
HT 7733	CMOS-IC	3.3V, 0.1A, Pfm Step-up DC/DC Converter	48	Hol	-	pdf pinout
HT 7733 A	CMOS-IC	3.3V, 0.2A, Pfm Step-up DC/DC Converter	48	Hol	-	pdf pinout
HT 7737	CMOS-IC	3.7V, 0.1A, Pfm Step-up DC/DC Converter	48	Hol	-	pdf pinout
HT 7750	CMOS-IC	5V, 0.1A, Pfm Step-up DC/DC Converter	48	Hol	-	pdf pinout
HT 7750 A	CMOS-IC	5V, 0.2A, Pfm Step-up DC/DC Converter	48	Hol	-	pdf pinout
HT 7818	LIN/Z-IC	1.8V, 0.5A TinyPowerTM LDO	48	Hol	-	pdf pinout
HT 7825	LIN/Z-IC	2.5V, 0.5A TinyPowerTM LDO	48	Hol	-	pdf pinout
HT 7827	LIN/Z-IC	2.7V, 0.5A TinyPowerTM LDO	48	Hol	-	pdf pinout
HT 7830	LIN/Z-IC	3V, 0.5A TinyPowerTM LDO	48	Hol	-	pdf pinout
HT 7833	LIN/Z-IC	3.3V, 0.5A TinyPowerTM LDO	48	Hol	-	pdf pinout
HT 7850	LIN/Z-IC	5V, 0.5A TinyPowerTM LDO	48	Hol	-	pdf pinout
HT 7937	IC	Built-in Ovp White LED Step-up Converter	GRID	Hol	-	pdf pinout
HT 81 R 09	Dig. Audio	Easyvoicetm Speech	DIP	Hol	-	pdf pinout
HT 82 V 26 A	CCD-IC	16-Bit CCD/CIS Analog Signal Processor	DIP	Hol	-	pdf pinout
HT 86 B 05 /SSOP24	Dig. Audio	Enhanced Voice Mask MCU	TSOP	Hol	-	pdf pinout
HT 86 B 10 /SSOP24	Dig. Audio	Enhanced Voice Mask MCU	TSOP	Hol	-	pdf pinout
HT 8950	Dig. Audio	Voice Modulator	DIP	Hol	-	pdf pinout
HT 8950 A	Dig. Audio	Voice Modulator	DIP	Hol	-	pdf pinout
HT 8970	Dig. Audio	Voice Echo	DIP	Hol	-	pdf pinout
HT 8972	Dig. Audio	Voice Echo	GRID	Hol	-	pdf pinout
HT 9020 B	LIN-IC	Call Progress Tone Detector	LLP	Hol	-	pdf pinout
HT 9032 C	LIN-IC	Calling Line Identification Receiver	LLP	Hol	-	pdf pinout
HT 9032 D	LIN-IC	Calling Line Identification Receiver	LLP	Hol	-	pdf pinout
HT 9033	LIN-IC	Cas Tone Detector	LLP	Hol	-	pdf pinout
HT 9170 B	LIN-IC	Dtmf Receiver	LLP	Hol	-	pdf pinout
HT 9170 D	LIN-IC	Dtmf Receiver	LLP	Hol	-	pdf pinout
HT 9172	LIN-IC	Dtmf Receiver	LLP	Hol	-	pdf pinout
HT 9200 A	LIN-IC	Dtmf Generators	LLP	Hol	-	pdf pinout
HT 9200 B	LIN-IC	Dtmf Generators	LLP	Hol	-	pdf pinout
HT 9274	LIN-IC	Quad Micropower Op Amp	TSOP	Hol	-	pdf pinout
HT 9302 A	LIN-IC	1-Memory/2-Memory Tone/pulse Dialer	LLP	Hol	-	pdf pinout
HT 9302 B	LIN-IC	1-Memory/2-Memory Tone/pulse Dialer	LLP	Hol	-	pdf pinout
HT 9302 C	LIN-IC	1-Memory/2-Memory Tone/pulse Dialer	LLP	Hol	-	pdf pinout
HT 9302 D	LIN-IC	1-Memory/2-Memory Tone/pulse Dialer	LLP	Hol	-	pdf pinout
HT 9302 G	LIN-IC	1-Memory/2-Memory Tone/pulse Dialer	LLP	Hol	-	pdf pinout
HT 9315 A	LIN-IC	15-Memory Tone/pulse Dialer	LLP	Hol	-	pdf pinout
HT 9315 AL	LIN-IC	15-Memory Tone/pulse Dialer	LLP	Hol	-	pdf pinout
HT 9315 B	LIN-IC	15-Memory Tone/pulse Dialer	LLP	Hol	-	pdf pinout
HT 9315 BL	LIN-IC	15-Memory Tone/pulse Dialer	LLP	Hol	-	pdf pinout
HT 9315 C	LIN-IC	15-Memory Tone/pulse Dialer	LLP	Hol	-	pdf pinout
HT 9315 CL	LIN-IC	15-Memory Tone/pulse Dialer	LLP	Hol	-	pdf pinout
HT 9315 D	LIN-IC	15-Memory Tone/pulse Dialer	LLP	Hol	-	pdf pinout
HT 9315 DL	LIN-IC	15-Memory Tone/pulse Dialer	LLP	Hol	-	pdf pinout
HT 9320 A	LIN-IC	22-Memory Tone/pulse Dialer	LLP	Hol	-	pdf pinout
HT 9320 B	LIN-IC	22-Memory Tone/pulse Dialer	LLP	Hol	-	pdf pinout
HT 9320 C	LIN-IC	22-Memory Tone/pulse Dialer	LLP	Hol	-	pdf pinout
HT 9320 H	LIN-IC	22-Memory Tone/pulse Dialer	LLP	Hol	-	pdf pinout
HT 9320 K	LIN-IC	22-Memory Tone/pulse Dialer	LLP	Hol	-	pdf pinout
HT 9320 L-X	LIN-IC	22-Memory Tone/pulse Dialer	LLP	Hol	-	pdf pinout
HT 9320 L	LIN-IC	22-Memory Tone/pulse Dialer	LLP	Hol	-	pdf pinout
HT 9480	LIN-IC	8-Bit Numerical Pager Controller MCU	LLP	Hol	-	pdf pinout
HT 95 A 10 P	LIN-IC	I/O Type Phone 8-Bit MCU	LLP	Hol	-	pdf pinout
HT 95 A 20P	LIN-IC	I/O Type Phone 8-Bit MCU	LLP	Hol	-	pdf pinout
HT 95 C 20 P	LIN-IC	CID Type Phone 8-Bit MCU	LLP	Hol	-	pdf pinout
HT 95 R 22	LIN-IC	I/O Type Phone 8-bit MCU	TSOP	Hol	-	pdf pinout
HT 95 A 30 P	LIN-IC	I/O Type Phone 8-Bit MCU	LLP	Hol	-	pdf pinout
HT 95 C 30 P	LIN-IC	CID Type Phone 8-Bit MCU	LLP	Hol	-	pdf pinout
HT 95 A 40 P	LIN-IC	I/O Type Phone 8-Bit MCU	LLP	Hol	-	pdf pinout
HT 95 C 40 P	LIN-IC	CID Type Phone 8-Bit MCU	LLP	Hol	-	pdf pinout
HT 16220	LIN-IC	RAM Mapping 32x8 LCD Controller, I/O MCU	DIP	Hol	-	pdf pinout
HT 16511	LIN-IC	1/8 to 1/16 Duty VFD Controller	DIP	Hol	-	pdf pinout
HT 16512	LIN-IC	1/4 to 1/11 Duty VFD Controller	DIP	Hol	-	pdf pinout
HT 16514	LIN-IC	Dot Character VFD Controller & Driver	DIP	Hol	-	pdf pinout
HT 16515	LIN-IC	1/4 to 1/12 Duty VFD Controller	DIP	Hol	-	pdf pinout
HT 16561	LIN-IC	VFD Digital Clock	DIP	Hol	-	pdf pinout
HT 16561 B	LIN-IC	VFD Digital Clock	DIP	Hol	-	pdf pinout
HT 16562	LIN-IC	1/2 Duty VFD Digital Clock	DIP	Hol	-	pdf pinout
HT 16562 B	LIN-IC	1/2 Duty VFD Digital Clock	DIP	Hol	-	pdf pinout
HT 16565	LIN-IC	24 Hour System VFD Digital Clock	DIP	Hol	-	pdf pinout
HT 16565 B	LIN-IC	24 Hour System VFD Digital Clock	DIP	Hol	-	pdf pinout
HT 16566	LIN-IC	1/2 Duty VFD Digital Clock	DIP	Hol	-	pdf pinout
HT 36 B 0 /SOP28	LIN-IC	Music Synthesizer 8-Bit MCU	TSOP	Hol	-	pdf pinout
HT 36 B 0 /QFP64	LIN-IC	Music Synthesizer 8-Bit MCU	TSOP	Hol	-	pdf pinout
HT 36 A 1 /SOP28	LIN-IC	Music Synthesizer 8-Bit MCU	TSOP	Hol	-	pdf pinout
HT 36 A 1 /SOP48	LIN-IC	Music Synthesizer 8-Bit MCU	TSOP	Hol	-	pdf pinout
HT 36 F 2/ 16SOP	LIN-IC	Music Synthesizer 8-Bit MCU	TSOP	Hol	-	pdf pinout
HT 36 A 2 /SOP28	LIN-IC	8-Bit Music Synthesizer MCU	TSOP	Hol	-	pdf pinout
HT 36 B 2 /SOP28	LIN-IC	Music Synthesizer 8-Bit MCU	TSOP	Hol	-	pdf pinout
HT 36 F 2 /28SOP	LIN-IC	Music Synthesizer 8-Bit MCU	TSOP	Hol	-	pdf pinout

Type	Device	Short Description	Fig.	Manu	Comparision Types	More at
HT 36 A 2 /SOP48	LIN-IC	8-Bit Music Synthesizer MCU	TSOP	Hol	-	pdf pinout
HT 36 B 2 /SOP56	LIN-IC	Music Synthesizer 8-Bit MCU	TSOP	Hol	-	pdf pinout
HT 36 A 3 /SOP28	LIN-IC	Music Synthesizer 8-Bit MCU	TSOP	Hol	-	pdf pinout
HT 36 A 3 /SOP48	LIN-IC	Music Synthesizer 8-Bit MCU	TSOP	Hol	-	pdf pinout
HT 36 F 6 /20SOP	LIN-IC	Music Synthesizer 8-Bit MCU	TSOP	Hol	-	pdf pinout
HT 36 F 6 /32SOP	LIN-IC	Music Synthesizer 8-Bit MCU	TSOP	Hol	-	pdf pinout
HT 81003	LIN-IC	EasyVoice TM 3-Second Speech	TSOP	Hol	-	pdf pinout
HT 81006	Dig. Audio	Easyvoicetm Speech	DIP	Hol	-	pdf pinout
HT 81009	Dig. Audio	Easyvoicetm Speech	DIP	Hol	-	pdf pinout
HT 81012	Dig. Audio	Easyvoicetm Speech	DIP	Hol	-	pdf pinout
HT 81018	Dig. Audio	Easyvoicetm Speech	DIP	Hol	-	pdf pinout
HT 82 V 731	Dig. Audio	16-Bit Stereo Audio D/A Converter	DIP	Hol	-	pdf pinout
HT 82 V 732	Dig. Audio	60mA Audio Power Amp	DIP	Hol	-	pdf pinout
HT 82 V 733	Dig. Audio	240mA Audio Power Amp	DIP	Hol	-	pdf pinout
HT 82 V 735	Dig. Audio	330mW Stereo Audio Power Amp, Shutdown	DIP	Hol	-	pdf pinout
HT 82 V 736	Dig. Audio	Stereo Headphone Driver with Mute	DIP	Hol	-	pdf pinout
HT 82 V 737	Dig. Audio	16-Bit Stereo Audio D/A Converter	DIP	Hol	-	pdf pinout
HT 82 V 738	Dig. Audio	24-Bit Stereo Audio D/A Converter	DIP	Hol	-	pdf pinout
HT 82 V 739	Dig. Audio	1200mW Audio Power Amp with Shutdown	DIP	Hol	-	pdf pinout
HT 82 V 805	CCD-IC	Ccd 4 Channel Vertical Driver	DIP	Hol	-	pdf pinout
HT 82 V 806	CCD-IC	Ccd 6 Channel Vertical Driver	DIP	Hol	-	pdf pinout
HT 82 V 842 A	IC	10- Bit 20MSPS Ccd Analog Signal Processor	DIP	Hol	-	pdf pinout
HT 86030	Dig. Audio	Voice Synthesizer 8-Bit MCU	DIP	Hol	-	pdf pinout
HT 86070	Dig. Audio	Voice Synthesizer 8-Bit MCU	DIP	Hol	-	pdf pinout
HT 86 R 192 /SOP	Dig. Audio	Voice Synthesizer 8-Bit Otp MCU	DIP	Hol	-	pdf pinout
HT 86 R 384	Dig. Audio	Voice Synthesizer 8-Bit MCU	DIP	Hol	-	pdf pinout
HT 93214 A	LIN-IC	1-memory Tone/pulse Dialer	LLP	Hol		pdf pinout
HT 93214 AT	LIN-IC	1-memory Tone/pulse Dialer	LLP	Hol		pdf pinout
HT 93214 B	LIN-IC	1-memory Tone/pulse Dialer	LLP	Hol		pdf pinout
HT 95 A 100	LIN-IC	I/O Type Phone 8-Bit MCU	LLP	Hol		pdf pinout
HT 95168	LIN-IC	HT95168 Caller Id Telephone IC	LLP	Hol		pdf pinout
HT 95168-A	LIN-IC	HT95168 Caller Id Telephone IC	LLP	Hol		pdf pinout
HT 95 A 200	LIN-IC	I/O Type Phone 8-Bit MCU	LLP	Hol		pdf pinout
HT 95 C 200	LIN-IC	CID Type Phone 8-Bit MCU	LLP	Hol		pdf pinout
HT 95 A 300	LIN-IC	I/O Type Phone 8-Bit MCU	LLP	Hol		pdf pinout
HT 95 C 300	LIN-IC	CID Type Phone 8-Bit MCU	LLP	Hol		pdf pinout
HT 95 A 400	LIN-IC	I/O Type Phone 8-Bit MCU	LLP	Hol		pdf pinout
HT 95 C 400	LIN-IC	CID Type Phone 8-Bit MCU	LLP	Hol		pdf pinout
HT 1087 /SOT89	LIN-IC	0.5A General Purpose LDO	LLP	Hol		pdf pinout
HT 1087 /TO92	LIN-IC	0.5A General Purpose LDO	LLP	Hol		pdf pinout
HT 1622 /44QFP	LIN-IC	RAM Mapping 32x8 LCD Controller,I/O MCU	DIP	Hol		pdf pinout
HT 1622 /52QFP	LIN-IC	RAM Mapping 32x8 LCD Controller,I/O MCU	DIP	Hol		pdf pinout
HT 3834 /16SOP	LIN-IC	36 Melody Music Generator	TSOP	Hol		pdf pinout
HT 3834 /28SOP	LIN-IC	36 Melody Music Generator	TSOP	Hol		pdf pinout
HT 6010 /18DIP	LIN-IC	3^12 Series of Encoders	TSOP	Hol		pdf pinout
HT 6010 /20SOP	LIN-IC	3^12 Series of Encoders	TSOP	Hol		pdf pinout
HT 82 V 24 /20 SOP	CCD-IC	16-Bit,15MSPS,3-CH CCD/CIS Analog Sig. Proc.	DIP	Hol	-	pdf pinout
HT 82 V 24 /28 SOP	CCD-IC	16-Bit,15MSPS,3-CH CCD/CIS Analog Sig. Proc.	DIP	Hol		pdf pinout
HT 86 B 10 /SSOP28	Dig. Audio	Enhanced Voice Mask MCU	TSOP	Hol	-	pdf pinout
HT 86 B 05 /SSOP28	Dig. Audio	Enhanced Voice Mask MCU	TSOP	Hol	-	pdf pinout
HT 86 B 05 /QFP44	Dig. Audio	Enhanced Voice Mask MCU	TSOP	Hol	-	pdf pinout
HT 86 B 10 /QFP44	Dig. Audio	Enhanced Voice Mask MCU	TSOP	Hol	-	pdf pinout
HT 86 B 20 /SSOP28	Dig. Audio	Enhanced Voice Mask MCU	TSOP	Hol	.	pdf pinout
HT 86 B 20 /QFP44	Dig. Audio	Enhanced Voice Mask MCU	TSOP	Hol	-	pdf pinout
HT 86 B 30 /SSOP28	Dig. Audio	Enhanced Voice Mask MCU	TSOP	Hol	-	pdf pinout
HT 86 B 30 /QFP44	Dig. Audio	Enhanced Voice Mask MCU	TSOP	Hol	-	pdf pinout
HT 86 B 40 /SSOP28	Dig. Audio	Enhanced Voice Mask MCU	TSOP	Hol	-	pdf pinout
HT 86 B 40 /QFP44	Dig. Audio	Enhanced Voice Mask MCU	TSOP	Hol	-	pdf pinout
HT 86 B 50 /SSOP28	Dig. Audio	Enhanced Voice Mask MCU	TSOP	Hol	-	pdf pinout
HT 86 B 50 /QFP44	Dig. Audio	Enhanced Voice Mask MCU	TSOP	Hol	-	pdf pinout
HT 86 B 60 /SSOP28	Dig. Audio	Enhanced Voice Mask MCU	TSOP	Hol	-	pdf pinout
HT 86 B 60 /QFP44	Dig. Audio	Enhanced Voice Mask MCU	TSOP	Hol	-	pdf pinout
HT 86 B 70 /QFP44	Dig. Audio	Enhanced Voice Mask MCU	TSOP	Hol	-	pdf pinout
HT 86 B 80 /QFP44	Dig. Audio	Enhanced Voice Mask MCU	TSOP	Hol	-	pdf pinout
HT 1015-1 /SOT-25	LIN-IC	1.5V Low Power LDO	LLP	Hol	-	pdf pinout
HT 1015-1 /SOT-89	LIN-IC	1.5V Low Power LDO	LLP	Hol	-	pdf pinout
HT 1015-1 /TO-92	LIN-IC	1.5V Low Power LDO	LLP	Hol	-	pdf pinout
HT 1086 /TO-220	LIN-IC	1.5A General Purpose LDO	LLP	Hol	-	pdf pinout
HT 1086 /SOT-223	LIN-IC	1.5A General Purpose LDO	LLP	Hol	-	pdf pinout
HT 1086 /TO-252	LIN-IC	1.5A General Purpose LDO	LLP	Hol	-	pdf pinout
HT 1086 /TO-263	LIN-IC	1.5A General Purpose LDO	LLP	Hol	-	pdf pinout
HT 6240-002 /20SOP	LIN-IC	Infrared Remote Encoder	TSOP	Hol	-	pdf pinout
HT 82 V805/20TSSOP	CCD-IC	Ccd 4 Channel Vertical Driver	DIP	Hol	-	pdf pinout
HT 83004 /20 SSOP	Dig. Audio	Q-voice TM	DIP	Hol	-	pdf pinout
HT 83004 /28 SOP	Dig. Audio	Q-voice TM	DIP	Hol	-	pdf pinout
HT 83007 /20 SSOP	Dig. Audio	Q-voice TM	DIP	Hol	-	pdf pinout
HT 83007 /28 SOP	Dig. Audio	Q-voice TM	DIP	Hol	-	pdf pinout
HT 83010 /20 SSOP	Dig. Audio	Q-voice TM	DIP	Hol	-	pdf pinout
HT 83010 /28 SOP	Dig. Audio	Q-voice TM	DIP	Hol	-	pdf pinout
HT 83020 /20 SSOP	Dig. Audio	Q-voice TM	DIP	Hol	-	pdf pinout
HT 83020 /28 SOP	Dig. Audio	Q-voice TM	DIP	Hol	-	pdf pinout
HT 83038 /20 SSOP	Dig. Audio	Q-voice TM	DIP	Hol	-	pdf pinout

Type	Device	Short Description	Fig.	Manu	Comparision Types	More at
HT 83038 /28 SOP	Dig. Audio	Q-voice TM	DIP	Hol	-	pdf pinout
HT 83050 /20 SSOP	Dig. Audio	Q-voice TM	DIP	Hol	-	pdf pinout
HT 83050 /28 SOP	Dig. Audio	Q-voice TM	DIP	Hol	-	pdf pinout
HT 83 R 074/20SSOP	Dig. Audio	Q-voice TM	DIP	Hol	-	pdf pinout
HT 83074 /20 SSOP	Dig. Audio	Q-voice TM	DIP	Hol	-	pdf pinout
HT 83 R 074 /28SOP	Dig. Audio	Q-voice TM	DIP	Hol	-	pdf pinout
HT 83074 /28 SOP	Dig. Audio	Q-voice TM	DIP	Hol	-	pdf pinout
HT 86072 /28 SOP	Dig. Audio	Voice Synthesizer 8-Bit MCU	DIP	Hol	-	pdf pinout
HT 86072 /44 QFP	Dig. Audio	Voice Synthesizer 8-Bit MCU	DIP	Hol	-	pdf pinout
HT 86144 /28 SOP	Dig. Audio	Voice Synthesizer 8-Bit MCU	DIP	Hol	-	pdf pinout
HT 86144 /44 QFP	Dig. Audio	Voice Synthesizer 8-Bit MCU	DIP	Hol	-	pdf pinout
HT 86192 /28 SOP	Dig. Audio	Voice Synthesizer 8-Bit MCU	DIP	Hol	-	pdf pinout
HT 86192 /44 QFP	Dig. Audio	Voice Synthesizer 8-Bit MCU	DIP	Hol	-	pdf pinout
HT 86 R 192 /44QFP	Dig. Audio	Voice Synthesizer 8-Bit Otp MCU	DIP	Hol	-	pdf pinout
HT 86384 /28 SOP	Dig. Audio	Voice Synthesizer 8-Bit MCU	DIP	Hol	-	pdf pinout
HT 86576 /32 SOP	Dig. Audio	Voice Synthesizer 8-Bit MCU	DIP	Hol	-	pdf pinout
HT 86 B 70 /QFP100	Dig. Audio	Enhanced Voice Mask MCU	TSOP	Hol	-	pdf pinout
HT 86768 /32 SOP	Dig. Audio	Voice Synthesizer 8-Bit MCU	DIP	Hol	-	pdf pinout
HT 86 B 80 /QFP100	Dig. Audio	Enhanced Voice Mask MCU	TSOP	Hol	-	pdf pinout
HT 86 B 90 /QFP100	Dig. Audio	Enhanced Voice Mask MCU	TSOP	Hol	-	pdf pinout
HT 86072 /100 QFP	Dig. Audio	Voice Synthesizer 8-Bit MCU	DIP	Hol	-	pdf pinout
HT 86144 /100 QFP	Dig. Audio	Voice Synthesizer 8-Bit MCU	DIP	Hol	-	pdf pinout
HT 86192 /100 QFP	Dig. Audio	Voice Synthesizer 8-Bit MCU	DIP	Hol	-	pdf pinout
HT 86 R 192/100QFP	Dig. Audio	Voice Synthesizer 8-Bit Otp MCU	DIP	Hol	-	pdf pinout
HT 86384 /100 QFP	Dig. Audio	Voice Synthesizer 8-Bit MCU	DIP	Hol	-	pdf pinout
HT 86576 /100 QFP	Dig. Audio	Voice Synthesizer 8-Bit MCU	DIP	Hol	-	pdf pinout
HT 86768 /100 QFP	Dig. Audio	Voice Synthesizer 8-Bit MCU	DIP	Hol	-	pdf pinout
HT 6240-002 /24SOP	LIN-IC	Infrared Remote Encoder	TSOP	Hol	-	pdf pinout
HTA 2003 R 3 S	IC	High Reliability DC/DC Converter	42	Inr	-	pdf pinout
HTA 20005 D	IC	High Reliability DC/DC Converter	42	Inr	-	pdf pinout
HTA 20005 S	IC	High Reliability DC/DC Converter	42	Inr	-	pdf pinout
HTA 20012 D	IC	High Reliability DC/DC Converter	42	Inr	-	pdf pinout
HTA 20012 S	IC	High Reliability DC/DC Converte.	42	Inr	-	pdf pinout
HTA 20015 D	IC	High Reliability DC/DC Converter	42	Inr	-	pdf pinout
HTA 20015 S	IC	High Reliability DC/DC Converter	42	Inr	-	pdf pinout
HTG 2150	LIN-IC	8-Bit 320 Pixel Dot Matrix LCD MCU	TSOP	Hol	-	pdf pinout
HTG 2190	LIN-IC	8-Bit 1024 Pixel Dot Matrix LCD MCU	TSOP	Hol	-	pdf pinout
HTLP912A	LED	am ø1.6mm I:600mcd Vf:2V If:20mA		Hrv		data pdf pinout
HTLP912AE	LED	yl ø1.6mm I:600mcd Vf:2V If:20mA		Hrv		data pdf pinout
HTLP912B	LED	bl ø1.6mm I:200mcd Vf:3.6V If:20mA		Hrv		data pdf pinout
HTLP912BG	LED	gr ø1.6mm I:200mcd Vf:3.6V If:20mA		Hrv		data pdf pinout
HTLP912E	LED	or ø1.6mm I:600mcd Vf:2V If:20mA		Hrv		data pdf pinout
HTLP912H	LED	rd ø1.66mm I:3.5mcd Vf:2.1V If:20mA		Hrv		data pdf pinout
HTLP912R	LED	rd ø1.6mm I:500mcd Vf:2V If:20mA		Hrv		data pdf pinout
HTLP912SR	LED	srd ø1:6mm I:400mcd Vf:2V If:20mA		Hrv		data pdf pinout
HTLP912TG	LED	gr ø1.6mm I:200mcd Vf:3.6V If:20mA		Hrv		data pdf pinout
HTLP912UR	LED	rd ø1.6mm I:1000mcd Vf:2.2V If:20mA		Hrv		data pdf pinout
HTLP912Y	LED	yl ø1.6mm I:500mcd Vf:2V If:20mA		Hrv		data pdf pinout
HTLP912YG	LED	gr ø1.6mm I:200mcd Vf:2.1V If:20mA		Hrv		data pdf pinout
HTLP9123	LED	yl ø1.66mm I:25mcd Vf:2.1V If:20mA		Hrv		data pdf pinout
HTLP9124	LED	gr ø1.66mm I:40mcd Vf:2.1V If:20mA		Hrv		data pdf pinout
HTLP9125	LED	or ø1.66mm I:60mcd Vf:2.1V If:20mA		Hrv		data pdf pinout
HTLP9129	LED	rd ø1.66mm I:130mcd Vf:1.7V If:20mA		Hrv		data pdf pinout
HTQ	Si-N	=2SC5378-Q (Typ-Code/Stempel/marking)	35(2mm)	Mat	→2SC5378	data
HTQ	Si-N	=2SC5379-Q (Typ-Code/Stempel/marking)	35(1,6mm)	Mat	→2SC5379	data
HTR	Si-N	=2SC5378-R (Typ-Code/Stempel/marking)	35(2mm)	Mat	→2SC5378	data
HTR	Si-N	=2SC5379-R (Typ-Code/Stempel/marking)	35(1,6mm)	Mat	→2SC5379	data
HTS	Si-N	=2SC5378-S (Typ-Code/Stempel/marking)	35(2mm)	Mat	→2SC5378	data
HTS	Si-N	=2SC5379-S (Typ-Code/Stempel/marking)	35(1,6mm)	Mat	→2SC5379	data
HTT 1115 E	Si-N	SMD, Includes 2 Different Transistors 4V, 50mA,	8-SMDIPbh4	Hit	HTT1213E,	data data
HTT 1115 S	Si-N	SMD, Includes 2 Different Transistors 4V, 50mA,	46bh4	Hit	HTT1213S,	data data
HTT 1213 E	Si-N	SMD, Includes 2 Transistors	8-SMDIPbh4	Hit		data
HTT 1213 S	Si-N	SMD, Includes 2 Transistors	46bh4	Hit		data
HU	Z-Di	=Z2 SMB-160 (Typ-Code/Stempel/marking)	71(5x3,5)	Fag	→Z2 SMB-160	data
HV	MOS-N-FET-d	=3SK285 (Typ-Code/Stempel/marking)	44	Mat	→3SK285	data
HV	C-Di	=SVC 252 (Typ-Code/Stempel/marking)	35	Say	→SVC 252	data
HV	Z-Di	=Z2 SMB-180 (Typ-Code/Stempel/marking)	71(5x3,5)	Fag	→Z2 SMB-180	data
HV-03 SS	Si-Di	kV-Gl, 3kV, 2mA <180ns	31a	Sak	-	data
HV-05	Si-Di	kV-Gl, 5kV, 1mA <500ns	31a	Sak	-	data
HV-06 SS	Si-Di	kV-Gl, 6kV, 2mA <180ns	31a	Sak	-	data
HV7A20	LED	am ø3mm I:800mcd Vf:2V If:20mA	T1	Hrv		data pdf pinout
HV7A40	LED	am ø3mm I:650mcd Vf:2V If:20mA	T1	Hrv		data pdf pinout
HV7B20	LED	bl ø3mm I:1300mcd Vf:3.6V If:20mA	T1	Hrv		data pdf pinout
HV7B40	LED	bl ø3mm I:900mcd Vf:3.6V If:20mA	T1	Hrv		data pdf pinout
HV7BG20	LED	gr ø3mm I:1200mcd Vf:3.6V If:20mA	T1	Hrv		data pdf pinout
HV7BG40	LED	gr ø3mm I:1500mcd Vf:3.6V If:20mA	T1	Hrv		data pdf pinout
HV7CB20	LED	bl ø3mm I:150mcd Vf:3.8V If:20mA	T1	Hrv		data pdf pinout
HV7CB40	LED	bl ø3mm I:700mcd Vf:3.6V If:20mA	T1	Hrv		data pdf pinout
HV7E20	LED	or ø3mm I:750mcd Vf:2V If:20mA	T1	Hrv		data pdf pinout
HV7E40	LED	or ø3mm I:650mcd Vf:2V If:20mA	T1	Hrv		data pdf pinout
HV7R20	LED	rd ø3mm I:550mcd Vf:2V If:20mA	T1	Hrv		data pdf pinout

pe	Device	Short Description	Fig.	Manu	Comparision Types	More at
R40	LED	rd ø3mm I:800mcd Vf:2V If:20mA	T1	Hrv		data pdf pinout
SR20	LED	srd ø3mm I:450mcd Vf:2V If:20mA	T1	Hrv		data pdf pinout
SR40	LED	srd ø3mm I:800mcd Vf:2V If:20mA	T1	Hrv		data pdf pinout
TG20	LED	gr ø3mm I:600mcd Vf:3.6V If:20mA	T1	Hrv		data pdf pinout
TG40	LED	gr ø3mm I:1700mcd Vf:3.6V If:20mA	T1	Hrv		data pdf pinout
UR40	LED	rd ø3mm I:1500mcd Vf:2.2V If:20mA	T1	Hrv		data pdf pinout
Y20	LED	yl ø3mm I:650mcd Vf:2V If:20mA	T1	Hrv		data pdf pinout
Y40	LED	yl ø3mm I:550mcd Vf:2V If:20mA	T1	Hrv		data pdf pinout
YG40	LED	gr ø3mm I:180mcd Vf:2.1V If:20mA	T1	Hrv		data pdf pinout
8	Si-Di	kV-Gl, 8kV, 1mA <500ns	31a	Sak	BY 617, BY 707, BY 717	data
A10	LED	am ø5mm I:1500mcd Vf:2V If:20mA	T1	Hrv		data pdf pinout
A20	LED	am ø5mm I:1650mcd Vf:2V If:20mA	T1	Hrv		data pdf pinout
A20F	LED	am ø5mm I:1900mcd Vf:2V If:20mA	T1	Hrv		data pdf pinout
B02F	LED	bl ø5mm I:1000mcd Vf:3.6V If:20mA	T1	Hrv		data pdf pinout
B10	LED	bl ø5mm I:900mcd Vf:3.6V If:20mA	T1	Hrv		data pdf pinout
B20	LED	bl ø5mm I:1500mcd Vf:3.6V If:20mA	T1	Hrv		data pdf pinout
BG10	LED	gr ø5mm I:3150mcd Vf:3.6V If:20mA	T1	Hrv		data pdf pinout
BG20	LED	gr ø5mm I:1700mcd Vf:3.6V If:20mA	T1	Hrv		data pdf pinout
BG20F	LED	gr ø5mm I:3000mcd Vf:3.6V If:20mA	T1	Hrv		data pdf pinout
CB10	LED	bl ø5mm I:200mcd Vf:3.8V If:20mA	T1	Hrv		data pdf pinout
CB20	LED	bl ø5mm I:250mcd Vf:3.8V If:20mA	T1	Hrv		data pdf pinout
E10	LED	or ø5mm I:1400mcd Vf:2V If:20mA	T1	Hrv		data pdf pinout
E20	LED	or ø5mm I:1150mcd Vf:2V If:20mA	T1	Hrv		data pdf pinout
E20F	LED	or ø5mm I:1900mcd Vf:2V If:20mA	T1	Hrv		data pdf pinout
O20F	LED	yl ø5mm I:2200mcd Vf:2V If:20mA	T1	Hrv		data pdf pinout
R10	LED	rd ø5mm I:1250mcd Vf:2V If:20mA	T1	Hrv		data pdf pinout
R20	LED	rd ø5mm I:1300mcd Vf:2V If:20mA	T1	Hrv		data pdf pinout
R20F	LED	rd ø5mm I:2500mcd Vf:2V If:20mA	T1	Hrv		data pdf pinout
SR10	LED	srd ø5mm I:950mcd Vf:2V If:20mA	T1	Hrv		data pdf pinout
SR20	LED	srd ø5mm I:1200mcd Vf:2V If:20mA	T1	Hrv		data pdf pinout
SR20F	LED	srd ø5mm I:1500mcd Vf:2V If:20mA	T1	Hrv		data pdf pinout
8 SS	Si-Di	kV-Gl, 8kV, 2mA <180ns	31a	Sak	BY 617, BY 707, BY 717	data
TG10	LED	gr ø5mm I:1500mcd Vf:3.6V If:20mA	T1	Hrv		data pdf pinout
TG20	LED	gr ø5mm I:1500mcd Vf:3.6V If:20mA	T1	Hrv		data pdf pinout
TG20F	LED	gr ø5mm I:3200mcd Vf:3.6V If:20mA	T1	Hrv		data pdf pinout
UR20F	LED	rd ø3mm I:4000mcd Vf:2.2V If:20mA	T1	Hrv		data pdf pinout
Y10	LED	yl ø5mm I:1500mcd Vf:2V If:20mA	T1	Hrv		data pdf pinout
Y20	LED	yl ø5mm I:1200mcd Vf:2V If:20mA	T1	Hrv		data pdf pinout
Y20F	LED	yl ø5mm I:2200mcd Vf:2V If:20mA	T1	Hrv		data pdf pinout
10	Si-Di	≈BA 127	31a		→BA 127	
13	Si-Di	≈BA 127	31a		→BA 127	
15	Ge-St	0,145V(1,5mA)	2	Hit	-	data
16	Ge-St	0,145V(2,3mA)	2	Hit	-	data
17(C)	Ge-St	0,145V(3,5mA)	2	Hit	-	data
18	Ge-St	0,145V(1mA)	2	Hit	-	data
23 D,E,F	Si-St	0,66...0,8V(3mA)	9c	Hit	BA 216, BA 314	data
23 G	Si-St	0,58...0,69V(3mA)	31a	Hit	BA 216, BA 314	data
23 GBL	Si-St	0,58...0,69V(3mA)	31a	Hit	BA 216, BA 314	data
23 GWT	Si-St	0,58...0,65V(3mA)	31a	Hit	BA 216, BA 314	data
23 GYL	Si-St	0,62...0,69V(3mA)	31a	Hit	BA 216, BA 314	data
25	Si-Di	≈BA 127	31a		→BA 127	
26	Si-St	0,68...0,75V(30mA)	31a	Hit	BA 220, BZ 102/0V7, BZX 55/C0V8	data
46	C-Di	≈BA 102	31a		→BA 102	
46(RD)	Si-St	1,17...1,4V(3mA)	31a	Hit	BZV 86/1V4, BZX 75/C1V4, ZTE 1,5	data
66 PG	MOS-IC	32-Ch. LCD Driver, Backplane Output	44-PLCC	Stx	-	pdf
70	Si-St	0,61...0,81V(0,5mA)	31a	Hit	BA 216, BA 314	data
75	Si-St	1A, 0,77...0,87V(100mA)	31a	Hit	ZY 1	data
80	Si-St	0,55...0,75V(3mA)	31a	Hit	BA 216, BA 314...315	data
82	MOS-N-FET-e	=STHV 82	18c§	Tho	→STHV 82	data
100	Si-St	70...115V(1mA)	31a	Hit	-	data
110 K4	LIN-I/O-IC	Power-over-ethernet Interf. PD Contr. IEEE802.3af(TM) Standard	87	Stx	-	pdf pinout
123(GBL...GYL)	Si-St	=HV 23(GBL...GYL)	31a	Hit	→HV 23...	data
202GC	LED	gr ø3mm I:90mcd Vf:2.1V If:20mA	T1	Hrv		data pdf pinout
202H	LED	rd ø3mm I:4mcd Vf:2.1V If:20mA	T1	Hrv		data pdf pinout
202HC	LED	rd ø3mm I:7mcd Vf:2.1V If:20mA	T1	Hrv		data pdf pinout
202O	LED	or ø3mm I:45mcd Vf:2.1V If:20mA	T1	Hrv		data pdf pinout
202OC	LED	or ø3mm I:70mcd Vf:2.1V If:20mA	T1	Hrv		data pdf pinout
202Y	LED	yl ø3mm I:30mcd Vf:2.1V If:20mA	T1	Hrv		data pdf pinout
202YC	LED	yl ø3mm I:55mcd Vf:2.1V If:20mA	T1	Hrv		data pdf pinout
202YG	LED	gr ø3mm I:22mcd Vf:2.1V If:20mA	T1	Hrv		data pdf pinout
203GC	LED	gr ø3mm I:45mcd Vf:2.1V If:20mA	T1	Hrv		data pdf pinout
203H	LED	rd ø3mm I:3mcd Vf:2.1V If:20mA	T1	Hrv		data pdf pinout
203HC	LED	rd ø3mm I:6mcd Vf:2.1V If:20mA	T1	Hrv		data pdf pinout
203O	LED	or ø3mm I:16mcd Vf:2.1V If:20mA	T1	Hrv		data pdf pinout
203OC	LED	or ø3mm I:50mcd Vf:2.1V If:20mA	T1	Hrv		data pdf pinout
203Y	LED	yl ø3mm I:12mcd Vf:2.1V If:20mA	T1	Hrv		data pdf pinout
203YC	LED	yl ø3mm I:36mcd Vf:2.1V If:20mA	T1	Hrv		data pdf pinout
203YG	LED	gr ø3mm I:13mcd Vf:2.1V If:20mA	T1	Hrv		data pdf pinout
209 FG	MOS-IC	SMD, 12 Ch. Analog Switch, hi-volt	48-MP	Stx		pdf

Type	Device	Short Description	Fig.	Manu	Comparision Types	More at
HV 400	LIN-IC	Mosfet Power Drv, 35V, 6A, 300kHz	8-DIP	Har	-	
HV 421 LG	MOS-IC	SMD, Telecom, Ring Generator, Udd<10V	8-MDIP	Stx	-	pdf
HV 430 WG	MOS-IC	SMD, Telecom, Ring Generator, Udd<325V	20-MDIP	Stx	-	pdf
HV 440 WG	MOS-IC	SMD, Telecom, Ring Generator, Udd<220V	16-MDIP	Stx	-	pdf
HV 441 PJ	MOS-IC	Telecom, Ring Generator, Udd<220V	44-PLCC	Stx	-	pdf
HV 450 WG	MOS-IC	SMD,Telecom, Ring Generator, Udd<-220V	16-MDIP	Stx	-	pdf
HV 506 PG	MOS-IC	SMD, 40-Ch. Row Driver, Uq<275V	64-MP	Stx	-	pdf
HV 507 PG	MOS-IC	SMD, 64-Ch. Ser.→Par. Conv., Uop<325V	80-MP	Stx	-	pdf
HV 508 LG	MOS-IC	SMD, High Volt. Liquid Crystal Driver	8-MDIP	Stx	-	pdf
HV 510 WG	MOS-IC	SMD, 12-Ch. Ser.→Par. Conv., Uop<240V	24-MDIP	Stx	-	pdf
HV 518 P	MOS-IC	32-Ch. Fluorescent Display Driver	40-DIP	Stx	-	pdf
HV 518 PJ	MOS-IC	32-Ch. Fluorescent Display Driver	44-PLCC		-	pdf
HV 574 PG	MOS-IC	SMD, 80-Ch. Ser.→Par. Conv., Uq<80V	100-MP	Stx	-	pdf
HV 623 PG	MOS-IC	32-Ch. Gray-shade Display Col. Driver	64-MP	Stx	-	pdf
HV 803 LG	MOS-IC	SMD, EL Lamp Driver, Udd<10V, Uq<120V	8-MDIP	Stx	-	pdf
HV 809 K2	MOS-IC	EL Lamp Driver, Udd<15V, Uin<200V	17/7Pin	Stx	-	pdf
HV 809 LG	MOS-IC	SMD,EL Lamp Driver, Udd<15V, Uin<200V	8-MDIP		-	pdf
HV 823 LG	MOS-IC	SMD, EL Lamp Driver, Udd<10V, Uq<120V	8-MDIP	Stx	-	pdf
HV 824 LG	MOS-IC	SMD, EL Lamp Driver, Udd<2V, Uq<120V	8-MDIP	Stx	-	pdf pinout
HV 825 LG	MOS-IC	SMD, EL Lamp Driver, Udd<2,5V, Uq<62V	8-MDIP	Stx	-	pdf
HV 825 MG	MOS-IC	SMD, EL Lamp Driver, Udd<2,5V, Uq<62V	8-SSMDIP		-	pdf
HV 826 LG	MOS-IC	SMD, EL Lamp Driver, Udd<4,5V, Uq<120V	8-MDIP	Stx	-	pdf
HV 826 MG	MOS-IC	SMD, EL Lamp Driver, Udd<4,5V, Uq<120V	8-SSMDIP		-	pdf
HV1085	LED	rd ø3mm I:4mcd Vf:2.1V If:20mA	T1	Hrv	-	data pdf pinou
HV-1205(-5,-9)	LIN-IC	Power Supply, 120V~-→+5..24V, 0,05A -5: 0...+75°, -9: -40...+85°	8-DIP	Har	-	data pdf pinou
HV1360	LED	or ø3mm I:50mcd Vf:2.1V If:20mA	T1	Hrv	-	data pdf pinou
HV1385	LED	or ø3mm I:35mcd Vf:2.1V If:20mA	T1	Hrv	-	data pdf pinou
HV1440	LED	yl ø3mm I:40mcd Vf:2.1V If:20mA	T1	Hrv	-	data pdf pinou
HV1485	LED	yl ø3mm I:24mcd Vf:2.1V If:20mA	T1	Hrv	-	data pdf pinou
HV 1516 P	MOS-IC	1 of 8 Decoded 8-Ch. Analog S, hi-volt	20-DIP	Stx	-	
HV1540	LED	gr ø3mm I:60mcd Vf:2.1V If:20mA	T1	Hrv	-	data pdf pinou
HV1585	LED	gr ø3mm I:25mcd Vf:2.1V If:20mA	T1	Hrv	-	data pdf pinou
HV 1616 P	MOS-IC	8-Ch. Analog S, hi-volt	24-DIP	Stx	-	
HV 1616 PJ	MOS-IC	8-Ch. Analog S, hi-volt	28-PLCC		-	
HV 1816 P	MOS-IC	8-Ch. Analog S, hi-volt	28-DIP	Stx	-	
HV 1816 PJ	MOS-IC	8-Ch. Analog S, hi-volt	28-PLCC		-	
HV 2116 PJ	MOS-IC	8 Ch. Analog Switch, hi-volt	28-PLCC	Stx	-	
HV 2216 P	MOS-IC	8 Ch. Analog Switch, hi-volt	28-DIP	Stx	-	
HV 2216 PJ	MOS-IC	8 Ch. Analog Switch, hi-volt	28-PLCC		-	
HV-2405(-5,-9)	LIN-IC	Power Supply, 264V~-→+5..24V, 0,05A	8-DIP	Har	-	
HV3050F	LED	rd ø5mm I:16mcd Vf:2.1V If:20mA	T1	Hrv	-	data pdf pinou
HV 3137 PG	MOS-IC	SMD, 64-Ch. Serial→Parallel Conv.	80-MP	Stx	-	pdf
HV3301F	LED	or ø5mm I:70mcd Vf:2.1V If:20mA	T1	Hrv	-	data pdf pinou
HV3350F	LED	yl ø5mm I:140mcd Vf:2.1V If:20mA	T1	Hrv	-	data pdf pinou
HV 3418 DG,PG	MOS-IC	SMD, 64-Ch. Serial→Parallel Conv.	80-MP	Stx	-	
HV3450F	LED	gr ø5mm I:220mcd Vf:2.1V If:20mA	T1	Hrv	-	data pdf pinou
HV 3527 PG	MOS-IC	SMD, 64-Ch. Serial→Parallel Conv.	80-MP	Stx	-	pdf pinout
HV3750F	LED	or ø5mm I:160mcd Vf:2.1V If:20mA	T1	Hrv	-	data pdf pinou
HV 3922 C	MOS-IC	High Voltage PIN Diode Driver, Uq<220V	20-DIC	Stx	-	pdf
HV 3922 DJ	MOS-IC	High Voltage PIN Diode Driver, Uq<220V	28-PLCC		-	pdf
HV 4522 DJ,PJ	MOS-IC	32-Ch. Ser.→Par. Conv., Uop<-220V	44-PLCC	Stx	-	
HV 4522 PG	MOS-IC	SMD,32-Ch. Ser.→Par. Conv.	44-MP		-	pdf
HV 4530 DJ,PJ	MOS-IC	32-Ch. Ser.→Par. Conv., Uop<-300V	44-PLCC	Stx	-	
HV 4522PG...4530PG	MOS-IC	=HV 4522DJ,PJ...4530DJ,PJ: SMD	44-MP		-	
HV 4530 PG	MOS-IC	SMD,32-Ch. Ser.→Par. Conv.	44-MP		-	pdf
HV 4622 DJ,PJ	MOS-IC	32-Ch. Ser.→Par. Conv., Uop<-220V	44-PLCC		-	
HV 4622 PG	MOS-IC	SMD,32-Ch. Ser.→Par. Conv.	44-MP		-	pdf
HV 4630 DJ,PJ	MOS-IC	32-Ch. Ser.→Par. Conv., Uop<-300V	44-PLCC	Stx	-	
HV 4622PG...4630PG	MOS-IC	=HV 4622DJ,PJ...4630DJ,PJ: SMD	44-MP		-	
HV 4630 PG	MOS-IC	SMD,32-Ch. Ser.→Par. Conv.	44-MP		-	pdf
HV 4937 PG	MOS-IC	64-Ch. Ser.→Par. Conv., Uq<-375V	80-MP	Stx	-	pdf pinout
HV 5122 DJ,PJ	MOS-IC	32-Ch. Ser.→Par. Conv., Uq<225V	44-PLCC	Stx	-	
HV 5122 PG	MOS-IC	SMD,32-Ch. Ser.→Par. Conv., Uq<225V	44-MP		-	pdf
HV 5222 DJ,PJ	MOS-IC	32-Ch. Ser.→Par. Conv., Uq<225V	44-PLCC	Stx	-	
HV 5222 PG	MOS-IC	SMD,32-Ch. Ser.→Par. Conv., Uq<225V	44-MP		-	pdf
HV 5308 DJ,PJ	MOS-IC	32-Ch. Serial→Parallel Converter	44-PLCC	Stx	-	
HV 5308 PG	MOS-IC	SMD,32-Ch. Serial→Parallel Convert.	44-MP		-	pdf
HV5336	LED	or/yl ø5mm I:11.7mcd Vf:2.1V If:20mA	T1	Hrv	-	data pdf pinout
HV5396	LED	or/yl ø5mm I:10mcd Vf:2.1V If:20mA	T1	Hrv	-	data pdf pinout
HV 5408 DJ,PJ	MOS-IC	32-Ch. Serial→Parallel Converter	44-PLCC	Stx	-	
HV 5408 PG	MOS-IC	SMD,32-Ch. Serial→Parallel Convert.	44-MP		-	pdf
HV5431	LED	gr/rd ø5mm I:2mcd Vf:2.1V If:20mA	T1	Hrv	-	data pdf pinout
HV5433	LED	gr/yl ø5mm I:8.3mcd Vf:2.1V If:20mA	T1	Hrv	-	data pdf pinout
HV5436	LED	or/gr ø5mm I:11.7mcd Vf:2.1V If:20mA	T1	Hrv	-	data pdf pinout
HV5491	LED	gr/rd ø5mm I:2mcd Vf:2.1V If:20mA	T1	Hrv	-	data pdf pinout
HV5493	LED	gr/yl ø5mm I:7mcd Vf:2.1V If:20mA	T1	Hrv	-	data pdf pinout
HV5496	LED	or/gr ø5mm I:10mcd Vf:2.1V If:20mA	T1	Hrv	-	data pdf pinout
HV 5522 DJ,PJ	MOS-IC	32-Ch. Ser.→Par. Conv., Uop<220V	44-PLCC	Stx	-	
HV 5522 PG	MOS-IC	SMD,32-Ch. Ser.→Par. Conv.,Uop<220V	44-MP		-	pdf
HV 5530 DJ,PJ	MOS-IC	32-Ch. Ser.→Par. Conv., Uop<300V	44-PLCC	Stx	-	
HV 5522PG...5530PG	MOS-IC	=HV 5522DJ,PJ...5530DJ,PJ: SMD	44-MP		-	
HV 5530 PG	MOS-IC	SMD,32-Ch. Ser.→Par. Conv.	44-MP		-	pdf
HV 5622 DJ,PJ	MOS-IC	32-Ch. Ser.→Par. Conv., Uop<220V	44-PLCC	Stx	-	
HV 5622 PG	MOS-IC	SMD,32-Ch. Ser.→Par. Conv.,Uop<220V	44-MP		-	pdf

Type	Device	Short Description	Fig.	Manu	Comparision Types	More at
HV 5630 DJ,PJ	MOS-IC	32-Ch. Ser.→Par. Conv., Uop<300V	44-PLCC	Stx	-	
HV 5622PG...5630PG	MOS-IC	=HV 5622DJ,PJ...5630DJ,PJ; SMD	44-MP		-	
HV 5630 PG	MOS-IC	SMD,32-Ch. Ser.→Par. Conv.	44-MP		-	pdf
HV 5812 P	MOS-IC	20-Ch. Fluorescent Display Driver	28-DIP	Stx	-	pdf
HV 5812 PJ	MOS-IC	20-Ch. Fluorescent Display Driver	28-PLCC	Stx	-	pdf
HV 6008 DJ,PJ	MOS-IC	32-Ch. Liquid Crystal Driver, Uq=±40V	44-PLCC	Stx	-	
HV 6008 PG	MOS-IC	SMD,32-Ch. Liquid Crystal Driver	44-MP		-	pdf
HV6050F	LED	rd ø5mm I:6mcd Vf:2.1V If:20mA	T1	Hrv	-	data pdf pinout
HV6336	LED	or/yl ø3mm I:16mcd Vf:2.1V If:20mA	T1	Hrv	-	data pdf pinout
HV6353F	LED	yl ø5mm I:60mcd Vf:2.1V If:20mA	T1	Hrv	-	data pdf pinout
HV6366	LED	or/yl ø3mm I:9.7mcd Vf:2.1V If:20mA	T1	Hrv	-	data pdf pinout
HV6433	LED	gr/yl ø3mm I:12mcd Vf:2.1V If:20mA	T1	Hrv	-	data pdf pinout
HV6436	LED	or/gr ø3mm I:16mcd Vf:2.1V If:20mA	T1	Hrv	-	data pdf pinout
HV6463	LED	gr/yl ø3mm I:7mcd Vf:2.1V If:20mA	T1	Hrv	-	data pdf pinout
HV6466	LED	or/gr ø3mm I:9.1mcd Vf:2.1V If:20mA	T1	Hrv	-	data pdf pinout
HV 6506 PJ	MOS-IC	32-Ch. LCD Driver, Backplane Output	44-PLCC	Stx	-	
HV 6506 PG	MOS-IC	SMD,32-Ch. LCD Drv, Backplane Output	44-MP		-	
HV 6810 P	MOS-IC	10-Ch. Latched Display Drv. Ser. Input	18-DIP	Stx	-	pdf
HV 6810 PJ	MOS-IC	10-Ch. Latched Display Drv. Ser. Input	20-PLCC		-	pdf
HV 6810 WG	MOS-IC	SMD,10-Ch. Latched Display Drv,Ser.Inp	20-MDIP		-	pdf
HV 7022 DJ-C, PJ-C	MOS-IC	34-Ch. Symmetric Row Drv, Uq<230V	44-PLCC	Stx	-	
HV 7224 DG	MOS-IC	SMD, 40-Ch. Symmetric Row Drv, Uq<240V	80-MP	Stx	-	pdf
HV 7224 PG	MOS-IC	SMD, 40-Ch. Symmetric Row Drv, Uq<240V	64-MP		-	pdf
HV7336	LED	or/yl ø3mm I:24.5mcd Vf:2.1V If:20mA	T1	Hrv	-	data pdf pinout
HV7376	LED	or/yl ø3mm I:14mcd Vf:2.1V If:20mA	T1	Hrv	-	data pdf pinout
HV7431	LED	gr/rd ø3mm I:4mcd Vf:2.1V If:20mA	T1	Hrv	-	data pdf pinout
HV7433	LED	gr/yl ø3mm I:18mcd Vf:2.1V If:20mA	T1	Hrv	-	data pdf pinout
HV7436	LED	or/gr ø3mm I:24.5mcd Vf:2.1V If:20mA	T1	Hrv	-	data pdf pinout
HV7471	LED	gr/rd ø3mm I:3mcd Vf:2.1V If:20mA	T1	Hrv	-	data pdf pinout
HV7473	LED	gr/yl ø3mm I:10mcd Vf:2.1V If:20mA	T1	Hrv	-	data pdf pinout
HV7476	LED	or/gr ø3mm I:14mcd Vf:2.1V If:20mA	T1	Hrv	-	data pdf pinout
HV 7620 DG	MOS-IC	SMD, 32-Ch. Ser.→Par. Conv., Uq<200V	80-MP	Stx	-	pdf
HV 7620 PG	MOS-IC	SMD, 32-Ch. Ser.→Par. Conv., Uq<200V	64-MP		-	pdf
HV 8051 LG	MOS-IC	SMD, EL Lamp Driver, Udd<4,5V, Uq>41V	8-MDIP	Stx	-	
HV 8053 LG	MOS-IC	SMD, EL Lamp Driver, Udd<4,5V, Uq>41V	8-MDIP	Stx	-	
HV 9100 P,C	SMPS-IC	SMPS Controller, MOSFET: 150V, <5Ω	14-DIP,DIC	Stx	-	
HV 9100 PJ	SMPS-IC	SMPS Controller, MOSFET: 150V, <5Ω	20-PLCC		-	pdf
HV 9102 P,C	SMPS-IC	SMPS Controller, MOSFET: 200V, <7Ω	14-DIP,DIC	Stx	-	
HV 9102 PJ	SMPS-IC	SMPS Controller, MOSFET: 150V, <5Ω	20-PLCC		-	pdf
HV 9103 P,C	SMPS-IC	SMPS Controller, MOSFET: 200V, <7Ω	14-DIP,DIC	Stx	-	
HV 9103 PJ	SMPS-IC	SMPS Controller, MOSFET: 150V, <5Ω	20-PLCC		-	pdf
HV 9105 P	SMPS-IC	SMPS Controller, MOSFET: 200V, <5Ω	14-DIP	Stx	-	pdf
HV 9105 PJ	SMPS-IC	SMPS Controller, MOSFET: 200V, <5Ω	20-PLCC		-	pdf
HV 9108 P	SMPS-IC	SMPS Controller, MOSFET: 200V, <5Ω	14-DIP	Stx	-	pdf
HV 9108 PJ	SMPS-IC	SMPS Controller, MOSFET: 200V, <5Ω	20-PLCC		-	pdf
HV 9110 NG	SMPS-IC	SMD,SMPS Ctrl., Current Mode	14-MDIP		-	pdf
HV 9110 P	SMPS-IC	SMPS Ctrl., Current Mode, Uin=10...120V	14-DIP	Stx	-	pdf
HV 9110 PJ	SMPS-IC	SMPS Ctrl., Current Mode, Uin=10...120V	20-PLCC		-	pdf
HV 9112 NG	SMPS-IC	SMD,SMPS Ctrl., Current Mode	14-MDIP		-	pdf
HV 9112 P	SMPS-IC	SMPS Ctrl., Current Mode, Uin=9...80V	14-DIP	Stx	-	pdf
HV 9112 PJ	SMPS-IC	SMPS Ctrl., Current Mode, Uin=9...80V	20-PLCC		-	pdf
HV 9113 NG	SMPS-IC	SMD,SMPS Ctrl., Current Mode	14-MDIP		-	pdf
HV 9113 P	SMPS-IC	SMPS Ctrl., Current Mode, Uin=10...120V	14-DIP	Stx	-	pdf
HV 9113 PJ	SMPS-IC	SMPS Ctrl., Current Mode, Uin=10...120V	20-PLCC		-	pdf
HV 9120 NG	SMPS-IC	SMD,SMPS Ctrl., Current Mode	16-MDIP		-	pdf
HV 9120 P	SMPS-IC	SMPS Ctrl., Current Mode, Uin=10...450V	16-DIP	Stx	-	pdf
HV 9120 PJ	SMPS-IC	SMPS Ctrl., Current Mode, Uin=10...450V	20-PLCC		-	pdf
HV 9123 NG	SMPS-IC	SMD,SMPS Ctrl., Current Mode	16-MDIP		-	pdf
HV 9123 P	SMPS-IC	SMPS Ctrl., Current Mode, Uin=10...450V	16-DIP	Stx	-	pdf
HV 9123 PJ	SMPS-IC	SMPS Ctrl., Current Mode, Uin=10...450V	20-PLCC		-	pdf
HV 9221 P	SMPS-IC	SMPS Ctrl., Current Mode, Multi-funct.	28-DIP	Stx	-	
HV 9221 WG	SMPS-IC	SMD,SMPS Ctrl., Current Mode	16-MDIP		-	
HV 9308 DJ,PJ	MOS-IC	32-Ch. Ser.→Par. Conv., Uop<80V	44-PLCC	Stx	-	
HV 9408 DJ,PJ	MOS-IC	32-Ch. Ser.→Par. Conv., Uop<80V	44-PLCC	Stx	-	
HV 9605 NG	SMPS-IC	SMD,SMPS Ctrl., Current Mode, f. ISDN	14-MDIP		-	pdf
HV 9605 P	SMPS-IC	SMPS Ctrl., Current Mode, f. Isdn Equ.	14-DIP	Stx	-	pdf
HV 9606 SP	LIN-IC	Current Mode PWM Contr., Supervisor	20-SSMDIP	Stx	-	pdf pinout
HV 9708 PJ	MOS-IC	32-Ch. Ser.→Par. Conv., Uq<80V	44-PLCC	Stx	-	pdf
HV 9808 PJ	MOS-IC	32-Ch. Ser.→Par. Conv., Uq<80V	44-PLCC	Stx	-	pdf
HV 20220 FG	MOS-IC	SMD, 8-Ch. Analog S,hi-volt, lo-charge	48-MP		-	pdf
HV 20220 P	MOS-IC	8-Ch. Analog S, hi-volt, lo-charge	28-DIP	Stx	-	pdf
HV 20220 PJ	MOS-IC	8-Ch. Analog S, hi-volt, lo-charge	28-PLCC		-	pdf
HV 20320 P	MOS-IC	8-Ch. Analog S, hi-volt, lo-charge	28-DIP	Stx	-	
HV 20320 PJ	MOS-IC	8-Ch. Analog S, hi-volt, lo-charge	28-PLCC		-	pdf
HV 20420 P	MOS-IC	8-Ch. Analog S, hi-volt, lo-charge	28-DIP	Stx	-	
HV 20420 PJ	MOS-IC	8-Ch. Analog S, hi-volt, lo-charge	28-PLCC		-	
HV 20620 P	MOS-IC	8-Ch. Analog S, hi-volt, lo-charge	28-DIP	Stx	-	
HV 20620 PJ	MOS-IC	8-Ch. Analog S, hi-volt, lo-charge	28-PLCC		-	
HV 20720 PJ	MOS-IC	Dual, 1 of 4 Analog Switch, hi-volt	28-PLCC	Stx	-	pdf
HV 20822 FG	MOS-IC	SMD, 16 Ch. Analog Switch, hi-volt	48-MP	Stx	-	pdf
HV 21716 PJ	MOS-IC	8-Ch. Analog S, hi-volt, lo-charge	28-PLCC	Stx	-	
HV 21816 PJ	MOS-IC	8-Ch. Analog S, hi-volt, lo-charge	28-PLCC	Stx	-	
HV 22716 PJ	MOS-IC	8-Ch. Analog S, hi-volt, lo-charge	28-PLCC	Stx	-	
HV 22816 PJ	MOS-IC	8-Ch. Analog S, hi-volt, lo-charge	28-PLCC	Stx	-	
HV 22816 WG	MOS-IC	SMD, 8-Ch. Analog S,hi-volt, lo-charge	28-MDIP		-	

Type	Device	Short Description	Fig.	Manu	Comparision Types	More at
HV 57009 DG,PG	MOS-IC	SMD,64-Ch. Ser.→Par. Conv., Uq<-85V	80-MP	Stx	-	
HV 57708 DG,PG	MOS-IC	SMD, 64-Ch. Ser.→Par. Conv., Uq<80V	80-MP	Stx	-	
HV 57908 DG,PG	MOS-IC	SMD, 64-Ch. Ser.→Par. Conv., Uq<80V	80-MP	Stx	-	
HV 62106 X,XW	MOS-IC	64-Ch. Gray-shade Display Col. Driver	Die	Stx	-	
HV 62208 PG	MOS-IC	32-Ch. 256 Gray-shade High Volt. Drv	64-MP	Stx	-	pdf
HV64531	LED	gr ø5mm I:150mcd Vf:2.1V If:20mA	T1	Hrv		data pdf pinout
HV 7800 K1	IC	High Side Current Monitor	45	Stx		pdf pinout
HV 998 2K 6-G	IC	Three-channel, Switch Mode LED Driver IC	QFN	Stx		pdf pinout
HV 3-2405 E-5	IC	World-wide Single Chip Power Supply	DIP	Har	-	pdf pinout
HV 3-2405 E-9	IC	World-wide Single Chip Power Supply	DIP	Har	-	pdf pinout
HVA23	HI-SPEED	Vs:±15V Vu:96dB Vo:±10V f:100Mc	12A	Ilc		data pinout
HVB 14 S	PIN-Di	=HVM 14S:	35o(2mm)	Hit	-	dat⁻
HVB 27 WK	C-Di	=HVM 27WK: 15V	35I(2mm)	Hit	-	data
HVC 131	PIN-Di	=HVU 131:	71a(1,3mm)	Hit	-	data
HVC 132	PIN-Di	=HVU 132:	71a(1,3mm)	Hit		data
HVC 133	PIN-Di	=HVU 133:	71a(1,3mm)	Hit		data
HVC 134	PIN-Di	Antenn.-S, 65V, <2Ω(10mA)	71a(1,3mm)	Hit	-	data
HVC 200 A	C-Di	=HVU 200A:	71a(1,3mm)	Hit	-	data
HVC 202 A,B	C-Di	=HVU 202A,B:	71a(1,3mm)	Hit		data
HVC 300 A,B	C-Di	=HVU 300A,B:	71a(1,3mm)	Hit	1SV278	data
HVC 306 A,B	C-Di	=HVU 306A,B:	71a(1,3mm)	Hit		data
HVC 308 A	C-Di	=HVU 308A:	71a(1,3mm)	Hit	1SV283	data
HVC 317 B	C-Di	SMD, UHF-Tuning, 10/0,7pF(1/25V)	71a(1,3mm)	Hit		data
HVC 321 B	C-Di	SMD, VHF-Tuning, 31/2,7pF(1/10V)	71a(1,3mm)	Hit		data
HVC 322 B	C-Di	SMD, VHF-Tuning, 15/2,2pF(2/25V)	71a(1,3mm)	Hit		data
HVC 350 B	C-Di	=HVU 350B:	71a(1,3mm)	Hit	1SV281	data
HVC 351	C-Di	=HVU 351:	71a(1,3mm)	Hit	1SV257, 1SV279	data
HVC 355 B	C-Di	=HVU 355B:	71a(1,3mm)	Hit		data
HVC 357	C-Di	=HVU 357:	71a(1,3mm)	Hit		data
HVC 358 B	C-Di	=HVU 358B:	71a(1,3mm)	Hit		data
HVC 359	C-Di	=HVU 359:	71a(1,3mm)	Hit		data
HVC 362	C-Di	=HVU 362:	71a(1,3mm)	Hit		data
HVC 363 A,B	C-Di	=HVU 363A,B:	71a(1,3mm)	Hit		data
HVC 365	C-Di	=HVU 365:	71a(1,3mm)	Hit		data
HVC 366	C-Di	SMD,VCO, 15V, 6,35/4,65pF(1/2V), <0,6Ω	71a(1,3mm)	Hit		data
HVC 368 B	C-Di	SMD, VCO, 15V, 15,7/5,5pF(1/3V), <1,1Ω	71a(1,3mm)	Hit	-	data
HVC 369 B	C-Di	SMD, VCO, 15V, 4,9/1,9pF(1/4V), <0,5Ω	71a(1,3mm)	Hit	1SV285	data
HVC 372 B	C-Di	SMD, VCO, 15V, 16/7,75pF(1/4V), <0,4Ω	71a(1,3mm)	Hit		data
HVC 373 B	C-Di	SMD, VCO, 10V, 11/5,3pF(1/3V), <0,8Ω	71a(1,3mm)	Hit		data
HVC 374 B	C-Di	SMD,VCO, 10V, 22,8/13,5pF(1/2V), <1,2Ω	71a(1,3mm)	Hit		data
HVC 375 B	C-Di	SMD,VCO, 10V, 15,7/3,75pF(1/4V), <1,1Ω	71a(1,3mm)	Hit		data
HVC 376 B	C-Di	SMD, VCO, 15V, 26,8/5,8pF(1/4V), <0,8Ω	71a(1,3mm)	Hit		data
HVC 377 B	C-Di	SMD, VCO, 15V, 6,9/2,2pF(1/4V), <0,8Ω	71a(1,3mm)	Hit		data
HVE 200 A	C-Di	=HVU 200A:	71a(1,3mm)	Hit	-	data
HVE 202 A	C-Di	=HVU 202A:	71a(1,3mm)	Hit		data
HVE 300 A	C-Di	=HVU 300A:	71a(1,3mm)	Hit	-	data
HVE 306 A	C-Di	=HVU 306A:	71a(1,3mm)	Hit		data
HVE 310	C-Di	UHF/SHF-Tuning, 35V, 2,2...3,3pF(1V)	Chip	Hit	-	data
HVE 350	C-Di	=HVU 350:	71a(1,3mm)	Hit	-	data
HVE 351	C-Di	=HVU 351:	71a(1,3mm)	Hit	1SV257, 1SV279	data
HVE 355	C-Di	=HVU 355:	71a(1,3mm)	Hit		data
HVE 358	C-Di	=HVU 358:	71a(1,3mm)	Hit	-	data
HVG 1	Si-Di	kV-Gl, 1kV, 0,3A, 2µs	31a	Gie	GP 02-20, MR 250-2, 1N1732(A)	data
HVG 2	Si-Di	=HVG 1: 2kV, 0,3A	31a	Gie	GP 02-20, MR 250-2, 1N1732(A)	data
HVG 3	Si-Di	=HVG 1: 3kV, 0,25A	31a	Gie	GP 02-30, MR 250-3, 1N1733(A)	data
HVG 4	Si-Di	=HVG 1: 4kV, 0,2A	31a	Gie	GP 02-40, MR 250-4, 1N1734(A)	data
HVG 5	Si-Di	=HVG 1: 5kV, 0,15A	31a	Gie	MR 250-5, 1N1734(A)	data
HVG 100	C-Di	FM-AFC, 12...19pf(8V)	31a	Hit	FC 54M, MA 345...346, SD 115	data
HVH3050	LED	rd ø5mm I:14mcd Vf:2.1V If:20mA	T1	Hrv		data pdf pinout
HVH3301	LED	or ø5mm I:70mcd Vf:2.1V If:20mA	T1	Hrv		data pdf pinout
HVH3350	LED	yl ø5mm I:130mcd Vf:2.1V If:20mA	T1	Hrv		data pdf pinout
HVH3450	LED	gr ø5mm I:220mcd Vf:2.1V If:20mA	T1	Hrv		data pdf pinout
HVH3750	LED	or ø5mm I:160mcd Vf:2.1V If:20mA	T1	Hrv		data pdf pinout
HVH6053	LED	rd ø5mm I:6mcd Vf:2.1V If:20mA	T1	Hrv		data pdf pinout
HVH6353	LED	yl ø5mm I:60mcd Vf:2.1V If:20mA	T1	Hrv		data pdf pinout
HVH63531	LED	gr ø5mm I:150mcd Vf:2.1V If:20mA	T1	Hrv		data pdf pinout
HVK 89(A,B)	C-Di	SMD, VHF/UHF-AFC, 3,3...5,7pF(10V)	72a(3,4mm)	Hit	MA 342	data
HVK105	LED	rd ø3mm I:80mcd Vf:1.7V If:20mA	T1	Hrv		data pdf pinout
HVM 11	C-Di	SMD, UHF/SHF-Tuning, 3,6...5,6pF(1V)	35e	Hit	MA 370	data
HVM 13	PIN-Di	SMD, HF-Abschw./attenuator, 60V, 50mA	35e	Hit		data
HVM 14	PIN-Di	SMD, VHF/UHF-Abschw./att., 50V, 50mA	35e	Hit	MA 551, 1SV128, 1SV2333	data
HVM 14 S	PIN-Di	=HVM 14: Dual	35o	Hit	1SV234	data
HVM 14 SR	PIN-Di	=HVM 14: Dual	35n	Hit	-	data
HVM 15	C-Di	=HVM 11: Dual	35I			data
HVM 16	C-Di	SMD, Dual, FM-Tuning, 43...48pF(2V)	35I	Hit	BB 404, BB 804, BB 814	data
HVM 17	C-Di	SMD, Dual, FM-Tuning, 50...85pF(1V)	35I	Hit		data
HVM 17 WA	C-Di	=HVM 17:	35m		-	data
HVM 25	C-Di	SMD, Dual, FM-Tuning, 36...45pF(3V)	35I	Hit	BB 404, BB 804, BB 814	data

Type	Device	Short Description	Fig.	Manu	Comparision Types	More at
HVM 27 WK	C-Di	SMD, Dual, FM-Tuning, 52...62pF(1V)	35l	Hit	1SV225	data
HVM 55	C-Di	SMD, Dual, FM-Tuner, 37...42pF(3V)	35l	Hit	-	data
HVM 89	C-Di	SMC, AFC, 3,6...5,9pF(10V)	35c1	Hit	-	data
HVM 100	C-Di	=HVR 100:	35c	Hit	-	data
HVM 121 WK	PIN-Di	SMD, HF-Abschw./atten., 100V, 0,1A	35l	Hit	-	data
HVM 131 S	PIN-Di	SMD,Dual, HF-Abschw./atten., 60V, 0,1A	35o	Hit	-	data
HVM 131 SR	PIN-Di	=HVM 131S:	35n	Hit	-	data
HVM 132	PIN-Di	SMD, HF-Abschw./attenuator, 60V, 0,1A	35e	Hit	-	data
HVM 132 WK	PIN-Di	=HVM 132: Dual	35l	Hit	-	data
HVM 187 S	PIN-Di	=HVM 13: Dual	35o	Hit	RN 719D	data
HVM 187 WK	PIN-Di	=HVM 13: Dual	35l	Hit	-	data
HVM 306	C-Di	SMD, Dual, Filter, Tuning, 32pF(2V)	35l	Hit	-	data
HVP 21	C-Di	130...170pF(1V)	31a	Hit	-	data
HVP 100	C-Di	AM-Tuning, 440...520pF(1V)	31a	Hit	BB 112, BB 130, BB 509, MV 1401, 1SV100+	data
HVR0001	LED	rd 7.1x7.5mm I:1.6mcd Vf:2.1V	T1	Hrv		data pdf pinout
HVR-1X-40B	Si-Di	kV-Gl, µW Oven, 9kV, 0,35A(Ta=60°)	31a	Sak	-	data
HVR 1X-70B	Si-Di	kV-Gl, 10kV, 0,4A(Tc=84°) µW-Herd/µWave Oven	31a	Sak	-	data
HVR 2X 062	Si-Di	kV-Gl, 6kV µW-Herd/µWave Oven Protection	31a	Sak	-	data
HVR5H123	LED	rd 5.0x7.0mm I:2mcd Vf:2.1V If:20mA	T1	Hrv		data pdf pinout
HVR 12	C-Di	SMD, UHF/SHF-Tuning, 3,6...5,6pF(1V)	71a(2,7mm)	Hit	1SV186	data
HVR 17	C-Di	SMD, FM-Tuning, 50...85pF(1V)	71a(2,7mm)	Hit		data
HVR 100	C-Di	SMD, AM-Tuning, 421...524pF(1V)	71a(2,7mm)	Hit	BB 512	data
HVR 300	C-Di	SMD, VHF-Tuning, 38,5..46,1pF(2V)	71a(2,7mm)	Hit	BB 419, BB 619...620, MA 329, MA 335	data
HVR0301	LED	or 7.1x7.5mm I:7mcd Vf:2.1V If:20mA	T1	Hrv		data pdf pinout
HVR0401	LED	yl 7.1x7.5mm I:5mcd Vf:2.1V If:20mA	T1	Hrv		data pdf pinout
HVR0504	LED	gr 7.1x7.5mm I:6mcd Vf:2.1V If:20mA	T1	Hrv		data pdf pinout
HVR5336RA	LED	or/yl 5.0x7.0mm Vf:2.1V	T1	Hrv		data pdf pinout
HVR5431RA	LED	gr/rd 5.0x7.0mm I:3mcd Vf:2.1V	T1	Hrv		data pdf pinout
HVR5433RA	LED	gr/yl 5.0x7.0mm I:6mcd Vf:2.1V	T1	Hrv		data pdf pinout
HVR5436RA	LED	or/gr 5.0x7.0mm I:8mcd Vf:2.1V	T1	Hrv		data pdf pinout
HVR53123	LED	yl 5.0x7.0mm I:4mcd Vf:2.1V If:20mA	T1	Hrv		data pdf pinout
HVR54123	LED	gr 5.0x7.0mm I:5mcd Vf:2.1V If:20mA	T1	Hrv		data pdf pinout
HVR57123	LED	or 5.0x7.0mm I:6mcd Vf:2.1V If:20mA	T1	Hrv		data pdf pinout
HVS 303	C-Di	VHF-Tuning, 35,8...42pF(1V)	31a	Hit	BB 405G, BB 505G, BB 809, 1SV124, ++	data
HVS 304	C-Di	VHFUHF-Tuning, 15...18pF(1V)	31a	Hit	BB 141...142, BB 405A,B, 1SV110...113, ++	data
HVT-15(SS)	Si-Di	kV-Gl, 11kV, 15mA, <180ns	31a	Sak	HS 10	data
HVT-18	Si-Di	kV-Gl, 12kV, 15mA, <500ns	31a	Sak	HS 10	data
HVT-20(SS)	Si-Di	kV-Gl, 14kV, 15mA, <180ns	31a	Sak	HS 13	data
HVT-22	Si-Di	kV-Gl, 14kV, 15mA, <500ns	31a	Sak	HS 13	data
HVT-25(SS)	Si-Di	kV-Gl, 17kV, 15mA, <180ns	31a	Sak	HS 13	data
HVT-30(E,S,SS)	Si-Di	kV-Gl, 18...24kV, 10mA, <200ns E=24kV, S=18kV, SS=20kV	31a	Sak	.	data
HVU 12	C-Di	=HVR 12:	71a(1,7mm)	Hit	MA 368, MA 373, 1SV245	data
HVU 17	C-Di	=HVR 17:	71a(1,7mm)	Hit	-	data
HVU 89	C-Di	=HVK 89:	71a(1,7mm)	Hit	MA 367	data
HVU 131	PIN-Di	Antenn.-S, 65V, 0,1A, <1Ω(10mA)	71a(1,7mm)	Hit	-	data
HVU 132	PIN-Di	Antenn.-S, 65V, 0,1A, <2Ω(10mA)	71a(1,7mm)	Hit	-	data
HVU 133	PIN-Di	Antenn.-S, 15V, <0,7Ω(2mA)	71a(1,7mm)	Hit	-	data
HVU 187	PIN-Di	=HVM 13:	71a(1,7mm)	Hit		data
HVU 200	C-Di	SMD, VHF-Tuning, 26,5...31,4pF(2V)	71a(1,7mm)	Hit	BB 439, MA 366, 1SV215, 1SV232	data
HVU 200 A	C-Di	=HVU 200: 27,7...31,8pF(2V)	71a(1,7mm)		BB 439, MA 366, 1SV215, 1SV232	data
HVU 202	C-Di	SMD, UHF-Tuning, 14...16,3pF(2V)	71a(1,7mm)	Hit	BB 535, MA 360, MA 372, 1SV214, 1SV223	data
HVU 202 A	C-Di	=HVU 202: 14,1...16,5pF(2V), <0,57Ω	71a(1,7mm)		BB 535, MA 360, MA 372, 1SV214, 1SV223	data
HVU 202 B	C-Di	=HVU 202: 14,2...15,8pF(2V), 0,57Ω	71a(1,7mm)		BB 535, MA 360, MA 372, 1SV214, 1SV223	data
HVU 300	C-Di	SMD, VHF-Tuning, 38,5...46,1pF(2V)	71a(1,7mm)	Hit	BB 639, MA 365, MA 374, 1SV231	data
HVU 300 A	C-Di	=HVU 300: 39,5...47,4pF(2V), <1,1Ω	71a(1,7mm)		BB 639	data
HVU 300 B	C-Di	=HVU 300: 47...53pF(2V), <1,1Ω	71a(1,7mm)		BB 639	data
HVU 305	C-Di	SMD, VHF-Tuning, 33,2...36,7pF(2V)	71a(1,7mm)	Hit	BB 439, BB 639, MA 361, MA 364	data
HVU 306	C-Di	SMD, VHF-Tuning, 29,3...33,2pF(2V)	71a(1,7mm)	Hit	BB 439, MA 355, MA 357, MA 366	data
HVU 306 A	C-Di	=HVU 306: 29,3...34,2pF(2V), <0,75Ω	71a(1,7mm)		BB 439, MA 361, MA 364	data
HVU 306 B	C-Di	=HVU 306: 29,5...33,5pF(2V), <0,75Ω	71a(1,7mm)		BB 439, MA 361, MA 364	data
HVU 307	C-Di	SMD, VHF-Tuning, 32,2...37,5pF(2V)	71a(1,7mm)	Hit	BB 439, BB 639, MA 371, 1SV217	data
HVU 308	C-Di	SMD, VHF-Tuning, 13,9...16,1pF(2V)	71a(1,7mm)	Hit	BB 535, MA 360, MA 363, MA 372	data
HVU 308 A	C-Di	=HVU 308: 13,7...15,9pF(2V)	71a(1,7mm)		BB 535, MA 360, MA 363, MA 372	data
HVU 314	C-Di	SMD, BS Tuning, 4,4...6,4pF(1V)	71a(1,7mm)	Hit	MA 368, 1SV245	data
HVU 315	C-Di	SMD, VHF-Tuning, 27,6...32,1pF(2V)	71a(1,7mm)	Hit	-	data
HVU 316	C-Di	SMD, BS/CS Tuning, 5,16...7,22pF(1V)	71a(1,7mm)	Hit	MA 368, MA 373, 1SV245	data
HVU 350(B)	C-Di	SMD, VCO, 15...17,5pF(1V)	71a(1,7mm)	Hit	1SV270	data
HVU 351	C-Di	SMD, VCO, 14...16pF(2V)	71a(1,7mm)	Hit	1SV229	data
HVU 352	C-Di	SMD, VCO ...1GHz, 6...8pF(1V)	71a(1,7mm)	Hit	-	data
HVU 354	C-Di	SMD, VCO, 16...18,5pF(1V)	71a(1,7mm)	Hit	-	data
HVU 355	C-Di	SMD, VCO, 6,4...7,4pF(1V)	71a(1,7mm)	Hit	-	data
HVU 355 B	C-Di	=HVU 355: 6,4...7,2pF(1V)	71a(1,7mm)		-	data
HVU 356	C-Di	SMD, VCO, BS-Tuner, 27,7...31,8pF(2V)	71a(1,7mm)	Hit	-	data

Type	Device	Short Description	Fig.	Manu	Comparision Types	More at
HVU 357	C-Di	SMD, VCO, 19,5..23,5pF(1V)	71a(1,7mm)	Hit	-	data
HVU 358	C-Di	SMD, VCO, 19...21pF(1V)	71a(1,7mm)	Hit	-	data
HVU 359	C-Di	SMD, VCXO Unit, 24,8...29,8pF(1V)	71a(1,7mm)	Hit	-	data
HVU 362	C-Di	SMD, VCXO, 41,6...49,9pF(1V)	71a(1,7mm)	Hit	-	data
HVU 363 A	C-Di	SMD, VHF-Tuning, 34,6...42,3pF(1V)	71a(1,7mm)	Hit	BB 639	data
HVU 363 B	C-Di	=HVU 363A: 36...42pF(1V)	71a(1,7mm)		BB 639	data
HVU 365	C-Di	SMD, VCO, 27...28,55pF(1V)	71a(1,7mm)	Hit	-	data
HW	MOS-N-FET-d	=3SK286 (Typ-Code/Stempel/marking)	44	Mat	→3SK286	data
HW	Z-Di	=Z2 SMB-200 (Typ-Code/Stempel/marking)	71(5x3,5)	Fag	→Z2 SMB-200	data
HWTS901	LED	wht 5.0x7.0mm I:700..800mcd If:20mA		Seo		data pdf pinout
HWTS902	LED	wht 5.0x7.0mm I:700..800mcd If:20mA		Seo		data pdf pinout
HWTS905	LED	wht 5.0x7.0mm I:700..800mcd If:20mA		Seo		data pdf pinout
HX	Si-P	=2SB1572-HX (Typ-Code/Stempel/marking)	39	Nec	→2SB1572	data
HX	MOS-N-FET-e	=2SK2922 (Typ-Code/Stempel/marking)	39	Hit	→2SK2922	data
HX0032	HI-SPEED	Vs:±15V Vu:60dB Vo:±15V Vi0:3mV		Hal		data pinout
HY	Si-P	=2SA1203-Y (Typ-Code/Stempel/marking)	39	Tos	→2SA1203	data
HY	Si-P	=2SB1572-HY (Typ-Code/Stempel/marking)	39	Nec	→2SB1572	data
HY	MOS-N-FET-e*	=2SK1772 (Typ-Code/Stempel/marking)	39	Hit	→2SK1772	data
HY	Si-P	=KTA1663-Y (Typ-Code/Stempel/marking)	39	Kec	→KTA 1663	data
HYB 4116 A-3	dRAM-IC	16384x1 Bit, 200ns	16-DIP	Sie	...4116-...	data
HYB 4116 A-4	dRAM-IC	=HYB 4116A-3: 250ns	16-DIP	Sie	...4116-...	data
HYB 4116-P2	dRAM-IC	16384x1 Bit, 150ns	16-DIP	Sie	...4116-...	data
HYB 4116-P3	dRAM-IC	=HYB 4116-P2: 200ns	16-DIP	Sie	...4116-...	data
HYB 4116 -P4,P-4	dRAM-IC	=HYB 4116-P2: 250ns	16-DIP	Sie	...4116-...	data
HYB 4164-P1	dRAM-IC	65536x1 Bit, 120ns	16-DIP	Sie	...4164-...	
HYB 4164-P2	dRAM-IC	=HYB 4164-P1: 150ns	16-DIP	Sie	...4164-...	
HYB 4164-P3	dRAM-IC	=HYB 4164-P1: 200ns	16-DIP	Sie	...4164-...	
HYB 41256-12	dRAM-IC	262144x1 Bit, 120ns	16-DIP	Sie	...41256-...	data
HYB 41256-15	dRAM-IC	=HYB 41256-12: 150ns	16-DIP	Sie	...41256-...	data
HYB 41256-20	dRAM-IC	=HYB 41256-12: 200ns	16-DIP	Sie	...41256-...	data
HYB 511000-12	dRAM-IC	=HYB 511000-10: 120ns	18-DIP	Sie	...41024-...	data
HYB 511000-10	dRAM-IC	1048576x1 Bit, 100ns, page mode	18-DIP	Sie	...41024-...	data
HYB 514171BJ-50	dRAM-IC	256Kx16-Bit Dram w. Self Refresh RAS 50ns	MDIP	Sie	-	pdf pinout
HYB 514171BJ-60	dRAM-IC	256Kx16-Bit Dram w. Self Refresh RAS 60ns	MDIP	Sie	-	pdf pinout
HYB 514171BJ-70	dRAM-IC	256Kx16-Bit Dram w. Self Refresh RAS 70ns	MDIP	Sie	-	pdf pinout
HYB 514171BJL-50	dRAM-IC	256Kx16-Bit Dram w. Self Refresh RAS 50ns low power	MDIP	Sie	-	pdf pinout
HYB 514171BJL-60	dRAM-IC	256Kx16-Bit Dram w. Self Refresh RAS 60ns low power	MDIP	Sie	-	pdf pinout
HYB 514171BJL-70	dRAM-IC	256Kx16-Bit Dram w. Self Refresh RAS 70ns low power	MDIP	Sie	-	pdf pinout
HZ	Si-P	=2SB1572-HZ (Typ-Code/Stempel/marking)	39	Nec	→2SB1572	data
HZ 1 C3	Z-Di	1,5...1,7V(5mA), 0,5W	31a	Hit	-	data
HZ 2...36(H) ...	Z-Di	2...36V, 0,5W, A,B,C,1,2,3≈-15...+8%	31a	Hit	BZV 85/..., BZX 55/..., BZX 83/..., ZPD...,+	data
HZ 2.7...39E(B1..4)	Z-Di	2,7...39V, 0,5W, B1..B4≈-8...+3%	31a	Hit	BZV 85/..., BZX 55/..., BZX 83/..., ZPD...,+	data
HZ 6...36(...)L	Z-Di	ra, 6...36V, 0,4W, A,B,C,1,2,3≈-12...+8%	31a	Hit	BZV 39/..., 1N4099...4122	data
HZ 2...5(A,B,C)LL	Z-Di	ra, 2...5V, 0,25W, A,B,C≈-12...+10%	31a	Hit	BZV 39/..., 1N4615...25	data
HZ 2.0...36(B,C)P	Z-Di	2,0...36V, 0,8W, BP≈6%, CP≈0...+12%	31a	Hit	BZV 85/..., BZW 22/..., BZX 61/..., ZPY...,+	data
HZF 2.0...36(B,C)P	Z-Di	=HZ 2.0...36(B,C)P: SMD, 0,9W	71a(5x2,5)	Hit	BZG 03-..., MA 1Z..., PTZ..., RD...FM	data
HZK 2...36 A,B,C	Z-Di	=HZ 2...36(H)A,B,C: SMD, 0,5W	72a(3,4mm)	Hit	BZD 27/..., BZV 55/..., RLZ ...(J)	data
HZK 6...36(A,B,C)L	Z-Di	=HZ 6...36(A,B,C)L: SMD, 0,4W	72a(3,4mm)	Hit		data
HZK 2...5(A,B,C)LL	Z-Di	=HZ 2...5(A,B,C)LL: SMD, 0,25W	72a(3,4mm)	Hit		data
HZM 4.3 FA	Z-Di	SMD, 4x Z, TAZ, 4,04...4,48V	45(KAKKK)	Hit	-	data
HZM 6.8 FA	Z-Di	SMD, 4x Z, TAZ, 6,47...7V	45(KAKKK)	Hit	-	data
HZM 27 FA	Z-Di	SMD, 4x Z, TAZ, 25,1...28,9V	45(KAKKK)	Hit	-	data
HZM 2.0...36N(B...)	Z-Di	SMD, 2,0...36V, B...B3≈-5...+5%	35e	Hit	BZX 84/..., RD ...M	data
HZM 6.2 ZWA	Z-Di	SMD, 2x Z, TAZ, 5,9...6,5V	35m	Hit	-	data
HZM 6.8 WA	Z-Di	SMD, 2x Z, TAZ, 6,47...7V	35m	Hit	-	data
HZM 27 WA	Z-Di	SMD, 2x Z, TAZ, 25,1...28,9V	35m	Hit	-	data
HZS 2...36 ...	Z-Di	=HZ 2...36(H)...: 0,4W	31a	Hit	→HZ 2...36(H)...	data
HZS 2.7...39E...	Z-Di	=HZ 2.7...39E...: 0,4W	31a	Hit	→HZ 2.7...39E...	data
HZS 5.6...36J(B...)	Z-Di	5,6...36V, 0,4W, B1...B3≈-6...+5%	31a	Hit	BZV 85/..., BZX 55/..., BZX 83/..., ZPD...,+	data
HZS 6...36 L...	Z-Di	=HZ 6...36L...: 0,4W	31a	Hit	→HZ 6...36L...	data
HZS 2...5 LL...	Z-Di	=HZ 2...5LL...: 0,25W	31a	Hit	→HZ 2...5LL...	data
HZS 2.0...39N(B...)	Z-Di	2,0...39V, 0,4W, B1...B4≈-8...+3%	31a	Hit	BZV 85/..., BZX 55/..., BZX 83/..., ZPD...,+	data
HZS 5.6...36S...	Z-Di	=HZS 5.6...36J...: 0,4W	31a	Hit	→HZS 5.6...36J...	data
HZT 7 A1	Ref-Di	6,5...7V, 20mA, ±0,01%/°C, <20Ω(10mA)	31a	Hit		data
HZT 7 A2	Ref-Di	=HZT 7A1: ±0,005%/°C, <20Ω(10mA)	31a			data
HZT 7 A3	Ref-Di	=HZT 7A1: ±0,002%/°C, <20Ω(10mA)	31a			data
HZT 7 B1	Ref-Di	=HZT 7A1: ±0,01%/°C, <25Ω(5mA)	31a			data
HZT 7 B2	Ref-Di	=HZT 7A1: ±0,005%/°C, <25Ω(5mA)	31a			data
HZT 7 B3	Ref-Di	=HZT 7A1: ±0,002%/°C, <25Ω(5mA)	31a			data
HZT 7 C1	Ref-Di	=HZT 7A1: ±0,01%/°C, <100Ω(1mA)	31a			data
HZT 7 C2	Ref-Di	=HZT 7A1: ±0,005%/°C, <100Ω(1mA)	31a			data

ype	Device	Short Description	Fig.	Manu	Comparision Types	More at
T 9	Z-IC, Ref-Di	Tuner-Stab., 9...10V, 20mA, 0,2W	31h	Hit	ZTK 9	
T 22	Z-IC, Ref-Di	Tuner-Stab., 21,5...23V, 20mA, 0,2W	31h	Hit	ZTK 22	
T 31	Z-IC, Ref-Di	Tuner-Stab., 30...32V, 20mA, 0,2W	31h	Hit	TDA 1550, ZTK 33	
T 33	Z-IC, Ref-Di	Tuner-Stab., 31...35V, 20mA, 0,4W	31h	Hit	TDA 1550, ZTK 33	pdf
U 2.0...36...	Z-Di	=HZM 2.0...36N...:	71a(1,7mm)	Hit	DTZ ..., RD ...S	data
U 2LL...5LL...	Z-Di	=HZ 2LL...5LL:	71a(1,7mm)	Hit	DTZ ..., RD ...S	data
	Si-P	=2SB1073 (Typ-Code/Stempel/marking)	39	Mat	→2SB1073	data
	Si-P	=2SB1220 (Typ-Code/Stempel/marking)	35(2mm)	Mat	→2SB1220	data
	Si-P	=2SB1463 (Typ-Code/Stempel/marking)	35(1,6mm)	Mat	→2SB1463	data
	Si-P	=2SB792 (Typ-Code/Stempel/marking)	35	Mat	→2SB792	data
	GaAs-N-FET	=2SK1196 (Typ-Code/Stempel/marking)	51	Mat	→2SK1196	data
	C-Di	=BBY 51-02W (Typ-Code/Stempel/marking)	71(1,3mm)	Sie	→BBY 51-02W	data
C	Si-N+R	=UN 7231 (Typ-Code/Stempel/marking)	39	Mat	→UN 7231	data
	Si-Di	=1SS336 (Typ-Code/Stempel/marking)	35	Tos	→1SS336	data
103	dRAM-IC	1024 Bit	18-DIP	Int	U 253D	
302	ROM-IC	2048 Bit	24-DIP	Int	U 501D	
602 A	PROM-IC	256x8 Bit	24-DIP	Int	U 551D	
D 1700 S/P	-IC	Single-Chip Voice Record & Playback Devices	QIP	Wbd	-	pdf pinout
702 A	EPROM-IC	256x8 Bit	24-DIP	Int	U 552C	
102 A	sRAM-IC	1024 Bit	16-DIP	Int	U 202D	
114	sRAM-IC	1024x4 Bit	18-DIP	Int	...2114..., U 214D	
115	sRAM-IC	1024x1 Bit	16-DIP	Int	U 215D	
124	sRAM-IC	1024x4 Bit	18-DIP	Int	U 224D, UL 224D, US 224D, VL 224D	
125	sRAM-IC	1024x1 Bit	16-DIP	Int	U 225D	
2147 H	sRAM-IC	4096x1 Bit	18-DIP	Int	...2147...	
2148	sRAM-IC	1024x4 Bit	18-DIP	Int	...2148..., U 2148C	
2308	ROM-IC	8192 Bit, mask.-progr.	24-DIP	Int	U 505D	
2364	ROM-IC	8192x8 Bit	28-DIP	Int	MN 2364, U 2364D	
2616	PROM-IC	2048x8 Bit	24-DIP	Int	U 2616D	
2708	EPROM-IC	1024x8 Bit	24-DIP	Int	U 555C	
2716	EPROM-IC	2048x8 Bit	24-DIP	Int	...2716..., U 2716C	
2764	EPROM-IC	8192x8 Bit, 200ns	28-DIP	Int	...2764...	
3245	I/O-IC	=µA 9645		Int	→µA 9645	
4016	NMOS-sRAM-IC	2048x8 Bit	24-DIP	Int	MN 4216, MN 4316, ... 4016...	
8008	µP-IC	8 Bit-CPU	18-DIP	Int	U 808D	
8205	TTL-IC	Schottky, 1/8-Binärdecoder/binary dec.	16-DIP	Int	DS 8205D	
8212	TTL-IC	Schottky, 8 Bit Bustreiber/bus driver	24-DIP	Int	DS 8212D	
8216	TTL-IC	Schottky, 4 Bit Bustreiber/bus driver	16-DIP	Int	DS 8216D	
8272	I/O-IC	Floppy-disk-controller	40-DIP	Int	U 8272D	
8282	TTL-IC	Schottky, 8 Bit Bustreiber/bus driver	20-DIP	Int	DS 8282D	
8283	TTL-IC	Schottky, 8 Bit Bustreiber/bus driver	20-DIP	Int	DS 8283D	
8286	TTL-IC	Schottky, 8 Bit Bustreiber/bus driver	20-DIP	Int	DS 8286D	
8287	TTL-IC	Schottky, 8 Bit Bustreiber/bus driver	20-DIP	Int	DS 8287D	
23128	ROM-IC	16384x8 Bit, 250ns	28-DIP	Int	MN 23128	
82720	MOS-IC	Grafic-display-controller		Int	U 82720D	
A	MOS-N-FET-e	=µPA602T (Typ-Code/Stempel/marking)	46	Nec	→µPA602T	data
A 6805 E2-PDW	µP-IC	Microprocessor Unit	DIP	Ina	-	pdf pinout
A 6805 E2-PLC	µP-IC	Microprocessor Unit	PLCC	Ina	-	pdf pinout
A 8044-PDW	µP-Periph	SDLC Communications Ctrl.	DIP	Ina	-	pdf pinout
A 8044-PLC	µP-Periph	Sdlc Communications Ctrl.	PLCC	Ina	-	pdf pinout
A 8344-PDW	µP-Periph	Sdlc Communications Ctrl.	DIP	Ina	-	pdf pinout
A 8344-PLC	µP-Periph	Sdlc Communications Ctrl.	PLCC	Ina	-	pdf pinout
B	Si-N	=XP 06531 (Typ-Code/Stempel/marking)	46(2mm)	Mat	→XP 06531	data
B	MOS-N-FET-e*	=µPA 611TA (Typ-Code/Stempel/marking)	46	Nec	→µPA 611TA	data
BB 1488 B	OP-IC	Dual, Serie 158	8-DIP		→MC 1458	data
BM 25 PPC 750 GX	MOS-µP-IC	IBM PowerPC 750GX RISC Microprocessor	GRID	Ibm	-	pdf pinout
IC 16 F 627 A/P	CMOS-IC	Flash-Based 8-Bit CMOSMicrocon-trollers	DIP	Mcp	-	pdf pinout
IC 16 F 628 A/P	CMOS-IC	Flash-Based 8-Bit CMOSMicrocon-trollers	DIP	Mcp	-	pdf pinout
IC 16 F 648 A/P	CMOS-IC	Flash-Based 8-Bit CMOSMicrocon-trollers	DIP	Mcp	-	pdf pinout
IC-	Si-N	=2SC3016 (Typ-Code/Stempel/marking)	35	Hit	→2SC3016	data
ICB 1 FL 02 G	IC	Smart Ballast Control IC,Fluorescent Lamp	QFP	Inf	-	pdf pinout
ICB8000	COMP	Vu:83dB Vo:7/0.2V Vi0:35mV	10A	Isi		data pinout
ICB8001	COMP	Vu:63dB Vo:7/0.2V Vi0:6mV	10A	Isi		data pinout
ICB8008	UNI	Vs:±15V Vu:86dB Vo:±12V Vi0:7.5mV	8A	Isi		data pinout
ICB8741	UNI	Vs:±15V Vu:86dB Vo:±12V Vi0:7.5mV	8A	Isi		pdf pinout
ICE 1 PCS01	SMPS-IC	Power Factor Correction (PFC) Contr., CCM	DIP	Inf	-	pdf pinout
ICE 1 PCS01G	SMPS-IC	SMD, Power Factor Corr. (pfc) Contr., CCM	MDIP	Inf	-	pdf pinout
ICE 1 PCS02	LIN-IC	Standalone Power Factor Correction (pfc)	DIC	Inf	-	pdf pinout
ICE 1 PCS02G	LIN-IC	Standalone Power Factor Correction (pfc)	MDIP	Inf	-	pdf pinout
ICE 1PD265G	SMPS-IC	SMD, Power Factor Contr. & Coolmos	MDIP	Inf	-	pdf pinout
ICE 1 QS01	LIN/MOS-IC	Controller for Switch Mode Power Supplies; Pfc	DIC	Inf	-	pdf pinout
ICE 1 QS01G	LIN/MOS-IC	Controller for Switch Mode Power Supplies; Pfc	DIC	Inf	-	pdf pinout
ICE 2(A,B) S01	SMPS-IC	SMPS Current Mode Contr., 100/67kHz Standby, Soft Start, +8.5...21V	8-DIP	Inf	-	pdf pinout
ICE 2(A,B) S01G	SMPS-IC	SMD, SMPS Curr. Mode Contr., 100/67kHz	8-MDIP	Inf		pdf pinout

Type	Device	Short Description	Fig.	Manu	Comparision Types	More at
		Standby, Soft Start, +8.5..21V				
ICE 2A0565	IC	SMPS Current Mode Controller, 650V, 23W	DIP	Inf	-	pdf pinout
ICE 2A0565 G	IC	SMPS Current Mode Controller, 650V, 23W	MDIP	Inf	-	pdf pinout
ICE 2A0565 Z	IC	SMPS Current Mode Controller, 650V, 23W	DIP	Inf	-	pdf pinout
ICE 2A165	IC	SMPS Current Mode Controller, 650V/800V, 31W	DIP	Inf	-	pdf pinout
ICE 2A180 Z	IC	SMPS Current Mode Controller, 800V, 29W	DIP	Inf	-	pdf pinout
ICE 2A265	IC	SMPS Current Mode Controller, 650V, 52W	DIP	Inf	-	pdf pinout
ICE 2A280 Z	IC	SMPS Current Mode Controller, 800V, 50W	DIP	Inf	-	pdf pinout
ICE 2A365	IC	SMPS Current Mode Controller, 650V, 76W	DIP	Inf	-	pdf pinout
ICE 2A765 I	IC	SMPS Current Mode Controller, 650V, 240W	86	Inf	-	pdf pinout
ICE 2A765 P2	IC	SMPS Current Mode Controller, 650V, 240W	86	Inf	-	pdf pinout
ICE 2B0565	IC	SMPS Current Mode Controller, 650V, 23W	DIP	Inf	-	pdf pinout
ICE 2B165	IC	SMPS Current Mode Controller, 650V, 31W	DIP	Inf	-	pdf pinout
ICE 2B265	IC	SMPS Current Mode Controller, 650V, 52W	DIP	Inf	-	pdf pinout
ICE 2B365	IC	SMPS Current Mode Controller, 650V, 76W	DIP	Inf	-	pdf pinout
ICE 2B765 I	IC	SMPS Current Mode Controller, 650V, 240W	86	Inf	-	pdf pinout
ICE 2B765 P2	IC	SMPS Current Mode Controller, 650V, 240W	86	Inf	-	pdf pinout
ICE 2 PCS01	SMPS-IC	Standalone Power Factor Correction (PFC), CCM	DIC	Inf	-	pdf pinout
ICE 2 PCS01G	SMPS-IC	Standalone Power Factor Correction (pfc), CCM	MDIP	Inf	-	pdf pinout
ICE 2 PCS02	SMPS-IC	Standalone Power Factor Correction (pfc) Cntrl. in Ccm,	DIC	Inf	-	pdf pinout
ICE 2 PCS02G	SMPS-IC	Standalone Power Factor Correction (pfc) Cntrl. in Ccm,	MDIP	Inf	-	pdf pinout
ICE 2 QS01	SMPS-IC	Quasi-Resonant PWM Controller	DIC	Inf	-	pdf pinout
ICE 3A0365	SMPS-IC	SMPS, Current Mode Contr., 650V 22W	DIP	Inf	-	pdf pinout
ICE 3A0565	SMPS-IC	SMPS, Current Mode Contr., 650V 25W	DIP	Inf	-	pdf pinout
ICE 3A0565Z	SMPS-IC	SMPS, Current Mode Contr., 650V 25W	DIP	Inf	-	pdf pinout
ICE 3A1065	SMPS-IC	SMPS, Current Mode Contr., 650V 32W	DIP	Inf	-	pdf pinout
ICE 3A1565	SMPS-IC	SMPS, Current Mode Contr., 650V 42W	DIP	Inf	-	pdf pinout
ICE 3A2065	SMPS-IC	SMPS, Current Mode Contr., 650V 57W	DIP	Inf	-	pdf pinout
ICE 3A2065I	SMPS-IC	SMPS, Current Mode Contr., 650V 102W	86	Inf	-	pdf pinout
ICE 3A2065P	SMPS-IC	SMPS, Current Mode Contr., 650V 102W	86	Inf	-	pdf pinout
ICE 3A2065Z	SMPS-IC	SMPS, Current Mode Contr., 650V 57W	DIP	Inf	-	pdf pinout
ICE 3A2565	SMPS-IC	SMPS, Current Mode Contr., 650V 68W	DIP	Inf	-	pdf pinout
ICE 3A3065I	SMPS-IC	SMPS, Current Mode Contr., 650V 128W	86	Inf	-	pdf pinout
ICE 3A3065P	SMPS-IC	SMPS, Current Mode Contr., 650V 128W	86	Inf	-	pdf pinout
ICE 3A3565I	SMPS-IC	SMPS, Current Mode Contr., 650V 170W	86	Inf	-	pdf pinout
ICE 3A3565P	SMPS-IC	SMPS, Current Mode Contr., 650V 170W	86	Inf	-	pdf pinout
ICE 3A5065I	SMPS-IC	SMPS, Current Mode Contr., 650V 220W	86	Inf	-	pdf pinout
ICE 3A5065P	SMPS-IC	SMPS, Current Mode Contr., 650V 220W	86	Inf	-	pdf pinout
ICE 3A5565I	SMPS-IC	SMPS, Current Mode Contr., 650V 240W	86	Inf	-	pdf pinout
ICE 3A5565P	SMPS-IC	SMPS, Current Mode Contr., 650V 240W	86	Inf	-	pdf pinout
ICE 3 AS02	SMPS-IC	Off-Line SMPS Current Mode Cntrl., int. 500V Startup Cell, 100kHz	DIC	Inf	-	pdf pinout
ICE 3 AS02G	SMPS-IC	Off-Line SMPS Current Mode Cntrl., int. 500V Startup Cell, 100kHz	MDIP	Inf	-	pdf pinout
ICE 3B0365	SMPS-IC	SMPS, Current Mode Contr., 650V 22W	DIP	Inf	-	pdf pinout
ICE 3B0565	SMPS-IC	SMPS, Current Mode Contr., 650V 25W	DIP	Inf	-	pdf pinout
ICE 3B1065	SMPS-IC	SMPS, Current Mode Contr., 650V 32W	DIP	Inf	-	pdf pinout
ICE 3B1565	SMPS-IC	SMPS, Current Mode Contr., 650V 42W	DIP	Inf	-	pdf pinout
ICE 3B2065	SMPS-IC	SMPS, Current Mode Contr., 650V 57W	DIP	Inf	-	pdf pinout
ICE 3B2065I	SMPS-IC	SMPS, Current Mode Contr., 650V 102W	86	Inf	-	pdf pinout
ICE 3B2065P	SMPS-IC	SMPS, Current Mode Contr., 650V 102W	86	Inf	-	pdf pinout
ICE 3B2565	SMPS-IC	SMPS, Current Mode Contr., 650V 68W	DIP	Inf	-	pdf pinout
ICE 3B3065I	SMPS-IC	SMPS, Current Mode Contr., 650V 128W	86	Inf	-	pdf pinout
ICE 3B3065P	SMPS-IC	SMPS, Current Mode Contr., 650V 128W	86	Inf	-	pdf pinout
ICE 3B3565I	SMPS-IC	SMPS, Current Mode Contr., 650V 170W	86	Inf	-	pdf pinout
ICE 3B3565P	SMPS-IC	SMPS, Current Mode Contr., 650V 170W	86	Inf	-	pdf pinout
ICE 3B5065I	SMPS-IC	SMPS, Current Mode Contr., 650V 220W	86	Inf	-	pdf pinout
ICE 3B5065P	SMPS-IC	SMPS, Current Mode Contr., 650V 220W	86	Inf	-	pdf pinout
ICE 3B5565I	SMPS-IC	SMPS, Current Mode Contr., 650V 240W	86	Inf	-	pdf pinout
ICE 3B5565P	SMPS-IC	SMPS, Current Mode Contr., 650V 240W	86	Inf	-	pdf pinout
ICE 3 BS02	SMPS-IC	Off-Line SMPS Current Mode Cntrl., int. 500V Startup Cell, 67kHz	DIC	Inf	-	pdf pinout
ICE 3 BS02G	SMPS-IC	Off-Line SMPS Current Mode Cntrl., int. 500V Startup Cell, 67kHz	DIC	Inf	-	pdf pinout
ICE 3 BS02L	SMPS-IC	Off-Line SMPS Current Mode Cntrl., integ. 500V Startup Cell	DIC	Inf	-	pdf pinout
ICE 3 DS01L	SMPS-IC	Off-Line SMPS Current Mode Controller, integrated 500V Startup Cell	DIC	Inf	-	pdf pinout
ICE 3 DS01LG	SMPS-IC	Off-Line SMPS Current Mode Controller, integrated 500V Startup Cell	MDIP	Inf	-	pdf pinout
ICE 3 B 0365 J	SMPS-IC	SMPS, Current Mode Contr., 650V 22W	DIP	Inf	-	pdf pinout
ICE 3 B 0365 JG	SMPS-IC	SMPS, Current Mode Contr., 650V 22W	DIP	Inf	-	pdf pinout
ICE 3 B 0365 L	SMPS-IC	SMPS, Current Mode Contr., 650V 22W	DIP	Inf	-	pdf pinout
ICE 3 B 0565 J	SMPS-IC	SMPS, Current Mode Contr., 650V 22W	DIP	Inf	-	pdf pinout
ICE 3 B 0565 JG	SMPS-IC	SMPS, Current Mode Contr., 650V 22W	DIP	Inf	-	pdf pinout
ICE 3 A 1065 L	SMPS-IC	SMPS, Current Mode Contr., 650V 22W	DIP	Inf	-	pdf pinout
ICE 3 A 1065 LJ	SMPS-IC	SMPS, Current Mode Contr., 650V 32W	DIP	Inf	-	pdf pinout
ICE 3 A 1565 L	SMPS-IC	SMPS, Current Mode Contr., 650V 22W	DIP	Inf	-	pdf pinout
ICE 3 B 1565 J	SMPS-IC	SMPS, Current Mode Contr., 650V 22W	DIP	Inf	-	pdf pinout
ICE 3 BR 4765 J	SMPS-IC	Off-Line SMPS Current Mode Controller	DIP	Inf	-	pdf pinout

Type	Device	Short Description	Fig.	Manu	Comparision Types	More at
ICH8500	UNI	Vs:±15V Vu:86dB Vo:±12V Vio:20mV	8A	Isi		data pinout
ICL101A	UNI	Vs:±20V Vu:87dB Vo:±12V Vi0:3mV		Isi		data pinout
ICL108	LO-POWER	Vs:±20V Vu:87dB Vo:±13V Vio:3mV	8A	Isi		data pinout
ICL 108...	OP-IC	→µA 108...		Isi	→µA 108...	
ICL 232 ...	I/O-IC	Dual RS-232 Transmitter/Receiver			→MAX 232 ...	pdf pinout
ICL301A	UNI	Vs:±15V Vu:83dB Vo:±12V Vi0:10mV		Isi		data pinout
ICL308	LO-POWER	Vs:±15V Vu:83dB Vo:±13V Vi0:10mV	8A	Isi		data pinout
ICL741	UNI	Vs:±22V Vu:106dB Vo:±14V Vi0:1mV	8A	Isi		data pinout
ICL 741...	OP-IC	→µA 741...		Isi	→µA 741...	
ICL741C	UNI	Vs:±18V Vu:100dB Vo:±14V Vi0:2mV	8A	Isi		data pinout
ICL748	UNI	Vs:±22V Vu:103dB Vo:±14V Vi0:1mV	8A	Isi		data pinout
ICL748C	UNI	Vs:±22V Vu:100dB Vo:±14V Vi0:2mV	8A	Isi		data pinout
ICL 7106	A/D-IC	3,5 Digit, A/D-Conv., LCD Drv		Tsc	TSC 7106A	pdf
ICL 7106(C)PL,JL	CMOS-A/D-IC	3,5 Digit, LCD Driver, 0...+70°	40-DIP,DIC	Max, Tsc	MAX 130..., TSC 7106A	
ICL 7106(C)PL,JL	CMOS-A/D-IC	3,5 Digit, A/D-Conv., LCD Drv, 0...+70°	40-DIP,DIC	Max,Tsc	MAX 130..., TSC 7106A	pdf
ICL 7106(C)QH	CMOS-A/D-IC	=ICL 7106...PL,JL:	44-PLCC			
ICL 7106(C)QH	CMOS-A/D-IC	3,5 Digit, A/D-Conv., LCD Drv, 0...+70°	44-PLCC	Max	-	pdf
ICL 7107(C)PL,JL	CMOS-A/D-IC	3,5 Digit, LED Driver, 0...+70°	40-DIP,DIC	Max	-	
ICL 7107(C)PL,JL	CMOS-A/D-IC	3,5 Digit, A/D-Conv., LED Drv, 0...+70°	40-DIP,DIC	Max	-	pdf
ICL 7107(C)QH	CMOS-A/D-IC	=ICL 7107...PL,JL:	44-PLCC			
ICL 7107(C)QH	CMOS-A/D-IC	3,5 Digit, A/D-Conv., LED Drv, 0...+70°	44-PLCC	Max	-	pdf
ICL 7109 CQH	CMOS-A/D-IC	12 Bit A/D-Converter, µP-Interface 3-State Binary Out, 0...+70°	44-PLCC	Max	-	pdf
ICL 7109(...)PL,JL		12 Bit, µP Interface, 3-State Output C: 0...+70°, I:-25...+85°, M: -55...+125°	40-DIP,DIC	Max	-	
ICL 7109(...)PL,JL	CMOS-A/D-IC	12Bit A/D-Conv., µP-Interf.,Binary Out C: 0...+70°, I:-25...+85°, M: -55...+125°	40-DIP,DIC	Max	-	pdf
ICL 7116(C)PL,JL	CMOS-A/D-IC	3,5 Digit, LCD Driver, 0...+70°	40-DIP,DIC	Max	-	
ICL 7116(C)PL,JL	CMOS-A/D-IC	3,5 Digit, A/D-Conv., LCD+Hold,0...+70°	40-DIP,DIC	Max	-	pdf
ICL 7116(C)QH	CMOS-A/D-IC	=ICL 7116...PL,JL:	44-PLCC			
ICL 7116(C)QH	CMOS-A/D-IC	3,5 Digit, A/D-Conv., LCD+Hold,0...+70°	44-PLCC	Max	-	pdf
ICL 7117(C)PL,JL	CMOS-A/D-IC	3,5 Digit, LED Driver, 0...+70°	40-DIP,DIC	Max	-	
ICL 7117(C)PL,JL	CMOS-A/D-IC	3,5 Digit, A/D-Conv., LED+Hold,0...+70°	40-DIP,DIC	Max	-	pdf
ICL 7117(C)QH	CMOS-A/D-IC	=ICL 7117...PL,JL:	44-PLCC			
ICL 7117(C)QH	CMOS-A/D-IC	3,5 Digit, A/D-Conv., LED+Hold,0...+70°	44-PLCC	Max	-	pdf
ICL 7126(C)PL,JL	CMOS-A/D-IC	3,5 Digit,lo-power,LCD Driver, 0...+70°	40-DIP,DIC	Max	-	
ICL 7126(C)PL,JL	CMOS-A/D-IC	3,5 Digit, A/D-Conv., Low Power LCD Driver, 0...+70°	40-DIP,DIC	Max	-	pdf
ICL 7126(C)QH	CMOS-A/D-IC	=ICL 7126...PL,JL:	44-PLCC			
ICL 7126(C)QH	CMOS-A/D-IC	3,5 Digit, A/D-Conv., Low Power LCD Driver, 0...+70°	44-PLCC	Max	-	pdf
ICL 7129A(C)PL,JL	CMOS-A/D-IC	4,5 Digit, LCD Driver, 0...+70°	40-DIP,DIC	Max	MAX 7129...	
ICL 7129A(C)PL,JL	CMOS-A/D-IC	4,5 Digit, A/D-Conv., LCD Drv, 0...+70°	40-DIP,DIC	Max	MAX 7129...	pdf
ICL 7129A(C)QH	CMOS-A/D-IC	=ICL 7129A...PL,JL:	44-PLCC			
ICL 7129A(C)QH	CMOS-A/D-IC	4,5 Digit, A/D-Conv., LCD Drv, 0...+70°	44-PLCC	Max	MAX 7129...	pdf
ICL 7135(C)PI,JI	CMOS-A/D-IC	4,5 Digit, MPX BCD Output, 0...+70°	28-DIP,DIC	Max	-	
ICL 7135(C)PI,JI	CMOS-A/D-IC	4,5 Digit, A/D-Converter, 0...+70° µp-interface, MPX BCD Output	28-DIP,DIC	Max	-	pdf
ICL 7135(C)QI	CMOS-A/D-IC	=ICL 7135...PI,JI:	28-PLCC			
ICL 7135(C)QI	CMOS-A/D-IC	4,5 Digit, A/D-Converter, 0...+70° µP-interface, MPX BCD Output	28-PLCC	Max	-	pdf
ICL 7136	A/D-IC	3,5-Digit A/D-Converter	40-DIP	Sie	C 7136D	pdf
ICL 7136 CMH	CMOS-A/D-IC	3,5 Digit, A/D-Conv., Low Power, LCD Driver, 0...+70°	44-QFP	Max		pdf
ICL 7136(C)PL,JL	CMOS-A/D-IC	3,5 Digit,lo-power,LCD Driver, 0...+70°	40-DIP,DIC	Max	C 7136D, MAX 131...	
ICL 7136(C)PL,JL	CMOS-A/D-IC	3,5 Digit, A/D-Conv., Low Power, LCD Driver, 0...+70°	40-DIP,DIC	Max	C 7136D, MAX 131...	pdf
ICL 7136(C)QH	CMOS-A/D-IC	=ICL 7136...PL,JL:	44-PLCC			
ICL 7136(C)QH	CMOS-A/D-IC	3,5 Digit, A/D-Conv., Low Power, LCD Driver, 0...+70°	44-PLCC	Max		pdf
ICL 7137(C)PL,JL	CMOS-A/D-IC	3,5 Digit,lo-power,LED Driver, 0...+70°	40-DIP,DIC	Max	-	
ICL 7137(C)PL,JL	CMOS-A/D-IC	3,5 Digit, A/D-Conv., Low Power, LED Driver, 0...+70°	40-DIP,DIC	Max		pdf
ICL 7137(C)QH	CMOS-A/D-IC	=ICL 7137...PL,JL:	44-PLCC			
ICL 7137(C)QH	CMOS-A/D-IC	3,5 Digit, A/D-Conv., Low Power, LED Driver, 0...+70°	44-PLCC	Max		pdf
ICL7600	OP	Vs:±18V Vu:105dB Vo:±4.8V Vi0:5µµV	14D	Isi		data pinout
ICL7601	OP	Vs:±18V Vu:105dB Vo:±4.8V Vi0:5µµV	14D	Isi		data pinout
ICL 7605	OP	Vs:±18V Vu:105dB Vo:±4.8V Vi0:±2mV	18D	Isi		data pinout
ICL 7605 CJN	CMOS-OP-IC	Auto-Zero Amp., int.Comp.,18V, 0...+70°	18-DIC	Isi		data
ICL 7605 IJN	CMOS-OP-IC	=ICL 7605CJN: -25...+85°	18-DIC			data
ICL 7605 MJN	CMOS-OP-IC	=ICL 7605CJN: -55...+125°	18-DIC			data
ICL7606	OP	Vs:±18V Vu:105dB Vo:±4.8V Vi0:±2mV	18D	Isi		data pinout
ICL 7606 ...	CMOS-OP-IC	=ICL 7605...: externe Compensation	18-DIC	Isi		data
ICL7611A	LO-POWER	Vs:±9V Vu:104dB Vo:>±4.9V Vi0:<2mV	8DSA	Har		data pdf pinout
ICL7611B	LO-POWER	Vs:±9V Vu:104dB Vo:>±4.9V Vi0:<5mV	8DSA	Har		data pdf pinout
ICL 7611(A...E)CPA	CMOS-OP-IC	lo-power, 18(±9)V, Ucmr>±4,4V, 0...+70° Offs A:<2, B:<5, C:<10, D:<15, E:<20mV	8-DIP	Isi, Max		data
ICL7611D	LO-POWER	Vs:±9V Vu:83dB Vo:>±4.5V Vi0:<15mV	d,8SDA	Max		data pinout
ICL 7611 ... M...	CMOS-OP-IC	=ICL 7611 ... C...: -55...+125°			-	data
ICL 7611.... ...BA,SA	CMOS-OP-IC	=ICL 7611... ... PA: SMD	8-MDIP			data
ICL 7611 ... ... TV	CMOS-OP-IC	=ICL 7611... ... PA:	81			data
ICL 7611(A...E)CPA	CMOS-OP-IC	OP, Low Power, ±1...8V(+2...16)V,0...+70°	8-DIP	Max		data pdf pinout

469

Type	Device	Short Description	Fig.	Manu	Comparision Types	More at
		Offs A:<2, B:<5, C:<10, D:<15, E:<20mV				
ICL 7611(A...E)CSA	CMOS-OP-IC	SMD, OP, Low Power, ±1..8V(+2...16)V Offs A:<2, B:<5, C:<10, D:<15, E:<20mV	8-MDIP	Max		data pdf pinout
ICL 7611(A...E)CTV	CMOS-OP-IC	OP, Low Power, ±1..8V(+2...16)V,0..+70° Offs A:<2, B:<5, C:<10, D:<15, E:<20mV	81	Max	-	data pdf pinout
ICL 7611(A...E)MTV	CMOS-OP-IC	OP, Low Power, ±1..8V(+2...16)V Offs A:<2, B:<5, C:<10, D:<15, E:<20mV	81	Max		data pdf pinout
ICL 7612	CMOS-OP-IC	=ICL 7611.........: Ucmr>±5,3V		Isi, Max		data
ICL7612A	LO-POWER	Vs:±9V Vu:104dB Vo:>±4.9V Vi0:<2mV	8DSA	Har		data pdf pinout
ICL7612B	LO-POWER	Vs:±9V Vu:104dB Vo:>±4.9V Vi0:<5mV	8DSA	Har		data pdf pinout
ICL7612D	LO-POWER	Vs:±9V Vu:83dB Vo:>±4.5V Vi0:<15mV	d,8SDA	Max		data pinout
ICL 7612(A...E)CPA	CMOS-OP-IC	OP, Low Power, ±1..8V(+2...16)V,0..+70° Offs A:<2, B:<5, C:<10, D:<15, E:<20mV	8-DIP	Max		data pdf pinout
ICL 7612(A...E)CSA	CMOS-OP-IC	SMD, OP, Low Power, ±1..8V(+2...16)V Offs A:<2, B:<5, C:<10, D:<15, E:<20mV	8-MDIP	Max		data pdf pinout
ICL 7612(A...E)CTV	CMOS-OP-IC	OP, Low Power, ±1..8V(+2...16)V,0..+70° Offs A:<2, B:<5, C:<10, D:<15, E:<20mV	81	Max		data pdf pinout
ICL 7612(A...E)MTV	CMOS-OP-IC	OP, Low Power, ±1..8V(+2...16)V Offs A:<2, B:<5, C:<10, D:<15, E:<20mV	81	Max		data pdf pinout
ICL7613A	CMOS	Vs:±18V	8DA	Isi		data pinout
ICL7613B	CMOS	Vs:±18V	8DA	Isi		data pinout
ICL7613D	CMOS	Vs:±18V	8DA	Isi		data pinout
ICL 7614	CMOS-OP-IC	=ICL 7611.........: Extern. Compensated		Isi, Max	-	data
ICL7614A	LO-VOLT	Vs:±9V Vu:102dB Vo:>±4.9V Vi0:<2mV	8DSA	Isi,Max		data pdf pinout
ICL7614B	LO-VOLT	Vs:±9V Vu:102dB Vo:>±4.9V Vi0:<5mV	8DSA	Isi,Max		data pdf pinout
ICL7614C	LO-VOLT	Vs:±9V Vu:102dB Vo:>±4.9V Vi0:<15mV	8S	Max		data pdf pinout
ICL7614D	LO-VOLT	Vs:±9V Vu:102dB Vo:>±4.9V Vi0:<15mV	d,8DA	Max		data pinout
ICL 7614(A...E)CPA	CMOS-OP-IC	OP, Low Power, ±1..8V(+2...16)V,0..+70° Offs A:<2, B:<5, C:<10, D:<15, E:<20mV	8-DIP	Max		data pdf pinout
ICL 7614(A...E)CSA	CMOS-OP-IC	SMD, OP, Low Power, ±1..8V(+2...16)V Offs A:<2, B:<5, C:<10, D:<15, E:<20mV	8-MDIP	Max		data pdf pinout
ICL 7614(A...E)CTV	CMOS-OP-IC	OP, Low Power, ±1..8V(+2...16)V,0..+70° Offs A:<2, B:<5, C:<10, D:<15, E:<20mV	81	Max		data pdf pinout
ICL 7614(A...E)MTV	CMOS-OP-IC	OP, Low Power, ±1..8V(+2...16)V Offs A:<2, B:<5, C:<10, D:<15, E:<20mV	81	Max	-	data pdf pinout
ICL7615A	CMOS	Vs:±9V f:0.48Mc	8DA	Isi		data pinout
ICL7615B	CMOS	Vs:±9V f:0.48Mc	8DA	Isi		data pinout
ICL7615D	CMOS	Vs:±9V f:0.48Mc	8DA	Isi		data pinout
ICL 7611	CMOS-OP-IC	=ICL 7611.........: Ucmr>±5,3V		Isi, Max		data
ICL7616A	LO-VOLT	Vs:±9V Vu:83dB Vo:>±4.5V Vi0:<2mV	8DSA	Isi,Max		data pdf pinout
ICL7616B	LO-VOLT	Vs:±9V Vu:83dB Vo:>±4.5V Vi0:<5mV	8DSA	Isi,Max		data pdf pinout
ICL7616D	LO-VOLT	Vs:±9V Vu:83dB Vo:>±4.5V Vi0:<15mV	d,8DSA	Max		data pinout
ICL 7616(A...E)CPA	CMOS-OP-IC	OP, Low Power, ±1..8V(+2...16)V,0..+70° Offs A:<2, B:<5, C:<10, D:<15, E:<20mV	8-DIP	Max		data pdf pinout
ICL 7616(A...E)CSA	CMOS-OP-IC	SMD, OP, Low Power, ±1..8V(+2...16)V Offs A:<2, B:<5, C:<10, D:<15, E:<20mV	8-MDIP	Max		data pdf pinout
ICL 7616(A...E)CTV	CMOS-OP-IC	OP, Low Power, ±1..8V(+2...16)V,0..+70° Offs A:<2, B:<5, C:<10, D:<15, E:<20mV	81	Max	-	data pdf pinout
ICL 7616(A...E)MTV	CMOS-OP-IC	OP, Low Power, ±1..8V(+2...16)V Offs A:<2, B:<5, C:<10, D:<15, E:<20mV	81	Max		data pdf pinout
ICL 7621	CMOS-OP-IC	=ICL 7611.........: Dual		Isi, Max		data
ICL7621	2xLO-POWER	Vs:±9V Vu:104dB Vo:>±4.9V Vi0:<2mV	8S	Har		data pinout
ICL7621A	2xLO-POWER	Vs:±9V Vu:104dB Vo:>±4.9V Vi0:<2mV	8DSA	Har		data pdf pinout
ICL7621B	2xLO-POWER	Vs:±9V Vu:104dB Vo:>±4.9V Vi0:<5mV	8DSA	Har		data pdf pinout
ICL7621D	2xLO-VOLT	Vs:±9V Vu:102dB Vo:>±4.9V Vi0:<15mV	d,8DSA	Max		data pinout
ICL7621DCPA	2xLO-POWER	Vs:±9V Vu:104dB Vo:>±4.9V Vi0:<15mV	8D	Har		data pdf pinout
ICL 7621(A.,E)CPA	CMOS-OP-IC	2xOP,Low Power,±1..8V(+2...16)V,0..+70° Offs A:<2, B:<5, C:<10, D:<15, E:<20mV	8-DIP	Max		data pdf pinout
ICL 7621(A...E)CSA	CMOS-OP-IC	SMD, 2xOP, Low Power,±1..8V(+2...16)V Offs A:<2, B:<5, C:<10, D:<15, E:<20mV	8-MDIP	Max	-	data pdf pinout
ICL 7621(A...E)CTV	CMOS-OP-IC	2xOP, Low Power, ±1..8V(+2...16)V Offs A:<2, B:<5, C:<10, D:<15, E:<20mV	81	Max		data pdf pinout
ICL 7621(A...E)MTV	CMOS-OP-IC	2xOP, Low Power,±1..8V(+2...16)V Offs A:<2, B:<5, C:<10, D:<15, E:<20mV	81	Max	-	data pdf pinout
ICL 7622	CMOS-OP-IC	=ICL 7612.........: Dual		Isi, Max		data
ICL7622A	2xLO-VOLT	Vs:±9V Vu:102dB Vo:>±4.9V Vi0:<2mV	14DS	Isi,Max		data pdf pinout
ICL7622B	2xLO-VOLT	Vs:±9V Vu:102dB Vo:>±4.9V Vi0:<5mV	14DS	Isi,Max		data pdf pinout
ICL7622D	2xLO-VOLT	Vs:±9V Vu:102dB Vo:>±4.9V Vi0:<15mV	d,14DS	Max		data pinout
ICL 7622(A...E)CJD	CMOS-OP-IC	2xOP, Low Power,±1..8V(+2...16)V Offs A:<2, B:<5, C:<10, D:<15, E:<20mV	14-DIC	Max	-	data pdf pinout
ICL 7622(A...E)CPD	CMOS-OP-IC	2xOP, Low Power,±1..8V(+2...16)V Offs A:<2, B:<5, C:<10, D:<15, E:<20mV	14-DIP	Max		data pdf pinout
ICL 7622(A...E)CSA	CMOS-OP-IC	SMD, 2xOP, Low Power,±1..8V(+2...16)V Offs A:<2, B:<5, C:<10, D:<15, E:<20mV	14-MDIP	Max	-	data pdf pinout
ICL 7622(A...E)CSD	CMOS-OP-IC	SMD, 2xOP, Low Power,±1..8V(+2...16)V Offs A:<2, B:<5, C:<10, D:<15, E:<20mV	14-MDIP	Max		data pdf pinout
ICL 7622(A...E)MJD	CMOS-OP-IC	2xOP, Low Power,±1..8V(+2...16)V Offs A:<2, B:<5, C:<10, D:<15, E:<20mV	14-DIC	Max	-	data pdf pinout
ICL7631B	3xLO-VOLT	Vs:±9V Vu:98dB Vo:>±4.5V Vi0:<5mV	16DS	Isi,Max		data pdf pinout
ICL7631C	3xLO-VOLT	Vs:±9V Vu:98dB Vo:>±4.5V Vi0:<10mV	16DS	Isi,Max		data pdf pinout
ICL7631EC	3xLO-VOLT	Vs:±9V Vu:98dB Vo:>±4.5V Vi0:<20mV	d,16DS	Max		data pinout
ICL 7631......PE,JE	CMOS-OP-IC	=ICL 7611.........: Triple	16-DIP,DIC	Isi, Max	-	data
ICL 7631(A...E)CPE	CMOS-OP-IC	3xOP, Low Power, ±1..8V(+2...16)V Offs A:<2, B:<5, C:<10, D:<15, E:<20mV	16-DIP	Max		data pdf pinout
ICL7632B	3xLO-VOLT	Vs:±9V Vu:98dB Vo:>±4.5V Vi0:<5mV	16DS	Isi,Max		data pdf pinout

Type	Device	Short Description	Fig.	Manu	Comparision Types	More at
CL7632C	3xLO-VOLT	Vs:±9V Vu:98dB Vo:>±4.5V Vi0:<5mV	16DS	Isi,Max		data pdf pinout
CL7632E	3xLO-VOLT	Vs:±9V Vu:98dB Vo:>±4.5V Vi0:<5mV	d,16DS	Max		data pdf pinout
CL 7632... ...PE,JE	CMOS-OP-IC	=ICL 7612... ... .: Triple	16-DIP,DIC	Isi, Max	-	data
CL 7632(A...E)CPE	CMOS-OP-IC	3xOP, Low Power, ±1...8V(+2...16)V Offs A:<2, B:<5, C:<10, D:<15, E:<20mV	16-DIP	Max	-	data pdf pinout
CL7641B	4xLO-VOLT	Vs:±9V Vu:98dB Vo:>±4.5V Vi0:<5mV	14D,16S	Isi,Max		data pdf pinout
CL7641C	4xLO-VOLT	Vs:±9V Vu:98dB Vo:>±4.5V Vi0:<10mV	14D,16S	Isi,Max		data pdf pinout
CL7641E	4xLO-VOLT	Vs:±9V Vu:98dB Vo:>±4.5V Vi0:<20mV	d,14D,16S	Max		data pdf pinout
CL 7641... ...PD,JD	CMOS-OP-IC	=ICL 7611... ... .: Quad	14-DIP,DIC	Isi, Max	-	data
CL 7641(A...E)CJD	CMOS-OP-IC	4xOP, Low Power, ±1...8V(+2...16)V Offs A:<2, B:<5, C:<10, D:<15, E:<20mV	14-DIC	Max	-	data pdf pinout
CL 7641(A...E)CPD	CMOS-OP-IC	4xOP, Low Power, ±1...8V(+2...16)V Offs A:<2, B:<5, C:<10, D:<15, E:<20mV	14-DIP	Max	-	data pdf pinout
CL 7641(A...E)CWE	CMOS-OP-IC	SMD, 4xOP, Low Power, ±1...8V(+2...16)V Offs A:<2, B:<5, C:<10, D:<15, E:<20mV	16-MDIP	Max	-	data pdf pinout
CL 7641(A...E)MJD	CMOS-OP-IC	4xOP, Low Power, ±1...8V(+2...16)V Offs A:<2, B:<5, C:<10, D:<15, E:<20mV	14-DIC	Max	-	data pdf pinout
CL7642B	4xLO-VOLT	Vs:±9V Vu:104dB Vo:>±4.9V Vi0:<5mV	14D,16S	Isi,Max		data pdf pinout
CL7642C	4xLO-VOLT	Vs:±9V Vu:104dB Vo:>±4.9V Vi0:<10mV	14D,16S	Isi,Max		data pdf pinout
CL7642E	4xLO-VOLT	Vs:±9V Vu:104dB Vo:>±4.9V Vi0:<20mV	d,14D,16S	Max		data pinout
CL 7642... ...PD,JD	CMOS-OP-IC	=ICL 7612... ... .: Quad	14-DIP,DIC	Isi, Max		data
CL 7642(A...E)CJD	CMOS-OP-IC	4xOP, Low Power, ±1...8V(+2...16)V Offs A:<2, B:<5, C:<10, D:<15, E:<20mV	14-DIC	Max		data pdf pinout
CL 7642(A...E)CPD	CMOS-OP-IC	4xOP, Low Power, ±1...8V(+2...16)V Offs A:<2, B:<5, C:<10, D:<15, E:<20mV	14-DIP	Max		data pdf pinout
CL 7642(A...E)CWE	CMOS-OP-IC	SMD, 4xOP, Low Power, ±1...8V(+2...16)V Offs A:<2, B:<5, C:<10, D:<15, E:<20mV	16-MDIP	Max		data pdf pinout
CL 7642(A...E)MJD	CMOS-OP-IC	4xOP, Low Power, ±1...8V(+2...16)V Offs A:<2, B:<5, C:<10, D:<15, E:<20mV	14-DIC	Max		data pdf pinout
CL7650	CHOPPER	Vs:±9V Vu:120dB Vo:±4.85V Vi0:0.7mV	8A,14D	Isi		data pinout
CL7650B	CHOPPER	Vs:18V Vo:±4.85V Vi0:±1mV f:2Mc	d,8D,14DSS	Max		data pinout
CL7650BCTV1	CHOPPER	Vs:±9V Vu:120dB Vo:±4.85V Vi0:1mV	8A	Isi		data pinout
CL7650C	CHOPPER	Vs:18V Vo:±4.85V Vi0:0.7mV f:2Mc	d,14D,8SS	Max		data pinout
CL 7650(B,S)CPA-1	CMOS-OP-IC	Chopper-Stabilized, 18(±9)V, 0...+70°	8-DIP	Isi, Max		data
CL7650CPA-1	CHOPPER	Vs:18V Vo:±4.85V Vi0:0.7mV f:2Mc	8D	Max		data pinout
CL 7650 SCPA-1	CMOS-OP-IC	Chopper-Stabilized, 18(±9)V, 0...+70°	8-DIP	Isi		data pinout
CL 7650...CPD,CJD	CMOS-OP-IC	=ICL 7650...CPA-1:	14-DIP,DIC		U 7650 DD	data
CL 7650 SCPD,CJD	CMOS-OP-IC	=ICL 7650SCPA-1: Fig.→	14-DIP,DIC		U 7650 DD	data pinout
CL 7650...CSA-1	CMOS-OP-IC	=ICL 7650...CPA-1: SMD	8-MDIP		-	
CL 7650...CSD	CMOS-OP-IC	=ICL 7650...CPA-1: SMD	14-MDIP			
CL 7650...CTV-1	CMOS-OP-IC	=ICL 7650...CPA-1:	81		-	data
CL7650CTV-1	CHOPPER	Vs:18V Vo:±4.85V Vi0:0.7mV f:2Mc	8A	Max		data pinout
CL 7650 SCTV-1	CMOS-OP-IC	=ICL 7650SCPA-1: Fig.→	81		-	data pinout
CL 7650 ...I...	CMOS-OP-IC	=ICL 7650...C...: -25...+85°			-	data
CL7650I	CHOPPER	Vs:18V Vo:±4.85V Vi0:0.7mV f:2Mc	14D	Max		data pdf pinout
CL 7650 SI ...	CMOS-OP-IC	=ICL 7650SC...: -25...+85°				data pinout
CL7650IJA-1	CHOPPER	Vs:18V Vo:±4.85V Vi0:0.7mV f:2Mc	8D	Max		data pinout
CL7650IJD-1	CHOPPER	Vs:±9V Vu:120dB Vo:±4.85V Vi0:0.7mV	14D	Isi		data pinout
CL 7650 ...M ...	CMOS-OP-IC	=ICL 7650...M...: -55...+125°			·	data
CL7650M	CHOPPER	Vs:18V Vo:±4.85V Vi0:0.7mV f:2Mc	14D	Max		data pdf pinout
CL 7650 SM ...	CMOS-OP-IC	=ICL 7650SM...: -55...+125°				data pinout
CL7650MTV-1	CHOPPER	Vs:18V Vo:±4.85V Vi0:0.7mV f:2Mc	8A	Max		data pinout
CL7650S	CHOPPER	Vs:±9V Vu:150dB Vo:±4.85V Vi0:0.7mV	8S,14SDDA	Har		data pinout
CL7652B	CHOPPER	Vs:±9V Vu:150dB Vo:±4.85V Vi0:2µµV	8D,14DA	Isi		data pinout
CL7652C	CHOPPER	Vs:18V Vu:150dB Vo:±4.85V f:0.45Mc	d,14D,8A	Max		data pinout
CL7652CPA	CHOPPER	Vs:18V Vu:150dB Vo:±4.85V f:0.45Mc	8D	Max		data pdf pinout
CL7652(B,S)CPD,JD	CMOS-OP-IC	Chopper-Stabilized, 18(±9)V, 0...+70°	14-DIP,DIC	Isi, Max	-	data
CL 7650 SCPD	CMOS-OP-IC	Chopper-Stabilized, 18(±9)V, 0...+70°	14-DIP	Isi		data pinout
CL7652I	CHOPPER	Vs:18V Vu:150dB Vo:±4.85V f:0.45Mc	14D,8A	Max		data pdf pinout
CL7652IJA	CHOPPER	Vs:18V Vu:150dB Vo:±4.85V f:0.45Mc	8D	Max		data pdf pinout
CL 7652(B,S)IPD	CMOS-OP-IC	=ICL 7652...CPD,CJD: -25...+85°	14-DIP		-	data
CL 7650 SiPD	CMOS-OP-IC	=ICL 7652SCPD: -25...+85°	14-DIP			data pinout
CL7652IT	CHOPPER	Vs:18V Vu:150dB Vo:±4.85V f:0.45Mc	8A	Max		data pdf pinout
CL 7652... ...PA,JA	CMOS-OP-IC	=ICL 7652...CPD,CJD:	8-DIP,DIC		-	data
CL 7652... ...TV	CMOS-OP-IC	=ICL 7652...CPD,CJD:	81		-	data
CL 7652... ...WE	CMOS-OP-IC	=ICL 7652...CPD,CJD: SMD	8-MDIP		-	data
CL 7660 AMTV	CMOS-IC	Voltage Converter, +5V→±5V, 0...+70°	81	Max	ICL 7660S..., U 7660 DC	pdf pinout
CL 7660 CBA	CMOS-IC	=ICL 7660CPA: SMD	8-MDIP		ICL 7660S...	pdf
CL 7660 CPA,CJA	CMOS-IC	Voltage Converter, +5V→±5V, 0...+70°	8-DIP,DIC	Isi, Max	ICL 7660S..., U 7660 DC	
CL 7660 (C/E)PA	CMOS-IC	Voltage Converter, +5V→±5V, 0...+70°	8-DIP	Max	ICL 7660S..., U 7660 DC	pdf pinout
CL 7660 (C/E)SA	CMOS-IC	SMD,Voltage Converter, +5V→±5V	8-MDIP	Max	ICL 7660S...	pdf pinout
CL 7660 CTV	CMOS-IC	=ICL 7660CPA: Fig.→	81		ICL 7660S...	
CL 7660 MTV	CMOS-IC	=ICL 7660CPA: -55...+125°, Fig.→	81		ICL 7660S...	pdf
CL 7660 S...	CMOS-IC	=ICL 7660...: verbessert/improved			U 7660 DG	
CL 7660 SI...	CMOS-IC	=ICL 7660S...: -25...+85°				
CL 7662 CPA	CMOS-IC	Voltage Converter, +15V→-15V, 0...+70°	8-DIP	Isi	-	pdf
CL 7662 CTV	CMOS-IC	=ICL 7662CPA: Fig.→	81			
CL 7662 MTV	CMOS-IC	=ICL 7662CPA: -55...+125°, Fig.→	81			
CL7663A	ER+	Io=40mA Vo:1...16V P:300mW	8DSA	Max		data pdf pinout
CL7663BC	ER+	Io=40mA Vo:1...16V	8SD	Rca		data pinout
CL7663BC/D	ER+	Io=40mA Vo:1.5...10V Vin:9V P:200mW		Max		data pinout
CL7663BCPA	ER+	Io=40mA Vo:1...16V P:200mW	8D	Rca		data pdf pinout
CL7663BCTV	ER+	Io=40mA Vo:1...16V Vin:9V P:300mW	8A	Rca		data pinout
CL7663BI	ER+	Io=40mA Vo:1.5...10V Vin:9V P:300mW	8DA	Max		data pdf pinout
CL7663C	ER+	Io=40mA Vo:1...16V	8SD	Rca		data pinout

471

Type	Device	Short Description	Fig.	Manu	Comparision Types	More at
ICL7663C/D	ER+	Io=40mA Vo:1.5...16V Vin:9V P:200mW		Max		data pinout
ICL7663CPA	ER+	Io=40mA Vo:1...16V P:200mW	8D	Rca		data pdf pinout
ICL 7663 CPA,CJA	CMOS-IC	1,3...16V, 0,04A,Uref:1,2...1,4V,0...+70°	8-DIP,DIC	Isi, Max	MAX 663...	data
ICL7663CSA	ER+	Io=40mA Vo:1.5...16V Vin:9V P:200mW	8S	Max		data pdf pinout
ICL 7663 CBA,CSA	CMOS-IC	=ICL 7663CPA: SMD	8-MDIP			data
ICL7663CTV	ER+	Io=40mA Vo:1...16V P:300mW	8A	Rca		data
ICL 7663 CTV	CMOS-IC	=ICL 7663CPA:	81			data
ICL7663I	ER+	Io=40mA Vo:1.5...16V Vin:9V P:300mW	8DA	Max		data pdf pinout
ICL7663S	ER+	Io=40mA Vo:1.3...16V P:300mW	8DAS	Rca		data pinout
ICL 7663 S...	CMOS-Z-IC	Uref:1,275...1,305V:verbessert/improved				
ICL 7663 (S)A...	CMOS-IC	=ICL 7663(S)...: Uref:1,275...1,305V				data
ICL 7663 SA...	CMOS-Z-IC	=ICL 7663S...: Uref=1,275...1,305V				data pinout
ICL 7663 SCBA	CMOS-Z-IC	=ICL 7663SCPA: SMD	8-MDIP			data pinout
ICL 7663 SCPA,SCJA	CMOS-Z-IC	1,3...16V, 0,04A,Uref=1,2...1,4V,0...+70°	8-DIP,DIC	Isi		data pinout
ICL 7663 S(A)I...	CMOS-Z-IC	=ICL 7663S(A)...: -25...+85°				data pinout
ICL7664	ER-	Io=25mA Vo:-1.3...-16V Vin:-9V	8DSA	Max		data
ICL 7664 CPA,CJA	CMOS-IC	-1,3ō-16V, 25mA, Uref=1,261,4V,0...+70°	8-DIP,DIC	Isi, Max		data
ICL 7664 CSA	CMOS-IC	=ICL 7664CPA,CJA: SMD	8-MDIP			data
ICL 7664 CTV	CMOS-IC	=ICL 7664CPA,CJA:	81			
ICL 7665 CPA,CJA	CMOS-IC	Voltage Detector, 1,15...1,45V, 0...+70°	8-DIP,DIC	Isi, Max		
ICL 7665 CBA,CSA	CMOS-IC	SMD,Voltage Detector, 1,15...1,45V	8-MDIP			
ICL 7665 CTV	CMOS-IC	Voltage Detector, 1,15...1,45V, 0...+70°	81			pdf
ICL 7665 S...	CMOS-Z-IC	Voltage Detector: verbessert/improved				
ICL 7665 SA...	CMOS-IC	=ICL 7665S...: Utrip=1,275...1,325V				pdf
ICL 7665 SCBA	CMOS-IC	=ICL 7665SCPA: SMD	8-MDIP			
ICL 7665 SCPA,SCJA	CMOS-IC	Voltage Detector, 1,15...1,45V, 0...+70°	8-DIP,DIC	Isi		pdf
ICL 7665 S(A)I...	CMOS-IC	=ICL 7665S(A)...: -25...+85°				
ICL 7667 CBA	CMOS-IC	=ICL 7667CPA: SMD	8-MDIP			pdf
ICL 7667 CPA,CJA	CMOS-IC	2x MOS-FET Driver, Ucc=4,5...15V	8-DIP,DIC	Har, Isi		
ICL 7667 CTV	CMOS-IC	=ICL 7667CPA: Fig.→	81			pdf
ICL 7667 MJA,MTV	CMOS-IC	=ICL 7667CPA: -55...+125°				pdf
ICL 7673 CBA	CMOS-IC	=ICL 7673CPA: SMD	8-MDIP			pdf
ICL 7673 CPA	CMOS-IC	Autom. Battery Backup Switch, 0...+70°	8-DIP	Isi		
ICL 7675 CPA,CJA	SMPS-IC	SMPS Controller(+ICL 7676), 0...+70°	8-DIP,DIC	Isi		
ICL 7675 IPA	SMPS-IC	=ICL 7675CPA: -25...+85°	8-DIP			
ICL 7675 MPA	SMPS-IC	=ICL 7675CPA: -55...+125°	8-DIC			
ICL 7676 CPA	SMPS-IC	SMPS Controller(+ICL 7675), 0...+70°	8-DIP	Isi		
ICL 7676 MPA	SMPS-IC	=ICL 7676CPA: -55...+125°	8-DIC			
ICL 7680 CPE	SMPS-IC	2xS-Reg, +15V/0,1A, -15V/50mA, 0...+70°	16-DIP	Isi		
ICL 7680 IDE,IPE	SMPS-IC	=ICL 7680CPE: -25...+85°	16-DIC,DIP			
ICL 7680 MDE	SMPS-IC	=ICL 7680CPE: -55...+125°	16-DIC			
ICL8001C	COMP	Vu:84dB Vo:7/0.2V Vi0:6mV	10A	Isi		data pinout
ICL8001M	COMP	Vu:84dB Vo:7/0.2V Vi0:4mV	10A	Isi		data pinout
ICL8007A	FET	Vs:±18V Vo:±13V Vi0:15mV f:1Mc	8A	Isi		data pinout
ICL8007C	UNI	Vs:±15V Vu:94dB Vo:±12V Vi0:2mV	8A	Isi		data pinout
ICL8007C3	UNI	Vs:±15V Vu:94dB Vo:±12V Vi0:4mV	8A	Isi		data pinout
ICL8007M	FET	Vs:±18V Vo:±13V Vi0:20mV f:1Mc	8A	Isi		data pinout
ICL8007M2	UNI	Vs:±15V Vu:94dB Vo:±12V Vi0:3.5mV	8A	Isi		data pinout
ICL8007M5	UNI	Vs:±15V Vu:94dB Vo:±12V Vi0:11mV	8A	Isi		data pinout
ICL8007MTA	HI-OHM	Vs:±15V Vu:93dB Vo:±12V Vi0:30mV	8A	Isi		data pinout
ICL8008C	UNI	Vs:±15V Vu:86dB Vo:±12V Vi0:7.5mV	8DA	Isi		data pinout
ICL8008M	UNI	Vs:±15V Vu:86dB Vo:±12V Vi0:6mV	8A	Isi		data pinout
ICL 8013(A,B,C)CTX	LIN-IC	1MHz, 4-Quadrant Analog Multiplier, ±18V, 0...+70°	82	Isi		pdf pinout
ICL 8013(A,B,C)MTX	LIN-IC	=ICL 8013(..)CTX: -55...+125°	82			
ICL8017C	HI-SPEED	Vs:±15V Vu:87dB Vo:±10V Vi0:7.5mV	10A	Isi		data pinout
ICL8017M	HI-SPEED	Vs:±15V Vu:87dB Vo:±10V Vi0:6mV	10A	Isi		data pinout
ICL8021C	LO-VOLT	Vs:±6V Vu:96dB Vi0:7.5mV f:0.25Mc	8A	Isi		data
ICL 8021 CBA	OP-IC	=ICL 8021CJA,CPA: SMD	8-MDIP			data
ICL 8021 CJA,CPA	OP-IC	Io-power, ±18V, 0,27MHz, 0...+70°	8-DIC,DIP	Isi	→Serie 741	data
ICL 8021 CTY	OP-IC	=ICL 8021CJA,CPA: Fig.→	81			data pinout
ICL8021M	LO-VOLT	Vs:±6V Vu:100dB Vi0:4mV f:0.3Mc	8A	Isi		data pinout
ICL 8021 M ...	OP-IC	=ICL 8021C...: -55...+125°				data pinout
ICL8022C	3xOP	Vs:±6V Vu:96dB Vi0:7.5mV f:0.25Mc	14D	Isi		data pinout
ICL8022M	3xOP	Vs:±6V Vu:100dB Vi0:4mV f:0.3Mc	14D	Isi		data pinout
ICL 8023 ... JE	OP-IC	=ICL 8021...: Triple	16-DIC	Isi		data
ICL8023M	3xOP	Vs:±18V Vu:200V/mV Vo:±13V Vi0:2mV	16D	Isi		data pinout
ICL 8023 ... PE	OP-IC	=ICL 8021...: Triple, SMD	16-MDIC	Isi		data
ICL 8038(A,B,C)CPD	LIN-IC	Waveform Gen., VCO, 36V, 0...+70° Drift: A: 120, B: 180, C: 250ppm/°C	14-DIP	Isi		pdf pinout
ICL 8038(A,B)MJD	LIN-IC	Waveform Gen., VCO, 36V, -55...+125° Drift: A:<250ppm/°C, B: <350ppm/°C	14-DIC	Har		pdf pinout
ICL8043C	HI-OHM	Vs:±15V Vu:86dB Vo:±12V Vi0:60mV		Isi		data pinout
ICL8043M	HI-OHM	Vs:±15V Vu:100dB Vo:±12V Vi0:30mV		Isi		data pinout
ICL 8048(B,C)CJE	LIN-IC	Log Amp., Dual JFET Inp, ±18V, 0...+70° Error: B: <30mV, C: <60mV	16-DIC	Isi		
ICL 8049(B,C)CJE	LIN-IC	=ICL 8048C...: Antilog Amp. Error: B: <15mV, C: <25mV	16-DIC	Isi		
ICL8069	Z+	Io=10mA Vo:1.20...1.25V P:int	2A,3N	Rca		data pdf pinou
ICL 8069 ... SQ2	Ref-Z-IC	Prec. Voltage Ref., +1,2V, <0,01%/°C	2d	Max		data
ICL 8069 ... ZQ2	Ref-Z-IC	=ICL 8069SQ2:	7d	Max		data
ICL 8211 CBA	LIN-IC	=ICL 8211CPA: SMD	8-MDIP			pdf
ICL 8211 CPA	LIN-IC	Voltage Detector, Utrip=1,15V, 0...+70°	8-DIP	Isi		pdf
ICL 8211 CTY	LIN-IC	=ICL 8211CPA: Fig.→	81			pdf
ICL 8211 MTY	LIN-IC	=ICL 8211CPA: -55...+125°, Fig.→	81			pdf

Type	Device	Short Description	Fig.	Manu	Comparision Types	More at
CL 8212 CBA	LIN-IC	=ICL 8212CPA: SMD	8-MDIP		-	pdf
CL 8212 CPA	LIN-IC	Voltage Detector, Utrip=1,15V, 0...+70°	8-DIP	Isi	-	pdf
CL 8212 CTY	LIN-IC	=ICL 8212CPA: Fig.→	81		-	pdf
CL 8212 MTY	LIN-IC	=ICL 8212CPA: -55..+125°, Fig.→	81		-	pdf
CM 7207 AIPD	CMOS-Logic	Cmos Timebase generator 5,24288 MHz	DIP	Isi	-	pdf pinout
CM 7207 IPD	CMOS-Logic	Cmos Timebase generator 6,5536 MHz	DIP	Isi	-	pdf pinout
CM 7208 IPI	CMOS-Logic	Cmos 7 decade counter,decoder	DIP	Isi	-	pdf pinout
CM 7211(A,M,AM)	CMOS-IC	8 Digit LCD Decoder/Drv, 4 Bit Input A: Code B Output, M: µP Interface	40-DIP	Isi, Max	-	pdf
CM 7212(A,M,AM)	CMOS-IC	8 Digit LED Decoder/Drv, 4 Bit Input A: Code B Output, M: µP Interface	40-DIP	Isi, Max	-	pdf
CM 7216(A,B,D)	CMOS-IC	8-Digit, Freq.Counter/Timer, f<10MHz 7-Segm.& DP LED, com. Cathode, +5V	28-DIP	Isi	-	pdf
CM 7217(A,B,C)	CMOS-IC	4Digit LED Presettable Up/Down Counter ICM 7217,B: Common Anode, A,C: C.Cath. ICM 7217,A: max 9999, B,C: max 5959	28-DIP	Isi, Max	-	pdf
CM 7218(A..D)PI	CMOS-IC	8 Digit Multiplexed LED Display Driver A,C: Common Anode, B,D: Common Cathode A,B: Serial Input, C,D: Parallel Input	28-DIP	Isi, Max	-	pdf
CM 7218(A..D)Q	CMOS-IC	SMD,8 Digit Multiplexed LED Display Dr	28-MP		-	pdf
CM 7224(C)PL	CMOS-IC	4 1/2 Digit LCD Counter/Decoder/Driver -25...+85°, C: 0...+70°	40-DIP	Isi, Max	-	pdf
CM 7224(C)Q	CMOS-IC	4 1/2 Digit LCD Counter/Decoder/Drive	44-PLCC		-	pdf
CM 7225(C)PL	CMOS-IC	4 1/2 Digit LED Counter/Decoder/Driver	40-DIP	Isi, Max	-	pdf
CM 7225(C)Q	CMOS-IC	4 1/2 Digit LED Counter/Decoder/Driver	44-PLCC		-	pdf
CM 7228 ...	CMOS-IC	8 Digit Multiplexed LED Display Driver		Isi	ICM 7218...	pdf
CM 7228(A..D)PI	CMOS-IC	8 Digit Multiplexed LED Display Driver A,C: Common Anode, B,D: Common Cathode A,B: Serial Input, C,D: Parallel Input	28-DIP	Isi	-	pdf
CM 7240 IPE,IJE	CMOS-IC	Progr. Timer/Counter, 18V, -25...+85°	16-DIP,DIC	Isi, Max	-	pdf
CM 7242 IPA,IJA	CMOS-IC	Long-Range Timer, 18V, -25...+85°	8-DIP,DIC	Isi, Max	-	pdf
CM 7242 CBA	CMOS-IC	=ICM 7242IPA: SMD, 0...+70°	8-MDIP		-	pdf
CM 7242 IPA	CMOS-IC	Long-Range Timer, 18V, -25...+85°	8-DIP	Isi	-	pdf
CM 7243 BIPL	IC	8-Character,LED Display Decoder Driver	DIP	Isi	-	pdf pinout
CM 7243 BIPLZ	IC	8-Character,LED Display Decoder Driver	DIP	Isi	-	pdf pinout
CM 7250 IPE,IJE	CMOS-IC	Progr. Timer/Counter, 18V, -25...+85°	16-DIP,DIC	Isi, Max	-	pdf
CM 7260 IPE,IJE	CMOS-IC	Progr. Timer/Counter, 18V, -25...+85°	16-DIP,DIC	Isi, Max	-	pdf
CM 7555 CBA,CD	CMOS-IC	=ICM 7555CFE,CN: SMD	8-MDIP	Isi, Phi	-	pdf pinout
CM 7555 CFE,CN	CMOS-IC	Timer, Uni, +18V, Iq=100mA, 0...+70°	8-DIC,DIP	Phi	KS 555N	pdf pinout
CM 7555 ID,ISA	CMOS-IC	SMD,Timer, Uni, +18V, Iq=100mA	8-MDIP		-	pdf
CM 7555 ID	CMOS-IC	=ICM 7555CFE,CN: SMD, -40...+85°	8-MDIP	Phi	-	pdf pinout
CM 7555IPA,IFE,IN	CMOS-IC	=ICM 7555CFE,CN: -40...85°	8-DIP	Isi, Phi	-	pdf pinout
CM 7555 ITV	CMOS-IC	=ICM 7555CFE,CN: -25...+85°	81	Isi	-	pdf
CM 7555 MFE,MN	CMOS-IC	=ICM 7555CFE,CN: -55...+125°	8-DIC,DIP	Phi	-	pdf pinout
CM 7555 MTV	CMOS-IC	=ICM 7555CFE,CN: -55...+125°	81	Isi	-	pdf
ICM 7556 IPD	CMOS-IC	=ICM 7555CFE,CN: Dual, -25...+85°	14-DIP	Isi	KS 556N	pdf pinout
ICM 7556 ISD	CMOS-IC	SMD,Timer, Uni, +18V, Iq=100mA	16-MDIP		-	pdf
ICM 7556 MJD	CMOS-IC	=ICM 7555CFE,CN: Dual, -55...+125°	14-DIC	Isi	-	pdf
ICM 7243 AIM 44 Z	IC	8-Character,LED Display Decoder Driver	DIP	Isi	-	pdf pinout
ICM 7243 AIM 44 ZT	IC	8-Character,LED Display Decoder Driver	DIP	Isi	-	pdf pinout
ICPL2502	OC	LED/PD+NPN Viso:3000V CTR:>15%		Iso		data pinout
ICPL2503	OC	LED/PD+NPN Viso:3000V CTR:>15%		Iso		data pinout
ICPL2530	DOC	LED/PD+NPN CTR:>7% Ibs:16mA		Iso		data pdf pinout
ICPL2531	DOC	LED/PD+NPN Viso:3000V CTR:>19%		Iso		data pdf pinout
ICPL2533	DOC	LED/PD+NPN Viso:3000V CTR:>12%		Iso		data pdf pinout
ICPL2601	OC	LED/PD+IC Viso:3000V CTR:700%		Iso		data pdf pinout
ICPL2602	OC	LED/PD+IC Viso:3000V CTR:700%		Iso		data pinout
ICPL2630	DOC	LED/PD+IC Viso:3000V CTR:700%		Iso		data pdf pinout
ICPL2631	DOC	LED/PD+IC Viso:3000V CTR:700%		Iso		data pdf pinout
ICPL2730	DOC	LED/PD+Dar Viso:3000V CTR:>300%		Iso		data pdf pinout
ICPL2731	DOC	LED/PD+Dar Viso:3000V CTR:>400%		Iso		data pdf pinout
ICPL3700	OC	LED/PD+Dar Viso:3000V Ibe:4.2mA		Iso		data pinout
ICS 9107 C-21	LIN-IC	Frequency Generator for Fibre Channel Systems, 106.25 MHz clock	DIC	Ics	AV9107C-19	pdf pinout
ICS 9107 C-22	LIN-IC	Frequency Generator for Fibre Channel Systems, 106.25 MHz clock	DIC	Ics	AV9107C-20	pdf pinout
ICS 9154 A-39	LIN-IC	Low Cost 16-Pin Frequency Generator	DIC	Ics	-	pdf pinout
ICS 307 G-03	CMOS-IC	Serially Programmable Clock source	TSOP	Idt	-	pdf pinout
ICS 307 G-03 T	CMOS-IC	Serially Programmable Clock source	TSOP	Idt	-	pdf pinout
ICTE-5...45	Z-Di	TAZ, 5...45V, 1,5kW (=1N6373...6381)	31a	Mot	1N5629...5649, 1N6267...6387	data
ICTE-5C...-45C	Z-Di	=ICTE-5...-45: bidirektional	31a	Mot	1N6036...55, 1N6138...58	data
ICX 098 BQ	CCD-IC	Diagonal 4.5mm (Type 1/4) Progressive Scan Ccd Image Sensor	DIP	Son	-	pdf pinout
ICX 205 AK	CCD-IC	Diagonal 8mm Progressive Scan Ccd Image Sensor	DIC	Son	-	pdf pinout
ICX 205 AL	CCD-IC	Diagonal 8mm Progressive Scan Ccd Image Sensor	DIC	Son	-	pdf pinout
ICX 206AK	Si-Sensor	Ccd Image Sensor for NTSC Color Video Cameras, effective pixels: 510(H)/492(V)	MDIP	Son	-	pdf pinout
ICX 206 AKB	Si-Sensor	Ccd Image Sensor for NTSC Color Video Cameras	MDIP	Son	-	pdf pinout
ICX 206 AL	Si-Sensor	Ccd Image Sensor for EIA B/w Video Cameras	MDIP	Son	-	pdf pinout
ICX 207 AK	Si-Sensor	Ccd Image Sensor for Pal Color Video Cameras	MDIP	Son	-	pdf pinout

Type	Device	Short Description	Fig.	Manu	Comparision Types	More at
ICX 207 AKB	Si-Sensor	Ccd Image Sensor for Pal Color Video Cameras	MDIP	Son	-	pdf pinout
ICX 207 AL	Si-Sensor	Ccd Image Sensor for Ccir B/w Video Cameras	MDIP	Son	-	pdf pinout
ICX 226 AK	CCD-IC	Diagonal 4.5mm (Type 1/4) Ccd Image Sensor	DIC	Son	-	pdf pinout
ICX 226 AL	Si-Sensor	Ccd Image Sensor for Eia B/w Video Cameras	MDIP	Son	-	pdf pinout
ICX 227 AK	CCD-IC	Diagonal 4.5mm (Type 1/4) Ccd Image Sensor	DIC	Son	-	pdf pinout
ICX 227 AL	Si-Sensor	Ccd Image Sensor for Ccir B/w Video Cameras	MDIP	Son	-	pdf pinout
ICX 228 AK	CCD-IC	Diagonal 4.5mm (Type 1/4) Ccd Image Sensor	DIC	Son	-	pdf pinout
ICX 228 AL	Si-Sensor	Ccd Image Sensor for Eia B/w Video Cameras	MDIP	Son	-	pdf pinout
ICX 229 AK	CCD-IC	Diagonal 4.5mm (Type 1/4) Ccd Image Sensor for Pal Color Video Cameras	DIC	Son	-	pdf pinout
ICX 229 AL	Si-Sensor	Ccd Image Sensor for Ccir B/w Video Cameras	MDIP	Son	-	pdf pinout
ICX 238 AKE	CCD-IC	Diagonal 3mm (Type 1/6) Ccd Image Sensor f. NTSC Color Cameras	LCC	Son		pdf pinout
ICX 239 AKE	CCD-IC	Diagonal 3mm (Type 1/6) Ccd Image Sensor f. Pal Color Cameras	LCC	Son		pdf pinout
ICX 254 AK	CCD-IC	Diagonal 6mm (Type 1/3) Ccd Image Sensor for NTSC Color Video Cameras	DIC	Son		pdf pinout
ICX 254 AL	CCD-IC	Diagonal 6mm (Type 1/3) Ccd Image Sensor f. Eia B/w Video Cameras	DIC	Son		pdf pinout
ICX 255 AK	CCD-IC	Diagonal 6mm (Type 1/3) Ccd Image Sensor f. Pal Color Video Cameras	DIC	Son		pdf pinout
ICX 255 AL	CCD-IC	Diagonal 6mm (Type 1/3) Ccd Image Sensor f. Ccir B/w Video Cameras	DIC	Son		pdf pinout
ICX 258 AK	CCD-IC	Diagonal 6mm (Type 1/3) Ccd Image Sensor f. NTSC Color Video Cameras	DIC	Son	-	pdf pinout
ICX 258 AL	Si-Sensor	Ccd Image Sensor for Eia B/w Video Cameras	MDIP	Son	-	pdf pinout
ICX 259 AL	Si-Sensor	Ccd Image Sensor for Ccir B/w Video Cameras	MDIP	Son	-	pdf pinout
ICX 278 AK	CCD-IC	Diagonal 4.5mm (Type 1/4) Ccd Image Sensor f. NTSC Color Video Cameras	DIC	Son	-	pdf pinout
ICX 278 AL	CCD-IC	Diagonal 4.5mm (Type 1/4) Ccd Image Sensor for Eia B/w Video Cameras	DIC	Son	-	pdf pinout
ICX 279 AK	CCD-IC	Diagonal 4.5mm (Type 1/4) Ccd Image Sensor for Pal Color Video Cameras	DIC	Son	-	pdf pinout
ICX 279 AL	CCD-IC	Diagonal 4.5mm (Type 1/4) Ccd Image Sensor for Ccir B/w Video Cameras	DIC	Son	-	pdf pinout
ICX 285 AL	CCD-IC	Diagonal 11mm(Type 2/3) Progressive Scan Ccd Image Sensor w. Square Pixel	DIC	Son	-	pdf pinout
ICX 285 AQ	CCD-IC	Diagonal 11mm (Type 2/3) Progressive Scan Ccd Image Sensor w. Square Pixel	DIC	Son	-	pdf pinout
ICX 404 AK	CCD-IC	Diagonal 6mm (Type 1/3) Ccd Image Sensor	DIC	Son	-	pdf pinout
ICX 404 AL	Si-Sensor	Ccd Image Sensor for Eia B/w Video Cameras	MDIP	Son	-	pdf pinout
ICX 405 AK	Si-Sensor	Ccd Image Sensor for Pal Color Video Cameras	MDIP	Son	-	pdf pinout
ICX 405 AL	CCD-IC	Diagonal 6mm (Type 1/3) Ccd Image Sensor for Ccir B/w Video Cameras	DIC	Son	-	pdf pinout
ICX 408 AK	CCD-IC	Diagonal 6mm (Type 1/3) Ccd Image Sensor for NTSC Color Video Cameras	DIC	Son	-	pdf pinout
ICX 408 AL	Si-Sensor	Ccd Image Sensor for Eia B/w Color Video Cameras	MDIP	Son	-	pdf pinout
ICX 409 AK	CCD-IC	Diagonal 6mm (Type 1/3) Ccd Image Sensor for Pal Color Video Cameras	DIC	Son	-	pdf pinout
ICX 409 AL	CCD-IC	Diagonal 6mm (Type 1/3) Ccd Image Sensor for Ccir B/w Video Cameras	DIC	Son	-	pdf pinout
ICX 414 AL	CCD-IC	Diagonal 8mm(Type 1/2) Progres. Scan Ccd Solid-state Image Sens., Square Pixel	DIC	Son	-	pdf pinout
ICX 414 AQ	CCD-IC	Diagonal 8mm (Type 1/2) Progressive Scan Ccd Image Sensor w. Square Pixel	DIC	Son	-	pdf pinout
ICX 415 AL	CCD-IC	Diagonal 8mm(Type 1/2) Progres. Scan Ccd Solid-state Image Sens., Square Pixel	DIC	Son	-	pdf pinout
ICX 415 AQ	CCD-IC	Diagonal 8mm(Type 1/2) Progres. Scan Ccd Image Sensor, Square Pixel	DIC	Son	-	pdf pinout
ICX 418 AKB	CCD-IC	Diagonal 8mm (Type 1/2) Ccd Image Sensor for NTSC Color Video Cameras	DIC	Son	-	pdf pinout
ICX 418 AKL	CCD-IC	Diagonal 8mm (Type 1/2) Ccd Image Sensor for NTSC Color Video Cameras	DIC	Son	-	pdf pinout
ICX 418 ALB	CCD-IC	Diagonal 8mm (Type 1/2) Ccd Image Sensor for Eia B/w Video Cameras	DIC	Son	-	pdf pinout
ICX 418 ALL	CCD-IC	Diagonal 8mm (Type 1/2) Ccd Image Sensor for Eia B/w Video Cameras	DIC	Son	-	pdf pinout
ICX 419 AKB	CCD-IC	Diagonal 8mm (Type 1/2) Ccd Image Sensor for Pal Color Video Cameras	DIC	Son	-	pdf pinout
ICX 419 AKL	CCD-IC	Diagonal 8mm (Type 1/2) Ccd Image Sensor for Pal Color Video Cameras	DIC	Son		pdf pinout

Type	Device	Short Description	Fig.	Manu	Comparision Types	More at
ICX 419 ALB	CCD-IC	Diagonal 8mm (Type 1/2) Ccd Image Sensor for Ccir B/w Video Cameras	DIC	Son	-	pdf pinout
ICX 419 ALL	CCD-IC	Diagonal 8mm (Type 1/2) Ccd Image Sensor for Ccir B/w Video Cameras	DIC	Son	-	pdf pinout
ICX 422 AL	CCD-IC	Diagonal 11mm (Type 2/3) Ccd Image Sensor	DIC	Son	-	pdf pinout
ICX 423 AL	CCD-IC	Diagonal 11mm (Type 2/3) Ccd Image Sensor	DIC	Son	-	pdf pinout
ICX 424 AL	CCD-IC	Diagonal 6mm(Type 1/3) Progres. Scan Ccd Solid-state Image Sensor	DIC	Son	-	pdf pinout
ICX 424 AQ	CCD-IC	Diagonal 6mm(Type 1/3) Progressive Scan Ccd Image Sensor, Square Pixel	DIC	Son	-	pdf pinout
ICX 428 AKL	CCD-IC	Diagonal 8mm (Type 1/2) Ccd Image Sensor	DIC	Son	-	pdf pinout
ICX 428 ALB	CCD-IC	Diagonal 8mm (Type 1/2) Ccd Image Sensor	DIC	Son	-	pdf pinout
ICX 428 ALL	CCD-IC	Diagonal 8mm (Type 1/2) Ccd Image Sensor	DIC	Son	-	pdf pinout
ICX 429 AKL	CCD-IC	Diagonal 8mm (Type 1/2) Ccd Image Sensor	DIC	Son	-	pdf pinout
ICX 429 ALB	CCD-IC	Diagonal 8mm (Type 1/2) Ccd Image Sensor for Ccir B/w Video Cameras	DIC	Son	-	pdf pinout
ICX 429 ALL	CCD-IC	Diagonal 8mm (Type 1/2) Ccd Image Sensor	DIC	Son	-	pdf pinout
ICX 259 AK	CCD-IC	Diagonal 6mm (Type 1/3) Ccd Image Sensor f. Pal Color Video Cameras	DIC	Son	-	pdf pinout
ID 82C55A-5	µP-Periph	CMOS, Ppi, 3x8 Programmable I/O Pins, TTL Compatible, 5MHz, -40...+85°C	DIC	Har	...82C55...	pdf pinout
ID 82C55A	µP-Periph	CMOS, Ppi, 3x8 Programmable I/O Pins, TTL Compatible, 8MHz, -40...+85°C	DIC	Har	...82C55...	pdf pinout
ID 100	Si-Di	Dual, Pico-A, 30V, 20mA, Ir<10pA(10V)	81	Isi	-	data
ID 101	Si-Di	Dual, Pico-A, 30V, 20mA, Ir<10pA(10V)	81	Isi	-	data
ID4250	LO-VOLT	Vs:±6V Vu:100dB Vi0:3mV f:0,25Mc		Sol	-	data pinout
ID4251	LO-VOLT	Vs:±1.5V Vu:96dB Vo:0.5V Vi0:6mV		Sol	-	data pinout
ID4301A	UNI	Vs:±15V Vu:88dB Vo:±10V Vi0:7.5mV		Sol	-	data pinout
ID4741	UNI	Vs:±15V Vu:86dB Vo:±10V Vi0:6mV		Sol	-	data pinout
ID-	Si-N	=2SC3127 (Typ-Code/Stempel/marking)	35	Hit	→2SC3127	data
IDT 75 C19...	CMOS-A/D-IC	9Bit, Video, Ecl Level, C19x70: 70Msps ...100: 70Msps, ...125: 100Msps	24-DIP			
IDT 75 C58	CMOS-A/D-IC	8 Bit, Video, >20 Msps, TTL Level	28-LCC	\N		
IDT 75 C58	CMOS-A/D-IC	8 Bit, Video, >20 Msps, TTL Level	28-DIP			
IDTVS 330 DCG	IC	low On-resist. wideb./video Quad 2-CH	DIP	Idt	-	pdf pinout
IDTVS 330 QG	IC	low On-resist. wideb./video Quad 2-CH	DIP	Idt	-	pdf pinout
IE	Si-N	=2SC3722K-E (Typ-Code/Stempel/marking)	35	Rhm	→2SC3722K	data
IFB	Si-P	=2SB973-B (Typ-Code/Stempel/marking)	35	Hit	→2SB973	data
IFC	Si-P	=2SB973-C (Typ-Code/Stempel/marking)	35	Hit	→2SB973	data
IGD	MOS-N-FET-d	=2SK1215-D (Typ-Code/Stempel/marking)	35(2mm)	Hit	→2SK1215	data
IGD	MOS-N-FET-d	=2SK360-D (Typ-Code/Stempel/marking)	35	Hit	→2SK360	data
IGE	MOS-N-FET-d	=2SK1215-E (Typ-Code/Stempel/marking)	35(2mm)	Hit	→2SK1215	data
IGE	MOS-N-FET-d	=2SK360-E (Typ-Code/Stempel/marking)	35	Hit	→2SK360	data
IGF	MOS-N-FET-d	=2SK1215-F (Typ-Code/Stempel/marking)	35(2mm)	Hit	→2SK1215	data
IGF	MOS-N-FET-d	=2SK360-F (Typ-Code/Stempel/marking)	35	Hit	→2SK360	data
IGT 4 D10...11	MOS-N-IGBT	L, 400V, 18A, 75W, 250/4500ns	17id	Gen	BUK 856-400IZ, MGP 20N45..N50	data
IGT 4 E10...11	MOS-N-IGBT	=IGT 4D10...11: 500V	17id	Gen	BUP 400, IGTP 20N50, MGP 20N50	data
IGT 5 E10CS	MOS-N-IGBT	Current Sensing, 500V, 10A, 75W	17/5Pin (GSensCEE)	Gen	-	data
IGT 6 D20..21	MOS-N-IGBT	L, 400V, 32A, 125W, 300/5650ns	23ib	Gen	GN 4530C	data
IGT 6 E10...11	MOS-N-IGBT	=IGT 4 D/E10...11:	23ib	Gen	GT 25J101, IGTH 20N40...N50	data
IGT 6 E20...21	MOS-N-IGBT	=IGT 6D20...21: 500V	23ib	Gen	GN 6030C	data
IGT 7 E20CS	MOS-N-IGBT	Current Sensing, 500V, 25A, 125W	18/5Pin (GSensCEE)	Gen	-	data
IGT 7 E50CS	MOS-N-IGBT	Current Sensing, 500V, 50A, 200W	18/5Pin (GSensCEE)	Gen	-	data
IGT 8 D/E20...21	MOS-N-IGBT	=IGT 6 D/E20...21:	16id	Gen	GN 4530C, GN 6030C	data
IGTH 10 N40	MOS-N-IGBT	L, 400V, 10A, 75W, 100/1400ns	18id	Gen	GT 15J101	data
IGTH 10 N50	MOS-N-IGBT	=IGTH 10N40: 500V	18id		GT 15J101	data
IGTH 20 N40	MOS-N-IGBT	L, 400V, 20A, 100W, 100/1400ns	18id	Gen	GN 6020C, GT 25J101	data
IGTH 20 N50	MOS-N-IGBT	=IGTH 20N40: 500V	18id		GN 6020C, GT 25J101	data
IGTM 10 N...	MOS-N-IGBT	=IGTH 10N...:	23ib	Gen	GT 15J101	data
IGTM 20 N...	MOS-N-IGBT	=IGTH 20N...:	23ib	Gen	GN 6030C, GT 25J101	data
IGTP 10 N...	MOS-N-IGBT	=IGTH 10N...:	17id	Gen	GN 6010A	data
IGTP 20 N...	MOS-N-IGBT	=IGTH 20N...:	17id	Gen	BUP 400, MGP 20N45...N50	data
IGT... 10 N..A	MOS-N-IGBT	Energy Loss=400µJ			-	data
IGT... 10 N..(A)D	MOS-N-IGBT	int. FW-Diode			-	data
IGT... 20 N..A	MOS-N-IGBT	Energy Loss=1070µJ			-	data
IGT... 20 N..(A)D	MOS-N-IGBT	int. FW-Diode			-	data
IH 5040(...)PE,JE	CMOS-IC	Analog Switch, ±4,5...18V, <30mA, <130Ω C: 0...+70°, M: -55...+125°	16-DIP,DIC	Max		
IH 5040(...)WE	CMOS-IC	SMD,Analog Switch, ±4,5...18V, <30mA	16-MDIP			
IH 5041(...)PE,JE	CMOS-IC	2x Analog Sw., ±4,5...18V, <30mA, <130Ω	16-DIP,DIC	Max		
IH 5041(...)TW	CMOS-IC	2x Analog Sw., ±4,5...18V, <30mA, <130Ω	82			
IH 5041(...)WE	CMOS-IC	SMD,2x Analog Sw., ±4,5...18V, <30mA	16-MDIP			
IH 5042(...)PE,JE	CMOS-IC	Analog Switch, ±4,5...18V, <30mA, <130Ω	16-DIP,DIC	Max		
IH 5042(...)WE	CMOS-IC	SMD,Analog Switch, ±4,5...18V, <30mA	16-MDIP			
IH 5043(...)PE,JE	CMOS-IC	2x Analog Sw., ±4,5...18V, <30mA, <130Ω	16-DIP,DIC	Max		
IH 5043(...)WE	CMOS-IC	SMD,2x Analog Sw., ±4,5...18V, <30mA	16-MDIP			

Type	Device	Short Description	Fig.	Manu	Comparision Types	More at
IH 5044(...)PE,JE	CMOS-IC	Analog Switch, ±4,5...18V, <30mA, <130Ω	16-DIP,DIC	Max	-	
IH 5044(...)WE	CMOS-IC	SMD,Analog Switch, ±4,5...18V, <30mA	16-MDIP		-	
IH 5045(...)PE,JE	CMOS-IC	2x Analog Sw., ±4,5...18V, <30mA, <130Ω	16-DIP,DIC	Max	-	
IH 5045(...)WE	CMOS-IC	SMD,2x Analog Sw., ±4,5...18V, <30mA	16-MDIP		-	
IH 5048(...)PE,JE	CMOS-IC	2x Analog Switch, 36V, 30mA, <75Ω C: 0...+70°, M: -55...+125°	16-DIP,DIC	Max	-	
IH 5048(...)WE	CMOS-IC	SMD,2x Analog Switch, 36V, 30mA, <75Ω	16-MDIP		-	
IH 5049(...)PE,JE	CMOS-IC	2x Analog Switch, 36V, 30mA, <75Ω	16-DIP,DIC	Max	-	
IH 5049(...)WE	CMOS-IC	SMD,2x Analog Switch, 36V, 30mA, <75Ω	16-MDIP		-	
IH 5050(...)PE,JE	CMOS-IC	Analog Switch, 36V, 30mA, <75Ω	16-DIP,DIC	Max	-	
IH 5050(...)WE	CMOS-IC	SMD,Analog Switch, 36V, 30mA, <75Ω	16-MDIP		-	
IH 5051(...)PE,JE	CMOS-IC	2x Analog Switch, 36V, 30mA, <75Ω	16-DIP,DIC	Max	-	
IH 5051(...)WE	CMOS-IC	SMD,2x Analog Switch, 36V, 30mA, <75Ω	16-MDIP		-	
IH 5108 ...	CMOS-IC	Fault Protected 8 Ch. Multiplexer		Isi	-	
IH 5108(...)PE,JE	CMOS-IC	Fault Protected 8 Ch. Multiplexer C: 0...+70°, E:-40...+85°, M: -55...+125°	16-DIP,DIC	Isi	-	
IH 5108(...)WE	CMOS-IC	SMD,Fault Protected 8 Ch. Multiplexer	16-MDIP	Isi	-	
IH5140	SPST	Vs:±4.5...±18V Ron:Ohm at 15/-15V	d,16DS	Max		data pinout
IH 5140(...)PE,JE	CMOS-IC	Analog Switch, ±4,5...18V, <30mA, <100Ω lo-power, C: 0...+70°, M: -55...+125°	16-DIP,DIC	Max	-	
IH 5140(...)WE	CMOS-IC	SMD,Analog Switch, ±4,5...18V, <30mA	16-MDIP		-	
IH5141	SPST	Vs:±4.5...±18V Ron:Ohm at 15/-15V	d,16D,10AS	Max		data pinout
IH 5141(...)PE,JE	CMOS-IC	2x Analog Sw., ±4,5...18V, <30mA, <100Ω	16-DIP,DIC	Max	-	
IH 5141(...)TW	CMOS-IC	2x Analog Sw., ±4,5...18V, <30mA, <100Ω	82		-	
IH 5141(...)WE	CMOS-IC	SMD,2x Analog Sw., ±4,5...18V, <30mA	16-MDIP		-	
IH5142	SPDT	Vs:±4.5...±18V Ron:Ohm at 15/-15V	d,16DS	Max		data pinout
IH 5142(...)PE,JE	CMOS-IC	Analog Switch, ±4,5...18V, <30mA, <100Ω	16-DIP,DIC	Max	-	
IH 5142(...)WE	CMOS-IC	SMD,Analog Switch, ±4,5...18V, <30mA	16-MDIP		-	
IH5143	SPDT	Vs:±4.5...±18V Ron:Ohm at 15/-15V	d,16DS	Max		data pinout
IH 5143(...)PE,JE	CMOS-IC	2x Analog Sw., ±4,5...18V, <30mA, <100Ω	16-DIP,DIC	Max	-	
IH 5143(...)WE	CMOS-IC	SMD,2x Analog Sw., ±4,5...18V, <30mA	16-MDIP		-	
IH5144	DPST	Vs:±4.5...±18V Ron:Ohm at 15/-15V	d,16DS	Max		data pinout
IH 5144(...)PE,JE	CMOS-IC	Analog Switch, ±4,5...18V, <30mA, <100Ω	16-DIP,DIC	Max	-	
IH 5144(...)WE	CMOS-IC	SMD,Analog Switch, ±4,5...18V, <30mA	16-MDIP		-	
IH5145	DPST	Vs:±4.5...±18V Ron:Ohm at 15/-15V	d,16DS	Max		data pinout
IH 5145(...)PE,JE	CMOS-IC	2x Analog Sw., ±4,5...18V, <30mA, <100Ω	16-DIP,DIC	Max	-	
IH 5145(...)WE	CMOS-IC	SMD,2x Analog Sw., ±4,5...18V, <30mA	16-MDIP		-	
IH 5208 ...	CMOS-IC	Fault Protected 4 Ch. Multiplexer		Isi	-	
IH 5208(...)PE,JE	CMOS-IC	Fault Protected 4 Ch. Multiplexer	16-DIP,DIC	Isi	-	
IH 5208(...)WE	CMOS-IC	SMD,Fault Protected 4 Ch. Multiplexer	16-MDIP	Isi	-	
IH 5341(...)PD,JD	CMOS-IC	2x RF/Video Sw., ±17V, 50mA, >100MHz C: 0...+70°,I: -20...+85°, M: -55...+125°	14-DIP,DIC	Max	-	
IH 5341(...)TW	CMOS-IC	2x RF/Video Sw., ±17V, 50mA, >100MHz	82		-	
IH 5352(...)PE,JE	CMOS-IC	4x RF/Video Sw., ±17V, 50mA, >100MHz	16-DIP,DIC	Max	-	
IH 6108 ...	CMOS-IC	8-Ch. Analog Multiplexer, -55...+125°		Isi	-	
IH 6108 ACWE	CMOS-IC	SMD,8-Ch. Analog Multiplexer, 0...+70°	16-MDIP	Isi	-	
IH 6116 ...	CMOS-IC	16-Ch. Analog Multiplexer, -55...+125°		Isi	-	
IH 6116 ACWI	CMOS-IC	SMD,16-Ch. Analog Multiplexer, 0...+70°	28-MDIP	Isi	-	
IH 6208 ...	CMOS-IC	2x4-Ch. Analog Multiplexer, -55...+125°		Isi	-	
IH 6208 ACWE	CMOS-IC	SMD,2x4-Ch. Analog Multiplexer,0...+70°	16-MDIP	Isi	-	
IH 6216 ...	CMOS-IC	2x8-Ch. Analog Multiplexer, -55...+125°		Isi	-	
IH 6216 ACWI	CMOS-IC	SMD,2x8-Ch. Analog Multiplexer,0...+70°	28-MDIP	Isi	-	
IHC	N-FET	=2SK431-C (Typ-Code/Stempel/marking)	35	Hit	→2SK431	data
IHD	N-FET	=2SK431-D (Typ-Code/Stempel/marking)	35	Hit	→2SK431	data
IHE	N-FET	=2SK431-E (Typ-Code/Stempel/marking)	35	Hit	→2SK431	data
IHF	N-FET	=2SK431-F (Typ-Code/Stempel/marking)	35	Hit	→2SK431	data
IJ-	Si-N	=2SC3374 (Typ-Code/Stempel/marking)	35	Hit	→2SC3374	data
IK	Si-P	=2SA1463-IK (Typ-Code/Stempel/marking)	39	Nec	→2SA1463	data
IK	Si-N	=2SD2210 (Typ-Code/Stempel/marking)	39	Mat	→2SD2210	data
IK 2101	IC	LED Driver IC	DIP	Iks		pdf pinout
IK 2102 DW	IC	LED Driver IC	TSOP	Iks		pdf pinout
IK 2102 TSD	IC	LED Driver IC	TSOP	Iks		pdf pinout
IK 2103 SD	IC	LED Driver IC	TSOP	Iks		pdf pinout
IK 2107	IC	LED Driver IC	DIP	Iks		pdf pinout
IK 2802	IC	High-Power LED Driver	TSOP	Iks		pdf pinout
IK 2804	IC	High-Power LED Driver	TSOP	Iks		pdf pinout
IK 2816	IC	Constant-current LED Driver	DIP	Iks		pdf pinout
IK 3051 AD	IC	Constant Voltage/current Controller	TSOP	Iks		pdf pinout
IK 3051 AL	IC	Constant Voltage/current Controller	TSOP	Iks		pdf pinout
IK 3051 D	IC	Constant Voltage/current Controller	TSOP	Iks		pdf pinout
IK 3051 L	IC	Constant Voltage/current Controller	TSOP	Iks		pdf pinout
IK 3101 D	IC	Low Drop Regulator,Signal Interf. Logic Circuit	DIP	Iks		pdf pinout
IK 3301	IC	Triple DC/DC Converter Control IC	DIP	Iks		pdf pinout
IK 3401 D	IC	Green Mode PWM Controller	QFN	Iks		pdf pinout
IK 3401 N	IC	Green Mode PWM Controller	QFN	Iks		pdf pinout
IK 6418 QN	IC	FM Stereo Transmitter IC	QFN	Iks		pdf pinout
IK 6486	IC	33 Output LED Driver	DIP	Iks		pdf pinout
IK 7437	IC	Digitally controlled audio processor	TSOP	Iks		pdf pinout
IK 7437 T	IC	Digitally controlled audio processor	TSOP	Iks		pdf pinout
IK 62083 DW	IC	8CH Darlington Sink Driver	DIP	Iks		pdf pinout
IK 62083 N	IC	8CH Darlington Sink Driver	DIP	Iks		pdf pinout
IK 62084 DW	IC	8CH Darlington Sink Driver	DIP	Iks		pdf pinout
IK 62084 N	IC	8CH Darlington Sink Driver	DIP	Iks		pdf pinout
IKB	Si-N	=2SC3521-B (Typ-Code/Stempel/marking)	35	Hit	→2SC3521	data

Type	Device	Short Description	Fig.	Manu	Comparision Types	More at
KC	Si-N	=2SC3521-C (Typ-Code/Stempel/marking)	35	Hit	→2SC3521	data
-	Si-N	=2SC3493 (Typ-Code/Stempel/marking)	35	Hit	→2SC3493	data
_1	OC	LED/NPN Viso:5300V CTR:50-400%		Sie,Vis	IS-1,TIL116,TIL125,TIL154,4N25	data pdf pinout
_2	OC	LED/NPN Viso:7500V CTR:>20% Ibs:10mA		Sie		data pinout
_9	OC	LED/NPN Viso:5300V CTR:>20% Ibs:10mA		Sie		data pinout
_10	OC	LED/NPN Viso:5300V CTR:>20% Ibs:10mA		Sie		data pinout
_11	OC	LED/NPN Viso:5300V CTR:>50% Ibs:10mA		Sie		data pinout
_15	OC	LED/NPN Viso:1500V CTR:>6% Vbs:1.3V		Fch	OPI2251	data pinout
_16	OC	LED/NPN Viso:1500V CTR:>6% Vbs:1.3V		Fch	OPI2251	data pinout
_30	OC	LED/Dar Viso:5300Vrms CTR:>500%		Sie,Vis	IS-1,TIL116,TIL125,TIL154,4N25	data pdf pinout
_66	QOC	LED/Dar Viso:5300Vrms CTR:500-750%		Vis		data pdf pinout
_66B-1	OC	LED/Dar Viso:5300Vrms CTR:>200%		Sie,Vis		data pdf pinout
_ 70 (...)	IC	Voltage Detector	TSOP	Iks	-	pdf pinout
_ 70 (...) A	IC	Voltage Detector	TSOP	Iks	-	pdf pinout
_74	OC	LED/NPN Viso:5300Vrms CTR:35%		Fch,Sie,Vis	IS74,MCT276,TIL124,TIL153,IL1	data pdf pinout
_ 75 (...) LF	IC	Voltage Detector	TSOP	Iks	-	pdf pinout
_ 75 (...) LS	IC	Voltage Detector	TSOP	Iks	-	pdf pinout
_ 75 (...) PT	IC	Voltage Detector	TSOP	Iks	-	pdf pinout
_ 75 (...) ST	IC	Voltage Detector	TSOP	Iks	-	pdf pinout
_100	OC	LED/PD+IC Viso:2500V CTR:1000%		Iso	6N137	data pinout
_101	OC	LED/PD+IC Viso:6000V CTR:1000%		Iso,Sie		data pdf pinout
_ 135 Z	IC	Integrated circuit of temperature sensor	TSOP	Iks		pdf pinout
_201	OC	LED/NPN Viso:5300Vrms CTR:75-150%		Sie,Vis	IS201	data pdf pinout
_202	OC	LED/NPN Viso:5300Vrms CTR:125-250%		Sie,Vis	IS202	data pdf pinout
_203	OC	LED/NPN Viso:5300Vrms CTR:225-450%		Sie,Vis	IS203	data pdf pinout
_205	OC	LED/NPN Viso:2500Vrms CTR:150-320%		Sie	IS201	data pdf pinout
_211	OC	LED/NPN Viso:2500Vrms CTR:>100% Ibe:5mA		Sie		data pinout
_215	OC	LED/NPN Viso:2500V CTR:130% Vbs:1V		Sie		data pinout
_221	OC	LED/NPN Viso:2500V CTR:>500% Vbs:1V		Sie,Vis		data pinout
_250	OC	bid. LED/NPN Viso:7500V CTR:>100%		Sie,Vis	CNY35,IS-604,PC733,CNY71	data pdf pinout
_256	OC	bid. LED/NPN Viso:2500V CTR:>20%		Sie		data pinout
_300	OC	LED/2 PD Viso:5300Vrms Vbs:1.2V		Sie		data pinout
L 311 D	IC	High Performance Voltage Comparators	DIP	Iks	-	pdf pinout
L 311 N	IC	High Performance Voltage Comparators	DIP	Iks	-	pdf pinout
_329	DOC	LED/2 LED/MOSFET Viso:2500Vrms		Sie		data pinout
L 339 D	IC	Quad Single Supply Comparator	DIP	Iks	-	pdf pinout
L 339 N	IC	Quad Single Supply Comparator	DIP	Iks	-	pdf pinout
_350	OC	LED/2 PD Viso:2500Vrms Vbs:1.8V		Sie		data pinout
_352	OC	LED/NPN Viso:2500Vrms CTR:>100%		Sie		data pinout
_356	OC	LED/MOSFET Viso:2500Vrms Vbs:1.26V		Sie		data pinout
L 386 D	Dig. Audio	Low Voltage Audio Power Amp	DIP	Iks		pdf pinout
L 386 N	Dig. Audio	Low Voltage Audio Power Amp	DIP	Iks		pdf pinout
L 393 D	IC	Low Offset Voltage Dual Comparators	DIP	Iks	-	pdf pinout
L 393 N	IC	Low Offset Voltage Dual Comparators	DIP	Iks	-	pdf pinout
_400	OC	LED/SCR Viso:5300Vrms Vbs:1.2V		Sie,Vis		data pdf pinout
_410	OC	LED/Triac Viso:5300Vrms Vbs:1.16V		Sie,Vis		data pdf pinout
_420	OC	LED/Triac Viso:5300Vrms Vbs:1.16V		Sie,Vis		data pdf pinout
_440	OC	LED/Triac Viso:5300Vrms Vbs:1.25V		Sie		data pdf pinout
_485	OC	LED/PD Viso:2500Vrms Vbs:1.26V		Sie		data pdf pinout
L 494 AD	IC	PWM Control Circuit	DIP	Iks	-	pdf pinout
L 494 AN	IC	PWM Control Circuit	DIP	Iks	-	pdf pinout
L 494 D	IC	PWM Control Circuit	DIP	Iks	-	pdf pinout
L 494 N	IC	PWM Control Circuit	DIP	Iks	-	pdf pinout
L709	UNI	Vs:±18V Vu:93dB Vo:±14V Vi0:2mV	14D	Riz		data pinout
L755	DOC	bid. LED/Dar Viso:5300Vrms CTR:>500%		Sie,Vis		data pdf pinout
L 1705 D	IC	Micro Monitor Supply Control	DIP	Iks	-	pdf pinout
L 1705 N	IC	Micro Monitor Supply Control	DIP	Iks	-	pdf pinout
L 2410 D	LIN-IC	Tone Ringer	DIP	Iks	-	pdf pinout
L 2410 N	LIN-IC	Tone Ringer	DIP	Iks	-	pdf pinout
L 2411 D	LIN-IC	Tone Ringer	DIP	Iks	-	pdf pinout
L 2411 N	LIN-IC	Tone Ringer	DIP	Iks	-	pdf pinout
L 2418 D	IC	Telephone Tone Ringer, Bridge Diode	DIP	Iks	-	pdf pinout
L 2418 N	IC	Telephone Tone Ringer, Bridge Diode	DIP	Iks	-	pdf pinout
L 2533 D	IC	Multistanda. Cmos IC,Single Chip phone	DIP	Iks	-	pdf pinout
L 2533 N	IC	Multistanda. Cmos IC,Single Chip phone	DIP	Iks	-	pdf pinout
L 2901 D	IC	Quad Single Supply Comparator	DIP	Iks	-	pdf pinout
L 2901 N	IC	Quad Single Supply Comparator	DIP	Iks	-	pdf pinout
L 2903 D	IC	Low Offset Voltage Dual Comparators	DIP	Iks	-	pdf pinout
L 2903 N	IC	Low Offset Voltage Dual Comparators	DIP	Iks	-	pdf pinout
L 34 C 86 D	IC	Differential line receiver	TSOP	Iks	-	pdf pinout
L 34 C 86 N	IC	Differential line receiver	TSOP	Iks	-	pdf pinout
L 34 C 87 D	CMOS-IC	Cmos Quad Tristate Differential Line Driver	TSOP	Iks	-	pdf pinout
L 34 C 87 N	CMOS-IC	Cmos Quad Tristate Differential Line Driver	TSOP	Iks		pdf pinout
L 3726 ANN	IC	Univ. Speech Circuit,Dialler Interface	DIP	Iks		pdf pinout
L 3842	IC	Fixed Frequency Current Mode PWM Controller	TSOP	Iks		pdf pinout
L 3843	IC	Fixed Freq. Current Mode PWM Controller	DIP	Iks		pdf pinout
L4108	OC	LED/Triac Viso:5300Vrms Vbs:1.16V		Sie,Vis		data pdf pinout
L4116	OC	LED/Triac Viso:5300Vrms Vbs:1.3V		Sie,Vis		data pdf pinout
L4208	OC	LED/Triac Viso:5300Vrms Vbs:1.16V		Sie,Vis		data pdf pinout
L4216	OC	LED/Triac Viso:5300Vrms Vbs:1.3V		Sie,Vis		data pdf pinout
L 5009 D	IC	Speaker Amplifier	TSOP	Iks	-	pdf pinout
L 5009 N	IC	Speaker Amplifier	TSOP	Iks	-	pdf pinout

Type	Device	Short Description	Fig.	Manu	Comparision Types	More at
IL 5767 QFP	IC	FM Stereo Radio IC	30	Iks	-	pdf pinout
IL 5767 QN	IC	FM Stereo Radio IC	30	Iks	-	pdf pinout
IL 5851 N	IC	Pulse Dialer with Redial	DIP	Iks	-	pdf pinout
IL 6083 N	IC	PWM Power Control IC, Interference Suppression	TSOP	Iks	-	pdf pinout
IL 6840 D	IC	Telephone Tone Ringer, Bridge Diode	DIP	Iks	-	
IL 6840 N	IC	Telephone Tone Ringer, Bridge Diode	DIP	Iks	-	pdf pinout
IL 6965 DW	IC	Telephone Speech Network, Dialler Interface	DIP	Iks	-	pdf pinout
IL 6965 N	IC	Telephone Speech Network, Dialler Interface	DIP	Iks	-	pdf pinout
IL 70 (....)/SOT89	IC	Voltage Detector	TSOP	Iks	-	pdf pinout
IL 33035 D	IC	brushless DC Motor Controller	DIP	Iks	-	pdf pinout
IL 33035 N	IC	brushless DC Motor Controller	DIP	Iks	-	pdf pinout
IL 33091 AD	MOS-IC	Powerful High-grade Fet driver	DIP	Iks	-	pdf pinout
IL 33091 AN	MOS-IC	Powerful High-grade Fet driver	DIP	Iks	-	pdf pinout
IL 33153 D	IC	Single IGBT Gate Driver	DIP	Iks	-	pdf pinout
IL 33153 N	IC	Single IGBT Gate Driver	DIP	Iks	-	pdf pinout
IL 34063 AD	IC	DC-to-DC Converter Control Circuits	DIP	Iks	-	pdf pinout
IL 34063 AN	IC	DC-to-DC Converter Control Circuits	DIP	Iks	-	pdf pinout
IL 34118 DW	IC	Voice Switched Speakerphone Circuit	DIP	Iks	-	pdf pinout
IL 34118 N	IC	Voice Switched Speakerphone Circuit	DIP	Iks	-	pdf pinout
IL 34119 D	Dig. Audio	Low Power Audio Amplifier	DIP	Iks	-	pdf pinout
IL 34119 N	Dig. Audio	Low Power Audio Amplifier	DIP	Iks	-	pdf pinout
IL 34262 N	IC	Power Factor Controllers	DIP	Iks	-	pdf pinout
IL 44608 N	IC	SMPS Controller	DIP	Iks	-	pdf pinout
IL 91214 AD	IC	Tone/pulse Dialer with Flash Function	DIP	Iks	-	pdf pinout
IL 91214 AN	IC	Tone/pulse Dialer with Flash Function	DIP	Iks	-	pdf pinout
IL 91214 BD	IC	Tone/pulse Dialer with Flash Function	DIP	Iks	-	pdf pinout
IL 91214 BN	IC	Tone/pulse Dialer with Flash Function	DIP	Iks	-	pdf pinout
IL 91260 CN	CMOS-IC	10 Memory Tone/pulse Dialler	DIP	Iks	-	pdf pinout
IL 91350 AN	CMOS-IC	13 Memory Tone/pulse Dialler	DIP	Iks	-	pdf pinout
IL 91350 ADW	CMOS-IC	13 Memory Tone/pulse Dialler	DIP	Iks	-	pdf pinout
IL 9151-3 N	IC	Pulse Dialer	DIP	Iks	-	pdf pinout
IL 91531 N	IC	parallel Input tone/pulse Dialer, Cmos	DIP	Iks	-	pdf pinout
IL 145567 N	IC	Pcm codec - Filter	DIP	Iks	-	pdf pinout
IL-	Si-P	=2SA1463-IL (Typ-Code/Stempel/marking)		Nec	→2SA1463	data
IL-	Si-N	=2SC4263 (Typ-Code/Stempel/marking)	35(2mm)	Hit	→2SC4263	data
ILA 1062 AD	IC	Telephone Speech Network, Dialer Interface	DIP	Iks	-	pdf pinout
ILA 1062 AN	IC	Telephone Speech Network, Dialer Interface	DIP	Iks	-	pdf pinout
ILA 1062 D	IC	Telephone Speech Network, Dialer Interface	DIP	Iks	-	pdf pinout
ILA 1062 N	IC	Telephone Speech Network, Dialer Interface	DIP	Iks	-	pdf pinout
ILA 1185 AD	IC	Triac Phase Angle Controller	DIP	Iks	-	pdf pinout
ILA 1185 AN	IC	Triac Phase Angle Controller	DIP	Iks	-	pdf pinout
ILA 1519 B	Dig. Audio	2 x 6 WATT Stereo Power Amplifier	DIP	Iks	-	pdf pinout
ILA 2003	IC	10W Audio Amplifier	DIP	Iks	-	pdf pinout
ILA 2616	IC	2x12W Hi-fi Audio Power Amp. with Mute	DIP	Iks	-	pdf pinout
ILA 3354 N	IC	IC for Electronic Ballasts	DIP	Iks	-	pdf pinout
ILA 3654	IC	Vert. deflection, guard Circuit (110°)	DIP	Iks	-	pdf pinout
ILA 4661 N	IC	Baseband Delay Line	DIP	Iks	-	pdf pinout
ILA 6107 Q	IC	Triple video output amplifier	DIP	Iks	-	pdf pinout
ILA 7052 N	Dig. Audio	mono Output Amplifier	DIP	Iks	-	pdf pinout
ILA 7053 N	Dig. Audio	2x1W portable/mains-fed stereo power amp.	DIP	Iks	-	pdf pinout
ILA 7056	IC	3 W Mono BTL Audio Output Amplifier	DIP	Iks	-	pdf pinout
ILA 7056 B	IC	5W mono BTL Audio Amp, DC Volume Control	DIP	Iks	-	pdf pinout
ILA 7056 SH	IC	3W Mono BTL Audio Output Amplifier	DIP	Iks	-	pdf pinout
ILA 8133 A	IC	Dual +5.1V +8V Regulator,disable,reset	DIP	Iks	-	pdf pinout
ILA 8137 A	IC	Dual 5.1V Regulator, disable, reset	DIP	Iks	-	pdf pinout
ILA 8138 A	IC	5.1V +12V Regulator, disable, reset	QFN	Iks	-	pdf pinout
ILA 8351	IC	Dc-coupled Vertical Deflection Circuit	DIP	Iks	-	pdf pinout
ILA 8356	IC	Dc-coupled Vertical Deflection Circuit	DIP	Iks	-	pdf pinout
ILA 8362 ANS	IC	Pal and PAL/NTSC TV processor	DIP	Iks	-	pdf pinout
ILA 8362 WNS	IC	Single-chip TV Processor IC	DIP	Iks	-	pdf pinout
ILA 8395 N	IC	SECAM decoder	DIP	Iks	-	pdf pinout
ILA 8842 NS	IC	I²C Bus Controlled Single Chip TV-RECEIVER	DIP	Iks	-	pdf pinout
ILA 1519 B 1	IC	2 x 6 WATT Stereo Power Amplifier	DIP	Iks	-	pdf pinout
ILA 1519 B 1 Q	IC	2 x 6 WATT Stereo Power Amplifier	DIP	Iks	-	pdf pinout
ILA 4605-2 N	SMPS-IC	Control IC for SMP Supplies	DIP	Iks	-	pdf pinout
ILC 555 D	CMOS-IC	Cmos general purpose timer	DIP	Iks	-	pdf pinout
ILC 555 N	CMOS-IC	Cmos general purpose timer	DIP	Iks	-	pdf pinout
ILC 556 D	IC	General purpose dual timer	DIP	Iks	-	pdf pinout
ILC 556 N	IC	General purpose dual timer	DIP	Iks	-	pdf pinout
ILCA2-30	OC	LED/Dar Viso:6000V CTR:>100%		Sie	4N32	data pinout
ILCA2-55	OC	LED/Dar Viso:6000V CTR:>100%		Sie		data pinout
ILCT6	DOC	LED/NPN Viso:5300Vrms CTR:50%		Sie,Vis	MCT6	data pdf pinout
ILD1	DOC	LED/NPN Viso:5300Vrms CTR:50-400%		Sie,Vis	ISD-1,TLP504,CNY74-2	data pdf pinout
ILD3	DOC	LED/NPN Viso:5300Vrms CTR:>100%		Sie		data pinout
ILD30	DOC	LED/Dar Viso:5300Vrms CTR:100-400%		Sie,Vis		data pdf pinout
ILD55	DOC	LED/Dar Viso:5300Vrms CTR:100-400%		Sie,Vis	H11B255,H11B1	data pdf pinout

Type	Device	Short Description	Fig.	Manu	Comparision Types	More at
ILD66	DOC	LED/Dar Viso:5300Vrms CTR:500-750%		Sie,Vis		data pdf pinout
ILD74	DOC	LED/NPN Viso:5300Vrms CTR:35%		Fch,Sie,Vis		data pdf pinout
ILD205	DOC	LED/NPN Viso:2500V CTR:130% Vbs:1.3V		Sie,Vis		data pdf pinout
ILD221	DOC	LED/Dar Viso:2500V CTR:>500% Vbs:1V		Sie		data pdf pinout
ILD250	DOC	bid. LED/NPN Viso:7500V CTR:>100%		Sie,Vis	CNY35,IS-604,PC733,CNY71	data pdf pinout
ILD255	DOC	bid. LED/NPN Viso:2500Vrms CTR:>50%		Sie		data pdf pinout
ILD256	DOC	bid. LED/NPN Viso:2500V CTR:>20%		Sie,Vis		data pdf pinout
ILD610	QOC	LED/NPN Viso:5300Vrms CTR:160-320%		Sie,Vis		data pdf pinout
ILD615	DOC	LED/NPN Viso:5300Vrms CTR:160-320%		Sie,Vis		data pdf pinout
ILD620	QOC	bid. LED/NPN Viso:5300Vrms Vbs:1.15V		Sie		data pdf pinout
ILD621	QOC	LED/NPN Viso:5300Vrms CTR:100-600%		Sie,Vis		data pdf pinout
ILN 2003 AD	IC	Darlington Transistor Arrays	DIP	Iks	-	pdf pinout
ILN 2003 AN	IC	Darlington Transistor Arrays	DIP	Iks	-	pdf pinout
ILQ1	QOC	LED/NPN Viso:5300Vrms CTR:20-300%		Sie,Vis	ISQ1,TLP504,CNY74-4	data pdf pinout
ILQ2	QOC	LED/NPN Viso:5300Vrms CTR:100-500%		Sie,Vis	ISD-1,TLP504,CNY74-2	data pdf pinout
ILQ3	QOC	LED/NPN Viso:5300Vrms CTR:>100%		Sie		data pinout
ILQ5	QOC	LED/NPN Viso:5300Vrms CTR:20-300%		Sie,Vis		data pdf pinout
ILQ30	QOC	LED/Dar Viso:5300Vrms CTR:200-400%		Sie,Vis		data pdf pinout
ILQ32	QOC	LED/Dar Viso:5300Vrms CTR:200-400%		Sie		data pdf pinout
ILQ55	QOC	LED/Dar Viso:5300Vrms CTR:100-400%		Sie,Vis		data pdf pinout
ILQ66	QOC	LED/Dar Viso:5300Vrms CTR:500-750%		Sie,Vis		data pdf pinout
ILQ74	QOC	LED/NPN Viso:5300Vrms CTR:35%		Sie,Vis	ISQ74	data pdf pinout
ILX 208 DW	IC	Interface IC of Data Serial Transfer	TSOP	Iks		pdf pinout
ILX 232 D	IC	Interface transceiver of RS-232 standard	TSOP	Iks		pdf pinout
ILX 232 N	IC	Interface transceiver of RS-232 standard	TSOP	Iks		pdf pinout
ILX 485 D	IC	Slew- Rate- Limited RS- 485/ RS- 422 Transceivers	TSOP	Iks		pdf pinout
ILX 485 N	IC	Slew- Rate- Limited RS- 485/ RS- 422 Transceivers	TSOP	Iks		pdf pinout
ILX 3232 D	IC	Interface Transcei., RS-232 Standard	TSOP	Iks		pdf pinout
ILX 3232 DW	IC	Interface Transcei., RS-232 Standard	TSOP	Iks		pdf pinout
ILX 3232 N	IC	Interface Transcei., RS-232 Standard	TSOP	Iks		pdf pinout
ILX 3232 TSD	IC	Interface Transcei., RS-232 Standard	TSOP	Iks		pdf pinout
ILX 3485 D	IC	Slew- Rate- Limited RS- 485/ RS- 422 Transceivers	QFN	Iks		pdf pinout
ILX 3485 N	IC	Slew- Rate- Limited RS- 485/ RS- 422 Transceivers	QFN	Iks		pdf pinout
IMB 1(A)	Si-P+R	SMD, Dual, Rb=Rbe=22kΩ, 50V, 100mA	46bh1	Rhm	-	data
IMB 1A	Si-P+R	SMD, 2x DTA124E, Dual Digital Transistor	46bh1	Rhm	=UMB1N	data data
IMB 2(A)	Si-P+R	SMD, Dual, Rb=Rbe=47kΩ, 50V, 30mA	46bh1	Rhm	-	data
IMB 2A	Si-P+R	SMD, 2x DTA144E, Dual Digital Transistor 100mA,	46bh1	Rhm	=UMB2N	data data
IMB 3(A)	Si-P+R	SMD, Dual, Rb=4,7kΩ, 50V, 100mA	46bh1	Rhm	-	data
IMB 3A	Si-P+R	SMD, 2x DTA143T, Dual Digital Transistor 50V, 100mA,	46bh1	Rhm	IMB4A,	data data
IMB 4(A)	Si-P+R	SMD, Dual, Rb=10kΩ, 50V, 100mA	46bh1	Rhm	-	data
IMB 4A	Si-P+R	SMD, 2x DTA114T, Dual Digital Transistor 50V, 100mA,	46bh1	Rhm	IMB3A,	data data
IMB 5(A)	Si-P+R	SMD, Dual, Rb=Rbe=22kΩ, 50V, 100mA	46bl1	Rhm	-	data
IMB 5A	Si-P+R	SMD, 2x DTA124E, Dual Digital Transistor	46bl1	Rhm	=UMB5N	data data
IMB 6(A)	Si-P+R	SMD, Dual, Rb=Rbe=47kΩ, 50V, 30mA	46bl1	Rhm	-	data
IMB 6A	Si-P+R	SMD, 2x DTA144E, Dual Digital Transistor	46bl1	Rhm	=UMB6N	data data
IMB 7(A)	Si-P+R	SMD, Dual, Rb=4,7kΩ, 50V, 100mA	46bl1	Rhm	-	data
IMB 7A	Si-P+R	SMD, 2x DTA143T, Dual Digital Transistor 50V, 100mA,	46bl1	Rhm	IMB8A,	data data
IMB 8(A)	Si-P+R	SMD, Dual, Rb=10kΩ, 50V, 100mA	46bl1	Rhm	-	data
IMB 8A	Si-P+R	SMD, 2x DTA114T, Dual Digital Transistor 50V, 100mA,	46bl1	Rhm	IMB7A,	data data
IMB 9 A	Si-P+R	SMD,Dual, Rb=10k, Rbe=47kΩ, 50V, 100mA	46bh1	Rhm	-	data
IMB 10 A	Si-P+R	SMD,Dual,Rb=2,2k, Rbe=47kΩ, 50V, 100mA	46bh1	Rhm	-	data
IMB 11 A	Si-P+R	SMD, Dual, Rb=Rbe=10kΩ, 50V, 100mA	46bh1	Rhm	-	data
IMB 14 A	Si-P+R	SMD, Rb=47kΩ, 50V, 100mA	46bl1	Rhm	-	data
IMB 16	Si-P+R	SMD,Dual,Rb=4,7k, Rbe=10kΩ, 50V, 500mA	46bh1	Rhm	-	data
IMB 17 A	Si-P+R	SMD, Dual, Rb=1k, Rbe=10kΩ, 50V, 100mA	46bl1	Rhm	-	data
IMBD 4148	Si-Di	SMD, SS, 100V, 0,15A, <4pF(0V), <4ns	35e	Gsi	MMBD 914, MMBD 4148	data
IMBD 4448	Si-Di	SMD, SS, 100V, 0,15A, <4pF(0V), <4ns	35e	Gsi	MMBD 914, MMBD 4148	data
IMBT 3903..3904	Si-N	=2N3903..3904: SMD	35a	Itt	BC 846, BCV 71..72, 2SC4209	data
IMBT 3905..3906	Si-P	=2N3905..3906: SMD	35a	Itt	BC 856..857, BCX 71, 2SA1620	data
IMD 1 A	Si-N/P+R	SMD, N+P, Rb=22kΩ, 50V, 100mA, 250MHz	46bh1	Rhm	-	data
IMD 2(A)	Si-N/P+R	SMD, N+P, Rb=Rbe=22kΩ, 50V, 30mA	46bh1	Rhm	-	data
IMD 2A	Si-P/N+R	SMD, DTA124E + DTC124E, Dual Digital Transistor	46bh1	Rhm	=UMD2N	data data
IMD 3(A)	Si-N/P+R	SMD, N+P, Rb=Rbe=10kΩ, 50V, 100mA	46bh1	Rhm	-	data
IMD 3A	Si-P/N+R	SMD, DTA114E + DTC114E, Dual Digital Transistor	46bh1	Rhm	=UMD3N	data data
IMD 6(A)	Si-N/P+R	SMD, N+P, Rb=4,7kΩ, 50V, 100mA	46bh1	Rhm	-	data
IMD 6A	Si-P/N+R	SMD, DTA143T + DTC143T, Dual Digital Transistor 50V, 100mA,	46bh1	Rhm	IMD8A,	data data
IMD 8(A)	Si-N/P	SMD, T1=NPN T2=PNP, Rb=47k 50V, 100mA,	46bh1	Rhm	IMD3(A), IMD6(A), IMD9(A), IMD14,	data data
IMD 8 A	Si-N/P+R	SMD, N+P, Rb=47kΩ, 50V, 100mA	46bh1	Rhm	-	data

Type	Device	Short Description	Fig.	Manu	Comparision Types	More at
IMD 9(A)	Si-N/P	SMD, T1=N, T2=P Rb=10k, Rbe=47k Rbe/Rb=3,7...5,7 50V, 100mA	46bh1	Rhm	IMD3(A), IMD6(A), IMD8(A), IMD14,	data data
IMD 9 A	Si-N/P+R	SMD, N+P, Rb=10k, Rbe=47kΩ, 50V, 100mA	46bh1	Rhm	-	data
IMD 10 A	Si-P/N+R	SMD, P+N, Rb=100, Rbe=10kΩ, 50V P: 500mA, N: 100mA	46bh1	Rhm	-	data
IMD 14	Si-N/P+R	SMD, N+P,Rb=220Ω, Rbe=10kΩ, 50V, 500mA	46bh1	Rhm	-	data
IMD 16 A	Si-P/N+R	SMD, P+N, Rb=2,2k, Rbe=22kΩ, 50V P: 500mA, N: 100mA	46bh1	Rhm	-	data
IMF 5564	N-FET	=2N5564: dielectrically isolated	81bx	Isi	→2N5564	data
IMF 5565	N-FET	=2N5565: dielectrically isolated	81bx	Isi	→2N5565	data
IMF 5566	N-FET	=2N5566: dielectrically isolated	81bx	Isi	→2N5566	data
IMF 5911	N-FET	=2N5911: dielectrically isolated	81bx	Isi	→2N5911	data
IMF 5912	N-FET	=2N5912: dielectrically isolated	81bx	Isi	→2N5912	data
IMF 6485	N-FET	=2N6485: dielectrically isolated	81bx	Isi	→2N6485	data
IMH 1(A)	Si-N+R	SMD, Dual, Rb=Rbe=22kΩ, 50V, 30mA	46bh1	Rhm	-	data
IMH 1A	Si-N+R	SMD, 2x DTC124E, Dual Digital Transistor	46bh1	Rhm	=UMH1N	data
IMH 2(A)	Si-N+R	SMD, Dual, Rb=Rbe=47kΩ, 50V, 30mA	46bh1	Rhm	-	data
IMH 2A	Si-N+R	SMD, 2x DTC144E, Dual Digital Transistor 100mA,	46bh1	Rhm	=UMH2N	data
IMH 3(A)	Si-N+R	SMD, Dual, Rb=4,7kΩ, 50V, 100mA	46bh1	Rhm	-	data
IMH 3A	Si-N+R	SMD, 2x DTC143T, Dual Digital Transistor 50V, 100mA,	46bh1	Rhm	IMH4A, IMH15A,	data
IMH 4(A)	Si-N+R	SMD, Dual, Rb=10kΩ, 50V, 100mA	46bh1	Rhm	-	data
IMH 4A	Si-N+R	SMD, 2x DTC114T, Dual Digital Transistor 50V, 100mA,	46bh1	Rhm	IMH3A, IMH15A,	data
IMH 5(A)	Si-N+R	SMD, Dual, Rb=Rbe=22kΩ, 50V, 30mA	46bl1	Rhm	-	data
IMH 5A	Si-N+R	SMD, 2x DTC124E, Dual Digital Transistor	46bl1	Rhm	=UMH5N	data
IMH 6(A)	Si-N+R	SMD, Dual, Rb=Rbe=47kΩ, 50V, 30mA	46bl1	Rhm	-	data
IMH 6A	Si-N+R	SMD, 2x DTC144E, Dual Digital Transistor	46bl1	Rhm	=UMH6N	data
IMH 7(A)	Si-N+R	SMD, Dual, Rb=4,7kΩ, 50V, 100mA	46bl1	Rhm	-	data
IMH 7A	Si-N+R	SMD, 2x DTC143T, Dual Digital Transistor 50V, 100mA,	46bl1	Rhm	IMH8A, IMH14A,	data
IMH 8(A)	Si-N+R	SMD, Dual, Rb=10kΩ, 50V, 100mA	46bl1	Rhm	-	data
IMH 8A	Si-N+R	SMD, 2x DTC114T, Dual Digital Transistor 50V, 100mA,	46bl1	Rhm	IMH7A, IMH14A,	data
IMH 9 A	Si-N+R	SMD,Dual, Rb=10k, Rbe=47kΩ, 50V, 100mA	46bh1	Rhm	-	data
IMH 10(A)	Si-N	SMD, Dual, int. Rb=2,2k,Rbe=47k Rbe/Rb=17...26 50V, 100mA,	46bh1	Rhm	IMH3(A), IMH4(A), IMH9(A), IMH11(A),	data
IMH 10 A	Si-N+R	SMD,Dual,Rb=2,2k, Rbe=47kΩ, 50V, 100mA	46bh1	Rhm	-	data
IMH 11(A)	Si-N	SMD, Dual, int. Rb=Rbe=10k˘ Rbe/Rb=0,8...1,2 50V, 100mA,	46bh1	Rhm	IMH3(A), IMH4(A), IMH9(A), IMH10(A),	data
IMH 11 A	Si-N+R	SMD, Dual, Rb=Rbe=10kΩ, 50V, 100mA	46bh1	Rhm	-	data
IMH 14 A	Si-N+R	SMD, Dual, Rb=47kΩ, 50V, 100mA	46bl1	Rhm	-	data
IMH 15 A	Si-N+R	SMD, Dual, Rb=47kΩ, 50V, 100mA	46bh1	Rhm	-	data
IMS 1403	sRAM-IC	Sram 16k x 1Bit, TTL compatible, +5V	DIP		-	pdf
IMS 1424	sRAM-IC	Sram 4k x 4Bit, TTL compatible, +5V	DIP		-	pdf
IMSA 100	DIG-IC	16 Bit Image Processor, Filter	84-PGA	Tho	-	
IMSA 110	DIG-IC	Image Processor, Delay Lines Proc.	100-PGA	Tho	-	pdf
IMSA 113	CMOS-IC	Delay-Line, digital, ...20MHz	44-PLCC	Tho	-	
IMSA 121	DIG-IC	Image Processor, Loop Filter	44-PLCC	Tho	-	
IMSG 171	DIG-IC	μcomp, Color Look-up Table, 6 Bit VGA	28-DIP	Tho	-	
IMSG 176	DIG-IC	μcomp, Color Look-up Table, 6 Bit VGA	28-DIP	Tho	-	
IMSG 176 L	DIG-IC	=IMSG 176: lo-power	28-DIP	Tho	-	
IMSG 178	DIG-IC	μcomp,Color Look-up Table, 6/8 Bit VGA	32/44-PLCC	Tho	-	
IMSG 300	DIG-IC	μcomp, Graphics Controller, 66...110ns	84-PGA	Tho	-	
IMSG 332	DIG-IC	μcomp, Graphics Controller, 85...110MHz	100-PGA	Tho	-	
IMSG 364	DIG-IC	μcomp, Graphics Controller, 85...110MHz	132-PGA	Tho	-	
IMT 1(A)	Si-P	SMD, Dual, 50V, 0,1A, 140MHz	46bh1	Rhm	-	data
IMT 1A	Si-P	SMD, 2x 2SA1037AK	46bh1	Rhm	IMT17,	data
IMT 2(A)	Si-P	SMD, Dual, 50V, 0,1A, 140MHz	46bl1	Rhm	-	data
IMT 3(A)	Si-P	SMD, Dual, 50V, 0,1A, 140MHz	46bh2	Rhm	-	data
IMT 4	Si-P	SMD, Dual, 120V, 0,05A, 140MHz	46bh1	Rhm	-	data
IMT 5 A	Si-P	SMD, Dual, -/25V, 0,1A	46bh1	Rhm	-	data
IMT 17	Si-P	SMD, Dual, 60V, 0,5A, 200MHz	46bh1	Rhm	-	data
IMX 1	Si-N	SMD, Dual, 60V, 0,15A, 180MHz	46bh1	Rhm	-	data
IMX 2	Si-N	SMD, Dual, 60V, 0,15A, 180MHz	46bl1	Rhm	-	data
IMX 3	Si-N	SMD, Dual, 60V, 0,15A, 180MHz	46bh2	Rhm	-	data
IMX 4	Si-N	SMD, Dual, HF, 30V, 50mA, >600MHz	46bh2	Rhm	-	data
IMX 5	Si-N	SMD, Dual, HF, 20V, 50mA, >1400MHz	46bh2	Rhm	-	data
IMX 6	Si-N	SMD, Dual, HF, 40V, 50mA, >150MHz	46bl1	Rhm	-	data
IMX 7	Si-N	SMD, Dual, 40V, 0,5A	46bh1	Rhm	-	data
IMX 8	Si-N	SMD, Dual, 120V, 0,05A, 140MHz	46bh2	Rhm	-	data
IMX 9	Si-N	SMD, Dual, hi-beta, 25V, 0,5A, B>560	46bh1	Rhm	-	data
IMX 11	Si-N	SMD, Dual, HF, 25V, 20mA	46bh2	Rhm	-	data
IMX 17	Si-N	SMD, Dual, 60, 0,5A, 250MHz	46bh1	Rhm	-	data
IMZ 1(A)	Si-N/P	SMD, N+P, 50V, 0,1A, 180+140MHz	46bh1	Rhm	-	data
IMZ 1A	Si-PN	SMD, 2SA1037AK + 2SC2412K, Power Management 50V, 150mA	46bh1	Rhm	=UMZ1N	data data
IMZ 2(A)	Si-P/N	SMD, P+N, 50V, 0,1A, 140+180MHz	46bh2	Rhm	-	data
IMZ 2A	Si-PN	SMD, 2SA1037AK + 2SC2412K, Power	46bh2	Rhm	UMZ2N,	data

Type	Device	Short Description	Fig.	Manu	Comparision Types	More at
		Management 50V, 150mA,				data
IMZ 3 A	Si-N/P	SMD, N+P, 60+30V, 0,1A, 250+140MHz	46bh1	Rhm	-	data
IMZ 4	Si-N/P	SMD, N+P, =2SC2411K+2SB1036K	46bh1	Rhm	-	data
IN 472 DW	IC	Liquid Crystal Display Controller	QFN	Iks	-	pdf pinout
IN 472 N	IC	Liquid Crystal Display Controller	QFN	Iks	-	pdf pinout
IN 555 D	IC	Timing Circuit	DIP	Iks	-	pdf pinout
IN 555 N	IC	Timing Circuit	DIP	Iks	-	pdf pinout
IN 556 D	CMOS-IC	Timing Circuit	DIP	Iks	-	pdf pinout
IN 556 N	CMOS-IC	Timing Circuit	DIP	Iks	-	pdf pinout
IN 1232 D	IC	Power Supply Control, Watchdog Timer	DIP	Iks	-	pdf pinout
IN 1232 N	IC	Power Supply Control, Watchdog Timer	DIP	Iks	-	pdf pinout
IN 1307 D	IC	Cmos IC of Real Time Watch Serial Inter.	TSOP	Iks	-	pdf pinout
IN 1307 N	IC	Cmos IC of Real Time Watch Serial Inter.	TSOP	Iks	-	pdf pinout
IN 1488 D	IC	Quadruple Line Drivers	DIP	Iks	-	pdf pinout
IN 1488 N	IC	Quadruple Line Drivers	DIP	Iks	-	pdf pinout
IN 1489 AN	IC	Quadruple Line Drivers	DIP	Iks	-	pdf pinout
IN 1489 AD	IC	Quadruple Line Drivers	DIP	Iks	-	pdf pinout
IN 24 LC 04 BD	CMOS-IC	4K/8K 2.5V Cmos Serial EEPROMs	DIP	Iks	-	pdf pinout
IN 24 LC 04 BN	CMOS-IC	4K/8K 2.5V Cmos Serial EEPROMs	DIP	Iks	-	pdf pinout
IN 24 LC 08 BD	CMOS-IC	4K/8K 2.5V Cmos Serial EEPROMs	DIP	Iks	-	pdf pinout
IN 24 LC 08 BN	CMOS-IC	4K/8K 2.5V Cmos Serial EEPROMs	DIP	Iks	-	pdf pinout
IN 24 LC 16 BD	LIN-IC	16K Serial EEPROM with I²C bus	DIP	Iks	-	pdf pinout
IN 24 LC 16 BN	IC	16K Serial EEPROM with I²C bus	DIP	Iks	-	pdf pinout
IN 4728A...4761A	Z-Di	=1N4728A...4761A: 1W	31a	Aeg	→1N4728...4761	data
IN 82 C 55	IC	Chmos Programmable Peripheral Interface	TSOP	Iks		pdf pinout
IN 16 C 554 PL	IC	asynchronous communications element	TSOP	Iks		pdf pinout
IN 16 C 554 TQ	IC	asynchronous communications element	TSOP	Iks		pdf pinout
IN 44780	IC	dot Matrix LCD Controller & Driver	QFN	Iks		pdf pinout
IN 75232	IC	EIA-232-D Interface 1 Chip IC	TSOP	Iks		pdf pinout
IN 16 C 1054 PL	IC	Quard Uart with 256-Byte Fifo	TSOP	Iks		pdf pinout
IN 16 C 1054 TQ	IC	Quard Uart with 256-Byte Fifo	TSOP	Iks		pdf pinout
INA101	INA	Vs:±20V Vo:±12.5V Vi0:±25µµV f:0.3Mc	14D,10A	Bub		data pdf pinout
INA 101(A,C,S)G	OP-IC	Gain-set(1...1000), ±20V, 2,5...300kHz	14-DIC	Bub		
INA 101 HP	OP-IC	=INA 101...G:	20-DIP	Bub		
INA 101(A,C,S)M	OP-IC	=INA 101...G:	82			
INA 102(A,C)G	OP-IC	Gain-set(1...1000), ±18V, 0,3...300kHz	16-DIC	Bub		
INA104	INA	Vs:±18V Vo:±14.5V Vi0:<±0.3mV f:3kc	16D,18Y	Bub		data pinout
INA 104(H,J,K)P	OP-IC	Gain-set(1...1000), ±20V, 2,5...300kHz	18-DIP	Bub		
INA 105(K)P	OP-IC	Unity-gain, ±18V, 1MHz	8-DIP	Bub		pdf
INA 105(A,B,K)M	OP-IC	=INA 105...P:	81			
INA 106(K)P	OP-IC	Unity-gain, ±18V, 5MHz	8-DIP	Bub		pdf
INA 106(A,B,K)M	OP-IC	=INA 106...P:	81			
INA 110(A,B)G	OP-IC	Gain-set(1...500), ±18V, 0,1...2,5MHz	16-DIC	Bub		
INA141	INA	Vs:±18V Vi0:±50µµV f:1Mkc	8DS	Bub		data pdf pinout
INA141M	INA	Vs:±18V Vi0:mV	8DS	Bub		data pdf pinout
INA159	HI-PREC	Vs:5.5V Vu:200V/mV Vi0:±0.1mV	8S	Bub		data pdf pinout
INA 1238 NS	IC	FM/AM Stereo Radioreceiver	30	Iks		pdf pinout
INA 2586	NMOS-IC	1024x8- Bit n- MOS Eeprom, I²c- Bus Interface	QFN	Iks		pdf pinout
INA 3010 DW	IC	Infrared Remote Control Transmitter	QFN	Iks		pdf pinout
INA 3010 N	IC	Infrared Remote Control Transmitter	QFN	Iks		pdf pinout
INA 8583 N	IC	Cmos timer, RAM, I2C-bus control	DIP	Iks		pdf pinout
INA 84 C 641	IC	Z8-BIT Microcontrollers, OCD VST	DIP	Iks		pdf pinout
INA84C640ANS-(...)	IC	Z8-BIT Microcontrollers, OCD VST	DIP	Iks		pdf pinout
INB	Si-P	=2SA1468-B (Typ-Code/Stempel/marking)	35	Hit	→2SA1468	data
INC	Si-P	=2SA1468-C (Typ-Code/Stempel/marking)	35	Hit	→2SA1468	data
INF 8574	CMOS-IC	8-bit quasi-bidirect. Port,I²C interf.	TSOP	Iks		pdf pinout
INF 8594 E	IC	512x8- bit Cmos EEPROMS, I2C- bus Interface	DIP	Iks		pdf pinout
INS 8255	µP-Periph	Ppi, 3x8 Programmable I/O Pins, TTL Compatible	DIP	Nsc	...8255...	pdf pinout
IO	Si-P	=2SA1298-O (Typ-Code/Stempel/marking)	35	Tos	→2SA1298	data
IO	Si-N	=2SC3515-O (Typ-Code/Stempel/marking)	39	Tos	→2SC3515	data
IO	Si-N	=2SC3515-O (Typ-Code/Stempel/marking)	39	Tos	→2SC3515	data
IO	Si-N	=2SC5262-O (Typ-Code/Stempel/marking)	44	Tos	→2SC5262	
IO	Si-P	=KTA1298-O (Typ-Code/Stempel/marking)	35	Kec	→KTA 1298	data
IOB 1680	NMOS-IC	Input/output Buffer f. µComp.	40-DIC	Itt		
IP	Si-P	=2SB1073-P (Typ-Code/Stempel/marking)	39	Mat	→2SB1073	data
IP 82C55A-5	µP-Periph	CMOS, Ppi, 3x8 Programmable I/O Pins, TTL Compatible, 5MHz, -40...+85°C	DIP	Har	...82C55...	pdf pinout
IP 82C55A	µP-Periph	CMOS, Ppi, 3x8 Programmable I/O Pins, TTL Compatible, 8MHz, -40...+85°C	DIP	Har	...82C55...	pdf pinout
IP 100	F-Thy	30V, 0,8A, Igt/Ih<0,2/<5mA	7e	Uni	BR103, BRX44...45, BRX50, TAG60Y, TAG62Y+	data
IP 101	F-Thy	=IP100: 60V	7e	Uni	BRX46...47, BRX51...52, TAG60A, TAG62A,++	data
IP 102	F-Thy	=IP100: 100V	7e	Uni	BRX46...47, BRX51...52, TAG60A, TAG62A,++	data
IP 103	F-Thy	=IP100: 150V	7e	Uni	BRX 47...49, BRX 52...54	data
IP 104	F-Thy	=IP100: 200V	7e	Uni	BRX 47...49, BRX 52...54	data
IP 105	F-Thy	=IP100: 300V	7e	Uni	BRX 48...49, BRX 53...56	data
IP 106	F-Thy	=IP100: 400V	7e	Uni	BRX49, BRX54...56	data
iP 1001	CMOS-Z-IC	Synchr. Buck Power Block	GRID	Inr		pdf pinout
iP 1201	DIG-IC	Synchronous Buck Power Block	GRID	Inr		pdf pinout
iP 1202	CMOS-IC	15A, 2Ph Synchr. Buck Power Block	GRID	Inr		pdf
iP 1203	IC	Full Function Synchr Buck Power Block	QFN	Inr		pdf pinout
IP 1525 AJ	IC	Regulating Pulse Width Modulators	DIC	Seb		pdf pinout

Type	Device	Short Description	Fig.	Manu	Comparision Types	More at
IP 1527 AJ	IC	Regulating Pulse Width Modulators	DIC	Seb	-	pdf pinout
iP 2001	CMOS-Z-IC	20A, Synchr. Buck Multiph. Power Block	GRID	Inr	-	pdf pinout
iP 2002	IC	Synchr Buck Multiphase Power Block	GRID	Inr	-	pdf pinout
iP 2003 A	CMOS-Z-IC	40A, Synchr. Buck Multiph. Power Block	LGA	Inr	-	pdf pinout
IP 2005 APBF	LIN-IC	Synchronous Buck Optimized Lga Power Stage	20-DIP	Inr	-	pdf pinout
IP 3525 AD-16	IC	Regulating Pulse Width Modulators	DIC	Seb	-	pdf pinout
IP 3525 AJ	IC	Regulating Pulse Width Modulators	DIC	Seb	-	pdf pinout
IP 3525 AN	IC	Regulating Pulse Width Modulators	DIC	Seb	-	pdf pinout
IP 3527 AJ	IC	Regulating Pulse Width Modulators	DIC	Seb	-	pdf pinout
IP 3527 AN	IC	Regulating Pulse Width Modulators	DIC	Seb	-	pdf pinout
IP 3527 AD-16	IC	Regulating Pulse Width Modulators	DIC	Seb	-	pdf pinout
IP-	Si-N	=2SC3793 (Typ-Code/Stempel/marking)	35	Hit	→2SC3793	data
IP-	Si-N	=2SC4262 (Typ-Code/Stempel/marking)	35(2mm)	Hit	→2SC4262	data
IPS 021	MOS-N-FET-e	PROFET, 50V, 5A, <0,15Ω, 1,5ms	17c§	Inr	-	
IPS 021 L	MOS-N-FET-e	=IPS 021: SMD	48c§		-	
IPS 021 S	MOS-N-FET-e	=IPS 021:	30c§		-	
IPS 022 G	MOS-N-FET-e	SMD, Dual, PROFET, 50V, 5A, <0,15Ω	8-MDIP	Inr	-	
IPS 024 G	MOS-N-FET-e	SMD, Quad, PROFET, 50V, 5A, <0,15Ω	16-MDIP	Inr	-	
IPS 031	MOS-N-FET-e	PROFET, 50V, 12A, <0,06Ω, 1,5ms	17c§	Inr	-	
IPS 031 G	MOS-N-FET-e	SMD, PROFET, 50V, 12A, <0,07Ω, 1,5ms	8-MDIP	Inr	-	
IPS 031 S	MOS-N-FET-e	=IPS 031:	30c§		-	
IPS 032 G	MOS-N-FET-e	SMD, Dual, PROFET, 50V, 12A, <0,07Ω	8-MDIP	Inr	-	
IPS 041 L	MOS-N-FET-e	SMD, PROFET, 50V, 2A, <0,5Ω, 1,5ms	48c§	Inr	-	
IPS 042 G	MOS-N-FET-e	SMD, Dual, PROFET, 50V, 2A, <0,5Ω	8-MDIP	Inr	-	
IPS 151	MOS-N-FET-e	PROFET, 50V, 35A, <25mΩ, 1,5µs	17c§	Inr	-	
IPS 151 S	MOS-N-FET-e	=IPS 151:	30c§		-	
IPS 511	MOS-N-FET-e	PROFET, High Side Sw., 50V, 5A, <0,13Ω	86/5Pin	Inr	-	
IPS 511 G	MOS-N-FET-e	=IPS 511: SMD, <0,15Ω	8-MDIP		-	
IPS 511 S	MOS-N-FET-e	=IPS 511:	30/5Pin		-	
IPS 512 G	MOS-N-FET-e	=IPS 511: SMD, Dual, <0,15Ω	16-MDIP		-	
IPS 521	MOS-N-FET-e	PROFET, High Side Sw., 50V, 10A, <80mΩ	86/5Pin	Inr	-	
IPS 521 G	MOS-N-FET-e	=IPS 521: SMD	8-MDIP		-	
IPS 521 G	MOS-N-FET-e	=IPS 521: SMD, <0,1Ω	8-MDIP		-	
IPS 521 S	MOS-N-FET-e	=IPS 521:	30/5Pin		-	
IPS 0551 T	MOS-N-FET-e	PROFET, 40V, 100A, <5mΩ, 4µs	78c	Inr	-	
IPS 1021 PBF	MOS	Intelligent Power low Side Switch	17	Inr	-	pdf pinout
IPS 1021 RPBF	MOS	Intelligent Power low Side Switch	17	Inr	-	pdf pinout
IPS 1021 SPBF	MOS	Intelligent Power low Side Switch	17	Inr	-	pdf pinout
IPS 1031 PBF	MOS	Intelligent Power low Side Switch	17	Inr	-	pdf pinout
IPS 1031 RPBF	MOS	Intelligent Power low Side Switch	17	Inr	-	pdf pinout
IPS 1031 SPBF	MOS	Intelligent Power low Side Switch	17	Inr	-	pdf pinout
IPS 1041 LPBF	MOS	Single/dual Ch Intelligent Power low Side Switch	48	Inr	-	pdf pinout
IPS 1041 RPBF	MOS	Single/dual Ch Intelligent Power low Side Switch	48	Inr	-	pdf pinout
IPS 1042 GPBF	MOS	Single/dual Ch Intelligent Power low Side Switch	48	Inr	-	pdf pinout
IPS 1051 LPBF	MOS	Single/dual-ch Intelligent Power low Side Switch	48	Inr	-	pdf pinout
IPS 1052 GPBF	MOS	Single/dual-ch Intelligent Power low Side Switch	48	Inr	-	pdf pinout
IPS 2041 LPBF	MOS	Intelligent Power low Side Switch	48	Inr	-	pdf pinout
IPS 2041 RPBF	MOS	Intelligent Power low Side Switch	48	Inr	-	pdf pinout
IPS 5451	MOS-N-FET-e	PROFET, High Side Sw., 50V, 35A, <25mΩ	86/5Pin	Inr	-	
IPS 5451 S	MOS-N-FET-e	=IPS 5451:	30/5Pin		-	
IPS 5551 T	MOS-N-FET-e	PROFET, High Side Sw., 40V, 100A, <5mΩ	78	Inr	-	
IPS 6011 PBF	MOS	Intelligent Power High Side Switch	17	Inr	-	pdf pinout
IPS 6011 RPBF	MOS	Intelligent Power High Side Switch	17	Inr	-	pdf pinout
IPS 6011 SPBF	MOS	Intelligent Power High Side Switch	17	Inr	-	pdf pinout
IPS 6021 PBF	MOS	Intelligent Power High Side Switch	17	Inr	-	pdf pinout
IPS 6021 RPBF	MOS	Intelligent Power High Side Switch	17	Inr	-	pdf pinout
IPS 6021 SPBF	MOS	Intelligent Power High Side Switch	17	Inr	-	pdf pinout
IPS 6031 PBF	MOS	Intelligent Power High Side Switch	17	Inr	-	pdf pinout
IPS 6031 RPBF	MOS	Intelligent Power High Side Switch	17	Inr	-	pdf pinout
IPS 6031 SPBF	MOS	Intelligent Power High Side Switch	17	Inr	-	pdf pinout
IPS 6041 GPBF	MOS	Intelligent Power High Side Switch	17	Inr	-	pdf pinout
IPS 6041 PBF	MOS	Intelligent Power High Side Switch	17	Inr	-	pdf pinout
IPS 6041 RPBF	MOS	Intelligent Power High Side Switch	17	Inr	-	pdf pinout
IPS 6041 SPBF	MOS	Intelligent Power High Side Switch	17	Inr	-	pdf pinout
IPS 6044 GPBF	MOS	Intelligent Power High Side Switch	17	Inr	-	pdf pinout
IPS 7071 GPBF	DIG-IC	Protected High Side Switch	17	Inr	-	pdf pinout
IPS 7081 PBF	DIG-IC	Intelligent Power High Side Switch	17	Inr	-	pdf pinout
IPS 7081 RPBF	DIG-IC	Intelligent Power High Side Switch	17	Inr	-	pdf pinout
IPS 7081 SPBF	DIG-IC	Intelligent Power High Side Switch	17	Inr	-	pdf pinout
IPS 7091 GPBF	DIG-IC	Intelligent Power High Side Switch	17	Inr	-	pdf pinout
IPS 7091 PBF	DIG-IC	Intelligent Power High Side Switch	17	Inr	-	pdf pinout
IPS 7091 SPBF	DIG-IC	Intelligent Power High Side Switch	17	Inr	-	pdf pinout
IQ	Si-P	=2SB1073-Q (Typ-Code/Stempel/marking)	39	Mat	→2SB1073	data
IQ	Si-P	=2SB1220-Q (Typ-Code/Stempel/marking)	35(2mm)	Mat	→2SB1220	data
IQ	Si-P	=2SB792-Q (Typ-Code/Stempel/marking)	35	Mat	→2SB792	data

Type	Device	Short Description	Fig.	Manu	Comparision Types	· More at
▪R	Si-P	=2SB1073-R (Typ-Code/Stempel/marking)	39	Mat	→2SB1073	data
▪R	Si-P	=2SB1220-R (Typ-Code/Stempel/marking)	35(2mm)	Mat	→2SB1220	data
▪R	Si-P	=2SB1463-R (Typ-Code/Stempel/marking)	35(1,6mm)	Mat	→2SB1463	data
▪R	Si-P	=2SB792-R (Typ-Code/Stempel/marking)	35	Mat	→2SB792	data
▪R	Si-N	=2SC3722K-R (Typ-Code/Stempel/marking)	35	Rhm	→2SC3722K	data
▪R	Si-N	=2SC5262-R (Typ-Code/Stempel/marking)	44	Tos	→2SC5262	
▪R 01 H420	Hybrid-IC	Mosfet Half Bridge, 500V, on=3Ω	9-SIP	Inr	-	
▪R 2 E02	LIN-IC	→KA 2288	16-DIP	Sha	KA 2288	pinout
▪R 02 H420	Hybrid-IC	Mosfet Half Bridge, 500V, on=3Ω	9-SIP	Inr	-	pdf
▪R3F01	2xOP	Vs:36V Vu:106dB Vo:±13.5V Vi0:2mV	8DS			data pinout
▪R3F02	2xOP	Vs:36V Vu:110dB Vo:±14V Vi0:0.5mV	8D			data pinout
▪R 03 H420	Hybrid-IC	Mosfet Half Bridge, 500V, on=3Ω	9-SIP	Inr	-	
▪R 3 K02:ECL	D/A-IC	8 Bit, Video, >20Msps, Ecl Level	40-DIP	Sha		
IR 3 K02:TTL	D/A-IC	8 Bit, Video, >20Msps, TTL Level	40-DIP	Sha		
IR 3 K03A	A/D-IC	8 Bit, Video, 20Msps, Ecl Level	28-DIP	Sha		
IR 3 K04A	A/D-IC	8 Bit, Video, 20Msps, Ecl Level	28-DIP	Sha		
IR 3 K06	CMOS-A/D-IC	8 Bit, Video, >20Msps, CMOS/TTL Level	28-DIP	Sha		
IR 3 K07	D/A-IC	8 Bit, Video, 40Msps, TTL Level	20-DIP	Sha		
IR 3 M01	LIN-IC	Sw-Reg, +35V, 0.1A, 5..200kHz	16-DIP	Sha	-	pdf
IR 3 M02(A)	LIN-IC	S-Reg, +41V, 0,25A, 1...300kHz	16-DIP	Sha	µPC 494	pinout
IR 3 M02N,AN	LIN-IC	=IR 3M02(A): SMD, 0,5W	16-MDIP		-	pinout
▪R 3 M03A	SMPS-IC	DC-DC Converter, Uout=1,25...40V, 1,8A	8-DIP	Sha	-	
IR 3 M04	SMPS-IC	S-Reg, -20...+20mA, 100...500kHz	8-DIP	Sha	-	
IR 3 M04N	SMPS-IC	=IR 3M04: SMD	14-MDIP			
IR 3 M05	SMPS-IC	SMD, S-Reg, 50/150/150mA, 1...300kHz	24-MDIP	Sha	-	
IR 3 N05	LIN-IC	=LM 567CN	8-DIP	Sha	→LM 567CN	pinout
IR 3 N06	LIN-IC	FM-ZF + Limiter + Squelch	16-DIP	Sha	-	
IR 3 P04	LIN-IC	Ctv, NTSC-Video + Chroma + Synchr.	42-DIP	Sha	KA 2153, TA 7644	
IR 3 R10	LIN-IC	Mikrof.-vorverst./pre-amp. + Agc	9-SIP	Sha		
IR 3 R13	LIN-IC	Recorder Alc	9-SIP	Sha		
IR 3 R14	LIN-IC	2x Kopfh.-Verst./head phone amp.	16-DIP	Sha		
IR 3 R15	LIN-IC	NF-V, Ucc=±3..20V	9-SIP	Sha		
IR 3 R16	LIN-IC	2x Recorder-Kopfverst./head amplifier	9-SIP	Sha		
IR 3 R17	LIN-IC	2x Recorder-Verst./amplifier + Alc	16-DIP	Sha		
IR 3 R18	LIN-IC	2x Vorverst./preamplifier + Equalizer	9-SIP	Sha		
IR 3 R21	LIN-IC	2x Vorverst./preamplifier + Agc	16-DIP	Sha		
IR 3 R22	LIN-IC	NF-E, 8V, >70mW(3V/4Ω)	8-DIP	Sha		
IR 3 R24	LIN-IC	→KA 2230	22-DIP	Sha	KA 2230	
IR 04 H420	Hybrid-IC	Mosfet Half Bridge, 500V, on=3Ω	9-SIP	Inr	-	
IR 9 K08	LIN-IC		8-DIP	Sha		
IR 51 H214	Hybrid-IC	Mosfet Half Bridge, 250V, on=2Ω	9-SIP	Inr	-	pdf
IR 51 H224	Hybrid-IC	Mosfet Half Bridge, 250V, on=1,1Ω	9-SIP	Inr	-	pdf
IR 51 H310	Hybrid-IC	Mosfet Half Bridge, 400V, on=3,6Ω	9-SIP	Inr	-	pdf
IR 51 H320	Hybrid-IC	Mosfet Half Bridge, 400V, on=1,8Ω	9-SIP	Inr	-	pdf
IR 51 H420	Hybrid-IC	Mosfet Half Bridge, 500V, on=3Ω	9-SIP	Inr	-	pdf
IR 51 H737	Hybrid-IC	Mosfet Half Bridge, 300V, on=0,8Ω	9-SIP	Inr	-	pdf
IR 51 HD214	Hybrid-IC	Mosfet Half Bridge, 250V, on=2Ω	9-SIP	Inr	-	pdf
IR 51 HD224	Hybrid-IC	Mosfet Half Bridge, 250V, on=3,6Ω	9-SIP	Inr	-	pdf
IR 51 HD310	Hybrid-IC	Mosfet Half Bridge, 400V, on=1,8Ω	9-SIP	Inr	-	pdf
IR 51 HD320	Hybrid-IC	Mosfet Half Bridge, 400V, on=1,8Ω	9-SIP	Inr	-	pdf
IR 51 HD420	Hybrid-IC	Mosfet Half Bridge, 500V, on=3Ω	9-SIP	Inr	-	pdf
IR 51 HD737	Hybrid-IC	Mosfet Half Bridge, 300V, on=0,8Ω	9-SIP	Inr	-	pdf
IR 062 HD4C10U-P2	Hybrid-IC	IGBT Half Bridge, 600V, Uceon=1,5V	9-SIP	Inr	-	
IR 082 HD4C10U-P2	Hybrid-IC	IGBT Half Bridge, 600V, Uceon=1,5V	9-SIP	Inr	-	
IR 3 Y 29 AM	IC	IC,NTSC/PAL,TFT,Lcd,color monitors	QFP	Say	-	pdf pinout
IR 1110	LIN-IC	SMD, Soft Start Controller, ±5V, ±5mA	64-MP	Inr	-	
IR 1150	CMOS-IC	µPFC 1 Cycle Cntrl. Pfc IC	DIC	Inr	-	pdf pinout
IR 1150 I	CMOS-IC	µPFC 1 Cycle Cntrl. Pfc IC	DIC	Inr	-	pdf pinout
IR 1150 IS	CMOS-IC	µPFC 1 Cycle Cntrl. Pfc IC	MDIP	Inr	-	pdf pinout
IR 1150 S	CMOS-IC	µPFC 1 Cycle Cntrl. Pfc IC	MDIP	Inr	-	pdf pinout
IR 1166 SPbF	IC	Smartrectifier Control IC	MDIP	Inr	-	pdf pinout
IR 1167 ASPbF	IC	Smartrectifier Control IC	MDIP	Inr	-	pdf pinout
IR 1167 BSPbF	IC	Smartrectifier Control IC	MDIP	Inr	-	pdf pinout
IR 1210	LIN-IC	SMD, Dual Low Side Driver, Io=±1.5A	8-MDIP	Inr	-	pdf
IR 2085 S	IC	100V Self Oscillat. Half-Bridge Driver	MDIP	Inr	-	pdf pinout
IR 2101	CMOS/LIN-IC	Mosfet Gate Drv, Uoffs=600V, 0,1/0,21A	8-DIP	Inr	-	pdf
IR 2101 S	CMOS/LIN-IC	SMD,Mosfet Gate Drv, Uoffs=600V	8-MDIP		-	pdf
IR 2102	CMOS/LIN-IC	Mosfet Gate Drv, Uoffs=600V, 0,1/0,21A	8-DIP	Inr	-	pdf
IR 2102 S	CMOS/LIN-IC	SMD,Mosfet Gate-Drv, Uoffs=600V	8-MDIP		-	pdf
IR 2103	CMOS/LIN-IC	Mosfet Gate Drv, Uoffs=600V, 0,1/0,21A	8-DIP	Inr	-	pdf
IR 2103 S	CMOS/LIN-IC	SMD,Mosfet Gate Drv, Uoffs=600V	8-MDIP		-	pdf
IR 2104	CMOS/LIN-IC	Mosfet Gate Drv, Uoffs=600V, 0,1/0,21A	8-DIP	Inr	-	pdf
IR 2104 S	CMOS/LIN-IC	SMD,Mosfet Gate Drv, Uoffs=600V	8-MDIP		-	pdf
IR 2105	CMOS/LIN-IC	Mosfet Gate Drv,Uoffs=600V, 0,13/0,27A	8-DIP	Inr	-	pdf
IR 2105 S	CMOS/LIN-IC	SMD,Mosfet Gate Drv,Uoffs=600V	8-MDIP		-	pdf
IR 2106	CMOS/LIN-IC	Mosfet Gate Drv, Uoffs=600V, 0,12/0,25A	8-DIP	Inr	-	pdf
IR 2106 S	CMOS/LIN-IC	SMD,Mosfet Gate Drv,Uoffs=600V	8-MDIP		-	pdf
IR 2107	CMOS/LIN-IC	Mosfet Gate Drv,Uoffs=600V, 0,12/0,25A	8-DIP	Inr	-	
IR 2107 S	CMOS/LIN-IC	SMD,Mosfet Gate Drv,Uoffs=600V	8-MDIP		-	
IR 2108	CMOS/LIN-IC	Mosfet Gate Drv,Uoffs=600V, 0,12/0,25A	DIC	Inr	-	pdf pinout
IR 2108 S	CMOS/LIN-IC	SMD,Mosfet Gate Drv,Uoffs=600V	8-MDIP		-	pdf pincut
IR 2109	CMOS/LIN-IC	Mosfet Gate Drv,Uoffs=600V, 0,12/0,25A	8-DIP	Inr	-	pdf pinout
IR 2109 S	CMOS/LIN-IC	SMD,Mosfet Gate Drv,Uoffs=600V	8-MDIP		-	pdf pinout
IR 2110	CMOS/LIN-IC	Mosfet Gate Driver, Uoffs=-4...500V	14-DIP	Inr	-	pdf
IR 2110 E4	CMOS/LIN-IC	2A High & Low Side Drv	LCC	Inr	-	pdf pinout

Type	Device	Short Description	Fig.	Manu	Comparision Types	More at
IR 2110 L4	CMOS/LIN-IC	High and Low Side Driver	DIC	Inr	-	pdf pinout
IR 2110 S	CMOS/LIN-IC	SMD,Mosfet Gate Driver, Uoffs=-4...500V	16-MDIP	Inr	-	pdf
IR 2111	CMOS/LIN-IC	Mosfet Gate Driver, Uoffs=-5...600V	8-DIP	Inr	-	pdf
IR 2111 S	CMOS/LIN-IC	SMD,Mosfet Gate Driver, Uoffs=-5...600V	8-MDIP		-	pdf
IR 2112	CMOS/LIN-IC	Mosfet Gate Driver, Uoffs=-5...600V	14-DIP	Inr	-	pdf
IR 2112 S	CMOS/LIN-IC	SMD,Mosfet Gate Driver, Uoffs=-5...600V	16-MDIP			
IR 2113	CMOS/LIN-IC	Mosfet Gate Driver, Uoffs=-5...600V	14-DIP	Inr	-	pdf
IR 2113 E6	CMOS/LIN-IC	High & Low Side Drv.	LCC	Inr	-	pdf pinout
IR 2113 L6	CMOS/LIN-IC	High & Low Side Drv.	DIP	Inr	-	pdf pinout
IR 2113 S	CMOS/LIN-IC	SMD,Mosfet Gate Driver, Uoffs=-5...600V	16-MDIP	Inr	-	pdf
IR 2114 SSPbF	CMOS/LIN-IC	600V Half-Bridge Gate Driver IC	TSOP	Inr	-	pdf pinout
IR 2117	CMOS/LIN-IC	Mosfet Gate Driver, Uoffs=-5...600V	8-DIP	Inr	-	pdf
IR 2117 S	CMOS/LIN-IC	SMD,Mosfet Gate Driver, Uoffs=-5...600V	8-MDIP		-	
IR 2118	CMOS/LIN-IC	Mosfet Gate Drv, Uoffs=600V, 0,2/0,42A	8-DIP	Inr	-	pdf
IR 2118 S	CMOS/LIN-IC	SMD,Mosfet Gate Drv, Uoffs=600V	8-MDIP		-	
IR 2121	CMOS/LIN-IC	Mosfet Gate Drv, Uoffs=5V, 1/2A	8-DIP	Inr	-	pdf
IR 2122	CMOS/LIN-IC	Mosfet Gate Drv,Uoffs=600V, 0,11/0,11A Current Sensing Single Channel Driver	8-DIP	Inr	-	pdf
IR 2122 S	CMOS/LIN-IC	SMD,Mosfet Gate Drv,Uoffs=600V	8-MDIP		-	pdf
IR 2125	CMOS/LIN-IC	Mosfet Gate Driver, Uoffs=-5...500V	8-DIP	Inr	-	pdf
IR 2127	CMOS/LIN-IC	Mosfet Gate Drv, Uoffs=600V, 0,2/0,42A	8-DIP	Inr	-	pdf
IR 2128	CMOS/LIN-IC	Mosfet Gate Drv, Uoffs=600V, 0,2/0,42A	8-DIP	Inr	-	pdf
IR 2130	CMOS/LIN-IC	Mosfet Gate Driver, Uoffs=-5...600V	28-DIP	Inr	-	pdf
IR 2130 D	CMOS/LIN-IC	3-Ph. Driver, +400V	DIP	Inr	-	pdf pinout
IR 2131	CMOS/LIN-IC	Mosfet Gate Driver, Uoffs=-5...600V	28-DIP	Inr	-	pdf
IR 2132	CMOS/LIN-IC	Mosfet Gate Drv, Uoffs=600V, 0,2/0,42A	28-DIP	Inr	-	pdf
IR 2133	CMOS/LIN-IC	Mosfet Gate Drv, Uoffs=600V, 0,2/0,42A	44-MP	Inr	-	pdf
IR 2135	CMOS/LIN-IC	Mosfet Gate Drv, Uoffs=600V, 0,2/0,42A	44-MP	Inr	-	pdf
IR 2136	CMOS/LIN-IC	Mosfet Gate Drv,Uoffs=600V, 0,12/0,25A 3-Phase Bridge Driver	28-DIP	Inr	-	pdf pinout
IR 2136 J	CMOS/LIN-IC	Mosfet Gate Drv,Uoffs=600V, 0,12/0,25A 3-Phase Bridge Driver	PLCC	Inr	-	pdf pinout
IR 2136 S	CMOS/LIN-IC	Mosfet Gate Drv,Uoffs=600V, 0,12/0,25A 3-Phase Bridge Driver	MDIP	Inr	-	pdf pinout
IR 2137	CMOS/LIN-IC	Mosfet Gate Drv, Uoffs=600V, 0,2/0,46A 3-Phase Bridge Driver	68-PLCC	Inr	-	pdf
IR 2151	CMOS/LIN-IC	Mosfet Gate Drv, Uoffs=600V, 0,1/0,21A	8-DIP	Inr	-	pdf
IR 2152	CMOS/LIN-IC	Mosfet Gate Drv, Uoffs=600V, 0,1/0,21A	8-DIP	Inr	-	pdf
IR 2153	CMOS/LIN-IC	Mosfet Gate Drv, Uoffs=600V, 0,2/0,4A	8-DIP	Inr	-	pdf
IR 2154	CMOS/LIN-IC	Mosfet Gate Drv, Uoffs=600V	8-DIP	Inr	-	pdf
IR 2154 S	CMOS/LIN-IC	SMD,Mosfet Gate Drv, Uoffs=600V	8-MDIP		-	
IR 2155	CMOS/LIN-IC	Mosfet Gate Driver, Uoffs=-5...600V	8-DIP	Inr	-	pdf
IR 2156 PbF	IC	Ballast Cntrl. IC	DIC	Inr	-	pdf pinout
IR 2156 SPbF	IC	Ballast Cntrl. IC	MDIP	Inr	-	pdf pinout
IR 2157	LIN-IC	600V Ballast Control	16-DIP	Inr	-	
IR 2157 S	LIN-IC	SMD,600V Ballast Control	16-MDIP		-	
IR 2159	LIN-IC	600V Dimming Ballast Control	16-DIP	Inr	-	
IR 2159 S	LIN-IC	SMD,600V Dimming Ballast Control	16-MDIP		-	pdf
IR 2161	IC	Halogen Conv. Cntrl. IC	DIC	Inr	-	pdf pinout
IR 2161 S	IC	Halogen Conv. Cntrl. IC	MDIP	Inr	-	pdf pinout
IR 2166 PbF	IC	Pfc & Ballast Control IC	DIC	Inr	-	pdf pinout
IR 2166 SPbF	IC	Pfc & Ballast Control IC	MDIP	Inr	-	pdf pinout
IR 2175	LIN-IC	linear Current Sensing IC	DIP	Inr	-	pdf pinout
IR 2175 S	LIN-IC	linear Current Sensing IC	DIP	Inr	-	pdf pinout
IR 2181	IC	High and low Side Driver	DIP	Inr	-	pdf pinout
IR 2181 S	IC	High and low Side Driver	DIP	Inr	-	pdf pinout
IR 2183	CMOS-IC	Half-bridge Driver	DIP	Inr	-	pdf pinout
IR 2183 S	CMOS-IC	Half-bridge Driver	TSOP	Inr	-	pdf pinout
IR 2184	CMOS-IC	600V Half-Bridge Driver	DIP	Inr	-	pdf pinout
IR 2184 S	CMOS-IC	600V Half-Bridge Driver	MDIP	Inr	-	pdf pinout
IR 2213	CMOS/LIN-IC	Mosfet Gate Drv, Uoffs=1200V, 1,7/2A	14-DIP	Inr	-	pdf pinout
IR 2213 S	LIN-IC	SMD,Mosfet Gate Drv, Uoffs=1200V	16-MDIP	Inr	-	pdf pinout
IR 2214 SSPbF	CMOS/LIN-IC	600V Half-Bridge Gate Driver IC	TSOP	Inr	-	pdf
IR 2233	CMOS/LIN-IC	Mosfet Gate Drv,Uoffs=1200V, 0,2/0,42A	44-MP	Inr	-	pdf
IR 2235	CMOS/LIN-IC	Mosfet Gate Drv,Uoffs=1200V, 0,2/0,42A	44-MP	Inr	-	pdf
IR 2301	CMOS/LIN-IC	600V High & Low Side Drv	DIC	Inr	-	pdf pinout
IR 2301 S	CMOS/LIN-IC	600V High & Low Side Drv	MDIP	Inr	-	pdf pinout
IR 2302	CMOS/LIN-IC	600V Half-Bridge Driver	DIC	Inr	-	pdf pinout
IR 2302 S	CMOS/LIN-IC	600V Half-Bridge Driver	MDIP	Inr	-	pdf pinout
IR 2304	CMOS/LIN-IC	600V Half-Bridge Driver	DIC	Inr	-	pdf pinout
IR 2304 S	CMOS/LIN-IC	600V Half-Bridge Driver	MDIP	Inr	-	pdf pinout
IR 2308	CMOS/LIN-IC	600V Half-Bridge Drv	DIC	Inr	-	pdf pinout
IR 2308 S	CMOS/LIN-IC	600V Half-Bridge Drv	MDIP	Inr	-	pdf pinout
IR 2403-2	IC					
IR 2520 D	IC	Adaptive Ballast Cntrl. IC	DIC	Inr	-	pdf pinout
IR 2520 DS	IC	Adaptive Ballast Cntrl. IC	MDIP	Inr	-	pdf pinout
IR 3080 M	LIN-IC	xphase VRD10 Control IC with VCCVID & Overtemp detect	MLP	Inr	-	pdf pinout
IR 3081 AM	LIN-IC	Xphase VR 10 Control IC	MLP	Inr	-	pdf pinout
IR 3082 AMPBF	LIN-IC	9.6V AMD Xphase Control IC	MLP	Inr	-	pdf pinout
IR 3082 MPBF	LIN-IC	Xphase AMD Opteron/Athlon 64 Control IC	MLP	Inr	-	pdf pinout
IR 3084 AMPBF	LIN-IC	Xphase VR 10/11 Control IC	MLP	Inr	-	pdf pinout
IR 3084 UMPBF	LIN-IC	Xphase VR10, VR11 & Opteron/Athlon64 Control IC	MLP	Inr	-	pdf pinout
IR 3086 AM(TR)	LIN-IC	Xphase Phase IC,Ovp,Fault/overtempdetect	MLP	Inr	-	pdf pinout

Type	Device	Short Description	Fig.	Manu	Comparision Types	More at
IR 3087 (M)TR	LIN-IC	Xphase Phase IC with Opti-phase, Ovp	MLP	Inr	-	pdf pinout
R 3088 AM(TR)	LIN-IC	Xphase Phase IC with Fault, Overtemp DETECT	MLP	Inr	-	pdf pinout
IR 3094 PBF	MOS	3 Phase PWM Controller For point of load	MLP	Inr	-	pdf pinout
IR 3101	CMOS/LIN-IC	1.6A, 500V, Half-Bridge FredFET Drv	SIP	Inr	-	pdf pinout
R 3103	CMOS/LIN-IC	0.75A, 500V, Half-Bridge FredFET Drv	SIP	Inr	-	pdf pinout
IR 3220 S /8	IC	Protected H-bridge For D.c. Motor	TSOP	Inr	-	pdf pinout
IR 3313 PBF	DIG-IC	Programmable Current Sense High Side SWITCH	17	Inr	-	pdf pinout
IR 3313 SPBF	DIG-IC	Programmable Current Sense High Side SWITCH	17	Inr	-	pdf pinout
IR 3314 PBF	DIG-IC	Programmable Current Sense High Side SWITCH	17	Inr	-	pdf pinout
R 3315 PBF	DIG-IC	low Emi Current Sense High Side Switch	17	Inr	-	pdf pinout
R 3315 SPBF	DIG-IC	low Emi Current Sense High Side Switch	17	Inr	-	pdf pinout
R 3316 PBF	DIG-IC	low Emi Current Sense High Side Switch	87	Inr	-	pdf pinout
R 3316 S	DIG-IC	low Emi Current Sense High Side Switch	87	Inr	-	pdf pinout
R 3316 SPBF	DIG-IC	low Emi Current Sense High Side Switch	87	Inr	-	pdf pinout
R3404	OP	Vs:20V Vu:110dB Vi0:2mV	9I			data pinout
R 3500 M(TR)PBF	LIN-IC	Xphase3 VR11.0 & AMD PVID Control IC	MLP	Inr	-	pdf pinout
R 3502 A	LIN-IC	Xphase3 Control IC	MLP	Inr	-	pdf
R 3505 MPBF	LIN-IC	Xphase3 Phase IC	MLP	Inr	-	pdf pinout
R 3507 M(TR)PBF	LIN-IC	Xphase3 Phase IC	MLP	Inr	-	pdf pinout
R 3513 MPBF	LIN-IC	Xphase3 pol Control IC	MLP	Inr	-	pdf pinout
R 3519 MPBF	MOS	Synchronous Mosfet Gate Driver IC	MLP	Inr	-	pdf pinout
R 3519 SPBF	MOS	Synchronous Mosfet Gate Driver IC	MLP	Inr	-	pdf pinout
R 3621 F	LIN-IC	synchronous PWM Controller	MLP	Inr	-	pdf pinout
R 3621 M	LIN-IC	synchronous PWM Controller	MLP	Inr	-	pdf pinout
R 3622 AMPBF	LIN-IC	synchronous step Down Controller	MLP	Inr	-	pdf pinout
R 3622 MPBF	LIN-IC	synchronous step Down Controller	MLP	Inr	-	pdf pinout
R 3623 MPBF	MOS	synchronous step Down Controller	MLP	Inr	-	pdf pinout
R 3624 MPBF	LIN-IC	synchronous PWM buck Controller	MLP	Inr	-	pdf pinout
R 3628 MPBF	LIN-IC	synchronous PWM buck Controller	MLP	Inr	-	pdf pinout
R 3629 AMPBF	LIN-IC	synchronous PWM buck Controller	MLP	Inr	-	pdf pinout
R 3629 MPBF	LIN-IC	synchronous PWM buck Controller	MLP	Inr	-	pdf pinout
R 3637 ASPBF	LIN-IC	1% accurate synchronous PWM Controller	MLP	Inr	-	pdf pinout
R 3637 SPBF	LIN-IC	1% accurate synchronous PWM Controller	MLP	Inr	-	pdf pinout
R 3638 SPBF	LIN-IC	synchronous PWM buck Controller	MLP	Inr	-	pdf pinout
R 3651 SPBF	LIN-IC	synchronous PWM buck Controller	MLP	Inr	-	pdf pinout
R3702	4xOP	Vs:±36V Vu:95dB Vo:6/7V Vi0:2mV	14DS		-	data pinout
R 3721 MPBF	LIN-IC	Power Monitor IC, Analog Output	20-DIP	Inr	-	pdf pinout
R3741	OP	Vs:±18V Vu:106dB Vo:±14V Vi0:1mV	8D		-	data pinout
R 3810 MTRPBF	LIN-IC	12A, synchronous buck regulator	20-DIP	Inr	-	pdf pinout
R 3811 MTRPBF	LIN-IC	7A, synchronous buck regulator	20-DIP	Inr	-	pdf pinout
R 3820 MTRPBF	MOS-IC	12A, synchronous buck regulator	20-DIP	Inr	-	pdf pinout
R 3821 AMTRPBF	MOS-IC	9A, synchronous buck regulator	20-DIP	Inr	-	pdf pinout
R 3821 MTRPBF	MOS-IC	7A, synchronous buck regulator	20-DIP	Inr	-	pdf pinout
R 3822 AMTRPBF	MOS-IC	6A, synchronous buck regulator	20-DIP	Inr	-	pdf pinout
R 3822 MTRPBF	MOS-IC	synchronous buck regulator	20-DIP	Inr	-	pdf pinout
R 4426	CMOS/LIN-IC	1.5A, 2x Low Side Drv	DIP	Inr	-	pdf pinout
R 4426 S	CMOS/LIN-IC	1.5A, 2x Low Side Drv	MDIP	Inr	-	pdf pinout
R 4427	CMOS/LIN-IC	1.5A, 2x Low Side Drv	DIP	Inr	-	pdf pinout
R 4427 S	CMOS/LIN-IC	1.5A, 2x Low Side Drv	MDIP	Inr	-	pdf pinout
R 4428	CMOS/LIN-IC	1.5A, 2x Low Side Drv	DIP	Inr	-	pdf pinout
R 4428 S	CMOS/LIN-IC	1.5A, 2x Low Side Drv	MDIP	Inr	-	pdf pinout
R 5001 S	MOS-IC	universal Active Oring Controller	20-DIP	Inr	-	pdf pinout
R 5001 STR	MOS-IC	universal Active Oring Controller	20-DIP	Inr	-	pdf pinout
R 6000	DMOS-IC	High-Side Power Switch, 5...35V, 3,5A		Inr	-	
R 6210	MOS-IC	Intelligent High Side Mosfet Power Sw. 5...50V, 28W, on=0,2Ω, Ilim=5A	86/5Pin	Inr		pdf
R 6216	MOS-IC	Intelligent High Side Mosfet Power Sw. 5...50V, 28W, on=0,2Ω, Ilim=10A	86/5Pin	Inr		pdf
R 6220	MOS-IC	Intelligent High Side Mosfet Power Sw. 5...50V, 28W, on=0,1Ω, Ilim=10A	86/5Pin	Inr		pdf
R 6224	MOS-IC	Intelligent High Side Mosfet Power Sw. 5...50V, 28W, on=0,1Ω, Ilim=20A	86/5Pin	Inr		pdf
R 6226	MOS-IC	Intelligent High Side Mosfet Power Sw. 5...50V, 28W, on=0,1Ω, Ilim=20A	86/5Pin	Inr		pdf
R 6311 G	MOS-IC	Intelligent High Side Mosfet Power Sw. 5...50V, on=0,15Ω, Ilim=5A	8-MDIP	Inr		
R 6320 G	MOS-IC	Intelligent High Side Mosfet Power Sw. 5...50V, on=85mΩ, Ilim=10A	8-MDIP	Inr		
R9022	2xOP	Vs:36V Vu:80dB Vo:26/0V Vi0:1mV	8D			data pinout
R 9161	KOP-IC	Quad, ±18V, 0...+70°	16-DIP	Sha		
R 9161 N	KOP-IC	=IR 9161: SMD	16-MDIP			
R 9331	LIN-IC	V/F-Converter	8-DIP	Sha		
R 9331 N	LIN-IC	SMD, V/f-converter	8-MDIP			
R 9431	LIN-IC	Regler/shunt reg., Uref=2,5V	8-DIP	Sha	(TL431CP)	
R 9494	LIN-IC	S-Reg, +41V, 0,25A, 1...300kHz	16-DIP	Sha		
R 9494 N	LIN-IC	=IR 9494: SMD	16-MDIP			
R9725	OP	Vs:36V Vu:130dB Vo:±13V Vi0:0.5mV	8DS			data pinout
R 21014	CMOS/LIN-IC	Mosfet Gate Drv, Uoffs=600V, 0,1/0,21A	14-DIP	Inr	-	pdf
R 21014 S	CMOS/LIN-IC	SMD,Mosfet Gate Drv, Uoffs=600V	14-MDIP		-	
R 21024	CMOS/LIN-IC	Mosfet Gate Drv, Uoffs=600V, 0,1/0,21A	14-DIP	Inr	-	pdf
R 21024 S	CMOS/LIN-IC	SMD,Mosfet Gate Drv, Uoffs=600V	14-MDIP		-	

Type	Device	Short Description	Fig.	Manu	Comparision Types	More at
IR 21034	CMOS/LIN-IC	Mosfet Gate Drv, Uoffs=600V, 0,1/0,21A	14-DIP	Inr	-	
IR 21034 S	CMOS/LIN-IC	SMD,Mosfet Gate Drv, Uoffs=600V	14-MDIP		-	
IR 21044	CMOS/LIN-IC	Mosfet Gate Drv, Uoffs=600V, 0,1/0,21A	14-DIP	Inr	-	
IR 21044 S	CMOS/LIN-IC	SMD,Mosfet Gate Drv, Uoffs=600V	14-MDIP			
IR 21064	CMOS/LIN-IC	Mosfet Gate Drv,Uoffs=600V, 0,12/0,25A	14-DIP	Inr	-	pdf
IR 21064 S	CMOS/LIN-IC	SMD,Mosfet Gate Drv,Uoffs=600V	14-MDIP		-	pdf
IR 21074	CMOS/LIN-IC	Mosfet Gate Drv,Uoffs=600V, 0,12/0,25A	14-DIP	Inr	-	
IR 21074 S	CMOS/LIN-IC	SMD,Mosfet Gate Drv,Uoffs=600V	14-MDIP		-	
IR 21084	CMOS/LIN-IC	Mosfet Gate Drv,Uoffs=600V, 0,12/0,25A	14-DIP	Inr	-	pdf
IR 21084 S	CMOS/LIN-IC	SMD,Mosfet Gate Drv,Uoffs=600V	14-MDIP		-	pdf pinout
IR 21091	CMOS/LIN-IC	600V Half-Bridge Driver	DIC	Inr	-	pdf pinout
IR 21091 S	CMOS/LIN-IC	600V Half-Bridge Driver	MDIP	Inr	-	pdf pinout
IR 21094	CMOS/LIN-IC	Mosfet Gate Drv,Uoffs=600V, 0,12/0,25A	14-DIP	Inr	-	pdf pinout
IR 21094 S	CMOS/LIN-IC	SMD,Mosfet Gate Drv,Uoffs=600V	14-MDIP	Inr	-	pdf pinout
IR 21141 SSPbF	CMOS/LIN-IC	Half-Bridge Gate Driver IC	TSOP	Inr	-	pdf pinout
IR 21362	CMOS/LIN-IC	Mosfet Gate Drv,Uoffs=600V, 0,12/0,25A 3-Phase Bridge Driver	28-DIP	Inr	-	pdf pinout
IR 21362 J	CMOS/LIN-IC	Mosfet Gate Drv,Uoffs=600V, 0,12/0,25A 3-Phase Bridge Driver	PLCC	Inr	-	pdf pinout
IR 21362 S	CMOS/LIN-IC	Mosfet Gate Drv,Uoffs=600V, 0,12/0,25A 3-Phase Bridge Driver	MDIP	Inr	-	pdf pinout
IR 21363	CMOS/LIN-IC	Mosfet Gate Drv,Uoffs=600V, 0,12/0,25A 3-Phase Bridge Driver	28-DIP	Inr	-	pdf pinout
IR 21365	CMOS/LIN-IC	Mosfet Gate Drv,Uoffs=600V, 0,12/0,25A 3-Phase Bridge Driver	28-DIP	Inr	-	pdf pinout
IR 21366	CMOS/LIN-IC	Mosfet Gate Drv,Uoffs=600V, 0,12/0,25A 3-Phase Bridge Driver	28-DIP	Inr	-	pdf pinout
IR 21367	CMOS/LIN-IC	Mosfet Gate Drv,Uoffs=600V, 0,12/0,25A 3-Phase Bridge Driver	28-DIP	Inr	-	pdf pinout
IR 21368	CMOS/LIN-IC	Mosfet Gate Drv,Uoffs=600V, 0,12/0,25A 3-Phase Bridge Driver	28-DIP	Inr	-	pdf pinout
IR 21381 Q	LIN-IC	3-Ph. AC Motor Cntrl. IC	QFP	Inr	-	pdf pinout
IR 21531(D)	CMOS/LIN-IC	Mosfet Gate Drv, Uoffs=600V	8-DIP	Inr	-	
IR 21531(D)S	CMOS/LIN-IC	SMD,Mosfet Gate Drv, Uoffs=600V	8-MDIP			
IR 21541	CMOS/LIN-IC	Mosfet Gate Drv, Uoffs=600V	8-DIP	Inr	-	
IR 21541 S	CMOS/LIN-IC	SMD,Mosfet Gate Drv, Uoffs=600V	8-MDIP			
IR 21571	LIN-IC	Fully Integrated Ballast Control IC	DIC	Inr	-	pdf pinout
IR 21571 S	LIN-IC	Fully Integrated Ballast Control IC	TSOP	Inr	-	pdf pinout
IR 21592	IC	Dimming Ballast Cntrl. IC	DIC	Inr	-	pdf pinout
IR 21592 S	IC	Dimming Ballast Cntrl. IC	MDIP	Inr	-	pdf pinout
IR 21593	IC	Dimming Ballast Cntrl. IC	DIC	Inr	-	pdf pinout
IR 21593 S	IC	Dimming Ballast Cntrl. IC	MDIP.	Inr	-	pdf pinout
IR 21771 S	LIN-IC	Phase Current Sensor IC,AC motor control	DIP	Inr	-	pdf pinout
IR 21814	IC	High and low Side Driver	DIP	Inr	-	pdf pinout
IR 21814 S	IC	High and low Side Driver	DIP	Inr	-	pdf pinout
IR 21834	CMOS-IC	Half-bridge Driver	DIP	Inr	-	pdf pinout
IR 21834 S	CMOS-IC	Half-bridge Driver	TSOP	Inr	-	pdf pinout
IR 21844	CMOS-IC	600V Half-Bridge Driver	DIP	Inr	-	pdf pinout
IR 21844 S	CMOS-IC	600V Half-Bridge Driver	MDIP	Inr	-	pdf pinout
IR 22381 Q	LIN-IC	3Ph. AC Motor Cntrl. IC	DIP	Inr	-	pdf pinout
IR 22771 S	LIN-IC	Phase Current Sensor IC,AC motor control	DIP	Inr	-	pdf pinout
IR 53 H(D) 420	MOS	Self-oscillating Half Bridge	SIP	Inr	-	pdf pinout
IR91308	OP	Vs:15V Vu:70V/mV Vo:±5V Vi0:3mV	8D			data pinout
IR91458	2xOP	Vs:±18V Vu:104dB Vo:±14V Vi0:1mV	8D			data pinout
IR93403	4xOP	Vs:±18V Vu:106dB Vo:±13.5V Vi0:2mV	14D			data pinout
IR94558	2xOP	Vs:36V Vu:110dB Vo:±14V Vi0:0.5mV	8D			data pinout
IR94559	2xOP	Vs:36V Vu:110dB Vo:±14V Vi0:0.5mV	8D			data pinout
IR 3220 S /20	IC	Protected H-bridge For D.c. Motor	TSOP	Inr	-	pdf pinout
IRAM 109-015 SD	IC	Integrated Power Hybrid IC	DIP	Inr	-	pdf pinout
IRAM 136-0461 G	IC	Plug N Drive Integrated Power Module	SIL	Inr	-	pdf pinout
IRAM 136-1060 BS	IC	Integrated Power Hybrid IC	SIL	Inr	-	pdf pinout
IRAM 136-3023 B	IC	Integrated Power Hybrid IC	DIP	Inr	-	pdf pinout
IRAM 136-3063 B	IC	Integrated Power Hybrid IC	SIL	Inr	-	pdf pinout
IRAMS 06 UP 60 B	IC	Plug N Drivetm Integrated Power Module	DIP	Inr	-	pdf pinout
IRAMS 10 UP 60 B	IC	Integrated Power Hybrid IC	DIP	Inr	-	pdf pinout
IRAMX 16 UP 60 A	IC	Plug N Drive(tm) Integrated Power Module	DIP	Inr	-	pdf pinout
IRAMX 20 UP 60 A	IC	Integrated Power Hybrid IC	DIP	Inr	-	pdf pinout
IRAMY 20 UP 60 B	IC	Integrated Power Hybrid IC	DIP	Inr	-	pdf pinout
IRC 150	MOS-N-FET-e	V-MOS, L, Current Sense, 100V, 30A	23/4Pin	Inr	-	data
IRC 250	MOS-N-FET-e	V-MOS, L, Current Sense, 200V, 29A	23/4Pin	Inr	-	data
IRC 254	MOS-N-FET-e	V-MOS, L, Current Sense, 250V, 22,2A	23/4Pin	Inr	-	data
IRC 350	MOS-N-FET-e	V-MOS, L, Current Sense, 400V, 14,5A	23/4Pin	Inr	-	data
IRC 450	MOS-N-FET-e	V-MOS, L, Current Sense, 500V, 12,2A	23/4Pin	Inr	-	data
IRC 530	MOS-N-FET-e	V-MOS, Current Sense, 100V, 14A, 79W	86/5Pin	Inr	-	data
IRC 531	MOS-N-FET-e	=IRC 530: 80V	86/5Pin	Inr	-	data
IRC 540	MOS-N-FET-e	V-MOS, Current Sense, 100V, 28A, 150W	86/5Pin	Inr	-	data
IRC 630	MOS-N-FET-e	V-MOS, Current Sense, 200V, 9A, 74W	86/5Pin	Inr	-	data
IRC 634	MOS-N-FET-e	V-MOS, Current Sense, 250V, 8,1A, 74W	86/5Pin	Inr	-	data
IRC 640	MOS-N-FET-e	V-MOS, Current Sense, 250V, 18A, 125W	86/5Pin	Inr	-	data
IRC 644	MOS-N-FET-e	V-MOS, Current Sense, 250V, 14A, 125W	86/5Pin	Inr	-	data
IRC 730	MOS-N-FET-e	V-MOS, Current Sense, 400V, 5,5A, 74W	86/5Pin	Inr	-	data
IRC 740	MOS-N-FET-e	V-MOS, Current Sense, 400V, 10A, 125W	86/5Pin	Inr	-	data
IRC 830	MOS-N-FET-e	V-MOS, Current Sense, 500V, 4,5A, 74W	86/5Pin	Inr	-	data

Type	Device	Short Description	Fig.	Manu	Comparision Types	More at
IRC 832	MOS-N-FET-e	V-MOS, Current Sense, 500V, 4A	86/5Pin	Inr	-	data
IRC 840	MOS-N-FET-e	V-MOS, Current Sense, 500V, 8A, 125W	86/5Pin	Inr	-	data
IRCP 054	MOS-N-FET-e	V-MOS, Current Sense, 60V, 70A, 230W	85/5Pin	Inr	-	data
IRCZ 24	MOS-N-FET-e	V-MOS, Current Sense, 60V, 17A, 60W	86/5Pin	Inr	-	data
IRCZ 34	MOS-N-FET-e	V-MOS, Current Sense, 60V, 30A, 88W	86/5Pin	Inr	-	data
IRCZ 44	MOS-N-FET-e	V-MOS, Current Sense, 60V, 50A, 150W	86/5Pin	Inr	-	data
IRD	Si-P	=2SA1484-D (Typ-Code/Stempel/marking)	35	Hit	→2SA1484	data
IRE	Si-P	=2SA1484-E (Typ-Code/Stempel/marking)	35	Hit	→2SA1484	data
IRF 9 Z10	MOS-P-FET-e	V-MOS, 50V, 4,7A, 20W, <0,5Ω(2,5A)	17c§	Inr,Sam	BUZ 171...172, IRF 9521, 2SJ102, 2SJ153	data
IRF 9 Z12	MOS-P-FET-e	=IRF 9Z10: 4A, <0,7Ω(2,5A)	17c§	Inr,Sam	→IRF 9Z10	data
IRF 9 Z14	MOS-P-FET-e	V-MOS, 60V, 6,7A, 43W, <0,5Ω(4A)	17c§	Inr,Sam	IRF 9630...33, MTP 8P08, 2SJ247	data
IRF 9 Z14L,S	MOS-P-FET-e	=IRF 9Z14:	30c§		2SJ266	data
IRF 9 Z15	MOS-P-FET-e	=IRF 9Z14: 4A, <0,7Ω(2,5A)	17c§	Inr,Sam	→IRF 9Z14	data
IRF 9 Z20	MOS-P-FET-e	V-MOS, 50V, 9,7A, 40W, <0,28Ω(5,6A)	17c§	Inr,Sam	IRF 9531, IRF 9533, 2SJ122...123, 2SJ169+	data
IRF 9 Z22	MOS-P-FET-e	=IRF 9Z20: 8,9A, <0,33Ω(5,6A)	17c§	Inr,Sam	→IRF 9Z20	data
IRF 9 Z24	MOS-P-FET-e	V-MOS, 60V,·11A, 60W, <0,28Ω(6,6A)	17c§	Inr,Sam	IRF 9530...31, 2SJ122...123, 2SJ169, ++	data
IRF 9 Z24L,S	MOS-P-FET-e	=IRF 9Z24:	30c§		2SJ219, 2SJ268, 2SJ296, 2SJ384	data
IRF 9 Z24N	MOS-P-FET-e	V-MOS, 55V, 12A, 45W, <0,175Ω(7,2A)	17c§	Inr	IRF 9530...31, 2SJ122, 2SJ140, 2SJ169, ++	data
IRF 9 Z24NL,NS	MOS-P-FET-e	=IRF 9Z24N:	30c§		2SJ219, 2SJ268, 2SJ296, 2SJ384	data
IRF 9 Z25	MOS-P-FET-e	=IRF 9Z24: 8,9A, <0,33Ω(5,6A)	17c§	Inr,Sam	→IRF 9Z24	data
IRF 9 Z30	MOS-P-FET-e	V-MOS, 50V, 18A, 74W, <0,14Ω(9,3A)	17c§	Inr,Sam	BUZ 271, IRF 9541, 2SJ174, 2SJ291	data
IRF 9 Z32	MOS-P-FET-e	=IRF 9Z30: 15A, <0,21Ω(9,3A)	17c§	Inr,Sam	→IRF 9Z30	data
IRF 9 Z34	MOS-P-FET-e	V-MOS, 60V, ↑8A, 88W, <0,14Ω(11A)	17c§	Inr,Sam	IRF 9540...41, 2SJ174, 2SJ291	data
IRF 9 Z34N	MOS-P-FET-e	V-MOS, 55V, 19A, 68W, <0,1Ω(10A)	17c§	Inr	IRF 9540...41, 2SJ174, 2SJ291	data
IRF 9 Z34(N)L;S	MOS-P-FET-e	=IRF 9Z34..:	30c§		2SJ220, 2SJ241, 2SJ297, 2SJ401	data
IRF 9 Z35	MOS-P-FET-e	=IRF 9Z34: 15A, <0,21Ω(9,3A)	17c§	Inr,Sam	→IRF 9Z34	data
IRF 034	MOS-N-FET-e	V-MOS, 60V, 30A, 90W, <0,05Ω(18A)	23a§	Inr	BUZ 24, 2N6763, 2SK561, 2SK906, 2SK1429+	data
IRF 035	MOS-N-FET-e	V-MOS, 60V, 25A, 90W, <0,07Ω(18A)	23a§	Inr	BUZ 24, 2N6763, 2SK561, 2SK906, 2SK1429+	data
IRF 044	MOS-N-FET-e	V-MOS, 60V, 30A, 150W, <0,028Ω(33A)	23a§	Inr	BUZ 24, 2N6764, 2SK561, 2SK906, 2SK1429+	data
IRF 045	MOS-N-FET-e	V-MOS, 60V, 30A, 150W, <0,035Ω(33A)	23a§	Inr	BUZ 24, 2N6764, 2SK561, 2SK906, 2SK1429+	data
IRF 120	MOS-N-FET-e	V-MOS, 100V, 9,2A, 60W, <0,27Ω(5,6A)	23a§	Inr,Fch,Sam	MTM 12N10, 2N6765, 2SK398...399, 2SK405++	data
IRF 121	MOS-N-FET-e	=IRF 120: 80V	23a§	Inr,Fch,Sam	MTM 12N08, 2N6765, 2SK398...399, 2SK405++	data
IRF 122	MOS-N-FET-e	V-MOS, 100V, 8A, 60W, <0,36Ω(5,6A)	23a§	Inr,Fch,Sam	MTM 12N10, 2N6765, 2SK398...399, 2SK405++	data
IRF 123	MOS-N-FET-e	=IRF 122: 80V	23a§	Inr,Fch,Sam	MTM 12N08, 2N6765, 2SK398...399, 2SK405++	data
IRF 130	MOS-N-FET-e	V-MOS, 100V, 14A, 79W, <0,16Ω(8,3A)	23a§	Inr,Sam++	BUZ 36, MTM 12N10, 2N6765, 2SK572	data
IRF 131	MOS-N-FET-e	=IRF 130: 80V	23a§	Inr,Sam++	BUZ 36, MTM 12N08, 2N6765, 2SK572	data
IRF 132	MOS-N-FET-e	V-MOS, 100V, 12A, 79W, <0,23Ω(8,3A)	23a§	Inr,Sam++	BUZ 36, MTM 12N10, 2N6765, 2SK572	data
IRF 133	MOS-N-FET-e	=IRF 132: 80V	23a§	Inr,Sam++	BUZ 36, MTM 12N08, 2N6765, 2SK572	data
IRF 140	MOS-N-FET-e	V-MOS, 100V, 28A, 150W, <0,077Ω(17A)	23a§	Inr,Sam++	BUZ 24, BUZ 349, 2N6764, 2SK906, 2SK1433	data
IRF 141	MOS-N-FET-e	=IRF 140: 80V	23a§	Inr,Sam++	BUZ 24, BUZ 349, 2N6764, 2SK906, 2SK1433	data
IRF 142	MOS-N-FET-e	V-MOS, 100V, 25A, 150W, <0,1Ω(17A)	23a§	Inr,Sam++	BUZ 24, BUZ 349, 2N6764, 2SK906, 2SK1433	data
IRF 143	MOS-N-FET-e	=IRF 142: 80V	23a§	Inr,Sam++	BUZ 24, BUZ 349, 2N6764, 2SK906, 2SK1433	data
IRF 150	MOS-N-FET-e	V-MOS, 100V, 30A, 150W, <0,055Ω(20A)	23a§	Inr,Sam++	BUZ 24, BUZ 349, 2N6764, 2SK906, 2SK1433	data
IRF 151	MOS-N-FET-e	=IRF 150: 60V	23a§	Inr,Sam++	BUZ 24, BUZ 349, 2N6763, 2SK906, 2SK1433	data
IRF 152	MOS-N-FET-e	V-MOS, 100V, 30A, 150W, <0,08Ω(20A)	23a§	Inr,Sam++	BUZ 24, BUZ 349, 2N6764, 2SK906, 2SK1433	data
IRF 153	MOS-N-FET-e	=IRF 152: 60V	23a§	Inr,Sam++	BUZ 24, BUZ 349, 2N6763, 2SK906	data
IRF 220	MOS-N-FET-e	V-MOS, 200V, 5A, 40W, <0,8Ω(2,5A)	23a§	Inr,Fch,Sam	BUZ 63, MTM 8N20, 2SK400	data
IRF 221	MOS-N-FET-e	=IRF 220: 150V	23a§	Inr,Fch,Sam	BUZ 63, MTM 8N18, 2SK400	data
IRF 222	MOS-N-FET-e	V-MOS, 200V, 4A, 40W, <1,2Ω(2,5A)	23a§	Inr,Fch,Sam	BUZ 63, MTM 8N20, 2SK400	data
IRF 223	MOS-N-FET-e	=IRF 222: 150V	23a§	Inr,Fch,Sam	BUZ 63, MTM 8N18, 2SK400	data
IRF 224	MOS-N-FET-e	V-MOS, 250V, 3,8A, 40W, <1,1Ω(2,1A)	23a§	Inr	BUZ 63, 2SK259...260, 2SK2264	data
IRF 225	MOS-N-FET-e	V-MOS, 250V, 3,3A, 40W, <1,5Ω(2,1A)	23a§	Inr	BUZ 63, 2SK259...260, 2SK2264	data
IRF 230	MOS-N-FET-e	V-MOS, 200V, 9A, 75W, <0,4Ω(5A)	23a§	Inr,Sam++	MTM 8N20, 2SK400...401, 2SK412	data
IRF 231	MOS-N-FET-e	=IRF 230: 150V	23a§	Inr,Sam++	MTM 8N18, 2SK400...401, 2SK412	data
IRF 232	MOS-N-FET-e	V-MOS, 200V, 8A, 75W, <0,6Ω(5A)	23a§	Inr,Sam++	MTM 8N20, 2SK400...401, 2SK412	data
IRF 233	MOS-N-FET-e	=IRF 232: 150V	23a§	Inr,Sam++	MTM 8N18, 2SK400...401, 2SK412	data
IRF 234	MOS-N-FET-e	V-MOS, 250V, 8,1A, 75W, <0,45Ω(4,1A)	23a§	Inr	BUZ 64, BUZ 325, 2SK293, 2SK501, ++	data
IRF 235	MOS-N-FET-e	V-MOS, 250V, 6,5A, 75W, <0,68Ω(4,1A)	23a§	Inr	BUZ 64, BUZ 325, 2SK293, 2SK501, ++	data
IRF 236	MOS-N-FET-e	=IRF 234: 275V	23a§	Inr	BUZ 64, BUZ 325, 2SK293, 2SK501, ++	data
IRF 237	MOS-N-FET-e	=IRF 235: 275V	23a§	Inr	BUZ 64, BUZ 325, 2SK293, 2SK501, ++	data
IRF 240	MOS-N-FET-e	V-MOS, 200V, 18A, 125W, <0,18Ω(10A)	23a§	Inr,Sam++	BUZ 36, BUZ 350, MTM 15N20, 2SK901	data
IRF 241	MOS-N-FET-e	=IRF 240: 150V	23a§	Inr,Sam++	BUZ 36, BUZ 350, MTM 15N20, 2SK901	data
IRF 242	MOS-N-FET-e	V-MOS, 200V, 16A, 125W, <0,22Ω(10A)	23a§	Inr,Sam++	BUZ 36, BUZ 350, MTM 15N20, 2SK901	data
IRF 243	MOS-N-FET-e	=IRF 242: 150V	23a§	Inr,Sam++	BUZ 36, BUZ 350, MTM 15N20, 2SK901	data
IRF 244	MOS-N-FET-e	V-MOS, 250V, 14A, 125W, <0,28Ω(8A)	23a§	Inr	BUZ 323, 2SK559...560, 2SK573, 2SK1135,++	data
IRF 245	MOS-N-FET-e	V-MOS, 250V, 13A, 125W, <0,34Ω(8A)	23a§	Inr	BUZ 323, 2SK559...560, 2SK573, 2SK1135,++	data
IRF 246	MOS-N-FET-e	=IRF 244: 275V	23a§	Inr	BUZ 323, 2SK559...560, 2SK868, 2SK1401,++	data
IRF 247	MOS-N-FET-e	=IRF 245: 275V	23a§	Inr	BUZ 323, 2SK559...560, 2SK868, 2SK1401	data
IRF 250	MOS-N-FET-e	V-MOS, 200V, 30A, 150W, <85mΩ(16A)	23a§	Inr,Sam++	BUZ 341, 2N6766, 2SK851, 2SK902, 2SK1492	data
IRF 251	MOS-N-FET-e	=IRF 250: 150V	23a§	Inr,Sam++	BUZ 341, 2N6766, 2SK851, 2SK902, 2SK1492	data
IRF 252	MOS-N-FET-e	V-MOS, 200V, 25A, 150W, <0,12Ω(16A)	23a§	Inr,Sam++	BUZ 341, 2N6766, 2SK851, 2SK902, 2SK1492	data
IRF 253	MOS-N-FET-e	=IRF 250: 150V	23a§	Inr,Sam++	BUZ 341, 2N6766, 2SK851, 2SK902, 2SK1492	data
IRF 254	MOS-N-FET-e	V-MOS, 250V, 22A, 150W, <0,14Ω(12A)	23a§	Inr,Sam	2SK623, 2SK901, 2SK944, 2SK1641, 2SK2007	data
IRF 255	MOS-N-FET-e	V-MOS, 250V, 20A, 150W, <0,17Ω(12A)	23a§	Inr,Sam	2SK623, 2SK901, 2SK944, 2SK1641, 2SK2007	data
IRF 256	MOS-N-FET-e	=IRF 254: 275V	23a§	Inr	2SK868(A), 2SK1169...70, 2SK1497...1500,++	data
IRF 257	MOS-N-FET-e	=IRF 255: 275V	23a§	Inr	2SK868(A), 2SK1169...70, 2SK1497...1500,++	data

Type	Device	Short Description	Fig.	Manu	Comparision Types	More at
IRF 320	MOS-N-FET-e	V-MOS, 400V, 3,3A, 50W, <1,8Ω(1,8A)	23a§	Inr,Fch,Sam	MTM 5N40, 2SK635	data
IRF 321	MOS-N-FET-e	=IRF 320: 350V	23a§	Inr,Fch,Sam	MTM 5N35, 2SK635	data
IRF 322	MOS-N-FET-e	V-MOS, 400V, 2,8A, 50W, <2,5Ω(1,8A)	23a§	Inr,Fch,Sam	MTM 5N40, 2SK635	data
IRF 323	MOS-N-FET-e	=IRF 322: 350V	23a§	Inr,Fch,Sam	MTM 5N35, 2SK635	data
IRF 330	MOS-N-FET-e	V-MOS, 400V, 5,5A, 75W, <1Ω(3A)	23a§	Inr,Sam++	BUZ 63, MTM 5N40, 2SK260, 2SK2264	data
IRF 331	MOS-N-FET-e	=IRF 320: 350V	23a§	Inr,Sam++	BUZ 63, MTM 5N35, 2SK259...260, 2SK2264	data
IRF 332	MOS-N-FET-e	V-MOS, 400V, 4,5A, 75W, <46/80ns	23a§	Inr,Sam++	BUZ 63, MTM 5N40, 2SK260, 2SK2264	data
IRF 333	MOS-N-FET-e	=IRF 322: 350V	23a§	Inr,Sam++	BUZ 63, MTM 5N35, 2SK259...260, 2SK2264	data
IRF 340	MOS-N-FET-e	V-MOS, 400V, 10A, 125W, <0,55Ω(5,2A)	23a§	Inr,Sam++	BUZ 64, BUZ 325...326, 2SK642, 2SK896,++	data
IRF 341	MOS-N-FET-e	=IRF 340: 350V	23a§	Inr,Sam++	BUZ 64, BUZ 325...326, 2SK642, 2SK896,++	data
IRF 342	MOS-N-FET-e	V-MOS, 400V, 8,3A, 125W, <0,8Ω(5,2A)	23a§	Inr,Sam++	BUZ 64, BUZ 325...326, 2SK642, 2SK896,++	data
IRF 343	MOS-N-FET-e	=IRF 342: 350V	23a§	Inr,Sam++	BUZ 64, BUZ 325...326, 2SK642, 2SK896,++	data
IRF 350	MOS-N-FET-e	V-MOS, 400V, 15A, 150W, <0,3Ω(8A)	23a§	Inr,Sam++	BUZ 323, MTM 15N40, 2SK559...60, 2SK899++	data
IRF 351	MOS-N-FET-e	=IRF 350: 350V	23a§	Inr,Sam++	BUZ 323, MTM 15N35, 2SK559...60, 2SK899++	data
IRF 352	MOS-N-FET-e	V-MOS, 400V, 13A, 150W, <0,4Ω(8A)	23a§	Inr,Sam++	BUZ 323, MTM 15N40, 2SK559...60, 2SK899++	data
IRF 353	MOS-N-FET-e	=IRF 352: 350V	23a§	Inr,Sam++	BUZ 323, MTM 15N35, 2SK559...60, 2SK899++	data
IRF 360	MOS-N-FET-e	V-MOS, 400V, 25A, 300W, <0,2Ω(14A)	23a§	Inr	-	data
IRF 362	MOS-N-FET-e	V-MOS, 400V, 22A, 300W, <0,25Ω(14A)	23a§	Inr	-	data
IRF 420	MOS-N-FET-e	V-MOS, 500V, 2,5A, 50W, <3Ω(1,4A)	23a§	Inr,Fch,Sam	BUZ 308, MTM 3N55, 2SK635	data
IRF 421	MOS-N-FET-e	=IRF 420: 450V	23a§	Inr,Fch,Sam	BUZ 308, MTM 3N55, 2SK635	data
IRF 422	MOS-N-FET-e	V-MOS, 500V, 2,2A, 50W, <4Ω(1,4A)	23a§	Inr,Fch,Sam	BUZ 308, MTM 3N55, 2SK635	data
IRF 423	MOS-N-FET-e	=IRF 422: 450V	23a§	Inr,Fch,Sam	BUZ 308, MTM 3N55, 2SK635	data
IRF 430	MOS-N-FET-e	V-MOS, 500V, 4,5A, 75W, <1,5Ω(2,5A)	23a§	Inr,Fch,Gen	BUZ 84, BUZ 355...356, MTM 4N50, 2SK351	data
IRF 431	MOS-N-FET-e	=IRF 430: 450V	23a§	Inr,Fch,Gen	BUZ 84, BUZ 355...356, MTM 4N45, 2SK351	data
IRF 432	MOS-N-FET-e	V-MOS, 500V, 4A, 75W, <2Ω(2,5A)	23a§	Inr,Fch,Gen	BUZ 84, BUZ 355...356, MTM 4N50, 2SK351	data
IRF 433	MOS-N-FET-e	=IRF 432: 450V	23a§	Inr,Fch,Gen	BUZ 84, BUZ 355...356, MTM 4N45, 2SK351	data
IRF 440	MOS-N-FET-e	V-MOS, 500V, 8A, 125W, <0,85Ω(4,4A)	23a§	Inr,Sam++	BUZ 94, BUZ 330...331, MTM 7N50, 2SK636	data
IRF 441	MOS-N-FET-e	=IRF 440: 450V	23a§	Inr,Sam++	BUZ 94, BUZ 330...331, MTM 7N45, 2SK636	data
IRF 442	MOS-N-FET-e	V-MOS, 500V, 7A, 125W, <1,1Ω(4,4A)	23a§	Inr,Sam++	BUZ 94, BUZ 330...331, MTM 7N50, 2SK636	data
IRF 443	MOS-N-FET-e	=IRF 442: 450V	23a§	Inr,Sam++	BUZ 94, BUZ 330...331, MTM 7N45, 2SK636	data
IRF 448	MOS-N-FET-e	V-MOS, 500V, 9,6A, 130W, <0,6Ω(5,4A)	23a§	Inr	BUZ 45, BUZ 339, 2SK512, 2SK683, 2SK724+	data
IRF 449	MOS-N-FET-e	V-MOS, 500V, 8,6A, 130W, <0,75Ω(5,4A)	23a§	Inr	BUZ 45, BUZ 339, 2SK512, 2SK683, 2SK724+	data
IRF 450	MOS-N-FET-e	V-MOS, 500V, 13A, 150W, <0,4Ω(7,2A)	23a§	Inr,Sam++	BUZ 323, BUZ 338, MTM 15N50, 2SK899, ++	data
IRF 451	MOS-N-FET-e	=IRF 450: 450V	23a§	Inr,Sam++	BUZ 323, BUZ 338, MTM 15N45, 2SK899, ++	data
IRF 452	MOS-N-FET-e	V-MOS, 500V, 11A, 150W, <0,5Ω(7,2A)	23a§	Inr,Sam++	BUZ 323, BUZ 338, MTM 15N50, 2SK899, ++	data
IRF 453	MOS-N-FET-e	=IRF 452: 450V	23a§	Inr,Sam++	BUZ 323, BUZ 338, MTM 15N45, 2SK899, ++	data
IRF 460	MOS-N-FET-e	V-MOS, 500V, 21A, 300W, <0,27Ω(12A)	23a§	Inr	-	data
IRF 462	MOS-N-FET-e	V-MOS, 500V, 19A, 300W, <0,35Ω(12A)	23a§	Inr	-	data
IRF 510	MOS-N-FET-e	V-MOS, 100V, 5,6A, 43W, <0,54Ω(3,4A)	17c§	Inr, Sam++	BUZ 20, BUZ 72, 2SK628, 2SK917, 2SK920++	data
IRF 510 A	MOS-N-FET-e	V-MOS, 100V, 5,6A, 33W, <0,4Ω(2,8A)	17c§	Sam	→IRF 510	data
IRF 511	MOS-N-FET-e	=IRF 510: 80V	17c§	Inr, Sam++	BUZ 20, BUZ 72, 2SK628, 2SK917, 2SK920++	data
IRF 512	MOS-N-FET-e	V-MOS, 100V, 4,9A, 43W, <0,74Ω(3,4A)	17c§	Inr, Sam++	BUZ 20, BUZ 72, 2SK628, 2SK917, 2SK920++	data
IRF 513	MOS-N-FET-e	=IRF 510: 80V	17c§	Inr, Sam++	BUZ 20, BUZ 72, 2SK628, 2SK917, 2SK920++	data
IRF 520	MOS-N-FET-e	V-MOS, 100V, 9,2A, 60W, <0,27Ω(5,5A)	17c§	Inr, Sam++	BUZ 20, BUZ 72, 2SK918, 2SK921, 2SK1427+	data
IRF 520 A	MOS-N-FET-e	V-MOS, 100V, 9,2A, 45W, <0,2Ω(4,6A)	17c§	Sam	→IRF 520	data
IRF 520 N	MOS-N-FET-e	V-MOS, 100V, 9,5A, 47W, <0,2Ω(5,7A)	17c§	Inr	→IRF 520	data
IRF 520(N)L,S	MOS-N-FET-e	=IRF 520..:	30c§		2SK1741, 2SK1907	data
IRF 521	MOS-N-FET-e	=IRF 520: 80V	17c§	Inr, Sam++	BUZ 20, BUZ 72, 2SK918, 2SK921, 2SK1427+	data
IRF 522	MOS-N-FET-e	V-MOS, 100V, 8A, 60W, <0,36Ω(5,6A)	17c§	Inr, Sam++	BUZ 20, BUZ 72, 2SK918, 2SK921, 2SK1427+	data
IRF 523	MOS-N-FET-e	=IRF 520: 80V	17c§	Inr, Sam++	BUZ 20, BUZ 72, 2SK918, 2SK921, 2SK1427+	data
IRF 520...523FI,P	MOS-N-FET-e	Iso, ≈75%Imax, 35W	17c	Sgs	BUZ 72(A)F, BUK 443-100, BUK 543-100	data
IRF 530	MOS-N-FET-e	V-MOS, 100V, 14A, 88W, <0,16Ω(8,4A)	17c§	Inr, Sam++	BUZ21, 2SK919, 2SK922, 2SK1301, 2SK1559+	data
IRF 530 A	MOS-N-FET-e	V-MOS, 100V, 14A, 55W, <0,11Ω(7A)	17c§	Sam	→IRF 530	data
IRF 530 N	MOS-N-FET-e	V-MOS, 100V, 17A, 79W, <0,11Ω(9A)	17c§	Inr	→IRF 530	data
IRF 530(N)L,S	MOS-N-FET-e	=IRF 530..:	30c§		IRFW 530A, 2SK1559S, 2SK1908, 2SK1927	data
IRF 531	MOS-N-FET-e	=IRF 530: 80V	17c§	Inr, Sam++	BUZ21, 2SK919, 2SK922, 2SK1301, 2SK1559+	data
IRF 532	MOS-N-FET-e	V-MOS, 100V, 12A, 88W, <0,23Ω(8,3A)	17c§	Inr, Sam++	BUZ21, 2SK919, 2SK922, 2SK1301, 2SK1559+	data
IRF 533	MOS-N-FET-e	=IRF 532: 80V	17c§	Inr, Sam++	BUZ21, 2SK919, 2SK922, 2SK1301, 2SK1559+	data
IRF 530...533FI,P	MOS-N-FET-e	Iso, ≈67%Imax, 40W	17c	Sgs	BUK 545-100, 2SK1230, 2SK1305, 2SK1430++	data
IRF 540	MOS-N-FET-e	V-MOS, 100V, 28A, 150W, <77mΩ(17A)	17c§	Inr, Sam++	BUK 456-100, BUZ 22	data
IRF 540 A	MOS-N-FET-e	V-MOS, 100V, 28A, 107W, <52mΩ(14A)	17c§	Sam	→IRF 540	data
IRF 540 N	MOS-N-FET-e	V-MOS, 100V, 27A, 94W, <52mΩ(16A)	17c§	Inr	→IRF 540	data
IRF 540(N)S	MOS-N-FET-e	=IRF 540....	30c§		MTB 33N10E	data
IRF 541	MOS-N-FET-e	=IRF 540: 80V	17c§	Inr,Sam++	BUK 456-100, BUZ 22	data
IRF 542	MOS-N-FET-e	V-MOS, 100V, 25A, 150W, <0,1Ω(17A)	17c§	Inr,Sam++	BUK 456-100, BUZ 22	data
IRF 543	MOS-N-FET-e	=IRF 542: 80V	17c§	Inr,Sam++	BUK 456-100, BUZ 22	data
IRF 540...543FI,P	MOS-N-FET-e	Iso, ≈55%Imax, 45W	17c	Sgs	BUK 445-100, 2SK1306...07, 2SK1318	data
IRF 550 A	MOS-N-FET-e	V-MOS, 100V, 40A, 167W, <40mΩ(20A)	17c§	Sam	RFP 40N10	data
IRF 610	MOS-N-FET-e	V-MOS, 200V, 3,3A, 36W, <1,5Ω(2A)	17c§	Inr, Sam++	MTP 5N20, 2SK923...924, 2SK1391, 2SK1921	data
IRF 610 A	MOS-N-FET-e	V-MOS, 200V, 3,3A, 38W, <1,5Ω(1,65A)	17c§	Sam	→IRF 610	data
IRF 610 S	MOS-N-FET-e	=IRF 610..:	30c§		IRFW 610A	data
IRF 611	MOS-N-FET-e	=IRF 610: 150V	17c§	Inr, Sam++	MTP 5N18, 2SK923...924, 2SK1391, 2SK1921	data
IRF 612	MOS-N-FET-e	V-MOS, 200V, 2,6A, 36W, <2,4Ω(1,6A)	17c§	Inr, Sam++	MTP 5N20, 2SK923...924, 2SK1391, 2SK1921	data
IRF 613	MOS-N-FET-e	=IRF 612: 150V	17c§	Inr, Sam++	MTP 5N18, 2SK923...924, 2SK1391, 2SK1921	data
IRF 614	MOS-N-FET-e	V-MOS, 250V, 2,7A, 36W, <2Ω(1,6A)	17c§	Inr, Sam	BUZ 76, 2SK310, 2SK923	data
IRF 614 A	MOS-N-FET-e	V-MOS, 250V, 2,8A, 40W, <2Ω(1,4A)	17c§	Sam	→IRF 614	data
IRF 614 S	MOS-N-FET-e	=IRF 614..:	30c§		IRFW 624A, 2SK2790	data
IRF 615	MOS-N-FET-e	V-MOS, 250V, 1,8A, 36W, <3Ω(1A)	17c§	Inr, Sam	BUZ 74, 2SK382, 2SK892	data

Type	Device	Short Description	Fig.	Manu	Comparision Types	More at
IRF 620	MOS-N-FET-e	V-MOS, 200V, 5,2A, 50W, <0,8Ω(3,1A)	17c§	Inr, Sam++	BUZ60, MTP 5N20, 2SK358, 2SK924, 2SK1391	data
IRF 620 A	MOS-N-FET-e	V-MOS, 200V, 5A, 47W, <0,8Ω(2,5A)	17c§	Sam	→IRF 620	data
IRF 620 S	MOS-N-FET-e	=IRF 620...:	30c§		IRFW 620A, 2SK782	data
IRF 621	MOS-N-FET-e	=IRF 620: 150V	17c§	Inr, Sam++	BUZ60, MTP 5N18, 2SK357, 2SK924, 2SK1391	data
IRF 622	MOS-N-FET-e	V-MOS, 200V, 4A, 50W, <1,2Ω(2,5A)	17c§	Inr, Sam++	BUZ60, MTP 5N20, 2SK358, 2SK924, 2SK1391	data
IRF 623	MOS-N-FET-e	=IRF 622: 150V	17c§	Inr, Sam++	BUZ60, MTP 5N18, 2SK357, 2SK924, 2SK1391	data
IRF 620..623FI,P	MOS-N-FET-e	Iso, ≈80%Imax, 30W	17c	Sgs	BUZ 73(A)F, BUK 444-200	data
IRF 624	MOS-N-FET-e	V-MOS, 250V, 4,4A, 50W, <1,1Ω(2,6A)	17c§	Inr, Sam	BUZ 60, BUZ 41A, 2SK358, 2SK924, 2SK1391	data
IRF 624 A	MOS-N-FET-e	V-MOS, 250V, 4,1A, 49W, <1,1Ω(2,05A)	17c§	Sam	→IRF 624	data
IRF 624 S	MOS-N-FET-e	=IRF 624..:	30c§ *		IRFW 624A, 2SK2790	data
IRF 625	MOS-N-FET-e	V-MOS, 250V, 3,3A, 50W, <1,5Ω(2,1A)	17c§	Inr, Sam	BUZ 60, BUZ 41A, 2SK358, 2SK924, 2SK1391	data
IRF 626	MOS-N-FET-e	=IRF 624: 275V	17c§	Inr	BUZ 60, BUZ 41A, 2SK553	data
IRF 627	MOS-N-FET-e	=IRF 625: 275V	17c§	Inr	BUZ 60, BUZ 41A, 2SK553	data
IRF 630	MOS-N-FET-e	V-MOS, 200V, 9A, 74W, <0,4Ω(5,4A)	17c§	Inr, Sam++	BUZ 31..32, 2SK459, 2SK925, 2SK1393,++	data
IRF 630 A	MOS-N-FET-e	V-MOS, 200V, 9A, 72W, <0,4Ω(4,5A)	17c§	Sam	→IRF 630	data
IRF 630 FI	MOS-N-FET-e	=IRF 630: Iso, 6A, 35W	17c	Sgs	2SK1478, 2SK1578, 2SK1957, 2SK2425, ++	data
IRF 630 S	MOS-N-FET-e	=IRF 630..:	30c§		IRFW 630A, 2SK2088	data
IRF 631	MOS-N-FET-e	=IRF 630: 150V	17c§ *	Inr, Sam++	BUZ 31..32, 2SK740, 2SK925, 2SK1393,++	data
IRF 632	MOS-N-FET-e	V-MOS, 200V, 8A, 74W, <0,6Ω(5A)	17c§	Inr, Sam++	BUZ 31..32, 2SK459, 2SK925, 2SK1393,++	data
IRF 633	MOS-N-FET-e	=IRF 632: 150V	17c§	Inr, Sam++	BUZ 31..32, 2SK740, 2SK925, 2SK1393,++	data
IRF 634	MOS-N-FET-e	V-MOS, 250V, 8,1A, 74W, <0,45Ω(5,1A)	17c§	Inr, Sam	2SK741, 2SK925, 2SK1221, 2SK1393,++	data
IRF 634 A	MOS-N-FET-e	V-MOS, 250V, 8,1A, 74W, <0,45Ω(4,05A)	17c§	Sam	→IRF 634	data
IRF 634 S	MOS-N-FET-e	=IRF 634..:	30c§		MTB 9N25E, IRFW 634A, 2SK1621, 2SK2252	data
IRF 635	MOS-N-FET-e	V-MOS, 250V, 6,5A, 74W, <0,68Ω(4,1A)	17c§	Inr, Sam	2SK741, 2SK925, 2SK1221, 2SK1393,++	data
IRF 636	MOS-N-FET-e	V-MOS, 275V, 8,1A, 75W, 0,5Ω(4,1A)	17c§	Inr, Sam	2SK554...555, 2SK894, 2SK1495...96, ++	data
IRF 636 A	MOS-N-FET-e	V-MOS, 275V, 8,1A, 74W, 0,5Ω(4,05A)	17c§	Sam	2SK554...555, 2SK894, 2SK1495...96, ++	data
IRF 637	MOS-N-FET-e	=IRF 634: 275V	17c§	Inr	2SK554...555, 2SK894, 2SK1495...96	data
IRF 640	MOS-N-FET-e	V-MOS, 200V, 18A, 125W, <0,18Ω(11A)	17c§	Inr, Sam++	BUK 456-200, BUZ 30A, 2SK2136	data
IRF 640 A	MOS-N-FET-e	V-MOS, 200V, 18A, 139W, <0,18Ω(9A)	17c§	Sam	→IRF 640	data
IRF 640 FI	MOS-N-FET-e	=IRF 640: Iso, 10A, 40W	17c	Sgs	2SK526, 2SK1036, 2SK1762	data
IRF 640 L,S	MOS-N-FET-e	=IRF 640..:	30c§		MTB 20N20E, 2SK2107, 2SK2136-Z	data
IRF 641	MOS-N-FET-e	=IRF 640: 150V	17c§	Inr, Sam++	BUZ 30A, PRF 540, PRF 542, 2SK2136	data
IRF 642	MOS-N-FET-e	V-MOS, 200V, 16A, 125W, <0,22Ω(10A)	17c§	Inr, Sam++	BUZ 30A, 2SK2136	data
IRF 643	MOS-N-FET-e	=IRF 642: 150V	17c§	Inr, Sam++	BUZ 30A, PRF 540, PRF 542, 2SK2136	data
IRF 644	MOS-N-FET-e	V-MOS, 250V, 14A, 125W, <0,28Ω(8,4A)	17c§	Inr, Sam	IRF 654A	data
IRF 644 A	MOS-N-FET-e	V-MOS, 250V, 14A, 139W, <0,28Ω(7A)	17c§	Sam	→IRF 644	data
IRF 644 S	MOS-N-FET-e	=IRF 644..:	30c§		2SK2065	data
IRF 645	MOS-N-FET-e	V-MOS, 250V, 13A, 125W, <0,34Ω(8A)	17c§	Inr, Sam	IRF 654A	data
IRF 646	MOS-N-FET-e	=IRF 644: 275V	17c§	Inr	IRF 750A	data
IRF 647	MOS-N-FET-e	=IRF 645: 275V	17c§	Inr	IRF 750A	data
IRF 650 A	MOS-N-FET-e	V-MOS, 200V, 28A, 156W, <85mΩ(14A)	17c§	Sam	-	data
IRF 654 A	MOS-N-FET-e	V-MOS, 250V, 21A, 156W, <0,14Ω(10,5A)	17c§	Sam	-	data
IRF 710	MOS-N-FET-e	V-MOS, 400V, 2A, 36W, <3,6Ω(1,2A)	17c§	Inr, Sam++	BUZ 74, MTP 2N40, 2SK382, 2SK892	data
IRF 710 A	MOS-N-FET-e	V-MOS, 400V, 2A, 36W, <3,6Ω(1A)	17c§	Sam	→IRF 710	data
IRF 710 S	MOS-N-FET-e	=IRF 710..:	30c§		IRFW 710A, MTB 2N40E	data
IRF 711	MOS-N-FET-e	=IRF 710: 350V	17c§	Inr, Sam++	BUZ 74, MTP 2N35, 2SK382, 2SK892	data
IRF 712	MOS-N-FET-e	V-MOS, 400V, 1,7A, 36W, <3,6Ω(1,1A)	17c§	Inr, Sam++	BUZ 74, MTP 2N40, 2SK382, 2SK892	data
IRF 713	MOS-N-FET-e	=IRF 712: 350V	17c§	Inr, Sam++	BUZ 74, MTP 2N35, 2SK382, 2SK892	data
IRF 720	MOS-N-FET-e	V-MOS, 400V, 3,3A, 50W, <1,8Ω(2A)	17c§	Inr, Sam++	BUZ 76, MTP 4N45, 2SK310...311, 2SK1244++	data
IRF 720 A	MOS-N-FET-e	V-MOS, 400V, 3,3A, 49W, <1,8Ω(1,65A)	17c§	Sam	→IRF 720	data
IRF 720 S	MOS-N-FET-e	=IRF 720..:	30c§		IRFW 720A, 2SK1493Z, 2SK1690	data
IRF 721	MOS-N-FET-e	=IRF 720: 350V	17c§	Inr, Sam++	BUZ 76, MTP 4N45, 2SK310...311, 2SK1244++	data
IRF 722	MOS-N-FET-e	V-MOS, 400V, 3,3A, 50W, <2,5Ω(1,8A)	17c§	Inr, Sam++	BUZ 76, MTP 4N45, 2SK310...311, 2SK1244++	data
IRF 723	MOS-N-FET-e	=IRF 722: 350V	17c§	Inr, Sam++	BUZ 76, MTP 4N45, 2SK310...311, 2SK1244++	data
IRF 720...723FI,P	MOS-N-FET-e	Iso, ≈71%Imax, 35W	17c	Sgs	BUK 445-450, 2SK503, 2SK1862...63	data
IRF 730	MOS-N-FET-e	V-MOS, 400V, 5,5A, 74W, <1Ω(3,3A)	17c§	Inr, Sam++	BUZ 60, 2SK552...553, 2SK1156, 2SK1246,++	data
IRF 730 A	MOS-N-FET-e	V-MOS, 400V, 5,5A, 73W, <1Ω(2,75A)	17c§	Inr, Sam	→IRF 730	data
IRF 730 S	MOS-N-FET-e	=IRF 730..:	30c§		IRFW 730A, 2SK1308, 2SK2838	data
IRF 731	MOS-N-FET-e	=IRF 730: 350V	17c§	Inr, Sam++	BUZ 60, 2SK552...553, 2SK1156, 2SK1246,++	data
IRF 732	MOS-N-FET-e	V-MOS, 400V, 4,5A, 74W, <1,5Ω(3A)	17c§	Inr, Sam++	BUZ 60, 2SK552...553, 2SK1156, 2SK1246,++	data
IRF 733	MOS-N-FET-e	=IRF 732: 350V	17c§	Inr, Sam++	BUZ 60, 2SK552...553, 2SK1156, 2SK1246,++	data
IRF 730...733FI,P	MOS-N-FET-e	Iso, ≈64%Imax, 35W	17c	Sgs	BUK 445-400, 2SK1231...32, 2SK1608,++	data
IRF 734	MOS-N-FET-e	V-MOS, 450V, 4,9A, 74W, <1,2Ω(2,9A)	17c§	Inr	BUZ 42, 2SK552...53, 2SK1156, 2SK1246, ++	data
IRF 737 LC	MOS-N-FET-e	V-MOS, lo Gate Charge, 300V, 6,1A, 74W, <0,75Ω(3,7A)	17c§	Inr	-	data
IRF 740	MOS-N-FET-e	V-MOS, 400V, 10A, 125W, <0,55Ω(6A)	17c§	Inr, Sam++	BUZ 61, MTP 10N40, 2SK1378, 2SK1442, ++	data
IRF 740 A	MOS-N-FET-e	V-MOS, 400V, 10A, 134W, <0,55Ω(5A)	17c§	Sam	→IRF 740	data
IRF 740 LC	MOS-N-FET-e	V-MOS, lo Gate Charge, 400V, 10A 125W, <0,55Ω(6A)	17c§	Inr	-	data
IRF 740 S	MOS-N-FET-e		30c§		IRFW 740A, MTB 10N40E, 2SK2949	data
IRF 741	MOS-N-FET-e	=IRF 740: 350V	17c§	Inr, Sam++	BUZ 61, MTP 10N35, 2SK1378, 2SK1442, ++	data
IRF 742	MOS-N-FET-e	V-MOS, 400V, 8,3A, 125W, <0,8Ω(5,2A)	17c§	Inr, Sam++	BUZ 61, MTP 10N40, 2SK554...555, 2SK894++	data
IRF 743	MOS-N-FET-e	=IRF 742: 350V	17c§	Inr, Sam++	BUZ 61, MTP 10N35, 2SK554...555, 2SK894++	data
IRF 740...743FI,P	MOS-N-FET-e	Iso, ≈55%Imax, 40W	17c	Sgs	BUK 445-400, 2SK1231...32, 2SK1566,++	data
IRF 744	MOS-N-FET-e	V-MOS, 450V, 8,8A, 125W, <0,63Ω(5,3A)	17c§	Inr	-	data
IRF 750 A	MOS-N-FET-e	V-MOS, 400V, 15A, 156W, <0,3Ω(7,5A)	17c§	Sam	-	data
IRF 820	MOS-N-FET-e	V-MOS, 500V, 2,5A, 50W, <3Ω(1,5A)	17c§	Inr, Sam++	BUZ74, MTP 3N50, 2SK382, 2SK892, 2SK1244	data
IRF 820 A	MOS-N-FET-e	V-MOS, 500V, 2,5A, 49W, <3Ω(1,25A)	17c§	Sam	→IRF 820	data
IRF 820 S	MOS-N-FET-e	=IRF 820..:	30c§	..	IRFW 820A, MTB 2N60E, 2SK2509	data
IRF 821	MOS-N-FET-e	=IRF 820: 450V	17c§	Inr, Sam++	BUZ74, MTP 3N45, 2SK382, 2SK892, 2SK1244	data
IRF 822	MOS-N-FET-e	V-MOS, 500V, 2,2A, 50W, <4Ω(1,4A)	17c§	Inr, Sam++	BUZ74, MTP 3N50, 2SK382, 2SK892, 2SK1244	data

Type	Device	Short Description	Fig.	Manu	Comparision Types	More at
IRF 823	MOS-N-FET-e	=IRF 822: 450V	17c§	Inr, Sam++	BUZ74, MTP 3N45, 2SK382, 2SK982, 2SK1244	data
IRF 820...823FI,P	MOS-N-FET-e	Iso, ≈70%Imax, 35W	17c	Sgs	BUK 444-500, BUK 445-600, 2SK1833	data
IRF 830	MOS-N-FET-e	V-MOS, 500V, 4,5A, 74W, <1,5Ω(2,7A)	17c§	Inr, Sam++	BUZ 41A..42, MTP 6N55, 2SK553, 2SK893,++	data
IRF 830 A	MOS-N-FET-e	V-MOS, 500V, 4,5A, 73W, <1,5Ω(2,25A)	17c§	Inr, Sam	→IRF 830	data
IRF 830 S	MOS-N-FET-e	=IRF 830..:	30c§		IRFW 830A, 2SK1703S, 2SK1751Z, 2SK2356Z+	data
IRF 831	MOS-N-FET-e	=IRF 830: 450V	17c§	Inr, Sam++	BUZ 41A..42, MTP 6N55, 2SK553, 2SK893,++	data
IRF 832	MOS-N-FET-e	V-MOS, 500V, 4A, 74W, <2Ω(2,5A)	17c§	Inr, Sam++	BUZ 41A..42, MTP 6N55, 2SK553, 2SK893,++	data
IRF 833	MOS-N-FET-e	=IRF 832: 450V	17c§	Inr, Sam++	BUZ 41A..42, MTP 6N55, 2SK553, 2SK893,++	data
IRF 830...833FI,P	MOS-N-FET-e	Iso, ≈63%Imax, 35W	* 17c	Sgs	BUK 445-500, 2SK1232, 2SK1608, 2SK1627++	data
IRF 840	MOS-N-FET-e	V-MOS, 500V, 8A, 125W, <0,85Ω(4,8A)	17c§	Inr, Sam++	MTP 8N50, 2SK555, 2SK894, 2SK1496,++	data
IRF 840 A	MOS-N-FET-e	V-MOS, 500V, 8A, 134W, <0,85Ω(4A)	17c§	Inr, Sam	→IRF 840	data
IRF 840 LC	MOS-N-FET-e	V-MOS, lo Gate Charge, 500V, 8A, 125W <0,85Ω(4,8A)	17c§	Inr	-	data
IRF 840 S,AS	MOS-N-FET-e	=IRF 840..:	30c§		IRFW 830A, 2SK1496Z, 2SK2360Z, 2SK1864++	data
IRF 841	MOS-N-FET-e	=IRF 840: 450V	17c§	Inr, Sam++	MTP 8N45, 2SK555, 2SK894, 2SK1495...96,++	data
IRF 842	MOS-N-FET-e	V-MOS, 500V, 7A, 125W, <1,1Ω(4,4A)	17c§	Inr, Sam++	MTP 8N50, 2SK555, 2SK894, 2SK1496,++	data
IRF 843	MOS-N-FET-e	=IRF 842: 450V	17c§	Inr, Sam++	MTP 8N45, 2SK555, 2SK894, 2SK1495...96,++	data
IRF 840...843FI,P	MOS-N-FET-e	Iso, ≈56%Imax, 40W	17c	Inr, Sam++	2SK1232, 2SK1608, 2SK1567, 2SK1627,++	data
IRF 1010	MOS-N-FET-e	V-MOS, 55V, 75A, 150W, <14mΩ(45A)	17c§	Inr	IRL 2505, IRL 3705N	data
IRF 1010 E	MOS-N-FET-e	V-MOS, 60V, 81A, 170W, <12mΩ(43A)	17c§	Inr	-	data
IRF 1010 N	MOS-N-FET-e	V-MOS, 55V, 84A, 170W, <11mΩ(43A)	17c§	Inr	IRL 2505, IRL 3705N	data
IRF 1010 NL,NS,S	MOS-N-FET-e	=IRF 1010..:	30c§		IRL 2505L,S, IRL 3705NL,NS	data
IRF 1104	MOS-N-FET-e	V-MOS, 40V, 100A, 170W, <9mΩ(60A)	17c§	Inr	IRL 2505	data
IRF 1310 N	MOS-N-FET-e	V-MOS, 100V, 36A, 120W, <36mΩ(22A)	17c§	Inr	IRF 540A, RFP 40N10E	data
IRF 1310 NL,NS	MOS-N-FET-e	=IRF 1310N:	30c§		IRFW 550A	data
IRF 1310 S	MOS-N-FET-e	V-MOS, 100V, 41A, 170W, <40mΩ(25A)	30c§	Inr	IRFW 550A	data
IRF 1404	MOS-N-FET-e	V-MOS, 40V, 162A, 200W, <4mΩ(95A)	17c§	Inr	-	data
IRF 2525	MOS-N-FET-e	V-MOS, 250V, 20A, 125W, <125mΩ(12A)	17c§	Inr	IRF 654A	data
IRF 2807	MOS-N-FET-e	V-MOS, 75V, 71A, 150W, <13mΩ(43A)	17c§	Inr	-	data
IRF 2807 L,S	MOS-N-FET-e	=IRF 2807:	30c§		-	data
IRF 3205	MOS-N-FET-e	V-MOS, 55V, 110A, 200W, <8mΩ(59A)	17c§	Inr	-	data
IRF 3205 L,S	MOS-N-FET-e	=IRF 3205L,S:	30c§		-	data
IRF 3315	MOS-N-FET-e	V-MOS, 150V, 21A, 94W, <82mΩ(12A)	17c§	Inr	BUZ 30A, IRF 650A, PRF 540, PRF 542	data
IRF 3315 L,S	MOS-N-FET-e	=IRF 3315:	30c§		MTB 20N20E, 2SK2107	data
IRF 3415	MOS-N-FET-e	V-MOS, 150V, 37A, 150W, <42mΩ(22A)	17c§	Inr	-	data
IRF 3415 L,S	MOS-N-FET-e	=IRF 3415:	30c§		IRF 3515S	data
IRF 3515 S	MOS-N-FET-e	V-MOS, 150V, 41A, 200W, <45mΩ(25A)	30c§	Inr	-	data
IRF 3710	MOS-N-FET-e	V-MOS, 100V, 49A, 150W, <25mΩ(28A)	17c§	Inr	-	data
IRF 3710 S	MOS-N-FET-e	=IRF 3710:	30c§		-	data
IRF 4905	MOS-P-FET-e	V-MOS, 55V, 74A, 200W, <20mΩ(38A)	17c§	Inr	-	data
IRF 4905 L,S	MOS-P-FET-e	=IRF 4905L,S:	30c§		-	data
IRF 5210	MOS-P-FET-e	V-MOS, 100V, 35A, 150W, <60mΩ(21A)	17c§	Inr	-	data
IRF 5210 L,S	MOS-P-FET-e	=IRF 5210:	30c§		-	data
IRF 5305	MOS-P-FET-e	V-MOS, 55V, 31A, 110W, <60mΩ(16A)	17c§	Inr	MTP 30P06V	data
IRF 5305 L,S	MOS-P-FET-e	=IRF 5305L,S:	30c§		MTB 30P06V	data
IRF 6215	MOS-P-FET-e	V-MOS, 150V, 11A, 80W, <0,29Ω(6,6A)	17c§	Inr	IRF 9641, 2SJ127	data
IRF 7101	MOS-N-FET-e	SMD, V-MOS, Dual, 20V, 3,5A, <0,1Ω(1,8A)	8-MDIP/ca3	Inr	-	data
IRF 7102	MOS-N-FET-e	SMD, V-MOS, Dual, 50V, 2A, <0,3Ω(1,5A)	8-MDIP	Inr	-	data
IRF 7103	MOS-N-FET-e	SMD, V-MOS, Dual, 50V, 3A, <0,13Ω(3A)	8-MDIP/ca3	Inr	-	data
IRF 7104	MOS-P-FET-e	SMD, V-MOS, Dual, 20V, 2,3A, <0,25Ω(1A)	8-MDIP/ca3	Inr	-	data
IRF 7105	MOS-NP-FET-e	SMD, V-MOS, T1=N, 25V, 3,5A, <0,1Ω(1A) T2=P, 25V, 2,3A, <0,25Ω(1A)	8-MDIP/ca3	Inr	-	data
IRF 7106	MOS-NP-FET-e	SMD, V-MOS, T1=N, 20V, 3A, <0,125Ω(1A) T2=P, 20V, 2,5A, <0,2Ω(1A)	8-MDIP/ca3	Inr	-	data
IRF 7107	MOS-NP-FET-e	SMD, V-FET, T1=N, 20V, 3A, <0,125Ω(3A) T2=P, 20V, 2,8A, <0,16Ω(3A)	8-MDIP/ca3	Inr	-	data
IRF 7201	MOS-N-FET-e	SMD, V-MOS, 30V, 7A, <30mΩ(7A)	8-MDIP/ca1	Inr	-	data
IRF 7204	MOS-P-FET-e	SMD, V-MOS, 20V, 5,3A, <60mΩ(5,3A)	8-MDIP/ca1	Inr	-	data
IRF 7205	MOS-P-FET-e	SMD, V-MOS, 30V, 4,6A, <70mΩ(4,6A)	8-MDIP/ca1	Inr	-	data
IRF 7207	MOS-P-FET-e	SMD,V-MOS,LogL, 20V, 5,4A, <60mΩ(5,4A)	8-MDIP/ca1	Inr	-	data
IRF 7210	MOS-P-FET-e	SMD, V-MOS, LogL, 12V, ±16A, <7mΩ(16A)	8-MDIP/ca1	Inr	-	data
IRF 7220	MOS-P-FET-e	SMD, V-MOS,LogL, 14V, ±11A, <12mΩ(11A)	8-MDIP/ca1	Inr	-	data
IRF 7233	MOS-P-FET-e	SMD, V-MOS, LogL, 12V, ±9,5A, <20mΩ	8-MDIP/ca1	Inr	-	data
IRF 7301	MOS-N-FET-e	SMD, V-MOS, Dual, LogL, 20V, 5,2A <50mΩ(2,6A), 51/83ns	8-MDIP/ca3	Inr	-	data
IRF 7303	MOS-N-FET-e	SMD, V-MOS, Dual, LogL, 30V, 5,3A <50mΩ(2,4A), 28/30ns	8-MDIP/ca3	Inr	-	data
IRF 7304	MOS-P-FET-e	SMD, V-MOS, Dual, LogL, 20V, 4,3A <90mΩ(2,2A), 34,4/84ns	8-MDIP/ca3	Inr	-	data
IRF 7306	MOS-P-FET-e	SMD, V-MOS, Dual, LogL, 30V, 4A <0,1Ω(1,8A), 28/43ns	8-MDIP/ca3	Inr	-	data
IRF 7307	MOS-NP-FET-e	=IRF 7301: 5A + IRF 7304: 4A	8-MDIP	Inr	-	data
IRF 7309	MOS-NP-FET-e	=IRF 7303: 4A + IRF 7306: 3A	8-MDIP	Inr	-	data

Type	Device	Short Description	Fig.	Manu	Comparision Types	More at
IRF 7311	MOS-N-FET-e	SMD, V-MOS, Dual, LogL, 20V, 6,6A <29mΩ(6A), 25,1/69ns	8-MDIP/ca3	Inr	-	data
IRF 7313	MOS-N-FET-e	SMD, V-MOS, Dual, LogL, 30V, 6,5A <29mΩ(5,8A), 17/43ns	8-MDIP/ca3	Inr	-	data
IRF 7314	MOS-P-FET-e	SMD, V-MOS, Dual, LogL, 20V, 5,3A <58mΩ(2,9A), 55/91ns	8-MDIP/ca3	Inr	-	data
IRF 7316	MOS-P-FET-e	SMD, V-MOS, Dual, LogL, 30V, 4,9A <58mΩ(4,9A), 26/66ns	8-MDIP/ca3	Inr	-	data
IRF 7317	MOS-NP-FET-e	=IRF 7311 + IRF 7314	8-MDIP	Inr	-	data
IRF 7319	MOS-NP-FET-e	=IRF 7313 + IRF 7316	8-MDIP	Inr	-	data
IRF 7321 D2	MOS-P-FET-e	SMD, V-MOS, 30V, 4,9A, <98mΩ(3,6A) int. Schottky-Di (30V, 2,8A)	8-MDIP (AASGDDKK)	Inr		data
IRF 7322 D1	MOS-P-FET-e	SMD,V-MOS,LogL, 20V, 5,3A, <62mΩ(2,9A) int. Schottky-Di (20V, 2,7A)	8-MDIP (AASGDDKK)	Inr		data
IRF 7324 D1	MOS-P-FET-e	SMD, V-MOS, LogL, 20V, 2,9A, <180mΩ int. Schottky-Di (20V, 1,7A)	8-MDIP (AASGDDKK)	Inr		data
IRF 7333	MOS-N-FET-e	SMD, V-MOS, Dual, LogL, T1: 30V, 3,5A <0,1Ω(2,2A), T2: 30V, 4,9A, <0,05Ω	8-MDIP/ca3	Inr	-	data
IRF 7341	MOS-N-FET-e	SMD, V-MOS, Dual, LogL, 55V, 4,7A <65mΩ(3,8A), 11,5/45ns	8-MDIP/ca3	Inr	-	data
IRF 7342	MOS-P-FET-e	SMD, V-MOS, Dual, LogL, 55V, 3,4A <0,17Ω(2,7A), 24/65ns	8-MDIP/ca3	Inr	-	data
IRF 7343	MOS-NP-FET-e	SMD, V-MOS, T1=N, 55V, 4,7A, <0,05Ω T2=P, 55V, 3,4A, <0,105Ω(3,1A)	8-MDIP/ca3	Inr	-	data
IRF 7353 D1	MOS-N-FET-e	SMD, V-MOS, 30V, 6,5A, <46mΩ(4,7A) int. Schottky-Di (30V, 2,7A)	8-MDIP (AASGDDKK)	Inr	-	data
IRF 7379	MOS-NP-FET-e	SMD,V-FET,LogL, T1=N, 30V, 5,7A, <75mΩ T2=P, 30V, 4,3A, <0,18Ω(3,7A)	8-MDIP/ca3	Inr	-	data
IRF 7389	MOS-NP-FET-e	SMD,V-FET,LogL, T1=N, 30V, 7,3A, <46mΩ T2=P, 30V, 5,3A, <98mΩ(3,6A)	8-MDIP/ca3	Inr	-	data
IRF 7401	MOS-N-FET-e	SMD,V-MOS,LogL, 20V, 8,7A, <22mΩ(4,1A)	8-MDIP/ca1	Inr	-	data
IRF 7403	MOS-N-FET-e	SMD, V-MOS, LogL, 30V, 8,5A, <22mΩ(4A)	8-MDIP/ca1	Inr	-	data
IRF 7404	MOS-P-FET-e	SMD,V-MOS,LogL, 20V, 6,7A, <40mΩ(3,2A)	8-MDIP/ca3	Inr	-	data
IRF 7406	MOS-P-FET-e	SMD,V-MOS,LogL, 30V, 5,8A, <45mΩ(2,8A)	8-MDIP/ca3	Inr	-	data
IRF 7413	MOS-N-FET-e	SMD,V-MOS, LogL, 30V, 13A, <11mΩ(7,3A)	8-MDIP/ca1	Inr	-	data
IRF 7413 A	MOS-N-FET-e	=IRF 7413: 12A, <13,5mΩ(6,6A)	8-MDIP/ca1	Inr	-	data
IRF 7416	MOS-P-FET-e	SMD,V-MOS,LogL, 30V, 10A, <20mΩ(5,6A)	8-MDIP/ca1	Inr	-	data
IRF 7421 D1	MOS-N-FET-e	SMD,V-MOS,LogL, 30V, 5,8A, <35mΩ(4,1A) int. Schottky-Di (30V, 2A)	8-MDIP (ASSGDDDD)	Inr	-	data
IRF 7422 D2	MOS-P-FET-e	SMD,V-MOS,LogL, 20V, 4,1A, <90mΩ(2,3A) int. Schottky-Di (20V, 3A)	8-MDIP (ASSGDDDD)	Inr	-	data
IRF 7501	MOS-N-FET-e	SMD, V-MOS, Dual, LogL, 20V, 2,4A <0,135Ω(1,7A), 29,7/31ns	8-SSMDIP /ca3	Inr	-	data
IRF 7503	MOS-N-FET-e	SMD, V-MOS, Dual, LogL, 30V, 2,4A <0,135Ω(1,7A), 14,7/17,3ns	8-SSMDIP /ca3	Inr	-	data
IRF 7504	MOS-P-FET-e	SMD, V-MOS, Dual, LogL, 20V, 1,7A <0,27Ω(1,2A), 44,1/81ns	8-SSMDIP /ca3	Inr	-	data
IRF 7506	MOS-P-FET-e	SMD, V-MOS, Dual, LogL, 30V, 1,7A <0,27Ω(1,2A), 21,7/28,3ns	8-SSMDIP /ca3	Inr	-	data
IRF 7507	MOS-PN-FET-e	=IRF 7504 + IRF 7501	8-SSMDIP	Inr	-	data
IRF 7509	MOS-PN-FET-e	=IRF 7506 + IRF 7503	8-SSMDIP	Inr	-	data
IRF 7521 D1	MOS-N-FET-e	SMD, V-MOS, 20V, 2,4A, <135mΩ(1,7A) int. Schottky-Di (20V, 1,9A)	8-MDIP (AASGDDKK)	Inr	-	data
IRF 7523 D1	MOS-N-FET-e	SMD, V-MOS, 30V, 2,4A, <222mΩ(0,85A) int. Schottky-Di (30V, 1,9A)	8-MDIP (AASGDDKK)	Inr	-	data
IRF 7524 D1	MOS-P-FET-e	SMD, V-MOS, 20V, 1,7A, <0,27Ω(1,2A) int. Schottky-Di (20V, 1,9A)	8-MDIP (AASGDDKK)	Inr	-	data
IRF 7526 D1	MOS-P-FET-e	SMD, V-MOS, 30V, 1,7A, <0,45Ω(0,6A) int. Schottky-Di (30V, 1,9A)	8-MDIP (AASGDDKK)	Inr	-	data
IRF 7555	MOS-P-FET-e	SMD, V-MOS, Dual, LogL, 20V, 4,3A <55mΩ(4,3A), 56/124ns	8-SSMDIP /ca3	Inr	-	data
IRF 7601	MOS-N-FET-e	SMD,V-MOS,LogL, 20V, 5,7A, <35mΩ(3,8A) 97,1/56ns	8-SSMDIP /ca1	Inr	-	data
IRF 7603	MOS-N-FET-e	SMD,V-MOS,LogL, 30V, 5,6A, <35mΩ(3,7A) 33,7/30ns	8-SSMDIP /ca1	Inr	-	data
IRF 7604	MOS-P-FET-e	SMD,V-MOS,LogL, 20V, 3,6A, <90mΩ(2,4A) 70/69ns	8-SSMDIP /ca1	Inr	-	data
IRF 7606	MOS-P-FET-e	SMD,V-MOS,LogL, 30V, 3,6A, <90mΩ(2,4A) 33/82ns	8-SSMDIP /ca1	Inr	-	data
IRF 7663	MOS-P-FET-e	SMD, V-MOS, LogL, 20V, 8,2A, <20mΩ(7A) 111/197ns	8-SSMDIP /ca1	Inr	-	data
IRF 7805	MOS-N-FET-e	SMD, V-MOS, LogL, 30V, 13A, <11mΩ(7A)	8-MDIP/ca1	Inr	-	data
IRF 7807	MOS-N-FET-e	SMD, V-MOS, LogL, 30V, 8,3A, <25mΩ(7A)	8-MDIP/ca1	Inr	-	data

Type	Device	Short Description	Fig.	Manu	Comparision Types	More at
IRF 9120...9123	MOS-P-FET-e	=IRF 9520...9523:	23a§	Sam	2SJ115, 2SJ118...119	data
IRF 9130	MOS-P-FET-e	V-MOS, 100V, 12A, 75W, <0,3Ω(6,5A)	23a§	Inr, Sam	MTM 12P10, 2SJ112...113	data
IRF 9131	MOS-P-FET-e	=IRF 9130: 60V	23a§	Inr, Sam	MTM 12P06, 2SJ112...113	data
IRF 9132	MOS-P-FET-e	V-MOS, 100V, 10A, 75W, <0,4Ω(6,5A)	23a§	Inr, Sam	MTM 12P10, 2SJ112...113	data
IRF 9133	MOS-P-FET-e	=IRF 9132: 60V	23a§	Inr, Sam	MTM 12P06, 2SJ112...113	data
IRF 9140	MOS-P-FET-e	V-MOS, 100V, 19A, 125W, <0,2Ω(10A)	23a§	Inr, Sam	MTH 20P10	data
IRF 9141	MOS-P-FET-e	=IRF 9140: 60V	23a§	Inr, Sam	MTH 20P08	data
IRF 9142	MOS-P-FET-e	V-MOS, 100V, 15A, 125W, <0,3Ω(10A)	23a§	Inr, Sam	MTH 20P10	data
IRF 9143	MOS-P-FET-e	=IRF 9142: 60V	23a§	Inr, Sam	MTH 20P08	data
IRF 9150	MOS-P-FET-e	V-MOS, 100V, 25A, 150W, <0,15Ω(10A)	23a§	Inr	-	data
IRF 9151	MOS-P-FET-e	=IRF 9150: 60V	23a§	Inr	MTH 25P06	data
IRF 9220...9223	MOS-P-FET-e	=IRF 9620...9623:	23a§	Sam	-	data
IRF 9230	MOS-P-FET-e	V-MOS, 200V, 6,5A, 75W, <0,8Ω(3,5A)	23a§	Inr, Sam	MTM 8P20, 2SJ56, 2SJ114, 2SJ351...352	data
IRF 9231	MOS-P-FET-e	=IRF 9230: 150V	23a§	Inr, Sam	MTM 8P18, 2SJ55...56, 2SJ114, 2SJ351...352	data
IRF 9232	MOS-P-FET-e	V-MOS, 200V, 5,5A, 75W, <1,2Ω(3,5A)	23a§	Inr, Sam	MTM 8P20, 2SJ56, 2SJ114, 2SJ351...352	data
IRF 9233	MOS-P-FET-e	=IRF 9232: 150V	23a§	Inr, Sam	MTM 8P18, 2SJ55...56, 2SJ114, 2SJ351...352	data
IRF 9240	MOS-P-FET-e	V-MOS, 200V, 11A, 125W, <0,5Ω(6A)	23a§	Inr, Sam	2SJ131	data
IRF 9241	MOS-P-FET-e	=IRF 9240: 150V	23a§	Inr, Sam	2SJ131, 2SJ200	data
IRF 9242	MOS-P-FET-e	V-MOS, 200V, 9A, 125W, <0,7Ω(6A)	23a§	Inr, Sam	2SJ131	data
IRF 9243	MOS-P-FET-e	=IRF 9242: 150V	23a§	Inr, Sam	2SJ131, 2SJ200	data
IRF 9410	MOS-N-FET-e	SMD, V-MOS, LogL, 30V, 7A, <30mΩ(7A)	8-SSMDIP	Inr	-	data
IRF 9510	MOS-P-FET-e	V-MOS, 100V, 4A, 43W, <1,2Ω(2,4A)	17§	Inr, Sam	BUZ 172, MTP 5P18...25, 2SJ134	data
IRF 9510 S	MOS-P-FET-e	=IRF 9510:	30c§		SFW 9520	data
IRF 9511	MOS-P-FET-e	=IRF 9510: 60V	17c§	Inr, Sam	BUZ 172, MTP 5P18...25, 2SJ102, 2SJ134	data
IRF 9512	MOS-P-FET-e	V-MOS, 100V, 2,5A, 43W, <1,5Ω(1,5A)	17c§	Inr, Sam	BUZ 172, MTP 5P18...25, 2SJ134	data
IRF 9513	MOS-P-FET-e	=IRF 9512: 60V	17c§	Inr, Sam	BUZ 172, MTP 5P18...25, 2SJ102, 2SJ134	data
IRF 9520	MOS-P-FET-e	V-MOS, 100V, 6,8A, 60W, <0,6Ω(4,1A)	17c§	Inr, Sam	MTP 8P10, 2SJ247	data
IRF 9520 N	MOS-P-FET-e	V-MOS, 100V, 6,7A, 47W, <0,48Ω(4A)	17c§	Inr	MTP 8P10, 2SJ247	data
IRF 9520 NL,NS	MOS-P-FET-e	=IRF 9520N:	30c§		SFW 9520	data
IRF 9520 S	MOS-P-FET-e	=IRF 9520:	30c§		SFW 9520	data
IRF 9521	MOS-P-FET-e	=IRF 9520: 60V	17c§	Inr, Sam	MTP 8P08, 2SJ247	data
IRF 9522	MOS-P-FET-e	V-MOS, 100V, 5A, 60W, <0,8Ω(3,5A)	17c§	Inr, Sam	MTP 8P10, 2SJ247	data
IRF 9523	MOS-P-FET-e	=IRF 9522: 60V	17c§	Inr, Sam	MTP 8P08, 2SJ247	data
IRF 9530	MOS-P-FET-e	V-MOS, 100V, 12A, 88W, <0,3Ω(7,2A)	17c§	Inr,Sam,Six	MTP 12P10, 2SJ127, 2SJ138	data
IRF 9530 NL,NS	MOS-P-FET-e	V-MOS, 100V, 14A, 75W, <0,2Ω(8,4A)	30c§	Inr	SFW 9530	data
IRF 9530S	MOS-P-FET-e	=IRF 9530S:	30c§		SFW 9530	data
IRF 9531	MOS-P-FET-e	=IRF 9530: 60V	17c§	Inr, Sam	MTP 12P06, 2SJ123, 2SJ127, 2SJ138	data
IRF 9532	MOS-P-FET-e	V-MOS, 100V, 10A, 88W, <0,4Ω(6,5A)	17c§	Inr, Sam	MTP 12P10, 2SJ127, 2SJ138	data
IRF 9533	MOS-P-FET-e	=IRF 9532: 60V	17c§	Inr, Sam	MTP 12P06, 2SJ123, 2SJ127, 2SJ138	data
IRF 9540	MOS-P-FET-e	V-MOS, 100V, 19A, 150W, <0,2Ω(11A)	17c§	Inr, Sam	2SJ221	data
IRF 9540 N	MOS-P-FET-e	V-MOS, 100V, 19A, 94W, <0,117Ω(11A)	17c§	Inr	2SJ221	data
IRF 9540(N)L,S	MOS-P-FET-e	=IRF 9540S...:	30c§		SFW 9540	data
IRF 9541	MOS-P-FET-e	=IRF 9540: 60V	17c§	Inr, Sam	MTP 20P06, 2SJ174, 2SJ221, 2SJ291	data
IRF 9542	MOS-P-FET-e	V-MOS, 100V, 15A, 150W, <0,3Ω(10A)	17c§	Inr, Sam	2SJ221	data
IRF 9543	MOS-P-FET-e	=IRF 9542: 60V	17c§	Inr, Sam	MTP 20P06, 2SJ174, 2SJ221, 2SJ291	data
IRF 9610	MOS-P-FET-e	V-MOS, 200V, 1,8A, 20W, <3Ω(0,9A)	17c§	Inr, Sam	BUZ 173, MTP 3P25	data
IRF 9610S	MOS-P-FET-e	=IRF 9610S:	30c§		SFW 9610	data
IRF 9611	MOS-P-FET-e	=IRF 9610: 150V	17c§	Inr, Sam	BUZ 173, MTP 3P25	data
IRF 9612	MOS-P-FET-e	V-MOS, 200V, 1,5A, 20W, <5,5Ω(0,9A)	17c§	Inr, Sam	BUZ 173, MTP 3P25	data
IRF 9613	MOS-P-FET-e	=IRF 9612: 150V	17c§	Inr, Sam	BUZ 173, MTP 3P25	data
IRF 9620	MOS-P-FET-e	V-MOS, 200V, 3,5A, 40W, <1,5Ω(1,5A)	17c§	Inr, Sam	BUZ 173, MTP 3P25	data
IRF 9620S	MOS-P-FET-e	=IRF 9620S:	30c§		SFW 9620	data
IRF 9621	MOS-P-FET-e	=IRF 9620: 150V	17c§	Inr, Sam	BUZ 173, MTP 3P25	data
IRF 9622	MOS-P-FET-e	V-MOS, 200V, 3A, 40W, <2,4Ω(1,5A)	17c§	Inr, Sam	BUZ 173, MTP 3P25	data
IRF 9623	MOS-P-FET-e	=IRF 9622: 150V	17c§	Inr, Sam	BUZ 173, MTP 3P25	data
IRF 9630	MOS-P-FET-e	V-MOS, 200V, 6,5A, 74W, <0,8Ω(3,9A)	17c§	Inr, Sam++	MTP 6P20E	data
IRF 9630S	MOS-P-FET-e	=IRF 9630S:	30c§		SFW 9630	data
IRF 9631	MOS-P-FET-e	=IRF 9630: 150V	17c§	Inr, Sam	MTP 6P20E	data
IRF 9632	MOS-P-FET-e	V-MOS, 200V, 5,5A, 74W, <1,2Ω(3,5A)	17c§	Inr, Sam	MTP 6P20E	data
IRF 9633	MOS-P-FET-e	=IRF 9632: 150V	17c§	Inr, Sam	MTP 6P20E	data
IRF 9640	MOS-P-FET-e	V-MOS, 200V, 11A, 125W, <0,5Ω(6,6A)	17c§	Inr, Sam	-	data
IRF 9640S	MOS-P-FET-e	=IRF 9640S:	30c§		SFW 9640	data
IRF 9641	MOS-P-FET-e	=IRF 9640: 150V	17c§	Inr, Sam	2SJ127	data
IRF 9642	MOS-P-FET-e	V-MOS, 200V, 9A, 125W, <0,7Ω(6A)	17c§	Inr, Sam	-	data
IRF 9643	MOS-P-FET-e	=IRF 9642: 150V	17c§	Inr, Sam	2SJ127	data
IRF 9952	MOS-NP-FET-e	=IRF 9956 + IRF 9953	8-MDIP	Inr	-	data
IRF 9953	MOS-P-FET-e	SMD, V-MOS, Dual, LogL, 30V, 2,3A <0,25Ω(1A), 23,7/26,9ns	8-MDIP	Inr	-	data
IRF 9956	MOS-N-FET-e	SMD, V-MOS, Dual, LogL, 30V, 3,5A <0,1Ω(2,2A), 15/16ns	8-MDIP	Inr	-	data
IRFA 1 Z0	MOS-N-FET-e	V-MOS, 100V, 0,5A, 1W, <2,4Ω(0,25A)	14	Sam	-	data
IRFA 1 Z3	MOS-N-FET-e	V-MOS, 60V, 0,4A, 1W, <3,2Ω(0,25A)	14	Sam	-	data
IRFAC 30	MOS-N-FET-e	V-MOS, 600V, 3,6A, 74W, <2,2Ω(2A)	23a§	Inr	BUZ 84, BUZ 307, 2SK351, 2SK954, 2SK1213	data
IRFAC 32	MOS-N-FET-e	V-MOS, 600V, 3,2A, 74W, <2,7Ω(2A)	23a§	Inr	BUZ 84, BUZ 307, 2SK351, 2SK954, 2SK1213	data
IRFAC 40	MOS-N-FET-e	V-MOS, 600V, 6,2A, 125W, <1,2Ω(3,4A)	23a§	Inr	BUZ 94, BUZ 332A, 2SK684, 2SK1032,++	data
IRFAC 42	MOS-N-FET-e	V-MOS, 600V, 5,4A, 125W, <1,5Ω(3,4A)	23a§	Inr	BUZ 94, BUZ 332A, 2SK684, 2SK1032,++	data
IRFAE 40	MOS-N-FET-e	V-MOS, 800V, 4,8A, 125W, <2Ω(2,8A)	23a§	Inr	BUZ 84, BUZ 355...356, 2SK351, 2SK695,++	data

Type	Device	Short Description	Fig.	Manu	Comparision Types	More at
IRFAE 42	MOS-N-FET-e	V-MOS, 800V, 4,4A, 125W, <2,4Ω(2,8A)	23a§	Inr	BUZ 84, BUZ 355...356, 2SK351, 2SK695	data
IRFAE 50	MOS-N-FET-e	V-MOS, 800V, 7,1A, 150W, <1,2Ω(4,2A)	23a§	Inr	BUK 638-800, BUZ 305, 2SK684, 2SK1032,++	data
IRFAE 52	MOS-N-FET-e	V-MOS, 800V, 6,6A, 150W, <1,4Ω(4,2A)	23a§	Inr	BUK 638-800, BUZ 305, 2SK684, 2SK1032,++	data
IRFAF 40	MOS-N-FET-e	V-MOS, 900V, 4,3A, 125W, <2,5Ω(2,5A)	23a§	Inr	BUZ 54, BUZ 357...358, 2SK727, 2SK1461,++	data
IRFAF 42	MOS-N-FET-e	V-MOS, 900V, 3,9A, 125W, <3Ω(2,5A)	23a§	Inr	BUZ 54, BUZ 357...358, 2SK727, 2SK1461,++	data
IRFAF 50	MOS-N-FET-e	V-MOS, 900V, 6,2A, 150W, <1,6Ω(3,6A)	23a§	Inr	2SK1358, 23SK1502, 2SK1692, 2SK1795...96+	data
IRFAF 52	MOS-N-FET-e	V-MOS, 900V, 5,7A, 150W, <1,9Ω(3,6A)	23a§	Inr	BUZ 54, BUZ 357...358, 2SK685, 2SK727,++	data
IRFAG 40	MOS-N-FET-e	V-MOS, 1000V, 3,9A, 125W, <3,5Ω(2,3A)	23a§	Inr	BUZ 54, BUZ 357...358, 2SK685, 2SK1359,++	data
IRFAG 42	MOS-N-FET-e	V-MOS, 1000V, 3,6A, 125W, <4,2Ω(2,3A)	23a§	Inr	BUZ 54, BUZ 357...358, 2SK685, 2SK1359,++	data
IRFAG 50	MOS-N-FET-e	V-MOS, 1000V, 5,6A, 150W, <2Ω(3,2A)	23a§	Inr	BUZ 54, BUZ 357, 2SK685, 2SK1120,++	data
IRFAG 52	MOS-N-FET-e	V-MOS, 1000V, 5,1A, 150W, <2,4Ω(3,2A)	23a§	Inr	BUZ 54, BUZ 357...358, 2SK685, 2SK1120,++	data
IRFB 9 N30A	MOS-N-FET-e	V-MOS, 300V, 9,3A, 96W, <0,45Ω(5,6A)	17c§	Inr	BUZ 61, IRF 740, 2SK1378, 2SK1442, ++	data
IRFB 9 N60A	MOS-N-FET-e	V-MOS, 600V, 9,2A, 170W, <0,75Ω(5,5A)	17c§	Inr	-	data
IRFB 9 N65A	MOS-N-FET-e	V-MOS, 650V, 8,5A, 167W, <0,93Ω(5,1A)	17c§	Inr	-	data
IRFB 11 N50A	MOS-N-FET-e	V-MOS, 500V, 11A, 170W, <0,52Ω(6,6A)	17c§	Inr	-	data
IRFBC 20	MOS-N-FET-e	V-MOS, 600V, 2,2A, 50W, <4,4Ω(1,3A)	17c§	Inr	BUK 456-800, BUZ 80, 2SK858, 2SK1323,++	data
IRFBC 20 L,S	MOS-N-FET-e	=IRFBC 20:	30c§		MTB 2N60E, 2SK1746	data
IRFBC 30	MOS-N-FET-e	V-MOS, 600V, 3,6A, 74W, <2,2Ω(2,2A)	17c§	Inr, Sgs	BUZ 90, 2SK1117, 2SK1402, 2SK1809	data
IRFBC 30 A	MOS-N-FET-e	V-MOS, 600V, 3,6A, 74W, <2,2Ω(2,2A)	17c§	Inr	→IRFBC 30	data
IRFBC 30 L,S,AS	MOS-N-FET-e	=IRFBC 30...:	30c§		2SK1624, 2SK1915, 2SK2777	data
IRFBC 32	MOS-N-FET-e	V-MOS, 600V, 3,2A, 74W, <2,7Ω(2A)	17c§	Inr	BUZ 90, 2SK1117, 2SK1402, 2SK1809	data
IRFBC 40	MOS-N-FET-e	V-MOS, 600V, 6,2A, 125W, <1,2Ω(3,7A)	17c§	Inr	BUK 657-600, BUZ 91	data
IRFBC 40 A	MOS-N-FET-e	V-MOS, 600V, 6,2A, 125W, <1,2Ω(3,7A)	17c§	Inr	→IRFBC 40	data
IRFBC 40 L,S,AS	MOS-N-FET-e	=IRFBC 40...:	30c§		SSW 7N60A, 2SK2060, 2SK2140Z, 2SK2908	data
IRFBC 40 LC	MOS-N-FET-e	V-MOS, lo Gate Charge,600V, 6,2A, 125W <1,2Ω(3,7A)	17c§	Inr	-	data
IRFBC 42	MOS-N-FET-e	V-MOS, 600V, 5,4A, 125W, <1,6Ω(3,4A)	17c§	Inr	BUK 657-600, BUZ 91	data
IRFBE 20	MOS-N-FET-e	V-MOS, 800V, 1,8A, 54W, <6,5Ω(1,1A)	17c§	Inr	BUZ 80, 2SK1199, 2SK1324, 2SK1456, ++	data
IRFBE 30	MOS-N-FET-e	V-MOS, 800V, 4,1A, 125W, <3Ω(2,5A)	17c§	Inr	BUZ 81, 2SK1639, 2SK1501, 2SK1807, ++	data
IRFBF 20	MOS-N-FET-e	V-MOS, 900V, 1,7A, 54W, <8Ω(1A)	17c§	Inr	BUZ 50, 2SK1199, 2SK1324, 2SK1456, ++	data
IRFBF 20 L,S	MOS-N-FET-e	=IRFBF 20:	30c§		2SK1647	data
IRFBF 30	MOS-N-FET-e	V-MOS, 900V, 3,6A, 125W, <3,7Ω(2,2A)	17c§	Inr	2SK1501, 2SK1639, 2SK1643, 2SK1807	data
IRFBG 20	MOS-N-FET-e	V-MOS, 1000V, 1,4A, 54W, <11Ω(0,84A)	17c§	Inr	BUK 456-1000, BUZ 51	data
IRFBG 30	MOS-N-FET-e	V-MOS, 1000V, 3,1A, 125W, <5Ω(1,9A)	17c§	Inr	BUK 456-1000, BUZ 51	data
IRFD 1 Z0	MOS-N-FET-e	V-MOS, 100V, 0,5A,1W, <2,4Ω(0,3A)	4-DIP	Inr,Gen,Mot	-	data
IRFD 1 Z1	MOS-N-FET-e	=IRFD 1Z0: 60V	4-DIP	Inr,Gen	-	data
IRFD 1 Z2	MOS-N-FET-e	V-MOS, 100V, 0,4A,1W, <3,2Ω(0,25A)	4-DIP	Inr,Gen	-	data
IRFD 1 Z3	MOS-N-FET-e	=IRFD 1Z2: 60V	4-DIP	Inr,Gen,Mot	-	data
IRFD 2 Z0	MOS-N-FET-e	V-MOS, 200V, 0,32A,1W, <5Ω(0,15A)	4-DIP	Inr,Gen	-	data
IRFD 2 Z1	MOS-N-FET-e	=IRFD 2Z0: 150V	4-DIP	Inr,Gen	-	data
IRFD 2 Z2	MOS-N-FET-e	V-MOS, 200V, 0,3A, 1W, <6,5Ω(0,15A),25	4-DIP	Gen	-	data
IRFD 2 Z3	MOS-N-FET-e	=IRFD 2Z2: 150V	4-DIP	Inr,Gen	-	data
IRFD 010	MOS-N-FET-e	V-MOS, 50V, 1,7A, 1W,<0,2Ω(0,86A)	4-DIP	Inr	-	data
IRFD 012	MOS-N-FET-e	V-MOS, 50V, 1,4A, 1W, <0,3Ω(0,86A)	4-DIP	Inr	-	data
IRFD 014	MOS-N-FET-e	V-MOS, 60V, 1,7A, 1,3W, <0,2Ω(1A)	4-DIP	Inr	-	data
IRFD 020	MOS-N-FET-e	V-MOS, 50V, 2,4A, 1W, <0,1Ω(1,4A)	4-DIP	Inr, Six	-	data
IRFD 022	MOS-N-FET-e	V-MOS, 50V, 2,2A, 1W, <0,12Ω(1,4A)	4-DIP	Inr	-	data
IRFD 024	MOS-N-FET-e	V-MOS, 60V, 2,5A, 1,3W, <0,1Ω(1,5A)	4-DIP	Inr	-	data
IRFD 110	MOS-N-FET-e	V-MOS, 100V, 1A, 1,3W, <0,54Ω(0,6A)	4-DIP	Inr,Gen,Mot	-	data
IRFD 111	MOS-N-FET-e	=IRFD 110: 60V	4-DIP	Inr,Gen	-	data
IRFD 112	MOS-N-FET-e	V-MOS, 100V, 0,8A, 1,3W, <0,8Ω(0,8A)	4-DIP	Inr,Gen	-	data
IRFD 113	MOS-N-FET-e	=IRFD 112: 60V	4-DIP	Inr,Gen,Mot	-	data
IRFD 120	MOS-N-FET-e	V-MOS, 100V, 1,3A, 1,3W, <0,27Ω(0,78A)	4-DIP	Inr,Gen,Mot	-	data
IRFD 121	MOS-N-FET-e	=IRFD 120: 60V	4-DIP	Inr,Gen,Mot	-	data
IRFD 122	MOS-N-FET-e	V-MOS, 100V, 1,1A, 1,3W, <0,4Ω(0,6A)	4-DIP	Inr,Gen,Mot	-	data
IRFD 123	MOS-N-FET-e	=IRFD 122: 60V	4-DIP	Inr,Gen,Mot	-	data
IRFD 210	MOS-N-FET-e	V-MOS, 200V, 0,6A, 1W, <1,5Ω(0,36A)	4-DIP	Inr,Gen,Mot	-	data
IRFD 211	MOS-N-FET-e	=IRFD 210: 150V	4-DIP	Inr,Gen,Mot	-	data
IRFD 212	MOS-N-FET-e	V-MOS, 200V, 0,45A, 1W, <2,4Ω(0,3A)	4-DIP	Inr,Gen,Mot	-	data
IRFD 213	MOS-N-FET-e	=IRFD 212: 150V	4-DIP	Inr,Gen,Mot	-	data
IRFD 214	MOS-N-FET-e	V-MOS, 250V, 0,45A, 1W, <2Ω(0,27A)	4-DIP	Inr	-	data
IRFD 220	MOS-N-FET-e	V-MOS, 200V, 0,8A, 1W, <0,8Ω(0,48A)	4-DIP	Inr,Gen,Mot	-	data
IRFD 221	MOS-N-FET-e	=IRFD 220: 150V	4-DIP	Inr,Gen,Mot	-	data
IRFD 222	MOS-N-FET-e	V-MOS, 200V, 0,7A, 1W, <1,2Ω(0,4A)	4-DIP	Inr,Gen,Mot	-	data
IRFD 223	MOS-N-FET-e	=IRFD 222: 150V	4-DIP	Inr,Gen,Mot	-	data
IRFD 224	MOS-N-FET-e	V-MOS, 250V, 0,63A, 1W, <1,1Ω(0,38A)	4-DIP	Inr	-	data
IRFD 310	MOS-N-FET-e	V-MOS, 400V, 0,35A, 1W, <3,6Ω(0,21A)	4-DIP	Gen	-	data
IRFD 311	MOS-N-FET-e	=IRFD 310: 350V	4-DIP	Gen	-	data
IRFD 312	MOS-N-FET-e	V-MOS, 400V, 0,3A, 1W, <5Ω(0,2A)	4-DIP	Gen	-	data
IRFD 313	MOS-N-FET-e	=IRFD 312: 350V	4-DIP	Gen	-	data
IRFD 320	MOS-N-FET-e	V-MOS, 400V, 0,49A, 1W, <1,8Ω(0,21A)	4-DIP	Inr, Gen	-	data
IRFD 321	MOS-N-FET-e	=IRFD 320: 350V	4-DIP	Inr,Gen	-	data
IRFD 322	MOS-N-FET-e	V-MOS, 400V, 0,4A, 1W, <2,5Ω(0,25A)	4-DIP	Gen	-	data
IRFD 323	MOS-N-FET-e	=IRFD 322: 350V	4-DIP	Gen	-	data
IRFD 420	MOS-N-FET-e	V-MOS, 500V, 0,37A, 1W, <3Ω(0,22A)	4-DIP	Inr	-	data
IRFD 9010	MOS-P-FET-e	V-MOS, 50V, 1,1A, 1W, <0,5Ω(0,58A)	4-DIP	Inr	-	data
IRFD 9012	MOS-P-FET-e	V-MOS, 50V, 0,91A, 1W, <0,7Ω(0,58A)	4-DIP	Inr	-	data
IRFD 9014	MOS-P-FET-e	V-MOS, 60V, 1,1A, 1,3W, <0,5Ω(0,66A)	4-DIP	Inr	-	data

Type	Device	Short Description	Fig.	Manu	Comparision Types	More at
IRFD 9020	MOS-P-FET-e	V-MOS, 50V, 1,6A, 1W, <0,28Ω(1,1A)	4-DIP	Inr, Six	-	data
IRFD 9022	MOS-P-FET-e	V-MOS, 50V, 1,4A, 1W, <0,33Ω(1,1A)	4-DIP	Inr	-	data
IRFD 9024	MOS-P-FET-e	V-MOS, 60V, 1,6A, 1,3W, <0,28Ω(0,96A)	4-DIP	Inr	-	data
IRFD 9110	MOS-P-FET-e	V-MOS, 100V, 0,7A, 1,3W, <1,2Ω(0,48A)	4-DIP	Inr	-	data
IRFD 9113	MOS-P-FET-e	V-MOS, 60V, 0,6A, 1,3W, <1,6Ω(0,3A)	4-DIP	Inr	-	data
IRFD 9120	MOS-P-FET-e	V-MOS, 100V, 1A, 1,3W, <0,6Ω(0,6A)	4-DIP	Inr,Mot,Six	-	data
IRFD 9121	MOS-P-FET-e	=IRFD 9120: 60V	4-DIP	Inr	-	data
IRFD 9122	MOS-P-FET-e	=IRFD 9120: 0,8A, <0,8Ω(0,8A)	4-DIP	Inr	-	data
IRFD 9123	MOS-P-FET-e	=IRFD 9120: 60V, 0,8A, 0,8Ω(0,8A)	4-DIP	Inr,Mot,Six	-	data
IRFD 9210	MOS-P-FET-e	V-MOS, 200V, 0,4A, 1W, <3Ω(0,24A)	4-DIP	Inr	-	data
IRFD 9213	MOS-P-FET-e	V-MOS, 150V, 0,3A, 1W, <4,5Ω(0,3A)	4-DIP	Inr	-	data
IRFD 9220	MOS-P-FET-e	V-MOS, 200V, 0,56A, 1W, <1,5Ω(0,34A)	4-DIP	Inr	-	data
IRFD 9223	MOS-P-FET-e	V-MOS, 150V, 0,45A, 1W, <2,4Ω(0,3A)	4-DIP	Inr	-	data
IRFDC 20	MOS-N-FET-e	V-MOS, 600V, 0,32A, 1W, <4,4Ω(0,19A)	4-DIP	Inr	-	data
IRFE 110	MOS-FET	= 4x IRFD 110	4x 4-DIP	Inr	-	data
IRFE 111	MOS-FET	= 4x IRFD 111	4x 4-DIP	Inr	-	data
IRFE 112	MOS-FET	= 4x IRFD 112	4x 4-DIP	Inr	-	data
IRFE 113	MOS-FET	= 4x IRFD 113	4x 4-DIP	Inr	-	data
IRFE 5110	MOS-FET	= 2x IRFD 110 + 2x IRFD 9120	4x 4-DIP	Inr	-	data
IRFE 5111	MOS-FET	= 2x IRFD 111 + 2x IRFD 9121	4x 4-DIP	Inr	-	data
IRFE 5112	MOS-FET	= 2x IRFD 112 + 2x IRFD 9122	4x 4-DIP	Inr	-	data
IRFE 5113	MOS-FET	= 2x IRFD 113 + 2x IRFD 9123	4x 4-DIP	Inr	-	data
IRFE 9120	MOS-FET	= 4x IRFD 9120	4x 4-DIP	Inr	-	data
IRFE 9121	MOS-FET	= 4x IRFD 9121	4x 4-DIP	Inr	-	data
IRFE 9122	MOS-FET	= 4x IRFD 9122	4x 4-DIP	Inr	-	data
IRFE 9123	MOS-FET	= 4x IRFD 9123	4x 4-DIP	Inr	-	data
IRFF 110	MOS-N-FET-e	V-MOS, 100V, 3,5A, 15W(Tc=25°), <0,6Ω	2a§	Inr	(2SK782, 2SK2790)[4]	data
IRFF 111	MOS-N-FET-e	=IRFF 110: 60V	2a§	Inr	(2SK782, 2SK2790)[4]	data
IRFF 112	MOS-N-FET-e	V-MOS, 100V, 3A, 15W(Tc=25°), <0,8Ω	2a§	Inr	(2SK782, 2SK2790)[4]	data
IRFF 113	MOS-N-FET-e	=IRFF 112: 60V	2a§	Inr	(2SK782, 2SK2790)[4]	data
IRFF 110R...113R	MOS-N-FET-e	=IRFF 110...113: Avalanche Energy Rated	2a§		-	data
IRFF 120	MOS-N-FET-e	V-MOS, 100V, 6A, 20W(Tc=25°), <0,3Ω	2a§	Inr	(MTB 9N25E, 2SK1621, 2SK2252)[4]	data
IRFF 121	MOS-N-FET-e	=IRFF 120: 60V	2a§	Inr	(MTB 9N25E, 2SK1621, 2SK2252)[4]	data
IRFF 122	MOS-N-FET-e	V-MOS, 100V, 5A, 20W(Tc=25°), <0,4Ω	2a§	Inr	(MTB 9N25E, 2SK1621, 2SK2252)[4]	data
IRFF 123	MOS-N-FET-e	=IRFF 122: 60V	2a§	Inr	(MTB 9N25E, 2SK1621, 2SK2252)[4]	data
IRFF 120R...123R	MOS-N-FET-e	=IRFF 120...123: Avalanche Energy Rated	2a§		-	data
IRFF 130	MOS-N-FET-e	V-MOS, 100V, 8A, 25W(Tc=25°), <0,18Ω	2a§	Inr	(MTB 9N25E, 2SK1621, 2SK2252)[4]	data
IRFF 131	MOS-N-FET-e	=IRFF 130: 60V	2a§	Inr	(MTB 9N25E, 2SK1621, 2SK2252)[4]	data
IRFF 132	MOS-N-FET-e	V-MOS, 100V, 7A, 25W(Tc=25°), <0,25Ω	2a§	Inr	(MTB 9N25E, 2SK1621, 2SK2252)[4]	data
IRFF 133	MOS-N-FET-e	=IRFF 132: 60V	2a§	Inr	(MTB 9N25E, 2SK1621, 2SK2252)[4]	data
IRFF 130R...133R	MOS-N-FET-e	=IRFF 130...133: Avalanche Energy Rated	2a§		-	data
IRFF 210	MOS-N-FET-e	V-MOS, 200V, 2,2A, 15W(Tc=25°), <1,5Ω	2a§	Inr	(2SK782, 2SK2790)[4]	data
IRFF 211	MOS-N-FET-e	=IRFF 210: 150V	2a§	Inr	(2SK782, 2SK2790)[4]	data
IRFF 212	MOS-N-FET-e	V-MOS, 200V, 1,8A, 15W(Tc=25°), <2,4Ω	2a§	Inr	(2SK782, 2SK2790)[4]	data
IRFF 213	MOS-N-FET-e	=IRFF 212: 150V	2a§	Inr	(2SK782, 2SK2790)[4]	data
IRFF 210R..213R	MOS-N-FET-e	=IRFF 210...213: Avalanche Energy Rated	2a§		-	data
IRFF 220	MOS-N-FET-e	V-MOS, 200V, 3,5A, 20W(Tc=25°), <0,8Ω	2a§	Inr	(2SK782, 2SK2790)[4]	data
IRFF 221	MOS-N-FET-e	=IRFF 220: 150V	2a§	Inr	(2SK782, 2SK2790)[4]	data
IRFF 222	MOS-N-FET-e	V-MOS, 200V, 3A, 20W(Tc=25°), <2Ω(2A)	2a§	Inr	(2SK782, 2SK2790)[4]	data
IRFF 223	MOS-N-FET-e	=IRFF 222: 150V	2a§	Inr	(2SK782, 2SK2790)[4]	data
IRFF 220R...223R	MOS-N-FET-e	=IRFF 220...223: Avalanche Energy Rated	2a§		-	data
IRFF 230	MOS-N-FET-e	V-MOS, 200V, 5,5A, 25W(Tc=25°), <0,4Ω	2a§	Inr	(2SK782, 2SK2790)[4]	data
IRFF 231	MOS-N-FET-e	=IRFF 230: 150V	2a§	Inr	(2SK782, 2SK2790)[4]	data
IRFF 232	MOS-N-FET-e	V-MOS, 200V, 4,5A, 25W(Tc=25°), <0,6Ω	2a§	Manu	(2SK782, 2SK2790)[4]	data
IRFF 233	MOS-N-FET-e	=IRFF 232: 150V	2a§	Manu	(2SK782, 2SK2790)[4]	data
IRFF 230R...233R	MOS-N-FET-e	=IRFF 230...233: Avalanche Energy Rated	2a§		-	data
IRFF 310	MOS-N-FET-e	V-MOS, 400V, 1,35A, 15W(Tc=25°), <3,6Ω	2a§	Inr	(MTB 2N40, 2SK2509)[4]	data
IRFF 311	MOS-N-FET-e	=IRFF 310: 350V	2a§	Inr	(MTB 2N40, 2SK2509)[4]	data
IRFF 312	MOS-N-FET-e	V-MOS, 400V, 1,15A, 15W(Tc=25°), <5Ω	2a§	Inr	(MTB 2N40, 2SK2509)[4]	data
IRFF 313	MOS-N-FET-e	=IRFF 312: 350V	2a§	Inr	(MTB 2N40, 2SK2509)[4]	data
IRFF 310R..313R	MOS-N-FET-e	=IRFF 310...313: Avalanche Energy Rated	2a§		-	data
IRFF 320	MOS-N-FET-e	V-MOS, 400V, 2,5A, 20W(Tc=25°), <1,8Ω	2a§	Inr	(2SK1493Z, 2SK1690)[4]	data
IRFF 321	MOS-N-FET-e	=IRFF 320: 350V	2a§	Inr	(2SK1493Z, 2SK1690)[4]	data
IRFF 322	MOS-N-FET-e	V-MOS, 400V, 2A, 20W(Tc=25°), <2,5Ω	2a§	Inr	(2SK1493Z, 2SK1690)[4]	data
IRFF 323	MOS-N-FET-e	=IRFF 322: 350V	2a§	Inr	(2SK1493Z, 2SK1690)[4]	data
IRFF 320R...323R	MOS-N-FET-e	=IRFF 320...323: Avalanche Energy Rated	2a§		-	data
IRFF 330	MOS-N-FET-e	V-MOS, 400V, 3,5A, 25W(Tc=25°), <1Ω	2a§	Inr	(2SK1493Z, 2SK1690)[4]	data
IRFF 331	MOS-N-FET-e	=IRFF 330: 350V	2a§	Inr	(2SK1493Z, 2SK1690)[4]	data
IRFF 332	MOS-N-FET-e	V-MOS, 400V, 3A, 25W(Tc=25°), <1,5Ω	2a§	Inr	(2SK1493Z, 2SK1690)[4]	data
IRFF 333	MOS-N-FET-e	=IRFF 332: 350V	2a§	Inr	(2SK1493Z, 2SK1690)[4]	data
IRFF 330R...333R	MOS-N-FET-e	=IRFF 330...333: Avalanche Energy Rated	2a§		-	data
IRFF 420	MOS-N-FET-e	V-MOS, 500V, 1,6A, 20W(Tc=25°), <3Ω	2a§	Inr	(MTB 2N60E, 2SK1746, 2SK2509)[4]	data
IRFF 421	MOS-N-FET-e	=IRFF 420: 450V	2a§	Inr	(MTB 2N60E, 2SK1746, 2SK2509)[4]	data
IRFF 422	MOS-N-FET-e	V-MOS, 500V, 1,4A, 20W(Tc=25°), <4Ω	2a§	Inr	(MTB 2N60E, 2SK1746, 2SK2509)[4]	data
IRFF 423	MOS-N-FET-e	=IRFF 422: 450V	2a§	Inr	(MTB 2N60E, 2SK1746, 2SK2509)[4]	data
IRFF 420R..423R	MOS-N-FET-e	=IRFF 420...423: Avalanche Energy Rated	2a§		-	data
IRFF 430	MOS-N-FET-e	V-MOS, 500V, 2,75A, 25W(Tc=25°), <1,5Ω	2a§	Inr	(2SK1494Z, 2SK1721)[4]	data
IRFF 431	MOS-N-FET-e	=IRFF 430: 450V	2a§	Inr	(2SK1494Z, 2SK1721)[4]	data

Type	Device	Short Description	Fig.	Manu	Comparision Types	More at
IRFF 432	MOS-N-FET-e	V-MOS, 500V, 2,25A, 25W(Tc=25°), <2Ω	2a§	Inr	(2SK1494Z, 2SK1721)[4]	data
IRFF 433	MOS-N-FET-e	=IRFF 432: 450V	2a§	Inr	(2SK1494Z, 2SK1721)[4]	data
IRFF 430R...433R	MOS-N-FET-e	=IRFF 430...433: Avalanche Energy Rated	2a§			data
IRFF 9110	MOS-P-FET-e	V-MOS, 100V, 2,6A, 15W(Tc=25°), <1,2Ω	2a§	Inr	(2SJ275)[4]	data
IRFF 9111	MOS-P-FET-e	=IRFF 9110: 60V	2a§	Inr	(2SJ275)[4]	data
IRFF 9112	MOS-P-FET-e	V-MOS, 100V, 2,3A, 15W(Tc=25°), <1,6Ω	2a§	Inr	(2SJ275)[4]	data
IRFF 9113	MOS-P-FET-e	=IRFF 9112: 60V	2a§	Inr	(2SJ275)[4]	data
IRFF 9120	MOS-P-FET-e	V-MOS, 100V, 4A, 20W(Tc=25°), <0,6Ω	2a§	Inr	(2SJ275)[4]	data
IRFF 9121	MOS-P-FET-e	=IRFF 9120: 60V	2a§	Inr	(2SJ275)[4]	data
IRFF 9122	MOS-P-FET-e	V-MOS, 100V, 3,5A, 20W(Tc=25°), <0,8Ω	2a§	Inr	(2SJ275)[4]	data
IRFF 9123	MOS-P-FET-e	=IRFF 9122: 60V	2a§	Inr	(2SJ275)[4]	data
IRFF 9130	MOS-P-FET-e	V-MOS, 100V, 6,5A, 25W(Tc=25°), <0,3Ω	2a§	Inr	(2SJ275)[4]	data
IRFF 9131	MOS-P-FET-e	=IRFF 9130: 60V	2a§	Inr	(2SJ275)[4]	data
IRFF 9132	MOS-P-FET-e	V-MOS, 100V, 5,5A, 25W(Tc=25°), <0,4Ω	2a§	Inr	(2SJ275)[4]	data
IRFF 9133	MOS-P-FET-e	=IRFF 9132: 60V	2a§	Inr	(2SJ275)[4]	data
IRFF 9210	MOS-P-FET-e	V-MOS, 200V, 1,6A, 15W(Tc=25°), <3Ω	2a§	Inr	(MTB 2P50E)[4]	data
IRFF 9211	MOS-P-FET-e	=IRFF 9210: 150V	2a§	Inr	(MTB 2P50E)[4]	data
IRFF 9212	MOS-P-FET-e	V-MOS, 200V, 1,3A, 15W(Tc=25°), <4,5Ω	2a§	Inr	(MTB 2P50E)[4]	data
IRFF 9213	MOS-P-FET-e	=IRFF 9212: 150V	2a§	Inr	(MTB 2P50E)[4]	data
IRFF 9220	MOS-P-FET-e	V-MOS, 200V, 2,5A, 20W(Tc=25°), <1,5Ω	2a§	Inr	(MTB 2P50E)[4]	data
IRFF 9221	MOS-P-FET-e	=IRFF 9220: 150V	2a§	Inr	(MTB 2P50E)[4]	data
IRFF 9222	MOS-P-FET-e	V-MOS, 200V, 2A, 20W(Tc=25°), <2,4Ω	2a§	Inr	(MTB 2P50E)[4]	data
IRFF 9223	MOS-P-FET-e	=IRFF 9222: 150V	2a§	Inr	(MTB 2P50E)[4]	data
IRFF 9230	MOS-P-FET-e	V-MOS, 200V, 4A, 25W(Tc=25°), <0,8Ω	2a§	Inr	-	data
IRFF 9231	MOS-P-FET-e	=IRFF 9230: 150V	2a§	Inr	-	data
IRFF 9232	MOS-P-FET-e	V-MOS, 200V, 3,5A, 25W(Tc=25°), <1,2Ω	2a§	Inr	-	data
IRFF 9233	MOS-P-FET-e	=IRFF 9232: 150V	2a§	Inr	-	data
IRFG 1 Z0	MOS-N-FET-e	4x V-MOS, 100V, 0,4A, 1,8W, <3Ω(0,2A)	14-DIC	Inr		data
IRFG 1 Z3	MOS-N-FET-e	4xV-MOS, 60V, 0,35A, 1,8W, <3,6Ω(0,2A)	14-DIC	Inr		data
IRFG 110	MOS-N-FET-e	4xV-MOS, 100V, 0,95A, 2,5W,<0,8Ω(0,5A)	14-DIC	Inr	-	data
IRFG 113	MOS-N-FET-e	4x V-MOS, 60V, 0,85A, 2,5W, <1Ω(0,5A)	14-DIC	Inr	-	data
IRFG 6110	MOS-FET-e	= 1/2 IRFG 110 + 1/2 IRFG 9110	14-DIC	Inr		data
IRFG 6113	MOS-FET-e	= 1/2 IRFG 113 + 1/2 IRFG 9113	14-DIC	Inr		data
IRFG 9110	MOS-P-FET-e	4xV-MOS, 100V, 0,75A, 2,5W,<1,4Ω(0,3A)	14-DIC	Inr		data
IRFG 9113	MOS-P-FET-e	4xV-MOS, 60V, 0,65A, 2,5W, <1,8Ω(0,3A)	14-DIC	Inr		data
IRFH 150	MOS-N-FET-e	V-MOS, 100V, 30A, 150W, <135/225ns	49a (Iso)	Inr		data
IRFH 250	MOS-N-FET-e	V-MOS, 200V, 30A, 150W, <135/225ns	49a (Iso)	Inr		data
IRFH 350	MOS-N-FET-e	V-MOS, 400V, 15A, 150W, <100/225ns	49a (Iso)	Inr		data
IRFH 450	MOS-N-FET-e	V-MOS, 500V, 13A, 150W, <85/220ns	49a (Iso)	Inr		data
IRFI 9 Z14G	MOS-P-FET-e	V-MOS, 60V, 5,3A, 27W, <0,5Ω(3,1A)	17c	Inr	2SJ248, 2SJ350	data
IRFI 9 Z24G	MOS-P-FET-e	V-MOS, 60V, 8,5A, 37W, <0,28Ω(5,1A)	17c	Inr	2SJ137, 2SJ147, 2SJ175, 2SJ236, 2SJ390	data
IRFI 9 Z24N	MOS-P-FET-e	V-MOS, 55V, 9,5A, 29W, <0,175Ω(5,4A)	17c	Inr	2SJ137, 2SJ147, 2SJ175, 2SJ236, 2SJ390	data
IRFI 9 Z34G	MOS-P-FET-e	V-MOS, 60V, 12A, 42W, <0,14Ω(7,2A)	17c	Inr	2SJ143, 2SJ176, 2SJ293, 2SJ321, 2SJ329++	data
IRFI 9 Z34N	MOS-P-FET-e	V-MOS, 55V, 14A, 37W, <0,1Ω(7,8A)	17c	Inr	2SJ143, 2SJ176, 2SJ293, 2SJ321, 2SJ329++	data
IRFI 510 A	MOS-N-FET-e	=IRFW 510A	30c§	Sam	→IRFW 510A	data
IRFI 510 G	MOS-N-FET-e	V-MOS, 100V, 4,5A, 27W, <0,54Ω(2,7A)	17c	Inr	BUK 442-100, BUK 542-100, 2SK1260,++	data
IRFI 520 A	MOS-N-FET-e	=IRFW 520A	30c§	Sam	→IRFW 520A	data
IRFI 520 G	MOS-N-FET-e	V-MOS, 100V, 7,2A, 37W, <0,27Ω(4,3A)	17c	Inr	BUK 443-100, BUZ 72(A)F, 2SK1261, ++	data
IRFI 520 N	MOS-N-FET-e	V-MOS, 100V, 7,2A, 27W, <0,2Ω(4,3A)	17c	Inr	→IRFI 520G	data
IRFI 530 A	MOS-N-FET-e	=IRFW 530A	30c§	Sam	→IRFW 530A	data
IRFI 530 G	MOS-N-FET-e	V-MOS, 100V, 9,7A, 42W, <0,16Ω(5,8A)	17c	Inr	BUK 545-100, 2SK1230, 2SK1305, 2SK1430++	data
IRFI 530 N	MOS-N-FET-e	V-MOS, 100V, 11A, 33W, <0,11Ω(10A)	17c	Inr	→IRFI 530G	data
IRFI 540 A	MOS-N-FET-e	=IRFW 540A	30c§	Sam	→IRFW 540A	data
IRFI 540 G	MOS-N-FET-e	V-MOS, 100V, 17A, 48W, <77mΩ(10A)	17c	Inr	2SK1307, 2SK1318, 2SK1348, 2SK1432, ++	data
IRFI 540 N	MOS-N-FET-e	V-MOS, 100V, 17A, 48W, <77mΩ(10A)	17c	Inr	→IRFI 540G	data
IRFI 550 A	MOS-N-FET-e	=IRFW 550A	30c§	Sam	→IRFW 550A	data
IRFI 610 A	MOS-N-FET-e	=IRFW 610A	30c§	Sam	→IRFW 610A	data
IRFI 614 A	MOS-N-FET-e	=IRFW 614A	30c§	Sam	→IRFW 614A	data
IRFI 614 G	MOS-N-FET-e	V-MOS, 250V, 2,1A, 23W, <2Ω(1,3A)	17c	Inr	BUK 444-400, 2SK1833	data
IRFI 620 A	MOS-N-FET-e	=IRFW 620A	30c§	Sam	→IRFW 620A	data
IRFI 620 G	MOS-N-FET-e	V-MOS, 200V, 4,1A, 30W, <0,8Ω(2,5A)	17c	Inr	BUK 444-200, BUZ 73(A)F	data
IRFI 624 A	MOS-N-FET-e	=IRFW 624A	30c§	Sam	→IRFW 624A	data
IRFI 624 G	MOS-N-FET-e	V-MOS, 250V, 3,4A, 30W, <1,1Ω(2A)	17c	Inr	BUK 445-400, 2SK530, 2SK1377	data
IRFI 630 A	MOS-N-FET-e	=IRFW 630A	30c§	Sam	→IRFW 630A	data
IRFI 630 G	MOS-N-FET-e	V-MOS, 200V, 5,9A, 35W, <0,4Ω(3,5A)	17c	Inr	BUK 445-200, 2SK1478, 2SK1668, 2SK1957++	data
IRFI 634 A	MOS-N-FET-e	=IRFW 634A	30c§	Sam	→IRFW 634A	data
IRFI 634 G	MOS-N-FET-e	V-MOS, 250V, 5,6A, 35W, <0,45Ω(3,4A)	17c	Inr	2SK1478, 2SK1568, 2SK1668	data
IRFI 640 A	MOS-N-FET-e	=IRFW 640A	30c§	Sam	→IRFW 640A	data
IRFI 640 G	MOS-N-FET-e	V-MOS, 200V, 9,8A, 40W, <0,18Ω(5,9A)	17c	Inr	BUK 426-200, 2SK526, 2SK1036, 2SK2212,++	data
IRFI 644 A	MOS-N-FET-e	=IRFW 644A	30c§	Sam	→IRFW 644A	data
IRFI 644 G	MOS-N-FET-e	V-MOS, 250V, 7,9A, 40W, <0,28Ω(4,7A)	17c	Inr	2SK1478, 2SK1568, 2SK1668...69	data
IRFI 710 A	MOS-N-FET-e	=IRFW 710A	30c§	Sam	→IRFW 710A	data
IRFI 720 A	MOS-N-FET-e	=IRFW 720A	30c§	Sam	→IRFW 720A	data
IRFI 720 G	MOS-N-FET-e	V-MOS, 400V, 2,6A, 30W, <1,8Ω(1,6A)	17c	Inr	BUK 445-450, 2SK503, 2SK1862...63	data
IRFI 730 A	MOS-N-FET-e	=IRFW 730A	30c§	Sam	→IRFW 730A	data
IRFI 730 G	MOS-N-FET-e	V-MOS, 400V, 3,7A, 35W, <1Ω(2,7A)	17c	Inr	BUK 445-400, 2SK530, 2SK1231...32, ++	data
IRFI 734 G	MOS-N-FET-e	V-MOS, 450V, 3,4A, 35W, <1,2mΩ(2A)	17c	Inr	BUK 445-450, 2SK1231...32, 2SK2353...54,++	data

Type	Device	Short Description	Fig.	Manu	Comparision Types	More at
IRFI 740 A	MOS-N-FET-e	=IRFW 740A	30c§	Sam	→IRFW 740A	data
IRFI 740 G	MOS-N-FET-e	V-MOS, 400V, 5,4A, 40W, <0,55Ω(3,2A)	17c	Inr	BUK 445-400, 2SK530, 2SK1231...32, ++	data
IRFI 740 GLC	MOS-N-FET-e	V-MOS, Io Gate Charge, 400V, 5,7A, 40W <0,55Ω(3,4A)	17c	Inr	-	data
IRFI 744 G	MOS-N-FET-e	V-MOS, 450V, 4,9A, 40W, <0,63Ω(2,9A)	17c	Inr	2SK1231...32, 2SK2353...54, 2SK2114...15,++	data
IRFI 820 A	MOS-N-FET-e	=IRFW 820A	30c§	Sam	→IRFW 820A	data
IRFI 820 G	MOS-N-FET-e	V-MOS, 500V, 2,1A, 30W, <3Ω(1,3A)	17c	Inr	BUK 444-500, 2SK1833	data
IRFI 830 A	MOS-N-FET-e	=IRFW 830A	30c§	Sam	→IRFW 830A	data
IRFI 830 G	MOS-N-FET-e	V-MOS, 500V, 3,1A, 35W, <1,5Ω(1,9A)	17c	Inr	BUK 445-500, 2SK1863	data
IRFI 840 A	MOS-N-FET-e	=IRFW 840A	30c§	Sam	→IRFW 840A	data
IRFI 840 G	MOS-N-FET-e	V-MOS, 500V, 4,6A, 40W, <0,85Ω(2,8A)	17c	Inr	2SK1232, 2SK1608, 2SK1627, 2SK2115, ++	data
IRFI 840 GLC	MOS-N-FET-e	V-MOS, Io Gate Charge, 500V, 4,5A, 40W <0,85Ω(2,7A)	17c	Inr	-	data
IRFI 1010 N	MOS-N-FET-e	V-MOS, 55V, 49A, 58W, <12mΩ(26A)	17c	Inr	2SK1421, 2SK2120, 2SK2312, 2SK2498, ++	data
IRFI 1310 G	MOS-N-FET-e	V-MOS, 100V, 22A, 48W, <40mΩ(13A)	17c	Inr	2SK1292, 2SK1349, 2SK1432, 2SK1906, ++	data
IRFI 1310 N	MOS-N-FET-e	V-MOS, 100V, 22A, 45W, <36mΩ(13A)	17c	Inr	→IRFI 1310G	data
IRFI 3205	MOS-N-FET-e	V-MOS, 55V, 64A, 63W, <8mΩ(34A)	17c	Inr	-	data
IRFI 3710 N	MOS-N-FET-e	V-MOS, 100V, 28A, 48W, <25mΩ(16A)	17c	Inr	2SK1262, 2SK2466, 2SK2577, 2SK2954	data
IRFI 4905	MOS-P-FET-e	V-MOS, 55V, 41A, 63W, <20mΩ(22A)	17c	Inr	-	data
IRFI 9520 G	MOS-P-FET-e	V-MOS, 100V, 5,2A, 37W, <0,6Ω(3,1A)	17c	Inr	2SJ248, 2SJ350	data
IRFI 9530 G	MOS-P-FET-e	V-MOS, 100V, 7,7A, 42W, <0,3Ω(4,6A)	17c	Inr	2SJ139, 2SJ274, 2SJ380	data
IRFI 9540 G	MOS-P-FET-e	V-MOS, 100V, 11A, 48W, <0,2Ω(6,6A)	17c	Inr	2SJ139, 2SJ274, 2SJ380	data
IRFI 9540 N	MOS-P-FET-e	V-MOS, 100V, 13A, 42W, <0,117Ω(7,8A)	17c	Inr	→IRFI 9540G	data
IRFI 9620 G	MOS-P-FET-e	V-MOS, 200V, 2A, 30W, <1,5Ω(1,8A)	17c	Inr	2SJ403, 2SJ407	data
IRFI 9630 G	MOS-P-FET-e	V-MOS, 200V, 4,3A, 30W, <0,8Ω(2,6A)	17c	Inr	2SJ403, 2SJ407	data
IRFI 9634	MOS-P-FET-e	V-MOS, 250V, 4,1A, 35W, <1Ω(2,5A)	17c	Inr	2SJ320, 2SJ448, 2SJ512	data
IRFI 9640 G	MOS-P-FET-e	V-MOS, 200V, 6,1A, 40W, <0,5Ω(3,7A)	17c	Inr	2SJ404, 2SJ405, 2SJ410	data
IRFIB 5 N65A	MOS-N-FET-e	V-MOS, 650V, 5,1A, 60W, <0,93Ω(3,1A)	17c	Inr	2SK2101	data
IRFIB 6 N60A	MOS-N-FET-e	V-MOS, 600V, 5,5A, 60W, <0,75Ω(3,3A)	17c	Inr	2SK1118, 2SK2045, 2SK2118, 2SK2139, ++	data
IRFIB 7 N50A	MOS-N-FET-e	V-MOS, 500V, 6,6A, 60W, <52mΩ(4A)	17c	Inr	2SK1352, 2SK2117, 2SK2237, 2SK2364, ++	data
IRFIBC 20 G	MOS-N-FET-e	V-MOS, 600V, 1,7A, 30W, <4,4Ω(1A)	17c	Inr	BUK 445-600, 2SK1758, 2SK1834, 2SK1953++	data
IRFIBC 30 G	MOS-N-FET-e	V-MOS, 600V, 2,5A, 35W, <2,2Ω(1,5A)	17c	Inr	2SK1572, 2SK1767, 2SK2144, 2SK2325, ++	data
IRFIBC 40 G	MOS-N-FET-e	V-MOS, 600V, 3,5A, 40W, <1,2Ω(2,1A)	17c	Inr	2SK1572, 2SK1767, 2SK2144, 2SK2325, ++	data
IRFIBC 40 GLC	MOS-N-FET-e	V-MOS, Io Gate Charge, 600V, 3,5A, 40W <1,2Ω(2,1A)	17c	Inr	-	data
IRFIBE 20 G	MOS-N-FET-e	V-MOS, 800V, 1,4A, 30W, <6,5Ω(0,84A)	17c	Inr	BUK 446-800, 2SK1459, 2SK1834	data
IRFIBE 30 G	MOS-N-FET-e	V-MOS, 800V, 2,1A, 35W, <3Ω(1,3A)	17c	Inr	BUK 446-800, 2SK1459, 2SK1834	data
IRFIBF 20 G	MOS-N-FET-e	V-MOS, 900V, 1,2A, 30W, <8Ω(0,72A)	17c	Inr	BUK 446-1000, 2SK808A	data
IRFIBF 30 G	MOS-N-FET-e	V-MOS, 900V, 1,9A, 35W, <3,7Ω(1,1A)	17c	Inr	2SK1143, 2SK1275, 2SK1459	data
IRFIP 044	MOS-N-FET-e	V-MOS, 60V, 43A, 100W, <28mΩ(26A)	18c	Inr	2SK1298, 2SK1424, 2SK1666, 2SK2203	data
IRFIP 054	MOS-N-FET-e	V-MOS, 60V, 64A, 120W, <14mΩ(38A)	18c	Inr	2SK1425	data
IRFIP 140	MOS-N-FET-e	V-MOS, 100V, 23A, 100W, <77mΩ(14A)	18c	Inr	BUK 426-100, 2SK1435	data
IRFIP 150	MOS-N-FET-e	V-MOS, 100V, 31A, 120W, <55mΩ(19A)	18c	Inr	2SK1435...36	data
IRFIP 240	MOS-N-FET-e	V-MOS, 200V, 14A, 83W, <0,18Ω(8,4A)	18c	Inr	2SK2008	data
IRFIP 244	MOS-N-FET-e	V-MOS, 250V, 11A, 83W, <0,28Ω(6,6A)	18c	Inr	-	data
IRFIP 250	MOS-N-FET-e	V-MOS, 200V, 22A, 96W, <85mΩ(13A)	18c	Inr	-	data
IRFIP 254	MOS-N-FET-e	V-MOS, 250V, 17A, 96W, <0,14Ω(10A)	18c	Inr	-	data
IRFIP 340	MOS-N-FET-e	V-MOS, 400V, 8A, 83W, <0,55Ω(4,8A)	18c	Inr	2SK1206, 2SK1225, 2SK1328, 2SK1831,++	data
IRFIP 350	MOS-N-FET-e	V-MOS, 400V, 11A, 96W, <0,3Ω(6,6A)	18c	Inr	2SK1206, 2SK1225, 2SK1328, 2SK1831,++	data
IRFIP 440	MOS-N-FET-e	V-MOS, 500V, 6,4A, 83W, <0,85Ω(3,8A)	18c	Inr	2SK1206, 2SK1329, 2SK1523, 2SK1696,++	data
IRFIP 448	MOS-N-FET-e	V-MOS, 500V, 7,4A, 89W, <0,6Ω(4,4A)	18c	Inr	2SK1206, 2SK1329, 2SK1523, 2SK1696,++	data
IRFIP 450	MOS-N-FET-e	V-MOS, 500V, 10A, 96W, <0,4Ω(6A)	18c	Inr	2SK1206, 2SK1329, 2SK1523, 2SK1696,++	data
IRFIP 9140	MOS-P-FET-e	V-MOS, 100V, 15A, 100W, <0,2Ω(9A)	18c	Inr	-	data
IRFIP 9240	MOS-P-FET-e	V-MOS, 200V, 8,9A, 83W, <0,5Ω(5,3A)	18c	Inr	-	data
IRFIZ 14 A	MOS-N-FET-e	=IRFWZ 14A	30c§	Sam	→IRFWZ 14A	data
IRFIZ 14 G	MOS-N-FET-e	V-MOS, 60V, 8A, 27W, <0,2Ω(4,8A)	17c	Inr	BUK 442-60, 2SK1033, 2SK1099, 2SK1344,++	data
IRFIZ 24 A	MOS-N-FET-e	=IRFWZ 24A	30c§	Sam	→IRFWZ 24A	data
IRFIZ 24 E	MOS-N-FET-e	V-MOS, 60V, 14A, 29W, <71mΩ(7,8A)	17c	Inr	2SK1094, 2SK1286, 2SK1419, 2SK1896	data
IRFIZ 24 G	MOS-N-FET-e	V-MOS, 60V, 14A, 37W, <0,1Ω(8,4A)	17c	Inr	2SK1094, 2SK1286, 2SK1419, 2SK1896	data
IRFIZ 24 N	MOS-N-FET-e	V-MOS, 55V, 14A, 29W, <70mΩ(7,8A)	17c	Inr	2SK1094, 2SK1286, 2SK1419, 2SK1896	data
IRFIZ 34 A	MOS-N-FET-e	=IRFWZ 34A	30c§	Sam	→IRFWZ 34A	data
IRFIZ 34 E	MOS-N-FET-e	V-MOS, 60V, 21A, 37W, <42mΩ(11A)	17c	Inr	2SK943, 2SK1095, 2SK1345...46, 2SK1420,++	data
IRFIZ 34 G	MOS-N-FET-e	V-MOS, 60V, 20A, 42W, <50mΩ(12A)	17c	Inr	2SK943, 2SK1095, 2SK1345...46, 2SK1420,++	data
IRFIZ 34 N	MOS-N-FET-e	V-MOS, 55V, 21A, 37W, <40mΩ(11A)	17c	Inr	2SK943, 2SK1095, 2SK1345...46, 2SK1420,++	data
IRFIZ 44 A	MOS-N-FET-e	=IRFWZ 44A	30c§	Sam	→IRFWZ 44A	data
IRFIZ 44 G	MOS-N-FET-e	V-MOS, 60V, 30A, 48W, <28mΩ(18A)	17c	Inr	2SK1262, 2SK2466	data
IRFIZ 44 N	MOS-N-FET-e	V-MOS, 55V, 31A, 45W, <24mΩ(17A)	17c	Inr	2SK1262, 2SK2466	data
IRFIZ 46 N	MOS-N-FET-e	V-MOS, 55V, 33A, 45W, <20mΩ(19A)	17c	Inr	2SK1262, 2SK2466	data
IRFIZ 48 G	MOS-N-FET-e	V-MOS, 60V, 37A, 50W, <18mΩ(22A)	17c	Inr	2SK1257, 2SK1653, 2SK1952, 2SK2312, ++	data
IRFIZ 48 N	MOS-N-FET-e	V-MOS, 55V, 40A, 45W, <16mΩ(22A)	17c	Inr	2SK1257, 2SK1653, 2SK1952, 2SK2312, ++	data
IRFJ 120	MOS-N-FET-e	V-MOS, L, 100V, 8A, 40W, <0,3Ω(4A)	22a§	Inr	BUZ 20, BUZ 72, 2SK470, 2SK740, 2SK918	data
IRFJ 121	MOS-N-FET-e	=IRFJ 120: 60V	22a§	Inr	BUZ 20, BUZ 72, 2SK470, 2SK740, 2SK918	data
IRFI 122	MOS-N-FET-e	V-MOS, L, 100V, 7A, 40W, <0,4Ω(4A)	22a§	Inr	BUZ 20, BUZ 72, 2SK470, 2SK740, 2SK918	data
IRFJ 123	MOS-N-FET-e	=IRFJ 122: 60V	22a§	Inr	BUZ 20, BUZ 72, 2SK470, 2SK740, 2SK918	data
IRFJ 130	MOS-N-FET-e	V-MOS, L, 100V, 10A, 50W, <0,18Ω(6A)	22a§	Inr	BUZ 20, BUZ 72, 2SK740, 2SK918, 2SK921	data

Type	Device	Short Description	Fig.	Manu	Comparision Types	More at.
IRFJ 131	MOS-N-FET-e	=IRFJ 130: 60V	22a§	Inr	BUZ 20, BUZ 72, 2SK740, 2SK918, 2SK921	data
IRFJ 132	MOS-N-FET-e	V-MOS, L, 100V, 10A, 50W, <0,25Ω(6A)	22a§	Inr	BUZ 20, BUZ 72, 2SK740, 2SK918, 2SK921	data
IRFJ 133	MOS-N-FET-e	=IRFJ 132: 60V	22a§	Inr	BUZ 20, BUZ 72, 2SK740, 2SK918, 2SK921	data
IRFJ 140	MOS-N-FET-e	V-MOS, L, 100V, 15A, 70W, <85mΩ(10A)	22a§	Inr	BUZ 21..22, 2SK919, 2SK922	data
IRFJ 141	MOS-N-FET-e	=IRFJ 140: 60V	22a§	Inr	BUZ 21..22, 2SK919, 2SK922	data
IRFJ 142	MOS-N-FET-e	V-MOS, L, 100V, 10A, 70W, <0,11Ω(10A)	22a§	Inr	BUZ 20..21, 2SK918...919, 2SK921...922	data
IRFJ 143	MOS-N-FET-e	=IRFJ 142: 60V	22a§	Inr	BUZ 20..21, 2SK918...919; 2SK921...22	data
IRFJ 220	MOS-N-FET-e	V-MOS, L, 200V, 5A, 40W, <0,8Ω(2,5A)	22a§	Inr	BUZ 30, BUZ 73, 2SK477, 2SK741, 2SK924	data
IRFJ 221	MOS-N-FET-e	=IRFJ 220: 150V	22a§	Inr	BUZ 30, BUZ 73, 2SK477, 2SK741, 2SK924	data
IRFJ 222	MOS-N-FET-e	V-MOS, L, 200V, 4A, 40W, <1,2Ω(2,5A)	22a§	Inr	BUZ 30, BUZ 73, 2SK477, 2SK741, 2SK924	data
IRFJ 223	MOS-N-FET-e	=IRFJ 222: 150V	22a§	Inr	BUZ 30, BUZ 73, 2SK477, 2SK741, 2SK924	data
IRFJ 230	MOS-N-FET-e	V-MOS, L, 200V, 8A, 50W, <0,4Ω(4A)	22a§	Inr	BUZ 30, BUZ 73, 2SK477, 2SK741	data
IRFJ 231	MOS-N-FET-e	=IRFJ 230: 150V	22a§	Inr	BUZ 30, BUZ 73, 2SK477, 2SK741	data
IRFJ 232	MOS-N-FET-e	V-MOS, L, 200V, 6,5A, 50W, <0,6Ω(4A)	22a§	Inr	BUZ 30, BUZ 73, 2SK477, 2SK741	data
IRFJ 233	MOS-N-FET-e	=IRFJ 232: 150V	22a§	Inr	BUZ 30, BUZ 73, 2SK477, 2SK741	data
IRFJ 240	MOS-N-FET-e	V-MOS, L, 200V, 13A, 70W, <0,18Ω(7A)	22a§	Inr	BUK 456-200, BUZ 30A, IRF 640, 2SK891	data
IRFJ 241	MOS-N-FET-e	=IRFJ 240: 150V	22a§	Inr	BUK 456-200, BUZ 30A, IRF 640, 2SK891	data
IRFJ 242	MOS-N-FET-e	V-MOS, L, 200V, 11A, 70W, <0,22Ω(7A)	22a§	Inr	BUZ 31..32, 2SK459, 2SK925	data
IRFJ 243	MOS-N-FET-e	=IRFJ 242: 150V	22a§	Inr	BUZ 31..32, 2SK459, 2SK470, 2SK925	data
IRFJ 320	MOS-N-FET-e	V-MOS, L, 400V, 3A, 40W, <1,8Ω(1,6A)	22a§	Inr	BUZ 76, IRF 720, IRF 722, 2SK310...311	data
IRFJ 321	MOS-N-FET-e	=IRF 320: 350V	22a§	Inr	BUZ 76, IRF 720, IRF 722, 2SK310...311	data
IRFJ 322	MOS-N-FET-e	V-MOS, L, 400V, 2,5A, 40W, <2,5Ω(1,6A)	22a§	Inr	BUZ 76, IRF 720, IRF 722, 2SK310...311	data
IRFJ 323	MOS-N-FET-e	=IRF 322: 350V	22a§	Inr	BUZ 76, IRF 720, IRF 722, 2SK310...311	data
IRFJ 330	MOS-N-FET-e	V-MOS, L, 400V, 4,5A, 50W, <1Ω(2,5A)	22a§	Inr	BUZ 41A...42, IRF 830, IRF 832, 2SK553	data
IRFJ 331	MOS-N-FET-e	=IRFJ 330: 350V	22a§	Inr	BUZ 41A...42, IRF 830, IRF 832, 2SK553	data
IRFJ 332	MOS-N-FET-e	V-MOS, L, 400V, 4A, 50W, <1,5Ω(2,5A)	22a§	Inr	BUZ 41A...42, IRF 830, IRF 832, 2SK553	data
IRFJ 333	MOS-N-FET-e	=IRFJ 332: 350V	22a§	Inr	BUZ 41A...42, IRF 830, IRF 832, 2SK553	data
IRFJ 340	MOS-N-FET-e	V-MOS, L, 400V, 7,5A, 70W,<0,55Ω(6,2A)	22a§	Inr	IRF 840, IRF 842, 2SK554...555	data
IRFJ 341	MOS-N-FET-e	=IRFJ 340: 350V	22a§	Inr	IRF 840, IRF 842, 2SK554...555	data
IRFJ 342	MOS-N-FET-e	V-MOS, L, 400V, 6A, 70W, <0,8Ω(6,2A)	22a§	Inr	IRF 840, IRF 842, 2SK554...555	data
IRFJ 343	MOS-N-FET-e	=IRFJ 342: 350V	22a§	Inr	IRF 840, IRF 842, 2SK554...555	data
IRFJ 420	MOS-N-FET-e	V-MOS, L, 500V, 2,5A, 40W, <3Ω(1A)	22a§	Inr	BUZ 74, IRF 820, 2SK382	data
IRFJ 421	MOS-N-FET-e	=IRFJ 420: 450V	22a§	Inr	BUZ 74, IRF 820, 2SK382	data
IRFJ 422	MOS-N-FET-e	V-MOS, L, 500V, 2A, 40W, <4Ω(1A)	22a§	Inr	BUZ 74, IRF 820, IRF 822, 2SK382	data
IRFJ 423	MOS-N-FET-e	=IRFJ 422: 450V	22a§	Inr	BUZ 74, IRF 820, IRF 822, 2SK382	data
IRFJ 430	MOS-N-FET-e	V-MOS, L, 500V, 4A, 50W, <1,5Ω(2,1A)	22a§	Inr	BUZ 41A...42, IRF 830, IRF 832, 2SK553	data
IRFJ 431	MOS-N-FET-e	=IRFJ 430: 450V	22a§	Inr	BUZ 41A...42, IRF 830, IRF 832, 2SK553	data
IRFJ 432	MOS-N-FET-e	V-MOS, L, 500V, 3A, 50W, <2Ω(2,1A)	22a§	Inr	BUZ 41A...42, IRF 830, IRF 832, 2SK553	data
IRFJ 433	MOS-N-FET-e	=IRFJ 432: 450V	22a§	Inr	BUZ 41A...42, IRF 830, IRF 832, 2SK553	data
IRFJ 440	MOS-N-FET-e	V-MOS, L, 500V, 6A, 70W, <0,85Ω(3,3A)	22a§	Inr	IRF 840, IRF 842, 2SK555	data
IRFJ 441	MOS-N-FET-e	=IRFJ 440: 450V	22a§	Inr	IRF 840, IRF 842, 2SK555	data
IRFJ 442	MOS-N-FET-e	V-MOS, L, 500V, 5A, 70W, <1,1Ω(3,3A)	22a§	Inr	BUZ 41A...42, IRF 830, 2SK553	data
IRFJ 443	MOS-N-FET-e	=IRFJ 440: 450V	22a§	Inr	BUZ 41A...42, IRF 830, 2SK553	data
IRFK ...	MOS	Leistungsmodule/Power Modules		Inr		
IRFL 014	MOS-N-FET-e	SMD, V-MOS, 60V, 2,7A, <0,2Ω(1,6A)	48c§	Inr	BUK 483-60A, BUK 583-60A	data
IRFL 024 N	MOS-N-FET-e	SMD, V-MOS, 55V, 4A, <75mΩ(2,8A)	48c§	Inr		data
IRFL 110	MOS-N-FET-e	SMD, V-MOS,100V, 1,5A,<0,54Ω(0,9A)	48c§	Inr	BUK 482-100A, BUK 582-100A	data
IRFL 210	MOS-N-FET-e	SMD, V-MOS, 200V, 0,96A, <1,5Ω(0,58A)	48c§	Inr	-	data
IRFL 214	MOS-N-FET-e	SMD, V-MOS, 250V, 0,79A, <2Ω(0,47A)	48c§	Inr	-	data
IRFL 1006	MOS-N-FET-e	SMD, V-MOS, 60V, 2,3A, <0,22Ω(1,6A)	48c§	Inr	BUK 483-60A, BUK 583-60A, IRFL 014	data
IRFL 4105	MOS-N-FET-e	SMD, V-MOS, 55V, 3,7A, <45mΩ(3,7A)	48c§	Inr	IRFL 024N	data
IRFL 4310	MOS-N-FET-e	SMD, V-MOS, 100V, 2,2A, <0,2Ω(1,6A)	48c§	Inr	-	data
IRFL 9014	MOS-P-FET-e	SMD, V-MOS, 60V, 1,8A, <0,5Ω(1,1A)	48c§	Inr	-	data
IRFL 9110	MOS-P-FET-e	SMD, V-MOS, 100V, 1,1A, <1,2Ω(0,66A)	48c§	Inr	-	data
IRFM 014 A	MOS-N-FET-e	SMD, V-MOS, 60V, 2,8A, <0,14Ω(1,4A)	48c§	Sam	BUK 483-60A.	data
IRFM 040	MOS-N-FET-e	V-MOS, L, 50V, 25A, 96W, <0,04Ω(23A)	18f (Iso)	Inr	-	data
IRFM 110 A	MOS-N-FET-e	SMD, V-MOS, 100V, 1,5A, <0,4Ω(0,75A)	48c§	Sam	BUK 481-100A, BUK 482-100A	data
IRFM 120 A	MOS-N-FET-e	SMD, V-MOS, 100V, 2,3A, <0,2Ω(1,15A)	48c§	Sam	IRFL 4310	data
IRFM 140	MOS-N-FET-e	V-MOS, L, 100V, 23A, 96W, <0,1Ω(15A)	18f (Iso)	Inr	-	data
IRFM 150	MOS-N-FET-e	V-MOS, L, 100V, 25A, 125W, <65mΩ(21A)	18f (Iso)	Inr	-	data
IRFM 210 A	MOS-N-FET-e	SMD, V-MOS, 200V, 0,77A, <1,5Ω(0,39A)	48c§	Sam	IRFL 210	data
IRFM 214 A	MOS-N-FET-e	SMD, V-MOS, 250V, 0,64A, <2Ω(0,32A)	48c§	Sam	IRFL 214	data
IRFM 220 A	MOS-N-FET-e	SMD, V-MOS, 200V, 1,13A, <0,8Ω(0,57A)	48c§	Sam	BUK 482-200A	data
IRFM 224 A	MOS-N-FET-e	SMD, V-MOS, 250V, 0,92A, <1,1Ω(0,46A)	48c§	Sam	-	data
IRFM 240	MOS-N-FET-e	V-MOS, L, 200V, 15A, 96W, <0,2Ω(9A)	18f (Iso)	Inr	-	data
IRFM 250	MOS-N-FET-e	V-MOS, L, 200V, 24A, 125W, <0,1Ω(16A)	18f (Iso)	Inr	-	data
IRFM 340	MOS-N-FET-e	V-MOS,L, 400V, 8,5A, 96W, <0,56Ω(5,4A)	18f (Iso)	Inr	-	data
IRFM 350	MOS-N-FET-e	V-MOS,L, 400V, 14A, 125W, <0,31Ω(8,6A)	18f (Iso)	Inr	-	data
IRFM 440	MOS-N-FET-e	V-MOS,L, 500V, 6,9A, 96W, <0,86Ω(4,3A)	18f (Iso)	Inr	-	daia
IRFM 450	MOS-N-FET-e	V-MOS,L, 500V, 11A, 125W, <0,42Ω(7,2A)	18f (Iso)	Inr	-	data
IRFM 9140	MOS-P-FET-e	V-MOS, L, 100V, 15A, 96W, <0,21Ω(9,8A)	18f (Iso)	Inr		data
IRFM 9240	MOS-P-FET-e	V-MOS,L, 200V, 9,3A, 96W, <0,51Ω(5,9A)	18f (Iso)	Inr		data

Type	Device	Short Description	Fig.	Manu	Comparision Types	More at
IRFP 22 N50A	MOS-N-FET-e	V-MOS, 500V, 22A, 277W, <0,23Ω(13A)	16c§	Inr	MTW 20N50E, 2SK1411, 2SK1680	data
IRFP 040	MOS-N-FET-e	V-MOS, 40V, 40A, 150W, <28mΩ(32A)	16c§	Inr	BUZ 347...348, 2SK849, 2SK905, 2SK1297,++	data
IRFP 042	MOS-N-FET-e	V-MOS, 40V, 40A, 150W, <35mΩ(32A)	16c§	Inr	BUZ 347...348, 2SK849, 2SK905, 2SK1297,++	data
IRFP 044	MOS-N-FET-e	V-MOS, 60V, 57A, 180W, <28mΩ(34A)	16c§	Inr	BUK 438-60, 2SK1422, 2SK2121, 2SK2173,++	data
IRFP 044 N	MOS-N-FET-e	V-MOS, 55V, 53A, 120W, <20mΩ(29A)	16c§	Inr	BUK 438-60, 2SK1422, 2SK2121, 2SK2173,++	data
IRFP 045	MOS-N-FET-e	V-MOS, 60V, 40A, 180W, <35mΩ(36A)	16c§	Inr	2SK849, 2SK857, 2SK1124, 2SK1297,++	data
IRFP 048	MOS-N-FET-e	V-MOS, 60V, 70A, 190W, <18mΩ(44A)	16c§	Inr	(2SK1423)[7]	data
IRFP 048 N	MOS-N-FET-e	V-MOS, 55V, 64A, 140W, <16mΩ(37A)	16c§	Inr	2SK1423	data
IRFP 054	MOS-N-FET-e	V-MOS, 60V, 70A, 230W, <14mΩ(54A)	16c§	Inr	(2SK1423)[7]	data
IRFP 054 N	MOS-N-FET-e	V-MOS, 55V, 81A, 170W, <12mΩ(43A)	16c§	Inr	(2SK1423)[7]	data
IRFP 064	MOS-N-FET-e	V-MOS, 60V, 70A, 300W, <9mΩ(78A)	16c§	Inr	-	data
IRFP 064 N	MOS-N-FET-e	V-MOS, 55V, 110A, 200W, <8mΩ(59A)	16c§	Inr	-	data
IRFP 120...123	MOS-N-FET-e	=IRF 520...523:	18c§	Sam	2SK631, 2SK1529	data
IRFP 130...133	MOS-N-FET-e	=IRF 530...533:	18c§	Sam	2SK572	data
IRFP 140	MOS-N-FET-e	V-MOS, 100V, 31A, 180W, <77mΩ(19A)	16c§,18c§	Inr,Nsc,Sam	MTW 35N15E	data
IRFP 140 A	MOS-N-FET-e	V-MOS, 100V, 31A, 131W, <52mΩ(15,5A)	18c§	Sam	BUZ 349, 2SK850, 2SK906	data
IRFP 140 N	MOS-N-FET-e	V-MOS, 100V, 27A, 94W, <52mΩ(16A)	16c§	Inr	2SK850, 2SK906, 2SK1303, 2SK1433	data
IRFP 141	MOS-N-FET-e	=IRFP 140: 80V	16c§,18c§	Inr,Nsc,Sam	MTW 35N15E	data
IRFP 142	MOS-N-FET-e	V-MOS, 100V, 27A, 180W, <99mΩ(19A)	16c§,18c§	Inr,Nsc,Sam	MTW 35N15E	data
IRFP 143	MOS-N-FET-e	=IRFP 142: 80V	16c§,18c§	Inr,Nsc,Sam	MTW 35N15E	data
IRFP 150	MOS-N-FET-e	V-MOS, 100V, 41A, 230W, <55mΩ(25A)	16c§,18c§	Inr,Sam,++	MTW 45N10E	data
IRFP 150 A	MOS-N-FET-e	V-MOS, 100V, 43A, 193W, <40mΩ(21,5A)	18c§	Sam	BUZ 345, MTW 45N10E	data
IRFP 150 N	MOS-N-FET-e	V-MOS, 100V, 39A, 140W, <36mΩ(23A)	16c§	Inr	BUZ 345, MTW 45N10E	data
IRFP 151	MOS-N-FET-e	=IRFP 150: 60V	16c§,18c§	Inr, Sam	MTW 45N10E	data
IRFP 152	MOS-N-FET-e	V-MOS, 100V, 34A, 230W, <80mΩ(22A)	16c§,18c§	Inr, Sam	MTW 45N10E	data
IRFP 153	MOS-N-FET-e	=IRFP 152: 60V	16c§,18c§	Inr, Sam	MTW 45N10E	data
IRFP 150...153FI,P	MOS-N-FET-e	=MRFP 150...153: Iso, ≈62%Imax, 65W	18c	Sgs	2SK1435	data
IRFP 220...223	MOS-N-FET-e	=IRF 620...623:	18c§	Sam	BUK 426-200	data
IRFP 230...233	MOS-N-FET-e	=IRF 630...633:	18c§	Sam	BUK 426-200	data
IRFP 240	MOS-N-FET-e	V-MOS, 200V, 20A, 150W, <0,18Ω(12A)	16c§,18c§	Inr,Nsc,Sam	BUZ 350, 2SK944, 2SK1641, 2SK1673, ++	data
IRFP 240 A	MOS-N-FET-e	V-MOS, 200V, 20A, 180W, <0,18Ω(10A)	18c§	Sam	2SK1673...74	data
IRFP 240 FI	MOS-N-FET-e	=IRFP 240: Iso, 12A, 55W	18c	Sgs	2SK820	data
IRFP 241	MOS-N-FET-e	=IRFP 240: 150V	16c§,18c§	Inr,Nsc,Sam	BUZ 350, 2SK944, 2SK1641, 2SK1673, ++	data
IRFP 242	MOS-N-FET-e	V-MOS, 200V, 18A, 150W, <0,22Ω(11A)	16c§,18c§	Inr, Sam	BUZ 350, 2SK944, 2SK1641, 2SK1673, ++	data
IRFP 243	MOS-N-FET-e	=IRFP 242: 150V	16c§,18c§	Inr, Sam	BUZ 350, 2SK944, 2SK1641, 2SK1673, ++	data
IRFP 244	MOS-N-FET-e	V-MOS, 250V, 15A, 150W, <0,28Ω(9A)	16c§,18c§	Inr, Sam	BUZ 323, MTW 16N40E, 2SK1673...74	data
IRFP 244 A	MOS-N-FET-e	V-MOS, 250V, 16A, 180W, <0,28Ω(8A)	18c§	Sam	BUZ 323, MTW 16N40E, 2SK1673...74	data
IRFP 245	MOS-N-FET-e	V-MOS, 250V, 14A, 150W, <0,34Ω(8A)	16c§,18c§	Inr, Sam	BUZ 323, MTW 16N40E, 2SK1673...74	data
IRFP 246	MOS-N-FET-e	=IRFP 244: 275V	16c§	Inr	BUZ 323, MTW 16N40E, 2SK1674	data
IRFP 247	MOS-N-FET-e	=IRFP 245: 275V	16c§	Inr	BUZ 323, MTW 16N40E, 2SK1674	data
IRFP 250	MOS-N-FET-e	V-MOS, 200V, 30A, 190W, <85mΩ(18A)	16c§,18c§	Inr,Gen,Sam	BUZ 341, MTW 32N20E, 2SK1675	data
IRFP 250 A	MOS-N-FET-e	V-MOS, 200V, 32A, 204W, <85mΩ(16A)	18c§	Sam	BUZ 341, MTW 32N20E, 2SK1675	data
IRFP 251	MOS-N-FET-e	=IRFP 250: 150V	16c§,18c§	Inr,Gen,Sam	BUZ 341, MTW 32N20E, 2SK1675	data
IRFP 252	MOS-N-FET-e	V-MOS, 200V, 27A, 190W, <0,12Ω(17A)	16c§,18c§	Inr,Gen,Sam	BUZ 341, MTW 32N20E, 2SK1675	data
IRFP 253	MOS-N-FET-e	=IRFP 252: 150V	16c§,18c§	Inr,Gen,Sam	BUZ 341, MTW 32N20E, 2SK1675	data
IRFP 254	MOS-N-FET-e	V-MOS, 250V, 23A, 190W, <0,14Ω(14A)	16c§,18c§	Inr,Sam	MTW 32N25E, 2SK1675...76	data
IRFP 254 A	MOS-N-FET-e	V-MOS, 250V, 25A, 210W, <0,14Ω(12,5A)	18c§	Sam	MTW 32N25E, 2SK1675...76	data
IRFP 255	MOS-N-FET-e	V-MOS, 250V, 21A, 190W, <0,17Ω(13A)	16c§,18c§	Inr, Sam	MTW 32N25E, 2SK1675...76	data
IRFP 256	MOS-N-FET-e	=IRFP 254: 275V	16c§	Inr	2SK1674, 2SK1676	data
IRFP 257	MOS-N-FET-e	=IRFP 256: 275V	16c§	Inr	2SK1674, 2SK1676	data
IRFP 260	MOS-N-FET-e	V-MOS, 200V, 46A, 280W, <55mΩ(28A)	16c§	Inr	-	data
IRFP 264	MOS-N-FET-e	V-MOS, 200V, 38A, 280W, <75mΩ(23A)	16c§	Inr	-	data
IRFP 320...323	MOS-N-FET-e	=IRF 720...723:	18c§	Sam	2SK635	data
IRFP 330...333	MOS-N-FET-e	=IRF 730...733:	18c§	Sam	2SK501, 2SK2264	data
IRFP 340	MOS-N-FET-e	V-MOS, 400V, 11A, 150W, <0,55Ω(6,6A)	16c§,18c§	Inr,Nsc,Sam	BUZ 325...326, MTW 16N40E	data
IRFP 340 A	MOS-N-FET-e	V-MOS, 400V, 11A, 162W, <0,55Ω(5,5A)	18c§	Sam	BUZ 325...326, MTW 16N40E	data
IRFP 341	MOS-N-FET-e	=IRFP 340: 350V	16c§,18c§	Inr,Nsc,Sam	BUZ 325...326, MTW 16N40E	data
IRFP 342	MOS-N-FET-e	V-MOS, 400V, 8,7A, 150W, <0,8Ω(5,5A)	16c§,18c§	Inr, Sam	BUZ 325...326, MTW 16N40E	data
IRFP 343	MOS-N-FET-e	=IRFP 342: 350V	16c§,18c§	Inr, Sam	BUZ 325...326, MTW 16N40E	data
IRFP 344	MOS-N-FET-e	V-MOS, 450V, 9,5A, 150W, <0,63Ω(5,7A)	16c§	Inr	BUK 637-500, BUZ 334	data
IRFP 350	MOS-N-FET-e	V-MOS, 400V, 16A, 190W, <0,3Ω(9,6A)	16c§,18c§	Inr,Sam,++	BUZ 323, MTW 16N40E, 2SK1677...78	data
IRFP 350 A	MOS-N-FET-e	V-MOS, 400V, 17A, 202W, <0,3Ω(8,5A)	18c§	Sam	BUZ 323, MTW 16N40E, 2SK1677...78	data
IRFP 350 FI	MOS-N-FET-e	=IRFP 350: Iso	18c	Sgs	2SK1268...69, 2SK1331, 2SK1524	data
IRFP 350 LC	MOS-N-FET-e	V-MOS, lo Gate Charge, 400V, 16A, 190W <0,3Ω(9,6A)	16c§	Inr	-	data
IRFP 351	MOS-N-FET-e	=IRFP 350: 350V	16c§,18c§	Inr,Gen,Sam	BUZ 323, MTW 16N40E, 2SK1677...78	data
IRFP 352	MOS-N-FET-e	V-MOS, 400V, 14A, 190W, <0,4Ω(8,9A)	16c§,18c§	Inr,Gen,Sam	BUZ 323, MTW 16N40E, 2SK1677...78	data
IRFP 353	MOS-N-FET-e	=IRFP 352: 350V	16c§,18c§	Inr,Gen,Sam	BUZ 323, MTW 16N40E, 2SK1677...78	data
IRFP 354	MOS-N-FET-e	V-MOS, 450V, 14A, 190W, <0,35Ω(8,4A)	16c§	Inr	BUK 638-500, BUZ 338, 2SK1677...78	data
IRFP 360	MOS-N-FET-e	V-MOS, 400V, 23A, 280W, <0,2Ω(14A)	16c§	Inr	MTW 24N40E, 2SK1409, 2SK1411	data
IRFP 360 LC	MOS-N-FET-e	V-MOS, lo Gate Charge, 400V, 23A, 280W <0,2Ω(14A)	16c§	Inr	-	data
IRFP 362	MOS-N-FET-e	V-MOS, 400V, 20A, 280W, <0,25Ω(13A)	16c§	Inr	MTW 24N40E, 2SK1409, 2SK1411	data
IRFP 420...423	MOS-N-FET-e	=IRF 820...823:	18c§	Sam	2SK635	data
IRFP 430...433	MOS-N-FET-e	=IRF 830...833:	18c§	Sam	-	data
IRFP 440	MOS-N-FET-e	V-MOS, 500V, 8,8A, 150W, <0,85Ω(6,3A)	16c§,18c§	Inr,Nsc,Sam	BUZ 339, IRFPC 50, 2SK1723, 2SK2699, ++	data
IRFP 440 A	MOS-N-FET-e	V-MOS, 500V, 8,5A, 162W, <0,85Ω(4,25A)	18c§	Sam	BUZ 339, IRFPC 50, 2SK1723, 2SK2699, ++	data
IRFP 441	MOS-N-FET-e	=IRFP 440: 450V	16c§,18c§	Inr,Nsc,Sam	BUZ 339, IRFPC 50, 2SK1723, 2SK2699, ++	data

Type	Device	Short Description	Fig.	Manu	Comparision Types	More at
IRFP 442	MOS-N-FET-e	V-MOS, 500V, 7,7A, 150W, <1,1Ω(4,9A)	16c§,18c§	Inr, Sam	BUZ 330...331, BUZ 339, IRFPC 50	data
IRFP 443	MOS-N-FET-e	=IRFP 442: 450V	16c§,18c§	Inr, Sam	BUZ 330...331, BUZ 339, IRFPC 50	data
IRFP 448	MOS-N-FET-e	V-MOS, 500V, 11A, 180W, <0,6Ω(6,6A)	16c§	Inr	BUZ 339, IRFPC 50, 2SK1723, 2SK2699, ++	data
IRFP 450	MOS-N-FET-e	V-MOS, 500V, 14A, 190W, <0,4Ω(8,4A)	16c§,18c§	Inr,Sam,++	BUK 638-500, BUZ 338, 2SK1678	data
IRFP 450 A	MOS-N-FET-e	V-MOS, 500V, 14A, 205W, <0,4Ω(7A)	16c§,18c§	Inr, Sam	→IRFP 450	data
IRFP 450 LC	MOS-N-FET-e	V-MOS, Io Gate Charge, 500V, 14A, 190W <0,4Ω(8,4A)	16c§	Inr	-	data
IRFP 451	MOS-N-FET-e	=IRFP 450: 450V	16c§,18c§	Inr,Sam,++	BUK 638-500, BUZ 338, 2SK1677...78	data
IRFP 452	MOS-N-FET-e	V-MOS, 500V, 12A, 190W, <0,5Ω(7,9A)	16c§,18c§	Inr,Sam,++	BUK 638-500, BUZ 338, 2SK1677...78	data
IRFP 453	MOS-N-FET-e	=IRFP 452: 450V	16c§,18c§	Inr,Sam	BUK 638-500, BUZ 338, 2SK1677...78	data
IRFP 450...453Fl,P	MOS-N-FET-e	=IRFP 450...453: Iso, ≈64%Imax, 70W	18c	Sgs	2SK1206, 2SK1268...1269, 2SK1523...24,++	data
IRFP 460	MOS-N-FET-e	V-MOS, 500V, 20A, 280W, <0,27Ω(12A)	16c§	Inr	MTW 20N50E, 2SK1411, 2SK1680	data
IRFP 460 A	MOS-N-FET-e	V-MOS, 500V, 20A, 280W, <0,27Ω(12A)	16c§	Inr	→IRFP 460	data
IRFP 460 LC	MOS-N-FET-e	V-MOS, Io Gate Charge, 500V, 20A, 280W <0,27Ω(12A)	16c§	Inr	-	data
IRFP 462	MOS-N-FET-e	V-MOS, 500V, 17A, 280W, <0,35Ω(11A)	16c§	Inr	2SK1411, 2SK1680	data
IRFP 2410	MOS-N-FET-e	V-MOS, 100V, 61A, 230W, <25mΩ(37A)	16c§	Inr	BUZ 344	data
IRFP 3710	MOS-N-FET-e	V-MOS, 100V, 49A, 180W, <25mΩ(28A)	16c§	Inr	BUZ 345, MTW 45N10E	data
IRFP 9120...9123	MOS-P-FET-e	=IRF 9520...9523:	18c§	Sam	2SJ115, 2SJ118...119	data
IRFP 9130...9133	MOS-P-FET-e	=IRF 9530...9533:	18c§	Sam	2SJ113, 2SJ200	data
IRFP 9140	MOS-P-FET-e	V-MOS, 100V, 21A, 180W, <0,2Ω(13A)	16c§,18c§	Inr,Sam	-	data
IRFP 9140 N	MOS-P-FET-e	V-MOS, 100V, 21A, 120W, <0,117Ω(13A)	16c§	Inr	-	data
IRFP 9141	MOS-P-FET-e	=IRFP 9140: 60V	16c§,18c§	Inr,Sam	-	data
IRFP 9142	MOS-P-FET-e	V-MOS, 100V, 16A, 180W, <0,3Ω(10A)	16c§,18c§	Inr,Sam	-	data
IRFP 9143	MOS-P-FET-e	=IRFP 9142: 60V	16c§,18c§	Inr,Sam	-	data
IRFP 9150...9151	MOS-P-FET-e	=IRF 9150...9151:	16c§	Inr	2SJ215	data
IRFP 9220...9223	MOS-P-FET-e	=IRF 9620...9623:	17c§	Sam	→IRF 9620...9623	data
IRFP 9230...9233	MOS-P-FET-e	=IRF 9630...9633:	18c§	Sam	2SJ114, 2SJ351...352	data
IRFP 9240	MOS-P-FET-e	V-MOS, 200V, 12A, 150W, <0,5Ω(7,2A)	16c§,18c§	Inr,Sam	2SJ131	data
IRFP 9241	MOS-P-FET-e	=IRFP 9240: 150V	16c§,18c§	Inr,Sam	2SJ131	data
IRFP 9242	MOS-P-FET-e	V-MOS, 200V, 10A, 150W, <0,7Ω(6,3A)	16c§,18c§	Inr,Sam	2SJ131	data
IRFP 9243	MOS-P-FET-e	=IRFP 9242: 150V	16c§,18c§	Inr,Sam	2SJ131	data
IRFPC 30	MOS-N-FET-e	V-MOS, 600V, 4,3A, 100W, <2,2Ω(2,6A)	16c§	Inr	2SK695, 2SK793, 2SK1213, 2SK1403,++	data
IRFPC 40	MOS-N-FET-e	V-MOS, 600V, 6,8A, 150W, <1,2Ω(4,1A)	16c§	Inr	BUZ 305, BUZ 332A, 2SK684, 2SK1032,++	data
IRFPC 42	MOS-N-FET-e	V-MOS, 600V, 5,9A, 150W, <1,6Ω(3,7A)	16c§	Inr	BUZ 305, BUZ 332A, 2SK684, 2SK1032,++	data
IRFPC 48	MOS-N-FET-e	V-MOS, 600V, 8,9A, 170W, <0,82Ω(5,3A)	16c§	Inr	BUK 637-600, BUZ 334	data
IRFPC 50	MOS-N-FET-e	V-MOS, 600V, 11A, 180W, <0,6Ω(6A)	16c§	Inr	BUZ 334	data
IRFPC 50 A	MOS-N-FET-e	V-MOS, 600V, 11A, 180W, <0,58Ω(6A)	16c§	Inr	→IRFBC 50	data
IRFPC 50 LC	MOS-N-FET-e	V-MOS, Io Gate Charge, 600V, 11A, 190W <0,6Ω(6,6A)	16c§	Inr	-	data
IRFPC 60	MOS-N-FET-e	V-MOS, 600V, 16A, 280W, <2,2Ω(2,6A)	16c§	Inr	2SK1573	data
IRFPC 60 LC	MOS-N-FET-e	V-MOS, Io Gate Charge, 600V, 16A, 280W <0,4Ω(9,6A)	16c§	Inr	-	data
IRFPE 30	MOS-N-FET-e	V-MOS, 800V, 4,1A, 125W, <2,2Ω(2,6A)	16c§	Inr	BUZ 355...356, 2SK534, 2SK604, 2SK695,++	data
IRFPE 40	MOS-N-FET-e	V-MOS, 800V, 5,4A, 150W, <2Ω(3,2A)	16c§	Inr	BUZ 355...356, 2SK534, 2SK604, 2SK695,++	data
IRFPE 42	MOS-N-FET-e	V-MOS, 800V, 5,3A, 150W, <2Ω(3A)	16c§	Inr	BUZ 355...356, 2SK534, 2SK604, 2SK695,++	data
IRFPE 50	MOS-N-FET-e	V-MOS, 800V, 7,8A, 190W, <1,2Ω(4,7A)	16c§	Inr	BUZ 305, 2SK684, 2SK1032, 2SK1358,++	data
IRFPE 52	MOS-N-FET-e	V-MOS, 800V, 7,8A, 190W, <1,4Ω(4,5A)	16c§	Inr	BUZ 305, 2SK684, 2SK1032, 2SK1358,++	data
IRFPF 30	MOS-N-FET-e	V-MOS, 900V, 3,6A, 125W, <3,7Ω(2,2A)	16c§	Inr	BUZ 357...358, 2SK685, 2SK727, 2SK1341,++	data
IRFPF 40	MOS-N-FET-e	V-MOS, 900V, 4,7A, 150W, <2,5Ω(2,8A)	16c§	Inr	BUZ 357...358, 2SK685, 2SK727, 2SK1341,++	data
IRFPF 42	MOS-N-FET-e	V-MOS, 900V, 4,3A, 150W, <3Ω(2,7A)	16c§	Inr	BUZ 357...358, 2SK685, 2SK727, 2SK1341,++	data
IRFPF 50	MOS-N-FET-e	V-MOS, 900V, 6,7A, 190W, <1,6Ω(4A)	16c§	Inr	-	data
IRFPF 52	MOS-N-FET-e	V-MOS, 900V, 6,2A, 190W, <1,9Ω(3,9A)	16c§	Inr	-	data
IRFPG 30	MOS-N-FET-e	V-MOS, 1000V, 3,1A, 125W, <5Ω(1,9A)	16c§	Inr	2SK696	data
IRFPG 40	MOS-N-FET-e	V-MOS, 1000V, 4,3A, 150W, <3,5Ω(2,6A)	16c§	Inr	BUZ 312, BUZ 357...358, 2SK685, 2SK1205++	data
IRFPG 42	MOS-N-FET-e	V-MOS, 1000V, 3,9A, 150W, <4,2Ω(2,5A)	16c§	Inr	BUZ 312, BUZ 357...358, 2SK685, 2SK1205++	data
IRFPG 50	MOS-N-FET-e	V-MOS, 1000V, 6,1A, 190W, <2Ω(3,6A)	16c§	Inr	2SK1120, 2SK1934	data
IRFPG 52	MOS-N-FET-e	V-MOS, 1000V, 5,5A, 190W, <2,4Ω(3,5A)	16c§	Inr	2SK1120, 2SK1934	data
IRFPS 37 N50A	MOS-N-FET-e	V-MOS, 500V, 36A, 446W, <0,13Ω(22A)	30c§	Inr	-	data
IRFR 1 N60A	MOS-N-FET-e	V-MOS, 600V, 1,4A, 36W, <7Ω(0,84A)	30c§	Inr	IRFRC 20(A)	data
IRFR 010	MOS-N-FET-e	V-MOS, 50V, 8,2A, 25W, <0,2Ω(4,2A)	30c§	Inr, Sam	2SK1472, 2SK1748	data
IRFR 012	MOS-N-FET-e	V-MOS, 50V, 6,7A, 25W, <0,3Ω(4,2A)	30c§	Inr, Sam	2SK1472, 2SK1748	data
IRFR 014	MOS-N-FET-e	V-MOS, 60V, 7,7A, 25W, <0,2Ω(4,6A)	30c§	Inr, Sam	MTD 9N10E, 2SK1472, 2SK1748, 2SK2415	data
IRFR 014 A	MOS-N-FET-e	V-MOS, 60V, 8,2A, 18W, <0,14Ω(4,1A)	30c§	Sam	→IRFR 014	data
IRFR 015	MOS-N-FET-e	V-MOS, 60V, 6,7A, 25W, <0,3Ω(4,2A)	30c§	Inr, Sam	MTD 9N10E, 2SK1472, 2SK1748, 2SK2415	data
IRFR 020	MOS-N-FET-e	V-MOS, 50V, 15A, 42W, <0,1Ω(8,7A)	30c§	Inr, Sam	MTD 15N06V, 2SK2334	data
IRFR 022	MOS-N-FET-e	V-MOS, 50V, 14A, 42W, <0,12Ω(8,7A)	30c§	Inr, Sam	MTD 15N06V, 2SK2334	data
IRFR 024	MOS-N-FET-e	V-MOS, 60V, 15A, 42W, <0,1Ω(8,4A)	30c§	Inr, Sam	MTD 15N06V, 2SK2334	data
IRFR 024 A	MOS-N-FET-e	V-MOS, 60V, 15A, 30W, <0,07Ω(7,5A)	30c§	Sam	MTD 15N06V, 2SK2334	data
IRFR 024 N	MOS-N-FET-e	V-MOS, 55V, 16A, 38W, <75mΩ(9,6A)	30c§	Inr	MTD 15N06V, 2SK2334	data
IRFR 025	MOS-N-FET-e	V-MOS, 60V, 14A, 42W, <0,12Ω(8,7A)	30c§	Inr, Sam	MTD 15N06V, 2SK2334	data
IRFR 34 A	MOS-N-FET-e	V-MOS, 60V, 23A, 40W, <0,04Ω(11,5A)	30c§	Sam	MTD 20N06HD	data
IRFR 110	MOS-N-FET-e	V-MOS, 100V, 4,3A, 25W, <0,54Ω(2,6A)	30c§	Inr, Sam	2SK1474, 2SK1804, 2SK2016	data
IRFR 110 A	MOS-N-FET-e	V-MOS, 100V, 4,7A, 20W, <0,4Ω(2,35A)	30c§	Sam	2SK1474, 2SK1804, 2SK2016	data
IRFR 111	MOS-N-FET-e	=IRFR 110: 80V	30c§		2SK1474, 2SK1804, 2SK2016	data
IRFR 120	MOS-N-FET-e	V-MOS, 100V, 7,7A, 42W, <0,27Ω(4,6A)	30c§	Inr, Sam	MTD 9N10E, 2SK1560	data
IRFR 120 A	MOS-N-FET-e	V-MOS, 100V, 8,4A, 32W, <0,2Ω(4,2A)	30c§	Sam	MTD 9N10E, 2SK1475, 2SK1557, 2SK1560	data
IRFR 120 N	MOS-N-FET-e	V-MOS, 100V, 9,1A, 39W, <0,21Ω(5,5A)	30c§	Inr	MTD 9N10E, 2SK1560	data

Type	Device	Short Description	Fig.	Manu	Comparision Types	More at
IRFR 121	MOS-N-FET-e	=IRFR 120: 80V	30c§	Inr, Sam	MTD 9N10E, 2SK1560	data
IRFR 130 A	MOS-N-FET-e	V-MOS, 100V, 13A, 41W, <0,11Ω(6,5A)	30c§	Sam		data
IRFR 210	MOS-N-FET-e	V-MOS, 200V, 2,6A, 25W, <1,5Ω(1,6A)	30c§	Inr, Sam	MTD 2N20, 2SK1335	data
IRFR 210 A	MOS-N-FET-e	V-MOS, 200V, 2,7A, 26W, <1,5Ω(1,35A)	30c§	Sam	MTD 2N20, 2SK1335	data
IRFR 212	MOS-N-FET-e	V-MOS, 200V, 2,1A, 25W, <2,4Ω(1,3A)	30c§	Inr, Sam	MTD 2N20, 2SK1335	data
IRFR 214	MOS-N-FET-e	V-MOS, 250V, 2,2A, 25W, <2Ω(1,3A)	30c§	Inr, Sam	MTD 2N40E, 2SK2235, 2SK2250	data
IRFR 214 A	MOS-N-FET-e	V-MOS, 250V, 2,2A, 25W, <2Ω(1,1A)	30c§	Sam	MTD 2N40E, 2SK2235, 2SK2250	data
IRFR 215	MOS-N-FET-e	V-MOS, 250V, 1,7A, 25W, <3Ω(1A)	30c§	Inr, Sam	MTD 2N40E, 2SK2235, 2SK2250	data
IRFR 220	MOS-N-FET-e	V-MOS, 200V, 4,8A, 42W, <0,8Ω(2,9A)	30c§	Inr, Sam	MTD 5N25E	data
IRFR 220 A	MOS-N-FET-e	V-MOS, 200V, 4,6A, 40W, <0,8Ω(2,3A)	30c§	Sam	MTD 5N25E	data
IRFR 221	MOS-N-FET-e	=IRFR 220: 150V	30c§	Inr	MTD 5N25E	data
IRFR 222	MOS-N-FET-e	V-MOS, 200V, 3,8A, 42W, <1,2Ω(2,4A)	30c§	Inr, Sam	MTD 4N20	data
IRFR 224	MOS-N-FET-e	V-MOS, 250V, 3,8A, 42W, <1,1Ω(2,3A)	30c§	Inr, Sam	MTD 5N25E	data
IRFR 224 A	MOS-N-FET-e	V-MOS, 250V, 3,8A, 42W, <1,1Ω(1,9A)	30c§	Sam	MTD 5N25E	data
IRFR 230 A	MOS-N-FET-e	V-MOS, 200V, 7,5A, 50W, <0,4Ω(3,75A)	30c§	Sam	-	data
IRFR 234 A	MOS-N-FET-e	V-MOS, 250V, 6,6A, 49W, <0,45Ω(3,3A)	30c§	Sam	- .	data
IRFR 310	MOS-N-FET-e	V-MOS, 400V, 1,7A, 25W, <3,6Ω(1A)	30c§	Inr, Sam	MTD 2N40E	data
IRFR 310 A	MOS-N-FET-e	V-MOS, 400V, 1,7A, 26W, <3,6Ω(0,85A)	30c§	Sam	MTD 2N40E	data
IRFR 311	MOS-N-FET-e	V-MOS, 400V, 1A, 25W, <5Ω(0,8A)	30c§	Inr, Sam	2SK579...580, 2SK1151...52, 2SK2017, ++	data
IRFR 320	MOS-N-FET-e	V-MOS, 400V, 3,1A, 42W, <1,8Ω(1,9A)	30c§	Inr, Sam	-	data
IRFR 320 A	MOS-N-FET-e	V-MOS, 400V, 3,1A, 41W, <1,8Ω(1,55A)	30c§	Sam	-	data
IRFR 321	MOS-N-FET-e	=IRFR 320: 350V	30c§	Inr	-	data
IRFR 322	MOS-N-FET-e	V-MOS, 400V, 2,3A, 42W, <2,5Ω(1,4A)	30c§	Inr, Sam	MTD 2N40E	data
IRFR 330 A	MOS-N-FET-e	V-MOS, 400V, 4,5A, 48W, <1Ω(2,25A)	30c§	Sam		data
IRFR 410	MOS-N-FET-e	V-MOS, 500V, 1,5A, 42W, <7Ω(1,5A)	30c§	Inr	MTD 2N50E	data
IRFR 420	MOS-N-FET-e	V-MOS, 500V, 2,4A, 42W, <3Ω(1,4A)	30c§	Inr, Sam		data
IRFR 420 A	MOS-N-FET-e	V-MOS, 500V, 2,3A, 41W, <3Ω(1,15A)	30c§	Sam		data
IRFR 421	MOS-N-FET-e	=IRFR 420: 450V	30c§	Inr		data
IRFR 422	MOS-N-FET-e	V-MOS, 500V, 2,2A, 42W, <4Ω(1,4A)	30c§	Inr, Sam	MTD 2N50E	data
IRFR 430 A	MOS-N-FET-e	V-MOS, 500V, 3,5A, 48W, <1,5Ω(1.75A)	30c§	Sam		data
IRFR 1205	MOS-N-FET-e	V-MOS, 55V, 42A, 89W, <27mΩ(22A)	30c§	Inr	-	data
IRFR 2605	MOS-N-FET-e	V-MOS, 55V, 19A, 50W, <85mΩ(11A)	30c§	Inr	-	data
IRFR 3303	MOS-N-FET-e	V-MOS, 30V, 33A, 57W, <31mΩ(18A)	30c§	Inr	-	data
IRFR 3910	MOS-N-FET-e	V-MOS, 100V, 15A, 52W, <0,11Ω(9A)	30c§	Inr	-	data
IRFR 4105	MOS-N-FET-e	V-MOS, 55V, 28A, 57W, <45mΩ(16A)	30c§	Inr	-	data
IRFR 5305	MOS-P-FET-e	V-MOS, 55V, 25A, 69W, <65mΩ(15A)	30c§	Inr	-	data
IRFR 5410	MOS-P-FET-e	V-MOS, 100V, 13A, 66W, <205mΩ(7,8A)	30c§	Inr	-	data
IRFR 5505	MOS-P-FET-e	V-MOS, 55V, 18A, 57W, <110mΩ(9,6A)	30c§	Inr	MTD 20P06HDL	data
IRFR 6215	MOS-P-FET-e	V-MOS, 150V, 13A, 110W, <295mΩ(6,6A)	30c§	Inr		data
IRFR 9010	MOS-P-FET-e	V-MOS, 50V, 5,3A, 25W, <0,5Ω(2,8A)	30c§	Inr, Sam	2SJ183, 2SJ239, 2SJ245, 2SJ279	data
IRFR 9012	MOS-P-FET-e	V-MOS, 50V, 4,5A, 25W, <0,7Ω(2,8A)	30c§	Inr, Sam	2SJ183, 2SJ239, 2SJ245, 2SJ279	data
IRFR 9014	MOS-P-FET-e	V-MOS, 60V, 5,1A, 25W, <0,5Ω(3,1A)	30c§	Inr, Sam	2SJ183, 2SJ239, 2SJ245, 2SJ279	data
IRFR 9015	MOS-P-FET-e	V-MOS, 60V, 4,5A, 25W, <0,7Ω(2,8A)	30c§	Inr, Sam	2SJ183, 2SJ239, 2SJ245, 2SJ279	data
IRFR 9020	MOS-P-FET-e	V-MOS, 50V, 9,9A, 42W, <0,28Ω(5,7A)	30c§	Inr, Sam	(2SJ389)[7]	data
IRFR 9022	MOS-P-FET-e	V-MOS, 50V, 9A, 42W, <0,33Ω(5,7A)	30c§	Inr, Sam	(2SJ389)[7]	data
IRFR 9024	MOS-P-FET-e	V-MOS, 60V, 8,8A, 42W, <0,28Ω(5,3A)	30c§	Inr, Sam	MTD 2955	data
IRFR 9024 N	MOS-P-FET-e	V-MOS, 55V, 11A, 38W, <175mΩ(6,6A)	30c§	Inr	MTD 2955, 2SJ389	data
IRFR 9025	MOS-P-FET-e	V-MOS, 60V, 9A, 42W, <0,33Ω(5,7A)	30c§	Inr, Sam	MTD 2955	data
IRFR 9110	MOS-P-FET-e	V-MOS, 100V, 3,1A, 25W, <1,2Ω(1,9A)	30c§	Inr, Sam	MTD 6P10E, SFR 9120, 2SJ195	data
IRFR 9111	MOS-P-FET-e	=IRFR 9110: 80V	30c§	Inr, Sam	MTD 6P10E, SFR 9120, 2SJ195	data
IRFR 9120	MOS-P-FET-e	V-MOS, 100V, 5,6A, 42W, <0,6Ω(3,4A)	30c§	Inr, Sam	MTD 6P10E, SFR 9120	data
IRFR 9120 N	MOS-P-FET-e	V-MOS, 100V, 6,5A, 39W, <0,48Ω(3,9A)	30c§	Inr	MTD 6P10E, SFR 9120, 2SJ195	data
IRFR 9121	MOS-P-FET-e	=IRFR 9120: 80V	30c§	Inr, Sam	MTD 6P10E, SFR 9120, 2SJ195	data
IRFR 9210	MOS-P-FET-e	V-MOS, 200V, 1,9A, 25W, <3Ω(1,1A)	30c§	Inr, Sam	SFR 9220, SFR 6224, 2SJ319	data
IRFR 9212	MOS-P-FET-e	V-MOS, 200V, 1,6A, 25W, <4,5Ω(1A)	30c§	Inr, Sam	SFR 9210, SFR 9214	data
IRFR 9214	MOS-P-FET-e	V-MOS, 250V, 2,7A, 50W, <3Ω(1,7A)	30c§	Inr	SFR 9224	data
IRFR 9220	MOS-P-FET-e	V-MOS, 200V, 3,6A, 42W, <1,5Ω(2,2A)	30c§	Inr, Sam	SFR 9220	data
IRFR 9222	MOS-P-FET-e	V-MOS, 200V, 2,8A, 42W, <2,4Ω(1,8A)	30c§	Inr, Sam	SFR 9220	data
IRFRC 20	MOS-N-FET-e	V-MOS, 600V, 2A, 42W, <4,4Ω(1,2A)	30c§	Inr	(2SK2040, 2SK2865)[7]	data
IRFS 1 Z0	MOS-N-FET-e	V-MOS, 100V, 0,82A, <2,4Ω(0,49A)	39b§	Inr	2SK1485	data
IRFS 1 Z3	MOS-N-FET-e	V-MOS, 80V, 0,75A, <3,2Ω(0,45A)	39b§	Inr	2SK1078...79, 2SK1485	data
IRFS 130	MOS-N-FET-e	V-MOS, 100V, 9,7A, 42W, <0,16Ω(8,3A)	18c	Sam	BUK 426-200	data
IRFS 131	MOS-N-FET-e	=IRFS 130: 80V	18c	Sam	BUK 426-200	data
IRFS 132	MOS-N-FET-e	V-MOS, 100V, 8,3A, 42W, <0,23Ω(8,3A)	18c	Sam	BUK 426-200	data
IRFS 133	MOS-N-FET-e	=IRFS 132: 80V	18c	Sam	BUK 426-200	data
IRFS 140	MOS-N-FET-e	V-MOS, 100V, 19,4A, 65W, <77mΩ(17A)	18c	Sam	BUK 426-100	data
IRFS 140 A	MOS-N-FET-e	V-MOS, 100V, 23A, 72W, <52mΩ(11,5A)	18c	Sam	BUK 426-100	data
IRFS 141	MOS-N-FET-e	=IRFS 140: 80V	18c	Sam	BUK 426-100	data
IRFS 142	MOS-N-FET-e	V-MOS, 100V, 17,3A, 65W, <0,1Ω(17A)	18c	Sam	BUK 426-100	data
IRFS 143	MOS-N-FET-e	=IRFS 142: 80V	18c	Sam	BUK 426-100	data
IRFS 150	MOS-N-FET-e	V-MOS, 100V, 27,7A, 70W, <55mΩ(22A)	18c	Sam	2SK1435	data
IRFS 150 A	MOS-N-FET-e	V-MOS, 100V, 31A, 100W, <40mΩ(15,5A)	18c	Sam	2SK1435	data
IRFS 151	MOS-N-FET-e	=IRFS 150: 80V	18c	Sam	2SK1435	data
IRFS 152	MOS-N-FET-e	V-MOS, 100V, 23,5A, 70W, <80mΩ(22A)	18c	Sam	2SK1435	data
IRFS 153	MOS-N-FET-e	=IRFS 152: 80V	18c	Sam	2SK1435	data

Type	Device	Short Description	Fig.	Manu	Comparision Types	More at
IRFS 230	MOS-N-FET-e	V-MOS, 200V, 6,2A, 42W, <0,4Ω(5A)	18c	Sam	BUK 426-200	data
IRFS 231	MOS-N-FET-e	=IRFS 230: 150V	18c	Sam	BUK 426-200	data
IRFS 232	MOS-N-FET-e	V-MOS, 200V, 5,5A, 42W, <0,6Ω(5A)	18c	Sam	BUK 426-200	data
IRFS 233	MOS-N-FET-e	=IRFS 232: 150V	18c	Sam	BUK 426-200	data
IRFS 240	MOS-N-FET-e	V-MOS, 200V, 12,5A, 65W, <0,18Ω(10A)	18c	Sam	2SK2008	data
IRFS 240 A	MOS-N-FET-e	V-MOS, 200V, 12,8A, 73W, <0,18Ω(6,4A)	18c	Sam	2SK2008	data
IRFS 241	MOS-N-FET-e	=IRFS 240: 150V	18c	Sam	2SK2008	data
IRFS 242	MOS-N-FET-e	V-MOS, 200V, 11A, 65W, <0,22Ω(10A)	18c	Sam	2SK2008	data
IRFS 243	MOS-N-FET-e	=IRFS 242: 150V	18c	Sam	2SK2008	data
IRFS 244 A	MOS-N-FET-e	V-MOS, 250V, 10,2A, 73W, <0,28Ω(5,1A)	18c	Sam	2SK820	data
IRFS 250	MOS-N-FET-e	V-MOS, 200V, 20,7A, 70W, <85mΩ(16A)	18c	Sam	2SK822, 2SK1549, 2SK2008	data
IRFS 250 A	MOS-N-FET-e	V-MOS, 200V, 21,3A, 90W, <85mΩ(10,6A)	18c	Sam	2SK822, 2SK1549, 2SK2008	data
IRFS 251	MOS-N-FET-e	=IRFS 250: 150V	18c	Sam	2SK822, 2SK1549, 2SK2008	data
IRFS 252	MOS-N-FET-e	V-MOS, 200V, 17,3A, 70W, <0,12Ω(16A)	18c	Sam	2SK822, 2SK1549, 2SK2008	data
IRFS 253	MOS-N-FET-e	=IRFS 252: 150V	18c	Sam	2SK822, 2SK1549, 2SK2008	data
IRFS 254 A	MOS-N-FET-e	V-MOS, 250V, 16A, 90W, <0,14Ω(8A)	18c	Sam	2SK822, 2SK1549, 2SK2008	data
IRFS 330	MOS-N-FET-e	V-MOS, 400V, 3,8A, 42W, <1Ω(3A)	18c	Sam	BUK 427-450	data
IRFS 331	MOS-N-FET-e	=IRFS 330: 350V	18c	Sam	BUK 427-450	data
IRFS 332	MOS-N-FET-e	V-MOS, 400V, 3,5A, 42W, <1,5Ω(3A)	18c	Sam	BUK 427-450	data
IRFS 333	MOS-N-FET-e	=IRFS 332: 350V	18c	Sam	BUK 427-450	data
IRFS 340	MOS-N-FET-e	V-MOS, 400V, 6,9A, 65W, <0,55Ω(5,2A)	18c	Sam	2SK1225, 2SK1328, 2SK1451, 2SK1831	data
IRFS 340 A	MOS-N-FET-e	V-MOS, 400V, 8A, 85W, <0,55Ω(4A)	18c	Sam	2SK1225, 2SK1328, 2SK1451, 2SK1831	data
IRFS 341	MOS-N-FET-e	=IRFS 340: 350V	18c	Sam	2SK1225, 2SK1328, 2SK1451, 2SK1831	data
IRFS 342	MOS-N-FET-e	V-MOS, 400V, 5,5A, 65W, <0,8Ω(5,2A)	18c	Sam	BUK 427-450, 2SK1451	data
IRFS 343	MOS-N-FET-e	=IRFS 342: 350V	18c	Sam	BUK 427-450, 2SK1451	data
IRFS 350	MOS-N-FET-e	V-MOS, 400V, 10,4A, 70W, <0,3Ω(8A)	18c	Sam	2SK1225, 2SK1328, 2SK1452, 2SK1831	data
IRFS 350 A	MOS-N-FET-e	V-MOS, 400V, 11,5A, 92W, <0,3Ω(5,75A)	18c	Sam	2SK1225, 2SK1328, 2SK1452, 2SK1831	data
IRFS 351	MOS-N-FET-e	=IRFS 350: 350V	18c	Sam	2SK1225, 2SK1328, 2SK1452, 2SK1831	data
IRFS 352	MOS-N-FET-e	V-MOS, 400V, 9A, 70W, <0,4Ω(8A)	18c	Sam	2SK1225, 2SK1328, 2SK1452, 2SK1831	data
IRFS 353	MOS-N-FET-e	=IRFS 352: 350V	18c	Sam	2SK1225, 2SK1328, 2SK1452, 2SK1831	data
IRFS 430	MOS-N-FET-e	V-MOS, 500V, 3,1A, 42W, <1,5Ω(2,5A)	18c	Sam	BUK 427-500	data
IRFS 431	MOS-N-FET-e	=IRFS 430: 450V	18c	Sam	BUK 427-500	data
IRFS 432	MOS-N-FET-e	V-MOS, 500V, 2,8A, 42W, <2Ω(2,5A)	18c	Sam	BUK 427-500	data
IRFS 433	MOS-N-FET-e	=IRFS 432: 450V	18c	Sam	BUK 427-500	data
IRFS 440	MOS-N-FET-e	V-MOS, 500V, 5,5A, 65W, <0,85Ω(4A)	18c	Sam	BUK 428-500	data
IRFS 440 A	MOS-N-FET-e	V-MOS, 500V, 6,2A, 85W, <0,85Ω(3,1A)	18c	Sam	BUK 428-500	data
IRFS 441	MOS-N-FET-e	=IRFS 440: 450V	18c	Sam	BUK 428-500	data
IRFS 442	MOS-N-FET-e	V-MOS, 500V, 4,8A, 65W, <1,1Ω(4A)	18c	Sam	BUK 427-500, BUK 428-500	data
IRFS 443	MOS-N-FET-e	=IRFS 442: 450V	18c	Sam	BUK 427-500, BUK 428-500	data
IRFS 450	MOS-N-FET-e	V-MOS, 500V, 9A, 70W, <0,4Ω(7A)	18c	Sam	2SK1206, 2SK1329, 2SK1523, 2SK1696, ++	data
IRFS 450 A	MOS-N-FET-e	V-MOS, 500V, 9,6A, 96W, <0,4Ω(4,8A)	18c	Sam	2SK1206, 2SK1329, 2SK1523, 2SK1696, ++	data
IRFS 451	MOS-N-FET-e	=IRFS: 450: 450V	18c	Sam	2SK1206, 2SK1329, 2SK1523, 2SK1696, ++	data
IRFS 452	MOS-N-FET-e	V-MOS, 500V, 8,3A, 70W, <Γ,5Ω(7A)	18c	Sam	2SK1206, 2SK1329, 2SK1523, 2SK1696, ++	data
IRFS 453	MOS-N-FET-e	=IRFS: 452: 450V	18c	Sam	2SK1206, 2SK1329, 2SK1523, 2SK1696, ++	data
IRFS 510 A	MOS-N-FET-e	V-MOS, 100V, 4,5A, 21W, <0,4Ω(2,25A)	17c	Sam	BUK442-100, 2SK992, 2SK1260, 2SK2581	data
IRFS 520	MOS-N-FET-e	V-MOS, 100V, 7A, 30W, <0,27Ω(5,6A)	17c	Sam	BUK443-100, BUZ72(A)F, 2SK1261, 2SK1556+	data
IRFS 520 A	MOS-N-FET-e	V-MOS, 100V, 7,2A, 28W, <0,2Ω(3,6A)	17c	Sam	BUK443-100, BUZ72(A)F, 2SK1261, 2SK1556+	data
IRFS 521	MOS-N-FET-e	=IRFS 520: 80V	17c	Sam	BUK443-100, BUZ72(A)F, 2SK1261, 2SK1556+	data
IRFS 522	MOS-N-FET-e	V-MOS, 100V, 6A, 30W, <0,36Ω(5,6A)	17c	Sam	BUK443-100, BUZ72(A)F, 2SK1261, 2SK1556+	data
IRFS 523	MOS-N-FET-e	=IRFS 521: 80V	17c	Sam	BUK443-100, BUZ72(A)F, 2SK1261, 2SK1556+	data
IRFS 530	MOS-N-FET-e	V-MOS, 100V, 9A, 35W, <0,16Ω(8,3A)	17c	Sam	BUK 545-100, 2SK1230, 2SK1305, 2SK1430++	data
IRFS 530 A	MOS-N-FET-e	V-MOS, 100V, 10,7A, 32W, <0,11Ω(5,35A)	17c	Sam	BUK 545-100, 2SK1230, 2SK1305, 2SK1430++	data
IRFS 531	MOS-N-FET-e	=IRFS 530: 80V	17c	Sam	BUK 545-100, 2SK1230, 2SK1305, 2SK1430++	data
IRFS 532	MOS-N-FET-e	V-MOS, 100V, 8A, 35W, <0,23Ω(8,3A)	17c	Sam	BUK 545-100, 2SK1230, 2SK1305, 2SK1430++	data
IRFS 533	MOS-N-FET-e	=IRFS 532: 80V	17c	Sam	BUK 545-100, 2SK1230, 2SK1305, 2SK1430++	data
IRFS 540	MOS-N-FET-e	V-MOS, 100V, 15A, 40W, <77mΩ(17A)	17c	Sam	2SK1034, 2SK1306, 2SK1431, 2SK1558, ++	data
IRFS 540 A	MOS-N-FET-e	V-MOS, 100V, 17A, 39W, <52mΩ(8,5A)	17c	Sam	2SK1034, 2SK1306, 2SK1431, 2SK1558, ++	data
IRFS 541	MOS-N-FET-e	=IRFS 540: 80V	17c	Sam	2SK1034, 2SK1306, 2SK1431, 2SK1558, ++	data
IRFS 542	MOS-N-FET-e	V-MOS, 100V, 14A, 40W, <0,1Ω(17A)	17c	Sam	BUK445-100, 2SK1306, 2SK1034, 2SK1431,++	data
IRFS 543	MOS-N-FET-e	=IRFS 542: 80V	17c	Sam	BUK445-100, 2SK1306, 2SK1034, 2SK1431,++	data
IRFS 550 A	MOS-N-FET-e	V-MOS, 100V, 21A, 46W, <40mΩ(10,5A)	17c	Sam	2SK1348, 2SK1432, 2SK1906, 2SK2391, ++	data
IRFS 610 A	MOS-N-FET-e	V-MOS, 200V, 2,5A, 22W, <1,5Ω(1,25A)	17c	Sam	BUK 444-400, 2SK2146, 2SK2538	data
IRFS 614 A	MOS-N-FET-e	V-MOS, 250V, 2,1A, 22W, <2Ω(1,05A)	17c	Sam	BUK 444-400, 2SK2146, 2SK2538	data
IRFS 620	MOS-N-FET-e	V-MOS, 200V, 4A, 30W, <0,8Ω(2,5A)	17c	Sam	BUK 444-200, BUZ 73(A)F, 2SK2010, ++	data
IRFS 620 A	MOS-N-FET-e	V-MOS, 200V, 4,1A, 30W, <0,8Ω(2,05A)	17c	Sam	BUK 444-200, BUZ 73(A)F, 2SK2010, ++	data
IRFS 621	MOS-N-FET-e	=IRFS 620: 150V	17c	Sam	BUK 444-200, BUZ 73(A)F, 2SK2010, ++	data
IRFS 622	MOS-N-FET-e	V-MOS, 200V, 3,5A, 30W, <1,2Ω(2,5A)	17c	Sam	BUK 444-200, BUZ 73(A)F, 2SK2010, ++	data
IRFS 623	MOS-N-FET-e	=IRFS 622: 150V	17c	Sam	BUK 444-200, BUZ 73(A)F, 2SK2010, ++	data
IRFS 624	MOS-N-FET-e	V-MOS, 250V, 3,3A, 30W, <1,1Ω(2,1A)	17c	Sam	BUK 445-400, 2SK2010, 2SK2108, 2SK2345	data
IRFS 624 A	MOS-N-FET-e	V-MOS, 250V, 3,4A, 34W, <1,1Ω(1,7A)	17c	Sam	BUK 445-400, 2SK2010, 2SK2108, 2SK2345	data
IRFS 625	MOS-N-FET-e	V-MOS, 250V, 2,9A, 30W, <1,5Ω(2,1A)	17c	Sam	BUK 445-400, 2SK2010, 2SK2108, 2SK2345	data
IRFS 630	MOS-N-FET-e	V-MOS, 200V, 6,1A, 35W, <0,4Ω(5A)	17c	Sam	BUK 445-200, 2SK1987, 2SK2160, 2SK2341++	data
IRFS 630 A	MOS-N-FET-e	V-MOS, 200V, 6,5A, 38W, <0,4Ω(3,25A)	17c	Sam	BUK 445-200, 2SK1987, 2SK2160, 2SK2341++	data
IRFS 631	MOS-N-FET-e	=IRFS 630: 150V	17c	Sam	BUK 445-200, 2SK1987, 2SK2160, 2SK2341++	data
IRFS 632	MOS-N-FET-e	V-MOS, 200V, 5,5A, 35W, <0,4Ω(5A)	17c	Sam	BUK 445-200, BUZ 73F, 2SK2010, 2SK2108++	data
IRFS 633	MOS-N-FET-e	=IRFS 632: 150V	17c	Sam	BUK 445-200, BUZ 73F, 2SK2010, 2SK2108++	data
IRFS 634	MOS-N-FET-e	V-MOS, 250V, 5,5A, 35W, <0,45Ω(4,1A)	17c	Sam	2SK1478, 2SK1568...69, 2SK1987, 2SK2341++	data
IRFS 634 A	MOS-N-FET-e	V-MOS, 250V, 5,8A, 38W, <0,45Ω(2,9A)	17c	Sam	2SK1478, 2SK1568...69, 2SK1987, 2SK2341++	data
IRFS 635	MOS-N-FET-e	V-MOS, 250V, 4,4A, 35W, <0,68Ω(4,1A)	17c	Sam	2SK1478, 2SK1568...69, 2SK1987, 2SK2341++	data

Type	Device	Short Description	Fig.	Manu	Comparision Types	More at
IRFS 640	MOS-N-FET-e	V-MOS, 200V, 10,2A, 40W, <0,18Ω(10A)	17c	Sam	2SK1036, 2SK1762, 2SK2011, 2SK2212, ++	data
IRFS 640 A	MOS-N-FET-e	V-MOS, 200V, 9,8A, 43W, <0,18Ω(4,9A)	17c	Sam	2SK1036, 2SK1762, 2SK2011, 2SK2212, ++	data
IRFS 641	MOS-N-FET-e	=IRFS 640: 150V	17c	Sam	2SK1036, 2SK1762, 2SK2011, 2SK2212, ++	data
IRFS 642	MOS-N-FET-e	V-MOS, 200V, 9,1A, 40W, <0,22Ω(10A)	17c	Sam	2SK1036, 2SK1762, 2SK2011, 2SK2212, ++	data
IRFS 643	MOS-N-FET-e	=IRFS 642: 150V	17c	Sam	2SK1036, 2SK1762, 2SK2011, 2SK2212, ++	data
IRFS 644	MOS-N-FET-e	V-MOS, 250V, 8,5A, 40W, <0,28Ω(8A)	17c	Sam	2SK1036, 2SK1478, 2SK1762, 2SK2011, ++	data
IRFS 644 A	MOS-N-FET-e	V-MOS, 250V, 7,9A, 43W, <0,28Ω(3,95A)	17c	Sam	2SK1036, 2SK1478, 2SK1762, 2SK2011, ++	data
IRFS 645	MOS-N-FET-e	V-MOS, 250V, 7,4A, 40W, <0,34Ω(8A)	17c	Sam	2SK1036, 2SK1478, 2SK1762, 2SK2011, ++	data
IRFS 650 A	MOS-N-FET-e	V-MOS, 200V, 15,8A, 50W, <85mΩ(7,9A)	17c	Sam	2SK1350, 2SK2135, 2SK2255, 2SK2382	data
IRFS 654 A	MOS-N-FET-e	V-MOS, 250V, 12A, 50W, <0,14Ω(6A)	17c	Sam	2SK2012, 2SK2255, 2SK2508	data
IRFS 710 A	MOS-N-FET-e	V-MOS, 400V, 1,6A, 23W, <3,6Ω(0,8A)	17c	Sam	2SK528...529, 2SK1443	data
IRFS 720	MOS-N-FET-e	V-MOS, 400V, 2,5A, 30W, <1,8Ω(1,8A)	17c	Sam	BUK 444-400, 2SK1833, 2SK1988...89	data
IRFS 720 A	MOS-N-FET-e	V-MOS, 400V, 2,8A, 33W, <1,8Ω(1,4A)	17c	Sam	BUK 444-400, 2SK1833, 2SK1988...89	data
IRFS 721	MOS-N-FET-e	=IRFS 720: 350V	17c	Sam	BUK 444-400, 2SK1833, 2SK1988...89	data
IRFS 722	MOS-N-FET-e	V-MOS, 400V, 2A, 30W, <2,5Ω(1,8A)	17c	Sam	BUK 444-400, 2SK1833, 2SK1988...89	data
IRFS 723	MOS-N-FET-e	=IRFS 722: 350V	17c	Sam	BUK 444-400, 2SK1833, 2SK1988...89	data
IRFS 730	MOS-N-FET-e	V-MOS, 400V, 3,5A, 35W, <1Ω(3A)	17c	Sam	BUK 445-450, 2SK503, 2SK1862..63	data
IRFS 730 A	MOS-N-FET-e	V-MOS, 400V, 3,9A, 38W, <1Ω(1,95A)	17c	Sam	BUK 445-450, 2SK503, 2SK1862..63	data
IRFS 731	MOS-N-FET-e	=IRFS 730: 350V	17c	Sam	BUK 445-450, 2SK503, 2SK1862..63	data
IRFS 732	MOS-N-FET-e	V-MOS, 400V, 3A, 35W, <1,5Ω(3A)	17c	Sam	BUK 445-450, 2SK503, 2SK1862..63	data
IRFS 733	MOS-N-FET-e	=IRFS 732: 350V	17c	Sam	BUK 445-450, 2SK503, 2SK1862..63	data
IRFS 740	MOS-N-FET-e	V-MOS, 400V, 5,5A, 40W, <0,55Ω(5,2A)	17c	Sam	BUK 445-400, 2SK1377, 2SK1992, 2SK2114++	data
IRFS 740 A	MOS-N-FET-e	V-MOS, 400V, 5,7A, 44W, <0,55Ω(2,85A)	17c	Sam	BUK 445-400, 2SK1377, 2SK1992, 2SK2114++	data
IRFS 741	MOS-N-FET-e	=IRFS 740: 350V	17c	Sam	BUK 445-400, 2SK1377, 2SK1992, 2SK2114++	data
IRFS 742	MOS-N-FET-e	V-MOS, 400V, 4,5A, 40W, <0,8Ω(5,2A)	17c	Sam	BUK 445-400, 2SK1377, 2SK1990, 2SK2114++	data
IRFS 743	MOS-N-FET-e	=IRFS 742: 350V	17c	Sam	BUK 445-400, 2SK1377, 2SK1990, 2SK2114++	data
IRFS 750 A	MOS-N-FET-e	V-MOS, 400V, 8,4A, 49W, <0,3Ω(4,2A)	17c	Sam	2SK1447, 2SK1566, 2SK1642, 2SK2363	data
IRFS 820	MOS-N-FET-e	V-MOS, 500V, 2A, 30W, <3Ω(1,4A)	17c	Sam	BUK 444-500, BUK 445-600, 2SK1758	data
IRFS 820 A	MOS-N-FET-e	V-MOS, 500V, 2,1A, 33W, <3Ω(1,05A)	17c	Sam	BUK 444-500, BUK 445-600, 2SK1758	data
IRFS 821	MOS-N-FET-e	=IRFS 820: 450V	17c	Sam	BUK 444-500, BUK 445-600, 2SK1758	data
IRFS 822	MOS-N-FET-e	V-MOS, 500V, 1,5A, 30W, <4Ω(1,4A)	17c	Sam	BUK 444-500, BUK 445-600, 2SK1758	data
IRFS 823	MOS-N-FET-e	=IRFS 822: 450V	17c	Sam	BUK 444-500, BUK 445-600, 2SK1758	data
IRFS 830	MOS-N-FET-e	V-MOS, 500V, 3A, 35W, <1,5Ω(2,5A)	17c	Sam	BUK 445-500, 2SK1572, 2SK1767, 2SK1863++	data
IRFS 830 A	MOS-N-FET-e	V-MOS, 500V, 3,1A, 38W, <1,5Ω(1,55A)	17c	Sam	BUK 445-500, 2SK1572, 2SK1767, 2SK1863++	data
IRFS 831	MOS-N-FET-e	=IRFS 830: 450V	17c	Sam	BUK 445-500, 2SK1572, 2SK1767, 2SK1863++	data
IRFS 832	MOS-N-FET-e	V-MOS, 500V, 2,5A, 35W, <2Ω(2,5A)	17c	Sam	BUK 445-500, 2SK1572, 2SK1767, 2SK1863++	data
IRFS 833	MOS-N-FET-e	=IRFS 832: 450V	17c	Sam	BUK 445-500, 2SK1572, 2SK1767, 2SK1863++	data
IRFS 840	MOS-N-FET-e	V-MOS, 500V, 4,5A, 40W, <0,85Ω(4A)	17c	Sam	2SK1351, 2SK1991, 2SK2115, 2SK2354, ++	data
IRFS 840 A	MOS-N-FET-e	V-MOS, 500V, 4,6A, 44W, <0,85Ω(2,3A)	17c	Sam	2SK1351, 2SK1991, 2SK2115, 2SK2354, ++	data
IRFS 841	MOS-N-FET-e	=IRFS 840: 450V	17c	Sam	2SK1351, 2SK1991, 2SK2115, 2SK2354, ++	data
IRFS 842	MOS-N-FET-e	V-MOS, 500V, 4A, 40W, <1,1Ω(4A)	17c	Sam	2SK1351, 2SK1991, 2SK2115, 2SK2354, ++	data
IRFS 843	MOS-N-FET-e	=IRFS 842: 450V	17c	Sam	2SK1351, 2SK1991, 2SK2115, 2SK2354, ++	data
IRFS 9130	MOS-P-FET-e	V-MOS, 100V, 8,3A, 42W, <0,3Ω(6,5A)	18c	Sam	-	data
IRFS 9131	MOS-P-FET-e	=IRFS 9130: 60V	18c	Sam	-	data
IRFS 9132	MOS-P-FET-e	V-MOS, 100V, 6,9A, 42W, <0,4Ω(6,5A)	18c	Sam	-	data
IRFS 9133	MOS-P-FET-e	=IRFS 9132: 60V	18c	Sam	-	data
IRFS 9140	MOS-P-FET-e	V-MOS, 100V, 13,2A, 65W, <0,2Ω(10A)	18c	Sam	-	data
IRFS 9141	MOS-P-FET-e	=IRFS 9140: 60V	18c	Sam	-	data
IRFS 9142	MOS-P-FET-e	V-MOS, 100V, 10,4A, 65W, <0,3Ω(10A)	18c	Sam	-	data
IRFS 9143	MOS-P-FET-e	=IRFS 9142: 60V	18c	Sam	-	data
IRFS 9230	MOS-P-FET-e	V-MOS, 200V, 4,5A, 42W, <0,8Ω(3,5A)	18c	Sam	-	data
IRFS 9231	MOS-P-FET-e	=IRFS 9230: 150V	18c	Sam	-	data
IRFS 9232	MOS-P-FET-e	V-MOS, 200V, 3,8A, 42W, <1,2Ω(3,5A)	18c	Sam	-	data
IRFS 9233	MOS-P-FET-e	=IRFS 9232: 150V	18c	Sam	-	data
IRFS 9240	MOS-P-FET-e	V-MOS, 200V, 7,6A, 65W, <0,5Ω(6A)	18c	Sam	-	data
IRFS 9241	MOS-P-FET-e	=IRFS 9240: 150V	18c	Sam	-	data
IRFS 9242	MOS-P-FET-e	V-MOS, 200V, 6,2A, 65W, <0,7Ω(6A)	18c	Sam	-	data
IRFS 9243	MOS-P-FET-e	=IRFS 9242: 150V	18c	Sam	-	data
IRFS 9520	MOS-P-FET-e	V-MOS, 100V, 5A, 30W, <0,6Ω(3A)	17c	Sam	2SJ248, 2SJ350	data
IRFS 9521	MOS-P-FET-e	=IRFS 9520: 60V	17c	Sam	2SJ248, 2SJ350	data
IRFS 9522	MOS-P-FET-e	V-MOS, 100V, 4,3A, 30W, <0,8Ω(3A)	17c	Sam	2SJ248, 2SJ350	data
IRFS 9523	MOS-P-FET-e	=IRFS 9522: 60V	17c	Sam	2SJ248, 2SJ350	data
IRFS 9530	MOS-P-FET-e	V-MOS, 100V, 8A, 35W, <0,3Ω(6,5A)	17c	Sam	2SJ139	data
IRFS 9531	MOS-P-FET-e	=IRFS 9530: 60V	17c	Sam	2SJ137, 2SJ139, 2SJ147, 2SJ175	data
IRFS 9532	MOS-P-FET-e	V-MOS, 100V, 5,8A, 35W, <0,4Ω(6,5A)	17c	Sam	2SJ248, 2SJ350	data
IRFS 9533	MOS-P-FET-e	=IRFS 9532: 60V	17c	Sam	2SJ248, 2SJ350	data
IRFS 9540	MOS-P-FET-e	V-MOS, 100V, 10,7A, 40W, <0,2Ω(10A)	17c	Sam	2SJ139	data
IRFS 9541	MOS-P-FET-e	=IRFS 9540: 60V	17c	Sam	2SJ137, 2SJ139, 2SJ147, 2SJ175	data
IRFS 9542	MOS-P-FET-e	V-MOS, 100V, 8,4A, 40W, <0,3Ω(10A)	17c	Sam	2SJ139	data
IRFS 9543	MOS-P-FET-e	=IRFS 9542: 60V	17c	Sam	2SJ137, 2SJ139, 2SJ147, 2SJ175	data
IRFS 9620	MOS-P-FET-e	V-MOS, 200V, 3A, 30W, <1,5Ω(1,5A)	17c	Sam	2SJ306	data
IRFS 9621	MOS-P-FET-e	=IRFS 9620: 150V	17c	Sam	2SJ159	data
IRFS 9622	MOS-P-FET-e	V-MOS, 200V, 2,6A, 30W, <2,4Ω(1,5A)	17c	Sam	2SJ306	data
IRFS 9623	MOS-P-FET-e	=IRFS 9622: 150V	17c	Sam	2SJ159	data
IRFS 9630	MOS-P-FET-e	V-MOS, 200V, 4,4A, 35W, <0,8Ω(3,5A)	17c	Sam	2SJ410	data
IRFS 9631	MOS-P-FET-e	=IRFS 9630: 150V	17c	Sam	2SJ410	data
IRFS 9632	MOS-P-FET-e	V-MOS, 200V, 3A, 35W, <1,2Ω(3,5A)	17c	Sam	2SJ410	data
IRFS 9633	MOS-P-FET-e	=IRFS 9632: 150V	17c	Sam	2SJ410	data

Type	Device	Short Description	Fig.	Manu	Comparision Types	More at
IRFS 9640	MOS-P-FET-e	V-MOS, 200V, 6,2A, 40W, <0,5Ω(6A)	17c	Sam	2SJ410	data
IRFS 9641	MOS-P-FET-e	=IRFS 9640: 150V	17c	Sam	2SJ410	data
IRFS 9642	MOS-P-FET-e	V-MOS, 200V, 5A, 40W, <0,7Ω(6A)	17c	Sam	2SJ410	data
IRFS 9643	MOS-P-FET-e	=IRFS 9642: 150V	17c	Sam	2SJ410	data
IRFSL 9 N60A	MOS-N-FET-e	V-MOS, 600V, 9,2A, 170W, <0,75Ω(5,5A)	30c§	Inr	SSW 10N60A	data
IRFSZ 14 A	MOS-N-FET-e	V-MOS, 60V, 8A, 19W, <0,14Ω(4A)	17c	Sam	BUK 442-60, 2SK1344, 2SK1895, 2SK1974,++	data
IRFSZ 20	MOS-N-FET-e	V-MOS, 50V, 12,9A, 30W, <0,1Ω(9A)	17c	Sam	BUK 443-50, 2SK1094, 2SK1419, 2SK1896	data
IRFSZ 22	MOS-N-FET-e	V-MOS, 50V, 12A, 30W, <0,12Ω(9A)	17c	,Sam	BUK 443-50, 2SK1094, 2SK1419, 2SK1896	data
IRFSZ 24	MOS-N-FET-e	=IRFSZ 20: 60V	17c	Sam	2SK1094, 2SK1419, 2SK1286, 2SK1896	data
IRFSZ 24 A	MOS-N-FET-e	V-MOS, 60V, 14A, 30W, <0,07Ω(7A)	17c	Sam	2SK1094, 2SK1419, 2SK1286, 2SK1896	data
IRFSZ 25	MOS-N-FET-e	=IRFSZ 22: 60V	17c	Sam	2SK1094, 2SK1419, 2SK1286, 2SK1896	data
IRFSZ 30	MOS-N-FET-e	V-MOS, 50V, 18,7A, 35W, <50mΩ(18A)	17c	Sam	BUK445-50, 2SK1290, 2SK1345, 2SK1420, ++	data
IRFSZ 32	MOS-N-FET-e	V-MOS, 50V, 15,6A, 35W, <70mΩ(18A)	17c	Sam	BUK445-50, 2SK1290, 2SK1345, 2SK1420, ++	data
IRFSZ 34	MOS-N-FET-e	=IRFSZ 30: 60V	17c	Sam	2SK1290, 2SK1345, 2SK1420, 2SK1882, ++	data
IRFSZ 34 A	MOS-N-FET-e	V-MOS, 60V, 20A, 34W, <40mΩ(10A)	17c	Sam	2SK1290, 2SK1345, 2SK1420, 2SK1882, ++	data
IRFSZ 35	MOS-N-FET-e	=IRFSZ 32: 60V	17c	Sam	2SK1290, 2SK1345, 2SK1420, 2SK1882, ++	data
IRFSZ 40	MOS-N-FET-e	V-MOS, 50V, 34A, 55W, <28mΩ(23A)	17c	Sam	2SK1257, 2SK1262, 2SK1421, 2SK1653, ++	data
IRFSZ 42	MOS-N-FET-e	V-MOS, 50V, 30A, 55W, <35mΩ(23A)	17c	Sam	2SK1257, 2SK1262, 2SK1421, 2SK1653, ++	data
IRFSZ 44	MOS-N-FET-e	=IRFSZ 40: 60V	17c	Sam	2SK1257, 2SK1262, 2SK1421, 2SK1653, ++	data
IRFSZ 44 A	MOS-N-FET-e	V-MOS, 60V, 30A, 45W, <24mΩ(15A)	17c	Sam	2SK1257, 2SK1262, 2SK1421, 2SK1653, ++	data
IRFSZ 45	MOS-N-FET-e	=IRFSZ 42: 60V	17c	Sam	2SK1257, 2SK1262, 2SK1421, 2SK1653, ++	data
IRFT ...	MOS-FET	Leistungsmodule/Power Modules		Inr		
IRFU 010...9310	MOS-N/P-FET	=IRFR 010...9310:	30c§	Inr, Sam	→IRFR 010...9310	data
IRFUC 20	MOS-N-FET-e	=IRFRC 20:	30c§	Inr	→IRFRC 20	data
IRFV 260	MOS-N-FET-e	V-MOS, 200V, 45A, 300W, <0,06Ω(29A)	18f	Inr	-	data
IRFW 450	MOS-N-FET-e	V-MOS, 500V, 14A, 180W, <0,4Ω	16c§	Sgs	2SK1411, 2SK1678	data
IRFW 510 A	MOS-N-FET-e	V-MOS, 100V, 5,6A, 33W, <0,4Ω(2,8A)	30c§	Sam	2SK782, 2SK2790	data
IRFW 520 A	MOS-N-FET-e	V-MOS, 100V, 9,2A, 45W, <0,2Ω(4,6A)	30c§	Sam	2SK1741, 2SK1620, 2SK1907, 2SK2088	data
IRFW 530 A	MOS-N-FET-e	V-MOS, 100V, 14A, 55W, <0,11Ω(7A)	30c§	Sam	2SK1559S, 2SK1561S, 2SK1908, 2SK1927	data
IRFW 540 A	MOS-N-FET-e	V-MOS, 100V, 28A, 107W, <52mΩ(14A)	30c§	Sam	2SK1623, 2SK1909, 2SK1928	data
IRFW 550 A	MOS-N-FET-e	V-MOS, 100V, 40A, 167W, <40mΩ(20A)	30c§	Sam		data
IRFW 610 A	MOS-N-FET-e	V-MOS, 200V, 3,3A, 38W, <1,5Ω(1,65A)	30c§	Sam	2SK782, 2SK2790	data
IRFW 614 A	MOS-N-FET-e	V-MOS, 250V, 2,8A, 40W, <2Ω(1,4A)	30c§	Sam	2SK1308, 2SK2790, 2SK2838	data
IRFW 620 A	MOS-N-FET-e	V-MOS, 200V, 5A, 47W, <0,8Ω(2,5A)	30c§	Sam	2SK1308, 2SK2790, 2SK2838	data
IRFW 624 A	MOS-N-FET-e	V-MOS, 250V, 4,1A, 49W, <1,1Ω(2,05A)	30c§	Sam	2SK1308, 2SK2790, 2SK2838	data
IRFW 630 A	MOS-N-FET-e	V-MOS, 200V, 9A, 72W, <0,4Ω(4,5A)	30c§	Sam	2SK2088, 2SK2062, 2SK2064	data
IRFW 634 A	MOS-N-FET-e	V-MOS, 250V, 8,1A, 74W, <0,45Ω(4,05A)	30c§	Sam	MTB 10N40E, 2SK2062, 2SK2064	data
IRFW 640 A	MOS-N-FET-e	V-MOS, 200V, 18A, 139W, <0,18Ω(9A)	30c§	Sam	MTB 16N25E, 2SK1636, 2SK2134-Z, 2SK2254	data
IRFW 644 A	MOS-N-FET-e	V-MOS, 250V, 14A, 139W, <0,28Ω(7A)	30c§	Sam	MTB 16N25E, 2SK1636, 2SK2133-Z, 2SK2254	data
IRFW 710 A	MOS-N-FET-e	V-MOS, 400V, 2A, 36W, <3,6Ω(1A)	30c§	Sam	MTB 2N40E	data
IRFW 720 A	MOS-N-FET-e	V-MOS, 400V, 3,3A, 49W, <1,8Ω(1,65A)	30c§	Sam	2SK1493Z...94Z, 2SK1690, 2SK1721	data
IRFW 730 A	MOS-N-FET-e	V-MOS, 400V, 5,5A, 73W, <1Ω(2,75A)	30c§	Sam	2SK1308, 2SK1313, 2SK1691, 2SK2838, ++	data
IRFW 740 A	MOS-N-FET-e	V-MOS, 400V, 10A, 134W, <0,55Ω(5A)	30c§	Sam	MTB 10N40E, 2SK1865, 2SK2366Z	data
IRFW 820 A	MOS-N-FET-e	V-MOS, 500V, 2,5A, 49W, <3Ω(1,25A)	30c§	Sam	MTB 2N60E, 2SK1746, 2SK2509	data
IRFW 830 A	MOS-N-FET-e	V-MOS, 500V, 4,5A, 80W, <1,5Ω(2,25A)	30c§	Sam	2SK1314, 2SK1703S, 2SK1722, 2SK1751Z, ++	data
IRFW 840 A	MOS-N-FET-e	V-MOS, 500V, 8A, 142W, <0,85Ω(4A)	30c§	Sam	2SK1316, 2SK1496Z, 2SK1541, 2SK2360Z, ++	data
IRFWZ 14 A	MOS-N-FET-e	V-MOS, 60V, 10A, 30W, <0,14Ω(5A)	30c§	Sam	2SK1741, 2SK1907, 2SK2041	data
IRFWZ 24 A	MOS-N-FET-e	V-MOS, 60V, 17A, 44W, <0,07Ω(8,5A)	30c§	Sam	2SK1622...23, 2SK1909, 2SK1918, 2SK2311++	data
IRFWZ 34 A	MOS-N-FET-e	V-MOS, 60V, 30A, 77W, <0,04Ω(15A)	30c§	Sam	MTB33N10E, 2SK1900, 2SK2411-Z, 2SK2789++	data
IRFWZ 44 A	MOS-N-FET-e	V-MOS, 60V, 50A, 126W, <24mΩ(25A)	30c§	Sam	MTB52N06V, 2SK2499-Z, 2SK2553	data
IRFZ 010	MOS-N-FET-e	V-MOS, 50V, 7,2A, 20W, <0,2Ω(3,7A)	17c§	Inr, Sam	BUZ 71, 2SK464, 2SK428, 2SK442, 2SK993	data
IRFZ 012	MOS-N-FET-e	V-MOS, 50V, 5,9A, 20W, <0,3Ω(3,7A)	17c§	Inr, Sam	BUZ 71, 2SK464, 2SK428, 2SK442, 2SK993	data
IRFZ 14	MOS-N-FET-e	V-MOS, 60V, 10A, 43W, <0,2Ω(6A)	17c§	Inr, Sam	BUZ 20, BUZ 72, 2SK970...71, 2SK1114, ++	data
IRFZ 14 A	MOS-N-FET-e	V-MOS, 60V, 10A, 30W, <0,14Ω(5A)	17c§	Sam	BUZ 20, BUZ 72, 2SK970...71, 2SK1114, ++	data
IRFZ 14 L,S	MOS-N-FET-e	=IRFZ 14:	30c§		2SK2041, 2SK2282	data
IRFZ 15	MOS-N-FET-e	V-MOS, 60V, 8,3A, 43W, <0,3Ω(5,8A)	17c§	Inr, Sam	BUZ 20, BUZ 72, 2SK970...71, 2SK1114, ++	data
IRFZ 20	MOS-N-FET-e	V-MOS, 50V, 15A, 40W, <0,1Ω(9A)	17c§	Inr, Sam++	BUZ 10, 2SK673, 2SK971, 2SK1416, 2SK2175	data
IRFZ 22	MOS-N-FET-e	V-MOS, 50V, 14A, 40W, <0,12Ω(9A)	17c§	Inr, Sam++	BUZ 10, 2SK673, 2SK971, 2SK1416, 2SK2175	data
IRFZ 20...22FI,P	MOS-N-FET-e	=IRFZ 20...22: Iso, ≈83%Imax, 30W	17c	Sgs	BUZ 71(A)F, BUK 442-50, BUK 543-50,++	
IRFZ 24	MOS-N-FET-e	V-MOS, 60V, 17A, 60W, <0,1Ω(10A)	17c§	Inr, Sam	BUZ 21, 2SK942, 2SK1115, 2SK1417, ++	data
IRFZ 24 A	MOS-N-FET-e	V-MOS, 60V, 17A, 44W, <0,07Ω(8,5A)	17c§	Sam	BUZ 21, 2SK942, 2SK1115, 2SK1417, ++	data
IRFZ 24 N	MOS-N-FET-e	V-MOS, 55V, 17A, 45W, <70mΩ(10A)	17c§	Inr	BUZ 21, 2SK942, 2SK1115, 2SK1417, ++	data
IRFZ 24 NL,L,NS,S	MOS-N-FET-e	=IRFZ 24...:	30c§		2SK1622, 2SK1918, 2SK1967, 2SK2311	data
IRFZ 25	MOS-N-FET-e	V-MOS, 60V, 14A, 60W, <0,12Ω(9A)	17c§	Inr, Sam	BUZ 21, 2SK673, 2SK971, 2SK1416,++	data
IRFZ 30	MOS-N-FET-e	V-MOS, 50V, 30A, 75W, <50mΩ(16A)	17c§	Inr,Mot,Sam	BUK 555-50, BUZ 11, 2SK1296, 2SK2411, ++	data
IRFZ 32	MOS-N-FET-e	V-MOS, 50V, 25A, 75W, <70mΩ(16A)	17c§	Inr,Mot,Sam	BUK 555-60, BUZ 11, 2SK972, 2SK1296, ++	data
IRFZ 34	MOS-N-FET-e	V-MOS, 60V, 30A, 88W, <50mΩ(18A)	17c§	Inr, Sam	BUK 555-60, 2SK856, 2SK1296, 2SK2411, ++	data
IRFZ 34 A	MOS-N-FET-e	V-MOS, 60V, 30A, 77W, <40mΩ(15A)	17c§	Sam	BUK 555-60, 2SK856, 2SK1296, 2SK2411, ++	data
IRFZ 34 E	MOS-N-FET-e	V-MOS, 60V, 28A, 68W, <42mΩ(16A)	17c§	Inr	BUK 555-60, 2SK856, 2SK1296, 2SK2411, ++	data
IRFZ 34 N	MOS-N-FET-e	V-MOS, 55V, 29A, 68W, <40mΩ(16A)	17c§	Inr	BUK 555-60, 2SK856, 2SK1296, 2SK2411, ++	data
IRFZ 34 NL,L,NS,S	MOS-N-FET-e	=IRFZ 34...:	30c§		MTB 30N06..., 2SK1900, 2SK2288, 2SK2411-Z	data
IRFZ 35	MOS-N-FET-e	V-MOS, 60V, 25A, 88W, <70mΩ(18A)	17c§	Inr, Sam	BUK 555-60, BUZ 21, 2SK942, 2SK1296,++	data
IRFZ 40	MOS-N-FET-e	V-MOS, 50V, 35A, 125W, <28mΩ(29A)	17c§	Inr, Sam,++	BUZ 11...12, PRFZ 42, 2SK1418, 2SK1542,++	data
IRFZ 40 FI	MOS-N-FET-e	=IRFZ 40: Iso, 27A, 45W	17c	Sgs	2SK1262	data
IRFZ 42	MOS-N-FET-e	V-MOS, 50V, 35A, 125W, <35mΩ(29A)	17c§	Inr, Sam,++	BUZ 11...12, PRFZ 42, 2SK1418, 2SK1542,++	data
IRFZ 44	MOS-N-FET-e	V-MOS, 60V, 50A, 150W, <28mΩ(31A)	17c§	Inr, Sam	BUK 556-60, 2SK2049, 2SK2499	data

Type	Device	Short Description	Fig.	Manu	Comparision Types	More at
IRFZ 44 A	MOS-N-FET-e	V-MOS, 60V, 50A, 126W, <24mΩ(25A)	17c§	Sam	BUK 556-60, 2SK2049, 2SK2499	data
IRFZ 44 E	MOS-N-FET-e	V-MOS, 60V, 48A, 110W, <23mΩ(25A)	17c§	Inr	BUK 556-60, 2SK2049, 2SK2499	data
IRFZ 44 N	MOS-N-FET-e	V-MOS, 55V, 49A, 110W, <22mΩ(25A)	17c§	Inr, Gsi	BUK 556-60, 2SK2049, 2SK2499	data
IRFZ 44 NL,L,NS,S	MOS-N-FET-e	=IRFZ 44..:	30c§		MTB 52N06..., 2SK2499-Z, 2SK2553	data
IRFZ 45	MOS-N-FET-e	V-MOS, 50V, 35A, 150W, <35mΩ(33A)	17c§	Inr, Sam	PRFZ42, 2SK856,2SK1418, 2SK1542, 2SK1911	data
IRFZ 46	MOS-N-FET-e	V-MOS, 50V, 50A, 150W, <24mΩ(32A)	17c§	Inr	BUK 456-50, BUK 556-60, BUZ 100, 2SK2049	data
IRFZ 46 N	MOS-N-FET-e	V-MOS, 55V, 53A, 120W, <20mΩ(28A)	17c§	Inr	BUK 456-50, BUK 556-60, BUZ 100, 2SK2049	data
IRFZ 46 NL,L,NS,S	MOS-N-FET-e	=IRFZ 46..:	30c§		MTB 52N06..., 2SK2499-Z, 2SK2553	data
IRFZ 48	MOS-N-FET-e	V-MOS, 60V, 50A, 190W, <18mΩ(43A)	17c§	Inr	(BUZ 100)[7]	data
IRFZ 48 N	MOS-N-FET-e	V-MOS, 55V, 64A, 140W, <16mΩ(32A)	17c§	Inr	(BUZ 100)[7]	data
IRFZ 48 NL,L,NS,S	MOS-N-FET-e	=IRFZ 48..:	30c§		MTB 52N06V	data
IRG 4 BC10 K	MOS-N-IGBT	L, >5kHz, 600V, 9A, 38W, 35/241ns	17id	Inr	GN 6010A	data
IRG 4 BC10 KD	MOS-N-IGBT	L, 600V, 9A, 38W, 77/237ns int. Damper-Di (4A, <42ns)	17id	Inr	BUP 410D	data
IRG 4 BC10 S	MOS-N-IGBT	L, 600V, 14A, 38W, 53/1340ns	17id	Inr	GN 6015A	data
IRG 4 BC10 SD	MOS-N-IGBT	L, 600V, 14A, 38W, 108/1535ns int. Damper-Di (4A, <42ns)	17id	Inr	BUP 400D	data
IRG 4 BC10 UD	MOS-N-IGBT	L, 8..40kHz, 600V, 8,5A, 38W, 56/227ns int. Damper-Di (4A, <42ns)	17id	Inr	BUP 410D	data
IRG 4 BC20 F	MOS-N-IGBT	L, 1...5kHz, 600V, 16A, 60W, 41/400ns	17id	Inr	BUP 400, GN 6015A	data
IRG 4 BC20 FD	MOS-N-IGBT	=IRG 4BC20F: 63/390ns, int. Damper-Di	17id	Inr	BUP 400D	data
IRG 4 BC20 K	MOS-N-IGBT	L, >5kHz, 600V, 16A, 60W, 55/250ns	17id	Inr	BUP 400, GN 6015A	data
IRG 4 BC20 KD	MOS-N-IGBT	=IRG 4BC20K: 88/252ns, int. Damper-Di	17id		BUP 400D	data
IRG 4 BC20 KD-S	MOS-N-IGBT	=IRG 4BC20KD:	30id		-	data
IRG 4 BC20 K-S	MOS-N-IGBT	=IRG 4BC20K:	30id		GT 15J103	data
IRG 4 BC20 S	MOS-N-IGBT	L, 600V, 19A, 60W, 36,7/970ns	17id	Inr	BUP 400	data
IRG 4 BC20 SD	MOS-N-IGBT	=IRG 4BC20S: 94/1170ns, int. Damper-Di	17id		BUP 400D	data
IRG 4 BC20 SD-S	MOS-N-IGBT	=IRG 4BC20SD:	30id		-	data
IRG 4 BC20 U	MOS-N-IGBT	L, 8..40kHz, 600V, 13A, 60W, 34/206ns	17id	Inr	BUK 854-800A, GN 6015A	data
IRG 4 BC20 UD	MOS-N-IGBT	=IRG 4BC20U: 54/203ns, int. Damper-Di	17id		BUP 410D	data
IRG 4 BC20 W	MOS-N-IGBT	L, SMPS, 600V, 13A, 60W, 36/174ns	17id	Inr	BUK 854-800A, GN 6015A	data
IRG 4 BC30 F	MOS-N-IGBT	L, 1...5kHz, 600V, 31A, 100W, 36/380ns	17id	Inr	BUP 401, BUP 402	data
IRG 4 BC30 FD	MOS-N-IGBT	=IRG 4BC30F: 68/390ns, int. Damper-Di	17id		-	data
IRG 4 BC30 K	MOS-N-IGBT	L, 600V, 28A, .100W, 54/250ns	17id	Inr	BUP 401, BUP 402	data
IRG 4 BC30 KD	MOS-N-IGBT	=IRG 4BC30K: 102/240ns, int. Damper-Di	17id		-	data
IRG 4 BC30 KD-S	MOS-N-IGBT	=IRG 4BC30KD:	30id		-	data
IRG 4 BC30 K-S	MOS-N-IGBT	=IRG 4BC30K:	30id		-	data
IRG 4 BC30 S	MOS-N-IGBT	L, 600V, 34A, 100W, 40/930ns	17id	Inr	BUP 402	data
IRG 4 BC30 U	MOS-N-IGBT	L, 8..40kHz, 600V, 23A, 100W, 27/175ns	17id	Inr	BUP 401	data
IRG 4 BC30 UD	MOS-N-IGBT	=IRG 4BC30U: 61/171ns, int. Damper-Di	17id		-	data
IRG 4 BC30 U-S	MOS-N-IGBT	=IRG 4BC30U:	30id		-	data
IRG 4 BC30 W	MOS-N-IGBT	L, SMPS, 600V, 23A, 100W, 41/166ns	17id	Inr	BUP 401	data
IRG 4 BC30 W-S	MOS-N-IGBT	=IRG 4BC30W:	30id		-	data
IRG 4 BC40 F	MOS-N-IGBT	L, 1...5kHz, 600V, 49A, 160W, 46/690ns	17id	Inr	-	data
IRG 4 BC40 K	MOS-N-IGBT	L, >5kHz, 600V, 42A, 160W, 45/280ns	17id	Inr	BUP 403	data
IRG 4 BC40 S	MOS-N-IGBT	L, 600V, 60A, 160W, 40/1030ns	17id	Inr	-	data
IRG 4 BC40 U	MOS-N-IGBT	L, 8..40kHz, 600V, 40A, 160W, 53/230ns	17id	Inr	BUP 403	data
IRG 4 BC40 W	MOS-N-IGBT	L, SMPS, 600V, 40A, 160W, 49/174ns	17id	Inr	BUP 403	data
IRG 4 CC10KB,SB,UB	MOS-N-IGBT	=IRG 4BC10..: Die in Wafer Form	6" Wafer	Inr	-	data
IRG 4 CC20 FB	MOS-N-IGBT	=IRG 4BC20F: Die in Wafer Form	6" Wafer	Inr	-	data
IRG 4 CC40 SB	MOS-N-IGBT	=IRG 4BC40S: Die in Wafer Form	6" Wafer	Inr	-	data
IRG 4 CC50 SB,WB	MOS-N-IGBT	=IRG 4BC50..: Die in Wafer Form	6" Wafer	Inr	-	data
IRG 4 CC71 KB	MOS-N-IGBT	=IRG 4BC71K: Die in Wafer Form	6" Wafer	Inr	-	data
IRG 4 CH40 SB	MOS-N-IGBT	=IRG 4PH40S: Die in Wafer Form	6" Wafer	Inr	-	data
IRG 4 IBC20 FD	MOS-N-IGBT	L, Iso, 600V, 14,3A, 34W, 63/390ns int. Damper-Di (6,5A, <55ns)	17ic	Inr	GT 15J102	data
IRG 4 IBC20 KD	MOS-N-IGBT	L, Iso, 600V, 11,5A, 34W, 88/252ns int. Damper-Di (6,3A, <55ns)	17ic	Inr	GT 15J102	data
IRG 4 IBC20 UD	MOS-N-IGBT	L, Iso, 600V, 11,4A, 34W, 54/203ns int. Damper-Di (6,5A, <55ns)	17ic	Inr	GT 15J102	data
IRG 4 IBC20 W	MOS-N-IGBT	L, Iso, 600V, 11,8A, 34W, 36/174ns	17ic	Inr	GT 15J102	data
IRG 4 IBC30 FD	MOS-N-IGBT	L, Iso, 600V, 11A, 45W, 68/390ns int. Damper-Di (8,5A, <60ns)	17ic	Inr	GT 15J102	data
IRG 4 IBC30 KD	MOS-N-IGBT	L, Iso, 600V, 17A, 45W, 102/240ns int. Damper-Di (9,2A, <60ns)	17ic	Inr	-	data
IRG 4 IBC30 UD	MOS-N-IGBT	L, Iso, 600V, 17A, 45W, 61/171ns int. Damper-Di (8,5A, <60ns)	17ic	Inr	-	data
IRG 4 IBC30 W	MOS-N-IGBT	L, Iso, 600V, 17A, 45W, 41/166ns	17ic	Inr	-	data
IRG 4 P254 S	MOS-N-IGBT	L, <10kHz, 250V, 98A, 200W, 84/780ns	16id	Inr	-	data
IRG 4 PC30 F	MOS-N-IGBT	L, 1...5kHz, 600V, 31A, 100W; 36/380ns	16id	Inr	GN 6030C	data
IRG 4 PC30 FD	MOS-N-IGBT	=IRG 4PC30F: 68/390ns, int. Damper-Di	16id		BUP 602D, MGW 20N60D	data
IRG 4 PC30 K	MOS-N-IGBT	L, 600V, 28A, 100W, 54/250ns	16id	Inr	GN 6030C	data
IRG 4 PC30 KD	MOS-N-IGBT	=IRG 4PC30K: 102/240ns, int. Damper-Di	16id		BUP 602D, MGW 20N60D	data
IRG 4 PC30 S	MOS-N-IGBT	L, 600V, 34A, 100W, 40/930ns	16id	Inr	-	data
IRG 4 PC30 U	MOS-N-IGBT	L, 600V, 23A, 100W, 33/320ns	16id	Inr	GN 6030C, GT 25J101	data
IRG 4 PC30 UD	MOS-N-IGBT	=IRG 4PC30U: 61/171ns, int. Damper-Di	16id		BUP 602D, MGW 20N60D	data
IRG 4 PC30 W	MOS-N-IGBT	L, 600V, 23A, 100W, 41/166ns	16id	Inr	GN 6030C, GT 25J101	data
IRG 4 PC40 F	MOS-N-IGBT	L, 1...5kHz, 600V, 49A, 160W, 44/410ns	16id	Inr	MGW 30N60	data
IRG 4 PC40 FD	MOS-N-IGBT	=IRG 4PC40F: 95/400ns, int. Damper-Di	16id		-	data
IRG 4 PC40 K	MOS-N-IGBT	L, >5kHz, 600V, 42A, 160W, 48/340ns	16id	Inr	MGW 30N60	data
IRG 4 PC40 KD	MOS-N-IGBT	=IRG 4PC40K: 86/210ns, int. Damper-Di	16id		BUP 603D	data

Type	Device	Short Description	Fig.	Manu	Comparision Types	More at
IRG 4 PC40 S	MOS-N-IGBT	L, 600V, 60A, 160W, 40/103ns	16id	Inr	BUP 604	data
IRG 4 PC40 U	MOS-N-IGBT	L, 600V, 40A, 160W, 53/230ns	16id	Inr	MGW 30N60	data
IRG 4 PC40 UD	MOS-N-IGBT	=IRG 4PC40U: 111/190ns, int. Damper-Di	16id		BUP 603D	data
IRG 4 PC40 W	MOS-N-IGBT	L, 600V, 40A, 160W, 49/174ns	16id	Inr	MGW 30N60	data
IRG 4 PC50 F	MOS-N-IGBT	L, 1..5kHz, 600V, 70A, 200W, 56/370ns	16id	Inr	BUP 604	data
IRG 4 PC50 FD	MOS-N-IGBT	=IRG 4PC50F: 80/380ns, int. Damper-Di	16id		-	data
IRG 4 PC50 K	MOS-N-IGBT	L, >5kHz, 600V, 52A, 200W, 72/239ns	16id	Inr	MGW 30N60	data
IRG 4 PC50 KD	MOS-N-IGBT	=IRG 4PC50K: 112/245ns, int. Damper-Di	16id		-	data
IRG 4 PC50 S	MOS-N-IGBT	L, 600V, 70A, 200W, 63/105ns	16id	Inr	BUP 604	data
IRG 4 PC50 U	MOS-N-IGBT	L, 8..40kHz, 600V, 55A, 200W, 52/258ns	16id	Inr	MGW 30N60	data
IRG 4 PC50 UD	MOS-N-IGBT	=IRG 4PC50U: 71/214ns, int. Damper-Di	16id		-	data
IRG 4 PC50 W	MOS-N-IGBT	L, SMPS, 600V, 55A, 200W, 90/177ns	16id	Inr	MGW 30N60	data
IRG 4 PF50 W	MOS-N-IGBT	L, SMPS, 900V, 51A, 200W, 55/260ns	16id	Inr	BUP 314	data
IRG 4 PF50 WD	MOS-N-IGBT	=IRG 4PF50W: 121/260ns, int. Damper-Di	16id			data
IRG 4 PH20 K	MOS-N-IGBT	L, 1200V, 11A, 60W, 49/363ns	16id	Inr	GN 12015C, GT 15Q101	data
IRG 4 PH20 KD	MOS-N-IGBT	=IRG 4PH20K: 80/350ns, int. Damper-Di	16id		-	data
IRG 4 PH30 K	MOS-N-IGBT	L, 1200V, 20A, 100W, 51/310ns	16id	Inr	MGW 12N120	data
IRG 4 PH30 KD	MOS-N-IGBT	=IRG 4PH30K: 123/310ns, int. Damper-Di	16id		MGW 12N120D	data
IRG 4 PH40 K	MOS-N-IGBT	L, 1200V, 30A, 160W, 52/350ns	16id	Inr	BUP 307	data
IRG 4 PH40 KD	MOS-N-IGBT	=IRG 4PH40K: 81/316ns, int. Damper-Di	16id		BUP 307D, BUP 313D	data
IRG 4 PH40 U	MOS-N-IGBT	L, <40kHz, 1200V, 30A, 160W, 45/280ns	16id	Inr	BUP 307	data
IRG 4 PH40 UD	MOS-N-IGBT	=IRG 4PH40U: 96/280ns, int. Damper-Di	16id		BUP 307D, BUP 313D	data
IRG 4 PH50 K	MOS-N-IGBT	L, 1200V, 45A, 200W, 63/330ns	16id	Inr	BUP 314	data
IRG 4 PH50 KD	MOS-N-IGBT	=IRG 4PH50K: 187/340ns, int. Damper-Di	16id		BUP 314D	data
IRG 4 PH50 S	MOS-N-IGBT	L, 1200V, 57A, 200W, 61/127ns	16id	Inr	BUP 314	data
IRG 4 PH50 U	MOS-N-IGBT	L, <40kHz, 1200V, 45A, 200W, 50/490ns	16id	Inr	BUP 314	data
IRG 4 PH50 UD	MOS-N-IGBT	=IRG 4PH50U: 71/290ns, int. Damper-Di	16id		BUP 314D	data
IRG 4 PSC71 K	MOS-N-IGBT	L, 600V, 85A, 350W, 88/340ns	30id	Inr	-	data
IRG 4 PSC71 KD	MOS-N-IGBT	=IRG 4PSC71K: 189/379ns, int.Damper-Di	30id		-	data
IRG 4 PSC71 U	MOS-N-IGBT	L, 8..40kHz, 600V, 85A, 350W, 84/142ns	30id	Inr	-	data
IRG 4 PSC71 UD	MOS-N-IGBT	=IRG 4PSC71U: 184/355ns, int.Damper-Di	30id		-	data
IRG 4 PSH71 K	MOS-N-IGBT	L, 1200V, 78A, 350W, 83/380ns	30id	Inr	-	data
IRG 4 PSH71 KD	MOS-N-IGBT	=IRG 4PSH71K: 151/360ns, int.Damper-Di	30id		-	data
IRG 4 RC10 K	MOS-N-IGBT	L, >5kHz, 600V, 9A, 38W, 35/241ns	30id	Inr	-	data
IRG 4 RC10 KD	MOS-N-IGBT	=IRG 4RC10K: 77/237ns, int. Damper-Di	30id		-	data
IRG 4 RC10 S	MOS-N-IGBT	L, 600V, 14A, 38W, 53/1340ns	30id	Inr	-	data
IRG 4 RC10 SD	MOS-N-IGBT	=IRG 4RC10S: 108/1535ns, Damper-Di	30id		-	data
IRG 4 RC10 U	MOS-N-IGBT	L, 8..40kHz, 600V, 8,5A, 38W, 30/197ns	30id	Inr	-	data
IRG 4 RC10 UD	MOS-N-IGBT	=IRG 4RC10U: 56/227ns, int. Damper-Di	30id		-	data
IRG 4 ZC70 UD	MOS-N-IGBT	SMD, 600V, 100A, 350W, 112/360ns int. Damper-Di (50A, <140ns)		Inr		data
IRG 4 ZC71 KD	MOS-N-IGBT	SMD, 600V, 100A, 350W, 203/608ns int. Damper-Di (50A, <140ns)		Inr		data
IRG 4 ZH50.KD	MOS-N-IGBT	SMD, 1200V, 54A, 210W, 153/350ns int. Damper-Di (16A, <135ns)		Inr		data
IRG 4 ZH70 UD	MOS-N-IGBT	SMD, 1200V, 78A, 350W, 128/460ns int. Damper-Di (42A, <160ns)		Inr		data
IRG 4 ZH71 KD	MOS-N-IGBT	SMD, 1200V, 78A, 350W, 125/435ns int. Damper-Di (42A, <160ns)		Inr		data
IRGB 420 U	MOS-N-IGBT	L, >5kHz, 500V, 14A, 60W, 39/168ns	17id	Inr	BUK 854-500IS, GN6015A	data
IRGB 420 UD2	MOS-N-IGBT	=IRGB 420U: 109/250ns, int. Damper-Di	17id		-	data
IRGB 430 U	MOS-N-IGBT	L, >5kHz, 500V, 14A, 60W, 39/168ns	17id	Inr	BUP 401	data
IRGB 430 UD2	MOS-N-IGBT	=IRGB 430U: 145/220ns, int. Damper-Di	17id		-	data
IRGB 440 U	MOS-N-IGBT	L, >5kHz, 500V, 40A, 160W, 40/156ns	17id	Inr	BUP 403	data
IRGBC 20 F	MOS-N-IGBT	L, 1..10kHz, 600V, 16A, 60W, 37/470ns	17id	Inr	BUP 400	data
IRGBC 20 FD2	MOS-N-IGBT	=IRGBC 20F: -/-ns, int. Damper-Di	17id		BUP 400D	data
IRGBC 20 K	MOS-N-IGBT	L, >5kHz, 600V, 10A, 60W, 47/178ns	17id	Inr	GN 6010A	data
IRGBC 20 KD2	MOS-N-IGBT	=IRGBC 20K: 97/190ns, int. Damper-Di	17id		BUP 410D	data
IRGBC 20 K-S,KD2-S	MOS-N-IGBT	=IRGBC 20K,KD2:	30id		GT 15J103	data
IRGBC 20 M	MOS-N-IGBT	L, 1..10kHz, 600V, 13A, 60W, 51/550ns	17id	Inr	GN 6015A	data
IRGBC 20 MD2	MOS-N-IGBT	=IRGBC 20M: 106/599ns, int. Damper-Di	17id		BUP 410D	data
IRGBC 20 M-S,MD2-S	MOS-N-IGBT	=IRGBC 20M,MD2:	30id		GT 15J103	data
IRGBC 20 S	MOS-N-IGBT	L, ..400Hz, 600V, 19A, 60W, 47/1730ns	17id	Inr	BUP 400	data
IRGBC 20 SD2	MOS-N-IGBT	=IRGBC 20S: 141/1730ns, int. Damper-Di	17id		BUP 400D	data
IRGBC 20 U	MOS-N-IGBT	L, >5kHz, 600V, 13A, 60W, 34/162ns	17id	Inr	GN 6015A	data
IRGBC 20 UD2	MOS-N-IGBT	=IRGBC 20U: 89/195ns, int. Damper-Di	17id		BUP 400D	data
IRGBC 30 F	MOS-N-IGBT	L, 1..10kHz, 600V, 31A, 100W, 46/510ns	17id	Inr	BUP 402	data
IRGBC 30 FD2	MOS-N-IGBT	=IRGBC 30F: 147/520ns, int. Damper-Di	17id		-	data
IRGBC 30 K	MOS-N-IGBT	L, >5kHz, 600V, 23A, 100W, 54/184ns	17id	Inr	BUP 401	data
IRGBC 30 KD2	MOS-N-IGBT	=IRGBC 30K: 187/230ns, int. Damper-Di	17id		-	data
IRGBC 30 K-S,KD2-S	MOS-N-IGBT	=IRGBC 30K,KD2:	30id		-	data
IRGBC 30 M	MOS-N-IGBT	L, 1..10kHz, 600V, 26A, 100W, 62/590ns	17id	Inr	BUP 401	data
IRGBC 30 MD2	MOS-N-IGBT	=IRGBC 30M: 460/640ns, int. Damper-Di	17id		-	data
IRGBC 30 M-S,MD2-S	MOS-N-IGBT	=IRGBC 30M,MD2:	30id		-.	data
IRGBC 30 S	MOS-N-IGBT	L, ..400Hz, 600V, 34A, 100W, 47/1730ns	17id	Inr	BUP 402	data
IRGBC 30 U	MOS-N-IGBT	L, >5kHz, 600V, 23A, 100W, 39/185ns	17id	Inr	BUP 401	data
IRGBC 30 UD2	MOS-N-IGBT	=IRGBC 30U: 112/380ns, int. Damper-Di	17id		-	data
IRGBC 40 F	MOS-N-IGBT	L, 1..10kHz, 600V, 49A, 160W, 63/470ns	17id	Inr	-	data

Type	Device	Short Description	Fig.	Manu	Comparision Types	More at
IRGBC 40 K	MOS-N-IGBT	L, >5kHz, 600V, 42A, 160W, 62/290ns	17id	Inr	BUP 403	data
IRGBC 40 K-S	MOS-N-IGBT	=IRGBC 40K:	30id		-	data
IRGBC 40 M	MOS-N-IGBT	L, 1...10kHz, 600V, 40A, 160W, 63/470ns	17id	Inr	BUP 403	data
IRGBC 40 M-S	MOS-N-IGBT	=IRGBC 40M:	30id		-	data
IRGBC 40 S	MOS-N-IGBT	L, ...400Hz, 600V, 50A, 160W, 78/1720ns	17id	Inr	-	data
IRGBC 40 U	MOS-N-IGBT	L, >5kHz, 600V, 40A, 160W, 46/139ns	17id	Inr	BUP 403	data
IRGBF 20 F	MOS-N-IGBT	L, 1...10kHz, 900V, 9A, 60W, 41/290ns	17id	Inr	BUP 202	data
IRGBF 30 F	MOS-N-IGBT	L, 1...10kHz, 900V, 20A, 100W, 37/300ns	17id	Inr	BUP 203	data
IRGCC 20...E	MOS-N-IGBT	=IRGBC 20...: Die in Wafer Form	5" Wafer	Inr	-	data
IRGCC 30...E	MOS-N-IGBT	=IRGBC 30...: Die in Wafer Form	5" Wafer	Inr	-	data
IRGCC 40...E	MOS-N-IGBT	=IRGBC 40...: Die in Wafer Form	5" Wafer	Inr	-	data
IRGCC 50...E	MOS-N-IGBT	=IRGBC 50...: Die in Wafer Form	5" Wafer	Inr	-	data
IRGCH 20...E	MOS-N-IGBT	=IRGPH 20...: Die in Wafer Form	5" Wafer	Inr	-	data
IRGCH 40...E	MOS-N-IGBT	=IRGPH 40...: Die in Wafer Form	5" Wafer	Inr	-	data
IRGCH 50...E	MOS-N-IGBT	=IRGPH 50...: Die in Wafer Form	5" Wafer	Inr	-	data
IRGCH 70...E	MOS-N-IGBT	=IRGPH 70...: Die in Wafer Form	5" Wafer	Inr	-	data
IRGIH 50 F	MOS-N-IGBT	L, Iso, 1200V, 25A, 200W, <94/810ns		Inr	-	data
IRGKIN 50 M12	MOS-N-IGBT	Power Module		Inr	-	data
IRGNI 200 F06	MOS-N-IGBT	Power Module		Inr	-	data
IRGP 420 U	MOS-N-IGBT	L, >5kHz, 500V, 14A, 60W, 39/168ns	16id	Inr	GT 15J101	data
IRGP 430 U	MOS-N-IGBT	L, >5kHz, 500V, 25A, 100W, 40/157ns	16id	Inr	GN 6030C1	data
IRGP 430 UD2	MOS-N-IGBT	=IRGP 430U: 145/220ns, int. Damper-Di	16id		MGW 20N60D	data
IRGP 440 U	MOS-N-IGBT	L, >5kHz, 500V, 40A, 160W, 40/156ns	16id	Inr	MGW 30N60	data
IRGP 440 UD2	MOS-N-IGBT	=IRGP 440U: 148/290ns, int. Damper-Di	16id		BUP 603D	data
IRGP 450 U	MOS-N-IGBT	L, >5kHz, 500V, 59A, 200W, 59/201ns	16id	Inr	BUP 604	data
IRGP 450 UD2	MOS-N-IGBT	=IRGP 450U: 59/201ns, int. Damper-Di	16id		-	data
IRGPC 20 F	MOS-N-IGBT	L, 1...10kHz, 600V, 16A, 60W, 37/470ns	16id	Inr	GT 15J101	data
IRGPC 20 K	MOS-N-IGBT	L, >5kHz, 600V, 10A, 60W, 47/178ns	16id	Inr	GT 15J101	data
IRGPC 20 KD2	MOS-N-IGBT	=IRGPC 20K: 97/190ns, int. Damper-Di	16id		-	data
IRGPC 20 M	MOS-N-IGBT	L, 1...10kHz, 600V, 13A, 60W, 51/550ns	16id	Inr	GT 15J101	data
IRGPC 20 MD2	MOS-N-IGBT	=IRGPC 20M: 106/590ns, int. Damper-Di	16id		-	data
IRGPC 20 U	MOS-N-IGBT	L, >5kHz, 600V, 13A, 60W, 34/162ns	16id	Inr	GT 15J101	data
IRGPC 30 F	MOS-N-IGBT	L, 1...10kHz, 600V, 31A, 100W, 46/510ns	16id	Inr	GN 6030C	data
IRGPC 30 FD2	MOS-N-IGBT	=IRGPC 30F: 147/520ns, int. Damper-Di	16id		BUP 602D, MGW 20N60D	data
IRGPC 30 K	MOS-N-IGBT	L, >5kHz, 600V, 23A, 100W, 54/184ns	16id	Inr	GT 25J101	data
IRGPC 30 KD2	MOS-N-IGBT	=IRGPC 30K: 187/204ns, int. Damper-Di	16id		BUP 602D, MGW 20N60D	data
IRGPC 30 M	MOS-N-IGBT	L, 1...10kHz, 600V, 26A, 100W, 62/590ns	16id	Inr	GT 25J101	data
IRGPC 30 MD2	MOS-N-IGBT	=IRGPC 30M: 198/640ns, int. Damper-Di	16id		BUP 602D, MGW 20N60D	data
IRGPC 30 S	MOS-N-IGBT	L, ...400Hz, 600V, 34A, 100W, 58/1540ns	16id	Inr	GN 6030C	data
IRGPC 30 U	MOS-N-IGBT	L, >5kHz, 600V, 23A, 100W, 39/185ns	16id	Inr	GT 25J101	data
IRGPC 30 UD2	MOS-N-IGBT	=IRGPC 30U: 123/310ns, int. Damper-Di	16id		BUP 602D, MGW 20N60D	data
IRGPC 40 F	MOS-N-IGBT	L, 1...10kHz, 600V, 49A, 160W, 62/470ns	16id	Inr	BUP 604	data
IRGPC 40 FD2	MOS-N-IGBT	=IRGPC 40F: 147/530ns, int. Damper-Di	16id		-	data
IRGPC 40 K	MOS-N-IGBT	L, >5kHz, 600V, 42A, 160W, 62/610ns	16id	Inr	MGW 30N60	data
IRGPC 40 KD2	MOS-N-IGBT	=IRGPC 40K: 162/310ns, int. Damper-Di	16id		BUP 306D	data
IRGPC 40 M	MOS-N-IGBT	L, 1...10kHz, 600V, 40A, 160W, 63/470ns	16id	Inr	MGW 30N60	data
IRGPC 40 MD2	MOS-N-IGBT	=IRGPC 40M: 63/470ns, int. Damper-Di	16id		BUP 603D	data
IRGPC 40 S	MOS-N-IGBT	L, ...400Hz, 600V, 50A, 160W, 78/1720ns	16id	Inr	BUP 604	data
IRGPC 40 U	MOS-N-IGBT	L, >5kHz, 600V, 40A, 160W, 46/139ns	16id	Inr	MGW 30N60	data
IRGPC 40 UD2	MOS-N-IGBT	=IRGPC 40U: 117/280ns, int. Damper-Di	16id		BUP 603D	data
IRGPC 50 F	MOS-N-IGBT	L, 1...10kHz, 600V, 70A, 200W, 74/480ns	16id	Inr	BUP 604	data
IRGPC 50 FD2	MOS-N-IGBT	=IRGPC 50F: 180/690ns, int. Damper-Di	16id		-	data
IRGPC 50 KD2	MOS-N-IGBT	L, >5kHz, 600V, 52A, 200W, 118/350ns int. Damper-Di (25A, <75ns)	16id	Inr	-	data
IRGPC 50 M	MOS-N-IGBT	L, 1...10kHz, 600V, 60A, 200W, 68/430ns	16id	Inr	BUP 604	data
IRGPC 50 MD2	MOS-N-IGBT	=IRGPC 50M: 188/605ns, int. Damper-Di	16id		-	data
IRGPC 50 S	MOS-N-IGBT	L, ...400Hz, 600V, 70A, 200W,111/1700ns	16id	Inr	BUP 604	data
IRGPC 50 UD2	MOS-N-IGBT	L, >5kHz, 600V, 55A, 200W, 144/360ns int. Damper-Di (25A, <75ns)	16id	Inr	-	data
IRGPC 60 K	MOS-N-IGBT	L, >5kHz, 600V, 70A, 280W, 66/240ns	16id	Inr	BUP 604	data
IRGPC 60 M	MOS-N-IGBT	L, 1...10kHz, 600V, 70A, 280W, 81/590ns	16id	Inr	BUP 604	data
IRGPF 20 F	MOS-N-IGBT	L, 1...10kHz, 600V, 9A, 60W, 41/290ns	16id	Inr	BUP 302	data
IRGPF 30 F	MOS-N-IGBT	L, 1...10kHz, 600V, 20A, 100W, 37/300ns	16id	Inr	BUP 303	data
IRGPF 40 F	MOS-N-IGBT	L, 1...10kHz, 600V, 31A, 160W, 40/280ns	16id	Inr	BUP 304	data
IRGPF 50 F	MOS-N-IGBT	L, 1...10kHz, 600V, 51A, 200W, 54/330ns	16id	Inr	BUP 314	data
IRGPH 20 M	MOS-N-IGBT	L,1...10kHz, 1200V, 4,5A, 60W, 39/473ns	16id	Inr	GT 8Q101	data
IRGPH 20 S	MOS-N-IGBT	L, 1...10kHz, 1200V, 10A, 60W, 39/473ns	16id	Inr	GN 12015C, GT 15Q101	data
IRGPH 30 MD2	MOS-N-IGBT	L,1...10kHz, 1200V, 15A, 100W, 12/410ns int. Damper-Di (6A, <80ns)	16id	Inr	GN 12015C, GT 15Q101	data
IRGPH 30 S	MOS-N-IGBT	L,...400Hz, 1200V, 22A, 100W, 50/1860ns	16id	Inr	MGW 20N120	data
IRGPH 40 F	MOS-N-IGBT	L,1...10kHz, 1200V, 29A, 160W, 50/460ns	16id	Inr	BUP 307	data
IRGPH 40 FD2	MOS-N-IGBT	=IRGPH 40F: 128/690ns, int. Damper-Di	16id		BUP 307D	data
IRGPH 40 K	MOS-N-IGBT	L, >5kHz, 1200V, 18A, 160W, 36/350ns	16id	Inr	MGW 12N120	data
IRGPH 40 KD2	MOS-N-IGBT	=IRGPH 40K: 77/392ns, int. Damper-Di	16id		MGW 12N120D	data
IRGPH 40 M	MOS-N-IGBT	L,1...10kHz, 1200V, 31A, 160W, 45/421ns	16id	Inr	BUP 307	data
IRGPH 40 MD2	MOS-N-IGBT	=IRGPH 40M: 138/1750ns, int. Damper-Di	16id		BUP 307D	data

Type	Device	Short Description	Fig.	Manu	Comparision Types	More at
IRGPH 40 S	MOS-N-IGBT	L,..400Hz, 1200V, 33A, 160W, 52/2080ns	16id	Inr	BUP 307	data
IRGPH 50 F	MOS-N-IGBT	L,1...10kHz, 1200V, 45A, 200W, 47/560ns	16id	Inr	BUP 314	data
IRGPH 50 FD2	MOS-N-IGBT	=IRGPH 50F: 152/680ns, int. Damper-Di	16id	Inr	BUP 314D	data
IRGPH 50 KD2	MOS-N-IGBT	L, >5kHz, 1200V, 36A, 200W, 138/390ns int. Damper-Di (16A, <135ns)	16id	Inr	BUP 307	data
IRGPH 50 M	MOS-N-IGBT	L,1...10kHz, 1200V, 42A, 200W, 74/470ns	16id	Inr	BUP 314	data
IRGPH 50 MD2	MOS-N-IGBT	=IRGPH 50M: 240/980ns, int. Damper-Di	16id	Inr	BUP 314D	data
IRGPH 50 S	MOS-N-IGBT	..400Hz, 1200V, 57A, 200W, 139/1980ns	16id	Inr	BUP 314	data
IRGS 14 B40L	MOS-N-IGBT	L, Voltage Clamped, 400V, 18A	30id	Inr	-	data
IRH 150	MOS-N-FET-e	V-MOS, RadH, 100V, 38A, 150W, <55mΩ	23a§	Inr	-	data
IRH 254	MOS-N-FET-e	V-MOS, RadH, 250V, 19A, 150W, <0,19Ω	23a§	Inr	-	data
IRH 450	MOS-N-FET-e	V-MOS, RadH, 500V, 10A, 150W, <0,6Ω	23a§	Inr	-	data
IRL80A	IRED	950nm ø1.52mm I:>0.4mW/Sr Vf:<1.5V		Osr,Sie		data pdf pinout
IRL81A	IRED	880nm ø1.52mm I:>1mW/Sr Vf:1.5V		Osr,Sie		data pdf pinout
IRL 510	MOS-N-FET-e	V-MOS, LogL, 100V, 5,6A, 43W, <0,54Ω	17c§	Inr	BUK 552-100, 2SK991	data
IRL 510 A	MOS-N-FET-e	V-MOS, LogL, 100V, 5,6A, 37W, <0,44Ω	17c§	Sam	BUK 552-100, 2SK991	data
IRL 510 L,S	MOS-N-FET-e	=IRL 510..:	30c§		IRLW 510A	data
IRL 520	MOS-N-FET-e	V-MOS, LogL, 100V, 9,2A, 60W, <0,27Ω	17c§	Inr	2SK1300	data
IRL 520 A	MOS-N-FET-e	V-MOS, LogL, 100V, 9,2A, 49W, <0,22Ω	17c§	Sam	2SK1300	data
IRL 520 N	MOS-N-FET-e	V-MOS, LogL, 100V, 10A, 47W, <0,22Ω	17c§	Inr	2SK1300	data
IRL 520 NL,L,NS,S	MOS-N-FET-e	=IRL 520..:	30c§		IRLW 520A	data
IRL 530	MOS-N-FET-e	V-MOS, LogL, 100V, 15A, 88W, <0,16(9A)	17c§	Inr	2SK1301, 2SK1559, 2SK1561	data
IRL 530 A	MOS-N-FET-e	V-MOS, LogL, 100V, 14A, 62W, <0,12(7A)	17c§	Sam	2SK1301, 2SK1559, 2SK1561	data
IRL 530 N	MOS-N-FET-e	V-MOS, LogL, 100V, 15A, 63W, <0,12(9A)	17c§	Inr	2SK1301, 2SK1559, 2SK1561	data
IRL 530 NL,L,NS,L	MOS-N-FET-e	=IRL 530..:	30c§		IRLW 530A, 2SK1559S, 2SK1561S, 2SK1908	data
IRL 540	MOS-N-FET-e	V-MOS, LogL, 100V, 28A, 150W, <77mΩ	17c§	Inr	-	data
IRL 540 A	MOS-N-FET-e	V-MOS, LogL, 100V, 28A, 121W, <58mΩ	17c§	Sam	-	data
IRL 540 N	MOS-N-FET-e	V-MOS, LogL, 100V, 30A, 94W, <53mΩ	17c§	Inr	-	data
IRL 540 NL,L,NS,L	MOS-N-FET-e	=IRL 540..:	30c§		IRLW 540A, 2SK1623, 2SK1909, 2SK1928	data
IRL 610 A	MOS-N-FET-e	V-MOS, LogL, 200V, 3,3A, 33W, <1,5Ω	17c§	Sam	-	data
IRL 620	MOS-N-FET-e	V-MOS, LogL, 200V, 5,2A, 50W, <0,8Ω	17c§	Inr	BUK 554-200	data
IRL 620 A	MOS-N-FET-e	V-MOS, LogL, 200V, 5A, 39W, <0,8Ω	17c§	Sam	BUK 554-200	data
IRL 620 S	MOS-N-FET-e	=IRL 620:	30c§		IRLW 620A	data
IRL 630	MOS-N-FET-e	V-MOS, LogL, 200V, 9A, 74W, <0,4Ω	17c§	Inr	BUK 554-200	data
IRL 630 A	MOS-N-FET-e	V-MOS, LogL, 200V, 9A, 69W, <0,4Ω	17c§	Sam	BUK 554-200	data
IRL 630 S	MOS-N-FET-e	=IRL 630:	30c§		IRLW 630A	data
IRL 640	MOS-N-FET-e	V-MOS, LogL, 200V, 17A, 125W, <0,18Ω	17c§	Inr	BUK 555-200	data
IRL 640 A	MOS-N-FET-e	V-MOS, LogL, 200V, 18A, 110W, <0,18Ω	17c§	Sam	BUK 555-200	data
IRL 640 S	MOS-N-FET-e	=IRL 640:	30c§		IRLW 640A	data
IRL 1004	MOS-N-FET-e	V-MOS, LogL, 40V, 130A, 200W, <9mΩ	17c§	Inr	-	data
IRL 1004 L,S	MOS-N-FET-e	V-MOS, LogL, 40V, 110A, 450W, <6,5mΩ	30c§	Inr	-	data
IRL 1104 L,S	MOS-N-FET-e	V-MOS, LogL, 40V, 104A, 167W, <12mΩ	30c§	Inr	-	data
IRL 2203 N	MOS-N-FET-e	V-MOS,LogL, 30V, 116A, 170W, <7mΩ(60A)	17c§	Inr	-	data
IRL 2203 NL,NS	MOS-N-FET-e	=IRL 2203N:	30c§	Inr	-	data
IRL 2310	MOS-N-FET-e	V-MOS,LogL, 100V, 40A, 170W, <50mΩ(20A)	17c§	Inr	-	data
IRL 2505	MOS-N-FET-e	V-MOS,LogL, 55V, 104A, 200W, <8mΩ(54A)	17c§	Inr	IRL 3705	data
IRL 2505 L,S	MOS-N-FET-e	=IRL 2205:	30c§		IRL 3705NL,NS	data
IRL 2703	MOS-N-FET-e	V-MOS, LogL, 30V, 24A, 45W, <40mΩ(14A)	17c§	Inr	2SK1115, 2SK1287, 2SK1910	data
IRL 2703 L,S	MOS-N-FET-e	=IRL 2703:	30c§		2SK1622, 2SK1918, 2SK2311	data
IRL 2910	MOS-N-FET-e	V-MOS,LogL, 100V, 48A, 150W,<30mΩ(29A)	17c§	Inr	-	data
IRL 2910 L,S	MOS-N-FET-e	=IRL 2910:	30c§		-	data
IRL 3102	MOS-N-FET-e	V-MOS, LogL, 20V, 61A, 89W, <15mΩ(37A)	17c§	Inr	IRL 2505, IRL 3705N	data
IRL 3102 S	MOS-N-FET-e	=IRL 3102:	30c§		-	data
IRL 3103	MOS-N-FET-e	V-MOS,LogL, 30V, 64A, 110W, <14mΩ(34A)	17c§	Inr	IRL 2505, IRL 3705N	data
IRL 3103 L,S	MOS-N-FET-e	=IRL 3103:	30c§		PHB 80N06LT, 2SK3070	data
IRL 3103 D1	MOS-N-FET-e	V-MOS, LogL, int. Schottky-Di (S→D) 30V, 54A, 70W, <14mΩ(32A), 219/74ns	17c§	Inr	-	data
IRL 3103 D1S	MOS-N-FET-e	=IRL 3103D1:	30c§			data
IRL 3202	MOS-N-FET-e	V-MOS, LogL, 20V, 48A, 69W, <19mΩ(39A)	17c§	Inr	BUZ 100L, MTP 52N06VL, 2SK2049, 2SK2499	data
IRL 3202 S	MOS-N-FET-e	=IRL 3202:	30c§		IRL 3102S	data
IRL 3215	MOS-N-FET-e	V-MOS, LogL, 150V, 12A, 80W, <184mΩ	17c§	Inr	BUK 555-200, IRL 640(A), 2SK1561	data
IRL 3302	MOS-N-FET-e	V-MOS, LogL, 20V, 39A, 57W, <23mΩ(23A)	17c§	Inr	BUZ 102AL, 2SK1542, 2SK1911, 2SK2513	data
IRL 3302 S	MOS-N-FET-e	=IRL 3302:	30c§		IRL 3202S	data
IRL 3303	MOS-N-FET-e	V-MOS, LogL, 30V, 38A, 68W, <26mΩ(20A)	17c§	Inr	BUZ 102AL, 2SK1542, 2SK1911, 2SK2513	data
IRL 3303 L,S	MOS-N-FET-e	=IRL 3303:	30c§		2SK1919, 2SK2164, 2SK2266, 2SK2376, ++	data
IRL 3402	MOS-N-FET-e	V-MOS,LogL, 20V, 85A, 110W, <10mΩ(51A)	17c§	Inr	IRL 2505, IRL 3705N	data
IRL 3402 S	MOS-N-FET-e	=IRL 3402:	30c§		IRL 2505L,S, IRL 3705NL,NS	data
IRL 3502	MOS-N-FET-e	V-MOS,LogL, 20V, 110A, 140W, <8mΩ(64A)	17c§	Inr	IRL 2505, IRL 3803	data
IRL 3502 S	MOS-N-FET-e	=IRL 3502:	30c§		IRL 1004L,S, IRL 2505L,S	data
IRL 3705 N	MOS-N-FET-e	V-MOS,LogL, 55V, 89A, 170W, <10mΩ(46A)	17c§	Inr	IRL 2505	data
IRL 3705 NL,NS	MOS-N-FET-e	=IRL 3705N:	30c§		IRL 2505L,S	data
IRL 3803	MOS-N-FET-e	V-MOS,LogL, 30V, 140A, 200W, <6mΩ(71A)	17c§	Inr	-	data
IRL 3803 L,S	MOS-N-FET-e	=IRL 3803:	30c§		-	data
IRL 5602 S	MOS-P-FET-e	V-MOS,LogL, 20V, 24A, 75W, <42mΩ(12A)	30c§	Inr	-	data
IRL 6903	MOS-P-FET-e	V-MOS,LogL, 30V, 105A, 200W,<20mΩ(46A)	17c§	Inr	-	data

Type	Device	Short Description	Fig.	Manu	Comparision Types	More at
IRL 6903 L,S	MOS-P-FET-e	=IRL 6903:	30c§		-	data
IRLBA 1304/P	MOS-N-FET-e	V-MOS, LogL, 40V, 95A, 300W, <6,5mΩ	30c§	Inr	-	data
IRLBA 3803/P	MOS-N-FET-e	V-MOS, LogL, 30V, 179A, 270W, <9mΩ	30c§	Inr	-	data
IRLBL 10 N60A	MOS-N-FET-e	V-MOS, 600V, 11A, 180W, <0,61Ω(6,5A)	30c§	Inr	-	data
IRLBL 1304	MOS-N-FET-e	V-MOS, LogL, 40V, 95A, 300W, <6,5mΩ	30c§	Inr	-	data
IRLC 1304	MOS-N-FET-e	=IRBL 1304: Die in Wafer Form	6" Wafer	Inr	-	data
IRLC 9024 N	MOS-N-FET-e	=IRLR 9024N: Die in Wafer Form	6" Wafer	Inr	-	data
IRLD 014	MOS-N-FET-e	V-MOS, LogL, 60V, 1,7A, 1,3W, <0,2Ω	4-DIP	Inr	-	data
IRLD 024	MOS-N-FET-e	V-MOS, LogL, 60V, 2,5A, 1,3W, <0,1Ω	4-DIP	Inr	-	data
IRLD 110	MOS-N-FET-e	V-MOS, LogL, 100V, 1A, 1,3W, <0,54Ω	4-DIP	Inr	-	data
IRLD 120	MOS-N-FET-e	V-MOS, LogL, 100V, 1,3A, 1,3W, <0,27Ω	4-DIP	Inr	-	data
IRLF 230	MOS-N-FET-e	V-MOS, LogL, 200V, 5,2A, 25W, <0,4Ω	2a§	Inr	(IRLW 620A)[4]	data
IRLI 510 A	MOS-N-FET-e	=IRLW 510A	30c§	Sam	→IRLW 510A	data
IRLI 520 A	MOS-N-FET-e	=IRLW 520A	30c§	Sam	→IRLW 520A	data
IRLI 520 G	MOS-N-FET-e	V-MOS, LogL, 100V, 7,2A, 37W, <0,27Ω	17c	Inr	BUK 543-100, 2SK1261, 2SK1556	data
IRLI 520 N	MOS-N-FET-e	V-MOS, LogL, 100V, 7,7A, 27W, <0,22Ω	17c	Inr	→IRLI 520G	data
IRLI 530 A	MOS-N-FET-e	=IRLW 530A	30c§	Sam	→IRLW 530A	data
IRLI 530 G	MOS-N-FET-e	V-MOS, LogL, 100V, 9,7A, 42W, <0,16Ω	17c	Inr	BUK 545-100, 2SK1035, 2SK1305, 2SK1904++	data
IRLI 530 N	MOS-N-FET-e	V-MOS, LogL, 100V, 11A, 33W, <0,12Ω	17c	Inr	→IRLI 530G	data
IRLI 540 A	MOS-N-FET-e	=IRLW 540A	30c§	Sam	→IRLW 540A	data
IRLI 540 G	MOS-N-FET-e	V-MOS, LogL, 100V, 17A, 48W, <77mΩ	17c	Inr	2SK1266, 2SK1307, 2SK1348...49, 2SK1906++	data
IRLI 540 N	MOS-N-FET-e	V-MOS, LogL, 100V, 20A, 42W, <53mΩ	17c	Inr	→IRLI 540G	data
IRLI 610 A	MOS-N-FET-e	=IRLW 610A	30c§	Sam	→IRLW 610A	data
IRLI 620 A	MOS-N-FET-e	=IRLW 620A	30c§	Sam	→IRLW 620A	data
IRLI 620 G	MOS-N-FET-e	V-MOS, LogL, 200V, 4A, 30W, <0,8Ω	17c	Inr	IRLS 620A, 2SK2108	data
IRLI 630 A	MOS-N-FET-e	=IRLW 630A	30c§	Sam	→IRLW 630A	data
IRLI 630 G	MOS-N-FET-e	V-MOS, LogL, 200V, 6,2A, 35W, <0,4Ω	17c	Inr	BUK 545-200, IRLS 630A, 2SK2161	data
IRLI 640 A	MOS-N-FET-e	=IRLW 640A	30c§	Sam	→IRLW 640A	data
IRLI 640 G	MOS-N-FET-e	V-MOS, LogL, 200V, 9,9A, 40W, <0,18Ω	17c	Inr	BUK 545-200, IRLS 640A, 2SK2161	data
IRLI 2203 G	MOS-N-FET-e	V-MOS, LogL, 30V, 52A, 48W, <15mΩ(26A)	17c	Inr	IRLI 3803	data
IRLI 2203 N	MOS-N-FET-e	V-MOS, LogL, 30V, 68A, 58W, <7mΩ(37A)	17c	Inr	IRLI 3803	data
IRLI 2505	MOS-N-FET-e	V-MOS, LogL, 55V, 58A, 63W, <8mΩ(31A)	17c	Inr		data
IRLI 2910	MOS-N-FET-e	V-MOS,LogL, 100V, 27A, 48W, <30mΩ(16A)	17c	Inr	2SK1262, 2SK2577, 2SK2954, 2SK3033	data
IRLI 3103	MOS-N-FET-e	V-MOS, LogL, 30V, 38A, 38W, <19mΩ(19A)	17c	Inr	IRLI 2203G, IRLI 3705	data
IRLI 3303	MOS-N-FET-e	V-MOS, LogL, 30V, 25A, 31W, <40mΩ(13A)	17c	Inr	BUK 545-50, 2SK2289, 2SK2507	data
IRLI 3705(N)	MOS-N-FET-e	V-MOS, LogL, 55V, 52A, 58W, <10mΩ(28A)	17c	Inr	IRLI 2505	data
IRLI 3803	MOS-N-FET-e	V-MOS, LogL, 30V, 76A, 63W, <6mΩ(40A)	17c	Inr		data
IRLIZ 14 A	MOS-N-FET-e	=IRLWZ 14A	30c§	Sam	→IRLWZ 14A	data
IRLIZ 14 G	MOS-N-FET-e	V-MOS, LogL, 60V, 8A, 27W, <0,2Ω(4,8A)	17c	Inr	BUK 542-60, 2SK1256, 2SK1344, 2SK1895,++	data
IRLIZ 24 A	MOS-N-FET-e	=IRLWZ 24A	30c§	Sam	→IRLWZ 24A	data
IRLIZ 24 G	MOS-N-FET-e	V-MOS, LogL, 60V, 14A, 37W, <0,1Ω(8,4A)	17c	Inr	2SK1286, 2SK1288, 2SK1306, 2SK1419, ++	data
IRLIZ 24 N	MOS-N-FET-e	V-MOS, LogL, 55V, 15A, 29W, <60mΩ(8,4A)	17c	Inr	2SK1286, 2SK1288, 2SK1306, 2SK1419, ++	data
IRLIZ 34 A	MOS-N-FET-e	=IRLWZ 34A	30c§	Sam	→IRLWZ 34A	data
IRLIZ 34 G	MOS-N-FET-e	V-MOS, LogL, 60V, 20A, 42W, <50mΩ(12A)	17c	Inr	2SK943, 2SK1095, 2SK1345, 2SK1951, ++	data
IRLIZ 34 N	MOS-N-FET-e	V-MOS, LogL, 55V, 22A, 37W, <35mΩ(12A)	17c	Inr	2SK943, 2SK1095, 2SK1345, 2SK1951, ++	data
IRLIZ 44 A	MOS-N-FET-e	=IRLWZ 44A	30c§	Sam	→IRLWZ 44A	data
IRLIZ 44 G	MOS-N-FET-e	V-MOS, LogL, 60V, 30A, 48W, <28mΩ(18A)	17c	Inr	2SK1257, 2SK1262, 2SK1653, 2SK1952, ++	data
IRLIZ 44 N	MOS-N-FET-e	V-MOS, LogL, 55V, 30A, 45W, <22mΩ(17A)	17c	Inr	2SK1257, 2SK1262, 2SK1653, 2SK1952, ++	data
IRLL 014	MOS-N-FET-e	SMD, V-MOS, LogL, 60V, 2,7A, <0,2Ω	48c§	Inr	BUK 583-60A	data
IRLL 014 N	MOS-N-FET-e	SMD, V-MOS, LogL, 55V, 2,8A, <0,14Ω	48c§	Inr	BUK 538-60A	data
IRLL 024 N	MOS-N-FET-e	SMD, V-MOS, LogL, 55V, 4,4A, <80mΩ	48c§	Inr	-	data
IRLL 110	MOS-N-FET-e	SMD, V-MOS, LogL, 100V, 1,5A, <0,54Ω	48c§	Inr	BUK 582-100	data
IRLL 2703	MOS-N-FET-e	SMD, V-MOS, LogL, 30V, 5,5A, <0,06Ω	48c§	Inr	BUK 9880-55, BUK 98150-55, IRLL 3303	data
IRLL 2705	MOS-N-FET-e	SMD, V-MOS, LogL, 55V, 5,2A, <0,04Ω	48c§	Inr	BUK 9880-55, BUK 98150-55	data
IRLL 3303	MOS-N-FET-e	SMD, V-MOS, LogL, 30V, 6,5A, <31mΩ	48c§	Inr	-	data
IRLL 6702	MOS-P-FET-e	SMD, V-MOS, LogL, 20V, 2,3A, <0,2Ω	48c§	Inr	-	data
IRLL 6802	MOS-P-FET-e	SMD, V-MOS, LogL, 20V, 5,6A, <0,05Ω	48c§	Inr	-	data
IRLM 014 A	MOS-N-FET-e	SMD, V-MOS, LogL, 60V, 2,8A, <155mΩ	48c§	Sam	BUK 483-60A	data
IRLM 110 A	MOS-N-FET-e	SMD, V-MOS, LogL, 100V, 1,5A, <0,44Ω	48c§	Sam	BUK 582-100A	data
IRLM 120 A	MOS-N-FET-e	SMD, V-MOS, LogL, 100V, 2,3A, <0,22Ω	48c§	Sam		data
IRLM 210 A	MOS-N-FET-e	SMD, V-MOS, LogL, 200V, 0,77A, <1,5Ω	48c§	Sam	-	data
IRLM 220 A	MOS-N-FET-e	SMD, V-MOS, LogL, 200V, 1,13A, <0,8Ω	48c§	Sam	-	data
IRLML 2402	MOS-N-FET-e	SMD, V-MOS, LogL, 20V, 1,2A, <0,25Ω	35a	Inr	-	data
IRLML 2803	MOS-N-FET-e	SMD, V-MOS, LogL, 30V, 1,2A, <0,25Ω	35a	Inr	-	data
IRLML 5103	MOS-P-FET-e	SMD, V-MOS, LogL, 30V, 0,76A, <0,6Ω	35a	Inr	-	data
IRLML 6302	MOS-P-FET-e	SMD, V-MOS, LogL, 20V, 0,78A, <0,6Ω	35a	Inr	-	data
IRLMS 1503	MOS-N-FET-e	SMD, V-MOS, LogL, 30V, 3,2A, <0,1Ω	46bn4	Inr	-	data
IRLMS 1902	MOS-N-FET-e	SMD, V-MOS, LogL, 20V, 3,2A, <0,1Ω	46bn4	Inr	-	data

Type	Device	Short Description	Fig.	Manu	Comparision Types	More at
IRLMS 5703	MOS-P-FET-e	SMD, V-MOS, LogL, 30V, 2,3A, <0,2Ω	46bn4	Inr	-	data
IRLMS 6702	MOS-P-FET-e	SMD, V-MOS, LogL, 20V, 2,3A, <0,2Ω	46bn4	Inr	-	data
IRLP 2505	MOS-N-FET-e	V-MOS,LogL, 55V, 90A, 150W, <10mΩ(54A)	16c§	Inr	-	data
IRLP 3803	MOS-N-FET-e	V-MOS,LogL,30V, 120A, 150W, <11mΩ(59A)	16c§	Inr	-	data
IRLR 014	MOS-N-FET-e	V-MOS, LogL, 60V, 7,7A, 25W, <0,2Ω	30c§	Inr	2SK1472, 2SK1748, 2SK2415	data
IRLR 014 A	MOS-N-FET-e	V-MOS, LogL, 60V, 8,2A, 18W, <155mΩ	30c§	Sam	2SK1472, 2SK1748, 2SK2415	data
IRLR 024	MOS-N-FET-e	V-MOS, LogL, 60V, 14A, 42W, <0,1Ω	30c§	Inr	MTD 15N06VL, 2SK2334	data
IRLR 024 A	MOS-N-FET-e	V-MOS, LogL, 60V, 15A, 30W, <75mΩ	30c§	Sam	MTD 15N06VL, 2SK2334	data
IRLR 024 N	MOS-N-FET-e	V-MOS, LogL, 55V, 17A, 38W, <65mΩ(10A)	30c§	Inr	MTD 15N06VL, 2SK2334	data
IRLR 034 A	MOS-N-FET-e	V-MOS, LogL, 60V, 23A, 42W, <46mΩ	30c§	Sam	MTD 20N06HDL, 2SK2334	data
IRLR 110	MOS-N-FET-e	V-MOS, LogL, 100V, 4,3A, 25W, <0,54Ω	30c§	Inr	2SK1474, 2SK1804, 2SK2016, 2SK2399	data
IRLR 110 A	MOS-N-FET-e	V-MOS, LogL, 100V, 4,7A, 22W, <0,44Ω	30c§	Inr, Sam	2SK1474, 2SK1804, 2SK2016, 2SK2399	data
IRLR 120	MOS-N-FET-e	V-MOS, LogL, 100V, 7,7A, 42W, <0,27Ω	30c§	Inr	2SK1475, 2SK1557, 2SK1560	data
IRLR 120 A	MOS-N-FET-e	V-MOS, LogL, 100V, 8,4A, 35W, <0,22Ω	30c§	Inr, Sam	2SK1475, 2SK1557, 2SK1560	data
IRLR 120 N	MOS-N-FET-e	V-MOS, LogL, 100V, 11A, 39W, <0,225Ω	30c§	Inr	(2SK3037)[7]	data
IRLR 130 A	MOS-N-FET-e	V-MOS, LogL, 100V, 13A, 46W, <0,12Ω	30c§	Inr, Sam		data
IRLR 210 A	MOS-N-FET-e	V-MOS, LogL, 200V, 2,7A, 21W, <1,5Ω	30c§	Inr, Sam	2SK1335	data
IRLR 220 A	MOS-N-FET-e	V-MOS, LogL, 200V, 4,6A, 33W, <0,8Ω	30c§	Inr, Sam	-	data
IRLR 230 A	MOS-N-FET-e	V-MOS, LogL, 200V, 7,5A, 48W, <0,4Ω	30c§	Inr, Sam	-	data
IRLR 2703	MOS-N-FET-e	V-MOS, LogL, 30V, 22A, 38W, <45mΩ(13A)	30c§	Inr	MTD 20N06HDL, 2SK2614	data
IRLR 2705	MOS-N-FET-e	V-MOS, LogL, 55V, 27A, 57W, <40mΩ(14A)	30c§	Inr	-	data
IRLR 2905	MOS-N-FET-e	V-MOS, LogL, 55V, 41A, 89W, <27mΩ(22A)	30c§	Inr	-	data
IRLR 3103	MOS-N-FET-e	V-MOS, LogL, 30V, 52A, 89W, <19mΩ(28A)	30c§	Inr	-	data
IRLR 3303	MOS-N-FET-e	V-MOS, LogL, 30V, 33A, 46W, <31mΩ(18A)	30c§	Inr	-	data
IRLR 3410	MOS-N-FET-e	V-MOS,LogL, 100V, 15A, 52W, <0,12Ω(9A)	30c§	Inr	(2SK3031)[7]	data
IRLS 510 A	MOS-N-FET-e	V-MOS, LogL, 100V, 4,5A, 23W, <0,44Ω	17c	Sam	BUK 542-100, 2SK992, 2SK1260, 2SK2581	data
IRLS 520 A	MOS-N-FET-e	V-MOS, LogL, 100V, 7,2A, 30W, <0,22Ω	17c	Sam	BUK 543-100, 2SK1261, 2SK1556	data
IRLS 530 A	MOS-N-FET-e	V-MOS, LogL, 100V, 10,7A, 36W, <0,12Ω	17c	Sam	BUK 545-100, 2SK1035, 2SK1305, 2SK1904	data
IRLS 540 A	MOS-N-FET-e	V-MOS, LogL, 100V, 17A, 44W, <58mΩ	17c	Sam	2SK1034, 2SK1288, 2SK1306, 2SK1558, ++	data
IRLS 610 A	MOS-N-FET-e	V-MOS, LogL, 200V, 2,5A, 19W, <1,5Ω	17c	Sam	-	data
IRLS 620 A	MOS-N-FET-e	V-MOS, LogL, 200V, 4,1A, 26W, <0,8Ω	17c	Sam	2SK2108	data
IRLS 630 A	MOS-N-FET-e	V-MOS, LogL, 200V, 6,5A, 36W, <0,4Ω	17c	Sam	BUK 545-200, 2SK2160...61	data
IRLS 640 A	MOS-N-FET-e	V-MOS, LogL, 200V, 9,8A, 40W, <0,18Ω	17c	Sam	BUK 545-200, 2SK2161	data
IRLSZ 14 A	MOS-N-FET-e	V-MOS, LogL, 60V, 8A, 21W, <155mΩ(4A)	17c	Sam	BUK 542-60, 2SK994	data
IRLSZ 24 A	MOS-N-FET-e	V-MOS, LogL, 60V, 14A, 32W, <75mΩ(7A)	17c	Sam	2SK1094, 2SK1286, 2SK1896	data
IRLSZ 34 A	MOS-N-FET-e	V-MOS, LogL, 60V, 20A, 40W, <46mΩ(10A)	17c	Sam	2SK1345, 2SK1882, 2SK1897, 2SK2232, ++	data
IRLSZ 44 A	MOS-N-FET-e	V-MOS, LogL, 60V, 30A, 49W, <25mΩ(15A)	17c	Sam	2SK1262, 2SK2289, 2SK2466	data
IRLU 014...3410	MOS-N-FET-e	=IRLR 014...3410:	30c§	Inr	→IRLR 014...3410	data
IRLW 510 A	MOS-N-FET-e	V-MOS, LogL, 100V, 5,6A, 37W, <0,44Ω	30c§	Sam	-	data
IRLW 520 A	MOS-N-FET-e	V-MOS, LogL, 100V, 9,2A, 49W, <0,22Ω	30c§	Sam	-	data
IRLW 530 A	MOS-N-FET-e	V-MOS, LogL, 100V, 14A, 62W, <0,12Ω	30c§	Sam	2SK1559S, 2SK1561S, 2SK1908	data
IRLW 540 A	MOS-N-FET-e	V-MOS, LogL, 100V, 28A, 121W, <58mΩ	30c§	Sam	-	data
IRLW 610 A	MOS-N-FET-e	V-MOS, LogL, 200V, 3,3A, 33W, <1,5Ω	30c§	Sam	-	data
IRLW 620 A	MOS-N-FET-e	V-MOS, LogL, 200V, 5A, 39W, <0,8Ω	30c§	Sam	-	data
IRLW 630 A	MOS-N-FET-e	V-MOS, LogL, 200V, 9A, 69W, <0,4Ω	30c§	Sam	-	data
IRLW 640 A	MOS-N-FET-e	V-MOS, LogL, 200V, 18A, 110W, <0,18Ω	30c§	Sam	-	data
IRLWZ 14 A	MOS-N-FET-e	V-MOS, LogL, 60V, 10A, 31W, <155mΩ(5A)	30c§	Sam	2SK1907, 2SK2041	data
IRLWZ 24 A	MOS-N-FET-e	V-MOS, LogL, 60V, 17A, 45W,<75mΩ(8,5A)	30c§	Sam	2SK1622, 2SK1918, 2SK1967, 2SK2311, ++	data
IRLWZ 34 A	MOS-N-FET-e	V-MOS, LogL, 60V, 30A, 83W, <46mΩ(15A)	30c§	Sam	MTB 35N06ZL, 2SK1900, 2SK2411-Z, ++	data
IRLWZ 44 A	MOS-N-FET-e	V-MOS, LogL, 60V, 50A, 125W,<25mΩ(25A)	30c§	Sam	MTB 52N06ZL, 2SK2499-Z, 2SK2553, ++	data
IRLZ 14	MOS-N-FET-e	V-MOS, LogL, 60V, 10A, 43W, <0,2Ω(6A)	17c§	Inr	2SK970, 2SK1114, 2SK1300, 2SK2175, ++	data
IRLZ 14 A	MOS-N-FET-e	V-MOS, LogL, 60V, 10A, 34W, <155mΩ(5A)	17c§	Sam	2SK970, 2SK1114, 2SK1300, 2SK2175, ++	data
IRLZ 14 L,S	MOS-N-FET-e	=IRLZ 14:	30c§		2SK2041, 2SK2282, 2SK3082	data
IRLZ 24	MOS-N-FET-e	V-MOS, LogL, 60V, 17A, 60W, <0,1Ω(10A)	17c§	Inr	BUK 552-60, 2SK971...972, 2SK1115, ++	data
IRLZ 24 A	MOS-N-FET-e	V-MOS, LogL, 60V, 17A, 47W,<75mΩ(8,5A)	17c§	Sam	BUK 552-60, 2SK971...972, 2SK1115, ++	data
IRLZ 24 N	MOS-N-FET-e	V-MOS, LogL, 55V, 18A, 45W, <60mΩ(11A)	17c§	Inr	BUK 552-60, 2SK971...972, 2SK1115, ++	data
IRLZ 24 NL,NS,L,S	MOS-N-FET-e	=IRLZ 24...:	30c§		2SK1622, 2SK1918, 2SK2175, 2SK2311	data
IRLZ 34	MOS-N-FET-e	V-MOS, LogL, 60V, 30A, 88W, <50mΩ(18A)	17c§	Inr	BUK 555-60, 2SK1291, 2SK1296, 2SK2411,++	data
IRLZ 34 A	MOS-N-FET-e	V-MOS, LogL, 60V, 30A, 91W, <46mΩ(15A)	17c§	Sam	BUK 555-60, 2SK1291, 2SK1296, 2SK2411,++	data
IRLZ 34 N	MOS-N-FET-e	V-MOS, LogL, 55V, 30A, 68W, <35mΩ(16A)	17c§	Inr	BUK 555-60, 2SK1291, 2SK1296, 2SK2411,++	data
IRLZ 34 NL,L,NS,S	MOS-N-FET-e	=IRLZ 34...:	30c§		MTB 30N06VL, 2SK1900, 2SK2411Z, 2SK2288Z	data
IRLZ 44	MOS-N-FET-e	V-MOS, LogL, 60V, 50A, 150W,<28mΩ(31A)	17c§	Inr	BUK 556-60, MTP 52N06VL, 2SK2049	data
IRLZ 44 A	MOS-N-FET-e	V-MOS, LogL, 60V, 50A, 137W,<25mΩ(25A)	17c§	Sam	BUK 556-60, MTP 52N06VL, 2SK2049	data
IRLZ 44 N	MOS-N-FET-e	V-MOS, LogL, 55V, 47A, 110W,<22mΩ(25A)	17c§	Inr	BUK 556-60, MTP 52N06VL, 2SK2049	data
IRLZ 44 NL,NS,S	MOS-N-FET-e	=IRLZ 44:	30c§		MTB 52N06VL, 2SK2499-Z, 2SK2553	data
IRMCF 311	IC	Dual Channel Sensorless Motor Control IC,Appliances	DIP	Inr	-	pdf pinout
IRMCF 312	IC	Dual Channel Sensorless Motor Control IC,Appliances	DIP	Inr	-	pdf pinout
IRMCF 341	IC	Sensorless Motor Control IC,Appliances	DIP	Inr	-	pdf pinout
IRMCF 343	IC	Sensorless Motor Control IC,Appliances	DIP	Inr	-	pdf pinout
IRMCF 371	IC	Sensorless Motor Control IC,Appliances	DIP	Inr	-	pdf pinout
IRMCK 203	IC	High Performance Sensorless Motion Control IC	DIP	Inr	-	pdf pinout
IRMCK 311	IC	Dual-CH Sensorless Motor Control IC ,Appliances	DIP	Inr	-	pdf pinout
IRMCK 312	IC	Dual- Ch Sensorless Motor Control IC, Appliances	DIP	Inr	-	pdf pinout

Type	Device	Short Description	Fig.	Manu	Comparision Types	More at
IRMCK 343	IC	Sensorless Motor Control iC, Appliances	DIP	Inr	-	pdf pinout
IRMCK 371	IC	Sensorless Motor Control IC, Appliances	DIP	Inr	-	pdf pinout
IRMCT 3UF1	IC	HiRel digital sensorless motor drive module, dig. speed/torque controller	FLP	Inr	-	pdf pinout
IRPT 2060A	LIN-IC	Power Modul,3 phase motor drives, 180V-240V AC, 2.2kW		Inr		pdf pinout
IRS 2001	CMOS/LIN-IC	200V High & Low Side Drv	DIC	Inr		pdf pinout
IRS 2001 S	CMOS/LIN-IC	200V High & Low Side Drv	MDIP	Inr		pdf pinout
IRS 2003	CMOS/LIN-IC	200V Half-Bridge Drv	DIC	Inr		pdf pinout
IRS 2003 S	CMOS/LIN-IC	200V Half-Bridge Drv	MDIP	Inr		pdf pinout
IRS 2004	CMOS/LIN-IC	200V Half-Bridge Drv	DIC	Inr		pdf pinout
IRS 2004 S	CMOS/LIN-IC	200V Half-Bridge Drv	MDIP	Inr		pdf pinout
IRS 2011	CMOS/LIN-IC	200V High & Low Side Drv	DIC	Inr		pdf pinout
IRS 2011 S	CMOS/LIN-IC	200V High & Low Side Drv	MDIP	Inr		pdf pinout
IRS 2092 PbF	IC	Protected Digital Audio Amp	DIC	Inr		pdf pinout
IRS 2092 SPbF	IC	Protected Digital Audio Amp	MDIP	Inr		pdf pinout
IRS 2101	CMOS/LIN-IC	600V High & Low Side Drv	DIC	Inr		pdf pinout
IRS 2101 S	CMOS/LIN-IC	600V High & Low Side Drv	MDIP	Inr		pdf pinout
IRS 2103	CMOS/LIN-IC	600V Half-Bridge Driver	DIC	Inr		pdf pinout
IRS 2103 S	CMOS/LIN-IC	600V Half-Bridge Driver	DIC	Inr		pdf pinout
IRS 2104	CMOS/LIN-IC	600V Half-Bridge Drv	DIC	Inr		pdf pinout
IRS 2104 S	CMOS/LIN-IC	600V Half-Bridge Drv	MDIP	Inr		pdf pinout
IRS 2106	CMOS/LIN-IC	600V High & Low Side Drv	DIC	Inr		pdf pinout
IRS 2106 S	CMOS/LIN-IC	600V High & Low Side Drv	MDIP	Inr		pdf pinout
IRS 2108	CMOS/LIN-IC	600V Half-Bridge Drv	DIC	Inr		pdf pinout
IRS 2108 S	CMOS/LIN-IC	600V Half-Bridge Drv	MDIP	Inr		pdf pinout
IRS 2109	CMOS/LIN-IC	600V Half Bridge Drv	DIC	Inr		pdf pinout
IRS 2109 S	CMOS/LIN-IC	600V Half Bridge Drv	MDIP	Inr		pdf pinout
IRS 2110 -1	CMOS/LIN-IC	600V High & Low Side Drv	DIC	Inr		pdf pinout
IRS 2110 -2	CMOS/LIN-IC	600V High & Low Side Drv	DIC	Inr		pdf pinout
IRS 2110	CMOS/LIN-IC	600V High & Low Side Drv	DIC	Inr		pdf pinout
IRS 2110 S	CMOS/LIN-IC	600V High & Low Side Drv	MDIP	Inr		pdf pinout
IRS 2111	CMOS/LIN-IC	600V Half-Bridge Drv	DIC	Inr		pdf pinout
IRS 2111 S	CMOS/LIN-IC	600V Half-Bridge Drv	MDIP	Inr		pdf pinout
IRS 2112 -1	CMOS/LIN-IC	600V High & Low Side Drv	DIC	Inr		pdf pinout
IRS 2112 -2	CMOS/LIN-IC	600V High & Low Side Drv	DIC	Inr		pdf pinout
IRS 2112	CMOS/LIN-IC	600V High & Low Side Drv	DIC	Inr		pdf pinout
IRS 2112 S	CMOS/LIN-IC	600V High & Low Side Drv	MDIP	Inr		pdf pinout
IRS 2113 -1	CMOS/LIN-IC	500V High & Low Side Drv	DIC	Inr		pdf pinout
IRS 2113 -2	CMOS/LIN-IC	500V High & Low Side Drv	DIC	Inr		pdf pinout
IRS 2113	CMOS/LIN-IC	500V High & Low Side Drv	DIC	Inr		pdf pinout
IRS 2113 S	CMOS/LIN-IC	500V High & Low Side Drv	MDIP	Inr		pdf pinout
IRS 2117	CMOS/LIN-IC	+600V Single High Side Drv	DIC	Inr		pdf pinout
IRS 2117 S	CMOS/LIN-IC	+600V Single High Side Drv	MDIP	Inr		pdf pinout
IRS 2118	CMOS/LIN-IC	+600V Single High Side Drv	DIC	Inr		pdf pinout
IRS 2118 S	CMOS/LIN-IC	+600V Single High Side Drv	MDIP	Inr		pdf pinout
IRS 2127 (1)	CMOS/LIN-IC	600V Current Sensing 1Ch. Drv	DIC	Inr		pdf pinout
IRS 2127 (1)S	CMOS/LIN-IC	600V Current Sensing 1Ch. Drv	MDIP	Inr		pdf pinout
IRS 2128 (1)	CMOS/LIN-IC	600V Current Sensing 1Ch. Drv	DIC	Inr		pdf pinout
IRS 2128 (1)S	CMOS/LIN-IC	600V Current Sensing 1Ch. Drv	MDIP	Inr		pdf pinout
IRS 2153 (1)DPbF	CMOS-IC	Self-oscillating Half-Bridge Driver IC	DIC	Inr		pdf pinout
IRS 2153 (1)DSPbF	CMOS-IC	Self-oscillating Half-Bridge Driver IC	MDIP	Inr		pdf pinout
IRS 2158 D	IC	600V ballast control IC	DIC	Inr		pdf pinout
IRS 2158 DS	IC	600V ballast control IC	MDIP	Inr		pdf pinout
IRS 2166 DPbF	IC	Pfc + 600V Ballast Cntrl. IC	DIC	Inr		pdf pinout
IRS 2166 DSPbF	IC	Pfc + 600V Ballast Cntrl. IC	MDIP	Inr		pdf pinout
IRS 2168 DPbF	IC	Pfc, 600V Ballast Cntrl. IC	DIC	Inr		pdf pinout
IRS 2168 DSPbF	IC	Pfc, 600V Ballast Cntrl. IC	MDIP	Inr		pdf pinout
IRS 2181	CMOS/LIN-IC	+600V High & Low Side Drv	DIC	Inr		pdf pinout
IRS 2181 S	CMOS/LIN-IC	+600V High & Low Side Drv	MDIP	Inr		pdf pinout
IRS 2183	CMOS/LIN-IC	+600V Half-Bridge Drv	DIC	Inr		pdf pinout
IRS 2183 S	CMOS/LIN-IC	+600V Half-Bridge Drv	MDIP	Inr		pdf pinout
IRS 2184 PbF	CMOS-IC	Half-bridge Driver	DIC	Inr		pdf pinout
IRS 2184 SPbF	CMOS-IC	Half-bridge Driver	MDIP	Inr		pdf pinout
IRS 2186	CMOS-IC	High and low Side Driver	DIP	Inr		pdf pinout
IRS 2186 S	CMOS-IC	High and low Side Driver	DIP	Inr		pdf pinout
IRS 2308	IC	Half-bridge Driver	DIP	Inr		pdf pinout
IRS 2308 S	IC	Half-bridge Driver	DIP	Inr		pdf pinout
IRS 2336	MOS-IC	High Voltage 3 Phase gate Driver IC	DIP	Inr		pdf pinout
IRS 2336 D	MOS-IC	High Voltage 3 Phase gate Driver IC	DIP	Inr		pdf pinout
IRS 2336 DJ	MOS-IC	High Voltage 3 Phase gate Driver IC	DIP	Inr		pdf pinout
IRS 2336 DM	MOS-IC	High Voltage 3 Phase gate Driver IC	DIP	Inr		pdf pinout
IRS 2336 DS	MOS-IC	High Voltage 3 Phase gate Driver IC	DIP	Inr		pdf pinout
IRS 2336 J	MOS-IC	High Voltage 3 Phase gate Driver IC	DIP	Inr		pdf pinout
IRS 2336 S	MOS-IC	High Voltage 3 Phase gate Driver IC	DIP	Inr		pdf pinout
IRS 2453 (1)D	IC	Self-oscillating Full-bridge Driver	DIC	Inr		pdf pinout
IRS 2453 (1)DS	IC	Self-oscillating Full-bridge Driver	MDIP	Inr		pdf pinout
IRS 2530 D	IC	DIM8TM DIMMING Ballast Control IC	MDIP	Inr		pdf pinout
IRS 2530 DS	IC	DIM8TM DIMMING Ballast Control IC	MDIP	Inr		pdf pinout
IRS 2540 PbF	IC	LED Buck Reg. Cnttrl. IC	DIC	Inr		pdf pinout
IRS 2540 SPbF	IC	LED Buck Reg. Cnttrl. IC	MDIP	Inr		pdf pinout
IRS 2541 PbF	IC	LED Buck Reg. Cnttrl. IC	DIC	Inr		pdf pinout
IRS 2541 SPbF	IC	LED Buck Reg. Cnttrl. IC	MDIP	Inr		pdf pinout
IRS 2552 D	IC	Ccfl/eefl Ballast Cntrl. IC	DIC	Inr		pdf pinout
IRS 2552 DS	IC	Ccfl/eefl Ballast Cntrl. IC	MDIP	Inr		pdf pinout

Type	Device	Short Description	Fig.	Manu	Comparision Types	More at
RS 2607 DS	CMOS/LIN-IC	600V High & Low Side Drv	MDIP	Inr	-	pdf pinout
RS 2608 DSPbF	CMOS-IC	600V Half-bridge Driver	MDIP	Inr	-	pdf pinout
RS 2609 DS	CMOS/LIN-IC	600V Half-Bridge Drv	MDIP	Inr	-	pdf pinout
RS 20124 SPbF	IC	Digital Audio Driver	MDIP	Inr	-	pdf pinout
RS 20954 SPbF	IC	Protected Digital Audio Driver	MDIP	Inr	-	pdf pinout
RS 20955 PbF	IC	Protected Digital Audio Driver	DIC	Inr	-	pdf pinout
RS 20955 SPbF	IC	Protected Digital Audio Driver	MDIP	Inr	-	pdf pinout
RS 20957 SPbF	IC	Protected Digital Audio Driver	MDIP	Inr	-	pdf pinout
RS 21064	CMOS/LIN-IC	600V High & Low Side Drv	DIC	Inr	-	pdf pinout
RS 21064 S	CMOS/LIN-IC	600V High & Low Side Drv	MDIP	Inr	-	pdf pinout
RS 21084	CMOS/LIN-IC	600V Half-Bridge Drv	DIC	Inr	-	pdf pinout
RS 21084 S	CMOS/LIN-IC	600V Half-Bridge Drv	MDIP	Inr	-	pdf pinout
RS 21091	CMOS/LIN-IC	600V Half-Bridge Drv	DIC	Inr	-	pdf pinout
RS 21091 S	CMOS/LIN-IC	600V Half-Bridge Drv	MDIP	Inr	-	pdf pinout
RS 21094	CMOS/LIN-IC	600V Half Bridge Drv	DIC	Inr	-	pdf pinout
RS 21094 S	CMOS/LIN-IC	600V Half Bridge Drv	MDIP	Inr	-	pdf pinout
RS 21171 S	CMOS/LIN-IC	+600V Single High Side Drv	MDIP	Inr	-	pdf pinout
RS 21571 DS	IC	Fully Integrated Ballast Cntrl. IC	MDIP	Inr	-	pdf pinout
RS 21814	CMOS/LIN-IC	+600V High & Low Side Drv	DIC	Inr	-	pdf pinout
RS 21814 S	CMOS/LIN-IC	+600V High & Low Side Drv	MDIP	Inr	-	pdf pinout
RS 21834	CMOS/LIN-IC	+600V Half-Bridge Drv	DIC	Inr	-	pdf pinout
RS 21834 S	CMOS/LIN-IC	+600V Half-Bridge Drv	MDIP	Inr	-	pdf pinout
RS 21844 PbF	CMOS-IC	Half-bridge Driver	DIC	Inr	-	pdf pinout
RS 21844 SPbF	CMOS-IC	Half-bridge Driver	MDIP	Inr	-	pdf pinout
RS 21850 S	CMOS-IC	Single High-Side Driver IC	MDIP	Inr	-	pdf pinout
RS 21853 SPBF	CMOS-IC	2x High Side Driver IC	MDIP	Inr	-	pdf pinout
RS 21864	CMOS-IC	High and low Side Driver	DIP	Inr	-	pdf pinout
RS 21864 S	CMOS-IC	High and low Side Driver	DIP	Inr	-	pdf pinout
RS 21952 SPBF	CMOS-IC	High Side & Dual low Side Driver IC	MDIP	Inr	-	pdf pinout
RS 21953 SPBF	CMOS-IC	High Side & Dual low Side Driver IC	TSOP	Inr	-	pdf pinout
RS 23364 D	MOS-IC	High Voltage 3 Phase gate Driver IC	DIP	Inr	-	pdf pinout
RS 23364 DJ	MOS-IC	High Voltage 3 Phase gate Driver IC	DIP	Inr	-	pdf pinout
RS 23364 DS	MOS-IC	High Voltage 3 Phase gate Driver IC	DIP	Inr	-	pdf pinout
RS 26302 DJ	CMOS/LIN-IC	600V Protected 3Ph Bridge 1 Gate Drv	PLCC	Inr	-	pdf pinout
RS 26310 DJ	CMOS/LIN-IC	600V Hi-volt 3Ph Gate Drv IC	PLCC	Inr	-	pdf pinout
RS 210614 S	CMOS/LIN-IC	600V High & Low Side Drv	MDIP	Inr	-	pdf pinout
RSF 3010	DMOS-IC	Protected Power Switch, 50V, 11A	17c	Inr	-	pdf
RSF 3011	DMOS-IC	Protected Power Switch, 50V, 5A, 200mΩ	17	Inr	-	pdf
RSF 3021	DMOS-IC	Protected Power Switch, 50V, 3A, 200mΩ	17	Inr	-	pdf
RSF 3031	DMOS-IC	Protected Power Sw., 50V, 1,8A, 200mΩ	17	Inr	-	pdf
RT 1250	CMOS-IC	IR-FB-sender/transmitter	24-DIP	Itt	-	
RT 1260	CMOS-IC	Ir-fb-sender/transmitter	24-DIP	Itt	-	
RU 3037 ACF	LIN-IC	8-PIN synchronous PWM Controller	20-DIP	Inr	-	pdf pinout
RU 3037 ACS	LIN-IC	8-PIN synchronous PWM Controller	20-DIP	Inr	-	pdf pinout
RU 3037 CF	LIN-IC	8-PIN synchronous PWM Controller	20-DIP	Inr	-	pdf pinout
RU 3037 CS	LIN-IC	8-PIN synchronous PWM Controller	20-DIP	Inr	-	pdf pinout
S	Si-P	=2SA1881 (Typ-Code/Stempel/marking)	35	Say	→2SA1881	data
S	Si-P	=2SB1220-S (Typ-Code/Stempel/marking)	35(2mm)	Mat	→2SB1220	data
S	Si-P	=2SB1463-S (Typ-Code/Stempel/marking)	35(1,6mm)	Mat	→2SB1463	data
S	Si-P	=2SB792-S (Typ-Code/Stempel/marking)	35	Mat	→2SB792	data
S	Si-N	=2SC3722K-S (Typ-Code/Stempel/marking)	35	Rhm	→2SC3722K	data
S1	OC	LED/NPN Viso:2500V CTR:>20% Ibs:16mA		Iso	IL1	data pinout
S1U60	PD	950nm Vb:4.7...5.3V Ib:2.8mA		Sha		data pdf pinout
S1U621	PD	950nm Vb:4.7...5.3V Ib:2.8mA		Sha		data pdf pinout
S5	OC	LED/NPN Viso:2500V CTR:>50% Ibs:16mA		Iso	IL1	data pinout
S 5	Si-P	=2SA1881-5 (Typ-Code/Stempel/marking)	35	Say	→2SA1881	data
S 6	Si-P	=2SA1881-6 (Typ-Code/Stempel/marking)	35	Say	→2SA1881	data
S 7	Si-P	=2SA1881-7 (Typ-Code/Stempel/marking)	35	Say	→2SA1881	data
S74	OC	LED/NPN Viso:1500V CTR:>2.5% Ibe:2mA		Iso	IL74	data pinout
IS 82C55A-5	µP-Periph	CMOS, Ppi, 3x8 Programmable I/O Pins, TTL Compatible, 5MHz, -40...+85°C	PLCC	Har	...82C55...	pdf pinout
IS 82C55A	µP-Periph	CMOS, Ppi, 3x8 Programmable I/O Pins, TTL Compatible, 8MHz, -40...+85°C	PLCC	Har	...82C55...	pdf pinout
IS201	OC	LED/NPN Viso:5000V CTR:>75% Ibs:10mA		Iso	IL201	data pdf pinout
IS202	OC	LED/NPN Viso:5000V CTR:125-250%		Iso	IL202	data pdf pinout
IS203	OC	LED/NPN Viso:5000V CTR:225-450%		Iso	IL203	data pdf pinout
IS431	PD	910nm Vb:5V Ib:2.3mA		Sha		data pinout
IS432	PD	910nm Vb:5V Ib:2.3mA		Sha		data pinout
IS435	PD	910nm Vb:5V Ib:2.5mA		Sha		data pinout
IS436	PD	910nm Vb:5V Ib:2.5mA		Sha		data pinout
IS437	PD	910nm Vb:5V Ib:2.5mA		Sha		data pinout
IS440	PD	900(940)nm Vb:4.5...16V Ib:3.5mA		Sha		data pinout
IS441	PD	900(940)nm Vb:4.5...16V Ib:3.5mA		Sha		data pinout
IS445	PD	700(830)nm Vb:<8V Ib:<1mA		Sha		data pinout
IS450	PD	900(940)nm Vb:4.5...9V Ib:<7mA		Sha		data pinout
IS455	2 PD	700(830)nm Vb:<8V Ib:<1mA		Sha		data pinout
IS456	PD	660(780)nm Vb:4.5...5.5V Ib:<6.6mA		Sha		data pinout
IS457	PD	660(780)nm Vb:4.5...5.5V Ib:<8.6mA		Sha		data pinout
IS471	PD	940nm Vb:4.5...16V Ib:<7mA		Sha		data pinout
IS485	PD	910nm Vb:5V Ib:<3.8mA		Sha		data pdf pinout
IS486	PD	910nm Vb:5V Ib:<3.8mA		Sha		data pdf pinout
IS487	PD	910nm Vb:5V Ib:<3.8mA		Sha		data pdf pinout
IS604	OC	bid. LED/NPN Viso:5000V CTR:>50%		Iso	IL250	data pinout
IS605	OC	LED/SCR Viso:5000V Ibs:11mA		Iso	IL400	data pinout
IS606	OC	LED/SCR Viso:5000V Ibs:14mA		Iso		data pinout

Type	Device	Short Description	Fig.	Manu	Comparision Types	More at
IS607	OC	LED/Triac Viso:4000V Ibs:10mA		Iso		data pdf pinout
IS608	OC	LED/Triac Viso:4000V Ibs:7mA		Iso		data pdf pinout
IS609	OC	LED/PD+IC Viso:5000V		Iso		data pinout
IS610	OC	LED/FET Viso:2500V Ibs:16mA		Iso		data pdf pinout
IS611	OC	LED/FET Viso:5000V Ibs:16mA		Iso		data pdf pinout
IS614	OC	LED/NPN Viso:4000V CTR:>15% Ibs:20mA		Iso		data pinout
IS615	OC	LED/NPN Viso:4000V CTR:>50% Ibs:20mA		Iso		data pinout
IS620	OC	LED/Triac Viso:7500V Ibs:30mA		Iso		data pdf pinout
IS621	OC	LED/Triac Viso:7500V Ibs:15mA		Iso		data pdf pinout
IS622	OC	LED/Triac Viso:7500V Ibs:10mA		Iso		data pdf pinout
IS623	OC	LED/Triac Viso:7500V Ibs:7mA		Iso		data pdf pinout
IS660	OC	LED/Dar Viso:5000V CTR:>1000%		Iso		data pinout
IS661	OC	LED/Dar Viso:5000V CTR:>1000%		Iso		data pinout
IS662	OC	LED/Dar Viso:5000V CTR:>1000%		Iso		data pinout
IS801	OC	LED/NPN Viso:2500V CTR:40-80%		Iso		data pinout
IS802	OC	LED/NPN Viso:2500V CTR:63-125%		Iso		data pinout
IS803	OC	LED/NPN Viso:2500V CTR:100-200%		Iso		data pinout
IS804	OC	LED/NPN Viso:2500V CTR:>20% Ibs:10mA		Iso		data pinout
IS805	OC	LED/NPN Viso:2500V CTR:>50% Ibs:10mA		Iso		data pinout
IS850	OC	LED/Dar Viso:2500V CTR:>500% Ibs:8mA		Iso		data pinout
IS851	OC	LED/Dar Viso:2500V CTR:>100% Ibs:8mA		Iso		data pinout
ISD 1700 E	-IC	Single-Chip Voice Record & Playback Devices	QIP	Wbd	-	pdf pinout
IS3020	OC	LED/Triac Viso:2500V Ibs:30mA		Iso		data pdf pinout
IS3021	OC	LED/Triac Viso:2500V Ibs:15mA		Iso		data pdf pinout
IS66WV25616BLL-55B	IC	4Mb, low Power pseudo Cmos static RAM	TSOP	Iss	-	pdf pinout
IS66WV25616BLL-55T	IC	4Mb, low Power pseudo Cmos static RAM	TSOP	Iss	-	pdf pinout
IS66WV25616ALL-70B	IC	4Mb, low Power pseudo Cmos static RAM	TSOP	Iss	-	pdf pinout
IS66WV25616ALL-70T	IC	4Mb, low Power pseudo Cmos static RAM	TSOP	Iss	-	pdf pinout
IS66WV51216BLL-55B	IC	8Mb, low Power pseudo Cmos static RAM	TSOP	Iss	-	pdf pinout
IS66WV51216BLL-55T	IC	8Mb, low Power pseudo Cmos static RAM	TSOP	Iss	-	pdf pinout
IS66WV51216ALL-70B	IC	8Mb, low Power pseudo Cmos static RAM	TSOP	Iss	-	pdf pinout
IS66WV51216ALL-70T	IC	8Mb, low Power pseudo Cmos static RAM	TSOP	Iss	-	pdf pinout
IS-	Si-N	=2SC3513 (Typ-Code/Stempel/marking)	35	Hit	→2SC3513	data
IS-	Si-N	=2SC4537 (Typ-Code/Stempel/marking)	35(2mm)	Hit	→2SC4537	data
ISB 9xxx...12xxx	DIG-IC	Gate Arrays		Tho	-	
ISB-A 27-0	MOS	N-channel Mosfet Array	GRID	Say	-	pdf pinout
ISB-A30-0	MOS	4 P-channel MOSFETs	GRID	Say	-	pdf pinout
ISB-E48-0	MOS	Charg. Circ. Volt. Sens.,3 MOSFETs	GRID	Say	-	pdf pinout
ISD1	DOC	LED/NPN Viso:2500V CTR:>20% Ibs:16mA		Iso	ILD1	data pinout
ISD5	DOC	LED/NPN Viso:2500V CTR:>50% Ibs:16mA		Iso	ILD1	data pinout
ISD74	DOC	LED/NPN Viso:1500V CTR:>2.5% Ibe:2mA		Iso	ILD74	data pinout
ISD201	DOC	LED/NPN Viso:5000V CTR:>75% Ibs:10mA		Iso		data pdf pinout
ISD202	DOC	LED/NPN Viso:5000V CTR:125-250%		Iso		data pdf pinout
ISD203	DOC	LED/NPN Viso:5000V CTR:225-450%		Iso		data pdf pinout
ISD 1110 P	CMOS-IC	Voice record / playback devices 10-Second duration	DIP	Wbd	-	pdf pinout
ISD 1110 S	CMOS-IC	Voice record / playback devices 10-Second duration	MDIP	Wbd	-	pdf pinout
ISD 1110 X	CMOS-IC	Voice record / playback devices 10-Second duration	CSP	Wbd	-	pdf pinout
ISD 1112 P	CMOS-IC	Voice record / playback devices 12-Second duration	DIP	Wbd	-	pdf pinout
ISD 1112 S	CMOS-IC	Voice record / playback devices 12-Second duration	MDIP	Wbd	-	pdf pinout
ISD 1112 X	CMOS-IC	Voice record / playback devices 12-Second duration	CSP	Wbd	-	pdf pinout
ISD 1210 P	CMOS-IC	Voice record / playback devices 10-Second duration	DIP	Wbd	-	pdf pinout
ISD 1210 S	CMOS-IC	Voice record / playback devices 10-Second duration	MDIP	Wbd	-	pdf pinout
ISD 1210 X	CMOS-IC	Voice record / playback devices 10-Second duration	CSP	Wbd	-	pdf pinout
ISD 1212 P	CMOS-IC	Voice record / playback devices 12-Second duration	DIP	Wbd	-	pdf pinout
ISD 1212 S	CMOS-IC	Voice record / playback devices 12-Second duration	MDIP	Wbd	-	pdf pinout
ISD 1212 X	CMOS-IC	Voice record / playback devices 12-Second duration	CSP	Wbd	-	pdf pinout
ISD 1416 P(I)	CMOS-IC	Voice record/playback devices 16-second DURATION	DIC	Wbd		pdf pinout
ISD 1416 S(I)	CMOS-IC	Voice record/playback devices 16-second DURATION	TSOP	Wbd		pdf pinout
ISD 1416 X	CMOS-IC	Voice record/playback devices 16-second DURATION	CSP	Wbd		pdf pinout
ISD 1420 P(I)	CMOS-IC	Voice record/playback devices 20-second DURATION	DIC	Wbd		pdf pinout
ISD 1420 S(I)	CMOS-IC	Voice record/playback devices 20-second DURATION	TSOP	Wbd		pdf pinout
ISD 1420 X	CMOS-IC	Voice record/playback devices 20-second DURATION	CSP	Wbd		pdf pinout
ISD 1610 P(I)	CMOS-IC	6.6-second duration Voice record/ playback device	DIC	Wbd		pdf pinout
ISD 1610 S(I)	CMOS-IC	6.6-second duration Voice record/ playback device	TSOP	Wbd		pdf pinout

Type	Device	Short Description	Fig.	Manu	Comparision Types	More at
SD 1610 X	CMOS-IC	6.6-second duration Voice record/ playback device	CSP	Wbd	-	pdf pinout
SD 1612 P(I)	CMOS-IC	8-second duration Voice record/ playback DEVICE	DIC	Wbd	-	pdf pinout
SD 1612 S(I)	CMOS-IC	8-second duration Voice record/ playback DEVICE	TSOP	Wbd	-	pdf pinout
SD 1612 X	CMOS-IC	8-second duration Voice record/ playback DEVICE	CSP	Wbd	-	pdf pinout
SD 1616 P(I)	CMOS-IC	10,6-second duration Voice record/ playback device	DIC	Wbd	-	pdf pinout
SD 1616 S(I)	CMOS-IC	10,6-second duration Voice record/ playback device	TSOP	Wbd	-	pdf pinout
SD 1616 X	CMOS-IC	10,6-second duration Voice record/ playback device	CSP	Wbd	-	pdf pinout
ISD 1620 P(I)	CMOS-IC	13,3-second duration Voice record/ playback device	DIC	Wbd	-	pdf pinout
ISD 1620 S(I)	CMOS-IC	13,3-second duration Voice record/ playback device	TSOP	Wbd	- -	pdf pinout
ISD 1620 X	CMOS-IC	13,3-second duration Voice record/ playback device	CSP	Wbd	-	pdf pinout
ISD 1806 P	CMOS-IC	Voice record/playback Device, 6- TO 12-second duration	DIC	Wbd	-	pdf pinout
ISD 1806 S	CMOS-IC	Voice record/playback Device, 6- TO 12-second duration	TSOP	Wbd	-	pdf pinout
ISD 1806 X	CMOS-IC	Voice record/playback Device, 6- TO 12-second duration	CSP	Wbd	-	pdf pinout
ISD 1810 P	CMOS-IC	Voice record/playback Device, 8- TO 16-second duration	DIC	Wbd	-	pdf pinout
ISD 1810 S	CMOS-IC	Voice record/playback Device, 8- TO 16-second duration	TSOP	Wbd	-	pdf pinout
ISD 1810 X	CMOS-IC	Voice record/playback Device, 8- TO 16-second duration	CSP	Wbd	-	pdf pinout
ISD 2532 E	CMOS-IC	Voice record/playback Device, 32-second DURATION	TSOP	Wbd	-	pdf pinout
ISD 2532 P	CMOS-IC	Voice record/playback Device, 32-second DURATION	DIC	Wbd	-	pdf pinout
ISD 2532 S	CMOS-IC	Voice record/playback Device, 32-second DURATION	MDIP	Wbd	-	pdf pinout
ISD 2532 X	CMOS-IC	Voice record/playback Device, 32-second DURATION	CSP	Wbd	-	pdf pinout
ISD 2540 E	CMOS-IC	Voice record/playback Device, 40-second DURATION	TSOP	Wbd	-	pdf pinout
ISD 2540 P	CMOS-IC	Voice record/playback Device, 40-second DURATION	DIC	Wbd	-	pdf pinout
ISD 2540 S	CMOS-IC	Voice record/playback Device, 40-second DURATION	MDIP	Wbd	-	pdf pinout
ISD 2540 X	CMOS-IC	Voice record/playback Device, 40-second DURATION	CSP	Wbd	-	pdf pinout
ISD 2548 E	CMOS-IC	Voice record/playback Device, 48-second DURATION	TSOP	Wbd	-	pdf pinout
ISD 2548 P	CMOS-IC	Voice record/playback Device, 48-second DURATION	DIC	Wbd	-	pdf pinout
ISD 2548 X	CMOS-IC	Voice record/playback Device, 48-second DURATION	CSP	Wbd	-	pdf pinout
ISD 2560 E	CMOS-IC	Voice record/playback Device, 60-second DURATION	TSOP	Wbd	-	pdf pinout
ISD 2560 P	CMOS-IC	Voice record/playback Device, 60-second DURATION	DIC	Wbd	-	pdf pinout
ISD 2560 S	CMOS-IC	Voice record/playback Device, 60-second DURATION	MDIP	Wbd	-	pdf pinout
ISD 2560 X	CMOS-IC	Voice record/playback Device, 60-second DURATION	CSP	Wbd	-	pdf pinout
ISD 2564 P	CMOS-IC	Voice record/playback Device, 64-second DURATION	DIC	Wbd	-	pdf pinout
ISD 2564 X	CMOS-IC	Voice record/playback Device, 64-second DURATION	CSP	Wbd	-	pdf pinout
ISD 2575 E	CMOS-IC	Voice record/playback Device, 75-second DURATION	TSOP	Wbd	-	pdf pinout
ISD 2575 P	CMOS-IC	Voice record/playback Device, 75-second DURATION	DIC	Wbd	-	pdf pinout
ISD 2575 S	CMOS-IC	Voice record/playback Device, 75-second DURATION	MDIP	Wbd	-	pdf pinout
ISD 2575 X	CMOS-IC	Voice record/playback Device, 75-second DURATION	CSP	Wbd	-	pdf pinout
ISD 2590 E	CMOS-IC	Voice record/playback Device, 90-second DURATION	TSOP	Wbd	-	pdf pinout
ISD 2590 P	CMOS-IC	Voice record/playback Device, 90-second DURATION	DIC	Wbd	-	pdf pinout
ISD 2590 S	CMOS-IC	Voice record/playback Device, 90-second DURATION	MDIP	Wbd	-	pdf pinout
ISD 2590 X	CMOS-IC	Voice record/playback Device, 90-second DURATION	CSP	Wbd	-	pdf pinout
ISD 5008 E(D/I)	CMOS-IC	Voice Record/Playback Device 4-, 5-, 6-, and 8-Minute Durations	TSOP	Wbd	-	pdf pinout
ISD 5008 P	CMOS-IC	Voice Record/Playback Device 4-, 5-, 6-,	DIC	Wbd		pdf pinout

Type	Device	Short Description	Fig.	Manu	Comparision Types	More at
		and 8-Minute Durations				
ISD 5008 S(D/I)	CMOS-IC	Voice Record/Playback Device 4-, 5-, 6-, and 8-Minute Durations	MDIP	Wbd	-	pdf pinout
ISD 5008 X	CMOS-IC	Voice Record/Playback Device 4-, 5-, 6-, and 8-Minute Durations	MLP	Wbd	-	pdf pinout
ISD 5008 Z(D/I)	CMOS-IC	Voice Record/Playback Device 4-, 5-, 6-, and 8-Minute Durations	CSP	Wbd	-	pdf pinout
ISD 5102 E(I)	CMOS-IC	1 TO 2 minutes duration, Voice record/ playback devices	TSOP	Wbd	-	pdf pinout
ISD 5102 S(I)	CMOS-IC	1 TO 2 minutes duration, Voice record/ playback devices	MDIP	Wbd	-	pdf pinout
ISD 5102 X	CMOS-IC	1 TO 2 minutes duration, Voice record/ playback devices	CSP	Wbd	-	pdf pinout
ISD 5104 E(I)	CMOS-IC	2 TO 4 minutes duration, Voice record/ playback devices	TSOP	Wbd	-	pdf pinout
ISD 5104 S(I)	CMOS-IC	2 TO 4 minutes duration, Voice record/ playback devices	MDIP	Wbd	-	pdf pinout
ISD 5104 X	CMOS-IC	2 TO 4 minutes duration, Voice record/ playback devices	CSP	Wbd	-	pdf pinout
ISD 5108 E(I)	CMOS-IC	4 TO 8 minutes duration, Voice record/ playback devices	TSOP	Wbd	-	pdf pinout
ISD 5108 S(I)	CMOS-IC	4 TO 8 minutes duration, Voice record/ playback devices	MDIP	Wbd	-	pdf pinout
ISD 5108 X	CMOS-IC	4 TO 8 minutes duration, Voice record/ playback devices	CSP	Wbd	-	pdf pinout
ISD 5116 E(I)	CMOS-IC	8 TO 16 minutes duration, Voice record/ playback devices	TSOP	Wbd	-	pdf pinout
ISD 5116 P	CMOS-IC	8 TO 16 minutes duration, Voice record/ playback devices	DIC	Wbd	-	pdf pinout
ISD 5116 S(I)	CMOS-IC	8 TO 16 minutes duration, Voice record/ playback devices	MDIP	Wbd	-	pdf pinout
ISD 5116 X	CMOS-IC	8 TO 16 minutes duration, Voice record/ playback devices	CSP	Wbd	-	pdf pinout
ISD 5216 E(D/I)	CMOS-IC	8 - 16 min. Voice record/playback Device, Integrated codec	TSOP	Wbd	-	pdf pinout
ISD 5216 P	CMOS-IC	8 - 16 min. Voice record/playback Device, Integrated codec	DIC	Wbd	-	pdf pinout
ISD 5216 S(I)	CMOS-IC	8 - 16 min. Voice record/playback Device, Integrated codec	MDIP	Wbd	-	pdf pinout
ISD 25120 P	CMOS-IC	Voice record/playback Device, 120-second DURATION	DIC	Wbd	-	pdf pinout
ISD 25120 S	CMOS-IC	Voice record/playback Device, 120-second DURATION	MDIP	Wbd	-	pdf pinout
ISD 25120 X	CMOS-IC	Voice record/playback Device, 120-second DURATION	CSP	Wbd	-	pdf pinout
ISD 400304M E(D/I)	CMOS-IC	Voice record/playback Devices, 4-minute DURATION	TSOP	Wbd	-	pdf pinout
ISD 400304M P	CMOS-IC	Voice record/playback Devices, 4-minute DURATION	DIC	Wbd	-	pdf pinout
ISD 400304M S(I)	CMOS-IC	Voice record/playback Devices, 4-minute DURATION	MDIP	Wbd	-	pdf pinout
ISD 400304M X	CMOS-IC	Voice record/playback Devices, 4-minute DURATION	CSP	Wbd	-	pdf pinout
ISD 400305M E(D/I)	CMOS-IC	Voice record/playback Devices, 5-minute DURATION	TSOP	Wbd	-	pdf pinout
ISD 400305M P	CMOS-IC	Voice record/playback Devices, 5-minute DURATION	DIC	Wbd	-	pdf pinout
ISD 400305M S(I)	CMOS-IC	Voice record/playback Devices, 5-minute DURATION	MDIP	Wbd	-	pdf pinout
ISD 400305M X	CMOS-IC	Voice record/playback Devices, 5-minute DURATION	CSP	Wbd	-	pdf pinout
ISD 400306M E(D/I)	CMOS-IC	Voice record/playback Devices, 6-minute DURATION	TSOP	Wbd	-	pdf pinout
ISD 400306M P	CMOS-IC	Voice record/playback Devices, 6-minute DURATION	DIC	Wbd	-	pdf pinout
ISD 400306M S(I)	CMOS-IC	Voice record/playback Devices, 6-minute DURATION	MDIP	Wbd	-	pdf pinout
ISD 400306M X	CMOS-IC	Voice record/playback Devices, 6-minute DURATION	CSP	Wbd	-	pdf pinout
ISD 400308M E(D/I)	CMOS-IC	Voice record/playback Devices, 8-minute DURATION	TSOP	Wbd	-	pdf pinout
ISD 400308M P	CMOS-IC	Voice record/playback Devices, 8-minute DURATION	DIC	Wbd	-	pdf pinout
ISD 400308M S(I)	CMOS-IC	Voice record/playback Devices, 8-minute DURATION	MDIP	Wbd	-	pdf pinout
ISD 400308M X	CMOS-IC	Voice record/playback Devices, 8-minute DURATION	CSP	Wbd	-	pdf pinout
ISD 400408M E(D/I)	CMOS-IC	Voice record/playback Devices, 8-minute DURATION	TSOP	Wbd	-	pdf pinout
ISD 400408M P	CMOS-IC	Voice record/playback Devices, 8-minute DURATION	DIC	Wbd	-	pdf pinout
ISD 400408M S(I)	CMOS-IC	Voice record/playback Devices, 8-minute DURATION	MDIP	Wbd	-	pdf pinout
ISD 400408M X	CMOS-IC	Voice record/playback Devices, 8-minute DURATION	CSP	Wbd	-	pdf pinout

Type	Device	Short Description	Fig.	Manu	Comparision Types	More at
ISD 400410M E(D/I)	CMOS-IC	Voice record/playback Devices, 10-minute DURATION	TSOP	Wbd	-	pdf pinout
ISD 400410M P	CMOS-IC	Voice record/playback Devices, 10-minute DURATION	DIC	Wbd	-	pdf pinout
ISD 400410M S(I)	CMOS-IC	Voice record/playback Devices, 10-minute DURATION	MDIP	Wbd	-	pdf pinout
ISD 400410M X	CMOS-IC	Voice record/playback Devices, 10-minute DURATION	CSP	Wbd	-	pdf pinout
ISD 400412M E(D/I)	CMOS-IC	Voice record/playback Devices, 12-minute DURATION	TSOP	Wbd	-	pdf pinout
ISD 400412M P	CMOS-IC	Voice record/playback Devices, 12-minute DURATION	DIC	Wbd	-	pdf pinout
ISD 400412M S(I)	CMOS-IC	Voice record/playback Devices, 12-minute DURATION	MDIP	Wbd	-	pdf pinout
ISD 400412M X	CMOS-IC	Voice record/playback Devices, 12-minute DURATION	CSP	Wbd	-	pdf pinout
ISD 400416M E(D/I)	CMOS-IC	Voice record/playback Devices, 16-minute DURATION	TSOP	Wbd	-	pdf pinout
ISD 400416M P	CMOS-IC	Voice record/playback Devices, 16-minute DURATION	DIC	Wbd	-	pdf pinout
ISD 400416M S(I)	CMOS-IC	Voice record/playback Devices, 16-minute DURATION	MDIP	Wbd	-	pdf pinout
ISD 400416M X	CMOS-IC	Voice record/playback Devices, 16-minute DURATION	CSP	Wbd	-	pdf pinout
ISD 4002120 E(D/I)	CMOS-IC	Voice record/ playback Devices, 120-Second duration	TSOP	Wbd	-	pdf pinout
ISD 4002120 P	CMOS-IC	Voice record/ playback Devices, 120-Second duration	DIC	Wbd	-	pdf pinout
ISD 4002120 S(I)	CMOS-IC	Voice record/ playback Devices, 120-Second duration	MDIP	Wbd	-	pdf pinout
ISD 4002120 X	CMOS-IC	Voice record/ playback Devices, 120-Second duration	CSP	Wbd	-	pdf pinout
ISD 4002150 E(D/I)	CMOS-IC	Voice record/ playback Devices, 150-Second duration	TSOP	Wbd	-	pdf pinout
ISD 4002150 P	CMOS-IC	Voice record/ playback Devices, 150-Second duration	DIC	Wbd	-	pdf pinout
ISD 4002150 S(I)	CMOS-IC	Voice record/ playback Devices, 150-Second duration	MDIP	Wbd	-	pdf pinout
ISD 4002150 X	CMOS-IC	Voice record/ playback Devices, 150-Second duration	CSP	Wbd	-	pdf pinout
ISD 4002180 E(D/I)	CMOS-IC	Voice record/ playback Devices, 180-Second duration	TSOP	Wbd	-	pdf pinout
ISD 4002180 P	CMOS-IC	Voice record/ playback Devices, 180-Second duration	DIC	Wbd	-	pdf pinout
ISD 4002180 S(I)	CMOS-IC	Voice record/ playback Devices, 180-Second duration	MDIP	Wbd	-	pdf pinout
ISD 4002180 X	CMOS-IC	Voice record/ playback Devices, 180-Second duration	CSP	Wbd	-	pdf pinout
ISD 4002240 E(D/I)	CMOS-IC	Voice record/ playback Devices, 240-Second duration	TSOP	Wbd	-	pdf pinout
ISD 4002240 P	CMOS-IC	Voice record/ playback Devices, 240-Second duration	DIC	Wbd	-	pdf pinout
ISD 4002240 S(I)	CMOS-IC	Voice record/ playback Devices, 240-Second duration	MDIP	Wbd	-	pdf pinout
ISD 4002240 X	CMOS-IC	Voice record/ playback Devices, 240-Second duration	CSP	Wbd	-	pdf pinout
ISH700	OC	LED/NPN Viso:500V CTR:>10% Ibs:16mA		Iso		data pinout
ISH701	OC	LED/NPN Viso:500V CTR:>20% Ibs:16mA		Iso		data pinout
ISH710	OC	LED/NPN Viso:2500V CTR:>20% Ibs:16mA		Iso		data pinout
ISH711	OC	LED/NPN Viso:2500V CTR:>50% Ibs:16mA		Iso		data pinout
ISL5120	SPST	Vs:2.7...12V Ron:Ohm Ton:28ns at 5/0V	8S	Isi		data pdf pinout
ISL5121	SPST	Vs:2.7...12V Ron:Ohm Ton:28ns at 5/0V	8S	Isi		data pdf pinout
ISL5122	SPST	Vs:2.7...12V Ron:Ohm Ton:28ns at 5/0V	8S	Isi		data pdf pinout
ISL5123	SPDT	Vs:2.7...12V Ron:Ohm Ton:28ns at 5/0V	8S	Isi		data pdf pinout
ISL 6719 ARZ	IC	100V Linear Bias Supply	DFN	Isi		pdf pinout
ISL8323	SPST	Vs:2.7...12V Ron:aOhm at 5/0V	8S	Isi		data pdf pinout
ISL8324	SPST	Vs:2.7...12V Ron:aOhm at 5/0V	8S	Isi		data pdf pinout
ISL8325	SPST	Vs:2.7...12V Ron:aOhm at 5/0V	8S	Isi		data pdf pinout
ISL8391	SPST	Vs:±2...±6V Ron:Ohm Ton:85ns at 5/0V	16S	Isi		data pdf pinout
ISL8392	SPST	Vs:±2...±6V Ron:Ohm Ton:85ns at 5/0V	16S	Isi		data pdf pinout
ISL8393	SPST	Vs:±2...±6V Ron:Ohm Ton:85ns at 5/0V	16S	Isi		data pdf pinout
ISL8394	SPDT	Vs:±2...±6V Ron:Ohm Ton:80ns at 5/0V	20S	Isi		data pdf pinout
ISL43110	SPST	Vs:2.7...12V Ron:Ohm Ton:45ns at 5/0V	8S,5S	Isi		data pdf pinout
ISL43111	SPST	Vs:2.7...12V Ron:Ohm Ton:45ns at 5/0V	8S,5S	Isi		data pdf pinout
ISL43112	SPST	Vs:±1.5...±6V Ron:Ohm at 5/-5V	8S,5S	Isi		data pdf pinout
ISL43113	SPST	Vs:±1.5...±6V Ron:Ohm at 5/-5V	8S,5S	Isi		data pdf pinout
ISL43120	SPST	Vs:2.7...12V Ron:Ohm Ton:28ns at 5/0V	8S	Isi		data pdf pinout
ISL43121	SPST	Vs:2.7...12V Ron:Ohm Ton:28ns at 5/0V	8S	Isi		data pdf pinout
ISL43122	SPST	Vs:2.7...12V Ron:Ohm Ton:28ns at 5/0V	8S	Isi		data pdf pinout
ISL43140	SPST	Vs:2...12V Ron:Ohm Ton:50ns at 5/0V	16SC	Isi		data pdf pinout
ISL43141	SPST	Vs:2...12V Ron:Ohm Ton:50ns at 5/0V	16SC	Isi		data pdf pinout
ISL43142	SPST	Vs:2...12V Ron:Ohm Ton:50ns at 5/0V	16SC	Isi		data pdf pinout
ISL43143	SPST	Vs:±2...±6V Ron:Ohm Ton:64ns at 5/0V	16CS	Isi		data pdf pinout
ISL43144	SPST	Vs:±2...±6V Ron:Ohm Ton:64ns at 5/0V	16CS	Isi		data pdf pinout
ISL43145	SPST	Vs:±2...±6V Ron:Ohm Ton:64ns at 5/0V	16CS	Isi		data pdf pinout

Type	Device	Short Description	Fig.	Manu	Comparision Types	More at
ISL43210	SPDT	Vs:2.7...12V Ron:Ohm Ton:28ns at 5/0V	6S	Isi		data pdf pjnout
ISL43231	SPDT	Vs:±2...±6V Ron:Ohm Ton:35ns at 5/-5V	20C	Isi		data pdf pinout
ISL43240	SPDT	Vs:±2...±6V Ron:Ohm Ton:64ns at 5/0V	20SC	Isi		data pdf pinout
ISL43410	DPDT	Vs:2...12V Ron:Ohm Ton:60ns at 5/0V	16C,10S	Isi		data pdf pinout
ISL43640	MULT4	Vs:2...12V Ron:Ohm Ton:60ns at 5/0V	16C,10S	Isi		data pdf pinout
ISL43681	MULT8	Vs:±2...±6V Ron:Ohm Ton:35ns at 5/-5V	20C	Isi		data pdf pinout
ISL43741	MULT4	Vs:±2...±6V Ron:Ohm Ton:35ns at 5/-5V	20C	Isi		data pdf pinout
ISL43840	MULT4	Vs:±2...±6V Ron:Ohm Ton:35ns at 5/-5V	20C	Isi		data pdf pinout
ISL43841	MULT4	Vs:±2...±6V Ron:Ohm at 5/-5V	20C	Isi		data pdf pinout
ISL84051	MULT8	Vs:±2...±6V Ron:Ohm Ton:50ns at 5/-5V	16S	Isi		data pdf pinout
ISL84052	MULT4	Vs:±2...±6V Ron:Ohm Ton:50ns at 5/-5V	16S	Isi		data pdf pinout
ISL84053	SPDT	Vs:±2...±6V Ron:Ohm Ton:50ns at 5/-5V	16S	Isi		data pdf pinout
ISL84514	SPST	Vs:2.7...12V Ron:aOhm at 5/0V	8S,5S	Isi		data pdf pinout
ISL84515	SPST	Vs:2.7...12V Ron:aOhm at 5/0V	8S,5S	Isi		data pdf pinout
ISL84516	SPST	Vs:±1.5...±6V Ron:Ohm at 5/-5V	8S,5S	Isi		data pdf pinout
ISL84517	SPST	Vs:±1.5...±6V Ron:Ohm at 5/-5V	8S,5S	Isi		data pdf pinout
ISL84521	SPST	Vs:±2...±6V Ron:Ohm Ton:50ns at 5/0V	16S	Isi		data pdf pinout
ISL84522	SPST	Vs:±2...±6V Ron:Ohm Ton:50ns at 5/0V	16S	Isi		data pdf pinout
ISL84523	SPST	Vs:±2...±6V Ron:Ohm Ton:50ns at 5/0V	16S	Isi		data pdf pinout
ISL84524	MULT4	Vs:2...12V Ron:aOhm Ton:90ns at 5/0V	10S,8S	Isi		data pdf pinout
ISL84525	DPDT	Vs:2...12V Ron:aOhm Ton:90ns at 5/0V	10S	Isi		data pdf pinout
ISL84541	SPST	Vs:2.7...12V Ron:Ohm Ton:35ns at 5/0V	8SD	Isi		data pdf pinout
ISL84542	SPST	Vs:2.7...12V Ron:Ohm Ton:35ns at 5/0V	8SD	Isi		data pdf pinout
ISL84543	SPST	Vs:2.7...12V Ron:Ohm Ton:35ns at 5/0V	8SD	Isi		data pdf pinout
ISL84544	SPDT	Vs:2.7...12V Ron:Ohm Ton:35ns at 5/0V	8SD	Isi		data pdf pinout
ISL84714	SPDT	Vs:1.65...3.6V Ron:Ohm at 3/0V	6S	Isi		data pdf pinout
ISL84715	SPDT	Vs:1.65...3.6V Ron:Ohm Ton:9ns at 3/0V	6S	Isi		data pdf pinout
ISL84716	SPDT	Vs:1.65...3.6V Ron:Ohm Ton:9ns at 3/0V	6S	Isi		data pdf pinout
ISL 97701	IC	Boost Regulator with Integrated Schottky	DFN	Isi	-	pdf pinout
ISL 97702	IC	Boost with Dual Reference Outputs	DFN	Isi	-	pdf pinout
ISL 95711 UIU 10 Z	CMOS-IC	Digitally Controlled Potentiometer	TSOP	Isi	-	pdf pinout
ISL 95711 WIU 10 Z	CMOS-IC	Digitally Controlled Potentiometer	TSOP	Isi	-	pdf pinout
ISM801	OC	LED/NPN Viso:2500V CTR:40-80%		Iso		data pinout
ISM802	OC	LED/NPN Viso:2500V CTR:63-125%		Iso		data pinout
ISM803	OC	LED/NPN Viso:2500V CTR:100-200%		Iso		data pinout
ISM804	OC	LED/NPN Viso:2500V CTR:>20% Ibs:10mA		Iso		data pinout
ISM805	OC	LED/NPN Viso:2500V CTR:>50% Ibs:10mA		Iso		data pinout
ISM850	OC	LED/Dar Viso:2500V CTR:>500% Ibs:8mA		Iso		data pinout
ISM851	OC	LED/Dar Viso:2500V CTR:>100% Ibs:8mA		Iso		data pinout
ISO253	PREC-BUFF	Vs:18V ViO:<±100mV f:50kkc	28Y	Bub		data pinout
ISO256	HI-PREC	Vs:18V ViO:60µµV f:50kkc	28Y	Bub		data pinout
IPS 1011 PBF	MOS	Intelligent Power low Side Switch	17	Inr	-	pdf pinout
IPS 1011 RPBF	MOS	Intelligent Power low Side Switch	17	Inr	-	pdf pinout
IPS 1011 SPBF	MOS	Intelligent Power low Side Switch	17	Inr	-	pdf pinout
ISP 1122 ABD	LIN-IC	Universal Serial Bus stand-alone hub	TSOP	Phi	-	pdf pinout
ISP 1122 AD	LIN-IC	Universal Serial Bus stand-alone hub	TSOP	Phi	-	pdf pinout
ISP 1122 ANB	LIN-IC	Universal Serial Bus stand-alone hub	TSOP	Phi	-	pdf pinout
ISP 1520 BD	IC	USB hub controller Rev. 2.0 480Mbit/s	64-QFP	Phi	-	pdf pinout
ISPD60	OC	LED/Dar Viso:2500V CTR:>100%		Iso		data pdf pinout
ISPD61	OC	LED/Dar Viso:2500V CTR:>500%		Iso		data pdf pinout
ISPD62	OC	LED/Dar Viso:2500V CTR:>1000%		Iso		data pdf pinout
ISPD63	OC	LED/Dar Viso:5000V CTR:>100%		Iso		data pdf pinout
ISPD64	OC	LED/Dar Viso:5000V CTR:>500%		Iso		data pdf pinout
ISPD65	OC	LED/Dar Viso:5000V CTR:>1000%		Iso		data pdf pinout
ISQ1	QOC	LED/NPN Viso:2500V CTR:>20% Ibs:16mA		Iso	ILQ-1	data pinout
ISQ5	QOC	LED/NPN Viso:2500V CTR:>50% Ibs:16mA		Iso		data pinout
ISQ74	QOC	LED/NPN Viso:1500V CTR:>2.5% Ibe:2mA		Iso	ILQ-74	data pinout
ISQ201	QOC	LED/NPN Viso:5000V CTR:>75% Ibs:10mA		Iso		data pdf pinout
ISQ202	QOC	LED/NPN Viso:5000V CTR:125-250%		Iso		data pdf pinout
ISQ203	QOC	LED/NPN Viso:5000V CTR:225-450%		Iso		data pdf pinout
ISS-53	Si-Di	≈BA 127	31a		→BA 127	
IT	Si-P	=2SB1220-T (Typ-Code/Stempel/marking)	35(2mm)	Mat	→2SB1220	data
IT	Si-P	=2SB1463-T (Typ-Code/Stempel/marking)	35(1,6mm)	Mat	→2SB1463	data
IT	Si-P	=2SB792-T (Typ-Code/Stempel/marking)	35	Mat	→2SB792	data
IT	Si-N/P+R	=XP 04387 (Typ-Code/Stempel/marking)	46(2mm)	Mat	→XP 04387	data
IT 404	Si-N	Log/Antilog 4-Trans.-Array, 20V, 20mA B>100(0,01...1mA)		Isi	-	data
IT 500	N-FET	Dual, 60V, Idss=0,7...7mA, ΔUgs<5mV	81bx	Isi	2N3954, 2N5045	data
IT 501	N-FET	=IT 500: ΔUgs<10mV	81bx	Isi	2N3955, 2N5046	data
IT 502	N-FET	=IT 500: ΔUgs<15mV	81bx	Isi	2N3956, 2N5047	data
IT 1700	MOS-P-FET-e	Chopper, 40V, Ugs=±125V, 50mA, <400Ω	5n	Isi	-	data
IT 1750	MOS-N-FET-e	Chopper, 25V, Ugs=±125V, 100mA, <50Ω	5g+	Isi	-	data
iT 3760 A-AS	MOS-IC	LCD Armbanduhr/lcd watch	Chip		U 130(X), U 1301XS	
iT 3812 A-AS	MOS-IC	LCD Armbanduhr/lcd watch	Chip		U 132(X)	
IT-	MOS-N-FET-d	=3SK162 (Typ-Code/Stempel/marking)	44	Hit	→3SK162	data
ITA 6 V1M3	LIN-IC, Z-Di	SMD,18x TAZ(6,1V) f.Centronics Interf.	20-MDIP	Tho	-	data
ITA 6 V1U1	LIN-IC, Z-Di	SMD, 6x TAZ(6,1V) f. RS422,485 Interf.	8-MDIP	Tho	-	data pdf
ITA 6 V1U3	LIN-IC, Z-Di	SMD, 8x TAZ(6,1V) f. RS422,485 Interf.	20-MDIP	Tho	-	data
ITA 6 V5B1	LIN-IC, Z-Di	SMD, 6x TAZ(6,5V) f. RS232,423 Interf.	8-MDIP	Tho	-	data pdf
ITA 6 V5B3	LIN-IC, Z-Di	SMD,10x TAZ(6,5V) f. RS232,423 Interf.	20-MDIP	Tho	-	data pdf
ITA 10 B1	LIN-IC, Z-Di	=ITA 6V5B1: 10V	8-MDIP	Tho		data pdf
ITA 10 B3	LIN-IC, Z-Di	=ITA 6V5B3: 10V	20-MDIP	Tho		data pdf
ITA 10 B4	LIN-IC, Z-Di	=ITA 6V5B3: 10V	20-DIP	Tho		data pdf

Type	Device	Short Description	Fig.	Manu	Comparision Types	More at
ITA 18 B1	LIN-IC, Z-Di	=ITA 6V5B1: 18V	8-MDIP	Tho	-	data pdf
ITA 18 B3	LIN-IC, Z-Di	=ITA 6V5B3: 18V	20-MDIP	Tho	-	data pdf
ITA 25 B1	LIN-IC, Z-Di	=ITA 6V5B1: 25V	8-MDIP	Tho	-	data pdf
ITA 25 B3	LIN-IC, Z-Di	=ITA 6V5B3: 25V	20-MDIP	Tho	-	data pdf
ITA 25 B4	LIN-IC, Z-Di	=ITA 6V5B3: 25V	20-DIP	Tho	-	
ITA 25 B5	LIN-IC, Z-Di	=ITA 6V5B3: 25V	16-MDIP	Tho	-	
ITA 100 U2	LIN-IC	Stepper Motor Drv. Protection, 100V	-MDIP	Tho	-	
ITE 4091	N-FET	=2N4091:	7c	Isi	→2N4091	data
ITE 4092	N-FET	=2N4092:	7c	Isi	→2N4092	data
ITE 4093	N-FET	=2N4093:	7c	Isi	→2N4093	data
ITE 4391	N-FET	=2N4391:	4b, 7c	Isi	→2N4391	data
ITE 4392	N-FET	=2N4392:	4b, 7c	Isi	→2N4392	data
ITE 4393	N-FET	=2N4393:	4b, 7c	Isi	→2N4393	data
ITE 4416	N-FET	=2N4416:	4b, 7c	Isi	→2N4416	data
ITR 17052	F-Thy+Di	=17052	17j§		→17052	
ITR 17053	F-Thy+Di	=17053	17j§		→17053	
ITT 90	LIN-IC	Telecom, Verstärker/amplifier	24-DIP	Itt	-	
ITT 91	CMOS-IC	Telecom, Controller f. ITT 90, ITT 92	24-DIP	Itt	-	
ITT 92	LIN-IC	Telecom, Lautspr.-verst./speaker amp.	18-DIP	Itt	-	
ITT 210	C-Di	≈BA 102	31a		→BA 102	
ITT 410	C-Di	≈BA 102	31a		→BA 102	
ITT 600	Opto	LED				
ITT 600(DPD)	Si-Di	SS, 75V, 0,2A, <6ns	31a	Itt	BAW 62, BAW 76, BAX 95, 1N4148...49, ++	data
ITT 601(DPD)	Si-Di	SS, 50V, 0,2A, <6ns	31a	Itt	BAW 62, BAW 76, BAX 95, 1N4148...49, ++	data
ITT 789(A)	Si-Di	≈1N4148	31a		→1N4148	
ITT 1451	PMOS-IC	Ir-fb Empf./receiver, 64 Codes	24-DIP	Itt	SAA1251	
ITT 2001	Si-Di	S, 100V, 0,15A, <50ns	31a	Itt	BA 195, BA 197...198, BAV 19..21, ++	data
ITT 2002	Si-Di	=ITT 2001: 200V	31a	Itt	BA 195, BA 197...198, BAV 20..21, ++	data
ITT 2003	Si-Di	=ITT 2001: 250V	31a	Itt	BA 198, BAV 21	data
ITT 2147-...	sRAM-IC	4096x1 Bit	18-DIP	Itt	...2147...	
ITT 2167 (D,N)-...	sRAM-IC	16384x1 Bit	20-DIP	Itt	...2167...	
ITT 2168 (D,N)-...	sRAM-IC	4096x4 Bit	20-DIP	Itt	...2168...	
ITT 2559	CMOS-IC	Telecom, Tastwahl/digital dialing	16-DIP	Itt	-	
ITT 3001	Si-Di	Uni, 70V, 0,1A, 250ns	31a	Itt	BA 157...159, BAX 17, BY 206..207, ++	data
ITT 3002	Si-Di	=ITT 3001: 150V	31a	Itt	BA 157...159, BAX 17, BY 206..207, ++	data
ITT 3003	Si-Di	=ITT 3001: 200V	31a	Itt	BA 157...159, BAX 17, BY 206..207, ++	data
ITT 3065 N	LIN-IC	TV, Ton-ZF, FM-ZF	14-DIP	Itt	MC 1358, LA 1365	
ITT 4027 (D,N)-...	dRAM-IC	4096x1 Bit	16-DIP	Itt	...4027...	
ITT 4116 (D,N)-...	dRAM-IC	16384x1 Bit	16-DIP	Itt	...4116...	
ITT 4164-...	dRAM-IC	65536x1 Bit	16-DIP	Itt	...4164...	
ITT 5101 S	sRAM-IC	256x4 Bit	22-DIP	Itt	-	
ITT 7163 (D,N)	PMOS-IC	5x Relaistreiber/relais driver	14-DIC,DIP	Itt	-	
ITT 7164 (D,N)	PMOS-IC	5xRelaistreiber/relais driver	16-DIP	Itt	-	
ITZ 5.6H	LIN-IC, Z-Di	9x Z-Di, common Cathode, 5,8. .5,91V	9-SIP	Rhm	-	data
IU	Si-P	=2SB1589 (Typ-Code/Stempel/marking)	39	Mat	→2SB1589	data
IU-	GaAs-N-FET	=2SK668 (Typ-Code/Stempel/marking)	44	Hit	→2SK668	data
IV-	MOS-N-FET-d	=3SK136 (Typ-Code/Stempel/marking)	44	Hit	→3SK136	data
IVN 5000 A...	MOS-N-FET-e	V-MOS, LogL, 0,7A, 2W, <2,5Ω(1A) AND=40V, ANE=60V, ANF=80V, ANH=100V	30e§	Gen, Isi	(2SK2014)[4]	data
IVN 5000 S...	MOS-N-FET-e	=IVN 5000A...: 0,9A, 3,13W(Tc=25°)	2a§		(2SK2014)[4]	data
IVN 5000 T...	MOS-N-FET-e	=IVN 5000A...: 1,2A, 6,25W(Tc=25°)	2a§		(2SK2014)[4]	data
IVN 5001 A...	MOS-N-FET-e	V-MOS, 0,7A, 2W, <2,5Ω(1A) AND=40V, ANE=60V, ANF=80V, ANH=100V	30e§	Gen, Isi	(2SK2014)[4]	data
IVN 5001 S...	MOS-N-FET-e	=IVN 5001A...: 0,9A, 3,13W(Tc=25°)	2a§		(2SK2014)[4]	
IVN 5001 T...	MOS-N-FET-e	=IVN 5001A...: 1,2A, 6,25W(Tc=25°)	2a§		(2SK2014)[4]	
IVN 5200 H...	MOS-N-FET-e	V-MOS, LogL, 5A, 30W, <0,5Ω(5A) HND=40V, HNE=60V, HNF=80V	22a§	Isi	IRL 510(A), 2SK702, 2SK991	data
IVN 5200 K...	MOS-N-FET-e	=IVN 5200H: 50W	23a§		2SK405	
IVN 5200 T...	MOS-N-FET-e	=IVN 5200H: 4A, 12,5W(Tc=25°)	2a§		(2SK1474, 2SK1804, 2SK2016, 2SK2504,++)[4]	data
IVN 5201 H...	MOS-N-FET-e	V-MOS, 5A, 30W, <0,5Ω(5A) HND=40V, HNE=60V, HNF=80V	22a§	Isi	IRF 510, 2SK702, 2SK917, 2SK920, 2SK991	data
IVN 5201 C...	MOS-N-FET-e	=IVN 5201H:	17b§		(IRF510, 2SK702, 2SK917, 2SK920, 2SK991)[5]	data
IVN 5201 K...	MOS-N-FET-e	=IVN 5201H: 50W	23a§		2SK400, 2SK405	data
IVN 5201 T...	MOS-N-FET-e	=IVN 5201H: 4A, 12,5W(Tc=25°)	2a§		(MTD 6N10, 2SK1060, 2SK2016, 2SK2504,++)[4]	data
IVN 6657	MOS-N-FET-e*	=2N6657: 2,4A, Gate Protection Diode	23a§	Isi	→2N6657	data
IVN 6658	MOS-N-FET-e*	=2N6658: 2,4A, Gate Protection Diode	23a§	Isi	→2N6658	data
IVN 6660	MOS-N-FET-e*	=2N6660: 1,2A, Gate Protection Diode	2a§	Isi	→2N6660	data
IVN 6661	MOS-N-FET-e*	=2N6661: 1,2A, Gate Protection Diode	2a§	Isi	→2N6661	data
IW	Si-N	=2SC5019 (Typ-Code/Stempel/marking)	39	Mat	→2SC5019	data
IW 8377	Si-N	≈BF 199	8a		→BF 199	
IW 9288	Si-N	≈BC 141	8a		→BC 141	
IW 10612	Si-N	≈BC 337	8a		→BC 337	
IW 12134	Si-N	≈BC 237	8a		→BC 237	
IW-	MOS-N-FET-d	=3SK137 (Typ-Code/Stempel/marking)	44	Hit	→3SK137	data

Type	Device	Short Description	Fig.	Manu	Comparision Types	More at
IX-	MOS-N-FET-d	=3SK138 (Typ-Code/Stempel/marking)	44	Hit	→3SK138	data
IY	Si-P	=2SA1298-Y (Typ-Code/Stempel/marking)	35	Tos	→2SA1298	data
IY	Si-N	=2SC3515-Y (Typ-Code/Stempel/marking)	39	Tos	→2SC3515	data
IY	Si-P	=KTA1298-Y (Typ-Code/Stempel/marking)	35	Kec	→KTA 1298	data
IY-	MOS-N-FET-d	=3SK194 (Typ-Code/Stempel/marking)	44	Hit	→3SK194	data
IZ	Si-N/P	=XN 04A88 (Typ-Code/Stempel/marking)	46	Mat	→XN 04A88	data
IZ 2	N-FET	=3SK167-2 (Typ-Code/Stempel/marking)	44	Say	→3SK167	data
IZ 3	N-FET	=3SK167-3 (Typ-Code/Stempel/marking)	44	Say	→3SK167	data
IZ 4	N-FET	=3SK167-4 (Typ-Code/Stempel/marking)	44	Say	→3SK167	data
IZ 5	N-FET	=3SK167-5 (Typ-Code/Stempel/marking)	44	Say	→3SK167	data
IZ 0065	IC	40 Ch Segment/Common Driver,Dot Matrix LCD	QFN	Iks	-	pdf pinout
IZ 0066	IC	dot Matrix LCD Controller &driver	QFN	Iks	-	pdf pinout
IZ 1503	CMOS-IC	Cmos Lsi LEDs	DIP	Iks	-	pdf pinout
IZ 1517	CMOS-IC	LED Driver	DIP	Iks	-	pdf pinout
IZ 1621	IC	RAM Mapping 32x4 LCD Controller,I/O µC	QFN	Iks	-	pdf pinout
IZ 6451	IC	Cmos dot Matrix LCD Driver	QFN	Iks	-	pdf pinout
IZ 6570	CMOS-IC	Cmos dot Matrix LCD Driver	QFN	Iks	-	pdf pinout
IZ 7065	IC	40CH Segment/Common Driver	QFN	Iks	-	pdf pinout
IZ 7066	IC	LCD Controller,Built-in ROM of Sign Generator	QFN	Iks	-	pdf pinout
IZ 8016	IC	Digit Thermometer	30	Iks	-	pdf pinout
IZ-	MOS-N-FET-d	=3SK154 (Typ-Code/Stempel/marking)	44	Hit	→3SK15	data
IZD 1520	IC	dot matrix LCD drivers	QFN	Iks	-	pdf pinout

J

Type	Device	Short Description	Fig.	Manu	Comparision Types	More at
J...	...-FET	→2SJ.., z.B./e.g. "J160" = 2SJ160→	Japantypen	JAP		
J	GaAs-N-FET	=2SK1687 (Typ-Code/Stempel/marking)	51	Mat	→2SK1687	data
J	C-Di	=HVC 357 (Typ-Code/Stempel/marking)	71(1,3mm)	Hit	→HVC 357	data
J	C-Di	=HVU 357 (Typ-Code/Stempel/marking)	71(1,7mm)	Hit	→HVU 357	data
J	Si-Di	=S 5566J (Typ-Code/Stempel/marking)	31	Tos	→S 5566J	data
J	Si-Di	=SB 07-03C (Typ-Code/Stempel/marking)	35	Say	→SB 07-03C	data
J 1	MOS-N-FET-e	=BSS 138 (Typ-Code/Stempel/marking)	35	Ons	→BSS 138	data
J 1	Si-P+R	=FMJ 1A (Typ-Code/Stempel/marking)	46	Rhm	→FMJ 1A	data
J 1	Si-Di	=HSM 402WK (Typ-Code/Stempel/marking)	35	Hit	→HSM 402WK	data
J 1	Z-Di	=MMSZ 5246B (Typ-Code/marking)	71(2,7mm)	Ons	→MMSZ 5246B	data
J 01	Si-P	=SO 2906R (Typ-Code/Stempel/marking)	35	Tho	→SO 2906R	data
J 1	C-Di	=ZC 830A (Typ-Code/Stempel/marking)	35	Fer	→ZC 830A	data
J 1 A	C-Di	=ZC 830A (Typ-Code/Stempel/marking)	35	Ztx	→ZC 830A	data
J 1 B	C-Di	=ZC 830B (Typ-Code/Stempel/marking)	35	Ztx	→ZC 830B	data
J 1 O	Si-P	=KSA 1298-O (Typ-Code/Stempel/marking)	35	Sam	→KSA 1298	data
J 1 S	C-Di	=ZC 830 (Typ-Code/Stempel/marking)	35	Ztx	→ZC 830	data
J 1 Y	Si-P	=KSA 1298-Y (Typ-Code/Stempel/marking)	35	Sam	→KSA 1298	data
J 2	Si-N	=2SC3142-2 (Typ-Code/Stempel/marking)	35	Say	→2SC3142	data
J 2	N-FET	=2SK2552-J2 (Typ-Code/Stempel/marking)	35(1,6mm)	Nec	→2SK2552	data
J 2	N-FET	=2SK67-J2 (Typ-Code/Stempel/marking)	≈35	Nec	→2SK67	data
J 2	N-FET	=2SK67-2 (Typ-Code/Stempel/marking)	35	Nec	→2SK67	data
J 2	Si-Di	=F 1J2 (Typ-Code/Stempel/marking)	71(5x2,5)	Org	→F 1J2	data
J 2	Z-Di	=MMSZ 5247B (Typ-Code/marking)	71(2,7mm)	Ons	→MMSZ 5247B	data
J 2	Z-Di	=UDZ 8.2B (Typ-Code/Stempel/marking)	71(1,7mm)	Rhm	→UDZ 2.0B...36B	data
J 2	Z-Di	=UDZS 8.2B (Typ-Code/Stempel/marking)	71(1,7mm)	Rhm	→UDZS 5.1B...10B	data
J 2	C-Di	=ZC 833A (Typ-Code/Stempel/marking)	35	Fer	→ZC 833A	data
J 2 A	C-Di	=ZC 833A (Typ-Code/Stempel/marking)	35	Ztx	→ZC 833A	data
J 2 B	C-Di	=ZC 833B (Typ-Code/Stempel/marking)	35	Ztx	→ZC 833B	data
J 2 S	C-Di	=ZC 833 (Typ-Code/Stempel/marking)	35	Ztx	→ZC 833	data
J 3	Si-Di	=HSM 402S (Typ-Code/Stempel/marking)	35	Hit	→HSM 402S	data
J 3	C-Di	=KDS 2236S (Typ-Code/Stempel/marking)	35	Kec	→KDS 2236S	data
J 3	Z-Di	=MMSZ 5248B (Typ-Code/marking)	71(2,7mm)	Ons	→MMSZ 5248B	data
J 03	Si-P	=SO 2907AR (Typ-Code/Stempel/marking)	35	Tho	→SO 2907AR	data
J 3	C-Di	=ZC 831A (Typ-Code/Stempel/marking)	35	Fer	→ZC 831A	data
J 3	Si-N	=2SC3142-3 (Typ-Code/Stempel/marking)	35	Say	→2SC3142	data
J 3	N-FET	=2SK2552-J3 (Typ-Code/Stempel/marking)	35(1,6mm)	Nec	→2SK2552	data
J 3	N-FET	=2SK67-J3 (Typ-Code/Stempel/marking)	≈35	Nec	→2SK67	data
J 3	N-FET	=2SK67-3 (Typ-Code/Stempel/marking)	35	Nec	→2SK67	data
J 3 A	C-Di	=ZC 831A (Typ-Code/Stempel/marking)	35	Ztx	→ZC 831A	data
J 3 B	C-Di	=ZC 831B (Typ-Code/Stempel/marking)	35	Ztx	→ZC 831B	data
J 3 S	C-Di	=ZC 831 (Typ-Code/Stempel/marking)	35	Ztx	→ZC 831	data
J 3 U	Si-Di	=SS 1J3U (Typ-Code/Stempel/marking)	31	Org	→SS 1J3U	data
J 3 U	Si-Di	=SS 3J3U (Typ-Code/Stempel/marking)	31	Org	→SS 3J3U	data
J 4	Si-N	=2SC3142-4 (Typ-Code/Stempel/marking)	35	Say	→2SC3142	data
J 4	Si-Di	=F 1J4 (Typ-Code/Stempel/marking)	71(5x2,5)	Org	→F 1J4	data
J 4	Si-Di	=F 2J4 (Typ-Code/Stempel/marking)	71(6,4mm)	Org	→F 2J4	data
J 4	Z-Di	=MMSZ 5249B (Typ-Code/marking)	71(2,7mm)	Ons	→MMSZ 5249B	data
J 4	Si-Di	=SS 1J4 (Typ-Code/Stempel/marking)	31	Org	→SS 1J4	data
J 4	Si-Di	=SS 1.5J4 (Typ-Code/Stempel/marking)	31	Org	→SS 1.5J4	data
J 4	Si-Di	=SS 3J4 (Typ-Code/Stempel/marking)	31	Org	→SS 3J4	data
J 4	C-Di	=ZC 832A (Typ-Code/Stempel/marking)	35	Fer	→ZC 832A	data
J 4	N-FET	=2SK2552-J4 (Typ-Code/Stempel/marking)	35(1,6mm)	Nec	→2SK2552	data
J 4	N-FET	=2SK67-J4 (Typ-Code/Stempel/marking)	≈35	Nec	→2SK67	data
J 4	N-FET	=2SK67-4 (Typ-Code/Stempel/marking)	35	Nec	→2SK67	data
J 4 A	C-Di	=ZC 832A (Typ-Code/Stempel/marking)	35	Ztx	→ZC 832A	data
J 4 B	C-Di	=ZC 832B (Typ-Code/Stempel/marking)	35	Ztx	→ZC 832B	data
J 4 S	Si-Di	=F 2J4S (Typ-Code/Stempel/marking)	71(6,4mm)	Org	→F 2J4S	data

Type	Device	Short Description	Fig.	Manu	Comparision Types	More at
S	C-Di	=ZC 832 (Typ-Code/Stempel/marking)	35	Ztx	→ZC 832	data
5...06	Si-Br	50...600V, 10A	70z	Mic	B...C10A	data
	N-FET	=2SK2552-J5 (Typ-Code/Stempel/marking)	35(1,6mm)	Nec	→2SK2552	data
	Z-Di	=MMSZ 5250B (Typ-Code/Stempel/marking)	71(2,7mm)	Ons	→MMSZ 5250B	
	Si-P	=SO 2907R (Typ-Code/Stempel/marking)	35	Tho	→SO 2907R	data
	Si-Di	=SS 1.5J5 (Typ-Code/Stempel/marking)	31	Org	→SS 1.5J5	data
	C-Di	=ZC 834A (Typ-Code/Stempel/marking)	35	Fer	→ZC 834A	data
	N-FET	=2SK67-J5 (Typ-Code/Stempel/marking)	≈35	Nec	→2SK67	data
	N-FET	=2SK67-5 (Typ-Code/Stempel/marking)	35	Nec	→2SK67	data
A	C-Di	=ZC 834A (Typ-Code/Stempel/marking)	35	Ztx	→ZC 834A	data
B	C-Di	=ZC 834B (Typ-Code/Stempel/marking)	35	Ztx	→ZC 834B	data
S	C-Di	=ZC 834 (Typ-Code/Stempel/marking)	35	Ztx	→ZC 834	data
	Si-Di	=F 1J6 (Typ-Code/Stempel/marking)	71(5x2,5)	Org	→F 1J6	data
	Si-Di	=F 2J6 (Typ-Code/Stempel/marking)	71(6,4mm)	Org	→F 2J6	data
	Si-Di	=M 1MA174 (Typ-Code/Stempel/marking)	35(2mm)	Ons	→M 1MA174	data
	Si-P	=SO 2894R (Typ-Code/Stempel/marking)	35	Tho	→SO 2894R	data
	Si-Di	=SS 1J6 (Typ-Code/Stempel/marking)	31	Org	→SS 1J6	data
	Si-Di	=SS 3J6 (Typ-Code/Stempel/marking)	31	Org	→SS 3J6	data
	C-Di	=ZC 835A (Typ-Code/Stempel/marking)	35	Fer	→ZC 835A	data
	N-FET	=2SK2552-J6 (Typ-Code/Stempel/marking)	35(1,6mm)	Nec	→2SK2552	data
	N-FET	=2SK67-J6 (Typ-Code/Stempel/marking)	≈35	Nec	→2SK67	data
	N-FET	=2SK67-6 (Typ-Code/Stempel/marking)	35	Nec	→2SK67	data
A	C-Di	=ZC 835A (Typ-Code/Stempel/marking)	35	Ztx	→ZC 835A	data
B	C-Di	=ZC 835B (Typ-Code/Stempel/marking)	35	Ztx	→ZC 835B	data
S	Si-Di	=F 2J6S (Typ-Code/Stempel/marking)	71(6,4mm)	Org	→F 2J6S	data
S	C-Di	=ZC 835 (Typ-Code/Stempel/marking)	35	Ztx	→ZC 835	data
	C-Di	=ZC 836A (Typ-Code/Stempel/marking)	35	Fer	→ZC 836A	data
	N-FET	=2SK2552-J7 (Typ-Code/Stempel/marking)	35(1,6mm)	Nec	→2SK2552	data
	N-FET	=2SK67-J7 (Typ-Code/Stempel/marking)	≈35	Nec	→2SK67	data
	N-FET	=2SK67-7 (Typ-Code/Stempel/marking)	35	Nec	→2SK67	data
A	C-Di	=ZC 836A (Typ-Code/Stempel/marking)	35	Ztx	→ZC 836A	data
B	C-Di	=ZC 836B (Typ-Code/Stempel/marking)	35	Ztx	→ZC 836B	data
S	C-Di	=ZC 836 (Typ-Code/Stempel/marking)	35	Ztx	→ZC 836	data
	Si-P	=BCX 71RJ (Typ-Code/Stempel/marking)	35	Fer	→BCX 71RJ	data
	N-FET	=2SK67-J8 (Typ-Code/Stempel/marking)	≈35	Nec	→2SK67	data
	N-FET	=2SK67-8 (Typ-Code/Stempel/marking)	35	Nec	→2SK67	data
	Si-Di	=F 1J9 (Typ-Code/Stempel/marking)	71(5x2,5)	Org	→F 1J9	data
	Si-Di	=F 2J9 (Typ-Code/Stempel/marking)	71(6,4mm)	Org	→F 2J9	data
	Si-Di	=SS 1J9 (Typ-Code/Stempel/marking)	31	Org	→SS 1J9	data
	Si-Di	=SS 2J9 (Typ-Code/Stempel/marking)	31	Org	→SS 2J9	data
	Si-P	=SO 2906AR (Typ-Code/Stempel/marking)	35	Tho	→SO 2906AR	data
	Si-P	=SO 679R (Typ-Code/Stempel/marking)	35	Tho	→SO 679R	data
	Si-Di	=SS 05J25 (Typ-Code/Stempel/marking)	31	Org	→SS 05J25	data
	Si-P	=SO 3906R (Typ-Code/Stempel/marking)	35	Tho	→SO 3906R	data
	Si-P	=SO 3905R (Typ-Code/Stempel/marking)	35	Tho	→SO 3905R	data
	Si-P	=SO 5400R (Typ-Code/Stempel/marking)	35	Tho	→SO 5400R	data
	Si-P	=SO 5401R (Typ-Code/Stempel/marking)	35	Tho	→SO 5401R	data
	Si-P	=SO 692R (Typ-Code/Stempel/marking)	35	Tho	→SO 692R	data
	Si-P	=SO 506R (Typ-Code/Stempel/marking)	35	Tho	→SO 506R	data
	Si-P	=SO 970R (Typ-Code/Stempel/marking)	35	Tho	→SO 970R	data
5	N-FET	=J 107: Idss>500mA, Up<10V, <3Ω	7d	Nsc, Six	-	data
06	N-FET	=J 107: Idss>200mA, Up<6V, <6Ω	7d	Nsc, Six	-	data
07	N-FET	Unj, 25V, Idss>100mA, Up<4,5V, 4/6ns	7d	Mot,Nsc,Six	-	data
08	N-FET	=J 107: Idss>80mA, Up<10V	7d	Nsc,Phi,Six	BFS 74, BFS 77, BSJ 108	data
9	N-FET	=J 107: Idss>40mA, Up<6V, <12Ω	7d	Nsc,Phi,Six	BSJ 109	data
0	N-FET	=J 107: Idss>10mA, Up<4V, <18Ω	7d	Nsc,Phi,Six	BFS 76, BFS 79, BSJ 110	data
1	N-FET	sym, 40V, Idss>20mA, Up<10V, 13/35ns	7d	Phi,Nsc,++	BSJ 111	data
2	N-FET	=J 111: Idss>5mA, Up<5V	7d	Phi,Nsc,++	BSJ 112	data
3	N-FET	=J 111: Idss>2mA, Up<3V	7d	Phi,Nsc,++	BSJ 113	data
74	P-FET	sym, 30V, Idss>20mA, Up<10V, 7/15ns	7c,a	Phi,Nsc++	BSJ 174	data
75	P-FET	sym, 30V, Idss>7mA, Up<6V, 15/30ns	7c,a	Phi,Nsc++	BSJ 175	data
76	P-FET	sym, 30V, Idss>2mA, Up<4V, 35/35ns	7c,a	Phi,Nsc++	BSJ 176	data
77	P-FET	sym,30V, Idss>1,5mA, Up<2,25V, 25/45ns	7c,a	Phi,Nsc++	BSJ 177	data
1	N-FET	NF, 40V, Idss=0,2...1mA, Up<1,5V	7d	Nsc,Six,++	BFW 13, 2N4338	data
2	N-FET	=J 201: Idss=0,9...4,5mA, Up<4V	7d	Nsc,Six,++	2N5246, 2N5360	data
03	N-FET	=J 201: Idss=4...20mA, Up<10V	7d	Nsc,Six,++	2N3368, 2SK113	data
04	N-FET	=J 201: Idss=0,2...3mA, Up<2V	7d	Six	-	data
0	N-FET	Uni, 25V, Idss=2...15mA, Up<3V	7d	Nsc, Six	BC 264A...B, BF 410B, 2SK192, 2SK370	data
1	N-FET	=J 210: Idss=7...20mA, Up<4,5V	7d	Nsc, Six	-	data
2	N-FET	=J 210: Idss=15...40mA, Up<6V	7d	Nsc, Six	BFT 10B	data
70	P-FET	Uni, 30V, Idss=2...15mA, Up<2V	7a,d	Nsc,Six,++	2SJ104, 2SJ107	data
71	P-FET	=J 270: Idss=6...50mA, Up<4,5V	7a,d	Nsc,Six,++	-	data
04	N-FET	HF, 30V, Idss=5...15mA, Up<6V	7d,c	Nsc,Six,++	BF 256A, 2N5398	data
05	N-FET	=J 305: Idss=1...8mA, Up<3V	7d,c	Nsc,Six,++	-	data
08	N-FET	AM...UHF, 25V, Idss=12...60mA, Up<6,5V	7d	Phi,Nsc,++	BF 256C, BF 348, BFT 10A...B	data
	N-FET	=J 308: Idss=12...30mA, Up<1...4V	7d	Phi,Nsc,++	BF 256C, BF 348, BFT 10A...B	data
0	N-FET	=J 308: Idss=24...60mA, Up=2...6,5V	7d	Phi,Nsc,++	BFT 10B...C, 2SK582	data
01	N-FET	Dual, Idss>10mA, Up<2,5V, ΔUgs<5mV	8-DIP	Nsc, Six	-	data
02	N-FET	=J 401: ΔUgs<10mV	8-DIP	Nsc, Six	-	data
03	N-FET	=J 401: ΔUgs<10mV	8-DIP	Nsc, Six	-	data
04	N-FET	=J 401: ΔUgs<15mV	8-DIP	Nsc, Six	-	data

Type	Device	Short Description	Fig.	Manu	Comparision Types	More at
J 405	N-FET	=J 401: ΔUgs<20mV	8-DIP	Nsc, Six	-	data
J 406	N-FET	=J 401: ΔUgs<40mV	8-DIP	Nsc, Six	-	data
J 500	Si-Di	FED, Strom-Stabi/current reg., 0,35W Uf=25V(240µA), >4MΩ, Ulim<1,2V (80% If)	7c	Six	1N5287..5289	data
J 501	Si-Di	=J500:Uf=25V(330µA), >2,2MΩ, Ulim<1,3V	7c	Six	1N5289..5291	data
J 502	Si-Di	=J500:Uf=25V(430µA), >1,5MΩ, Ulim<1,5V	7c	Six	1N5291..5293	data
J 503	Si-Di	=J500:Uf=25V(560µA), >1,2MΩ, Ulim<1,7V	7c	Six	1N5293..5295	data
J 504	Si-Di	=J500:Uf=25V(750µA), >800kΩ, Ulim<1,9V	7c	Six	1N5295..5297	data
J 505	Si-Di	=J 500: Uf=25V(1mA), >500kΩ, Ulim<2,1V	7c	Six	1N5298..5301	data
J 506	Si-Di	=J500:Uf=25V(1,4mA), >330kΩ, Ulim<2,5V	7c	Six	1N5302..5305	data
J 507	Si-Di	=J500:Uf=25V(1,8mA), >200kΩ, Ulim<2,8V	7c	Six	1N5306..5309	data
J 508	Si-Di	=J500:Uf=25V(2,4mA), >200kΩ, Ulim<3,1V	7c	Six	1N5308..5311	data
J 509	Si-Di	=J 500: Uf=25V(3mA), >150kΩ, Ulim<3,5V	7c	Six	1N5309..5313	data
J 510	Si-Di	=J500:Uf=25V(3,6mA), >150kΩ, Ulim<3,9V	7c	Six	1N5310..5314	data
J 511	Si-Di	=J500:Uf=25V(4,7mA), >120kΩ, Ulim<4,2V	7c	Six	1N5311..5314	data
J1360	LDR	610nm Ro:3MOhm Vb:2V		Hei		data pinout
JA	MOS-P-FET-e	=2SJ187 (Typ-Code/Stempel/marking)	39	Say	→2SJ187	data
JA	N-FET	=2SK2170 (Typ-Code/Stempel/marking)	35(1,6mm)	Say	→2SK2170	data
JA	Si-Di	=BAV 74 (Typ-Code/Stempel/marking)	35	Mot,Phi,Sie Fer,Tho	→BAV 74	data
JA	Si-Di	=BAV 74L (Typ-Code/Stempel/marking)	35	Ons	→BAV 74L	
JA	MOS-P-FET-e*	=CPH 6301 (Typ-Code/Stempel/marking)	46	Say	→CPH 6301	data
JA	MOS-N-FET-e*	=µPA603T (Typ-Code/Stempel/marking)	46	Nec	→µPA603T	data
JA 100	Si-P	Uni, 30V, 0,1A, 0,5W, 130MHz	7c	Itt,Phi	BC 213, BC 258, BC 308, BC 558, ++	data
JA 101	Si-P	=JA 100: 50V	7c	Itt,Phi	BC 212, BC 257, BC 307, BC 557, ++	data
JA 1350(B,W)	Si-N	≈BC 550			→BC 550	
JAN 7805 T	LIN/Z-IC	Positive Fixed Voltage Regulator, 5V	2	Lin	-	pdf pinout
JAN 7812 T	LIN/Z-IC	Positive Fixed Voltage Regulator, 12V	2	Lin	-	pdf pinout
JAN 7815 T	LIN/Z-IC	Positive Fixed Voltage Regulator, 15V	2	Lin	-	pdf pinout
JAN 7905 K	LIN/Z-IC	negative Fixed Voltage Regulator, 1.5A, -5V	22	Lin	SG7905K	pdf pinout
JAN 7905 T	LIN/Z-IC	negative Fixed Voltage Regulator, 1.5A, -5V	2	Lin	SG7905T	pdf pinout
JAN 7912 K	LIN/Z-IC	negative Fixed Voltage Regulator, 1.5A, -12V	22	Lin	SG7912K	pdf pinout
JAN 7912 T	LIN/Z-IC	negative Fixed Voltage Regulator, 1.5A, -12V	2	Lin	SG7912T	pdf pinout
JAN 7915 K	LIN/Z-IC	negative Fixed Voltage Regulator, 1.5A, -15V	22	Lin	SG7915K	pdf pinout
JAN 7915 T	LIN/Z-IC	negative Fixed Voltage Regulator, 1.5A, -15V	2	Lin	SG7915T	pdf pinout
JAp	Si-Di	=BAV 74 (Typ-Code/Stempel/marking)	35	Phi	→BAV 74	data
JAs	Si-Di	=BAV 74 (Typ-Code/Stempel/marking)	35	Sie	→BAV 74	data
JAt	Si-Di	=BAV 74 (Typ-Code/Stempel/marking)	35	Phi	→BAV 74	data
JB	Si-Di	=BAR 74 (Typ-Code/Stempel/marking)	35	Sie, Tho	→BAR 74	data
JB	MOS-P-FET-e*	=CPH 6302 (Typ-Code/Stempel/marking)	46	Say	→CPH 6302	data
JB	MOS-P-FET-e	=µPA 610TA (Typ-Code/Stempel/marking)	46	Nec	→µPA 610TA	data
JBs	Si-Di	=BAR 74 (Typ-Code/Stempel/marking)	35	Sie	→BAR 74	data
JC	Si-Di	=BAL 74 (Typ-Code/Stempel/marking)	35	Sie,Phi,Tho	→BAL 74	data
JC	Si-Di	=BAL 74W (Typ-Code/Stempel/marking)	35(2mm)	Phi	→BAL 74W	data
JC	MOS-P-FET-e*	=CPH 3303 (Typ-Code/Stempel/marking)	35	Say	→CPH 3303	data
JC	Si-N	=2SC2735 (Typ-Code/Stempel/marking)	35	Hit	→2SC2735	data
JC	Si-N	=2SC4265 (Typ-Code/Stempel/marking)	35(2mm)	Mit	→2SC4265	data
JC	P-FET	=2SJ125-C (Typ-Code/Stempel/marking)	35	Mit	→2SJ125	data
JC	P-FET	=2SJ145-C (Typ-Code/Stempel/marking)	35(2mm)	Mit	→2SJ145	data
JC	Si-Di	=U 1JC44 (Typ-Code/Stempel/marking)	71(5x2,5)	Tos	→U 1JC44	data
JC 327(A)...328	Si-P	=BC 327(A)...328:	7c	Phi	→BC 327(A)...328	data
JC 337(A)...338	Si-N	=BC 337(A)...338:	7c	Phi	→BC 337(A)...338	data
JC 500	Si-N	Uni, 30V, 0,1A, 0,5W, 130MHz	7c	Itt,Phi	BC 168, BC 183, BC 238, BC 548, ++	data
JC 501	Si-N	=JC 500: 50V	7c	Itt,Phi	BC 167, BC 182, BC 237, BC 547, ++	data
JC 546..550	Si-N	=BC 546..550:	7c	Phi	→BC 546..550	data
JC 556..560	Si-P	=BC 556..560:	7c	Phi	→BC 556..560	data
JCp	Si-Di	=BAL 74 (Typ-Code/Stempel/marking)	35	Phi	→BAL 74	data
JCs	Si-Di	=BAL 74 (Typ-Code/Stempel/marking)	35	Sie	→BAL 74	data
JD	P-FET	=2SJ125-D (Typ-Code/Stempel/marking)	35	Mit	→2SJ125	data
JD	P-FET	=2SJ145-D (Typ-Code/Stempel/marking)	35(2mm)	Mit	→2SJ145	data
JD	MOS-P-FET-e	=2SJ287 (Typ-Code/Stempel/marking)	39	Say	→2SJ287	data
JD	Si-Di	=BAW 56 (Typ-Code/Stempel/marking)	35	Sie,Tho,Gsi	→BAW 56	data
JD	MOS-P-FET-e*	=CPH 3304 (Typ-Code/Stempel/marking)	35	Say	→CPH 3304	data
JD 3	Si-Di	=BAT 54SL... (Typ-Code/Stempel/marking)	35	Ons	→BAT 54SL	
JDP 2 S01E,T,U	PIN-Di	SMD, VHF/UHF HF-Abschw., 30V, 50mA	71a(1,2mm)	Tos	-	
JDP 2 S02S,T	PIN-Di	SMD, VHF/UHF HF-Abschw., 30V, 50mA	71a(1,2mm)	Tos	-	
JDP 4 P02U	PIN-Di	SMD, VHF/UHF HF-Abschw., 30V, 50mA	44s4(2mm)	Tos	-	
JDV 2 S01...14...	C-Di	SMD, Tuning	71	Tos	-	
JE	P-FET	=2SJ125-E (Typ-Code/Stempel/marking)	35	Mit	→2SJ125	data
JE	P-FET	=2SJ145-E (Typ-Code/Stempel/marking)	35(2mm)	Mit	→2SJ145	data
JE	Si-Di	=BAV 99 (Typ-Code/Stempel/marking)	35	Aeg,Sie	→BAV 99	data
JE	MOS-P-FET-e*	=CPH 3305 (Typ-Code/Stempel/marking)	35	Say	→CPH 3305	data
JE 9011(D...J)	Si-N	=SS 9011(D...J)	7e	Nec	→SS 9011	data
JE 9012(D...H)	Si-P	=SS 9012(D...H)	7e	Nec	→SS 9012	data

Type	Device	Short Description ·	Fig.	Manu	Comparision Types	More at
9013(D..H)	Si-N	=SS 9013(D..H)	7e	Nec	→SS 9013	data
9014(A..D)	Si-N	=SS 9014(A..D)	7e	Nec	→SS 9014	data
9015(A...C)	Si-P	=SS 9015(A...C)	7e	Nec	→SS 9015	data
9016(D..I)	Si-N	=SS 9016(D..I)	7e	Nec	→SS 9016	data
9018(D..I)	Si-N	=SS 9018(D..I)	7e	Nec	→SS 9018	data
9100(A...G)	Si-N	Uni, 60V, 0,1A, 0,625W, 300MHz	7e	Nec	BC 174, BC 546, 2SC2240, 2SC2459, ++	data
9101(A..D)	Si-N	NF/S, 40V, 0,1A, 0,625W, 300MHz	7e	Nec	BC 167, BC 182, BC 237, BC 547, ++	data
T-XXX-10	IRED	rd 870nm ca. ø12mm I:>1mW/Sr Vf:2.1V		Roi		data pdf pinout
	Si-Di	=BAL 99 (Typ-Code/Stempel/marking)	35	Phi,Sie,Tho	→BAL 99	data
	Si-Di	=BAL 99W (Typ-Code/Stempel/marking)	35(2mm)	Phi	→BAL 99W	data
	MOS-P-FET-e*	=CPH 3306 (Typ-Code/Stempel/marking)	35	Say	→CPH 3306	data
494	Si-N	=BF 494:	7c	Phi	→BF 494	data
495	Si-N	=BF 495:	7c	Phi	→BF 495	data
	Si-Di	=BAL 99 (Typ-Code/Stempel/marking)	35	Phi	→BAL 99	data
	Si-Di	=BAL 99 (Typ-Code/Stempel/marking)	35	Phi	→BAL 99	data
	MOS-P-FET-e	=2SJ316 (Typ-Code/Stempel/marking)	39	Say	→2SJ316	data
	N-FET	=2SK208-G (Typ-Code/Stempel/marking)	35	Tos	→2SK208	data
	N-FET	=2SK879-GR (Typ-Code/Stempel/marking)	35(2mm)	Tos	→2SK879	data
	Si-Di	=BAR 99 (Typ-Code/Stempel/marking)	35	Sie, Tho	→BAR 99	data
	Si-Di	=BAR 99 (Typ-Code/Stempel/marking)	35	Sie	→BAR 99	data
	Si-P	=2SB1518-D (Typ-Code/Stempel/marking)	39	Hit	→2SB1518	data
	Si-Di	=BAV 70 (Typ-Code/Stempel/marking)	35	Sie,Aeg	→BAV 70	data
	Si-Di	=U 05JH44 (Typ-Code/Stempel/marking)	71(5x2,5)	Tos	→U 05JH44	data
	Si-P	=2SA1566-D (Typ-Code/Stempel/marking)	35	Hit	→2SA1566	data
	Si-P	=2SA1566-E (Typ-Code/Stempel/marking)	35	Hit	→2SA1566	data
	Si-P	=2SB1518-E (Typ-Code/Stempel/marking)	39	Hit	→2SB1518	data
	Si-Di	=BAV 70 (Typ-Code/Stempel/marking)	35	Sie, Gsi	→BAV 70	data
	Si-Di	=BAW 56 (Typ-Code/Stempel/marking)	35	Aeg	→BAW 56	data
	MOS-N-FET-d	=BF 1009 (Typ-Code/Stempel/marking)	44	Sie	→BF 1009	data
	Si-P	=2SA1508 (Typ-Code/Stempel/marking)	35	Say	→2SA1508	data
	MOS-P-FET-e*	=2SJ560 (Typ-Code/Stempel/marking)	39	Say	→2SJ560	data
101A	UNI	Vs:±22V Vu:>50V/mV Vo:±10V Vi0:±2mV	8DA,14F	Nsc		data pdf pinout
108	LO-POWER	Vs:±20V Vu:110dB Vo:±13V Vi0:0.3mV	14D,8A,10F	Nsc		data pdf pinout
111	COMP	Vs:36V Vu:>40V/mV Vi0:±3mV	14D,8A,10F	Nsc		data pdf pinout
118	HI-SPEED	Vs:±20V Vu:106dB Vo:±13V Vi0:2mV	14D,8A,10F	Nsc		data pdf pinout
119	2xCOMP	Vs:36V Vu:10.5V/mV Vo:TTL Vi0:±3.8mV	10A	Nsc		data pdf pinout
124	4xOP	Vs:32V Vu:100V/mV Vi0:3mV	14DFS	Nsc		data pdf pinout
139	4xCOMP	Vs:±18V Vu:>50V/mV Vo:TTL Vi0:±5mV	14DFS	Nsc		data pdf pinout
148	4xOP	Vs:±22V Vu:104dB Vo:±12V Vi0:1mV	14DFS	Has,Nsc,Ray		data pdf pinout
s	MOS-N-FET-d	=BF 1009S (Typ-Code/Stempel/marking)	44	Sie	→BF 1009S	data
s	MOS-N-FET-d	=BF 1009SW (Typ-Code/Stempel/marking)	44(2mm)	Sie	→BF 1009SW	
	Si-N	=2SC2059-JM (Typ-Code/Stempel/marking)	≳35	Rhm	→2SC2059	data
	Si-N	=2SC2059K-M (Typ-Code/Stempel/marking)	35	Rhm	→2SC2059K	data
	Si-N	=2SC4099-M (Typ-Code/Stempel/marking)	35(2mm)	Rhm	→2SC4099	data
	Si-N	=2SC4649-M (Typ-Code/Stempel/marking)	35(1,6mm)	Rhm	→2SC4649	data
	MOS-P-FET-e*	=2SJ561 (Typ-Code/Stempel/marking)	39	Say	→2SJ561	data
38510/10104	LO-CURR	Vs:±20V Vo:>16/0V Vi0:±0.5mV	8AD	Pmi		data pinout
38510/11401	JFET	Vs:±22V Vo:>±16V Vi0:±5mV	8AD	Pmi		data pinout
38510/11402	JFET	Vs:±22V Vo:>±16V Vi0:±5mV	8AD	Pmi		data pinout
38510/11403	JFET	Vs:±22V Vo:>±16V Vi0:±5mV	8AD	Pmi		data pinout
38510/11404	JFET	Vs:±22V Vo:>±16V Vi0:±2mV	8AD	Pmi		data pinout
38510/11405	JFET	Vs:±22V Vo:>±16V Vi0:±2mV	8AD	Pmi		data pinout
38510/11406	JFET	Vs:±22V Vo:>±16V Vi0:±2mV	8AD	Pmi		data pinout
38510/13501	LO-OFFSET	Vs:±22V Vu:>300V/mV Vo:±12V	8AD	Pmi		data pinout
38510/13502	LO-OFFSET	Vs:±22V Vu:>200V/mV Vo:±12V	8AD	Pmi		data pinout
38510/13503	LO-NOISE	Vs:±22V Vo:±11.5V Vi0:±25µµV	8AD	Pmi		data pinout
38510/10104...	OP-IC	=PM 108...: MIL-M-38510 Standard		Pmi	·	data
38510/10106...	OP-IC	=PM 2108...: MIL-M-38510 Standard		Pmi	·	data
38510/11004...	OP-IC	=PM 4136...: MIL-M-38510 Standard		Pmi	·	data
38510/11301...	D/A-IC	8 Bit, multipl., hi-speed: MIL-M-38510		Pmi	·	data
38510/11302...	A/D-IC	8 Bit, multiplying, hi-speed		Pmi	·	data
38510/11401...	OP-IC	=LF 155...: MIL-M-38510 Standard		Pmi	·	data
38510/11402...	OP-IC	=LF 156...: MIL-M-38510 Standard		Pmi	·	data
38510/11403...	OP-IC	=LF 157...: MIL-M-38510 Standard		Pmi	·	data
38510/11404...	OP-IC	=LF 155A...: MIL-M-38510 Standard		Pmi	·	data
38510/11405...	OP-IC	=LF 156A...: MIL-M-38510 Standard		Pmi	·	data
38510/11406...	OP-IC	=LF 157A...: MIL-M-38510 Standard		Pmi	·	data
38510/13501...	OP-IC	=OP-07A...: MIL-M-38510 Standard		Pmi	·	data
38510/13502...	OP-IC	=OP-07...: MIL-M-38510 Standard		Pmi	·	data
38510/13503...	OP-IC	=OP-27A...: MIL-M-38510 Standard		Pmi	·	data
38510/10106B	2xLO-CURR	Vs:±20V Vo:±16V Vi0:±0.5mV	16D	Pmi		data pinout
38510/11004B	4xUNI	Vs:±22V Vo:±16V Vi0:±5mV	14D	Pmi		data pinout
	Si-N	=2SC2059-JN (Typ-Code/Stempel/marking)	≳35	Rhm	→2SC2059	data
	Si-N	=2SC2059K-N (Typ-Code/Stempel/marking)	35	Rhm	→2SC2059K	data
	Si-N	=2SC4099-N (Typ-Code/Stempel/marking)	35(2mm)	Rhm	→2SC4099	data
	Si-N	=2SC4649-N (Typ-Code/Stempel/marking)	35(1,6mm)	Rhm	→2SC4649	data
	MOS-P-FET-e*	=2SJ562 (Typ-Code/Stempel/marking)	39	Say	→2SJ562	data
	Si-P	=2SA1384-O (Typ-Code/Stempel/marking)	39	Tos	→2SA1384	data
	MOS-P-FET-e*	=2SJ563 (Typ-Code/Stempel/marking)	39	Say	→2SJ563	data
	N-FET	=2SK208-O (Typ-Code/Stempel/marking)	35	Tos	→2SK208	data
	N-FET	=2SK879-O (Typ-Code/Stempel/marking)	35(2mm)	Tos	→2SK879	data
	Si-P	=KTA 1668-O (Typ-Code/Stempel/marking)	39	Kec	→KTA 1668	data
1006	Si-N	VHF-L, 60V, 12A, PQ=100W(175MHz)	57s	Mot		data

Type	Device	Short Description	Fig.	Manu	Comparision Types	More at
JO 2015	Si-N	UHF-L, 65V, 9A, PQ=50W(400MHz)	57s	Mot	MRF 327[Motorola]	data
JO 2017	Si-N	UHF-L, 65V, 15A, PQ=65W(440MHz)	57s	Mot		data
JO 3037	Si-N	UHF-L, 36V, 5A, PQ=37W(470MHz)	57s	Mot	MRF 646[Motorola]	data
JO 3501	Si-N	UHF-L, 50V, 2A, PQ=15W(470MHz)	61w	Mot	-	data
JO 3502	Si-N	UHF-L., 50V, 4A, PQ=35W(960MHz)	≈61w	Mot	-	data
JO 4036	Si-N	VHF-L, 36V, 6,5A, PQ=36W(175MHz)	57s	Mot	-	data
JO 4045	Si-N	VHF-L, 36V, 6,5A, PQ=45W(175MHz)	57s	Mot	-	data
JP	Si-N	=2SC2059-JP (Typ-Code/Stempel/marking)	≈35	Rhm	→2SC2059	data
JP	Si-N	=2SC2059K-P (Typ-Code/Stempel/marking)	35	Rhm	→2SC2059K	data
JP	Si-N	=2SC4099-P (Typ-Code/Stempel/marking)	35(2mm)	Rhm	→2SC4099	data
JP	Si-N	=2SC4649-P (Typ-Code/Stempel/marking)	35(1,6mm)	Rhm	→2SC4649	data
JP	MOS-P-FET-e*	=2SJ578 (Typ-Code/Stempel/marking)	39	Say	→2SJ578	data
JP	Si-Di	=BAS 19 (Typ-Code/Stempel/marking)	35	Phi, Sie	→BAS 19	data
JP	Si-Di	=BAW 101 (Typ-Code/Stempel/marking)	44	Sie	→BAW 101	data
JP	Si-Di	=BAW 56GT (Typ-Code/Stempel/marking)	44	Sie	→BAW 56GT	data
JPAD 5	Si-Di	Pico-Amp Diode, 35V, 10mA, Ir<5pA(20V)	7d	Six	-	data
JPAD 50	Si-Di	Pico-Amp Di, 35V, 10mA, Ir<50pA(20V)	7d	Six	-	data
JPp	Si-Di	=BAS 19 (Typ-Code/Stempel/marking)	35	Phi	→BAS 19	data
JPs	Si-Di	=BAS 19 (Typ-Code/Stempel/marking)	C5	Sie	→BAS 19	data
JPs	Si-Di	=BAW 101 (Typ-Code/Stempel/marking)	44	Sie	→BAW 101	data
JPt	Si-Di	=BAS 19 (Typ-Code/Stempel/marking)	35	Phi	→BAS 19	data
JQ	MOS-P-FET-e*	=2SJ579 (Typ-Code/Stempel/marking)	39	Say	→2SJ579	data
JR	Si-P	=2SA1384-R (Typ-Code/Stempel/marking)	39	Tos	→2SA1384	data
JR	Si-N	=2SC5631 (Typ-Code/Stempel/marking)	39	Hit	→2SC5631	
JR	N-FET	=2SK208-R (Typ-Code/Stempel/marking)	35	Tos	→2SK208	data
JR	N-FET	=2SK879-R (Typ-Code/Stempel/marking)	35(2mm)	Tos	→2SK879	data
JR	Si-Di	=BAS 20 (Typ-Code/Stempel/marking)	35	Phi, Sie	→BAS 20	data
JR	Si-Di	=BAS 20H... (Typ-Code/Stempel/marking)	71(1,7mm)	Ons	→BAS 20H	
JR-	Si-N	=2SC5773 (Typ-Code/Stempel/marking)	35	Hit	→2SC5773	
JRp	Si-Di	=BAS 20 (Typ-Code/Stempel/marking)	35	Phi	→BAS 20	data
JRs	Si-Di	=BAS 20 (Typ-Code/Stempel/marking)	35	Sie	→BAS 20	data
JRt	Si-Di	=BAS 20 (Typ-Code/Stempel/marking)	35	Phi	→BAS 20	data
JS	Si-P	=2SA1815 (Typ-Code/Stempel/marking)	35	Say	→2SA1815	data
JS	Si-N.	=2SD2532 (Typ-Code/Stempel/marking)	39	Hit	→2SD2532	data
JS	Si-Di	=BAS 21 (Typ-Code/Stempel/marking)	35	Phi, Sie	→BAS 21	data
JS	Si-Di	=BAS 21H... (Typ-Code/Stempel/marking)	71(1,7mm)	Ons	→BAS 21H	
JS	Si-Di	=BAS 21L... (Typ-Code/Stempel/marking)	35	Ons	→BAS 21L	
JS	Si-Di	=BAS 21SL... (Typ-Code/Stempel/marking)	35	Ons	→BAS 21SL	
JS	Si-Di	=BAS 221 (Typ-Code/Stempel/marking)	71(2mm)	Phi	→BAS 221	data
JS	Si-Di	=BAW 100 (Typ-Code/Stempel/marking)	44	Sie	→BAW 100	data
JS	Si-Di	=BAW 56G (Typ-Code/Stempel/marking)	44	Sie	→BAW 56G	data
JS	Si-Di	=MMSD 103 (Typ-Code/Stempel/marking)	71(2,7mm)	Ons	→MMSD 103	
JS 3	Si-P	=2SA1815-3 (Typ-Code/Stempel/marking)	35	Say	→2SA1815	data
JS 4	Si-P	=2SA1815-4 (Typ-Code/Stempel/marking)	35	Say	→2SA1815	data
JS 5	Si-P	=2SA1815-5 (Typ-Code/Stempel/marking)	35	Say	→2SA1815	data
JS601	IRED	870nm ca. ø12mm I:80mW/Sr Vf:1.5V	TO46	Oec		data pdf pinout
JS640	LED	rd ca. ø12mm I:90mW/Sr Vf:2V If:75mA	TO46	Oec		data pdf pinout
JS645	LED	rd ca. ø12mm I:80mW/Sr Vf:2V If:75mA	TO46	Oec		data pdf pinout
JS675	LED	rd ca. ø12mm I:90mW/Sr Vf:2V If:55mA	TO46	Oec		data pdf pinout
JS680	IRED	770nm ca. ø12mm I:110mW/Sr Vf:2V	TO46	Oec		data pdf pinout
JS770	IRED	770nm ca. ø12mm I:110mW/Sr Vf:2V	TO46	Oec		data pdf pinout
JS870	IRED	870nm ca. ø12mm I:14mW/Sr Vf:1.4V	TO46	Oec		data pdf pinout
JSp	Si-Di	=BAS 21 (Typ-Code/Stempel/marking)	35	Phi	→BAS 21	data
JSs	Si-Di	=BAS 21 (Typ-Code/Stempel/marking)	35	Sie	→BAS 21	data
JSs	Si-Di	=BAW 100 (Typ-Code/Stempel/marking)	44	Sie	→BAW 100	data
JSt	Si-Di	=BAS 21 (Typ-Code/Stempel/marking)	35	Phi	→BAS 21	data
JT	Si-N	=2SC4006 (Typ-Code/Stempel/marking)	35	Say	→2SC4006	data
JT	Si-N	=2SC4269 (Typ-Code/Stempel/marking)	35	Say	→2SC4269	data
JT	Si-Di	=BAS 28 (Typ-Code/Stempel/marking)	44	Phi, Sie	→BAS 28	data
JT 4 B25-AS(C)	BiMOS-IC	64 Bit Thermal Head Driver	Chip	Tos	-	
JT 1538-AS	LIN-IC	32 Bit Thermal Printer Head Driver	Chip	Tos	-	
JTA1205	OC	LED VR/2 Viso:4000Vrms Vbs:<8V		Cpc,The		data pinout
JTA1210	OC	LED VR/2 Viso:4000Vrms Vbs:<8V		Cpc,The		data pinout
JTA2405	OC	LED VR/2 Viso:4000Vrms Vbs:<8V		Cpc,The		data pinout
JTA2410	OC	LED VR/2 Viso:4000Vrms Vbs:<8V		Cpc,The		data pinout
JTA2415	OC	LED VR/2 Viso:4000Vrms Vbs:<8V		Cpc,The		data pinout
JTB1205	OC	LED VR/2 Viso:4000Vrms Vbs:<16V		Cpc,The		data pinout
JTB1210	OC	LED VR/2 Viso:4000Vrms Vbs:<16V		Cpc,The		data pinout
JTB2405	OC	LED VR/2 Viso:4000Vrms Vbs:<16V		Cpc,The		data pinout
JTB2410	OC	LED VR/2 Viso:4000Vrms Vbs:<16V		Cpc,The		data pinout
JTB2415	OC	LED VR/2 Viso:4000Vrms Vbs:<16V		Cpc,The		data pinout
JTC240H	OC	LED VR/2 Viso:4000Vrms Vbs:<26V		Cpc,The		data pinout
JTC1205	OC	LED VR/2 Viso:4000Vrms Vbs:<26V		Cpc,The		data pinout
JTC1210	OC	LED VR/2 Viso:4000Vrms Vbs:<26V		Cpc,The		data pinout
JTC2405	OC	LED VR/2 Viso:4000Vrms Vbs:<26V		Cpc,The		data pinout
JTC2410	OC	LED VR/2 Viso:4000Vrms Vbs:<26V		Cpc,The		data pinout
JTC2415	OC	LED VR/2 Viso:4000Vrms Vbs:<26V		Cpc,The		data pinout
JTp	Si-Di	=BAS 28 (Typ-Code/Stempel/marking)	44	Phi	→BAS 28	data
JTs	Si-Di	=BAS 28W (Typ-Code/Stempel/marking)	44(2mm)	Sie	→BAS 28W	data
JTs	Si-Di	=BAS 28 (Typ-Code/Stempel/marking)	44	Sie	→BAS 28	data
JU	Si-Di	=BAS 16 (Typ-Code/Stempel/marking)	35	Mot,Sie,Tho	→BAS 16	data
JU	Si-Di	=U 1JU44 (Typ-Code/Stempel/marking)	71(5x2,5)	Tos	→U 1JU44	data

Type	Device	Short Description	Fig.	Manu	Comparision Types	More at
JUCA 4532	MOS-IC	Thermo-druckkopf/printing head driver	40-	Tho	-	
JUCA 4632	IC	2x 16 Dioden, gem./common Anode, 30V	40-	Tho	-	
JV	Si-Di	=BAS 116 (Typ-Code/Stempel/marking)	35	Sie, Ons	→BAS 116	data
JV	Si-Di	=BAT 54H... (Typ-Code/Stempel/marking)	71(1,7mm)	Ons	→BAT 54H	
JV	PIN-Di	=1SV263 (Typ-Code/Stempel/marking)	35 (2mm)	Say	→1SV263	data
JV	PIN-Di	=1SV266 (Typ-Code/Stempel/marking)	35	Say	→1SV266	data
JV 3	Si-Di	=BAT 54L... (Typ-Code/Stempel/marking)	35	Ons	→BAT 54L	
JVp	Si-Di	=BAS 116 (Typ-Code/Stempel/marking)	35	Phi	→BAS 116	data
JVs	Si-Di	=BAS 116 (Typ-Code/Stempel/marking)	35	Sie	→BAS 116	data
JXp	Si-Di	=BAV 170 (Typ-Code/Stempel/marking)	35	Phi	→BAV 170	data
JXs	Si-Di	=BAV 170 (Typ-Code/Stempel/marking)	35	Sie	→BAV 170	data
JXt	Si-Di	=BAV 170 (Typ-Code/Stempel/marking)	35	Phi	→BAV 170	data
JY	Si-Di	=BAV 199L (Typ-Code/Stempel/marking)	35	Ons	→BAV 199L	data
JY-	Si-Di	=BAV 199W (Typ-Code/Stempel/marking)	35(2mm)	Phi	→BAV 199W	data
JY	Si-N	=2SC3770 (Typ-Code/Stempel/marking)	35	Say	→2SC3770	data
JY	MOS-P-FET-e	=2SJ244 (Typ-Code/Stempel/marking)	39	Hit	→2SJ244	data
JY	N-FET	=2SK208-Y (Typ-Code/Stempel/marking)	35	Tos	→2SK208	data
JY	N-FET	=2SK879-Y (Typ-Code/Stempel/marking)	35(2mm)	Tos	→2SK879	data
JY	Si-P	=KTA 1668-Y (Typ-Code/Stempel/marking)	39	Kec	→KTA 1668	data
JYp	Si-Di	=BAV 199 (Typ-Code/Stempel/marking)	35	Phi	→BAV 199	data
JYs	Si-Di	=BAV 199 (Typ-Code/Stempel/marking)	35	Sie	→BAV 199	data
JYt	Si-Di	=BAV 199 (Typ-Code/Stempel/marking)	35	Phi	→BAV 199	data
JYt	Si-Di	=BAV 199W (Typ-Code/Stempel/marking)	35(2mm)	Phi	→BAV 199W	data
JZ	Si-Di	=BAW 156 (Typ-Code/Stempel/marking)	35	Sie	→BAW 156	data
JZp	Si-Di	=BAW 156 (Typ-Code/Stempel/marking)	35	Phi	→BAW 156	data
JZs	Si-Di	=BAW 156 (Typ-Code/Stempel/marking)	35	Sie	→BAW 156	data
JZt	Si-Di	=BAW 156 (Typ-Code/Stempel/marking)	35	Phi	→BAW 156	data

K

Type	Device	Short Description	Fig.	Manu	Comparision Types	More at
K...	...-FET	→2SK.., z.B./e.g. "K727" = 2SK727→	Japantypen	JAP		
K...	...-FET	→KSK.., z.B./e.g. "K 65" = KSK 65→	Samsung	Sam		
K...	...-FET	→KTK.., z.B./e.g. "K821" = KTK821→	KEC	Kec		

Type	Device	Short Description	Fig.	Manu	Comparision Types	More at
K	Si-P	=2SB1208 (Typ-Code/Stempel/marking)	39	Mat	→2SB1208	data
K	Si-N	=2SC2778 (Typ-Code/Stempel/marking)	35	Mat	→2SC2778	data
K	Si-N	=2SC3936 (Typ-Code/Stempel/marking)	35(2mm)	Mat	→2SC3936	data
K	Si-N	=2SC4655 (Typ-Code/Stempel/marking)	35(1,6mm)	Mat	→2SC4655	data
K	GaAs-N-FET	=2SK1688 (Typ-Code/Stempel/marking)	51	Mat	→2SK1688	data
K	Si-Di	=BAT 68-03W (Typ-Code/Stempel/marking)	71(1,7mm)	Sie	→BAT 68-03W	data
K	C-Di	=BB 142 (Typ-Code/Stempel/marking)	71(1,3mm)	Phi	→BB 142	data
K	C-Di	=BBY 52-02W (Typ-Code/Stempel/marking)	71(1,3mm)	Sie	→BBY 52-02W	data
K	C-Di	=HVU 315 (Typ-Code/Stempel/marking)	71(1,7mm)	Hit	→HVU 315	data
K	Si-Di	=MA 2S304 (Typ-Code/Stempel/marking)	71(1,3mm)	Mat	→MA 2S304	data
K	Z-Di	=MAZS 039 (Typ-Code/Stempel/marking)	71(1,3mm)	Mat	→MAZS 039	data
K	Si-Di	=SB 007W03C (Typ-Code/Stempel/marking)	35	Say	→SB 007W03C	data
K	Si-Di	=SB 007W03Q (Typ-Code/Stempel/marking)	35(2mm)	Say	→SB 007W03Q	data
K 1	Si-N	=BCW 71 (Typ-Code/Stempel/marking)	35	Fer, Phi	→BCW 71	data
K 1	Z-Di	=MMSZ 5251B (Typ-Code/marking)	71(2,7mm)	Ons	→MMSZ 5251B	
K 1	MOS-N-FET-e*	=UM 5K1N (Typ-Code/Stempel/marking)	45(2mm)	Rhm	→UM 5K1N	data
K 1	MOS-N-FET-e*	=UM 6K1N (Typ-Code/Stempel/marking)	46(2mm)	Rhm	→UM 6K1N	data
K 1	MOS-N-FET-e	=VN 10KT (Typ-Code/Stempel/marking)	35	Six	→VN 10KT	data
K 1	MOS-N-FET-e	=2SK2741 (Typ-Code/Stempel/marking)	39	Tos	→2SK2741	data
K 1	N-FET	=2SK57-1 (Typ-Code/Stempel/marking)	35	Nec	→2SK57	data
K 1 O	Si-N	=KSC 3265-O (Typ-Code/Stempel/marking)	35	Sam	→KSC 3265	data
K 1 p	Si-N	=BCW 71 (Typ-Code/Stempel/marking)	35	Phi	→BCW 71	data
K 1 t	Si-N	=BCW 71 (Typ-Code/Stempel/marking)	35	Phi	→BCW 71	data

Type	Device	Short Description	Fig.	Manu	Comparision Types	More at
K 1 V 5	Diac	Ub=45..60V, Ib<0,5mA, Itsm=13A	31	Shi	-	data
K 1 V 6	Diac	=K1V5: Ub=55..65V	31	Shi	-	
K 1 V 8	Diac	Ub=77..95V, Ib<0,5mA, Itsm=20A	31	Shi	-	data
K 1 V 10	Diac	=K1V5: Ub=95..113V, Itsm=20A	31	Shi	-	data
K 1 V 11	Diac	=K1V5: Ub=104..118V, Itsm=20A	31	Shi	-	
K 1 V 12	Diac	=K1V5: Ub=110..125V, Itsm=20A	31	Shi	-	
K 1 V 14	Diac	Ub=125..150V, Ib<0,5mA, Itsm=20A	31	Shi	-	
K 1 V 16	Diac	=K1V14: Ub=145..170V	31	Shi	-	
K 1 V 18	Diac	=K1V14: Ub=165..200V	31	Shi	-	
K 1 V 22	Diac	Ub=200..230V, Ib<0,5mA, Itsm=20A	31	Shi	-	data
K 1 V 22W	Diac	Ub=200..230V, Ib<0,5mA, Itsm=16A	31	Shi	-	data
K 1 V 24	Diac	=K1V22: Ub=220..250V	31	Shi	-	
K 1 V 24W	Diac	=K1V22W: Ub=220..250V	31	Shi	-	
K 1 V 26	Diac	=K1V22: Ub=240..270V	31	Shi	-	
K 1 V 26W	Diac	=K1V22W: Ub=240..265V	31	Shi	-	data
K 1 V 33W	Diac	=K1V22W: Ub=309..355V, Itsm=13A	31	Shi	-	
K 1 V 34W	Diac	=K1V22W: Ub=320..360V, Itsm=13A	31	Shi	-	
K 1 V 36W	Diac	=K1V22W: Ub=340..380V, Itsm=13A	31	Shi	-	
K 1 V 38W	Diac	=K1V22W: Ub=360..400V, Itsm=13A	31	Shi	-	
K 1 V(A)10...16	Diac	=K1V10...16: Itsm=16A	31	Shi	-	data

Type	Device	Short Description	Fig.	Manu	Comparision Types	More at
K1Y	Si-N	=KSC 3265-Y (Typ-Code/Stempel/marking)	35	Sam	→KSC 3265	data

Type	Device	Short Description	Fig.	Manu	Comparision Types	More at
K 1,5/20	Ge-Di	≈OA 161	31a		→OA 161	

Type	Device	Short Description	Fig.	Manu	Comparision Types	More at
K 2	MOS-N-FET-e*	=2SK2742 (Typ-Code/Stempel/marking)	39	Tos	→2SK2742	data
K 2	N-FET	=2SK57-2 (Typ-Code/Stempel/marking)	35	Nec	→2SK57	data
K 2	Si-N	=BCW 72 (Typ-Code/Stempel/marking)	35	Fer,Mot,Phi	→BCW 72	data
K 2	MOS-NP-FET-e	=HN 1L02FU (Typ-Code/Stempel/marking)	46(2mm)	Tos	→HN 1L02FU	data

Type	Device	Short Description	Fig.	Manu	Comparision Types	More at
K 2	Z-Di	=MMSZ 5252B (Typ-Code/marking)	71(2,7mm)	Ons	→MMSZ 5252B	
K 2 C5	Diac	Ub=44...61V, Ib<0,5mA, Itsm=20A	31	Shi	K1V5	data
K 2 C6	Diac	=K2C5: Ub=56...67V	31	Shi	K1V6	data
K 2 F5	Diac	=K2C5:	34	Shi	K1V5	data
K 2 F6	Diac	=K2C6:	34	Shi	K1V6	data
K 2 p	Si-N	=BCW 72 (Typ-Code/Stempel/marking)	35	Phi	→BCW 72	data
K 2 t	Si-N	=BCW 72 (Typ-Code/Stempel/marking)	35	Phi	→BCW 72	data
K 2,5/15	Ge-Di	≈OA 161	31a		→OA 161	
K 3	Si-N	=2SC3143-3 (Typ-Code/Stempel/marking)	35	Say	→2SC3143	data
K 3	MOS-N-FET-e*	=2SK2839 (Typ-Code/Stempel/marking)	39	Tos	→2SK2839	
K 3	N-FET	=2SK57-3 (Typ-Code/Stempel/marking)	35	Nec	→2SK57	data
K 3	Si-N	=BCW 81 (Typ-Code/Stempel/marking)	35	Phi	→BCW 81	data
K 3	MOS-NP-FET-e	=HN 1L03FU (Typ-Code/Stempel/marking)	46(2mm)	Tos	→HN 1L03FU	
K 3	Z-Di	=MMSZ 5253B (Typ-Code/marking)	71(2,7mm)	Ons	→MMSZ 5253B	
K 3 A	C-Di	=KDV 804S (Typ-Code/Stempel/marking)	35	Kec	→KDV 804S	data
K 3 B	C-Di	=KDV 804S (Typ-Code/Stempel/marking)	35	Kec	→KDV 804S	data
K 3 C	C-Di	=KDV 804S (Typ-Code/Stempel/marking)	35	Kec	→KDV 804S	data
K 3 D	C-Di	=KDV 804S (Typ-Code/Stempel/marking)	35	Kec	→KDV 804S	data
K 3 E	C-Di	=KDV 804S (Typ-Code/Stempel/marking)	35	Kec	→KDV 804S	data
K 3 p	Si-N	=BCW 81 (Typ-Code/Stempel/marking)	35	Phi	→BCW 81	data
K 4	Si-N	=2SC3143-4 (Typ-Code/Stempel/marking)	35	Say	→2SC3143	
K 4	N-FET	=2SK160-4 (Typ-Code/Stempel/marking)	35	Nec	→2SK160	data
K 4	MOS-N-FET-e*	=2SK2836 (Typ-Code/Stempel/marking)	39	Tos	→2SK2836	
K 4	N-FET	=2SK853-K4 (Typ-Code/Stempel/marking)	35(2mm)	Nec	→2SK853	
K 4	Si-N	=BCW 71R (Typ-Code/Stempel/marking)	35	Val,Fer	→BCW 71R	data
K 4	Z-Di	=MMSZ 5254B (Typ-Code/marking)	71(2,7mm)	Ons	→MMSZ 5254B	
K 4	MOS-NP-FET-e	=SSM 6L05FU (Typ-Code/Stempel/marking)	46(2mm)	Tos	→SSM 6L05FU	
K 005...06	Si-Br	50...600V, 25A	70	Mic	B...C25A	data
K 5	Si-N	=2SC3143-5 (Typ-Code/Stempel/marking)	35	Say	→2SC3143	
K 5	N-FET	=2SK160-5 (Typ-Code/Stempel/marking)	35	Nec	→2SK160	data
K 5	N-FET	=2SK853-K5 (Typ-Code/Stempel/marking)	35(2mm)	Nec	→2SK853	
K 5	Si-N	=BCW 72R (Typ-Code/Stempel/marking)	35	Val	→BCW 72R	data
K 5	Z-Di	=MMSZ 5255B (Typ-Code/marking)	71(2,7mm)	Ons	→MMSZ 5255B	
K 5	MOS-NP-FET-e	=SSM 6L09FU (Typ-Code/Stempel/marking)	46(2mm)	Tos	→SSM 6L09FU	
K 5/6	Ge-Di	≈OA 161	31a		→OA 161	
K 6	N-FET	=2SK160-6 (Typ-Code/Stempel/marking)	35	Nec	→2SK160	data
K 6	N-FET	=2SK853-K6 (Typ-Code/Stempel/marking)	35(2mm)	Nec	→2SK853	data
K 6	Si-N	=BCV 71R (Typ-Code/Stempel/marking)	35	Fer	→BCV 71R	data
K 7	N-FET	=2SK160-7 (Typ-Code/Stempel/marking)	35	Nec	→2SK160	data
K 7	N-FET	=2SK853-K7 (Typ-Code/Stempel/marking)	35(2mm)	Nec	→2SK853	data
K 7	Si-N	=BCV 71 (Typ-Code/Stempel/marking)	35	Fer, Phi	→BCV 71	data
K 7 p	Si-N	=BCV 71 (Typ-Code/Stempel/marking)	35	Phi	→BCV 71	data
K 7 t	Si-N	=BCV 71 (Typ-Code/Stempel/marking)	35	Phi	→BCV 71	data
K 8	Si-N	=BCV 72 (Typ-Code/Stempel/marking)	35	Fer, Phi	→BCV 72	data
K 8 p	Si-N	=BCV 72 (Typ-Code/Stempel/marking)	35	Phi	→BCV 72	data
K 8 t	Si-N	=BCV 72 (Typ-Code/Stempel/marking)	35	Phi	→BCV 72	data
K 9	Si-Di	=1SS348 (Typ-Code/Stempel/marking)	35	Tos	→1SS348	data
K 9	Si-N	=BCF 81 (Typ-Code/Stempel/marking)	35	Phi	→BCF 81	data
K 9	Si-N	=BCV 72R (Typ-Code/Stempel/marking)	35	Fer	→BCV 72R	data
K 9 p	Si-N	=BCF 81 (Typ-Code/Stempel/marking)	35	Phi	→BCF 81	data
K 11	Si-N	=BCX 70G (Typ-Code/Stempel/marking)	35	Val	→BCX 70G	data
K 12	Si-N	=BCX 70H (Typ-Code/Stempel/marking)	35	Val	→BCX 70H	data
K 13	Si-N	=BCX 70J (Typ-Code/Stempel/marking)	35	Val	→BCX 70J	data
K 14	N-FET	=2SK238-K14 (Typ-Code/Stempel/marking)	35	Nec	→2SK238	data
K 14	Si-N	=BCX 70K (Typ-Code/Stempel/marking)	35	Val	→BCX 70K	data
K 15	N-FET	=2SK238-K15 (Typ-Code/Stempel/marking)	35	Nec	→2SK238	data
K 15 N	Diac	SMD, Ub=125...160V	71			
K 16	N-FET	=2SK238-K16 (Typ-Code/Stempel/marking)	35	Nec	→2SK238	data
K 17	N-FET	=2SK238-K17 (Typ-Code/Stempel/marking)	35	Nec	→2SK238	data
K 24	N-FET	=2SK160A-K24(Typ-Code/Stempel/marking)	35	Nec	→2SK160A	data
K 24	N-FET	=2SK853A-K24(Typ-Code/Stempel/marking)	35(2mm)	Nec	→2SK853A	data
K 25	N-FET	=2SK160A-K25(Typ-Code/Stempel/marking)	35	Nec	→2SK160A	data
K 25	N-FET	=2SK853A-K25(Typ-Code/Stempel/marking)	35(2mm)	Nec	→2SK853A	data
K 26	N-FET	=2SK160A-K26(Typ-Code/Stempel/marking)	35	Nec	→2SK160A	data
K 26	N-FET	=2SK853A-K26(Typ-Code/Stempel/marking)	35(2mm)	Nec	→2SK853A	data
K 27	N-FET	=2SK160A-K27(Typ-Code/Stempel/marking)	35	Nec	→2SK160A	data
K 27	N-FET	=2SK853A-K27(Typ-Code/Stempel/marking)	35(2mm)	Nec	→2SK853A	data
K 31	Si-N	=BCW 81R (Typ-Code/Stempel/marking)	35	Val	→BCW 81R	data
K 41	N-FET	=2SK520-K41 (Typ-Code/Stempel/marking)	35	Nec	→2SK520	data
K 42	N-FET	=2SK520-K42 (Typ-Code/Stempel/marking)	35	Nec	→2SK520	data
K 43	N-FET	=2SK520-K43 (Typ-Code/Stempel/marking)	35	Nec	→2SK520	data
K 44	N-FET	=2SK520-K44 (Typ-Code/Stempel/marking)	35	Nec	→2SK520	data
K 45	N-FET	=2SK520-K45 (Typ-Code/Stempel/marking)	35	Nec	→2SK520	data
K 51	N-FET	=2SK508-K51 (Typ-Code/Stempel/marking)	35	Nec	→2SK508	data
K 52	N-FET	=2SK508-K52 (Typ-Code/Stempel/marking)	35	Nec	→2SK508	data
K 53	N-FET	=2SK508-K53 (Typ-Code/Stempel/marking)	35	Nec	→2SK508	data
K 71	Si-N	=BCV 71R (Typ-Code/Stempel/marking)	35	Val	→BCV 71R	data
K 80	Si-Di	=BAS 19 (Typ-Code/Stempel/marking)	35	Aeg	→BAS 19	
K 81	Si-N	=BCV 72R (Typ-Code/Stempel/marking)	35	Val	→BCV 72R	data
K 81	Si-Di	=BAS 20 (Typ-Code/Stempel/marking)	35	Aeg	→BAS 20	
K 82	Si-Di	=BAS 21 (Typ-Code/Stempel/marking)	35	Aeg	→BAS 21	

Type	Device	Short Description	Fig.	Manu	Comparision Types	More at
91	Si-N	=BCF 81R (Typ-Code/Stempel/marking)	35	Val	→BCF 81R	data
102P	OC	LED/NP;j Viso:1000V CTR:50% Vbs:1.25V		Aeg,Tel	CNY57,SFH601G	data pinout
104P	OC	LED/NPN Viso:1500V CTR:25-400%		Aeg	SL5504	data pinout
109P	OC	LED/NPN Viso:1500V CTR:>20% lbs:20mA		Aeg	SL5511	data pinout
119P	OC	LED/NPN CTR:100-200%		Aeg		data pinout
133P	OC	LED/NPr! Viso:10000V CTR:25-300%		Aeg		data pinout
174 АФ1А	LIN-IC	TV,Ctv, Synchr.-signal Prozessor	16-DIP	GUS	-	
174 АФ4А	LIN-IC	Ctv, RGB Signal-prozessor, Ucc=12V	16-DIP	GUS	-	
174 АФ5	LIN-IC	Ctv, RGB Matrix, Ucc=12V	16-DIP	GUS	-	
174 KH1	LIN-IC	8-Kanal Schalter/8-Ch. Switch, Ucc=12V	16-DIP	GUS		
174 KH2	LIN-IC	8-Kanal Schalter/8-Ch. Switch, Ucc=12V	16-DIP	GUS		
174 КП1	LIN/OP-IC	Dual, Audio Select, 2x4 Inputs	16-DIP	GUS	TDA 1029	
174 КП3	LIN-IC	TV, Programmwahl/program Select Ctrl.	28-DIP	GUS		
174 XA1M	LIN-IC	Ctv, SECAM-Signal, Ucc=12V	16-DIP	GUS	-	
174 XA2	LIN-IC	AM-Empfänger/Receiver, Ucc=9V	16-DIP	GUS	A 244D, TCA 440	
174 XA3(А,Б)	LIN-IC	Dolby B Noise Processor, Ucc=15V	16-DIP	GUS	NE 545	
174 XA4	LIN-IC	AM/FM Synchron-Detector, Ucc=15V	16-DIP	GUS		
174 XA5	LIN-IC	FM Quadratur-Detector, Ucc=12V	18-DIP	GUS	-	
174 XA6	LIN-IC	FM-ZF, Dem., Afc, Ucc=12V	18-DIP	GUS	A 225D, TDA 1047	
174 XA7	LIN-IC	Radio, Phasendetector, Ucc=9V	16-DIP	GUS	-	
174 XA8	LIN-IC	Ctv, Pal/secam Synchr.-Demod., Ucc=12V	16-DIP	GUS	MCA 650, TCA 650	
174 XA9	LIN-IC	Ctv, Chroma PAL/SECAM, Ucc=12V	16-DIP	GUS	TCA 640	
174 XA10	LIN-IC	AM-V, AM/FM-ZF+NF, 0,3W, Ucc=9V	16-DIP	GUS	A 283D, HA 12402, KA 22424, TA 7613AP, U	
174 XA11	LIN-IC	TV, HA/VA-Synchr., HA-O, Ucc=12V	16-DIP	GUS	TDA 2591, TDA 2593	
174 XA14	LIN-IC	PLL Stereo-Decoder, Ucc=12V	24-DIP	GUS	-	
174 XA15	LIN-IC	FM-Tuner/front end, Ucc=9V	16-DIP	GUS	TDA 1062	
174 XA16	LIN-IC	Ctv, SECAM-Decoder, Ucc=12V	28-DIP	GUS	A 3520D, TDA 3520	
174 XA17	LIN-IC	Ctv, RGB, Video-E, Ucc=12V	28-DIP	GUS	A 3501D, TDA 3501	
174 XA19	LIN-IC	FM Tuning-Interface f.TDA 1062	16-DIP	GUS	TDA 1093B	
174 XA20	LIN/MOS-IC	TV-VHF-Tuner, Ucc=12V	16-DIP	GUS	TUA 2000	
174 XA26	LIN-IC	Schmalb.-/narrow band FM-ZF, Ucc=6V	18-DIP	GUS	MC 3359, NJM 3359...	
174 XA27	LIN-IC	Ctv, Farbkorrektur/color Correction	18-DIP	GUS	-	
174 XA28	LIN-IC	Ctv, PAL-Decoder, Ucc=12V	24-DIP	GUS	-	
174 XA31	LIN-IC	Ctv, SECAM-Decoder, Ucc=12V	28-DIP	GUS	-	
174 XA32	LIN-IC	Ctv,PAL/SECAM/NTSC-3,58/4,43-Dec.(neg)	28-DIP	GUS	A 4555DC, MDA 4555, TDA 4555	
174 XA33	LIN-IC	Ctv, RGB(neg. B-y,R-y), Video-e	28-DIP	GUS	MDA 3505, TDA 3505	
174 YB5	LIN/OP-IC	Breitb.-diff./wideband Diff. Amp.	14-DIP	GUS	-	
174 YH3	LIN-IC	NF-Vorverst./Audio Preamp., Ucc=6V	12-DIP+a	GUS	-	
174 YH4(А,Б)	LIN-IC	NF-E, A: 1W(4Ω), Б: 0,7W(4Ω), Ucc=9V	8-DIP+a	GUS	TAA 300	
174 YH5	LIN-IC	NF-E, >2W(4Ω), Ucc=12V, 1,45A	12-DIP+a	GUS	-	
174 YH7	LIN-IC	NF-E, >4,5W(4Ω), Ucc=15V, 1,8A	12-DIP+a	GUS	-	
174 YH8	LIN-IC	NF-E, 2W(4Ω), Ucc=12V, 1,09A	8-DIP+a	GUS	-	
174 YH9(А,Б,В)	LIN-IC	NF-V-E, А,Б: 7W(4Ω), В: 4,5W, Ucc=18V A: 0,04...20kHz, Б,В: 0,04...16kHz	12-DIP+a	GUS	TBA 810AT, TCA 940	
174 YH10(А,Б)	LIN-IC	2x NF-Klangrg./AF DC tone control A: Pnoise<50mW, Б: <100mW	16-DIP	GUS	A 274D, TCA 740	
K 174 YH11	LIN-IC	NF-E, ±18V, 2,4A, 15W(4Ω), Ucc=±15V	14-DIP	GUS	MDA 2020, TDA 2020	
K 174 YH12	LIN-IC	2x NF-Stereo-Potentiometer (Vol.+Bal.)	16-DIP	GUS	A 273D, TCA 730	
K 174 YH13	LIÑ-IC	A/w-verst. + Alc f. Recorder	16-DIP	GUS	A 202D, TDA 1002	
K 174 YH14	LIN-IC	NF-E, 16,5V, 5,5W(2Ω), Ucc=15V	86/5Pin	GUS	TDA 2003	
K 174 YH15	LIN-IC	2x NF-E, 16,5V, 2x9W, Ucc=15V	11-SQL	GUS	TDA 2004	
K 174 YH17	LIN-IC	SMD, 2xKopfh.-Vst./headph. amp, Ucc=3V	16-MDIP	GUS	TA 7688F	
K 174 YH18	LIN-IC	2x NF-E, 2x 1W(4Ω), Ucc=9V	17-SQL	GUS	-	
K 174 YH19	LIN-IC	NF-E, 15W(4Ω), Ucc=15V	86/5Pin	GUS	-	
K 174 YH20	LIN-IC	2x NF-E, 2x 2,25W, Ucc=12V	12-DIP+a	GUS	-	
K 174 YH24	LIN-IC	2x NF-E, 2x 2,2W, Ucc=6V	8-DIP	GUS	-	
K 174 YK1	LIN-IC	Ctv, Einstellkombi/adjust Control	16-DIP	GUS	-	
K 174 YP1	LIN-IC	ZF, Demod., NF-V, Ucc=12V	14-DIP	GUS	-	
K 174 YP2	LIN-IC	TV-ZF Dem., Synchr. Signal, Ucc=12V	16-DIP	GUS	-	
K 174 YP3	LIN-IC	TV, Ton-zf f. LC-filter	14-DIP	GUS	SN 76660, TBA 120	
K 174 YP4	LIN-IC	TV, Ton-ZF, Vc-signal	14-DIP	GUS	A 223D, SN 76622, TBA 120U, U 223	
K 174 YP5	LIN-IC	TV, Video-ZF, Agc (PNP Tuner)	16-DIP	GUS	A 241D, TDA 2541, TDA 3541, TDA 8341	
K 174 YP7	LIN-IC	FM-ZF, Demod., NF-V, Ucc=6V	16-DIP	GUS	TCA 770	
K 174 YP8	LIN-IC	TV, Quasi-parallelton/split-sound	16-DIP	GUS	TDA 2545	
K 174 YP10	LIN-IC	TV, Video/Ton-ZF; Ucc=12V	8-DIP	GUS	-	
K 174 YP11	LIN-IC	TV-Ton-ZF, NF-V, Ucc=12V	18-DIP	GUS	TDA 1236	
K 174 УП1	LIN-IC	Ctv, Einstellkombi/adjust Control	16-DIP	GUS	-	
K 174 ГФ1	LIN-IC	TV, HA-Generator, Ucc=12V	14-DIP	GUS	-	
K 174 ПС1	LIN-IC	Mischer/mixer bis/up to 200MHz, Ucc=9V	14-DIP	GUS	S 042P	
K241	OC	LED/NPN Viso:1500V CTR:>30% lbs:10mA		Aeg		data pinout
K258	OC	LED/NPN Viso:1500V CTR:25-400%		Aeg	SL5500	data pinout
K814P	OC	bid. LED/PT Viso:5000Vrms Vbs:1.25V		Vis		data pdf pinout
K815P	OC	LED/Dar Viso:5000Vrms CTR:800%		Vis		data pdf pinout
K817P	DOC	LED/NPN Viso:5300V CTR:50-600%		Vis		data pinout
K824P	DOC	bid. LED/PT Viso:5000Vrms Vbs:1.25V		Vis		data pdf pinout
K825P	DOC	LED/Dar Viso:5000Vrms CTR:800%		Vis		data pdf pinout
K827P	DOC	LED/NPN Viso:5300V CTR:50-600%		Aeg,Tel,Tem	PC627,PC827,TLP521	data pdf pinout
K844P	QOC	bid. LED/PT Viso:5000Vrms Vbs:1.25V		Vis		data pdf pinout
K845P	QOC	LED/Dar Viso:5000Vrms CTR:800%		Vis		data pdf pinout
K847P	QOC	LED/NPN Viso:2500V CTR:50-600%		Aeg,Tel	PC847,TLP521,TLP621	data pinout
K 1010	\N	\N		\N		
K3010	OC	LED/Triac Viso:3750Vrms Vbs:1.25V		Aeg,Tel,Vis	GE3009,H11J,MCP3009,MOC300	data pdf pinout
K3010G	OC	LED/Triac Viso:3750Vrms Vbs:1.25V		Aeg,Tel,Vis	GE3009,H11J,MCP3009,MOC300	data pdf pinout
K3010P	OC	LED/Triac Viso:3750Vrms Vbs:1.25V		Aeg,Tel,Vis	GE3009,H11J,MCP3009,MOC300	data pdf pinout
K3010PG	OC	LED/Triac Viso:3750Vrms Vbs:1.25V		Aeg,Tel,Vis	GE3009,H11J,MCP3009,MOC300	data pdf pinout

Type	Device	Short Description	Fig.	Manu	Comparision Types	More at
K3020P	OC	LED/Triac Viso:3750Vrms Vbs:1.25V		Aeg,Tel,Vis	GE3020,MCP3020,MOC3020,OPI3020	data pdf pinout
K3020PG	OC	LED/Triac Viso:3750Vrms Vbs:1.25V		Aeg,Tel,Vis	GE3020,MCP3020,MOC3020,OPI3020	data pdf pinout
K8013P	OC	LED/NPN Viso:5300V CTR:90% Vbs:1.25V		Aeg,Tel,Vis		data pdf pinout
K8031P	OC	LED/PD+IC Viso:5300V Vbs:1.25V		Aeg,Tel		data pinout
K8901	OC	LED/NPN		Aeg,Tel	OPI1264	data pinout
K8910	OC	LED/Dar		Aeg,Tel	OPI113	data pinout
K8920	OC	LED/NPN		Aeg,Tel	OPI120	data pinout
K8930	OC	LED/Dar		Aeg,Tel	OPI120	data pinout
K8940	OC	LED/NPN		Aeg,Tel	OPI102	data pinout
K8945	OC	LED/Dar		Aeg,Tel	OPI1264	data pinout

KA

Type	Device	Short Description	Fig.	Manu	Comparision Types	More at
KA-3022-4.5SF	LED	srd+gr+yl 2.5x1.7mm I>5mcd Vf:2.1V		Kin		data pinout
KA	Si-P	=2SA1965 (Typ-Code/Stempel/marking)	35(1,6mm)	Say	→2SA1965	data
KA	Si-N	=2SC4409 (Typ-Code/Stempel/marking)	39	Tos	→2SC4409	data
KA	MOS-N-FET-e	=2SK2103 (Typ-Code/Stempel/marking)	39	Rhm	→2SK2103	data
KA	Si-N	=BFT 75 (Typ-Code/Stempel/marking)	35	Sie	→BFT 75	data
KA	MOS-N-FET-e	=BSS 87 (Typ-Code/Stempel/marking)	39	Sie	→BSS 87	data
KA	MOS-N-FET-e*	=MCH 3401 (Typ-Code/Stempel/marking)	35(2mm)	Say	→MCH 3401	data
KA	Si-N+R	=RN 1441A (Typ-Code/Stempel/marking)	35	Tos	→RN 1441	
KA	Si-N	=µPA604T (Typ-Code/Stempel/marking)	46	Nec	→µPA604T	data
KA	Si-N	=µPA670T (Typ-Code/Stempel/marking)	46(2mm)	Nec	→µPA670T	data
KA 1 H0265R	SMPS-IC	S-Reg(SPS), 650V, 2A, 60W, 100kHz	86/4Pin	Sam	-	pdf
KA 1 H0280R	SMPS-IC	S-Reg(SPS), 800V, 2A, 20W, 100kHz	86/4Pin	Sam	-	
KA 1 H0365R	SMPS-IC	S-Reg(SPS), 650V, 3A, 60W, 100kHz	86/4Pin	Sam	-	pdf
KA 1 H0380R	SMPS-IC	S-Reg(SPS), 800V, 3A, 50W, 100kHz	86/4Pin	Sam	-	pdf
KA 1 H0565	SMPS-IC	S-Reg(SPS), 650V, 5A, 110W, 100kHz	86/4Pin	Sam	-	
KA 1 H0680	SMPS-IC	S-Reg(SPS), 800V, 6A, 100W, 100kHz	18/5Pin	Sam	-	pdf
KA 1 H0765	SMPS-IC	S-Reg(SPS), 650V, 7A, 170W, 100kHz	18/5Pin	Sam	-	
KA 1 H0965	SMPS-IC	S-Reg(SPS), 650V, 9A, 230W, 100kHz	18/5Pin	Sam	-	
KA 1 L0365R	SMPS-IC	S-Reg(SPS), 650V, 3A, 60W, 50kHz	86/4Pin	Sam	-	pdf
KA 1 L0380R	SMPS-IC	S-Reg(SPS), 800V, 3A, 50W, 50kHz	86/4Pin	Sam	-	
KA 1 L0765	SMPS-IC	S-Reg(SPS), 650V, 7A, 170W, 50kHz	18/5Pin	Sam	-	
KA 1 L0880	SMPS-IC	S-Reg(SPS), 800V, 8A, 150W, 50kHz	18/5Pin	Sam	-	pdf
KA 1 L0965	SMPS-IC	S-Reg(SPS), 650V, 9A, 230W, 50kHz	18/5Pin	Sam	-	pdf
KA 1 M0265R	SMPS-IC	S-Reg(SPS), 650V, 2A, 30W, 67kHz	86/4Pin	Sam	-	pdf
KA 1 M0280R	SMPS-IC	S-Reg(SPS), 800V, 2A, 20W, 67kHz	86/4Pin	Sam	-	
KA 1 M0365R	SMPS-IC	S-Reg(SPS), 650V, 3A, 60W, 67kHz	86/4Pin	Sam	-	pdf
KA 1 M0380R	SMPS-IC	S-Reg(SPS), 800V, 3A, 50W, 67kHz	86/4Pin	Sam	-	
KA 1 M0565	SMPS-IC	S-Reg(SPS), 650V, 5A, 110W, 67kHz	86/4Pin	Sam	-	
KA 1 M0680	SMPS-IC	S-Reg(SPS), 800V, 6A, 100W, 67kHz	18/5Pin	Sam	-	pdf
KA 1 M0765	SMPS-IC	S-Reg(SPS), 650V, 7A, 170W, 67kHz	18/5Pin	Sam	-	
KA 1 M0880	SMPS-IC	S-Reg(SPS), 800V, 8A, 150W, 67kHz	18/5Pin	Sam	-	pdf
KA 1 M0965	SMPS-IC	S-Reg(SPS), 650V, 9A, 230W, 67kHz	18/5Pin	Sam	-	
KA 2 S0565	SMPS-IC	S-Reg(SPS), 650V, 5A, 110W, 20...100kHz	18/5Pin	Sam	-	
KA 2 S0680	SMPS-IC	S-Reg(SPS), 800V, 6A, 100W, 20...100kHz	18/5Pin	Sam	-	
KA 2 S0765	SMPS-IC	S-Reg(SPS), 650V, 7A, 170W, 20...100kHz	18/5Pin	Sam	-	pdf
KA 2 S0880	SMPS-IC	S-Reg(SPS), 800V, 8A, 150W, 20...100kHz	18/5Pin	Sam	-	pdf
KA 2 S0965	SMPS-IC	S-Reg(SPS), 650V, 9A, 230W, 20...100kHz	18/5Pin	Sam	-	pdf
KA 3 S0680R	SMPS-IC	S-Reg(SPS), 800V, 6A, 100W, 20...100kHz	18/5Pin	Sam	-	pdf
KA 3 S0880R	SMPS-IC	S-Reg(SPS), 800V, 8A, 150W, 20...100kHz	18/5Pin	Sam	-	
KA 3 Z07	Z-Di	TAZ, >5,5V, Ifsm=30A(1ms), Ih=50mA	31a	Shi	-	data
KA 3 Z18	Z-Di	TAZ, >15,5V, Ifsm=30A(1ms), Ih=50mA	31a	Shi	-	data
KA5H0165R	SMPS-IC	Offline SMPS, Current Mode PWM with high volt. Power SenseFET, fosc=100kHz	86	Fcs	-	pdf pinout
KA5H0165RN	SMPS-IC	Offline SMPS, Current Mode PWM with high volt. Power SenseFET, fosc=100kHz	8-DIP	Fcs	-	pdf pinout
KA5H02659RN	SMPS-IC	Offline SMPS, Current Mode PWM with high volt. Power SenseFET, fosc=100kHz	8-DIP	Fcs	-	pdf pinout
KA5H0265R	SMPS-IC	Offline SMPS, Current Mode PWM with high volt. Power SenseFET, fosc=100kHz	86	Fcs	-	pdf pinout
KA5H0265RC	SMPS-IC	Offline SMPS, Current Mode PWM with high volt. Power SenseFET, fosc=100kHz	86	Fcs	-	pdf pinout
KA5H0280R	SMPS-IC	Offline SMPS, Current Mode PWM with high volt. Power SenseFET, fosc=100kHz	86	Fcs	-	pdf pinout
KA 5 H 0365 R	SMPS-IC	Offline SMPS, Current Mode PWM with high volt. Power SenseFET, fosc=100kHz	86	Fcs	-	pdf pinout
KA5H0380R	SMPS-IC	Offline SMPS, Current Mode PWM with high volt. Power SenseFET, fosc=100kHz	86	Fcs	-	pdf pinout
KA5L0165R	SMPS-IC	Offline SMPS, Current Mode PWM with high volt. Power SenseFET, fosc=50kHz	86	Fcs	-	pdf pinout
KA5L0165RN	SMPS-IC	Offline SMPS, Current Mode PWM with high volt. Power SenseFET, fosc=50kHz	8-DIP	Fcs	-	pdf pinout
KA5L0165RVN	SMPS-IC	Offline SMPS, Current Mode PWM with high volt. Power SenseFET, fosc=50kHz	8-DIP	Fcs	-	pdf pinout
KA5L0265R	SMPS-IC	Offline SMPS, Current Mode PWM with high volt. Power SenseFET, fosc=50kHz	86	Fcs	-	pdf pinout
KA5L0365R	SMPS-IC	Offline SMPS, Current Mode PWM with high volt. Power SenseFET, fosc=50kHz	86	Fcs	-	pdf pinout
KA5L0365RN	SMPS-IC	Offline SMPS, Current Mode PWM with high volt. Power SenseFET, fosc=50kHz	8-DIP	Fcs	-	pdf pinout
KA5L0380R	SMPS-IC	Offline SMPS, Current Mode PWM with high volt. Power SenseFET, fosc=50kHz	86	Fcs	KA5M0380R	pdf pinout

Type	Device	Short Description	Fig.	Manu	Comparision Types	More at
KA5M0165R	SMPS-IC	Offline SMPS, Current Mode PWM with high volt. Power SenseFET, fosc=67kHz	86	Fcs		pdf pinout
KA5M0165RN	SMPS-IC	Offline SMPS, Current Mode PWM with high volt. Power SenseFET, fosc=67kHz	8-DIP	Fcs		pdf pinout
KA5M02659RN	SMPS-IC	Offline SMPS, Current Mode PWM with high volt. Power SenseFET, fosc=67kHz	8-DIP	Fcs		pdf pinout
KA5M0265R	SMPS-IC	Offline SMPS, Current Mode PWM with high volt. Power SenseFET, fosc=67kHz	86	Fcs		pdf pinout
KA5M0280R	SMPS-IC	Offline SMPS, Current Mode PWM with high volt. Power SenseFET, fosc=67kHz	86	Fcs		pdf pinout
KA5M0365R	SMPS-IC	Offline SMPS, Current Mode PWM with high volt. Power SenseFET, fosc=67kHz	86	Fcs		pdf pinout
KA5M0365RN	SMPS-IC	Offline SMPS, Current Mode PWM with high volt. Power SenseFET, fosc=67kHz	8-DIP	Fcs		pdf pinout
KA5M0380R	SMPS-IC	Offline SMPS, Current Mode PWM with high volt. Power SenseFET, fosc=67kHz	86	Fcs	KA5H0380R	pdf pinout
KA5M0765RC	SMPS-IC	Offline SMPS, Current Mode PWM with high volt. Power SenseFET, fosc=67kHz	86	Fcs		pdf pinout
KA5M0965Q	SMPS-IC	Offline SMPS, Current Mode PWM with high volt. Power SenseFET, fosc=67kHz	86	Fcs		pdf pinout
KA5Q0740RTTU	SMPS-IC	high voltage power Sensefet and current mode PWM controller, 400V	86	Fcs		pdf pinout
KA5Q0740RTYDTU	SMPS-IC	high voltage power Sensefet and current mode PWM controller, 400V	86	Fcs		pdf pinout
KA5Q0765RTTU	SMPS-IC	high voltage power SenseFET and current mode PWM controller, 650V 7A	86	Fcs		pdf pinout
KA5Q0765RTYDTU	SMPS-IC	high voltage power Sensefet and current mode PWM controller, 650V 7A	86	Fcs		pdf pinout
KA5Q12656RTTU	SMPS-IC	high voltage power Sensefet and current mode PWM controller, 650V 12A	86	Fcs		pdf pinout
KA5Q12656RTYDTU	SMPS-IC	high voltage power Sensefet and current mode PWM controller, 650V 12A	86	Fcs		pdf pinout
KA5Q1265RFTU	SMPS-IC	high voltage power Sensefet and current mode PWM controller, 650V 12A	86	Fcs		pdf pinout
KA5Q1265RFYDTU	SMPS-IC	high voltage power Sensefet and current mode PWM controller, 650V 12A	86	Fcs		pdf pinout
KA5Q1565RFTU	SMPS-IC	high voltage power Sensefet and current mode PWM controller, 650V 15A	86	Fcs		pdf pinout
KA5Q1565RFYDTU	SMPS-IC	high voltage power Sensefet and current mode PWM controller, 650V 15A	86	Fcs		pdf pinout
KA 6	Si-Di	=BAS 16 (Typ-Code/Stempel/marking)	35	Aeg	→BAS 16	
KA 10 N14	Z-Di	TAZ, >130V, Ifsm=100A(1ms), Ih=100mA	31a	Shi	-	data
KA 10 R25	Z-Di	TAZ, >220V, Ifsm=100A(1ms), Ih=100mA	31a	Shi	-	data
KA33V	Z+	Io=10mA Vo:31...35V P:200mW	2N	Sam		data pinout
KA 33 V	Z-IC	Tuner-Stab., 31...35V, 10mA, (~TAA 550)	7j	Sam	TAA 550, TAA 940, ZTK 33, µPC574	data pdf pinout
KA 76 L05(Z)	Z-IC	lo-drop, +5V, 0,1A, -40...+125°	7d	Sam	TA 78DS05...P	pdf
KA 78 L05AD...L24AD	Z-IC	=KA 78L05AZ...L24AZ: SMD	8-MDIP	Sam		pdf
KA 78 L05A(Z)	Z-IC	+5V, 0,1A	7b	Sam	...78L05 (TO-92)	pdf
KA 78 L06A(Z)	Z-IC	+6V, 0,1A	7b	Sam	...78L06 (TO-92)	pdf
KA 78 L08A(Z)	Z-IC	+8V, 0,1A	7b	Sam	...78L08 (TO-92)	pdf
KA 78 L09A(Z)	Z-IC	+9V, 0,1A	7b	Sam	...78L09 (TO-92)	pdf
KA 78 L10A(Z)	Z-IC	+10V, 0,1A	7b	Sam	...78L10 (TO-92)	pdf
KA 78 L12A(Z)	Z-IC	+12V, 0,1A	7b	Sam	...78L12 (TO-92)	pdf
KA 78 L15A(Z)	Z-IC	+15V, 0,1A	7b	Sam	...78L15 (TO-92)	pdf
KA 78 L18A(Z)	Z-IC	+18V, 0,1A	7b	Sam	...78L18 (TO-92)	pdf
KA 78 L24A(Z)	Z-IC	+24V, 0,1A	7b	Sam	...78L24 (TO-92)	pdf
KA 78 M05	Z-IC	+5V, 0,5A, 4%, 0...+125°	17b	Sam	...78M05...(TO-220)	pdf
KA 78 M06	Z-IC	+6V, 0,5A, 4%, 0...+125°	17b	Sam	...78M06...(TO-220)	pdf
KA 78 M08	Z-IC	+8V, 0,5A, 4%, 0...+125°	17b	Sam	...78M08...(TO-220)	pdf
KA 78 M10	Z-IC	+10V, 0,5A, 4%, 0...+125°	17b	Sam	...78M10...(TO-220)	pdf
KA 78 M12	Z-IC	+12V, 0,5A, 4%, 0...+125°	17b	Sam	...78M12...(TO-220)	pdf
KA 78 M15	Z-IC	+15V, 0,5A, 4%, 0...+125°	17b	Sam	...78M15...(TO-220)	pdf
KA 78 M18	Z-IC	+18V, 0,5A, 4%, 0...+125°	17b	Sam	...78M18...(TO-220)	pdf
KA 78 M20	Z-IC	+20V, 0,5A, 4%, 0...+125°	17b	Sam	...78M20...(TO-220)	pdf
KA 78 M24	Z-IC	+24V, 0,5A, 4%, 0...+125°	17b	Sam	...78M24...(TO-220)	pdf
KA 78M05 I...M24 I	Z-IC	=KA 78M05...78M24: -40...+125°	17b	Sam		
KA 78M05 R...M24 Rö	Z-IC	=KA 78M05...78M24(I):	30b	Sam		
KA 78 R05	Z-IC	Iso, lo-drop, +5V, 1A, -20...+80°	86/4Pin	Sam	-	pdf
KA 78 R08	Z-IC	Iso, lo-drop, +8V, 1A, -20...+80°	86/4Pin	Sam	-	pdf
KA 78 R09	Z-IC	Iso, lo-drop, +9V, 1A, -20...+80°	86/4Pin	Sam	-	pdf
KA 78 R12	Z-IC	Iso, lo-drop, +12V, 1A, -20...+80°	86/4Pin	Sam	-	pdf
KA 78 R15	Z-IC	Iso, lo-drop, +15V, 1A, -20...+80°	86/4Pin	Sam	-	pdf
KA 78 R33	Z-IC	Iso, lo-drop, +3,3V, 1A, -20...+80°	86/4Pin	Sam	-	pdf
KA 78 S40(N)	SMPS-IC	Schaltregler/S-Reg., 1,3...40V, 1,5A	16-DIP	Sam	µA 78S40	
KA78T05	R+	Io=int Vo:4.8...5.2V Vin:10V P:int	3P	Sam		data pinout
KA78T05A	R+	Io=int Vo:4.9...5.1V Vin:10V P:int	3P	Sam		data pinout
KA 78 T05(A)CH	Z-IC	+5V, 3A, 4% (A=2%)	18b	Sam		data pinout
KA 78 T05(A)CT	Z-IC	+5V, 3A, 4% (A=2%)	17b	Sam	MC 78T05CT	data pinout
KA78T06	R+	Io=int Vo:5.75...6.25V Vin:11V P:int	3P	Sam		data pinout
KA 78 T06CH	Z-IC	+6V, 3A, 4%	18b	Sam		data pinout
KA 78 T06CT	Z-IC	+6V, 3A, 4%	17b	Sam	MC 78T06CT	data pinout
KA78T08	R+	Io=int Vo:7.7...8.3V Vin:14V P:int	3P	Sam		data pinout
KA 78 T08CH	Z-IC	+8V, 3A, 4%	18b	Sam	-	data pinout
KA 78 T08CT	Z-IC	+8V, 3A, 4%	17b	Sam	MC 78T08CT	data pinout

Type	Device	Short Description	Fig.	Manu	Comparision Types	More at
KA78T12	R+	Io=int Vo:11.5...12.5V Vin:19V P:int	3P	Sam		data pinout
KA78T12A	R+	Io=int Vo:11.75...12.25V Vin:19V P:int	3P	Sam		data pinout
KA 78 T12(A)CH	Z-IC	+12V, 3A, 4% (A=2%)	18b	Sam		data pinout
KA 78 T12(A)CT	Z-IC	+12V, 3A, 4% (A=2%)	17b	Sam	MC 78T12CT	data pinout
KA78T15	R+	Io=int Vo:14.4...15.6V Vin:23V P:int	3P	Sam		data pinout
KA78T15A	R+	Io=int Vo:14.7...15.3V Vin:23V P:int	3P	Sam		data pinout
KA 78 T15(A)CH	Z-IC	+15V, 3A, 4% (A=2%)	18b	Sam	-	data pinout
KA 78 T15(A)CT	Z-IC	+15V, 3A, 4% (A=2%)	17b	Sam	MC 78T15CT	data pinout
KA78T18	R+	Io=int Vo:17.3...18.7V Vin:27V P:int	3P	Sam		data pinout
KA 78 T18CH	Z-IC	+18V, 3A, 4%	18b	Sam		data pinout
KA 78 T18CT	Z-IC	+18V, 3A, 4%	17b	Sam	MC 78T18CT	data pinout
KA78T24	R+	Io=int Vo:23...25V Vin:33V P:int	3P	Sam		data pinout
KA 78 T24CH	Z-IC	+24V, 3A, 4%	18b	Sam		data pinout
KA 78 T24CT	Z-IC	+24V, 3A, 4%	17b	Sam	MC 78T24CT	data pinout
KA 79 L05A(Z)	Z-IC	-5V, 0,1A	7a	Sam	... 79L05 (TO-92)	pdf
KA 79 L12A(Z)	Z-IC	-12V, 0,1A	7a	Sam	... 79L12 (TO-92)	
KA 79 L15A(Z)	Z-IC	-15V, 0,1A	7a	Sam	... 79L15 (TO-92)	
KA 79 L18A(Z)	Z-IC	-18V, 0,1A	7a	Sam	... 79L18 (TO-92)	
KA 79 L24A(Z)	Z-IC	-24V, 0,1A	7a	Sam	... 79L24 (TO-92)	
KA 79 M05	Z-IC	-5V, 0,5A, 4%, 0...+125°	17c	Sam	... 79M05...(TO-220)	pdf
KA 79 M06	Z-IC	-6V, 0,5A, 4%, 0...+125°	17c	Sam	... 79M06...(TO-220)	pdf
KA 79 M08	Z-IC	-8V, 0,5A, 4%, 0...+125°	17c	Sam	... 79M08...(TO-220)	pdf
KA 79 M12	Z-IC	-12V, 0,5A, 4%, 0...+125°	17c	Sam	... 79M12...(TO-220)	pdf
KA 79 M15	Z-IC	-15V, 0,5A, 4%, 0...+125°	17c	Sam	... 79M15...(TO-220)	pdf
KA 79 M18	Z-IC	-18V, 0,5A, 4%, 0...+125°	17c	Sam	... 79M18...(TO-220)	pdf
KA 79 M24	Z-IC	-24V, 0,5A, 4%, 0...+125°	17c	Sam	... 79M24...(TO-220)	pdf
KA 79M05 R...M24 R	Z-IC	=KA 79M05...79M24:	30c	Sam		
KA201	UNI	Vs:±22V Vu:160V/mV Vo:±14V Vi0:0.5mV	8SD	Sam		data pinout
KA 201 AD	OP-IC	=KA 201AN: SMD	8-MDIP	Sam	... 101..., ... 201...	
KA 201 A(N)	OP-IC	Uni,Serie 101, ±22V, 10V/µs, -25...+85°	8-DIP	Sam	→Serie 101	data
KA219	2xCOMP	Vs:36V Vu:40V/mV Vi0:0.7mV	14DS	Sam		data pinout
KA 219...	KOP-IC	→LM 219...: -25...+85°	14-(M)DIP	Sam	→LM 219...	
KA224	4xHI-GAIN	Vs:±18V Vu:100V/mV Vo:28/0V	14DS	Sam		data pdf pinout
KA 224...	OP-IC	=LM 224...	14-(M)DIP	Sam	→LM 224...	
KA 236-...	Ref-Z-IC	=KA 336-...: -25...+85°	7(-+Reg)	Sam	LM 136, LM 236	
KA239	4xDIFF-COMP	Vs:±18V Vu:200V/mV Vo:TTL Vi0:±1.4mV	14DS	Sam		data pdf pinout
KA 239...	KOP-IC	=LM 239...	14-(M)DIP	Sam	→LM 239...	
KA239D	4xDIFF-COMP	Vs:±18V Vu:200V/mV Vo:TTL Vi0:±1.4mV	14S	Sam		data pdf pinout
KA248	4xHI-GAIN	Vs:±18V Vu:160V/mV Vo:±12V Vi0:1mV	14DS	Sam		data pdf pinout
KA 248...	OP-IC	=LM 248...	14-(M)DIP	Sam	→LM 248...	
KA258	2xHI-GAIN	Vs:±16V Vu:100V/mV Vo:28/0V	8DS,9I	Sam		data pdf pinout
KA 258...	OP-IC	=LM 258...	8-(M)DIP	Sam	→LM 258...	
KA 258 S	OP-IC	=LM 258...	9-SIP	Sam	→LM 258...	
KA 278 R05	Z-IC	Iso, lo-drop, +5V, 2A, -20...+80°	86/4Pin	Sam	-	pdf
KA 278 R09	Z-IC	Iso, lo-drop, +9V, 2A, -20...+80°	86/4Pin	Sam	-	pdf
KA 278 R12	Z-IC	Iso, lo-drop, +12V, 2A, -20...+80°	86/4Pin	Sam	-	pdf
KA 278 R33	Z-IC	Iso, lo-drop, +3,3V, 2A, -20...+80°	86/4Pin	Sam	-	pdf
KA293	2xDIFF-COMP	Vs:±18V Vu:200V/mV Vi0:±1mV	8DS,9I	Sam		data pdf pinout
KA 293...	KOP-IC	=LM 293...	8-(M)DIP	Sam	→LM 293...	
KA 293 S	KOP-IC	=LM 293...	9-SIP	Sam	→LM 293...	
KA301	UNI	Vs:±18V Vu:160V/mV Vo:±14V Vi0:2mV	8SD	Sam		data pinout
KA 301 AD	OP-IC	=KA 201AN: SMD, 0...70°	8-MDIP	Sam	... 101..., ... 201..., ... 301...	
KA 301 AN	OP-IC	=KA 201AN: 0...70°	8-DIP	Sam	... 101..., ... 201..., ... 301...	
KA311	COMP	Vs:36V Vu:200V/mV Vi0:1mV	8DS	Sam		data pdf pinout
KA 311...	KOP-IC	=LM 311...	8-(M)DIP	Sam	→LM 311...	
KA 317	Z-IC	+1,2...37V, 1A, 0...+125°	17I	Sam	LM 317...(TO-220)	pdf
KA 317 L(Z)	Z-IC	=KA 317: 0,1A	7o	Sam	LM 317...(TO-92)	pdf
KA 317 M	Z-IC	=KA 317: 0,5A	7o	Sam	LM 317...(TO-220)	pdf
KA319	2xCOMP	Vs:36V Vu:40V/mV Vi0:2mV	14DS	Sam		data pdf pinout
KA 319...	KOP-IC	→LM 319...	14-(M)DIP	Sam	→LM 319...	pdf
KA324	4xHI-GAIN	Vs:±18V Vu:100V/mV Vo:28/0V	14DS	Sam		data pdf pinout
KA 324...	OP-IC	=LM 324...	14-(M)DIP	Sam	→LM 324...	
KA 331	LIN-IC	V/F-Converter, 1Hz...100kHz, 0...+70°	8-DIP	Sam	LM 331, RC 4151	pdf pinout
KA 336-2.5(B)(Z)	Ref-Z-IC	2,5V(1mA), 0,3<1Ω, 0...70°	7(-+Reg)	Sam	-	
KA 336-5.0(B)(Z)	Ref-Z-IC	5V(1mA), 0,6<2Ω, 0...70°	7(-+Reg)	Sam	LM 336	
KA 337	Z-IC	-1,2...-37V, 1A, 0...+125°	17n	Sam	LM 337...(TO-220)	pdf
KA 337 L	Z-IC	=KA 337: Fig.→	7o	Sam	LM 337...(TO-92)	
KA339	4xDIFF-COMP	Vs:±18V Vu:200V/mV Vo:TTL Vi0:±1.4mV	14DS	Sam		data pdf pinout
KA 339...	KOP-IC	=LM 339...	14-(M)DIP	Sam	→LM 339...	
KA339D	4xDIFF-COMP	Vs:±18V Vu:200V/mV Vo:TTL Vi0:±1.4mV	14S	Sam		data pdf pinout
KA 340 T05	Z-IC	+5V, 1A	17b	Sam	LM 340-5, ... 7805...(TO-220)	
KA 340 T06	Z-IC	+6V, 1A	17b	Sam	LM 340-6, ... 7806...(TO-220)	
KA 340 T08	Z-IC	+8V, 1A	17b	Sam	LM 340-8, ... 7808...(TO-220)	
KA 340 T09	Z-IC	+9V, 1A	17b	Sam	LM 340-9, ... 7809...(TO-220)	
KA 340 T10	Z-IC	+10V, 1A	17b	Sam	LM 340-10, ... 7810...(TO-220)	
KA 340 T11	Z-IC	+11V, 1A	17b	Sam	... 7811...(TO-220)	
KA 340 T12	Z-IC	+12V, 1A	17b	Sam	LM 340-12, ... 7812...(TO-220)	
KA 340 T15	Z-IC	+15V, 1A	17b	Sam	LM 340-15, ... 7815...(TO-220)	
KA 340 T18	Z-IC	+18V, 1A	17b	Sam	LM 340-18, ... 7818...(TO-220)	
KA 340 T24	Z-IC	+24V, 1A	17b	Sam	LM 340-24, ... 7824...(TO-220)	
KA348	4xHI-GAIN	Vs:±18V Vu:160V/mV Vo:±12V Vi0:1mV	14DS	Sam		data pdf pinout
KA 348...	OP-IC	=LM 348...	14-(M)DIP	Sam	→LM 348...	
KA 350	Z-IC	+1,2...33V, 3A, 0...+125°	17I	Sam	LM 350T	pdf
KA350H	ER+	Io=int Vo:1.2...33V P:int	3P	Sam		data pinout
KA 350 H	Z-IC	=KA 350. Fig.→	18I	Sam		data pinout

Type	Device	Short Description	Fig.	Manu	Comparision Types	More at
KA358	2xHI-GAIN	Vs:±16V Vu:100V/mV Vo:28/0V	8DS,9I	Sam		data pdf pinout
KA 358 ...	OP-IC	=LM 358...	8-(M)DIP	Sam	→LM 358...	
KA 358 S	OP-IC	=LM 358...	9-SIP	Sam	→LM 358...	
KA 361 D	KOP-IC	=KA 361N: SMD	14-MDIP	Sam	→LM 361	
KA 361 N	KOP-IC	hi-speed, 0...70° (=LM 361)	14-DIP	Sam	→LM 361	
KA 378 R05	Z-IC	Iso, lo-drop, +5V, 3A, -20...+80°	86/4Pin	Sam	-	pdf
KA 378 R33	Z-IC	Iso, lo-drop, +3,3V, 3A, -20...+80°	86/4Pin	Sam	-	pdf
KA 385(Z)-1.2	Ref-Z-IC	1,235V, 1...2%, 0...+70°	7(AK-)	Sam	LM 385(B)Z-1,2	
KA 386(D,S)	LIN-IC	=LM 386...		Sam	→LM 386...	pinout
KA393	2xDIFF-COMP	Vs:±18V Vu:200V/mV Vi0:±1mV	8DS,9I	Sam		data pdf pinout
KA 393...	KOP-IC	=LM 393...	8-(M)DIP	Sam	→LM 393...	
KA 393 S	KOP-IC	=LM 393...	9-SIP	Sam	→LM 393...	
KA 431 (C)D	Ref-Z-IC	=KA 431..Z: SMD	8-MDIP	Sam	NJM 431M	
KA 431(CN)	Ref-Z-IC	=KA 431..Z: Fig.→	8-DIP	Sam	NJM 431D, TL 431CP,CJG	
KA 431(C)Z,(A)Z	Ref-Z-IC	2,5...36V, 1...100mA, 0...70°	7(KARef)	Sam	NJM 431L, TL 431CLP, µA 431	
KA 431 IN	Ref-Z-IC	=KA 431..ZF: -40...+85°	8-DIP	Sam	TL 431IP,ILP	
KA 431 IZ	Ref-Z-IC	=KA 431..Z: -40...+85°	7(KARef)	Sam	TL 431IP	
KA 555...	LIN-IC	=NE 555...	8-(M)DIP	Sam	→NE 555...	
KA 555 I...	LIN-IC	→NE 555..: -40...+85°	8-(M)DIP		→NE 555...	
KA 556...	LIN-IC	=NE 556...	14-(M)DIP	Sam	→NE 556...	
KA 556 I...	LIN-IC	→NE 556..: -40...+85°	14-(M)DIP		→NE 556...	
KA 558(B)...	LIN-IC	=NE 558...	16-(M)DIP		→NE 558...	pinout
KA 558(B)I...	LIN-IC	=NE 558..: -40...+85°	16-(M)DIP	Sam	→NE 558...	pinout
KA 567	LIN-IC	Pll-ton-/tone-decoder	8-DIP	Sam	→LM 567...	pdf pinout
KA 567D	LIN-IC	SMD, Pll-ton-/tone-decoder	8-MDIP	Sam	→LM 567...	pdf pinout
KA 601 D	LIN-IC	PCM-Repeater (isa-system)		Hfo	-	
KA 602 D	LIN-IC	Pcm-repeater (isa-system)		Hfo	-	
KA 610 D	LIN-IC	V10-Port (isa-system)		Hfo	-	
KA710	COMP	Vs:-7/14V Vi0:1.8mV	14DS	Sam		data pinout
KA 710(C)D	KOP-IC	=KA 710(C)N: SMD	14-MDIP	Sam	... 710..., ... 1710...	
KA 710(C)N	KOP-IC	Serie 110, hi-speed, +14/-7V, 0...70°	14-DIP	Sam	→Serie 110	data
KA710I	COMP	Vs:-7/14V Vi0:0.75mV	14DS	Sam		data pinout
KA 710 I,ID	KOP-IC	=KA 710(C)N: -25...+85V	14-(M)DIP	Sam	... 710..., ... 1710...	
KA 711	KOP-IC	Serie 711, hi-speed, +14/-7V, 0...70°	14-DIP	Sam	→Serie 711	data
KA 711D	KOP-IC	=KA 711: SMD	14-MDIP	Sam	... 711..., ... 1711...	
KA 711 I,ID	KOP-IC	=KA 711: -25...+85V	14-(M)DIP	Sam	... 711..., ... 1711...	
KA 723	Z-IC	Voltage Reg., +2...37V 0.15A, 0...70°	14-DIP	Fcs	... 723...	pdf
KA 723 D	Z-IC	SMD, Volt. Reg., +2...37V 0.15A, 0...70°	14-MDIP	Fcs	... 723...(SMD)	pdf
KA 723 I	Z-IC	Voltage Reg., +2...37V 0.15A, -25...+85°	14-DIP	Sam	... 723...	
KA 723 ID	Z-IC	SMD, Volt.Reg. +2...37V 0.15A,-25...+85°	14-MDIP	Sam	... 723...(SMD)	
KA 733 CD	LIN-IC	=KA 733CN: SMD	14-MDIP	Sam	µA 733SC	pinout
KA 733 CN	LIN-IC	Video-Verst./Amp., 90MHz, ±8V, 0...70°	14-DIP	Sam	LM 733CN, µA 733...	pinout
KA741	UNI	Vs:±22V Vu:>50V/mV Vo:>±16V f:1.5Mc	8SD	Sam		data pinout
KA 741	OP-IC	Uni, Serie 741, ±18V, 0...+70°	8-DIP	Sam	→Serie 741	data
KA741D	UNI	Vs:±18V Vu:200V/mV Vo:±14V Vi0:2mV	8S	Sam		data pdf pinout
KA 741 E	OP-IC	=KA 741: ±22V	8-DIP		... 741...	
KA 741 I	OP-IC	=KA 741: -40...+85°	8-DIP		... 741...	
KA 741 ...D	OP-IC	=KA 741..: SMD	8-MDIP		-	
KA 950(Y)	Si-N	≈2SC2120			→2SC2120	
KA 1170 N	LIN-IC	TV, VA-synchr. + Osc. + E	12-DIP+b	Sam	-	
KA 1180 P	LIN-IC	TV, Ha-synchr. + Osc.	16-DIP	Sam	LA 3160...61, M 51521...22L, MB 3105...06	pdf pinout
KA 1222	LIN-IC	2x NF-V-ra, Ucc=2,5...6V	8-SIP	Sam	LA 1201, UL 1211	
KA 1241	LIN-IC	AM/FM-ZF + Agc + AM-demod.	14-DIP	Sam		data pdf pinout
KA1458	2xUNI	Vs:±18V Vu:200V/mV Vo:±14V Vi0:2mV	8DS,9I	Sam	→LM 1458...	
KA 1458...	OP-IC	=LM 1458...	8-(M)DIP	Sam	→LM 1558...	
KA 1458...I	OP-IC	=LM 1458..: -25...+85°	8-(M)DIP	Sam	→LM 1458...	
KA 1458 S	OP-IC	=LM 1458...	9-SIP	Sam	AN 241, CA 3065, HA 1125, LA 1365, LM 30	pdf
KA 2101	LIN-IC	TV, Ton-ZF, NF-Tr, Ucc=24V	14-DIP	Sam	µPC 1353	pdf
KA 2102 A	LIN-IC	Tv-ton-zf + NF-V-E, 2,4W(18V/8Ω)	14-DIP+g	Sam		
KA 2103 L	LIN-IC	TV, Stummschaltg./Sound Muting	8-SIP	Sam		
KA 2104	LIN-IC	TV, Auto Power off, Muting	9-SIP	Sam	TA 7337	
KA 2105	LIN-IC	TV, Ton-ZF-begrenzer/limiter, Demod.	9-SIP	Sam		
KA 2106	LIN-IC	TV-dual-Ton-ZF	16-DIP	Sam	AN 5836	pdf pinout
KA 2107	LIN-IC	DC Ctrl. f. Volume, Tone, Balance	12-SIP	Sam	LA 1385, TA 7242, µPC 1031H2	
KA 2130 A	LIN-IC	TV, VA-Synchr./O/E, Ucc=9...18V	10-SIL	Sam	AN 5512	pdf pinout
KA 2131	LIN-IC	TV, VA-E, Ucc=24V(Pin 10 = frei/n.c.)	10-SIL	Sam	TDA 1044(U)	
KA 2132	LIN-IC	TV, VA-Synchr./Osc./E, Ucc=11...27V	12-DIP+b	Sam	µPC 1379	pdf
KA 2133	LIN-IC	TV, HA/VA-Synchr./O, VA-E, Ucc=24V	16-DIP+g	Sam	AN 5436	pinout
KA 2134	LIN-IC	Ctv, Ha/va Signal-prozessor, Ucc=12V	18-DIP	Sam	AN 5790	
KA 2135	LIN-IC	TV, Ha Signal-prozessor, Ucc=11V	12-SIP	Sam	TDA 1170N	
KA 2136	LIN-IC	TV, VA-Synchr./O/E, Ucc=30V	12-QIP+b	Sam	TDA 1180	
KA 2137	LIN-IC	CTV/TV, HA-Prozessor, Ucc=12V	16-DIP	Sam	LA 7851	pdf
KA 2138	LIN-IC	Monitor, HA/VA-Prozessor, Ucc=12V	20-DIP	Sam	-	pdf pinout
KA 2139	LIN-IC	Monitor, RGB-Video-Verst./Amp.,70MHz	28-DIP	Sam	-	pdf pinout
KA 2140B	LIN-IC	Monitor, RGB-Video-Amp. 130MHz, OSD	28-DIP	Sam	-	pdf pinout
KA 2141	LIN-IC	Monitor, RGB-Video-Amp. 85MHz	20-DIP	Sam	-	pdf pinout
KA 2142	LIN-IC	Ctv, VA-E, Ucc=35V	10-SIL	Sam	-	pdf
KA 2142C	LIN-IC	Vert. Deflection Output, Ucc=35V	10-SIL	Fcs	-	pdf pinout
KA 2143B	LIN-IC	Monitor, RGB-Video-Amp. 110MHz, OSD	24-DIP	Sam	-	pdf pinout
KA 2151	LIN-IC	Ctv, Pal Chroma-prozessor, Ucc=12V	24-DIP	Sam	TA 7193	
KA 2153	LIN-IC	Ctv, NTSC Video, Chroma, Synchr.	42-DIP	Sam	IR 3P04, TA 7644BP	pdf
KA 2154	LIN-IC	Ctv, PAL/NTSC Video, Chroma, Synchr.	42-DIP	Sam	TA 7698	pdf
KA 2155	LIN-IC	Ctv, NTSC Video, Chroma, Synchr.	30-SDIP	Sam	LA 7625	pdf
KA 2156	LIN-IC	=KA 2155: integr. Peak Clip Circuit	30-SDIP	Sam	LA 7626	pdf
KA 2161	LIN-IC	Ctv, NTSC/PAL Signal-prozessor	54-SDIP	Sam		pdf

Type	Device	Short Description	Fig.	Manu	Comparision Types	More at
KA 2163(B)	LIN-IC	CTV, NTSC Signal-Prozessor, I²C-Bus	56-SDIP	Sam	-	
KA 2163	LIN-IC	Ctv, NTSC Signal-prozessor, I²c-bus	56-SDIP	Sam		pdf
KA 2181	LIN-IC	IR-FB, Vorverst./preamplifier	8-SIP	Sam	µPC 1373H	pdf pinout
KA 2182	LIN-IC	IR-FB, Vorverst./preamplifier	8-SIP	Sam	µPC 1374H	
KA 2183	LIN-IC	=KA 2182: invert. Ausg./Output	8-SIP	Sam	µPC 1374H	
KA 2184-15	LIN-IC	IR-FB, Vorvst./Preamp., Pin 5: R=150kΩ	8-SIP	Sam	CX 20106A	pinout
KA 2184-18	LIN-IC	=KA 2184-15: Pin 5: R=180kΩ	8-SIP	Sam	CX 20106A	pinout
KA 2184-20	LIN-IC	=KA 2184-15: Pin 5: R=200kΩ	8-SIP	Sam	CX 20106A	pinout
KA 2184 D-01,02,03	LIN-IC	=KA 2184-15,18,20: SMD	8-MDIP	Sam		pinout
KA 2185	LIN-IC	Tv/vc, Tuning Frequ. Synthesizer(NTSC)	20-DIP	Sam	TD 6359P	
KA 2186	LIN-IC	TV, Video-Umschalter/Switch,Ucc=8...14V	8-DIP	Sam	TEA 2014	pdf
KA 2187	LIN-IC	Tv/vc, Tuning Frequ. Synthesizer(PAL)	20-DIP	Sam	TD 6358P	
KA 2192	LIN-IC	TV, Audio/video-umschalter/switch	30-SDIP	Sam	-	pdf
KA 2194 D	LIN-IC	Video, RGB-Encoder(NTSC/PAL), Ucc=5V	24-MDIP	Sam	-	pinout
KA 2195 D	LIN-IC	Video, Rgb-encoder (ntsc), Ucc=5V	24-MDIP	Sam	-	pdf pinout
KA 2196 D	LIN-IC	Video, Rgb-encoder (ntsc), Ucc=5V	24-MDIP	Sarn	-	pinout
KA 2197 D	LIN-IC	Video, Rgb-encoder (PAL), Ucc=5V	24-MDIP	Sam	-	pdf
KA 2198 BD	LIN-IC	Video, Rgb-encoder (pal/ntsc), Ucc=5V	24-MDIP	Sam	-	pdf
KA 2201	LIN-IC	NF-E, 14V, 1,5A, 1,2W(9V/8Ω)	8-DIP	Sam	TBA 820M	pdf pinout
KA 2201 A	LIN-IC	=KA 2201: lo Crossover Distortion	8-DIP	Sam	TBA 820M	pinout
KA 2201 B	LIN-IC	=KA 2201: integr. 56Ω Pin 6→7	8-DIP	Sam	TBA 820M	pinout
KA 2201 N	LIN-IC	=KA 2201: 16V, 1,5A, 2W(12V/8Ω)	8-DIP	Sam	TBA 820M	pinout
KA 2206	LIN-IC	2x Power Amp, 9V, 2x2.3W(9V/4Ohm)	12-DIP+b	Sam	KA 22061, LA 4182..4183, TEA 2025, NTE16	pdf pinout
KA 2206	LIN-IC	2x Power Amp, 9V, 2x2.3W(9V/4Ohm)	16-DIP	Fci	KA 22061, LA 4182..4183, TEA 2025, NTE16	
KA 2206B	LIN-IC	2x Power Amp, 9V, 2x2.3W(9V/4Ohm)	12-DIP+b	Sam	KA 22061, LA 4182..4183, TEA 2025, NTE16	
KA 2209	LIN-IC	2x Power Amp, 2x0.65W(6V/ 4Ohm), Vcc=1.8V~9V	8-DIP	Sam	TDA 2822M	
KA 2210	LIN-IC	2x NF-E, 18V, 3,5A, 2x5,5W(13V/4Ω)	12-SIL	Sam	LA 4445	
KA 2211	LIN-IC	2x NF-E, 18V, 3,5A, 2x5,8W(13V/4Ω)	12-SIL	Sam	(TA 7240AP)[4]	
KA 2212	LIN-IC	NF-E, 14V, 0,5W(6V/8Ω)	9-SIP	Sam	AN 7112, LA 4140, TA 7313(AP)	pdf
KA 2213	LIN-IC	NF-V+E, ALC, 13V, 2,2W(9V/4Ω)	14-DIP+g	Sam	LA 4160	pdf pinout
KA 2213 D	LIN-IC	=KA 2213: 0,5W(6V/8Ω)	16-DIP	Sam	-	
KA 2214	LIN-IC	2x NF-E, 18V, 2x1,2W(9V/8Ω)	14-DIP+g	Sam	µPC 1263C	pdf
KA 2220	LIN-IC	Recorder, NF-V + Equal. + ALC, Ucc=5V	9-SIP	Sam	BA 333, LA 3210, TA 7137	pdf pinout
KA 2221	LIN-IC	2x NF-V-ra, Ucc=12V	8-SIP	Sam	BA 328, LA 3161, M 5152L, TA 7375P	pdf pinout
KA 2222	LIN-IC	2x NF-V-ra, Ucc=4V	8-DIP	Sam	-	
KA 2223	LIN-IC	5 Band Graphic Equal., Ucc=5...13V	16-DIP	Sam	LA 3600, M 5226P, TA 7796P	pdf pinout
KA 2224(B)	LIN-IC	2x NF-V, Equal., ALC, Ucc=4...13V	14-DIP	Sam	LA 3220	pinout
KA 2225	LIN-IC	2x NF-V, Ucc=3V	16-DIP	Sam	TA 7709	pdf
KA 2225 D	LIN-IC	=KA 2225: SMD	16-MDIP	Sam	TA 7709F	pdf
KA 2226	LIN-IC	2x NF-V-ra, Equal., Alc	16-DIP	Sam	-	
KA 2227	LIN-IC	2x NF-V-ra, Equal., Alc	16-DIP	Sam	-	
KA 2228	LIN-IC	2x NF-V, ra, Mikrofon-V, ALC, Ucc=5V	21-SQP	Sam	TA 7417P	
KA 2229	LIN-IC	4x NF-V, ra	30-SDIP	Sam	-	
KA 2230	LIN-IC	Recorder-Suchlauf/Music Sel., 9 Progr.	22-DIP	Sam	IR 3R24	
KA 2231	LIN-IC	Recorder, NF-pegelsensor/level Sensor	9-SIP	Sam	LA 2010	
KA 2241	LIN-IC	AM/FM-ZF, Ucc=5V	16-DIP	Sam	-	
KA 2243(/N)	LIN-IC	AM/FM-ZF, Demod., Ucc=3...14V	16-DIP	Sam	BA 4220, HA 12413	
KA 2244	LIN-IC	FM-ZF f. Auto-/car Radio ,Ucc=8...15V	9-SIP	Sam	BA 404, TA 7303	pinout
KA 2245	LIN-IC	FM-ZF f. Auto-/car Radio, Ucc=8...14V	7-SIP	Sam	BA 403, LA 1150, TA 7130P, µPC1028H	pdf pinout
KA 2246	LIN-IC	AM-Tuner, Ucc=12V	16-DIP	Sam	-	
KA 2247	LIN-IC	AM-Tuner, FM-ZF, Ucc=3...8V	16-DIP	Sam	BA 4260, LA 1260	
KA 2248(A)	LIN-IC	AM-Tuner, FM-ZF, Ucc=3V	16-DIP	Sam	TA 7687(A)P	
KA 2248(A)D	LIN-IC	=KA 2248(A): SMD	20-MDIP	Sam	-	
KA 2249	LIN-IC	FM-Vorstufe/FM Front End, Ucc=2...7V	7-SIP	Sam	AN 7213	pdf pinout
KA 2249 D	LIN-IC	=KA 2249: SMD, Ucc=2...5V	8-MDIP	Sam		pdf pinout
KA 2250	CMOS-IC	Dual El. Lautst./volume Ctrl., Udd=12V	16-DIP	Sam	TC 9153	pdf pinout
KA 2261	LIN-IC	FM MPX Stereo-Decoder, Ucc=3...14V	16-DIP	Sam	AN7410, BA1330, HA11227, LA3361, TA7604	pdf
KA 2262	LIN-IC	FM MPX Stereo-Decoder, Ucc=6,5...14V	16-SQP	Sam	LA 3370	
KA 2263(/N)	LIN-IC	FM PLL MPX Stereo-Decoder, Ucc=3...12V	9-SIP	Sam	AN 7420, TA 7343(A)P	
KA 2264	LIN-IC	FM PLL MPX Stereo-Decoder, Ucc=1,8...5V	9-SIP	Sam	AN 7421, TA 7342	pdf
KA 2264 D	LIN-IC	=KA 2264: SMD	16-MDIP	Sam		pdf
KA 2265	LIN-IC	FM PLL MPX Stereo-Decoder,HiFi,Ucc=12V	16-DIP	Sam	LA 3410	pdf
KA 2266	LIN-IC	FM PLL MPX Stereo-Decoder,Ucc=6,5...14V	16-SQP	Sam	LA 3375	
KA 2268	LIN-IC	TV Multiplex-demod. (korea-system)	28-DIP	Sam	-	pinout
KA 2268 N	LIN-IC	~KA 2268	28-DIP	Sam		
KA 2270	LIN-IC	Noise-suppr., dbx-system	16-DIP	Sam	CXA 1011P	
KA 2271	LIN-IC	Dolby B Prozessor, Ucc=8...16V	16-DIP	Sam	CXA 1101P	pdf pinout
KA 2271B	LIN-IC	Dolby B Prozessor, Ucc=8...16V	16-DIP	Sam	CXA 1101P	pdf pinout
KA 2271BD	LIN-IC	SMD, Dolby B Prozessor, Ucc=8...16V	16-MDIP	Sam	CXA 1101M	pdf pinout
KA 2272	LIN-IC	FM Noise-Suppression, Ucc=8...15V	16-SQP	Sam	LA 2110	pdf
KA 2272D	LIN-IC	SMD, FM Noise-Suppression, Ucc=8...15V	16-MDIP	Sam	LA 2110	pdf pinout
KA 2281	LIN-IC	2x LED-Treiber/level meter, 2x5 LED	16-DIP	Sam	TA 7666	pdf
KA 2283	LIN-IC	2x LED-Treiber/level meter, 2x5 LED	16-DIP	Sam	TA 7667	
KA 2284(B)	LIN-IC	LED-Treiber/level meter, 15mA, 5 LED	9-SIP	Sam	AN 6884, BA 6124, LB 1403	pdf pinout
KA 2285(B)	LIN-IC	LED-Treiber/level meter, 7mA, 5 LED	9-SIP	Sam	BA 6137, LB 1423	pdf pinout
KA 2286	LIN-IC	LED-Treiber/level meter, 7mA, 5 LED	9-SIP	Sam	LB 1433	pinout
KA 2287(B)	LIN-IC	LED-Treiber/level meter, 15mA, 5 LED	9-SIP	Sam	BA 6125, LB 1413	pinout
KA 2288	LIN-IC	LED-Treiber/level meter, 10mA, 7 LED	16-DIP	Sam	IR 2 E02	pdf pinout
KA 2290	LIN-IC	AM/FM-Radio, Ucc=3...8V	24-SDIP	Sam	LA 1810	
KA 2292	LIN-IC	AM/FM-Tuner, Stereo-Decoder, Ucc=3V	24-SDIP	Sam	TA 8127, TA 8167	pdf
KA 2292 D	LIN-IC	=KA 2292: SMD	24-MDIP	Sam	TA 8127F	pdf
KA 2293	LIN-IC	AM/FM-Tuner, Stereo-Decoder, Ucc=3V	24-SDIP	Sam	TA 8122	pdf
KA 2293 D	LIN-IC	=KA 2293: SMD	24-MDIP	Sam	TA 8122AF	pdf
KA 2295(Q)	LIN-IC	AM/FM-Radio, MPX, f. DTS, Ucc=6...12V	48-MP	Sam	-	

Type	Device	Short Description	Fig.	Manu	Comparision Types	More at
KA 2297	LIN-IC	AM/FM-Radio, Ucc=1,8...7V	16-DIP	Sam	-	pdf
KA 2297 D	LIN-IC	=KA 2297: SMD	16-MDIP	Sam	-	
KA 2298 B	LIN-IC	AM/FM-Radio, Ucc=1,8...7V	24-SDIP	Sam	-	pdf
KA 2301	LIN-IC	Spielzeugsteuerung/toy Radio Control	9-SIP	Sam	-	
KA 2302	LIN-IC	Spielzeugsteuerung/toy Radio Control	9-SIP	Sam	-	
KA 2303	LIN-IC	Spielzeugsteuerung/toy Radio Control	9-SIP	Sam	-	
KA 2304	LIN-IC	Spielzeugsteuerung/toy Radio Control	9-SIP	Sam	-	
KA 2305 A	LIN-IC	Spielzeugsteuerung/toy Radio Control	12-SIP	Sam	-	
KA 2306 A	LIN-IC	Spielzeugsteuerung/toy Radio Control	14-DIP	Sam	-	
KA 2309	LIN-IC	Spielzeugsteuerung/toy Radio Control	16-DIP	Sam	-	
KA 2310	LIN-IC	Spielzeugsteuerung/toy Radio Control	9-SIP	Sam	-	
KA 2311	LIN-IC	Spielzeugsteuerung/toy Radio Control	16-DIP	Sam	-	
KA 2312	LIN-IC	Spielzeugsteuerung/toy Radio Control	9-SIP	Sam	-	
KA 2314	LIN-IC	Spielzeugsteuerung/toy Radio Control	9-SIP	Sam	-	
KA 2401	LIN-IC	Motorregler/dc Motor-Tr, Uref=1,27V	8-DIP	Sam	(µPC1470H)	pdf
KA 2402	LIN-IC	DC motor speed controller, Vcc=1.8~8V	8-DIP	Sam	LA 5521D	pdf pinout
KA 2402 D	LIN-IC	SMD, DC motor speed controller, Vcc=1.8~4.5V	8-MDIP	Sam	LA 5521M	pdf pinout
KA 2403	LIN-IC	=KA 2402: Uref=0,5V	8-DIP	Sam	-	pinout
KA 2404	LIN/Z-IC	Motorregler/dc Motor-Tr, Uref=1,27V	7d(Z-IC)	Sam	-	pdf
KA 2404 A	LIN/Z-IC	=KA 2404: Fig.→	14	Sam	(µPC 1470H)	pdf
KA 2407	LIN-IC	Motorregler/dc Motor-Tr, Uref=1V	85/4Pin	Sam	(AN 6651)	pdf
KA 2410	LIN-IC	Telefonwecker/tone Ringer	8-DIP	Sam	CS 8204, ML 8204, TA 31001	pdf
KA 2411	LIN-IC	Telefonwecker/tone Ringer	8-DIP	Sam	CS 8205, ML 8205, TA 31002	pdf
KA 2412 A	LIN-IC	Telefonsprechkreis/speech Circuit	14-DIP	Sam	CIC 9185, LS 285A	
KA 2412 B	LIN-IC	Telefonsprechkreis/speech Circuit	14-DIP	Sam	-	
KA 2412 C	LIN-IC	Telefonsprechkreis/speech Circuit	14-DIP	Sam	-	
KA 2413(B)	LIN-IC	Telefon, Zweiton-/dual Tone Generator	16-DIP	Sam	PBD 3535	
KA 2414	LIN-IC	Ein-chip-telefon/one Chip Telephone	40-DIP	Sam	MC 34011	
KA 2417	LIN-IC	Ein-chip-telefon/one Chip Telephone	40-DIP	Sam	MC 34010	
KA 2418	LIN-IC	Telefonwecker/tone Ringer	8-DIP	Sam	-	pdf pinout
KA 2419	LIN-IC	Telefonwecker/tone Ringer, I²l	8-DIP	Sam	-	
KA 2420 D	LIN-IC	=KA 2420N: SMD	28-MDIP	Sam	-	
KA 2420 N	LIN-IC	Telefon, Lautspr.-verst./speaker Amp.	28-DIP	Sam	-	
KA 2425 A,B	LIN-IC	Telefonsprechkreis/speech Network	18-DIP	Sam	-	
KA 2428	LIN-IC	Telefonwecker/tone Ringer	8-DIP	Sam	-	pdf
KA 2580 A	LIN-IC	8x Treiber/Driver, 50V	18-DIP	Sam	UDN 2580A	
KA 2588 A	LIN-IC	8x Treiber/Driver, 50V	20-DIP	Sam	UDN 2588A	
KA 2605	LIN-IC	TV, HA/VA-Synchr., Ucc=8...15V	9-SIP	Sam	-	
KA 2606	LIN-IC	TV, HA/VA-Synchr., Ucc=10...15V	9-SIP	Sam	-	pinout
KA 2611	LIN-IC	Digital Lautst./volume Ctrl.	8-DIP	Sam	-	
KA 2615	LIN-IC	4x LED/LAMP-Tr, 0,15A, Rin=-	9-SIP	Sam	TD 62551S	
KA 2616	LIN-IC	=KA 2615: Rb=2,7kΩ f. TTL Input	9-SIP	Sam	TD 62553S	
KA 2617	LIN-IC	=KA 2615: Rb=10,5kΩ f. PMOS/CMOS Inp.	9-SIP	Sam	TD 62554S	
KA 2618	LIN-IC	=KA 2615: Rb=20kΩ f. PMOS Input	9-SIP	Sam	TD 62555S	
KA 2619	LIN-IC	5 NPN Trans.-Array, 48/24V, 50mA	14-DIP	Sam	-	
KA 2619 D	LIN-IC	=KA 2619: SMD	14-MDIP	Sam	-	
KA 2651	LIN-IC	Flt Display-Tr., 8xNPN-Darl., 65V	18-DIP	Sam	UDN 6118, XR 6118	
KA 2653	LIN-I/O-IC	Interface f. RS232, V28	16-DIP	Sam	DS 75154, SN 75154, µA 75154	
KA 2654	LIN-I/O-IC	Line Driver & Tranceiver f. RS232C	8-DIP	Sam	SN 751701	
KA 2654 D	LIN-I/O-IC	=KA 2654: SMD	8-MDIP	Sam		
KA 2655(D)	LIN-IC	=ULN 2001A(D)		Sam	→ULN 2001A(D)	pdf
KA 2656(D)	LIN-IC	=ULN 2002A(D)		Sam	→ULN 2002A(D)	pdf
KA 2657(D)	LIN-IC	=ULN 2003A(D)		Sam	→ULN 2003A(D)	pdf
KA 2658(D)	LIN-IC	=ULN 2004A(D)		Sam	→ULN 2004A(D)	pdf
KA 2659(D)	LIN-IC	=ULN 2005A(D)		Sam	→ULN 2005A(D)	pdf
KA 2803(B)	LIN-IC	Fehlstromdetektor/earth Leakage Det.	8-DIP	Fcs	M 54123	pdf pinout
KA 2804	LIN-IC	Triac Nullspg.-schalter/zero Volt. Sw.	8-DIP	Sam	µPC 1701C	pdf pinout
KA 2807	LIN-IC	Fehlstromdetektor/earth Leakage Det.	8-DIP	Fcs	LM 1851	pdf pinout
KA 2808	LIN-IC	Fehlstromdetektor/earth Leakage Det.	8-DIP	Sam	-	
KA 2811 B,C	LIN-IC	SMD, Hard Disk, Motor Drv.	48-MP	Sam	-	
KA 2811 B	LIN-IC	SMD, Hard Disk, Motor Drv.	48-MP	Sam	-	
KA 2820 D	LIN-IC	SMD, Floppy Disk, Step Motor Drv.	20-MDIP	Sam	-	
KA 2821 D	LIN-IC	SMD, Floppy Disk, Step Motor Drv.	20-MDIP	Sam	-	
KA 2822 D	LIN-IC	SMD, Floppy Disk, Spindle Motor Drv.	28-SMDIP+b	Sam	-	pdf
KA2901	4xDIFF-COMP	Vs:±18V Vu:100V/mV Vo:TTL Vi0:2mV	14DS	Sam		data pdf pinout
KA 2901...	KOP-IC	=LM 2901...	14-(M)DIP	Sam	→LM 2901...	
KA 2902	4xHI-GAIN	Vs:±13V Vu:100V/mV Vo:24/0V	14DS,9l	Sam		data pdf pinout
KA 2902...	OP-IC	=LM 2902...	14-(M)DIP	Sam	→LM 2902..	
KA2903	2xDIFF-COMP	Vs:±18V Vu:100V/mV Vi0:±1mV	8DS	Sam		data pdf pinout
KA2903...	KOP-IC	=LM 2903...	8-(M)DIP	Sam	→LM 2903...	
KA 2903 S	KOP-IC	=LM 2903...	9-SIP	Sam	→LM 2903...	
KA2904	2xDIFF-COMP	Vs:±18V Vu:100V/mV Vi0:±1mV	9l,8DS	Sam		data pinout
KA 2904...	OP-IC	=LM 2904...	8-(M)DIP	Sam	→LM 2904...	
KA 2904 S	OP-IC	=LM 2904...	9-SIP	Sam	→LM 2904...	
KA 2911	LIN-IC	TV, Video-ZF, neg. Agc, (mosfet-tuner)	16-DIP	Sam	TA 7607, TA 7660, TDA 2544	pdf
KA 2912	LIN-IC	TV, Video-ZF, pos. Agc, Ucc=12V	14-DIP+g	Sam	µPC 1366	pdf
KA 2913 A	LIN-IC	TV, Video-, Ton-ZF, pos. Agc, Ucc=12V	16-DIP	Sam	TA 7678, µPC 1414	
KA 2914 A	LIN-IC	Ctv, Video-, Ton-ZF, neg. Agc, Ucc=12V	24-DIP	Sam	TA 7680	pdf pinout
KA 2915	LIN-IC	TV-Video-, Ton-ZF, pos. Agc, HA, VA	28-DIP	Sam	AN 5151	pdf pinout
KA 2916	LIN-IC	=KA 2911: f. NPN-Tuner, pos. Agc	16-DIP	Sam	TA 7611, TA 7661	pdf
KA 2917	LIN-IC	=KA 2913A: neg. Agc	16-DIP	Sam	TA 7675	
KA 2918	LIN-IC	=KA 2914A: pos. Agc	24-DIP	Sam	TA 7681	pdf pinout
KA 2919	LIN-IC	Ctv, Video-, Ton-ZF, Ucc=12V	30-SDIP	Sam	LA 7520	pdf
KA 2921	LIN-IC	TV, Video-, Ton-ZF, pos. Agc, Ha, VA	16-DIP	Sam	TA 7609	

Type	Device	Short Description	Fig.	Manu	Comparision Types	More at
KA 2922	LIN-IC	TV, Video-ZF, Ton-ZF, pos.AGC, Ucc=12V	20-DIP	Sam	LA 7535	
KA 2923	LIN-IC	=KA 2922: neg. Agc	20-DIP	Sam	LA 7530	
KA 2924	LIN-IC	Ctv, Video+quasi-ton-zf, Ucc=12V	20-SDIP	Sam	TA 8712	
KA 2944	LIN-IC	Vc, Video-a/w-verst./amplifier	28-DIP	Sam	HA 11744, µPC 1534C	
KA 2945	LIN-IC	Vc, Video-verst./amplifier	28-DIP	Sam	HA 11745, µPC 1524A	
KA 2981	LIN-IC	Vc, VHF-Modulator, Ucc=6,2V	16-DIP	Sam	TA 7673, µPC 1507(A)C	
KA 2983	LIN-IC	Vc, Audio-a/w-verst./ampl. + Logic	18-DIP	Sam	BA 5102	
KA 2984	LIN-IC	Vc, A/W-verst./amplifier	28-DIP	Sam	HA 11724	
KA 2985	LIN-IC	Vc, Video Signal-prozessor	28-DIP	Sam	HA 11725	
KA 2986	LIN-IC	Vc, Chroma-signal (pal,NTSC), Ucc=9V	28-DIP	Sam	HA 11726	
KA 2987	LIN-IC	Vc, Servo-controller	28-DIP	Sam	HA 11727	
KA 2988	LIN-IC	Vc, Chroma-signal-prozessor (pal,Ntsc)	28-DIP	Sam	HA 11741, µPC 1536C	
KA 3010 D	LIN-IC	SMD, Cd-rom, 4-Channel Motor Driver	28-SMDIP	Sam	-	pdf
KA 3011 BD	IC	3-Phase Bldc Motor Driver for CD	DIP_K	Fch	-	pdf
KA 3011 BDTF	IC	3-Phase Bldc Motor Driver for CD	DIP_K	Fch	-	pdf pinout
KA 3011 D	LIN-IC	SMD, Cd-rom, 3-Phase Spindle Motor Drv	28-SMDIP	Sam		pdf
KA 3012 D	LIN-IC	SMD, Cd-rom, 4-Channel Motor Driver	28-SMDIP	Sam	-	pdf
KA 3030 D	LIN-IC	SMD, Cd-rom, 6-Channel Motor Driver	28-SMDIP	Sam	-	pdf
KA 3080	LIN-IC	Vc, 3-Phase Capstan Motor Driver	32-SDIP	Sam	-	pdf
KA 3080 D	LIN-IC	=KA 3080: SMD	28-SMDIP			
KA 3081 D	LIN-IC	SMD, Vc, 3-Phase Drum Motor Driver	20-MDIP	Sam		
KA 3100 D	LIN-IC	SMD, Floppy Disk, Step Motor Drv.	16-MDIP	Sam	-	pdf
KA3302	4xDIFF-COMP	Vs:±14V Vu:30V/mV Vo:TTL Vi0:2mV	14DS	Sam		data pdf pinout
KA 3302...	KOP-IC	=LM 3302...	14-(M)DIP	Sam	→LM 3302...	
KA3303	4xHI-GAIN	Vs:±18V Vu:200V/mV Vo:12/0V	14DS	Sam		data pdf pinout
KA 3303...	OP-IC	=MC 3303...	14-(M)DIP	Sam	→MC 3303...	
KA 3361...	LIN-IC	=MC 3361...		Sam	→MC 3361...	
KA3403	4xHI-GAIN	Vs:±18V Vu:200V/mV Vo:±13V Vi0:1.5mV	14DS	Sam		data pdf pinout
KA 3403...	OP-IC	=MC 3403...	14-(M)DIP	Sam	→MC 3403...	
KA 3524	LIN-IC	S-Reg PWM Ctrl., 5V, ...100kHz, 0...+70°	16-DIP	Fcs	HA 17524, LM 3524, SG 3524, UC 3524A	pdf pinout
KA 3525 A	LIN-IC	S-Reg PWM Ctrl., 5V, 0,5A, 0...+70°	16-DIP	Fcs	SG 3525A, UC 3525A	pdf pinout
KA 3525 AD	LIN-IC	=KA 3525A: SMD	16-MDIP	Fcs	-	pdf pinout
KA 3526 B	LIN-IC	S-Reg PWM Ctrl., 5V, 0,2A, 0...+70°	18-DIP	Sam	SG 3526, UC 3526	pinout
KA 3825	LIN-IC	PWM Ctrl., hi-speed,+5V, 0,5A, 0...+70°	16-DIP	Sam	-	pinout
KA 3842 A	LIN-IC	PWM Ctrl., ...500kHz, 0...+70°	8-DIP	Fcs		pdf pinout
KA 3842 B	LIN-IC	PWM Ctrl., ...500kHz, 0...+70°	8-DIP	Fcs	UC 3842	pdf pinout
KA 3842 BD	LIN-IC	PWM Ctrl., ...500kHz, 0...+70°	TSOP	Fcs	-	pdf pinout
KA 3843 A	LIN-IC	PWM Ctrl., ...500kHz, 0...+70°	8-DIP	Fcs	UC 3843	pdf pinout
KA 3843 AD	LIN-IC	PWM Ctrl., ...500kHz, 0...+70°	MDIP	Fcs	UC 3843	pdf pinout
KA 3843 B	LIN-IC	PWM Ctrl., ...500kHz, 0...+70°	8-DIP	Fcs	UC 3843	pdf pinout
KA 3843 BD	LIN-IC	PWM Ctrl., ...500kHz, 0...+70°	TSOP	Fcs	-	pdf pinout
KA 3844 B	LIN-IC	PWM Ctrl., ...500kHz, 0...+70°	8-DIP	Fcs	UC 3844	pdf pinout
KA 3844 BD	LIN-IC	PWM Ctrl., ...500kHz, 0...+70°	TSOP	Fcs	-	pdf pinout
KA 3845 B	LIN-IC	PWM Ctrl., ...500kHz, 0...+70°	8-DIP	Fcs	UC 3845	pdf pinout
KA 3845BD...3845BD	LIN-IC	=KA 3842B...3845B: SMD	14-MDIP		UC 3842D...3845D	pdf pinout
KA 3845 BD	LIN-IC	PWM Ctrl., ...500kHz, 0...+70°	TSOP	Fcs	-	pdf pinout
KA 3846	LIN-IC	PWM Ctrl., 0,2A, ...500kHz, 0...+70°	16-DIP	Fcs	-	pdf pinout
KA 3882	LIN-IC	PWM Ctrl., 0,1A, ...500kHz, 0...+85°	8-DIP	Sam	-	pdf pinout
KA 3882 E	LIN-IC	PWM Ctrl., 0,2A, ...500kHz, 0...+70°	8-DIP	Fcs	-	pdf
KA 3883	LIN-IC	PWM Ctrl., 0,1A, ...500kHz, 0...+85°	8-DIP	Sam	-	pdf pinout
KA 3883 E	LIN-IC	PWM Ctrl., 0,2A, ...500kHz, 0...+70°	8-DIP	Fcs	-	pdf
KA 3884	LIN-IC	PWM Ctrl., 0,1A, ...500kHz, 0...+85°	8-DIP	Sam	-	pdf pinout
KA 3885	LIN-IC	PWM Ctrl., 0,1A, ...500kHz, 0...+85°	8-DIP	Sam	-	pdf pinout
KA 3882 D...3885 D	LIN-IC	=KA 3882...3885: SMD	8-MDIP		-	pdf pinout
KA 3902	LIN-IC	Car, Lüfter/fan Motor Driver	14-DIP	Sam	-	pdf
KA 3903	LIN-IC	Kfz. Fensterheber/automotive Window IC	16-DIP	Sam	-	pdf
KA4558	2xHI-GAIN	Vs:22V Vu:200V/mV Vo:13/0V Vi0:1mV	8DS,9I	Sam		data pinout
KA 4558...	OP-IC	=MC 4558...	8-(M)DIP	Sam	→MC 4558...	
KA 4558...I	OP-IC	=MC 4558...: -40...+85°	8-(M)DIP	Sam	→MC 4558...	
KA 4558 S	OP-IC	=MC 4558...	9-SIP	Sam	→MC 4558...	
KA5532	2xLO-NOISE	Vs:±22V Vu:2.2V/mV Vo:±13V Vi0:0.5mV	8D	Sam		data pdf pinout
KA 5532	OP-IC	=NE 5532...	8-DIP	Sam	→NE 5532...	
KA 6101	LIN-IC	Analog Interface f. Teletext-system	18-DIP	Sam	AN 5355	
KA 6102	LIN-IC	~KA 6101	18-DIP	Sam	AN 5356	
KA 6201	LIN-IC	Fdd, Lese-Verst./Read Amp., Ucc=5V	18-DIP	Sam	HA 16631P	
KA 6202	LIN-IC	Stepping Motor-Tr, 2 Phase, 12V, 0,33A	14-DIP	Sam	-	
KA 7226	LIN-IC	2x Recorder-A/W-Vst., ALC, Ucc=3...16V	14-DIP	Sam	TA 7658P	pdf
KA 7302 D	LIN-IC	SMD, Ccd Color Camera, Agc, Gamma Corr	30-SMDIP	Sam	-	pdf
KA 7305	LIN-IC	SMD, Ccd B/w Camera, Signal Prozessor	48-MP	Sam	-	pdf
KA 7307 D	LIN-IC	SMD, Ccd Color Camera, Agc, Cds	20-MDIP	Sam	-	pdf
KA 7405	LIN-IC	SMD, Camera, Film + Zoom Motor Drv.	22-MDIP	Sam	-	
KA 7406	LIN-IC	SMD, Camera, Strobo Trigger, Motor Drv	48-MP	Sam	-	
KA 7500(B)	SMPS-IC	S-Reg Ctrl., +5V, 0,25A, -20...+85°	16-DIP	Sam	TL 494, µPC 494	pdf pinout
KA 7500 C	SMPS-IC	SMPS Controller	DIP	Fch	-	pdf pinout
KA 7500 CD	SMPS-IC	SMPS Controller	DIP	Fch	-	pdf pinout
KA 7500(B)D	SMPS-IC	=KA 7500: SMD	16-MDIP		-	pdf pinout
KA 7511	SMPS-IC	Tv/vc, SMPS Controller, 0...+70°	10-SIL	Sam	-	pdf pinout
KA 7514	LIN-IC	Power Factor Controller, Uref=5%	8-DIP	Sam	-	
KA 7514 A	LIN-IC	=KA 7514: Uref=1%	8-DIP	Sam	-	
KA 7514 D,AD	LIN-IC	=KA 7514(A): SMD	8-MDIP	Sam	-	
KA 7515	SMPS-IC	SMPS Controller, -20...+85°	8-DIP	Sam	-	pdf
KA 7521	LIN-IC	Ballast Controller, Ucc=10V	16-DIP	Sam	-	
KA 7522	LIN-IC	Dimming Ballast Controller, Ucc=12V	22-SDIP	Sam	-	
KA 7522 D	LIN-IC	=KA 7522: SMD	20-MDIP	Sam	-	
KA 7524(B)	LIN-IC	Power Factor Controller, Uref=1%	8-DIP	Sam	-	pdf pinout

Type	Device	Short Description	Fig.	Manu	Comparision Types	More at
KA 7524(B)D	LIN-IC	=KA 7524: SMD	8-MDIP		-	pdf pinout
KA 7531	LIN-IC	Dimming Ballast Controller, Ucc=12V	22-SDIP	Sam	-	pinout
KA 7531 D	LIN-IC	=KA 7531: SMD	24-MDIP		-	pinout
KA 7533(Z)	LIN-IC	3,3V Spannungs-/Voltage Detector	7d	Sam	KIA 7033	
KA 7536	LIN-IC	3,6V Spannungs-/Voltage Detector	7d	Sam	KIA 7036	
KA 7539	LIN-IC	3,9V Spannungs-/Voltage Detector	7d	Sam	KIA 7039	
KA 7541	IC	Simple Ballast Controller	DIP	Fch	-	pdf pinout
KA 7541 D	IC	Simple Ballast Controller	DIP	Fch	-	pdf pinout
KA 7542	LIN-IC	4,2V Spannungs-/Voltage Detector	7d	Sam	KIA 7042	
KA 7545	LIN-IC	4,5V Spannungs-/Voltage Detector	7d	Sam	KIA 7045	
KA 7551	LIN-IC	SMD, Dual PWM Controller, -25...+85°	16-MDIP	Sam	-	pinout
KA 7552	LIN-IC	PWM Controller, -25...+85°	8-DIP	Sam	-	pdf pinout
KA 7552 A	SMPS-IC	SMPS Controller	DIP	Fch	-	pdf pinout
KA 7553	LIN-IC	PWM Controller, -25...+85°	8-DIP	Sam	-	pdf pinout
KA 7553 A	SMPS-IC	SMPS Controller	DIP	Fch	-	pdf pinout
KA 7560 D	LIN-IC	Bleibatterie/lead Battery Control	24-MDIP	Sam	AN 8360	pinout
KA 7561	LIN-IC	Reusable Alk./mangan Batt. Charge Ctrl	16-DIP	Sam	-	
KA 7577	SMPS-IC	SMPS PWM Ctrl., ±2A, -30...+85°	16-DIP	Sam	-	pdf
KA 7577 D	SMPS-IC	=KA 7577: SMD	20-MDIP		-	pdf
KA 7630	Z-IC	5.1V/0.5A, 8V/0.5A, 12V/1A	10-SIL	Sam	-	pdf pinout
KA 7631	Z-IC	5.1V/0.5A, 9V/0.5A, 12V/1A	10-SIL	Sam	-	pdf pinout
KA 7632	Z-IC	3.3V/0.5A, 8V/0.5A, 5.1V/1A	10-SIL	Sam	-	pdf pinout
KA 7633	Z-IC	3.3V/0.5A, 9V/0.5A, 5.1V/1A	10-SIL	Sam	-	pdf pinout
KA 7805(A)	Z-IC	+5V, 1A, 4% (A=2%), 0...+125°	17b	Sam	... 7805...(TO-220)	pdf
KA 7806(A)	Z-IC	+6V, 1A, 4% (A=2%), 0...+125°	17b	Sam	... 7806...(TO-220)	pdf
KA 7808(A)	Z-IC	+8V, 1A, 4% (A=2%), 0...+125°	17b	Sam	... 7808...(TO-220)	pdf
KA 7809(A)	Z-IC	+9V, 1A, 4% (A=2%), 0...+125°	17b	Sam	... 7809...(TO-220)	pdf
KA 7810(A)	Z-IC	+10V, 1A, 4% (A=2%), 0...+125°	17b	Sam	... 7810...(TO-220)	pdf
KA 7811(A)	Z-IC	+11V, 1A, 4% (A=2%), 0...+125°	17b	Sam	... 7811...(TO-220)	
KA 7812(A)	Z-IC	+12V, 1A, 4% (A=2%), 0...+125°	17b	Sam	... 7812...(TO-220)	pdf
KA 7815(A)	Z-IC	+15V, 1A, 4% (A=2%), 0...+125°	17b	Sam	... 7815...(TO-220)	pdf
KA 7818(A)	Z-IC	+18V, 1A, 4% (A=2%), 0...+125°	17b	Sam	... 7818...(TO-220)	pdf
KA 7824(A)	Z-IC	+24V, 1A, 4% (A=2%), 0...+125°	17b	Sam	... 7824...(TO-220)	pdf
KA 7805 I...7824 I	Z-IC	=KA 7805...7824: -40...+125°	17b	Sam		
KA 7805...7824 R(A)	Z-IC	=KA 7805...7824:	30b	Sam	-	pdf
KA 7805RI...7824RI	Z-IC	=KA 7805...7824: -40...+125°	30b	Sam	-	
KA 7905(A)	Z-IC	-5V, 1A, 4%, 0...+125°	17c	Sam	... 7905...(TO-220)	pdf
KA 7906	Z-IC	-6V, 1A, 4%, 0...+125°	17c	Sam	... 7906...(TO-220)	pdf
KA 7908	Z-IC	-8V, 1A, 4%, 0...+125°	17c	Sam	... 7908...(TO-220)	pdf
KA 7912(A)	Z-IC	-12V, 1A, 4%, 0...+125°	17c	Sam	... 7912...(TO-220)	pdf
KA 7915(A)	Z-IC	-15V, 1A, 4%, 0...+125°	17c	Sam	... 7915...(TO-220)	pdf
KA 7918	Z-IC	-18V, 1A, 4%, 0...+125°	17c	Sam	... 7918...(TO-220)	pdf
KA 7924	Z-IC	-24V, 1A, 4%, 0...+125°	17c	Sam	... 7924...(TO-220)	pdf
KA 8101	LIN-IC	Vc, Detailverbess./detail enhancer	9-SIP	Sam	LA 7308(P)	
KA 8102	LIN-IC	Vc, A/W-verst./amplifier	22-SDIP	Sam	µPC 2313CA	
KA 8103	LIN-IC	Vc, Luminanz Signal Prozessor	48-SDIP	Sam	µPC 2317CA	
KA 8104	LIN-IC	Vc, Chroma-prozessor	30-SDIP	Sam	µPC 2315CA	
KA 8106	LIN-IC	Vc, SVHS Sub-emphasis/sub-deempasis	18-DIP	Sam	VC 2031	
KA 8112	LIN-IC	Vc, VHS A/W-Verst./Amp. (f. 2 Heads)	22-SDIP	Sam	LA 7320	
KA 8113	LIN-IC	Vc, VHS-HQ Luminancesignal-prozesser	30-SDIP	Sam	LA 7323	
KA*8114	LIN-IC	Vc, SECAM Chroma Prozessor	30-SDIP	Sam	-	
KA 8301	LIN-IC	Vc, DC Motor Control, Loading Motor	10-SIL	Sam	BA 6209	pdf pinout
KA 8302	LIN-IC	Vc, Servo-controller	12-SIP	Sam	TA 8617S	
KA 8303	CMOS-IC	Vc, Digital Servo-controller	42-SDIP	Sam	TD 6362N	
KA 8304	LIN-IC	VC++,Motorreg./2-Phase DD Motor Driver	12-SIL	Sam	BA 6411	pdf pinout
KA 8305	LIN-IC	VC++, Motorregler/motor Driver	10-SIL	Sam	BA 6238A	
KA 8306	LIN-IC	Vc, Dual Br.-motor-tr/bridge Driver	10-SIL	Sam	TA 7288P	pdf pinout
KA 8307	LIN-IC			Sam	HD 49723	
KA 8308	CMOS-IC	Vc, Digital Servo Controller	30-SDIP	Sam	-	
KA 8309(B)	MOS-IC	SMD, Cd, Servo Signal-prozessor	48-MP	Sam	CXA 1082	pdf
KA 8310	LIN-IC	VC++, Motorregler/2-Phase DD Motor-Tr	20-SQP	Sam	M 51721ATL	pdf pinout
KA 8311	LIN-IC	Vc, Motorreg./3-Phase Brushl. Motor-Tr	32-SDIP	Sam	M 52440ASP	pinout
KA 8312	LIN-IC	Vc, Servo-controller	12-SIP	Sam	-	
KA 8316	CMOS-IC	Vc, Digital Servo Controller	56-SDIP	Sam	HD 49748	
KA 8319	CMOS-IC	Vc, Digital Servo Controller	56-SDIP	Sam	-	
KA 8320	CMOS-IC	SMD, Vc, Digital Servo Prozessor	60-MP	Sam	-	pdf
KA 8328 D	LIN-IC	SMD, Vc, 2-Phase Drum Motor Driver	20-MDIP	Sam	-	pdf
KA 8329 B	LIN-IC	Vc, 3-Phase Capstan Motor Driver	32-SDIP	Sam	-	
KA 8330	LIN-IC	Vc, Loading Motor Driver	10-SIP	Sam	-	
KA 8401	LIN-IC	Vc, A/W-verst./amp., Switchless	24-SQP	Sam	BA 7751LS	
KA 8402	LIN-IC	Vc, Audio+Video-S/2-Input Switch	7-SIP	Sam	TA 7347P	pinout
KA 8403	LIN-IC	Vc, Audio+Video-S/3-Input Switch	9-SIP	Sam	TA 7348P	pinout
KA 8500 A	LIN-IC	Telefonsprechkreis/speech Network	16-DIP	Sam	LS 156	
KA 8501 A	LIN-IC	Telefonsprechkreis/speech Network	16-DIP	Sam	LS 256	pdf
KA 8503	LIN-IC	Telefonsprechkr./speech Netw., lo-volt	18-DIP	Sam	-	pinout
KA 8504	LIN-IC	Telefonsprechkreis/speech Network	16-DIP	Sam	TA 31003	pdf pinout
KA 8507	LIN-IC	Telefon, Kompander/compandor	20-DIP	Sam	-	pinout
KA 8507B	LIN-IC	Automatic gain control system	DIP	Sam	-	pdf pinout
KA 8507BD	LIN-IC	SMD, aut. gain control system	MDIP	Sam	-	pdf pinout
KA 8507 D	LIN-IC	=KA 8507: SMD	20-MDIP	Sam	-	pinout
KA 8508	LIN-IC	Telefon, FM Sender/transmitter	16-DIP	Sam	MC 2833	
KA 8508 A	LIN-IC	=KA 8508: + HF-Verst./Amp. (Pin11-13)	16-DIP	Sam	-	
KA 8509	LIN-IC	SMD, Telefon, FM Empfänger/receiver	24-SDIP	Sam	MC 3362	pdf
KA 8601	LIN-IC	Telefon, Voice Switched Speaker Phone	48-SDIP	Sam	-	pdf
KA 8602 D	LIN-IC	=KA 8602: SMD	8-MDIP	Sam	-	pdf pinout

Type	Device	Short Description	Fig.	Manu	Comparision Types	More at
KA 8602 N	LIN-IC	NF-E, Ucc=2...16V, 0,25A,>0,25W(6V/32Ω)	8-DIP	Sam	-	pdf pinout
KA 9201 D	LIN-IC	=KA 9201M: SMD	30-MDIP	Sam	CXA 1081	pdf
KA 9201 M	LIN-IC	Cd, HF Signal-prozessor	30-SDIP	Sam	CXA 1081	pdf
KA 9201 Q	LIN-IC	=KA 9201M: SMD	32-MP	Sam	CXA 1271	pdf
KA 9220 B	IC	Rf Amp + Ssp for CD Player	MP	Sam	-	pdf pinout
KA 9220 C	BiCMOS-IC	SMD, Cd, Servo Signal-prozessor, HF-A.	80-MP	Sam		
KA 9221	MOS-IC	SMD, Cd, Servo Signal-prozessor	48-MP	Sam	CXA 1272	
KA 9250 D	LIN-IC	SMD, Cd-rom, Sled Motor Controller	24-MDIP	Sam		
KA 9255 D	LIN-IC	SMD, Cd, PWM Spindle-motor Ctrl.	22-MDIP		BA 6280	
KA 9256	OP-IC	Dual, hi-power, ±18V, 1A, 12,5W	10-SIL	Sam	TA 7256	pdf
KA 9257	OP-IC	Dual, hi-power, 18V, 0,5A, 15W	12-SIL	Sam	BA 6290*	pdf pinout
KA 9258 CD	IC	4-Channel Motor Driver	MSIP	Fcs		pdf pinout
KA 9258 CDTF	IC	4-Channel Motor Driver	MSIP	Fcs		pdf pinout
KA 9258 D	LIN-IC	SMD, Cd, 4-Channel Motor Driver	28-SMDIP	Sam	-	pdf
KA 9259 D	LIN-IC	SMD, Cd, 5-Channel Motor Driver	28-SMDIP	Sam	-	pdf
KA 9270	LIN-IC	Cd, Dual Audio Filter	20-DIP	Sam	-	pdf pinout
KA 9270 D	LIN-IC	=KA 9270: SMD	20-MDIP	Sam		pdf pinout
KA 9401	LIN-IC	SMD, Laser Disk, Audio Signal Proz.	80-MP	Sam		
KA 9410	BiCMOS-IC	Laser Disk, Video Signal Proz, (ntsc)	42-SDIP	Sam		
KA 9412	BiCMOS-IC	Laser Disk, Video Proz, (ntsc/pal)	42-SDIP	Sam		
KA 9413-01	BiCMOS-IC	=KA 9413: OSD Japanisch/Japanese	100-QFP	Sam		
KA 9413	BiCMOS-IC	Laser Disk, Video Proz, OSD Englisch	100-QFP	Sam		
KA 9414 D	BiCMOS-IC	SMD, Laser Disk, Video-Demodulator	30-MDIP	Sam		
KA 9421	BiCMOS-IC	SMD, Laser Disk, Servo Control	100-QFP	Sam		
KA 9430	BiCMOS-IC	Laser Disk, Time Base Corr, Motor Ctrl	42-SDIP	Sam		
KA 9431	BiCMOS-IC	Laser Disk, Time Base Corr, Motor Ctrl	48-SDIP	Sam		
KA 9490	LIN-IC	Laser Disk, Ccc Clock Driver	9-SIP	Sam		
KA 22061	LIN-IC	=KA 2206	12-DIP+b	Sam	KA 2206, LA 4182...4183, TEA 2025	pinout
KA 22062	LIN-IC	2x NF-E, 16V, 2x2,5A, 2x4,5W(12V/4Ω)	12-SIL	Sam	TA 7283AP	
KA 22063	LIN-IC	2x NF-E, 16V, 2x2,5A, 2x4,5W(12V/4Ω)	12-SIL	Sam	KA 22065, TA 8207K, TA 7282AP	
KA 22065	LIN-IC	2x NF-E, 20V, 2x2,5A, 2x4,6W(12V/4Ω)	12-SIL	Sam	TA 8207K	pdf pinout
KA 22069BN	LIN-IC	2x Power Amp, 9V, 2x2.3W(9V/4Ohm)	DIP	Sam	KA 22061, LA 4182...4183, TEA 2025, NTE16	pdf pinout
KA 22101	LIN-IC	NF-E, 18V, 9A, 23W(13V/4Ω) BTL	12-SQL	Sam	TA 7250BP	
KA 22102	LIN-IC	2x NF-E, 18V, 9A, 2x15W(13V/4Ω)	17-QILP	Sam	TA 8205	
KA 22103	LIN-IC	2x NF-E, 18V, 9A, 2x19W(13V/4Ω)	17-QILP	Sam	TA 8210	pdf
KA 22130	LIN-IC	=KA 2213: Pin 8+9→Masse/Ground	16-DIP_g	Sam	LA 4160	pdf pinout
KA 22131	LIN-IC	2x NF-V+E, Ucc=3V, 2x69mW(3V/16Ω)	24-MDIP	Sam	BA 3502F	
KA 22132	LIN-IC	Recorder, 2x Nf-v+e		Sam	AN 7108, CXA 1034P	
KA 22134	LIN-IC	Recorder, 2x NF-V, Ucc=1,8...6V	16-DIP	Sam	TA 8119	pinout
KA 22135	LIN-IC	Recorder, 2x NF-V, Motorreg./ctrl.	22-SDIP	Sam	LAG 637D	pinout
KA 22136	LIN-IC	Recorder, 2x NF-V, Motorreg./ctrl.	28-SDIP	Sam	LAG 665	
KA 22136 D	LIN-IC	=KA 22136: SMD	28-MDIP	Sam		
KA 22211	LIN-IC	~KA 2221: Ucc=9V	8-SIP	Sam	KA 2221, LA 3160, M 51521L	pdf pinout
KA 22231	LIN-IC	2x 5 Band Graphic Equal., Ucc=1,6...6V	28-MDIP	Sam	-	
KA 22232	LIN-IC	2x 3 Band Graphic Equal., Ucc=1,6...6V	20-MDIP	Sam	-	
KA 22233	LIN-IC	2x 3 Band Graphic Equal., Ucc=5...15V	22-DIP	Sam	-	pdf
KA 22234	LIN-IC	2x 5 Band Graphic Equal., Ucc=3,5...14V	24-SQP	Sam	BA 3822L	pdf pinout
KA 22235	LIN-IC	5 Band Graphic Equal., Ucc=3,5...16V	18-SQP	Sam	BA 3812L	
KA 22241(B)	LIN-IC	2x NF-V, Equal., ALC, Ucc=4,5...14V	9-SIP	Sam	BA 3308	
KA 22241(B)	LIN-IC	2x NF-V, Equal., ALC, Ucc=4,5...14V	9-SIP	Sam	BA 3308	pdf pinout
KA 22242	LIN-IC	2x NF-V, Equal., ALC, Ucc=4...12V	10-SIP	Sam	BA 3312N	pdf
KA 22261	LIN-IC	2x NF-V-ra, Alc	16-DIP	Sam	TA 7668BP	pdf pinout
KA 22291	LIN-IC	Recorder, 4x A/W-Verst., Ucc=4...12V	24-SDIP	Sam	-	pdf pinout
KA 22296	LIN-IC	Recorder, 4x A/W-E, Ucc=6...9V	30-SDIP	Sam	-	
KA 22421	LIN-IC	AM-Radio, NF-V+E, Ucc=2...5V	16-DIP	Sam	CIC 7641, TA 7641BP	
KA 22421 D	LIN-IC	=KA 22421: SMD	16-MDIP	Sam		
KA 22424	LIN-IC	AM/FM-Radio, NF-V+E, Ucc=3...12V	16-DIP	Sam	A 283D, HA 12402, TA 7613AP, TDA 1083, U	pinout
KA 22426(B)	LIN-IC	AM/FM-Radio, NF-V+E, Ucc=2...8,5V	28-DIP	Sam	CXA 1019	pinout
KA 22427(C)	LIN-IC	AM/FM-Radio, NF-V+E, Ucc=3...12V	16-DIP	Sam	→KA 22424	pinout
KA 22429 D	LIN-IC	SMD, FM-Radio, NF-V, Ucc=1,8...6V	16-MDIP	Sam	TDA 7021T	pdf pinout
KA 22441	LIN-IC	FM-ZF, Afc, Agc, Muting, Ucc=6...14V	16-SQP	Sam	BA 4110, LA 1140	pdf
KA 22461	LIN-IC	AM-Tuner, ZF, Agc, Ucc=8...15V	19-SQP	Sam	µPC 1215V	
KA 22471	LIN-IC	AM-Tuner, FM-ZF, Ucc=3...8V	16-DIP	Sam	TA 7640AP,P	pdf
KA 22491	LIN-IC	FM-Vorstufe/FM front end	9-SIP	Sam	-	
KA 22495	LIN-IC	FM-V, Front End, Ucc=1,6...6V	9-SIP	Sam	AN 7205, LA 1185, TA 7358AP	pdf
KA 22495 D	LIN-IC	=KA 22495: SMD	14-MDIP	Sam		pdf
KA 22496	LIN-IC	FM-Vorstufe/FM Front End, Ucc=1,6...6V	9-SIP	Sam	TA 7358P	
KA 22682	LIN-IC	TV Multiplex-demod. (korea-system)	28-DIP	Sam	-	pinout
KA 22711	LIN-IC	Dolby B Prozessbr, Ucc=5...16V	16-DIP	Sam	CXA 1163P	pdf pinout
KA 22712(B)	LIN-IC	Dolby B Prozessor, Ucc=6,5...16V	16-DIP	Sam	CXA 1102P	pinout
KA 22901	LIN-IC	AM/FM-Radio, MPX, Ucc=1,8...7V	24-SDIP	Sam		
KA 22901 D	LIN-IC	=KA 22901: SMD	24-MDIP			
KA 34063 A	SMPS-IC	DC-DC Converter Ctrl., Ucc=5V, 0...+70°	8-DIP	Fcs	MC 34063	pdf pinout
KA 34063 AD	LIN-IC	=KA 34063A: SMD	8-MDIP	Fch	-	pdf
KA 5 Q 0740 RT	SMPS-IC	Fairchild Power Switch(fps)	86	Fch		pdf
KA 75250	LIN-IC	2,5V Spannungs-/Voltage Detector	7d	Fch	KIA 7025	pdf pinout
KA 75270	LIN-IC	2,7V Spannungs-/Voltage Detector	7d	Fch	KIA 7027	pdf pinout
KA 75290	LIN-IC	2,9V Spannungs-/Voltage Detector	7d	Fch	KIA 7029	pinout
KA 75310	LIN-IC	3,1V Spannungs-/Voltage Detector	7d	Fch	KIA 7031	pinout
KA 75330	LIN-IC	3,3V Spannungs-/Voltage Detector	7d	Fch		pdf pinout
KA 75360	LIN-IC	3,6V Spannungs-/Voltage Detector	7d	Fch		pdf pinout
KA 75390	LIN-IC	3,9V Spannungs-/Voltage Detector	7d	Fch		pdf pinout
KA 75420	LIN-IC	3,9V Spannungs-/Voltage Detector	7d	Fch		pdf pinout
KA 75450	LIN-IC	3,9V Spannungs-/Voltage Detector	7d	Fch		pdf pinout
KAB 3403 T	IC	White LED Step-up Converter	TSOP	Kec		pdf pinout

Type	Device	Short Description	Fig.	Manu	Comparision Types	More at
KAB 3405 T	IC	Step Up Type DC/DC Converter, White LED	TSOP	Kec	-	pdf pinout
KAC 3301 QN	IC	1X/1. 5X Fractional Cha ge Pump	TSOP	Kec	-	pdf pinout
KAC 3302 QN	IC	White LED 1X/1. 5X Fractional Charge Pump	TSOP	Kec	-	pdf pinout
KAC 3303 DN	IC	1X/1. 5X Fractional Charge Pump	TSOP	Kec	-	pdf pinout
KAC 3305 DN	IC	700mA White LED Regulating Charge Pump	TSOP	Kec	-	pdf pinout
KAD 0206	A/D-IC	6 Bit, hi-speed, 20Msps	30-SDIP	Sam		
KAD 0206 D	A/D-IC	=KAD 0206: SMD	32-MDIP	Sam		
KAD 0228 D	CMOS-A/D-IC	SMD, 8 Bit, hi-speed, Video, 40Msps	24-MDIP	Sam		
KAD 0808 IN	A/D-IC	8 Bit, Multiplexer	28-DIP	Sam		pinout
KAD 0809 IN	A/D-IC	8 Bit, Multiplexer	28-DIP	Sam		
KAD 0816	A/D-IC	8 Bit, µcomp compat.; 16-Ch. MPX	40-DIP	Sam		
KAD 0820 AIN,BIN	A/D-IC	8 Bit, Track/Hold	20-DIP	Sam		
KAD 7001(CQ)	CMOS-A/D-IC	SMD, 3 3/4 Digit 3260-Count, Udd=3V	80-MP	Sam		
KAG	N-FET	=2SK368-GR (Typ-Code/Stempel/marking)	35	Tos	→2SK368	data
KAO	N-FET	=2SK368-O (Typ-Code/Stempel/marking)	35	Tos	→2SK368	data
KAY	N-FET	=2SK368-Y (Typ-Code/Stempel/marking)	35	Tos	→2SK368	data

KB

Type	Device	Short Description	Fig.	Manu	Comparision Types	More at
KB	Si-N	=2SC2778-B (Typ-Code/Stempel/marking)	35	Mat	→2SC2778	data
KB	Si-N	=2SC3936-B (Typ-Code/Stempel/marking)	35(2mm)	Mat	→2SC3936	data
KB	Si-N	=2SC4539 (Typ-Code/Stempel/marking)	39	Tos	→2SC4539	data
KB	Si-N	=2SC4655-B (Typ-Code/Stempel/marking)	35(1,6mm)	Mat	→2SC4655	data
KB	Si-N	=2SC4972-B (Typ-Code/Stempel/marking)	≈35	Mat	→2SC4972	data
KB	MOS-N-FET-e	=2SK1311 (Typ-Code/Stempel/marking)	39	Say	→2SK1311	data
KB	N-FET	=2SK323-B (Typ-Code/Stempel/marking)	35	Hit	→2SK323	data
KB	N-FET	=2SK433-B (Typ-Code/Stempel/marking)	35	Mit	→2SK433	data
KB	Si-N	=BFQ 29 (Typ-Code/Stempel/marking)	35	Sie	→BFQ 29	data
KB	MOS-P-FET	=BSS 192 (Typ-Code/Stempel/marking)	39	Val	→BSS 192	data
KB	MOS-N-FET-e*	=CPH 6402 (Typ-Code/Stempel/marking)	46	Say	→CPH 6402	data
KB	MOS-N-FET-e*	=MCH 3402 (Typ-Code/Stempel/marking)	35(2mm)	Say	→MCH 3402	data
KB	Si-N+R	=RN 1441B (Typ-Code/Stempel/marking)	35	Tos	→RN 1441	data
KB 102 А,В,Б,Д,Г	C-Di	Radio, 45V(Д=80V), 14…40pF(4V) Б=19…30, Д=19…30, Г=19…30pF(4V)	31a	GUS	-	data
KB 104 А,В,Е,Б,Д,Г	C-Di	Radio, 45V(Д,Г=80V), 90…192pF(4V) Е=95…143, Б=106…144, Д=128…192, Г=95…143pF(4V)	31a	GUS	-	data
KB 110 А,В,Е,Б,Д,Г	C-Di	Radio, 45V, 12…26,4pF(4V) Б=14,4…21,6, Д=14,4…21,6, Г=12…18pF	31a	GUS	-	data
KB 112 А-1,Б-1	C-Di	Min, 25V, 9,6…18pF		GUS		data
KB 114 А-1,Б-1	C-Di	Min, A-1=150V, Б-1=115V, 54,4…81,6pF	36a	GUS		data
KB 122 А1,В1,Б1	C-Di	=KB 122A9,В9,Б9:	31	GUS	-	data
KB 122 А9,В9,Б9,Г9	C-Di	SMD, TV Band-S, 30V, 1,9…3,1pF(25V) Б9=2…2,3, Г9=1,9…2,35pF(25V)	35	GUS	-	data
KB 123 A	C-Di	TV Band-S, 28V, 2,6…3,8pF(25V)	71b	GUS	-	data
KB 126 A-5	C-Di	TV,Radio Band-S, 28V, 2,6…3,8pF(25V)	Chip	GUS	-	data
KB 128 A	C-Di	Car Radio, FM, 12V, 22…28pF(1V)	31b	GUS	-	data
KB 129 A	C-Di	Frequ. Modulat., 25V, <10,8pF(5V)	31b	GUS	-	data
KB 132 A	C-Di	Radio, 12V, 26,4…39,6pF(2V)	71a	GUS	-	data
KB 136 А,Б	C-Di	Telefon, Quarz-Generator, 30V 17,1…24,2pF(4V)	31a	GUS	-	data
KB 138 А,Б	C-Di	FM Radio, 12V, 14…21pF(2V)	71	GUS	-	data
KB 139 A	C-Di	AM-Tuning, 16V, 500…620pF(1V)	40(2Pin)	GUS	-	data
KB 142 А,Б	C-Di	AM-Tuning, 32V, 230…320pF(1V)	71	GUS	-	data
KB 143 А,В,Б	C-Di	FM/VHF-Tuning, 18(B=28)V, <29,7pF(3V)	31	GUS	-	data
KB 144 А,В,Б,Г	C-Di	CATV-Tuning, 32V, 2,6…3,2pF(28V) Б,Г=2,8…3,2/33,5pF(26/1V)	31	GUS	-	data
KB 144 А1…Г1	C-Di	=KB 144A,В,Б,Г:	71		-	data
KB 146 A	C-Di	Video, 15V, 10…16pF(10V)	31a	GUS	-	data
KB 149 А,В,Б	C-Di	TV, UHF-Tuning, 30V, 1,8…2,7pF(28V) В=2,2…2,7, Б=1,8…2,4pF(28V)	31	GUS	-	data
KB-162	Si-St	0,59…0,65V(3mA)	31a	Unz	BA 216, BA 314	data
KB-162 C5B	Si-St	0,59…0,75V(20mA)	(36)	Unz		data
KB-165	Si-St	0,62…0,69V(3mA)	31a	Unz	BA 216, BA 314	data
KB-169	Si-St	0,66…0,72V(3mA)	31a	Unz	BA 216, BA 314	data
KB-262(C4,M)	Si-St	1,18…1,30V(3mA)	31a	Unz	BZV 86/1V4, BZX 75/C1V4, ZTE 1,5	data
KB-265(C4)	Si-St	1,24…1,38V(3mA)	31a	Unz	BZV 86/1V4, BZX 75/C1V4, ZTE 1,5	data
KB-269	Si-St	1,32…1,44V(3mA)	31a	Unz	BZV 86/1V4, BZX 75/C1V4, ZTE 1,5	data
KB-362(M)	Si-St	1,77…1,95V(3mA)	31a	Unz	BZ102/2V1, BZV86/2V0, BZX75/C2V1, ZTE 2	data
KB-365	Si-St	1,86…2,07V(3mA)	31a	Unz	BZ102/2V1, BZV86/2V0, BZX75/C2V1, ZTE 2	data
KB-369	Si-St	1,98…2,16V(3mA)	31a	Unz	BZ102/2V1, BZV86/2V0, BZX75/C2V1, ZTE 2	data
KB-462	Si-St	2,36…2,60V(3mA)	31a	Unz	BZ 102/2V8, BZV 86/2V6, BZX 85/C2V8	data
KB-465	Si-St	2,48…2,76V(3mA)	31a	Unz	BZ 102/2V8, BZV 86/2V6, BZX 75/C2V8	data
KB-469	Si-St	2,64…2,88V(3mA)	31a	Unz	BZ 102/2V8, BZV 86/2V6, BZX 75/C2V8	data
KB 2825	LIN-IC	SMD, Floppy Disk, Pre-amp., Controller	64-MP	Sam	-	
KB 4433 A	LIN-IC	2x NF-E, 6,8V, 2x0,15W(3V/4Ω)	16-MDIP	old	-	
KB 9223	BiCMOS-IC	SMD, Cd, Servo Signal Proz., Hf Amp.	80-MP	Sam		pdf
KB 22686	LIN-IC	TV, Sound MPX f. 2-Carrier System	32-SDIP	Sam	-	
KB 22688 B	LIN-IC	TV, Sound MPX f. 2-Carrier System	32-SDIP	Sam	-	
KBL 01	Si-Br	Gl-Br, 100V, 4A	33x1	Gsi	B70 C4000	data

Type	Device	Short Description	Fig.	Manu	Comparision Types	More at
KBL 02	Si-Br	Gl-Br, 200V, 4A	33x1	Gsi	B140 C4000	data
KBL 04	Si-Br	Gl-Br, 400V, 4A	33x1	Gsi	B280 C4000	data
KBL 005	Si-Br	Gl-Br, 50V, 4A	33x1	Gsi	B35 C4000	data
KBL 06	Si-Br	Gl-Br, 600V, 4A	33x1	Gsi	B420 C4000	data
KBL 08	Si-Br	Gl-Br, 800V, 4A	33x1	Gsi	B560 C4000	data
KBL 10	Si-Br	Gl-Br, 1000V, 4A	33x1	Gsi	B700 C4000	data
KBP 01(M)	Si-Br	Gl-Br, 100V, 1A	33x1	Gie	B70 C1000	data
KBP 02(M)	Si-Br	Gl-Br, 200V, 1A	33x1	Gie	B140 C1000	data
KBP 04(M)	Si-Br	Gl-Br, 400V, 1A	33x1	Gie	B280 C1000	data
KBP 005(M)	Si-Br	Gl-Br, 50V, 1A	33x1	Gie	B35 C1000	data
KBP 06(M)	Si-Br	Gl-Br, 600V, 1A	33x1	Gie	B420 C1000	data
KBP 08(M)	Si-Br	Gl-Br, 800V, 1A	33x1	Gie	B560 C1000	data
KBP 10(M)	Si-Br	Gl-Br, 1000V, 1A	33x1	Gie	B700 C1000	data
KBPC 10(-W,-D)-...	Si-Br	Gl-Br, 50...1000V, 10A 01=100, 02=200, 04=400, 005=50V 06=600, 08=800, 10=1000V	79x3	Gsi	GBPC 12-...	data
KBPC 12(-W,-D)-...	Si-Br	Gl-Br, 50...1000V, 12A 01=100, 02=200, 04=400, 005=50V 06=600, 08=800, 10=1000V	79x3	Gsi	GBPC 12-...	data
KBPC 15(-W,-D)-...	Si-Br	Gl-Br, 50...1000V, 15A 01=100, 02=200, 04=400, 005=50V 06=600, 08=800, 10=1000V	79x3	Gsi	GBPC 15-...	data
KBPC 25(-W,-D)-...	Si-Br	Gl-Br, 50...1000V, 25A 01=100, 02=200, 04=400, 005=50V 06=600, 08=800, 10=1000V	79x3	Gsi	GBPC 25-...	data
KBPC 35(-W,-D)-...	Si-Br	Gl-Br, 50...1000V, 35A 01=100, 02=200, 04=400, 005=50V 06=600, 08=800, 10=1000V	79x3	Gsi	GBPC 35-...	data
KBU 4 A...M	Si-Br	Gl-Br, 50...1000V, 4A A=50, B=100, D=200, G=400, J=600V K=800, M=1000V	33x1	Gsi	B35...700 C4000	data
KBU 6 A...M	Si-Br	Gl-Br, 50...1000V, 6A A=50, B=100, D=200, G=400, J=600V K=800, M=1000V	33x1	Gsi	B35...700 C6000	data
KBU 8 A...M	Si-Br	Gl-Br, 50...1000V, 8A A=50, B=100, D=200, G=400, J=600V K=800, M=1000V	33x1	Gsi	B35...700 C8000	data

KC

Type	Device	Short Description	Fig.	Manu	Comparision Types	More at
KC	Si-N	=2SC2778-C (Typ-Code/Stempel/marking)	35	Mat	→2SC2778	data
KC	Si-N	=2SC3936-C (Typ-Code/Stempel/marking)	35(2mm)	Mat	→2SC3936	data
KC	Si-N	=2SC4540 (Typ-Code/Stempel/marking)	39	Tos	→2SC4540	data
KC	Si-N	=2SC4655-C (Typ-Code/Stempel/marking)	35(1,6mm)	Mat	→2SC4655	data
KC	Si-N	=2SC4972-C (Typ-Code/Stempel/marking)	≈35	Mat	→2SC4972	data
KC	MOS-N-FET-e	=2SK1467 (Typ-Code/Stempel/marking)	39	Say	→2SK1467	data
KC	MOS-N-FET-e	=2SK2463 (Typ-Code/Stempel/marking)	39	Rhm	→2SK2463	data
KC	N-FET	=2SK323-C (Typ-Code/Stempel/marking)	35	Hit	→2SK323	data
KC	N-FET	=2SK433-C (Typ-Code/Stempel/marking)	35	Mit	→2SK433	data
KC	Si-N	=BFQ 29P (Typ-Code/Stempel/marking)	35	Sie	→BFQ 29P	data
KC	MOS-N-FET-e*	=CPH 3403 (Typ-Code/Stempel/marking)	35	Say	→CPH 3403	data
KC 133 A	Z-Di	3...3,7V, 3...81mA, 0,3W	34	GUS	BZD 10/C3V3, BZX 55/C3V3, 1N5226, ++	
KC 139 A	Z-Di	3,5...4,3V, 3...70mA, 0,3W	34	GUS	BZD 10/C3V9, BZX 55/C3V9, 1N5228, ++	
KC 147	Si-N	→BC 147	9a	Tes		data
KC 147 A	Z-Di	4,1...5,2V, 3...58mA, 0,3W	34	GUS	BZD 10/C4V7, BZX 55/C4V7, 1N5230, ++	
KC 148	Si-N	→BC 148	9a	Tes		data
KC 149	Si-N	→BC 149	9a	Tes		data
KC 156 A	Z-Di	5,6V ±10%, 3...55mA, 0,3W	34	GUS	BZD 10/C5V6, BZX 55/C5V6, 1N5232, ++	
KC 168 A	Z-Di	6,8V ±10%, 3...45mA, 0,3W	34	GUS	BZD 10/C6V8, BZX 55/C6V8, 1N5235, ++	
KC 507	Si-N	≈BC 547	2a	Tes	→BC 547	data
KC 508	Si-N	≈BC 548	2a	Tes	→BC 548	data
KC 509	Si-N	≈BC 549	2a	Tes	→BC 549	data
KC 510	Si-N	≈BC 550	2a	Tes	→BC 550	data
KC 620 A	Z-Di	120V±10%, 5...42mA, 5W	32	GUS	BZX 98/C120, 1N3008	
KC 630 A	Z-Di	130V±10%, 5...38mA, 5W	32	GUS	BZX 98/C130, 1N3009	
KC 650 A	Z-Di	150V±10%, 2,5...33mA, 5W	32	GUS	BZX 98/C150, 1N3011	
KC 680 A	Z-Di	180V±10%, 2,5...28mA, 5W	32	GUS	BZX 98/C180,-1N3014	
KC 73125-M,UB,UC	CCD-IC	SMD, Ccd Sensor f. 1/3' Camera	16-DIP	Sam	-	
KC 73129-M,UB,UC	CCD-IC	SMD, Ccd Sensor f. 1/3' Camera	16-DIP	Sam	-	
KCG	N-FET	=2SK625-G (Typ-Code/Stempel/marking)	35	Tos	→2SK625	data
KCO	N-FET	=2SK625-O (Typ-Code/Stempel/marking)	35	Tos	→2SK625	data
KCR	N-FET	=2SK625-R (Typ-Code/Stempel/marking)	35	Tos	→2SK625	data
KCY	N-FET	=2SK625-Y (Typ-Code/Stempel/marking)	35	Tos	→2SK625	data

Type	Device	Short Description	Fig.	Manu	Comparision Types	More at
KD...KE						
КД ...... КДС ...	-Di	→nach/after KZ ...		GUS		
KD	Si-N	=2SC4541 (Typ-Code/Stempel/marking)	39	Tos	→2SC4541	data
KD	MOS-N-FET-e	=2SK1470 (Typ-Code/Stempel/marking)	39	Say	→2SK1470	data
KD	N-FET	=2SK323-D (Typ-Code/Stempel/marking)	35	Hit	→2SK323	data
KD	N-FET	=2SK433-D (Typ-Code/Stempel/marking)	35	Mit	→2SK433	data
KD	N-FET	=2SK930-D (Typ-Code/Stempel/marking)	35(2mm)	Mit	→2SK930	data
KD	MOS-N-FET-e*	=CPH 6404 (Typ-Code/Stempel/marking)	46	Say	→CPH 6404	data
KD	Z-Di	=P6 SMB-6.8 (Typ-Code/Stempel/marking)	71(5x3,5)	Fag	→P6 SMB-6.8	data
KD	Z-Di	=SMBJ 5.0 (Typ-Code/Stempel/marking)	71(5x3,5)	Mop	→SMBJ ...	data
KD 1	Z-Di	=AZ 23C2V7 (Typ-Code/Stempel/marking)	35	Aeg	→AZ 23C2V7	
KD 2	Z-Di	=AZ 23C3V0 (Typ-Code/Stempel/marking)	35	Aeg	→AZ 23C3V0	
KD 3	Z-Di	=AZ 23C3V3 (Typ-Code/Stempel/marking)	35	Aeg	→AZ 23C3V3	
KD 4	Z-Di	=AZ 23C3V6 (Typ-Code/Stempel/marking)	35	Aeg	→AZ 23C3V6	
KD 5	Z-Di	=AZ 23C3V9 (Typ-Code/Stempel/marking)	35	Aeg	→AZ 23C3V9	
KD 6	Z-Di	=AZ 23C4V3 (Typ-Code/Stempel/marking)	35	Aeg	→AZ 23C4V3	
KD 7	Z-Di	=AZ 23C4V7 (Typ-Code/Stempel/marking)	35	Aeg	→AZ 23C4V7	
KD 8	Z-Di	=AZ 23C5V1 (Typ-Code/Stempel/marking)	35	Aeg	→AZ 23C5V1	
KD 9	Z-Di	=AZ 23C5V6 (Typ-Code/Stempel/marking)	35	Aeg	→AZ 23C5V6	
KD 501	Si-N	NF-L, -/40V, 20A, 150W, 2MHz	23a§	Tes	BD 249A, BD 368, 2N3772, 2N5886	data
KD 502	Si-N	NF-L, -/60V, 20A, 150W, 2MHz	23a§	Tes	BD 249B, BD 368, 2N3772, 2N5886	data
KD 503	Si-N	NF-L, -/80V, 20A, 150W, 2MHz	23a§	Tes	BD 249C, BD 368, 2N3772, 2N5886	data
KD 601	Si-N	NF-L, 40V, 10A, 35W, 10MHz	23a§	Tes	BD 245, BD 311, 2N5055, ++	data
KD 602	Si-N	NF-L, 110V, 40A, 35W, 0,5MHz	23a§	Tes	2N6032, 2N6274...6275	data
KD 605	Si-N	NF-L, -/40V, 10A, 70W, 2MHz	23a§	Tes	BD 245A, BD 311, BD 315, 2N5877...78, ++	data
KD 606	Si-N	NF-L, -/60V, 10A, 70W, 2MHz	23a§	Tes	BD 245B, BD 313, BD 315, 2N5878, ++	data
KD 607	Si-N	NF-L, -/80V, 10A, 70W, 2MHz	23a§	Tes	BC 245C, BD 317, BDY 53, 2N3055, ++	data
KD 610	Si-N-Darl	NF-L, -/40V, 10A, 35W	23a§	Tes	BDV 65(A..C), BDX 65(A..C), MJ 3000, ++	data
KD 1084 AV33	LIN-IC	5A, fix. 3.3V low drop positive voltage regulator	17	Stm	-	pdf pinout
KD 1084 (A)D2T18	LIN-IC	5A, fix. 1.8V low drop positive voltage regulator	17	Stm	-	pdf pinout
KD 1084 D2T25	LIN-IC	5A, fix. 2.5V low drop positive voltage regulator	17	Stm	-	pdf pinout
KD 1084 (A)DT25	LIN-IC	5A, fix. 2.5V low drop positive voltage regulator	17	Stm	-	pdf pinout
KD 1084 (A)DT3.3	LIN-IC	5A, fix. 3.3V low drop positive voltage regulator	17	Stm	-	pdf pinout
KD 1084 (A)DT	LIN-IC	5A, adj. low drop positive voltage regulator	17	Stm	-	pdf pinout
KDA	Z-Di	=AZ 23C6V2 (Typ-Code/Stempel/marking)	35	Aeg	→AZ 23C6V2	
KDA	MOS-N-FET-e*	=SSM 6K06FU (Typ-Code/Stempel/marking)	46(2mm)	Tos	→SSM 6K06FU	data
KDA 0316 LD,D	CMOS-D/A-IC	=KDA 0316LN,N: SMD	20-MDIP	Sam	-	pinout
KDA 0316 LN,N	CMOS-D/A-IC	Cd, 16 Bit, Dynamic Level Shift	20-DIP	Sam	LC 7881, CDX 1161	pinout
KDA 0318 D	CMOS-D/A-IC	SMD, 16/18 Bit, Digital Audio Stereo	32-MDIP	Sam	LC 7881, CDX 1161	
KDA 0340 D	CMOS-D/A-IC	SMD, 1 Bit, Digital Audio Stereo	28-MDIP	Sam	-	
KDA 0406 CD	D/A-IC	=KDA 0406CN: SMD	28-MDIP	Sam	-	
KDA 0406 CN	D/A-IC	Triple, 6 Bit, hi-speed, 20Msps, Video	28-SDIP	Sam	-	
KDA 0408	D/A-IC	Triple, 8 Bit, hi-speed, 60Msps, Video	42-SDIP	Sam	-	
KDA 0408 Q	D/A-IC	=KDA 0408: SMD	48-MP	Sam	-	
KDA 0471 BPL-...	CMOS-D/A-IC	RAMDAC, 6 Bit+Overlay f. µComp.	44-PLCC	Sam	-	
KDA 0476 BCN-...	CMOS-D/A-IC	RAMDAC, 6 Bit f. µComp.	28-DIP	Sam	-	
KDA 0476 BPJ-...	CMOS-D/A-IC	=KDA 476BCN-...: Fig.→	32-PLCC	Sam	-	
KDA 0476 BPL-...	CMOS-D/A-IC	=KDA 476BCN-...: Fig.→	44-PLCC	Sam	-	
KDA 0478 BPL-...	CMOS-D/A-IC	RAMDAC, 6 Bit+Overlay f. µComp.	44-PLCC	Sam	-	
KDA 0800 CD	D/A-IC	=KDA 0800CN: SMD	16-MDIP	Sam	-	
KDA 0800 CN	D/A-IC	8 Bit D/a-converter	16-DIP	Sam	-	
KDA 0801 CN	D/A-IC	8 Bit D/a-converter	16-DIP	Sam	-	
KDA 0802 CD	D/A-IC	=KDA 0802CN: SMD	16-MDIP	Sam	-	
KDA 0802 CN	D/A-IC	8 Bit D/a-converter	16-DIP	Sam	-	
KDA 0807 CD	D/A-IC	=KDA 0807CN: SMD	16-MDIP	Sam	-	
KDA 0807 CN	D/A-IC	8 Bit D/a-converter	16-DIP	Sam	-	
KDA 0808 CD	D/A-IC	=KDA 0808CN: SMD	16-MDIP	Sam	-	
KDA 0808 CN	D/A-IC	8 Bit D/a-converter	16-DIP	Sam	DAC 0808, MC 1408	
KDA 3310-8...-10	D/A-IC	10 Bit, hi-speed, 25Msps, Video	28-DIP	Sam	-	
KDA 3310 L-...	D/A-IC	=KDA 3310-...: Fig.→	32-PLCC	Sam	-	
KDB	Z-Di	=AZ 23C6V8 (Typ-Code/Stempel/marking)	35	Aeg	→AZ 23C6V8	
KDB	MOS-P-FET-e*	=SSM 6J06FU (Typ-Code/Stempel/marking)	46(2mm)	Tos	→SSM 6J06FU	data
KDC	Z-Di	=AZ 23C7V5 (Typ-Code/Stempel/marking)	35	Aeg	→AZ 23C7V5	
KDC	MOS-N-FET-e*	=SSM 6K08FU (Typ-Code/Stempel/marking)	46(2mm)	Tos	→SSM 6K08FU	
KDD	Z-Di	=AZ 23C8V2 (Typ-Code/Stempel/marking)	35	Aeg	→AZ 23C8V2	
KDD	MOS-P-FET-e*	=SSM 6J08FU (Typ-Code/Stempel/marking)	46(2mm)	Tos	→SSM 6J08FU	
KDE	Z-Di	=AZ 23C9V1 (Typ-Code/Stempel/marking)	35	Aeg	→AZ 23C9V1	
KDE	MOS-N-FET-e*	=SSM 6K07FU (Typ-Code/Stempel/marking)	46(2mm)	Tos	→SSM 6K07FU	data
KDF	Z-Di	=AZ 23C10 (Typ-Code/Stempel/marking)	35	Aeg	→AZ 23C10	
KDF	MOS-P-FET-e*	=SSM 6J07FU (Typ-Code/Stempel/marking)	46(2mm)	Tos	→SSM 6J07FU	data
KDG	Z-Di	=AZ 23C11 (Typ-Code/Stempel/marking)	35	Aeg	→AZ 23C11	
KDH	Z-Di	=AZ 23C12 (Typ-Code/Stempel/marking)	35	Aeg	→AZ 23C12	
KDH	MOS-P-FET-e*	=SSM 3J13T (Typ-Code/Stempel/marking)	35	Tos	→SSM 3J13T	
KDI	Z-Di	=AZ 23C13 (Typ-Code/Stempel/marking)	35	Aeg	→AZ 23C13	
KDJ	Z-Di	=AZ 23C15 (Typ-Code/Stempel/marking)	35	Aeg	→AZ 23C15	
KDJ	MOS-N-FET-e*	=SSM 3K12T (Typ-Code/Stempel/marking)	35	Tos	→SSM 3K12T	
KDK	Z-Di	=AZ 23C16 (Typ-Code/Stempel/marking)	35	Aeg	→AZ 23C16	

Type	Device	Short Description	Fig.	Manu	Comparision Types	More at
KDL	Z-Di	=AZ 23C18 (Typ-Code/Stempel/marking)	35	Aeg	→AZ 23C18	
KDM	Z-Di	=AZ 23C20 (Typ-Code/Stempel/marking)	35	Aeg	→AZ 23C20	
KDN	Z-Di	=AZ 23C22 (Typ-Code/Stempel/marking)	35	Aeg	→AZ 23C22	
KDO	Z-Di	=AZ 23C24 (Typ-Code/Stempel/marking)	35	Aeg	→AZ 23C24	
KDP	Z-Di	=AZ 23C27 (Typ-Code/Stempel/marking)	35	Aeg	→AZ 23C27	
KDQ	Z-Di	=AZ 23C30 (Typ-Code/Stempel/marking)	35	Aeg	→AZ 23C30	
KDR	Z-Di	=AZ 23C33 (Typ-Code/Stempel/marking)	35	Aeg	→AZ 23C33	
KDR 331	Si-Di	SMD, Dual, Schottky-Di, S, 15V, 0,05A	35l(2mm)	Kec	-	data
KDR 331 E	Si-Di	=KDR 331:	35l(1,6mm)		-	data
KDR 357	Si-Di	SMD, Schottky-Di, S, 45V, 0,1A	71a(1,7mm)	Kec	1SS357	data
KDR 367	Si-Di	SMD, Schottky-Di, S, 15V, 0,1A	71a(1,7mm)	Kec	1SS367	data
KDR 367 E	Si-Di	=KDR 367:	71a(1,2mm)		1SS373, 1SS389	data
KDR 368	Si-Di	SMD, Schottky-Di, S, 30V, 0,1A	71a(1,7mm)	Kec	BAS 140W	data
KDR 368 E	Si-Di	=KDR 368:	71a(1,2mm)		-	data
KDR 377	Si-Di	SMD, Schottky-Di, S, 40V, 30mA	71a(1,7mm)		-	data
KDR 378	Si-Di	SMD, Dual, Schottky-Di, S, 15V, 0,1A	35l(2mm)	Kec	1SS378	data
KDR 411	Si-Di	SMD, Schottky-Di, S, 40V, 0,5A	35e(2mm)	Kec	-	data
KDR 411 S	Si-Di	=KDR 411:	35e		-	data
KDR 412	Si-Di	=KDR 411:	71a(1,7mm)	Kec	BAT 165	data
KDR 728	Si-Di	SMD, Schottky-Di, S, 30V, 30mA	71a(1,7mm)	Kec	-	data
KDR 729	Si-Di	SMD, Schottky-Di, S, 30V, 0,2A	71a(1,7mm)	Kec	MA 729	data
KDR 731	Si-Di	SMD, Dual, Schottky, SS, 35V, 0,07A <10ns	35o(2mm)	Kec	-	data
KDR 731 S	Si-Di	=KDR 731:	35o		-	data
KDR 732	Si-Di	=KDR 731:	35l(2mm)	Kec	-	data
KDR 732 E	Si-Di	=KDR 731:	35l(1,6mm)		-	data
KDR 784	Si-Di	SMD, Schottky-Di, S, 30V, 0,1A	71a(1,7mm)	Kec	BAS 140W, MA 784	data
KDS	Z-Di	=AZ 23C36 (Typ-Code/Stempel/marking)	35	Aeg	→AZ 23C36	
KDS 112	C-Di	SMD, VHF-Band-S, 30V, <1,2pF(6V)	35l(2mm)	Kec	-	data
KDS 112 E	C-Di	=KDS 112:	35l(1,6mm)		-	data
KDS 113	C-Di	=KDS 112:	35m(2mm)	Kec	-	data
KDS 114	C-Di	=KDS 112:	71a(.,7mm)	Kec	-	data
KDS 114E	C-Di	=KDS 112:	71a(1,2mm)		-	data
KDS 115	C-Di	=KDS 112:	35m(2mm)	Kec	-	data
KDS 120	Si-Di	=1SS181:	35m(2mm)	Kec	→1SS181	data
KDS 120 E	Si-Di	=1SS181:	35m(1,6mm)		-	data
KDS 121	Si-Di	=1SS184:	35l(2mm)	Kec	→1SS184	data
KDS 121 E	Si-Di	=1SS184:	35l(1,6mm)		-	data
KDS 122	Si-Di	=1SS187:	35m(2mm)	Kec	→1SS187	data
KDS 160	Si-Di	=1SS184:	71a(1,7mm)	Kec	HSU 119, MA 111, 1SS352, 1SS353	data
KDS 160 E	Si-Di	=1SS184:	71a(1,2mm)		1SS368, 1SS387	data
KDS 181	Si-Di	=1SS181:	35m	Kec	→1SS181	data
KDS 184	Si-Di	=1SS184:	35l	Kec	→1SS184	data
KDS 187	Si-Di	=1SS187:	35f	Kec	→1SS187	data
KDS 190	Si-Di	=1SS190:	35d	Kec	→1SS190	data
KDS 193	Si-Di	=1SS193:	35e	Kec	→1SS193	data
KDS 196	Si-Di	=1SS196:	35c	Kec	→1SS196	data
KDS 200	Si-Di	=1SS200	7q	Kec	→1SS200	data
KDS 201	Si-Di	=1SS201	7q	Kec	→1SS201	data
KDS 226	Si-Di	=1SS226	35o	Kec	→1SS226	data
KDS 2236 M	Si-Di	=1S2236:	41d	Kec	→1S2236	data
KDS 2236 S	Si-Di	=1S2236: SMD	35		-	data
KDS 2236 S	Si-Di	=1S2236: SMD	35e			data
KDT	Z-Di	=AZ 23C39 (Typ-Code/Stempel/marking)	35	Aeg	→AZ 23C39	
KDU	Z-Di	=AZ 23C43 (Typ-Code/Stempel/marking)	35	Aeg	→AZ 23C43	
KDV	Z-Di	=AZ 23C47 (Typ-Code/Stempel/marking)	35	Aeg	→AZ 23C47	
KDV 147	Si-Di	=1SV147:	41d	Kec	→1SV147	data
KDV 149	Si-Di	=1SV149	41	Kec	→1SV149	data
KDV 152 M	C-Di	CB, Schnurlostelefon/cordless phone	41d	Kec	-	data
KDV 152 S	C-Di	=KDV 152M: SMD	35e		-	data
KDV 153	C-Di	SMD, VCO f. UHF-Band Radio	35e	Kec	-	data
KDV 154	C-Di	SMD, VCO f. UHF-Band Radio	71a(1,7mm)	Kec	-	data
KDV 154 E	C-Di	=KDV 154:	71a(1,2mm)		-	data
KDV 173	PIN-Di	SMD, VHF/UHF-Abschw./Attenuator, 50V	35o	Kec	-	data
KDV 174	PIN-Di	=KDV 173:	35m(2mm)	Kec	-	data
KDV 175	PIN-Di	=KDV 173:	71a(1,7mm)	Kec	-	data
KDV 175 E	PIN-Di	=KDV 173:	71a(1,2mm)		-	data
KDV 214	C-Di	SMD, TV-Tuning, 30V, 14...16pF(2V)	71a(1,7mm)	Kec	MA 360, MA 363, MA 372, 1SV214	data
KDV 214 E	C-Di	=KDV 214:	71a(1,2mm)		-	data
KDV 239	C-Di	=1SV239:	71a(1,7mm)	Kec	-	data
KDV 239 E	C-Di	=1SV239:	71a(1,2mm)		-	data
KDV 251 M	C-Di	CB, Schnurlostelefon/cordless phone	41d	Kec	-	data
KDV 251 S	C-Di	=KDV 251M: SMD	35e		-	data
KDV 262	C-Di	=1SV262:	71a(1,7mm)	Kec	→1SV262	data

Type	Device	Short Description	Fig.	Manu	Comparision Types	More at
KDV 262 E	C-Di	=1SV262:	71a(1,2mm)		-	data
KDV 269	C-Di	=1SV269:	71a(1,7mm)	Kec	→1SV269	data
KDV 287	C-Di	=1SV287:	71a(1,7mm)	Kec		data
KDV 350	C-Di	SMD, VCO, 15V, 15...17,5pF(1V)	71a(1,7mm)	Kec	1SV270	data
KDV 350 E	C-Di	=KDV 350:	71a(1,2mm)		-	data
KDV 804 M	Si-Di	=BB 304: 15V	41p	Kec	→BB 304	data
KDV 804 S	Si-Di	=BB 804: 15V	35l	Kec	→BB 804	data
KDV 1430(A...D)	C-Di	SMD,Dual, FM-Tuning, 18V, 69...77pF(2V)	35l	Kec	-	data
KDV 1470(A...C)	C-Di	SMD,Dual, FM-Tuning, 16V, 66...74pF(1V)	35l	Kec	-	data
KDV 1471	C-Di	SMD, FM-Tuning, 16V, 32...37pF(1V)	35c	Kec	-	data
KDV 1472(A...C)	C-Di	SMD, FM-Tuning, 16V, 30...41pF(1V)	71a(1,7mm)	Kec	-	data
KDW	Z-Di	=AZ 23C51 (Typ-Code/Stempel/marking)	35	Aeg	→AZ 23C51	
KDZ 2.0EV...24EV	Z-Di	SMD, 2,0...24V ±6%, 0,1W	71a(1,2mm)	Kec	-	data
KDZ 2.0M...24M	Z-Di	SMD, 2,0...24V ±6%, 0,15W	71a(1,7mm)	Kec	DTZ .., HZU .., RD ...S	data
KDZ 2.0V...24V	Z-Di	SMD, 2,0...24V ±6%, 0,15W	71a(1,7mm)	Kec	DTZ .., HZU .., RD ...S	data
KE	Z-Di	=1SMB 5.0A (Typ-Code/Stempel/marking)	71(5x3,5)	Ons	→1SMB 5.0A...170A	
KE	N-FET	=2SK1062 (Typ-Code/Stempel/marking)	35	Tos	→2SK1062	data
KE	MOS-N-FET-e	=2SK1473 (Typ-Code/Stempel/marking)	39	Say	→2SK1473	data
KE	MOS-N-FET-e*	=2SK3065 (Typ-Code/Stempel/marking)	39	Rhm	→2SK3065	data
KE	N-FET	=2SK323-E (Typ-Code/Stempel/marking)	35	Hit	→2SK323	data
KE	N-FET	=2SK433-E (Typ-Code/Stempel/marking)	35	Mit	→2SK433	data
KE	N-FET	=2SK930-E (Typ-Code/Stempel/marking)	35(2mm)	Mit	→2SK930	data
KE	Z-Di	=P6 SMB-6.8A(Typ-Code/Stempel/marking)	71(5x3,5)	Fag	→P6 SMB-6.8A	data
KE	Z-Di	=SMBJ 5.0A (Typ-Code/Stempel/marking)	71(5x3,5)	Mop	→SMBJ ...	data

KF

Type	Device	Short Description	Fig.	Manu	Comparision Types	More at
KF	P-FET	=2SJ168 (Typ-Code/Stempel/marking)	35	Tos	→2SJ168	data
KF	MOS-N-FET-e	=2SK1724 (Typ-Code/Stempel/marking)	39	Say	→2SK1724	data
KF	MOS-N-FET-e*	=CPH 3406 (Typ-Code/Stempel/marking)	35	Say	→CPH 3406	data
KF	Z-Di	=P6 SMB-7.5 (Typ-Code/Stempel/marking)	71(5x3,5)	Fag	→P6 SMB-7.5	data
KF	Z-Di	=SMBJ 6.0 (Typ-Code/Stempel/marking)	71(5x3,5)	Mop	→SMBJ ...	data
KF12B	R+	Io=1A Vo:1.225...1.275V Vin:3.3V P:int	8S,3P	Sgs		data pdf pinout
KF15B	R+	Io=1A Vo:1.47...1.53V Vin:3.5V P:int	8S,3P	Sgs		data pdf pinout
KF 15 BDT-TR	Z-IC	Very low drop voltage regulators with inhibit	45	Stm	-	pdf pinout
KF25B	R+	Io=1A Vo:1.47...1.53V Vin:3.5V P:int	8S,3P	Sgs		data pdf pinout
KF 25 BD-TR	Z-IC	Very low drop voltage regulators with inhibit	45	Stm	-	pdf pinout
KF 25 BDT-TR	Z-IC	Very low drop voltage regulators with inhibit	45	Stm	-	pdf pinout
KF27B	R+	Io=1A Vo:2.646...2.754V Vin:4.7V P:int	8S,3P	Sgs		data pdf pinout
KF30B	R+	Io=1A Vo:2.94...3.06V Vin:5V P:int	8S,3P	Sgs		data pdf pinout
KF33B	R+	Io=1A Vo:3.234...3.366V Vin:5.3V P:int	8S,3P	Sgs		data pdf pinout
KF 33 BD-TR	Z-IC	Very low drop voltage regulators with inhibit	45	Stm	-	pdf pinout
KF 33 BDT-TR	Z-IC	Very low drop voltage regulators with inhibit	45	Stm	-	pdf pinout
KF35B	R+	Io=1A Vo:3.43...3.57V Vin:5.5V P:int	8S,3P	Sgs		data pdf pinout
KF40B	R+	Io=1A Vo:3.92...4.08V Vin:6V P:int	8S,3P	Sgs		data pdf pinout
KF 40 BDT-TR	Z-IC	Very low drop voltage regulators with inhibit	45	Stm	-	pdf pinout
KF45B	R+	Io=1A Vo:4.41...4.59V Vin:6.5V P:int	8S,3P	Sgs		data pdf pinout
KF47B	R+	Io=1A Vo:4.606...4.794V Vin:6.7V P:int	8S,3P	Sgs		data pdf pinout
KF50B	R+	Io=1A Vo:4.9...5.1V Vin:7V P:int	8S,3P	Sgs		data pdf pinout
KF 50 BD-TR	Z-IC	Very low drop voltage regulators with inhibit	45	Stm	-	pdf pinout
KF 50 BDT-TR	Z-IC	Very low drop voltage regulators with inhibit	45	Stm	-	pdf pinout
KF52B	R+	Io=1A Vo:5.096...5.304V Vin:7.2V P:int	8S,3P	Sgs		data pdf pinout
KF55B	R+	Io=1A Vo:5.39...5.61V Vin:7.5V P:int	8S,3P	Sgs		data pdf pinout
KF60B	R+	Io=1A Vo:5.88...6.12V Vin:8V P:int	8S,3P	Sgs		data pdf pinout
KF80B	R+	Io=1A Vo:7.84...8.16V Vin:10V P:int	8S,3P	Sgs		data pdf pinout
KF 80 BDT-TR	Z-IC	Very low drop voltage regulators with inhibit	45	Stm		pdf pinout
KF85B	R+	Io=1A Vo:8.33...8.67V Vin:10.5V P:int	8S,3P	Sgs		data pdf pinout
KF120B	R+	Io=1A Vo:11.76...12.24V Vin:14V P:int	8S,3P	Sgs		data pdf pinout
KF 124	Si-N	≈BF 254	9a	Tes	→BF 254	data
KF 125	Si-N	≈BF 255	9a	Tes	→BF 255	data
KF 167	Si-N	≈BF 167	5k	Tes	→BF 167	data
KF 173	Si-N	≈BF 173	5k	Tes	→BF 173	data
KF 257	Si-N	≈BF 257	2a§		→BF 257	
KF 258	Si-N	≈BF 258	2a§		→BF 258	
KF 259	Si-N	≈BF 259	2a§		→BF 259	
KF 272	Si-P	≈BF 272	5g	Tes	→BF 272	data
KF347	4xJFET	Vs:±18V Vu:100V/mV Vo:±13.5V Vi0:5mV	14D	Sam		data pdf pinout
KF 347(A)	OP-IC	=LF 347...	14-DIP	Sam	→LF 347...	
KF347A	4xJFET	Vs:±18V Vu:100V/mV Vo:±13.5V Vi0:3mV	14D	Sam		data pinout
KF351	JFET	Vs:±18V Vu:100V/mV Vo:±13.5V Vi0:5mV	8SD	Sam		data pinout

Type	Device	Short Description	Fig.	Manu	Comparision Types	More at
KF 351 N,D	OP-IC	=LF 351...	8-(M)DIP	Sam	→LF 351...	
KF353	2xJFET	Vs:±18V Vu:100V/mV Vo:±13.5V Vi0:5mV	8DS,9I	Sam		data pdf pinout
KF 353 N,D	OP-IC	=LF 353...	8-(M)DIP	Sam	→LF 353...	
KF 353 S	OP-IC	=LF 353...	9-SIP	Sam	→LF 353...	
KF 442	2xJFET	Vs:±18V Vu:200V/mV Vo:±13V Vi0:1mV	8D	Sam		data pinout
KF 442(A)	OP-IC	=LF 442...	8-DIP	Sam	→LF 442...	
KF442A	2xJFET	Vs:±20V Vu:200V/mV Vo:±18V Vi0:0.5mV	8D,9I	Sam		data pinout
KF442S	2xJFET	Vs:±18V Vu:200V/mV Vo:±13V Vi0:1mV	9I	Sam		data pinout
KF 442(A)S	OP-IC	=LF 442...	9-SIP	Sam	→LF 442...	
KF 503	Si-N	Uni, -/60V, 0,05A, 0,7W, 150MHz	2a	Tes	BC 546, 2SC1890(A), 2SC2363, 2SD755, ++	data
KF 504	Si-N	Uni, -/100V, 0,05A, 0,7W, 150MHz	2a	Tes	2SC1890A, 2SC2363, 2SC3245, 2SD756(A),++	data
KF 506	Si-N	≈BC 141	2a	Tes	→BC 141	data
KF 507	Si-N	≈BC 140	2a	Tes	→BC 140	data
KF 508	Si-N	≈BC 141	2a	Tes	→BC 141	data
KF 517	Si-P	≈BC 160	2a	Tes	→BC 160	data
KF 520	MOS-N-FET		2e	Tes	-	data
KF 521	MOS-N-FET-d	Dual-Gate	5r	Tes	-	data
KF 522	MOS-P-FET-e	Dual	81bw	Tes	-	data
KF 524	Si-N	≈BF 254	9a	Tes	→BF 254	data
KF 525	Si-N	≈BF 255	9a	Tes	→BF 255	data

KG...KL

Type	Device	Short Description	Fig.	Manu	Comparision Types	More at
KG	Z-Di	=1SMB 6.0A (Typ-Code/Stempel/marking)	71(5x3,5)	Ons	→1SMB 5.0A...170A	
KG	MOS-N-FET-e	=2SK1726 (Typ-Code/Stempel/marking)	39	Say	→2SK1726	data
KG	N-FET	=2SK211-GR (Typ-Code/Stempel/marking)	35	Tos	→2SK211	data
KG	N-FET	=2SK881-GR (Typ-Code/Stempel/marking)	35(2mm)	Tos	→2SK881	data
KG	N-FET	=KTK 211-G (Typ-Code/Stempel/marking)	35	Kec	→KTK 211	data
KG	Z-Di	=P6 SMB-7.5A(Typ-Code/Stempel/marking)	71(5x3,5)	Fag	→P6 SMB-7.5A	data
KG	Z-Di	=SMBJ 6.0A (Typ-Code/Stempel/marking)	71(5x3,5)	Mop	→SMBJ ...	data
KG 70	Se-Di	Amplitudenbegrenzer/limiter		Hfo	-	
KG 10200...12000	CMOS-IC	Semi-custom Gate Arrays		Sam	-	
KG 30400...312000	CMOS-IC	Semi-custom Gate Arrays		Sam	-	
KGA 4115	IC	10Gbps EA Modulator Driver, -5.5 ...-5V		Son		pdf pinout
KH	MOS-N-FET-e	=2SK1728 (Typ-Code/Stempel/marking)	39	Say	→2SK1728	data
KH	MOS-N-FET-e	=2SK1826 (Typ-Code/Stempel/marking)	35	Tos	→2SK1826	data
KH	MOS-N-FET-e	=2SK1827 (Typ-Code/Stempel/marking)	35(2mm)	Tos	→2SK1827	data
KH	MOS-N-FET-e*	=HN 1K04FU (Typ-Code/Stempel/marking)	46(2mm)	Tos	→HN 1K04FU	
KH	Z-Di	=P6 SMB-8.2 (Typ-Code/Stempel/marking)	71(5x3,5)	Fag	→P6 SMB-8.2	data
KH	Z-Di	=SMBJ 6.5 (Typ-Code/Stempel/marking)	71(5x3,5)	Mop	→SMBJ ...	data
KI	MOS-N-FET-e	=2SK1828 (Typ-Code/Stempel/marking)	35	Tos	→2SK1828	data
KI	MOS-N-FET-e	=2SK1829 (Typ-Code/Stempel/marking)	35(2mm)	Tos	→2SK1829	data
KI	MOS-N-FET-e	=2SK1830 (Typ-Code/Stempel/marking)	35(1,6mm)	Tos	→2SK1830	data
KI	MOS-N-FET-e	=2SK2151 (Typ-Code/Stempel/marking)	39	Say	→2SK2151	data
KI	MOS-N-FET-e*	=HN 1K02FU (Typ-Code/Stempel/marking)	46(2mm)	Tos	→HN 1K02FU	data
KIA 78DL05P...15P	Z-IC	lo-drop, +5...+15V, 0,25A	17b	Kec	TA 78DLxxP	
KIA 78DL05PI...15PI	Z-IC	Iso, lo-drop, +5...+15V, 0,25A	17b	Kec	TA 78DLxxS	
KIA 78 DS 05PBV	Z-IC	lo-drop, 5V/30mA, 0,8W	7b(9mm)	Kec		
KIA 78L05BP...24BP	Z-IC	+5...+24V, 0,15A	7b(9mm)	Kec	... 78Lxx... (TO-92L)	pdf
KIA 78L05F...24F	Z-IC	SMD, +5...+24V, 0;15A	39b	Kec	... 78Lxx... (SOT-89)	
KIA 78 R 05PI	Z-IC	Iso, On/Off Control, 5V/1A, 15W	17/4Pin	Kec	-	pdf
KIA 78 R 09PI	Z-IC	Iso, On/Off Control, 9V/1A, 15W	17/4Pin	Kec	-	pdf
KIA 78 R 12PI	Z-IC	Iso, On/Off Control, 12V/1A, 15W	17/4Pin	Kec	-	pdf
KIA 78S05BP...24BP	Z-IC	+5...+24V, 0,1A	7b	Kec	... 78Lxx... (TO-92)	
KIA 79L05BP...24BP	Z-IC	-5...-24V, 0,15A	7a(9mm)	Kec	... 79Lxx... (TO-92L)	pdf
KIA 79L05F...24F	Z-IC	SMD, -5...-24V, 0,15A	39a	Kec	... 79Lxx... (SOT-89)	pdf
KIA 79S05BP...24BP	Z-IC	-5...-24V, 0,15A	7a	Kec	... 79Lxx... (TO-92)	
KIA 324 F	OP-IC	=KIA 342P: SMD	14-MDIP		-	pdf
KIA 324 P	OP-IC	Quad, Serie 124, ±18V	14-DIP	Kec	→Serie 124	data pdf
KIA 339 F	KOP-IC	=KIA 339P: SMD	14-MDIP		-	pdf
KIA 339 P	KOP-IC	Quad, Serie 139, ±18V	14-DIP	Kec	→Serie 139	data pdf
KIA 358 F	OP-IC	=KIA 358P: SMD	8-MDIP		-	pdf
KIA 358 P	OP-IC	Dual, Serie 158, ±18V	8-DIP	Kec	→Serie 158	data pdf
KIA 358 S	OP-IC	=KIA 358P: Fig.→	9-SIP		-	pdf
KIA 393 F	KOP-IC	=KIA 393P: SMD	8-MDIP		-	pdf
KIA 393 P	KOP-IC	Dual, Serie 193, ±18V	8-DIP	Kec	→Serie 193	data pdf
KIA 393 S	KOP-IC	=KIA 393P: Fig.→	9-SIP		-	pdf
KIA 431(A)	Ref-Z-IC	2,5...36V, 1...100mA, 0...+70°, A=±1%	7(KARef)	Kec	KA 431...Z, NJM 431..., µA 431	pdf
KIA 431 F,AF	Ref-Z-IC	=KIA 431: SMD	39		-	
KIA 494AF	LIN-IC	SMD, PWM Contr. 41V, 0.25A, -40...+85°C	16-MDIP	Kec	...494...	pdf pinout
KIA 494AP	LIN-IC	PWM Controller, 41V, 0.25A, -40...+85°C	16-DIP	Kec	...494...	pdf pinout
KIA 494 F,FV	LIN-IC	=KIA 494P: SMD	16-MDIP		-	
KIA 494 P,PV	LIN-IC	PWM Controller, 42V, 0,25A, 0...+70°	16-DIP	Kec	MB3759, KA7500, TL594, TA76494, µPC494	
KIA 555 F	LIN-IC	=KIA 555P: SMD	8-MDIP		-	
KIA 555 P	LIN-IC	Zeitgeber/Timer, 4,5...16V, 0...+70°	8-DIP	Kec	CA535..., LM555..., SE555..., TDB0555..., ++	pdf
KIA 556 F	LIN-IC	=KIA 556P: SMD	14-MDIP		-	
KIA 556 P	LIN-IC	Dual-Zeitgeber/Timer, 4,5...16V, 0...70°	14-DIP	Kec	LM556..., MC3456..., SE556..., TDB0556...	
KIA 574	Z-IC	Tuner-Stab., 31...35V, 10mA, (~TAA550)	7j	Kec	TAA 550, ZTK 33	
KIA 643 PV	LIN-IC	Car, Blinker-/Flasher Ctrl, Ucc=9...15V	8-DIP	Kec	-	
KIA 750 LPV	Z-IC	lo-drop, +5V/0,15A, 0,6W	7b	Kec	-	
KIA 750 V	Z-IC	lo-drop, +5V/0,15A, 20,8W	17b	Kec	-	
KIA 2017 FN	LIN-IC	SMD, Car, FM-ZF Diversity, Ucc=7...9V	24-SSMDIP	Kec	-	pdf

Type	Device	Short Description	Fig.	Manu	Comparision Types	More at
KIA 2022 AFN	LIN-IC	SMD,AM/FM-ZF, MPX(DTS), Ucc=0,95..2,2V	24-SSMDIP	Kec	-	pdf
KIA 2025 P	LIN-IC	Car, 2xPlay Amp.(Auto Rev), Ucc=6...16V	16-DIP	Kec	-	
KIA 2039 FN	LIN-IC	SMD,Car, FM Ant. Diversity, Ucc=7...10V	16-SSMDIP	Kec	-	pdf
KIA 2041 F	LIN-IC	=KIA 2041N: SMD	24-SMDIP		-	
KIA 2041 N	LIN-IC	Radio Cass., Surround Proc.,Ucc=4...12V	24-SDIP	Kec	-	
KIA 2042 F	LIN-IC	SMD, Car, 2xPlay Amp., Music Sensor	24-SMDIP	Kec	-	
KIA 2051 FN	LIN-IC	SMD, Car, Am Noise Cancel, Ucc=8±1V	24-SSMDIP	Kec	-	pdf
KIA 2057 N	LIN-IC	Radio Cass., AM/FM-ZF, Mpx(dts)	24-SDIP	Kec	-	pdf
KIA 2058 F	LIN-IC	SMD, Cd, 4xBTL Power Drv., Ucc=4...10V	20-SMDIP+b	Kec	-	
KIA 2062 F	LIN-IC	=KIA 2062P: SMD	16-SMDIP		-	
KIA 2062 P	LIN-IC	5-Band Graphic Equalizer, Ucc=4...16V	16-DIP	Kec	-	
KIA 2068 N	LIN-IC	Port. Cass., 2x Mic. Amp., Ucc=4...9V	24-SDIP	Kec	-	
KIA 2074 F	LIN-IC	SMD, Car, HF-Prozessor, Ucc=8±1V	48-MP	Kec	-	pdf
KIA 2078 P	LIN-IC	Car, 3Mode Preset Equal., Ucc=7,5...14V	16-DIP	Kec	-	pdf
KIA 2093 F	LIN-IC	SMD, Car, FM-ZF, MPX, Ucc=8±1V	48-MP	Kec	-	pdf
KIA 2575 FP XX	LIN-IC	150kHz 1A Step-down DC/DC Converter	87	Kec	-	pdf pinout
KIA 2575 PI XX	LIN-IC	150kHz 1A Step-down DC/DC Converter	86	Kec	-	pdf pinout
KIA 2576 FP XX	LIN-IC	52kHz 3A Step-down DC/DC Converter	87	Kec	-	pdf pinout
KIA 2576 PI XX	LIN-IC	52kHz 3A Step-down DC/DC Converter	86	Kec	-	pdf pinout
KIA 2596 FP XX	LIN-IC	150kHz 3A Step-down DC/DC Converter	87	Kec	-	pdf pinout
KIA 2596 PI XX	LIN-IC	150kHz 3A Step-down DC/DC Converter	86	Kec	-	pdf pinout
KIA 2951 F(A)30	LIN-IC	3.0V, 0.1A Micropower Voltage Reg.	MDIP	Kec	-	pdf pinout
KIA 2951 F(A)33	LIN-IC	3.3V, 0.1A Micropower Voltage Reg.	MDIP	Kec	-	pdf pinout
KIA 2951 F(A)50	LIN-IC	5.0V, 0.1A Micropower Voltage Reg.	MDIP	Kec	-	pdf pinout
KIA 2951 P(A)30	LIN-IC	3.0V, 0.1A Micropower Voltage Reg.	DIP	Kec	-	pdf pinout
KIA 2951 P(A)33	LIN-IC	3.3V, 0.1A Micropower Voltage Reg.	DIP	Kec	-	pdf pinout
KIA 2951 P(A)50	LIN-IC	5.0V, 0.1A Micropower Voltage Reg.	DIP	Kec	-	pdf pinout
KIA 33 C 15 F	IC	1A, 2-output low Drop Voltage regulator	TSOP	Kec	-	pdf pinout
KIA 33 C 18 F	IC	1A, 2-output low Drop Voltage regulator	TSOP	Kec	-	pdf pinout
KIA 33 M 18 FL	IC	1A, 2-output low Drop Voltage regulator	TSOP	Kec	-	pdf pinout
KIA 33 C 25 F	IC	1A, 2-output low Drop Voltage regulator	TSOP	Kec	-	pdf pinout
KIA 33 M 25 FL	IC	1A, 2-output low Drop Voltage regulator	TSOP	Kec	-	pdf pinout
KIA 3524 F	LIN-IC	=KIA 3524P: SMD	16-MDIP		-	
KIA 3524 P	LIN-IC	S-Reg PWM Ctrl., 8...35V, 0...+70°	16-DIP	Kec	HA 17524, LM 3524, SG 3524, UC 3524A,++	
KIA 3525 F	LIN-IC	=KIA 3525P: SMD	16-MDIP		-	
KIA 3525 P	LIN-IC	S-Reg PWM Ctrl., 8...35V, 0...+70°	16-DIP	Kec	KA 3525A, SG 3525, TL 3525A, UC 3525A	
KIA 3802 F	IC	LCD Backlight Inverter Controller	TSOP	Kec	-	pdf pinout
KIA 3820 F	IC	Gate Pulse Modulator	MDIP	Kec	-	pdf pinout
KIA 3820 FK	IC	Gate Pulse Modulator	MDIP	Kec	-	pdf pinout
KIA 3842 F	LIN-IC	=KIA 3842P: SMD	14-MDIP		-	pdf pinout
KIA 3842 P	LIN-IC	PWM Ctrl., ...500kHz, Ucc=7...40V	8-DIP	Kec	KA 3842, UC 3842	
KIA 4210 FV	LIN-IC	Lamp Failure Indicator	TSOP	Kec	-	pdf pinout
KIA 4210 SV	LIN-IC	Lampentest/lamp failure Indicator	9-SIP	Kec	-	pdf
KIA 4558 F	OP-IC	=KIA 4558P: SMD	8-MDIP		-	pdf
KIA 4558 P	OP-IC	Dual,lo-noise,Serie 158, ±18V, 0...+70°	8-DIP	Kec	→Serie 158	data pdf
KIA 4558 S	OP-IC	=KIA 4558P: Fig.→	9-SIP		-	pdf
KIA 4559 F	OP-IC	=KIA 4558P: SMD	8-MDIP		-	pdf
KIA 4559 P	OP-IC	=KIA 4558P: 2V/µs, 6MHz	8-DIP	Kec	→Serie 158	data pdf
KIA 4559 S	OP-IC	=KIA 4559P: Fig.→	9-SIP		-	pdf
KIA 6003 S	LIN-IC	Car, FM-ZF, Ucc=8...15V	9-SIP	Kec	-	pdf
KIA 6005 F	LIN-IC	SMD, AM/FM-Stereo-Tuner, Ucc=0,95..5V	24-MDIP	Kec	-	
KIA 6010 SN	LIN-IC	Car, Noise Cancel, Ucc=7...9V	12-SIP	Kec	-	pdf
KIA 6029 Z	LIN-IC	Car, FM-ZF System, Ucc=7...16V	16-SQP	Kec	-	pdf
KIA 6030 Z	LIN-IC	Car, FM Stereo MPX, Ucc=7...16V	16-SQP	Kec	-	pdf
KIA 6032 AF	LIN-IC	=KIA 6032AN: SMD	24-SMDIP		-	pdf
KIA 6032 AN	LIN-IC	Portable, AM/FM-ZF, MPX, Ucc=1,8..8V	24-SDIP	Kec	-	pdf
KIA 6035 P	LIN-IC	Car, AM-Tuner, Ucc=7,5...12V	20-DIP	Kec	-	pdf
KIA 6037 F	LIN-IC	=KIA 6037P: SMD	20-MDIP		-	
KIA 6037 P	LIN-IC	AM-Tuner, Ucc=7,5...10V	20-DIP	Kec	-	
KIA 6040 P	LIN-IC	Portable Cass., AM/FM-ZF, Ucc=3...8V	16-DIP	Kec	-	pdf
KIA 6043 S	LIN-IC	PLL Stereo-Decoder, Ucc=3,5...12V	9-SIP	Kec	AN 7420	pdf
KIA 6050 AF	LIN-IC	FM-System, Stereo-Decoder, Ucc=7...9V	44-MP	Kec	-	
KIA 6057 P	LIN-IC	Portable Cass., AM/FM-ZF, Ucc=1,7...6V	16-DIP	Kec	-	
KIA 6058 F	LIN-IC	=KIA 6058(A)S: SMD	8-MDIP		-	pdf
KIA 6058(A)S	LIN-IC	FM Front End, Ucc=1,6...6V	9-SIP	Kec	AN 7205, BA 4422S	pdf pinout
KIA 6067 F	LIN-IC	=KIA 6067N: SMD	24-MDIP		-	
KIA 6067 N	LIN-IC	AM/FM-Tuner, Ucc=1,8...7V	24-SDIP	Kec	-	
KIA 6072 AF	LIN-IC	SMD, Car, FM-System, MPX, Ucc=7...9V	44-MP	Kec	-	
KIA 6076 F	LIN-IC	SMD, FM Stereo MPX, Ucc=0,9...5V	16-SMDIP	Kec	-	pdf
KIA 6078 S	LIN-IC	FM Front End, Ucc=1,6...6V	9-SIP	Kec	-	
KIA 6092 F	LIN-IC	SMD, FM V, AM/FM-ZF, Ucc=0,95...5V	16-SMDIP	Kec	-	pdf
KIA 6200 K	LIN-IC	2x NF-E, 4,6W, Ucc=6...18V	12-SIL	Kec	-	
KIA 6201 F	LIN-IC	2x Kopfh./Headphone Amp, 5mW	8-MDIP	Kec	-	
KIA 6202 F	LIN-IC	=KIA 6202P: SMD	8-MDIP		-	
KIA 6202 P	LIN-IC	2x Power Amp 0.1W(3V/4Ohm), Ucc=1.8...6V	8-DIP	Kec	-	
KIA 6203 H	LIN-IC	2x NF-E, 2x 13W(28V/8Ω), Ucc=10...37V	12-	Kec	-	
KIA 6206 F	LIN-IC	SMD, Dual Kopfh./ Headph.(14mW), Ucc=0.9... 5V	16-SMDIP	Kec	-	pdf pinout
KIA 6210 AH	LIN-IC	2x Power Amp, 2x19W(13.2V/ 4Ohm), Ucc=9... 18V	17-QILP	Kec	-	pdf pinout
KIA 6213 S	LIN-IC	Audio Amp 0.5W(6V/8Ohm), Ucc=4...14V	9-SIP	Kec	-	pdf pinout
KIA 6216 H	LIN-IC	2xBTL Power Amp 15W+18W, Ucc=9...18V	17-QILP	Kec	-	pdf pinout
KIA 6220 H	LIN-IC	2xBTL Aud. PAmp, 30W+26W, Ucc=9...18V	17-QILP	Kec	-	pdf pinout
KIA 6221 AH	LIN-IC	2xBTL Power Amp, 30W+26W, Ucc=9...18V	17-QILP	Kec	-	pdf pinout
KIA 6221 H	LIN-IC	2x NF-E, 2x 26W(13,2V/2Ohm), Ucc=9...18V	17-QILP	Kec	-	

Type	Device	Short Description	Fig.	Manu	Comparision Types	More at
KIA 6225 P	LIN-IC	2x Vorverst./preamp, Ucc=6...16V	8-DIP	Kec	-	pdf pinout
KIA 6225 S	LIN-IC	2x Vorverst./preamp, Ucc=6...16V	9-SIP	Kec	AN 7310	pdf pinout
KIA 6238 P	LIN-IC	NF-V + E, Alc Ucc=3,5...9V	16-DIP	Kec	-	
KIA 6240K	LIN-IC	2x Power Amp, 2x6W(22V, 8Ohm), Vcc=8V... 26V	SIL	Kec	-	pdf pinout
KIA 6248 K	LIN-IC	2x Power Amp, 7.3W+6.4W, Ucc=6...18V	15-SIP	Kec	-	pdf pinout
KIA 6250 H	LIN-IC	NF-E, 30W(BTL), Ucc=9...18V	12-	Kec		
KIA 6259 P	LIN-IC	2x Preamp low noise, Ucc=±1,5...18V	8-DIP	Kec	-	pdf pinout
KIA 6259 S	LIN-IC	2x Preamp low noise, Ucc=±1,5...18V	9-SIP	Kec	-	pdf pinout
KIA 6266 BP	LIN-IC	4x Vorverst./preamp, ALC, Ucc=5...12V	16-DIP	Kec	-	
KIA 6268 P	LIN-IC	2x Rec./Play Preamp, ALC, Ucc=6...15V	16-DIP	Kec	-	pdf pinout
KIA 6269 P	LIN-IC	2x Power Amp, 1W(6V/4Ohm), Ucc=4.5V...9V	16-DIP	Kec	-	pdf pinout
KIA 6276 S	LIN-IC	2x NF-E, 2x0,1W(3V/4Ω), Ucc=1,8...6V	9-SIP	Kec	-	
KIA 6278 F	LIN-IC	SMD, 1W Aud. PAmp, Ucc=2...10V	8-MDIP	Kec	-	
KIA 6278 P	LIN-IC	SMD, 1W Aud. PAmp, Ucc=2...10V	8-DIP	Kec	-	pdf pinout
KIA 6278 S	LIN-IC	SMD, 1W Aud. PAmp, Ucc=2...10V	9-SIP	Kec	AN 7110, BA 527	pdf pinout
KIA 6280 H	LIN-IC	2x Aud. PAmp/BTL, 2x 5.8W, Ucc=9...18V	12-SQL	Kec	-	pdf pinout
KIA 6281 H	LIN-IC	2x Aud. PAmp/BTL, 2x 5.8W, Ucc=9...18V	12-SQL	Kec	-	pdf pinout
KIA 6282 K	LIN-IC	2x Audio amp, 4.6W(12V/4Ohm), Ucc=6...15V	12-SIL	Kec	-	pdf pinout
KIA 6283 K	LIN-IC	2x Audio amp, 4.6W(12V/4Ohm), Ucc=6...15V	12-SIL	Kec	BA 5406	pdf pinout
KIA 6289 N	LIN-IC	4x Vorverst./preamp, ALC, Ucc=4...13.5V	24-SDIP	Kec	-	pdf pinout
KIA 6347 S,SV	LIN-IC	Timer, Long Time, 250mA, Ucc=5...16V	9-SIP	Kec	-	
KIA 6401 F	LIN-IC	=KIA 6401: SMD	8-MDIP		-	pdf
KIA 6401 P	LIN-IC	Telefon, Wecker/tone ringer	8-DIP	Kec	-	pdf
KIA 6402 F	LIN-IC	=KIA 6402: SMD	8-MDIP			pdf
KIA 6402 P	LIN-IC	Telefon, Wecker/tone ringer	8-DIP	Kec	-	pdf
KIA 6419 F	LIN-IC	=KIA 6419P: SMD	8-MDIP			pdf
KIA 6419 P	LIN-IC	NF-E, 55mW, Ucc=2...16V	8-DIP	Kec	-	pdf
KIA 6451 F,AF	LIN-IC	=KIA 6451P,AP: SMD	16-MDIP	Kec		
KIA 6451 P,AP	LIN-IC	Audio Compander, Ucc=2,4...9V	16-DIP	Kec	-	
KIA 6500 S	Z-IC	5V, Watch Dog Timer, Ucc=7...17V	9-SIP	Kec	-	
KIA 6501 S	LIN-IC	Car, Drehzahlmesser/tacho Meter	9-SIP	Kec	-	
KIA 6801 K	LIN-IC	Bi-directional DC Motor Driver	SIP	Kec	-	pdf pinout
KIA 6900 P	LIN-IC	5 Band Graphic Equal., Ucc=3...16V	16-DIP	Kec	-	
KIA 6900 Z	LIN-IC	=KIA 6900P:	16-SQP			
KIA 6901 F	LIN-IC	=KIA 6401: SMD	8-MDIP		-	pdf
KIA 6901 P	LIN-IC	Motorregler/dc Motor Speed Ctrl.	8-DIP	Kec	-	pdf
KIA 6903 P	LIN-IC	Motorregler/dc Motor Speed Ctrl.	8-DIP	Kec	-	pdf
KIA 6924 S	LIN-IC	Portable Cass., Muting	9-SIP	Kec	-	pdf
KIA 6930 P	LIN-IC	Port. Cass., 2x Vol.-/Bal.-/Tone-Ctrl.	16-DIP	Kec	-	pdf
KIA 6941 S	LIN-IC	Prg. Sensor f. Cassette Tape Player	9-SIP	Kec	-	pdf
KIA 6966 S	LIN-IC	LED-Treiber/level meter, 5 LED	9-SIP	Kec	AN 6884	pdf pinout
KIA 6967 S	LIN-IC	5 Dot LED Level Meter	SIP	Kec	-	pdf pinout
KIA 6976 P	LIN-IC	LED-Treiber/level meter, 2x5 LED	16-DIP	Kec	(AN 6886)	
KIA 6987 P	LIN-IC	Toy Radio Control Dem.	DIC	Kec	-	pdf pinout
KIA 6988 P	LIN-IC	Cdi Control System IC	DIP	Kec	-	pdf pinout
KIA 7019 P	LIN-IC	Volt. Detection, lo reset, 1,9V	7	Kec	-	
KIA 7021 P	LIN-IC	=KIA 7019P: 2,1V	7	Kec	-	
KIA 7023 P	LIN-IC	=KIA 7019P: 2,3V	7	Kec	-	
KIA 7025 P	LIN-IC	=KIA 7019P: 2,5V	7	Kec	KA 75250	
KIA 7027 P	LIN-IC	=KIA 7019P: 2,7V	7	Kec	KA 75270	
KIA 7029 P	LIN-IC	=KIA 7019P: 2,9V	7	Kec	KA 75290	
KIA 7031 P	LIN-IC	=KIA 7019P: 3,1V	7	Kec	KA 75310	
KIA 7032 P	LIN-IC	=KIA 7019P: 3,2V	7	Kec		
KIA 7033 P	LIN-IC	=KIA 7019P: 3,3V	7	Kec	KA 7533	
KIA 7034 P	LIN-IC	=KIA 7019P: 3,4V	7	Kec		
KIA 7035 P	LIN-IC	=KIA 7019P: 3,5V	7	Kec		
KIA 7036 P	LIN-IC	=KIA 7019P: 3,6V	7	Kec	KA 7536	
KIA 7039 P	LIN-IC	=KIA 7019P: 3,9V	7	Kec	KA 7539	
KIA 7042 P	LIN-IC	=KIA 7019P: 4,2V	7	Kec	KA 7542	
KIA 7045 P	LIN-IC	=KIA 7019P: 4,5V	7	Kec	KA 7545	
KIA 7019F...7045F	LIN-IC	=KIA 7019P...7045P: SMD	39	Kec		
KIA 7125 FV	LIN-IC	SMD, CPU Reset, 1,25V Delay Time Ctrl.	8-MDIP	Kec	-	
KIA 7125 F	LIN-IC	SMD, CPU Reset, 1,25V Delay Time Ctrl.	8-MDIP	Kec	-	pdf pinout
KIA 7125 P	LIN-IC	SMD, CPU Reset, 1,25V Delay Time Ctrl.	8-MDIP	Kec	-	pdf pinout
KIA 7125 PV	LIN-IC	SMD, CPU Reset, 1,25V Delay Time Ctrl.	8-MDIP	Kec	-	pdf pinout
KIA 7217AP	LIN-IC	Aud. PAmp, 5.8W(13.2V/4Ohm), Ucc=9...18V	10-SIL	Kec	TA 7217AP	pdf pinout
KIA 7219F...7245F	LIN-IC	SMD,CPU Reset, Low Volt. Detection	39	Kec		
KIA 7219P...7245P	LIN-IC	CPU Reset, Low Volt. Detection	7	Kec		
KIA 7219S...7245S	LIN-IC	CPU Reset, Low Volt. Detection	35	Kec		
KIA 7282	LIN-IC	2x NF-E, 16V, 2,5A, 2x4,6W(12V/4Ω)	12-SIL	Kec	KA 22063	
KIA 7288P	IC	Bidir. DC Motor Drv. (2 Motors)	SIL	Kec	-	
KIA 7291	IC	Bidir. DC Motor Drv., various cases	SIL	Kec	-	pdf pinout
KIA 7419 P	LIN-IC	Volt. Detection, hi reset, 1,9V	7	Kec	-	pdf
KIA 7421 P	LIN-IC	=KIA 7419P: 2,1V	7	Kec	-	pdf
KIA 7423 P	LIN-IC	=KIA 7419P: 2,3V	7	Kec	-	pdf
KIA 7425 P	LIN-IC	=KIA 7419P: 2,5V	7	Kec	-	pdf
KIA 7427 P	LIN-IC	=KIA 7419P: 2,7V	7	Kec	-	pdf
KIA 7429 P	LIN-IC	=KIA 7419P: 2,9V	7	Kec	-	pdf
KIA 7431 P	LIN-IC	=KIA 7419P: 3,1V	7	Kec	-	pdf
KIA 7432 P	LIN-IC	=KIA 7419P: 3,2V	7	Kec	-	pdf
KIA 7433 P	LIN-IC	=KIA 7419P: 3,3V	7	Kec	-	pdf
KIA 7434 P	LIN-IC	=KIA 7419P: 3,4V	7	Kec	-	pdf
KIA 7435 P	LIN-IC	=KIA 7419P: 3,5V	7	Kec	-	pdf
KIA 7436 P	LIN-IC	=KIA 7419P: 3,6V	7	Kec	-	pdf

Type	Device	Short Description	Fig.	Manu	Comparision Types	More at
KIA 7439 P	LIN-IC	=KIA 7419P: 3,9V	7	Kec	-	pdf
KIA 7442 P	LIN-IC	=KIA 7419P: 4,2V	7	Kec	-	pdf
KIA 7445 P	LIN-IC	=KIA 7419P: 4,5V	7	Kec	-	pdf
KIA 7419F...7445F	LIN-IC	=KIA 7419P...7445P: SMD	39	Kec	-	
KIA 7719P...7745P	LIN-IC	CPU Reset, Low Volt. Detection	7	Kec		
KIA 7719S...7745S	LIN-IC	SMD,CPU Reset, Low Volt. Detection	35	Kec		
KIA 78 MR 05 PI	IC	4 terminal low Drop Voltage regulator	TSOP	Kec	-	pdf pinout
KIA 7805P...7824P	Z-IC	+5...+24V, 1A	17b	Kec	... 78xx... (TO-220)	
KIA 7805PI...7824PI	Z-IC	Iso, +5...+24V, 1A	17b	Kec	... 78xx... (TO-220 Iso)	
KIA 78 MR 06 PI	IC	4 terminal low Drop Voltage regulator	TSOP	Kec	-	pdf pinout
KIA 78 MR 08 PI	IC	4 terminal low Drop Voltage regulator	TSOP	Kec	-	pdf plhout
KIA 78 MR 09 PI	IC	4 terminal low Drop Voltage regulator	TSOP	Kec	-	pdf pinout
KIA 78 MR 10 PI	IC	4 terminal low Drop Voltage regulator	TSOP	Kec	-	pdf pinout
KIA 78 MR 12 PI	IC	4 terminal low Drop Voltage regulator	TSOP	Kec	-	pdf pinout
KIA 78 MR 15 PI	IC	4 terminal low Drop Voltage regulator	TSOP	Kec	-	pdf pinout
KIA 78 MR 18 PI	IC	4 terminal low Drop Voltage regulator	TSOP	Kec	-	pdf pinout
KIA 78 MR 25 PI	IC	4 terminal low Drop Voltage regulator	TSOP	Kec	-	pdf pinout
KIA 78 MR 30 PI	IC	4 terminal low Drop Voltage regulator	TSOP	Kec	-	pdf pinout
KIA 78 MR 33 PI	IC	4 terminal low Drop Voltage regulator	TSOP	Kec	-	pdf pinout
KIA 78 MR 35 PI	IC	4 terminal low Drop Voltage regulator	TSOP	Kec	-	pdf pinout
KIA 78 MR 37 PI	IC	4 terminal low Drop Voltage regulator	TSOP	Kec	-	pdf pinout
KIA 7905P...7924P	Z-IC	-5...-24V, 1A	17c	Kec	... 79xx... (TO-220)	
KIA 7905PI...7924PI	Z-IC	Iso, -5...-24V, 1A	17c	Kec	... 79xx... (TO-220 Iso)	
KIA 8000 S,SV	LIN-IC	Watch Dog Timer, Ucc=6...17V	9-SIP	Kec		pdf
KIA 8052 S,SV	LIN-IC	Full Bridge Driver, Ucc=8...16V	9-SIP	Kec		
KIA 8122 AF	LIN-IC	=KIA 8122AN: SMD	24-SMDIP			pdf
KIA 8122 AN	LIN-IC	Portable, AM/FM Tuner, Ucc=1,8...7V	24-NDIP	Kec		pdf
KIA 8145 FN	LIN-IC	SMD, 2x NF-E, 8mW(1,5V/16Ω), 0,9...2,2V	16-SSMDIP	Kec		
KIA 8155 FN	LIN-IC	SMD, 2x Vorverst/Preamp., 0,9...4V	24-SSMDIP	Kec		
KIA 8157 AFN	LIN-IC	SMD, 2x NF-E, 14mW, 0,9...2,2V	24-SSMDIP	Kec		pdf
KIA 8176 SN	LIN-IC	TV-FM Front End (DTS), Ucc=4...14V	12-SIP	Kec		
KIA 8182 FN	LIN-IC	SMD, TV-FM Front End (DTS), Ucc=4...14V	16-SSMDIP	Kec		pdf
KIA 8192 F	LIN-IC	SMD, Cd, 2x BTL Power Drv., Ucc=4...12V	20-SMDIP+b	Kec		
KIA 8207 K	LIN-IC	2x NF-E, 2x 4,6W(12V/4Ω), Ucc=6...15V	12-SIL	Kec		pdf
KIA 8224 H	LIN-IC	Power Supply,5V/0,1A, 5V/0,3A, 8V/1,2A	12-SQL	Kec		pdf
KIA 8231 L	LIN-IC	2x NF-E, 2x 37W(13,2V/2Ω), Ucc=9...18V	17-SILP	Kec		pdf
KIA 8244 H	LIN-IC	System Voltage Reg., Ucc=9,6...18V	12-SQL	Kec		
KIA 8246 H	LIN-IC	TV,2x NF-E, 2x 6W(20V/8Ω), Ucc=10...30V	12-SQL	Kec		pdf
KIA 8251 AH	LIN-IC	4x NF-E, 21W(14,4V/4Ω), Ucc=9...18V	25-QILP	Kec		pdf
KIA 8255 AH	LIN-IC	4x NF-E, 17W(14,4V/4Ω), Ucc=9...18V	25-QILP	Kec		
KIA 34063 A(F)	IC	DC/DC Converter Controller	TSOP	Kec		pdf pinout
KIA 78 R 000 F/PI	IC	4 terminal low Drop Voltage regulator	TSOP	Kec	-	pdf pinout
KIA 78 RM 000 F/PI	IC	4,5 terminal low Drop Voltage regulator	TSOP	Kec	-	pdf pinout
KIA 78 DM015S/F/PI	IC	3 terminal low Drop Voltage regulator	TSOP	Kec	-	pdf pinout
KIA 78 D 015 PI	IC	3 terminal low Drop Voltage regulator	TSOP	Kec	-	pdf pinout
KIA 78 D 015 S/F	IC	3 terminal low Drop Voltage regulator	TSOP	Kec	-	pdf pinout
KIA 78 R 015 F/PI	IC	4 terminal low Drop Voltage regulator	TSOP	Kec	-	pdf pinout
KIA 78 RM 015 F/PI	IC	4,5 terminal low Drop Voltage regulator	TSOP	Kec	-	pdf pinout
KIA 78 DM018S/F/PI	IC	3 terminal low Drop Voltage regulator	TSOP	Kec	-	pdf pinout
KIA 78 D 018 PI	IC	3 terminal low Drop Voltage regulator	TSOP	Kec	-	pdf pinout
KIA 78 D 018 S/F	IC	3 terminal low Drop Voltage regulator	TSOP	Kec	-	pdf pinout
KIA 78 R 018 F/PI	IC	4 terminal low Drop Voltage regulator	TSOP	Kec	-	pdf pinout
KIA 78 RM 018 F/PI	IC	4,5 terminal low Drop Voltage regulator	TSOP	Kec	-	pdf pinout
KIA 78 DM020S/F/PI	IC	3 terminal low Drop Voltage regulator	TSOP	Kec	-	pdf pinout
KIA 78 D 020 PI	IC	3 terminal low Drop Voltage regulator	TSOP	Kec	-	pdf pinout
KIA 78 D 020 S/F	IC	3 terminal low Drop Voltage regulator	TSOP	Kec	-	pdf pinout
KIA 78 R 020 F/PI	IC	4 terminal low Drop Voltage regulator	TSOP	Kec	-	pdf pinout
KIA 78 RM 020 F/PI	IC	4,5 terminal low Drop Voltage regulator	TSOP	Kec	-	pdf pinout
KIA 78 DM025S/F/PI	IC	3 terminal low Drop Voltage regulator	TSOP	Kec	-	pdf pinout
KIA 78 D 025 PI	IC	3 terminal low Drop Voltage regulator	TSOP	Kec	-	pdf pinout
KIA 78 D 025 S/F	IC	3 terminal low Drop Voltage regulator	TSOP	Kec	-	pdf pinout
KIA 78 R 025 F/PI	IC	4 terminal low Drop Voltage regulator	TSOP	Kec	-	pdf pinout
KIA 78 RM 025 F/PI	IC	4,5 terminal low Drop Voltage regulator	TSOP	Kec	-	pdf pinout
KIA 78 DM030S/F/PI	IC	3 terminal low Drop Voltage regulator	TSOP	Kec	-	pdf pinout
KIA 78 D 030 PI	IC	3 terminal low Drop Voltage regulator	TSOP	Kec	-	pdf pinout
KIA 78 D 030 S/F	IC	3 terminal low Drop Voltage regulator	TSOP	Kec	-	pdf pinout
KIA 78 R 030 F/PI	IC	4 terminal low Drop Voltage regulator	TSOP	Kec	-	pdf pinout
KIA 78 RM 030 F/PI	IC	4,5 terminal low Drop Voltage regulator	TSOP	Kec	-	pdf pinout
KIA 78 DM033S/F/PI	IC	3 terminal low Drop Voltage regulator	TSOP	Kec	-	pdf pinout
KIA 78 D 033 PI	IC	3 terminal low Drop Voltage regulator	TSOP	Kec	-	pdf pinout
KIA 78 D 033 S/F	IC	3 terminal low Drop Voltage regulator	TSOP	Kec	-	pdf pinout
KIA 78 R 033 F/PI	IC	4 terminal low Drop Voltage regulator	TSOP	Kec	-	pdf pinout
KIA 78 RM 033 F/PI	IC	4,5 terminal low Drop Voltage regulator	TSOP	Kec	-	pdf pinout
KIA 78 DM050S/F/PI	IC	3 terminal low Drop Voltage regulator	TSOP	Kec	-	pdf pinout
KIA 78 D 050 PI	IC	3 terminal low Drop Voltage regulator	TSOP	Kec	-	pdf pinout
KIA 78 D 050 S/F	IC	3 terminal low Drop Voltage regulator	TSOP	Kec	-	pdf pinout
KIA 78 R 050 F/PI	IC	4 terminal low Drop Voltage regulator	TSOP	Kec	-	pdf pinout
KIA 78 RM 050 F/PI	IC	4,5 terminal low Drop Voltage regulator	TSOP	Kec	-	pdf pinout
KIA 278 R 000 FP	IC	2A ADJ. Output low Drop Voltage Reg.	TSOP	Kec	-	pdf pinout
KIA 278 R 000 PI	IC	2A ADJ. Output low Drop Voltage Reg.	TSOP	Kec	-	pdf pinout
KIA 278 R 015 FP	IC	2A low dropout Voltage regulator	TSOP	Kec	-	pdf pinout
KIA 278 R 015 PI	IC	2A low dropout Voltage regulator	TSOP	Kec	-	pdf pinout
KIA 278 R 018 FP	IC	2A low dropout Voltage regulator	TSOP	Kec	-	pdf pinout
KIA 278 R 018 PI	IC	2A low dropout Voltage regulator	TSOP	Kec	-	pdf pinout
KIA 278 R 020 FP	IC	2A low dropout Voltage regulator	TSOP	Kec	-	pdf pinout

Type	Device	Short Description	Fig.	Manu	Comparision Types	More at
KIA 278 R 020 PI	IC	2A low dropout Voltage regulator	TSOP	Kec	-	pdf pinout
KIA 278 R 025 FP	IC	2A low dropout Voltage regulator	TSOP	Kec	-	pdf pinout
KIA 278 R 025 PI	IC	2A low dropout Voltage regulator	TSOP	Kec	-	pdf pinout
KIA 278 R 030 FP	IC	2A low dropout Voltage regulator	TSOP	Kec	-	pdf pinout
KIA 278 R 030 PI	IC	2A low dropout Voltage regulator	TSOP	Kec	-	pdf pinout
KIA 278 R 033 FP	IC	2A low dropout Voltage regulator	TSOP	Kec	-	pdf pinout
KIA 278 R 033 PI	IC	2A low dropout Voltage regulator	TSOP	Kec	-	pdf pinout
KIA 278 R 050 FP	IC	2A low dropout Voltage regulator	TSOP	Kec	-	pdf pinout
KIA 278 R 050 PI	IC	2A low dropout Voltage regulator	TSOP	Kec	-	pdf pinout
KIA 378 R 000 FP	IC	3A ADJ-TYPE low dropout Voltage Reg.	TSOP	Kec	-	pdf pinout
KIA 378 R 015 FP	IC	3.5A low dropout Voltage regulator	TSOP	Kec	-	pdf pinout
KIA 378 R 015 PI	IC	3.5A low dropout Voltage regulator	TSOP	Kec	-	pdf pinout
KIA 378 R 018 FP	IC	3.5A low dropout Voltage regulator	TSOP	Kec	-	pdf pinout
KIA 378 R 018 PI	IC	3.5A low dropout Voltage regulator	TSOP	Kec	-	pdf pinout
KIA 378 R 020 FP	IC	3.5A low dropout Voltage regulator	TSOP	Kec	-	pdf pinout
KIA 378 R 020 PI	IC	3.5A low dropout Voltage regulator	TSOP	Kec	-	pdf pinout
KIA 378 R 025 FP	IC	3.5A low dropout Voltage regulator	TSOP	Kec	-	pdf pinout
KIA 378 R 025 PI	IC	3.5A low dropout Voltage regulator	TSOP	Kec	-	pdf pinout
KIA 378 R 030 FP	IC	3.5A low dropout Voltage regulator	TSOP	Kec	-	pdf pinout
KIA 378 R 030 PI	IC	3.5A low dropout Voltage regulator	TSOP	Kec	-	pdf pinout
KIA 378 R 033 FP	IC	3.5A low dropout Voltage regulator	TSOP	Kec	-	pdf pinout
KIA 378 R 033 PI	IC	3.5A low dropout Voltage regulator	TSOP	Kec	-	pdf pinout
KIA 378 R 050 FP	IC	3.5A low dropout Voltage regulator	TSOP	Kec	-	pdf pinout
KIA 378 R 050 PI	IC	3.5A low dropout Voltage regulator	TSOP	Kec	-	pdf pinout
KIA 578 R 000 FP	IC	5A Adj. Output Low Drop Voltage Reg.	TSOP	Kec	-	pdf pinout
KIA 578 R 000 PI	IC	5A Adj. Output Low Drop Voltage Reg.	TSOP	Kec	-	pdf pinout
KIA 278 R 000 P5	IC	2.0A Adj. Output Low Drop Voltage Reg.	TSOP	Kec	-	pdf pinout
KIB 3401 F	IC	White LED Step-up Converter	TSOP	Kec	-	pdf pinout
KIB 3402 F	IC	White LED Step-up Converter	TSOP	Kec	-	pdf pinout
KIB 6990 T	LIN-IC	Load Swith, 2.3A	46	Kec	-	pdf pinout
KIC 3812 FT	IC	LCD Backlight Inverter Controller	TSOP	Kec	-	pdf pinout
KIC 9162 AF	CMOS-IC	=KIC 9162AN: SMD	28-MDIP			
KIC 9162 AN	CMOS-IC	Analog Switch Array, Udd=8...18V	28-SDIP	Kec		
KIC 9163 AF	CMOS-IC	=KIC 9163AN: SMD	28-MDIP			
KIC 9163 AN	CMOS-IC	Analog Switch Array, Udd=8...18V	28-SDIP	Kec		
KIC 9164 AF	CMOS-IC	=KIC 9164AN: SMD	28-MDIP			
KIC 9164 AN	CMOS-IC	Analog Switch Array, Udd=8...18V	28-SDIP	Kec		
KIC 9208 N	CMOS-IC	Analog Switch Array, Udd=8...18V	28-SDIP	Kec	-	
KIC 9209 N	CMOS-IC	Analog Switch Array, Udd=8...18V	28-SDIP	Kec	-	
KIC 9210 P	CMOS-IC	Electr. Volume Control, Udd=6...18V	16-DIP	Kec		
KIC 9211 P	CMOS-IC	Electr. Volume Control, Udd=6...18V	20-DIP	Kec		
KIC 9235 F	CMOS-IC	=KIC 9235P: SMD	16-MDIP			
KIC 9235 P	CMOS-IC	Electr. Volume Control, Udd=4,5...12V	16-DIP	Kec		
KIC 9243 F	CMOS-IC	SMD, Ir-fb Sender/Transmitter, 2...4V	20-MDIP	Kec		
KIC 9244 P	CMOS-IC	Ir-fb Empfänger/Receiver, 9 Command	20-DIP	Kec		
KIC 9256 F	CMOS-IC	=KIC 9256P: SMD	16-MDIP			
KIC 9256 P	CMOS-IC	PLL f. HiFi-Tuner(DTS), Udd=4,5...5,5V	16-DIP	Kec		
KIC 9257 F	CMOS-IC	=KIC 9257P: SMD	20-MDIP			
KIC 9257 P	CMOS-IC	PLL f. HiFi-Tuner(DTS), Udd=4,5...5,5V	20-DIP	Kec		
KIC 9259 N	CMOS-IC	Ir-fb Empfänger/Receiver, 17 Command	28-SDIP	Kec		
KIC 9260 F	CMOS-IC	=KIC 9260P: SMD	16-MDIP			
KIC 9260 P	CMOS-IC	Electr. Volume Control, Udd=6...18V	16-DIP	Kec		
KIC 9285 P	CMOS-IC	Ir-fb Empfänger/Receiver, 11 Command	20-DIP	Kec		
KIC 9289 N	CMOS-IC	Karaoke Echo IC, Udd=4,5...5,5V	24-SDIP	Kec		
KIC 9293 F	CMOS-D/A-IC	=KIC 9293N: SMD	24-SMDIP			
KIC 9293 FN	CMOS-D/A-IC	=KIC 9293N: SMD	24-SSMDIP			
KIC 9293 N	CMOS-D/A-IC	18 Bit, Analog Filter f. Dig. Audio	24-SDIP	Kec		
KIC 9297 F	CMOS-IC	SMD, LCD Drv., Key Scan, Udd=2,5...5,5V	60-MP	Kec		
KIC 9299 P	CMOS-IC	Electr. Volume Control, Udd=6...18V	16-DIP	Kec		
KIC 9409 BF-001	CMOS-IC	SMD, Karaoke IC, Key Ctrl, Mic Echo	44-MP	Kec		
KIC 9412 AF	CMOS-IC	=KIC 9412AP: SMD	24-SMDIP			
KIC 9412 AP	CMOS-IC	Electr. Volume Control	20-DIP	Kec		
KIC 9413 AP	CMOS-IC	Electr. Volume Control	16-DIP	Kec		
KIC 9421 F	CMOS-IC	SMD, Electr. Vol. Ctrl., Udd=4,5...5,5V	44-MP	Kec		
KIC 9432 F	CMOS-IC	SMD, Cd, Digital Servo, Sync. Sep.	100-MP	Kec		
KIC 9433 F	CMOS-IC	SMD, Cd, Digital Servo, Sync. Sep.	100-MP	Kec		
KIC 9459 F	CMOS-IC	=KIC 9459N: SMD	24-SMDIP			
KIC 9459 N	CMOS-IC	Electr. Volume Control	28-SDIP	Kec		
KIC 9463 F	CMOS-IC	=KIC 9463N: SMD	28-MDIP			
KIC 9463 N	CMOS-IC	Electr. Volume Control	28-SDIP	Kec		
KID	Si-N	=2SC4050-D (Typ-Code/Stempel/marking)	35	Hit	→2SC4050	data
KID 65001...65507AF	LIN-IC	=KID 65001...65004AP: SMD	16-MDIP	Kec	-	
KID 65001 AP	LIN-IC	7x Darl.-Tr +Di, 50V, 0,5A	16-DIP	Kec	TD 62001AP	pdf
KID 65002 AP	LIN-IC	7x Darl.-Tr +Di, 50V, 0,5A	16-DIP	Kec	TD 62002AP	pdf
KID 65003 AP	LIN-IC	7x Darl.-Tr +Di, 50V, 0,5A	16-DIP	Kec	TD 62003AP	pdf
KID 65004 AP	LIN-IC	7x Darl.-Tr +Di, 50V, 0,5A	16-DIP	Kec	TD 62004AP	pdf
KID 65501 P	LIN-IC	7x Single Driver, common E, 35V, 0,2A	16-DIP	Kec	TD 62501P	pdf
KID 65502 P	LIN-IC	7x Single Driver, common E, 35V, 0,2A	16-DIP	Kec	TD 62502P	pdf
KID 65503 P	LIN-IC	7x Single Driver, common E, 35V, 0,2A	16-DIP	Kec	TD 62503P	pdf
KID 65504 P	LIN-IC	7x Single Driver, common E, 35V, 0,2A	16-DIP	Kec	TD 62504P	pdf
KID 65505 P	LIN-IC	7x Single Driver, common C, 35V, 0,2A	16-DIP	Kec	TD 62505P	pdf
KID 65506 P	LIN-IC	7x Single Driver, common C, 35V, 0,2A	16-DIP	Kec	TD 62506P	pdf
KID 65507 P	LIN-IC	5x Single Driver, 35V, 0,2A	16-DIP	Kec	TD 62507P	pdf
KID 65551 S	LIN-IC	4x Single Driver, common E, 25V, 0,15A	9-SIP	Kec	TD 62551S	pdf
KID 65553 S	LIN-IC	4x Single Driver, common E, 25V, 0,15A	9-SIP	Kec	TD 62553S	pdf

Type	Device	Short Description	Fig.	Manu	Comparision Types	More at
KID 65554 S	LIN-IC	4x Single Driver, common E, 25V, 0,15A	9-SIP	Kec	TD 62554S	pdf
KID 65555 S	LIN-IC	4x Single Driver, common E, 25V, 0,15A	9-SIP	Kec	TD 62555S	pdf
KID 65603 P,PV	LIN-IC	Hex Threshold Free Driver	16-DIP	Kec	-	
KID 65783 AF	-IC	8CH High-voltage source Driver	DIP	Kec	-	pdf pinout
KID 65783 AP	-IC	8CH High-voltage source Driver	DIP	Kec	-	pdf pinout
KIE	Si-N	=2SC4050-E (Typ-Code/Stempel/marking)	35	Hit	→2SC4050	data
KJ	MOS-N-FET-e	=2SK1847 (Typ-Code/Stempel/marking)	35	Say	→2SK1847	data
KJ	MOS-N-FET-e	=2SK2036 (Typ-Code/Stempel/marking)	35	Tos	→2SK2036	data
KJ	MOS-N-FET-e	=2SK2037 (Typ-Code/Stempel/marking)	35(2mm)	Tos	→2SK2037	data
KJ	MOS-N-FET-e	=2SK2152 (Typ-Code/Stempel/marking)	39	Say	→2SK2152	data
KJ	MOS-N-FET-e*	=HN 1K06FU (Typ-Code/Stempel/marking)	46(2mm)	Tos	→HN 1K06FU	
KJD	Si-Di	=BAW 56 (Typ-Code/Stempel/marking)	35	Aeg	→BAW 56	
KJE	Si-Di	=BAV 99 (Typ-Code/Stempel/marking)	35	Aeg	→BAV 99	
KJF	Si-Di	=BAL 99 (Typ-Code/Stempel/marking)	35	Aeg	→BAL 99	
KJH	Si-Di	=BAV 70 (Typ-Code/Stempel/marking)	35	Aeg	→BAV 70	
KK	Z-Di	=1SMB 6.5A (Typ-Code/Stempel/marking)	71(5x3,5)	Ons	→1SMB 5.0A...170A	
KK	Si-P	=2SB805-KK (Typ-Code/Stempel/marking)	39	Nec	→2SB805	data
KK	MOS-N-FET-e*	=HN 1K05FU (Typ-Code/Stempel/marking)	46(2mm)	Tos	→HN 1K05FU	
KK	Z-Di	=P6 SMB-8.2A(Typ-Code/Stempel/marking)	71(5x3,5)	Fag	→P6 SMB-8.2A	data
KK	Z-Di	=SMBJ 6.5A (Typ-Code/Stempel/marking)	71(5x3,5)	Mop	→SMBJ ...	data
KL	Si-P	=2SA1510 (Typ-Code/Stempel/marking)	35	Say	→2SA1510	data
KL	Si-P	=2SB805-KL (Typ-Code/Stempel/marking)	39	Nec	→2SB805	data
KL	MOS-N-FET-e*	=2SK2731 (Typ-Code/Stempel/marking)	35	Rhm	→2SK2731	data
KL	N-FET	=KTK 211-L (Typ-Code/Stempel/marking)	35	Kec	→KTK 211	data
KL	Z-Di	=P6 SMB-9.1 (Typ-Code/Stempel/marking)	71(5x3,5)	Fag	→P6 SMB-9.1	data
KL	Z-Di	=SMBJ 7.0 (Typ-Code/Stempel/marking)	71(5x3,5)	Mop	→SMBJ ...	data
KL 3 L07	Z-Di	SMD,TAZ, >65V, Ifsm=30A(1ms), Ih=100mA	71a(5x2,5)	Shi	-	data
KL 3 LU06	Z-Di	SMD, TAZ, >48V, Ifsm=30A(1ms), Ih=50mA	71a(5x2,5)	Shi	-	data
KL 3 LU07	Z-Di	SMD,TAZ, >65V, Ifsm=30A(1ms), Ih=100mA	71a(5x2,5)	Shi	-	data
KL 3 LU08	Z-Di	SMD,TAZ,Ur<63V,Ifsm=30A(1ms), Ih=100mA	71a(5x2,5)	Shi	-	data
KL 3 N14	Z-Di	SMD,TAZ,>130V, Ifsm=30A(1ms), Ih=100mA	71a(5x2,5)	Shi	-	data
KL 3 R20	Z-Di	SMD,TAZ,>185V, Ifsm=30A(1ms), Ih=100mA	71a(5x2,5)	Shi	-	data
KL 3 R25	Z-Di	SMD,TAZ,>220V, Ifsm=30A(1ms), Ih=100mA	71a(5x2,5)	Shi	-	data
KL 3 RU25	Z-Di	SMD,TAZ,>220V, Ifsm=30A(1ms), Ih=100mA	71a(5x2,5)	Shi	-	data
KL 3 Z07	Z-Di	SMD,TAZ, >5,5V, Ifsm=30A(1ms), Ih=50mA	71a(5x2,5)	Shi	-	data
KL 3 Z18	Z-Di	SMD,TAZ,>15,5V, Ifsm=30A(1ms), Ih=50mA	71a(5x2,5)	Shi	-	data
KL 2822 MN	IC	Dual Low - Voltage Power Amplifier	TSOP	Iks	-	pdf pinout
KL 2822 MD	IC	Dual Low - Voltage Power Amplifier	TSOP	Iks	-	pdf pinout
KL 3340 P	IC	Electronic Attenuator	DIP	Iks	-	pdf pinout
KL 6270 N	Si	Gain Controlled Microphone Preamp./vogad	DIP	Iks	-	pdf pinout
KL 6270 D	Si	Gain Controlled Microphone Preamp./vogad	DIP	Iks	-	pdf pinout
KL 7496 LN	Dig. Audio	2W+2W amplifier, DC Volume Control	DIP	Iks	-	pdf pinout
KL 34018 DW	LIN-IC	Voice Switched Speaker phone Circuit	DIP	Iks	-	pdf pinout
KL 34018 N	LIN-IC	Voice Switched Speaker phone Circuit	DIP	Iks	-	pdf pinout

KM

Type	Device	Short Description	Fig.	Manu	Comparision Types	More at
KM	Z-Di	=1SMB 7.0A (Typ-Code/Stempel/marking)	71(5x3,5)	Ons	→1SMB 5.0A...170A	
KM	Si-P	=2SB805-KM (Typ-Code/Stempel/marking)	39	Nec	→2SB805	data
KM	MOS-N-FET-e	=2SK2009 (Typ-Code/Stempel/marking)	35	Tos	→2SK2009	data
KM	N-FET	=2SK2171 (Typ-Code/Stempel/marking)	39	Say	→2SK2171	data
KM	MOS-N-FET-e	=BST 80 (Typ-Code/Stempel/marking)	39	Val	→BST 80	data
KM	Z-Di	=P6 SMB-9.1A(Typ-Code/Stempel/marking)	71(5x3,5)	Fag	→P6 SMB-9.1A	data
KM	Z-Di	=SMBJ 7.0A (Typ-Code/Stempel/marking)	71(5x3,5)	Mop	→SMBJ ...	data
KM 3	N-FET	=2SK2171-3 (Typ-Code/Stempel/marking)	39	Say	→2SK2171	data
KM 4	N-FET	=2SK2171-4 (Typ-Code/Stempel/marking)	39	Say	→2SK2171	data
KM 5	N-FET	=2SK2171-5 (Typ-Code/Stempel/marking)	39	Say	→2SK2171	data
KM 28 C64	EEPROM-IC	CMOS, 8192 x 8 Bit, 5V, 150...250ns	28-DIP	Sam	...28C64...	data
KM 28 C65	EEPROM-IC	CMOS, 8192 x 8 Bit, 5V, 150...250ns	28-DIP	Sam	...28C65...	data
KM 41C1000P-10...12	CMOS-dRAM-IC	1048576 x 1 Bit, 100...120ns, Page Mode	18-DIP	Sam		
KM 41C1000Z-...	CMOS-dRAM-IC	=KM 41C1000P-...:	20-SQP			
KM 41 C1001 ...	CMOS-dRAM-IC	=KM 41 C1000...: Nibble Mode		Sam		
KM 41 C1002 ...	CMOS-dRAM-IC	=KM 41 C1000...: Static Column Mode		Sam		
KM 44 C256P-10...12	CMOS-dRAM-IC	262144 x 4 Bit, 100...120ns, Page Mode	20-DIP	Sam	...424256...	data
KM 44 C256Z-...	CMOS-dRAM-IC	=KM 44 C256P-...:	20-SQP			data
KM 44 C258 ...	CMOS-dRAM-IC	=KM 44 C256...: Static Column Mode		Sam	...424258...	data
KM 110 B(H)/...	Sensor	Magnetfeld-/Magnetic Field Sensor		Phi		
KM 901(D..H)	Si-N	AM, 140MHz	7e, 8a	Mic	BF 240...241, BF 254, BF 494, BF 594, ++	data
KM 904(D..J)	Si-N	NF-E, 25V, 0,5A, 0,5W, 200MHz	7e	Mic	BC 337...338, BC 635, BC 637, BC 639, ++	data
KM 905(D..J)	Si-P	NF-E, 25V, 0,5A, 0,5W, 120MHz	7e	Mic	BC 327...328, BC 636, BC 638, BC 640, ++	data
KM 917(D..H)	Si-N	AM/FM-ZF, 210MHz	7e, 8a	Mic	BF 240...241, BF 254, BF 494, BF 594, ++	data
KM 918(D..H)	Si-N	FM, 450MHz	7e, 8a	Mic	BF 240...241, BF 255, BF 495, BF 595, ++	data
KM 928(A,B)	Si-N	VHF/UHF, 800MHz	7e, 8a	Mic	BF 377...378, BF 689, BF 763, 2N2857, ++	data
KM 934(B..D)	Si-N	Uni, 35V, 0,5A, 0,5W, 180MHz	7e	Mic	BC 337...338, BC 635, BC 637, BC 639, ++	data
KM 935(B..D)	Si-P	Uni, 35V, 0,5A, 0,5W, 180MHz	7e	Mic	BC 327...328, BC 636, BC 638, BC 640, ++	data
KM 951	Si-P	≈BC 559	7		→BC 559	
KM 2816 A	EEPROM-IC	2048 x 8 Bit, 5V, 250...350ns	24-DIP	Sam	...2816...	data
KM 2817 A	EEPROM-IC	2048 x 8 Bit, 5V, 250...350ns	28-DIP	Sam	...2817...	data
KM 2864 A,AH	EEPROM-IC	8192 x 8 Bit, 5V, 200...300ns	28-DIP	Sam	...2864...	data
KM 2865 A,AH	EEPROM-IC	8192 x 8 Bit, 5V, 200...300ns	28-DIP	Sam	...2865...	data
KM 4164 A-12...-20	dRAM-IC	65536x1 Bit, 120...200ns	16-DIP	Sam	...4164...	

Type	Device	Short Description	Fig.	Manu	Comparision Types	More at
KM 4164 B-12	dRAM-IC	DRAM, 65536x1 Bit, 120ns	16-DIP	Sam	...4164...	data pdf pinout
KM 4164 B-15	dRAM-IC	DRAM, 65536x1 Bit, 150ns	16-DIP	Sam	...4164...	data pdf pinout
KM 4164 B-10	dRAM-IC	DRAM, 65536x1 Bit, 100ns	16-DIP	Sam	...4164...	data pdf pinout
KM 6264(A,AL,L)-...	CMOS-sRAM-IC	hi-speed, 8192 x 8 Bit, <100...150ns	28-DIP	Sam	...6164...	data
KM 9011	Si-N	Uni, -/20V, 0,1A, 0,3W, >150MHz	7e	Mic	BC 237, BC 547, BF 240...41, BF 254...55++	data
KM 9012	Si-P	NF-E, -/20V, 0,6A, 0,5W, 150MHz	7e	Mic	BC 327, BC 636, BC 638, 2SB909...910, ++	data
KM 9013	Si-N	NF-E, -/20V, 0,6A, 0,5W, 200MHz	7e	Mic	BC 337, BC 635, BC 637, 2SD1225...26, ++	data
KM 9014(A...D)	Si-N	Uni, 25V, 0,1A, 0,3W, 140MHz	7e, 8a	Mic	BC 168, BC 183, BC 238, BC 548, ++	data
KM 9015(A...D)	Si-P	Uni, 25V, 0,1A, 0,3W, 120MHz	7e, 8a	Mic	BC 213, BC 258, BC 308, BC 558, ++	data
KM 9016	Si-N	HF, -/20V, 0,05A, 0,3W, >500MHz	7e	Mic	BF 240...241, BF 254...255, BF 594...595,++	data
KM 9018	Si-N	HF, -/15V, 0,05A, 0,3W, >400MHz	7e	Mic	BF 225, BF 255, BF 314, BF 505, BF 507++	data
KM 41256 AJ-...	dRAM-IC	=KM 41256AP..:	18-PLCC		-	data
KM 41256AP-10...-15	dRAM-IC	262144x1 Bit, 100...150ns, page mode	16-DIP	Sam	...41256...	data
KM 41256 AZ-...	dRAM-IC	=KM 41256AP..:	16-SQP			data
KM 41257 ...	dRAM-IC	=KM 41257...: Nibble Mode		Sam	...41257...	data
KM 41464 AJ-...	dRAM-IC	=KM 41464AP..:	18-PLCC		...4464...	data
KM 41464AP-12...-15	dRAM-IC	65536 x 4 Bit, 120...150ns	18-DIP	Sam	...4464...	data
KM 41464 AZ-...	dRAM-IC	=KM 41464AP..:	20-SQP		...4464...	data
KM 62256P(LP)-...	CMOS-sRAM-IC	hi-speed, 32768 x 8 Bit, <100...150ns	28-DIP	Sam	...61256...	data
KM 416 C 1004 CJ	CMOS-IC	1M x 16Bit Cmos Dynamic RAM	TSOP	Sam	-	pdf pinout
KM 416 C 1004 CT	CMOS-IC	1M x 16Bit Cmos Dynamic RAM	TSOP	Sam	-	pdf pinout
KM 416 V 1004 CJ	CMOS-IC	1M x 16Bit Cmos Dynamic RAM	TSOP	Sam	-	pdf pinout
KM 416 V 1004 CT	CMOS-IC	1M x 16Bit Cmos Dynamic RAM	TSOP	Sam	-	pdf pinout
KM 416 C 1204 CJ	CMOS-IC	1M x 16Bit Cmos Dynamic RAM	TSOP	Sam	-	pdf pinout
KM 416 C 1204 CT	CMOS-IC	1M x 16Bit Cmos Dynamic RAM	TSOP	Sam	-	pdf pinout
KM 416 V 1204 CJ	CMOS-IC	1M x 16Bit Cmos Dynamic RAM	TSOP	Sam	-	pdf pinout
KM 416 V 1204 CT	CMOS-IC	1M x 16Bit Cmos Dynamic RAM	TSOP	Sam	-	pdf pinout
KMA 200	Sensor	Programmable Angle Sensor		Phi	-	
KMI 15/1...4	Sensor	Integrated Rotational Speed Sensor		Phi	-	
KMI 16/1	Sensor	Integrated Rotational Speed Sensor		Phi	-	
KMI 18/2...4	Sensor	Integrated Rotational Speed Sensor		Phi	-	
KMI 20/1...4	Sensor	Integrated Rotational Speed Sensor		Phi	-	
KMI 22/1	Sensor	Integrated Rotational Speed Sensor		Phi	-	
KMM 48256..591001	dRAM-IC	Speichermodule/memory Modules	SIP,SIMM	Sam	-	data
KMOC3053	OC	LED/Triac Viso:5000Vrms Vbs:1.2V		Cos		data pdf pinout
KMZ 10 A...C	Sensor	Magnetfeld-/Magnetic Field Sensor	4-SIP	Phi	-	
KMZ 41	Sensor	SMD, Magnetfeld-/Magnetic Field Sensor	8-MDIP	Phi	-	
KMZ 42	Sensor	SMD, Magnetfeld-/Magnetic Field Sensor	8-MDIP	Phi	-	
KMZ 51	Sensor	SMD, Magnetfeld-/Magnetic Field Sensor	8-MDIP	Phi	-	
KMZ 52	Sensor	SMD, Magnetfeld-/Magnetic Field Sensor	16-MDIP	Phi	-	

KN...KS

Type	Device	Short Description	Fig.	Manu	Comparision Types	More at
KN	Si-N	=2SC4983 (Typ-Code/Stempel/marking)	35	Say	→2SC4983	data
KN	MOS-P-FET-e	=2SJ305 (Typ-Code/Stempel/marking)	35	Tos	→2SJ305	data
KN	N-FET	=2SK2218 (Typ-Code/Stempel/marking)	39	Say	→2SK2218	data
KN	MOS-N-FET-e*	=2SK3018 (Typ-Code/Stempel/marking)	35(2mm)	Rhm	→2SK3018	data
KN	MOS-N-FET-e*	=2SK3019 (Typ-Code/Stempel/marking)	35(1,6mm)	Rhm	→2SK3019	data
KN	MOS-N-FET-e	=BST 84 (Typ-Code/Stempel/marking)	39	Val	→BST 84	data
KN	Z-Di	=P6 SMB-10 (Typ-Code/Stempel/marking)	71(5x3,5)	Fag	→P6 SMB-10	data
KN	Z-Di	=SMBJ 7.5 (Typ-Code/Stempel/marking)	71(5x3,5)	Mop	→SMBJ ...	data
KN 3	N-FET	=2SK2218-3 (Typ-Code/Stempel/marking)	39	Say	→2SK2218	data
KN 4	N-FET	=2SK2218-4 (Typ-Code/Stempel/marking)	39	Say	→2SK2218	data
KN 5	Si-N	=2SC4983-5 (Typ-Code/Stempel/marking)	35	Say	→2SC4983	data
KN 5	N-FET	=2SK2218-5 (Typ-Code/Stempel/marking)	39	Say	→2SK2218	data
KN 6	Si-N	=2SC4983-6 (Typ-Code/Stempel/marking)	35	Say	→2SC4983	data
KN 7	Si-N	=2SC4983-7 (Typ-Code/Stempel/marking)	35	Say	→2SC4983	data
KN 2222(A)	Si-N	=2N2222(A): 0,6A, 0,625W	7e	Kec	→2N2222(A)	data
KN 2222(A)S	Si-N	=2N2222(A): SMD, 0,6A	35a		→MMBT 2222(A)	data
KN 2907(A)	Si-P	=2N2907(A): 0,6A, 0,625W	7e	Kec	→2N2907(A)	data
KN 2907(A)S	Si-P	=2N2907(A): SMD, 0,6A	35a		→MMBT 2907(A)	data
KN 3903	Si-N	=2N3903:	7e	Kec	→2N3903	data
KN 3903 S	Si-N	=2N3903: SMD	35a		→MMBT 3903	data
KN 3904	Si-N	=2N3904:	7e	Kec	→2N3904	data
KN 3904 S	Si-N	=2N3904: SMD	35a		→MMBT 3904	data
KN 3905	Si-P	=2N3905:	7e	Kec	→2N3905	data
KN 3905 S	Si-P	=2N3905: SMD	35a		→MMBT 3905	data
KN 3906	Si-P	=2N3906:	7e	Kec	→2N3906	data
KN 3906 S	Si-P	=2N3906: SMD	35a		→MMBT 3906	data
KN 4400	Si-N	=2N4400:	7e	Kec	→2N4400	data
KN 4400 S	Si-N	=2N4400: SMD	35a		→MMBT 4400	data
KN 4401	Si-N	=2N4401:	7e	Kec	→2N4401	data
KN 4401 S	Si-N	=2N4401: SMD	35a		→MMBT 4401	data
KN 4402	Si-P	=2N4402:	7e	Kec	→2N4402	data
KN 4402 S	Si-P	=2N4402: SMD	35a		→MMBT 4402	data
KN 4403	Si-P	=2N4403:	7e	Kec	→2N4403	data
KN 4403 S	Si-P	=2N4403: SMD	35a		→MMBT 4403	data
KO	N-FET	=2SK211-O (Typ-Code/Stempel/marking)	35	Tos	→2SK211	data
KO	MOS-N-FET-e	=2SK2260 (Typ-Code/Stempel/marking)	39	Say	→2SK2260	data

Type	Device	Short Description	Fig.	Manu	Comparision Types	More at
KO	N-FET	=2SK881-O (Typ-Code/Stempel/marking)	35(2mm)	Tos	→2SK881	data
KO	MOS-N-FET-e	=BST 86 (Typ-Code/Stempel/marking)	39	Val	→BST 86	data
KO	Si-P	=KTA1001-O (Typ-Code/Stempel/marking)	39	Kec	→KTA 1001	data
KO	N-FET	=KTK 211-O (Typ-Code/Stempel/marking)	35	Kec	→KTK 211	data
KOM2033A	PD	840(850)nm S:0.56mA/mW P:15µA/mW/cm²		Sie		data pinout
KOM2033B	PD	870(850)nm S:0.64mA/mW P:16µA/mW/cm²		Sie		data pinout
KOM2045	PD	880(850)nm S:0.62mA/mW P:80nA/Lx	DIP-8	Sie		data pdf pinout
KOM2084	PD	880(850)nm S:0.62mA/mW P:80nA/Lx		Sie		data pinout
KOM2085	PD	920(850)nm S:0.62mA/mW P:180nA/Lx		Sie		data pinout
KOM2100A	PD	840(850)nm S:0.6mA/mW P:16µA/mW/cm²		Sie		data pinout
KOM2100AF	PD	880nm S:0.56mA/mW P:15µA/mW/cm²		Sie		data pinout
KOM2100B	PD	840(950)nm S:0.68mA/mW P:18µA/mW/cm²		Sie		data pdf pinout
KOM2100BF	PD	880(950)nm S:0.64mA/mW P:18µA/mW/cm²		Sie		data pdf pinout
KOM2108	PD	850nm S:0.59mA/mW P:2.2µA/mW/cm²	PLCC-28	Sie		data pinout
KOM2125	PD	850nm S:0.62mA/mW P:40/100nA/Lx		Sie		data pdf pinout
KP-1608	PT	940nm P:0.3µA/mW/cm²	T1	Kin		data pdf pinout
KP-2012	PT	940nm P:0.3µA/mW/cm²	T1	Kin		data pdf pinout
KP-3015	PT	940nm P:0.3µA/mW/cm²	T1	Kin		data pdf pinout
KP-3216	PT	940nm P:0.3µA/mW/cm²	T1	Kin		data pdf pinout
KP	Z-Di	=1SMB 7.5A (Typ-Code/Stempel/marking)	71(5x3,5)	Ons	→1SMB 5.0A...170A	
KP	Si-P	=2SB806-KP (Typ-Code/Stempel/marking)	39	Nec	→2SB806	data
KP	MOS-N-FET-e	=2SK2033 (Typ-Code/Stempel/marking)	35	Tos	→2SK2033	data
KP	MOS-N-FET-e	=2SK2034 (Typ-Code/Stempel/marking)	35(2mm)	Tos	→2SK2034	data
KP	MOS-N-FET-e	=2SK2035 (Typ-Code/Stempel/marking)	35(1,6mm)	Tos	→2SK2035	data
KP	MOS-N-FET-e	=2SK2316 (Typ-Code/Stempel/marking)	39	Say	→2SK2316	data
KP	MOS-N-FET-e*	=HN 1K03FU (Typ-Code/Stempel/marking)	46(2mm)	Tos	→HN 1K03FU	data
KP	MOS-N-FET-e	=HN 4K03JU (Typ-Code/Stempel/marking)	45(2mm)	Tos	→HN 4K03JU	data
KP	Z-Di	=P6 SMB-10A (Typ-Code/Stempel/marking)	71(5x3,5)	Fag	→P6 SMB-10A	data
KP	Z-Di	=SMBJ 7.5A (Typ-Code/Stempel/marking)	71(5x3,5)	Mop	→SMBJ ...	data
KP 4-5	Z-Di	TAZ, 50...60V, Ifsm=40A, 2W, Ih>50mA	31a	Shi	-	data
KP 4-8	Z-Di	TAZ, 75...98V, Ifsm=40A, 2W, Ih>95mA	31a	Shi	-	data
KP 4 L07	Z-Di	SMD,TAZ, >65V, Ifsm=40A(1ms), Ih=100mA	71a(7,6x4)	Shi	-	data
KP 4 L08	Z-Di	SMD,TAZ, >70V, Ifsm=40A(1ms), Ih=100mA	71a(7,6x4)	Shi	-	data
KP 4 N12	Z-Di	SMD,TAZ,>110V, Ifsm=40A(1ms), Ih=100mA	71a(7,6x4)	Shi	-	data
KP 4 N14	Z-Di	SMD,TAZ,>130V, Ifsm=40A(1ms), Ih=100mA	71a(7,6x4)	Shi	-	data
KP 4 R25	Z-Di	SMD,TAZ,>220V, Ifsm=40A(1ms), Ih=100mA	71a(7,6x4)	Shi	-	data
KP 8 LU07	Z-Di	SMD,TAZ, >65V, Ifsm=80A(1ms), Ih=100mA	71a(7,6x4)	Shi		data
KP 10 L06	Z-Di	SMD,TAZ,>55V, Ifsm=100A(1ms), Ih=100mA	71a(7,6x4)	Shi	-	data
KP 10 L07	Z-Di	SMD,TAZ, >65V, Ifsm=100A(1ms), Ih=100mA	71a(7,6x4)	Shi	-	data
KP 10 L08	Z-Di	SMD,TAZ,>70V, Ifsm=100A(1ms), Ih=100mA	71a(7,6x4)	Shi	-	data
KP 10 LU06	Z-Di	SMD,TAZ,>55V, Ifsm=100A(1ms), Ih=100mA	71a(7,6x4)	Shi	-	data
KP 10 LU07	Z-Di	SMD,TAZ,>63V, Ifsm=100A(1ms), Ih=100mA	71a(7,6x4)	Shi	-	data
KP 10 LU08	Z-Di	SMD,TAZ,>70V, Ifsm=100A(1ms), Ih=100mA	71a(7,6x4)	Shi	-	data
KP 10 N12	Z-Di	SMD,TAZ,>110V, Ifsm=100A(1mS), Ih=100mA	71a(7,6x4)	#hi	-	data
KP 10 N14	Z-Di	SMD,TAZ,>130V, Ifsm=100A(1ms), Ih=100mA	71a(7,6x4)	Shi	-	data
KP 10 R25	Z-Di	SMD,TAZ,>220V, Ifsm=100A(1ms), Ih=100mA	71a(7,6x4)	Shi	-	data
KP 10 R29	Z-Di	SMD,TAZ,<260V, Ifsm=100A(1ms), Ih=100mA	71a(7,6x4)	Shi	-	data
KP 15 L07	Z-Di	SMD,TAZ,>65V, Ifsm=150A(1ms), Ih=100mA	71a(7,6x4)	Shi	-	data
KP 15 L08	Z-Di	SMD,TAZ,>70V, Ifsm=150A(1ms), Ih=100mA	71a(7,6x4)	Shi	-	data
KP 15 N12	Z-Di	SMD,TAZ,>110V, Ifsm=150A(1ms), Ih=100mA	71a(7,6x4)	Shi	-	data
KP 15 N14	Z-Di	SMD,TAZ,>130V, Ifsm=150A(1ms), Ih=100mA	71a(7,6x4)	Shi	-	data
KP 15 R25	Z-Di	SMD,TAZ,>220V, Ifsm=150A(1ms), Ih=100mA	71a(7,6x4)	Shi	-	data
KP 15 R29	Z-Di	SMD,TAZ,>270V, Ifsm=150A(1ms), Ih=100mA	71a(7,6x4)	Shi	-	data
KP 823 C04	Si-Di	Dual, Schottky, S-L, 40V, 5A(Tc=87°)	30p§	Fjd	DE 5SC4M, MA 3U749, KS 823C04	data
KPA-3010	PT	940nm P:0.3µA/mW/cm²	T1	Kin		data pdf pinout
KQ	Si-P	=2SB806-KQ (Typ-Code/Stempel/marking)	39	Nec	→2SB806	data
KQ	MOS-P-FET-e	=2SJ343 (Typ-Code/Stempel/marking)	35	Tos	→2SJ343	data
KQ	MOS-P-FET-e	=2SJ344 (Typ-Code/Stempel/marking)	35(2mm)	Tos	→2SJ344	data
KQ	Z-Di	=P6 SMB-11 (Typ-Code/Stempel/marking)	71(5x3,5)	Fag	→P6 SMB-11	data
KQ	Z-Di	=SMBJ 8.0 (Typ-Code/Stempel/marking)	71(5x3,5)	Mop	→SMBJ ...	data
KR	Z-Di	=1SMB 8.0A (Typ-Code/Stempel/marking)	71(5x3,5)	Ons	→1SMB 5.0A...170A	
KR	Si-P	=2SB806-KR (Typ-Code/Stempel/marking)	39	Nec	→2SB806	data
KR	Z-Di	=P6 SMB-11A (Typ-Code/Stempel/marking)	71(5x3,5)	Fag	→P6 SMB-11A	data
KR	Z-Di	=SMBJ 8.0A (Typ-Code/Stempel/marking)	71(5x3,5)	Mop	→SMBJ ...	data
KRA 101 M	Si-P+R	S, Rb=Rbe=4,7kΩ, 50V, 0,1A, 200MHz	7c	Kec	AN 1L3M, DTA 143ES, RN 2001, UN 411L,++	data
KRA 101 S	Si-P+R	=KRA 101M: SMD	35a		FN 1L3M, DTA 143EK, RN 2401, UN 211L,++	data
KRA 102 M	Si-P+R	=KRA 101M: Rb=Rbe=10kΩ	7c	Kec	AN 1A4M, DTA 114ES, RN 2002, UN 4111,++	data
KRA 102 S	Si-P+R	=KRA 102M: SMD	35a		FN 1A4M, DTA 114EK, RN 2402, UN 2111,++	data
KRA 103 M	Si-P+R	=KRA 101M: Rb=Rbe=22kΩ	7c	Kec	AN 1F4M, DTA 124ES, RN 2003, UN 4112,++	data
KRA 103 S	Si-P+R	=KRA 103M: SMD	35a		FN 1F4M, DTA 124EK, RN 2403, UN 2112,++	data
KRA 104 M	Si-P+R	=KRA 101M: Rb=Rbe=47kΩ	7c	Kec	AN 1L4M, DTA 144ES, RN 2004, UN 4113,++	data
KRA 104 S	Si-P+R	=KRA 104M: SMD	35a		FN 1L4M, DTA 144EK, RN 2404, UN 2113,++	data
KRA 105 M	Si-P+R	=KRA 101M: Rb=2,2k, Rbe=47kΩ	7c	Kec	DTA 123JS, RN 2005	data
KRA 105 S	Si-P+R	=KRA 105M: SMD	35a		DTA 123JK, RN 2405	data
KRA 106 M	Si-P+R	=KRA 101M: Rb=4,7k, Rbe=47kΩ	7c	Kec	DTA 143ZS, RN 2006, 2SA1591, 2SA1616	data
KRA 106 S	Si-P+R	=KRA 106M: SMD	35a		DTA 143ZK, RN 2406, 2SA1597	data
KRA 107 M	Si-P+R	=KRA 101M: Rb=10k, Rbe=47kΩ	7c	Kec	AN 1A4P, DTA 114YS, RN 2007, UN 4114,++	data
KRA 107 S	Si-P+R	=KRA 107M: SMD	35a		FN 1A4P, DTA 114YK, RN 2407, UN 2114,++	data
KRA 108 M	Si-P+R	=KRA 101M: Rb=22k, Rbe=47kΩ	7c	Kec	AN 1F4N, DTA 124XS, KSR 2007, RN 2008	data
KRA 108 S	Si-P+R	=KRA 108M: SMD	35a		FN 1F4N, DTA 124XK, KSR 2107, RN 2408	data
KRA 109 M	Si-P+R	=KRA 101M: Rb=47k, Rbe=22kΩ	7c	Kec	AN 1L4L, DTA 144WS, RN 2009, UN 411E,++	data
KRA 109 S	Si-P+R	=KRA 109M: SMD	35a		FN 1L4L, DTA 144WK, RN 2409, UN 211E,++	data

Type	Device	Short Description	Fig.	Manu	Comparision Types	More at
KRA 110 M	Si-P+R	=KRA 101M: Rb=4,7kΩ, Rbe=-	7c	Kec	AN 1L3Z, DTA 143TS, RN 2010, UN 4116,++	data
KRA 110 S	Si-P+R	=KRA 110M: SMD	35a		FN 1L3Z, DTA 143TK, RN 2410, UN 2116,++	data
KRA 111 M	Si-P+R	=KRA 101M: Rb=10kΩ, Rbe=-	7c	Kec	AN 1A4Z, DTA 114TS, RN 2011, UN 4115,++	data
KRA 111 S	Si-P+R	=KRA 111M: SMD	35a		FN 1A4Z, DTA 114TK, RN 2411, UN 2115,++	data
KRA 112 M	Si-P+R	=KRA 101M: Rb=100kΩ, Rbe=-	7c	Kec	DTA 115TS	data
KRA 112 S	Si-P+R	=KRA 112M: SMD	35a		DTA 115TK	data
KRA 113 M	Si-P+R	=KRA 101M: Rb=22kΩ, Rbe=-	7c	Kec	AN 1F4Z, DTA 124TS, UN 4117, 2SA1590,++	data
KRA 113 S	Si-P+R	=KRA 113M: SMD	35a		FN 1F4Z, DTA 124TK, UN 2117, 2SA1589,++	data
KRA 114 M	Si-P+R	=KRA 101M: Rb=47kΩ, Rbe=-	7c	Kec	AN 1L4Z, DTA 144TS, UN 4110, 2SA1509,++	data
KRA 114 S	Si-P+R	=KRA 114M: SMD	35a		FN 1L4Z, DTA 144TK, UN 2110, 2SA1508,++	data
KRA 116 M	Si-P+R	=KRA 101M: Rb=1k, Rbe=10kΩ	7c	Kec	AN 1A3Q, DTA 113ZS, UN 4119	data
KRA 116 S	Si-P+R	=KRA 116M: SMD	35a		FN 1A3Q, DTA 113ZK, UN 2119	data
KRA 117 M	Si-P+R	=KRA 101M: Rb=Rbe=2,2kΩ	7c	Kec	DTA 123ES	data
KRA 117 S	Si-P+R	=KRA 117M: SMD	35a		DTA 123EK	data
KRA 118 M	Si-P+R	=KRA 101M: Rb=2,2k, Rbe=10kΩ	7c	Kec	DTA 123YS, UN 411H, 2SA1503	data
KRA 118 S	Si-P+R	=KRA 118M: SMD	35a		DTA 123YK, UN 211H, 2SA1502	data
KRA 119 M	Si-P+R	=KRA 101M: Rb=4,7k, Rbe=10kΩ	7c	Kec	AN 1L3N, DTA 143XS, KSR 2005, UN 411F,++	data
KRA 119 S	Si-P+R	=KRA 119M: SMD	35a		DTA 143XK, FN 1L3N, UN 211F, 2SA1653, ++	data
KRA 120 M	Si-P+R	=KRA 101M: Rb=10k, Rbe=4,7kΩ	7c	Kec	DTA 114WS, UN 411K	data
KRA 120 S	Si-P+R	=KRA 120M: SMD	35a		DTA 114WK, UN 211K	data
KRA 121 M	Si-P+R	=KRA 101M: Rb=47k, Rbe=10kΩ	7c	Kec	DTA 144VS, UN 411D	data
KRA 121 S	Si-P+R	=KRA 121M: SMD	35a		DTA 144VK, UN 211D	data
KRA 122 M	Si-P+R	=KRA 101M: Rb=100k, Rbe=100kΩ	7c	Kec	DTA 115ES	data
KRA 122 S	Si-P+R	=KRA 122M: SMD	35a		DTA 115EK	data
KRA 221 M	Si-P+R	S, Rb=Rbe=1kΩ, 50V, 0,8A, 200MHz	7c	Kec	-	data
KRA 222 M	Si-P+R	=KRA 221M: Rb=Rbe=2,2kΩ	7c	Kec	-	data
KRA 223 M	Si-P+R	=KRA 221M: Rb=Rbe=4,7kΩ	7c	Kec	-	data
KRA 224 M	Si-P+R	=KRA 221M: Rb=Rbe=10kΩ	7c	Kec	-	data
KRA 225 M	Si-P+R	=KRA 221M: Rb=1k, Rbe=10kΩ	7c	Kec	-	data
KRA 226 M	Si-P+R	=KRA 221M: Rb=2,2k, Rbe=10kΩ	7c	Kec	-	data
KRA 221S..226S	Si-P+R	=KRA 221..226M: SMD	35a		-	data
KRA 301	Si-P+R	=KRA 101M: SMD	35a(2mm)	Kec	DTA 143EU, RN 2301, UN 511L	data
KRA 301 E	Si-P+R	=KRA 101M: SMD	35a(1,6mm)	Kec	RN 2101, UN 911L	data
KRA 302	Si-P+R	=KRA 102M: SMD	35a(2mm)	Kec	DTA 114EU, RN 2302, UN 5111, 2SA1678	data
KRA 302 E	Si-P+R	=KRA 102M: SMD	35a(1,6mm)	Kec	RN 2102, UN 9111	data
KRA 303	Si-P+R	=KRA 103M: SMD	35a(2mm)	Kec	DTA 124EU, RN 2303, UN 5112, 2SA1677	data
KRA 303 E	Si-P+R	=KRA 103M: SMD	35a(1,6mm)	Kec	RN 2103, UN 9112	data
KRA 304	Si-P+R	=KRA 104M: SMD	35a(2mm)	Kec	DTA 144EU, RN 2304, UN 5113, 2SA1676	data
KRA 304 E	Si-P+R	=KRA 104M: SMD	35a(1,6mm)	Kec	RN 2104, UN 9113	data
KRA 305	Si-P+R	=KRA 105M: SMD	35a(2mm)	Kec	DTA 123JU, RN 2305	data
KRA 305 E	Si-P+R	=KRA 105M: SMD	35a(1,6mm)	Kec	RN 2105	data
KRA 306	Si-P+R	=KRA 106M: SMD	35a(2mm)	Kec	DTA 143ZU, RN 2306	data
KRA 306 E	Si-P+R	=KRA 106M: SMD	35a(1,6mm)	Kec	RN 2106	data
KRA 307	Si-P+R	=KRA 107M: SMD	35a(2mm)	Kec	DTA 114YU, RN 2307, UN 5114	data
KRA 307 E	Si-P+R	=KRA 107M: SMD	35a(1,6mm)	Kec	RN 2107, UN 9114	data
KRA 308	Si-P+R	=KRA 108M: SMD	35a(2mm)	Kec	DTA 124XU, RN 2308	data
KRA 308 E	Si-P+R	=KRA 108M: SMD	35a(1,6mm)	Kec	RN 2108	data
KRA 309	Si-P+R	=KRA 109M: SMD	35a(2mm)	Kec	DTA 144WU, RN 2309, UN 511E	data
KRA 309 E	Si-P+R	=KRA 109M: SMD	35a(1,6mm)	Kec	RN 2109, UN 911E	data
KRA 310	Si-P+R	=KRA 110M: SMD	35a(2mm)	Kec	DTA 143TU, RN 2310, UN 5116	data
KRA 310 E	Si-P+R	=KRA 110M: SMD	35a(1,6mm)	Kec	RN 2110, UN 9116	data
KRA 311	Si-P+R	=KRA 111M: SMD	35a(2mm)	Kec	DTA 114TU, RN 2311, UN 5115	data
KRA 311 E	Si-P+R	=KRA 111M: SMD	35a(1,6mm)	Kec	RN 2111, UN 9115	data
KRA 312	Si-P+R	=KRA 112M: SMD	35a(2mm)	Kec	DTA 115TU	data
KRA 312 E	Si-P+R	=KRA 112M: SMD	35a(1,6mm)	Kec	UNR 911BJ	data
KRA 313	Si-P+R	=KRA 113M: SMD	35a(2mm)	Kec	DTA 124TU, RN 2312, UN 5117	data
KRA 313 E	Si-P+R	=KRA 113M: SMD	35a(1,6mm)	Kec	RN 2112, UN 9117	data
KRA 314	Si-P+R	=KRA 114M: SMD	35a(2mm)	Kec	DTA 144TU, RN 2313, UN 5110	data
KRA 314 E	Si-P+R	=KRA 114M: SMD	35a(1,6mm)	Kec	RN 2113, UN 9110	data
KRA 316	Si-P+R	=KRA 116M: SMD	35a(2mm)	Kec	DTA 113ZU, UN 5119	data
KRA 316 E	Si-P+R	=KRA 116M: SMD	35a(1,6mm)	Kec	UN 9119	data
KRA 317	Si-P+R	=KRA 117M: SMD	35a(2mm)	Kec	DTA 123EU	data
KRA 317 E	Si-P+R	=KRA 117M: SMD	35a(1,6mm)	Kec	-	data
KRA 318	Si-P+R	=KRA 118M: SMD	35a(2mm)	Kec	DTA 123YU, UN 511H, 2SA1722	data
KRA 318 E	Si-P+R	=KRA 118M: SMD	35a(1,6mm)	Kec	UN 911H	data
KRA 319	Si-P+R	=KRA 119M: SMD	35a(2mm)	Kec	DTA 143XU, UN 511F	data
KRA 319 E	Si-P+R	=KRA 119M: SMD	35a(1,6mm)	Kec	UN 911F	data
KRA 320	Si-P+R	=KRA 120M: SMD	35a(2mm)	Kec	DTA 114WU, UN 511K	data
KRA 320 E	Si-P+R	=KRA 120M: SMD	35a(1,6mm)	Kec	UN 911K	data
KRA 321	Si-P+R	=KRA 121M: SMD	35a(2mm)	Kec	DTA 114VU, UN 511D	data
KRA 321 E	Si-P+R	=KRA 121M: SMD	35a(1,6mm)	Kec	UN 911D	data
KRA 322	Si-P+R	=KRA 122M: SMD	35a(2mm)	Kec	DTA 115EU	data
KRA 322 E	Si-P+R	=KRA 122M: SMD	35a(1,6mm)	Kec	UNR 911A	data
KRA 551 U	Si-P+R	=KRA 101M: SMD, Dual	45ba1(2mm)	Kec	-	data
KRA 552 U	Si-P+R	=KRA 102M: SMD, Dual	45ba1(2mm)	Kec	-	data
KRA 553 U	Si-P+R	=KRA 103M: SMD, Dual	45ba1(2mm)	Kec	-	data
KRA 554 U	Si-P+R	=KRA 104M: SMD, Dual	45ba1(2mm)	Kec	-	data
KRA 555 U	Si-P+R	=KRA 105M: SMD, Dual	45ba1(2mm)	Kec	-	data
KRA 556 U	Si-P+R	=KRA 106M: SMD, Dual	45ba1(2mm)	Kec	-	data
KRA 557 U	Si-P+R	=KRA 107M: SMD, Dual	45ba1(2mm)	Kec	-	data
KRA 558 U	Si-P+R	=KRA 108M: SMD, Dual	45ba1(2mm)	Kec	-	data
KRA 559 U	Si-P+R	=KRA 109M: SMD, Dual	45ba1(2mm)	Kec	-	data
KRA 560 U	Si-P+R	=KRA 110M: SMD, Dual	45ba1(2mm)	Kec	-	data
KRA 561 U	Si-P+R	=KRA 111M: SMD, Dual	45ba1(2mm)	Kec	-	data

Type	Device	Short Description	Fig.	Manu	Comparision Types	More at
KRA 562 U	Si-P+R	=KRA 112M: SMD, Dual	45ba1(2mm)	Kec	-	data
KRA 563 U	Si-P+R	=KRA 113M: SMD, Dual	45ba1(2mm)	Kec	-	data
KRA 564 U	Si-P+R	=KRA 114M: SMD, Dual	45ba1(2mm)	Kec	-	data
KRA 566 U	Si-P+R	=KRA 116M: SMD, Dual	45ba1(2mm)	Kec	-	data
KRA 567 U	Si-P+R	=KRA 117M: SMD, Dual	45ba1(2mm)	Kec	-	data
KRA 568 U	Si-P+R	=KRA 118M: SMD, Dual	45ba1(2mm)	Kec	-	data
KRA 569 U	Si-P+R	=KRA 119M: SMD, Dual	45ba1(2mm)	Kec	-	data
KRA 570 U	Si-P+R	=KRA 120M: SMD, Dual	45ba1(2mm)	Kec	-	data
KRA 571 U	Si-P+R	=KRA 121M: SMD, Dual	45ba1(2mm)	Kec	-	data
KRA 572 U	Si-P+R	=KRA 122M: SMD, Dual	45ba1(2mm)	Kec	-	data
KRA 751 U	Si-P+R	=KRA 101M: SMD, Dual	46bh1(2mm)	Kec	-	data
KRA 752 U	Si-P+R	=KRA 102M: SMD, Dual	46bh1(2mm)	Kec	-	data
KRA 753 U	Si-P+R	=KRA 103M: SMD, Dual	46bh1(2mm)	Kec	-	data
KRA 754 U	Si-P+R	=KRA 104M: SMD, Dual	46bh1(2mm)	Kec	-	data
KRA 755 U	Si-P+R	=KRA 105M: SMD, Dual	46bh1(2mm)	Kec	-	data
KRA 756 U	Si-P+R	=KRA 106M: SMD, Dual	46bh1(2mm)	Kec	-	data
KRA 757 U	Si-P+R	=KRA 107M: SMD, Dual	46bh1(2mm)	Kec	-	data
KRA 758 U	Si-P+R	=KRA 108M: SMD, Dual	46bh1(2mm)	Kec	-	data
KRA 759 U	Si-P+R	=KRA 109M: SMD, Dual	46bh1(2mm)	Kec	-	data
KRA 760 U	Si-P+R	=KRA 110M: SMD, Dual	46bh1(2mm)	Kec	-	data
KRA 761 U	Si-P+R	=KRA 111M: SMD, Dual	46bh1(2mm)	Kec	-	data
KRA 762 U	Si-P+R	=KRA 112M: SMD, Dual	46bh1(2mm)	Kec	-	data
KRA 763 U	Si-P+R	=KRA 113M: SMD, Dual	46bh1(2mm)	Kec	-	data
KRA 764 U	Si-P+R	=KRA 114M: SMD, Dual	46bh1(2mm)	Kec	-	data
KRA 766 U	Si-P+R	=KRA 116M: SMD, Dual	46bh1(2mm)	Kec	-	data
KRA 767 U	Si-P+R	=KRA 117M: SMD, Dual	46bh1(2mm)	Kec	-	data
KRA 768 U	Si-P+R	=KRA 118M: SMD, Dual	46bh1(2mm)	Kec	-	data
KRA 769 U	Si-P+R	=KRA 119M: SMD, Dual	46bh1(2mm)	Kec	-	data
KRA 770 U	Si-P+R	=KRA 120M: SMD, Dual	46bh1(2mm)	Kec	-	data
KRA 771 U	Si-P+R	=KRA 121M: SMD, Dual	46bh1(2mm)	Kec	-	data
KRA 772 U	Si-P+R	=KRA 122M: SMD, Dual	46bh1(2mm)	Kec	-	data
KRC 101 M	Si-N+R	S, Rb=Rbe=4,7kΩ, 50V, 0,1A, 200MHz	7c	Kec	AA 1L3M, DTC 143ES, RN 1001, UN 421L,++	data
KRC 101 S	Si-N+R	=KRC 101M: SMD	35a		FA 1L3M, DTC 143EK, RN 1401, UN 221L,++	data
KRC 102 M	Si-N+R	=KRC 101M: Rb=Rbe=10kΩ	7c	Kec	AA 1A4M, DTC 114ES, RN 1002, UN 4211,++	data
KRC 102 S	Si-N+R	=KRC 102M: SMD	35a		FA 1A4M, DTC 114EK, RN 1402, UN 2211,++	data
KRC 103 M	Si-N+R	=KRC 101M: Rb=Rbe=22kΩ	7c	Kec	AA 1F4M, DTC 124ES, RN 1003, UN 4212,++	data
KRC 103 S	Si-N+R	=KRC 103M: SMD	35a		FA 1F4M, DTC 124EK, RN 1403, UN 2212,++	data
KRC 104 M	Si-N+R	=KRC 101M: Rb=Rbe=47kΩ	7c	Kec	AA 1L4M, DTC 144ES, RN 1004, UN 4213,++	data
KRC 104 S	Si-N+R	=KRC 104M: SMD	35a		FA 1L4M, DTC 144EK, RN 1404, UN 2213,++	data
KRC 105 M	Si-N+R	=KRC 101M: Rb=2,2k, Rbe=47kΩ	7c	Kec	DTC 123JS, RN 1005	data
KRC 105 S	Si-N+R	=KRC 105M: SMD	35a		DTC 123JK, RN 1405	data
KRC 106 M	Si-N+R	=KRC 101M: Rb=4,7k, Rbe=47kΩ	7c	Kec	DTC 143ZS, RN 1006, 2SC4133, 2SC4195	data
KRC 106 S	Si-N+R	=KRC 106M: SMD	35a		DTC 143ZK, RN 1406, 2SC4146	data
KRC 107 M	Si-N+R	=KRC 101M: Rb=10k, Rbe=47kΩ	7c	Kec	AA 1A4P, DTC 114YS, RN 1007, UN 4214,++	data
KRC 107 S	Si-N+R	=KRC 107M: SMD	35a		FA 1A4P, DTC 114YK, RN 1407, UN 2214,++	data
KRC 108 M	Si-N+R	=KRC 101M: Rb=22k, Rbe=47kΩ	7c	Kec	AA 1F4N, DTC 124XS, KSR 1007, RN 1008	data
KRC 108 S	Si-N+R	=KRC 108M: SMD	35a		FA 1F4N, DTC 124XK, KSR 1107, RN 1408	data
KRC 109 M	Si-N+R	=KRC 101M: Rb=47k, Rbe=22kΩ	7c	Kec	AA 1L4L, DTC 144WS, RN 1009, UN 421E,++	data
KRC 109 S	Si-N+R	=KRC 109M: SMD	35a		FA 1L4L, DTC 144WK, RN 1409, UN 221E,++	data
KRC 110 M	Si-N+R	=KRC 101M: Rb=4,7kΩ, Rbe=-	7c	Kec	AA 1L3Z, DTC 143TS, RN 1010, UN 4216,++	data
KRC 110 S	Si-N+R	=KRC 110M: SMD	35a		FA 1L3Z, DTC 143T, RN 1410, UN 2216,++	data
KRC 111 M	Si-N+R	=KRC 101M: Rb=10kΩ, Rbe=-	7c	Kec	AA 1A4Z, DTC 114TS, RN 1011, UN 4215,++	data
KRC 111 S	Si-N+R	=KRC 111M: SMD	35a		FA 1A4Z, DTC 114TK, RN 1411, UN 2215,++	data
KRC 112 M	Si-N+R	=KRC 101M: Rb=100kΩ, Rbe=-	7c	Kec	DTC 115TS	data
KRC 112 S	Si-N+R	=KRC 112M: SMD	35a		DTC 115TK	data
KRC 113 M	Si-N+R	=KRC 101M: Rb=22kΩ, Rbe=-	7c	Kec	AA 1F4Z, DTC 124TS, KSR 1011, UN 4217,++	data
KRC 113 S	Si-N+R	=KRC 113M: SMD	35a		FA 1F4Z, DTC 124TK, KSR 1111, UN 2217,++	data
KRC 114 M	Si-N+R	=KRC 101M: Rb=47kΩ, Rbe=-	7c	Kec	AA 1L4Z, DTC 144TS, KSR 1012, UN 4210,++	data
KRC 114 S	Si-N+R	=KRC 114M: SMD	35a		FA 1L4Z, DTC 144TK, KSR 1112, UN 2210,++	data
KRC 116 M	Si-N+R	=KRC 101M: Rb=1k, Rbe=10kΩ	7c	Kec	AA 1A3Q, DTC 113ZS, UN 4219	data
KRC 116 S	Si-N+R	=KRC 116M: SMD	35a		FA 1A3Q, DTC 113ZK, UN 2219	data
KRC 117 M	Si-N+R	=KRC 101M: Rb=Rbe=2,2kΩ	7c	Kec	DTC 123ES	data
KRC 117 S	Si-N+R	=KRC 117M: SMD	35a		DTC 123EK	data
KRC 118 M	Si-N+R	=KRC 101M: Rb=2,2k, Rbe=10kΩ	7c	Kec	DTC 123YS, UN 421H, 2SC3864	data
KRC 118 S	Si-N+R	=KRC 118M: SMD	35a		DTC 123YK, UN 221H, 2SC3863	data
KRC 119 M	Si-N+R	=KRC 101M: Rb=4,7k, Rbe=10kΩ	7c	Kec	AA 1L3N, DTC 143XS, UN 421F, 2SC4361, ++	data
KRC 119 S	Si-N+R	=KRC 119M: SMD	35a		FA 1L3N, DTC 143XK, UN 221F, 2SC4360, ++	data
KRC 120 M	Si-N+R	=KRC 101M: Rb=10k, Rbe=4,7kΩ	7c	Kec	DTC 114WS, UN 421K	data
KRC 120 S	Si-N+R	=KRC 120M: SMD	35a		DTC 114WK, UN 221K	data
KRC 121 M	Si-N+R	=KRC 101M: Rb=47k, Rbe=10kΩ	7c	Kec	DTC 114VS, UN 421D	data
KRC 121 S	Si-N+R	=KRC 121M: SMD	35a		DTC 144VK, UN 221D	data
KRC 122 M	Si-N+R	=KRC 101M: Rb=100k, Rbe=100kΩ	7c	Kec	DTC 115ES	data
KRC 122 S	Si-N+R	=KRC 122M: SMD	35a		DTC 115EK	data
KRA 231 M	Si-N+R	S, Rb=2,2kΩ, Rbe=-, 50V, 0,6A, 200MHz	7c	Kec	DTC 323TS, DTD 123TS	data
KRA 231 S	Si-N+R	=KRC 231M: SMD	35a		DTC 323TK, DTD 123TK	data
KRA 232 M	Si-N+R	=KRC 231M: Rb=5,6kΩ, Rbe=-	7c	Kec		data
KRA 232 S	Si-N+R	=KRC 232M: SMD	35a			data
KRA 233 M	Si-N+R	=KRC 231M: Rb=10kΩ, Rbe=-	7c	Kec	DTC 314TS, DTD 114TS	data
KRA 233 S	Si-N+R	=KRC 233M: SMD	35a		DTC 314TK, DTD 114TK	data
KRA 234 M	Si-N+R	=KRC 231M: Rb=4,7kΩ, Rbe=-	7c	Kec	DTC 343TS, DTD 143TS	data
KRA 234 S	Si-N+R	=KRC 234M: SMD	35a		DTC 343TK, DTD 143TK	data
KRA 235 M	Si-N+R	=KRC 231M: Rb=6,8kΩ, Rbe=-	7c	Kec	DTC 363TS, DTD 163TS	data
KRA 235 S	Si-N+R	=KRC 235M: SMD	35a		DTC 363TK, DTD 163TK	data
KRC 241 M	Si-N+R	S, Rb=Rbe=1kΩ, 50V, 0,8A, 200MHz	7c	Kec	-	data

Type	Device	Short Description	Fig.	Manu	Comparision Types	More at
KRC 242 M	Si-N+R	=KRC 241M: Rb=Rbe=2,2kΩ	7c	Kec	-	data
KRC 243 M	Si-N+R	=KRC 241M: Rb=Rbe=4,7kΩ	7c	Kec	-	data
KRC 244 M	Si-N+R	=KRC 241M: Rb=Rbe=10kΩ	7c	Kec	-	data
KRC 245 M	Si-N+R	=KRC 241M: Rb=1k, Rbe=10kΩ	7c	Kec	-	data
KRC 246 M	Si-N+R	=KRC 241M: Rb=2,2k, Rbe=10kΩ	7c	Kec	-	data
KRA 241S..246S	Si-N+R	=KRC 241M..246M: SMD	35a			data
KRC 401	Si-N+R	=KRC 101M: SMD	35a(2mm)	Kec	DTC 143EU, RN 1301, UN 521L	data
KRC 401 E	Si-N+R	=KRC 101M: SMD	35a(1,6mm)	Kec	RN 1101, UN 921L	data
KRC 402	Si-N+R	=KRC 102M: SMD	35a(2mm)	Kec	DTC 114EU, RN 1302, UN 5211, 2SC4398	data
KRC 402 E	Si-N+R	=KRC 102M: SMD	35a(1,6mm)	Kec	RN 1102, UN 9211	data
KRC 403	Si-N+R	=KRC 103M: SMD	35a(2mm)	Kec	DTC 124EU, RN 1303, UN 5212, 2SC4397	data
KRC 403 E	Si-N+R	=KRC 103M: SMD	35a(1,6mm)	Kec	RN 1103, UN 9212	data
KRC 404	Si-N+R	=KRC 104M: SMD	35a(2mm)	Kec	DTC 144EU, RN 1304, UN 5213, 2SC4396	data
KRC 404 E	Si-N+R	=KRC 104M: SMD	35a(1,6mm)	Kec	RN 1104, UN 9213	data
KRC 405	Si-N+R	=KRC 105M: SMD	35a(2mm)	Kec	DTC 123JU, RN 1305	data
KRC 405 E	Si-N+R	=KRC 105M: SMD	35a(1,6mm)	Kec	RN 1105	data
KRC 406	Si-N+R	=KRC 106M: SMD	35a(2mm)	Kec	DTC 143ZU, RN 1306	data
KRC 406 E	Si-N+R	=KRC 106M: SMD	35a(1,6mm)	Kec	RN 1106	data
KRC 407	Si-N+R	=KRC 107M: SMD	35a(2mm)	Kec	DTC 114YU, RN 1307, UN 5214	data
KRC 407 E	Si-N+R	=KRC 107M: SMD	35a(1,6mm)	Kec	RN 1107, UN 9214	data
KRC 408	Si-N+R	=KRC 108M: SMD	35a(2mm)	Kec	DTC 124XU, RN 1308	data
KRC 408 E	Si-N+R	=KRC 108M: SMD	35a(1,6mm)	Kec	RN 1108	data
KRC 409	Si-N+R	=KRC 109M: SMD	35a(2mm)	Kec	DTC 144WU, RN 1309, UN 521E	data
KRC 409 E	Si-N+R	=KRC 109M: SMD	35a(1,6mm)	Kec	RN 1109	data
KRC 410	Si-N+R	=KRC 110M: SMD	35a(2mm)	Kec	DTC 143TU, RN 1310, UN 5216	data
KRC 410 E	Si-N+R	=KRC 110M: SMD	35a(1,6mm)	Kec	RN 1110, UN 9216	data
KRC 411	Si-N+R	=KRC 111M: SMD	35a(2mm)	Kec	DTC 114TU, RN 1311, UN 5215	data
KRC 411 E	Si-N+R	=KRC 111M: SMD	35a(1,6mm)	Kec	RN 1111, UN 9215	data
KRC 412	Si-N+R	=KRC 112M: SMD	35a(2mm)	Kec	DTC 115TU	data
KRC 412 E	Si-N+R	=KRC 112M: SMD	35a(1,6mm)	Kec	UNR 921BJ	data
KRC 413	Si-N+R	=KRC 113M: SMD	35a(2mm)	Kec	DTC 124TU, RN 1312, UN 5217	data
KRC 413 E	Si-N+R	=KRC 113M: SMD	35a(1,6mm)	Kec	RN 1112, UN 9217	data
KRC 414	Si-N+R	=KRC 114M: SMD	35a(2mm)	Kec	DTC 144TU, RN 1313, UN 5210	data
KRC 414 E	Si-N+R	=KRC 114M: SMD	35a(1,6mm)	Kec	RN 1113, UN 9210	data
KRC 416	Si-N+R	=KRC 116M: SMD	35a(2mm)	Kec	DTC 113ZU, UN 5219	data
KRC 416 E	Si-N+R	=KRC 116M: SMD	35a(1,6mm)	Kec	UN 9219	data
KRC 417	Si-N+R	=KRC 117M: SMD	35a(2mm)	Kec	DTC 123EU	data
KRC 417 E	Si-N+R	=KRC 117M: SMD	35a(1,6mm)	Kec		data
KRC 418	Si-N+R	=KRC 118M: SMD	35a(2mm)	Kec	DTC 123YU, UN 521H, 2SC4498	data
KRC 418 E	Si-N+R	=KRC 118M: SMD	35a(1,6mm)	Kec	UN 921H	data
KRC 419	Si-N+R	=KRC 119M: SMD	35a(2mm)	Kec	DTC 143XU, UN 521F	data
KRC 419 E	Si-N+R	=KRC 119M: SMD	35a(1,6mm)	Kec	UN 921F	data
KRC 420	Si-N+R	=KRC 120M: SMD	35a(2mm)	Kec	DTC 114WU, UN 521K	data
KRC 420 E	Si-N+R	=KRC 120M: SMD	35a(1,6mm)	Kec	UN 921K	data
KRC 421	Si-N+R	=KRC 121M: SMD	35a(2mm)	Kec	DTC 144VU, UN 521D	data
KRC 421 E	Si-N+R	=KRC 121M: SMD	35a(1,6mm)	Kec	UN 921D	data
KRC 422	Si-N+R	=KRC 122M: SMD	35a(2mm)	Kec	DTC 115EU	data
KRC 422 E	Si-N+R	=KRC 122M: SMD	35a(1,6mm)	Kec	UNR 921AJ	data
KRC 651 U	Si-N+R	=KRC 101M: SMD, Dual	45ba1(2mm)	Kec	-	data
KRC 652 U	Si-N+R	=KRC 102M: SMD, Dual	45ba1(2mm)	Kec	-	data
KRC 653 U	Si-N+R	=KRC 103M: SMD, Dual	45ba1(2mm)	Kec	-	data
KRC 654 U	Si-N+R	=KRC 104M: SMD, Dual	45ba1(2mm)	Kec	-	data
KRC 655 U	Si-N+R	=KRC 105M: SMD, Dual	45ba1(2mm)	Kec	-	data
KRC 656 U	Si-N+R	=KRC 106M: SMD, Dual	45ba1(2mm)	Kec	-	data
KRC 657 U	Si-N+R	=KRC 107M: SMD, Dual	45ba1(2mm)	Kec	-	data
KRC 658 U	Si-N+R	=KRC 108M: SMD, Dual	45ba1(2mm)	Kec	-	data
KRC 659 U	Si-N+R	=KRC 109M: SMD, Dual	45ba1(2mm)	Kec	-	data
KRC 660 U	Si-N+R	=KRC 110M: SMD, Dual	45ba1(2mm)	Kec	-	data
KRC 661 U	Si-N+R	=KRC 111M: SMD, Dual	45ba1(2mm)	Kec	-	data
KRC 662 U	Si-N+R	=KRC 112M: SMD, Dual	45ba1(2mm)	Kec	-	data
KRC 663 U	Si-N+R	=KRC 113M: SMD, Dual	45ba1(2mm)	Kec	-	data
KRC 664 U	Si-N+R	=KRC 114M: SMD, Dual	45ba1(2mm)	Kec	-	data
KRC 666 U	Si-N+R	=KRC 116M: SMD, Dual	45ba1(2mm)	Kec	-	data
KRC 667 U	Si-N+R	=KRC 117M: SMD, Dual	45ba1(2mm)	Kec	-	data
KRC 668 U	Si-N+R	=KRC 118M: SMD, Dual	45ba1(2mm)	Kec	-	data
KRC 669 U	Si-N+R	=KRC 119M: SMD, Dual	45ba1(2mm)	Kec	-	data
KRC 670 U	Si-N+R	=KRC 120M: SMD, Dual	45ba1(2mm)	Kec	-	data
KRC 671 U	Si-N+R	=KRC 121M: SMD, Dual	45ba1(2mm)	Kec	-	data
KRC 672 U	Si-N+R	=KRC 122M: SMD, Dual	45ba1(2mm)	Kec	-	data
KRC 851 U	Si-N+R	=KRC 101M: SMD, Dual	46bh1(2mm)	Kec	-	data
KRC 852 U	Si-N+R	=KRC 102M: SMD, Dual	46bh1(2mm)	Kec	-	data
KRC 853 U	Si-N+R	=KRC 103M: SMD, Dual	46bh1(2mm)	Kec	-	data
KRC 854 U	Si-N+R	=KRC 104M: SMD, Dual	46bh1(2mm)	Kec	-	data
KRC 855 U	Si-N+R	=KRC 105M: SMD, Dual	46bh1(2mm)	Kec	-	data
KRC 856 U	Si-N+R	=KRC 106M: SMD, Dual	46bh1(2mm)	Kec	-	data
KRC 857 U	Si-N+R	=KRC 107M: SMD, Dual	46bh1(2mm)	Kec	-	data
KRC 858 U	Si-N+R	=KRC 108M: SMD, Dual	46bh1(2mm)	Kec	-	data
KRC 859 U	Si-N+R	=KRC 109M: SMD, Dual	46bh1(2mm)	Kec	-	data
KRC 860 U	Si-N+R	=KRC 110M: SMD, Dual	46bh1(2mm)	Kec	-	data
KRC 861 U	Si-N+R	=KRC 111M: SMD, Dual	46bh1(2mm)	Kec	-	data
KRC 862 U	Si-N+R	=KRC 112M: SMD, Dual	46bh1(2mm)	Kec	-	data
KRC 863 U	Si-N+R	=KRC 113M: SMD, Dual	46bh1(2mm)	Kec	-	data
KRC 864 U	Si-N+R	=KRC 114M: SMD, Dual	46bh1(2mm)	Kec	-	data
KRC 866 U	Si-N+R	=KRC 116M: SMD, Dual	46bh1(2mm)	Kec	-	data
KRC 867 U	Si-N+R	=KRC 117M: SMD, Dual	46bh1(2mm)	Kec	-	data

Type	Device	Short Description	Fig.	Manu	Comparision Types	More at
KRC 868 U	Si-N+R	=KRC 118M: SMD, Dual	46bh1(2mm)	Kec	-	data
KRC 869 U	Si-N+R	=KRC 119M: SMD, Dual	46bh1(2mm)	Kec	-	data
KRC 870 U	Si-N+R	=KRC 120M: SMD, Dual	46bh1(2mm)	Kec	-	data
KRC 871 U	Si-N+R	=KRC 121M: SMD, Dual	46bh1(2mm)	Kec	-	data
KRC 872 U	Si-N+R	=KRC 122M: SMD, Dual	46bh1(2mm)	Kec	-	data
KRX 101 U	Si-N/P+R	=KRC 102M + KRA 102M: SMD	45ba4(2mm)	Kec	-	data
KRX 102 U	Si-N/P+R	=KRC 104M + KRA 107M: SMD	45ba4(2mm)	Kec	-	data
KRX 103 U	Si-N/P+R	=KRC 104M + KRA 103M: SMD	45ba4(2mm)	Kec	-	data
KRX 104 U	Si-N/P+R	=KRC 104M + KRA 106M: SMD	45ba4(2mm)	Kec	-	data
KRX 201 U	Si-N/P+R	=KRC 102M + KRA 102M: SMD	46bh1(2mm)	Kec	-	data
KRX 202 U	Si-N/P+R	=KRC 103M + KRA 103M: SMD	46bh1(2mm)	Kec	-	data
KRX 203 U	Si-N/P+R	=KRC 104M + KRA 104M: SMD	46bh1(2mm)	Kec	-	data
KRX 204 U	Si-N/P+R	=KRC 105M + KRA 105M: SMD	46bh1(2mm)	Kec	-	data
KS-2529	LED	srd+rd/gr+rd/yl+gr/yl 2.4x1.5mm		Kin		data pinout
KS-3020	LED	srd+gr+yl 2x1.6mm I>5mcd Vf:2.1V		Kin		data pinout
KS	Si-P	=2SA1813 (Typ.-Code/Stempel/marking)	35(2mm)	Say	→2SA1813	data
KS	Si-P	=2SA1814 (Typ-Code/Stempel/marking)	35	Say	→2SA1814	data
KS	Si-N	=2SD2533 (Typ-Code/Stempel/marking)	39	Hit	→2SD2533	data
KS	MOS-P-FET-e	=2SJ345 (Typ-Code/Stempel/marking)	35	Tos	→2SJ345	data
KS	MOS-P-FET-e	=2SJ346 (Typ-Code/Stempel/marking)	35(2mm)	Tos	→2SJ346	data
KS	MOS-P-FET-e	=2SJ347 (Typ-Code/Stempel/marking)	35(1,6mm)	Tos	→2SJ347	data
KS	MOS-N-FET-e*	=2SK3119 (Typ-Code/Stempel/marking)	39	Say	→2SK3119	data
KS	MOS-P-FET-e*	=HN 1J02FU (Typ-Code/Stempel/marking)	46(2mm)	Tos	→HN 1J02FU	data
KS	Z-Di	=P6 SMB-12 (Typ-Code/Stempel/marking)	71(5x3,5)	Fag	→P6 SMB-12	data
KS	Z-Di	=SMBJ 8.5 (Typ-Code/Stempel/marking)	71(5x3,5)	Mop	→SMBJ ...	data
KS 25 C02IN	CMOS-IC	8 Bit Successive Approximation Regist.	16-DIP	Sam	-	
KS 25 C03IN	CMOS-IC	8 Bit Successive Approximation Regist.	16-DIP	Sam	-	
KS 25 C04IN	CMOS-IC	12 Bit Successive Approximation Regist	24-SDIP	Sam	-	
KS 56 C820	CMOS-µC-IC	SMD, Cd, 4 Bit µComp., LCD-Tr, I/O	80-MP	Sam	-	
KS 58 C05	CMOS-IC	=KS 5851		Sam	→KS 5851	
KS 58 D05	CMOS-IC	=KS 5852		Sam	→KS 5852	
KS 58 E05	CMOS-IC	=KS 5853		Sam	→KS 5853	
KS 093	CMOS-IC	Telecom, Crosspoint Switch		Sam	GM 62093, M 093, MT 8812	
KS 0118 B,C	CMOS-IC	SMD, Generation Lock	80-MP	Sam	-	
KS 0118 B	CMOS-IC	SMD, Generation Lock	80-MP	Sam	-	
KS 0119	MOS-IC	SMD, NTSC Encoder f. Multimedia Video	80-MP	Sam	-	
KS 0119 Q2	MOS-IC	=KS 0119:	100-MP	Sam	-	
KS 0122	MOS-IC	SMD,Multistd. Video-Dec. f. Multimedia	100-MP	Sam	-	
KS 0123	MOS-IC	Set-top Box Encoder f. Multimedia	44-PLCC	Sam	-	pdf
KS 0125	MOS-IC	Video CD Encoder + OSD f. Multimedia	80-MP	Sam	-	pdf
KS 0127(B)	MOS-IC	SMD,Multistd. Video-Dec. f. Multimedia	100-MP	Sam	-	
KS 0127	MOS-IC	SMD,Multistd. Video-dec. f. Multimedia	100-MP	Sam	-	pdf
KS 0143	MOS-IC	Image Compression Proc. f. Multimedia	208-FLP	Sam		
KS 0144	MOS-IC	Motion Estimation Proc. f. Multimedia	208-FLP	Sam		
KS 0161(Q)	MOS-IC	SMD, Sound Synthesizer f. Multimedia	160-MP	Sam	-	
KS 0164	MOS-IC	SMD, Sound Synthesizer f. Multimedia	100-MP	Sam	-	
KS 0165	MOS-IC	SMD, Sound Synthesizer f. Multimedia	100-MP	Sam	-	
KS 0174-1M	MOS-IC	1 MB Mask-ROM f. KS 0164	44-MDIP	Sam	-	
KS 0174-2M	MOS-IC	2 MB Mask-ROM f. KS 0164	44-MDIP	Sam	-	
KS 272	CMOS-OP-IC	Dual, lo-power, Udd=3...16V	8-DIP	Sam	TLC 272...	
KS 274	CMOS-OP-IC	Quad, lo-power, Udd=3...16V	14-DIP	Sam	TLC 274...	
KS 555 D	CMOS-IC	=KS 555: SMD	8-MDIP	Sam	LMC 555CM	pinout
KS 555 HD	CMOS-IC	=KS 555: hi-speed, SMD, 0..+70°	8-MDIP	Sam	LMC 555M	pinout
KS 555 H(N)	CMOS-IC	=KS 555: hi-speed, 0..+70°	8-DIP	Sam	TLC 555	pinout
KS 555(N)	CMOS-IC	=NE 555: CMOS-Version, -20..+85°	8-DIP	Sam	ICM 7555, LMC 555CN	pinout
KS 556 D	CMOS-IC	=KS 556: SMD	14-MDIP	Sam		
KS 556(N)	CMOS-IC	=NE 556: CMOS-Version, 0..+70°	14-DIP	Sam	ICM 7556, TLC 556	
KS 823 C04	Si-Di	Dual, Schottky, S-L, 40V, 5A(Tc=87°)	30p§	Fjd	DE 5SC4M, MA 3U749, KS 823C04	data
KS 826 S04	Si-Di	Schottky, S-L, 40V, 5A(Tc=92°)	30f§	Fjd	PBYR 740D, SPB-G54S, STPS 1045B	data
KS 926 S2	Si-Di	S-L, 200V, 5A(Tc=106°), <35ns	30f§	Fjd	BYW 29ED-200	data
KS 1461	IC	Rf Signal Processor	QFP	Sam		pdf pinout
KS 2145 B	MOS-IC	Monitor, 7-Mode Selector & Power Save	24-SDIP	Sam	-	
KS 2146 B	MOS-IC	Monitor, 11-Mode Selector & Power Save	24-SDIP	Sam	-	
KS 5112	CMOS-IC	LCD Uhr/watch, 1,5V, 0,8µA		Sam	-	
KS 5113	CMOS-IC	LCD Uhr/watch, 1,5V, 1,5µA		Sam	-	
KS 5114	CMOS-IC	LCD Uhr/watch, 1,5V, 0,8µA		Sam	-	
KS 5116	MOS-IC	6 Digit Up/Down Counter, 1,5V, 9µA	Chip	Sam	-	
KS 5171	CMOS-IC	LCD Uhr, Alarm/clock, 1,5V, 1,5µA		Sam	-	
KS 5172	CMOS-IC	LCD Uhr, Alarm/clock, 1,5V, 1µA		Sam	-	
KS 5184	CMOS-IC	LCD Uhr, Alarm/watch, 1,5V, 1,5µA		Sam	-	
KS 5189	CMOS-IC	LCD Uhr, Alarm/watch, 1,5V, 1µA		Sam	-	
KS 5190	CMOS-IC	LCD Uhr, Alarm/watch, 1,5V, 1µA		Sam	-	
KS 5194	CMOS-IC	LCD Uhr, Alarm/watch, 1,5V, 1µA		Sam	-	
KS 5198	CMOS-IC	LCD Uhr/watch, 1,5V, 0,8µA		Sam	-	
KS 5199 A	CMOS-IC	LCD Uhr/watch, 1,5V, 0,8µA		Sam	-	
KS 5205	CMOS-IC	Analoguhr, Alarm/analog clock	14-DIP	Sam	-	
KS 5206	CMOS-IC	Analoguhr, Alarm/analog clock	8-DIP	Sam	-	
KS 5207	CMOS-IC	Analoguhr/analog clock	8-DIP	Sam	-	
KS 5209	CMOS-IC	Analoguhr, Alarm/analog clock	8-DIP	Sam	-	
KS 5210	CMOS-IC	Analoguhr, Alarm/analog clock	8-DIP	Sam	-	
KS 5211	CMOS-IC	Analoguhr/analog clock	8-DIP	Sam	-	
KS 5243	CMOS-IC	Analoguhr/analog watch		Sam	-	

Type	Device	Short Description	Fig.	Manu	Comparision Types	More at
KS 5310(A)	MOS-IC	Melodie/melody IC	Chip	Sam	-	
KS 5311(A)	MOS-IC	Melodie/melody IC	8-DIP	Sam	-	
KS 5313 N...T	CMOS-IC	Melodie/melody IC	8-,16-DIP	Sam	-	
KS 5314	CMOS-IC	Melodie/melody IC		Sam	-	
KS 5340	CMOS-IC	Melodie/melody IC		Sam	-	
KS 5380	CMOS-IC	Melodie/melody IC		Sam	-	
KS 5381	CMOS-IC	Melodie/melody IC	24-DIP	Sam	-	
KS 5401(A)	CMOS-IC	Melodie/melody IC	18-DIP	Sam	-	
KS 5410	CMOS-IC	SMD, IR-FB, Sender/transmitter	20-MDIP	Sam	TC 9012F	
KS 5512	CMOS-IC	TV, Screen Display Prozessor	32-SDIP	Sam	M 50554	
KS 5513	CMOS-IC	TV, On-screen Display, Synchr. Sep.	32-SDIP	Sam		
KS 5514 B-XX	CMOS-IC	TV, Screen Display Prozessor	SMDIP	Sam		pdf pinout
KS 5520-01	CMOS-IC	TV, Screen Display Prozessor	MDIP	Sam		pdf pinout
KS 5520 D-01	CMOS-IC	TV, Screen Display Prozessor	SMDIP	Sam		pdf pinout
KS 5706	CMOS-IC	Transceiver f. RS232		Sam	MC 145406	
KS 5788	CMOS-I/O-IC	Quad,Leitungstr./Line Driver f. RS232C	14-DIP	Sam	DS 14C88	
KS 5788 D	CMOS-I/O-IC	=KS 5788: SMD	14-MDIP	Sam	-	
KS 5789 A	CMOS-I/O-IC	4x Leitungsempf./Line Rcv. f. RS232C	14-DIP	Sam	DS 14C89A	
KS 5789 D	CMOS-I/O-IC	=KS 5789: SMD	14-MDIP	Sam	-	
KS 5803 A	CMOS-IC	IR-FB, Sender/Transmitter, 736 Codes	16-DIP	Sam	µPD 1913C	pinout
KS 5803 B	CMOS-IC	IR-FB, Sender/Transmitter, 8960 Codes	20-MDIP	Sam	µPD 1943G	pinout
KS 5804	CMOS-IC	Telefon, Impulswahl/dig. Pulse Dialer	16-DIP	Sam	LR 40981, MK 50981, TP 50981	
KS 5805 A	CMOS-IC	Telefon, Impulswahl/dig. Pulse Dialer	18-DIP	Sam	LR 40992, MK 50992, T 40992	pdf
KS 5805 B	CMOS-IC	Telefon, Impulswahl/dig. Pulse Dialer	18-DIP	Sam	LR 40993, MK 50993, T 40993	pdf
KS 5806	CMOS-IC	Telefon, Wählspeicher/dialer Memory	18-DIP	Sam	MK 5177	
KS 5808(A)	CMOS-IC	Telefon, Zweitonwahl/dual Tone Dialer	16-DIP	Sam	LR4089, MK5089, S25089, SBA5089, UM95089	
KS 5809	CMOS-IC	Telefon, Zweitonwahl/dual Tone Dialer	16-DIP	Sam	-	
KS 5810	CMOS-IC	Telefon, Zweitonwahl/dual Tone Dialer	16-DIP	Sam	-	
KS 5811	CMOS-IC	Telefon, Zweitonwahl/dual Tone Dialer	16-DIP	Sam	-	
KS 5812	CMOS-I/O-IC	Quad Uart	40-DIP	Sam	-	
KS 5314	MOS-IC	Melodie/melody IC	14-DIP	Sam	-	
KS 5814	MOS-IC	Melodie/melody IC	14-DIP	Sam	-	
KS 5815	MOS-IC	Fieberthermometer/clinical Thermometer	Chip	Sam	-	
KS 5819	CMOS-IC	Telefon-impulswahl/tone-pulse Dialer	22-(S)DIP	Sam	S 7230A/B, UM 91230	
KS 5820	CMOS-IC	Telefon-impulswahl/tone-pulse Dialer	18-DIP	Sam	LC 7360, UM 91210	
KS 5821	CMOS-IC	Telefon-impulswahl/tone-pulse Dialer	22-(S)DIP	Sam		
KS 5823	CMOS-IC	Telefon, Pulse/dual Tone Dialer	18-DIP	Sam		
KS 5824	I/O-IC	UART	24-DIP	Sam	HD 6350, MC 6850	
KS 5851	CMOS-IC	Telefon,Impulswahl/pulse Dialer+redial	18-DIP	Sam		
KS 5852	CMOS-IC	Telefon,Impulswahl/pulse Dialer+redial	18-DIP	Sam		
KS 5853	CMOS-IC	Telefon,Impulswahl/pulse Dialer+redial	16-DIP	Sam		
KS 5901 A	CMOS-IC	SMD, Sprachausgabe/speech synthesis	60-MP	Sam		
KS 5902	CMOS-IC	Sprachausgabe/speech synthesis	24-DIP	Sam		
KS 5911	CMOS-IC	SMD, Voice Recording & Reproducing	48-MP	Sam	-	pdf
KS 5912	CMOS-IC	Voice & Sound Generation, 64kB ROM	16-DIP	Sam	-	
KS 5990	CMOS-IC	SMD, Cd, Digital Signal-prozessor	80-MP	Sam	-	
KS 5991	CMOS-IC	SMD, Cd, Digital Signal-prozessor	80-MP	Sam	-	
KS 6022 A	MOS-IC	Taschenrechner/Calculator, 3V, 25µA	48-MP	Sam	-	
KS 6025	MOS-IC	Taschenrechner/Calculator, 1,5V, 6µA	48-MP	Sam	-	
KS 6026	MOS-IC	Taschenrechner/Calculator, 1,5V, 4µA	48-MP	Sam	-	
KS 6027	MOS-IC	Tischrechner/Calculator, 1,5V, 6µA		Sam	-	
KS 6028	MOS-IC	Taschenrechner/Calculator, 1,5V, 1,3µA	48-MP	Sam	-	
KS 6029	MOS-IC	Taschenrechner/Calculator, 1,5V, 1,5µA	48-MP	Sam	-	
KS 6041	MOS-IC	Taschenrechner/Calculator, 3V, 15µA	60-MP	Sam	-	
KS 7126 CN	A/D-IC	A/D-Converter, 3,5 digit	40-DIP	Sam	TSC 7126	
KS 7212	CMOS-IC	SMD, Ccd B/w Camera, Timing/sync. Gen.	48-MP	Sam	-	pdf
KS 7213	CMOS-IC	SMD, Ccd Col.camera, Timing/sync. Gen.	80-MP	Sam	-	
KS 7214	CMOS-IC	SMD, Ccd B/w Camera, Timing/sync. Gen.	40-MP	Sam	-	pdf
KS 7220	CMOS-IC	SMD, Ccd Camera, Vertical Driver	20-MDIP	Sam	-	
KS 7221 D	CMOS-IC	SMD, Ccd Camera, Vertical Driver	20-SMDIP	Sam	-	pdf
KS 7301 B	CMOS-IC	SMD, Dig. Camcorder, NTSC/PAL Proz.	160-MP	Sam	-	pdf
KS 7306	CMOS-IC	SMD, Ccd Color Camera, NTSC/PAL Proz.	100-MP	Sam	-	pdf
KS 7308	CMOS-IC	SMD, Camcorder, Frequ. Conver. Memory	48-MP	Sam	-	pdf
KS 7314	CMOS-IC	SMD, Ccd/camcorder, Digital Zoom	80-MP	Sam	-	pdf
KS 7719 NZ	CMOS-IC	1,9V Spg.-/Voltage Detector,nFET Outp	7	Sam		
KS 7720 CZ	CMOS-IC	2V Spg.-/Voltage Detector, Cmos Outp.	7	Sam		
KS 7721 NZ	CMOS-IC	2,1V Spg.-/Voltage Detector,nFET Outp.	7	Sam		
KS 7723 CZ	CMOS-IC	2,3V Spg.-/Voltage Detector,Cmos Outp.	7	Sam		
KS 7724 NZ	CMOS-IC	2,4V Spg.-/Voltage Detector,nFET Outp.	7	Sam		
KS 7725 CZ	CMOS-IC	2,5V Spg.-/Voltage Detector,Cmos Outp.	7	Sam		
KS 7727 CZ	CMOS-IC	2,7V Spg.-/Voltage Detector,Cmos Outp.	7	Sam		
KS 7727 NZ	CMOS-IC	2,7V Spg.-/Voltage Detector,nFET Outp.	7	Sam		
KS 7730 CZ	CMOS-IC	2,95V Spg.-/Voltage Detector,Cmos Outp	7	Sam		
KS 7733 CZ	CMOS-IC	3,25V Spg.-/Voltage Detector,Cmos Outp	7	Sam		
KS 7736 CZ	CMOS-IC	3,55V Spg.-/Voltage Detector,Cmos Outp	7	Sam		
KS 7736 NZ	CMOS-IC	3,55V Spg.-/Voltage Detector,nFET Outp	7	Sam		
KS 7740 CZ	CMOS-IC	4V Spg.-/Voltage Detector, Cmos Outp.	7	Sam		
KS 7740 NZ	CMOS-IC	4V Spg.-/Voltage Detector, nFET Outp.	7	Sam		
KS 7742 CZ	CMOS-IC	4,15V Spg.-/Voltage Detector,Cmos Outp	7	Sam		
KS 7745 CZ	CMOS-IC	4,45V Spg.-/Voltage Detector,Cmos Outp	7	Sam		
KS 7746 NZ	CMOS-IC	4,6V Spg.-/Voltage Detector,nFET Outp	7	Sam		
KS 7748 CZ	CMOS-IC	4,75V Spg.-/Voltage Detector,Cmos Outp	7	Sam		
KS 7751 CZ	CMOS-IC	5,10V Spg.-/Voltage Detector,Cmos Outp	7	Sam	-	
KS 8695 P	IC	Integrated Multi-port PCI Gateway Solution	GRID	Mcr	-	pdf

Type	Device	Short Description	Fig.	Manu	Comparision Types	More at
KS 8695 PX	IC	Integrated Multi-port PCI Gateway Solution	GRID	Mcr	-	pdf
KS 8695 X	IC	Integrated Multi-port High-performance Gateway Solution	MP	Mcr	-	pdf
KS 8721 B	IC	2.5V 10/100BasTX/FX MII Physical Layer Transceiver	TSOP	Mcr	-	pdf pinout
KS 8721 BL(I)	IC	3.3V Single Power Supply 10/100BASE-Tx/fx MII Physical Layer Transceiver	MP	Mcr	-	pdf pinout
KS 8721 BT	IC	2.5V 10/100BasTX/FX MII Physical Layer Transceiver	MP	Mcr	-	pdf pinout
KS 8721 CL	IC	3.3V Single Power Supply 10/100BASE-Tx/fx MII Physical Layer Transceiver	MP	Mcr	-	pdf pinout
KS 8721 SL(I)	IC	3.3V Single Power Supply 10/100BASE-Tx/fx MII Physical Layer Transceiver	TSOP	Mcr	-	pdf pinout
KS 8737	IC	3.3V 10/100BaseTX/FX MII Physical Layer Transceiver	MP	Mcr	-	pdf pinout
KS 8802 D	CMOS-IC	=KS 8802: SMD	16-MDIP	Sam	-	
KS 8802(N)	CMOS-IC	Telefon, Kanalwahl/channel Select(15)	16-DIP	Sam	-	
KS 8993 (I)	IC	3-Port 10/100 Igtegrated Switch	MP	Mcr	-	pdf pinout
KS 8993 F/FL	IC	Fast Ethernet Media Converter with TS-1000 OAM	MP	Mcr	-	pdf pinout
KS 8993 M(L/I)	IC	Integrated 3-Port 10/100 Managed Switch with Phys	MP	Mcr	-	pdf pinout
KS 8995 E	IC	5-Port 10/100 Integrated Switch with PHY and Frame Buffer	MP	Mcr	-	pdf pinout
KS 8995 M(I)	IC	Integrated 5-Port 10/100 Managed Switch	MP	Mcr	-	pdf pinout
KS 8995 MA/FQ(I)	CMOS-IC	Integrated 5-Port 10/100 Managed Switch	MP	Mcr	-	pdf pinout
KS 8995 X	CMOS-IC	Integrated 5-Port 10/100 QoS Switch	MP	Mcr	-	pdf pinout
KS 8995 XA	CMOS-IC	Integrated 5-Port 10/100 QoS Switch	MP	Mcr	-	pdf pinout
KS 8997	CMOS-IC	8-Port 10/100 Integrated Switch w. PHY & Frame Buffer	MP	Mcr	-	pdf pinout
KS 8999 (I)	CMOS-IC	Integrated 9-Port 10/100 Switch w. PHY & Frame Buffer	MP	Mcr	-	pdf
KS 9210	CMOS-IC	SMD, Cd, Digital Signal-prozessor	80-MP	Sam	CXD 1167	pdf
KS 9211(B)	CMOS-IC	SMD, Cd, Digital Signal-prozessor	80-MP	Sam	-	
KS 9241 B	CMOS-IC	SMD, Cd, Real-time Error Correction	80-MP	Sam	-	
KS 9282 B	CMOS-D/A-IC	SMD, Cd, Digital Audio Signal Proz.	80-MP	Sam	-	
KS 9283	CMOS-sRAM-IC	SMD, Cd, Digital Audio Signal Proz.	80-MP	Sam	-	
KS 9284	CMOS-sRAM-IC	SMD, Cd, Digital Audio Signal Proz.	80-MP	Sam	-	
KS 9286	CMOS-D/A-IC	SMD, Cd, Digital Audio Signal Proz.	80-MP	Sam	-	
KS 9801	CMOS-IC	Cd,Tv++, IR-FB-empf./receiver	16-DIP	Sam	-	pdf
KS 9802	CMOS-IC	Cd,Tv++, IR-FB-sender/transmitter	16-DIP	Sam	-	pdf
KS 9803	CMOS-IC	Cd,Tv++, Ir-fb-empf./receiver	22-DIP	Sam	-	pdf
KS 51000...51020	CMOS-µC-IC	4 Bit, 1024x4 Bit ROM, LED Drive	28-,40-DIP	Sam	-	
KS 51200	CMOS-µC-IC	4 Bit, 1024x4 Bit ROM, LED Drive	28-,40-DIP	Sam	-	
KS 51400...51404	CMOS-µC-IC	4 Bit, 2038x8 Bit ROM, 128x4Bit RAM	40-SDIP	Sam	-	
KS 51500...51504	CMOS-µC-IC	4 Bit, 2038x8 Bit ROM, 128x4Bit RAM	30-SDIP	Sam	-	
KS 52000...52036	CMOS-µC-IC	SMD, 4 Bit, 2268x4 Bit ROM, LCD Drive	60-MP	Sam	-	
KS 56000	µC-IC			Sam	µPD 75104	
KS 58002(E)	CMOS-IC	Telefon, Ton,Pulswahl/tone,Pulse Dial.	22-DIP	Sam	-	
KS 58006	CMOS-IC	Telefon, Ton,Pulswahl/tone,Pulse Dial.	18-DIP	Sam	-	pdf pinout
KS 58006 D	CMOS-IC	=KS 58006: SMD	20-MDIP	Sam	-	pinout
KS 58008	CMOS-IC	Telefon, Ton,Pulswahl/tone,Pulse Dial.	18-DIP	Sam	-	pinout
KS 58010	CMOS-IC	Telefon, Ton,Pulswahl/tone,Pulse Dial.	20-DIP	Sam	-	pinout
KS 58012	CMOS-IC	Telefon, Ton,Pulswahl/tone,Pulse Dial.	22-DIP	Sam	-	pinout
KS 58013	CMOS-IC	Telefon, Ton,Pulswahl/tone,Pulse Dial.	22-DIP	Sam	-	pinout
KS 58014	CMOS-IC	Telefon, Ton,Pulswahl/tone,Pulse Dial.	22-DIP	Sam	-	pinout
KS 58015	CMOS-IC	Telefon, Zweitonwahl/dual Tone Dialer	14-DIP	Sam	-	pdf pinout
KS 58015 D	CMOS-IC	=KS 58015: SMD	14-MDIP	Sam	-	pdf pinout
KS 58500	CMOS-IC	Telefon, Wahlwiederh./repertory Dialer	22-SDIP	Sam	-	
KS 58500 E	CMOS-IC	=KS 58500: Fig.→	22-DIP	Sam	-	
KS 58503 D	CMOS-IC	=KS 58503: SMD	20-MDIP	Sam	-	
KS 58503(N)	CMOS-IC	Telefon, Wahlwiederh./repertory Dialer	18-DIP	Sam	-	
KS 58505(N)	CMOS-IC	Telefon, 3 Notruf-Nr./Emergency Dialer	22-DIP	Sam	-	pinout
KS 58506(N)	CMOS-IC	Telefon, 3 Notruf-Nr./Emergency Dialer	18-DIP	Sam	-	
KS 58510(N)	CMOS-IC	Telefon, 3 Notruf-Nr./Emergency Dialer	18-DIP	Sam	-	
KS 58512(N)	CMOS-IC	Telefon, Wahlwiederh./repertory Dialer	18-DIP	Sam	-	
KS 58514(N)	CMOS-IC	Telefon, Wahlwiederh./repertory Dialer	18-DIP	Sam	-	
KS 58517(N)	CMOS-IC	Telefon, Wahlwiederh./repertory Dialer	20-DIP	Sam	-	
KS 58519(E)	CMOS-IC	Telefon, Wahlwiederh./repertory Dialer	22-DIP	Sam	-	
KS 58520(E)	CMOS-IC	Telefon, Wahlwiederh./repertory Dialer	22-DIP	Sam	-	
KS 58521(E)	CMOS-IC	Telefon, Wahlwiederh./repertory Dialer	22-DIP	Sam	-	
KS 58522(N)	CMOS-IC	Telefon, Wahlwiederh./repertory Dialer	20-DIP	Sam	-	
KS 58523(E)	CMOS-IC	Telefon, Wahlwiederh./repertory Dialer	22-DIP	Sam	-	
KS 58525(E)	CMOS-IC	Telefon, Wahlwiederh./repertory Dialer	22-DIP	Sam	-	
KS 58526(E)	CMOS-IC	Telefon, Wahlwiederh./repertory Dialer	22-DIP	Sam	-	
KS 58527(E)	CMOS-IC	Telefon, Wahlwiederh./repertory Dialer	22-DIP	Sam	-	
KS 58530(E)	CMOS-IC	Telefon, Wahlwiederh./repertory Dialer	22-DIP	Sam	-	
KS 58531 D	CMOS-IC	=KS 58531(N): SMD	28-MDIP	Sam	-	pinout
KS 58531(N)	CMOS-IC	Telefon, Wahlwiederh./repertory Dialer	28-DIP	Sam	-	pinout
KS 58535 D	CMOS-IC	=KS 58535(N): SMD	28-MDIP	Sam	-	pinout
KS 58535(N)	CMOS-IC	Telefon, Wahlwiederh./repertory Dialer	28-DIP	Sam	-	pinout
KS 58536 D	CMOS-IC	=KS 58536(N): SMD	28-MDIP	Sam	-	pinout
KS 58536(N)	CMOS-IC	Telefon, Wahlwiederh./repertory Dialer	28-DIP	Sam	-	pinout

Type	Device	Short Description	Fig.	Manu	Comparision Types	More at
KSA						
KSA-2529	LED	rd/gr+rd/yl+gr/yl 2.4x1.5mm Vf:2.2V		Kin		data pinout
KSA 473	Si-P	=2SA473	17j§	Sam	→2SA473	data
KSA 539	Si-P	=2SA539: 0,4W	7e	Sam	→2SA539	data
KSA 542	Si-P	Uni, 30V, 0,05A, 0,25W, 100MHz	7e	Sam	BC 214, BC 258, BC 308, BC 558, 2SB725++	data
KSA 545	Si-P	=2SA545:	7e	Sam	→2SA545	data
KSA 614	Si-P	NF/S-L, 80V, 3A, 25W	17j§	Sam	BD 242B, BD 538, BD 540B, 2SB690, ++	data
KSA 634	Si-P	=2SA634: 2A	13j§	Sam	→2SA634	data
KSA 636	Si-P	=2SA636: 2A	13j§	Sam	→2SA636	data
KSA 640	Si-P	=2SA640:	7e	Sam	→2SA640	data
KSA 642	Si-P	=2SA642:	7e	Sam	→2SA642	data
KSA 643	Si-P	=2SA643:	7e	Sam	→2SA643	data
KSA 707	Si-P	NF-Tr/E, 60V, 0,7A, 0,8W	7e	Sam	BC 327A, BC 638, BC 640, 2SB910, ++	data
KSA 708	Si-P	Uni, 80V, 0,7A, 0,8W, 50MHz	7e	Sam	BC 640, 2SB647, 2SB910, 2SB1116A, ++	data
KSA 709	Si-P	Uni, 160/50V, 0,7A, 0,8W, 50MHz	7e	Sam	2SA1013, 2SA1319, 2SB1212	data
KSA 733	Si-P	Uni, 60V, 0,15A, 0,25W, 180MHz	7e	Sam	BC 212, BC 256, BC 266, BC 556, 2SB725++	data
KSA 812	Si-P	=2SA812	35a	Sam	→2SA812	data
KSA 910	Si-P	NF/S, 150/150V, 0,05A, 0,8W, 100MHz	7c(9mm)	Sam	BF 423, 2SA1019, 2SA1281, 2SA1285A, ++	data
KSA 916	Si-P	NF-Tr/E, 120/120V, 0,8A, 0,9W, 120MHz	7c(9mm)	Sam	2SA965, 2SA1013, 2SB647, 2SB984, ++	data
KSA 928 A	Si-P	NF-Tr/E, 30V, 2A, 1W, 120MHz	7c(9mm)	Sam	MPS 750...751, 2SA1382, 2SB822, 2SB911,++	data
KSA 931	Si-P	NF/S, 80V, 0,7A, 1W, 100MHz	7c(9mm)	Sam	BC 640, 2SA965, 2SB647, 2SB984, ++	data
KSA 940	Si-P	=2SA940	17j§	Sam	→2SA940	data
KSA 952	Si-P	=2SA952	7c	Sam	→2SA952	data
KSA 953	Si-P	=2SA953	7c	Sam	→2SA953	data
KSA 954	Si-P	=2SA954	7c	Sam	→2SA954	data
KSA 992	Si-P	=2SA992	7c	Sam	→2SA992	data
KSA 1010	Si-P	=2SA1010	17j§	Sam	→2SA1010	data
KSA 1013	Si-P	=2SA1013	7c(9mm)	Sam	→2SA1013	data
KSA 1015	Si-P	=2SA1015	7c	Sam	→2SA1015	data
KSA 1142	Si-P	=2SA1142	14h§	Sam	→2SA1142	data
KSA 1150	Si-P	Uni, 40V, 0,5A, 0,3W	41c	Sam	BC 327...328, BC 636, BC 638, BC 640, ++	data
KSA 1156	Si-P	=2SA1156	14h§	Sam	→2SA1156	data
KSA 1174	Si-P	=2SA1174	41c	Sam	→2SA1174	data
KSA 1175	Si-P	=2SA1175	41c	Sam	→2SA1175	data
KSA 1182	Si-P	=2SA1182	35a	Sam	→2SA1182	data
KSA 1201	Si-P	=2SA1201	39b	Sam	→2SA1201	data
KSA 1203	Si-P	=2SA1203	39b	Sam	→2SA1203	data
KSA 1220(A)	Si-P	=2SA1220(A)	14h§	Sam	→2SA1220(A)	data
KSA 1241	Si-P	=2SA1241	30j§	Sam	→2SA1241	data
KSA 1242	Si-P	=2SA1242	30j§	Sam	→2SA1242	data
KSA 1243	Si-P	=2SA1243	30j§	Sam	→2SA1243	data
KSA 1244	Si-P	=2SA1244	30j§	Sam	→2SA1244	data
KSA 1298	Si-P	=2SA1298	35a	Sam	→2SA1298	data
KSA 1304	Si-P	=2SA1304	17c	Sam	→2SA1304	data
KSA 1370	Si-P	=2SA1370	7c(9mm)	Sam	→2SA1370	data
KSA 1378	Si-P	Uni, 30V, 0,3A, 0,3W	41c	Sam	BC 327...328, BC 636, BC 638, 2SB910, ++	data
KSA 1381	Si-P	=2SA1381	14h§	Sam	→2SA1381	data
KSA 1406	Si-P	=2SA1406	14h§	Sam	→2SA1406	data
KSA 1614	Si-P	NF-L, Reg, 80V, 3A, 20W	17c	Sam	BD 938F, BDT 32...F, 2SA1488A, 2SA1635,++	data
KSA 1625	Si-P	=2SA1625	7e	Sam	→2SA1625	data
KSB						
KSB 546	Si-P	=2SB546	17j§	Sam	→2SB546	data
KSB 564 A	Si-P	=2SB564	7e	SAm	→2SB564	data
KSB 596	Si-P	=2SB596	17j§	Sam	→2SB596	data
KSB 601	Si-P-Darl+Di	=2SB601	17j§	Sam	→2SB601	data
KSB 707	Si-P	=2SB707	17j§	Sam	→2SB707	data
KSB 708	Si-P	=2SB708	17j§	Sam	→2SB708	data
KSB 744(A)	Si-P	=2SB744(A)	14h§	Sam	→2SB744(A)	data
KSB 772	Si-P	=2SB772	14h§	Sam	→2SB772	data
KSB 794	Si-P-Darl+Di	=2SB794	14h§	Sam	→2SB794	data
KSB 795	Si-P-Darl+Di	=2SB795	14h§	Sam	→2SB795	data
KSB 798	Si-P	=2SB798	39b	Sam	→2SB798	data
KSB 810	Si-P	=2SB810	41c	Sam	→2SB810	data
KSB 811	Si-P	=2SB811	41c	Sam	→2SB811	data
KSB 817	Si-P	=2SB817	18j§	Sam	→2SB817	data
KSB 834	Si-P	=2SB834	17j§	Sam	→2SB834	data
KSB 906	Si-P	=2SB906	30j§	Sam	→2SB906	data
KSB 907	Si-P	=2SB907	30j§	Sam	→2SB907	data
KSB 1015	Si-P	=2SB1015	17c	Sam	→2SB1015	data
KSB 1017	Si-P	=2SB1017	17c	Sam	→2SB1017	data
KSB 1022	Si-P-Darl	=2SB1022	17c	Sam	→2SB1022	data
KSB 1023	Si-P-Darl	=2SB1023	17c	Sam	→2SB1023	data
KSB 1097	Si-P	=2SB1097	17c	Sam	→2SB1097	data
KSB 1098	Si-P-Darl-Di	=2SB1098	17c	Sam	→2SB1098	data
KSB 1116(A)	Si-P	=2SB1116(A)	7c	Sam	→2SB1116(A)	data
KSB 1121	Si-P	=2SB1121	39b	Sam	→2SB1121	data

Type	Device	Short Description	Fig.	Manu	Comparision Types	More at
KSB 1149	Si-P-Darl+Di	=2SB1149	14h§	Sam	→2SB1149	data
KSB 1150	Si-P-Darl+Di	=2SB1150	14h§	Sam	→2SB1150	data
KSB 1151	Si-P	=2SB1151	14h§	Sam	→2SB1151	data
KSB 1366	Si-P	NF-L, 60V, 3A, 25W, 9MHz	17c	Sam	2SA1440, 2SA1488, 2SA1635, 2SB1335, ++	data

KSC

Type	Device	Short Description	Fig.	Manu	Comparision Types	More at
KSC 184	Si-N	AM-V/M/ZF, 30V, 0,05A, 0,25W, 100MHz	7e	Sam	BF 240..241, BF 254..255, BF 594..595,++	data
KSC 388	Si-N	=2SC388ATM:	7e	Sam	→2SC388	data
KSC 815	Si-N	=2SC815: 0,4W	7e	Sam	→2SC815	data
KSC 838	Si-N	FM-V/M/O/ZF, 35V, 30mA, 0,25W, 250MHz	7e	Sam	BF 240..241, BF 254..255, BF 594..595,++	data
KSC 839	Si-N	AM/FM-V/M/O/ZF, 35V, 0,1A, 200MHz	7e	Sam	BF 240..241, BF 254..255, BF 594..595,++	data
KSC 853	Si-N	=2SC853:	7e	Sam	→2SC853	data
KSC 900	Si-N	=2SC900:	7e	Sam	→2SC900	data
KSC 921	Si-N	FM-V/O, 35V, 0,1A, 250MHz	7e	Sam	BFX 59, (BF 240..241, BF 254..255,++)[7]	data
KSC 945	Si-N	Uni, 60V, 0,15A, 0,25W, 300MHz	7e	Sam	BC 174, BC 182, BC 190, BC 546, 2SB725++	data
KSC 1008	Si-N	Uni, 80V, 0,7A, 0,8W, 50MHz	7e	Sam	BC 639, 2SD667, 2SD1226, 2SD1616A, ++	data
KSC 1009	Si-N	Uni, 160/140V, 0,7A, 0,8W, 50MHz	7e	Sam	2SC2383, 2SC3228, 2SC3332, 2SD1812	data
KSC 1070(-1,-2)	Si-N	=2SC1070	25u	Sam	→2SC1070	data
KSC 1072	Si-N	NF-Tr/E, 60V, 0,7A, 0,8W	7e	Sam	BC 337A, BC 637, BC 639, 2SD1226, ++	data
KSC 1096	Si-N	=2SC1096: 2A	13j§	Sam	→2SC1096	data
KSC 1098	Si-N	=2SC1098: 2A	13j§	Sam	→2SC1098	data
KSC 1173	Si-N	=2SC1173	17j§	Sam	→2SC1173	data
KSC 1187	Si-N	=2SC1187:	7e	Sam	→2SC1187	data
KSC 1188	Si-N	=2SC1188:	7e, 7f	Sam	→2SC1188	data
KSC 1222	Si-N	=2SC1222	7e	Sam	→2SC1222	data
KSC 1330	Si-N	=2SC1330:	7e, 7f	Sam	→2SC1330	data
KSC 1393	Si-N	=2SC1393:	7f	Sam	→2SC1393	data
KSC 1394	Si-N	=2SC1394:	7f	Sam	→2SC1394	data
KSC 1395	Si-N	=2SC1395:	7e	Sam	→2SC1395	data
KSC 1506	Si-N	Vid, 300/300V, 0,1A, 0,75W, 80MHz	7e	Sam	BF 299, BF 393, BF 420A, 2SC3468, ++	data
KSC 1507	Si-N	=2SC1505	17j§	Sam	→2SC1505	data
KSC 1520	Si-N	=2SC1519	13j§	Sam	→2SC1519	data
KSC 1520 A	Si-N	=2SC1520	13j§	Sam	→2SC1520	data
KSC 1623	Si-N	=2SC1623	35a	Sam	→2SC1623	data
KSC 1674	Si-N	=2SC1674:	7e	Sam	→2SC1674	data
KSC 1675	Si-N	=2SC1675:	7e	Sam	→2SC1675	data
KSC 1730	Si-N	=2SC1730	7c	Sam	→2SC1730	data
KSC 1815	Si-N	=2SC1815	7e	Sam	→2SC1815	data
KSC 1845	Si-N	=2SC1845	7c	Sam	→2SC1845	data
KSC 1983	Si-N	=2SC1983	17j§	Sam	→2SC1983	data
KSC 2001	Si-N	=2SC2001	7c	Sam	→2SC2001	data
KSC 2002	Si-N	=2SC2002	7c	Sam	→2SC2002	data
KSC 2003	Si-N	=2SC2003	7c	Sam	→2SC2003	data
KSC 2073	Si-N	=2SC2073	17j§	Sam	→2SC2073	data
KSC 2223	Si-N	=2SC2223	35a	Sam	→2SC2223	data
KSC 2233	Si-N	=2SC2233	17j§	Sam	→2SC2233	data
KSC 2258(A)	Si-N	=2SC2258(A):	14h§	Sam	→2SC2258(A)	data
KSC 2310	Si-N	NF/S, 200/150V, 0,05A, 0,8W, 100MHz	7c(9mm)	Sam	BF 391..393, BF 420, BF 422	data
KSC 2316	Si-N	NF-Tr/E, 120V, 0,8A, 0,9W, 120MHz	7c(9mm)	Sam	2SC2235, 2SC2383, 2SD667, 2SD1665, ++	data
KSC 2328 A	Si-N	NF-Tr/E, 30V, 2A, 1W, 120MHz	7c(9mm)	Sam	MPS650...51, 2SD1055, 2SD1100, 2SD1227,++	data
KSC 2330	Si-N	Vid, 300/300V, 0,1A, 1W, 50MHz	7c(9mm)	Sam	BF 393, BF 420A, 2SC3249, 2SC3468, ++	data
KSC 2330 A	Si-N	=KSC 2330: 400/400V	7c(9mm)	Sam	2SC2267, 2SC3469, 2SC4166, 2SD1385	data
KSC 2331	Si-N	NF/S, 80V, 0,7A, 1W, 50MHz	7c(9mm)	Sam	BC 639, 2SC2235, 2SD667, 2SD1665, ++	data
KSC 2333	Si-N	=2SC2333:	17j§	Sam	→2SC2333	data
KSC 2334	Si-N	=2SC2334	17j§	Sam	→2SC2334	data
KSC 2335	Si-N	=2SC2335	17j§	Sam	→2SC2335	data
KSC 2335 F	Si-N	=2SC2335: Iso	17c	Sam	BU 306F, 2SC3574, 2SC3890, 2SC4130, ++	data
KSC 2340	Si-N	Vid, 350/350V, 0,1A, 1W, >50MHz	7c(9mm)	Sam	BF 485, BF 487, 2N6517, 2SC3469, ++	data
KSC 2383	Si-N	=2SC2383	7c(9mm)	Sam	→2SC2383	data
KSC 2500	Si-N	=2SC2500	7c(9mm)	Sam	→2SC2500	data
KSC 2517	Si-N	=2SC2517	17j§	Sam	→2SC2517	data
KSC 2518	Si-N	=2SC2518	17j§	Sam	→2SC2518	data
KSC 2669	Si-N	=2SC2669	41c	Sam	→2SC2669	data
KSC 2682	Si-N	=2SC2682	14h§	Sam	→2SC2682	data
KSC 2688	Si-N	=2SC2688	14h§	Sam	→2SC2688	data
KSC 2690(A)	Si-N	=2SC2690(A)	14h§	Sam	→2SC2690(A)	data
KSC 2710	Si-N	Uni, 40V, 0,5A, 0,3W	41c	Sam	BC 337..338, BC 635, BC 637, 2SC2710, ++	data
KSC 2715	Si-N	=2SC2715	35a	Sam	→2SC2715	data
KSC 2734	Si-N	=2SC2734	35a	Sam	→2SC2734	data
KSC 2749	Si-N	=2SC2749	18j§	Sam	→2SC2749	data
KSC 2751	Si-N	=2SC2751	18j§	Sam	→2SC2751	data
KSC 2752	Si-N	=2SC2752	14h§	Sam	→2SC2752	data
KSC 2753	Si-N	=2SC2753	7f	Sam	→2SC2753	data
KSC 2755	Si-N	=2SC2755	35a	Sam	→2SC2755	data
KSC 2756	Si-N	=2SC2756	35a	Sam	→2SC2756	data

Type	Device	Short Description	Fig.	Manu	Comparision Types	More at
KSC 2757	Si-N	=2SC2757	35a	Sam	→2SC2757	data
KSC 2758	Si-N	=2SC2758	35a	Sam	→2SC2758	data
KSC 2759	Si-N	=2SC2759	35a	Sam	→2SC2759	data
KSC 2784	Si-N	=2SC2784	41c	Sam	→2SC2784	data
KSC 2785	Si-N	=2SC2785	41c	Sam	→2SC2785	data
KSC 2786	Si-N	=2SC2786	41c	Sam	→2SC2786	data
KSC 2787	Si-N	=2SC2787	41c	Sam	→2SC2787	data
KSC 2859	Si-N	=2SC2859	35a	Sam	→2SC2859	data
KSC 2881	Si-N	=2SC2881	39b	Sam	→2SC2881	data
KSC 2883	Si-N	=2SC2883	39b	Sam	→2SC2883	data
KSC 2982	Si-N	=2SC2982	39b	Sam	→2SC2982	data
KSC 3073	Si-N	=2SC3073:	30j§	Sam	→2SC3073	data
KSC 3074	Si-N	=2SC3074:	30j§	Sam	→2SC3074	data
KSC 3076	Si-N	=2SC3076:	30j§	Sam	→2SC3076	data
KSC 3120	Si-N	=2SC3120	35a	Sam	→2SC3120	data
KSC 3123	Si-N	=2SC3123	35a	Sam	→2SC3123	data
KSC 3125	Si-N	=2SC3125	35a	Sam	→2SC3125	data
KSC 3233	Si-N	=2SC3233:	30j§	Sam	→2SC3233	data
KSC 3265	Si-N	=2SC3265	35a	Sam	→2SC3265	data
KSC 3296	Si-N	=2SC3296:	17c	Sam	→2SC3296	data
KSC 3488	Si-N	Uni, 30V, 0,3A, 0,3W	41c	Sam	BC 337...338, BC 635, BC 637, 2SC3488, ++	data
KSC 3502	Si-N	=2SC3502	14h§	Sam	→2SC3502	data
KSC 3503	Si-N	=2SC3503	14h§	Sam	→2SC3503	data
KSC 3552	Si-N	=2SC3552	18j§	Sam	→2SC3552	data
KSC 3569	Si-N	=2SC3569	17c	Sam	→2SC3569	data
KSC 3953	Si-N	=2SC3953	14h§	Sam	→2SC3953	data
KSC 5019	Si-N	lo-sat, 30V, 2A, 0,75W, 150MHz	7c	Sam	2SC4484, 2SD1246, 2SD1835, 2SD2177,++	data
KSC 5020	Si-N	Ballast,800/500V, 3A, 40W, <500/3300ns	17j§	Sam	BUT 11(A), MJE 13071, 2SC3086, 2SD841,++	data
KSC 5020 F	Si-N	=KSC 5020: Iso, 30W	17c		BUT 11(A)F, 2SC3749, 2SC4517, 2SC4908,++	data
KSC 5021	Si-N	Ballast,800/500V, 5A, 50W, <500/3300ns	17j§		BUT 11(A), MJE 13071, 2SC3047, 2SC3087++	data
KSC 5021 F	Si-N	=KSC 5021: Iso, 40W	17c		BUT 11(A)F, 2SC3750, 2SC4518, 2SC4638,++	data
KSC 5022	Si-N	S-L, -/500V, 4A, 60W, <500/3300ns	18j§	Sam	BUW 11(A), BUW 131(A), BUV 82...83	data
KSC 5023	Si-N	S-L, 800/500V, 7A, 80W, <500/3300ns	18j§	Sam	BUP 22B...C, 2SC3089, 2SC3449, 2SC3636,++	data
KSC 5024	Si-N	S-L, 800/500V, 10A, 90W, <500/3300ns	18j§	Sam	BUV 47(A), BUW 12(A), 2SC3450, 2SC3637++	data
KSC 5025	Si-N	S-L, 800/500V, 15A, 100W, <500/3300ns	18j§	Sam	BUV 48(A), BUW 13(A), 2SC3451, 2SC3638++	data
KSC 5026	Si-N	S-L, 1100/800V, 1,5A, 40W, <500/3300ns	17j§	Sam	MJE 8500, 2SC3178, 2SC3456, 2SC4230	data
KSC 5027	Si-N	S-L, 1100/800V, 3A, 50W, <500/3300ns	17j§	Sam	BUL 213, MJE 8502. 2SC3050, 2SC3457	data
KSC 5027 F	Si-N	=KSC 5027: Iso	17c		MJF 18204	data
KSC 5028	Si-N	S-L, -/800V, 3A, 80W, <500/3300ns	18j§		2SC3387, 2SC3459, 2SC4236	data
KSC 5029	Si-N	S-L, 1100/800V, 4,5A, 90W, <500/3300ns	18j§	Sam	2SC3387, 2SC3459, 2SC4236	data
KSC 5030	Si-N	S-L, 1100/800V, 6A, 100W, <500/3300ns	18j§	Sam	BUV 89, 2SC3466, 2SC3643	data
KSC 5030 F	Si-N	=KSC 5030: Iso, 60W	18c		2SC4428, 2SC4584	data
KSC 5031	Si-N	S-L, -/800V, 8A, 140W, <500/3300ns	18j§	Sam	BUV 70...71, 2SC3461, 2SC4237	data
KSC 5039	Si-N	Ballast, 800/400V, 5A, 70W, <1/3,8µs	17j§	Sam	BUT 11(A), MJE 13071, 2SC3047, 2SC3087++	data
KSC 5039 D	Si-N+Di	=KSC 5039: int. Damper-Diode	17j§		(BUT 11(A), MJE 13071, 2SC3047, 2SC3087)[2]	data
KSC 5039 F	Si-N	=KSC 5039: Iso, 30W	17c		BUT 11(A)F, 2SC3750, 2SC3795, 2SC4518,++	data
KSC 5042	Si-N	Dyn. Focus, 1500/900V, 0,1A, 10W	17j§	Sam	2SC3675	data
KSC 5042 F	Si-N	=KSC 5042: Iso, 6W	17c		2SC4630, 2SC4686	data
KSC 5042 M	Si-N	=KSC 5042: 4W	14h§		-	data
KSC 5047	Si-N	S-L, hi-Ueb, lo-sat, 100V, 15A, 100W	18j§	Sam	-	data
KSC 5054	Si-N	S-L, 500/400V, 0,5A, 10W, <1/3,5µs	30j§	Sam	2SC3129, 2SC3588, 2SC4499	data
KSC 5060	Si-N	S-L, 800/500V, 3A, 40W, <500/3300ns	17j§	Sam	BUT 11(A), MJE 13071, 2SC3086, 2SD841,++	data
KSC 5061	Si-N	S-L, 800/500V, 5A, 50W, <500/3300ns	17j§	Sam	BUT 11(A), MJE 13071, 2SC3047, 2SC3087++	data
KSC 5086	Si-N+Di	CTV-HA, hi-res, 1500/800V, 7A, 50W	18c		2SC4123, 2SC4769, 2SC5041	data
KSC 5086 F	Si-N+Di	=KSC 5086: 40W	17c		-	data
KSC 5088	Si-N	CTV-HA, hires, 1500/800V, 8A, 50W	18c	Sam	BU 2522AF, 2SC3886A, 2SC3896, 2SC4758	data
KSC 5089	Si-N	CTV-HA, hi-res, 1500/800V, 8A, 150W	18j§	Sam	BU 2522A, 2SC3687	data
KSC 5321	Si-N	Ballast,800/500V, 5A, 100W,<500/3300ns	17j§	Sam	BUF 636A, BUT 11, BUT 18, BUV 46	data
KSC 5321 F	Si-N	=KSC 5321: Iso, 40W	17c		BUT 11AF, 2SC4518A, 2SC4898	data
KSC 5337	Si-N	S-L, 700/400V, 8A, 100W, <200/2500ns	17j§	Sam	BUF 644A, BUL 57, BUL 147, BUT 22B, ++	data
KSC 5337 F	Si-N	=KSC 5337: Iso, 40W	17c		BU 307F, BUL 57PI, BUL 147F, BUT 22BF,++	data
KSC 5338	Si-N	Ball.,1000/450V, 5A, 100W, <200/2500ns	17j§	Sam	BUF 636A, BUT 11A, BUT 18A, BUV 46A, ++	data
KSC 5338 D	Si-N+Di	=KSC 5338: int. Damper-Diode	17j§		(BUF 636A, BUT 11A, BUT 18A, BUV 46A)[2]	data
KSC 5338 F	Si-N	=KSC 5338: Iso, 40W	17c		BUT 11AF, 2SC4518A, 2SC4898	data
KSC 5367	Si-N	Ball., 1600/800V, 3A, 80W, <500/4500ns	17j§	Sam	BU 1706A, BUL 310, BUL 416	data
KSC 5367 F	Si-N	=KSC 5367: Iso, 40W	17c		BU 1706AX, BUL 310PI	data
KSC 5386	Si-N+Di	Monitor-HA, 48kHz, 1500/800V, 7A, 50W	18c	Sam	BU 2508DF, 2SC3892A, 2SC4123, 2SC4916	data

KSD...KSP

Type	Device	Short Description	Fig.	Manu	Comparision Types	More at
KSD 73	Si-N	NF/S-L, 100V, 5A, 30W, 20MHz	17j§	Sam	BD 243C, BD 539C, BD 953, 2SD613, ++	data
KSD 227	Si-N	=2SD227	7e	Sam	→2SD227	data
KSD 261	Si-N	=2SD261:	7e	Sam	→2SD261	data
KSD 288	Si-N	NF/S-L, 80V, 3A, 25W	17j§	Sam	BD 241B, BD 537, BD 539B, 2SD726, ++	data
KSD 362	Si-N	TV-HA, 150/70V, 5A, 40W, 10MHz	17j§	Sam	BU 406...409, 2SD823, 2SD1150, 2SD1163,++	data
KSD 363	Si-N	TV-HA, 300/120V, 6A, 40W, 10MHz	17j§	Sam	BU 406...409, 2SC3590, 2SD1069, 2SD1163++	data
KSD 401	Si-N	=2SD401	17j§	Sam	→2SD401	data

Type	Device	Short Description	Fig.	Manu	Comparision Types	More at
KSD 471 A	Si-N	=2SD471A	7e	Sam	→2SD471A	data
KSD 526	Si-N	=2SD526	17j§	Sam	→2SD526	data
KSD 560	Si-N-Darl+Di	=2SD560	17j§	Sam	→2SD560	data
KSD 568	Si-N	=2SD568	17j§	Sam	→2SD568	data
KSD 569	Si-N	=2SD569	17j§	Sam	→2SD569	data
KSD 794(A)	Si-N	=2SD794(A)	14h§	Sam	→2SD794(A)	data
KSD 818	Si-N	CTV-HA, 1500/600V, 2,5A, 50W	18j§	Sam	BU 705, BU 706, 2SC3483, 2SD1493...94, ++	data
KSD 819	Si-N	CTV-HA, 1500/600V, 3,5A, 50W	18j§	Sam	BU 706, 2SC3484, 2SD1495	data
KSD 820	Si-N	CTV-HA, 1500/600V, 5A, 50W	18j§	Sam	BU 2508(A), BU 908, 2SD1496...97, ++	data
KSD 821	Si-N	CTV-HA, 1500/600V, 6A, 50W	18j§	Sam	BU 2508(A), BU 908, 2SD1496...97, ++	data
KSD 868	Si-N+Di	CTV-HA, 1500/600V, 2,5A, 50W	18j§	Sam	BU 706D, 2SC3479, 2SD1290...1291, ++	data
KSD 869	Si-N+Di	CTV-HA, 1500/600V, 3,5A, 50W	18j§	Sam	BU 706D, 2SC3480, 2SD 1729	data
KSD 870	Si-N+Di	CTV-HA, 1500/600V, 5A, 50W	18j§	Sam	BU 2508D, 2SC3481...82, 2SC3681...82, ++	data
KSD 871	Si-N+Di	CTV-HA, 1500/600V, 6A, 50W	18j§	Sam	BU 2508D, 2SC3482, 2SC3681...83, ++	data
KSD 880	Si-N	=2SD880	17j§	Sam	→2SD880	data
KSD 882	Si-N	=2SD882	14h§	Sam	→2SD882	data
KSD 985	Si-N-Darl+Di	=2SD985	14h§	Sam	→2SD985	data
KSD 986	Si-N-Darl+Di	=2SD986	14h§	Sam	→2SD986	data
KSD 1020	Si-N	=2SD1020	41c	Sam	→2SD1020	data
KSD 1021	Si-N	=2SD1021	41c	Sam	→2SD1021	data
KSD 1047	Si-N	=2SD1047	18j§	Sam	→2SD1047	data
KSD 1221	Si-N	=2SD1221	30j§	Sam	→2SD1221	data
KSD 1222	Si-N	=2SD1222	30j§	Sam	→2SD1222	data
KSD 1273	Si-N	=2SD1273	17c	Sam	→2SD1273	data
KSD 1362	Si-N	B/W TV-HA, 150/70V, 5A, 20W	17c	Sam	BU 407F, 2SD1404	data
KSD 1406	Si-N	=2SD1406	17c	Sam	→2SD1406	data
KSD 1408	Si-N	=2SD1408	17c	Sam	→2SD1408	data
KSD 1413	Si-N-Darl	=2SD1413	17c	Sam	→2SD1413	data
KSD 1417	Si-N-Darl	=2SD1417	17c	Sam	→2SD1417	data
KSD 1588	Si-N	=2SD1588	17c	Sam	→2SD1588	data
KSD 1589	Si-N-Darl+Di	=2SD1589	17c	Sam	→2SD1589	data
KSD 1616(A)	Si-N	=2SD1616(A)	7c	Sam	→2SD1616(A)	data
KSD 1621	Si-N	=2SD1621	39b	Sam	→2SD1621	data
KSD 1691	Si-N	=2SD1691	14h§	Sam	→2SD1691	data
KSD 1692	Si-N-Darl+Di	=2SD1692	14h§	Sam	→2SD1692	data
KSD 1693	Si-N	=2SD1693	14h§	Sam	→2SD1693	data
KSD 1943	Si-N	=2SD1943	17j§	Sam	→2SD1943	data
KSD 1944	Si-N	=2SD1944	17c	Sam	→2SD1944	data
KSD 2012	Si-N	NF-L, 60V, 3A, 25W, 3MHz	17c	Sam	2SC3690, 2SD1762, 2SD1913, 2SD1985, ++	data
KSD 2058	Si-N	NF-L, 60V, 3A, 25W, >0,4MHz	17c	Sam	2SC3690, 2SD1762, 2SD1913, 2SD1985, ++	data
KSD 5000	Si-N+Di	TV-HA, 1500/800V, 2,5A, 80W	18j§	Sam	BU 706D, 2SC3479, 2SD1290...1291, ++	data
KSD 5001	Si-N+Di	TV-HA, 1500/800V, 3,5A, 80W	18j§	Sam	BU 706D, 2SC3480, 2SD1729	data
KSD 5002	Si-N+Di	TV-HA, 1500/800V, 5A, 120W	18j§	Sam	BU 706D, 2SC3481...82, 2SC4291...92, ++	data
KSD 5003	Si-N+Di	TV-HA, 1500/800V, 6A, 120W	18j§	Sam	BU 2508D, 2SC3482, 2SC4292, 2SD1731, ++	data
KSD 5004	Si-N	TV-HA, 1500/800V, 2,5A, 80W	18j§	Sam	BU 705, 2SC3483, 2SD1493...94	data
KSD 5005	Si-N	TV-HA, 1500/800V, 3,5A, 80W	18j§	Sam	BU 706, 2SC3484, 2SD1495	data
KSD 5006	Si-N	TV-HA, 1500/800V, 5A, 120W	18j§	Sam	BU 706, BU 2508A, 2SC3485...86, 2SC3685	data
KSD 5007	Si-N	TV-HA, 1500/800V, 6A, 120W	18j§	Sam	BU 2508A, 2SC3486, 2SC3685	data
KSD 5010	Si-N+Di	=KSD 5000: Iso, 50W	18c	Sam	BU 705DF, 2SD1553, 2SD1649, 2SD1876, ++	data
KSD 5011	Si-N+Di	=KSD 5001: Iso, 50W	18c	Sam	BU 706DF, 2SD1554, 2SD1650, 2SD1877, ++	data
KSD 5012	Si-N+Di	=KSD 5002: Iso, 60W	18c	Sam	BU 706DF, 2SD1555, 2SD1651, 2SD1878, ++	data
KSD 5013	Si-N+Di	=KSD 5003: Iso, 60W	18c	Sam	BU 2508DF, 2SD1556, 2SD1652, 2SD1879, ++	data
KSD 5014	Si-N	=KSD 5004: Iso, 50W	18c	Sam	BU 705F, 2SD1543, 2SD1653, 2SD1882	data
KSD 5015	Si-N	=KSD 5005: Iso, 50W	18c	Sam	BU 706F, 2SD1544, 2SD1654, 2SD1883	data
KSD 5016	Si-N	=KSD 5006: Iso, 60W	18c	Sam	BU 706F, 2SD1545, 2SD1655, 2SD1884, ++	data
KSD 5017	Si-N	=KSD 5007: Iso, 60W	18c	Sam	BU 2508AF, 2SD1546, 2SD1656, 2SD1885, ++	data
KSD 5018	Si-N-Darl	S-L, int. Rbe, 600/275V, 4A, 40W	17j§	Sam	2SC3579, 2SD798	data
KSD 5041	Si-N	NF-E, Blitzg./strobo, 40V, 5A, 0,75W	7c	Sam	2SC3671, 2SD1961...62, 2SD2097, 2SD2179++	data
KSD 5060	Si-N+Di	CTV-HA, 1500/800V, 2,5A, 80W	18j§	Sam	BU 705D, 2SD1290...91, 2SD1728	data
KSD 5061	Si-N+Di	CTV-HA, 1500/800V, 3,5A, 80W	18j§	Sam	BU 706D, 2SC3480, 2SD1729	data
KSD 5062	Si-N+Di	CTV-HA, 1500/800V, 5A, 120W	18j§	Sam	BU 706D, BU 2508D, 2SC3681, ++	data
KSD 5064	Si-N	CTV-HA, 1500/800V, 2,5A, 80W	18j§	Sam	2SD705, 2SC3483, 2SD1493...94	data
KSD 5065	Si-N	CTV-HA, 1500/800V, 3,5A, 80W	18j§	Sam	2SD706, 2SC3484, 2SD1495	data
KSD 5066	Si-N	CTV-HA, 1500/800V, 5A, 120W	18j§	Sam	2SD706, BU 2508A, 2SC3685, 2SD1496...97++	data
KSD 5068	Si-N	CTV-HA, 1500/800V, 8A, 150W	18j§	Sam	BU 2508A, BU 2520A, 2SC3687...88, ++	data
KSD 5070	Si-N+Di	CTV-HA, 1500/800V, 2,5A, 50W	18c	Sam	BU 705DF, 2SD1553, 2SD1649, 2SD1876, ++	data
KSD 5071	Si-N+Di	CTV-HA, 1500/800V, 3,5A, 50W	18c	Sam	BU 706DF, 2SD1650, 2SD1877	data
KSD 5072	Si-N+Di	CTV-HA, 1500/800V, 5A, 60W	18c	Sam	BU 706DF, BU 2508DF, 2SC4122, 2SD1878,++	data
KSD 5074	Si-N	CTV-HA, 1500/800V, 2,5A, 50W	18c	Sam	BU 705F, 2SD1653, 2SD1882	data
KSD 5075	Si-N	CTV-HA, 1500/800V, 3,5A, 50W	18c	Sam	BU 706F, 2SC4142, 2SD1654, 2SD1883	data
KSD 5075 T	Si-N	=KSD 5075: 75W	17j§		BU 505, MJE 12007, 2SD1575	data
KSD 5076	Si-N	CTV-HA, 1500/800V, 5A, 60W	18c	Sam	BU 706F, BU 2508AF, 2SC4142, 2SD1655	data
KSD 5078	Si-N	CTV-HA, 1500/800V, 8A, 70W	18c	Sam	BU 2508AF, 2SC3896, 2SC5067, 2SD1886, ++	data
KSD 5079	Si-N	CTV-HA, 1500/800V, 10A, 70W	18c	Sam	BU 2520AF, 2SC3897, 2SC4924, 2SC5068, ++	data
KSD 5080	Si-N+Di	CTV-HA, 1500/800V, 8A, 70W	18c	Sam	BU 2508DF, 2SC4124, 2SD1880	data
KSD 5090	Si-N+Di	CTV-HA, 1500/800V, 8A, 150W	18j§	Sam	BU 2508D, BU 2520D, 2SC3683	data
KSD 5338 D	Si-N+Di	Ballast, 1000/450V, 5A, 75W	17j§	Sam	MJE18004D	data
KSD 5702	Si-N+Di	CTV-HA, 1500V, 6A, 70W	18c	Sam	2SC4122, 2SC4294, 2SC4764, 2SD1556, ++	data
KSE 44 H...	Si-N	=D 44H...	17j§	Sam	→D 44H...	data

Type	Device	Short Description	Fig.	Manu	Comparision Types	More at
KSE 45 H...	Si-P	=D 45H...	17j§	Sam	→D 45H...	data
KSE 170	Si-P	=MJE 170	14h§	Sam	→MJE 170	data
KSE 171	Si-P	=MJE 171	14h§	Sam	→MJE 171	data
KSE 172	Si-P	=MJE 172	14h§	Sam	→MJE 172	data
KSE 180	Si-N	=MJE 180	14h§	Sam	→MJE 180	data
KSE 181	Si-N	=MJE 181	14h§	Sam	→MJE 181	data
KSE 182	Si-N	=MJE 182	14h§	Sam	→MJE 182	data
KSE 200	Si-N	=MJE 200	14h§	Sam	→MJE 200	data
KSE 210	Si-P	=MJE 210	14h§	Sam	→MJE 210	data
KSE 340	Si-N	=MJE 340	14h§	Sam	→MJE 340	data
KSE 350	Si-P	=MJE 350	14h§	Sam	→MJE 350	data
KSE 700	Si-P-Darl+Di	=MJE 700	14h§	Sam	→MJE 700	data
KSE 701	Si-P-Darl+Di	=MJE 701	14h§	Sam	→MJE 701	data
KSE 702	Si-P-Darl+Di	=MJE 702	14h§	Sam	→MJE 702	data
KSE 703	Si-P-Darl+Di	=MJE 703	14h§	Sam	→MJE 703	data
KSE 800	Si-N-Darl+Di	=MJE 800	14h§	Sam	→MJE 800	data
KSE 801	Si-N-Darl+Di	=MJE 801	14h§	Sam	→MJE 801	data
KSE 802	Si-N-Darl+Di	=MJE 802	14h§	Sam	→MJE 802	data
KSE 803	Si-N-Darl+Di	=MJE 803	14h§	Sam	→MJE 803	data
KSE 2955 T	Si-P	=TIP 2955	18j§	Sam	→MJ, MJE, PH, TIP 2955	data
KSE 3055 T	Si-N	=TIP 3055	18j§	Sam	→MJ, MJE, PH, TIP 3055	data
KSE 5740	Si-N-Darl+Di	=MJE 5740	17j§	Sam	→MJE 5740	data
KSE 5741	Si-N-Darl+Di	=MJE 5741	17j§	Sam	→MJE 5741	data
KSE 5742	Si-N-Darl+Di	=MJE 5742	17j§	Sam	→MJE 5742	data
KSE 13003	* Si-N	=MJ 4360: 700/400V, 20W	14h§	Sam	BUD 44, BUX 99	data
KSE 13004	Si-N	=MJE 13004	17j§	Sam	→MJE 13004	data
KSE 13005	Si-N	=MJE 13005	17j§	Sam	→MJE 13005	data
KSE 13006	Si-N	=MJE 13006	17j§	Sam	→MJE 13006	data
KSE 13007	Si-N	=MJE 13007	17j§	Sam	→MJE 13007	data
KSE 13008	Si-N	=MJE 13008	17j§	Sam	→MJE 13008	data
KSE 13009	Si-N	=MJE 13009	17j§	Sam	→MJE 13009	data
KSE 13005...13009 F	Si-N	=MJE 13005F...13009F: Iso	17c			data
KSE 15002	Si-N+Di	Ballast, 800V, 2A, 40W, <500/3300ns	17j§	Sam	BUV 36(A), BUX 84...85	data
KSH 29 ...	Si-N	=TIP 29...: 15W	30j§	Sam	2SC4134, 2SD1074...75, 2SD1733	data
KSH 30 ...	Si-P	=TIP 30...: 15W	30j§	Sam	2SA1592, 2SB844...845, 2SB1181	data
KSH 31 ...	Si-N	=TIP 31...: 15W	30j§	Sam	2SC4978, 2SD1815...16	data
KSH 32 ...	Si-P	=TIP 32...: 15W	30j§	Sam	2SB1215...16	data
KSH 41 ...	Si-N	=MJD 41...	30j§	Sam	→MJD 41...	data
KSH 42 ...	Si-P	=MJD 42...	30j§	Sam	→MJD 42...	data
KSH 44 ...	Si-N	=MJD 44...	30j§	Sam	→MJD 44...	data
KSH 45 ...	Si-P	=MJD 45...	30j§	Sam	→MJD 45...	data
KSH 47	Si-N	=MJD 47	30j§	Sam	→MJD 47	data
KSH 50	Si-N	=MJD 50	30j§	Sam	→MJD 50	data
KSH 112	Si-N-Darl+Di	=MJD 112	30j§	Sam	→MJD 112	data
KSH 117	Si-P-Darl+Di	=MJD 117	30j§	Sam	→MJD 117	data
KSH 122	Si-N-Darl+Di	=MJD 122	30j§	Sam	→MJD 122	data
KSH 127	Si-P-Darl+Di	=MJD 127	30j§	Sam	→MJD 127	data
KSH 200	Si-N	=MJD 200	30j§	Sam	→MJD 200	data
KSH 210	Si-P	=MJD 210	30j§	Sam	→MJD 210	data
KSH 340	Si-N	=MJD 340	30j§	Sam	→MJD 340	data
KSH 350	Si-P	=MJD 350	30j§	Sam	→MJD 350	data
KSH 2955	Si-P	=MJD 2955	30j§	Sam	→MJD 2955	data
KSH 3055	Si-N	=MJD 3055	30j§	Sam	→MJD 3055	data
KSH 13003	Si-N	=MJE 13003: 40W	30j§	Sam	-	data
K4S640432H-TC(L)75	CMOS-IC	SDRAM 64Mb H-die (x4, x8, x16)	TSOP	Sam	-	pdf pinout
K4S640432H-UC(L)75	CMOS-IC	SDRAM 64Mb H-die (x4, x8, x16)	TSOP	Sam	-	pdf pinout
K4S640832H-TC(L)75	CMOS-IC	SDRAM 64Mb H-die (x4, x8, x16)	TSOP	Sam	-	pdf pinout
K4S640832H-UC(L)75	CMOS-IC	SDRAM 64Mb H-die (x4, x8, x16)	TSOP	Sam	-	pdf pinout
K4S641632H-TC(L)60	CMOS-IC	SDRAM 64Mb H-die (x4, x8, x16)	TSOP	Sam		pdf pinout
K4S641632H-UC(L)60	CMOS-IC	SDRAM 64Mb H-die (x4, x8, x16)	TSOP	Sam		pdf pinout
K4S641632H-TC(L)70	CMOS-IC	SDRAM 64Mb H-die (x4, x8, x16)	TSOP	Sam		pdf pinout
K4S641632H-UC(L)70	CMOS-IC	SDRAM 64Mb H-die (x4, x8, x16)	TSOP	Sam	-	pdf pinout
K4S641632H-TC(L)75	CMOS-IC	SDRAM 64Mb H-die (x4, x8, x16)	TSOP	Sam	-	pdf pinout
K4S641632H-UC(L)75	CMOS-IC	SDRAM 64Mb H-die (x4, x8, x16)	TSOP	Sam	-	pdf pinout
KSK 30	N-FET	≈2SK30	7e	Sam	→2SK30	data
KSK 65	N-FET	≈2SK65	41e	Sam	→2SK65	data
KSK 117	N-FET	≈2SK117	7a	Sam	→2SK117	data
KSK 123	N-FET	≈2SK123	35b	Sam	→2SK123	data
KSK 161	N-FET	≈2SK161	41d	Sam	→2SK161	data
KSK 211	N-FET	≈2SK211	35c	Sam	→2SK211	data
KSK 596	N-FET	Kondens.-Mikr.-V, 20V, Idss>0,1...0,8mA	41e	Sam	2SK 596	data
KSP 5	Si-N	=MPSA 05	7e	Sam	→MPSA 05	data
KSP 6	Si-N	=MPSA 06	7e	Sam	→MPSA 06	data
KSP 10	Si-N	=MPSH 10	7f	Sam	→MPSH 10	data
KSP 11	Si-N	=MPSH 11	7f	Sam	→MPSH 11	data
KSP 12	Si-N-Darl	=MPSA 12	7e	Sam	→MPSA 12	data
KSP 13	Si-N-Darl	=MPSA 13	7e	Sam	→MPSA 13	data
KSP 14	Si-N-Darl	=MPSA 14	7e	Sam	→MPSA 14	data

Type	Device	Short Description	Fig.	Manu	Comparision Types	More at
KSP 17	Si-N	=MPSH 17	7f	Sam	→MPSH 17	data
KSP 20	Si-N	=MPSH 20	7f	Sam	→MPSH 20	data
KSP 24	Si-N	=MPSH 24	7f	Sam	→MPSH 24	data
KSP 25	Si-N-Darl	=MPSA 25	7e	Sam	→MPSA 25	data
KSP 26	Si-N-Darl	=MPSA 26	7e	Sam	→MPSA 26	data
KSP 27	Si-N-Darl	=MPSA 27	7e	Sam	→MPSA 27	data
KSP 42	Si-N	=MPSA 42	7e	Sam	→MPSA 42	data
KSP 43	Si-N	=MPSA 43	7e	Sam	→MPSA 43	data
KSP 44	Si-N	=MPSA 44	7e	Sam	→MPSA 44	data
KSP 45	Si-N	=MPSA 45	7e	Sam	→MPSA 45	data
KSP 55	Si-P	=MPSA 55	7e	Sam	→MPSA 55	data
KSP 56	Si-P	=MPSA 56	7e	Sam	→MPSA 56	data
KSP 62	Si-P-Darl	=MPSA 62	7e	Sam	→MPSA 62	data
KSP 63	Si-P-Darl	=MPSA 63	7e	Sam	→MPSA 63	data
KSP 64	Si-P-Darl	=MPSA 64	7e	Sam	→MPSA 64	data
KSP 75	Si-P-Darl	=MPSA 75	7e	Sam	→MPSA 75	data
KSP 76	Si-P-Darl	=MPSA 76	7e	Sam	→MPSA 76	data
KSP 77	Si-P-Darl	=MPSA 77	7e	Sam	→MPSA 77	data
KSP 92	Si-P	=MPSA 92	7e	Sam	→MPSA 92	data
KSP 93	Si-P	=MPSA 93	7e	Sam	→MPSA 93	data
KSP 94	Si-P	S, Vid, 400/400V, 0,3A, 0,625W	7e	Sam	2SA1625, 2SB1074, 2SB1488	data
KSP 2222(A)	Si-N	=2N2222(A): 0,6A, 0,625W	7e	Sam	→2N2222(A)	data
KSP 2907(A)	Si-P	=2N2906(A): 0,625W	7e	Sam	→2N2907(A)	data
KSP 5172	Si-N	=2N5172: 0,625W	7e	Sam	→2N5172	data
KSP 5179	Si-N	=2N5179: 0,2W	7e	Sam	→2N5179	data
KSP 6520	Si-N	=MPS 6520	7e	Sam	→MPS 6520	data
KSP 6521	Si-N	=MPS 6521	7e	Sam	→MPS 6521	data
KSP 8097	Si-N	=MPS 8097: 0,625W	7e	Sam	→MPS 8097	data
KSP 8098	Si-N	=MPS 8098	7e	Sam	→MPS 8098	data
KSP 8099	Si-N	=MPS 8099	7e	Sam	→MPS 8099	data
KSP 8598	Si-P	=MPS 8598	7e	Sam	→MPS 8598	data
KSP 8599	Si-P	=MPS 8599	7e	Sam	→MPS 8599	data

KSR...KSZ

Type	Device	Short Description	Fig.	Manu	Comparision Types	More at
KSR 1001	Si-N+R	S, Rb=Rbe=4,7kΩ, 50V, 0,1A, 0,3W	7c	Sam	AA 1L3M, DTC 143ES, RN 1001, UN 421L,++	data
KSR 1002	Si-N+R	=KSR 1001: Rb=Rbe=10kΩ	7c	Sam	AA 1A4M, DTC 114ES, RN 1002, UN 4211,++	data
KSR 1003	Si-N+R	=KSR 1001: Rb=Rbe=22kΩ	7c	Sam	AA 1F4M, DTC 124ES, RN 1009, UN 4212,++	data
KSR 1004	Si-N+R	=KSR 1001: Rb=Rbe=47kΩ	7c	Sam	AA 1L4M, DTC 144ES, RN 1004, UN 4213,++	data
KSR 1005	Si-N+R	=KSR 1001: Rb=4,7kΩ, Rbe=10kΩ	7c	Sam	AA 1L3N, DTC 143XS, UN 421F, 2SC4361	data
KSR 1006	Si-N+R	=KSR 1001: Rb=10kΩ, Rbe=47kΩ	7c	Sam	AA 1A4P, DTC 114YS, RN 1007, UN 4214,++	data
KSR 1007	Si-N+R	=KSR 1001: Rb=22kΩ, Rbe=47kΩ	7c	Sam	AA 1F4N, DTC 124XS, RN 1008	data
KSR 1008	Si-N+R	=KSR 1001: Rb=47kΩ, Rbe=22kΩ	7c	Sam	AA 1L4L, DTC 144WS, RN 1009, UN 421E,++	data
KSR 1009	Si-N+R	=KSR 1001: 40V, Rb=4,7kΩ, Rbe=-	7c	Sam	AA 1L3Z, DTC 143TS, RN 1010, UN 4216,++	data
KSR 1010	Si-N+R	=KSR 1001: 40V, Rb=10kΩ, Rbe=-	7c	Sam	AA 1A4Z, DTC 114TS, RN 1011, UN 4215,++	data
KSR 1011	Si-N+R	=KSR 1001: 40V, Rb=22kΩ, Rbe=-	7c	Sam	AA 1F4Z, DTC 124TS, UN 4217, 2SC4121	data
KSR 1012	Si-N+R	=KSR 1001: 40V, Rb=47kΩ, Rbe=-	7c	Sam	AA 1L4Z, DTC 144TS, UN 4210, 2SC3899	data
KSR 1013	Si-N+R	=KSR 1001: Rb=2,2kΩ, Rbe=47kΩ	7c	Sam	DTC 123JS, RN 1005	data
KSR 1014	Si-N+R	=KSR 1001: Rb=4,7kΩ, Rbe=47kΩ	7c	Sam	DTC 143ZS, RN 1006, 2SC4133, 2SC4195	data
KSR 1101	Si-N+R	=KSR 1001: SMD	35a	Sam	FA 1L3M, DTC 143EK, RN 1401, UN 221L,++	data
KSR 1102	Si-N+R	=KSR 1002: SMD	35a	Sam	FA 1A4M, DTC 114EK, RN 1402, UN 2211,++	data
KSR 1103	Si-N+R	=KSR 1003: SMD	35a	Sam	FA 1F4M, DTC 124EK, RN 1403, UN 2212,++	data
KSR 1104	Si-N+R	=KSR 1004: SMD	35a	Sam	FA 1L4M, DTC 144EK, RN 1404, UN 2213,++	data
KSR 1105	Si-N+R	=KSR 1005: SMD	35a	Sam	FA 1L3N, DTC 143XK, UN 221F, 2SC4360	data
KSR 1106	Si-N+R	=KSR 1006: SMD	35a	Sam	FA 1A4P, DTC 114YK, RN 1407, UN 2214,++	data
KSR 1107	Si-N+R	=KSR 1007: SMD	35a	Sam	FA 1F4N, DTC 124XK, RN 1408	data
KSR 1108	Si-N+R	=KSR 1008: SMD	35a	Sam	FA 1L4L, DTC 144WK, RN 1409, UN 221E,++	data
KSR 1109	Si-N+R	=KSR 1009: SMD	35a	Sam	FA 1L3Z, DTC 143TK, RN 1410, UN 2216,++	data
KSR 1110	Si-N+R	=KSR 1010: SMD	35a	Sam	FA 1A4Z, DTC 114TK, RN 1411, UN 2215,++	data
KSR 1111	Si-N+R	=KSR 1011: SMD	35a	Sam	FA 1F4Z, DTC 124TK, UN 2217, 2SC4120	data
KSR 1112	Si-N+R	=KSR 1012: SMD	35a	Sam	FA 1L4Z, DTC 144TK, UN 2210, 2SC3898	data
KSR 1113	Si-N+R	=KSR 1013: SMD	35a	Sam	DTC 123JK, RN 1405	data
KSR 1114	Si-N+R	=KSR 1014: SMD	35a	Sam	DTC 143ZK, RN 1406, 2SC4146	data
KSR 1201...1214	Si-N+R	=KSR 1001...1014:	41c	Sam	→KSR 1001...1014	data
KSR 2001	Si-P+R	S, Rb=Rbe=4,7kΩ, 50V, 0,1A, 0,3W	7c	Sam	AN 1L3M, DTA 143ES, RN 2001, UN 411L,++	data
KSR 2002	Si-P+R	=KSR 2001: Rb=Rbe=10kΩ	7c	Sam	AN 1A4M, DTA 114ES, RN 2002, UN 4111,++	data
KSR 2003	Si-P+R	=KSR 2001: Rb=Rbe=22kΩ	7c	Sam	AN 1F4M, DTA 124ES, RN 2003, UN 4112,++	data
KSR 2004	Si-P+R	=KSR 2001: Rb=Rbe=47kΩ	7c	Sam	AN 1L4M, DTA 144ES, RN 2004, UN 4113,++	data
KSR 2005	Si-P+R	=KSR 2001: Rb=4,7kΩ, Rbe=10kΩ	7c	Sam	AN 1L3N, DTA 143XS, UN 411F, 2SA1654	data
KSR 2006	Si-P+R	=KSR 2001: Rb=10kΩ, Rbe=47kΩ	7c	Sam	AN 1A4P, DTA 114YS, RN 2007, UN 4114,++	data
KSR 2007	Si-P+R	=KSR 2001: Rb=22kΩ, Rbe=47kΩ	7c	Sam	AN 1F4N, DTA 124XS, RN 2008	data
KSR 2008	Si-P+R	=KSR 2001: Rb=47kΩ, Rbe=22kΩ	7c	Sam	AN 1L4L, DTA 144WS, RN 2009, UN 411E,++	data
KSR 2009	Si-P+R	=KSR 2001: 40V, Rb=4,7kΩ, Rbe=-	7c	Sam	AN 1L3Z, DTA 143TS, RN 2010, UN 4116,++	data
KSR 2010	Si-P+R	=KSR 2001: 40V, Rb=10kΩ, Rbe=-	7c	Sam	AN 1A4Z, DTA 114TS, RN 2011, UN 4115,++	data
KSR 2011	Si-P+R	=KSR 2001: 40V, Rb=22kΩ, Rbe=-	7c	Sam	AN 1F4Z, DTA 124TS, UN 4117, 2SA1590	data
KSR 2012	Si-P+R	=KSR 2001: 40V, Rb=47kΩ, Rbe=-	7c	Sam	AN 1L4Z, DTA 144TS, UN 4110, 2SA1509	data
KSR 2013	Si-P+R	=KSR 2001: Rb=2,2kΩ, Rbe=47kΩ	7c	Sam	DTA 123JS, RN 2005	data
KSR 2014	Si-P+R	=KSR 2001: Rb=4,7kΩ, Rbe=47kΩ	7c	Sam	DTA 143ZS, RN 2006, 2SA1591, 2SA1616	data
KSR 2101	Si-P+R	=KSR 2001: SMD	35a	Sam	FN 1L3M, DTA 143EK, RN 2401, UN 211L,++	data
KSR 2102	Si-P+R	=KSR 2002: SMD	35a	Sam	FN 1A4M, DTA 114EK, RN 2402, UN 2111,++	data
KSR 2103	Si-P+R	=KSR 2003: SMD	35a	Sam	FN 1F4M, DTA 124EK, RN 2403, UN 2112,++	data
KSR 2104	Si-P+R	=KSR 2004: SMD	35a	Sam	FN 1L4M, DTA 144EK, RN 2404, UN 2113,++	data
KSR 2105	Si-P+R	=KSR 2005: SMD	35a	Sam	FN 1L3N, DTA 143XK, UN 211F, 2SA1653	data

Type	Device	Short Description	Fig.	Manu	Comparision Types	More at
KSR 2106	Si-P+R	=KSR 2006: SMD	35a	Sam	FN 1A4P, DTA 114YK, RN 2407, UN 2114,++	data
KSR 2107	Si-P+R	=KSR 2007: SMD	35a	Sam	FN 1F4N, DTA 124XK, RN 2408	data
KSR 2108	Si-P+R	=KSR 2008: SMD	35a	Sam	FN 1L4L, DTA 144WK, RN 2409, UN 211E,++	data
KSR 2109	Si-P+R	=KSR 2009: SMD	35a	Sam	FN 1L3Z, DTA 143TK, RN 2410, UN 2116,++	data
KSR 2110	Si-P+R	=KSR 2010: SMD	35a	Sam	FN 1A4Z, DTA 114TK, RN 2411, UN 2115,++	data
KSR 2111	Si-P+R	=KSR 2011: SMD	35a	Sam	FN 1F4Z, DTA 124TK, UN 2117, 2SA1589	data
KSR 2112	Si-P+R	=KSR 2012: SMD	35a	Sam	FN 1L4Z, DTA 144TK, UN 2210, 2SA1508	data
KSR 2113	Si-P+R	=KSR 2013: SMD	35a	Sam	DTA 123JK, RN 2405	data
KSR 2114	Si-P+R	=KSR 2014: SMD	35a	Sam	DTA 143ZK, RN 2406, 2SA1597	data
KSR 2201...2214	Si-P+R	=KSR 2001...2014:	41c	Sam	→KSR 2001...2014	data
KST 05	Si-N	=MPSA 05: SMD	35a	Sam	→MMBTA 05	data
KST 06	Si-N	=MPSA 06: SMD	35a	Sam	→MMBTA 06	data
KST 10	Si-N	=MPSH 10: SMD	35a	Sam	→MMBTH 10	data
KST 13	Si-N-Darl	=MPSA 13: SMD	35a	Sam	→MMBTA 13	data
KST 14	Si-N-Darl	=MPSA 14: SMD	35a	Sam	→MMBTA 14	data
KST 17	Si-N	=MPSH 17: SMD	35a	Sam	→MMBTH 17	data
KST 20	Si-N	=MPSA 20: SMD	35a	Sam	→MMBTA 20	data
KST 24	Si-N	=MPSH 24: SMD	35a	Sam	→MMBTH 24	data
KST 42	Si-N	=MPSA 42: SMD	35a	Sam	→MMBTA 42	data
KST 43	Si-N	=MPSA 43: SMD	35a	Sam	→MMBTA 43	data
KST 55	Si-P	=MPSA 55: SMD	35a	Sam	→MMBTA 55	data
KST 56	Si-P	=MPSA 56: SMD	35a	Sam	→MMBTA 56	data
KST 63	Si-P-Darl	=MPSA 63: SMD	35a	Sam	→MMBTA 63	data
KST 64	Si-P-Darl	=MPSA 64: SMD	35a	Sam	→MMBTA 64	data
KST 92	Si-P	=MPSA 92: SMD	35a	Sam	→MMBTA 92	data
KST 93	Si-P	=MPSA 93: SMD	35a	Sam	→MMBTA 93	data
KST 812	Si-P	≈2SA812: SMD	35a	Sam	→2SA812	data
KST 1009	Si-N	≈2SC1009: SMD	35a	Sam	→2SC1009	data
KST 1623	Si-N	≈2SC1623: SMD	35a	Sam	→2SC1623	data
KST 2222(A)	Si-N	=2N2222(A): SMD	35a	Sam	→MMBT 2222(A)	data
KST 2484	Si-N	=2N2484: SMD	35a	Sam	→MMBT 2484	data
KST 2907(A)	Si-P	=2N2907(A): SMD	35a	Sam	→MMBT 2907(A)	data
KST 3903	Si-N	=2N3903: SMD	35a	Sam	→MMBT 3903	data
KST 3904	Si-N	=2N3904: SMD	35a	Sam	→MMBT 3904	data
KST 3905	Si-P	=2N3905: SMD	35a	Sam	→MMBT 3905	data
KST 3906	Si-P	=2N3906: SMD	35a	Sam	→MMBT 3906	data
KST 4123	Si-N	=2N4123: SMD	35a	Sam	→MMBT 4123	data
KST 4124	Si-N	=2N4124: SMD	35a	Sam	→MMBT 4124	data
KST 4125	Si-P	=2N4125: SMD	35a	Sam	→MMBT 4125	data
KST 4126	Si-P	=2N4126: SMD	35a	Sam	→MMBT 4126	data
KST 4401	Si-N	=2N4401: SMD	35a	Sam	→MMBT 4401	data
KST 4403	Si-P	=2N4403: SMD	35a	Sam	→MMBT 4403	data
KST 5086	Si-P	=2N5086: SMD	35a	Sam	→MMBT 5086	data
KST 5087	Si-P	=2N5087: SMD	35a	Sam	→MMBT 5087	data
KST 5088	Si-N	=2N5088: SMD	35a	Sam	→MMBT 5088	data
KST 5089	Si-N	=2N5089: SMD	35a	Sam	→MMBT 5089	data
KST 5179	Si-N	=2N5179: SMD	35a	Sam	→MMBT 5179	data
KST 5401	Si-P	=2N5401: SMD	35a	Sam	→MMBT 5401	data
KST 5550	Si-N	=2N5550: SMD	35a	Sam	→MMBT 5550	data
KST 5551	Si-N	=2N5551: SMD	35a	Sam	→MMBT 5551	data
KST 6428	Si-N	=2N6428: SMD	35a	Sam	→MMBT 6428	data
KSV 3100 ACN-...	A/D-D/A-IC	8 Bit A/D- + 10 Bit D/a-converter	40-DIP	Sam	-	
KSV 3110(CN-...)	A/D-D/A-IC	8 Bit A/D- + 10 Bit D/A-Con., Video	40-DIP	Sam	-	
KSV 3208(CN)	A/D-IC	8 Bit, hi-speed, TV, Video, 20Msps	28-DIP	Sam	-	
KSV 3310	D/A-IC	10 Bit, TV, Video, 20MHz	28-DIP	Sam	-	
KSV 3404	D/A-IC	4 Bit, Triple, TV, Video, 100MHz	28-DIC	Sam	-	
KSY 13	GaAs-Sensor	SMD, Hall-Sensor	44	Sie	-	
KSZ 8041 FTL(I)	CMOS-IC	10Base-T/100Base-TX/100Base-FX Physical Layer Transceiver	MP	Mcr	-	pdf
KSZ 8041 NL(I)	IC	10Base-T/100Base-TX Physical Layer Transceiver	QFN	Mcr	-	pdf pinout
KSZ 8041 TL(I)	CMOS-IC	10Base-T/100Base-TX/100Base-FX Physical Layer Transceiver	MP	Mcr	-	pdf
KSZ 8721 BL(I)	IC	3.3V Single Power Supply 10/100BASE-Tx/fx MII Physical Layer Transceiver	MP	Mcr	-	pdf pinout
KSZ 8721 CL	IC	3.3V Single Power Supply 10/100BASE-Tx/fx MII Physical Layer Transceiver	MP	Mcr	-	pdf pinout
KSZ 8721 SL(I)	IC	3.3V Single Power Supply 10/100BASE-Tx/fx MII Physical Layer Transceiver	TSOP	Mcr	-	pdf pinout
KSZ 8841 -16MBL(I)	CMOS-IC	Single-port Ethernet Mac Controller with Non-PCI Interface	GRID	Mcr	-	pdf pinout
KSZ 8841 -16MQL	CMOS-IC	Single-port Ethernet Mac Controller with Non-PCI Interface	MP	Mcr	-	pdf pinout
KSZ 8841 -16MVL(I)	CMOS-IC	Single-port Ethernet Mac Controller with Non-PCI Interface	MP	Mcr	-	pdf pinout
KSZ 8841 -32MQL	CMOS-IC	Single-port Ethernet Mac Controller with Non-PCI Interface	MP	Mcr	-	puf pinout
KSZ 8841 -32MVL(I)	CMOS-IC	Single-port Ethernet Mac Controller with Non-PCI Interface	MP	Mcr	-	pdf pinout
KSZ 8841 -PMQL(I)	CMOS-IC	Single-port Ethernet Mac Controller with PCI Interface	MP	Mcr	-	pdf pinout
KSZ 8842 -16MBL(I)	CMOS-IC	2-Port Ethernet Switch with Non-PCI Interface	MP	Mcr	-	pdf pinout

Type	Device	Short Description	Fig.	Manu	Comparision Types	More at
KSZ 8842 -16MQL	CMOS-IC	2-Port Ethernet Switch with Non-PCI Interface	MP	Mcr	-	pdf pinout
KSZ 8842 -16MVL(I)	CMOS-IC	2-Port Ethernet Switch with Non-PCI Interface	MP	Mcr	-	pdf pinout
KSZ 8842 -32MQL	CMOS-IC	2-Port Ethernet Switch with Non-PCI Interface	MP	Mcr	-	pdf pinout
KSZ 8842 -32MVL(I)	CMOS-IC	2-Port Ethernet Switch with Non-PCI Interface	MP	Mcr	-	pdf pinout
KSZ 8842 -PMBL(AM)	CMOS-IC	2-Port Ethernet Switch with Non-PCI Interface	GRID	Mcr	-	pdf pinout
KSZ 8842 -PMQL(I)	CMOS-IC	2-Port Ethernet Switch with Non-PCI Interface	MP	Mcr	-	pdf pinout
KSZ 8862 -16MQL-FL	CMOS-IC	2-Port(1x 10BASE-FL) Ethernet Switch w. Non-PCI Interface & Fiber Support	MP	Mcr	-	pdf pinout
KSZ 8862 -16MQL-FX	CMOS-IC	2-Port(1x 100BASE-FX) Ethernet Switch w. Non-PCI Interface & Fiber Support	MP	Mcr	-	pdf pinout
KSZ 8862 -16MQL-SX	CMOS-IC	2-Port(1x 100BASE-SX) Ethernet Switch w. Non-PCI Interface & Fiber Support	MP	Mcr	-	pdf pinout
KSZ 8862 -32MQL-FL	CMOS-IC	2-Port(1x 10BASE-FL) Ethernet Switch w. Non-PCI Interface & Fiber Support	MP	Mcr	-	pdf pinout
KSZ 8862 -32MQL-FX	CMOS-IC	2-Port(1x 100BASE-FX) Ethernet Switch w. Non-PCI Interface & Fiber Support	MP	Mcr	-	pdf pinout
KSZ 8862 -32MQL-SX	CMOS-IC	2-Port(1x 100BASE-SX) Ethernet Switch w. Non-PCI Interface & Fiber Support	MP	Mcr	-	pdf pinout
KSZ 8893 FQL-FX -	CMOS-IC	Single-Chip 3-Port(1x 100Base-FX) Switch w. Fiber Support	MP	Mcr	-	pdf pinout
KSZ 8893 FQL	CMOS-IC	Single-Chip 3-Port Switch w. Fiber Support	MP	Mcr	-	pdf pinout
KSZ 8893 MBL	IC	Integrated 3-Port 10/100 Managed Switch with Phys	GRID	Mcr	-	pdf pinout
KSZ 8893 MQL(I)	IC	Integrated 3-Port 10/100 Managed Switch with Phys	MP	Mcr	-	pdf pinout
KSZ 8993	IC	3-Port 10/100 Integrated Switch	MP	Mcr	-	pdf pinout
KSZ 8993 M(L/I)	IC	Integrated 3-Port 10/100 Managed Switch with Phys	MP	Mcr	-	pdf pinout
KSZ 8995 M(I)	IC	Integrated 5-Port 10/100 Managed Switch	MP	Mcr	-	pdf pinout
KSZ 8995 X	CMOS-IC	Integrated 5-Port 10/100 QoS Switch	MP	Mcr	-	pdf pinout
KSZ 8995 XA	CMOS-IC	Integrated 5-Port 10/100 QoS Switch	MP	Mcr	-	pdf pinout
KSZ 8997	CMOS-IC	8-Port 10/100 Integrated Switch w. PHY & Frame Buffer	MP	Mcr	-	pdf pinout
KSZ 8999 (I)	CMOS-IC	Integrated 9-Port 10/100 Switch w. PHY & Frame Buffer	MP	Mcr	-	pdf

KT

Type	Device	Short Description	Fig.	Manu	Comparision Types	More at
KT	Z-Di	=1SMB 8.5A (Typ-Code/Stempel/marking)	71(5x3,5)	Ons	→1SMB 5.0A...170A	
KT	Si-N	=2SC4007 (Typ-Code/Stempel/marking)	35	Say	→2SC4007	data
KT	Si-N	=2SC4270 (Typ-Code/Stempel/marking)	35	Say	→2SC4270	data
KT	MOS-N-FET-e*	=2SK3120 (Typ-Code/Stempel/marking)	39	Say	→2SK3120	data
KT	Z-Di	=P6 SMB-12A (Typ-Code/Stempel/marking)	71(5x3,5)	Fag	→P6 SMB-12A	data
KT	Z-Di	=SMBJ 8.5A (Typ-Code/Stempel/marking)	71(5x3,5)	Mop	→SMBJ ...	data
KT 10 L07	Z-Di	TAZ, >65V, Ifsm=100A(1ms), Ih=100mA	78(10x9x4)	Shi	-	data
KT 10 L08	Z-Di	TAZ, >70V, Ifsm=100A(1ms), Ih=100mA	78(10x9x4)	Shi	-	data
KT 10 N12	Z-Di	TAZ, >110V, Ifsm=100A(1ms), Ih=100mA	78(10x9x4)	Shi	-	data
KT 10 N14	Z-Di	TAZ, >130V, Ifsm=100A(1ms), Ih=100mA	78(10x9x4)	Shi	-	data
KT 10 R25	Z-Di	TAZ, >220V, Ifsm=100A(1ms), Ih=100mA	78(10x9x4)	Shi	-	data
KT 10 R29	Z-Di	TAZ, <260V, Ifsm=100A(1ms), Ih=100mA	78(10x9x4)	Shi	-	data
KT 15 N12	Z-Di	TAZ, >110V, Ifsm=150A(1ms), Ih=100mA	78(10x9x4)	Shi	-	data
KT 15 N14	Z-Di	TAZ, >130V, Ifsm=150A(1ms), ih=100mA	78(10x9x4)	Shi	-	data
KT 15 R25	Z-Di	TAZ, >220V, Ifsm=150A(1ms), Ih=100mA	78(10x9x4)	Shi	-	data
KT 40 N14	Z-Di	TAZ, >130V, Ifsm=400A(1ms), Ih=200mA	78(10x9x4)	Shi	-	data
KT 40 R23	Z-Di	TAZ, >200V, Ifsm=400A(1ms), Ih=200mA	78(10x9x4)	Shi	-	data
KT 208(A...V)	Si-P	Uni, 20...60V, 0,15A, 0,2W	2d	GUS	BC 212, BC 556, 2SA970, 2SA1049, ++	
KT 209(A...V)	Si-P	Uni, 15...60V, 0,3A, 0,2W	7a	GUS	BC 327...328, BC 636, BC 638, BC 640, ++	
KT 301 А,Б,В	Si-N	Uni, 20(Б,В=30)V, 10mA, 0,15W, 30MHz, ß>20, А: >40, Б: >10, В: >20	2f	GUS	BC 108, BC 183, BC 238, BC 548, ++	
KT 301 Г,Д,Е,Ж	Si-N	Uni, 20V, 10mA, 0,15W, 60MHz, Г: ß>10, Д: >20, Е: >40, Ж: 80	2f	GUS	BC 108, BC 183, BC 238, BC 548, ++	
KT 312 А,Б,В	Si-N	Uni, 15(Б=30)V, 30mA, 0,225W, >80MHz, А: B=12...100, Б: 25...100, В: 50...250	2f	GUS	BC 108, BC 183, BC 238, BC 548, ++	
KT 370 А9,Б9	Si-P	SMD, S, 15V, 30mA, А9: >10, Б9: >12MHz	35a	GUS	BSR 12, 2SA1764	data
KT 393 А-1, Б-1	Si-P	Min, Dual, А-1=10V, Б-1=15V, >500MHz	36(6Pin)	GUS	-	data
KT 396 А-2	Si-N	Min, UHF, 15V, 40mA, >2100MHz	36c	GUS		data
KT 396 А9	Si-N	=KT 396A-2: SMD	35a		BFR 53, BFT 75, 2SC3014, 2SC3773	data
KT 397 А-2	Si-N	Min, HF, 40V, 10mA, >500MHz	36c	GUS	-	data
KT 398 А-1, Б-1	Si-N	Min, Dual, HF, 10V, 10mA, >1GHz	36(6Pin)	GUS	-	data
KT 398 А94, Б94	Si-N	=KT 398A-1,Б-1: SMD	6-MDIP		-	data
KT 399 АМ	Si-N	VHF/UHF, 15V, 20mA, 0,15W, >1800MHz	7a	GUS	BF 377...78, BF 630, BF 689K, 2SC2570, ++	data
KT 502 А	Si-P	Uni, 40V, 0,15A, 0,35W, >5MHz, B>40	7a	GUS	BC 257, BC 212, BC 307, BC 557, 2SB725++	data

561

Type	Device	Short Description	Fig.	Manu	Comparision Types	More at
KT 502 B	Si-P	=KT 502A: 60V	7a		BC 256, BC 212, BC 556, 2SA933A, 2SB725	data
KT 502 E	Si-P	=KT 502A: 90V	7a		2SA970, 2SA1049, 2SA1136, 2SA1335	data
KT 502 Б	Si-P	=KT 502A: B>80	7a		BC 257, BC 212, BC 307, BC 557, 2SB725++	data
KT 502 Д	Si-P	=KT 502A: 80V	7a		2SA970, 2SA1049, 2SA1136, 2SA1335	data
KT 502 Г	Si-P	=KT 502A: 60V, B>80	7a		BC 251, BC 212, BC 556, 2SA933A, 2SB725	data
KT 503 A	Si-N	Uni, 40V, 0,15A, 0,35W, >5MHz, B>40	7a	GUS	BC 167, BC 182, BC 237, BC 547, 2SD767++	data
KT 503 B	Si-N	=KT 503A: 60V	7a		BC 174, BC 182, BC 546, 2SC1815, 2SD767	data
KT 503 E	Si-N	=KT 503A: 90V	7a		2SC2240, 2SC2459, 2SC2674, 2SC3378	data
KT 503 Б	Si-N	=KT 503A: B>80	7a		BC 167, BC 182, BC 237, BC 547, 2SD767++	data
KT 503 Д	Si-N	=KT 503A: 80V	7a		2SC2240, 2SC2459, 2SC2674, 2SC3378	data
KT 503 Г	Si-N	=KT 503A: 60V, B>80	7a		BC 174, BC 182, BC 546, 2SC1815, 2SD767	data
KT 506 A	Si-N	S, 800V, 2A, 0,8W, >10MHz	2a§	GUS	(BUV 36, BUV 94...95, BUX 84...85, ++)[6]	data
KT 506 Б	Si-N	=KT 506A: 600/600V	2a§		(BUV 36, BUV 94...95, BUX 84...85, ++)[6]	data
KT 601 A	Si-N	HF, TV, 100V, 30mA, 0,5W, 40MHz	2b§	GUS	BC 141, BCX 40, 2N1990, 2N2102, ++	data
KT 601 AM	Si-N	HF, TV, 100V, 30mA, 0,5W, 40MHz	14b	GUS	BF 457...459, BF 469, BF 471, 2SC3423, ++	data
KT 602 А,Б,В,Г	Si-N	120/100(B,Г=80)V, 75mA, 2,8W, >150MHz	2b§	GUS	(BF457...459, 2SC3416, 2SD1609, 2SD2491+)[4]	data
KT 602 АМ,БМ	Si-N	HF, TV, 120/100V, 75mA, 0,85W, >150MHz AM: B=20...80, B: B=15...80, БМ,Г: B>50	14b	GUS	BF 457...459, 2SC3416, 2SD1609, 2SD2491++	data
KT 604 АМ,БМ	Si-N	HF, S, 300/250V, 0,2A, 0,8W, >40MHz	14b	GUS	BF 417, 2SC2621, 2SC4217	data
KT 605 АМ,БМ	Si-N	HF, Vid, 300/250V, 0,1A, 0,4W, >40MHz	14b	GUS	BF 459, 2SC3417, 2SC3789...90, 2SC4147,++	data
KT 606 А,Б	Si-N	VHF-E, 60V, 0,4A, A: >350, Б: >300MHz	49	GUS	-	data
KT 611 АМ,БМ	Si-N	HF, S, 200/180V, 0,1A, 0,8W, >60MHz	14b	GUS	BF 415, BF 458...459, 2SC3788, 2SC3955,++	data
KT 626 A	Si-P	NF/HF-L, 45V, 1,5A, 1W, >75MHz	14b	GUS	2SB1142	data
KT 626 B	Si-P	=KT 626A: 80V	14b		2SB1144	data
KT 626 Б	Si-P	=KT 626A: 60V	14b		2SB1142	data
KT 626 Д	Si-P	=KT 626A: 20V, 45MHz	14b		2SB891, 2SB1142	data
KT 626 Г	Si-P	=KT 626A: 20V, 45MHz	14b		2SB891, 2SB1142	data
KT 632 В1,Б1	Si-P	Uni, 110V, 0,1A, 0,5W, >200MHz	7a	GUS	2SA970, 2SA1049, 2SA1136, 2SA1335, ++	data
KT 632 Б	Si-P	Uni, 100V, 0,1A, 0,5W, >200MHz	2a		2SA970, 2SA1049, 2SA1136, 2SA1335, ++	data
KT 639 А,В,Б	Si-P	NF/HF-L, 45V, 1,5A, 1W, >80MHz	14b	GUS	2SB1142	data
KT 639 E	Si-P	=KT 639A,В,Б: 100V	14b		2SB1144	data
KT 639 Д	Si-P	=KT 639A,В,Б: 60V	14b		2SB1142	data
KT 639 Ж	Si-P	=KT 639A,В,Б: 100V	14b		2SB1144	data
KT 639 И	Si-P	=KT 639A,В,Б: 30V	14b		2SB1142	data
KT 639 Г	Si-P	=KT 639A,В,Б: 60V	14b		2SB1142	data
KT 640 A-2,A1-2	Si-N	UHF/SHF, 25V, 60mA, PQ>0,1W(7GHz)	25zf	GUS	BFQ 82, BFQ 196, 2SC3603	data
KT 640 B-2	Si-N	=KT 640A-2,A1-2: PQ>0,08W(7GHz)	25zf		BFQ 82, BFQ 196, 2SC3603	data
KT 643 A-2	Si-N	UHF/SHF, 25V, 1,1W, PQ>0,5W(7GHz)	25zf	GUS	BFQ 196, 2SC3603, 2SC3358	data
KT 644 А,Б	Si-P	NF/HF-L, 60V, 0,6A, 1W, >200MHz	14b	GUS	(2SB889, 2SB1309)[8]	data
KT 644 В,Г	Si-P	=KT 644A,Б: 40V	14b		(2SB889, 2SB1309)[8]	data
KT 645 A	Si-N	Uni, 60V, 0,3A, 0,5W	7a	GUS	BC 337A, 2SC2002...03, 2SC1627, 2SD1226	data
KT 645 Б	Si-N	=KT 645A: 40V *	7a		BC 337(A), 2SC2002...03, 2SC1627, 2SD1226	data
KT 646 А,Б	Si-N	NF/HF-L, 60V, 1A, 1W, >200MHz	14b	GUS	(2SC3421, 2SD1200, 2SD1381)[8]	data
KT 659 A	Si-N	S, 60V, 12A, 1W, >300MHz, <40/80ns	2a§	GUS		data
KT 660 A	Si-N	Uni, 50V, 0,8A, 0,5W, >200MHz	7a	GUS	BC 637, BC 639, 2SD1616, 2SD1768, ++	data
KT 660 Б	Si-N	=KT 660A: 30V	7a		BC 635, BC 637, 2SD1225, 2SD1616, ++	data
KT 661 A	Si-P	Uni,60V, 0,3A, 0,4W, >200MHz, <-/150ns	2a	GUS	BSW 24, 2N2907(A), 2SA1153, 2SA1218	data
KT 662 A	Si-P	Uni,60V, 0,4A, 0,6W, >200MHz, <-/150ns	2a§	GUS	BSW 23, 2N2905(A), 2SN4030...33	data
KT 664 A9	Si-P	SMD, Uni, 120V, 1A	39b	GUS	2SA1416, 2SB1025...26	data
KT 664 Б9	Si-P	=KT 664A9: 100V	39b		BCX 53, 2SA1416, 2SB803...04, 2SB1025...26	data
KT 665 A9	Si-N	SMD, Uni, 120V, 1A	39b		2SC3646, 2SD1418...19, 2SD1615A	data
KT 665 Б9	Si-N	=KT 665A9: 100	39b		BCX 56, 2SD1005, 2SD1368, 2SD1422	data
KT 668 А,В,Б	Si-P	Uni, 50V, 0,1A, 0,5W, >200MHz	7a	GUS	BC 212, BC 257, BC 307, BC 557, 2SB725++	data
KT 680 A	Si-N	Uni, 30V, 0,6A, 0,35W, >120MHz	7a	GUS	BC337...338, 2SD1226, 2SD1578, 2SD2322S...	data
KT 681 A	Si-P	Uni, 30V, 0,6A, 0,35W, >120MHz	7a	GUS	BC 327...328, 2SA1426, 2SB909...910, ++	data
KT 683 A	Si-N	NF-L, 150V, 1A, 8W, B=40...120	14b	GUS	2SC2690A, 2SC3117, 2SC3621, 2SD1563A	data
KT 683 B	Si-N	=KT 683A: 120V	14b		2SC2690(A), 2SC3621, 2SD1563(A), ++	data
KT 683 E	Si-N	=KT 683A: 60V, B=160...480	14b		BD 137, BD 228, 2SC2690(A), 2SD1563(A)++	data
KT 683 Б	Si-N	=KT 683A: 120V, B=80...240	14b		2SC2690(A), 2SC3621, 2SD1563(A), ++	data
KT 683 Д	Si-N	=KT 683A: 60V, B=80...240	14b		BD 137, BD 228, 2SC2690(A), 2SD1563(A)++	data
KT 683 Г	Si-N	=KT 683A: 100V	14b		BD 139, BD 230, 2SC2690(A), 2SD1563(A)++	data
KT 684 A	Si-P	Uni, 45V, 1A, 0,8W	7b	GUS	BC 636, BC 638, 2SA1705, 2SB1116, ++	data
KT 684 B	Si-P	=KT 684A: 100V	7b		BC 640, 2SA1708, 2SB984, 2SB1437, ++	data
KT 684 Б	Si-P	=KT 684A: 60V	7b		BC 638, BC 640, 2SA1705, 2SB1116, ++	data
KT 684 Г	Si-P	=KT 684A: 30V, 1,5A	7b		2SA966, 2SA1703, 2SB819, 2SB978, ++	data
KT 685 А,В	Si-P	Uni, 60V, 0,6A, 0,6W, >200MHz	7a	GUS	BC 327A, BC 638, 2SA1705, 2SB910, ++	data
KT 685 Е,Д,Ж	Si-P	=KT 685A,B: 30V	7a		BC 327...328, BC 636, 2SA1426, 2SB909, ++	data
KT 685 Б,Г	Si-P	=KT 685A,B: -/60V	7a		BC 640, 2SA965, 2SA1425, 2SB910, ++	data
KT 686 А,В	Si-P	Uni, 50V, 0,8A, 0,625W, >100MHz	7a	GUS	BC 327, BC 638, 2SA1705, 2SB1116, ++	data
KT 686 Д,Ж,Г	Si-P	=KT 686A,B: 30V	7a		BC 327...328, BC 636, 2SA1705, 2SB909, ++	data
KT 712 A	Si-P-Darl	L, 200V, 10A, 50W, B=500...10k	17j§	GUS	-	data
KT 712 Б	Si-P-Darl	=KT 712A: 160V, B=400...1000	17j§		-	data
KT 719 A	Si-N	NF/S-L, 120V, 1,5A, 10W, <-/2,5µs	14h§	GUS	2SC2690(A), 2SC3621, 2SD1563(A), ++	data
KT 720 A	Si-P	NF/S-L, 120V, 1,5A, 10W, <-/2,5µs	14h§	GUS	2SA1220(A), 2SA1249, 2SB1086(A), ++	data
KT 721 A	Si-N	NF/S-L, 120V, 3A, 25W, <-/2,5µs	14h§	GUS	BD 443(A)	data

Type	Device	Short Description	Fig.	Manu	Comparision Types	More at
KT 722 A	Si-P	NF/S-L, 120V, 3A, 25W, <-/2,5µs	14h§	GUS	-	data
KT 723 A	Si-N	NF/S-L, 120V, 10A, 60W, <-/3µs	17h§	GUS	(2SC2527, 2SC4330)[5]	data
KT 724 A	Si-P	NF/S-L, 120V, 10A, 60W, <-/3µs	17h§	GUS	(2SA1077, 2SA1646)[5]	data
KT 801 A,Б	Si-N	L, A=-/80, Б=-/60V, 2A, 5W, >10MHz A: B=13...50, Б: B=20...100	2c§	GUS	(BD 519, BD 527, 2SD1079, 2SD1281...82++)[4]	data
KT 802 A	Si-N	L, 150/130V, 5A, 50W, >10MHz	2e§	GUS	(BD 243D, MJE 5182, 2SC4329...30,++)[4]	data
KT 805 A,Б	Si-N	=KT 805 AM,БM:	2e§	GUS	(MJE 15030, 2SC4329, 2SC4330,++)[4]	
KT 805 AM	Si-N	NF/S-L, 160/60V, 5A, 30W, >20MHz	17h§	GUS	(2SC4329, 2SC4330, 2SD772A,B)[5]	data
KT 805 BM,БM	Si-N	=KT 805AM: 135/60V	17h§	GUS	(2SC4329, 2SC4330, 2SD772(A,B))[5]	data
KT 814 A,B,Б,Г	Si-P	NF/S-L, 40...100V, 1,5A, 10W, 3MHz A=40, B=70, Б=50, Г=100V	14h§	GUS	BD 140, BD 238, BD 380, 2SC1220(A), ++	
KT 815 A,B,Б,Г	Si-N	NF/S-L, 40...100V, 1,5A, 10W, 3MHz A=40, B=70V, Б=50, Г=100V	14h§	GUS	BD 139, BD 237, BD 379, 2SC2690(A), ++	
KT 816 A,B,Б,Г	Si-P	NF/S-L, 40...100V, 3A, 25W A=40, B=60, Б=45, Г=100V	14h§	GUS	BD 190, BD 724, BD 726	
KT 817 A,B,Б,Г	Si-N	NF/S-L, 40...100V, 3A, 25W A=40, B=60V, Б=45, Г=100V	14h§	GUS	BD 189, BD 723, BD 725, BD 443	
KT 818 A,B,Б,Г	Si-P	NF/S-L, 40...90V, 10A, 60W A=40, B=70, Б=50, Г=90V	17h§	GUS	(BD 806, BD 808, BD 810, BDT 96)[5]	data
KT 818 AM,BM,БM,ГM	Si-P	=KT 818A,B,Б,Г: 15A, 100W	23a§		BD 318, BD 746A...F, BDW 52(A...C)	data
KT 819 A,B,Б,Г	Si-N	NF/S-L, 40...100V, 10A, 60W A=40, B=70, Б=50, Г=100V	17h§	GUS	(BD 805, BD 807, BD 809, BDT 95)[5]	data
KT 819 AM,BM,БM,ГM	Si-N	=KT 819A,B,Б,Г: 15A, 100W	23a§		BD 317, BD 745A...F, BDW 51(A...C)	data
KT 822 A-1,B-1,Б-1	Si-P	Min, Uni, -/45...100V, 2A, 20W(Tc=25°) A-1=-/45, B-1=-/100, Б-1=-/60V	36	GUS	-	data
KT 823 A-1,B-1,Б-1	Si-N	Min, Uni, -/45...100V, 2A, 20W(Tc=25°) A-1=-/45, B-1=-/100, Б-1=-/60V	36	GUS	-	data
KT 825 Е,Д,Г	Si-P-Darl	NF/S-L, 30..90V, 20A, 125W, B>750 E=30V, Д=60V, Е=90V	23a§	GUS	BDX 68(A,B), 2N6285...87, 2SB1079	data
KT 827 A,B,Б	Si-P-Darl	NF/S-L, 60...100V, 20A, 125W, B>750 A=100V, B=60V, Б=80V	23a§	GUS	BDX 68(A,B), 2N6285...87, 2SB1079	data
KT 829 A,B,Б,Г	Si-N-Darl	NF/S-L, 45...100V, 8A, 60W, B>750 A=100V, B=60V, Б=80V, Г=45V	17j§	GUS	BD 649, BD 901, BDW 73A...D), 2SD1196, ++	data
KT 834 A,B,Б	Si-N-Darl	S-L, 400...500V, 15A, 100W, B>150 A=500, B=400V, Б=450V	23a§	GUS	BU 932(P), 2SD1314, 2SD1466	data
KT 835 A,Б	Si-P	NF-L, A=30V, Б=45V, 7,5A, 25W, >1MHz	17h§	GUS	(BD 544(A...D), BD 796, BD 806, ++)[5]	data
KT 837 A,B,Б	Si-P	NF-L, hi-Ueb, 80/60/15V, 7,5A, 30W	17h§	GUS	-	data
KT 837 Е,Д,Г	Si-P	NF-L, hi-Ueb, 60/45/15V, 7,5A, 30W	17h§	GUS	-	data
KT 837 К,Ж,И	Si-P	NF-L, hi-Ueb, 45/30/15V, 7,5A, 30W	17h§	GUS	-	data
KT 837 М,Л,Н	Si-P	NF-L, 80/60/5V, 7,5A, 30W	17h§	GUS	(BD 544B...D, BD 800, BD 810, BDT 96, ++)[5]	data
KT 837-Р,П,С	Si-P	NF-L, 60/45/5V, 7,5A, 30W	17h§	GUS	(BD 544A...D, BD 798, BD 808, BDT 96, ++)[5]	data
KT 837 Т,У,Ф	Si-P	NF-L, 45/30/5V, 7,5A, 30W	17h§	GUS	(BD 544A...D, BD 796, BD 806, BDT 96, ++)[5]	data
KT 838 A,Б	Si-N	TV, Video, A=1500V, Б=1200V, 0,1A, 52W	23a§	GUS	BU 208, BU 706, 2SC2928, 2SD350(A)	data
KT 840 A,B,Б	Si-N	S-L, 750...900V, 6A, 60W, <-/4,1µs A=900/400, B=800/375, Б=750/350V	23a§	GUS	BUW 132A, BUX 47A, 2SC2834A, 2SC3636, ++	data
KT 846 A,B,Б	Si-N	TV-HA, 1500V(Б=1200V), 5A, 52W	23a§	GUS	BU 208A, BU 706, 2SC2928, 2SD350A, ++	data
KT 847 A	Si-N	S-L, 650V, 15A, 125W, <-/4,5µs	23a§	GUS	BUS 13, BUS 23B, BUW 13, BUW 45...46, ++	data
KT 848 A	Si-N	S-L, 520/400V, 15A, 87W	23a§	GUS	BUP 23A,B, BUS 23(A), BUW 44...46, ++	data
KT 850 A	Si-N	S-L, 250/200V, 2A, 25W(Tc=100°)	17j§	GUS	D 44Q3,Q7, TIP 75, 2SC2023, 2SD610	data
KT 850 B	Si-N	=KT 850A: 180/150V	17j§		BF 239F, D 44Q1, 2SC2660, 2SD1138	data
KT 850 Б	Si-N	=KT 850A: 300/250V	17j§		D 44Q7, TIP 75, 2SC2023, 2SC3968	data
KT 851 A	Si-P	S-L, 250/200V, 2A, 25W(Tc=100°)	17j§	GUS	2SA1009(A), 2SA1236	data
KT 851 B	Si-P	=KT 851A: 180/150V	17j§		BF 240F, 2SA1133, 2SB861, 2SB720	data
KT 851 Б	Si-P	=KT 851A: 300/250V	17j§		2SA1009(A), 2SA1236	data
KT 852 A,B,Б,Г	Si-P-Darl	L, 45...100V, 2,5A, 50W, B>500 A=100V, B=60V, Б=80V, Г=45V	17j§	GUS	BDT 60(A...C, BDW 54(A...D), 2SB751(A,B)++	data
KT 853 A,B,Б,Г	Si-P-Darl	L, 45...100V, 8A, 60W, B>750 A=100V, B=60V, Б=80V, Г=45V	17j§	GUS	BDW 47(A...D), BDX 54(A...D), 2SB886, ++	data
KT 854 A	Si-N	S-L, 600/500V, 10A, 60W, >10MHz	17j§	GUS	BUF 650, MJE 13008...09	data
KT 854 Б	Si-N	=KT 854A: 400/300V	17j§		BUT 62, BUV 28(A), 2SC3562, 2SC4107	data
KT 855 A	Si-N	S-L, 250V, 5A, 40W, >5MHz	17j§	GUS	MJE 51T, 2N6497, 2SC2767	data
KT 855 B,Б	Si-N	=KT 855A: 150V	17j§		MJE 51T, 2N6497, 2SC2767	data
KT 857 A,B	Si-N	S-L, 250V, 7A, 60W(B=40W), >100MHz	17j§	GUS	BU 409, TIP 150, 2SC4848	data
KT 858 A,Б	Si-N	S-L, 400V, 7A, 60W(Б=40W), >100MHz	17j§	GUS	TIP 152, 2SC2335, 2SC2427, 2SC4106, ++	data
KT 859 A	Si-N	S-L, 800V, 3A, 40W, >10MHz	17j§	GUS	BUT 11(A), 2SC3086, 2SC3148, 2SC3490, ++	data
KT 862 Б	Si-N	S-L, 500/250V, 15A, 50W, <400/1250ns	≈62/b§	GUS	-	data
KT 862 B,Г	Si-N	S-L, 600V, 10A, 50W, <500/2500ns B=600/350V, Г=600/400V	≈62/b§	GUS	-	data
KT 863 A,Б	Si-N	lo-sat, 30V, 10A, 50W, >4MHz	17j§	GUS	2SD1062, 2SD1443(A)	data
KT 863 B	Si-N	lo-sat, 160V, 10A, 50W, >4MHz	17j§	GUS	2SC4330	data
KT 864 A	Si-N	NF/S-L, 200V, 10A, 100W, >15MHz	23a§	GUS	BD 245F, MJ 15015, 2SC3263	data
KT 865 A	Si-P	NF/S-L, 200V, 10A, 100W, >15MHz	23a§	GUS	BD 246F, MJ 15016, 2SA1294	data
KT 872 A,Б	Si-N	TV-HA, 1500/700V, 8A, 100W	18j§	GUS	BU 508A, BU 2508A, 2SC3687, 2SC5076, ++	data

Type	Device	Short Description	Fig.	Manu	Comparision Types	More at
KT 874 А,Б	Si-N	S-L, 150V, 30A, 75W, <600/700ns A=150/100V, Б=150/120V	≈62/b§	GUS	-	data
KT 877 A-5	Si-P-Darl	80V, 20A, B>750, >100MHz, <400/2300ns	Chip	GUS	-	data
KT 888 A-5	Si-P	S, 900V, 0,1A, >15MHz, <-/3600ns	Chip	GUS	-	data
KT 888 Б-5	Si-P	=KT 888A-5: 600V	Chip		-	data
KT 896 A	Si-N-Darl	L, 90V, 20A, 125W, B>750, <1/4,5µs	18j§	GUS	2SD1559, 2SD2083, 2SD2256	data
KT 896 Б	Si-N-Darl	=KT 896A: 60V	18j§		2SD1559, 2SD2083, 2SD2256	data
KT 897 A	Si-N-Darl	L, 350V, 20A, 150W, B>400, >10MHz	23a§	GUS	MJ 10000...01, MJ 10004...05, 2SD831	data
KT 897 Б	Si-N-Darl	=KT 897A: 200V	23a§		MJ 10000...01, MJ 10004...05, 2SD831	data
KT 898 А,Б	Si-N-Darl	=KT 897A,Б: 125W	18j§		BUT 13P	data
KT 914 A	Si-N	VHF/UHF-L, 65V, 0,8A, PQ>2,5W(400MHz)	49a	GUS	2N4440	data
KT 919 A	Si-N	UHF-L, 45V, 0,7A, PQ>4,4W(200MHz)	62a&	GUS	-	data
KT 919 B	Si-N	=KT 919A: 0,2A, PQ>1W(200MHz)	62a&		-	data
KT 919 Б	Si-N	=KT 919A: 0,35A, PQ>2W(200MHz)	62a&		-	data
KT 920 A	Si-N	FM/VHF-Tr/E, 36V, 0,5A, 5W, >400MHz	55r	GUS	-	data
KT 920 B	Si-N	=KT 920A: 3A, 25W	55r		-	data
KT 920 Б	Si-N	=KT 920A: 1A, 10W	55r		-	data
KT 920 Г	Si-N	=KT 920A: 3A, 25W, >350MHz	55r		-	data
KT 921 A	Si-N	AM/FM-L, 70V, >90MHz, PQ>12,5W(50MHz)	49a	GUS	-	data
KT 921 B	Si-N	=KT 921A: 50V, >60MHz	49a		-	data
KT 922 A	Si-N	AM/FM-Tr/E, 65V, 0,8A, 8W, >300MHz	55r	GUS	-	data
KT 922 B	Si-N	=KT 922A: 3A, 40W	55r		-	data
KT 922 Б	Si-N	=KT 922A: 1,5A, 20W	55r		-	data
KT 922 Д	Si-N	=KT 922A: 3A, 40W, >250MHz	55r		-	data
KT 922 Г	Si-N	=KT 922A: 1,5A, 20W	55r		-	data
KT 925 A	Si-N	VHF-Tr/E, 36V, 0,5A, PQ>2W(300MHz)	55r	GUS	-	data
KT 925 B	Si-N	=KT 925A: 3,3A, PQ>5W(300MHz)	55r		-	data
KT 925 Б	Si-N	=KT 925A: 1A, PQ>20W(300MHz)	55r		-	data
KT 925 Г	Si-N	=KT 925A: 3,3A, PQ>15W(300MHz)	55r		-	data
KT 929 A	Si-N	FM/VHF-Tr/E, 30V, 0,8A, PQ>2W(175MHz)	55r	GUS	-	data
KT 930 A	Si-N	VHF/UHF-L, 6A, >450MHz, PQ>40W(400MHz)	57s	GUS	2N6439, 2SC2796	data
KT 930 Б	Si-N	=KT 930A: 10A, PQ>75W(400MHz)	57s		MRF 327, 2SC2186	data
KT 931 A	Si-N	FM/VHF-L, 15A, >2,5GHz, PQ>80W(175MHz)	57s	GUS	2SC2609, 2SC2891	data
KT 932 А,В,Б	Si-P	NF/S-L, 40...80V, 2A, 20W, >30MHz A=80V, B=40V, Б=60V	23a§	GUS	2N6226, 2SA1693, 2SD193, 2SB775	data
KT 933 А,Б	Si-P	L, A=90V, Б=60V, 0,5A, 6,5W, >75MHz	49a	GUS	-	data
KT 934 A	Si-N	VHF/UHF-Tr/E, 60V, 0,6A, PQ>3W(400MHz)	55r	GUS	-	data
KT 934 B	Si-N	=KT 934A: 20A, PQ>25W(400MHz)	55r			data
KT 934 Б	Si-N	=KT 934A: 10A, PQ>12W(400MHz)	55r			data
KT 934 Д	Si-N	=KT 934A: 20A, PQ>20W(400MHz)	55r			data
KT 934 Г	Si-N	=KT 934A: 10A, PQ>10W(400MHz)	55r			data
KT 940 A	Si-N	TV, Vid, 300V, 0,1A, 10W, >90MHz	14h§	GUS	BF 459, 2SC3417, 2SC3503, 2SC4147,++	data
KT 940 B	Si-N	=KT 940A: 160V	14h§		BF 457, 2SC3416, 2SC3502, 2SC3600,++	data
KT 940 Б	Si-N	=KT 940A: 250V	14h§		BF 458, 2SC3417, 2SC3503, 2SC4147,++	data
KT 944 A	Si-N	AM-L, 100V, 12,5A, PQ>100W(30MHz)	49a	GUS	-	data
KT 945 Б	Si-N	S-L, 150V, 15A, 50W(Tc=50°), >51MHz	23a§	GUS	BDW 12, MJE 4343, 2SD551, 2SD1047, ++	data
KT 947 A	Si-N	HF-L, 100V, 20A, PQ>250W(1,5MHz)	49a	GUS	-	data
KT 948 A	Si-N	UHF-L, 45V, 2,5A, PQ>15W(2GHz)	62a&	GUS	-	data
KT 948 Б	Si-N	=KT 948A: 1,25A, PQ>8W(2GHz)	62a&		-	data
KT 955 A	Si-N	AM-L, 70V, 2A, PQ>10W(30MHz)	59r	GUS	-	data
KT 956 A	Si-N	AM-L, 100V, 15A, PQ>80W(30MHz)	55r	GUS	-	data
KT 957 A	Si-N	AM-L, 60V, 20A, PQ>150W(30MHz)	55r	GUS	-	data
KT 958 A	Si-N	FM/VHF-L, 36V, 10A, PQ>40W(175MHz)	57s	GUS	-	data
KT 960 A	Si-N	VHF/UHF-L, 36V, 10A, PQ>40W(400MHz)	57s	GUS	-	data
KT 961 А,В,Б	Si-N	NF/S-L, 60...100V, 1,5A, 12,5W, >50MHz A=100V, B=60V, Б=80V	14h§	GUS	BD 139, BD 230, 2SD1177...78, 2SD1563, ++	data
KT 962 A	Si-N	UHF-L, 50V, 1,5A, PQ>10W(1GHz)	55v	GUS	-	data
KT 962 B	Si-N	=KT 962A: 4A, PQ>40W(1GHz)	55v		-	data
KT 962 Б	Si-N	=KT 962A: 2,5A, PQ>20W(1GHz)	55v		-	data
KT 965 A	Si-N	AM-L, 36V, 4A, PQ>10W(30MHz)	59r	GUS	MRF 449	data
KT 966 A	Si-N	AM-L, 36V, 8A, PQ>40W(30MHz)	59r	GUS	MRF 450	data
KT 967 A	Si-N	AM-L, 36V, 15A, PQ>90W(30MHz)	55r	GUS	MRF 454A	data
KT 969 A	Si-N	TV, Vid, 300V, 0,1A, 6W, >60MHz	14h§	GUS	BF 459, 2SC3417, 2SC3503, 2SC4147,++	data
KT 970 A	Si-N	VHF/UHF-L, 50V, 13A, PQ>100W(400MHz)	≈57s	GUS	-	data
KT 971 A	Si-N	FM/VHF-L, 50V, 17A, PQ>150W(175MHz)	≈57s	GUS	-	data
KT 972 А,Б	Si-N-Darl	L, 60(Б=45)V, 2A, 8W, >200MHz, B>2000	14h§	GUS	BD 677, BD 679, 2SC4341, 2SD1438	data
KT 973 А,Б	Si-P-Darl	L, 60(Б=45)V, 2A, 8W, >200MHz, B>2000	14h§	GUS	BD 678, BD 680, 2SB795, 2SB1034	data
KT 976 A	Si-N	UHF-L, 50V, 6A, PQ>60W(1GHz)	55v	GUS	-	data
KT 981 A	Si-N	AM/FM-L, 36V, 10A, PQ>50W(80MHz)	55r	GUS	-	data
KT 983 A	Si-N	VHF/UHF-L, 40V, 0,5A, PQ>0,5W(860MHz)	55r	GUS	-	data

Type	Device	Short Description	Fig.	Manu	Comparision Types	More at
KT 983 B	Si-N	=KT 983A: 2A, PQ>3,5W(860MHz)	55r		-	data
KT 983 Б	Si-N	=KT 983A: 1A, PQ>1W(860MHz)	55r		-	data
KT 984 A	Si-N	UHF-L, 65V, 7A, PQ>75W(10µs)(820MHz)	62a&	GUS	-	data
KT 984 Б	Si-N	=KT 984A: 16A, PQ>250W(10µs)(820MHz)	62a&		-	data
KT 985 AC	Si-N	Dual, UHF-L, 50V, 17A, PQ>125W(400MHz)	≈61(8Pin)	GUS	-	data
KT 991 AC	Si-N	Dual,UHF-L, 50V, 3,75A, PQ>55W(700MHz)	≈61(4Pin)	GUS	-	data
KT 997 A,Б	Si-N	NF/S-L, -/45V, 10A, 50W, >51MHz	17j§	GUS	2SC3345...46, 2SD1212, 2SD1443A	data
KT 2206B	LIN-IC	2x Power Amp, 9V, 2x2.3W(9V/4Ohm)	12-DIP+b		KA2206, KA 22061, LA 4182...4183, TEA 202	pinout
KT 3030	CMOS-IC	Telecom, µ-Law Combo codec	16-DIC	Sam	ETC 5054, TP 3054, µA 3054, 2916	pinout
KT 3031	CMOS-IC	Telecom, µ-Law Combo codec	20-DIC	Sam	ETC 5064, KT 3064, TP 3064	
KT 3032	CMOS-IC	Telecom, A-law Combo codec	16-DIC	Sam	ETC 5057, TP 3057, µA 3057, 2917	pinout
KT 3033	CMOS-IC	Telecom, A-law Combo codec	20-DIC	Sam	ETC 5067, TP 3067	
KT 3040 J	CMOS-IC	Telecom, Pcm Codec-filter	16-DIC	Sam	ETC 5040, M 5912, TP 3040, µA 5912, 2912	
KT 3054 J	CMOS-IC	Telecom, µ-Law Combo codec	16-DIC	Sam	→KT 3030	
KT 3064 J	CMOS-IC	Telecom, µ-Law Combo codec	20-DIC	Sam	→KT 3031	
KT 3102 АМ,БМ,ЖМ, ИМ	Si-N	Uni, 50V, 0,2A, 0,25W, >200MHz	7a	GUS	BC 167, BC 182, BC 237, BC 547, 2SC1740	data
KT 3102 ВМ,КМ	Si-N	=KT 3102AM...: 30V	7a		BC 168, BC 183, BC 238, BC 548, 2SC1740	data
KT 3102 ЕМ	Si-N	=KT 3102AM...: ra, 20V, >300MHz	7a		BC 184, BC 239, BC 549, 2SC3112...13	data
KT 3102 ДМ	Si-N	=KT 3102AM...: ra, 30V	7a		BC 167, BC 184, BC 239, BC 549, 2SC2675	data
KT 3102 ГМ	Si-N	=KT 3102AM...: 20V, >300MHz	7a		BC 168, BC 183, BC 238, BC 548, 2SC3069	data
KT 3103 A1,Б1	Si-	Dual, 15V, 20mA, 0,3W, >600MHz	8-DIP	GUS	-	data
KT 3106 A-2	Si-N	Min, VHF/UHF, 15V, 20mA, >1000MHz	36c	GUS	-	data
KT 3107 A,B,И	Si-P	Uni, 50V, 0,1A, 0,3W, >250MHz	7a	GUS	BC 212, BC 307, BC 557, 2SA1015, ++	data
KT 3107 В,К,Д,Г	Si-P	=KT 3107A...: 30V	7a		BC 213, BC 308, BC 558, 2SA1015, ++	data
KT 3107 Е,Ж,Л	Si-P	=KT 3107A...: ra, 25V	7a		BC 214, BC 309, BC 559, 2SA1137, ++	data
KT 3109 A,B,Б	Si-P	VHF/UHF, 25...30V, 50mA, >1100MHz	24e	GUS	BF 679...680, BF 779...780, BF 967...970	data
KT 3117 A1	Si-N	Uni, lo-sat, 60V, 0,4A, 0,5W, >200MHz	7a		2SC4414, 2SC4485, 2SD1616(A), 2SD2181	data
KT 3117 Б	Si-N	=KT 3117A1: 75V, 0,3W	2a		2SC4414, 2SC4488, 2SD1616A, 2SD2181	data
KT 3121 A-6	Si-N	UHF, 12V, 10mA, F<2dB(1GHz)	Chip	GUS	-	data
KT 3122 A,Б	Si-N	SS, 35V, 0,1A, 0,15W, tf<1...1,5ns	7c	GUS	BSV 91...92, BSX 92...93, 2SC3811	data
KT 3123 АМ,ВМ,БМ	Si-P	UHF, 15V, 30mA, >3..4GHz	24e	GUS	BFQ 23, BFQ 56, BFQ 75...76, 2SA967, ++	data
KT 3126 A,Б	Si-P	VHF, 30V, 30mA, 0,15W, >500MHz	7a	GUS	BF 324, BF 414, BF 506, BF 914, BF 936++	data
KT 3126 A9	Si-P	=KT 3126A,Б: SMD, 35V, >450MHz	35a		BF 568...569	data
KT 3127 A	Si-P	VHF, 20V, 25mA, 0,1W, >600MHz	5g	GUS	BF 324, BF 414, BF 506, BF 914, BF 936++	data
KT 3128 A	Si-P	VHF, 40V, 20mA, 0,1W, >700MHz	5g	GUS	BF 324, BF 414, BF 506, BF 914, BF 936++	data
KT 3128 A9	Si-P	=KT 3128A: SMD, 35V, >650MHz	35a		BF 568...569	data
KT 3129 A9,Б9	Si-P	SMD, Uni, 50V, 0,1A, >200MHz	35a	GUS	BC 856...857, BCW 69...70, BCW 89	data
KT 3129 В9,Г9	Si-P	=KT 3129A9,Б9: 30V	35a		BC 856...858, BCW 29...30, BCW 61	data
KT 3129 Д9	Si-P	=KT 3129A9,Б9: 20V	35a		BC 856...858, BCW 29...30, ECW 61	data
KT 3130 A9,Б9	Si-N	SMD, Uni, 50V, 0,1A, >15MHz	35a	GUS	BC 846...847, BCW 71...72, BCV 71...72	data
KT 3130 В9,Ж9	Si-N	=KT 3130A9,Б9: 30V	35a		BC 846...848, BCW 31...33, BCW 81, ++	data
KT 3130 Е9,Г9	Si-N	=KT 3130A9,Б9: 20V	35a		BC 846...848, BCW 31...33, BCW 81, 2SC3689	data
KT 3130 Д9	Si-N	=KT 3130A9,Б9: ra, 30V	35a		BC 849...850, BCF 32...33, BCF 81, 2SC3323	data
KT 3142 A	Si-N	S, VHF/UHF, 40V, 0,2A, >500MHz	2a	GUS	BSS 11, 2N2368...69(A), 2SC3732	data
KT 3150 Б-2	Si-P	Min, VHF/UHF, 40V, 50mA, >1200MHz	36c	GUS	-	data
KT 3151 A9,Б9	Si-N	SMD, Uni, 80V, 0,1A, >100MHz	35a	GUS	BC 846, BCV 71...72, 2SC1622A, 2SC4050	data
KT 3151 В9	Si-N	=KT 3151A9,Б9: 60V	35a		BC 846, BCV 71...72, 2SC3323, 2SC4076	data
KT 3151 Е9	Si-N	=KT 3151A9,Б9: 20V	35a		BC 846...848, BCW 31...33, 2SC1622, ++	data
KT 3151 Д9	Si-N	=KT 3151A9,Б9: 30V	35a		BC 846...848, BCW 31...33, 2SC1622, ++	data
KT 3151 Г9	Si-N	=KT 3151A9,Б9: 40V	35a		BC 846...847, BCW 71...72, BCW 81, ++	data
KT 3153 A9	Si-N	SMD, Uni, 60V, 0,4A, >250MHz	35a	GUS	BCW 65, MBTA 05...06, 2SC4519	data
KT 3157 A	Si-P	Uni, Vid, 250V, 30mA, 0,2W	7d	GUS	BF 421, BF 423, BF 484, 2SA1371, ++	data
KT 3165 A	Si-P	VHF/UHF, 40V, 30mA, >750MHz	24a	GUS	BF 679...680, BF 779...780, BF 967...970	data
KT 3170 J,N	CMOS-IC	Telecom, DTMF-Empf./receiver	18-DIC,DIP	Sam	G 8870, MT 8870	pinout
KT 3172 A9	Si-N	SMD, Uni, 20V, 0,2A	35a	GUS	BC 846...48, BCW 60, BCX 70, 2SC3052, ++	data
KT 3174 AC-2	Si-N	Dual, 10V, 7,5mA, 0,15W	8-MDIP	GUS		data
KT 5116 J	CMOS-IC	Telecom, µ-Law Companding codec	16-DIC	Sam	M 5116, MK 5116, TP 5116, µA 5116	
KT 8101 A	Si-N	NF/S-L, 200V, 16A, 150W	18j§	GUS	BD 249F, 2SC3263, 2SC3856	data
KT 8101 Б	Si-N	=KT 8101A: 160V	18j§		BD 745E...F, MJ 4343, 2SC3263, 2SC3856	data
KT 8102 A	Si-P	NF/S-L, 200V, 16A, 150W	18j§	GUS	BD 250F, 2SA1294	data
KT 8102 Б	Si-P	=KT 8102A: 160V	18j§		BD 746E...F, MJ 4353, 2SA1294, 2SA1386(A)	data
KT 8107 A	Si-N+Di	TV-HA, 1500V, 8A, 100W, sat<1V(4,5A)	18j§	GUS	BU 508D, BU 2508D, 2SC3683	data
KT 8107 B	Si-N	=KT 8107A: ohne/wo. Damper-Di	18j§		BU 508A, BU 2508A, 2SC3687	data
KT 8107 Б	Si-N+Di	=KT 8107A: sat<3V(4,5A)	18j§		BU 508D, BU 2508D, 2SC3683	data
KT 8107 A1,Б1	Si-N+Di	=KT 8107A1,Б1: Iso	18c		BU 508DF, BU 2508DF, 2SC4763, 2SC5149,++	data
KT 8107 Г	Si-N	=KT 8107A: ohne/wo. Damper-Di, sat<3V	18j§		BU 508A, BU 2508A, 2SC3687	data
KT 8107 B1,Г1	Si-N	=KT 8107B1,Г1: Iso	18c		BU 508AF, BU 2508AF, 2SC4758, 2SC5067,++	data
KT 8109 A	Si-N-Darl	S-L, 350V, 5A, 40W, B>180	17h§	GUS	(2SD978, 2SD987, 2SD1072)[5]	data
KT 8110 A	Si-N	S-L, 700V, 7A, 60W, 20MHz, <700/3200ns A=700/400V, B=700/350V, Б=700/400V	17j§	GUS	BUT 12, BUT 54, BUT 56, MJE 13007, ++	data
KT 8518	CMOS-IC	8 Bit Addressable Latched Driver	16-DIP	Sam		

Type	Device	Short Description	Fig.	Manu	Comparision Types	More at
KT 8520	CMOS-IC	Telecom, µ-Law codec		Sam	TP 3020, 2910	
KT 8521	CMOS-IC	Telecom, A-law codec		Sam	TP 3021, 2911	
KT 8554(J)	CMOS-IC	Telecom, 1 Chip Codec, µ-Law	16-DIC	Sam	ETC 5054, TP 3054, µA 3054, 2916	pinout
KT 8555(J)	CMOS-IC	Telecom, Time Slot Assignment Circuit	20-DIC	Sam	TP 3155	pdf pinout
KT 8557(J)	CMOS-IC	Telecom, 1 Chip Codec, A-law	16-DIC	Sam	ETC 5057, TP 3057, µA 3057, 2917	pinout
KT 8592(N)	CMOS-IC	Telecom, 4x4 Crosspoint Switch, Memory	16-DIP	Sam	-	
KT 8593(N)	CMOS-IC	Telecom,12x8 Crosspoint Switch, Memory	40-DIP	Sam	-	pinout
KT 9101 AC	Si-N	Dual, UHF-L, 50V, 7A, PQ>100W(700MHz)	≈61(4Pin)	GUS	•	data
KT 9104 A	Si-N	UHF-L, 50V, 1,5A, PQ>5W(700MHz)	≈62a&	GUS	-	data
KT 9104 Б	Si-N	=KT 9104A: 5A, PQ: 20W(700MHz)	≈62a&	GUS	-	data
KT 9115 A	Si-P	TV, Vid-E, 300V, 0,1A, 10W	14h§	GUS	2SA1353, 2SA1381	data
KT 9115 Б	Si-P	=KT 9115A: 150V	14h§	GUS	2SA1352...53, 2SA1380...81, 2SB1109, ++	data
KT 9120 A	Si-P	NF/S-L, 45V, 12A, 50W, >50MHz	17j§	GUS	2SA1328...29, 2SB903	data
KT 9133 A	Si-N	VHF-L, 55V, 16A, PQ>30W(225MHz)	≈57s	GUS	MRF 327, 2SC2186, 2SC2896...97	data
KT 9141 A	Si-N	HF, 120V, 0,3A, 3W(Tc=85°), >1020MHz	2a§	GUS	-	data
KT 9141 A1	Si-N	=KT 9141A: 0,4A, 5W(Tc=85°)	55v	GUS	-	data
KT 9142 A	Si-N	Dual, UHF-L, 55V, 15A, PQ>50W(860MHz)	≈61(4Pin)	GUS	-	data
KT 9144 A9	Si-P	SMD, S, 500V, 50mA, >30MHz	39b	GUS	2SA1583	data
KT 9145 A9	Si-N	SMD, S, 500V, 50mA, >50MHz	39b	GUS	2SC4114, 2SD1473	data
KT 9150 A	Si-N	Dual, UHF-L, 40V, 5A, PQ>8W(860MHz)	≈61(4Pin)	GUS	-	data
KT 9151 A	Si-N	Dual, VHF-L, 55V, 3A, PQ>200W(230MHz)	≈61(4Pin)	GUS	-	data
KT 9152 A	Si-N	Dual, UHF-L, 55V, 24A, PQ>100W(860MHz)	≈61(4Pin)	GUS	-	data
KT 9157 A	Si-N	Foto, L, 30V, 5A, 10W	14h§	GUS	2SC2270, 2SD826, 2SD1691, 2SD2147	data
KT 9160 A,B,Б	Si-N	L, 140V, 467W(Tc=50°), PQ>700W(1,5MHz)		GUS	-	data
KT 9166 A	Si-N	L, lo-sat, 60V, 15A, 50W, >90MHz	17j§	GUS	-	data

KTA

Type	Device	Short Description	Fig.	Manu	Comparision Types	More at
KTA 200	Si-P	Uni, 60V, 0,5A, 0,625W, 200MHz	7c	Kec	BC 327A, BC 638, BC 640, 2SB910, ++	data
KTA 473	Si-P	=2SA473	17j§	Kec	→2SA473	data
KTA 501 U	Si-P	SMD, Dual, 50V, 0,15A, >80MHz	45ba1(2mm)	Kec	-	data
KTA 701 U	Si-P	SMD, Dual, 50V, 0,15A, >80MHz	46bh1(2mm)	Kec	-	data
KTA 708	Si-P	Uni, 80V, 1A, 0,625W, 150MHz	7c	Kec	BC 640, 2SB647, 2SB1041, 2SB1437, ++	data
KTA 940	Si-P	=2SA940	17j§	Kec	→2SA940	data
KTA 950	Si-P	=2SA950	7c(9mm)	Kec	→2SA950	data
KTA 968(A)	Si-P	=2SA968(A)	17j§	Kec	→2SA968	data
KTA 1001	Si-P	SMD, Blitzg./Strobo, 35V, 3A, 170MHz	39b	Kec	2SA1730, 2SA1736, 2SB1124	data
KTA 1015	Si-P	≈2SA1015	7c	Kec	→2SA1015	data
KTA 1021	Si-P	Uni, 35V, 0,5A, 0,4W, 200MHz	41c	Kec	BC 327...328, BC 638, BC 640, 2SB910, ++	data
KTA 1023	Si-P	Uni, 120/120V, 0,8A, 1W, 120MHz	7c(9mm)	Kec	2SA1013, 2SA1275, 2SB647(A), 2SB1456, ++	data
KTA 1024	Si-P	Uni, 150/150V, 0,05A, 1W, 120MHz	7c(9mm)	Kec	2SA1016K, 2SA1123..24, 2SA1482	data
KTA 1036	Si-P	NF/S-L, 60V, 3A, 30W, 30MHz	17j§	Kec	BD 242A, BD 536, BD 936, 2SA1288, ++	data
KTA 1038	Si-P	NF/S-L, 100V, 5A, 40W, 30MHz	17j§	Kec	BD 244C, BD 544C, BD 802, 2SB633	data
KTA 1040 D,L	Si-P	NF/S-L, 60V, 3A, 20W, 30MHz	30j§	Kec	2SA1244, 2SB1184, 2SB1202...03	data
KTA 1042 D,L	Si-P	NF/S-L, 100V, 5A, 20W, 30MHz	30j§	Kec	2SA1648, 2SB1535	data
KTA 1045 D,L	Si-P	NF/S-L, lo-sat, 120V, 1A, 8W, 110MHz	30j§	Kec	(2SA1592, 2SB844, 2SB845, 2SB959,2SB960)[17]	data
KTA 1046	Si-P	NF-L, 60V, 3A, 25W, 30MHz, 400/2200ns	17c	Kec	BD 936F, BDT 32...F, 2SB1187, 2SB1375, ++	data
KTA 1049	Si-P	NF-L, 100V, 5A, 30W, 30MHz	17c	Kec	BD 954F, 2SA1441, 2SB1016, 2SB1294, ++	data
KTA 1070	Si-P	Vid, 200/200V, 0,1A, 1W, 150MHz	7c(9mm)	Kec	BF 421A, BF 423A, 2SA1370...71	data
KTA 1099	Si-P	Vid-L, -/120V, 0,3A, 8W	17c	Kec	(2SA1799)[3]	data
KTA 1225 D,L	Si-P	NF/S-L, 160V, 1,5A, 10W, 100MHz	30j§	Kec	2SA1225, 2SA1552, 2SB840..842, 2SB1275	data
KTA 1241	Si-P	Blitzg./Strobo, 35V, 5A, 1W, 170MHz	7c(9mm)	Kec	2SA1050, 2SA1431, 2SB1288, 2SB1306, ++	data
KTA 1242 D,L	Si-P	S-L, lo-sat, 35V, 5A, 170MHz	30j§	Kec	2SA1242, 2SA1244, 2SB1203...1204	data
KTA 1243	Si-P	=KTA 1241: 0,625W	7c	Kec	2SA1050, 2SA1431, 2SB1288, 2SB1306, ++	data
KTA 1266	Si-P	Uni, 55V, 0,15A, 0,625W, >80MHz	7c	Kec	BC 556, BC 212, BC 256, 2SB725, ++	data
KTA 1266 L	Si-P	=KTA 1266: ra	7c	Kec	BC 416, BC 560, 2SA970, 2SA1136...37, ++	data
KTA 1267(L)	Si-P	=KTA 1266(L): 0,4W	41c	Kec	→KTA 1266(L)	data
KTA 1268	Si-P	Uni,ra, 120/120V, 0,1A, 0,625W, 100MHz	7c	Kec	2SA1016K, 2SA1123..24, 2SA1482, ++	data
KTA 1270	Si-P	Uni, 35V, 0,5A, 0,625W, 200MHz	7c	Kec	BC 327...328, BC 636, BC 638, 2SB910, ++	data
KTA 1271	Si-P	NF-E, 35V, 0,8A, 0,625W, 120MHz	7c	Kec	BC 327...328, BC 636, BC 638, 2SB910, ++	data
KTA 1272	Si-P	=KTA 1271: 0,4W	41c	Kec	→KTA 1271	data
KTA 1273	Si-P	NF-E, 30V, 2A, 1W, 120MHz	7c(9mm)	Kec	2SA1382, 2SB892, 2SB1312, 2SB1434, ++	data
KTA 1274	Si-P	Uni, 80V, 0,4A, 1W, 100MHz	7c(9mm)	Kec	BC 640, 2SB647, 2SB1041, 2SB1437, ++	data
KTA 1275	Si-P	TV-NF-E, 160/160V, 1A, 1W, 50MHz	7c(9mm)	Kec	2SA1013, 2SA1275, 2SA1770, 2SB1212	data
KTA 1276	Si-P	NF-L, 30V, 3A, 10W, 100MHz	17j§	Kec	2SA1012, 2SA1288, 2SB1064, 2SB1273	data
KTA 1277	Si-P	S, S-Reg, 400/400V, 0,5A, 1W, <1/5µs	7c(9mm)	Kec	2SA1776, 2SA1780, 2SA1972, 2SB1488	data
KTA 1279	Si-P	Telefon, 300/300V, 0,5A, 0,625W, >50MHz	7c	Kec	BF 421A, BF 493, MPSA 92, 2N6519	data
KTA 1281	Si-P	NF-E, S, 50V, 2A, 1W, 100MHz	7c(9mm)	Kec	2SA1382, 2SB892, 2SB1312, 2SB1434, ++	data

Type	Device	Short Description	Fig.	Manu	Comparision Types	More at
KTA 1282	Si-P	Uni, 30V, 2A, 0,625W, 120MHz	7c	Kec	2SA1382, 2SB892, 2SB1312, 2SB1434, ++	data
KTA 1283	Si-P	Uni, 40V, 1,5A, 0,625W, 200MHz	7c	Kec	2SA1382, 2SB892, 2SB1312, 2SB1434, ++	data
KTA 1297	Si-P	Uni, lo-sat, 20V, 2A, 0,4W, 120MHz	41c	Kec	2SB892, 2SB927, 2SB1433, 2SB1434, ++	data
KTA 1298	Si-P	=2SA1298	35a	Kec	→2SA1298	data
KTA 1302	Si-P	=2SA1302	77j§	Kec	→2SA1302	data
KTA 1381	Si-P	Vid-L, hi-def, 300V, 0,1A, 7W, 150MHz	14h§	Kec	2SA1353, 2SA1381	data
KTA 1385 D,L	Si-P	NF/S-L, 60V, 5A, 15W, <1/3,5µs	30j§	Kec	2SA1244, 2SA1385, 2SA1401, 2SB1203...04	data
KTA 1504(S)	Si-P	SMD, Uni, 50V, 0,15A, >80MHz	35a	Kec	BC 856...857, BCF 70, BCW 69...70	data
KTA 1505(S)	Si-P	SMD, Uni, 35V, 0,5A, 200MHz	35a	Kec	BC 807, BCW 67...68, BCX 17, 2SA1313, ++	data
KTA 1517	Si-P	SMD, Uni, ra, 120/120V, 0,1A, 100MHz	35a	Kec	2SA1312	data
KTA 1657	Si-P	Monitor-VA, 150/150V, 1,5A, 20W, 4MHz	17c	Kec	2SA1306(A,B), 2SA1332, 2SA1606, 2SB1186A	data
KTA 1658	Si-P	NF-L, 30/30V, 3A, 15W, 100MHz	17c	Kec	2SA1440, 2SB1185, 2SB1274, 2SB1314	data
KTA 1659	Si-P	NF/S-L, 160/160V, 1,5A, 20W, 100MHz	17c	Kec	2SA1306(A,B), 2SA1332, 2SA1606, 2SA1859A	data
KTA 1659 A	Si-P	=KTA 1659: 180/180V	17c	Kec	2SA1306A,B, 2SA1606, 2SA1859A, 2SA1930	data
KTA 1660	Si-P	SMD, Uni, 150/150V, 0,05A, 120MHz	39b	Kec	2SA1200, 2SA1660, 2SB807, 2SB1046...47	data
KTA 1661	Si-P	SMD, Uni, 120/120V, 0,8A, 120MHz	39b	Kec	2SA1201, 2SA1661, 2SB806, 2SB1025...26	data
KTA 1662	Si-P	SMD, Uni, 80/80V, 0,4A, 120MHz	39b	Kec	2SA1202, 2SA1368, 2SA1662, 2SB767, ++	data
KTA 1663	Si-P	SMD, Uni, 30V, 1,5A, 120MHz	39b	Kec	2SA1213, 2SA1663, 2SB1313, 2SB1519, ++	data
KTA 1664	Si-P	SMD, Uni, 35V, 0,8A, 120MHz	39b	Kec	BCX 51, 2SA1204, 2SA1664, 2SA1734, ++	data
KTA 1666	Si-P	SMD, S, lo-sat, 50V, 2A, 120MHz	39b	Kec	2SA1681, 2SB1123, 2SB1313, 2SB1519, ++	data
KTA 1668	Si-P	SMD, Uni, 80V, 1A, 150MHz	39b	Kec	2SA1859, 2SB1115A, 2SB1260	data
KTA 1700	Si-P	NF/S-L, 160/160V, 1,5A, 10W, 100MHz	14h§	Kec	2SA1249, 2SB649, 2SB1086A	data
KTA 1703	Si-P	S-Reg, 400/400V, 0,5A, 200/2200ns	14h§	Kec	2SA1156, 2SA1775	data
KTA 1704	Si-P	NF/HF-L, 120/120V, 1,2A, 20W, 175MHz	14h§	Kec	2SA1220, 2SA1249, 2SB649, 2SB1086	data
KTA 1705	Si-P	Uni-L, 30V, 3A, 10W, 100MHz	14h§	Kec	BD786, MJE250...254, 2SB744(A), 2SB772,++	data
KTA 1709	Si-P	lo-sat, 35V, 5A, 10W, 170MHz	14h§	Kec	2SA1357	data
KTA 1715	Si-P	NF/S-L, lo-sat, 50V, 2A, 10W, 100MHz	14h§	Nec	2SB986, 2SB1143, 2SB1217	data
KTA 1718 D,L	Si-P	lo-sat, 50V, 2A, 10W, 100/110ns	30j§	Kec	2SA1241, 2SA1731, 2SB1201, 2SB1202	data
KTA 1726	Si-P	NF-L, 80V, 6A, 50W, 20MHz	17j§	Kec	BD 244B...F, BD 544B...C, BD 810, 2SA1726+	data
KTA 1940	Si-P	NF-HiFi-E, 120/120V, 8A, 80W, 30MHz	18j§	Tos	BD 246C, BD 546D, 2SA1694, 2SA1940, ++	data
KTA 1943	Si-P	NF-HiFi-E, 230/230V, 15A, 150W, 25MHz	77j§	Kec	2SA1553, 2SA1943	data
KTA 1962	Si-P	NF-HiFi-E, 230/230V, 15A, 150W, 30MHz	18j§	Kec	2SA1294, 2SA1962	data
KTA 2014	Si-P	=KTA 1504:	35a(2mm)	Kec	BC 856W...857W, 2SA1576, 2SA1603, 2SA1644	data
KTA 2015	Si-P	=KTA 1505:	35a(2mm)	Kec	BC 807W, 2SB1219A	data
KTA 2017	Si-P	=KTA 1517:	35a(2mm)	Kec	2SA1587	data
KTA 2400	Si-P	Diff. Amp., 120/120V, 0,1A, 100MHz	7c	Kec	2SA970, 2SA1049, 2SA1136, 2SA1335	data

KTB

Type	Device	Short Description	Fig.	Manu	Comparision Types	More at
KTB 595	Si-P	=2SB595	17j§	Kec	→2SB595	data
KTB 596	Si-P	=2SB596	17j§	Kec	→2SB596	data
KTB 598	Si-P	NF,Uni, 30V, 1A, 0,625W, 180MHz	7c	Kec	BC 636, BC 638, 2SB909, 2SB1116(A), ++	data
KTB 631 K	Si-P	=2SB631K	14h§	Kec	→2SB631K	data
KTB 688	Si-P	=2SB688: 10A	18j§	Kec	→2SB688	data
KTB 688 A	Si-P	=2SB688:	18j§	Kec	→2SB688	data
KTB 764	Si-P	Uni, 60V, 1A, 1W, 150MHz	7c(9mm)	Kec	BC 640, 2SA1705, 2SB647, 2SB1041, ++	data
KTB 772	Si-P	NF/S-L, lo-sat, 40V, 3A, 10W, 80MHz	14h§	Kec	BD 786, MJE 250...254, 2SB744(A), 2SB772	data
KTB 778	Si-P	NF-L, 120/120V, 10A, 80W, 10MHz	18c	Kec	2SA1672, 2SA1805, 2SA1909, 2SB1860	data
KTB 817	Si-P	=2SB817	18j§	Kec	→2SB817	data
KTB 834	Si-P	=2SB834	17j§	Kec	→2SB834	data
KTB 985	Si-P	S, Tr, lo-sat, 60V, 3A, 1W, 150MHz	7c(9mm)	Kec	2SA1707, 2SA1926, 2SB985	data
KTB 988	Si-P	Uni-L, 60/60V, 3A, 30W, 9MHz (≈2SB988)	17j§	Kec	BD 242A...F, BD 538, BD 938, 2SB690, ++	data
KTB 989	Si-P	Uni-L, 80/80V, 4A, 30W, >3MHz(≈2SB989)	17j§	Kec	BD 244A...F, BD 538, BD 952, 2SB690, ++	data
KTB 1124	Si-P	SMD, S, lo-sat, 60V, 3A, 150MHz	39b	Kec	2SA1736, 2SB1124	data
KTB 1151	Si-P	lo-sat, 60V, 5A, 20W, <1/3,5µs	14h§	Kec	2SB1151	data
KTB 1241	Si-P	Uni, 80V, 1A, 1W, 100MHz	7c(9mm)	Kec	BC 640, 2SB647, 2SB1041, 2SB1437, ++	data
KTB 1260	Si-P	SMD, Uni, 80V, 1A, 100MHz	39b	Kec	BCX 53, 2SA1890, 2SB1115A, 2SB1260, ++	data
KTB 1366(V)	Si-P	NF-L, 60/60V, 3A, 25W, 9MHz (≈2SB1366) V: Industrial Version	17c	Kec	BDT 32(A...C)F, 2SA1440, 2SA1635, 2SB1335	data
KTB 1367(V)	Si-P	NF-L,100/100V, 5A, 30W, 5MHz(≈2SB1367)	17c	Kec	BD 956F, 2SA1741, 2SB1016, 2SB1294, ++	data
KTB 1368(V)	Si-P	NF-L, 80/80V, 4A, 25W, 9MHz (≈2SB1368)	17c	Kec	BD 954F, 2SA1741, 2SB1095, 2SB1294, ++	data
KTB 1369	Si-P	Monitor-VA-E, 200/180V, 2A, 20W, 50MHz	17c	Kec	2SA1306A, 2SA1606, 2SA1859A, 2SA1930	data
KTB 1370	Si-P	NF/S-L, lo-sat, 100V, 7A, 30W, 10MHz	17c	Kec	2SA1442, 2SA1742, 2SB946	data
KTB 1423	Si-P-Darl+Di	Motor-Tr, 120/120V, 5A, 30W, B>1000	17c	Kec	2SB1252, 2SB1340	data
KTB 1424	Si-P-Darl	Uni-L, 80V, 3A, 25W, B>3000	17c	Kec	2SA1718, 2SB1098, 2SB1250, 2SB1430	data
KTB 1469	Si-P	TV, VA,NF-E, 200/180V, 2A, 25W, 100MHz	17j§	Kec	2SA1133, 2SB720	data
KTB 1772	Si-P	S,NF-E, 40V, 3A, 0,625W, 80MHz	7c	Kec	2SB985, 2SB1505, 2SB1592	data

Type	Device	Short Description	Fig.	Manu	Comparision Types	More at
KTB 2510	Si-P-Darl	NF-E, 160/150V, 10A, 100W, B>5000	18j§	Kec	2SB1560	data
KTB 2955	Si-P	NF-L, 120/120V, 10A, 40W, 10MHz	17j§	Kec	BD 744D...F, 2SA1077	data

KTC

Type	Device	Short Description	Fig.	Manu	Comparision Types	More at
KTC 200	Si-N	Uni, 60V, 0,5A, 0,625W, 300MHz	7c	Kec	BC 337A, BC 637, BC 639, 2SD1226, ++	data
KTC 388(A)	Si-N	≈2SC388(A)	7c	Kec	→2SC388(A)	
KTC 601 U	Si-N	SMD, Dual, 50V, 0,15A, >80MHz	45ba1(2mm)	Kec	-	data
KTC 801 U	Si-N	SMD, Dual, 60V, 0,15A, >80MHz	46bh1(2mm)	Kec	-	data
KTC 941	Si-N	≈BF 254		Kec	→BF 254	
KTC 1001	Si-N	Vid-L, 300/300, 0,15A, 12,5W, 100MHz	17c	Kec	2SC3565, 2SC3946, 2SC4075	data
KTC 1003	Si-N	TV-HA, 200/60, 4A, 30W, 8MHz	17c	Kec	2SD1958, (2SD1136, 2SD1159)[3]	data
KTC 1006	Si-N	CB, HF-Tr, 80/80, 0,8A, 1W, 150MHz	7c(9mm)	Kec	-	data
KTC 1008	Si-N	Reg,Driver, 80, 1A, 0,625W, 150MHz	7c	Kec	BC 639, 2SC4488, 2SD1616A, 2SD1768, ++	data
KTC 1020	Si-N	Uni, 35, 0,5A, 0,4W, 300MHz	41c	Kec	BC 337, BC 635, BC 637, 2SD1225, ++	data
KTC 1026	Si-N	Uni, 200/180, 0,1A, 1W, >50MHz	7c(9mm)	Kec	2SD2031	data
KTC 1027	Si-N	Uni, 120/120, 0,8A, 1W, 120MHz	7c(9mm)	Kec	2SC2235, 2SC2383, 2SC3228, 2SD2184, ++	data
KTC 1094	Si-N	Vid-E, -/60, 50mA, 1W	7c(9mm)	Kec	2SC3526, 2SC4829	data
KTC 1095	Si-N	Vid-E, -/20, 0,5A, 5W	17c	Kec	2SC4605, (2SC4201...02, 2SC4411)[3]	data
KTC 1170	Si-N	Vid-E,hi-def,200/200, 0,1A, 5W, 150MHz	17c	Kec	2SC3565, 2SC4075	data
KTC 1173	Si-N	=2SC1173	17j§	Kec	→2SC1173	data
KTC 1199	Si-N	Vid-E, -/120, 0,3A, 8W	17c	Kec	(2SC3612, 2SC4714)[3]	data
KTC 1815	Si-N	≈2SC1815	7c	Kec	→2SC1815	
KTC 1932	Si-N	≈BF 255		Kec	→BF 255	
KTC 1959	Si-N	≈2SC1959	7c	Kec	→2SC1959	
KTC 1969	Si-N	=2SC1969	17j§	Kec	→2SC1969	
KTC 2016	Si-N	NF/S-L, 60V, 3A, 30W, 30MHz	17j§	Kec	BD 241A, BD 535, BD 935, 2SC2562, ++	data
KTC 2018	Si-N	NF/S-L, 100V, 5A, 40W, 30MHz	17j§	Kec	BD 243C, BD 543C, BD 801, 2SD613	data
KTC 2020 D,L	Si-N	NF/S-L, 60V, 3A, 20W, 30MHz	30j§	Kec	2SC4522, 2SD1221, 2SD1760, 2SD1802	data
KTC 2022 D,L	Si-N	NF/S-L, 100V, 5A, 20W, 30MHz	30j§	Kec	2SC3303, 2SC4332, 2SC5103	data
KTC 2025 D,L	Si-N	NF/S-L, lo-sat, 120V, 1A, 8W, 130MHz	30j§	Kec	(2SC4134, 2SD1074, 2SD1075)[17]	data
KTC 2026	Si-N	NF-L, 60V, 3A, 25W, 30MHz, 650/1950ns	17c	Kec	BD 935F, BDT 31...F, 2SD1406, 2SD1585, ++	data
KTC 2028	Si-N	NF-L, 100V, 5A, 30W, 30MHz	17c	Kec	BD 953F, 2SC3540, 2SD1407, 2SD1940, ++	data
KTC 2068	Si-N	≈2SC2068	13j§	Kec	→2SC2068	
KTC 2073	Si-N	=2SC2073	17j§	Kec	→2SC2073	data
KTC 2078	Si-N	CB, HF-E, 50/80, 4A, 10W, >100MHz	17j§	Kec	2SC1816, 2SC2075, 2SC2078, 2SC2119	data
KTC 2120	Si-N	=2SC2120	7c(9mm)	Kec	→2SC2120	
KTC 2235	Si-N	≈2SC2235		Kec	→2SC2235	
KTC 2236(A)	Si-N	=2SC2236(A): 2A, 1W	7c(9mm)	Kec	→2SC2236	data
KTC 2238(A)	Si-N	=2SC2238(A)	17j§	Kec	→2SC2238	data
KTC 2334	Si-N	=2SC2334	17j§	Kec	→2SC2334	data
KTC 2347	Si-N	=2SC2347	7c	Kec	→2SC2347	data
KTC 2553	Si-N	=2SC2553	17j§	Kec	→2SC2553	data
KTC 2800	Si-N	NF/S-L, 160/160V, 1,5A, 10W, 100MHz	14h§	Kec	2SC3117, 2SD669, 2SC2690A	data
KTC 2801	Si-N	Chroma-E, 300/300V, 0,1A, 50MHz	14h§	Kec	2SC3417, 2SC3503, 2SC4147	data
KTC 2803	Si-N	NF/HF-L, 120/120V, 1,2A, 20W, 155MHz	14h§	Kec	2SC2690, 2SC3117, 2SD669, 2SD1563	data
KTC 2804	Si-N	Uni-L, 30V, 3A, 10W, 100MHz	14h§	Kec	BD785, MJE240..244, 2SD794(A), 2SB882,++	data
KTC 2814	Si-N	NF/S-L, lo-sat, 50V, 2A, 10W, 100MHz	14h§	Nec	2SD1348, 2SD1818	data
KTC 2815 D,L	Si-N	lo-sat, 50V, 2A, 10W, 100/1100ns	30j§	Kec	2SC3076, 2SD1801, 2SD1802	data
KTC 2874	Si-N	Muting,hi-Ueb,25V, 0,3A, 0,625W, 30MHz	7c	Kec	2SC2878, 2SC3792, 2SC3836	data
KTC 2875	Si-N	=KTC 2874: SMD	35a	Kec	2SC3326	data
KTC 2983 D,L	Si-N	NF/S-L, 160V, 1,5A, 10W, 100MHz	30j§	Kec	2SC2983, 2SD1080...1082, 2SD1918	data
KTC 3072	Si-N	NF/S-L, lo-sat, 40V, 5A, 100MHz	30j§	Kec	2SC3072, 2SC3074, 2SD1803, 2SD1805	data
KTC 3112	Si-N	=2SC3112: 0,625W	7c	Kec	→2SC3112	data
KTC 3113	Si-N	=2SC3113	41c	Kec	→2SC3113	data
KTC 3120	Si-N	=2SC3120	35a	Kec	* →2SC3120	data
KTC 3121	Si-N	=2SC3121	35a	Kec	→2SC3121	data
KTC 3190	Si-N	=2SC3190	7c	Kec	→2SC3190	data
KTC 3191	Si-N	=2SC3191	41c	Kec	→2SC3191	data
KTC 3192	Si-N	=2SC3192	7c	Kec	→2SC3192	data
KTC 3193	Si-N	=2SC3193	41c	Kec	→2SC3193	data
KTC 3194	Si-N	=2SC3194	7c	Kec	→2SC3194	data
KTC 3195	Si-N	=2SC3195	41c	Kec	→2SC3195	data
KTC 3197	Si-N	=2SC3197	7c	Kec	→2SC3197	data
KTC 3198	Si-N	Uni, 60V, 0,15A, 0,625W, >80MHz	7c	Kec	BC 546, BC 174, BC 182, 2SD767, ++	data
KTC 3198 A	Si-N	=KTC 3198: 50V	7c		BC 546...547, BC 174, BC 182, 2SD767, ++	data
KTC 3198 L	Si-N	=KTC 3198: ra	7c		BC 141, BC 550, 2SC2240, 2SC2674...75, ++	data
KTC 3199...	Si-N	=KTC 3198...: 0,4W	41c	Kec	→KTC 3198...	data
KTC 3200	Si-N	Uni,ra, 120/120V, 0,1A, 0,625W, 100MHz	7c	Kec	2SC2362K, 2SC2631...32, 2SC3800, ++	data
KTC 3202	Si-N	Uni, 35V, 0,5A, 0,625W, 300MHz	7c	Kec	BC 337...338, BC 635, BC 637, 2SD1226, ++	data

Type	Device	Short Description	Fig.	Manu	Comparision Types	More at
KTC 3203	Si-N	NF-E, 35V, 0,8A, 0,625W, 120MHz	7c	Kec	BC 337...338, BC 635, BC 637, 2SC1226, ++	data
KTC 3204	Si-N	=KTC 3203: 0,4W	41c	Kec	→KTC 3203	data
KTC 3205	Si-N	NF-E, 30V, 2A, 1W, 120MHz	7c(9mm)	Kec	2SC4486, 2SD1207, 2SD2069, 2SD2484, ++	data
KTC 3206	Si-N	S, Vid-E, 200/150V, 0,05A, 1W, 120MHz	7c(9mm)	Kec	2SC3249, 2SC3467...68, 2SC4490, 2SC4650++	data
KTC 3207	Si-N	S, Vid-E, 300/300V, 0,1A, 1W, >50MHz	7c(9mm)	Kec	2SC3249, 2SC3468...69, 2SC4218, 2SC4490++	data
KTC 3208	Si-N	Vid-E, 300/300V, 0,15A, 12,5W, 100MHz	17j§	Kec	2SC1505...07, 2SC1755...57, 2SC1905	data
KTC 3209	Si-N	NF-E, S, 50V, 2A, 1W, 100MHz	7c(9mm)	Kec	2SC4486, 2SD1207, 2SD2069, 2SD2484, ++	data
KTC 3210	Si-N	Uni, 30V, 2A, 0,625W, 120MHz	7c	Kec	2SC3205, 2SD1207, 2SD1055, 2SD1227, ++	data
KTC 3211	Si-N	Uni, 40V, 1,5A, 0,625W, 190MHz	7c	Kec	2SD1207, 2SD1051, 2SD1055, 2SD2177, ++	data
KTC 3217	Si-N	CTV, 450/450V, 0,1A, 1W, >50MHz	7c(9mm)	Kec	MPSA 44, 2SD1573[6]	data
KTC 3226	Si-N	Blitzg./Strobo Flash, lo-sat, 30V, 2A, 1W, 150MHz, B>200	7c(9mm)	Kec	2SC5039, 2SC3225...26, 2SD1247, 2SD2159++	data
KTC 3227	Si-N	Uni, 80V, 0,4A, 1W, 100MHz	7c(9mm)	Kec	BC 639, 2SD667, 2SD1292, 2SD2181, ++	data
KTC 3228	Si-N	TV-NF-E, 160/160V, 1A, 1W, 20MHz	7c(9mm)	Kec	2SC2383, 2SC3228, 2SC4614, 2SD1812	data
KTC 3229	Si-N	Vid-E, 300/300V, 0,1A, 95MHz	17c	Kec	2SC3565, 2SC3942, 2SC4075, 2SC4544	data
KTC 3230	Si-N	NF-L, 30V, 3A, 10W, 100MHz	17j§	Kec	2SC2562, 2SC3253, 2SD1505, 2SD1912	data
KTC 3231	Si-N	TV-HA, 200/60V, 4A, 40W, >8MHz	17j§	Kec	2SD823, 2SD1136, 2SD1159	data
KTC 3245	Si-N	Uni, 400/350V, 0,3A, 0,625W	7c	Kec	MPSA45, 2SD1350	data
KTC 3265	Si-N	=2SC3265	35a	Kec	→2SC3265	data
KTC 3267	Si-N	Uni, lo-sat, 20V, 2A, 0,4W, 120MHz	41c	Kec	2SC4484, 2SD1207, 2SD1835, 2SD2177, ++	data
KTC 3281	Si-N	=2SC3281	77j§	Kec	→2SC3281	data
KTC 3295	Si-N	=2SC3295	35a	Kec	→2SC3295	data
KTC 3400	Si-N	Diff.-Amp., 120/120V, 0,1A, 100MHz	7c	Kec	2SC2240, 2SC2674, 2SC2459, 2SC3378	data
KTC 3467	Si-N	=2SC3467	7c(9mm)	Kec	→2SC3467	data
KTC 3502	Si-N	Vid-L, hi-def, 200V, 0,1A, 5W, 150MHz	14h§	Kec	2SC3416, 2SC3417, 2SC3502, 2SC3503	data
KTC 3503	Si-N	Vid-L, hi-def, 300V, 0,1A, 7W, 150MHz	14h§	Kec	2SC3417, 2SC3503	data
KTC 3875(S)	Si-N	SMD, Uni, 60V, 0,15A, >80MHz	35a	Kec	BC 846, BCV 71...72, 2SC3323, 2SC4076	data
KTC 3876(S)	Si-N	SMD, Uni, 35V, 0,5A, 300MHz	35a	Kec	BC 817, BCW 65...66, BCX 19, 2SC3325, ++	data
KTC 3878(S)	Si-N	SMD, HF(AM), 35V, 0,1A, 120MHz	35a	Kec	BC 850, BCF 81, BF 840...841, 2SC3323, ++	data
KTC 3879(S)	Si-N	SMD, AM/FM, 35V, 0,05A, >100MHz	35a	Kec	BF 554, BF 840...841, BFS 18...19, 2SC2778	data
KTC 3880(S)	Si-N	SMD, FM, 40V, 0,02A, 550MHz	35a	Kec	BF 599, BFS 20, 2SC3015, 2SC3125, ++	data
KTC 3881(S)	Si-N	SMD, VHF, 30V, 0,05A, 600MHz	35a	Kec	BF 599, BFS 17, BFS 20, 2SC3015...16, ++	data
KTC 3882	Si-N	SMD, VHF-O, 30V, 0,05A, 1100MHz	35a	Kec	BF 599, BFS 17, BFS 20, 2SC3015...16, ++	data
KTC 3883	Si-N	SMD, VHF, 30V, 0,2A, >500MHz	35a	Kec	-	data
KTC 3911(S)	Si-N	SMD, ra, 120/120V, 0,1A, 100MHz	35a	Kec	2SC3324	data
KTC 3920	Si-N	SMD, Uni, 35V, 0,5A, 300MHz	35a	Kec	BC 817, BCW 65, BCX 19, 2SC3325, ++	data
KTC 4021	Si-N	=2SC3121	35a(2mm)	Kec	BF 547W, BFS 17W, 2SC4780	data
KTC 4075	Si-N	=KTC 3875:	35a(2mm)	Kec	BC 846W, 2SC4081, 2SC4128, 2SA4141	data
KTC 4076	Si-N	=KTC 3876:	35a(2mm)	Kec	BC 817W, 2SC4173, 2SD1820A	data
KTC 4077	Si-N	=KTC 3911:	35a(2mm)	Kec	2SC4117	data
KTC 4079	Si-N	=KTC 3879:	35a(2mm)	Kec	2SC3931, 2SC3936, 2SC5020	data
KTC 4080	Si-N	=KTC 3880:	35a(2mm)	Kec	2SC3931, 2SC3935, 2SC5020	data
KTC 4081	Si-N	=KTC 3881:	35a(2mm)	Kec	BF 547W, BFS 17W, 2SC3931...33, 2SC4780	data
KTC 4082	Si-N	=KTC 3882:	35a(2mm)	Kec	BF 547W, BFS 17W, 2SC3931...33, 2SC4780	data
KTC 4368	Si-N	Monitor-VA, 150/150V, 1,5A, 20W, 4MHz	17c	Kec	2SC3289(A,B), 2SC3364, 2SC4159, 2SD1763A	data
KTC 4369	Si-N	NF-E, 30V, 3A, 15W, 100MHz	17c	Kec	2SC3690, 2SC4935, 2SD1762, 2SD1913	data
KTC 4370	Si-N	NF/S-L, 160/160V, 1,5A, 20W, 100MHz	17c	Kec	2SC3298(A,B), 2SC4370, 2SC3364, 2SD1763A	data
KTC 4370 A	Si-N	=KTC 4370: 180/180V	17c	Kec	2SC3298A,B, 2SC4883A, 2SC4159, 2SC5171	data
KTC 4371	Si-N	S-Reg, Ballast, 500/400V, 5A, 30W	17c	Kec	2SC4026, 2SC4418, 2SC4662, 2SC5206, ++	data
KTC 4372	Si-N	SMD, Uni, 200/150V, 0,05A, 120MHz	39b	Kec	2SC2880, 2SC4372	data
KTC 4373	Si-N	SMD, Uni, 120/120V, 0,8A, 120MHz	39b	Kec	2SC2881, 2SC4373, 2SD1007, 2SD1418...19	data
KTC 4374	Si-N	SMD, Uni, 80/80V, 0,4A, 100MHz	39b	Kec	2SC2882, 2SC3438, 2SC4374, 2SD875, ++	data
KTC 4375	Si-N	SMD, Uni, 30V, 1,5A, 120MHz	39b	Kec	2SC2873, 2SC4377, 2SD2457, 2SD2532, ++	data
KTC 4376	Si-N	SMD, Uni, 35V, 0,8A, 120MHz	39b	Kec	BCX 54, 2SC2884, 2SC4376, 2SC4539, ++	data
KTC 4377	Si-N	SMD, Blitzg./Strobo Flash, lo-sat, 30V, 2A, 150MHz, B=140...600	39b	Kec	2SC2982, 2SD1621, 2SD1766	data
KTC 4378	Si-N	SMD, Reg/Driver, 80/60V, 1A, 150MHz	39b	Kec	BCX 56, 2SC5026, 2SD1005, 2SD1368, ++	data
KTC 4379	Si-N	SMD, S, lo-sat, 50V, 2A, 120MHz	39b	Kec	2SC4409, 2SD1623, 2SD2391, 2SD2532, ++	data
KTC 4419	Si-N	S-Reg, 500/400V, 5A, 30W, <1/3µs	17c	Kec	2SC4026, 2SC4418, 2SC4662, 2SC5206, ++	data
KTC 4512	Si-N	NF-L, 80V, 6A, 50W, 20MHz	17j§	Kec	BD 243B...F, BD 543B...C, BD 809, 2SC4512+	data
KTC 4520	Si-N	S-L, S-Reg, 800/500V, 3A, 40W, 18MHz <500/3300ns	17j§	Kec	BUT 11(A), BUV 46(A), 2SC3047, 2SC3086 2SC3490, 2SC3150	data
KTC 4520 F	Si-N	=KTC 4520: Iso, 30W	17c		BUT11(A)F, BUV46(A)FI, 2SC3352, 2SC3794	data
KTC 4521	Si-N	S-L, S-Reg, 800/500V, 5A, 50W, 18MHz <500/3300ns	17j§	Kec	BUT 11(A), BUV 46(A), 2SC3047, 2SC3087 BUT 54	data
KTC 4521 F	Si-N	=KTC 4521: Iso, 40W	17c		BUT11(A)F, BUV46(A)FI, 2SC3353, 2SC3795	data
KTC 4666	Si-N	SMD,ra,lo-sat,hi-beta,60V,0,15A,250MHz	35a(2mm)	Kec	2SC4666	data
KTC 5103 D,L	Si-N	NF/S-L, 60V, 5A, 15W, <1/3,5µs	30j§	Kec	2SC3074, 2SD1803, 2SD1804, 2SD1805	data
KTC 5197	Si-N	NF-HiFi-E, 120/120V, 8A, 80W, 30MHz	18j§	Kec	BD 245C, BD 545D, 2SC3181, 2SC5197, ++	data
KTC 5200	Si-N	NF-HiFi-E, 230/230V, 15A, 150W, 30MHz	77j§	Kec	2SC4029, 2SC5200	data

Type	Device	Short Description	Fig.	Manu	Comparision Types	More at
KTC 5242	Si-N	HF-HiFi-E, 230/230V, 15A, 150W, 30MHz	18j§	Kec	2SC3263	data
KTC 8050	Si-N	Uni, 35, 0,8A, 0,625W, 120MHz	7e	Kec	BC 337, BC 635, BC 637, 2SD1225, ++	data
KTC 8550	Si-P	Uni, 35, 0,8A, 0,625W, 120MHz	7e	Kec	BC 327, BC 636, BC 636, 2SB909, ++	data
KTC 9011	Si-N	AM, FM-ZF, 35V, 50mA, 0,625W, >100MHz	7e	Kec	BF 240...41, BF 254...55, BF 594..595, ++	data
KTC 9012	Si-P	NF-E, 40V, 0,5A, 0,625W, >150MHz	7e	Kec	BC 327, BC 636, BC 638, 2SB909...910, ++	data
KTC 9013	Si-N	NF-E, 40V, 0,5A, 0,625W, >140MHz	7e	Kec	BC 337, BC 635, BC 637, 2SD1225...26, ++	data
KTC 9014	Si-N	Uni, 50V, 0,15A, 0,625W, >60MHz	7e	Kec	BC 414, BC 550, 2SC2240, 2SC2675, ++	data
KTC 9015	Si-P	Uni, 50V, 0,15A, 0,625W, >60MHz	7e	Kec	BC 416, BC 560, 2SA970, 2SA1137, ++	data
KTC 9016	Si-N	FM-V/M/O, 40V, 20mA, 0,625W, >260MHz	7e	Kec	BF 240...241, BF 254...255, BF 594...595,++	data
KTC 9018	Si-N	FM-V/M/O, 40V, 20mA, 0,625W, >260MHz	7e	Kec	BF 225, BF 255, BF 314, BF 505, BF 507++	data

KTD

Type	Device	Short Description	Fig.	Manu	Comparision Types	More at
KTD 525	Si-N	=2SD525	17j§	Kec	→2SD525	data
KTD 526	Si-N	=2SD526	17j§	Kec	→2SD526	data
KTD 545	Si-N	NF,Uni, 30V, 1A, 0,625W, 180MHz	7c	Kec	BC 635, BC 637, 2SC4485, 2SD1616, ++	data
KTD 600 K	Si-N	=2SD600K	14h§	Kec	→2SD600K	data
KTD 686	Si-N-Darl+Di	L,100/80V, 4A, 30W, B>2000, 200/2100ns	17j§	Kec	BDW 23C, BDW 63C..D, 2SD1147, 2SD1195	data
KTD 718	Si-N	=2SD718: 10A	18j§	Kec	→2SD718	data
KTD 718 A	Si-N	=2SD718:	18j§	Kec	→2SD718	data
KTD 863	Si-N	=2SD863	7c(9mm)	Kec	→2SD863	data
KTD 880	Si-N	≈2SD880	17j§	Kec	→2SD880	data
KTD 882	Si-N	NF/S-L, lo-sat, 40V, 3A, 10W, 90MHz	14h§	Kec	BD 785, MJE 240...244, 2SD882, 2SD1818	data
KTD 998	Si-N	NF-L, 120/120V, 10A, 80W, 10MHz	18c	Kec	2SC3762, 2SC4690, 2SC4886	data
KTD 1003	Si-N	=KTD 1028: SMD	39b	Kec	2SC3727, 2SC5209	data
KTD 1028(V)	Si-N	hi-beta, lo-sat, 60/50V, 1A, B>800 V: Industrial Version	7c(9mm)	Kec	2SC3247	data
KTD 1047	Si-N	=2SD1047	18j§	Kec	→2SD1047	data
KTD 1145	Si-N	=2SD1145	7c(9mm)	Kec	→2SD1145	data
KTD 1146	Si-N	Blitzg./Strobo Flash, lo-sat, 40V, 5A 0,625W, 100MHz	7c(9mm)	Kec	2SB1446, 2SB1483	data
KTD 1302	Si-N	Muting,hi-Ueb,25V, 0,3A, 0,625W, 60MHz	7c	Kec	2SC3068...69, 2SC3576, 2SD2144...45, ++	data
KTD 1303	Si-N	=KTD 1302: 0,4W	41c	Kec	→KTD 1302	data
KTD 1304	Si-N	=KTD 1302: SMD	35a	Kec	2SD2114K	data
KTD 1347	Si-N	S, Tr, lo-sat, 60V, 3A, 1W, 150MHz	7c(9mm)	Kec	2SC4487, 2SD1347	data
KTD 1351	Si-N	Uni-L, 60/60V, 3A, 30W, 3MHz(≈2SD1351)	17j§	Kec	BD 241A..F, BD 537, BD 937, 2SD712, ++	data
KTD 1352	Si-N	Uni-L, 80/80V, 4A, 30W, 8MHz(≈2SD1352)	17j§	Kec	BD 243A..F, BD 537, BD 951, 2SD712, ++	data
KTD 1411	Si-N-Darl	L, 80V, 4A, 15W, B>3000	14h§	Kec	(BD 679, BD 779, 2N6039)[13]	data
KTD 1413	Si-N-Darl+Di	Motor-Tr, 150/100V, 5A, 25W, B>2000	17c	Kec	2SD1315, 2SD1336(A)	data
KTD 1414	Si-N-Darl+Di	=2SD1414: 25W	17c	Kec	→2SD1414	data
KTD 1415(V)	Si-N-Darl+Di	Motor-Tr, 100/100V, 7A, 30W, B>2000 V: Industrial Version	17c	Kec	BD 647F, 2SD1415, 2SD1590, 2SD2162, ++	data
KTD 1510	Si-N-Darl	NF-E, 160/150V, 10A, 100W, B>5000	18j§	Kec	2SD2390	data
KTD 1554 A	Si-N+Di	=2SD1554	18c	Kec	→2SD1554	data
KTD 1555 A	Si-N+Di	=2SD1555	18c	Kec	→2SD1555	data
KTD 1624	Si-N	SMD, S, lo-sat, 60V, 3A, 150MHz	39b	Kec	2SC4521, 2SD1624	data
KTD 1691	Si-N	lo-sat, 60V, 5A, 20W, <1/3,5μs	14h§	Kec	2SD1691	data
KTD 1863	Si-N	Uni, 80V, 1A, 1W, 100MHz	7c(9mm)	Kec	BC 639, 2SD667, 2SD1292, 2SD1616A, ++	data
KTD 1882	Si-N	S,NF-E, 40V, 3A, 0,625W, 90MHz	7c	Kec	2SC4487, 2SD1507, 2SD1347, 2SD2152, ++	data
KTD 1898	Si-N	SMD, Uni, 100V, 1A, 100MHz	39b	Kec	BCX 56, 2SC5026, 2SD1005, 2SD1368, ++	data
KTD 1937	Si-N+Di	hi-beta, lo-sat, 100/80V, 10A, 40W 70MHz, B=500...1500	17c	Kec	2SC4024, 2SD1947	data
KTD 1945	Si-N	lo-sat, -/60V, 3A, 25W	17c	Kec	2SC3540, 2SC3746, 2SC4549, 2SC4596	data
KTD 2058(V)	Si-N	NF-L, 60/60V, 3A, 25W, 3MHz (≈2SD2058) V: Industrial Version	17c	Kec	BDT 31(A..C)F, 2SC3691, 2SC3851, 2SD2000	data
KTD 2059(V)	Si-N	NF-L,100/100V, 5A, 30W, 12MHz(≈2SD2059)	17c	Kec	BD 955F, 2SC4549, 2SC3540, 2SD1407, ++	data
KTD 2060	Si-N	NF-L, 80/80V, 4A, 25W, 8MHz (≈2SD2060)	17c	Kec	BD 953F, 2SC4549, 2SD1586, 2SD1407, ++	data
KTD 2061	Si-N	Monitor-VA-E, 200/180V, 2A, 20W,100MHz	17c	Kec	2SC3298A, 2SC4159, 2SC4883A, 2SC5171	data
KTD 2066(V)	Si-N+Di	hi-beta, lo-sat, 100/80V, 5A, 30W 130MHz, B=500...1500, V: Industrial V.	17c	Kec	2SC4553, 2SC4558, 2SD1474, 2SD2076, 2SD1947, 2SD2165	data
KTD 2092	Si-N+Di	=2SD2092	17c	Kec	→2SD2092	data
KTB 2161	Si-N	TV, VA,NF-E, 200/180V, 2A, 25W, 100MHz	17j§	Kec	2SC2660, 2SD760	data
KTD 2424	Si-N-Darl	Uni-L, 80V, 3A, 25W, B>3000	17c	Kec	2SD1891...92	data
KTD 2499	Si-N+Di	CTV-HA, 1500/600V, 6A, 50W	18c	Kec	2SC4764, 2SD2499	data
KTD 3055	Si-N	NF-L, 120/120V, 10A, 40W, 12MHz	17j§	Kec	BD 743D..F, 2SC2527	data

Type	Device	Short Description	Fig.	Manu	Comparision Types	More at
KTK						
KTK 24	MOS-N-FET-e	V-MOS, S-L, 60V, 16A, 60W, <65mΩ(8A)	17c§	Kec	BUK 552-60, 2SK971...72, 2SK1115, 2SK1416	data
KTK 117	N-FET	=2SK117	7a	Kec	→2SK117	data
KTK 123	N-FET	Uni, 15V, 0,625W, Idss=5...14mA	7	Kec	-	data
KTK 127	N-FET	=2SK123: 0,4W	41	Kec	-	data
KTK 136	N-FET	=2SK123: SMD	35	Kec	-	data
KTK 161	N-FET	=2SK161: 0,4W	41d	Kec	→2SK161	data
KTK 184	N-FET	=2SK184: 0,4W	41	Kec	→2SK184	data
KTK 192 A	N-FET	=2SK192A:	41a	Kec	→2SK192A	data
KTK 209	N-FET	=2SK209	35	Kec	→2SK209	data
KTK 211	N-FET	=2SK211:	35c	Kec	→2SK211	data
KTK 596	N-FET	=2SK596: 0,4W	41	Kec	→2SK596	data
KTK 740	MOS-N-FET-e	V-MOS, S-L, 450V, 10A, 125W, <0,7Ω(5A)	17c§	Kec	2SK1442, 2SK1503	data
KTK 821	MOS-N-FET-e	V-MOS, S-L, 450V, 3A, 40W, <2,6Ω(1A)	17c§	Kec	2SK1154, 2SK1244, 2SK1439, 2SK1493...94	data
KTK 831	MOS-N-FET-e	V-MOS, S-L, 450V, 5A, 75W, <1,6Ω(3A)	17c§	Kec	BUZ 41...42A, IRF 830, 2SK1156, 2SK1246++	data
KTK 840	MOS-N-FET-e	V-MOS, S-L, 450V, 8A, 125W, <0,75Ω(8A)	17c§	Kec	IRF 840, 2SK894, 2SK1495...96, 2SK1574	data
KTK 930	MOS-N-FET-e	V-MOS, S-L, 900V, 5A, 150W, <3Ω(5A)	77c§	Kec	2SK539, 2SK1465	data
KTK 1021	MOS-N-FET-e	=KTK 821: Iso	17c	Kec	BUK 445-450, 2SK1444, 2SK1862...63,++	data
KTK 1031	MOS-N-FET-e	=KTK 831: Iso	17c	Kec	2SK1231, 2SK1351, 2SK1445, 2SK1608,++	data
KTK 1040	MOS-N-FET-e	=KTK 840: Iso	17c	Kec	2SK1352, 2SK1446, 2SK1609, 2SK1566...67++	data
KTK 1140	MOS-N-FET-e	=KTK 740: Iso	17c	Kec	-	data
KTK 2312	MOS-N-FET-e*	V-MOS, LogL, 60V, 45A, 45W, <25mΩ(25A) (=2SK2312)	17c	Kec	2SK1257, 2SK1653, 2SK1952, 2SK2120	data
KTK 2314	MOS-N-FET-e*	V-MOS, LogL,100V, 27A, 75W, <130mΩ(15A) (=2SK2314)	17c§	Kec	2SK1293, 2SK2050, 2SK2314	data
KTK 2385	MOS-N-FET-e*	V-MOS, LogL, 60V, 36A, 40W, <55mΩ(15A) (=2SK2385)	17c	Kec	2SK1257, 2SK1952, 2SK2120, 2SK2312, ++	data
KTK 2542	MOS-N-FET-e*	V-MOS, 500V, 8A, 80W, <0,85Ω(4A) (=2SK2542)	17c§	Kec	IRF 840, 2SK1496, 2SK1574, 2SK2360, ++	data
KTK 2608	MOS-N-FET-e*	V-MOS, 900V, 3A, 100W, <4,3Ω(1,5A) (=2SK2608)	17c§	Kec	BUK456-1000, 2SK1456, 2SK1638, 2SK1793++	data
KTK 2610	MOS-N-FET-e*	V-MOS, 900V, 5A, 125W, <2,5Ω(3A) (=2SK2610)	18c§	Kec	2SK727, 2SK1341, 2SK1649...50, 2SK1794,++	data
KTK 2611	MOS-N-FET-e*	V-MOS, 900V, 9A, 150W, <1,4Ω(4A) (=2SK2611)	18c§	Kec	2SK1796, 2SK1933, 2SK2488, 2SK2676, ++	data
KTK 2661	MOS-N-FET-e*	V-MOS, 500V, 5A, 75W, <1,5Ω(2,5A) (=2SK2661)	17c§	Kec	BUZ 41A...42, IRF 830, 2SK1751, 2SK2356++	data
KTK 2662	MOS-N-FET-e*	=KTK 2661: Iso, 35W, (=2SK2662)	17c	Kec	BUK 445-500, 2SK1993, 2SK2115, 2SK2354++	data
KTK 2679	MOS-N-FET-e*	V-MOS, 400V, 5,5A, 35W, <1,2Ω(3A) (=2SK2679)	17c	Kec	BUK 445-450, 2SK1377, 2SK1445, 2SK1990++	data
KTK 2700	MOS-N-FET-e*	V-MOS, 900V, 3A, 40W, <4,3Ω(1,5A) (=2SK2700)	17c	Kec	2SK1356, 2SK1460, 2SK1995, 2SK2480, ++	data
KTK 2749	MOS-N-FET-e*	V-MOS, 900V, 7A, 150W, <2Ω(3,5A) (=2SK2749)	18c§	Kec	2SK1342, 2SK1502, 2SK1795, 2SK2611, ++	data
KTK 2841	MOS-N-FET-e*	V-MOS, 400V, 10A, 80W, <0,55Ω(5A) (=2SK2841)	17c§	Kec	BUZ 61, 2SK1378, 2SK1442, 2SK2365	data
KTN...KTZ						
KTN 2222(A)	Si-N	=2N2222(A): 0,625W	7e	Kec	→2N2222	data
KTN 2222(A)S	Si-N	=2N2222(A): SMD	35a	Kec	BSR14, BSS79, PMBT2222(A), SMBT2222(A)	data
KTN 2222(A)U	Si-N	=2N2222(A): SMD	35a(2mm)	Kec	2SC4173	data
KTN 2369(A)	Si-N	=2N2369(A): 0,625W	7e	Kec	→2N2369	data
KTN 2369(A)S	Si-N	=2N2369(A): SMD	35a	Kec	2SC4168, 2SC4453, PMBT 2369(A)	data
KTN 2369(A)U	Si-N	=2N2369(A): SMD	35a(2mm)	Kec	2SC4175...76, 2SC4452	data
KTN 2907(A)	Si-P	=2N2907(A): 0,625W	7e	Kec	→2N2907	data
KTN 2907(A)S	Si-P	=2N2907(A): SMD	35a	Kec	BSR15...16,BSS80,PMBT2907(A),SMBT2907(A)	data
KTN 2907(A)U	Si-P	=2N2907(A): SMD	35a(2mm)	Kec	2SA1608	data
KTQ 127	MOS-N-FET-d	=3SK127:	44v	Kec	→3SK127	data
KTX 101 U	Si-N+P	SMD, =KTA 701U + KTC 801U	46bh1(2mm)	Kec	-	data
KTX 301 U	Si-P	SMD, int. Diode(85V, 0,1A), 50V, 0,15A	45 (2mm)	Kec	-	data
KTX 401 U	Si-N	SMD, int. Diode(85V, 0,1A), 60V, 0,15A	45 (2mm)	Kec	-	data
KTY 13	Si-Sensor	SMD, Temperatur-Sensor	35	Sie	-	
KTY 81-110...-152	Si-Sensor	Temperatur-Sensor, 1kΩ(1mA/25°)	7	Phi	-	
KTY 81-210...-252	Si-Sensor	Temperatur-Sensor, 2kΩ(1mA/25°)	7	Phi	-	
KTY 82-...	Si-Sensor	=KTY 81-...: SMD	35	Phi	-	
KTY 83-110...-152	Si-Sensor	Temperatur-Sensor, 1kΩ(1mA/25°)	31	Phi	-	
KTY 84-130...-152	Si-Sensor	Temperatur-Sensor, 1kΩ(2mA/100°)	31	Phi	-	
KTY 85-110...-152	Si-Sensor	SMD, Temperatur-Sensor, 1kΩ(1mA/25°)	72(3,4mm)	Phi	-	
KTY 86-205	Si-Sensor	Temperatur-Sensor, 2kΩ(0,1mA/25°)		Phi	-	
KTY 87-205	Si-Sensor	Temperatur-Sensor, 2kΩ(0,1mA/25°)		Phi	-	

Type	Device	Short Description	Fig.	Manu	Comparision Types	More at
KU						
KU	MOS-N-FET-e*	=2SK3121 (Typ-Code/Stempel/marking)	39	Say	→2SK3121	data
KU	Z-Di	=P6 SMB-13 (Typ-Code/Stempel/marking)	71(5x3,5)	Fag	→P6 SMB-13	data
KU	Z-Di	=SMBJ 9.0 (Typ-Code/Stempel/marking)	71(5x3,5)	Mop	→SMBJ ...	data
KU	MOS-N-FET-e*	=SSM 3K02F (Typ-Code/Stempel/marking)	35	Tos	→SSM 3K02F	data
KU163	PS	LED/NPN Vbs:1.1V Ibs:5mA Ibe:0.3mA		Stl		data pdf pinout
KU 601	Si-N	S-L, 60V, 2A, 10W(Tc=105°), 15MHz	23a§	Tes	BUX 16, 2SC2809	data
KU 602	Si-N	S-L, 100V, 2A, 10W(Tc=105°), 15MHz	23a§	Tes	BUX 16, 2SC2809	data
KU 605	Si-N	S-L, 200V, 10A, 50W(Tc=80°), 5MHz	23a§	Tes	BU 109, BUX 17, TIP 160, 2SC2769, ++	data
KU 606	Si-N	S-L, 120V, 8A, 50W(Tc=80°), 5MHz	23a§	Tes	BU 109, BUX 17, TIP 160, 2SC2769, ++	data
KU 607	Si-N	S-L, 210V, 10A, 70W(Tc=45°)	23a§	Tes	BU 109, BUX 17, TIP 160, 2SC2769, ++	data
KU 608	Si-N	250V, 10A, 70W(Tc=45°), 9MHz	23a§	Tes	BU 109, BUX 17, TIP 160, 2SC2769, ++	data
KU 611	Si-N	S-L, 60V, 3A, 10W(Tc=105°), 30MHz	22a§	Tes	BD 241B, BD 243B, BU 409, 2SD422...423,++	data
KU 612	Si-N	S-L, 120V, 3A, 10W(Tc=105°), 30MHz	22a§	Tes	BD 241C, BD 243C, BU 409, 2SD422...423,++	data
KV...KZ						
KV	Z-Di	=1SMB 9.0A (Typ-Code/Stempel/marking)	71(5x3,5)	Ons	→1SMB 5.0A...170A	
KV	PIN-Di	=1SV264 (Typ-Code/Stempel/marking)	35 (2mm)	Say	→1SV264	data
KV	PIN-Di	=1SV267 (Typ-Code/Stempel/marking)	35	Say	→1SV267	data
KV	Z-Di	=P6 SMB-13A (Typ-Code/Stempel/marking)	71(5x3,5)	Fag	→P6 SMB-13A	data
KV	Z-Di	=SMBJ 9.0A (Typ-Code/Stempel/marking)	71(5x3,5)	Mop	→SMBJ ...	data
KV 1	Z-Di	=DZ 23C2V7 (Typ-Code/Stempel/marking)	35	Aeg	→DZ 23C2V7	
KV 2	Z-Di	=DZ 23C3V0 (Typ-Code/Stempel/marking)	35	Aeg	→DZ 23C3V0	
KV 3	Z-Di	=DZ 23C3V3 (Typ-Code/Stempel/marking)	35	Aeg	→DZ 23C3V3	
KV 4	Z-Di	=DZ 23C3V6 (Typ-Code/Stempel/marking)	35	Aeg	→DZ 23C3V6	
KV 5	Z-Di	=DZ 23C3V9 (Typ-Code/Stempel/marking)	35	Aeg	→DZ 23C3V9	
KV 6	Z-Di	=DZ 23C4V3 (Typ-Code/Stempel/marking)	35	Aeg	→DZ 23C4V3	
KV 7	Z-Di	=DZ 23C4V7 (Typ-Code/Stempel/marking)	35	Aeg	→DZ 23C4V7	
KV 8	Z-Di	=DZ 23C5V1 (Typ-Code/Stempel/marking)	35	Aeg	→DZ 23C5V1	
KV 9	Z-Di	=DZ 23C5V6 (Typ-Code/Stempel/marking)	35	Aeg	→DZ 23C5V6	
KVA	Z-Di	=DZ 23C6V2 (Typ-Code/Stempel/marking)	35	Aeg	→DZ 23C6V2	
KVB	Z-Di	=DZ 23C6V8 (Typ-Code/Stempel/marking)	35	Aeg	→DZ 23C6V8	
KVC	Z-Di	=DZ 23C7V5 (Typ-Code/Stempel/marking)	35	Aeg	→DZ 23C7V5	
KVD	Z-Di	=DZ 23C8V2 (Typ-Code/Stempel/marking)	35	Aeg	→DZ 23C8V2	
KVE	Z-Di	=DZ 23C9V1 (Typ-Code/Stempel/marking)	35	Aeg	→DZ 23C9V1	
KVF	Z-Di	=DZ 23C10 (Typ-Code/Stempel/marking)	35	Aeg	→DZ 23C10	
KVG	Z-Di	=DZ 23C11 (Typ-Code/Stempel/marking)	35	Aeg	→DZ 23C11	
KVH	Z-Di	=DZ 23C12 (Typ-Code/Stempel/marking)	35	Aeg	→DZ 23C12	
KVI	Z-Di	=DZ 23C13 (Typ-Code/Stempel/marking)	35	Aeg	→DZ 23C13	
KVJ	Z-Di	=DZ 23C15 (Typ-Code/Stempel/marking)	35	Aeg	→DZ 23C15	
KVK	Z-Di	=DZ 23C16 (Typ-Code/Stempel/marking)	35	Aeg	→DZ 23C16	
KVL	Z-Di	=DZ 23C18 (Typ-Code/Stempel/marking)	35	Aeg	→DZ 23C18	
KVM	Z-Di	=DZ 23C20 (Typ-Code/Stempel/marking)	35	Aeg	→DZ 23C20	
KVN	Z-Di	=DZ 23C22 (Typ-Code/Stempel/marking)	35	Aeg	→DZ 23C22	
KVO	Z-Di	=DZ 23C24 (Typ-Code/Stempel/marking)	35	Aeg	→DZ 23C24	
KVP	Z-Di	=DZ 23C27 (Typ-Code/Stempel/marking)	35	Aeg	→DZ 23C27	
KVQ	Z-Di	=DZ 23C30 (Typ-Code/Stempel/marking)	35	Aeg	→DZ 23C30	
KVR	Z-Di	=DZ 23C33 (Typ-Code/Stempel/marking)	35	Aeg	→DZ 23C33	
KVS	Z-Di	=DZ 23C36 (Typ-Code/Stempel/marking)	35	Aeg	→DZ 23C36	
KVT	Z-Di	=DZ 23C39 (Typ-Code/Stempel/marking)	35	Aeg	→DZ 23C39	
KVU	Z-Di	=DZ 23C43 (Typ-Code/Stempel/marking)	35	Aeg	→DZ 23C43	
KVV	Z-Di	=DZ 23C47 (Typ-Code/Stempel/marking)	35	Aeg	→DZ 23C47	
KVW	Z-Di	=DZ 23C51 (Typ-Code/Stempel/marking)	35	Aeg	→DZ 23C51	
KW	MOS-N-FET-e*	=2SK3122 (Typ-Code/Stempel/marking)	39	Say	→2SK3122	
KW	Z-Di	=P6 SMB-15 (Typ-Code/Stempel/marking)	71(5x3,5)	Fag	→P6 SMB-15	data
KW	Z-Di	=SMBJ 10 (Typ-Code/Stempel/marking)	71(5x3,5)	Mop	→SMBJ ...	data
KW	MOS-N-FET-e*	=SSM 3K01F (Typ-Code/Stempel/marking)	35	Tos	→SSM 3K01F	data
KW	MOS-N-FET-e*	=SSM 3K01T (Typ-Code/Stempel/marking)	35	Tos	→SSM 3K01T	
KWT824	LED	wht 2.4x1.5mm I=4cd Vf:3.2 #)V		Seo		data pdf pinout
KX	Z-Di	=1SMB 10A (Typ-Code/Stempel/marking)	71(5x3,5)	Ons	→1SMB 5.0A...170A	
KX	MOS-N-FET-e*	=2SK3291 (Typ-Code/Stempel/marking)	39	Say	→2SK3291	data
KX	Z-Di	=P6 SMB-15A (Typ-Code/Stempel/marking)	71(5x3,5)	Fag	→P6 SMB-15A	data
KX	MOS-N-FET-e	=PMBF 170 (Typ-Code/Stempel/marking)	35	Phi	→PMBF 170	data
KX	Z-Di	=SMBJ 10A (Typ-Code/Stempel/marking)	71(5x3,5)	Mop	→SMBJ ...	data
KX-1	Si-Di	≈BAV 20			→BAV 20	
KXC	Z-Di	=1SMB 10CA (Typ-Code/Stempel/marking)	71(5x3,5)	Ons	→1SMB 10CA...78CA	
KY	Si-N	=2SC3771 (Typ-Code/Stempel/marking)	35	Say	→2SC3771	data
KY	MOS-N-FET-e*	=2SK1764 (Typ-Code/Stempel/marking)	39	Hit	→2SK1764	data
KY	N-FET	=2SK211-Y (Typ-Code/Stempel/marking)	35	Tos	→2SK211	data
KY	MOS-N-FET-e*	=2SK3292 (Typ-Code/Stempel/marking)	39	Say	→2SK3292	data
KY	N-FET	=2SK881-Y (Typ-Code/Stempel/marking)	35(2mm)	Tos	→2SK881	data
KY	Si-P	=KTA1001-Y (Typ-Code/Stempel/marking)	39	Kec	→KTA 1001	data
KY	N-FET	=KTK 211-Y (Typ-Code/Stempel/marking)	35	Kec	→KTK 211	data
KY	Z-Di	=P6 SMB-16 (Typ-Code/Stempel/marking)	71(5x3,5)	Fag	→P6 SMB-16	data
KY	Z-Di	=SMBJ 11 (Typ-Code/Stempel/marking)	71(5x3,5)	Mop	→SMBJ ...	data
KY 1	Z-Di	=BZX 84C11 (Typ-Code/Stempel/marking)	35	Aeg	→BZX 84C11	
KY 2	Z-Di	=BZX 84C12 (Typ-Code/Stempel/marking)	35	Aeg	→BZX 84C12	
KY 3	Z-Di	=BZX 84C13 (Typ-Code/Stempel/marking)	35	Aeg	→BZX 84C13	
KY 4	Z-Di	=BZX 84C15 (Typ-Code/Stempel/marking)	35	Aeg	→BZX 84C15	
KY 5	Z-Di	=BZX 84C16 (Typ-Code/Stempel/marking)	35	Aeg	→BZX 84C16	
KY 6	Z-Di	=BZX 84C18 (Typ-Code/Stempel/marking)	35	Aeg	→BZX 84C18	
KY 7	Z-Di	=BZX 84C20 (Typ-Code/Stempel/marking)	35	Aeg	→BZX 84C20	

Type	Device	Short Description	Fig.	Manu	Comparision Types	More at
KY 8	Z-Di	=BZX 84C22 (Typ-Code/Stempel/marking)	35	Aeg	→BZX 84C22	
KY 9	Z-Di	=BZX 84C24 (Typ-Code/Stempel/marking)	35	Aeg	→BZX 84C24	
KYA	Z-Di	=BZX 84C27 (Typ-Code/Stempel/marking)	35	Aeg	→BZX 84C27	
KYB	Z-Di	=BZX 84C30 (Typ-Code/Stempel/marking)	35	Aeg	→BZX 84C30	
KYC	Z-Di	=BZX 84C33 (Typ-Code/Stempel/marking)	35	Aeg	→BZX 84C33	
KYD	Z-Di	=BZX 84C36 (Typ-Code/Stempel/marking)	35	Aeg	→BZX 84C36	
KYE	Z-Di	=BZX 84C39 (Typ-Code/Stempel/marking)	35	Aeg	→BZX 84C39	
KYF	Z-Di	=BZX 84C43 (Typ-Code/Stempel/marking)	35	Aeg	→BZX 84C43	
KYG	Z-Di	=BZX 84C47 (Typ-Code/Stempel/marking)	35	Aeg	→BZX 84C47	
KYH	Z-Di	=BZX 84C51 (Typ-Code/Stempel/marking)	35	Aeg	→BZX 84C51	
КЦ ...	-Di	→nach/after КД ...		GUS		
KZ	Z-Di	=1SMB 11A (Typ-Code/Stempel/marking)	71(5x3,5)	Ons	→1SMB 5.0A...170A	
KZ	MOS-N-FET-e*	=2SK3293 (Typ-Code/Stempel/marking)	39	Say	=2SK3293	data
KZ	Z-Di	=KA 3Z07 (Typ-Code/Stempel/marking)	31	Shi	→KA 3Z07	data
KZ	Z-Di	=KA 3Z18 (Typ-Code/Stempel/marking)	31	Shi	→KA 3Z18	data
KZ	Z-Di	=P6 SMB-16A (Typ-Cods/Stempel/marking)	71(5x3,5)	Fag	→P6 SMB-16A	data
KZ	Z-Di	=SMBJ 11A (Typ-Code/Stempel/marking)	71(5x3,5)	Mop	→SMBJ ...	data
KZ 1	Z-Di	=BZX 84C4V7 (Typ-Code/Stempel/marking)	35	Aeg	→BZX 84C4V7	
KZ 2	Z-Di	=BZX 84C5V1 (Typ-Code/Stempel/marking)	35	Aeg	→BZX 84C5V1	
KZ 3	Z-Di	=BZX 84C5V6 (Typ-Code/Stempel/marking)	35	Aeg	→BZX 84C5V6	
KZ 4	Z-Di	=BZX 84C6V2 (Typ-Code/Stempel/marking)	35	Aeg	→BZX 84C6V2	
KZ 5	Z-Di	=BZX 84C6V8 (Typ-Code/Stempel/marking)	35	Aeg	→BZX 84C6V8	
KZ 6	Z-Di	=BZX 84C7V5 (Typ-Code/Stempel/marking)	35	Aeg	→BZX 84C7V5	
KZ 7	Z-Di	=BZX 84C8V2 (Typ-Code/Stempel/marking)	35	Aeg	→BZX 84C8V2	
KZ 8	Z-Di	=BZX 84C9V1 (Typ-Code/Stempel/marking)	35	Aeg	→BZX 84C9V1	
KZ 9	Z-Di	=BZX 84C10 (Typ-Code/Stempel/marking)	35	Aeg	→BZX 84C10	
KZC	Z-Di	=1SMB 11CA (Typ-Code/Stempel/marking)	71(5x3,5)	Ons	→1SMB 10CA...78CA	
KZC	Z-Di	=BZX 84C2V7 (Typ-Code/Stempel/marking)	35	Aeg	→BZX 84C2V7	
KZD	Z-Di	=BZX 84C3V0 (Typ-Code/Stempel/marking)	35	Aeg	→BZX 84C3V0	
KZE	Z-Di	=BZX 84C3V3 (Typ-Code/Stempel/marking)	35	Aeg	→BZX 84C3V3	
KZF	Z-Di	=BZX 84C3V6 (Typ-Code/Stempel/marking)	35	Aeg	→BZX 84C3V6	
KZG	Z-Di	=BZX 84C3V9 (Typ-Code/Stempel/marking)	35	Aeg	→BZX 84C3V9	
KZH	Z-Di	=BZX 84C4V3 (Typ-Code/Stempel/marking)	35	Aeg	→BZX 84C4V3	

КД...КЦ

Type	Device	Short Description	Fig.	Manu	Comparision Types	More at
КД 102 А,Б	Si-Di	Uni, А=250V, Б=300V, 0,1A	31a	GUS	BA 147/300, BAY 21, BAY 46, BAY 88, ++	data
КД 103 А,Б	Si-Di	Uni, 50V, 0,1A, <4µs	31a	GUS	BA 147/50, BAY 18, BAY 73, 1N5194, ++	data
КД 110 А	GaAs-Di	SS, 30V, 10mA, <10ns	Koax	GUS	-	data
КД 202 Б,Г,Е,И Л,Н,С	Si-Di	GI, 1A, Б=50, Г=100, Е=200, И=300V Л=400, Н=500, С=600V	32	GUS	BY 249-...	
КД 202 А,В,Д,Ж	Si-Di	GI, 3A, А=50, В=100, Д=200, Ж=300V К=400, М=500, Р=600V	32	GUS	BY 249-...	
КД 203 А,В,Б,Д,Г	Si-Di	GI-L, 10A(Б,Г=5A)(Tc=100°) А=600V, В,Б=800V, Д,Г=1000V	≈32	GUS	BYW 88/.., BYX 42/.., 1N4508...4511	data
КД 206 А,В,Б	Si-Di	GI-L, 400...600V, 10A(Tc=70°) А=400V, В=600V, Б=500V	32	GUS	BYW 88/.., BYX 42/.., 1N4507...4511	data
КД 213 А-5,В-5, Б-5,Г-5	Si-Di	GI,S, 200V(Г=100V), 10A(Tc=70°) А-5:<300,В-5:<500,Б-5:<170,Г-5:<300ns	Cell	GUS	-	data
КД 222 А-5,В-5,Б-5	Si-Di	Schottky, 20...40V, 2A А-5=20V, В-5=40V, Б-5=30V	Cell	GUS	-	data
КД 226 А,В,Б,Д,Г	Si-Di	GI, S, 100...800V, 2A, <250ns А=100V, В=400V, Б=200V, Д=800V, Г=600V	31a	GUS	BY 218/.., BYW 32...36, BYT 77...78, ++	data
КД 235 А,Б	Si-Di	Schottky, А=30V, Б=40V, 1A	31a	GUS	BYV 10-.., SB 130...140, 1N5818...19	data
КД 238 АС,ВС,БС	Si-Di	Dual, Schottky, 25...45V, 2x7,5A АС=25V, ВС=45V, БС=35V	17p§	GUS	BYV 18-.., BYV 118-.., MBR 735...760	data
КД 243 А,В,Е,Б, Д,Ж,Г	Si-Di	GI, Uni, 100...1000V, 1A, А=100, В=200V Е=800, Б=100, Д=600, Ж=1000, Г=400V	31a	GUS	BY 126...127, BY 133...135, 1N4002...07, ++	data
КД 244 А,В,Б,Г	Si-Di	GI/S-L, 100...200V, 6A(Tc=100°) А,Б=100V, В,Г=200V, А,В:<50, Б,Г:<35ns	17	GUS	BYR 29-.., BYT 08-.., BYW 29-...	data
КД 247 А,В,Е,Б, Д,Г	Si-Di	GI, S, 50...800V, 1A, <150ns, Д:<250ns А=100,В=400,Е=50,Б=200,Д=800,Г=600V	31a	GUS	BYT 52A...M, BYV 12...16, RGP 10A...M, ++	data
КД 249 А,В,Б	Si-Di	Schottky-GI, А=40, В=20, Б=30V, 3A	31a	GUS	BYS 26-.., MBR 320...340, SB 320...340	data
КД 257 А	Si-Di	GI, S, 200V, 3A, <250ns	31	GUS	BY 318/200, BYW 16/200, RGP 30D, ++	data
КД 258 А	Si-Di	GI, S, 200V, 1,5A, <250ns	31	GUS	BY 218/200, BYV 27/200, BYV 36A, ++	data
КД 304 А,В,Е,Б,Д, Ж,И,Г (1-1)	Si-Di	Magnet-Diode, Sensor	12(2Pin)	GUS	-	
КД 401 А,Б	Si-Di	Uni, 75V, 92mA	31	GUS	BA 147/100, BAS 34, BAY 19, 1N5194, ++	
КД 407 А	Si-Di	HF, S, 24V, 50mA, <1Ω	31	GUS	-	data
КД 409 А9,Б9	C-Di	SMD, TV Band-S, 40V, А9:<1, Б9:<1,5pF	35p	GUS	-	data
КД 413 А,Б	PIN-Di	HF-Reg, 24V, 20mA, А=30...60, Б=40...80Ω	31	GUS	-	data
КД 419 А,В,Б,Г	Si-Di	Schottky, 15...50V, 10mA, <1,5pF(0V) А=15V, В=50V, Б=30V, Г=15V	31	GUS	BAT 81...83, 1N5711...12	data

Type	Device	Short Description	Fig.	Manu	Comparision Types	More at
КД 424 А,В,Г	Si-Di	Uni, 150..250V, 0,35A, <1µs A=250V, B=200V, Б=150V	31a	GUS	BA 157...159, BA 199/..., BAS 11	data
КД 425 А,Б	GaAs-Di	Impuls-Di, A=600, Б=400V, 2A, <2µs	32a§	GUS	-	data
КД 503 А,Б	Si-Di	S, 30V, 20mA, 10ns	31	GUS	BA 217, BA 218, BAX 13, 1N4148, ++	data
КД 504 А	Si-Di	Uni, 40V, 80mA	31	GUS	BA 128, BA 222, BAS 33, 1N5194, ++	data
КД 510 А	Si-Di	Uni, S. 50V, 0,2A	31	GUS	BAW 62, BAX 95, BAY 61, 1N4148..49, ++	data
КД 512 А,А1,Б	Si-Di	SS, 20V, 20mA	31	GUS	BAW 62, BAX 95, BAY 74, 1N4148..49, ++	data
КД 514 А,А1	Si-Di	SS, 10V, 20mA	31	GUS	BAW 62, BAX 95, BAY 74, 1N4148..49, ++	data
КД 521 А,В	Si-Di	Uni, S, A=75V, B=50V, 0,05A	31	GUS	BAW 62, BAX 95, BAY 61, 1N4148..49, ++	data
КД 522 Б	Si-Di	Uni, S, 50V, 0,1A	31	GUS	BAW 62, BAX 95, BAY 61, 1N4148..49, ++	data
КД 532 А	Si-Di	VC, Uni, S, 30V, 0,1A, <250ns	31	GUS	BAX 90, BAX 94, BAY 42...43, 1SS84, ++	data
КД 629 AC9	Si-Di	SMD, Dual, S, 90V, 0,2A, <100ns	35o	GUS	BAS 31	data
КД 704 AC9	Si-Di	SMD, Dual, SS, 70V, 0,1A, <6ns	35l	GUS	BAV 70, HSM 2838C, 1SS184	data
КД 708 АВБ	Si-Di	SS, 200V(B=100V), 1A, <10ns A: <10ns, B: <25ns, Б: <15ns	31a	GUS	-	data
КД 710 A	Si-Di	VC, SS, 35V, 0,1A, <6ns	31	GUS	BAW 62, BAY 71, BAY 95, 1N4148..49, ++	data
КД 711 А	Si-Di	VC, SS, 35V, 0,1A, <10ns	31	GUS	BAW 62, BAY 71, BAY 95, 1N4148..49, ++	data
КД 805 А	Si-Di	TV, SS, 75V, 0,2A, <4ns	31	GUS	BAW 62, BAW 76, BAX 95, 1N4148..49, ++	data
КД 808 А	Si-Di	Schottky, 25V, 0,2A, <5ns	31	GUS	BAT 42...43, BAT 85, MA 723, MA 777	data
КД 810 А	Si-Di	VC, SS, 3V, 10mA, <2ns	31	GUS	-	data
КД 901 А-1	Si-Di	Min, SS, 10V, 5mA, <20ns	36b(2Pin)	GUS	-	data
КД 901 В-1	Si-Di	=КД 901А-1: 3 Di	36(4Pin)		Pin-Code: KAAA	data
КД 901 Б-1	Si-Di	=КД 901А-1: 2 Di	36(3Pin)		Pin-Code: KAA	data
КД 901 Г-1	Si-Di	=КД 901А-1: 4 Di	36(5Pin)		Pin-Code: KAAAA	data
КД 904 А-1	Si-Di	Min, SS, 10V, 5mA, <10ns	36b(2Pin)	GUS	-	data
КД 904 В-1	Si-Di	=КД 904А-1: 3 Di	36(4Pin)		Pin-Code: KAAA	data
КД 904 Б-1	Si-Di	=КД 904А-1: 2 Di	36(3Pin)		Pin-Code: KAA	data
КД 904 Г-1	Si-Di	=КД 904А-1: 4 Di	36(5Pin)		Pin-Code: KAAAA	data
КД 908 АМ	Si-Di	Array, S, 40V, 0,2A, <20ns	14-DIP	GUS	-	data
КД 910 А-1	Si-Di	Min, SS, 5V, 10mA, <5ns	36a(2Pin)	GUS	-	data
КД 910 В-1	Si-Di	=КД 910А-1: 3 Di	36(4Pin)		Pin-Code: AKKK	data
КД 910 Б-1	Si-Di	=КД 910А-1: 2 Di	36(3Pin)		Pin-Code: AKK	data
КД 917 АМ	Si-Di	Array, S, 40V, 0,2A, <40ns	14-DIP	GUS	-	data
КД 922 А	Si-Di	Schottky, 18V, 50mA, <1pF(0V)	31b	GUS	BAT 45, MA 774	data
КД 922 В	Si-Di	=КД 922А: 10V, 10mA	31b		BAR 10...11, BAT 19, 1N5712, 1SS281	data
КД 922 Б	Si-Di	=КД 922А: 21V, 35mA	31b		BAT 45, MA 774	data
КД 923 А	Si-Di	Schottky, 14V, 0,1A, <3,6pF(0V)	31	GUS	BAT 42...43, BAT 85, MA 774	data
КД 927 А	Si-Di	Schottky, 35V, 10mA, <0,5pF(1V)	31	GUS	BAT 81...82, MA 776	data
КД 2994 А	Si-Di	GI/S-L, 100V, 8A(Tc=100°), <35ns	17	GUS	BYT 79-300...-500, BYV 79-100...-200	data
КД 2995 А,В,Е,Б, Д,Г	Si-Di	GI/S-L, 50...200V, 10A(Tc=125°), <50ns A=50, B=100, E=100(<100ns), Б=70V Д=200, Г=150V	32	GUS	BYW 13/..., MUR 2505...2520	data
КД 2997 А,В,Б	Si-Di	GI, S, 50...200V, 30A(Tc=85°), <200ns A=200V, B=50V, Б=100V	31	GUS	-	data
КД 2998 А,В,Б,Д,Г	Si-Di	Schottky-GI, 15...35V, 30A(Tc=85°) A=15V, B=25V, Б=20V, Д=30V, Г=35V	32	GUS	BYS 31-..., BYS 32, BYS 41-..., MBR3535	data
КД 2999 А,В,Б	Si-Di	GI, S, 50...200V, 20A(Tc=95°), <200ns A=200V, B=50V, Б=100V	31	GUS	-	data
КДС 523 АР,ВР	Si-Di	AP=Dual, BP=Quad, SS, 50V, 20mA, <1ns	31°	GUS	-	data
КДС 628 АМ	Si-Di	Array, S, 50V, 0,3A, <40ns	14-DIP	GUS	-	data
КП 303 А,Б,В,Г,Д, Е,Ж,И	N-FET	HF, 25V, 20mA, А,Б: Idss>0,5, В: >1,5 Г,Д: >3, Е: >5, Ж: >0,3, И: >1,5mA А,Б: Ugs<3, В: <4, Г,Д,Е: <8, Ж: <3V И: <2V	5h	GUS	BF 245, BFW 61, BFS 72, 2N3819, 2N3823++	data
КЦ 108 А	Si-Di	kV-GI, 2kV, 0,1A, 50kHz	31a	GUS	GP 02-20, HVG 2	data
КЦ 108 В	Si-Di	=КЦ 108А: 6kV	31a		BYX 90	data
КЦ 108 Б	Si-Di	=КЦ 108А: 4kV	31a		GP 02-40, HVG 4	data
КЦ 113 А-1	Si-Di	kV-GI, 1,6kV, 0,5A, 20kHz	31a	GUS	BYD 43-20	data
КЦ 114 А	Si-Di	kV-GI, 4kV, 50mA, 0...10kHz	31a	GUS	BV 6, HVG 4	data
КЦ 114 Б	Si-Di	=КЦ 114А: 6kV	31a		BV 6, BYX 90	data
КЦ 401 А	Si-Br	GI, 500V, 0,5A	42/3Pin	GUS	-	data
КЦ 401 В	Si-Br	GI-Br, 400V, 0,5A	42/8Pin	GUS	-	data
КЦ 401 Б	Si-Br	GI-Br, 500V, 0,5A	42/8Pin	GUS	-	data
КЦ 407 А	Si-Br	GI-Br, 300V, 0,5A, <5µs	6-DIP	GUS	-	data

Type	Device	Short Description	Fig.	Manu	Comparision Types	More at
L						
L	Si-N	=2SD1821A (Typ-Code/Stempel/marking)	35(2mm)	Mat	→2SD1821A	data
L	Si-N	=2SD2240A (Typ-Code/Stempel/marking)	35(1,6mm)	Mat	→2SD2240A	data
L	Si-N	=2SD814A (Typ-Code/Stempel/marking)	35	Mat	→2SD814A	data
L	Si-Di	=BAT 62-02W (Typ-Code/Stempel/marking)	71(1,3mm)	Sie	→BAT 62-02W	data
L	Si-Di	=BAT 62-03W (Typ-Code/Stempel/marking)	71(1,7mm)	Sie	→BAT 62-03W	data
L	C-Di	=BBY 53-02W (Typ-Code/Stempel/marking)	71(1,3mm)	Sie	→BBY 53-02W	data
L	C-Di	=HVU 356 (Typ-Code/Stempel/marking)	71(1,7mm)	Hit	→HVU 356	data
L	Si-Di	=MA 2S372 (Typ-Code/Stempel/marking)	71(1,3mm)	Mat	→MA 2S372	data
L	Z-Di	=MAZS 043 (Typ-Code/Stempel/marking)	71(1,3mm)	Mat	→MAZS 043	data
L	Si-Di	=SB 02W03C (Typ-Code/Stempel/marking)	35	Say	→SB 02W03C	data
L 0	Si-Di	=SST 510 (Typ-Code/Stempel/marking)	35	Six	→SST 510	data
L-1002	LED	gr+rd+srd+yl 1.1x3.4mm I>0.5mcd		Kin		data pdf pinout
L-103	LED	or+gr+rd+srd 2x5mm I>2mcd Vf:2.1V		Kin		data pdf pinout
L-1034	LED	gr+rd+srd+yl ø2mm I>2mcd Vf:2.1V		Kin		data pdf pinout
L-1043	LED	gr+srd+yl 5.8x1.3mm I>1mcd Vf:2.1V		Kin		data pdf pinout
L-1053	LED	gr+rd+srd+yl 1x5mm I>2mcd Vf:2.1V		Kin		data pdf pinout
L-1060	LED	gr+rd+srd+yl ø1.8mm I<5mcd		Kin		data pdf pinout
L-113	LED	or+gr+rd+srd+am+rd/gr+yl 2x5mm		Kin		data pdf pinout
L-1154	LED	gr+rd+srd+or+yl ø3mm I>10mcd Vf:2.1V	T1	Kin		data pdf pinout
L-115V	LED	rd/gr+rd/yl+gr/yl ø3mm I>8mcd		Kin		data pdf pinout
L-117	LED	rd/gr+rd/yl+gr/yl 2x5mm I>3mcd		Kin		data pdf pinout
L-119	LED	rd/gr 2x5mm I>0mcd Vf:2.2V If:20mA		Kin		data pdf pinout
L-13	LED	gr+rd+srd+yl ø2mm I>3.2mcd Vf:2.1V		Kin		data pdf pinout
L-130WCP	LED	rd/gr+rd/yl+gr/yl ø2.9mm I>8mcd		Kin		data pdf pinout
L-130WDT	LED	rd/gr+rd/yl+gr/yl ø2.9mm I>8mcd		Kin		data pdf pinout
L-131EB	LED	gr+srd+yl 1.8x5.3mm I>1mcd		Kin		data pdf pinout
L-132BR	LED	gr+rd+srd+yl ø3mm I>8mcd Vf:2.1V	T1	Kin		data pdf pinout
L-132CB	LED	gr+rd+yl 1.8x7mm I>1mcd Vf:2.1V		Kin		data pdf pinout
L-132X	LED	gr+rd+srd+or+yl ø3mm I>10mcd Vf:2.1V	T1	Kin		data pdf pinout
L-1334	LED	srd+gr+or+yl ø3mm I>1mcd Vf:2.1V		Kin		data pdf pinout
L-1384AD	LED	gr+srd+yl ø3mm I>8mcd Vf:2.2V		Kin		data pdf pinout
L-1384AL	LED	gr+srd+yl ø3mm I>8mcd Vf:2.2V		Kin		data pdf pinout
L-1394	LED	gr+rd+srd+yl ø2mm I>1.8mcd Vf:2.1V		Kin		data pdf pinout
L-1413	LED	gr+rd+srd+yl ø4mm I>1.2mcd Vf:2.1V		Kin		data pdf pinout
L-144	LED	or+gr+rd+srd+yl 1.9x3.9mm I>2mcd		Kin		data pdf pinout
L-144FI	LED	gr+srd+yl ø2.7x1.9mm I>1mcd Vf:2.1V		Kin		data pdf pinout
L-144FJ	LED	gr+srd+yl ø2.7x1.9mm I>1mcd Vf:2.1V		Kin		data pdf pinout
L-1503	LED	srd+gr+yl ø5mm I>20mcd Vf:2.1V	T1¾	Kin		data pdf pinout
L-1503CB	LED	gr+srd+yl ø5mm I>5mcd Vf:2.1V		Kin		data pdf pinout
L-1503EB	LED	gr+srd+yl ø5mm I>5mcd Vf:2.1V		Kin		data pdf pinout
L-1503S	LED	gr+rd ø5mm I>100mcd Vf:1.85V If:20mA	T1¾	Kin		data pdf pinout
L-150LVS	LED	rd/gr+rd/yl+gr/yl ø5mm I>20mcd		Kin		data pdf pinout
L-1513	LED	srd+gr+yl ø5mm I>20mcd Vf:2.1V	T1¾	Kin		data pdf pinout
L-1513S	LED	gr+rd ø5mm I>100mcd Vf:1.85V If:20mA	T1¾	Kin		data pinout
L-1513SUR	LED	rd ø5mm I>1300mcd Vf:1.9V If:20mA	T1¾	Kin		data pinout
L-153	LED	or+gr+rd+srd+yl 2.3x7mm I>1.3mcd		Kin		data pdf pinout
L-1533BQ/1ID	LED	gr+srd+yl ø4.7mm I>5mcd Vf:2.1V		Kin		data pdf pinout
L-1543S	LED	gr+rd ø5mm I>1000mcd Vf:1.85V	T1¾	Kin		data pdf pinout
L-1543SUR	LED	rd ø5mm I>400mcd Vf:1.9V If:20mA	T1¾	Kin		data pdf pinout
L-1553	LED	or+gr+rd+srd+yl 5x5mm I>2mcd Vf:2.1V		Kin		data pdf pinout
L-159	LED	gr+rd+srd+yl ø5mm I>60mcd Vf:2.1V	T1¾	Kin		data pdf pinout
L-1593	LED	srd+gr+yl ø5mm I>20mcd Vf:2.1V	T1¾	Kin		data pdf pinout
L-1619U	LED	rd 5.6x5mm I>480mcd Vf:1.85V If:20mA		Kin		data pdf pinout
L-161T	LED	srd+gr+yl 5.6x5mm I>100mcd Vf:2.1V		Kin		data pdf pinout
L-1633S	LED	rd ø5mm I>700mcd Vf:1.9V If:20mA	T1¾	Kin		data pdf pinout
L-169	LED	gr+rd+srd+yl 2x4mm I>5mcd Vf:2.1V		Kin		data pdf pinout
L-1703S	LED	rd ø5mm I>300mcd Vf:1.9V If:20mA	T1¾	Kin		data pdf pinout
L-173	LED	or+gr+rd+srd+yl 2.5x5mm I>2mcd		Kin		data pdf pinout
L-174	LED	gr+rd+srd+yl ø3.2mm I>8mcd Vf:2.1V		Kin		data pdf pinout
L-1773	LED	or+gr+rd+srd+yl 1.75x4mm I>3mcd		Kin		data pdf pinout
L-1799	LED	rd ø7.5mm I>480mcd Vf:1.85V If:20mA		Kin		data pdf pinout
L-2060	LED	gr+srd+yl ø1.8mm I>3mcd Vf:2.1V		Kin		data pdf pinout
L-21	LED	or+gr+srd+yl ø2mm I>1.2mcd Vf:2.1V		Kin		data pdf pinout
L-233	LED	am+or+rd/gr+srd+yl 2x5mm I>2mcd		Kin		data pinout
L-239	LED	rd/gr ø5.1mm I>8mcd Vf:2.2V If:10mA		Kin		data pdf pinout
L-2523S	LED	rd ø5mm I>1300mcd Vf:1.9V If:20mA	T1¾	Kin		data pdf pinout
L-301	LED	or+gr+rd+srd+yl 2x5x5mm I>0.8mcd		Kin		data pinout
L-311	LED	gr+rd+srd+yl 2x4.5x4.5mm I>0.8mcd		Kin		data pinout
L-32	LED	am+gr+rd+srd+yl ø3mm I<12.5mcd	T1	Kin		data pinout
L-32	PT	940nm P:0.3µA/mW/cm²	T1	Kin		data pdf pinout
L-323	LED	or+gr+rd+srd+yl 5.6x4.9x4.9mm I>2mcd		Kin		data pinout
L-32SL	LED	gr+rd+srd+yl ø3mm I<12.5mcd Vf:2.1V	T1	Kin		data pinout
L-34	LED	am+gr+rd+srd+yl ø3mm I>5mcd Vf:2.1V	T1	Kin		data pinout
L-34SF	IRED	850nm ø3mm I:§7mW/Sr Vf:1.4V If:20mA	T1	Kin		data pdf pinout
L-34SL	LED	gr+rd+yl ø3mm I>5mcd Vf:2.1V If:10mA	T1	Kin		data pinout
L-362	LED	gr+rd+srd+yl 3x4.5x4.5mm I>1.25mcd		Kin		data pdf pinout
L-36B	LED	gr+rd+srd+yl ø3mm I>5mcd Vf:2.1V	T1	Kin		data pdf pinout
L-383	LED	or+gr+rd+srd+yl 2.5x5mm I>2mcd		Kin		data pdf pinout
L-3V	LED	rd/gr+rd/yl+gr/yl ø3mm I>8mcd	T1	Kin		data pdf pinout
L-3W	LED	rd/gr ø3mm I>36mcd Vf:2.1V If:20mA	T1	Kin		data pdf pinout
L-4060VH2	LED	gr+srd+yl ø1.8mm I>5mcd Vf:2.1V		Kin		data pdf pinout
L-424	LED	or+gr+rd+srd+yl ø3mm I>1.3mcd	T1	Kin		data pdf pinout
L-433	LED	or+gr+rd+srd+yl ø5mm I>2mcd Vf:2.1V		Kin		data pdf pinout
L-44	LED	gr+rd+srd+yl ø4mm I>3.2mcd Vf:2.1V	T1	Kin		data pdf pinout

Type	Device	Short Description	Fig.	Manu	Comparision Types	More at
L-443	LED	or+gr+rd+srd+yl ø2.4mm I>1mcd		Kin		data pdf pinout
L-453	LED	gr+rd+srd+yl ø5mm I>5mcd Vf:2.1V	T1¾	Kin		data pdf pinout
L-456B	LED	gr+rd+srd+yl ø5mm I>5mcd Vf:2.1V	T1¾	Kin		data pdf pinout
L-469	LED	rd/gr+gr/yl ø3mm I>10mcd Vf:2.1V	T1	Kin		data pdf pinout
L-474	LED	srd+gr+yl ø2.9mm I>2mcd Vf:2.1V	T1	Kin		data pdf pinout
L-483	LED	or+gr+rd+srd+rd/gr+yl ø5mm I>1.3mcd	T1¾	Kin		data pdf pinout
L-493	LED	srd+gr+yl ø5mm I>2mcd Vf:2.1V	T1¾	Kin		data pdf pinout
L-503	LED	gr+rd+srd 5x5mm I>2mcd Vf:2.1V		Kin		data pdf pinout
L-51	LED	or+gr+rd+srd+yl ø5mm I>3.2mcd	T1¾	Kin		data pinout
L-513	LED	or+gr+rd+srd+yl 2.5x5.2mm I>2mcd		Kin		data pdf pinout
L-52	LED	am+gr+rd+srd+yl ø5mm I>3.2mcd	T1¾	Kin		data pinout
L-53	LED	am+srd+or+gr+yl ø5mm I>20mcd Vf:2.1V	T1¾	Kin		data pdf pinout
L-53	PT	940nm P:0.3µA/mW/cm²	T1¾	Kin		data pinout
L-53B	LED	bl ø5mm I>3mcd Vf:3.8V If:20mA	T1¾	Kin		data pinout
L-53BR	LED	gr+srd+yl ø5mm I>5mcd Vf:2.2V	T1¾	Kin		data pdf pinout
L-53BT	LED	gr+srd+yl ø5mm I>5mcd Vf:2.1V	T1¾	Kin		data pdf pinout
L-53L	LED	gr+srd+yl ø5mm I>0.8mcd Vf:2.1V	T1¾	Kin		data pdf pinout
L-53M	LED	gr ø5mm I>480mcd Vf:2.1V If:20mA	T1¾	Kin		data pdf pinout
L-53MB	LED	bl ø5mm I>40mcd Vf:4V If:20mA	T1¾	Kin		data pdf pinout
L-53MW	LED	wht ø5mm I>80mcd Vf:3.8V If:20mA	T1¾	Kin		data pdf pinout
L-53PB	LED	bl ø5mm I>650mcd Vf:3.7V If:20mA	T1¾	Kin		data pdf pinout
L-53RES	LED	gr+srd+yl ø5mm I>5mcd Vf:5V If:13mA	T1¾	Kin		data pdf pinout
L-53S	LED	or+gr+rd+yl ø5mm I>2800mcd Vf:4V	T1¾	Kin		data pdf pinout
L-53SF	IRED	850nm ø5mm I:§10mW/Sr Vf:1.4V	T1¾	Kin		data pdf pinout
L-53SRSGW	LED	rd/gr ø5mm I>12.5mcd Vf:2.2V If:20mA	T1¾	Kin		data pdf pinout
L-54P	LED	wht ø5mm I>500mcd Vf:3.6V If:20mA	T1¾	Kin		data pdf pinout
L-56B	LED	gr+rd+srd+yl ø5mm I>5mcd Vf:2.1V	T1¾	Kin		data pdf pinout
L-57	LED	rd/gr+rd/yl+gr/yl+srd ø5mm I>5mcd	T1¾	Kin		data pdf pinout
L-59	LED	rd/gr+rd/yl+gr/yl ø5mm I>70mcd	T1¾	Kin		data pdf pinout
L-59BL1	LED	rd/gr+rd/yl+gr/yl ø5mm I>20mcd	T1¾	Kin		data pdf pinout
L-59CL1	LED	rd/gr+rd/yl+gr/yl ø5mm I>20mcd	T1¾	Kin		data pdf pinout
L-5L	LED	rd/gr+rd/yl ø5mm I>20mcd Vf:2.1V	T1¾	Kin		data pdf pinout
L-610MP	PT	940nm P:0.5µA/mW/cm²		Kin		data pdf pinout
L-616B	LED	gr+rd+srd+yl ø3mm I>5mcd Vf:2.1V	T1	Kin		data pdf pinout
L-63	LED	gr+srd+yl ø5mm I>20mcd Vf:2.1V	T1¾	Kin		data pdf pinout
L-631	LED	gr+srd+yl ø5mm I>20mcd Vf:2.1V	T1¾	Kin		data pinout
L-631SRC	LED	rd ø5mm I>200mcd Vf:2.25V If:20mA	T1¾	Kin		data pinout
L-631SRD	LED	rd ø5mm I>100mcd Vf:2.25V If:20mA	T1¾	Kin		data pinout
L-63*SRT	LED	rd ø5mm I>200mcd Vf:2.25V If:20mA	T1¾	Kin		data pinout
L-704	LED	or+gr+rd+srd+yl 3x3mm I>1mcd Vf:2.1V		Kin		data pdf pinout
L-7104	LED	srd+or+gr+yl ø3mm I>8mcd Vf:2.1V	T1	Kin		data pdf pinout
L-7104MB	LED	bl ø3mm I>10mcd Vf:4V If:20mA	T1	Kin		data pdf pinout
L-7104PB	LED	bl ø3mm I>110mcd Vf:3.65V If:20mA	T1	Kin		data pdf pinout
L-7104PW	LED	wht ø3mm I>200mcd Vf:3.6V If:20mA	T1	Kin		data pdf pinout
L-7104S	LED	gr+or+rd+yl ø3mm I>1500mcd Vf:4V	T1	Kin		data pdf pinout
L-7104W	LED	wht ø3mm I>50mcd Vf:3.8V If:20mA	T1	Kin		data pdf pinout
L-793	LED	or+gr+srd+yl ø8mm I:20mcd Vf:2.1V		Kin		data pdf pinout
L-793S	LED	rd ø8mm I>650mcd Vf:1.85V If:20mA		Kin		data pdf pinout
L-796B	LED	gr+srd+yl ø8mm I>20mcd Vf:2.1V		Kin		data pdf pinout
L-799	LED	rd/gr ø8mm I>50mcd Vf:2.2V If:20mA	T1¾	Kin		data pdf pinout
L-803	LED	gr+srd+yl 8x8mm I>20mcd Vf:2.1V		Kin		data pdf pinout
L-813	LED	or+gr+srd+yl ø10mm I>20mcd Vf:2.1V		Kin		data pdf pinout
L-813S	LED	rd ø10mm I>650mcd Vf:1.85V If:20mA		Kin		data pdf pinout
L-816B	LED	gr+srd+yl ø8mm I>20mcd Vf:2.1V		Kin		data pdf pinout
L-819	LED	rd/gr+srd+yl ø10mm I>12.5mcd Vf:2.1V		Kin		data pdf pinout
L-934	LED	srd+or+gr+yl ø3mm I<50mcd Vf:2.1V	T1	Kin		data pdf pinout
L-934B	LED	bl ø3mm I>3mcd Vf:3.8V If:20mA	T1	Kin		data pinout
L-934L	LED	gr+srd+yl ø3mm I>0.8mcd Vf:2.1V	T1	Kin		data pdf pinout
L-934RES	LED	gr+srd+yl ø3mm I>8mcd Vf:5V If:13mA	T1	Kin		data pdf pinout
L-937	LED	rd/gr+rd/yl+gr/yl+srd ø3mm I>5mcd	T1	Kin		data pdf pinout
L-93W	LED	rd/gr+yl/gr+gr/yl ø3mm I>20mcd	T1	Kin		data pinout
L	Si-N	=2SC4211 (Typ-Code/Stempel/marking)	35(2mm)	Say	→2SC4211	data
L	GaAs-N-FET	=2SK1689 (Typ-Code/Stempel/marking)	51	Mat	→2SK1689	data
L	C-Di	=BB 143 (Typ-Code/Stempel/marking)	71(1,3mm)	Phi	→BB 143	data
L 0 p	Si-Di	=BAT 721S (Typ-Code/Stempel/marking)	35	Phi	→BAT 721S	data
L 0 t	Si-Di	=BAT 721S (Typ-Code/Stempel/marking)	35	Phi	→BAT 721S	data
L 1	Si-N	=BFS 36 (Typ-Code/Stempel/marking)	35(2mm)	Fer	→BFS 36	data
L 1	Si-P	=BSS 65 (Typ-Code/Stempel/marking)	35	Fer	→BSS 65	data
L 1	Si-P	=NSL 35T (Typ-Code/Stempel/marking)	35(1,6mm)	Ons	→NSL 35T	
L 1	Si-Di	=SST 511 (Typ-Code/Stempel/marking)	35	Six	→SST 511	data
L 1	Si-P+Di	=UML 1N (Typ-Code/Stempel/marking)	45(2mm)	Rhm	→UML 1N	data
L 1 G	Si-N	=HN 2C01FU-GR(Typ-Code/Stempel/mark.)	46(2mm)	Tos	→HN 2C01FU	data
L 1 N06C	MOS-N-FET-e	=MLD 1N06CL (Typ-Code/Stempel/marking)	30	Ons	→MLD 1N06CL	
L 1 N06CL	MOS-N-FET-e	=MLP 1N06CL (Typ-Code/Stempel/marking)	17	Ons	→MLP 1N06CL	
L 1 X	N-FET	=KSK 123: (Typ-Code/Stempel/marking)	35	Sam	→2SK123	data
L 1 Y	Si-N	=HN 2C01FU-Y (Typ-Code/Stempel/mark.)	46(2mm)	Tos	→HN 2C01FU	data
L 2	Si-Di	=BAS40-06L…(Typ-Code/Stempel/marking)	35	Ons	→BAS 40-06L	
L 2	Si-Di	=BAT 42W (Typ-Code/Stempel/marking)	71(2,7mm)	Gsi	→BAT 42W	data
L 2	Si-N	=BFS 36A (Typ-Code/Stempel/marking)	35(2mm)	Fer	→BFS 36A	
L 2	Si-P	=BSS 69 (Typ-Code/Stempel/marking)	35	Fer	→BSS 69	
L 2	Si-P	=NSL 12T (Typ-Code/Stempel/marking)	35(1,6mm)	Ons	→NSL 12T	
L 2	Si-Di	=SST 502 (Typ-Code/Stempel/marking)	35	Six	→SST 502	data
L 2	Z-Di	=UDZ 9.1B (Typ-Code/Stempel/marking)	71(1,7mm)	Rhm	→UDZ 2.0B…36B	data
L 2	Z-Di	=UDZS 9.1B (Typ-Code/Stempel/marking)	71(1,7mm)	Rhm	→UDZS 5.1B…10B	data
L 2	Si-N+Di	=UML 2N (Typ-Code/Stempel/marking)	45(2mm)	Rhm	→UML 2N	data

Type	Device	Short Description	Fig.	Manu	Comparision Types	More at
L 2 N06CL	MOS-N-FET-e	=MLP 2N06CL (Typ-Code/Stempel/marking)	17	Ons	→MLP 2N06CL	
L 2 X	N-FET	=KSK 211: (Typ-Code/Stempel/marking)	35	Sam	→2SK211	data
L 3	Si-N	=2SC1623-L3 (Typ-Code/Stempel/marking)	35	Nec	→2SC1623	data
L 3	Si-Di	=BAT 43W (Typ-Code/Stempel/marking)	71(2,7mm)	Gsi	→BAT 43W	data
L 3	Si-P	=BFS 37 (Typ-Code/Stempel/marking)	35(2mm)	Fer	→BFS 37	data
L 3	Si-P	=BSS 70 (Typ-Code/Stempel/marking)	35	Fer	→BSS 70	data
L 3	Si-N	=KST 1623L3 (Typ-Code/Stempel/marking)	35	Sam	→KST 1623	data
L 3	Si-P	=NSL 5T (Typ-Code/Stempel/marking)	35(1,6mm)	Ons	→NSL 5T	
L 3	Si-Di	=SST 503 (Typ-Code/Stempel/marking)	35	Six	→SST 503	data
L 3 A	C-Di	=KDV 1430A (Typ-Code/Stempel/marking)	35	Kec	→KDV 1430	data
L 3 B	C-Di	=KDV 1430B (Typ-Code/Stempel/marking)	35	Kec	→KDV 1430	data
L 3 C	C-Di	=KDV 1430C (Typ-Code/Stempel/marking)	35	Kec	→KDV 1430	data
L 3 D	C-Di	=KDV 1430D (Typ-Code/Stempel/marking)	35	Kec	→KDV 1430	data
L 4	Si-Di	=BAT 254 (Typ-Code/Stempel/marking)	71(2mm)	Phi	→BAT 254	data
L 4	Si-Di	=BAT 54 (Typ-Code/Stempel/marking)	35	Phi	→BAT 54	data
L 4	Si-Di	=BAT 54W (Typ-Code/Stempel/marking)	35(2mm)	Phi	→BAT 54W	data
L 4	Si-Di	=BAT 54W (Typ-Code/Stempel/marking)	71(2,7mm)	Gsi	→BAT 54W	data
L 4	Si-P	=BFS 37A (Typ-Code/Stempel/marking)	35(2mm)	Fer	→BFS 37A	data
L 4	Si-P	=BSS 65R (Typ-Code/Stempel/marking)	35	Fer	→BSS 65R	data
L 4	Si-Di	=D 1FL40 (Typ-Code/Stempel/marking)	71(5x2,5)	Shi	→D 1FL40	data
L 4	Si-N	=KST 1623L4 (Typ-Code/Stempel/marking)	35	Sam	→KST 1623	data
L 4	Si-P	=KTC 601U-Y (Typ-Code/Stempel/marking)	45(2mm)	Kec	→KTC 601U	data
L 4	Si-N	=KTC 801U-Y (Typ-Code/Stempel/marking)	46(2mm)	Kec	→KTC 801U	data
L 4	Si-Di	=SST 504 (Typ-Code/Stempel/marking)	35	Six	→SST 504	data
L 4	Si-N	=2SC1623-L4 (Typ-Code/Stempel/marking)	35	Nec	→2SC1623	data
L 4	Si-N	=2SC2812-L4 (Typ-Code/Stempel/marking)	35	Say	→2SC2812	data
L 4	Si-N	=2SC4177-L4 (Typ-Code/Stempel/marking)	35(2mm)	Nec	→2SC4177	data
L 4	Si-N	=2SC4783-L4 (Typ-Code/Stempel/marking)	35(1,6mm)	Nec	→2SC4783	data
L 4	Si-Di	=D 1NL40 (Typ-Code/Stempel/marking)	31	Shi	→D 1NL40	data
L 4 p	Si-Di	=BAT 54 (Typ-Code/Stempel/marking)	35	Phi	→BAT 54	data
L 4 t	Si-Di	=BAT 54 (Typ-Code/Stempel/marking)	35	Phi	→BAT 54	data
L 4 Z	Si-Di	=BAT 54 (Typ-Code/Stempel/marking)	35	Ztx	→BAT 54	data
L 5	Si-Di	=SST 505 (Typ-Code/Stempel/marking)	35	Six	→SST 505	data
L 5	Si-N	=2SC1623-L5 (Typ-Code/Stempel/marking)	35	Nec	→2SC1623	data
L 5	Si-N	=2SC2812-L5 (Typ-Code/Stempel/marking)	35	Say	→2SC2812	data
L 5	Si-N	=2SC4177-L5 (Typ-Code/Stempel/marking)	35(2mm)	Nec	→2SC4177	data
L 5	Si-N	=2SC4783-L5 (Typ-Code/Stempel/marking)	35(1,6mm)	Nec	→2SC4783	data
L 005(T1)	Z-IC	+5V, 0,93A	23a	Sgs	... 7805... (TO-3)	data pinout
L 5	Si-Di	=BAS 55 (Typ-Code/Stempel/marking)	35	Phi	→BAS 55	data
L 5	Si-N	=BFT 27 (Typ-Code/Stempel/marking)	35(2mm)	Fer	→BFT 27	data
L 5	Si-P	=BSS 69R (Typ-Code/Stempel/marking)	35	Fer	→BSS 69R	data
L 5	Si-N	=KST 1623L5 (Typ-Code/Stempel/marking)	35	Sam	→KST 1623	data
L 5 p	Si-Di	=BAS 55 (Typ-Code/Stempel/marking)	35	Phi	→BAS 55	data
L005T1	R+	Io=930mA Vo:4.75...5.25V Vin:7.5..20V	3O	Sgs		data pinout
L 6	Si-Di	=BAT 46W (Typ-Code/Stempel/marking)	71(2,7mm)	Gsi	→BAT 46W	data
L 6	Si-P	=KTC 601U-GR(Typ-Code/Stempel/marking)	45(2mm)	Kec	→KTC 601U	data
L 6	Si-N	=KTC 801U-GR(Typ-Code/Stempel/marking)	46(2mm)	Kec	→KTC 801U	data
L 6	Si-Di	=SST 506 (Typ-Code/Stempel/marking)	35	Six	→SST 506	data
L 6	Si-N	=2SC1623-L6 (Typ-Code/Stempel/marking)	35	Nec	→2SC1623	data
L 6	Si-N	=2SC2812-L6 (Typ-Code/Stempel/marking)	35	Say	→2SC2812	data
L 6	Si-N	=2SC4177-L6 (Typ-Code/Stempel/marking)	35(2mm)	Nec	→2SC4177	data
L 6	Si-N	=2SC4783-L6 (Typ-Code/Stempel/marking)	35(1,6mm)	Nec	→2SC4783	data
L 6	PIN-Di	=BAR 17 (Typ-Code/Stempel/marking)	35	Sie	→BAR 17	data
L 6	Si-N	=BSS 47 (Typ-Code/Stempel/marking)	35(2mm)	Fer	→BSS 47	data
L 6	Si-P	=BSS 70R (Typ-Code/Stempel/marking)	35	Fer	→BSS 70R	data
L 6	Si-N	=KST 1623L6 (Typ-Code/Stempel/marking)	35	Sam	→KST 1623	data
L 6 p	Si-Di	=BAT 720 (Typ-Code/Stempel/marking)	35	Phi	→BAT 720	data
L 7	Si-Di	=BAT 17 (Typ-Code/Stempel/marking)	35	Gsi	→BAT 17	data
L 7	Si-Di	=BAT 17W (Typ-Code/Stempel/marking)	71(2,7mm)	Gsi	→BAT 17W	data
L 7	Si-Di	=BAT 17WS (Typ-Code/Stempel/marking)	71(1,7mm)	Gsi	→BAT 17WS	data
L 7	Si-Di	=SST 507 (Typ-Code/Stempel/marking)	35	Six	→SST 507	data
L 7	Si-N	=2SC1623-L7 (Typ-Code/Stempel/marking)	35	Nec	→2SC1623	data
L 7	Si-N	=2SC2812-L7 (Typ-Code/Stempel/marking)	35	Say	→2SC2812	data
L 7	Si-N	=2SC4177-L7 (Typ-Code/Stempel/marking)	35(2mm)	Nec	→2SC4177	data
L 7	Si-N	=2SC4783-L7 (Typ-Code/Stempel/marking)	35(1,6mm)	Nec	→2SC4783	data
L 7	PIN-Di	=BAR 14-1 (Typ-Code/Stempel/marking)	35	Sie	→BAR 14	data
L 7	Si-N	=BSS 56 (Typ-Code/Stempel/marking)	35(2mm)	Fer	→BSS 56	data
L 7	Si-N	=KST 1623L7 (Typ-Code/Stempel/marking)	35	S.m	→KST 1623	data
L 7 p	Si-Di	=BAT 721 (Typ-Code/Stempel/marking)	35	Phi	→BAT 721	data
L 7 t	Si-Di	=BAT 721 (Typ-Code/Stempel/marking)	35	Phi	→BAT 721	data
L 8	C-Di	=BA 792 (Typ-Code/Stempel/marking)	71(2mm)	Phi	→BAT 792	data
L 8	PIN-Di	=BAR 15-1 (Typ-Code/Stempel/marking)	35	Sie	→BAR 15	data
L 8	Si-Di	=SST 508 (Typ-Code/Stempel/marking)	35	Six	→SST 508	data
L 8 p	Si-Di	=BAT 721A (Typ-Code/Stempel/marking)	35	Phi	→BAT 721A	data
L 8 t	Si-Di	=BAT 721A (Typ-Code/Stempel/marking)	35	Phi	→BAT 721A	data
L 9	Si-Di	=SST 509 (Typ-Code/Stempel/marking)	35	Six	→SST 509	data
L 9	Si-Di	=1SS349 (Typ-Code/Stempel/marking)	35	Tos	→1SS349	data
L 9	PIN-Di	=BAR 16-1 (Typ-Code/Stempel/marking)	35	Sie	→BAR 16	data
L 9 p	Si-Di	=BAT 721C (Typ-Code/Stempel/marking)	35	Phi	→BAT 721C	data
L 9 t	Si-Di	=BAT 721C (Typ-Code/Stempel/marking)	35	Phi	→BAT 721C	data
L14C	PT	940nm P:>50nµA/mW/cm²	TO18	Har,Qtc,Rca		data pdf pinout
L14F	PT	850nm P:>5µA/mW/cm²	TO18	Har,Qtc,Rca		data pdf pinout
L14G	PT	780(950)nm P:>3µA/mW/cm²	TO18	Har,Qtc,Rca		data pdf pinout
L14N	PT	770nm P:2µA/mW/cm²	TO18	Har,Qtc,Rca		data pdf pinout
L14P	PT	840nm P:11µA/mW/cm²	TO18	Har,Qtc,Rca		data pdf pinout

Type	Device	Short Description	Fig.	Manu	Comparision Types	More at
L14Q	PT	840nm P:0.8µA/mW/cm²	•	Har,Qtc,Rca		data pdf pinout
L14R	PT	840nm P:>18µA/mW/cm²		Har,Qtc,Rca		data pdf pinout
L 15	Si-N	=2SC3624A-L15 (Typ-Code/Stempel/mark.)	35	Nec	→2SC3624A	data
L 15	Si-N	=2SC4181A-L15 (Typ-Code/Stempel/mark.)	35(2mm)	Nec	→2SC4181A	data
L 16	Si-N	=2SC3624A-L16 (Typ-Code/Stempel/mark.)	35	Nec	→2SC3624A	data
L 16	Si-N	=2SC4181A-L16 (Typ-Code/Stempel/mark.)	35(2mm)	Nec	→2SC4181A	data
L 17	Si-N	=2SC3624-L17(Typ-Code/Stempel/marking)	35	Nec	→2SC3624	data
L 17	Si-N	=2SC4181-L17(Typ-Code/Stempel/marking)	35(2mm)	Nec	→2SC4181	data
L 18	Si-N	=2SC3624-L18(Typ-Code/Stempel/marking)	35	Nec	→2SC3624	data
L 18	Si-N	=2SC4181-L18(Typ-Code/Stempel/marking)	35(2mm)	Nec	→2SC4181	data
L 20	Si-Di	=BAS 29 (Typ-Code/Stempel/marking)	35	Val	→BAS 29	data
L 21	Si-Di	=BAS 31 (Typ-Code/Stempel/marking)	35	Val	→BAS 31	data
L 22	Si-Di	=BAS 35 (Typ-Code/Stempel/marking)	35	Val	→BAS 35	data
L 30	Si-Di	=BAV 23 (Typ-Code/Stempel/marking)	44	Phi	→BAV 23	data
L 30	Si-N	=FA 1L4L (Typ-Code/Stempel/marking)	35	Nec	→FA 1...	data
L 30	Si-N	=GA 1L4L (Typ-Code/Stempel/marking)	35(2mm)	Nec	→GA 1...	data
L 31	Si-Di	=BAV 23S (Typ-Code/Stempel/marking)	35	Phi	→BAV 23S	data
L 31	Si-N	=FA 1L4M (Typ-Code/Stempel/marking)	35	Nec	→FA 1...	data
L 31	Si-N	=GA 1L4M (Typ-Code/Stempel/marking)	35(2mm)	Nec	→GA 1...	data
L 32	Si-N	=FA 1F4M (Typ-Code/Stempel/marking)	35	Nec	→FA 1...	data
L 32	Si-N	=GA 1F4M (Typ-Code/Stempel/marking)	35(2mm)	Nec	→GA 1...	data
L 33	Si-N	=FA 1A4M (Typ-Code/Stempel/marking)	35	Nec	→FA 1...	data
L 33	Si-N	=GA 1A4M (Typ-Code/Stempel/marking)	35(2mm)	Nec	→GA 1...	data
L 34	Si-N	=FA 1A4P (Typ-Code/Stempel/marking)	35	Nec	→FA 1...	data
L 34	Si-N	=GA 1A4P (Typ-Code/Stempel/marking)	35(2mm)	Nec	→GA 1...	data
L 35	Si-N	=FA 1F4N (Typ-Code/Stempel/marking)	35	Nec	→FA 1...	data
L 35	Si-N	=GA 1F4N (Typ-Code/Stempel/marking)	35(2mm)	Nec	→GA 1...	data
L 36	Si-N	=FA 1L3Z-O (Typ-Code/Stempel/marking)	35	Nec	→FA 1...	data
L 36	Si-N	=GA 1L3Z-O (Typ-Code/Stempel/marking)	35(2mm)	Nec	→GA 1...	data
L036T1	R+	Io=750mA Vo:11.4...12.6V P:1.25W	3O	Sgs		data pinout
L 036(T1)	Z-IC	+12V, 1A	23a	Sgs	... 7812... (TO-3)	data pinout
L 37	Si-N	=FA 1L3Z-P (Typ-Code/Stempel/marking)	35	Nec	→FA 1...	data
L 37	Si-N	=GA 1L3Z-P (Typ-Code/Stempel/marking)	35(2mm)	Nec	→GA 1...	data
L037T1	R+	Io=680mA Vo:14.25...15.75V P:1.25W	3O	Sgs		data pinout
L 037(T1)	Z-IC	+15V, 0,9A	23a	Sgs	... 7815... (TO-3)	data pinout
L 38	Si-N	=FA 1L3Z-K (Typ-Code/Stempel/marking)	35	Nec	→FA 1...	data
L 38	Si-N	=GA 1L3Z-K (Typ-Code/Stempel/marking)	35(2mm)	Nec	→GA 1...	data
L 41	Si-Di	=BAT 74 (Typ-Code/Stempel/marking)	44	Phi	→BAT 74	data
L 42	Si-Di	=BAT 54A (Typ-Code/Stempel/marking)	35	Phi, Ztx	→BAT 54A	data
L 42	Si-Di	=BAT 54AW (Typ-Code/Stempel/marking)	35(2mm)	Phi	→BAT 54AW	data
L 43	Si-Di	=BAT 54C (Typ-Code/Stempel/marking)	35	Phi, Ztx	→BAT 54C	data
L 43	Si-Di	=BAT 54CW (Typ-Code/Stempel/marking)	35(2mm)	Phi	→BAT 54CW	data
L 44	Si-Di	=BAT 54S (Typ-Code/Stempel/marking)	35	Phi, Ztx	→BAT 54S	data
L 44	Si-Di	=BAT 54SW (Typ-Code/Stempel/marking)	35(2mm)	Phi	→BAT 54SW	data
L045T9	UNI	Vs:±18V Vu:100dP nV	8A	Sgs		data pinout
L 51	Si-Di	=BAS 56 (Typ-Code/Stempel/marking)	44	Phi	→BAS 56	data
L 52	Si-Di	=BAS 678 (Typ-Code/Stempel/marking)	35	Phi	→BAS 678	data
L 61	Si-N	=FA 1L4Z-Q (Typ-Code/Stempel/marking)	35	Nec	→FA 1...	data
L 61	Si-N	=GA 1L4Z-Q (Typ-Code/Stempel/marking)	35(2mm)	Nec	→GA 1...	data
L 62	Si-N	=FA 1L4Z-P (Typ-Code/Stempel/marking)	35	Nec	→FA 1...	data
L 62	Si-N	=GA 1L4Z-P (Typ-Code/Stempel/marking)	35(2mm)	Nec	→GA 1...	data
L 63	Si-N	=FA 1L4Z-K (Typ-Code/Stempel/marking)	35	Nec	→FA 1...	data
L 63	Si-N	=GA 1L4Z-K (Typ-Code/Stempel/marking)	35(2mm)	Nec	→GA 1...	data
L 64	Si-N	=FA 1F4Z-Q (Typ-Code/Stempel/marking)	35	Nec	→FA 1...	data
L 64	Si-N	=GA 1F4Z-Q (Typ-Code/Stempel/marking)	35(2mm)	Nec	→GA 1...	data
L 65	Si-N	=FA 1F4Z-P (Typ-Code/Stempel/marking)	35	Nec	→FA 1...	data
L 65	Si-N	=GA 1F4Z-P (Typ-Code/Stempel/marking)	35(2mm)	Nec	→GA 1...	data
L 66	Si-N	=FA 1F4Z-K (Typ-Code/Stempel/marking)	35	Nec	→FA 1...	data
L 66	Si-N	=GA 1F4Z-K (Typ-Code/Stempel/marking)	35(2mm)	Nec	→GA 1...	data
L 67	Si-N	=FA 1A4Z-Q (Typ-Code/Stempel/marking)	35	Nec	→FA 1...	data
L 67	Si-N	=GA 1A4Z-Q (Typ-Code/Stempel/marking)	35(2mm)	Nec	→GA 1...	data
L 68	Si-N	=FA 1A4Z-P (Typ-Code/Stempel/marking)	35	Nec	→FA 1...	data
L 68	Si-N	=GA 1A4Z-P (Typ-Code/Stempel/marking)	35(2mm)	Nec	→GA 1...	data
L 69	Si-N	=FA 1A4Z-K (Typ-Code/Stempel/marking)	35	Nec	→FA 1...	data
L 69	Si-N	=GA 1A4Z-K (Typ-Code/Stempel/marking)	35(2mm)	Nec	→GA 1...	data
L 72	Si-Di	=BAT 17DS (Typ-Code/Stempel/marking)	35	Gsi	→BAT 17DS	data
L 78LxxABD,ABZ	Z-IC	=L 78LxxCD,CZ: ±5%, -40...+125°		Sgs	•	
L 78LxxACD,ACZ	Z-IC	=L 78LxxCD,CZ: ±5%		Sgs		
L 78L05CD...15CD	Z-IC	=L 78L...CZ: SMD	8-MDIP	Sgs	... 78Lxx... (TO-92)	
L 78L05CZ...15CZ	Z-IC	+5...+15V, ±10%, 0,1A, 0...+125°	7b	Sgs	... 78Lxx... (TO-92)	
L78L05AB	R+	Io=100mA Vo:4.8...5.2V Vin:10V P:int	4S,8S,3N	Sgs		data pdf pinout
L78L05AC	R+	Io=100mA Vo:4.8...5.2V Vin:10V P:int	4S,8S,3N	Sgs		data pdf pinout
L78L05C	R+	Io=100mA Vo:4.6...5.4V Vin:10V P:int	8S,3N	Sgs		data pdf pinout
L78L06AB	R+	Io=100mA Vo:5.76...6.24V Vin:12V P:int	8S,4S,3N	Sgs		data pdf pinout
L78L06AC	R+	Io=100mA Vo:5.76...6.24V Vin:12V P:int	8S,4S,3N	Sgs		data pdf pinout
L78L06C	R+	Io=100mA Vo:5.52...6.48V Vin:12V P:int	8S,3N	Sgs		data pdf pinout
L78L08AB	R+	Io=100mA Vo:7.68...8.42V Vin:14V P:int	8S,4S,3N	Sgs		data pdf pinout
L78L08AC	R+	Io=100mA Vo:7.68...8.42V Vin:14V P:int	8S,4S,3N	Sgs		data pdf pinout
L78L08C	R+	Io=100mA Vo:7.68...8.42V Vin:14V P:int	3N,8S	Sgs		data pdf pinout
L78L09AB	R+	Io=100mA Vo:8.64...9.36V Vin:15V P:int	8S,4S,3N	Sgs		data pdf pinout
L78L09AC	R+	Io=100mA Vo:8.64...9.36V Vin:15V P:int	8S,4S,3N	Sgs		data pdf pinout
L78L09C	R+	Io=100mA Vo:8.28...9.72V Vin:15V P:int	8S,3N	Sgs		data pdf pinout
L78L12AB	R+	Io=100mA Vo:11.5...12.5V Vin:19V P:int	8S,4S,3N	Sgs		data pdf pinout
L78L12AC	R+	Io=100mA Vo:11.5...12.5V Vin:19V P:int	8S,4S,3N	Sgs		data pdf pinout

Type	Device	Short Description	Fig.	Manu	Comparision Types	More at
L78L12C	R+	Io=100mA Vo:11.1...12.9V Vin:19V P:int	8S,3N	Sgs		data pdf pinout
L78L15AB	R+	Io=100mA Vo:14.4...15.6V Vin:19V P:int	8S,4S,3N	Sgs		data pdf pinout
L78L15AC	R+	Io=100mA Vo:14.4...15.6V Vin:19V P:int	8S,4S,3N	Sgs		data pdf pinout
L78L15C	R+	Io=100mA Vo:13.8...16.2V Vin:23V P:int	8S,3N	Sgs		data pdf pinout
L78L18AB	R+	Io=100mA Vo:17.3...18.7V Vin:27V P:int	8S,4S,3N	Sgs		data pdf pinout
L78L18AC	R+	Io=100mA Vo:17.3...18.7V Vin:27V P:int	8S,4S,3N	Sgs		data pdf pinout
L78L18C	R+	Io=100mA Vo:16.6...19.4V Vin:27V P:int	8S,3N	Sgs		data pdf pinout
L78L24AB	R+	Io=100mA Vo:23...25V Vin:27V P:int	8S,3N	Sgs		data pdf pinout
L78L24AC	R+	Io=100mA Vo:23...25V Vin:27V P:int	8S,3N	Sgs		data pdf pinout
L78L24C	R+	Io=100mA Vo:22.1...25.9V Vin:33V P:int	8S,3N	Sgs		data pdf pinout
L78L33	R+	Io=100mA Vo:3.268...3.432V Vin:8.3V	4S	Sgs		data pdf pinout
L78L33AB	R+	Io=100mA Vo:3.268...3.432V Vin:8.3V	4S,8S,3N	Sgs		data pdf pinout
L78L33AC	R+	Io=100mA Vo:3.268...3.432V Vin:8.3V	8S,3N	Sgs		data pdf pinout
L78LR05	R+	Io=150mA Vo:4.75...5.25V Vin:7...20V	5P	Say		data pdf pinout
L 78 LR05 B...H	Z-IC	+5V, 150mA, Reset	30/5-Pin	Say	-	
L78M0	R+	Io=500mA Vo:4.75...5.25V Vin:7...20V	3P	Say		data pdf pinout
L 78 M05..24 AB(V)	Z-IC	=L 78MxxCV: hi-prec, 2%, -40...125°	17b	Sgs		data pinout
L 78 M05..24 CP	Z-IC	=L 78MxxCV: Iso	17b	Sgs	...78Mxx... (TO-220 Iso)	
L 78 M05..24 CV	Z-IC	+5...24V, 0,5A	17b	Say,Sgs	...78Mxx... (TO-220)	data pinout
L 78 M05..24 CX	Z-IC	=L 78MxxCV: Fig.→	≈14b	Sgs		data pinout
L 78 M05..24 ML	Z-IC	=L 78MxxCV: Iso	17b	Say	...78Mxx... (TO-220 Iso)	data pinout
L78M05	R+	Io=500mA Vo:4.8...5.2V Vin:10V P:1.75W	3P	Say		data pdf pinout
L78M05AB	R+	Io=int Vo:4.9...5.1V Vin:10V P:int	3P	Sgs		data pdf pinout
L78M05AC	R+	Io=int Vo:4.9...5.1V Vin:10V P:int	3P	Sgs		data pdf pinout
L78M06	R+	Io=500mA Vo:5.7...6.3V Vin:8...21V P:1W	3P	Say		data pdf pinout
L78M06AB	R+	Io=int Vo:5.88...6.12V Vin:11V P:int	3P	Sgs		data pdf pinout
L78M06AC	R+	Io=int Vo:5.88...6.12V Vin:11V P:int	3P	Sgs		data pdf pinout
L78M07	R+	Io=500mA Vo:6.6...7.4V Vin:9...22V P:1W	3P	Say		data pdf pinout
L78M08	R+	Io=500mA Vo:7.6...8.4V Vin:10.5...23V	3P	Say		data pdf pinout
L78M08AB	R+	Io=int Vo:7.84...8.16V Vin:14V P:int	3P	Sgs		data pdf pinout
L78M08AC	R+	Io=int Vo:7.84...8.16V Vin:14V P:int	3P	Sgs		data pdf pinout
L78M09	R+	Io=500mA Vo:8.5...9.5V Vin:11.5...24V	3P	Say		data pdf pinout
L78M09AB	R+	Io=int Vo:8.82...9.18V Vin:14V P:int	3P	Sgs		data pdf pinout
L78M09AC	R+	Io=int Vo:8.82...9.18V Vin:14V P:int	3P	Sgs		data pdf pinout
L78M10	R+	Io=500mA Vo:9.5...10.5V Vin:12.5...25V	3P	Say		data pdf pinout
L78M10AB	R+	Io=int Vo:9.8...10.2V Vin:16V P:int	3P	Sgs		data pdf pinout
L78M10AC	R+	Io=int Vo:9.8...10.2V Vin:16V P:int	3P	Sgs		data pdf pinout
L78M12	R+	Io=500mA Vo:11.4...12.6V P:1W	3P	Say		data pdf pinout
L78M12AB	R+	Io=int Vo:11.75...12.25V Vin:19V P:int	3P	Sgs		data pdf pinout
L78M12AC	R+	Io=int Vo:11.75...12.25V Vin:19V P:int	3P	Sgs		data pdf pinout
L78M15	R+	Io=500mA Vo:14.25...15.75V P:1W	3P	Say		data pdf pinout
L78M15AB	R+	Io=int Vo:14.7...15.3V Vin:23V P:int	3P	Sgs		data pdf pinout
L78M15AC	R+	Io=int Vo:14.7...15.3V Vin:23V P:int	3P	Sgs		data pdf pinout
L78M18	R+	Io=500mA Vo:17.1...18.9V Vin:21...33V	3P	Say		data pdf pinout
L78M18AB	R+	Io=int Vo:17.64...18.36V Vin:26V P:int	3P	Sgs		data pdf pinout
L78M18AC	R+	Io=int Vo:17.64...18.36V Vin:26V P:int	3P	Sgs		data pdf pinout
L78M20	R+	Io=500mA Vo:19...21V Vin:23...35V P:1W	3P	Say		data pdf pinout
L78M20AB	R+	Io=int Vo:19.6...20.4V Vin:29V P:int	3P	Sgs		data pd' pinout
L78M20AC	R+	Io=int Vo:19.6...20.4V Vin:29V P:int	3P	Sgs		data pdf pinout
L78M24	R+	Io=500mA Vo:22.8...25.8V Vin:27...35V	3P	Say		data pdf pinout
L78M24AB	R+	Io=int Vo:23.5...24.5V Vin:33V P:int	3P	Sgs		data pdf pinout
L78M24AC	R+	Io=int Vo:23.5...24.5V Vin:33V P:int	3P	Sgs		data pdf pinout
L78MG	ER+	Io=500mA Vo:5...30V Vin:10V P:1.2W	4P	Say		data pdf pinout
L 78 MG	Z-IC	+7,5...35V, 0,5A	15/4Pin	Say		data pinout
L 78 MR05...12	Z-IC	Volt. Reg. +5...12V, 0,5A, Reset	17/5Pin	Say	-	pdf
L78MR05	R+	Io=500mA Vo:4.75...5.25V Vin:7...20.5V	5P	Say		data pdf pinout
L78MR06	R+	Io=500mA Vo:5.7...6.3V Vin:8...21.5V	5P	Say		data pdf pinout
L78MR08	R+	Io=500mA Vo:7.6...8.4V P:1.75W	5P	Say		data pdf pinout
L78MR09	R+	Io=500mA Vo:8.5...9.5V Vin:11.5...24V	5P	Say		data pdf pinout
L78MR12	R+	Io=500mA Vo:11.4...12.6V P:1.75W	5P	Say		data pdf pinout
L 78 M05T...M24T	Z-IC	=L 78Mxx(CV): Fig.→	30b	Say		data pinout
L 78 N05..24	Z-IC	+5...24V, 0,5A	14b	Say	...78Nxx... (TO-126)	data pinout
L78N05	R+	Io=500mA Vo:4.8...5.2V Vin:10V P:1.2W	3P	Say		data pinout
L78N06	R+	Io=500mA Vo:5.75...6.25V Vin:11V	3P	Say		data pinout
L78N08	R+	Io=500mA Vo:7.6...8.4V Vin:10.5...23V	3P	Say		data pinout
L78N09	R+	Io=500mA Vo:8.5...9.5V Vin:11.5...24V	3P	Say		data pinout
L78N10	R+	Io=500mA Vo:9.5...10.5V Vin:12.5...25V	3P	Say		data pinout
L78N12	R+	Io=500mA Vo:11.4...12.6V P:1.2W	3P	Say		data pinout
L78N15	R+	Io=500mA Vo:14.25...15.75V P:1.2W	3P	Say		data pinout
L78N18	R+	Io=500mA Vo:17.1...18.9V Vin:21...33V	3P	Say		data pinout
L78N20	R+	Io=500mA Vo:19...21V Vin:23...35V	3P	Say		data pinout
L78N24	R+	Io=500mA Vo:22.8...25.8V Vin:27...35V	3P	Say		data pinout
L 78 S05..24 CV	Z-IC	+5...24V, 2A	17b	Sgs	L 2005..2024	data pinout
L 78 S05..24 T,CT	Z-IC	+5...24V, 2A	23a	Sgs	L 2005..2024	data pinout
L78S05	R+	Io=int Vo:4.8...5.2V Vin:10V P:int	3OP	Sgs		data pdf pinout
L78S09	R+	Io=int Vo:8.65...9.35V Vin:14V P:int	3OP	Sgs		data pdf pinout
L78S10	R+	Io=int Vo:9.5...10.5V Vin:15V P:int	3OP	Sgs		data pdf pinout
L78S12	R+	Io=int Vo:11.5...12.5V Vin:19V P:int	3OP	Sgs		data pdf pinout
L78S15	R+	Io=int Vo:14.4...15.6V Vin:23V P:int	3OP	Sgs		data pdf pinout
L78S18	R+	Io=int Vo:17.1...18.9V Vin:26V P:int	3OP	Sgs		data pdf pinout
L78S24	R+	Io=int Vo:23...25V Vin:33V P:int	3OP	Sgs		data pdf pinout
L78S75	R+	Io=int Vo:7.15...7.9V Vin:12.5V P:int	3OP	Sgs		data pdf pinout
L 79 LxxABD	Z-IC	-5...-15V, Negative voltage regulators,	MDIP	Stm	-	pdf pinout
L 79 LxxABUT	Z-IC	-5...-15V, Negative voltage regulators,	48	Stm	-	pdf pinout
L 79 LxxABZ	Z-IC	-5...-15V, Negative voltage regulators	7	Stm	-	pdf pinout

Type	Device	Short Description	Fig.	Manu	Comparision Types	More at
L 79 LxxACD	Z-IC	-5...-15V, Negative voltage regulators	MDIP	Stm	L 79LxxCD	pdf pinout
L 79 LxxACUT	Z-IC	-5...-15V, Negative voltage regulators	48	Stm	L 79LxxCD	pdf pinout
L 79 LxxACZ	Z-IC	-5...-15V, Negative voltage regulators	7	Stm	L 79LxxCD	pdf pinout
L 79 L05CD...15CD	Z-IC	-5...-15V, Negative voltage regulators	MDIP	Stm	L 79L...CZ	pdf pinout
L 79 L05CZ...15CZ	Z-IC	-5...-15V, ±10%, 0,1A, 0...+125°	7b	Stm	... 79Lxx... (TO-92)	pdf pinout
L79L05AB	R-	Io=100mA Vo:-4.8...-5.2V Vin:-10V	8S,4S,3N	Sgs		data pdf pinout
L79L05AC	R-	Io=100mA Vo:-4.8...-5.2V Vin:-10V	8S,4S,3N	Sgs		data pdf pinout
L79L05C	R-	Io=100mA Vo:-4.6...-5.4V Vin:-10V	8S,3N	Sgs		data pdf pinout
L79L06AB	R-	Io=100mA Vo:-5.76...-6.24V Vin:-12V	8S,4S,3N	Sgs		data pdf pinout
L79L06AC	R-	Io=100mA Vo:-5.76...-6.24V Vin:-12V	8S,4S,3N	Sgs		data pdf pinout
L79L06C	R-	Io=100mA Vo:-5.52...-6.48V Vin:-12V	8S,3N	Sgs		data pdf pinout
L79L08AB	R-	Io=100mA Vo:-7.68...-8.32V Vin:-14V	8S,4S,3N	Sgs		data pdf pinout
L79L08AC	R-	Io=100mA Vo:-7.68...-8.32V Vin:-14V	8S,4S,3N	Sgs		data pdf pinout
L79L08C	R-	Io=100mA Vo:-7.36...-8.64V Vin:-14V	8S,3N	Sgs		data pdf pinout
L79L09AB	R-	Io=100mA Vo:-8.64...-9.36V Vin:-15V	8S,4S,3N	Sgs		data pdf pinout
L79L09AC	R-	Io=100mA Vo:-8.64...-9.36V Vin:-15V	8S,4S,3N	Sgs		data pdf pinout
L79L09C	R-	Io=100mA Vo:-8.28...-9.72V Vin:-15V	8S,3N	Sgs		data pdf pinout
L79L12AB	R-	Io=100mA Vo:-11.5...-12.5V Vin:-19V	8S,4S,3N	Sgs		data pdf pinout
L79L12AC	R-	Io=100mA Vo:-11.5...-12.5V Vin:-19V	8S,4S,3N	Sgs		data pdf pinout
L79L12C	R-	Io=100mA Vo:-11.1...-12.9V Vin:-19V	8S,3N	Sgs		data pdf pinout
L79L15AB	R-	Io=100mA Vo:-14.4...-15.6V Vin:-23V	8S,3N	Sgs		data pdf pinout
L79L15AC	R-	Io=100mA Vo:-14.4...-15.6V Vin:-23V	8S,3N	Sgs		data pdf pinout
L79L15C	R-	Io=100mA Vo:-13.8...-16.2V Vin:-23V	8S,3N	Sgs		data pdf pinout
L 79 M05..24	Z-IC	-5...-24V, 0,5A	17c	Say	... 79Mxx... (TO-220)	data pinout
L 79 M05ML...M24ML	Z-IC	=L 79 Mxx: Iso	17c	Say	... 79Mxx... (TO-220 Iso)	data pinout
L 79 M05T...M24T	Z-IC	=L 79 Mxx: Fig.→	30c	Say		data pinout
L79M05	R-	Io=500mA Vo:-4.75...-5.25V P:1W	3P	Say		data pinout
L79M06	R-	Io=500mA Vo:-5.7...-6.3V Vin:-8...-25V	3P	Say		data pinout
L79M08	R-	Io=500mA Vo:-7.6...-8.4V P:1W	3P	Say		data pinout
L79M09	R-	Io=500mA Vo:-8.5...-9.5V P:1W	3P	Say		data pinout
L79M10	R-	Io=500mA Vo:-9.5...-10.5V P:1W	3P	Say		data pinout
L79M12	R-	Io=500mA Vo:-11.4...-12.6V P:1W	3P	Say		data pinout
L79M15	R-	Io=500mA Vo:-14.25...-15.75V P:1W	3P	Say		data pinout
L79M20	R-	Io=500mA Vo:-19...-21V Vin:-23...-35V	3P	Say		data pinout
L79M24	R-	Io=500mA Vo:-22.8...-25.2V P:1W	3P	Say		data pinout
L 81	Si-N	=FA 1L3M (Typ-Code/Stempel/marking)	35	Nec	→FA 1...	data
L 81	Si-N	=GA 1L3M (Typ-Code/Stempel/marking)	35(2mm)	Nec	→GA 1...	data
L 82	Si-N	=FA 1L3N (Typ-Code/Stempel/marking)	35	Nec	→FA 1...	data
L 82	Si-N	=GA 1L3N (Typ-Code/Stempel/marking)	35(2mm)	Nec	→GA 1...	data
L 83	Si-N	=FA 1A3Q (Typ-Code/Stempel/marking)	35	Nec	→FA 1...	data
L 83	Si-N	=GA 1A3Q (Typ-Code/Stempel/marking)	35(2mm)	Nec	→GA 1...	data
L 88 MS04...12T	Z-IC	Strobe, +4...12V, 0,5A	30/5Pin	Say	-	
L 88 R05 C	Z-IC	Reset, +5V, 1A, Threshold=4,3...4,7V	86/5Pin	Say	-	pdf
L 88 R05 D	Z-IC	Reset, +5V, 1A, Threshold=4...4,4V	86/5Pin	Say	-	pdf
L 88 R05 E	Z-IC	Reset, +5V, 1A, Threshold=3,7...4,1V	86/5Pin	Say	-	pdf
L 93 PI	IC	Quad low Side Driver	TSOP	Stm	-	pdf pinout
L 103 T1	LIN-IC	FM-ZF		Stm	SN76603	
L115T1	HI-SPEED	Vs:±18V Vu:90dB Vo:±13V Vi0:2mV	10A	Sgs		data pinout
L 120 A(AB)(B1)	LIN-IC	Thyristor-ansteuerung/phase control	16-DIP	Sgs	-	
L 120 AB	LIN-IC	Zero-Cross. Trig.-Phase Control., 12V	16-DIP	Sgs	ECG936,NTE936	
L 121 (B1)	LIN-IC	Thyristoransteuerg./burst control	16-DIP	Sgs	-	
L123	ER+	Io=150mA Vo:2...37V P:520mW	14D,10A	Sgs		data pinout
L 123(C)T,T1,T2	Z-IC	+2...37V, 0,15A	82	Sgs	...723... (TO-100)	data pinout
L 123(C)TB	Z-IC	=L 123 (C)T	82	Sgs	...723... (TO-100)	data pinout
L 123 B1,CB,D1	Z-IC	+2...37V, 0,15A	14-DIP	Sgs	...723... (DIP)	data pinout
L129	R+	Io=930mA Vo:4.75...5.25V Vin:7.5...20V	3P	Sgs		data pinout
L 129	Z-IC	+5V, 0,93A	14b	Sgs	...7805... (TO-126)	data pinout
L130	R+	Io=750mA Vo:11.4...12.6V	3P	Sgs		data pinout
L 130	Z-IC	+12V, 0,75A	14b	Sgs	...7812... (TO-126)	data pinout
L131	R+	Io=680mA Vo:14.25...15.75V	3P	Sgs		data pinout
L 131	Z-IC	+15V, 0,6A	14b	Sgs	...7815... (TO-126)	data pinout
L 133 T	LIN-IC	NF-E, 30V, 1,5A, 5W(18V/8Ω)	23/8pin	Sgs	-	
L141	UNI	Vs:±18V Vu:100dB Vo:±14V Vi0:2mV	14D,8A	Sgs		data pinout
L 141 B1	OP-IC	Uni, Serie 741, ±18V, 0...+70°	14-DIP	Sgs	→Serie 741	
L 141 T1	OP-IC	=L 141B1: Fig.→	82	Sgs	→Serie 741	
L141T2	UNI	Vs:±22V Vu:100dB Vo:±14V Vi0:1mV	8A	Sgs		
L144	3xOP	Vs:±15V Vu:66dB Vo:±10V Vi0:5mV	14FD	Six		
L144CJ	3xOP	Vs:±15V Vu:66dB Vo:±10V Vi0:20mV	14D	Six		
L146	ER+	Io=150mA Vo:2...77V P:520mW	14D,10A	Sgs		data pinout
L 146 CB	Z-IC	+2...7V, 0,15A	14-DIP	Sgs	-	data pinout
L 146 (C)T,(C)TB	Z-IC	=L 146 CB: Fig.→	82	Sgs		data pinout
L147B1	2xOP	Vs:±18V Vu:100dB Vo:±14V Vi0:2mV	14D	Sgs		
L 147 B1,CB	OP-IC	Dual, Serie 747, ±18V, 0...+70°	14-DIP	Sgs	→Serie 747	
L 148 C,CT,T1,T2	OP-IC	Serie 748	81	Sgs	→Serie 748	data pinout
L148T1	UNI	Vs:±22V Vu:100dB Vo:±14V Vi0:2mV	8A	Sgs		data pinout
L148T2	UNI	Vs:±22V Vu:103dB Vo:±14V Vi0:1mV	8A	Sgs		data pinout
L 149(V)	LIN-IC	NF-E, ±20V, 3A, 20W(±16V/4Ω)	86/5Pin	Sgs	(TDA 1420)	pdf
L 150(D)	LIN-IC	60dB-Expander	14-DIP	Sgs	-	
L 165(V)	OP-IC	hi-power, ±18V, 3,5A, 18W(±15V/4Ω)	86/5Pin	Sgs	-	pdf
L165V	POWER-OP	Vs:±18V Vu:80dB Vo:24V Vi0:2mV	5P	Sgs		data pinout
L 192/5	Z-IC	+5V, 0,25A	86/5Pin	Sgs		
L 192/12	Z-IC	+12V, 0,25A	86/5Pin	Sgs		
L 192/15	Z-IC	+15V, 0,25A	86/5Pin	Sgs		
L 194- 5(V)	Z-IC	+ Gleichr./rectifier, +5V, 0,5A	86/5Pin	Sgs		data pinout
L194-12V	R+	Io=int Vo:11.4...12.6V Vin:22V P:int	5P	Sgs		data pinout

Type	Device	Short Description	Fig.	Manu	Comparision Types	More at
L194-15V	R+	Io=int Vo:14.25...15.75V Vin:25V P:int	5P	Sgs		data pinout
L194-5V	R+	Io=int Vo:4.75...5.25V Vin:15V P:int	5P	Sgs		data pinout
L 194-12(V)	Z-IC	+ Gleichr./rectifier, +12V, 0,5A	86/5Pin	Sgs	-	data pinout
L 194-15(V)	Z-IC	+ Gleichr./rectifier, +15V, 0,5A	86/5Pin	Sgs	-	data pinout
L200	ER+	Io=int Vo:2.85...36V P:int	5PO	Sgs		data pdf pinout
L 200 C,CV,SP	Z-IC	+2,85...36V, 2A	86/5Pin	Sgs	-	data pdf pinout
L 200 CH		=L 200 CV	86/5Pin	Sgs		data pdf pinout
L 200 CT	Z-IC	Adjustable Voltage,Current Reg.	86/5Pin	Sgs		data pdf pinout
L 200 CV	Z-IC	=L 200 CV	86/5Pin	Sgs		data pdf pinout
L 200 C,CT,T	Z-IC	=L 200 CV: Fig.→	23/4Pin	Sgs		data pdf pinout
L 201(B,B-4)	LIN-IC	7x NPN-Darlington-Array, 30V, 0,5A	16-DIP	Sgs		
L 202(B,B-4)	LIN-IC	7x NPN-Darlington-Array, 30V, 0,5A	16-DIP	Sgs		
L 203(B,B-4)	LIN-IC	7x NPN-Darlington-Array, 30V, 0,5A	16-DIP	Sgs		
L 204(B,B-4)	LIN-IC	7x NPN-Darlington-Array, 30V, 0,5A	16-DIP	Sgs		
L272	2xPOWER-OP	Vs:28V Vu:70dB Vio:<10mV f:1,2Mc	14D,8D,16D	Sgs		data pdf pinout
L 272(B)	OP-IC	Dual, hi-power, 28V, 1A, 1V/µs	16-DIP	Sgs	-	data pdf pinout
L 272 D	OP-IC	=L 272(B): SMD	16-MDIP	Sgs		pdf pinout
L 272 M(MB)	OP-IC	=L 272(B): Fig.→	8-DIP	Sgs	-	data pdf pinout
L 282	LIN-IC	Vc, Capstan Head Wheel Interface	18-DIP	Tho		
L 290(B)	LIN-IC	Schrittmotor-steuerg./tachometer conv.	16-DIP	Sgs		pdf
L 291(B)	LIN-IC	Schrittmotor-Steuerg./5 Bit D/A conv.	16-DIP	Sgs		
L 292(V,VH)	LIN-IC	Schrittmotor-treiber/switch mode dr.	15-SQL	Sgs,Uni		pdf
L 293(B)	LIN-IC	4-Kanal/Channel Motor-Tr., 1A	16-DIP	Sgs,Uni	LM 18293	data pdf pinout
L 293 C	LIN-IC	4-Kanal/Channel Motor-tr.	20-DIP	Tho	-	pdf
L 293 D	LIN-IC	=L 293B: 0,6A, + int. Dioden	16-DIP	Sgs,Uni	-	pdf pinout
L 293 E	LIN-IC	4-Kanal/Channel Motor-tr.	20-DIP	Sgs	-	data pdf pinout
L 294	LIN-IC	Magnet-treiber/driver, switch mode	11-SQL	Sgs	-	data pdf pinout
L 295(V,VH)	LIN-IC	Dual-leistungstreiber/solenoid driver	15-SQL	Sgs,Uni	-	pdf
L 296(HT,V,VH)	Z-IC, LIN-IC	S-Reg, +5,1...40V, 4A	15-SQL	Sgs,Uni	B 2960VG	pdf
L 296 P,PHT	Z-IC	=L 296(HT): + extern.limiting current	15-SQL	Sgs	-	pdf
L 297	LIN-IC	Schrittmotor-strg./stepper motor cont.	20-DIP	Sgs	-	pdf
L 297 A	LIN-IC	=L 297: + step pulse doubler	20-DIP	Sgs	-	pdf
L 298(V,VH)	LIN-IC	Dual, Leistungstreiber/power Driver	15-SQL	Sgs,Uni	LM 18298	pdf
L 298 D	LIN-IC	=L 298: 1A, + int. Dioden	15-SQL	Uni	-	pdf
L 300(N)	LIN-IC	Telecom, Line Interface (slic)	28-DIP	Tho	-	
L 343 M	LIN-IC	=TDA 3420: SMD	16-MDIP	Sgs		
L387	R+	Io=500mA Vo:4.8...5.2V Vin:12V	5P	Sgs		data pdf pinout
L 387(A)	Z-IC	lo-drop, +5V, ±4%, 0,5A	86/5Pin	Sgs		data pdf pinout
L 465(A)	OP-IC	±20V, 4A, 20W	86/5Pin	Sgs		
L465A	POWER-OP	Vs:±20V Vu:80dB Vo:25V Vi0:2mV	5P	Sgs		data pdf pinout
L 482(B)	LIN-IC	Hall-effekt Zündstrg./ignition control	16-DIP	Tho	-	pdf
L 482 D1	LIN-IC	=L 482(B): SMD	16-MDIP	Tho		pdf
L 483	LIN-IC	Benzineinspritzg./fuel Injection Drv.	86/5Pin	Tho	-	
L 484(B)	LIN-IC	Magnetic Zündstrg./ignition control	16-DIP	Tho		pdf
L 484 D1	LIN-IC	=L 484(B): SMD	16-MDIP	Tho		pdf
L 485	LIN-IC	Spg.-regler/voltage Reg., hi-prec	16-DIP	Sgs		
L 486	LIN-IC	Kfz-blinkgeber/car Direction Indic. D.	8-DIP	Sgs	-	pdf
L487	R+	Io=500mA Vo:4.8...5.2V Vin:14.4V	5P	Sgs		data pinout
L 487(B)	Z-IC	lo-drop, Reset, +5V, 0,5A	86/5Pin	Sgs		data pinout
L 497(B)	LIN-IC	Hall-effekt Zündstrg./ignition control	16-DIP	Tho	-	pdf
L 497 D1	LIN-IC	=L 497(B): SMD	16-MDIP	Tho		pdf
L 530	LIN-IC	Electron. Zündung/ignition Interface	16-DIP	Tho		
L 530 D1	LIN-IC	=L 530: SMD	16-MDIP	Tho		
L 583	LIN-IC	Benzineinspritzg./fuel Injection Ctrl.	16-DIP	Sgs		
L 584	LIN-IC	Multifunction Injection Interface	16-DIP	Tho		pdf
L 585	LIN-IC	Kfz-regler/car Alternator Regulator	16-DIP	Tho		pdf
L 585 D1	LIN-IC	=L 585: SMD	16-MDIP	Tho		pdf
L 601(B)	LIN-IC	8x NPN-Darlington-Array, 90V, 0,4A	18-DIP	Sgs		
L 602(B)	LIN-IC	8x NPN-Darlington-Array, 90V, 0,4A	18-DIP	Sgs		
L 603(B)	LIN-IC	8x NPN-Darlington-Array, 90V, 0,4A	18-DIP	Sgs		
L 604(B)	LIN-IC	8x NPN-Darlington-Array, 90V, 0,4A	18-DIP	Sgs		
L 702 B,DP	LIN-IC	4x NPN-Darlington-Array, 90V, 2A, 4W	16-DIP	Sgs		pdf
L 702 N,SP	LIN-IC	=L 702B,DP: 20W	11-SQL	Sgs		
L 780 S05..24	Z-IC	+5..24V, 1A, Strobe	86/5Pin	Say		data pinout
L780S05	R+	Io=1A Vo:4.8...5.2V Vin:10V P:1.75W	5P	Say		data pdf pinout
L780S06	R+	Io=1A Vo:5.7...6.3V Vin:8...21V	5P	Say		data pdf pinout
L780S07	R+	Io=1A Vo:6.65...7.35V Vin:9...22V	5P	Say		data pdf pinout
L780S08	R+	Io=1A Vo:7.7...8.3V Vin:15V P:1.75W	5P	Say		data pdf pinout
L780S085	R+	Io=1A Vo:8.16...8.84V Vin:11...24V	5P	Say		data pinout
L780S09	R+	Io=1A Vo:8.64...9.36V Vin:16V P:1.75W	5P	Say		data pdf pinout
L780S10	R+	Io=1A Vo:9.6...10.4V Vin:17V P:1.75W	5P	Say		data pdf pinout
L780S12	R+	Io=1A Vo:11.5...12.5V Vin:19V P:1.75W	5P	Say		data pdf pinout
L780S15	R+	Io=1A Vo:14.4...15.6V Vin:23V P:1.75W	5P	Say		data pdf pinout
L780S18	R+	Io=1A Vo:17.3...18.7V Vin:27V P:1.75W	5P	Say		data pdf pinout
L780S20	R+	Io=1A Vo:19.2...20.8V Vin:29V P:1.75W	5P	Say		data pdf pinout
L780S24	R+	Io=1A Vo:22.8...25.2V Vin:27...35V	5P	Say		data pdf pinout
L 2005(C)	Z-IC	+5V, 2A	17b	Sgs	-	data
L2005	R+	Io=int Vo:4.8...5.2V Vin:10V P:int	3OP	Sgs		data pinout
L 2005(C)	Z-IC	+5V, 2A	17b	Sgs		data pinout
L 2005(C)	Z-IC	+5V, 2A	23a	Sgs	-	data pdf
L 2009(C)	Z-IC	+9V, 2A	17b	Sgs		data
L2009	R+	Io=int Vo:8.65...9.35V Vin:14V P:int	3OP	Sgs		data pinout
L 2009(C)	Z-IC	+9V, 2A	17b	Sgs		data pdf
L 2009(C)	Z-IC	+9V, 2A	23a	Sgs	-	data pdf pinout
L2010	R+	Io=int Vo:9.5...10.5V Vin:15V P:int	3OP	Sgs		data pinout

Type	Device	Short Description	Fig.	Manu	Comparision Types	More at
L 2012(C)	Z-IC	+12V, 2A	17b	Sgs	-	data
L2012	R+	Io=int Vo:11.5...12.5V Vin:19V P:int	3OP	Sgs	-	data pinout
L 2012(C)	Z-IC	+12V, 2A	17b	Sgs	-	data pinout
L 2012(C)	Z-IC	+12V, 2A	23a	Sgs	-	data
L 2015(C)	Z-IC	+15V, 2A	17b	Sgs	-	data
L2015	R+	Io=int Vo:14.4...15.6V Vin:23V P:int	3OP	Sgs	-	data pinout
L 2015(C)	Z-IC	+15V, 2A	17b	Sgs	-	data pinout
L 2015(C)	Z-IC	+15V, 2A	23a	Sgs	-	data
L 2018(C)	Z-IC	+18V, 2A	17b	Sgs	-	data
L2018	R+	Io=int Vo:17.1...18.9V Vin:26V P:int	3OP	Sgs	-	data pinout
L 2018(C)	Z-IC	+18V, 2A	17b	Sgs	-	data pinout
L 2018(C)	Z-IC	+18V, 2A	23a	Sgs	-	data
L 2024(C)	Z-IC	+24V, 2A	17b	Sgs	-	data
L2024	R+	Io=int Vo:23...25V Vin:33V P:int	3OP	Sgs	-	data pinout
L 2024(C)	Z-IC	+24V, 2A	17b	Sgs	-	data pinout
L 2024(C)	Z-IC	+24V, 2A	23a	Sgs	-	data
L 2075(C)	Z-IC	+7,5V, 2A	17b	Sgs	-	data
L2075	R+	Io=int Vo:7.15...7.9V Vin:12.5V P:int	3OP	Sgs	-	data pinout
L 2075(C)	Z-IC	+7,5V, 2A	17b	Sgs	-	data pinout
L 2075(C)	Z-IC	+7,5V, 2A	23a	Sgs	-	data
L2605V	R+	Io=500mA Vo:4.8...5.2V Vin:12...16V	3P	Sgs	-	data pdf pinout
L 2605 V	Z-IC	+5V, 0,5A	17b	Sgs	... 78M05... (TO-220)	data pdf pinout
L 2605 X	Z-IC	=L 2605V: Fig.→	≈14b	Sgs	-	data pdf pinout
L2610V	R+	Io=500mA Vo:9.55...10.45V Vin:12...16V	3P	Sgs	-	data pdf pinout
L 2610 V	Z-IC	+10V, 0,5A	17b	Sgs	... 78M10... (TO-220)	data pdf pinout
L 2610 X	Z-IC	=L 2610V: Fig.→	≈14b	Sgs	-	data pdf pinout
L2685V	R-	Io=500mA Vo:-8.15...-8.85V P:int	3P	Sgs	-	data pdf pinout
L 2685 V	Z-IC	+8,5V, 0,5A	17b	Sgs	... 78M85... (TO-220)	data pdf pinout
L 2685 X	Z-IC	=L 2610V: Fig.→	≈14b	Sgs	-	data pdf pinout
L 2720	OP-IC	Dual, lo-drop, hi-power, 28V, 1A	16-DIP	Sgs	-	data pdf pinout
L 2722	OP-IC	=L 2720: Fig.→	8-DIP	Sgs	-	data pdf pinout
L2724	2xPOWER-OP	Vs:28V Vu:70dB Vi0:<10mV f:1.2Mc	9I	Sgs	-	data pdf pinout
L 2724	OP-IC	=L 2720: Fig.→	9-SIL	Sgs	-	pdf pinout
L2726	2xPOWER-OP	Vs:28V Vu:70dB Vi0:<10mV f:1.2Mc	20S	Sgs	-	pdf pinout
L 2726	OP-IC	=L 2720: SMD	20-MDIP	Tho	-	pdf pinout
L2750	2xPOWER-OP	Vs:18V Vu:85dB Vi0:<5mV f:10Mc	11P	Sgs	-	data pdf pinout
L 2750	OP-IC	Dual, lo-drop, hi-power, 28V, 4A	11-SQL	Tho	-	pdf pinout
L 3000	LIN-IC	Telecom, Line Interface, hi-volt	28-DIC	Tho	-	pdf pinout
L 3000 N(SO)	LIN-IC	=L 3000: SMD	28-MDIP			pinout
L 3010	LIN-IC	Telecom, Control Unit, Ser. Interface	28-DIC	Tho	-	
L 3025	LIN-IC	Telecom, Control Unit, Ser. Interface	28-DIC	Tho	-	
L 3030(N)	LIN-IC	Telecom, Line Interface (slic)	28-DIC,DIP	Tho	-	
L 3030(N)	LIN-IC	Telecom, Line Interface (slic)	28,44-PLCC	Tho	-	pinout
L 3035	LIN-IC	Telecom, Line Interface (slic)	44-PLCC	Tho	-	
L 3036	LIN-IC	Telecom, Line Interface (slic)	44-PLCC	Tho	-	
L 3037(FN)	LIN-IC	Telecom, Line Interface (slic)	44-PLCC	Tho	L 3035, L 3036	pdf pinout
L 3037 QN	LIN-IC	=L 3037(FN): SMD	44-MP		-	pdf pinout
L 3090	LIN-IC	Telecom, Ctrl. Unit, Par. Interface	28-DIC,DIP	Tho	-	
L 3091	LIN-IC	Telecom, Ctrl. Unit, Par. Interface	28-PLCC	Tho	-	
L 3091	LIN-IC	Telecom, Ctrl. Unit, Par. Interface	28-DIP	Tho	-	
L 3092(N)	LIN-IC	Telecom, Line Interface (slic)	28-DIP	Tho	-	pinout
L 3092 FN	LIN-IC	=L 3092(N): Fig.→	28-PLCC		-	pinout
L 3100(B,B1)	LIN-IC	Telecom, Overvolt.+overcurrent Prot.	8-DIP	Sgs	-	
L 3101(B1)	LIN-IC	Telecom, Overvolt.+overcurrent Prot.	8-DIP	Sgs	-	
L 3121(B)	LIN-IC	Telecom, Overvoltage Protection	4-SIP	Sgs	-	pdf
L 3211	LIN-IC	Telefon, Sprechkreis/speech Circuit	16-DIP	Sgs	-	
L 3234	LIN-IC	Telecom, Klingeleinspeisung/ring Inj.	86/7Pin	Tho	-	pdf pinout
L 3235	LIN-IC	Telecom, Line Interface	28-PLCC	Tho	-	pdf pinout
L 3240(B1)	LIN-IC	Telefon, Wecker/2-Tone Ringer	8-DIP	Tho	(LS 1240)	pdf pinout
L 3240 D,D1	LIN-IC	=L 3240: SMD	8-MDIP	Tho	-	
L 3280(A,B)	LIN-IC	Telefon, Sprechkreis/speech Circuit	14-DIP	Tho	-	pinout
L 3281(AB)	LIN-IC	Telefon, Sprechkreis/speech Circuit	14-DIP	Tho	-	pdf pinout
L 3281 D,AD1	LIN-IC	=L 3281: SMD	14-MDIP	Tho	-	
L 3654B	LIN-IC	Druckeranst.-tr./printer solenoid drv.	16-DIP	Sgs	-	pdf pinout
L 3654S	LIN-IC	Druckeranst.-tr./printer solenoid drv.	16-DIP	Sgs	-	pdf pinout
L 3845	LIN-IC	Telefon, Line Interface (trunk)	8-DIP	Tho	-	pdf pinout
L 3845 D	LIN-IC	=L 3845: SMD	8-MDIP		-	pdf pinout
L 3913	BiMOS-IC	Telefon, Monochip with Loud Heading	44-PLCC	Tho	-	pinout
L 3914(AN,AD)	LIN-IC	Telefon, Sprechen/speech + Dialer	28-(M)DIP	Tho	L 3916A, L 3924A, L 3934A	pinout
L 3916(AN,AD)	LIN-IC	Telefon, Sprechen/speech + Dialer	28-(M)DIP	Tho	L 3914A, L 3926A. L 3936A	pinout
L 3924 AN,AD	LIN-IC	Telefon, Sprechen/speech + Dialer	28-(M)DIP	Tho	L 3914A, L 3916A, L 3934A	pinout
L 3926 AN,AD	LIN-IC	Telefon, Sprechen/speech + Dialer	28-(M)DIP	Tho	L 3914A, L 3916A, L 3936A	pinout
L 3934 AN,AD	LIN-IC	Telefon, Sprechen/speech + Dialer	28-(M)DIP	Tho	L 3914A, L 3916A, L 3924A	pinout
L 3936 AN,AD	LIN-IC	Telefon, Sprechen/speech + Dialer	28-(M)DIP	Tho	L 3914A, L 3916A, L 3926A	pinout
L 4620	LIN-IC	Flüssigkeitspegelsensor/liquid level	8-DIP	Tho	-	pdf
L4705CV	R+	Io=500mA Vo:4.8...5.2V Vin:14.4V	3P	Sgs		data pdf pinout
L 4705 CV	Z-IC	lo-drop, +5V, 0,5A	17b	Sgs		data pdf pinout
L4710CV	R+	Io=500mA Vo:9.6...10.4V Vin:14.4V	3P	Sgs		data pdf pinout
L 4710 CV	Z-IC	lo-drop, +10V, 0,5A	17b	Sgs		data pdf pinout
L4785CV	R+	Io=500mA Vo:8.16...8.54V Vin:14.4V	3P	Sgs		data pdf pinout
L 4785 CV	Z-IC	lo-drop, +8,5V, 0,5A	17b	Sgs		data pdf pinout
L4805CV	R+	Io=400mA Vo:4.8...5.2V Vin:14.4V	3P	Sgs		data pdf pinout
L 4805 CV	Z-IC	lo-drop, +5V, 0,4A	17b	Sgs		data pinout
L 4805 CX	Z-IC	=L 4805CV: Fig.→	≈14b	Sgs		pdf
L4810CV	R+	Io=400mA Vo:9.6...10.4V Vin:14.4V	3P	Sgs		data pdf pinout

Type	Device	Short Description	Fig.	Manu	Comparision Types	More at
L 4810 CV	Z-IC	lo-drop, +10V, 0,4A	17b	Sgs	-	
L 4810 CX	Z-IC	=L 4810CV: Fig.→	≈14b	Sgs	-	data pinout
L 4812 CV	Z-IC	lo-drop, +12V, 0,4A	17b	Sgs	-	pdf
L 4812 CX	Z-IC	=L 4812CV: Fig.→	≈14b	Sgs	-	pdf
L4885CV	R+	Io=400mA Vo:8.16...8.84V Vin:14.4V	3P	Sgs	-	pdf
L 4885 CV	Z-IC	lo-drop, +8,5V, 0,4A	17b	Sgs	-	data pdf pinout
L 4885 CX	Z-IC	=L 4885CV: Fig.→	≈14b	Sgs	-	data pinout
L 4892 CV	Z-IC	lo-drop, +9,2V, 0,4A	17b	Tho	-	pdf
L 4892 CX	Z-IC	=L 4892CV: Fig.→	≈14b	Sgs	-	pdf
L 4901	Z-IC	Dual, Reset, +5V, 0,3+0,4A	86/7Pin	Sgs	-	pdf
L4901A	R+	Io=>400mA Vo:4.95...5.15V Vin:7...18V	7P	Sgs		
L 4901 A	Z-IC	=L 4901: +5V, ±2%, 0,4+0,4A	86/7Pin	Tho		data pdf pinout
L 4902	Z-IC	Dual, Reset, Disable, +5V, 0,3+0,4A	86/7Pin	Sgs		data pdf pinout
L4902A	R+	Io=>300mA Vo:4.95...5.15V Vin:7...24V	7P	Sgs		
L 4902 A	Z-IC	=L 4902: +5V, ±2%, 0,3+0,3A	86/7Pin	Tho		data pdf pinout
L4903	R+	Io=>100mA Vo:4.95...5.15V Vin:7...18V	8D	Sgs		data pdf pinout
L 4903	Z-IC	Dual,Reset,Disable, +5V±2%, 0,05+0,1A	8-DIP	Sgs		data pdf pinout
L4904	R+	Io=100mA Vo:4.95...5.15V Vin:7...18V	8D	Sgs		data pdf pinout
L 4904	Z-IC	Dual, Reset, +5V±2%, 0,05+0,1A	8-DIP	Sgs		data pdf pinout
L 4904 A	Z-IC	~L 4904	8-DIP	Tho		
L 4905	R+	Io=>300mA Vo:5.0...5.1V Vin:7...24V	7P	Sgs		data pdf pinout
L 4905	Z-IC	Dual, Reset, +5V±1%, 0,2+0,3A	86/7Pin	Tho		data pinout
L4915	ER+	Io=250mA Vin:6...18V P:int	8D	SGS		data pinout
L 4915	Z-IC	+ Filter, +4...+11V, 0,2A	8-DIP	Sgs		data pdf pinout
L4916	R+	Io=250mA Vo:8.1...8.9V Vin:12...18V	8D	Sgs		data pdf pinout
L 4916	Z-IC	+ Filter, +8,5V, 0,25A	8-DIP	Sgs		data pdf pinout
L4918	R+	Io=250mA Vo:8.1...8.9V Vin:12...18V	5P	Sgs		data pdf pinout
L 4918	Z-IC	+ Filter, +8,5V, 0,25A	86/5Pin	Sgs		data pdf pinout
L 4920	Z-IC	+1,25...+20V, 0,4A	86/5Pin	Sgs	-	data pdf pinout
L 4921	Z-IC	=L 4920: Fig.→	8-DIP	Sgs		pdf
L 4922	Z-IC	lo-drop, Reset, +5V, ±4%, 1A	86/5Pin	Tho		pdf
L 4923	Z-IC	lo-drop, Reset, Inhibit, +5V, ±4%, 1A	86/5Pin	Tho		pdf
L 4925	Z-IC	lo-drop, +5V, ±2%, 0,5A	86/5Pin	Tho	-	pdf
L 4926	Z-IC	Dual,lo-drop, +5V±2%/0,05A,+5-20V/0,5A	11-SQL	Tho	-	pdf
L 4927	Z-IC	Dual, lo-drop, +5V±2%/0,05A, +5V/0,5A	86/5Pin	Tho		
L 4928	Z-IC	=L 4926: Fig.→	16-DIP	Tho		
L 4930	Z-IC	Dual, lo-drop, +5V/0,1A, +9,5V/0,5A	86/5Pin	Tho		
L4931ABX12	R+	Io=300mA Vo:1.238...1.263V Vin:3.3V	8S,5P,3PN	Sgs		
L4931ABX120	R+	Io=300mA Vo:11.88...12.12V Vin:14V	8S,5P,3PN	Sgs		data pdf pinout
L4931ABX15	R+	Io=300mA Vo:1.485...1.515V Vin:3.5V	8S,5P,3PN	Sgs		data pdf pinout
L4931ABX25	R+	Io=300mA Vo:2.475...2.525V Vin:4.5V	8S,5P,3PN	Sgs		data pdf pinout
L4931ABX27	R+	Io=300mA Vo:2.673...2.727V Vin:4.7V	8S,5P,3PN	Sgs		data pdf pinout
L4931ABX30	R+	Io=300mA Vo:2.97...3.03V Vin:5V P:int	8S,5P,3PN	Sgs		data pdf pinout
L4931ABX33	R+	Io=300mA Vo:3.267...3.333V Vin:5.3V	8S,5P,3PN	Sgs		data pdf pinout
L4931ABX35	R+	Io=300mA Vo:3.465...3.535V Vin:5.5V	8S,5P,3PN	Sgs		data pdf pinout
L4931ABX40	R+	Io=300mA Vo:3.96...4.04V Vin:6V P:int	8S,5P,3PN	Sgs		data pdf pinout
L4931ABX45	R+	Io=300mA Vo:4.455...4.545V Vin:6.5V	8S,5P,3PN	Sgs		data pdf pinout
L4931ABX47	R+	Io=300mA Vo:4.653...4.747V Vin:6.7V	8S,5P,3PN	Sgs		data pdf pinout
L4931ABX50	R+	Io=300mA Vo:4.95...5.05V Vin:7V P:int	8S,5P,3PN	Sgs		data pdf pinout
L4931ABX52	R+	Io=300mA Vo:5.148...5.252V Vin:7.2V	8S,5P,3PN	Sgs		data pdf pinout
L4931ABX55	R+	Io=300mA Vo:5.445...5.444V Vin:7.5V	8S,5P,3PN	Sgs		data pdf pinout
L4931ABX60	R+	Io=300mA Vo:5.94...6.06V Vin:8V P:int	8S,5P,3PN	Sgs		data pdf pinout
L4931ABX80	R+	Io=300mA Vo:7.92...8.08V Vin:10V P:int	8S,5P,3PN	Sgs		data pdf pinout
L4931CX12	R+	Io=300mA Vo:1.225...1.275V Vin:3.3V	8S,5P,3PN	Sgs		data pdf pinout
L4931CX120	R+	Io=300mA Vo:11.76...12.24V Vin:14V	8S,5P,3PN	Sgs		data pdf pinout
L4931CX15	R+	Io=300mA Vo:1.47...1.53V Vin:3.5V	8S,5P,3PN	Sgs		data pdf pinout
L4931CX25	R+	Io=300mA Vo:2.45...2.55V Vin:4.5V	8S,5P,3PN	Sgs		data pdf pinout
L4931CX27	R+	Io=300mA Vo:2.646...2.808V Vin:4.7V	8S,5P,3PN	Sgs		data pdf pinout
L4931CX30	R+	Io=300mA Vo:2.94...2.06V Vin:5V P:int	8S,5P,3PN	Sgs		data pdf pinout
L4931CX33	R+	Io=300mA Vo:3.234...3.366V Vin:5.3V	8S,5P,3PN	Sgs		data pdf pinout
L4931CX35	R+	Io=300mA Vo:3.43...3.57V Vin:5.5V	8S,5P,3PN	Sgs		data pdf pinout
L4931CX40	R+	Io=300mA Vo:3.92...4.08V Vin:6V P:int	8S,5P,3PN	Sgs		data pdf pinout
L4931CX45	R+	Io=300mA Vo:4.41...4.59V Vin:6.5V	8S,5P,3PN	Sgs		data pdf pinout
L4931CX47	R+	Io=300mA Vo:4.606...4.888V Vin:6.7V	8S,5P,3PN	Sgs		data pdf pinout
L4931CX50	R+	Io=300mA Vo:4.9...5.0V Vin:7V P:int	8S,5P,3PN	Sgs		data pdf pinout
L4931CX52	R+	Io=300mA Vo:5.096...5.304V Vin:7.2V	8S,5P,3PN	Sgs		data pdf pinout
L4931CX55	R+	Io=300mA Vo:5.59...5.61V Vin:7.5V	8S,5P,3PN	Sgs		data pdf pinout
L4931CX60	R+	Io=300mA Vo:5.88...6.12V Vin:8V P:int	8S,5P,3PN	Sgs		data pdf pinout
L4931CX80	R+	Io=300mA Vo:7.84...8.16V Vin:10V P:int	8S,5P,3PN	Sgs		data pdf pinout
L 4934	Z-IC	Dual, lo-drop, +5V/0,05A, +5V/0,1A	8-DIP	Tho	-	data pdf pinout ·
L 4935	Z-IC	=L 4934: Fig.→	16-DIP	Tho	-	
L 4936	Z-IC	Dual Volt. Reg., Power-On Reset	11-SQL	Tho	-	
L 4937	Z-IC	Dual, Multifunction	86/7Pin	Tho	-	pdf pinout
L 4938 E	Z-IC	Adv. Voltage Reg.	12+4-DIP	Stm	-	pdf
L 4938 ED	Z-IC	Adv. Voltage Reg.	MDIP	Stm	-	pdf pinout
L 4938 EPD	Z-IC	Adv. Voltage Reg.	TSOP	Stm	-	pdf pinout
L 4938 N	Z-IC	Dual Voltage Reg.	12+4-DIP	Stm	-	pdf pinout
L 4938 ND	Z-IC	Dual Voltage Reg.	MDIP.	Stm	-	pdf pinout
L 4938 NPD	Z-IC	Dual Voltage Reg.	TSOP	Stm	-	pdf pinout
L4940	R+	Io=int Vo:8.3...8.7V Vin:10.5V	3SP	Sgs		pdf pinout
L 4940V5	Z-IC	low drop Volt. Reg., +5V, 2%, 1.5A	17b	Sgs		data pdf pinout
L 4940V10	Z-IC	low drop Volt. Reg., +10V, 2%, 1.5A	17b	Sgs		pdf pinout
L 4940V12	Z-IC	low drop Volt. Reg., +12V, 2%, 1.5A	17b	Sgs		pdf pinout
L 4940V85	Z-IC	low drop Volt. Reg., +8.5V, 2%, 1.5A	17b	Sgs		pdf pinout
L 4940D2T10	Z-IC	low drop Volt. Reg., +10V, 2%, 1.5A	87	Sgs		pdf pinout

583

Type	Device	Short Description	Fig.	Manu	Comparision Types	More at
L 4940D2T12	Z-IC	low drop Volt. Reg., +12V, 2%, 1.5A	87	Sgs	-	pdf pinout
L 4940D2T5	Z-IC	low drop Volt. Reg., +5V, 2%, 1.5A	87	Sgs	-	pdf pinout
L 4940D2T85	Z-IC	low drop Volt. Reg., +8.5V, 2%, 1.5A	87	Sgs	-	pdf pinout
L 4941(BV)	Z-IC	lo-drop, +5V, 2%, 1A	17b	Sgs	-	pdf
L4941B	R+	Io=int Vo:4.8...5.2V Vin:7V P:int12W	3SP	Sgs	-	data pdf pinout
L 4941 X	Z-IC	=L 4941(BV): Fig.→	≈14b	Tho	-	pdf
L 4943	LIN-IC	Multifunction Voltage regulator For car RADIO	SQL	Stm	-	pdf pinout
L 4945	Z-IC	lo-drop, +5V, ±4%, 0,5A	17b	Tho	-	pdf
L 4947	Z-IC	lo-drop, Reset, +5V, ±4%, 0,5A	86/5Pin	Tho	-	pdf
L 4948	Z-IC	Quad, Inhibit, Reset	11-SQL	Tho	-	pdf
L 4949	Z-IC	lo-drop, Multifunction	8-(M)DIP	Tho	-	pdf
L 4950	Z-IC	lo-drop, +8,5V, 0,5A	17	Tho	-	
L 4951	Z-IC	lo-drop, +10V, 0,5A	17	Tho	-	
L 4952	LIN/Z-IC	Voltage regulator	MDIP	Stm	-	pdf pinout
L 4953 G	LIN-IC	Multifunction Voltage regulator For car RADIO	SQL	Stm	-	pdf pinout
L 4957 AD1.5	LIN/Z-IC	1.5V, up to 5A ULDO linear regulator	30	Stm	-	pdf pinout
L 4957 AD1.8	LIN/Z-IC	1.8V, up to 5A ULDO linear regulator	30	Stm	-	pdf pinout
L 4957 AD2.5	LIN/Z-IC	2.5V, up to 5A ULDO linear regulator	30	Stm	-	pdf pinout
L 4957 AD3.3	LIN/Z-IC	3.3V, up to 5A ULDO linear regulator	30	Stm	-	pdf pinout
L 4959	Z-IC	2x12V/1,3A/0,8A, 8,6V/0,6A, 5,6V/0,25A	11-SQL	Tho	-	pdf
L 4960(H,V)	Z-IC	Schaltregler/sw. reg., 5,1...40V, 2,5A	86/7Pin	Sgs	-	pdf
L 4962	Z-IC	Schaltregler/sw. reg., 5,1...40V, 1,5A	16-DIP	Sgs	-	pdf
L 4962 E,EH	Z-IC	=L 4962: Fig.→	86/7Pin	Sgs	-	pdf
L 4963(W)	Z-IC	Schaltregler/sw. reg., 5,1...36V, 1,5A	18-DIP	Tho	-	pdf
L 4963 D	Z-IC	=L 4963(W): SMD	20-MDIP	Tho	-	pdf
L 4964(HT)	Z-IC	Schaltregler/sw. reg., 5,1...28V, 4A	15-SQL	Sgs	-	pdf
L 4970(A)	SMPS-IC	Schaltregler/sw. reg., 10A	15-SQL	Tho	-	pdf pinout
L 4971	LIN/Z-IC	1.5A step Down Switching regulator	DIC	Stm	-	pdf pinout
L 4971 D	LIN/Z-IC	1.5A step Down Switching regulator	MDIP	Stm	-	pdf pinout
L 4972 A	Z-IC	Schaltregler/sw. reg., 5,1...40V, 2A	20-DIP	Tho	-	pdf pinout
L 4972 AD	Z-IC	=L 4972A: SMD	20-MDIP	Tho	-	pdf pinout
L 4973 V3.3	LIN/Z-IC	3.5A step down switching regulator, 3.3V, 0.5V-50V	DIC	Stm	-	pdf pinout
L 4973 D3.3	LIN/Z-IC	3.5A step down switching regulator, 3.3V, 0.5V-50V	MDIP	Stm	-	pdf pinout
L 4973 D5.1	LIN/Z-IC	3.5A step down switching regulator, 5.1V, 5.1V-50V	MDIP	Stm	-	pdf pinout
L 4973 V5.1	LIN/Z-IC	3.5A step down switching regulator, 5.1V, 5.1V-50V	DIC	Stm	-	pdf pinout
L 4974(A)	SMPS-IC	Schaltregler/sw. reg., 3,5A	20-DIP	Tho	-	pdf pinout
L 4975(A)	SMPS-IC	Schaltregler/sw. reg., 5A	15-SQL	Tho	-	pdf pinout
L 4976	LIN/Z-IC	1A step down switching regulator	DIC	Stm	-	pdf pinout
L 4976 D	LIN/Z-IC	1A step down switching regulator	MDIP	Stm	-	pdf pinout
L 4977(A)	SMPS-IC	Schaltregler/sw. reg., 7A	15-SQL	Tho	-	pdf pinout
L 4978	SMPS-IC	Step down switching reg., 2A/3.3V...50V	8-DIP	Sgs	-	pdf pinout
L 4978D	SMPS-IC	SMD, step down switch.reg.,2A/3.3...50V	16-MDIP	Sgs	-	pdf pinout
L 4979 D	LIN/Z-IC	Low Drop Voltage Reg	MDIP	Stm	-	pdf pinout
L 4979 MD	LIN/Z-IC	Low Drop Voltage Reg	MDIP	Stm	-	pdf pinout
L 4981 A	SMPS-IC	Smps-controller	20-DIP	Tho	-	pdf
L 4981 B	SMPS-IC	Smps-controller	20-DIP	Tho	-	pdf
L 4981 AD,BD	SMPS-IC	=L 4981A,B: SMD	20-MDIP		-	
L 4989	LIN/Z-IC	Low Power Voltage Reg	TSOP	Stm	-	pdf pinout
L 4989 D	LIN/Z-IC	Low Power Voltage Reg	MDIP	Stm	-	pdf pinout
L 4993 D	LIN/Z-IC	Low drop voltage reg	MDIP	Stm	-	pdf pinout
L 4993 MD	LIN/Z-IC	Low drop voltage reg	MDIP	Stm	-	pdf pinout
L 5431	Z-IC	2,5...36V, 1...100mA, hi-prec	7(KARef)	Say	-	
L 5450(B7)	MOS-IC	LED-Display-Tr, 35 Segments	40-DIP	Tho	-	
L5630	Z+	Io=10mA Vo:31...35V P:200mW	2N	Say	-	data pinout
L 5630	Z-IC	Tuner-Stab., 31...35V, 10mA (~ZTK 33)	7j	Say	TAA 550, TDA 1057, MVS 460, µPC 574,++	data pdf pinout
L5631	Z+	Io=10mA Vo:31...35V P:400mW	2N	Say	-	data pinout
L 5631	Z-IC	Tuner-Stab., 31...35V, 10mA	7j	Say	TAA 550, TDA 1057, MVS 460, µPC574,++	data pdf pinout
L 5832	LIN-IC	Ventiltreiber/solenoid-controller	16-DIP	Sgs	-	
L 5950	LIN-IC	Multiple Multifunction Voltage regulator For car Radio	SQL	Stm	-	pdf pinout
L 5952	LIN-IC	Multiple Multifunction Voltage regulator	SILP	Stm	-	pdf pinout
L 5953	LIN-IC	Multiple switching voltage regulator	MDIP	Sln	-	pdf pinout
L 5955	LIN-IC	Multiple multifunction voltage regulator for car radio	QILP	Stm	-	pdf pinout
L 5956	LIN-IC	Multifunction voltage regulator for car radio	SQL	Stm	-	pdf pinout
L 5956 PD(TR)	LIN-IC	Multifunction voltage regulator for car radio	MDIP	Stm	-	pdf pinout
L 5957	LIN/Z-IC	Multifunction voltage reg	SQL	Stm	-	pdf pinout
L 5957 PD	LIN/Z-IC	Multifunction voltage reg	TSOP	Stm	-	pdf pinout
L 5959	LIN-IC	Multifunction voltage regulator for car radio	SQL	Stm	-	pdf pinout
L 5961	CMOS-IC	Most power management device (PMD)	MDIP	Stm	-	pdf
L 5970 AD	LIN/Z-IC	1.5A Switch step Down Switching regulator	MDIP	Stm	-	pdf pinout
L 5970 D	LIN/Z-IC	Up to 1A step down switching regulator	MDIP	Stm	-	pdf pinout
L 5972 D	LIN/Z-IC	2A switch step down switching regulator	MDIP	Stm	-	pdf pinout
L 5973 AD	LIN/Z-IC	2A switch step down switching regulator	MDIP	Stm	-	pdf pinout
L 5973 D	LIN/Z-IC	2.5A switch step down switching	MDIP	Stm	-	pdf pinout

Type	Device	Short Description	Fig.	Manu	Comparision Types	More at
		regulator				
L 5980	LIN/Z-IC	0.7A step-down switching regulator	MLP	Stm	-	pdf pinout
L 5981	LIN/Z-IC	1A step-down switching regulator	MLP	Stm	-	pdf pinout
L 5983	LIN/Z-IC	1.5A step-down switching regulator	MLP	Stm	-	pdf pinout
L 5985	LIN/Z-IC	2A step-down switching regulator	MLP	Stm	-	pdf pinout
L 5988 D	LIN/Z-IC	4A step-down switching regulator with synchronous rectification	TSOP	Stm	-	pdf pinout
L 5994 (A)	LIN-IC	Adj. triple out. power supply contrl.	MP	Stm	-	pdf pinout
L 6000	LIN-IC	SMD,Fdd, Schreib-lese/read-write Ctrl.	64-MP	Tho	-	pdf
L 6100	MOS-IC	Dmos Power Switch, 100V, 1A	18-DIP	Sgs	-	
L 6101	MOS-IC	=L 6100: Fig.→	11-SQL	Sgs	-	
L 6102	MOS-IC	=L 6100: Fig.→	15-SQL	Sgs	-	
L 6114	MOS-IC	4x Dmos Power Switch, TTL Input	20-DIP	Tho	-	pdf
L 6115	MOS-IC	4x Dmos Power Switch, TTL Input	15-SQL	Tho	-	pdf
L 6122	MOS-IC	Dmos Switch, 100V	20-DIP	Tho	-	pdf
L 6123	MOS-IC	Dmos Switch, 100V	15-SQL	Tho	-	pdf
L 6201	LIN-IC	Dmos Full Bridge Driver	20-MDIP	Sgs	-	pdf pinout
L 6201 PS	LIN-IC	Dmos Full Bridge Driver	20-MDIP	Sgs	-	pdf pinout
L 6202	LIN-IC	Dmos Full Bridge Driver	18-DIP	Sgs	-	pdf pinout
L 6202 D	LIN-IC	=L 6202: SMD	20-MDIP		-	pdf
L 6203	LIN-IC	Dmos Full Bridge Driver	11-SQL	Sgs	-	pdf pinout
L 6204	LIN-IC	2x Brücken-Tr./bridge driver	24-DIP	Sgs	-	pdf
L 6204 D	LIN-IC	=L 6204: SMD	20-MDIP		-	pdf
L 6208	LIN-IC	Dual, Leistungstreiber/power Driver	15-SQL	Sgs,Uni	-	pdf
L 6210	LIN-IC	2x Schottky-Br., 50V, 2A	16-DIP	Sgs	-	pdf
L 6212	SMPS-IC	Magnet-treiber/solenoid driver, SMPS	15-SQL	Sgs	-	
L 6213	SMPS-IC	Magnet-treiber/solenoid driver, SMPS	16+4-DIP	Sgs	-	data pdf pinout
L 6217(A)	LIN-IC	Schrittmotor-tr/stepper motor driver	44-PLCC	Sgs	-	
L 6218	LIN-IC	Schrittmotor-tr/stepper motor driver	16+4-DIP	Sgs	-	
L 6219	LIN-IC	Schrittmotor-tr/stepper motor driver, 10...46V, 2 Brücken/bridges a 750mA	20+4-DIP	Sgs	-	data pdf pinout
L 6219 D	LIN-IC	Schrittmotor-tr/stepper motor driver 10...46V, 2 Brücken/bridges a 750mA	44-PLCC	Sgs	-	data pdf pinout
L 6219 DS	LIN-IC	SMD, Schrittmot.-tr/stepper motor drv. 10...46V, 2 Brücken/bridges a 750mA	20+4-MDIP	Sgs	-	data pdf pinout
L 6220	LIN-IC	4x Darl.-Schalter/switch	16-DIP	Sgs	-	pdf
L 6220 N	LIN-IC	=L 6220: Fig.→	15-SQL	Sgs	-	
L 6221 A	LIN-IC	4x Darl.-Schalter/switch, 50V, 1,8A	16-DIP	Sgs	-	pdf
L 6221 C	LIN-IC	4x Darl.-Schalter/switch	16-DIP	Sgs	-	pdf
L 6221 N,CN	LIN-IC	=L 6221A: Fig.→	15-SQL	Sgs	D 6221VC	
L 6222	LIN-IC	4x Schalter/switch, 50V, 1,2A	16-DIP	Sgs	-	
L 6223(A)	LIN-IC	Schrittmotor-tr/stepper motor driver	20-DIP	Sgs	-	
L 6227	LIN-IC	4x Dmos Switch with Patches	16+4-DIP	Sgs	-	pdf
L 6230(A)	LIN-IC	Motor-tr./dc motor driver, bidirekt.	15-SQL	Sgs	-	
L 6231	LIN-IC	Motor-tr./dc motor driver	15-SQL	Sgs	-	
L 6232 A	DMOS-IC	Spindelmotor-tr/spindle Driver	44-PLCC	Tho	-	
L 6232 B	DMOS-IC	Spindelmotor-tr/spindle Driver	28-PLCC	Tho	-	
L 6233	LIN-IC	Phase locked frequency controller	16-DIP	Sgs	-	pdf pinout
L 6233 P	LIN-IC	Phase locked frequency controller	20-PLCC	Sgs	-	pdf pinout
L 6234	MOS-IC	3-Ph.-Motor-Tr/driver	16+4-DIP	Sgs	-	pdf
L 6235	LIN-IC	R-dat, 3-ph. brushless DC motor drv.	20-PLCC	Sgs	-	pdf pinout
L 6235D	LIN-IC	SMD, R-dat, 3-ph. brushl. DC motor drv	MDIP	Sgs	-	pdf pinout
L 6235N	LIN-IC	R-dat, 3-ph. brushless DC motor drv.	DIP	Sgs	-	pdf pinout
L 6235PD	LIN-IC	SMD, R-dat, 3-ph. brushl. DC motor drv	SSMDIP	Sgs	-	pdf pinout
L 6236	LIN-IC	Brushl. 3-ph. DC motor drv, bidirect.	20-PLCC	Sgs	-	pdf
L 6237	MOS-IC	SMD,Spindelmotor-tr/spindle Driver, 5V	64-MP	Sgs	-	
L 6238	MOS-IC	Spindelmotor-Tr./Spindle Drv, 12V 2,5W	44-PLCC	Sgs	-	data pdf pinout
L 6238 S	MOS-IC	Spindelmotor-Tr./Spindle Drv, 12V 2,3W	44-PLCC	Sgs	-	data pdf pinout
L 6238 SQA	MOS-IC	Spindelmotor-Tr./Spindle Drv, 12V 1,3W	44-QFP	Sgs	-	data pdf pinout
L 6238 SQT	MOS-IC	Spindelmotor-Tr./Spindle Drv, 12V 1,7W	64-QFP	Sgs	-	data pdf pinout
L 6242	DIG-IC	SMD, Voice Coil Motor Driver	20-MDIP	Sgs	-	pdf
L 6243	DIG-IC	Voice Coil Motor Driver (hd Head Act.)	44-PLCC	Sgs	-	pdf
L 6243 D	DIG-IC	=L 6243: SMD	28-MDIP		-	pdf
L 6245	MOS-IC	SMD, Hard-disk Drive Power Combo, 5V	64-MP	Sgs	-	pdf
L 6280	MOS-IC	3-Channel Multipower System	44-PLCC	Sgs	-	pdf
L 6285	MOS-IC	3-Channel Multipower System	44-PLCC	Sgs	-	pdf
L 6286	MOS-IC	3-Channel Multipower System	44-PLCC	Sgs	-	pdf
L 6384	CMOS-IC	Hi-volt Half Bridge Drv.	DIP	Stm	-	pdf pinout
L 6384 D	CMOS-IC	Hi-volt Half Bridge Drv.	MDIP	Stm	-	pdf pinout
L 6503	LIN-IC	Hammer Solenoid Controller	20-DIP	Sgs	-	pdf
L 6504	LIN-IC	Solenoid Controller	14-DIP	Sgs	-	pdf
L 6505	LIN-IC	Stromstg./current ctrl.f.stepp.motors	18-DIP	Sgs	-	pdf
L 6506	LIN-IC	Stromstg./current ctrl.f.stepp.motors	18-DIP	Sgs	-	pdf
L 6506 D	LIN-IC	=L 6506: SMD	18-MDIP		-	pdf
L 6515	LIN-IC	Dual DC Motor Positioning System	44-PLCC	Sgs	-	pdf
L 6560	SMPS-IC	Smps-controller	8-DIP	Sgs	-	pdf
L 6560 D	SMPS-IC	=L 6560: SMD	8-MDIP		-	pdf
L 6562 D	LIN-IC	Transition-mode Pfc Controller	DIP	Stm	-	pdf pinout
L 6562 N	LIN-IC	Transition-mode Pfc Controller	DIP	Stm	-	pdf pinout
L 6570 A,B	LIN-IC	Fdd, Schreib-lese/read-write Ctrl.	28-DIP	Tho	-	pdf
L 6603	I/O-IC	Memory Card Interface (smart Card)	28-DIP	Tho	-	pdf
L 6604	I/O-IC	=L 6603: Fig.→	28PLCC	Tho	-	pdf
L 6605	I/O-IC	Memory Card Interface (smart Card)	18-DIP	Tho	-	pdf
L 6615 D	LIN-IC	High/low Side Load Share Ctrl.	MDIP	Stm	-	pdf pinout
L 6615 N	LIN-IC	High/low Side Load Share Ctrl.	DIP	Stm	-	pdf pinout

Type	Device	Short Description	Fig.	Manu	Comparision Types	More at
L 6701	LIN-IC	3 Phase Controller for VR10, VR9 and K8 CPUs	TSOP	Stm	-	pdf pinout
L 6711	LIN-IC	3 Phase controller with dynamic VID and selectable DACs	MP	Stm	-	pdf pinout
L 6712 (A)D	LIN-IC	Two-phase interleaved DC/DC controller	MDIP	Stm	-	pdf pinout
L 6712 (A)Q	LIN-IC	Two-phase interleaved DC/DC controller	QFN	Stm	-	pdf pinout
L 6713 A	CMOS-IC	2/3 Phase cntrl. with embedded drivers f. Intel VR10, VR11 and AMD 6 bit CPUs	MP	Stm	-	pdf pinout
L 6714	LIN-IC	4 phase cntrl. with embedded drivers f. Intel VR10, VR11 and AMD 6Bit CPUs	MP	Stm	-	pdf pinout
L 6720	LIN-IC	Minitel Interface	20-DIP	Tho	-	
L 6721	LIN-IC	Minitel Interface	20-DIP	Tho	-	
L 6722	LIN-IC	3 Phase controller for DC/DC converters	QFN	Stm	-	
L 6725 (A)	LIN-IC	Voltage mode PWM cntrl., bootstrap anti-discharging system	MDIP	Stm	-	pdf pinout
L 6726 A	LIN-IC	Single phase PWM controller	MDIP	Stm	-	pdf pinout
L 6727	LIN-IC	Single phase PWM controller	MDIP	Stm	-	pdf pinout
L 6728	IC	Single phase PWM controller	DFN	Stm	-	pdf pinout
L 6730 (B)	BiCMOS-IC	Adj. step-down cntrl. with synchronous rectification	TSOP	Stm	-	pdf pinout
L 6730 C/D	BiCMOS-IC	Adj. step-down cntrl. with synchronous rectification	TSOP	Stm	-	pdf pinout
L 6730 CQ	BiCMOS-IC	Adj. step-down cntrl. with synchronous rectification	LCC	Stm	-	pdf pinout
L 6731 D	LIN-IC	Adj. step-down cntrl. w. synchronous rectification	TSOP	Stm	-	pdf pinout
L 6732	LIN-IC	Adj. step-down cntrl. w. synchronous rectification	TSOP	Stm	-	pdf pinout
L 6740 L	IC	Hybrid controller (4+1) for AMD SVID and PVID processors	MP	Stm	-	pdf pinout
L 6741	IC	High current Mosfet driver	MDIP	Stm	-	pdf pinout
L 6743	IC	High current Mosfet driver	MDIP	Stm	-	pdf pinout
L 6743 B	IC	High current Mosfet driver	MLP	Stm	-	pdf pinout
L 6743 Q	IC	High current Mosfet driver	MLP	Stm	-	pdf pinout
L 6750	IC	2 - 5 phase cntrl. w. LTB technology®	QFN	Stm	-	pdf
L 6756	IC	2/3/4 phase buck cntrl.	QFN	Stm	-	pdf
L 6902 D	LIN/Z-IC	Up to 1A switching regulator with adj. current limit	MDIP	Stm	-	pdf pinout
L 6910	LIN/Z-IC	Adj. step-down cntrl. w. synchronous rectification	MDIP	Stm	-	pdf pinout
L 6910 A	LIN/Z-IC	Adj. step-down cntrl. w. synchronous rectification	TSOP	Stm	-	pdf pinout
L 6910 G	LIN/Z-IC	Adj. step-down cntrl. w. synchronous rectification	MDIP	Stm	-	pdf pinout
L 6911 C	LIN-IC	5 bit programmable step-down cntrl. w. synchr. rectification	MDIP	Stm	-	pdf pinout
L 6911 D	LIN-IC	5 bit programmable step-down cntrl. w. synchr. rectification	MDIP	Stm	-	pdf pinout
L 6911 E	LIN-IC	5 bit programmable step-down cntrl. w. synchr. rectification	MDIP	Stm	-	pdf pinout
L 6917 BD	IC	5 bit programmable dual-phase cntrl.	MDIP	Stm	-	pdf pinout
L 6919 E	IC	5 bit programmable dual-phase cntrl. w. dynamic VID management	MDIP	Stm	-	pdf pinout
L 6920 DB	LIN-IC	Synchronous rectifier step-up converter	TSOP	Stm	-	pdf pinout
L 6920 DC	LIN-IC	Synchronous rectifier step up converter	TSOP	Stm	-	pdf pinout
L 6924 D	LIN-IC	Battery charger system with integrated power switch for Li-ion/li-polymer	QFN	Stm	-	pdf pinout
L 6925 D	LIN-IC	High efficiency monolitic synchronous step-down regulator	MDIP	Stm	-	pdf
L 6926	LIN/Z-IC	High efficiency monolithic synchronous step down regulator	MDIP	Stm	-	pdf pinout
L 6926 D1/Q1	LIN/Z-IC	High efficiency monolithic synchronous step down regulator	MLP	Stm	-	pdf pinout
L 6928 D	LIN-IC	High efficiency monolithic synchronous step down regulator	MDIP	Stm	-	pdf pinout
L 6928 Q1	LIN-IC	High efficiency monolithic synchronous step down regulator	MLP	Stm	-	pdf pinout
L 6932 D1.2	LIN/Z-IC	2A, ADJ. 1.2V - 5V, High performance ULDO linear regulator	MDIP	Stm	-	pdf pinout
L 6932 D1.5	LIN/Z-IC	2A, fix. 1.5V, High performance ULDO linear regulator	MDIP	Stm	-	pdf pinout
L 6932 D1.8	LIN/Z-IC	2A, fix. 1.8V, High performance ULDO linear regulator	MDIP	Stm	-	pdf pinout
L 6932 D2.5	LIN/Z-IC	2A, fix. 2.5V, High performance ULDO linear regulator	MDIP	Stm	-	pdf pinout
L 6932 H1.2	LIN/Z-IC	Adj. 1.2V - 5V, 2A High performance ULDO linear regulator	MDIP	Stm	-	pdf pinout
L 6933 H1.2	LIN/Z-IC	Adj. 1.2V - 5V, 2A high performance ULDO linear regulator w. Soft Start	MDIP	Stm	-	pdf pinout
L 6997 S	LIN-IC	step Down Controller For low Voltage Operations	TSOP	Stm	-	pdf pinout
L 7150	LIN-IC	4x NPN-Darlington-E, 50V, 1,5A	15-SQL	Sgs	-	
L 7152	LIN-IC	4x NPN-Darlington-E, 50V, 1,5A	15-SQL	Sgs	-	
L 7180	LIN-IC	=L 7150: 80V, 1,5A	15-SQL	Sgs	-	
L 7182	LIN-IC	=L 7152: 80V, 1,5A	15-SQL	Sgs	-	

Type	Device	Short Description	Fig.	Manu	Comparision Types	More at
L7805	R+	Io=int Vo:4.8...5.2V Vin:10V P:int	3PO	Sgs		data pdf pinout
L7805AB	R+	Io=int Vo:4.9...5.1V Vin:10V P:int	3SP	Sgs		data pdf pinout
L 7805 ABP	Z-IC	Precision 1 A regulators	45	Stm	-	pdf pinout
L 7805 ABV	Z-IC	Precision 1 A regulators	45	Stm		pdf pinout
L7805AC	R+	Io=int Vo:4.9...5.1V Vin:10V P:int	3SP	Sgs		data pdf pinout
L 7805 ACP	Z-IC	Precision 1 A regulators	45	Stm	-	pdf pinout
L 7805 ACV	Z-IC	Precision 1 A regulators	45	Stm		pdf pinout
L7805C	R+	Io=int Vo:4.8...5.2V Vin:10V P:int	3SPO	Sgs		data pdf pinout
L 7805 CDT-TR	Z-IC	Positive voltage regulators	17b	Stm	...7824... (TO-220)	data pdf pinout
L 7805 CT	Z-IC	+5V, 1,5A	23a	Stm	...7805... (TO-3)	data pinout
L 7805 CV,ACV	Z-IC	+5V, 1,5A, 5%, (A=2%)	17b	Stm	...7805... (TO-220)	data pdf pinout
L7805ML	R+	Io=1A Vo:4.75...5.25V Vin:7...20V P:15W	3P	Say		data pinout
L 7805 ML	Z-IC	=L 7805CV,ACV: Iso, 1A	17b	Say	...7808... (TO-220 Iso)	data pinout
L 7805 T	Z-IC	Positive voltage regulators	23a	Stm	...7809... (TO-3)	data pdf pinout
L7806	R+	Io=int Vo:5.75...6.25V Vin:11V P:int	3PO	Sgs		data pdf pinout
L7806AB	R+	Io=int Vo:4.9...5.1V Vin:10V P:int	3SP	Sgs		data pdf pinout
L 7806 ABV	Z-IC	Precision 1 A regulators	45	Stm	-	pdf pinout
L7806AC	R+	Io=int Vo:5.88...6.12V Vin:11V P:int	3SP	Sgs		data pdf pinout
L 7806 ACV	Z-IC	Precision 1 A regulators	45	Stm	-	pdf pinout
L7806C	R+	Io=int Vo:5.75...6.25V Vin:11V P:int	3SPO	Sgs		data pdf pinout
L 7806 CT	Z-IC	+6V, 1,5A	23a	Stm	...7806... (TO-3)	data pdf pinout
L 7806 CV,ACV	Z-IC	+6V, 1,5A, 5%, (A=2%)	17b	Stm	...7806... (TO-220)	data pdf pinout
L7806ML	R+	Io=1A Vo:5.7...6.3V Vin:8...21V P:15W	3P	Say		data pinout
L 7806 ML	Z-IC	=L 7806CV,ACV: Iso, 1A	17b	Say	...7806... (TO-220 Iso)	data pinout
L7807	R+	Io=1A Vo:6.72...7.28V Vin:12V	3P	Say		data pinout
L 7807	Z-IC	+7V, 1,5A, 5%	17b	Say	...7807... (TO-220)	data pinout
L 7807 CV,ACV	Z-IC	+7V, 1,5A, 5%	17b	Say	...7807... (TO-220)	data
L7807ML	R+	Io=1A Vo:6.6...7.4V Vin:9...22V P:15W	3P	Say		data pinout
L 7807 ML	Z-IC	=L 7807CV,ACV: Iso, 1A	17b	Say	...7807... (TO-220 Iso)	data pinout
L7808	R+	Io=int Vo:7.7...8.3V Vin:14V P:int	3PO	Sgs		data pdf pinout
L7808AB	R+	Io=int Vo:7.84...8.16V Vin:10V P:int	3SP	Sgs		data pdf pinout
L 7808 ABV	Z-IC	Precision 1 A regulators	45	Stm		pdf pinout
L7808AC	R+	Io=int Vo:7.84...8.16V Vin:10V P:int	3SP	Sgs		data pdf pinout
L 7808 ACV	Z-IC	Precision 1 A regulators	45	Stm	-	pdf pinout
L7808C	R+	Io=int Vo:7.7...8.3V Vin:14V P:int	3SPO	Sgs		data pdf pinout
L 7808 CP	Z-IC	=L 7805CV...7885CV: Iso	17b	Stm	...78xx... (TO-220 Iso)	pdf pinout
L 7808 CT	Z-IC	+8V, 1,5A	23a	Sgs	...7808... (TO-3)	data pdf
L 7808 CV,ACV	Z-IC	+8V, 1,5A, 5%, (A=2%)	17b	Stm	...7808... (TO-220)	data pdf pinout
L7808ML	R+	Io=1A Vo:7.6...8.4V Vin:10.5...23V	3P	Say		data pinout
L 7808 ML	Z-IC	=L 7808CV,ACV: Iso, 1A	17b	Say	...7808... (TO-220 Iso)	data pinout
L7809	R+	Io=1A Vo:8.6...9.4V Vin:16V P:1.75AmW	3P	Say		data pinout
L7809AB	R+	Io=int Vo:8.82...9.18V Vin:15V P:int	3SP	Sgs		data pdf pinout
L 7809 ABV	Z-IC	Precision 1 A regulators	45	Stm	-	pdf pinout
L7809AC	R+	Io=int Vo:8.82...9.18V Vin:15V P:int	3SP	Sgs		data pdf pinout
L 7809 ACV	Z-IC	Precision 1 A regulators	45	Stm	-	pdf pinout
L7809C	R+	Io=int Vo:8.65...9.35V Vin:15V P:int	3SPO	Sgs		data pdf pinout
L 7809 CP,CT	Z-IC	+9V, 1,5A	23a	Stm	...7809... (TO-3)	data pdf pinout
L 7809 CV	Z-IC	+9V, 1,5A	17b	Stm	...7809... (TO-220)	data pdf pinout
L7809ML	R+	Io=1A Vo:8.5...9.5V Vin:11.5...24V	3P	Say		data pinout
L 7809 ML	Z-IC	=L 7809CV,ACV: Iso, 1A	17b	Say	...7809... (TO-220 Iso)	data pinout
L7810	R+	Io=1A Vo:9.6...10.4V Vin:17V	3P	Say		data pinout
L 7810 CV,ACV	Z-IC	+10V, 1,5A, 5%	17b	Say	...7810... (TO-220)	data pinout
L7810C	R+	Io=1A Vo:9.5...10.5V Vin:15V P:int	3OP	Sgs		data pdf pinout
L7810ML	R+	Io=1A Vo:9.5...10.5V Vin:12.5...25V	3P	Say		data pinout
L 7810 ML	Z-IC	=L 7810CV,ACV: Iso, 1A	17b	Say	...7810... (TO-220 Iso)	data pinout
L7810T	R+	Io=int Vo:9.5...10.5V Vin:15V P:int	3O	Sgs		data pinout
L7812	R+	Io=int Vo:11.5...12.5V Vin:19V P:int	3PO	Sgs		data pdf pinout
L7812AB	R+	Io=int Vo:11.75...12.25V Vin:19V P:int	3SP	Sgs		data pdf pinout
L 7812 ABV	Z-IC	Precision 1 A regulators	45	Stm		pdf pinout
L7812AC	R+	Io=int Vo:11.75...12.25V Vin:19V P:int	3SP	Sgs		data pdf pinout
L 7812 ACP	Z-IC	Precision 1 A regulators	45	Stm	-	pdf pinout
L 7812 ACV	Z-IC	Precision 1 A regulators	45	Stm	-	pdf pinout
L7812C	R+	Io=int Vo:11.5...12.5V Vin:19V P:int	3SPO	Sgs		data pdf pinout
L 7812 CP	Z-IC	Positive voltage regulators	23a	Stm	...7809... (TO-3)	data pdf pinout
L 7812 CT	Z-IC	+12V, 1,5A	23a	Sgs	...7812... (TO-3)	data pinout
L 7812 CV,ACV	Z-IC	+12V, 1,5A, 5%, (A=2%)	17b	Sgs	...7812... (TO-220)	data pdf pinout
L7812ML	R+	Io=1A Vo:11.4...12.6V Vin:14.5...27V	3P	Say		data pinout
L 7812 ML	Z-IC	=L 7812CV,ACV: Iso, 1A	17b	Say	...7812... (TO-220 Iso)	data pinout
L7815	R+	Io=int Vo:14.4...15.6V Vin:23V P:int	3PO	Sgs		data pdf pinout
L7815AB	R+	Io=int Vo:14.7...15.3V Vin:23V P:int	3SP	Sgs		data pdf pinout
L 7815 ABV	Z-IC	Precision 1 A regulators	45	Stm	-	pdf pinout
L7815AC	R+	Io=int Vo:14.7...15.3V Vin:23V P:int	3SP	Sgs		data pdf pinout
L 7815 ACV	Z-IC	Precision 1 A regulators	45	Stm	-	pdf pinout
L7815C	R+	Io=int Vo:14.5...15.6V Vin:23V P:int	3SPO	Sgs		data pdf pinout
L 7815 CP	Z-IC	Positive voltage regulators	23a	Stm	...7809... (TO-3)	data pdf pinout
L 7815 CT	Z-IC	+15V, 1,5A	23a	Sgs	...7815... (TO-3)	data pinout
L 7815 CV,ACV	Z-IC	+15V, 1,5A, 5%, (A=2%)	17b	Stm	...7815... (TO-220)	data pdf pinout
L7815ML	R+	Io=1A Vo:14.25...15.75V Vin:17.5...30V	3P	Say		data pinout
L 7815 ML	Z-IC	=L 7815CV,ACV: Iso, 1A	17b	Say	...7815... (TO-220 Iso)	data pinout
L7818	R+	Io=int Vo:17.3...18.7V Vin:26V P:int	3PO	Sgs		data pdf pinout
L7818AB	R+	Io=int Vo:17.64...18.36V Vin:27V P:int	3SP	Sgs		data pdf pinout
L 7818 ABV	Z-IC	Precision 1 A regulators	45	Stm	-	pdf pinout
L7818AC	R+	Io=int Vo:17.64...18.36V Vin:27V P:int	3SP	Sgs		data pdf pinout
L 7818 ACV	Z-IC	Precision 1 A regulators	45	Stm	-	pdf pinout
L7818C	R+	Io=int Vo:17.3...18.7V Vin:26V P:int	3SPO	Sgs		data pdf pinout

Type	Device	Short Description	Fig.	Manu	Comparision Types	More at
L 7818 CT	Z-IC	+18V, 1,5A	23a	Sgs	... 7818... (TO-3)	data pinout
L 7818 CV,ACV	Z-IC	+18V, 1,5A, 5%, (A=2%)	17b	Stm	... 7818... (TO-220)	data pdf pinout
L7818ML	R+	Io=1A Vo:17.1...18.9V Vin:21...33V	3P	Say		data pinout
L 7818 ML	Z-IC	=L 7818CV,ACV: Iso, 1A	17b	Say	... 7818... (TO-220 Iso)	data pinout
L7820	R+	Io=int Vo:19.2...20.8V Vin:28V P:int	3PO	Sgs		data pdf pinout
L7820AB	R+	Io=int Vo:19.6...20.4V Vin:28V P:int	3SP	Sgs		data pdf pinout
L7820AC	R+	Io=int Vo:19.6...20.4V Vin:28V P:int	3SP	Sgs		data pdf pinout
L7820C	R+	Io=int Vo:19.2...20.8V Vin:28V P:int	3SPO	Sgs		data pdf pinout
L 7820 CV	Z-IC	+20V, 1,5A	17b	Sgs	... 7820... (TO-220)	data pdf pinout
L7820ML	R+	Io=1A Vo:19...21V Vin:23...35V P:15W	3P	Say		data pinout
L 7820 ML	Z-IC	=L 7820CV,ACV: Iso, 1A	17b	Say	... 7820... (TO-220 Iso)	data pinout
L7824	R+	Io=int Vo:23...25V Vin:33V P:int	3PO	Sgs		data pinout
L7824AB	R+	Io=int Vo:23.5...24.5V Vin:33V P:int	3SP	Sgs		data pdf pinout
L 7824 ABV	Z-IC	Precision 1 A regulators	45	Stm	-	pdf pinout
L7824AC	R+	Io=int Vo:23.5...24.5V Vin:33V P:int	3SP	Sgs		data pdf pinout
L 7824 ACV	Z-IC	Precision 1 A regulators	45	Stm	-	pdf pinout
L7824C	R+	Io=int Vo:23...25V Vin:33V P:int	3SPO	Sgs		data pdf pinout
L 7824 CP	Z-IC	Positive voltage regulators	23a	Stm	... 7809... (TO-3)	data pinout
L 7824 CT	Z-IC	+24V, 1,5A	23a	Sgs	... 7824... (TO-3)	data pinout
L 7824 CV,ACV	Z-IC	+24V, 1,5A, 5%, (A=2%)	17b	Stm	... 7824... (TO-220)	data pdf pinout
L7824ML	R+	Io=1A Vo:22.8...25.2V Vin:27...35V	3P	Say		data pinout
L 7824 ML	Z-IC	=L 7824CV,ACV: Iso, 1A	17b	Say	... 7824... (TO-220 Iso)	data pinout
L7852C	R+	Io=int Vo:5.0...5.4V Vin:10V P:int	3SPO	Sgs		data pdf pinout
L7875	R+	Io=int Vo:7.15...7.9V Vin:12.5V P:int	3OP	Sgs		data pinout
L 7875 CT	Z-IC	+7,5V, 1,5A	23a	Sgs	... 7875... (TO-3)	data pinout
L 7875 CV	Z-IC	+7,5V, 1,5A	17b	Sgs	... 7875... (TO-220)	data pinout
L 7885 CV	Z-IC	+8,5V, 1,5A	17b	Stm	... 7885... (TO-220)	pdf pinout
L7885C	R+	Io=int Vo:8.2...8.8V Vin:14.5V P:int	3SPO	Sgs		data pdf pinout
L 7805CP...7885CP	Z-IC	=L 7805CV...7885CV: Iso	17b	Stm	... 78xx... (TO-220 Iso)	pdf pinout
L 7885 CT	Z-IC	Positive voltage regulators	23a	Stm	... 7806... (TO-3)	data pdf pinout
L7900AC	R-	Io=int Vo:-4.9...-5.1V Vin:-10V P:int	3SP	Sgs		data pdf pinout
L 7905 ACD2T	Z-IC	-5V, 1.5A, 5%, (A=2%)	48	Stm		pdf pinout
L7905C	R-	Io=int Vo:-4.8...-5.2V Vin:-10V P:int	3SPO	Sgs		data pdf pinout
L 7905 CD2T	Z-IC	Negative voltage regulators, -5V, 1.5A	48	Stm		pdf pinout
L 7905 CT	Z-IC	-5V, 1,5A	23d	Sgs	... 7905... (TO-3)	pdf
L 7905 ACV	Z-IC	-5V, 1,5A, 5%, (A=2%)	17	Stm	... 7905... (TO-220)	pdf pinout
L 7905 CV/CP	Z-IC	Negative voltage regulators, -5V, 1.5A	17	Stm		pdf pinout
L7906AC	R-	Io=int Vo:-5.88...-6.12V Vin:-11V	3SP	Sgs		data pdf pinout
L7906C	R-	Io=int Vo:-5.75...-6.25V Vin:-11V	3SPO	Sgs		data pdf pinout
L7908AC	R-	Io=int Vo:-7.84...-8.16V Vin:-14V	3SP	Sgs		data pdf pinout
L7908C	R-	Io=int Vo:-7.7...-8.3V Vin:-14V P:int	3SPO	Sgs		data pdf pinout
L 7908 CT	Z-IC	-8V, 1,5A	23d	Sgs	... 7908... (TO-3)	pdf
L 7908 CV,ACV	Z-IC	-8V, 1,5A, 5%, (A=2%)	17c	Sgs	... 7908... (TO-220)	
L 7908 CV	Z-IC	Negative voltage regulators, -8V, 1.5A	17	Stm		pdf pinout
L7912AC	R-	Io=int Vo:-11.75...-12.25V Vin:-19V	3SP	Sgs		data pdf pinout
L 7912 ACD2T	Z-IC	-12V, 1,5A, 5%, (A=2%)	48	Stm		pdf pinout
L7912C	R-	Io=int Vo:-11.5...-12.5V Vin:-19V	3SPO	Sgs		data pdf pinout
L 7912 CD2T	Z-IC	Negative voltage regulators, -12V, 1.5A	48	Stm		pdf pinout
L 7912 CT	Z-IC	-12V, 1,5A	23d	Sgs	... 7912... (TO-3)	pdf
L 7912 ACV	Z-IC	-12V, 1,5A, 5%, (A=2%)	17	Stm	... 7912... (TO-220)	pdf pinout
L 7912 CV/CP	Z-IC	Negative voltage regulators, -12V, 1.5A	17	Stm		pdf pinout
L7915AC	R-	Io=int Vo:-14.7...-15.3V Vin:-23V	3SP	Sgs		pdf pinout
L7915C	R-	Io=int Vo:-14.4...-15.6V Vin:-23V	3P	Sgs		data pdf pinout
L 7915 CD2T	Z-IC	Negative voltage regulators, -15V, 1.5A	48	Stm		pdf pinout
L 7915 CT	Z-IC	-15V, 1,5A	23d	Sgs	... 7915... (TO-3)	pdf
L 7915 ACV	Z-IC	-15V, 1,5A, 5%, (A=2%)	17	Stm	... 7915... (TO-220)	pdf pinout
L 7915 CV/CP	Z-IC	Negative voltage regulators, -15V, 1.5A	17	Stm		pdf pinout
L7918AC	R-	Io=int Vo:-17.64...-18.36V Vin:-27V	3SP	Sgs		data pdf pinout
L 7918 CT	Z-IC	-18V, 1,5A	23d	Sgs	... 7918... (TO-3)	pdf
L 7918 CV,ACV	Z-IC	-18V, 1,5A, 5%, (A=2%)	17c	Sgs	... 7918... (TO-220)	
L7920AC	R-	Io=int Vo:-19.6...-20.4V Vin:-29V	3SP	Sgs		
L 7920 CD2T	Z-IC	Negative voltage regulators, -20V, 1.5A	48	Stm		data pdf pinout
L 7920 CT	Z-IC	-20V, 1,5A	23d	Sgs	... 7920... (TO-3)	pdf
L 7920 CV	Z-IC	-20V, 1,5A, 5%, (A=2%)	17	Sgs	... 7920... (TO-220)	pdf pinout
L7922AC	R-	Io=int Vo:-23.5...-24.5V Vin:-31V	3SP	Sgs		data pdf pinout
L7924AC	R-	Io=int Vo:-23.5...-24.5V Vin:-33V	3SP	Sgs		data pdf pinout
L 7924 CT	Z-IC	-24V, 1,5A	23d	Sgs	... 7924... (TO-3)	pdf
L 7924 CV,ACV	Z-IC	-24V, 1,5A, 5%, (A=2%)	17c	Sgs	... 7924... (TO-220)	
L 7952 CT	Z-IC	-5,2V, 1,5A	23d	Sgs	... 7952... (TO-3)	pdf
L 7952 CV	Z-IC	-5,2V, 1,5A	17c	Sgs	... 7952... (TO-220)	pdf
L7952AC	R-	Io=int Vo:-5.1...-5.3V Vin:-10V P:int	3SP	Sgs		data pdf pinout
L7952C	R-	Io=int Vo:-5.0...-5.4V Vin:-10V P:int	3SOP	Sgs		data pdf pinout
L 7905CP...7924CP	Z-IC	=L 7905CV...7924CV: Iso	17c	Sgs	... 79xx... (TO-220 Iso)	
L 9222	LIN-IC	4x invert. Transistor-Schalter/switch	16-DIP	Tho	-	pdf
L 9305(C)	LIN-IC	Dual, Relais-Tr/Relay Dr., 3,5...18V,1A	16-DIP	Sgs	-	
L 9306	LIN-IC	Dual, Relais-/Relay Dr., 3,5...18V,0,3A	8-DIP	Sgs	-	pdf
L 9307	LIN-IC	Dual, Low Side Driver, hi-current	11-SQL	Tho	-	
L 9308	LIN-IC	Dual, Low Side Driver	8-DIP	Tho	-	pdf
L 9309	LIN-IC	Dual, Low Side Driver, hi-current	10-SIL	Tho	-	
L 9322	LIN-IC	4x Trans.-Schalter/switch	16-DIP	Tho	-	
L 9324	LIN-IC	Window Lift Controller	20-DIP	Tho	-	
L 9326	LIN-IC	SMD, Dual IPS	24-MDIP	Sgs	-	
L 9332	IC	Quad low Side Driver	TSOP	Sgs	-	
L 9333 MD	IC	Quad low Side Driver	TSOP	Stm	-	pdf pinout
L 9335	LIN-IC	Injector Driver, 2,4A+0,6A	86/5Pin	Sgs	-	pdf pinout

Type	Device	Short Description	Fig.	Manu	Comparision Types	More at
9336	LIN-IC	Injector Driver, 4A+1A	86/5Pin	Sgs	-	
9337 MD	IC	triple low Side Driver	TSOP	Stm	-	pdf pinout
9338 D	IC	Quad low Side Driver	TSOP	Stm	-	pdf pinout
9338 MD	IC	Quad low Side Driver	TSOP	Stm	-	pdf pinout
9339	IC	Quad low Side Driver	TSOP	Stm	-	pdf pinout
9339 MD	IC	Quad low Side Driver	TSOP	Stm	-	pdf pinout
9341	LIN-IC	Quad, Low Side Driver	13-SQL	Sgs	-	pdf
9342	LIN-IC	Multipoint Fuel Injector Driver	24-DIP	Sgs	-	
9347 LF	IC	Quad (2x5A/2x2.5A) low-side switch	TSOP	Stm	-	pdf pinout
9348	IC	Quad low-side drv	TSOP	Stm	-	pdf pinout
9350	LIN-IC	High Side Driver	86/5Pin	Sgs	-	
9351	LIN-IC	High Side Driver	86/5Pin	Sgs	-	pdf
9352	LIN-IC	High Side Driver	86/5Pin	Sgs	-	pdf
9355	LIN-IC	Switchmode High Side Driver, 6A	86/5Pin	Sgs	-	
9360	LIN-IC	Dual Injection Driver	11-SQL	Sgs	-	
9362	IC	Quad low Side Driver	TSOP	Stm	-	pdf pinout
9363	LIN-IC	Quad, Low Side Driver	15-SQL	Sgs	-	
9374	IC	4-Channel valve drv	TSOP	Stm	-	pdf pinout
9374 XP	IC	4-Channel valve drv	TSOP	Stm	-	pdf pinout
9375 (XP)	CMOS/LIN-IC	8-Ch. valve driver	TSOP	Stm	-	pdf pinout
9380	MOS-IC	triple High-side Mosfet Driver	MDIP	Stm	-	pdf pinout
9386 MD	IC	Dual Intelligent Power low Side Switch	TSOP	Stm	-	pdf pinout
9390	CMOS/LIN-IC	12 Ch. valve drv	MP	Stm	-	pdf pinout
9407 F	LIN/Z-IC	Voltage Reg.	SIL	Stm	-	pdf pinout
9444 VB	LIN-IC	Kfz-regler/car Alternator Regulator	17	Sgs	-	pdf pinout
9448 VB	LIN-IC	Kfz-regler/car Alternator Regulator	17	Sgs	-	pdf
9466 N	LIN/Z-IC	Car Alternator Voltage Reg	SIL	Stm	-	pdf pinout
9468 N	LIN/Z-IC	All Silicon Voltage Reg.	SIL	Stm	-	pdf pinout
9473 J	LIN/Z-IC	Car Alternator Voltage Regulator	SIL	Stm	-	pdf pinout
9474	LIN/Z-IC	All Silicon Voltage Reg.	SIL	Stm	-	pdf pinout
9480 VB	LIN-IC	Kfz-regler/car Alternator Regulator	17b	Sgs	-	pdf
9484	LIN/Z-IC	Car Alternator Voltage Reg.	SIL	Stm	-	pdf pinout
9610 C	LIN-IC	=L 9611C: SMD	16-MDIP	Tho	-	pdf
9611 C	LIN-IC	PWM-controller f. MOS-V-FETs	16-DIP	Tho	-	pdf
9613 B	IC	Data Interface	MDIP	Stm	-	pdf pinout
9615	IC	Can Bus Transceiver	MDIP	Stm	-	pdf pinout
9616	IC	High Speed Can Bus Transceiver	MDIP	Stm	-	pdf pinout
9637 D	CMOS-IC	ISO 9141 Interface	MDIP	Stm	-	pdf pinout
9638	IC	Lin Bus Transceiver	MDIP	Stm	-	pdf pinout
9651	IC	Smart Quad Switch	TSOP	Stm	-	pdf pinout
9686	LIN-IC	Kfz-blinkgeber/car Direction indicator	8-DIP	Tho	-	pdf
9700	LIN-IC	Hex Limiter, hi-prec	8-DIP	Tho	-	pdf
9703	LIN-IC	8x Erdungsüberw./ground contact monit.	20-DIP	Tho	-	pdf
9703 D	LIN-IC	=L 9703: SMD	20-MDIP		-	pdf
9704	LIN-IC	8x Spg.-Überw./voltage contact monit.	20-DIP	Tho	-	pdf
9704 D	LIN-IC	=L 9704: SMD	20-MDIP		-	pdf
9705	LIN-IC	2x4 Spg. Überw./contact interface	20-DIP	Tho	-	pdf
9705 D	LIN-IC	=L 9705: SMD	20-MDIP		-	pdf
9733	CMOS/LIN-IC	8x self configuring low/high side drv	MDIP	Stm	-	pdf pinout
9733 XP	CMOS/LIN-IC	8x self configuring low/high side drv	TSOP	Stm	-	pdf pinout
9801	LIN-IC	High Side Driver, 25A	86/5Pin	Tho	-	
9803	LIN-IC	Super smart power motor drv.	MP	Stm	-	pdf pinout
9805 E	LIN-IC	Super smart power motor drv	MP	Stm	-	pdf pinout
9811	LIN-IC	DC & PWM High Side Driver	86/7Pin	Tho	-	
9813	µP-IC,MOS-IC	Super Smart Mirror, Embeded MCU	MP	Stm	-	pdf pinout
9820	IC	High Side Driver	DIC	Stm	-	pdf pinout
9820 D	IC	High Side Driver	MDIP	Stm	-	pdf pinout
9821	LIN-IC	High Side Driver, 25A	86/5Pin	Tho	-	pdf
9822	LIN-IC	Octal Serial Ventil-Tr/solenoid Drv.	15-SQL	Tho	-	
9822 E	LIN-IC	8-Bit Serial Solenoid driver	SQL	Stm	-	pdf pinout
9822 EPD	LIN-IC	8-Bit Serial Solenoid driver	TSOP	Stm	-	pdf pinout
9822 N	LIN-IC	SMD, 8x Serial Ventil-Tr./Solenoid Drv	20-MDIP	Sgs	-	data pdf pinout
9823	IC	Octal Low-side Driver, serial input control	TSOP	Stm	-	pdf pinout
9825	IC	Octal Low-side Driver, Serial/parallel input control	TSOP	Stm	-	pdf pinout
9826	IC	Octal Low-side Driver, Serial/parallel input control	TSOP	Stm	-	pdf pinout
9830	LIN-IC	PWM Dimmer	86/7Pin	Tho	-	pdf
9842	LIN-IC	Octal Parallel Low Side Driver	20-DIP	Tho	-	
9842 D	LIN-IC	=L 9842: SMD	20-MDIP	Tho	-	
9848	CMOS/LIN-IC	Octal Config. Low/high Side Drv.	MDIP	Stm	-	pdf pinout
9856	IC	High voltage high-side driver	MDIP	Stm	-	pdf pinout
9857	CMOS/LIN-IC	High voltage high-side driver	MDIP	Stm	-	pdf pinout
9903	CMOS/LIN-IC	Motor Bridge Cntrl.	MDIP	Stm	-	pdf pinout
9904	CMOS/LIN-IC	Motor Bridge Cntrl.	MDIP	Stm	-	pdf pinout
9907	LIN-IC	Motor Bridge Daver	8-DIP	Tho	-	pdf
9907 D	LIN-IC	=L 9907: SMD	20-MDIP		-	
9911	LIN/Z-IC	Car alternator smart voltage reg.	SIL	Stm	-	pdf pinout
9930	LIN-IC	Dual, Brücken-tr/full Bridge	11-SQL	Tho	-	pdf
9935	LIN-IC	2-phase stepper motor drv	MDIP	Stm	-	pdf pinout
9936	LIN-IC	Motor Brücken-tr./half bridge driver	8-SQL	Tho	-	
9937	LIN-IC	Motor Brücken-tr./full bridge driver	11-SQL	Tho	-	pdf
9942 XP1	LIN-IC	Stepper Motor drv	TSOP	Stm	-	pdf pinout
9946	LIN-IC	Motor Brücken-tr./half bridge driver	15-SQL	Tho	-	

Type	Device	Short Description	Fig.	Manu	Comparision Types	More at
L 9947	LIN-IC	Quad Brücken-tr./half bridge driver	15-SQL	Tho	-	pdf
L 9949	LIN-IC	Door Actuatro Drv	TSOP	Stm		pdf pinout
L 9950	LIN-IC	Door actuator driver	MDIP	Stm		pdf pinout
L 9950 XP	LIN-IC	Door actuator driver	TSOP	Stm		pdf pinout
L 9951	LIN-IC	Rear door actuator driver	MDIP	Stm		pdf pinout
L 9951 XP	LIN-IC	Rear door actuator driver	TSOP	Stm		pdf pinout
L 9953	LIN-IC	Door actuator driver	MDIP	Stm		pdf pinout
L 9953 XP	LIN-IC	Door actuator driver	TSOP	Stm		pdf pinout
L 9954	LIN-IC	Door actuator driver	MDIP	Stm		pdf pinout
L 9954 XP	LIN-IC	Door actuator driver	TSOP	Stm		pdf pinout
L 9997 ND	CMOS/LIN-IC	2x Half Bridge Driver	MDIP	Stm		pdf pinout
L 7805 ABD 2 T-TR	Z-IC	Precision 1 A regulators	45	Stm	-	pdf pinout
L 7805 ACD 2 T-TR	Z-IC	Precision 1 A regulators	45	Stm	-	pdf pinout
L 7805 CD 2 T-TR	Z-IC	Positive voltage regulators	17b	Stm	... 7824... (TO-220)	data pdf pinout
L 7806 ABD 2 T-TR	Z-IC	Precision 1 A regulators	45	Stm	-	pdf pinout
L 7806 ACD 2 T-TR	Z-IC	Precision 1 A regulators	45	Stm	-	pdf pinout
L 7806 CD 2 T-TR	Z-IC	Positive voltage regulators	17b	Stm	... 7824... (TO-220)	data pdf pinout
L 7808 ABD 2 T-TR	Z-IC	Precision 1 A regulators	45	Stm	-	pdf pinout
L 7808 ACD 2 T-TR	Z-IC	Precision 1 A regulators	45	Stm	-	pdf pinout
L 7808 CD 2 T-TR	Z-IC	Positive voltage regulators	17b	Stm	... 7824... (TO-220)	data pdf pinout
L 7809 ABD 2 T-TR	Z-IC	Precision 1 A regulators	45	Stm	-	pdf pinout
L 7809 ACD 2 T-TR	Z-IC	Precision 1 A regulators	45	Stm	-	pdf pinout
L 7809 CD 2 T-TR	Z-IC	Positive voltage regulators	17b	Stm	... 7824... (TO-220)	data pdf pinout
L 7812 ABD 2 T-TR	Z-IC	Precision 1 A regulators	45	Stm	-	pdf pinout
L 7812 ACD 2 T-TR	Z-IC	Precision 1 A regulators	45	Stm	-	pdf pinout
L 7812 CD 2 T-TR	Z-IC	Positive voltage regulators	17b	Stm	... 7824... (TO-220)	data pdf pinout
L 7815 ABD 2 T-TR	Z-IC	Precision 1 A regulators	45	Stm	-	pdf pinout
L 7815 ACD 2 T-TR	Z-IC	Precision 1 A regulators	45	Stm	-	pdf pinout
L 7815 CD 2 T-TR	Z-IC	Positive voltage regulators	17b	Stm	... 7824... (TO-220)	data pdf pinout
L 7818 CD 2 T-TR	Z-IC	Positive voltage regulators	17b	Stm	... 7824... (TO-220)	data pdf pinout
L 7824 ABD 2 T-TR	Z-IC	Precision 1 A regulators	45	Stm	-	pdf pinout
L 7824 CD 2 T-TR	Z-IC	Positive voltage regulators	17b	Stm	... 7824... (TO-220)	data pdf pinout
L 7885 CD 2 T-TR	Z-IC	Positive voltage regulators	17b	Stm	... 7824... (TO-220)	data pdf pinout
L 4931 ABPT 25 TR	Z-IC	Very low drop voltage regulators with inhibit	45	Stm	-	pdf pinout
L 4931 CDT 25-TR	Z-IC	Very low drop voltage regulators with inhibit	45	Stm	-	pdf pinout
L 4931 CPT 25-TR	Z-IC	Very low drop voltage regulators with inhibit	45	Stm	-	pdf pinout
L 4931 ABPT 27 TR	Z-IC	Very low drop voltage regulators with inhibit	45	Stm	-	pdf pinout
L 4931 CD 27-TR	Z-IC	Very low drop voltage regulators with inhibit	45	Stm	-	pdf pinout
L 4931 CD 27-TRY	Z-IC	Very low drop voltage regulators with inhibit	45	Stm	-	pdf pinout
L 4931 CPT 27-TR	Z-IC	Very low drop voltage regulators with inhibit	45	Stm	-	pdf pinout
L 4931 ABD 33-TR	Z-IC	Very low drop voltage regulators with inhibit	45	Stm	-	pdf pinout
L 4931 ABDT 33-TR	Z-IC	Very low drop voltage regulators with inhibit	45	Stm	-	pdf pinout
L 4931 ABV 33	Z-IC	Very low drop voltage regulators with inhibit	45	Stm	-	pdf pinout
L 4931 CD 33-TR	Z-IC	Very low drop voltage regulators with inhibit	45	Stm	-	pdf pinout
L 4931 CD 33-TRY	Z-IC	Very low drop voltage regulators with inhibit	45	Stm	-	pdf pinout
L 4931 CDT 33-TR	Z-IC	Very low drop voltage regulators with inhibit	45	Stm	-	pdf pinout
L 4931 CPT 33-TR	Z-IC	Very low drop voltage regulators with inhibit	45	Stm	-	pdf pinout
L 4931 CZ 33-AP	Z-IC	Very low drop voltage regulators with inhibit	45	Stm	-	pdf pinout
L 4931 ABD 35-TR	Z-IC	Very low drop voltage regulators with inhibit	45	Stm	-	pdf pinout
L 4931 ABD 35-TRY	Z-IC	Very low drop voltage regulators with inhibit	45	Stm	-	pdf pinout
L 4931 ABDT 35 TR	Z-IC	Very low drop voltage regulators with inhibit	45	Stm	-	pdf pinout
L 4931 CD 35-TR	Z-IC	Very low drop voltage regulators with inhibit	45	Stm	-	pdf pinout
L 4931 CDT 35-TR	Z-IC	Very low drop voltage regulators with inhibit	45	Stm	-	pdf pinout
L 4931 ABDT 50-TR	Z-IC	Very low drop voltage regulators with inhibit	45	Stm	-	pdf pinout
L 4931 CD 50-TR	Z-IC	Very low drop voltage regulators with inhibit	45	Stm	-	pdf pinout
L 4931 CDT 50-TR	Z-IC	Very low drop voltage regulators with inhibit	45	Stm	-	pdf pinout
L 4931 CPT 50-TR	Z-IC	Very low drop voltage regulators with inhibit	45	Stm	-	pdf pinout
L 4931 CZ 50-AP	Z-IC	Very low drop voltage regulators with inhibit	45	Stm	-	pdf pinout
L 4931 ABDT 80-TR	Z-IC	Very low drop voltage regulators with inhibit	45	Stm	-	pdf pinout

Type	Device	Short Description	Fig.	Manu	Comparision Types	More at
4931 ABPT 80 TR	Z-IC	Very low drop voltage regulators with inhibit	45	Stm	-	pdf pinout
4931 CD 80-TR	Z-IC	Very low drop voltage regulators with inhibit	45	Stm	-	pdf pinout
4931 CDT 80-TR	Z-IC	Very low drop voltage regulators with inhibit	45	Stm	-	pdf pinout
4931 CPT 80-TR	Z-IC	Very low drop voltage regulators with inhibit	45	Stm	-	pdf pinout
4931 ABD 120 TR	Z-IC	Very low drop voltage regulators with inhibit	45	Stm	-	pdf pinout
4931 ABPT 120 R	Z-IC	Very low drop voltage regulators with inhibit	45	Stm	-	pdf pinout
4931 CD 120-TR	Z-IC	Very low drop voltage regulators with inhibit	45	Stm	-	pdf pinout
4931 CDT 120-TR	Z-IC	Very low drop voltage regulators with inhibit	45	Stm	-	pdf pinout
4931 CPT 120-TR	Z-IC	Very low drop voltage regulators with inhibit	45	Stm	-	pdf pinout

A

Type	Device	Short Description	Fig.	Manu	Comparision Types	More at
A	Si-P	=2SA1681 (Typ-Code/Stempel/marking)	39	Tos	→2SA1681	data
A	Si-P	=BF 550 (Typ-Code/Stempel/marking)	35	Phi, Sie	→BF 550	data
A	Si-N	=HD 2A3M (Typ-Code/Stempel/marking)	39	Nec	→HD 2...	data
A	Si-N	=KTD1003-A (Typ-Code/Stempel/marking)	39	Kec	→KTD 1003	data
A	Si-N+R	=RN 1442A (Typ-Code/Stempel/marking)	35	Tos	→RN 1442	
A	Si-Di	=SB 005W03 (Typ-Code/Stempel/marking)	35(1,6mm)	Say	→SB 005W03	data
A	MOS-P-FET-e	=Si 1303DL (Typ-Code/Stempel/marking)	35(2mm)	Six	→Si 1303DL	data
A	Si-P	=µPA605T (Typ-Code/Stempel/marking)	46	Nec	→µPA605T	data
A	Si-P	=µPA671T (Typ-Code/Stempel/marking)	46(2mm)	Nec	→µPA671T	data
A543B	LED	am ø5mm l:1120...4500mcd Vf:2V	T1¾	Osr		data pdf pinout
A 733 P	Si-P		7c	Ons	2SA608N, 2SA608N-F, 2SA608N-G, 2SA659	data
A 733 Q	Si-P		7c	Ons	2SA608N, 2SA608N-F, 2SA608N-G, 2SA659,	data
A 1050	HF-IC	HF, Buffer, Rf Amp, Am Detektor Vcc=1.2... 1.6V, Iccmax=1.2mA, fc=1MHz	7	Say	-	pdf pinout
A 1060	HF-IC	FM-antenn.-diversity-schalter/switch	9-SIP	Say	-	pdf pinout
A 1061 M	HF-IC	SMD, Antenna Switch. Contr., Vs=7...12V	14-SMDIP	Say	-	pdf pinout
A 1065 M	HF-IC	SMD, 2 Tuner Diversity, Car FM Tuner Vcc=7...12V, Icc=16mA, -30...80°C	20-MDIP	Say	-	pdf pinout
A 1111(A)P	LIN-IC	TV-Ton-ZF, FM-ZF	8-DIP	Say	-	pdf pinout
A 1130	LIN-IC	AM-Tuner, ZF, Agc	16-SQP	Say	16-SQP	pdf pinout
A 1132	LIN-IC	Am-tuner + Zf	16-SQP	Say	-	pdf pinout
A 1135	LIN-IC	AM-Tuner, ZF, Agc	20-DIP	Say	-	pdf pinout
A 1135 M	LIN-IC	SMD, AM-Tuner, ZF, Agc	20-MDIP	Say	-	pdf pinout
A 1136(N)	LIN-IC	AM-Tuner, ZF, AGC, AM-Stereo. Ucc=8V	24-SDIP	Say	-	
A 1136	LIN-IC	AM-Tuner, ZF, Agc, AM-Stereo	24-DIP	Say	-	pdf
A 1136 M,NM	LIN-IC	=LA 1136: SMD	24-SMDIP	Say	-	
A 1136 M	LIN-IC	SMD, AM-Tuner, ZF, Agc, AM-Stereo	24-MDIP	Say	-	pdf
A 1136 N	HF-IC	AM-Tuner, ZF, Agc, AM-Stereo	24-DIP	Say	-	pdf pinout
A 1136 NM	HF-IC	SMD, AM-Tuner, ZF, Agc, AM-Stereo	24-SMDIP	Say	-	pdf pinout
A 1137(N)	LIN-IC	AM-Tuner, ZF, AGC, Ucc=8V)	20-DIP	Say	-	
A 1137	LIN-IC	AM-Tuner, ZF, Agc	20-DIP	Say	-	pdf
A 1137 M,NM	LIN-IC	=LA 1137: SMD	20-MDIP	Say	-	
A 1137 M	LIN-IC	SMD, AM-Tuner, ZF, Agc	20-MDIP	Say	-	pdf
A 1137 N	HF-IC	AM-Tuner, ZF, Agc	20-DIP	Say	-	pdf pinout
A 1137 NM	HF-IC	SMD, AM-Tuner, ZF, Agc	20-MDIP	Say	-	pdf pinout
A 1140	LIN-IC	FM-ZF, Agc, Afc, Quadratur-dem.	16-SQP	Say	BA 4110, KA 22441	pdf pinout
A 1143	LIN-IC	FM-ZF, Agc, Afc, Quadratur-dem.	16-DIP	Say	-	pdf pinout
A 1145	LIN-IC	FM-ZF, Afc, Quadratur-dem.	18-SQP	Say	-	pdf pinout
A 1145 M	LIN-IC	=LA 1145: SMD	20-MDIP	Say	-	pdf pinout
A 1150	LIN-IC	FM-ZF+Demod.(Pin1 fehlt/without pin1)	8-SIP	Say	BA 403, KA 2245, TA 7130, µPC 1028H	pdf pinout
A 1150 N	LIN-IC	FM-ZF, Dem.(mit/with Pin1)	8-SIP	Say	-	pdf
A 1160	LIN-IC	FM-ZF, Agc	16-DIP	Say	-	pdf pinout
A 1165	LIN-IC	FM-Tuner	16-SQP	Say	-	pdf pinout
A 1170	LIN-IC	FM-frontend, Fm-tuner	16-SQP	Say	-	pdf pinout
A 1175	LIN-IC	Fm-frontend, Fm-tuner	16-SQP	Say	-	pdf pinout
A 1175 M	LIN-IC	SMD, Fm-frontend, Fm-tuner	16-MDIP	Say	-	pdf pinout
A 1177	LIN-IC	FM Frontend, FM-Tuner, Ucc=8...9V	9-SIP	Say	-	pdf pinout
A 1178 M	LIN-IC	FM Frontend, Fm-tuner	10-MDIP	Say	-	pdf pinout
A 1180	LIN-IC	FM-Tuner/front end, Ucc=1,6...6V	9-SIP	Say	-	pdf pinout
A 1185	LIN-IC	FM-Tuner/Front end, Ucc=1,5...8V	9-SIP	Say	AN 7205, KA 22495, TA 7358AP	pdf pinout
A 1186 N	LIN-IC	FM-Tuner/Front end, Ucc=1,8...7,5V	9-SIP	Say	-	pdf pinout
A 1188 A	LIN-IC	FM-Tuner/Front end, Ucc=3,5...7,5V	9-SIP	Say	-	pdf pinout
A 1193 M,V	HF-IC	SMD, Fm-frontend, Mixer, Agc, IF-Amp	20-MDIP	Say	-	pdf pinout
A 1201(FM)	LIN-IC	AM/FM-ZF, AGC, AM-Demod.	14-DIP	Say	KA 1241, UL 1211	
A 1201	LIN-IC	AM/FM-ZF, Agc, Am-demod.	14-DIP	Say	KA 1241, UL 1211	pdf pinout
A 1205	LIN-IC	AM/FM-ZF, Agc	16-DIP	Say	-	pdf pinout
A 1207	LIN-IC	AM/FM-ZF, Agc	16-DIP	Say	-	pdf pinout
A 1210	LIN-IC	AM/FM-ZF, Dem., Agc	16-DIP	Say	-	pdf pinout
A 1221	LIN-IC	FM-ZF, Vccmax=24V	5	Say	-	pdf pinout
A 1222	LIN-IC	FM-ZF, Vccmax=20V	8-DIP	Say	-	pdf pinout
A 1225 M	LIN-IC	SMD, FM If/zf Detector, Vcc=1.8...8V	10-SMDIP	Say	-	pdf pinout
A 1230	LIN-IC	FM-ZF, Agc	16-DIP	Say	-	pdf pinout
A 1231(N)	LIN-IC	=LA 1230: + mute level control(Pin16)	16-DIP	Say	-	pdf pinout

Type	Device	Short Description	Fig.	Manu	Comparision Types	More at
LA 1232	LIN-IC	FM-ZF, Agc, Afc, 9...14V	16-DIP	Say	-	pdf pinout
LA 1235	LIN-IC	FM-ZF, Afc, Ucc=10...14V	16-DIP	Say	-	pdf pinout
LA 1240	LIN-IC	AM-Tuner	16-DIP	Say	-	pdf pinout
LA 1245	LIN-IC	AM-Tuner, Vccmax=16V	20-DIP	Say	-	pdf pinout
LA 1247	LIN-IC	AM-Tuner, Vccmax=16V	20-DIP	Say	-	pdf
LA 1260	LIN-IC	FM/AM-Tuner, Ucc=3...8V	16-DIP	Say	BA 4260, KA 2247	pdf
LA 1261	LIN-IC	FM/AM-Tuner, AM/FM-ZF, Vcc=9V	16-DIP	Say	-.	pdf
LA 1265	LIN-IC	FM/AM-Tuner, AM/FM-ZF, Ucc=6...14V	22-SDIP	Say	-	pdf pinout
LA 1266	LIN-IC	AM/FM-Tuner, ...-ZF, NF-V, Ucc=6...14V	24-SDIP	Say	-	pdf pinout
LA 1267	LIN-IC	FM/AM-Tuner, ...-ZF, NF-V, Ucc=6...14V	24-SDIP	Say	-	pdf
LA 1270	LIN-IC	FM/AM-Tuner, AM/FM-ZF, Ucc=3V	20-DIP	Say	-	pdf pinout
LA 1320	LIN-IC	Tv-ton-zf	14-DIP	Say	-	pdf
LA 1342	LIN-IC	FM-ZF, TV-Ton-ZF, Nf-tr.	14-DIC	Say	-	pdf
LA 1352	LIN-IC	TV-Video-ZF, Agc(pos.)	14-DIP	Say	MC 1352, TA 7074	pdf
LA 1353	LIN-IC	Tv-video-zf, Agc(neg.)	14-DIP	Say	MC 1353	pdf
LA 1354	LIN-IC	TV, Video-Dem, Vcc=12V	8-DIP	Say	-	pdf
LA 1357(N)	LIN-IC	TV, Video-ZF, AGC, AFT	22-DIP	Say	-	pdf
LA 1357(N)	LIN-IC	TV, Video-ZF, Agc, Aft	22-DIP	Say	-	pdf
LA 1362	LIN-IC	Vc/tv, Lautst.-reg./dc volume control	9-SIP	Say	-	pdf
LA 1363	LIN-IC	TV-Ton-ZF, Dem	14-DIP	Say	→LA 1365	pdf
LA 1364	LIN-IC	TV, AFT	14-DIP	Say	HA 1126, M 5135, TA 7070	pdf
LA 1365	LIN-IC	TV-Ton-ZF, Dem, + integr.Stab.(Pin5)	14-DIP	Say	AN 241, CA 3065, HA 1125, KA 2101, LM 30	pdf
LA 1366 N	LIN-IC	Ctv, NTSC Chroma	16-DIP	Say	-	
LA 1367	LIN-IC	Ctv, NTSC Chroma	16-DIP	Say	-	
LA 1368	LIN-IC	Ctv, NTSC Chroma	14-DIP	Say	-	pdf
LA 1369	LIN-IC	Ctv, Chroma-dem	14-DIP	Say	-	
LA 1373	LIN-IC	Ctv, Color Bandpass	14-DIP	Say	-	
LA 1374	LIN-IC	Ctv, NTSC Chroma	16-DIP	Say	-	
LA 1375	LIN-IC	Ctv, NTSC Chroma-dem	14-DIP	Say	-	
LA 1376	LIN-IC	Ctv, NTSC Chroma-dem	14-DIP	Say	-	
LA 1381	LIN-IC	TV-HA/VA-synchr.-kombi	14-DIP	Say	-	pdf
LA 1382	LIN-IC	Tv-ha/va-synchr.-kombi	14-DIP	Say	-	pdf
LA 1383	LIN-IC	TV-VA-O	14-DIP	Say	-	pdf
LA 1384	LIN-IC	TV-HA-O	8-DIP	Say	-	pdf
LA 1385	LIN-IC	TV/CTV-VA-O, VA-E, Vccmax=20V	10-SIL	Say	KA 2130A, TA 7242, µPC 1031H2	pdf
LA 1387	LIN-IC	Ctv, Synchr. (PAL), Vccmax=14V	24-DIP	Say	(LA 7800)	pdf
LA 1388	LIN-IC	Ctv, Synchr. (ntsc), Vccmax=14V	24-DIP	Say	-	pdf
LA 1390	LIN-IC	Ctv, NTSC Chroma, Vccmax=22V	16-DIP	Say	-	pdf
LA 1460	LIN-IC	Ctv, HA/VA-Synchr., Vcc=14V	22-DIP	Say	-	pdf
LA 1463	LIN-IC	Ctv, Ha/va-synchr. (PAL), Vcc=14V	24-DIP	Say	-	pdf
LA 1464	LIN-IC	Ctv, HA/VA-Synchr.(NTSC), Vcc=14V	24-DIP	Say	-	pdf
LA 1503	LIN-IC	Telecom, FM-ZF, Ucc=2,5...9V	16-DIP	Say	-	pdf
LA 1600	LIN-IC	AM-Radio, Ucc=1,8...6V	9-SIP	Say	-	pdf
LA 1610	LIN-IC	AM-Radio, NF-V+E, Ucc=1,8...6V	14-DIP	Say	-	
LA 1650	LIN-IC	DCF77-Empf./Receiver, Vcc=1,2...6,5V	18-DIP	Say	-	pdf
LA 1780 M	LIN-IC	SMD, Single-Chip Tuner, Car Radio	64-QFP	Say	-	pdf
LA 1781 M	LIN-IC	SMD, Single-Chip Tuner, Car Radio	64-QFP	Say	-	pdf
LA 1784 M	LIN-IC	SMD, Single-Chip Tuner, Car Radio	64-QFP	Say	-	pdf
LA 1787 M	LIN-IC	SMD, Single-Chip Tuner, Car Radio	64-QFP	Say	-	pdf
LA 1800	LIN-IC	AM/FM-Radio, Kopfh./Earphone-E, Ucc=3V	22-SDIP	Say	-	pdf
LA 1801	LIN-IC	AM/FM-Radio, Ucc=3V		Say		
LA 1805	LIN-IC	AM/FM-Radio, Stereo-Decoder, USA-band	24-SDIP	Say	-	pdf
LA 1806	LIN-IC	AM/FM-Radio, Stereo-Dec., Japan-band	24-SDIP	Say	-	
LA 1810	LIN-IC	AM/FM-Radio, Stereo-Dec., Usa-band	24-SDIP	Say	KA 2290	pdf
LA 1811	LIN-IC	AM/FM-Radio, Stereo-Dec., Japan-band	24-SDIP	Say	-	
LA 1815	LIN-IC	AM/FM-Tuner, AM-ZF	20-DIP	Say	-	pdf
LA 1816	LIN-IC	AM/FM-Tuner, Stereo-Decoder, Usa-band	24-SDIP	Say	-	pdf
LA 1816 M	LIN-IC	SMD,AM/FM-Tuner, Stereo-Dec., Usa-band	24-SMDIP	Say	-	pdf
LA 1817	LIN-IC	AM/FM-Tuner, Stereo-Dec., Japan-band	24-SDIP	Say	-	pdf
LA 1817 M	LIN-IC	SMD, AM/FM-Tuner, Stereo-Dec., Japan	24-SMDIP	Say	-	pdf
LA 1823	LIN-IC	Single-Chip Tuner, FM/AM, Vcc=1.8...6V	24-SDIP	Say	-	pdf
LA 1824	LIN-IC	Single-Chip Tuner, FM/AM, Vcc=1.8..6V Manual Tuning	24-SDIP	Say	-	pdf
LA 1826	LIN-IC	AM/FM-Radio, Stereo-Decoder, Ucc=3V	24-SDIP	Say	-	pdf
LA 1827 M	LIN-IC	Single-Chip Tuner, Vcc=1.8...6V	24-SMDIP	Say	-	pdf
LA 1828	LIN-IC	Single-Chip Tuner, FM/AM, Vcc=2.5...6V Manual Tuning	24-SDIP	Say	-	pdf
LA 1831	LIN-IC	AM/FM-Radio, Stereo-Decoder, Ucc=5V	24-SDIP	Say	-	pdf
LA 1831 M	LIN-IC	SMD, AM/FM-Radio, Stereo-Dec., Ucc=5V	24-SMDIP	Say	-	pdf
LA 1832	LIN-IC	AM/FM-Radio, Stereo-Decoder, Ucc=5V	24-SDIP	Say	-	pdf
LA 1832 M	LIN-IC	SMD, AM/FM-Radio, Stereo-Dec., Ucc=5V	24-SMDIP	Say	-	pdf
LA 1833 N	LIN-IC	SMD, AM/FM Tuner IC, Vcc=4...8V	24-SMDIP	Say	-	pdf
LA 1833 NM	LIN-IC	AM/FM Tuner IC, Home Stereo, Vcc=4...8V	24-SDIP	Say	-	pdf
LA 1835	LIN-IC	AM/FM-Radio, Stereo-Decoder, Ucc=7V	30-SDIP	Say	-	pdf
LA 1835 M	LIN-IC	=LA 1835: SMD	30-SMDIP	Say	-	
LA 1836	LIN-IC	AM/FM-Radio, Stereo-Decoder, Ucc=7V	30-SDIP	Say	-	pdf
LA 1836 M	LIN-IC	=LA 1836: SMD	30-SMDIP	Say	-	
LA 1837	LIN-IC	Single-Chip Home Stereo IC, Vcc=9V	30-SDIP	Say	-	pdf
LA 1838	LIN-IC	Single-Chip Home Stereo IC, Vcc=9V	30-SDIP	Say	-	pdf
LA 1845 N	LIN-IC	Single-Chip Home Stereo IC, Vcc=8V	24-SDIP	Say	-	pdf
LA 1851 N	LIN-IC	Dts Single-Chip AM/FM Tuner, Ucc=7V	30-SDIP	Say	-	pdf pinout
LA 1851 NM	LIN-IC	SMD, DTS, AM/FM Tuner, Ucc=7V	30-SMDIP	Say	-	pdf pinout
LA 1860 M	LIN-IC	SMD, FM-ZF, Stereo-Decoder, Ucc=9V	36-MDIP	Say	-	
LA 1861 M	LIN-IC	SMD, FM Tuner, Car Radio, Vccmax=8V	36-HSMDIP	Say	-	pdf
LA 1862 M	LIN-IC	SMD, FM Tuner, Car Radio, Vccmax=10V	36-HSMDIP	Say	-	pdf

Type	Device	Short Description	Fig.	Manu	Comparision Types	More at
LA 1867 NM	LIN-IC	SMD, FM Tuner, Car Radio, Vccmax=9,2V	44-QFP	Say	-	pdf
LA 1875 M	LIN-IC	SMD, AM/FM Tuner, Car Radio,Vcc=7...10V	36-HSMDIP	Say	-	pdf
LA 1883 M	LIN-IC	FM/AM Tuner chip 7,5 to 9,2V supply	QIP	Say	-	pdf
LA 1888 NM	LIN-IC	SMD, Car Stereo Tuner, Vcc=7,5...9,2V	64-QFP	Say	-	pdf
LA 1895 M	LIN+D/A-IC	SMD, Car Radio Tuner, Vcc=+5V & +8,5V Data Bus Interface, D/A Converter	80-QFP	Say	-	pdf
LA 2000	LIN-IC	Recorder, Audio Level Sensor, 3.5...14V	9-SIP	Say	-	pdf pinout
LA 2000 M	LIN-IC	SMD, Recorder, Audio Level Sensor	8-MDIP	Say	-	pdf pinout
LA 2010	LIN-IC	Recorder, Audio Level Sensor, 3.5...14V	9-SIP	Say	KA 2231	pdf pinout
LA 2030	LIN-IC	Recorder, Audio Level Sensor	16-DIP	Say	-	pdf pinout
LA 2100	LIN-IC	FM Noise-suppr.	16-DIP	Say	-	pdf pinout
LA 2101	LIN-IC	FM Noise-suppr.	16-DIP	Say	-	pdf pinout
LA 2110	LIN-IC	FM Noise-suppr.	16-SQP	Say	KA 2272	pdf pinout
LA 2110 M	LIN-IC	=LA 2110: SMD	20-MDIP	Say	-	pdf pinout
LA 2113	LIN-IC	FM Noise-suppr.	16-DIP	Say	-	pdf pinout
LA 2200	LIN-IC	Ari Senderkennung/station code	16-DIP	Say	-	pdf pinout
LA 2200 M	LIN-IC	SMD, Ari Senderkennung/station code	16-MDIP	Say	-	pdf pinout
LA 2205	LIN-IC	Ari Senderkennung/station code	16-SQP	Say	-	pdf pinout
LA 2210	LIN-IC	ARI Durchsagekenn./inform. code	22-DIP	Say	-	pdf pinout
LA 2211	LIN-IC	Ari Durchsagekenn./inform. code	22-DIP	Say	-	pdf pinout
LA 2220	LIN-IC	Ari Senderkennung/station code	22-SQP	Say	-	pdf pinout
LA 2230	LIN-IC	FM-RDS-Decoder, 57kHz DSB, SD, DK	24-SDIP	Say	-	pdf
LA 2230 M	LIN-IC	SMD, FM-RDS-Decoder, 57kHz DSB, SD, DK	24-SMDIP	Say	-	pdf pinout
LA 2231	LIN-IC	FM-RDS-Decoder, 57kHz DSB, SD, DK	24-SDIP	Say	-	pdf pinout
LA 2231 M	LIN-IC	SMD, FM-RDS-Decoder, 57kHz DSB, SD, DK	24-SMDIP	Say	-	pdf pinout
LA 2232	LIN-IC	FM-RDS-Decoder, Ucc=5V	24-SDIP	Say	-	pdf pinout
LA 2232 M	LIN-IC	SMD, FM-RDS-Decoder, Ucc=5V	24-MDIP	Say	-	pdf pinout
LA 2300	LIN-IC	FM-ZF-Peripherie	22-DIP	Say	-	pdf pinout
LA 2351 M	I/O-IC	SMD,7.5Mbps Automotive Lan Transceiver	30-SMDIP	Say	-	data pinout
LA2400	COMP	Vs:22V Vi0:50mV	8I	Say	-	data pdf pinout
LA 2400	KOP-IC	Uni, 22V, 30mA, -20...+75°	8-SIP	Say	-	pdf pinout
LA 2500	LIN-IC	Arbeitspkt.-gener.f.nf-e(a-class bias)	9-SIP	Say	-	pdf pinout
LA 2600	LIN-IC	Dual, Lautst.-reg./electronic volume	14-DIP	Say	-	pdf pinout
LA 2610	LIN-IC	TV, PC, Analog Surround Processor	12-SIP	Say	-	pdf pinout
LA 2615	LIN-IC	TV, PC, Analog Surround, AViSS(TM)3D	16-DIP	Say	-	pdf pinout
LA 2615 M	LIN-IC	SMD, TV, PC, Analog Surr., AViSS(TM)3D	16-MDIP	Say	-	pdf pinout
LA 2650	LIN-IC	TV, Radio Casette, Bass Boost IC 8 Bit µP-Interface	20-DIP	Say	-	pdf pinout
LA 2655 V	LIN-IC	SMD, Clear Sound Control IC, Vcc=9V	20-SSMDIP	Say	-	pdf pinout
LA 2730	LIN-IC	Dolby-B System, mono	16-DIP	Say	-	pdf
LA 2735	LIN-IC	Dolby-B,C Prozessor	20-DIP	Say	-	pdf
LA 2746	LIN-IC	Dolby-B System, stereo	24-DIP	Say	-	pdf pinout
LA 2775	LIN-IC	Dolby Pro-Logic Noise Sequencer, 12V	14-DIP	Say	-	pdf pinout
LA 2800 N	LIN-IC	Telecom, Anrufbeant./answering Machine	30-SDIP	Say	-	pdf pinout
LA 2805	LIN-IC	Telecom, Anrufbeant./answering machine	24-SDIP	Say	-	pdf pinout
LA 2805 M	LIN-IC	=LA 2805: SMD	24-MDIP	Say	-	pdf pinout
LA 2806 M	LIN-IC	Telecom, Anrufbeant./answering machine	30-SMDIP	Say	-	pdf pinout
LA 2900 M	LIN-IC	SMD, 2 Chan. Line Amplifier, Car Audio	20-MDIP	Say	-	pdf pinout
LA 2901 V	LIN-IC	SMD, 4 Chan. Line Amplifier, Car Audio	24-SSMDIP	Say	-	pdf pinout
LA 2902 V	LIN-IC	SMD, 4 Chan. Line Amplifier, Car Audio	24-SSMDIP	Say	-	pdf pinout
LA 3020	LIN-IC	NF/ZF-Verstärker/amplifier		Say	-	pdf pinout
LA 3100	LIN-IC	2x Nf-v/tr.	14-DIP	Say	-	pdf pinout
LA 3101	LIN-IC	2x Nf-v/tr.	14-DIP	Say	-	pdf pinout
LA 3110	LIN-IC	NF-V, (Ucc=15V)	8-SIP	Say	LA 3120	pdf pinout
LA 3115	LIN-IC	2x NF-V, (Ucc=15V)	14-DIP	Say	LA 3122	pdf
LA 3120	LIN-IC	=LA 3110: Ucc=20V	8-SIP	Say	-	pdf pinout
LA 3122	LIN-IC	=LA 3115: Ucc=20V	14-DIP	Say	-	pdf
LA 3130	LIN-IC	NF-V-ra, (Ucc=36V)	8-SIP	Say	(LA 3150)	pdf pinout
LA 3133	LIN-IC	2x NF-V-ra	14-DIP	Say	-	
LA 3148	LIN-IC	4x NF	16-DIP	Say	-	
LA 3150	LIN-IC	NF-V, ra, (Ucc=15V)	8-SIP	Say	(LA 3130)	pdf pinout
LA 3155	LIN-IC	2x NF-V	16-DIP	Say	-	pdf pinout
LA 3160	LIN-IC	2x NF-V, Ucc=9V	8-SIP	Say	KA 1222, M 51521...22L, MB 3105...06	data pdf pinout
LA3161	OP	Vs:18V Vu:78dB	8I	Say	-	pdf pinout
LA 3161	LIN-IC	=LA 3160 +Stabi	8-SIP	Say	BA 328, KA 2221, M 5152L, TA 7375P	pdf pinout
LA 3170	LIN-IC	Breitbandverst./wide-band amplifier	8-SIP	Say	-	pdf pinout
LA 3180	LIN-IC	Dual, NF-equalizer	16-SQP	Say	-	pdf pinout
LA 3181	LIN-IC	2x Equalizer, Ucc=9V	16-SQP	Say	-	pdf pinout
LA 3190	LIN-IC	Recorder, A/W-Verst., Auto-reverse	18-DIP	Say	-	pdf pinout
LA 3200	LIN-IC	Recorder, NF-V, Alc	14-DIP	Say	-	pdf pinout
LA 3201	LIN-IC	NF-V, Alc	8-DIP	Say	-	pdf pinout
LA 3210	LIN-IC	NF-V, Alc	9-SIP	Say	BA 333, KA 2220, TA 7137	pdf pinout
LA 3220	LIN-IC	Recorder,2x Equalizer, ALC, Ucc=5...13V	14-DIP	Say	KA 2224	pdf pinout
LA 3225(T)	LIN-IC	Recorder, 2x NF-V, ALC	10-SIP	Say	-	
LA 3225 T	LIN-IC	Recorder, 2x NF-V, Alc	10-SIP	Say	-	pdf pinout
LA 3226(T)	LIN-IC	Recorder, 2x NF-V, ALC	10-SIP	Say	-	
LA 3226 T	LIN-IC	Recorder, 2x NF-V, Alc	10-SIP	Say	-	pdf pinout
LA 3230 M	LIN-IC	SMD, 2x NF-V, Ucc=3V	14-MDIP	Say	-	pdf pinout
LA 3235 W	LIN-IC	Recorder, Audio System, Kopfh./headph.	48-MP	Say	-	pdf pinout
LA 3240	LIN-IC	Recorder, 2x NF-V, ALC, 6x el. Switch	24-SDIP	Say	-	pdf
LA 3241	LIN-IC	Recorder, 2x NF-V, ALC, 6x el. Switch	24-SDIP	Say	-	pdf pinout
LA 3242	LIN-IC	Recorder, 2x NF-V, ALC, 6x el. Switch	24-SDIP	Say	-	pdf pinout
LA 3245	LIN-IC	Recorder, NF-V, Auto-reverse	20-DIP	Say	-	pdf
LA 3246	LIN-IC	Stereo Preamp.,Compact Double Cassette	20-DIP	Say	-	pdf pinout
LA 3300	LIN-IC	FM-Dem, Stereo-decoder	14-DIP	Say	LA 3301	

Type	Device	Short Description	Fig.	Manu	Comparision Types	More at
LA 3301	LIN-IC	Fm-dem, Stereo-decoder	14-DIP	Say	-	
LA 3310(N)	LIN-IC	Stereo-decoder	16-DIP	Say		pdf
LA 3311	LIN-IC	Stereo-Decoder, Nf-treiber/driver	16-DIP	Say		
LA 3330	LIN-IC	FM PLL MPX Stereo-decoder	16-DIP	Say	-	
LA 3330 M	LIN-IC	SMD, FM PLL MPX Stereo-decoder	20-MDIP	Say		pdf
LA 3335	LIN-IC	FM PLL MPX Stereo-Decoder, Ucc=3V	9-SIP	Say		pdf
LA 3335 M	LIN-IC	SMD, FM PLL MPX Stereo-Decoder, Ucc=3V	10-SMDIP	Say		pdf
LA 3350(B)	LIN-IC	FM PLL MPX Stereo-decoder	16-DIP	Say	-	pdf
LA 3361	LIN-IC	FM PLL MPX Stereo-decoder, Ucc=6V	16-DIP	Say	-	pdf
LA 3365	LIN-IC	FM PLL MPX Stereo-Decoder, Ucc=6V	16-SQP	Say	AN7410, BA1330, HA11227, KA2261, TA7604	pdf
LA3366	LED	am ø3mm I:180...450mcd Vf:2V If:20mA	T1	Osr	-	pdf
LA 3370	LIN-IC	FM PLL MPX Stereo-Decoder, Car Stereo	16-SQP	Say	-	data pdf pinout
LA 3373	LIN-IC	FM PLL MPX Stereo-Decoder, Car Stereo	16-DIP	Say	KA 2262	pdf
LA 3375	LIN-IC	PLL FM MPX Stereo-Decoder, Car Stereo	16-SQP	Say	-	pdf
LA 3376	LIN-IC	FM PLL MPX Stereo-Decoder, Car Stereo	16-SQP	Say	KA 2266	pdf
LA 3376 M	LIN-IC	SMD, FM PLL MPX Stereo-Dec.,Car Stereo	MDIP	Say	-	pdf pinout
LA 3380(A)	LIN-IC	Stereo-Decoder	20-DIP	Say	LA 3381	pdf pinout
LA 3380	LIN-IC	PLL FM Stereo-decoder	20-DIP	Say	LA 3381	
LA 3380 A	LIN-IC	PLL FM Stereo-decoder	20-DIP	Say	LA 3381	pdf
LA 3381	LIN-IC	PLL FM Stereo-decoder	20-DIP	Say		pdf
LA 3390	LIN-IC	FM PLL MPX Stereo-decoder	20-DIP	Say		pdf
LA 3400(N)	LIN-IC	FM PLL MPX Stereo-Decoder, HiFi	22-DIP	Say		pdf
LA 3400	LIN-IC	FM PLL MPX Stereo-Decoder, HiFi	22-DIP	Say		
LA 3400 N	LIN-IC	FM PLL MPX Stereo-Decoder, HiFi	22-DIP	Say		pdf
LA 3401	LIN-IC	FM PLL MPX Stereo-Decoder, HiFi	22-SDIP	Say		pdf
LA 3410	LIN-IC	FM PLL MPX Stereo-Decoder, HiFi	16-DIP	Say	KA 2265	pdf
LA 3430	LIN-IC	FM PLL MPX Stereo-decoder	16-SQP	Say		pdf
LA 3430 M	LIN-IC	=LA 3430: SMD	16-MDIP	Say		pdf
LA 3433	LIN-IC	FM PLL MPX Stereo-decoder	16-DIP	Say		pdf
LA 3440	LIN-IC	FM/FMX Stereo-Decoder, Ucc=7,5...10V	30-SDIP	Say		pdf
LA 3450	LIN-IC	Stereo-Decoder, VCO, Ucc=13V	28-SDIP	Say		pdf
LA 3510 M	LIN-IC	Recorder, Verst./amplifier system, 3V	20-MDIP	Say		pdf
LA 3550 M	LIN-IC	Recorder, Kopfh./headph. Amp.,Ucc=1,5V	14-SMDIP	Say		pdf
LA 3600	LIN-IC	5-Band Graphic Equalizer, Ucc=8V	16-DIP	Say		pdf
LA 3605	LIN-IC	5-Band Graphic Equalizer, Ucc=8V	16-DIP	Say	KA 2223, M 5226P, TA 7796P	pdf pinout
LA 3607	LIN-IC	7-Band Graphic Equalizer	20-DIP	Say	-	pdf
LA 3607 M	LIN-IC	7-Band Graphic Equalizer, Ucc=8V	20-DIP	Say	-	pdf pinout
LA 3610 M	LIN-IC	=LA 3607: SMD	20-MDIP	Say	-	pdf pinout
LA 3800	LIN-IC	5-Band Graphic Equalizer	16-MDIP	Say		pdf
LA 3801	LIN-IC	TV, MPX-demodulator (japan-system)		Say		
LA 3805	LIN-IC	TV, Mpx-demodulator (japan-system)		Say		
LA 4000	LIN-IC	TV, Mpx-demodulator (japan-system)	30-SDIP	Say		
	LIN-IC	NF-V/E, 3,5W(13V/4Ω)		Say		
LA 4030(P)	LIN-IC	=LA 4032: 16V, 1A, 1W(11V/8Ω)	8-DIP+a	Say	LA 4031, LA 4032	
LA 4031(P)	LIN-IC	=LA 4032: 18V, 1,5A, 2W(13V/4Ω)	8-DIP+a	Say	LA 4032	
LA 4032(P)	LIN-IC	NF-E, 25V, 1,5A, 3W(18V/8Ω)	8-DIP+a	Say	(LA 4200, LA 4201)[10]	
LA 4033	LIN-IC	NF-E, Ucc=19...23V, 5W		Say		
LA 4050 P	LIN-IC	AF/NF-Amp, 16V, 1A, 1W(11V/8Ohm)	8-DIP+a	Say	LA 4051	
LA 4051 P	LIN-IC	AF/NF-Amp, 23V, 1,5A, 2,5W(17V/8Ohm)	8-DIP+a	Say	(LA 4200, LA 4201)[10]	pdf
LA 4070	LIN-IC	Telefon, NF-E, 0,5W(9V/8Ω)	14-DIP+g	Say	-	pdf
LA 4100	LIN-IC	Power Amp, 9V, 1W(6V/4Ohm)	14-DIP+g	Say	-	pdf
LA 4101	LIN-IC	Power Amp, 11V, 1.5W(7.5V/4Ohm)	14-DIP+g	Say	LA 4101, LA 4102	pdf pinout
LA 4102	LIN-IC	Power Amp, 13V, 2.1W(9V/4Ohm)	14-DIP+g	Say	LA 4102	pdf pinout
LA 4110	LIN-IC	=LA 4112: 1W(6V/4Ω)	14-DIP+g	Say	-	pdf pinout
LA 4112	LIN-IC	NF/AF-Amp, 2.3W(9V/4Ohm)	DIP_g	Say	LA 4112	pdf
LA 4120	LIN-IC	=LA 4125T: 2x1W(6V/2...8Ω)	20-DILP	Say	-	pdf
LA 4125	LIN-IC	=LA 4125T: 2x2,4W(9V/2...8Ω)	20-DILP	Say	LA 4125(T), LA 4126(T)	pdf
LA 4125 T	LIN-IC	2x NF-E, 2x4,2W(12V/4...8Ω)	20-DILP	Say	LA 4126(T)	pdf
LA 4126	LIN-IC	2x NF/AF-Amp, 2x2,4W(9V/4Ohm)	DILP	Say	LA 4126 T	pdf
LA 4126 T	LIN-IC	2x NF/AF-Amp, 2x4,2W(12V/4Ohm)	DILP	Say	LA 4125(T)	pdf
LA 4135	LIN-IC	NF-E, 9V, 2,5W	14-DIP+g	Say	LA 4125 T	pdf
LA 4137	LIN-IC	NF-E, 11V, 1,8W(7,5V/3,2Ω)	14-DIP+g	Say	-	pdf
LA 4138	LIN-IC	=LA 4137: 13V, 2,7W(9V/3,2Ω)	14-DIP+g	Say	LA 4138	pinout
LA 4140	LIN-IC	NF-E, 0,5W(6V/8Ω)	9-SIP	Say	-	pdf pinout
LA 4142(A)	LIN-IC	NF-E, 0,5W(6V/8Ω)	9-SIP	Say	AN 7112, KA 2212, TA 7313(AP)	pdf
LA 4145	LIN-IC	Radio Cassette Recorder, AF PAmp 9V, 0.5A, 0.9W(6V/4Ohm), Icc0=10mA	9-SIP	Say	LA 4146	pdf
LA 4146	LIN-IC	Radio Cassette Recorder, AF PAmp 9V, 0.5A, 0.9W(6V/4Ohm), Icc0=5mA	9-SIP	Say	(LA 4145)	pdf
LA 4147	LIN-IC	Radio Cassette Recorder, AF PAmp 9V, 0.5A, 0.9W(6V/4Ohm)	9-SIP	Say	-	pdf
LA 4160	LIN-IC	AF/NF-PreAmp, AF/NF-Amp, ALC, 13V, 2,2W(9V/4Ohm)	14-DIP+g	Say	KA 2213	pdf pinout
LA 4162	LIN-IC	AF/NF-PreAmp,AF/NF-Amp, ALC, 11V, 0,5W(6V/8Ohm)	16-DIP	Say	-	pdf
LA 4165 M	LIN-IC	Recorder, AF/ NF- Pre- / Amp, ALC, 0, 2W(3V/ 4Ohm)	24-MDIP	Say	-	pdf
LA 4166 M	LIN-IC	Recorder, AF/ NF- Pre- / Amp, ALC, 0, 2W(3V/ 4Ohm)	24-MDIP	Say	-	pdf
LA 4167 M	LIN-IC	Recorder, AF/ NF- Pre- / Amp, ALC, 2x0, 12W(3V/ 4Ohm)	30-SMDIP	Say	-	pdf
LA 4168 M	LIN-IC	Recorder, AF/ NF- Pre- / Amp, ALC, 2x0, 12W(3V/ 4Ohm)	30-SMDIP	Say	-	pdf
LA 4170	LIN-IC	2x NF-E(Kopfh./headphone),14V/8...200Ω	14-DIP+g	Say	(LA 4175)	
LA 4175	LIN-IC	2x NF-E(Kopfh./headphone),12V/39Ω	14-DIP+g	Say	(LA 4170)	
LA 4177	LIN-IC	2x NF-E(Kopfh./headphone), 6V/32Ω	14-DIP+h	Say	(LA 4175)	

Type	Device	Short Description	Fig.	Manu	Comparision Types	More at
LA 4178	LIN-IC	2x NF-E(Kopfh./headphone), 6V/32Ω	14-DIP+h	Say	-	
LA 4180	LIN-IC	=LA 4182: 2x1W(6V/4Ω)	12-DIP+b	Say	LA 4182, TEA 2025	
LA 4182	LIN-IC	2x NF-E, 2x2,3W(9V/4Ω)	12-DIP+b	Say	KA 2206, KA 22061, LA 4183, TEA 2025	pdf pinout
LA 4183	LIN-IC	=LA 4182	12-DIP+b	Say	KA 2206, KA 22061, LA 4182, TEA 2025	pdf pinout
LA 4185	LIN-IC	2x NF-E, 13V, 3A, 2x2,4W(9V/4Ω)	14-SILP	Say	LA 4185T	pdf pinout
LA 4185 T	LIN-IC	=LA 4185: 18V, 3A, 2x4,2W(12V/4Ω)	14-SILP	Say		pdf pinout
LA 4190	LIN-IC	=LA 4192: 9V, 2x1W(6V/4Ω)	12-DIP+b	Say	LA 4192	pdf
LA 4192	LIN-IC	2x Preamp+Power Amp, 11V, 2x2.3W(9V/4Ohm)	12-DIP+b	Say	-	pdf pinout
LA 4195	LIN-IC	2x NF-E, 13V, 2,25A, 2x2,4W(9V/4Ω)	20-DILP	Say	LA 4195T	pdf
LA 4195 T	LIN-IC	=LA 4195: 18V, 2,25A, 2x4,2W(12V/4Ω)	20-DILP	Say		
LA 4200	LIN-IC	=LA 4201: 16V, 1,5W(11V/8Ω)	14-DIP+g	Say	LA 4201	pdf
LA 4201	LIN-IC	NF-E, 20V, 2,5W(14V/8Ω)	14-DIP+g	Say	-	
LA 4210	LIN-IC	NF-E, 22V, 3,5W(16V/8Ω)	12-SIL	Say	(LA 4220)[10]	pdf
LA 4220	LIN-IC	NF-E, 22V, 3,5W(16V/8Ω)	10-SIL	Say	-	pdf
LA 4227	LIN-IC	2-channel AF amplifier 2.5W typ., standby SW function	12-DIP	Say	-	pdf pinout
LA 4230	LIN-IC	NF-E, 31V, 1,95A, 6W(22V/8Ω)	12-SIL	Say	LA 4250	
LA 4250	LIN-IC	=LA 4230: 35V, 2,2A, 8W(25V/8Ω)	12-SIL	Say	-	pdf pinout
LA 4260	LIN-IC	Home Stereos and Music Centers, 2x AF PAmp 22V, 2A, 2x 2.5W(14V/8Ohm)	10-SIL	Say	LA 4261	pdf
LA 4261	LIN-IC	Home Stereos and Music Centers, 2x AF PAmp 25V, 2A, 2x 3.5W(16V/8Ohm)	10-SIL	Say		pdf
LA 4265	LIN-IC	NF-E, 25V, 2A, 3,5W(16V/8Ω)	10-SIL	Say	-	pdf
LA 4266	IC	NF/AF PAmp, 25V, 1,5A, 3W(16V/8Ohm), mute, therm.+ overvolt. protection	SIL	Say	LA4267, LA4268	pdf pinout
LA 4267	IC	NF/AF PAmp, 35V, 2,5A, 5W(20V/8Ohm), mute, therm.+ overvolt. protection	SIL	Say	LA4268	pdf pinout
LA 4268	IC	NF/AF PAmp, 35V, 2,5A, 10W(28V/8Ohm), mute, therm.+ overvolt. protection	SIL	Say	-	pdf pinout
LA 4270	LIN-IC	2x NF-E, 35V, 3,5A, 2x6W(25V/8Ω)	10-SILP	Say	-	pdf
LA 4275	LIN-IC	NF-E, 25V, 3,5A, 6W(25V/8Ω)	7-SIL	Say	-	pdf
LA 4276	IC	2x NF/AF PAmp, 25V, 2x3W(16V/8Ohm), mute, therm.+ overvolt. protection	SIL	Say	LA4277, LA4278	pdf
LA 4277	IC	2x NF/AF PAmp, 25V, 2x5W(20V/8Ж), mute, therm.+ overvolt. protection	SIL	Say	LA4278	pdf pinout
LA 4278	IC	2x NF-E, 35V, 2.5A, 2x10W(28V/8Ohm), mute, therm.+ overvolt. protection	SILP	Say	-	pdf pinout
LA 4280	LIN-IC	2x NF-E, 32V, 4A, 2x10W(32V/8Ω)	14-SILP	Say	-	pdf pinout
LA 4282	LIN-IC	2x NF-E, 32V, 2x10W(32V/8Ω)	12-SILP	Say	-	pdf pinout
LA 4400	LIN-IC	NF-E, 18V, 4,5W(13V/4Ω)	12-SIL	Say	-	
LA 4410	LIN-IC	NF-E, 18V, 3,5W(13V/4Ω)	12-SIL	Say	-	pdf
LA 4420	LIN-IC	NF-E, 18V, 5,5W(13V/4Ω)	10-SIL	Say	-	pdf
LA 4422	LIN-IC	NF-E, 18V, 5,8W(13V/4Ω)	10-SIL	Say	-	pdf
LA 4425 A	LIN-IC	NF-E, 18V, 3,3A, 5W(13,2V/4Ω)	14/5Pin	Say	-	
LA 4430	LIN-IC	=LA 4420: 4,5W(13V/4Ω)	10-SIL	Say	LA 4420	
LA 4440	LIN-IC	2x NF-E, 18V, 2x6W(13V/4Ω)	14-SILP	Say	-	pdf
LA 4445	LIN-IC	2x NF-E, 18V, 3,5A, 2x5,5W(13V/4Ω)	12-SILP	Say	KA 2210	pdf
LA 4446	LIN-IC	2x NF-E, 18V, 3,5A, 2x5,5W(13V/4Ω)	13-SIL	Say	-	pdf
LA 4450	LIN-IC	2x Power Amp, 37V 4A, 2x20W(26.4V/4Ohm)	14-SILP	Say	-	pdf pinout
LA 4460(N)	LIN-IC	NF-E, 18V, 4,5A, 12W(13V/4Ω)	10-SILP	Say	-	pdf
LA 4461(N)	LIN-IC	=LA4460: spiegelb.Pinbel./rev.pinning	10-SILP	Say	-	pdf
LA 4465	LIN-IC	NF-E, 18V, 4A, 12W(13V/4Ω)	10-SILP	Say	-	pdf
LA 4466	LIN-IC	=LA4465: spiegelb.Pinbel./rev.pinning	10-SILP	Say	-	pdf
LA 4467	LIN-IC	NF-E, 18V, 4A, 12W(13V/4Ω)	12-SILP	Say	-	pdf
LA 4468	LIN-IC	=LA4467: spiegelb.Pinbel./rev.pinning	12-SILP	Say	-	
LA 4470	LIN-IC	NF-E, 18V, 4A, 20W(13V/4Ω)	14-SILP	Say	LA 4490	pdf
LA 4471	LIN-IC	=LA4470: spiegelb.Pinbel./rev.pinning	14-SILP	Say	LA 4491	pdf
LA 4475	LIN-IC	NF-E, 18V, 4A, 12W(20V/4Ω)	14-QILP	Say	LA 4495, LA 4497	pdf
LA 4476	LIN-IC	=LA4475: spiegelb.Pinbel./rev.pinning	14-QILP	Say	LA 4496, LA 4498	pdf
LA 4480	LIN-IC	2x NF-E, 18V, 4A, 2x4W(13V/4Ω)	12-SILP	Say	-	
LA 4485	LIN-IC	2x NF-E, 16V, 3,3A, 2x5W(13V/4Ω)	13-SILP	Say	-	
LA 4490	LIN-IC	=LA4470: + Standby (Pin 9)	14-SILP	Say	-	
LA 4491	LIN-IC	=LA4490: spiegelb.Pinbel./rev.pinning	14-SILP	Say	-	
LA 4495	LIN-IC	=LA 4475: + Standby (Pin9)	14-QILP	Say	-	pdf
LA 4496	LIN-IC	=LA4495: spiegelb.Pinbel./rev.pinning	14-QILP	Say	-	pdf
LA 4497	LIN-IC	=LA4475: + Standby (Pin 9)	14-QILP	Say	-	
LA 4498	LIN-IC	=LA4497: spiegelb.Pinbel./rev.pinning	14-QILP	Say	-	
LA 4500	LIN-IC	2x NF-E, 24V, 2,5A, 2x5,3W(12V/3Ω)	20-DILP	Say	LA 4505	pdf
LA 4505	LIN-IC	=LA 4500: 2x8,5W(15V/3Ω)	20-DILP	Say	-	pdf
LA 4507	LIN-IC	2x NF-E, 24V, 2,5A, 2x4,5W(16V/8Ω)	14-SILP	Say	-	
LA 4508	LIN-IC	2x NF-E, 24V, 2,5A, 2x8,5W(15V/3Ω)	14-SILP	Say	-	
LA 4510	LIN-IC	Headph. Power Amp, 3V...4.5V 0.57A, 0.24W(3V/4Ohm)	9-SIP	Say	-	pdf pinout
LA 4520	LIN-IC	Recorder, 2x Verst./ampl., Ucc=4,5...6V	20-DIP	Say	-	
LA 4525	LIN-IC	NF-E, 15V, 1W(9V/8Ω)	8-DIP	Say	-	
LA 4530 M	LIN-IC	=LA 4530S: SMD	20-MDIP	Say	-	
LA 4530 S	LIN-IC	2x NF-E, 6V, 0,2A, 2x36mW(3V/32Ω)	16-SQP	Say	-	
LA 4533 M	LIN-IC	Kopfh.-Vst./Headphone Ampl., Ucc=3V	8-SMDIP	Say	-	
LA 4534 M	LIN-IC	Kopfh.-Vst./Headphone Ampl., Ucc=3V	8-SMDIP	Say	-	
LA 4535 M	LIN-IC	Kopfh.-Vst./Headphone Ampl., Ucc=1,5V	8-MDIP	Say	-	
LA 4537 M	LIN-IC	Kopfh.-Vst./Headphone Ampl., Ucc=1,5V	8-SMDIP	Say	-	
LA 4538 M	LIN-IC	Kopfh.-Vst./Headphone Ampl., Ucc=1,5V	14-SMDIP	Say	-	pdf
LA 4540	LIN-IC	NF-E, 15V, 18W(15V/4Ω)		Say		pdf
LA 4550	LIN-IC	2x NF-E, 13V, 2x1W(9V/4Ω)	12-DIP+b	Say	LA 4555	pdf

Type	Device	Short Description	Fig.	Manu	Comparision Types	More at
LA 4555	LIN-IC	=LA 4500: 2x2,3W(9V/4Ω)	12-DIP+b	Say	-	
LA 4557	LIN-IC	2x NF-E, 12V, 2x2,4W(9V/4Ω)	12-DIP+b	Say	-	
LA 4558	LIN-IC	2x NF-E, 12V, 2x2,4W(9V/4Ω)	12-DIP+b	Say	-	
LA 4560 M	LIN-IC	Recorder, Verstärker/Amp., Ucc=4,5...6V	24-MDIP	Say	-	pdf
LA 4570	LIN-IC	Kopfh.-Vst./Headphone Amp., Ucc=3V	16-DIP	Say	-	
LA 4570 M	LIN-IC	LA 4570: SMD	16-MDIP	Say	-	
LA 4571 MB	LIN-IC	Kopfh.-Vst./Headphone Amp., Ucc=3V	20-MDIP	Say	-	
LA 4575	LIN-IC	Kopfh.-Vst./Headphone Ampl., Ucc=3V	16-DIP	Say	-	
LA 4575 M	LIN-IC	LA 4575: SMD	16-MDIP	Say	-	
LA 4580 M	LIN-IC	Kopfh.-Vst./Headphone Ampl., Ucc=3V	20-MDIP	Say	-	pdf
LA 4581 MB	LIN-IC	Kopfh.-Vst./Headphone Amp., Ucc=3V	24-SMDIP	Say	-	pdf
LA 4582 CM,NM	LIN-IC	Kopfh.-Vst./Headphone Amp., Ucc=3V	36-MP	Say	-	
LA 4582 NM	LIN-IC	Kopfh.-Vst./Headphone Amp., Ucc=3V	36-MP	Say	-	
LA 4583 M	LIN-IC	Recorder, Audio System, Kopfh./headph.	44-MP	Say	-	
LA 4590 W	LIN-IC	Audio System, Kopfh./headph. Amp.,1,5V	48-MP	Say	-	
LA 4597	LIN-IC	2x NF-E, 18V, 2x2,9W(9V/3,2Ω)	13-SIL	Say	-	pdf pinout
LA 4598	LIN-IC	2x NF-E, 18V, 2x2,9W(9V/3,2Ω)	14-QILP	Say	-	
LA 4620	LIN-IC	2x NF-E, 24V, 2x17W(15V/4Ω)	23-QILP	Say	-	
LA 4630 N	LIN-IC	2x NF-E, 20V, 2x5W(12V/3Ω)	18-SILP	Say	-	pdf
LA 4631	LIN-IC	5W 2-Ch. Audio AF Power Amp.	SIL	Say	-	pdf pinout
LA 4700(N)	LIN-IC	2x NF-E, 18V, 4A, 2x12W(13V/4Ω)	18-SILP	Say	-	pdf pinout
LA 4705	LIN-IC	2x NF-E, 24V, 2x15W(13V/4Ω)	18-SILP	Say	-	pdf
LA 4742		4x NF-E, 26V, 4.5A, 4x40W(13.7V/4Ж BTL)	SILP	Say		pdf
LA 4800 V	LIN-IC	Kopfh.-Vst./Headphone Amp., Ucc=3V	16-SSMDIP	Say		
LA 4805 V	LIN-IC	Kopfh.-Vst./Headphone Amp., Ucc=3V	30-SSMDIP	Say		
LA 4810 M	IC	Stereo/monaural BTL Power Amplifier	DIP	Say		pdf pinout
LA5002	R+	Io=60mA Vo:1.85...2.15V Vin:4V	8S,4P	Say		data pdf pinout
LA 5002	Z-IC	Io-drop, +2V, 0,06A, 5%, -20...+80°	15/4Pin	Say		data pinout
LA 5002 M	Z-IC	=LA 5002: SMD	8-MDIP	Say		data pinout
LA5003	R+	Io=60mA Vo:2.8...3.2V Vin:4V P:560mW	8S,4P	Say		data pdf pinout
LA 5003	Z-IC	Io-drop, +3V, 0,06A, 5%, -20...+80°	15/4Pin	Say	-	data pinout
LA 5003 M	Z-IC	=LA 5003: SMD	8-MDIP	Say		data pdf pinout
LA5004	R+	Io=60mA Vo:3.75...4.25V Vin:5V	8S,4P	Say		data pdf pinout
LA 5004	Z-IC	Io-drop, +4V, 0,06A, 5%, -20...+80°	15/4Pin	Say		data pinout
LA 5004 M	Z-IC	=LA 5004: SMD	8-MDIP	Say		data pinout
LA5005	R+	Io=60mA Vo:4.75...5.25V Vin:6V	8S,4P	Say		data pdf pinout
LA 5005	Z-IC	Io-drop, +5V, 0,06A, 5%, -20...+80°	15/4Pin	Say	-	data pinout
LA 5005 M	Z-IC	=LA 5005: SMD	8-MDIP	Say		data pinout
LA 5006	Z-IC	+6V, 0,06, 5%, -20...+80°	15/4Pin	Say		data pdf pinout
LA5006M	R+	Io=60mA Vo:5.7...6.3V Vin:7V P:300mW	8S	Say		data pinout
LA 5006 M	Z-IC	=LA 5006: SMD	8-MDIP	Say		data pinout
LA 5008	Z-IC	+8V, 0,06, 5%, -20...+80°	15/4Pin	Say		data pinout
LA5008M	R+	Io=60mA Vo:7.6...8.4V Vin:9V P:300mW	8S	Say		data pinout
LA 5008 M	Z-IC	=LA 5008: SMD	8-MDIP	Say	-	data pinout
LA 5009	Z-IC	+9V, 0,06, 5%, -20...+80°	15/4Pin	Say		data pinout
LA5009M	R+	Io=60mA Vo:8.55...9.45V Vin:10V	8S	Say		data pinout
LA 5009 M	Z-IC	=LA 5009: SMD	8-MDIP	Say	-	data pinout
LA 5010	Z-IC	+10V, 0,06, 5%, -20...+80°	15/4Pin	Say		data pinout
LA5010M	R+	Io=60mA Vo:9.4...10.6V Vin:11V	8S	Say		data pinout
LA 5010 M	Z-IC	=LA 5010: SMD	8-MDIP	Say		data pinout
LA 5100	LIN-IC	Ctv, 12V-Stromversorg./power supply		Say		
LA 5110	SMPS-IC	Ctv, Smps-controller	9-SIP	Say		pdf
LA 5112(N)	SMPS-IC	Ctv, Smps-controller	9-SIP	Say		pdf
LA 5200	LIN-IC	24V-Stromversorgung/power supply		Say		pdf
LA 5310 M	LIN-IC	Spg.-Teiler/Volt. Div. f. LCD-matrix	8-MDIP	Say		pdf
LA 5311 M	LIN-IC	Variable Divided Voltage Generator for LCD	8-MDIP	Say	-	pdf pinout
LA 5315 M	LIN-IC	Stromvers./power Supply f. Lcd-matrix	20-MDIP	Say		pdf
LA 5316 M	LIN-IC	Stromvers./power Supply f. Lcd-matrix	20-MDIP	Say		pdf
LA 5317 M	LIN-IC	Stromvers./power Supply f. Lcd-matrix	30-MDIP	Say		pdf
LA 5318 M	LIN-IC	Stromvers./power Supply f. Lcd-matrix	16-MDIP	Say		pdf
LA 5511	LIN-IC	Motorregler/dc motor speed control	85/4Pin	Say	(LA 5512)	pdf
LA 5512	LIN-IC	Motorregler/dc motor speed control	85/4Pin	Say	(LA 5511)	pdf
LA 5515	LIN-IC	Motorregler/dc motor speed control		Say		pdf
LA 5521(D)	LIN-IC	Motorregler/dc motor speed control	8-DIP	Say	KA 2402	pdf pinout
LA 5521 M	LIN-IC	=LA 5521(D): SMD	8-MDIP	Say		pdf pinout
LA 5522	LIN-IC	Motorregler/dc motor speed control	85/5Pin	Say		pdf
LA 5523	LIN-IC	Motorregler/dc motor speed control	5-SIP	Say		pdf
LA 5524	LIN-IC	Motorregler/dc motor speed control	8-DIP	Say		pdf
LA 5524 M	LIN-IC	=LA 5524: SMD	8-MDIP	Say		pdf
LA 5525	LIN-IC	Motorregler/dc motor speed control	85/5Pin	Say		pdf
LA 5527	LIN-IC	Motorreg./DC Speed Ctrl., Ucc=1,8...10V	6-DIP	Say		pdf
LA 5527 M	LIN-IC	=LA 5527: SMD	8-MDIP	Say		pdf
LA 5528(N)	LIN-IC	Motorreg./DC Speed Ctrl., Ucc=1,8...10V	8-DIP	Say		pdf
LA 5528(N)M	LIN-IC	=LA 5528: SMD	8-MDIP	Say		pdf
LA 5536	LIN-IC	Motorregler/dc Motor Speed Control	85/5Pin	Say		pdf
LA 5536 N	LIN-IC	=LA 5536: Fig.→	85/4Pin	Say		pdf
LA 5537(N)	LIN-IC	Motorregler/dc Motor Speed Control	5-SIP	Say		pdf
LA 5540	LIN-IC	Motorreg./motor dr., regulator, brake	85/4Pin	Say		pdf
LA 5550	LIN-IC	Motorregler/dc Motor Controller	16-DIP	Say		pdf
LA 5550 M	LIN-IC	=LA 5550: SMD	20-MDIP	Say		pdf
LA 5555	LIN-IC	Motorreg./DC Motor Controller, Ucc=3V	22-SDIP	Say		pdf pinout
LA 5562 D	LIN-IC	Motorregler/dc Motor Speed Controller		Say		pdf
LA 5562 S	LIN-IC	Motorregler/dc Motor Speed Controller	5-SIP	Say		pdf
LA 5586	LIN-IC	Motorregler/dc Motor Speed Controller	30/5-Pin	Say		pdf

Type	Device	Short Description	Fig.	Manu	Comparision Types	More ...
LA 5587	LIN-IC	Motorregler/dc Motor Speed Controller	5-SIP	Say	-	pdf
LA 5588	LIN-IC	Motorregler/dc Motor Speed Controller	5-SIP	Say	-	pdf
LA 5601	Z-IC	Vc, Reset, 5,2V, 250mA	10-DIP	Say	-	pdf
LA 5603	Z-IC	Reset, 2x 5,2V, ±7,5, 4,8V	18-SILP	Say	-	pdf
LA 5605	Z-IC	Cd, 7,5V/0,5A, 9V/60mA	13-SIL	Say	-	pdf
LA 5606(N)	Z-IC	BS/CS Tuner, 15,7V, 12V, 9V, 5V	14-SILP	Say	-	
LA 5606	Z-IC	Bs/cs Tuner, 15,7V, 12V, 9V, 5V	14-SILP	Say	-	pdf
LA 5607(N)	Z-IC	BS/CS Tuner, 15,7V, 12V, 9V, 5V	14-SILP	Say	-	
LA 5607	Z-IC	Bs/cs Tuner, 15,7V, 12V, 9V, 5V	14-SILP	Say	-	pdf
LA 5630	Z-IC	→L 5630	7j	Say	→L 5630	
LA 5640(N)	Z-IC	Stabi/voltage Regulator f. LCDs	7b	Say	-	pdf
LA 5655	Z-IC	Stabi/voltage regulator f. Flt	10-SIL	Say	-	pdf
LA 5658	Z-IC	30V/0,05A, 15,5V/0,35A, 5,9V/0,1A	12-SILP	Say	-	pdf
LA 5659	Z-IC	+5V/1A, +5V/35mA, Logic Control	86/5Pin	Say	-	pdf
LA 5660 M	LIN-IC	Single-ch Switching Regulator IC	MDIP	Say	-	pdf pinout
LA 5665	Z-IC	Reset, +15,5V/350mA, +5,6V/100mA	7-SIL	Say	-	pdf
LA 5666	Z-IC	+13,0V/350mA, +5,6V/100mA	7-SIL	Say	-	pdf
LA 5667	Z-IC	+13,0V/350mA, +5,6V/100mA	7-SIL	Say	-	pdf
LA 5668	Z-IC	+5,5V/100rnA, +5,0V/100mA	10-SIL	Say	-	pdf
LA 5670 M	SMPS-IC	SMD, TV, Schaltregler/switching reg.	16-MDIP	Say	-	pdf
LA 5677 M	SMPS-IC	SMD, Dual, S-Reg Control, 3,6...18V	16-MDIP	Say	-	pdf
LA 5685 N	Z-IC	3x 8,5V, 5,2V, 5,1V	10-SIL	Say	-	pdf
LA 5690 D	Z-IC	Reset, Watchdog, 5V	8-DIP	Say	-	pdf
LA 5690 S	Z-IC	=LA 5690D: Fig.→	9-SIP	Say	-	pdf
LA 5691 D	Z-IC	Reset, Watchdog, 5V	8-DIP	Say	-	pdf
LA 5691 S	Z-IC	=LA 5691D: Fig.→	9-SIP	Say	-	pdf
LA 5692 D	Z-IC	Reset, Watchdog, 5V	8-DIP	Say	-	pdf
LA 5692 M	Z-IC	=LA 5692D: SMD	8-MDIP	Say	-	pdf
LA 5692 S	Z-IC	=LA 5692: Fig.→	9-SIP	Say	-	pdf
LA 5693 D	Z-IC	Reset, Watchdog, 5V	8-DIP	Say	-	pdf
LA 5693 M	Z-IC	=LA 5693D: SMD	8-MDIP	Say	-	pdf
LA 5693 S	Z-IC	=LA 5693D: Fig.→	9-SIP	Say	-	pdf
LA 5700	LIN-IC	Abstimmspg.-erz./tuning volt. supply	14-DIP+g	Say	-	pdf pinout
LA6082	2xJFET	Vs:±18V Vu:200V/mV Vo:±12V Vi0:5mV	8DS,9I	Say	-	data pdf pinout
LA 6082 D	OP-IC	Dual, JFET, Uni, ±18V, -30...+85°	8-DIP	Say	→Serie 080	data pdf pinout
LA 6082 M	OP-IC	=LA 6082D: SMD	8-MDIP	Say	-	data pdf pinout
LA 6082 S	OP-IC	=LA 6082D: Fig.→	9-SIP	Say	-	data pdf pinout
LA6083	2xJFET	Vs:±18V Vu:200V/mV Vo:±12V Vi0:5mV	14DS	Say	-	data pdf pinout
LA 6083 D	OP-IC	Dual, JFET, Uni, ±18V, -30...+85°	14-DIP	Say	→Serie 080	data pdf pinout
LA 6083 M	OP-IC	=LA 6083D: SMD	14-MDIP	Say	-	data pdf pinout
LA6324	4xHI-PERF	Vs:32V Vu:100V/mV Vo:<30.5V ViJ:±2mV	14DS	Say	-	data pdf pinout
LA 6324(N)	OP-IC	Quad, Serie 124, 32V, -30...+85°	14-DIP	Say	→Serie 124	data pdf pinout
LA 6324(N)M	OP-IC	=LA 6324: SMD	14-MDIP	Say	-	data pdf pinout
LA6339	4xCOMP	Vs:36V Vo:<34.5V Vi0:±2mV	14DS	Say	-	data pdf pinout
LA 6339	KOP-IC	Quad, Serie 139, 36V, -30...+85°	14-DIP	Say	→Serie 139	data pinout
LA 6339 M	KOP-IC	=LA 6339: SMD	14-MDIP	Say.	-	data pinout
LA6358	2xHI-PERF	Vs:32V Vu:100V/mV Vo:<30.5V Vi0:±2mV	8DS,9I	Say	-	data pdf pinout
LA 6358(N)	OP-IC	Dual, Serie 158, 32V, -30...+85°	8-DIP	Say	→Serie 158	data pdf pinout
LA 6358(N)M	OP-IC	=LA 6358: SMD	8-MDIP	Say	-	data pdf pinout
LA 6358(N)S	OP-IC	=LA 6358: Fig.→	9-SIP	Say	-	data pdf pinout
LA6393	2xCOMP	Vs:36V Vo:<34.5V Vi0:±1mV	8DS,9I	Say	-	data pdf pinout
LA 6393 D	KOP-IC	Dual, 36V, -30...+85°	8-DIP	Say	→Serie 193	data pinout
LA 6393 M	KOP-IC	=LA 6393: SMD	8-MDIP	Say	-	data pinout
LA 6393 S	KOP-IC	=LA 6393D: Fig.→	9-SIP	Say	-	data pinout
LA6458	2xHI-PERF	Vs:±18V Vo:±13V Vi0:0.5mV	8DS,9I	Say	-	data pinout
LA 6458 D	OP-IC	Dual, Serie 158, ±18V, -20...+75°	8-DIP	Say	→Serie 158	data pinout
LA 6458 M	OP-IC	=LA 6458: SMD	8-MDIP	Say	-	data pinout
LA 6458 S	OP-IC	=LA 6458D: Fig.→	9-SIP	Say	-	data pinout
LA6462	2xHI-PERF	Vs:±18V Vo:±13V Vi0:0.3mV f:6Mc	8DS,9I	Say	-	data pinout
LA 6462 D	OP-IC	Dual, NF, Uni, ±18V, -20...+75°	8-DIP	Say	-	data pinout
LA 6462 M	OP-IC	=LA 6462D: SMD	8-MDIP	Say	-	data pinout
LA 6462 S	OP-IC	=LA 6462: Fig.→	9-SIP	Say	-	data pinout
LA6500	HI-GAIN	Vs:±18V Vo:±13V Vi0:2mV	5P	Say	-	data pdf pinout
LA 6500	OP-IC	hi-power, ±18V, 1A, -20...+75°	86/5Pin	Say	-	data pdf pinout
LA6501	HI-GAIN	Vs:±18V Vo:±13V Vi0:2mV	5P	Say	-	data pinout
LA 6501	OP-IC	=LA 6500: Fig.→	30/5Pin	Say	-	data pinout
LA6510	2xPOWER-OP	Vs:±18V Vo:±13V Vi0:2mV	10I,18S	Say	-	data pdf pinout
LA6510	OP-IC	Dual, hi-power, ±18V, 1A, -20...+75°	10-SIL	Say.	-	data pinout
LA6512	2xOP	Vs:±30V Vu:100V/mV Vo:±13V Vi0:2mV	10I	Say	-	data pdf pinout
LA 6512	OP-IC	Dual, hi-power, ±30V, 1A, -20...+75°	10-SIL	Say	-	
LA6513	2xOP	Vs:±30V Vo:±12V Vi0:6mV	10I	Say	-	data pdf pinout
LA 6513	OP-IC	Dual, hi-power, ±30V, 1A, -20...+75°	10-SIL	Say	-	
LA6515	2xPOWER-OP	Vs:±18V Vo:±13V Vi0:2mV	10I	Say	-	data pdf pinout
LA 6515	OP-IC	=LA 6510: Fig.→	10-SIP	Say	-	data pinout
LA6517	2xOP	Vs:±18V Vu:85dB Vo:±12V Vi0:2mV	8D	Say	-	data pdf pinout
LA 6517	OP-IC	Dual, hi-power, ±18V, 0,5A, -20...+75°	8-DIP	Say	-	data pdf pinout
LA 6517 M	OP-IC	=LA 6517: SMD	16-MDIP		-	pdf
LA6520	3xPOWER-OP	Vs:±18V Vo:±12V Vi0:2mV	14D	Say	-	data pdf pinout
LA 6520	OP-IC	Triple, hi-power, ±18V, 0,5A, -20...75°	12-DIP+b	Say	-	data pdf pinout
LA 6530	OP-IC	Quad, hi-power, 16V, 0,7A, -20...75°	16-DIP	Say	-	pdf
LA 6531	OP-IC	Quad, hi-power, 16V, 0,7A, -20...75°	16-DIP	Say	-	pdf
LA 6532 M	OP-IC	SMD, CD, Quad, hi-power, 9V, -20...75°	30-MDIP	Say	-	pdf
LA 6533	OP-IC	Quad, hi-power, 16V, 0,7A, -20...75°	16-DIP	Say	-	pdf
LA 6534	OP-IC	Quad, hi-power, 16V, 0,7A, -20...75°	16-DIP	Say	-	pdf
LA 6535 M	OP-IC	SMD, Cd, Quad, hi-power, 0,7A	30-MDIP	Say	-	pdf

597

Type	Device	Short Description	Fig.	Manu	Comparision Types	More at
LA 6536 M	OP-IC	SMD, Cd, Quad, hi-power, 0,7A	30-MDIP	Say	-	pdf
LA 6800 M	LIN-IC	piezoel. Lautspr.-/speaker-tr.	8-MDIP	Say	-	pdf
LA 6805 M	LIN-IC	Leistungsverst./power ampl., 7V	14-MDIP	Say	-	pdf
LA 7000	LIN-IC	Vc, Luminanz-signal (beta)	28-DIP	Say	-	pdf pinout
LA 7000 S	LIN-IC	=LA 7000: Fig.→	28-SDIP	Say	-	pdf pinout
LA 7003	LIN-IC	Vc, Chroma-signal (beta)	28-DIP	Say	-	pdf pinout
LA 7003 S	LIN-IC	=LA 7003: Fig.→	28-SDIP	Say	-	pdf pinout
LA 7005	LIN-IC	Vc, Luminanz-signal (beta)	28-DIP	Say	-	pdf pinout
LA 7005 S	LIN-IC	=LA 7005: Fig.→	28-SDIP	Say	-	pdf pinout
LA 7007	LIN-IC	Vc, Playback (beta)	28-DIP	Say	-	pdf pinout
LA 7007 S	LIN-IC	=LA 7007: Fig.→	28-SDIP	Say	-	pdf pinout
LA 7009	LIN-IC	Vc, Chroma-synchr. (beta)	24-DIP	Say	-	pdf pinout
LA 7009 S	LIN-IC	=LA 7007: Fig.→	24-SDIP	Say	-	pdf pinout
LA 7011	LIN-IC	Vc, Bandlauf-sensor/tape sensor (beta)	14-DIP	Say	-	pdf pinout
LA 7013	LIN-IC	Vc, Noise-suppr. (beta)	16-DIP	Say	-	pdf pinout
LA 7015 N	LIN-IC	Vc, VHF-Modulator	16-SQP	Say	-	pdf pinout
LA 7016	LIN-IC	Vc, Analogschalter/analog switch	8-SIP	Say	-	pdf pinout
LA 7017	LIN-IC	Vc, Audio, Multiplex-encoder	16-SQP	Say	-	pdf pinout
LA 7018	LIN-IC	Vc, Analogschalter/analog switch	8-DIP	Say	-	pdf pinout
LA 7019	LIN-IC	Vc, Analogschalter/analog switch1	8-DIP	Say	-	pdf pinout
LA 7020	LIN-IC	Vc, Pal-prozessor	28-SDIP	Say	-	pdf pinout
LA 7025	LIN-IC	Vc, Pal-prozessor	28-SDIP	Say	-	pdf pinout
LA 7027	LIN-IC	Vc, Prozessor	28-SDIP	Say	-	pdf pinout
LA 7031	LIN-IC	Vc, VHS-Prozessor	30-SDIP	Say	-	pdf pinout
LA 7032	LIN-IC	Vc, VHS-Prozessor	30-SDIP	Say	-	pdf pinout
LA 7033	LIN-IC	Vc, VHS-Prozessor	30-SDIP	Say.	-	pdf pinout
LA 7034	LIN-IC	Vc, VHS-Prozessor	30-SDIP	Say	-	pdf
LA 7035	LIN-IC	Vc, VHS-Prozessor	30-SDIP	Say	-	pdf pinout
LA 7040	LIN-IC	Vc, Audio-prozessor	18-DIP	Say	-	pdf pinout
LA 7042	LIN-IC	Vc, Verstärker, Schalter/ampl., switch	20-DIP	Say	-	pdf pinout
LA 7043 M	LIN-IC	Vc, Verstärker, Schalter/ampl., switch	20-MDIP	Say	-	pdf pinout
LA 7045	LIN-IC	Vc, Verstärker, Schalter/ampl., switch	20-DIP	Say	-	pdf pinout
LA 7046	LIN-IC	Vc, Verstärker, Schalter/ampl., switch	20-DIP	Say	-	pdf pinout
LA 7048	LIN-IC	Vc, Verstärker, Schalter/ampl., switch	20-DIP	Say	-	pdf pinout
LA 7050	LIN-IC	Vc, Video-,Audio-proz. f.UHF-modulator	9-SIP	Say	-	pdf pinout
LA 7051	LIN-IC	Vc, Video-,Audio-proz. f.uhf-modulator	9-SIP	Say	-	pdf pinout
LA 7053	LIN-IC	Vc, Video-,Audio-proz. f.uhf-modulator	9-SIP	Say	-	pdf pinout
LA 7054	LIN-IC	Vc, Video-,Audio-proz. f.uhf-modulator	9-SIP	Say	-	pdf
LA 7054 Z	LIN-IC	=LA 7054: Fig.→	9-SQP	Say	-	pdf pinout
LA 7055	LIN-IC	Vc, Audio-,Video-Proz. f.uhf-modulator	15-SQP	Say	-	pdf pinout
LA 7056	LIN-IC	=LA 7055: Fig.→	16-DIP	Say	-	pdf pinout
LA 7058	LIN-IC	Vc, Audio-,Video-proz. f.uhf-modulator	9-SIP	Say	-	pdf pinout
LA 7058 R	LIN-IC	=LA 7058: Fig.→	9-SQP	Say	▪	pdf pinout
LA 7060	LIN-IC	Vc, Kopfverstärker/head amplifier	18-DIP	Say	-	pdf pinout
LA 7061	LIN-IC	Vc, Kopfverstärker/head amplifier	16-SQP	Say	-	pdf pinout
LA 7070	LIN-IC	Vc, Luminanz-signal, (beta-system)	30-SDIP	Say	-	pdf pinout
LA 7071	LIN-IC	Vc, Prozessor (beta-ntsc)	30-DIP	Say	-	pdf pinout
LA 7072	LIN-IC	Vc, Noise-suppr.	18-DIP	Say	-	pdf pinout
LA 7073	LIN-IC	Vc, Prozessor (beta-pal)	30-DIP	Say	-	pdf pinout
LA 7074	LIN-IC	Vc, Prozessor (beta-pal)	30-DIP	Say	-	pdf pinout
LA 7090	LIN-IC	Vc, Verstärker, Schalter/ampl., switch	22-DIP	Say	-	pdf pinout
LA 7095	LIN-IC	Vc, Prozessor, Ucc=12V	30-SDIP	Say	-	
LA 7096	LIN-IC	Vc, Prozessor, Ucc=12V	30-SDIP	Say	-	pdf pinout
LA 7097	LIN-IC	Vc, Prozessor, Ucc=12V	30-SDIP	Say	-	
LA 7098	LIN-IC	Vc, Prozessor, Ucc=12V	30-SDIP	Say	-	
LA 7110	LIN-IC	Vc, Servo-interface	42-DIP	Say	-	pdf pinout
LA 7112 M	LIN-IC	Vc, Trommel-/drum Servo Interface	14-DIP	Say	-	
LA 7116	LIN-IC	Vc, Servo-interface	24-SDIP	Say	-	pdf pinout
LA 7117	LIN-IC	Vc, Servo-interface	28-SDIP	Say	-	pdf pinout
LA 7122	LIN-IC	Vc, Servo-interface	30-SDIP	Say	-	
LA 7140	LIN-IC	Vc, S-VHS, Video-schalter/switch	30-SDIP	Say	-	pdf pinout
LA 7150	LIN-IC	Vc, PAL-AV-Schalter/switch, Ucc=12V	22-SDIP	Say	-	pdf pinout
LA 7151	LIN-IC	Vc,Camera, AV-schalter/switch	12-SIP	Say	-	pdf pinout
LA 7151 M	LIN-IC	=LA 7151: SMD	10-MDIP	Say	-	pdf pinout
LA 7152	LIN-IC	Vc, Video-Schalter/switch, Ucc=5V	9-SIP	Say	-	pdf pinout
LA 7155 M	LIN-IC	SCART Audio-Schalter/switching, ±5V	24-MDIP	Say	-	pdf pinout
LA 7156	LIN-IC	SCART Audio/video Switch, 8...13V	24-SDIP	Say	-	odf pinout
LA 7205	LIN-IC	Vc, Bandlauf/tape control	9-SIP	Say	-	
LA 7205 M	LIN-IC	=LA 7205: SMD	14-MDIP	Say	-	pdf pinout
LA 7210	LIN-IC	Vc, Synchr.-Detector, Ucc=7...13V	10-SIP	Say	-	pdf pinout
LA 7212	LIN-IC	Vc, Synchr.-Detector, Ucc=7...13V	14-DIP	Say	-	pdf pinout
LA 7213	LIN-IC	Vc, Synchr.-Separator, Ucc=5V	5-SIP	Say	-	pdf pinout
LA 7214	LIN-IC	Vc, Synchr.-Separator, Video-buffer	9-SIP	Say	-	pdf pinout
LA 7215	LIN-IC	Vc, Synchr.-Detector, Tsg	15-SQP	Say	-	pdf pinout
LA 7217	LIN-IC	Vc, TV, Camera, Afc, Synchr., Ucc=5V	14-DIP	Say	-	pdf pinout
LA 7217 M	LIN-IC	=LA 7217: SMD	14-MDIP	Say	-	pdf pinout
LA 7218 M	LIN-IC	SMD, AFC	14-MDIP	Say	-	pdf pinout
LA 7220	LIN-IC	Vc, NF-Umschalter/audio switch	16-DIP	Say	-	pdf pinout
LA 7220 M	LIN-IC	=LA 7220: SMD	24-MDIP	Say	-	pdf pinout
LA 7221	LIN-IC	Vc, Umschalter/electronic switch, 5V	9-SIP	Say	-	pdf pinout
LA 7222	LIN-IC	Vc, AV-Umschalter/switch, Ucc=12V	12-SIP	Say	-	pdf pinout
LA 7223	LIN-IC	Vc, Pal AV-umschalter/switch	24-SDIP	Say	-	pdf pinout
LA 7224	LIN-IC	FB-empf.-vorverst./pre amplifier	8-SIP	Say	-	pdf pinout
LA 7225	LIN-IC	FB-empf., Vorverst./pre amplifier	9-SIP	Say	-	pdf pinout
LA 7227	LIN-IC	IR-FB, Vorverstärker/pre amplifier	7-SIP	Say	-	pdf pinout

Type	Device	Short Description	Fig.	Manu	Comparision Types	More at
LA 7227 M	LIN-IC	=LA 7227: SMD	8-MDIP	Say	-	pdf pinout
LA 7228	LIN-IC	IR-FB, Vorverstärker/pre amplifier	9-SIP	Say	-	pdf pinout
LA 7228 M	LIN-IC	=LA 7228: SMD	10-MDIP	Say	-	pdf pinout
LA 7230	LIN-IC	HiFi Vc, Eingangsschalter/input select	30-SDIP	Say	-	
LA 7235	LIN-IC	HiFi Vc, Eingangsschalter/input select	16-DIP	Say	-	
LA 7236	LIN-IC	HiFi Vc, Eingangsschalter/input select	24-SDIP	Say	-	
LA 7236 M	LIN-IC	=LA 7236: SMD	24-MDIP	Say	-	
LA 7237	LIN-IC	HiFi Vc, Eingangsschalter/input select	32-SQP	Say	-	
LA 7245	LIN-IC	Vc, 2x Muting, Ucc=5V	9-SIP	Say		pdf
LA 7250	LIN-IC	HiFi Vc, Ucc Reg, HiFi Mute	32-SQP	Say		
LA 7256	LIN-IC	HiFi Vc, Audio Signal Processing	24-SDIP	Say	-	pdf
LA 7261 M	LIN-IC	SMD, Camera, Ccd Signal-proz., Ucc=5V	48-MP	Say	-	pdf
LA 7262 M	LIN-IC	Camera, Ccd Signal-prozessor (color)		Say		
LA 7264 M	LIN-IC	Camera, Ccd Signal-prozessor (color)		Say	-	
LA 7265 W	LIN-IC	SMD, B/w Camera, Ccd Signal-prozessor	48-MP	Say		pdf
LA 7266 M	LIN-IC	Camera, Ccd Signal-encoder (color)		Say		pdf
LA 7270	LIN-IC	Vc, VHS, HiFi A/w-verst.	22-SDIP	Say	-	pdf
LA 7270 M	LIN-IC	=LA 7270: SMD	24-MDIP	Say		pdf
LA 7282	LIN-IC	Vc, Audio-Prozessor, Ucc=12V	24-SDIP	Say	-	pdf
LA 7282 M	LIN-IC	=LA 7282: SMD	24-MDIP	Say	-	pdf
LA 7285	LIN-IC	Vc, Audio-Prozessor, Ucc=12V	24-SDIP	Say	-	pdf
LA 7285 M	LIN-IC	=LA 7285: SMD	24-MDIP	Say	-	pdf
LA 7286	LIN-IC	Vc, Audio-Prozessor, Ucc=12V	24-SDIP	Say	-	pdf
LA 7291	LIN-IC	Vc, Prozessor, Ucc=9V	30-SDIP	Say	-	
LA 7292	LIN-IC	Vc, Prozessor, Ucc=9V	30-SDIP	Say	-	
LA 7293 M	LIN-IC	Camera, Mikrofon-verst./amplifier	8-MDIP	Say	-	pdf
LA 7294	LIN-IC	Vc, Audio-Prozessor, Ucc=12V	30-SQP	Say	-	pdf
LA 7295	LIN-IC	Vc, Audio-Prozessor, Ucc=12V	30-SQP	Say	-	pdf
LA 7296	LIN-IC	Vc, Audio-Prozessor, Ucc=12V	30-SQP	Say	-	pdf
LA 7297	LIN-IC	Vc, Audio-Prozessor, Ucc=9V	30-SQP	Say	-	pdf
LA 7300	LIN-IC	Vc, VHS A/W-Verst,Kopfverst./head Amp.	22-SDIP	Say	-	
LA 7300 M	LIN-IC	=LA 7300: SMD	24-MDIP		-	pdf
LA 7301	LIN-IC	Vc, VHS A/W-Verst,Kopfverst./head Amp.	30-SDIP	Say	-	pdf
LA 7301 M	LIN-IC	=LA 7301: SMD	30-SMDIP		-	pdf
LA 7305 N	LIN-IC	Vc, VHS Signal-prozessor, Ucc=5V	42-SDIP	Say	-	
LA 7306 N	LIN-IC	Vc, VHS Signal-prozessor, Ucc=5V	42-SDIP	Say	-	
LA 7307 M	LIN-IC	SMD, Vc, VHS Signal-prozessor, Ucc=5V	48-MP	Say	-	
LA 7308(P)	LIN-IC	Vc, Detailverbess./detail enhancer	9-SIP	Say	KA 8101	
LA 7308 (M)	LIN-IC	=LA 7308: SMD	8-MDIP		-	
LA 7309	LIN-IC	Vc, Prozessor, Noise Suppr., Ucc=5V	20-SDIP	Say	-	
LA 7311	LIN-IC	Vc, PAL/SECAM/S-VHS-Diskr, Ucc=5V	16-SQP	Say	-	pdf
LA 7315	LIN-IC	Vc, VHS Chroma, Signal-prozessor	42-SDIP	Say	-	
LA 7316(A-N)	LIN-IC	VC, VHS(NTSC) Chroma, Signal Processor	24-SDIP	Say	-	
LA 7316	LIN-IC	Vc, VHS(NTSC) Chroma, Signal-prozessor	24-SDIP	Say	-	pdf
LA 7316 M,AM	LIN-IC	=LA 7316: SMD	24-SMDIP		-	
LA 7316 M	LIN-IC	=LA 7316: SMD	24-SMDIP		-	
LA 7317	LIN-IC	Vc, Vhs(ntsc) Chroma, Signal-prozessor	24-SDIP	Say	-	pdf
LA 7318	LIN-IC	Vc, Vhs(ntsc) Chroma, Signal-prozessor	24-SDIP	Say	-	
LA 7318 M	LIN-IC	=LA 7318: SMD	24-SMDIP		-	
LA 7319	LIN-IC	Vc, Vhs(ntsc) Chroma, Signal-prozessor	24-SDIP	Say	-	
LA 7319 M	LIN-IC	=LA 7319: SMD	24-SMDIP		-	
LA 7320	LIN-IC	Vc, VHS A/W-Verst,Kopfverst./head Amp.	22-SDIP	Say	KA 8112	pdf
LA 7320 M	LIN-IC	=LA 7320: SMD	24-MDIP		-	pdf
LA 7321	LIN-IC	Vc, VHS A/W-Verst,Kopfverst./head Amp.	30-SDIP	Say	-	pdf
LA 7321 M	LIN-IC	=LA 7321: SMD	30-MDIP	Say	-	
LA 7323	LIN-IC	Vc, VHS-HQ, Luminanzsignal-prozessor	30-SDIP	Say	KA 8113	pdf
LA 7323 M	LIN-IC	=LA 7323: SMD	30-MDIP		-	pdf
LA 7324	LIN-IC	Vc, Vhs-hq, Luminanzsignal-prozessor	30-SDIP	Say	-	
LA 7324 M	LIN-IC	=LA 7324: SMD	30-MDIP		-	
LA 7327	LIN-IC	Vc, Chroma, Noise Suppr., Ucc=5V	8-SIP	Say	-	pdf
LA 7328	LIN-IC	Vc, Kammfilter/comb filter, Ucc=5V	15-SQP	Say	-	pdf
LA 7330	LIN-IC	Vc, VHS Chroma, Signal-prozessor	24-SDIP	Say	-	pdf
LA 7330 M	LIN-IC	=LA 7330: SMD	24-MDIP	Say	-	pdf
LA 7331	LIN-IC	Vc, VHS Chroma, Signal-prozessor	24-SDIP	Say	-	
LA 7356	LIN-IC	Vc, PAL/SECAM(MESECAM)-Diskr, Ucc=5V	10-SDIP	Say	-	pdf
LA 7356 M	LIN-IC	SMD,Vc, PAL/SECAM(MESECAM)-Diskr	10-SMDIP		-	pdf
LA 7357	LIN-IC	Pal/secam Discrimination Circuit f. VHS VCRs	DIC	Say	-	pdf pinout
LA 7358 M	LIN-IC	SECAM Chroma Signal Processor	TSOP	Say	-	pdf pinout
LA 7371 A	LIN-IC	Vc, Vhs, A/W-Control, Ucc=5V	24-SDIP	Say	-	pdf
LA 7375	LIN-IC	Vc, Vhs, A/W-Control, Ucc=5V	16-DIP	Say	-	pdf
LA 7383(N)	LIN-IC	VC, NTSC HQ-Processor, Ucc=5V	36-SDIP	Say	-	
LA 7383	LIN-IC	Vc, NTSC HQ-Prozessor, Ucc=5V	36-SDIP	Say	-	pdf
LA 7390(N)	LIN-IC	Vc(vhs), PAL/SECAM-Prozessor, Ucc=5V	36-SDIP	Say	-	pdf
LA 7391 A(AN)	LIN-IC	Vc(vhs), NTSC/PAL/SECAM-Prozessor, 5V	42-SDIP	Say	-	pdf pinout
LA 7400 M	LIN-IC	SMD, Digital Vc, Pip-prozessor	30-SMDIP	Say	-	pdf
LA 7401 M	LIN-IC	SMD, Digital Vc, Pip-prozessor	30-SMDIP	Say	-	
LA 7402 M	LIN-IC	SMD, Digital Vc, Pip-prozessor	48-MP	Say	-	pdf
LA 7403	LIN-IC	TV, Pip-prozessor (ntsc)	48-SDIP	Say	-	
LA 7411	LIN-IC	Vc, Vhs, A/w-verst./amp., Ucc=5V	24-SDIP	Say	-	pdf
LA 7411 M	LIN-IC	SMD,Vc, Vhs, A/w-verst./amp., Ucc=5V	24-SMDIP		-	pdf,
LA 7416	LIN-IC	Vc, Vhs, A/w verst./amp., Ucc=5V	30-SDIP	Say	-	pdf
LA 7416 M	LIN-IC	SMD,Vc, Vhs, A/w-verst./amp., Ucc=5V	30-SMDIP		-	pdf
LA 7425	LIN-IC	Vc, Vhs, NTSC Signal-proz., Ucc=5V	36-SDIP	Say	-	pdf
LA 7440	LIN-IC	Vc, Vhs, Video Signal-proz., Ucc=5V	48-SDIP	Say	-	pdf

Type	Device	Short Description	Fig.	Manu	Comparision Types	More at
LA 7440 M	LIN-IC	SMD,Vc, Vhs, Video Signal-proz.,Ucc=5V	48-MP		-	pdf
LA 7449	LIN-IC	Vc, Vhs, Video Signal-proz., Ucc=5V	48-SDIP	Say	-	pdf
LA 7449 M	LIN-IC	SMD,Vc, Vhs, Video Signal-proz.,Ucc=5V	48-MP		-	pdf
LA 7451 M	LIN-IC	SMD, Vc, PCM-Prozessor(2Kanal/channel)	30-MDIP	Say	-	pdf
LA 7454 W	LIN-IC	SMD, Vc, FM-Audio-Prozessor, Ucc=4,75V	48-MP	Say	-	pdf
LA 7456 M	LIN-IC	SMD, Vc, Stereo-Matrix, Ucc=4,75V	36-MP	Say	-	pdf
LA 7461 W	LIN-IC	SMD, Vc, NTSC/PAL-Prozessor (Hi8)	64-MP	Say	-	
LA 7470 M	LIN-IC	Camera, Mikrofon-Verst./amp.(2channel)	24-MDIP	Say	-	pdf
LA 7470 V	LIN-IC	=LA 7470M:	24-SMDIP	Say	-	pdf
LA 7471 M	LIN-IC	SMD,Camera, 2-Ch. Mikrofon-Verst./amp.	36-MP	Say	-	pdf
LA 7505	LIN-IC	Tv-video-zf, Agc, AFT, f. Saw-filter	16-DIP	Say	-	pdf
LA 7506	LIN-IC	TV, Video-zf	16-DIP	Say	-	
LA 7507	LIN-IC	TV, Video-ZF, Agc, f. Saw-filter	16-DIP	Say	-	pdf
LA 7508	LIN-IC	TV, Video-ZF, Agc, Aft, f. Saw-filter	16-DIP	Say	-	
LA 7509	LIN-IC	TV, Video-ZF, Agc, Aft, f. Saw-filter	16-DIP	Say	-	pdf
LA 7510	LIN-IC	TV, Vc, Quasi-parallel-ton-zf	9-SIP	Say	-	pdf
LA 7520(N)	LIN-IC	Ctv, Video-ZF, Ton-zf	30-SDIP	Say	KA 2919	pdf pinout
LA 7521 N	LIN-IC	Ctv, Video-ZF, Ton-zf	30-SDIP	Say	-	
LA 7522	LIN-IC			Say	-	
LA 7530 N	LIN-IC	Ctv/vc, Video-ZF, Ton-ZF, Ucc=9...12V	20-DIP	Say	KA 2923	pdf
LA 7531	LIN-IC	Ctv, Video-ZF, Ton-ZF, Ucc=9...12V	20-DIP	Say	-	
LA 7533	LIN-IC	Ctv/vc, Video-ZF, Ton-ZF, Ucc=9V	20-DIP	Say	-	pdf
LA 7535	LIN-IC	TV, Video-ZF, Ton-ZF, pos. Agc,Ucc=12V	20-DIP	Say	KA 2922	pdf
LA 7545	LIN-IC	TV, Vc, Video-ZF, Ton-ZF, Ucc=12V	16-DIP	Say	-	pdf
LA 7550	LIN-IC	TV, Vc, Video-ZF, Ton-ZF, Ucc=12V	24-SDIP	Say	-	pdf
LA 7555	LIN-IC	TV, Vc, Video-ZF, Ton-ZF, Ucc=12V	24-SDIP	Say	-	
LA 7556	LIN-IC	TV, Vc, Video-ZF, Ton-ZF, Ucc=12V	24-SDIP	Say	-	
LA 7570	LIN-IC	TV, Vc, PLL-video-ZF, Ton-zf	24-SDIP	Say	-	pdf
LA 7577(N)	LIN-IC	TV, Vc, Pll-video-zf, Ton-zf	24-SDIP	Say	-	pdf
LA 7578 N	LIN-IC	TV, Vc, Pll-video-zf, Ton-zf	24-SDIP	Say	-	pdf
LA 7580	LIN-IC	TV, Vc, Pll-video-zf, Ton-zf	24-SDIP	Say	-	
LA 7583	LIN-IC	TV, Vc, Pll-video-zf, Ton-zf	24-SDIP	Say	-	pdf
LA 7590	LIN-IC	TV, VHF-M/O/ZF	14-DIP	Say	-	
LA 7600	LIN-IC	Vc, NTSC Video+chroma (beta)	28-DIP	Say	-	pdf
LA 7601(W)	LIN-IC	Vc, NTSC Video+chroma (beta, Usa)	28-DIP	Say	-	pdf
LA 7605 M	-IC	Video Signal Processor, Pal and NTSC	PLCC	Say	-	pdf pinout
LA 7620	LIN-IC	Ctv, Video-, Chroma-prozessor (ntsc)	30-SDIP	Say	-	pdf
LA 7621	LIN-IC	Ctv, Video-, Chroma-prozessor (ntsc)	30-SDIP	Say	-	pdf
LA 7625	LIN-IC	Ctv, Video-, Chroma-prozessor (ntsc)	30-SDIP	Say	KA 2155	pdf
LA 7626	LIN-IC	Ctv, Video-, Chroma-prozessor (ntsc)	30-SDIP	Say	KA 2156	pdf
LA 7629	LIN-IC	Ctv, Video-, Chroma-prozessor (ntsc)	30-SDIP	Say	-	pdf
LA 7640 N	LIN-IC	Ctv, SECAM Chroma System	24-SDIP	Say	-	pdf
LA 7651 N	LIN-IC	Ctv, Signal-prozessor (ntsc), Ucc=9V	42-SDIP	Say	-	
LA 7652 N	LIN-IC	Ctv, Signal-prozessor (ntsc), Ucc=9V	42-SDIP	Say	-	
LA 7655 N	LIN-IC	Ctv, Signal-prozessor (ntsc), Ucc=9V	42-SDIP	Say	-	pdf
LA 7668 N	LIN-IC	Ctv, Video-signal Auto-white Process.	24-SDIP	Say	-	pdf
LA 7672	LIN-IC	Ctv, Signal-prozessor (ntsc)	52-SDIP	Say	-	pdf
LA 7674	DIG-IC	CTV signal processor NTSC	MDIP	Say	-	pdf
LA 7680	LIN-IC	Ctv, Signal-prozessor (pal/ntsc)	48-SDIP	Say	-	pdf
LA 7681	LIN-IC	Ctv, Signal-prozessor (pal/ntsc)	48-SDIP	Say	-	pdf
LA 7685	LIN-IC	Ctv, NTSC/PAL/SECAM-Prozessor	64-SDIP	Say	-	
LA 7685 J	LIN-IC	Ctv, NTSC/PAL Signal-prozessor	64-SDIP	Say	-	pdf
LA 7688	LIN-IC	Vif, Sif, video, chrominance, and deflection processing for PAL/NTSC	SDIP	Say	-	pdf pinout
LA 7690	LIN-IC	Ctv, Color Limiter	9-SIP	Say	-	
LA 7696	LIN-IC	Ctv, On-screen Display Interface, 12V	20-DIP	Say	-	pdf
LA 7698	LIN-IC	Ctv, Chroma-prozessor, Ucc=9V	24-SDIP	Say	-	pdf
LA 7710	LIN-IC	TV, Vc, Quasi-parallel-ton, SECAM/PAL	16-DIP	Say	-	pdf
LA 7750	LIN-IC	Tv-stereo	28-DIP	Say	-	
LA 7751	LIN-IC	Tv-stereo	28-DIP	Say	-	pdf pinout
LA 7755	LIN-IC	Tv-stereo	14-DIP	Say	-	pdf pinout
LA 7760	LIN-IC	Ctv, Audio Multiplex-decoder (ntsc)	30-SDIP	Say	-	pdf pinout
LA 7761	LIN-IC	Ctv, Audio Multiplex-decoder (ntsc)	28-SDIP	Say	-	pdf pinout
LA 7770	LIN-IC	Catv, FSK-empfänger/receiver	20-DIP	Say	-	pdf
LA 7780 M	LIN-IC	QPSK Downconverter for CATV Systems	TSOP	Say	-	pdf pinout
LA 7790 M	LIN-IC	QPSK Transmitter for Cable TV	TSOP	Say	-	pdf pinout
LA 7800	LIN-IC	Ctv, HA/VA-O, Synchr.	16-DIP	Say	-	pdf
LA 7801	LIN-IC	Ctv, Ha/va-o, Synchr.	16-DIP	Say	-	pdf pinout
LA 7802	LIN-IC	Ctv, Ha/va-o, Synchr.	18-DIP	Say	-	pdf
LA 7806	LIN-IC	B/w(s/w) TV Synchr., Deflection	16-DIP	Say	-	pdf pinout
LA 7808	LIN-IC	B/w(s/w) TV Synchr., Deflection	9-SIP	Say	-	pdf pinout
LA 7810	LIN-IC	Ctv, Ha/va-o, Synchr.	22-DIP	Say	-	pdf
LA 7811	LIN-IC	Ctv, Ha/va-o, Synchr.	22-DIP	Say	-	pdf
LA 7820	LIN-IC	Color TV, Synchr., Deflection	18-DIP	Say	-	pdf pinout
LA 7822	LIN-IC	Color TV, Synchr., Deflection	16-DIP	Say	-	pdf pinout
LA 7823	LIN-IC	Color TV, Synchr., Deflection, Drv.	16-DIP	Say	-	pdf pinout
LA 7824	LIN-IC	Color TV, Synchr., Deflection, Drv.	16-DIP	Say	-	pdf pinout
LA 7830	LIN-IC	Color TV, Vert. Deflect. Output, Ucc=18...27V	7-SIL	Say	-	pdf pinout
LA 7831	LIN-IC	Ctv, VA-Tr, VA-E, Ucc=18...27V	10-SILP	Say	-	pdf pinout
LA 7832	LIN-IC	TV, HA-Tr, Ha-e		Say	-	pdf
LA 7833	LIN-IC	TV, VA-Tr, VA-E	SIP	Say	-	pdf pinout
LA 7835	LIN-IC	TV, Ha-tr, HA-E, max. Iq=1,8A	13-SIL	Say	LA 7836	pdf
LA 7836	LIN-IC	=LA 7835: max. Iq=2,2A	13-SIL	Say	-	pdf
LA 7837	LIN-IC	TV, Ha-tr, HA-E, max. Iq=1,8A	13-SIL	Say	-	pdf pinout

Type	Device	Short Description	Fig.	Manu	Comparision Types	More at
LA 7838	LIN-IC	=LA 7837: max. Iq=2,2A	13-SIL	Say	-	pdf pinout
LA 7840	LIN-IC	Ctv, VA-Tr, VA-E, max. Iq=1,8A	7-SIL	Say	-	pdf pinout
LA 7841	LIN-IC	Ctv, VA-Tr, VA-E, max. Iq=2,2A	7-SIL	Say	-	pdf pinout
LA 7845	LIN-IC	Vert. Deflection Output Circuit	7-SIL	Say	LA7845N	pdf pinout
LA 7845 N	LIN-IC	Vert. Deflection Output Circuit	7-SIL	Say	-	pdf pinout
LA 7846 N	LIN-IC	Ctv, VA-Tr, VA-E, max. Iq=3A	10-SILP	Say	-	pdf pinout
LA 7847	LIN-IC	TV Vertical Output, E/W Drive	SIL	Say	-	pdf pinout
LA 7849	LIN-IC	TV Vertical Output, E/w Drive	SILP	Say	-	pdf
LA 7850	LIN-IC	Crt Display Synchr., Deflection	20-DIP	Say	LA7855, LA7851	pdf pinout
LA 7851	LIN-IC	Crt Display Synchr., Deflection	20-DIP	Say	LA7856, KA2138	pdf pinout
LA 7852	LIN-IC	Crt Display Synchr. Deflection IC	22-SDIP	Say	LA7857, LA7853	pdf pinout
LA 7853	LIN-IC	Crt Display Synchr. Deflection IC	22-SDIP	Say	LA7858	pdf pinout
LA 7855	LIN-IC	Crt Display Synchr., Deflection	20-DIP	Say	-	pdf pinout
LA 7856	LIN-IC	Crt Display Synchronization, Deflection	20-DIP	Say	-	pdf pinout
LA 7857	LIN-IC	Crt Display Synchr., Deflection	22-SDIP	Say	-	pdf pinout
LA 7858	LIN-IC	Crt Display Synchr., Deflection	22-SDIP	Say	-	pdf pinout
LA 7860	LIN-IC	Multisync Display Deflection Proc.	30-SDIP	Say	-	pdf pinout
LA 7860 M	LIN-IC	SMD, Multisync Display Deflection Proc.	30-SSMDIP	Say	-	pdf pinout
LA 7875N	IC	Vert. Deflection Output, 2.2App	SILP	Say	-	pdf pinout
LA 7876N	IC	Vert. Deflection Output, 3App	SILP	Say	-	pdf pinout
LA 7890	LIN-IC	Crt, RGB Cutoff Adjustment	10-SIP	Say	-	pdf pinout
LA 7900	LIN-IC	TV, Tuner Band-selector	9-SIP	Say	-	pdf
LA 7905	LIN-IC	TV, Tuner Band-selector	9-SIP	Say	-	pdf
LA 7906	LIN-IC	TV, Tuner Band-selector	9-SIP	Say	-	pdf
LA 7910	LIN-IC	TV, Tuner Band-selector	9-SIP	Say	-	pdf
LA 7911	LIN-IC	TV, Tuner Controller	16-DIP	Say	-	pdf
LA 7912	LIN-IC	CTV, AFT	9-SIP	Say	-	pdf
LA 7913	LIN-IC	TV, Tuner Controller	16-DIP	Say	-	pdf
LA 7915	LIN-IC	Vc, TV, Tuning-system	20-DIP	Say	-	pdf
LA 7916	LIN-IC	Vc, TV, Tuning-system	20-DIP	Say	-	pdf
LA 7920	LIN-IC	TV, Tuner Band-selector	9-SIP	Say	-	pdf
LA 7925	LIN-IC	Vc, TV, Tuning-system	20-DIP	Say	-	pdf pinout
LA 7926	LIN-IC	Vc, TV, Tuning-system	16-DIP	Say	-	pdf
LA 7930	LIN-IC	Vc, TV, Tuning-system-control	30-SDIP	Say	-	pdf pinout
LA 7933	LIN-IC	Vc, TV, Tuning-system-control	20-DIP	Say	-	pdf pinout
LA 7935	LIN-IC	Vc, TV, Tuning-system-control	22-SDIP	Say	-	pdf
LA 7936	LIN-IC	Vc, TV, Tuning-system-control	30-SDIP	Say	-	pdf
LA 7937	LIN-IC	Vc, TV, Tuning-system-control	22-SDIP	Say	-	pdf
LA 7938	LIN-IC	Vc, TV, Tuning-system-control	22-SDIP	Say	-	pdf
LA 7940	LIN-IC	Ctv, Vc, Kanalwahl/channel select(16x)		Say	-	
LA 7945(N)	LIN-IC	TV, Synchr. (US Standard), Ucc=5V	22-SDIP	Say	-	pdf
LA 7950	LIN-IC	Ctv, Frequ.-diskr., PAL/NTSC-Select.	10-SIP	Say	-	pdf
LA 7951	LIN-IC	TV, Vc, Video-Schalter/switch, Ucc=12V	14-DIP	Say	-	pdf
LA 7952	LIN-IC	TV, Vc, Video-Schalter/switch, Ucc=12V	9-SIP	Say	-	pdf
LA 7953	LIN-IC	TV, Audio-Controller, Ucc=12V	30-SDIP	Say	-	pdf
LA 7954	LIN-IC	TV, Vc, Video-schalter/switch	9-SIP	Say	-	pdf
LA 7955	LIN-IC	TV, Vc, Video-Schalter/switch, Ucc=12V	20-DIP	Say	-	pdf
LA 7956	LIN-IC	TV, Vc, Video-Schalter/switch, Ucc=12V	9-SIP	Say	-	pdf
LA 7957	LIN-IC	TV, Vc, Video-Schalter/switch, Ucc=12V	20-DIP	Say	-	pdf
LA 7970	LIN-IC	Catv, Standard Base Band Interface		Say	-	
LA 7975	LIN-IC	Ctv, PAL-Signal, Ton-ZF-converter	5-SIP	Say	-	pdf
LA 7976	LIN-IC	Pal Sif Converter Circuit f. TV & Vcr Multi-system	MSIP	Say	-	pdf pinout
LA 7995 M	LIN-IC	Charge Pump Type DC-DC Converter	TSOP	Say	-	pdf pinout
LA 8200 M	LIN-IC	Floppy-disk,Verstärker/read-write amp.	44-FLP	Say	-	pdf
LA 8202 M	LIN-IC	Floppy-disk,Verstärker/read-write amp.		Say	-	
LA 8205	LIN-IC	Floppy-disk,Verstärker/read-write amp.	22-DIP	Say	-	pdf
LA 8205 M	LIN-IC	=LA 8205: SMD	24-MDIP	Say	-	pdf
LA 8206 M	LIN-IC	Floppy-disk,Verstärker/read-write amp.	24-MDIP	Say	-	pdf
LA 8500	LIN-IC	Telecom, Wecker/tone ringer	8-DIP	Say	-	
LA 8501(-P)	LIN-IC	Telecom, Wecker/tone ringer	8-DIP	Say	-	
LA 8501	LIN-IC	Telecom, Wecker/tone ringer	8-DIP	Say	-	
LA 8510	LIN-IC	Telecom, Sprechkreis/speech network	20-DIP	Say	-	pdf
LA 8515(N)	LIN-IC	Telecom, Sprechkreis/speech network	20-DIP	Say	-	
LA 8515	LIN-IC	Telecom, Sprechkreis/speech network	20-DIP	Say	-	
LA 8517 M	LIN-IC	Telecom, Schnurlossystem/cordless Base	64-MP	Say	-	pdf
LA 8518 NM	LIN-IC	Signal Processor for Cordless Telephone Base Sets	PLCC	Say	-	pdf pinout
LA 8519 M	LIN-IC	I/O Switch/voice Signal-processing IC f. Cordless Telephones	PLCC	Say	-	pdf pinout
LA 8520 M	LIN-IC	Audio Signal-processing IC with I/O Switching	TSOP	Say	-	pdf pinout
LA 8522 M	LIN-IC	Audio Signal-processing IC with I/O Switching	TSOP	Say	-	pdf pinout
LA 8580 W	LIN-IC	Compander + I/O Switching Telephone Audio Signal-processing IC	MP	Say	-	pdf pinout
LA 8601 M	LIN-IC	Telecom, Schnurlossystem/cordless Base	64-MP	Say	-	
LA 8602 M	LIN-IC	Telecom,Schnurlossystem/cordless Handy	64-MP	Say	-	
LA 8604 M	LIN-IC	SMD, Telecom, FM-ZF, Ucc=2,4...6V	24-SMDIP	Say	-	pdf
LA 8606 M	LIN-IC	Telecom,Schnurlossystem/cordless Phone	44-MP	Say	-	pdf
LA 8608 V	LIN-IC	SMD, Telecom, FM-ZF, Ucc=1,8..6V	24-SSMDIP	Say	-	pdf
LA 8610 M	LIN-IC	SMD, Telecom, Pager, FM-ZF	20-MDIP	Say	-	pdf
LA 8615 M	LIN-IC	SMD, Telecom, Pager, FM-ZF	16-SMDIP	Say	-	pdf
LA 8615 V	LIN-IC	=LA 8615M:	16-SSMDIP	Say	-	pdf
LA 8630	LIN-IC	Telecom, Kompander f. Cordless Phone	16-DIP	Say	-	pdf

Type	Device	Short Description	Fig.	Manu	Comparision Types	More at
LA 8630 M	LIN-IC	=LA 8630: SMD	16-MDIP		-	pdf
LA 8632	LIN-IC	Telecom, Kompander f. Cordless Phone	24-SDIP	Say	-	pdf
LA 8632 M	LIN-IC	=LA 8632: SMD	24-SMDIP		-	pdf
LA 8633 V	LIN-IC	SMD,Telecom, Kompander f. Cordless Ph.	24-SSMDIP	Say	-	pdf
LA 8640 W	LIN-IC	Telecom, Kompander f. Cordless Phone	48-MP	Say	-	
LA 8645 W	LIN-IC	Telecom, Kompander f. Cordless Phone	64-MP	Say	-	
LA 9000 N	LIN-IC	f. Dad-player	16-DIP	Say	-	pdf pinout
LA 9010	LIN-IC	f. Dad-player	16-DIP	Say	-	pdf pinout
LA 9100	LIN-IC	f. Dad-player	28-DIP	Say	-	pdf pinout
LA 9200 NM	LIN-IC	SMD, Cd, Analogsignal-prozessor	48-MP	Say	-	pdf
LA 9201 M	LIN-IC	SMD, Cd, Analogsignal-prozessor	48-MP	Say	-	pdf
LA 9210 M	LIN-IC	SMD, Cd, Analogsignal-prozessor	80-MP	Say	-	pdf
LA 9215	LIN-IC	CD, Logic, Audio Output, D/A-Converter	24-SDIP	Say	-	
LA 9215 M	LIN-IC	Cd, Logic, Verstärker/ampl.	24-SDIP	Say	-	
LA 9233 M	LIN-IC	Analog Signal Processor (asp) for CD Players	MP	Say	-	pdf pinout
LA 9240 M	LIN-IC	Analog Signal Processor (asp) for CD Players	MP	Say	-	pdf pinout
LA 9241 M	LIN-IC	Analog Signal Processor (asp) for CD players	MP	Say	-	pdf pinout
LA 9242 M	LIN-IC	Analog Signal Processor (asp) for CD players	MP	Say	-	pdf pinout
LA 9250 M	LIN-IC	CD Player Analog Signal Processor (asp)	MP	Say	-	pdf pinout
LA 9251 M	LIN-IC	CD Player Analog Signal Processor (asp)	MP	Say	-	pdf pinout
LA 9400 M	LIN-IC	SMD, LD Player, Servo	64-MP	Say	-	
LA 9410	LIN-IC	Ld Player	28-SDIP	Say	-	pdf
LA 9450 CL	BiCMOS-IC	Laser Diode Pulse Driver IC	MLP	Say	-	pdf pinout
LA 9511 W	LIN-IC	AV Remote Coupler Transmitter	MP	Say	-	pdf pinout
LA 9520 V	LIN-IC	AV Coupler Receiver	TSOP	Say	-	pdf pinout
LA 9605 W	LIN-IC	MD Player Rf and Matrix Signal-Processing IC	MP	Say	-	pdf pinout
LA 9702 W	LIN-IC	DVD Player Front End Processor	MP	Say	-	pdf pinout
LA 42031	LIN-IC	5W Monaural Audio Power Amp	SIL	Say	-	pdf pinout
LA 42032	LIN-IC	5W ×2ch Audio Power Amp	SIL	Say	-	pdf pinout
LA 42052	LIN-IC	5W ×2ch Audio Power Amplifier	SIL	Say	-	pdf pinout
LA 42072	LIN-IC	7W ×2ch Audio Power Amp	SIL	Say	-	pdf pinout
LA 42101	LIN-IC	10W ×1ch Audio Power Amp	SIL	Say	-	pdf pinout
LA 42102	LIN-IC	10W ×2ch Audio Power Amp	SIL	Say	-	pdf pinout
LA 42105	LIN-IC	BTL 5W ×1ch Audio Power Amp	QILP	Say	-	pdf pinout
LA 42152	LIN-IC	15W ×2ch Audio Power Amp	SILP	Say	-	pdf pinout
LA 42201	LIN-IC	BTL 20W Monaural Audio Power Amp	86	Say	-	pdf pinout
LA 42205	LIN-IC	5W ×2ch Audio Power Amp	QILP	Say	-	pdf pinout
LA 42210	LIN-IC	BTL 10W ×2ch Audio Power Amp	QILP	Say	-	pdf pinout
LA 42352	LIN-IC	5W 2Ch. AF Power Amp., DC Volume Cntrl.	SIL	Say	-	pdf pinout
LA 47501	LIN-IC	BTL Power Amp 50W×4, 43W×4(Vcc=14.4V, 4Ohm, 1kHz)	QILP	Say	-	pdf
LA 47510	LIN-IC	4-ch. 50W Car stereo. Vcc=14,4 V	QILP	Say	-	pdf pinout
LA 47511	CMOS-IC	4-CH BTL Power Amp IC,Car Stereo Systems	128-MP	Say	-	pdf pinout
LA 47512	Dig. Audio	4-CH BTL Power Amp IC,Car Stereo Systems	QILP	Say	-	pdf pinout
LA 47516	Dig. Audio	4-CH BTL Power Amp IC,Car Stereo Systems	QILP	Say	-	pdf pinout
LA 47536	IC	4x NF/AF PAmp, 26V, 4x45W(14V/4Ohm) BTL mute, therm.+ overvolt. prot.	SILP	Say	-	pdf pinout
LA 70100 M	LIN-IC	SECAM Chroma-signal Processor IC for Vcr	TSOP	Say	-	pdf pinout
LA 72702 NV	IC	Btsc Decoder	16-DIP	Say	-	pdf pinout
LA 73024 AV	-IC	Double Scart Interface IC	FLP	Say	-	pdf pinout
LA 73054	IC	Video Signal Driver	TSOP	Say	-	pdf pinout
LA 73073 CL	LIN-IC	Video Driver for DSC	SQL	Say	-	pdf pinout
LA 76810	IC	Ntsc/pal/secam TV signal proc., I2C		Say	-	pdf
LA 76810A	IC	Ntsc/pal/secam TV signal proc., I2C		Say	-	pdf pinout
LA 78040	IC	TV,Crt Display Vert. Output IC	SILP	Say	-	pdf pinout
LA 78040 N	LIN-IC	TV,Crt Display Vert. Output IC	SILP	Say	-	pdf pinout
LA 78041	IC	Crt, HDTV, vertical deflection output	SQL	Say	-	pdf pinout
LA 78045	IC	Crt, HDTV, vertical deflection output, bus control support	SQL	Say	-	pdf pinout
LAA110	DOC	LED/MOSFET Viso:3750Vrms Vbs:1.2V		Cpc		data pinout
LAA676	LED	am ø5mm I:180...355mcd Vf:2V If:20mA		Osr		data pdf pinout
LAD 010	Hybrid-IC	Nf-treiber/driver	≈SIP	Mts		
LAD 050	Hybrid-IC	Nf-treiber/driver	≈SIP	Mts		
LAD 070	Hybrid-IC	Nf-treiber/driver	≈SIP	Mts	-	
LAE63B	LED	am ø5mm I:5600mcd Vf:2.2V If:50mA	PLCC-4	Osr		data pdf pinout
LAE65	LED	am ø5mm I>2800mcd Vf:2.3V If:50mA	PLCC-4	Osr		data pdf pinout
LAE67B	LED	am ø2.55mm I:560...1400mcd Vf:2.2V	PLCC-4	Osr		data pdf pinout
LAE655	LED	am ø2.55mm I:900...1800mcd Vf:2.1V	PLCC-4	Osr		data pdf pinout
LAE675	LED	am ø2.55mm I:450...900mcd Vf:2.1V	PLCC-4	Osr		data pdf pinout
LAG 637 D	LIN-IC	Recorder, 2x Nf-v+e			KA 22135	
LAL896	LED	am ø2.55mm I:90...180mcd Vf:2V	PLCC-4	Osr		data pdf pinout
LAM67B	LED	am ø2.55mm I:280...710mcd Vf:2.1V		Osr		data pdf pinout
LAM676	LED	am ø2.55mm I:140...355mcd Vf:2V		Osr		data pdf pinout
LAM776	LED	am ø2.55mm I:440mlm Vf:2V If:20mA		Osr		data pdf pinout
LAp	Si-P	=BF 550 (Typ-Code/Stempel/marking)	35	Phi	→BF 550	data
LAP 040	Hybrid-IC	Nf-verstärker/amplifier	≈SIP	Mts	-	
LAP 170	Hybrid-IC	NF-Vorverstärker/pre-amplifier	≈SIP	Mts	-	
LAP 191	Hybrid-IC	Nf-verstärker/amplifier	≈SIP	Mts	-	
LAS 14 CB	Z-IC	+14V, 3A, 30W(Tc=80°), -55...+150°	23a	Lbd		
LAS14U	ER+	Io=3A Vo:2.6..30V P:30W	5O	Lam,Pmi		data pinout

Type	Device	Short Description	Fig.	Manu	Comparision Types	More at
LAS 14 U	Z-IC	+2,65...30V, 3A, 30W(Tc=70°),-55..+150°	23/4Pin	Lbd	-	data
LAS 15 A05	Z-IC	+5V(±2%),1,5A,15W(Tc=105°),-55...+150°	23a	Lbd	... 7805... (TO-3)	
LAS 15 A12	Z-IC	+12V(±2%),1,5A,15W(Tc=105°),-55...+150°	23a	Lbd	... 7812... (TO-3)	
LAS 15 A15	Z-IC	+15V(±2%),1,5A,15W(Tc=105°),-55...+150°	23a	Lbd	... 7815... (TO-3)	
LAS15U	ER+	Io=1.5A Vo:4...30V P:15W	5O	Lam,Pmi		data pinout
LAS 15 U	Z-IC	+4...+30V,1,5A, 15W(Tc=105°),-55...+150°	23/4Pin	Lbd	-	data pdf
LAS16CB	R+	Io=2A Vo:13.3...14.7V	3O	Lam		data pinout
LAS 16 CB	Z-IC	+14V, 2A, 20W(Tc=80°), -55...+150°	23a	Lbd	-	data pdf
LAS16U	ER+	Io=2A Vo:4...30V P:20W	5O	Lam		data pinout
LAS 16 U	Z-IC	+4...+30V, 2A, 20W(Tc=80°), -55...+150°	23/4Pin	Lbd		data pdf
LAS 18 A05	Z-IC	-5V(±2%),1,5A,15W(Tc=105°),-55...+150°	23a	Lbd	... 7905... (TO-3)	pdf
LAS 18 A12	Z-IC	-12V(±2%),1,5A,15W(Tc=105°),-55...+150°	23a	Lbd	... 7912... (TO-3)	pdf
LAS 18 A15	Z-IC	-15V(±2%),1,5A,15W(Tc=105°),-55...+150°	23a	Lbd	... 7915... (TO-3)	pdf
LAS18U	ER-	Io=1.5A Vo:-2.6..-30V P:15W	5O	Pmi		data pinout
LAS 18 U	Z-IC	-4...-30V,1,5A, 15W(Tc=105°),-55...+150°	23/4Pin	Lbd	-	data pdf
LAS 19 A05...A12(P)	Z-IC	=LAS 1905...1912(P): ±2%		Uni		
LAS19U	R+	Io=5A Vo:11.4...12.6V P:50W	3O	Lam		data pinout
LAS 19 U	Z-IC	+4...+30V, 5A	23/4Pin	Uni		data pinout
LAS 39 A05(K)	Z-IC	=LAS 3905(K): ±2%		Uni		
LAS39U	ER+	Io=8A Vo:4...16V P:80W	5O	Uni		data pinout
LAS 39 U	Z-IC	+4...+16V, 8A	23/4Pin	Uni		data pinout
LAS723	ER+	Io=150mA Vo:2...38V P:800mW	10A	Pmi		data pinout
LAS 723 A	Z-IC	+2...38V, 0,15A, 0,8W, -55...+150°	82	Lbd	-	data
LAS723B	ER+	Io=150mA Vo:2...48V P:800mW	10A	Pmi		data pinout
LAS 723 B	Z-IC	+2...48V, 0,15A, 0,8W, -55...+150°	82	Lbd	-	data
LAS1000	ER+	Io=150mA Vo:2...38V P:800mW	10A	Pmi		data pinout
LAS 1000	Z-IC	=LAS 723A: integr. shot down	82	Lbd		data
LAS1100	ER+	Io=150mA Vo:2...48V P:800mW	10A	Pmi		data pinout
LAS 1100	Z-IC	=LAS 723B: integr. shot down	82	Lbd		data
LAS1405	R+	Io=3A Vo:4.5...5.5V P:30W	3O	Lam		data pinout
LAS 1405	Z-IC	+5V, 3A, 30W(Tc=80°), -55...+150°	23a	Lbd	KA 78T05CH, MC 78T05(A)CK	data
LAS1406	R+	Io=3A Vo:5.7...6.3V P:30W	3O	Lam,Pmi		data pinout
LAS 1406	Z-IC	+6V, 3A, 30W(Tc=80°), -55...+150°	23a	Lbd	KA 78T06CH, MC 78T06(A)CK	data
LAS1408	R+	Io=3A Vo:7.6...8.4V P:30W	3O	Lam,Pmi		data pinout
LAS 1408	Z-IC	+8V, 3A, 30W(Tc=80°), -55...+150°	23a	Lbd	KA 78T08CH, MC 78T08(A)CK	data
LAS1410	R+	Io=3A Vo:9.5...10.5V	3O	Pmi		data pinout
LAS 1410	Z-IC	+10V, 3A, 30W(Tc=80°), -55...+150°	23a	Lbd	KA 78T10CH, MC 78T10(A)CK	data
LAS1412	R+	Io=3A Vo:11.4...12.6V P:30W	3O	Lam,Pmi		data pinout
LAS 1412	Z-IC	+12V, 3A, 30W(Tc=80°), -55...+150°	23a	Lbd	KA 78T12CH, MC 78T12(A)CK	data
LAS1415	R+	Io=3A Vo:14.25...15.75V	3O	Lam,Pmi		data pinout
LAS 1415	Z-IC	+15V, 3A, 30W(Tc=80°), -55...+150°	23a	Lbd	KA 78T15CH, MC 78T15(A)CK	data
LAS1505	R+	Io=1.5A Vo:4.5...5.5V P:15W	3O	Lam		data
LAS 1505	Z-IC	+5V, 1,5A, 15W(Tc=105°), -55...+150°	23a	Lbd	... 7805... (TO-3)	data
LAS1506	R+	Io=1.5A Vo:5.7...6.3V P:15W	3O	Lam,Pmi		data pinout
LAS 1506	Z-IC	+6V, 1,5A, 15W(Tc=105°), -55...+150°	23a	Lbd	... 7806... (TO-3)	data
LAS1508	R+	Io=1.5A Vo:7.6...8.4V P:15W	3O	Lam,Pmi		data
LAS 1508	Z-IC	+8V, 1,5A, 15W(Tc=105°), -55...+150°	23a	Lbd	... 7808... (TO-3)	data
LAS1510	R+	Io=1.5A Vo:9.5...10.5V	3O	Pmi		data
LAS 1510	Z-IC	+10V, 1,5A, 15W(Tc=105°), -55...+150°	23a	Lbd	... 7810... (TO-3)	data
LAS1512	R+	Io=1.5A Vo:11.4...12.6V P:15W	3O	Lam,Pmi		data
LAS 1512	Z-IC	+12V, 1,5A, 15W(Tc=105°), -55...+150°	23a	Lbd	... 7812... (TO-3)	data
LAS1515	R+	Io=1.5A	3O	Lam,Pmi		data pinout
LAS 1515	Z-IC	+15V, 1,5A, 15W(Tc=105°), -55...+150°	23a	Lbd	... 7815... (TO-3)	data
LAS1518	R+	Io=1.5A Vo:17.1...18.9V P:15W	3O	Lam,Pmi		data
LAS 1518	Z-IC	+18V, 1,5A, 15W(Tc=105°), -55...+150°	23a	Lbd	... 7818... (TO-3)	data
LAS 1520	R+	Io=1.5A Vo:19...21V P:15W	3O	Pmi		data pinout
LAS 1520	Z-IC	+20V, 1,5A, 15W(Tc=105°), -55...+150°	23a	Lbd	... 7820... (TO-3)	data
LAS1524	R+	Io=1.5A Vo:22.8...25.2V P:15W	3O	Lam,Pmi		data pinout
LAS 1524	Z-IC	+24V, 1,5A, 15W(Tc=105°), -55...+150°	23a	Lbd	... 7824... (TO-3)	data
LAS1528	R+	Io=1.5A Vo:26.6...29.4V P:15W	3O	Pmi		data pinout
LAS 1528	Z-IC	+28V, 1,5A, 15W(Tc=105°), -55...+150°	23a	Lbd	... 7828... (TO-3)	data
LAS1605	R+	Io=2A Vo:4.75...5.25V P:20W	3O	Lam		data pinout
LAS 1605	Z-IC	+5V, 2A, 20W(Tc=80°), -55...+150°	23a	Lbd	KA 78T05CH, MC 78T05(A)CK	data pdf
LAS1606	R+	Io=2A Vo:5.7...6.3V P:20W	3O	Lam		data pinout
LAS 1606	Z-IC	+6V, 2A, 20W(Tc=80°), -55...+150°	23a	Lbd	KA 78T06CH, MC 78T06(A)CK	data
LAS1608	R+	Io=2A Vo:7.6...8.4V P:20W	3O	Lam		data pinout
LAS 1608	Z-IC	+8V, 2A, 20W(Tc=80°), -55...+150°	23a	Lbd	KA 78T08CH, MC 78T08(A)CK	data pdf
LAS 1610	Z-IC	+10V, 2A, 20W(Tc=80°), -55...+150°	23a	Lbd	KA 78T10CH, MC 78T10(A)CK	data
LAS1612	R+	Io=2A Vo:11.4...12.6V P:20W	3O	Lam		data pinout
LAS 1612	Z-IC	+12V, 2A, 20W(Tc=80°), -55...+150°	23a	Lbd	KA 78T12CH, MC 78T12(A)CK	data pdf
LAS1615	R+	Io=2A Vo:14.25...15.75V	3O	Lam		data pinout
LAS 1615	Z-IC	+15V, 2A, 20W(Tc=80°), -55...+150°	23a	Lbd	KA 78T15CH, MC 78T15(A)CK	data pdf
LAS1802	R-	Io=1.5A Vo:-1.9...-2.1V P:15W	3O	Pmi		data pinout
LAS 1802	Z-IC	-2V, 1,5A, 15W(Tc=105°), -55...+150°	23d	Lbd	... 7902... (TO-3)	data pdf
LAS1805	R-	Io=1.5A Vo:-4.75...-5.25V P:15W	3O	Lam,Pmi		data pinout
LAS 1805,2	Z-IC	-5,2V, 1,5A, 15W(Tc=105°), -55...+150°	23d	Lbd	... 7952... (TO-3)	data pdf
LAS 1805	Z-IC	-5V, 1,5A, 15W(Tc=105°), -55...+150°	23d	Lbd	... 7905... (TO-3)	data
LAS1806	R-	Io=1.5A Vo:-5.7...-6.3V P:15W	3O	Pmi		data pinout
LAS 1806	Z-IC	-6V, 1,5A, 15W(Tc=105°), -55...+150°	23d	Lbd	... 7906... (TO-3)	data pdf
LAS1808	R-	Io=1.5A Vo:-7.6...-8.4V P:15W	3O	Pmi		data pinout
LAS 1808	Z-IC	-8V, 1,5A, 15W(Tc=105°), -55...+150°	23d	Lbd	... 7908... (TO-3)	data pdf
LAS1810	R-	Io=1.5A Vo:-9.5...-10.5V P:15W	3O	Pmi		data pinout
LAS 1810	Z-IC	-10V, 1,5A, 15W(Tc=105°), -55...+150°	23d	Lbd	... 7910... (TO-3)	data pdf
LAS 1812	Z-IC	-12V, 1,5A, 15W(Tc=105°), -55...+150°	23d	Lbd	... 7912... (TO-3)	data pdf
LAS1815	R-	Io=1.5A Vo:-14.25...-15.75V P:15W	3O	Lam,Pmi		data pinout

Type	Device	Short Description	Fig.	Manu	Comparision Types	More at
LAS 1815	Z-IC	-15V, 1,5A, 15W(Tc=105°), -55...+150°	23d	Lbd	... 7915... (TO-3)	data pdf
LAS1818	R-	Io=1.5A Vo:-17.1...-18.9V P:15W	3O	Lam,Pmi		data pinout
LAS 1818	Z-IC	-18V, 1,5A, 15W(Tc=105°), -55...+150°	23d	Lbd	... 7918... (TO-3)	data pdf
LAS1820	R-	Io=1.5A Vo:-19...-21V P:15W	3O	Pmi		data pinout
LAS 1820	Z-IC	-20V, 1,5A, 15W(Tc=105°), -55...+150°	23d	Lbd	... 7920... (TO-3)	data pdf
LAS1824	R-	Io=1.5A Vo:-22.8...-25.2V P:15W	3O	Lam,Pmi		data pinout
LAS 1824	Z-IC	-24V, 1,5A, 15W(Tc=105°), -55...+150°	23d	Lbd	... 7924... (TO-3)	data pdf
LAS1828	R-	Io=1.5A Vo:-26.6...-29.4V P:15W	3O	Pmi		data pinout
LAS 1828	Z-IC	-28V, 1,5A, 15W(Tc=105°), -55...+150°	23d	Lbd	... 7928... (TO-3)	data
LAS1905	R+	Io=5A Vo:4.75...5.25V P:50W	3O	Lam		data pinout
LAS 1905(B)	Z-IC	+5V, 5A, ±5%, B=+5/-3%	23a	Uni	-	data pdf pinout
LAS 1905(B)P	Z-IC	=LAS 1905(B): Fig.→	16b	Uni	-	data pinout
LAS1906	R+	Io=5A Vo:5.7...6.3V P:50W	3O	Lam		data pinout
LAS 1906	Z-IC	+6V, 5A, ±5%, 50W(Tc=90°), -55...+150°	23a	Lbd	-	data
LAS1908	R+	Io=5A Vo:7.6...8.4V P:50W	3O	Lam		data pinout
LAS 1908	Z-IC	+8V, 5A, ±5%, 50W(Tc=90°), -55...+150°	23a	Lbd	-	data
LAS 1910	Z-IC	+10V, 5A, ±5%, 50W(Tc=90°), -55...+150°	23a	Lbd	-	data
LAS1912	R+	Io=5A Vo:11.4...12.6V P:50W	3O	Lam		data pinout
LAS 1912(B)	Z-IC	+12V, 5A, ±5%, B=+5/-3%	23a	Uni	-	data pinout
LAS 1912(B)P	Z-IC	=LAS 1912(B): Fig.→	16b	Uni	-	data pinout
LAS1915	R+	Io=5A Vo:14.25...15.75V	3O	Lam		data pinout
LAS 1915	Z-IC	+15V, 5A, ±5%, 50W(Tc=90°), -55...+150°	23a	Lbd	-	data
LAS3905	R+	Io=8A Vo:0.95...1.05V P:80W	3O,5O	Lam		data pinout
LAS 3905	Z-IC	+5V, 8A, ±5%, Sense Pin	23/4Pin	Uni	-	data pinout
LAS 3905 K	Z-IC	+5V, 8A, ±5%	23a	Uni	-	data pinout
LAS 6350	SMPS-IC	S-Reg, 5A, Fixed Current Limit	23/8Pin	Uni	-	data pinout
LAS 6350 P1	SMPS-IC	=LAS 6350: Fig.→	9-SIL	Uni	-	data pinout
LAS 6351	SMPS-IC	S-Reg, 5A, Adjustable Current Limit	23/8Pin	Uni	-	data pinout
LAS 6351 P1	SMPS-IC	=LAS 6351: Fig.→	9-SIL	Uni	-	data pinout
LAS 6380	SMPS-IC	S-Reg, 8A, Fixed Current Limit	23/8Pin	Uni	-	data pinout
LAS 6380 P1	SMPS-IC	=LAS 6380: Fig.→	9-SIL	Uni	-	data pinout
LAS 6381	SMPS-IC	S-Reg, 8A, Adjustable Current Limit	23/8Pin	Uni	-	data pinout
LAS 6381 P1	SMPS-IC	=LAS 6381: Fig.→	9-SIL	Uni	-	data pinout
Lat	Si-P	=BF 550 (Typ-Code/Stempel/marking)	35	Phi	→BF 550	data
LAT68B	LED	am ø2.55mm I:355...900mcd Vf:2.1V		Osr		data pdf pinout
LAT676	LED	am ø2.55mm I:180...355mcd Vf:2V	PLCC-2	Osr		data pdf pinout
LAT776	LED	am ø2.55mm I:180...355mcd Vf:2V		Osr		data pdf pinout
LATBG66B	LED	am/gr/bl ø2.55mm I<1120 *)mcd	PLCC-4	Osr		data pdf pinout
LATBT66B	LED	am/gr/bl ø2.55mm I<450 *)mcd	PLCC-4	Osr		data pdf pinout
LATBT686	LED	am/gr/bl ø2.55mm I<280 *)mcd	PLCC-4	Osr		data pdf pinout
LAW57B	LED	am ø2.55mm I=13...24cd Vf:2.2V		Osr		data pdf pinout
LAY876	LED	am ø2.55mm I:140...355mcd Vf:2V		Osr		data pdf pinout

LB

Type	Device	Short Description	Fig.	Manu	Comparision Types	More at
LB	Si-P	=2SA1734 (Typ-Code/Stempel/marking)	39	Tos	→2SA1734	data
LB	Si-N	=2SC2462-B (Typ-Code/Stempel/marking)	35	Hit	→2SC2462	data
LB	MOS-N-FET	=BF 999 (Typ-Code/Stempel/marking)	35	Aeg,Sie	→BF 999	data
LB	Si-N	=HD 2F3P (Typ-Code/Stempel/marking)	39	Nec	→HD 2...	data
LB	Si-N	=KTD1003-B (Typ-Code/Stempel/marking)	39	Kec	→KTD 1003	data
LB	Si-N+R	=RN 1442B (Typ-Code/Stempel/marking)	35	Tos	→RN 1442	
LB	MOS-N-FET-d*	=S 525T (Typ-Code/Stempel/marking)	35	Aeg	→S 525T	data
LB	MOS-P-FET-e	=Si 1305DL (Typ-Code/Stempel/marking)	35(2mm)	Six	→Si 1305DL	data
LB 1006	LIN-IC	Telefon, Anrufdetektor/ringing detect.	8-DIP	Tho		
LB 1009	LIN-IC	Telefon, Ein-chip Telefon (µproc.)	20-DIP	Tho		
LB 1011	LIN-IC	Telecom, Line Battery Feed	8-DIP	Tho		
LB 1013	LIN-IC	Telecom, 2x OP-Amp, 85V	15-SQL	Tho		
LB 1020	LIN-IC	Telefon, Speakerphone Kit	24-DIP	Tho		
LB 1021	LIN-IC	Telefon, Speakerphone Kit	18-DIP	Tho		
LB 1026	LIN-IC	Telefon, Voice Frequ. Level Expander	8-DIP	Tho		
LB 1027	LIN-IC	Telefon, Electret Preamplifier	8-DIP	Tho		
LB 1100	DIG-IC	9-Dioden-Matrix	20-DIP	Say		
LB 1101	DIG-IC	4-Dioden-Matrix	20-DIP	Say		
LB 1103	DIG-IC	Dioden-matrix	20-DIP	Say		
LB 1105 M	DIG-IC	SMD,4x6 Dioden-Array, gem./comm. Anode	30-MDIP	Say		pdf
LB 1200	DIG-IC	8x PNP+NPN-Trans.-Array, 25V, 0,1A	20-DIP	Say	-	pdf pinout
LB 1205	DIG-IC	4x Darl.-Tr, 62V, 1,5A	16-DIP	Say	-	pdf
LB 1206	DIG-IC	4x Darl.-Tr, 62V, 1,5A	16-DIP	Say	-	pdf
LB 1211	DIG-IC	7x Trans.-Array, com. Emitt.,50V, 0,2A	16-DIP	Say	-	pdf
LB 1212	DIG-IC	=LB 1211: Rb=10,5, Rbe=10kΩ, int. Z-Di	16-DIP	Say	-	pdf
LB 1213	DIG-IC	=LB 1211: Rb=2,7kΩ, Rbe=10kΩ	16-DIP	Say	-	pdf
LB 1213 M	DIG-IC	=LB 1213: SMD, 0,1A	16-MDIP	Say	-	pdf
LB 1214	DIG-IC	=LB 1211: Rb=10,5kΩ, Rbe=10kΩ	16-DIP	Say	-	pdf
LB 1215	DIG-IC	7x Trans.-Array, com. Coll., 50V, 0,2A	16-DIP	Say	-	pdf
LB 1216	DIG-IC	=LB 1215: Rb=2,7kΩ	16-DIP	Say	-	pdf
LB 1217	DIG-IC	5x Trans.-Array, 50V, 0,2A	16-DIP	Say	-	pdf
LB 1221	DIG-IC	2x Flip-flop, Ucc=12V		Say	-	
LB 1231	DIG-IC	7x Darl.-Trans.-Array, 50V, 0,5A	16-DIP	Say	-	pdf
LB 1232	DIG-IC	=LB 1231: Rb=10,5kΩ, int. Z-Diode	16-DIP	Say	-	pdf
LB 1233	DIG-IC	=LB 1231: Rb=2,7kΩ	16-DIP	Say	-	pdf
LB 1234	DIG-IC	=LB 1231: Rb=10,5kΩ	16-DIP	Say	-	pdf
LB 1235	DIG-IC	4x Darl.-Tr, 65V, 1,5A	16-DIP	Say	-	pdf
LB 1240	DIG-IC	FLT-Tr, 8x NPN-Darlington, 55V, 30mA	18-DIP	Say	-	pdf
LB 1241	DIG-IC	FLT-Tr, 8x NPN-Darlington, 45V, 30mA	18-DIP	Say	-	pdf

Type	Device	Short Description	Fig.	Manu	Comparision Types	More at
LB 1245	DIG-IC	FLT-Tr, 8x NPN-Darlington, 55V, 30mA	20-DIP	Say	-	pdf
LB 1246	DIG-IC	Druckertreiber/printer driver, 8xDarl.	18-DIP	Say	-	pdf
LB 1247	DIG-IC	8x NPN-Darlington, 7V, 0,4A	20-DIP	Say	-	pdf
LB 1248	DIG-IC	9x NPN-Darlington, 7V, 0,6A	20-DIP	Say	-	pdf pinout
LB 1249	DIG-IC	8x NPN-Darlington, 7V, 0,4A	20-DILP	Say	-	pdf
LB 1252	DIG-IC	Druckertreiber/printer driver	14-DIP	Say	-	pdf pinout
LB 1253	DIG-IC	Druckertreiber/printer driver	16-DIP	Say	-	pdf pinout
LB 1254	DIG-IC	7x Treiber/driver		Say	-	
LB 1255	DIG-IC	Druckertreiber/printer driver	18-DIP	Say		pdf pinout
LB 1256	DIG-IC	Druckertreiber/printer driver	18-DIP	Say	-	pdf
LB 1256 M	DIG-IC	=LB 1256: SMD	20-MDIP	Say		pdf
LB 1257	DIG-IC	8x NPN-Darlington, 7V, 0,4A	18-DIP	Say	-	pdf
LB 1258	DIG-IC	7x NPN-Darlington, 7V, 0,5A	16-DIP	Say	-	pdf
LB 1259	DIG-IC	Druckertreiber/printer driver	20-DILP	Say		pdf pinout
LB 1260	DIG-IC	Druckertreiber/printer driver	18-DIP	Say	M 54535P	pdf pinout
LB 1261	DIG-IC	Druckertreiber/printer driver	16-DIP	Say	-	pdf pinout
LB 1262	DIG-IC	7x NPN-Darlington, 20V, 0,05A	18-DIP	Say	-	pdf pinout
LB 1264	DIG-IC	Druckertreiber/printer driver	16-DIP	Say	-	pdf
LB 1265	DIG-IC	8x Treiber/driver, 25V, 0,1A	20-DIP	Say	-	pdf
LB 1265 M	DIG-IC	=LB 1265: SMD	20-MDIP	Say	-	pdf
LB 1267	DIG-IC	2x Magnet-treiber/driver, 8V, 2A	8-DIP	Say	-	pdf
LB 1267 M	DIG-IC	=LB 1267: SMD	8-MDIP	Say	-	pdf
LB 1268	DIG-IC	3x NPN-Darlington, 8V, 1A	8-DIP	Say	-	pdf pinout
LB 1269	DIG-IC	6x Druckertreiber/printer driver	16-DIP	Say	-	pdf pinout
LB 1270	DIG-IC	6x NPN-Darlington	14-DIP	Say	-	
LB 1271	DIG-IC	6x NPN-Darlington, 40V, 0,15A	14-DIP	Say	M 54527P	
LB 1272	DIG-IC	6x NPN-Darlington, 20V, 0,1A	14-DIP	Say	-	pdf
LB 1273 R	DIG-IC	6x NPN-Darlington, 20V, 0,15A	14-DIP+h	Say	-	pdf
LB 1274	DIG-IC	6x NPN-Darlington, 20V, 0,1A	14-DIP+h	Say	-	pdf
LB 1275	DIG-IC	7x NPN-Darlington, 20V, 0,1A	16-DIP	Say	-	pdf
LB 1276	DIG-IC	7x NPN-Darlington-Array, f. LED	16-DIP	Say	-	pdf
LB 1281	DIG-IC	2x Leistungs-Tr./power driver, Ucc=24V		Say	-	
LB 1282	DIG-IC	2x Leistungs-Tr./power driver, Ucc=12V		Say	-	
LB 1283	DIG-IC	Timer, Relais-Tr., Ucc=12V		Say	-	
LB 1284 F	DIG-IC	Darlington-Array, 20V, 0,4A	14-DIP+h	Say	-	
LB 1285 F	DIG-IC	5x NPN-Darlington, 20V, 0,4A	14-DIP+h	Say	-	
LB 1286 F	DIG-IC	=LB 1285F: 30V	14-DIP+h	Say	-	
LB 1287	DIG-IC	5x NPN-Darlington, 30V, 0,5A	14-DIP+h	Say	-	pdf
LB 1288	DIG-IC	=LB 1287: 20V, 0,5A	14-DIP+h	Say	M 54516P	pdf
LB 1290	DIG-IC	8x NPN-Darlington, 55V, 30mA	18-DIP	Say	-	pdf pinout
LB 1291	DIG-IC	8x NPN-Darlington, 55V, 30mA	18-DIP	Say	-	pdf
LB 1292	DIG-IC	6x NPN-Darlington, 55V, 30mA	16-DIP	Say	-	pdf
LB 1293	DIG-IC	6x NPN-Darlington, 55V, 30mA	16-DIP	Say	-	pdf
LB 1294	DIG-IC	6x NPN-Darlington, 60V, 60mA	16-DIP	Say	-	pdf
LB 1331	DIG-IC	Ctv, NTSC VIR-Signal	16-DIP	Say	-	
LB 1332	DIG-IC	Ctv, NTSC Vir-signal	16-DIP	Say	-	pdf
LB 1403(N)	DIG-IC	5-LED Pegelanz.-Tr./level meter	9-SIP	Say	AN 6884, BA 6124, KA 2284	pdf pinout
LB 1405	DIG-IC	5-LED Pegelanz.-Tr./level meter	16-DIP	Say	-	pdf
LB 1407	DIG-IC	7-LED Pegelanz.-Tr./level meter, log.	14-DIP	Say	-	pdf
LB 1408	DIG-IC	7-LED Pegelanz.-Tr./level meter	16-DIP	Say	-	pdf
LB 1409	DIG-IC	9-LED Pegelanz.-Tr./level meter	16-DIP	Say	-	pdf pinout
LB 1409 M	DIG-IC	=LB 1409: SMD	16-MDIP	Say	-	
LB 1410	DIG-IC	10-LED Pegelanz.-Tr./level meter	18-DIP	Say	-	pdf
LB 1411	DIG-IC	10-LED Pegelanz.-Tr./level meter	16-DIP	Say	-	pdf
LB 1412	DIG-IC	12-LED Anz.-Tr./level meter, peak hold	22-DIP	Say	-	pdf
LB 1412 M	DIG-IC	=LB 1412: SMD	24-MDIP	Say	-	pdf
LB 1413(N)	DIG-IC	5-LED Pegelanz.-Tr./level meter	9-SIP	Say	KA 2287	pdf pinout
LB 1415	DIG-IC	5-LED Pegelanz.-Tr./level meter	16-DIP	Say	-	pdf
LB 1416	DIG-IC	5-LED Pegelanz.-Tr./level meter, log.	14-DIP+g	Say	-	pdf
LB 1417	DIG-IC	=LB 1407: linear	14-DIP	Say	-	pdf
LB 1419	DIG-IC	9-LED Pegelanz.-Tr./level meter	16-DIP	Say	-	pdf pinout
LB 1423(N)	DIG-IC	5-LED Pegelanz.-Tr./level meter	9-SIP	Say	BA 6137, KA 2285	pdf pinout
LB 1426	DIG-IC	5-LED Pegelanz.-Tr./level meter, lin.	14-DIP+g	Say	-	pdf
LB 1433(N)	DIG-IC	5-LED Pegelanz.-Tr./level meter	9-SIP	Say	BA 6125, KA 2286	pdf pinout
LB 1436	DIG-IC	5-LED Pegelanz.-Tr./level meter, log.	14-DIP+g	Say	-	pdf
LB 1443(N)	DIG-IC	5-LED Pegelanz.-Tr./level meter	9-SIP	Say	-	pdf
LB 1450	KOP-IC	FM LED-Tuning-indicator	9-SIP	Say	-	pdf
LB 1460	DIG-IC		14-DIP	Say	-	
LB 1470	DIG-IC	16-LED AM/FM-Frequ.-Anz.-Tr./display	22-DIP	Say	-	
LB 1473	DIG-IC	16-LED AM/FM-Frequ.-Anz.-Tr./display	22-DIP	Say	-	pdf pinout
LB 1475	DIG-IC	2-Draht-FB/wired remote control	20-DIP	Say	-	pdf pinout
LB 1475 M	DIG-IC	=LB 1475: SMD	24-MDIP	Say	-	
LB 1493	DIG-IC	5-LED Pegelanz.-Tr./level meter	9-SIP	Say	-	
LB 1494	DIG-IC	10-Dot Pegelanz.-Tr./level meter f.FLT	16-DIP	Say	-	
LB 1500	DIG-IC	Tv-kanalwahl/channel selector	16-DIP	Say	-	pdf
LB 1501	DIG-IC	Tv-kanalwahl/channel selector	16-DIP	Say	-	
LB 1515	DIG-IC	Tv-kanalwahl/channel selector	16-DIP	Say	-	pdf
LB 1530	DIG-IC	Tv-kanalwahl/channel selector	16-DIP	Say	-	
LB 1550	DIG-IC	Tv-kanalwahl/channel selector		Say	-	
LB 1551	DIG-IC	TV-sensor	16-DIP	Say	-	pdf
LB 1601	DIG-IC	Motorregler/dc motor speed control	14-DIP+g	Say	-	pdf pinout
LB 1609	DIG-IC	Motorregler/dc Motor Driver	16-DIP	Say	-	
LB 1615	DIG-IC	3-Phasen Motor-Tr./DD motor driver	20-DILP	Say	-	
LB 1616 N	DIG-IC	3-Phasen Motor-Tr./DD motor driver	20-DILP	Say	-	
LB 1619 M	DIG-IC	SMD, Vc, 3-Phase DD-Motor-Tr., 1,5A	30-SDIP	Say		data pdf pinout

Type	Device	Short Description	Fig.	Manu	Comparision Types	More at
LB 1620	DIG-IC	Vc, 3-Phase DD-Motor-Tr., 1,5A	20-DILP	Say	-	data pdf pinout
LB 1620 M	DIG-IC	=LB 1620: SMD	30-MDIP	Say	-	
LB 1622	DIG-IC	3-Phasen Motor-Tr./DD motor driver	30-SDIP	Say	-	pdf pinout
LB 1623	DIG-IC	3-Phasen Motor-Tr./DD motor driver	20-DILP	Say	-	pdf pinout
LB 1624 D	DIG-IC	3-Phasen Motor-Tr./DD motor driver	30-SDIP	Say	-	
LB 1624 M	DIG-IC	=LB 1624D: SMD	30-MDIP	Say	-	
LB 1625	DIG-IC	3-Phasen Motor-Tr/DD motor driver		Say	-	
LB 1630	DIG-IC	Motorreg./motor driver, bidirectional	8-DIP	Say	-	pdf
LB 1630 M	DIG-IC	=LB 1630: SMD	20-MDIP	Say	-	pdf
LB 1631 D	DIG-IC	Motorreg./motor diver, bidirectional		Say	-	
LB 1631 M	DIG-IC	=LB 1631D: SMD	16+4-MDIP	Say	-	
LB 1633 M	DIG-IC	Motorreg./motor driver, bidirectional	16+4-MDIP	Say	-	pdf
LB 1634 M	DIG-IC	Motorreg.-/motor driver, bidirectional	16+4-SMDIP	Say	-	pdf
LB 1635 M	DIG-IC	Motorreg.-/motor driver, bidirectional	10-SMDIP	Say	-	pdf
LB 1636 M	DIG-IC	Motorreg./motor driver, bidirectional	16-MDIP	Say	-	pdf
LB 1638	DIG-IC	Motorreg./motor driver, bidirectional	10-DIP	Say	-	pdf
LB 1638 M	DIG-IC	=LB 1638: SMD	10-SMDIP	Say	-	pdf
LB 1640 N	DIG-IC	Forw./rev. Motor Driver with Brake	10-SIL	Say	-	pdf pinout
LB 1641	DIG-IC	Motorreg./motor drv., bidirect., brake	10-SIP	Say	-	pdf pinout
LB 1642	DIG-IC	Motor-Tr., bidirectional, brake, ±0,1A	10-SIP	Say	-	data pdf pinout
LB 1642 B	DIG-IC	=LB 1642: ±0,3mA	10-SIP	Say	-	
LB 1643	DIG-IC	Motorreg./motor drv., bidirect., brake	10-SIP	Say	-	pdf
LB 1644	DIG-IC	Dual, Motorreg./motor driver,bidirect.	16-DIP	Say	-	pdf
LB 1645 N	DIG-IC	Motorregler/motor driver, bidirect.	10-SIL	Say	-	pdf
LB 1646	DIG-IC	Vc, Motorreg./reel motor controller	14-SILP	Say	-	pdf
LB 1648	DIG-IC	Dual, Motorreg./motor driver,bidirect.	12-DIP+b	Say	-	pdf
LB 1649	DIG-IC	Dual, Motorreg./motor driver,bidirect.	12-DIP+b	Say	-	pdf
LB 1650	DIG-IC	Dual, Motorreg./motor driver,bidirect.	16-DIP	Say	-	pdf
LB 1651	DIG-IC	Dual, Motorreg./motor driver,bidirect.	20-DILP	Say	-	pdf
LB 1656	DIG-IC	Fdd, 2-Phase Stepping Motor Driver	16-DIP	Say	-	pdf
LB 1656 M	DIG-IC	=LB 1656: SMD	16+4-SMDIP	Say	-	pdf
LB 1657 M	DIG-IC	SMD,Fdd, 2-Phase Stepping Motor Driver	16+4-SMDIP	Say	-	pdf
LB 1660 N	DIG-IC	2-Ph. Brushless DC Fan-Motor-Tr, 1,5A	8-DIP	Say	-	pdf
LB 1661	DIG-IC	2-Ph. Brushless DC Fan-Motor-Tr, 1,5A	8-DIP	Say	-	pdf
LB 1662 D	DIG-IC	2-Ph. Brushless DC Fan-Motor-Tr, 1,5A	10-DIP	Say	-	pdf
LB 1662 M	DIG-IC	=LB 1662D: SMD	16+4-SMDIP	Say	-	pdf
LB 1663	DIG-IC	2-Ph. Brushless DC Fan-Motor-Tr, 1,5A	10-DIP	Say	-	pdf
LB 1663 M	DIG-IC	=LB 1663: SMD	16+4-SMDIP	Say	-	pdf
LB 1664 N	DIG-IC	2-Ph. Brushless DC Fan-Motor-Tr, 1,5A	16-DIP	Say	-	pdf
LB 1665	DIG-IC	2-Ph. Brushless DC Fan-Motor-Tr, 1,5A	16-DIP	Say	-	pdf
LB 1666	DIG-IC	2-Ph. Brushless DC Fan-Motor-Tr, 1,5A	16-DIP	Say	-	pdf
LB 1667	DIG-IC	2-Ph. Brushless DC Fan-Motor-Tr, 1,5A	10-DIP	Say	-	pdf
LB 1667 M	DIG-IC	=LB 1667: SMD	14-SMDIP	Say	-	pdf
LB 1668	DIG-IC	2-Ph. Brushless DC Fan-Motor-Tr, 1,5A	10-DIP	Say	-	pdf
LB 1668 M	DIG-IC	=LB 1668: SMD	14-SMDIP	Say	-	pdf
LB 1669	DIG-IC	2-Ph. Brushless DC Fan-Motor-Tr, 1,5A	10-DIP	Say	-	pdf
LB 1669 M	DIG-IC	=LB 1669: SMD	10-SMDIP	Say	-	pdf
LB 1670 M	DIG-IC	SMD, Cd, 3-Phase Motor-Tr., Ucc=3V	30-SMDIP	Say	-	pdf
LB 1671 M	DIG-IC	SMD, Cd, 3-Phase Motor-Tr., Ucc=1,5V	30-MDIP	Say	-	
LB 1673 M	DIG-IC	SMD, Cd, 3-Phase Motor-Tr., Ucc=1,5V	24-SMDIP	Say	-	data pdf pinout
LB 1674 M	DIG-IC	SMD, Cd, 3-Phase Motor-Tr., Ucc=1,5V	24-SMDIP	Say	-	data pdf pinout
LB 1674 V	DIG-IC	Brushless, Sensorless Motor Driver	24-SSMDIP	Say	-	pdf pinout
LB 1676 M	DIG-IC	SMD, Fdd, 3-Phase Motor-Tr., Ucc=5V	16+4-SMDIP	Say	-	
LB 1677 M	DIG-IC	SMD, Fdd, 3-Phase Motor-Tr., Ucc=5V	16+4-SMDIP	Say	-	
LB 1682	DIG-IC	3-Phase DD-Motor-Tr	30-SDIP	Say	-	
LB 1684	DIG-IC	Vc, 3-Phase DD-Motor-Tr., 1,5A	20-DILP	Say	-	data pdf pinout
LB 1686 D	DIG-IC	3-Phase DD-Motor-Tr	20-DILP	Say	-	
LB 1686 M	DIG-IC	=LB 1686D: SMD	30-MDIP	Say	-	
LB 1687	DIG-IC	Vc, 3-Phase DD-Motor-Tr., 1,5A	30-SDIP	Say	-	data pdf pinout
LB 1687 M	DIG-IC	=LB 1687D: SMD	30-SMDIP	Say	-	pdf
LB 1688	DIG-IC	Vc, 3-Phase DD-Motor-Tr., 1,5A	30-SDIP	Say	-	pdf
LB 1689 D	DIG-IC	Vc, 3-Phase DD-Motor-Tr, 1,5A	30-SDIP	Say	-	pdf
LB 1689 M	DIG-IC	=LB 1689D: SMD	30-SMDIP	Say	-	pdf
LB 1690	DIG-IC	3-Phase DD Fan-Motor-Tr., Ucc=5V, 2,5A	20-DILP	Say	-	data pdf pinout
LB 1692	DIG-IC	3-Phase DD Fan-Motor-Tr., Ucc=5V, 2,5A	20-DILP	Say	-	data pdf pinout
LB 1693	DIG-IC	3-Phase DD(Fan)Motor-Tr., Ucc=5V, 2,5A	20-DILP	Say	-	data pdf pinout
LB 1710	DIG-IC	7x NPN-Darlington, 50V, 0,4A	16-DIP	Say	-	pdf
LB 1720	DIG-IC	8x NPN-Tr.,Rb=4,7, Rbe=47kΩ, 20V, 0,2A	18-DIP	Say	-	pdf
LB 1721 M	DIG-IC	=LB 1720: SMD	20-MDIP	Say	-	pdf
LB 1730	DIG-IC	4x NPN-Darlington, 85V, 1,5A	16-DIP	Say	-	pdf
LB 1731	DIG-IC	4x NPN-Darlington, 82V, 1,5A	16-DIP	Say	-	pdf
LB 1740	DIG-IC	8x NPN-Darlington, 50V, 0,5A	18-DIP	Say	-	pdf
LB 1745	DIG-IC	8x NPN-Darlington, 50V, 0,5A	18-DIP	Say	-	pdf
LB 1760	DIG-IC	6x NPN-Transistor, 20V, 0,32A	16-DIP	Say	-	pdf pinout
LB 1807	DIG-IC	VC		Say	-	
LB 1810 M	DIG-IC	SMD, Fdd, 3,5", 3-Phase DD-Motor-Tr	44-MP	Say	-	pdf
LB 1813 M	DIG-IC	SMD, Fdd, 3,5", 3-Phase DD-Motor-Tr.	36-SMDIP	Say	-	data pdf pinout
LB 1816	DIG-IC	Fdd, 3,5/5,25", 3-Phase DD-Motor-Tr	28-DILP	Say	-	
LB 1817 M	DIG-IC	SMD, Fdd, 3-Phase DD-Motor-Tr.	36-SMDIP	Say	-	data pdf pinout
LB 1820	DIG-IC	3-Phase DD-Motor-Tr., 2,5A	28-DILP	Say	-	data pdf pinout
LB 1822	DIG-IC	3-Phase DD-Motor-Tr., 30mA	28-DILP	Say	-	data pdf pinout
LB 1823	DIG-IC	3-Phase DD-Motor-Tr., 30mA	30-SDIP	Say	-	data pdf pinout
LB 1823 M	DIG-IC	=LB 1823: SMD	30-SMDIP	Say	-	pdf
LB 1824	DIG-IC	3-Phase DD-Motor-Tr., 2,5A	28-DILP	Say	-	data pdf pinout
LB 1825	DIG-IC	3-Phase DD-Motor-Tr., 2A	28-DILP	Say	-	data pdf pinout

Type	Device	Short Description	Fig.	Manu	Comparision Types	More at
LB 1830 M	DIG-IC	Motorreg./motor driver, bidirectional	10-SMDIP	Say	-	pdf
LB 1831 M	DIG-IC	Motorreg./motor driver, bidirectional	16+4-SMDIP	Say	-	pdf
LB 1832 V	DIG-IC	Motorreg./motor driver, bidirectional	24-SSMDIP	Say	-	pdf
LB 1833 M	DIG-IC	Motorreg./motor driver, bidirectional	16+4-SMDIP	Say	-	pdf
LB 1834 M	DIG-IC	Motorreg./motor driver, bidirectional	16+4-SMDIP	Say	-	pdf
LB 1836 M	DIG-IC	Motorreg./motor driver, bidirectional	14-SMDIP	Say	-	pdf pinout
LB 1837 M	DIG-IC	Motorreg./motor driver, bidirectional	14-SMDIP	Say	-	pdf
LB 1838 M	DIG-IC	Motorreg./motor driver, bidirectional	14-SMDIP	Say	-	pdf
LB 1839 M	DIG-IC	Motorreg./motor driver, bidirectional	14-SMDIP	Say	-	pdf
LB 1841 V	DIG-IC	Motorreg./motor driver, bidirectional	20-SSMDIP	Say	-	pdf
LB 1843 V	DIG-IC	Motorreg./motor driver, bidirectional	20-SSMDIP	Say	-	pdf
LB 1847	DIG-IC	PWM Stepping Motor Driver	MP	Say	-	pdf pinout
LB 1851 M	DIG-IC	SMD, Vc, 3-Phase DD-Motor-Tr., 1,5A	30-SMDIP	Say	-	data pdf pinout
LB 1854 M	DIG-IC	SMD, Vc, 3-Phase DD-Motor-Tr., 1,5A	30-SMDIP	Say	-	data pdf pinout
LB 1855 M,NM	DIG-IC	SMD, VC, 3-Phase DD Motor Drv, 1,2A	16+4-SMDIP	Say	-	
LB 1855 M	DIG-IC	SMD, Vc, 3-Phase DD-Motor-Tr, 1,2A	16+4-SMDIP	Say	-	
LB 1857 M	DIG-IC	SMD, Vc, 3-Phase DD-Motor-Tr, 1,5A	30-SMDIP	Say	-	data pdf pinout
LB 1860	DIG-IC	2-Ph. Brushless DC Fan-Motor-Tr, 1,5A	10-DIP	Say	-	pdf
LB 1860 M	DIG-IC	=LB 1860: SMD	14-SMDIP	Say	-	pdf
LB 1861	DIG-IC	2-Ph. Brushless DC Fan-Motor-Tr, 1,5A	10-DIP	Say	-	pdf
LB 1861 M	DIG-IC	=LB 1861: SMD	14-SMDIP	Say	-	pdf
LB 1863 M	DIG-IC	2-Ph. Brushless DC Fan-Motor-Tr, 1,5A	14-SMDIP	Say	-	pdf
LB 1869 M	DIG-IC	2-Ph. Brushless DC Fan-Motor-Tr, 1,5A	14-SMDIP	Say	-	pdf
LB 1870	DIG-IC	3-Phase DD-Motor-Tr., 1A	28-DILP	Say	-	data pdf pinout
LB 1870 M	DIG-IC	=LB 1870: SMD	36-SMDIP	Say	-	data pdf pinout
LB 1871	DIG-IC	3-Phase DD-Motor-Tr., 1A	28-DILP	Say	-	data pdf pinout
LB 1871 M	DIG-IC	=LB 1871: SMD	36-SMDIP	Say	-	data pdf pinout
LB 1877 M	DIG-IC	Brushless Motor Driver	SIP	Say	-	pdf pinout
LB 1877 V	DIG-IC	Brushless Motor Driver	SSMDIP	Say	-	pdf pinout
LB 1880 M	DIG-IC	SMD, VC(8mm), 3-Ph. DD-Motor-Tr, 0,8A	36-SMDIP	Say	-	
LB 1881 M	DIG-IC	SMD, Vc, 3-Phase DD-Motor-Tr., 1A	30-SMDIP	Say	-	data pdf pinout
LB 1881 V	DIG-IC	=LB 1881M:	24-SSMDIP	Say	-	pdf
LB 1882 V	DIG-IC	SMD, CD/DAT, 3-Phase DD-Motor-Tr., 1A	24-SSMDIP	Say	-	data pdf pinout
LB 1885 M	DIG-IC	SMD, VC(8mm), 3-Ph. DD-Motor-Tr, 0,8A	36-SMDIP	Say	-	pdf
LB 1886 V	DIG-IC	SMD, Vc, 3-Phase DD-Motor-Tr., 1A	24-SSMDIP	Say	-	data pdf pinout
LB 1890 M	DIG-IC	SMD, Fdd, 3,5", 3-Phase DD-Motor-Tr.	36-SMDIP	Say	-	pdf
LB 1893	DIG-IC	SMD, Cd-rom, 3-Phase DD-Spindle-Motor	34-MP	Say	-	pdf
LB 1894 M	DIG-IC	SMD,Cd-rom, 3-Ph. DD-Spindle-Motor-Tr.	30-SMDIP	Say	-	data pdf pinout
LB 1920	DIG-IC	3-Phase DD-Motor-Tr., 3,1A	28-DILP	Say	-	data pdf pinout
LB 1927	DIG-IC	Three-phase Brushless Motor Driver	8-SIP	Say	-	pdf pinout
LB 1928	DIG-IC	Three-phase Brushless Motor Driver	8-SIP	Say	-	pdf pinout
LB 1929	DIG-IC	Three-phase Brushless Motor Driver	MP	Say	-	pdf pinout
LB 1930 M	DIG-IC	Single-channel Motor Driver	DIP	Say	-	pdf pinout
LB 1945 H	IC	PWM Cur.contr.,Stepping Motor Driver	TSOP	Sca	-	pdf pinout
LB 1947	DIG-IC	PWM DC Motor Driver	8-SIP	Say	-	pdf pinout
LB 1948 M	DIG-IC	forward/reverse 12 V motor driver	MDIP	Say	-	pdf pinout
LB 1971 V	DIG-IC	Three-phase Spindle Motor Driver	SSMDIP	Say	-	pdf pinout
LB 3120	DIG-IC			Say	-	
LB3331	LED	bl ø5mm I:3mcd Vf:3.1V If:20mA	T1¾	Sie		data pinout
LB3333	LED	bl ø3.0mm I:112...450mcd Vf:3.5V	T1	Osr		data pdf pinout
LB3336	LED	bl ø3mm I:71...112mcd Vf:3.5V If:10mA	T1	Osr		data pdf pinout
LB3356	LED	bl ø3mm I:18...28mcd Vf:3.5V If:10mA	T1	Osr		data pdf pinout
LB 3500	DIG-IC	FM-Frequ.-Teiler/divider, 1:8	9-SIP	Say	-	pdf
LB5413	LED	bl ø5mm I:710...2800mcd Vf:3.5V	T1¾	Osr		data pdf pinout
LB5416	LED	bl ø5mm I:180...280mcd Vf:3.5V	T1¾	Osr		data pdf pinout
LB5433	LED	bl ø5.0mm I:71...710mcd Vf:3.5V	T1¾	Osr		data pdf pinout
LB5456	LED	bl ø5mm I:11.2...18mcd Vf:3.5V	T1¾	Osr		data pdf pinout
LB 8015	DIG-IC	Schmitt-Trigger	8-DIP	Say	-	pdf pinout
LB 8016	DIG-IC	Schmitt-trigger, Timer		Say	-	
LB 8050	DIG-IC	Kfz Scheibenw.-strg./car window wiper	8-DIP	Say	-	pdf
LB 8106 M	DIG-IC	SMD, Cd, Actuator Driver, Ucc=2...4V	44-MP	Say	-	pdf
LB 8107 M	DIG-IC	SMD, Cd, Actuator Driver, Ucc=2...4V	44-MP	Say	-	pdf
LB 8108 M	DIG-IC	SMD, Cd, Actuator Driver, Ucc=2...4V	44-MP	Say	-	pdf
LB 8555 D	DIG-IC	Timer, Ucc=4,5...16V, ±200mA	8-DIP	Say	→NE 555 N	pdf
LB 8555 M	DIG-IC	=LB 8555D: SMD	8-MDIP	Say	-	pdf
LB 8555 S	DIG-IC	=LB 8555D: Fig.→	8-SIP	Say	-	pdf
LB 8620 M	DIG-IC	Motorreg./motor driver, bidirectional	30-SMDIP	Say	-	pdf
LB 8700	DIG-IC	Thermodrucker-tr./thermal Printer Drv.	9-SIP	Say	-	pdf pinout
LB 8900 M	DIG-IC	Clock Driver f. Ccd-sensor	24-MDIP	Say	-	
LB 9010 C	DIG-IC	Foto-interrupter		Say	-	
LB 9050	DIG-IC	Hall-IC, Schalter/switch	≈3-SIP	Say	-	
LB 9051	DIG-IC	Hall-IC, Schalter/switch	≈3-SIP	Say	-	pdf
LB 9052	DIG-IC	Hall-IC, Schalter/switch	≈3-SIP	Say	-	pdf
LB 11847	DIG-IC	PWM Stepping Motor Driver	SIP	Say	-	pdf pinout
LB 11920	Dig. Audio	Three-phase Brushless Motor Driver	24-SSMDIP	Say	-	pdf pinout
LB 11922	DIG-IC	Three-phase Brushless Motor Driver	24-SSMDIP	Say	-	pdf pinout
LB 11923 V	DIG-IC	Three-phase Brushless Motor Driver	24-SSMDIP	Say	-	pdf pinout
LB 11995 M	DIG-IC	Three-ph. Brushless Motor Driver	DIP	Say	-	pdf pinout
LBA67C	LED	bl ø5mm I:56...140mcd Vf:3.6V If:20mA		Osr		data pdf pinout
LBA110	DOC	LED/MOSFET Viso:3750Vrms Vbs:1.2V		Cpc		data pinout
LBA673	LED	bl ø5mm I:45...112mcd Vf:3.5V If:20mA		Osr		data pdf pinout
LBA676	LED	bl ø5mm I:9...22.4mcd Vf:3.5V If:10mA		Osr		data pdf pinout
LBB110	DOC	LED/MOSFET Viso:3750Vrms Vbs:1.2V		Cpc		data pinout
LBBT670	LED	bl ø5mm I:1.8mcd Vf:2V If:10mA	PLCC-4	Sie		data pinout
LBC 546...550	Si-N	→BC 546...550				

607

Type	Device	Short Description	Fig.	Manu	Comparision Types	More at
LBC 556..560	Si-P	→BC 556..560				
LBE63C	LED	bl ø5mm I:90...224mcd Vf:3.9V If:30mA	PLCC-4	Osr		data pdf pinout
LBE673	LED	bl ø5mm I:45...112mcd Vf:3.7V If:30mA	PLCC-4	Osr		data pdf pinout
LBK376	LED	bl ø3mm I:28...35.5mcd Vf:3.5V	T1	Osr		data pdf pinout
LBL89C	LED	bl ø3mm I:28...71mcd Vf:3.6V If:20mA	SOD80	Osr		data pdf pinout
LBL89S	LED	bl ø3mm I:11.2...28mcd Vf:3.6V	SOD80	Osr		data pdf pinout
LBM47C	LED	bl ø3mm I:71...180mcd Vf:3.6V If:20mA		Osr		data pdf pinout
LBM670	LED	bl ø3mm I>0.35mcd Vf:3.1V If:20mA	PLCC-2	Sie		data pinout
LBM673	LED	bl ø3mm I:35...90mcd Vf:3.5V If:30mA		Osr		data pdf pinout
LBM676	LED	bl ø3mm I:7.1...18mcd Vf:3.5V If:10mA		Osr		data pdf pincut
LBQ99A	LED	bl ø3mm I:7.1...18mcd Vf:3.5V If:10mA		Osr		data pdf pinout
LBQ993	LED	bl ø3mm I:7.1...28mcd Vf:3.5V If:10mA		Osr		data pdf pinout
LBR99A	LED	bl ø3mm I:2.8...10mcd Vf:3.8V If:10mA		Osr		data pdf pinout
LBT67C	LED	bl ø2.4mm I:56...140mcd Vf:3.6V	PLCC-2	Osr		data pdf pinout
LBT68C	LED	bl ø2.4mm I:56...140mcd Vf:3.6V	PLCC-2	Osr		data pdf pinout
LBT670	LED	bl ø2.4mm I:0.35mcd Vf:3.1V If:20mA	PLCC-2	Sie		data pinout
LBT673	LED	bl ø2.4mm I:45...112mcd Vf:3.5V	PLCC-2	Osr		data pdf pinout
LBT676	LED	bl ø2.4mm I:9...22.4mcd Vf:3.5V	PLCC-2	Osr		data pdf pinout
LBT770	LED	bl ø2.4mm I:0.35mcd Vf:3.1V If:20mA	PLCC-2	Sie		data pinout
LBT773	LED	bl ø2.4mm I:45...112mcd Vf:3.5V	PLCC-2	Osr		data pdf pinout
LBT776	LED	bl ø2.4mm I:9...22.4mcd Vf:3.5V		Osr		data pdf pinout
LBW5SG	LED	bl ø2.4mm I=5.2...11.2cd Vf:3.8V		Osr		data pdf pinout
LBY87C	LED	bl ø2.4mm I:90...224mcd Vf:3.6V		Osr		data pdf pinout
LBY876	LED	bl ø2.4mm I:7.1...18mcd Vf:3.5V		Osr		data pdf pinout

LC

Type	Device	Short Description	Fig.	Manu	Comparision Types	More at
LC	Si-P	=2SA1735 (Typ-Code/Stempel/marking)	39	Tos	→2SA1735	data
LC	Si-N	=2SC2462-C (Typ-Code/Stempel/marking)	35	Hit	→2SC2462	data
LC	Si-N	=2SC4357-C (Typ-Code/Stempel/marking)	39	Mit	→2SC4357	data
LC	Si-N	=HD 2L3N (Typ-Code/Stempel/marking)	39	Nec	→HD 2...	data
LC	Si-N	=KTD1003-C (Typ-Code/Stempel/marking)	39	Kec	→KTD 1003	data
LC	MOS-P-FET-e	=Si 1307DL (Typ-Code/Stempel/marking)	35(2mm)	Six	→Si 1307DL	data
LC307P	LED	gr+bl ø3.1rnm I=3.2cd Vf:3.4V If:30mA	T1	Cot		data pdf pinout
LC307T	LED	rd+or+yl ø3.1mm I=1.8cd Vf:2.3V	T1	Cot		data pdf pinout
LC374TWN	LED	wht ø3.1mm I=2.6cd Vf:3.4V	T1	Cot		data pdf pinout
LC377PWH	LED	wht ø3mm I=1cd Vf:4V		Cot		data pdf pinout
LC503	LED	gr+bl ø5mm I=7.5cd Vf:3.4V If:30mA	T1¾	Cot		data pdf pinout
LC503A	LED	rd+yl ø5mm I=4.5cd Vf:2.1V If:50mA	T1¾	Cot		data pdf pinout
LC503TWN	LED	wht ø5mm I=18cd Vf:3.4V	T1¾	Cot		data pdf pinout
LC503TX	LED	wht ø5mm I=7cd Vf:3.4V	T1¾	Cot		data pdf pinout
LC512TX	LED	wht ø5mm I=7.2cd Vf:3.4V	T1¾	Cot		data pdf pinout
LC810	OP	Vs:6V Vu:1000V/mV Vi0:2.5mV f:0.6Mc	8D	Gen		data pdf pinout
LC 3100	CMOS-ROM-IC	Telecom, 128 Bit Mask ROM	24-DIP,div	Say	-	pdf
LC 3101	CMOS-ROM-IC	Telecom, 128-kBit Mask ROM	24-DIP	Say	-	pdf
LC 3514(D,E,L)	CMOS-sRAM-IC	1024 x 4 Bit, <200ns	18-DIP	Say	-	
LC 3516(D)	CMOS-sRAM-IC	2048 x 8 Bit, <250ns	24-DIP	Say	-	
LC 3517(D)	CMOS-sRAM-IC	2048 x 8 Bit, <250ns	24-DIP	Say	-	
LC 3518 ...	CMOS-sRAM-IC	2048 x 8 Bit	24-DIP	Say	-	
LC 3564 ...	CMOS-sRAM-IC	8kB x 8 Bit	28-DIP	Say	-	
LC 3664 ..	CMOS-sRAM-IC	8kB x 8 Bit	28-DIP	Say	-	
LC 3764 P	CMOS-sRAM-IC	8kB x 8 Bit Mask ROM	28-DIP	Say	-	pdf
LC 4913	CMOS-Logic	Dual, D-Flip-Flop, Schmitt Clock Input	14-DIP	Say	-	
LC 4913 M	CMOS-Logic	=LC 4913: SMD	14-MDIP		-	
LC 4966	CMOS-Logic	Quad Bilateral Digital or Analog Sw.	14-DIP	Say	... 4066...	pdf pinout
LC 4969	CMOS-Logic	Triple Inverter	8-DIP	Say	-	
LC 5613	CMOS-IC	LCD-digitaluhr/digital watch	Chip	Say	-	pdf pinout
LC 5621	CMOS-IC	Lcd-digitaluhr/digital watch	Chip	Say	-	pdf
LC 5631	CMOS-IC	Lcd-digitaluhr/digital watch + alarm	Chip	Say	-	pdf
LC 5632	CMOS-IC	Lcd-digitaluhr/digital watch + alarm	Chip	Say	-	pdf
LC 5633	CMOS-IC	Lcd-digitaluhr/digital watch	Chip	Say	-	pdf
LC 5641	CMOS-IC	LCD Alarm-Digital-Uhr/Watch, 1,5V	Chip	Say	-	pdf pinout
LC 5642	CMOS-IC	Alarm-digital-uhr/watch		Say	-	
LC 5643	CMOS-IC	LCD Digital-Uhr/Watch, 1,5V	Chip	Say	-	pdf pinout
LC 5644	CMOS-IC	Digital-uhr/watch		Say	-	
LC 5645 N	CMOS-IC	LCD Digital-Uhr/Watch, 1,5V	Chip	Say	-	pdf
LC 5646	CMOS-IC	LCD Digital-Uhr/Watch, 1,5V	Chip	Say	-	pdf pinout
LC 5700 N	CMOS-µC-IC	4 Bit, 1392x14Bit ROM, 84x8Bit RAM	MP, Chip	Say	-	
LC 5701	CMOS-IC	Alarm-digital-uhr/clock + Kalender		Say	-	
LC 5732(H)	CMOS-µC-IC	SMD,4 Bit,2kB ROM, 48x4Bit RAM, LCD-Tr	64-MP	Say		pdf
LC 5733(H)	CMOS-µC-IC	SMD, 4 Bit, 2kB ROM, 128x4Bit RAM, LCD	64-MP	Say		pdf
LC 5734(H)	CMOS-µC-IC	SMD, 4 Bit, 2kB ROM, 128x4Bit RAM, LCD	64-MP	Say		pdf
LC 5800	CMOS-µC-IC	SMD, 4 Bit, 2kB ROM, 152x4Bit RAM, LCD	80-MP	Say		
LC 5805	CMOS-µC-IC	4 Bit, 1kB ROM, 128x4Bit RAM, LCD-Tr	Chip	Say		pdf
LC 5812	CMOS-µC-IC	SMD, 4 Bit, 2kB ROM, 152x4Bit RAM, LCD	80-MP	Say		pdf
LC 5813 C	CMOS-µC-IC	SMD, 4 Bit, 2kB ROM, 152x4Bit RAM, LCD	64-MP	Say		
LC 5814 H	CMOS-µC-IC	SMD, 4 Bit, 2kB ROM, 152x4Bit RAM, LCD	64-MP	Say		
LC 5816 H	CMOS-µC-IC	SMD, 4 Bit, 2kB ROM, 248x4Bit RAM, LCD	64-MP	Say		
LC 5850	CMOS-µC-IC	SMD, 4 Bit, 1kB ROM, 64x4Bit RAM, LCD	64-MP	Say		
LC 5851(H)	CMOS-µC-IC	SMD, 4 Bit, 1kB ROM, 64x4Bit RAM, LCD	64-MP	Say		
LC 5863 H	CMOS-µC-IC	4 Bit, 3kB ROM, 256x4Bit RAM, LCD-Tr	80-FLP	Say		
LC 5864 H	CMOS-µC-IC	4 Bit, 4kB ROM, 256x4Bit RAM, LCD-Tr	80-FLP	Say		pdf
LC 5866 H	CMOS-µC-IC	4 Bit, 6kB ROM, 256x4Bit RAM, LCD-Tr	80-FLP	Say		
LC 5868 H	CMOS-µC-IC	4 Bit, 8kB ROM, 256x4Bit RAM, LCD-Tr	80-FLP	Say		

Type	Device	Short Description	Fig.	Manu	Comparision Types	More at .
.C 6502(B,C,D)	CMOS-µC-IC	4 Bit, 2kB ROM, 128x4Bit RAM	42-(S)DIP	Say	-	
.C 6505(B,C,D)	CMOS-µC-IC	4 Bit, 1kB ROM, 64x4Bit RAM	42-(S)DIP	Say	-	
.C 6510 C	CMOS-µC-IC	4 Bit, 4096x8Bit ROM, 256x4Bit RAM	42-SDIP	Say	-	
.C 6511 C	CMOS-µC-IC	4 Bit, 4096x8Bit ROM, 256x4Bit RAM	42-SDIP	Say	-	
.C 6512 D	CMOS-µC-IC	4 Bit, 2048x8Bit ROM, 128x4Bit RAM	42-SDIP	Say	-	
.C 6513 D	CMOS-µC-IC	4 Bit, 1024x8Bit ROM, 64x4Bit RAM	42-SDIP	Say	-	
.C 6514 B	CMOS-µC-IC	4 Bit, 4096x8Bit ROM, 256x4Bit RAM	42-SDIP	Say	-	pdf
.C 6515 B	CMOS-µC-IC	4 Bit, 4096x8Bit ROM, 256x4Bit RAM	42-SDIP	Say	-	
.C 6520 C	CMOS-µC-IC	4 Bit, 4096x8Bit ROM, 256x4Bit RAM	42-SDIP	Say	-	pdf
.C 6520 H	CMOS-µC-IC	4 Bit, 4096x8Bit ROM, 256x4Bit RAM	42-SDIP	Say	-	pdf
.C 6522 C	CMOS-µC-IC	4 Bit, 2048x8Bit ROM, 128x4Bit RAM	42-SDIP	Say	-	pdf
.C 6522 H	CMOS-µC-IC	4 Bit, 2048x8Bit ROM, 128x4Bit RAM	42-SDIP	Say	-	pdf
.C 6523 C,H	CMOS-µC-IC	4 Bit, 2048x8Bit ROM, 128x4Bit RAM	30-SDIP	Say	-	
.C 6526 C,H	CMOS-µC-IC	4 Bit, 1024x8Bit ROM, 64x4Bit RAM	30-SDIP	Say	-	
.C 6527 C,H	CMOS-µC-IC	4 Bit, 1kB ROM, 64x4Bit RAM	18-DIP	Say	-	
.C 6528 C,H	CMOS-µC-IC	4 Bit, 0,5kB ROM, 32x4Bit RAM	18-DIP	Say	-	
.C 6530 C	CMOS-µC-IC	4 Bit, 4kB ROM, 256x4Bit RAM, D/A Conv	42-SDIP	Say	-	
.C 6532 C	CMOS-µC-IC	4 Bit, 2kB ROM, 128x4Bit RAM, D/A Conv	42-SDIP	Say	-	
.C 6538 D	CMOS-µC-IC	4 Bit, 8kB ROM, 448x4Bit RAM, FLT-Tr	64-SDIP	Say	-	pdf
.C 6543 C,H	CMOS-µC-IC	4 Bit, 2kB ROM, 128x4Bit RAM	30-SDIP	Say	-	
.C 6546 C,H	CMOS-µC-IC	4 Bit, 1kB ROM, 64x4Bit RAM	30-SDIP	Say	-	
.C 6554 D,H	CMOS-µC-IC	4 Bit, 4kB ROM, 256x4Bit RAM, Flt	64-SDIP	Say	-	
.C 6568 D,H	CMOS-µC-IC	4 Bit, 8kB ROM, 256x4Bit RAM, Flt	64-SDIP	Say	-	
LC 7000	CMOS-IC	AM/FM Tuning-system	20-DIP	Say	-	pdf
LC 7010	CMOS-IC	PLL Frequ.-Synth. Controller f. LM7000	42-SDIP	Say	-	pdf pinout
LC 7011	CMOS-IC	PLL Frequ.-Synth. Controller f. LM7000	42-SDIP	Say	-	pdf
LC 7020(A,B)K,T.	CMOS-IC	PLL Frequ.-Synth. Controller f. LM7000	64-FLP	Say	-	
LC 7030	CMOS-IC	FM-Tuning Controller (f. LM 7000)	42-SDIP	Say	-	pdf pinout
LC 7031	CMOS-IC	AM/FM-Tuning Controller (f. LM 7000)	42-SDIP	Say	-	
LC 7060	CMOS-IC	Graphic Equalizer Controller	42-SDIP	Say	-	pdf pinout
LC 7060 AN	CMOS-IC	=LC 7060: SMD	64-MP	Say	-	
LC 7070 N	CMOS-IC	Rds Data System	18-DIP	Say	-	pdf
LC 7070 NM	CMOS-IC	=LC 7070N: SMD	18-MDIP	Say	-	pdf
LC 7071 NM	CMOS-IC	SMD, Rds Data System	18-MDIP	Say	-	pdf
LC 7073	CMOS-IC	FM-RDS-Decoder	18-DIP	Say	-	pdf
LC 7073 M	CMOS-IC	=LC 7073: SMD	18-MDIP	Say	-	pdf
LC 7074	CMOS-IC	Fm-rds-decoder	18-DIP	Say	-	pdf
LC 7074 M	CMOS-IC	=LC 7074: SMD	18-MDIP	Say	-	pdf
LC 7110	CMOS-IC	Cb, PLL Frequ.-synthesizer	20-DIP	Say	-	
LC 7112	CMOS-IC	Cb, PLL Frequ.-synthesizer	20-DIP	Say	-	
LC 7113	CMOS-IC	Cb, PLL Frequ.-synthesizer	16-DIP	Say	-	
LC 7120	CMOS-IC	Cb, 27MHz Synthesizer	20-DIP	Say	-	pdf
LC 7130	MOS-IC	Cb, 27MHz PLL Frequ.-synthesizer	20-DIP	Say	-	
LC 7131	CMOS-IC	Cb, 27MHz Synthesizer, 40kanal./chann.	20-DIP	Say	-	pdf
LC 7132	CMOS-IC	Cb, 27MHz Synthesizer	20-DIP	Say	-	pdf
LC 7135	CMOS-IC	Cb, 27MHz Sythesizer, 22kanal./chann.	20-DIP	Say	-	pdf
LC 7136	CMOS-IC	Cb, 27MHz Synthesizer, 40kanal./chann.	20-DIP	Say	-	pdf
LC 7137	CMOS-IC	Cb, PLL Frequ.-synthesizer	20-DIP	Say	-	pdf
LC 7150	CMOS-IC	Telecom, PLL Synthesizer f. Cordl. Ph.	18-DIP	Say	-	pdf
LC 7151	CMOS-IC	Telecom, PLL Synthesizer f. Cordl. Ph.	18-DIP	Say	-	pdf
LC 7152	CMOS-IC	Telecom, PLL Synthesizer f. Cordl. Ph.	24-SDIP	Say	-	pdf
LC 7152 M,KM,NM	CMOS-IC	=LC 7152: SMD	24-SMDIP		-	
LC 7152 M	CMOS-IC	=LC 7152: SMD	24-SMDIP		-	pdf
LC 7153	CMOS-IC	Telecom, PLL Synthesizer f. Cordl. Ph.	24-SDIP	Say	-	pdf
LC 7153 M	CMOS-IC	=LC 7153: SMD	24-SMDIP		-	pdf
LC 7180	CMOS-IC	Cb, 27MHz Kanalwahl/channel selector	20-DIP	Say	-	
LC 7181	CMOS-IC	Cb, 27MHz Kanalwahl/channel selector.	20-DIP	Say	-	pdf pinout
LC 7184	CMOS-IC	Cb, 27MHz Kanalwahl/channel selector	20-DIP	Say	-	
LC 7185-8750	CMOS-IC	Cb, PLL-Synthesizer, Controller	30-SDIP	Say	-	
LC 7190	CMOS-IC	Cb, 27MHz Kanalwahl/channel selector	20-DIP	Say	-	
LC 7191	CMOS-IC	Cb, 27MHz Kanalwahl/channel selector	20-DIP	Say	-	pdf pinout
LC 7200	CMOS-IC	AM/FM Tuning-System, 5+5 Station.	42-DIP	Say	-	
LC 7201	CMOS-IC	AM/FM Tuning-System, 6+6 Station.	42-DIP	Say	-	
LC 7203	CMOS-IC	AM/FM Tuning-System, 6+6 Station.	42-DIP	Say	-	
LC 7207	CMOS-IC	AM/FM Tuning-System, 7+7 Station.	42-DIP	Say	-	
LC 7210	CMOS-IC	AM/FM Tuning-system	28-DIP	Say	-	pdf pinout
LC 7215	CMOS-IC	Am PLL Frequ.-synthesizer (0,5..13MHz)	14-DIP	Say	-	pdf
LC 7215 F	CMOS-IC	=LC 7215: 0,5..20MHz	14-DIP	Say	-	pdf
LC 7215 FM	CMOS-IC	=LC 7215: SMD, 0,5..20MHz	14-MDIP	Say	-	pdf
LC 7216 M	CMOS-IC	SMD, AM/FM PLL Frequ.-synthesizer	20-MDIP	Say	-	pdf
LC 7217	CMOS-IC	AM/FM/TV PLL Frequ.-synthesizer	42-SDIP	Say	-	
LC 7218	CMOS-IC	AM/FM PLL Frequ.-synthesizer	24-SDIP	Say	-	pdf
LC 7218 M	CMOS-IC	=LC 7218: SMD	24-MDIP	Say	-	pdf
LC 7219	CMOS-IC	PLL Frequ.-synthesizer, Car Stereo	24-SDIP	Say	-	pdf
LC 7219 M	CMOS-IC	=LC 7219: SMD	24-MDIP	Say	-	pdf
LC 7220	CMOS-IC	AM/FM Tuning-system (usa)	42-SDIP	Say	-	pdf pinout
LC 7221	CMOS-IC	AM/FM Tuning-system (japan)	42-SDIP	Say	-	pdf pinout
LC 7222	CMOS-IC	AM/FM Tuning-system (europa)	42-SDIP	Say	-	pdf pinout
LC 7223	CMOS-IC	AM/FM, PLL Frequ.-synthesizer	42-SDIP	Say	-	pdf pinout
LC 7225	CMOS-IC	=LC 7220: SMD	64-MP	Say	-	pdf pinout
LC 7226	CMOS-IC	=LC 7221: SMD	64-MP	Say	-	pdf pinout
LC 7227	CMOS-IC	=LC 7222: SMD	64-MP	Say	-	pdf pinout
LC 7230-8221	CMOS-IC	SMD, PLL, Controller, ROM, Lcd-encoder	80-MP	Say	-	
LC 7230-8272	CMOS-IC	SMD, PLL, Controller, ROM, Lcd-encoder	80-MP	Say	-	
LC 7230	CMOS-IC	SMD, PLL, Controller, ROM, LCD-encoder	80-MP	Say	-	pdf

Type	Device	Short Description	Fig.	Manu	Comparision Types	More at
LC 7232-8291	CMOS-IC	SMD, PLL, Controller, ROM, Lcd-encoder	80-MP	Say	-	
LC 7232-8377	CMOS-IC	SMD, PLL, Controller, ROM, Lcd-encoder	80-MP	Say	-	
LC 7232	CMOS-IC	SMD, PLL, Controller, ROM, Lcd-encoder	80-MP	Say	-	
LC 7233(N)	CMOS-IC	SMD, PLL, Controller, ROM, Lcd-encoder	64-MP	Say		pdf
LC 7234-8460	CMOS-IC	SMD, PLL, Controller, ROM, Lcd-encoder	64-MP	Say		
LC 7234	CMOS-IC	SMD, PLL, Controller, ROM, Lcd-encoder	64-MP	Say		pdf
LC 7250	CMOS-IC	AM/FM Frequ.-display (f. LED)	42-DIP	Say		pdf pinout
LC 7253	MOS-IC	AM/FM Frequ.-disp.+uhr/clock	42-DIP	Say	-	
LC 7257	CMOS-IC	AM/FM-Frequenzanz./frequ. display	42-DIP	Say	-	
LC 7258	CMOS-IC	AM/FM Frequenzanz./frequ. display	42-DIP	Say	-	
LC 7259	CMOS-IC	AM/FM Tuning Frequ.-display (f. Led)	42-DIP	Say	-	
LC 7265	CMOS-IC	AM/FM Frequ.-display (f. Led)	42-SDIP	Say		pdf pinout
LC 7266	CMOS-IC	AM/FM Frequ.-display (f. FLT)	42-SDIP	Say		pdf pinout
LC 7267	CMOS-IC	AM/FM Frequ.-display (f. Led)	42-SDIP	Say		pdf
LC 7268	CMOS-IC	AM/FM Frequ.-display (f. Flt)	42-SDIP	Say		pdf pinout
LC 7350	CMOS-IC	Telecom, Impulswahl/pulse dialer	18-DIP	Say		pdf pinout
LC 7351	CMOS-IC	Telecom, Impulswahl/pulse dialer	20-DIP	Say		pdf pinout
LC 7352	CMOS-IC	Telecom, Impulswahl/pulse dialer	18-DIP	Say		pdf pinout
LC 7353	CMOS-IC	Telecom, Impulswahl/pulse dialer	18-DIP	Say		pdf pinout
LC 7354	CMOS-IC	Telecom, Impulswahl/pulse dialer	20-DIP	Say		pdf pinout
LC 7355	CMOS-IC	Telecom, Impulswahl/pulse dialer	18-DIP	Say		pdf pinout
LC 7356	CMOS-IC	Telecom, Impulswahl/pulse dialer	20-DIP	Say		pdf pinout
LC 7360	CMOS-IC	→KS 5820	18-DIP	Say	KS 5820, UM 91210	
LC 7363(J)	CMOS-IC	Telecom, DTMF/impulswahl/pulse dialer	22-SDIP	Say	-	pdf
LC 7363(J)M	CMOS-IC	=LC 7363: SMD	30-SMDIP		-	pdf
LC 7364(J)	CMOS-IC	Telecom, Dtmf/impulswahl/pulse dialer	22-SDIP	Say	-	pdf
LC 7365(N)	CMOS-IC	Telecom, Dtmf-tonwahl/tone dialer	16-DIP	Say	-	pdf
LC 7365(N)M	CMOS-IC	=LC 7365: SMD	20-MDIP		-	pdf
LC 7366(N)	CMOS-IC	Telecom, Dtmf-tonwahl/tone dialer	16-DIP	Say	-	pdf
LC 7366(N)M	CMOS-IC	=LC 7366: SMD	20-MDIP		-	pdf
LC 7367(J)	CMOS-IC	Telecom, Dtmf/impulswahl/pulse dialer	22-SDIP	Say	-	pdf
LC 7367(J)M	CMOS-IC	=LC 7367: SMD	30-SMDIP		-	pdf
LC 7368(J)	CMOS-IC	Telecom, Dtmf/impulswahl/pulse dialer	22-SDIP	Say	-	pdf
LC 7370	CMOS-IC	Telecom, Impulswahl/pulse dialer	42-DIP	Say	-	pdf pinout
LC 7371	CMOS-IC	Telecom, Impulswahl/pulse dialer	30-SDIP	Say	-	
LC 7385	CMOS-IC	Telecom, Dtmf-empfänger/receiver	18-DIP	Say	-	pdf
LC 7385 M	CMOS-IC	=LC 7385: SMD	18-MDIP		-	pdf
LC 7387 M	CMOS-IC	SMD, Telecom, Dtmf-empfänger/receiver	14-SMDIP	Say	-	pdf
LC 7410	CMOS-IC	Vc(vhs,Beta), Servo-digital-controller	30-SDIP	Say	-	pdf pinout
LC 7412(N)	CMOS-IC	VC, Servo Digital Controller	30-SDIP	Say	-	
LC 7412	CMOS-IC	Vc, Servo-digital-controller	30-SDIP	Say	-	pdf
LC 7413(N)	CMOS-IC	VC, Servo Digital Controller	42-SDIP	Say	-	
LC 7413	CMOS-IC	Vc, Servo-digital-controller	42-SDIP	Say	-	pdf
LC 7414	CMOS-IC	SMD, Vc, Servo-digital-controller	48-MP	Say	-	
LC 7415	CMOS-IC	SMD, Vc(vhs,Beta), Servo-digital-ctrl.	48-MP	Say	-	
LC 7431	CMOS-IC	Vc, VHS Chroma-prozessor (ntsc)	16-SQP	Say	-	pdf
LC 7431 M	CMOS-IC	=LC 7431: SMD	16-MDIP	Say	-	pdf
LC 7432	CMOS-IC	Vc, VHS Chroma-prozessor (ntsc)	16-SQP	Say	-	pdf
LC 7432 M	CMOS-IC	=LC 7432: SMD	16-MDIP	Say	-	
LC 7441 N	CMOS-IC	Ctv,Vc, NTSC/PAL Pip Controller	64-SDIP	Say	-	pdf
LC 7441 NE	CMOS-IC	SMD,Ctv,Vc, NTSC/PAL Pip Controller	64-MP	Say	-	pdf
LC 7442 N	CMOS-IC	Ctv,Vc, NTSC/PAL Pip Controller	64-SDIP	Say	-	
LC 7442 NE	CMOS-IC	SMD,Ctv,Vc, NTSC/PAL Pip Controller	64-MP		-	
LC 7444	CMOS-IC	Ctv,Vc, Dual Vco	14-DIP	Say	-	pdf
LC 7450 H,N	CMOS-IC	PLL f. CATV Hf Modulator	16-DIP	Say	-	pdf pinout
LC 7450 M	CMOS-IC	=LC 7450H,N: SMD	-MDIP		-	pdf pinout
LC 7460	CMOS-IC	IR-FB, Sender/transmitter (8 Codes)	16-DIP	Say	-	pdf pinout
LC 7461 M	CMOS-IC	SMD, IR-FB, Sender/transmitter	24-MDIP	Say	-	pdf
LC 7462 M	CMOS-IC	SMD, IR-FB, Sender/transmitter	20-MDIP	Say	-	pdf
LC 7463 M	CMOS-IC	SMD, IR-FB, Sender/transmitter	20-MDIP	Say	-	pdf
LC 7464 M	CMOS-IC	SMD, IR-FB, Sender/transmitter	24-MDIP	Say	-	pdf
LC 7465 M	CMOS-IC	SMD, IR-FB, Sender/transmitter	30-MDIP	Say	-	pdf
LC 7471	CMOS-IC	Vc, NTSC On-screen Display Controller	22-SDIP	Say	-	pdf
LC 7472 M	CMOS-IC	Vc, NTSC On-screen Display Controller	24-MDIP	Say	-	pdf
LC 7475	CMOS-IC	Vc, Pal On-screen Display Controller	22-SDIP	Say	-	pdf
LC 7480	CMOS-A/D-IC	Video, 6 Bit 3-Input Multipl., <20Msps	24-SDIP	Say	-	pdf
LC 7480 M	CMOS-A/D-IC	SMD,Video,6Bit 3-Input Multipl.,<20Msp	24-SMDIP		-	pdf
LC 7500	CMOS-IC	NF-lautst.-reg./el. volume control	16-DIP	Say	-	pdf pinout
LC 7500 M	CMOS-IC	=LC 7500: SMD	-MDIP		-	pdf pinout
LC 7510	CMOS-IC	Recorder,Suchlauf/autom. music select.	18-DIP	Say	-	pdf pinout
LC 7512	CMOS-IC	Recorder,Suchlauf/autom. music select.	18-DIP	Say	-	pdf pinout
LC 7515	CMOS-IC	Recorder,Suchlauf/autom. music select.	20-DIP	Say	-	pdf pinout
LC 7517	CMOS-IC	Recorder,Suchlauf/autom. music select.	22-DIP	Say	-	pdf pinout
LC 7520	CMOS-IC	Graphic Equalizer Controller	28-DIP	Say	-	pdf
LC 7522	CMOS-IC	Graphic Equalizer Controller, Udd=14V	28-SDIP	Say	-	pdf
LC 7523	CMOS-IC	Graphic Equalizer Controller, Udd=8V	28-SDIP	Say	-	pdf pinout
LC 7523 M	CMOS-IC	=LC 7523: SMD	30-MDIP		-	
LC 7527 E	CMOS-IC	SMD, Graphic Equalizer System, Udd=5V	64-MP	Say	-	pdf
LC 7530	CMOS-IC	Stereo Lautst.-reg./el. volume control	20-DIP	Say	-	pdf pinout
LC 7531	CMOS-IC	Mono Lautst.-Reg./el.volume ctrl.(12V)	16-SQP	Say	-	
LC 7532 M	CMOS-IC	Mono Lautst.-Reg./el.volume ctrl.(3V)	16-MDIP	Say	-	
LC 7533	CMOS-IC	Lautst.-reg./el. volume control (3V)	16-DIP	Say	-	pdf pinout
LC 7534 M	CMOS-IC	Lautst.-Reg./el volume control (3V)	20-MDIP	Say	-	pdf pinout
LC 7535	CMOS-IC	Lautst.+bal.-reg./el. vol.+bal. ctrl.	22-DIP	Say	-	
LC 7536	CMOS-IC	Serially Controlled Electronic Volume	30-SDIP	Say	-	pdf pinout

Type	Device	Short Description	Fig.	Manu	Comparision Types	More at
		Control that Handles High Voltages				
LC 7537 AN	CMOS-IC	=LC 7537N: SMD	48-MP	Say	-	pdf
LC 7537N	CMOS-IC	NF-Klangreg./electr. tone ctrl.	42-SDIP	Say	-	pdf
LC 7537 NE,AN	CMOS-IC	=LC 7537N: SMD	48-MP	Say	-	
LC 7550	CMOS-IC	SMD,2xPegelanz./level mtr. + peak hold	64-MP	Say	-	pdf pinout
LC 7555	CMOS-IC	2x Pegelanz./level mtr. + peak hold	42-DIP	Say	-	pdf pinout
LC 7556	CMOS-IC	2x Pegelanz./level mtr. + peak hold	42-DIP	Say	-	pdf
LC 7560	CMOS-IC	SMD, LCD-treiber/drv f. Graph. Equal.	64-MP	Say	-	pdf pinout
LC 7561	CMOS-IC	SMD, Lcd-treiber/drv f. Graph. Equal.	64-MP	Say	-	
LC 7565	CMOS-IC	Flt-treiber/driver f. Graph. Equal.	42-SDIP	Say	-	pdf
LC 7565 A,B	CMOS-IC	=LC 7565: SMD	48-MP	Say	-	
LC 7566	CMOS-IC	FLT-Tr. f. Spectrum Analyzer	42-SDIP	Say	-	
LC 7570	CMOS-IC	Tuning Display-Tr (f. Flt)	42-SDIP	Say	-	pdf
LC 7572	CMOS-IC	Tuning Display-tr (f. Led)	18-DIP	Say	-	pdf pinout
LC 7573(N)	CMOS-IC	Display Driver (f. FLT)	30-SDIP	Say	-	
LC 7573	CMOS-IC	Display-tr (f. Flt)	30-SDIP	Say	-	pdf
LC 7573 M,MN	CMOS-IC	=LC 7573: SMD	30-SMDIP			
LC 7573 M	CMOS-IC	=LC 7573: SMD	30-MDIP			
LC 7574 E,NE,NW	CMOS-IC	SMD, Display Driver (f. FLT)	48-MP	Say	-	
LC 7574 E	CMOS-IC	SMD, Display-tr (f. Flt)	48-MP	Say	-	
LC 7580	CMOS-IC	SMD, LCD-Treiber/driver, max. 53 Segm.	64-MP	Say	-	pdf pinout
LC 7581 C	CMOS-IC	LCD-Treiber/driver, max. 53 Segm.	Chip	Say	-	
LC 7582(A,B,E,W)	CMOS-IC	SMD,Frequ.-Display, LCD-treiber/driver	64-MP	Say	-	
LC 7582(E,W)	CMOS-IC	SMD,Frequ.-display, LCD-treiber/driver	64-MP	Say	-	
LC 7583 N(A,B)	CMOS-IC	SMD,Frequ.-Display, LCD-Treiber/driver	64-MP	Say	-	
LC 7583 N	CMOS-IC	SMD,Frequ.-display, Lcd-treiber/driver	64-MP	Say	-	pdf
LC 7584 N	CMOS-IC	SMD,LCD-Tr., Zeichen-/character Gener.	80-MP	Say	-	pdf
LC 7600	CMOS-IC	LCD Digitaluhr/digital clock	FLP	Say	-	pdf pinout
LC 7650	CMOS-IC	Alarm-analog-uhr/clock	8-DIP	Say	-	pdf pinout
LC 7651	CMOS-IC	Alarm-analog-uhr/clock	8-DIP	Say	-	pdf pinout
LC 7652	CMOS-IC	Alarm-analog-uhr/clock	8-DIP	Say	-	pdf pinout
LC 7653	CMOS-IC	Alarm-analog-uhr/clock	8-DIP	Say	-	pdf pinout
LC 7660	CMOS-IC	Step Motor Driver f. Analog Clock	8-DIP	Say	-	
LC 7662	CMOS-IC	Step Motor Driver f. Analog Clock	8-DIP	Say	-	
LC 7665	CMOS-IC	Step Motor Driver f. Analog Clock	8-DIP	Say	-	
LC 7667	CMOS-IC	Step Motor Driver f. Analog Clock	8-DIP	Say	-	
LC 7730	CMOS-IC	Timer, Udd=1,8..5,5V, -10...+75°	16-DIP	Say	-	pdf
LC 7730 M	CMOS-IC	=LC 7730: SMD	16-MDIP	Say	-	pdf
LC 7800(N)	CMOS-IC	µComputer Input-expander	28-DIP	Say	-	pdf pinout
LC 7815	CMOS-IC	4x Analog-Schalter/Switch, Udd=15...18V	28-SDIP	Say	-	pdf
LC 7815 H	CMOS-IC	=LC 7815: Ucc=15...23V	28-SDIP	Say	-	pdf
LC 7816	CMOS-IC	4x Analog-Schalter/Switch	28-SDIP	Say	-	pdf
LC 7817	CMOS-IC	4x Analog-Schalter/Switch	28-DIP	Say	-	pdf
LC 7818	CMOS-IC	5x Analog-Schalter/Switch	30-SDIP	Say	-	pdf
LC 7820	CMOS-IC	4x Analog-Schalter/Switch	28-DIP	Say	-	pdf pinout
LC 7821N	CMOS-IC	8x Analog-Schalter/Switch	30-SDIP	Say	-	pdf pinout
LC 7822N	CMOS-IC	8x Analog-Schalter/Switch	30-SDIP	Say	-	pdf pinout
LC 7823N	CMOS-IC	7x Analog-Schalter/Switch	30-SDIP	Say	-	pdf pinout
LC 7824	CMOS-IC	7x Analog Function Switch, Udd=±9V	16-DIP	Say	-	
LC 7824	CMOS-IC	Analog-Switch, 6 Inputs	16-DIP	Say	-	pdf pinout
LC 7860 N,KA	CMOS-IC	SMD, Cd, Digitalsignal-prozessor	80-MP	Say	-	
LC 7861 KE,NE	CMOS-IC	SMD, CD, Digitalsignal Processor	64-MP	Say	-	
LC 7861 NE	CMOS-IC	SMD, Cd, Digitalsignal-prozessor	64-MP	Say	-	
LC 7863 KA	CMOS-IC	SMD, Cd, Digitalsignal-prozessor	80-MP	Say	-	pdf
LC 7865 E	CMOS-IC	SMD, Cd, Digitalsignal-prozessor	64-MP	Say	-	
LC 7867 E	CMOS-IC	SMD, Cd, Digitalsignal-prozessor	64-MP	Say	-	pdf
LC 7868 E	CMOS-IC	SMD, Cd, Digitalsignal-prozessor	64-MP	Say	-	pdf
LC 7869 E,KE	CMOS-IC	SMD, CD, Digitalsignal Processor	64-MP	Say	-	
LC 7869 E	CMOS-IC	SMD, Cd, Digitalsignal-prozessor	64-MP	Say	-	pdf
LC 7870 E,NE	CMOS-IC	SMD, Cd, G-decoder	100-MP	Say	-	
LC 7880	CMOS-D/A-IC	16 Bit, Audio	20-DIP	Say	-	
LC 7880 M	CMOS-D/A-IC	=LC 7880: SMD	20-MDIP	Say	-	
LC 7881	CMOS-D/A-IC	16 Bit, Audio	20-DIP	Say	KDA 0316, CDX 1161	pdf pinout
LC 7881 M	CMOS-D/A-IC	=LC 7881: SMD	20-MDIP	Say	-	pdf
LC 7883 K	CMOS-D/A-IC	16 Bit, Audio, Filter	28-DIP	Say	-	pdf
LC 7883 KM	CMOS-D/A-IC	=LC 7883K: SMD	28-MDIP	Say	-	pdf
L 7885 CP	Z-IC	=L 7805CV...7885CV: Iso	17b	Stm	... 78xx... (TO-220 Iso)	pdf pinout
LC 7886	CMOS-A/D-IC	18 Bit, Audio	24-DIP	Say	-	pdf
LC 7886 M	CMOS-A/D-IC	=LC 7886: SMD	24-MDIP	Say	-	pdf
LC 7900	CMOS-IC	System Clock Generator	8-SIP	Say	-	pdf pinout
LC 7910	A/D-D/A-IC	CMOS, 6 Bit, 64 Step	16-DIP	Say	-	pdf pinout
LC 7930 N	CMOS-IC	SMD, LDC Display-tr	60-MP	Say	-	pdf
LC 7930 NW	CMOS-IC	=LC 7930N: Fig.→	64-MP	Say	-	pdf
LC 7931 D	CMOS-IC	SMD, Ldc Display-Tr, 80 Kanal/Channel	100-MP	Say	-	pdf
LC 7932	CMOS-IC	16 Bit LED Driver, Udd=5V, 5MHz	30-SDIP	Say	-	pdf
LC 7932 M	CMOS-IC	=LC 7932: SMD	30-SMDIP	Say	-	pdf
LC 7935(AN)	CMOS-IC	32Bit Thermal Printer Head Driver	64-MP	Say	-	
LC 7935	CMOS-IC	32Bit Thermal Printer Head Driver	64-FLP	Say	-	
LC 7936(A,B)	CMOS-IC	32Bit PPC LED Erasing Head Driver	64-MP	Say	-	
LC 7936	CMOS-IC	Thermal Printer Head Driver		Say	-	pdf
LC 7938 C	CMOS-IC	Image Sensor Driver, Udd=5V	Chip	Say	-	pdf
LC 7940 A,ND,H,YD	CMOS-IC	SMD, Dot Matrix LCD Display Driver	100-MP	Say	-	
LC 7940 A,ND,H	CMOS-IC	SMD, Dotmatrix Ldc Display-tr	100-MP	Say	-	
LC 7941 A,ND,H,YD	CMOS-IC	SMD, Dot Matrix LCD Display Driver	100-MP	Say	-	

Type	Device	Short Description	Fig.	Manu	Comparision Types	More at
LC 7941 A,ND,H	CMOS-IC	SMD, Dotmatrix Ldc Display-tr	100-MP	Say	-	
LC 7942 A,ND,YD	CMOS-IC	SMD, Dot Matrix LCD Display Driver	80-MP	Say	-	
LC 7942 A,ND	CMOS-IC	SMD, Dotmatrix Ldc Display-tr	30-MP	Say	-	
LC 7943 D	CMOS-IC	SMD, Dotmatrix Ldc Display-tr	80-MP	Say	-	pdf
LC 7960	CMOS-IC	BCD-Zähler/Counter (3 Digits)	16-DIP	Say	-	
LC 7961	CMOS-IC	BCD-Zähler/Counter (4 Digits)	20-DIP	Say	-	pdf pinout
LC 7970	CMOS-IC	Key Scan Detector f. capacit. Keyboard	28-SDIP	Say	-	
LC 7980	CMOS-IC	SMD, LCD Dotmatrix Grafik Controller	64-MP	Say	-	pdf
LC 7981	CMOS-IC	SMD, LCD Dotmatrix Grafik Controller	60-MP	Say	-	pdf
LC 7985 NA,ND	CMOS-IC	SMD, Ldc Display-tr/controller	80-MP	Say	-	
LC 7986 C	CMOS-IC	Ldc Display-tr/controller	128-Pad	Say	-	pdf
LC 7990	CMOS-IC	Digital Servo f. Hard Disk Drive(HDD)	22-SDIP	Say	-	
LC 7990 M	CMOS-IC	=LC 7990: SMD	30-MDIP	Say	-	
LC 7991	CMOS-IC	Motor Controller, Digital Servo	16-DIP	Say	-	pdf
LC 7991 M	CMOS-IC	=LC 7991: SMD	20-MDIP	Say	-	pdf
LC 7995	CMOS-IC	Key Logic, Mechanic Ctrl., LED Driver	42-SDIP	Say	-	
LC 8100	CMOS-IC	Telecom, Sprachausg./speech Synthesiz.	28-DIP,div	Say	-	pdf pinout
LC 8200	CMOS-IC	Telecom, Sprachausg./speech Synthesiz.	40-DIP	Say	-	
LC 8200 M	CMOS-IC	=LC 8200: SMD	48-MP	Say	-	
LC 8200 S	CMOS-IC	=LC 8200:	42-SDIP		-	
LC 8210	CMOS-IC	SMD, Telecom, Fax, Digital Copy, Scan	64-MP	Say	-	
LC 8211	CMOS-IC	SMD, Telecom, Fax, Digital Copy, Scan	100-MP	Say	-	pdf
LC 8213(K)	CMOS-IC	SMD,Telecom, FAX, G3/G4 Data Processor	80-MP	Say	-	
LC 8213	CMOS-IC	SMD,Telecom, Fax, G3/G4 Data Prozessor	80-MP	Say	-	pdf
LC 8220	CMOS-IC	SMD, Jpeg (ISO 10918) Color Prozessor	160-MP	Say	-	pdf
LC 8230	CMOS-IC	SMD, MPEG (ISO/IEC 11172-3) Audio Dec.	100-MP	Say	-	pdf
LC 8390 M	A/D-D/A-IC	16 Bit, Audio	30-SMDIP	Say	-	pdf
LC 8500 E	CMOS-IC	SMD, Telecom	80-MP	Say	-	
LC 8910	CMOS-IC	Telecom, FB, Home Automation	24-DIP	Say	-	pdf
LC 8912	CMOS-IC	Telecom, Fb, Home Automation	28-DIP	Say	-	pdf
LC 8913	CMOS-IC	Telecom, Fb, Home Automation	40-DIP	Say	-	pdf
LC 8920	CMOS-IC	SMD, Telecom, Fax Data Prozessor	100-MP	Say	-	pdf
LC 8930	CMOS-IC	Telecom, Scrambler	18-DIP	Say	-	
LC 8931	CMOS-IC	Telecom, Scrambler	28-MDIP	Say	-	
LC 8935	CMOS-IC	SMD, Telecom, Fax Data Prozessor	80-MP	Say	-	
LC 8950	CMOS-IC	SMD, CD-ROM Controller f. Cd-i	80-MP	Say	-	
LC 8953	CMOS-IC	SMD, 68000 MPU Interface	160-MP	Say	-	pdf
LC 8954	CMOS-IC	SMD, CD-ROM Controller f. Cd-i	128-MP	Say	-	pdf
LC 8955	CMOS-IC	SMD, ADPCM Decoder	80-MP	Say	-	pdf
LC 8991	CMOS-IC	Camera, CCD, Delay Line (ntsc)	8-DIP	Say	-	pdf
LC 8992	CMOS-IC	Camera, CCD, Delay Line (pal)	8-DIP	Say	-	pdf
LC 8993	CMOS-IC	Camera, CCD, Linear Image Sensor	12-DIC	Say	-	
LC 9103 B	CMOS-IC	Gate Array (340), 2ns, 3...5,5V		Say	-	
LC 9105 B	CMOS-IC	Gate Array (588), 2ns, 3...5,5V		Say	-	
LC 9108 B	CMOS-IC	Gate Array (858), 2ns, 3...5,5V		Say	-	
LC 9111 A	CMOS-IC	Gate Array (1190), 2ns, 3...5,5V		Say	-	
LC 9116 A	CMOS-IC	Gate Array (1692), 2ns, 3...5,5V		Say	-	
LC 9123 A	CMOS-IC	Gate Array (2394), 2ns, 3...5,5V		Say	-	
LC 9130 A	CMOS-IC	Gate Array (3013), 2ns, 3...5,5V		Say	-	
LC 9140 A	CMOS-IC	Gate Array (4082), 2ns, 3...5,5V		Say	-	
LC 9153 A	CMOS-IC	Gate Array (5365), 2ns, 3...5,5V		Say	-	
LC 9600	CMOS-IC	Gate Array (1k...10k), 2ns, 3...5,5V		Say	-	
LC 9901..9904	CCD-IC	Camera, CCD, Timing	64-QIP	Say	-	
LC 9911..9977	CCD-IC	Camera, Bild-/image-sensor		Say	-	
LC 9931 B-10	CMOS-IC	Ccd-camera, 1/3", monochrome	≈20-SDIP	Say	-	
LC 9945	CCD-IC	CCD-Videocamera, CCIR, monochrome,1/6"	≈20-SDIP	Say	-	pdf
LC 9946	CCD-IC	CCD-Videocamera, EIA, monochrome, 1/6"	≈20-SDIP	Say	-	pdf
LC 9947 G	CCD-IC	CCD-Videocamera, EIA, monochrome, 1/5"	≈20-SDIP	Say	-	pdf
LC 9997 G	CCD-IC	CCD-Videocamera, Color (ntsc), 1/5"	≈20-SDIP	Say	-	pdf
LC 37256 P	CMOS-sRAM-IC	32kB x 8 Bit Mask ROM	28-DIP	Say	-	
LC 65102 A	CMOS-µC-IC	4 Bit, 2kBx8Bit ROM, 128x4Bit RAM	30-SDIP	Say	-	pdf
LC 65104 A	CMOS-µC-IC	4 Bit, 4kBx8Bit ROM, 256x4Bit RAM	30-SDIP	Say	-	pdf
LC 65204 A	CMOS-µC-IC	4 Bit, 4kBx8Bit ROM, 256x4Bit RAM, Flt	52-SDIP	Say	-	pdf
LC 65506 B	CMOS-µC-IC	4 Bit, 6kB ROM, 512x4Bit RAM	64-SDIP	Say	-	
LC 65508 B	CMOS-µC-IC	4 Bit, 8kB ROM, 512x4Bit RAM	64-SDIP	Say	-	
LC 65512 A	CMOS-µC-IC	4 Bit, 12kB ROM, 512x4Bit RAM	64-SDIP	Say	-	
LC 65516 A	CMOS-µC-IC	4 Bit, 16kB ROM, 512x4Bit RAM	64-SDIP	Say	-	
LC 67216 A	CMOS-µC-IC	4 Bit, 16kBx8Bit ROM, 608x4Bit RAM	64-SDIP	Say	-	
LC 72144 M	A/D-IC	PLL Frequency Synthesizer	7-SIL	Say	-	pdf pinout
LC 72151 V	CMOS-IC	PLL Freq. Synthesizer, elec.tuning	MP	Say	-	pdf pinout
LC 72191	CMOS-IC	PLL Frequ.-synthesizer, Car Stereo	24-SDIP	Say	-	pdf
LC 72191 M	CMOS-IC	=LC 72191: SMD	24-MDIP	Say	-	pdf
LC 72711 LW	CMOS-IC	Mobile Fm, Receiver IC	8-SIP	Say	-	pdf pinout
LC 72711 W	CMOS-IC	Mobile Fm, Receiver IC	8-SIP	Say	-	pdf pinout
LC 72713 W	CMOS-IC	Mobile FM Multiplex Broadcast IC	20-SSMDIP	Say	-	pdf pinout
LC 72714 W	CMOS-IC	Mobile FM Multiplex Broadcast IC	MP	Say	-	pdf pinout
LC 72722	CMOS-IC	Single-Chip Rds	8-SIP	Say	-	pdf pinout
LC 72722 M	CMOS-IC	Single-Chip Rds	8-SIP	Say	-	pdf pinout
LC 72722 PM	CMOS-IC	Single-Chip Rds	8-SIP	Say	-	pdf pinout
LC 72723	CMOS-IC	Rds Demodulation IC	DIP	Say	-	pdf pinout
LC 72723M	CMOS-IC	Rds Demodulation IC	DIP	Say	-	pdf pinout
LC 72725 M	CMOS-IC	Rds Demodulation IC	MDIP	Say	-	pdf pinout
LC 72725 NV	CMOS-KOP-IC	Rds Demodulation IC	MDIP	Say	-	pdf pinout
LC 72725 V	CMOS-KOP-IC	Rds Demodulation IC	MDIP	Say	-	pdf pinout
LC 73711 N	CMOS-IC	Telecom, memory DTMF/pulse dialer	30-SDIP	Say	-	pdf

Type	Device	Short Description	Fig.	Manu	Comparision Types	More at
LC 73720	CMOS-IC	Telecom, memory Dtmf/pulse dialer	36-SDIP	Say	-	pdf
LC 73721	CMOS-IC	Telecom, memory Dtmf/pulse dialer	36-SDIP	Say	-	
LC 73721 M	CMOS-IC	=LC 73721: SMD	36-MDIP		-	
LC 73860	CMOS-IC	Telecom, Dtmf-empfänger/receiver	8-DIP	Say	-	pdf
LC 73861	CMOS-IC	Telecom, Dtmf-empfänger/receiver	8-DIP	Say	-	pdf
LC 74401 E	CMOS-IC	SMD, Ctv,Vc, NTSC/PAL Pip Controller	80-MP	Say	-	pdf
LC 74402	CMOS-IC	Ctv,Vc, NTSC/PAL Pip Controller	64-SDIP	Say	-	pdf
LC 74402 E	CMOS-IC	SMD,Ctv, Vc, NTSC/PAL Pip Controller	64-MP		-	pdf
LC 74711	CMOS-IC	Vc, NTSC On-screen Display Controller	22-SDIP	Say	-	pdf
LC 74721	CMOS-IC	OSD Controller, 24 Chars. x 10 Lines f. NTSC, Pal-m	24-SDIP	Say	-	pdf
LC 74721 M	CMOS-IC	SMD,OSD Controller, 24 Chars x 10 Line	24-MDIP		-	pdf
LC 74723	CMOS-IC	OSD Controller, 24 Chars. x 10 Lines	24-SDIP	Say	-	pdf
LC 74723 M	CMOS-IC	SMD,OSD Controller, 24 Chars x 10 Line	24-MDIP		-	pdf
LC 74730 M	CMOS-IC	SMD, OSD Ctrl., 24 Chars. x 10 Lines	30-SMDIP	Say	-	pdf
LC 74760	CMOS-IC	OSD Controller, 24 Chars. x 12 Lines	30-SDIP	Say	-	pdf
LC 74760 M	CMOS-IC	SMD,OSD Controller, 24 Chars x 12 Line	30-MDIP		-	pdf
LC 74761	CMOS-IC	OSD Controller, 24 Chars. x 12 Lines	30-SDIP	Say	-	pdf
LC 74761 M	CMOS-IC	SMD,OSD Controller, 24 Chars x 12 Line	30-MDIP		-	pdf
LC 74763	CMOS-IC	OSD Controller, 24 Chars. x 12 Lines	30-SDIP	Say	-	pdf
LC 74763 M	CMOS-IC	SMD,OSD Controller, 24 Chars x 12 Line	30-MDIP		-	pdf
LC 74770	CMOS-IC	Camcorder,On-screen Display Controller	24-MDIP	Say	-	pdf
LC 74770 M	CMOS-IC	SMD, OSD Ctrl., 24 Chars. x 12 Rows f. Camcorder Viewfinder	24-MDIP	Say	-	pdf
LC 74780	CMOS-IC	Vc, OSD Ctrl., 24 Chars. x 12 Lines	24-SDIP	Say	-	pdf
LC 74780 M	CMOS-IC	SMD,Vc, OSD Ctrl., 24 Chars x 12 Line	24-MDIP		-	pdf
LC 74781	CMOS-IC	Vc, OSD Ctrl., 24 Chars. x 12 Lines	24-SDIP	Say	-	pdf
LC 74781 M	CMOS-IC	SMD,Vc, OSD Ctrl., 24 Chars x 12 Line	24-MDIP		-	pdf
LC 74782	CMOS-IC	Vc, OSD Ctrl., 24 Chars. x 12 Lines	24-SDIP	Say	-	pdf
LC 74782 M	CMOS-IC	SMD,Vc, OSD Ctrl., 24 Chars x 12 Line	24-MDIP		-	pdf
LC 74783	CMOS-IC	Vc, OSD Ctrl., 24 Chars. x 12 Lines	24-SDIP	Say	-	pdf
LC 74783 M	CMOS-IC	SMD,Vc, OSD Ctrl., 24 Chars x 12 Line	24-MDIP		-	pdf
LC 74784	CMOS-IC	Vc, OSD Ctrl., 24 Chars. x 12 Lines	24-SDIP	Say	-	pdf
LC 74784 M	CMOS-IC	SMD,Vc, OSD Ctrl., 24 Chars x 12 Line	24-MDIP		-	pdf
LC 74986 NWF	CMOS-IC	LCD TV Scan Converter IC	MDIP	Say	-	pdf pinout
LC 74986 NWV	CMOS-IC	LCD TV Scan Converter IC	MDIP	Say	-	pdf pinout
LC 75010 W	CMOS-IC	Car Audio Dsp	24-SSMDIP	Say	-	pdf pinout
LC 75341	IC	volume/tone control system, Vdd=5...10V	SDIP	Say	-	pdf pinout
LC 75341M	IC	SMD, vol/tone control system, Vdd=5...10V	SMDIP	Say	-	pdf pinout
LC 75392	CMOS-IC	Single-Chip Electronic Volume Control System	30-SDIP	Say	-	pdf pinout
LC 75392 M	CMOS-IC	Single-Chip Electronic Volume Control System	30-SDIP	Say	-	pdf pinout
LC 75396 NE	CMOS-IC	Single-Chip Volume Control	MP	Say	-	pdf pinout
LC 75410 E	CMOS-IC	Electronic Volume Controller	MP	Say	-	pdf pinout
LC 75410 W	CMOS-IC	Electronic Volume Controller	MP	Say	-	pdf pinout
LC 75412 E	CMOS-IC	Electronic Volume Controller	MP	Say	-	pdf pinout
LC 75412 W	CMOS-IC	Electronic Volume Controller	MP	Say	-	pdf pinout
LC 75710 E,NE	CMOS-IC	SMD, Display Driver (f. FLT)	64-MP	Say	-	
LC 75710 E	CMOS-IC	SMD, Display-tr (f. Flt)	64-MP	Say	-	
LC 75711 E,NE	CMOS-IC	SMD, Display Driver (f. FLT)	64-MP	Say	-	
LC 75711 E	CMOS-IC	SMD, Display-tr (f. Flt)	64-MP	Say	-	
LC 75821 E,W	CMOS-IC	SMD, LCD Display-Tr., 104 Segments	64-MP	Say	-	
LC 75822 E,W	CMOS-IC	SMD, LCD Display-Tr., 104 Segments	64-MP	Say	-	
LC 75823 E,W	CMOS-IC	SMD, LCD Display-Tr., 156 Segments	64-MP	Say	-	
LC 75850 E,W	CMOS-IC	SMD, LCD Display Tr., 156 Segments	64-MP	Say	-	
LC 75853 (N)E,W	CMOS-IC	SMD, LCD Display Driver, 126 Segments	64-MP	Say	-	
LC 75853 E,W	CMOS-IC	SMD,LCD Display-Tr, 126Segm., Key Scan	64-MP	Say	-	
LC 78211	IC	Function switching, serial data control	SDIP	Say	-	pdf pinout
LC 78212	IC	Function switching, serial data control	SDIP	Say	-	pdf pinout
LC 78213	IC	Function switching, serial data control	SDIP	Say	-	pdf pinout
LC 78648 E	CMOS-IC	Compact Disc Player IC	MDIP	Say	-	pdf pinout
LC 78681 E,KE(-L)	CMOS-IC	SMD, Cd, Digitalsignal-prozessor	64-MP	Say	-	
LC 78684	CMOS-IC	MP3 Decoder for CDPlayers	24-SSMDIP	Say	-	pdf pinout
LC 78815	CMOS-D/A-IC	16 Bit, Audio	20-DIP	Say	-	
LC 78815 M	CMOS-D/A-IC	=LC 78815: SMD	20-MDIP	Say	-	pdf
LC 78816	CMOS-D/A-IC	16 Bit, Audio	20-DIP	Say	-	pdf
LC 78816 M	CMOS-D/A-IC	=LC 78816: SMD	20-MDIP	Say	-	
LC 78816 V	CMOS-D/A-IC	=LC 78816: SMD	24-SSMDIP	Say	-	
LC 78820	CMOS-D/A-IC	18 Bit, Audio	20-DIP	Say	-	
LC 78820 M	CMOS-D/A-IC	=LC 78820: SMD	20-MDIP	Say	-	
LC 78835 K	CMOS-D/A-IC	18 Bit, Audio, Filter	24-DIP	Say	-	pdf
LC 78835 M,KM	CMOS-D/A-IC	=LC 78835: SMD	24-MDIP	Say	-	
LC 78840 M	CMOS-IC	SMD, Digital Audio, Filter	24-MDIP	Say	-	pdf
LC 78855 M,KM	CMOS-D/A-IC	SMD, 1 Bit, Audio, Filter	28-MDIP	Say	-	
LC 78865 M	CMOS-A/D-IC	SMD, 16 Bit, Audio	20-DIP	Say	-	pdf
LC 79400 D	CMOS-IC	SMD, Dotmatrix Ldc Display-tr	100-MP	Say	-	pdf
LC 79401 D	CMOS-IC	SMD, Dotmatrix Ldc Display-tr	100-MP	Say	-	pdf
LC 79430 D	CMOS-IC	SMD, Dotmatrix Ldc Display-tr	100-MP	Say	-	pdf
LC 79431 D	CMOS-IC	SMD, Dotmatrix Ldc Display-tr	100-MP	Say	-	pdf
LC 81096	CMOS-IC	SMD, Telecom	28-MDIP	Say	-	
LC 81192	CMOS-IC	SMD, Telecom	28-MDIP	Say	-	
LC 83010 E	CMOS-IC	=LC 83010N: SMD	80-MP	Say	-	
LC 83010 N	CMOS-IC	Audio, Digitalsignal-prozessor	64-SDIP	Say	-	pdf
LC 83010 NE	CMOS-IC	=LC 83010N: SMD	80-MP	Say	-	

Type	Device	Short Description	Fig.	Manu	Comparision Types	More at
LC 83015 E,JE	CMOS-IC	SMD, Audio, Digital Signal Processor	80-MP	Say	-	
LC 83015 E	CMOS-IC	SMD, Audio, Digitalsignal-prozessor	80-MP	Say	-	pdf
LC 83020 E (-86...)	CMOS-IC	SMD, Audio, Digital Signal Processor	100-MP	Say	-	
LC 83020 E	CMOS-IC	SMD, Audio, Digitalsignal-prozessor	100-MP	Say	-	
LC 86104 A,C	CMOS-µC-IC	8 Bit, 4kx8Bit ROM, 168x8Bit RAM	100-FLP	Say		
LC 86108 A,C	CMOS-µC-IC	8 Bit, 8kx8Bit ROM, 168x8Bit RAM	100-FLP	Say		
LC 86208 A	CMOS-µC-IC	8 Bit, 8kx8Bit ROM, 336x8Bit RAM		Say		
LC 86212 A	CMOS-µC-IC	8 Bit, 12kx8Bit ROM, 336x8Bit RAM	-	Say		
LC 86216 A	CMOS-µC-IC	8 Bit, 16kx8Bit ROM, 336x8Bit RAM	-	Say		
LC 89058 W-E	CMOS-IC	Digital Audio Interface Receiver	MP	Say	-	pdf pinout
LC 89060	CMOS-D/A-IC	6 Bit, Video, 30 Msps	16-DIP	Say	-	pdf
LC 89060 M	CMOS-D/A-IC	=LC 89060: SMD	20-MDIP		-	pdf
LC 89061	CMOS-A/D-IC	6 Bit, Video, 15 Msps	16-DIP	Say	-	
LC 89061 M	CMOS-A/D-IC	=LC 89061: SMD	20-MDIP		-	
LC 89066	CMOS-A/D-IC	6 Bit, Video, >15 Msps	16-DIP	Say	-	pdf
LC 89066 M	CMOS-A/D-IC	SMD,6 Bit, Video, >15 Msps	20-MDIP		-	pdf
LC 89080	CMOS-D/A-IC	3 Bit, Video 3-Channel, >30 Msps	42-SDIP	Say	-	pdf
LC 89080 Q	CMOS-D/A-IC	SMD,3 Bit, Video 3-Channel, >30 Msps	48-MP		-	pdf
LC 89512(K)	CMOS-IC	SMD, CD-ROM Error Corr., SCSI Interf.	128-MP	Say	-	
LC 89512 K	CMOS-IC	SMD, CD-ROM Prozessor, Scsi-bus	128-MP	Say	-	
LC 89512 W	CMOS-IC	=LC 89512K:	100-MP	Say	-	pdf
LC 89513 K	CMOS-IC	SMD, CD-ROM Prozessor, Host Control	64-MP	Say	-	pdf
LC 89515(K)	CMOS-IC	SMD, CD-ROM Prozessor	80-MP	Say	-	pdf
LC 89517 K	CMOS-IC	SMD, CD-ROM Proz., Sub-code Interface	100-MP	Say	-	pdf
LC 89560	CMOS-IC	SMD, CD-ROM Proz., XA, ADPCM Decoder	100-MP	Say	-	
LC 89562	CMOS-IC	SMD, CD-ROM Proz., Xa, ADPCM Decoder	100-MP	Say	-	
LC 89581	CMOS-IC	SMD, Cd-rom/cd-i Proz., Format Encoder	80-MP	Say	-	pdf
LC 89582	CMOS-IC	SMD, Cd-mo, Cd-wo, Atip Decoder	48-MP	Say	-	pdf
LC 89583	CMOS-IC	SMD, Cd-mo, Cd-wo, CD Encoder	100-MP	Say	-	pdf
LC 89915	CMOS-IC	CCD, NTSC 1H Delay Line, Ucc=5V	8-DIP	Say	-	pdf
LC 89915 M	CMOS-IC	SMD,CCD, NTSC 1H Delay Line, Ucc=5V	8-MDIP		-	pdf
LC 89925	CMOS-IC	CCD, Pal 1H Delay Line, Ucc=5V	8-DIP	Say	-	pdf
LC 89925 M	CMOS-IC	SMD,CCD, Pal 1H Delay Line, Ucc=5V	8-MDIP		-	pdf
LC 89960	CMOS-IC	CCD, NTSC 1H Delay Line, Ucc=5V	14-DIP	Say	-	pdf
LC 89960 M	CMOS-IC	SMD,CCD, NTSC 1H Delay Line, Ucc=5V	14-MDIP		-	pdf
LC 89970	NMOS-IC	CCD, Pal 1H Delay Line, Ucc=5V	22-SDIP	Say	-	pdf
LC 89970 M	NMOS-IC	SMD,CCD, Pal 1H Delay Line, Ucc=5V	24-MDIP		-	pdf
LC 89971	NMOS-IC	CCD, Multisystem 1H Delay Line, Ucc=5V	22-SDIP	Say	-	pdf
LC 89971 M	NMOS-IC	SMD,CCD, Multisystem 1H Delay Line	24-MDIP		-	pdf
LC 89972 M	NMOS-IC	SMD, CCD, Pal 1H Delay Line, Ucc=5V	24-MDIP	Say	-	pdf
LC 89973 M	NMOS-IC	SMD, CCD, Pal 1H Delay Line, Ucc=5V	24-MDIP	Say	-	pdf
LC 92007 B	CMOS-IC	Gate Array (708), 1,4ns, 3...5,5V		Say	-	
LC 92011 B	CMOS-IC	Gate Array (1110), 1,4ns, 3...5,5V		Say	-	
LC 92018 B	CMOS-IC	Gate Array (1881), 1,4ns, 3...5,5V		Say	-	
LC 92032 A	CMOS-IC	Gate Array (3216), 1,4ns, 3...5,5V		Say	-	
LC 92041 A	CMOS-IC	Gate Array (4185), 1,4ns, 3...5,5V		Say	-	
LC 92060 A	CMOS-IC	Gate Array (6016), 1,4ns, 3...5,5V		Say	-	
LC 92080 A	CMOS-IC	Gate Array (8028), 1,4ns, 3...5,5V		Say	-	
LC 92100 A	CMOS-IC	Gate Array (10023), 1,4ns, 3...5,5V		Say	-	
LC 97000	CMOS-IC	Gate Array (1k..15k), 2ns, 3...5,5V		Say	-	
24 LC 1025 P	EEPROM-IC	1024K I2C Cmos Serial EEPROM	DIP	Mcp	-	pdf pinout
24 LC 1025 SM	EEPROM-IC	1024K I2C Cmos Serial EEPROM	DIP	Mcp	-	pdf pinout
LC 371000 Q	CMOS-sRAM-IC	128kB x 8 Bit Mask ROM	28-DIP	Say	-	
LC 374000 P	CMOS-sRAM-IC	512kB x 8 Bit Mask ROM	40-DIP	Say	-	
LC 737231	CMOS-IC	Telecom, memory Dtmf/pulse dialer	36-SDIP	Say	-	
LC 737231 M	CMOS-IC	=LC 737231: SMD	36-MDIP		-	
LC 823208	CMOS-IC	SMD, DCT	44-MP	Say	-	
LC 823240	CMOS-IC	SMD, MPEG (ISO/IEC 11172-3) Video Dec.	100-MP	Say	-	
LC 823400	CMOS-IC	SMD, MPEG (ISO/IEC 11172-3) Video Dec.	120-MP	Say	-	
LC 864616A	µP-IC	OSD controller (192 char),caption data slicer, 6x 7Bit PWM output ...	DIP	Say	-	pdf pinout
LC 895170 W	CMOS-IC	SMD, CD-ROM Prozessor, Buffer RAM	100-MP	Say	-	pdf
LC 895199 K	CMOS-IC	32x CD-ROM Decoder	MP	Say	-	pdf pinout
LCA110	OC	LED/MOSFET Viso:3750Vrms Vbs:1.2V		Cpc		data pinout
LCA131	OC	LED/MOSFET Viso:2500Vrms Vbs:1.2V		Hpa,The		data pinout
LCA142	OC	LED/MOSFET Viso:2500Vrms Vbs:1.2V		The		data pinout
LCA210	OC	LED/2 MOSFET Viso:2500Vrms Vbs:1.2V		The		data pdf pinout
LCB110	OC	LED/MOSFET Viso:3750Vrms Vbs:1.2V		Cpc,The		data pinout
LCB111	OC	LED/MOSFET Viso:2500Vrms Vbs:1.2V		The		data pdf pinout
LCB112	OC	LED/MOSFET Viso:2500Vrms Vbs:1.2V		The		data pinout
LCB113	OC	LED/MOSFET Viso:2500Vrms Vbs:1.2V		The		data pinout
LCB114	OC	LED/MOSFET Viso:2500Vrms Vbs:1.2V		The		data pinout
LCC110	OC	LED/2 MOSFET Viso:3750Vrms Vbs:1.2V		Cpc,The		data pinout
LCP 150 S	LIN-IC	Telecom, Dual, Progr. Taz f. Clic	4-SIP	Tho	-	
LCP 1511	LIN-IC	SMD,Telecom, Dual, Progr. Taz f. Clic	8-MDIP	Tho	-	
LCP 1512	LIN-IC	Telecom, Dual, Progr. Taz f. Clic	8-DIP	Tho	-	
LCTA 6 V1M3	LIN-IC, Z-Di	SMD,18x TAZ(6,1V) f.Centronics Interf. Low Capacitance	20-MDIP	Tho	-	data

Type	Device	Short Description	Fig.	Manu	Comparision Types	More at
LD...LE						
LD-001	LED	or+gr+rd+yl 9.5x4.5mm I:4mcd Vf:2.1V		Roh		data pdf pinout
LD-002	LED	or+gr+rd+yl 6.5x1.3mm I:0.63mcd		Roh		data pdf pinout
LD-003	LED	or+gr+rd+yl 6.5x1.3mm I:0.63mcd		Roh		data pinout
LD-006	LED	or+gr+rd+yl 6.5x4mm I:10mcd Vf:2.1V		Roh		data pinout
LD-010	LED	or+gr+rd+yl 1.62x58mm I:2.5mcd		Roh		data pinout
LD-101	LED	or+gr+rd+yl 5x5mm I:2.5mcd Vf:2.1V		Roh		data pinout
LD-1203	LED	or+rd+gr+yl 28.8x6.8mm I:6.3mcd		Roh		data pdf pinout
LD-201	LED	or+gr+rd+yl 5.6x3.2mm I:2.5mcd		Roh		data pinout
LD-404	LED	or+gr+rd+yl 10x10mm I:2.5mcd Vf:2.1V		Roh		data pdf pinout
LD-603	LED	or+gr+rd+yl 13.8x6.8mm I:6.3mcd		Roh		data pinout
LD-605	LED	rd/gr 15x12mm I:25mcd Vf:2.1V		Roh		data pdf pinout
LD-701	LED	or+gr+rd+yl 20x4.5mm I:6.3mcd		Roh		data pinout
LD-701AX	LED	bl+gr+wht ø5.6mm I:35lm Vf:3.6V		Cot		data pdf pinout
LD-701CX	LED	rd+am ø5.6mm I:35lm Vf:2.4V If:450mA		Cot		data pdf pinout
LD-706	LED	or+gr+rd+yl 18x13mm I:10mcd Vf:4.2V		Roh		data pdf pinout
LD...	Opto	LED				data pinout
LD	Si-P	=2SA1736 (Typ-Code/Stempel/marking)	39	Tos	→2SA1736	data
LD	Si-N	=2SC2462-D (Typ-Code/Stempel/marking)	35	Hit	→2SC2462	data
LD	Si-N	=2SC4357-D (Typ-Code/Stempel/marking)	39	Mit	→2SC4357	data
LD	MOS-N-FET-d	=BF 543 (Typ-Code/Stempel/marking)	35	Sie	→BF 543	data
LD	Si-N	=HD 2A4M (Typ-Code/Stempel/marking)	39	Nec	→HD 2...	data
LD	Z-Di	=P6 SMB-18 (Typ-Code/Stempel/marking)	71(5x3,5)	Fag	→P6 SMB-18	data
LD	Z-Di	=SM 6T 6V8C (Typ-Code/Stempel/marking)	71(6x4mm)	Tho	→SM 6T...	data
LD	Z-Di	=SMBJ 12 (Typ-Code/Stempel/marking)	71(5x3,5)	Mop	→SMBJ ...	data
LD 1	Si-N	=FA 3L4Z (Typ-Code/Stempel/marking)	35	Nec	→FA 3...	data
LD30	LED	rd ø3mm I>0.3mcd Vf:1.6V If:20mA	T1	Sie		data pinout
LD32	LED	rd ø3mm I>0.8mcd Vf:2.4V If:10mA	T1	Sie		data pinout
LD35	LED	yl ø3mm I>0.6mcd Vf:2.4V If:20mA	T1	Sie		data pinout
LD36	LED	yl ø3mm I>0.6mcd Vf:2.4V If:10mA	T1	Sie		data pinout
LD37	LED	gr ø3mm I>0.5mcd Vf:1.6V If:20mA	T1	Sie		data pinout
LD41	LED	rd ø5mm I>0.3mcd Vf:1.6V If:20mA	T1¾	Sie		data pinout
LD41S	LED	rd ø5mm I>2mcd Vf:1.6V If:20mA		Sie		data pinout
LD50	LED	rd ø5mm I>1mcd Vf:1.6V If:20mA	T1¾	Sie		data pinout
LD52	LED	rd ø5mm I>0.8mcd Vf:2.4V If:10mA	T1¾	Sie		data pinout
LD52C	LED	rd ø5.1mm I>9mcd Vf:2.4V If:20mA	T1¾	Sie		data pinout
LD52S	LED	rd ø5mm I>2mcd Vf:2.4V If:10mA		Sie		data pinout
LD55	LED	yl ø5mm I>0.8mcd Vf:2.4V If:20mA	T1¾	Sie		data pinout
LD56	LED	yl ø5mm I>0.6mcd Vf:2.4V If:10mA	T1¾	Sie		data pinout
LD56C	LED	yl ø5.1mm I>6mcd Vf:2.4V If:20mA	T1¾	Sie		data pinout
LD56S	LED	yl ø5mm I>1.6mcd Vf:2.4V If:10mA		Sie		data pinout
LD57	LED	gr ø5mm I>0.6mcd Vf:2.4V If:20mA	T1¾	Sie		data pinout
LD57C	LED	gr ø5.1mm I>12mcd Vf:2.4V If:20mA	T1¾	Sie		data pinout
LD57S	LED	gr ø5mm I>4mcd Vf:2.4V If:20mA		Sie		data pinout
LD80	LED	rd 2.4x4.8mm I>0.6mcd Vf:1.6		Sie		data pinout
LD82	LED	rd 2.4x4.8mm I>0.6mcd Vf:2.4V		Sie		data pinout
LD86	LED	yl 2.4x4.8mm I>0.6mcd Vf:2.4V		Sie		data pinout
LD87	LED	gr 2.4x4.8mm I>0.6mcd Vf:2.4V		Sie		data pinout
LD121	LED	rd+yl+gr 0.7x1mm I>0.63mcd Vf:2.4V		Sie		data pinout
LD242	IRED	950nm ø4.15mm I:<3.2mW/Sr Vf:1.3V	TO18	Osr		data pinout
LD260	IRED	950nm 2.54x2.54mm I:<8mW/Sr Vf:1.25V		Sie		data pdf pinout
LD261	IRED	950nm ø1.8mm I:>5mW/Sr Vf:1.25V		Sie		data pinout
LD262	IRED	950nm 2.54x2.54mm I:<5mW/Sr Vf:1.25V		Osr		data pinout
LD263	IRED	950nm 2.54x2.54mm I:<5mW/Sr Vf:1.25V		Osr		data pdf pinout
LD264	IRED	950nm 2.54x2.54mm I:<5mW/Sr Vf:1.25V		Osr		data pdf pinout
LD265	IRED	950nm 2.54x2.54mm I:<8mW/Sr Vf:1.25V		Sie		data pdf pinout
LD266	IRED	950nm 2.54x2.54mm I:<8mW/Sr Vf:1.25V		Sie		data pinout
LD267	IRED	950nm 2.54x2.54mm I:<8mW/Sr Vf:1.25V		Sie		data pinout
LD268	IRED	950nm 2.54x2.54mm I:<8mW/Sr Vf:1.25V		Sie		data pinout
LD269	IRED	950nm 2.54x2.54mm I:<8mW/Sr Vf:1.25V		Sie		data pinout
LD271	IRED	950nm ø5.1mm I:>16mW/Sr Vf:1.3V	T1¾	Sie		data pinout
LD271L	IRED	950nm ø5.1mm I:>16mW/Sr Vf:1.3V	T1¾	Sie		data pdf pinout
LD273	IRED	950nm ø5mm I:>25mW/Sr Vf:1.3V	T1¾	Sie		data pdf pinout
LD274	IRED	950nm ø5mm I:60mW/Sr Vf:1.3V	T1¾	Sie		data pdf pinout
LD275	IRED	950nm ø5.1mm I:>80mW/Sr Vf:1.3V	T1¾	Sie		data pdf pinout
LD301	LED	rd+wht ø2.4mm I:7lm Vf:2.8V If:100mA		Cot		data pinout
LD461	LED	rd 2.4x2.8mm I>0.6mcd Vf:1.6V		Sie		data pdf pinout
LD471	LED	gr 2.4x2.8mm I>0.6mcd Vf:2.4V		Sie		data pinout
LD481	LED	yl 2.4x2.8mm I>0.6mcd Vf:2.4V		Sie		data pinout
LD491	LED	yl 2.4x2.8mm I>0.6mcd Vf:2.4V		Sie		data pinout
LD 1020	Hybrid-IC	Cb, 27MHz Osc., Am-zf	5-SQP	Say		
LD 1041	Hybrid-IC	AM-ZF	9-SQP	Say	-	
LD 1084 V12	LIN-IC	5A low drop positive voltage regulator fixed 12V	17	Stm	-	pdf pinout
LD 1084 V25	LIN-IC	5A low drop positive voltage regulator fixed 2.5V	17	Stm	-	pdf pinout
LD 1084 V50	LIN-IC	5A low drop positive voltage regulator fixed 5.0V	17	Stm	-	pdf pinout
LD 1084 V	LIN-IC	5A low drop positive voltage regulator adj.	17	Stm	-	pdf pinout
LD 1085 CDT	LIN/Z-IC	3A low-drop, adj. positive voltage regulator	30	Stm	-	pdf pinout
LD 1085 D2M18	LIN/Z-IC	3A low drop positive voltage regulator fixed 1.8V	30	Stm		pdf pinout

Type	Device	Short Description	Fig.	Manu	Comparision Types	More at
LD 1085 D2M25	LIN/Z-IC	3A low drop positive voltage regulator fixed 2.5V	30	Stm		pdf pinout
LD 1085 D2M33	LIN/Z-IC	3A low drop positive voltage regulator fixed 3.3V	30	Stm		pdf pinout
LD 1085 D2M	LIN/Z-IC	3A low drop positive voltage regulator adjustable	30	Stm		pdf pinout
LD 1085 D2T18	LIN/Z-IC	3A low drop positive voltage regulator fixed 1.8V	30	Stm		pdf pinout
LD 1085 D2T33	LIN/Z-IC	3A low drop positive voltage regulator fixed 3.3V	30	Stm		pdf pinout
LD 1085 D2T	LIN/Z-IC	3A low drop positive voltage regulator adjustable	30	Stm		pdf pinout
LD 1085 DT15	LIN/Z-IC	3A low drop positive voltage regulator fixed 1.5V	30	Stm		pdf pinout
LD 1085 DT18	LIN/Z-IC	3A low drop positive voltage regulator fixed 1.8V	30	Stm		pdf pinout
LD 1085 DT25	LIN/Z-IC	3A low drop positive voltage regulator fixed 2.5V	30	Stm		pdf pinout
LD 1085 DT33	LIN/Z-IC	3A low drop positive voltage regulator fixed 3.3V	30	Stm		pdf pinout
LD 1085 V25	LIN/Z-IC	3A low drop positive voltage regulator fixed 2.5V	17	Stm		pdf pinout
LD 1085 V50	LIN/Z-IC	3A low drop positive voltage regulator fixed 5.0V	17	Stm		pdf pinout
LD 1085 V/P	LIN/Z-IC	3A low drop positive voltage regulator adjustable	17	Stm		pdf pinout
LD1085X	ER+	Io=int Vo:1.25...28V Vin:2.85...30V	3SP	Sgs		data pdf pinout
LD1085X120	R+	Io=int Vo:11.76...12.24V P:int	3SP	Sgs		data pdf pinout
LD1085X15	R+	Io=int Vo:1.47...1.53V Vin:3.1...30V	3SP	Sgs		data pdf pinout
LD1085X18	R+	Io=int Vo:1.764...1.836V Vin:3.4...30V	3SP	Sgs		data pdf pinout
LD1085X25	R+	Io=int Vo:2.45...2.55V Vin:4.1...30V	3SP	Sgs		data pdf pinout
LD1085X285	R+	Io=int Vo:2.793...2.907V Vin:4.5...30V	3SP	Sgs		data pdf pinout
LD1085X33	R+	Io=int Vo:3.234...3.366V Vin:4.9...30V	3SP	Sgs		data pdf pinout
LD1085X36	R+	Io=int Vo:3.528...3.572V Vin:5.2...30V	3SP	Sgs		data pdf pinout
LD1085X50	R+	Io=int Vo:4.9...5.1V Vin:6.6...30V	3SP	Sgs		data pdf pinout
LD1085X80	R+	Io=int Vo:7.84...8.16V Vin:9.8...30V	3SP	Sgs		data pdf pinout
LD1085X90	R+	Io=int Vo:8.82...9.18V Vin:11...30V	3SP	Sgs		data pdf pinout
LD 1086 D2M33	LIN-IC	1.5A low drop positive voltage regulator fixed 3.3V	30	Stm	-	pdf pinout
LD 1086 D2M	LIN-IC	1.5A low drop positive voltage regulator adj.	30	Stm	-	pdf pinout
LD 1086 D2T12	LIN-IC	1.5A low drop positive voltage regulator fixed 12V	30	Stm	-	pdf pinout
LD 1086 D2T15	LIN-IC	1.5A low drop positive voltage regulator fixed 1.5V	30	Stm	-	pdf pinout
LD 1086 D2T18	LIN-IC	1.5A low drop positive voltage regulator fixed 1.8V	30	Stm	-	pdf pinout
LD 1086 D2T25	LIN-IC	1.5A low drop positive voltage regulator fixed 2.5V	30	Stm	-	pdf pinout
LD 1086 D2T33	LIN-IC	1.5A low drop positive voltage regulator fixed 3.3V	30	Stm	-	pdf pinout
LD 1086 D2T50	LIN-IC	1.5A low drop positive voltage regulator fixed 5V	30	Stm	-	pdf pinout
LD 1086 D2T80	LIN-IC	1.5A low drop positive voltage regulator fixed 8V	30	Stm	-	pdf pinout
LD 1086 D2T	LIN-IC	1.5A low drop positive voltage regulator adj.	30	Stm	-	pdf pinout
LD 1086 DT18	LIN-IC	1.5A low drop positive voltage regulator fixed 1.8V	30	Stm	-	pdf pinout
LD 1086 DT15	LIN-IC	1.5A low drop positive voltage regulator fixed 1.5V	30	Stm	-	pdf pinout
LD 1086 DT25	LIN-IC	1.5A low drop positive voltage regulator fixed 2.5V	30	Stm	-	pdf pinout
LD 1086 DT33	LIN-IC	1.5A low drop positive voltage regulator fixed 3.3V	30	Stm	-	pdf pinout
LD 1086 DT50	LIN-IC	1.5A low drop positive voltage regulator fixed 5V	30	Stm	-	pdf pinout
LD 1086 DT80	LIN-IC	1.5A low drop positive voltage regulator fixed 8V	30	Stm	-	pdf pinout
LD 1086 DT	LIN-IC	1.5A low drop positive voltage regulator adj.	30	Stm	-	pdf pinout
LD 1086 V12	LIN-IC	1.5A low drop positive voltage regulator fixed 12V	17	Stm	-	pdf pinout
LD 1086 V18	LIN-IC	1.5A low drop positive voltage regulator fixed 1.8V	17	Stm	-	pdf pinout
LD 1086 V25	LIN-IC	1.5A low drop positive voltage regulator fixed 2.5V	17	Stm	-	pdf pinout
LD 1086 V33	LIN-IC	1.5A low drop positive voltage regulator fixed 3.3V	17	Stm	-	pdf pinout
LD 1086 V36	LIN-IC	1.5A low drop positive voltage regulator fixed 3.6V	17	Stm	-	pdf pinout
LD 1086 V80	LIN-IC	1.5A low drop positive voltage regulator fixed 8V	17	Stm	-	pdf pinout
LD 1086 V	LIN-IC	1.5A low drop positive voltage regulator adj.	17	Stm	-	pdf pinout

Type	Device	Short Description	Fig.	Manu	Comparision Types	More at
LD1086X	ER+	Io=int Vo:1.25...28V Vin:2.85...30V	3SP	Sgs		data pdf pinout
LD1086X120	R+	Io=int Vo:11.76...12.24V P:int	3SP	Sgs		data pdf pinout
LD1086X15	R+	Io=int Vo:1.47...1.53V Vin:3.1...30V	3SP	Sgs		data pdf pinout
LD1086X18	R+	Io=int Vo:1.764...1.836V Vin:3.4...30V	3SP	Sgs		data pdf pinout
LD1086X25	R+	Io=int Vo:2.45...2.55V Vin:4.1...30V	3SP	Sgs		data pdf pinout
LD1086X285	R+	Io=int Vo:2.793...2.907V Vin:4.5...30V	3SP	Sgs		data pdf pinout
LD1086X33	R+	Io=int Vo:3.234...3.366V Vin:4.9...30V	3SP	Sgs		data pdf pinout
LD1086X36	R+	Io=int Vo:3.528...3.572V Vin:5.2...30V	3SP	Sgs		data pdf pinout
LD1086X50	R+	Io=int Vo:4.9...5.1V Vin:6.6...30V	3SP	Sgs		data pdf pinout
LD1086X80	R+	Io=int Vo:7.84...8.16V Vin:9.8...30V	3SP	Sgs		data pdf pinout
LD1086X90	R+	Io=int Vo:8.82...9.18V Vin:11...30V	3SP	Sgs		data pdf pinout
LD1117	ER+	Vo:1.25...15V Vin:3.2V P:12W	8S,3P,4S	Sgs		data pdf pinout
LD 1117 AD2M	LIN-IC	1A, adj. 1.25 to 15V, low drop positive voltage regulators	30	Stm	-	pdf pinout
LD 1117 ADT12	LIN-IC	1A, fix. 1.2V, low drop positive voltage regulators	30	Stm	-	pdf pinout
LD 1117 ADT18	LIN-IC	1A, fix. 1.8V, low drop positive voltage regulators	30	Stm	-	pdf pinout
LD 1117 ADT25	LIN-IC	1A, fix. 2.5V, low drop positive voltage regulators	30	Stm	-	pdf pinout
LD 1117 ADT28	LIN-IC	1A, fix. 2.85V, low drop positive voltage regulators	30	Stm	-	pdf pinout
LD 1117 ADT33	LIN-IC	1A, fix. 3.3V, low drop positive voltage regulators	30	Stm	-	pdf pinout
LD 1117 ADT50	LIN-IC	1A, fix. 5V, low drop positive voltage regulators	30	Stm	-	pdf pinout
LD 1117 ADT	LIN-IC	1A, adj. 1.25 to 15V, low drop positive voltage regulators	30	Stm	-	pdf pinout
LD 1117 AS12	LIN-IC	1A, fix. 1.2V, low drop positive voltage regulators	48	Stm	-	pdf pinout
LD 1117 AS18	LIN-IC	1A, fix. 1.8V, low drop positive voltage regulators	48	Stm	-	pdf pinout
LD 1117 AS25	LIN-IC	1A, fix. 2.5V, low drop positive voltage regulators	48	Stm	-	pdf pinout
LD 1117 AS28	LIN-IC	1A, fix. 2.85V, low drop positive voltage regulators	48	Stm	-	pdf pinout
LD 1117 AS33	LIN-IC	1A, fix. 3.3V, low drop positive voltage regulators	48	Stm	-	pdf pinout
LD 1117 AS50	LIN-IC	1A, fix. 5V, low drop positive voltage regulators	48	Stm	-	pdf pinout
LD 1117 AS	LIN-IC	1A, adj. 1.25 to 15V, low drop positive voltage regulators	48	Stm	-	pdf pinout
LD 1117 AV12	LIN-IC	1A, fix. 1.2V, low drop positive voltage regulators	17	Stm	-	pdf pinout
LD 1117 AV18	LIN-IC	1A, fix. 1.8V, low drop positive voltage regulators	17	Stm	-	pdf pinout
LD 1117 AV25	LIN-IC	1A, fix. 2.5V, low drop positive voltage regulators	17	Stm	-	pdf pinout
LD 1117 AV28	LIN-IC	1A, fix. 2.85V, low drop positive voltage regulators	17	Stm	-	pdf pinout
LD 1117 AV33	LIN-IC	1A, fix. 3.3V, low drop positive voltage regulators	17	Stm	-	pdf pinout
LD 1117 AV50	LIN-IC	1A, fix. 5V, low drop positive voltage regulators	17	Stm	-	pdf pinout
LD 1117 AV	LIN-IC	1A, adj. 1.25 to 15V, low drop positive voltage regulators	17	Stm	-	pdf pinout
LD1117AX	ER+	Io=int Vo:1.25...8V P:12W	3P,4S	Sgs		data pdf pinout
LD1117AX18	R+	Io=int Vo:1.764...1.836V Vin:3.8...9V	3P,4S	Sgs		data pdf pinout
LD1117AX25	R+	Io=int Vo:2.45...2.55V Vin:3.9...8V	3P,4S	Sgs		data pdf pinout
LD1117AX28	R+	Io=int Vo:2.793...2.907V Vin:4.85V	3P,4S	Sgs		data pdf pinout
LD1117AX33	R+	Io=int Vo:3.234...3.366V Vin:5.3V	3P,4S	Sgs		data pdf pinout
LD1117AX50	R+	Io=int Vo:4.9...5.1V Vin:7V P:12W	3P,4S	Sgs		data pdf pinout
LD1117X12	R+	Vo:1.176...1.224V Vin:2V P:12W	8S,3P,4S	Sgs		data pdf pinout
LD1117X18	R+	Vo:1.76...1.84V Vin:3.8V P:12W	8S,3P,4S	Sgs		data pdf pinout
LD1117X25	R+	Vo:2.45...2.55V Vin:4.5V P:12W	8S,3P,4S	Sgs		data pdf pinout
LD1117X28	R+	Vo:2.82...2.88V Vin:4.85V P:12W	8S,3P,4S	Sgs		data pdf pinout
LD1117X30	R+	Vo:2.97...3.03V Vin:5V P:12W	8S,3P,4S	Sgs		data pdf pinout
LD1117X33	R+	Vo:3.24...3.36V Vin:5.3V P:12W	8S,3P,4S	Sgs		data pdf pinout
LD 1120	Hybrid-IC	f.drahtlose Mikrofone/wireless micr.	9-SQP	Say		
LD 1207	IC	LED Driver IC	DIP	Iks	-	pdf pinout
LD 1207 L	IC	LED Driver IC	DIP	Iks	-	pdf pinout
LD 1207 VL	IC	LED Driver IC	DIP	Iks	-	pdf pinout
LD 1585 CD2M15	LIN/Z-IC	5A low dropout fast response positive voltage regulator fixed 1.5V	30	Stm	-	pdf pinout
LD 1585 CD2M25	LIN/Z-IC	5A low dropout fast response positive voltage regulator fixed 2.5V	30	Stm	-	pdf pinout
LD 1585 CD2M28	LIN/Z-IC	5A low dropout fast response positive voltage regulator fixed 2.85V	30	Stm	-	pdf pinout
LD 1585 CD2M50	LIN/Z-IC	5A low dropout fast response positive voltage regulator fixed 5.0V	30	Stm	-	pdf pinout
LD 1585 CD2M80	LIN/Z-IC	5A low dropout fast response positive voltage regulator fixed 8.0V	30	Stm	-	pdf pinout
LD 1585 CD2M90	LIN/Z-IC	5A low dropout fast response positive voltage regulator fixed 9.0V	30	Stm	-	pdf pinout
LD 1585 CV50	LIN/Z-IC	5A low dropout fast response positive	17	Stm	-	pdf pinout

Type	Device	Short Description	Fig.	Manu	Comparision Types	More at
		voltage regulator fixed 5.0V				
LD 1585 CV	LIN/Z-IC	5A low dropout fast response positive voltage regulator adj.	17	Stm		pdf pinout
LD2980ABX15	R+	Io=int Vo:1.489...1.511V Vin:2.5V	5S,4S	Sgs		data pdf pinout
LD2980ABX18	R+	Io=int Vo:1.786...1.814V Vin:2.8V	5S,4S	Sgs		data pdf pinout
LD2980ABX25	R+	Io=int Vo:2.481...2.519V Vin:3.5V	5S,4S	Sgs		data pdf pinout
LD2980ABX28	R+	Io=int Vo:2.828...2.871V Vin:3.85V	5S,4S	Sgs		data pdf pinout
LD2980ABX30	R+	Io=int Vo:2.978...3.023V Vin:4.0V	5S,4S	Sgs		data pdf pinout
LD2980ABX32	R+	Io=int Vo:3.176...3.224V Vin:4.2V	5S,4S	Sgs		data pdf pinout
LD2980ABX33	R+	Io=int Vo:3.275...3.325V Vin:4.3V	5S,4S	Sgs		data pdf pinout
LD2980ABX36	R+	Io=int Vo:3.573...3.627V Vin:4.6V	5S,4S	Sgs		data pdf pinout
LD2980ABX38	R+	Io=int Vo:3.781...3.819V Vin:4.8V	5S,4S	Sgs		data pdf pinout
LD2980ABX40	R+	Io=int Vo:3.97...4.03V Vin:5.0V P:int	5S,4S	Sgs		data pdf pinout
LD2980ABX47	R+	Io=int Vo:4.665...4.735V Vin:5.7V	5S,4S	Sgs		data pdf pinout
LD2980ABX48	R+	Io=int Vo:4.814...4.886V Vin:5.85V	5S,4S	Sgs		data pdf pinout
LD2980ABX50	R+	Io=int Vo:4.963...5.038V Vin:6.0V	5S,4S	Sgs		data pdf pinout
LD2980CX15	R+	Io=int Vo:1.477...1.523V Vin:2.5V	5S,4S	Sgs		data pdf pinout
LD2980CX18	R+	Io=int Vo:1.773...1.827V Vin:2.8V	5S,4S	Sgs		data pdf pinout
LD2980CX25	R+	Io=int Vo:2.463...2.538V Vin:3.5V	5S,4S	Sgs		data pdf pinout
LD2980CX28	R+	Io=int Vo:2.807...2.893V Vin:3.85V	5S,4S	Sgs		data pdf pinout
LD2980CX30	R+	Io=int Vo:2.955...3.045V Vin:4.0V	5S,4S	Sgs		data pdf pinout
LD2980CX32	R+	Io=int Vo:3.152...3.248V Vin:4.2V	5S,4S	Sgs		data pdf pinout
LD2980CX33	R+	Io=int Vo:3.251...3.35V Vin:4.3V P:int	5S,4S	Sgs		data pdf pinout
LD2980CX36	R+	Io=int Vo:3.546...3.664V Vin:4.6V	5S,4S	Sgs		data pdf pinout
LD2980CX38	R+	Io=int Vo:3.76...3.838V Vin:4.8V P:int	5S,4S	Sgs		data pdf pinout
LD2980CX40	R+	Io=int Vo:3.94...4.06V Vin:5.0V P:int	5S,4S	Sgs		data pdf pinout
LD2980CX47	R+	Io=int Vo:4.63...4.771V Vin:5.7V P:int	5S,4S	Sgs		data pdf pinout
LD2980CX48	R+	Io=int Vo:4.777...4.923V Vin:5.85V	5S,4S	Sgs		data pdf pinout
LD2980CX50	R+	Io=int Vo:4.925...5.075V Vin:6.0V	5S,4S	Sgs		data pdf pinout
LD2981ABX15	R+	Io=int Vin:2.5V P:int	5S,4S	Sgs		data pdf pinout
LD2981ABX18	R+	Io=int Vo:1.782...1.818V Vin:2.8V	5S,4S	Sgs		data pdf pinout
LD2981ABX25	R+	Io=int Vo:2.475...2.525V Vin:3.5V	5S,4S	Sgs		data pdf pinout
LD2981ABX28	R+	Io=int Vo:2.822...2.878V Vin:3.8V	5S,4S	Sgs		data pdf pinout
LD2981ABX30	R+	Io=int Vo:2.970...3.030V Vin:4.0V	5S,4S	Sgs		data pdf pinout
LD2981ABX32	R+	Io=int Vo:3.168...3.232V Vin:4.2V	5S,4S	Sgs		data pdf pinout
LD2981ABX33	R+	Io=int Vo:3.267...3.333V Vin:4.3V	5S,4S	Sgs		data pdf pinout
LD2981ABX36	R+	Io=int Vo:3.564...3.636V Vin:4.6V	5S,4S	Sgs		data pdf pinout
LD2981ABX38	R+	Io=int Vo:3.762...3.838V Vin:4.8V	5S,4S	Sgs		data pdf pinout
LD2981ABX40	R+	Io=int Vo:3.96...4.04V Vin:5.0V P:int	5S,4S	Sgs		data pdf pinout
LD2981ABX47	R+	Io=int Vo:4.653...4.747V Vin:5.7V	5S,4S	Sgs		data pdf pinout
LD2981ABX48	R+	Io=int Vo:4.801...4.899V Vin:5.8V	5S,4S	Sgs		data pdf pinout
LD2981ABX50	R+	Io=int Vo:4.95...5.050V Vin:6.0V P:int	5S,4S	Sgs		data pdf pinout
LD2981CX15	R+	Io=int Vo:1.470...1.519V Vin:2.5V	5S,4S	Sgs		data pdf pinout
LD2981CX18	R+	Io=int Vo:1.764...1.836V Vin:2.8V	5S,4S	Sgs		data pdf pinout
LD2981CX25	R+	Io=int Vo:2.45...2.55V Vin:3.5V P:int	5S,4S	Sgs		data pdf pinout
LD2981CX28	R+	Io=int Vo:2.793...2.907V Vin:3.85V	5S,4S	Sgs		data pdf pinout
LD2981CX30	R+	Io=int Vo:2.94...3.06V Vin:4.0V P:int	5S,4S	Sgs		data pdf pinout
LD2981CX32	R+	Io=int Vo:3.136...3.264V Vin:4.2V	5S,4S	Sgs		data pdf pinout
LD2981CX33	R+	Io=int Vo:3.234...3.366V Vin:4.3V	5S,4S	Sgs		data pdf pinout
LD2981CX36	R+	Io=int Vo:3.528...3.672V Vin:4.6V	5S,4S	Sgs		data pdf pinout
LD2981CX38	R+	Io=int Vo:3.724...3.876V Vin:4.8V	5S,4S	Sgs		data pdf pinout
LD2981CX40	R+	Io=int Vo:3.92...4.08V Vin:5.0V P:int	5S,4S	Sgs		data pdf pinout
LD2981CX47	R+	Io=int Vo:4.606...4.794V Vin:5.7V	5S,4S	Sgs		data pdf pinout
LD2981CX48	R+	Io=int Vo:4.753...4.947V Vin:5.85V	5S,4S	Sgs		data pdf pinout
LD2981CX50	R+	Io=int Vo:4.9...5.1V Vin:6.0V P:int	5S,4S	Sgs		data pdf pinout
LD 3001	Hybrid-IC	Recorder, Verst./amplifier	9-SQP	Say	-	
LD 3030	Hybrid-IC	Nf-verst./af amplifier	9-SQP	Say	-	
LD 3040	Hybrid-IC	Nf-verst./af amplifier	5-SQP	Say	-	
LD 3050	Hybrid-IC	Recorder, Nf-verst./af amplifier	13-SQP	Say	-	
LD 3061	Hybrid-IC	Orgel-master-o + Puffer/buffer	9-SQP	Say	-	
LD 3070	Hybrid-IC	Verst./amplifier f. Bell Boy	5-SQP	Say	-	
LD 3080	Hybrid-IC		13-SQP	Say	-	
LD 3100	Hybrid-IC	Recorder, Verst./amplifier	9-SQP	Say	-	
LD 3115	Hybrid-IC	NF-Equal., rauscharm/low noise	9-SQP	Say	-	
LD 3120	Hybrid-IC	NF-Equal., rauscharm/low noise	9-SQP	Say	-	
LD 3130	Hybrid-IC	NF-Equal., rauscharm/low noise	9-SQP	Say	-	
LD 3141	Hybrid-IC	NF-Equal., rauscharm/low noise	9-SQP	Say	-	
LD 3150	Hybrid-IC	NF-Equal., rauscharm/low noise	9-SQP	Say	-	
LD 3180	Hybrid-IC	NF-Equal., rauscharm/low noise	9-SQP	Say	-	
LD 3200	Hybrid-IC	VC, VA-O	9-SQP	Say	-	
LD 3210	Hybrid-IC	VC, HA-O	9-SQP	Say	-	
LD 5100	Hybrid-Z-IC	Z-IC, 1,15V	5-SQP	Say	-	
LD 6010	Hybrid-IC	Batt.-Überladeschutz/over charg.prot.	5-SIP	Say	-	
LD 6020	Hybrid-IC		5-SQP	Say	-	
LD 6040	Hybrid-IC	Recorder, Alc	5-SQP	Say	-	
LD 7552 IN/BN	SMPS-IC	Green-mode PWM Ctrl.	DIP	Lea	-	pdf pinout
LD 7552 IS/BS	SMPS-IC	Green-mode PWM Ctrl.	MDIP	Lea	-	pdf pinout
LD 1580 P 2 T-R	Z-IC	7 A positive voltage regulator adj.	45	Stm	-	pdf pinout
LD 29080 PTR	IC	800mA Fixed and adjustable output very low drop voltage regulator	48	Stm		pdf pinout
LD 29150 PTR	IC	1.5 A, very low drop voltage regulators	48	Stm	-	pdf pinout
LD 2979 M 30 TR	Z-IC	Very low drop voltage regulators with inhibit	45	Stm	-	pdf pinout
LD 2979 M 33 TR	Z-IC	Very low drop voltage regulators with inhibit	45	Stm	-	pdf pinout

Type	Device	Short Description	Fig.	Manu	Comparision Types	More at
LD 2979 M 38 TR	Z-IC	Very low drop voltage regulators with inhibit	45	Stm	-	pdf pinout
LD 2979 M 50 TR	Z-IC	Very low drop voltage regulators with inhibit	45	Stm	-	pdf pinout
LD 2980 CM 18 TR	Z-IC	Ultra low drop voltage regulators	45	Stm	-	pdf pinout
LD 2980 CU 18 TR	Z-IC	Ultra low drop voltage regulators	45	Stm	-	pdf pinout
LD 2980 ABM 25 TR	Z-IC	Ultra low drop voltage regulators	45	Stm	-	pdf pinout
LD 2980 ABU 25 TR	Z-IC	Ultra low drop voltage regulators	45	Stm	-	pdf pinout
LD 2980 CM 25 TR	Z-IC	Ultra low drop voltage regulators	45	Stm	-	pdf pinout
LD 2980 CU 25 TR	Z-IC	Ultra low drop voltage regulators	45	Stm	-	pdf pinout
LD 2980 ABM 30 TR	Z-IC	Ultra low drop voltage regulators	45	Stm	-	pdf pinout
LD 2980 ABU 30 TR	Z-IC	Ultra low drop voltage regulators	45	Stm	-	pdf pinout
LD 2980 CM 30 TR	Z-IC	Ultra low drop voltage regulators	45	Stm	-	pdf pinout
LD 2980 CU 30 TR	Z-IC	Ultra low drop voltage regulators	45	Stm	-	pdf pinout
LD 2980 ABM 33 TR	Z-IC	Ultra low drop voltage regulators	45	Stm	-	pdf pinout
LD 2980 ABU 33 TR	Z-IC	Ultra low drop voltage regulators	45	Stm	-	pdf pinout
LD 2980 CM 33 TR	Z-IC	Ultra low drop voltage regulators	45	Stm	-	pdf pinout
LD 2980 ABM 36 TR	Z-IC	Ultra low drop voltage regulators	45	Stm	-	pdf pinout
LD 2980 ABU 36 TR	Z-IC	Ultra low drop voltage regulators	45	Stm	-	pdf pinout
LD 2980 CM 36 TR	Z-IC	Ultra low drop voltage regulators	45	Stm	-	pdf pinout
LD 2980 CU 36 TR	Z-IC	Ultra low drop voltage regulators	45	Stm	-	pdf pinout
LD 2980 ABM 50 TR	Z-IC	Ultra low drop voltage regulators	45	Stm	-	pdf pinout
LD 2980 ABU 50 TR	Z-IC	Ultra low drop voltage regulators	45	Stm	-	pdf pinout
LD 2980 CM 50 TR	Z-IC	Ultra low drop voltage regulators	45	Stm	-	pdf pinout
LD 2980 CU 50 TR	Z-IC	Ultra low drop voltage regulators	45	Stm	-	pdf pinout
LD 2981 CM 15 TR	IC	voltage regulators,inhibit low Esr output capacitors compatible	48	Stm	-	pdf pinout
LD 2981 CM 18 TR	IC	voltage regulators,inhibit low Esr output capacitors compatible	48	Stm	-	pdf pinout
LD 2981 CU 18 TR	IC	voltage regulators,inhibit low Esr output capacitors compatible	48	Stm	-	pdf pinout
LD 2981 ABM 25 TR	IC	voltage regulators,inhibit low ESR output capacitors compatible	48	Stm	-	pdf pinout
LD 2981 ABU 25 TR	IC	voltage regulators,inhibit low Esr output capacitors compatible	48	Stm	-	pdf pinout
LD 2981 CM 25 TR	IC	voltage regulators,inhibit low Esr output capacitors compatible	48	Stm	-	pdf pinout
LD 2981 CU 25 TR	IC	voltage regulators,inhibit low Esr output capacitors compatible	48	Stm	-	pdf pinout
LD 2981 ABM 30 TR	IC	voltage regulators,inhibit low Esr output capacitors compatible	48	Stm	-	pdf pinout
LD 2981 ABU 30 TR	IC	voltage regulators,inhibit low Esr output capacitors compatible	48	Stm	-	pdf pinout
LD 2981 CM 30 TR	IC	voltage regulators,inhibit low Esr output capacitors compatible	48	Stm	-	pdf pinout
LD 2981 CU 30 TR	IC	voltage regulators,inhibit low Esr output capacitors compatible	48	Stm	-	pdf pinout
LD 2981 ABM 33 TR	IC	voltage regulators,inhibit low Esr output capacitors compatible	48	Stm	-	pdf pinout
LD 2981 ABU 33 TR	IC	voltage regulators,inhibit low Esr output capacitors compatible	48	Stm	-	pdf pinout
LD 2981 CM 33 TR	IC	voltage regulators,inhibit low Esr output capacitors compatible	48	Stm	-	pdf pinout
LD 2981 CU 33 TR	IC	voltage regulators,inhibit low Esr output capacitors compatible	48	Stm	-	pdf pinout
LD 2981 ABM 50 TR	IC	voltage regulators,inhibit low Esr output capacitors compatible	48	Stm	-	pdf pinout
LD 2981 ABU 50 TR	IC	voltage regulators,inhibit low Esr output capacitors compatible	48	Stm	-	pdf pinout
LD 2981 CM 50 TR	IC	voltage regulators,inhibit low Esr output capacitors compatible	48	Stm	-	pdf pinout
LD 2981 CU 50 TR	IC	voltage regulators,inhibit low Esr output capacitors compatible	48	Stm	-	pdf pinout
LD 29080 DT 15 R	IC	800mA Fixed and adjustable output very low drop voltage regulator	48	Stm	-	pdf pinout
LD 29080 PT 15 R	IC	800mA Fixed and adjustable output very low drop voltage regulator	48	Stm	-	pdf pinout
LD 29080 DT 18 R	IC	800mA Fixed and adjustable output very low drop voltage regulator	48	Stm	-	pdf pinout
LD 29080 PT 18 R	IC	800mA Fixed and adjustable output very low drop voltage regulator	48	Stm	-	pdf pinout
LD 29080 DT 25 R	IC	800mA Fixed and adjustable output very low drop voltage regulator	48	Stm	-	pdf pinout
LD 29080 PT 25 R	IC	800mA Fixed and adjustable output very low drop voltage regulator	48	Stm	-	pdf pinout
LD 29080 DT 33 R	IC	800mA Fixed and adjustable output very low drop voltage regulator	48	Stm	-	pdf pinout
LD 29080 PT 33 R	IC	800mA Fixed and adjustable output very low drop voltage regulator	48	Stm	-	pdf pinout
LD 29080 DT 50 R	IC	800mA Fixed and adjustable output very low drop voltage regulator	48	Stm	-	pdf pinout
LD 29080 PT 50 R	IC	800mA Fixed and adjustable output very low drop voltage regulator	48	Stm	-	pdf pinout
LD 29080 DT 80 R	IC	800mA Fixed and adjustable output very low drop voltage regulator	48	Stm	-	pdf pinout

Type	Device	Short Description	Fig.	Manu	Comparision Types	More at
LD 29080 PT 80 R	IC	800mA Fixed and adjustable output very low drop voltage regulator	48	Stm	-	pdf pinout
LD 29150 DT 15 R	IC	1.5 A, very low drop voltage regulators	48	Stm	-	pdf pinout
LD 29150 PT 15 R	IC	1.5 A, very low drop voltage regulators	48	Stm	-	pdf pinout
LD 29150 DT 18 R	IC	1.5 A, very low drop voltage regulators	48	Stm	-	pdf pinout
LD 29150 PT 18 R	IC	1.5 A, very low drop voltage regulators	48	Stm	-	pdf pinout
LD 29150 DT 25 R	IC	1.5 A, very low drop voltage regulators	48	Stm	-	pdf pinout
LD 29150 PT 25 R	IC	1.5 A, very low drop voltage regulators	48	Stm	-	pdf pinout
LD 29150 DT 33 R	IC	1.5 A, very low drop voltage regulators	48	Stm	-	pdf pinout
LD 29150 PT 33 R	IC	1.5 A, very low drop voltage regulators	48	Stm	-	pdf pinout
LD 29150 DT 50 R	IC	1.5 A, very low drop voltage regulators	48	Stm	-	pdf pinout
LD 29150 PT 50 R	IC	1.5 A, very low drop voltage regulators	48	Stm	-	pdf pinout
LDA100	OC	bid. LED/NPN Viso:3750Vrms CTR:100%		Cpc		data pinout
LDA101	OC	LED/NPN Viso:3750Vrms CTR:100%		Cpc		data pinout
LDA110	OC	bid. LED/Dar Viso:3750Vrms CTR:1000%		Cpc		data pinout
LDA111	OC	LED/Dar Viso:3750Vrms CTR:1000%		Cpc		data pinout
LDA200	DOC	LED/Dar Viso:3750Vrms CTR:1000%		Cpc,The		data pinout
LDD 080	Hybrid-IC	Differentialverstärker/diff. amplifier		Mts	-	
LDD 100	Hybrid-IC	Differentialverstärker/diff. amplifier		Mts	-	
LDP 24 AS	LIN-IC	TAZ(25...32V), Load Dump Protection	20-MDIP	Tho		data
LDR 1825	LIN/Z-IC	1.8V/2.5V, Very low drop dual voltage regulator	87	Stm	-	pdf pinout
LDR 1833	LIN/Z-IC	1.8V/3.3V, Very low drop dual voltage regulator	87	Stm	-	pdf pinout
LDR 2518	LIN/Z-IC	2.5V/1.8V, Very low drop dual voltage regulator	87	Stm	-	pdf pinout
LDR 2533	LIN/Z-IC	2.5V/3.3V, Very low drop dual voltage regulator	87	Stm	-	pdf pinout
LDR 3318	LIN/Z-IC	3.3V/1.8V, Very low drop dual voltage regulator	87	Stm	-	pdf pinout
LDR 3325	LIN/Z-IC	3.3V/2.5V, Very low drop dual voltage regulator	87	Stm	-	pdf pinout
LDs	MOS-N-FET-d	=BF 543 (Typ-Code/Stempel/marking)	35	Sie	→BF 543	data
LE-0503-03	LED	bl+gr+rd+uv+wht+yl ø9.1mm I:8600mcd	T3¾	Jkl		data pdf pinout
LE-0504-01	LED	bl+gr+rd+uv+wht+yl ø4.5mm I:400mcd		Jkl		data pdf pinout
LE-0504-06	LED	wht ø4.5mm I:900mcd Vf:6V If:20mA		Jkl		data pdf pinout
LE-0509-01	LED	bl+gr+rd+wht+yl ø4.5mm I:400mcd		Jkl		data pdf pinout
LE-0509-02	LED	bl+gr+rd+wht+yl ø4.5mm I:3000mlm		Jkl		data pdf pinout
LE-0603-02	LED	bl+gr+wht+yl ø4.5mm I:4200mcd		Jkl		data pdf pinout
LE-0603-04	LED	bl+gr+rd+wht+yl ø4.5mm Vf:24V		Jkl		data pdf pinout
LE-BA9S-12W	LED	wht ø4.5mm I:800mcd Vf:12...14V		Jkl		data pdf pinout
LE	Z-Di	=1SMB 12A (Typ-Code/Stempel/marking)	71(5x3,5)	Ons	→1SMB 5.0A...170A	
LE	Si-N	=2SC2412KLN-E (Typ-Code/Stempel/mark.)	35	Rhm	→2SC2412KLN	data
LE	Si-N	=2SC2412L-LE(Typ-Code/Stempel/marking)	≈35	Rhm	→2SC2412L	data
LE	Si-N	=2SC3052-E (Typ-Code/Stempel/marking)	35	Mit	→2SC3052	data
LE	Si-N	=2SC4154-E (Typ-Code/Stempel/marking)	35(2mm)	Mit	→2SC4154	data
LE	Si-N	=2SC4357-E (Typ-Code/Stempel/marking)	39	Mit	→2SC4357	data
LE	Si-P	=BF 660 (Typ-Code/Stempel/marking)	35	Phi, Sie	→BF 660	data
LE	Si-N	=HD 2L2Q (Typ-Code/Stempel/marking)	39	Nec	→HD 2...	data
LE	Z-Di	=P6 SMB-18A (Typ-Code/Stempel/marking)	71(5x3,5)	Fag	→P6 SMB-18A	data
LE	Z-Di	=SM 6T 6V8CA(Typ-Code/Stempel/marking)	71(6x4mm)	Tho	→SM 6T...	data
LE	Z-Di	=SMBJ 12A (Typ-Code/Stempel/marking)	71(5x3,5)	Mop	→SMBJ ...	data
LE12AB	R+	Io=150mA Vo:1.225...1.275V Vin:3.3V	8S,3N	Sgs		data pdf pinout
LE12C	R+	Io=150mA Vo:1.225...1.275V Vin:3.3V	8S,3N	Sgs		data pdf pinout
LE15AB	R+	Io=150mA Vo:1.47...1.53V Vin:3.5V	8S,3N	Sgs		data pdf pinout
LE15C	R+	Io=150mA Vo:1.47...1.53V Vin:3.5V	8S,3N	Sgs		data pdf pinout
LE25AB	R+	Io=150mA Vo:2.475...2.525V Vin:4.5V	8S,3N	Sgs		data pdf pinout
LE25C	R+	Io=150mA Vo:2.45...2.55V Vin:4.5V	8S,3N	Sgs		data pdf pinout
LE 25 LB1282 M	CMOS-IC	128Kbit, Serial SPI EEPROM (SPI Bus)	MDIP	Say	-	pdf pinout
LE 25 LB1282 TT	CMOS-IC	128Kbit, Serial SPI EEPROM (SPI Bus)	TSOP	Say	-	pdf pinout
LE27AB	R+	Io=150mA Vo:2.673...2.727V Vin:4.7V	8S,3N	Sgs		data pdf pinout
LE27C	R+	Io=150mA Vo:2.646...2.754V Vin:4.7V	8S,3N	Sgs		data pdf pinout
LE30AB	R+	Io=150mA Vo:2.97...3.03V Vin:5V	8S,3N	Sgs		data pdf pinout
LE30C	R+	Io=150mA Vo:7.84...8.16V Vin:10V	8S,3N	Sgs		data pdf pinout
LE 37 C01P	CMOS-PROM-IC	128 x 8 Bit, Serial Input	8-DIP	Say		
LE120AB	R+	Io=150mA Vo:11.88...12.12V Vin:14V	8S,3N	Sgs		data pdf pinout
LE120C	R+	Io=150mA Vo:11.76...12.24V Vin:14V	8S,3N	Sgs		data pdf pinout
LEC	Z-Di	=1SMB 12CA (Typ-Code/Stempel/marking)	71(5x3,5)	Ons	→1SMB 10CA...78CA	
LED47W-66-16100	LED	wht ø4.5mm I:80lm Vf:13V If:1.2AmA		Roi		data pdf pinout
LED55	IRED	940nm ø4.7mm I=7cd Vf:1.4V If:100mA	TO18	Fcs,Har,Qtc		data pdf pinout
LED55F	IRED	940nm ø4.7mm I=7cd Vf:1.4V If:100mA	TO18	Fcs,Har,Qtc		data pdf pinout
LED370-66-60-110	LED	uv ø4.5mm I:17.5mcd Vf:18V If:200mA	TO66	Roi		data pdf pinout
LED375-66-60-110	LED	uv ø4.5mm I:17.5mcd Vf:17V If:200mA	TO66	Roi		data pdf pinout
LED385-66-60-110	LED	uv ø4.5mm I:17.5mcd Vf:17V If:240mA	TO66	Roi		data pdf pinout
LED395-4X4PC66	LED	uv ø4.5mm I:17.5mcd Vf:14.4V If:1AmA	TO66	Roi		data pdf pinout
LED395-66-60	LED	uv ø4.5mm I:17.5mcd Vf:18V If:240mA	TO66	Roi		data pdf pinout
LED395-66-60-110	LED	uv ø4.5mm I:90mcd Vf:18V If:240mA	TO66	Roi		data pdf pinout
LED405-66-60	LED	uv ø4.5mm I:90mcd Vf:18V If:240mA	TO66	Roi		data pdf pinout
LED405-66-60-110	LED	uv ø4.5mm I:70mcd Vf:18V If:240mA	TO66	Mts		data pdf pinout
LED430-66-60	LED	bl ø4.5mm I:360mcd Vf:19.5V If:240mA	TO66	Roi		data pdf pinout
LED435-66-60	LED	bl ø4.5mm I:360mcd Vf:18V If:240mA	TO66	Roi		data pdf pinout
LED450-66-60	LED	bl ø4.5mm I:450mcd Vf:18V If:240mA	TO66	Roi		data pdf pinout
LED470-4X4PC66	LED	bl ø4.5mm I=11cd Vf:14V If:1AmA	TO66	Roi		data pdf pinout
LED470-66-60	LED	bl ø4.5mm I=18cd Vf:17.5V If:240mA	TO66	Roi		data pdf pinout
LED505-4X4PC66	LED	cy ø4.5mm I=45cd Vf:12.5V If:1AmA	TO66	Roi		data pdf pinout

Type	Device	Short Description	Fig.	Manu	Comparision Types	More at
LED505-66-60	LED	cy ø4.5mm I:10mW/Sr Vf:18.5V	TO66	Roi		data pdf pinout
LED525-4X4PC66	LED	gr ø4.5mm I=45cd Vf:12.5V If:1AmA	TO66	Roi		data pdf pinout
LED525-66-60	LED	gr ø4.5mm I=7cd Vf:18.5V If:240mA	TO66	Roi		data pdf pinout
LED565-66-60	LED	gr ø4.7mm I=45cd Vf:11V If:240mA	TO66	Roi		data pdf pinout
LED570-66-60	LED	gr ø4.7mm I=45cd Vf:11V If:240mA	TO66	Roi		data pdf pinout
LED590-66-60	LED	yl ø4.7mm I=7cd Vf:10V If:240mA	TO66	Roi		data pdf pinout
LED625-4X4PC66	LED	rd ø4.7mm I=20cd Vf:7.8V If:1AmA	TO66	Roi		data pdf pinout
LED630-66-60	LED	rd ø4.7mm I=6cd Vf:10V If:240mA	TO66	Roi		data pdf pinout
LED645-66-60	LED	rd ø4.7mm I=7cd Vf:10V If:240mA	TO66	Roi		data pdf pinout
LED660-66-16100	LED	rd ø4.7mm I=7cd Vf:8V If:1.4AmA	TO66	Roi		data pdf pinout
LED660-66-60	LED	rd ø4.7mm I=3.5cd Vf:10V If:240mA	TO66	Roi		data pdf pinout
LED680-66-16100	LED	rd ø4.7mm I=3.5cd Vf:7.7V If:1.4AmA	TO66	Roi		data pdf pinout
LED680-66-60	LED	rd ø4.7mm I=3.5cd Vf:9.8V If:240mA	TO66	Roi		data pdf pinout
LED690-66-60	LED	rd ø4.7mm I=3.5cd Vf:9.8V If:240mA	TO66	Roi		data pdf pinout
LED700-66-60	LED	rd ø4.7mm I=3.5cd Vf:9.8V If:240mA	TO66	Roi		data pdf pinout
LED720-66-16100	IRED	720nm ø4.7mm I=3.5cd Vf:7.4V	TO66	Roi		data pdf pinout
LED730-66-16100	IRED	730nm ø4.7mm I=3.5cd Vf:7.4V	TO66	Roi		data pdf pinout
LED735-66-60	IRED	735nm ø4.7mm I:§450mW/Sr Vf:9V	TO66	Roi		data pdf pinout
LED740-66-16100	IRED	740nm ø4.7mm I:§450mW/Sr Vf:5.5V	TO66	Roi		data pdf pinout
LED740-66-60	IRED	740nm ø4.7mm I:§450mW/Sr Vf:9V	TO66	Roi		data pdf pinout
LED750-66-60	IRED	750nm ø4.7mm I:§450mW/Sr Vf:9V	TO66	Roi		data pdf pinout
LED760-66-60	IRED	760nm ø4.7mm I:§450mW/Sr Vf:9V	TO66	Roi		data pdf pinout
LED780-66-60	IRED	780nm ø4.7mm I:§450mW/Sr Vf:9V	TO66	Roi		data pdf pinout
LED810-4X4PC66	IRED	810nm ø4.7mm I:§450mW/Sr Vf:6.2V	TO66	Roi		data pdf pinout
LED810-66-60	IRED	810nm ø4.7mm I:§450mW/Sr Vf:9V	TO66	Roi		data pdf pinout
LED850-66-60	IRED	850nm ø4.7mm I:400mW/Sr Vf:7.5V	TO66	Roi		data pdf pinout
LED870-4X4PC66	IRED	870nm ø4.7mm I:400mW/Sr Vf:5.8V	TO66	Roi		data pdf pinout
LED870-66-60	IRED	870nm ø4.7mm I:400mW/Sr Vf:7.5V	TO66	Roi		data pdf pinout
LED890-66-60	IRED	890nm ø4.7mm I:400mW/Sr Vf:7V	TO66	Roi		data pdf pinout
LED910-66-60	IRED	910nm ø4.7mm I:400mW/Sr Vf:7.1V	TO66	Roi		data pdf pinout
LED920-66-16100	IRED	950nm ø4.7mm I:400mW/Sr Vf:5.2V	TO66	Roi		data pdf pinout
LED940-4X4PC66	IRED	940nm ø4.7mm I:400mW/Sr Vf:5.2V	TO66	Roi		data pdf pinout
LED940-66-60	IRED	940nm ø4.7mm I:400mW/Sr Vf:7.1V	TO66	Roi		data pdf pinout
LED940-66-60-520	IRED	940nm ø4.7mm I:400mW/Sr Vf:7.1V	TO66	Roi		data pdf pinout
LED950-66-16100	IRED	920nm ø4.7mm I:400mW/Sr Vf:5.5V	TO66	Roi		data pdf pinout
LED970-66-60	IRED	970nm ø4.7mm I:400mW/Sr Vf:6.5V	TO66	Roi		data pdf pinout
LED1050-66-60	IRED	1050nm ø4.5mm I:7000mcd Vf:7V	TO66	Roi		data pdf pinout
LED1200-66-60	IRED	1200nm ø4.5mm I:7000mcd Vf:6.6V	TO66	Roi		data pdf pinout
LED1300-66-60	IRED	1300nm ø4.5mm I:7000mcd Vf:6.6V	TO66	Roi		data pdf pinout
LED1450-66-60	IRED	1450nm ø4.5mm I:7000mcd Vf:6V	TO66	Roi		data pdf pinout
LED1550-66-60	IRED	1550nm ø4.5mm I:7000mcd Vf:6.6V	TO66	Roi		data pdf pinout
LED 7706	LIN-IC	6- rows 30mA LEDs driver w. boost regulator	QFN	Stm		pdf pinout
LEDW47-66-60-120	LED	wht ø4.7mm I=13cd Vf:19V If:240mA	TO3	Roi		data pdf pinout
LEp	Si-P	=BF 660 (Typ-Code/Stempel/marking)	35	Phi	→BF 660	data
LEs	Si-P	=BF 660W (Typ-Code/Stempel/marking)	35(2mm)	Sie	→BF 660W	data
LEs	Si-P	=BF 660 (Typ-Code/Stempel/marking)	35	Sie	→BF 660	data

LF...LL

Type	Device	Short Description	Fig.	Manu	Comparision Types	More at
LF	Si-N	=2SC3052-F (Typ-Code/Stempel/marking)	35	Mit	→2SC3052	data
LF	Si-N	=2SC4154-F (Typ-Code/Stempel/marking)	35(2mm)	Mit	→2SC4154	data
LF	Si-N	=HD 2F2Q (Typ-Code/Stempel/marking)	39	Nec	→HD 2...	data
LF	Z-Di	=P6 SMB-20 (Typ-Code/Stempel/marking)	71(5x3,5)	Fag	→P6 SMB-20	data
LF	Z-Di	=SM 6T 7V5C (Typ-Code/Stempel/marking)	71(6x4mm)	Tho	→SM 6T...	data
LF	Z-Di	=SMBJ 13 (Typ-Code/Stempel/marking)	71(5x3,5)	Mop	→SMBJ ...	data
LF5W	LED	rgb ø5mm I:12mcd Vf:2-4V If:20mA	T1¾	Kin		data pdf pinout
LF12XXX	R+	Io=1A Vo:1.238...1.263V Vin:3.3V P:int	3P,5P	Sgs		data pdf pinout
LF15XXX	R+	Io=1A Vo:1.485...1.515V Vin:3.5V P:int	3P,5P	Sgs		data pdf pinout
LF18XXX	R+	Io=1A Vo:1.782...1.818V Vin:3.3V P:int	3P,5P	Sgs		data pdf pinout
LF25XXX	R+	Io=1A Vo:2.475...2.525V Vin:4.5V P:int	3P,5P	Sgs		data pdf pinout
LF27XXX	R+	Io=1A Vo:2.673...2.727V Vin:4.7V P:int	3P,5P	Sgs		data pdf pinout
LF30XXX	R+	Io=1A Vo:2.97...3.03V Vin:5V P:int	3P,5P	Sgs		data pdf pinout
LF33XXX	R+	Io=1A Vo:3.267...3.333V Vin:5.3V P:int	3P,5P	Sgs		data pdf pinout
LF35XXX	R+	Io=1A Vo:3.465...3.535V Vin:5.5V P:int	3P,5P	Sgs		data pdf pinout
LF40XXX	R+	Io=1A Vo:3.96...4.04V Vin:6V P:int	3P,5P	Sgs		data pdf pinout
LF45XXX	R+	Io=1A Vo:4.455...4.545V Vin:6.5V P:int	3P,5P	Sgs		data pdf pinout
LF47XXX	R+	Io=1A Vo:4.653...4.747V Vin:6.7V P:int	3P,5P	Sgs		data pdf pinout
LF50XXX	R+	Io=1A Vo:4.95...5.05V Vin:7V P:int	3P,5P	Sgs		data pdf pinout
LF52XXX	R+	Io=1A Vo:5.148...5.252V Vin:7.2V P:int	3P,5P	Sgs		data pdf pinout
LF55XXX	R+	Io=1A Vo:5.445...5.555V Vin:7.5V P:int	3P,5P	Sgs		data pdf pinout
LF59	LED	rgb ø5mm I:1.2-20mcd Vf:2-3.8V	T1¾	Kin		data pinout
LF60XXX	R+	Io=1A Vo:5.94...6.06V Vin:8V P:int	3P,5P	Sgs		data pdf pinout
LF80XXX	R+	Io=1A Vo:7.92...8.08V Vin:10V P:int	3P,5P	Sgs		data pdf pinout
LF81W	LED	rgb ø5mm I:12mcd Vf:2-4V If:20mA		Kin		data pdf pinout
LF85XXX	R+	Io=1A Vo:8.415...8.585V Vin:10.5V	3P,5P	Sgs		data pdf pinout
LF90XXX	R+	Io=1A Vo:8.91...9.09V Vin:11V P:int	3P,5P	Sgs		data pdf pinout
LF111	FET	Vs:±18V Vu:106dB Vo:±14V Vi0:0.7mV	14D,10F,8A	Amd,Nsc		data pinout
LF 111 D,J	KOP-IC	=LF 111H: Fig.→	14-DIC	Nsc	→Serie 111	data pinout
LF 111 F	KOP-IC	=LF 111H: Min	10-FLP	Nsc	→Serie 111	data pinout
LF 111 H	KOP-IC	JFET Inp, Serie 111, ±18V, -55...+125°	81	Nsc,Phi	→Serie 111	data pinout
LF120XXX	R+	Io=1A Vo:11.88...12.12V Vin:14V P:int	3P,5P	Sgs		data pinout
LF147	4xFET	Vs:±22V Vu:100dB Vo:±13.5V Vi0:1mV	14D	Nsc		data pinout
LF147A	4xFET	Vs:±18V Vu:200V/mV Vo:±12V Vi0:1mV	20C,14D	Sgs		data pinout

Type	Device	Short Description	Fig.	Manu	Comparision Types	More at
LF147B	4xFET	Vs:±18V Vu:200V/mV Vo:±12V Vi0:3mV	20C,14D	Sgs		data pinout
LF147D	4xFET	Vs:±22V Vu:100dB Vo:±13.5V Vi0:1mV	14D	Nsc		data pinout
LF 147(A,B)D,J	OP-IC	Quad, JFET, ±18V, 13V/µs, -55...+125°	14-DIC	Nsc,Tho	→Serie 124	data pdf pinout
LF147GC	4xFET	Vs:±18V Vu:200V/mV Vo:±12V Vi0:3mV	20C	Sgs		data pinout
LF 147(A,B)GC	OP-IC	=LF 147(A.B)J: Fig.→	20-(P)LCC	Tho	→Serie 124	data pdf pinout
LF147J	4xFET	Vs:±18V Vu:200V/mV Vo:±12V Vi0:3mV	14SD	Sgs		data pdf pinout
LF147J/883	4xFET	Vs:±22V Vu:100dB Vo:±13.5V Vi0:1mV	14D	Nsc		data pinout
LF151	FET	Vs:±18V Vu:200V/mV Vo:±12V Vi0:3mV	20C,8A	Sgs		data pdf pinout
LF151A	FET	Vs:±18V Vu:200V/mV Vo:±12V Vi0:2mV	20C,8A	Sgs		data pinout
LF151B	FET	Vs:±18V Vu:200V/mV Vo:±12V Vi0:5mV	20C,8A	Sgs		data pinout
LF151D	JFET	Vs:±18V Vu:200V/mV Vo:±12V Vi0:3mV	8S	Sgs		data pinout
LF 151(A,B)GC	OP-IC	=LF 151(A.B)H: Fig.→	20-(P)LCC	Tho	→Serie 155	data pdf pinout
LF 151(A,B)H	OP-IC	JFET, ±18V, 13V/µs, -55...+125°	81	Tho	→Serie 155	data pdf pinout
LF151N	JFET	Vs:±18V Vu:200V/mV Vo:±12V Vi0:3mV	8D	Sgs		data pinout
LF153	2xFET	Vs:±18V Vu:200V/mV Vo:±12V Vi0:3mV	20C,8A	Sgs		data pinout
LF153A	2xFET	Vs:±18V Vu:200V/mV Vo:±12V Vi0:2mV	20C,8A	Sgs		data pinout
LF153B	2xFET	Vs:±18V Vu:200V/mV Vo:±12V Vi0:5mV	20C,8A	Sgs		data pinout
LF153D	2xJFET	Vs:±18V Vu:200V/mV Vo:±12V Vi0:3mV	8S	Sgs		data pinout
LF 153(A,B)GC	OP-IC	=LF 153(A,B)H: Fig.→	20-(P)LCC	Tho	→Serie 153	data pdf pinout
LF 153(A,B)H	OP-IC	Dual, JFET, ±18V, 13V/µs, -55...+125°	81	Tho	→Serie 153	data pdf pinout
LF153N	2xJFET	Vs:±18V Vu:200V/mV Vo:±12V Vi0:3mV	8D	Sgs		data pinout
LF155	FET	Vs:±22V Vu:106dB Vo:±12V Vi0:3mV	8AD	Amd,Ltc		data pinout
LF155A	JFET	Vs:±22V Vu:200V/mV Vo:±13.2V f:6Mc	8D,20CdA	Ray		data pinout
LF155AH/883	JFET	Vs:±22V Vu:>50V/mV Vo:±10V Vi0:±2mV	8A	Nsc		data pinout
LF155D	JFET	Vs:±22V Vu:200V/mV Vo:±13.2V f:5.7Mc	8D	Ray		data pinout
LF 155(A)D,DG	OP-IC	=LF 155(A)H: Fig.→	.8-DIP	Tho	→Serie 155	data pinout
LF155GC	FET	Vs:±22V Vu:200V/mV Vo:±12V Vi0:3mV	20Cd	Sgs		data pinout
LF 155(A)GC	OP-IC	=LF 155(A)H: Fig.→	20-(P)LCC	Tho	→Serie 155	data pinout
LF155H	FET	Vs:±22V Vu:200V/mV Vo:±12V Vi0:3mV	8A	Sgs		data pdf pinout
LF155H/883	JFET	Vs:±22V Vu:>50V/mV Vo:±10V Vi0:±5mV	8A	Nsc		data pinout
LF 155(A)H	OP-IC	Serie 155, JFET,±22V,-55...+125°, 5V/µs	81	Nsc,Tho	→Serie 155	data pinout
LF155N	FET	Vs:±22V Vu:200V/mV Vo:±12V Vi0:3mV	8D	Sgs		data pdf pinout
LF155T	FET	Vs:±22V Vu:106dB Vo:±12V Vi0:3mV	8A	Mul,Phi,Val		data pinout
LF156	FET	Vs:±22V Vu:106dB Vo:±12V Vi0:3mV	d,8AD	Tho		data pinout
LF156A	JFET	Vs:±22V Vu:200V/mV Vo:±13.2V f:8Mc	8D,20CdA	Ray		data pinout
LF156D	JFET	Vs:±22V Vu:200V/mV Vo:±13.2V f:7.6Mc	8D	Ray		data pdf pinout
LF 156(A)DG	OP-IC	=LF 155(A)H: 12V/µs	8-DIP	Tho	→Serie 155	data pinout
LF156GC	FET	Vs:±22V Vu:200V/mV Vo:±12V Vi0:3mV	20C	Sgs		data pinout
LF 156(A)GC	OP-IC	=LF 156(A)H: Fig.→	20-(P)LCC	Tho	→Serie 155	data pinout
LF156H	FET	Vs:±22V Vu:200V/mV Vo:±12V Vi0:3mV	8A	Sgs		data pdf pinout
LF 156(A)H	OP-IC	=LF 155(A)H: 12V/µs	81	Nsc,Tho	→Serie 155	data pinout
LF156N	FET	Vs:±22V Vu:200V/mV Vo:±12V Vi0:3mV	8D	Sgs		data pdf pinout
LF156T	FET	Vs:±22V Vu:106dB Vo:±12V Vi0:3mV	8A	Mul,Phi,Val		data pinout
LF157	FET	Vs:±22V Vu:106dB Vo:±12V Vi0:3mV	d,8AD	Tho		data pinout
LF157A	JFET	Vs:±22V Vu:200V/mV Vo:±13.2V f:30Mc	8D,20CdA	Ray		data pinout
LF157D	JFET	Vs:±22V Vu:200V/mV Vo:±13.2V f:28Mc	8D	Ray		data pdf pinout
LF 157(A)DG	OP-IC	=LF 155(A)H: 50V/µs	8-DIP	Tho	→Serie 155	data pinout
LF157GC	FET	Vs:±22V Vu:200V/mV Vo:±12V Vi0:3mV	20C	Sgs		data pinout
LF 157(A)GC	OP-IC	=LF 157(A)H: Fig.→	20-(P)LCC	Tho	→Serie 155	data pinout
LF157H	FET	Vs:±22V Vu:200V/mV Vo:±12V Vi0:3mV	8A	Sgs		data pdf pinout
LF 157(A)H	OP-IC	=LF 155(A)H: 50V/µs	81	Nsc,Tho	→Serie 155	data pinout
LF157N	FET	Vs:±22V Vu:200V/mV Vo:±12V Vi0:3mV	8D	Sgs		data pdf pinout
LF157T	FET	Vs:±22V Vu:106dB Vo:±12V Vi0:3mV	8A	Mul,Phi,Val		data pinout
LF 198 FE	LIN-IC	=LF 198(A)H: Fig.→	8-DIC	Phi		pdf pinout
LF 198(A)H	LIN-IC	Bi-FET Sample & Hold, -55...+125°	81	Nsc,Phi,Tho	-	pdf
LF211	COMP	Vs:±18V Vu:106dB Vo:±14V Vi0:0.7mV	14D,10F,8A	Nsc		data pinout
LF 211 D,F,H	KOP-IC	=LF 111H: -25...+85°		Nsc	→Serie 111	data pinout
LF247	4xFET	Vs:±18V Vu:200V/mV Vo:±12V Vi0:3mV	14SD	Sgs		data pinout
LF247A	4xFET	Vs:±18V Vu:200V/mV Vo:±12V Vi0:1mV	14SD	Sgs		data pinout
LF 247(A,B)D	OP-IC	=LF 147(A.B)J: SMD, -40...+105°	14-MDIP	Tho	→Serie 124	data pdf pinout
LF 247(A,B)N	OP-IC	=LF 147(A.B)J: -40...+105°	14-DIP	Tho	→Serie 124	data pdf pinout
LF251	JFET	Vs:±18V Vu:200V/mV Vo:±12V Vi0:3mV	8SD	Sgs		data pdf pinout
LF251A	FET	Vs:±18V Vu:200V/mV Vo:±12V Vi0:2mV	8SD	Sgs		data pinout
LF251B	FET	Vs:±18V Vu:200V/mV Vo:±12V Vi0:5mV	8SD	Sgs		data pinout
LF 251(A,B)D	OP-IC	=LF 151(A.B)H: SMD, -40...+105°	8-MDIP	Tho	→Serie 155	data pdf pinout
LF 251(A,B)N	OP-IC	=LF 151(A.B)H: -40...+105°	8-DIP	Tho	→Serie 155	data pdf pinout
LF253A	2xFET	Vs:±18V Vu:200V/mV Vo:±12V Vi0:2mV	8SD	Sgs		data pinout
LF253B	2xFET	Vs:±18V Vu:200V/mV Vo:±12V Vi0:5mV	8SD	Sgs		data pinout
LF253D	2xJFET	Vs:±18V Vu:200V/mV Vo:±12V Vi0:3mV	8S	Sgs		data pinout
LF 253(A,B)D	OP-IC	=LF 153(A,B)H: SMD, -40...+105°	8-MDIP	Tho	→Serie 153	data pdf pinout
LF253N	2xFET	Vs:±18V Vu:200V/mV Vo:±12V Vi0:3mV	8D	Sgs		data pinout
LF 253(A,B)N	OP-IC	=LF 153(A,B)H: -40...+105°	8-DIP	Tho	→Serie 153	data pdf pinout
LF255	FET	Vs:±22V Vu:106dB Vo:±12V Vi0:3mV	8AD	Amd,Isi,Nsc		data pdf pinout
LF255D	FET	Vs:±22V Vu:200V/mV Vo:±12V Vi0:3mV	8S	Sgs		data pdf pinout
LF 255 D	OP-IC	=LF 155(A)H: SMD, -40...+105°	8-MDIP	Tho	→Serie 155	data pinout
LF 255 H	OP-IC	=LF 155(A)H: -25...+85°	81	Nsc,Tho	→Serie 155	data pinout
LF255N	FET	Vs:±22V Vu:200V/mV Vo:±12V Vi0:3mV	8D	Sgs		data pdf pinout
LF 255 N	OP-IC	=LF 155(A)H: -40...+105°	8-DIP	Tho	→Serie 155	data pinout
LF256	FET	Vs:±22V Vu:106dB Vo:±12V Vi0:3mV	8AD	Amd,Isi,Nsc		data pinout
LF256D	FET	Vs:±22V Vu:200V/mV Vo:±12V Vi0:3mV	8S	Sgs		data pdf pinout
LF 256 D	OP-IC	=LF 156(A)H: SMD, -40...+105°	8-MDIP	Tho	→Serie 155	data pinout
LF 256 H	OP-IC	=LF 156(A)H: -25...+85°	81	Nsc,Tho	→Serie 155	data pinout
LF256N	FET	Vs:±22V Vu:200V/mV Vo:±12V Vi0:3mV	8D	Sgs		data pinout
LF 256 N	OP-IC	=LF 156(A)H: -40...+105°	8-DIP	Tho	→Serie 155	data pinout
LF257	FET	Vs:±22V Vu:106dB Vo:±12V Vi0:3mV	8AD	Isi,Nsc		data pdf pinout

Type	Device	Short Description	Fig.	Manu	Comparision Types	More at
LF257D	FET	Vs:±22V Vu:200V/mV Vo:±12V Vi0:3mV	8S	Sgs		data pdf pinout
LF 257 D	OP-IC	=LF 157(A)H: SMD, -40...+105°	8-MDIP	Tho	→Serie 155	data pinout
LF 257 H	OP-IC	=LF 155(A)H: -25...+85°	81	Nsc,Tho	→Serie 155	data pinout
LF257N	FET	Vs:±22V Vu:200V/mV Vo:±12V Vi0:3mV	8D	Sgs		data pdf pinout
LF 257 N	OP-IC	=LF 157(A)H: -40...+105°	8-DIP	Tho	→Serie 155	data pinout
LF 298 FE,N	LIN-IC	=LF 198(A)H: -25...+85°	8-DIC,DIP	Phi	-	pinout
LF 298(A)H	LIN-IC	=LF 198H: -25...+85°	81	Nsc,Phi,Tho	-	
LF311	COMP	Vs:±18V Vu:106dB Vo:±14V Vi0:2mV	14D,10F,8A	Nsc		data pinout
LF 311 D,F,H	KOP-IC	=LF 111H: 0...+70°		Nsc	→Serie 111	data pinout
LF347A	4xFET	Vs:±18V Vu:200V/mV Vo:±12V Vi0:1mV	14SD	Sgs		data pinout
LF347B	4xFET	Vs:±18V Vu:200V/mV Vo:±12V Vi0:3mV	14SD	Sgs,Tix		data pdf pinout
LF347D	4xFET	Vs:±18V Vu:100dB Vo:±13.5V Vi0:5mV	14DS	Nsc		data pinout
LF 347(A,B)D	OP-IC	=LF 147(A,B)H: 0...+70°	14-MDIP		→Serie 124	data pdf pinout
LF347J	4xFET	Vs:±18V Vu:100dB Vo:±13V Vi0:5mV	14D	Has,Nsc		data pinout
LF347M	4xFET	Vs:±18V Vu:100dB Vo:±13.5V Vi0:3mV	14S	Nsc		data pinout
LF347N	4xFET	Vs:±18V Vu:100dB Vo:±13.5V Vi0:5mV	14D	Has,Mot,Nsc		data pinout
LF 347(A,B)N,DP,J	OP-IC	=LF 147(A,B)J: 0...+70°	14-DIP,DIC	Nsc, Tho	→Serie 124	data pinout
LF347WM	4xFET	Vs:±18V Vu:100dB Vo:±13.5V Vi0:3mV	14S	Nsc		data pinout
LF351	FET	Vs:±18V Vu:100dB Vo:±13V Vi0:5mV	8ADS	Nsc		data pinout
LF351A	FET	Vs:±18V Vu:200V/mV Vo:±12V Vi0:2mV	8SAD	Sgs		data pinout
LF351B	FET	Vs:±18V Vu:200V/mV Vo:±12V Vi0:5mV	8SAD	Sgs		data pinout
LF351D	FET	Vs:±18V Vu:100dB Vo:±13V Vi0:5mV	8S	Mot,Tix		data pinout
LF 351(A,B)D	OP-IC	=LF 151(A,B)H: SMD, 0...+70°	8-MDIP	Nsc,Tho	→Serie 155	data pdf pinout
LF 351 H	OP-IC	=LF 151(A,B)H: 0...+70°	81	Nsc,Tho	→Serie 155	data pdf pinout
LF351N	JFET	Vs:±18V Vu:200V/mV Vo:±12V Vi0:3mV	8D	Sgs		data pinout
LF 351(A,B)N,DP	OP-IC	=LF 151A,B)H: 0...+70°	8-DIP	Mot,Nsc,Tho	→Serie 155	data pdf pinout
LF351P	2xJFET	Vs:±18V Vu:100dB Vo:±13V Vi0:5mV	8D	Tix		data pinout
LF353A	2xFET	Vs:±18V Vu:200V/mV Vo:±12V Vi0:2mV	8SAD	Sgs		data pinout
LF353B	2xFET	Vs:±18V Vu:200V/mV Vo:±12V Vi0:5mV	8SAD	Sgs		data pinout
LF353D	2xFET	Vs:±18V Vu:100dB Vo:±13V Vi0:5mV	8S	Mot		data pinout
LF 353(A,B)D	OP-IC	=LF 153(A,B)H: SMD, 0...+70°	8-MDIP	Nsc,Tho	→Serie 153	data pdf pinout
LF353H	2xFET	Vs:±18V Vu:100dB Vo:±13V Vi0:5mV	8A	Har		data pdf pinout
LF 353 H	OP-IC	=LF 153(A,B)H: 0...+70°	81	Nsc,Tho	→Serie 153	data pdf pinout
LF353J	2xFET	Vs:±18V Vu:100dB Vo:±13V Vi0:5mV	8D	Nsc		data pinout
LF353M	2xJFET	Vs:±18V Vu:100V/mV Vo:±13.5V Vi0:5mV	8S	Nsc		data pinout
LF353N	2xFET	Vs:±18V Vu:100dB Vo:±13V Vi0:5mV	8D	Has,Mot		data pinout
LF 353(A,B)N,DP	OP-IC	=LF 153(A,B)H: 0...+70°	8-DIP	Mot,Nsc,Tho	→Serie 153	data pdf pinout
LF353P	2xJFET	Vs:±18V Vu:200V/mV Vo:±12V Vi0:3mV	8D	Tix		data pdf pinout
LF354A	2xFET	Vs:±18V Vu:100dB Vo:±13V Vi0:1mV	14D	Nsc		data pinout
LF354B	2xFET	Vs:±18V Vu:100dB Vo:±13V Vi0:3mV	14D	Nsc		data pinout
LF354N	2xFET	Vs:±18V Vu:100dB Vo:±13V Vi0:5mV	14D	Nsc		data pinout
LF355	FET	Vs:±18V Vu:106dB Vo:±12V Vi0:3mV	8DA,14DS	Tho		data pinout
LF355A	FET	Vs:±18V Vu:200V/mV Vo:±12V Vi0:1mV	8SDA	Sgs		data pinout
LF355B	FET	Vs:±22V Vu:106dB Vo:±12V Vi0:1mV	8A,14DSD	Har,Mot,Nsc		data pinout
LF355D	FET	Vs:±18V Vu:200V/mV Vo:±12V Vi0:3mV	8S	Sgs		data pdf pinout
LF 355(A)D	OP-IC	=LF 155(A)H: SMD, 0...+70°	8-MDIP	Tho	→Serie 155	data pdf pinout
LF 355(A,B)DP,J,N	OP-IC	=LF 155(A)H: 0...+70°	8-DIP,DIC	Mot,Nsc,Tho	→Serie 155	data pdf pinout
LF 355(A,B)H	OP-IC	=LF 155(A)H: 0...+70°	81	Mot,Nsc,Tho	→Serie 155	data pdf pinout
LF355N	FET	Vs:±18V Vu:200V/mV Vo:±12V Vi0:3mV	8D	Sgs		data pdf pinout
LF356	FET	Vs:±18V Vu:106dB Vo:±12V Vi0:3mV	8DA,14DS	Tho		data pinout
LF356A	FET	Vs:±18V Vu:200V/mV Vo:±12V Vi0:1mV	8SDA	Sgs		data pinout
LF356B	FET	Vs:±22V Vu:106dB Vo:±12V Vi0:3mV	8ADS	Mot,Nsc		data pdf pinout
LF356D	FET	Vs:±18V Vu:200V/mV Vo:±12V Vi0:3mV	8S	Sgs		data pdf pinout
LF 356(A)	OP-IC	=LF 156(A): SMD, 0...70°	8-MDIP	Phi,Tho	→Serie 155	data pdf pinout
LF 356(A,B)DP,J,N	OP-IC	=LF 155(A)H: 0...+70°	8-DIP,DIC	Mot,Nsc,Tho	→Serie 155	data pdf pinout
LF 356(A,B)H .	OP-IC	=LF 155(A)H: 0...+70°	81	Mot,Nsc,Tho	→Serie 155	data pdf pinout
LF356N	FET	Vs:±18V Vu:200V/mV Vo:±12V Vi0:3mV	8D	Sgs		data pdf pinout
LF357	FET	Vs:±18V Vu:106dB Vo:±12V Vi0:3mV	8DA,14DS	Tho		data pinout
LF357A	FET	Vs:±18V Vu:200V/mV Vo:±12V Vi0:1mV	8SDA	Sgs		data pinout
LF357B	FET	Vs:±22V Vu:106dB Vo:±12V Vi0:3mV	8A,14DD	Mot,Nsc		data pinout
LF357D	FET	Vs:±18V Vu:200V/mV Vo:±12V Vi0:3mV	8S	Sgs		data pdf pinout
LF 357(A)D	OP-IC	=LF 157(A)H: 0...+70°	8-MDIP	Tho	→Serie 155	data pdf pinout
LF 357(A,B)DP,J,N	OP-IC	=LF 155(A)H: 0...+70°	8-DIP,DIC	Mot,Nsc,Tho	→Serie 155	data pdf pinout
LF 357(A)H	OP-IC	=LF 155(A)H: 0...+70°	81	Mot,Nsc,Tho	→Serie 155	data pdf pinout
LF357N	FET	Vs:±18V Vu:200V/mV Vo:±12V Vi0:3mV	8D	Sgs		data pinout
LF 398 D	LIN-IC	=LF 198H: SMD, 0...+70°	14-MDIP	Phi	-	pdf pinout
LF 398(A)DP,N	LIN-IC	=LF 198(A)H: 0...+70°	8-DIP	Nsc,Phi,Tho	-	pinout
LF 398 FE	LIN-IC	=LF 198H: 0...+70°	8-DIC	Phi	-	pdf pinout
LF 398(A)H	LIN-IC	=LF 198(A)H: 0...+70°	81	Nsc,Phi,Tho	-	pdf
LF400A	FET	Vs:±18V Vu:106dB Vo:±12V Vi0:<0.5mV	8A	Nsc		data pinout
LF400CH	FET	Vs:±18V Vu:106dB Vo:±12V Vi0:4mV	8A	Nsc		data pinout
LF 400(A)CH	OP-IC	JFET, ±18V, 57V/µs, 18MHz, 0...+70°	81	Nsc		data pinout
LF 400(A)CN	OP-IC	=LF 400(A)CH: Fig.→	-DIP	Nsc		data
LF400MH	JFET	Vs:±18V Vu:50V/mV Vo:±11.5V Vi0:±3mV	8A	Nsc		data pinout
LF401A	JFET	Vs:±18V Vu:300V/mV Vo:±12V f:16Mc	14D	Nsc		data pinout
LF401CD	JFET	Vs:±18V Vu:300V/mV Vo:±12V f:16Mc	14D	Nsc		data pinout
LF411A	FET	Vs:±22V Vu:106dB Vo:±12V Vi0:0.3mV	8AD	Nsc		data pdf pinout
LF411CD	JFET	Vs:±18V Vu:80V/mV Vo:±13.5V f:3Mc	8S	Mot,Tix		data pinout
LF 411 CD	OP-IC	=LF 411(A)CH: SMD	8-MDIP	Nsc	-	data pinout
LF411CH	2xFET	Vs:±18V Vu:106dB Vo:±12V Vi0:1mV	8A	Nsc		data pinout
LF 411(A)CH	OP-IC	JFET, ±18V, 15V/µs, 4MHz, 0...+70°	81	Nsc	-	data pinout
LF411CN	JFET	Vs:±18V Vu:80V/mV Vo:±12V Vi0:0.5mV	8D	Mot,Nsc		data pdf pinout
LF 411(A)CN,CJ	OP-IC	=LF 411(A)CH: Fig.→	8-DIP,DIC	Nsc		data pinout
LF411CP	JFET	Vs:±18V Vu:80V/mV Vo:±13.5V f:3Mc	8D	Tix		data pinout
LF411M	FET	Vs:±18V Vu:106dB Vo:±12V Vi0:0.8mV	8AD	Nsc		data pinout

Type	Device	Short Description	Fig.	Manu	Comparision Types	More at
LF411MH/883	JFET	Vs:±18V Vu:25V/mV Vo:±12V Vi0:±2mV	8A	Nsc		data pinout
LF 411 MH	OP-IC	=LF 411(A)CH: -55...+125°	81	Nsc	-	data pinout
LF412	2xFET	Vu:106dB Vo:±12V Vi0:1mV f.4Mc	8AD	Ltc		data pdf pinout
LF 412(A)...	OP-IC	=LF 411(A): Dual		Nsc		data pinout
LF412A	2xFET	Vs:±22V Vu:106dB Vo:±12V Vi0:0.5mV	8AD	Nsc		data pdf pinout
LF412CD	2xJFET	Vs:±18V Vu:200V/mV Vo:±13.5V Vi0:1mV	8S	Mot,Tix		data pdf pinout
LF412CH	2xJFET	Vs:±18V Vu:200V/mV Vo:±13.5V Vi0:1mV	8A	Nsc		data pdf pinout
LF412CJ	2xFET	Vu:106dB Vo:±12V Vi0:1mV f.4Mc	8D	Nsc		data pinout
LF412CN	2xFET	Vu:106dB Vo:±12V Vi0:1mV f.4Mc	8D	Mot		data pdf pinout
LF412CP	2xJFET	Vs:±18V Vu:200V/mV Vo:±13.5V Vi0:1mV	8D	Tix		data pdf pinout
LF412MH	2xJFET	Vs:±18V Vu:200V/mV Vo:±13.5V Vi0:1mV	8A	Nsc		data pdf pinout
LF412MH/883	2xJFET	Vs:±18V Vu:>±25V/mV Vo:±12V Vi0:±3mV	8A	Nsc		data pinout
LF412MJ/883	2xJFET	Vs:±18V Vu:>±25V/mV Vo:±12V Vi0:±3mV	8D	Nsc		data pinout
LF422CN	2xLO-POWER	Vs:±18V Vu:25V/mV Vo:±12.7V Vi0:<5mV	8D	Mot		data pinout
LF441A	FET	Vs:±22V Vu:100dB Vo:±12V Vi0:0.3mV	8AD	Ltc,Nsc		data pinout
LF441C	JFET	Vs:±18V Vu:60V/mV Vo:±12V Vi0:3mV	8DS	Nsc		data pinout
LF441CD	JFET	Vs:±18V Vu:60V/mV Vo:±12V Vi0:3mV	8D	Mot		data pdf pinout
LF 441 CD	OP-IC	=LF 441(A)CH: SMD	8-MDIP	Nsc		data pinout
LF441CH	FET	Vs:±18V Vu:100dB Vo:±12V Vi0:1mV	8A	Nsc		data pinout
LF 441(A)CH	OP-IC	JFET, ±18V, 1V/µs, 1MHz, 0...+70°	81	Nsc		data pinout
LF 441(A)CN,CJ	OP-IC	=LF 441(A)CH: Fig.→	8-DIP,DIC	Nsc		data pinout
LF441MH/883	FET	Vs:±18V Vu:100dB Vo:±12V Vi0:1mV	8A	Nsc		data pinout
LF 441 MH	OP-IC	=LF 441(A)CH: -55...+125°	81	Nsc		data pinout
LF 442(A)...	OP-IC	=LF 441(A): Dual		Nsc		data pinout
LF442A	2xFET	Vs:±22V Vu:100dB Vo:±12V Vi0:0.5mV	8AD	Nsc		data pinout
LF442C	2xFET	Vs:±18V Vu:60V/mV Vo:±12V Vi0:3mV	8D	Nsc		data pinout
LF442CD	2xFET	Vs:±18V Vu:60V/mV Vo:±12V Vi0:3mV	8D	Mot		data pinout
LF442CH	2xFET	Vs:±18V Vu:100dB Vo:±12V Vi0:1mV	8A	Nsc		data pinout
LF442MH	2xFET	Vs:±18V Vu:100dB Vo:±12V Vi0:1mV	8A	Nsc		data pinout
LF 444(A)...	OP-IC	=LF 441(A): Quad	14-...	Nsc		data pinout
LF444A	4xJFET	Vs:±22V Vu:100mV, Vo:±13V Vi0:2mV	14D	Nsc		data pdf pinout
LF444C	4xFET	Vs:±18V Vu:100dB Vo:±12V Vi0:3mV	14D	Nsc		data pdf pinout
LF444CD	4xFET	Vs:±18V Vu:60V/mV Vo:±12V Vi0:3mV	14D	Mot		data pdf pinout
LF444CM	4xJFET	Vs:±18V Vu:100V/mV Vo:±13V Vi0:3mV	14S	Nsc		data pdf pinout
LF444CN	4xFET	Vs:±18V Vu:60V/mV Vo:±12V Vi0:3mV	14D	Mot		data pdf pinout
LF444CWM	4xFET	Vs:±18V Vu:100dB Vo:±12V Vi0:3mV	14S	Nsc		data pinout
LF444MD/883	4xJFET	Vs:±18V Vu:>25V/mV Vo:±12V Vi0:±10mV	14D	Nsc		data pinout
LF451CM	WIDEBAND	Vs:±18V Vu:200V/mV Vo:±13.5V f:4Mc	8S	Nsc		data pinout
LF453CM	2xWIDEBAND	Vs:±18V Vu:200V/mV Vo:±13.5V Vi0:5mV	8S	Nsc		data pdf pinout
LF455	FET	Vu:106dB Vi0:0.25mV f:3Mc	8AD	Nsc		data pinout
LF456	FET	Vu:106dB Vi0:0.25mV f:5Mc	8AD	Nsc		data pinout
LF457	FET	Vu:106dB Vi0:0.25mV f:20Mc	8AD	Nsc		data pinout
LF819	LED	rgb ø10mm I:1.2-20mcd Vf:2-3.8V		Kin		data pinout
LF2155	2xFET	Vs:±22V Vu:106dB Vo:±12V Vi0:3mV	8DA	Tix		data pinout
LF2155A	2xFET	Vs:±22V Vu:106dB Vo:±12V Vi0:1mV	8DA	Tix		data pinout
LF2156	2xFET	Vs:±22V Vu:106dB Vo:±12V Vi0:3mV	8DA	Tix		data pinout
LF2156A	2xFET	Vs:±22V Vu:106dB Vo:±12V Vi0:1mV	8DA	Tix		data pinout
LF2157	2xFET	Vs:±22V Vu:106dB Vo:±12V Vi0:3mV	8DA	Tix		data pinout
LF2157A	2xFET	Vs:±22V Vu:106dB Vo:±12V Vi0:1mV	8DA	Tix		data pinout
LF2255	2xFET	Vs:±22V Vu:106dB Vo:±12V Vi0:3mV	8DA	Tix		data pinout
LF2256	2xFET	Vs:±22V Vu:106dB Vo:±12V Vi0:3mV	8DA	Tix		data pinout
LF2257	2xFET	Vs:±22V Vu:106dB Vo:±12V Vi0:3mV	8DA	Tix		data pinout
LF2355	2xFET	Vs:±18V Vu:106dB Vo:±12V Vi0:3mV	8DA	Tix		data pinout
LF2355A	2xFET	Vs:±18V Vu:106dB Vo:±12V Vi0:1mV	8DA	Tix		data pinout
LF2356	2xFET	Vs:±18V Vu:106dB Vo:±12V Vi0:3mV	8DA	Tix		data pinout
LF2356A	2xFET	Vs:±18V Vu:106dB Vo:±12V Vi0:1mV	8DA	Tix		data pinout
LF2357	2xFET	Vs:±18V Vu:106dB Vo:±12V Vi0:3mV	8DA	Tix		data pinout
LF2357A	2xFET	Vs:±18V Vu:106dB Vo:±12V Vi0:1mV	8DA	Tix		data pinout
LF 11201 D	LIN-IC	4x JFET Analog S, 36V, 20mA,-55...+125°	16-DIC	Nsc	MB 47201	pdf
LF 11202 D	LIN-IC	4x JFET Analog S, 36V, 20mA,-55...+125°	16-DIC	Nsc	-	pdf
LF 11331 D	LIN-IC	4x JFET Analog S, 36V, 20mA,-55...+125°	16-DIC	Nsc	-	pdf
LF 11332 D	LIN-IC	4x JFET Analog S, 36V, 20mA,-55...+125°	16-DIC	Nsc	-	pdf
LF 11333 D	LIN-IC	4x JFET Analog S, 36V, 20mA,-55...+125°	16-DIC	Nsc	-	pdf
LF 11508 D	LIN-IC	8x JFET Analog S, 36V, 10mA,-55...+125°	16-DIC	Nsc		
LF 11509 D	LIN-IC	4x JFET Analog S, 36V, 10mA,-55...+125°	16-DIC	Nsc		
LF 13201 D,N	LIN-IC	=LF 11201D: 0...+70°	16-DIC,DIP	Nsc	LF 11201	
LF 13202 D,N	LIN-IC	=LF 11202D: 0...+70°	16-DIC,DIP	Nsc	LF 11202	
LF 13300 D	A/D-IC	Bi-FET A/D Analog Building Block	18-DIC	Nsc		
LF 13331 D,N	LIN-IC	=LF 11331D: 0...+70°	16-DIC,DIP	Nsc	LF 11331	
LF 13332 D,N	LIN-IC	=LF 11332D: 0...+70°	16-DIC,DIP	Nsc	LF 11332	
LF 13333 D,N	LIN-IC	=LF 11333D: 0...+70°	16-DIC,DIP	Nsc	LF 11333	
LF 13508 D,N	LIN-IC	=LF 11508D: 0...+70°	16-DIC,DIP	Nsc		
LF 13509 D,N	LIN-IC	=LF 11509D: 0...+70°	16-DIC,DIP	Nsc		
LF13741	FET	Vs:±18V Vu:100dB Vo:±12V Vi0:5mV	8AD	Nsc		data pinout
LF 13741 H	OP-IC	JFET, ±18V, 0,5V/µs, 1MHz, 0...+70°	81	Nsc	-	data pinout
LF 13741 N	OP-IC	=LF 13741H: Fig.→	8-DIP	Nsc		data pinout
LFB 01(L)	Si-Di	SMD, 80V, 0,15A, L=50V		Say	-	data
LFs	Si-P	=BF 770 (Typ-Code/Stempel/marking)	35	Sie	→BF 770	data
LFT155	FET	Vs:±22V Vu:106dB Vo:±12V Vi0:0.5mV	8A	Nsc		data pinout
LFT156	FET	Vs:±22V Vu:106dB Vo:±12V Vi0:0.5mV	8A	Nsc		data pinout
LFT355	FET	Vs:±18V Vu:106dB Vo:±12V Vi0:0.5mV	8A	Nsc		data pinout
LFT356	FET	Vs:±18V Vu:106dB Vo:±12V Vi0:0.5mV	8A	Nsc		data pinout
LG	Z-Di	=1SMB 13A (Typ-Code/Stempel/marking)	71(5x3,5)	Ons	→1SMB 5.0A...170A	
LG	Si-N	=2SC2712-GR (Typ-Code/Stempel/marking)	35	Tos	→2SC2712	data
LG	Si-N	=2SC3052-G (Typ-Code/Stempel/marking)	35	Mit	→2SC3052	data

Type	Device	Short Description	Fig.	Manu	Comparision Types	More at
LG	Si-N	=2SC4116-GR (Typ-Code/Stempel/marking)	35(2mm)	Tos	→2SC4116	data
LG	Si-N	=2SC4154-G (Typ-Code/Stempel/marking)	35(2mm)	Mit	→2SC4154	data
LG	Si-N	=2SC4207-GR (Typ-Code/Stempel/marking)	45	Tos	→2SC4207	data
LG	Si-N	=2SC4738-GR (Typ-Code/Stempel/marking)	35(1,6mm)	Tos	→2SC4738	data
LG	Si-N	=2SC4944-GR (Typ-Code/Stempel/marking)	45(2mm)	Tos	→2SC4944	data
LG	Si-P	=BF 767 (Typ-Code/Stempel/marking)	35	Sie	→BF 767	data
LG	Si-N	=BF 775A (Typ-Code/Stempel/marking)	35	Sie	→BF 775A	data
LG,47K	LED	gr ø5mm I:2.8…7.1mcd Vf:1.8V If:2mA		Osr		data pdf pinout
LG	Si-N	=KTC 4075-GR(Typ-Code/Stempel/marking)	35(2mm)	Kec	→KTC 4075	data
LG	Z-Di	=P6 SMB-20A (Typ-Code/Stempel/marking)	71(5x3,5)	Fag	→P6 SMB-20A	data
LG	Z-Di	=SM 6T 7V5CA(Typ-Code/Stempel/marking)	71(6x4mm)	Tho	→SM 6T…	data
LG	Z-Di	=SMBJ 13A (Typ-Code/Stempel/marking)	71(5x3,5)	Mop	→SMBJ …	data
LG3330	LED	gr ø3mm I:45…71mcd Vf:2V If:10mA	T1	Osr,Sie		data pdf pinout
LG3341	LED	gr ø3mm I:45…71mcd Vf:2V If:10mA	T1	Osr		data pdf pinout
LG3360	LED	gr ø3mm I:11.2…18mcd Vf:2V If:10mA	T1	Osr		data pdf pinout
LGA670	LED	gr ø3mm I:11.2…28mcd Vf:2V If:10mA		Osr		data pdf pinout
LGA672	LED	gr ø3mm I:71…180mcd Vf:2.6V If:50mA		Osr		data pdf pinout
LGA676	LED	gr ø3mm I:56…112mcd Vf:2V If:20mA		Osr		data pdf pinout
LGA679	LED	gr ø3mm I:1.12…2.8mcd Vf:1.9V		Osr		data pdf pinout
LGC	Z-Di	=SMB 13CA (Typ-Code/Stempel/marking)	71(5x3,5)	Ons	→1SMB 10CA…78CA	
LGK380	LED	gr ø3mm I:71…112mlm Vf:2.1V If:15mA	T1	Osr		data pdf pinout
LGK389	LED	gr ø3mm I:1.2…7.2mcd Vf:1.9V If:2mA	T1	Osr		data pdf pinout
LGL890	LED	gr ø3mm I:11.2…28mcd Vf:2.2V		Osr		data pdf pinout
LGM47K	LED	gr ø3mm I:1.8…3.55mcd Vf:1.8V		Osr		data pdf pinout
LGM67K	LED	gr ø3mm I:2.8…7.1mcd Vf:1.8V		Osr		data pdf pinout
LGM470	LED	gr ø3mm I:9…22.4mcd Vf:2V If:10mA		Osr		data pdf pinout
LGM670	LED	gr ø3mm I:9…22.4mcd Vf:2V If:10mA	PLCC-2	Osr		data pdf pinout
LGM676	LED	gr ø3mm I:45…90mcd Vf:2V If:20mA		Osr		data pdf pinout
LGM770	LED	gr ø3mm I:9…22.4mcd Vf:2V If:20mA		Osr		data pdf pinout
LGM779	LED	gr ø3mm I:1mcd Vf:1.9V If:2mA		Osr		data pdf pinout
LGN971	LED	gr ø3mm I:7.1…10mcd Vf:2.2V If:20mA	1206	Osr		data pdf pinout
LGQ971	LED	gr ø3mm I:7.1…10mcd Vf:2.2V If:20mA	0603	Osr		data pdf pinout
LGR971	LED	gr ø3mm I:7.1…28mcd Vf:2.2V If:20mA	0805	Osr		data pdf pinout
LGT67K	LED	gr ø3mm I:3.55…9mcd Vf:1.8V If:2mA	PLCC-2	Osr		data pdf pinout
LGT670	LED	gr ø3mm I:11.2…28mcd Vf:2V If:10mA	PLCC-2	Osr		data pdf pinout
LGT671	LED	gr ø3mm I:14…35.5mcd Vf:2V If:10mA	PLCC-2	Osr		data pdf pinout
LGT672	LED	gr ø3mm I:71…180mcd Vf:2.6V If:50mA	PLCC-2	Osr		data pdf pinout
LGT676	LED	gr ø3mm I:56…112mcd Vf:2V If:20mA	PLCC-2	Osr		data pdf pinout
LGT679	LED	gr ø3mm I:1.12…2.8mcd Vf:1.9V	PLCC-2	Osr		data pdf pinout
LGT770	LED	gr ø3mm I:11.2…28mcd Vf:2V If:10mA	PLCC-2	Osr		data pdf pinout
LGY870	LED	gr ø3mm I:9…22.4mcd Vf:2V If:10mA		Osr		data pdf pinout
LH	Si-P	=BF 569 (Typ-Code/Stempel/marking)	35	Aeg,Phi,Sie Tho	→BF 569	data
LH	Z-Di	=P6 SMB-22 (Typ-Code/Stempel/marking)	71(5x3,5)	Fag	→P6 SMB-22	data
LH	Z-Di	=SMBJ 14 (Typ-Code/Stempel/marking)	71(5x3,5)	Mop	→SMBJ …	data
LH0001	LO-POWER	Vs:±20V Vu:95dB Vo:±11.5V Vio:0.2mV	10A	Nsc		data pinout
LH0001A	LO-POWER	Vs:±20V Vu:95dB Vo:±11.5V Vio:0.2mV	14D,10A	Nsc		data pinout
LH0002	FOLLOWER	Vs:±22V Vo:±11V f:50Mc	20C,8A	Nsc		data pinout
LH0002C	FOLLOWER	Vs:±22V Vo:±11V f:50Mc	8A,10D	Nsc		data pdf pinout
LH0002CH	BUFF	Vs:±22V Vo:±11V Vio:±10mV f:50Mc	8A	Alp		data pinout
LH 0002 CH	Hybrid-OP-IC	=LH 0002H: 0…+85°	81	Nsc	-	data pdf pinout
LH 0002 CN	Hybrid-OP-IC	=LH 0002H: 0…+85°	10-DIP	Nsc	-	data pdf pinout
LH0002CP	BUFF	Vs:±22V Vo:±11V Vio:±10mV f:50Mc	10D	Alp		data pinout
LH 0002 E	Hybrid-OP-IC	=LH 0002H: SMD	20-(P)LCC	Nsc	-	data pinout
LH0002H	BUFF	Vs:±22V Vo:±11V Vi0:±10mV f:50Mc	8A	Alp		data pdf pinout
LH 0002 H	Hybrid-OP-IC	hi-current, ±22V, ±0,1A, -55…+125°	81	Nsc	-	data pinout
LH0003	HI-SPEED	Vs:±20V Vu:97dB Vo:±12V Vi0:0.4mV	10A	Nsc		data pinout
LH 0003(C)H	Hybrid-OP-IC	hi-speed, ±20V, 30MHz, -55…+125°	82	Nsc	-	data pdf pinout
LH0004	HI-VOLT	Vs:±45V Vu:95dB Vo:±33V Vi0:0.3mV	10A	Nsc		data pinout
LH 0004(C)H	Hybrid-OP-IC	hi-volt, ±45V, -55…+125°(C=0…+85°)	82	Nsc	-	data pinout
LH0005A	HI-SPEED	Vs:±20V Vu:75dB Vo:±5V Vi0:1mV	10A	Nsc		data
LH 0005 AH	Hybrid-OP-IC	=LH 0005H: verbess./improved Version	82	Nsc	-	data pdf pinout
LH0005CH	HI-SPEED	Vs:±20V Vu:74dB Vo:±6V Vi0:3mV	10A	Nsc		data pinout
LH 0005 CH	Hybrid-OP-IC	=LH 0005H: 0…+85°	82	Nsc	-	data pinout
LH0005H	HI-SPEED	Vs:±20V Vu:72dB Vo:±5V Vi0:5mV	10A	Nsc		data pinout
LH 0005 H	Hybrid-OP-IC	hi-speed, ±20V, ±50mA, -55…+125°	82	Nsc	-	data pinout
LH0020CG	UNI+FOLL	Vs:±22V Vu:83dB Vo:±14.5V Vi0:1mV	12A	Nsc		data pinout
LH0020G	UNI+FOLL	Vs:±22V Vu:109dB Vo:±14.5V Vi0:1mV	12A	Nsc		data pinout
LH0021CK	POWER-OP	Vs:±18V Vu:106dB Vo:±14V Vi0:3mV	8O	Nsc		data pdf pinout
LH0021K	POWER-OP	Vs:±18V Vu:106dB Vo:±14V Vi0:1mV	8O	Nsc		data pdf pinout
LH 0021(C)K	Hybrid-OP-IC	Power, ±18V, 1A,-55…+125(C=-25…+85)°.	23/8-Pin	Nsc	-	data pinout
LH0022	FET	Vs:±22V Vu:106dB Vo:±12.5V Vi0:2mV	14D,8A	Nsc		data pinout
LH0022C	FET	Vs:±22V Vu:104dB Vo:±12V Vi0:3.5mV	14D,8A	Nsc		data pinout
LH 0022(C)D	Hybrid-OP-IC	=LH 0022(C)H: Fig.→	14-DIC	Nsc	-	data pinout
LH 0022(C)H	Hybrid-OP-IC	FET Inp, ±22V, -55…+125(C=-25…+85)°	81	Nsc	-	data pinout
LH 0023G	Hybrid-IC	Sample&Hold, Vcc=±15V, -55…+125°C	83	Nsc	-	pdf
LH 0023CG	Hybrid-IC	Sample&Hold, Vcc=±15V, -25…+85°C	83	Nsc	-	pdf
LH0024CH	HI-SPEED	Vs:±18V·Vu:72dB Vo:±13V Vi0:5mV	8A	Nsc		data pdf pinout
LH0024H	HI-SPEED	Vs:±18V Vu:74dB Vo:±13V Vi0:2mV	8A	Nsc		data pdf pinout
LH 0024(C)H	Hybrid-OP-IC	±22V, 500V/µs, -55…+125(C=-25…+85)°	81	Nsc	-	data pdf pinout
LH0032	HI-SPEED	Vs:±18V Vu:70dB Vo:±13.5V Vi0:2mV	48C,12A	Nsc		data pdf pinout
LH0032A	HI-SPEED	Vs:±18V Vu:70dB Vo:±13.5V Vi0:2mV	12A	Nsc		data
LH0032CG	HI-SPEED	Vs:±18V Vo:±13V Vi0:2mV	12A	Alp		data pinout
LH0032G	HI-SPEED	Vs:±18V Vo:±13V Vi0:2mV	12A	Alp		data pdf pinout
LH 0032(C)G	Hybrid-OP-IC	FET,±18V,500V/µs,-55…+125(C=-25…+85)	83	Nsc	-	data pinout

Type	Device	Short Description	Fig.	Manu	Comparision Types	More at
LH0033	FOLLOWER	Vs:±20V Vu:-0.2dB Vo:±13V Vi0:5mV	12A,8D	Max,Nsc	✓	data pdf pinout
LH0033A	FOLLOWER	Vs:±20V Vu:-0.2dB Vo:±13V Vi0:6mV	12A	Max,Nsc		data pinout
LH0033C	FOLLOWER	Vs:±20V Vu:-0.2dB Vo:±13V Vi0:12mV	12A,8D	Max,Nsc		data pdf pinout
LH0033CG	FAST-BUFF	Vs:±20V Vo:>±12V Vi0:12mV f:100Mc	12A	Alp		data pdf pinout
LH0033E	BUFF	Vs:±20V Vu:-0.2dB Vo:±13V Vi0:5mV	48C	Nsc		data pinout
LH0033G	FOLLOWER	Vs:±20V Vu:-0.2dB Vo:±13V Vi0:5mV	12A	Nsc		data pinout
LH 0033(A)(C)G	Hybrid-OP-IC	±20V, 1500V/µs, -55...+125(C=-25..+85)	83	Nsc, Max	-	data
LH 0033(C)G	Hybrid-OP-IC	±20V, 1500V/µs, -55...+125(C=-25..+85)	83	Nsc	-	data pinout
LH 0033(C)J	Hybrid-OP-IC	=LH 0033(C)G: Fig.→	22x12mm	Nsc	-	data pinout
LH 0033(C)K	Hybrid-OP-IC	=LH 0033(C)G: Fig.→	23/8Pin	Nsc	-	data
LH0036	OP	Vs:±18V Vo:±13.5V Vi0:0.5mV	12A	Nsc		data pinout
LH0036C	OP	Vs:±18V Vo:±13.5V Vi0:1mV	12A	Nsc		data pinout
LH 0036(C)G	Hybrid-OP-IC	lo-power, ±18V, -55...+125(C=-25..+85)°	83	Nsc		data pinout
LH 0038	OP	Vs:±18V Vo:±12V Vi0:25µµV f:1.6kc	16Y	Nsc		
LH0038C	OP	Vs:±18V Vo:±12V Vi0:30µµV f:1.6kc	16Y	Nsc		data pinout
LH 0038(C)D	Hybrid-OP-IC	hi-prec, ±18V, -55...+125(C=-25..+85)°	16-DIC	Nsc		data pinout
LH0041	POWER-OP	Vs:±18V Vu:106dB Vo:±14V Vi0:3mV	12A,8D,48C	Nsc		data pdf pinout
LH 0041 CJ	Hybrid-OP-IC	=LH 0041(C)G: Fig.→	22x12mm	Nsc	-	data pinout
LH0041G	POWER-OP	Vs:±18V Vu:106dB Vo:±14V Vi0:1mV	12A	Nsc		data pinout
LH 0041(C)G	Hybrid-OP-IC	Power,±18V, 0,2A,-55...+125(C=-25...85)°	83	Nsc	-	data pinout
LH0042	FET	Vs:±22V Vu:100dB Vo:±12V Vi0:6mV	14D,8A	Nsc		data pinout
LH0042D	FET	Vs:±22V Vu:103dB Vo:±12.5V Vi0:5mV	14D	Nsc		data pinout
LH 0042(C)D	Hybrid-OP-IC	=LH 0042(C)H: Fig.→	14-DIC	Nsc	-	data pinout
LH 0042(C)H	Hybrid-OP-IC	FET Inp, ±22V, -55...+125(C=-25..+85)°	81	Nsc	-	data pinout
LH 0043G	Hybrid-IC	Sample & Hold, ±15V, -55...+125°C	83	Nsc	-	pdf
LH 0043CG	Hybrid-IC	Sample & Hold, ±15V, -25...+85°C	83	Nsc	-	pdf
LH0044	LO-NOISE	Vs:±20V Vu:140dB Vo:±13.5V Vi0:12µµV	8A	Nsc		data pinout
LH0044A	LO-NOISE	Vs:±20V Vu:145dB Vo:±13.7V Vi0:8µµV	8A	Nsc		data pinout
LH 0044(A)BH,CH	Hybrid-OP-IC	=LH 0044(A)H: -25...+85°	81	Nsc	-	data pinout
LH 0044(A)H	Hybrid-OP-IC	lo-noise,±20V, -55...+125(A: <InOffset)	81	Nsc	-	data pinout
LH0045CG	TRANS	Vs:50V Vi0:2mV	12A	Nsc		data pinout
LH0045G	TRANS	Vs:50V Vi0:0.7mV	12A	Nsc		data pinout
LH 0045(C)G	Hybrid-OP-IC	+50V, 50mA, -55...+125(C=-25...85)°	83	Nsc	-	data pinout
LH 0045(C)K	Hybrid-OP-IC	=LH 0045(C)G: Fig.→	23/8Pin	Nsc	-	data pinout
LH0052	FET	Vs:±22V Vu:106dB Vo:±12.5V Vi0:0.1mV	14D,8A	Nsc		data pinout
LH0052C	FET	Vs:±22V Vu:104dB Vo:±12V Vi0:0.2mV	14D,8A	Nsc		data pinout
LH 0052(C)D	Hybrid-OP-IC	=LH 0052(C)H: Fig.→	14-DIC	Nsc	-	data pinout
LH 0052(C)H	Hybrid-OP-IC	FET Inp, ±22V, -55...+125(C=-25..+85)°	81	Nsc	-	data pinout
LH 0053(C)G	Hybrid-IC	Sample & Hold, ±15V, -55...+125(C=-25..+85)°	83	Nsc	-	
LH0061CK	WIDEBAND	Vs:±18V Vu:94dB Vo:±12V Vi0:3mV	8O	Nsc		data pinout
LH0061K	WIDEBAND	Vs:±18V Vu:100dB Vo:±12V Vi0:1mV	8O	Nsc		data pinout
LH 0061(C)K	Hybrid-OP-IC	±18V, 0,5A, -55...+125(C=-25..+85)°	23/8Pin	Nsc	-	data pinout
LH0062	FET	Vs:±20V Vu:106dB Vo:±12.5V Vi0:2mV	14D,8A	Nsc		data pinout
LH0062C	FET	Vs:±20V Vu:104dB Vo:±12V Vi0:5mV	14D,8A	Nsc		data pinout
LH 0062(C)D	Hybrid-OP-IC	=LH 0062(C)H: Fig.→	14-DIC	Nsc	-	data pinout
LH 0062(C)H	Hybrid-OP-IC	FET Inp, ±20V, -55...+125(C=-25..+85)°	81	Nsc	-	data pinout
LH0063	FAST-BUFF	Vs:±20V Vo:±13V Vi0:10mV f:200Mc	12A,8O	Nsc		data pinout
LH 0063(C)G	Hybrid-OP-IC	±20V, 6000V/µs, -55...+125(C=-25..+85)	83	Nsc	-	
LH 0063(C)J	Hybrid-OP-IC	=LH 0063(C)G: Fig.→	22x12mm	Nsc	-	
LH 0063(C)K	Hybrid-OP-IC	=LH 0063(C)G: Fig.→	23/8-Pin	Nsc	-	data pinout
LH0070	R+	Io=±mA P:600mW	3A	Ltc		data pinout
LH 0070-0H,-1H,-2H	Ref-Z-IC	10,000V, ±0,02%, -55...+125°	2e	Nsc	-	data pinout
LH007,1	R+	Io=±20mA P:600mW	3A	Nsc		data pinout
LH 0071-0H,-1H,-2H	Ref-Z-IC	10,240V, ±0,02%, -55...+125°	2e	Nsc	-	data pinout
LH0075	ER+	Io=1mA Vo:5...27V Vin:32V	12A	Nsc		data pinout
LH 0075(G),C(G)	Ref-Z-IC	0...+27V, 0,2A, -55...+125°, C: 0...+70°	83	Nsc	-	data pinout
LH0076	ER-	Io=1mA Vo:-3...-27V Vin:-32V	12A	Nsc		data pinout
LH 0076(G),C(G)	Ref-Z-IC	0...-27V, 0,2A, -55...+125°, C:-25..+85°	83	Nsc	-	data pinout
LH 0080	NMOS-µC-IC	Z80-Family, 8 Bit, 5V, 2.5MHz, Z80 clone	40-DIP	Sha	Z80	pdf pinout
LH0082CD	PRE-AMP	Vs:15V Vo:4/0V	14D	Nsc		
LH0084	OP	Vs:±18V Vo:±12V Vi0:7mV f:3.25Mc	16DY	Nsc		data pinout
LH0084C	OP	Vs:±18V Vo:±12V Vi0:13mV f:3.25Mc	16DY	Nsc		data pinout
LH 0084(C)D	Hybrid-OP-IC	progr., ±18V, -55...+125(C=-25..+85)°	16-DIC	Nsc	-	
LH0086	PROG-GAIN	Vs:±18V Vo:±12V Vi0:0.3mV f:3Mc	14D	Nsc		data pinout
LH 0086(C)D	Hybrid-OP-IC	±18V, 10V/µs, -55...+125(C=-25..+85)°	14-DIC	Nsc	-	data pinout
LH 0091(C)D	Hybrid-IC	RMS→DC Converter, Ucc=±15V	16-DIC	Nsc	-	
LH 0094(C)D	Hybrid-IC	Multifunction Converter, Ucc=±15V	16-DIC	Nsc	-	pdf
LH 0101	POWER-OP	Vs:±22V Vu:106dB Vo:±12.5V Vi0:5mV	8OA	Max		data pdf pinout
LH 0101(A)CK	Hybrid-OP-IC	=LH 0101(A)K: -25...+125°	23/8-Pin	Nsc	-	data pinout
LH0101D	UNI	Vs:±20V Vu:93dB Vo:±12V Vi0:6mV	14D	Ray		data pinout
LH0132	FET	Vs:±18V Vu:70dB Vo:±13.5V Vi0:2mV	12A	Nsc		data pinout
LH0132C	FET	Vs:±18V Vu:70dB Vo:±13V Vi0:2mV	12A	Nsc		data pinout
LH0201D	UNI	Vs:±20V Vu:86dB Vo:±12V Vi0:10mV	14D	Ray		data pinout
LH0201H	UNI	Vs:±22V Vu:103dB Vo:±14V Vi0:2mV	8A	Ray		data pinout
LH0740A	HI-OHM	Vs:±22V Vu:100dB Vo:±14V Viu:10mV	8A	Nsc		data pinout
LH 740 AH(C)	Hybrid-OP-IC	Fet Inp, ±22V, -55...+125, C=0...+85°	81	Nsc	-	data pinout
LH 1028	LIN-IC	Telefon, Interface Circuit	8-DIP	Tho	-	
LH1056	DOC	LED/MOSFET Viso:3750Vrms Vbs:1.25V		Sie		
LH 1056	LIN-IC	Telecom, Opto-coupled AC/DC Relay	6-DIP	Tho	-	data pdf pinout
LH 1061	LIN-IC	Telecom, Opto-coupled Ac/dc Relay	DIP	Tho	-	
LH1298	DOC	LED/MOSFET Viso:3750Vrms Vbs:1.25V		Sie		data pdf pinout
LH1529	DOC	LED/2 LED/MOSFET Viso:3750Vrms		Sie		data pinout
LH1540	DOC	LED/MOSFET Viso:3750Vrms Vbs:1.25V		Sie		data pdf pinout
LH1550	DOC	LED/MOSFET Viso:3750Vrms Vbs:1.25V		Sie		data pdf pinout
LH 1605 CK	Z-IC	=LH 1605K: 0...+70°	23/8Pin	Nsc	-	

Type	Device	Short Description	Fig.	Manu	Comparision Types	More at
LH 1605 K	Z-IC	S-Reg., +3..30V, 5A, ...100kHz	23/8Pin	Nsc		
LH2003	VIDEO-BUFF	Vs:±18V Vo:±11V Vi0:5mV f:100Mc	8AD	Nsc		data pinout
LH2011	2xOP	Vs:±20V Vu:109.5dB Vi0:0.1mV	16DF	Nsc		data pinout
LH2011B	2xOP	Vs:±20V Vu:109.5dB Vi0:0.2mV	16DF	Nsc		data pinout
LH2011CD	2xOP	Vs:±20V Vu:109.5dB Vi0:0.5mV	16D	Nsc		data pinout
LH 2011 CD	Hybrid-OP-IC	=LH 2011(B)D: -25...+85°	16-DIC	Nsc	-	data pinout
LH 2011(B)D	Hybrid-OP-IC	Dual, ±20V, -55...+125°, B: <InOffset	16-DIC	Nsc		data pinout
LH 2011(B)F	Hybrid-OP-IC	=LH 2011(B)D: Min	16-FLP	Nsc		data pinout
LH2033	VIDEO-BUFF	Vs:±18V Vo:±11V Vi0:5mV f:100Mc	8D	Nsc		data pinout
LH2101A	2xOP	Vs:±22V Vu:94dB Vo:±12V Vi0:2mV	16DF	Nsc		data pinout
LH 2101 AD,AJ	Hybrid-OP-IC	Dual, ±22V, 10V/µs, -55...+125°	16-DIC	Nsc, Ray	-	data pinout
LH 2101 AF	Hybrid-OP-IC	=LH 2101AD,AJ: Min	16-FLP	Nsc		data pinout
LH2101D	UNI	Vs:±22V Vu:160V/mV Vo:±13V Vi0:0.7mV	16D	Ray		data pinout
LH2108	2xOP	Vs:±20V Vu:300V/mV Vo:±14V Vi0:0.7mV	16DF	Ltc,Ray		data pinout
LH2108A	2xOP	Vs:±20V Vu:300V/mV Vo:±14V Vi0:0.3mV	16DF	Ltc,Ray	-	data pinout
LH2108D	2xOP	Vs:±20V Vu:>88dB Vo:±13V Vi0:<2mV	16D	Nsc		data pinout
LH 2108(A)D,J	Hybrid-OP-IC	Dual, ±20V, -55...+125°, A: <InOffset	16-DIC	Nsc, Ray		data pinout
LH2108F	2xOP	Vu:94dB Vo:±13V Vi0:3mV	16D	Mul,Phi,Sig		data pinout
LH 2108(A)F	Hybrid-OP-IC	=LH 2108(A)D,J: Min	16-FLP	Nsc		data pinout
LH2110	2xFOLLOWER	Vs:±18V Vo:>±10V Vi0:<4mV	16DF	Nsc		data pinout
LH 2110 D,J	Hybrid-OP-IC	Dual Follower, ±18V, 30V/µs,-55...+125°	16-DIC	Nsc		data pinout
LH 2110 F	Hybrid-OP-IC	=LH 2110D,J: Min	16-FLP	Nsc		data pinout
LH2111	2xCOMP	Vs:±18V Vu:106dB Vi0:3mV	16DF	Nsc		data pinout
LH 2111 D,J	Hybr.-KOP-IC	Dual, ±18V, 50mA, -55...+125°	16-DIC	Nsc, Ray		data pinout
LH 2111 F	Hybr.-KOP-IC	=LH 2111D,J: Min	16-FLP	Nsc		data pinout
LH2201A	2xOP	Vs:±22V Vu:94dB Vo:±12V Vi0:2mV	16DF	Nsc		data pinout
LH 2201 A...	Hybrid-OP-IC	=LH 2101...: -25...+85°	16-...	Nsc	-	data pinout
LH2208	2xOP	Vs:±20V Vu:>88dB Vo:±13V Vi0:<2mV	16FD	Nsc		data pinout
LH 2208(A)...	Hybrid-OP-IC	=LH 2108...: -25...+85°	16-...	Nsc	-	data pinout
LH2208A	2xOP	Vs:±20V Vu:>92dB Vo:±13V Vi0:<0.5mV	16DF	Nsc		data pinout
LH2208D	2xOP	Vs:±20V Vu:>88dB Vo:±13V Vi0:<2mV	16D	Nsc		data pinout
LH2208F	2xOP	Vu:94dB Vo:±13V Vi0:3mV	16D	Mul,Phi,Sig		data pinout
LH2210	2xFOLLOWER	Vs:±18V Vo:>±10V Vi0:<4mV	16DF	Nsc		data pinout
LH 2210	Hybrid-OP-IC	=LH 2110...: -25...+85°	16-...	Nsc	-	data pinout
LH 2211	Hybr.-KOP-IC	=LH 2211...: -25...+85°	16-...	Nsc	-	data pinout
LH2211D	2xCOMP	Vs:±18V Vu:106dB Vi0:3mV	16D	Nsc		data pinout
LH2211F	2xCOMP	Vi0:4mV	16DF	Mul,Phi,Sig		data pinout
LH2211J	2xCOMP	Vs:±18V Vu:106dB Vi0:3mV	16D	Nsc		data pinout
LH2301A	2xOP	Vs:±22V Vu:>84dB Vo:±12V Vi0:7.5mV	16DF	Nsc		data pinout
LH 2301 A...	Hybrid-OP-IC	=LH 2101...: -25...+85°	16-...	Nsc	-	data pinout
LH 2308(A)...	Hybrid-OP-IC	=LH 2108...: 0...+70°	16-...	Nsc	-	data pinout
LH2308A	2xOP	Vs:±20V Vu:>96dB Vo:±13V Vi0:<0.5mV	16DF	Nsc		data pinout
LH2308D	2xOP	Vs:±20V Vu:88dB Vo:±13V Vi0:7.5mV	16D	Nsc		data pinout
LH2308F	2xOP	Vu:88dB Vo:±13V Vi0:10mV	16DF	Mul,Phi,Sig		data pinout
LH2310	2xFOLLOWER	Vs:±18V Vo:>±10V Vi0:<7.5mV	16DF	Nsc		data pinout
LH 2310	Hybrid-OP-IC	=LH 2110...: 0...+70°	16-...	Nsc	-	data pinout
LH2311	2xCOMP	Vs:±18V Vu:106dB Vi0:7.5mV	16FD	Amd,Nsc,Val		data pinout
LH 2311	Hybr.-KOP-IC	=LH 2311...: 0...+70°	16-...	Nsc	-	data pdf pinout
LH2311D	2xCOMP	Vs:±18V Vu:106dB Vi0:7.5mV	16D	Nsc		data pinout
LH2311F	2xCOMP	Vu:106dB Vi0:10mV	16D	Mul,Phi,Sig		data pinout
LH 2422(J)	Hybrid-IC	Video Driver/Amp, ~10MHz, Ucc=60V	≈9-SIL	Nsc	-	pdf
LH 2424	Hybrid-IC	Video Driver/Amp, 175MHz, Ucc=60V	≈9-SIL	Nsc	-	
LH 2426	Hybrid-IC	3x Video Driver/Amp, 100MHz, Ucc=80V	12-SIL	Nsc		
LH 2440(A)	Hybrid-IC	Video Driver/Amp, 90MHz, Ucc=60V	12-SIL	Nsc		pdf
LH 2464-10 ... 20	dRAM-IC	64k x 4Bit DRAM, 100ns, +5V	DIP	Sha		pdf pinout
LH3344	LED	rd ø3.0mm I:224...560mcd Vf:1.9V	T1	Osr		data pdf pinout
LH3364	LED	srd ø3.0mm I:56...140mcd Vf:1.9V	T1	Osr		data pdf pinout
LH4001CN	WIDE-BUFF	Vs:±22V Vo:±11V	10D	Nsc		data pdf pinout
LH 4001 N	Hybrid-OP-IC	Buffer, ±22V, ±0,1A, 0...+70°	10-DIP	Nsc		data
LH4002	WIDE-BUFF	Vs:±6V Vo:±2.4V Vi0:20mV f:150Mc	8AD,10D	Nsc		data pdf pinout
LH 4002 N	Hybrid-OP-IC	Video Buffer, ±6V, ±60mA, -55...+125°	10-DIP	Nsc		data pdf
LH 4002 E	Hybrid-OP-IC	=LN 4002N:	20-(P)LCC			data
LH 4002 H	Hybrid-OP-IC	=LN 4002N:	81			data
LH4003CD	PREC-BUFF	Vs:±8V Vo:±3V Vi0:5mV f:250Mc	24D	Nsc		data pinout
LH4003D	PREC-BUFF	Vs:±8V Vo:±3V Vi0:2mV f:250Mc	24D	Nsc		data pinout
LH4004	WIDE-BUFF	Vs:±15V Vo:±9.6V Vi0:8mV f:70Mkc	24D	Nsc		data pinout
LH4006CD	PREC-BUFF	Vs:±8V Vo:±3V Vi0:5mV f:350Mc	24D	Nsc		data pinout
LH4006D	PREC-BUFF	Vs:±8V Vo:±3V Vi0:2mV f:350Mc	24D	Nsc		data pinout
LH4008	FAST-BUFF	Vs:40V Vo:11.9/0V Vi0:10mV f:180Mc	8O,11P	Nsc		data pinout
LH 4008(C)K	Hybrid-OP-IC	Buffer, 40V, ±0,2A, 130MHz, -55...+125°C, C: -25...+85°C	23/8Pin	Nsc		data
LH 4008(C)T	Hybrid-OP-IC	=LN 4008...K:	11-SQL			data
LH 4009	FAST-BUFF	Vs:40V Vo:±11V Vi0:10mV f:190Mc	8O,11P	Nsc		data pinout
LH 4009(C)K	Hybrid-OP-IC	Buffer, 40V, ±0,2A, 150MHz, -55...+125°C, C: -25...+85°C	23/8Pin	Nsc		data
LH 4009(C)T	Hybrid-OP-IC	=LN 4009K...:	11-SQL			data
LH4010	FAST-BUFF	Vs:40V Vo:±9V Vi0:5mV f:140Mc	12A,16D	Nsc		data pinout
LH 4010(C)K	Hybrid-OP-IC	Buffer, 40V, ±0,1A, 20MHz, -55...+125°C, C: -25...+85°C	2/12Pin	Nsc		data
LH 4010(C)N	Hybrid-OP-IC	=LN 4010G...:	16-DIP			data
LH4011	FAST-BUFF	Vs:40V Vo:11.4/0V Vi0:10mV f:160Mc	8O,11P	Nsc		data pinout
LH 4011(C)K	Hybrid-OP-IC	Buffer, 40V, ±0,2A, 80MHz, -55...+125°C, C: -25...+85°C	23/8Pin	Nsc		data
LH 4011(C)T	Hybrid-OP-IC	=LN 4011K...:	11-SQL			data
LH4012	WIDE-BUFF	Vs:40V Vo:11.4/0V Vi0:±20mV f:460Mc	8O	Nsc		data pinout

Type	Device	Short Description	Fig.	Manu	Comparision Types	More at
LH 4012(C)K	Hybrid-OP-IC	Buffer, 40V, ±0,2A, 230MHz, -55...+125°C, C: -25...+85°C	23/8Pin	Nsc	-	data pdf
LH4033	FAST-BUFF	Vs:40V Vo:±12V Vi0:12mV f:100Mc	16D,11P	Nsc		data pinout
LH 4033 CN	Hybrid-OP-IC	Buffer, 40V, ±0,1A, 100MHz, -25...+85°	16-DIP	Nsc	-	data
LH 4063 CT	Hybrid-OP-IC	Buffer, 40V, ±0,1A, 100MHz, -25...+85°	11-SQL	Nsc		data
LH4101	WIDEBAND	Vs:±17V Vu:65dB Vo:±13.5V Vi0:<15mV	24D	Nsc		data pinout
LH4104	POWER-OP	Vs:±18V Vu:106dB Vo:±13V Vi0:2mV	12A,16D	Nsc		data pinout
LH4105	POWER-OP	Vs:±18V Vu:106dB Vo:±13V Vi0:0.2mV	12A	Nsc		data
LH4106	HI-SPEED	Vs:±7.5V Vu:65dB Vo:3/2.6V Vi0:5mV	10A,16D	Nsc		data pinout
LH4117	HI-PREC	Vs:±18V Vo:±11V Vi0:15mV f:40Mkc	24D	Nsc		data pinout
LH4118	WIDEBAND	Vs:±18V Vo:±13V Vi0:±2mV	12A	Nsc		data pinout
LH4124CN	HI-SLEW	Vs:±18V Vu:5V/mV Vo:±13V Vi0:5mV	16D	Nsc		data pinout
LH4141CN	POWER-OP	Vs:±18V Vo:±14V Vi0:3mV	16D	Nsc		data pinout
LH4161	HI-SPEED	Vs:36V Vo:14/13V Vi0:0.5mV f:40Mc	10A,8D	Nsc		data pinout
LH4162	2xHI-SPEED	Vs:36V Vo:14/13V Vi0:0.5mV f:40Mc	10A,16D	Nsc		data pinout
LH4200	UNI	Vs:12V Vo:3/0V	24D,12A	Nsc		data pinout
LH 4266(C)D	Hybrid-IC	Video Analog Switch, ...150MHz, ±18V -55...+125°C, C: -25...+85°	≈24-DIC	Nsc		
LH5424	LED	rd ø5mm I>63mcd Vf:1.75V If:10mA	T1¾	Sie		data pinout
LH5464	LED	rd ø5mm I:71...112mcd Vf:1.75V	T1¾	Osr		data pdf pinout
LH6121	FAST-BUFF	Vs:±18V Vo:13.5/0V Vi0:15mV f:50Mc	8A	Nsc		data
LH6125	FAST-BUFF	Vs:±18V Vo:13.5/0V Vi0:15mV f:50Mc	8A	Nsc		data
LH6221	FAST-BUFF	Vs:±18V Vo:13.5/0V Vi0:15mV f:50Mc	8AD	Nsc		data
LH6225	FAST-BUFF	Vs:±18V Vo:13.5/0V Vi0:15mV f:50Mc	8A,14D	Nsc		data
LH6321	FAST-BUFF	Vs:±18V Vo:13.5/0V Vi0:15mV f:50Mc	8D	Nsc		data
LH6325	FAST-BUFF	Vs:±18V Vo:13.5/0V Vi0:15mV f:50Mc	14D	Nsc		data
LH7001	ER±	Io=200mA Vo:±1.2...±37V Vin:±3...±40V	8DA	Nsc		data
LH24250	2xμ-POWER	Vs:±18V Vo:>±12V Vi0:<3mV	16DF	Nsc		data
LH24250C	2xμ-POWER	Vs:±18V Vo:>±12V Vi0:<5mV	16DF	Nsc		data
LH 24250(C)D,J	Hybrid-OP-IC	Dual μpower, ±18V, -55...+125(C=0...70)°	16-DIC	Nsc	-	data pinout
LH 24250(C)F	Hybrid-OP-IC	=LH 24250(C)D,J: Min	16-FLP	Nsc	-	data pinout
LHA674	LED	rd ø5mm I:28...71mcd Vf:1.9V If:10mA	T1	Osr		data pdf pinout
LHGBT686	LED	rd/gr/bl ø5mm I:4.5...11.2mcd Vf:3.5V	PLCC-4	Osr		data pdf pinout
LHGT680	LED	rd/gr ø5mm I:7.1...28mcd Vf:2V	PLCC-4	Osr		data pdf pinout
LHI807	PYRODET	8μnm Resp:0.64V/mW Vb:>2V	TO5	Hei		data pinout
LHI807TC	PYRODET	11μnm Resp:0.32V/mW Vb:>2V	TO5	Hei		data pinout
LHI878	PYRODET	8μnm Resp:4V/mW Vb:>2V	TO5	Hei,Sie		data pdf pinout
LHI907	PYRODET	8μnm Resp:3.6V/mW Vb:>2V	TO5	Hei		data pdf pinout
LHI978	PYRODET	8μnm Resp:3.5V/mW Vb:>2V	TO5	Hei		data pinout
LHI1148	PYRODET	8μnm Resp:4.8V/mW Vb:>2V	TO5	Hei		data pdf pinout
LHL376	LED	rd ø3mm I:180...280mcd Vf:1.9V	T1	Osr		data pdf pinout
LMH 6515 SQ	IC	600 MHz, Dig.Contr., Variable Gain Amp.	LLP	Nsc		pdf pinout
LHN974	LED	rd ø3mm I:7.1...15mcd Vf:1.8V If:20mA	1206	Osr		data pdf pinout
LHp	Si-P	=BF 569 (Typ-Code/Stempel/marking)	35	Phi	→BF 569	data
LHR974	LED	rd ø3mm I:11.2...15mcd Vf:1.8V	1206	Osr		data pdf pinout
LHs	Si-P	=BF 569 (Typ-Code/Stempel/marking)	35	Sie	→BF 569	data
LHT674	LED	rd ø3mm I:28...71mcd Vf:1.9V If:10mA	* PLCC-2	Osr		data pdf pinout
LHT676	LED	rd ø3mm I:71...180mcd Vf:1.85V	PLCC-2	Osr		data pdf pinout
LHT774	LED	rd ø3mm I:28...71mcd Vf:1.9V If:10mA	PLCC-2	Osr		data pdf pinout
LIS 302DL	IC	SMD, 3 axes linear accelerometer, sensing element, I²c/spi interface	LGA	Sgs	-	pdf pinout
LIU-01 FP,FQ	LIN-IC	Telecom, Serial Data Receiver, Ucc=5V	16-DIP,DIC	Pmi	-	
LIU-01 FS	LIN-IC	SMD,Telecom, Serial Data Receiver	16-MDIP		-	
LIU-01 FP,FW	LIN-IC	Telecom, Serial Data Transcvr., Ucc=5V	24-DIP,DIC	Pmi	-	
LIU-02 FS	LIN-IC	SMD, Ser. Data Transcvr., Ucc=5V	24-MDIP			
LJ	MOS-N-FET-e	=2SK1848 (Typ-Code/Stempel/marking)	35	Say	→2SK1848	data
LJ	Si-P	=BF 579 (Typ-Code/Stempel/marking)	35	Phi, Sie	→BF 579	data
LJp	Si-P	=BF 579 (Typ-Code/Stempel/marking)	35	Phi	→BF 579	data
LK	Z-Di	=1SMB 14A (Typ-Code/Stempel/marking)	71(5x3,5)	Ons	→1SMB 5.0A...170A	
LK	Si-N	=2SD1000-LK (Typ-Code/Stempel/marking)	39	Nec	→2SD1000	data
LK	Si-P	=BF 568 (Typ-Code/Stempel/marking)	35	Sie	→BF 568	data
LK	Si-N	=BF 799 (Typ-Code/Stempel/marking)	35	Sie	→BF 799	data
LK	Z-Di	=P6 SMB-22A (Typ-Code/Stempel/marking)	71(5x3,5)	Fag	→P6 SMB-22A	
LK	Z-Di	=SMBJ 14A (Typ-Code/Stempel/marking)	71(5x3,5)	Mop	→SMBJ ...	
LK112	R+	Io=150mA Vo:7.840...8.160V Vin:9V	5S	Sgs		data pdf pinout
LK112S	R+	Io=150mA Vo:7.840...8.160V Vin:9V	5S	Sgs		data pdf pinout
LKC	Z-Di	=1SMB 14CA (Typ-Code/Stempel/marking)	71(5x3,5)	Ons	→1SMB 10CA...78CA	
LKs	Si-N	=BF 799W (Typ-Code/Stempel/marking)	35(2mm)	Sie	→BF 799W	data
LL	Si-P	=2SA1518 (Typ-Code/Stempel/marking)	35	Say	→2SA1518	
LL	Si-N	=2SC2712-BL (Typ-Code/Stempel/marking)	35	Tos	→2SC2712	data
LL	Si-N	=2SC4116-BL (Typ-Code/Stempel/marking)	35(2mm)	Tos	→2SC4116	
LL	Si-N	=2SC4207-BL (Typ-Code/Stempel/marking)	45	Tos	→2SC4207	data
LL	Si-N	=2SC4738-BL (Typ-Code/Stempel/marking)	35(1,6mm)	Tos	→2SC4738	
LL	Si-N	=2SD1000-LL (Typ-Code/Stempel/marking)	39	Nec	→2SD1000	
LL	Si-N	=KTC 4075-BL(Typ-Code/Stempel/marking)	35(2mm)	Kec	→KTC 4075	
LL	Z-Di	=P6 SMB-24 (Typ-Code/Stempel/marking)	71(5x3,5)	Fag	→P6 SMB-24	data
LL	Z-Di	=SMBJ 15 (Typ-Code/Stempel/marking)	71(5x3,5)	Mop	→SMBJ ...	data
LL 41	Si-Di	=BAT 41: SMD	72(3,4mm)	Gsi	TMMBAT 41, TMMBAT 46	data
LL 42	Si-Di	=BAT 42: SMD	72(3,4mm)	Gsi	TMMBAT 42, TMMBAT 43	data
LL 43	Si-Di	=BAT 43: SMD	72(3,4mm)	Gsi	TMMBAT 42, TMMBAT 43	data
LL 46	Si-Di	=BAT 46: SMD	72(3,4mm)	Gsi	TMMBAT 46	data
LL 48	Si-Di	=BAT 48: SMD	72(3,4mm)	Gsi	TMMBAT 48	data

Type	Device	Short Description	Fig.	Manu	Comparision Types	More at
LL 101 A,B,C	Si-Di	SMD, Schottky, SS, 40...60V, 15mA, <1ns A=60V, B=50V, C=40C	72(3,4mm)	Aeg, Gsi	BAS 83...81	data
LL 103 A,B,C	Si-Di	SMD,Schottky, SS, 20...40V, 0,2A, <10ns A=40V, B=30V, C=20C	72(3,4mm)	Aeg, Gsi	BAS 86...85	data
LL 4148	Si-Di	=1N4148: SMD	72a(3,4mm)	Aeg,Itt,Die Gsi	BAS 32, PMLL 4148	data
LL 4150	Si-Di	=1N4150: SMD	72a(3,4mm)	Aeg,Die,Gsi	BAS 32, PMLL 4150	data
LL 4151	Si-Di	=1N4151: SMD	72a(3,4mm)	Aeg, Die	BAS 32, PMLL 4151	data
LL 4154	Si-Di	=1N4154: SMD	72a(3,4mm)	Aeg	BAS 32	data
LL 4448	Si-Di	=1N4448: SMD	72a(3,4mm)	Aeg,Itt,Die Gsi	BAS 32, PMLL 4448	data
LL 5711	Si-Di	SMD, Schottky, VHF/UHF-Dem, 70V, 15mA	72(3,4mm)	Gsi	TMM 1N5711	data
LL 6263	Si-Di	SMD, Schottky, VHF/UHF-Dem, 60V, 15mA	72(3,4mm)	Gsi	TMM 1N6273	data
LLM338	ER+	Io=5A Vo:1.2...32V P:int	30	Uni		data pinout
LLM 338	Z-IC	+1,2...+32V, 5A, 0...+125°	23k	Uni		data pinout

LM

Type	Device	Short Description	Fig.	Manu	Comparision Types	More at
LM	Z-Di	=1SMB 15A (Typ-Code/Stempel/marking)	71(5x3,5)	Ons	→1SMB 5.0A...170A	
LM	Si-N	=2SD1000-LM (Typ-Code/Stempel/marking)	39	Nec	→2SD1000	data
LM	Si-P	=BF 569R (Typ-Code/Stempel/marking)	35	Aeg	→BF 569R	data
LM	MOS-P-FET-e	=BST 120 (Typ-Code/Stempel/marking)	39	Val	→BST 120	data
LM	Z-Di	=P6 SMB-24A (Typ-Code/Stempel/marking)	71(5x3,5)	Fag	→P6 SMB-24A	data
LM	Z-Di	=SMBJ 15A (Typ-Code/Stempel/marking)	71(5x3,5)	Mop	→SMBJ ...	data
LM1-AHR-01	LED	rd ø3mm I:500mcd Vf:2.1V If:50mA		Cot		data pdf pinout
LM1-AHR-11	LED	rd ø3mm I:1000mcd Vf:2.4V If:70mA		Cot		data pdf pinout
LM1-NRR1	LED	rd/gr/bl ø3mm I:70mcd Vf:3.4V		Cot		data pdf pinout
LM1-PBL-01	LED	bl ø3mm I:160mcd Vf:3.4V If:25mA		Cot		data pdf pinout
LM1-PBL-11	LED	bl ø3mm I:200mcd Vf:3.6V If:30mA		Cot		data pdf pinout
LM1-PPG-01	LED	gr ø3mm I:560mcd Vf:3.4V If:25mA		Cot		data pdf pinout
LM1-PPG-11	LED	gr ø3mm I:700mcd Vf:3.6V If:30mA		Cot		data pdf pinout
LM1-PPP1	LED	rd/gr/bl ø3mm I:110mcd Vf:3.4V		Cot		data pdf pinout
LM1-PWN1-01	LED	wht ø3mm I:600mcd Vf:3.4V		Cot		data pdf pinout
LM1-PWN1-11	LED	wht ø3mm I:780mcd Vf:3.6V		Cot		data pdf pinout
LM1-PWR1-11	LED	wht ø3mm I:700mcd Vf:3.6V		Cot		data pdf pinout
LM1-PYL-01	LED	gr+yl ø3mm I:450mcd Vf:2.1V If:50mA		Cot		data pdf pinout
LM1-PYL-11	LED	gr+yl ø3mm I:1100mcd Vf:2.4V If:70mA		Cot		data pdf pinout
LM10	OP+V-REF	Vs:45V Vio:±2mV	8A	Nsc		data pdf pinout
LM10B	OP+V-REF	Vs:45V Vu:102dB Vio:0.3mV	8AD	Ltc,Nsc		data pdf pinout
LM 10 B... ...	OP-IC	=LM 10 H..: -25...+85°		Nsc		data pinout
LM10C	OP+V-REF	Vs:45V Vu:102dB Vio:0.5mV	8AD,14S	Ltc,Nsc		data pdf pinout
LM 10 C... ...	OP-IC	=LM 10H..: 0...+70°		Nsc	-	data pinout
LM 10(B,C)H(L)	OP-IC	integr. Volt.Ref., 45(L=7)V,-55...+125°	81	Nsc	-	data pdf pinout
LM 10(B,C)N(L)	OP-IC	=LM 10H..: Fig.→	8-DIP	Nsc	-	data pdf pinout
LM 10(C)WM	OP-IC	=LM 10H..: SMD	8-MDIP	Nsc	-	data pdf pinout
LM11	OP	Vs:40V Vu:109.5dB Vio:0.2mV	8DAF,14D	Nsc		data pinout
LM 11(L)CDP,C(L)N	OP-IC	=LM 11CH: Fig.→	8-DIP	Mot,Nsc,Tho	-	data pinout
LM 11 C(L)H	OP-IC	Uni, 40V, InOffset<0,6(L<5)mV, 0...+70°	81	Mot,Nsc,Tho	-	data pdf pinout
LM 11 C(L)J,D	OP-IC	=LM 11CH: Fig.→	14-DIC	Mot,Nsc	-	data pinout
LM 11 C(L)J-8	OP-IC	=LM 11CH: Fig.→	8-DIC	Mot	-	data pinout
LM 11 C(L)N-14	OP-IC	=LM 11CH: Fig.→	14-DIP	Mot,Nsc	-	data pinout
LM 11 H,MH	OP-IC	=LM 11CH: -55...+125°	81	Mot,Tho	-	data pdf pinout
LM 11 J	OP-IC	=LM 11CH: -55...+125°	14-DIC	Mot	-	data pinout
LM 11 MDG,J-8	OP-IC	=LM 11CH: -55...+125°	8-DIP,DIC	Mot,Tho	-	data pinout
LM12	HI-GAIN	Vs:100V Vu:100V/mV Vio:2mV	5O	Nsc		data pinout
LM 19 CIZ	LIN-IC	Temp. Sensor, 2.4V/10µA, -55...130°C	7	Nsc		pdf pinout
LM 20 (B,C)IM7	LIN-IC	SMD, Temp. Sensor, 2.4V/10µA -55...130°C, BIM7: ±2.5°C CIM7: ±5°C	45	Nsc		pdf pinout
LM 20 SI(BP,TL)	LIN-IC	SMD, Temp.Sensor, ±3.5°C, -40...125°C	4-GRID	Nsc		pdf
LM 34 ..H	LIN-IC	Precision Fahrenheit Temp. Sensor	2	Nsc	-	pdf pinout
LM 34 ..M	LIN-IC	Precision Fahrenheit Temp. Sensor	8-MDIP	Nsc	-	pdf pinout
LM 34 ..Z	LIN-IC	Precision Fahrenheit Temp. Sensor	7	Nsc	-	pdf pinout
LM34D	4xPROG	Vs:±22V Vu:1000V/mV Vo:±14V f:1Mc	16S	Sgs		data pdf pinout
LM 35 ..H	LIN-IC	Precision Centigrade Temp. Sensor	2	Nsc	-	pdf pinout
LM 35 ..M	LIN-IC	Precision Centigrade Temp. Sensor	8-MDIP	Nsc	-	pdf pinout
LM 35 ...T	LIN-IC	Precision Centigrade Temp. Sensor	17	Nsc	-	pdf pinout
LM 35 ..Z	LIN-IC	Precision Centigrade Temp. Sensor	7	Nsc	-	pdf pinout
LM 45 (B,C)IM3	LIN-IC	Centigrade Temp. Sensor, -20...100°C	35	Nsc	-	pdf pinout
LM 50 (B,C)IM3	LIN-IC	Precision Centigrade Temp. Sensor	35	Nsc	-	pdf pinout
LM 60 (B,C)IM3	LIN-IC	Precision Centigrade Temp.Sensor, 2.7V	35	Nsc	-	pdf pinout
LM 60 (B,C)IZ	LIN-IC	Precision Centigrade Temp. Sensor, 2.7V	7	Nsc	-	pdf pinout
LM 61 (B,C)IM3	LIN-IC	Precision Centigrade Temp.Sensor, 2.7V	35	Nsc	-	pdf pinout
LM 61 (B,C)IZ	LIN-IC	Precision Centigrade Temp.Sensor, 2.7V	7	Nsc	-	pdf pinout
LM 62 (B,C)IM3	LIN-IC	Centigrade Temp.Sensor, 0...+90°C, 2.7V	35	Nsc	-	pdf pinout
LM78GCP	ER+	Io=1A Vo:5..30V P:int	4P	Nsc		data pinout
LM 78 Lxx(A)CH	Z-IC	=LM 78Lxx(A)CZ: Fig.→	2e	Nsc	...78Lxx... (TO-5)	data pinout
LM 78 L05(A)CZ	Z-IC	+5V, 100mA, ±10%, A=±5%	7b	Nsc	...78L05... (TO-92)	data pinout
LM 78 L06(A)CZ	Z-IC	+6V, 100mA, ±10%, A=±5%	7b	Nsc	...78L06... (TO-92)	data pinout
LM 78 L08(A)CZ	Z-IC	+8V, 100mA, ±10%, A=±5%	7b	Nsc	...78L08... (TO-92)	data pinout
LM 78 L09(A)CZ	Z-IC	+9V, 100mA, ±10%, A=±5%	7b	Nsc	...78L09... (TO-92)	data pinout
LM 78 L10(A)CZ	Z-IC	+10V, 100mA, ±10%, A=±5%	7b	Nsc	...78L10... (TO-92)	data pinout
LM 78 L12(A)CZ	Z-IC	+12V, 100mA, ±10%, A=±5%	7b	Nsc	...78L12... (TO-92)	data pinout

Type	Device	Short Description	Fig.	Manu	Comparision Types	More at
LM 78 L15(A)CZ	Z-IC	+15V, 100mA, ±10%, A=±5%	7b	Nsc	...78L15... (TO-92)	data pinout
LM 78 L18(A)CZ	Z-IC	+18V, 100mA, ±10%, A=±5%	7b	Nsc	...78L18... (TO-92)	data pinout
LM 78 L24(A)CZ	Z-IC	+24V, 100mA, ±10%, A=±5%	7b	Nsc	...78L24... (TO-92)	data pinout
LM78L62(A)CZ	Z-IC	+6,2V, 100mA, ±10%, A=±5%	7b	Nsc	...78L62... (TO-92)	data pinout
LM78L82(A)CZ	Z-IC	+8,2V, 100mA, ±10%, A=±5%	7b	Nsc	...78L82... (TO-92)	data pinout
LM78LXX	R+	Io=100mA Vo:4.5...5.5V	3ANP	Nsc		data pinout
LM78LXXA	R+	Io=100mA Vo:4.75...5.25V Vin:7...20V	3NA,8S,4S	Fcs		data pdf pinout
LM 78MxxCP	Z-IC	=LM 78MxxCT: Fig.→	13b	Nsc	...78Mxx... (TO-202)	data pinout
LM 78 M05CT	Z-IC	+5V, >0,5A	17b	Nsc	...78M05... (TO-220)	data pinout
LM 78 M06CT	Z-IC	+6V, >0,5A	17b	Nsc	...78M06... (TO-220)	data pinout
LM 78 M08CT	Z-IC	+8V, >0,5A	17b	Nsc	...78M08... (TO-220)	data pinout
LM 78 M10CT	Z-IC	+10V, >0,5A	17b	Nsc	...78M10... (TO-220)	data pinout
LM 78 M12CT	Z-IC	+12V, >0,5A	17b	Nsc	...78M12... (TO-220)	data pinout
LM 78 M15CT	Z-IC	+15V, >0,5A	17b	Nsc	...78M15... (TO-220)	data pinout
LM 78 M18CT	Z-IC	+18V, >0,5A	17b	Nsc	...78M18... (TO-220)	data pinout
LM 78 M24CT	Z-IC	+24V, >0,5A	17b	Nsc	...78M24... (TO-220)	data pinout
LM78MXX	R+	Io=500mA Vo:4.75...5.25V Vin:7.5...20V	4S,3APP	Nsc		data pdf pinout
LM79GCP	ER-	Io=1A Vo:-2.55...-30V P:int	4P	Nsc		data pdf pinout
LM79L05A	R-	Io=100mA Vo:-4.8...-5.2V Vin:-10V	8S,3N	Nsc		data pdf pinout
LM 79 L05(A)CZ	Z-IC	-5V, 100mA, ±10%, A=±5%	7a	Nsc	...79L05... (TO-92)	data pinout
LM 79 L06(A)CZ	Z-IC	-6V, 100mA, ±10%, A=±5%	7a	Nsc	...79L06... (TO-92)	data pinout
LM79L12A	R-	Io=100mA Vo:-11.4...-12.6V P:int	8S,3N	Nsc		data pdf pinout
LM 79 L12(A)CZ	Z-IC	-12V, 100mA, ±10%, A=±5%	7a	Nsc	...79L12... (TO-92)	data pinout
LM79L15A	R-	Io=100mA Vo:-14.4...-15.6V Vin:-20V	8S,3N	Nsc		data pdf pinout
LM 79 L15(A)CZ	Z-IC	-15V, 100mA, ±10%, A=±5%	7a	Nsc	...79L15... (TO-92)	data pinout
LM79L18A	R-	Io=100mA Vo:-17.3...-18.7V Vin:-23V	3N	Nsc		data pdf pinout
LM 79 L18(A)CZ	Z-IC	-18V, 100mA, ±10%, A=±5%	7a	Nsc	...79L18... (TO-92)	data pinout
LM79L24A	R-	Io:100mA Vo:-23...-25V Vin:-29V	3N	Nsc		data pdf pinout
LM 79 L24(A)CZ	Z-IC	-24V, 100mA, ±10%, A=±5%	7a	Nsc	...79L24.. (TO-92)	data pinout
LM79M05	R-	Io=500mA Vo:-4.8...-5.2V Vin:-10V	3P	Nsc		data pdf pinout
LM79M05A	R-	Io=1.5A Vo:-4.8...-5.2V Vin:-10V	3AP	Nsc		data pinout
LM 78 Mxx(A)CH	Z-IC	=LM 79Mxx(A)CP: Fig.→	2f	Nsc	...79Mxx... (TO-5)	data
LM 79 Mxx(A)CH	Z-IC	=LM79Mxx(A)CP: Fig.→	2f	Nsc	...79Mxx... (TO-5)	data
LM 79 M05(A)CP	Z-IC	-5V, 0,5A, ±10%, A=±5%	13c	Nsc	...79M05... (TO-202)	data pinout
LM79M06A	R-	Io=500mA Vo:-5.75...-6.25V Vin:-11V	3AP	Nsc		data pinout
LM 79 M06(A)CP	Z-IC	-6V, 0,5A, ±10%, A=±5%	13c	Nsc	...79M06... (TO-202)	data pinout
LM79M08A	R-	Io=500mA Vo:-7.7...-8.3V Vin:-14V	3AP	Nsc		data pinout
LM 79 M08(A)CP	Z-IC	-8V, 0,5A, ±10%, A=±5%	13c	Nsc	...79M08... (TO-202)	data pinout
LM79M12	R-	Io=500mA Vo:-11.4...-12.6V P:int	3P	Nsc		data pdf pinout
LM 79 M12(A)CP	Z-IC	-12, 0,5A, ±10%, A=±5%	13c	Nsc	...79M12... (TO-202)	data pinout
LM79M15	R-	Io=500mA Vo:-14.25...-15.75V P:int	3P	Nsc		data pinout
LM79M15A	R-	Io=500mA Vo:-14.4...-15.6V Vin:-23V	3AP	Nsc		data pinout
LM 79 M15(A)CP	Z-IC	-15, 0,5A, ±10%, A=±5%	13c	Nsc	...79M15... (TO-202)	data pinout
LM 79 M18(A)CP	Z-IC	-18, 0,5A, ±10%, A=±5%	13c	Nsc	...79M18... (TO-202)	data pinout
LM79M24A	R-	Io=500mA Vo:-23...-25V Vin:-33V	3AP	Nsc		data pinout
LM 79 M24(A)CP	Z-IC	-24, 0,5A, ±10%, A=±5%	13c	Nsc	...79M24... (TO-202)	data pinout
LM79MGCT	ER-	Io=500mA Vo:-2.2...-30V P:int	4P	Nsc		data pinout
LM100	ER+	Io=12mA Vu:2...30V P:800mW	10F,8A	Isi,Nsc		data pinout
LM 100 ...	Z-IC	~LM 105...	81	Nsc	→LM 105...	data pdf pinout
LM101	UNI	Vs:±22V Vu:104dB Vo:±13V Vi0:1mV	14D,10F,8A	Sig		data pinout
LM101A	UNI	Vs:±22V Vu:160V/mV Vo:±13V Vi0:0.7mV	8DS,14D,20	Ray		data pinout
LM 101(A)D,JG,P	OP-IC	=LM 101(A)H: Fig.→	8-DIP,DIC	Tix	→Serie 101	data pinout
LM 101(A)DG,J	OP-IC	=LM 101(A)H: Fig.→	8-DIP,DIC	Mot,Tho	→Serie 101	data pdf pinout
LM 101(A)F,U	OP-IC	=LM 101(A)H: SMD	10-FLP	Nsc,Tix	-	data pdf pinout
LM 101(A)FK	OP-IC	=LM 101(A)H: SMD	20-(P)LCC	Tix		data pdf pinout
LM 101(A)GC	OP-IC	=LM 101(A)H: SMD	20-(P)LCC	Tho		data pdf pinout
LM 101(A)H,L	OP-IC	Uni, Serie 101, ±22V, -55...+125°	81	Mot,Nsc,Tho	→Serie 101	data pdf pinout
LM 101(A)J-14,N-14	OP-IC	=LM 101(A)H: Fig.→	14-DIC,DIP	Nsc	→Serie 101	data pdf pinout
LM 101(A)W	OP-IC	=LM 101(A)H: SMD	14-FLP	Tix		data pdf pinout
LM102	FOLLOWER	Vs:±18V Vo:±13V Vi0:2mV	8A	Nsc		data pinout
LM 102 H	OP-IC	Volt.Follower, ±18V, -55...+125°	81	Nsc	→Serie 102	data pinout
LM103-1.8	Z+	Vo:1.8...V P:250mW	2A	Nsc		data pinout
LM103-2.0	Z+	Vo:2...V P:250mW	2A	Nsc		data pinout
LM103-2.2	Z+	Vo:2.2...V P:250mW	2A	Nsc		data pinout
LM103-2.4	Z+	Vo:2.4...V P:250mW	2A	Nsc		data pinout
LM103-2.7	Z+	Vo:2.7...V P:250mW	2A	Nsc		data pinout
LM103-3.0	Z+	Vo:3...V P:250mW	2A	Nsc		data pinout
LM103-3.3	Z+	Vo:3.3...V P:250mW	2A	Nsc		data pinout
LM103-3.6	Z+	Vo:3.6...V P:250mW	2A	Nsc		data pinout
LM103-3.9	Z+	Vo:3.9...V P:250mW	2A	Nsc		data pinout
LM103-4.3	Z+	Vo:4.3...V P:250mW	2A	Nsc		data pinout
LM103-4.7	Z+	Vo:4.7...V P:250mW	2A	Nsc		data pinout
LM103-5.1	Z+	Vo:5.1...V P:250mW	2A	Nsc		data pinout
LM103-5.6	Z+	Vo:5.6...V P:250mW	2A	Nsc		data pinout
LM 103(H)-1,8..5,6	Ref-Z-IC	1,8...5,6V, ±10%, 0,25W, 5<25Ω	2d	Nsc	-	data pdf pinout
LM104	ER-	Vo:-0.015...-40V P:500mW	10FA	Nsc		data pdf pinout
LM 104 H,L	Z-IC	-0,015...-40V, 20mA, -55...+125	82	Nsc,Tix		data pinout
LM104J	ER-	Io=20mA Vo:-0.015...-40V P:800mW	14D,10A	Tix		data pinout
LM 104 J	Z-IC	=LM 104H,L: Fig.→	14-DIC	Tix		data pinout
LM105	ER+	Io=12mA Vo:4.5...40V P:800mW	8A,10FD	Tix		data pinout
LM 105 H,L	Z-IC	+4,5...40V, 12mA, -55...+125°	81	Nsc,Tho,Tix	µA 105H	data pinout
LM 105 JG	Z-IC	=LM 105H,L: Fig.→	8-DIC	Tix		data pinout
LM106	COMP	Vs:±15V Vu:92dB Vo:TTL Vi0:0.5mV	14F,8ADD	Nsc		data pinout
LM 106 D	KOP-IC	=LM 106H: SMD	8-MDIP	Tix	→Serie 106	data pdf pinout
LM 106 F,U,W	KOP-IC	=LM 106H: Min	14-FLP	Nsc,Tix	→Serie 106	data pdf pinout

Type	Device	Short Description	Fig.	Manu	Comparision Types	More at
LM 106 H,L	KOP-IC	Serie 106, ±15V, 0,1A, -55...+125°	81	Nsc,Tix	→Serie 106	data pdf pinout
LM 106 J,N	KOP-IC	=LM 106H: Fig.→	14-DIC,DIP	Tix	→Serie 106	data pdf pinout
LM 106 JG,P	KOP-IC	=LM 106H: Fig.→	8-DIC,DIP	Tix	→Serie 106	data pdf pinout
LM107	UNI	Vs:±22V Vu:104dB Vo:±13V Vi0:0.7mV	14D,10F,8A	Nsc		data pinout
LM 107 D,J,J-14	OP-IC	=LM 107H: Fig.→	14-DIP,DIC	Nsc,Tix	→Serie 107	data pinout
LM 107 F,U	OP-IC	=LM 107H: Fig.→	10-FLP	Nsc,Tix	-	data pdf pinout
LM 107 H	OP-IC	Uni, Serie 107, ±22V, 0...+70°	81	Nsc	→Serie 107	data pdf pinout
LM 107 J,JG,N	OP-IC	=LM 107H: Fig.→	8-DIC,DIP	Nsc,Tix	→Serie 107	data pdf pinout
LM 107 W	OP-IC	=LM 107H: Fig.→	14-FLP	Tix	-	data pdf pinout
LM108	LO-POWER	Vs:±20V Vu:200V/mV Vo:±18V Vi0:1mV	8D,14DS,10	Ray		data pinout
LM108A	LO-POWER	Vs:±20V Vu:200V/mV Vo:±18V Vi0:0.4mV	8D,14DS,10	Ray		data pinout
LM 108(A)D,J	OP-IC	=LM 108(A)H: Fig.→	14-DIP,DIC	Mot,Nsc	→Serie 108	data pdf pinout
LM 108(A)DG,J-8	OP-IC	=LM 108(A)H: Fig.→	8-DIP,DIC	Mot,Nsc,Tho	→Serie 108	data pdf pinout
LM 108(A)F	OP-IC	=LM 108(A)H: Fig.→	10-FLP	Nsc	-	data pdf pinout
LM 108(A)GC	OP-IC	=LM 108(A)H: SMD	20-(P)LCC	Tho	→Serie 108	data pdf pinout
LM 108(A)H,T	OP-IC	Serie 108, lo-power, ±20V, -55...+125°	81	Mot,Nsc,Tho	→Serie 108	data pdf pinout
LM109	R+	Io=500mA Vo:4.7...5.3V P:600mW	3A	Tix	µA109H	data pinout
LM109DA	R+	Io=1A Vo:4.8...5.2V	3O	Sig		data pinout
LM109DB	R+	Io=200mA Vo:4.8...5.2V	3A	Sig		data pinout
LM 109 H,LA	Z-IC	+5V, 1A, 2W, -55...+150°	2e	Nsc,Tho++	µA 109HM	data pdf pinout
LM109K	R+	Io=int Vo:4.7...5.3V Vin:10V P:int	3O	Tho	A109KM,µA109K	data pdf pinout
LM109K03	R+	Io=1.5A Vo:4.85...5.15V P:20W	3O	Ray	A109KM,µA109K	data pinout
LM 109 K	Z-IC	=LM 109H,LA: 20W	23a	Nsc,Tho++	µA 109KM	data pdf pinout
LM110	FOLLOWER	Vs:±18V Vo:±10V Vi0:1.5mV f:250kc	14D,10F,8A	Isi,Nsc		data pinout
LM 110 D,J	OP-IC	=LM 110H: Fig.→	14-DIC	Nsc	→Serie 110	data pinout
LM 110 F	OP-IC	=LM 110H: Fig.→	10-FLP	Nsc	-	data pinout
LM 110 H	OP-IC	Volt.Follower, ±18V, -55...+125°	81	Nsc	→Serie 110	data pinout
LM 110 J-8	OP-IC	=LM 110H: Fig.→	8-DIC	Nsc	→Serie 110	data
LM111	COMP	Vs:36V Vu:200V/mV Vi0:0.7mV	8D,14DS,20	Ray		data pinout
LM 111 D	KOP-IC	=LM 111H: SMD	8-MDIP	Mot,Tix	→Serie 111	data pdf pinout
LM 111DG,FE,J-8,JG	KOP-IC	=LM 111H: Fig.→	8-DIC	Mot,Nsc,++	→Serie 111	data pdf pinout
LM 111 F,U	KOP-IC	=LM 111H: Min	10-FLP	Nsc,Tix	→Serie 111	data pdf pinout
LM 111 GC,FK	KOP-IC	=LM 111H: SMD	20-(P)LCC	Tho,Tix	→Serie 111	data pdf pinout
LM 111 H,L,T	KOP-IC	Serie 111, ±18V, 50mA, -55...+125°	81	Mot,Nsc,++	→Serie 111	data pdf pinout
LM 111 J	KOP-IC	=LM 111H: Fig.→	14-DIC	Nsc,Tix	→Serie 111	data pdf pinout
LM112	LO-POWER	Vs:±20V Vu:110dB Vo:±13V Vi0:0.7mV	14D,10F,8A	Nsc,Ray		data pdf pinout
LM 112 H	OP-IC	Serie 112, lo-power, ±20V, -25...+125°	81	Nsc	→Serie 112	data pinout
LM113-1H	Z+	Io=50mA Vo:1.21...1.232V P:100mW	2A	Nsc		data pinout
LM 113-2H	Z+	Io=50mA Vo:1.195...1.245V P:100mW	2A	Nsc		data pinout
LM 113H, -1H, -2H	Ref-Z-IC	1,220V, <5%(-1H<1%, -2H<2%), -55...+125°	2d	Nsc		data pdf pinout
LM113H	Z+	Io=50mA Vo:1.16...1.28V P:100mW	2A	Nsc		data pdf pinout
LM117	ER+	Io=int Vo:1.2...37V P:int	3OA	Mot		data pinout
LM117AH	ER+	Io=0.8A Vo:1.2...37V Vin:3...40V P:2W	3A	Nsc		data pinout
LM117AK	ER+	Io=2.2A Vo:1.2...37V Vin:3...40V P:20W	3O	Nsc		data pinout
LM117E	ER+	Io=1.5A Vo:1.2...37V P:20W	20C	Nsc		data pinout
LM117H	ER+	Io=1.5A Vo:1.2...57V	3AO	Nsc		data pinout
LM 117 H	Z-IC	+1,2...37V, >0,5A, -55...+150°	2k	Mot,Nsc,Tho	-	data pdf pinout
LM 117 HV(H,K)	Z-IC	=LM 117H: +1,2...57V		Nsc	-	data pdf pinout
LM117K	ER+	Io=1.5A Vo:1.2...37V P:int	3O	Nsc		data pdf pinout
LM 117 K	Z-IC	=LM 117H: >1,5A	23k	Mot,Nsc,Tho	µA 117K...	data pdf pinout
LM 117 LH	Z-IC	=LM 117H: >0,1A	2k	Mot		data pdf pinout
LM 117 MT	Z-IC	=LM 117H: >0,5A	17I	Mot	µA 117U...	data pdf
LM118	HI-SREED		d,8D,14DS,	Tho		data pinout
LM118A	HI-SPEED	Vs:±20V Vu:86dB Vo:±13V Vi0:1mV	8AD	Har		data pinout
LM 118 DG,J,J-8	OP-IC	=LM 118H: Fig.→	8-DIP,DIC	Nsc,Tho,Tix	→Serie 118	data pinout
LM 118 GC	OP-IC	=LM 118H: SMD	20-(P)LCC	Tho	→Serie 118	data pinout
LM 118 H,L	OP-IC	Serie 118, hi-speed, ±20V, -55...+125°	81	Nsc,Tho,Tix	→Serie 118	data pinout
LM 118 J	OP-IC	=LM 118H: Fig.→	14-DIC	Nsc	→Serie 118	data pinout
LM119	2xCOMP	Vs:36V Vu:40V/mV Vo:TTL Vi0:0.7mV	10A,14DS	Sgs		data pinout
LM 119 D,DG,F,J	KOP-IC	=LM 119H: Fig.→	14-DIC	Nsc,Phi,Tho	→Serie 119	data pinout
LM 119 F	KOP-IC	=LM 119H: Min	10-FLP	Nsc,Phi	→Serie 119	data pinout
LM 119 GC	KOP-IC	=LM 119H: SMD	20-(P)LCC	Tho	→Serie 119	data pinout
LM 119 H	KOP-IC	Dual, ±18V, -55...+125°	82	Nsc,Phi,Tho	→Serie 119	data pinout
LM120H-12	R-	Io=200mA Vo:-11.7...-12.3V Vin:-35V	3A	Nsc		data pinout
LM120H-15	R-	Io=200mA Vo:-14.7...-15.3V Vin:-20V	3A	Nsc		data pinout
LM120H-18	R-	Io=200mA Vo:-17.6...-18.4V Vin:-23V	3A	Nsc		data pinout
LM120H-24	R-	Io=200mA Vo:-23.5...-24.5V Vin:-29V	3A	Nsc		data pinout
LM120H-5.0	R-	Io=500mA Vo:-4.9...-5.1V Vin:-10V	3A	Nsc		data pinout
LM120H-5.2	R-	Io=500mA Vo:-5.1...-5.3V Vin:-10V	3A	Nsc		data pinout
LM120H-6	R-	Io=500mA Vo:-5.85...-6.15V Vin:-11V	3A	Nsc		data pinout
LM120H-8	R-	Io=200mA Vo:-7.8...-8.2V Vin:-13V	3A	Nsc		data pinout
LM120H-9	R-	Io=200mA Vo:-8.8...-9.2V Vin:-14V	3A	Nsc		data pinout
LM 120 H-...	Z-IC	=LM 320H-... -55...+150°	2f	Nsc		data pinout
LM120K-12	R-	Io=1A Vo:-11.7...-12.3V Vin:-35V P:int	3O	Nsc		data pinout
LM120K-15	R-	Io=1A Vo:-14.7...-15.3V Vin:-20V P:int	3O	Nsc		data pinout
LM120K-18	R-	Io=1A Vo:-17.6...-18.4V Vin:-23V	3O	Nsc		data pinout
LM120K-24	R-	Io=1A Vo:-23.5...-24.5V Vin:-29V	3O	Nsc		data pinout
LM120K-5.0	R-	Io=1.5A Vo:-4.9...-5.1V Vin:-10V P:int	3O	Nsc		data pinout
LM120K-5.2	R-	Io=1.5A Vo:-5.1...-5.3V Vin:-10V	3O	Nsc		data pinout
LM120K-6	R-	Io=1.5A Vo:-5.85...-6.15V Vin:-11V	3O	Nsc		data pinout
LM120K-8	R-	Io=1.5A Vo:-7.8...-8.2V Vin:-13V	3O	Nsc		data pinout
LM120K-9	R-	Io=1A Vo:-8.8...-9.2V Vin:-14V	3O	Nsc		data pinout
LM 120 K-...	Z-IC	=LM 320K-... -55...+150°	23d	Nsc		data pinout
LM121	PRE-AMP	Vs:±20V Vu:26dB Vi0:0.7mV	14D,10F,8A	Nsc		data pinout
LM121A	PRE-AMP	Vs:±20V Vu:26dB Vi0:0.2mV	14D,10F,8A	Nsc		data pinout

Type	Device	Short Description	Fig.	Manu	Comparision Types	More at
LM 121(A)D	OP-IC	=LM 121(A)H: Fig.→	14-DIP	Nsc	-	data pinout
LM 121(A)F	OP-IC	=LM 121(A)H: Min	10-FLP	Nsc	-	data pinout
LM 121(A)H	OP-IC	Vorverst./Preamp, ±20V, -55...+125°	81	Nsc	-	data pinout
LM 122(H)	LIN-IC	Timer, Ucc:4,5...40V, -55...+125°	82	Nsc		pdf pinout
LM123	R+	Io=int Vo:4.7...5.3V Vin:7.5V P:int	3PO	Tho		data pdf pinout
LM123AK	R+	Io=3A Vo:4.7...5.3V Vin:7.5V P:int	3O	Nsc		data pinout
LM 123 ISP-3	Z-IC	=LM 123(A)K: -40...+150°	18b	Tho		data pdf pinout
LM 123(A)K	Z-IC	+5V, 3A, 6%(A=2%), -55...+150°	23a	Mot,Nsc,Tho	-	data pdf pinout
LM124	4xOP	Vs:32V Vu:100V/mV Vi0:3mV	14DSF,20Cd	Nsc		data pinout
LM124A	4xOP	Vs:32V Vu:50V/mV Vi0:2mV	14DS,20CF	Tix		data pinout
LM 124(A)D	OP-IC	=LM 124(A)DP,N: SMD	14-MDIP	Tho,Tix	→Serie 124	data pinout
LM 124(A)DG,F,J	OP-IC	=LM 124(A)DP,N: Fig.→	14-DIC	Mot,Nsc,++	→Serie 124	data pinout
LM 124(A)DP,N	OP-IC	Quad, Serie 124, ±16V, -55...+125°	14-DIP	Tho,Tix	→Serie 124	data pinout
LM 124(A)FK	OP-IC	=LM 124(A)DP,N: SMD	20-(P)LCC	Tix	→Serie 124	data pinout
LM 124(A)GC	OP-IC	=LM 124(A)DP,N: SMD	20-(P)LCC	Tho	→Serie 124	data pinout
LM125H	R±	Io=100mA Vo:±14.8...±15.2V P:2W	10A	Nsc		data pinout
LM 125 H	Z-IC	±15V, 0,1A, -55...+125°	82	Nsc	-	data pinout
LM126H	R±	Io=100mA Vo:±11.8...±12.2V P:int	10A	Nsc		data pinout
LM 126 H	Z-IC	±12V, 0,1A, -55...+125°	82	Nsc	-	data pinout
LM127H	R-+	Vo:-11.8...-12.2V		Nsc		data pinout
LM129	Z+	Vo:6.7...7.2V	2A	Nsc		data pdf pinout
LM 129AH,BH,CH,DH	Ref-Z-IC	6,9V, 3%, <0,001...0,01%/°C, -55...+125°	2d	Nsc		data pinout
LM 131(A)H	LIN-IC	V/F-Converter, 1Hz...100kHz, -55...+125°	81	Nsc,Tho	-	
LM133K	ER-	Io=3A Vo:-1.2...-32V P:int	3O	Nsc		data pdf pinout
LM 134 H-3	LIN-IC	Temp. Sensor, ±3° Temp. Error	2(Reg+-)	Nsc		pdf pinout
LM 134 H-6	LIN-IC	Temp. Sensor, ±6° Temp. Error	2(Reg+-)	Nsc		pdf pinout
LM 134 H	LIN-IC	Adj. Current Source, 1µ...10mA, ±3%	2(Reg+-)	Nsc		pdf pinout
LM_134 Z(-...)	LIN-IC	Adj. Current Source, 1µ...10mA, ±3%	7(-Reg+)	Tho		pdf pinout
LM 135(A)D	LIN-IC	=LM 135(A)H: SMD	8-MDIP	Tho		pdf
LM 135(A)H	LIN-IC	Temp. Sensor, 0,4...5mA, -55...+150°	2(Reg+-)	Nsc	-	pdf
LM 135(A)Z	LIN-JC	=LM 135(A)H: Fig.→	7(-+Reg)	Tho	-	pdf
LM136AH-2.5	Z+	Io=10mA Vo:2.465...2.515V	3A	Nsc		data pinout
LM136AH-5.0	Z+	Io=10mA Vo:4.95...5.05V	3A	Nsc		data pinout
LM136H-2.5	Z+	Io=10mA Vo:2.44...2.54V	3A	Nsc		data pinout
LM 136(A)H-2.5	Ref-Z-IC	2,5V(1mA), ±2%(A=±1%), -55...+150°	2(RegKA)	Nsc	-	data pinout
LM136H-5.0	Z+	Io=10mA Vo:4.9...5.1V	3A	Nsc		data pinout
LM 136(A)H-5.0	Ref-Z-IC	5,0V(1mA), ±2%(A=±1%), -55...+150°	2(RegKA)	Nsc	-	data pinout
LM 136(A)Z	Ref-Z-IC	=LM 136H-5.0: Fig.→	7(AKReg)	Tho		data
LM137	ER-	Io=int Vo:-1.2...-37V P:int	3O	Mot		data pinout
LM137H	ER-	Io=int Vo:-1.2...-37V P:int	3A	Tho		data pdf pinout
LM 137 H	Z-IC	-1,2...-37V, >0,5A, 2W, -55...+150°	2I	Mot,Nsc,Tho	-	data pdf pinout
LM 137 HV(H,K)	Z-IC	=LM 137H: -1,2...-47V		Nsc		data pdf pinout
LM137HVH	ER-	Io=200mA Vo:-1.2...-47V P:int	3A	Nsc		data pdf pinout
LM137HVK	ER-	Io=1.5A Vo:-1.2...-47V P:int	3O	Nsc		data pinout
LM137K	ER-	Io=1.5A Vo:-1.2...-37V P:20W	3O	Mot,Nsc		data pdf pinout
LM 137 K	Z-IC	=LM 137H: >1,5A, 20W	23m	Mot,Nsc,Tho	-	data pdf pinout
LM 137 MR	Z-IC	=LM 137H:	22m	Mot		data pdf pinout
LM 137 MT	Z-IC	=LM 137H: >0,5A	17n	Mot	-	data pdf
LM138AK	ER+	Io=8A Vo:1.2...32V P:int	3O	Nsc		data pdf pinout
LM138K	ER+	Io=int Vo:1.2...32V P:int	3O	Nsc,Tho		data pdf pinout
LM 138 K	Z-IC	+1,2...32V, >5A, -55...+150°	23k	Nsc,Tho	µA 138K...	data pdf pinout
LM139	4xSS-COMP	Vs:36V Vu:200V/mV Vi0:<2mV	14DS,20CF	Ray		data pinout
LM139A	4xSS-COMP	Vs:36V Vu:200V/mV Vi0:<5mV	14DS,20CF	Ray		data pinout
LM 139(A)D	KOP-IC	=LM 139(A)DG,F,J,N: SMD	14-MDIP	Mot,Tho,Tix	→Serie 139	data pdf pinout
LM 139(A)DG,F,J,N	KOP-IC	Quad, Serie 139, ±18V, -55...+125°	14-DIC,DIP	Mot,Nsc,++	→Serie 139	data pdf pinout
LM 139(A)F,W	KOP-IC	=LM 139(A)DG,F,J,N: Min	14-FLP	Nsc,Tix	→Serie 139	data pdf pinout
LM 139(A)GC,FK	KOP-IC	=LM 139(A)DG,F,J,N: SMD	20-(P)LCC	Tho	→Serie 139	data pdf pinout
LM140AK-10	R+	Io=2.4A Vo:9.6...10.4V Vin:17V	3O	Nsc		data pinout
LM140AK-12	R+	Io=1.5A Vo:11.75...12.25V Vin:19V	3O	Mot,Nsc		data pinout
LM140AK-15	R+	Io=1.5A Vo:14.7...15.3V Vin:23V P:int	3O	Mot,Nsc		data pinout
LM140AK-18	R+	Io=2.4A Vo:17.3...18.7V Vin:27V	3O	Nsc		data pinout
LM140AK-24	R+	Io=2.4A Vo:23...25V Vin:33V	3O	Nsc		data pinout
LM140AK-5	R+	Io=1.5A Vo:4.8...5.2V Vin:7.5...20V	3O	Nsc		data pinout
LM140AK-5.0	R+	Io=1.5A Vo:4.9...5.1V P:int	3O	Mot,Nsc		data pinout
LM140AK-6	R+	Io=2.4A Vo:5.76...6.24V Vin:11V	3O	Nsc		data pinout
LM140AK-8	R+	Io=2.4A Vo:7.7...8.3V Vin:14V	3O	Nsc		data pinout
LM140AT-10	R+	Io=2.4A Vo:9.6...10.4V Vin:17V	3P	Nsc		data pinout
LM140AT-12	R+	Io=2.4A Vo:11.5...12.5V Vin:19V	3P	Nsc		data pinout
LM140AT-15	R+	Io=2.4A Vo:14.4...15.6V Vin:23V	3P	Nsc		data pinout
LM140AT-18	R+	Io=2.4A Vu:17.3...18.7V Vin:27V	3P	Nsc		data pinout
LM140AT-24	R+	Io=2.4A Vo:23...25V Vin:33V	3P	Nsc		data pinout
LM140AT-5	R+	Io=2.4A Vo:4.8...5.2V Vin:10V	3P	Nsc		data pinout
LM140AT-6	R+	Io=2.4A Vo:5.76...6.24V Vin:11V	3P	Nsc		data pinout
LM140AT-8	R+	Io=2.4A Vo:7.7...8.3V Vin:14V	3P	Nsc		data pinout
LM140H-12	R+	Io=100mA Vo:11.5...12.5V P:int	3A	Nsc		data pinout
LM140H-15	R+	Io=100mA Vo:14.4...15.6V P:int	3A	Nsc		data pinout
LM140H-24	R+	Io=100mA Vo:23...25V Vin:33V	3A	Nsc		data pinout
LM140H-5.0	R+	Io=100mA Vo:4.9...5.1V Vin:7.2...20V	3A	Nsc		data pinout
LM140H-6.0	R+	Io=100mA Vo:5.76...6.24V Vin:11V	3A	Nsc		data pinout
LM140H-8.0	R+	Io=100mA Vo:7.7...8.3V Vin:14V	3A	Nsc		data pinout
LM140K-10	R+	Io=2.4A Vo:9.5...10.5V Vin:17V	3O	Nsc		data pinout
LM140K-12	R+	Io=int Vo:11.5...12.5V Vin:19V P:int	3O	Mot		data pinout
LM140K-12/883	R+	Io=1.5A Vo:11.5...12.5V Vin:19V P:int	3O	Nsc		data pinout
LM140K-15	R+	Io=int Vo:14.4...15.6V Vin:23V P:int	3O	Mot		data pinout
LM140K-15/883	R+	Io=1.5A Vo:14.4...15.6V Vin:23V P:int	3O	Nsc		data pinout

Type	Device	Short Description	Fig.	Manu	Comparision Types	More at
LM140K-18	R+	Io=2.4A Vo:17.1...18.9V Vin:27V	3O	Nsc		data pinout
LM140K-24	R+	Io=2.4A Vo:22.8...25.2V Vin:33V	3O	Nsc		data pinout
LM140K-5	R+	Io=1.5A Vo:4.75...5.25V Vin:8...20V	3O	Nsc		data pinout
LM140K-5.0	R+	Io=1.5A Vo:4.8...5.2V Vin:10V P:int	3O	Mot,Nsc		data pinout
LM140K-6	R+	Io=2.4A Vo:5.7...6.3V Vin:11V	3O	Nsc		data pinout
LM140K-6.0	R+	Io=int Vo:5.75...6.25V Vin:11V P:int	3O	Mot		data pinout
LM140K-8	R+	Io=2.4A Vo:7.6...8.4V Vin:14V	3O	Nsc		data pinout
LM140K-8.0	R+	Io=int Vo:7.7...8.3V Vin:14V P:int	3O	Mot		data pinout
LM 140(A)K-...	Z-IC	=LM 340(A)K-...: -55...+150°	23a	Mot,Nsc		data pinout
LM140LAH-10	R+	Io=100mA Vo:9.6...10.4V Vin:16V	3A	Nsc		data pinout
LM140LAH-12	R+	Io=100mA Vo:11.5...12.5V P:int	3A	Nsc		data pinout
LM140LAH-15	R+	Io=100mA Vo:14.4...15.6V P:int	3A	Nsc		data pinout
LM140LAH-18	R+	Io=100mA Vo:17.3...18.7V Vin:27V	3A	Nsc		data pinout
LM140LAH-24	R+	Io=100mA Vo:23...25V Vin:33V	3A	Nsc		data pinout
LM140LAH-5	R+	Io=100mA Vo:4.8...5.2V Vin:7.2...20V	3A	Nsc		data pinout
LM140LAH-5.0	R+	Io=100mA Vo:4.9...5.1V Vin:7.2...20V	3A	Nsc		data pinout
LM140LAH-6	R+	Io=100mA Vo:5.76...6.24V Vin:11V	3A	Nsc		data pinout
LM140LAH-8	R+	Io=100mA Vo:7.7...8.3V Vin:14V	3A	Nsc		data pinout
LM 140 LAH-...	Z-IC	=LM 340 LAH-...: -55...+150°	2e	Nsc	-	data pinout
LM140T-10	R+	Io=2.4A Vo:9.5...10.5V Vin:17V	3P	Nsc		data pinout
LM140T-12	R+	Io=2.4A Vo:11.4...12.6V Vin:19V	3P	Nsc		data pinout
LM140T-15	R+	Io=2.4A Vo:14.25...15.75V Vin:23V	3P	Nsc		data pinout
LM140T-18	R+	Io=2.4A Vo:17.1...18.9V Vin:27V	3P	Nsc		data pinout
LM140T-24	R+	Io=2.4A Vo:22.8...25.2V Vin:33V	3P	Nsc		data pinout
LM140T-5	R+	Io=2.4A Vo:4.75...5.25V Vin:10V	3P	Nsc		data pinout
LM140T-6	R+	Io=2.4A Vo:5.7...6.3V Vin:11V	3P	Nsc		data pinout
LM140T-8	R+	Io=2.4A Vo:7.6...8.4V Vin:14V	3P	Nsc		data pinout
LM 140(A)T-...	Z-IC	=LM 340(A)T-...: -55...+150°	17b	Nsc	-	data pinout
LM143	HI-VOLT	Vs:±40V Vu:105dB Vo:±35V Vi0:2mV	14D,10F,8A	Nsc		data pinout
LM143A	HI-VOLT	Vs:±55V Vu:106dB Vo:>±35V Vi0:2mV	8AD	Har		data pinout
LM 143 H	OP-IC	Serie 136, hi-volt, ±40V, -55...+125°	81	Nsc	→Serie 136	data pdf pinout
LM144	HI-VOLT	Vs:±40V Vu:105dB Vo:±25V Vi0:2mV	14D,10F,8A	Nsc		data pinout
LM 144 H	OP-IC	hi-volt, ±40V, 30V/µs, -55...+125°	81	Nsc		data pinout
LM145K-5.0	R-	Io=3A Vo:-4.9...-5.1V Vin:-7.5V P:int	3O	Nsc		data pinout
LM145K-5.2	R-	Io=3A Vo:-5.1...-5.3V Vin:-7.5V P:int	3O	Nsc		data pinout
LM 145 K-5.2	Z-IC	-5,2V, 3A, 25W, -55...+150°	23d	Nsc	-	data pinout
LM 145 K-5.0	Z-IC	-5V, 3A, 25W, -55...+150°	23d	Nsc	-	data pinout
LM146	4xPROG	Vs:±22V Vu:1000V/mV Vo:±14V f:1.2Mc	16DS,20Cd	Mit		data pinout
LM 146 DG,J	OP-IC	Quad, progr., ±22V, 1,2MHz, -55...+125°	16-DIP,DIC	Nsc,Tho		data pdf pinout
LM 146 GC	OP-IC	=LM 146DG,J: SMD	20-(P)LCC	Tho	-	data pdf pinout
LM148	4xOP	Vs:±22V Vu:104dB Vo:±12V Vi0:1mV	14DS,20CFd	Isi		data pinout
LM 148 DG,J	OP-IC	Quad, Serie 124, ±22V, 1MHz,-55...+125°	14-DIC	Nsc,Mot,++	→Serie 124	data pdf pinout
LM 148 FK	OP-IC	=LM 148DG,J: SMD	20-(P)LCC	Tix	→Serie 124	data pdf pinout
LM 148 GC	OP-IC	=LM 148DG,J: SMD	20-(P)LCC	Tho	→Serie 124	data pdf pinout
LM149	4xOP	Vs:±22V Vu:104dB Vo:±12V Vi0:1mV	14DFd	Isi		data pinout
LM 149 D,DG,J	OP-IC	Quad, Serie 124, ±22V, 4MHz,-55...+125°	14-DIC	Nsc,Tho	→Serie 124	data pinout
LM150	ER+	Io=3A Vo:1.2...33V P:int	3O	Nsc		data pdf pinout
LM150K	ER+	Io=int Vo:1.2...33V P:int	3O	Mot		data pdf pinout
LM 150(A)K	Z-IC	+1,2...33V, 3A, -55...+150°	23k	Mot,Nsc	µA 150K...	data pdf pinout
LM151	2xCOMP	Vs:-7/14V Vu:65dB Vo:TTL Vi0:0.6mV	14D	Nsc		data pinout
LM155	FET	Vs:±22V Vu:106dB Vo:±12V Vi0:3mV	8A	Nsc		data pinout
LM156	FET	Vs:±22V Vu:106dB Vo:±12V Vi0:3mV	8A	Nsc		data pinout
LM156A	FET	Vs:±22V Vu:106dB Vo:±12V Vi0:1mV	8A	Nsc		data pinout
LM157	FET	Vs:±22V Vu:106dB Vo:±12V Vi0:3mV	8A	Nsc		data pinout
LM157A	FET	Vs:±22V Vu:106dB Vo:±12V Vi0:1mV	8A	Nsc		data pinout
LM158	2xLO-POWER	Vs:32V Vu:100V/mV Vo:3.5/0V Vi0:2mV	8SD,20CdA	Sgs		data pdf pinout
LM158A	2xLO-POWER	Vs:32V Vu:100V/mV Vo:3.5/0V Vi0:1mV	8S,20CAD	Sgs		data pdf pinout
LM 158 AWG/883	OP-IC	SMD, Dual, Low Power OPAmp -55...125°V, 1MHz, ±1.5/+3...±16/+32V	14-MDIP	Nsc	-	pdf
LM 158(A)D,M	OP-IC	=LM 158(A)H: SMD	8-MDIP	Nsc,Tho,Tix	→Serie 158	data pinout
LM 158(A)DG,N,P	OP-IC	=LM 158(A)H: Fig.→	8-DIP	Tho,Tix	→Serie 158	data pinout
LM 158(A)FE,J,JG	OP-IC	=LM 158(A)H: Fig.→	8-DIC	Mot,++	→Serie 158	data pinout
LM 158 FK	OP-IC	=LM 158(A)H: SMD	20-(P)LCC	Tix	→Serie 158	data pinout
LM 158(A)GC	OP-IC	=LM 158(A)H: SMD	20-(P)LCC	Tho	→Serie 158	data pinout
LM 158(A)H,L	OP-IC	Dual, Serie 158, ±16V, 1MHz,-55...+125°	81	Mot	→Serie 158	data pinout
LM 158(A)U	OP-IC	=LM 158(A)H: Fig.→	10-FLP	Tix	→Serie 158	data pinout
LM 158 W(D/S)	IC	Low power dual operational amplifiers, -55 to +125	MDIP	Stm	-	pdf pinout
LM 158 WN	IC	Low power dual operational amplifiers, -55 to +125	DIC	Stm	-	pdf pinout
LM 158 WP	IC	Low power dual operational amplifiers, -55 to +125	TSOP	Stm	-	pdf pinout
LM 158 (A)H	OP-IC	Dual, Low Power OPAmp -55...125°C, 1MHz, ±1.5/+3...±16/+32V	81	Nsc	-	pdf
LM 158 (A)J	OP-IC	Dual, Low Power OPAmp -55...125°C, 1MHz, ±1.5/+3...±16/+32V	8-DIC	Nsc	-	pdf
LM159	2xHI-SPEED	Vs:±11V Vu:72dB Vo:10.3/0V f:400Mc	14DS	Nsc		data pinout
LM 159 J	OP-IC	Dual, hi-speed, 22V, 400MHz,-55...+125°	14-DIC	Nsc		data pinout
LM160	FAST-COMP	Vs:±8V Vo:TTL Vi0:2mV	14DF,8ADS	Nsc		data pinout
LM 160 D,J-14	KOP-IC	=LM 160H: Fig.→	14-DIC	Nsc	µA 760...	data pinout
LM 160 F	KOP-IC	=LM 160H: Min	14-FLP	Nsc	µA 760...	data pinout
LM 160 H	KOP-IC	hi-speed, ±8V, <20ns, -55...+125°	81	Nsc	µA 760...	data pinout
LM161	FAST-COMP	Vs:±16V Vu:70dB Vo:TTL Vi0:1mV	14DF,10AS	Nsc		data pinout
LM 161 D,J	KOP-IC	=LM 161H: Fig.→	14-DIC	Nsc	SE 529...	data pinout
LM 161 F	KOP-IC	=LM 161H: Min	14-FLP	Nsc		data pinout

Type	Device	Short Description	Fig.	Manu	Comparision Types	More at
LM 161 H	KOP-IC	hi-speed, ±16V, <20ns, -55...+125°	81	Nsc	SE 529...	data pinout
LM 163(A)D	OP-IC	=LM 163(A)H: Fig.→	16-DIC	Nsc	-	
LM163(A)H-10...1000	OP-IC	hi-prec, ±18V, -55...+125°, A: <InOffs.	81	Nsc	-	
LM168B	R+	Io=50mA P:600mW	8A	Nsc		data pinout
LM169	R+	Io=50mA P:600mW	8A	Nsc		data pdf pinout
LM185BH	EZ+	Io=20mA Vo:1.24...5.3V	3A	Nsc		data pdf pinout
LM185BXH	EZ+	Io=20mA Vo:1.24...5.3V	3A	Nsc		data pdf pinout
LM185BXH1.2	Z+	Io=20mA Vo:1.223...1.247V	2A	Nsc		data pinout
LM185BXH2.5	Z+	Io=20mA Vo:2.462...2.538V	2A	Nsc		data pinout
LM185BYH	EZ+	Io=20mA Vo:1.24...5.3V	3A	Nsc		data pinout
LM185BYH1.2	Z+	Io=20mA Vo:1.223...1.247V	2A	Nsc		data pinout
LM185BYH2.5	Z+	Io=20mA Vo:2.462...2.538V	2A	Nsc		data pinout
LM185D-1.2	Z+	Io=20mA Vo:1.22...1.24V P:int	8S	Tix		data pinout
LM185D-2.5	Z+	Io=20mA Vo:2.46...2.53V P:int	8S	Tix		data pinout
LM185H-1.2	Z+	Io=20mA Vo:1.223...1.247V	2A	Nsc		data pinout
LM 185 Z-1.2	Ref-Z-IC	1,235V, 1%, 0,2<0,6Ω, -55...+125°	2(K-A)	Nsc	-	data pinout
LM185H-2.5	Z+	Io=20mA Vo:2.462...2.538V	2A	Nsc		data pinout
LM 185 Z-2.5	Ref-Z-IC	2,5V, 1%, 0,2<0,6Ω, -55...+125°	2(K-A)	Nsc	-	data pinout
LM185LP-1.2	Z+	Io=20mA Vo:1.22...1.24V P:int	3N	Tix		data pinout
LM185LP-2.5	Z+	Io=20mA Vo:2.46...2.53V P:int	3N	Tix		data pinout
LM192	OP	Vs:±16V Vu:100dB Vo:±14V Vi0:2mV	8AD	Nsc		data pinout
LM 192 H	OP/KOP-IC	1x OP + 1x KOP, ±16V, -55...+125°	81	Nsc	-	data pinout
LM 192 J	OP/KOP-IC	=LM 192H: Fig.→	8-DIC	Nsc	-	data pinout
LM193	2xCOMP	Vs:36V Vu:200V/mV Vi0:2mV	8ADS,20C	Tix		data pinout
LM193A	2xCOMP	Vu:200V/mV Vo:3.5/0V Vi0:<2mV	8SD,20CA	Sgs		data pinout
LM 193(A)D	KOP-IC	=LM 193(A)H: SMD	8-MDIP	Tho	→Serie 193	data pdf pinout
LM193(A)J/DG,F,J,N	KOP-IC	=LM 193(A)H: Fig.→	8-DIC,DIP	Mot,Nsc,++	→Serie 193	data pdf pinout
LM 193(A)GC,FK	KOP-IC	=LM 193(A)H: SMD	20-(P)LCC	Tho,Tix	→Serie 193	data pdf pinout
LM 193(A)H,L	KOP-IC	Dual, lo-power, ±18V, -55...+125°	81	Mot,Nsc,++	→Serie 193	data pdf pinout
LM 193(A)W	KOP-IC	=LM 193(A)H: Min	10-FLP	Tix	→Serie 193	data pinout
LM 194(H)	LIN-IC	2xNPN-T., 40V, 20mA, B>500, -55...+125°	81	Nsc	-	pdf
LM 195 H	LIN-IC	NPN-Darl, 42V, 1A, 500ns, -55...+150°	2a*	Nsc	-	pdf
LM 195 K	LIN-IC	=LM 195H: Fig.→	23e*	Nsc	-	pdf
LM196K	ER+	Io=10A Vo:1.25...15V P:int	3O	Nsc		data pdf pinout
LM 196 K	Z-IC	+1,25...15V, 10A, 70W, -55...+150	23l	Nsc		data pinout
LM199	Z+	Io=10mA Vo:6.8...7.1V	4A	Ltc		data pinout
LM199AH	Z+	Io=10mA Vo:6.8...7.1V P:300mW	4A	Nsc		data pinout
LM 199(A)H(-20)	Ref-Z-IC	6,95V, +1/-2%, 0,5Ω, -55...+125°	5(-+AK)	Nsc	-	data pinout
LM200	ER+	Io=12mA Vo:2..30V P:800mW	10F,8A	Isi,Nsc		data pinout
LM 200 ...	Z-IC	~LM 205...	81	Nsc	→LM 205...	data pinout
LM201	UNI	Vs:±22V Vu:103.5dB Vo:±13V Vi0:2mV	14D,10F,8A	Nsc,Sig		data pinout
LM201A	UNI	Vs:±22V Vu:104dB Vo:±13V Vi0:0.7mV	8DS,14D	Tho		data pinout
LM 201(A)D,JG,P	OP-IC	=LM 101(A)H: -25...+85°	8-DIP,DIC	Tix	→Serie 101	data pinout
LM 201(A)D	OP-IC	=LM 101(A)H: SMD, -25...+85°	8-MDIP	Mot,Tho	→Serie 101	data pinout
LM 201(A)DP,J,N	OP-IC	=LM 101(A)H: -25...+85°	8-DIP,DIC	Mot,Nsc,Tho	→Serie 101	data pinout
LM 201(A)H,L	OP-IC	=LM 101(A)H: -25...+85°	81	Mot,Nsc,Tho	→Serie 101	data pinout
LM 201(A)J-14,N-14	OP-IC	=LM 101(A)H: -25...+85°	14-DIC,DIP	Nsc	→Serie 101	data pinout
LM 201(A)W	OP-IC	=LM 101(A)H: -25...+85°	14-FLP	Tix	-	data pinout
LM202	FOLLOWER	Vs:±18V Vo:±13V Vi0:3mV	8A	Nsc		data pinout
LM 202 H	OP-IC	=LM 102H: -25...+85°	81	Nsc	→Serie 102	data pinout
LM204	ER-	Vo:-0.015...-40V P:500mW	10FA	Nsc		data pinout
LM 204 H,L	Z-IC	=LM 104H,L: -25...+85°	82	Nsc,Tix	LM 104...	data pinout
LM204J	ER-	Io=20mA Vo:-0.015...-40V P:1W	14D,10A	Tix		data pinout
LM 204 J,N	Z-IC	=LM 104H,L: -25...+85°	14-DIC,DIP	Tix	LM 104...,	data pinout
LM205	ER+	Io=12mA Vo:4.5...40V P:800mW	8A,10FD	Nsc		data pinout
LM205F	ER+	Io=12mA Vo:4.5..40V P:800mW	10F	Tdy		data pinout
LM205H	ER+	Io=375mA Vo:4.5..40V P:500mW	8A	Tho		data pinout
LM 205 H,L	Z-IC	=LM 105H: -25...85° (Tho: +100°)	81	Nsc,Tho,Tix	LM 105...	data pinout
LM 205 JG,P	Z-IC	=LM 105H: -25...+85°	8-DIC,DIP	Tix	LM 105...	data pinout
LM206	COMP	Vs:±15V Vu:92dB Vo:TTL Vi0:0.5mV	14F,8ADD	Nsc		data pinout
LM 206 D...W	KOP-IC	=LM 106..: -25...+85°	81	Nsc,Tix	→Serie 106	data pinout
LM207	UNI	Vs:±22V Vu:200V/mV Vi0:0.6mV	8S,14D,10F	Tix		data pinout
LM 207 D,J,J-14	OP-IC	=LM 107H: -25...+85°	14-DIP,DIC	Nsc,Tix	→Serie 107	data pdf pinout
LM 207 F,U	OP-IC	=LM 107H: -25...+85°	10-FLP	Nsc,Tix	-	data pinout
LM 207 H,T	OP-IC	=LM 107H: -25...+85°	81	Nsc,Phi	→Serie 107	data pinout
LM 207 J,JG,N,P	OP-IC	=LM 107H: -25...+85°	8-DIC,DIP	Nsc,Tix	→Serie 107	data pdf pinout
LM 207 W	OP-IC	=LM 107H: -25...+85°	14-FLP	Tix	-	pdf
LM208	LO-POWER	Vs:±20V Vu:110dB Vo:±13V Vi0:0.7mV	10F,14D,8S	Ray		data pinout
LM208A	LO-POWER	Vs:±20V Vu:110dB Vo:±13V Vi0:0.3mV	10F,14D,8S	Ray		data pinout
LM 208(A)D	OP-IC	=LM 108(A)H: SMD, -25...+85°	8-MDIP	Mot	→Serie 108	data pdf pinout
LM 208(A)H,T	OP-IC	=LM 108(A)H: -25...+85°	81	Mot,Nsc,Tho	→Serie 108	data pinout
LM 208(A)N,J-8	OP-IC	=LM 108(A)H: -25...+85°	8-DIP,DIC	Mot,Nsc	→Serie 108	data pinout
LM209DA	R+	Io=1A Vo:4.8...5.2V	3O	Nsc		data pinout
LM209DB	R+	Io=200mA Vo:4.8...5.2V	3A	Nsc		data pinout
LM209H	R+	Io=int Vo:4.7...5.3V Vin:10V P:int	3A	Sig,Tho	µA109H	data pinout
LM 209 H,LA	Z-IC	=LM 109H,LA: -25...+150°	2e	Nsc,Tho++	LM 109.., µA 209HM	data pinout
LM209K	R+	Io=int Vo:4.7...5.3V Vin:10V P:int	3O	Sig,Tho	µA109K,µA209K	data pinout
LM 209 K	Z-IC	=LM 109H,LA: 20W, -25...+150°	23a	Nsc,Tho++	LM 109.., µA 209KM	data pinout
LM209LA	R+	Io=500mA Vo:4.7...5.3V P:600mW	3A	Tix	µA109H	data pinout
LM210	FOLLOWER	Vs:±18V Vo:±10V Vi0:1.5mV f:250kc	14D,10F,8A	Isi,Nsc		data pinout
LM 210 D,J	OP-IC	=LM 110H: -25...+85°	14-DIC	Nsc	→Serie 110	data pinout
LM 210 F	OP-IC	=LM 110H: -25...+85°	10-FLP	Nsc	-	data pinout
LM 210 H	OP-IC	=LM 110H: -25...+85°	81	Nsc	→Serie 110	data pinout
LM 210 J-8	OP-IC	=LM 110H: -25...+85°	8-DIC	Nsc	→Serie 110	data
LM211	COMP	Vs:36V Vu:200V/mV Vo:TTL Vi0:0.7mV	14D,8S,10F	Phi		data pinout

Type	Device	Short Description	Fig.	Manu	Comparision Types	More at
LM 211 D,J,N14	KOP-IC	=LM 111H: -25...+85°	8-DIP	Nsc,Tix,++	→Serie 111	data pdf pinout
LM 211 D	KOP-IC	=LM 111H: SMD, -25...+85°	8-MDIP	Mot,Nsc,++	→Serie 111	data pdf pinout
LM 211 F,U	KOP-IC	=LM 111H: Min, -25...+85°	10-FLP	Nsc,Tix	→Serie 111	data pdf pinout
LM 211 FE,J-8,JG	KOP-IC.	=LM 111H: -25...+85°	8-DIC	Mot,Nsc,++	→Serie 111	data pdf pinout
LM 211 H,L,T	KOP-IC	=LM 111H: -25...+85°	81	Mot,Nsc,++	→Serie 111	data pdf pinout
LM 211 N,P	KOP-IC	=LM 111H: -25...+85°	8-DIP	Thc,Tix,++	→Serie 111	data pdf pinout
LM212	LO-POWER	Vs:±20V Vu:110dB Vo:±13V Vi0:0.7mV	14D,10F,8A	Nsc		data pdf pinout
LM 212 H	OP-IC	=LM 112H: -25...+85°	81	Nsc	→Serie 112	data pinout
LM216	LO-BIAS	Vs:±20V Vu:86dB Vo:±12V Vi0:10mV	8A	Nsc		data pinout
LM216A	LO-BIAS	Vs:±20V Vu:92dB Vo:±12V Vi0:3mV	8A	Nsc		data pinout
LM 216(A)H	OP-IC	Io-bias, ±20V, -25...+85°	81	Nsc		data pinout
LM217	ER+	Io=int Vo:1.2...37V P:int	3AO	Mot		data pinout
LM217H	ER+	Io=int Vo:1.2...37V P:int	3A	Tho		data pinout
LM 217 H	Z-IC	=LM 117H: -25...+150°	2k	Mot,Nsc,Tho	LM 117...	data pinout
LM 217 HV(H,K)	Z-IC	=LM 117H: +1,2...57V		Nsc	LM 117HV...	data pinout
LM217K	ER+	Io=int Vo:1.2...37V P:int	3OP	Tho		data pinout
LM 217 K	Z-IC	=LM 117H: -25...+150°	23k	Mot,Nsc,Tho	LM 117..., µA 217K...	data pinout
LM 217 KC	Z-IC	=LM 117H: >1,5A, -25...+150°	17I	Tix	LM 117..., µA 217U...	data pinout
LM217LA	ER+	Io=500mA Vo:1.2...37V P:600mW	3A	Tix		data pinout
LM 217 LH	Z-IC	=LM 117LH: -25...+150°	2k	Mot	LM 117...	data pinout
LM 217 MDT	Z-IC	1.2 to 37 V adjustable voltage regulator	17I	Mot		data pdf pinout
LM 217 MT	Z-IC	=LM 117MT: -25...+150°	17I	Mot	LM 117..., µA 217U...	data pdf
LM218	HI-SPEED	Vs:±20V Vu:106dB Vo:±13V Vi0:2mV	14D,8S,10F	Nsc		data pinout
LM 218 D,J	OP-IC	=LM 118H: Fig.→	14-DIC	Nsc	→Serie 118	data pdf pinout
LM 218 D	OP-IC	=LM 118H: SMD, -40...+105°	8-MDIP	Tho,Tix	→Serie 118	data pdf pinout
LM 218 H,L	OP-IC	=LM 118: -25...+85°	81	Nsc,Tho,Tix	→Serie 118	data pdf pinout
LM 218 JG,J-8,N,P	OP-IC	=LM 118H: -25...+85°(Tho: -40...+105°)	8-DIP	Nsc,Tho,Tix	→Serie 118	data pdf pinout
LM219	2xCOMP	Vs:36V Vu:40V/mV Vo:TTL Vi0:0.7mV	14DS,10FA	Sgs		data pdf pinout
LM 219 D	OP-IC	=LM 119H: SMD, -40...+105°	14-MDIP	Tho	→Serie 119	data pdf pinout
LM 219 DP,N,F,J	KOP-IC	=LM 119H: -25...+85°	14-DIP,DIC	Nsc,Phi,Tho	→Serie 119	data pinout
LM 219 F	KOP-IC	=LM 119H: -25...+85°	10-FLP	Nsc,Phi	→Serie 119	data pinout
LM 219 H,L	KOP-IC	=LM 119: -25...+85°	82	Nsc,Tho	→Serie 119	data pinout
LM220H-15	R-	Io=200mA Vo:-14.7...-15.3V Vin:-20V	3A	Nsc		data pinout
LM220H-18	R-	Io=200mA Vo:-17.6...-18.4V Vin:-23V	3A	Nsc		data pinout
LM220H-24	R-	Io=200mA Vo:-23.5...-24.5V Vin:-29V	3A	Nsc		data pinout
LM220H-5.0	R-	Io=500mA Vo:-4.9...-5.1V Vin:-10V	3A	Nsc		data pinout
LM220H-5.2	R-	Io=500mA Vo:-5.1...-5.3V Vin:-10V	3A	Nsc		data pinout
LM220H-6	R-	Io=500mA Vo:-5.85...-6.15V Vin:-11V	3A	Nsc		data pinout
LM220H-8	R-	Io=200mA Vo:-7.8...-8.2V Vin:-13V	3A	Nsc		data pinout
LM220H-9	R-	Io=200mA Vo:-8.8...-9.2V Vin:-14V	3A	Nsc		data pinout
LM 220 H-...	Z-IC	=LM 320H-...: -25...+150°	2f	Nsc	LM 120H-...	data pinout
LM220K-15	R-	Io=1A Vo:-14.7...-15.3V Vin:-20V	3O	Nsc		data pinout
LM220K-18	R-	Io=1A Vo:-17.6...-18.4V Vin:-23V	3O	Nsc		data pinout
LM220K-24	R-	Io=1A Vo:-23.5...-24.5V Vin:-29V	3O	Nsc		data pinout
LM220K-5.0	R-	Io=1.5A Vo:-4.9...-5.1V Vin:-10V	3O	Nsc		data pinout
LM220K-5.2	R-	Io=1.5A Vo:-5.1...-5.3V Vin:-10V	3O	Nsc		data pinout
LM220K-6	R-	Io=1.5A Vo:-5.85...-6.15V Vin:-11V	3O	Nsc		data pinout
LM220K-8	R-	Io=1.5A Vo:-7.8...-8.2V Vin:-13V	3O	Nsc		data pinout
LM220K-9	R-	Io=1A Vo:-8.8...-9.2V Vin:-14V	3O	Nsc		data pinout
LM 220 K-...	Z-IC	=LM 320K-...: -25...+150°	23d	Nsc	LM 120K-...	data pinout
LM221	PRE-AMP	Vs:±20V Vu:26dB Vi0:0.2mV	14D,10F,8A	Nsc		data pinout
LM 221(A)...	OP-IC	=LM 121...: -25...+85°		Nsc	-	data pinout
LM 222(H)	LIN-IC	=LM 122(H): -25...+85°	82	Nsc	-	
LM223K	R+	Io=int Vo:4.9...5.1V Vin:7.5V P:int	3O	Mot		data pinout
LM 223(A)K	Z-IC	=LM 123(A)K: -25...+150°	23a	Mot,Nsc,Tho	LM 123...	data pinout
LM224	4xOP	Vs:32V Vu:100V/mV Vi0:3mV	14SDF	Tix		data pinout
LM 224(A)D,DG,F,J	OP-IC	=LM 124(A)DP,N: -25...+85°	14-DIC	Mot,Nsc,++	→Serie 124	data pdf pinout
LM 224(A)D	OP-IC	=LM 124(A)DP,N: SMD, -25...+85°	14-MDIP	Mot,Tho,Tix	→Serie 124	data pdf pinout
LM 224(A)DP,N	OP-IC	=LM 124(A)DP,N: -25...+85°	14-DIP	Mot,Tho,Tix	→Serie 124	data pdf pinout
LM224A	4xOP	Vs:32V Vu:100V/mV Vi0:2mV	14DSF	Tix		data pinout
LM 224 D	OP-IC	SMD, Quad Op Amp, =LM124, +3...32V	14-MDIP	Ons	-	pdf pinout
LM 224 DR2	OP-IC	SMD, Quad Op Amp, =LM124, +3...32V	14-MDIP	Ons	-	pdf pinout
LM 224 DTB	OP-IC	SMD, Quad Op Amp, =LM124, +3...32V	14-SSMDIP	Ons	-	pdf pinout
LM 224 DTBR2	OP-IC	SMD, Quad Op Amp, =LM124, +3...32V	14-SSMDIP	Ons	-	pdf pinout
LM 224 N	OP-IC	Quad Op Amp, =LM124, +3...32V	14-DIP	Ons	-	pdf pinout
LM225H	R±	Io=100mA Vo:±14.8...±15.2V	10A	Nsc		data pinout
LM 225 H	Z-IC	=LM 125H: -25...+85°	82	Nsc		data pinout
LM226H	R±	Io=100mA Vo:±11.8...±12.2V	10A	Nsc		data pinout
LM 226 H	Z-IC	=LM 126H: -25...+85°	82	Nsc		data pinout
LM227H	R-+	Vo:-11.8...-12.2V		Nsc		data pinout
LM 231(A)H	LIN-IC	=LM 131(A)H: -25...+85°	81	Nsc,Tho	-	
LM 231(A)N	LIN-IC	=LM 131(A)H: -25...+85°	8-DIP	Nsc,Tho	-	pdf
LM 234 H-...,Z-...	LIN-IC	=LM 134...: -25...+85°		Nsc,Tho		
LM 234 H-(3,6)	LIN-IC	Temp. Sensor, ±3/6° Temp. Error	2(Reg+-)	Nsc		pdf pinout
LM 234 H	LIN-IC	Adj. Current Source, 1µ...10mA, ±3%	2(Reg+-)	Nsc		pdf pinout
LM 234 Z-(3,6)	LIN-IC	Temp. Sensor, ±3/6° Temp. Error	7(-Reg+)	Nsc		pdf pinout
LM 235(A)D	LIN-IC	=LM 135(A)H: SMD, -40...+125°	8-MDIP	Tho		pdf
LM 235(A)H,Z	LIN-IC	=LM 135(A)H: -40...+125°		Nsc,Tho		pdf
LM236AD-2.5	Z+	Io=10mA Vo:2.46...2.51V P:int	8S	Tix		data pinout
LM236AH-2.5	Z+	Io=10mA Vo:2.465...2.515V	3A	Nsc		data pinout
LM236AH-5.0	Z+	Io=10mA Vo:4.95...5.05V	3N	Nsc		data pinout
LM236ALP-2.5	Z+	Io=10mA Vo:2.46...2.51V P:int	3N	Tix		data pinout
LM236D-2.5	Z+	Io=10mA Vo:2.44...2.54V P:int	8S	Tix		data pinout
LM 236(A)D	Ref-Z-IC	=LM 236(A)Z: SMD	8-MDIP	Tho	-	data pinout
LM236H-2.5	Z+	Io=10mA Vo:2.44...2.54V	3A	Nsc		data pinout

Type	Device	Short Description	Fig.	Manu	Comparision Types	More at
LM236H-5.0	Z+	Io=10mA Vo:4.9..5.1V	3A	Nsc		data pinout
LM 236(A)H-...	Ref-Z-IC	=LM 136H-... : -25...+85°	2(RegKA)	Tho	KA 236-...	data pinout
LM236LP-2.5	Z+	Io=10mA Vo:2.44..2.54V P:int	3N	Tix		data pinout
LM 236(A)Z	Ref-Z-IC	=LM 136H-5.0: -25...+150°	7(AKReg)	Tho	KA 236-5.0	data pinout
LM237	ER-	Io:-int Vo:-1.2..-37V P:int	3O	Mot		data pinout
LM237H	ER-	Io=int Vo:-1.2..-37V P:int	3A	Tho		data pdf pinout
LM 237 H	Z-IC	=LM 137H: -25...+150°	2I	Mot,Nsc,Tho	LM 137...	data pdf pinout
LM 237 HV(H,K)	Z-IC	=LM 137H: -1,2...-47V, -25...+150°		Nsc	LM 137HV...	data pinout
LM237K	ER-	Io=1.5A Vo:-1.2..-37V P:20W	3O	Mot,Nsc		data pdf pinout
LM 237 K	Z-IC	=LM 137H: 1,5A, 20W, -25...+150°	23m	Mot,Nsc,Tho	LM 137...	data pdf pinout
LM237KC	ER-	Io=1.5A Vo:-1.2..-37V P:2W	3P	Tix		data pdf pinout
LM 237 KC,SP	Z-IC	=LM 137H: >1,5V, -25..+150	17n	Tix,Tho	LM 137...	data pinout
LM 237 MR	Z-IC	=LM 137H: -25..+150	22m	Mot	LM 137...	data pinout
LM 237 MT	Z-IC	=LM 137H: >0,5A, -25...+150°	17n	Mot	LM 137...	data
LM238K	ER+	Io=int Vo:1.2..32V P:int	3O	Tho		data pdf pinout
LM 238 K	Z-IC	=LM 138K: -25...+150°	23k	Nsc,Tho	LM 138.., µA 238K...	data pinout
LM239	4xCOMP	Vs:36V Vu:500V/mV Vi0:±2mV	14DSF	Tix		data pinout
LM239A	4xCOMP	Vs:±18V Vu:106dB Vi0:2mV	14DSF	Mul,Phi,Sig		data pinout
LM 239(A)CM,D,FP	KOP-IC	=LM 139..: SMD, -25...+85°	14-MDIP	Mot,Tho,Tix	→Serie 139	data pdf pinout
LM 239(A)DP,N,F,J	KOP-IC	=LM 139..: -25...+85°	14-DIP,DIC	Mot,Nsc,++	→Serie 139	data pdf pinout
LM 239(A)F,W	KOP-IC	=LM 139.. : -25...+85°	14-DIC,DIP	Nsc,Tix	→Serie 139	data pdf pinout
LM240LAH-10	R+	Io=100mA Vo:9.6..10.4V Vin:16V	3A	Nsc		data pinout
LM240LAH-12	R+	Io=100mA Vo:11.5..12.5V Vin:19V	3A	Nsc		data pinout
LM240LAH-15	R+	Io=100mA Vo:14.4..15.6V Vin:23V	3A	Nsc		data pinout
LM240LAH-18	R+	Io=100mA Vo:17.3..18.7V Vin:27V	3A	Nsc		data pinout
LM240LAH-24	R+	Io=100mA Vo:23..25V Vin:33V	3A	Nsc		data pinout
LM240LAH-6	R+	Io=100mA Vo:5.76..6.24V Vin:11V	3A	Nsc		data pinout
LM240LAH-8	R+	Io=100mA Vo:7.7..8.3V Vin:14V	3A	Nsc		data pinout
LM 240 LAH-...	Z-IC	=LM 340(A)K-..: 0,1A, 2W, -25...+85°	2e	Nsc		data pinout
LM240LAZ-10	R+	Io=100mA Vo:9.6..10.4V Vin:16V		Nsc		data pinout
LM240LAZ-12	R+	Io=100mA Vo:11.5..12.5V Vin:19V		Nsc		data pinout
LM240LAZ-15	R+	Io=100mA Vo:14.4..15.6V Vin:23V		Nsc		data pinout
LM240LAZ-18	R+	Io=100mA Vo:17.3..18.7V Vin:27V		Nsc		data pinout
LM240LAZ-24	R+	Io=100mA Vo:23..25V Vin:33V		Nsc		data pinout
LM240LAZ-6	R+	Io=100mA Vo:5.76..6.24V Vin:11V		Nsc		data pinout
LM240LAZ-8	R+	Io=100mA Vo:7.7..8.3V Vin:14V		Nsc		data pinout
LM 240 LAZ-...	Z-IC	=LM 340(A)K-..: 0,1A, 1,2W, -25...+85°	7b	Nsc		data pinout
LM245K-5.0	R-	Io=3A Vo:-4.9..-5.1V	3O	Nsc		data pinout
LM245K-5.2	R-	Io=3A Vo:-5.1..-5.3V	3O	Nsc		data pinout
LM 245 K-...	Z-IC	=LM 145K-.. : -25...+150°	23d	Nsc	LM 145K-...	data pinout
LM246	4xPROG	Vs:±18V Vu:1000V/mV Vo:±14V f:1.2Mc	16DS,14D	Mit		data pinout
LM 246 D	OP-IC	=LM 146DG,J: SMD, -40...+105°	10-MDIP	Tho	LM 146...	data pinout
LM 246 DG,DP,J,N	OP-IC	=LM 146DG,J: ±18V, -25...+85°	16-DIP,DIC	Nsc,Tho	LM 146...	data pinout
LM248	4xHI-GAIN	Vs:±22V Vu:160V/mV Vo:±12V Vi0:1mV	14SDd	Sgs		data pdf pinout
LM 248 D	OP-IC	=LM 148DG,J: SMD, -40...+105°	14-MDIP	Tho,Tix	→Serie 124	data pdf pinout
LM 248 DP,J,N	OP-IC	=LM 148DG: ±18V, -25...+85°	14-DIP,DIC	Mot,Nsc,++	→Serie 124	data pdf pinout
LM 248 GC	OP-IC	=LM 148DG,J: SMD, -40...+105°	20-(P)LCC	Tho	→Serie 124	data pdf
LM249	4xOP	Vs:±18V Vu:104dB Vo:±12V Vi0:1mV	14Dd	Nsc		data pinout
LM 249 D,DP,J	OP-IC	=LM 149DG: ±18V, -25...+85°	14-DIP	Nsc,Tho	→Serie 124	data pinout
LM 249 GC	OP-IC	=LM 149DG,J: SMD, -40...+105°	20-(P)LCC	Tho	→Serie 124	data
LM250K	ER+	Io=3A Vo:1.2...33V	3O	Mot,Nsc		data pdf pinout
LM 250 K	Z-IC	=LM 150K: -25...+150°	23k	Mot,Nsc	LM 150K, µA 250K...	data pinout
LM258	2xLO-POWER	Vs:32V Vu:100V/mV Vo:3.5/0V Vi0:2mV	8SDA	Sgs		data pdf pinout
LM258A	2xLO-POWER	Vs:32V Vu:100V/mV Vo:3.5/0V Vi0:1mV	8SAD	Sgs		data pdf pinout
LM 258(A)D	OP-IC	=LM 158(A)H: SMD, -25...+85°	8-MDIP	Mot,Sam,Tix	→Serie 158	data pinout
LM 258(A)DP,N,P	OP-IC	=LM 158(A)H: -25...+85°	8-DIP	Nsc,Mot,++	→Serie 158	data pinout
LM 258(A)FE,J,JG	OP-IC	=LM 158(A)H: -25...+85°	8-DIC	Mot,Nsc,++	→Serie 158	data pinout
LM 258(A)H	OP-IC	=LM 158(A)H: -25...+85°	81	Mot,++	→Serie 158	data pinout
LM 258 H	OP-IC	Dual, Low Power OPAmp -25..85°C, 1MHz, ±1.5/+3..±16/+32V	81	Nsc	-	pdf pinout
LM 258 S.	OP-IC	=LM 158(A)H: -25...+85°	9-SIP	Sam	→Serie 158	data
LM 258 W(D/S)	IC	Low power dual operational amplifiers, -40 to +105	MDIP	Stm	-	pdf pinout
LM 258 WN	IC	Low power dual operational amplifiers, -40 to +105	DIC	Stm	-	pdf pinout
LM 258 WP	IC	Low power dual operational amplifiers, -40 to +105	TSOP	Stm	-	pdf pinout
LM260	HI-SPEED	Vs:±8V Vo:TTL Vi0:2mV	14D,8A	Nsc		data pinout
LM 260 D,J-14	KOP-IC	=LM 160H: -25...+85°	14-DIC	Nsc	µA 760...	data pinout
LM 260 H	KOP-IC	=LM 160H: -25...+85°	81	Nsc	µA 760...	data pinout
LM261	HI-SPEED	Vs:±16V Vu:70dB Vo:TTL Vi0:1mV	14D,10A	Nsc		data pinout
LM 261 D,J	KOP-IC	=LM 161H: -25...+85°	14-DIC	Nsc	SE 529...	data pinout
LM 261 H	KOP-IC	=LM 161H: -25...+85°	81	Nsc	SE 529...	data pinout
LM268-BYH5.0	R+	Io=50mA P:600mW	8AD	Nsc		data pinout
LM268BYH-10	R+	Io=50mA P:600mW	8A	Nsc		data pinout
LM285B	Z+	Io=20mA Vo:2.462..2.538V	3N,2A	Nsc		data pinout
LM285BEOA-1.2	Z+	Io=20mA Vo:1.223..1.235V	8S	Tcs		data pinout
LM285BEOA-2.5	Z+	Io=20mA Vo:2.462..2.538V	8S	Tcs		data pinout
LM285BEZB-1.2	Z+	Io=20mA Vo:1.223..1.247V	3N	Tcs		data pinout
LM285BXZ	EZ+	Io=20mA Vo:1.24..5.3V	3N	Nsc		data pdf pinout
LM285BXZ1.2	Z+	Io=20mA Vo:1.223..1.247V	3N	Nsc		data pinout
LM285BXZ2.5	Z+	Io=20mA Vo:2.462..2.538V	3N	Nsc		data pinout
LM285BYH1.2	Z+	Io=20mA Vo:1.223..1.247V	2A	Nsc		data pinout
LM285BYH2.5	Z+	Io=20mA Vo:2.462..2.538V	2A	Nsc		data pinout
LM285BYZ	EZ+	Io=20mA Vo:1.24..5.3V	3N	Nsc		data pdf pinout

Type	Device	Short Description	Fig.	Manu	Comparision Types	More at
LM285BYZ1.2	Z+	Io=20mA Vo:1.223...1.247V	3N	Nsc		data pinout
LM285BYZ2.5	Z+	Io=20mA Vo:2.462...2.538V	3N	Nsc		data pinout
LM285D-1.2	Z+	Io=20mA Vo:1.22...1.24V P:int	8S	Tix		data pinout
LM285D-2.5	Z+	Io=20mA Vo:2.46...2.53V P:int	8S	Tix		data pinout
LM285EOA-1.2	Z+	Io=20mA Vo:1.205...1.270V	8S	Tcs		data pinout
LM285EOA-2.5	Z+	Io=20mA Vo:2.425...2.575V	8S	Tcs		data pinout
LM285EZB-1.2	Z+	Io=20mA Vo:1.205...1.270V	3N	Tcs		data pinout
LM285EZB-2.5	Z+	Io=20mA Vo:2.425...2.575V	3N	Tcs		data pinout
LM285H-1.2	Z+	Io=20mA Vo:1.223...1.247V	2A	Nsc		data pinout
LM285H-2.5	Z+	Io=20mA Vo:2.462...2.538V	2A	Nsc		data pinout
LM 285 H-...	Ref-Z-IC	=LM 185H-...: -40..+85°	2(K-A)	Nsc		data pinout
LM285LP-1.2	Z+	Io=20mA Vo:1.22...1.24V P:int	3N	Tix		data pinout
LM285LP-2.5	Z+	Io=20mA Vo:2.46...2.53V P:int	3N	Tix		data pinout
LM285M	EZ+	Io=20mA Vo:1.24...5.3V	8S	Nsc		data pdf pinout
LM285M-1.2	Z+	Io=20mA Vo:1.223...1.247V	8S	Nsc		data pinout
LM285M-2.5	Z+	Io=20mA Vo:2.462...2.538V	8S	Nsc		data pinout
LM285Z	EZ+	Io=20mA Vo:1.24...5.3V	3N	Nsc		data pdf pinout
LM285Z-1.2	Z+	Io=20mA Vo:1.223...1.247V	3N	Nsc		data pinout
LM285Z-2.5	Z+	Io=20mA Vo:2.462...2.538V	3N	Nsc		data pinout
LM 285 Z-...	Ref-Z-IC	=LM 185H-...: -40..+85°	7(AK-)	Mot		data pinout
LM292	OP	Vs:±16V Vu:100dB Vo:±14V Vio:2mV	8AD	Nsc		data pinout
LM292	R+	Io=750mA Vo:4.75...5.25V Vin:6...26V	5P	Nsc		data pdf pinout
LM 292 H	OP/KOP-IC	=LM 192H: -25...+85°	81	Nsc	LM 192...	data pinout
LM 292 J	OP/KOP-IC	=LM 192H: -25...+85°	8-DIC	Nsc	LM 192...	data pinout
LM293	2xCOMP	Vs:36V Vu:200V/mV Vio:2mV	8SDA	Tix		data pinout
LM293A	2xCOMP	Vs:36V Vu:200V/mV Vio:1mV	8SDA	Tix		data pdf pinout
LM 293(A)D	KOP-IC	=LM 139(A)H: SMD, -25...+85°	8-MDIP	Sam,Tho,Tix	→Serie 193	data pdf pinout
LM 293(A)FE,J,JG	KOP-IC	=LM 193(A)H: -25...+85°	8-DIC	Mot,Nsc,++	→Serie 193	data pdf pinout
LM 293(A)H,L	KOP-IC	=LM 193(A)H: -25...+85°	81	Mot,Nsc,++	→Serie 193	data pdf pinout
LM 293(A)N,P	KOP-IC	=LM 139(A)H: -25...+85°	8-DIP	Mot,Nsc,++	→Serie 193	data pdf pinout
LM 293 S	KOP-IC	=LM 139(A)H: -25...+85°	9-SIP	Sam	→Serie 193	data pdf
LM 295 H	LIN-IC	=LM 195H: -25...+150°	2a*	Nsc	-	data pinout
LM 295 K	LIN-IC	=LM 195H: -25...+150°	23e*	Nsc	-	data pinout
LM299	Z+	Io=10mA Vo:6.8...7.1V P:300mW	4A	Nsc		data pinout
LM 299(A)H(-20)	Ref-Z-IC	=LM 199...: -25...+85°	5(-+AK)	Nsc	-	pdf
LM300	ER+	Io=12mA Vo:2...20V P:500mW	10F,8A	Isi,Nsc		data pinout
LM 300 ...	Z-IC	~LM 305...	81	Nsc	→LM 305...	data pinout
LM301A	UNI	Vs:±18V Vu:104dB Vo:±13V Vio:2mV	8DS,14DAF	Tho		data pinout
LM 301(A)D,FP	OP-IC	=LM 101(A)H: SMD, ±18V, 0...+70°	8-MDIP	Nsc,Tho	→Serie 101	data pdf pinout
LM 301(A)D,JG,P	OP-IC	=LM 101(A)H: ±18V, 0...+70°	8-DIP,DIC	Tix	→Serie 101	data pinout
LM 301(A)DP,J,N	OP-IC	=LM 101(A)H: ±18V, 0...+70°	8-DIP,DIC	Mot,Nsc,Tho	→Serie 101	data pinout
LM 301(A)H,L	OP-IC	=LM 101(A)H: ±18V, 0...+70°	81	Mot,Nsc,Tho	→Serie 101	data pdf pinout
LM 301(A)J-14,N-14	OP-IC	=LM 101(A)H: 0...+70°	14-DIC,DIP	Nsc	→Serie 101	data pinout
LM 301(A)W	OP-IC	=LM 101(A)H: ±18V, 0...+70°	14-FLP	Tix	-	data pinout
LM302	FOLLOWER	Vs:±18V Vo:±13V Vio:5mV	8A	Nsc		data pinout
LM 302 H	OP-IC	=LM 102H: 0...+70°	81	Nsc	→Serie 102	data pinout
LM304F	ER-	Io=20mA Vo:-0.035...-30V P:500mW	10F	Nsc		data pinout
LM304H	ER-	Vo:-0.035...-30V P:500mW	10A	Nsc		data pdf pinout
LM 304 H,L	Z-JC	=LM 104H,L: 0...+70°	82	Nsc	LM 104..., LM 204...	data pdf pinout
LM304J	ER-	Io=20mA Vo:-0.035...-30V P:1W	14D,10A	Tix		data pinout
LM 304 J,N	Z-IC	=LM 104H,L: -0.035...30V, 0...+70°	14-DIC,DIP	Tix	LM 104..., LM 204...	data pinout
LM305	ER+	Io=12mA Vo:4.5...30V P:500mW	14D,8A,10F	Tix		data pinout
LM305(A)(H,JG,L,P)	Z-IC	=LM 105H,L: 0...+70°, A=45mA		Nsc,Tix	µA 305...	data pdf
LM305A	ER+	Io=375mA Vo:4.5...30V P:500mW	8AD	Tho		data pdf pinout
LM 305 H,L	Z-IC	=LM 105H,L: +4,5...30V, 0...+70°	81	Nsc,Tho,Tix	LM 105..., LM 205...	data pinout
LM 305 JG,P	Z-IC	=LM 105H,L: +4,5...30V, 0...+70°	8-DIC,DIP	Nsc,Tho,Tix	LM 105..., LM 205...	data pinout
LM306	COMP	Vs:±15V Vu:92dB Vo:TTL Vio:1.6mV	8SA,14DDF	Tix		data pinout
LM 306 D...W	KOP-IC	=LM 106...: 0...+70°		Nsc,Tix	→Serie 106	data pinout
LM307	UNI	Vs:±18V Vu:200V/mV Vio:2mV	8S,14D,10F	Tix		data pinout
LM 307 D,J,J-14	OP-IC	=LM 107H: 0...+70°	14-DIP,DIC	Nsc,Tix	→Serie 107	data pdf pinout
LM 307 F,U	OP-IC	=LM 107H: 0...+70°	10-FLP	Nsc,Tix		data pinout
LM 307 H,L	OP-IC	=LM 107H: 0...+70°	81	Nsc,Tix	→Serie 107	data pinout
LM 307 J,JG,N,P	OP-IC	=LM 107H: 0...+70°	8-DIC,DIP	Mot,Nsc,Tix	→Serie 107	data pinout
LM 307 M	OP-IC	=LM 107H: SMD, 0...+70°	8-MDIP	Nsc	→Serie 107	data pinout
LM 307 W	OP-IC	=LM 107H: 0...+70°	14-FLP	Tix	-	pdf
LM308	LO-POWER	Vs:±18V Vu:110dB Vo:±13V Vio:2mV	10F,8D,14D	Ray		data pinout
LM308A	LO-POWER	Vs:±18V Vu:110dB Vo:±13V Vio:0.3mV	10F,8D,14D	Ray		data pinout
LM 308(A)D,FP,M	OP-IC	=LM 108(A)H: ±18V, 0...+70°	8-MDIP	Mot,Nsc,Tho	→Serie 108	data pinout
LM 308(A)DP,N,J-8	OP-IC	=LM 108(A)H: ±18V, 0...+70°	8-DIP,DIC	Mot,Nsc,Tho	→Serie 108	data pinout
LM308(A)H(-1,-2),T	OP-IC	=LM 108(A)H: ±18V, 0...+70°	81	Mot,Nsc,Tho	→Serie 108	data pinout
LM309DA	R+	Io=1.5A Vo:4.8...5.2V P:8W	3O	Sig		data pinout
LM 309 DA,K	Z-IC	=LM 109L,LA: 20W, 0...+125°	23a	Nsc,Tho++	LM 109..., LM 209..., ... 7805... (TO-3)	data pdf pinout
LM309DB	R+	Io=500mA Vo:4.8...5.2V P:8W	3A	Sig		data pinout
LM309H	R+	Io=200mA Vo:4.8...5.2V Vin:10V P:2W	3A	Nsc		data pdf pinout
LM 309 H,LA	Z-IC	=LM 109H,LA: 0...+125°	2e	Nsc,Tho++	LM 109..., LM 209..., ... 7805...(TO-5)	data pdf pinout
LM309K	R+	Io=int Vo:4.8...5.2V Vin:10V P:int	3O	Tho		data pinout
LM309LA	R+	Io=500mA Vo:4.8...5.2V Vin:10V	3A	Tix		data pinout
LM310	FOLLOWER	Vs:±18V Vo:±10V Vio:2.5mV	14D,10F,8A	Isi,Nsc		data pinout
LM 310 D,J	OP-IC	=LM 110H: 0...+70°	14-DIC	Nsc	→Serie 110	data pinout
LM 310 F	OP-IC	=LM 110H: 0...+70°	10-FLP	Nsc		data pinout
LM 310 H	OP-IC	=LM 110H: 0...+70°	81	Nsc	→Serie 110	data pinout
LM 310 J-8,N	OP-IC	=LM 110H: 0...+70°	8-DIC,DIP	Nsc	→Serie 110	data pinout
LM 310 M	OP-IC	=LM 110H: SMD, 0...+70°	8-MDIP	Nsc	→Serie 110	data pinout
LM311	COMP	Vs:36V Vu:200V/mV Vo:TTL Vio:2mV	14D,8SD	Phi		data pdf pinout
LM 311 D,FP,M	KOP-IC	=LM 111H: SMD, 0...+70°	8-MDIP	Nsc,Tho,Phi	→Serie 111	data pdf pinout

Type	Device	Short Description	Fig.	Manu	Comparision Types	More at
LM 311 D,J,N,N-14	KOP-IC	=LM 111H: 0...+70°	14-DIC,DIP	Nsc,Tix,Phi	→Serie 111	data pdf pinout
LM 311 DP,N,N-8,P	KOP-IC	=LM 111H: 0...+70°	8-DIP	Mot,Tix,++	→Serie 111	data pdf pinout
LM 311 F,U	KOP-IC	=LM 111H: Min, 0...+70°	10-FLP	Nsc,Tix,Phi	→Serie 111	data pdf pinout
LM 311 FE,J-8,JG	KOP-IC	=LM 111H: 0...+70°	8-DIC	Mot,Nsc,++	→Serie 111	data pdf pinout
LM 311 H,L,T	KOP-IC	=LM 111H: 0...+70°	81	Mot,Nsc,++	→Serie 111	data pdf pinout
LM312	LO-POWER	Vs:±18V Vu:110dB Vo:13V Vio:2mV	14D,10F,8A	Nsc		data pdf pinout
LM 312 H	OP-IC	=LM 112H: ±18V, 0...+70°	81	Nsc	→Serie 112	data pdf pinout
LM 313(H)	Ref-Z-IC	=LM 113(H): 0...+70°	2d	Nsc	-	data pinout
LM313H	Z+	Io=50mA Vo:1.16...1.28V P:100mW	2A	Nsc		data pdf pinout
LM 314	IC			Nsc		
LM316	LO-BIAS	Vs:±20V Vu:86dB Vo:±12V Vi0:10mV	8A	Nsc		data pinout
LM316A	LO-BIAS	Vs:±20V Vu:92dB Vo:±12V Vi0:3mV	8A	Nsc		data pinout
LM 316(A)H	OP-IC	=LM 216(H): 0...+70°	81	Nsc	LM 216...	data pinout
LM317	ER+	Io=1.5A Vo:1.2...37V P:15W	3SP	Mot		data pinout
LM317A	ER+	Io=2.2A Vo:1.2...37V P:20W	3P	Nsc		data pinout
LM317AH	ER+	Io=500mA Vo:1.2...37V P:2W	3A	Mot,Nsc		data pinout
LM317AK	ER+	Io=2.2A Vo:1.2...37V P:20W	3O	Nsc		data pinout
LM317BT	ER+	Io=1.5A Vo:1.2...37V P:int	3P	Mot		data pinout
LM317H	ER+	Io=500mA Vo:1.2...37V P:int	3A	Nsc		data pinout
LM 317 H	Z-IC	=LM 117H: 0...+125°	2k	Mot,Nsc,Tho	LM 117..., LM 217...	data pinout
LM317HV	ER+	Io=1.5A Vo:1.2...57V P:int	3AOP	Nsc		data pinout
LM 317 HV(H,K)	Z-IC	=LM 117H: +1,2...57V		Nsc	LM 117HV..., LM 217HV...	data pinout
LM317ISP-3	ER+	Io=int Vo:1.2...37V P:int	3P	Tho		data pinout
LM317K	ER+	Io=int Vo:1.2...37V P:int	3OP	Tho		data pinout
LM 317 K	Z-IC	=LM 117K: 0...+125°	23k	Mot,Nsc,Tho	LM 117..., LM 217..., µA 317K...	data pinout
LM317L	ER+	Io=100mA Vo:1.2...37V P:int	8S,3N	Nsc		data pinout
LM317LA	ER+	Io=500mA Vo:1.2...37V P:600mW	3A	Tix		data pinout
LM317LH	ER+	Io=int Vo:1.2...37V P:int	3A	Mot		data pinout
LM 317 LH	Z-IC	=LM 117LH: 0...+125°	2k	Mot	LM 117..., LM 217...	data pinout
LM317LZ	ER+	Io=int Vo:1.2...37V P:int	3N	Mot		data pinout
LM 317 LZ	Z-IC	=LM 117H: >0,1A, >0,62W, 0...+125°	7o	Mot,Nsc	KA 317L, TL 317CLP	data pinout
LM 317 MDT	Z-IC	1.2 to 37 V adjustable voltage regulator	17I	Mot		data pinout
LM317MP	ER+	Io=800mA Vo:1.2...37V P:20W	3P	Nsc		data pinout
LM 317 MP	Z-IC	=LM 117H: >0,5A, 7,5W, 0...+125°	13I	Nsc	LM 117..., LM 217...	pdf
LM317MR	ER+	Io=int Vo:1.2...37V P:int	3O	Mot		data pinout
LM317MT	ER+	Io=500mA Vo:1.2...37V P:int	3P	Mot		data pinout
LM 317 MT	Z-IC	=LM 117MT: 0...+125°	17I	Mot	KA 317M, LM 117..., LM 217..., µA 317U...	data pinout
LM317SP	ER+	Io=int Vo:1.2...37V P:int	3P	Tho		data pinout
LM 317 SP3	Z-IC	=LM 117H: >1,5A, 0...+125°	18I	Tho		data pinout
LM 317 SP	Z-IC	=LM 117H: >1,5A, 0...+125°	17I	Tho	KA 317, LM 117..., LM 217..., µA 317U...	data pinout
LM317T	ER+	Io=2.2A Vo:1.2...37V P:20W	3P	Nsc		data pinout
LM 317 T,KC	Z-IC	=LM 117H: >1,5A, 0...+125°	17I	Mot,Nsc,Tho	KA 317, LM 117..., LM 217..., µA 317U	data pinout
LM318	HI-SPEED	Vs:±20V Vu:106dB Vo:13V Vi0:4mV	14D,8SDA	Nsc		data pinout
LM318A	HI-SPEED	Vs:±20V Vu:15V/mV Vo:13V Vi0:5mV	8AD	Har		data pinout
LM 318 D,FP,M	OP-IC	=LM 118H: SMD, 0...+70°	8-MDIP	Nsc,Tho,Tix	→Serie 118	data pinout
LM 318 DP,N,N8,P	OP-IC	=LM 118H: 0...+70°	8-DIP	Nsc,Tho,Tix	→Serie 118	data pinout
LM 318 H,L	OP-IC	=LM 118H: 0...+70°	81	Tho	→Serie 118	data pinout
LM 318 JG,J-8	OP-IC	=LM 118H: 0...+70°	8-DIC	Nsc,Tix	→Serie 118	data pinout
LM319	2xCOMP	Vs:36V Vu:40V/mV Vo:TTL Vi0:2mV	14DS,10FA	Phi		data pinout
LM319A	2xCOMP	Vs:36V Vu:40V/mV Vo:TTL Vi0:0.5mV	10A,14DS	Nsc		data pdf pinout
LM 319 D,FP,M	KOP-IC	=LM 119H: SMD, 0...+70°	14-MDIP	Nsc,Phi,Tho	→Serie 119	data pinout
LM 319 DP,N,F,J	KOP-IC	=LM 119H: 0...+70°	14-DIP,DIC	Nsc,Phi,Tho	→Serie 119	data pinout
LM 319 F	KOP-IC	=LM 119H: 0...+70°	10-FLP	Phi	→Serie 119	data pinout
LM 319 H	KOP-IC	=LM 119H: 0...+70°	82	Nsc,Phi,Tho	→Serie 119	data pinout
LM320H-12	R-	Io=200mA Vo:-11.6...-12.4V Vin:-35V	3A	Nsc		data pinout
LM320H-15	R-	Io=200mA Vo:-14.6...-15.4V Vin:-20V	3A	Nsc		data pinout
LM320H-18	R-	Io=200mA Vo:-17.4...-18.6V Vin:-23V	3A	Nsc		data pinout
LM320H-24	R-	Io=200mA Vo:-23.2...-24.8V Vin:-29V	3A	Nsc		data pinout
LM320H-5.0	R-	Io=500mA Vo:-4.8...-5.2V Vin:-10V	3A	Nsc		data pinout
LM320H-5.2	R-	Io=500mA Vo:-5...-5.4V Vin:-10V	3A	Nsc		data pinout
LM320H-6	R-	Io=500mA Vo:-5.75...-6.25V Vin:-11V	3A	Nsc		data pinout
LM320H-8	R-	Io=200mA Vo:-7.7...-8.3V Vin:-13V	3A	Nsc		data pinout
LM320H-9	R-	Io=200mA Vo:-8.65...-9.35V Vin:-14V	3A	Nsc		data pinout
LM 320 H-...	Z-IC	=LM 320T-...: 0,5A, 2W	2f	Nsc	...79xx... (TO-5)	data pinout
LM320K-12	R-	Io=1A Vo:-11.6...-12.4V Vin:-35V P:int	3O	Nsc		data pinout
LM320K-15	R-	Io=1A Vo:-14.6...-15.4V Vin:-20V P:int	3O	Nsc		data pinout
LM320K-18	R-	Io=1A Vo:-17.4...-18.6V Vin:-23V	3O	Nsc		data pinout
LM320K-24	R-	Io=1A Vo:-23.2...-24.8V Vin:-29V	3O	Nsc		data pinout
LM320K-5.0	R-	Io=1.5A Vo:-4.8...-5.2V Vin:-10V P:int	3O	Nsc		data pinout
LM320K-5.2	R-	Io=1.5A Vo:-5...-5.4V Vin:-10V	3O	Nsc		data pinout
LM320K-6	R-	Io=1.5A Vo:-5.75...-6.25V Vin:-11V	3O	Nsc		data pinout
LM320K-8	R-	Io=1.5A Vo:-7.7...-8.3V Vin:-13V	3O	Nsc		data pinout
LM320K-9	R-	Io=1A Vo:-8.65...-9.35V Vin:-14V	3O	Nsc		data pinout
LM 320 K,KC-...	Z-IC	=LM 320T-...: 1,5A, 20W	23d	Nsc	...79xx... (TO-3)	data pinout
LM320KC-15	R-	Io=1A Vo:-14.6...-15.4V Vin:-20V P:int	3O	Nsc		data pinout
LM320KC-5.0	R-	Io=1.5A Vo:-4.8...-5.2V Vin:-10V P:int	3O	Nsc		data pinout
LM 320 KC-...	Z-IC	=LM 320T-...	17c	Tix	...79xx... (TO-220)	data pinout
LM320LZ-10	R-	Io=100mA Vo:-9.6...-10.4V Vin:-15V	3N	Nsc		data pinout
LM320LZ-12	R-	Io=100mA Vo:-11.5...-12.5V Vin:-10V	3N	Nsc		data pinout
LM320LZ-15	R-	Io=100mA Vo:-14.4...-15.6V Vin:-20V	3N	Nsc		data pinout
LM320LZ-18	R-	Io=100mA Vo:-17.3...-18.7V Vin:-23V	3N	Nsc		data pinout
LM320LZ-24	R-	Io=100mA Vo:-23...-25V Vin:-29V	3N	Nsc		data pinout
LM320LZ-5.0	R-	Io=100mA Vo:-4.8...-5.2V Vin:-10V	3N	Nsc		data pinout
LM 320 LZ-...	Z-IC	=LM 320T-...: 0,1A, 1,2W	7a	Nsc	...79Lxx... (TO-92)	data pinout
LM320MLP-10	R-	Io=250mA Vo:-9.6...-10.4V Vin:-15V	3P	Nsc		data pinout

Type	Device	Short Description	Fig.	Manu	Comparision Types	More at
LM320MLP-15	R-	Io=250mA Vo:-14.4...-15.6V Vin:-20V	3P	Nsc		data pinout
LM320MLP-18	R-	Io=250mA Vo:-17.3...-18V Vin:-23V	3P	Nsc		data pinout
LM320MLP-24	R-	Io=250mA Vo:-23...-25V Vin:-29V	3P	Nsc		data pinout
LM320MLP-5.0	R-	Io=250mA Vo:-4.8...-5.2V Vin:-10V	3P	Nsc		data pinout
LM 320 MLP-...	Z-IC	=LM 320T-...: 0,25A, 7,5W	13c	Nsc	... 79xx... (TO-202)	data pinout
LM320MP-12	R-	Io=500mA Vo:-11.5...-12.5V Vin:-35V	3P	Nsc		data pinout
LM320MP-15	R-	Io=500mA Vo:-14.4...-15.6V Vin:-20V	3P	Nsc		data pinout
LM320MP-18	R-	Io=500mA Vo:-17.4...-18.6V Vin:-23V	3P	Nsc		data pinout
LM320MP-24	R-	Io=500mA Vo:-23.2...-24.8V Vin:-29V	3P	Nsc		data pinout
LM320MP-5.0	R-	Io=500mA Vo:-4.8...-5.2V Vin:-10V	3P	Nsc		data pinout
LM320MP-5.2	R-	Io=500mA Vo:-5...-5.4V Vin:-10V	3P	Nsc		data pinout
LM320MP-6	R-	Io=500mA Vo:-5.75...-6.25V Vin:-11V	3P	Nsc		data pinout
LM320MP-8	R-	Io=500mA Vo:-7.7...-8.3V Vin:-13V	3P	Nsc		data pinout
LM320MP-9	R-	Io=500mA Vo:-8.65...-9.35V Vin:-14V	3P	Nsc		data pinout
LM 320 MP-...	Z-IC	=LM 320T-...: 0,5A, 7,5W	13c	Nsc	... 79xx... (TO-202)	data pinout
LM 320 T-5.0	Z-IC	-5V, 1,5A, 15W, 0...+125°	17c	Nsc	... 7905... (TO-220)	data pinout
LM 320 T-5.2	Z-IC	-5,2V, 1,5A, 15W, 0...+125°	17c	Nsc	... 7952... (TO-220)	data pinout
LM 320 T-6.0	Z-IC	-6V, 1,5A, 15W, 0...+125°	17c	Nsc	... 7906... (TO-220)	data pinout
LM 320 T-8.0	Z-IC	-8V, 1,5A, 15W, 0...+125°	17c	Nsc	... 7908... (TO-220)	data pinout
LM 320 T-9.0	Z-IC	-9V, 1,5A, 15W, 0...+125°	17c	Nsc	... 7909... (TO-220)	data pinout
LM320T-12	R-	Io=1A Vo:-11.6...-12.4V Vin:-35V P:int	3P	Nsc		data pinout
LM 320 T-12	Z-IC	-12V, 1,5A, 15W, 0...+125°	17c	Nsc	... 7912... (TO-220)	data pinout
LM320T-15	R-	Io=1A Vo:-14.5...-15.5V Vin:-20V P:int	3P	Nsc		data pinout
LM 320 T-15	Z-IC	-15V, 1,5A, 15W, 0...+125°	17c	Nsc	... 7915... (TO-220)	data pinout
LM320T-18	R-	Io=1A Vo:-17.4...-18.6V Vin:-23V	3P	Nsc		data pinout
LM 320 T-18	Z-IC	-18V, 1,5A, 15W, 0...+125°	17c	Nsc	... 7918... (TO-220)	data pinout
LM320T-24	R-	Io=1A Vo:-23.2...-24.8V Vin:-29V	3P	Nsc		data pinout
LM 320 T-24	Z-IC	-24V, 1,5A, 15W, 0...+125°	17c	Nsc	... 7924... (TO-220)	data pinout
LM320T-5.0	R-	Io=1.5A Vo:-4.8...-5.2V Vin:-10V	3P	Nsc		data pinout
LM320T-5.2	R-	Io=1.5A Vo:-5...-5.4V Vin:-10V	3P	Nsc		data pinout
LM320T-6	R-	Io=1.5A Vo:-5.75...-6.25V Vin:-11V	3P	Nsc		data pinout
LM320T-8	R-	Io=1.5A Vo:-7.7...-8.3V Vin:-13V	3P	Nsc		data pinout
LM320T-9	R-	Io=1A Vo:-8.65...-9.35V Vin:-14V	3P	Nsc		data pinout
LM321	LO-POWER	Vs:32V Vu:100V/mV Vo:26V Vio:2mV	14D,10F,8A	Nsc		data pdf pinout
LM 321(A)...	OP-IC	=LM 121...: 0...+70°		Nsc	-	data pinout
LM 322(H)	LIN-IC	=LM 122(H): -0...+70°	82	Nsc	-	pdf pinout
LM 322(N)	LIN-IC	=LM 122(H): -0...+70°	14-DIP	Nsc	-	pdf pinout
LM323	R+	Io=int Vo:4.8...5.2V Vin:7.5V P:int	3OP	Mot		data pdf pinout
LM323AK	R+	Io=3A Vo:4.9...5.1V Vin:7.5V P:int	3O	Mot,Nsc		data pdf pinout
LM323AT	R+	Io=int Vo:4.9...5.1V Vin:7.5V P:int	3P	Mot		data pdf pinout
LM 323 H,SP3	Z-IC	=LM 123(A)K: 0...+125°	18b	Sam,Tho	LM 123...	data pinout
LM323K	R+	Io=3A Vo:4.8...5.2V Vin:7.5V P:int	3O	Nsc		data pdf pinout
LM 323(A)K	Z-IC	=LM 123(A)K: 0...+125°	23a	Mot	LM 123..., LM 223...	data pinout
LM 323(A)T	Z-IC	=LM 123(A)K: 0...+125°	17b	Mot	-	data pinout
LM324	4xOP	Vs:32V Vu:100V/mV Vio:3mV	14SDd	Tix		data pinout
LM 324(A)CM,D,FP,M	OP-IC	=LM 124(A)DP,N: SMD, 0...+70°	14-MDIP	Mot,Nsc,++	→Serie 124	data pdf pinout
LM 324(A)DG,F,J	OP-IC	=LM 124(A)DP,N: 0...+70°	14-DIC	Mot,Nsc,++	→Serie 124	data pdf pinout
LM 324(A)DP,N	OP-IC	=LM 124(A)DP,N: 0...+70°	14-DIP	Sgs	→Serie 124	data pdf pinout
LM324A	4xOP	Vs:32V Vu:100V/mV Vio:2mV	14DS	Tix		data pinout
LM 324 AD	OP-IC	SMD, Quad Op Amp, =LM124, +3...32V	14-MDIP	Ons	-	pdf pinout
LM 324 ADR2	OP-IC	SMD, Quad Op Amp, =LM124, +3...32V	14-MDIP	Ons	-	pdf pinout
LM 324 ADTB	OP-IC	SMD, Quad Op Amp, =LM124, +3...32V	14-SSMDIP	Ons	-	pdf pinout
LM 324 ADTBR2	OP-IC	SMD, Quad Op Amp, =LM124, +3...32V	14-SSMDIP	Ons	-	pdf pinout
LM 324 AN	OP-IC	Quad Op Amp, =LM124, +3...32V	14-DIP	Ons	-	pdf pinout
LM 324 D	OP-IC	SMD, Quad Op Amp, =LM124, +3...32V	14-MDIP	Ons	-	pdf pinout
LM 324 DR2	OP-IC	SMD, Quad Op Amp, =LM124, +3...32V	14-MDIP	Ons	-	pdf pinout
LM 324 DTB	OP-IC	SMD, Quad Op Amp, =LM124, +3...32V	14-SSMDIP	Ons	-	pdf pinout
LM 324 DTBR2	OP-IC	SMD, Quad Op Amp, =LM124, +3...32V	14-SSMDIP	Ons	-	pdf pinout
LM 324 N	OP-IC	Quad Op Amp, =LM124, +3...32V	14-DIP	Ons	-	pdf pinout
LM324x2	8xOP	Vs:32V Vu:100V/mV Vio:3mV	30S	Tix		data pinout
LM325	R±	Io=100mA Vo:±14.5...±15.5V P:5W	14Ds,10A	Nsc		data pinout
LM 325 H	Z-IC	=LM 125H: 0...+70°	82	Nsc	-	data pdf pinout
LM 325(A)N	Z-IC	=LM 125H: 0...+70°, ±2%, A=±1%	14-DIP	Nsc	-	data pdf pinout
LM 325(A)S	Z-IC	=LM 125H: 0...+70°	14-DIP+d	Nsc	-	data pdf pinout
LM326	R±	Io=100mA Vo:±11.5...±12.5V P:5W	10A,14Ds	Nsc		data pinout
LM 326 H	Z-IC	=LM 126H: 0...+70°	82	Nsc	-	data pinout
LM 326 N	Z-IC	=LM 126H: 0...+70°	14-DIP	Nsc	-	data pinout
LM 326 S	Z-IC	=LM 126H: 0...+70°	14-DIP+d	Nsc	-	data pinout
LM327	R-+	Vo:-11.8...-12.2V		Nsc		data pinout
LM329AH	Z+	Vo:6.6...7.2V	2A	Nsc		data pdf pinout
LM329AZ	Z+	Vo:6.6...7.25V	3N	Ltc		data pdf pinout
LM329B	Z+	Vo:6.6...7.2V	2A,3N	Nsc		data pdf pinout
LM329BZ	Z+	Vo:6.6...7.25V	3N	Ltc		data pdf pinout
LM 329 BZ,CZ,DZ	Ref-Z-IC	=LM 129...: ±5%, 0...+70°	7a	Nsc		data pinout
LM329C	Z+	Vo:6.6...7.2V	2A,3N	Nsc		data pdf pinout
LM329CZ	Z+	Vo:6.6...7.25V	3N	Ltc		data pdf pinout
LM329D	Z+	Vo:6.6...7.2V	2A,3N	Nsc		data pdf pinout
LM329DZ	Z+	Vo:6.6...7.25V	3N	Ltc		data pdf pinout
LM 330 KC	Z-IC	=LM 330T-5.0	17b	Tix	-	
LM330T-5.0	R+	Io=150mA Vo:4.8...5.2V Vin:14V P:int	3P	Nsc		data pinout
LM 330 T-5.0	Z-IC	Io-drop, +5V, 0,15A, 0...+70	17b	Nsc	-	data pinout
LM 331(A)DP,N	LIN-IC	=LM 131(A)H: 0...+70°	8-DIP	Nsc,Tho	KA 331, RC 4151	pinout
LM 331(A)H	LIN-IC	=LM 131H: 0...+70°	81	Nsc,Tho	-	
LM333	ER-	Io=3A Vo:-1.2...-32V P:int	3OP	Nsc		data pdf pinout
LM 334 H-...,Z-...	LIN-IC	=LM 134..: 0...+70°		Nsc,Tho	-	

Type	Device	Short Description	Fig.	Manu	Comparision Types	More at
LM 334 H	LIN-IC	Adj. Current Source, 1µ...10mA, ±3%	2(Reg+-)	Nsc	-	pdf pinout
LM 334 M(X)	LIN-IC	SMD, Adj. Curr. Source, 1µ...10mA, ±3%	8-MDIP	Nsc		pdf pinout
LM 334 SM(X)	LIN-IC	SMD, Adj. Curr. Source, 1µ...10mA, ±3%	8-MDIP	Nsc		pdf pinout
LM 334 Z	LIN-IC	Adj. Current Source, 1µ...10mA, ±3%	7(-Reg+)	Nsc	-	pdf pinout
LM 335(A)D	LIN-IC	=LM 135(A)H: SMD, -40..+100°	8-MDIP	Tho		pdf
LM 335Z	LIN-IC	=LM 135(A)H: -40..+100°		Nsc,Tho		pdf
LM336BD-2.5	Z+	Io=10mA Vo:2.44..2.54V P:int	8S	Tix		data pinout
LM336BH-2.5	Z+	Io=10mA Vo:2.44..2.54V	3A	Ltc		data pinout
LM336BLP-2.5	Z+	Io=10mA Vo:2.44..2.54V P:int	3N	Tix		data pinout
LM336BM-2.5	Z+	Io=10mA Vo:2.44..2.54V P:int	8S	Nsc		data pinout
LM336BM-5.0	Z+	Io=10mA Vo:4.9..5.1V	8S	Nsc		data pinout
LM336BZ-2.5	Z+	Io=10mA Vo:2.44..2.54V	3N	Nsc		data pdf pinout
LM336BZ-5.0	Z+	Io=10mA Vo:4.9..5.1V	3N	Nsc		data pinout
LM336D-2.5	Z+	Io=10mA Vo:2.39..2.59V P:int	8S	Tix		data pinout
LM 336(B)D	Ref-Z-IC	=LM 336(B)Z: SMD	8-MDIP	Tho		data pinout
LM336H-2.5	Z+	Io=10mA Vo:2.39..2.59V	3A	Nsc		data pinout
LM 336(B)H-...	Ref-Z-IC	=LM 136(A)H..: ±4%(B=±2%), 0...+70°	7(RegKA)	Nsc	KA 336-...	data pinout
LM336LP-2.5	Z+	Io=10mA Vo:2.39..2.59V P:int	3N	Tix		data pinout
LM336M-2.5	Z+	Io=10mA Vo:2.39..2.59V	8S	Nsc		data pinout
LM336M-5.0	Z+	Io=10mA Vo:4.8..5.2V	8S	Nsc		data pinout
LM336Z-2.5	Z+	Io=10mA Vo:2.39..2.59V	3N	Nsc		data pinout
LM336Z-5.0	Z+	Io=10mA Vo:4.8..5.2V	3N	Nsc		data pinout
LM 336(B)Z-...	Ref-Z-IC	=LM 136(A)Z: ±4%(B=±2%), 0...+70°	7(AKReg)	Nsc,Tho	KA 336-...	data pinout
LM337H	ER-	Io=int Vo:-1.2..-37V P:int	3A	Tho		data pdf pinout
LM 337 H	Z-IC	=LM 137H: 0...+125°	2I	Mot,Nsc,Tho	LM 137..., LM 237...	data pdf pinout
LM 337 HV(H,K)	Z-IC	=LM 137H: -1,2..-47V, -0..+125°		Nsc	LM 137HV..., LM 237HV...	data pdf pinout
LM337HVH	ER-	Io=200mA Vo:-1.2..-47V P:int	3A	Nsc		data pinout
LM337HVK	ER-	Io=1.5A Vo:-1.2..-47V P:int	3O	Nsc		data pdf pinout
LM337ISP-3	ER-	Io=int Vo:-1.2..-37V P:int	3P	Tho		data pinout
LM337K	ER-	Io=int Vo:-1.2..-37V P:int	3O	Tho		data pdf pinout
LM 337 K	Z-IC	=LM 137H: >1,5A, 20W, 0...+125°	23m	Mot,Nsc,Tho	LM 137..., LM 237...	data pdf pinout
LM337KC	ER-	Io=1.5A Vo:-1.2..-37V P:2W	3P	Tix		data pdf pinout
LM 337 KC,SP,T	Z-IC	=LM 137H: >1,5A, 15W, 0...+125°	17n	Nsc,Tix,Tho	KA 337, LM 137..., LM 237...	data pdf pinout
LM337L	ER-	Io=100mA Vo:-1.2..-37V P:int	8S,3N	Nsc		data pdf pinout
LM 317 LZ	Z-IC	=LM 137H: >0,1A, >0,62W, 0..+125°	7o	Nsc	-	data
LM 337 LZ	Z-IC	=LM 137H: >0,1A, >0,62W, 0..+125°	7o	Nsc	-	data pdf
LM337MP	ER-	Io=500mA Vo:-1.2..-37V P:7.5W	3P	Nsc		data pinout
LM 337 MP	Z-IC	=LM 137H: >0,5A, 7,5W, 0..+125°	13n	Nsc	LM 137..., LM 237...	pdf
LM337MR	ER-	Io=int Vo:-1.2..-37V P:int	3O	Mot		data pinout
LM 337 MR	Z-IC	=LM 137H: 0..+125	22m	Mot	LM 137..., LM 237...	data pdf pinout
LM337MT	ER-	Io=500mA Vo:-1.2..-37V P:int	3P	Mot		data pdf pinout
LM 337 MT	Z-IC	=LM 137H: >0,5A, 0..+125°	17n	Mot	KA 337, LM 137..., LM 237...	data pdf pinout
LM337SP	ER-	Io=int Vo:-1.2..-37V P:int	3P	Tho		data pinout
LM 337 SP3	Z-IC	=LM 137H: >1,5A, 20W, 0..+125°	18n	Tho	-	data pinout
LM337T	ER-	Io=1.5A Vo:-1.2..-37V P:15W	3P	Mot,Nsc		data pinout
LM338A	ER+	Io=8A Vo:1.2..32V P:int	3OP	Nsc		data pdf pinout
LM338K	ER+	Io=int Vo:1.2..32V P:int	3O	Nsc,Tho		data pdf pinout
LM 338 K	Z-IC	=LM 138K: 0..+125°	23k	Nsc,Tho	LM 138.., LM 238.., µA 338K...	data pdf pinout
LM338T	ER+	Io=8A Vo:1.2..32V P:int	3P	Nsc		data pinout
LM 338 T	Z-IC	=LM 138K: 0..+125°	17l	Nsc	µA 338U...	data pinout
LM339	4xCOMP	Vs:36V Vi0:2mV	14SDd	Har		data pinout
LM339A	4xCOMP	Vs:36V Vi0:1mV	14DSd	Har		data pinout
LM 339(A)CM,D,FP,M	KOP-IC	=LM 139..: SMD, 0...+70°	14-MDIP	Nsc,Tho,Tix	→Serie 139	data pdf pinout
LM 339(A)DP,N	KOP-IC	=LM 139..: 0..+70°	14-DIP	Mot,Nsc,++	→Serie 139	data pdf pinout
LM 339(A)F,J	KOP-IC	=LM 139..: 0..+70°	14-DIC	Phi	→Serie 139	data pinout
LM339x2	8xCOMP	Vs:36V Vu:200V/mV Vi0:2mV	30S	Tix		data pinout
LM340AK-10	R+	Io=2.4A Vo:9.6..10.4V Vin:17V	3O	Nsc		data pinout
LM340AK-12	R+	Io=2.4A Vo:11.75..12.25V Vin:19V	3O	Mot,Nsc		data pinout
LM340AK-15	R+	Io=2.4A Vo:14.7..15.3V Vin:23V P:int	3O	Mot,Nsc		data pinout
LM340AK-18	R+	Io=2.4A Vo:17.3..18.7V Vin:27V	3O	Nsc		data pinout
LM340AK-24	R+	Io=2.4A Vo:23..25V Vin:33V	3O	Nsc		data pinout
LM340AK-5	R+	Io=2.4A Vo:4.8..5.2V Vin:7.5..20V	3O	Nsc		data pinout
LM340AK-5.0	R+	Io=1.5A Vo:4.9..5.1V Vin:10V P:int	3O	Mot,Nsc		data pinout
LM340AK-6	R+	Io=2.4A Vo:5.76..6.24V Vin:11V	3O	Nsc		data pinout
LM340AK-8	R+	Io=2.4A Vo:7.7..8.3V Vin:14V	3O	Nsc		data pinout
LM340AKC-10	R+	Io=2.4A Vo:9.6..10.4V Vin:17V	3O	Nsc		data pinout
LM340AKC-12	R+	Io=2.4A Vo:11.5..12.5V Vin:19V	3O	Tix		data pinout
LM340AKC-15	R+	Io=2.4A Vo:14.4..15.6V Vin:23V	3O	Nsc		data pinout
LM340AKC-18	R+	Io=2.4A Vo:17.3..18.7V Vin:27V	3O	Nsc		data pinout
LM340AKC-24	R+	Io=2.4A Vo:23..25V Vin:33V	3O	Nsc		data pinout
LM340AKC-5	R+	Io=2.4A Vo:4.8..5.2V Vin:10V	3O	Nsc		data pinout
LM340AKC-6	R+	Io=2.4A Vo:5.76..6.24V Vin:11V	3O	Nsc		data pinout
LM340AKC-8	R+	Io=2.4A Vo:7.7..8.3V Vin:14V	3O	Nsc		data pinout
LM340AT-10	R+	Io=2.4A Vo:9.6..10.4V Vin:17V	3P	Nsc		data pinout
LM340AT-12	R+	Io=1.5A Vo:11.75..12.25V Vin:19V	3P	Mot,Nsc,Ons		data pdf pinout
LM340AT-15	R+	Io=1.5A Vo:14.7..15.3V Vin:23V P:int	3P	Mot,Nsc		data pinout
LM340AT-18	R+	Io=2.4A Vo:17.3..18.7V Vin:27V	3P	Nsc		data pinout
LM340AT-24	R+	Io=2.4A Vo:23..25V Vin:33V	3P	Nsc		data pinout
LM340AT-5	R+	Io=1.5A Vo:4.8..5.2V Vin:10V P:int	3P	Mot,Ons		data pdf pinout
LM340AT-5.0	R+	Io=2.4A Vo:4.8..5.2V Vin:7.5..20V	3P	Nsc		data pinout
LM340AT-6.0	R+	Io=2.4A Vo:5.76..6.24V Vin:11V	3P	Nsc		data pinout
LM340AT-8.0	R+	Io=2.4A Vo:7.7..8.3V Vin:14V	3P	Nsc		data pinout
LM 340 DA-5	Z-IC	+5V, 1A, 15W, 0...+70°	23a	Mot,Phi	... 7805... (TO-3)	data pdf pinout
LM340DA-12	R+	Io=1A Vo:11.5..12.5V P:15W	3O	Sig		data pinout
LM 340 DA-12	Z-IC	+12V, 1A, 15W, 0...+70°	23a	Mot,Phi	... 7812... (TO-3)	data pdf pinout

Type	Device	Short Description	Fig.	Manu	Comparision Types	More at
LM340DA-15	R+	Io=1A Vo:14.4...15.6V	3O	Sig		data pinout
LM340DA-18	R+	Io=1A Vo:17.3...18.7V	3O	Sig		data pinout
LM340DA-24	R+	Io=1A Vo:23...25V	3O	Sig		data pinout
LM340DA-5	R+	Io=1A Vo:4.8...5.2V P:15W	3O	Sig		data pinout
LM340DA-6	R+	Io=1A Vo:5.75...6.25V P:15W	3O	Sig		data pinout
LM340DA-8	R+	Io=1A Vo:7.7...8.3V P:15W	3O	Sig		data pinout
LM 340(A)K,KC-5	Z-IC	+5V, 1,5A, 20W, 4%(A=2%), 0...+125°	23a	Mot,Nsc	...7805... (TO-3)	data pdf pinout
LM 340(A)K,KC-6	Z-IC	+6V, 1,5A, 20W, 4%(A=2%), 0...+125°	23a	Mot,Nsc	...7806... (TO-3)	data pdf pinout
LM 340(A)K,KC-8	Z-IC	+8V, 1,5A, 20W, 4%(A=2%), 0...+125°	23a	Mot,Nsc	...7808... (TO-3)	data pdf pinout
LM340K-10	R+	Io=2.4A Vo:9.5...10.5V Vin:17V	3O	Nsc		data pinout
LM 340(A)K,KC-10	Z-IC	+10V, 1,5A, 20W, 4%(A=2%), 0...+125°	23a	Mot,Nsc	...7810... (TO-3)	data pdf pinout
LM340K-12	R+	Io=int Vo:11.5...12.5V Vin:19V P:int	3O	Mot		data pinout
LM 340(A)K,KC-12	Z-IC	+12V, 1,5A, 20W, 4%(A=2%), 0...+125°	23a	Mot,Nsc	...7812... (TO-3)	data pdf pinout
LM340K-15	R+	Io=int Vo:14.4...15.6V Vin:23V P:int	3O	Mot		data pinout
LM 340(A)K,KC-15	Z-IC	+15V, 1,5A, 20W, 4%(A=2%), 0...+125°	23a	Mot,Nsc	...7815... (TO-3)	data pdf pinout
LM340K18	R+	Io=2.4A Vo:17.1...18.9V Vin:27V	3O	Nsc		data pinout
LM 340(A)K,KC-18	Z-IC	+18V, 1,5A, 20W, 4%(A=2%), 0...+125°	23a	Mot,Nsc	...7818... (TO-3)	data pdf pinout
LM340K-24	R+	Io=2.4A Vo:22.8...25.2V Vin:33V	3O	Nsc		data pinout
LM 340(A)K,KC-24	Z-IC	+24V, 1,5A, 20W, 4%(A=2%), 0...+125°	23a	Mot,Nsc	...7824... (TO-3)	data pdf pinout
LM340K-5	R+	Io=2.4A Vo:4.75...5.25V Vin:7.5...20V	3O	Nsc		data pinout
LM340K-5.0	R+	Io=1.5A Vo:4.8...5.2V Vin:10V P:int	3O	Mot,Nsc		data pinout
LM340K-6	R+	Io=2.4A Vo:5.7...6.3V Vin:11V	3O	Nsc		data pinout
LM340K-6.0	R+	Io=int Vo:5.75...6.25V Vin:11V P:int	3O	Mot		data pinout
LM340K-8	R+	Io=2.4A Vo:7.6...8.4V Vin:14V	3O	Nsc		data pinout
LM340K-8.0	R+	Io=int Vo:7.7...8.3V Vin:14V P:int	3O	Mot,Nsc		data pinout
LM 340(A)KC,T,U-5	Z-IC	+5V, 1,5A, 15W, 4%(A=2%), 0...+125°	17b	Mot,Nsc++	...7805... (TO-220)	data pdf pinout
LM 340(A)KC,T,U-6	Z-IC	+6V, 1,5A, 15W, 4%(A=2%), 0...+125°	17b	Mot,Nsc++	...7806... (TO-220)	data pdf pinout
LM 340(A)KC,T,U-8	Z-IC	+8V, 1,5A, 15W, 4%(A=2%), 0...+125°	17b	Mot,Nsc++	...7808... (TO-220)	data pdf pinout
LM 340(A)KC,T,U-10	Z-IC	+10V, 1,5A, 15W, 4%(A=2%), 0...+125°	17b	Mot,Nsc++	...7810... (TO-220)	data pdf pinout
LM 340(A)KC,T,U-12	Z-IC	+12V, 1,5A, 15W, 4%(A=2%), 0...+125°	17b	Mot,Nsc++	...7812... (TO-220)	data pdf pinout
LM 340(A)KC,T,U-15	Z-IC	+15V, 1,5A, 15W, 4%(A=2%), 0...+125°	17b	Mot,Nsc++	...7815... (TO-220)	data pdf pinout
LM 340(A)KC,T,U-18	Z-IC	+18V, 1,5A, 15W, 4%(A=2%))	17b	Mot,Nsc++	...7818... (TO-220)	data pdf pinout
LM 340(A)KC,T,U-24	Z-IC	+24V, 1,5A, 15W, 4%(A=2%), 0...+125°	17b	Mot,Nsc++	...7824... (TO-220)	data pdf pinout
LM340KC-10	R+	Io=2.4A Vo:9.5...10.5V Vin:17V	3O	Nsc		data pinout
LM340KC-12	R+	Io=1.5A Vo:11.5...12.5V Vin:19V P:int	3O	Nsc		data pinout
LM340KC-15	R+	Io=1.5A Vo:14.4...15.6V Vin:23V P:int	3O	Nsc		data pinout
LM340KC-18	R+	Io=2.4A Vo:17.1...18.9V Vin:27V	3O	Nsc		data pinout
LM340KC-24	R+	Io=2.4A Vo:22.8...25.2V Vin:33V	3O	Nsc		data pinout
LM340KC-5	R+	Io=2.4A Vo:4.75...5.25V Vin:7.5...20V	3O	Nsc		data pinout
LM340KC-5.0	R+	Io=1.5A Vo:4.8...5.2V Vin:10V P:int	3O	Nsc		data pinout
LM340KC-6	R+	Io=2.4A Vo:5.7...6.3V Vin:11V	3O	Nsc		data pinout
LM340KC-8	R+	Io=2.4A Vo:7.6...8.4V Vin:14V	3O	Nsc		data pinout
LM340LAH-10	R+	Io=100mA Vo:9.7...10.3V Vin:16V	3A	Nsc		data pinout
LM340LAH-12	R+	Io=100mA Vo:11.5...12.5V P:int	3A	Nsc		data pinout
LM340LAH-15	R+	Io=100mA Vo:14.55...15.45V P:int	3A	Nsc		data pinout
LM340LAH-18	R+	Io=100mA Vo:17.45...18.55V Vin:27V	3A	Nsc		data pinout
LM340LAH-24	R+	Io=100mA Vo:23.3...24.7V Vin:33V	3A	Nsc		data pinout
LM340LAH-5	R+	Io=100mA Vo:4.85...5.15V Vin:7...20V	3A	Nsc		data pinout
LM340LAH-5.0	R+	Io=100mA Vo:4.9...5.1V Vin:7.2...20V	3A	Nsc		data pinout
LM340LAH-6	R+	Io=100mA Vo:5.82...6.18V Vin:11V	3A	Nsc		data pinout
LM340LAH-8	R+	Io=100mA Vo:7.76...8.24V Vin:14V	3A	Nsc		data pinout
LM 340 LAH-...	Z-IC	=LM 340(A)K,KC-...: 0,1A, 2W	2e	Nsc	...78xx... (TO-5)	data pdf pinout
LM340LAZ-10	R+	Io=100mA Vo:9.7...10.3V Vin:16V	3N	Nsc		data pinout
LM340LAZ-12	R+	Io=100mA Vo:11.65...12.35V P:int	3N	Nsc		data pinout
LM340LAZ-15	R+	Io=100mA Vo:14.55...15.45V P:int	3N	Nsc		data pinout
LM340LAZ-18	R+	Io=100mA Vo:17.45...18.55V Vin:27V	3N	Nsc		data pinout
LM340LAZ-24	R+	Io=100mA Vo:23.3...24.7V Vin:33V	3N	Nsc		data pinout
LM340LAZ-5	R+	Io=100mA Vo:4.85...5.15V Vin:7...20V	3N	Nsc		data pinout
LM340LAZ-5.0	R+	Io=100mA Vo:4.8...5.2V Vin:7.2...20V	3N	Nsc		data pinout
LM340LAZ-6	R+	Io=100mA Vo:5.82...6.18V Vin:11V	3N	Nsc		data pinout
LM340LAZ-8	R+	Io=100mA Vo:7.76...8.24V Vin:14V	3N	Nsc		data pinout
LM 340 LAZ-...	Z-IC	=LM 340(A)K-...: 0,1A, 1,2W	7b	Nsc	...78Lxx... (TO-92)	data pdf pinout
LM340T-10	R+	Io=2.4A Vo:9.5...10.5V Vin:17V	3P	Nsc		data pinout
LM340T-12	R+	Io=1.5A Vo:11.5...12.5V Vin:19V P:int	3P	Mot,Nsc,Ons		data pdf pinout
LM340T-15	R+	Io=1.5A Vo:14.4...15.6V Vin:23V P:int	3P	Mot,Nsc		data pinout
LM340T-18	R+	Io=2.4A Vo:17.1...18.9V Vin:27V	3P	Mot,Nsc		data pinout
LM340T-24	R+	Io=2.4A Vo:22.8...25.2V Vin:33V	3P	Mot,Nsc		data pinout
LM340T-5	R+	Io=2.4A Vo:4.75...5.25V Vin:7.5...20V	3P	Nsc,Ons		data pdf pinout
LM340T-5.0	R+	Io=1.5A Vo:4.8...5.2V Vin:10V P:int	3P	Mot,Nsc		data pinout
LM340T-6	R+	Io=2.4A Vo:5.7...6.3V Vin:11V	3P	Nsc		data pinout
LM340T-6.0	R+	Io=1.5A Vo:5.75...6.25V Vin:11V P:int	3P	Mot,Nsc		data pinout
LM340T-8	R+	Io=2.4A Vo:7.6...8.4V Vin:14V	3P	Nsc		data pinout
LM340T-8.0	R+	Io=1A Vo:7.7...8.3V Vin:14V P:int	3P	Mot,Nsc		data pinout
LM340U-12	R+	Io=1A Vo:11.5...12.5V P:15W	3P	Sig		data pinout
LM340U-15	R+	Io=1A Vo:14.4...15.6V	3P	Sig		data pinout
LM340U-18	R+	Io=1A Vo:17.3...18.7V	3P	Sig		data pinout
LM340U-24	R+	Io=1A Vo:23...25V	3P	Sig		data pinout
LM340U-5	R+	Io=1A Vo:4.8...5.2V P:15W	3P	Sig		data pinout
LM340U-6	R+	Io=1A Vo:5.75...6.25V P:15W	3P	Sig		data pinout
LM340U-8	R+	Io=1A Vo:7.7...8.3V P:15W	3P	Sig		data pinout
LM 341 P-5	Z-IC	+5V, 0,5A, 7,5W, 0...+70°	13b	Nsc	...7805... (TO-202)	data pinout
LM 341 P-6	Z-IC	+6V, 0,5A, 7,5W, 0...+70°	13b	Nsc	...7806... (TO-202)	data pinout
LM 341 P-8	Z-IC	+8V, 0,5A, 7,5W, 0...+70°	13b	Nsc	...7808... (TO-202)	data pinout
LM341P-10	R+	Io=500mA Vo:9.6...10.4V Vin:17V P:7.5W	3P	Nsc		data pinout
LM 341 P-10	Z-IC	+10V, 0,5A, 7,5W, 0...+70°	13b	Nsc	...7810... (TO-202)	data pinout

Type	Device	Short Description	Fig.	Manu	Comparision Types	More at
LM341P-12	R+	Io=500mA Vo:11.5...12.5V Vin:19V P:int	3P	Nsc		data pinout
LM 341 P-12	Z-IC	+12V, 0,5A, 7,5W, 0...+70°	13b	Nsc	... 7812... (TO-202)	data pinout
LM341P-15	R+	Io=200mA Vo:14.4...15.6V	3P	Nsc		data pinout
LM 341 P-15	Z-IC	+15V, 0,5A, 7,5W, 0...+70°	13b	Nsc	... 7815... (TO-202)	data pinout
LM341P-18	R+	Io=500mA Vo:17.25...18.75V	3P	Nsc		data pinout
LM 341 P-18	Z-IC	+18V, 0,5A, 7,5W, 0...+70°	13b	Nsc	... 7818... (TO-202)	data pinout
LM341P-24	R+	Io=500mA Vo:23..25V	3P	Nsc		data pinout
LM 341 P-24	Z-IC	+24V, 0,5A, 7,5W, 0...+70°	13b	Nsc	... 7824... (TO-202)	data pinout
LM341P-5	R+	Io=500mA Vo:4.8...5.2V P:7.5W	3P	Nsc		data pinout
LM341P-5.0	R+	Io=500mA Vo:4.8...5.2V Vin:10V P:int	3P	Nsc		data pinout
LM341P-6	R+	Io=500mA Vo:5.75...6.25V P:7.5W	3P	Nsc		data pinout
LM341P-8	R+	Io=500mA Vo:7.7...8.3V P:7.5W	3P	Nsc		data pinout
LM341T-12	R+	Io=500mA Vo:11.5...12.5V Vin:19V P:int	3P	Nsc		data pinout
LM341T-15	R+	Io=500mA Vo:14.4...15.6V Vin:23V P:int	3P	Nsc		data pinout
LM341T-5.0	R+	Io=500mA Vo:4.8...5.2V Vin:10V P:int	3P	Nsc		data pinout
LM 341 T-...	Z-IC	=LM 341P-...	17b	Nsc	... 78xx... (TO-220)	data pinout
LM 342 P-5	Z-IC	+5V, 0,25A, 7,5W, 0...+70°	13b	Nsc	... 7805... (TO-202)	data pinout
LM 342 P-6	Z-IC	+6V, 0,25A, 7,5W, 0...+70°	13b	Nsc	... 7806... (TO-202)	data pinout
LM 342 P-8	Z-IC	+8V, 0,25A, 7,5W, 0...+70°	13b	Nsc	... 7808... (TO-202)	data pinout
LM342P-10	R+	Io=200mA Vo:9.6...10.4V	3P	Nsc		data pinout
LM 342 P-10	Z-IC	+10V, 0,25A, 7,5W, 0...+70°	13b	Nsc	... 7810... (TO-202)	data pinout
LM342P-12	R+	Io=250mA Vo:11.5...12.5V P:int	3P	Nsc		data pinout
LM 342 P-12	Z-IC	+12V, 0,25A, 7,5W, 0...+70°	13b	Nsc	... 7812... (TO-202)	data pinout
LM342P-15	R+	Io=250mA Vo:14.25...15.75V P:int	3P	Nsc		data pinout
LM 342 P-15	Z-IC	+15V, 0,25A, 7,5W, 0...+70°	13b	Nsc	... 7815... (TO-202)	data pinout
LM342P-18	R+	Io=200mA Vo:17.25...18.75V	3P	Nsc		data pinout
LM 342 P-18	Z-IC	+18V, 0,25A, 7,5W, 0...+70°	13b	Nsc	... 7818... (TO-202)	data pinout
LM342P-24	R+	Io=200mA Vo:23..25V	3P	Nsc		data pinout
LM 342 P-24	Z-IC	+24V, 0,25A, 7,5W, 0...+70°	13b	Nsc	... 7824... (TO-202)	data pinout
LM342P-5	R+	Io=200mA Vo:4.8...5.2V	3P	Nsc		data pinout
LM342P-5.0	R+	Io=250mA Vo:4.75...5.25V Vin:7.5...20V	3P	Nsc		data pinout
LM342P-6	R+	Io=200mA Vo:5.75...6.25V	3P	Nsc		data pinout
LM342P-8	R+	Io=200mA Vo:7.7...8.3V	3P	Nsc		data pinout
LM343	HI-VOLT	Vs:±40V Vu:105dB Vo:±25V Vio:2mV	8AD	Has,Nsc		data pdf pinout
LM343A	HI-VOLT	Vs:±55V Vu:106dB Vo:±35V Vio:2mV	8AD	Har		data pinout
LM 343 H	OP-IC	=LM 143H: ±34V, 0...+70°	81	Nsc	→Serie 136	data pinout
LM344	HI-VOLT	Vs:±34V Vu:105dB Vo:±25V Vio:2mV	8A	Nsc		data pdf pinout
LM 344 H	OP-IC	=LM 144H: ±34V, 0...+70°	81	Nsc	LM 144...	data pinout
LM345K-5.0	R-	Io=3A Vo:-4.8...-5.2V Vin:-7.5V P:int	3O	Nsc		data pinout
LM345K-5.2	R-	Io=3A Vo:-5...-5.4V Vin:-7.5V P:int	3O	Nsc		data pinout
LM 345 K-...	Z-IC	=LM 145K-...: 0...+125°	23d	Nsc	LM 145K-..., LM 245K...	data pinout
LM346	4xPROG	Vs:±18V Vu:120dB Vo:±14V Vio:0.5mV	16DS,14D	Tho		data pdf pinout
LM346-2	4xPROG	Vs:±18V Vu:1000V/mV Vo:±14V f:1.2Mc	16D	Mit		data pinout
LM 346 D,FP,M	OP-IC	=LM 146DG: SMD, ±18V, 0...+70°	16-MDIP	Nsc,Tho	LM 146..., LM 246...	data pdf pinout
LM 346 DG,DP,J,N	OP-IC	=LM 146DG: ±18V, 0...+70°	16-DIP,DIC	Tho	LM 146..., LM 246...	data pinout
LM348	4xHI-GAIN	Vs:±22V Vu:160V/mV Vo:±12V Vio:1mV	14SD	Sgs		data pdf pinout
LM348A	4xOP	Vs:±20V Vu:108dB Vo:±12V Vio:0.5mV	14D	Har		data pinout
LM 348 D,DP,J,N	OP-IC	=LM 148DG: ±18V, 0...+70°	14-DIP,DIC	Mot,Nsc,++	→Serie 124	data pinout
LM 348 D,FP,M	OP-IC	=LM 148DG: SMD, ±18V, 0...+70°	14-MDIP	Mot,Nsc,++	→Serie 124	data pinout
LM349	4xOP	Vs:±18V Vu:83.5dB Vo:±12V Vio:1mV	14DS	Nsc		data pinout
LM 349 D,FP	OP-IC	=LM 149DG: SMD, ±18V, 0...+70°	14-MDIP	Tho,Tix	→Serie 124	data pinout
LM 349 DG,DP,J,N	OP-IC	=LM 149DG: ±18V, 0...+70°	14-DIP	Nsc,Tho	→Serie 124	data pinout
LM350	ER+	Io=3A Vo:1.2...33V P:int	3OP	Nsc		data pdf pinout
LM350K	ER+	Io=3A Vo:1.2...33V P:int	3OP	Nsc		data pdf pinout
LM 350(A)K	Z-IC	=LM 150K: 0...+125°	23k	Mot,Nsc	KA 350H, LM 150K, LM 250K..., µA 350K...	data pdf pinout
LM 350(A)T,KC	Z-IC	=LM 150K: 0...+125°	17I	Mot,Nsc,Tix	KA 350, µA 350U...	data pdf pinout
LM358	2xLO-POWER	Vs:32V Vu:100V/mV Vo:3.5/0V Vio:2mV	8GSDAd	Sgs		data pdf pinout
LM 358	LIN/OP-IC	Low Power Dual Op Amp	MDIP	Mtm	-	pdf pinout
LM358A	2xLO-POWER	Vs:32V Vu:100V/mV Vo:3.5/0V Vio:1mV	8SDA	Sgs		data pinout
LM 358 BP	OP-IC	SMD, Dual, Low Power OPAmp 0...70°C, 1MHz, ±1.5/+3...±16/+32V	8-GRID	Nsc	-	pdf
LM 358 H	OP-IC	Dual, Low Power OPAmp 0...70°C, 1MHz, ±1.5/+3...±16/+32V	81	Nsc	-	pdf pinout
LM 358(A)CM,D,FP,M	OP-IC	=LM 158(A)H: SMD, 0...+70°	8-MDIP	Tho,Tix	→Serie 158	data pdf pinout
LM 358(A)DP,N,P	OP-IC	=LM 158(A)H: 0...+70°	8-DIP	Mot,++	→Serie 158	data pdf pinout
LM 358(A)FE,J,JG	OP-IC	=LM 158(A)H: 0...+70°	8-DIC	Mot,Nsc,++	→Serie 158	data pdf pinout
LM 358(A)H,L	OP-IC	=LM 158(A)H: 0...+70°	81	Mot,++	→Serie 158	data pdf pinout
LM 358 S	OP-IC	=LM 158H: 0...70°	9-SIP	Sam	→Serie 158	data pdf pinout
LM 358 W(D/S)	IC	Low power dual operational amplifiers, 0 to +70	MDIP	Stm	-	pdf pinout
LM 358 WN	IC	Low power dual operational amplifiers, 0 to +70	DIC	Stm	-	pdf pinout
LM 358 WP	IC	Low power dual operational amplifiers, 0 to +70	TSOP	Stm	-	pdf pinout
LM 358 (A)M	OP-IC	SMD, Dual, Low Power OPAmp 0...70°C, 1MHz, ±1.5/+3...±16/+32V	8-MDIP	Nsc	-	pdf
LM 358 (A)N	OP-IC	Dual, Low Power OPAmp 0...70°C, 1MHz, ±1.5/+3...±16/+32V	8-DIP	Nsc	-	pdf
LM359	2xHI-SPEED	Vs:±11V Vu:72dB Vo:10.3/0V f:400Mc	14DS	Nsc		data pdf pinout
LM 359 J,N	OP-IC	=LM 159J: 0...70°	14-DIC,DIP	Nsc	LM 159...	data pdf pinout
LM 359 M	OP-IC	=LM 159J: SMD, 0...70°	14-MDIP	Nsc	LM 159...	data pdf pinout
LM360	FAST-COMP	Vs:±8V Vo:TTL Vio:2mV	14D,8ADS	Nsc		data pinout
LM 360 D,J-14,N-14	KOP-IC	=LM 160H: 0...+70°	14-DIC,DIP	Nsc	µA 760...	data pinout
LM 360 H	KOP-IC	=LM 160H: 0...+70°	81	Nsc	µA 760...	data pinout
LM 360 M	KOP-IC	=LM 160H: SMD, 0...+70°	14-DIC,DIP	Nsc		data pdf pinout

Type	Device	Short Description	Fig.	Manu	Comparision Types	More at
LM 360 N	KOP-IC	=LM 160H: 0...+70°	8-DIP	Nsc	NJM 360D, µA 760...	data pinout
LM361	HI-SPEED	Vs:±16V Vu:70dB Vo:TTL Vi0:1mV	14D,10AS	Nsc		data pinout
LM 361 D,J,N	KOP-IC	=LM 161H: 0...+70°	14-DIC,DIP	Nsc	NE 529..., SE 529...	data pinout
LM 361 H	KOP-IC	=LM 161H: 0...+70°	81	Nsc	NE 529..., SE 529...	data pinout
LM 361 M	KOP-IC	=LM 161H: 0...+70°	14-MDIP	Nsc	-	data pdf pinout
LM363	GAIN-SET	Vs:±18V Vu:0.1V/mV Vi0:2.5mV	16D,8A	Nsc		data pdf pinout
LM 363(A)D	OP-IC	=LM 163(A)H: 0...+70°	16-DIC	Nsc	-	data pinout
LM 363(A)H	OP-IC	=LM 163(A)H: 0...+70°	81	Nsc	-	data pinout
LM368H-10	R+	Io=50mA P:600mW	8A	Nsc		data pinout
LM368H-2.5	ER+	Vo:1.9...5.2V Vin:4.9...10.5V P:600mW	8A	Nsc		data pinout
LM368H-5.0	R+	Io=50mA P:600mW	8A	Nsc		data pinout
LM368M-2.5	ER+	Vo:1.9...5.2V Vin:4.9...10.5V P:600mW	8S	Nsc		data pinout
!LM368M-5.0	R+	Io=50mA P:600mW	8S	Nsc		data pinout
LM368N-2.5	ER+	Vo:1.9...5.2V Vin:4.9...10.5V P:600mW	8D	Nsc		data pinout
LM368N-5.0	R+	Io=50mA P:600mW	8D	Nsc		data pinout
LM368YH-10	R+	Io=50mA P:600mW	8A	Nsc		data pinout
LM368YH-2.5	ER+	Vo:1.9...5.2V Vin:4.9...10.5V P:600mW	8A	Nsc		data pinout
LM368YH-5.0	R+	Io=50mA P:600mW	8A	Nsc		data pinout
LM369	R+	Io=50mA P:600mW	8ADS,3N	Nsc		data pinout
LM 370(N)	LIN-IC		14-DIP	Nsc	-	
LM 376	ER+	Io=25mA Vo:5...37V P:800mW	8DA	Tix		data pinout
LM 376 JG,N,P	Z-IC	+5...37V, 25mA, 0...+70°	8-DIC,DIP	Nsc,Tix	µA 376	data pinout
LM 376 L	Z-IC	=LM 376JG,N,P: Fig.→	81	Tix		data pinout
LM 377(N)	LIN-IC	2x NF-E, 10...26V, 2x>2W(20V/8Ω)	14-DIP	Nsc	LM 1877	
LM 378(N)	LIN-IC	2x NF-E, 10...35V, 2x>4W(24V/8Ω,16Ω)	14-DIP	Nsc	-	
LM 379(N)	LIN-IC	2x NF-E, 10...35V, 2x>6W(28V/8Ω)	14-DIP+d	Nsc		
LM380	AUDIO-P (5W)	Vs:22V Vu:34dB Vo:14V f:0.1Mkc	8D,14D	Nsc		data pdf pinout
LM 380	LIN-IC	NF-E, 10...22V, 1.3A, >2,5W(18V/8Ohm	14-DIP	Nsc		pdf pinout
LM 380 N-8	LIN-IC	=LM 380(N): Fig.→	8-DIP	Nsc		pdf pinout
LM 380 N	LIN-IC	NF-E, 10...22V, 1.3A, >2,5W(18V/8Ohm	14-DIP	Nsc		pdf
LM381	2xOP	Vs:40V f:15Mc f:75kc	14D	Nsc		data pdf pinout
LM 381(N),A(N)	LIN-IC	2x NF-V, ra, Ur<1(A<0,7)µV,	14-DIP	Nsc		
LM382	2xOP	Vs:40V f:15Mc f:75kc	14D	Nsc		data pinout
LM 382(N)	LIN-IC	2x NF-V, ra, Ucc=9...40V	14-DIP	Nsc	-	
LM383	AUDIO-P (10W	Vs:25V Vu:40dB f:30kc	5P	Nsc		data pdf pinout
LM 383(T),A(T)	LIN-IC	NF-E, 5...20V, 3,5A, >7W(14V/2Ω)	86/5Pin	Nsc	LM 2002(A), TDA 2002(A), TDA 2003	pdf pinout
LM383A	AUDIO-P (10W	Vs:40V Vu:40dB f:30kc	5P	Nsc		data pdf pinout
LM384	AUDIO-P (10W	Vs:28V Vu:&0.05dB f:450kc	14D	Nsc		data pdf pinout
LM 384(N)	LIN-IC	NF-E, 12...26V, >5W(22V/8Ω)	14-DIP	Nsc	-	pinout
LM385BCOA-1.2	Z+	Io=20mA Vo:1.223...1.235V	8S	Tcs		data pinout
LM385BCOA-2.5	Z+	Io=20mA Vo:2.462...2.538V	8S	Tcs		data pinout
LM385BCZB-1.2	Z+	Io=20mA Vo:1.223...1.235V	3N	Tcs		data pinout
LM385BCZB-2.5	Z+	Io=20mA Vo:2.462...2.538V	3N	Tcs		data pinout
LM385BD-1.2	Z+	Io=20mA Vo:1.22...1.24V P:int	8S	Tix		data pinout
LM385BD-2.5	Z+	Io=20mA Vo:2.62...2.53V P:int	8S	Tix		data pinout
LM385BH-1.2	Z+	Io=10mA Vo:1.22...1.25V	2A	Ltc		data pinout
LM385BH-2.5	Z+	Io=10mA Vo:2.46...2.54V	2A	Ltc		data pinout
LM385BLP-1.2	Z+	Io=20mA Vo:1.22...1.24V P:int	3N	Tix		data pinout
LM385BLP-2.5	Z+	Io=20mA Vo:2.62...2.53V P:int	3N	Tix		data pinout
LM385BM-1.2	Z+	Io=20mA Vo:1.223...1.247V	8S	Nsc		data pinout
LM385BM-2.5	Z+	Io=20mA Vo:2.462...2.538V	8S	Nsc		data pinout
LM385BXZ	EZ+	Io=20mA Vo:1.24...5.3V	3N	Nsc		data pdf pinout
LM385BXZ1.2	Z+	Io=20mA Vo:1.223...1.247V	3N	Nsc		data pinout
LM385BXZ2.5	Z+	Io=20mA Vo:2.462...2.538V	3N	Nsc		data pinout
LM385BYZ	EZ+	Io=20mA Vo:1.24...5.3V	3N	Nsc		data pdf pinout
LM385BYZ1.2	Z+	Io=20mA Vo:1.223...1.247V	3N	Nsc		data pinout
LM385BYZ2.5	Z+	Io=20mA Vo:2.462...2.538V	3N	Nsc		data pinout
LM385BZ-1.2	Z+	Io=20mA Vo:1.223...1.247V	3N	Nsc		data pinout
LM385BZ-2.5	Z+	Io=20mA Vo:2.462...2.538V	3N	Nsc		data pinout
LM385COA-1.2	Z+	Io=20mA Vo:1.205...1.270V	8S	Tcs		data pinout
LM385COA-2.5	Z+	Io=20mA Vo:2.425...2.575V	8S	Tcs		data pinout
LM385CZB-1.2	Z+	Io=20mA Vo:1.205...1.270V	3N	Tcs		data pinout
LM385CZB-2.5	Z+	Io=20mA Vo:2.425...2.575V	3N	Tcs		data pinout
LM385D-1.2	Z+	Io=20mA Vo:1.21...1.26V P:int	8S	Tix		data pinout
LM385D-2.5	Z+	Io=20mA Vo:2.42...2.57V P:int	8S	Tix		data pinout
LM385H-1.2	Z+	Io=10mA Vo:1.2...1.26V	2A	Ltc		data pinout
LM385H-2.5	Z+	Io=10mA Vo:2.43...2.58V	2A	Ltc		data pinout
LM 385(B)H-...	Ref-Z-IC	=LM 185H-...: ±3%(B=±1...1,5%), 0...+70°	2(K-A)	Nsc	KA 385-...	data pinout
LM385LP-1.2	Z+	Io=20mA Vo:1.21...1.26V P:int	3N	Tix		data pinout
LM385LP-2.5	Z+	Io=20mA Vo:2.42...2.57V P:int	3N	Tix		data pinout
LM385M	EZ+	Io=20mA Vo:1.24...5.3V	8S	Nsc		data pdf pinout
LM385M-1.2	Z+	Io=20mA Vo:1.205...1.260V	8S	Nsc		data pinout
LM385M-2.5	Z+	Io=20mA Vo:2.425...2.575V	8S	Nsc		data pinout
LM385S8-1.2	Z+	Io=20mA Vo:1.20...1.26V	8S	Ltc		data pinout
LM385S8-2.5	Z+	Io=20mA Vo:2.43...2.58V	8S	Ltc		data pinout
LM385Z	EZ+	Io=20mA Vo:1.24...5.3V	3N	Nsc		data pdf pinout
LM385Z-1.2	Z+	Io=20mA Vo:1.205...1.260V	3N	Nsc		data pinout
LM385Z-2.5	Z+	Io=20mA Vo:2.425...2.575V	3N	Nsc		data pinout
LM 385(B)Z-...	Ref-Z-IC	=LM 185H-...: ±3%(B=1...1,5%), 0...+70°	7(AK-)	Mot,Nsc	KA 385-...	data pinout
LM386	AUDIO-P (1W)	Vs:15V f:300kc	8SD	Nsc		data pdf pinout
LM 386 D,M-1	LIN-IC	=LM 386(N-1): SMD	8-MDIP	Nsc	KA 386D	pdf pinout
LM 386(N,N-1)	LIN-IC	NF-E, 4...12V, 0,325W(6V/8Ω)	8-DIP	Nac	NJM 386, KA 386	pdf pinout
LM 386 N-3	LIN-IC	=LM 386N-1: 4...12V, 0,7W(9V/8Ω)	8-DIP	Nsc	-	pdf pinout
LM 386 N-4	LIN-IC	=LM 386N-1: 5...18V, 1W(16V/32Ω)	8-DIP	Nsc	-	pdf pinout
LM 386 S	LIN-IC	=LM 386(N-1): Fig.→	9-SIP	Sam	KA 386S	pdf

Type	Device	Short Description	Fig.	Manu	Comparision Types	More at
LM387A	2xOP	Vs:40V f:75kc	8D	Nsc		data pdf pinout
LM 387 A(N)	LIN-IC	=LM 387N: Ucc=9...40V	8-DIP	Nsc	NJM 387	pdf pinout
LM387N	2xOP	Vs:30V f:75kc	8D	Nsc		data pdf pinout
LM 387 N	LIN-IC	2x NF-V, ra, Ucc=9...30V	8-DIP	Nsc	NJM 387	pdf pinout
LM388	AUDIO-P (5W)	Vs:15V Vu:26dB f:300kc	14D	Nsc		data pdf pinout
LM 388(N,N-1)	LIN-IC	NF-E, 4...12V, 2,2W(12V/8Ω)	14-DIP	Nsc	-	pinout
LM 388 N-2	LIN-IC	=LM 388N-1: 4...12V, 0,9W(6V/4Ω)	14-DIP	Nsc	-	pinout
LM 388 N-3	LIN-IC	=LM 388N-1: 5...18V, 3,8W(16V/8Ω)	14-DIP	Nsc	-	pinout
LM 389(N)	LIN-IC	NF-E, 4...12V, 0,5W(9V/16Ω), +3x NPN-T.	18-DIP	Nsc	-	pdf pinout
LM390	AUDIO-P (1W)	Vs:10V Vu:dB f:300kc	14D	Nsc		data pdf pinout
LM 390(A)	Si-P	≈BC 328	7e		→BC 328	data pdf pinout
LM 390(N)	LIN-IC	NF-E, 4...9V, 1W(6V/4Ω)	14-DIP	Nsc		pinout
LM391	AUDIO-OP	Vs:100V Vu:5.5V/mV Vi0:5mV f:25Mc	16D	Nsc		data pdf pinout
LM 391(N)	LIN-IC	NF-Tr, N-60=±30, N-80=±40, N-100=±50V	16-DIP	Nsc		pdf pinout
LM392	OP	Vs:±16V Vu:100dB Vo:±14V Vi0:2mV	8ADS	Nsc		data pinout
LM 392 H	OP/KOP-IC	=LM 192H: 0...70°	81	Nsc	LM 192..., LM 292...	data pinout
LM 392 J,N	OP/KOP-IC	=LM 192H: 0...70°	8-DIC,DIP	Nsc	LM 192..., ... LM 292...	data pinout
LM 392 M	OP/KOP-IC	=LM 192H: SMD, 0...70°	8-MDIP	Nsc	-	data pinout
LM393	2xCOMP	Vs:36V Vu:200V/mV Vi0:2mV	8SDAGd	Tix		data pinout
LM393A	2xCOMP	Vs:36V Vu:200V/mV Vo:3.5/0V Vi0:1mV	8SDA	Sgs		data pinout
LM 393(A)CM,D,FP,M	KOP-IC	=LM 193(A)H: SMD, 0...+70°	8-MDIP	Nsc,Tho,Tix	→Serie 193	data pdf pinout
LM 393(A)DP,N,P	KOP-IC	=LM 193(A)H: 0...+70°	8-DIP	Mot,Nsc,++	→Serie 193	data pdf pinout
LM 393(A)FE,J,JG	KOP-IC	=LM 193(A)H: 0...+70°	8-DIC	Nsc,Tix,++	→Serie 193	data pdf pinout
LM 393(A)H,L	KOP-IC	=LM 193(A)H: 0...+70°	81	Nsc,Mot,++	→Serie 193	data pdf pinout
LM 393 S	KOP-IC	=LM 193(A)H: Fig.→	9-SIP	Sam	→Serie 193	data pdf
LM 394(H)	LIN-IC	=LM 194(H): -25...+85°	81	Nsc	-	pdf
LM 394 B(H)	LIN-IC	=LM 194(H): B>225, -25...+85°	81	Nsc	-	pdf
LM 394 C(H)	LIN-IC	=LM 194(H): 20V, B>225, -25...+85°	81	Nsc	-	pdf
LM 394 BN,CN	LIN-IC	=LM 394B,C(H): Fig.→	8-DIP	Nsc	-	pdf
LM 395 H	LIN-IC	=LM 195H: 36V, 0...+125°	2a*	Nsc	-	pdf
LM 395 K	LIN-IC	=LM 195H: 36V, 0...+125°	23e*	Nsc	-	pdf
LM 395 P	LIN-IC	=LM 195H: 36V, 0...+125°	13q	Nsc	-	pdf
LM 395 T	LIN-IC	=LM 195H: 36V, 0...+125°	17q	Nsc	-	pdf
LM396K	ER+	Io=10A Vo:1.25...15V P:int	3O	Nsc		data pinout
LM 396 K	Z-IC	=LM 196K: 0...+125°	23l	Nsc	LM 196K	data pinout
LM397	UNI	Vs:30V Vu:120V/mV Vi0:2mV	5S	Nsc		data pdf pinout
LM 399(A)	Si-N	≈BC 338	7e		→BC 338	
LM399A	Z+	Io=10mA Vo:6.6...7.3V P:300mW	4A	Nsc		data pdf pinout
LM 399(A)H(-50)	Ref-Z-IC	=LM 199...: ±5%, 0...+70°	5(-+AK)	Nsc	-	data pdf pinout
LM431A	EZ+	Io=0.4mA Vo:2.5...36V P:780mW	3S,8SN	Nsc		data pdf pinout
LM431B	EZ+	Io=0.4mA Vo:2.5...36V P:780mW	3S,8SN	Nsc		data pdf pinout
LM431C	EZ+	Io=0.4mA Vo:2.5...36V P:780mW	3S,8SN	Nsc		data pdf pinout
LM 433 MA	IC	Dual Op Amp with On-chip Fixed 2.5V Reference	47	Nsc	-	pdf pinout
LM529	COMP	Vs:±16V Vu:70dB Vi0:3mV	10A	Nsc		data pinout
LM529C	COMP	Vs:±16V Vu:70dB Vi0:5mV	10A,14D	Nsc		data pinout
LM 555 CH	LIN-IC	=LM 555H: 0...+70°	81	Nsc	→NE 555...	pinout
LM 555 CJ,CN	LIN-IC	=LM 555H: 0...+70°	8-DIC,DIP	Nsc	→NE 555...	pinout
LM 555 CM	LIN-IC	=LM 555H: SMD, 0...+70°	8-MDIP	Nsc	→NE 555...	pdf pinout
LM 555 H	LIN-IC	Zeitgeber/Timer, 5...15V, -25...+125°	81	Nsc	→NE 555...	pinout
LM 556 CJ,CN	LIN-IC	2x Zeitgeber/Timer, 5...15V, 0...+70°	14-DIC,DIP	Nsc	→NE 556...	pinout
LM 556 CM	LIN-IC	SMD, 2x Zeitgeber/Timer, 5...15V	14-MDIP	Nsc	→NE 556...	pinout
LM 556 J	LIN-IC	2x Zeitgeber/Timer, 5...15V, -25...+125°	14-DIC	Nsc	→NE 556...	pinout
LM 565 CH	LIN-IC	PLL, ±5..±12V, 500kHz, 0...+70°	82	Nsc	→NE 565...	pinout
LM 565 CN	LIN-IC	PLL, ±5..±12V, 500kHz, 0...+70°	14-DIP	Nsc	→NE 565...	pinout
LM 565 H	LIN-IC	PLL, ±5..±12V, 500kHz, -55...+125°	82	Nsc	→NE 565...	pdf pinout
LM 566(N)	LIN-IC	VCO, Function/Signal Gen., -55...+125°	8-DIP	Nsc	NE 566...	pinout
LM 566 C(N)	LIN-IC	=LM 566(N): 0...+70°	8-DIP	Nsc	NE 566...	pinout
LM 567CD	LIN-IC	SMD, Pll-ton-/tone-decoder	8-MDIP	Nsc	KA 567D	pinout
LM 567CH	LIN-IC	Pll-ton-/tone-decoder	81	Nsc		pdf pinout
LM 567CM	LIN-IC	SMD, Pll-ton-/tone-decoder	8-MDIP	Nsc	BA 567, NE 567, SE 567, IR 3N05, XR 567	pdf pinout
LM 567CN	LIN-IC	Pll-ton-/tone-decoder	8-DIP	Nsc	BA 567, NE 567, SE 567, IR 3N05, XR 567	pdf pinout
LM 567H	LIN-IC	Pll-ton-/tone-decoder	81	Nsc		pdf pinout
LM 567 LD	LIN-IC	SMD, PLL-Ton-/Tone-Decoder, lo-power	8-MDIP	Sam	KA 567LD	pinout
LM 567 LN	LIN-IC	PLL-Ton-/Tone-Decoder, lo-power	8-DIP	Sam	KA 567L, XRL 567	pinout
LM 578	Ge-P	≈AF 127			→AF 127	
LM 592(N)	LIN-IC	Video Verst./Amp., 90MHz, 0...+70°	14-DIP	Nsc	NE 592F,N, NJM 592D, µA592D,P	data pinout
LM 592 M	LIN-IC	=LM 592(N): SMD	14-MDIP	Nsc	NE 592D, NJM 592M	data pinout
LM 596	Si-N	≈BC 548			→BC 548	
LM604	MUX-OP	Vs:±18V Vu:200V/mV Vo:±12.6V Vi0:1mV	20S,18D	Nsc		data pinout
LM604A	MUX-OP	Vs:±18V Vu:200V/mV Vo:±12.6V f:7Mc	20S,18D	Nsc		data pinout
LM607A	HI-PREC	Vs:±22V Vu:160dB Vo:±13.8V Vi0:15µV	8AD	Nsc		data pinout
LM607B	HI-PREC	Vs:±22V Vu:160dB Vo:±13.8V Vi0:15µµV	8AD	Nsc		data pinout
LM607C	HI-PREC	Vs:±22V Vu:160dB Vo:±13.8V Vi0:15µµV	8ADS	Nsc		data pinout
LM611	OP+V-REF	Vs:36V Vo:1/1.6V Vi0:±2mV f:0.52Mc	8D,14S	Nsc		data pdf pinout
LM611A	OP+V-REF	Vs:36V Vo:1/1.6V Vi0:±2mV f:0.52Mc	8D	Nsc		data pdf pinout
LM611C	OP+V-REF	Vs:32V Vo:1/1.6V Vi0:±2mV f:0.52Mc	14S,8D	Nsc		data pdf pinout
LM612	2xCOMP	Vs:36V Vu:100V/mV Vi0:2mV	8DS	Nsc		data pdf pinout
LM613	2xOP+COMP	Vs:36V Vu:500V/mV Vo:0.8/1.V f:0.8Mc	16DS	Nsc		data pdf pinout
LM613A	2xOP+COMP	Vs:36V Vu:500V/mV Vo:0.8/1.V f:0.8Mc	16D,20C	Nsc		data pdf pinout
LM614	2xOP+V-REF	Vs:36V Vo:1/1.6V Vi0:±2mV f:0.52Mc	16DS	Nsc		data pdf pinout
LM614C	2xOP+V-REF	Vs:32V Vo:1/1.6V Vi0:±2mV f:0.52Mc	16DS	Nsc		data pdf pinout
LM615	4xOP+V-REF	Vs:36V Vu:100V/mV Vi0:2mV	16DS	Nsc		data pdf pinout
LM 621(N)	LIN-IC	3/4-Phase Brushless DC Motor-Tr	18-DIP	Nsc	-	pdf

Type	Device	Short Description ·	Fig.	Manu	Comparision Types	More at
LM 622(N)	LIN-IC	PWM Motor-Tr., ±4,5...±20V	20-DIP	Nsc	-	
LM627A	HI-PREC	Vs:±22V Vu:7000V/mV Vo:±13.8V f:14Mc	8AD	Nsc		data pinout
LM627B	HI-PREC	Vs:±22V Vu:7000V/mV Vo:±13.8V f:14Mc	8AD	Nsc		data pinout
LM627C	HI-PREC	Vs:±22V Vu:7000V/mV Vo:+13.8V f:14Mc	8AD	Nsc		data pinout
LM637A	HI-PREC	Vs:±22V Vu:7000V/mV Vo:±13.8V f:65Mc	8AD	Nsc		data pinout
LM637B	HI-PREC	Vs:±22V Vu:7000V/mV Vo:±13.8V f:65Mc	8AD	Nsc		data pinout
LM637C	HI-PREC	Vs:±22V Vu:7000V/mV Vo:±13.8V f:65Mc	8AD	Nsc		data pinout
LM675	POWER-OP	Vs:±30V Vo:±21V Vi0:1mV f:5.5Mc	5P	Nsc		data pdf pinout
LM688	OP	Vs:18V Vo:±4.7V Vi0:±10mV f:>1Mc	8D,14D	Nsc		data pinout
LM688A	OP	Vs:18V Vo:±4.7V Vi0:±5mV f:>1Mc	8D,14D	Nsc		data pinout
LM 703	IC	Limiter	8-DIP	Nsc		
LM709	UNI	Vs:±18V Vu:<70V/mV Vo:±10V Vi0:<5mV	8A,14D,10F	Nsc		data pinout
LM709A	UNI	Vs:±18V Vu:<70V/mV Vo:±10V Vi0:<2mV	8A,14D,10F	Nsc		data pinout
LM709C	UNI	Vs:±18V Vu:93dB Vo:±13V Vi0:2mV	8A,14DD	Nsc		data pdf pinout
LM 709 CH	OP-IC	=LM 709(A)H: 0...+70°	81	Nsc	→Serie 109	data pdf pinout
LM 709 CN-8	OP-IC	Operational Amplifier	8-DIP	Nsc		data pdf pinout
LM 709 CN	OP-IC	Operational Amplifier	8-DIP	Nsc		data pdf pinout
LM 709(A)H	OP-IC	Uni, Serie 109, ±18V, -55...+125°	81	Nsc	→Serie 109	data pdf pinout
LM 709(A)J	OP-IC	=LM 709(A)C: 0...+70°	14-DIC	Nsc	→Serie 109	data pinout
LM710	COMP	Vs:-7/14V Vu:65dB Vo:TTL Vi0:0.6mV	8A,14D	Nsc		data pinout
LM710A	COMP	Vs:-7/14V Vu:65dB Vo:TTL Vi0:0.6mV	8A,14D,10F	Nsc		data pinout
LM710C	COMP	Vs:-7/14V Vu:64dB Vo:TTL Vi0:1.6mV	8A,14D	Nsc		data pdf pinout
LM 710(C)H	KOP-IC	Serie 110, ±14/-7V,-55...+125(C=0...70)°	81	Nsc	→Serie 110	data pinout
LM 710(C)N	KOP-IC	=LM 710(C)H: Fig.→	14-DIP	Nsc	→Serie 110	data pinout
LM711	2xCOMP	Vs:-7/14V Vu:64dB Vo:TTL Vi0:1mV	10A,14D	Nsc		data pinout
LM 711(C)H	KOP-IC	Serie 711, ±14/-7V,-55...+125° (C=0...70)°	82	Nsc	→Serie 711	data pinout
LM 711(C)N	KOP-IC	=LM 711(C)H: Fig.→	14-DIP	Nsc	→Serie 711	data pinout
LM712A	HI-SPEED	Vs:±15V f:200Mc	5S	Nsc		data pinout
LM715	HI-SPEED	Vs:±18V Vu:30V/mV Vo:±13V Vi0:2mV	10A,14D	Nsc		data pdf pinout
LM723	ER+	Io=150mA Vo:2...37V P:660mW	14S,10AD	Nsc		data pinout
LM 723 CD,CM	Z-IC	=LM 723(C)H: SMD	14-MDIP	Nsc,Sam,Tho	...723...	data pinout
LM 723(C)H	Z-IC	+2...37V, 0,15A, -55...+125° (C=0...70°)	82	Nsc,Tho	...723...	data pinout
LM 723(C)J,(C)N	Z-IC	=LM 723(C)H: Fig.→	14-DIC,DIP	Nsc,Sam,Tho	...723...	data pinout
LM 723 ID	Z-IC	=LM 723(C)H: SMD, -25...+85°	14-MDIP	Sam	...723...	data pinout
LM 723 IN	Z-IC	=LM 723(C)H: -25...+85°	14-DIP	Sam	...723...	data pinout
LM725	HI-GAIN	Vs:±22V Vu:130dB Vo:±13.5V Vi0:0.5mV	8AD,14D	Nsc		data pinout
LM725A	HI-GAIN	Vs:±22V Vo:±18V Vi0:±0.5mV	8AD	Nsc		data pdf pinout
LM 725 CH	OP-IC	=LM 725(A)H: 0...+70°	81	Nsc	→Serie 725	data pdf pinout
LM 725 CJ,CN	OP-IC	=LM 725(A)H: 0...+70°	8-DIC,DIP	Nsc	→Serie 725	data pinout
LM 725 D	OP-IC	=LM 725(A)H: 0...+70°	14-DIP	Nsc	→Serie 725	data pinout
LM 725(A)H	OP-IC	Uni, Serie 725, ±22V, -55...+125°	81	Nsc	→Serie 725	data pinout
LM 725(A)J	OP-IC	=LM 725(A)C: 0...+70°	8-DIC	Nsc	→Serie 725	data pinout
LM 733 CH	LIN-IC	=LM 733H: 0...+70°	82	Nsc	µA 733C,CH	pdf
LM 733 CN	LIN-IC	=LM 733H: 0...+70°	14-DIP	Nsc	KA 733CN, µA733...	pdf pinout
LM 733 H	LIN-IC	Diff. Video Verst./Amp., -55...+125°	82	Nsc	µA 733HM	pdf
LM741	HI-GAIN	Vs:44V Vu:200kV/mV Vo:±13V Vi0:1mV	14D,8DA	Har		data pdf pinout
LM741A	UNI	Vs:±22V Vu:106dB Vo:±15V Vi0:0.8mV	14D,10F,8A	Nsc ·		data pdf pinout
LM741C	II-GAIN	Vs:36V Vu:200kV/mV Vo:±13V Vi0:2mV	14D,8DAS	Har		data pdf pinout
LM 741 CH,EH	OP-IC	=LM 741(A)H: 0...+70°, C=±18V, E=±22V	81	Nsc,Tho	→Serie 741	data pinout
LM 741 CM	IC	Single Operational Amplifier	DIP	Fch	-	pdf pinout
LM 741 CN	IC	Single Operational Amplifier	DIP	Fch	-	pdf pinout
LM 741(E)D,CM	OP-IC	=LM 741..H: SMD	8-MDIP	Nsc,Sam	→Serie 741	data pinout
LM 741(A)H	OP-IC	Uni, Serie 741, ±22V, -55...+125°	81	Nsc,Tho	→Serie 741	data pinout
LM 741 ID	OP-IC	=LM 741(A)H: SMD, -40...+85°	8-MDIP	Sam	→Serie 741	data pinout
LM 741 IN	OP-IC	Single Operational Amplifier	8-DIP	Fcs		data pdf pinout
LM 741(C,E)J,N	OP-IC	=LM 741..H: Fig.→	8-DIC,DIP	Nsc,Tho	→Serie 741	data pinout
LM 741..J-14,N-14	OP-IC	=LM 741..H: Fig.→	14-DIC,DIP	Nsc	→Serie 741	data pinout
LM747_1	2xOP	Vs:±18V Vu:106dB Vo:±13V Vi0:2mV	14D,10A	Nsc		data pinout
LM747	2xOP	Vs:±18V Vu:106dB Vo:±13V Vi0:2mV	14DF,10A	Nsc		data pinout
LM747_1A	2xOP	Vs:±22V Vu:106dB Vo:±15V Vi0:0.8mV	14D,10A	Nsc		data pinout
LM747A	2xOP	Vs:±22V Vu:106dB Vo:±15V Vi0:0.8mV	14D,10A	Nsc		data pinout
LM 747 CH,EH	OP-IC	=LM 741(A)H: ±18V, 0...+70°	82	Nsc	→Serie 747	data pinout
LM 747(A)H	OP-IC	Dual, Serie 747, ±22V, -55...+125°	82	Nsc	→Serie 747	data pinout
LM 747(A,C,E)J,N	OP-IC	=LM 747..H: Fig.→	14-DIC,DIP	Nsc	→Serie 747	data pinout
LM748	UNI	Vs:±22V Vu:100dB Vo:±13V Vi0:2mV	8AD	Nsc,Sgs		data pinout
LM 748(C)H	OP-IC	Serie 748, ±22V, -55...+125° (C=0...+70°)	81	Nsc,Tho	→Serie 748	data pinout
LM 748(C)J,N	OP-IC	=LM 748(C)H: Fig.→	8-DIC,DIP	Nsc,Tho	→Serie 748	data pinout
LM759	POWER-OP	Vs:±18V Vu:200V/mV Vo:±12.5V Vi0:1mV	8A,4P	Nsc		data pdf pinout
LM760	COMP	Vs:±8V Vu:70dB Vo:TTL Vi0:1mV	14D,8ADF	Nsc		data pinout
LM 760...	KOP-IC	→µA 760...		Nsc	→µA 760...	data pinout
LM776	PROG	Vs:±18V Vu:400V/mV Vo:±13V Vi0:2mV	8AD	Nsc	•	data pinout
LM 776...	OP-IC	→µA 776...		Nsc	→µA 776...	data pinout
LM 831(N)	LIN-IC	2x NF-E, 1,8...6V, 2x0,4W(5V/4Ω)	16-(M)DIP	Nsc	-	pinout
LM 832(N)	LIN-IC	Rauschunterdr./noise Red. System DNR	14-DIP	Nsc		pdf
LM 832 M	LIN-IC	=LM 832(N): SMD	14-MDIP	Nsc		pdf
LM833	2xAUDIO-OP	Vs:36V Vu:110dB Vo:±13.4V Vi0:0.3mV	8SD	Nsc		data pdf pinout
LM 833 D,M	OP-IC	=LM 833N: SMD	8-MDIP	Nsc	-	data pinout
LM 833 N	OP-IC	Dual, lo-noise, ±18V, -40...+105°	8-DIP	Nsc, Tho	MC 33078...	data pdf pinout
LM837	4xLO-NOISE	Vs:±18V Vu:110dB Vo:±13.5V Vi0:0.3mV	14SD	Nsc		data pdf pinout
LM 837 M	OP-IC	=LM 837N: SMD	14-MDIP	Nsc	-	data pdf pinout
LM 837 N	OP-IC	Quad, lo-noise, ±18V, -40...+105°	14-DIP	Nsc	-	data pdf pinout
LM 903(N)	LIN-IC	Flüssigk.-pegel/fluid Level Detector	16-DIP	Nsc		pdf pinout
LM 909(N)	LIN-IC	Fb, Empfänger/Receiver, ...40MHz, 9V	18-DIP	Nsc		
LM 1005	IC			Nsc		pdf
LM 1011	LIN-IC	Dolby B	16-DIP	Nsc	(NE 645, NE 650)	pdf

Type	Device	Short Description	Fig.	Manu	Comparision Types	More at
LM 1014	Si-N	≈BF 273	7f		→BF 273	
LM 1014(A)N	LIN-IC	Motorregler/Speed Control, 5...20V	10-DIP	Nsc	-	pdf
LM 1015	Si-N	≈BF 240	7f		→BF 240	
LM 1017(N)	LIN-IC	7-Segment Display-Decoder, Ucc=5...12V	16-DIP	Nsc	SN 29764	pdf
LM 1019(N)	LIN-IC	TV, Digital Tuning Station Detector	16-DIP	Nsc		pdf
LM 1020 N	LIN-IC	Ctv, Video/chroma Signal	16-DIP	Nsc	TDA 2560	pdf
LM 1035(N)	LIN-IC	2x Audio DC Balance,Volume/Tone Ctrl.	20-DIP	Nsc	-	pdf
LM 1036(N)	LIN-IC	2x Audio DC Balance,Volume,Tone Ctrl.	20-(M)DIP	Nsc	-	pdf pinout
LM 1037(N)	LIN-IC	2x 4-Kanal/Channel Analog Switch	18-DIP	Nsc	-	pdf
LM 1038(N)	LIN-IC	2x 4-Kanal/Channel Analog Audio Switch	18-DIP	Nsc	-	pdf
LM 1040(N)	LIN-IC	2x Audio DC Balance,Volume/Tone Ctrl.	24-DIP	Nsc	-	pdf
LM 1042(N)	LIN-IC	Flüssigk.-pegel/fluid Level Detector	16-DIP	Nsc	-	pdf
LM 1044(N)	LIN-IC	3-Ch. Analog Video Switch, Ucc=8...16V	24-DIP	Nsc	-	pdf
LM1084IS-12	R+	Io=5A Vo:11.76...12.24V Vin:13.5...25V	3S	Nsc		data pinout
LM1084IS-3.3	R+	Io=5A Vo:3.235...3.365V Vin:4.8...15V	3S	Nsc		data pinout
LM1084IS-5.0	R+	Io=5A Vo:4.9...5.1V Vin:6.5...20V P:int	3S	Nsc		data pinout
LM1084IS-ADJ	ER+	Io=5A Vo:1.2...15V P:int	3S	Nsc		data pinout
LM1084ISX-12	R+	Io=5A Vo:11.76...12.24V Vin:13.5...25V	3S	Nsc		data pinout
LM1084ISX-3.3	R+	Io=5A Vo:3.235...3.365V Vin:4.8...15V	3S	Nsc		data pinout
LM1084ISX-5.0	R+	Io=5A Vo:4.9...5.1V Vin:6.5...20V P:int	3S	Nsc		data pinout
LM1084ISX-ADJ	ER+	Io=5A Vo:1.2...15V P:int	3S	Nsc		data pinout
LM1084IT-12	R+	Io=5A Vo:11.76...12.24V Vin:13.5...25V	3P	Nsc		data pinout
LM1084IT-3.3	R+	Io=5A Vo:3.235...3.365V Vin:4.8...15V	3P	Nsc		data pinout
LM1084IT-5.0	R+	Io=5A Vo:4.9...5.1V Vin:6.5...20V P:int	3P	Nsc		data pinout
LM1084IT-ADJ	ER+	Io=5A Vo:1.2...15V P:int	3P	Nsc		data pinout
LM1085IS-12	R+	Io=3A Vo:11.76...12.24V Vin:13.5...25V	3S	Nsc		data pinout
LM1085IS-3.3	R+	Io=3A Vo:3.235...3.365V Vin:4.8...15V	3S	Nsc		data pinout
LM1085IS-5.0	R+	Io=3A Vo:4.9...5.1V Vin:6.5...20V P:int	3S	Nsc		data pinout
LM1085IS-ADJ	ER+	Io=3A Vo:1.2...15V Vin:1.5...15V P:int	3S	Nsc		data pinout
LM1085ISX-12	R+	Io=3A Vo:11.76...12.24V Vin:13.5...25V	3S	Nsc		data pinout
LM1085ISX-3.3	R+	Io=3A Vo:3.235...3.365V Vin:4.8...15V	3S	Nsc		data pinout
LM1085ISX-5.0	R+	Io=3A Vo:4.9...5.1V Vin:6.5...20V P:int	3S	Nsc		data pinout
LM1085ISX-ADJ	ER+	Io=3A Vo:1.2...15V Vin:1.5...15V P:int	3S	Nsc		data pinout
LM1085IT-12	R+	Io=3A Vo:11.76...12.24V Vin:13.5...25V	3P	Nsc		data pinout
LM1085IT-3.3	R+	Io=3A Vo:3.235...3.365V Vin:4.8...15V	3P	Nsc		data pinout
LM1085IT-5.0	R+	Io=3A Vo:4.9...5.1V Vin:6.5...20V P:int	3P	Nsc		data pinout
LM1085IT-ADJ	ER+	Io=3A Vo:1.2...15V Vin:1.5...15V P:int	3P	Nsc		data pinout
LM1086CS-2.85	R+	Io=1.5A Vo:2.79...2.91V Vin:4.35...18V	3S	Nsc		data pinout
LM1086CS-3.3	R+	Io=1.5A Vo:3.235...3.365V P:int	3S	Nsc		data pinout
LM1086CS-5.0	R+	Io=1.5A Vo:4.9...5.1V Vin:6.5...20V	3S	Nsc		data pinout
LM1086CS-ADJ	ER+	Io=1.5A Vo:1.2...15V Vin:1.5...15V	3S	Nsc		data pinout
LM1086CSX-2.85	R+	Io=1.5A Vo:2.79...2.91V Vin:4.35...18V	3S	Nsc		data pinout
LM1086CSX-3.3	R+	Io=1.5A Vo:3.235...3.365V P:int	3S	Nsc		data pinout
LM1086CSX-5.0	R+	Io=1.5A Vo:4.9...5.1V Vin:6.5...20V	3S	Nsc		data pinout
LM1086CSX-ADJ	ER+	Io=1.5A Vo:1.2...15V Vin:1.5...15V	3S	Nsc		data pinout
LM1086CT-2.85	R+	Io=1.5A Vo:2.79...2.91V Vin:4.35...18V	3P	Nsc		data pinout
LM1086CT-3.3	R+	Io=1.5A Vo:3.235...3.365V P:int	3P	Nsc		data pinout
LM1086CT-5.0	R+	Io=1.5A Vo:4.9...5.1V Vin:6.5...20V	3P	Nsc		data pinout
LM1086CT-ADJ	ER+	Io=1.5A Vo:1.2...15V Vin:1.5...15V	3P	Nsc		data pinout
LM1086IS-2.85	R+	Io=1.5A Vo:2.79...2.91V Vin:4.35...18V	3S	Nsc		data pinout
LM1086IS-3.3	R+	Io=1.5A Vo:3.235...3.365V P:int	3S	Nsc		data pinout
LM1086IS-5.0	R+	Io=1.5A Vo:4.9...5.1V Vin:6.5...20V	3S	Nsc		data pinout
LM1086IS-ADJ	ER+	Io=1.5A Vo:1.2...15V Vin:1.5...15V	3S	Nsc		data pinout
LM1086ISX-2.85	R+	Io=1.5A Vo:2.79...2.91V Vin:4.35...18V	3S	Nsc		data pinout
LM1086ISX-3.3	R+	Io=1.5A Vo:3.235...3.365V P:int	3S	Nsc		data pinout
LM1086ISX-5.0	R+	Io=1.5A Vo:4.9...5.1V Vin:6.5...20V	3S	Nsc		data pinout
LM1086ISX-ADJ	ER+	Io=1.5A Vo:1.2...15V Vin:1.5...15V	3S	Nsc		data pinout
LM1086IT-2.85	R+	Io=1.5A Vo:2.79...2.91V Vin:4.35...18V	3P	Nsc		data pinout
LM1086IT-3.3	R+	Io=1.5A Vo:3.235...3.365V P:int	3P	Nsc		data pinout
LM1086IT-5.0	R+	Io=1.5A Vo:4.9...5.1V Vin:6.5...20V	3P	Nsc		data pinout
LM1086IT-ADJ	ER+	Io=1.5A Vo:1.2...15V Vin:1.5...15V	3P	Nsc		data pinout
LM 1098	Si-N	≈BF 255	7e		→BF 255	
LM 1111(N)	LIN-IC	~LM 1112	16-DIP	Nsc	LM 1112	pdf
LM 1112 A,B,C(N)	LIN-IC	Dolby B, A,B,C: Encode Characteristic	16-DIP	Nsc		pdf
LM1117DT-2.85	R+	Io=800mA Vo:2.79...2.91V P:int	4S	Nsc		data pinout
LM1117DT-3.3	R+	Io=800mA Vo:3.235...3.365V P:int	4S	Nsc		data pinout
LM1117DT-5.0	R+	Io=800mA Vo:4.9...5.1V Vin:6.5...12V	4S	Nsc		data pinout
LM1117DT-ADJ	ER+	Io=800mA Vo:1.25...13.8V Vin:1.4...10V	4S	Nsc		data pinout
LM1117DTX-2.85	R+	Io=800mA Vo:2.79...2.91V P:int	4S	Nsc		data pinout
LM1117DTX-3.3	R+	Io=800mA Vo:3.235...3.365V P:int	4S	Nsc		data pinout
LM1117DTX-5.0	R+	Io=800mA Vo:4.9...5.1V Vin:6.5...12V	4S	Nsc		data pinout
LM1117DTX-ADJ	ER+	Io=800mA Vo:1.25...13.8V Vin:1.4...10V	4S	Nsc		data pinout
LM1117MP-2.85	R+	Io=800mA Vo:2.79...2.91V P:int	4S	Nsc		data pinout
LM1117MP-3.3	R+	Io=800mA Vo:3.235...3.365V P:int	4S	Nsc		data pinout
LM1117MP-5.0	R+	Io=800mA Vo:4.9...5.1V Vin:6.5...12V	4S	Nsc		data pinout
LM1117MP-ADJ	ER+	Io=800mA Vo:1.25...13.8V Vin:1.4...10V	4S	Nsc		data pinout
LM1117MPX-2.85	R+	Io=800mA Vo:2.79...2.91V P:int	4S	Nsc		data pinout
LM1117MPX-3.3	R+	Io=800mA Vo:3.235...3.365V P:int	4S	Nsc		data pinout
LM1117MPX-5.0	R+	Io=800mA Vo:4.9...5.1V Vin:6.5...12V	4S	Nsc		data pinout
LM1117MPX-ADJ	ER+	Io=800mA Vo:1.25...13.8V Vin:1.4...10V	4S	Nsc		data pinout
LM1117T-2.85	R+	Io=800mA Vo:2.79...2.91V P:int	3P	Nsc		data pinout
LM1117T-3.3	R+	Io=800mA Vo:3.235...3.365V P:int	3P	Nsc		data pinout
LM1117T-5.0	R+	Io=800mA Vo:4.9...5.1V Vin:6.5...12V	3P	Nsc		data pinout
LM1117T-ADJ	ER+	Io=800mA Vo:1.25...13.8V Vin:1.4...10V	3P	Nsc		data pinout

Type	Device	Short Description	Fig.	Manu	Comparision Types	More at
LM 1121 A,B,C(N)	LIN-IC	Dolby B, A,B,C: Encode Characteristic	16-DIP	Nsc	-	
LM 1131 A,B,C(N)	LIN-IC	Dolby B, A,B,C: Encode Characteristic	20-DIP	Nsc		pinout
LM 1141(N)	LIN-IC	Dolby B/C, Ucc=5...16V	28-DIP	Nsc		
LM 1141 V	LIN-IC	=LM 1141(N): SMD	28-PLCC	Nsc	-	
LM 1181	Si-N	≈BC 239	7e		→BC 239	
LM 1201(N)	LIN-IC	Video-verst./amp. System, 100MHz	16-DIP	Nsc		pdf pinout
LM 1203	LIN-IC	3x RGB Video Amplifier System, 70MHz	28-DIP	Nsc	≈ TLS1233, ≈ LM1282	pdf pinout
LM 1203A	LIN-IC	3x RGB Video-verst./amp. System,150MHz	28-DIP	Nsc	≈ TLS1233, ≈ LM1282	pdf pinout
LM 1203AN	LIN-IC	3x RGB Video-verst./amp. System,150MHz	28-DIP	Nsc	≈ TLS1233, ≈ LM1282	pdf pinout
LM 1203B	LIN-IC	3x RGB Video-verst./amp. System,100MHz	28-DIP	Nsc	≈ TLS1233, ≈ LM1282	pdf pinout
LM 1203BN	LIN-IC	3x RGB Video-verst./amp. System,100MHz	28-DIP	Nsc	≈ TLS1233, ≈ LM1282	pdf pinout
LM 1203N	LIN-IC	3x RGB Video Amplifier System, 70MHz	28-DIP	Nsc	≈ TLS1233, ≈ LM1282, ≈ ECG7081	pdf pinout
LM 1205N	IC	RGB Video Amp. (130MHz), Blanking	DIP	Nsc	-	pdf pinout
LM 1207N	IC	RGB Video Amp. (85MHz), Blanking	DIP	Nsc		pdf pinout
LM 1211(N)	LIN-IC	Breitb./Wideb. Dem. System, 20...80MHz	20-DIP	Nsc	-	pdf pinout
LM 1237	D/A-IC	I²C,RGB Preamp.,Internal 254 Character OSD,4 DACs	DIP	Nsc	-	pdf pinout
LM 1267	D/A-IC	I²C RGB Video Amp System, OSD, DACs	DIP	Nsc	-	pdf pinout
LM 1279 A	IC	110 MHz RGB Video Amplifier System, OSD	DIP	Nsc	-	pdf pinout
LM 1281 N	IC	RGB Video Amp. System (85MHz),OSD	DIP	Nsc	-	pdf pinout
LM 1282N	IC	RGB Video Amp. System (110MHz),OSD	DIP	Nsc	-	pdf pinout
LM 1283 N	IC	RGB Video Amp. System (140MHz),OSD	DIP	Nsc	-	pdf pinout
LM 1290 N	IC	Autosync Horizontal Deflection Processor	DIP	Nsc	-	pdf pinout
LM 1295 N	IC	DC Controlled Geometry Correction System	DIP	Nsc	-	pdf pinout
LM 1303(N,P)	OP-IC	Dual	14-DIP	Nsc	SN 76131, TBA 231(A)	
LM 1304(N)	LIN-IC			Nsc	MC 1304, SN 76104, µA 732	
LM 1305(N)	LIN-IC	Stereo-decoder		Nsc	MC 1305, SN 76105, ULN 2122	
LM 1307(N)	LIN-IC				MC 1307	
LM 1310(N,P)	LIN-IC	PLL FM Stereo-Decoder, Ucc=10...18V	14-DIP	Nsc	A 290D, CA 1310, MC 1310, SN 76115	
LM 1391(N)	LIN-IC	TV, PLL, HA-Synchr., Ha-o	8-DIP	Nsc		pdf pinout
LM 1403	Si-N	≈BC 337	7e		→BC 337	
LM 1404	Si-P	≈BC 327	7e		→BC 327	
LM 1410 K	Si-P	≈BC 307	7e		→BC 307	
LM1414	2xCOMP	Vs:-7/14V Vu:64dB Vo:TTL Vi0:1mV	14D	Nsc		data pinout
LM 1414 J,N,P	KOP-IC	Dual, hi-speed, ±14/-7V, 30ns, 0...+70°	14-DIC,DIP	Nsc	LM 1514...	data pinout
LM 1427	Si-N	≈BC 337	7e		→BC 337	
LM 1428	Si-P	≈BC 327	7e		→BC 327	
LM1458	2xOP	Vs:36V Vu:106dB Vo:±13V Vi0:2mV	8DSA	Har		data pdf pinout
LM1458A	2xOP	Vs:±18V Vu:92dB Vo:±13V Vi0:1mV	8AD	Har		data pdf pinout
LM 1458 DG,DP,J,N	OP-IC	=LM 1458H: Fig.→	8-DIC,DIP	Nsc,Tho	→Serie 158	data pinout
LM 1458 FP,M	OP-IC	=LM 1458H: SMD	8-MDIP	Nsc,Tho	→Serie 158	data pinout
LM 1458 H	OP-IC	Dual, Serie 158, ±18V, 0...+70°	81	Nsc,Tho	→Serie 158	data pinout
LM 1496 H	LIN-IC	=LM 1496N: Fig.→	82	Nsc	LM 1596H, MC 1496G, MC 1596G	pdf pinout
LM 1496 M	LIN-IC	=LM 1496N: SMD	14-MDIP	Nsc	MC 1496D	pdf pinout
LM 1496 N	LIN-IC	Modulator/demodulator,...100MHz,0...+70°	14-DIP	Nsc	MC 1496N, MC 1596N	pdf pinout
LM 1514 J	KOP-IC	=LM 1414...: -55...+125°	14-DIC	Nsc		data pinout
LM 1524 J	SMPS-IC	SMPS/PWM Controller, 50mA, -25...+125°	16-DIC	Nsc	→SG 1524...	pinout
LM1558	2xOP	Vs:±18V Vu:100dB Vo:±13V Vi0:2mV	8SDdA	Nsc		data pdf pinout
LM1558A	2xOP	Vs:±22V Vu:92dB Vo:±13V Vi0:1mV	8AD	Har		data pinout
LM 1558 DG,J,N	OP-IC	=LM 1458H: ±22V, -55...+125°	8-DIC,DIP	Nsc,Tho	→Serie 158	data pinout
LM 1558 H	OP-IC	=LM 1458H: ±22V, -55...+125°	81	Nsc,Tho	→Serie 158	data pinout
LM1575J-12	S+	Vin:8...40V P:int	16D	Nsc		data pinout
LM1575J-15	S+	Vin:8...40V P:int	16D	Nsc		data pinout
LM1575J-3.3	S+	Vin:8...40V P:int	16D	Nsc		data pinout
LM1575J-5.0	S+	Vin:8...40V P:int	16D	Nsc		data pinout
LM1575J-ADJ	ES+	Vo:1.23...37V Vin:40V P:int	16D	Nsc		data pinout
LM 1575J-3.3...-QML	SMPS-IC	1A Step-down Voltage Reg. Family 3.3V ... 15V and adjustable versions	16-DIC	Nsc		data pinout pdf
LM1577K-12	S+	Io=3A Vin:3.5...40V P:int	5O	Nsc		
LM1577K-15	S+	Io=3A Vin:3.5...40V P:int	5O	Nsc		data pinout
LM1577K-ADJ	ES+	Io=3A Vo:1.23...37V Vin:40V P:int	5O	Nsc		data pinout
LM 1578 H	SMPS-IC	=LM 1578N:	81	Nsc		data pinout
LM 1578 N	SMPS-IC	S-Reg, +2...40V, 0,75A, -55...+125°	8-DIP	Nsc	-	
LM 1596 H	LIN-IC	=LM 1496N: -55...+125°	82	Nsc	MC 1596G	pdf pinout
LM 1737	Si-N	≈BC 337	7b		→BC 337	
LM 1738	Si-P	≈BC 327	7b		→BC 327	
LM 1771 (S,T,U)MMX	SMPS-IC	SMD, synchr. buck contr. ton=0.5/1/2µs	SSMDIP	Nsc	-	pdf pinout
LM 1771 (S,T,U)SDX	SMPS-IC	SMD, synchr. buck contr. ton=0.5/1/2µs	LLP	Nsc	-	pdf pinout
LM 1800(N)	LIN-IC	PLL FM Stereo-Decoder, Ucc=10...18V	16-DIP	Nsc	CA 758, MC 1311, ULX 2244, µA 758 C	pdf
LM 1801(N)	LIN-IC	Power KOP, Gas Detector, Ucc=8...14V	14-DIP	Nsc	-	
LM 1812(N)	LIN-IC	Us Transceiver, PQ=12W(Peak)	18-DIP	Nsc		
LM 1815(N)	LIN-IC	Sense Amp. f. Motor Ctrl., Ucc=2...12V	14-DIP	Nsc		
LM 1818(N)	LIN-IC	Recorder, A/W-Verst.,ALC, Ucc=3,5...18V	20-DIP	Nsc,Phi	A 1818D	pdf
LM 1819(N)	LIN-IC	Air-core Meter Driver/tachometer	14-DIP	Nsc		pdf
LM 1820(N)	LIN-IC	~LM 3820	14-DIP	Nsc	LM 3830	pdf
LM 1821 S	LIN-IC	Ctv, Video-zf PLL Synchr. Detector	16-DIP+c	Nsc	-	
LM 1823(N)	LIN-IC	TV, Video-ZF, PLL-dem., Agc(neg.), Afc	28-DIP	Nsc	-	
LM 1828(N)	LIN-IC	Ctv, Chroma Demodulator, Ucc=24V	14-DIP	Nsc		
LM 1830 H	LIN-IC	=LM 1830N: Fig.→	82	Nsc		
LM 1830 N	LIN-IC	Flüssigkeit/fluid Detector	14-DIP	Nsc		pdf
LM 1837N	LIN-IC	2x Autoreverse-Recorder PreAmp, 4...18V	18-DIP	Nsc	-	pdf pinout
LM 1837M	LIN-IC	SMD, 2x Autorev.-Rec. PreAmp, 4...18V	16-MDIP	Nsc	-	pdf pinout
LM 1841	LIN-IC	=CA 2136A	14-DIP	Nsc	CA 2136A, ULN 2136A	
LM 1848(N)	LIN-IC	Ctv, Chroma Demodulator, Ucc=24V	14-DIP	Nsc		

Type	Device	Short Description	Fig.	Manu	Comparision Types	More at
LM 1851(N)	LIN-IC	Fehlstromdetektor/earth Leakage Det.	8-DIP	Nsc, Ray	KA 2807	pdf pinout
LM 1851 M	LIN-IC	=LM 1851(N): SMD	8-MDIP	Ray	-	pdf
LM 1863(M)	LIN-IC	Am Radio f. Electr. Tuning, Ucc=7...16V	20-MDIP	Nsc	-	pdf
LM 1865(N)	LIN-IC	FM-ZF f. Electr. Tuning, Agc(neg.)	20-DIP	Nsc	-	pdf pinout
LM 1866(N)	LIN-IC	AM/FM Radio, Ucc=3...15V	20-DIP	Nsc	-	
LM 1868(N)	LIN-IC	AM/FM Radio, 4,5...15V,NF-E,0,7W(9V/8Ω)	20-DIP	Nsc	-	pdf
LM 1870(N)	LIN-IC	PLL FM Stereo-Decoder, Blend Ctrl.	20-DIP	Nsc,Phi	-	
LM 1870 M	LIN-IC	=LM 1870(N): SMD	20-MDIP	Nsc	-	
LM 1871N	LIN-IC	27/49MHz Fernstrg./Radio Ctrl.Transm.	18-DIP	Nsc	-	pdf pinout
LM 1872N	LIN-IC	27/49/72MHz Fernstrg./Radio Ctrl. Rcv.	18-DIP	Nsc	-	pdf pinout
LM1875	AUDIO-OP	Vs:60V Vi0:±15mV f:5.5Mc	5P	Nsc	-	data pinout
LM 1875(T)	LIN-IC	NF-E, 20...60V, 25W(±25V/8Ω)	86/5Pin	Nsc	-	pdf pinout
LM 1876T	LIN-IC	2x Audio PAmp, Mute, Standby, 20...64V, 20W(±22V/8Ohm)	15-SQL	Nsc	-	pdf pinout
LM 1876TF	LIN-IC	2x Audio PAmp, Mute, Standby, 20...64V, 20W(±22V/8Ohm)	15-SQL	Nsc	-	pdf pinout
LM1877	2xAUDIO-OP	Vs:26V Vu:70dB Vi0:15mV f:65kc	14SD	Nsc		data pinout
LM 1877N-9	LIN-IC	2x PAmp, 6...24V, 2x >2W(20V/8Ohm)	14-DIP	Nsc	-	pdf pinout
LM 1877M-9	LIN-IC	SMD, 2x PAmp, 6...24V, 2x >2W(20V/8Ohm)	14-MDIP	Nsc	-	pdf pinout
LM 1880(N)	LIN-IC	TV, HA-/VA-prozessor	14-DIP	Nsc	-	
LM 1881N/M	LIN-IC	Video Sync Separator	8-DIP	Nsc	-	pdf pinout
LM 1882 L	TTL-IC	=LM 1882: SMD	20-(P)LCC			
LM 1882 P	TTL-IC	Monitor, μcomp, progr. Video	20-DIP	Nsc		
LM 1882 S	TTL-IC	=LM 1882: SMD	20-MDIP			
LM 1884(N)	LIN-IC	TV, Stereo-Decoder, Ucc=12V	16-DIP	Nsc		
LM 1886(N)	LIN-IC	TV, Video Matrix, Digital→Analog	20-DIP	Nsc	-	pdf
LM 1889(N)	LIN-IC	TV, Video Modulator, Ucc=12...18V	18-DIP	Nsc	-	pdf pinout
LM 1893(N)	LIN-IC	Bi-line Carrier-current Transceiver	18-DIP	Nsc	-	pdf pinout
LM 1894(N)	LIN-IC	Rausch-/noise Reduction System Dnr	14-DIP	Nsc	-	
LM 1895(N)	LIN-IC	NF-E, 3...10V, 1,1W(6V/4Ω)	8-DIP	Nsc	-	data pdf pinout
LM1896	2xAUDIO-OP	Vs:12V Vi0:5mV f:20kc	14D	Nsc	-	pdf pinout
LM 1896(N-1,N-2)	LIN-IC	=LM 1895: Dual, 2x1,1W(6V/4Ω), 1x1,8W	14-DIP	Nsc	-	
LM 1897(N)	LIN-IC	2x Recorder-V, Ucc=4...18V	16-DIP	Nsc	-	
LM1900	4xOP	Vs:±16V Vu:70dB Vo:±14V f:2.5Mc	14D	Nsc,Ray	-	data pinout
LM 1921(T)	LIN-IC	Leistungsschalter/Power Switch, 1A	86/5Pin	Nsc	-	pdf
LM 1946(N)	LIN-IC	Over/under Current Limit Diagnostic	20-DIP	Nsc	-	pdf
LM 1949(N)	LIN-IC	Fuel Injector Drive Controller	8-DIP	Nsc	-	pdf
LM 1950 T	LIN-IC	750 mA High Side Switch	8-DIP	Nsc	-	pdf pinout
LM 1951(T)	LIN-IC	Leistungsschalter/Power Switch, 1A	86/5Pin	Nsc	-	pdf pinout
LM 1964(V)	LIN-IC	Car Sensor Interface Amp.	20-PLCC	Nsc	-	
LM 1965(N)	LIN-IC	=LM 1865(N): f. Manual Tuning	20-DIP	Nsc	-	
LM 1971	LIN-IC	1-Ch. Volume Ctrl., Ucc=4,5...12V	8-(M)DIP	Nsc	-	pdf pinout
LM 1972	LIN-IC	2-Ch. Volume Ctrl., Ucc=4,5...12V	20-(M)DIP	Nsc	-	pdf pinout
LM 1973	LIN-IC	3-Ch. Volume Ctrl., Ucc=4,5...12V	20-(M)DIP	Nsc	-	pdf pinout
LM 2002(T),A(T)	LIN-IC	NF-E, 5...20V, 3,5A, 10W(16V/2Ω)	86/5Pin	Nsc	TDA 2002(A), TDA 2003	pinout
LM 2005(T-M,T-S)	LIN-IC	2x NF-E, 8...18V, 3,5V, 2x10W(14V/4Ω)	11-SQL	Nsc	TDA 2005	
LM 2065(N)	LIN-IC	=LM 1865(N): pos. Agc	20-DIP	Nsc		data pinout
LM2101A	2xOP	Vs:±22V	16FD	Six		data pinout
LM2108	LO-POWER	Vs:±20V Vu:200V/mV Vo:±18V Vi0:1mV	16D	Ray		data pinout
LM2108A	LO-POWER	Vs:±20V Vu:200V/mV Vo:±18V Vi0:0.4mV	16D	Ray		data pinout
LM2111	COMP	Vs:36V Vu:200V/mV Vi0:0.7mV	16D	Ray		pdf
LM 2111(N)	LIN-IC	Tv-ton-zf	14-DIP	Nsc	CA 2111A, MC 1357, ULN 2111A	pdf pinout
LM 2402T	IC	Triple 3ns Crt Driver	SQL	Nsc		pdf pinout
LM 2403 T	IC	Triple 4.5ns Crt Driver	SQL	Nsc		pdf pinout
LM 2405 T	IC	Triple 7ns Crt Driver	SQL	Nsc	-	pdf pinout
LM 2406T	IC	Triple 9ns Crt Driver	SQL	Nsc		pdf pinout
LM 2407 T	IC	Triple 7.5ns Crt Driver	SQL	Nsc	-	pdf pinout
LM 2408T	IC	Triple 4.5ns Crt Driver	SQL	Nsc		pdf pinout
LM 2409T	IC	Triple 9.5ns Crt Driver	SQL	Nsc	-	pdf pinout
LM 2412T	IC	Triple 2.8ns Crt Driver	SQL	Nsc	-	pdf pinout
LM 2413T	IC	Triple 4ns Crt Driver	SQL	Nsc		pdf pinout
LM 2415T	IC	Triple 5.5ns Crt Driver	SQL	Nsc	-	pdf pinout
LM 2416T	LIN-IC	Triple 50MHz Crt Driver, fb=42MHz	11-SIL	Nsc	-	pdf pinout
LM 2416CT	LIN-IC	Triple 50MHz Crt Driver, fb=35MHz	11-SIL	Nsc	-	pdf pinout
LM 2419T	IC	Triple 65MHz Crt Driver	SQL	Nsc	-	pdf pinout
LM 2421AT	IC	Triple Channel HDTV Driver	SQL	Nsc		pdf pinout
LM 2422	IC	220V Triple Chan. 30MHz Crt DTV Drv.	SQL	Nsc	-	pdf pinout
LM 2423TE	IC	220V Triple Chan. 15MHz Crt DTV Drv.	SQL	Nsc	-	pdf pinout
LM 2425TE	IC	220V Triple Chan. 10MHz Crt DTV Drv.	SQL	Nsc	-	pdf pinout
LM 2426TE	IC	Triple Chan. 30MHz Crt DTV Drv.	SQL	Nsc	-	pdf pinout
LM 2427T	IC	Triple 80MHz Crt Driver	SIL	Nsc	-	pdf pinout
LM 2429	IC	Triple Chan. 15MHz DTV Drv.	SQL	Nsc	-	pdf pinout
LM 2432TE	IC	220V 37MHz Hdtv Crt Drv.	SQL	Nsc	-	pdf pinout
LM 2463 TA	IC	Monolithic Triple 4 ns Crt Driver	SQL	Nsc	-	pdf pinout
LM 2465 TA	IC	Monolithic Triple 5.5 hs High Gain Crt Driver	SQL	Nsc	-	pdf pinout
LM 2467 T	IC	Monolithic Triple 7.5 ns Crt Driver	SQL	Nsc		pdf pinout
LM 2468 TA	IC	Monolothic Triple 14nS High Gain Crt Driver	SQL	Nsc		pdf pinout
LM 2480 NA	IC	80V Triple Bias Clamp	SQL	Nsc		pdf pinout
LM 2501 J	Si-N	≈BC 337	7e		→BC 337	
LM 2502 J	Si-P	≈BC 327	7e		→BC 327	
LM 2524 J,N	SMPS-IC	=LM 1524: 0...+70°	16-DIC,DIP	Nsc	→SG 2524...	pinout
LM 2574 BN	LIN-IC	52kHz Simple 0.5A Buck Regulator	DIP	Mcr	-	pdf pinout
LM2574HVM-12	S+	Io=500mA Vin:12V P:int	14S	Nsc		data pinout

Type	Device	Short Description	Fig.	Manu	Comparision Types	More at
LM2574HVM-15	S+	Io=500mA Vin:12V P:int	14S	Nsc		data pinout
LM2574HVM-3.3	S+	Io=500mA Vo:3.3...V Vin:12V P:int	14S	Nsc		data pinout
LM2574HVM-5.0	S+	Io=500mA Vin:12V P:int	14S	Nsc		data pinout
LM2574HVM-ADJ	ES+	Io=500mA Vo:1.23...57V Vin:12V P:int	14S	Nsc		data pinout
LM2574HVN-12	R+	Io=500mA Vo:11.40...12.66V P:int	8D	Nsc		data pinout
LM2574HVN-15	R+	Io=500mA Vo:14.25...15.83V P:int	8D	Nsc		data pinout
LM2574HVN-3.3	R+	Io=500mA Vo:3.135...3.482V P:int	8D	Nsc		data pinout
LM2574HVN-5.0	R+	Io=500mA Vo:4.750...5.275V Vin:7...60V	8D	Nsc		data pinout
LM2574HVN-ADJ	ES+	Io=500mA Vo:1.23...57V Vin:12V P:int	8D	Nsc		data pinout
LM2574M-12	S+	Io=500mA Vin:12V P:int	14S	Nsc		data pinout
LM2574M-15	S+	Io=500mA Vin:12V P:int	14S	Nsc		data pinout
LM2574M-3.3	S+	Io=500mA Vo:3.3...V Vin:12V P:int	14S	Nsc		data pinout
LM2574M-5.0	S+	Io=500mA Vin:12V P:int	14S	Nsc		data pinout
LM2574M-ADJ	ES+	Io=500mA Vo:1.23...37V Vin:12V P:int	14S	Nsc		data pinout
LM2574N-12	S+	Io=500mA Vin:12V P:int	8D	Nsc		data pinout
LM2574N-15	S+	Io=500mA Vin:12V P:int	8D	Nsc		data pinout
LM2574N-3.3	S+	Io=500mA Vin:12V P:int	8D	Nsc		data pinout
LM2574N-5.0	S+	Io=500mA Vin:12V P:int	8D	Nsc		data pinout
LM2574N-ADJ	ES+	Io=500mA Vo:1.23...37V Vin:12V P:int	8D	Nsc		data pinout
LM 2574 YN	LIN-IC	52kHz Simple 0.5A Buck Regulator	DIP	Mcr	-	pdf pinout
LM 2575 BN	LIN-IC	52kHz Simple 1A Buck Regulator	DIP	Mcr	-	pdf pinout
LM 2575 BT	LIN-IC	52kHz Simple 1A Buck Regulator	DIP	Mcr	-	pdf pinout
LM 2575 BU	LIN-IC	52kHz Simple 1A Buck Regulator	DIP	Mcr	-	pdf pinout
LM 2575 BWM	LIN-IC	52kHz Simple 1A Buck Regulator	DIP	Mcr	-	pdf pinout
LM 2575HV ...	SMPS-IC	1A Step-down Voltage Reg. Family high input voltage rating (60V)	17/5Pin	Nsc	-	pdf
LM2575HVM-12	S+	Vin:8...60V P:int	24S	Nsc		data pinout
LM2575HVM-15	S+	Vin:8...60V P:int	24S	Nsc		data pinout
LM2575HVM-3.3	S+	Vin:8...60V P:int	24S	Nsc		data pinout
LM2575HVM-5.0	S+	Vin:8...60V P:int	24S	Nsc		data pinout
LM2575HVM-ADJ	ES+	Vo:1.23...57V Vin:60V P:int	24S	Nsc		data pinout
LM 2575HVM-3.3...15	SMPS-IC	1A Step-down Voltage Reg. Family 3.3V ... 15V and adjustable versions	24-MDIP	Nsc	-	pdf
LM2575HVN-12	S+	Vin:8...60V P:int	16D	Nsc		data pinout
LM2575HVN-15	S+	Vin:8...60V P:int	16D	Nsc		data pinout
LM2575HVN-3.3	S+	Vin:8...60V P:int	16D	Nsc		data pinout
LM2575HVN-5.0	S+	Vin:8...60V P:int	16D	Nsc		data pinout
LM2575HVN-ADJ	ES+	Vo:1.23...57V Vin:60V P:int	16D	Nsc		data pinout
LM 2575HVN-3.3...15	SMPS-IC	1A Step-down Voltage Reg. Family 3.3V ... 15V and adjustable versions	16-DIP	Nsc	-	pdf
LM2575HVS-12	S+	Vin:8...60V P:int	5S	Nsc		data pinout
LM2575HVS-15	S+	Vin:8...60V P:int	5S	Nsc		data pinout
LM2575HVS-3.3	S+	Vin:8...60V P:int	5S	Nsc		data pinout
LM2575HVS-5.0	S+	Vin:8...60V P:int	5S	Nsc		data pinout
LM2575HVS-ADJ	ES+	Vo:1.23...57V Vin:60V P:int	5S	Nsc		data pinout
LM 2575HVS-3.3...15	SMPS-IC	1A Step-down Voltage Reg. Family 3.3V ... 15V and adjustable versions	30/5pin	Nsc	-	pdf
LM2575HVT-12	S+	Vin:8...60V P:int	5P	Nsc		data pinout
LM2575HVT-15	S+	Vin:8...60V P:int	5P	Nsc		data pinout
LM2575HVT-3.3	S+	Vin:8...60V P:int	5P	Nsc		data pinout
LM2575HVT-5.0	S+	Vin:8...60V P:int	5P	Nsc		data pinout
LM2575HVT-ADJ	ES+	Vo:1.23...57V Vin:60V P:int	5P	Nsc		data pinout
LM 2575HVT-3.3...15	SMPS-IC	1A Step-down Voltage Reg. Family 3.3V ... 15V and adjustable versions	17/5pin	Nsc	-	pdf
LM2575M-12	S+	Vin:8...40V P:int	24S	Nsc		data pinout
LM2575M-15	S+	Vin:8...40V P:int	24S	Nsc		data pinout
LM2575M-3.3	S+	Vin:8...40V P:int	24S	Nsc		data pinout
LM2575M-5.0	S+	Vin:8...40V P:int	24S	Nsc		data pinout
LM2575M-ADJ	ES+	Vo:1.23...37V Vin:40V P:int	24S	Nsc		data pinout
LM 2575M-3.3...15	SMPS-IC	1A Step-down Voltage Reg. Family 3.3V ... 15V and adjustable versions	24-MDIP	Nsc	-	pdf
LM2575N-12	S+	Vin:8...40V P:int	16D	Nsc		data pinout
LM2575N-15	S+	Vin:8...40V P:int	16D	Nsc		data pinout
LM2575N-3.3	S+	Vin:8...40V P:int	16D	Nsc		data pinout
LM2575N-5.0	S+	Vin:8...40V P:int	16D	Nsc		data pinout
LM2575N-ADJ	ES+	Vo:1.23...37V Vin:40V P:int	16D	Nsc		data pinout
LM 2575N-3.3...15	SMPS-IC	1A Step-down Voltage Reg. Family 3.3V ... 15V and adjustable versions	16-DIP	Nsc	-	pdf
LM2575S-12	S+	Vin:8...40V P:int	5S	Nsc		data pinout
LM2575S-15	R+	Vo:14.25...15.75V Vin:18...40V P:int	5S	Nsc		data pinout
LM2575S-3.3	S+	Vin:8...40V P:int	5S	Nsc		data pinout
LM2575S-5.0	S+	Vin:8...40V P:int	5S	Nsc		data pinout
LM2575S-ADJ	ES+	Vo:1.23...37V Vin:40V P:int	5S	Nsc		data pinout
LM 2575S-3.3...15	SMPS-IC	1A Step-down Voltage Reg. Family 3.3V ... 15V and adjustable versions	30/5pin	Nsc	-	pdf
LM2575T-12	S+	Vin:8...40V P:int	5P	Nsc		data pinout
LM2575T-15	S+	Vin:8...40V P:int	5P	Nsc		data pinout
LM2575T-3.3	S+	Vin:8...40V P:int	5P	Nsc		data pinout
LM2575T-5.0	S+	Vin:8...40V P:int	5P	Nsc		data pinout
LM2575T-ADJ	ES+	Vo:1.23...37V Vin:40V P:int	5P	Nsc		data pinout
LM 2575T-3.3...15	SMPS-IC	1A Step-down Voltage Reg. Family 3.3V ... 15V and adjustable versions	17/5pin	Nsc	-	pdf
LM 2575T ... Flow	SMPS-IC	1A Step-down Voltage Reg. Family	86/5pin	Nsc	-	pdf
LM 2575 WT	LIN-IC	52kHz Simple 1A Buck Regulator	DIP	Mcr	-	pdf pinout
LM 2575 WU	LIN-IC	52kHz Simple 1A Buck Regulator	DIP	Mcr	-	pdf pinout

Type	Device	Short Description	Fig.	Manu	Comparision Types	More at
LM 2575 YWM	LIN-IC	52kHz Simple 1A Buck Regulator	DIP	Mcr	-	pdf pinout
LM 2576 BT	LIN-IC	52kHz Simple 3A Buck Regulator	DIP	Mcr	-	pdf pinout
LM 2576 BU	LIN-IC	52kHz Simple 3A Buck Regulator	DIP	Mcr	-	pdf pinout
LM2576HVS-12	S+	Io=3A Vin:12V P:int	5S	Nsc		data pinout
LM2576HVS-15	S+	Io=3A Vin:12V P:int	5S	Nsc		data pinout
LM2576HVS-3.3	S+	Io=3A Vin:12V P:int	5S	Nsc		data pinout
LM2576HVS-5.0	S+	Io=3A Vin:12V P:int	5S	Nsc		data pinout
LM2576HVS-ADJ	ES+	Io=3A Vo:1.23...57V Vin:12V P:int	5S	Nsc		data pinout
LM2576HVSX-12	S+	Io=3A Vin:12V P:int	5S	Nsc		data pinout
LM2576HVSX-15	S+	Io=3A Vin:12V P:int	5S	Nsc		data pinout
LM2576HVSX-3.3	S+	Io=3A Vin:12V P:int	5S	Nsc		data pinout
LM2576HVSX-5.0	S+	Io=3A Vin:12V P:int	5S	Nsc		data pinout
LM2576HVSX-ADJ	ES+	Io=3A Vo:1.23...57V Vin:12V P:int	5S	Nsc		data pinout
LM2576HVT-12	S+	Io=3A Vin:12V P:int	5P	Nsc		data pinout
LM2576HVT-15	S+	Io=3A Vin:12V P:int	5P	Nsc		data pinout
LM2576HVT-3.3	S+	Io=3A Vin:12V P:int	5P	Nsc		data pinout
LM2576HVT-5.0	S+	Io=3A Vin:12V P:int	5P	Nsc		data pinout
LM2576HVT-ADJ	ES+	Io=3A Vo:1.23...57V Vin:12V P:int	5P	Nsc		data pinout
LM2576S-12	S+	Io=3A Vin:12V P:int	5S	Nsc		data pinout
LM2576S-15	S+	Io=3A Vin:12V P:int	5S	Nsc		data pinout
LM2576S-3.3	S+	Io=3A Vin:12V P:int	5S	Nsc		data pinout
LM2576S-5.0	S+	Io=3A Vin:12V P:int	5S	Nsc		data pinout
LM2576S-ADJ	ES+	Io=3A Vo:1.23...37V Vin:12V P:int	5S	Nsc		data pinout
LM2576SX-12	S+	Io=3A Vin:12V P:int	5S	Nsc		data pinout
LM2576SX-15	S+	Io=3A Vin:12V P:int	5S	Nsc		data pinout
LM2576SX-3.3	S+	Io=3A Vin:12V P:int	5S	Nsc		data pinout
LM2576SX-5.0	S+	Io=3A Vin:12V P:int	5S	Nsc		data pinout
LM2576SX-ADJ	ES+	Io=3A Vo:1.23...37V Vin:12V P:int	5S	Nsc		data pinout
LM2576T-12	S+	Io=3A Vin:12V P:int	5P	Nsc		data pinout
LM2576T-15	S+	Io=3A Vin:12V P:int	5P	Nsc		data pinout
LM2576T-3.3	S+	Io=3A Vin:12V P:int	5P	Nsc		data pinout
LM2576T-5.0	S+	Io=3A Vin:12V P:int	5P	Nsc		data pinout
LM2576T-ADJ	ES+	Io=3A Vo:1.23...37V Vin:12V P:int	5P	Nsc		data pinout
LM 2576 WT	LIN-IC	52kHz Simple 3A Buck Regulator	DIP	Mcr	-	pdf pinout
LM 2576 WU	LIN-IC	52kHz Simple 3A Buck Regulator	DIP	Mcr	-	pdf pinout
LM2577M-12	R+	Io=6A Vo:11.40...12.60V Vin:5...10V	24S	Nsc		data pinout
LM2577M-15	R+	Io=6A Vo:14.25...15.75V Vin:5...12V	24S	Nsc		data pinout
LM2577M-ADJ	S+	Io=6A Vo:12.0...V Vin:3.5...40V P:int	24S	Nsc		data pinout
LM2577N-12	R+	Io=6A Vo:11.40...12.60V Vin:5...10V	16D	Nsc		data pinout
LM2577N-15	R+	Io=6A Vo:14.25...15.75V Vin:5...12V	16D	Nsc		data pinout
LM2577N-ADJ	S+	Io=6A Vo:12.0...V Vin:3.5...40V P:int	16D	Nsc		data pinout
LM2577S-12	R+	Io=6A Vo:11.40...12.60V Vin:5...10V	5S	Nsc		data pinout
LM2577S-15	R+	Io=6A Vo:14.25...15.75V Vin:5...12V	5S	Nsc		data pinout
LM2577S-ADJ	S+	Io=6A Vo:15.0...V Vin:3.5...40V P:int	5S	Nsc		data pinout
LM2577T-12	R+	Io=6A Vo:11.40...12.60V Vin:5...10V	5P	Nsc		data pinout
LM2577T-15	R+	Io=6A Vo:14.25...15.75V Vin:5...12V	5P	Nsc		data pinout
LM2577T-ADJ	S+	Io=6A Vo:12.0...V Vin:3.5...40V P:int	5P	Nsc		data pinout
LM 2578 H	SMPS-IC	=LM 1578N: -40...+85°	81	Nsc	-	
LM 2578 N	SMPS-IC	=LM 1578N: -40...+85°	8-DIP	Nsc	-	
LM 2579 T	IC	Switching Regulator	SIL	Nsc	-	pdf pinout
LM2585S-12	R+	Io=3A Vo:11.40...12.60V Vin:4...10V	5S	Nsc		data pinout
LM2585S-3.3	R+	Io=3A Vo:3.14...3.46V Vin:4...12V P:int	5S	Nsc		data pinout
LM2585S-5.0	R+	Io=3A Vo:4.75...5.25V Vin:4...12V P:int	5S	Nsc		data pinout
LM2585S-ADJ	S+	Io=3A Vin:4...40V P:int	5S	Nsc		data pinout
LM2585SX-12	R+	Io=3A Vo:11.40...12.60V Vin:4...10V	5S	Nsc		data pinout
LM2585SX-3.3	R+	Io=3A Vo:3.14...3.46V Vin:4...12V P:int	5S	Nsc		data pinout
LM2585SX-5.0	R+	Io=3A Vo:4.75...5.25V Vin:4...12V P:int	5S	Nsc		data pinout
LM2585SX-ADJ	S+	Io=3A Vin:4...40V P:int	5S	Nsc		data pinout
LM2585T-12	R+	Io=3A Vo:11.40...12.60V Vin:4...10V	5P	Nsc		data pinout
LM2585T-3.3	R+	Io=3A Vo:3.14...3.46V Vin:4...12V P:int	5P	Nsc		data pinout
LM2585T-5.0	R+	Io=3A Vo:4.75...5.25V Vin:4...12V P:int	5P	Nsc		data pinout
LM2585T-ADJ	S+	Io=3A Vin:4...40V P:int	5P	Nsc		data pinout
LM2586S-12	S+	Io=3A Vin:4...10V P:int	7S	Nsc		data pinout
LM2586S-3.3	S+	Io=3A Vin:4...10V P:int	7S	Nsc		data pinout
LM2586S-5.0	S+	Io=3A Vin:4...10V P:int	7S	Nsc		data pinout
LM2586S-ADJ	ES+	Io=3A Vo:1.23...37V Vin:10V P:int	7S	Nsc		data pinout
LM2586SX-12	S+	Io=3A Vin:4...10V P:int	7S	Nsc		data pinout
LM2586SX-3.3	S+	Io=3A Vin:4...10V P:int	7S	Nsc		data pinout
LM2586SX-5.0	S+	Io=3A Vin:4...10V P:int	7S	Nsc		data pinout
LM2586SX-ADJ	ES+	Io=3A Vo:1.23...37V Vin:10V P:int	7S	Nsc		data pinout
LM2586T-12	S+	Io=3A Vin:4...10V P:int	7P	Nsc		data pinout
LM2586T-3.3	S+	Io=3A Vin:4...10V P:int	7P	Nsc		data pinout
LM2586T-5.0	S+	Io=3A Vin:4...10V P:int	7P	Nsc		data pinout
LM2586T-ADJ	ES+	Io=3A Vo:1.23...37V Vin:10V P:int	7P	Nsc		data pinout
LM2587S-12	S+	Io=5A Vin:4...10V P:int	5S	Nsc		data pinout
LM2587S-3.3	S+	Io=5A Vin:4...10V P:int	5S	Nsc		data pinout
LM2587S-5.0	S+	Io=5A Vin:4...10V P:int	5S	Nsc		data pinout
LM2587S-ADJ	ES+	Io=5A Vo:1.23...37V Vin:10V P:int	5S	Nsc		data pinout
LM2587SX-12	R+	Io=5A Vo:11.40...12.60V Vin:4...10V	5S	Nsc		data pinout
LM2587SX-3.3	R+	Io=5A Vo:3.14...3.46V Vin:4...12V P:int	5S	Nsc		data pinout
LM2587SX-5.0	R+	Io=5A Vo:4.75...5.25V Vin:4...12V P:int	5S	Nsc		data pinout
LM2587SX-ADJ	ER+	Io=5A Vo:3.14...3.46V Vin:4...12V P:int	5S	Nsc		data pinout
LM2587T-12	S+	Io=5A Vin:4...10V P:int	5P	Nsc		data pinout
LM2587T-3.3	S+	Io=5A Vin:4...10V P:int	5P	Nsc		data pinout
LM2587T-5.0	S+	Io=5A Vin:4...10V P:int	5P	Nsc		data pinout

Type	Device	Short Description	Fig.	Manu	Comparision Types	More at
LM2587T-ADJ	ES+	Io=5A Vo:1.23...37V Vin:10V P:int	5P	Nsc		data pinout
LM2588S-12	S+	Io=5A Vin:4...10V P:int	7S	Nsc		data pinout
LM2588S-3.3	S+	Io=5A Vin:4...10V P:int	7S	Nsc		data pinout
LM2588S-5.0	S+	Io=5A Vin:4...10V P:int	7S	Nsc		data pinout
LM2588S-ADJ	ES+	Io=5A Vo:1.23...37V Vin:10V P:int	7S	Nsc		data pinout
LM2588SX-12	S+	Io=5A Vin:4...10V P:int	7S	Nsc		data pinout
LM2588SX-3.3	S+	Io=5A Vin:4...10V P:int	7S	Nsc		data pinout
LM2588SX-5.0	S+	Io=5A Vin:4...10V P:int	7S	Nsc		data pinout
LM2588SX-ADJ	ES+	Io=5A Vo:1.23...37V Vin:10V P:int	7S	Nsc		data pinout
LM2588T-12	S+	Io=5A Vin:4...10V P:int	7P	Nsc		data pinout
LM2588T-3.3	S+	Io=5A Vin:4...10V P:int	7P	Nsc		data pinout
LM2588T-5.0	S+	Io=5A Vin:4...10V P:int	7P	Nsc		data pinout
LM2588T-ADJ	ES+	Io=5A Vo:1.23...37V Vin:10V P:int	7P	Nsc		data pinout
LM2594HVM-12	S+	Io=500mA Vin:4.5...60V P:int	8S	Nsc		data pinout
LM2594HVM-3.2	S+	Io=500mA Vin:4.5...60V P:int	8S	Nsc		data pinout
LM2594HVM-3.3	R+	Io=500mA Vo:3.135...3.465V P:int	8S	Nsc		data pinout
LM2594HVM-5.0	S+	Io=500mA Vin:4.5...60V P:int	8S	Nsc		data pinout
LM2594HVM-ADJ	ES+	Io=500mA Vo:1.2...57V Vin:60V P:int	8S	Nsc		data pinout
LM2594HVN-12	S+	Io=500mA Vin:4.5...60V P:int	8D	Nsc		data pinout
LM2594HVN-3.3	S+	Io=500mA Vin:4.5...60V P:int	8D	Nsc		data pinout
LM2594HVN-5.0	S+	Io=500mA Vin:4.5...60V P:int	8D	Nsc		data pinout
LM2594HVN-ADJ	ES+	Io=500mA Vo:1.2...57V Vin:60V P:int	8D	Nsc		data pinout
LM2594M-12	S+	Io=500mA Vin:4.5...40V P:int	8S	Nsc		data pinout
LM2594M-3.3	S+	Io=500mA Vin:4.5...40V P:int	8S	Nsc		data pinout
LM2594M-5.0	S+	Io=500mA Vin:4.5...40V P:int	8S	Nsc		data pinout
LM2594M-ADJ	ES+	Io=500mA Vo:1.2...37V Vin:40V P:int	8S	Nsc		data pinout
LM2594N-12	S+	Io=500mA Vin:4.5...40V P:int	8D	Nsc		data pinout
LM2594N-3.3	S+	Io=500mA Vin:4.5...40V P:int	8D	Nsc		data pinout
LM2594N-5.0	S+	Io=500mA Vin:4.5...40V P:int	8D	Nsc		data pinout
LM2594N-ADJ	ES+	Io=500mA Vo:1.2...37V Vin:40V P:int	8D	Nsc		data pinout
LM2595J-12-QML	S+	Io=1mA Vin:4.5...40V P:int	16D	Nsc		data pinout
LM2595J-3.3-QML	S+	Io=1mA Vin:4.5...40V P:int	16D	Nsc		data pinout
LM2595J-5.0-QML	S+	Io=1mA Vin:4.5...40V P:int	16D	Nsc		data pinout
LM2595J-ADJ-QML	ES+	Io=1mA Vo:1.2...37V Vin:40V P:int	16D	Nsc		data pinout
LM2595S-12	S+	Io=1mA Vin:4.5...40V P:int	5S	Nsc		data pinout
LM2595S-3.3	S+	Io=1mA Vin:4.5...40V P:int	5S	Nsc		data pinout
LM2595S-5.0	S+	Io=1mA Vin:4.5...40V P:int	5S	Nsc		data pinout
LM2595S-ADJ	ES+	Io=1mA Vo:1.2...37V Vin:40V P:int	5S	Nsc		data pinout
LM2595T-12	S+	Io=1mA Vin:4.5...40V P:int	5P	Nsc		data pinout
LM2595T-3.3	S+	Io=1mA Vin:4.5...40V P:int	5P	Nsc		data pinout
LM2595T-5.0	S+	Io=1mA Vin:4.5...40V P:int	5P	Nsc		data pinout
LM2595T-ADJ	ES+	Io=1mA Vo:1.2...37V Vin:40V P:int	5P	Nsc		data pinout
LM2596S-12	S+	Io=3mA Vin:4.5...40V P:int	5S	Nsc		data pinout
LM2596S-3.3	S+	Io=3mA Vin:4.5...40V P:int	5S	Nsc		data pinout
LM2596S-5.0	S+	Io=3mA Vin:4.5...40V P:int	5S	Nsc		data pinout
LM2596S-ADJ	ES+	Io=3mA Vo:1.2...37V Vin:40V P:int	5S	Nsc		data pinout
LM 2596 S-ADJ	SMPS-IC	Pow.Conv. 150kHz 3A Step-down Vol.Reg.	86	Nsc		pdf pinout
LM2596T-12	S+	Io=3mA Vin:4.5...40V P:int	5P	Nsc		data pinout
LM2596T-3.3	S+	Io=3mA Vin:4.5...40V P:int	5P	Nsc		data pinout
LM2596T-5.0	S+	Io=3mA Vin:4.5...40V P:int	5P	Nsc		data pinout
LM2596T-ADJ	ES+	Io=3mA Vo:1.2...37V Vin:40V P:int	5P	Nsc		data pinout
LM 2596 T-ADJ	SMPS-IC	Pow.Conv. 150kHz 3A Step-down Vol.Reg.	86	Nsc		pdf pinout
LM2597HVM-12	S+	Io=0.5mA Vin:4.5...60V P:int	8S	Nsc		data pinout
LM2597HVM-3.3	S+	Io=0.5mA Vin:4.5...60V P:int	8S	Nsc		data pinout
LM2597HVM-5.0	S+	Io=0.5mA Vin:4.5...60V P:int	8S	Nsc		data pinout
LM2597HVM-ADJ	ES+	Io=0.5mA Vo:1.2...57V Vin:60V P:int	8S	Nsc		data pinout
LM2597HVN-12	S+	Io=0.5mA Vin:4.5...60V P:int	8D	Nsc		data pinout
LM2597HVN-3.3	S+	Io=0.5mA Vin:4.5...60V P:int	8D	Nsc		data pinout
LM2597HVN-5.0	S+	Io=0.5mA Vin:4.5...60V P:int	8D	Nsc		data pinout
LM2597HVN-ADJ	ES+	Io=0.5mA Vo:1.2...57V Vin:60V P:int	8D	Nsc		data pinout
LM2597M-12	S+	Io=0.5mA Vin:4.5...40V P:int	8S	Nsc		data pinout
LM2597M-3.3	S+	Io=0.5mA Vin:4.5...40V P:int	8S	Nsc		data pinout
LM2597M-5.0	S+	Io=0.5mA Vin:4.5...40V P:int	8S	Nsc		data pinout
LM2597M-ADJ	ES+	Io=0.5mA Vo:1.2...37V Vin:40V P:int	8S	Nsc		data pinout
LM2597N-12	S+	Io=0.5mA Vin:4.5...40V P:int	8D	Nsc		data pinout
LM2597N-3.3	S+	Io=0.5mA Vin:4.5...40V P:int	8D	Nsc		data pinout
LM2597N-5.0	S+	Io=0.5mA Vin:4.5...40V P:int	8D	Nsc		data pinout
LM2597N-ADJ	ES+	Io=0.5mA Vo:1.2...37V Vin:40V P:int	8D	Nsc		data pinout
LM2598S-12	S+	Io=1mA Vin:4.5...40V P:int	7S	Nsc		data pinout
LM2598S-3.3	S+	Io=1mA Vin:4.5...40V P:int	7S	Nsc		data pinout
LM2598S-5.0	S+	Io=1mA Vin:4.5...40V P:int	7S	Nsc		data pinout
LM2598S-ADJ	ES+	Io=1mA Vo:1.2...37V Vin:40V P:int	7S	Nsc		data pinout
LM2598T-12	S+	Io=1mA Vin:4.5...40V P:int	7P	Nsc		data pinout
LM2598T-3.3	S+	Io=1mA Vin:4.5...40V P:int	7P	Nsc		data pinout
LM2598T-5.0	S+	Io=1mA Vin:4.5...40V P:int	7P	Nsc		data pinout
LM2598T-ADJ	ES+	Io=1mA Vo:1.2...37V Vin:40V P:int	7P	Nsc		data pinout
LM2599S-12	S+	Io=3mA Vin:4.5...40V P:int	7S	Nsc		data pinout
LM2599S-3.3	S+	Io=3mA Vin:4.5...40V P:int	7S	Nsc		data pinout
LM2599S-5.0	S+	Io=3mA Vin:4.5...40V P:int	7S	Nsc		data pinout
LM2599S-ADJ	ES+	Io=3mA Vo:1.2...37V Vin:40V P:int	7S	Nsc		data pinout
LM2599T-12	S+	Io=3mA Vin:4.5...40V P:int	7P	Nsc		data pinout
LM2599T-3.3	S+	Io=3mA Vin:4.5...40V P:int	7P	Nsc		data pinout
LM2599T-5.0	S+	Io=3mA Vin:4.5...40V P:int	7P	Nsc		data pinout
LM2599T-ADJ	ES+	Io=3mA Vo:1.2...37V Vin:40V P:int	7P	Nsc		data pinout
LM 2636	Si-N	≈BC 337	2a		→BC 337	data pinout

Type	Device	Short Description	Fig.	Manu	Comparision Types	More at
LM 2638 M	IC	Motherboard Power Supply Solution	SQL	Nsc		pdf pinout
LM2651MTC-1.8	S+	Io=900mA Vo:1.719...1.854V Vin:10V	16S	Nsc		data pinout
LM2651MTC-2.5	S+	Io=900mA Vo:2.388...2.575V Vin:10V	16S	Nsc		data pinout
LM2651MTC-3.3	S+	Io=900mA Vo:3.201...3.399V Vin:10V	16S	Nsc		data pinout
LM2651MTCX-1.8	S+	Io=900mA Vo:1.719...1.854V Vin:10V	16S	Nsc		data pinout
LM2651MTCX-2.5	S+	Io=900mA Vo:2.388...2.575V Vin:10V	16S	Nsc		data pinout
LM2651MTCX-3.3	S+	Io=900mA Vo:3.201...3.399V Vin:10V	16S	Nsc		data pinout
LM2653MTC-ADJ	ES+	Io=900mA Vo:-1.5...-5V P:933mW	16S	Nsc		data pinout
LM2670S-12	S+	Io=3mA Vo:11.64...12.36V Vin:15...40V	7S	Nsc		data pinout
LM2670S-3.3	S+	Io=3mA Vo:3.201...3.399V Vin:8...40V	7S	Nsc		data pinout
LM2670S-5.0	S+	Io=3mA Vo:4.85...5.15V Vin:8...40V	7S	Nsc		data pinout
LM2670S-ADJ	ES+	Io=3mA Vo:-1.2...-37V Vin:40V P:int	7S	Nsc		data pinout
LM2670T-12	S+	Io=3mA Vo:11.64...12.36V Vin:15...40V	7P	Nsc		data pinout
LM2670T-3.3	S+	Io=3mA Vo:3.201...3.399V Vin:8...40V	7P	Nsc		data pinout
LM2670T-5.0	S+	Io=3mA Vo:4.85...5.15V Vin:8...40V	7P	Nsc		data pinout
LM2670T-ADJ	ES+	Io=3mA Vo:1.2...37V Vin:40V P:int	7P	Nsc		data pinout
LM2671M-12	S+	Io=0.5mA Vo:11.64...12.36V P:int	8S	Nsc		data pinout
LM2671M-3.3	S+	Io=0.5mA Vo:3.201...3.399V Vin:8...40V	8S	Nsc		data pinout
LM2671M-5.0	S+	Io=0.5mA Vo:4.85...5.15V Vin:8...40V	8S	Nsc		data pinout
LM2671M-ADJ	ES+	Io=0.5mA Vo:1.2...37V Vin:40V P:int	8S	Nsc		data pinout
LM2671N-12	S+	Io=0.5mA Vo:11.64...12.36V P:int	8D	Nsc		data pinout
LM2671N-3.3	S+	Io=0.5mA Vo:3.201...3.399V Vin:8...40V	8D	Nsc		data pinout
LM2671N-5.0	S+	Io=0.5mA Vo:4.85...5.15V Vin:8...40V	8D	Nsc		data pinout
LM2671N-ADJ	ES+	Io=0.5mA Vo:1.2...37V Vin:40V P:int	8D	Nsc		data pinout
LM2672M-12	S+	Io=1mA Vo:11.64...12.36V Vin:15...40V	8S	Nsc		data pinout
LM2672M-3.3	S+	Io=1mA Vo:3.201...3.399V Vin:8...40V	8S	Nsc		data pinout
LM2672M-5.0	S+	Io=1mA Vo:4.85...5.15V Vin:8...40V	8S	Nsc		data pinout
LM2672M-ADJ	ES+	Io=1mA Vo:1.2...37V Vin:40V P:int	8S	Nsc		data pinout
LM2672N-12	S+	Io=1mA Vo:11.64...12.36V Vin:15...40V	8D	Nsc		data pinout
LM2672N-3.3	S+	Io=1mA Vo:3.201...3.399V Vin:8...40V	8D	Nsc		data pinout
LM2672N-5.0	S+	Io=1mA Vo:4.85...5.15V Vin:8...40V	8D	Nsc		data pinout
LM2672N-ADJ	ES+	Io=1mA Vo:1.2...37V Vin:40V P:int	8D	Nsc		data pinout
LM2673S-12	S+	Io=3mA Vo:11.64...12.36V Vin:15...40V	7S	Nsc		data pinout
LM2673S-3.3	S+	Io=3mA Vo:3.201...3.399V Vin:8...40V	7S	Nsc		data pinout
LM2673S-5.0	S+	Io=3mA Vo:4.85...5.15V Vin:8...40V	7S	Nsc		data pinout
LM2673S-ADJ	ES+	Io=3mA Vo:1.2...37V Vin:40V P:int	7S	Nsc		data pinout
LM2673T-12	S+	Io=3mA Vo:11.64...12.36V Vin:15...40V	7P	Nsc		data pinout
LM2673T-3.3	S+	Io=3mA Vo:3.201...3.399V Vin:8...40V	7P	Nsc		data pinout
LM2673T-5.0	S+	Io=3mA Vo:4.85...5.15V Vin:8...40V	7P	Nsc		data pinout
LM2673T-ADJ	ES+	Io=3mA Vo:1.2...37V Vin:40V P:int	7P	Nsc		data pinout
LM2674M-12	S+	Io=0.5mA Vo:11.64...12.36V P:int	8S	Nsc		data pinout
LM2674M-3.3	S+	Io=0.5mA Vo:3.201...3.399V Vin:8...40V	8S	Nsc		data pinout
LM2674M-5.0	S+	Io=0.5mA Vo:4.85...5.15V Vin:8...40V	8S	Nsc		data pinout
LM2674M-ADJ	ES+	Io=0.5mA Vo:1.2...37V Vin:40V P:int	8S	Nsc		data pinout
LM2674N-12	S+	Io=0.5mA Vo:11.64...12.36V P:int	8D	Nsc		data pinout
LM2674N-3.3	S+	Io=0.5mA Vo:3.201...3.399V Vin:8...40V	8D	Nsc		data pinout
LM2674N-5.0	S+	Io=0.5mA Vo:4.85...5.15V Vin:8...40V	8D	Nsc		data pinout
LM2674N-ADJ	ES+	Io=0.5mA Vo:1.2...37V Vin:40V P:int	8D	Nsc		data pinout
LM2675M-12	S+	Io=1mA Vo:11.64...12.36V Vin:15...40V	8S	Nsc		data pinout
LM2675M-3.3	S+	Io=1mA Vo:3.201...3.399V Vin:8...40V	8S	Nsc		data pinout
LM2675M-5.0	S+	Io=1mA Vo:4.85...5.15V Vin:8...40V	8S	Nsc		data pinout
LM2675M-ADJ	ES+	Io=1mA Vo:1.2...37V Vin:40V P:int	8S	Nsc		data pinout
LM2675N-12	S+	Io=1mA Vo:11.64...12.36V Vin:15...40V	8D	Nsc		data pinout
LM2675N-3.3	S+	Io=1mA Vo:3.201...3.399V Vin:8...40V	8D	Nsc		data pinout
LM2675N-5.0	S+	Io=1mA Vo:4.85...5.15V Vin:8...40V	8D	Nsc		data pinout
LM2675N-ADJ	ES+	Io=1mA Vo:1.2...37V Vin:40V P:int	8D	Nsc		data pinout
LM2676S-12	S+	Io=3mA Vo:11.64...12.36V Vin:15...40V	7S	Nsc		data pinout
LM2676S-3.3	S+	Io=3mA Vo:3.201...3.399V Vin:8...40V	7S	Nsc		data pinout
LM2676S-5.0	S+	Io=3mA Vo:4.85...5.15V Vin:8...40V	7S	Nsc		data pinout
LM2676S-ADJ	ES+	Io=3mA Vo:1.2...37V Vin:40V P:int	7S	Nsc		data pinout
LM2676T-12	S+	Io=3mA Vo:11.64...12.36V Vin:15...40V	7P	Nsc		data pinout
LM2676T-3.3	S+	Io=3mA Vo:3.201...3.399V Vin:8...40V	7P	Nsc		data pinout
LM2676T-5.0	S+	Io=3mA Vo:4.85...5.15V Vin:8...40V	7P	Nsc		data pinout
LM2676T-ADJ	ES+	Io=3mA Vo:1.2...37V Vin:40V P:int	7P	Nsc		data pinout
LM2678S-12	S+	Io=5mA Vo:11.64...12.36V Vin:15...40V	7S	Nsc		data pinout
LM2678S-3.3	S+	Io=5mA Vo:3.201...3.399V Vin:8...40V	7S	Nsc		data pinout
LM2678S-5.0	S+	Io=5mA Vo:4.85...5.15V Vin:8...40V	7S	Nsc		data pinout
LM2678S-ADJ	ES+	Io=5mA Vo:1.2...37V Vin:40V P:int	7S	Nsc		data pinout
LM2678T-12	S+	Io=5mA Vo:11.64...12.36V Vin:15...40V	7P	Nsc		data pinout
LM2678T-3.3	S+	Io=5mA Vo:3.201...3.399V Vin:8...40V	7P	Nsc		data pinout
LM2678T-5.0	S+	Io=5mA Vo:4.85...5.15V Vin:8...40V	7P	Nsc		data pinout
LM2678T-ADJ	ES+	Io=5mA Vo:1.2...37V Vin:40V P:int	7P	Nsc		data pinout
LM2679S-12	S+	Io=5mA Vo:11.64...12.36V Vin:15...40V	7S	Nsc		data pinout
LM2679S-3.3	S+	Io=5mA Vo:3.201...3.399V Vin:8...40V	7S	Nsc		data pinout
LM2679S-5.0	S+	Io=5mA Vo:4.85...5.15V Vin:8...40V	7S	Nsc		data pinout
LM2679S-ADJ	ES+	Io=5mA Vo:1.2...37V Vin:40V P:int	7S	Nsc		data pinout
LM2679T-12	S+	Io=5mA Vo:11.64...12.36V Vin:15...40V	7P	Nsc		data pinout
LM2679T-3.3	S+	Io=5mA Vo:3.201...3.399V Vin:8...40V	7P	Nsc		data pinout
LM2679T-5.0	S+	Io=5mA Vo:4.85...5.15V Vin:8...40V	7P	Nsc		data pinout
LM2679T-ADJ	ES+	Io=5mA Vo:1.2...37V Vin:40V P:int	7P	Nsc		data pinout
LM 2747MTCX	SMPS-IC	SMD, synchr. buck controller, f<1MHz	SSMDIP	Nsc	-	pdf pinout
LM 2750 LD-...	SMPS-IC	Switched Capacitor Boost Reg., 0.12A low noise,...-ADJ: 3.8...5.2V, ...-5.0:5V	10-LLP	Nsc	-	pdf pinout
LM 2754SQ	IC	SMD, switched capacitor DC/DC conv., four reg. current sinks, max. 800mA	LLP	Nsc	-	pdf pinout

Type	Device	Short Description	Fig.	Manu	Comparision Types	More at
LM 2808(N)	LIN-IC	TV, Ton-/Sound ZF, NF-E, 2,6W(16V/8Ω)	18-DIP	Nsc	-	
LM 2825 N3.3	SMPS-IC	DC-DC Converter, Uout=3,3V, 1A	24-DIP	Nsc	-	
LM 2825 N5.0	LIN-IC	=LM 2825N3.3: Uout=5V	24-DIP	Nsc	-	
LM 2830 XMFX	SMPS-IC	SMD, high frequency, PWM step-down DC/DC45 converter, 1.0A, 1.6MHz		Nsc	-	pdf pinout
LM 2830 ZMFX	SMPS-IC	SMD, high frequency, PWM step-down DC/DC45 converter, 1.0A, 3MHz		Nsc	-	pdf pinout
LM 2830 ZSDX	SMPS-IC	SMD, high frequency, PWM step-down DC/DCLLP converter, 1.0A, 3MHz		Nsc	-	pdf pinout
LM 2831 (X,Y,Z)MFX	SMPS-IC	SMD, high frequency, PWM step-down DC/DC45 converter, 1.5A, 1.6/0.55/3MHz		Nsc	-	pdf pinout
LM 2832 (X,Y,Z)MYX	SMPS-IC	SMD, high frequency, PWM step-down DC/DCSSMDIP converter, 2.0A, 1.6/0.55/3MHz		Nsc	-	pdf pinout
LM 2832 (X,Y,Z)SDX	SMPS-IC	SMD, high frequency, PWM step-down DC/DCLLP converter, 2.0A, 1.6/0.55/3MHz		Nsc	-	pdf pinout
LM 2876	LIN-IC	NF-E, 20...60V, 50W(±30V/8Ω)	11-SQL	Nsc	-	pdf pinout
LM2877	2xAUDIO-OP	Vs:26V Vi0:15mV f:65kc	11P	Nsc	-	data pdf pinout
LM 2877(P)	LIN-IC	2x NF-E, 6..24V, 2x4,5W(20V/8Ω)	11-SIL	Nsc	-	data pdf pinout
LM2878	2xAUDIO-OP	Vs:35V Vi0:10mV f:65kc	11P	Nsc	-	data pdf pinout
LM 2878(P)	LIN-IC	2x NF-E, 6...32V, 2x5,5W(22V/8Ω)	11-SIL	Nsc	-	pinout
LM2879	2xAUDIO-OP	Vs:35V Vu:70dB Vi0:10mV f:65kc	11P	Nsc	-	data pdf pinout
LM 2879(T)	LIN-IC	2x NF-E, 6...32V, 2x8W(28V/8Ω)	11-SQL	Nsc	-	pinout
LM 2889(N)	LIN-IC	TV, Video Modulator, Ucc=10...16V	18-DIP	Nsc	-	pdf
LM 2893(N)	LIN-IC	Bi-line Carrier-current Transceiver	18-DIP	Nsc	-	pdf pinout
LM 2895(P)	LIN-IC	NF-E, 3...15V, 4,3W(12V/4Ω)	11-SIL	Nsc	-	
LM2896	2xAUDIO-OP	Vs:15V Vi0:5mV f:20kc	11P	Nsc	-	data pdf pinout
LM 2896(P-1,P-2)	LIN-IC	=LM 2895: Dual, 2x2,5W(12V/8Ω), 1x9W	11-SIL	Nsc	-	data pdf pinout
LM2900	4xOP	Vs:36V Vu:2.8V/mV f:2.5Mc	14D	Tix	-	data pdf pinout
LM 2900 D,J,N	OP-IC	Quad, ±16V, 50mA, -40...+85°	14-DIC,DIP	Mot,Nsc,Tix	NJM 2900..., (LM 3900...)[16]	data pdf pinout
LM2901	4xCOMP	Vs:36V Vu:100V/mV Vo:TTL Vi0:2mV	14SDF	Har	-	data pdf pinout
LM 2901 D,FP,M	KOP-IC	=LM 2901DP,N: SMD	14-MDIP	Nsc,Tho	→Serie 139	data pdf pinout
LM 2901(A)DP,F,J,N	KOP-IC	Quad, Serie 139, ±18V, -40...+85°	14-DIP,DIC	Mot,Nsc,++	→Serie 139	data pdf pinout
LM 2901 GC	KOP-IC	=LM 2901(A)DP,J,N: SMD, -40...+105°	20-(P)LCC	Tho	→Serie 139	data pdf
LM 2901 W	KOP-IC	=LM 2901(A)DP,J,N: Min	14-FLP	Tix	→Serie 139	data pinout
LM2902	4xOP	Vs:26V Vu:100V/mV Vi0:3mV	14SDdF	Tix		data pinout
LM 2902 CM,D,FP,M	OP-IC	=LM 124(A)DP,N: SMD, ±13V, -40...+85°	14-MDIP	Nsc,Tho,Tix	→Serie 124	data pdf pinout
LM 2902 DP,J,N	OP-IC	=LM 124(A)DP,N: 26V, -40...+85°	14-DIP,DIC	Mot,Nsc,++	→Serie 124	data pdf pinout
LM 2902 D	OP-IC	SMD, Quad Op Amp, =LM124, +3..26V	14-MDIP	Ons	-	pdf pinout
LM 2902 DR2	OP-IC	SMD, Quad Op Amp, =LM124, +3..26V	14-MDIP	Ons	-	pdf pinout
LM 2902 DTB	OP-IC	SMD, Quad Op Amp, =LM124, +3..26V	14-SSMDIP	Ons	-	pdf pinout
LM 2902 DTBR2	OP-IC	SMD, Quad Op Amp, =LM124, +3..26V	14-SSMDIP	Ons	-	pdf pinout
LM 2902 N	OP-IC	Quad Op Amp, =LM124, +3..26V	14-DIP	Ons	-	pdf pinout
LM 2902 VD	OP-IC	SMD, Quad Op Amp, =LM124, +3..26V	14-MDIP	Ons	-	pdf pinout
LM 2902 VDR2	OP-IC	SMD, Quad Op Amp, =LM124, +3..26V	14-MDIP	Ons	-	pdf pinout
LM 2902 VDTB	OP-IC	SMD, Quad Op Amp, =LM124, +3..26V	14-SSMDIP	Ons	-	pdf pinout
LM 2902 VDTBR2	OP-IC	SMD, Quad Op Amp, =LM124, +3..26V	14-SSMDIP	Ons	-	pdf pinout
LM 2902 VN	OP-IC	Quad Op Amp, =LM124, +3..26V	14-DIP	Ons	-	pdf pinout
LM2903	2xCOMP	Vs:36V Vu:100V/mV Vi0:2mV	8SAGD,10F	Tix		data pinout
LM 2903 CM,D,FP,M	KOP-IC	~LM 193(A)H: SMD, -40...+85°	8-MDIP	Nsc,Tho,Tix	→Serie 193	data pdf pinout
LM 2903DP,J,JG,N,P	KOP-IC	~LM 193(A)H: -40...+85°	8-DIP,DIC	Mot,Nsc,++	→Serie 193	data pdf pinout
LM 2903 H,L	KOP-IC	~LM 193(A)H: -40...+85°	81	Phi,Tix	→Serie 193	data pinout
LM 2903 U	KOP-IC	~LM 193(A)H: -40...+85°	10-FLP	Tix	→Serie 193	data pinout
LM2904	2xLO-POWER	Vs:32V Vu:100V/mV Vo:3.5/0V Vi0:2mV	8SDAG,10F	Sgs		data pinout
LM 2904 CM,D,FP,M	OP-IC	=LM 158(A)H: SMD, ±13V, -40...+85°	8-MDIP	Tho,Tix	→Série 158	data pinout
LM 2904DP,GJ,J,N,P	OP-IC	=LM 158(A)H: ±13V, -40...+85°	8-DIP,DIC	Mot,++	→Serie 158	data pinout
LM 2904 H,L	OP-IC	=LM 158(A)H: -40...+85°	81	Nsc,Mot,Tix	→Serie 158	data pinout
LM 2904 IBP	OP-IC	SMD, Dual, Low Power OPAmp -40...85°C, 1MHz, ±1.5/+3...±13/+26V	8-GRID	Nsc	-	pdf
LM 2904 M	OP-IC	SMD, Dual, Low Power OPAmp -40...85°C, 1MHz, ±1.5/+3...±13/+26V	8-MDIP	Nsc	-	pdf pinout
LM 2904 N	OP-IC	Dual, Low Power OPAmp -40...85°C, 1MHz, ±1.5/+3...±13/+26V	8-DIP	Nsc	-	pdf pinout
LM 2904 U	OP-IC	=LM 158(A)H: ±13V, -40...+85°	10-FLP	Tix	-	data pinout
LM 2904 WHD	IC	Low power dual operational amplifier	MDIP	Stm	-	pdf
LM 2905(N)	LIN-IC	=LM 122(H): -40...+85°	8-DIP	Nsc	-	pdf
LM 2907 N-8,P	LIN-IC	=LM 2907N: Fig.→	8-DIP	Nsc,Tix	-	pdf
LM 2907 N	LIN-IC	F/V-Converter, Ucc<28V, -40...+85°	14-DIP	Nsc,Tix	-	pdf
LM2908	4xOP	Vs:±18V Vu:104dB Vo:±12V Vi0:6mV	14D	Har	-	data pinout
LM 2917 N-8	LIN-IC	=LM 2917N: Fig.→	8-DIP	Nsc,Tix	-	pdf
LM 2917 N	LIN-IC	F/V-Converter, Ucc<28V, -40...+85°	14-DIP	Nsc,Tix	-	pdf
LM2924	2xOP+COMP	Vs:±13V Vu:100V/mV Vi0:2mV	8DS	Nsc	-	data pinout
LM 2924 N	OP/KOP-IC	=LM 192H: ±13V, -40...+85°	8-DIP	Nsc	LM 192..., ... LM 292...	data pinout
LM2930KC-5	R+	Io=150mA Vo:4.5...5.5V Vin:6...26V P:2W	3P	Tix		data pdf pinout
LM2930KC-8	R+	Io=150mA Vo:7.2...8.8V Vin:6...26V P:2W	3P	Tix		data pdf pinout
LM2930LP-5	R+	Io=150mA Vo:4.5...5.5V Vin:6...26V	3N	Tix		data pinout
LM2930LP-8	R+	Io=150mA Vo:7.2...8.8V Vin:6...26V	3N	Tix		data pinout
LM 2930 LP-...	Z-IC	=LM 2930T-...:	7b	Tix	NJM 2930L-xx	data pdf pinout
LM2930S-5.0	R+	Io=150mA Vo:4.7...5.3V Vin:14V P:int	3S	Nsc		data pdf pinout
LM2930S-8.0	R+	Io=150mA Vo:7.5...8.5V Vin:14V P:int	3S	Nsc		data pdf pinout
LM2930T-5.0	R+	Io=150mA Vo:4.7...5.3V Vin:14V P:int	3P	Nsc		data pdf pinout
LM 2930(A)KC,T-5.0	Z-IC	Io-drop, +5V, 0,4>0,15A	17b	Nsc,Sgs,Tix	NJM 2930-05	data pdf pinout
LM2930T-8.0	R+	Io=150mA Vo:7.5...8.5V Vin:14V P:int	3P	Nsc		data pdf pinout
LM 2930 KC,T-8.0	Z-IC	Io-drop, +8V, 0,4>0,15A	17b	Nsc,Tix	NJM 2930-08	data pdf pinout
LM2931	R+	Io=100mA Vo:4.81...5.19V P:int	8S,3SP	Nsc		data pinout
LM2931A	R+	Io=100mA Vo:4.81...5.19V Vin:14V P:int	8S,3SP	Nsc		data pinout

Type	Device	Short Description	Fig.	Manu	Comparision Types	More at
LM2931AD-5.0	R+	Io=100mA Vo:4.75...5.25V Vin:6...26V	8S	Mot		data pinout
LM 2931(A)D-5.0	Z-IC	=LM 2930T-...: SMD	8-MDIP	Mot	-	data pdf pinout
LM2931AT-5.0	R+	Io=100mA Vo:4.75...5.25V Vin:6...26V	5P	Mot		data pinout
LM 2931(A)T-5.0	Z-IC	lo-drop, +5V, 0,4>0,15A	17b	Mot,Nsc,Sgs	-	data pdf pinout
LM 2931(A)Z,LP-5.0	Z-IC	=LM 2930T-...:	7b	Mot,Nsc,Tix	-	data pdf pinout
LM2931AZ-5.0	R+	Io=100mA Vo:4.81...5.19V Vin:14V P:int	3N	Nsc		data pdf pinout
LM2931C	ER+	Io=100mA Vo:3...24V Vin:14V P:int	8S,5SP	Nsc		data pinout
LM2931CD	ER+	Io=100mA Vo:2.7...29.5V Vin:14V P:int	8S,5P	Mot		data pdf pinout
LM2931D	R+	Io=100mA Vo:4.75...5.25V Vin:6...26V	8S	Mot		data pdf pinout
LM2931T	ER+	Io=150mA Vo:3...24V P:int	5P	Nsc		data pinout
LM2931T-5.0	R+	Io=100mA Vo:4.75...5.25V Vin:6...26V	5P	Mot		data pinout
LM 2931 T	Z-IC	=LM 2930T-...: 3...24V	86/5Pin	Mot,Nsc	-	data pdf pinout
LM2931Z-5.0	R+	Io=100mA Vo:4.75...5.25V Vin:14V P:int	3N	Nsc		data pinout
LM 2935	Z-IC	Dual, lo-drop, 5V, 0,75+0,01A	86/5Pin	Sgs	-	data pdf
LM2935T	R+	Io=750mA Vo:4.75...5.25V Vin:6...26V	5P	Nsc		data pinout
LM2936M-5.0	R+	Io=50mA Vo:4.85...5.15V Vin:5.5...26V	8S	Nsc		data pdf pinout
LM 2936 M-5.0	Z-IC	=LM 2936Z-...:	8-MDIP		-	data pdf pinout
LM2936Z	R+	Io=50mA Vo:4.9...5.1V Vin:5.5...26V	3N	Nsc		data pinout
LM2936Z-5.0	R+	Io=50mA Vo:4.85...5.15V Vin:5.5...26V	3N	Nsc		data pdf pinout
LM 2936 Z-5.0	Z-IC	+5V, 0,05A	7b	Nsc	... 78L05 (TO-92)	data pdf pinout
LM2937ES-XX	R+	Io=500mA Vo:9.5...10.5V Vin:6...26V	3S	Nsc		data pdf pinout
LM2937ET-XX	R+	Io=500mA Vo:9.5...10.5V Vin:6...26V	3P	Nsc		data pinout
LM2937IMP-XX	R+	Io=500mA Vo:9.5...10.5V Vin:6...26V	4S	Nsc		data pdf pinout
LM2940CS-12	R+	Io=1A Vo:11.64...12.36V Vin:13.6...26V	3S	Nsc		data pinout
LM2940CS-15	R+	Io=1A Vo:14.55...15.45V P:int	3S	Nsc		data pinout
LM2940CS-5.0	R+	Io=1A Vo:4.85...5.15V Vin:6.25...26V	3S	Nsc		data pinout
LM2940CS-9.0	R+	Io=1A Vo:8.55...9.45V Vin:10.5...26V	3S	Nsc		data pinout
LM2940CT-12	R+	Io=1A Vo:11.64...12.36V Vin:13.6...26V	3P	Nsc		data pinout
LM2940CT-15	R+	Io=1A Vo:14.55...15.45V P:int	3P	Nsc		data pinout
LM2940CT-5.0	R+	Io=1A Vo:4.85...5.15V Vin:6.25...26V	3P	Nsc		data pinout
LM2940CT-9.0	R+	Io=1A Vo:8.55...9.45V Vin:10.5...26V	3P	Nsc		data pinout
LM2940IMP-10	R+	Io=1A Vo:9.7...10.3V Vin:11.5...26V	4S	Nsc		data pinout
LM2940IMP-12	R+	Io=1A Vo:11.64...12.36V Vin:13.6...26V	4S	Nsc		data pinout
LM2940IMP-15	R+	Io=1A Vo:14.55...15.45V P:int	4S	Nsc		data pinout
LM2940IMP-5.0	R+	Io=1A Vo:4.85...5.15V Vin:6.25...26V	4S	Nsc		data pinout
LM2940IMP-8.0	R+	Io=1A Vo:7.76...8.24V Vin:9.4...26V	4S	Nsc		data pinout
LM2940IMP-9.0	R+	Io=1A Vo:8.73...9.27V Vin:10.5...26V	4S	Nsc		data pinout
LM2940J-12/883	R+	Io=1A Vo:11.64...12.36V Vin:13.6...26V	16D	Nsc		data pinout
LM2940J-15/883	R+	Io=1A Vo:14.55...15.45V P:int	16D	Nsc		data pinout
LM2940J-5.0/883	R+	Io=1A Vo:4.85...5.15V Vin:6.25...26V	16D	Nsc		data pinout
LM2940J-8.0/883	R+	Io=1A Vo:7.76...8.24V Vin:9.4...26V	16D	Nsc		data pinout
LM2940K-12/883	R+	Io=1A Vo:11.4...12.6V Vin:13.6...26V	3O	Nsc		data pinout
LM2940K-15/883	R+	Io=1A Vo:14.25...15.75V P:int	3O	Nsc		data pinout
LM2940K-5.0/883	R+	Io=1A Vo:4.75...5.25V Vin:6.25...26V	3O	Nsc		data pinout
LM2940K-8.0/883	R+	Io=1A Vo:7.6...8.4V Vin:9.4...26V P:int	3O	Nsc		data pinout
LM2940S-10	R+	Io=1A Vo:9.7...10.3V Vin:11.5...26V	3S	Nsc		data pinout
LM2940S-12	R+	Io=1A Vo:11.4...12.6V Vin:13.6...26V	3S	Nsc		data pinout
LM2940S-15	R+	Io=1A Vo:14.25...15.75V P:int	3S	Nsc		data pinout
LM2940S-5.0	R+	Io=1A Vo:4.85...5.15V Vin:6.25...26V	3S	Nsc		data pinout
LM2940S-8.0	R+	Io=1A Vo:7.76...8.24V Vin:9.4...26V	3S	Nsc		data pinout
LM2940S-9.0	R+	Io=1A Vo:8.55...9.45V Vin:10.5...26V	3S	Nsc		data pinout
LM2940T-10	R+	Io=1A Vo:9.7...10.3V Vin:11.5...26V	3P	Nsc		data pinout
LM2940T-12	R+	Io=1A Vo:11.4...12.6V Vin:13.6...26V	3P	Nsc		data pinout
LM2940T-15	R+	Io=1A Vo:14.25...15.75V P:int	3P	Nsc		data pinout
LM2940T-5.0	R+	Io=1A Vo:4.85...5.15V Vin:6.25...26V	3P	Nsc		data pinout
LM2940T-8.0	R+	Io=1A Vo:7.76...8.24V Vin:9.4...26V	3P	Nsc		data pinout
LM2940T-9.0	R+	Io=1A Vo:8.55...9.45V Vin:10.5...26V	3P	Nsc		data pinout
LM 2940(C)T-12	Z-IC	+5V, 1A, lo-drop	17b	Nsc		data pinout
LM2940WG	R+	Io=1A Vo:4.85...5.15V Vin:6.25...26V	16S	Nsc		data pinout
LM2941	ER+	Io=1A Vo:5...20V P:int	5SPO	Nsc		data pdf pinout
LM2941J	ER+	Io=1A Vo:1.24...1.31V Vin:5...20V P:int	16DS	Nsc		data pdf pinout
LM2984	R+	Io=500mA Vo:4.85...5.15V Vin:6...26V	11P	Nsc		data pdf pinout
LM2990J-12	R-	Io=1A Vo:-11.4...-12.6V Vin:-1...-26V	16D	Nsc		data pinout
LM2990J-15	R-	Io=1A Vo:-14.25...-15.75V P:int	16D	Nsc		data pinout
LM2990J-5.0	R-	Io=1A Vo:-4.75...-5.25V Vin:-1...-26V	16D	Nsc		data pinout
LM2990S-12	R-	Io=1A Vo:-11.4...-12.6V Vin:-1...-26V	3S	Nsc		data pinout
LM2990S-15	R-	Io=1A Vo:-14.25...-15.75V P:int	3S	Nsc		data pinout
LM2990S-5.0	R-	Io=1A Vo:-4.75...-5.25V Vin:-1...-26V	3S	Nsc		data pinout
LM2990T-12	R-	Io=1A Vo:-11.4...-12.6V Vin:-1...-26V	3P	Nsc		data pinout
LM2990T-15	R-	Io=1A Vo:-14.25...-15.75V P:int	3P	Nsc		data pinout
LM2990T-5.0	R-	Io=1A Vo:-4.75...-5.25V Vin:-1...-26V	3P	Nsc		data pinout
LM2990T-5.2	R-	Io=1A Vo:-4.94...-5.46V Vin:-1...-26V	3P	Nsc		data pinout
LM2990WG-5.	R-	Io=1A Vo:-4.75...-5.25V Vin:-1...-26V	16S	Nsc		data pinout
LM2991	ER-	Io=1A Vo:-2...-25V Vin:-1...-26V P:int	5SP	Nsc		data pdf pinout
LM2991J	ER-	Io=1A Vo:-2...-25V Vin:-1...-26V P:int	16D	Nsc		data pinout
LM2991WG	ER-	Io=1A Vo:-2...-25V Vin:-1...-26V P:int	16S	Nsc		data pinout
LM 3009	IC			Nsc	-	
LM 3011(H)	LIN-IC	Breitb./Wideband Amp., Ucc<15V	82	Nsc	CA 3011	pdf
LM3026	HI-SPEED	Vs:±9V Vu:32dB Vo:±6V Vi0:5mV	12A	Nsc		data pinout
LM3028	OP	Vs:±12V Vu:30dB Vo:±8V Vi0:2mV	8A	Nsc		data pinout
LM3028A	OP	Vs:±12V Vu:30dB Vi0:5mV	8A	Nsc		data pinout
LM 3045 J,N	LIN-IC	5xNPN-Trans, 20V, 0,05A, -55...+125°	14-DIP,DIC	Nsc,Sgs	CA 3045, TBA 331, µA 3045	
LM3046	OP	f:120Mc	14S	Nsc		data pdf pinout
LM 3046 J,N	LIN-IC	=LM 3045: -40...+85°	14-DIP,DIC	Nsc,Sgs	CA 3046, TBA 331, µA 3046	
LM 3046 D,M	LIN-IC	=LM 3046J,N: SMD	14-MDIP	Nsc,Sgs	-	

Type	Device	Short Description	Fig.	Manu	Comparision Types	More at
LM3053	OP	Vs:±23V	8A	Nsc		data pinout
LM3054	HI-SPEED	Vs:±9V Vu:32dB Vo:±6V Vi0:5mV	14D	Nsc		data pinout
LM 3064(N,N-01)	LIN-IC	TV, Aft, Agc, Ucc=30V	14-DIP	Nsc	-	
LM 3065(N)	LIN-IC	TV, Ton-ZF, FM-ZF	14-DIP	Nsc	AN 241, CA 3065, HA 1125, KA 2101, LA 13	
LM 3075(N,N-01)	LIN-IC	FM Detector/Limiter, NF-V, Ucc>12,5V	14-DIP	Nsc	CA 3075	pdf
LM3080	OTA	Vs:±18V Vo:±14V Vi0:0.4mV f:2Mc	8DS	Nsc		data pinout
LM3080A	OTA	Vs:±22V Vo:±14V Vi0:0.4mV f:2Mc	8D	Nsc		data pinout
LM 3080(A)J,N	OP-IC	OTA, ±18(A=±22)V, 0...+70(A=-55...+125)°	8-DIC,DIP	Nsc		data pdf pinout
LM 3086(N)	LIN-IC	5x NPN-Trans, 20V, 0,05A, -40...+85°	14-DIP	Nsc	CA 3086, TBA 331, µA 3086	pdf
LM 3089(N)	LIN-IC	TV-Ton-ZF, FM-ZF, Agc, Afc, Ucc=12V	16-DIP	Nsc	CA 3089, TCA 3089, TDA 1200(A)	pdf
LM 3100MHX	SMPS-IC	SMD, synchr. rectified Buck Conv., 1.5A, integr. dual N-chan. switches	SSMDIP	Nsc		pdf pinout
LM 3146 M	LIN-IC	=LM 3146(N): SMD	14-MDIP	Nsc	-	pdf
LM 3146(N)	LIN-IC	5x NPN-Trans, 40V, 0,05A, -40...+85°	14-DIP	Nsc	-	pdf
LM 3150	IC	simple switcher® Controller, 42V Sync.step- Down	TSOP	Nsc	-	pdf pinout
LM 3151	IC	simple switcher® Controller, 42V Sync.step- Down	TSOP	Nsc	-	pdf pinout
LM 3152	IC	simple switcher® Controller, 42V Sync.step- Down	TSOP	Nsc	-	pdf pinout
LM 3153	IC	simple switcher® Controller, 42V Sync.step- Down	TSOP	Nsc	-	pdf pinout
LM 3189 N	LIN-IC	FM-ZF, Dem, Afc, AGC(progr.), 8,5...16V	16-DIP	Nsc	CA 3189E, TCA 3189	pdf
LM 3211	PMOS-IC	6 Bit Frequ.-T/divider f.Orgel/organ		Say		pdf
LM 3216	PMOS-IC	6-Stage Divider	14-DIP	Say	M3216	pdf pinout
LM3301	4xOP	Vs:±14V Vu:69dB Vo:±13.5V	14D	Nsc		data pdf pinout
LM 3301 N	OP-IC	=LM 2900... : ±14V	14-DIP	Nsc	LM 2900...	data pdf pinout
LM3302	4xCOMP	Vs:36V Vu:30V/mV Vo:TTL Vi0:1mV	14SD	Har		data pdf pinout
LM 3302 DP,J,L,N,P	KOP-IC	Quad, Serie 139, ±14V, -40...+85°	14-DIP,DIC	Mot,Nsc,++	→Serie 139	data pdf pinout
LM3303	4xOP	Vs:±18V Vu:200V/mV Vo:>±12V Vi0:2mV	14DS	Nsc	-	data pinout
LM 3351	IC	Switched Capacitor Voltage Converter	SQL	Nsc	-	pdf pinout
LM 3354 MM-x.x	SMPS-IC	SMD, Reg. Buck-Boost DC-DC, +x.xV/90mA	10-VSMDIP	Nsc		pdf pinout
LM 3361(A)N	LIN-IC	FM-ZF, Schmalb./Narrow Band, Ucc=2...9V	16-DIP	Nsc	KA 3361, MC 3361	pdf
LM 3364 K	NMOS-dRAM-IC	64kB x 1 Bit, 150ns access time	16-DIP	Say		
LM3401	4xOP	Vs:±9V	14D	Nsc		data pinout
LM 3401 N	OP-IC	=LM 2900... : ±9V, 0...+75°	14-DIP	Nsc	LM 2900..., LM 3301..., LM 3900...	data pdf pinout
LM 3402HVMM	SMPS-IC	SMD, 0.5A constant current buck reg. for driving power LEDs, HiVolt version	SSMDIP	Nsc		pdf pinout
LM 3402MM	SMPS-IC	SMD, 0.5A constant current buck reg. for driving high power LEDs	SSMDIP	Nsc	-	pdf pinout
LM3403	4xOP	Vs:±18V Vu:200V/mV Vo:>±12V Vi0:2mV	14DS	Nsc		data pdf pinout
LM 3404	SMPS-IC	SMD, 1A constant current buck reg. for driving power LEDs	VSMDIP	Nsc	-	pdf pinout
LM 3404HV	SMPS-IC	SMD, 1A constant current buck reg. for driving power LEDs, Hivolt version	VSMDIP	Nsc	-	pdf pinout
LM 3405-X	SMPS-IC	SMD, 1A constant current buck reg. for driving high power LEDs, 1.6MHz	46	Nsc		pdf pinout
LM 3405-Y	SMPS-IC	SMD, 1A constant current buck reg. for driving high power LEDs, 550kHz	46	Nsc		pdf pinout
LM3411AM5-3.3	R+	Io=20mA Vo:3.267...3.333V P:300mW	5S	Nsc		data pinout
LM3411AM5-5.0	R+	Io=20mA Vo:4.95...5.05V P:300mW	5S	Nsc		data pinout
LM3411AM5X-3.3	R+	Io=20mA Vo:3.267...3.333V P:300mW	5S	Nsc		data pinout
LM3411AM5X-5.0	R+	Io=20mA Vo:4.95...5.05V P:300mW	5S	Nsc		data pinout
LM3411M5-3.3	R+	Io=20mA Vo:3.234...3.366V P:300mW	5S	Nsc		data pinout
LM3411M5-5.0	R+	Io=20mA Vo:4.9...5.1V P:300mW	5S	Nsc		data pinout
LM3411M5X-3.3	R+	Io=20mA Vo:3.234...3.366V P:300mW	5S	Nsc		data pinout
LM3411M5X-5.0	R+	Io=20mA Vo:4.9...5.1V P:300mW	5S	Nsc		data pinout
LM3420-12.6	R+	Io=20mA Vo:12.35...12.85V Vin:12.6V	5S	Nsc		data pinout
LM3420-4.2	R+	Io=20mA Vo:4.116...4.284V Vin:4.2V	5S	Nsc		data pinout
LM3420-8.4	R+	Io=20mA Vo:8.232...8.568V Vin:8.4V	5S	Nsc		data pinout
LM3420A-12.6	R+	Io=20mA Vo:12.47...12.73V Vin:12.6V	5S	Nsc		data pinout
LM3420A-4.2	R+	Io=20mA Vo:4.158...4.242V Vin:4.2V	5S	Nsc		data pinout
LM3420A-8.4	R+	Io=20mA Vo:8.316...8.484V Vin:8.4V	5S	Nsc		data pinout
LM3420AM5-12.6	R+	Io=20mA Vo:12.47...12.72V P:300mW	5S	Nsc		data pinout
LM3420AM5-16.8	R+	Io=20mA Vo:16.63...16.96V P:300mW	5S	Nsc		data pinout
LM3420AM5-4.2	R+	Io=20mA Vo:4.158...4.242V P:300mW	5S	Nsc		data pinout
LM3420AM5-8.2	R+	Io=20mA Vo:8.118...8.282V P:300mW	5S	Nsc		data pinout
LM3420AM5-8.4	R+	Io=20mA Vo:8.316...8.484V P:300mW	5S	Nsc		data pinout
LM3420M5-12.6	R+	Io=20mA Vo:12.34...12.85V P:300mW	5S	Nsc		data pinout
LM3420M5-16.8	R+	Io=20mA Vo:16.46...17.13V P:300mW	5S	Nsc		data pinout
LM3420M5-4.2	R+	Io=20mA Vo:4.116...4.284V P:300mW	5S	Nsc		data pinout
LM3420M5-8.2	R+	Io=20mA Vo:8.036...8.364V P:300mW	5S	Nsc		data pinout
LM3420M5-8.4	R+	Io=20mA Vo:8.232...8.568V P:300mW	5S	Nsc		data pinout
LM3460M5-1.2	R+	Io=7A Vo:1.19...1.24V P:300mW	5S	Nsc		data pinout
LM3460M5-1.5	R+	Io=7A Vo:1.47...1.53V P:300mW	5S	Nsc		data pinout
LM3460M5X-1.2	R+	Io=7A Vo:1.19...1.24V P:300mW	5S	Nsc		data pinout
LM3460M5X-1.5	R+	Io=7A Vo:1.47...1.53V P:300mW	5S	Nsc		data pinout
LM 3478	IC	Low-side N-channel Controller, Switching Regulator	SQL	Nsc		pdf pinout
LM3480iM3-12	R+	Vo:11.40...12.60V Vin:1.5...30V	3S	Nsc		data pinout
LM3480iM3-15	R+	Vo:14.25...15.75V Vin:1.5...30V	3S	Nsc		data pinout
LM3480IM3-3.3	R+	Vo:3.14...3.46V Vin:1.5...30V P:250mW	3S	Nsc		data pinout
LM3480IM3-5.0	R+	Vo:4.75...5.25V Vin:1.5...30V P:250mW	3S	Nsc		data pinout
LM3480IM3X-12	R+	Vo:11.40...12.60V Vin:1.5...30V	3S	Nsc		data pinout
LM3480IM3X-15	R+	Vo:14.25...15.75V Vin:1.5...30V	3S	Nsc		data pinout

Type	Device	Short Description	Fig.	Manu	Comparision Types	More at
LM3480IM3X-3.3	R+	Vo:3.14...3.46V Vin:1.5...30V P:250mW	3S	Nsc		data pinout
LM3480IM3X-5.0	R+	Vo:4.75...5.25V Vin:1.5...30V P:250mW	3S	Nsc		data pinout
LM3490IM5	R+	Vo:4.75...5.25V Vin:1.5...30V P:300mW	5S	Nsc		data pinout
LM3490IM5-12	R+	Vo:11.40...12.60V Vin:1.5...30V	5S	Nsc		data pinout
LM3490IM5-15	R+	Vo:14.25...15.75V Vin:1.5...30V	5S	Nsc		data pinout
LM3490IM5X-12	R+	Vo:11.40...12.60V Vin:1.5...30V	5S	Nsc		data pinout
LM3490IM5X-15	R+	Vo:14.25...15.75V Vin:1.5...30V	5S	Nsc		data pinout
LM3490IM5X-3.3	R+	Vo:3.14...3.46V Vin:1.5...30V P:300mW	5S	Nsc		data pinout
LM3490IM5X-5.0	R+	Vo:4.75...5.25V Vin:1.5...30V P:300mW	5S	Nsc		data pinout
LM3503	4xOP	Vs:±18V Vu:200V/mV Vo:>±13.5V f:1Mc	14D	Nsc		data pinout
LM 3524	SMPS-IC	=SG 3524	16-DIP	Nsc	→SG 3524	pdf pinout
LM 3524 J,N	SMPS-IC	=LM 1524: 0...+70°	16-DIC,DIP	Nsc	→SG 3524...	pdf pinout
LM 3578 H	SMPS-IC	=LM 1578N: 0...+70°	81	Nsc	-	pdf
LM 3578 N	SMPS-IC	=LM 1578N: 0...+70°	8-DIP	Nsc	-	pdf
LM3622AM-4.1	R+	Vo:4.07...4.13V Vin:4.5V P:350mW	8S	Nsc		data pinout
LM3622AM-4.2	R+	Vo:4.17...4.23V Vin:4.5V P:350mW	8S	Nsc		data pinout
LM3622AM-8.2	R+	Vo:8.14...8.26V Vin:4.5V P:350mW	8S	Nsc		data pinout
LM3622AM-8.4	R+	Vo:8.34...8.46V Vin:4.5V P:350mW	8S	Nsc		data pinout
LM3622AMX-4.1	R+	Vo:4.07...4.13V Vin:4.5V P:350mW	8S	Nsc		data pinout
LM3622AMX-4.2	R+	Vo:4.17...4.23V Vin:4.5V P:350mW	8S	Nsc		data pinout
LM3622AMX-8.2	R+	Vo:8.14...8.26V Vin:4.5V P:350mW	8S	Nsc		data pinout
LM3622AMX-8.4	R+	Vo:8.34...8.46V Vin:4.5V P:350mW	8S	Nsc		data pinout
LM3622M-4.1	R+	Vo:4.05...4.15V Vin:4.5V P:350mW	8S	Nsc		data pinout
LM3622M-4.2	R+	Vo:4.15...4.25V Vin:4.5V P:350mW	8S	Nsc		data pinout
LM3622M-8.2	R+	Vo:8.1...8.3V Vin:4.5V P:350mW	8S	Nsc		data pinout
LM3622M-8.4	R+	Vo:8.3...8.5V Vin:4.5V P:350mW	8S	Nsc		data pinout
LM3622MX-4.1	R+	Vo:4.05...4.15V Vin:4.5V P:350mW	8S	Nsc		data pinout
LM3622MX-4.2	R+	Vo:4.15...4.25V Vin:4.5V P:350mW	8S	Nsc		data pinout
LM3622MX-8.2	R+	Vo:8.1...8.3V Vin:4.5V P:350mW	8S	Nsc		data pinout
LM3622MX-8.4	R+	Vo:8.3...8.5V Vin:4.5V P:350mW	8S	Nsc		data pinout
LM 3700	ROM-IC	4096x10Bit	28-DIP	Say	-	pdf
LM 3708	IC	Microprocessor Supervisory Circuits,Low Line Output	DIP	Nsc	-	pdf pinout
LM 3709	IC	Microprocessor Supervisory Circuits,Low Line Output	DIP	Nsc	-	pdf pinout
LM 3710	ROM-IC	8192x10Bit	28-DIP	Say	-	pdf
LM 3743 MM-300	SMPS-IC	SMD, voltage mode PWM buck controller, 300kHz, Hiccup mode protection	VSMDIP	Nsc	-	pdf pinout
LM 3743 MM-1000	SMPS-IC	SMD, voltage mode PWM buck controller, 1MHz, Hiccup mode protection	VSMDIP	Nsc	-	pdf pinout
LM 3764	EPROM-IC	NMOS, 8192 x 8 Bit, <250...450ns	28-DIP	Say	-	pdf
LM 3820(N)	LIN-IC	Am Radio System, Ucc=6V	14-DIP	Nsc	-	pdf
LM3875	AUDIO-OP	Vs:±84V Vu:120dB Vi0:1mV f:8Mc	11P	Nsc		data pdf pinout
LM 3875	LIN-IC	NF-E, 20...84V, 70W(±35V/8Ω)	11-SQL	Nsc	-	pdf pinout
LM 3876	LIN-IC	NF-E, 20...84V, 70W(±35V/8Ω)	11-SQL	Nsc	-	pdf pinout
LM 3880MFX	IC	SMD, Power Sequencer, power up/down control, 3 open drain outputs	46	Nsc	-	pdf pinout
LM 3886	LIN-IC	NF-E, 20...84V, 78W(±35V/8Ω)	11-SQL	Nsc	-	pdf pinout
LM3900	4xOP	Vs:36V Vu:2.8V/mV f:2.5Mc	14SD	Tix	-	data pdf pinout
LM 3900 D,M	OP-IC	=LM 2900...: SMD, 0...+70°	14-MDIP	Mot,Nsc,Tix	NJM 2900..., NJM 3900...	data pdf pinout
LM 3900 J,N	OP-IC	=LM 2900...: 0...+70°	14-DIC,DIP	Mot,Nsc	LM 2900...	data pdf pinout
LM 3905(N)	LIN-IC	=LM 122(H): 0...+70°	8-DIP	Nsc	-	pdf pinout
LM 3909(N)	LIN-IC	LED Flasher/Oscillator, Ucc=1...5V	8-DIP	Nsc	-	pdf pinout
LM 3911(N)	LIN-IC	Temp. Sensor, Volt. Ref., Op-amp.	8-DIP	Nsc	-	pdf
LM 3914 J	LIN-IC	10 LED Level Meter linear, bar/dot	18-DIC	Nsc	-	pdf pinout
LM 3914 N	LIN-IC	10 LED Level Meter linear, bar/dot	18-DIP	Nsc	-	pdf
LM 3914 V	LIN-IC	10 LED Level Meter linear, bar/dot	20-PLCC	Nsc	-	pdf pinout
LM 3915 J	LIN-IC	10 LED Level Meter log., bar/dot	18-DIC	Nsc	-	pdf pinout
LM 3915 N	LIN-IC	10 LED Level Meter log., bar/dot	18-DIP	Nsc	-	pdf
LM 3916 J	LIN-IC	10 LED Level VU-Meter, bar/dot	18-DIC	Nsc	-	pdf pinout
LM 3916 N	LIN-IC	10 LED Level VU-Meter, bar/dot	18-DIP	Nsc	-	pdf
LM3940	R+	Io=1A Vo:3.13...3.47V Vin:5V P:int	4S,3SP,16D	Nsc		data pinout
LM3999	Z+	Io=10mA Vo:6.6...7.3V P:300mW	3N	Nsc		data pinout
LM 3999 Z	Ref-Z-IC	6,95V, ±5%, 0,6<2,2Ω, 0...+70°	7(AK+)	Nsc		pdf
LM4040AIM3-10	Z+	Io=20mA Vo:10.0...V P:306mW	3S	Nsc		data pinout
LM4040AIM3-2.5	Z+	Io=20mA Vo:2.5...V P:306mW	3S	Nsc		data pinout
LM4040AIM3-4.1	Z+	Io=20mA Vo:4.1...V P:306mW	3S	Nsc		data pinout
LM4040AIM3-5.0	Z+	Io=20mA Vo:5.0...V P:306mW	3S	Nsc		data pinout
LM4040AIM3-8.2	Z+	Io=20mA Vo:8.2...V P:306mW	3S	Nsc		data pinout
LM4040AIZ-10	Z+	Io=20mA Vo:10.0...V P:550mW	3N	Nsc		data pinout
LM4040AIZ-2.5	Z+	Vo:20mA Vo:2.5...V P:550mW	3N	Nsc		data pinout
LM4040AIZ-4.1	Z+	Io=20mA Vo:4.1...V P:550mW	3N	Nsc		data pinout
LM4040AIZ-5.0	Z+	Io=20mA Vo:5.0...V P:550mW	3N	Nsc		data pinout
LM4040AIZ-8.2	Z+	Io=20mA Vo:8.2...V P:550mW	3N	Nsc		data pinout
LM4040BIM3-10	Z+	Io=20mA Vo:10.0...V P:306mW	3S	Nsc		data pinout
LM4040BIM3-2.5	Z+	Io=20mA Vo:2.5...V P:306mW	3S	Nsc		data pinout
LM4040BIM3-4.1	Z+	Io=20mA Vo:4.1...V P:306mW	3S	Nsc		data pinout
LM4040BIM3-5.0	Z+	Io=20mA Vo:5.0...V P:306mW	3S	Nsc		data pinout
LM4040BIM3-8.2	Z+	Io=20mA Vo:8.2...V P:306mW	3S	Nsc		data pinout
LM4040BIZ-10	Z+	Io=20mA Vo:10.0...V P:550mW	3N	Nsc		data pinout
LM4040BIZ-2.5	Z+	Io=20mA Vo:2.5...V P:550mW	3N	Nsc		data pinout
LM4040BIZ-4.1	Z+	Io=20mA Vo:4.1...V P:550mW	3N	Nsc		data pinout
LM4040BIZ-5.0	Z+	Io=20mA Vo:5.0...V P:550mW	3N	Nsc		data pinout
LM4040BIZ-8.2	Z+	Io=20mA Vo:8.2...V P:550mW	3N	Nsc		data pinout
LM4040CEM3-2.5	Z+	Io=20mA Vo:2.5...V P:306mW	3S	Nsc		data pinout

Type	Device	Short Description	Fig.	Manu	Comparision Types	More at
LM4040CEM3-5.0	Z+	Io=20mA Vo:5.0...V P:306mW	3S	Nsc		data pinout
LM4040CIM3-10	Z+	Io=20mA Vo:10.0...V P:306mW	3S	Nsc		data pinout
LM4040CIM3-2.5	Z+	Io=20mA Vo:2.5...V P:306mW	3S	Nsc		data pinout
LM4040CIM3-4.1	Z+	Io=20mA Vo:4.1...V P:306mW	3S	Nsc		data pinout
LM4040CIM3-5.0	Z+	Io=20mA Vo:5.0...V P:306mW	3S	Nsc		data pinout
LM4040CIM3-8.2	Z+	Io=20mA Vo:8.2...V P:306mW	3S	Nsc		data pinout
LM4040CIZ-10	Z+	Io=20mA Vo:10.0...V P:550mW	3N	Nsc		data pinout
LM4040CIZ-2.5	Z+	Io=20mA Vo:2.5...V P:550mW	3N	Nsc		data pinout
LM4040CIZ-4.1	Z+	Io=20mA Vo:4.1...V P:550mW	3N	Nsc		data pinout
LM4040CIZ-5.0	Z+	Io=20mA Vo:5.0...V P:550mW	3N	Nsc		data pinout
LM4040CIZ-8.2	Z+	Io=20mA Vo:8.2...V P:550mW	3N	Nsc		data pinout
LM4040DEM3-2.5	Z+	Io=20mA Vo:2.5...V P:306mW	3S	Nsc		data pinout
LM4040DEM3-5.0	Z+	Io=20mA Vo:5.0...V P:306mW	3S	Nsc		data pinout
LM4040DIM3-10	Z+	Io=20mA Vo:10.0...V P:306mW	3S	Nsc		data pinout
LM4040DIM3-2.5	Z+	Io=20mA Vo:2.5...V P:306mW	3S	Nsc		data pinout
LM4040DIM3-4.1	Z+	Io=20mA Vo:4.1...V P:306mW	3S	Nsc		data pinout
LM4040DIM3-5.0	Z+	Io=20mA Vo:5.0...V P:306mW	3S	Nsc		data pinout
LM4040DIM3-8.2	Z+	Io=20mA Vo:8.2...V P:306mW	3S	Nsc		data pinout
LM4040DIZ-10	Z+	Io=20mA Vo:10.0...V P:550mW	3N	Nsc		data pinout
LM4040DIZ-2.5	Z+	Io=20mA Vo:2.5...V P:550mW	3N	Nsc		data pinout
LM4040DIZ-4.1	Z+	Io=20mA Vo:4.1...V P:550mW	3N	Nsc		data pinout
LM4040DIZ-5.0	Z+	Io=20mA Vo:5.0...V P:550mW	3N	Nsc		data pinout
LM4040DIZ-8.2	Z+	Io=20mA Vo:8.2...V P:550mW	3N	Nsc		data pinout
LM4040E	Z+	Io=20mA Vo:2.5...V P:550mW	3SN	Nsc		data pinout
LM4041	Z+	Io=20mA Vo:1.22...V P:306mW	3SN	Nsc		data pinout
LM4041CEM3-ADJ	EZ+	Io=20mA Vo:1.225...10V P:306mW	3S	Nsc		data pinout
LM4041CIM3-1.2	Z+	Io=20mA Vo:1.22...V P:306mW	3S	Nsc		data pinout
LM4041CIM3-ADJ	EZ+	Io=20mA Vo:1.225...10V P:306mW	3S	Nsc		data pinout
LM4041CIZ-1.2	Z+	Io=20mA Vo:1.225...V P:550mW	3N	Nsc		data pinout
LM4041CIZ-ADJ	EZ+	Io=20mA Vo:1.225...10V P:550mW	3N	Nsc		data pinout
LM4041DEM3-1.2	Z+	Io=20mA Vo:1.22...V P:306mW	3S	Nsc		data pinout
LM4041DEM3-ADJ	EZ+	Io=20mA Vo:1.225...10V P:306mW	3S	Nsc		data pinout
LM4041DIM3-1.2	Z+	Io=20mA Vo:1.22...V P:306mW	3S	Nsc		data pinout
LM4041DIM3-ADJ	EZ+	Io=20mA Vo:1.225...10V P:306mW	3S	Nsc		data pinout
LM4041DIZ-1.2	Z+	Io=20mA Vo:1.22...V P:550mW	3N	Nsc		data pinout
LM4041DIZ-ADJ	EZ+	Io=20mA Vo:1.225...10V P:550mW	3N	Nsc		data pinout
LM4041E	Z+	Io=20mA Vo:1.22...V P:550mW	3SN	Nsc		data pinout
LM4050AIM3-10	Z+	Io=15mA Vo:10...V P:280mW	3S	Nsc		data pinout
LM4050AIM3-2.5	Z+	Io=15mA Vo:2.5...V P:280mW	3S	Nsc		data pinout
LM4050AIM3-4.1	Z+	Io=15mA Vo:4.1...V P:280mW	3S	Nsc		data pinout
LM4050AIM3-5.0	Z+	Io=15mA Vo:5...V P:280mW	3S	Nsc		data pinout
LM4050AIM3-8.2	Z+	Io=15mA Vo:8.2...V P:280mW	3S	Nsc		data pinout
LM4050AIM3X-10	Z+	Io=15mA Vo:10...V P:280mW	3S	Nsc		data pinout
LM4050AIM3X-2.5	Z+	Io=15mA Vo:2.5...V P:280mW	3S	Nsc		data pinout
LM4050AIM3X-4.1	Z+	Io=15mA Vo:4.1...V P:280mW	3S	Nsc		data pinout
LM4050AIM3X-5.0	Z+	Io=15mA Vo:5...V P:280mW	3S	Nsc		data pinout
LM4050AIM3X-8.2	Z+	Io=15mA Vo:8.2...V P:280mW	3S	Nsc		data pinout
LM4050BIM3-10	Z+	Io=15mA Vo:10...V P:280mW	3S	Nsc		data pinout
LM4050BIM3-2.5	Z+	Io=15mA Vo:2.5...V P:280mW	3S	Nsc		data pinout
LM4050BIM3-4.1	Z+	Io=15mA Vo:4.1...V P:280mW	3S	Nsc		data pinout
LM4050BIM3-5.0	Z+	Io=15mA Vo:5...V P:280mW	3S	Nsc		data pinout
LM4050BIM3-8.2	Z+	Io=15mA Vo:8.2...V P:280mW	3S	Nsc		data pinout
LM4050BIM3X-10	Z+	Io=15mA Vo:10...V P:280mW	3S	Nsc		data pinout
LM4050BIM3X-2.5	Z+	Io=15mA Vo:2.5...V P:280mW	3S	Nsc		data pinout
LM4050BIM3X-4.1	Z+	Io=15mA Vo:4.1...V P:280mW	3S	Nsc		data pinout
LM4050BIM3X-5.0	Z+	Io=15mA Vo:5...V P:280mW	3S	Nsc		data pinout
LM4050BIM3X-8.2	Z+	Io=15mA Vo:8.2...V P:280mW	3S	Nsc		data pinout
LM4050CIM3-10	Z+	Io=15mA Vo:10...V P:280mW	3S	Nsc		data pinout
LM4050CIM3-2.5	Z+	Io=15mA Vo:2.5...V P:280mW	3S	Nsc		data pinout
LM4050CIM3-4.1	Z+	Io=15mA Vo:4.1...V P:280mW	3S	Nsc		data pinout
LM4050CIM3-5.0	Z+	Io=15mA Vo:5...V P:280mW	3S	Nsc		data pinout
LM4050CIM3-8.2	Z+	Io=15mA Vo:8.2...V P:280mW	3S	Nsc		data pinout
LM4050CIM3X-10	Z+	Io=15mA Vo:10...V P:280mW	3S	Nsc		data pinout
LM4050CIM3X-2.5	Z+	Io=15mA Vo:2.5...V P:280mW	3S	Nsc		data pinout
LM4050CIM3X-4.1	Z+	Io=15mA Vo:4.1...V P:280mW	3S	Nsc		data pinout
LM4050CIM3X-5.0	Z+	Io=15mA Vo:5...V P:280mW	3S	Nsc		data pinout
LM4050CIM3X-8.2	Z+	Io=15mA Vo:8.2...V P:280mW	3S	Nsc		data pinout
LM4051AIM3-1.2	Z+	Io=20mA Vo:1.22...V P:280mW	3S	Nsc		data pinout
LM4051AIM3-ADJ	EZ+	Io=20mA Vo:1.22...10V P:280mW	3S	Nsc		data pinout
LM4051BIM3-1.2	Z+	Io=20mA Vo:1.22...V P:280mW	3S	Nsc		data pinout
LM4051BIM3-ADJ	EZ+	Io=20mA Vo:1.22...10V P:280mW	3S	Nsc		data pinout
LM4051CIM3-1.2	Z+	Io=20mA Vo:1.22...V P:280mW	3S	Nsc		data pinout
LM4051CIM3-ADJ	EZ+	Io=20mA Vo:1.24...10V P:280mW	3S			data pinout
LM4136	4xOP	Vs:±18V Vu:300V/mV Vo:±14V Vi0:0.5mV	14DS	Nsc		data pinout
LM4250	LO-POWER	Vs:±18V Vu:96dB Vo:±12V Vi0:6mV	8S,10FAD	Nsc		data pdf pinout
LM4250C	LO-POWER	Vs:±18V Vu:96dB Vo:±12V Vi0:6mV	8ADS	Has,Nsc		data pdf pinout
LM 4250(C)H	OP-IC	lo-power, ±18V, -55...+125°(C=0...+70°)	81	Nsc,Isi	→Serie 250	data pdf pinout
LM 4250(C)J,N	OP-IC	=LM 4250(C): Fig.→	8-DIC,DIP		→Serie 250	data pdf pinout
LM 4250(C)M	OP-IC	=LM 4250(C): SMD	8-MDIP		→Serie 250	data pdf pinout
LM4431M3-2.5	Z+	Io=20mA Vo:2.50...V P:306mW	3S	Nsc		data pinout
LM 4500 A	LIN-IC	PLL Stereo-Decoder, Blend	16-DIP	Nsc	TCA 4500A, UL 1261N	pdf
LM 4560	Dig. Audio	Advanced PCI Audio Accelerator	DIP	Nsc	-	pdf pinout
LM 4610	Dig. Audio	Dual DC Operated Tone/volume/balance Circuit,National 3-D Sound	DIP	Nsc	-	pdf pinout
LM 4675SD	IC	SMD, 2.65W, Class D audio power amp, low	LLP	Nsc	-	pdf pinout

Type	Device	Short Description	Fig.	Manu	Comparision Types	More at
LM 4675TL	IC	EMI, filterless, Spread Spectrum SMD, 2.65W, Class D audio power amp, low EMI, filterless, Spread Spectrum	GRID	Nsc	-	pdf pinout
LM 4700	LIN-IC	NF-E, 20...64V, 42W(±28V/8Ω)	11-SQL	Nsc	-	pdf pinout
LM 4700 T(F)	LIN-IC	Aud. PAmp, 20...66V, 30W(±28V/8Ω) Mute, Standby, TF:isolated package	11-SQL	Nsc	-	pdf pinout
LM 4834 MS	LIN-IC	SMD, Aud. PAmp Boomer(R), +5V Mic PreAmp, DC Vol.Control,1.5W(5V/8Ω)	28-SSMDIP	Nsc	-	pdf pinout
LM 4857 ITL	LIN-IC	Stereo Audio Sub-System,3DEnhanc.,1.2W Loudspk.-, Headph.-, Earpiece-amp, I²C	30-GRID	Nsc	-	pdf
LM 4857 SP	LIN-IC	Stereo Audio Sub-System,3DEnhanc.,1.2W Loudspk.-, Headph.-, Earpiece-amp, I²C	28-LLP	Nsc	-	pdf
LM 4860	LIN-IC	SMD, NF-E, 2,7...5,5V, 1,45W(5V/8Ω)	16-MDIP	Nsc	-	pdf pinout
LM 4860 M	LIN-IC	SMD, Aud. PAmp Boomer(r) Shutdown, 2.7...5.5V, 1.45W(5V/8Ω)	16-MDIP	Nsc	-	pdf pinout
LM 4861	LIN-IC	SMD, NF-E, 2,7...5,5V, 1,45W(5V/8Ω)	8-MDIP	Nsc	-	pdf pinout
LM 4861 M	LIN-IC	SMD, Aud. PAmp Boomer(r) Shutdown, 2.7...5.5V, 1.45W(5V/8Ω)	8-MDIP	Nsc	-	pdf pinout
LM 4862	LIN-IC	NF-E, 2,7...5,5V, 0,83W(5V/8Ω)	8-(M)DIP	Nsc	-	pdf pinout
LM 4862 M	LIN-IC	SMD, Aud. PAmp Boomer(r) Shutdown, 2.7...5.5V, 0.83W(5V/8Ω)	8-MDIP	Nsc	-	pdf pinout
LM 4862 N	LIN-IC	Aud. PAmp Boomer(r) Shutdown, 2.7...5.5V, 0.83W(5V/8Ω)	8-DIP	Nsc	-	pdf pinout
LM 4880	LIN-IC	NF-E, 2,7...5,5V, 0,33W(5V/8Ω)	8-(M)DIP	Nsc	-	pdf pinout
LM 4880 M	LIN-IC	SMD, Stereo Aud. PAmp Boomer(r) Shutdown, 2.7...5.5V, 0.33W(5V/ 8Ω)	8-MDIP	Nsc	-	pdf pinout
LM 4880 N	LIN-IC	Stereo Aud. PAmp Boomer(r) Shutdown, 2.7...5.5V, 0.33W(5V/8Ω)	8-DIP	Nsc	-	pdf pinout
LM 5005MH	SMPS-IC	SMD, high voltage switching reg., 75V N-channel buck switch, 2.5A	SSMDIP	Nsc	-	pdf pinout
LM 5007 MM	SMPS-IC	SMD, Step Down Switch. Reg., 80V	8-SSMDIP	Nsc	-	pdf pinout
LM 5007 SD	SMPS-IC	High Volt.(80V) Step Down Switch. Reg.	8-LLP	Nsc	-	pdf pinout
LM 5009MM	SMPS-IC	SMD, 100V step-down switching reg., 150mA, integr. N-channel Mosfet	SSMDIP	Nsc	-	pdf pinout
LM 5009SDC	SMPS-IC	SMD, 100V step-down switching reg., 150mA, integr. N-channel Mosfet	LLP	Nsc	-	pdf pinout
LM 5069MMX	IC	SMD, pos. high voltage hot swap/inrush current controller, power limiting	VSMDIP	Nsc	-	pdf pinout
LM 5100AM...CM	IC	SMD, high voltage, high- and low-side gate driver, 3/2/1A, input Cmos	MDIP	Nsc	-	pdf pinout
LM 5100ASD...BSD	IC	SMD, high voltage, high- and low-side gate driver, 3/2/1A, input Cmos	LLP	Nsc	-	pdf pinout
LM 5101AM...CM	IC	SMD, high voltage, high- and low-side gate driver, 3/2/1A, input TTL	MDIP	Nsc	-	pdf pinout
LM 5101ASD...CSD	IC	SMD, high voltage, high- and low-side gate driver, 3/2/1A, input TTL	LLP	Nsc	-	pdf pinout
LM 5102MM	IC	SMD, high voltage, high- and low-side gate driver	VSMDIP	Nsc	-	pdf pinout
LM 5102SD	IC	SMD, high voltage, high- and low-side gate driver	LLP	Nsc	-	pdf pinout
LM6104	4xFB-AMP	Vs:24V Vo:6.5/0V Vi0:10mV f:7.5Mc	14S	Nsc		data pdf pinout
LM6118	2xFAST-SET	Vs:42V Vu:>150V/mV Vi0:±1mV f:>14Mc	20C,8AD	Nsc		data pinout
LM6121	FAST-BUFF	Vs:±18V Vo:±13.5V Vi0:15mV f:50Mc	8AD	Nsc		data pdf pinout
LM 6121 H	OP-IC	Buffer, ±18V, 50MHz, -55...+125°	81	Nsc	-	data
LM 6121 N	OP-IC	=LM 6121H:	8-DIP	Nsc		data
LM6125	FAST-BUFF	Vs:±18V Vo:±13.5V Vi0:15mV f:50Mc	8A	Nsc		data pdf pinout
LM 6125 H	OP-IC	Buffer, ±18V, 50MHz, -55...+125°	81	Nsc	-	data pdf
LM 6125 N	OP-IC	=LM 6125H:	14-DIP	Nsc		data
LM6132	2xRR-OP	Vs:24V Vu:100V/mV Vo:4.99/0V f:10Mc	8SD	Nsc		data pdf pinout
LM6134	4xRR-OP	Vs:24V Vu:100V/mV Vo:4.99/0V f:10Mc	14SD	Nsc		data pdf pinout
LM6142	2xRR-OP	Vs:24V Vu:270V/mV Vo:4.97V Vi0:0.3mV	8SD,14S	Nsc		data pdf pinout
LM6142A	2xRR-OP	Vs:24V Vu:270V/mV Vo:4.97V Vi0:0.3mV	8SD	Nsc		data pdf pinout
LM 6142 M	OP-IC	=LM 6142N: SMD	8-MDIP	Nsc	-	pinout
LM 6142 N	OP-IC	Dual, lo-power, ±24V, 10MHz	8-DIP	Nsc		pinout
LM6144	4xRR-OP	Vs:24V Vu:270V/mV Vo:4.97V Vi0:0.3mV	14SD	Nsc		data pinout
LM6144A	4xRR-OP	Vs:24V Vu:270V/mV Vo:4.97V Vi0:0.3mV	14SD	Nsc		data pdf pinout
LM 6144 M	OP-IC	=LM 6144N: SMD	14-MDIP	Nsc		pinout
LM 6144 N	OP-IC	Quad, lo-power, ±24V, 10MHz	8-DIP	Nsc		pinout
LM6152A	2xRR-OP	Vs:24V Vu:214V/mV Vo:4.89/0V f:75Mc	8SD	Nsc		data pdf pinout
LM6152B	2xRR-OP	Vs:24V Vu:214V/mV Vo:4.89/0V f:75Mc	8SD	Nsc		data pdf pinout
LM6154A	4xRR-OP	Vs:24V Vu:214V/mV Vo:4.89/0V f:75Mc	14SD	Nsc		data pdf pinout
LM6154B	4xRR-OP	Vs:24V Vu:214V/mV Vo:4.89/0V f:75Mc	14SD	Nsc		data pdf pinout
LM6161	HI-SPEED	Vs:36V Vo:±13.5V Vi0:±7mV f:>40Mc	20C,8D,10F	Nsc		data pdf pinout
LM6162	HI-SPEED	Vs:36V Vo:±14V Vi0:±3mV f:100Mc	20C,8D,10F	Nsc		data pinout
LM6164	HI-SPEED	Vs:36V Vu:2.5V/mV Vo:14.2V Vi0:2mV	20C,8D,10F	Nsc		data pinout
LM6165	HI-SPEED	Vs:36V Vu:10.5V/mV Vo:14.2V Vi0:1mV	20C,8D,10F	Nsc		data pinout
LM6171A	HI-SPEED	Vs:36V Vu:90dB Vo:±13.3V Vi0:1.5mV	8SD	Nsc		data pdf pinout
LM6171B	HI-SPEED	Vs:36V Vu:90dB Vo:±13.3V Vi0:1.5mV	8SD	Nsc		data pdf pinout
LM6172	2xHI-SPEED	Vs:36V Vu:86dB Vo:±13.2V Vi0:0.4mV	8DS	Nsc		data pinout
LM6172A	2xHI-SPEED	Vs:36V Vu:>75dB Vo:±12.5V Vi0:<1.5mV	8D	Nsc		data pinout
LM6181A	HI-SPEED	Vs:±18V Vo:±12V Vi0:2mV f:100Mc	8S,16SD	Nsc		data pdf pinout
LM6181I	HI-SPEED	Vs:±18V Vo:±12V Vi0:3.5mV f:100Mc	8S,16SD	Nsc		data pdf pinout
LM6182	2xFB-AMP	Vs:±18V Vo:±12V Vi0:2mV f:100Mc	8D,16S	Nsc		data pdf pinout
LM6182A	2xFB-AMP	Vs:±18V Vo:±12V Vi0:2mV f:100Mc	16S,8D,14D	Nsc		data pdf pinout

Type	Device	Short Description	Fig.	Manu	Comparision Types	More at
LM6211	LO-POWER	Vs:25V Vu:120dB Vi0:0.25mV f:20Mc	5S	Nsc		data pdf pinout
LM6218	2xFAST-SET	Vs:±21V Vu:500V/mV Vi0:0.2mV f:17Mc	8AD,14S	Nsc		data pinout
LM6221	FAST-BUFF	Vs:±18V Vo:13.5/0V Vi0:15mV f:50Mc	8AD	Nsc		data pdf pinout
LM 6221 ...	OP-IC	=LM 6121...: -40...+85°				data pinout
LM6225	FAST-BUFF	Vs:±18V Vo:±13.5V Vi0:15mV f:50Mc	14D	Nsc		data pdf pinout
LM 6225 ...	OP-IC	=LM 6125...: -40...+85°				data pinout
LM6261	HI-SPEED	Vs:36V Vu:14.2V Vi0:5mV f:50Mc	8DS	Nsc		data pinout
LM6262	HI-SPEED	Vs:36V Vo:±14V Vi0:±3mV f:100Mc	8SD	Nsc		data pinout
LM6264	HI-SPEED	Vs:36V Vu:2.5V/mV Vo:14.2V Vi0:2mV	8D	Nsc		data pdf pinout
LM6265	HI-SPEED	Vs:36V Vu:10.5V/mV Vo:14.2V Vi0:1mV	8D	Nsc		data pinout
LM6310I	OP	Vs:22V Vo:3.5/0V Vi0:1mV f:90Mc	8SD	Nsc		data pinout
LM6311I	OP	Vs:±6V Vu:3.16V/mV Vo:±3.4V f:110Mc	8SD	Nsc		data pinout
LM6313	HI-SPEED	Vs:±18V Vo:±13.1V Vi0:5mV f:35Mc	16D	Nsc		data pinout
LM6317I	OP	Vs:±6V Vu:3.16V/mV Vo:±3.2V f:120Mc	8SD	Nsc		data pinout
LM6321	FAST-BUFF	Vs:±18V Vo:13.5/0V Vi0:15mV f:50Mc	14S,8D	Nsc		data pinout
LM 6321 ...	OP-IC	=LM 6121...: -20...+80°			-	data pdf pinout
LM6325	FAST-BUFF	Vs:±18V Vo:±13.5V Vi0:15mV f:50Mc	14D	Nsc		data pinout
LM 6325 ...	OP-IC	=LM 6125...: -20...+80°				data pinout
LM6361	HI-SPEED	Vs:36V Vu:14.2V Vi0:5mV f:50Mc	8DS	Nsc		data pinout
LM6362	HI-SPEED	Vs:36V Vo:±14V Vi0:±3mV f:100Mc	8SD	Nsc		data pinout
LM6364	HI-SPEED	Vs:36V Vu:2.5V/mV Vo:14.2V Vi0:2mV	8SD	Nsc		data pinout
LM6365	HI-SPEED	Vs:36V Vu:10.5V/mV Vo:14.2V Vi0:1mV	8SD	Nsc		data pinout
LM 6402(A,H,L)	µC-IC	4 Bit, 2kB ROM, 128x4Bit RAM	42-DIP	Say	-	
LM 6405(A,H,L)	µC-IC	4 Bit, 1kB ROM, 64x4Bit RAM	42-DIP	Say	-	
LM 6413 E	µC-IC	4 Bit, 1kB ROM , 64x4Bit RAM	28-DIP	Say	-	
LM 6416 E	µC-IC	4 Bit, 1kB ROM , 64x4Bit RAM	28-DIP	Say	-	
LM 6417 E	µC-IC	4 Bit	22-DIP	Say	-	
LM 6497	µC-IC		Chip	Say	-	
LM 6499	µC-IC		Chip	Say	-	
LM6511	COMP	Vs:36V Vu:40V/mV Vi0:1.5mV	8SD	Nsc		data pdf pinout
LM6584	4xRR-OP	Vs:18V Vu:108dB Vo:0.1...15.9V	14S	Nsc		data pdf pinout
LM6685	FAST-COMP	Vs:±7V Vi0:0.3mV	14S,16D	Nsc		data pinout
LM6687	2xFAST-COMP	Vs:±7V Vi0:0.6mV	16SD	Nsc		data pinout
LM 7000(N)	NMOS-IC	AM/FM-Tuning Pll-frequ.-synthesizer	20-DIP	Say	-	pdf pinout
LM 7001	NMOS-IC	AM/FM-Tuning Pll-frequ.-synthesizer	16-DIP	Say	-	pdf pinout
LM 7001 M,JM	NMOS-IC	=LC 7001(J): SMD	20-MDIP		-	
LM 7001 M	NMOS-IC	=LM 7001: SMD	20-MDIP	Say	-	pinout
LM 7005	NMOS-IC	AM...UHF Tuning Pll-frequ.-synthesizer	24-SDIP	Say	-	pdf
LM 7006(H)	NMOS-IC	Telecom, PLL Synthesizer f. Cordl. Ph.	20-DIP	Say	-	pdf
LM 7007 M, MH	NMOS-IC	Telecom, PLL Synthesizer f. Cordl. Ph.	24-SMDIP	Say	-	
LM 7008 M, MH	NMOS-IC	Telecom, PLL Synthesizer f. Cordl. Ph.	24-SMDIP	Say	-	
LM7121A	HI-SPEED	Vs:±15V f:200Mc	5S,8SD	Nsc		data pinout
LM7121B	HI-SPEED	Vs:±15V f:200Mc	5S,8SD	Nsc		data pinout
LM7121I	FB-AMP	Vs:33V Vu:72dB Vo:±13.4V Vi0:0.9mV	5S,8SD	Nsc		data pdf pinout
LM7131A	HI-SPEED	Vs:12V Vu:60dB Vo:2.6/0V Vi0:0.02mV	5S,8SD	Nsc		data pdf pinout
LM7131B	HI-SPEED	Vs:12V Vu:60dB Vo:2.6/0V Vi0:0.02mV	5S,8SD	Nsc		data pdf pinout
LM7171A	HI-SPEED	Vs:36V Vu:85dB Vo:±13.3V Vi0:0.2mV	8SD,16S	Nsc		data pdf pinout
LM7171B	HI-SPEED	Vs:36V Vu:85dB Vo:±13.3V Vi0:0.2mV	8SD,16S	Nsc		data pdf pinout
LM7301I	OP	Vs:35V Vu:71V/mV Vo:4.93/0V f:4Mc	5S,8S	Nsc		data pinout
LM7805	R+	Io=1A Vo:4.8...5.2V Vin:10V P:int	3OPS	Nsc		data pinout
LM 7805CK...7824CK	Z-IC	=LM 7805CT,CU...LM 7824CT,CU:	23a	Nsc	... 78xx... (TO-3)	data pinout
LM 7805 CT,CU	Z-IC	+5V, >1A	17b	Nsc	... 7805... (TO-220)	data pinout
LM7806	R+	Io=1A Vo:5.76...6.24V P:2W	3OP	Nsc		data pinout
LM 7806 CT,CU	Z-IC	+6V, >1A	17b	Nsc	... 7806... (TO-220)	data pinout
LM7808	R+	Io=1.5A Vo:7.6...8.4V P:2.5W	3OP	Nsc		data pinout
LM 7808 CT,CU	Z-IC	+8V, >1A	17b	Nsc	... 7808... (TO-220)	data pinout
LM7810	R+	Io=1A Vo:9.6...10.4V Vin:17V	3OP	Nsc		data pinout
LM 7810 CT,CU	Z-IC	+10V, >1A	17b	Nsc	... 7810... (TO-220)	data pinout
LM7812	R+	Io=1A Vo:11.5...12.5V Vin:19V P:int	3OPS	Nsc		data pinout
LM 7812 CT,CU	Z-IC	+12V, >1A	17b	Nsc	... 7812... (TO-220)	data pinout
LM7815	R+	Io=1A Vo:14.4...15.6V P:int	3OP	Nsc		data pinout
LM 7815 CT,CU	Z-IC	+15V, >1A	17b	Nsc	... 7815... (TO-220)	data pinout
LM7818	R+	Io=1.5A Vo:17.3...18.7V P:2W	3OP	Nsc		data pinout
LM 7818 CT,CU	Z-IC	+18V, >1A	17b	Nsc	... 7818... (TO-220)	data pinout
LM7824	R+	Io=1A Vo:23...25V P:2W	3OP	Nsc		data pinout
LM 7824 CT,CU	Z-IC	+24V, >1A	17b	Nsc	... 7824... (TO-220)	data pinout
LM 78 S 40	IC	Universal Switching Regulator Subsystem	DIP	Nsc		pdf pinout
LM7905.2	R-	Io=1.5A Vo:-5...-5.4V Vin:-10V	3OP	Nsc		data pdf pinout
LM7905	R-	Io=1.5A Vo:-4.8...-5.2V Vin:-10V P:int	3OP	Nsc		data pdf pinout
LM 7905 CT,CU	Z-IC	-5V, 1A	17c	Nsc	... 7905... (TO-220)	data pinout
LM7906	R-	Io=1.5A Vo:-5.75...-6.25V Vin:-11V	3OP	Nsc		data pinout
LM7908	R-	Io=1A Vo:-7.6...-8.4V Vin:-11.5...-23V	3OP	Nsc		data pinout
LM7909	R-	Io=1.5A Vo:-8.65...-9.35V Vin:-15V	3OP	Nsc		data pinout
LM7912	R-	Io=1.5A Vo:-11.4...-12.6V P:int	3OP	Nsc		data pdf pinout
LM 7912 CT,CU	Z-IC	-12V, 1A	17c	Nsc	... 7912... (TO-220)	data pinout
LM7915	R-	Io=1.5A Vo:-14.4...-15.6V Vin:-23V	3OP	Nsc		data pinout
LM7915C	R-	Io=int Vo:-14.4...-15.6V Vin:-23V	3SPO	Sgs		data pdf pinout
LM 7915 CT,CU	Z-IC	-15V, 1A	17c	Nsc	... 7915... (TO-220)	data pinout
LM7918	R-	Io=1.5A Vo:-17.3...-18.7V Vin:-27V	3OP	Nsc		data pinout
LM7918C	R-	Io=int Vo:-17.3...-18.3V Vin:-27V	3SPO	Sgs		data pdf pinout
LM7920C	R-	Io=int Vo:-19.2...-20.8V Vin:-29V	3SPO	Sgs		data pdf pinout
LM7922C	R-	Io=int Vo:-21.1...-22.9V Vin:-31V	3SOP	Sgs		data pdf pinout
LM7924	R-	Io=1.5A Vo:-23...-25V Vin:-33V	3OP	Nsc		data pinout
LM7924C	R-	Io=int Vo:-23...-24.5V Vin:-33V P:int	3SPO	Sgs		data pdf pinout
LM 8050 I	Si-N		7e	Ons	CX908, KTC8050, LM8050J, LM9013G,	data

Type	Device	Short Description	Fig.	Manu	Comparision Types	More at
LM 8050 J	Si-N		7e	Ons	CX908, KTC8050, LM8050I, LM9013G,	data
LM 8061	MOS-IC	Digital Clock, Timer	24-DIP	Say	-	
LM 8062	MOS-IC	Digital Clock, Timer	24-DIP	Say	-	
LM 8063	MOS-IC	Digital Clock, Timer	24-DIP	Say	-	
LM 8064	MOS-IC	Digital Clock, Timer	24-DIP	Say	-	
LM 8065	MOS-IC	Digital Clock, Timer	24-DIP	Say	-	
LM 8071	PMOS-IC	Orgel-tongenerator/tone gener.f.organ	24-DIP	Say	-	pdf pinout
LM 8072	MOS-IC	Generator f. el. Orgel/organ	24-DIP	Say	-	
LM 8081	MOS-IC	Radio/TV-kanalwahl/channel selector	24-DIP	Say	-	
LM 8162	MOS-IC	Alarm-digital-uhr/clock	24-DIP	Say	-	
LM 8164	MOS-IC	Alarm-digital-uhr/clock	24-DIP	Say	-	
LM 8165	MOS-IC	Alarm-digital-uhr/clock	24-DIP	Say	-	
LM 8168	MOS-IC	Alarm-digital-uhr/clock	24-DIP	Say	-	
LM 8360	PMOS-IC	Alarm-digital-uhr/clock + blanking	40-DIP	Say	MD 8009, MM 5316	pdf
LM 8361	PMOS-IC	Alarm-digital-uhr/clock (no blanking)	40-DIP	Say	MM 5387	pdf·pinout
LM 8362	PMOS-IC	Alarm-digital-uhr/clock	40-DIP	Say	-	pdf pinout
LM 8363(N,DH,S)	PMOS-IC	2x Alarm-Digital-Uhr/clock, Udd=-12V	42-DIP	Say	-	pdf pinout
LM 8364(N,DH,S)	PMOS-IC	2x Alarm-Digital-Uhr/clock, Udd=-12V	42-DIP	Say	-	pdf pinout
LM 8365 N,DH,S	PMOS-IC	=LM 8364: spiegelb.Anschl./rev. pin.	42-DIP	Say	-	pdf pinout
LM 8365 MF	CMOS-IC	Micropower Undervoltage Sensing IC w. Programmable Out. Delay	45	Nsc	MC33465	pdf pinout
LM 8368	PMOS-IC	Alarm-Digital-Uhr/clock, Udd=-6,5...16V	40-DIP	Say	-	pdf pinout
LM 8369	MOS-IC	Timer	20-DIP	Say	-	pdf
LM 8371	PMOS-IC	Rhythmusgenerator/rhythm generator	16-DIP	Say	-	pdf
LM 8372	PMOS-IC	Rhythmusgenerator/rhythm generator	16-DIP	Say	-	
LM 8460	PMOS-IC	Alarm-uhr/clock, Clock Radio	42-SDIP	Say	-	pdf pinout
LM 8471	PMOS-IC	Rhythmusgenerator/rhythm generator	28-DIP	Say	-	pdf pinout
LM 8521	MOS-IC	Recorder, Bandzählwerk/tape counter	42-DIP	Say	-	
LM 8523	PMOS-IC	Recorder, Bandzählwerk/tape counter	42-DIP	Say	-	pdf pinout
LM 8560(B,D,J,N)	PMOS-IC	Alarm-Digital-Uhr/Clock, Udd=-12V	28-SDIP	Say	TMS 3450	
LM 8560	PMOS-IC	Alarm-Digital-Uhr/clock, Udd=-12V	28-SDIP	Say	TMS 3450	pdf pinout
LM 8561	PMOS-IC	Alarm-Digital-Uhr/clock, Flt Display	28-SDIP	Say	-	
LM 8562	PMOS-IC	Alarm-Digital-Uhr/clock, Udd=-14...-8V	28-SDIP	Say	-	pdf
LM 8569	PMOS-IC	Alarm-Digital-Uhr/clock,Udd=-14..-6,5V	28-SDIP	Say	-	
LM 8802	µC-IC	8 Bit, 2kx8Bit ROM, 128x8Bit RAM	42-(S)DIP	Say	-	
LM 8804	µC-IC	8 Bit, 4kx8Bit ROM, 256x8Bit RAM	42-(S)DIP	Say	-	
LM 8852	µC-IC	8 Bit, 2kx8Bit ROM, 128x8Bit RAM	64-SDIP	Say	-	
LM 8854	µC-IC	8 Bit, 4kx8Bit ROM, 256x8Bit RAM	64-SDIP	Say	-	
LM 8899	µC-IC	8 Bit, 8kx8Bit ext. ROM, 256x8Bit RAM	100-PGA	Say	-	
LM 8942	PMOS-IC	Inverter	16-DIP	Say	-	pdf pinout
LM 8948	MOS-IC	7x Buffer	16-DIP	Say	-	
LM 8972	PMOS-IC	Rhythmusgenerator/rhythm generator	16-DIP	Say	-	pdf pinout
LM 9013 G	Si-N		7e	Ons	CX908, LM9013H, MPS6601, MPS8050,	data
LM 9013 H	Si-N		7e	Ons	CX908, LM9013G, MPS6601, MPS8050,	data
LM 9036 DT-3.3	LIN-IC	3.3V 50mA Voltage Regulator	87	Nsc	-	pdf pinout
LM 9036 DT-5.0	LIN-IC	5.0V 50mA Voltage Regulator	87	Nsc	-	pdf pinout
LM 9036 M-3.3	LIN-IC	3.3V 50mA Voltage Regulator	MDIP	Nsc	-	pdf pinout
LM 9036 M-5.0	LIN-IC	5.0V 50mA Voltage Regulator	MDIP	Nsc	-	pdf pinout
LM 9036 MM-3.3	LIN-IC	3.3V 50mA Voltage Regulator	TSOP	Nsc	-	pdf pinout
LM 9036 MM-5.0	LIN-IC	5.0V 50mA Voltage Regulator	TSOP	Nsc	-	pdf pinout
LM 9071 S	LIN-IC	Low-Dropout System Voltage Reg.,Delayed Reset	DIP	Nsc	-	pdf pinout
LM 9071 T	LIN-IC	Low-Dropout System Voltage Reg.,Delayed Reset	DIP	Nsc	-	pdf pinout
LM 9072 S	LIN-IC	Dual Tracking Low- Dropout System Regulator	DIP	Nsc	-	pdf pinout
LM 9072 T	LIN-IC	Dual Tracking Low- Dropout System Regulator	DIP	Nsc	-	pdf pinout
LM11600	2xOTA	Vs:±22V Vo:14.2/0V Vi0:0.4mV f:2Mc	16D	Nsc	-	data pinout
LM 11600 AJ	OP-IC	=LM 13600N: ±22V, -55...+125°	16-DIC	Nsc	-	
LM 11700 AJ	OP-IC	=LM 13700N: ±22V, -55...+125°	16-DIC	Nsc	-	pdf
LM 12(H)454 CIV	A/D-IC	4 Kan./Chan., Data Acquisition System 8/12 Bit + Vorz./Sign, Watchdog-Mode H: High Speed Version	44-PLCC	Nsc	-	pdf
LM 12(H,L)458 CIV	A/D-IC	8 Kan./Chan., Data Acquisition System 8/12 Bit + Vorz./Sign, Watchdog-Mode L: 3.3V version, H: High Speed Version	44-PLCC	Nsc	-	pdf
LM 12(H)458 CIVF	A/D-IC	8 Kan./Chan., Data Acquisition System 8/12 Bit + Vorz./Sign, Watchdog-Mode H: High Speed Version	44-QFP	Nsc	-	pdf
LM 12(H)458 MEL	A/D-IC	8 Kan./Chan., Data Acquisition System 8/12 Bit + Vorz./Sign, Watchdog-Mode H: High Speed Version	44-PLCC	Nsc	-	pdf
LM13080	POWER-OP	Vs:15V Vu:80dB Vi0:3mV	8D,11I	Nsc	-	data pinout
LM 13080 N	OP-IC	Power, ±7,5, 0,25A, 1W, 0...+70°	8-DIP	Nsc	-	data pinout
LM 13080 P	OP-IC	=LM 13080N: 1,9W	11-SIL	Nsc	-	data
LM13600	2xOTA	Vs:±18V Vi0:0.4mV f:2Mc	16SD	Nsc	-	data pdf pinout
LM13600A	2xOTA	Vs:±22V Vi0:0.4mV f:2Mc	16D	Nsc	-	data pdf pinout
LM 13600 AN	OP-IC	=LM 13600N: ±22V	16-DIP	Nsc,Phi	-	data pdf pinout
LM 13600 D	OP-IC	=LM 13600N: SMD	16-MDIP	Phi	-	pdf
LM 13600(A)J	OP-IC	=LM 13600N: -55...+125°	16-DIC	Nsc	-	pdf
LM 13600 N	OP-IC	Dual, Transconductance, ±18V, 0...+70°	16-DIP	Nsc,Phi	-	data pdf pinout
LM13700	2xOTA	Vs:±18V Vi0:0.4mV f:2Mc	16SD	Nsc	-	data pdf pinout
LM13700A	2xOTA	Vs:±22V Vi0:0.4mV f:2Mc	16D	Nsc	-	data pdf pinout
LM 13700 AN	OP-IC	=LM 13700N: ±22V	16-DIP	Nsc	-	data pdf pinout

Type	Device	Short Description	Fig.	Manu	Comparision Types	More at
LM 13700(A)J	OP-IC	=LM 13700N: -55...+125°	16-DIC	Nsc	-	pdf
LM 13700 N	OP-IC	Dual, Transconductance, ±18V, 0...+70°	16-DIP	Nsc		data pdf pinout
LM 18293(N)	LIN-IC	4-Channel Push-Pull (Motor) Driver, 1A	16-DIP	Nsc	L 293 B	data pdf pinout
LM 18298(T)	LIN-IC	Dual, Brückentreiber/Bridge Drv, 4A	15-SQL	Nsc	L 298	pdf
LM 22671	IC	500 mA simple switcher®, Step-down Voltage Reg.	TSOP	Nsc	-	pdf pinout
LM 22672	IC	1A simple switcher®, Step-down Voltage Reg.	TSOP	Nsc		pdf pinout
LM 22673 M	IC	3A simple switcher®, Step-down Voltage Reg.	TSOP	Nsc		pdf pinout
LM 22673 T	IC	3A simple switcher®, Step-down Voltage Reg.	TSOP	Nsc		pdf pinout
LM 22674 M	IC	500 mA simple switcher®, Step-down Voltage Reg.	TSOP	Nsc		pdf pinout
LM 22675 M	IC	1A simple switcher®, Step-down Voltage Reg.	TSOP	Nsc		pdf pinout
LM 22676 M	IC	3A simple switcher®, Step-down Voltage Reg.	TSOP	Nsc		pdf pinout
LM 22676 T	IC	3A simple switcher®, Step-down Voltage Reg.	TSOP	Nsc		pdf pinout
LM 22677 T	IC	5A simple switcher®, Step-down Voltage Reg.	TSOP	Nsc		pdf pinout
LM 22678 T	IC	5A simple switcher®, Step-down Voltage Reg.	TSOP	Nsc		pdf pinout
LM 22679 T	IC	5A simple switcher®, Step-down Voltage Reg.	TSOP	Nsc		pdf pinout
LM 22680 MR	IC	2A simple switcher®, Step-down Voltage Reg.	TSOP	Nsc		pdf pinout
LM24250	2xLO-POWER	Vs:±18V Vu:96dB Vo:±12V Vi0:6mV	16DF	Nsc		data pinout
LM 33256 K,N	NMOS-dRAM-IC	256kB x 1 Bit, 150ns access time	16-DIP	Say		
LM 48510 SD	LIN-IC	Boosted Class D Audio Power Amp., 1.2W	MLP	Nsc		pdf pinout
LM 48511 SQ	Dig. Audio	3W, Mono, Class D Audio Power Amp.	QFN	Nsc		pdf pinout
LM77000	POWER-OP	Vs:±18V Vu:200V/mV Vo:±12.5V Vi0:1mV	4P	Nsc		data pdf pinout
LM 2574-3.3 BN	LIN-IC	52kHz Simple 0.5A Buck Regulator	DIP	Mcr	-	pdf pinout
LM 2574-3.3 BY	LIN-IC	52kHz Simple 0.5A Buck Regulator	DIP	Mcr	-	pdf pinout
LM 2574-5.0 BN	LIN-IC	52kHz Simple 0.5A Buck Regulator	DIP	Mcr	-	pdf pinout
LM 2574-5.0 YN	LIN-IC	52kHz Simple 0.5A Buck Regulator	DIP	Mcr	-	pdf pinout
LM 2575-12 BN	LIN-IC	52kHz Simple 1A Buck Regulator	DIP	Mcr	-	pdf pinout
LM 2575-12 BT	LIN-IC	52kHz Simple 1A Buck Regulator	DIP	Mcr	-	pdf pinout
LM 2575-12 BU	LIN-IC	52kHz Simple 1A Buck Regulator	DIP	Mcr	-	pdf pinout
LM 2575-12 BWM	LIN-IC	52kHz Simple 1A Buck Regulator	DIP	Mcr	-	pdf pinout
LM 2575-12 WT	LIN-IC	52kHz Simple 1A Buck Regulator	DIP	Mcr	-	pdf pinout
LM 2575-12 WU	LIN-IC	52kHz Simple 1A Buck Regulator	DIP	Mcr	-	pdf pinout
LM 2575-12 YWM	LIN-IC	52kHz Simple 1A Buck Regulator	DIP	Mcr	-	pdf pinout
LM 2575-3.3 BN	LIN-IC	52kHz Simple 1A Buck Regulator	DIP	Mcr	-	pdf pinout
LM 2575-3.3 BT	LIN-IC	52kHz Simple 1A Buck Regulator	DIP	Mcr	-	pdf pinout
LM 2575-3.3 BU	LIN-IC	52kHz Simple 1A Buck Regulator	DIP	Mcr	-	pdf pinout
LM 2575-3.3 BWM	LIN-IC	52kHz Simple 1A Buck Regulator	DIP	Mcr	-	pdf pinout
LM 2575-3.3 WU	LIN-IC	52kHz Simple 1A Buck Regulator	DIP	Mcr	-	pdf pinout
LM 2575-3.3 YWM	LIN-IC	52kHz Simple 1A Buck Regulator	DIP	Mcr	-	pdf pinout
LM 2575-5.0 BN	LIN-IC	52kHz Simple 1A Buck Regulator	DIP	Mcr	-	pdf pinout
LM 2575-5.0 BT	LIN-IC	52kHz Simple 1A Buck Regulator	DIP	Mcr	-	pdf pinout
LM 2575-5.0 BU	LIN-IC	52kHz Simple 1A Buck Regulator	DIP	Mcr	-	pdf pinout
LM 2575-5.0 BWM	LIN-IC	52kHz Simple 1A Buck Regulator	DIP	Mcr	-	pdf pinout
LM 2575-5.0 WT	LIN-IC	52kHz Simple 1A Buck Regulator	DIP	Mcr	-	pdf pinout
LM 2575-5.0 WU	LIN-IC	52kHz Simple 1A Buck Regulator	DIP	Mcr	-	pdf pinout
LM 2575-5.0 YN	LIN-IC	52kHz Simple 1A Buck Regulator	DIP	Mcr	-	pdf pinout
LM 2575-5.0 YWM	LIN-IC	52kHz Simple 1A Buck Regulator	DIP	Mcr	-	pdf pinout
LM 2576-12 BT	LIN-IC	52kHz Simple 3A Buck Regulator	DIP	Mcr	-	pdf pinout
LM 2576-12 BU	LIN-IC	52kHz Simple 3A Buck Regulator	DIP	Mcr	-	pdf pinout
LM 2576-12 WT	LIN-IC	52kHz Simple 3A Buck Regulator	DIP	Mcr	-	pdf pinout
LM 2576-12 WU	LIN-IC	52kHz Simple 3A Buck Regulator	DIP	Mcr	-	pdf pinout
LM 2576-3.3 BT	LIN-IC	52kHz Simple 3A Buck Regulator	DIP	Mcr	-	pdf
LM 2576-3.3 BU	LIN-IC	52kHz Simple 3A Buck Regulator	DIP	Mcr	-	pdf pinout
LM 2576-3.3 WT	LIN-IC	52kHz Simple 3A Buck Regulator	DIP	Mcr	-	pdf pinout
LM 2576-3.3 WU	LIN-IC	52kHz Simple 3A Buck Regulator	DIP	Mcr	-	pdf pinout
LM 2576-5.0 BT	LIN-IC	52kHz Simple 3A Buck Regulator	DIP	Mcr	-	pdf pinout
LM 2576-5.0 BU	LIN-IC	52kHz Simple 3A Buck Regulator	DIP	Mcr	-	pdf pinout
LM 2576-5.0 WT	LIN-IC	52kHz Simple 3A Buck Regulator	DIP	Mcr	-	pdf pinout
LM 2576-5.0 WU	LIN-IC	52kHz Simple 3A Buck Regulator	DIP	Mcr	-	pdf pinout
LM 2596 S-12	SMPS-IC	Pow.Conv. 150kHz 3A Step-down Vol.Reg.	86	Nsc	-	pdf pinout
LM 2596 T-12	SMPS-IC	Pow.Conv. 150kHz 3A Step-down Vol.Reg.	86	Nsc	-	pdf pinout
LM 2596 S-3.3	SMPS-IC	Pow.Conv. 150kHz 3A Step-down Vol.Reg.	86	Nsc	-	pdf pinout
LM 2596 T-3.3	SMPS-IC	Pow.Conv. 150kHz 3A Step-down Vol.Reg.	86	Nsc	-	pdf pinout
LM 2596 S -5.0	SMPS-IC	Pow.Conv. 150kHz 3A Step-down Vol.Reg.	86	Nsc	-	pdf pinout
LM 2596 T-5.0	SMPS-IC	Pow.Conv. 150kHz 3A Step-down Vol.Reg.	86	Nsc	-	pdf pinout
LMC	Z-Di	=1SMB 15CA (Typ-Code/Stempel/marking)	71(5x3,5)	Ons	→1SMB 10CA...78CA	
LMC272C	2xRR-OP	Vs:16V Vu:88dB Vo:2.64/0V Vi0:1.4mV	8SD	Nsc		data pdf pinout
LMC 555 CH	CMOS-IC	=LMC 555CN: Fig.→	81	Nsc	-	pdf pinout
LMC 555 CM	CMOS-IC	=LMC 555CN: SMD	8-MDIP	Nsc	-	pdf pinout
LMC 555 CN	CMOS-IC	Zeitgeber/Timer, Ucc<15V, -40...+85°	8-DIP	Nsc	ICM 7555, KS 555	pdf pinout
LMC 567 CM	CMOS-IC	=LMC 567CN SMD	8-MDIP	Nsc	-	pdf pinout
LMC 567 CN	CMOS-IC	PLL-Ton-/Tone-Decoder(=LM 567: CMOS)	8-DIP	Nsc	-	pdf pinout
LMC 568 CM	CMOS-IC	=LMC 568CN: SMD	8-MDIP	Nsc	-	pdf pinout
LMC 568 CN	CMOS-IC	PLL, 2...9V, 500kHz, -25...+125°	8-DIP	Nsc	-	pdf pinout

Type	Device	Short Description	Fig.	Manu	Comparision Types	More at
LMC660	4xOP	Vs:16V Vo:4.87V Vi0:1mV f:1.4Mc	14SD	Nsc		data pdf pinout
LMC660A	4xOP	Vs:16V Vo:4.87V Vi0:1mV f:1.4Mc	14DS	Nsc		data pinout
LMC662	2xCMOS	Vs:16V Vo:4.87/0V Vi0:1mV f:1.4Mc	8SD	Nsc		data pdf pinout
LMC662A	2xCMOS	Vs:16V Vo:4.87/0V Vi0:1mV f:1.4Mc	8DS	Nsc		data pinout
LMC668	CHOPPER	Vs:18V Vu:134dB Vo:±4.85V Vi0:1μμV	8D	Nsc		data pinout
LMC 835(N)	CMOS-IC	Digital Graphic Equalizer, 14 Band	28-DIP	Nsc	.	pdf pinout
LMC 1982	CMOS-IC	Audio Tone/volume/balance Ctrl., 2Inp.	28-DIP	Nsc		pdf pinout
LMC 1983	CMOS-IC	Audio Tone/volume/balance Ctrl., 3Inp.	28-DIP	Nsc		pdf pinout
LMC 1992(N)	CMOS-IC	Car Audio Tone/Volume/Fader Ctrl.,4Inp	28-DIP	Nsc		pinout
LMC 1993(N)	CMOS-IC	=LMC 1992: 3 Inp., Loudness	28-DIP	Nsc	. -	
LMC2001A	RR-OP	Vs:5.25V Vu:137dB Vo:4.975V f:6Mc	5S,8S	Nsc		data pdf pinout
LMC6001	LO-NOISE	Vs:16V Vu:1400V/mV Vo:4,87/0V	8AD	Nsc		data pdf pinout
LMC6001A	LO-NOISE	Vs:16V Vu:1400V/mV Vo:4.87/0V	8AD	Nsc		data pdf pinout
LMC6022	2xCMOS	Vs:16V Vu:1000V/mV Vo:4.98/0V	8S	Nsc		data pdf pinout
LMC6022I	2xCMOS	Vs:16V Vu:1000V/mV Vo:4.98/0V	8SD	Nsc		data pdf pinout
LMC6024	4xCMOS	Vs:16V Vu:1000V/mV Vo:4.98/0V	14S	Nsc		data pdf pinout
LMC6024I	4xCMOS	Vs:16V Vu:1000V/mV Vo:4.98/0V	14SD	Nsc		data pdf pinout
LMC6032	2xCMOS	Vs:16V Vu:2000V/mV Vo:4.87/0V	8S	Nsc		data pdf pinout
LMC6032I	2xCMOS	Vs:16V Vu:2000V/mV Vo:4.87/0V	8SD	Nsc		data pdf pinout
LMC6034	4xCMOS	Vs:16V Vu:2000V/mV Vo:4.87/0V	14S	Nsc		data pdf pinout
LMC6034I	4xCMOS	Vs:16V Vu:2000V/mV Vo:4.87/0V	14SD	Nsc		data pdf pinout
LMC6035	2xCMOS	Vs:16V Vu:2000V/mV Vo:14.8/0V	8G	Nsc		data pdf pinout
LMC6035I	2xCMOS	Vs:16V Vu:2000V/mV Vo:14.8/0V	8S	Nsc		data pdf pinout
LMC6036	4xCMOS	Vs:16V Vu:2000V/mV Vo:14.8/0V	8G,14S	Nsc		data pdf pinout
LMC6036i	4xCMOS	Vs:16V Vu:2000V/mV Vo:14.8/0V	8Y,14S	Nsc		data pinout
LMC6041	CMOS	Vs:16V Vu:1000V/mV Vo:4.98/0V	8DS	Nsc		data pinout
LMC6042	2xCMOS	Vs:16V Vu:1000V/mV Vo:4.98/0V	8DS	Nsc		data pdf pinout
LMC6044	4xCMOS	Vs:16V Vu:1000V/mV Vo:4.98/0V	14SD	Nsc		data pdf pinout
LMC6061	CMOS	Vs:16V Vu:4000V/mV Vo:4.99/0V	8SD	Nsc		data pdf pinout
LMC6061A	CMOS	Vs:16V Vu:4000V/mV Vo:4.99/0V	8SD	Nsc		data pdf pinout
LMC6062	2xCMOS	Vs:16V Vu:4000V/mV Vo:4.99/0V	8DS	Nsc		data pdf pinout
LMC6062A	2xCMOS	Vs:16V Vu:4000V/mV Vo:4.99/0V	8SD	Nsc		data pdf pinout
LMC6064	4xCMOS	Vs:16V Vu:4000V/mV Vo:4.99/0V	14SD	Nsc		data pdf pinout
LMC6081	CMOS	Vs:16V Vu:1400V/mV Vo:4.87/0V	8SD	Nsc		data pdf pinout
LMC6082	2xCMOS	Vs:16V Vu:1400V/mV Vo:4.87/0V	8SD	Nsc		data pdf pinout
LMC 6082	CMOS-KOP-IC	Precision Cmos Dual Operational Amp.	DIP	Nsc		pdf pinout
LMC6084	4xCMOS	Vs:16V Vu:1400V/mV Vo:4.99/0V	14SD	Nsc		data pinout
LMC6442	2xCMOS	Vs:16V Vu:141V/mV Vo:4.99/0V	8SD,10S	Nsc		data pdf pinout
LMC6462	2xCMOS	Vs:16V Vu:3000V/mV Vo:4.995V	8D	Nsc		data pdf pinout
LMC6462A	2xCMOS	Vs:16V Vu:3000V/mV Vo:4.995V	8SD	Nsc		data pdf pinout
LMC6462B	2xCMOS	Vs:16V Vu:3000V/mV Vo:4.995V Vi0:3mV	8SD	Nsc		data pdf pinout
LMC6464	4xCMOS	Vs:16V Vu:3000V/mV Vo:4.995V	14DS	Nsc		data pdf pinout
LMC6464A	4xCMOS	Vs:16V Vu:3000V/mV Vo:4.995V	14SD	Nsc		data pdf pinout
LMC6464B	4xCMOS	Vs:16V Vu:3000V/mV Vo:4.995V Vi0:3mV	14SD	Nsc		data pdf pinout
LMC6482	2xCMOS	Vs:16V Vu:666V/mV Vo:4.9/0V f:1.5Mc	8SD	Nsc		data pdf pinout
LMC6482A	2xCMOS	Vs:16V Vu:666V/mV Vo:4.9/0V f:1.5Mc	8SD	Nsc		data pdf pinout
LMC6484	4xCMOS	Vs:16V Vu:666V/mV Vo:4.9/0V Vi0:3mV	14SD	Nsc		data pdf pinout
LMC6484A	4xCMOS	Vs:16V Vu:666V/mV Vo:4.9/0V f:1.5Mc	14SD	Nsc		data pdf pinout
LMC6492A	2xCMOS	Vs:16V Vu:300V/mV Vo:4.9/0V Vi0:3mV	8SD	Nsc		data pdf pinout
LMC6492B	2xCMOS	Vs:16V Vu:300V/mV Vo:4.9/0V Vi0:6mV	8SD	Nsc		data pdf pinout
LMC6494A	4xCMOS	Vs:16V Vu:300V/mV Vo:4.9/0V Vi0:3mV	14SD	Nsc		data pdf pinout
LMC6494B	4xCMOS	Vs:16V Vu:300V/mV Vo:4.9/0V Vi0:6mV	14SD	Nsc		data pdf pinout
LMC6572A	2xLO-VOLT	Vs:12V Vu:1000V/mV Vo:2.695V Vi0:3mV	8SD	Nsc		data pdf pinout
LMC6572B	2xLO-VOLT	Vs:12V Vu:1000V/mV Vo:2.695V Vi0:7mV	8SD	Nsc		data pdf pinout
LMC6574A	4xLO-VOLT	Vs:12V Vu:1000V/mV Vo:2.695V Vi0:3mV	14SD	Nsc		data pdf pinout
LMC6574B	4xLO-VOLT	Vs:12V Vu:1000V/mV Vo:2.695V Vi0:7mV	14SD	Nsc		data pdf pinout
LMC6582A	2xOP	Vs:12V Vu:1000V/mV Vo:4.9/0V f:1.2Mc	8SD	Nsc		data pdf pinout
LMC6582B	2xOP	Vs:12V Vu:1000V/mV Vo:4.9/0V f:1.2Mc	8SD	Nsc		data pdf pinout
LMC6584A	4xOP	Vs:12V Vu:1000V/mV Vo:4.9/0V f:1.2Mc	14SD	Nsc		data pdf pinout
LMC6584B	4xOP	Vs:12V Vu:1000V/mV Vo:4.9/0V f:1.2Mc	14SD	Nsc		data pdf pinout
LMC6681	RR-OP	Vs:12V Vu:70V/mV Vo:2.95/0V f:1.2Mc	8SD	Nsc		data pdf pinout
LMC6682	2xRR-OP	Vs:12V Vu:70V/mV Vo:2.95/0V f:1.2Mc	14SD	Nsc		data pdf pinout
LMC6684	4xRR-OP	Vs:12V Vu:70V/mV Vo:2.95/0V f:1.2Mc	16SD	Nsc		data pdf pinout
LMC6762A	2xCOMP	Vs:16V Vu:100dB Vi0:3mV	8SD	Nsc		data pdf pinout
LMC6762B	2xCOMP	Vs:16V Vu:100dB Vi0:3mV	8SD	Nsc		data pdf pinout
LMC6772A	2xCOMP	Vs:16V Vu:100dB Vi0:3mV	8SD	Nsc		data pdf pinout
LMC6772B	2xCOMP	Vs:16V Vu:100dB Vi0:3mV	8SD	Nsc		data pdf pinout
LMC7101A	LO-POWER	Vs:16V Vo:2.45/0V Vi0:6mV f:0.6Mc	5S,8D	Nsc		data pdf pinout
LMC7101B	LO-POWER	Vs:16V Vo:2.45/0V Vi0:9mV f:0.6Mc	5S,8D	Nsc		data pdf pinout
LMC7111A	RR-OP	Vs:11V Vu:400V/mV Vo:2.69/0V	8D	Nsc		data pdf pinout
LMC7111B	RR-OP	Vs:11V Vu:400V/mV Vo:2.69/0V	5S,8D	Nsc		data pdf pinout
LMC7211A	COMP	Vs:16V Vu:100dB Vo:4.8/0V Vi0:5mV	5S,8SD	Nsc		data pdf pinout
LMC7211B	COMP	Vs:16V Vu:100dB Vo:4.8/0V Vi0:15mV	5S,8SD	Nsc		data pdf pinout
LMC7215I	COMP	Vs:10V Vu:140dB Vi0:1mV	5S,8S	Nsc		data pdf pinout
LMC7221	COMP	Vs:16V Vu:100dB Vi0:3mV	5S,8SD	Nsc		data pdf pinout
LMC7225	COMP	Vs:10V Vu:140dB Vi0:1mV	5S,8S	Nsc		data pdf pinout
LMC8101	RR-OP	Vs:12V Vu:138V/mV Vo:±4.9V f:1.3Mc	8YS	Nsc		data pdf pinout
LMC 13204 J	CMOS-IC	4x Analog S, 22V, 20mA, -55...+125°	16-DIC	Nsc		
LMC 13334 J	CMOS-IC	4x Analog S, 22V, 20mA, -55...+125°	16-DIC	Nsc		
LMC 13335 J	CMOS-IC	4x Analog S, 22V, 20mA, -55...+125°	20-DIC	Nsc		
LMD 18200	IC	3A, 55V H-Bridge	DIP	Nsc	.	pdf pinout
LMD 18201	IC	3A, 55V H-Bridge	DIP	Nsc		pdf pinout
LMD 18245	IC	3A, 55V Dmos Full-bridge Motor Driver	DIP	Nsc	.	pdf pinout
LMD 18400	IC	Quad High Side Driver	DIP	Nsc	.	pdf pinout
LME49713	AUDIO-OP	Vs:±19V Vo:±12.1V Vi0:50μμV f:30Mkc	8S	Nsc		data pdf pinout

Type	Device	Short Description	Fig.	Manu	Comparision Types	More at
LME49720	2xAUDIO-OP	Vs:±18V Vu:140dB Vo:±13.6V f:55Mc	8AS	Nsc		data pdf pinout
LME49721	2xAUDIO-OP	Vs:±6V Vu:118dB Vi0:0.3mV f:20Mc	8S	Nsc		data pdf pinout
LME49722	2xAUDIO-OP	Vs:±19V Vu:140dB Vo:±14V Vi0:0.7mV	8S	Nsc		data pdf pinout
LME49723	2xAUDIO-OP	Vs:±18V Vu:105dB Vo:±14V Vi0:±0.3mV	8S	Nsc		data pdf pinout
LME49725	2xAUDIO-OP	Vs:±19V Vu:135dB Vo:±3.8V f:40Mc	8S	Nsc		data pdf pinout
LME49740	4xAUDIO-OP	Vs:±18V Vu:140dB Vo:±13.6V f:55Mc	14DS	Nsc		data pdf pinout
LME49743	4xAUDIO-OP	Vs:±18V Vu:110dB Vo:±13V Vi0:±0.15mV	14S	Nsc		data pdf pinout
LMH 1981MTX	IC	SMD, multi-format sync separator, Macrovision compatible, +3.3V...5V	SSMDIP	Nsc		pdf pinout
LMH6321	FAST-BUFF	Vs:±18V Vo:13.5/0V Vi0:15mV f:50Mkc	8S,7P	Nsc		data pdf pinout
LMH6502	WIDEBAND	Vs:12.6V Vu:72dB Vo:±3.2V f:50Mkc	14S	Nsc		data pdf pinout
LMH 6502,6503 MA	LIN-IC	SMD,Wideband linear-in-db var.gain Amp 6502: 130MHz, 6503:135MHz	14-MDIP	Nsc		pdf pinout
LMH6503	WIDEBAND	Vs:12.6V Vu:79dB Vo:±3.2V f:50Mkc	14S	Nsc		data pdf pinout
LMH6504	WIDEBAND	Vs:12.6V Vu:80dB Vo:±2.2V f:58Mkc	8S	Nsc		data pdf pinout
LMH 6504 M(A,M)	LIN-IC	SMD,Wideband var.gain Amp, 150MHz	8-(SS)MDIP	Nsc		pdf pinout
LMH6505	WIDEBAND	Vs:12.6V Vu:80dB Vo:±2.4V f:38Mkc	8S	Nsc		data pdf pinout
LMH6514	PROG-GAIN	Vs:5.5V Vu:30dB Vo:5.5V f:600Mkc	16H	Nsc		data pdf pinout
LMH6515	PROG-GAIN	Vs:5.5V Vu:24.2dB Vo:5.5V f:600Mkc	16H	Nsc		data pdf pinout
LMH 6515 SQX	IC	600 MHz, Dig.Contr., Variable Gain Amp.	LLP	Nsc		pdf pinout
LMH6518	PROG-GAIN	Vu:38.84dB Vo:0.8V f:825Mkc	16H	Nsc		pdf pinout
LMH6550	DIFF-DRV	Vs:13.2V Vu:997V/mV Vo:±3.8V Vi0:1mV	8S	Nsc		data pdf pinout
LMH6551	DIFF-DRV	Vs:13.2V Vu:997V/mV Vo:±3.8V	8S	Nsc		data pdf pinout
LMH6552	DIFF-DRV	Vs:13.2V Vu:1000V/mV Vo:±1.25V	8SH	Nsc		data pdf pinout
LMH 6552 MA	IC	1.5 GHz Fully Differential Amplifier	LLP	Nsc		pdf pinout
LMH 6552 MAX	IC	1.5 GHz Fully Differential Amplifier	LLP	Nsc		pdf pinout
LMH 6552 SD	IC	1.5 GHz Fully Differential Amplifier	LLP	Nsc		pdf pinout
LMH 6552 SDX	IC	1.5 GHz Fully Differential Amplifier	LLP	Nsc		pdf pinout
LMH 6555 SQ	IC	Low Distortion 1.2 GHz Differential Driver	LLP	Nsc		pdf pinout
LMH 6555 SQX	IC	Low Distortion 1.2 GHz Differential Driver	LLP	Nsc		pdf pinout
LMH6559	FAST-BUFF	Vs:13V Vu:995V/mV Vo:0.93..2.07V	8S,5S	Nsc		data pdf pinout
LMH6560	FAST-BUFF	Vs:13V Vu:995V/mV Vo:±3.34V Vi0:2mV	14S	Nsc		data pdf pinout
LMH6601	FB-OP	Vs:6V Vu:65dB Vi0:±1mV f:73Mkc	6S	Nsc		data pdf pinout
LMH6609	FB-OP	Vs:±6.6V Vo:±2V Vi0:±0.8mV f:110Mkc	8S,5S	Nsc		data pdf pinout
LMH6611	RR-OP	Vs:12V Vu:103dB Vi0:0.074mV f:135Mc	6S	Nsc		data pdf pinout
LMH6612	2xRR-OP	Vs:12V Vu:103dB Vi0:0.095mV f:130Mc	8S	Nsc		data pdf pinout
LMH6618	RR-OP	Vs:12V Vu:95dB Vi0:0.1mV f:65Mc	6S	Nsc		data pdf pinout
LMH6619	2xRR-OP	Vs:12V Vu:95dB Vi0:0.1mV f:65Mc	8S	Nsc		data pdf pinout
LMH6622	2xWIDEBAND	Vs:13.2V Vu:82dB Vo:-1.4...1.5V	8S	Nsc		data pdf pinout
LMH6624	2xWIDEBAND	Vs:13.2V Vu:81dB Vo:±4.4V Vi0:±0.1mV	8S,5S	Nsc		data pdf pinout
LMH6626	2xWIDEBAND	Vs:13.2V Vu:79dB Vo:±1.5V f:80Mkc	8S	Nsc		data pdf pinout
LMH6628	2xWIDEBAND	Vs:13.5V Vu:63dB Vo:±3.5...1.5V	8S	Nsc		data pdf pinout
LMH6639	RR-OP	Vs:13.5V Vu:100dB Vo:±4.97V f:90Mc	8S,6S	Nsc		data pdf pinout
LMH6640	RR-OP	Vs:18V Vu:95dB Vo:0.1...15.9V Vi0:1mV	5S	Nsc		data pdf pinout
LMH6642	RR-OP	Vs:13.5V Vu:96dB Vo:±4.96V Vi0:±1mV	8S,5S	Nsc		data pdf pinout
LMH6643	2xRR-OP	Vs:13.5V Vu:96dB Vo:±4.96V Vi0:±1mV	8S	Nsc		data pdf pinout
LMH6644	4xRR-OP	Vs:13.5V Vu:96dB Vo:±4.96V Vi0:±1mV	14S	Nsc		data pdf pinout
LMH6645	RR-OP	Vs:12.6V Vu:85dB Vo:-4.92...4.95V	8S,5S	Nsc		data pdf pinout
LMH6646	2xRR-OP	Vs:12.6V Vu:85dB Vo:-4.92...4.95V	8S	Nsc		data pdf pinout
LMH6647	RR-OP	Vs:12.6V Vu:85dB Vo:-4.92...4.95V	8S,6S	Nsc		data pdf pinout
LMH6654	LO-POWER	Vs:13.2V Vu:67dB Vo:-3.6...3.4V	8S,5S	Nsc		data pdf pinout
LMH6655	2xLO-POWER	Vs:13.2V Vu:67dB Vo:-3.6...3.4V	8S	Nsc		data pdf pinout
LMH6657	2xOP	Vs:12.6V Vu:90dB Vo:-4.05...4.2V	5S	Nsc		data pdf pinout
LMH6658	2xOP	Vs:12.6V Vu:90dB Vo:-4.05...4.2V	8S	Nsc		data pdf pinout
LMH6682	2xOP	Vs:12.6V Vu:85dB Vo:0.87..4.19V	8S	Nsc		data pdf pinout
LMH6683	2xOP	Vs:12.6V Vu:90dB Vo:4.05...4.25V	14S	Nsc		data pdf pinout
LMH 7324 SQ	IC	Quad 700 PS High Speed Comparator	LLP	Nsc		pdf pinout
LMP 7300 MA	IC	Comparator, Reference, adj. Hysteresis	TSOP	Nsc		pdf pinout
LMP 7300 MM	IC	Comparator, Reference, adj. Hysteresis	TSOP	Nsc		pdf pinout
LMP 7731 MF	IC	2.9 nV/sqrt(Hz) Low Noise, RRIO, Op Amp.	TSOP	Nsc		pdf pinout
LMP 7732 MA	IC	2.9 nV/sqrt(Hz) Low Noise, RRIO Amp.	TSOP	Nsc		pdf pinout
LMS1585ACS-3.3	R+	Io=5A Vo:3.235...3.365V Vin:4.75...7V	3S	Nsc		data pinout
LMS1585ACS-ADJ	ER+	Io=5A Vo:2...6V Vin:1.5...5.75V P:int	3S	Nsc		data pinout
LMS1585ACSX-3.3	R+	Io=5A Vo:3.235...3.365V Vin:4.75...7V	3S	Nsc		data pinout
LMS1585ACSX-ADJ	ER+	Io=5A Vo:2...6V Vin:1.5...5.75V P:int	3S	Nsc		data pinout
LMS1585ACT-3.3	R+	Io=5A Vo:3.235...3.365V Vin:4.75...7V	3P	Nsc		data pinout
LMS1585ACT-ADJ	ER+	Io=5A Vo:2...6V Vin:1.5...5.75V P:int	3P	Nsc		data pinout
LMS1585AIS-3.3	R+	Io=5A Vo:3.235...3.365V Vin:4.75...7V	3S	Nsc		data pinout
LMS1585AIS-ADJ	ER+	Io=5A Vo:2...6V Vin:1.5...5.75V P:int	3S	Nsc		data pinout
LMS1585AISX-3.3	R+	Io=5A Vo:3.235...3.365V Vin:4.75...7V	3S	Nsc		data pinout
LMS1585AISX-ADJ	ER+	Io=5A Vo:2...6V Vin:1.5...5.75V P:int	3S	Nsc		data pinout
LMS1585AIT-3.3	R+	Io=5A Vo:3.235...3.365V Vin:4.75...7V	3P	Nsc		data pinout
LMS1585AIT-ADJ	ER+	Io=5A Vo:2...6V Vin:1.5...5.75V P:int	3P	Nsc		data pinout
LMS1587CS-3.3	R+	Io=3A Vo:3.235...3.365V Vin:4.75...7V	3S	Nsc		data pinout
LMS1587CS-ADJ	ER+	Io=3A Vo:2...6V Vin:1.5...5.75V P:int	3S	Nsc		data pinout
LMS1587CSX-3.3	R+	Io=3A Vo:3.235...3.365V Vin:4.75...7V	3S	Nsc		data pinout
LMS1587CSX-ADJ	ER+	Io=3A Vo:2...6V Vin:1.5...5.75V P:int	3S	Nsc		data pinout
LMS1587CT-3.3	R+	Io=3A Vo:3.235...3.365V Vin:4.75...7V	3P	Nsc		data pinout
LMS1587CT-ADJ	ER+	Io=3A Vo:2...6V Vin:1.5...5.75V P:int	3P	Nsc		data pinout
LMS1587CTX-3.3	R+	Io=3A Vo:3.235...3.365V Vin:4.75...7V	3P	Nsc		data pinout
LMS1587CTX-ADJ	ER+	Io=3A Vo:2...6V Vin:1.5...5.75V P:int	3P	Nsc		data pinout
LMS1587IS-3.3	R+	Io=3A Vo:3.235...3.365V Vin:4.75...7V	3S	Nsc		data pinout
LMS1587IS-ADJ	ER+	Io=3A Vo:2...6V Vin:1.5...5.75V P:int	3S	Nsc		data pinout

Type	Device	Short Description	Fig.	Manu	Comparision Types	More at
LMS1587ISX-3.3	R+	Io=3A Vo:3.235...3.365V Vin:4.75...7V	3S	Nsc		data pinout
LMS1587ISX-ADJ	ER+	Io=3A Vo:2...6V Vin:1.5...5.75V P:int	3S	Nsc		data pinout
LMS1587IT-3.3	R+	Io=3A Vo:3.235...3.365V Vin:4.75...7V	3P	Nsc		data pinout
LMS1587IT-ADJ	ER+	Io=3A Vo:2...6V Vin:1.5...5.75V P:int	3P	Nsc		data pinout
LMS1587ITX-3.3	R+	Io=3A Vo:3.235...3.365V Vin:4.75...7V	3P	Nsc		data pinout
LMS1587ITX-ADJ	ER+	Io=3A Vo:2...6V Vin:1.5...5.75V P:int	3P	Nsc		data pinout
LMS4684	SPDT	Vs:1.8...5.5V Ron:Ohm at 2.7...3.3/0V	10H,12G	Nsc		data pdf pinout
LMS 33460	LIN-IC	3V Under Voltage Detector	45	Nsc		pdf pinout
LMV321	RR-OP	Vs:5.5V Vu:100V/mV Vo:120V Vi0:1.7mV	5S	Nsc		data pdf pinout
LMV 321 ILT	IC	Low cost low power input/ output rail- to- rail operational amplifiers	45	Stm	-	pdf pinout
LMV 321 RILT	IC	Low cost low power input/ output rail- to- rail operational amplifiers	45	Stm	-	pdf pinout
LMV324	4xRR-OP	Vs:5.5V Vu:100V/mV Vo:120V Vi0:1.7mV	14S	Nsc		data pdf pinout
LMV 324 ID(T)	IC	Low cost low power input/ output rail- to-rail op amp, Quad	MDIP	Stm	-	pdf pinout
LMV 324 IP(T)	IC	Low cost low power input/ output rail- to-rail op amp, Quad	TSOP	Stm	-	pdf pinout
LMV331	COMP	Vs:5.5V Vu:50V/mV Vi0:1.7mV	5S	Nsc		data pdf pinout
LMV339	4xCOMP	Vs:5.5V Vu:50V/mV Vi0:1.7mV	14S	Nsc		data pdf pinout
LMV358	2xRR-OP	Vs:5.5V Vu:100V/mV Vo:120V Vi0:1.7mV	8S	Nsc		data pdf pinout
LMV 358 ID(T)	IC	Low cost low power input/ output rail- to- rail op amp, Dual	MDIP	Stm	-	pdf pinout
LMV 358 IP(T)	IC	Low cost low power input/ output rail- to- rail op amp, Dual	TSOP	Stm	-	pdf pinout
LMV393	2xCOMP	Vs:5.5V Vu:50V/mV Vi0:1.7mV	8S	Nsc		data pdf pinout
LMV431	EZ+	Io=0.1mA Vo:1.24...6V P:780mW	5S,3N	Nsc		data pdf pinout
LMV721	LO-NOISE	Vs:5.5V Vu:84dB Vo:2.177V Vi0:0.02mV	5S	Nsc		data pdf pinout
LMV722	2xLO-NOISE	Vs:5.5V Vu:84dB Vo:2.177V Vi0:0.02mV	8S	Nsc		data pdf pinout
LMV751	LO-NOISE	Vs:5.5V Vu:120dB Vo:4.89/0V f:5Mc	5S	Nsc		data pdf pinout
LMV821	RR-OP	Vs:5.5V Vu:105dB Vo:4.9/0V Vi0:1mV	5S	Nsc		data pdf pinout
LMV822	2xRR-OP	Vs:5.5V Vu:105dB Vo:4.9/0V Vi0:1mV	8S	Nsc		data pdf pinout
LMV824	4xRR-OP	Vs:5.5V Vu:105dB Vo:4.9/0V Vi0:1mV	14S	Nsc		data pdf pinout
LMV921	RR-OP	Vs:5.5V Vu:108dB Vo:4.965V Vi0:1.5mV	5S	Nsc		data pdf pinout

LN...LR

Type	Device	Short Description	Fig.	Manu	Comparision Types	More at
LN...	Opto					
LN	Si-N	=2SC5226 (Typ-Code/Stempel/marking)	35(2mm)	Say	→2SC5226	data
LN	Si-N	=2SC5227 (Typ-Code/Stempel/marking)	35	Say	→2SC5227	data
LN	Si-N	=2SC5228 (Typ-Code/Stempel/marking)	44	Say	→2SC5228	data
LN	Si-N	=2SC5488 (Typ-Code/Stempel/marking)	35(1,4mm)	Say	→2SC5488	data
LN	Si-N	=2SC5501 (Typ-Code/Stempel/marking)	44(2mm)	Say	→2SC5501	data
LN	MOS-P-FET-e	=BST 122 (Typ-Code/Stempel/marking)	39	Val	→BST 122	data
LN	Z-Di	=P6 SMB-27 (Typ-Code/Stempel/marking)	71(5x3,5)	Fag	→P6 SMB-27	data
LN	Z-Di	=SM 6T 10C (Typ-Code/Stempel/marking)	71(6x4mm)	Tho	→SM 6T...	data
LN	Z-Di	=SMBJ 16 (Typ-Code/Stempel/marking)	71(5x3,5)	Mop	→SMBJ ...	data
LN 1 VB60	Si-Br	Gl-Br, 600V, 1,2A(Ta=25°), <5µs	33x1	Shi	B380 C1500	data
LN 1 WBA60	Si-Br	Gl-Br, 600V, 1,1A(Ta=25°), <5µs	4-DIP	Shi	A 0580	data
LN 2 SB60	Si-Br	Gl-Br, 600V, 1,6A(Ta=25°), <5µs	33x1	Shi	B380 C1500	data
LN 3	Si-N	=2SC5226-3 (Typ-Code/Stempel/marking)	35(2mm)	Say	→2SC5226	data
LN 3	Si-N	=2SC5227-3 (Typ-Code/Stempel/marking)	35	Say	→2SC5227	data
LN 3	Si-N	=2SC5228-3 (Typ-Code/Stempel/marking)	44	Say	→2SC5228	data
LN 4	Si-N	=2SC5226-4 (Typ-Code/Stempel/marking)	35(2mm)	Say	→2SC5226	data
LN 4	Si-N	=2SC5227-4 (Typ-Code/Stempel/marking)	35	Say	→2SC5227	data
LN 4	Si-N	=2SC5228-4 (Typ-Code/Stempel/marking)	44	Say	→2SC5228	data
LN 4	Si-N	=2SC5501-4 (Typ-Code/Stempel/marking)	44(2mm)	Say	→2SC5501	data
LN 4 SB60	Si-Br	Gl-Br, 600V, 4A(Tc=111°), <5µs	33x1	Shi	B380 C4000	data
LN 5	Si-N	=2SC5226-5 (Typ-Code/Stempel/marking)	35(2mm)	Say	→2SC5226	data
LN 5	Si-N	=2SC5227-5 (Typ-Code/Stempel/marking)	35	Say	→2SC5227	data
LN 5	Si-N	=2SC5228-5 (Typ-Code/Stempel/marking)	44	Say	→2SC5228	data
LN 5	Si-N	=2SC5501-5 (Typ-Code/Stempel/marking)	44(2mm)	Say	→2SC5501	data
LN 6 SB60	Si-Br	Gl-Br, 600V, 6A(Tc=111°), <5µs	33x1	Shi	B380 C6000	data
LN51	IRED	950nm ø4.6mm I:4mcd Vf:1.25V	TO18	Pan		data pdf pinout
LN54	IRED	950nm ø2.2mm I:4mcd Vf:1.5V If:50mA		Pan		data pdf pinout
LN55	IRED	950nm ø3.5mm I:4mcd Vf:<1.5V If:50mA		Pan		data pdf pinout
LN57	IRED	950nm ø1.8mm I:4mcd Vf:1.25V If:50mA		Pan		data pinout
LN58	IRED	950nm ø2.4mm I:4mcd Vf:1.25V If:50mA		Pan		data pdf pinout
LN59	IRED	940nm ø2.5mm I:4mcd Vf:1.3V If:50mA		Pan		data pdf pinout
LN62S	IRED	950nm ø3mm I:4mcd Vf:1.2V If:50mA		Pan		data pinout
LN64	IRED	950nm ø4.4mm I:4mcd Vf:1.3V If:50mA		Pan		data pinout
LN65	IRED	950nm ø3.5mm I:4mcd Vf:1.3V If:100mA		Pan		data pdf pinout
LN66	IRED	950nm I:4mcd Vf:1.3V If:50mA		Pan		data pinout
LN68	IRED	880nm ø5mm I:4mcd Vf:1.5V If:50mA	T1¾	Pan		data pinout
LN122	LED	rd ø4.6mm I:1100mcd Vf:1.8V If:20mA	TO18	Pan		data pinout
LN123	LED	rd ø4.6mm I:1100mcd Vf:1.8V If:20mA	TO18	Pan		data pinout
LN124	LED	rd ø4.4mm I:1100mcd Vf:2V If:20mA		Pan		data pinout
LN125	LED	rd ø4.4mm I:1100mcd Vf:1.8V If:20mA		Pan		data pinout
LN126	LED	rd ø3mm I:9mcd Vf:1.8V If:20mA		Pan		data pinout
LN151	IRED	950nm ø4.6mm I:4.5mcd Vf:1.3V	TO18	Pan		data pdf pinout

Type	Device	Short Description	Fig.	Manu	Comparision Types	More at
LN152	IRED	950nm ø4.6mm I:4.5mcd Vf:1.3V	TO18	Pan		data pdf pinout
LN155	IRED	940nm ø4.6mm I:4.5mcd Vf:1.3V		Pan		data pinout
LN162S	IRED	950nm ø5mm I:4.5mcd Vf:1.2V If:50mA		Pan		data pdf pinout
LN166	IRED	950nm ø5mm I:4.5mcd Vf:1.35V If:50mA	T1¾	Pan		data pinout
LN172	IRED	900nm ø4.2mm I:4.5mcd Vf:1.4V	TO18	Pan		data pdf pinout
LN174	IRED	900nm ø4.4mm I:4.5mcd Vf:1.4V		Pan		data pinout
LN175	IRED	900nm ø4.4mm I:4.5mcd Vf:1.4V		Pan		data pdf pinout
LN176	IRED	900nm ø5mm I:4.5mcd Vf:1.5V If:50mA	T1¾	Pan		data pinout
LN181	IRED	880nm ø3mm I:4.5mcd Vf:1.8V If:100mA		Pan		data pinout
LN181L	IRED	880nm ø3mm I:4.5mcd Vf:1.8V If:100mA		Pan		data pinout
LN182	IRED	880nm ø4.6mm I:4.5mcd Vf:1.7V		Pan		data pinout
LN183	IRED	880nm ø4.6mm I:4.5mcd Vf:1.7V		Pan		data pinout
LN191	IRED	1300nm ø3mm I:5mcd Vf:1.2V If:50mA		Pan		data pinout
LN193	IRED	1300nm ø3mm I:5mcd Vf:1.2V If:100mA		Pan		data pinout
LN1251	LED	rd ø4.4mm I:9mcd Vf:<2.6V If:10mA		Roe		data pdf pinout
LN1261	LED	rd ø3mm I:12mcd Vf:<2.6V If:20mA		Roe		data pdf pinout
LN1351	LED	gr ø3mm I:5mcd Vf:<2.8V If:10mA		Roe		data pdf pinout
LN1361	LED	gr ø3mm I:7.5mcd Vf:<2.8V If:20mA		Roe		data pdf pinout
LN1451	LED	yl ø3mm I:2.2mcd Vf:<2.8V If:10mA		Roe		data pdf pinout
LN1461	LED	yl ø3mm I:4.5mcd Vf:<2.8V If:20mA		Roe		data pdf pinout
LN1851	LED	ord ø4.6mm I:3.5mcd Vf:<2.8V If:10mA		Roe		data pdf pinout
LN1861	LED	ord ø4.6mm I:5mcd Vf:<2.8V If:20mA		Roe		data pdf pinout
LN2152C13	LED	rd/gr ø3mm I:4mcd Vf:<2.8V If:10mA		Roe		data pinout
LN2162C13	LED	rd/gr ø3mm I:4mcd Vf:<2.8V If:10mA		Roe		data pdf pinout
LN 9014(B...D)	Si-N	Uni, ra, 30V, 0,1A, 0,3W, 120MHz	7e	Mic	BC 169, BC 184, BC 239, BC 549, ++	data
LN 9015(B...D)	Si-P	Uni, ra, 30V, 0,1A, 0,3W, 120MHz	7e	Mic	BC 214, BC 259, BC 309, BC 559, ++	data
LN9705	IR LD+MD	788nm ø2mm I:4mcd Vf:1.75V If:50mA		Pan		data pinout
LN9707	IR LD+MD	788nm ø2mm I:4mcd Vf:2V If:50mA		Pan		data pinout
LN9710	IR LD+MD	830nm ø2mm I:4mcd Vf:2.2V If:130mA		Pan		data pinout
LNBK 20 D2	LIN-IC	Lnb supply & cntrl. voltage reg.	MDIP	Stm	-	pdf pinout
LNBP 8 K7	LIN-IC	LNB Supply & control voltage regulator	87	Stm	-	pdf pinout
LNBP 9 K7	LIN-IC	Lnb Supply & control voltage regulator	87	Stm	-	pdf pinout
LNBP 10SP...16SP	IC	SMD, LNBP supply and control voltage regulator (parallel interface)	MDIP	Sgs	-	pdf pinout
LNBP 20CR	IC	Lnbp supply and control voltage regulator (parallel interface).	SIL	Sgs	-	pdf pinout
LNBP 20PD	IC	SMD, Lnbp supply and control voltage regulator (parallel interface)	MDIP	Sgs	-	pdf pinout
LND 150 N3	MOS-N-FET-d	V-MOS, 500V, 30mA, 0,74W, <1kΩ(0,5mA) Low Drive	7e	Stx	-	data
LND 150 N8	MOS-N-FET-d	=LND 150N3: SMD	39b			data
LND 250 K1	MOS-N-FET-d	=LND 150N3: SMD, 13mA	35e		-	data
LNE 250 K1	MOS-N-FET-e	V-MOS, SMD, 500V, 3mA, <1kΩ(0,5mA) Low Drive	35e	Stx	-	data
LNK 302 D	IC	Off-Line Switcher IC, 63mA/80mA	MDIP	Pwi	-	pdf pinout
LNK 302 P/G	IC	Off-Line Switcher IC, 63mA/80mA	DIC	Pwi	-	pdf pinout
LNK 304 D	IC	Off-Line Switcher IC, 120mA/170mA	MDIP	Pwi	-	pdf pinout
LNK 304 P/G	IC	Off-Line Switcher IC, 120mA/170mA	DIC	Pwi	-	pdf pinout
LNK 305 D	IC	Off-Line Switcher IC, 175mA/280mA	MDIP	Pwi	-	pdf pinout
LNK 305 P/G	IC	Off-Line Switcher IC, 175mA/280mA	DIC	Pwi	-	pdf pinout
LNK 306 D	IC	Off-Line Switcher IC, 225mA/360mA	MDIP	Pwi	-	pdf pinout
LNK 306 P/G	IC	Off-Line Switcher IC, 225mA/360mA	DIC	Pwi	-	pdf pinout
LNK 353 G	IC	Low Power Off-Line Switcher IC, 3W	47	Pwi	-	pdf pinout
LNK 353 P	IC	Low Power Off-Line Switcher IC, 3W	DIC	Pwi	-	pdf pinout
LNK 354 G	IC	Low Power Off-Line Switcher IC, 3.5W	47	Pwi	-	pdf pinout
LNK 354 P	IC	Low Power Off-Line Switcher IC, 3.5W	DIC	Pwi	-	pdf pinout
LNK 362 D	IC	Low Power Off-Line Switcher IC, 2.8W	MDIP	Pwi	-	pdf pinout
LNK 362 G	IC	Low Power Off-Line Switcher IC, 2.8W	47	Pwi	-	pdf pinout
LNK 362 P	IC	Low Power Off-Line Switcher IC, 2.8W	DIC	Pwi	-	pdf pinout
LNK 363 D	IC	Low Power Off-Line Switcher IC, 5W	MDIP	Pwi	-	pdf pinout
LNK 363 G	IC	Low Power Off-Line Switcher IC, 5W	47	Pwi	-	pdf pinout
LNK 363 P	IC	Low Power Off-Line Switcher IC, 5W	DIC	Pwi	-	pdf pinout
LNK 364 D	IC	Low Power Off-Line Switcher IC, 5.5W	MDIP	Pwi	-	pdf pinout
LNK 364 G	IC	Low Power Off-Line Switcher IC, 5.5W	47	Pwi	-	pdf pinout
LNK 364 P	IC	Low Power Off-Line Switcher IC, 5.5W	DIC	Pwi	-	pdf pinout
LNK 500 G	IC	CV or CV/CC Switcher, 4W	47	Pwi	-	pdf pinout
LNK 500 P	IC	CV or CV/CC Switcher, 4W	DIC	Pwi	-	pdf pinout
LNK 501 G	IC	CV/CC Switcher f. Very Low Cost Chargers & Adapters, 4W	47	Pwi	-	pdf pinout
LNK 501 P	IC	CV/CC Switcher f. Very Low Cost Chargers & Adapters, 4W	DIC	Pwi	-	pdf pinout
LNK 520 G	IC	CV or CV/CC Switcher f. Very Low Cost Adapters & Chargers, 4W	47	Pwi	-	pdf pinout
LNK 520 P	IC	CV or CV/CC Switcher f. Very Low Cost Adapters & Chargers, 4W	DIC	Pwi	-	pdf pinout
LNK 562 D	IC	Off- Line Switcher IC f. Linear Transformer Replacement, 1.9W	MDIP	Pwi	-	pdf pinout
LNK 562 G	IC	Off- Line Switcher IC f. Linear Transformer Replacement, 1.9W	47	Pwi	-	pdf pinout
LNK 562 P	IC	Off- Line Switcher IC f. Linear Transformer Replacement, 1.9W	DIC	Pwi	-	pdf pinout
LNK 563 D	IC	Off- Line Switcher IC f. Linear Transformer Replacement, 2.5W	MDIP	Pwi	-	pdf pinout
LNK 563 G	IC	Off- Line Switcher IC f. Linear	47	Pwi	-	pdf pinout

Type	Device	Short Description	Fig.	Manu	Comparision Types	More at
LNK 563 P	IC	Transformer Replacement, 2.5W Off- Line Switcher IC f. Linear	DIC	Pwi	-	pdf pinout
LNK 564 D	IC	Transformer Replacement, 2.5W Off- Line Switcher IC f. Linear	MCiP	Pwi	-	pdf pinout
LNK 564 G	IC	Transformer Replacement, 3W Off- Line Switcher IC f. Linear	47	Pwi	-	pdf pinout
LNK 564 P	IC	Transformer Replacement, 3W Off- Line Switcher IC f. Linear	DIC	Pwi	-	pdf pinout
LNK 603 D	IC	Transformer Replacement, 3W Accurate CV/CC Switcher, 2.5 - 3.3W	MDIP	Pwi	-	pdf pinout
LNK 603 P	IC	Accurate CV/CC Switcher, 2.5 - 3.3W	DIC	Pwi	-	pdf pinout
LNK 604 D	IC	Accurate CV/CC Switcher, 3.5 - 4.1W	MDIP	Pwi	-	pdf pinout
LNK 604 P	IC	Accurate CV/CC Switcher, 3.5 - 4.1W	DIC	Pwi	-	pdf pinout
LNK 605 D	IC	Accurate CV/CC Switcher, 4.5 - 5.1W	MDIP	Pwi	-	pdf pinout
LNK 605 P	IC	Accurate CV/CC Switcher, 4.5 - 5.1W	DIC	Pwi	-	pdf pinout
LNK 606 D	IC	Accurate CV/CC Switcher, 5.5 - 6.1W	MDIP	Pwi	-	pdf pinout
LNK 606 P	IC	Accurate CV/CC Switcher, 5.5 - 6.1W	DIC	Pwi	-	pdf pinout
LNK 613 DG	IC	Accurate CV/CC Switcher, 2.5 - 3.3W	MDIP	Pwi	-	pdf pinout
LNK 613 PG	IC	Accurate CV/CC Switcher, 2.5 - 3.3W	DIC	Pwi	-	pdf pinout
LNK 614 DG	IC	Accurate CV/CC Switcher, 3.5 - 4.1W	MDIP	Pwi	-	pdf pinout
LNK 614 PG	IC	Accurate CV/CC Switcher, 3.5 - 4.1W	DIC	Pwi	-	pdf pinout
LNK 615 DG	IC	Accurate CV/CC Switcher, 4.5 - 5.1W	MDIP	Pwi	-	pdf pinout
LNK 615 PG	IC	Accurate CV/CC Switcher, 4.5 - 5.1W	DIC	Pwi	-	pdf pinout
LNK 616 DG	IC	Accurate CV/CC Switcher, 5.5 - 6.1W	MDIP	Pwi	-	pdf pinout
LNK 616 PG	IC	Accurate CV/CC Switcher, 5.5 - 6.1W	DIC	Pwi	-	pdf pinout
LO	Si-N	=2SC2712-O (Typ-Code/Stempel/marking)	35	Tos	→2SC2712	data
LO	Si-N	=2SC4116-O (Typ-Code/Stempel/marking)	35(2mm)	Tos	→2SC4116	data
LO	Si-N	=BF 775 (Typ-Code/Stempel/marking)	35	Sie	→BF 775	data
LO	P-FET	=BSR174 (Typ-Code/Stempel/marking)	35	Phi,Val	→BSR174	data
LO	Si-N	=KTC 4075-O (Typ-Code/Stempel/marking)	35(2mm)	Kec	→KTC 4075	data
LO5SMARO4	LED	or ø5mm I:1100mcd Vf:2.1V If:50mA		Cot		data pdf pinout
LO5SMNHR4	LED	rd ø5mm I:350mcd Vf:2V If:50mA		Cot		data pdf pinout
LO5SMPBL4	LED	bl ø5mm I:350mcd Vf:3.4V If:30mA		Cot		data pdf pinout
LO5SMTPG4	LED	gr ø5mm I:1750mcd Vf:3.4V If:30mA		Cot		data pdf pinout
LO494X	LED	rd+bl+gr+or ø3mm I:930mcd Vf:2.3V		Cot		data pdf pinout
LO543B	LED	or ø5mm I:1800...7100mcd Vf:2V	T1¾	Osr		data pdf pinout
LO566AHR4	LED	rd ø5mm I:1300mcd Vf:2.1V If:50mA		Cot		data pdf pinout
LO566AYL4	LED	yl ø5mm I:1500mcd Vf:2.1V If:50mA		Cot		data pdf pinout
LO566TBL4	LED	bl ø5mm I:880mcd Vf:3.4V If:30mA		Cot		data pdf pinout
LO566TPG4	LED	gr ø5mm I:2800mcd Vf:3.4V If:30mA		Cot		data pdf pinout
LO3336	LED	or ø3.0mm I:450...1120mcd Vf:1.85V	T1	Osr		data pdf pinout
LO3360	LED	or ø3mm I:11.2...18mcd Vf:2V If:10mA	T1	Osr		data pdf pinout
LO3366	LED	or ø3mm I:180...450mcd Vf:2V If:20mA	T1	Osr		data pdf pinout
LO5436	LED	or ø5.0mm I:450...2800mcd Vf:2V	T1¾	Osr		data pdf pinout
LOA67B	LED	or ø5mm I:355...900mcd Vf:1.8V		Osr		data pdf pinout
LOA670	LED	or ø5mm I:5.6...18mcd Vf:2V If:10mA		Osr		data pdf pinout
LOA676	LED	or ø5mm I:180...355mcd Vf:2V If:20mA		Osr		data pdf pinout
LOE63B	LED	or ø5mm I:5000mcd Vf:2.2V If:50mA	PLCC-4	Osr		data pdf pinout
LOE67B	LED	or ø5mm I:560...1400mcd Vf:2.2V	PLCC-4	Osr		data pdf pinout
LOG 112 AID	IC	Logarithmic and Log Ratio Amplifiers	TSOP	Bub		pdf pinout
LOG 2112 AIDW	IC	Logarithmic and Log Ratio Amplifiers	TSOP	Bub		pdf pinout
LOGA671	LED	or/gr ø5mm I:7.1...28mcd Vf:2V		Osr		data pdf pinout
LOGT671	LED	or/gr ø5mm I:7.1...28mcd Vf:2V	PLCC-4	Osr		data pdf pinout
LOK376	LED	or ø3.0mm I:450...1120mcd Vf:1.85V	T1	Osr		data pdf pinout
LOK380	LED	or ø3mm I:71.:.112mlm Vf:2.1V If:15mA	T1	Osr		data pdf pinout
LOL89K	LED	or ø3mm I:5.6...14mcd Vf:1.7V If:2mA		Osr		data pdf pinout
LOL896	LED	or ø3mm I:90...180mcd Vf:2V If:50mA	PLCC-4	Osr		data pdf pinout
LOM67K	LED	or ø3mm I:7.1...18mcd Vf:1.8V If:2mA		Osr		data pdf pinout
LOM670	LED	or ø3mm I:5.6...14mcd Vf:2V If:10mA	PLCC-2	Osr		data pdf pinout
LOM676	LED	or ø3mm I:140...355mcd Vf:2V If:20mA		Osr		data pdf pinout
LOM770	LED	or ø3mm I:5.6...14mcd Vf:2V If:20mA		Osr		data pdf pinout
LOM776	LED	or ø3mm I:560mlm Vf:2V If:20mA		Osr		data pdf pinout
LOPT670	LED	or/gr ø3mm I<11,2/7.1mcd Vf:2V	PLCC-4	Osr		data pdf pinout
LOQ976	LED	or ø3mm I:70mcd Vf:2V If:20mA	0603	Osr		data pdf pinout
LOQ996	LED	or ø3mm I:70mcd Vf:2V If:20mA		Osr		data pdf pinout
LOR971	LED	or ø3mm I:6mcd Vf:2.1V If:20mA	0805	Osr		data pdf pinout
LOR976	LED	or ø3mm I:70mcd Vf:2V If:20mA	0805	Osr		data pdf pinout
LOs	Si-N	=BF 775 (Typ-Code/Stempel/marking)	35	Sie	→BF 775	data
LOs	Si-N	=BF 775W (Typ-Code/Stempel/marking)	35(2mm)	Sie	→BF 775W	data
LOT67K	LED	or ø3mm I:9...22.4mcd Vf:1.8V If:2mA	PLCC-2	Osr		data pdf pinout
LOT77K	LED	or ø3mm I:3...22.4mcd Vf:1.8V If:2mA		Osr		data pdf pinout
LOT670	LED	or ø3mm I:7.1...18mcd Vf:2V If:10mA	PLCC-2	Osr		data pdf pinout
LOT676	LED	or ø3mm I:180...355mcd Vf:2V If:20mA	PLCC-2	Osr		data pdf pinout
LOT770	LED	or ø3mm I:7.1...18mcd Vf:2V If:10mA	PLCC-2	Osr		data pdf pinout
LOT776	LED	or ø3mm I:180...355mcd Vf:2V If:20mA		Osr		data pdf pinout
LOY876	LED	or ø3mm I:140...355mcd Vf:2V If:20mA		Osr		data pdf pinout
LP	Z-Di	=1SMB 16A (Typ-Code/Stempel/marking)	71(5x3,5)	Ons	→1SMB 5.0A...170A	data
LP	P-FET	=BSR 175 (Typ-Code/Stempel/marking)	35	Phi,Val	→BSR 175	data
LP	Si-N	=HD 1A3M (Typ-Code/Stempel/marking)	39	Nec	→HD 1...	data
LP	Z-Di	=P6 SMB-27A (Typ-Code/Stempel/marking)	71(5x3,5)	Fag	→P6 SMB-27A	data
LP	Z-Di	=SM 6T 10CA (Typ-Code/Stempel/marking)	71(6x4mm)	Tho	→SM 6T...	data
LP	Z-Di	=SMBJ 16A (Typ-Code/Stempel/marking)	71(5x3,5)	Mop	→SMBJ ...	data
LP6-PWXX	LED	wht ø5mm I=3.1cd Vf:4.5V		Cot		data pdf pinout
LP6-TPP1	LED	rd/gr/bl ø5mm I:280mcd Vf:4V If:30mA		Cot		data pdf pinout
LP111	COMP	Vs:±18V Vu:100V/mV Vi0:2mV	20C,8D	Tix		data pdf pinout

Type	Device	Short Description	Fig.	Manu	Comparision Types	More at
LP124	4xLO-POWER	Vs:±16V Vu:100dB Vo:3.6/0V Vi0:1mV	14D	Nsc		data pdf pinout
LP165	4xCOMP	Vs:36V Vu:500V/mV Vi0:1mV	16D	Ray		data pinout
LP 165 D	KOP-IC	4x progr, +36(±18)V, -55...+125°	8-DIP	Ray		data pinout
LP211	COMP	Vs:±18V Vu:100V/mV Vi0:2mV	8SD	Tix		data pdf pinout
LP239	4xCOMP	Vs:36V Vu:500V/mV Vi0:±2mV	14SD	Tix		data pdf pinout
LP265N	4xCOMP	Vs:±18V Vu:500V/mV Vi0:1mV	16D	Ray		data pinout
LP 265 N	KOP-IC	Quad, progr, 36(±18)V, -40...+85°	16-DIP	Nsc		data
LP311	COMP	Vs:±18V Vu:200V/mV Vi0:2mV	8SD	Nsc		data pdf pinout
LP311D	COMP	Vs:±18V Vu:100V/mV Vi0:2mV	8S	Tix		data pdf pinout
LP311H	COMP	Vs:±18V Vu:200V/mV Vi0:2mV	8A	Nsc		data pdf pinout
LP311JG	COMP	Vs:±18V Vu:100V/mV Vi0:2mV	8D	Tix		data pdf pinout
LP 311 N	KOP-IC	36(±18)V, -20...+80°	8-DIP	Nsc		data pdf
LP311P	COMP	Vs:±18V Vu:100V/mV Vi0:2mV	8D	Tix		data pdf pinout
LP324	4xLO-POWER	Vs:±16V Vu:100dB Vo:3.6/0V Vi0:1mV	14DS	Nsc		data pdf pinout
LP339	4xCOMP	Vs:36V Vu:500V/mV Vi0:±2mV	14SD	Tix		data pdf pinout
LP339N	4xCOMP	Vs:36V Vu:500V/mV Vi0:±2mV	14D	Tix		data pdf pinout
LP 339 N	KOP-IC	Quad, 36(±18)V, -20...+80°	14-DIP	Nsc		data pdf
LP365	4xCOMP	Vs:36V Vu:500V/mV Vi0:2mV	16SD	Ray		data pdf pinout
LP365A	4xCOMP	Vs:36V Vu:500V/mV Vi0:1mV	16D	Ray		data pdf pinout
LP 365(A)M	KOP-IC	=LP 365...: SMD	16-MDIP	Nsc		data
LP 365(A)N	KOP-IC	=LP 165...: 0...+70°, Offs<6(A:<6)mV	8-DIC	Ray		data pdf pinout
LP377AHR1	LED	rd ø3mm I=4.5cd Vf:2.5V If:70mA		Cot		data pdf pinout
LP377AYL1	LED	am ø3mm I=5cd Vf:2.5V If:70mA		Cot		data pdf pinout
LP377HR1	LED	rd ø3mm I=4.5cd Vf:2.6V If:70mA		Cot		data pdf pinout
LP377PBG1	LED	gr ø3mm I=2cd Vf:3.6V If:30mA		Cot		data pdf pinout
LP377PBL1	LED	bl ø3mm I:850mcd Vf:3.6V If:30mA		Cot		data pdf pinout
LP377PPG1	LED	gr ø3mm I=2.5cd Vf:3.6V If:30mA		Cot		data pdf pinout
LP377TRO1	LED	or ø3mm I=4.5cd Vf:2.6V If:70mA		Cot		data pdf pinout
LP377TYL1	LED	am ø3mm I=3.2cd Vf:2.6V If:70mA		Cot		data pdf pinout
LP379AX	LED	rd+am ø3mm I=6cd Vf:2.5V If:70mA		Cot		data pdf pinout
LP379PX	LED	gr+bl ø3mm I=2cd Vf:3.6V If:30mA		Cot		data pdf pinout
LP 0701 N3	MOS-P-FET-e	V-MOS, LogL, 16,5V, 0,5A, 1W, <1,5Ω	7e	Stx	2SJ228	data
LP 0701 LG	MOS-P-FET-e	=LP 0701N3: SMD, 0,7A	8-MDIP (--SGDDDD)	Stx		data
LP 0801 K1	MOS-P-FET-e	V-MOS, SMD, LogL, 16,5V, 0,1A, <12Ω	35a	Stx	BSS 84, BSS 284, 2SJ185, 2SJ461	data
LP2901	4xCOMP	Vs:36V Vu:500V/mV Vi0:±2mV	14SD	Tix		data pdf pinout
LP2902	4xLO-POWER	Vs:±13V Vu:100dB Vo:3.6/0V Vi0:1mV	14SD	Nsc		data pdf pinout
LP2945AGC	LED	gr ø3mm I:40mcd Vf:2.1V If:10mA	T1	Lst		data pdf pinout
LP2945AGT	LED	gr ø3mm I:40mcd Vf:2.1V If:10mA	T1	Lst		data pdf pinout
LP2945AHD	LED	rd ø3mm I:4mcd Vf:2V If:10mA	T1	Lst		data pdf pinout
LP2945AHT	LED	rd ø3mm I:1.6mcd Vf:2V If:10mA	T1	Lst		data pdf pinout
LP2945AIT	LED	rd ø3mm I:18mcd Vf:2V If:10mA	T1	Lst		data pdf pinout
LP2945ASDRC	LED	rd ø3mm I:70mcd Vf:2V If:20mA	T1	Lst		data pdf pinout
LP2945ASRD	LED	rd ø3mm I:32mcd Vf:1.7V If:20mA	T1	Lst		data pdf pinout
LP2945ASUGC	LED	or ø3mm I:360mcd Vf:2V If:20mA	T1	Lst		data pdf pinout
LP2945ASURC	LED	or ø3mm I:360mcd Vf:2V If:20mA	T1	Lst		data pdf pinout
LP2945ASURD	LED	rd ø3mm I:120mcd Vf:2V If:20mA	T1	Lst		data pdf pinout
LP2945ASYGC	LED	gr ø3mm I:620mcd Vf:2V If:20mA	T1	Lst		data pdf pinout
LP2945AUBC	LED	bl ø3mm I:110mcd Vf:3.8V If:20mA	T1	Lst		data pdf pinout
LP2945AUBD	LED	bl ø3mm I:160mcd Vf:3.8V If:20mA	T1	Lst		data pdf pinout
LP2945AURC	LED	rd ø3mm I:500mcd Vf:1.7V If:20mA	T1	Lst		data pdf pinout
LP2945AURD	LED	rd ø3mm I:240mcd Vf:1.7V If:20mA	T1	Lst		data pdf pinout
LP2945AUYC	LED	yl ø3mm I:340mcd Vf:2.1V If:20mA	T1	Lst		data pdf pinout
LP2945AVGC	LED	gr ø3mm I:63mcd Vf:2.1V If:20mA	T1	Lst		data pdf pinout
LP2945AVGD	LED	gr ø3mm I:70mcd Vf:2.1V If:20mA	T1	Lst		data pdf pinout
LP2945AYT	LED	yl ø3mm I:16mcd Vf:2V If:10mA	T1	Lst		data pdf pinout
LP2950ACZ-3.0	R+	Io=100mA Vo:2.964...3.036V Vin:6V	3N	Nsc		data pinout
LP2950ACZ-3.3	R+	Io=100mA Vo:3.26...3.34V Vin:6V	3N	Nsc		data pinout
LP2950ACZ-5.0	R+	Io=100mA Vo:4.975...5.025V Vin:6V	3N	Nsc		data pinout
LP2950CZ-3.0	R+	Io=100mA Vo:2.94...3.06V Vin:6V	3N	Nsc		data pinout
LP2950CZ-3.3	R+	Io=100mA Vo:3.234...3.366V Vin:6V	3N	Nsc		data pinout
LP2950CZ-5.0	R+	Io=100mA Vo:4.95...5.05V Vin:6V P:int	3N	Nsc		data pinout
LP2951	ER+	Io=100mA Vo:1.24...29V Vin:6V P:int	8D,20CA	Nsc		data pinout
LP2951ACJ	ER+	Io=100mA Vo:1.24...29V Vin:6V P:int	8D	Nsc		data pinout
LP2951ACM	ER+	Io=100mA Vo:1.24...29V Vin:6V P:int	8S	Nsc		data pdf pinout
LP2951ACM-3.0	R+	Io=100mA Vo:2.964...3.036V Vin:6V	8S	Nsc		data pinout
LP2951ACM-3.3	R+	Io=100mA Vo:3.26...3.34V Vin:6V P:int	8S	Nsc		data pinout
LP2951ACMM-3.0	R+	Io=100mA Vo:2.964...3.036V Vin:6V	8S	Nsc		data pinout
LP2951ACMM-3.3	R+	Io=100mA Vo:3.26...3.34V Vin:6V P:int	8S	Nsc		data pinout
LP2951ACN-3.0	R+	Io=100mA Vo:2.964...3.036V Vin:6V	8D	Nsc		data pinout
LP2951ACN-3.3	R+	Io=100mA Vo:3.26...3.34V Vin:6V P:int	8D	Nsc		data pinout
LP2951CM	ER+	Io=100mA Vo:1.24...29V Vin:6V P:int	8S	Fcs,Nsc		data pdf pinout
LP2951CM-3.0	R+	Io=100mA Vo:2.94...3.06V Vin:6V P:int	8S	Fcs,Nsc		data pinout
LP2951CM-3.3	R+	Io=100mA Vo:3.234...3.366V Vin:6V	8S	Fcs,Nsc		data pinout
LP2951CMM-3.0	R+	Io=100mA Vo:2.94...3.06V Vin:6V P:int	8S	Fcs,Nsc		data pinout
LP2951CMM-3.3	R+	Io=100mA Vo:3.234...3.366V Vin:6V	8S	Fcs,Nsc		data pinout
LP2951CN-3.0	R+	Io=100mA Vo:2.94...3.06V Vin:6V P:int	8D	Nsc		data pinout
LP2951CN-3.3	R+	Io=100mA Vo:3.234...3.366V Vin:6V	8D	Nsc		data pinout
LP2952AIM	ER+	Io=250mA Vo:1.23...29V Vin:6V P:int	16S	Nsc		data pinout
LP2952AIM-3.3	R+	Io=250mA Vo:3.254...3.346V Vin:4.3V	16S	Nsc		data pinout
LP2952AIM-5.0	R+	Io=250mA Vo:4.93...5.07V Vin:6V P:int	16S	Nsc		data pdf pinout
LP2952AIN	ER+	Io=250mA Vo:1.23...29V Vin:6V P:int	14D	Nsc		data pinout
LP2952AIN-3.3	R+	Io=250mA Vo:3.254...3.346V Vin:4.3V	14D	Nsc		data pinout
LP2952AIN-5.0	R+	Io=250mA Vo:4.93...5.07V Vin:6V P:int	14D	Nsc		data pdf pinout
LP2952IM	ER+	Io=250mA Vo:1.23...29V Vin:6V P:int	16S	Nsc		data pdf pinout

Type	Device	Short Description	Fig.	Manu	Comparision Types	More at
LP2952IM-3.3	R+	Io=250mA Vo:3.221...3.379V Vin:4.3V	16S	Nsc		data pinout
LP2952IM-5.0	R+	Io=250mA Vo:4.88...5.12V Vin:6V P:int	16S	Nsc		data pinout
LP2952IN	ER+	Io=250mA Vo:1.23...29V Vin:6V P:int	14D	Nsc		data pdf pinout
LP2952IN-3.3	R+	Io=250mA Vo:3.221...3.379V Vin:4.3V	14D	Nsc		data pinout
LP2952IN-5.0	R+	Io=250mA Vo:4.88...5.12V Vin:6V P:int	14D	Nsc		data pinout
LP2953A	ER+	Io=250mA Vo:1.23...29V Vin:6V P:int	16DS	Nsc		data pinout
LP2953AIM	ER+	Io=250mA Vo:1.23...29V Vin:6V P:int	16S	Nsc		data pdf pinout
LP2953AIM-3.3	R+	Io=250mA Vo:3.254...3.346V Vin:4.3V	16S	Nsc		data pinout
LP2953AIM-5.0	R+	Io=250mA Vo:4.93...5.07V Vin:6V P:int	16S	Nsc		data pinout
LP2953AIN-3.3	R+	Io=250mA Vo:3.254...3.346V Vin:4.3V	16D	Nsc		data pinout
LP2953AIN-5.0	R+	Io=250mA Vo:4.93...5.07V Vin:6V P:int	16D	Nsc		data pinout
LP2953IM	ER+	Io=250mA Vo:1.23...29V Vin:6V P:int	16S	Nsc		data pdf pinout
LP2953IM-3.3	R+	Io=250mA Vo:3.221...3.379V Vin:4.3V	16S	Nsc		data pinout
LP2953IM-5.0	R+	Io=250mA Vo:4.88...5.12V Vin:6V P:int	16S	Nsc		data pinout
LP2953IN	ER+	Io=250mA Vo:1.23...29V Vin:6V P:int	16D	Nsc		data pdf pinout
LP2953IN-3.3	R+	Io=250mA Vo:3.221...3.379V Vin:4.3V	16D	Nsc		data pinout
LP2953IN-5.0	R+	Io=250mA Vo:4.88...5.12V Vin:6V P:int	16D	Nsc		data pinout
LP2954	R+	Io=250mA Vo:4.9...5.1V Vin:6V P:int	3SP	Nsc		data pdf pinout
LP2954A	R+	Io=250mA Vo:4.94...5.06V Vin:6V P:int	3SP	Nsc		data pdf pinout
LP2954AIM	ER+	Io=250mA Vo:1.23...29V Vin:6V P:int	8S	Nsc		data pdf pinout
LP2954IM	ER+	Io=250mA Vo:1.23...29V Vin:6V P:int	8S	Nsc		data pdf pinout
LP2956	ER+	Io=250mA Vo:1.23...29V Vin:6...30V	16SD	Nsc		data pdf pinout
LP2957	R+	Io=250mA Vo:4.93...5.07V Vin:6...30V	5SP	Nsc		data pdf pinout
LP2960AIM	ER+	Io=500mA Vo:1.23...29V Vin:6V P:int	16S	Nsc		data pinout
LP2960AIM-3.3	R+	Io=500mA Vo:3.254...3.346V Vin:6V	16S	Nsc		data pinout
LP2960AIM-5.0	R+	Io=500mA Vo:4.93...5.07V Vin:6V P:int	16S	Nsc		data pinout
LP2960AIN	ER+	Io=500mA Vo:1.23...29V Vin:6V P:int	16D	Nsc		data pinout
LP2960AIN-3.3	R+	Io=500mA Vo:3.254...3.346V Vin:6V	16D	Nsc		data pinout
LP2960AIN-5.0	R+	Io=500mA Vo:4.93...5.07V Vin:6V P:int	16D	Nsc		data pinout
LP2960IM	ER+	Io=500mA Vo:1.23...29V Vin:6V P:int	16S	Nsc		data pinout
LP2960IM-3.3	R+	Io=500mA Vo:3.221...3.379V Vin:6V	16S	Nsc		data pinout
LP2960IM-5.0	R+	Io=500mA Vo:4.88...5.12V Vin:6V P:int	16S	Nsc		data pinout
LP2960IN	ER+	Io=500mA Vo:1.23...29V Vin:6V P:int	16D	Nsc		data pinout
LP2960IN-3.3	R+	Io=500mA Vo:3.221...3.379V Vin:6V	16D	Nsc		data pinout
LP2960IN-5.0	R+	Io=500mA Vo:4.88...5.12V Vin:6V P:int	46D	Nsc		data pinout
LP2966IMM-1818	R++	Io=500mA Vo:1.782...1.818V Vin:1...7V	8S	Nsc		data pinout
LP2966IMM-1828	R++	Io=500mA Vo:1.782...1.818V Vin:1...7V	8S	Nsc		data pinout
LP2966IMM-2525	R++	Io=500mA Vo:2.475...2.525V Vin:1...7V	8S	Nsc		data pinout
LP2966IMM-2828	R++	Io=500mA Vo:2.772...2.828V Vin:1...7V	8S	Nsc		data pinout
LP2966IMM-2830	R++	Io=500mA Vo:2.772...2.828V Vin:1...7V	8S	Nsc		data pinout
LP2966IMM-3030	R++	Io=500mA Vo:2.97...3.03V Vin:1...7V	8S	Nsc		data pinout
LP2966IMM-3325	R++	Io=500mA Vo:2.475...2.525V Vin:1...7V	8S	Nsc		data pinout
LP2966IMM-3333	R++	Io=500mA Vo:3.267...3.633V Vin:1...7V	8S	Nsc		data pinout
LP2966IMM-3336	R++	Io=500mA Vo:3.267...3.633V Vin:1...7V	8S	Nsc		data pinout
LP2966IMM-3636	R++	Io=500mA Vo:3.564...3.636V Vin:1...7V	8S	Nsc		data pinout
LP2966IMM-5050	R++	Io=500mA Vo:4.95...5.05V Vin:1...7V	8S	Nsc		data pinout
LP2966IMMX-1818	R++	Io=500mA Vo:1.782...1.818V Vin:1...7V	8S	Nsc		data pinout
LP2966IMMX-1828	R++	Io=500mA Vo:1.782...1.818V Vin:1...7V	8S	Nsc		data pinout
LP2966IMMX-2525	R++	Io=500mA Vo:2.475...2.525V Vin:1...7V	8S	Nsc		data pinout
LP2966IMMX-2828	R++	Io=500mA Vo:2.772...2.828V Vin:1...7V	8S	Nsc		data pinout
LP2966IMMX-2830	R++	Io=500mA Vo:2.772...2.828V Vin:1...7V	8S	Nsc		data pinout
LP2966IMMX-3030	R++	Io=500mA Vo:2.97...3.03V Vin:1...7V	8S	Nsc		data pinout
LP2966IMMX-3325	R++	Io=500mA Vo:2.475...2.525V Vin:1...7V	8S	Nsc		data pinout
LP2966IMMX-3333	R++	Io=500mA Vo:3.267...3.633V Vin:1...7V	8S	Nsc		data pinout
LP2966IMMX-3336	R++	Io=500mA Vo:3.267...3.633V Vin:1...7V	8S	Nsc		data pinout
LP2966IMMX-3636	R++	Io=500mA Vo:3.564...3.636V Vin:1...7V	8S	Nsc		data pinout
LP2966IMMX-5050	R++	Io=500mA Vo:4.95...5.05V Vin:1...7V	8S	Nsc		data pinout
LP2975AIMM-12	R+	Vo:11.64...12.36V Vin:12.5...24V P:int	8S	Nsc		data pinout
LP2975AIMM-3.3	R+	Vo:3.201...3.399V Vin:3.8...24V P:int	8S	Nsc		data pinout
LP2975AIMM-5.0	R+	Vo:4.85...5.15V Vin:5.5...24V P:int	8S	Nsc		data pinout
LP2975AIMMX-12	R+	Vo:11.64...12.36V Vin:12.5...24V P:int	8S	Nsc		data pinout
LP2975AIMMX-3.3	R+	Vo:3.201...3.399V Vin:3.8...24V P:int	8S	Nsc		data pinout
LP2975AIMMX-5.0	R+	Vo:4.85...5.15V Vin:5.5...24V P:int	8S	Nsc		data pinout
LP2975IMM-12	R+	Vo:11.52...12.48V Vin:12.5...24V P:int	8S	Nsc		data pinout
LP2975IMM-3.3	R+	Vo:3.168...3.432V Vin:3.8...24V P:int	8S	Nsc		data pinout
LP2975IMM-5.0	R+	Vo:4.8...5.2V Vin:5.5...24V P:int	8S	Nsc		data pinout
LP2975IMMX-12	R+	Vo:11.52...12.48V Vin:12.5...24V P:int	8S	Nsc		data pinout
LP2975IMMX-3.3	R+	Vo:3.168...3.432V Vin:3.8...24V P:int	8S	Nsc		data pinout
LP2975IMMX-5.0	R+	Vo:4.8...5.2V Vin:5.5...24V P:int	8S	Nsc		data pinout
LP2978	R+	Io=400mA Vo:3.743...3.857V Vin:1...16V	5S	Nsc		data pinout
LP2978A	R+	Io=400mA Vo:3.762...3.838V Vin:1...16V	5S	Nsc		data pinout
LP2978I	R+	Io=400mA Vo:3.743...3.857V Vin:1...16V	5S	Nsc		data pinout
LP2980AI-3.0	R+	Io=150mA Vo:2.925...3.075V Vin:6V	5S	Nsc		data pinout
LP2980AI-3.3	R+	Io=150mA Vo:3.217...3.383V Vin:6V	5S	Nsc		data pinout
LP2980AI-5.0	R+	Io=150mA Vo:4.875...5.125V Vin:6V	5S	Nsc		data pinout
LP2980AIBP-3.3	R+	Io=150mA Vo:3.3...V Vin:1...16V P:int	5Y	Nsc		data pinout
LP2980AIBP-5.0	R+	Io=150mA Vo:5...V Vin:1...16V P:int	5Y	Nsc		data pinout
LP2980AIBPX-3.3	R+	Io=150mA Vo:3.3...V Vin:1...16V P:int	5Y	Nsc		data pinout
LP2980AIBPX-5.0	R+	Io=150mA Vo:5...V Vin:1...16V P:int	5Y	Nsc		data pinout
LP2980AIM5-2.5	R+	Io=150mA Vo:2.5...V Vin:1...16V P:int	5S	Nsc		data pinout
LP2980AIM5-2.6	R+	Io=150mA Vo:2.6...V Vin:1...16V P:int	5S	Nsc		data pinout
LP2980AIM5-2.7	R+	Io=150mA Vo:2.7...V Vin:1...16V P:int	5S	Nsc		data pinout
LP2980AIM5-2.8	R+	Io=150mA Vo:2.8...V Vin:1...16V P:int	5S	Nsc		data pinout
LP2980AIM5-2.9	R+	Io=150mA Vo:2.9...V Vin:1...16V P:int	5S	Nsc		data pinout
LP2980AIM5-3.0	R+	Io=150mA Vo:3...V Vin:1...16V P:int	5S	Nsc		data pinout

Type	Device	Short Description	Fig.	Manu	Comparision Types	More at
LP2980AIM5-3.1	R+	Io=150mA Vo:3.1...V Vin:1...16V P:int	5S	Nsc		data pinout
LP2980AIM5-3.2	R+	Io=150mA Vo:3.2...V Vin:1...16V P:int	5S	Nsc		data pinout
LP2980AIM5-3.3	R+	Io=150mA Vo:3.3...V Vin:1...16V P:int	5S	Nsc		data pinout
LP2980AIM5-3.5	R+	Io=150mA Vo:3.5...V Vin:1...16V P:int	5S	Nsc		data pinout
LP2980AIM5-3.6	R+	Io=150mA Vo:3.6...V Vin:1...16V P:int	5S	Nsc		data pinout
LP2980AIM5-3.8	R+	Io=150mA Vo:3.8...V Vin:1...16V P:int	5S	Nsc		data pinout
LP2980AIM5-4.0	R+	Io=150mA Vo:4...V Vin:1...16V P:int	5S	Nsc		data pinout
LP2980AIM5-4.5	R+	Io=150mA Vo:4.5...V Vin:1...16V P:int	5S	Nsc		data pinout
LP2980AIM5-4.7	R+	Io=150mA Vo:4.7...V Vin:1...16V P:int	5S	Nsc		data pinout
LP2980AIM5-5.0	R+	Io=150mA Vo:5...V Vin:1...16V P:int	5S	Nsc		data pinout
LP2980AIM5X-2.5	R+	Io=150mA Vo:2.5...V Vin:1...16V P:int	5S	Nsc		data pinout
LP2980AIM5X-2.6	R+	Io=150mA Vo:2.6...V Vin:1...16V P:int	5S	Nsc		data pinout
LP2980AIM5X-2.7	R+	Io=150mA Vo:2.7...V Vin:1...16V P:int	5S	Nsc		data pinout
LP2980AIM5X-2.8	R+	Io=150mA Vo:2.8...V Vin:1...16V P:int	5S	Nsc		data pinout
LP2980AIM5X-2.9	R+	Io=150mA Vo:2.9...V Vin:1...16V P:int	5S	Nsc		data pinout
LP2980AIM5X-3.0	R+	Io=150mA Vo:3...V Vin:1...16V P:int	5S	Nsc		data pinout
LP2980AIM5X-3.1	R+	Io=150mA Vo:3.1...V Vin:1...16V P:int	5S	Nsc		data pinout
LP2980AIM5X-3.2	R+	Io=150mA Vo:3.2...V Vin:1...16V P:int	5S	Nsc		data pinout
LP2980AIM5X-3.3	R+	Io=150mA Vo:3.3...V Vin:1...16V P:int	5S	Nsc		data pinout
LP2980AIM5X-3.5	R+	Io=150mA Vo:3.5...V Vin:1...16V P:int	5S	Nsc		data pinout
LP2980AIM5X-3.6	R+	Io=150mA Vo:3.6...V Vin:1...16V P:int	5S	Nsc		data pinout
LP2980AIM5X-3.8	R+	Io=150mA Vo:3.8...V Vin:1...16V P:int	5S	Nsc		data pinout
LP2980AIM5X-4.0	R+	Io=150mA Vo:4...V Vin:1...16V P:int	5S	Nsc		data pinout
LP2980AIM5X-4.5	R+	Io=150mA Vo:4.5...V Vin:1...16V P:int	5S	Nsc		data pinout
LP2980AIM5X-4.7	R+	Io=150mA Vo:4.7...V Vin:1...16V P:int	5S	Nsc		data pinout
LP2980AIM5X-5.0	R+	Io=150mA Vo:5...V Vin:1...16V P:int	5S	Nsc		data pinout
LP2980I-3.0	R+	Io=150mA Vo:2.895...3.105V Vin:6V	5S	Nsc		data pinout
LP2980I-3.3	R+	Io=150mA Vo:3.184...3.416V Vin:6V	5S	Nsc		data pinout
LP2980I-5.0	R+	Io=150mA Vo:4.825...5.175V Vin:6V	5S	Nsc		data pinout
LP2980IBP-3.3	R+	Io=150mA Vo:3.3...V Vin:1...16V P:int	5Y	Nsc		data pinout
LP2980IBP-5.0	R+	Io=150mA Vo:5...V Vin:1...16V P:int	5Y	Nsc		data pinout
LP2980IBPX-3.3	R+	Io=150mA Vo:3.3...V Vin:1...16V P:int	5Y	Nsc		data pinout
LP2980IBPX-5.0	R+	Io=150mA Vo:5...V Vin:1...16V P:int	5Y	Nsc		data pinout
LP2980IM5-2.5	R+	Io=150mA Vo:2.5...V Vin:1...16V P:int	5S	Nsc		data pinout
LP2980IM5-2.6	R+	Io=150mA Vo:2.6...V Vin:1...16V P:int	5S	Nsc		data pinout
LP2980IM5-2.7	R+	Io=150mA Vo:2.7...V Vin:1...16V P:int	5S	Nsc		data pinout
LP2980IM5-2.8	R+	Io=150mA Vo:2.8...V Vin:1...16V P:int	5S	Nsc		data pinout
LP2980IM5-2.9	R+	Io=150mA Vo:2.9...V Vin:1...16V P:int	5S	Nsc		data pinout
LP2980IM5-3.0	R+	Io=150mA Vo:3...V Vin:1...16V P:int	5S	Nsc		data pinout
LP2980IM5-3.1	R+	Io=150mA Vo:3.1...V Vin:1...16V P:int	5S	Nsc		data pinout
LP2980IM5-3.2	R+	Io=150mA Vo:3.2...V Vin:1...16V P:int	5S	Nsc		data pinout
LP2980IM5-3.3	R+	Io=150mA Vo:3.3...V Vin:1...16V P:int	5S	Nsc		data pinout
LP2980IM5-3.5	R+	Io=150mA Vo:3.5...V Vin:1...16V P:int	5S	Nsc		data pinout
LP2980IM5-3.6	R+	Io=150mA Vo:3.6...V Vin:1...16V P:int	5S	Nsc		data pinout
LP2980IM5-3.8	R+	Io=150mA Vo:3.8...V Vin:1...16V P:int	5S	Nsc		data pinout
LP2980IM5-4.0	R+	Io=150mA Vo:4...V Vin:1...16V P:int	5S	Nsc		data pinout
LP2980IM5-4.5	R+	Io=150mA Vo:4.5...V Vin:1...16V P:int	5S	Nsc		data pinout
LP2980IM5-4.7	R+	Io=150mA Vo:4.7...V Vin:1...16V P:int	5S	Nsc		data pinout
LP2980IM5-5.0	R+	Io=150mA Vo:5...V Vin:1...16V P:int	5S	Nsc		data pinout
LP2980IM5-ADJ	ER+	Io=150mA Vo:1.2...15V Vin:1...16V P:int	5S	Nsc		data pinout
LP2980IM5X-2.5	R+	Io=150mA Vo:2.5...V Vin:1...16V P:int	5S	Nsc		data pinout
LP2980IM5X-2.6	R+	Io=150mA Vo:2.6...V Vin:1...16V P:int	5S	Nsc		data pinout
LP2980IM5X-2.7	R+	Io=150mA Vo:2.7...V Vin:1...16V P:int	5S	Nsc		data pinout
LP2980IM5X-2.8	R+	Io=150mA Vo:2.8...V Vin:1...16V P:int	5S	Nsc		data pinout
LP2980IM5X-2.9	R+	Io=150mA Vo:2.9...V Vin:1...16V P:int	5S	Nsc		data pinout
LP2980IM5X-3.0	R+	Io=150mA Vo:3...V Vin:1...16V P:int	5S	Nsc		data pinout
LP2980IM5X-3.1	R+	Io=150mA Vo:3.1...V Vin:1...16V P:int	5S	Nsc		data pinout
LP2980IM5X-3.2	R+	Io=150mA Vo:3.2...V Vin:1...16V P:int	5S	Nsc		data pinout
LP2980IM5X-3.3	R+	Io=150mA Vo:3.3...V Vin:1...16V P:int	5S	Nsc		data pinout
LP2980IM5X-3.5	R+	Io=150mA Vo:3.5...V Vin:1...16V P:int	5S	Nsc		data pinout
LP2980IM5X-3.6	R+	Io=150mA Vo:3.6...V Vin:1...16V P:int	5S	Nsc		data pinout
LP2980IM5X-3.8	R+	Io=150mA Vo:3.8...V Vin:1...16V P:int	5S	Nsc		data pinout
LP2980IM5X-4.0	R+	Io=150mA Vo:4...V Vin:1...16V P:int	5S	Nsc		data pinout
LP2980IM5X-4.5	R+	Io=150mA Vo:4.5...V Vin:1...16V P:int	5S	Nsc		data pinout
LP2980IM5X-4.7	R+	Io=150mA Vo:4.7...V Vin:1...16V P:int	5S	Nsc		data pinout
LP2980IM5X-5.0	R+	Io=150mA Vo:5...V Vin:1...16V P:int	5S	Nsc		data pinout
LP2980IM5X-ADJ	ER+	Io=150mA Vo:1.2...15V Vin:1...16V P:int	5S	Nsc		data pinout
LP2980LVAIM5-1.5	R+	Io=150mA Vo:1.46...1.54V P:int	5S	Nsc		data pinout
LP2980LVAIM5-1.8	R+	Io=150mA Vo:1.76...1.84V P:int	5S	Nsc		data pinout
LP2980LVAIM5X-1.5	R+	Io=150mA Vo:1.46...1.54V P:int	5S	Nsc		data pinout
LP2980LVAIM5X-1.8	R+	Io=150mA Vo:1.76...1.84V P:int	5S	Nsc		data pinout
LP2980LVIM5-1.5	R+	Io=150mA Vo:1.45...1.55V P:int	5S	Nsc		data pinout
LP2980LVIM5-1.8	R+	Io=150mA Vo:1.74...1.86V P:int	5S	Nsc		data pinout
LP2980LVIM5X-1.5	R+	Io=150mA Vo:1.45...1.55V P:int	5S	Nsc		data pinout
LP2980LVIM5X-1.8	R+	Io=150mA Vo:1.74...1.86V P:int	5S	Nsc		data pinout
LP2981AIBP-2.5	R+	Io=400mA Vo:2.5...V Vin:1...16V P:int	5Y	Nsc		data pinout
LP2981AIBP-3.2	R+	Io=400mA Vo:3.2...V Vin:1...16V P:int	5Y	Nsc		data pinout
LP2981AIBP-3.3	R+	Io=400mA Vo:3.3...V Vin:1...16V P:int	5Y	Nsc		data pinout
LP2981AIBPX-2.5	R+	Io=400mA Vo:2.5...V Vin:1...16V P:int	5Y	Nsc		data pinout
LP2981AIBPX-3.2	R+	Io=400mA Vo:3.2...V Vin:1...16V P:int	5Y	Nsc		data pinout
LP2981AIBPX-3.3	R+	Io=400mA Vo:3.3...V Vin:1...16V P:int	5Y	Nsc		data pinout
LP2981AIM5-2.5	R+	Io=400mA Vo:2.5...V Vin:1...16V P:int	5S	Nsc		data pinout
LP2981AIM5-2.7	R+	Io=400mA Vo:2.7...V Vin:1...16V P:int	5S	Nsc		data pinout
LP2981AIM5-2.8	R+	Io=400mA Vo:2.8...V Vin:1...16V P:int	5S	Nsc		data pinout
LP2981AIM5-2.9	R+	Io=400mA Vo:2.9...V Vin:1...16V P:int	5S	Nsc		data pinout

Type	Device	Short Description	Fig.	Manu	Comparision Types	More at
LP2981AIM5-3.0	R+	Io=400mA Vo:3...V Vin:1...16V P:int	5S	Nsc		data pinout
LP2981AIM5-3.1	R+	Io=400mA Vo:3.1...V Vin:1...16V P:int	5S	Nsc		data pinout
LP2981AIM5-3.2	R+	Io=400mA Vo:3.2...V Vin:1...16V P:int	5S	Nsc		data pinout
LP2981AIM5-3.3	R+	Io=400mA Vo:3.3...V Vin:1...16V P:int	5S	Nsc		data pinout
LP2981AIM5-3.6	R+	Io=400mA Vo:3.6...V Vin:1...16V P:int	5S	Nsc		data pinout
LP2981AIM5-3.8	R+	Io=400mA Vo:3.8...V Vin:1...16V P:int	5S	Nsc		data pinout
LP2981AIM5-4.0	R+	Io=400mA Vo:4...V Vin:1...16V P:int	5S	Nsc		data pinout
LP2981AIM5-4.7	R+	Io=400mA Vo:4.7...V Vin:1...16V P:int	5S	Nsc		data pinout
LP2981AIM5-5.0	R+	Io=400mA Vo:5...V Vin:1...16V P:int	5S	Nsc		data pinout
LP2981AIM5X-2.5	R+	Io=400mA Vo:2.5...V Vin:1...16V P:int	5S	Nsc		data pinout
LP2981AIM5X-2.7	R+	Io=400mA Vo:2.7...V Vin:1...16V P:int	5S	Nsc		data pinout
LP2981AIM5X-2.8	R+	Io=400mA Vo:2.8...V Vin:1...16V P:int	5S	Nsc		data pinout
LP2981AIM5X-2.9	R+	Io=400mA Vo:2.9...V Vin:1...16V P:int	5S	Nsc		data pinout
LP2981AIM5X-3.0	R+	Io=400mA Vo:3...V Vin:1...16V P:int	5S	Nsc		data pinout
LP2981AIM5X-3.1	R+	Io=400mA Vo:3.1...V Vin:1...16V P:int	5S	Nsc		data pinout
LP2981AIM5X-3.2	R+	Io=400mA Vo:3.2...V Vin:1...16V P:int	5S	Nsc		data pinout
LP2981AIM5X-3.3	R+	Io=400mA Vo:3.3...V Vin:1...16V P:int	5S	Nsc		data pinout
LP2981AIM5X-3.6	R+	Io=400mA Vo:3.6...V Vin:1...16V P:int	5S	Nsc		data pinout
LP2981AIM5X-3.8	R+	Io=400mA Vo:3.8...V Vin:1...16V P:int	5S	Nsc		data pinout
LP2981AIM5X-4.0	R+	Io=400mA Vo:4...V Vin:1...16V P:int	5S	Nsc		data pinout
LP2981AIM5X-4.7	R+	Io=400mA Vo:4.7...V Vin:1...16V P:int	5S	Nsc		data pinout
LP2981AIM5X-5.0	R+	Io=400mA Vo:5...V Vin:1...16V P:int	5S	Nsc		data pinout
LP2981IBP-2.5	R+	Io=400mA Vo:2.5...V Vin:1...16V P:int	5Y	Nsc		data pinout
LP2981IBP-3.2	R+	Io=400mA Vo:3.2...V Vin:1...16V P:int	5Y	Nsc		data pinout
LP2981IBP-3.3	R+	Io=400mA Vo:3.3...V Vin:1...16V P:int	5Y	Nsc		data pinout
LP2981IBPX-2.5	R+	Io=400mA Vo:2.5...V Vin:1...16V P:int	5Y	Nsc		data pinout
LP2981IBPX-3.2	R+	Io=400mA Vo:3.2...V Vin:1...16V P:int	5Y	Nsc		data pinout
LP2981IBPX-3.3	R+	Io=400mA Vo:3.3...V Vin:1...16V P:int	5Y	Nsc		data pinout
LP2981IM5-2.5	R+	Io=400mA Vo:2.5...V Vin:1...16V P:int	5S	Nsc		data pinout
LP2981IM5-2.7	R+	Io=400mA Vo:2.7...V Vin:1...16V P:int	5S	Nsc		data pinout
LP2981IM5-2.8	R+	Io=400mA Vo:2.8...V Vin:1...16V P:int	5S	Nsc		data pinout
LP2981IM5-2.9	R+	Io=400mA Vo:2.9...V Vin:1...16V P:int	5S	Nsc		data pinout
LP2981IM5-3.0	R+	Io=400mA Vo:3...V Vin:1...16V P:int	5S	Nsc		data pinout
LP2981IM5-3.1	R+	Io=400mA Vo:3.1...V Vin:1...16V P:int	5S	Nsc		data pinout
LP2981IM5-3.2	R+	Io=400mA Vo:3.2...V Vin:1...16V P:int	5S	Nsc		data pinout
LP2981IM5-3.3	R+	Io=400mA Vo:3.3...V Vin:1...16V P:int	5S	Nsc		data pinout
LP2981IM5-3.6	R+	Io=400mA Vo:3.6...V Vin:1...16V P:int	5S	Nsc		data pinout
LP2981IM5-3.8	R+	Io=400mA Vo:3.8...V Vin:1...16V P:int	5S	Nsc		data pinout
LP2981IM5-4.0	R+	Io=400mA Vo:4...V Vin:1...16V P:int	5S	Nsc		data pinout
LP2981IM5-4.7	R+	Io=400mA Vo:4.7...V Vin:1...16V P:int	5S	Nsc		data pinout
LP2981IM5-5.0	R+	Io=400mA Vo:5...V Vin:1...16V P:int	5S	Nsc		data pinout
LP2981IM5X-2.5	R+	Io=400mA Vo:2.5...V Vin:1...16V P:int	5S	Nsc		data pinout
LP2981IM5X-2.7	R+	Io=400mA Vo:2.7...V Vin:1...16V P:int	5S	Nsc		data pinout
LP2981IM5X-2.8	R+	Io=400mA Vo:2.8...V Vin:1...16V P:int	5S	Nsc		data pinout
LP2981IM5X-2.9	R+	Io=400mA Vo:2.9...V Vin:1...16V P:int	5S	Nsc		data pinout
LP2981IM5X-3.0	R+	Io=400mA Vo:3...V Vin:1...16V P:int	5S	Nsc		data pinout
LP2981IM5X-3.1	R+	Io=400mA Vo:3.1...V Vin:1...16V P:int	5S	Nsc		data pinout
LP2981IM5X-3.2	R+	Io=400mA Vo:3.2...V Vin:1...16V P:int	5S	Nsc		data pinout
LP2981IM5X-3.3	R+	Io=400mA Vo:3.3...V Vin:1...16V P:int	5S	Nsc		data pinout
LP2981IM5X-3.6	R+	Io=400mA Vo:3.6...V Vin:1...16V P:int	5S	Nsc		data pinout
LP2981IM5X-3.8	R+	Io=400mA Vo:3.8...V Vin:1...16V P:int	5S	Nsc		data pinout
LP2981IM5X-4.0	R+	Io=400mA Vo:4...V Vin:1...16V P:int	5S	Nsc		data pinout
LP2981IM5X-4.7	R+	Io=400mA Vo:4.7...V Vin:1...16V P:int	5S	Nsc		data pinout
LP2981IM5X-5.0	R+	Io=400mA Vo:5...V Vin:1...16V P:int	5S	Nsc		data pinout
LP2982AIBP-2.8	R+	Io=150mA Vo:2.8...V Vin:1...16V P:int	5Y	Nsc		data pinout
LP2982AIBP-3.0	R+	Io=150mA Vo:3...V Vin:1...16V P:int	5Y	Nsc		data pinout
LP2982AIBPX-2.8	R+	Io=150mA Vo:2.8...V Vin:1...16V P:int	5Y	Nsc		data pinout
LP2982AIBPX-3.0	R+	Io=150mA Vo:3...V Vin:1...16V P:int	5Y	Nsc		data pinout
LP2982AIM5-2.5	R+	Io=150mA Vo:2.5...V Vin:1...16V P:int	5S	Nsc		data pinout
LP2982AIM5-2.8	R+	Io=150mA Vo:2.8...V Vin:1...16V P:int	5S	Nsc		data pinout
LP2982AIM5-3.0	R+	Io=150mA Vo:3...V Vin:1...16V P:int	5S	Nsc		data pinout
LP2982AIM5-3.3	R+	Io=150mA Vo:3.3...V Vin:1...16V P:int	5S	Nsc		data pinout
LP2982AIM5-3.6	R+	Io=150mA Vo:3.6...V Vin:1...16V P:int	5S	Nsc		data pinout
LP2982AIM5-3.8	R+	Io=150mA Vo:3.8...V Vin:1...16V P:int	5S	Nsc		data pinout
LP2982AIM5-4.0	R+	Io=150mA Vo:4...V Vin:1...16V P:int	5S	Nsc		data pinout
LP2982AIM5-4.5	R+	Io=150mA Vo:4.5...V Vin:1...16V P:int	5S	Nsc		data pinout
LP2982AIM5-4.7	R+	Io=150mA Vo:4.7...V Vin:1...16V P:int	5S	Nsc		data pinout
LP2982AIM5-5.0	R+	Io=150mA Vo:5...V Vin:1...16V P:int	5S	Nsc		data pinout
LP2982AIM5X-2.5	R+	Io=150mA Vo:2.5...V Vin:1...16V P:int	5S	Nsc		data pinout
LP2982AIM5X-2.8	R+	Io=150mA Vo:2.8...V Vin:1...16V P:int	5S	Nsc		data pinout
LP2982AIM5X-3.0	R+	Io=150mA Vo:3...V Vin:1...16V P:int	5S	Nsc		data pinout
LP2982AIM5X-3.3	R+	Io=150mA Vo:3.3...V Vin:1...16V P:int	5S	Nsc		data pinout
LP2982AIM5X-3.6	R+	Io=150mA Vo:3.6...V Vin:1...16V P:int	5S	Nsc		data pinout
LP2982AIM5X-3.8	R+	Io=150mA Vo:3.8...V Vin:1...16V P:int	5S	Nsc		data pinout
LP2982AIM5X-4.0	R+	Io=150mA Vo:4...V Vin:1...16V P:int	5S	Nsc		data pinout
LP2982AIM5X-4.5	R+	Io=150mA Vo:4.5...V Vin:1...16V P:int	5S	Nsc		data pinout
LP2982AIM5X-4.7	R+	Io=150mA Vo:4.7...V Vin:1...16V P:int	5S	Nsc		data pinout
LP2982AIM5X-5.0	R+	Io=150mA Vo:5...V Vin:1...16V P:int	5S	Nsc		data pinout
LP2982IBP-2.8	R+	Io=150mA Vo:2.8...V Vin:1...16V P:int	5Y	Nsc		data pinout
LP2982IBP-3.0	R+	Io=150mA Vo:3...V Vin:1...16V P:int	5Y	Nsc		data pinout
LP2982IBPX-2.8	R+	Io=150mA Vo:2.8...V Vin:1...16V P:int	5Y	Nsc		data pinout
LP2982IBPX-3.0	R+	Io=150mA Vo:3...V Vin:1...16V P:int	5Y	Nsc		data pinout
LP2982IM5-2.5	R+	Io=150mA Vo:2.5...V Vin:1...16V P:int	5S	Nsc		data pinout
LP2982IM5-2.8	R+	Io=150mA Vo:2.8...V Vin:1...16V P:int	5S	Nsc		data pinout
LP2982IM5-3.0	R+	Io=150mA Vo:3...V Vin:1...16V P:int	5S	Nsc		data pinout

Type	Device	Short Description	Fig.	Manu	Comparision Types	More at
LP2982IM5-3.3	R+	Io=150mA Vo:3.3...V Vin:1...16V P:int	5S	Nsc		data pinout
LP2982IM5-3.6	R+	Io=150mA Vo:3.6...V Vin:1...16V P:int	5S	Nsc		data pinout
LP2982IM5-3.8	R+	Io=150mA Vo:3.8...V Vin:1...16V P:int	5S	Nsc		data pinout
LP2982IM5-4.0	R+	Io=150mA Vo:4...V Vin:1...16V P:int	5S	Nsc		data pinout
LP2982IM5-4.5	R+	Io=150mA Vo:4.5...V Vin:1...16V P:int	5S	Nsc		data pinout
LP2982IM5-4.7	R+	Io=150mA Vo:4.7...V Vin:1...16V P:int	5S	Nsc		data pinout
LP2982IM5-5.0	R+	Io=150mA Vo:5...V Vin:1...16V P:int	5S	Nsc		data pinout
LP2982IM5X-2.5	R+	Io=150mA Vo:2.5...V Vin:1...16V P:int	5S	Nsc		data pinout
LP2982IM5X-2.8	R+	Io=150mA Vo:2.8...V Vin:1...16V P:int	5S	Nsc		data pinout
LP2982IM5X-3.0	R+	Io=150mA Vo:3...V Vin:1...16V P:int	5S	Nsc		data pinout
LP2982IM5X-3.3	R+	Io=150mA Vo:3.3...V Vin:1...16V P:int	5S	Nsc		data pinout
LP2982IM5X-3.6	R+	Io=150mA Vo:3.6...V Vin:1...16V P:int	5S	Nsc		data pinout
LP2982IM5X-3.8	R+	Io=150mA Vo:3.8...V Vin:1...16V P:int	5S	Nsc		data pinout
LP2982IM5X-4.0	R+	Io=150mA Vo:4...V Vin:1...16V P:int	5S	Nsc		data pinout
LP2982IM5X-4.5	R+	Io=150mA Vo:4.5...V Vin:1...16V P:int	5S	Nsc		data pinout
LP2982IM5X-4.7	R+	Io=150mA Vo:4.7...V Vin:1...16V P:int	5S	Nsc		data pinout
LP2982IM5X-5.0	R+	Io=150mA Vo:5...V Vin:1...16V P:int	5S	Nsc		data pinout
LP2985AIBP-2.5	R+	Io=350mA Vo:2.5...V Vin:1...16V P:int	5Y	Nsc		data pinout
LP2985AIBP-2.8	R+	Io=350mA Vo:2.8...V Vin:1...16V P:int	5Y	Nsc		data pinout
LP2985AIBP-3.3	R+	Io=350mA Vo:3.3...V Vin:1...16V P:int	5Y	Nsc		data pinout
LP2985AIBPX-2.5	R+	Io=350mA Vo:2.5...V Vin:1...16V P:int	5Y	Nsc		data pinout
LP2985AIBPX-2.8	R+	Io=350mA Vo:2.8...V Vin:1...16V P:int	5Y	Nsc		data pinout
LP2985AIBPX-3.3	R+	Io=350mA Vo:3.3...V Vin:1...16V P:int	5Y	Nsc		data pinout
LP2985AIM5-2.5	R+	Io=350mA Vo:2.5...V Vin:1...16V P:int	5S	Nsc		data pinout
LP2985AIM5-2.8	R+	Io=350mA Vo:2.8...V Vin:1...16V P:int	5S	Nsc		data pinout
LP2985AIM5-3.0	R+	Io=350mA Vo:3...V Vin:1...16V P:int	5S	Nsc		data pinout
LP2985AIM5-3.1	R+	Io=350mA Vo:3.1...V Vin:1...16V P:int	5S	Nsc		data pinout
LP2985AIM5-3.2	R+	Io=350mA Vo:3.2...V Vin:1...16V P:int	5S	Nsc		data pinout
LP2985AIM5-3.3	R+	Io=350mA Vo:3.3...V Vin:1...16V P:int	5S	Nsc		data pinout
LP2985AIM5-3.5	R+	Io=350mA Vo:3.5...V Vin:1...16V P:int	5S	Nsc		data pinout
LP2985AIM5-3.6	R+	Io=350mA Vo:3.6...V Vin:1...16V P:int	5S	Nsc		data pinout
LP2985AIM5-4.0	R+	Io=350mA Vo:4...V Vin:1...16V P:int	5S	Nsc		data pinout
LP2985AIM5-5.0	R+	Io=350mA Vo:5...V Vin:1...16V P:int	5S	Nsc		data pinout
LP2985AIM5X-2.5	R+	Io=350mA Vo:2.5...V Vin:1...16V P:int	5S	Nsc		data pinout
LP2985AIM5X-2.8	R+	Io=350mA Vo:2.8...V Vin:1...16V P:int	5S	Nsc		data pinout
LP2985AIM5X-3.0	R+	Io=350mA Vo:3...V Vin:1...16V P:int	5S	Nsc		data pinout
LP2985AIM5X-3.1	R+	Io=350mA Vo:3.1...V Vin:1...16V P:int	5S	Nsc		data pinout
LP2985AIM5X-3.2	R+	Io=350mA Vo:3.2...V Vin:1...16V P:int	5S	Nsc		data pinout
LP2985AIM5X-3.3	R+	Io=350mA Vo:3.3...V Vin:1...16V P:int	5S	Nsc		data pinout
LP2985AIM5X-3.5	R+	Io=350mA Vo:3.5...V Vin:1...16V P:int	5S	Nsc		data pinout
LP2985AIM5X-3.6	R+	Io=350mA Vo:3.6...V Vin:1...16V P:int	5S	Nsc		data pinout
LP2985AIM5X-4.0	R+	Io=350mA Vo:4...V Vin:1...16V P:int	5S	Nsc		data pinout
LP2985AIM5X-5.0	R+	Io=350mA Vo:5...V Vin:1...16V P:int	5S	Nsc		data pinout
LP2985IBP-2.5	R+	Io=350mA Vo:2.5...V Vin:1...16V P:int	5Y	Nsc		data pinout
LP2985IBP-2.8	R+	Io=350mA Vo:2.8...V Vin:1...16V P:int	5Y	Nsc		data pinout
LP2985IBP-3.3	R+	Io=350mA Vo:3.3...V Vin:1...16V P:int	5Y	Nsc		data pinout
LP2985IBPX-2.5	R+	Io=350mA Vo:2.5...V Vin:1...16V P:int	5Y	Nsc		data pinout
LP2985IBPX-2.8	R+	Io=350mA Vo:2.8...V Vin:1...16V P:int	5Y	Nsc		data pinout
LP2985IBPX-3.3	R+	Io=350mA Vo:3.3...V Vin:1...16V P:int	5Y	Nsc		data pinout
LP2985IM5-2.5	R+	Io=350mA Vo:2.5...V Vin:1...16V P:int	5S	Nsc		data pinout
LP2985IM5-2.8	R+	Io=350mA Vo:2.8...V Vin:1...16V P:int	5S	Nsc		data pinout
LP2985IM5-3.0	R+	Io=350mA Vo:3...V Vin:1...16V P:int	5S	Nsc		data pinout
LP2985IM5-3.1	R+	Io=350mA Vo:3.1...V Vin:1...16V P:int	5S	Nsc		data pinout
LP2985IM5-3.2	R+	Io=350mA Vo:3.2...V Vin:1...16V P:int	5S	Nsc		data pinout
LP2985IM5-3.3	R+	Io=350mA Vo:3.3...V Vin:1...16V P:int	5S	Nsc		data pinout
LP2985IM5-3.5	R+	Io=350mA Vo:3.5...V Vin:1...16V P:int	5S	Nsc		data pinout
LP2985IM5-3.6	R+	Io=350mA Vo:3.6...V Vin:1...16V P:int	5S	Nsc		data pinout
LP2985IM5-4.0	R+	Io=350mA Vo:4...V Vin:1...16V P:int	5S	Nsc		data pinout
LP2985IM5-5.0	R+	Io=350mA Vo:5...V Vin:1...16V P:int	5S	Nsc		data pinout
LP2985IM5X-2.5	R+	Io=350mA Vo:2.5...V Vin:1...16V P:int	5S	Nsc		data pinout
LP2985IM5X-2.8	R+	Io=350mA Vo:2.8...V Vin:1...16V P:int	5S	Nsc		data pinout
LP2985IM5X-3.0	R+	Io=350mA Vo:3...V Vin:1...16V P:int	5S	Nsc		data pinout
LP2985IM5X-3.1	R+	Io=350mA Vo:3.1...V Vin:1...16V P:int	5S	Nsc		data pinout
LP2985IM5X-3.2	R+	Io=350mA Vo:3.2...V Vin:1...16V P:int	5S	Nsc		data pinout
LP2985IM5X-3.3	R+	Io=350mA Vo:3.3...V Vin:1...16V P:int	5S	Nsc		data pinout
LP2985IM5X-3.5	R+	Io=350mA Vo:3.5...V Vin:1...16V P:int	5S	Nsc		data pinout
LP2985IM5X-3.6	R+	Io=350mA Vo:3.6...V Vin:1...16V P:int	5S	Nsc		data pinout
LP2985IM5X-4.0	R+	Io=350mA Vo:4...V Vin:1...16V P:int	5S	Nsc		data pinout
LP2985IM5X-5.0	R+	Io=350mA Vo:5...V Vin:1...16V P:int	5S	Nsc		data pinout
LP2986AIM-3.0	R+	Io=400mA Vo:2.946...3.054V Vin:1...16V	8S	Nsc		data pinout
LP2986AIM-3.3	R+	Io=400mA Vo:3.241...3.359V Vin:1...16V	8S	Nsc		data pinout
LP2986AIM-5.0	R+	Io=400mA Vo:4.91...5.09V Vin:1...16V	8S	Nsc		data pinout
LP2986AIMM-3.0	R+	Io=400mA Vo:2.946...3.054V Vin:1...16V	8S	Nsc		data pinout
LP2986AIMM-3.3	R+	Io=400mA Vo:3.241...3.359V Vin:1...16V	8S	Nsc		data pinout
LP2986AIMM-5.0	R+	Io=400mA Vo:4.91...5.09V Vin:1...16V	8S	Nsc		data pinout
LP2986AIMMX-3.0	R+	Io=400mA Vo:2.946...3.054V Vin:1...16V	8S	Nsc		data pinout
LP2986AIMMX-3.3	R+	Io=400mA Vo:3.241...3.359V Vin:1...16V	8S	Nsc		data pinout
LP2986AIMMX-5.0	R+	Io=400mA Vo:4.91...5.09V Vin:1...16V	8S	Nsc		data pinout
LP2986AIMX-3.0	R+	Io=400mA Vo:2.946...3.054V Vin:1...16V	8S	Nsc		data pinout
LP2986AIMX-3.3	R+	Io=400mA Vo:3.241...3.359V Vin:1...16V	8S	Nsc		data pinout
LP2986AIMX-5.0	R+	Io=400mA Vo:4.91...5.09V Vin:1...16V	8S	Nsc		data pinout
LP2986IM-3.0	R+	Io=400mA Vo:2.916...3.084V Vin:1...16V	8S	Nsc		data pinout
LP2986IM-3.3	R+	Io=400mA Vo:3.208...3.392V Vin:1...16V	8S	Nsc		data pinout
LP2986IM-5.0	R+	Io=400mA Vo:4.86...5.14V Vin:1...16V	8S	Nsc		data pinout
LP2986IMM-3.0	R+	Io=400mA Vo:2.916...3.084V Vin:1...16V	8S	Nsc		data pinout

Type	Device	Short Description	Fig.	Manu	Comparision Types	More at
LP2986IMM-3.3	R+	Io=400mA Vo:3.208...3.392V Vin:1...16V	8S	Nsc		data pinout
LP2986IMM-5.0	R+	Io=400mA Vo:4.86...5.14V Vin:1...16V	8S	Nsc		data pinout
LP2986IMMX-3.0	R+	Io=400mA Vo:2.916...3.084V Vin:1...16V	8S	Nsc		data pinout
LP2986IMMX-3.3	R+	Io=400mA Vo:3.208...3.392V Vin:1...16V	8S	Nsc		data pinout
LP2986IMMX-5.0	R+	Io=400mA Vo:4.86...5.14V Vin:1...16V	8S	Nsc		data pinout
LP2986IMX-3.0	R+	Io=400mA Vo:2.916...3.084V Vin:1...16V	8S	Nsc		data pinout
LP2986IMX-3.3	R+	Io=400mA Vo:3.208...3.392V Vin:1...16V	8S	Nsc		data pinout
LP2986IMX-5.0	R+	Io=400mA Vo:4.86...5.14V Vin:1...16V	8S	Nsc		data pinout
LP2987AIM-2.8	R+	Io=400mA Vo:2.8...V Vin:1...16V P:int	8S	Nsc		data pinout
LP2987AIM-3.0	R+	Io=400mA Vo:3...V Vin:1...16V P:int	8S	Nsc		data pinout
LP2987AIM-3.2	R+	Io=400mA Vo:3.2...V Vin:1...16V P:int	8S	Nsc		data pinout
LP2987AIM-3.3	R+	Io=400mA Vo:3.3...V Vin:1...16V P:int	8S	Nsc		data pinout
LP2987AIM-3.8	R+	Io=400mA Vo:3.8...V Vin:1...16V P:int	8S	Nsc		data pinout
LP2987AIM-5.0	R+	Io=400mA Vo:5...V Vin:1...16V P:int	8S	Nsc		data pinout
LP2987AIMM-2.8	R+	Io=400mA Vo:2.8...V Vin:1...16V P:int	8S	Nsc		data pinout
LP2987AIMM-3.0	R+	Io=400mA Vo:3...V Vin:1...16V P:int	8S	Nsc		data pinout
LP2987AIMM-3.2	R+	Io=400mA Vo:3.2...V Vin:1...16V P:int	8S	Nsc		data pinout
LP2987AIMM-3.3	R+	Io=400mA Vo:3:3...V Vin:1...16V P:int	8S	Nsc		data pinout
LP2987AIMM-3.8	R+	Io=400mA Vo:3.8...V Vin:1...16V P:int	8S	Nsc		data pinout
LP2987AIMM-5.0	R+	Io=400mA Vo:5...V Vin:1...16V P:int	8S	Nsc		data pinout
LP2987AIMMX-2.8	R+	Io=400mA Vo:2.8...V Vin:1...16V P:int	8S	Nsc		data pinout
LP2987AIMMX-3.0	R+	Io=400mA Vo:3...V Vin:1...16V P:int	8S	Nsc		data pinout
LP2987AIMMX-3.2	R+	Io=400mA Vo:3.2...V Vin:1...16V P:int	8S	Nsc		data pinout
LP2987AIMMX-3.3	R+	Io=400mA Vo:3.3...V Vin:1...16V P:int	8S	Nsc		data pinout
LP2987AIMMX-3.8	R+	Io=400mA Vo:3.8...V Vin:1...16V P:int	8S	Nsc		data pinout
LP2987AIMMX-5.0	R+	Io=400mA Vo:5...V Vin:1...16V P:int	8S	Nsc		data pinout
LP2987AIMX-2.8	R+	Io=400mA Vo:2.8...V Vin:1...16V P:int	8S	Nsc		data pinout
LP2987AIMX-3.0	R+	Io=400mA Vo:3...V Vin:1...16V P:int	8S	Nsc		data pinout
LP2987AIMX-3.2	R+	Io=400mA Vo:3.2...V Vin:1...16V P:int	8S	Nsc		data pinout
LP2987AIMX-3.3	R+	Io=400mA Vo:3.3...V Vin:1...16V P:int	8S	Nsc		data pinout
LP2987AIMX-3.8	R+	Io=400mA Vo:3.8...V Vin:1...16V P:int	8S	Nsc		data pinout
LP2987AIMX-5.0	R+	Io=400mA Vo:5...V Vin:1...16V P:int	8S	Nsc		data pinout
LP2987IM-2.8	R+	Io=400mA Vo:2.8...V Vin:1...16V P:int	8S	Nsc		data pinout
LP2987IM-3.0	R+	Io=400mA Vo:3...V Vin:1...16V P:int	8S	Nsc		data pinout
LP2987IM-3.2	R+	Io=400mA Vo:3.2...V Vin:1...16V P:int	8S	Nsc		data pinout
LP2987IM-3.3	R+	Io=400mA Vo:3.3...V Vin:1...16V P:int	8S	Nsc		data pinout
LP2987IM-3.8	R+	Io=400mA Vo:3.8...V Vin:1...16V P:int	8S	Nsc		data pinout
LP2987IM-5.0	R+	Io=400mA Vo:5...V Vin:1...16V P:int	8S	Nsc		data pinout
LP2987IMM-2.8	R+	Io=400mA Vo:2.8...V Vin:1...16V P:int	8S	Nsc		data pinout
LP2987IMM-3.0	R+	Io=400mA Vo:3...V Vin:1...16V P:int	8S	Nsc		data pinout
LP2987IMM-3.2	R+	Io=400mA Vo:3.2...V Vin:1...16V P:int	8S	Nsc		data pinout
LP2987IMM-3.3	R+	Io=400mA Vo:3.3...V Vin:1...16V P:int	8S	Nsc		data pinout
LP2987IMM-3.8	R+	Io=400mA Vo:3.8...V Vin:1...16V P:int	8S	Nsc		data pinout
LP2987IMM-5.0	R+	Io=400mA Vo:5...V Vin:1...16V P:int	8S	Nsc		data pinout
LP2987IMMX-2.8	R+	Io=400mA Vo:2.8...V Vin:1...16V P:int	8S	Nsc		data pinout
LP2987IMMX-3.0	R+	Io=400mA Vo:3...V Vin:1...16V P:int	8S	Nsc		data pinout
LP2987IMMX-3.2	R+	Io=400mA Vo:3.2...V Vin:1...16V P:int	8S	Nsc		data pinout
LP2987IMMX-3.3	R+	Io=400mA Vo:3.3...V Vin:1...16V P:int	8S	Nsc		data pinout
LP2987IMMX-3.8	R+	Io=400mA Vo:3.8...V Vin:1...16V P:int	8S	Nsc		data pinout
LP2987IMMX-5.0	R+	Io=400mA Vo:5...V Vin:1...16V P:int	8S	Nsc		data pinout
LP2987IMX-2.8	R+	Io=400mA Vo:2.8...V Vin:1...16V P:int	8S	Nsc		data pinout
LP2987IMX-3.0	R+	Io=400mA Vo:3...V Vin:1...16V P:int	8S	Nsc		data pinout
LP2987IMX-3.2	R+	Io=400mA Vo:3.2...V Vin:1...16V P:int	8S	Nsc		data pinout
LP2987IMX-3.3	R+	Io=400mA Vo:3.3...V Vin:1...16V P:int	8S	Nsc		data pinout
LP2987IMX-3.8	R+	Io=400mA Vo:3.8...V Vin:1...16V P:int	8S	Nsc		data pinout
LP2987IMX-5.0	R+	Io=400mA Vo:5...V Vin:1...16V P:int	8S	Nsc		data pinout
LP2988AIM-2.8	R+	Io=400mA Vo:2.8...V Vin:1...16V P:int	8S	Nsc		data pinout
LP2988AIM-3.0	R+	Io=400mA Vo:3...V Vin:1...16V P:int	8S	Nsc		data pinout
LP2988AIM-3.2	R+	Io=400mA Vo:3.2...V Vin:1...16V P:int	8S	Nsc		data pinout
LP2988AIM-3.3	R+	Io=400mA Vo:3.3...V Vin:1...16V P:int	8S	Nsc		data pinout
LP2988AIM-3.8	R+	Io=400mA Vo:3.8...V Vin:1...16V P:int	8S	Nsc		data pinout
LP2988AIM-5.0	R+	Io=400mA Vo:5...V Vin:1...16V P:int	8S	Nsc		data pinout
LP2988AIMM-2.8	R+	Io=400mA Vo:2.8...V Vin:1...16V P:int	8S	Nsc		data pinout
LP2988AIMM-3.0	R+	Io=400mA Vo:3...V Vin:1...16V P:int	8S	Nsc		data pinout
LP2988AIMM-3.2	R+	Io=400mA Vo:3.2...V Vin:1...16V P:int	8S	Nsc		data pinout
LP2988AIMM-3.3	R+	Io=400mA Vo:3.3...V Vin:1...16V	8S	Nsc		data pinout
LP2988AIMM-3.8	R+	Io=400mA Vo:3.8...V Vin:1...16V P:int	8S	Nsc		data pinout
LP2988AIMM-5.0	R+	Io=400mA Vo:5...V Vin:1...16V P:int	8S	Nsc		data pinout
LP2988AIMMX-2.8	R+	Io=400mA Vo:2.8...V Vin:1...16V P:int	8S	Nsc		data pinout
LP2988AIMMX-3.0	R+	Io=400mA Vo:3...V Vin:1...16V P:int	8S	Nsc		data pinout
LP2988AIMMX-3.2	R+	Io=400mA Vo:3.2...V Vin:1...16V P:int	8S	Nsc		data pinout
LP2988AIMMX-3.3	R+	Io=400mA Vo:3.3...V Vin:1...16V P:int	8S	Nsc		data pinout
LP2988AIMMX-3.8	R+	Io=400mA Vo:3.8...V Vin:1...16V P:int	8S	Nsc		data pinout
LP2988AIMMX-5.0	R+	Io=400mA Vo:5...V Vin:1...16V P:int	8S	Nsc		data pinout
LP2988AIMX-2.8	R+	Io=400mA Vo:2.8...V Vin:1...16V P:int	8S	Nsc		data pinout
LP2988AIMX-3.0	R+	Io=400mA Vo:3...V Vin:1...16V P:int	8S	Nsc		data pinout
LP2988AIMX-3.2	R+	Io=400mA Vo:3.2...V Vin:1...16V P:int	8S	Nsc		data pinout
LP2988AIMX-3.3	R+	Io=400mA Vo:3.3...V Vin:1...16V P:int	8S	Nsc		data pinout
LP2988AIMX-3.8	R+	Io=400mA Vo:3.8...V Vin:1...16V P:int	8S	Nsc		data pinout
LP2988AIMX-5.0	R+	Io=400mA Vo:5...V Vin:1...16V P:int	8S	Nsc		data pinout
LP2988IM-2.8	R+	Io=400mA Vo:2.8...V Vin:1...16V P:int	8S	Nsc		data pinout
LP2988IM-3.0	R+	Io=400mA Vo:3...V Vin:1...16V P:int	8S	Nsc		data pinout
LP2988IM-3.2	R+	Io=400mA Vo:3.2...V Vin:1...16V P:int	8S	Nsc		data pinout
LP2988IM-3.3	R+	Io=400mA Vo:3.3...V Vin:1...16V P:int	8S	Nsc		data pinout
LP2988IM-3.8	R+	Io=400mA Vo:3.8...V Vin:1...16V P:int	8S	Nsc		data pinout

Type	Device	Short Description	Fig.	Manu	Comparision Types	More at
LP2988IM-5.0	R+	Io=400mA Vo:5...V Vin:1...16V P:int	8S	Nsc		data pinout
LP2988IMM-2.8	R+	Io=400mA Vo:2.8...V Vin:1...16V P:int	8S	Nsc		data pinout
LP2988IMM-3.0	R+	Io=400mA Vo:3...V Vin:1...16V P:int	8S	Nsc		data pinout
LP2988IMM-3.2	R+	Io=400mA Vo:3.2...V Vin:1...16V P:int	8S	Nsc		data pinout
LP2988IMM-3.3	R+	Io=400mA Vo:3.3...V Vin:1...16V P:int	8S	Nsc		data pinout
LP2988IMM-3.8	R+	Io=400mA Vo:3.8...V Vin:1...16V P:int	8S	Nsc		data pinout
LP2988IMM-5.0	R+	Io=400mA Vo:5...V Vin:1...16V P:int	8S	Nsc		data pinout
LP2988IMMX-2.8	R+	Io=400mA Vo:2.8...V Vin:1...16V P:int	8S	Nsc		data pinout
LP2988IMMX-3.0	R+	Io=400mA Vo:3...V Vin:1...16V P:int	8S	Nsc		data pinout
LP2988IMMX-3.2	R+	Io=400mA Vo:3.2...V Vin:1...16V P:int	8S	Nsc		data pinout
LP2988IMMX-3.3	R+	Io=400mA Vo:3.3...V Vin:1...16V P:int	8S	Nsc		data pinout
LP2988IMMX-3.8	R+	Io=400mA Vo:3.8...V Vin:1...16V P:int	8S	Nsc		data pinout
LP2988IMMX-5.0	R+	Io=400mA Vo:5...V Vin:1...16V P:int	8S	Nsc		data pinout
LP2988IMX-2.8	R+	Io=400mA Vo:2.8...V Vin:1...16V P:int	8S	Nsc		data pinout
LP2988IMX-3.0	R+	Io=400mA Vo:3...V Vin:1...16V P:int	8S	Nsc		data pinout
LP2988IMX-3.2	R+	Io=400mA Vo:3.2...V Vin:1...16V P:int	8S	Nsc		data pinout
LP2988IMX-3.3	R+	Io=400mA Vo:3.3...V Vin:1...16V P:int	8S	Nsc		data pinout
LP2988IMX-3.8	R+	Io=400mA Vo:3.8...V Vin:1...16V P:int	8S	Nsc		data pinout
LP2988IMX-5.0	R+	Io=400mA Vo:5...V Vin:1...16V P:int	8S	Nsc		data pinout
LP3052BGT	LED	gr ø3mm I:44mcd Vf:2V If:20mA	T1	Lst		data pdf pinout
LP3052BIT	LED	rd ø3mm I:44mcd Vf:2V If:20mA	T1	Lst		data pdf pinout
LP3052BSRT	LED	rd ø3mm I:135mcd Vf:1.7V If:10mA	T1	Lst		data pdf pinout
LP3052BSUGC	LED	gr ø3mm I=4.5cd Vf:3.1V If:20mA	T1	Lst		data pdf pinout
LP3052BSYGC	LED	yl ø3mm I:340mcd Vf:2V If:20mA	T1	Lst		data pdf pinout
LP3052BUBC	LED	bl ø3mm I=1cd Vf:3.6V If:20mA	T1	Lst		data pdf pinout
LP3052BUYC	LED	yl ø3mm I:1380mcd Vf:2V If:20mA	T1	Lst		data pdf pinout
LP3052BUYD	LED	yl ø3mm I:270mcd Vf:2V If:10mA	T1	Lst		data pdf pinout
LP3052BVRD	LED	rd ø3mm I:60mcd Vf:2V If:20mA	T1	Lst		data pdf pinout
LP3052BYT	LED	yl ø3mm I:44mcd Vf:2V If:20mA	T1	Lst		data pdf pinout
LP3340	LED	gr ø3mm I:7.1...28mcd Vf:2V If:10mA	T1	Osr		data pdf pinout
LP3341	LED	gr ø3mm I:18..28mcd Vf:2V If:10mA	T1	Osr		data pdf pinout
LP3360	LED	gr ø3mm I:4.5...7.1mcd Vf:2V If:10mA	T1	Osr		data pdf pinout
LP3470IM5-2.63	R+	Io=10mA P:300mW	5S	Nsc		data pinout
LP3470IM5-2.93	R+	Io=10mA P:300mW	5S	Nsc		data pinout
LP3470IM5-3.08	R+	Io=10mA P:300mW	5S	Nsc		data pinout
LP3470IM5-4.00	R+	Io=10mA P:300mW	5S	Nsc		data pinout
LP3470IM5-4.38	R+	Io=10mA P:300mW	5S	Nsc		data pinout
LP3470IM5-4.63	R+	Io=10mA P:300mW	5S	Nsc		data pinout
LP3470IM5X-2.63	R+	Io=10mA P:300mW	5S	Nsc		data pinout
LP3470IM5X-2.93	R+	Io=10mA P:300mW	5S	Nsc		data pinout
LP3470IM5X-3.08	R+	Io=10mA P:300mW	5S	Nsc		data pinout
LP3470IM5X-4.00	R+	Io=10mA P:300mW	5S	Nsc		data pinout
LP3470IM5X-4.38	R+	Io=10mA P:300mW	5S	Nsc		data pinout
LP3470IM5X-4.63	R+	Io=10mA P:300mW	5S	Nsc		data pinout
LP3470M5-2.63	R+	Io=10mA P:300mW	5S	Nsc		data pinout
LP3470M5-2.93	R+	Io=10mA P:300mW	5S	Nsc		data pinout
LP3470M5-3.08	R+	Io=10mA P:300mW	5S	Nsc		data pinout
LP3470M5-4.00	R+	Io=10mA P:300mW	5S	Nsc		data pinout
LP3470M5-4.38	R+	Io=10mA P:300mW	5S	Nsc		data pinout
LP3470M5-4.63	R+	Io=10mA P:300mW	5S	Nsc		data pinout
LP3470M5X-2.63	R+	Io=10mA P:300mW	5S	Nsc		data pinout
LP3470M5X-2.93	R+	Io=10mA P:300mW	5S	Nsc		data pinout
LP3470M5X-3.08	R+	Io=10mA P:300mW	5S	Nsc		data pinout
LP3470M5X-4.00	R+	Io=10mA P:300mW	5S	Nsc		data pinout
LP3470M5X-4.38	R+	Io=10mA P:300mW	5S	Nsc		data pinout
LP3470M5X-4.63	R+	Io=10mA P:300mW	5S	Nsc		data pinout
LP 3905SD-00	IC	Power Management Unit, 2x 600mA DC/DC buck regulators, 2x 150mA linear reg.	LLP	Nsc	-	pdf pinout
LP 3905SD-30	IC	Power Management Unit, 2x 600mA DC/DC buck regulators, 2x 150mA linear reg., Buck1/2: Fixed PWM	LLP	Nsc	-	pdf pinout
LP 3905SD-A3	IC	Power Management Unit, 2x 600mA DC/DC buck regulators, 2x 150mA linear reg., Buck1/2 adjustable	LLP	Nsc	-	pdf pinout
LP 3905SDX-00	IC	Power Management Unit, 2x 600mA DC/DC buck regulators, 2x 150mA linear reg.	LLP	Nsc	-	pdf pinout
LP 3905SDX-30	IC	Power Management Unit, 2x 600mA DC/DC buck regulators, 2x 150mA linear reg., Buck1/2: Fixed PWM	LLP	Nsc	-	pdf pinout
LP 3905SDX-A3	IC	Power Management Unit, 2x 600mA DC/DC buck regulators, 2x 150mA linear reg., Buck1/2 adjustable	LLP	Nsc	-	pdf pinout
LP 3906SQ DJXI	IC	2x step-down DC/DC 1.5A, 2x linear regulator with I2C, SW1:0.9V, SW2:1.8V, LDO1:3.3V, LDO2:1.8V	LLP	Nsc	-	pdf pinout
LP 3906SQ JXXI	IC	2x step-down DC/DC 1.5A, 2x linear regulator with I2C, SW1:1.2V, SW2:3.3V, LDO1:3.3V, LDO2:1.8V	LLP	Nsc	-	pdf pinout
LP 3945,3946 ILD	LIN-IC	Battery Charge Managment, Li-ion,Ni-MH LP3946: nur Li-ion, I²C	14-LLP	Nsc		pdf pinout
LP 3971SQ-B410	IC	Power Management Unit, 6 low dropout linear reg., 3 DC/DC buck reg., back- up battery charger, serial interface	LLP	Nsc	-	pdf pinout
LP 3971SQ-D510	IC	Power Management Unit, 6 low dropout	LLP	Nsc	-	pdf pinout

Type	Device	Short Description	Fig.	Manu	Comparision Types	More at
		linear reg., 3 DC/DC buck reg., back- up battery charger, serial interface				
LP 3971SQ-F211	IC	Power Management Unit, 6 low dropout linear reg., 3 DC/DC buck reg., back- up battery charger, serial interface	LLP	Nsc		pdf pinout
LP5086A4UWC	LED	wht ø5mm I=10cd Vf:3.4V If:20mA	T1¾	Lst		data pdf pinout
LP5086AAD	LED	am ø5mm I:16mcd Vf:2V If:20mA	T1¾	Lst		data pdf pinout
LP5086AED	LED	or ø5mm I:17mcd Vf:2V If:20mA	T1¾	Lst		data pdf pinout
LP5086AGD	LED	gr ø5mm I:11mcd Vf:2.1V If:20mA	T1¾	Lst		data pdf pinout
LP5086AGT	LED	gr ø5mm I:33mcd Vf:2.1V If:20mA	T1¾	Lst		data pdf pinout
LP5086AHD	LED	rd ø5mm I:1mcd Vf:2V If:10mA	T1¾	Lst		data pdf pinout
LP5086AHRC	LED	rd ø5mm I:660mcd Vf:2V If:10mA	T1¾	Lst		data pdf pinout
LP5086AHRD	LED	rd ø5mm I:300mcd Vf:2V If:20mA	T1¾	Lst		data pdf pinout
LP5086AHT	LED	rd ø5mm I:9mcd Vf:2V If:10mA	T1¾	Lst		data pdf pinout
LP5086AID	LED	rd ø5mm I:11mcd Vf:2V If:20mA	T1¾	Lst		data pdf pinout
LP5086AIT	LED	rd ø5mm I:33mcd Vf:2V If:20mA	T1¾	Lst		data pdf pinout
LP5086ASDRC	LED	rd ø5mm I:600mcd Vf:2V If:20mA	T1¾	Lst		data pdf pinout
LP5086ASDRD	LED	rd ø5mm I:125mcd Vf:2V If:20mA	T1¾	Lst		data pdi pinout
LP5086ASRC	LED	rd ø5mm I:100mcd Vf:1.7V If:20mA	T1¾	Lst		data pdf pinout
LP5086ASRD	LED	rd ø5mm I:150mcd Vf:1.7V If:20mA	T1¾	Lst		data pdf pinout
LP5086ASUBC	LED	bl ø5mm I=1.5cd Vf:3.5V If:20mA	T1¾	Lst		data pdf pinout
LP5086ASUGC	LED	gr ø5mm I=8cd Vf:3.5V If:20mA	T1¾	Lst		data pdf pinout
LP5086ASURC	LED	ord ø5mm I=4cd Vf:2V If:20mA	T1¾	Lst		data pdf pinout
LP5086ASURD	LED	rd ø5mm I:180mcd Vf:2V If:20mA	T1¾	Lst		data pdf pinout
LP5086ASYGC	LED	yl ø5mm I=4.5cd Vf:2V If:20mA	T1¾	Lst		data pdf pinout
LP5086ASYGD	LED	gr ø5mm I:90mcd Vf:2V If:20mA	T1¾	Lst		data pdf pinout
LP5086AUBC	LED	bl ø5mm I:190mcd Vf:3.8V If:20mA	T1¾	Lst		data pdf pinout
LP5086AUBT	LED	bl ø5mm I:150mcd Vf:3.8V If:20mA	T1¾	Lst		data pdf pinout
LP5086AUBW	LED	bl ø5mm I:222mcd Vf:3.8V If:20mA	T1¾	Lst		data pdf pinout
LP5086AUGC	LED	yl ø5mm I:700mcd Vf:2.1V If:20mA	T1¾	Lst		data pdf pinout
LP5086AUGD	LED	gr ø5mm I:175mcd Vf:2.1V If:10mA	T1¾	Lst		data pdf pinout
LP5086AURC	LED	rd ø5mm I=1.2cd Vf:1.7V If:20mA	T1¾	Lst		data pdf pinout
LP5086AURD	LED	rd ø5mm I:440mcd Vf:1.7V If:20mA	T1¾	Lst		data pdf pinout
LP5086AURT	LED	rd ø5mm I=1.1cd Vf:1.7V If:20mA	T1¾	Lst		data pdf pinout
LP5086AUSOC	LED	or ø5mm I=4cd Vf:2V If:20mA	T1¾	Lst		data pdf pinout
LP5086AUSRC	LED	rd ø5mm I=3.5cd Vf:2V If:20mA	T1¾	Lst		data pdf pinout
LP5086AUYC	LED	yl ø5mm I:400mcd Vf:2V If:20mA	T1¾	Lst		data pdf pinout
LP5086AUYD	LED	yl ø5mm I:520mcd Vf:2V If:20mA	T1¾	Lst		data pdf pinout
LP5086AUYOC	LED	or ø5mm I=3cd Vf:2V If:20mA	T1¾	Lst		data pdf pinout
LP5086AUYW	LED	yl ø5mm I=2.2cd Vf:2V If:20mA	T1¾	Lst		data pdf pinout
LP5086AVGC	LED	yl ø5mm I:190mcd Vf:2.1V If:20mA	T1¾	Lst		data pdf pinout
LP5086AVGD	LED	gr ø5mm I:65mcd Vf:2.1V If:20mA	T1¾	Lst		data pdf pinout
LP5086AVGT	LED	gr ø5mm I:180mcd Vf:2.1V If:20mA	T1¾	Lst		data pdf pinout
LP5086AVRC	LED	rd ø5mm I=3.8cd Vf:2V If:20mA	T1¾	Lst		data pdf pinout
LP5086AVRD	LED	rd ø5mm I=1.2cd Vf:2V If:20mA	T1¾	Lst		data pdf pinout
LP5086AVYC	LED	yl ø5mm I:180mcd Vf:2V If:20mA	T1¾	Lst		data pdf pinout
LP5086AVYD	LED	yl ø5mm I:50mcd Vf:2V If:20mA	T1¾	Lst		data pdf pinout
LP5086AYD	LED	yl ø5mm I:11mcd Vf:2.1V If:20mA	T1¾	Lst		data pdf pinout
LP5086AYGT	LED	gr ø5mm I:450mcd Vf:2.1V If:20mA	T1¾	Lst		data pdf pinout
LP5086AYT	LED	yl ø5mm I:33mcd Vf:2V If:20mA	T1¾	Lst		data pdf pinout
LP5086BSUBC	LED	bl+yl ø5mm I:190mcd Vf:3.5V If:20mA	T1¾	Lst		data pdf pinout
LP5086BSUGC	LED	gr ø5mm I=10cd Vf:3.6V If:20mA	T1¾	Lst		data pdf pinout
LP5086BSURC	LED	or ø5mm I:>3cd Vf:1.7V If:20mA	T1¾	Lst		data pdf pinout
LP5086BSURT	LED	rd ø5mm I=2cd Vf:2V If:20mA	T1¾	Lst		data pdf pinout
LP5086BUBC	LED	bl ø5mm I:700mcd Vf:3.8V If:20mA	T1¾	Lst		data pdf pinout
LP5086BUBGC	LED	gr+or ø5mm I:9mcd Vf:3.5V If:20mA	T1¾	Lst		data pdf pinout
LP5086BUGC	LED	yl ø5mm I=4cd Vf:2.3V If:20mA	T1¾	Lst		data pdf pinout
LP5086BURC	LED	rd ø5mm I=1.1cd Vf:1.7V If:20mA	T1¾	Lst		data pdf pinout
LP5086BUSO	LED	or ø5mm I=11.78cd Vf:2V If:20mA	T1¾	Lst		data pdf pinout
LP5086BUYC	LED	yl ø5mm I=10.58cd Vf:2V If:20mA	T1¾	Lst		data pdf pinout
LP5086BUYT	LED	yl ø5mm I=2cd Vf:2V If:20mA	T1¾	Lst		data pdf pinout
LP5086BVGT	LED	gr ø5mm I:400mcd Vf:2.1V If:20mA	T1¾	Lst		data pdf pinout
LP 5550SQ	IC	PWI 1.0 compliant Energy Manag. System	LLP	Nsc	-	pdf pinout
LP5900	R+	Io=100mA P:int	4G	Nsc		data pdf pinout
LP 5951MFX-...	LIN-IC	SMD, micro power, 150mA low dropout Cmos voltage reg., range 1.3V...3.3V	45	Nsc	-	pdf pinout
LP 5951MGX-...	LIN-IC	SMD, micro power, 150mA low dropout Cmos voltage reg., range 1.3V...3.3V	45	Nsc		pdf pinout
LPA670	LED	gr ø5mm I:2.8...7.1mcd Vf:2V If:10mA		Osr		data pdf pinout
LPA672	LED	gr ø5mm I:22.4...56mcd Vf:2.6V	PLCC-2	Osr		data pdf pinout
LPA675	LED	gr ø5mm I:28...71mcd Vf:2V If:30mA		Osr		data pdf pinout
LPA676	LED	gr ø5mm I:11.2...28mcd Vf:2V If:20mA		Osr		data pdf pinout
LPC	Z-Di	=1SMB 16CA (Typ-Code/Stempel/marking)	71(5x3,5)	Ons	→1SMB 10CA...78CA	
LPC660	4xCMOS	Vs:16V Vu:1000V/mV Vo:4.99/0V	14SD	Nsc		data pdf pinout
LPC661	CMOS	Vs:16V Vu:1000V/mV Vo:4.99/0V	8SD	Nsc		data pdf pinout
LPC662	2xCMOS	Vs:16V Vu:1000V/mV Vo:4.99/0V	8SD	Nsc		data pdf pinout
LPC 2387 FBD 100	A/D-D/A-IC	Single-chip 16-bit/32-bit MCU; 512kB	16-DIP	Nxp		pdf pinout
LPD 4101	LIN-IC	2x NPN-Trans., 2A, Ucc=5V, 2xAND Input	23a/8Pin	Lbd	-	data
LPD 4104	LIN-IC	2x NPN-Trans., 2A, Ucc=5V, 2x or Input	23a/8Pin	Lbd	-	data
LPD 4106	LIN-IC	2x NPN-Trans., 2A,Ucc=5V, Buffer Input	23a/8Pin	Lbd		data
LPE675	LED	gr ø5mm I:45...112mcd Vf:2V If:50mA	PLCC-4	Osr		data pdf pinout
LPK376	LED	gr ø3.0mm I:56...140mcd Vf:2V If:20mA	T1	Osr		data pdf pinout
LPK380	LED	gr ø3mm I:18...28mlm Vf:2.1V If:15mA	T1	Osr		data pdf pinout
LPK382	LED	gr ø3.0mm I:112...180mcd Vf:2.6V	T1	Osr		data pdf pinout
LPM67K	LED	gr ø3.0mm I:0.9...2.24mcd Vf:1.8V		Osr		data pdf pinout

Type	Device	Short Description	Fig.	Manu	Comparision Types	More at
LPM670	LED	gr ø3.0mm I:2.24...5.6mcd Vf:2V	PLCC-2	Osr		data pdf pinout
LPM675	LED	gr ø3.0mm I:22.4...56mcd Vf:2V		Osr		data pdf pinout
LPM676	LED	gr ø3.0mm I:9...22.4mcd Vf:2V If:20mA		Osr		data pdf pinout
LPM770	LED	gr ø3.0mm I:2.24...5.6mcd Vf:2V		Osr		data pdf pinout
LPPT672	LED	gr ø3.0mm I:50mcd Vf:2V If:30mA	PLCC-4	Osr,Sie		data pdf pinout
LPT67K	LED	gr ø3.0mm I:1.12...2.8mcd Vf:1.8V	PLCC-2	Osr		data pdf pinout
LPT80A	PT	870(950)nm P:>0.4µA/mW/cm²		Sie		data
LPT655	LED	gr ø3.0mm I:71...180mcd Vf:2V If:30mA		Osr		data pdf pinout
LPT670	LED	gr ø3.0mm I:2.8...7.1mcd Vf:2V	PLCC-2	Osr		data pdf pinout
LPT672	LED	gr ø3.0mm I:22.4...56mcd Vf:2.6V		Osr		data pdf pinout
LPT675	LED	gr ø3.0mm I:28...71mcd Vf:2V If:30mA		Osr		data pdf pinout
LPT676	LED	gr ø3.0mm I:11.2...28mcd Vf:2V	PLCC-2	Osr		data pdf pinout
LPT770	LED	gr ø3.0mm I:2.8...7.1mcd Vf:2V	PLCC-2	Osr		data pdf pinout
LPV321	RR-OP	Vs:5.5V Vu:100V/mV Vi0:1.5mV	5S	Nsc		data pdf pinout
LPV324	4xRR-OP	Vs:5.5V Vu:100V/mV Vi0:1.5mV	14S	Nsc		data pdf pinout
LPV358	2xRR-OP	Vs:5.5V Vu:100V/mV Vi0:1.5mV	8S	Nsc		data pdf pinout
LQ	Si-N	=2SD1821A-Q (Typ-Code/Stempel/marking)	35(2mm)	Mat	→2SD1821A	data
LQ	Si-N	=2SD2240A-Q (Typ-Code/Stempel/marking)	35(1,6mm)	Mat	→2SD2240A	data
LQ	Si-N	=2SD814A-Q (Typ-Code/Stempel/marking)	35	Mat	→2SD814A	data
LQ	P-FET	=BSR 176 (Typ-Code/Stempel/marking)	35	Phi,Val	→BSR 176	data
LQ	Si-N	=HD 1F3P (Typ-Code/Stempel/marking)	39	Nec	→HD 1...	data
LQ	Z-Di	=P6 SMB-30 (Typ-Code/Stempel/marking)	71(5x3,5)	Fag	→P6 SMB-30	data
LQ	Z-Di	=SMBJ 17 (Typ-Code/Stempel/marking)	71(5x3,5)	Mop	→SMBJ ...	data
LQ18S3	LED	wht ø3.0mm I:7.1...28mcd Vf:3.3V	0603	Osr		data pdf pinout
LR	Z-Di	=1SMB 17A (Typ-Code/Stempel/marking)	71(5x3,5)	Ons	→1SMB 5.0A...170A	data
LR	Si-N	=2SC2412KLN-R (Typ-Code/Stempel/mark.)	35	Rhm	→2SC2412KLN	data
LR	Si-N	=2SC2412L-LR(Typ-Code/Stempel/marking)	≈35	Rhm	→2SC2412L	data
LR	Si-N	=2SD1821A-R (Typ-Code/Stempel/marking)	35(2mm)	Mat	→2SD1821A	data
LR	Si-N	=2SD2240A-R (Typ-Code/Stempel/marking)	35(1,6mm)	Mat	→2SD2240A	data
LR	Si-N	=2SD814A-R (Typ-Code/Stempel/marking)	35	Mat	→2SD814A	data
LR		=BF 517 (Typ-Code/Stempel/marking)	35	Sie	→BF 517	data
LR	P-FET	=BSR 177 (Typ-Code/Stempel/marking)	35	Phi,Val	→BSR 177	data
LR	Si-N	=HD 1L3N (Typ-Code/Stempel/marking)	39	Nec	→HD 1...	data
LR	Z-Di	=P6 SMB-30A (Typ-Code/Stempel/marking)	71(5x3,5)	Fag	→P6 SMB-30A	data
LR	Z-Di	=SMBJ 17A (Typ-Code/Stempel/marking)	71(5x3,5)	Mop	→SMBJ ...	data
LR 645 LG	Z-IC	SMD, 8...12V ±7%, 3mA, Uin=15...450V	8-MDIP	Stx	-	pdf
LR 645 N3	Z-IC	10V ±7%, 3mA, Uin=15...450V	7d	Stx	-	pdf
LR 645 N5	Z-IC	10V ±7%, 3mA, Uin=15...450V	17b	Stx	-	pdf
LR 645 N8	Z-IC	SMD, 10V ±7%, 3mA, Uin=15...450V	39d	Stx	-	pdf
LR 745 N3	Z-IC	22V, 3mA, Uin=25...450V	7d	Stx	-	pdf
LR 745 N8	Z-IC	SMD, 22V, 3mA, Uin=25...450V	39d	Stx	-	pdf
LR3360	LED	gr+or+srd+yl ø3mm I:7.1...11.2mcd	T1	Osr,Sie		data pdf pinout
LR 4089	CMOS-IC	→KS 5808	16-DIP	Sha	KS5808, MK5089, S25089, SBA5089, UM95089	
LR5360	LED	gr+rd+srd+yl ø5mm I:11.2...18mcd	T1¾	Osr,Sie		data pdf pinout
LR 40981	CMOS-IC	→KS 5804	16-DIP	Sha	KS 5804, MK 50981, TP 50981	
LR 40992	CMOS-IC	Telefon, Impulswahl/dig. Pulse Dialer	18-DIP	Sha	KS 5805A, MK 50992, T 40992	
LR 40993	CMOS-IC	→KS 5805B	18-DIP	Sha	KS 5805B, MK 50993, T 40993	pdf
LRB480	LED	gr+rd+srd+rd/gr+yl 2.5x5mm I>4mcd		Sie		data pinout
LRC	Z-Di	=1SMB 17CA (Typ-Code/Stempel/marking)	71(5x3,5)	Ons	→1SMB 10CA...78CA	
LRJ380	LED	gr+rd+srd+yl ø4mm I>4mcd Vf:2V	T1¾	Sie		data pinout
LRs	Si-N	=BF 517 (Typ-Code/Stempel/marking)	35	Sie	→BF 517	data

LS

Type	Device	Short Description	Fig.	Manu	Comparision Types	More at
LS	Si-P	=2SA1838 (Typ-Code/Stempel/marking)	35(2mm)	Say	→2SA1838	data
LS	Si-P	=2SA1839 (Typ-Code/Stempel/marking)	35	Say	→2SA1839	data
LS	Si-N	=2SC2412KLN-S (Typ-Code/Stempel/mark.)	35	Rhm	→2SC2412KLN	data
LS	Si-N	=2SC2412L-LS(Typ-Code/Stempel/marking)	≈35	Rhm	→2SC2412L	data
LS	Si-N	=2SD1821A-S (Typ-Code/Stempel/marking)	35(2mm)	Mat	→2SD1821A	data
LS	Si-N	=2SD2240A-S (Typ-Code/Stempel/marking)	35(1,6mm)	Mat	→2SD2240A	data
LS	Si-N	=2SD814A-S (Typ-Code/Stempel/marking)	35	Mat	→2SD814A	data
LS	Si-N	=BF 770A (Typ-Code/Stempel/marking)	35	Sie	→BF 770A	data
LS	Si-N	=HD 1A4M (Typ-Code/Stempel/marking)	39	Nec	→HD 1...	data
LS	Z-Di	=P6 SMB-33 (Typ-Code/Stempel/marking)	71(5x3,5)	Fag	→P6 SMB-33	data
LS	Z-Di	=SMBJ 18 (Typ-Code/Stempel/marking)	71(5x3,5)	Mop	→SMBJ ...	data
LS 025(T)	LIN-IC	Telefon, Modulator-demodulator	82	Sgs	-	
LS 045(T)	LIN-IC	Telefon, Kanalverst./channel amplifier	81	Sgs	-	
LS 101 A,B,C	Si-Di	=LL 101A,B,C:	72(3,5mm)	Aeg	BAS 83...81	
LS 101(A)T,TB	OP-IC	Uni, Serie 101, ±22V, -55...+125°	81	Sgs	→Serie 101	data pinout
LS101A	UNI	Vs:±22V Vu:104dB Vo:±13V Vi0:0.7mV	8A	Sgs		data pinout
LS101T	UNI	Vs:±22V Vu:104dB Vo:±13V Vi0:1mV	8A	Sgs		data pinout
LS 103 A,B,C	Si-Di	=LL 103A,B,C:	72(3,5mm)	Aeg	BAS 86...85	
LS107T	UNI	Vs:±22V Vu:104dB Vo:±13V Vi0:0.7mV	8A	Sgs		data pinout
LS 107 T,TB	OP-IC	Uni, Serie 107, ±22V, -55...+125°	81	Sgs	→Serie 107	data pinout
LS 141(A,C)T,TB	OP-IC	Uni, Serie 741	81	Sgs	→Serie 741	data pinout
LS141A	UNI	Vs:±22V Vu:106dB Vo:±13V Vi0:0.8mV	8A	Sgs		data pinout
LS141C	UNI	Vs:±18V Vu:106dB Vo:±13V Vi0:2mV	8DSA	Sgs		data pinout
LS 141 CB	OP-IC	=LS 141(A,C)T: Fig.→	8-DIP	Sgs	→Serie 741	data pinout
LS 141 CM	OP-IC	=LS 141(A,C)T: SMD	8-MDIP	Sgs	→Serie 741	data pinout
LS141T	UNI	Vs:±22V Vu:106dB Vo:±13V Vi0:1mV	8A	Sgs		data pinout
LS147	2xOP	Vs:±18V Vu:106dB Vo:±13V Vi0:1mV	14D	Sgs		data pinout
LS 148(A,C)T,TB	OP-IC	Uni, Serie 748	81	Sgs	→Serie 748	data pinout
LS 148 CB	OP-IC	=LS 148(A,C)T: Fig.→	8-DIP	Sgs	→Serie 748	data pinout

Type	Device	Short Description	Fig.	Manu	Comparision Types	More at
LS 148 CM	OP-IC	=LS 148(A,C)T: SMD	8-MDIP	Sgs	→Serie 748	data pinout
LS148CT	UNI	Vs:±18V Vu:106dB Vo:±13V Vi0:1mV	8A	Sgs		data pinout
LS148T	UNI	Vs:±22V Vu:106dB Vo:±13V Vi0:1mV	8A	Sgs		data pinout
LS 150 C(CD,CB)	LIN-IC	80dB Kompander	14-DIP	Sgs		
LS 156(B)	LIN-IC	Telefon, Sprechkreis/speech Circuit	16-DIP	Sgs	KA 8500A	
LS 159 M	LIN-IC	5x NPN-Transistor-Array, 20V, 50mA	14-MDIP	Sgs	-	
LS 185	LIN-IC	Telefon, Sprechkreis/speech Circuit	14-DIP	Sgs		
LS 188	LIN-IC	Telecom, Mikrof.-vst./micro Amp.	8-DIP	Sgs	-	pdf
LS201	UNI	Vs:±22V Vu:103.5dB Vo:±13V Vi0:2mV	8DSA	Sgs		data pinout
LS 201(A)B,TB	OP-IC	=LS 201(A)T: Fig.→	8-DIP	Sgs	→Serie 101	data pinout
LS 201(A)M	OP-IC	=LS 201(A)T: SMD	8-MDIP	Sgs	→Serie 101	data pinout
LS 201(A)T	OP-IC	Uni, Serie 101, ±22V	81	Sgs	→Serie 101	data pinout
LS201A	UNI	Vs:±22V Vu:104dB Vo:±13V Vi0:0.7mV	8A	Sgs		data pinout
LS204	2xOP	Vs:±18V Vu:100dB Vo:>±13V Vi0:0.5mV	8AD	Sgs		data pinout
LS 204 CB	OP-IC	=LS 204CTB: Fig.→	8-DIP	Sgs	TEB 1033, TF 1033	data-pinout
LS204CD	2xHI-PERF	Vs:±18V Vu:100dB Vo:>±13V Vi0:0.5mV	8S	Sgs		data pdf pinout
LS204CM	2xOP	Vs:±18V Vu:100dB Vo:>±13V Vi0:0.5mV	8S	Sgs		data pinout
LS204CN	2xHI-PERF	Vs:±18V Vu:100dB Vo:>±13V Vi0:0.5mV	8D	Sgs		data pdf pinout
LS204I	2xHI-PERF	Vs:±18V Vu:100dB Vo:>±13V Vi0:0.5mV	8SD	Sgs		data pdf pinout
LS204M	2xOP	Vs:±18V Vu:100dB Vo:>±13V Vi0:0.5mV	8SD	Sgs		data pinout
LS 204(C)M	OP-IC	=LS 204CTB,TB: SMD	8-MDIP	Sgs	-	data pinout
LS 204(A,C)TB	OP-IC	Dual, ±18V	81	Sgs		data pinout
LS207T	UNI	Vs:±22V Vu:104dB Vo:±13V Vi0:0.7mV	8A	Sgs		data pinout
LS 207 T,TB	OP-IC	Uni, Serie 107, ±22V, -25...+85°	81	Sgs	→Serie 107	data pinout
LS 256	LIN-IC	Telefon, Sprechkreis/speech Circuit	16-DIP	Sgs	KA 8501A	pinout
LS 285(A)B	LIN-IC	Telefon, Sprechkreis/speech Circuit	14-DIP	Sgs	KA 2412A, CIC 9185	
LS 285(B)	LIN-IC	Telefon, Sprechkreis/speech Circuit	16-DIP	Sgs	-	
LS 288	LIN-IC	Telefon, Sprechkreis/speech Circuit	16-DIP	Sgs		
LS301	UNI	Vs:±18V Vu:104dB Vo:±13V Vi0:2mV	8DSA	Sgs		data pinout
LS 301(A)B	OP-IC	=LS 301(A)T: Fig.→	8-DIP	Sgs	→Serie 101	data pinout
LS 301(A)M	OP-IC	=LS 301(A)T: SMD	8-MDIP	Sgs	→Serie 101	data pinout
LS 301(A)T,TB	OP-IC	Uni, Serie 101, 0...+85°	81	Sgs	→Serie 101	data pinout
LS307	UNI	Vs:±18V Vu:104dB Vo:±13V Vi0:2mV	8DSA	Sgs		data pinout
LS 307 B	OP-IC	=LS 307 T: Fig.→	8-DIP	Sgs	→Serie 107	data pinout
LS 307 M	OP-IC	=LS 307 T: SMD	8-MDIP	Sgs	→Serie 107	data pinout
LS 307 T,TB	OP-IC	Uni, Serie 107, ±18V, 0...+70°	81	Sgs	→Serie 107	data pinout
LS336K	LED	srd ø3mm I:7.1...18mcd Vf:1.8V If:2mA	T1	Osr		data pdf pinout
LS 342(D)	LIN-IC	Telefon, Interface	8-DIP	Sgs		
LS 346	LIN-IC	Telecom, Verpolschutz/polarity Guard	8-DIP	Sgs		
LS 348	LIN-IC	Telefon, Sprechkreis/speech Circuit	20-DIP	Sgs		
LS 356(B)	LIN-IC	Telefon, Sprechkreis/speech Circuit	16-DIP	Sgs		pdf
LS 388	LIN-IC	Telefon, Sprechkreis/speech Circuit	16-DIP	Sgs		
LS404	4xHI-PERF	Vs:±18V Vu:100dB Vo:>±13V Vi0:0.7mV	14SD	Sgs		data pdf pinout
LS 404B	OP-IC	Quad OP, ±18V	14-DIP	Sgs	TEB 4033, TEF 4033	pdf pinout
LS404C	4xOP	Vs:±18V Vu:100dB Vo:>±10V Vi0:1mV	14DS	Sgs		data pinout
LS 404CB	OP-IC	Quad OP, ±18V	14-DIP	Sgs	TEB 4033, TEF 4033	pdf pinout
LS404CM	OP-IC	SMD, Quad, ±18V	14-MDIP	Sgs	-	data pdf pinout
LS404D1	4xOP	Vs:±18V Vu:100dB Vo:>±10V Vi0:1mV	14S	Sgs		data pinout
LS 404M	OP-IC	SMD, Quad, ±18V	14-MDIP	Sgs	-	data pdf pinout
LS 454	LIN-IC	Recorder, NF-V + Alc	16-MDIP	Sgs		
LS 496	LIN-IC	Telecom, 4x Relais-Tr./Relay Driver	16-DIP	Sgs		
LS 588	LIN-IC	Telefon, Sprechkreis/speech Circuit	16-DIP	Sgs		
LS0603AUBC	LED	bl &1.6x0.8x0.8mm I:28...43mcd	0603	Lst		data pdf pinout
LS0603DRC	LED	rd &1.6x0.8x0.6mm I:14...29mcd Vf:2V	0603	Lst		data pdf pinout
LS0603PBC	LED	gr &1.6x0.8x0.8mm I:2.4...3.8mcd	0603	Lst		data pdf pinout
LS0603PGC	LED	gr &1.6x0.8x0.8mm I:1...2mcd Vf:2.1V	0603	Lst		data pdf pinout
LS0603RC	LED	rd &1.6x0.8x0.6mm I:9.5...17mcd	0603	Lst		data pdf pinout
LS0603SDRC	LED	rd &1.6x0.8x0.8mm I:15...28mcd Vf:2V	0603	Lst		data pdf pinout
LS0603SOC	LED	or &1.6x0.8x0.8mm I:4...6mcd Vf:2V	0603	Lst		data pdf pinout
LS0603SRC	LED	rd &1.6x0.8x0.8mm I:11...20mcd	0603	Lst		data pdf pinout
LS0603SRD	LED	rd &1.6x0.8x0.8mm I:10...20mcd	0603	Lst		data pdf pinout
LS0603SUBC	LED	bl &1.6x0.8x0.6mm I:17...30mcd	0603	Lst		data pdf pinout
LS0603SUGC	LED	gr &1.6x0.8x0.8mm I:87...148mcd	0603	Lst		data pdf pinout
LS0603SURC	LED	rd &1.6x0.8x0.8mm I:51...76mcd Vf:2V	0603	Lst		data pdf pinout
LS0603SYGC	LED	gr &1.6x0.8x0.6mm I:14.5...22mcd	0603	Lst		data pdf pinout
LS0603UBC	LED	bl &1.6x0.8x0.8mm I:26...44mcd	0603	Lst		data pdf pinout
LS0603UBG	LED	gr &1.6x0.8x0.6mm I:33...60mcd	0603	Lst		data pdf pinout
LS0603UBW	LED	bl &1.6x0.8x0.6mm I:12...20mcd	0603	Lst		data pdf pinout
LS0603UGC	LED	gr &1.6x0.8x0.8mm I:7.5...12.5mcd	0603	Lst		data pdf pinout
LS0603URC	LED	rd &1.6x0.8x0.6mm I:60...109mcd Vf:2V	0603	Lst		data pdf pinout
LS0603USOC	LED	or &1.6x0.8x0.8mm I:53...88mcd Vf:2V	0603	Lst		data pdf pinout
LS0603UYC	LED	am &1.6x0.8x0.6mm I:78...118mcd Vf:2V	0603	Lst		data pdf pinout
LS0603UYOC	LED	or &1.6x0.8x0.8mm I:16...45mcd Vf:2V	0603	Lst		data pdf pinout
LS0603VGC	LED	gr &1.6x0.8x0.8mm I:6...11mcd Vf:2.1V	0603	Lst		data pdf pinout
LS0603VRC	LED	rd &1.6x0.8x0.8mm I:4...8mcd Vf:2V	0603	Lst		data pdf pinout
LS0603VYC	LED	am &1.6x0.8x0.8mm I:2.5...4mcd Vf:2V	0603	Lst		data pdf pinout
LS 656(B,AB)	LIN-IC	Telefon, Sprechkreis/speech Circuit	16-DIP	Sgs		pinout
LS 656 D(AD1)	LIN-IC	=LS 656...: SMD	20-MDIP	Sgs	(LS 256)	pinout
LS709A	UNI	Vs:±18V Vu:93dB Vo:±13V Vi0:0.6mV	8A	Sgs		data pinout
LS709C	UNI	Vs:±18V Vu:93dB Vo:±13V Vi0:2mV	14D,8A	Sgs		data pinout
LS 709 CB	OP-IC	=LS 709CTB: Fig.→	14-DIP	Sgs	→Serie 709	data pinout
LS709T	UNI	Vs:±18V Vu:93dB Vo:±13V Vi0:1mV	8A	Sgs		data pinout
LS 709(A,C)TB	OP-IC	Uni, Serie 709, ±18V	81	Sgs	→Serie 709	data pinout
LS776	LO-POWER	Vs:±18V Vu:112dB Vi0:2mV	8DSA	Sgs		data pinout
LS 776(C)T,TB	OP-IC	lo-power, Serie 250, ±18V,	81	Sgs	→Serie 250	data pinout

Type	Device	Short Description	Fig.	Manu	Comparision Types	More at
LS 776 CB	OP-IC	=LS 776(C)T: Fig.→	8-DIP	Sgs	→Serie 250	data pinout
LS 776 CM	OP-IC	=LS 776(C)T: SMD	8-MDIP	Sgs	→Serie 250	data pinout
LS0805PGC	LED	gr &2x1.25x1.1mm I:1...3mcd Vr:2.1V	0805	Lst		data pdf pinout
LS0805SDRC	LED	rd &2x1.25x0.8mm I:10...22mcd Vf:2V	0805	Lst		data pdf pinout
LS0805SOC	LED	or &2x1.25x1.1mm I:3...6mcd Vf:2V	0805	Lst		data pdf pinout
LS0805SRC	LED	rd &2x1.25x1.1mm I:13...21mcd Vf:1.7V	0805	Lst		data pdf pinout
LS0805SURC	LED	rd &2x1.25x1.1mm I:81...138mcd Vf:2V	0805	Lst		data pdf pinout
LS0805SYG	LED	gr &2x1.25x0.8mm I:12...19mcd Vf:2V	0805	Lst		data pdf pinout
LS0805SYGC	LED	gr &2x1.25x0.8mm I:16...22mcd Vf:2V	0805	Lst		data pdf pinout
LS0805UBC	LED	bl &2x1.25x1.1mm I:10...20mcd Vf:4V	0805	Lst		data pdf pinout
LS0805UBGC	LED	gr &2x1.25x0.8mm I:27...51mcd Vf:3.5V	0805	Lst		data pdf pinout
LS0805UGC	LED	bl+gr &2x1.25x0.8mm I:65...103mcd	0805	Lst		data pdf pinout
LS0805USOC	LED	or &2x1.25x1.1mm I:17...47mcd Vf:2V	0805	Lst		data pdf pinout
LS0805UYC	LED	am &2x1.25x0.8mm I:44...74mcd Vf:2V	0805	Lst		data pdf pinout
LS0805UYOC	LED	or &2x1.25x1.1mm I:85...130mcd Vf:2V	0805	Lst		data pdf pinout
LS0805VGC	LED	gr &2x1.25x1.1mm I:6...12mcd Vf:2.1V	0805	Lst		data pdf pinout
LS0805VRC	LED	rd &2x1.25x1.1mm I:5...10mcd Vf:2V	0805	Lst		data pdf pinout
LS0805VYC	LED	am &2x1.25x1.1mm I:5...10mcd Vf:2V	0805	Lst		data pdf pinout
LS830	UNI	Vs:±18V Vu:104dB Vo:±13V Vi0:2mV	8S	Sgs		data pinout
LS0907PGC	LED	gr 3.2x2.4x2.6mm I:70...122mcd	0907	Lst		data pdf pinout
LS0907SUGC	LED	gr 3.2x2.4x2.6mm I:600...1100mcd	0907	Lst		data pdf pinout
LS0907SYGC	LED	gr 3.2x2.4x2.6mm I:228...380mcd Vf:2V	0907	Lst		data pdf pinout
LS0907UBC	LED	bl 3.2x2.4x2.6mm I:118...160mcd	0907	Lst		data pdf pinout
LS0907UBGC	LED	gr 3.2x2.4x2.6mm I:350...675mcd	0907	Lst		data pdf pinout
LS0907UYC	LED	rd+am 3.2x2.4x2.6mm I:171...460mcd	0907	Lst		data pdf pinout
LS0907VGC	LED	gr 3.2x2.4x2.6mm I:70...135mcd	0907	Lst		data pdf pinout
LS1008DRC	LED	rd ø1.9mm I:198...297mcd Vf:2V		Lst		data pdf pinout
LS1008GC	LED	gr ø1.9mm I:14...23mcd Vf:2V If:20mA		Lst		data pdf pinout
LS1008SRC	LED	rd ø1.9mm I:70...120mcd Vf:1.7V		Lst		data pdf pinout
LS1008SUGC	LED	gr ø1.9mm I:835...1560mcd Vf:3.5V		Lst		data pdf pinout
LS1008SYGC	LED	gr ø1.9mm I:132...220mcd Vf:2V		Lst		data pdf pinout
LS1008UBC	LED	bl ø1.9mm I:213...300mcd Vf:3.5V		Lst		data pdf pinout
LS1008UBGC	LED	gr ø1.9mm I:407...781mcd Vf:3.5V		Lst		data pdf pinout
LS1008URC	LED	rd ø1.9mm I:800...1370mcd Vf:2V		Lst		data pdf pinout
LS1008UYC	LED	am ø1.9mm I:198...496mcd Vf:2V		Lst		data pdf pinout
LS1008UYOC	LED	or ø1.9mm I:790...1280mcd Vf:2V		Lst		data pdf pinout
LS1008VGC	LED	gr ø1.9mm I:30...50mcd Vf:2.1V		Lst		data pdf pinout
LS1008VRC	LED	rd ø1.9mm I:48...88mcd Vf:2V If:10mA		Lst		data pdf pinout
LS1008VYC	LED	am ø1.9mm I:18...33mcd Vf:2V If:20mA		Lst		data pdf pinout
LS1206PGC	LED	gr &3x1.5x1.5mm I:2...4mcd Vf:2.1V	1206	Lst		data pdf pinout
LS1206PGD	LED	gr &3x1.5x1.5mm I:3...4.5mcd Vf:2.1V	1206	Lst		data pdf pinout
LS1206SDRC	LED	rd &3x1.5x1.5mm I:28...38mcd Vf:2V	1206	Lst		data pdf pinout
LS1206SGC	LED	gr &3x1.5x1.5mm I:4...6mcd Vf:2V	1206	Lst		data pdf pinout
LS1206SOC	LED	or &3x1.5x1.5mm I:9...17mcd Vf:2V	1206	Lst		data pdf pinout
LS1206SUBC	LED	bl &3x1.5x1.5mm I:57...85mcd Vf:3.5V	1206	Lst		data pdf pinout
LS1206SURC	LED	rd &3x1.5x1.5mm I:135...290mcd Vf:2V	1206	Lst		data pdf pinout
LS1206SUYC	LED	am &3x1.5x1.5mm I:68...118mcd Vf:2V	1206	Lst		data pdf pinout
LS1206SVYC	LED	am &3x1.5x1.5mm I:10...20mcd Vf:2V	1206	Lst		data pdf pinout
LS1206SYGC	LED	gr &3x1.5x1.5mm I:57...90mcd Vf:2V	1206	Lst		data pdf pinout
LS1206UBC	LED	bl &3x1.5x1.5mm I:80...145mcd Vf:3.5V	1206	Lst		data pdf pinout
LS1206UYOC	LED	or &3x1.5x1.5mm I:29...80mcd Vf:2V	1206	Lst		data pdf pinout
LS1206VGC	LED	gr &3x1.5x1.5mm I:21...38mcd Vf:2.1V	1206	Lst		data pdf pinout
LS1206VRC	LED	rd &3x1.5x1.5mm I:4...10mcd Vf:2V	1206	Lst		data pdf pinout
LS 1240	LIN-IC	Telecom, Telefonklingel/2 Tone Ringer	8-DIP	Sgs	PSB 6520	pdf pinout
LS 1240 A	LIN-IC	=LS 1240: f. Low Impedance Transducer	8-DIP	Sgs	-	pdf pinout
LS 1241	LIN-IC	Telecom, Telefonklingel/2 Tone Ringer	8-DIP	Sgs	PSB 6521	pdf pinout
LS1265SUGC	LED	gr &3.2x1.5x1.1mm I:65...120mcd		Lst		data pdf pinout
LS1265UBC	LED	bl &3.2x1.5x1.1mm I:14...20mcd		Lst		data pdf pinout
LS1265UGC	LED	gr &3.2x1.5x1.1mm I:6...120mcd		Lst		data pdf pinout
LS1306PGC	LED	gr &3,2x1.6x1.1mm I:1.5...3mcd	1306	Lst		data pdf pinout
LS1306SOC	LED	or &3,2x1.6x1.1mm I:3...6mcd Vf:2V	1306	Lst		data pdf pinout
LS1306SRC	LED	rd &3,2x1.6x1.1mm I:14...24mcd	1306	Lst		data pdf pinout
LS1306UBC	LED	bl &3,2x1.6x1.1mm I:19...27mcd	1306	Lst		data pdf pinout
LS1306UGC	LED	gr &3,2x1.6x1.1mm I:8...17mcd Vf:2.1V	1306	Lst		data pdf pinout
LS1306USOC	LED	or &3,2x1.6x1.1mm I:1.6...45mcd Vf:2V	1306	Lst		data pdf pinout
LS1306UYC	LED	am &3,2x1.6x1.1mm I:16...45mcd Vf:2V	1306	Lst		data pdf pinout
LS1306UYOC	LED	or &3,2x1.6x1.1mm I:1.6...45mcd Vf:2V	1306	Lst		data pdf pinout
LS1306VGC	LED	gr &3,2x1.6x1.1mm I:8...13mcd Vf:2.1V	1306	Lst		data pdf pinout
LS1306VRC	LED	rd &3,2x1.6x1.1mm I:5.5...9mcd Vf:2V	1306	Lst		data pdf pinout
LS1306VYC	LED	am &3,2x1.6x1.1mm I:3.5...6mcd Vf:2V	1306	Lst		data pdf pinout
LS2138SRC	LED	rd ø1.9mm I:60...110mcd Vf:1.7V		Lst		data pdf pinout
LS2138SYGC	LED	gr ø1.9mm I:132...220mcd Vf:2V		Lst		data pdf pinout
LS2138UBC	LED	bl ø1.9mm I:210...300mcd Vf:3.5V		Lst		data pdf pinout
LS2138URC	LED	rd ø1.9mm I:400...600mcd Vf:2V		Lst		data pdf pinout
LS2138UYC	LED	am ø1.9mm I:190...590mcd Vf:2V		Lst		data pdf pinout
LS2138UYOC	LED	or ø1.9mm I:200...550mcd Vf:2V		Lst		data pdf pinout
LS2138VGC	LED	gr ø1.9mm I:30...55mcd Vf:2.1V		Lst		data pdf pinout
LS2138VRC	LED	rd ø1.9mm I:48...88mcd Vf:2V If:20mA		Lst		data pdf pinout
LS2138VYC	LED	am ø1.9mm I:18...33mcd Vf:2V If:20mA		Lst		data pdf pinout
LS 2805 D	IC	Radiation Hardened DC/DC Converter	42	Inr	-	pdf pinout
LS 2805 S	IC	Radiation Hardened DC/DC Converter	42	Inr	-	pdf pinout
LS 2812 D	IC	Radiation Hardened DC/DC Converter	42	Inr	-	pdf pinout
LS 2812 S	IC	Radiation Hardened DC/DC Converter	42	Inr	-	pdf pinout
LS 2815 D	IC	Radiation Hardened DC/DC Converter	42	Inr	-	pdf pinout
LS 2815 S	IC	Radiation Hardened DC/DC Converter	42	Inr	-	pdf pinout

Type	Device	Short Description	Fig.	Manu	Comparision Types	More at
LS3226SURSUBSUG	LED	rd/gr/bl ø1.9mm I:32...60mcd Vf:3.5V		Lst		data pdf pinout
LS3226SURSUBUBC	LED	rd/gr/bl ø1.9mm I:20...45mcd Vf:3.5V		Lst		data pdf pinout
LS3226SURUGUBC	LED	rd/gr/bl ø1.9mm I:45...80mcd Vf:2.1V		Lst		data pdf pinout
LS3226UYOSUGSUB	LED	or/gr/bl ø1.9mm I:73...100mcd Vf:2V		Lst		data pdf pinout
LS3228SURC	LED	rd ø1.9mm I:285...430mcd Vf:2V	PLCC-4	Lst		data pdf pinout
LS 3240	LIN-IC	Telecom, Telefonklingel/2 Tone Ringer	8-DIP	Sgs		
LS3336	LED	srd ø3.0mm I:280...710mcd Vf:1.8V	T1	Osr		data pdf pinout
LS3340	LED	or+srd+yl ø3mm I:28...45mcd Vf:2V	T1	Osr		data pdf pinout
LS3341	LED	gr+srd+yl ø3mm I:45...71mcd Vf:2V	T1	Osr		data pdf pinout
LS3360	LED	srd ø3mm I:11.2...18mcd Vf:2V If:10mA	T1	Osr		data pdf pinout
LS3366	LED	srd ø3mm I:112...280mcd Vf:2V If:20mA	T1	Osr		data pdf pinout
LS3369	LED	gr+srd+yl ø3mm I:1.12...7.1mcd Vf:2V	T1	Osr		data pdf pinout
LS3380	LED	gr+srd+rd/gr+yl ø3mm I:7.1...11.2mcd	T1	Osr,Sie		data pdf pinout
LS3386	LED	srd ø3.0mm I:71...180mcd Vf:2V	T1	Osr		data pdf pinout
LS 4148	Si-Di	=1N4148: SMD	72a(3,4mm)	Aeg	BAS 32	data
LS 4150	Si-Di	=1N4150: SMD	72a(3,4mm)	Aeg	BAS 32	data
LS 4151	Si-Di	=1N4151: SMD	72a(3,4mm)	Aeg	BAS 32	data
LS 4154	Si-Di	=1N4154: SMD	72a(3,4mm)	Aeg	BAS 32	data
LS 4448	Si-Di	=1N4448: SMD	72a(3,4mm)	Aeg	BAS 32	data
LS4558N	2xOP	Vs:±18V Vu:100dB Vo:±13V Vio:0.5mV	8DS	Sgs		data pinout
LS 4558 NB	OP-IC	Dual, Single/Split Supply, Serie 158	8-DIP	Sgs	→Serie 158	data pdf pinout
LS 4558 NM	OP-IC	SMD, Dual, Single/Split Supply	8-MDIP	Sgs	→Serie 158	data pdf pinout
LS 5018(B)	LIN-IC	Telecom, Overvolt. Prot., bidirekt.	8-DIP	Sgs	-	
LS 5060(B)	LIN-IC	Telecom, Overvolt. Prot., bidirekt.	8-DIP	Sgs	-	
LS 5120(B,B1)	LIN-IC	Telecom, Overvolt. Prot., bidirekt.	8-DIP	Sgs	-	
LS5380	LED	gr+srd+yl ø5mm I>4mcd Vf:2V If:10mA	T1¾	Osr,Sie		data pdf pinout
LS5420	LED	gr+srd+yl ø5mm I:112...180mcd Vf:2V	T1¾	Osr		data pdf pinout
LS5421	LED	gr+srd+yl ø5mm I:180...280mcd Vf:2V	T1¾	Osr,Sie		data pdf pinout
LS5436	LED	srd ø5.0mm I:180...1120mcd Vf:2V	T1¾	Osr		data pdf pinout
LS5460	LED	gr+rd+srd+yl ø5mm I:11.2...18mcd	T1¾	Osr,Sie		data pdf pinout
LS5469	LED	gr+srd+yl ø5mm I:1.12...7.1mcd Vf:2V	T1¾	Osr		data pdf pinout
LS5480	LED	gr+rd+srd+yl ø5mm I:11.2...18mcd	T1¾	Osr,Sie		data pdf pinout
LS 7083	CMOS-IC	Quadrature Clock Converter	DIP	Lsi	-	pdf pinout
LS 7084	CMOS-IC	Quadrature Clock Converter	DIP	Lsi	-	pdf pinout
LS 7220	IC	Keyless Lock Circuit	DIP	Lsi	-	pdf pinout
LS 7232 ND	MOS	Touch Control Lamp Dimmer	DIP	Lsi	-	pdf pinout
LS 7232 ND-S	MOS	Touch Control Lamp Dimmer	DIP	Lsi	-	pdf pinout
LS 8025 M	LIN-IC	Telefon, Modulator-demodulator	14-MDIP	Sgs	-	
LS 8045 M	LIN-IC	Telefon, Kanalverst./channel amplifier	8-MDIP	Sgs	-	
LS8101	UNI	Vs:±22V Vu:104dB Vo:±13V Vi0:1mV	8S	Sgs		data pinout
LS 8101 (A)M	OP-IC	=LS 101(A)T: hi-rel, SMD	8-MDIP	Sgs	→Serie 101	data pinout
LS8101A	UNI	Vs:±22V Vu:104dB Vo:±13V Vi0:0.7mV	8S	Sgs		data pinout
LS8107M	UNI	Vs:±22V Vu:104dB Vo:±13V Vi0:0.7mV	8S	Sgs		data pinout
LS 8107 M	OP-IC	=LS 107 T: hi-rel, SMD	8-MDIP	Sgs	→Serie 107	data pinout
LS 8141 (A,C)M	OP-IC	=LS 141(A,C)T: hi-rel, SMD	8-MDIP	Sgs	→Serie 741	data pinout
LS8141A	UNI	Vs:±22V Vu:106dB Vo:±13V Vi0:0.8mV	8S	Sgs		data pinout
LS8141C	UNI	Vs:±18V Vu:106dB Vo:±13V Vi0:2mV	8S	Sgs		data pinout
LS8141M	UNI	Vs:±22V Vu:106dB Vo:±13V Vi0:1mV	8S	Sgs		data pinout
LS 8148 (A,C)M	OP-IC	=LS 148(A,C)T: hi-rel, SMD	8-MDIP	Sgs	→Serie 748	data
LS 8159 M	LIN-IC	=LS 159 M: hi-rel, SMD	14-MDIP	Sgs	-	
LS8201	UNI	Vs:±22V Vu:103.5dB Vo:±13V Vi0:2mV	8S	Sgs		data pinout
LS 8201 (A)M	OP-IC	=LS 201(A)T: hi-rel, SMD	8-MDIP	Sgs	→Serie 101	data pinout
LS8201A	UNI	Vs:±22V Vu:104dB Vo:±13V Vi0:0.7mV	8S	Sgs		data pinout
LS 8204 (A,C)M	LIN-IC	=LS 204(A,C)T: hi-rel, SMD	8-MDIP	Sgs	-	data
LS 8204 A,C	LIN-IC	=LS 204(A,C)T: hi-rel	81	Sgs	-	data
LS8207	UNI	Vs:±22V Vu:104dB Vo:±13V Vi0:0.7mV	8S	Sgs		data pinout
LS 8207 M	OP-IC	=LS 207 T: hi-rel, SMD	8-MDIP	Sgs	→Serie 107	data pinout
LS 8301 (A)M	OP-IC	=LS 301(A)T: hi-rel, SMD	8-MDIP	Sgs	→Serie 101	data pinout
LS 8307 M	OP-IC	=LS 307 T: hi-rel, SMD	8-MDIP	Sgs	→Serie 107	data pinout
LS8404	4xHI-PERF	Vs:±18V Vu:100dB Vo:>±10V Vi0:1mV	14S	Sgs		data pinout
LS 8404 CM	OP-IC	SMD, Quad, ±18V, hi-rel	14-MDIP	Sgs	-	pdf pinout
LS 8404 M	OP-IC	SMD, Quad, ±18V, hi-rel	14-MDIP	Sgs	-	pdf pinout
LS 8776 (C)M	OP-IC	=LS 776(C)T: hi-rel, SMD	8-MDIP	Sgs	→Serie 250	data
LS12065PGC	LED	rd &3.2x1.5x1.1mm I:0.63...1mcd Vf:2V		Lst		data pdf pinout
LS12065SDRC	LED	rd &3.2x1.5x1.1mm I:1.25...2mcd Vf:2V		Lst		data pdf pinout
LS12065SRC	LED	rd &3.2x1.5x1.1mm I:6...15mcd Vf:1.7V		Lst		data pdf pinout
LS12065SURC	LED	rd &3.2x1.5x1.1mm I:72...123mcd Vf:2V		Lst		data pdf pinout
LS12065SYGC	LED	gr &3.2x1.5x1.1mm I:15...23mcd Vf:2V		Lst		data pdf pinout
LS12065UGC	LED	gr &3.2x1.5x1.1mm I:7.5...15mcd		Lst		data pdf pinout
LS12065UYC	LED	am &3.2x1.5x1.1mm I:57...89mcd Vf:2V		Lst		data pdf pinout
LS12065UYOC	LED	or &3.2x1.5x1.1mm I:15...43mcd Vf:2V		Lst		data pdf pinout
LS12065VRC	LED	rd &3.2x1.5x1.1mm I:4...10mcd Vf:2V		Lst		data pdf pinout
LS12065VYC	LED	am &3.2x1.5x1.1mm I:4...10mcd Vf:2V		Lst		data pdf pinout
LS13064UBC	LED	bl &3.2x1.6x1.1mm I:15...18mcd		Lst		data pdf pinout
LS 2801 R 5 S	IC	Radiation Hardened DC/DC Converter	42	Inr	-	pdf pinout
LS 2802 R 5 S	IC	Radiation Hardened DC/DC Converter	42	Inr	-	pdf pinout
LS 2803 R 5 S	IC	Radiation Hardened DC/DC Converter	42	Inr	-	pdf pinout
LSA67B	LED	srd ø5mm I:224...560mcd Vf:1.8V		Osr		data pdf pinout
LSA67K	LED	srd ø5mm I:4.5...11.2mcd Vf:1.8V		Osr		data pdf pinout
LSA670	LED	gr+or+srd+yl ø5mm I>6.3mcd Vf:2V		Sie		data pinout
LSA672	LED	gr+or+srd+yl ø5mm I>40mcd Vf:2.2V		Sie		data pinout
LSA674	LED	rd ø5mm I>16mcd Vf:1.75V If:10mA		Sie		data pinout
LSA676	LED	or+srd+yl ø5mm I>160mcd Vf:2V		Sie		data pinout
LSA679	LED	gr+srd+yl ø5mm I:1mcd Vf:2V If:2mA		Sie		data pinout
LSBT676	LED	rd/bl ø5mm I:28...71mcd Vf:3.5V	PLCC-4	Osr		data pdf pinout

Type	Device	Short Description	Fig.	Manu	Comparision Types	More at
LSC870	LED	gr+or+srd+yl ⌀5mm I>6.3mcd Vf:2V		Sie		data pinout
LSE 12 A	Si-N	≈BUW 12A	18j§	Mot	→BUW 12A	data pinout
LSE63B	LED	srd ⌀5mm I:3200mcd Vf:2.2V	PLCC-4	Osr		data pdf pinout
LSE67B	LED	srd ⌀2.55mm I:355...900mcd Vf:2.2V	PLCC-4	Osr		data pdf pinout
LSE655	LED	srd ⌀2.55mm I:710...1800mcd Vf:2.1V	PLCC-4	Osr		data pdf pinout
LSE675	LED	srd ⌀2.55mm I:280...710mcd Vf:2.1V	PLCC-4	Osr		data pdf pinout
LSF3128UWC	LED	wht ⌀2.55mm I:85...122mcd Vf:3.5V	PLCC-2	Lst		data pdf pinout
LSF3427SURSYG	LED	rd/gr ⌀2.55mm I:11...19mcd Vf:2V		Lst		data pdf pinout
LSG3331	LED	rd/gr ⌀3mm I>4mcd Vf:2V If:10mA	T1	Sie		data pdf pinout
LSG3351	LED	rd/gr ⌀3mm I>2.5mcd Vf:2V If:10mA	T1	Sie		data pinout
LSGA671	LED	rd/gr ⌀3mm I:7.1...28mcd Vf:2V		Osr		data pinout
LSGK370	LED	or/gr+rd/gr ⌀3mm I:>32mlm Vf:2V	T1	Sie		data pdf pinout
LSGK372	LED	rd/gr ⌀3mm I:>100mlm Vf:2.6V If:50mA	T1	Sie		data pinout
LSGT670	LED	or/gr+rd/gr+rd/yl+yl/gr ⌀3mm Vf:2V	PLCC-4	Sie		data pdf pinout
LSGT671	LED	rd/gr ⌀3mm I:7.1...28mcd Vf:2V	PLCC-4	Osr		data pdf pinout
LSGT677	LED	rd/gr ⌀3mm I:4.5...18mcd Vf:2V	PLCC-4	Osr		data pdf pinout
LSGT770	LED	rd/gr ⌀3mm I:4.5...18mcd Vf:2V	PLCC-4	Osr		data pdf pinout
LSK376	LED	srd ⌀3.0mm I:280...710mcd Vf:1.8V	T1	Osr		data pdf pinout
LSK380	LED	gr+or+srd+yl ⌀3mm I:45...71mlm	T1	Osr,Sie		data pdf pinout
LSK382	LED	gr+or+srd+yl ⌀3mm I:160mlm Vf:2.4V	T1	Sie		data pdf pinout
LSK389	LED	gr+srd+yl ⌀3mm I:3.2mlm Vf:2V If:2mA	T1	Sie		data pinout
LSL89K	LED	srd ⌀3mm I:2.8...7.1mcd Vf:1.7V		Osr		data pinout
LSL896	LED	srd ⌀3mm I:56...112mcd Vf:2V If:50mA	PLCC-4	Osr		data pdf pinout
LSM67K	LED	srd ⌀3mm I:3.55...9mcd Vf:1.8V If:2mA		Osr		data pdf pinout
LSM670	LED	gr+or+srd+yl ⌀3mm I>6.3mcd Vf:2V		Sie		data pinout
LSM676	LED	or+srd+yl ⌀3mm I>160mcd Vf:2V	PLCC-2	Sie		data pdf pinout
LSM770	LED	gr+or+srd+yl ⌀3mm I>4mcd Vf:2V	PLCC-2	Sie		data pinout
LSM776	LED	srd ⌀3mm I:280mlm Vf:2V If:20mA		Sie		data pinout
LSM779	LED	srd ⌀3mm I:1.12...2.8mcd Vf:1.8V		Osr		data pdf pinout
LSPBT670	LED	rd/gr/bl ⌀3mm I:>4mW/Sr Vf:2/3.1V	PLCC-4	Sie		data pdf pinout
LSPT670	LED	rd/gr ⌀3mm I<11.2/7.1mcd Vf:2V	PLCC-4	Osr		data pdf pinout
LSQ976	LED	srd ⌀3mm I:50mcd Vf:2V If:20mA	0603	Osr		data pdf pinout
LSR976	LED	srd ⌀3mm I:50mcd Vf:2V If:20mA	0805	Osr		data pdf pinout
LSRA5SURC	LED	rd &2.1x0.6x1mm I:90...154mcd Vf:2V		Lst		data pdf pinout
LSRA5SYGC	LED	gr &2.1x0.6x1mm I:28...33mcd Vf:2V		Lst		data pdf pinout
LSRA5UBC	LED	bl &2.1x0.6x1mm I:17...26mcd Vf:3.5V		Lst		data pdf pinout
LSRA5UGC	LED	gr &2.1x0.6x1mm I:63...110mcd Vf:3.5V		Lst		data pdf pinout
LSRA5USOC	LED	or &2.1x0.6x1mm I:20...54mcd Vf:2V		Lst		data pdf pinout
LSRA5UYC	LED	am &2.1x0.6x1mm I:20...54mcd Vf:2V		Lst		data pdf pinout
LSRA5UYOC	LED	or &2.1x0.6x1mm I:20...49mcd Vf:2V		Lst		data pdf pinout
LSs	Si-N	=BF 770A (Typ-Code/Stempel/marking)	35	Sie	→BF 770A	data pinout
LSS260	LED	gr+srd+yl &2.1x0.6x1mm I>0.4mcd	SOT23	Sie		data
LSS269	LED	gr+srd+yl &2.1x0.6x1mm I>0.16mcd	SOT23	Sie		data pinout
LSST670	LED	gr+or+srd+yl &2.1x0.6x1mm I>8mcd	PLCC-4	Sie		data pinout
LSST672	LED	gr+or/gr+or/yl+rd/gr+srd+yl/gr &2.1x0.	PLCC-4	Sie		data pinout
LST67B	LED	srd ⌀1.9mm I:224...560mcd Vf:2.1V		Osr		data
LST67K	LED	srd ⌀1.9mm I:4.5...11.2mcd Vf:1.8V		Osr		data pdf pinout
LST670	LED	gr+or+srd+yl ⌀1.9mm I>4mcd Vf:2V	PLCC-2	Sie		data pdf pinout
LST672	LED	gr+or+srd+yl ⌀1.9mm I>40mcd Vf:2.2V	PLCC-2	Sie		data pdf pinout
LST675	LED	srd ⌀1.9mm I:140...355mcd Vf:2.1V	PLCC-2	Osr		data pdf pinout
LST676	LED	or+srd+yl ⌀1.9mm I>160mcd Vf:2V	PLCC-2	Osr		data pdf pinout
LST679	LED	gr+srd+yl ⌀1.9mm I>1mcd Vf:2V If:2mA	PLCC-2	Sie		data pdf pinout
LST770	LED	gr+or+srd+yl ⌀1.9mm I>6.3mcd Vf:2V	PLCC-2	Sie		data pdf pinout
LST776	LED	or+srd+yl ⌀1.9mm I>160mcd Vf:2V	PLCC-2	Sie		data pinout
LST2138UYC	LED	am ⌀1.9mm I:198...496mcd Vf:2V		Lst		data pinout
LSU260	LED	gr+srd+yl ⌀1mm I>0.63mcd Vf:2V		Sie		data pdf pinout
LSY876	LED	srd ⌀1mm I:90...224mcd Vf:2V If:20mA		Osr		data pdf pinout
LSYA676	LED	rd/yl ⌀1mm I>25mcd Vf:2V If:20mA		Sie		data pdf pinout
LSYT67B	LED	rd/yl ⌀1mm I:180...450mcd Vf:2.1V	PLCC-4	Osr		data pdf pinout
LSYT670	LED	rd/yl ⌀1mm I<18mcd Vf:2V If:10mA	PLCC-4	Osr		data pdf pinout
LSYT675	LED	rd/yl ⌀1mm I:180...450mcd Vf:2V	PLCC-4	Osr		data pdf pinout
LSYT676	LED	rd/yl ⌀1mm I:56...140mcd Vf:2V	PLCC-4	Osr		data pdf pinout

LT...LZ

Type	Device	Short Description	Fig.	Manu	Comparision Types	More at
LT	Z-Di	=1SMB 18A (Typ-Code/Stempel/marking)	71(5x3,5)	Ons	→1SMB 5.0A...170A	
LT	Si-N	=2SD1821A-T (Typ-Code/Stempel/marking)	35(2mm)	Mat	→2SD1821A	
LT	Si-N	=2SD2028 (Typ-Code/Stempel/marking)	35	Say	→2SD2028	data
LT	Si-N	=2SD2240A-T (Typ-Code/Stempel/marking)	35(1,6mm)	Mat	→2SD2240A	data
LT	Si-N	=2SD814A-T (Typ-Code/Stempel/marking)	35	Mat	→2SD814A	data
LT	Si-N	=HD 1L2Q (Typ-Code/Stempel/marking)	39	Nec	→HD 1...	data
LT	Z-Di	=P6 SMB-33A (Typ-Code/Stempel/marking)	71(5x3,5)	Fag	→P6 SMB-33A	data
LT	Z-Di	=SMBJ 18A (Typ-Code/Stempel/marking)	71(5x3,5)	Mop	→SMBJ ...	data
LT1X11A	LED	srd+gr+or ⌀2mm I:12mcd Vf:1.75V		Sha		data pdf pinout
LT1X21A	LED	srd+gr+yl+or ⌀2mm I:12mcd Vf:1.75V		Sha		data pdf pinout
LT1X51A	LED	srd+gr+yl+or 1.5x2.9mm I:9mcd		Sha		data pdf pinout
LT1X52A	LED	rd 1.5x2.9mm I:9mcd Vf:1.75V If:20mA	SOT143	Sha		data pinout
LT1X53A	LED	srd+gr+yl+or 1.5x2.9mm I:7.8mcd	SOT143	Sha		data pinout
LT1X73A	LED	srd+gr+yl+or 1.45x1.45mm I:7mcd		Sha		data pinout
LT1X82A	LED	srd+gr+yl+or 1.5x2.9mm I:54mcd		Sha		data pinout
LT1X83A	LED	srd+gr+or 1.5x2.9mm I:11.7mcd Vf:2V		Sha		data pdf pinout
LT1X92A	LED	srd+gr 1.8x3mm I:32mcd Vf:1.95V		Sha		data pinout
LT1XX53A	LED	gr/rd+yl/rd 1.5x2.9mm I:1.1mcd		Sha		data pinout
LT1XX82A	LED	gr/rd+gr/or 1.5x2.9mm I:10.5mcd		Sha		data pinout

Type	Device	Short Description	Fig.	Manu	Comparision Types	More at
		gr+rd 125x9.6mm I:1500lxmcd Vf:12V		Sha		data pinout
LT46XXX	LED					data pdf pinout
LT101	2xCOMP	Vs:40V Vi0:0.4mV	8AD,16SS	Ltc		data pdf pinout
LT111A	COMP	Vs:36V Vu:500V/mV Vi0:0.5mV	8AD	Ltc		data pdf pinout
LT117A	ER+	Io=1.5A Vo:1.2...15V Vin:3...40V P:int	3AO	Ltc		data pdf pinout
LT117H	ER+	Io=1.5A Vo:1.2...15V Vin:3...40V P:int	3AO	Ltc		data pdf pinout
LT118A	HI-SPEED	Vs:±20V Vu:500V/mV Vo:±13V Vi0:0.5mV	8AD	Ltc		data pdf pinout
LT123	R+	Io=3A Vo:4.6...5.4V Vin:7.5...15V P:int	3O	Ltc		data pdf pinout
LT137	ER-	Io=1.5A Vo:-1.2...-37V P:int	3OA	Ltc		data pdf pinout
LT137A	ER-	Io=1.5A Vo:-1.2...-37V P:int	3A	Ltc		data pinout
LT137AHV	ER-	Io=1.5A Vo:-1.2...-47V P:int	3AO	Ltc		data pinout
LT138	ER+	Io=5A Vo:1.2...32V P:int	3O	Ltc		data pinout
LT150	ER+	Io=3A Vo:1.2...32V P:int	3O	Ltc		data pdf pinout
LT162	RR-OP	Vs:7V Vu:80dB Vi0:±5mV	16S,8S	Ltc		data pdf pinout
LT311A	COMP	Vs:±18V Vu:500V/mV Vi0:0.5mV	8AD	Ltc		data pdf pinout
LT317	ER+	Io=1.5A Vo:1.2...37V Vin:3...40V P:int	3AOP	Ltc		data pdf pinout
LT317A	ER+	Io=1.5A Vo:1.2...37V Vin:3...40V P:int	3AOP	Ltc		data pdf pinout
LT318A	HI-SPEED	Vs:±20V Vu:500V/mV Vo:±13V Vi0:0.5mV	8AD	Ltc		data pdf pinout
LT323	R+	Io=3A Vo:4.75...5.25V Vin:7.5...15V	3OP	Ltc		data pdf pinout
LT323A	R+	Io=3A Vo:4.85...5.15V Vin:7.5...15V	3OP	Ltc		data pdf pinout
LT337	ER-	Io=1.5A Vo:-1.2...-37V P:int	3AOP	Ltc		data pinout
LT337AHV	EH-	Io=1.5A Vo:-1.2...-47V P:int	3AO	Ltc		data pinout
LT338	ER+	Io=5A Vo:1.2...32V P:int	3O	Ltc		data pinout
LT350	ER+	Io=3A Vo:1.2...32V P:int	3OP	Ltc		data pdf pinout
LT580	R+	Vo:2.49...2.51V	3A	Ltc		data pdf pinout
LT580J	R+	Vo:2.425...2.575V	3A	Ltc		data pdf pinout
LT580K	R+	Vo:2.475...2.525V	3A	Ltc		data pdf pinout
LT580S	R+	Vo:2.475...2.525V	3A	Ltc		data pdf pinout
LT581J	R+	Vo:9.97...10.03V	3A	Ltc		data pdf pinout
LT581K	R+	Vo:9.99...10.01V	3A	Ltc		data pdf pinout
LT581L	R+	Vo:9.995...10.00V	3A	Ltc		data pdf pinout
LT685	HI-SPEED	Vs:±7V Vi0:1mV	10A,16D	Ltc		data pinout
LT955XCU	LED	gr/rd ø7.4mm I:250mcd Vf:1.8V	T1¾	Sha		data pdf pinout
LT1001	OC	LED/LDR Viso:3000V Ibs:20mA		Hei		data
LT 1001 A	Si-N	VHF/UHF-A, 40V, 0,2A	2a§	Mot	MRF 586[Motorola]	
LT1001A	HI-PREC	Vs:±22V Vu:800V/mV Vo:±14V Vi0:10μμV	8ADS,20C	Ltc		data pdf pinout
LT1001ACN8	HI-PREC	Vs:±22V Vu:800V/mV Vo:±14V Vi0:10μμV	8D	Max		data pinout
LT1001ACS8	HI-PREC	Vs:±22V Vu:800V/mV Vo:±14V Vi0:10μμV	8S	Max		data pinout
LT1001C	HI-PREC	Vs:±22V Vu:800V/mV Vo:±14V Vi0:18μμV	8ADS	Ltc		data pdf pinout
LT1001CM	HI-PREC	Vs:±22V Vu:2000V/mV Vo:±14V f:0.8Mc	8S	Ray		data pinout
LT-1001(A)CM	OP-IC	=LT-1001(A)CN: SMD	8-MDIP			data pinout
LT1001CN	HI-PREC	Vs:±22V Vu:2000V/mV Vo:±14V f:0.8Mc	8D	Ray		data pinout
LT1001CN8	HI-PREC	Vs:±22V Vu:800V/mV Vo:±14V Vi0:18μμV	8D	Max		data pinout
LT-1001(A)CN	OP-IC	hi- prec, ±22V, 0, 8MHz, 0... +70° Offs.<0, 16mV, A: <0, 06mV	8-DIP	Ray		
LT1001CS8	HI-PREC	Vs:±22V Vu:800V/mV Vo:±14V Vi0:18μμV	8S	Max		data pdf pinout
LT1001M	HI-PREC	Vs:±22V Vu:800V/mV Vo:±14V Vi0:18μμV	8AD,20C	Ltc		data pdf pinout
LT1001MD	HI-PREC	Vs:±22V Vu:2000V/mV Vo:±14V f:0.8Mc	8D	Ray		data pinout
LT-1001(A)MD	OP-IC	=LT-1001(A)CN: -55...+125°	8-DIC			data pinout
LT-1001(A)ML	OP-IC	=LT-1001(A)CN: -55...+125°	20-(P)LCC			data pinout
LT-1001(A)MT	OP-IC	=LT-1001(A)CN: -55...+125°	81			data pinout
LT1002	OC	LED/LDR Viso:3000V Ibs:20mA		Hei		data pdf pinout
LT1002	2xHI-PREC	Vs:±22V Vu:800V/mV Vo:±14V Vi0:25μμV	14D	Ltc		data pdf pinout
LT1002A	2xHI-PREC	Vs:±22V Vu:800V/mV Vo:±14V Vi0:20μμV	14D	Ltc		data pinout
LT1003	OC	LED/LDR Viso:3000V Ibs:20mA		Hei		data pdf pinout
LT1003	R+	Io=5A Vo:4.8...5.2V Vin:7.5...15V P:int	3O	Ltc		pdf pinout
LT 1004(C,I)S8-1.2	Z-Ref-IC	SMD, 1.2V Micropower Volt. Ref., ±4mV C: 0...70°C / I:-40...85°C	8-MDIP	Lic	REF 1004 (C,I)-1.2	pdf pinout
LT 1004(C,I)S8-2.5	Z-Ref-IC	SMD, 2.5V Micropower Volt. Ref., ±20mV C: 0...70°C / I:-40...85°C	8-MDIP	Lic	REF 1004 (C,I)-2.5	pdf pinout
LT 1004(C,I)Z-1.2	Z-Ref-IC	1.2V Micropower Volt. Ref., ±4mV C: 0...70°C / I:-40...85°C	7	Lic		pdf pinout
LT 1004(C,I)Z-2.5	Z-Ref-IC	2.5V Micropower Volt. Ref., ±20mV C: 0...70°C / I:-40...85°C	7	Lic		pdf pinout
LT 1004(C,M)H-1.2	Z-Ref-IC	1.2V Micropower Volt. Ref., ±4mV C: 0...70°C / M:-55...125°C	2/2Pin	Lic		pdf pinout
LT 1004(C,M)H-2.5	Z-Ref-IC	2.5V Micropower Volt. Ref., ±20mV C: 0...70°C / M:-55...125°C	2/2Pin	Lic		pdf pinout
LT1004CD-1.2	Z+	Io=20mA Vo:1.22...1.24V P:int	8S	Tix		data pinout
LT1004CD-2.5	Z+	Io=20mA Vo:2.47...2.53V P:int	8S	Tix		data pinout
LT1004CLP-1.2	Z+	Io=20mA Vo:1.22...1.24V P:int	3N	Tix		data pinout
LT1004CLP-2.5	Z+	Io=20mA Vo:2.47...2.53V P:int	3N	Tix		data pinout
LT1004CS8-1.2	Z+	Io=10mA Vo:1.22...1.24V	8S	Ltc		data pinout
LT1004CS8-2.5	Z+	Io=10mA Vo:2.47...2.53V	8S	Ltc		data pinout
LT1004MD-1.2	Z+	Io=20mA Vo:1.22...1.24V P:int	8S	Tix		data pinout
LT1004MD-2.5	Z+	Io=20mA Vo:2.46...2.53V P:int	8S	Tix		data pinout
LT1004MLP-1.2	Z+	Io=20mA Vo:1.22...1.24V P:int	3N	Tix		data pinout
LT1004MLP-2.5	Z+	Io=20mA Vo:2.46...2.53V P:int	3N	Tix		data pinout
LT1005	Z+	Io=1A Vo:4.8...5.2V Vin:7.4...20V P:int	5OP	Ltc		data pdf pinout
LT1006	SS-OP	Vs:±22V Vu:2μV/mV Vo:4.4/0V	8ADS	Ltc		data pdf pinout
LT1006A	SS-OP	Vs:±22V Vu:2.5μV/mV Vo:4.4/0V	8AD	Ltc		data pdf pinout
LT1007	LO-NOISE	Vs:±22V Vu:20μV/mV Vo:±13.5V f:8Mc	8ADS,16S	Ltc		data pdf pinout
LT1007A	LO-NOISE	Vs:±22V Vu:20μV/mV Vo:±13.8V f:8Mc	8AD	Ltc		data pdf pinout
LT1007CDW	LO-NOISE	Vs:±22V Vu:20V/mV Vo:±13.5V	16S	Tix		data pinout
LT1007CP	LO-NOISE	Vs:±22V Vu:20V/mV Vo:±13.5V	8D	Tix		data pinout
LT1007M	LO-NOISE	Vs:±22V Vu:20V/mV Vo:±13.5V	8DA	Tix		data pinout

Type	Device	Short Description	Fig.	Manu	Comparision Types	More at
LT1008	LO-NOISE	Vs:±20V Vu:600V/mV Vo:±14V Vi0:30µµV	8AD	Ltc		data pdf pinout
LT1009	Z+	Vo:2.49...2.50V P:1.37W	20C,8D,3NS	Tix		data pinout
LT1009C	Z+	Vo:2.49...2.50V P:500mW	8D,3N,4S	Tix		data pinout
LT1009CD	Z+	Vo:2.49...2.51V P:725mW	8S	Tix		data pdf pinout
LT1009CF	Z+	Vo:2.49...2.50V P:1.37W	20C	Tix		data pinout
LT1009CH	Z+	Io=10mA Vo:2.49...2.51V	3A	Ltc		data pdf pinout
LT1009CZ	Z+	Io=10mA Vo:2.49...2.51V	3N	Ltc		data pdf pinout
LT1009I	Z+	Vo:2.49...2.51V P:725mW	8S	Tix		data pdf pinout
LT1009M	Z+	Vo:2.49...2.50V P:775mW	8D,3N	Tix		data pinout
LT1009MH	Z+	Io=10mA Vo:2.49...2.51V	3A	Ltc		data pdf pinout
LT1009S8	Z+	Io=10mA Vo:2.49...2.51V	8S	Ltc		data pdf pinout
LT1010	HI-SPEED	Vs:40V f:20Mc	4A,5O,8DP	Ltc		data pdf pinout
LT1011	OC	LED/LDR Viso:3000V Ibs:20mA		Hei		data pinout
LT1011	COMP	Vs:36V Vu:300V/mV Vi0:0.6mV	8AD	Ltc		data pdf pinout
LT1011A	COMP	Vs:36V Vu:300V/mV Vi0:0.3mV	8AD	Ltc		data pdf pinout
LT1012	LO-NOISE	Vs:±20V Vu:2000V/mV Vo:±14V	8DAS	Ltc		data pdf pinout
LT1012A	LO-NOISE	Vs:±20V Vu:2000V/mV Vo:±14V Vi0:8µV	8AD	Ltc		data pdf pinout
LT1012C	LO-NOISE	Vs:±20V Vu:2000V/mV Vo:±14V	8AD	Ltc		data pdf pinout
LT-1012 CN	OP-IC	Io-power,hi-prec, ±22V, 0,8MHz,0...+70°	8-DIP	Ray		data pinout
LT-1012 CT	OP-IC	=LT-1012CN:	81			data pinout
LT1012D	LO-NOISE	Vs:±20V Vu:2000V/mV Vo:±14V	8AD	Ltc		data pdf pinout
LT1012MT	LO-POWER	Vs:±22V Vu:1000V/mV Vo:±14V Vi0:7µµV	8A	Ray		data pdf pinout
LT-1012 MT	OP-IC	=LT-1012CN: -55...+125°	81			data pinout
LT1013	2xHI-PREC	Vs:±22V Vu:8µV/mV Vo:±14V Vi0:40µµV	8DS	Ltc		data pinout
LT1013A	2xHI-PREC	Vc⁺:±22V Vu:8µV/mV Vo:±14V Vi0:40µµV	8SD,20C	Ltc		data pinout
LT1013C	2xHI-PREC	Vs:±22V Vu:7µV/mV Vo:±14V Vi0:60µµV	8AD	Ltc		data pinout
LT1013D	2xOP	Vs:±22V Vu:7µV/mV Vo:±14V Vi0:200µµV	8SD	Tix		data pinout
LT1013DP	2xOP	Vs:±22V Vu:7µV/mV Vo:±14V Vi0:60µµV	8D	Tix		data pinout
LT1013IP	2xOP	Vs:±22V Vu:7µV/mV Vo:±14V Vi0:60µµV	8D	Tix		data pinout
LT1013M	2xOP	Vs:±22V Vu:7µV/mV Vo:±14V Vi0:60µµV	8D	Tix		data pinout
LT1013MFK	2xOP	Vs:±22V Vu:7µV/mV Vo:±14V Vi0:60µµV	20C	Tix		data pinout
LT1013Y	2xOP	Vs:±22V Vu:7dB Vo:±14V Vi0:200µµV	d	Tix		data pinout
LT1014	4xHI-PREC	Vs:±22V Vu:8µV/mV Vo:±14V Vi0:50µµV	14D,16S	Ltc		data pinout
LT1014A	4xHI-PREC	Vs:±22V Vu:8µV/mV Vo:±14V Vi0:50µµV	14D,20C	Ltc		data pinout
LT1014CN	4xHI-PREC	Vs:±22V Vo:±14V Vi0:60µµV	14D	Tix		data pinout
LT1014D	4xHI-PREC	Vs:±22V Vo:±14V Vi0:200µµV	16S,14D	Tix		data pinout
LT1014DN	4xHI-PREC	Vs:±22V Vo:±14V Vi0:200µµV	14D	Tix		data pinout
LT1014IN	4xHI-PREC	Vs:±22V Vo:±14V Vi0:60µµV	14D	Tix		data pinout
LT1014M	4xHI-PREC	Vs:±22V Vo:±14V Vi0:60µµV	14D	Tix		data pinout
LT1014MFK	4xHI-PREC	Vs:±22V Vo:±14V Vi0:60µµV	20C	Tix		data pinout
LT1015	FAST-COMP	Vs:7V Vi0:1mV	8D	Ltc		data pinout
LT 1016	Si-N	≈BF 255	7e		→BF 255	data pinout
LT1016	COMP	Vs:±7V Vi0:1mV	10A,8DS	Ltc		data pdf pinout
LT1019ACH-10	R+	Io=10mA	8A	Ltc		data pinout
LT1019ACH-2.5	R+	Io=10mA	8A	Ltc		data pinout
LT1019ACH-4.5	R+	Io=10mA	8A	Ltc		data pinout
LT1019ACH-5	R+	Io=10mA	8A	Ltc		data pinout
LT1019ACN8-10	R+	Io=10mA	8D	Ltc		data pinout
LT1019ACN8-2.5	R+	Io=10mA	8D	Ltc		data pinout
LT1019ACN8-4.5	R+	Io=10mA	8D	Ltc		data pinout
LT1019ACN8-5	R+	Io=10mA	8D	Ltc		data pinout
LT1019AMH-10	R+	Io=10mA	8A	Ltc		data pinout
LT1019AMH-2.5	R+	Io=10mA	8A	Ltc		data pinout
LT1019AMH-4.5	R+	Io=10mA	8A	Ltc		data pinout
LT1019AMH-5	R+	Io=10mA	8A	Ltc		data pinout
LT1019CH-10	R+	Io=10mA	8A	Ltc		data pinout
LT1019CH-2.5	R+	Io=10mA	8A	Ltc		data pinout
LT1019CH-4.5	R+	Io=10mA	8A	Ltc		data pinout
LT1019CH-5	R+	Io=10mA	8A	Ltc		data pinout
LT1019CN8-10	R+	Io=10mA	8D	Ltc		data pinout
LT1019CN8-2.5	R+	Io=10mA	8D	Ltc		data pinout
LT1019CN8-4.5	R+	Io=10mA	8D	Ltc		data pinout
LT1019CN8-5	R+	Io=10mA	8D	Ltc		data pinout
LT1019MH-10	R+	Io=10mA	8A	Ltc		data pinout
LT1019MH-2.5	R+	Io=10mA	8A	Ltc		data pinout
LT1019MH-4.5	R+	Io=10mA	8A	Ltc		data pinout
LT1019MH-5	R+	Io=10mA	8A	Ltc		data pinout
LT1020	R+	Io=>0.1mA Vo:2.46..2.54V P:int	14D,16S	Ltc		data pdf pinout
LT1021ACH-10	R+	Io=10mA Vo:9.95...10.05V	8A	Ltc		data pinout
LT1021ACH-7	R+	Io=10mA Vo:6.95...7.05V	8A	Ltc		data pinout
LT1021AMH-10	R+	Io=10mA Vo:9.95...10.05V	8A	Ltc		data pinout
LT1021AMH-7	R+	Io=10mA Vo:6.95...7.05V	8A	Ltc		data pinout
LT1021BCH-10	R+	Io=10mA Vo:9.95...10.05V	8A	Ltc		data pinout
LT1021BCH-5	R+	Io=10mA Vo:4.95...5.05V	8A	Ltc		data pinout
LT1021BCH-7	R+	Io=10mA Vo:6.95...7.05V	8A	Ltc		data pinout
LT1021BCN8-10	R+	Io=10mA Vo:9.95...10.05V	8D	Ltc		data pinout
LT1021BCN8-5	R+	Io=10mA Vo:4.95...5.05V	8D	Ltc		data pinout
LT1021BCN8-7	R+	Io=10mA Vo:6.95...7.05V	8D	Ltc		data pinout
LT1021BMH-10	R+	Io=10mA Vo:9.95...10.05V	8A	Ltc		data pinout
LT1021BMH-5	R+	Io=10mA Vo:4.95...5.05V	8A	Ltc		data pinout
LT1021BMH-7	R+	Io=10mA Vo:6.95...7.05V	8A	Ltc		data pinout
LT1021CCH-10	R+	Io=10mA Vo:9.95...10.05V	8A	Ltc		data pinout
LT1021CCH-5	R+	Io=10mA Vo:4.95...5.05V	8A	Ltc		data pinout
LT1021CCN8-10	R+	Io=10mA Vo:9.95...10.05V	8D	Ltc		data pinout
LT1021CCN8-5	R+	Io=10mA Vo:4.95...5.05V	8D	Ltc		data pinout

Type	Device	Short Description	Fig.	Manu	Comparision Types	More at
LT1021CMH-10	R+	Io=10mA Vo:9.95...10.05V	8A	Ltc		data pinout
LT1021CMH-5	R+	Io=10mA Vo:4.95...5.05V	8A	Ltc		data pinout
LT1021CMH-7	R+	Io=10mA Vo:6.95...7.05V	8A	Ltc		data pinout
LT1021DCH-10	R+	Io=10mA Vo:9.95...10.05V	8A	Ltc		data pinout
LT1021DCH-5	R+	Io=10mA Vo:4.95...5.05V	8A	Ltc		data pinout
LT1021DCH-7	R+	Io=10mA Vo:6.95...7.05V	8A	Ltc		data pinout
LT1021DCN8-10	R+	Io=10mA Vo:9.95...10.05V	8D	Ltc		data pinout
LT1021DCN8-5	R+	Io=10mA Vo:4.95...5.05V	8D	Ltc		data pinout
LT1021DCN8-7	R+	Io=10mA Vo:6.95...7.05V	8D	Ltc		data pinout
LT1021DCS8-10	R+	Io=10mA Vo:9.95...10.05V	8S	Ltc		data pinout
LT1021DCS8-5	R+	Io=10mA Vo:4.95...5.05V	8S	Ltc		data pinout
LT1021DCS8-7	R+	Io=10mA Vo:6.95...7.05V	8S	Ltc		data pinout
LT1021DMH-10	R+	Io=10mA Vo:9.95...10.05V	8A	Ltc		data pinout
LT1021DMH-5	R+	Io=10mA Vo:4.95...5.05V	8A	Ltc		data pinout
LT1021DMH-7	R+	Io=10mA Vo:6.95...7.05V	8A	Ltc		data pinout
LT1022	HI-SPEED	Vs:±20V Vu:400V/mV Vo:±13.2V f:8Mc	8AD	Ltc		data pdf pinout
LT1022A	HI-SPEED	Vs:±20V Vu:400V/mV Vo:±13.2V f:8.5Mc	8A	Ltc		data pdf pinout
LT1024	2xLO-NOISE	Vs:±20V Vu:2000V/mV Vo:±14V	14D	Ltc		data pinout
LT1024A	2xLO-NOISE	Vs:±20V Vu:2000V/mV Vo:±14V	14D	Ltc		data pinout
LT 1025	LIN-IC	Micropower Thermocouple Cold Junction Compensator	DIP	Lic	-	pdf pinout
LT 1026 CH	IC	Voltage Converter	QFN	Lic	-	pdf pinout
LT 1026 MH	IC	Voltage Converter	QFN	Lic	-	pdf pinout
LT1028	LO-NOISE	Vs:±22V Vu:30µV/mV Vo:±13V Vio:20µµV	8ADS,16S	Ltc		data pdf pinout
LT1028A	LO-NOISE	Vs:±22V Vu:30µV/mV Vo:±13V Vio:10µµV	8AD	Ltc		data pdf pinout
LT1029	R+	Io=10mA Vo:4.95...5.05V	3AN	Ltc		data pdf pinout
LT1029A	R+	Io=10mA Vo:4.99...5.01V	3AN	Ltc		data pdf pinout
LT1031B	R+	Io=±mA Vo:9.995...10.00V P:600mW	3A	Ltc		data pdf pinout
LT1031C	R+	Io=±mA Vo:9.99...10.1V P:600mW	3A	Ltc		data pdf pinout
LT1031D	R+	Io=±mA Vo:9.98...10.02V P:600mW	3A	Ltc		data pdf pinout
LT1033	ER-	Io=3A Vo:-1.2...-32V Vin:-3...-35V	3OP	Ltc		data pdf pinout
LT1034	R+	Io=20mA Vo:2.43...2.57V	3N,8S	Ltc		data pinout
LT1034BCH-1.2	R+	Io=20mA Vo:1.205...1.245V	3A	Ltc		data pinout
LT1034BCH-2.5	R+	Io=20mA Vo:2.43...2.57V	3A	Ltc		data pinout
LT1034BCZ-1.2	R+	Io=20mA Vo:1.205...1.245V	3N	Ltc		data pinout
LT1034BCZ-2.5	R+	Io=20mA Vo:2.43...2.57V	3N	Ltc		data pinout
LT1034BMH-1.2	R+	Io=20mA Vo:1.205...1.245V	3A	Ltc		data pinout
LT1034BMH-2.5	R+	Io=20mA Vo:2.43...2.57V	3A	Ltc		data pinout
LT1034CH-1.2	R+	Io=20mA Vo:1.205...1.245V	3A	Ltc		data pinout
LT1034CH-2.5	R+	Io=20mA Vo:2.43...2.57V	3A	Ltc		data pinout
LT1034CS8-1.2	R+	Io=20mA Vo:1.205...1.245V	8S	Ltc		data pinout
LT1034CS8-2.5	R+	Io=20mA Vo:2.43...2.57V	8S	Ltc		data pinout
LT1034CZ-1.2	R+	Io=20mA Vo:1.205...1.245V	3N	Ltc		data pinout
LT1034IZ-1.2	R+	Io=20mA Vo:1.205...1.245V	3N	Ltc		data pinout
LT1034IZ-2.5	R+	Io=20mA Vo:2.43...2.57V	3N	Ltc		data pinout
LT1034MH-1.2	R+	Io=20mA Vo:1.205...1.245V	3A	Ltc		data pinout
LT1034MH-2.5	R+	Io=20mA Vo:2.43...2.57V	3A	Ltc		data pinout
LT1035	R+	Io=3A Vo:4.8...5.2V Vin:7.7...20V P:int	5OP	Ltc		data pdf pinout
LT1036	R+	Io=3A Vo:11.52...12.48V Vin:15...30V	5OP	Ltc		data pdf pinout
LT1037	LO-NOISE	Vs:±22V Vu:20µV/mV Vo:±13.5V f:60Mc	8ADS,16S	Ltc		data pdf pinout
LT1037A	LO-NOISE	Vs:±22V Vu:20µV/mV Vo:±13.8V f:60Mc	8AD	Ltc		data pdf pinout
LT1037CDW	LO-NOISE	Vs:±22V Vu:20V/mV Vo:±13.5V	16S	Tix		data pinout
LT1037CP	LO-NOISE	Vs:±22V Vu:20V/mV Vo:±13.5V	8D	Tix		data pinout
LT1037M	LO-NOISE	Vs:±22V Vu:20V/mV Vo:±13.5V	8DA	Tix		data pinout
LT1038	ER+	Io=10A Vo:1.2...32V Vin:3...35V P:int	3O	Ltc		data pdf pinout
LT1054	R-	Io=100mA Vo:-4.7...-5.2V Vin:-7V	16S,8D	Tix		data pdf pinout
LT1055	PREC-JFET	Vs:±20V Vu:400V/mV Vo:±13.2V f:4.5Mc	8ADS	Ltc		data pdf pinout
LT1055A	PREC-JFET	Vs:±20V Vu:400V/mV Vo:±13.2V f:5Mc	8A	Ltc		data pdf pinout
LT1056	PREC-JFET	Vs:±20V Vu:400V/mV Vo:±13.2V f:5.5Mc	8ADS	Ltc		data pdf pinout
LT1056A	PREC-JFET	Vs:±20V Vu:400V/mV Vo:±13.2V f:6.5Mc	8A	Ltc		data pdf pinout
LT1057	2xJFET	Vs:±20V Vu:300V/mV Vo:±13V Vio:0.3mV	16S,8S	Ltc		data pdf pinout
LT1057A	2xJFET	Vs:±20V Vu:350V/mV Vo:±13V f:5Mc	8AD	Ltc		data pdf pinout
LT1057C	2xJFET	Vs:±20V Vu:300V/mV Vo:±13V f:5Mc	8AD	Ltc		data pdf pinout
LT1057M	2xJFET	Vs:±20V Vu:300V/mV Vo:±13V f:5Mc	8AD	Ltc		data pdf pinout
LT1058	4xJFET	Vs:±20V Vu:300V/mV Vo:±13V f:5Mc	14D,16S	Ltc		data pinout
LT1058A	4xJFET	Vs:±20V Vu:350V/mV Vo:±13V f:5Mc	14D	Ltc		data pinout
LT1070	ES+	Vo:3.0...40V Isw:10A	5PO	Ltc		data pdf pinout
LT1071	ES+	Vo:...40V Isw:2.5A P:1W	5PO,8D	Tix		data pinout
LT1072	ES+	Vo:3.0...40V Isw:3A	8D,5OS,16S	Ltc		data pdf pinout
LT1072H	ES+	Vo:...60V Isw:2.5A P:2W	5P	Tix		data pinout
LT1073	ES+	Io=0.8mA Vo:...50V P:500mW	8DS	Ltc		data pdf pinout
LT1073-12	S+	Vo:11.4...12.6V Vin:<8V P:500mW	8DS	Ltc		data pdf pinout
LT1073-5	S+	Vo:4.75...5.25V Vin:<8V P:500mW	8DS	Ltc		data pdf pinout
LT1074	ES+	Vo:2.2...40V Isw:11A	5O,7PP	Ltc		data pdf pinout
LT 1074 CK	LIN-IC	Step-down Switching Regulator	DIP	Lic	-	pdf pinout
LT 1074 CT	LIN-IC	Step-down Switching Regulator	DIP	Lic	-	pdf pinout
LT 1074 CY	LIN-IC	Step-down Switching Regulator	DIP	Lic	-	pdf pinout
LT 1074 IT	LIN-IC	Step-down Switching Regulator	DIP	Lic	-	pdf pinout
LT 1074 MK	LIN-IC	Step-down Switching Regulator	DIP	Lic	-	pdf pinout
LT1076	ES+	Vo:2.2...40V Isw:3.2A	5OP,7P	Ltc		data pdf pinout
LT1076-5	S+	Io=1.8A Vo:4.85...5.15V Vin:8...40V	7P,5P	Ltc		data pdf pinout
LT 1076 CK	LIN-IC	Step-down Switching Regulator	DIP	Lic	-	pdf pinout
LT 1076 CQ	LIN-IC	Step-down Switching Regulator	DIP	Lic	-	pdf pinout
LT 1076 CR	LIN-IC	Step-down Switching Regulator	DIP	Lic	-	pdf pinout
LT 1076 CT	LIN-IC	Step-down Switching Regulator	DIP	Lic	-	pdf pinout

Type	Device	Short Description	Fig.	Manu	Comparision Types	More at
LT 1076 CY	LIN-IC	Step-down Switching Regulator	DIP	Lic	-	pdf pinout
LT 1076 IT	LIN-IC	Step-down Switching Regulator	DIP	Lic	-	pdf pinout
LT 1076 MK	LIN-IC	Step-down Switching Regulator	DIP	Lic	-	pdf pinout
LT1077	SS-OP	Vs:±22V Vu:8000V/mV Vo:±14V	8ADS	Ltc		data pdf pinout
LT1077A	SS-OP	Vs:±22V Vu:8000V/mV Vo:±14V	8AD	Ltc		data pdf pinout
LT1078	2xSS-OP	Vs:±22V Vu:1000V/mV Vo:3.9/0V	8AD,16SS	Ltc		data pinout
LT1078A	2xSS-OP	Vs:±22V Vu:1000V/mV Vo:3.9/0V	8A,14DD	Ltc		data pdf pinout
LT1079	4xSS-OP	Vs:±22V Vu:1000V/mV Vo:3.9/0V	14D,16S	Ltc		data pinout
LT1079A	4xSS-OP	Vs:±22V Vu:1000V/mV Vo:3.9/0V	14D	Ltc		data pinout
LT1082	ES+	Vo:3.0...75V Isw:2.4A	8D,5P	Ltc		data pdf pinout
LT1083CK	ER+	Io=7.5A Vo:1.2...30V Vin:1.5...25V	3O	Ltc		data pinout
LT1083CK-12	R+	Io=7.5A Vo:11.76...12.24V P:int	3O	Ltc		data pinout
LT1083CK-5	R+	Io=7.5A Vo:4.9...5.1V Vin:6.5...20V	3O	Ltc		data pinout
LT1083CP	ER+	Io=7.5A Vo:1.2...30V Vin:1.5...25V	3P	Ltc		data pinout
LT1083CP-12	R+	Io=7.5A Vo:11.76...12.24V P:int	3P	Ltc		data pinout
LT1083CP-5	R+	Io=7.5A Vo:4.9...5.1V Vin:6.5...20V	3P	Ltc		data pinout
LT1083MK	ER+	Io=7.5A Vo:1.2...35V Vin:1.5...25V	3O	Ltc		data pinout
LT1083MK-12	R+	Io=7.5A Vo:11.76...12.24V P:int	3O	Ltc		data pinout
LT1083MK-5	R+	Io=7.5A Vo:4.9...5.1V Vin:6.5...20V	3O	Ltc		data pinout
LT1084	ER+	Io=5A Vo:1.2...30V Vin:1.5...25V P:int	3P	Ltc		data pinout
LT1084C	ER+	Io=5A Vo:1.2...30V Vin:1.5...25V P:int	3POS	Ltc		data pinout
LT1084CK-12	R+	Io=5A Vo:11.76...12.24V Vin:13.5...25V	3O	Ltc		data pinout
LT1084CK-5	R+	Io=5A Vo:4.9...5.1V Vin:6.5...20V P:int	3O	Ltc		data pinout
LT1084CP	ER+	Io=5A Vo:1.2...30V Vin:1.5...25V P:int	3P	Ltc		data pinout
LT1084CP-12	R+	Io=5A Vo:11.76...12.24V Vin:13.5...25V	3P	Ltc		data pinout
LT1084CP-5	R+	Io=5A Vo:4.9...5.1V Vin:6.5...20V P:int	3P	Ltc		data pinout
LT1084CT-12	R+	Io=5A Vo:11.76...12.24V Vin:13.5...25V	3P	Ltc		data pinout
LT1084CT-3.3	R+	Io=5A Vo:3.235...3.365V Vin:4.8...15V	3P	Ltc		data pinout
LT1084CT-5	R+	Io=5A Vo:4.9...5.1V Vin:6.5...20V P:int	3P	Ltc		data pinout
LT1084MK	ER+	Io=5A Vo:1.2...35V Vin:1.5...25V P:int	3O	Ltc		data pinout
LT1084MK-12	R+	Io=5A Vo:11.76...12.24V Vin:13.5...25V	3O	Ltc		data pinout
LT1084MK-5	R+	Io=5A Vo:4.9...5.1V Vin:6.5...20V P:int	3O	Ltc		data pinout
LT1085	ER+	Io=3A Vo:1.2...30V Vin:1.5...25V P:int	3P	Ltc		data pinout
LT1085CK	ER+	Io=3A Vo:1.2...30V Vin:1.5...25V P:int	3O	Ltc		data pinout
LT1085CK-12	R+	Io=3A Vo:11.76...12.24V Vin:13.5...25V	3O	Ltc		data pinout
LT1085CK-5	R+	Io=3A Vo:4.9...5.1V Vin:6.5...20V P:int	3O	Ltc		data pinout
LT1085CM	ER+	Io=3A Vo:1.2...30V Vin:1.5...25V P:int	3S	Ltc		data pinout
LT1085CM-3.3	R+	Io=3A Vo:3.235...3.365V Vin:4.8...15V	3S	Ltc		data pinout
LT1085CM-3.6	R+	Io=3A Vo:3.5...3.672V Vin:5...15V P:int	3S	Ltc		data pinout
LT1085CT-12	R+	Io=3A Vo:11.76...12.24V Vin:13.5...25V	3P	Ltc		data pinout
LT1085CT-3.3	R+	Io=3A Vo:3.235...3.365V Vin:4.8...15V	3P	Ltc		data pinout
LT1085CT-3.6	R+	Io=3A Vo:3.5...3.672V Vin:5...15V P:int	3P	Ltc		data pinout
LT1085CT-5	R+	Io=3A Vo:4.9...5.1V Vin:6.5...20V P:int	3P	Ltc		data pinout
LT1085MK	ER+	Io=3A Vo:1.2...35V Vin:1.5...25V P:int	3O	Ltc		data pdf pinout
LT1085MK-12	R+	Io=3A Vo:11.76...12.24V Vin:13.5...25V	3O	Ltc		data pinout
LT1085MK-5	R+	Io=3A Vo:4.9...5.1V Vin:6.5...20V P:int	3O	Ltc		data pinout
LT1086CH	ER+	Io=500mA Vo:1.2...30V Vin:1.5...15V	3A	Ltc		data pdf pinout
LT1086CK	ER+	Io=1.5A Vo:1.2...30V Vin:1.5...15V	3O	Ltc		data pinout
LT1086CK-12	R+	Io=1.5A Vo:11.76...12.24V P:int	3O	Ltc		data pinout
LT1086CK-5	R+	Io=1.5A Vo:4.9...5.2V Vin:1.5...15V	3O	Ltc		data pinout
LT1086CM	ER+	Io=1.5A Vo:1.2...30V Vin:1.5...15V	3S	Ltc		data pdf pinout
LT1086CM-3.3	R+	Io=1.5A Vo:3.235...3.365V P:int	3S	Ltc		data pinout
LT1086CM-3.6	R+	Io=1.5A Vo:3.5...3.672V Vin:4.75...18V	3S	Ltc		data pinout
LT1086CT	ER+	Io=1.5A Vo:1.2...30V Vin:1.5...15V	3P	Ltc		data pinout
LT1086CT-12	R+	Io=1.5A Vo:11.76...12.24V P:int	3P	Ltc		data pinout
LT1086CT-2.85	R+	Io=1.5A Vo:2.79...2.91V Vin:4.35...18V	3P	Ltc		data pinout
LT1086CT-3.3	R+	Io=1.5A Vo:3.235...3.365V P:int	3P	Ltc		data pinout
LT1086CT-3.6	R+	Io=1.5A Vo:3.5...3.672V Vin:4.75...18V	3P	Ltc		data pinout
LT1086CT-5	R+	Io=1.5A Vo:4.9...5.2V Vin:1.5...15V	3P	Ltc		data pinout
LT1086MH	ER+	Io=500mA Vo:1.2...30V Vin:1.5...15V	3A	Ltc		data pdf pinout
LT1086MK	ER+	Io=1.5A Vo:1.2...35V Vin:1.5...15V	3O	Ltc		data pdf pinout
LT1086MK-12	R+	Io=1.5A Vo:11.76...12.24V P:int	3O	Ltc		data pinout
LT1086MK-5	R+	Io=1.5A Vo:4.9...5.2V Vin:1.5...15V	3O	Ltc		data pinout
LT 1088 CD	LIN-IC	Wideband RMS-DC Converter Building Block	DIP	Lic	-	pdf pinout
LT 1088 CN	LIN-IC	Wideband Rms-dc Converter Building Block	DIP	Lic	-	pdf pinout
LT 1089 K	LIN-IC	High Side Switch	43	Lic	-	pdf pinout
LT 1089 T	LIN-IC	High Side Switch	43	Lic	-	pdf pinout
LT1097	LO-POWER	Vs:±20V Vu:1000V/mV Vo:±13V f:0.7Mc	8DS	Ltc		data pdf pinout
LT 1101 H	LIN-IC	Single Supply Instrumental Amp (fixed Gain=10 or 100)	43	Lic	-	pdf pinout
LT 1101 J	LIN-IC	Single Supply Instrumental Amp (fixed Gain=10 or 100)	43	Lic		pdf pinout
LT 1101 N	LIN-IC	Single Supply Instrumental Amp (fixed Gain=10 or 100)	43	Lic		pdf pinout
LT 1101 S	LIN-IC	Single Supply Instrumental Amp (fixed Gain=10 or 100)	43	Lic		pdf pinout
LT 1102 H	LIN-IC	JFET Input Instrumental Amp (fixed Gain=10 or 100)	43	Lic	-	pdf pinout
LT1106	S++	Vo:4.75...5.25V Isw:0.95A P:500mW	20S	Ltc		data pdf pinout
LT1107	ES+	Io=800mA Vo:1.25...10V Vin:30V	8DS	Ltc		data pdf pinout
LT1108	ES+	Io=0.8mA Vo:...30V Isw:1.5A P:500mW	8DS	Ltc		data pdf pinout
LT1109	ES+	Vo:...50V Isw:1.2A P:300mW	3N,8DS	Ltc		data pdf pinout
LT1109A	ES+	Vo:...50V Isw:2A P:300mW	8DS	Ltc		data pdf pinout
LT1110	ES+	Vo:...30V Isw:1.5A P:500mW	8DS	Ltc		data pdf pinout
LT1111	ES+	Io=800mA Vo:1.25...10V Vin:30V	8DS	Ltc		data pdf pinout

Type	Device	Short Description	Fig.	Manu	Comparision Types	More at
LT1112	2xLO-POWER	Vs:±20V Vu:1300V/mV Vo:±12.4V	8DS	Ltc		data pdf pinout
LT1112A	2xLO-POWER	Vs:±20V Vu:1500V/mV Vo:±12.4V	8D	Ltc		data pdf pinout
LT1113	2xLO-NOISE	Vs:±20V Vu:4500V/mV Vo:±13.8V	8DS	Ltc		data pdf pinout
LT1113A	2xLO-NOISE	Vs:±20V Vu:4800V/mV Vo:±13.8V	8D	Ltc		data pdf pinout
LT1114	4xLO-POWER	Vs:±20V Vu:1300V/mV Vo:±12.4V	14D,16S	Ltc		data pdf pinout
LT1114A	4xLO-POWER	Vs:±20V Vu:1500V/mV Vo:±12.4V	14D	Ltc		data pdf pinout
LT1115	LO-NOISE	Vs:±22V Vu:20V/mV Vo:±15V Vi0:0.05mV	8D,16S	Ltc		data pinout
LT1116	SS-COMP	Vs:7V Vi0:1mV	8DS	Ltc		data pdf pinout
LT1117CM	ER+	Io=800mA Vo:1.2...15V Vin:1.4...10V	3S	Ltc		data pdf pinout
LT1117CM-2.85	R+	Io=800mA Vo:2.79...2.91V P:int	3S	Ltc		data pinout
LT1117CM-3.3	R+	Io=800mA Vo:3.235...3.365V P:int	3S	Ltc		data pinout
LT1117CM-5	R+	Io=800mA Vo:4.9...5.1V Vin:6.5...12V	3S	Ltc		data pinout
LT1117CST	ER+	Io=800mA Vo:1.2...15V Vin:1.4...10V	4S	Ltc		data pdf pinout
LT1117CST-2.85	R+	Io=800mA Vo:2.79...2.91V P:int	4S	Ltc		data pinout
LT1117CST-3.3	R+	Io=800mA Vo:3.235...3.365V P:int	4S	Ltc		data pinout
LT1117CST-5	R+	Io=800mA Vo:4.9...5.1V Vin:6.5...12V	4S	Ltc		data pinout
LT 1118	LIN-IC	LIN/S-Reg, 2,5V, 800mA		Lic		pdf
LT1120	ER+	Io=125mA Vo:2.46...2.54V Vin:4.5...36V	8DSA	Ltc		data pdf pinout
LT1121ACS8	ER+	Io=150mA Vo:3.8...20V P:int	8S	Ltc		data pdf pinout
LT1121ACS8-3.3	R+	Io=150mA Vo:3.2...3.4V Vin:4.3...20V	8S	Ltc		data pinout
LT1121ACS8-5	R+	Io=150mA Vo:4.85...5.15V Vin:6...20V	8S	Ltc		data pinout
LT1121AIS8	ER+	Io=150mA Vo:3.8...20V P:int	8S	Ltc		data pdf pinout
LT1121AIS8-3.3	R+	Io=150mA Vo:3.2...3.4V Vin:4.3...20V	8S	Ltc		data pinout
LT1121AIS8-5	R+	Io=150mA Vo:4.85...5.15V Vin:6...20V	8S	Ltc		data pinout
LT1121CN8	ER+	Io=150mA Vo:3.8...20V P:int	8D	Ltc		data pdf pinout
LT1121CN8-3.3	R+	Io=150mA Vo:3.2...3.4V Vin:4.3...20V	8D	Ltc		data pinout
LT1121CN8-5	R+	Io=150mA Vo:4.85...5.15V Vin:6...20V	8D	Ltc		data pinout
LT1121CS8	ER+	Io=150mA Vo:3.8...20V P:int	8S	Ltc		data pdf pinout
LT1121CS8-3.3	R+	Io=150mA Vo:3.2...3.4V Vin:4.3...20V	8S	Ltc		data pinout
LT1121CS8-5	R+	Io=150mA Vo:4.85...5.15V Vin:6...20V	8S	Ltc		data pinout
LT1121CST-3.3	R+	Io=150mA Vo:3.2...3.4V Vin:4.3...20V	4S	Ltc		data pinout
LT1121CST-5	R+	Io=150mA Vo:4.85...5.15V Vin:6...20V	4S	Ltc		data pinout
LT1121CZ-3.3	R+	Io=150mA Vo:3.2...3.4V Vin:4.3...20V	3N	Ltc		data pinout
LT1121CZ-5	R+	Io=150mA Vo:4.85...5.15V Vin:6...20V	3N	Ltc		data pinout
LT1121IN8	ER+	Io=150mA Vo:3.8...20V P:int	8D	Ltc		data pdf pinout
LT1121IN8-3.3	R+	Io=150mA Vo:3.2...3.4V Vin:4.3...20V	8D	Ltc		data pinout
LT1121IN8-5	R+	Io=150mA Vo:4.85...5.15V Vin:6...20V	8D	Ltc		data pinout
LT1121IS8	ER+	Io=150mA Vo:3.8...20V P:int	8S	Ltc		data pdf pinout
LT1121IS8-3.3	R+	Io=150mA Vo:3.2...3.4V Vin:4.3...20V	8S	Ltc		data pinout
LT1121IS8-5	R+	Io=150mA Vo:4.85...5.15V Vin:6...20V	8S	Ltc		data pinout
LT1121IST-3.3	R+	Io=150mA Vo:3.2...3.4V Vin:4.3...20V	4S	Ltc		data pinout
LT1121IST-5	R+	Io=150mA Vo:4.85...5.15V Vin:6...20V	4S	Ltc		data pinout
LT1121IZ-3.3	R+	Io=150mA Vo:3.2...3.4V Vin:4.3...20V	3N	Ltc		data pinout
LT1121IZ-5	R+	Io=150mA Vo:4.85...5.15V Vin:6...20V	3N	Ltc		data pinout
LT1122	JFET	Vs:±20V Vu:500V/mV Vo:±12.5V f:14Mc	8DS	Ltc		data pdf pinout
LT 1123 CZ	LIN-IC	Low Dropout Regulator Driver	43	Lic		pdf pinout
LT1124	2xHI-SPEED	Vs:±22V Vo:±13.8V Vi0:25µµV f:12.5Mc	8DS	Ltc		data pdf pinout
LT1124A	2xHI-SPEED	Vs:±22V Vo:±13.8V Vi0:20µµV f:12.5Mc	8DS	Ltc		data pdf pinout
LT1125	4xHI-SPEED	Vs:±22V Vo:±13.8V Vi0:30µµV f:12.5Mc	14D,16S	Ltc		data pdf pinout
LT1125A	4xHI-SPEED	Vs:±22V Vo:±13.8V Vi0:25µµV f:12.5Mc	14D	Ltc		data pdf pinout
LT1126	2xLO-NOISE	Vs:±22V Vo:±13.8V Vi0:25µµV f:65Mc	8DS	Ltc		data pdf pinout
LT1126A	2xLO-NOISE	Vs:±22V Vo:±13.8V Vi0:20µµV f:65Mc	8D	Ltc		data pdf pinout
LT1127	4xLO-NOISE	Vs:±22V Vo:±13.8V Vi0:30µµV f:65Mc	14D,16S	Ltc		data pdf pinout
LT1127A	4xLO-NOISE	Vs:±22V Vo:±13.8V Vi0:25µµV f:65Mc	14D	Ltc		data pdf pinout
LT1128	LO-NOISE	Vu:5000V/mV Vi0:<80µµV f:13Mc	8DS	Ltc		data pdf pinout
LT1128A	LO-NOISE	Vu:5000V/mV Vi0:<80µµV f:13Mc	8D	Ltc		data pdf pinout
LT1128M	LO-NOISE	Vu:7000V/mV Vi0:<40µµV f:13Mc	8D	Ltc		data pdf pinout
LT1129CQ	ER+	Io=700mA Vo:3.8...30V P:int	5S	Ltc		data pdf pinout
LT1129CQ-3.3	R+	Io=700mA Vo:3.2...3.4V Vin:4.3...20V	5S	Ltc		data pinout
LT1129CQ-5	R+	Io=700mA Vo:4.85...5.15V Vin:6...20V	5S	Ltc		data pinout
LT1129CS8	ER+	Io=700mA Vo:3.8...30V P:int	8S	Ltc		data pdf pinout
LT1129CS8-3.3	R+	Io=700mA Vo:3.2...3.4V Vin:4.3...20V	8S	Ltc		data pinout
LT1129CS8-5	R+	Io=700mA Vo:4.85...5.15V Vin:6...20V	8S	Ltc		data pinout
LT1129CST-3.3	R+	Io=700mA Vo:3.2...3.4V Vin:4.3...20V	4S	Ltc		data pinout
LT1129CST-5	R+	Io=700mA Vo:4.85...5.15V Vin:6...20V	4S	Ltc		data pinout
LT1129CT	ER+	Io=700mA Vo:3.8...30V P:int	5P	Ltc		data pdf pinout
LT1129CT-3.3	R+	Io=700mA Vo:3.2...3.4V Vin:4.3...20V	5P	Ltc		data pinout
LT1129CT-5	R+	Io=700mA Vo:4.85...5.15V Vin:6...20V	5P	Ltc		data pinout
LT1129IQ	ER+	Io=700mA Vo:3.8...30V P:int	5S	Ltc		data pdf pinout
LT1129IQ-3.3	R+	Io=700mA Vo:3.2...3.4V Vin:4.3...20V	5S	Ltc		data pinout
LT1129IQ-5	R+	Io=700mA Vo:4.85...5.15V Vin:6...20V	5S	Ltc		data pinout
LT1129IS8	ER+	Io=700mA Vo:3.8...30V P:int	8S	Ltc		data pdf pinout
LT1129IS8-3.3	R+	Io=700mA Vo:3.2...3.4V Vin:4.3...20V	8S	Ltc		data pinout
LT1129IS8-5	R+	Io=700mA Vo:4.85...5.15V Vin:6...20V	8S	Ltc		data pinout
LT1129IST-3.3	R+	Io=700mA Vo:3.2...3.4V Vin:4.3...20V	4S	Ltc		data pinout
LT1129IST-5	R+	Io=700mA Vo:4.85...5.15V Vin:6...20V	4S	Ltc		data pinout
LT1129IT	ER+	Io=700mA Vo:3.8...30V P:int	5P	Ltc		data pdf pinout
LT1129IT-3.3	R+	Io=700mA Vo:3.2...3.4V Vin:4.3...20V	5P	Ltc		data pinout
LT1129IT-5	R+	Io=700mA Vo:4.85...5.15V Vin:6...20V	5P	Ltc		data pinout
LT 1158 (C/I)N	MOS	Half Bridge N-channel Power Mosfet Driver	DIC	Lic	-	pdf pinout
LT 1158 (C/I)S	MOS	Half Bridge N-channel Power Mosfet Driver	TSOP	Lic	-	pdf pinout
LT 1160 (C/I)N	MOS	Half-Bridge N-channel Power Mosfet Drivers	DIC	Lic	-	pdf pinout

Type	Device	Short Description	Fig.	Manu	Comparision Types	More at
LT 1160 (C/I)S	MOS	Half-Bridge N-channel Power Mosfet Drivers	TSOP	Lic	-	pdf pinout
LT 1161 (C/I)N	MOS	Quad Protected High-Side Mosfet Driver	DIC	Lic	-	pdf pinout
LT 1161 (C/I)SW	MOS	Quad Protected High-Side Mosfet Driver	TSOP	Lic	-	pdf pinout
LT 1162 (C/I)N	MOS	Full-bridge N-channel Power Mosfet Drivers	DIC	Lic	-	pdf pinout
LT 1162 (C/I)SW	MOS	Full-bridge N-channel Power Mosfet Drivers	TSOP	Lic	-	pdf pinout
LT 1166	LIN-IC	Automatic Class AB Gain Control		Lic		pdf
LT1169	2xLO-NOISE	Vs:±20V Vu:4500V/mV Vo:±13.8V	8DS	Ltc		data pdf pinout
LT1169A	2xLO-NOISE	Vs:±20V Vu:4800V/mV Vo:±13.8V	8D	Ltc		data pdf pinout
LT1170	ES+	Vo:3.0...65V Isw:10A	5OP	Ltc		data pdf pinout
LT1171	ES+	Vo:3.0...65V Isw:5A	5OP	Ltc		data pdf pinout
LT1172	ES+	Vo:3.0...40V Isw:3A	8D,5OPS	Ltc		data pdf pinout
LT1173	ES+	Io=0.8mA Vo:...30V Isw:1.5A P:500mW	8DS	Ltc		data pdf pinout
LT 1175 (C/I) Q	LIN-IC	500mA Negative Low Dropout Micropower Regulator	TSOP	Lic		pdf pinout
LT 1175 (C/I) T	LIN-IC	500mA Negative Low Dropout Micropower Regulator	TSOP	Lic		pdf pinout
LT1176	ES+	Io=960mA Vo:2.2...32V Vin:35V	8D,20S	Ltc		data pdf pinout
LT 1176 CS	LIN-IC	Step-down Switching Regulator	TSOP	Lic		pdf pinout
LT1178	2xSS-OP	Vs:±22V Vu:700V/mV Vo:3.8/0V	8ADS,16S	Ltc		data pdf pinout
LT1178A	2xSS-OP	Vs:±22V Vu:700V/mV Vo:3.8/0V	8AD	Ltc		data pdf pinout
LT1179	4xSS-OP	Vs:±22V Vu:700V/mV Vo:3.8/0V	14D,16S	Ltc		data pdf pinout
LT1179A	4xSS-OP	Vs:±22V Vu:700V/mV Vo:3.8/0V	14D	Ltc		data pdf pinout
LT 1185 (C/I)T	LIN-IC	Low Dropout Regulator	TSOP	Lic	-	pdf pinout
LT 1185 MK	LIN-IC	Low Dropout Regulator	TSOP	Lic		pdf pinout
LT1187	VIDEO-AMP	Vs:18V Vo:±4V Vi0:2mV f:53Mc	8DS	Ltc		data pdf pinout
LT1189	VIDEO-AMP	Vs:18V Vo:±4V Vi0:1mV f:35Mc	8DS	Ltc		data pdf pinout
LT1190	HI-SPEED	Vs:18V Vu:22V/mV Vo:±4V Vi0:3mV	8DS	Ltc		data pdf pinout
LT1191	HI-SPEED	Vs:18V Vu:45V/mV Vo:±4V Vi0:1mV	8DS	Ltc		data pdf pinout
LT1192	HI-SPEED	Vs:18V Vu:180V/mV Vo:±4V Vi02:0.2mV	8DS	Ltc		data pdf pinout
LT1193	VIDEO-AMP	Vs:18V Vo:±4V Vi0:2mV f:9Mc	8DS	Ltc		data pdf pinout
LT1194	VIDEO-AMP	Vs:18V Vo:±4V Vi0:1mV f:35Mc	8DS	Ltc		data pdf pinout
LT1195	LO-POWER	Vs:18V Vu:7.5V/mV Vo:±4V Vi0:3mV	8DS	Ltc		data pdf pinout
LT1200	LO-POWER	Vs:±18V Vu:6V/mV Vo:±13.8V Vi0:0.5mV	8DS	Ltc		data pinout
LT1201	2xLO-POWER	Vs:±18V Vu:6V/mV Vo:±13.8V Vi0:0.7mV	8DS	Ltc		data pinout
LT1202	4xLO-POWER	Vs:±18V Vu:6V/mV Vo:±13.8V Vi0:0.7mV	14D,16S	Ltc		data pinout
LT1203	MUX-OP	Vs:±18V Vi0:10mV f:47.7Mc	8DS	Ltc		data pdf pinout
LT1204	4xMUX-OP	Vs:±18V Vu:73dB Vo:±13.5V Vi0:5mV	16DS	Ltc		data pdf pinout
LT1205	2xMUX-OP	Vs:±18V Vi0:10mV f:47.7Mc	16S	Ltc		data pdf pinout
LT1206	FB-AMP	Vs:±18V Vu:71dB Vo:±12.5V Vi0:±3mV	8D,7PS	Ltc		data pdf pinout
LT1207	2xFB-AMP	Vs:±18V Vu:71dB Vo:±12.5V Vi0:±3mV	16S	Ltc		data pdf pinout
LT1208	2xHI-SPEED	Vs:±18V Vu:7V/mV Vo:±13.3V Vi0:0.5mV	8DS	Ltc		data pdf pinout
LT1209	4xHI-SPEED	Vs:±18V Vu:7V/mV Vo:±13.3V Vi0:0.5mV	14D,16S	Ltc		data pdf pinout
LT1210	FB-AMP	Vs:±18V Vu:71dB Vo:±11.5V Vi0:±3mV	7S,16SP	Ltc		data pdf pinout
LT1211	2xSS-OP	Vs:36V Vu:560V/mV Vo:4.4/0V f:13Mc	8DS	Ltc		data pdf pinout
LT1211A	2xSS-OP	Vs:36V Vu:560V/mV Vo:4.4/0V f:13Mc	8D	Ltc		data pdf pinout
LT1212	4xSS-OP	Vs:36V Vu:560V/mV Vo:4.4/0V f:13Mc	14D,16S	Ltc		data pdf pinout
LT1213	2x8S-OP	Vs:36V Vu:850V/mV Vo:4.39/0V f:26Mc	8DS	Ltc		data pdf pinout
LT1213A	2xSS-OP	Vs:36V Vu:850V/mV Vo:4.39/0V f:26Mc	8D	Ltc		data pdf pinout
LT1214	4xSS-OP	Vs:36V Vu:850V/mV Vo:4.39/0V f:26Mc	14D,16S	Ltc		data pdf pinout
LT1215	2xSS-OP	Vs:36V Vu:600V/mV Vo:4.39/0V f:23Mc	8DS	Ltc		data pdf pinout
LT1215A	2xSS-OP	Vs:36V Vu:600V/mV Vo:4.39/0V f:23Mc	8D	Ltc		data pdf pinout
LT1216	4xSS-OP	Vs:36V Vu:600V/mV Vo:4.39/0V f:23Mc	14D,16S	Ltc		data pdf pinout
LT1217	FB-AMP	Vs:±18V Vo:±13V Vi0:±1mV f:10Mc	8DS	Ltc		data pdf pinout
LT1218	RR-OP	Vs:±18V Vu:4000V/mV Vi0:85µµV	8D	Ltc		data pdf pinout
LT1218L	RR-OP	Vs:±8V Vu:2800V/mV Vi0:35µµV f:0.3Mc	8DS	Ltc		data pdf pinout
LT1219	RR-OP	Vs:±18V Vu:4000V/mV Vi0:85µµV	8DS	Ltc		data pdf pinout
LT1219L	RR-OP	Vs:±8V Vu:2800V/mV Vi0:35µµV	8DS	Ltc		data pdf pinout
LT1220	HI-SPEED	Vs:36V Vu:50V/mV Vo:±13V Vi0:0.5mV	8D	Ltc		data pdf pinout
LT1221	HI-SPEED	Vs:36V Vu:100V/mV Vo:±13V Vi0:0.5mV	8DS	Ltc		data pdf pinout
LT1222	LO-NOISE	Vs:36V Vu:200V/mV Vo:±13V Vi0:0.3mV	8DS	Ltc		data pdf pinout
LT1223	FB-AMP	Vs:±18V Vu:89dB Vo:±12V Vi0:±1mV	8DS	Ltc		data pdf pinout
LT1224	HI-SPEED	Vs:±18V Vu:7V/mV Vo:±13.3V Vi0:0.5mV	8DS	Ltc		data pdf pinout
LT1225	HI-SPEED	Vs:±18V Vu:20V/mV Vo:±13.3V f:150Mc	8DS	Ltc		data pdf pinout
LT1226	LO-NOISE	Vs:±18V Vu:150V/mV Vo:±13.3V f:1GMc	8DS	Ltc		data pdf pinout
LT1227	FB-AMP	Vs:±18V Vu:72dB Vo:±13.5V Vi0:±3mV	8DS	Ltc		data pdf pinout
LT1228	FB-AMP	Vs:±18V Vu:65dB Vo:±13.5V Vi0:±3mV	8DS	Ltc		data pdf pinout
LT1229	2xFB-AMP	Vs:±18V Vu:65dB Vo:±13.5V Vi0:±3mV	8DS	Ltc		data pdf pinout
LT1230	4xFB-AMP	Vs:±18V Vu:65dB Vo:±13.5V Vi0:±3mV	14DS	Ltc		data pdf pinout
LT1236ACN8-10	R+	Vo:9.995...10.00V Vin:15V	8D	Ltc		data pinout
LT1236ACN8-5	R+	Vo:4.998...5.002V Vin:10V	8D	Ltc		data pinout
LT1236ACS8-10	R+	Vo:9.995...10.00V Vin:15V	8S	Ltc		data pinout
LT1236ACS8-5	R+	Vo:4.998...5.002V Vin:10V	8S	Ltc		data pinout
LT1236AIN8-10	R+	Vo:9.995...10.00V Vin:15V	8D	Ltc		data pinout
LT1236AIN8-5	R+	Vo:4.998...5.002V Vin:10V	8D	Ltc		data pinout
LT1236AIS8-10	R+	Vo:9.995...10.00V Vin:15V	8S	Ltc		data pinout
LT1236AIS8-5	R+	Vo:4.998...5.002V Vin:10V	8S	Ltc		data pinout
LT1236BCN8-10	R+	Vo:9.990...10.01V Vin:15V	8D	Ltc		data pinout
LT1236BCN8-5	R+	Vo:4.995...5.005V Vin:10V	8D	Ltc		data pinout
LT1236BCS8-10	R+	Vo:9.990...10.01V Vin:15V	8S	Ltc		data pinout
LT1236BCS8-5	R+	Vo:4.995...5.005V Vin:10V	8S	Ltc		data pinout
LT1236BIN8-10	R+	Vo:9.990...10.01V Vin:15V	8D	Ltc		data pinout
LT1236BIN8-5	R+	Vo:4.995...5.005V Vin:10V	8D	Ltc		data pinout

Type	Device	Short Description	Fig.	Manu	Comparision Types	More at
LT1236BIS8-10	R+	Vo:9.990...10.01V Vin:15V	8S	Ltc		data pinout
LT1236BIS8-5	R+	Vo:4.995...5.005V Vin:10V	8S	Ltc		data pinout
LT1236CCN8-10	R+	Vo:9.990...10.01V Vin:15V	8D	Ltc		data pinout
LT1236CCN8-5	R+	Vo:4.995...5.005V Vin:10V	8D	Ltc		data pinout
LT1236CCS8-10	R+	Vo:9.990...10.01V Vin:15V	8S	Ltc		data pinout
LT1236CCS8-5	R+	Vo:4.995...5.005V Vin:10V	8S	Ltc		data pinout
LT1236CIN8-10	R+	Vo:9.990...10.01V Vin:15V	8D	Ltc		data pinout
LT1236CIN8-5	R+	Vo:4.995...5.005V Vin:10V	8D	Ltc		data pinout
LT1236CIS8-10	R+	Vo:9.990...10.01V Vin:15V	8S	Ltc		data pinout
LT1236CIS8-5	R+	Vo:4.995...5.005V Vin:10V	8S	Ltc		data pinout
LT 1241..45 CJ8	SMPS-IC	Current Mode PWM Controller, 0...+100°C	8-DIC	Lic	UC1842	pdf pinout
LT 1241..45 CN8	SMPS-IC	Current Mode PWM Controller, 0...+100°C	8-DIP	Lic	UC1842	pdf pinout
LT 1241..45 CS8	SMPS-IC	SMD, Current Mode PWM Controller, 0...+100°C	8-MDIP	Lic	-	pdf pinout
LT 1241..45 IN8	SMPS-IC	Current Mode PWM Controller, -40...+100°C	8-DIP	Lic	UC1842	pdf pinout
LT 1241..45 IS8	SMPS-IC	SMD, Current Mode PWM Controller, -40...+100°C	8-MDIP	Lic	-	pdf pinout
LT 1241..45 MJ8	SMPS-IC	Current Mode PWM Controller, -55...+125°C	8-DIC	Lic	UC1842	pdf pinout
LT 1246 CN8	LIN-IC	1MHz Off-Line Current Mode PWM and DC/DC Converter	DIC	Lic	UC1842	pdf pinout
LT 1246 CS8	LIN-IC	1MHz Off-Line Current Mode PWM and DC/DC Converter	TSOP	Lic	UC1842	pdf pinout
LT 1247 CN8	LIN-IC	1MHz Off-Line Current Mode PWM and DC/DC Converter	DIC	Lic	UC1843	pdf pinout
LT 1247 CS8	LIN-IC	1MHz Off-Line Current Mode PWM and DC/DC Converter	TSOP	Lic	UC1843	pdf pinout
LT 1248	SMPS-IC	Smps-controller	16-DIP			pdf
LT 124 XC (J/N/S)8	LIN-IC	High Speed Current Mode Pulse Width Modulators	TSOP	Lic	-	pdf pinout
LT 124 XI (N/S) 8	LIN-IC	High Speed Current Mode Pulse Width Modulators	TSOP	Lic	-	pdf pinout
LT 124 XMJ 8	LIN-IC	High Speed Current Mode Pulse Width Modulators	TSOP	Stm	-	pdf pinout
LT 1249 (C/I)N	LIN-IC	Power Factor Controller	DIC	Lic	-	pdf pinout
LT 1249 (C/I)S	LIN-IC	Power Factor Controller	TSOP	Lic	-	pdf pinout
LT1251	FB-AMP	Vs:±18V Vu:93dB Vo:±3.5V Vi0:2mV	14DS	Ltc		data pdf pinout
LT1252	VIDEO-AMP	Vs:±14V Vo:±10.5V Vi0:5mV	8DS	Ltc		data pdf pinout
LT1253	2xVIDEO-AMP	Vs:±14V Vo:±10.5V Vi0:5mV	8DS	Ltc		data pdf pinout
LT1254	4xVIDEO-AMP	Vs:±14V Vo:±10.5V Vi0:5mV	14DS	Ltc		data pdf pinout
LT1256	FB-AMP	Vs:±18V Vu:93dB Vo:±3.5V Vi0:2mV	14DS	Ltc		data pdf pinout
LT1259	2xFB-AMP	Vs:±18V Vu:72dB Vo:±14V Vi0:2mV	14DS	Ltc		data pdf pinout
LT1260	3xFB-AMP	Vs:±18V Vu:72dB Vo:±14V Vi0:2mV	16DS,14DS	Ltc		data pdf pinout
LT1268	ES+	Vo:3.0...60V Isw:15A	5P	Ltc		data pdf pinout
LT1268B	ES+	Vo:3.0...60V Isw:15A	5P	Ltc		data pdf pinout
LT 1268 (B) CQ	LIN-IC	7.5A, 150kHz Switching Regulators	TSOP	Lic	-	pdf pinout
LT 1268 (B) CT	LIN-IC	7.5A, 150kHz Switching Regulators	TSOP	Lic	-	pdf pinout
LT1269	ES+	Vo:...60V Isw:8A	5P,20S	Ltc		data pdf pinout
LT1270	ES+	Vo:...60V Isw:16A	5P	Ltc		data pdf pinout
LT1271	ES+	Vo:...60V Isw:8A	5P	Ltc		data pdf pinout
LT1300	S+	Vo:4.80...5.20V Isw:1.25A P:500mW	8DS	Ltc		data pdf pinout
LT 1300 CN8	LIN-IC	Micropower High Efficiency 3.3/5V Step-up DC/DC Converter	DIC	Lic	-	pdf pinout
LT 1300 CS8	LIN-IC	Micropower High Efficiency 3.3/5V Step-up DC/DC Converter	MDIP	Lic	-	pdf pinout
LT1301	S+	Vo:4.75...5.25V Isw:1.25A P:500mW	8DS	Ltc		data pdf pinout
LT 1301 CN8	LIN-IC	Micropower High Efficiency 5V/12V Step-up DC/DC Converter	DIC	Lic	-	pdf pinout
LT 1301 (C/I)S8	LIN-IC	Micropower High Efficiency 5V/12V Step-up DC/DC Converter	MDIP	Lic	-	pdf pinout
LT1302	ES+	Vo:...25V Isw:3.9A P:700mW	8DS	Ltc		data pdf pinout
LT 1302 CN8-5	LIN-IC	Micropower High Output Current Step-up Fixed 5V DC/DC Converter	DIC	Lic	-	pdf pinout
LT 1302 CN8	LIN-IC	Micropower High Output Current Step-up Adjustable DC/DC Converter	DIC	Lic	-	pdf pinout
LT 1302 CS8-5	LIN-IC	Micropower High Output Current Step-up Fixed 5V DC/DC Converter	MDIP	Lic	-	pdf pinout
LT 1302 CS8	LIN-IC	Micropower High Output Current Step-up Adjustable DC/DC Converter	MDIP	Lic	-	pdf pinout
LT1303	ES+	Vo:...50V Isw:1.25A P:500mW	8DS	Ltc		data pdf pinout
LT1304	ES+	Vo:...25V Isw:1.2A P:500mW	8S	Ltc		data pdf pinout
LT1305	ES+	Vo:...25V Isw:2.5A P:500mW	8S	Ltc		data pdf pinout
LT1306	ES+	Vo:...5V Isw:2A P:500mW	8S	Ltc		data pdf pinout
LT1307	ES+	Vo:...30V Isw:1.25A	8SD	Ltc		data pdf pinout
LT1307B	ES+	Vo:...30V Isw:1.25A	8S	Ltc		data pdf pinout
LT1308	ES+	Vo:...30V Isw:2.5A	8S	Ltc		data pdf pinout
LT 1308 A,B	LIN-IC	SMD, PWM S-Reg, Uin=1...10V, 600kHz	8-MDIP	Lic	-	
LT1309	ES+	Vo:...30V Isw:0.6A P:500mW	8S	Ltc		data pdf pinout
LT1310	ES+	Vo:...35V Isw:2.8A	10S	Ltc		data pdf pinout
LT1311	4xHI-PREC	Vs:±18V Vu:100V/mV Vi0:±150µµV	14S	Ltc		data pdf pinout
LT 1313	LIN-IC	Dual Pcmcia VPP Driver/regulator	MDIP	Lic	-	pdf pinout
LT1317	ES+	Vo:...30V Isw:1.35A	8S	Ltc		data pdf pinout
LT1317B	ES+	Vo:...30V Isw:1.35A	8S	Ltc		data pdf pinout
LT 1336 (C/I)N	MOS-IC	Half-Bridge N-channel Power Mosfet Driver with Boost Regulator	DIC	Lic	-	pdf pinout
LT 1336 (C/I)S	MOS-IC	Half-Bridge N-channel Power Mosfet	TSOP	Lic		pdf pinout

Type	Device	Short Description	Fig.	Manu	Comparision Types	More at
		Driver with Boost Regulator				
LT 1339 (C/I)N	MOS-IC	High Power Synchronousm DC/DC Controller	DIC	Lic	-	pdf pinout
LT 1339 (C/I)SW	MOS-IC	High Power Synchronousm DC/DC Controller	TSOP	Lic	-	pdf pinout
LT1351	LO-POWER	Vs:36V Vu:80V/mV Vo:±13.8V Vi0:0.2mV	8DS	Ltc		data pdf pinout
LT1352	2xLO-POWER	Vs:36V Vu:80V/mV Vo:±13.8V Vi0:0.2mV	8DS	Ltc		data pdf pinout
LT1353	4xLO-POWER	Vs:36V Vu:80V/mV Vo:±13.8V Vi0:0.2mV	14S	Ltc		data pdf pinout
LT1354	LO-POWER	Vs:±18V Vu:36V/mV Vo:±13.8V f:12Mc	8DS	Ltc		data pdf pinout
LT1355	2xLO-POWER	Vs:±18V Vu:36V/mV Vo:±13.8V f:12Mc	8DS	Ltc		data pdf pinout
LT1356	4xLO-POWER	Vs:±18V Vu:36V/mV Vo:±13.8V f:12Mc	14D,16S	Ltc		data pdf pinout
LT1357	HI-SPEED	Vs:±18V Vu:65V/mV Vo:±13.8V f:25Mc	8DS	Ltc		data pdf pinout
LT1358	2xHI-SPEED	Vs:±18V Vu:65V/mV Vo:±13.8V f:25Mc	8DS	Ltc		data pdf pinout
LT1359	4xHI-SPEED	Vs:±18V Vu:65V/mV Vo:±13.8V f:25Mc	14D,16S	Ltc		data pdf pinout
LT1360	HI-SPEED	Vs:±18V Vu:9V/mV Vo:±13.9V Vi0:0.3mV	8DS	Ltc		data pdf pinout
LT1361	2xHI-SPEED	Vs:±18V Vu:9V/mV Vo:±13.9V Vi0:0.3mV	8DS	Ltc		data pdf pinout
LT1362	4xHI-SPEED	Vs:±18V Vu:9V/mV Vo:±13.9V Vi0:0.3mV	14D,16S	Ltc		data pdf pinout
LT1363	HI-SPEED	Vs:±18V Vu:9V/mV Vo:±14V Vi0:0.5mV	8DS	Ltc		data pdf pinout
LT1364	2xHI-SPEED	Vs:±18V Vu:9V/mV Vo:±14V Vi0:0.5mV	8DS	Ltc		data pdf pinout
LT1365	4xHI-SPEED	Vs:±18V Vu:9V/mV Vo:±14V Vi0:0.5mV	14D,16S	Ltc		data pdf pinout
LT1366	2xHI-PREC	Vs:±15V Vu:10kV/mV Vi0:200µµV	8DS	Ltc		data pdf pinout
LT1367	4xHI-PREC	Vs:±15V Vu:10kV/mV Vi0:200µµV	16S	Ltc		data pdf pinout
LT1368	2xHI-PREC	Vs:±15V Vu:10kV/mV Vi0:200µµV	8DS	Ltc		data pdf pinout
LT1369	4xHI-PREC	Vs:±15V Vu:10kV/mV Vi0:200µµV	16S	Ltc		data pdf pinout
LT1370	ES+	Vo:...35V Isw:10A	7P	Ltc		data pdf pinout
LT 1370 (C/I)R	LIN-IC	500kHz High Efficiency 6A Switching Regulator	87	Lic	-	pdf pinout
LT 1370 (C/I)T7	LIN-IC	500kHz High Efficiency 6A Switching Regulator	86	Lic	-	pdf pinout
LT 1370 HV(C/I)R	LIN-IC	500kHz High Efficiency 6A Switching Regulator	87	Lic	-	pdf pinout
LT 1370 HV(C/I)T7	LIN-IC	500kHz High Efficiency 6A Switching Regulator	86	Lic	-	pdf pinout
LT1371	ES+	Vo:...35V Isw:5.4A	7P,20S	Ltc		data pdf pinout
LT 1371 (C,HV,I)R	LIN-IC	500kHz, 3A Current Mode Switch	87	Lic	-	pdf pinout
LT 1371(C,HV,I,)SW	LIN-IC	500kHz, 3A Current Mode Switch	86	Lic	-	pdf pinout
LT 1371 (C,HV,I)T7	LIN-IC	500kHz, 3A Current Mode Switch	86	Lic	-	pdf pinout
LT1372	ES+	Vo:...35V Isw:2.7A	8DS	Ltc		data pdf pinout
LT 1372 HV(C/I)N8	LIN-IC	500kHz High Efficiency 1.5A Switching Regulators	DIC	Lic	-	pdf pinout
LT 1372 HV(C/I)S8	LIN-IC	500kHz High Efficiency 1.5A Switching Regulators	MDIP	Lic	-	pdf pinout
LT 1372 (C/I)N8	LIN-IC	500kHz High Efficiency 1.5A Switching Regulators	DIC	Lic	-	pdf pinout
LT 1372 (C/I)S8	LIN-IC	500kHz High Efficiency 1.5A Switching Regulators	MDIP	Lic	-	pdf pinout
LT1373	ES+	Vo:...35V Isw:2.7A	8DS	Ltc		data pdf pinout
LT 1373 CN8	LIN-IC	250kHz Low Supply Current High Efficiency 1.5A Switching Regulator	DIC	Lic	-	pdf pinout
LT 1373 CS8	LIN-IC	250kHz Low Supply Current High Efficiency 1.5A Switching Regulator	MDIP	Lic	-	pdf pinout
LT1374	ES+	Io=3.6A Vo:2.42...23V Vin:25V	16S,7P,8S	Ltc		data pdf pinout
LT1374-5	S+	Io=3.6A Vo:4.94...5.06V Vin:15V	7P,8S	Ltc		data pdf pinout
LT 1374 CR,IR	Z-IC	3V-15V, 4,5A, 500kHz Step-down Reg	87	Lic	-	pdf pinout
LT 1374 CS8,IS8	Z-IC	Z-IC 3V-15V, 4,5A, 500kHz Step-down	SMDIP	Lic	-	pdf pinout
LT 1374 CT7,IT7	Z-IC	3V-15V, 4,5A, 500kHz Step-down Reg	86/7Pin	Lic	-	pdf pinout
LT1375	ES+	Io=1.2A Vo:2.42...23V Vin:25V Isw:3A	8DS	Ltc		data pdf pinout
LT1375-5	S+	Io=1.2A Vo:4.94...5.06V Vin:15V	8DS	Ltc		data pdf pinout
LT1376	ES+	Io=1.2A Vo:2.42...23V Vin:25V Isw:3A	8DS,16S	Ltc		data pdf pinout
LT1376-5	S+	Io=1.2A Vo:4.94...5.06V Vin:15V	8DS	Ltc		data pdf pinout
LT1377	ES+	Vo:...35V Isw:2.7A	8S	Ltc		data pdf pinout
LT 1377 (C/I)S8	LIN-IC	1MHz High Efficiency 1.5A Switching Regulators	MDIP	Lic	-	pdf pinout
LT 1389 CN	IC	Appletalk® Peripheral Interface Transceiver	DIC	Lic	-	pdf pinout
LT 1389 CSW	IC	Appletalk® Peripheral Interface Transceiver	TSOP	Lic	-	pdf pinout
LT 1394 CS8	CMOS-OP-IC	7ns, Low Power, Single Supply, Ground- Sensing Comparator	MDIP	Lic	-	pdf pinout
LT1395	FB-OP	Vs:±6V Vu:65dB Vo:4.2V Vi0:±12mV	5S,6S,8S	Ltc		data pinout
LT1396	2xFB-OP	Vs:±6V Vu:65dB Vo:4.2V Vi0:±12mV	8S	Ltc		data pinout
LT1397	4xFB-OP	Vs:±6V Vu:65dB Vo:4.2V Vi0:±12mV	16S,14S	Ltc		data pinout
LT1398	2xFB-AMP	Vs:±6V Vu:65dB Vo:±4.2V Vi0:1.5mV	16S	Ltc		data pinout
LT1399	3xFB-AMP	Vs:±6V Vu:65dB Vo:±4.2V Vi0:1.5mV	16S	Ltc		data pinout
LT1399HV	3xFB-AMP	Vs:±7.8V Vu:65dB Vo:±4.2V Vi0:1.5mV	16S	Ltc		data pinout
LT1413	2xSS-OP	Vs:±22V Vu:1000V/mV Vi0:60µµV	8DS	Ltc		data pinout
LT1413A	2xSS-OP	Vs:±22V Vu:1000V/mV Vi0:50µµV	8D	Ltc		data pinout
LT1424-5	S+	Io=400mA Vo:5.23...5.37V Isw:1.95A	8DS	Ltc		data pdf pinout
LT1424-9	S+	Io=200mA Vo:9...9.38V Isw:1.95A	8DS	Ltc		data pdf pinout
LT1425	ES+	Vo:2.8...35V Isw:1.9A	16S	Ltc		data pdf pinout
LT 1425 (C/I)S	IC	Isolated Flyback Switching Regulator	MDIP	Lic	-	pdf pinout
LT 1431 MJ8	Z-IC, LIN-IC	Programmable Reference, Vout 2.5V - 36V.	DIC	Lic	-	pdf pinout
LT 1431 (C/I)N8	Z-IC, LIN-IC	Programmable Reference, Vout 2.5V - 36V.	DIC	Lic	-	pdf pinout
LT 1431 (C/I)S8	Z-IC, LIN-IC	Programmable Reference, Vout 2.5V - 36V.	MDIP	Lic	-	pdf pinout
LT 1431 (C/I)Z	Z-IC, LIN-IC	Programmable Reference, Vout 2.5V - 36V.	7	Lic	-	pdf pinout
LT1457	2xPREC-JFET	Vs:±20V Vu:300V/mV Vo:±13V f:1.7Mc	8DS	Ltc		data pdf pinout
LT1457A	2xPREC-JFET	Vs:±20V Vu:350V/mV Vo:±13V f:1.7Mc	8D	Ltc		data pdf pinout

687

Type	Device	Short Description	Fig.	Manu	Comparision Types	More at
LT 1460-...	LIN-IC	SMD, 2,5...10V Reference Voltage		Lic	-	
LT 1461-2.5 ...	LIN-IC	SMD, 2,5V Reference Voltage		Lic	-	
LT1462	2xJFET	Vs:±20V Vu:500V/mV Vo:±11.8V Vi0:1mV	8DS	Ltc		data pdf pinout
LT1462A	2xJFET	Vs:±20V Vu:600V/mV Vo:±12.4V	8DS	Ltc		data pdf pinout
LT1463	4xJFET	Vs:±20V Vu:500V/mV Vo:±11.8V Vi0:1mV	14DS	Ltc		data pdf pinout
LT1464	2xJFET	Vs:±20V Vu:400V/mV Vo:±13.2V Vi0:1mV	8DS	Ltc		data pdf pinout
LT1464A	2xJFET	Vs:±20V Vu:1000V/mV Vo:±13.5V f:1Mc	8DS	Ltc		data pdf pinout
LT1465	4xJFET	Vs:±20V Vu:400V/mV Vo:±13.2V Vi0:1mV	14DS	Ltc		data pdf pinout
LT1466L	2xRR-OP	Vs:±8V Vu:450V/mV Vi0:120μμV	8DS	Ltc		data pdf pinout
LT1467L	4xRR-OP	Vs:±8V Vu:450V/mV Vi0:120μμV	16S	Ltc		data pdf pinout
LT1468	HI-PREC	Vs:±36V Vu:96.5dB Vo:±13.6V f:90Mc	8DS	Ltc		data pinout
LT 1468 CS8	BiMOS-OP-IC	90MHz, 22V/ ms 16- Bit Accurate Operational Amplifier	MDIP	Lic		pdf pinout
LT1469	2xHI-PREC	Vs:±36V Vu:96.5dB Vo:±13.6V f:90Mc	8DS	Ltc		data pinout
LT1490	2xRR-OP	Vs:44V Vu:250V/mV Vo:±14.5V f:0.2Mc	8SD	Ltc		data pdf pinout
LT1490A	2xRR-OP	Vs:44V Vu:250V/mV Vo:±14.6V f:0.2Mc	8SD	Ltc		data pinout
LT1491	4xRR-OP	Vs:44V Vu:250V/mV Vo:±14.5V f:0.2Mc	14DS	Ltc		data pdf pinout
LT1491A	2xRR-OP	Vs:44V Vu:250V/mV Vo:±14.6V f:0.2Mc	14DS	Ltc		data pinout
LT1492	2xHI-PREC	Vs.36V Vu:6500V/mV Vi0:120μμV f:5Mc	8DS	Ltc		data pdf pinout
LT1493	4xHI-PREC	Vs:36V Vu:6500V/mV Vi0:120μμV f:5Mc	16S	Ltc		data pdf pinout
LT1494	2xRR-OP	Vs:36V Vu:360V/mV Vo:±14.85V	8SD	Ltc		data pinout
LT1495	2xRR-OP	Vs:36V Vu:360V/mV Vo:±14.85V	8DS	Ltc		data pdf pinout
LT1496	4xRR-OP	Vs:36V Vu:360V/mV Vo:±14.85V	14DS	Ltc		data pdf pinout
LT1498	2xRR-OP	Vs:36V Vu:5200V/mV Vi0:200μμV	8DS	Ltc		data pdf pinout
LT1499	4xRR-OP	Vs:36V Vu:5200V/mV Vi0:200μμV	14S	Ltc		data pdf pinout
LT1506	ES+	Io=3.6A Vo:2.42...14V Vin:15V	7P,8S	Ltc		data pdf pinout
LT1506-3.3	S+	Io=3.6A Vo:3.2...3.4V Vin:5V Isw:8.5A	7P,8S	Ltc		data pdf pinout
LT1507	ES+	Io=1.2A Vo:2.42...14V Vin:15V Isw:3A	8DS	Ltc		data pdf pinout
LT1507-3.3	S+	Io=1.2A Vo:3.25...3.35V Vin:15V	8DS	Ltc		data pdf pinout
LT 1509 (C/I)N	LIN-IC	Power Factor and PWM Controller	DIC	Lic	-	pdf pinout
LT 1509 (C/I)SW	LIN-IC	Power Factor and PWM Controller	TSOP	Lic	-	pdf pinout
LT 1510 CGN	IC	Constant-voltage/-current Battery Charger	43	Lic	-	pdf pinout
LT 1510 CN	IC	Constant-voltage/-current Battery Charger	43	Lic	-	pdf pinout
LT 1510 CS	IC	Constant-voltage/-current Battery Charger	43	Lic	-	pdf pinout
LT 1510 IGN	IC	Constant-voltage/-current Battery Charger	43	Lic	-	pdf pinout
LT 1510 IN	IC	Constant-voltage/-current Battery Charger	43	Lic	-	pdf pinout
LT 1510 IS	IC	Constant-voltage/-current Battery Charger	43	Lic	-	pdf pinout
LT 1513	IC	Sepic Constant-,Prog.-cur./const-volt. Battery Charger	87	Lic	-	pdf pinout
LT 1521 CS8-3.3	Z-IC, LIN-IC	0.3A, 3.3V Low Dropout Regulators with Micropower Quiescent Current	MDIP	Lic		pdf pinout
LT 1521 CS8-3	Z-IC, LIN-IC	0.3A, 3V Low Dropout Regulators with Micropower Quiescent Current	MDIP	Lic		pdf pinout
LT 1521 CS8-5	Z-IC, LIN-IC	0.3A, 5V Low Dropout Regulators with Micropower Quiescent Current	MDIP	Lic		pdf pinout
LT 1521 CS8	Z-IC, LIN-IC	0.3A Low Dropout Regulators with Micropower Quiescent Current	MDIP	Lic		pdf pinout
LT 1521 CST-3.3	Z-IC, LIN-IC	0.3A, 3.3V Low Dropout Regulators with Micropower Quiescent Current	48	Lic	-	pdf pinout
LT 1521 CST-3	Z-IC, LIN-IC	0.3A, 3V Low Dropout Regulators with Micropower Quiescent Current	48	Lic	-	pdf pinout
LT 1521 CST-5	Z-IC, LIN-IC	0.3A, 5V Low Dropout Regulators with Micropower Quiescent Current	48	Lic	-	pdf pinout
LT 1528 CQ	Z-IC, LIN-IC	3A Low Dropout Regulator for Microprocessor Applications	87	Lic	-	pdf pinout
LT 1528 CT	Z-IC, LIN-IC	3A Low Dropout Regulator for Microprocessor Applications	86	Lic	-	pdf pinout
LT 1529 CQ-3.3	Z-IC, LIN-IC	3A, 3.3V Low Dropout Regulators with Micropower Quiescent Current	87	Lic	-	pdf pinout
LT 1529 CQ-5	Z-IC, LIN-IC	3A, 5V Low Dropout Regulators with Micropower Quiescent Current	87	Lic	-	pdf pinout
LT 1529 CQ	Z-IC, LIN-IC	3A Low Dropout Regulators with Micropower Quiescent Current	87	Lic	-	pdf pinout
LT 1529 CT-3.3	Z-IC, LIN-IC	3A, 3.3V Low Dropout Regulators with Micropower Quiescent Current	86	Lic	-	pdf pinout
LT 1529 CT-5	Z-IC, LIN-IC	3A, 5V Low Dropout Regulators with Micropower Quiescent Current	86	Lic	-	pdf pinout
LT 1529 CT	Z-IC, LIN-IC	3A Low Dropout Regulators with Micropower Quiescent Current	86	Lic	-	pdf pinout
LT1533	ES+	Io=1A Vo:...100V	16S	Ltc		data pdf pinout
LT 1533 (C/I)S	Z-IC, LIN-IC	Ultralow Noise 1A Switching Regulator	MDIP	Lic	-	pdf pinout
LT1534	ES+	Io=2A Vo:5...100V	16S	Ltc		data pdf pinout
LT1534-1	ES+	Io=2A Vo:5...100V	16S	Ltc		data pdf pinout
LT 1537 G	LIN-IC	5V RS232 Transceiver,Small Capacitors	TSOP	Lic		pdf pinout
LT 1537 SW	LIN-IC	5V RS232 Transceiver,Small Capacitors	TSOP	Lic		pdf pinout
LT 1567 MS8	IC	1.4nV/√Hz 180MHz Filter Building Block	TSOP	Lic		pdf pinout
LT 1568 GN	IC	Very Low Noise, High Frequency Active RC, Filter Building Block	TSOP	Lic	-	pdf pinout

Type	Device	Short Description	Fig.	Manu	Comparision Types	More at .
LT1572	ES+	Vo:3.0...40V Isw:3A	16S	Ltc		data pdf pinout
LT 1572 CS	IC	100kHz, 1.25A Switching Regulator with Catch Diode	MDIP	Lic		pdf pinout
LT 1573 CS8-2.5	Z-IC, LIN-IC	Low Dropout Regulator Driver, 2.5V	MDIP	Lic	-	pdf pinout
LT 1573 CS8-2.8	Z-IC, LIN-IC	Low Dropout Regulator Driver, 2.8V	MDIP	Lic		pdf pinout
LT 1573 CS8-3.3	Z-IC, LIN-IC	Low Dropout Regulator Driver, 3.3V	MDIP	Lic		pdf pinout
LT 1573 CS8	Z-IC, LIN-IC	Low Dropout Regulator Driver, 1.27V - 6.8V	MDIP	Lic		pdf pinout
LT 1575 CN8-1.5	LIN-IC	Ultrafast Transient Response, Low Dropout Regulator, 1.5V Fixed	DIC	Lic		pdf pinout
LT 1575 CN8-2.8	LIN-IC	Ultrafast Transient Response, Low Dropout Regulator, 2.8V Fixed	DIC	Lic		pdf pinout
LT 1575 CN8-3.3	LIN-IC	Ultrafast Transient Response, Low Dropout Regulator, 3.3V Fixed	DIC	Lic		pdf pinout
LT 1575 CN8-3.5	LIN-IC	Ultrafast Transient Response, Low Dropout Regulator, 3.5V Fixed	DIC	Lic		pdf pinout
LT 1575 CN8-5	LIN-IC	Ultrafast Transient Response, Low Dropout Regulator, 5V Fixed	DIC	Lic		pdf pinout
LT 1575 CN8	LIN-IC	Ultrafast Transient Response, Low Dropout Regulator, Adjustable	DIC	Lic		pdf pinout
LT 1575 CS8-1.5	LIN-IC	Ultrafast Transient Response, Low Dropout Regulator, 1.5V Fixed	MDIP	Lic		pdf pinout
LT 1575 CS8-2.8	LIN-IC	Ultrafast Transient Response, Low Dropout Regulator, 2.8V Fixed	MDIP	Lic		pdf pinout
LT 1575 CS8-3.3	LIN-IC	Ultrafast Transient Response, Low Dropout Regulator, 3.3V Fixed	MDIP	Lic		pdf pinout
LT 1575 CS8-3.5	LIN-IC	Ultrafast Transient Response, Low Dropout Regulator, 3.5V Fixed	MDIP	Lic		pdf pinout
LT 1575 CS8-5	LIN-IC	Ultrafast Transient Response, Low Dropout Regulator, 5V Fixed	MDIP	Lic		pdf pinout
LT 1575 CS8	LIN-IC	Ultrafast Transient Response, Low Dropout Regulator, Adjustable	MDIP	Lic		pdf pinout
LT1576	ES+	Io=1.2A Vo:2.42...23V Vin:25V	8S	Ltc		data pdf pinout
LT1576-5	S+	Io=1.2A Vo:4.94...5.06V Vin:15V	8S	Ltc		data pdf pinout
LT 1577 CS-3.3/2.8	LIN-IC	Ultrafast Transient Response, Dual Low Dropout Regulator, 3.3V/2.8V	MDIP	Lic		pdf pinout
LT 1577 CS-3.3/ADJ	LIN-IC	Ultrafast Transient Response, Dual Low Dropout Regulator, 3.3V/Adj	MDIP	Lic		pdf pinout
LT 1577 CS	LIN-IC	Ultrafast Transient Response, Dual Low Dropout Regulator, Adj/adj	MDIP	Lic		pdf pinout
LT1578	ES+	Io=1.2A Vo:1.21...14V Vin:15V	8S	Ltc		data pdf pinout
LT1578-2.5	S+	Io=1.2A Vo:2.46...2.54V Vin:5V	8S	Ltc		data pdf pinout
LT 1579 CGN-3.3	Z-IC, LIN-IC	0.3A Dual Input Smart Battery Backup Regulator, 3.3V	TSOP	Lic		pdf pinout
LT 1579 CGN-3	Z-IC, LIN-IC	0.3A Dual Input Smart Battery Backup Regulator, 3V	TSOP	Lic	-	pdf pinout
LT 1579 CGN-5	Z-IC, LIN-IC	0.3A Dual Input Smart Battery Backup Regulator, 5V	TSOP	Lic	-	pdf pinout
LT 1579 CGN	Z-IC, LIN-IC	0.3A Dual Input Smart Battery Backup Regulator, 1.5V - 20V	TSOP	Lic		pdf pinout
LT 1579 CS-3.3	Z-IC, LIN-IC	0.3A Dual Input Smart Battery Backup Regulator, 3.3V	MDIP	Lic		pdf pinout
LT 1579 CS-3	Z-IC, LIN-IC	0.3A Dual Input Smart Battery Backup Regulator, 3V	MDIP	Lic	-	pdf pinout
LT 1579 CS-5	Z-IC, LIN-IC	0.3A Dual Input Smart Battery Backup Regulator, 5V	MDIP	Lic		pdf pinout
LT 1579 CS8-3.3	Z-IC, LIN-IC	0.3A Dual Input Smart Battery Backup Regulator, 3.3V	MDIP	Lic		pdf pinout
LT 1579 CS8-3	Z-IC, LIN-IC	0.3A Dual Input Smart Battery Backup Regulator, 3V	MDIP	Lic	-	pdf pinout
LT 1579 CS8-5	Z-IC, LIN-IC	0.3A Dual Input Smart Battery Backup Regulator, 5V	MDIP	Lic		pdf pinout
LT 1579 CS8	Z-IC, LIN-IC	0.3A Dual Input Smart Battery Backup Regulator, 1.5V - 20V	MDIP	Lic		pdf pinout
LT 1579 CS	Z-IC, LIN-IC	0.3A Dual Input Smart Battery Backup Regulator, 1.5V - 20V	MDIP	Lic	-	pdf pinout
LT 1580 CT-2.5	Z-IC, LIN-IC	7A, Very Low Dropout Regulator, 2.5V	86	Lic		pdf pinout
LT 1580 CT	Z-IC, LIN-IC	7A, Very Low Dropout Regulator, Adjustable Output	86	Lic		pdf pinout
LT 1581 CT7-2.5	Z-IC, LIN-IC	10A, Very Low Dropout Regulator, 2.5V	86	Lic		pdf pinout
LT 1581 CT7	Z-IC, LIN-IC	10A, Very Low Dropout Regulator, Adjustable	86	Lic		pdf pinout
LT1610	ES+	Vo:...30V Isw:0.9A	8S	Ltc		data pdf pinout
LT 1610 CMS8	LIN-IC	1.7MHz, Single Cell Micropower DC/DC Converter	TSOP	Lic		pdf pinout
LT 1610 CS8	LIN-IC	1.7MHz, Single Cell Micropower DC/DC Converter	MDIP	Lic		pdf pinout
LT1611	ES-	Vo:...-34V Vin:10V Isw:0.8A	5S	Ltc		data pdf pinout
LT1612	ES+	Io=480mA Vo:0.62...5V Vin:5.5V	8S	Ltc		data pdf pinout
LT1613	ES+	Vo:..34V Isw:0.8A	5S	Ltc		data pdf pinout
LT1614	ES-	Vo:...-24V Vin:10V Isw:1.2A	8S	Ltc		data pdf pinout
LT1615	ES+	Vo:..34V Isw:0.4A	5S	Ltc		data pdf pinout
LT1615-1	ES+	Vo:..34V Isw:0.125A	5S	Ltc		data pdf pinout
LT 1615	SMPS-IC	SMD, DC-DC Up Converter, Uin=2,5...4,2V	45	Lic		pdf
LT1616	ES+	Io=500mA Vo:1.25...21.75V Vin:25V	6S	Ltc		data pdf pinout

Type	Device	Short Description	Fig.	Manu	Comparision Types	More at
LT1617	ES-	Vo:...-34V Isw:0.4A	5S	Ltc		data pdf pinout
LT1617-1	ES-	Vo:...-34V Isw:0.125A	5S	Ltc		data pdf pinout
LT1618	ES+	Vo:...35V Isw:2.8A	10HS	Ltc		data pdf pinout
LT 1618 EDD	LIN-IC	Constant-Current/Constant-Voltage 1.4MHz Step-up DC/DC Converter	MLP	Lic	-	pdf pinout
LT 1618 EMS	LIN-IC	Constant-Current/Constant-Voltage 1.4MHz Step-up DC/DC Converter	TSOP	Lic		pdf pinout
LT 1619 EMS8	LIN-IC	SMD, Low Voltage Current Mode PWM Controller	8-MDIP	Lic	-	pdf pinout
LT 1619 ES8	LIN-IC	SMD, Low Voltage Current Mode PWM Controller	8-MDIP	Lic	-	pdf pinout
LT 1630 CN8	LIN/OP-IC	30MHz, 10V/ms, Dual Rail-to-Rail Input and Output Precision Op Amps	DIC	Lic	-	pdf pinout
LT 1630 CS8	LIN/OP-IC	30MHz, 10V/ms, Dual Rail-to-Rail Input and Output Precision Op Amps	MDIP	Lic		pdf pinout
LT 1631 CS8	LIN/OP-IC	30MHz, 10V/ms, Quad Rail-to-Rail Input and Output Precision Op Amps	MDIP	Lic	-	pdf pinout
LT 1634 AS8-1.25	IC	Micropower Precision Shunt Voltage Reference, 1.25V	MDIP	Lic	-	pdf pinout
LT 1634 AS8-2.5	IC	Micropower Precision Shunt Voltage Reference, 2.5V	MDIP	Lic	-	pdf pinout
LT 1634 AS8-4.096	IC	Micropower Precision Shunt Voltage Reference, 4.096V	MDIP	Lic	-	pdf pinout
LT 1634 AS8-5	IC	Micropower Precision Shunt Voltage Reference, 5V	MDIP	Lic	-	pdf pinout
LT 1634 BS8-1.25	IC	Micropower Precision Shunt Voltage Reference, 1.25V	MDIP	Lic	-	pdf pinout
LT 1634 BS8-2.5	IC	Micropower Precision Shunt Voltage Reference, 2.5V	MDIP	Lic	-	pdf pinout
LT 1634 BS8-4.096	IC	Micropower Precision Shunt Voltage Reference, 4.096V	MDIP	Lic	-	pdf pinout
LT 1634 BS8-5	IC	Micropower Precision Shunt Voltage Reference, 5V	MDIP	Lic	-	pdf pinout
LT 1634 CZ-1.25	IC	Micropower Precision Shunt Voltage Reference, 1.25V	7	Lic	-	pdf pinout
LT 1634 CZ-2.5	IC	Micropower Precision Shunt Voltage Reference, 2.5V	7	Lic	-	pdf pinout
LT 1634 CZ-4.096	IC	Micropower Precision Shunt Voltage Reference, 4.096V	7	Lic	-	pdf pinout
LT 1634 CZ-5	IC	Micropower Precision Shunt Voltage Reference, 5V	7	Lic	-	pdf pinout
LT 1634 BMS8-1.25	IC	Micropower Precision Shunt Voltage Reference, 1.25V	TSOP	Lic	-	pdf pinout
LT 1634 BMS8-2.5	IC	Micropower Precision Shunt Voltage Reference, 2.5V	TSOP	Lic	-	pdf pinout
LT 1635 (C/I)N8	LIN/OP-IC	Micropower Rail-to-Rail Op Amp and Reference	DIC	Lic	-	pdf pinout
LT 1635 (C/I)S8	LIN/OP-IC	Micropower Rail-to-Rail Op Amp and Reference	MDIP	Lic	-	pdf pinout
LT 1637	OP-IC	SMD, 2,7...44V, 1,1MHz	8-SMDIP	Lic	-	pdf
LT 1641	LIN-IC	SMD, Hot-swap Power Supply Controller	8-MDIP	Lic	-	pdf
LT1676	ES+	Io=440mA Vo:1.24...54V Vin:60V Isw:1A	8DS	Ltc		data pdf pinout
LT 1676 (C/I)N8	LIN-IC	Wide Input Range, High Efficiency, Step-down Switching Regulator	DIC	Lic	-	pdf pinout
LT 1676 (C/I)S8	LIN-IC	Wide Input Range, High Efficiency, Step-down Switching Regulator	MDIP	Lic		pdf pinout
LT 1680 (C/I)N	MOS-IC	High Power DC/DC Step-up Controller	DIC	Lic	-	pdf pinout
LT 1680 (C/I)SW	MOS-IC	High Power DC/DC Step-up Controller	TSOP	Lic	-	pdf pinout
LT 1719	KOP-IC	hi-speed		Lic	-	pdf
LT 1720	KOP-IC	Dual, hi-speed		Lic	-	pdf
LT 1721	KOP-IC	Quad, hi-speed		Lic	-	pdf
LT 1739	Si-N	HF, Vid-Tr, hi-res, 100V, 0,3A,>900MHz	2a§	Mot	-	data
LT 1761 ES5-1.5	Z-IC, LIN-IC	100mA, Low Noise, LDO Micropower Regulators, 1.5V	45	Lic	-	pdf pinout
LT 1761 ES5-1.8	Z-IC, LIN-IC	100mA, Low Noise, LDO Micropower Regulators, 1.8V	45	Lic	-	pdf pinout
LT 1761 ES5-2.5	Z-IC, LIN-IC	100mA, Low Noise, LDO Micropower Regulators, 2.5V	45	Lic	-	pdf pinout
LT 1761 ES5-2.8	Z-IC, LIN-IC	100mA, Low Noise, LDO Micropower Regulators, 2.8V	45	Lic	-	pdf pinout
LT 1761 ES5-2	Z-IC, LIN-IC	100mA, Low Noise, LDO Micropower Regulators, 2V	45	Lic	-	pdf pinout
LT 1761 ES5-3.3	Z-IC, LIN-IC	100mA, Low Noise, LDO Micropower Regulators, 3.3V	45	Lic	-	pdf pinout
LT 1761 ES5-3	Z-IC, LIN-IC	100mA, Low Noise, LDO Micropower Regulators, 3V	45	Lic	-	pdf pinout
LT 1761 ES5-5	Z-IC, LIN-IC	100mA, Low Noise, LDO Micropower Regulators, 5V	45	Lic	-	pdf pinout
LT 1761 ES5-BYP	Z-IC, LIN-IC	100mA, Low Noise, LDO Micropower Regulators	45	Lic	-	pdf pinout
LT 1761 ES5-SD	Z-IC, LIN-IC	100mA, Low Noise, LDO Micropower Regulators	45	Lic	-	pdf pinout
LT 1762	Z-IC	SMD, 2,5/3,3/5V or 1,22...20V, 150mA	8-MDIP	Lic	-	
LT 1763	Z-IC	SMD, 2,5/3,3/5V or 1,22...20V, 500mA	8-MDIP	Lic	-	pdf
LT1765	ES+	Io=2.4A Vo:1.2...22.5V Vin:25V	16S,8S	Ltc		data pdf pinout

Type	Device	Short Description	Fig.	Manu	Comparision Types	More at
LT1765-1.8	S+	Io=2.4A Vo:1.76...1.82V Vin:15V	16S	Ltc		data pdf pinout
LT1765-2.5	S+	Io=2.4A Vo:2.45...2.55V Vin:15V	16S	Ltc		data pdf pinout
LT1765-3.3	S+	Io=2.4A Vo:3.234...3.366V Vin:15V	16S	Ltc		data pdf pinout
LT1765-5	S+	Io=2.4A Vo:4.9...5.1V Vin:15V	16S	Ltc		data pdf pinout
LT1766	ES+	Io=1.2A Vo:1.2...54V Vin:60V Isw:3A	16S	Ltc		data pdf pinout
LT1766-5	S+	Io=1.2A Vo:4.94...5.06V Vin:15V	16S	Ltc		data pdf pinout
LT1767	ES+	Io=1.2A Vo:2.42...22.5V Vin:25V	8S	Ltc		data pdf pinout
LT1767-1.8	S+	Io=1.2A Vo:1.764...1.836V Vin:15V	8S	Ltc		data pdf pinout
LT1767-2.5	S+	Io=1.2A Vo:2.45...2.55V Vin:15V	8S	Ltc		data pdf pinout
LT1767-3.3	S+	Io=1.2A Vo:3.234...3.366V Vin:15V	8S	Ltc		data pdf pinout
LT1767-5	S+	Io=1.2A Vo:4.9...6.1V Vin:15V Isw:3A	8S	Ltc		data pdf pinout
LT 1769 CFE	IC	Const-Cur./ Const-Vol. 2A Battery Charger	87	Lic	-	pdf pinout
LT 1769 CGN	IC	Const-Cur./ Const-Vol. 2A Battery Charger	87	Lic	-	pdf pinout
LT 1769 IFE	IC	Const-Cur./ Const-Vol. 2A Battery Charger	87	Lic	-	pdf pinout
LT 1769 IGN	IC	Const-Cur./ Const-Vol. 2A Battery Charger	87	Lic	-	pdf pinout
LT1776	ES+	Io=550mA Vo:1.24...36V Vin:40V Isw:1A	8DS	Ltc		data pdf pinout
LT 1776 (C/I)N8	LIN-IC	Wide Input Range, High Efficiency, Step-down Switching Regulator	DIC	Lic	-	pdf pinout
LT 1776 (C/I)S8	LIN-IC	Wide Input Range, High Efficiency, Step-down Switching Regulator	MDIP	Lic	-	pdf pinout
LT1777	ES+	Io=430mA Vo:1.24...43.2V Vin:48V	16S	Ltc		data pdf pinout
LT 1780 N	LIN-IC	5V RS232 Dual Driver/Receiver, ±15kV ESD Protec.	TSOP	Lic	-	pdf pinout
LT 1780 SW	LIN-IC	5V RS232 Dual Driver/Receiver, ±15kV ESD Protec.	TSOP	Lic		pdf pinout
LT 1781 N	LIN-IC	5V RS232 Dual Driver/Receiver, ±15kV ESD Protec.	TSOP	Lic		pdf pinout
LT 1781 S	LIN-IC	5V RS232 Dual Driver/Receiver, ±15kV ESD Protec.	TSOP	Lic		pdf pinout
LT 1781 SW	LIN-IC	5V RS232 Dual Driver/Receiver, ±15kV ESD Protec.	TSOP	Lic	-	pdf pinout
LT 1782	OP-IC	SMD, 2,7...18V, InOffs.<0,8mV, 200kHz	45	Lic	-	pdf
LT 1783	OP-IC	SMD, 2,7...18V, InOffs.<0,8mV, 1,25MHz	45	Lic	-	pdf
LT 1787	LIN-IC	SMD, Current Monitoring Amp., 2,5...36V	8-(S)MDIP	Lic	-	pdf
LT 1791 AN	LIN-IC	60V Fault Protected RS485/ RS422 Transceivers	TSOP	Lic	-	pdf pinout
LT 1791 AS	LIN-IC	60V Fault Protected RS485/ RS422 Transceivers	TSOP	Lic		pdf pinout
LT 1791 N	LIN-IC	60V Fault Protected RS485/ RS422 Transceivers	TSOP	Lic		pdf pinout
LT 1791 S	LIN-IC	60V Fault Protected RS485/ RS422 Transceivers	TSOP	Lic		pdf pinout
LT1792	JFET	Vs:±20V Vu:4500V/mV Vo:±13.2V	8DS	Ltc		data pdf pinout
LT1792A	JFET	Vs:±20V Vu:4800V/mV Vo:±13.2V	8DS	Ltc		data pdf pinout
LT1793	JFET	Vs:±20V Vu:4400V/mV Vo:±13.2V	8DS	Ltc		data pdf pinout
LT1793A	JFET	Vs:±20V Vu:4500V/mV Vo:±13.2V	8DS	Ltc		data pdf pinout
LT 1796	A/D-IC	Overvol.fault Protected Can Transceiv.	QFN	Lic	-	pdf pinout
LT 1814	Si-N	HF, Vid-Tr, hi-res, 120V, 0,4A, >1GHz	55r	Mot	-	data
LT 1817	Si-N	HF, Vid-Tr, hi-res, 120V, 0,4A, >1GHz	55v	Mot	-	data
LT 1839	Si-N	HF, Vid-Tr, hi-res, 120V, 0,3A, >1GHz	2a§	Mot	-	data
LT 1910	MOS-IC	Protected High Side Mosfet Driver	MDIP	Ltc		pdf pinout
LT1930	ES+	Vo:1.24...34V Vin:16V Isw:2.5A	5S	Ltc		data pdf pinout
LT1930A	ES+	Vo:1.24...34V Vin:16V Isw:2.5A	5S	Ltc		data pdf pinout
LT1931	ES-	Vo:...-34V Vin:16V Isw:2.5A	5S	Ltc		data pdf pinout
LT1931A	ES-	Vo:...-34V Vin:16V Isw:2.5A	5S	Ltc		data pdf pinout
LT1932	ES+	Io=45mA Vo:...34V Vin:10V Isw:0.78A	6S	Ltc		data pdf pinout
LT1934	ES+	Vo:1.25...30V Vin:34V Isw:0.49A	6SH	Ltc		data pdf pinout
LT1937	ES+	Vo:...-34V Vin:10V	5S,6S	Ltc		data pdf pinout
LT1940	E2S	Io=1.4A Vo:1.25...22.5V Vin:25V	16S	Ltc		data pdf pinout
LT1940L	E2S	Io=1.4A Vo:1.25...5.6V Vin:7V	16S	Ltc		data pdf pinout
LT 1942 EUF	LIN-IC	Quad DC/DC Converter for Triple Output TFT Supply Plus LED Driver	QFN	Lic		pdf pinout
LT1943	E4S	Vo:...40V Isw:0.7A	28S	Ltc		data pdf pinout
LT1944	E2S	Vo:...34V Vin:15V Isw:0.4A	10S	Ltc		data pdf pinout
LT1944-1	E2S	Vo:...34V Vin:15V Isw:225mA	10S	Ltc		data pdf pinout
LT1945	E2S	Vo:...-34V Vin:15V Isw:0.4A	10S	Ltc		data pdf pinout
LT1946	ES+	Vo:...34V Vin:16V Isw:3.1A	8S	Ltc		data pdf pinout
LT1946A	ES+	Vo:...34V Vin:16V Isw:3.1A	8S	Ltc		data pdf pinout
LT1947	E3S	Vo:...34V Isw:1A	10S	Ltc		data pdf pinout
LT1949	ES+	Io=1.2A Vo:1.2...54V Vin:60V Isw:3A	8S,16S	Ltc		data pdf pinout
LT 1949	SMPS-IC	SMD, PWM DC-DC Converter, Uin=1,5...10V	8-SMDIP	Lic		pdf
LT1959	ES+	Io=3.6A Vo:1.2...14V Vin:15V Isw:8.5A	7P,8S	Ltc		data pdf pinout
LT1961	ES+	Vo:1.2...34V Vin:25V Isw:3A	8S	Ltc		data pdf pinout
LT 1962 EMS8-1.5	LIN-IC	300mA, Low Noise, Micropower LDO Regulators, 1.5V	TSOP	Lic	-	pdf pinout
LT 1962 EMS8-1.8	LIN-IC	300mA, Low Noise, Micropower LDO Regulators, 1.8V	TSOP	Lic	-	pdf pinout
LT 1962 EMS8-2.5	LIN-IC	300mA, Low Noise, Micropower LDO Regulators, 2.5V	TSOP	Lic	-	pdf pinout
LT 1962 EMS8-3.3	LIN-IC	300mA, Low Noise, Micropower LDO Regulators, 3.3V	TSOP	Lic	-	pdf pinout

Type	Device	Short Description	Fig.	Manu	Comparision Types	More at
LT 1962 EMS8-3	LIN-IC	300mA, Low Noise, Micropower LDO Regulators, 3V	TSOP	Lic	-	pdf pinout
LT 1962 EMS8-5	LIN-IC	300mA, Low Noise, Micropower LDO Regulators, 5V	TSOP	Lic	-	pdf pinout
LT 1962 EMS8	LIN-IC	300mA, Low Noise, Micropower LDO Regulators, 1.22V - 20V	TSOP	Lic	-	pdf pinout
LT1976	ES+	Io=1.2mA Vo:1.2...54V Vin:60V	16S	Ltc		data pdf pinout
LT1976B	ES+	Io=1.2A Vo:1.2...54V Vin:60V Isw:4A	16S	Ltc		data pdf pinout
LT 1994 DD	IC	Fully Differential I/O Amplifier/driver	TSOP	Lic		pdf pinout
LT 2001	Si-N	UHF-A, 40V, 0,2A, 3GHz	55r	Mot	MRF 587[Motorola]	data
LT2001	OC	LED/2 LDR Viso:2500V Ibs:20mA		Hei		data pinout
LT2002	OC	LED/2 LDR Viso:2500V Ibs:20mA		Hei		data pinout
LT2011	OC	LED/2 LDR Viso:2500V Ibs:20mA		Hei		data pinout
LT2078	2xμ-POWER	Vs:±22V Vu:5000V/mV Vo:±13.2V	8S	Ltc		data pdf pinout
LT2078A	2xμ-POWER	Vs:±22V Vu:5000V/mV Vo:±13.2V	8S	Ltc		data pdf pinout
LT2079	4xμ-POWER	Vs:±22V Vu:5000V/mV Vo:±13.2V	14S	Ltc		data pdf pinout
LT2079A	4xμ-POWER	Vs:±22V Vu:5000V/mV Vo:±13.2V	14S	Ltc		data pdf pinout
LT2178	2xHI-PREC	Vs:±22V Vu:1000¹/mV Vo:±12.7V	8S	Ltc		data pdf pinout
LT2178A	2xHI-PREC	Vs:±22V Vu:1200V/mV Vo:±12.7V	8S	Ltc		data pdf pinout
LT2179	4xHI-PREC	Vs:±22V Vu:1000V/mV Vo:±12.7V	14S	Ltc		data pdf pinout
LT2179A	4xHI-PREC	Vs:±22V Vu:1200V/mV Vo:±12.7V	14S	Ltc		data pdf pinout
LT 3003	LIN-IC	3-Channel LED Ballaster with PWM	TSOP	Lic		pdf pinout
LT 3005	Si-N	UHF-A, 40V, 0,2A, 3GHz	55r	Mot	MRF 587[Motorola]	data
LT3011	OC	LED/LDR Viso:3000V Ibs:20mA		Hei		data pinout
LT 3014	Si-N	UHF-A, 40V, 0,2A, 3GHz	55r	Mot	MRF 587[Motorola]	data
LT 3046	Si-N	UHF-A, 40V, 0,15A, 3GHz	2a	Mot	BRF 96, MRF 965[Motorola]	data
LT3333	LED	gr ø3.0mm I:710...2800mcd Vf:3.3V	T1	Osr		data pdf pinout
LT3405	ES+	Io=300mA Vo:0.8...5.5V Vin:5.5V	5S	Ltc		data pdf pinout
LT3430	ES+	Io=2.75A Vo:1.2...57.6V Vin:60V	16S	Ltc		data pdf pinout
LT3431	ES+	Io=2.75A Vo:1.2...57.6V Vin:60V	16S	Ltc		data pdf pinout
LT3433	ES+	Io=400mA Vo:3.3...20V Isw:0.9A	16S	Ltc		data pdf pinout
LT3436	ES+	Vo:1.2...34V Vin:25V Isw:6A	16S	Ltc		data pdf pinout
LT3460	ES+	Vo:...36V Isw:0.6A	6S,5S	Ltc		data pdf pinout
LT3461	ES+	Vo:..38V Isw:0.6A	6S	Ltc		data pdf pinout
LT3461A	ES+	Vo:..38V Isw:0.6A	6S	Ltc		data pdf pinout
LT3462	ES-	Vo:..-38V Isw:0.6A	6S	Ltc		data pdf pinout
LT3462A	ES-	Vo:..-38V Isw:0.6A	6S	Ltc		data pdf pinout
LT3463	E2S	Vo:..40V Isw:0.32A	10H	Ltc		data pdf pinout
LT3463A	E2S	Vo:..40V Isw:0.46A	10H	Ltc		data pdf pinout
LT3464	ES+	Vo:...34V Isw:140mA	8S	Ltc		data pdf pinout
LT3465	ES+	Vo:...-30V	6S	Ltc		data pdf pinout
LT 3465 (A)	LIN-IC	1.2MHz/2.4MHz White LED Drivers with Built-in Schottky	46	Lic		pdf pinout
LT3466	E2S	Vo:...39.5V Vin:15V	10H,16S	Ltc		data pdf pinout
LT 3466 -1	LIN-IC	White LED Driver and Boost Converter,Up to 10 White LEDs	MLP	Lic		pdf pinout
LT 3466 EDD	LIN-IC	Dual Full Function White LED Step-up Converter w. Built-In Schottky Diodes	MLP	Lic		pdf pinout
LT 3466 EFE	LIN-IC	Dual Full Function White LED Step-up Converter w. Built-In Schottky Diodes	TSOP	Lic	-	pdf pinout
LT3467	ES+	Vo:...40V Isw:2.5A	6S	Ltc		data pdf pinout
LT 3474 (-1)	LIN-IC	Step-down 1A LED Driver	TSOP	Lic	-	pdf pinout
LT 3475 (-1)	LIN-IC	Dual Step-down 1.5A LED Driver	TSOP	Lic	-	pdf pinout
LT 3476	LIN-IC	High Current Quad Output LED Driver, 4x 1.5A, 36V	QFN	Lic		pdf pinout
LT 3477 EFE	LIN-IC	3A, DC/ DC Converter with Dual Rail-to- Rail Current Sense	TSOP	Lic		pdf pinout
LT 3477 EUF	LIN-IC	3A, DC/ DC Converter with Dual Rail-to- Rail Current Sense	QFN	Lic		pdf pinout
LT 3478 (-1)	LIN-IC	4.5A Monolithic LED Drivers with True Color PWM Dimming	TSOP	Lic	-	pdf pinout
LT 3486 EDHC	LIN-IC	Dual 1.3A White LED Step-up Converters with Wide Dimming	MLP	Lic		pdf pinout
LT 3486 EFE	LIN-IC	Dual 1.3A White LED Step-up Converters with Wide Dimming	TSOP	Lic		pdf pinout
LT 3491 EDC	LIN-IC	White LED Driver with Integrated Schottky	MLP	Lic		pdf pinout
LT 3491 ESC8	LIN-IC	White LED Driver with Integrated Schottky	47	Lic		pdf pinout
LT 3496	LIN-IC	Triple Output LED Driver	QFN	Lic	-	pdf pinout
LT 3497	LIN-IC	Dual Full Function White LED Driver with Integrated Schottky Diodes	MLP	Lic	-	pdf pinout
LT 3498	LIN-IC	20mA LED Driver and Oled Driver with Integrated Schottky	MLP	Lic		pdf pinout
LT 3517 EFE	LIN-IC	Full-Featured LED Driver with 1.5A Switch Current	TSOP	Lic	-	pdf pinout
LT 3517 EUF	LIN-IC	Full-Featured LED Driver with 1.5A Switch Current	QFN	Lic	-	pdf pinout
LT 3518	LIN-IC	Full-Featured LED Driver with 2.3A Switch Current	QFN	Lic		pdf pinout
LT 3590 EDC	LIN-IC	48V Buck Mode LED Driver	MLP	Lic	-	pdf pinout
LT 3590 ESC8	LIN-IC	48V Buck Mode LED Driver	47	Lic	-	pdf pinout
LT 3591	LIN-IC	White LED Driver with Integrated Schottky	MLP	Lic	-	pdf pinout

Type	Device	Short Description	Fig.	Manu	Comparision Types	More at
LT 3595	LIN-IC	16 Channel Buck Mode LED Driver	QFN	Lic	-	pdf pinout
LT 4217	Si-N	UHF-A, 20V, 0,4A, 5,5GHz	55r	Mot		data
LT 4220 GN	IC	Dual Supply Hot Swap Controller	86	Lic	-	pdf pinout
LT 4239	Si-N	=LT 4217: 5GHz	2a§	Mot		data
LT 4254 GN	IC	Positive High Volt. Hot Swap Controller	TSOP	Ltc		pdf pinout
LT 4356 DE-1	LIN-IC	Overvoltage Protection Regulator and Inrush Limiter	MLP	Lic	-	pdf pinout
LT 4356 MS-1	LIN-IC	Overvoltage Protection Regulator and Inrush Limiter	TSOP	Lic	-	pdf pinout
LT 4700	Si-N	UHF, 20V, 50mA, 6GHz	52r	Mot	MRF 572[Motorola]	data
LT 4746	Si-N	UHF, 20V, 50mA, 6GHz	2a§	Mot	MRF 905[Motorola]	data
LT 4772	Si-N	=LT 4746	5g	Mot	MRF 914[Motorola]	data
LT5433	LED	gr ø5.0mm I:280..2800mcd Vf:3.3V	T1¾	Osr		data pdf pinout
LT 5500 EGN	CMOS-IC	1.8-2.7GHz Receiver Front End	TSOP	Lic	-	pdf pinout
LT 5502 EGN	IC	400MHz Quadrature IF Demodulator with RSSI	TSOP	Lic	-	pdf pinout
LT 5503 EFE	IC	1.2-2.7GHz Direct IQ Modulator & Mixer	TSOP	Lic	-	pdf pinout
LT 5506	IC	40MHz-500MHz Quadrature Dem., VGA	QFN	Lic		pdf pinout
LT 5511	IC	High Signal Level Upconverting Mixer	TSOP	Lic		pdf pinout
LT 5512 EUF	IC	1kHz-3GHz High Signal Level Active Mixer	QFN	Lic		pdf pinout
LT 5514 FE	IC	IF Amp/adc Driver, Digitally Controlled Gain	TSOP	Lic		pdf pinout
LT 5516	IC	800MHz- 1.5GHz Direct Conversion Quadrature Dem.	QFN	Lic		pdf pinout
LT 5518	IC	1.5GHz–2.4GHz Hi Lin Direct Quadrature Modulator	QFN	Lic		pdf pinout
LT 5519	IC	0.7GHz to 1.4GHz High Linearity Upconverting Mixer	QFN	Lic		pdf pinout
LT 5521	IC	Very High Linearity Active Mixer	QFN	Lic		pdf pinout
LT 5522 EUF	IC	400MHz to 2.7GHz High Signal Level Downconverting Mixer	QFN	Lic		pdf pinout
LT 5524 FE	IC	IF Amp/adc Driver, Digitally Controlled Gain	TSOP	Lic		pdf pinout
LT 5525 EUF	IC	High Linearity, Low Power Downconverting Mixer	QFN	Lic		pdf pinout
LT 5526 EUF	IC	High Linearity, Low Power Downconverting Mixer	QFN	Lic		pdf pinout
LT 5527 EUF	IC	400MHz to 3.7GHz High Signal Level Downconverting Mixer	QFN	Lic		pdf pinout
LT 5528	IC	1.5GHz-2.4GHz Hi Line Direct Quadrature Modulator	QFN	Lic		pdf pinout
LT 5546	IC	40MHz-500MHz VGA & I/q Dem.	QFN	Lic		pdf pinout
LT 5557 EUFPBF	A/D-D/A-IC	Switched Capacitor Voltage Converter	81	Lic		pdf pinout
LT 5557 EUFPBF	IC	400MHz to 3.8GHz 3.3V High Signal Level Downconverting Mixer	QFN	Lic		pdf pinout
LT 5558	IC	600MHz-1100MHz Hi Lin Direct Quadrature Modulator	QFN	Lic		pdf pinout
LT 5560	IC	0.01MHz to 4GHz Low Power Active Mixer	MLP	Lic		pdf pinout
LT 5568 -2	IC	GSM/Edge Optimized, Hi Lin Direct I/q Modulator	QFN	Lic		pdf pinout
LT 5570 DD	LIN-IC	40MHz to 2.7GHz Mean-Squared Power Detector	TSOP	Lic		pdf pinout
LT 5571	IC	620MHz–1100MHz Hi Lin Direct I/ Q Modulator	QFN	Lic		pdf pinout
LT 5572	IC	1.5GHz-2.5GHz Direct I/q Modulator	QFN	Lic		pdf pinout
LT 5575	IC	800MHz-2.7GHz Hi Lin Direct Conversion Quadrature Demo	QFN	Lic		pdf pinout
LT 5817	Si-P	HF, Vid-Tr, hi-res, 80V, 0,4A, >1,5GHz	55v	Mot		data
LT 5839	Si-P	HF, Vid-Tr, hi-res, 80V, 0,3A, >1,5GHz	2a§	Mot		data
LT6600	LED	gr/rd ø26mm I:1800mcd Vf:9.3V		Sha		data pdf pinout
LT 6600 S8-10	IC	Very Low Noise, Differential Amplifier and 10MHz Lowpass Filter	TSOP	Lic		pdf pinout
LT 6600 DF-10	IC	Very Low Noise, Differential Amplifier and 10MHz Lowpass Filter	MLP	Lic		pdf pinout
LT 6600 S8-15	IC	Very Low Noise, Differential Amplifier and 15MHz Lowpass Filter	TSOP	Lic		pdf pinout
LT 6600 S8-2.5	IC	Very Low Noise, Differential Amplifier and 2.5MHz Lowpass Filter	TSOP	Lic		pdf pinout
LT 6600 S8-20	IC	Very Low Noise, Differential Amplifier and 20MHz Lowpass Filter	TSOP	Lic		pdf pinout
LT 6600 DF-2.5	IC	Very Low Noise, Differential Amplifier and 2.5MHz Lowpass Filter	MLP	Lic		pdf pinout
LT6650	LED	gr/rd/bl ø26mm I<2000mcd Vf:<9.2V		Sha		data pinout
LT6701	LED	gr/rd ø52mm I=7cd Vf:<18.8V If:60mA		Sha		data pinout
LT6710	LED	gr/rd ø52mm I=6cd Vf:18.5V If:60mA		Sha		data pdf pinout
LT6720	LED	gr/rd ø52mm I=6cd Vf:18.5V If:80mA		Sha		data pinout
LT6725	LED	gr/rd ø52mm I=4.5cd Vf:13.2V If:80mA		Sha		data pinout
LT9010	LED	rd+gr+yl 19.9x7mm I:5mcd Vf:1.9V		Sha		data pinout
LT9040	LED	rd+gr+yl 17x13mm I:112mcd Vf:2.05V		Sha		data pinout
LT9200	LED	rd+yl 10x4mm I:4mcd Vf:1.9V If:10mA		Sha		data pinout
LT9210	LED	rd+yl 11.4x:3.9mm I:9mcd Vf:1.9V		Sha		data pinout
LT9230	LED	rd+yl 16x7mm I:9mcd Vf:1.9V If:10mA		Sha		data pinout
LT9310	LED	rd+gr+yl 45x18mm I:6.4mcd Vf:1.9V		Sha		data pinout
LT9323	LED	rd+gr+yl 65x17.1mm I:8.5mcd Vf:2V		Sha		data pinout

Type	Device	Short Description	Fig.	Manu	Comparision Types	More at
LT9325	LED	gr 30x17mm I:30mcd Vf:2.2V If:40mA		Sha		data pinout
LT9400	LED	rd+yl+yl/rd 9x7mm I:35mcd Vf:2V	DIP-8	Sha		data pinout
LT9525	LED	srd+gr+yl+or ø20mm I:80mcd Vf:2V		Sha		data pinout
LT9526	LED	srd+gr+yl ø20mm I:250mcd Vf:2V		Sha		data pinout
LT9550	LED	gr+gr/rd ø7.5mm I:700mcd Vf:2.2V	T1¾	Sha		data pdf pinout
LT9552	LED	gr+rd ø7.5mm I:200mcd Vf:1.75V	T1¾	Sha		data pinout
LT9560	LED	srd+gr+yl+or ø10mm I:8000mcd If:20mA		Sha		data pdf pinout
LT9562	LED	rd ø10mm I:1400mcd Vf:1.75V If:20mA		Sha		data pinout
LT 1026 CJ 8	IC	Voltage Converter	QFN	Lic	-	pdf pinout
LT 1026 CN 8	IC	Voltage Converter	QFN	Lic	-	pdf pinout
LT 1026 CS 8	IC	Voltage Converter	QFN	Lic	-	pdf pinout
LT 1026 IS 8	IC	Voltage Converter	QFN	Lic	-	pdf pinout
LT 1026 MJ 8	IC	Voltage Conve.ter	QFN	Lic	-	pdf pinout
LT 1027 CH-5	LIN-IC	Precision 5V Reference	DIP	Lic	-	pdf pinout
LTC 1041 (MJ/CN) 8	LIN-IC	Bang-bang Controller	TSOP	Lic	-	pdf pinout
LT 1073 CN 8	LIN-IC	Micropower DC-DC Converter Adjustable and Fixed 5V, 12V	DIP	Lic	-	pdf pinout
LT 1073 CS 8	LIN-IC	Micropower DC-DC Converter Adjustable and Fixed 5V, 12V	DIP	Lic	-	pdf pinout
LT 1102 J 8	LIN-IC	JFET Input Instrumental Amp (fixed Gain=10 or 100)	43	Lic	-	pdf pinout
LT 1102 N 8	LIN-IC	JFET Input Instrumental Amp (fixed Gain=10 or 100)	43	Lic	-	pdf pinout
LT 1107 CN 8	LIN-IC	DC/DC Converter Adj. and Fixed 5V, 12V	43	Lic	-	pdf pinout
LT 1107 CS 8	LIN-IC	DC/DC Converter Adj. and Fixed 5V, 12V	43	Lic	-	pdf pinout
LT 1107 MJ 8	LIN-IC	DC/DC Converter Adj. and Fixed 5V, 12V	43	Lic	-	pdf pinout
LT 1108 CN 8	LIN-IC	DC/DC Converter Adjustable and Fixed 5V, 12V	43	Lic	-	pdf pinout
LT 1108 CS 8	LIN-IC	DC/DC Converter Adjustable and Fixed 5V, 12V	43	Lic	-	pdf pinout
LT 1111 CN 8	LIN-IC	DC/DC Converter Adjustable and Fixed 5V, 12V	43	Lic	-	pdf pinout
LT 1111 CS 8	LIN-IC	DC/DC Converter Adjustable and Fixed 5V, 12V	43	Lic	-	pdf pinout
LT 1111 MJ 8	LIN-IC	DC/DC Converter Adjustable and Fixed 5V, 12V	43	Lic	-	pdf pinout
LT 1167 (AC/AI) S8	LIN-IC	Gain Programmable, Instrumentation Amp.	TSOP	Lic	-	pdf pinout
LT 1167 (C/I) S8	LIN-IC	Gain Programmable, Instrumentation Amp.	TSOP	Lic	-	pdf pinout
LT 1173 CN 8	LIN-IC	DC/DC Converter Adjustable and Fixed 5V, 12V	TSOP	Lic	-	pdf pinout
LT 1173 CS 8	LIN-IC	DC/DC Converter Adjustable and Fixed 5V, 12V	TSOP	Lic	-	pdf pinout
LT 1175 (C/I) ST-5	LIN-IC	500mA Negative Low Dropout Micropower Regulator	TSOP	Lic	-	pdf pinout
LT 1175 (C/I)N 8	LIN-IC	500mA Negative Low Dropout Micropower Regulator	TSOP	Lic	-	pdf pinout
LT 1175 (C/I) S 8	LIN-IC	500mA Negative Low Dropout Micropower Regulator	TSOP	Lic	-	pdf pinout
LT 1176 CN 8	LIN-IC	Step-down Switching Regulator	TSOP	Lic	-	pdf pinout
LT 1510-5 CGN	IC	Constant-voltage/-current Battery Charger	43	Lic	-	pdf pinout
LT 1510-5 IGN	IC	Constant-voltage/-current Battery Charger	43	Lic	-	pdf pinout
LT 1510 CS 8	IC	Constant-voltage/-current Battery Charger	43	Lic	-	pdf pinout
LT 1510 IS 8	IC	Constant-voltage/-current Battery Charger	43	Lic	-	pdf pinout
LT 1513-2	IC	Sepic Constant-,Prog.-cur./const-volt. Battery Charger	87	Lic	-	pdf pinout
LT 1571 EGN-1	IC	Sepic Const-,Progr.-cur./const-vol. Battery Charger	87	Lic	-	pdf pinout
LT 1571 EGN-2	IC	Sepic Const-,Progr.-cur./const-vol. Battery Charger	87	Lic	-	pdf pinout
LT 1571 EGN-5	IC	Sepic Const-,Progr.-cur./const-vol. Battery Charger	87	Lic	-	pdf pinout
LT 1640 AHN 8	IC	Negative Voltage Hot Swap Controller	86	Lic		pdf pinout
LT 1640 AHS 8	IC	Negative Voltage Hot Swap Controller	86	Lic		pdf pinout
LT 1640 ALN 8	IC	Negative Voltage Hot Swap Controller	86	Lic		pdf pinout
LT 1640 ALS 8	IC	Negative Voltage Hot Swap Controller	86	Lic		pdf pinout
LT 1640 HN 8	IC	Negative Voltage Hot Swap Controller	86	Lic		pdf pinout
LT 1640 HS 8	IC	Negative Voltage Hot Swap Controller	86	Lic		pdf pinout
LT 1640 LN 8	IC	Negative Voltage Hot Swap Controller	86	Lic		pdf pinout
LT 1640 LS 8	IC	Negative Voltage Hot Swap Controller	86	Lic		pdf pinout
LT 1785 AN 8	LIN-IC	60V Fault Protected RS485/ RS422 Transceivers	TSOP	Lic		pdf pinout
LT 1785 AS 8	LIN-IC	60V Fault Protected RS485/ RS422 Transceivers	TSOP	Lic		pdf pinout
LT 1785 N 8	LIN-IC	60V Fault Protected RS485/ RS422 Transceivers	TSOP	Lic		pdf pinout
LT 1785 S 8	LIN-IC	60V Fault Protected RS485/ RS422 Transceivers	TSOP	Lic		pdf pinout
LT 1993-2 UD	A/D-IC	Differential Amp/ ADC Driver (AV = 2V/V)	TSOP	Lic		pdf pinout
LT 1993-4 UD	A/D-IC	Differential Amp/ ADC Driver (AV = 4V/V)	TSOP	Lic		pdf pinout
LT 1994 MS 8	IC	Fully Differential I/O Amplifier/driver	TSOP	Lic		pdf pinout
LT 4250 HN 8	IC	Negative 48V Hot Swap Controller	TSOP	Ltc		

Type	Device	Short Description	Fig.	Manu	Comparision Types	More at
LT 4250 HS 8	IC	Negative 48V Hot Swap Controller	TSOP	Ltc	-	pdf pinout
LT 4250 LN 8	IC	Negative 48V Hot Swap Controller	TSOP	Ltc	-	pdf pinout
LT 4250 LS 8	IC	Negative 48V Hot Swap Controller	TSOP	Ltc	-	pdf pinout
LT 4256-3 GN	IC	Positive High Volt. Hot Swap Controller	TSOP	Ltc	-	pdf pinout
LT 4356-1 MS	IC	Overvolt. Protection Regulator, Inrush Limiter	TSOP	Ltc	-	pdf pinout
LT 5504 MS 8	LIN-IC	800MHz to 2.7GHz Rf Measuring Receiver	TSOP	Lic		pdf pinout
LT 5534 SC 6	LIN-IC	50MHz to 3GHz Rf Power Detector 60dB Dynamic Range	TSOP	Lic		pdf pinout
LT 6402-6 UD	A/D-IC	Differential Amp/ ADC Driver (AV = 6dB)	TSOP	Lic		pdf pinout
LT 6703 CDC-2	IC	Comparator with 400mV Reference	TSOP	Lic	-	pdf pinout
LT 6703 HDC-2	IC	Comparator with 400mV Reference	TSOP	Lic		pdf pinout
LT 6703 IDC-2	IC	Comparator with 400mV Reference	TSOP	Lic		pdf pinout
LT 6703 CDC-3	IC	Comparator with 400mV Reference	TSOP	Lic		pdf pinout
LT 6703 HDC-3	IC	Comparator with 400mV Reference	TSOP	Lic		pdf pinout
LT 6703 IDC-3	IC	Comparator with 400mV Reference	TSOP	Lic		pdf pinout
LT 1027 CN 8-5	LIN-IC	Precision 5V Reference	DIP	Lic		pdf pinout
LT 1027 DCS 8-5	LIN-IC	Precision 5V Reference	DIP	Lic		pdf pinout
LT 1176 CN 8-5	LIN-IC	Step-down Switching Regulator	TSOP	Lic		pdf pinout
LT 1641-1 S 8	IC	Posit. High Volt. Hot Swap Controllers	86	Lic		pdf pinout
LT 1641-2 S 8	IC	Posit. High Volt. Hot Swap Controllers	86	Lic		pdf pinout
LT 1993-10 UD	A/D-IC	Differential Amp/ ADC Driver (AV =10V/V)	TSOP	Lic		pdf pinout
LT 6402-12 UD	A/D-IC	Differential Amp/ ADC Driver (AV = 12dB)	TSOP	Lic		pdf pinout
LT 6402-20	A/D-IC	Differential Amplifier/adc Driver	QFN	Lic		pdf pinout
LT 6600-10 DF	IC	Differential Amp, 10MHz Lowpass Filter	DFN	Lic		pdf pinout
LT 6600-2.5 DF	IC	Differential Amp, 2.5MHz Lowpass Filter	DFN	Lic		pdf pinout
LT 6600-5 S 8	IC	Differential Amp, 5MHz Lowpass Filter	DFN	Lic		pdf pinout
LT 4356-1 DE 12	IC	Overvolt. Protection Regulator, Inrush Limiter	TSOP	Ltc		pdf pinout
LT 6600-10 S 8	IC	Differential Amp, 10MHz Lowpass Filter	DFN	Lic		pdf pinout
LT 6600-15 S 8	IC	Differential Amp, 15MHz Lowpass Filter	DFN	Lic		pdf pinout
LT 6600-20 S 8	IC	Differential Amp, 20MHz Lowpass Filter	DFN	Lic		pdf pinout
LT 6600-2.5 S 8	IC	Differential Amp, 2.5MHz Lowpass Filter	DFN	Lic		pdf pinout
LTA-1000R	LED	or+gr+srd+yl 1.52x58mm I:2mcd		Lit		data pinout
LTA67C	LED	gr 1.52x58mm I:224...560mcd Vf:3.5V		Osr		data pdf pinout
LTA673	LED	gr 1.52x58mm I:140...355mcd Vf:3.3V		Osr		data pdf pinout
LTA709	UNI	Vs:±15V Vu:81dB Vo:±12V Vi0:10mV	d	Amp		data pinout
LTA741	UNI	Vs:±15V Vu:86dB Vo:±12V Vi0:6mV	d	Amp		data pinout
LTA747	2xOP	Vs:±15V Vu:87dB Vo:±12V Vi0:5mV	d	Amp		data pinout
LTC	Z-Di	=1SMB 18CA (Typ-Code/Stempel/marking)	71(5x3,5)	Ons	→1SMB 10CA...78CA	
LTC144	2xCOMP	Vs:±5V Vi0:±3mV	8DS,16DS	Ltc		data pdf pinout
LTC 221MJ/CJ/CN/CS	CMOS-IC	Quad Cmos Analog Switches, Data Latches	TSOP	Lic	-	pdf pinout
LTC 222MJ/CJ/CN/CS	CMOS-IC	Quad Cmos Analog Switches, Data Latches	TSOP	Lic	-	pdf pinout
LTC 691 N	LIN-IC	Microprocessor Supervisory Circuits	TSOP	Lic	-	pdf pinout
LTC 691 SW	LIN-IC	Microprocessor Supervisory Circuits	TSOP	Lic	-	pdf pinout
LTC 695 N	LIN-IC	Microprocessor Supervisory Circuits	TSOP	Lic	-	pdf pinout
LTC 695 SW	LIN-IC	Microprocessor Supervisory Circuits	TSOP	Lic	-	pdf pinout
LTC1040	2xCOMP	Vs:±9V Vi0:±0.3mV	18D	Ltc		data pdf pinout
LTC1042	COMP	Vs:18V Vo:TTL Vi0:±0.3mV	8D	Ltc		data pdf pinout
LTC 1043 (CN/MD)	LIN-IC	Dual Precision Instrumentation Switched-Capacitor Building Block	TSOP	Lic		pdf pinout
LTC 1044 CH	CMOS-IC	Switched Capacitor Voltage Converter	81	Lic	-	pdf pinout
LTC 1044 MH	CMOS-IC	Switched Capacitor Voltage Converter	81	Lic	-	pdf pinout
LTC1047	2xCHOPPER	Vs:16V Vu:150dB Vo:±4.95V Vi0:±3μμV	8D,16S	Ltc		data pdf pinout
LTC1049	CHOPPER	Vs:18V Vu:160dB Vo:±4.97V Vi0:±2μμV	8DS	Ltc		data pdf pinout
LTC1050	CHOPPER	Vs:±9V Vu:160V/mV Vo:±4.85V f:2.5Mc	8AD,14DS	Ltc		data pdf pinout
LTC1051	2xCHOPPER	Vs:16.5V Vu:160dB Vo:±4.85V f:2.5Mc	8D,16S	Ltc		data pdf pinout
LTC1052	CHOPPER	Vs:±9V Vu:150dB Vo:±4.85V f:1.2Mc	8AD,14D	Ltc		data pdf pinout
LTC1052CD	CHOPPER	Vs:16V Vu:150dB Vo:4.95/0V f:1.2Mc	8S	Tix		data pinout
LTC1052CP	CHOPPER	Vs:16V Vu:150dB Vo:4.95/0V f:1.2Mc	8D	Tix		data pinout
LTC1052MD	CHOPPER	Vs:16V Vu:150dB Vo:4.95/0V f:1.2Mc	8S	Tix		data pinout
LTC1052MP	CHOPPER	Vs:16V Vu:150dB Vo:4.95/0V f:1.2Mc	8D	Tix		data pinout
LTC1053	4xCHOPPER	Vs:16.5V Vu:160dB Vo:±4.85V f:2.5Mc	14D,18S	Ltc		data pdf pinout
LTC 1059 AC (N/J)	IC	High Performance Switched Capacitor Universal Filter	DIC	Lic		pdf pinout
LTC 1059 AMJ	IC	High Performance Switched Capacitor Universal Filter	DIC	Lic		pdf pinout
LTC 1059 C (N/J)	IC	High Performance Switched Capacitor Universal Filter	DIC	Lic		pdf pinout
LTC 1059 MJ	IC	High Performance Switched Capacitor Universal Filter	DIC	Lic		pdf pinout
LTC 1059 N	IC	High Performance Switched Capacitor Universal Filter	DIC	Lic		pdf pinout
LTC 1059 S	IC	High Performance Switched Capacitor Universal Filter	MDIP	Lic		pdf pinout
LTC 1060 (AC/AM) J	IC	Universal Dual Filter Building Block	TSOP	Lic		pdf pinout
LTC 1060 (AC/C) N	IC	Universal Dual Filter Building Block	TSOP	Lic		pdf pinout
LTC 1060 (C/M) J	IC	Universal Dual Filter Building Block	TSOP	Lic		pdf pinout
LTC 1060 (A)N	IC	Universal Dual Filter Building Block	DIC	Lic		pdf pinout
LTC 1060 SW	IC	Universal Dual Filter Building Block	TSOP	Lic		pdf pinout
LTC 1061 (A)N	IC	High Performance Triple Universal Filter Building Block	DIC	Lic		pdf pinout
LTC 1061 SW	IC	High Performance Triple Universal Filter Building Block	TSOP	Lic		pdf pinout
LTC 1062 N8	CMOS-IC	5th Order Lowpass Filter	DIC	Lic	-	pdf pinout

Type	Device	Short Description	Fig.	Manu	Comparision Types	More at
LTC 1062 SW	CMOS-IC	5th Order Lowpass Filter	TSOP	Lic		pdf pinout
LTC 1063 CN 8	D/A-IC	DC Accurate,5th Order Butterworth Lowpass Filter	TSOP	Lic		pdf pinout
LTC 1063 CSW	D/A-IC	DC Accurate,5th Order Butterworth Lowpass Filter	TSOP	Lic		pdf pinout
LTC 1064	IC	Quad Universal Filter Building Block	TSOP	Lic		pdf pinout
LTC 1065 CSW	LIN-IC	Linear Phase 5th Order Bessel Lowpass Filter	DIP	Lic		pdf pinout
LTC 1065 ISW	LIN-IC	Linear Phase 5th Order Bessel Lowpass Filter	DIP	Lic		pdf pinout
LTC 1067	LIN-IC	Universal Dual Filter Building Block	DIP	Lic		pdf pinout
LTC 1068 G	LIN-IC	Filter Building Blocks	DIP	Lic		pdf pinout
LTC 1068 N	LIN-IC	Filter Building Blocks	DIP	Lic		pdf pinout
LTC 1091	IC	1/2/6/8-CH, 10-Bit Serial I/O Data Acquisition Sys.	87	Lic		pdf pinout
LTC 1092	IC	1/2/6/8-CH, 10-Bit Serial I/O Data Acquisition Sys.	87	Lic		pdf pinout
LTC 1093	IC	1/2/6/8-CH, 10-Bit Serial I/O Data Acquisition Sys.	87	Lic		pdf pinout
LTC 1094	IC	1/2/6/8-CH, 10-Bit Serial I/O Data Acquisition Sys.	87	Lic		pdf pinout
LTC 1096	IC	8-Bit Serial I/O A/D Converters	87	Lic		pdf pinout
LTC 1096 L	IC	8-Bit Serial I/O A/D Converters	87	Lic		pdf pinout
LTC 1098	IC	8-Bit Serial I/O A/D Converters	87	Lic		pdf pinout
LTC 1098 L	IC	8-Bit Serial I/O A/D Converters	87	Lic		pdf pinout
LTC 1099	IC	8-Bit A/D Converter, Sample-and-Hold	87	Lic		pdf pinout
LTC 1099 CSW	IC	8-Bit A/D Converter, Sample-and-Hold	87	Lic		pdf pinout
LTC 1100 (AC/C) S	LIN-IC	Chopper-stabilized Instrumentation Amp.	TSOP	Lic		pdf pinout
LTC1150	CHOPPER	Vs:36V Vu:180dB Vo:±14.5V f:2.5Mc	8DS	Ltc		data pdf pinout
LTC1151	2xCHOPPER	Vs:36V Vu:140dB Vo:±14.5V Vi0:±5µµV	8D,16S	Ltc		data pdf pinout
LTC1152	LO-DRIFT	Vs:14V Vu:130dB Vo:4.4/0V Vi0:±1µµV	8DS	Ltc		data pdf pinout
LTC 1153 CN8	MOS-IC	Auto-reset Electronic Circuit Breaker	DIC	Lic		pdf pinout
LTC 1153 CS8	MOS-IC	Auto-reset Electronic Circuit Breaker	TSOP	Lic		pdf pinout
LTC 1154 N8	MOS-IC	High Side Micropower Mosfet Driver	DIC	Lic		pdf pinout
LTC 1154 S8	MOS-IC	High Side Micropower Mosfet Driver	TSOP	Lic		pdf pinout
LTC 1155 N8/J8	MOS-IC	Dual High Side Micropower Mosfet Driver	DIC			pdf pinout
LTC 1155 S8	MOS-IC	Dual High Side Micropower Mosfet Driver	TSOP			pdf pinout
LTC 1156 N	MOS-IC	Quad High Sidemicropower Mosfet Driver with Internal Charge Pump	DIC	Lic		pdf pinout
LTC 1156 S	MOS-IC	Quad High Sidemicropower Mosfet Driver with Internal Charge Pump	TSOP	Lic		pdf pinout
LTC 1157 CN8	MOS-IC	3.3V Dual Micropower High-Side/Low-Side Mosfet Driver	DIC	Lic		pdf pinout
LTC 1157 CS8	MOS-IC	3.3V Dual Micropower High-Side/Low-Side Mosfet Driver	TSOP	Lic		pdf pinout
LTC 1163 CN8	MOS-IC	Triple 1.8V to 6V High-Side Mosfet Drivers	DIC	Lic		pdf pinout
LTC 1163 CS8	MOS-IC	Triple 1.8V to 6V High-Side Mosfet Drivers	TSOP	Lic		pdf pinout
LTC 1164 CS	LIN-IC	Quad Universal Filter Building Block	DIP	Lic		pdf pinout
LTC 1164 J	LIN-IC	Quad Universal Filter Building Block	DIP	Lic		pdf pinout
LTC 1164 N	LIN-IC	Quad Universal Filter Building Block	DIP	Lic		pdf pinout
LTC 1164 SW	LIN-IC	Quad Universal Filter Building Block	DIP	Lic		pdf pinout
LTC 1165 CN8	MOS-IC	Triple 1.8V to 6V High-Side Mosfet Drivers	DIC	Lic		pdf pinout
LTC 1165 CS8	MOS-IC	Triple 1.8V to 6V High-Side Mosfet Drivers	TSOP	Lic		pdf pinout
LTC 1196	IC	8-Bit,1MSPS ADCs,Auto-Shutdown Options	87	Lic		pdf pinout
LTC 1197	A/D-IC	10-Bit, 500ksps ADCs, Auto Shutdown	87	Lic		pdf pinout
LTC 1197 L	A/D-IC	10-Bit, 500ksps ADCs, Auto Shutdown	87	Lic		pdf pinout
LTC 1198	IC	8-Bit,1MSPS ADCs,Auto-Shutdown Options	87	Lic		pdf pinout
LTC 1199	A/D-IC	10-Bit, 500ksps ADCs, Auto Shutdown	87	Lic		pdf pinout
LTC 1199 L	A/D-IC	10-Bit, 500ksps ADCs, Auto Shutdown	87	Lic		pdf pinout
LTC 1232	LIN-IC	Microprocessor Supervisory Circuits	MDIP	Lic		pdf pinout
LTC1250	CHOPPER	Vs:18V Vu:170dB Vo:±4.95V Vi0:±5µµV	8DS	Ltc		data pdf pinout
LTC 1255 (C/I)N8	MOS-IC	Dual 24V High-Side Mosfet Driver	DIC	Lic		pdf pinout
LTC 1255 (C/I)S8	MOS-IC	Dual 24V High-Side Mosfet Driver	TSOP	Lic		pdf pinout
LTC 1257	A/D-IC	Single Supply 12-Bit Volt. Output DAC	QFN	Lic		pdf pinout
LTC 1261 CS	LIN-IC	Switched Capacitor Regulated Voltage Inverter	TSOP	Lic		pdf pinout
LTC 1264 N	LIN-IC	Quad Universal Filter Building Block	DIP	Lic		pdf pinout
LTC 1264 S	LIN-IC	Quad Universal Filter Building Block	DIP	Lic		pdf pinout
LTC 1264 SW	LIN-IC	Quad Universal Filter Building Block	DIP	Lic		pdf pinout
LTC1265	ES+	Io=960mA Vo:1.25...13V Vin:13V	14S	Ltc		data pdf pinout
LTC1265-3.3	S+	Io=960mA Vo:3.4..3.22V Vin:10V	14S	Ltc		data pdf pinout
LTC1265-5	S+	Io=960mA Vo:4.9..5.2V Vin:10V	14S	Ltc		data pdf pinout
LTC 1272	IC	12-Bit, 3µs, 250kHz Sampling A/D Conv.	87	Lic		pdf pinout
LTC 1273	IC	12-Bit, 300ksps Sampling A/D Conv., Reference	87	Lic		pdf pinout
LTC 1274	A/D-IC	12-Bit,10mW,100ksps ADCs,1µA Shutdown	87	Lic		pdf pinout
LTC 1275	IC	12-Bit, 300ksps Sampling A/D Conv., Reference	87	Lic		pdf pinout
LTC 1276	IC	12-Bit, 300ksps Sampling A/D Conv., Reference	87	Lic		pdf pinout
LTC 1277	A/D-IC	12-Bit,10mW,100ksps ADCs,1µA Shutdown	87	Lic		pdf pinout

Type	Device	Short Description	Fig.	Manu	Comparision Types	More at
LTC 1278	A/D-IC	12-Bit, 500ksps Sampling A/D Conv., Shutdown	87	Lic	-	pdf pinout
LTC 1279	A/D-IC	12-Bit, 600ksps Sampling A/D Conv., Shutdown	87	Lic	-	pdf pinout
LTC 1282	A/D-IC	12-Bit Sampling A/D Conv., Reference	87	Lic	-	pdf pinout
LTC 1283	A/D-IC	3V 10-Bit Data Acquisition System	87	Lic	-	pdf pinout
LTC 1285	A/D-IC	3V Sampling 12-Bit A/D Converters	87	Lic	-	pdf pinout
LTC 1286	A/D-IC	Sampling 12-Bit A/D Converters	87	Lic	-	pdf pinout
LTC 1287	A/D-IC	3V 12-Bit Data Acquisition System	87	Lic	-	pdf pinout
LTC 1288	A/D-IC	3V Sampling 12-Bit A/D Converters	87	Lic	-	pdf pinout
LTC 1289	A/D-IC	3V Sampling 12-Bit A/D Converters	87	Lic	-	pdf pinout
LTC 1290	A/D-IC	1 chip,12-Bit Data Acquisition System	87	Lic	-	pdf pinout
LTC 1291	A/D-IC	1 chip,12-Bit Data Acquisition System	87	Lic	-	pdf pinout
LTC 1292	A/D-IC	1 chip,12-Bit Data Acquisition System	87	Lic	-	pdf pinout
LTC 1293	A/D-IC	1 chip,12-Bit Data Acquisition System	87	Lic	-	pdf pinout
LTC 1294	A/D-IC	1 chip,12-Bit Data Acquisition System	87	Lic	-	pdf pinout
LTC 1296	A/D-IC	1 chip,12-Bit Data Acquisition System	87	Lic	-	pdf pinout
LTC 1297	A/D-IC	1 chip,12-Bit Data Acquisition System	87	Lic	-	pdf pinout
LTC 1298	A/D-IC	Sampling 12-Bit A/D Converters	87	Lic	-	pdf pinout
LTC 1314 CS	IC	Pcmcia Switching Matrix, N-charinel VCC Switch Drivers	TSOP	Lic	-	pdf pinout
LTC 1315 CG	IC	Pcmcia Switching Matrix, N-channel Vcc Switch Drivers	TSOP	Lic	-	pdf pinout
LTC 1323 CG	LIN-IC	Single 5V Appletalk® Transceiver	TSOP	Lic	-	pdf pinout
LTC 1323 CS	LIN-IC	Single 5V Appletalk® Transceiver	TSOP	Lic	-	pdf pinout
LTC 1323 CSW	LIN-IC	Single 5V Appletalk® Transceiver	TSOP	Lic	-	pdf pinout
LTC 1324 C (N/SW)	LIN-IC	Single Supply Localtalk® Transceiver	TSOP	Lic	-	pdf pinout
LTC 1326 MS8-2.5	LIN-IC	Micropower Precision Triple Supply Monitors; 2.5V, 3.3V and ADJ	TSOP	Lic	-	pdf pinout
LTC 1326 CMS8	LIN-IC	Micropower Precision Triple Supply Monitors; 5V, 3.3V and ADJ	TSOP	Lic	-	pdf pinout
LTC 1326 S8-2.5	LIN-IC	Micropower Precision Triple Supply Monitors; 2.5V, 3.3V and ADJ	MDIP	Lic	-	pdf pinout
LTC 1326 CS8	LIN-IC	Micropower Precision Triple Supply Monitors; 5V, 3.3V and ADJ	MDIP	Lic	-	pdf pinout
LTC 1343 CGW	LIN-IC	Software-selectable Multiprotocol Transceiver	TSOP	Lic	-	pdf pinout
LTC 1344 CG	LIN-IC	Software-selectable Cable Terminator	TSOP	Lic	-	pdf pinout
LTC 1345 C (NW/SW)	LIN-IC	Single Supply V.35 Transceiver	TSOP	Lic	-	pdf pinout
LTC 1345 I (NW/SW)	LIN-IC	Single Supply V.35 Transceiver	TSOP	Lic	-	pdf pinout
LTC 1348 CG	LIN-IC	RS232 3-Driver/5-Receiver Transceiver	TSOP	Lic	-	pdf pinout
LTC 1348 IG	LIN-IC	RS232 3-Driver/5-Receiver Transceiver	TSOP	Lic	-	pdf pinout
LTC 1348 ISW	LIN-IC	RS232 3-Driver/5-Receiver Transceiver	TSOP	Lic	-	pdf pinout
LTC 1348 CSW	LIN-IC	RS232 3-Driver/5-Receiver Transceiver	TSOP	Lic	-	pdf pinout
LTC 1380 (C/I)GN	CMOS-IC	Single- Ended 8- Channel Analog Multiplexer with SMBus Interface	TSOP	Lic	-	pdf pinout
LTC 1380 (C/I)S	CMOS-IC	Single- Ended 8- Channel Analog Multiplexer with SMBus Interface	MDIP	Lic	-	pdf pinout
LTC 1387 (C/I)G	CMOS-IC	Single 5V RS232/RS485 Multiprotocol Transceiver	TSOP	Lic	-	pdf pinout
LTC 1387 (C/I)SW	CMOS-IC	Single 5V RS232/RS485 Multiprotocol Transceiver	TSOP	Lic	-	pdf pinout
LTC 1390 CN	CMOS-IC	8-Channel Analog Multiplexer with Serial Interface	DIC	Lic	-	pdf pinout
LTC 1390 CS	CMOS-IC	8-Channel Analog Multiplexer with Serial Interface	MDIP	Lic	-	pdf pinout
LTC 1393 (C/I)GN	CMOS-IC	Differential 4- Channel Analog Multiplexer with SMBus Interface	TSOP	Lic	-	pdf pinout
LTC 1393 (C/I)S	CMOS-IC	Differential 4- Channel Analog Multiplexer with SMBus Interface	MDIP	Lic	-	pdf pinout
LTC 1400 (C,I)S8	A/D-IC	12-Bit, 400ksps ADC, Shutdown	8-MDIP	Ltc	-	pdf pinout
LTC 1420	A/D-IC	12-Bit, 10Msps, Sampling ADC	28-SSMDIP	Lic	-	pdf pinout
LTC 1421 G	IC	Hot Swap Controller	TSOP	Ltc	-	pdf pinout
LTC 1421 SW	IC	Hot Swap Controller	TSOP	Ltc	-	pdf pinout
LTC 1426	D/A-IC	Micropower Dual 6-Bit PWM DAC	QFN	Lic	-	pdf pinout
LTC 1433 GN	Z-IC, LIN-IC	450mA, Low Noise Current Mode Step-down DC/DC Converter	TSOP	Lic	-	pdf pinout
LTC 1434 GN	Z-IC, LIN-IC	450mA, Low Noise Current Mode Step-down DC/DC Converter	TSOP	Lic	-	pdf pinout
LTC 1435 (C/I)G	LIN-IC	High Efficiency Low Noise Synchronous Step-down Switching Regulator	TSOP	Lic	-	pdf pinout
LTC 1435 (C/I)S	LIN-IC	High Efficiency Low Noise Synchronous Step-down Switching Regulator	MDIP	Lic	-	pdf pinout
LTC1440	COMP	Vs:±5V Vi0:±3mV	8DS	Ltc	-	data pdf pinout
LTC 1446	D/A-IC	Dual 12-Bit Rail-to-Rail DACs in SO-8	QFN	Lic	-	pdf pinout
LTC 1446 L	D/A-IC	Dual 12-Bit Rail-to-Rail DACs in SO-8	QFN	Lic	-	pdf pinout
LTC 1448	D/A-IC	SMD, Dual, 12 Bit	8-MDIP	Lic	-	pdf
LTC 1450	D/A-IC	Parallel Input, 12-Bit Rail-to-Rail DACs	QFN	Lic	-	pdf pinout
LTC 1450 L	D/A-IC	Parallel Input, 12-Bit Rail-to-Rail DACs	QFN	Lic	-	pdf pinout
LTC 1451	D/A-IC	12-Bit Rail-to-Rail Micropower DACs	QFN	Lic	-	pdf pinout
LTC 1452	D/A-IC	12-Bit Rail-to-Rail Micropower DACs	QFN	Lic	-	pdf pinout
LTC 1453	D/A-IC	12-Bit Rail-to-Rail Micropower DACs	QFN	Lic	-	pdf pinout
LTC 1454	D/A-IC	Dual 12-Bit Rail-to-Rail Micropower DACs	QFN	Lic	-	pdf pinout
LTC 1454 L	D/A-IC	Dual 12-Bit Rail-to-Rail Micropower DACs	QFN	Lic	-	pdf pinout
LTC 1456	D/A-IC	12-Bit Rail-to-Rail DAC, Clear Input	QFN	Lic	-	pdf pinout

Type	Device	Short Description	Fig.	Manu	Comparision Types	More at
LTC 1458	D/A-IC	Quad 12-Bit Rail-to-Rail Micropower DACs	QFN	Lic	-	pdf pinout
LTC 1458 L	D/A-IC	Quad 12-Bit Rail-to-Rail Micropower DACs	QFN	Lic	-	pdf pinout
LTC 1470 (C/E)S8	LIN-IC	Single Pcmcia Protected 3.3V/5V Vcc Switches	TSOP	Lic	-	pdf pinout
LTC 1471 CS	LIN-IC	Dual Pcmcia Protected 3.3V/5V Vcc Switches	TSOP	Lic	-	pdf pinout
LTC 1472 CS	LIN-IC	Protected Pcmcia Vcc and Vpp Switching Matrix	TSOP	Lic	-	pdf pinout
LTC 1473 (C/I)GN	MOS-IC	Dual Powerpathtm Switch Driver	TSOP	Lic	-	pdf pinout
LTC 1473 LCGN	MOS-IC	Dual Low Voltage Powerpathtm Switch Driver	TSOP	Lic	-	pdf pinout
LTC 1477 CS8	MOS-IC	Single Protected High Side Switches	TSOP	Lic	-	pdf pinout
LTC 1478 CS	MOS-IC	Dual Protected High Side Switches	TSOP	Lic	-	pdf pinout
LTC 1479 (C/I)G	MOS-IC	Powerpath Controller for Dual Battery Systems	TSOP	Lic	-	pdf pinout
LTC 1503	CMOS-IC	SMD, Down S-Reg, 2,4...6V, 100mA	8-(SS)MDIP	Lic		
LTC1504	ES+	Io=800mA Vo:1.27...9V Vin:10V	8S	Ltc		data pdf pinout
LTC1504-3.3	S+	Io=500mA Vo:3.2...3.4V Vin:5V	8S	Ltc		data pdf pinout
LTC 1504	CMOS-IC	LIN/S-Reg, 2,5V, 500mA		Lic		pdf
LTC1504A	ES+	Io=800mA Vo:1.27...9V Vin:10V	8S	Ltc		data pdf pinout
LTC1504A-3.3	S+	Io=500mA Vo:3.2...3.4V Vin:5V	8S	Ltc		data pdf pinout
LTC 1518 S	LIN-IC	52Mbps Precision Delay RS485 Quad Line Receivers	TSOP	Lic	-	pdf pinout
LTC 1519 S	LIN-IC	52Mbps Precision Delay RS485 Quad Line Receivers	TSOP	Lic	-	pdf pinout
LTC 1520 S	LIN-IC	50Mbps Precision Quad Line Receiver	TSOP	Lic	-	pdf pinout
LTC 1522 CMS8	Z-IC, LIN-IC	Micropower, Regulated 5V Charge Pump DC/DC Converter	TSOP	Lic	-	pdf pinout
LTC 1522 CS8	Z-IC, LIN-IC	Micropower, Regulated 5V Charge Pump DC/DC Converter	MDIP	Lic	-	pdf pinout
LTC1531	4xDIFF-COMP	Vs:7V Vi0:2mV	28S	Ltc		data pdf pinout
LTC 1535 SW	LIN-IC	Isolated RS485 Transceiver	TSOP	Lic	-	pdf pinout
LTC 1536 MS8	IC	Precision Triple Supply Monitor for PCI Applications	TSOP	Lic	-	pdf pinout
LTC 1536 S8	IC	Precision Triple Supply Monitor for PCI Applications	MDIP	Lic	-	pdf pinout
LTC 1540 CMS8	LIN/OP-IC	Nanopower Comparator with Reference	TSOP	Lic	-	pdf pinout
LTC 1540 (C/I)S8	LIN/OP-IC	Nanopower Comparator with Reference	MDIP	Lic	-	pdf pinout
LTC 1541 CMS8	LIN/OP-IC	Micropower Op Amp, Comparator and Reference	TSOP	Lic	-	pdf pinout
LTC 1541 (C/I)S8	LIN/OP-IC	Micropower Op Amp, Comparator and Reference	MDIP	Lic	-	pdf pinout
LTC 1542 CMS8	LIN/OP-IC	Micropower Op Amp and Comparator	TSOP	Lic	-	pdf pinout
LTC 1542 (C/I)S8	LIN/OP-IC	Micropower Op Amp and Comparator	MDIP	Lic	-	pdf pinout
LTC 1543 CG	IC	Software-selectable Multiprotocol Transceiver	TSOP	Lic	-	pdf pinout
LTC 1544 CG	IC	Software-selectable Multiprotocol Transceiver	TSOP	Lic	-	pdf pinout
LTC 1550 CGN	IC	Switched Capacitor Reg. Voltage Inverter	TSOP	Lic	-	pdf pinout
LTC 1550 (C/I)GN	LIN-IC	Low Noise, Switched Capacitor Regulated Voltage Inverters	TSOP	Lic	-	pdf pinout
LTC 1550 IGN	IC	Switched Capacitor Reg. Voltage Inverter	TSOP	Lic	-	pdf pinout
LTC 1550 LCGN	IC	Switched Capacitor Reg. Voltage Inverter	TSOP	Lic	-	pdf pinout
LTC 1550 LIGN	IC	Switched Capacitor Reg. Voltage Inverter	TSOP	Lic	-	pdf pinout
LTC 1550 CS8-4.1	LIN-IC	Low Noise, Switched Capacitor Regulated Voltage Inverters	MDIP	Lic	-	pdf pinout
LTC 1551 CS8-4.1	LIN-IC	Low Noise, Switched Capacitor Regulated Voltage Inverters	MDIP	Lic	-	pdf pinout
LTC 1555 GN	LIN-IC	SIM Power Supply and Level Translator	TSOP	Lic	-	pdf pinout
LTC 1556 GN	LIN-IC	SIM Power Supply and Level Translator	TSOP	Lic	-	pdf pinout
LTC 1562 G	LIN-IC	Active RC Quad Universal Filter	DIP	Lic	-	pdf pinout
LTC 1562 N	LIN-IC	Active RC Quad Universal Filter	DIP	Lic	-	pdf pinout
LTC 1564 G	LIN-IC	Antialiasing Filter and 4-Bit P.G.A.	DIP	Lic	-	pdf pinout
LTC 1569-6,-7	CMOS-IC	SMD, Low Pass Filter, <300kHz	8-MDIP	Lic	-	pdf pinout
LTC 1574 S-3.3	LIN-IC	High Efficiency Step- Down DC/ DC Converters w. Internal Schottky Diode	MDIP	Lic	-	pdf pinout
LTC 1574 S-5	LIN-IC	High Efficiency Step- Down DC/ DC Converters w. Internal Schottky Diode	MDIP	Lic	-	pdf pinout
LTC 1574 S	LIN-IC	High Efficiency Step- Down DC/ DC Converters w. Internal Schottky Diode	MDIP	Lic	-	pdf pinout
LTC 1588	D/A-IC	12-/14-/16-Bit SoftSpan DACs, Output Range	QFN	Lic	-	pdf pinout
LTC 1589	D/A-IC	12-/14-/16-Bit SoftSpan DACs, Output Range	QFN	Lic	-	pdf pinout
LTC 1590	D/A-IC	Dual Serial 12-Bit Multiplying DAC	QFN	Lic	-	pdf pinout
LTC 1591	D/A-IC	14-Bit/16-Bit Parallel Multiplying DACs	QFN	Lic	-	pdf pinout
LTC 1592	D/A-IC	12-/14-/16-Bit SoftSpan DACs, Output Range	QFN	Lic	-	pdf pinout
LTC 1594 L(C/I)S	LIN-IC	4-Channel, 3V Micropower Sampling 12-Bit Serial I/O A/D Converters	MDIP	Lic	-	pdf pinout
LTC 1594 (C/I)S	LIN-IC	4-Channel, Micropower Sampling 12-Bit Serial I/O A/D Converters	MDIP	Lic	-	pdf pinout
LTC 1595	D/A-IC	Serial 16-Bit Multiplying DACs	QFN	Lic	-	pdf pinout
LTC 1596	D/A-IC	Serial 16-Bit Multiplying DACs	QFN	Lic	-	pdf pinout
LTC 1597	D/A-IC	14-Bit/16-Bit Parallel Multiplying DACs	QFN	Lic	-	pdf pinout

Type	Device	Short Description	Fig.	Manu	Comparision Types	More at
LTC 1598 (C/I)G	LIN-IC	8-Channel, Micropower Sampling 12-Bit Serial I/O A/D Converters	TSOP	Lic	-	pdf pinout
LTC 1598 L(C/I)G	LIN-IC	8-Channel, 3V Micropower Sampling 12-Bit Serial I/O A/D Converters	TSOP	Lic	-	pdf pinout
LTC 1599	A/D-IC	SMD, 16 Bit, Multiplying, Double Buff.	24-SSMDIP	Lic	-	pdf
LTC 1623 (M)S8	MOS-IC	SMBus Dual High Side Switch Controller	TSOP	Lic	-	pdf pinout
LTC 1629	SMPS-IC	SMD, DC-DC Current Mode Controller		Lic	-	pdf
LTC 1642	CMOS-IC	SMD, Hot Swap Controller, 3,3...15V	16-SSMDIP	Lic	-	pdf
LTC 1642 AGN	CMOS-IC	Hot Swap Controller	16-SSMDIP	Lic	-	pdf pinout
LTC 1643 AHGN	CMOS-IC	Pci-bus Hot Swap Controller	16-SSMDIP	Lic	-	pdf pinout
LTC 1643 ALGN	CMOS-IC	PCI-bus Hot Swap Controller	16-SSMDIP	Lic	-	pdf pinout
LTC 1644 GN	CMOS-IC	Compactpci Bus Hot Swap Controller	16-SSMDIP	Lic	-	pdf pinout
LTC 1645 S	CMOS-IC	2-CH Hot Swap Controller/Power Sequencer	16-SSMDIP	Lic	-	pdf pinout
LTC 1646 GN	CMOS-IC	Compactpci Dual Hot Swap Controller	16-SSMDIP	Lic	-	pdf pinout
LTC 1649 S	LIN-IC	3.3V Input High Power Step- Down Switching Regulator Controller	MDIP	Lic	-	pdf pinout
LTC 1650	D/A-IC	Low Glitch 16-Bit Voltage Output DAC	QFN	Lic	-	pdf pinout
LTC 1654	D/A-IC	Dual 14-Bit Rail-to-Rail DAC	TSOP	Lic	-	pdf pinout
LTC 1655 L	D/A-IC	SMD, 16 Bit	8-MDIP	Lic	-	pdf
LTC 1657	D/A-IC	Parallel 16-Bit Rail-to-Rail DAC	DIP	Ltc	-	pdf pinout
LTC 1657 L	D/A-IC	Parallel 16-Bit Rail-to-Rail DAC	DIP	Lic	-	pdf pinout
LTC 1658	D/A-IC	SMD, 14 Bit	8-MDIP	Lic	-	pdf
LTC 1659	D/A-IC	SMD, 12 Bit	8-MDIP	Lic	-	pdf
LTC 1660	D/A-IC	Micropower Octal 8-Bit and 10-Bit DACs	TSOP	Lic	-	pdf pinout
LTC 1663	D/A-IC	SMD, 10 Bit	6-MDIP	Lic	-	pdf
LTC 1664	D/A-IC	Micropower Quad 10-Bit DAC	TSOP	Lic	-	pdf pinout
LTC 1665	D/A-IC	Micropower Octal 8-Bit and 10-Bit DACs	TSOP	Lic	-	pdf pinout
LTC 1666	D/A-IC	12-Bit, 14-Bit, 16-Bit, 50Msps DACs	TSOP	Lic	-	pdf pinout
LTC 1667	D/A-IC	12-Bit, 14-Bit, 16-Bit, 50Msps DACs	TSOP	Lic	-	pdf pinout
LTC 1668	D/A-IC	12-Bit, 14-Bit, 16-Bit, 50Msps DACs	TSOP	Lic	-	pdf pinout
LTC 1687 S	LIN-IC	52Mbps Precision Delay RS485 Fail-Safe Transcei.	TSOP	Lic	-	pdf pinout
LTC 1688 S	LIN-IC	100Mbps RS485 Hot Swapable Quad Drivers	TSOP	Lic	-	pdf pinout
LTC 1689 S	LIN-IC	100Mbps RS485 Hot Swapable Quad Drivers	TSOP	Lic	-	pdf pinout
LTC 1693 -1(C/I)S8	MOS-IC	High Speed Dual N-channel Mosfet Drivers	TSOP	Lic	-	pdf pinout
LTC 1693 -2(C/I)S8	MOS-IC	High Speed Dual N-channel Mosfet Drivers	TSOP	Lic	-	pdf pinout
LTC 1693 -3CMS8	MOS-IC	High Speed Single N-channel Mosfet Drivers	TSOP	Lic	-	pdf pinout
LTC 1693 -5CMS8	MOS-IC	High Speed Single P-channel Mosfet Driver	TSOP	Lic	-	pdf pinout
LTC 1696 ES6	MOS-IC	Overvoltage Protection Controller	47	Lic	-	pdf pinout
LTC1701	ES+	Io=720mA Vo:1.25...5.5V Vin:5.5V	5S	Ltc	-	data pdf pinout
LTC1701B	ES+	Io=640mA Vo:0.8..8.5V Vin:8.5V	5S,8S	Ltc	-	data pdf pinout
LTC 1702	CMOS-IC	SMD, 2-Phase Down S-Reg, 550kHz	24-SSMDIP	Lic	-	pdf
LTC 1703	SMPS-IC	SMD, DC-DC Converter, Dual Output	28-SSMDIP	Lic	-	pdf
LTC 1710	IC	SMBus Dual Monolithic High Side Switch	SSMDIP	Lic	-	pdf pinout
LTC 1726 MS8-2.5	LIN-IC	Triple Supply Monitor, µP Supervisor, Watchdog Timer; 2.5V, 3.3V and ADJ	TSOP	Lic	-	pdf pinout
LTC 1726 MS8-5	LIN-IC	Triple Supply Monitor, µP Supervisor, Watchdog Timer; 5V, 3.3V and ADJ	TSOP	Lic	-	pdf pinout
LTC 1726 S8-2.5	LIN-IC	Triple Supply Monitor, µP Supervisor, Watchdog Timer; 2.5V, 3.3V and ADJ	MDIP	Lic	-	pdf pinout
LTC 1726 S8-5	LIN-IC	Triple Supply Monitor, µP Supervisor, Watchdog Timer; 5V, 3.3V and ADJ	MDIP	Lic	-	pdf pinout
LTC 1727 MS8-2.5	LIN-IC	Micropower Precision Triple Supply Monitors; 2.5V, 3.3V and ADJ	TSOP	Lic	-	pdf pinout
LTC 1727 MS8-5	LIN-IC	Micropower Precision Triple Supply Monitors; 5V, 3.3V and ADJ	TSOP	Lic	-	pdf pinout
LTC 1727 S8-2.5	LIN-IC	Micropower Precision Triple Supply Monitors; 2.5V, 3.3V and ADJ	MDIP	Lic	-	pdf pinout
LTC 1727 S8-5	LIN-IC	Micropower Precision Triple Supply Monitors; 5V, 3.3V and ADJ	MDIP	Lic	-	pdf pinout
LTC 1728 S5-1.8	LIN-IC	Micropower Precision Triple Supply Monitors; 3V, 1.8V and ADJ	45	Lic	-	pdf pinout
LTC 1728 S5-2.5	LIN-IC	Micropower Precision Triple Supply Monitors; 2.5V, 3.3V and ADJ	45	Lic	-	pdf pinout
LTC 1728 S5-3.3	LIN-IC	Micropower Precision Triple Supply Monitors; 3.3V, 1.8V and ADJ	45	Lic	-	pdf pinout
LTC 1728 S5-5	LIN-IC	Micropower Precision Triple Supply Monitors; 5V, 3.3V and ADJ	45	Lic	-	pdf pinout
LTC 1733	IC	Lithium-ion Battery Charger,Thermal Regulation	87	Lic	-	pdf pinout
LTC 1734	IC	Lithium-ion Linear Battery Charger	87	Lic	-	pdf pinout
LTC 1734 L	IC	Lithium-ion Linear Battery Charger	87	Lic	-	pdf pinout
LTC 1740 (C/I)G	IC	14-Bit, 6Msps, Sampling ADC	TSOP	Lic	-	pdf pinout
LTC 1741 (C/I)FW	IC	12-Bit, 65Msps Low Noise ADC	TSOP	Lic	-	pdf pinout
LTC 1742 (C/I) FW	IC	14-Bit, 65Msps Low Noise ADC	TSOP	Lic	-	pdf pinout
LTC 1742 FW	IC	14-Bit, 65Msps Low Noise ADC	TSOP	Lic	-	pdf pinout
LTC 1743 FW	IC	12-Bit, 50Msps ADC	TSOP	Lic	-	pdf pinout
LTC 1744 (C/I) FW	IC	14-Bit, 50Msps ADC	TSOP	Lic	-	pdf pinout
LTC 1745 (C/I) FW	IC	Low Noise, 12-Bit, 25Msps ADC	TSOP	Lic	-	pdf pinout
LTC 1746 (C/I) FW	IC	Low Power, 14-Bit, 25Msps ADC	TSOP	Lic	-	pdf pinout
LTC 1747 FW	IC	12-Bit, 80Msps Low Noise ADC	TSOP	Lic	-	pdf pinout
LTC 1748 (C/I) FW	IC	14-Bit, 80Msps Low Noise ADC	TSOP	Lic	-	pdf pinout
LTC 1749 FW	IC	12-Bit, 80Msps Wide Bandwidth ADC	TSOP	Lic	-	pdf pinout

Type	Device	Short Description	Fig.	Manu	Comparision Types	More at
LTC 1754-5	SMPS-IC	SMD, DC-DC Converter, 3V→5V, 50mA	46	Lic	-	
LTC 1757 A-1	IC	Single Rf Power Controllers	TSOP	Lic	-	pdf pinout
LTC 1757 A-2	IC	Dual Band Rf Power Controllers	TSOP	Lic	-	pdf pinout
LTC 1758 -1	IC	Rf Power Controllers with 250kHz Control Loop Bandwidth	TSOP	Lic	-	pdf pinout
LTC 1758 -2	IC	Rf Power Controllers with 250kHz Control Loop Bandwidth	TSOP	Lic	-	pdf pinout
LTC 1759	CMOS-IC	SMD, Battery Charger, SMBus compatible	SSMDIP	Lic	-	pdf
LTC1779	ES+	Io=250mA Vo:0.8..9.8V Vin:9.8V	6S	Ltc		data pdf pinout
LTC 1795	CMOS-IC	SMD, Dual, Adsl Line Driver, 900V/µs	20-MDIP	Lic	-	
LTC 1813	CMOS-OP-IC	SMD, Dual, ±5V, Iq=±40mA, >500V/µs	8-SMDIP	Lic	-	
LTC 1821	D/A-IC	16-Bit, Ultra Precise, Fast Settling VOUT DAC	TSOP	Lic	-	pdf pinout
LTC 1840	IC	Dual Fan Controller, 2-Wire Interface	TSOP	Lic	-	pdf pinout
LTC 1850	LIN-IC	8-CH, 10-Bit/12-Bit, 1.25Msps Samp. ADCs	MDIP	Lic	-	pdf pinout
LTC 1851	LIN-IC	8-CH, 10-Bit/12-Bit, 1.25Msps Samp. ADCs	MDIP	Lic	-	pdf pinout
LTC 1852	LIN-IC	8-CH,10-Bit/12-Bit,400ksps,Sampl. ADCs	MDIP	Lic	-	pdf pinout
LTC 1853	LIN-IC	8-CH,10-Bit/12-Bit,400ksps,Sampl. ADCs	MDIP	Lic	-	pdf pinout
LTC 1854	A/D-IC	8-CH, ±10V 12-/14-/16-Bit, 100ksps ADC ,Shutdown	MDIP	Lic	-	pdf pinout
LTC 1855	A/D-IC	8-CH, ±10V 12-/14-/16-Bit, 100ksps ADC ,Shutdown	MDIP	Lic	-	pdf pinout
LTC 1856	A/D-IC	8-CH, ±10V 12-/14-/16-Bit, 100ksps ADC ,Shutdown	MDIP	Lic	-	pdf pinout
LTC 1857	A/D-IC	8-CH, 12-/14-/16-Bit, 100ksps SoftSpan ADC, Shutdown	MDIP	Lic	-	pdf pinout
LTC 1858	A/D-IC	8-CH, 12-/14-/16-Bit, 100ksps SoftSpan ADC, Shutdown	MDIP	Lic	-	pdf pinout
LTC 1859	A/D-IC	8-CH, 12-/14-/16-Bit, 100ksps SoftSpan ADC, Shutdown	MDIP	Lic	-	pdf pinout
LTC 1860	A/D-IC	12-Bit, 250ksps 1- and 2-CH ADCs	MDIP	Lic	-	pdf pinout
LTC 1860 L	A/D-IC	3V, 12-Bit, 150ksps, 1- and 2-CH ADCs	MDIP	Lic	-	pdf pinout
LTC 1861 LMS	A/D-IC	12-Bit, 250ksps 1- and 2-CH ADCs	MDIP	Lic	-	pdf pinout
LTC 1861 MS	A/D-IC	12-Bit, 250ksps 1- and 2-CH ADCs	MDIP	Lic	-	pdf pinout
LTC 1863	A/D-IC	12-/16-Bit, 8-Channel 200ksps ADCs	MDIP	Lic	-	pdf pinout
LTC 1863 L	A/D-IC	3V, 12-/16-Bit, 8-Channel 175ksps ADCs	MDIP	Lic	-	pdf pinout
LTC 1867	A/D-IC	12-/16-Bit, 8-Channel 200ksps ADCs	MDIP	Lic	-	pdf pinout
LTC 1867 L	A/D-IC	3V, 12-/16-Bit, 8-Channel 175ksps ADCs	MDIP	Lic	-	pdf pinout
LTC1875	ES+	Io=151mA Vo:0.8..6V Vin:6V Isw:3A	16S	Ltc		data pdf pinout
LTC1877	ES+	Io=600mA Vo:0.8..10V Vin:10V	8S	Ltc		data pdf pinout
LTC1878	ES+	Io=600mA Vo:0.8..6V Vin:6V Isw:1.5A	8S	Ltc		data pdf pinout
LTC1879	ES+	Io=1.2A Vo:0.8..10V Vin:10V Isw:3A	16S	Ltc		data pdf pinout
LTC 1923 EGN	IC	High Efficiency Thermoelectric Cooler Controller	SSMDIP	Lic	-	pdf pinout
LTC 1923 EUH	IC	High Efficiency Thermoelectric Cooler Controller	SSMDIP	Lic	-	pdf pinout
LTC 1929	SMPS-IC	SMD, DC-DC Current Mode Controller		Lic	-	pdf
LTC 1957 -1	IC	Single Band Rf Power Controllers with 40dB Dynamic Range	TSOP	Lic	-	pdf pinout
LTC 1957 -2	IC	Dual Band Rf Power Controllers with 40dB Dynamic Range	TSOP	Lic	-	pdf pinout
LTC 1985 S5-1.8	LIN-IC	Micropower Precision Triple Supply Monitor with Push-Pull Reset Output	45	Lic	-	
LTC 1998	IC	Comparator,Voltage Reference,Battery Monitoring	87	Lic	-	pdf pinout
LTC 2202	A/D-IC	16-Bit, 25Msps/10Msps ADCs	MDIP	Lic	-	pdf pinout
LTC 2203	A/D-IC	16-Bit, 25Msps/10Msps ADCs	MDIP	Lic	-	pdf pinout
LTC 2204	A/D-IC	16-Bit, 65Msps/40Msps ADCs	MDIP	Lic	-	pdf pinout
LTC 2205	A/D-IC	16-Bit, 65Msps/40Msps ADCs	MDIP	Lic	-	pdf pinout
LTC 2206	A/D-IC	16-Bit, 105Msps/80Msps ADCs	MDIP	Lic	-	pdf pinout
LTC 2207	A/D-IC	16-Bit, 105Msps/80Msps ADCs	MDIP	Lic	-	pdf pinout
LTC 2208	A/D-IC	16-Bit, 130Msps ADC	MDIP	Lic	-	pdf pinout
LTC 2209	A/D-IC	16-Bit, 160Msps ADC	MDIP	Lic	-	pdf pinout
LTC 2220 UP	IC	12-Bit, 170Msps ADCs	QFN	Lic	-	pdf pinout
LTC 2221 UP	IC	12-Bit, 135Msps ADCs	QFN	Lic	-	pdf pinout
LTG 2222 -11	IC	11-Bit, 105Msps ADCs	QFN	Lic	-	pdf pinout
LTC 2222 UK	IC	12-Bit, 105Msps ADCs	QFN	Lic	-	pdf pinout
LTC 2223 UK	IC	12-Bit, 80Msps ADCs	QFN	Lic	-	pdf pinout
LTC 2224 UK	IC	12-Bit, 135Msps ADC	QFN	Lic	-	pdf pinout
LTC 2225 UH	IC	12-Bit, 10Msps Low Power 3V ADC	QFN	Lic	-	pdf pinout
LTC 2226 UH	IC	12-Bit, 25Msps Low Power 3V ADCs	QFN	Lic	-	pdf pinout
LTC 2227 UH	IC	12-Bit, 40Msps Low Power 3V ADCs	QFN	Lic	-	pdf pinout
LTC 2228 UH	IC	12-Bit, 65Msps Low Power 3V ADCs	QFN	Lic	-	pdf pinout
LTC 2230 UP	IC	10-Bit,170Msps ADCs	QFN	Lic	-	pdf pinout
LTC 2231 UP	IC	10-Bit, 135Msps ADCs	QFN	Lic	-	pdf pinout
LTC 2232 UK	IC	10-Bit, 105Msps ADCs	QFN	Lic	-	pdf pinout
LTC 2233 UK	IC	10-Bit, 80Msps ADCs	QFN	Lic	-	pdf pinout
LTC 2234 UK	IC	10-Bit, 135Msps ADC	QFN	Lic	-	pdf pinout
LTC 2236 UH	IC	10-Bit, 25Msps Low Noise 3V ADCs	QFN	Lic	-	pdf pinout
LTC 2237 UH	IC	10-Bit, 40Msps Low Noise 3V ADCs	QFN	Lic	-	pdf pinout
LTC 2238 UH	IC	10-Bit, 65Msps Low Noise 3V ADCs	QFN	Lic	-	pdf pinout
LTC 2239 UH	IC	10-Bit, 80Msps Low Noise 3V ADC	QFN	Lic	-	pdf pinout
LTC 2245	A/D-IC	14-Bit, 10Msps Low Power 3V ADC	QFN	Lic	-	pdf pinout
LTC 2246	A/D-IC	14-Bit, 65/40/25Msps Low-Power 3V ADCs	QFN	Lic	-	pdf pinout
LTC 2247	A/D-IC	14-Bit, 65/40/25Msps Low Power 3V ADCs	QFN	Lic	-	pdf pinout

Type	Device	Short Description	Fig.	Manu	Comparision Types	More at
LTC 2248	A/D-IC	14-Bit, 65/40/25Msps Low Power 3V ADCs	QFN	Lic	-	pdf pinout
LTC 2249	A/D-IC	14-Bit, 80Msps Low Power 3V ADC	QFN	Lic	-	pdf pinout
LTC 2250	A/D-IC	10-Bit, 125/105Msps Low Noise 3V ADCs	QFN	Lic	-	pdf pinout
LTC 2251	A/D-IC	10-Bit, 125/105Msps Low Noise 3V ADCs	QFN	Lic	-	pdf pinout
LTC 2252	A/D-IC	12-Bit, 125/105Msps Low Power 3V ADCs	QFN	Lic	-	pdf pinout
LTC 2253	A/D-IC	12-Bit, 125/105Msps Low Power 3V ADCs	QFN	Lic	-	pdf pinout
LTC 2254	A/D-IC	14-Bit, 125/105Msps Low Power 3V ADCs	QFN	Lic	-	pdf pinout
LTC 2255	A/D-IC	14-Bit, 125/105Msps Low Power 3V ADCs	QFN	Lic	-	pdf pinout
LTC 2280	A/D-IC	Dual 10-Bit, 105Msps Low Noise 3V ADC	QFN	Lic	-	pdf pinout
LTC 2281	A/D-IC	Dual 10-Bit, 125Msps Low Power 3V ADC	QFN	Lic	-	pdf pinout
LTC 2282	A/D-IC	Dual 12-Bit, 105Msps Low Power 3V ADC	QFN	Lic	-	pdf pinout
LTC 2283	A/D-IC	Dual 12-Bit, 125Msps Low Power 3V ADC	QFN	Lic	-	pdf pinout
LTC 2285	A/D-IC	Dual 14-Bit, 125Msps Low Power 3V ADC	QFN	Lic	-	pdf pinout
LTC 2286	A/D-IC	Dual 10-Bit, 65/40/25Msps 3V ADCs	QFN	Lic	-	pdf pinout
LTC 2287	A/D-IC	Dual 10-Bit, 65/40/25Msps 3V ADCs	QFN	Lic	-	pdf pinout
LTC 2288	A/D-IC	Dual 10-Bit, 65/40/25Msps 3V ADCs	QFN	Lic	-	pdf pinout
LTC 2289	A/D-IC	Dual 10-Bit, 80Msps Low Noise 3V ADC	QFN	Lic	-	pdf pinout
LTC 2290	A/D-IC	Dual 12-Bit, 10Msps Low Power 3V ADC	QFN	Lic	-	pdf pinout
LTC 2291	A/D-IC	Dual 12-Bit, 65/40/25Msps 3V ADCs	QFN	Lic	-	pdf pinout
LTC 2292	A/D-IC	Dual 12-Bit, 65/40/25Msps 3V ADCs	QFN	Lic	-	pdf pinout
LTC 2293	A/D-IC	Dual 12-Bit, 65/40/25Msps 3V ADCs	QFN	Lic	-	pdf pinout
LTC 2294	A/D-IC	Dual 12-Bit, 80Msps Low Power 3V ADC	QFN	Lic	-	pdf pinout
LTC 2295	A/D-IC	Dual 14-Bit, 10Msps Low Power 3V ADC	QFN	Lic	-	pdf pinout
LTC 2296	A/D-IC	Dual 14-Bit, 65/40/25Msps 3V ADCs	QFN	Lic	-	pdf pinout
LTC 2297	A/D-IC	Dual 14-Bit, 65/40/25Msps 3V ADCs	QFN	Lic	-	pdf pinout
LTC 2298	A/D-IC	Dual 14-Bit, 65/40/25Msps 3V ADCs	QFN	Lic	-	pdf pinout
LTC 2299	A/D-IC	Dual 14-Bit, 80Msps Low Power 3V ADC	QFN	Lic	-	pdf pinout
LTC 2308	A/D-IC	500ksps, 8-Channel, 12-Bit ADC	QFN	Lic	-	pdf pinout
LTC 2400	CMOS-IC	SMD, Diff. Preamp f. A/D Conv., 5/±5V	16-SSMDIP	Lic	-	pdf
LTC 2408	A/D-IC	24 Bit, 8-Ch. Multiplex Input		Lic	-	pdf
LTC 2450	IC	Easy-to-Use, Ultra-Tiny 16-Bit ΔΣ ADC	87	Lic	-	pdf pinout
LTC 2453	IC	Differe., 16-Bit ΔΣ ADC, I²C Interface	87	Lic	-	pdf pinout
LTC 2600	D/A-IC	Octal 16-/14-/12-Bit Rail-to-Rail DACs	TSOP	Lic	-	pdf pinout
LTC 2601	D/A-IC	16-/14-/12-Bit Rail-to-Rail DACs	TSOP	Lic	-	pdf pinout
LTC 2602	D/A-IC	Dual 16-/14-/12-Bit Rail-to-Rail DACs	TSOP	Lic	-	pdf pinout
LTC 2604	D/A-IC	Quad 16-Bit Rail-to-Rail DACs	TSOP	Lic	-	pdf pinout
LTC 2605	D/A-IC	Octal 16-/14-/12-Bit Rail-to-Rail DACs	TSOP	Lic	-	pdf pinout
LTC 2606	D/A-IC	16-/14-/12-Bit Rail-to-Rail DACs, I²C Interface	TSOP	Lic	-	pdf pinout
LTC 2607	D/A-IC	16-/14-/12-Bit Dual Rail-to-Rail DACs, I²C Interface	TSOP	Lic	-	pdf pinout
LTC 2609	D/A-IC	Quad 16-/14-/12-Bit Rail-to-Rail DACs, I²C Interface	TSOP	Lic	-	pdf pinout
LTC 2610	D/A-IC	Octal 16-/14-/12-Bit Rail-to-Rail DACs	TSOP	Lic	-	pdf pinout
LTC 2611	D/A-IC	16-/14-/12-Bit Rail-to-Rail DACs	TSOP	Lic	-	pdf pinout
LTC 2612	D/A-IC	Dual 16-/14-/12-Bit Rail-to-Rail DACs	TSOP	Lic	-	pdf pinout
LTC 2614	D/A-IC	Quad 16-Bit Rail-to-Rail DACs	TSOP	Lic	-	pdf pinout
LTC 2615	D/A-IC	Octal 16-/14-/12-Bit Rail-to-Rail DACs	TSOP	Lic	-	pdf pinout
LTC 2616	D/A-IC	16-/14-/12-Bit Rail-to-Rail DACs, I²C Interface	TSOP	Lic	-	pdf pinout
LTC 2617	D/A-IC	16-/14-/12-Bit Dual Rail-to-Rail DACs, I²C Interface	TSOP	Lic	-	pdf pinout
LTC 2619	D/A-IC	Quad 16-/14-/12-Bit Rail-to-Rail DACs, I²C Interface	TSOP	Lic	-	pdf pinout
LTC 2620	D/A-IC	Octal 16-/14-/12-Bit Rail-to-Rail DACs	TSOP	Lic	-	pdf pinout
LTC 2621	D/A-IC	16-/14-/12-Bit Rail-to-Rail DACs	TSOP	Lic	-	pdf pinout
LTC 2622	D/A-IC	Dual 16-/14-/12-Bit Rail-to-Rail DACs	TSOP	Lic	-	pdf pinout
LTC 2624	D/A-IC	Quad 16-Bit Rail-to-Rail DACs	TSOP	Lic	-	pdf pinout
LTC 2625	D/A-IC	Octal 16-/14-/12-Bit Rail-to-Rail DACs	TSOP	Lic	-	pdf pinout
LTC 2626	D/A-IC	16-/14-/12-Bit Rail-to-Rail DACs, I²C Interface	TSOP	Lic	-	pdf pinout
LTC 2627	D/A-IC	16-/14-/12-Bit Dual Rail-to-Rail DACs, I²C Interface	TSOP	Lic	-	pdf pinout
LTC 2629	D/A-IC	Quad 16-/14-/12-Bit Rail-to-Rail DACs, I²C Interface	TSOP	Lic	-	pdf pinout
LTC 2630	D/A-IC	Single 12-/10-/8-Bit Rail-to-Rail DACs, Reference	TSOP	Lic	-	pdf pinout
LTC 2641	D/A-IC	16-/14-/12-Bit Vout DACs	TSOP	Lic	-	pdf pinout
LTC 2642	D/A-IC	16-/14-/12-Bit Vout DACs	TSOP	Lic	-	pdf pinout
LTC 2704	D/A-IC	Quad 12-/14-/16-Bit Voltage Output SoftSpan DACs, Readback	TSOP	Lic	-	pdf pinout
LTC 2751-12	D/A-IC	Current Output 12-/14-/16-Bit DACs, Parallel I/O	TSOP	Lic	-	pdf pinout
LTC 2801 DE	LIN-IC	1.8V...5.5V RS-232 Single/Dual Transcei.	TSOP	Lic	-	pdf pinout
LTC 2802 DE	LIN-IC	1.8V...5.5V RS-232 Single/Dual Transcei.	TSOP	Lic	-	pdf pinout
LTC 2803 DHC	LIN-IC	1.8V...5.5V RS-232 Single/Dual Transcei.	TSOP	Lic	-	pdf pinout
LTC 2803 GN	LIN-IC	1.8V...5.5V RS-232 Single/Dual Transcei.	TSOP	Lic	-	pdf pinout
LTC 2804 DHC	LIN-IC	1.8V...5.5V RS-232 Single/Dual Transcei.	TSOP	Lic	-	pdf pinout
LTC 2804 GN	LIN-IC	1.8V...5.5V RS-232 Single/Dual Transcei.	TSOP	Lic	-	pdf pinout
LTC 2854 DE	LIN-IC	3.3V 20Mbps RS485/RS422 Transceivers	TSOP	Lic	-	pdf pinout
LTC 2855 DE	LIN-IC	3.3V 20Mbps RS485/RS422 Transceivers	TSOP	Lic	-	pdf pinout
LTC 2855 GN	LIN-IC	3.3V 20Mbps RS485/RS422 Transceivers	TSOP	Lic	-	pdf pinout
LTC 2859 DD	LIN-IC	20Mbps RS485 Transceivers, Switchable Termination	DFN	Lic	-	pdf pinout
LTC 2861 DE	LIN-IC	20Mbps RS485 Transceivers, Switchable	DFN	Lic	-	pdf pinout

Type	Device	Short Description	Fig.	Manu	Comparision Types	More at
		Termination				
LTC 2861 GN	LIN-IC	20Mbps RS485 Transceivers, Switchable Termination	DFN	Lic	-	pdf pinout
LTC 2900 -1DD	IC	Programmable Quad Supply Monitor with Adjustable Reset Timer	MLP	Lic	-	pdf pinout
LTC 2900 -1MS	IC	Programmable Quad Supply Monitor with Adjustable Reset Timer	TSOP	Lic	-	pdf pinout
LTC 2900 -2DD	IC	Programmable Quad Supply Monitor with Adjustable Reset Timer	MLP	Lic	-	pdf pinout
LTC 2900 -2MS	IC	Programmable Quad Supply Monitor with Adjustable Reset Timer	TSOP	Lic	-	pdf pinout
LTC 2901 -1GN	IC	Programmable Quad Supply Monitor with Adjustable Reset and Watchdog Timers	TSOP	Lic	-	pdf pinout
LTC 2901 -2GN	IC	Programmable Quad Supply Monitor with Adjustable Reset and Watchdog Timers	TSOP	Lic	-	pdf pinout
LTC 2901 -3GN	IC	Programmable Quad Supply Monitor with Adjustable Reset and Watchdog Timers	TSOP	Lic	-	pdf pinout
LTC 2901 -4GN	IC	Programmable Quad Supply Monitor with Adjustable Reset and Watchdog Timers	TSOP	Lic	-	pdf pinout
LTC 2902 -1/2	IC	Programmable Quad Supply Monitor with Adj. Reset Timer and Supply Tolerance	TSOP	Lic	-	pdf pinout
LTC 2903 -A1	IC	Precision Quad Supply Monitor: 3.3V, 2.5V, 1.8V, ADJ	46	Lic	-	pdf pinout
LTC 2903 -B1	IC	Precision Quad Supply Monitor: 5V, 3.3V, 2.5V, 1.8V	46	Lic	-	pdf pinout
LTC 2903 -C1	IC	Precision Quad Supply Monitor: 5V, 3.3V, 1.8V, −5.2V	46	Lic	-	pdf pinout
LTC 2903 -D1	IC	Precision Quad Supply Monitor: 3.3V, ADJ, ADJ, ADJ	46	Lic	-	pdf pinout
LTC 2903 -E1	IC	Precision Quad Supply Monitor: 5V, ADJ, ADJ, ADJ	46	Lic	-	pdf pinout
LTC 2906 DDB 8	LIN-IC	Dual Supply Monitors,Pin-selectable Threshold,Adjustable Input	DFN	Lic	-	pdf pinout
LTC 2907 DDB 8	LIN-IC	Dual Supply Monitors,Pin-selectable Threshold,Adjustable Input	DFN	Lic	-	pdf pinout
LTC 2908 DDBC	LIN-IC	Precision Six Input Supply Monitor	DFN	Lic	-	pdf pinout
LTC 2909 DDB	LIN-IC	Triple/dual Input UV, OV, Neg. Voltage Monitor	DFN	Lic	-	pdf pinout
LTC 2915 DDB	LIN-IC	Low Volt. Supervisor, 27 Selectable Thresholds	TSOP	Lic	-	pdf pinout
LTC 2916 DDB	LIN-IC	Low Volt. Supervisor, 27 Selectable Thresholds	TSOP	Lic	-	pdf pinout
LTC 2917 DDB	LIN-IC	Supervisor, 27 Selectable Thresholds, Watchdog Timer	TSOP	Lic	-	pdf pinout
LTC 2917 MS	LIN-IC	Supervisor, 27 Selectable Thresholds, Watchdog Timer	TSOP	Lic	-	pdf pinout
LTC 2918 DDB	LIN-IC	Supervisor, 27 Selectable Thresholds, Watchdog Timer	TSOP	Lic	-	pdf pinout
LTC 2918 MS	LIN-IC	Supervisor, 27 Selectable Thresholds, Watchdog Timer	TSOP	Lic	-	pdf pinout
LTC 2921 GN	LIN-IC	Power Supply Tracker, Input Monitors	TSOP	Lic	-	pdf pinout
LTC 2922 F	LIN-IC	Power Supply Tracker, Input Monitors	TSOP	Lic	-	pdf pinout
LTC 2923 MS	LIN-IC	Power Supply Tracking Controller	TSOP	Lic	-	pdf pinout
LTC 2924 GN	LIN-IC	Quad Power Supply Sequencer	TSOP	Lic	-	pdf pinout
LTC 2925 GN	LIN-IC	Multiple Power Supply Tracking Controller	TSOP	Lic	-	pdf pinout
LTC 2925 UF	LIN-IC	Multiple Power Supply Tracking Controller	TSOP	Lic	-	pdf pinout
LTC 2926 GN	LIN-IC	MOSFET-controlled Power Supply Tracker	TSOP	Lic	-	pdf pinout
LTC 2926 UFD	LIN-IC	Mosfet-controlled Power Supply Tracker	TSOP	Lic	-	pdf pinout
LTC 2927 DDB	LIN-IC	Power Supply Tracking Controller	TSOP	Lic	-	pdf pinout
LTC 2928 G	LIN-IC	Multi- Ch Power Supply Sequencer/ Supervisor	TSOP	Lic	-	pdf pinout
LTC 2928 UHF	LIN-IC	Multi- Ch Power Supply Sequencer/ Supervisor	TSOP	Lic	-	pdf pinout
LTC 2952 /QFN	IC	Push Button Powerpath Controller with Supervisor	SSMDIP	Lic	-	pdf pinout
LTC 2952 /TSSOP	IC	Push Button Powerpath Controller with Supervisor	SSMDIP	Lic	-	pdf pinout
LTC 2953 DD	LIN-IC	Push Button On/Off Controller, Voltage Monitoring	DFN	Lic	-	pdf pinout
LTC 2954 DDB	LIN-IC	Push Button On/Off Controller, µP Interrupt	DFN	Lic	-	pdf pinout
LTC 2970 UFD	LIN-IC	Dual I²C Power Supply Monitor, Margining Controller	DFN	Lic	-	pdf pinout
LTC 3200 -5	LIN-IC	Low Noise, Regulated Charge Pump DC/DC Converters	46	Lic	-	pdf pinout
LTC 3200	LIN-IC	Low Noise, Regulated Charge Pump DC/DC Converters	TSOP	Lic	-	pdf pinout
LTC 3201	LIN-IC	100mA Ultralow Noise Charge Pump LED Supply with Output Current Adjust	TSOP	Lic	-	pdf pinout
LTC 3202 EDD	LIN-IC	Low Noise, High Efficiency Charge Pump for White LEDs, up to 125mA	MLP	Lic	-	pdf pinout
LTC 3202 EMS	LIN-IC	Low Noise, High Efficiency Charge Pump for White LEDs, up to 125mA	TSOP	Lic	-	pdf pinout

Type	Device	Short Description	Fig.	Manu	Comparision Types	More at
LTC 3203 BEDD	IC	500mA Output Current Low Noise Dual Mode Step-up Charge Pumps	DIP	Lic	-	pdf pinout
LTC 3203 EDD	IC	500mA Output Current Low Noise Dual Mode Step-up Charge Pumps	DIP	Lic	-	pdf pinout
LTC 3205	LIN-IC	Multidisplay LED Controller	QFN	Lic	-	pdf pinout
LTC 3206	LIN-IC	I2C Multidisplay LED Controller	QFN	Lic	-	pdf pinout
LTC 3207 (-1)	LIN-IC	0,6A Universal Multi-Output LED/CAM Driver	QFN	Lic	-	pdf pinout
LTC 3208	LIN-IC	High Current Software Confi gurable Multidisplay LED Controller	QFN	Lic	-	pdf pinout
LTC 3209 -1	LIN-IC	600mA Main/Camera LED Controller	QFN	Lic	-	pdf pinout
LTC 3209 -2	LIN-IC	600mA Main/Camera LED Controller	QFN	Lic	-	pdf pinout
LTC 3210 -1	LIN-IC	main/CAM LED Controller with 64-Step Brightness Control	QFN	Lic	-	pdf pinout
LTC 3210 -2	LIN-IC	main/CAM LED Controllers with 32-Step Brightness Control	QFN	Lic	-	pdf pinout
LTC 3210 -3	LIN-IC	main/CAM LED Controllers with 32-Step Brightness Control	QFN	Lic	-	pdf pinout
LTC 3210	LIN-IC	main/CAM LED Controller, Up to 0,5A	QFN	Lic	-	pdf pinout
LTC 3212	LIN-IC	RGB LED Driver and Charge Pump	MLP	Lic	-	pdf pinout
LTC 3214	LIN-IC	500mA Camera LED Charge Pump	MLP	Lic	-	pdf pinout
LTC 3215	LIN-IC, Z-Di	700mA Low Noise High Current LED Charge Pump	MLP	Lic	-	pdf pinout
LTC 3216	LIN-IC	1A Low Noise High Current LED Charge Pump	MLP	Lic	-	pdf pinout
LTC 3217	LIN-IC	0,6A Low Noise Multi-LED Camera Light Charge Pump	QFN	Lic	-	pdf pinout
LTC 3218	LIN-IC	0,4A Single Wire Camera LED Charge Pump	MLP	Lic	-	pdf pinout
LTC 3219	LIN-IC	250mA Universal Nine Channel LED Driver	QFN	Lic	-	pdf pinout
LTC 3221 EDC	IC	Micropower, Regulated Charge Pump in 2 × 2 DFN	DIP	Lic	-	pdf pinout
LTC 3230	LIN-IC	5-LED Main/Sub Display Driver with Dual LDO	QFN	Lic	-	pdf pinout
LTC 3250 ES6-1.2	LIN-IC	Step-down DC/DC Converter	46	Lic	-	pdf pinout
LTC 3250 ES6-1.5	LIN-IC	Step-down DC/DC Converter	46	Lic	-	pdf pinout
LTC 3251 -1.2	LIN-IC	0.5A, Step-down DC/DC Converter	TSOP	Lic	-	pdf pinout
LTC 3251 -1.5	LIN-IC	0.5A, Step-down DC/DC Converter	TSOP	Lic	-	pdf pinout
LTC 3251	LIN-IC	0.5A, Step-down DC/DC Converter	TSOP	Lic	-	pdf pinout
LTC 3252	LIN-IC	Dual, Step-down DC/DC Converter	MLP	Lic	-	pdf pinout
LTC3400	ES+	Vo:2.5...5.0V Isw:0.85A	5S	Ltc		data pdf pinout
LTC3401	ES+	Vo:2.6...5.5V Isw:1.6A	10S	Ltc		data pdf pinout
LTC3402	ES+	Vo:2.6...5.5V Isw:2.5A	10S	Ltc		data pdf pinout
LTC3403	ES+	Io=600mA Vo:0.3...3.6V Isw:1.3A	8H	Ltc		data pdf pinout
LTC3404	ES+	Io=600mA Vo:0.8...6V Isw:1.25A	10S	Ltc		data pdf pinout
LTC3405A	S+	Io=300mA Vo:1.746...1.854V Isw:0.625A	5S	Ltc		data pdf pinout
LTC3406	S+	Io=600mA Vo:1.746...1.854V Isw:1.25A	5S	Ltc		data pdf pinout
LTC3406B	ES+	Io=600mA Vo:0.6...5.5V Isw:1.25A	5S	Ltc		data pdf pinout
LTC3407	ES+	Io=600mA Vo:0.6...5.5V Isw:1.25A	10HS	Ltc		data pdf pinout
LTC3407-2	ES+	Io=800mA Vo:0.6...5.5V Isw:1.6A	10HS	Ltc		data pdf pinout
LTC3408	ES+	Io=600mA Vo:0.3...3.6V Isw:1.3A	8H	Ltc		data pdf pinout
LTC3410	ES+	Io=300mA Vo:1.2...5.5V Vin:5.5V	6S	Ltc		data pdf pinout
LTC3410B	ES+	Io=300mA Vo:1.2...5.5V Vin:5.5V	6S	Ltc		data pdf pinout
LTC3411	ES+	Io=1.25A Vo:0.8...5.0V Vin:5.5V	10H,8S	Ltc		data pdf pinout
LTC3412	ES+	Io=2.5A Vo:0.8...5.0V Vin:5V Isw:6.5A	16SH	Ltc		data pdf pinout
LTC3412A	ES+	Io=3A Vo:0.8...5.0V Isw:6A	16SH	Ltc		data pdf pinout
LTC3413	ES+	Io=3A Vo:0.8...5.0V Vin:5V Isw:7.2A	16S	Ltc		data pdf pinout
LTC3414	ES+	Io=4A Vo:0.8...5.0V Vin:5V Isw:9.5A	20S	Ltc		data pdf pinout
LTC3421	ES+	Vo:2.4...5.25V Vin:3.3V	24H	Ltc		data pdf pinout
LTC3423	ES+	Vo:1.5...5.5V Isw:1.6A	10S	Ltc		data pdf pinout
LTC3424	ES+	Vo:1.5...5.5V Isw:2.8A	10S	Ltc		data pdf pinout
LTC3425	ES+	Vo:2.4...5.25V Isw:7A	32H	Ltc		data pdf pinout
LTC3426	ES+	Vo:2.25...5.5V Isw:2.3A	5S	Ltc		data pdf pinout
LTC3440	ES+	Io=600mA Vo:2.5...5.5V	10HS	Ltc		data pdf pinout
LTC3441	ES+	Io=1.2A Vo:2.4...5.5V	12H	Ltc		data pdf pinout
LTC 3452	LIN-IC	Synchronous Buck-Boost Main/camera White LED Driver	QFN	Lic	-	pdf pinout
LTC 3453	LIN-IC	Synchronous Buck-Boost High Power White LED Driver	QFN	Lic	-	pdf pinout
LTC 3454	LIN-IC	1A Synchronous Buck-Boost High Current LED Driver	MLP	Lic	-	pdf pinout
LTC 3456	IC	2-Cell, DC/DC Converter,USB Power Manager	87	Lic	-	pdf pinout
LTC 3490 EDD	LIN-IC	Single Cell 350mA LED Driver	MLP	Lic	,	pdf pinout
LTC 3490 ES8	LIN-IC	Single Cell 350mA LED Driver	TSOP	Lic	-	pdf pinout
LTC 3550	IC	Li-Ion Battery Charger,600mA Buck Conv.	43	Lic	-	pdf pinout
LTC 3555	IC	USB Power Manager, 3x Step-down DC/DC	43	Lic	-	pdf pinout
LTC 3557	IC	USB Power Manager, Li- Ion Charger, 3xStep- Down Reg.	43	Lic	-	pdf pinout
LTC 3559	IC	Linear USB Battery Charger with Dual Buck Regulators	DFN	Lic	-	pdf pinout
LTC 3783 EDHD	LIN-IC	PWM LED Driver and Boost, Flyback and Sepic Controller	MLP	Lic	-	pdf pinout
LTC 3783 EFE	LIN-IC	PWM LED Driver and Boost, Flyback and Sepic Controller	TSOP	Lic	-	pdf pinout
LTC 4001	IC	2A Synchronous Buck Li-Ion Charger	DFN	Lic	-	pdf pinout

Type	Device	Short Description	Fig.	Manu	Comparision Types	More at
LTC 4010	IC	battery fast charge IC, Nicd, NiMH	DFN	Lic	-	pdf pinout
LTC 4011	IC	battery fast charge IC, Nicd, NiMH	DFN	Lic		pdf pinout
LTC 4050	IC	Li-Ion Battery Charger,Thermistor Interface	DFN	Lic		pdf pinout
LTC 4055	IC	USB Power Controller,Li-Ion Linear Charger	DFN	Lic		pdf pinout
LTC 4059	IC	900mA Linear Li-Ion Battery Chargers, Therm. Reg.	DFN	Lic		pdf pinout
LTC 4059 A	IC	900mA Linear Li-Ion Battery Chargers, Therm. Reg.	DFN	Lic		pdf pinout
LTC 4060 EDHC	IC	Linear NiMH/NiCd Fast Battery Charger	DFN	Lic		pdf pinout
LTC 4060 EFE	IC	Linear Nimh/nicd Fast Battery Charger	DFN	Lic		pdf pinout
LTC 4061	IC	Linear Li-Ion Battery Charger Thermistor Input	DFN	Lic		pdf pinout
LTC 4062	IC	Li-Ion Battery Charger, Comparator	DFN	Lic		pdf pinout
LTC 4063	IC	Li-Ion Charger,Low Dropout,Lin. Regul.	DFN	Lic		pdf pinout
LTC 4064	IC	Monolithic Linear Charger,Back-up Li-Ion Batt.	DFN	Lic		pdf pinout
LTC 4065 L	IC	250mA Li-Ion Battery Charger	DFN	Lic		pdf pinout
LTC 4065 LX	IC	250mA Li-Ion Battery Charger	DFN	Lic		pdf pinout
LTC 4067	IC	USB Power Manager,Ovp,Li-ion/polymer Charger	DFN	Lic		pdf pinout
LTC 4069	IC	750mA Li-Ion Battery Charger, NTC Thermistor Input	43	Lic		pdf pinout
LTC 4075	IC	Dual Input Usb/ac Adapter Li-Ion Battery Chargers	DFN	Lic		pdf pinout
LTC 4075 HVX	IC	2 Input Li-ion/polymer Battery Charger	DFN	Lic		pdf pinout
LTC 4075 X	IC	Dual Input Usb/ac Adapter Li-Ion Battery Chargers	DFN	Lic		pdf pinout
LTC 4076	IC	Dual Input Li-Ion Battery Charger	DFN	Lic		pdf pinout
LTC 4077	IC	Dual Input Li-Ion Battery Charger	DFN	Lic		pdf pinout
LTC 4078	IC	Dual Input Li-Ion Battery Charger Overvolt. Prot.	DFN	Lic		pdf pinout
LTC 4078 X	IC	Dual Input Li-Ion Battery Charger Overvolt. Prot.	DFN	Lic		pdf pinout
LTC 4080 EDD		500mA Li-Ion Charger,300mA Sync. Buck	DFN	Lic		pdf pinout
LTC 4080 EMSE		500mA Li-Ion Charger,300mA Sync. Buck	DFN	Lic		pdf pinout
LTC 4080 XEDD	IC	500mA Li-Ion Charger,300mA Sync. Buck	DFN	Lic		pdf pinout
LTC 4080 XEMSE	IC	500mA Li-Ion Charger,300mA Sync. Buck	DFN	Lic		pdf pinout
LTC 4081	IC	500mA Li-Ion Charger,NTC Input,300mA Sync. Buck	DFN	Lic		pdf pinout
LTC 4085	IC	USB Power Manager, Ideal Diode Contr., Li- Ion Charg.	DFN	Lic		pdf pinout
LTC 4088	IC	Battery Charger/usb Power Manager	DFN	Lic		pdf pinout
LTC 4090	IC	USB Power Manager,2A Bat-Track Buck Regulator	DFN	Lic		pdf pinout
LTC 4095	IC	USB Li-ion/polymer Battery Charger	DFN	Lic		pdf pinout
LTC 4096	IC	Dual Input Li-Ion Battery Chargers	DFN	Lic		pdf pinout
LTC 4096 X	IC	Dual Input Li-Ion Battery Chargers	DFN	Lic		pdf pinout
LTC 4097	IC	USB/Wall Adapter Li-ion/polymer Bat. Charg.	DFN	Lic		pdf pinout
LTC 4101	IC	Smart Battery Charger Controller	DFN	Lic		pdf pinout
LTC 4211 MS	CMOS-IC	Hot Swap Contr., Multifunction Current Control	46	Ltc		pdf pinout
LTC 4212 MS	CMOS-IC	Hot Swap Controller, Power-up Timeout	46	Ltc		pdf pinout
LTC 4213	IC	No RsenseÖ Electronic Circuit Breaker	MLP	Lic		pdf pinout
LTC 4214 -1	IC	Negative Low Voltage Hot Swap Controllers, Latch Off After Fault	TSOP	Lic		pdf pinout
LTC 4214 -2	IC	Negative Low Voltage Hot Swap Cntrl., Automatic Retry After Fault	TSOP	Lic	-	pdf pinout
LTC 4215 GN	IC	Hot Swap Controller with I2C Compatible Monitoring	TSOP	Lic		pdf pinout
LTC 4215 UFD	IC	Hot Swap Controller with I2C Compatible Monitoring	QFN	Lic		pdf pinout
LTC 4216 DE	IC	Ultralow Voltage Hot Swap Controller	MLP	Lic		pdf pinout
LTC 4216 MS	IC	Ultralow Voltage Hot Swap Controller	TSOP	Lic		pdf pinout
LTC 4221	IC	Dual Hot Swap Cntrl./ Power Sequencer with Dual Speed	TSOP	Lic		pdf pinout
LTC 4223 -1DHD	IC	Dual Supply Hot Swap Controller for Advanced Mezzanine Card	MLP	Lic		pdf pinout
LTC 4223 -1GN	IC	Dual Supply Hot Swap Controller for Advanced Mezzanine Card	TSOP	Lic		pdf pinout
LTC 4223 -2DHD	IC	Dual Supply Hot Swap Controller for Advanced Mezzanine Card	MLP	Lic		pdf pinout
LTC 4223 -2GN	IC	Dual Supply Hot Swap Controller for Advanced Mezzanine Card	TSOP	Lic	-	pdf pinout
LTC 4242 G	IC	Dual Slot Hot Swap Controller for PCI Express	TSOP	Lic		pdf pinout
LTC 4242 UHF	IC	Dual Slot Hot Swap Controller for PCI Express	QFN	Lic		pdf pinout
LTC 4251 -1	IC	Negative Voltage Hot Swap Controllers, operating range to –36V to –72V	46	Lic	-	pdf pinout
LTC 4251 -2	IC	Negative Voltage Hot Swap Controllers, UV threshold of –43V only.	46	Lic		pdf pinout
LTC 4251	IC	Negative Voltage Hot Swap Controllers,	46	Lic	-	pdf pinout

Type	Device	Short Description	Fig.	Manu	Comparision Types	More at
		operating range of –43V to –75V				
LTC 4252 -1MS8	IC	Negative Voltage Hot Swap Controllers	TSOP	Lic	-	pdf pinout
LTC 4252 (A)-1MS	IC	Negative Voltage Hot Swap Controllers	TSOP	Lic	-	pdf pinout
LTC 4252 -2MS8	IC	Negative Voltage Hot Swap Controllers, Automatic Retry After Fault	TSOP	Lic	-	pdf pinout
LTC 4252 (A)-2MS	IC	Negative Voltage Hot Swap Controllers, Automatic Retry After Fault	TSOP	Lic	-	pdf pinout
LTC 4253 AUF-ADJ	IC	–48V Hot Swap Controller with Sequencer	QFN	Lic	-	pdf pinout
LTC 4253 (A)	IC	–48V Hot Swap Controllers with Sequencer	TSOP	Lic	-	pdf pinout
LTC 4253 AGN-ADJ	IC	–48V Hot Swap Controller with Sequencer	TSOP	Lic	-	pdf pinout
LTC 4255	IC	Quad Network Power Controller with I2C Compatible Interface	TSOP	Lic	-	pdf pinout
LTC 4257 DD	LIN-IC	IEEE 802.3af Pd Power over Ethernet Interface Contr.	QFN	Lic	--	pdf pinout
LTC 4258	IC	Quad IEEE 802.3af Power over Ethernet Controller with Integrated Detection	TSOP	Lic	-	pdf pinout
LTC 4259 A-1	IC	Quad IEEE 802.3af Power over Ethernet Controller with AC Disconnect	TSOP	Lic	-	pdf pinout
LTC 4259 AGW	LIN-IC	Quad IEEE 802.3af Power over Ethernet Contr.	TSOP	Lic	-	pdf pinout
LTC 4260 GN	IC	Positive High Voltage Hot Swap Controller	TSOP	Lic	-	pdf pinout
LTC 4260 SW	IC	Positive High Voltage Hot Swap Controller	MDIP	Lic	-	pdf pinout
LTC 4260 UH	IC	Positive High Voltage Hot Swap Controller	QFN	Lic	-	pdf pinout
LTC 4261 GN-2	IC	Negative Voltage Hot Swap Controllers with ADC and I2C Monitoring	TSOP	Lic	-	pdf pinout
LTC 4261 GN	IC	Negative Voltage Hot Swap Controllers with ADC and I2C Monitoring	TSOP	Lic	-	pdf pinout
LTC 4261 UFD-2	IC	Negative Voltage Hot Swap Controllers with ADC and I2C Monitoring	QFN	Lic	-	pdf pinout
LTC 4261 UFD	IC	Negative Voltage Hot Swap Controllers with ADC and I2C Monitoring	QFN	Lic	-	pdf pinout
LTC 4263 DE-1	IC	High Power Single Pse Controller with Internal Switch	MLP	Lic	-	pdf pinout
LTC 4263 DE	IC	Single IEEE 802.3af Compliant Pse Controller with Internal Switch	MLP	Lic	-	pdf pinout
LTC 4263 S	IC	Single IEEE 802.3af Compliant Pse Controller with Internal Switch	MDIP	Lic	-	pdf pinout
LTC 4264	IC	High Power Pd Interface Controller with 750mA Current Limit	MLP	Lic	-	pdf pinout
LTC 4267 -1	IC	Power over Ethernet IEEE 802.3af Pd Interface	TSOP	Lic	-	pdf pinout
LTC 4268 -1	IC	High Power Pd with Synchronous Noopto Flyback Controller	MLP	Lic	-	pdf pinout
LTC 4303 DD	CMOS-IC	Hot Swappable 2-Wire Bus Buffer,Stuck Bus Recovery	46	Ltc	-	pdf pinout
LTC 4304 DD	CMOS-IC	Hot Swappable 2-Wire Bus Buffer,Stuck Bus Recovery	46	Ltc	-	pdf pinout
LTC 4305 DHP	CMOS-IC	2-Ch, 2-Wire Bus Multiplexer	46	Ltc	-	pdf pinout
LTC 4305 GN	CMOS-IC	2-Ch, 2-Wire Bus Multiplexer	46	Ltc	-	pdf pinout
LTC 4306 GN	CMOS-IC	4-Ch, 2-Wire Bus Multiplexer	46	Ltc	-	pdf pinout
LTC 4306 UFD	CMOS-IC	4-Ch, 2-Wire Bus Multiplexer	46	Ltc	-	pdf pinout
LTC 4307 DD	LIN-IC	Low Offset Hot Swappable 2-Wire Bus Buffer	DFN	Lic	-	pdf pinout
LTC 4308 DD	LIN-IC	Level Shifting Hot Swappable 2-Wire Bus Buffer	DFN	Lic	-	pdf pinout
LTC 4309 GN	LIN-IC	Low Offset Hot Swappable 2-Wire Bus Buffer	DFN	Lic	-	pdf pinout
LTC 4350	IC	Hot Swappable Load Share Controller	SSMDIP	Lic	-	pdf pinout
LTC 4355 S	IC	Positive High Voltage Ideal Diode-or with Input Supply and Fuse Monitors	SSMDIP	Lic	-	pdf pinout
LTC 4357 DCB	IC	Pos. High Volt. Ideal Diode Controller	SSMDIP	Lic	-	pdf pinout
LTC 4401 -1	IC	Rf Power Controllers with 250kHz Loop BW and 45dB Dynamic Range	46	Lic	-	pdf pinout
LTC 4401 -2	IC	Rf Power Controllers with 250kHz Loop BW and 45dB Dynamic Range	TSOP	Lic	-	pdf pinout
LTC 4402 -1	IC	Multiband Rf Power Controllers for Edge/tdma	TSOP	Lic	-	pdf pinout
LTC 4402 -2	IC	Multiband Rf Power Controllers for Edge/tdma	TSOP	Lic	-	pdf pinout
LTC 4403 -1	IC	Multiband Rf Power Controllers for Edge/tdma	TSOP	Lic	-	pdf pinout
LTC 4403 -2	IC	Multiband Rf Power Controllers for Edge/tdma	TSOP	Lic	-	pdf pinout
LTC 4414	IC	36V, Low Loss Powerpath Controller for Large PFETs	SSMDIP	Lic	-	pdf pinout
LTC 4416	IC	36V,Dual Controllers for Large PFETs	SSMDIP	Lic	-	pdf pinout
LTC 4441 MSE	IC	N-channel Mosfet Gate Driver	SSMDIP	Lic	-	pdf pinout
LTC 4444	IC	High Volt.sync.n-channel Mosfet Driver	SSMDIP	Lic	-	pdf pinout
LTC 5100 UF	DIG/LIN-IC	3.3V, 3.2Gbps VCSEL Driver	QFN	Lic	-	pdf pinout
LTC 6406 UD	IC	3GHz, Rail-to-Rail Input Differential Amp/driver	DFN	Lic	-	pdf pinout
LTC 6416	CMOS-IC	2GHz Low Noise Diff. 16-Bit ADC Buffer	MLP	Lic	-	pdf pinout

Type	Device	Short Description	Fig.	Manu	Comparision Types	More at
LTC 660 C(N/S) 8	CMOS-IC	100mA Cmos Voltage Converter	TSOP	Lic	-	pdf pinout
LTC 690 J 8	LIN-IC	Microprocessor Supervisory Circuits	TSOP	Lic	-	pdf pinout
LTC 690 N 8	LIN-IC	Microprocessor Supervisory Circuits	TSOP	Lic	-	pdf pinout
LTC 690 S 8	LIN-IC	Microprocessor Supervisory Circuits	TSOP	Lic	-	pdf pinout
LTC 694 N 8	LIN-IC	Microprocessor Supervisory Circuits	TSOP	Lic	-	pdf pinout
LTC 694 S 8	LIN-IC	Microprocessor Supervisory Circuits	TSOP	Lic	-	pdf pinout
LTC 699 N 8	LIN-IC	Microprocessor Supervisory Circuits	TSOP	Lic	-	pdf pinout
LTC 699 S 8	LIN-IC	Microprocessor Supervisory Circuits	TSOP	Lic	-	pdf pinout
LTC 7541 A	D/A-IC	Cmos 12-Bit Multiplying DAC	TSOP	Lic	-	pdf pinout
LTC 7543	D/A-IC	Serial 12-Bit Multiplying DACs	TSOP	Lic	-	pdf pinout
LTC 7545 A	D/A-IC	Serial 12-Bit Multiplying DACs	TSOP	Lic	-	pdf pinout
LTC7650	CHOPPER	Vs:±9V Vu:150dB Vo:±4.85V f:1.2Mc	8D,14D	Ltc	-	data pinout
LTC7652	CHOPPER	Vs:±9V Vu:150dB Vo:±4.85V f:1.2Mc		Ltc	-	data pdf pinout
LTC 8043	D/A-IC	Serial 12-Bit Multiplying DAC	TSOP	Lic	-	pdf pinout
LTC 8143	D/A-IC	Serial 12-Bit Multiplying DACs	TSOP	Lic	-	pdf pinout
LTC 1044 ACN 8	CMOS-IC	12V Cmos Voltage Converter	TSOP	Lic	-	pdf pinout
LTC 1044 ACS 8	CMOS-IC	12V Cmos Voltage Converter	TSOP	Lic	-	pdf pinout
LTC 1044 AIN 8	CMOS-IC	12V Cmos Voltage Converter	TSOP	Lic	-	pdf pinout
LTC 1044 AIS 8	CMOS-IC	12V Cmos Voltage Converter	TSOP	Lic	-	pdf pinout
LTC 1044 CJ 8	CMOS-IC	Switched Capacitor Voltage Converter	81	Lic	-	pdf pinout
LTC 1044 CN 8	CMOS-IC	Switched Capacitor Voltage Converter	81	Lic	-	pdf pinout
LTC 1044 CS 8	CMOS-IC	Switched Capacitor Voltage Converter	TSOP	Lic	-	pdf pinout
LTC 1044 MJ 8	CMOS-IC	Switched Capacitor Voltage Converter	81	Lic	-	pdf pinout
LTC 1064-1 ACN	IC	Clock Sweepable Elliptic Lowpass Filter	TSOP	Lic	-	pdf pinout
LTC 1064-1 CN	IC	Clock Sweepable Elliptic Lowpass Filter	TSOP	Lic	-	pdf pinout
LTC 1064-1 CSW	IC	Clock Sweepable Elliptic Lowpass Filter	TSOP	Lic	-	pdf pinout
LTC 1064-2 CN	IC	8th Order Butterworth Lowpass Filter	TSOP	Lic	-	pdf pinout
LTC 1064-2 CSW	IC	8th Order Butterworth Lowpass Filter	TSOP	Lic	-	pdf pinout
LTC 1064-3 CN	IC	8th Order Linear Phase Lowpass Filter	TSOP	Lic	-	pdf pinout
LTC 1064-3 CSW	IC	8th Order Linear Phase Lowpass Filter	TSOP	Lic	-	pdf pinout
LTC 1064-4 CN	CMOS-IC	Clock Sweepable Cauer Lowpass Filter	DIP	Lic	-	pdf pinout
LTC 1064-4 CSW	CMOS-IC	Clock Sweepable Cauer Lowpass Filter	DIP	Lic	-	pdf pinout
LTC 1064-7 CN	LIN-IC	Linear Phase, 8th Order Lowpass Filter	DIP	Lic	-	pdf pinout
LTC 1064-7 CSW	LIN-IC	Linear Phase, 8th Order Lowpass Filter	DIP	Lic	-	pdf pinout
LTC 1065 CN 8	LIN-IC	Linear Phase 5th Order Bessel Lowpass Filter	DIP	Lic	-	pdf pinout
LTC 1065 IN 8	LIN-IC	Linear Phase 5th Order Bessel Lowpass Filter	DIP	Lic	-	pdf pinout
LTC 1066-1 CSW	LIN-IC	14-Bit Clock-Tunable Phase Lowpass Filter	DIP	Lic	-	pdf pinout
LTC 1069-1	LIN-IC	Progressive Elliptic, Lowpass Filter	DIP	Lic	-	pdf pinout
LTC 1069-6	LIN-IC	Single Supply Elliptic Lowpass Filter	DIP	Lic	-	pdf pinout
LTC 1069-7	LIN-IC	Linear Phase 8th Order Lowpass Filter	DIP	Lic	-	pdf pinout
LTC 1100 (AC/C) N8	LIN-IC	Chopper-stabilized Instrumentation Amp.	TSOP	Lic	-	pdf pinout
LTC 1100 AM J8	LIN-IC	Chopper-stabilized Instrumentation Amp.	TSOP	Lic	-	pdf pinout
LTC 1100 (C/M) J8	LIN-IC	Chopper-stabilized Instrumentation Amp.	TSOP	Lic	-	pdf pinout
LTC 1164-5 CN	LIN-IC	Butterworth or Bessel Lowpass Filter	DIP	Lic	-	pdf pinout
LTC 1164-5 CSW	LIN-IC	Butterworth or Bessel Lowpass Filter	DIP	Lic	-	pdf pinout
LTC 1164-6 N	LIN-IC	Ellipt. or Linear Phase Lowpass Filter	DIP	Lic	-	pdf pinout
LTC 1164-6 SW	LIN-IC	Ellipt. or Linear Phase Lowpass Filter	DIP	Lic	-	pdf pinout
LTC 1164-7 N	LIN-IC	Linear Phase 8th Order Lowpass Filter	DIP	Lic	-	pdf pinout
LTC 1164-7 SW	LIN-IC	Linear Phase 8th Order Lowpass Filter	DIP	Lic	-	pdf pinout
LTC 1174 HVCN 8	LIN-IC	Step-down and Inverting DC/DC Converter	DIP	Lic	-	pdf pinout
LTC 1174 HVCS 8	LIN-IC	Step-down and Inverting DC/DC Converter	DIP	Lic	-	pdf pinout
LTC 1174 IN 8	LIN-IC	Step-down and Inverting DC/DC Converter	DIP	Lic	-	pdf pinout
LTC 1174 IS 8	LIN-IC	Step-down and Inverting DC/DC Converter	DIP	Lic	-	pdf pinout
LTC 1261 LCMS 8	IC	Switched Capacitor Reg. Voltage Inverter	TSOP	Lic	-	pdf pinout
LTC 1261 LCS 8	IC	Switched Capacitor Reg. Voltage Inverter	TSOP	Lic	-	pdf pinout
LTC 1262 C (N/S) 8	LIN-IC	12V, 30mA Flash Memory Programming Supply	TSOP	Lic	-	pdf pinout
LTC 1262 IS 8	LIN-IC	12V, 30mA Flash Memory Programming Supply	TSOP	Lic	-	pdf pinout
LTC 1263 CS 8	LIN-IC	12V, 60mA Flash Memory Programming Supply	TSOP	Lic	-	pdf pinout
LTC 1422 N 8	IC	Hot Swap Controller	TSOP	Ltc	-	pdf pinout
LTC 1422 S 8	IC	Hot Swap Controller	TSOP	Ltc	-	pdf pinout
LTC 1480 N 8	LIN-IC	3.3V Ultralow Power RS485 Transceiver	TSOP	Lic	-	pdf pinout
LTC 1480 S 8	LIN-IC	3.3V Ultralow Power RS485 Transceiver	TSOP	Lic	-	pdf pinout
LTC 1481 N 8	LIN-IC	Low Power RS485 Transceiver, Shutdown	TSOP	Lic	-	pdf pinout
LTC 1481 S 8	LIN-IC	Low Power RS485 Transceiver, Shutdown	TSOP	Lic	-	pdf pinout
LTC 1482 MS 8	LIN-IC	RS485 Transcei., Carrier Detect Receiver Fail-Safe	TSOP	Lic	-	pdf pinout
LTC 1482 N 8	LIN-IC	RS485 Transcei., Carrier Detect Receiver Fail-Safe	TSOP	Lic	-	pdf pinout
LTC 1482 S 8	LIN-IC	RS485 Transcei., Carrier Detect Receiver Fail-Safe	TSOP	Lic	-	pdf pinout
LTC 1483 N 8	LIN-IC	Power RS485 Low Emi Transceiver, Shutdown	TSOP	Lic	-	pdf pinout
LTC 1483 S 8	LIN-IC	Power RS485 Low Emi Transceiver, Shutdown	TSOP	Lic	-	pdf pinout
LTC 1484 MS 8	LIN-IC	Low Power RS485 Transceiver, Receiver Fail-Safe	TSOP	Lic	-	pdf pinout
LTC 1484 N 8	LIN-IC	Low Power RS485 Transceiver, Receiver Fail-Safe	TSOP	Lic	-	pdf pinout
LTC 1484 S 8	LIN-IC	Low Power RS485 Transceiver, Receiver	TSOP	Lic	-	pdf pinout

Type	Device	Short Description	Fig.	Manu	Comparision Types	More at
		Fail-Safe				
LTC 1487 N 8	LIN-IC	RS485 Low EMI, Shutdown, High Input Impedance	TSOP	Lic	-	pdf pinout
LTC 1487 S 8	LIN-IC	Low Power RS485 Low EMI,Shutdown,High Input Impedance	TSOP	Lic	-	pdf pinout
LTC 1550 LCMS 8	LIN-IC	Switched Capacitor Reg. Voltage Inverter	TSOP	Lic	-	pdf pinout
LTC 1550 LCS 8	IC	Switched Capacitor Reg. Voltage Inverter	TSOP	Lic	-	pdf pinout
LTC 1551 LCMS 8	LIN-IC	Switched Capacitor Reg. Voltage Inverter	TSOP	Lic	-	pdf pinout
LTC 1551 LCS 8	IC	Switched Capacitor Reg. Voltage Inverter	TSOP	Lic	-	pdf pinout
LTC 1562-2 G	LIN-IC	Active RC Quad Universal Filter	DIP	Lic	-	pdf pinout
LTC 1563-2 GN	LIN-IC	4th Order Lowpass Filter Family	DIP	Lic	-	pdf pinout
LTC 1563-3 GN	LIN-IC	4th Order Lowpass Filter Family	DIP	Lic	-	pdf pinout
LTC 1596-1	D/A-IC	Serial 16-Bit Multiplying DACs	QFN	Lic	-	pdf pinout
LTC 1643 AL-1 GN	CMOS-IC	Pci-bus Hot Swap Controller	16-SSMDIP	Lic	-	pdf pinout
LTC 1645 S 8	CMOS-IC	2-CH Hot Swap Controller/Power Sequencer	16-SSMDIP	Lic	-	pdf pinout
LTC 1647-3 GN	CMOS-IC	Dual Hot Swap Controllers	16-SSMDIP	Lic	-	pdf pinout
LTC 1682 CMS 8	LIN-IC	Doubler Charge Pumps, Low Noise Linear Regulator	TSOP	Lic	-	pdf pinout
LTC 1682 CS 8	LIN-IC	Doubler Charge Pumps, Low Noise Linear Regulator	TSOP	Lic	-	pdf pinout
LTC 1682 IMS 8	LIN-IC	Doubler Charge Pumps, Low Noise Linear Regulator	TSOP	Lic	-	pdf pinout
LTC 1682 IS 8	LIN-IC	Doubler Charge Pumps, Low Noise Linear Regulator	TSOP	Lic	-	pdf pinout
LTC 1686 S 8	LIN-IC	52Mbps Precision Delay RS485 Fail-Safe Transcei.	TSOP	Lic	-	pdf pinout
LTC 1690 MS 8	LIN-IC	Differential Driver, Receiver Pair, Fail- Safe Receiver Output	TSOP	Lic	-	pdf pinout
LTC 1690 N 8	LIN-IC	Differential Driver, Receiver Pair, Fail- Safe Receiver Output	TSOP	Lic	-	pdf pinout
LTC 1690 S 8	LIN-IC	Differential Driver, Receiver Pair, Fail- Safe Receiver Output	TSOP	Lic	-	pdf pinout
LTC 1694 S 5	CMOS-IC	Smbus/i²c Accelerator	46	Ltc	-	pdf pinout
LTC 1729 CMS 8	IC	Li-Ion Battery Charger Termination Contr.	87	Lic	-	pdf pinout
LTC 1729 CS 8	IC	Li-Ion Battery Charger Termination Contr.	87	Lic	-	pdf pinout
LTC 1730 EGN-4	IC	Lithium-ion Battery Pulse Charger, Overcur. Protection	87	Lic	-	pdf pinout
LTC 1732-4	IC	Lithium-ion Linear Battery Charger Controller	87	Lic	-	pdf pinout
LTC 1751 EMS 8	LIN-IC	Regulated Charge Pump DC/DC Converters	TSOP	Lic	-	pdf pinout
LTC 1842 (C/I) S8	LIN-IC	Ultralow Power Dual Comparator Reference	TSOP	Lic	-	pdf pinout
LTC 1861 LS 8	A/D-IC	12-Bit, 250ksps 1- and 2-CH ADCs	MDIP	Lic	-	pdf pinout
LTC 1861 S 8	A/D-IC	12-Bit, 250ksps 1- and 2-CH ADCs	MDIP	Lic	-	pdf pinout
LTC 1981 ES 5	IC	Single,Dual Micropower High Side Switch Controllers	SSMDIP	Lic	-	pdf pinout
LTC 1982 ES 6	IC	Single,Dual Micropower High Side Switch Controllers	SSMDIP	Lic	-	pdf pinout
LTC 2220-1	A/D-IC	12-Bit, 185Msps ADC	87	Lic	-	pdf pinout
LTC 2450-1	IC	Easy-to-Use, Ultra-Tiny 16-Bit ΛE ADC	87	Lic	-	pdf pinout
LTC 2904 DDB 8	LIN-IC	Dual Supply Monitors,Pin-selectable Thresholds	DFN	Lic	-	pdf pinout
LTC 2904 TS 8	LIN-IC	Dual Supply Monitors,Pin-selectable Thresholds	DFN	Lic	-	pdf pinout
LTC 2905 DDB 8	LIN-IC	Dual Supply Monitors,Pin-selectable Thresholds	DFN	Lic	-	pdf pinout
LTC 2905 TS 8	LIN-IC	Dual Supply Monitors,Pin-selectable Thresholds	DFN	Lic	-	pdf pinout
LTC 2906 TS 8	LIN-IC	Dual Supply Monitors,Pin-selectable Threshold,Adjustable Input	DFN	Lic	-	pdf pinout
LTC 2907 TS 8	LIN-IC	Dual Supply Monitors,Pin-selectable Threshold,Adjustable Input	DFN	Lic	-	pdf pinout
LTC 2908 DDB 8	LIN-IC	Precision Six Input Supply Monitor	DFN	Lic	-	pdf pinout
LTC 2909 TS 8	LIN-IC	Triple/dual Input Uv, Ov, Neg. Voltage Monitor	DFN	Lic	-	pdf pinout
LTC 2912 DDB-1	LIN-IC	Single Uv/ov Voltage Monitor	DFN	Lic	-	pdf pinout
LTC 2912 DDB-2	LIN-IC	Single Uv/ov Voltage Monitor	DFN	Lic	-	pdf pinout
LTC 2912 DDB-3	LIN-IC	Single Uv/ov Voltage Monitor	DFN	Lic	-	pdf pinout
LTC 2913 DD-1	LIN-IC	Dual Uv/ov Voltage Monitor	TSOP	Lic	-	pdf pinout
LTC 2913 MS-1	LIN-IC	Dual Uv/ov Voltage Monitor	TSOP	Lic	-	pdf pinout
LTC 2913 DD-2	LIN-IC	Dual Uv/ov Voltage Monitor	TSOP	Lic	-	pdf pinout
LTC 2913 MS-2	LIN-IC	Dual Uv/ov Voltage Monitor	TSOP	Lic	-	pdf pinout
LTC 2914 DHC-1	LIN-IC	Quad Uv/ov Positive/negative Voltage Monitor	TSOP	Lic	-	pdf pinout
LTC 2914 GN-1	LIN-IC	Quad Uv/ov Positive/negative Voltage Monitor	TSOP	Lic	-	pdf pinout
LTC 2914 DHC-2	LIN-IC	Quad Uv/ov Positive/negative Voltage Monitor	TSOP	Lic	-	pdf pinout
LTC 2914 GN-2	LIN-IC	Quad Uv/ov Positive/negative Voltage Monitor	TSOP	Lic	-	pdf pinout
LTC 2915 TS 8	LIN-IC	Low Volt. Supervisor, 27 Selectable Thresholds	TSOP	Lic	-	pdf pinout
LTC 2916 TS 8	LIN-IC	Low Volt. Supervisor, 27 Selectable Thresholds	TSOP	Lic	-	pdf pinout

Type	Device	Short Description	Fig.	Manu	Comparision Types	More at
LTC 2927 TS 8	LIN-IC	Power Supply Tracking Controller	TSOP	Lic	-	pdf pinout
LTC 2950-1 DDB	LIN-IC	Push Button On/Off Controller	TSOP	Lic	-	pdf pinout
LTC 2950-2 DDB	LIN-IC	Push Button On/Off Controller	TSOP	Lic	-	pdf pinout
LTC 2954 TS 8	LIN-IC	Push Button On/Off Controller, µP Interrupt	DFN	Lic	-	pdf pinout
LTC 2970-1 UFD	LIN-IC	Dual I²C Power Supply Monitor, Margining Controller	DFN	Lic	-	pdf pinout
LTC 3203 BEDD-1	IC	500mA Output Current Low Noise Dual Mode Step-up Charge Pumps	DIP	Lic	-	pdf pinout
LTC 3203 EDD-1	IC	500mA Output Current Low Noise Dual Mode Step-up Charge Pumps	DIP	Lic	-	pdf pinout
LTC 3204 BEDC-5	IC	Low Noise Regulated Charge Pump in 2 x 2 DFN	DIP	Lic	-	pdf pinout
LTC 3204 EDC-5	IC	Low Noise Regulated Charge Pump in 2 x 2 DFN	DIP	Lic	-	pdf pinout
LTC 3221 EDC-5	IC	Micropower, Regulated Charge Pump in 2 x 2 DFN	DIP	Lic	-	pdf pinout
LTC 3550-1	IC	Li-Ion Battery Charger,600mA Buck Conv.	43	Lic	-	pdf pinout
LTC 3552-1	IC	Li-Ion Battery Charger,2xBuck Convert.	43	Lic	-	pdf pinout
LTC 3557-1	IC	USB Power Manager, Li- Ion Charger, 3xStep- Down Reg.	43	Lic	-	pdf pinout
LTC 4001-1	IC	2A Synchronous Buck Li-Ion Charger	DFN	Lic	-	pdf pinout
LTC 4085-1	IC	USB Power Manager,Diode Controller, 4.1V Li-Ion Charg.	DFN	Lic	-	pdf pinout
LTC 4088-1	IC	Battery Charger/usb Power Manager	DFN	Lic	-	pdf pinout
LTC 4088-2	IC	Battery Charger/usb Power Manager	DFN	Lic	-	pdf pinout
LTC 4211 MS 8	CMOS-IC	Hot Swap Contr., Multifunction Current Control	46	Ltc	-	pdf pinout
LTC 4211 S 8	CMOS-IC	Hot Swap Contr., Multifunction Current Control	46	Ltc	-	pdf pinout
LTC 4257 S8	DIG/LIN-IC	IEEE 802.3af Pd Power over Ethernet Interface Contr.	QFN	Lic	-	pdf pinout
LTC 4300 A-3 DD	CMOS-IC	Level Shifting Hot Swappable 2-Wire Bus Buffer	46	Ltc	-	pdf pinout
LTC 4307-1 DD	LIN-IC	Hdmi Level- Shifting 2-Wire Bus Buffer	DFN	Lic	-	pdf pinout
LTC 4307 MS 8	LIN-IC	Low Offset Hot Swappable 2-Wire Bus Buffer	DFN	Lic	-	pdf pinout
LTC 4308 MS 8	LIN-IC	Level Shifting Hot Swappable 2-Wire Bus Buffer	DFN	Lic	-	pdf pinout
LTC 4354 DDB 8	IC	Negative Voltage Diode-or Controller and Monitor	SSMDIP	Lic	-	pdf pinout
LTC 4354 S 8	IC	Negative Voltage Diode-or Controller and Monitor	SSMDIP	Lic	-	pdf pinout
LTC 4357 MS 8	IC	Pos. High Volt. Ideal Diode Controller	SSMDIP	Lic	-	pdf pinout
LTC 4413-1	IC	Dual 2.6A, 2.5V to 5.5V Ideal Diodes	SSMDIP	Lic	-	pdf pinout
LTC 4413-2	IC	Dual 2.6A, 2.5V to 5.5V Ideal Diodes	SSMDIP	Lic	-	pdf pinout
LTC 4416-1	IC	36V,Dual Controllers for Large PFETs	SSMDIP	Lic	-	pdf pinout
LTC 4442 MS 8 E	IC	Synchronous N-channel Mosfet Drivers	SSMDIP	Lic	-	pdf pinout
LTC 6406 MS 8 E	IC	3GHz, Rail-to-Rail Input Differential Amp/driver	DFN	Lic	-	pdf pinout
LTC 6410-6 UD	IC	Differential IF Amp,Configurable Input Impedance	DFN	Lic	-	pdf pinout
LTC 1067-50	LIN-IC	Universal Dual Filter Building Block	DIP	Lic	-	pdf pinout
LTC 1068-25 G	LIN-IC	Filter Building Blocks	DIP	Lic	-	pdf pinout
LTC 1068-50 G	LIN-IC	Filter Building Blocks	DIP	Lic	-	pdf pinout
LTC 1261 LCMS 8-4	IC	Switched Capacitor Reg. Voltage Inverter	TSOP	Lic	-	pdf pinout
LTC 1261 LCS 8-4	IC	Switched Capacitor Reg. Voltage Inverter	TSOP	Lic	-	pdf pinout
LTC 1329-10	A/D-IC	8-Bit Current Output D/A Converter	QFN	Lic	-	pdf pinout
LTC 1329-50	A/D-IC	8-Bit Current Output D/A Converter	QFN	Lic	-	pdf pinout
LTC 1329 A-50	A/D-IC	8-Bit Current Output D/A Converter	QFN	Lic	-	pdf pinout
LTC 1421-2.5 G	IC	Hot Swap Controller	TSOP	Ltc	-	pdf pinout
LTC 1421-2.5 SW	IC	Hot Swap Controller	TSOP	Ltc	-	pdf pinout
LTC 1427-50	D/A-IC	10-Bit Current Out DAC, SMBus Ser. Interface	QFN	Lic	-	pdf pinout
LTC 1428-50	D/A-IC	8-Bit Current Sink Output DAC	QFN	Lic	-	pdf pinout
LTC 1550 LCMS 8-2	LIN-IC	Switched Capacitor Reg. Voltage Inverter	TSOP	Lic	-	pdf pinout
LTC 1550 LCS 8-2	IC	Switched Capacitor Reg. Voltage Inverter	TSOP	Lic	-	pdf pinout
LTC 1560-1 S 8	LIN-IC	1MHz/500kHz, Lowpass Elliptic Filter	DIP	Lic	-	pdf pinout
LTC 1566-1 S 8	LIN-IC	2.3MHz Continuous Time Lowpass Filter	DIP	Lic	-	pdf pinout
LTC 1647-1 S8	CMOS-IC	Dual Hot Swap Controllers	16-SSMDIP	Lic	-	pdf pinout
LTC 1647-2 S8	CMOS-IC	Dual Hot Swap Controllers	16-SSMDIP	Lic	-	pdf pinout
LTC 1682 CMS 8-5	LIN-IC	Doubler Charge Pumps, Low Noise Linear Regulator	TSOP	Lic	-	pdf pinout
LTC 1682 CS 8-5	LIN-IC	Doubler Charge Pumps, Low Noise Linear Regulator	TSOP	Lic	-	pdf pinout
LTC 1682 IMS 8-5	LIN-IC	Doubler Charge Pumps, Low Noise Linear Regulator	TSOP	Lic	-	pdf pinout
LTC 1682 IS 8-5	LIN-IC	Doubler Charge Pumps, Low Noise Linear Regulator	TSOP	Lic	-	pdf pinout
LTC 1694-1 S 5	CMOS-IC	Smbus/i²c Accelerator	46	Ltc	-	pdf pinout
LTC 1732-4.2	IC	Lithium-ion Linear Battery Charger Controller	87	Lic	-	pdf pinout
LTC 1732-8.4	IC	Lithium-ion Linear Battery Charger Controller	87	Lic	-	pdf pinout
LTC 1751 EMS 8-5	LIN-IC	Regulated Charge Pump DC/DC Converters	TSOP	Lic	-	pdf pinout

Type	Device	Short Description	Fig.	Manu	Comparision Types	More at
LTC 1754 ES 6-5	LIN-IC	Regulated 3.3V/5V Charge Pump, Shutdown	TSOP	Lic	-	pdf pinout
LTC 1928 E S 6-5	LIN-IC	Doubler Charge Pump with Low Noise Linear Regulator in Thinsot	TSOP	Lic	-	pdf pinout
LTC 1983 ES 6-3	LIN-IC	100mA Regulated Charge-pump Inverters in ThinSOT	TSOP	Lic	-	pdf pinout
LTC 1983 ES 6-5	LIN-IC	100mA Regulated Charge-pump Inverters in ThinSOT	TSOP	Lic	-	pdf pinout
LTC 2205-14	A/D-IC	14-Bit, 65Msps ADC	MDIP	Lic	-	pdf pinout
LTC 2206-14	A/D-IC	14-Bit, 105Msps/80Msps ADCs	MDIP	Lic	-	pdf pinout
LTC 2207-14	A/D-IC	14-Bit, 105Msps/80Msps ADCs	MDIP	Lic	-	pdf pinout
LTC 2208-14	A/D-IC	14-Bit, 130Msps ADC	MDIP	Lic	-	pdf pinout
LTC 2240-10	A/D-IC	10-Bit, 170Msps ADC	QFN	Lic	-	pdf pinout
LTC 2240-12	A/D-IC	12-Bit, 170Msps ADC	QFN	Lic	-	pdf pinout
LTC 2241-10	A/D-IC	10-Bit, 210Msps ADC	QFN	Lic	-	pdf pinout
LTC 2241-12	A/D-IC	12-Bit, 210Msps ADC	QFN	Lic	-	pdf pinout
LTC 2351-12	IC	6 Channel, 12-Bit, 1.5Msps Simultaneous Sampling ADC Shutdown	87	Lic	-	pdf pinout
LTC 2351-14	IC	6 Channel, 14-Bit, ADC, Shutdown	87	Lic	-	
LTC 2355-12	IC	Serial 12-Bit/14-Bit, ADCs, Shutdown	87	Lic	-	pdf pinout
LTC 2355-14	IC	Serial 12-Bit/14-Bit, ADCs, Shutdown	87	Lic	-	pdf pinout
LTC 2356-12	IC	Serial 12-Bit/14-Bit, ADCs, Shutdown	87	Lic	-	pdf pinout
LTC 2356-14	IC	Serial 12-Bit/14-Bit, ADCs, Shutdown	87	Lic	-	pdf pinout
LTC 2751-14	D/A-IC	Current Output 12-/14-/16-Bit DACs, Parallel I/O	TSOP	Lic	-	pdf pinout
LTC 2751-16	D/A-IC	Current Output 12-/14-/16-Bit DACs, Parallel I/O	TSOP	Lic	-	pdf pinout
LTC 2753-12	D/A-IC	Dual Curr. Out. 12- / 14- / 16- Bit DACs, Parallel I/. O	TSOP	Lic	-	pdf pinout
LTC 2753-14	D/A-IC	Dual Current Output 12-/14-/16-Bit DACs,Parallel I/O	TSOP	Lic	-	pdf pinout
LTC 2753-16	D/A-IC	Dual Current Output 12-/14-/16-Bit DACs,Parallel I/O .	TSOP	Lic	-	pdf pinout
LTC 2908 TS 8 -A1	LIN-IC	Precision Six Input Supply Monitor	DFN	Lic	-	pdf pinout
LTC 2908 TS 8 -B1	LIN-IC	Precision Six Input Supply Monitor	DFN	Lic	-	pdf pinout
LTC 2908 TS 8 -C1	LIN-IC	Precision Six Input Supply Monitor	DFN	Lic	-	pdf pinout
LTC 2910 DHC 16	LIN-IC	Octal Positive/negative Voltage Monitor	DFN	Lic	-	pdf pinout
LTC 2910 GN 16	LIN-IC	Octal Positive/negative Voltage Monitor	DFN	Lic	-	pdf pinout
LTC 2912 TS 8-1	LIN-IC	Single UV/OV Voltage Monitor	DFN	Lic	-	pdf pinout
LTC 2912 TS 8-2	LIN-IC	Single Uv/ov Voltage Monitor	DFN	Lic	-	pdf pinout
LTC 2912 TS 8-3	LIN-IC	Single Uv/ov Voltage Monitor	DFN	Lic	-	pdf pinout
LTC 2920-1 S 5	LIN-IC	Single/Dual Power Supply Margining Controller	TSOP	Lic	-	pdf pinout
LTC 2920-2 MS 8	LIN-IC	Single/Dual Power Supply Margining Controller	TSOP	Lic	-	pdf pinout
LTC 2923 DE 12	LIN-IC	Power Supply Tracking Controller	TSOP	Lic	-	
LTC 2950-1 TS 8	LIN-IC	Push Button On/Off Controller	TSOP	Lic	-	pdf pinout
LTC 2950-2 TS 8	LIN-IC	Push Button On/Off Controller	TSOP	Lic	-	pdf pinout
LTC 2951-1 DDB 8	LIN-IC	Push Button On/Off Controller	DFN	Lic	-	pdf pinout
LTC 2951-1 TS 8	LIN-IC	Push Button On/Off Controller	DFN	Lic	-	pdf pinout
LTC 2951-2 DDB 8	LIN-IC	Push Button On/Off Controller	DFN	Lic	-	pdf pinout
LTC 2951-2 TS 8	LIN-IC	Push Button On/Off Controller	DFN	Lic	-	pdf pinout
LTC 3204 BEDC-3.3	IC	Low Noise Regulated Charge Pump in 2 x 2 DFN	DIP	Lic	-	pdf pinout
LTC 3204 EDC-3.3	IC	Low Noise Regulated Charge Pump in 2 x 2 DFN	DIP	Lic	-	pdf pinout
LTC 3221 EDC-3.3	IC	Micropower, Regulated Charge Pump in 2 x 2 DFN	DIP	Lic	-	pdf pinout
LTC 3240 EDC-2.5	IC	Step-down Charge Pump DC/DC Converter	DIP	Lic	-	pdf pinout
LTC 3240 EDC-3.3	IC	Step-down Charge Pump DC/DC Converter	DIP	Lic	-	pdf pinout
LTC 4052-4.2	IC	Lithium-ion Battery Pulse Charger	DFN	Lic	-	pdf pinout
LTC 4053 EDD-4.2	IC	USB Compatible Lithium-ion Battery Charger	DFN	Lic	-	pdf pinout
LTC 4053 EMSE-4.2	IC	USB Compatible Lithium-ion Battery Charger	DFN	Lic	-	
LTC 4056-4.2	IC	Linear Li-Ion Charger with Termination	DFN	Lic	-	
LTC 4057-4.2	IC	Lin. Li-Ion Battery Charger,Thermal Regulation	DFN	Lic	-	pdf pinout
LTC 4058-4.2	IC	Li-Ion Battery Charger,Thermal Regula.	DFN	Lic	-	pdf pinout
LTC 4058 X-4.2	IC	Li-Ion Battery Charger,Thermal Regula.	DFN	Lic	-	pdf pinout
LTC 4061-4.4	IC	Linear Li-Ion Battery Charger Thermistor Input	DFN	Lic	-	pdf pinout
LTC 4065-4.4	IC	750mA Li-Ion Battery Charger	DFN	Lic	-	pdf pinout
LTC 4068-4.2	IC	Li-Ion Battery Charger Programmable Termination	DFN	Lic	-	pdf pinout
LTC 4068 X-4.2	IC	Li-Ion Battery Charger Programmable Termination	DFN	Lic	-	pdf pinout
LTC 4069-4.4	IC	750mA Li-Ion Battery Charger, NTC Thermistor Input	43	Lic	-	pdf pinout
LTC 4210-1 S 6	CMOS-IC	Hot Swap Controller	46	Ltc	-	pdf pinout
LTC 4210-2 S 6	CMOS-IC	Hot Swap Controller	46	Ltc	-	pdf pinout
LTC 4210-3 S 6	CMOS-IC	Hot Swap Controller	46	Ltc	-	pdf pinout
LTC 4210-4 S 6	CMOS-IC	Hot Swap Controller	46	Ltc	-	pdf pinout
LTC 4300 A-1 MS 8	CMOS-IC	Hot Swappable 2-Wire Bus Buffers	46	Ltc	-	pdf pinout
LTC 4300 A-2 MS 8	CMOS-IC	Hot Swappable 2-Wire Bus Buffers	46	Ltc	-	pdf pinout
LTC 4300 A-3 MS 8	CMOS-IC	Level Shifting Hot Swappable 2-Wire Bus	46	Ltc	-	pdf pinout

Type	Device	Short Description	Fig.	Manu	Comparision Types	More at
		Buffer				
LTC 4304 MS 10	CMOS-IC	Hot Swappable 2-Wire Bus Buffer,Stuck Bus Recovery	46	Ltc	-	pdf pinout
LTC 4307-1 MS 8	LIN-IC	Hdmi Level- Shifting 2-Wire Bus Buffer	DFN	Lic		pdf pinout
LTC 4309 DE 12	LIN-IC	Low Offset Hot Swappable 2-Wire Bus Buffer	DFN	Lic		pdf pinout
LTC 4355 DE 14	IC	Positive High Voltage Ideal Diode-or with Input Supply and Fuse Monitors	SSMDIP	Lic		pdf pinout
LTC 4440 S 6-5	IC	High Speed, High Voltage, High Side Gate Driver	SSMDIP	Lic		pdf pinout
LTC 4440 MS 8 E-5	IC	High Speed, High Voltage, High Side Gate Driver	SSMDIP	Lic		pdf pinout
LTC 4441 S 8-1	IC	N-channel Mosfet Gate Driver	SSMDIP	Lic		pdf pinout
LTC 4442 MS 8 E-1	IC	Synchronous N-channel Mosfet Drivers	SSMDIP	Lic		pdf pinout
LTC 6400-20 UD	IC	1.8GHz Differential ADC Driver,300MHz IF	DFN	Lic		pdf pinout
LTC 6401-20 UD	IC	1.3GHz Differential ADC Driver,140MHz IF	DFN	Lic		pdf pinout
LTC 1068-200 G	LIN-IC	Filter Building Blocks	DIP	Lic		pdf pinout
LTC 1159CG-(3.3/5)	LIN-IC	Synchronous Step-down Switching Regulat.	TSOP	Lic		pdf pinout
LTC 1159CN-(3.3/5)	LIN-IC	Synchronous Step-down Switching Regulat.	TSOP	Lic		pdf pinout
LTC 1159CS-(3.3/5)	LIN-IC	Synchronous Step-down Switching Regulat.	TSOP	Lic		pdf pinout
LTC 1261 LCMS8-4.5	IC	Switched Capacitor Reg. Voltage Inverter	TSOP	Lic		pdf pinout
LTC 1261 LCS 8-4.5	IC	Switched Capacitor Reg. Voltage Inverter	TSOP	Lic		pdf pinout
LTC 1265CS-(5/3.3)	LIN-IC	1.2A, Step-down DC/DC Converter	TSOP	Lic		pdf pinout
LTC 1550 LCMS8-2.5	LIN-IC	Switched Capacitor Reg. Voltage Inverter	TSOP	Lic		pdf pinout
LTC 1550 LCS 8-2.5	IC	Switched Capacitor Reg. Voltage Inverter	TSOP	Lic		pdf pinout
LTC 1550 LCMS8-4.1	LIN-IC	Switched Capacitor Reg. Voltage Inverter	TSOP	Lic		pdf pinout
LTC 1550 LCS 8-4.1	IC	Switched Capacitor Reg. Voltage Inverter	TSOP	Lic		pdf pinout
LTC 1551 LCMS8-4.1	LIN-IC	Switched Capacitor Reg. Voltage Inverter	TSOP	Lic		pdf pinout
LTC 1551 LCS 8-4.1	IC	Switched Capacitor Reg. Voltage Inverter	TSOP	Lic		pdf pinout
LTC 1565-31 S8	LIN-IC	650kHz Cont. Time, Lin. Phase Lowpass Filter	DIP	Lic	-	pdf pinout
LTC 1682 CMS 8-3.3	LIN-IC	Doubler Charge Pumps, Low Noise Linear Regulator	TSOP	Lic	-	pdf pinout
LTC 1682 CS 8-3.3	LIN-IC	Doubler Charge Pumps, Low Noise Linear Regulator	TSOP	Lic	-	pdf pinout
LTC 1682 IMS 8-3.3	LIN-IC	Doubler Charge Pumps, Low Noise Linear Regulator	TSOP	Lic	-	pdf pinout
LTC 1682 IS 8-3.3	LIN-IC	Doubler Charge Pumps, Low Noise Linear Regulator	TSOP	Lic	-	pdf pinout
LTC 1730 ES 8-4.2	IC	Lithium-ion Battery Pulse Charger, Overcur. Protection	87	Lic	-	pdf pinout
LTC 1731 EMS 8-4.1	IC	Lithium-ion Battery Charger Controlle.	87	Lic	-	pdf pinout
LTC 1731 ES 8-4.1	IC	Lithium-ion Battery Charger Controlle.	87	Lic	-	pdf pinout
LTC 1731 EMS 8-4.2	IC	Lithium-ion Battery Charger Controlle.	87	Lic	-	pdf pinout
LTC 1731 ES 8-4.2	IC	Lithium-ion Battery Charger Controlle.	87	Lic	-	pdf pinout
LTC 1731 EMS 8-8.2	IC	Lithium-ion Linear Battery Charger Controller	87	Lic	-	pdf pinout
LTC 1731 ES 8-8.2	IC	Lithium-ion Linear Battery Charger Controller	87	Lic	-	pdf pinout
LTC 1731 EMS 8-8.4	IC	Lithium-ion Linear Battery Charger Controller	87	Lic	-	pdf pinout
LTC 1731 ES 8-8.4	IC	Lithium-ion Linear Battery Charger Controller	87	Lic	-	pdf pinout
LTC 1751 EMS 8-3.3	LIN-IC	Regulated Charge Pump DC/DC Converters	TSOP	Lic	-	pdf pinout
LTC 1754 ES 6-3.3	LIN-IC	Regulated 3.3V/5V Charge Pump, Shutdown	TSOP	Lic	-	pdf pinout
LTC 1911 EMS 8-1.5	LIN-IC	Low Noise, High Efficiency, Inductorless Step-down DC/DC Converter	TSOP	Lic	-	pdf pinout
LTC 1911 EMS 8-1.8	LIN-IC	Low Noise, High Efficiency, Inductorless Step-down DC/DC Converter	TSOP	Lic	-	pdf pinout
LTC1174CN8-(3.3/5)	LIN-IC	Step-down and Inverting DC/DC Converter	DIP	Lic	-	pdf pinout
LTC1174CS8-(3.3/5)	LIN-IC	Step-down and Inverting DC/DC Converter	DIP	Lic	-	pdf pinout
LTC1261CS8-(4/4.5)	LIN-IC	Switched Capacitor Regulated Voltage Inverter	TSOP	Lic	-	pdf pinout
LTE-209	IRED	940nm ø3mm I:3mW/Sr Vf:1.2V If:20mA	T1	Lit		data pinout
LTE-2871	IRED	940nm ø5mm I:12mW/Sr Vf:1.2V If:20mA	T1¾	Lit		data pinout
LTE-2872U	IRED	940nm ø5mm I:10.5mW/Sr Vf:1.2V	T1¾	Lit		data pinout
LTE-302	IRED	940nm ø1.5mm I:1.13mW/Sr Vf:1.2V		Lit		data pinout
LTE-306	IRED	940nm ø1.6mm I:1.5mW/Sr Vf:1.2V		Lit		data pinout
LTE-3270	IRED	940nm ø5mm I:10.5mW/Sr Vf:1.35V	T1¾	Lit		data pinout
LTE-3271	IRED	940nm ø5mm I:12mW/Sr Vf:1.25V	T1¾	Lit		data pinout
LTE-4206	IRED	940nm ø3mm I:5.26mW/Sr Vf:1.2V	T1	Lit		data pinout
LTE-4238	IRED	940nm ø5mm I:5.26mW/Sr Vf:1.3V	T1¾	Lit		data pinout
LTE-5238	IRED	940nm ø5mm I:7.5mW/Sr Vf:1.3V	T1¾	Lit		data pinout
LTE63C	LED	gr ø5mm I:900...1800mcd Vf:3.8V	PLCC-4	Osr		data pdf pinout
LTE67C	LED	gr ø5mm I:355...900mcd Vf:3.8V	PLCC-4	Osr		data pdf pinout
LTE633	LED	bl+gr ø5mm I:710...1800mcd Vf:3.5V	PLCC-4	Osr		data pdf pinout
LTE673	LED	gr ø5mm I:180...450mcd Vf:3.7V	PLCC-4	Osr		data pdf pinout
LTH-1550	PS	LED/NPN CTR:4-8% Vbs:1.2V Ibs:20mA		Lit		data pinout
LTH-1650	PS	LED/NPN CTR:2-6% Vbs:1.2V Ibs:20mA		Lit		data pinout
LTH-209	PS	LED/NPN CTR:2% Vbs:1.2V Ibs:20mA		Lit		data pinout
LTH-301	PI	LED/NPN CTR:5% Vbs:1.2V Ibs:20mA		Lit		data pinout
LTH-301-W	PI	LED/NPN CTR:>10% Vbs:1.2V Ibs:20mA		Lit		data pinout
LTH-306	PI	LED/NPN CTR:2.5% Vbs:1.2V Ibs:20mA		Lit		data pinout
LTH-306-W	PI	LED/NPN CTR:10% Vbs:1.2V Ibs:20mA		Lit		data pinout
LTH860	PI	LED/NPN CTR:>2.5% Vbs:1.2V Ibs:20mA		Lit	MSTXXXX	data pinout

Type	Device	Short Description	Fig.	Manu	Comparision Types	More at
LTH861	PI	LED/NPN CTR:>10% Vbs:1.2V Ibs:10mA		Lit	MSTXXXX	data pinout
LTH862	PI	LED/NPN CTR:>6% Vbs:1.2V Ibs:20mA		Lit	MSTXXXX	data pinout
LTH865	PI	LED/NPN CTR:>2.5% Vbs:1.2V Ibs:20mA		Lit	MSTXXXX	data pinout
LTH866	PI	LED/NPN CTR:>10% Vbs:1.2V Ibs:10mA		Lit	MSTXXXX	data pinout
LTH867	PI	LED/NPN CTR:>9% Vbs:1.2V Ibs:20mA		Lit	MSTXXXX	data pinout
LTH870	PI	LED/NPN CTR:>2.5% Vbs:1.2V Ibs:20mA		Lit		data pinout
LTH871	PI	LED/NPN CTR:>2.5% Vbs:1.2V Ibs:20mA		Lit		data pinout
LTH872	PI	LED/NPN CTR:>9% Vbs:1.2V Ibs:20mA		Lit		data pinout
LTH873	PI	LED/NPN CTR:>1% Vbs:1.2V Ibs:20mA		Lit		data pinout
LTH875	PI	LED/NPN CTR:>2.5% Vbs:1.2V Ibs:20mA		Lit		data pinout
LTH876	PI	LED/NPN CTR:>2.5% Vbs:1.2V Ibs:20mA		Lit		data pinout
LTH877	PI	LED/NPN CTR:>9% Vbs:1.2V Ibs:20mA		Lit		data pinout
LTH878	PI	LED/NPN CTR:>1% Vbs:1.2V Ibs:20mA		Lit		data pinout
LTJ-811HR	LED	gr+srd+yl ø20mm I:25mcd Vf:2.1V		Lit		data pinout
LTL-10203	LED	rd+srd+gr+yl ø5mm I:12.6mcd Vf:2.1V	T1¾	Lit		data pinout
LTL-10C	LED	yl/gr+rd+or/gr ø3mm I:5.6mcd Vf:2.1V	T1	Lit		data pinout
LTL-1214A	LED	rd+gr+yl+or ø2mm I:2.5mcd Vf:2V		Lit		data pinout
LTL-14C	LED	or/gr+yl ø3mm I:4.8mcd Vf:2.1V	T1	Lit		data pinout
LTL-16K	LED	srd+gr+yl ø3mm I:12.6mcd Vf:2.1V	T1	Lit		data pinout
LTL-16KE	LED	srd+gr+yl ø3mm I:19mcd Vf:2.1V	T1	Lit		data pinout
LTL-201	LED	rd+gr+yl+or ø3mm I:5.6mcd Vf:2V	T1	Lit		data pinout
LTL-209	LED	rd+srd+gr+yl+or ø3mm I:8.7mcd Vf:2V	T1	Lit		data pinout
LTL-2211AL	LED	rd+srd+gr+yl ø3mm I:1.5mcd Vf:2.1V		Lit		data pinout
LTL-2214RT	LED	rd+srd+gr+yl ø4.8mm I:1.7mcd Vf:2.1V		Lit		data pinout
LTL-2214WC	LED	rd+gr+yl+or ø4.8mm I:2.5mcd Vf:2V		Lit		data pinout
LTL-2300	LED	srd+yl+gr 8.89x3.81mm I:4.2mcd		Lit		data pinout
LTL-23214A	LED	rd+gr+yl 6x2mm I:2.5mcd Vf:2.1V		Lit		data pinout
LTL-2350	LED	srd+yl+gr 195x3.81mm I:8mcd Vf:2.1V		Lit		data pinout
LTL-2600	LED	srd+yl+gr 8.89x3.81mm I:4.2mcd		Lit		data pinout
LTL-2620	LED	srd+yl+gr 8.89x3.81mm I:4.2mcd		Lit		data pinout
LTL-2655	LFD	srd+yl+gr 8.89x8.89mm I:8mcd Vf:2.1V		Lit		data pinout
LTL-2685	LED	srd+yl+gr 8.89x195mm I:16mcd Vf:2.1V		Lit		data pinout
LTL-293	LED	rd/gr ø5mm I:12mcd Vf:2.1V If:20mA	T1¾	Lit		data pinout
LTL-298	LED	yl/gr+srd+rd/gr+or/gr ø5mm I:8.7mcd	T1¾	Lit		data pinout
LTL-307CK	LED	rd ø5mm I:840mcd Vf:1.8V If:20mA	T1¾	Lit		data pinout
LTL-307ELC	LED	srd+gr+yl ø5mm I:3.7mcd Vf:1.9V	T1¾	Lit		data pinout
LTL-31	LED	rd/gr ø5mm I:13mcd Vf:2.1V If:20mA	T1¾	Lit		data pinout
LTL-3201A	LED	rd+srd+gr+yl+or 5x2mm I:3.7mcd Vf:2V	T1¾	Lit		data pinout
LTL-3213A	LED	rd+srd+gr+yl+or 7.1x2.4mm I:3.7mcd		Lit		data pinout
LTL-3217A	LED	rd+gr+yl+am+or 4.8x1mm I:1.7mcd		Lit		data pinout
LTL-3218A	LED	rd+gr+yl+am 5x1mm I:1.7mcd Vf:2.1V		Lit		data pinout
LTL-327P	LED	or+gr+srd+yl ø5mm I:12.6mcd Vf:2.1V	T1¾	Lit		data pinout
LTL-33221AA	LED	srd+gr+yl 5.6x3.15mm I:8.7mcd		Lit		data pinout
LTL-353BJ	LED	bl ø5mm I:19mcd Vf:3V If:20mA	T1¾	Lit		data pinout
LTL-353CK	LED	rd ø5mm I:2300mcd Vf:1.8V If:20mA	T1¾	Lit		data pinout
LTL-387URK	LED	rd ø5mm I:840mcd Vf:1.8V If:20mA	T1¾	Lit		data pinout
LTL-403	LED	srd+gr+yl 4x2mm I:2.5mcd Vf:2.1V		Lit		data pinout
LTL-4201	LED	rd+srd+gr+yl+or ø3mm I:12.6mcd Vf:2V	T1	Lit		data pinout
LTL-4201N	LED	rd+srd+gr+yl+or ø3mm I:8.7mcd Vf:2V	T1	Lit		data pinout
LTL-4203	LED	rd+srd+gr+yl+or ø5mm I:29mcd Vf:2V	T1¾	Lit		data pinout
LTL-4221NLC	LED	srd+gr+yl ø3mm I:1.1mcd Vf:1.9V	T1	Lit		data pinout
LTL-4268	LED	rd ø5mm I:1200mcd Vf:1.8V If:20mA	T1¾	Lit		data pinout
LTL-427	LED	srd+gr+yl 4.85x4.95mm I:1.7mcd		Lit		data pinout
LTL-42B5	LED	bl ø3mm I:12.6mcd Vf:3V If:20mA	T1	Lit		data pinout
LTL-433	LED	gr+or+rd+yl 5x2mm I:1.7mcd Vf:2.1V		Lit		data pinout
LTL-503-11	LED	rd+gr+yl ø4.8mm I:12.6mcd Vf:2.1V		Lit		data pinout
LTL-503-14	LED	rd+gr+yl ø4.8mm I:12.6mcd Vf:2.1V		Lit		data pinout
LTL-52	LED	gr+or/gr+rd/gr ø5mm I:8.7mcd Vf:2.1V	T1¾	Lit		data pinout
LTL-5203	LED	rd+srd+gr+yl ø5mm I:4mcd Vf:2.1V	T1¾	Lit		data pinout
LTL-57173	LED	yl+gr 12.7x6.35mm I:4.2mcd Vf:2.1V		Lit		data pinout
LTL-57173HB	LED	srd 12.7x6.35mm I:4.2mcd Vf:2V		Lit		data pinout
LTL-613	LED	rd+srd+gr ø4.75mm I:40mcd Vf:2.1V		Lit		data pinout
LTL-6203LN	LED	rd+srd+gr+yl+or 4.9x2.4mm I:12.6mcd	T1¾	Lit		data pinout
LTL-709P	LED	srd+gr+yl ø1.8mm I:8.7mcd Vf:2.1V		Lit		data pinout
LTL-717	LED	gr+srd+yl 1.8x1.8mm I:2.5mcd Vf:2.1V		Lit		data pinout
LTL-767	LED	srd+gr+yl 2x5mm I:1.7mcd Vf:2.1V		Lit		data pinout
LTL-815K	LED	srd+gr+yl ø3mm I:3.7mcd Vf:2.1V		Lit		data pinout
LTL-833	LED	srd+gr+yl ø2mm I:2.5mcd Vf:2.1V		Lit		data pinout
LTL-907CK	LED	rd ø2mm I:3.5mcd Vf:1.8V If:20mA	SOT23	Lit		data pinout
LTL-907HK	LED	or/gr ø2mm I:2.8mcd Vf:2.1V If:20mA	SOT23	Lit		data pinout
LTL-907JK	LED	or ø2mm I:3.7mcd Vf:2V If:20mA	SOT23	Lit		data pinout
LTL-907NK	LED	rd ø2mm I:12.6mcd Vf:1.8V If:20mA	SOT23	Lit		data pinout
LTL-907PK	LED	or+gr+rd+yl ø2mm I:2.8mcd Vf:2.1V	SOT23	Lit		data pinout
LTL-9212A	LED	rd+srd+gr+yl+or 3.8x3mm I:2.5mcd		Lit		data pinout
LTL-9213A	LED	rd+srd+gr+yl+or 5x6mm I:3.7mcd Vf:2V		Lit		data pinout
LTL-93BA	LED	gr+rd+yl ø1.6mm I:8.7mcd Vf:2.1V		Lit		data pinout
LTL-93BCA	LED	rd ø1.6mm I:80mcd Vf:1.8V If:20mA		Lit		data pinout
LTL-93BCK	LED	rd ø1.6mm I:200mcd Vf:1.8V If:20mA		Lit		data pinout
LTL-93BK	LED	or+gr+yl ø1.6mm I:8.7mcd Vf:2.1V		Lit		data pinout
LTL307	LED	srd+gr+yl ø5mm I:29mcd Vf:2.1V	T1¾	Lit		data pinout
LTL307E	LED	srd+gr+yl ø5mm I:60mcd Vf:2.1V	T1¾	Lit		data pinout
LTM673	LED	gr ø5mm I:112..280mcd Vf:3.3V		Osr		data pinout
LTM 2220-AA	IC	12-Bit, 170Msps ADC	87	Lic		data pdf pinout
LTM 4600EV	SMPS-IC	10A, DC/DC step down power sup., <20V inp., controller, power Fets, inductor	LGA	Lic		pdf pinout

Type	Device	Short Description	Fig.	Manu	Comparision Types	More at
LTM 4600HVEV	SMPS-IC	10A, DC/DC step down power sup., <28V inp., controller, power FETs, inductor	LGA	Lic	-	pdf pinout
LTM 4600HVIV	SMPS-IC	10A, DC/DC step down power sup., <28V inp., controller, power Fets, inductor	LGA	Lic	-	pdf pinout
LTM 4600IV	SMPS-IC	10A, DC/DC step down power sup., <20V inp., controller, power Fets, inductor	LGA	Lic	-	pdf pinout
LTOP-2230	PD	950nm Vb:5V Ib:<1.5mA		Lit		data pdf pinout
LTOP-4830	PD	950nm Vb:5V Ib:<1.5mA		Lit		data pdf pinout
LTR-1650	PT	940nm P:>6.4µA/mW/cm²	T1¾	Lit		data pdf pinout
LTR-208E	PT	940nm P:>2.40µA/mW/cm²	T1¾	Lit		data pdf pinout
LTR-209	PT	940nm P:>6.4µA/mW/cm²	T1	Lit		data pdf pinout
LTR-301	PT	940nm P:1µA/mW/cm²		Lit		data pdf pinout
LTR-305	PT	940nm P:0.4µA/mW/cm²		Har,Lit		data pinout
LTR-306	PT	940nm P:>1.6µA/mW/cm²		Lit		data pdf pinout
LTR-309	PT	940nm P:>1.6µA/mW/cm²		Lit		data pdf pinout
LTR-311	PT	940nm P:>6.4µA/mW/cm²		Lit		data pdf pinout
LTR-3208	PT	940nm P:>6.4µA/mW/cm²	T1¾	Lit		data pdf pinout
LTR-323DB	PD	900nm P:130µA/mW/cm²	T1¾	Lit		data pdf pinout
LTR-4206	PT	940nm P:>6.4µA/mW/cm²	T1	Har,Lit		data pdf pinout
LTR-516	PD	900nm P:20µA/mW/cm²		Lit		data pdf pinout
LTR-516AD	PD	900nm P:20µA/mW/cm²		Lit		data pdf pinout
LTR-526	PD	900nm P:20µA/mW/cm²		Lit		data pdf pinout
LTR-53	PD	900nm P:15µA/mW/cm²		Lit		data pdf pinout
LTR-536	PD	950nm P:20µA/mW/cm²		Lit		data pdf pinout
LTR-546	PD	900nm P:20µA/mW/cm²		Lit		data pdf pinout
LTR-5576	PT	940nm P:>0.7µA/mW/cm²		Lit		data pdf pinout
LTR-5579	PT	940nm P:>0.4µA/mW/cm²		Lit		data pdf pinout
LTR-5586	PT	940nm P:>0.7µA/mW/cm²		Lit		data pdf pinout
LTT67C	LED	gr ø5mm I:224..560mcd Vf:3.5V	PLCC-2	Osr		data pinout
LTT68C	LED	gr ø5mm I:224..560mcd Vf:3.5V	PLCC-2	Osr		data pdf pinout
LTT773	LED	gr ø5mm I:140...355mcd Vf:3.3V	PLCC-2	Osr		data pinout
LTV4N25	OC	LED/NPN Viso:500Vrms CTR:>10%		Lit		data pinout
LTV4N32	OC	LED/Dar Viso:1500Vrms CTR:>500%		Lit		data pinout
LTV4N35	OC	LED/NPN Viso:3550Vrms CTR:>100%		Lit		data pinout
LTV4N37	OC	LED/NPN Viso:1500Vrms CTR:>100%		Lit		data pinout
LTV6N135	OC	LED/PD+NPN Viso:2500Vrms CTR:35%		Lit		data pinout
LTV6N137	OC	LED/PD+IC Viso:2500Vrms CTR:1500%		Lit		data pinout
LTV6N138	OC	LED/PD+Dar Viso:2500Vrms CTR:1500%		Lit		data pinout
LTV702F	OC	LED/NPN Viso:5000Vrms CTR:30-320%		Lit		data pinout
LTV702V	OC	LED/NPN Viso:5000Vrms CTR:30-320%		Lit		data pinout
LTV703F	OC	LED/NPN Viso:5000Vrms CTR:30-320%		Lit		data pinout
LTV703V	OC	LED/NPN Viso:5000Vrms CTR:30-320%		Lit		data pinout
LTV713F	OC	LED/NPN Viso:5000Vrms CTR:50-600%		Lit		data pinout
LTV713V	OC	LED/NPN Viso:5000Vrms CTR:50-600%		Lit		data pinout
LTV725	OC	LED/Dar Viso:5000Vrms Vbs:1.2V		Lit		data pinout
LTV814	OC	bid. LED/NPN Viso:5000Vrms Vbs:1.2V		Lit		data pinout
LTV815	OC	LED/Dar Viso:5000Vrms CTR:600-7500%		Lit		data pinout
LTV817	OC	LED/NPN Viso:5000Vrms CTR:50-600%		Lit		data pinout
LTV819	OC	LED/NPN Viso:5000Vrms CTR:50-400%		Lit		data pinout
LTV824	QOC	bid. LED/NPN Viso:5000Vrms Vbs:1.2V		Lit		data pinout
LTV825	DOC	LED/Dar Viso:5000Vrms CTR:600-7500%		Lit		data pinout
LTV827	DOC	LED/NPN Viso:5000Vrms CTR:50-600%		Lit		data pinout
LTV829	DOC	LED/NPN Viso:5000Vrms CTR:50-400%		Lit		data pinout
LTV844	QOC	bid. LED/NPN Viso:5000Vrms Vbs:1.2V		Lit		data pinout
LTV845	QOC	LED/Dar Viso:5000Vrms CTR:600-7500%		Lit		data pinout
LTV847	QOC	LED/NPN Viso:5000Vrms CTR:50-600%		Lit		data pinout
LTV849	QOC	LED/NPN Viso:5000Vrms CTR:50-400%		Lit		data pinout
LTV8141	OC	bid. LED/Dar Viso:5000Vrms Vbs:1.2V		Lit		data pinout
LTV8241	DOC	bid. LED/Dar Viso:5000Vrms Vbs:1.2V		Lit		data pinout
LTV8441	QOC	bid. LED/Dar Viso:5000Vrms Vbs:1.2V		Lit		data pinout
LTVG 5	Si-Di	kV-Gl, 5kV, 0,02A	31a	Gie	HS 6	data
LTVG 10	Si-Di	kV-Gl, 10kV, 0,02A	31a	Gie	HS 8	data
LTVG 11	Si-Di	kV-Gl, 11kV, 0,02A	31a	Gie	HS 10	data
LTZ 1000 (A)	LIN-IC	Ultra Precision Reference	81	Lic		pdf pinout
LU	Si-N	=HD 1F2Q (Typ-Code/Stempel/marking)	39	Nec	→HD 1...	data
LU	Z-Di	=P6 SMB-36 (Typ-Code/Stempel/marking)	71(5x3,5)	Fag	→P6 SMB-36	data
LU	Z-Di	=SMBJ 20 (Typ-Code/Stempel/marking)	71(5x3,5)	Mop	→SMBJ ...	data
LU7615PG	LED	gr ø3mm I:60mlm Vf:3.5V If:20mA		Lst		data pdf pinout
LU7615SUB	LED	gr ø3mm I:445mlm Vf:3.5V If:20mA		Lst		data pdf pinout
LU7615SUG	LED	gr ø3mm I:450mlm Vf:3.5V If:20mA		Lst		data pdf pinout
LU7615SUR	LED	or ø3mm I:1.45cdmlm Vf:2.6V If:70mA		Lst		data pdf pinout
LU7615UBG	LED	gr ø3mm I:1.7cdmlm Vf:3.5V If:20mA		Lst		data pdf pinout
LU7615USO	LED	or ø3mm I:2.75cdmlm Vf:2.3V If:70mA		Lst		data pdf pinout
LU7615USR	LED	ord ø3mm I:500mlm Vf:2.6V If:70mA		Lst		data pdf pinout
LU7615UWC	LED	wht ø3mm I:1120mlm Vf:3.5V If:20mA		Lst		data pdf pinout
LU7615UYC	LED	am ø3mm I:1.5cdmlm Vf:2.3V If:70mA		Lst		data pdf pinout
LU7615UYO	LED	or ø3mm I:1.45cdmlm Vf:2.3V If:70mA		Lst		data pdf pinout
LU7619UWC	LED	wht ø3mm I:1240mlm Vf:3.5V If:20mA		Lst		data pdf pinout
LUB371	LED	rd/gr 2.5x5mm I>1.6mcd Vf:2V If:10mA		Sie		data pinout
LUS250	LED	rd/gr 2.5x5mm I>0.4mcd Vf:2V If:10mA	SOT23	Sie		data pinout
LV	Z-Di	=1SMB 20A (Typ-Code/Stempel/marking)	71(5x3,5)	Ons	→1SMB 5.0A...170A	
LV	PIN-Di	=1SV265 (Typ-Code/Stempel/marking)	45	Say	→1SV265	data
LV	Z-Di	=P6 SMB-36A (Typ-Code/Stempel/marking)	71(5x3,5)	Fag	→P6 SMB-36A	data
LV	Z-Di	=SMBJ 20A (Typ-Code/Stempel/marking)	71(5x3,5)	Mop	→SMBJ ...	data
LV 1035 M	Dig. Audio	Dolby Prologic Decoder	QFP	Say		pdf pinout

Type	Device	Short Description	Fig.	Manu	Comparision Types	More at
LV 1115 / M	LIN-IC	SMD, Bi-cmos LSI, Surround Processor, e-Volume	SSMDIP	Say	-	pdf
LV 2100 V	BiCMOS-IC	Telecom, PLL Synthesizer f. Cordl. Ph.	24-SSMDIP	Say	-	
LV 2101 V	BiCMOS-IC	Telecom, PLL Synthesizer f. Cordl. Ph.	24-SMDIP	Say	-	
LV 2205	C-Di	FM/VHF-Tuning, 25V, 13,5...16,5pF(4V)	7d	Ons	-	
LV 2209	C-Di	FM/VHF-Tuning, 25V, 29,7...36,3pF(4V)	7d	Ons	-	
LV 3100 M	CMOS-IC	SMD, Parametric Equalizer System	64-MP	Say	-	pdf
LV3333	LED	gr ø3.0mm l:180...1800mcd Vf:3.3V	T1	Osr		data pdf pinout
LV 4001 W	CMOS-IC	SMD, Vc, Chroma-prozessor (Hi8)	48-MP	Say	-	
LV 4101 W	CMOS-IC	SMD, Camera-prozessor	64-MP	Say	-	
LV 5044 V	CMOS-IC	DC-DC Converter Controller	SQL	Say	-	pdf pinout
LV5413	LED	gr ø5mm l:710...2800mcd Vf:3.3V	T1¾	Osr		data pdf pinout
LV5433	LED	gr ø5.0mm l:280...2800mcd Vf:3.3V	T1¾	Osr		data pdf pinout
LV 5803 NM	BiCMOS-IC	Step-down Switching Reg.	MDIP	Say	-	pdf pinout
LV 7980	BiCMOS-IC	3 in 1 RGB Driver	SQL	Say	-	pdf pinout
LV 8804 V	BiCMOS-IC	3-Ph., PWM PC & Server Fan Motor Drv.	TSOP	Say	-	pdf pinout
LV 23000 M	IC	Single-Chip Tuner IC	TSOP	Say	-	pdf pinout
LVA67C	LED	gr ø5.0mm l:180...450mcd Vf:3.5V		Osr		data pdf pinout
LVA673	LED	gr ø5.0mm l:112...280mcd Vf:3.3V		Osr		data pdf pinout
LVC	Z-Di	=1SMB 20CA (Typ-Code/Stempel/marking)	71(5x3,5)	Ons	→1SMB 10CA...78CA	
LVE63C	LED	gr ø5.0mm l:355...900mcd Vf:3.8V	PLCC-4	Osr		data pdf pinout
LVE633	LED	gr ø5.0mm l:710...1800mcd Vf:3.5V	PLCC-4	Osr		data pdf pinout
LVE673	LED	gr ø5.0mm l:140...355mcd Vf:3.7V	PLCC-4	Osr		data pdf pinout
LVI 030	Hybrid-IC	FM-ZF		Mts	-	
LVI 060	Hybrid-IC	Verstärker/amplifier		Mts	-	
LVM673	LED	gr ø5.0mm l:90...224mcd Vf:3.3V		Osr		data pdf pinout
LVS250	LED	srd+gr ø5.0mm l>0.4mcd Vf:2V If:10mA	SOT23	Sie		data pinout
LVT 3V3	Z-Di	TAZ, 3,3V, 600W(1ms)	31a	Tho	-	data
LVT67C	LED	gr ø5.0mm l:224...560mcd Vf:3.5V	PLCC-2	Osr		data pdf pinout
LVT673	LED	gr ø5.0mm l:112...280mcd Vf:3.3V	PLCC-2	Osr		data pdf pinout
LVT773	LED	gr ø5.0mm l:112...280mcd Vf:3.3V	PLCC-2	Osr		data pdf pinout
LVW5SG	LED	gr ø5.0mm l=18...33cd Vf:3.8V	PLCC-2	Osr		data pdf pinout
LW	Z-Di	=P6 SMB-39 (Typ-Code/Stempel/marking)	71(5x3,5)	Fag	→P6 SMB-39	data
LW	Z-Di	=SM 6T 15C (Typ-Code/Stempel/marking)	71(6x4mm)	Tho	→SM 6T...	data
LW	Z-Di	=SMBJ 22 (Typ-Code/Stempel/marking)	71(5x3,5)	Mop	→SMBJ ...	data
LW541C	LED	wht ø3.0mm l=2.8...7.1cd Vf:3.6V	T1¾	Osr		data pinout
LW3333	LED	wht ø3.0mm l:355...900mcd Vf:3.5V	T1	Osr		data pdf pinout
LWA67C	LED	wht ø3.0mm l:355...710mcd Vf:3.5V		Osr		data pdf pinout
LWA673	LED	wht ø3.0mm l:355...710mcd Vf:3.6V		Osr		data pdf pinout
LWA676	LED	wht ø3.0mm l:28...71mcd Vf:3.5V		Osr		data pdf pinout
LWE67C	LED	wht ø3.0mm l:560...1120mcd Vf:3.9V	PLCC-4	Osr		data pdf pinout
LWE673	LED	wht ø3.0mm l:224...5600mcd Vf:3.7V	PLCC-4	Osr		data pdf pinout
LWL88C	LED	wht ø3.0mm l:112...280mcd Vf:3.5V	SCD80	Osr		data pdf pinout
LWM67C	LED	wht ø3.0mm l:280...560mcd Vf:3.6V		Osr		data pdf pinout
LWM673	LED	wht ø3.0mm l:140...280mcd Vf:3.5V		Osr		data pdf pinout
LWT67C	LED	wht ø3.0mm l:355...710mcd Vf:3.6V	PLCC-2	Osr		data pdf pinout
LWT673	LED	wht ø3.0mm l:180...355mcd Vf:3.5V	PLCC-2	Osr		data pdf pinout
LWT676	LED	wht ø3.0mm l:28...71mcd Vf:3.5V	PLCC-2	Osr		data pdf pinout
LWT773	LED	wht ø3.0mm l:180...355mcd Vf:3.5V		Osr		data pdf pinout
LWW5SG	LED	wht ø3.0mm l=21...39cd Vf:3.8V		Osr		data pdf pinout
LWY87C	LED	wht ø3.0mm l:280...560mcd Vf:3.6V		Osr		data pdf pinout
LWY87S	LED	wht ø3.0mm l:45...112mcd Vf:3.4V		Osr		data pdf pinout
LX	Z-Di	=1SMB 22A (Typ-Code/Stempel/marking)	71(5x3,5)	Ons	→1SMB 5.0A...170A	
LX	Si-N	=HD 1A4A (Typ-Code/Stempel/marking)	39	Nec	→HD 1...	data
LX	Z-Di	=P6 SMB-39A (Typ-Code/Stempel/marking)	71(5x3,5)	Fag	→P6 SMB-39A	data
LX	Z-Di	=SM 6T 15CA (Typ-Code/Stempel/marking)	71(6x4mm)	Tho	→SM 6T...	data
LX	Z-Di	=SMBJ 22A (Typ-Code/Stempel/marking)	71(5x3,5)	Mop	→SMBJ ...	data
LX038	OP	Vs:±18V Vi0:<0.1mV f:0.002Mc f:1.6kc	16Y	Crb		data pinout
LX038C	OP	Vs:±18V Vi0:<0.2mV f:0.002Mc f:1.6kc	16Y	Crb		data pinout
LX061	SPECIAL	Vs:±18V Vu:60dB Vo:±10V	12A	Twa		data pinout
LX063	SPECIAL	Vs:±16V Vo:±10V	12A	Crb		data pinout
LX0101	HYBRID	Vs:±18V Vu:106dB Vo:±12.5V Vi0:±5mV	8O	Crb		data pinout
LX107	SPECIAL	Vs:±18V Vu:>100dB Vi0:2mV f:1Mc	7Y	Crb		data pinout
LX108	SPECIAL	Vs:±18V Vu:116dB Vi0:2mV f:1.5Mc	7Y	Crb		data pinout
LX109	SPECIAL	Vs:±26V Vu:116dB Vi0:2mV f:2Mc	7Y	Crb		data pinout
LX118	HYBRID	Vs:±18V Vu:116dB Vi0:±3.5mV f:2Mc	7Y	Crb		data pinout
LX119	HYBRID	Vs:±27V Vu:116dB Vi0:±3.5mV f:2.5Mc	7Y	Crb		data pinout
LX124	FET	Vs:±18V Vu:>106dB Vo:±10V f:0.75Mc	7Y	Crb		data pinout
LX127	OP	Vs:±18V Vo:±10V f:1Mc f:100kc	7Y	Crb		data pinout
LX133	OP	Vs:±18V Vo:±10V f:1.5Mc f:20kc	7Y	Crb		data pinout
LX143	OP	Vs:±18V Vu:>96dB Vo:±10V f:1.5Mc	6Y	Crb		data pinout
LX163	FET	Vs:±18V Vu:>100dB Vo:±1V f:0.03Mc	6Y	Crb		data pinout
LX207	HYBRID	Vs:±18V Vu:<80dB Vi0:±50mV f:20kc	9Y	Crb		data pinout
LX208	HYBRID	Vs:±18V Vu:<80dB Vi0:±50mV f:100kc	9Y	Crb		data pinout
LX401	OP	Vs:±18V Vo:±10V Vi0:<0.1mV f:1Mc	7Y	Crb		data pinout
LX401F	OP	Vs:±18V Vo:±10V Vi0:<0.25mV f:1Mc	7Y	Crb		data pinout
LX404	OP	Vu:140dB Vo:±10V Vi0:<±50µV f:0.2Mc	7Y	Crb		data pinout
LX404F	OP	Vu:140dB Vo:±10V Vi0:<±0.1mV f:0.2Mc	7Y	Crb		data pinout
LX412	OP	Vs:±18V Vu:160dB Vo:±10V Vi0:<±20µV	7Y	Crb		data pinout
LX412J	OP	Vs:±18V Vu:160dB Vo:±10V Vi0:<±50µV	7Y	Crb		data pinout
LX431	OP	Vs:±18V Vo:±10V Vi0:±2mV f:10Mc	7Y	Crb		data pinout
LX484	OP	Vs:±18V f:10kkc f:10kc	11Y	Crb		data pinout
LX485	OP	Vs:±18V Vo:±10V f:10kc	11Y	Crb		data pinout
LX605	OP	Vs:±18V f:6kkc f:5kc	14Y	Crb		data pinout

Type	Device	Short Description	Fig.	Manu	Comparision Types	More at
LX606	HYBRID	Vs:±18V Vo:±12V Vi0:±1mV f:0.6Mc	32Y	Crb		data pinout
LX703	FET	Vs:±18V Vu:>86dB Vo:±10V Vi0:<±20mV	8A	Crb		data pinout
LX704	FET	Vs:±18V Vo:>±10V Vi0:<±20mV f:1Mc	8A	Crb		data pinout
LX706	FET	Vs:±18V Vo:>±10V Vi0:±20mV f:1Mc	8A	Crb		data pinout
LX708	FET	Vs:±18V Vu:>90dB Vo:±10V Vi0:<±1mV	8A	Crb		data pinout
LX714A	HI-VOLT	Vs:±40V Vu:86dB Vo:±35V Vi0:60mV	8A	Twa		data pinout
LX714B	HI-VOLT	Vs:±40V Vu:86dB Vo:±35V Vi0:30mV	8A	Twa		data pinout
LX714C	HI-VOLT	Vs:±40V Vu:86dB Vo:±35V Vi0:20mV	8A	Twa		data pinout
LX715A	HI-VOLT	Vs:±40V Vu:86dB Vo:±32V Vi0:60mV	12A	Twa		data pinout
LX715B	HI-VOLT	Vs:±40V Vu:86dB Vo:±32V Vi0:30mV	12A	Twa		data pinout
LX715C	HI-VOLT	Vs:±40V Vu:86dB Vo:±32V Vi0:20mV	12A	Twa		data pinout
LX716	HYBRID	Vs:±18V Vu:>116dB Vi0:±2mV f:1.5Mc	12A	Crb		data pinout
LX718	OP	Vs:±18V Vu:>110dB Vi0:2mV f:1Mc	12A	Crb		data pinout
LX719	OP	Vs:±18V Vu:>126dB Vi0:±2mV f:1.5Mc	12A	Crb		data pinout
LX721	FET	Vs:±18V Vo:±10V f:0.5Mc f:10kc	10O	Crb		data pinout
LX722	FET	Vs:±18V Vo:±10V f:0.5Mc f:10kc	10O	Crb		data pinout
LX724	OP	Vs:±40V Vo:±30V f:2Mc f:50kc	10O	Crb		data pinout
LX727A	HI-SPEED	Vs:±18V Vu:80dB Vo:±10V Vi0:50mV	8A	Twa		data pinout
LX727B	HI-SPEED	Vs:±18V Vu:80dB Vo:±10V Vi0:20mV	8A	Twa		data pinout
LX728	HI-SPEED	Vs:±18V Vu:77dB Vo:±10V f:12Mc	8A	Twa		data pinout
LX729	OP	Vs:±18V Vo:±10V f:5Mc f:50kc	10O	Crb		data pinout
LX771	OP	Vs:±16V Vo:±10V Vi0:<±20mV f:70Mc	12A	Crb		data pinout
LX1020	OP	Vs:±18V Vu:100dB Vo:±12V Vi0:2mV	10Y	Crb		data pinout
LX1022	2xOP	Vs:±18V Vu:100dB Vo:±12V Vi0:2mV	14Y	Crb		data pinout
LX1024	4xOP	Vs:±18V Vu:100dB Vo:±12V Vi0:2mV	24Y	Crb		data pinout
LX1030	OP	Vs:±18V Vu:96dB Vo:±10V Vi0:2mV	10Y	Crb		data pinout
LX1032	OP	Vs:±18V Vu:96dB Vo:±10V Vi0:2mV	14Y	Crb		data pinout
LX1034	OP	Vs:±18V Vu:96dB Vo:±10V Vi0:2mV	24Y	Crb·		data pinout
LX 1562 IDM	IC	Second- Generation Power Factor Controller	TSOP	Lin		pdf pinout
LX 1562 IM	IC	Second- Generation Power Factor Controller	TSOP	Lin		pdf pinout
LX 1563 IDM	IC	Second- Generation Power Factor Controller	TSOP	Lin		pdf pinout
LX 1563 IM	IC	Second- Generation Power Factor Controller	TSOP·	Lin		pdf pinout
LX2108	HYBRID	Vs:±18V Vu:>116dB Vi0:±2mV f:1.5Mc	14Y	Crb		data pinout
LX7030	FET	Vs:±18V Vo:>±10V Vi0:<±20mV f:1Mc	8A	Crb		data pinout
LX7031	FET	Vs:±18V Vo:>±10V Vi0:<±20mV f:1Mc	8A	Crb		data pinout
LX7032	FET	Vs:±18V Vo:>±10V Vi0:<±20mV f:1Mc	8A	Crb		data pinout
LX7192	OP	Vs:±18V Vu:>126dB Vi0:±2mV f:1.5Mc	24D	Crb		data pinout
LXC	Z-Di	=1SMB 22CA (Typ-Code/Stempel/marking)	71(5x3,5)	Ons	→1SMB 10CA...78CA	
LXHL-BB1X	LED	bl ø6mm l:16lm Vf:3.42V If:350mA		Lux		data pdf pinout
LXHL-BD1X	LED	rd ø6mm l:44lm Vf:2.95V If:350mA		Lux		data pdf pinout
LXHL-BE1X	LED	cy ø6mm l:45lm Vf:3.42V If:350mA		Lux		data pdf pinout
LXHL-BH1X	LED	or ø6mm l:55lm Vf:2.95V If:350mA		Lux		data pdf pinout
LXHL-BL1X	LED	am ø6mm l:42lm Vf:2.95V If:350mA		Lux		data pdf pinout
LXHL-BM1X	LED	gr ø6mm l:53lm Vf:3.42V If:350mA		Lux		data pdf pinout
LXHL-BR1X	LED	bl ø6mm l:53lm Vf:3.42V If:350mA		Lux		data pdf pinout
LXHL-BW1X	LED	wht ø6mm l:45lm Vf:3.42V If:350mA		Lux		data pdf pinout
LXHL-DB1X	LED	bl ø6mm l:14.5lm Vf:3.42V If:350mA		Lux		data pdf pinout
LXHL-DBXX	LED	bl ø6mm l:13.9lm Vf:3.7V If:700mA		Lux		data pdf pinout
LXHL-DDXX	LED	rd ø6mm l:125lm Vf:2.95V If:1400mA		Lux		data pdf pinout
LXHL-DE1X	LED	cy ø6mm l:40.5lm Vf:3.42V If:350mA		Lux		data pdf pinout
LXHL-DH1X	LED	or ø6mm l:50lm Vf:2.95V If:350mA		Lux		data pdf pinout
LXHL-DHXX	LED	or ø6mm l:170lm Vf:2.95V If:1400mA		Lux		data pdf pinout
LXHL-DL1X	LED	am ø6mm l:38lm Vf:2.95V If:350mA		Lux		data pdf pinout
LXHL-DLXX	LED	or ø6mm l:100lm Vf:2.95V If:1400mA		Lux		data pdf pinout
LXHL-DM1X	LED	gr ø6mm l:48lm Vf:3.42V If:350mA		Lux		data pdf pinout
LXHL-DMXX	LED	gr ø6mm l:58lm Vf:3.7V If:700mA		Lux		data pdf pinout
LXHL-DR1X	LED	bl ø6mm l:58lm Vf:3.42V If:350mA		Lux		data pdf pinout
LXHL-DW1X	LED	wht ø6mm l:40.5lm Vf:3.42V If:350mA		Lux		data pdf pinout
LXHL-DWXX	LED	wht ø6mm l:58lm Vf:3.7V If:700mA		Lux		data pdf pinout
LXHL-FE·X	LED	bl ø4mm l:8.2lm Vf:3.42V If:350mA		Lux,Roi		data pdf pinout
LXHL-FB5X	LED	bl ø4mm l:18.1lm Vf:6.84V If:700mA		Lux,Roi		data pdf pinout
LXHL-FBXX	LED	bl ø4mm l:13.9lm Vf:3.9V If:1000mA		Lux		data pdf pinout
LXHL-FD1X	LED	rd ø4mm l:30.6lm Vf:2.95V If:350mA		Lux,Roi		data pdf pinout
LXHL-FE1X	LED	cy ø4mm l:30.6lm Vf:3.42V If:350mA		Lux,Roi		data pdf pinout
LXHL-FE5X	LED	cy ø4mm l:67.2lm Vf:6.84V If:700mA		Lux,Roi		data pdf pinout
LXHL-FH1X	LED	or ø4mm l:39.8lm Vf:2.95V If:350mA		Lux,Roi		data pdf pinout
LXHL-FL1X	LED	am ø4mm l:23.5lm Vf:2.95V If:350mA		Lux,Roi		data pdf pinout
LXHL-FM1X	LED	gr ø4mm l:30.6lm Vf:3.42V If:350mA		Lux,Roi		data pdf pinout
LXHL-FM5X	LED	gr ø4mm l:67.2lm Vf:6.84V If:700mA		Lux,Roi		data pdf pinout
LXHL-FMXX	LED	gr ø4mm l:51.7lm Vf:3.9V If:1000mA		Lux		data pdf pinout
LXHL-FR1X	LED	bl ø4mm l:51.7lm Vf:3.42V If:350mA		Lux,Roi		data pdf pinout
LXHL-FR5X	LED	bl ø4mm l:51.7lm Vf:6.84V If:700mA		Lux,Roi		data pdf pinout
LXHL-FW1X	LED	wht ø4mm l:30.6lm Vf:3.42V If:350mA		Lux,Roi		data pdf pinout
LXHL-FW5X	LED	wht ø4mm l:87.4lm Vf:6.84V If:700mA		Lux,Roi		data pdf pinout
LXHL-FWXX	LED	wht ø4mm l:61.7lm Vf:3.7V If:700mA		Lux		data pdf pinout
LXHL-LB3X	LED	bl ø4mm l:13.9lm Vf:3.9V If:1000mA		Lux,Roi		data pdf pinout
LXHL-LB5X	LED	bl ø4mm l:18.1lm Vf:6.84V If:700mA		Lux,Roi		data pdf pinout
LXHL-LE3X	LED	cy ø4mm l:51.7lm Vf:3.9V If:1000mA		Lux,Roi		data pdf pinout
LXHL-LE5X	LED	cy ø4mm l:67.2lm Vf:6.84V If:700mA		Lux,Roi		data pdf pinout
LXHL-LM3X	LED	gr ø4mm l:51.7lm Vf:3.9V If:1000mA		Lux,Roi		data pdf pinout
LXHL-LM5X	LED	gr ø4mm l:67.2lm Vf:6.84V If:700mA		Lux,Roi		data pdf pinout

Type	Device	Short Description	Fig.	Manu	Comparision Types	More at
LXHL-LR3X	LED	bl ø4mm I:67.2lm Vf:3.9V If:1000mA		Lux,Roi		data pdf pinout
LXHL-LR5X	LED	bl ø4mm I:67.2lm Vf:6.84V If:700mA		Lux,Roi		data pdf pinout
LXHL-LW3X	LED	wht ø4mm I:60lm Vf:3.7V If:700mA		Lux,Roi		data pdf pinout
LXHL-LW5X	LED	wht ø4mm I:87.4lm Vf:6.84V If:700mA		Lux,Roi		data pdf pinout
LXHL-MB1X	LED	bl ø4mm I:8.2lm Vf:3.42V If:350mA		Lux,Roi		data pdf pinout
LXHL-MD1X	LED	rd ø4mm I:30.6.9lmlm Vf:2.85V		Lux,Roi		data pdf pinout
LXHL-MDXX	LED	rd ø4mm I:30.6lm Vf:2.85V If:350mA		Lux		data pdf pinout
LXHL-ME1X	LED	cy ø4mm I:30.6lm Vf:3.42V If:350mA		Lux,Roi		data pdf pinout
LXHL-MH1X	LED	or ø4mm I:39.8lm Vf:2.95V If:350mA		Lux,Roi		data pdf pinout
LXHL-MHXX	LED	am ø4mm I:23.5lm Vf:2.95V If:350mA		Lux		data pdf pinout
LXHL-ML1X	LED	am ø4mm I:23.5lm Vf:2.85V If:350mA		Lux,Roi		data pdf pinout
LXHL-MLXX	LED	am ø4mm I:30.6lm Vf:2.95V If:350mA		Lux		data pdf pinout
LXHL-MM1X	LED	gr ø4mm I:30.6lm Vf:3.42V If:350mA		Lux,Roi		data pdf pinout
LXHL-MR1X	LED	bl ø4mm I:30.6lm Vf:3.42V If:350mA		Lux,Roi		data pdf pinout
LXHL-MW1X	LED	wht ø4mm I:13.9lm Vf:3.42V If:350mA		Lux,Roi		data pdf pinout
LXHL-MWXX	LED	wht ø4mm I:20lm Vf:3.42V If:350mA		Lux		data pdf pinout
LXHL-NB1X	LED	bl ø4mm I:8.2lm Vf:3.42V If:350mA		Lux,Roi		data pdf pinout
LXHL-ND1X	LED	rd ø4mm I:13.9lm Vf:2.85V If:350mA		Lux,Roi		data pdf pinout
LXHL-NE1X	LED	cy ø4mm I:30.6lm Vf:3.42V If:350mA		Lux,Roi		data pdf pinout
LXHL-NH1X	LED	or ø4mm I:39.8lm Vf:2.95V If:350mA		Lux,Roi		data pdf pinout
LXHL-NL1X	LED	am ø4mm I:10.7lm Vf:2.85V If:350mA		Lux,Roi		data pdf pinout
LXHL-NM1X	LED	gr ø4mm I:30.6lm Vf:3.42V If:350mA		Lux,Roi		data pdf pinout
LXHL-NR1X	LED	bl ø4mm I:30.6lm Vf:3.42V If:350mA		Lux,Roi		data pdf pinout
LXHL-NW1X	LED	wht ø4mm I:13.9lm Vf:3.42V If:350mA		Lux,Roi		data pdf pinout
LXHL-NWXX	LED	wht ø4mm I:17lm Vf:3.42V If:350mA		Lux		data pdf pinout
LXHL-PB1X	LED	bl ø6mm I:16lm Vf:3.42V If:350mA		Lux		data pdf pinout
LXHL-PBXX	LED	bl ø6mm I:23lm Vf:3.7V If:700mA		Lux		data pdf pinout
LXHL-PDXX	LED	rd ø6mm I:140lm Vf:2.95V If:1400mA		Lux		data pdf pinout
LXHL-PE1X	LED	cy ø6mm I:45lm Vf:3.42V If:350mA		Lux		data pdf pinout
LXHL-PEXX	LED	cy ø6mm I:64lm Vf:3.7V If:700mA		Lux		data pdf pinout
LXHL-PH1X	LED	or ø6mm I:55lm Vf:2.95V If:350mA		Lux		data pdf pinout
LXHL-PHXX	LED	or ø6mm I:190lm Vf:2.95V If:1400mA		Lux		data pdf pinout
LXHL-PL1X	LED	am ø6mm I:42lm Vf:2.95V If:350mA		Lux		data pdf pinout
LXHL-PLXX	LED	or ø6mm I:110lm Vf:2.95V If:1400mA		Lux		data pdf pinout
LXHL-PM1X	LED	gr ø6mm I:53lm Vf:3.42V If:350mA		Lux		data pdf pinout
LXHL-PMXX	LED	gr ø6mm I:64lm Vf:3.7V If:700mA		Lux		data pdf pinout
LXHL-PR1X	LED	bl ø6mm I:64lm Vf:3.42V If:350mA		Lux		data pdf pinout
LXHL-PRXX	LED	bl ø6mm I:64lm Vf:3.7V If:700mA		Lux		data pdf pinout
LXHL-PW1X	LED	wht ø6mm I:45lm Vf:3.42V If:350mA		Lux		data pdf pinout
LXHL-PWXX	LED	wht ø6mm I:65lm Vf:3.7V If:700mA		Lux		data pdf pinout
LXK2-PBXX	LED	bl ø6mm I:35lm Vf:3.72V If:1000mA		Lux		data pdf pinout
LXK2-PDXX	LED	rd ø6mm I:60lm Vf:2.95V If:350mA		Lux		data pdf pinout
LXK2-PEXX	LED	cy ø6mm I:100lm Vf:3.72V If:1000mA		Lux		data pdf pinout
LXK2-PHXX	LED	or ø6mm I:60lm Vf:2.95V If:350mA		Lux		data pdf pinout
LXK2-PLXX	LED	am ø6mm I:45lm Vf:2.95V If:350mA		Lux		data pdf pinout
LXK2-PMXX	LED	gr ø6mm I:45lm Vf:3.42V If:350mA		Lux		data pdf pinout
LXK2-PRXX	LED	bl ø6mm I:45lm Vf:3.72V If:1000mA		Lux		data pdf pinout
LXK2-PWXX	LED	wht ø6mm I:120lm Vf:3.72V If:1000mA		Lux		data pdf pinout
LY	Si-N	=2SC2712-Y (Typ-Code/Stempel/marking)	35	Tos	→2SC2712	data
LY	Si-N	=2SC3772 (Typ-Code/Stempel/marking)	35	Say	→2SC3772	data
LY	Si-N	=2SC4003 (Typ-Code/Stempel/marking)	35	Say	→2SC4003	data
LY	Si-N	=2SC4116-Y (Typ-Code/Stempel/marking)	35(2mm)	Tos	→2SC4116	data
LY	Si-N	=2SC4207-Y (Typ-Code/Stempel/marking)	45	Tos	→2SC4207	data
LY	Si-N	=2SC4738-Y (Typ-Code/Stempel/marking)	35(1,6mm)	Tos	→2SC4738	data
LY	Si-N	=2SC4944-Y (Typ-Code/Stempel/marking)	45(2mm)	Tos	→2SC4944	data
LY	Si-N	=KTC 4075-Y (Typ-Code/Stempel/marking)	35(2mm)	Kec	→KTC 4075	data
LY	Z-Di	=P6 SMB-43 (Typ-Code/Stempel/marking)	71(5x3,5)	Fag	→P6 SMB-43	data
LY	Z-Di	=SMBJ 24 (Typ-Code/Stempel/marking)	71(5x3,5)	Mop	→SMBJ ...	data
LY336K	LED	yl ø3mm I:9..22.4mcd Vf:1.8V If:2mA	T1	Osr		data pdf pinout
LY543B	LED	yl ø5.0mm I:1120...4500mcd Vf:2V	T1¾	Osr		data pdf pinout
LY3336	LED	yl ø3.0mm I:355...900mcd Vf:1.9V	T1	Osr		data pdf pinout
LY3360	LED	yl ø3mm I:11.2...18mcd Vf:2V If:10mA	T1	Osr		data pdf pinout
LY3366	LED	yl ø3mm I:180...450mcd Vf:2V If:20mA	T1	Osr		data pdf pinout
LY5436	LED	yl ø5.0mm I:280...1800mcd Vf:2V	T1¾	Osr		data pdf pinout
LYA67B	LED	yl ø5.0mm I:355...900mcd Vf:1.8V		Osr		data pdf pinout
LYA67K	LED	gr ø5.0mm I:7.1...18mcd Vf:<2.2V		Osr		data pdf pinout
LYA670	LED	yl ø5.0mm I:5.6...18mcd Vf:2V If:10mA		Osr		data pdf pinout
LYA676	LED	yl ø5.0mm I:180...355mcd Vf:2V		Osr		data pdf pinout
LYA679	LED	yl ø5.0mm I:0.9...2.24mcd Vf:2V		Osr		data pdf pinout
LYE63B	LED	yl ø5.0mm I:4400mcd Vf:2.2V If:50mA	PLCC-4	Osr		data pdf pinout
LYE67B	LED	yl ø2.55mm I:560...1400mcd Vf:2.2V	PLCC-4	Osr		data pdf pinout
LYE655	LED	yl ø2.55mm I:900...1800mcd Vf:2.1V	PLCC-4	Osr		data pdf pinout
LYE675	LED	yl ø2.55mm I:450...1120mcd Vf:2.1V	PLCC-4	Osr		data pdf pinout
LYG67B	LED	yl ø2.55mm I=1.4...3.55cd Vf:2.2V	PLCC-6	Osr		data pdf pinout
LYGT670	LED	yl/gr ø2.55mm I<18mcd Vf:2V If:10mA	PLCC-4	Osr		data pdf pinout
LYK376	LED	yl ø3.0mm I:450...1120mcd Vf:1.9V	T1	Osr		data pdf pinout
LYK389	LED	yl ø3mm I:1.2...7.2mcd Vf:2V If:2mA	T1	Osr		data pdf pinout
LYL89K	LED	yl ø3mm I:3.55...9mcd Vf:1.7V If:2mA		Osr		data pdf pinout
LYL896	LED	yl ø3mm I:90...180mcd Vf:2V If:50mA	PLCC-4	Osr		data pdf pinout
LYM67K	LED	yl ø3mm I:5.6...14mcd Vf:1.8V If:2mA		Osr		data pdf pinout
LYM670	LED	yl ø3mm I:5.6...14mcd Vf:2V If:10mA	PLCC-2	Osr		data pdf pinout
LYM676	LED	yl ø3mm I:140...355mcd Vf:2V If:20mA		Osr		data pdf pinout
LYM770	LED	yl ø3mm I:7.1...18mcd Vf:2V If:20mA		Osr		data pdf pinout
LYM776	LED	yl ø3mm I:440mlm Vf:2V If:20mA		Osr		data pdf pinout
LYM779	LED	yl ø3mm I:1.12...2.24mcd Vf:2V If:2mA		Osr		data pdf pinout

Type	Device	Short Description	Fig.	Manu	Comparision Types	More at
LYN971	LED	yl ø3mm I:2.8...6mcd Vf:2.2V If:20mA		Osr		data pdf pinout
LYQ971	LED	yl ø3mm I:2.8...6mcd Vf:2.2V If:20mA	0603	Osr		data pdf pinout
LYQ976	LED	yl ø3mm I:60mcd Vf:2V If:20mA	0603	Osr		data pdf pinout
LYR971	LED	yl ø3mm I:6mcd Vf:2.2V If:20mA	0805	Osr		data pdf pinout
LYR976	LED	yl ø3mm I:60mcd Vf:2V If:20mA	0805	Osr		data pdf pinout
LYT67K	LED	yl ø3mm I:7.1...18mcd Vf:1.8V If:2mA	PLCC-2	Osr		data pdf pinout
LYT68B	LED	yl ø3mm I:355...900mcd Vf:2.1V		Osr		data pdf pinout
LYT670	LED	yl ø3mm I:7.1...18mcd Vf:2V If:10mA	PLCC-2	Osr		data pdf pinout
LYT673	LED	gr ø3mm I:140...355mcd Vf:3.3V	PLCC-2	Osr		data pdf pinout
LYT676	LED	yl ø3mm I:180...355mcd Vf:2V If:20mA	PLCC-2	Osr		data pdf pinout
LYT679	LED	yl ø3mm I:1.4...2.8mcd Vf:2V If:2mA	PLCC-2	Osr		data pdf pinout
LYT686	LED	yl ø3mm I:180...355mcd Vf:2V If:20mA	PLCC-2	Osr		data pdf pinout
LYT770	LED	yl ø3mm I:7.1.18mcd Vf:2V If:10mA	PLCC-2	Osr		data pdf pinout
LYT776	LED	yl ø3mm I:180...355mcd Vf:2V If:20mA		Osr		data pdf pinout
LYW57B	LED	yl ø3mm I=13...24cd Vf:2.2V If:400mA		Osr		data pdf pinout
LYY876	LED	yl ø3mm I:140...355mcd Vf:2V If:20mA		Osr		data pdf pinout
LZ	Z-Di	=1SMB 24A (Typ-Code/Stempel/marking)	71(5x3,5)	Ons	→1SMB 5.0A...170A	
LZ	Z-Di	=P6 SMB-43A (Typ-Code/Stempel/marking)	71(5x3,5)	Fag	→P6 SMB-43A	data
LZ	Z-Di	=SMBJ 24A (Typ-Code/Stempel/marking)	71(5x3,5)	Mop	→SMBJ ...	data
LZC	Z-Di	=1SMB 24CA (Typ-Code/Stempel/marking)	71(5x3,5)	Ons	→1SMB 10CA...78CA	
LZR180	LED	gr+rd+yl ø2mm I>0.25mcd Vf:2V		Sie		data pinout

Continuation
M to Z
See VRT Volume- 1b

Pin Assignments

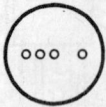

Fig.1 (von unten/bottom view)

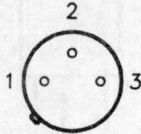

Fig.2 (von unten/bottom view)

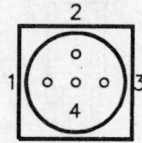

Fig.3 (von unten/bottom view)

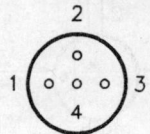

Fig.4 (von unten/bottom view)

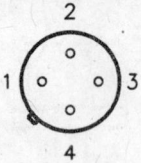

Fig.5 (von unten/bottom view)

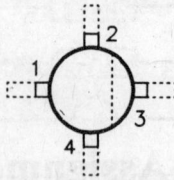

Fig.6 (von unten/bottom view)

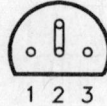

Fig.7 (von unten/bottom view)

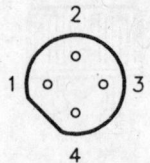

Fig.8 (von unten/bottom view)

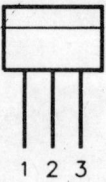

Fig.9

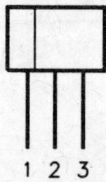

Fig.10

Fig.11

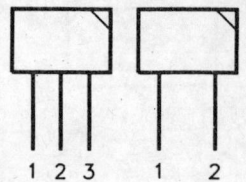

Fig.12

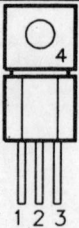

Fig.13 (von oben/top view)

Fig.14 (von oben/top view)

Fig.15 (von oben/top view)

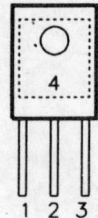

Fig.16 (von oben/top view)

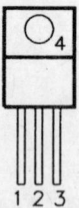

Fig.17 (von oben/top view)

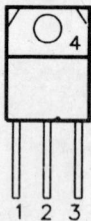

Fig.18 (von oben/top view)

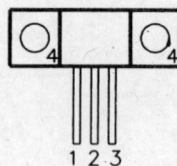

Fig.19 (von oben/top view)

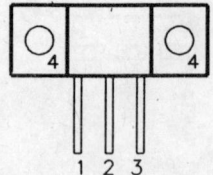

Fig.20 (von oben/top view)

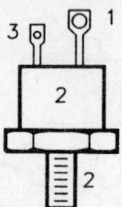

Fig.21

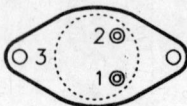

Fig.22 (von unten/bottom view)

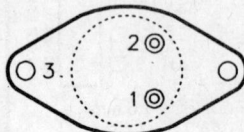

Fig.23 (von unten/bottom view)

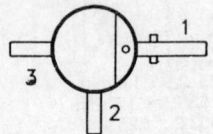

Fig.24 (von oben/top view)

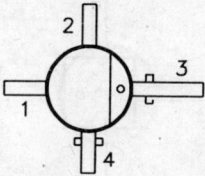

Fig.25 (von oben/top view)

Fig.26 (von oben/top view)

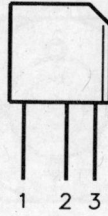

Fig.27

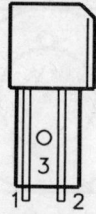

Fig.28

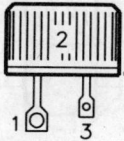

Fig.29

Fig.30 (von oben/top view)

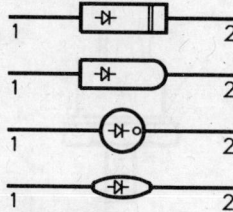

Fig.31

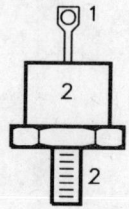

Fig.32

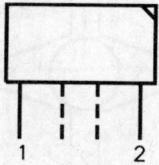

Fig.33

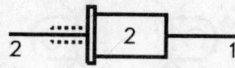

Fig.34

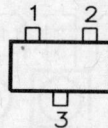

SOT−23: 2,9 x 1,5mm
SOT−323: 2 x 1,25mm
SS Mini: 1,6 x 0,8mm

Fig.35 (von oben/top view)

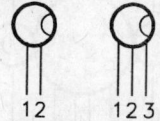

Fig.36

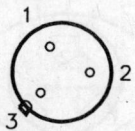

Fig.37 (von unten/bottom view)

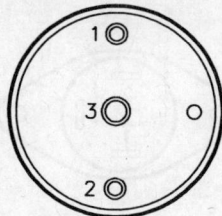

Fig.38 (von unten/bottom view)

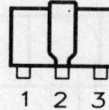

SOT−89

Fig.39

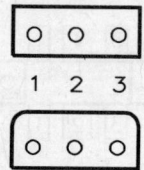

Fig.40 (von unten/bottom view)

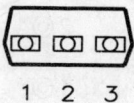

Fig.41 (von unten/bottom view)

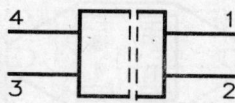

Fig.42 (von oben/top view)

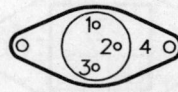

Fig.43 (von unten/bottom view)

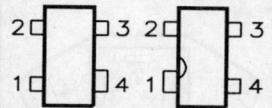

SOT−143(R): 2,9 x 1,5mm
SOT−343: 2 x 1,25mm

Fig.44 (von oben/top view)

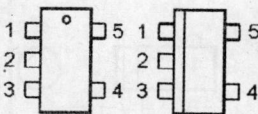

SOT-23 2,9 x 1,6 mm
SOT-153 2,9 x 1,5 mm
SOT-353 2 x 1,25 mm

Fig.45 (von oben/top view)

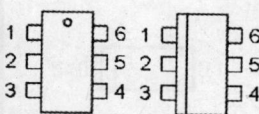

SOT-23 2,9 x 1,6 mm
SOT-163 2,9 x 1,5 mm
SOT-363 2 x 1,25 mm

Fig.46 (von oben/top view)

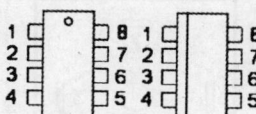

SOT-23 2,9 x 1,6 mm
SOT-163 2,9 x 1,5 mm
SOT-363 2 x 1,25 mm

Fig.47 (von oben/top view)

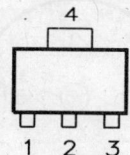

SOT−223

Fig.48 (von oben/top view)

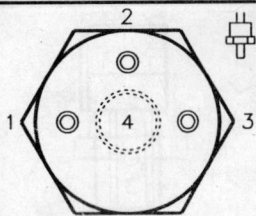

Fig.49 (von oben/ top view)

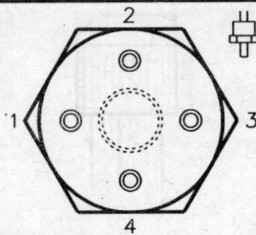

Fig.50 (von oben/ top view)

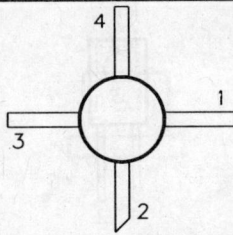

Fig.51 (von oben/ top view)

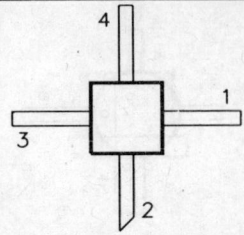

Fig.52 (von oben/ top view)

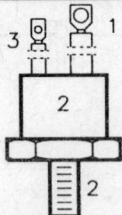

Fig.53

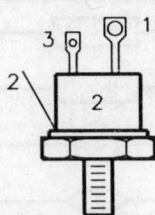

Fig.54

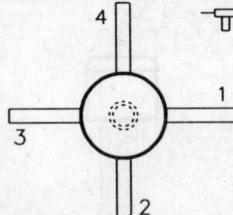

Fig.55 (von oben/ top view)

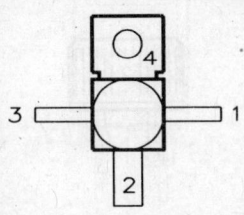

Fig.56 (von oben/ top view)

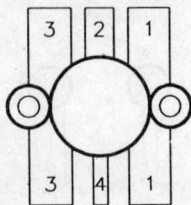

Fig.57 (von oben/ top view)

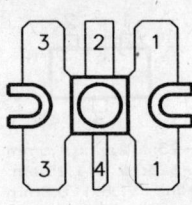

Fig.58 (von oben/ top view)

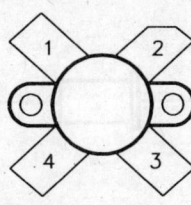

Fig.59 (von oben/ top view)

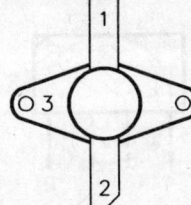

Fig.60 (von oben/ top view)

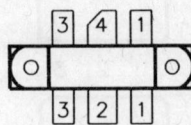

Fig.61 (von oben/ top view)

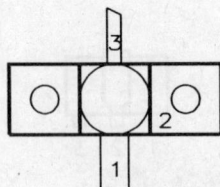

Fig.62 (von oben/ top view)

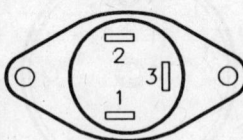

Fig.63 (von oben/ top view)

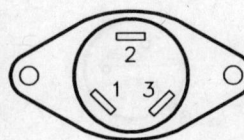

Fig.64 (von oben/ top view)

Fig.65 (von oben/ top view)

Fig.66 (von oben/ top view)

Fig.67 (von oben/ top view)

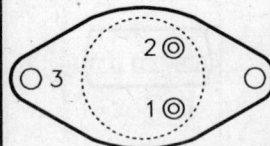

Fig.68 (von unten/ bottom view)

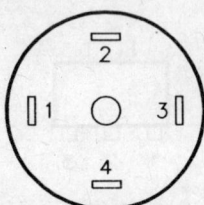

Fig.69 (von oben/ top view)

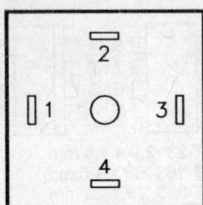

Fig.70 (von oben/ top view)

```
1 ▭ 2

SOD-6:    6,4 x 4,2mm
SOD-15:   8 x 5,2mm
SOD-123: 2,7 x 1,55mm
SOD-323: 1,7 x 1,25mm
```

Fig.71 (von oben/ top view)

```
1 ▭ 2    ◯

MELF:      5 x 2,5mm Ø
MINIMELF: 3,4 x 1,6mm Ø
DO-213:   3,5 x 1,4mm □
```

Fig.72

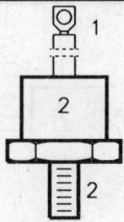

Fig.73

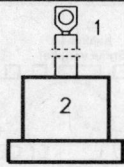

Fig.74

Fig.75

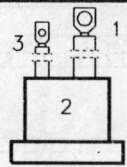

Fig.76

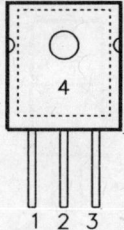

Fig.77 (von oben/ top view)

Fig.78 (von oben/ top view)

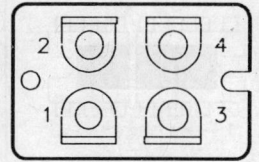

Fig.80 (von oben/ top view)

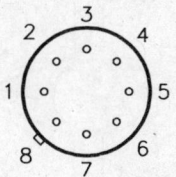

Fig.81 (von unten/ bottom view)

TO-70,71,76,77,78,99

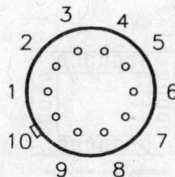

Fig.82 (von unten/ bottom view)

TO-96,100

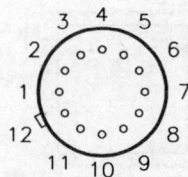

Fig.83 (von unten/ bottom view)

TO-101

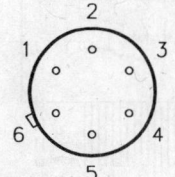

Fig.84 (von unten/ bottom view)

Fig.85 (von oben/ top view)

TO-126/.... TO-247/...

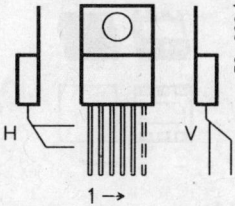

Fig.86 (von oben/ top view)

TO-220/...

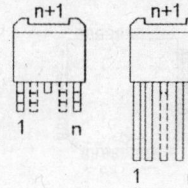

Fig.87 (von oben/ top view)

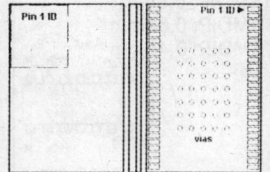

CSP

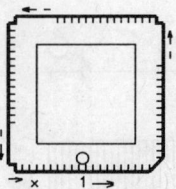

CCC (von oben/ top view)

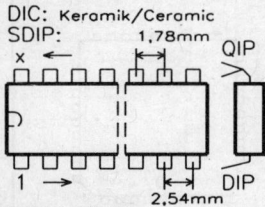

DIC: Keramik/Ceramic
SDIP: 1,78mm
2,54mm

**DIC,DIP
QIP,SDIP** (von oben/ top view)

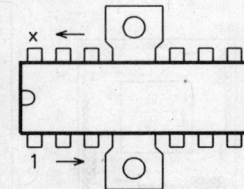

**DIP/
QIP+a** (von oben/ top view)

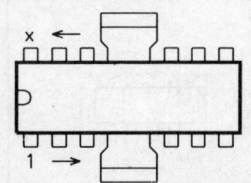

**DIP/
QIP+b** (von oben/ top view)

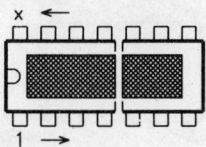

**DIP/
QIP+c** (von oben/ top view)

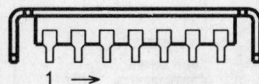

**DIP/
QIP+d**

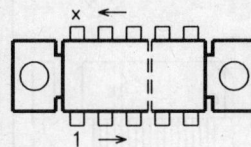

**DIP/
QIP+f** (von oben/ top view)

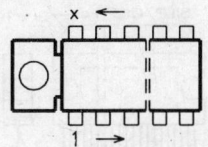

**DIP/
QIP+g** (von oben/ top view)

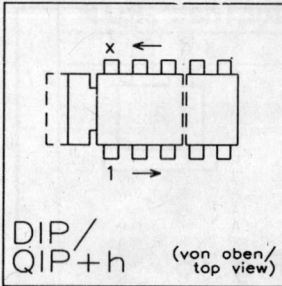

DIP/
QIP+h   (von oben/ top view)

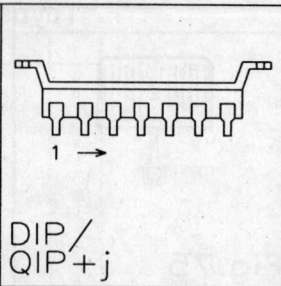

DIP/
QIP+j

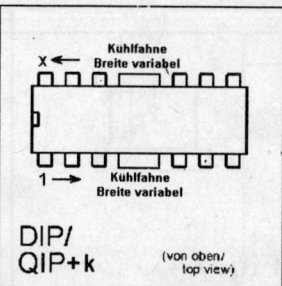

Kuhlfahne
Breite variabel

Kuhlfahne
Breite variabel

DIP/
QIP+k   (von oben/ top view)

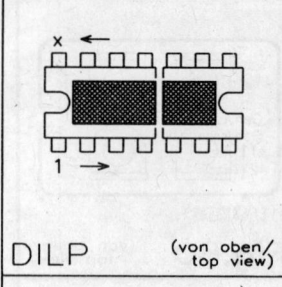

DILP   (von oben/ top view)

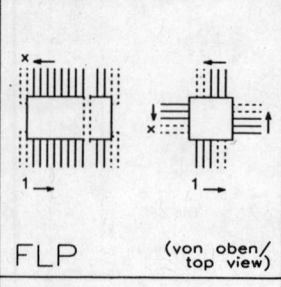

FLP   (von oben/ top view)

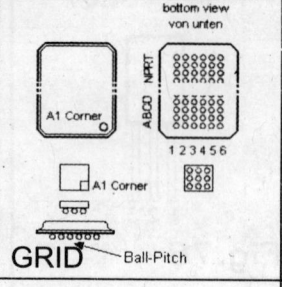

bottom view
von unten

A1 Corner

ABCD NPRT

1 2 3 4 5 6

A1 Corner

GRID   Ball-Pitch

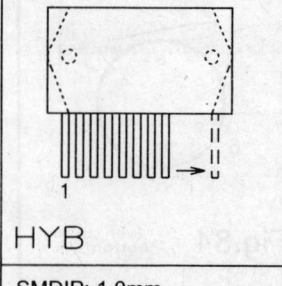

HYB

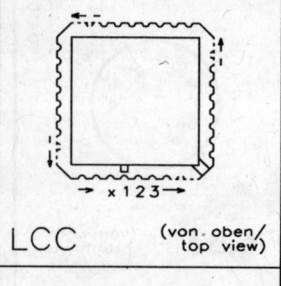

LCC   (von oben/ top view)

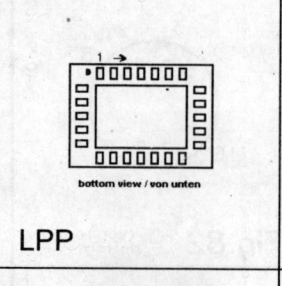

bottom view / von unten

LPP

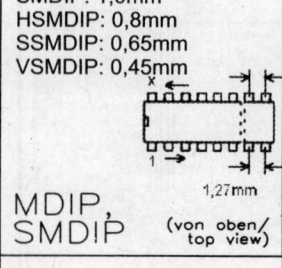

SMDIP: 1,0mm
HSMDIP: 0,8mm
SSMDIP: 0,65mm
VSMDIP: 0,45mm

1,27mm

MDIP,
SMDIP   (von oben/ top view)

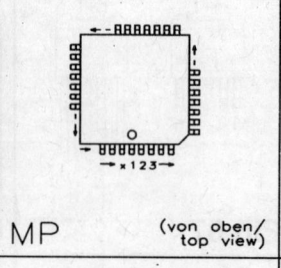

MP   (von oben/ top view)

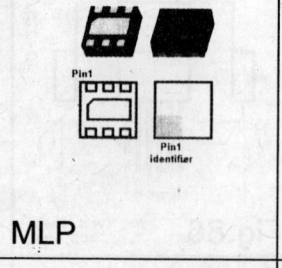

Pin1

Pin1
identifier

MLP

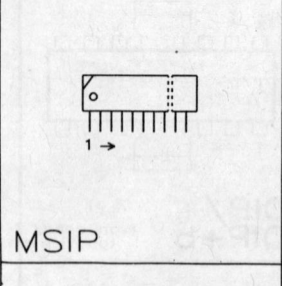

MSIP

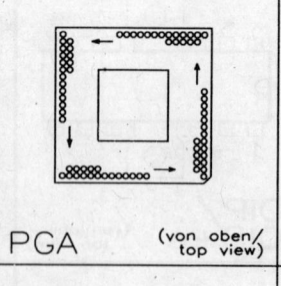

PGA   (von oben/ top view)

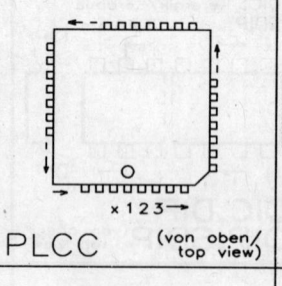

PLCC   (von oben/ top view)

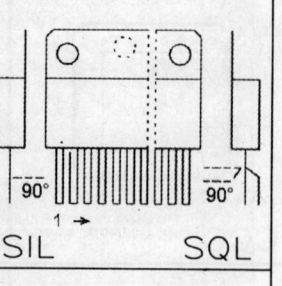

90°   90°

SIL   SQL

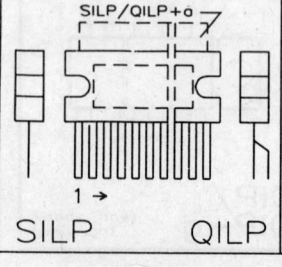

SILP/QILP+a

SILP   QILP

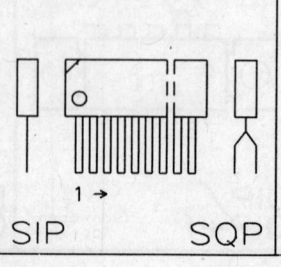

SIP   SQP

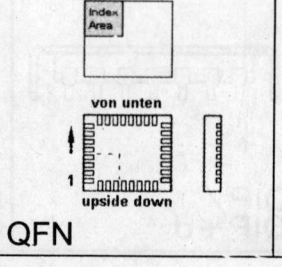

Index
Area

von unten

upside down

QFN

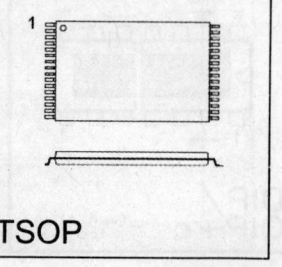

TSOP

Diode

	1	2	3	4
a	A	K		
aa	K1	-	K2	A1-2
ab	A1	-	A2	K1-2
ac	A1	-	K2	K1/A2
al	A	K	A	K
b	K	A		
b1	A1/K2	A2/K1		
b2	K	A	K	
c	A	-	K	
c1	A	-	K	
c2	A1/K2		A2/K1	
c3	K2	-	A1	
d	K	-	A	
d1	K	K	A	
e	-	A	K	
f		K	A	
g	A	K	Geh	
h	K	A	Geh	
j	A	K	A	
j3	A	K	-	K
j4	-	A	K	A
j5	A1	A2	A1	A2
j6	A1/A2	K1	K2	
j7	K1/K2	A1	A2	
j8	A1/K2	A1	A2	
l	A1	A2	K1-2	
l1	K1-2	A2	A1	
l2	A1	-	A2	K1-2
m	K1	K2	A1-2	
m1	A1-2	K1		
m2	K1	A1/2	K2	
n	A1	K2	A2/K1	
n1	A1	A2/K1	K2	
n2	A2/K1	A1	A2	
o	K1	A2	A1/K2	
o1	K1	A1/K2	A2	
o2	K2	A1	A2/K1	
o4	K1	A1/K2	A2	A1/K2
o5	A1	A1/K2	A2	
p	A2	K1-2	A1	
q	K2	A1-2	K1	
r	A1	A2	A3	K1-3
r1	A1	A2	K1-3	A3
r2	K1-3	A1	A2	A3
r3	A1-3	K1	K2	K3
s	K1	K2	K3	A1-3
s0	A1	K1	K2	A2
s1	K1	A2	A1-3	K3
s2	A1	A2	K2	K1
s3	A1	A1	A2	K2
s4	K1	K2	A2	A1
s5	K1	K2	A1	A2
s6	K1	A2	K2	A1
s7	A1	A2	K1	K2
s8	A1	K2	K1	A2
s9	A1	K2	A2	K1
t	K	-	A	Geh
t1	A	-	K	Geh
t2	n.c.	-	A	K
t3	n.c.	K	A	
t4	-	A	A	K
t5	K	-	A	K
u	A2/K1	A1	Geh	K2
u1	A2/K1	K2	Geh	A1
v	A	K	K	
v1	K	K	A	
w	~	-	~	+
x	-	~	+	~
x1	+	~	-	~
x2	-	+	~	~
x3	+	~	-	
x4	-	~	~	+
x5	-	+	~	
x6	~	+	-	
x7	~	-	+	-
y	A1/K4	A2/K1	A3/K2	A4/K3
z	Aufdruck/Imprint Impr.			

↗Weitere Pincode linke Seite
↗Continuation on the left side

Thy
Pin-Code: ba

GTO-Thy

Pin-Code: x, x1, x2, x3, x4, bn

Tetrode

Thy + Di (ITR)

Pin-Code: r, r1, r2, cc, ce, cg, cj, da, dc

Triac

Trigger-Di

Pin-Code: y

Diac

SAS, SBS

Pin-Code: s, s1, cd, cf, ch, ck, db

SUS

Pin-Code: n, n1, o, o1, u, u1, bm, cm, bo

UJT

Pin-Code: m, m1, q

PUT

Pin-Code: p, l, l1

Thyristor, Triac*, PUT

	1	2	3	4
a	K	G	A	
b	K	A	G	
c	G	A	K	
d	G	A	K	
e	A	G	K	
f	A	K	G	
g	K	G	A	Geh
h	K	G		Geh
h1	K	A	G	A
j	G	A	K	Geh
k	G	K	A	Geh
l	A	G	K	Geh
m	A	K	G	Geh
n	K	-	G	A
o	K	-	A	G
p	A	A	G	K
q	T1	T2	G	T2
r	A1	G	A2	
s	A1	A2	G	
§	A + Geh./case/boîter/invol.			
&	K + Geh./case/boîter/invol.			
*	K= A1 (Triac)			
	A= A2 (Triac)			

Tetroden

	1	2	3	4
a	K	Gk	Ga	A
b	A	K	Ga	Gk
c	Gk	K	Ga	A
d	K	Gk	A	Ga
e	Gk	K	A	Ga
f	K	Ga	Gk	A
§	Ga + Geh./case/boîter/invol.			

Trigger-Di, Diac**, UJT

	1	2	3	4
a	A	K		
b	K	A		
c	A	-	K	
d	K	-	A	
e	A2	G	A1	
f	A2	A1	G	
g	K	G	A	
h	A	K	G	
m	B1	-	B2	E
n	B1	E	B2	
o	B2	E	B1	
p	E	-	B1	B2
q	B1	B2	E	
r	E	B1	-	B2
s	E	B2	B1	
t	B2	B1	E	
§	B2 + Geh./case/boîter/invol.			
&	G + Geh./case/boîter/invol.			
**	K= A1 (Diac)			
	A= A2 (Diac)			

Z-IC

	1	2	3	4
a	output	input	Gnd	
b	input	Gnd	output	(Gnd)
c	Gnd	input	output	(input)
d	output	Gnd	input	
e	input	output	Gnd	
f	Gnd	output	input	
g	A	-	K	
h	A	K		
j	K		A	
k	input	Reg	output	
l	reg	output	input	
m	output	Reg	input	
n	reg	input	output	
o	input	output	Reg	

Reg = Regelung/adjust

Diode

	1	2	3	4	5	6	7	8	9	10	11	12	13	14	15	16	
am	A	K	-	-	K												am
an	A1	-	A2	K2	K1												an
az	A1	K1-2	A2	A3	K3												az
ba	A1	K1-3	A2	K1-3	A3												ba
bb	K1	A1-4	K2	K3	K4												bb
bc	A1	K1-4	A2	A3	A4												bc
bd	K1-4	A1	A2	A3	A4												bd
be	A1-4	K1	K2	K3	K4												be
bf	-	~	+	+	~	-											bf
bg	K1	K2	K3	A3	A2	A1											bg
bh	A1	A2	A3-4	K3	K4	K1-2											bh
bi	K1-5	A1	A2	A3	A4	A5											bi
bj	K1	K2	A3-4	K3	K4	A1-2											bj
bk	A1	A2	K3-4	A3	A4	K1-2											bk
bl	A1	A2	A3	K3	K2	K1											bl
bm	A1	K2	A3/K4	A4	K3	K1/A2											bm
bn	+	n.c.	-	~	n.c.	~											bn
bo	A1	K1/A2	K2	A3	K3/A4	K4											bo
bp	A1-2	K4	K3	A3-4	K2	K1											bp
bq	K1-2	A4	A3	K3-4	A2	A1											bq
br	A4	K1-4	A3	A2	K1-4	A1											br
bs	A1/K2	K4	A3	K3/A4	A2	K1											bs
bt	K4	K3	K1-2	A2	A1	A3-4											bt
bu	A4	A3	K1-2	A2	A1	K3-4											bu
bv	K4	K3	A1-2	K2	K1	A3-4											bv
bw	A1	K1-3	A2	A3	K1-3	-											bw
bx	A1	K1-4	A2	A3	K1-4	A4											bx
by	A1	A2	A3	A4	A5	K6											by
bz	A1-5	K1	K2	K3	K4	K5											bz
ca	A1	K2	K3	A4	K4	A3	A2	K1									ca
cb	A1	A2	A3	A4	K3	K2	K1										cb
cc	A1	-	A2	K1-4	A3	-	A4										cc
cd	K1	-	K2	A1-4	K3	-	K4										cd
ce	A1	-	A2	K1-5	A3	-	A4	A5									ce
cf	K1	-	K2	A1-5	K3	-	K4	K5									cf
cg	A1	A2	A3	K1-6	A4	-	A5	A6									cg
ch	K1	K2	K3	A1-6	K4	-	K5	K6									ch
cj	A1	A2	A3	K1-7	A4	A5	A6	A7									cj
ck	K1	K2	K3	A1-7	K4	K5	K6	K7									ck
cm	A1	K2	A3/K4	-	A4	K3	K1/A2										cm
cn	A1	K1-4	A2	K1-4	A3	K1-4	A4										cn
co	A1	A2	A3	K1-6	A4	A5	A6										co
cp	A1	A2	A3	A4	K4	K3		A2									cp
cq	K1	K2	K3	A1-6	K4	K5	K6										cq
cr	K1	K2	K3	K4	A4	A3	A2	A1									cr
cv	K1	A1	K2	A2	K3	A4	A4	-									cv
cw	K1	K2	K3	K4	A1-8	K5	K6	K7	K8								cw
cx	A1-8	K1	K2	K3	K4	K5	K6	K7	K8								cx
cy	K1-8	A1	A2	A3	A4	A5	A6	A7	A8								cy
cz	A1	A2	A3	A4	K1-8	A5	A6	A7	A8								cz
da	A1	A2	A3	K1-8	A4	A5	A6	A7	A8								da
db	K1	K2	K3	A1-8	K4	K5	K6	K7	K8								db
dc	-	A1	A2	A3	A4	A5	A6	A7	A8	K1-8							dc
de	-	K1	K2	K3	K4	K5	K6	K7	K8	A1-8							de
eb	K1	A1-2	K2	K3	A3-4	K4	K5	A5-6	K6	K7	A7-8	K8					eb
fa	-	-	A4	A3	A2	A1	-	-	A5	A6	A7	A8	-	K1-8			fa
fb	-	-	K4	K3	K2	K1	-	-	K5	K6	K7	K8	-	A1-8			fb
fc	-	A1	A2	-	A3	-	A4	A5	A6	-	A7	A8	-	K1-8			fc
fd	-	K1	K2	-	K3	-	K4	K5	K6	-	K7	K8	-	A1-8			fd
fe	A1	A2	A3	A4	A5	A6	K1-13	A7	A8	A9	A10	A11	A12	A13			fe
ff	K1	K2	K3	K4	K5	K6	A1-13	K7	K8	K9	K10	K11	K12	K13			ff
fg	K1	K2	K3	K4	K5	K6	K7	A7	A6	A5	A4	A3	A2	A1			fg
ga	A1	A2	A3	A4	A5	A6	A7	A8	K8	K7	K6	K5	K4	K3	K2	K1	ga
gc	K1	K2	K3	K4	K5	K6	K7	K8	A8	A7	A6	A5	A4	A3	A2	A1	gc
j2	A1	-	K2	A2	-	K1											j2
o3	K1	-	A1/K2	-	A2												o3

Transistor

	1	2	3	4	5	6	7	8	9	10	11	12	13	14	15	16	
ba	B2	C2	E1/2	C1	B1	-	-	-	-	-	-						ba
ba1	B1	E1/2	B2	C2	C1	-	-	-	-	-	-						ba1
ba2	D2	S1/2	D1	G1	G2	-	-	-	-	-	-						ba2
ba3	E1	B1/2	E2	C2	C1	-	-	-	-	-	-						ba3
ba4	E1	B1	E2	C2	B2/C1												ba4
ba5	B2	E1/2	B1	C1	C2												ba5
ba6	E1	B1	B2	C2	C1/E2												ba6
ba7	E1	E2/B1	B2	C2	C1												ba7
ba8	B1	E1	E2	C2	B2/C1												ba8
ba9	B1	E2	B2	C2	E1	C1											ba9
bb	B1	C1	C2	B2	E1/2	-	-	-	-	-	-						bb
bb1	G1	S	G2	D2	D1												bb1
bb2	G	S	A	K	D												bb2
bb3	D1	D2	S1/2	G2	G1												bb3
bb4	G2	S2	D1	S1/D2	G1												bb4
bb5	G1	D1	S1/S2	D2	G2												bb5
bb6	G	E	K	C/A	-												bb6
bb7	G1	E1	C2	C1/E2	G2												bb7
bb8	A1/K2	A3/K4	K1/K3/CA2/A4/E		G												bb8
bc	E2	B2	B1	E1	C1/2	-	-	-	-	-	-						bc
bc1	B	E	A	K	C												bc1
bc2	B	E	A	K+C	C+K												bc2
bc3	S	E/D	B	C	G												bc3
bc4	A	E	B	C	K												bc4
bc5	B	E	K	C/A	-												bc5
bc6	B/A	K	C	E/Re	Re												bc6
bc7	G	S	A	D	D	D											bc7
bc8	Re	E/Re	C	A	B/K												bc8
bd	E2	C2	B1/2	C1	E1	-	-	-	-	-							bd
bd1	input	US2	output	US1	Gnd	-	-	-	-	-							bd1
bd2	output	US	US	input	Gnd	-	-	-	-	-	-						bd2
be	E1/2	B1	B2	E1/2	C2	C1	-	-	-	-	-						be
bf	E1	B1	C1	E2	B2	C2	-	-	-	-	-						bf
bf1	B1	B2	-	E1	E2	C1/2	-	-	-	-	-						bf1
bf2	C1	B1	E1	E2	B2	C2	-	-	-	-	-						bf2
bf3	B1	E1	C1	C2	E2	B2	-	-	-	-	-						bf3
bf4	B1	E1/2	C1	C2	E1/2	B2	-	-	-	-	-						bf4
bf5	B	C	E	B	C	E	-	-	-	-	-						bf5
bg	B1	C1	E1	B2	C2	E2	-	-	-	-	-						bg
bg1	E2	C2	B2	B1	C1	E1											bg1
bh	E1	B1	B2	E2	C2	C1	-	-	-	-	-						bh
bh0	C	C	B	E	C	C											bh0
bh1	E1	B1	C2	B2	B2	C1	-	-	-	-	-						bh1
bh2	E1	B1	B2	C2	E2	C1	-	-	-	-	-						bh2
bh3	C1	E1	C2	E2	B2	B1	-	-	-	-	-						bh3
bh4	C1	E1	C2	B2	E2	B1	-	-	-	-	-						bh4
bh5	B1	E1	C2	B2	E2	C1	-	-	-	-	-						bh5
bh6	B1	E1	C2	C2	B2	C1	-	-	-	-	-						bh6
bh7	E1	B1	E2	C2	B2	C1	-	-	-	-	-						bh7
bh8	C1	B2	C2	E2	E1	B1	-	-	-	-	-						bh8
bh9	B1	E1	B2	B2	C2	C1	-	-	-	-	-						bh9
bi0	C	C	B	n.c.	C	E											bh10
bj	E1	C1	B1	E2	C2	B2	-	-	-	-	-						bj
bj1	E1	C1	B2	E2	C2	B1	-	-	-	-	-						bj1
bj2	B1	C1	E1	E2	C2	B2	-	-	-	-	-						bj2
bk	E1/2	C1	B2	E1/2	C2	B1	-	-	-	-	-						bk
bl	E1	C1	C2	E2	B2	B1	-	-	-	-	-						bl
bl1	E1	E2	B2	C2	B1	C1	-										bl1
bl2	E1	E2	C2	B2	B1	C1	-										bl2
bm	E1/2	C1	C2	E1/2	B2	B1	-										bm
bn	S1	D1	G1	S2	D2	G2	-	-	-	-	-						bn
bn1	S1	G1	D1	D2	G2	S2	-	-	-	-	-						bn1
bn2	S1	S2	G2	D2	G1	D1	-	-	-	-	-						bn2
bn3	S1	G1	D2	S2	G2	D1	-	-	-	-	-						bn3
bn4	D	D	G	S	D	D	-	-	-	-	-						bn4
bn5	E	B	A	C	C	C											bn5
bn6	G1	S1	S2	G2	D2	D1											bn6
bn7	D	D	G	S	D	D											bn7
bn8	G1	S2	G2	D2	S1	D1											bn8
bn9	S1	D2	D2	S2	G1	D1/G2											bn9
bn10	A	S	G	D	-	K											bn10
bn11	S	G	D	S	D												bn11
bn12	D	D	G	S	D	S											bn12
bn13	-	G	S	-	D	D											bn13
bo	D1	G1	S1	S2	G2	D2	-	-	-	-	-						bo
bo1	input	Gnd	US	US	Gnd	output	-	-	-	-	-						bo1
bp	D1	G1	S1	(Sub)1	D2	G2	S2	-	-	-	-						bp
bp1	D1	G1	S1	Sub	S2	G2	D2	-	-	-	-						bp1
bp2	B1	E1	C1	B2	E2	C2	Geh	-	-	-	-						bp2
bp3	B1	E1/2	C1	B2	E1/2	C2	Geh	-	-	-	-						bp3
bp4	B1	C1	E1	-	E2	C2	Geh	-	-	-	-						bp4
bp5	E	C1	C2	E	B2	B1	Geh	-	-	-	-						bp5
bq	E1	B1	C1	-	E2	B2	C2	-	-	-	-						bq
br	E1	E2	C2	-	B2	B1	C1	-	-	-	-						br
br1	A	A	S	G	D	D	K	K									br1
bs	C1	B1	E1	-	E2	B2	C2	-									bs
bt	E1	B1	C1	-	C2	B2	E2	-	-	-	-						bt
bu	E1	-	B1	C1/2	B2	-	E2	-	-	-	-						bu
bv	E2	-	E1	B1	C1/2	B2	-	-	-	-	-						bv
bv1	E1-3	B1	C1	B2	C2	B3	C3	E1-3	-	-	-						bv1
bw	D1	S1	G1	Sub	S2	G2	D2	(Sub)	-	-	-						bw
bx	S1	D1	G1	(Sub)	S2	G2	D2	(Sub)	-	-	-						bx
by	S1	G1	D1	-	D2	G2	S2	(Sub)	-	-	-						by
bz	D1	-	G1	Sub	G2	-	D2	S1/2	-	-	-						bz
bu1	input	Gnd	US1	Gnd	US2	Gnd	output	Geh	-	-	-						bu1
ca	D1	G1	-	Sub	-	G2	D2	S1/2									ca

Transistor

	1	2	3	4	5	6	7	8	9	10	11	12	13	14	15	16	
ca1	S	S	S	G	D	D	D	D	-	-	-	-					ca1
ca2	-	S	S	D	D	D	D	D	-	-	-	-					ca2
ca3	S1	G1	S2	G2	D2	D2	D1	D1	-	-	-	-					ca3
ca4	D1	S1	S1	G1	G2	S2	S2	D2	-	-	-	-					ca4
ca5	D	G	S	G	D	S	S	D									ca5
ca6	E	E	E	G	C	C	C	C									ca6
ca7	S	S	S	G	S	S	S	S									ca7
ca8	S2	G2	S1	G1	D1	D1	D2	D2									ca8
ca9	S1	G1	S2	G2	D	D	D	D									ca9
ca10	G1	S1	G2	S2	D2	D2	D1	D1									ca10
ca11	D	S	S	G	A	A	A	K									ca11
ca12	D1/D2	S1	S1	G1	G2	S2	S2	D1/D2									ca12
ca13	D1	S1	S1	G1	G2	S2	S2	D2									ca13
ca14	S1	G1	S2	G2	D2	D2	D1	D1									ca14
cb	S1	D1	-	G1	S2	D2	-	G2									cb
cb1	S	S	S	G													cb1
cb2	D	D	D	G	D	D	D	S									cb2
cb3	S1/D2	G1	S2	G2	D2/S1	D2/S1	D1	D1									cb3
cb4	D	D	D	G	S	D	D	D									cb4
cc	S1	D1	G1	Sub	G2	D2	S2		-	-	-	-					cc
cd	E1	C1	B1	E2	C2	B2	E3	C3	B3	-	-	-	-				cd
ce	E1/4	C1	B1	B2	C2	E2/3	C3	B3	B4	C4	-	-	-				ce
cf	E1	B1	C1/B2	E2	C2	C4	E4	C3/B4	B3	E3	-	-	-				cf
cg	input	Gnd	Gnd	Gnd	US	-	Gnd	Gnd	output	Geh	-	-	-				cg
cg1	input1	Gnd	Gnd	outp.1	US	input2	Gnd	Gnd	output.2	Geh	-	-	-				cg1
ch	B1	E1	-	E2	B2	-	C2	-	C1	-	-	-					ch
cj	C1-4	B1	E1	B2	E2	E3	B3	E4	B4	C1-4	-	-	-				cj
ck	C1	B1	E1-4	B2	C2	C3	B3	E1-4	B4	C4	-	-	-				ck
ck1	E1-4	B1	C1	B2	C2	B3	C3	B4	C4	E1-4	-	-	-				ck1
cl	D1	S1	-	S2	D2	-	G2	-	G1	-	-						cl
cm	E1	B1	C1	-	C2	B2	E2	E3	B3	C3	-	C4	B4	E4	-		cm
cn	-	C1	C2	C3	C4	C5	C6	C7	C8	E5-8	B4/8	B3/7	B2/6	B1/5	E1-4	-	cn
co	B1	C1	C2	C3	C4	C5	-	E1-5	B4	B3	B2	E1-5					co
cp	C1-4	E1	B1	-	B2	E2	-	E3	-	B4	E4	-	-				cp
cq	E1-4	C1	B1	-	B2	C2	-	C3	-	B4	C4	-					cq
cr	-	B1	C1/B2	C3	-	-	-	-	E2/C3	E1/3	-	-					cr
cs	C1	B1	E1	-	E2	B2	C2	C3	B3	E3	-	E4	B4	C4	-		cs
ct	C1	C1	C2	C2	C3	C3	C4	C4	B4	E4	B3	E3	B2	E2	B1	E1	ct
ct1	Sub	D3	G3	S3	S4	G4	D4	Sub	D1	G1	S1	S2	G2	D2	-	-	ct1
cu	C	B	E	-	E	B	C	C	B	E	-	E	B	C			cu
cv	D	S	G	-	G	S	D	D	S	G	-	G	S	D	-		cv
z1	G1	S2	G2	D2	S1	D1	-										z1
zz	D1/2	G2	S2/4/6	D3/4	G4	S2/4/6	G6	D5/6	D5/6	G5	S1/3/5	G5	D3/4	S1/3/5	G1/3	D1/2	zz

	Transistor				Darlington				FET				
	1	2	3	4	1	2	3	4	1	2	3	4	
a	E	B	C		E2	B1	C1/2		S	G	D		a
b	E	C	B		E2	C1/2	B1		S	D	G		b
c	B	C	E		B1	C1/2	E2		G	D	S		c
d	B	E	C		-				G	S	D		d
e	C	B	E		C1/2	B1	E2		D	G	S		e
f	C	E	B		-	-	-		D	S	G		f
g	E	B	C	Geh	E2	B1	C1/2	Geh	S	G	D	Sub	g
h	E	C	B	Geh	E2	C1/2	B1	Geh	S	D	G	Sub	h
j	B	C	E	Geh	B1	C1/2	E2	Geh	G	D	S	Sub	j
k	B	E	C	Geh	-	-	-	-	-	-	-	-	k
l	C	B	E	Geh	-	-	-	-	D	G	S	Sub	l
m	C	B	E	C	-	-	-	-	D	S	G	Sub	m
n	E	B	Geh	C	C1	E1/2	E2		D	G	Sub	S	n
o	E	C	Geh	B	-	-	-		D	S	Sub	G	o
p	C	Geh	E	B	-	-	-	-	G	Sub	S	D	p
q	C	C	E	B	-	-	-	-	S	D	G	G	q
r	E	C	E	B	E2	B1	C1/2	E1/2	S	G1	D	G2	r
s	E	B	E	C	E2	B1	E1/2	C1/2	S	G1	G2	D	s
t	E	C	B	C	E2	E1/2	B1	C1/2	D	G2	G1	S/Sub	t
u	E	B	C	B	B2	E1/2	C2/B1	C1	G1	G2	D	S/Sub	u
v	B	C	B	E	C1/2	E1/2	B1	E2	S	D	G2	G1	v
w	B	E	B	C	B2	E1/2	C1	C2/B1	G1	G2	S	D	w
x	E	B	C	E	E2	B1	C1/2	E2	D	S	G1	G2	x
y	E1	B	E2	C	-	-	-	-	S	G	S	D	y
z	E1	B	C	E2					S	D	S	G	z
za	B	E	C	E					G	S	D	S	za
zb	E/K	C	A	B					G		D	S	zb
zc	E2	E1	C1	B1/2 C2					G2	D	S	G1	zc
zd	input	Us	output	Gnd					G2	D	G1	S	zd
ze	C	E	B	E					S	S	G	D	ze
zf	C	B	E	B					G	-	S	D	zf
zg	C	B	E						G	D	S	D	zg
zh	B	E	C						G	S	S	D	zh
zj	E1/2	C2	B1/2	C1									zj
zk	E	E	B	C									zk
zm	output	Us2	Us1	input									zm
zn	E	B	C										zn

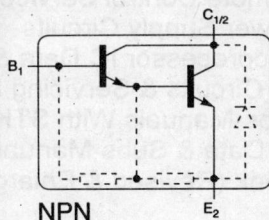

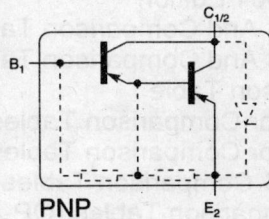

NPN

PNP

§ = C + case.
& = B + case.
* = E + case.
Geh = case
Gnd = ground

§ = D + case.
& = G + case.
* = S + case.
+ = Sub + case.
= = G2 + case.
Sub = bulk

OTHER TITLES OF INTEREST

300 Circuits
301 Circuits
302 Circuits
303 Circuits
304 Circuits
305 Circuits
306 Circuits
307 Circuits
308 Circuits
309 Circuits
Build Your Own Electronic Test Instruments
Designing Audio Circuits
Build Your Own Audio Valve Amplifiers
Build Your Own High End Audio Equipment
PIC In Practical/Diskette
Short Course 8051/8052 Microcontroller & Assembler
Hand Book For Sound Technicians
SMT Projects
Modern DVD Player Servicing
Modern Digital Colour TV Remote Control Service Codes
Modern Colour TV SMPS Power Supply Circuits
Modern Remote Control Microprocessor IC Data & Subs-Manual
Modern Latest Mobile Phone Circuits & Servicing Diagram
Modern Sound IC Data & Subs-Manuals With STK IC Data & App-Manual
Modern Japanese Transistor Data & Subs-Manual
Microprocessor Data Handbook (Revised & Enlarged Edition)
Microcontroller Basics
NTE/ECG Semiconductor Technical Cross Ref GDE
IC Master (Three Volume) 2001 Edition
Up-To-Date CMOS 4000 ICS And Comparison Tables
Up-To-Date CMOS 7400 ICS And Comparison Tables
Up-To-Date Data & Comparison Table
Up-To-Date World's Transistor Comparison Tables (O-U) Vol. 1
Up-To-Date World's Transistor Comparison Tables (O-U) Vol. 2
Up-To-Date TTL 7400 Data & Comparison Tables 7400....7450729 ICs
Up-To-Date linear ICs & Comparison Tables (OP-AMP & Comparators)
Up-To-Date Diodes Data & Comparison Tables
Up-To-Date Memory IC Data & Comparison Tables 8X350....882048
Up-To-Date Emitters & Optocouplers Data Comparison Tables
Up-To-Date Thyristor, Triac, Diac, UJT Data & Comparison Tables
Up-To-Date World's Transistors Diodes Thyristors And IC's A....Z Vol. 1
Up-To-Date World's Transistors Diodes Thyristors And IC's
60000....μ Vol. 2